CHILTON®

ASIAN SERVICE MANUAL 2010 EDITION VOLUME I ACURA HONDA

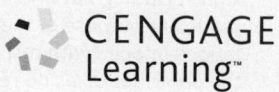

CENGAGE Learning™

Australia • Brazil • Japan • Korea • Mexico • Singapore • Spain • United Kingdom • United States

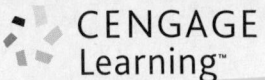

CENGAGE Learning™

CHILTON®
Asian Service Manual
2010 Edition
Volume I
Acura and Honda

Vice President, Technology Professional Business Unit:
Gregory L. Clayton

Publisher, Technology Professional Business Unit:
David Koontz

Director of Marketing:
Beth A. Lutz

Director Education Production:
Carolyn Miller

Marketing Manager:
Jennifer Barbic

Marketing Coordinator:
Rachael Torres

Chilton Content Specialist:
Paula Baillie

Graphical Designer:
Melinda Possinger

Art Director:
Benjamin Gleeksman

Sr. Content Project Manager:
Elizabeth C. Hough

Senior Editor:
Christine L. Sheeky

Editors:
Dennis L. Bailey
Jim Bailey
Eugene F. Hannon Jr., A.S.E.

ISBN-13: 978-1-1110-3764-2
ISBN-10: 1-1110-3764-7
ISSN: 1939-621X

Chilton
5 Maxwell Drive
Clifton Park, NY 12065-2919
USA

Chilton products are represented in Canada by Nelson Education, Ltd.

NOTICE TO THE READER

Publisher does not warrant or guarantee any of the products described herein or perform any independent analysis in connection with any of the product information contained herein. Publisher does not assume, and expressly disclaims, any obligation to obtain and include information other than that provided to it by the manufacturer.

The reader is expressly warned to consider and adopt all safety precautions that might be indicated by the activities described herein and to avoid all potential hazards. By following the instructions contained herein, the reader willingly assumes all risks in connection with such instructions.

The publisher makes no representations or warranties of any kind, including but not limited to, the warranties of fitness for particular purpose or merchantability, nor are any such representations implied with respect to the material set forth herein, and the publisher takes no responsibility with respect to such material. The publisher shall not be liable for any special, consequential, or exemplary damages resulting, in whole or part, from the readers' use of, or reliance upon, this material.

Printed in the United States of America
1 2 3 4 5 6 7 14 13 12 11 10

Contents

Sections

1 MDX

2 RDX

3 RL

4 TL

5 TSX

6 ZDX

7 Acura Diagnostic Trouble Codes

8 Accord

9 Civic

10 Civic Hybrid

11 CR-V

12 Element

13 Fit

14 Insight Hybrid

15 Odyssey

16 Pilot

17 Ridgeline

18 S2000

19 Honda Diagnostic Trouble Codes

Model Index

Model	Section No.
A	
Accord	8-1
C	
Civic	9-1
Civic Hybrid	10-1
CR-V	11-1
E	
Element	12-1
F	
Fit	13-1
I	
Insight Hybrid	14-1
M	
MDX	1-1
O	
Odyssey	15-1
P	
Pilot	16-1
R	
RDX	2-1
Ridgeline	17-1
RL	3-1
S	
S2000	18-1
T	
TL	4-1
TSX	5-1
Z	
ZDX	6-1

USING THIS INFORMATION

Organization

To find where a particular model section or procedure is located, look in the Table of Contents. Main topics are listed with the page number on which they may be found. Following the main topics is an alphabetical listing of all of the procedures within the section and their page numbers.

Manufacturer and Model Coverage

This product covers 2009–2010 Asian models that are produced in sufficient quantities to warrant coverage, and which have technical content available from the vehicle manufacturers before our publication date. Although this information is as complete as possible at the time of publication, some manufacturers may make changes which cannot be included here. While striving for total accuracy, the publisher cannot assume responsibility for any errors, changes, or omissions that may occur in the compilation of this data.

Part Numbers and Special Tools

Part numbers and special tools are recommended by the publisher and vehicle manufacturer to perform specific jobs. Before substituting any part or tool for the one recommended, you must be completely satisfied that neither your personal safety, nor the performance of the vehicle will be endangered.

ACKNOWLEDGEMENT

Portions of materials contained herein have been reprinted under license from American Honda Corporation, License Agreement 07201AH.

PRECAUTIONS

Before servicing any vehicle, please be sure to read all of the following precautions, which deal with personal safety, prevention of component damage, and important points to take into consideration when servicing a motor vehicle:

- Always wear safety glasses or goggles when drilling, cutting, grinding or prying.
- Steel-toed work shoes should be worn when working with heavy parts. Pockets should not be used for carrying tools. A slip or fall can drive a screwdriver into your body.
- Work surfaces, including tools and the floor should be kept clean of grease, oil or other slippery material.
- When working around moving parts, don't wear loose clothing. Long hair should be tied back under a hat or cap, or in a hair net.
- Always use tools only for the purpose for which they were designed. Never pry with a screwdriver.
- Keep a fire extinguisher and first aid kit handy.
- Always properly support the vehicle with approved stands or lift.
- Always have adequate ventilation when working with chemicals or hazardous material.
- Carbon monoxide is colorless, odorless and dangerous. If it is necessary to operate the engine with vehicle in a closed area such as a garage, always use an exhaust collector to vent the exhaust gases outside the closed area.
- When draining coolant, keep in mind that small children and some pets are attracted by ethylene glycol antifreeze, and are quite likely to drink any left in an open container, or in puddles on the ground. This will prove fatal in sufficient quantity. Always drain the coolant into a sealable container.
- To avoid personal injury, do not remove the coolant pressure relief cap while the engine is operating or hot. The cooling system is under pressure; steam and hot liquid can come out forcefully when the cap is loosened slightly. Failure to follow these instructions may result in personal injury. The coolant must be recovered in a suitable, clean container for reuse. If the coolant is contaminated it must be recycled or disposed of correctly.
- When carrying out maintenance on the starting system be aware that heavy gauge leads are connected directly to the battery. Make sure the protective caps are in place when maintenance is completed. Failure to follow these instructions may result in personal injury.
- Do not remove any part of the engine emission control system. Operating the engine without the engine emission control system will reduce fuel economy and engine ventilation. This will weaken engine performance and shorten engine life. It is also a violation of Federal law.
- Due to environmental concerns, when the air conditioning system is drained, the refrigerant must be collected using refrigerant recovery/recycling equipment. Federal law requires that refrigerant be recovered into appropriate recovery equipment and the process be conducted by qualified technicians who have been certified by an approved organization, such as MACS, ASI, etc. Use of a recovery machine dedicated to the appropriate refrigerant is necessary to reduce the possibility of oil and refrigerant incompatibility concerns. Refer to the instructions provided by the equipment manufacturer when removing refrigerant from or charging the air conditioning system.
- Always disconnect the battery ground when working on or around the electrical system.
- Batteries contain sulfuric acid. Avoid contact with skin, eyes, or clothing. Also, shield your eyes when working near batteries to protect against possible splashing of the acid solution. In case of acid contact with skin or eyes, flush immediately with water for a minimum of 15 minutes and get prompt medical attention. If acid is swallowed, call a physician immediately. Failure to follow these instructions may result in personal injury.
- Batteries normally produce explosive gases. Therefore, do not allow flames, sparks or lighted substances to come near the battery. When charging or working near a battery, always shield your face and protect your eyes. Always provide ventilation. Failure to follow these instructions may result in personal injury.

• When lifting a battery, excessive pressure on the end walls could cause acid to spew through the vent caps, resulting in personal injury, damage to the vehicle or battery. Lift with a battery carrier or with your hands on opposite corners. Failure to follow these instructions may result in personal injury.

• Observe all applicable safety precautions when working around fuel. Whenever servicing the fuel system, always work in a well-ventilated area. Do not allow fuel spray or vapors to come in contact with a spark, open flame, or excessive heat (a hot drop light, for example). Keep a dry chemical fire extinguisher near the work area. Always keep fuel in a container specifically designed for fuel storage; also, always properly seal fuel containers to avoid the possibility of fire or explosion. Do not smoke or carry lighted tobacco or open flame of any type when working on or near any fuel-related components.

• Fuel injection systems often remain pressurized, even after the engine has been turned OFF. The fuel system pressure must be relieved before disconnecting any fuel lines. Failure to do so may result in fire and/or personal injury.

• The evaporative emissions system contains fuel vapor and condensed fuel vapor. Although not present in large quantities, it still presents the danger of explosion or fire. Disconnect the battery ground cable from the battery to minimize the possibility of an electrical spark occurring, possibly causing a fire or explosion if fuel vapor or liquid fuel is present in the area. Failure to follow these instructions can result in personal injury.

• The EPA warns that prolonged contact with used engine oil may cause a number of skin disorders, including cancer! You should make every effort to minimize your exposure to used engine oil. Protective gloves should be worn when changing oil. Wash your hands and any other exposed skin areas as soon as possible after exposure to used engine oil. Soap and water, or waterless hand cleaner should be used.

• Some vehicles are equipped with an air bag system, often referred to as a Supplemental Restraint System (SRS) or Supplemental Inflatable Restraint (SIR) system. The system must be disabled before performing service on or around system components, steering column, instrument panel components, wiring and sensors. Failure to follow safety and disabling procedures could result in accidental air bag deployment, possible personal injury and unnecessary system repairs.

• Always wear safety goggles when working with, or around, the air bag system. When carrying a non-deployed air bag, be sure the bag and trim cover are pointed away from your body. When placing a non-deployed air bag on a work surface, always face the bag and trim cover upward, away from the surface. This will reduce the motion of the module if it is accidentally deployed.

• Electronic modules are sensitive to electrical charges. The ABS module can be damaged if exposed to these charges.

• Brake pads and shoes may contain asbestos, which has been determined to be a cancer-causing agent. Never clean brake surfaces with compressed air. Avoid inhaling brake dust. Clean all brake surfaces with a commercially available brake cleaning fluid.

• When replacing brake pads, shoes, discs or drums, replace them as complete axle sets.

• When servicing drum brakes, disassemble and assemble one side at a time, leaving the remaining side intact for reference.

• Brake fluid often contains polyglycol ethers and polyglycols. Avoid contact with the eyes and wash your hands thoroughly after handling brake fluid. If you do get brake fluid in your eyes, flush your eyes with clean, running water for 15 minutes. If eye irritation persists, or if you have taken brake fluid internally, immediately seek medical assistance.

• Clean, high quality brake fluid from a sealed container is essential to the safe and proper operation of the brake system. You should always buy the correct type of brake fluid for your vehicle. If the brake fluid becomes contaminated, completely flush the system with new fluid. Never reuse any brake fluid. Any brake fluid that is removed from the system should be discarded. Also, do not allow any brake fluid to come in contact with a painted or plastic surface; it will damage the paint.

• Never operate the engine without the proper amount and type of engine oil; doing so will result in severe engine damage.

• Timing belt maintenance is extremely important! Many models utilize an interference- type, non freewheeling engine. If the timing belt breaks, the valves in the cylinder head may strike the pistons, causing potentially serious (also time-consuming and expensive) engine damage.

• Disconnecting the negative battery cable on some vehicles may interfere with the functions of the on-board computer system(s) and may require the computer to undergo a relearning process once the negative battery cable is reconnected.

• Steering and suspension fasteners are critical parts because they affect performance of vital components and systems and their failure can result in major service expense. They must be replaced with the same grade or part number or an equivalent part if replacement is necessary. Do not use a replacement part of lesser quality or substitute design. Torque values must be used as specified during reassembly.

ACURA

MDX

1

AUXILIARY HEATING & AIR CONDITIONING SYSTEM 1-50

Blower Motor 1-50
Removal & Installation....1-50
Heater Core 1-50
Removal & Installation....1-50

BRAKES 1-9

ANTI-LOCK BRAKE SYSTEM (ABS)....1-10
Wheel Speed Sensors 1-10
Removal & Installation....1-10
BLEEDING THE BRAKE SYSTEM 1-9
Bleeding Procedure....1-9
Bleeding Procedure 1-9
FRONT DISC BRAKES 1-11
Brake Caliper....1-11
Removal & Installation....1-11
Disc Brake Pads....1-11
Removal & Installation....1-11
INFORMATION AND PRECAUTIONS....1-9
Anti-lock Systems....1-9
Disc and Drum Systems 1-9
PARKING BRAKE 1-13
Parking Brake Cables....1-13
Adjustment 1-13
Parking Brake Shoes....1-13
Removal & Installation....1-13
REAR DISC BRAKES 1-12
Brake Caliper....1-12
Removal & Installation....1-12
Disc Brake Pads....1-12
Removal & Installation....1-12

CHASSIS ELECTRICAL....1-15

AIR BAG (SUPPLEMENTAL RESTRAINT SYSTEM)....1-15
General Information....1-15
Arming the System 1-15
Clockspring Centering....1-15
Disarming the System....1-15
Service Precautions 1-15

DRIVE TRAIN....1-15

Front Halfshaft....1-15
Overhaul 1-15
Removal & Installation....1-15
Rear Halfshafts 1-16
Removal & Installation....1-16
Rear Pinion Seal....1-16
Removal & Installation....1-16
Transfer Case Assembly....1-18
Removal & Installation....1-18

ENGINE COOLING 1-18

Engine Fan 1-18
Removal & Installation....1-18
Radiator....1-18
Removal & Installation....1-18
Thermostat 1-19
Removal & Installation....1-19
Water Pump 1-20
Removal & Installation....1-20

ENGINE ELECTRICAL....1-20

CHARGING SYSTEM 1-20
Alternator 1-20
Removal & Installation....1-20
IGNITION SYSTEM 1-21
Firing Order....1-21
Ignition Coil 1-21
Removal & Installation....1-21
Ignition Timing....1-21
Adjustment 1-21
Inspection 1-21
Spark Plugs....1-21
Removal & Installation....1-21
STARTING SYSTEM 1-21
Starter 1-21
Removal & Installation....1-21
Solenoid Replacement 1-22

ENGINE MECHANICAL 1-23

Accessory Drive Belts 1-23
Accessory Belt Routing....1-23
Adjustment 1-23
Inspection 1-23
Removal & Installation....1-23
Camshaft and Valve Lifters....1-23
Inspection 1-23
Removal & Installation....1-24
Catalytic Converter....1-25
Removal & Installation....1-25
Crankshaft Damper....1-27
Removal & Installation....1-27
Crankshaft Front Seal....1-28
Removal & Installation....1-28
Cylinder Head 1-28
Removal & Installation....1-28
Driveplate....1-29
Removal & Installation....1-29
Intake Manifold 1-30
Removal & Installation....1-30
Oil Pan 1-30
Removal & Installation....1-30
Oil Pump....1-31
Inspection 1-32
Removal & Installation....1-31
Piston and Ring....1-32
Positioning 1-32
Rear Main Seal....1-32
Removal & Installation....1-32
Rocker Arms....1-32
Removal & Installation....1-32
Timing Belt & Sprockets 1-33
Removal & Installation....1-33
Timing Belt Front Cover 1-33
Removal & Installation....1-33
Timing Belt Rear Cover 1-35
Removal & Installation....1-35
Valve (Clearance)
Lash....1-36
Adjustment 1-36
Valve Covers 1-35
Removal & Installation....1-35

ENGINE PERFORMANCE & EMISSION CONTROLS....1-37

Accelerator Pedal Position (APP) Sensor 1-37
Location....1-37
Removal & Installation....1-37

Camshaft Position (CMP) Sensor 1-37
Location....1-37
Removal & Installation....1-37
Crankshaft Position (CKP) Sensor 1-38
Location....1-38
Removal & Installation....1-38
Engine Coolant Temperature (ECT) Sensor 1-38
Location....1-38
Removal & Installation....1-38
Evaporative Emissions (EVAP) Canister 1-38
Location....1-38
Removal & Installation....1-39
Heated Oxygen (HO2S) Sensor 1-39
Location....1-39
Removal & Installation....1-39
Intake Air Temperature (IAT) Sensor 1-40
Location....1-40
Removal & Installation....1-40
Knock Sensor (KS)....1-40
Location....1-40
Removal & Installation....1-40
Malfunction Indicator Light (MIL)....1-40
Reset Procedure....1-40
Manifold Absolute Pressure (MAP) Sensor....1-40
Removal & Installation....1-40
Mass Air Flow (MAF) Sensor....1-40
Location....1-40
Removal & Installation....1-40
Output Shaft Speed (OSS) Sensor 1-40
Location....1-40
Removal & Installation....1-41
Positive Crankcase Ventilation (PCV) Valve....1-41
Location....1-41
Removal & Installation....1-41
Powertrain Control Module (PCM)....1-41
Location....1-41
Removal & Installation....1-41
CKP Pattern Learn Procedure....1-43
PCM Idle Learn Procedure....1-42
PCM Update....1-43
Throttle Position Sensor (TPS)...1-43
Location....1-43
Removal & Installation....1-43

FUEL....1-44

GASOLINE FUEL INJECTION SYSTEM....1-44
Fuel Filter....1-45
Removal & Installation....1-45
Fuel Level Sending Unit....1-45
Location....1-45
Removal & Installation....1-45
Fuel Pressure Regulator....1-46
Removal & Installation....1-46
Fuel Rail and Injector....1-47
Removal & Installation....1-47
Fuel System Pressure....1-44
Relieving....1-44
Fuel System Service Precautions....1-44
Fuel Tank....1-47
Draining....1-47
Removal & Installation....1-47
Fuel Tank Unit (Fuel Pump) 1-45
Removal & Installation....1-45
Idle Speed....1-48
Adjustment....1-48
Throttle Body....1-48
Removal & Installation....1-48

HEATING, VENTILATION & AIR CONDITIONING SYSTEM 1-49
Blower Motor....1-49
Removal & Installation....1-49
Heater Core....1-49
Removal & Installation....1-49

SPECIFICATIONS AND MAINTENANCE CHARTS 1-3
Brake Specifications....1-7
Camshaft and Bearing Specifications....1-5
Capacities....1-4
Crankshaft and Connecting Rod Specifications....1-5
Engine and Vechicle Identification....1-3
Gasoline Engine Tune-Up Specifications....1-00
Fluid Specifications....1-4
General Engine Specifications....1-3
Piston and Ring Specifications....1-5
Scheduled Maintenance Intervals....1-8
Tire, Wheel and Ball Joint Specifications....1-7
Torque Specifications....1-6
Valve Specifications....1-4
Wheel Alignment....1-7

STEERING....1-51
Power Rack & Pinion Steering Gear....1-51
Removal & Installation....1-51
Power Steering Pump....1-55
Bleeding....1-55
Removal & Installation....1-55

SUSPENSION....1-55

FRONT SUSPENSION....1-55
Control Links....1-55
Removal & Installation....1-55
Lower Ball Joint....1-55
Removal & Installation....1-55
Lower Control Arm....1-56
Removal & Installation....1-56
Stabilizer Bar....1-60
Removal & Installation....1-60
Steering Knuckle....1-57
Removal & Installation....1-57
Strut....1-57
Overhaul....1-58
Removal & Installation....1-57
Wheel Hub & Bearing....1-60
Removal & Installation....1-60

REAR SUSPENSION....1-60
Ball Joint....1-61
Removal & Installation....1-61
Coil Spring....1-61
Removal & Installation....1-61
Control Arms/Links....1-62
Removal & Installation....1-62
Shock Absorber (Damper)....1-65
Removal & Installation....1-65
Testing....1-65
Stabilizer Bar....1-65
Removal & Installation....1-65
Wheel Hub & Bearing....1-66
Adjustment....1-66
Removal & Installation....1-66

SPECIFICATIONS AND MAINTENANCE CHARTS

ENGINE AND VEHICLE IDENTIFICATION

Engine						
Code	Liters (cc)	Cu. In.	Cyl.	Fuel Sys.	Engine Type	Eng. Mfg.
J37A1	3.7 (3664)	224	6	PGM-FI	SOHC	Honda

Model Year	
Code ①	Year
9	2009
0	2010

PGM-FI: Programmed Fuel Injection

SOHC: Single Overhead Camshaft

① 10th digit of the Vehicle Identification Number (VIN)

37647_AMDX_C0001

GENERAL ENGINE SPECIFICATIONS

Year	Model	Engine Displacement Liters	Engine ID	Net Horsepower @ rpm	Net Torque @ rpm (ft. lbs.)	Bore x Stroke (in.)	Com-pression Ratio	Oil Pressure @ rpm
2009	MDX	3.7	J37A1	300@6000	275@5000	3.54x3.78	11.0:1	71@3000
2010	MDX	3.7	J37A1	300@6000	275@5000	3.54x3.78	11.0:1	71@3000

37647_AMDX_C0002

GASOLINE ENGINE TUNE-UP SPECIFICATIONS

Year	Engine Displacement Liters	Engine ID/VIN	Spark Plug Gap (in.)	Ignition Timing (deg.)		Fuel Pump (psi)	Idle Speed (rpm)		Valve Clearance	
				MT	AT		MT	AT	In.	Ex.
2009	3.7	J37A1	0.039-0.043	—	8-12B	57-64 ①	—	660-760	0.008-0.009	0.011-0.013
2010	3.7	J37A1	0.039-0.043	—	8-12B	57-64 ①	—	660-760	0.008-0.009	0.011-0.013

NOTE: The Vehicle Emission Control Information label reflects specification changes during production and must be used if they differ from this chart.

B: Before Top Dead Center

HYD: Hydraulic

① At idle, pressure regulator vacuum hose disconnected

37647_AMDX_C0003

CAPACITIES

Year	Model	Engine Displacement Liters	Engine ID/VIN	Engine Oil with Filter (qts.)	Transmission (pts.) 5-Spd	6-Spd	Auto.	Drive Axle Front (pts.)	Rear (pts.)	Fuel Tank (gal.)	Cooling System (qts.)
2009	MDX	3.7	J37A1	4.5	—	—	3.0	—	5.34	21.0	7.7
2010	MDX	3.7	J37A1	4.5	—	—	3.0	—	5.34	21.0	7.7

NOTE: All capacities are approximate. Add fluid gradually and ensure a proper fluid level is obtained.

NOTE: Capacities given are service, not overhaul capacities

37647_AMDX_C0004

FLUID SPECIFICATIONS

Year	Model	Engine Displ. Liters	Engine Oil	Man. Trans.	Auto. Trans.	Drive Axle Front	Rear	Transfer Case	Power Steering Fluid	Brake Master Cylinder	Cooling System
2009	MDX	3.7	5W-20 Acura	Acura MTF	Acura ATF-Z1	—	—	①	Acura PS Fluid	Acura DOT 3	②
2010	MDX	3.7	5W-20 Acura	Acura MTF	Acura ATF-Z1	—	—	①	Acura PS Fluid	Acura DOT 3	②

① Hypoid gear oil SAE 90 or SAE 80W-90 viscosity, API classified GL4 or GL5 only

② Acura Long Life Antifreeze/Coolant-Type2

37647_AMDX_C0005

VALVE SPECIFICATIONS

Year	Engine Displacement Liters	Engine ID/VIN	Seat Angle (deg.)	Face Angle (deg.)	Spring Test Pressure (lbs. @ in.)	Spring Installed Height (in.)	Stem-to-Guide Clearance (in.) Intake	Exhaust	Stem Diameter (in.) Intake	Exhaust
2009	3.7	J37A1	45	45	NA	0.835-0.874	0.0008-0.0018	0.0022-0.0031	0.2159-0.2163	0.2146-0.2150
2010	3.7	J37A1	45	45	NA	0.811-0.85	0.0008-0.0018	0.0022-0.0031	0.2159-0.2163	0.2146-0.2150

NA: Not Available

37647_AMDX_C0006

CAMSHAFT AND BEARING SPECIFICATIONS

All measurements are given in inches.

Year	Engine Displacement Liters	Engine VIN	Journal Diameter	Brg. Oil Clearance	Shaft End-play	Runout	Journal Bore	Lobe Height Intake	Lobe Height Exhaust
2009	3.7	J37A1	NA	0.0020-0.0035	0.0020-0.0080	0.0010	NA	①	1.4326
2010	3.7	J37A1	NA	0.0020-0.0035	0.0020-0.0080	0.0010	NA	①	1.4326

NA: Information not available

① Primary: 1.3824 inches

Mid: 1.4328 inches

Secondary: 1.3824 inches

37647_AMDX_C0009

CRANKSHAFT AND CONNECTING ROD SPECIFICATIONS

All measurements are given in inches.

Year	Engine Displacement Liters	Engine ID/VIN	Crankshaft Main Brg. Journal Dia.	Crankshaft Main Brg. Oil Clearance	Crankshaft Shaft End-play	Crankshaft Thrust on No.	Connecting Rod Journal Diameter	Connecting Rod Oil Clearance	Connecting Rod Side Clearance
2009	3.7	J37A1	2.8337-2.8346	0.0007-0.0018	0.0040-0.0140	3	2.2431-2.2441	0.0008-0.0017	0.0060-0.0140
2010	3.7	J37A1	2.8337-2.8346	0.0007-0.0018	0.0040-0.0140	3	2.2431-2.2441	0.0008-0.0017	0.0060-0.0140

37647_AMDX_C0007

PISTON AND RING SPECIFICATIONS

All measurements are given in inches

Year	Engine Displacement Liters	Engine ID/VIN	Piston Clearance	Ring Gap Top Compression	Ring Gap Bottom Compression	Ring Gap Oil Control	Ring Side Clearance Top Compression	Ring Side Clearance Bottom Compression	Ring Side Clearance Oil Control
2009	3.7	J37A1	0.0002-0.0013	0.0120-0.0160	0.0160-0.0220	0.0080-0.0280	0.0022-0.0033	0.0012-0.0024	NA
2010	3.7	J37A1	0.0002-0.0013	0.0120-0.0160	0.0160-0.0220	0.0080-0.0280	0.0022-0.0033	0.0012-0.0024	NA

NA; Not Available

37647_AMDX_C0008

TORQUE SPECIFICATIONS

All readings in ft. lbs.

Year	Engine Displacement Liters	Engine ID/VIN	Cylinder Head Bolts	Main Bearing Bolts	Rod Bearing Bolts	Crankshaft Damper Bolts	Flywheel Bolts	Manifold Intake	Manifold Exhaust	Spark Plugs	Oil Pan Drain Plug
2009	3.7	J37A1	①	②	③	④	54	16	40	13	29
2010	3.7	J37A1	①	②	③	④	54	16	40	13	29

① Step 1: 22 ft. lbs.
Step 2: Rotate 90 degrees
Step 3: Rotate an additional 90 degrees
Step 4 (new bolts only): additional 90 degrees

② Step 1: Cap bolts 56 ft. lbs.
Step 2: Side bolts 36 ft. lbs.

③ Step 1: 14 ft. lbs.
Step 2: Rotate 90 degrees

④ Step 1: 47 ft. lbs.
Step 2: Rotate 60 degrees

37647_AMDX_C0010

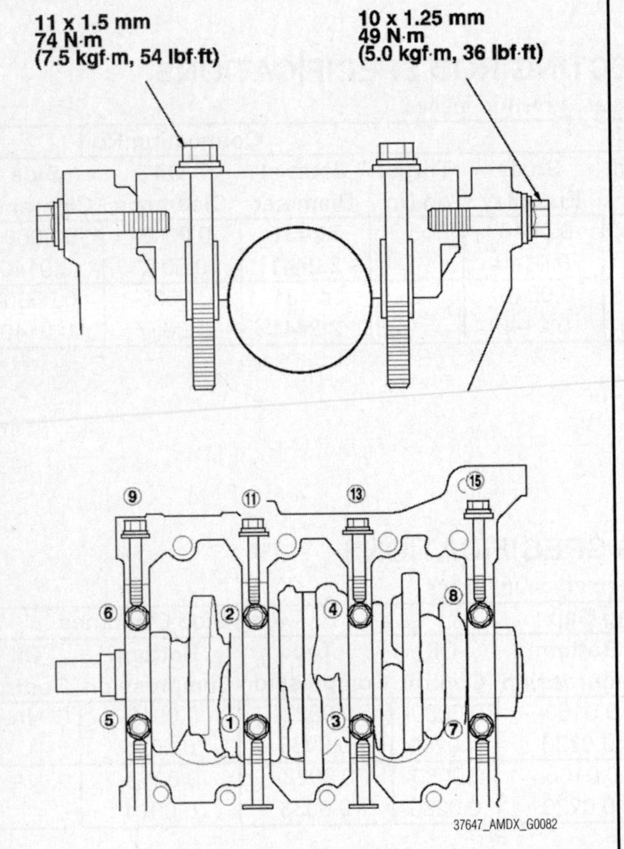

11 x 1.5 mm
74 N·m
(7.5 kgf·m, 54 lbf·ft)
10 x 1.25 mm
49 N·m
(5.0 kgf·m, 36 lbf·ft)
9
11
13
15
6
2
4
8
5
1
3
7
37647_AMDX_G0082

Fig. 1 Main bearing torque sequence—3.7L engine

WHEEL ALIGNMENT

Year	Model		Caster Range (+/-Deg.)	Caster Preferred Setting (Deg.)	Camber Range (+/-Deg.)	Camber Preferred Setting (Deg.)	Toe-in (in.)
2009	MDX	F	±35	+4.12	1.00	-0.30	0 +/- 0.08
		R	—	—	0.45	-0.30	0.08 +/- 0.08
2010	MDX	F	±35	+4.12	1.00	-0.30	0 +/- 0.08
		R	—	—	0.45	-0.30	0.08 +/- 0.08

37647_AMDX_C0011

TIRE, WHEEL AND BALL JOINT SPECIFICATIONS

Year	Model	OEM Tires Standard	OEM Tires Optional	Tire Pressures (psi) Front	Tire Pressures (psi) Rear	Wheel Size	Ball Joint Inspection	Lug Nut (ft. lbs.)
2009	MDX	P255/55R18	None	NA	NA	NA	NS	80
2010	MDX	P255/55R18	None	NA	NA	NA	NS	80

OEM: Original Equipment Manufacturer

PSI: Pounds Per Square Inch

NS: Not Specified by manufacturer

NA: Not available

37647_AMDX_C0012

BRAKE SPECIFICATIONS

All measurements in inches unless noted

Year	Model		Brake Disc Original Thickness	Brake Disc Minimum Thickness	Brake Disc Maximum Runout	Brake Drum Diameter Original Inside Diameter	Brake Drum Diameter Max. Wear Limit	Brake Drum Diameter Maximum Machine Diameter	Minimum Lining Thickness Front	Minimum Lining Thickness Rear	Brake Caliper Bracket Bolts (ft. lbs.)	Brake Caliper Mounting Bolts (ft. lbs.)
2009	MDX	F	1.100-1.111	1.020	0.004	—	—	—	0.06	—	101	53
		R	0.430-0.440	0.350	0.004	—	—	—	—	0.06	65	27
2010	MDX	F	1.100-1.111	1.020	0.004	—	—	—	0.06	—	101	53
		R	0.430-0.440	0.350	0.004	—	—	—	—	0.06	65	27

NA: Not Available

F: Front

R: Rear

37647_AMDX_C0013

SCHEDULED MAINTENANCE INTERVALS
Acura MDX

TO BE SERVICED	TYPE OF SERVICE	SERVICE INTERVALS						
		Symbol A	Symbol B	1	2	3	4	5
Air cleaner element	Replace				✓			
Brake fluid	Inspect		✓					
Brake hoses and lines	Inspect		✓					
Clutch fluid (if equipped)	Inspect		✓					
Drive belt	Inspect				✓			
Driveshaft boots	Inspect		✓					
Dust & pollen filter	Replace				✓			
Engine coolant	Inspect		✓					
Engine coolant	Replace							✓
Engine oil	Replace	✓						
Engine oil & filter	Replace		✓					
Exhaust system	Inspect		✓					
Front & Rear brakes	Inspect		✓					
Fuel lines	Inspect		✓					
Parking brake adjustment	Inspect		✓					
Power steering fluid	Inspect		✓					
Spark plugs	Replace						✓	
Suspension components	Inspect		✓					
Tie-rod ends, steering gearbox and boots	Inspect		✓					
Timing belt	Replace						✓	
Tires	Rotate			✓				
Transmission fluid	Inspect		✓					
Transmission fluid	Replace					✓		
Valve clearance	Inspect						✓	
Windshield washer fluid	Inspect		✓					

Symbol A and B represent the maintenance symbol that will appear on the dash maintenance reminder.

Numbers 1through 5 indicate when the service is performed. See the owners manual for specifics.

37647_AMDX_C0014

BRAKES

INFORMATION AND PRECAUTIONS

ANTI-LOCK SYSTEMS

- Certain components within the ABS system are not intended to be serviced or repaired individually.
- Do not use rubber hoses or other parts not specifically specified for and ABS system. When using repair kits, replace all parts included in the kit. Partial or incorrect repair may lead to functional problems and require the replacement of components.
- Lubricate rubber parts with clean, fresh brake fluid to ease assembly. Do not use shop air to clean parts; damage to rubber components may result.
- Use only DOT 3 brake fluid from an unopened container.
- If any hydraulic component or line is removed or replaced, it may be necessary to bleed the entire system.
- A clean repair area is essential. Always clean the reservoir and cap thoroughly before removing the cap. The slightest amount of dirt in the fluid may plug an orifice and impair the system function. Perform repairs after components have been thoroughly cleaned; use only denatured alcohol to clean components. Do not allow ABS components to come into contact with any substance containing mineral oil; this includes used shop rags.
- The Anti-Lock control unit is a microprocessor similar to other computer units in the vehicle. Ensure that the ignition switch is **OFF** before removing or installing controller harnesses. Avoid static electricity discharge at or near the controller.
- If any arc welding is to be done on the vehicle, the control unit should be unplugged before welding operations begin.

DISC AND DRUM SYSTEMS

CAUTION

Dust and dirt accumulating on brake parts during normal use may contain asbestos fibers from production or aftermarket brake linings. Breathing excessive concentrations of asbestos fibers can cause serious bodily harm. Exercise care when servicing brake parts. Do not sand or grind brake lining unless equipment used is designed to contain the dust residue. Do not clean brake parts with compressed air or by dry brushing. Cleaning should be done by dampening the brake components with a fine mist of water, then wiping the brake components clean with a dampened cloth. Dispose of cloth and all residue containing asbestos fibers in an impermeable container with the appropriate label. Follow practices prescribed by the Occupational Safety and Health Administration (OSHA) and the Environmental Protection Agency (EPA) for the handling, processing, and disposing of dust or debris that may contain asbestos fibers.

BRAKES

BLEEDING THE BRAKE SYSTEM

BLEEDING PROCEDURE

BLEEDING PROCEDURE

See Figure 2.

➡Do not reuse the drained fluid. Use only clean Acura DOT 3 Brake Fluid from an unopened container. Using a non-Acura brake fluid can cause corrosion and shorten the life of the system.

Make sure no dirt or other foreign matter is allowed to contaminate the brake fluid.

Do not spill brake fluid on the vehicle, it may damage the paint; if brake fluid does contact the paint, wash it off immediately with water.

The reservoir connected to the master cylinder must be at the MAX (upper) level mark at the start of the bleeding procedure and checked after bleeding each wheel. Add fluid as required.

1. Make sure the brake fluid level in the reservoir is at the MAX (upper) level line.
2. Have someone slowly pump the brake pedal several times, then apply steady pressure.
3. Start the bleeding at the driver's side of the front brake system.

➡Bleed the calipers or the wheel cylinders in the sequence shown.

4. Attach a length of clear drain tube to the bleed screw, then, loosen the bleed screw to allow air to escape from the system. Then tighten the bleed screw securely.

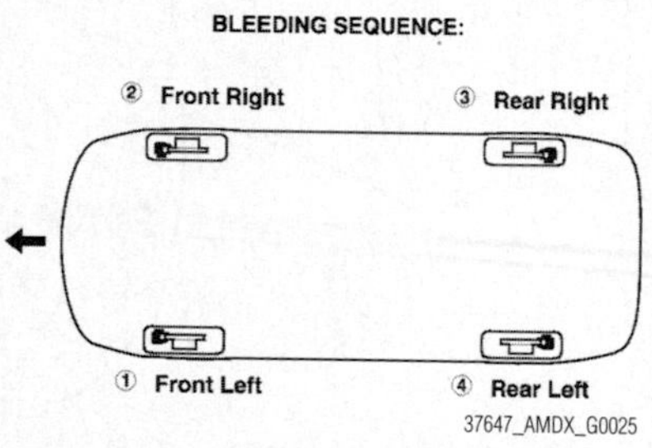
BLEEDING SEQUENCE:
② Front Right
③ Rear Right
① Front Left
④ Rear Left
37647_AMDX_G0025

Fig. 2 Bleed the calipers or the wheel cylinders in sequence

5. Refill the master cylinder reservoir to the MAX (upper) level line.
6. Repeat the procedure for each brake circuit until there are no air bubbles are in the fluid.

WHEEL SPEED SENSORS

REMOVAL & INSTALLATION

Front

See Figure 3.

1. Turn the ignition switch to LOCK (0).
2. Remove the clip, then disconnect the wheel speed sensor connector.
3. Remove the bolts and the wheel speed sensor.

To install:

4. Install the wheel speed sensor in the reverse order of removal.

✻✻ CAUTION

Install the sensor carefully to avoid twisting the wires.

➡If the wheel speed sensor comes in contact with the hub bearing unit, it is faulty.

5. Start the engine, and check that the ABS, the VSA indicators, and the trailer stability assist warning goes off.
6. Test-drive the vehicle, and check that the ABS, the VSA indicators, and the trailer stability assist warning do not come on.

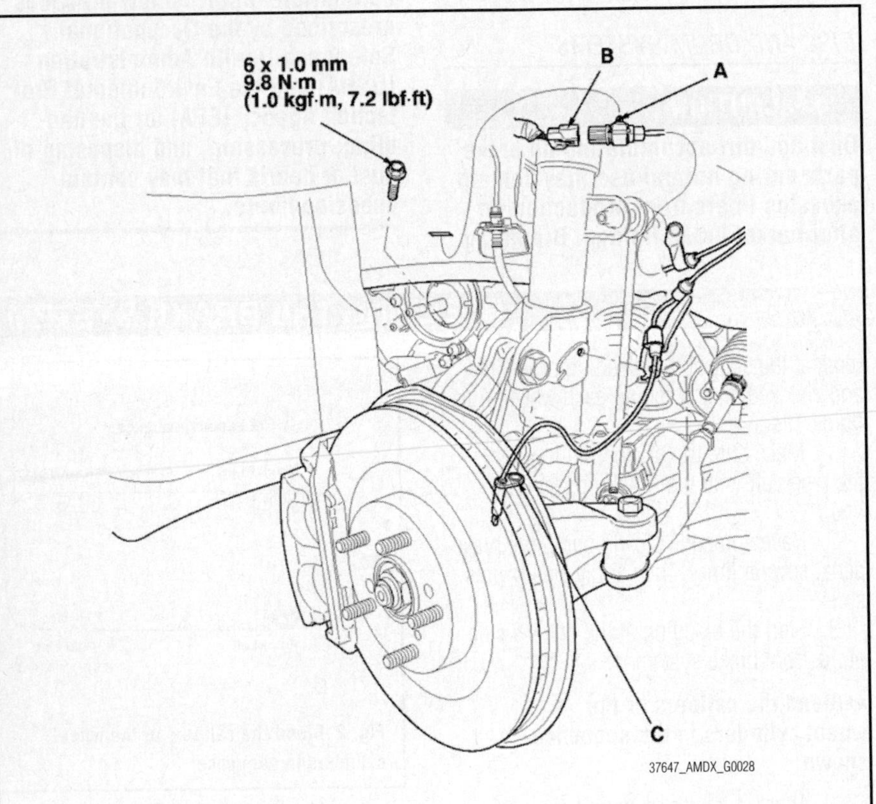
6 x 1.0 mm
9.8 N·m
(1.0 kgf·m, 7.2 lbf·ft)
B
A
C
37647_AMDX_G0028

Fig. 3 Remove the front wheel speed sensor (C)

Rear

See Figure 4.

1. Turn the ignition switch to LOCK (0).
2. Disconnect the wheel speed sensor connector.
3. Remove the clips, the bolt, and the wheel speed sensor.

To install:

4. Install the wheel speed sensor in the reverse order of removal.

✻✻ CAUTION

Install the sensor carefully to avoid twisting the wires.

➡If the wheel speed sensor comes in contact with the hub bearing unit, it is faulty.

5. Start the engine, and check that the ABS, the VSA indicators, and the trailer stability assist warning goes off.
6. Test-drive the vehicle, and check that the ABS, the VSA indicators, and the trailer stability assist warning do not come on.

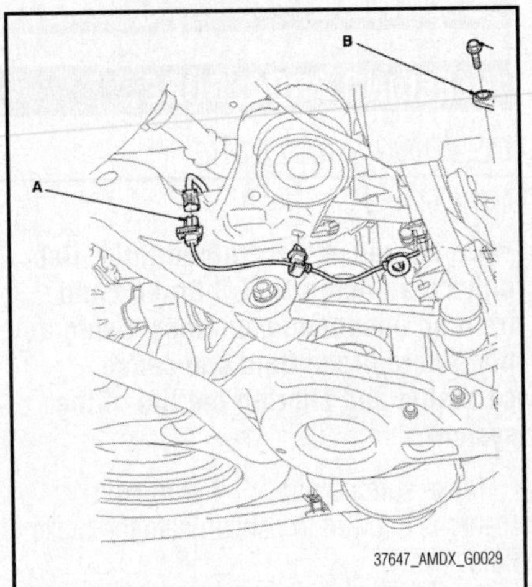
B
A
37647_AMDX_G0029

Fig. 4 Remove the rear wheel speed sensor

BRAKES

BRAKE CALIPER

REMOVAL & INSTALLATION

See Figure 5.

1. Raise and safely support the vehicle.
2. Remove the front wheel.
3. Remove the brake hose bracket mounting bolt and banjo bolt to disconnect the brake hose from the caliper. Plug the brake hose to prevent excessive fluid loss.
4. Remove the brake caliper bracket mounting bolts, then remove the caliper assembly from the knuckle.

To install:

5. Install the caliper assembly and tighten the mounting bolts to 101 ft. lbs. (137 Nm).
6. Connect the brake hose using new sealing washers and tighten the banjo bolt to 25 ft. lbs. (34 Nm).
7. The remainder of the installation is the reverse order of removal.

DISC BRAKE PADS

REMOVAL & INSTALLATION

See Figure 6.

1. Remove some brake fluid from the master cylinder.
2. Raise and safely support the vehicle
3. Remove the front wheel.
4. Remove the flange bolt, and pivot the caliper up out of the way.
5. Remove the pad shims and brake pads.
6. Remove the pad retainers.

➡**The upper and lower pad retainers are different. During installation, make sure the pad retainers are in proper positions.**

To install:

7. Install the pad retainers. Wipe excess assembly paste off the retainers. Keep the assembly paste off the discs and pads.
8. Mount the brake caliper piston compressor tool on the caliper body.
9. Press in the piston with the brake caliper piston compressor tool so the caliper will fit over the brake pads. Make sure the piston boot is in position to prevent damaging it when pivoting the caliper down.

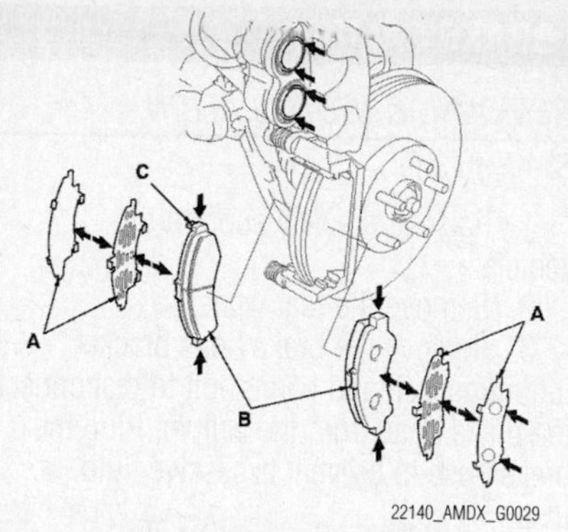
C
A
A
B
22140_AMDX_G0029

Fig. 6 Apply assembly to the pad side of the shims (A) and back of the brake pads (B). Install the pad with the wear indicator (C) on the upper side

➡**Be careful when pressing in the piston; brake fluid might overflow from the master cylinder's reservoir.**

10. Remove the brake caliper piston compressor tool.
11. Apply a thin coat of M-77 assembly paste (P/N 08798-9010) to the pad side of the shims, the back of the brake pads and the other areas indicated by the arrows. Wipe excess assembly paste off the pad shims and brake pads. Contaminated brake discs or brake pads reduce stopping ability. Keep grease and assembly paste off the brake discs and brake pads.
12. Install the brake pads and pad shims correctly. Install the brake pad with the wear indicator on the upper inside. If you are reusing the brake pads, always reinstall the brake pads in their original positions to prevent a momentary loss of braking efficiency.
13. Pivot the caliper down into position. Install the flange bolt and tighten it to 53 ft. lbs. (72 Nm).
14. Clean the mating surfaces of the brake disc and the inside of the wheel, then install the front wheels.
15. Press the brake pedal several times to make sure the brakes work.
16. Add brake fluid as needed.
17. After installation, check for leaks at hose and line joints or connections, and retighten if necessary.
18. Test-drive the vehicle, then recheck for leaks.

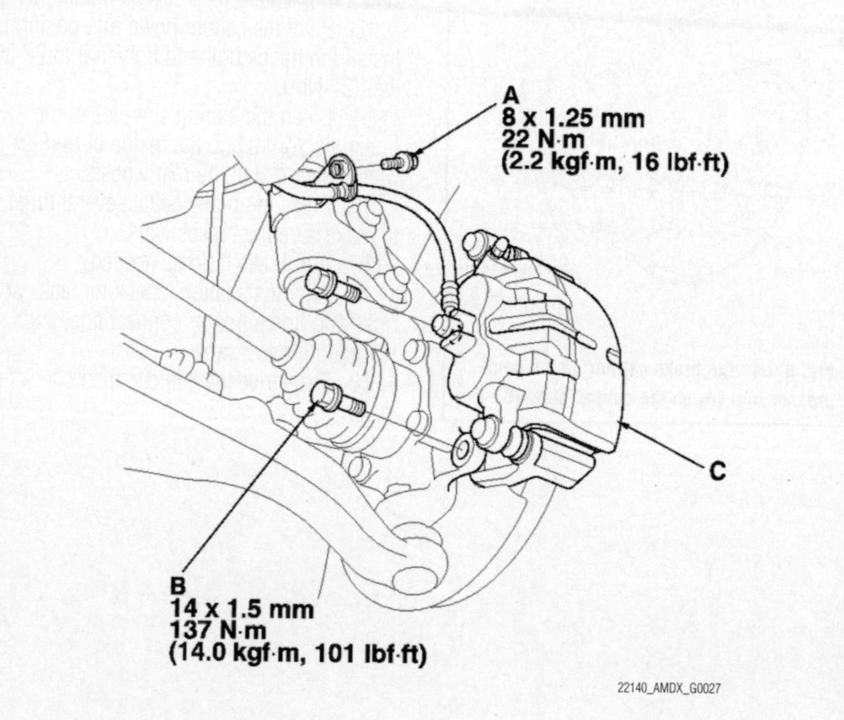
A
8 x 1.25 mm
22 N·m
(2.2 kgf·m, 16 lbf·ft)
C
B
14 x 1.5 mm
137 N·m
(14.0 kgf·m, 101 lbf·ft)
22140_AMDX_G0027

Fig. 5 Remove the caliper mounting bolts (B) to remove the caliper assembly (C) from the vehicle

BRAKE CALIPER

REMOVAL & INSTALLATION

See Figure 7.

1. Raise and safely support the vehicle.
2. Remove the rear wheel.
3. Remove the brake hose bracket mounting bolt and banjo bolt to disconnect the brake hose from the caliper. Plug the brake hose to prevent excessive fluid loss.
4. Remove the brake caliper bracket mounting bolts, then remove the caliper assembly from the knuckle.
5. Install the caliper assembly to the knuckle. Tighten the mounting bolts to 65 ft. lbs. (88 Nm).
6. Connect the brake hose using new sealing washers and tighten the banjo bolt to 25 ft. lbs. (34 Nm).
7. The remainder of the installation is the reverse order of removal.

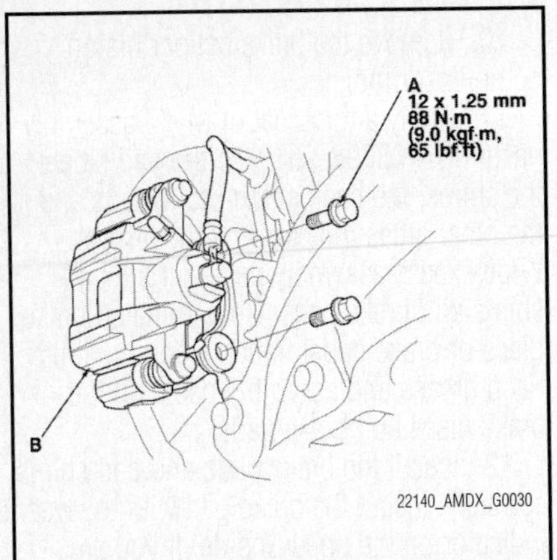
A
12 x 1.25 mm
88 N·m
(9.0 kgf·m,
65 lbf·ft)
B
22140_AMDX_G0030

Fig. 7 Remove the brake caliper mounting bolts (A) to remove the caliper assembly

DISC BRAKE PADS

REMOVAL & INSTALLATION

See Figure 8.

1. Remove some brake fluid from the master cylinder.
2. Raise and safely support the vehicle.
3. Remove the rear wheel.
4. Remove the flange bolt, and pivot the caliper up out of the way.
5. Remove the pad shims and brake pads.
6. Remove the pad retainers.

To install:

7. Install the pad retainers. Wipe excess assembly paste off the retainers. Keep the assembly paste off the discs and pads.
8. Mount the brake caliper piston compressor tool on the caliper body.
9. Press in the piston with the brake caliper piston compressor tool so the caliper will fit over the brake pads. Make sure the piston boot is in position to prevent damaging it when pivoting the caliper down.

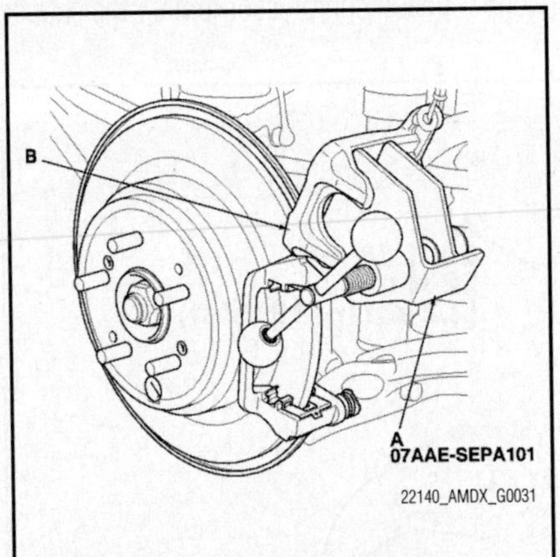
B
A
07AAE-SEPA101
22140_AMDX_G0031

Fig. 8 Use the brake caliper piston compressor tool (A) on the caliper body (B)

CAUTION

Be careful when pressing in the piston; brake fluid might overflow from the master cylinder's reservoir. If brake fluid gets on any painted surface, wash it off immediately with water.

10. Remove the brake caliper piston compressor tool.
11. Apply a thin coat of M-77 assembly paste to the pad side of the shims, the back of the brake pads, and the other areas indicated by the arrows. Wipe excess assembly paste off the pad shims and brake pads. Contaminated brake disc/drums or brake pads reduce stopping ability. Keep grease and assembly paste off the brake disc/drums and brake pads.
12. Install the brake pads and pad shims correctly. Install the brake pad with the wear indicator on the bottom inside. If you are reusing the brake pads, always reinstall the brake pads in their original positions to prevent a momentary loss of braking efficiency.
13. Pivot the caliper down into position. Install the flange bolt and tighten it to 27 ft. lbs. (37 Nm).
14. Clean the mating surfaces of the brake disc/drum and the inside of the wheel, then install the rear wheels.
15. Press the brake pedal several times to make sure the brakes work.
16. Add brake fluid as needed.
17. After installation, check for leaks at hose and line joints or connections, and retighten if necessary.
18. Test-drive the vehicle, then check for leaks.

PARKING BRAKE CABLES

ADJUSTMENT

See Figure 9.

1. Raise the rear of the vehicle, and make sure it is securely supported. Remove the rear wheels.
2. Release the parking brake, and back off the adjusting nut.
3. Remove the access plug.
4. Turn up the ratchet teeth or the adjuster assembly with a flat-tip screwdriver until the shoes lock against the drum. Then back off ten clicks, and install the access plug.
5. Do the minor adjustment procedure.
6. Install the rear wheels.

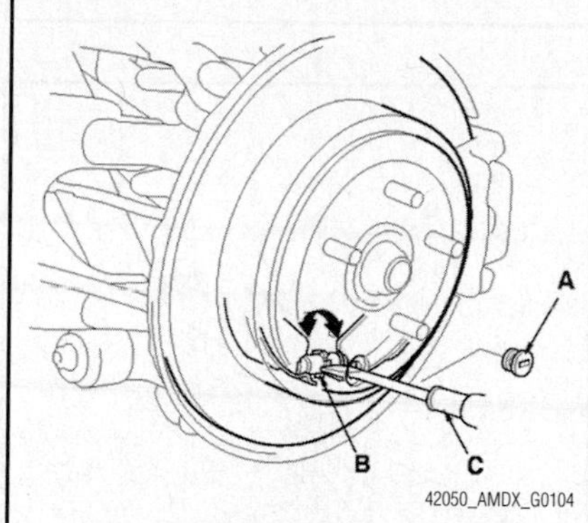
A
B
C
42050_AMDX_G0104

Fig. 9 Remove the access plug (A). Turn up the ratchet teeth (B) or the adjuster assembly with a flat-tip screwdriver (C) until the shoes lock against the drum. Then back off ten clicks, and install the access plug.

PARKING BRAKE SHOES

REMOVAL & INSTALLATION

See Figures 10 through 18.

1. Raise the rear of the vehicle, and make sure it is securely supported. Remove the rear wheels.
2. Release the parking brake, and remove the rear brake caliper and brake disc/drum.
3. Disconnect and remove the upper return springs.
4. Remove the tension pins by pushing and turning the retainers.
5. Remove the connecting rod.
6. Lower the parking brake shoe assembly.
7. Remove the forward brake shoe by removing the lower return spring and adjuster assembly.

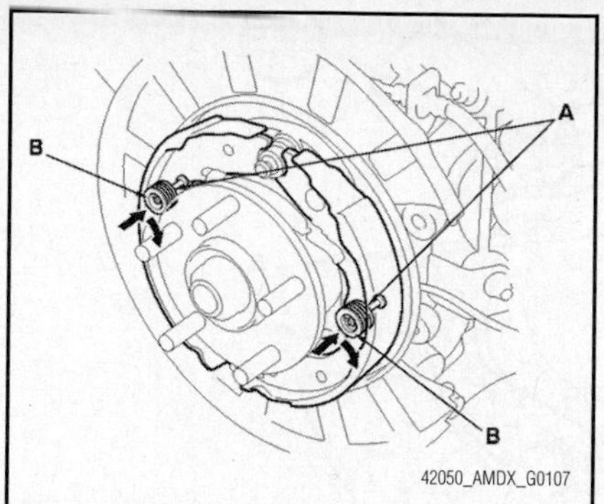
A
B
B
42050_AMDX_G0107

Fig. 10 Remove the tension pins (A) by pushing and turning the retainers (B)

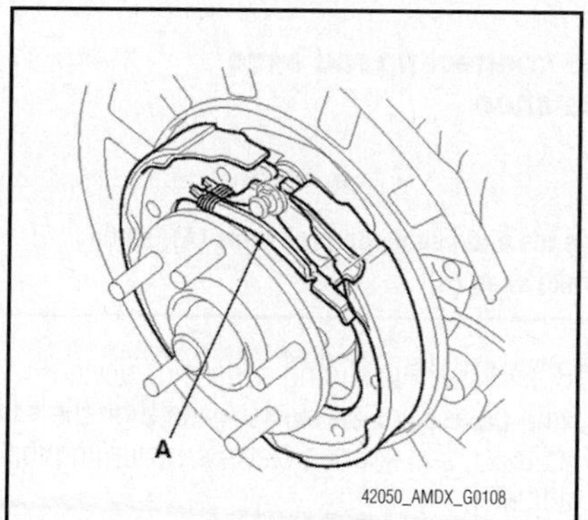
A
42050_AMDX_G0108

Fig. 11 Remove the connecting rod (A)

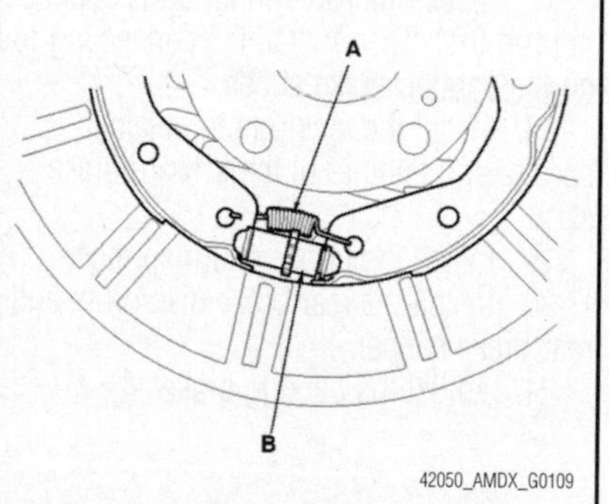
A
B
42050_AMDX_G0109

Fig. 12 Remove the forward brake shoe by removing the lower return spring (A) and adjuster assembly (B)

8. Remove the rearward brake shoe by disconnecting the parking brake cable from the parking brake lever.
9. Remove the U-clip, wave washer, and parking brake lever from the brake shoe.

To install:

10. Apply Molykote® 44 MA grease to the sliding surface of the pivot pin of the rearward brake shoe.
11. Install the parking brake lever and

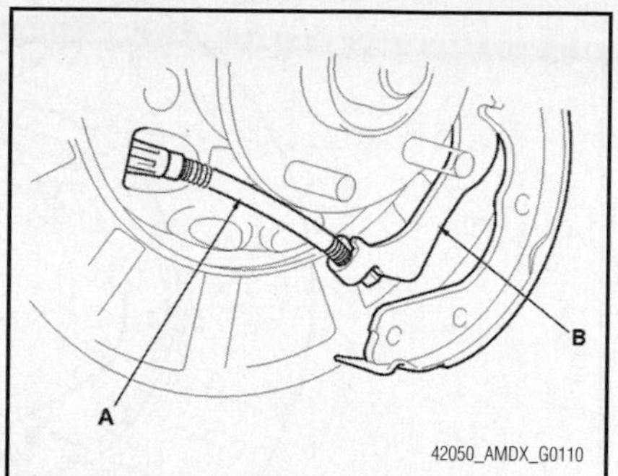
A
B
42050_AMDX_G0110

Fig. 13 Remove the rearward brake shoe by disconnecting the parking brake cable (A) from the parking brake lever (B)

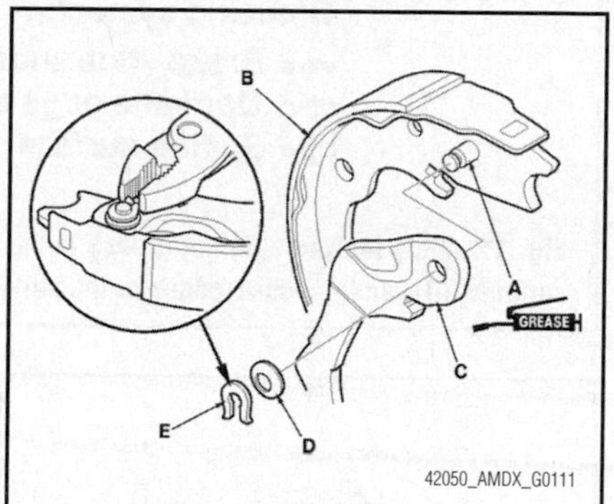
B
A
GREASE
C
E
D
42050_AMDX_G0111

Fig. 14 Apply Molykote® 44 MA grease to the sliding surface of the pivot pin (A) of the rearward brake shoe (B). Install the parking brake lever (C) and wave washer on the pivot pin, and secure with a new U-clip

wave washer on the pivot pin, and secure with a new U-clip, noting the following:

a. Install the wave washer with its convex side facing out.
b. Pinch the U-clip securely to prevent the parking brake lever from coming out from the brake shoe.

12. Connect the parking brake cable to the parking brake lever.
13. Apply Molykote® 44 MA grease to the shoe ends and connecting rod ends, sliding surfaces, and opposite edges of the parking brake shoe as shown. Wipe off any excess. Keep grease off the brake linings.
14. Install the tension pin, retainer spring, and retainer of the rearward brake shoe. Make sure the tension pin does not contact the parking brake lever.
15. Clean the threaded portions of clevis A, and coat the threads of clevis A with grease. Clean the sliding surface of clevis

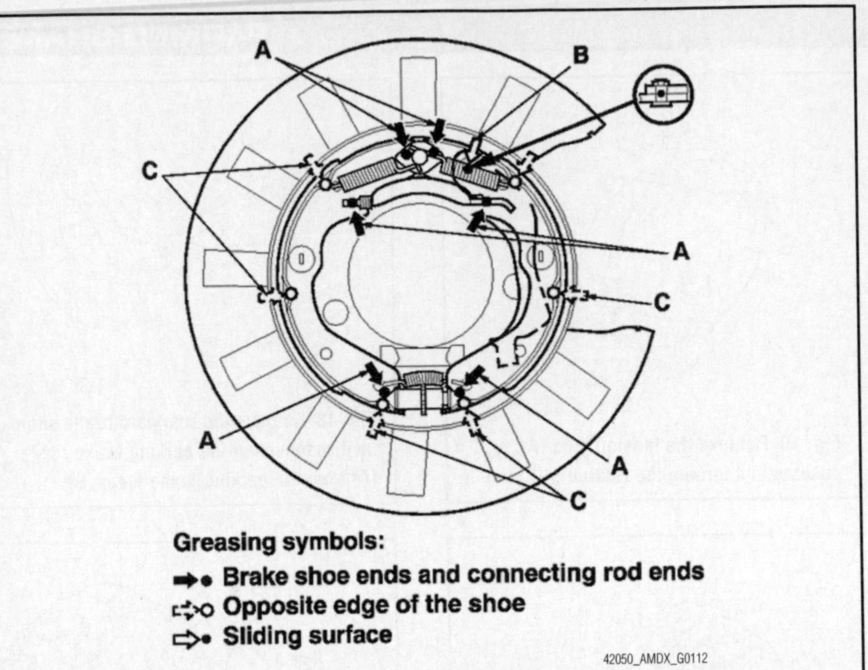
A
B
C
A
C
A
A
C
Greasing symbols:
Brake shoe ends and connecting rod ends
Opposite edge of the shoe
Sliding surface
42050_AMDX_G0112

Fig. 15 Apply Molykote 44 MA grease to the shoe ends and connecting rod ends (A), sliding surfaces (B), and opposite edges of the parking brake shoe (C)

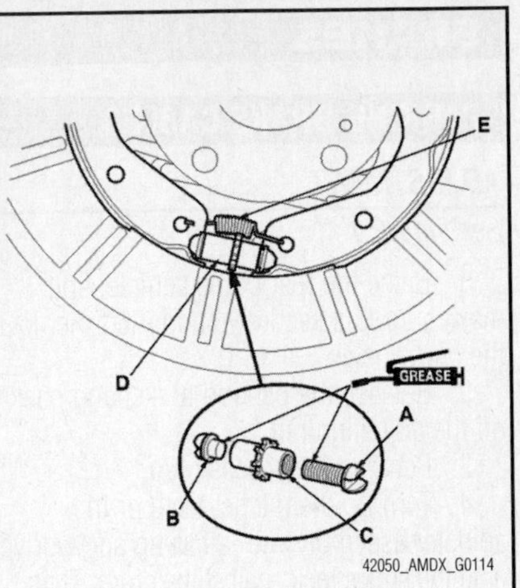
E
D
GREASE
A
B
C
42050_AMDX_G0114

Fig. 17 View of clevis (A & B), adjuster (C), adjuster assembly and lower return spring

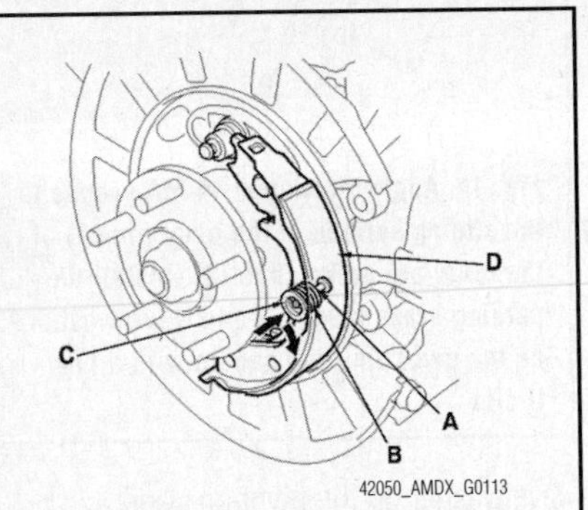
D
C
A
B
42050_AMDX_G0113

Fig. 16 Install the tension pin (A), retainer spring (B), and retainer (C) of the rearward brake shoe. Make sure the tension pin does not contact the parking brake lever

B, and coat the sliding surface of clevis B with grease. Install clevis A and B on the adjuster, and shorten clevis A by turning the adjuster.

16. Reinstall the brake shoe adjuster assembly, and hook the lower return spring on the parking brake shoes.

17. Install the rod spring to the connecting rod first. Then install the connecting rod on the parking brake shoes.

18. Install the tension pin, retainer spring, and retainer of the forward brake shoe.

19. Install the upper return springs.

20. Install the rear brake disc/drum and rear brake caliper.

21. Adjust the parking brake.

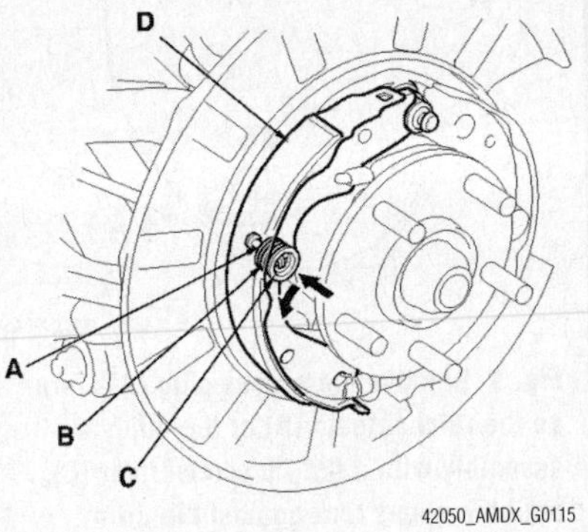
D
A
B
C
42050_AMDX_G0115

Fig. 18 Install the tension pin (A), retainer spring (B), and retainer (C) of the forward brake shoe

CHASSIS ELECTRICAL

AIR BAG (SUPPLEMENTAL RESTRAINT SYSTEM)

GENERAL INFORMATION

**** CAUTION**

These vehicles are equipped with an air bag system. The system must be disarmed before performing service on, or around, system components, the steering column, instrument panel components, wiring and sensors. Failure to follow the safety precautions and the disarming procedure could result in accidental air bag deployment, possible injury and unnecessary system repairs.

SERVICE PRECAUTIONS

Disconnect and isolate the battery negative cable before beginning any airbag system component diagnosis, testing, removal, or installation procedures. Allow system capacitor to discharge for two minutes before beginning any component service. This will disable the airbag system. Failure to disable the airbag system may result in accidental airbag deployment, personal injury, or death.

DISARMING THE SYSTEM

Disconnect and isolate the negative battery cable. Wait 3 minutes for the system capacitor to discharge before performing any service.

ARMING THE SYSTEM

Connect the negative battery cable.

CLOCKSPRING CENTERING

See Figure 19.

1. Rotate the cable reel clockwise until it stops. Then rotate it counterclockwise (about three turns) until the arrow mark on the cable reel label points straight up.

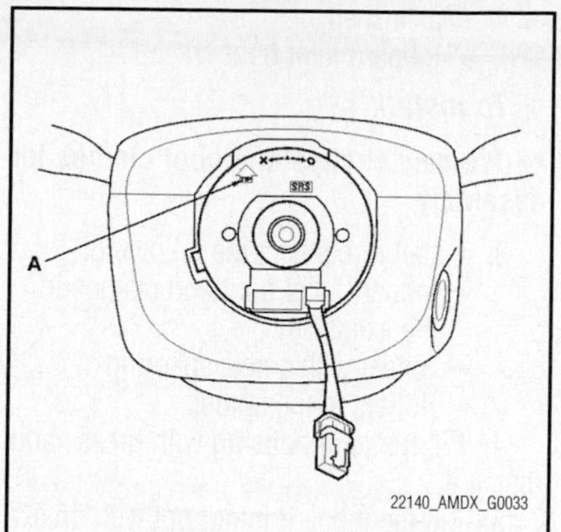
A
SRS
22140_AMDX_G0033

Fig. 19 Rotate the cable reel counterclockwise until the arrow mark (A) on the cable reel label points straight up

DRIVE TRAIN

FRONT HALFSHAFT

REMOVAL & INSTALLATION

See Figures 20 and 21.

1. Raise and safely support the vehicle.
2. Remove the front wheel
3. Pry up the locking tab on the spindle nut and remove the nut.
4. Remove the splash shield.
5. Drain the transaxle fluid.
6. Remove the suspension stroke sensor, if equipped.
7. Separate the lower ball joint from the wheel knuckle. For additional information, refer to the following section, "Lower Ball Joint, Removal & Installation."
8. Loosen the halfshaft outboard joint from the front hub using a plastic hammer.
9. Pull the wheel knuckle outward and separate the outboard joint from the front hub.
10. If removing the left halfshaft, pry the inboard joint from the differential with a prybar. Remove the halfshaft as an assembly.
11. If removing the right halfshaft, drive the inboard joint off of the intermediate shaft using a drift and a hammer. Remove the halfshaft as an assembly.

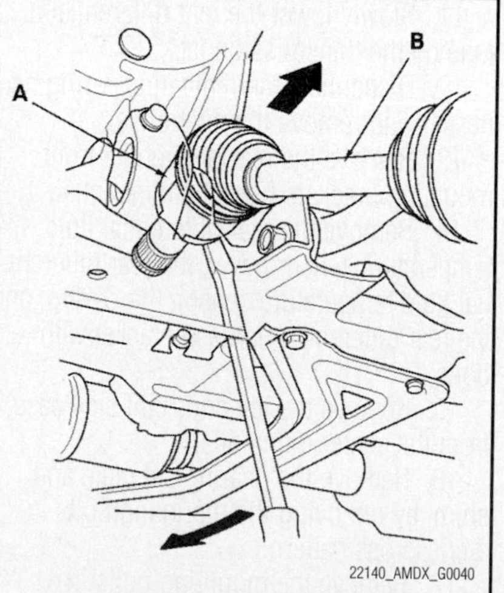
A
B
22140_AMDX_G0040

Fig. 20 To remove the left halfshaft (B), pry the inboard joint (A) from the differential with a prybar

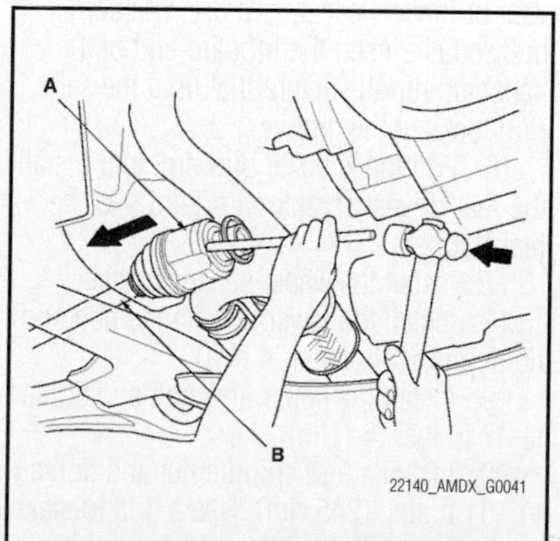
A
B
22140_AMDX_G0041

Fig. 21 If removing the right halfshaft (B), drive the inboard joint (A) off of the intermediate shaft using a drift and a hammer

To install:

➡Use new circlips and cotter pins during installation.

12. Insert the inboard end of the halfshaft into the differential (or intermediate shaft) until the set ring locks in the groove.
13. Install the outboard joint into the front hub.
14. Install the lower ball joint and tighten the nut to 76–83 ft. lbs. (103–113 Nm).
15. Install the suspension stroke sensor, if equipped.
16. Install the splash shield.
17. Install the new hub nut and tighten to 242 ft. lbs. (328 Nm) and stake the nut.
18. Install the wheel and lower the vehicle.
19. Refill the automatic transaxle to the correct level.
20. Test drive the vehicle and check for leaks.

OVERHAUL

Front

Inboard Joint

1. Before servicing the vehicle, refer to the precautions section.
2. Remove or disconnect the following:
   - Axle halfshaft from the vehicle.
   - Inboard joint boot clamps and push the boot back
   - Inboard joint housing from the axle

- Rollers from the spider
- Snapring and the spider from the axle shaft
- Inboard joint boot

To install:

➡Use new circlips and boot clamps for assembly.

3. Install or connect the following:
   - Inboard joint boot and clamps to the axle shaft
   - Spider with a new snapring
   - Rollers to the spider
4. Fill the joint housing with grease and install it.
5. Fill the inboard joint boot with grease and install the boot clamps.
6. Install the axle halfshaft to the vehicle.

Outboard Joint

1. Before servicing the vehicle, refer to the precautions section.
2. Remove or disconnect the following:
   - Axle halfshaft from the vehicle and place it in a vise
   - Outboard joint boot clamps and push the boot back
   - Outboard joint by driving it off the axle shaft with a brass drift and hammer
   - Outboard joint boot

To install:

➡Use new circlips and boot clamps for assembly.

3. Install the outboard joint boot and clamps to the axle shaft.
4. Fill the outboard joint with grease. Install the outboard joint to the axle shaft. Tap the stub shaft with a brass hammer to seat the circlip.
5. Fill the outboard joint boot with grease and install the boot clamps.
6. Install the axle halfshaft to the vehicle.

REAR HALFSHAFTS

REMOVAL & INSTALLATION

See Figure 22.

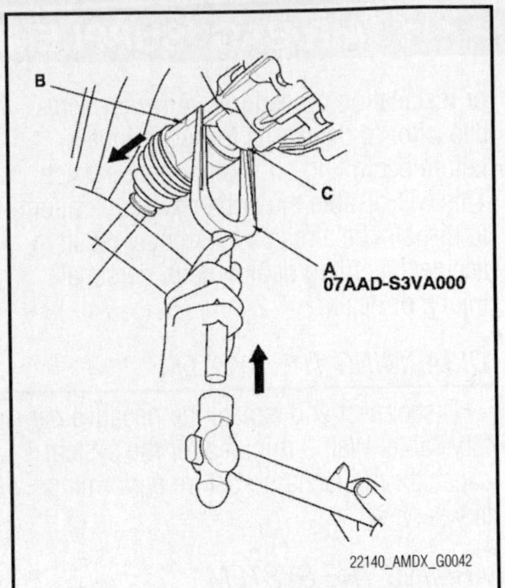
B
C
A
07AAD-S3VA000
22140_AMDX_G0042

Fig. 22 Using the driveshaft remover (A) and the hammer, pry out the inboard joint (B) from the rear differential (C)

1. Raise and safely support the vehicle.
2. Remove the rear wheel.
3. Pry up the locking tab and remove the hub nut.
4. Remove the wheel sensor and harness clip.
5. Remove the lock pin from the upper arm ball joint castle nut, and remove the nut.
6. Separate the ball joint from the upper arm with the ball joint remover.
7. Remove the lower control arm.
8. Place a transmission jack under lower arm and remove the flange bolt.
9. Loosen the rear driveshaft outboard joint from the rear hub using a plastic hammer.
10. Pull the knuckle outward, then separate the rear driveshaft outboard joint.
11. Using the driveshaft remover and the hammer, pry out the inboard joint from the rear differential.
12. Remove the rear halfshaft, then remove the set ring.

To install:

13. Apply grease to the whole splined surface of the halfshaft.
14. Install a new set ring in the set ring groove of the differential.
15. Clean the areas where the driveshaft contacts the differential thoroughly with solvent or brake cleaner, and dry with compressed air. Insert the inboard end of the halfshaft into the differential until the set ring locks in the groove.
16. Pull the knuckle outward, and install the rear driveshaft outboard joint into the rear hub.
17. Install the wheel sensor bracket.
18. Install the lower arm flange bolt and tighten to 54 ft. lbs. (74 Nm).
19. Install the upper arm bolt and tighten to 47 ft. lbs. (64 Nm).
20. Install a new spindle nut and tighten to 181 ft. lbs. (245 Nm). Use a drift to stake the spindle nut shoulder once the nut is tightened to specification.
21. Install the wheel
22. Turn the wheel to make sure there is no binding between the driveshaft and wheel.
23. Refill the differential until the fluid level is at the bottom of the fill hole. A complete oil change would require 2.79 quarts of VTM–4 differential fluid. Install the plug and tighten to 35 ft. lbs. (47 Nm).
24. Check and adjust the wheel alignment.

REAR PINION SEAL

REMOVAL & INSTALLATION

See Figures 23 through 29.

1. Drain the rear differential fluid.
2. Remove the exhaust system.
3. Remove the right rear halfshaft. For additional information, refer to the following section, "Halfshafts, Removal & Installation, Rear."
4. Matchmark the driveshaft to the rear differential companion flange. Separate the driveshaft from the rear differential.
5. Place the transmission jack under the rear differential.
6. Using the driveshaft remover and a hammer, and disconnect the left rear inboard joint from the rear differential.
7. Remove the rear differential mounting bolts.
8. Lower the rear differential a little on the transmission jack, then remove the left rear driveshaft inboard joint from the rear differential.
9. Disconnect the right solenoid wiring harness and the rear differential fluid temperature sensor wiring harness, then remove the harness clip.
10. Disconnect the breather tube from the breather pipe.
11. Slowly lower the rear differential a little on the transmission jack.
12. Disconnect the remaining wiring harnesses and remove the differential.
13. Remove the rear differential front mounting brackets from the differential.
14. Remove the rear differential fluid temperature sensor cover, the rear differential fluid temperature sensor, the O-ring, and the rear differential harness bracket with bolts.
15. Remove the left and right side case from the center differential.
16. Remove the bearing set plate and shim, by removing the mounting bolts in a crisscross pattern.
17. Remove the mounting bolts in a crisscross pattern and remove the differential housing assembly.
18. Remove the ring gear assembly, the

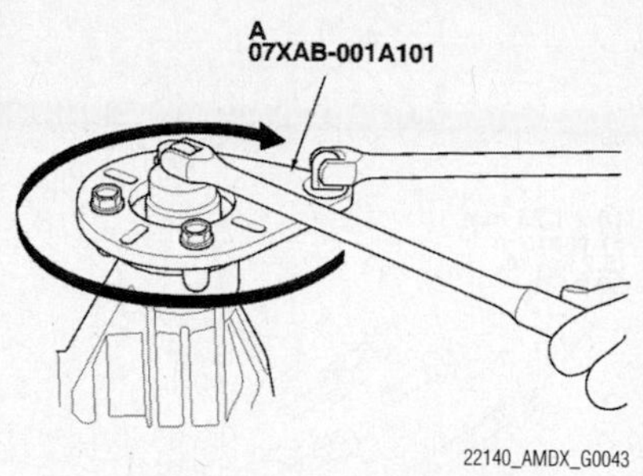
A
07XAB-001A101
22140_AMDX_G0043

Fig. 23 Install the companion flange holder (A) on the companion flange to loosen the locknut

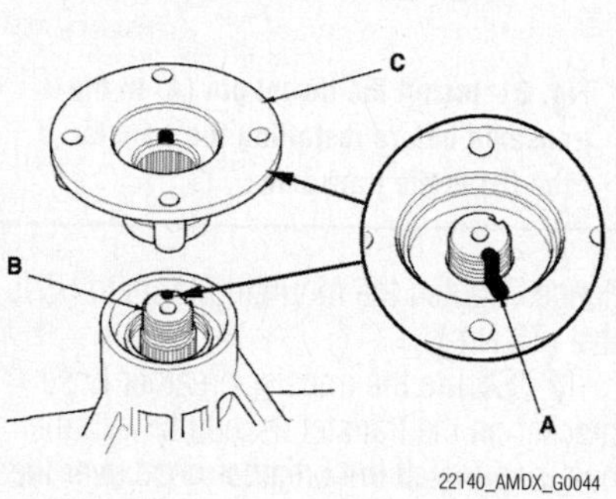
C
B
A
22140_AMDX_G0044

Fig. 24 Make a reference mark (A) across the input shaft (B) and companion flange (C), before removing the companion flange.

bearing outer races, the 75 mm shim, and the 8 x 14 mm dowel pins.

19. Raise the locknut tab from the groove of the input shaft, making sure that the tab completely clears the groove to prevent damaging the input shaft.

20. Install the companion flange holder on the companion flange. Loosen the 27 mm locknut (left-hand threads).

21. Loosen the locknut clockwise so that its tab comes out from the groove in the input shaft.

22. Tighten the locknut until its tab aligns with the groove.

23. Remove any dirt from inside of the groove in the input shaft, then loosen the locknut.

24. Remove the 27 mm locknut, the 27 mm spring washer, the 28 mm back-up ring, and the 28 mm O-ring.

25. Make a reference mark across the input shaft and companion flange, then remove the companion flange.

26. Remove the drive pinion, pinion spacer, and the thrust washer by tapping on the drive pinion with a plastic hammer.

27. Remove the pinion oil seal with a commercially available seal remover.

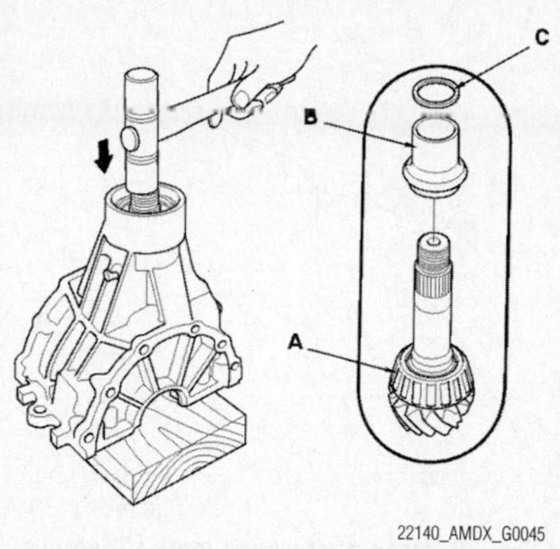
C
B
A
22140_AMDX_G0045

Fig. 25 Remove the drive pinion (A), pinion spacer (B), and the thrust washer (C) by tapping on the drive pinion with a plastic hammer.

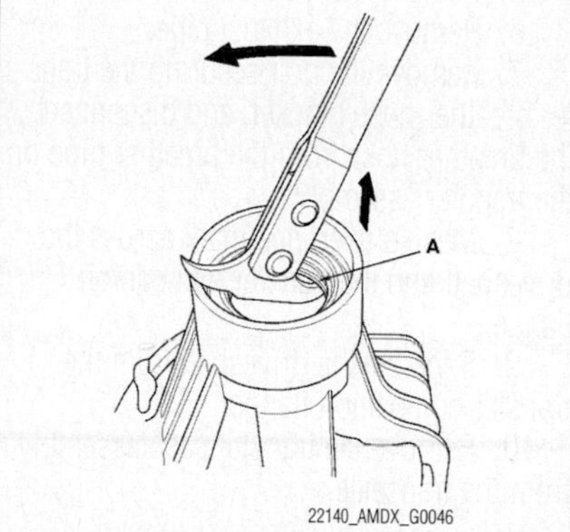
A
22140_AMDX_G0046

Fig. 26 Remove the front case oil seal (A) with a commercially available seal remover.

To install:

28. Apply ATF to the tapered roller bearing, then install the new front case oil seal with the dust seal driver and the 39 x 47.3 mm fork seal driver weight.

29. Apply ATF to the tapered roller bearing, then install the drive pinion, pinion spacer, and the thrust washer into the differential carrier.

30. Align the match mark and install the companion flange on the input shaft.

31. Apply ATF to the new 28 mm O-ring, then install the new 28 mm O-ring, the 28 mm backup ring, the 27 mm spring washer, and the new 27 mm locknut.

32. Install the companion flange holder to the companion flange, then tighten the locknut (left-hand threads) to 53 ft. lbs. (72 Nm).

33. Rotate the drive pinion several times to assure proper tapered roller bearing contact. Measure the drive pinion turning torque. Torque should be 69.0–134.0 ft. lbs. (93.2–181.4 Nm).

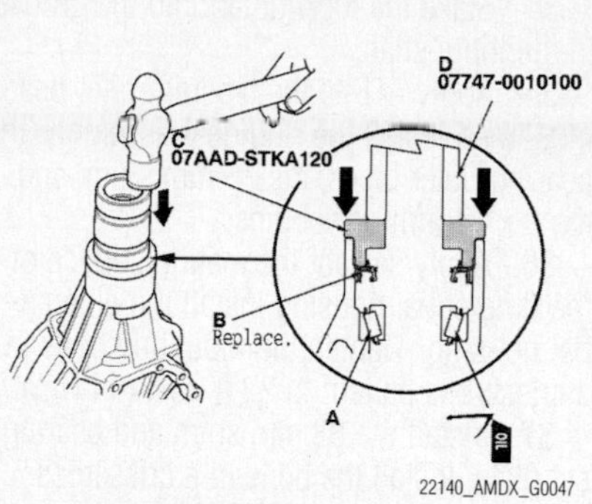
D
07747-0010100
C
07AAD-STKA120
B
Replace.
A
22140_AMDX_G0047

Fig. 27 Apply ATF to the tapered roller bearing (A), then install the new front case oil seal (B) with the dust seal driver (C) and the 39 x 47.3 mm fork seal driver weight.

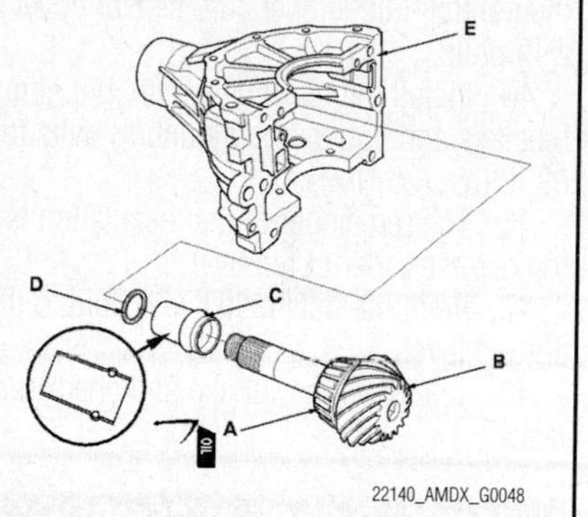
E
D
C
B
A
22140_AMDX_G0048

Fig. 28 Apply ATF to the tapered roller bearing (A), then install the drive pinion (B), pinion spacer (C), and the thrust washer into the differential carrier

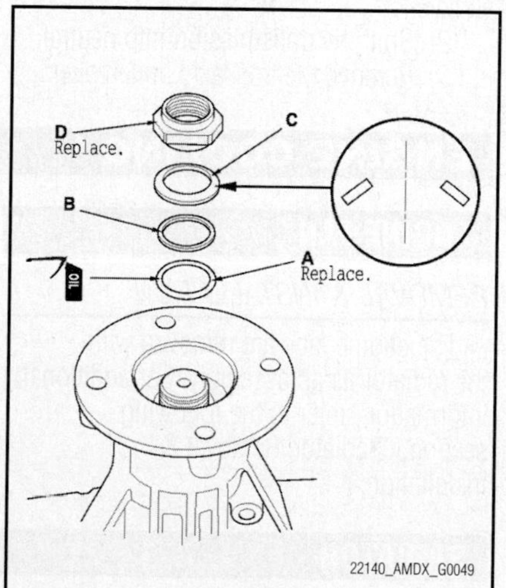
D
Replace.
C
B
A
Replace.
22140_AMDX_G0049

Fig. 29 Apply ATF to the new 28 mm O-ring (A), then install the new 28 mm O-ring, the 28 mm backup ring (B), the 27 mm spring washer (C), and the new 27 mm locknut

34. Stake the locknut tab into the groove in the input shaft.

35. Apply ATF to the tapered roller bearings, then install the ring gear assembly, the bearing outer races, the 75 mm shim, and the 8 x 14 mm dowel pins.

36. Apply sealant the mating surface of the differential housing install the differential housing. Tighten the mounting bolts in a crisscross pattern to 32 ft. lbs. (44 Nm).

37. Install the 83 mm shim and bearing set plate. Install the bolts in a crisscross pattern to 20 ft. lbs. (27 Nm).

38. Apply sealant the mating surfaces of the left and right cases. Install the left and right cases and tighten the mounting bolts in a crisscross pattern to 20 ft. lbs. (27 Nm).

39. Install the rear differential fluid temperature sensor, a new O-ring, the rear differential harness bracket, and the rear differential fluid temperature sensor cover with bolts.

40. Install the rear differential mounting brackets and tighten the mounting bolts to 63 ft. lbs. (85 Nm).

41. The remainder of the installation is the reverse order of removal.

42. Refill the differential with fluid to the correct level.

43. Test drive the vehicle and check for leaks.

TRANSFER CASE ASSEMBLY

REMOVAL & INSTALLATION

See Figures 30 and 31.

1. Raise and safely support the vehicle.
2. Shift the transmission into neutral.
3. Remove the transaxle undercover.

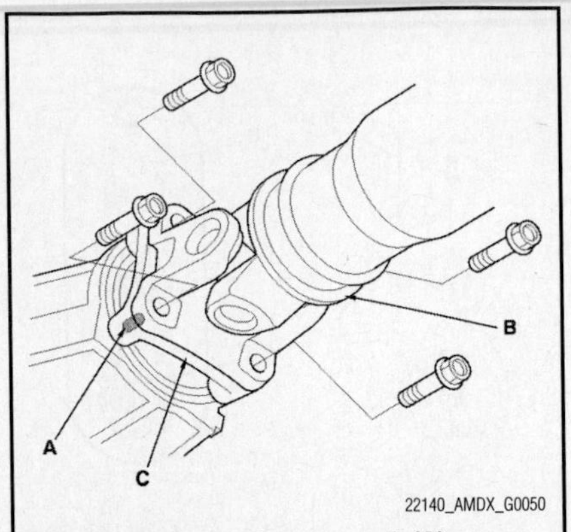
A
B
C
22140_AMDX_G0050

Fig. 30 Make a reference mark (A) across the driveshaft (B) and the transfer companion flange (C) before separating

4. Drain the transaxle fluid and replace the drain plug.
5. Remove the front subframe stiffener.
6. Remove the exhaust pipe.
7. Remove the bolt securing the transfer breather hose bracket, and disconnect the breather hose from the breather pipe on the transfer assembly.
8. Make a reference mark across the driveshaft and the transfer companion flange.
9. Separate the driveshaft from the transfer companion flange.
10. Remove the transfer case assembly from the transaxle.

To install:

11. Install the dowel pin in the transaxle, and install the transfer case assembly on the transaxle. Tighten the mounting bolts to 38 ft. lbs. (51 Nm).
12. Align the matchmarks and install the driveshaft to the transfer case companion flange. Tighten the mounting bolts to 53 ft. lbs. (72 Nm).

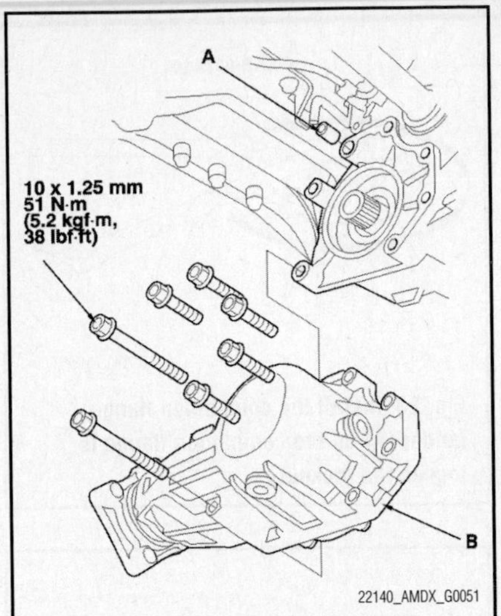
A
10 x 1.25 mm
51 N·m
(5.2 kgf·m,
38 lbf·ft)
B
22140_AMDX_G0051

Fig. 31 Install the dowel pin (A) in the transaxle before installing the transfer case (B) to the transaxle

13. Secure the transfer breather hose bracket on the transfer assembly with the bolt, and install the breather hose over the breather pipe with the dot facing out.
14. Install the exhaust pipe, using new nuts and gaskets.
15. Install the front subframe stiffener with new mounting bolts. Tighten the mounting bolts to 40 ft. lbs. (54 Nm).
16. Refill the transfer case with hypoid gear oil to the correct level.
17. Refill the transaxle with fluid to the correct level.
18. Install the transaxle undercover and tighten the mounting bolts to 87 inch lbs. (9.8 Nm).

ENGINE COOLING

ENGINE FAN

REMOVAL & INSTALLATION

The engine fans are removed with the radiator as an assembly. For additional information, refer to the following section, "Radiator, Removal & Installation."

RADIATOR

REMOVAL & INSTALLATION

See Figure 32.

1. Make sure you have the anti-theft code for the radio and the navigation system, then write down the frequencies for the radio's preset buttons.

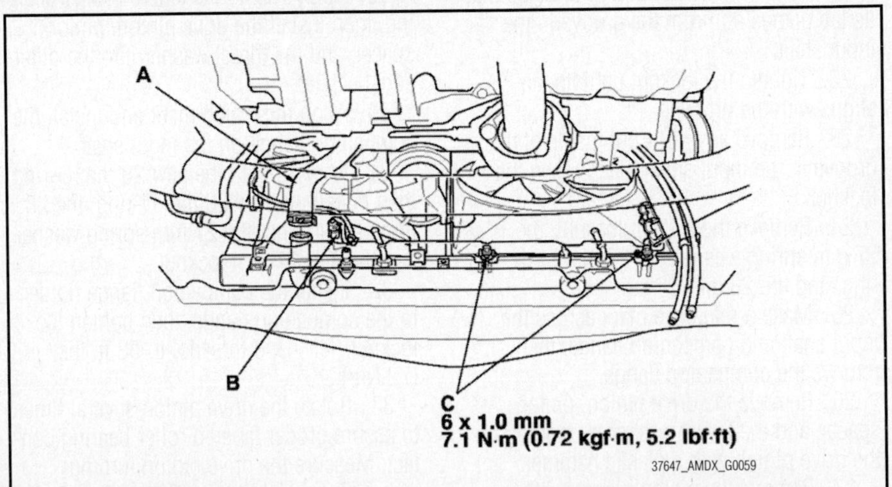
A
B
C
6 x 1.0 mm
7.1 N·m (0.72 kgf·m, 5.2 lbf·ft)
37647_AMDX_G0059

Fig. 32 Remove the lower radiator hose (A) and disconnect the ECT sensor 2 connector (B)

2. Disconnect the negative cable from the battery first, then disconnect the positive cable. Wait at least 3 minutes before proceeding.
3. Remove the battery and battery tray.
4. Remove the air intake assembly.
5. Remove the front grille cover, then remove the front grille.
6. Disconnect the fan motor connectors, harness clamp, Engine Coolant Temperature (ECT) sensor 2 sub-harness connector and coolant reservoir hose.
7. Drain the engine coolant.
8. Remove the upper radiator hose.
9. Unclamp the hose clamps and remove the ATF cooler hoses, then plug the line and hoses.
10. Remove the ATF cooler pipe mount bolts, then remove the ATF cooler pipe.
11. Remove the lower radiator hose and disconnect the ECT sensor 2 connector and loosen the two bolts.
12. Disconnect the hood latch switch connector, clamps, then remove the hood latch.
13. Remove the bulkhead bracket, and radiator and A/C condenser mount upper bracket/cushion, then remove the fan shroud mount bolts.
14. Move the fan shroud assemblies toward the engine, then pull up the radiator assembly.
15. Remove the fan shroud assemblies.

To install:

16. Install the radiator and fans in the reverse order of removal.
17. Install the bulkhead bracket in the reverse order of removal. Apply body paint to the bulkhead bracket mounting bolts.
18. Set the upper and lower cushions securely.
19. Fill the transmission with ATF.
20. Install the battery.
21. Do the battery terminal reconnection procedure.
22. Fill the radiator with engine coolant and bleed the air from the system.
23. Adjust the engine hood latch.

THERMOSTAT

REMOVAL & INSTALLATION

See Figure 33.

➡Make sure you have the anti-theft codes for the radio and navigation system, then write down the audio presets. Make sure the ignition switch is OFF.

1. Disconnect the negative cable from the battery first, then the positive cable.

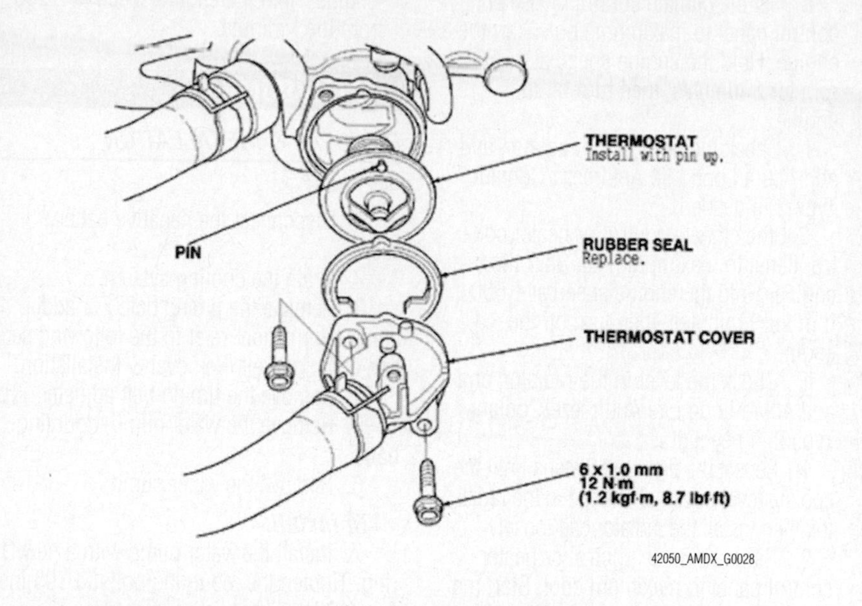
THERMOSTAT
Install with pin up.
PIN
RUBBER SEAL
Replace.
THERMOSTAT COVER
6 x 1.0 mm
12 N·m
(1.2 kgf·m, 8.7 lbf·ft)
42050_AMDX_G0028

Fig. 33 Exploded view of the thermostat and related components

2. Remove the battery.
3. Drain the engine coolant as follows:
   a. Turn the ignition switch ON (II). Set the displayed temperature on the climate control system to 90°F (32°C), then turn the ignition switch OFF. Make sure the engine and radiator are cool to the touch.
   b. Remove the radiator cap.
   c. Loosen the drain plug, and drain the coolant.
   d. Install a rubber hose on the drain bolt located at the rear of the cylinder block, then loosen the drain bolt.
   e. When the coolant stops draining, tighten the drain bolt.
   f. Tighten the radiator drain plug securely.
   g. Remove, drain, and reinstall the coolant reservoir.
4. Remove the thermostat cover, then remove the thermostat.

To install:

5. Install the thermostat with a new rubber seal.
6. Install the battery. Clean the battery posts and cable terminals with sandpaper, then assemble them and apply grease to prevent corrosion.
7. Fill the radiator with engine coolant and bleed the air from the system, as follows:
   a. Pour Acura Long Life Antifreeze/Coolant Type 2 (P/N 0L999-9001) into the radiator up to the base of the filler neck.

➡Always use Acura Long Life Antifreeze/Coolant Type 2 (P/N OL999-9001). Using a non-Honda coolant can result in corrosion, causing the cooling system to malfunction or fail. Acura Long Life Antifreeze/Coolant Type 2 is a mixture of 50 % antifreeze and 50 % water. Do not add water.

➡The engine coolant capacities (including the reserve tank capacity of 0.6L (0.16 US gal)): After coolant change: 7.1L (1.88 US gal). After engine overhaul: 9.0L (2.38 US gal).

   b. Start the engine. Hold the engine speed at 1,500 rpm until it warms up (the radiator fan comes on at least twice). Make sure the thermostat is open.
   c. Turn off the engine. Check the level in the radiator, and add Acura Long Life Antifreeze/Coolant Type 2, if needed.
   d. Set the climate control or heater control panel to maximum cool. Start the engine. Hold the engine speed at 1,500 rpm for 5 minutes, then turn off the engine.
   e. Check the level in the radiator, and add Acura Long Life Antifreeze/Coolant Type 2, if needed.
   f. Set the climate control or heater control panel to maximum heat. Start the engine. Hold the engine speed at 1,500 rpm for 5 minutes, then turn off the engine.
   g. Check the level in the radiator, and add Acura Long Life Antifreeze/Coolant Type 2, If needed.

h. Set the climate control or heater control panel to maximum cool. Start the engine. Hold the engine speed at 1,500 rpm for 3 minutes, then turn off the engine.

i. Check the level in the radiator, and add Acura Long Life Antifreeze/Coolant Type 2, if needed.

Set the climate control or heater control panel to maximum heat. Start the engine. Hold the engine speed at 1,500 rpm for 3 minutes, then turn off the engine.

j. Check the level in the radiator, and add Acura Long Life Antifreeze/Coolant Type 2, if needed.

k. Repeat the previous 3 steps until the coolant level does not change in the radiator, then install the radiator cap loosely.

l. Set the climate control or heater control panel to maximum cool. Start the engine. Hold the engine speed at 2,500 rpm for 1 minute.

WATER PUMP

REMOVAL & INSTALLATION

See Figure 34.

1. Disconnect the negative battery cable.
2. Drain the cooling system.
3. Remove the timing belt. For additional information, refer to the following section, "Timing Belt, Removal & Installation."
4. Remove the timing belt adjuster.
5. Remove the water pump mounting bolts.
6. Remove the water pump.

To install:

7. Install the water pump with a new O-ring. Tighten the mounting bolts to 105 inch lbs. (12 Nm).

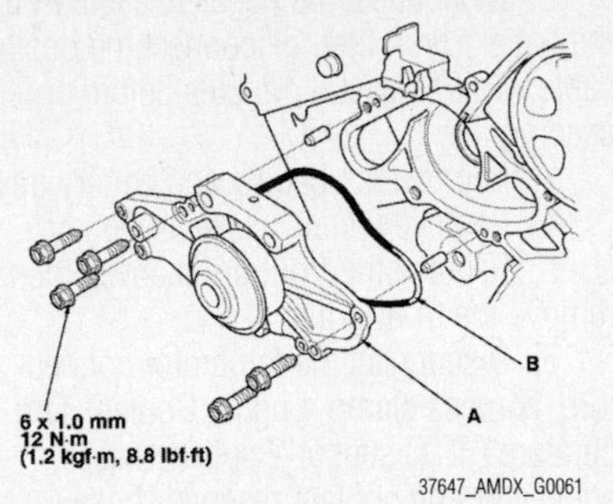
B
A
6 x 1.0 mm
12 N·m
(1.2 kgf·m, 8.8 lbf·ft)
37647_AMDX_G0061

Fig. 34 Exploded view of water pump mounting assembly

8. The remainder of the installation is the reverse order of removal.
9. Refill the cooling system to the correct level.
10. Start the engine and check for leaks.

ENGINE ELECTRICAL

CHARGING SYSTEM

ALTERNATOR

REMOVAL & INSTALLATION

See Figure 35.

1. Make sure you have the anti-theft code for the radio and the navigation system, then write down the frequencies for the radio's preset buttons.
2. Disconnect the negative cable from the battery first, then disconnect the positive cable. Wait at least 3 minutes before proceeding.
3. Remove the engine appearance cover.
4. Remove the accessory drive belt.
5. Remove the A/C suction line from the brackets.
6. Remove the coolant reservoir, then remove power steering fluid reservoir from the bracket.
7. Remove the harness bracket, then disconnect the A/C compressor clutch wiring harness.
8. Disconnect the alternator wiring harness and the BLK wire from the alternator.

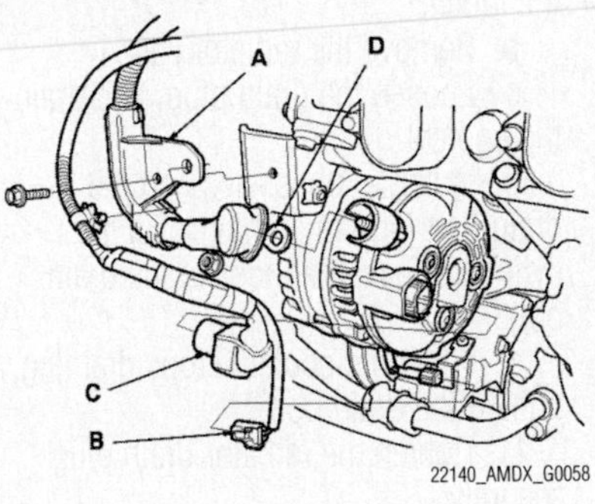
A
D
C
B
22140_AMDX_G0058

Fig. 35 Remove the following connectors to remove the alternator

9. Remove the mounting bolt, the bracket mounting bolt, then remove the alternator.

To install:

10. Install the alternator. Tighten the mounting bolt to 33 ft. lbs. (45 Nm) and bracket mounting bolt to 16 ft. lbs. (22 Nm).
11. Connect the alternator wiring harness and BLK wire to the alternator.
12. Install the harness bracket, then connect the A/C compressor clutch wiring harness
13. Install the power steering fluid reservoir to the bracket, then install the coolant reservoir.
14. Install the A/C suction line to the brackets.
15. Install the accessory drive belt.
16. Install the engine appearance cover.
17. Connect the negative battery cable.

ENGINE ELECTRICAL

IGNITION SYSTEM

FIRING ORDER

The firing order is 1–4–2–5–3–6.

IGNITION COIL

REMOVAL & INSTALLATION

See Figure 36.

1. Disconnect the negative battery cable.
2. Remove the engine appearance cover.
3. Disconnect the ignition coil wiring harness.

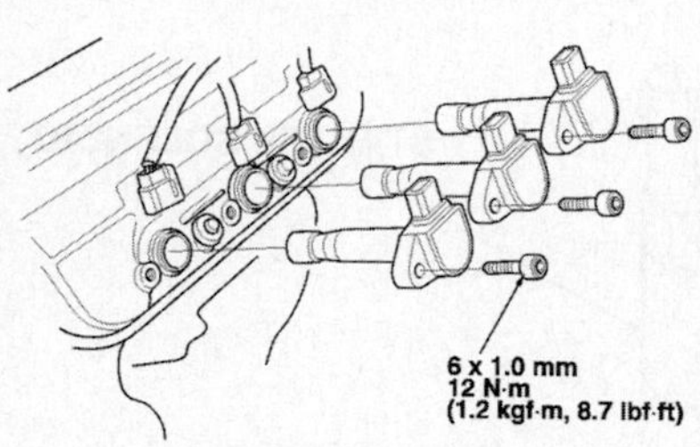
6 x 1.0 mm
12 N·m
(1.2 kgf·m, 8.7 lbf·ft)
42050_AMDX_G0004

Fig. 36 Disconnect the ignition coil connector (A), then remove the ignition coils (B)

4. Remove the ignition coil.

To install:

5. Installation is the reverse order of removal.
6. Tighten the ignition coil mounting nut to 108 inch lbs. (12 Nm).

IGNITION TIMING

INSPECTION

1. Connect the Honda Diagnostic System (HDS) to the data link connector (DLC).
2. Turn the ignition switch to ON (II).
3. Make sure the HDS communicates with the vehicle and the Powertrain Control Module (PCM). If it doesn't communicate, troubleshoot the DLC circuit.
4. Check for DTCs. If a DTC is present, diagnose and repair the cause before inspecting the ignition timing.
5. Start the engine. Hold the engine at 3,000 rpm with no load (in P or N) until the radiator fan comes on, then let it idle.
6. Check the idle speed.
7. Jump the SCS line with the HDS.
8. Connect the timing light to the service loop.
9. Aim the light toward the pointer on the timing belt cover. Check the ignition timing under a no load condition (headlights, blower fan, rear window defogger, and air conditioner are turned off).

ADJUSTMENT

Ignition timing is control by the Powertrain Control Module (PCM) and cannot be adjusted. If timing is out of specification and the rest of the ignition system work properly, the PCM must be replaced.

SPARK PLUGS

REMOVAL & INSTALLATION

1. Disconnect the negative battery cable.
2. Remove the ignition coil. For additional information, refer to the following section, "Ignition Coil, Removal & Installation."
3. Remove the spark plug.
4. Inspect the spark plug.

To install:

5. Install the spark plug and tighten to 13 ft. lbs. (18 Nm).
6. Install the ignition coil.
7. Connect the negative battery cable.

ENGINE ELECTRICAL

STARTING SYSTEM

STARTER

REMOVAL & INSTALLATION

See Figure 37.

1. Make sure you have the anti-theft codes for the radio, and the navigation system, then write down the audio presets. Make sure the ignition switch is OFF.
2. Disconnect the negative cable from the battery first, then disconnect the positive cable. Wait at least 3 minutes before proceeding.
3. Remove the air intake assembly.
4. Remove the battery and battery tray.
5. Remove the starter wiring harness clamp.
6. Disconnect the positive starter cable from the B terminal. Disconnect the wiring harness from the S terminal.
7. Remove the starter mounting bolts and remove the starter.

To install:

8. Install the starter using a new gasket, then install the harness clamp, and positive starter cable and S terminal connector.

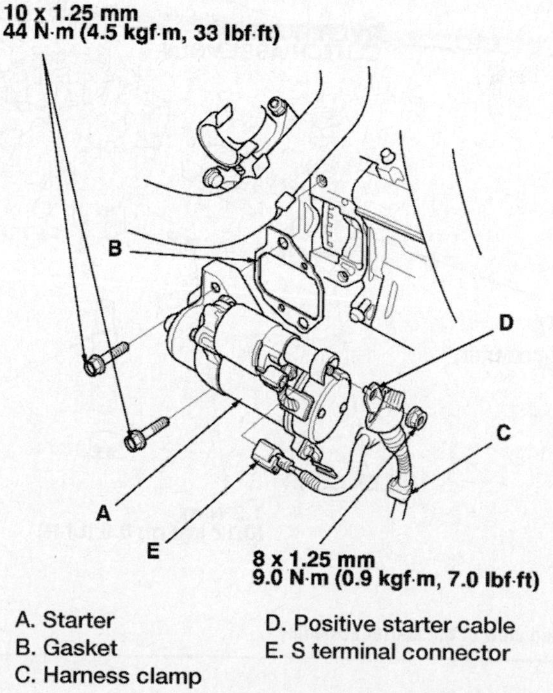
10 x 1.25 mm
44 N·m (4.5 kgf·m, 33 lbf·ft)
B
D
C
A
E
8 x 1.25 mm
9.0 N·m (0.9 kgf·m, 7.0 lbf·ft)
A. Starter
B. Gasket
C. Harness clamp
D. Positive starter cable
E. S terminal connector
37647_AMDX_G0065

Fig. 37 Install the starter

Make sure the crimped side of the B terminal is facing out, away from the starter.

9. Install the battery base and battery.
10. Install the air cleaner assembly.
11. Do the battery terminal reconnection procedure.
12. Start the engine to make sure the starter works properly.

SOLENOID REPLACEMENT

See Figure 38.

1. Remove the starter. For additional information, refer to the following section, "Starter, Removal or Installation."
2. Disassemble the starter as shown to remove the solenoid.
3. Installation is the reverse order of disassembly.

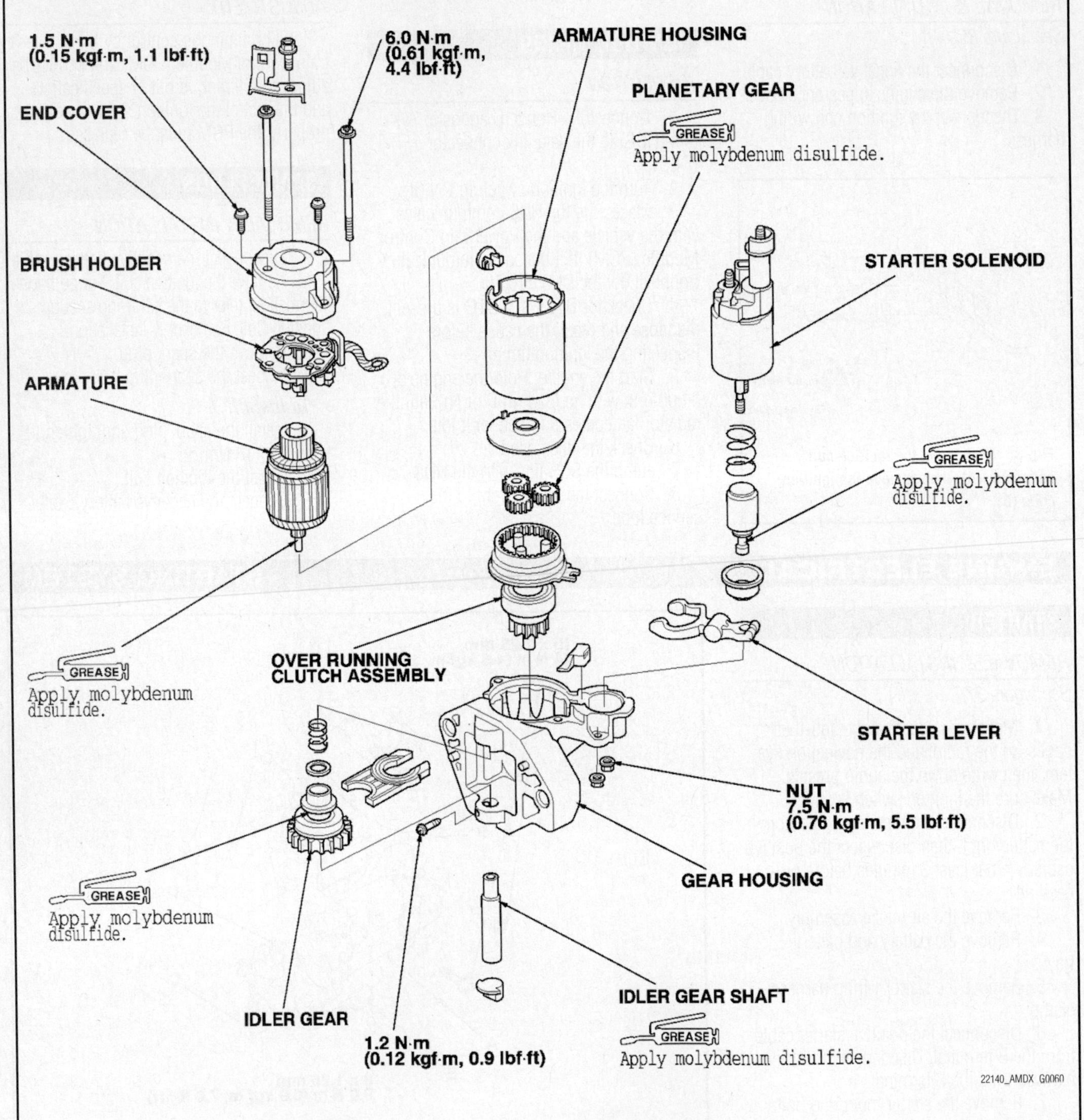
1.5 N·m
(0.15 kgf·m, 1.1 lbf·ft)
6.0 N·m
(0.61 kgf·m,
4.4 lbf·ft)
ARMATURE HOUSING
PLANETARY GEAR
GREASE
Apply molybdenum disulfide.
END COVER
BRUSH HOLDER
STARTER SOLENOID
ARMATURE
GREASE
Apply molybdenum disulfide.
GREASE
Apply molybdenum disulfide.
OVER RUNNING CLUTCH ASSEMBLY
STARTER LEVER
NUT
7.5 N·m
(0.76 kgf·m, 5.5 lbf·ft)
GEAR HOUSING
GREASE
Apply molybdenum disulfide.
IDLER GEAR SHAFT
IDLER GEAR
1.2 N·m
(0.12 kgf·m, 0.9 lbf·ft)
GREASE
Apply molybdenum disulfide.
22140_AMDX G0060

Fig. 38 Exploded view of the starter assembly

ENGINE MECHANICAL

➡Disconnecting the negative battery cable may interfere with the functions of the on board computer systems and may require the computer to undergo a relearning process, once the negative battery cable is reconnected.

ACCESSORY DRIVE BELTS

ACCESSORY BELT ROUTING

See Figure 39.

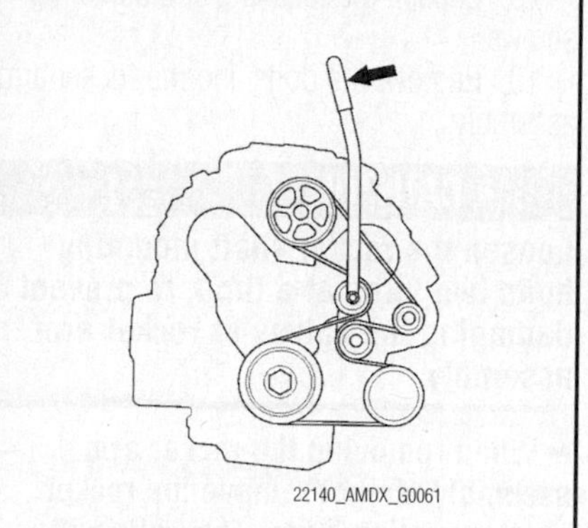
22140_AMDX_G0061

Fig. 39 Accessory drive belt routing

INSPECTION

See Figure 40.

1. Inspect the belt for cracks and damage. If the belt is cracked or damaged, replace it.
2. Check that the auto-tensioner indicator is within the standard range. If it is out of the standard range, replace the drive belt.

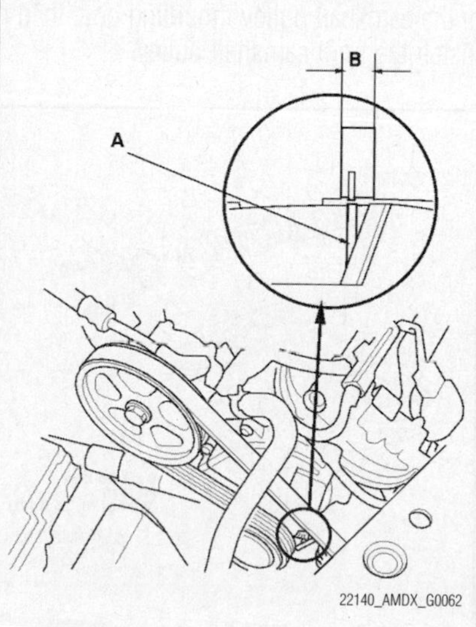
B
A
22140_AMDX_G0062

Fig. 40 Check that the auto-tensioner indicator (A) is within the standard range (B)

ADJUSTMENT

Belt tension is automatically maintained by a belt tensioner. No adjustments are necessary.

REMOVAL & INSTALLATION

See Figure 41.

➡For this procedure, you will need a Belt tension release tool (Snap-on YA9317) or equivalent.

1. Move the auto-tensioner with the belt tension release tool to relieve tension from the drive belt, then remove the drive belt.
2. Install the new belt in the reverse order of removal.

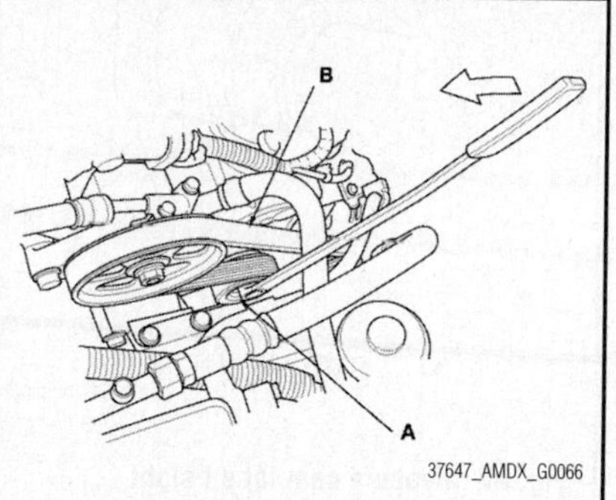
B
A
37647_AMDX_G0066

Fig. 41 Move the auto-tensioner with the belt tension release tool

CAMSHAFT AND VALVE LIFTERS

INSPECTION

See Figures 42 through 47.

1. Remove the cylinder head.
2. Remove the rocker arms.
3. Put the rocker shafts on the cylinder head, then tighten the bolts to the specified torque, as follows:
   a. Apply engine oil to the bolt threads and flange.
   b. Tighten the 8 x 1.255mm bolts to 17 ft. lbs. (24 Nm).
4. Seat the camshaft by pushing it toward the rear of the cylinder head.
5. Zero the dial indicator against the end of the camshaft. Push the camshaft back and forth and read the end play. If the end play is beyond the service limit, replace the thrust cover and recheck. If it is still beyond the service limit, replace the camshaft:

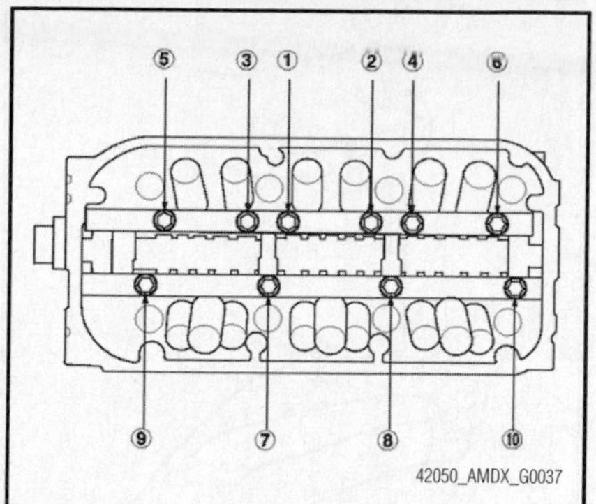
⑤ ③ ① ② ④ ⑥
⑨ ⑦ ⑧ ⑩
42050_AMDX_G0037

Fig. 42 Rocker shaft bolt tightening sequence

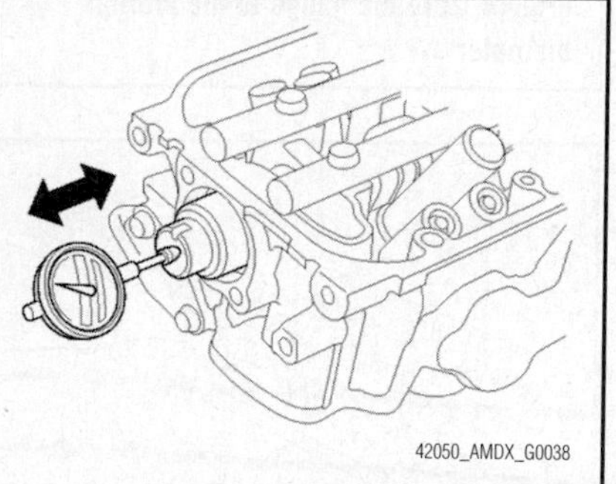
42050_AMDX_G0038

Fig. 43 Zero the dial indicator against the end of the camshaft. Push the camshaft back and forth and read the end play

   a. Standard (new): 0.002–0.008 inches (0.05–0.20mm)
   b. Service Limit: 0.008 inches (0.20mm)
6. Remove the camshaft thrust cover, then pull out the camshaft.
7. Wipe the camshaft clean, then inspect the lift ramps. Replace the camshaft if any lobes are pitted, scored or excessively worn.
8. Use a micrometer to measure the diameter of each camshaft journal.
9. Zero the gauge to the journal diameter.
10. Clean the camshaft bearing surfaces in the cylinder head. Measure the inside diameter of each camshaft bearing surface, and check for an out-of-round condition:
    - If the camshaft-to-holder clearance is within limits of 0.0020–0.0035 inches (0.050–0.089mm), go to step 11.
    - If the camshaft-to-holder clearance is beyond the service limit of 0.006 inches (0.15mm), and the camshaft has been replaced, replace the cylinder head.
    - If the camshaft-to-holder clearance is beyond the service limit of 0.006

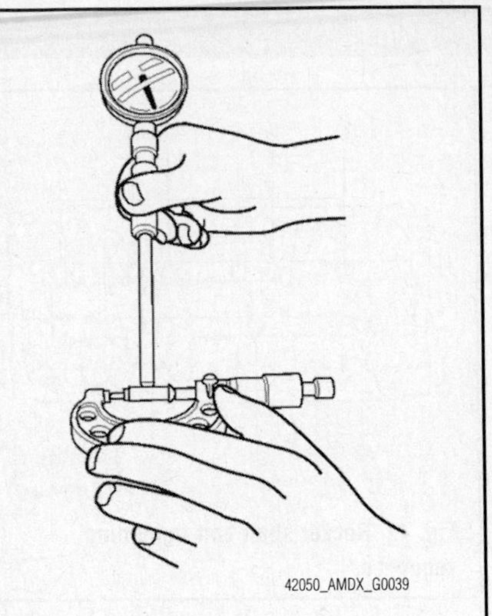
42050_AMDX_G0039

Fig. 44 Zero the gauge to the journal diameter

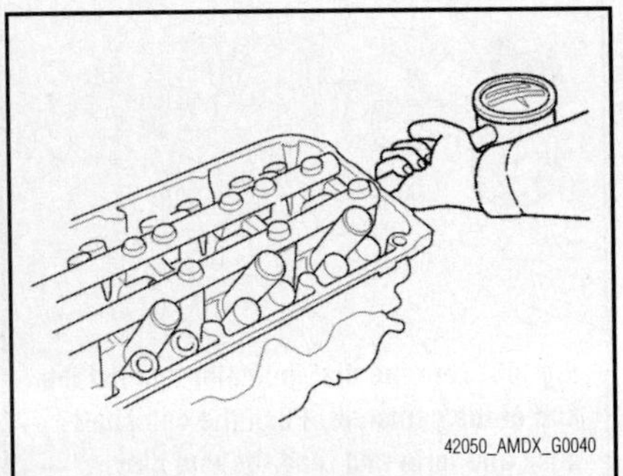
42050_AMDX_G0040

Fig. 45 Measure the inside diameter of each camshaft bearing surface, and check for an out-of-round condition

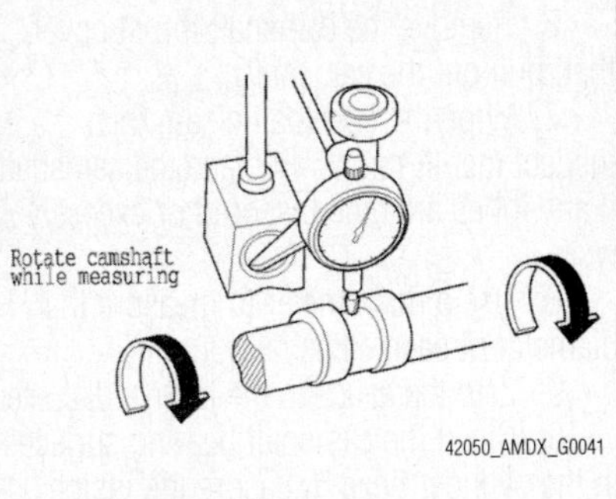
Rotate camshaft
while measuring
42050_AMDX_G0041

Fig. 46 Check total runout with the camshaft supported on V-blocks

inches (0.15mm), and the camshaft has not been replaced, go to next step.

11. Check total runout with the camshaft supported on V-blocks:
    - If the total runout of the camshaft is within the service limit of 0.001 inches (0.03mm) max, replace the cylinder head.

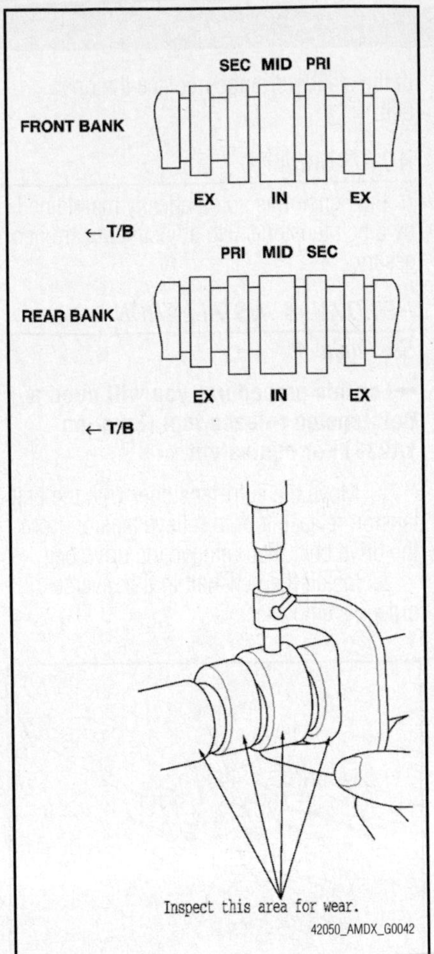
SEC MID PRI
FRONT BANK
EX IN EX
← T/B
PRI MID SEC
REAR BANK
EX IN EX
← T/B
Inspect this area for wear.
42050_AMDX_G0042

Fig. 47 Measure cam lobe height

- If the total runout is beyond the service limit 0.002 inches (0.04mm), replace the camshaft and recheck the oil clearance. If the oil clearance is still out of tolerance, replace the cylinder head.

12. Measure cam lobe height and compare with the following cam lobe height standard (new):
    - PRI Intake: 1.3824 inches (35.112 mm)
    - MID Intake: 1.4328 inches (36.394 mm)
    - SEC Intake: 1.3824 inches (35.112 mm)
    - Exhaust: 1.4326 inches (36.389 mm)

REMOVAL & INSTALLATION

Front Camshaft

See Figure 48.

1. Disconnect the negative and positive battery cable.
2. Remove the battery and battery box.
3. Drain the engine cooling system.
4. Disconnect the upper radiator hose.
5. Remove the exhaust gas recirculation valve.
6. Remove the timing belt. For additional information, refer to the following section, "Timing Belt, Removal & Installation."
7. Remove the intake manifold. For additional information, refer to the following section, "Intake Manifold, Removal & Installation."
8. Remove the cylinder head cover. For additional information, refer to the following section, "Cylinder Head, Removal & Installation."
9. Loosen the locknuts and adjusting screws.
10. Remove the bolts and the rocker arm assembly.

CAUTION

Loosen the rocker shaft mounting bolts two turns at a time, to prevent damaging the valves or rocker arm assembly.

➡When removing the rocker arm assembly, do not remove the rocker shaft mounting bolts. The bolts will keep the springs and the rocker arms on the shafts.

11. Remove the front camshaft pulley.
12. Remove the thrust cover, the remove the camshaft.

To install:

13. Apply clean engine oil to the journals and cam lobes of the camshaft. Install the camshaft using a new O-ring. Tighten the thrust plate to 16 ft. lbs. (22 Nm).
14. Apply clean engine oil to the threads of the camshaft pulley mounting bolt, then install the front camshaft pulley.

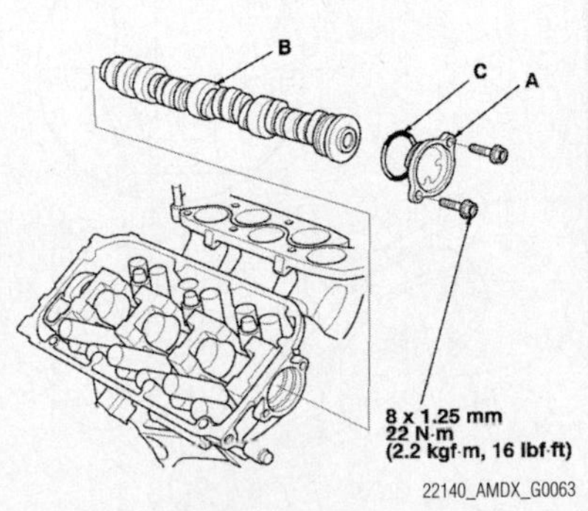
B
C
A
8 x 1.25 mm
22 N·m
(2.2 kgf·m, 16 lbf·ft)
22140_AMDX_G0063

Fig. 48 Remove the thrust cover (A) to remove the camshaft (B). Replace the O-ring (C) when reinstalling

15. Install the rocker arm assembly. Tighten the mounting bolts two turns at a time in the sequence shown to 17 ft. lbs. (24 Nm).
16. Install the timing belt.
17. Adjust the valve clearance. For additional information, refer to the following section, "Valve Lash, Adjustment."
18. The remainder of the installation is the reverse order of removal.
19. Refill the cooling system to the correct level.

Rear Camshaft

1. Disconnect the negative battery cable.
2. Drain the cooling system.
3. Remove the intake manifold. For additional information, refer to the following section, "Intake Manifold, Removal & Installation."
4. Remove the purge joint.
5. Remove the brake lines from the master cylinder.
6. Remove the timing belt. For additional information, refer to the following section, "Timing Belt, Removal & Installation."
7. Remove the cylinder head cover. For additional information, refer to the following section, "Cylinder Head, Removal & Installation."
8. Loosen the locknuts and adjusting screws.
9. Remove the bolts and the rocker arm assembly.

CAUTION

Loosen the rocker shaft mounting bolts two turns at a time, to prevent damaging the valves or rocker arm assembly.

➡When removing the rocker arm assembly, do not remove the rocker shaft mounting bolts. The bolts will keep the springs and the rocker arms on the shafts.

10. Remove the rear camshaft pulley.
11. Remove the thrust cover, then remove the camshaft.

To install:

12. Apply clean engine oil to the journals and cam lobes of the camshaft. Install the camshaft using a new O-ring. Tighten the thrust plate to 16 ft. lbs. (22 Nm).
13. Apply clean engine oil to the threads of the camshaft pulley mounting bolt, then install the rear camshaft pulley.
14. Install the rocker arm assembly. Tighten the mounting bolts two turns at a time in the sequence shown to 17 ft. lbs. (24 Nm).
15. Install the timing belt.
16. Adjust the valve clearance. For additional information, refer to the following section, "Valve Lash, Adjustment."
17. Install the brake lines to the master cylinder and bleed the brake system.
18. The remainder of the installation is the reverse order of removal.
19. Refill the cooling system to the correct level.

CATALYTIC CONVERTER

REMOVAL & INSTALLATION

Front

See Figure 49.

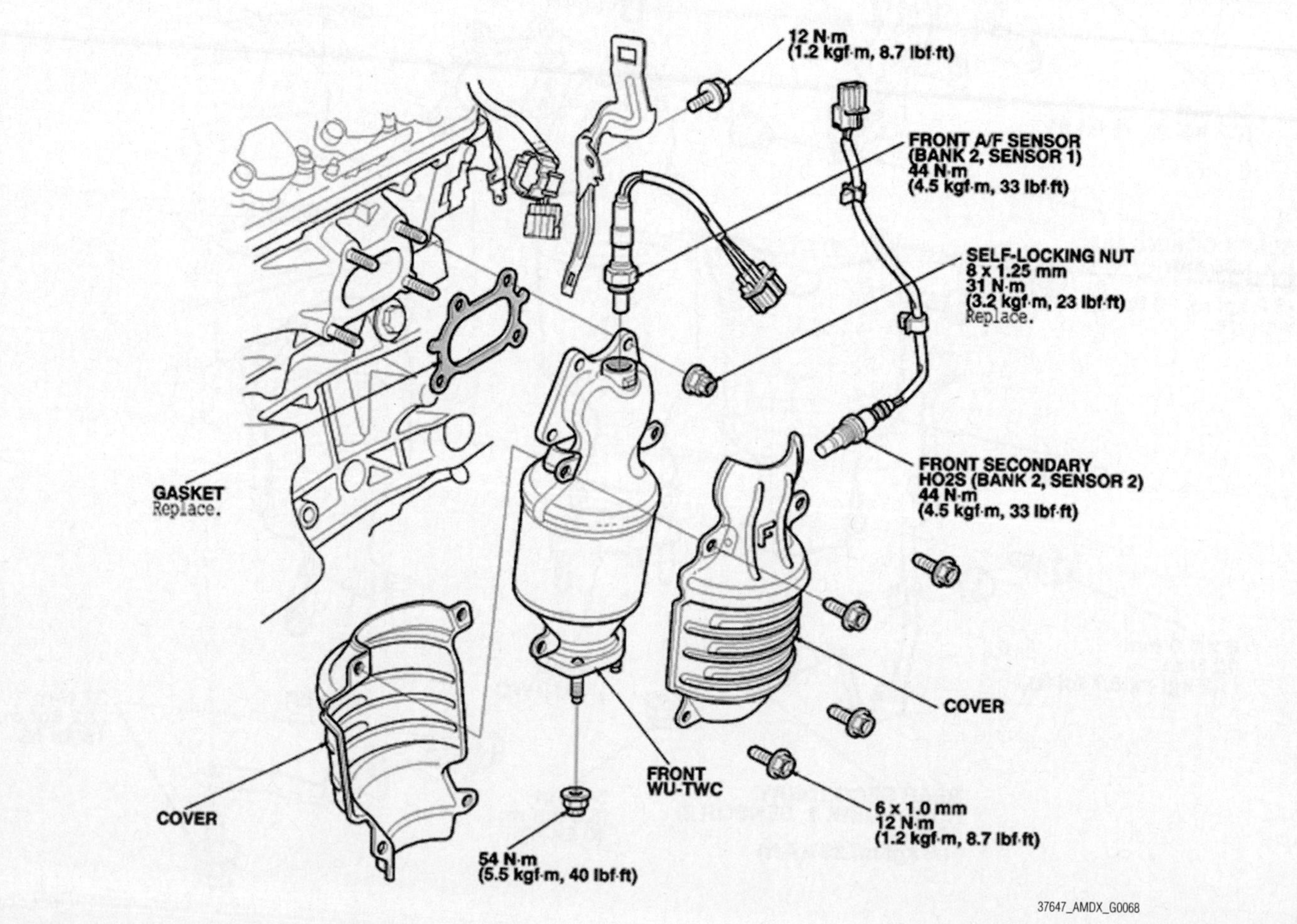
12 N·m
(1.2 kgf·m, 8.7 lbf·ft)
FRONT A/F SENSOR
(BANK 2, SENSOR 1)
44 N·m
(4.5 kgf·m, 33 lbf·ft)
SELF-LOCKING NUT
8 x 1.25 mm
31 N·m
(3.2 kgf·m, 23 lbf·ft)
Replace.
GASKET
Replace.
FRONT SECONDARY
HO2S (BANK 2, SENSOR 2)
44 N·m
(4.5 kgf·m, 33 lbf·ft)
COVER
FRONT
WU-TWC
COVER
6 x 1.0 mm
12 N·m
(1.2 kgf·m, 8.7 lbf·ft)
54 N·m
(5.5 kgf·m, 40 lbf·ft)
37647_AMDX_G0068

Fig. 49 Exploded view of front Warm Up-Three Way Catalyst (WU-TWC)

1. Remove the radiator and A/C condenser fan assemblies.
2. Disconnect the front Air Fuel Ratio (A/F) sensor connector and front secondary heated oxygen sensor (secondary HO2S) connector.
3. Carefully remove the front WU-TWC.
4. Carefully install the front WU-TWC, and tighten the nuts in a crisscross pattern in two or three steps.
5. Install the parts in the reverse order of removal.

Rear

See Figure 50.

1. Remove exhaust pipe A.
2. Remove the intermediate shaft.
3. Disconnect the rear Air Fuel Ratio (A/F) sensor connector and the rear secondary heated oxygen sensor (secondary HO2S) connector.
4. Remove the rear WU-TWC bracket, then remove the rear WU-TWC.
5. Install the rear WU-TWC, and tighten the nuts in a crisscross pattern in two or three steps.
6. Install the parts in the reverse order of removal.

Exhaust System Three Way Catalyst (TWC)

See Figures 51 and 52.

1. Remove the under-floor TWC.
2. Remove the converter cover.
3. Install the parts in the reverse order of removal with new gaskets and new self-locking nuts.

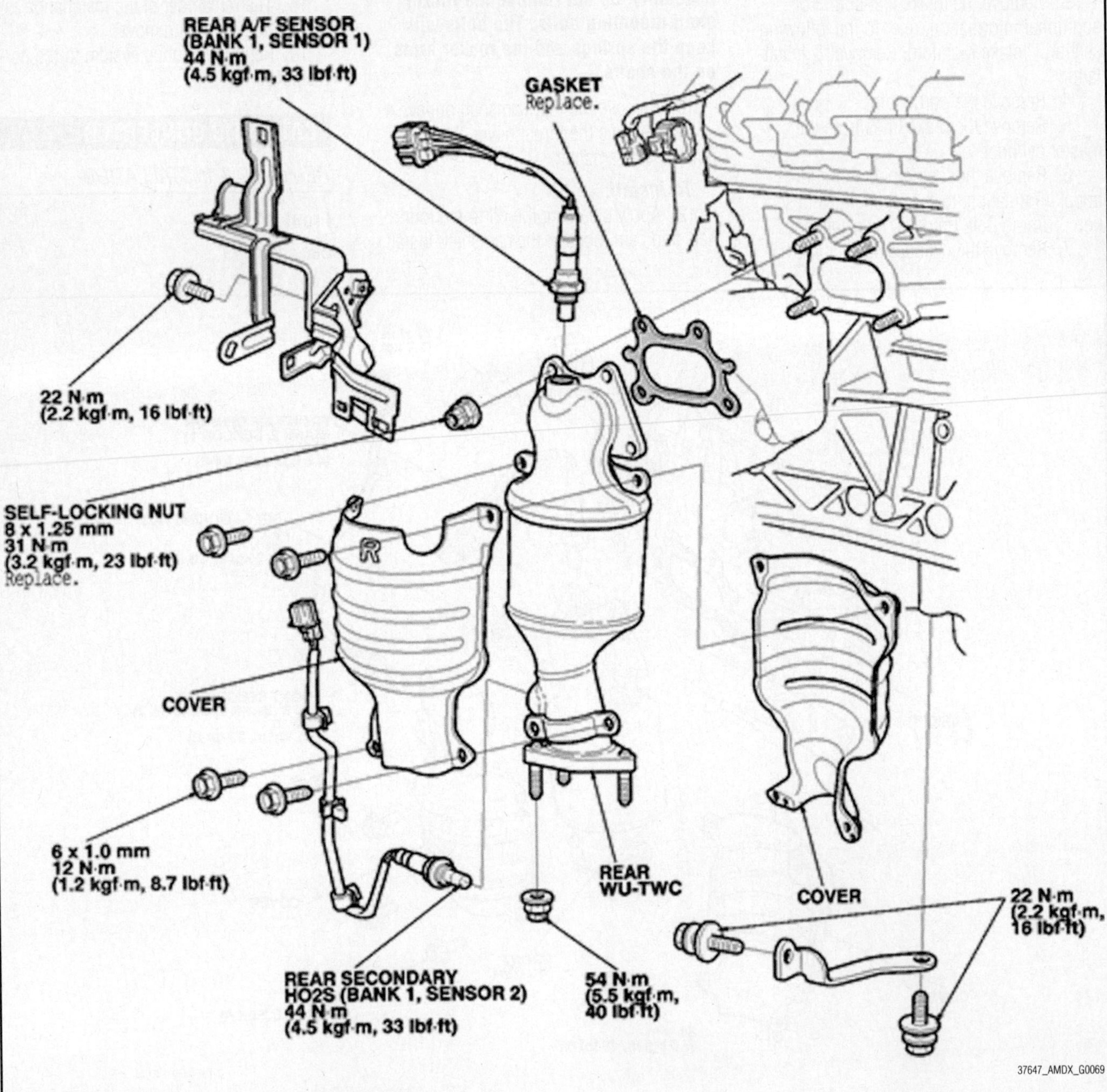
REAR A/F SENSOR
(BANK 1, SENSOR 1)
44 N·m
(4.5 kgf·m, 33 lbf·ft)
GASKET
Replace.
22 N·m
(2.2 kgf·m, 16 lbf·ft)
SELF-LOCKING NUT
8 x 1.25 mm
31 N·m
(3.2 kgf·m, 23 lbf·ft)
Replace.
COVER
6 x 1.0 mm
12 N·m
(1.2 kgf·m, 8.7 lbf·ft)
REAR
WU-TWC
COVER
22 N·m
(2.2 kgf·m,
16 lbf·ft)
REAR SECONDARY
HO2S (BANK 1, SENSOR 2)
44 N·m
(4.5 kgf·m, 33 lbf·ft)
54 N·m
(5.5 kgf·m,
40 lbf·ft)
37647_AMDX_G0069

Fig. 50 Exploded view of rear Warm Up-Three Way Catalyst (WU-TWC)

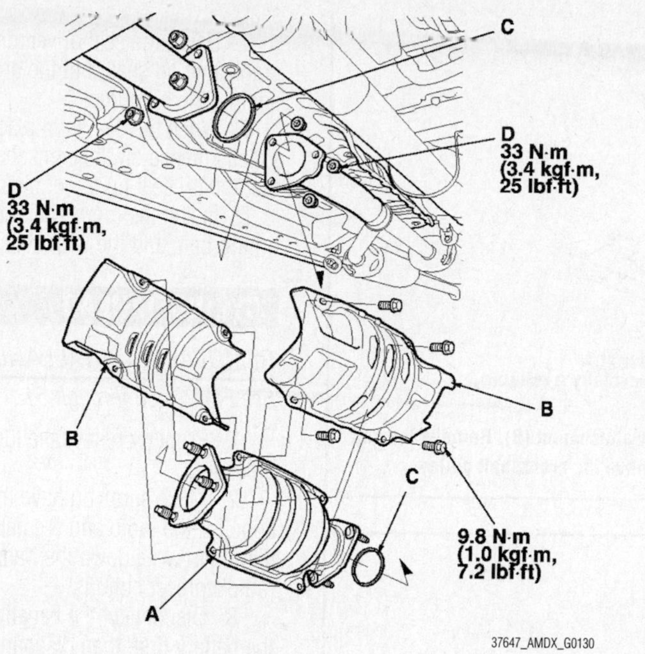
C
D
33 N·m
(3.4 kgf·m,
25 lbf·ft)
D
33 N·m
(3.4 kgf·m,
25 lbf·ft)
B
B
C
9.8 N·m
(1.0 kgf·m,
7.2 lbf·ft)
A
37647_AMDX_G0130

Fig. 51 Exploded view of under-floor TWC

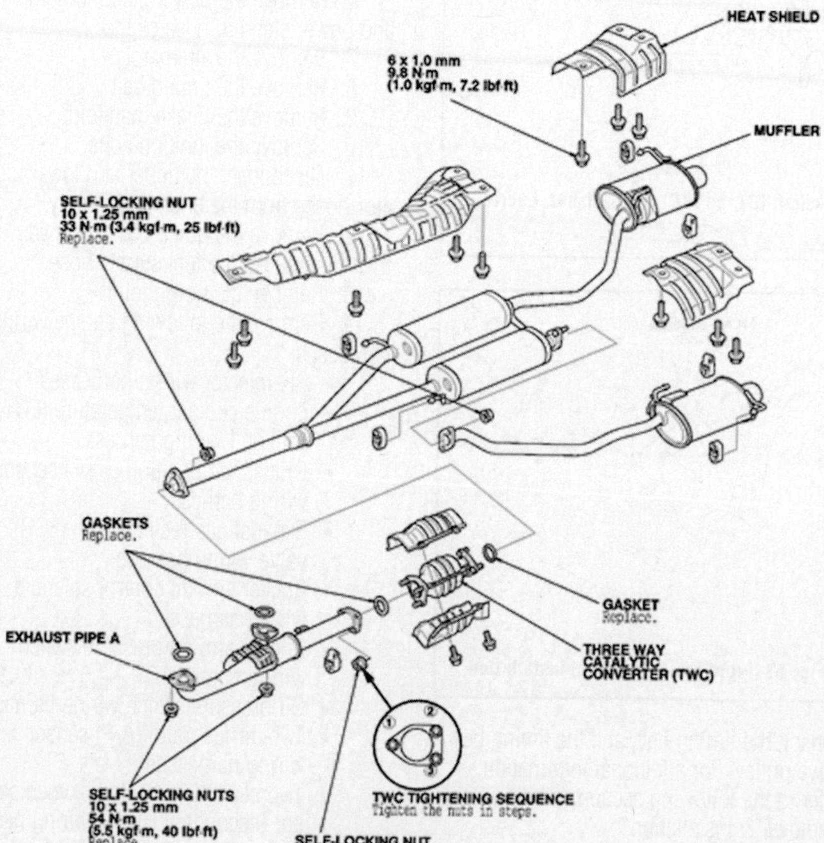
HEAT SHIELD
6 x 1.0 mm
9.8 N·m
(1.0 kgf·m, 7.2 lbf·ft)
MUFFLER
SELF-LOCKING NUT
10 x 1.25 mm
33 N·m (3.4 kgf·m, 25 lbf·ft)
Replace.
GASKETS
Replace.
GASKET
Replace.
EXHAUST PIPE A
THREE WAY
CATALYTIC
CONVERTER (TWC)
TWC TIGHTENING SEQUENCE
Tighten the nuts in steps.
SELF-LOCKING NUTS
10 x 1.25 mm
54 N·m
(5.5 kgf·m, 40 lbf·ft)
Replace.
SELF-LOCKING NUT
10 x 1.25 mm
33 N·m (3.4 kgf·m, 25 lbf·ft)
Replace.
37647_AMDX_G0070

Fig. 52 Exploded view of exhaust system

CRANKSHAFT DAMPER

REMOVAL & INSTALLATION

See Figures 53 through 56.

➡This procedure requires the use of the following special tools or their equivalents:

- Holder handle 07JAB-001020A
- Holder attachment, 50mm, offset 07MAB-PY3010A
- Socket, 19mm 07JAA-001020A or a commercially available 19mm socket

1. Remove the right front wheel.
2. Remove the splash shield.
3. Remove the drive belt.
4. Hold the pulley with the holder handle and holder attachment.
5. Remove the bolt with a heavy duty 19mm socket and breaker bar, then remove the crankshaft pulley.

To install:

6. Remove any oil or clean the pulleys, crankshaft, bolt, and washer. Lubricate new engine oil as shown in the accompanying illustration.
7. Install the crankshaft pulley, and tighten the bolt as follows. Do not use an impact wrench.
   a. Hold the pulley with the handle and holder attachment, then tighten the bolt to 47 ft. lbs. (64 Nm) with a torque wrench and 19mm socket.
   b. Mark the bolt head and crankshaft pulley as shown, then tighten the bolt an additional 60 degrees. (The mark on the bolt head line up with the mark on the crankshaft pulley).
8. Install the drive belt.
9. Install the splash shield.
10. Install the right front wheel.

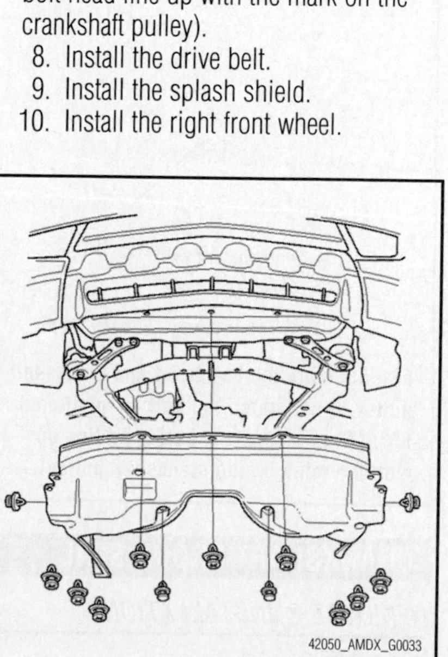
42050_AMDX_G0033

Fig. 53 Remove the splash shield

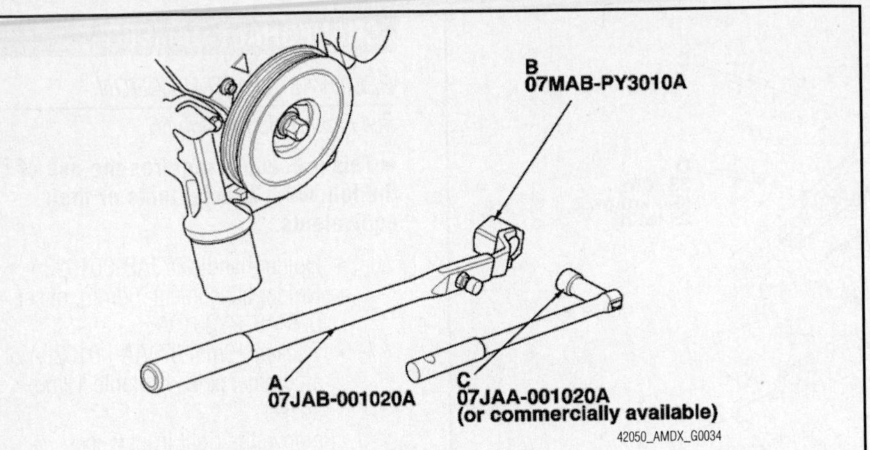
B
07MAB-PY3010A
A
07JAB-001020A
C
07JAA-001020A
(or commercially available)
42050_AMDX_G0034

Fig. 54 Hold the pulley with the holder handle (A) and holder attachment (B). Remove the bolt with a heavy duty 19mm socket (C) and breaker bar, then remove the crankshaft pulley

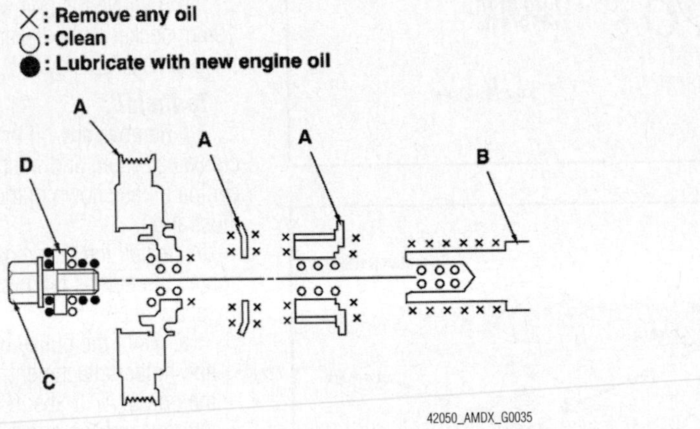
×: Remove any oil
○: Clean
●: Lubricate with new engine oil
A
A
A
B
D
C
42050_AMDX_G0035

Fig. 55 Remove any oil or clean the pulleys (A), crankshaft (B), bolt (C), and washer. Lubricate new engine oil

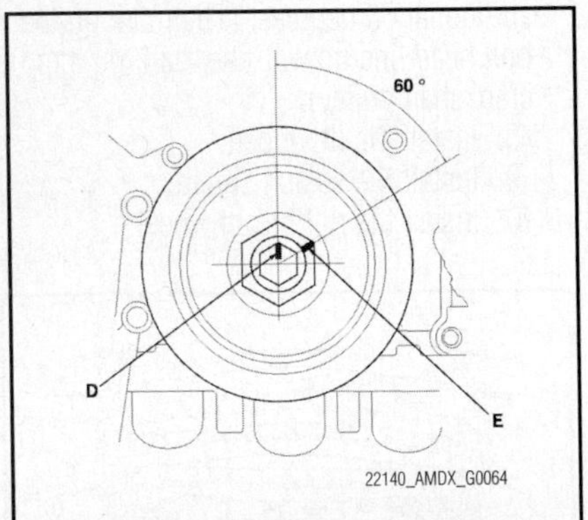
60 °
D
E
22140_AMDX_G0064

Fig. 56 Mark the bolt head and crankshaft pulley, then tighten the bolt an additional 60°. (The mark on the bolt head line up with the mark on the crankshaft pulley).

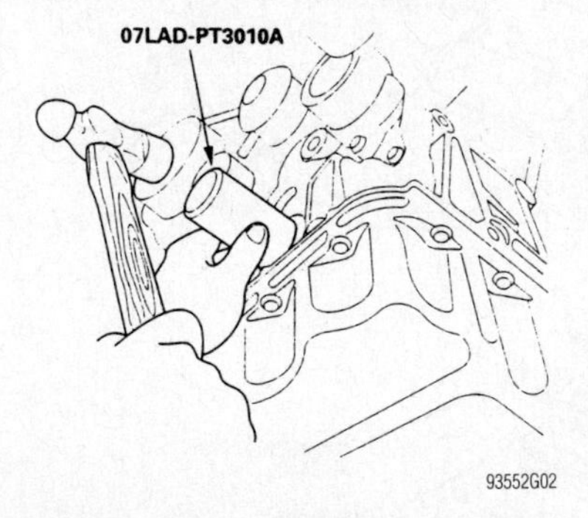
07LAD-PT3010A
93552G02

Fig. 57 Front crankshaft seal installation

CRANKSHAFT FRONT SEAL

REMOVAL & INSTALLATION

See Figure 57.

1. Remove the crankshaft position (CKP) sensor, the timing belt, and the timing belt drive pulley. For additional information, refer to the following section, "Timing Belt, Removal & Installation."
2. Remove the crankshaft front seal.

To install:

3. Clean and dry the crankshaft oil seal housing.
4. Apply a light coat of multipurpose grease to the crankshaft and to the lip of the seal.
5. Using the seal driver, drive in the crankshaft oil seal until the driver bottoms against the oil pump.
6. When the seal is in place, clean any excess grease off the crankshaft, and check that the oil seal lip is not distorted.
7. Install the timing belt drive pulley, the timing belt, and the CKP sensor.

CYLINDER HEAD

REMOVAL & INSTALLATION

See Figures 58 through 61.

1. Properly relieve the fuel system pressure.
2. Make sure you have the anti-theft code for the radio and the navigation system, then write down the frequencies for the radio's preset buttons.
3. Disconnect the negative cable from the battery first, then disconnect the positive cable. Wait at least 3 minutes before proceeding.
4. Drain the engine cooling system.
5. Remove the accessory drive belt.
6. Remove the power steering pump and power steering hose clamp.
7. Remove the alternator.
8. Remove the timing belt.
9. Remove the intake manifold.
10. Remove the ignition coils.
11. Disconnect the upper and lower radiator hoses from the engine assembly.
12. Remove the quick-connect cover, then disconnect the fuel supply hose.
13. Remove the purge joint.
14. Remove the following engine wiring harnesses:
    - Six injector wiring harnesses
    - Engine coolant temperature (ECT) sensor 1 wiring harness
    - Crankshaft position (CKP) sensor wiring harness
    - Exhaust gas recirculation (EGR) valve wiring harness
    - Rocker arm oil control solenoid wiring harness
    - Rocker arm oil pressure switch wiring harness
    - Oil pressure switch wiring harness
    - Two air fuel ratio (A/F) sensor wiring harnesses
    - Two secondary heated oxygen sensor (secondary HO2S) wiring harnesses
15. Remove the front warm up three way catalytic converter (front WU-TWC) and rear warm up three way catalytic converter (rear WU-TWC).

16. Remove the connector bracket from the front cylinder head.
17. Remove the bracket from the rear cylinder head.
18. Remove the fuel rails.
19. Remove the water passage.
20. Remove the front and rear camshaft pulleys and back covers.
21. Remove the cylinder head cover as follows:
  a. Remove the dipstick.
  b. Remove the two bolts securing the harness holder, and disconnect the front air fuel ratio (A/F) sensor connector, front secondary heated oxygen sensor (secondary HO2S) connector, exhaust gas recirculation (EGR) valve connector and engine coolant temperature (ECT) sensor 1 connector.
  c. Disconnect the three injector connectors from the injectors on the rear side cylinder head.
  d. Remove the power steering hose bracket mounting bolt, the harness holder mounting bolts, and the engine ground cable bolt.
  e. Remove the engine ground cable and disconnect the breather hose.
  f. Remove the cylinder head covers.
22. Loosen the cylinder head bolts in sequence and ⅓ turns until all bolts are loose.
23. Remove the cylinder head.

To install:

24. Clean the cylinder head and engine block surface.
25. Clean and install the oil control orifices with new O-rings.
26. Install the dowel pins and new cylinder head gaskets.
27. Clean the timing belt pulleys, timing belt guide plate, and the upper and lower covers.
28. Align the crankshaft and camshaft sprocket TDC marks as shown.
29. Put the cylinder head onto the engine block.
30. Apply clean engine oil to the cylinder head bolt threads and flanges.
31. Tighten the cylinder head bolts in sequence as follows:
  a. Step 1: 22 ft. lbs. (29 Nm)
  b. Step 2: Plus 90 degrees

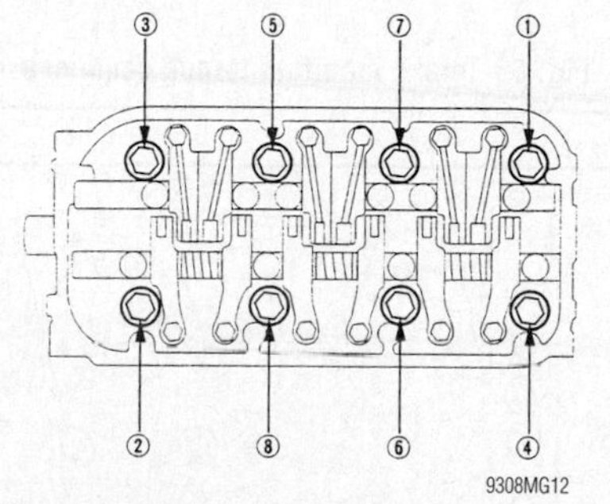
③
⑤
⑦
①
②
⑧
⑥
④
9308MG12

Fig. 58 Cylinder head bolt loosening sequence

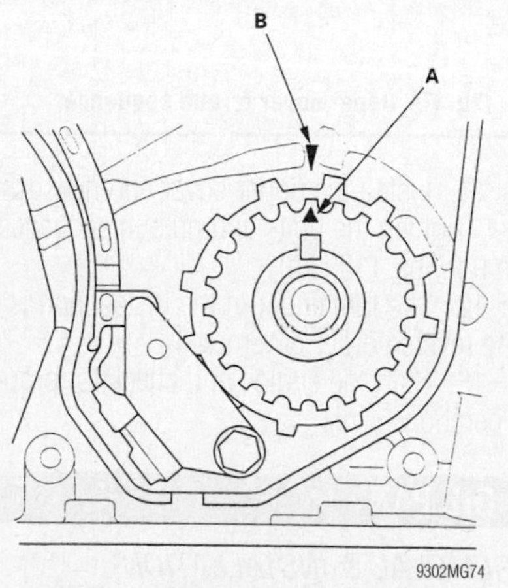
B
A
9302MG74

Fig. 59 Crankshaft timing belt sprocket TDC marks. Align sprocket mark (A) with pointer (B)

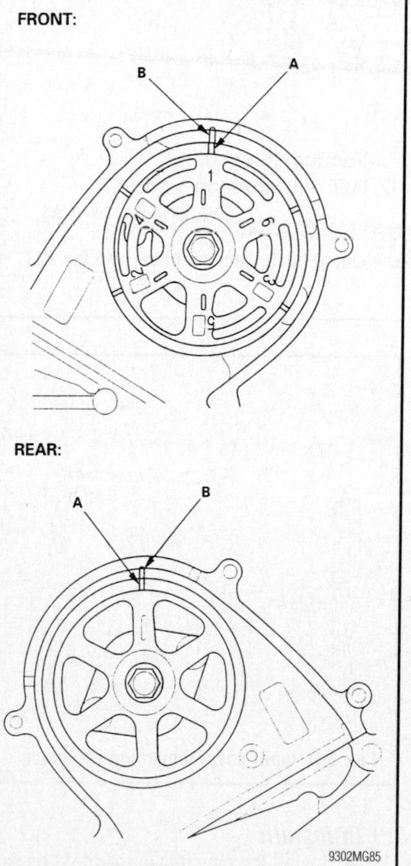
FRONT:
B
A
REAR:
A
B
9302MG85

Fig. 60 Camshaft TDC marks. Align sprocket mark (A) with the back cover pointer (B)—MDX

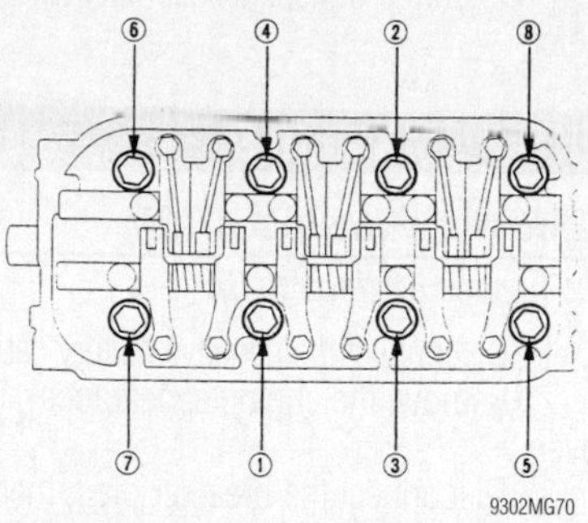
⑥
④
②
⑧
⑦
①
③
⑤
9302MG70

Fig. 61 Cylinder head bolt tightening sequence

  c. Step 3: Plus an additional 90 degrees
32. Install the timing belt.
33. Adjust the valve clearance.
34. Install the cylinder head covers and tighten to 108 inch lbs. (12 Nm).
35. Install the water passage.
36. Install the front warm up three way catalytic converter (front WU-TWC) and rear warm up three way catalytic converter (rear WU-TWC).
37. Install the fuel rails.
38. Install the connector bracket to the front cylinder head.
39. Install the bracket to the rear cylinder head.
40. Install the purge joint and tighten the mounting nuts to 109 inch lbs. (12 Nm).
41. Connect the fuel supply hose and replace the quick-connect fitting cover.
42. Reconnect the engine wiring harnesses that were previously removed.
43. Connect the upper and lower radiator hoses.
44. Install the intake manifold.
45. Install the alternator.
46. Install the power steering pump and hose bracket.
47. Install the accessory drive belt.
48. Reconnect the battery cables.
49. After installation, check that all tubes, hoses and connectors are installed correctly.
50. Refill the engine cooling system to the correct level.
51. Start the engine and check for leaks.

DRIVEPLATE

REMOVAL & INSTALLATION

1. Remove the transaxle assembly.
2. Remove the driveplate and washer from the engine.

To install:

3. Install the driveplate and washer on the crankshaft. Tighten the bolts in a criss-cross pattern to 54 ft. lbs. (74 Nm).

4. Install the transaxle assembly into the vehicle.

INTAKE MANIFOLD

REMOVAL & INSTALLATION

See Figures 62 through 67.

1. Disconnect the negative battery cable.
2. Remove the engine appearance cover.
3. Disconnect the breather pipe, then remove the intake air duct.
4. Disconnect the Positive Crankcase Ventilation (PCV) hose, the brake booster vacuum hose, vacuum hose, and the Intake Manifold Tuning (IMT) actuator connector.
5. Disconnect the Evaporative Emission (EVAP) canister hose, EVAP canister purge valve connector, throttle actuator connector, and Manifold Absolute Pressure (MAP) sensor connector.
6. Disconnect and plug the water bypass hoses.
7. Remove the upper cover mounting bolts and nuts sequentially in three steps, then remove the upper cover.
8. Remove the intake manifold mounting bolts and nuts sequentially in three steps, then remove the intake manifold.

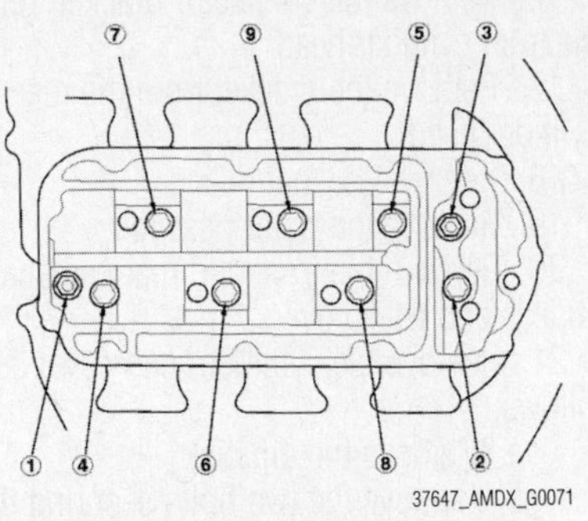
37647_AMDX_G0071

Fig. 65 Intake manifold loosening sequence

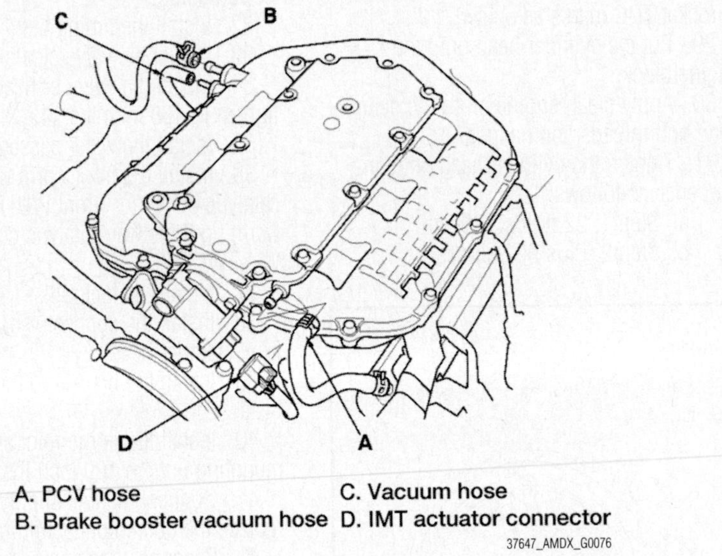
C
B
D
A
A. PCV hose
B. Brake booster vacuum hose
C. Vacuum hose
D. IMT actuator connector
37647_AMDX_G0076

Fig. 62 Disconnect the PCV hose, the brake booster vacuum hose, vacuum hose, and the IMT actuator connector

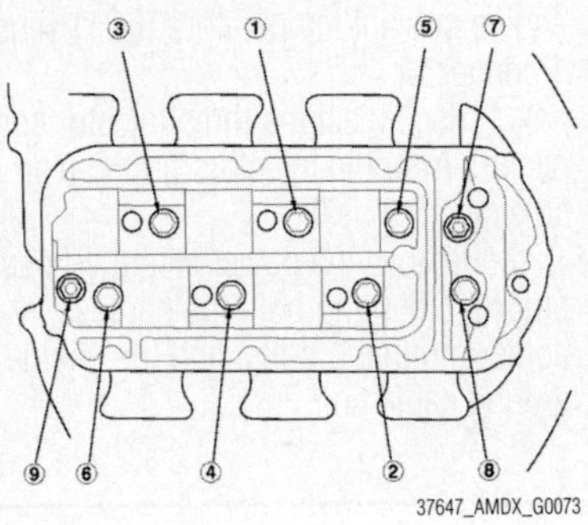
37647_AMDX_G0073

Fig. 66 Intake manifold torque sequence

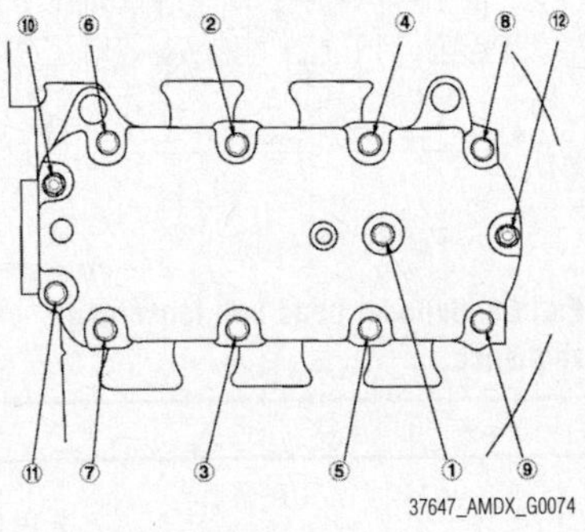
37647_AMDX_G0074

Fig. 67 Upper cover torque sequence

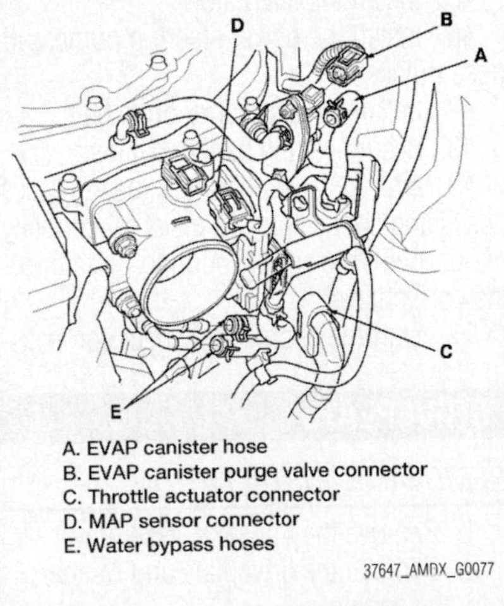
D
B
A
C
E
A. EVAP canister hose
B. EVAP canister purge valve connector
C. Throttle actuator connector
D. MAP sensor connector
E. Water bypass hoses
37647_AMDX_G0077

Fig. 63 Disconnect the EVAP canister hose, EVAP canister purge valve connector, throttle actuator connector, and MAP sensor connector

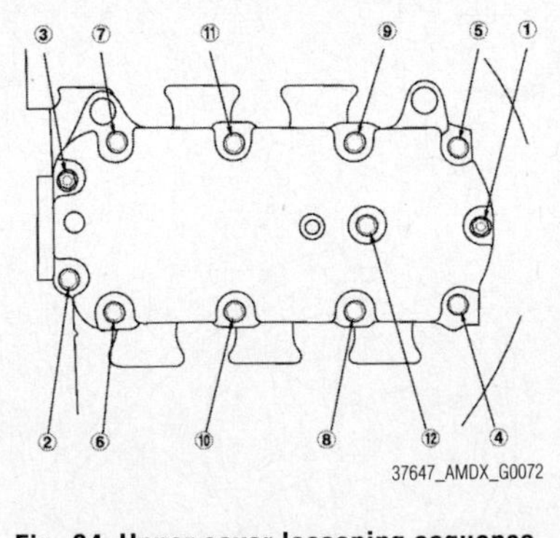
37647_AMDX_G0072

Fig. 64 Upper cover loosening sequence

To install:

9. Install the new intake manifold gasket.
10. Install the intake manifold and tighten the bolts in sequence to 16 ft. lbs. (22 Nm).
11. Install the upper cover and new gasket. Tighten the bolts and nuts in sequence to 9 ft. lbs. (12 Nm).
12. The remainder of the installation is the reverse order of removal.
13. Start the engine and check for proper operation.

OIL PAN

REMOVAL & INSTALLATION

See Figures 68 through 72.

1. Drain the engine oil.
2. Disconnect the negative battery cable.

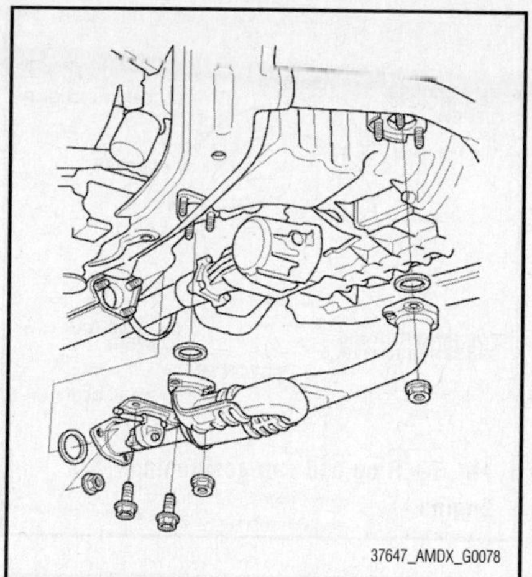
37647_AMDX_G0078
Fig. 68 Remove the exhaust front pipe

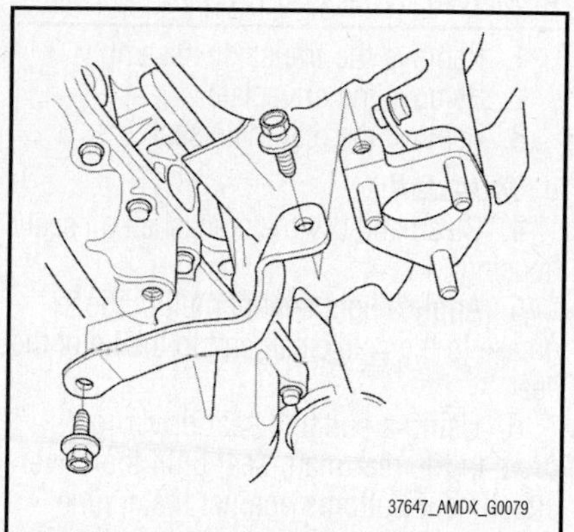
37647_AMDX_G0079
Fig. 69 Remove the rear WU-TWC bracket

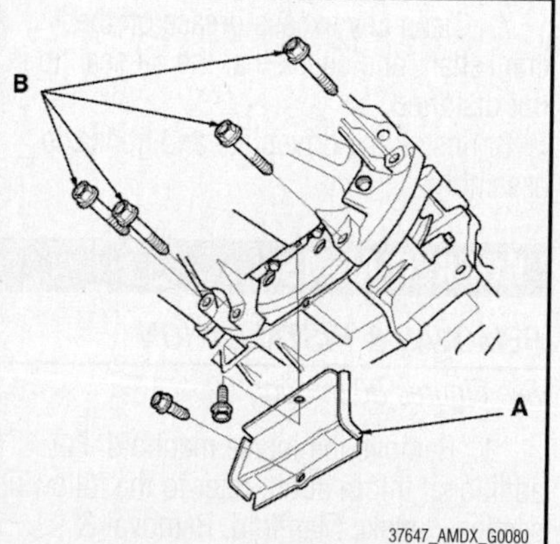
B
A
37647_AMDX_G0080
Fig. 70 Remove the torque converter cover (A) and the four bolts (B) securing the transaxle

3. Remove the front splash shield.
4. Remove the front exhaust pipe.
5. Remove the rear Warm Up Three Way Catalytic Converter (rear WU-TWC) bracket.
6. Remove the torque converter cover and the four bolts securing the transaxle.
7. Remove the oil pan mounting bolts.
8. Use a flat bladed screwdriver to separate the oil pan from the block.

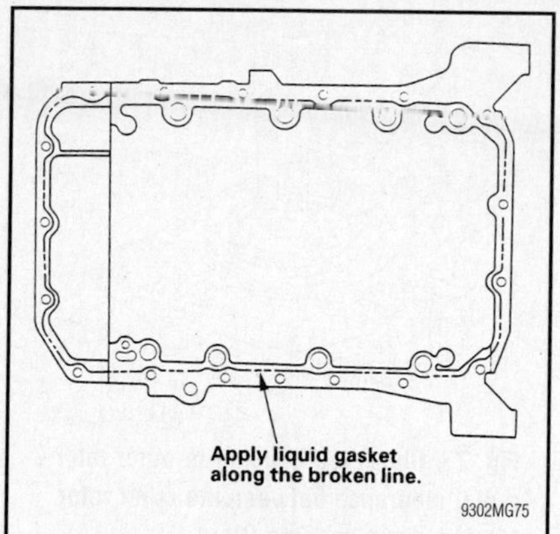
Apply liquid gasket along the broken line.
9302MG75
Fig. 71 Apply liquid gasket to the inner threads of the bolt holes and the engine block along the area indicated by the broken line

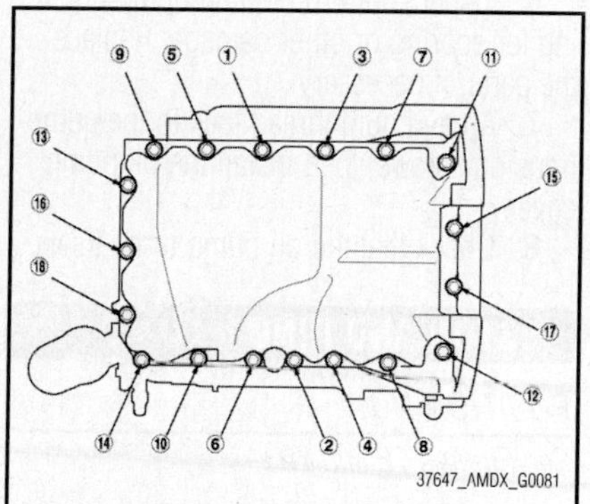
37647_AMDX_G0081
Fig. 72 Oil pan tightening sequence

To install:

9. Remove all of the old liquid gasket material.
10. Apply liquid gasket to the mating surfaces of the oil pan as shown.
11. Install the oil pan on the engine block. Tighten the mounting bolts in sequence to 9 ft. lbs. (12 Nm).
12. Tighten the four bolts securing the transmission. Torque the bolts to 54 ft. lbs. (74 N m).
13. Install the torque converter cover.
14. Install the rear Warm Up Three Way Catalytic Converter (rear WU-TWC) bracket.
15. Install the front exhaust pipe using new gaskets and new self-locking nuts.
16. Install the splash shield.
17. Connect the negative battery cable.

➡Wait at least 30 minutes before adding oil to the engine.

18. Refill the engine with oil to the correct level.

OIL PUMP

REMOVAL & INSTALLATION

See Figures 73 and 74.

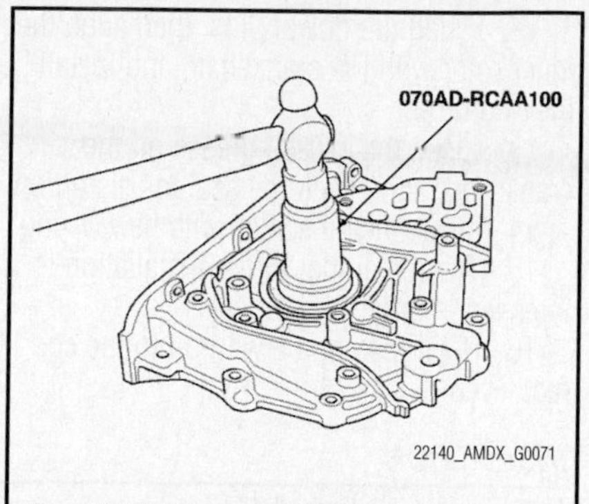
070AD-RCAA100
22140_AMDX_G0071
Fig. 73 Using a suitable driver tool, gently tap in a new oil seal

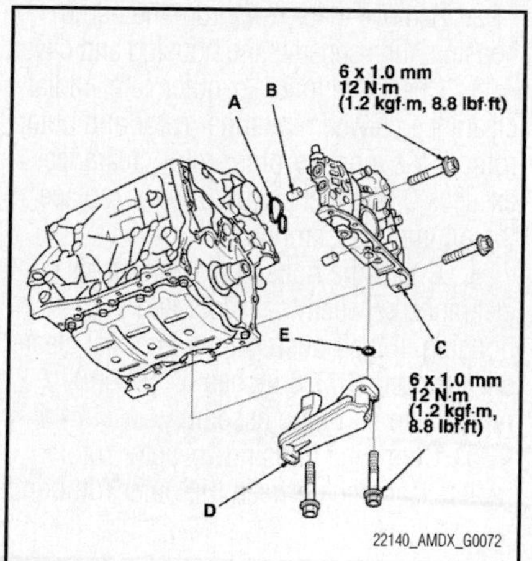
6 x 1.0 mm
12 N·m
(1.2 kgf·m, 8.8 lbf·ft)
A
B
C
D
E
6 x 1.0 mm
12 N·m
(1.2 kgf·m,
8.8 lbf·ft)
22140_AMDX_G0072
Fig. 74 Oil pump assembly

1. Remove the timing belt. For additional information, refer to the following section, "Timing Belt, Removal & Installation."
2. Remove the crankshaft position sensor.
3. Remove the rocker arm control solenoid/oil filter assembly.
4. Remove the oil pan.
5. Remove the oil screen before removing the oil pump.

To install:

6. Remove the old oil seal from the oil pump.
7. Gently tap in the new oil seal until the oil seal driver bottoms on the pump.
8. Remove all of the old liquid gasket from the oil pump mating surfaces, bolts, and bolt holes.
9. Clean and dry the oil pump mating surfaces.
10. Apply liquid gasket evenly to the engine block mating surface of the oil pump. Install the component within 5 minutes of applying the liquid gasket.
11. Grease the lip of the oil seal, and apply oil to the new O-ring.

12. Install the dowel pins, then align the inner rotor with the crankshaft, and install the oil pump.
13. Clean the excess grease off the crankshaft, and check the seal for distortion.
14. Install the oil screen with new O-ring.
15. The remainder of the installation is the reverse order of removal.
16. Refill the engine with oil to the correct level.

INSPECTION

See Figures 75 through 77.

1. Remove the oil pump assembly from the engine.
2. Remove the screws from the pump housing, then separate the housing and cover.
3. Check the inner-to-outer rotor radial clearance between the inner rotor and outer rotor. If the inner-to-outer rotor clearance exceeds 0.008 inches (0.20 mm), replace the oil pump assembly.
4. Check the housing-to-rotor axial clearance between the rotors and pump housing. If the housing-to-rotor axial clearance exceeds 0.005 inches (0.12 mm), replace the oil pump assembly.
5. Check the housing-to-outer rotor radial clearance between the outer rotor and pump housing. If the housing-to-outer rotor radial clearance exceeds 0.008 inches (0.20 mm), replace the oil pump assembly.
6. Inspect both rotors and pump housing for scoring or other damage. Replace the parts, if necessary.
7. Apply liquid thread lock to the pump housing screws, then install the oil pump cover.
8. Check that the oil pump turns freely.

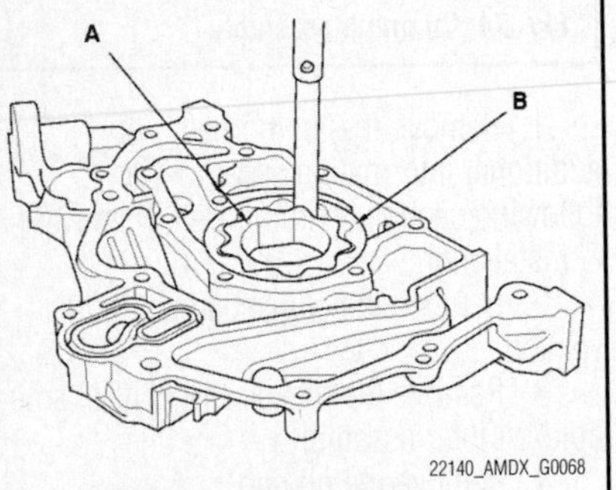
A
B
22140_AMDX_G0068

Fig. 75 Check the inner-to-outer rotor radial clearance between the inner rotor (A) and outer rotor (B)

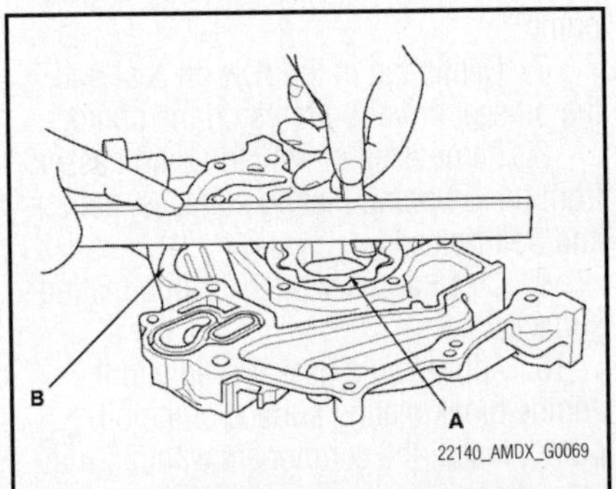
B
A
22140_AMDX_G0069

Fig. 76 Check the housing-to-rotor axial clearance between the rotors (A) and pump housing (B)

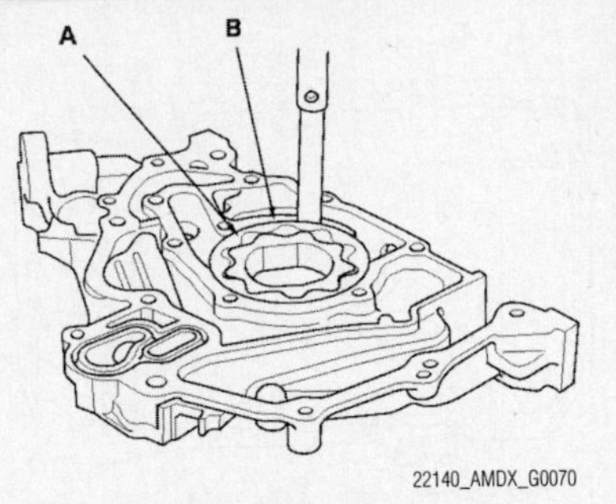
A
B
22140_AMDX_G0070

Fig. 77 Check the housing-to-outer rotor radial clearance between the outer rotor (A) and pump housing (B)

PISTON AND RING

POSITIONING

See Figures 78 and 79.

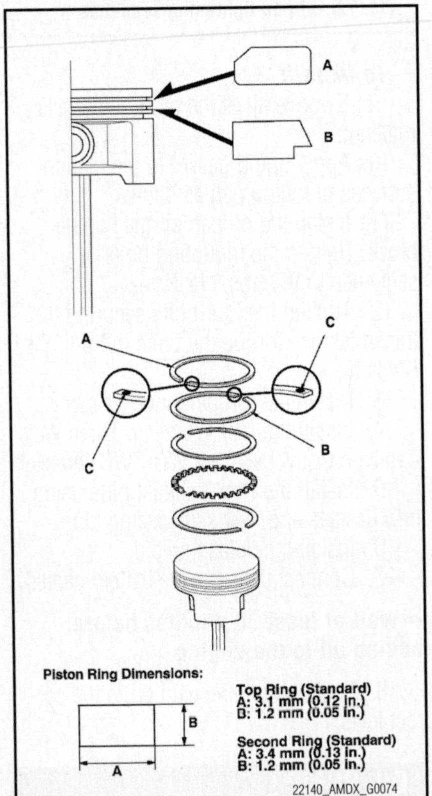
A
B
A
C
C
B
Piston Ring Dimensions:
B
A
Top Ring (Standard)
A: 3.1 mm (0.12 in.)
B: 1.2 mm (0.05 in.)
Second Ring (Standard)
A: 3.4 mm (0.13 in.)
B: 1.2 mm (0.05 in.)
22140_AMDX_G0074

Fig. 78 Piston ring installation—3.7L Engine

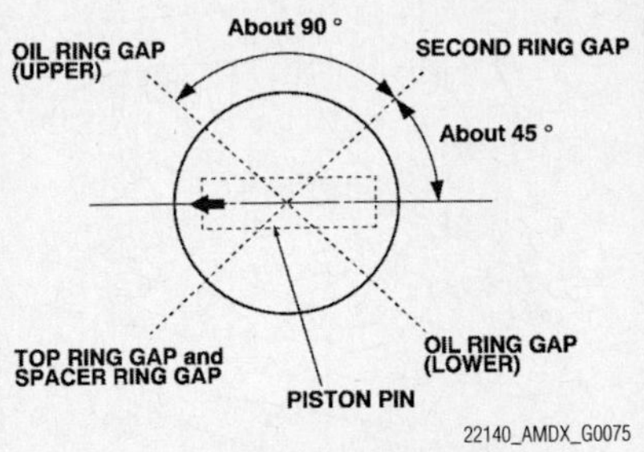
About 90 °
OIL RING GAP (UPPER)
SECOND RING GAP
About 45 °
TOP RING GAP and SPACER RING GAP
OIL RING GAP (LOWER)
PISTON PIN
22140_AMDX_G0075

Fig. 79 Ring end gap positioning—3.7 Engine

REAR MAIN SEAL

REMOVAL & INSTALLATION

1. Remove the transaxle assembly.
2. Remove the driveplate.
3. Remove the rear main seal.

To install:

4. Clean and dry the crankshaft oil seal housing.
5. Apply a light coat of multipurpose grease to the crankshaft and to the lip of the seal.
6. Using a suitable seal driver tool, drive in the rear main seal until the driver attachment bottoms against the engine block end cover. Align the hole in the driver attachment with the pin on the crankshaft.
7. Clean any excess grease off the crankshaft, and check that the oil seal lip is not distorted.
8. Install the driveplate and transaxle assembly

ROCKER ARMS

REMOVAL & INSTALLATION

See Figures 80 through 82.

1. Remove the intake manifold. For additional information, refer to the following section, "Intake Manifold, Removal & Installation."
2. Remove the cylinder head cover. For additional information, refer to the following section, "Cylinder Head, Removal & Installation."
3. Loosen the locknuts and adjusting screws.
4. Remove the rocker arm assembly as follows:
   a. Unscrew the rocker shaft bolts 2 turns at a time in a criss-cross pattern to avoid damaging the vales or rocker assembly.

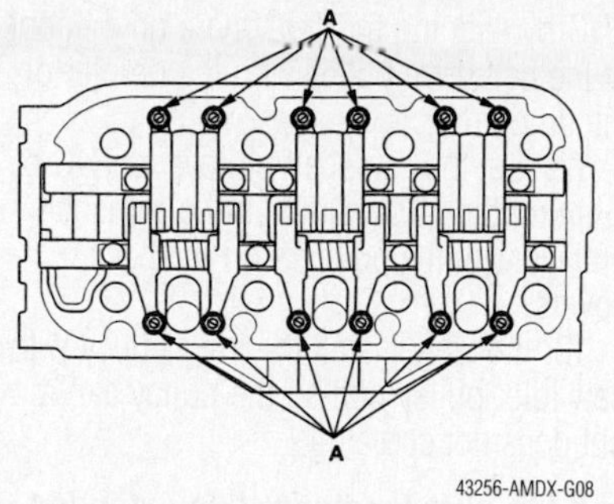
A
A
43256-AMDX-G08

Fig. 80 Rocker arm shaft adjusting screw locations

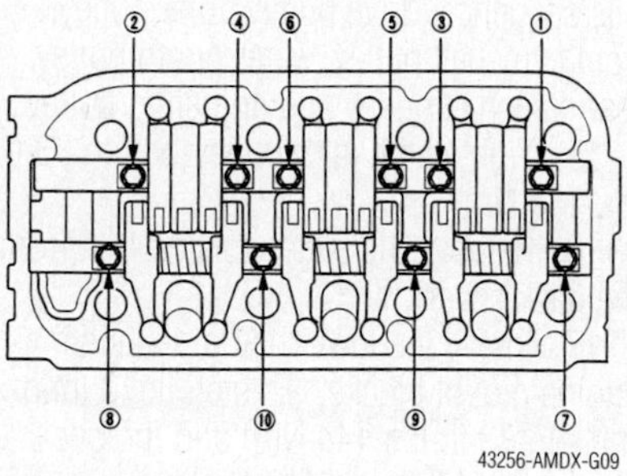
43256-AMDX-G09

Fig. 81 Rocker arm shaft loosening sequence

b. Do not remove the rocker shaft bolts. These bolts keep the springs and rocker arms on the shafts.

5. Remove the rocker arms and shafts from the vehicle as an assembly.

➡Keep all valve train components in order for assembly.

6. Remove the rocker arms and springs from the rocker arm shafts.

To install:

7. Assemble the rocker arms and springs to the rocker arm shafts in their original positions.

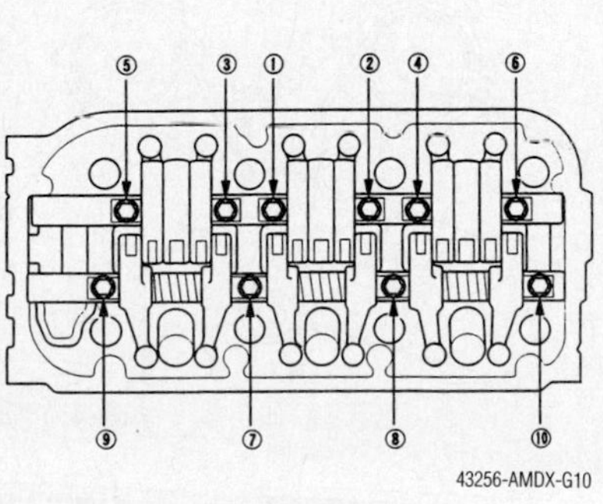
43256-AMDX-G10

Fig. 82 Rocker shaft tightening sequence

8. Install the rocker arm assemblies. Tighten the bolts in sequence and in multiple passes to 17 ft. lbs. (24 Nm).
9. Adjust the valve clearance.
10. Install the cylinder head cover.
11. Install the intake manifold.

TIMING BELT FRONT COVER

REMOVAL & INSTALLATION

See Figures 83 and 84.

1. Disconnect the negative battery cable.
2. Remove the splash shield.
3. Remove the accessory drive belt.
4. Remove the accessory drive belt auto-tensioner.
5. Support the engine assembly with a suitable jack.
6. Remove the ground cable, then remove the upper half of the side engine mount bracket.
7. Remove the front upper and rear upper cover.
8. Remove the crankshaft pulley.
9. Remove the lower cover.

To install:

10. Install the lower cover and tighten the mounting bolts to 9 ft. lbs. (12 Nm).
11. Install the front upper and rear upper cover and tighten the mounting bolts to 9 ft. lbs. (12 Nm).
12. Install the crankshaft pulley.
13. Install the accessory drive belt auto-tensioner.
14. Install the accessory drive belt.
15. Install the splash shield.
16. Connect the negative battery cable.
17. Start the engine and check for leaks.

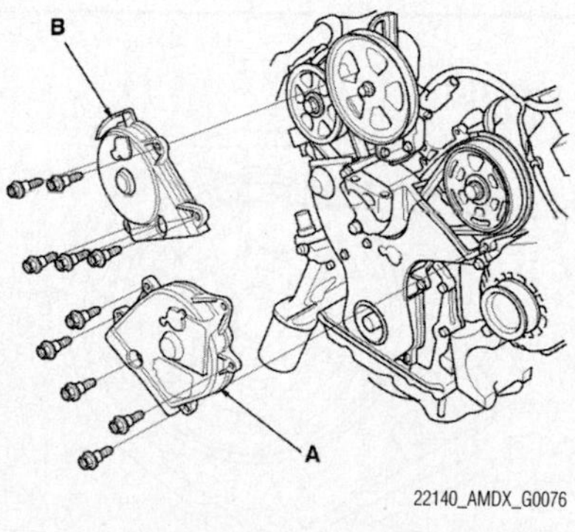
B
A
22140_AMDX_G0076

Fig. 83 Removing the front upper (A) and rear upper (B) cover

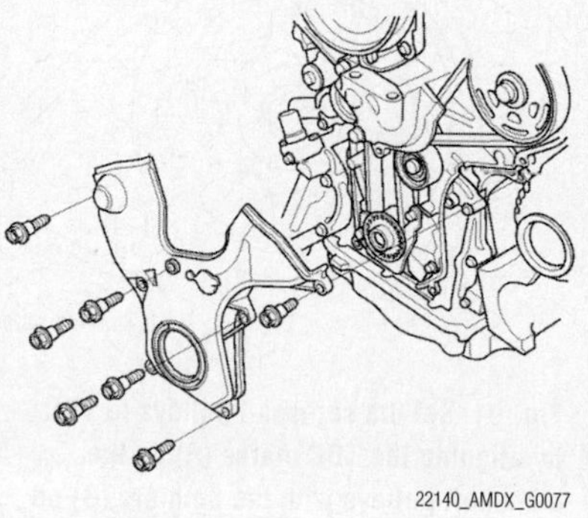
22140_AMDX_G0077

Fig. 84 Removing the lower cover

TIMING BELT & SPROCKETS

REMOVAL & INSTALLATION

See Figures 85 through 93.

1. Turn the crankshaft so the white mark aligns with the pointer.
2. Check that the No. 1 piston top dead center (TDC) mark on the front camshaft pulley and the pointer on the front upper cover are aligned.

➡If the marks are not aligned, rotate the crankshaft 360 degrees, and recheck the camshaft pulley mark.

3. Raise and safely support the vehicle, then remove the right front wheel.

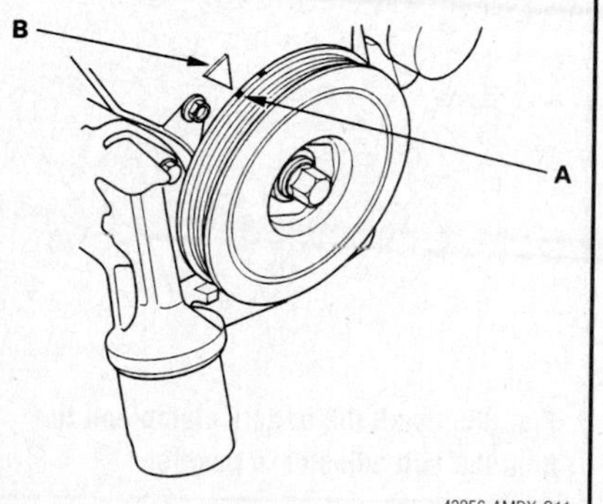
B
A
43256-AMDX-G11

Fig. 85 Turn the crankshaft so the white mark (A) aligns with the pointer (B)

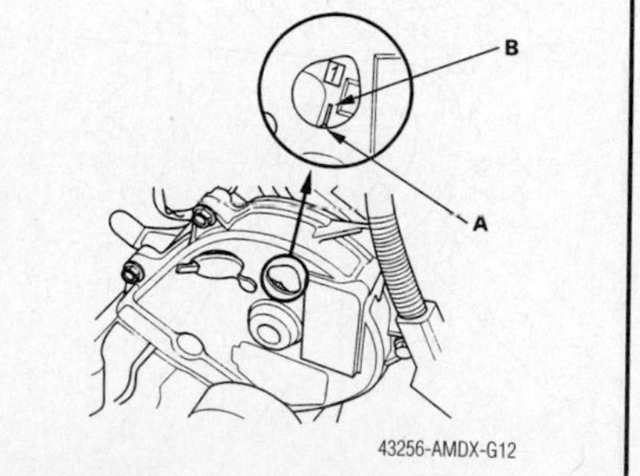
B
A
43256-AMDX-G12

Fig. 86 Make sure the number 1 piston is at top dead center (A) on the front camshaft pulley and pointer (B)

4. Remove the timing belt front covers. For additional information, refer to the following section, "Timing Belt Front Cover, Removal & Installation."

5. Remove one of the battery clamp bolts from the battery tray, and grind the end of it flat.

6. Thread in the battery clamp bolt as shown to hold the timing belt adjuster in its current position. Tighten it by hand; do not use a wrench.

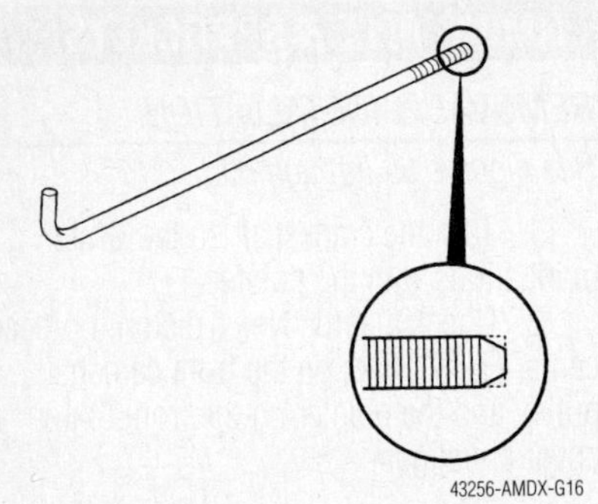
43256-AMDX-G16

Fig. 87 Remove a battery clamp bolt and grind the end

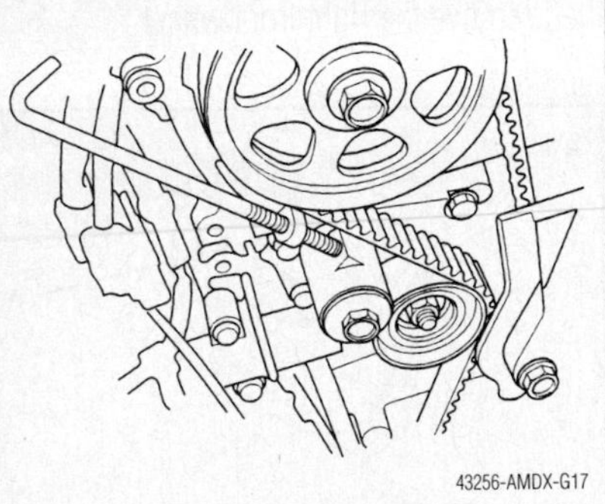
43256-AMDX-G17

Fig. 88 Install the battery clamp bolt to hold the belt adjuster in position

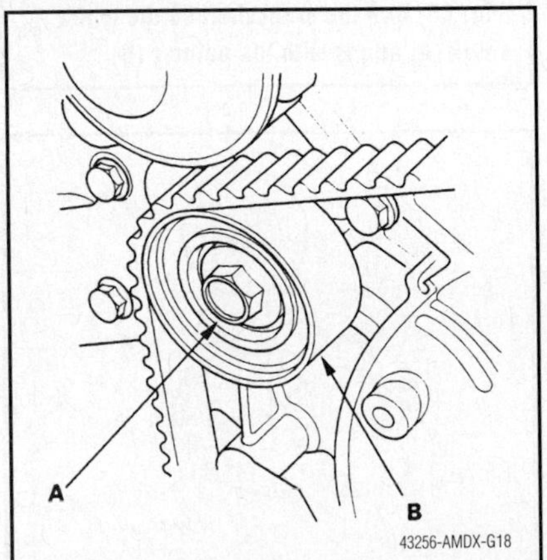
A
B
43256-AMDX-G18

Fig. 89 Remove the idler pulley bolt (A), pulley (B) and the timing belt

7. Remove the timing belt guide plate.

8. Remove the lower half of the side engine mount bracket.

9. Remove the idler pulley bolt and idler pulley, then remove the timing belt. Discard the idler pulley bolt.

To install:

10. Clean the pulleys, belt guide plate and the upper and lower covers.

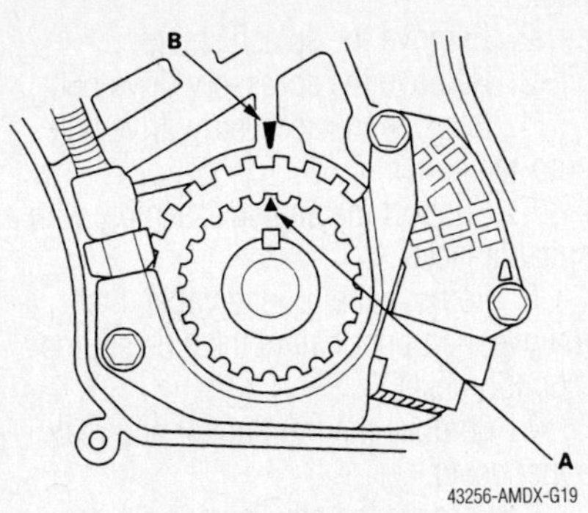
B
A
43256-AMDX-G19

Fig. 90 Set the timing belt pulley to TDC by aligning the TDC mark (A) on the tooth of the belt pulley with the pointer (B) on the oil pump

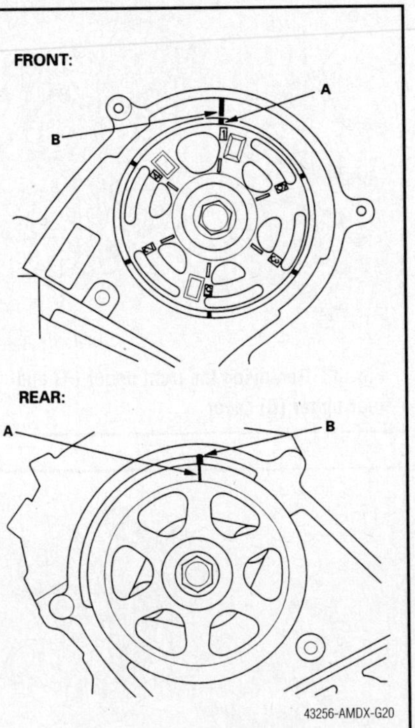
FRONT:
A
B
REAR:
A
B
43256-AMDX-G20

Fig. 91 Set the camshaft pulleys to TDC by aligning the TDC marks (A) on the camshaft pulleys with the pointers (B) on the back covers

11. Set the timing belt drive pulley to TDC by aligning the TDC mark on the tooth of the belt drive pulley with the pointer on the oil pump.

12. Set the camshaft pulleys to TDC by aligning the TDC marks on the camshaft pulleys with the pointers on the back covers.

13. Loosely install the idler pulley with a new idler pulley bolt so the pulley can move but does not come off.

➡If the auto-tensioner has extended and the timing belt cannot be installed, do the timing belt replacement procedure.

14. Install the belt over the pulleys in this sequence; drive pulley, idler pulley, front camshaft pulley, water pump pulley, rear camshaft pulley and adjusting pulley.

15. Tighten the idler pulley bolt to 33 ft. lbs. (44 Nm).

16. Remove the battery clamp bolt from the back cover.

17. Install the lower half of the side engine mount bracket. Tighten the 3 long bolts to 33 ft. lbs. (44 Nm) and the one short bolt to 9 ft. lbs. (12 Nm).

18. Install the timing belt guide plate as shown.

19. Install the timing belt front covers.

20. Install the crankshaft pulley.

21. Rotate the crankshaft pulley about six turns clockwise so the timing belt positions itself on the pulleys.

22. Turn the crankshaft pulley so the white mark lines up with the pointer.

23. Check the camshaft pulley marks are aligned. If the marks are aligned, proceed to

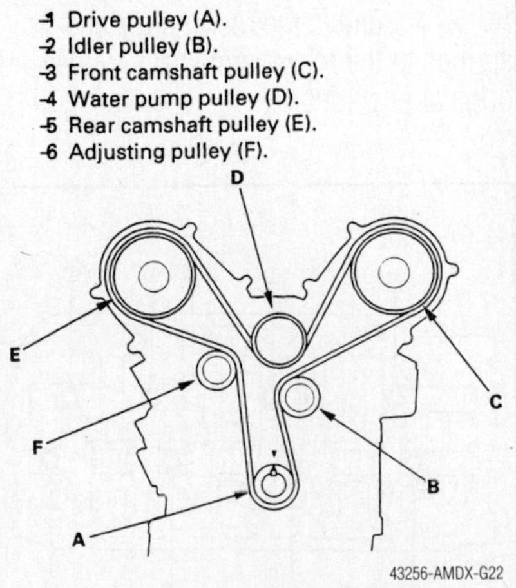
1 Drive pulley (A).
2 Idler pulley (B).
3 Front camshaft pulley (C).
4 Water pump pulley (D).
5 Rear camshaft pulley (E).
6 Adjusting pulley (F).
D
E
C
F
B
A
43256-AMDX-G22

Fig. 92 Route the belt in the sequence listed

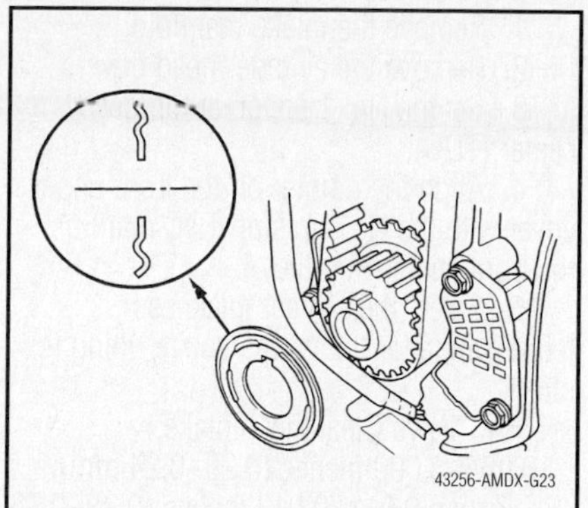
43256-AMDX-G23

Fig. 93 Install the timing belt guide plate

the next step. If the marks are not aligned, remove the timing belt and reinstall using the steps outlined before this step.

24. Install the upper half of the side engine mount bracket, and tighten the new mounting bolts to 33 ft. lbs. (44 Nm), then tighten the mass damper mounting bolt to 40 ft. lbs. (54 Nm).
25. Install the accessory drive belt and auto-tensioner.
26. Install the splash shield.
27. Install the right front wheel.
28. Start the engine and check for leaks.

TIMING BELT REAR COVER

REMOVAL & INSTALLATION

1. Remove the timing belt. For additional information, refer to the following section, "Timing Belt, Removal & Installation."
2. Remove the camshaft pulley.
3. Remove the rear cover.

To install:

4. Install the rear cover.
5. Install the camshaft pulley.
6. Install the timing belt.
7. Start the engine and check for leaks.

VALVE COVERS

REMOVAL & INSTALLATION

See Figures 94 through 98.

1. Remove the intake manifold.
2. Remove the six ignition coils.
3. Remove the dipstick.
4. Remove the two bolts securing the harness holder, and disconnect the front Air Fuel Ratio (A/F) sensor connector, front secondary heated oxygen sensor (secondary HO2S) connector, Exhaust Gas Recirculation (EGR) valve connector and Engine Coolant Temperature (ECT) sensor 1 connector.

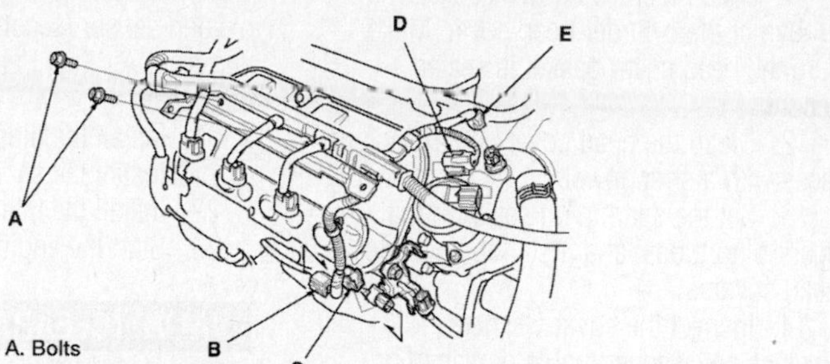
D
E
A
B
C
A. Bolts
B. A/F sensor connector
C. Secondary HO2S connector
D. EGR valve connector
E. ECT sensor 1 connector
37647_AMDX_G0083

Fig. 94 Remove the two bolts securing the harness holder, and disconnect the front A/F sensor connector, front secondary heated oxygen sensor connector, EGR valve connector and ECT sensor 1 connector

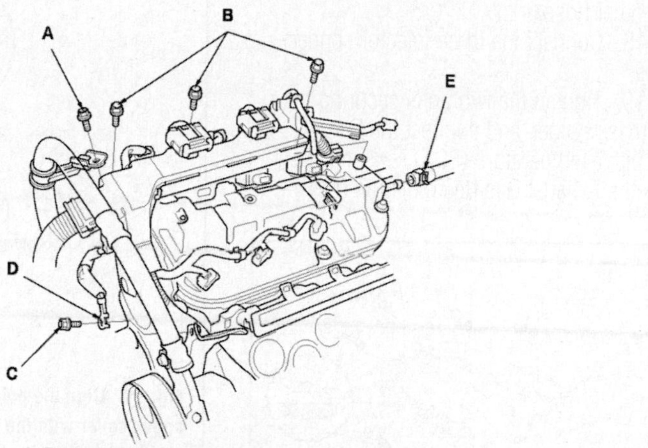
A
B
E
D
C
A. Power steering hose bracket mounting bolt
B. Harness holder mounting bolts
C. Engine ground cable bolt
D. Engine ground cable
E. Breather hose
37647_AMDX_G0084

Fig. 95 Remove the power steering hose bracket mounting bolt, the harness holder mounting bolts, and the engine ground cable bolt

5. Disconnect the three injector connectors from the injectors on the rear side cylinder head.
6. Remove the power steering hose bracket mounting bolt, the harness holder mounting bolts, and the engine ground cable bolt.
7. Remove the engine ground cable and disconnect the breather hose.
8. Remove the cylinder head covers.

To install:

9. Check the spark plug seals and head cover gasket for damage and deterioration. If any seal or gasket is damaged, replace it.
10. Thoroughly clean the head cover gasket and the groove of the cylinder cover.

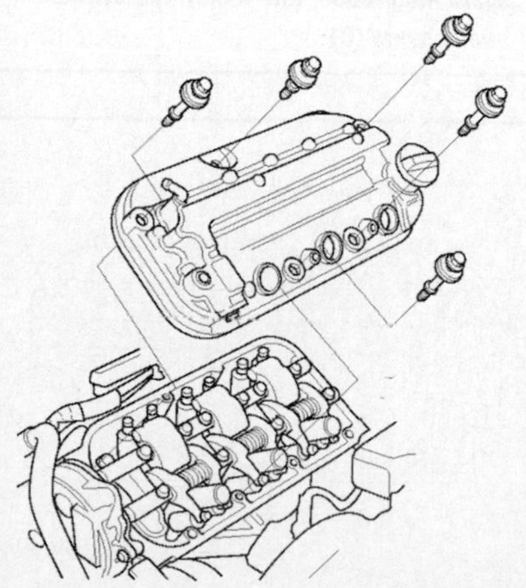
37647_AMDX_G0085

Fig. 96 Remove the cylinder head covers

11. Install the head cover gasket in the groove of the cylinder head cover. Make sure the head cover gasket is seated securely.
12. Clean the head cover contacting surfaces with a shop towel.
13. Set the spark plug seals on the spark plug tubes, and install the cylinder head covers.
14. Inspect the cover washers. Replace any washer that is damaged or deteriorated.
15. Tighten the cylinder head cover mounting bolts in sequence to 9 ft. lbs. (12 Nm).
16. Tighten the harness holder mounting bolts and the power steering hose bracket mounting bolt to 9 ft. lbs. (12 Nm).
17. Install the engine ground cable and breather hose.
18. Connect the three injector connectors.
19. Tighten the two bolts securing the harness holder, and connect the Engine Coolant Temperature (ECT) sensor 1 connector, Exhaust Gas Recirculation (EGR) valve connector, front secondary heated oxygen sensor (secondary HO2S) connector and front Air Fuel Ratio (A/F) sensor connector.
20. Install the dipstick.
21. Install the six ignition coils.
22. Install the intake manifold.
23. Start the engine to check for leaks.

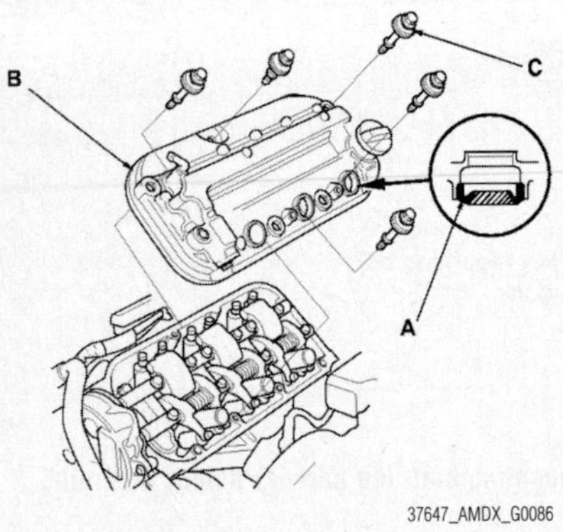
B
C
A
37647_AMDX_G0086

Fig. 97 Set the spark plug seals (A) on the spark plug tubes, and install the cylinder head covers (B)

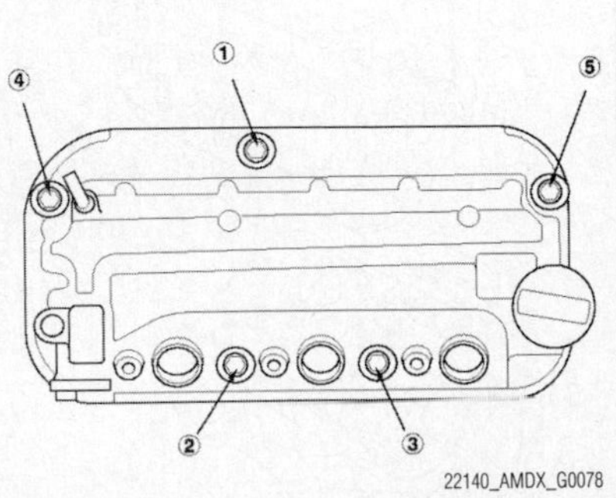
①
④
⑤
②
③
22140_AMDX_G0078

Fig. 98 Cylinder head cover mounting bolt tightening sequence

VALVE (CLEARANCE) LASH

ADJUSTMENT

See Figures 99 through 103.

➡Adjust the valves only when the cylinder head temperature is less than 100°F (38°C).

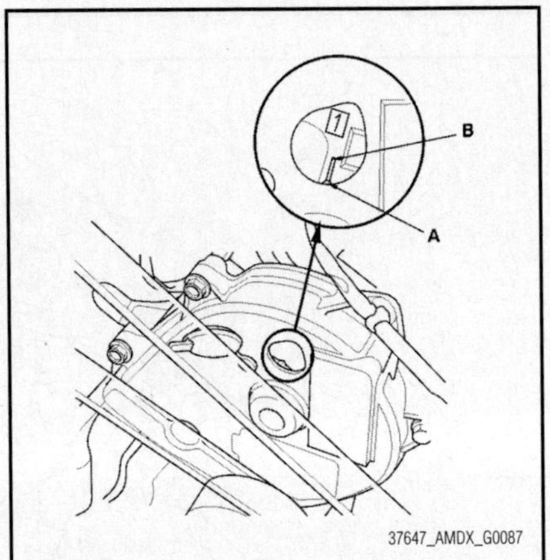
B
A
37647_AMDX_G0087

Fig. 99 Align the pointer (A) on the front upper cover with the No. 1 piston TDC mark (B) on the front camshaft pulley

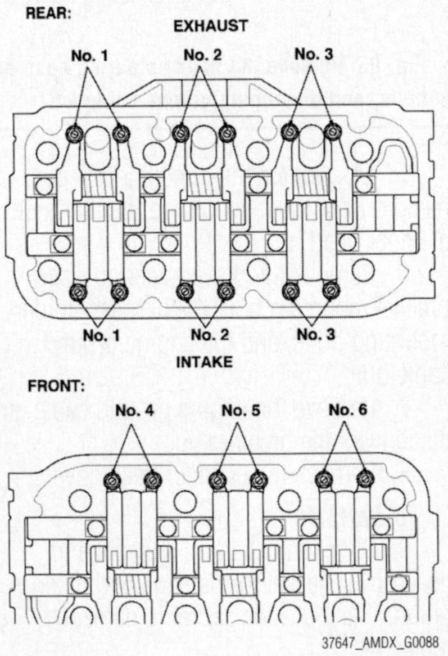
REAR:
EXHAUST
No. 1
No. 2
No. 3
No. 1
No. 2
No. 3
INTAKE
FRONT:
No. 4
No. 5
No. 6
37647_AMDX_G0088

Fig. 100 Intake and exhaust valve locations

1. Remove the intake manifold.
2. Remove the cylinder head covers.
3. Set the No. 1 piston at top dead center (TDC).
4. Align the pointer on the front upper cover with the No. 1 piston TDC mark on the front camshaft pulley.
5. Select the correct thickness feeler gauge for the valves you're going to check.
   a. Valve Clearance: Intake: 0.008–0.009 inches (0.20–0.24 mm); Exhaust: 0.011–0.013 inches (0.28–0.32 mm)
6. Insert the feeler gauge between the adjusting screw and the end of the valve stem on No. 1 cylinder and slide it back and forth; you should feel a slight amount of drag.
7. If you feel too much or too little drag, loosen the locknut, and turn the adjusting screw until the drag on the feeler gauge is correct.
8. Tighten the locknut and recheck the clearance.
9. Repeat the adjustment, if necessary.

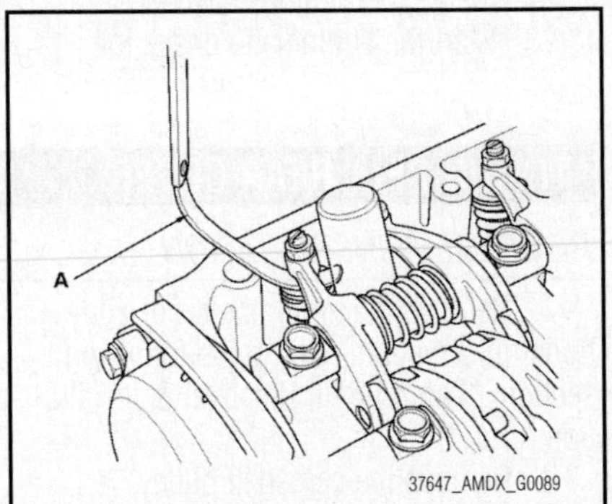
A
37647_AMDX_G0089

Fig. 101 Insert the feeler gauge (A) between the adjusting screw and the end of the valve stem on No. 1

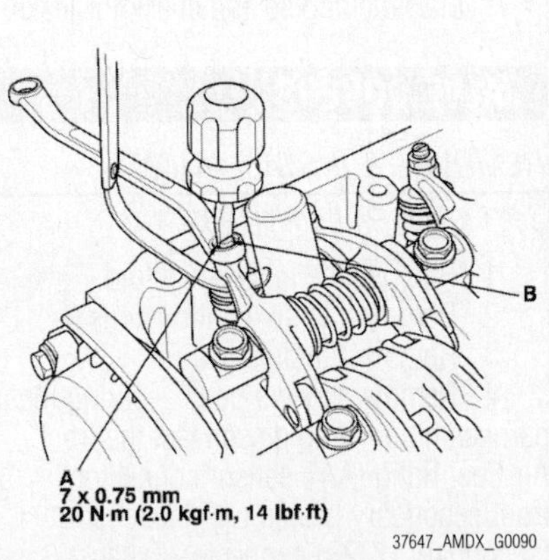
B
A
7 x 0.75 mm
20 N·m (2.0 kgf·m, 14 lbf·ft)
37647_AMDX_G0090

Fig. 102 Loosen the locknut (A), and turn the adjusting screw (B)

10. Rotate the crankshaft clockwise. Align the pointer on the front upper cover with the No. 4 piston TDC mark on the front camshaft pulley.

11. Check and, if necessary, adjust the valve clearance on No. 4 cylinder.

12. Rotate the crankshaft clockwise. Align the pointer on the front upper cover with the No. 2 piston TDC mark on the front camshaft pulley.

13. Check and, if necessary, adjust the valve clearance on No. 2 cylinder.

14. Rotate the crankshaft clockwise. Align the pointer on the front upper cover with the No. 5 piston TDC mark on the front camshaft pulley.

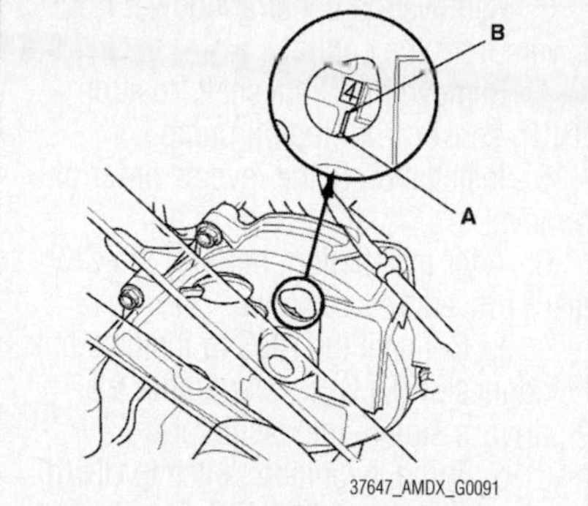
B
A
37647_AMDX_G0091

Fig. 103 Align the pointer (A) on the front upper cover with the No. 4 piston TDC mark (B) on the front camshaft pulley

15. Check and, if necessary, adjust the valve clearance on No. 5 cylinder.

16. Rotate the crankshaft clockwise. Align the pointer on the front upper cover with the No. 3 piston TDC mark on the front camshaft pulley.

17. Check and, if necessary, adjust the valve clearance on No. 3 cylinder.

18. Rotate the crankshaft clockwise. Align the pointer on the front upper cover with the No. 6 piston TDC mark on the front camshaft pulley.

19. Check and, if necessary, adjust the valve clearance on No. 6 cylinder.

20. Install the cylinder head covers.

21. Install the intake manifold.

ENGINE PERFORMANCE & EMISSION CONTROLS

ACCELERATOR PEDAL POSITION (APP) SENSOR

LOCATION

The Accelerator Pedal Position (APP) sensor is located on the accelerator pedal module. The accelerator pedal position sensor is an integrated part of the accelerator pedal module. If the sensor is faulty, the entire module must be replaced.

REMOVAL & INSTALLATION

See Figure 104.

1. Disconnect the Accelerator Pedal Position (APP) sensor 6P connector.
2. Remove the accelerator pedal module.
3. Installation is the reverse order of removal.

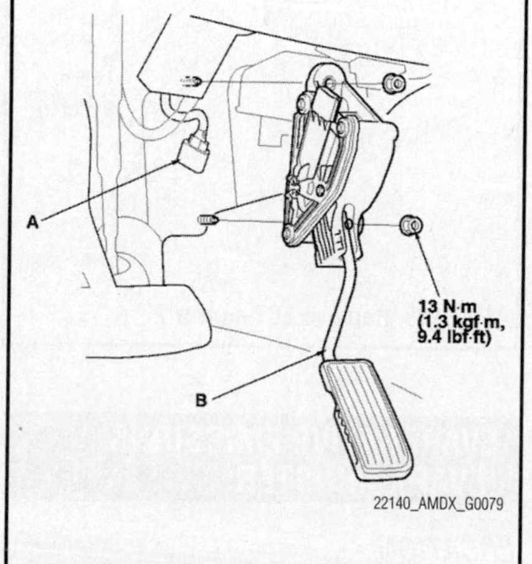
A
13 N·m
(1.3 kgf·m,
9.4 lbf·ft)
B
22140_AMDX_G0079

Fig. 104 Disconnect the APP sensor wiring harness (A) to remove the accelerator pedal module (B)

CAMSHAFT POSITION (CMP) SENSOR

LOCATION

The Camshaft Position (CMP) sensor is located on the timing belt cover.

REMOVAL & INSTALLATION

See Figures 105 and 106.

1. Remove the timing belt.
2. Remove the front camshaft pulley (CMP sensor pulse plate).
3. Disconnect the Camshaft Position Sensor (CMP), then remove the back cover.
4. Remove the CMP from the back cover.
5. Installation is the reverse order of removal.

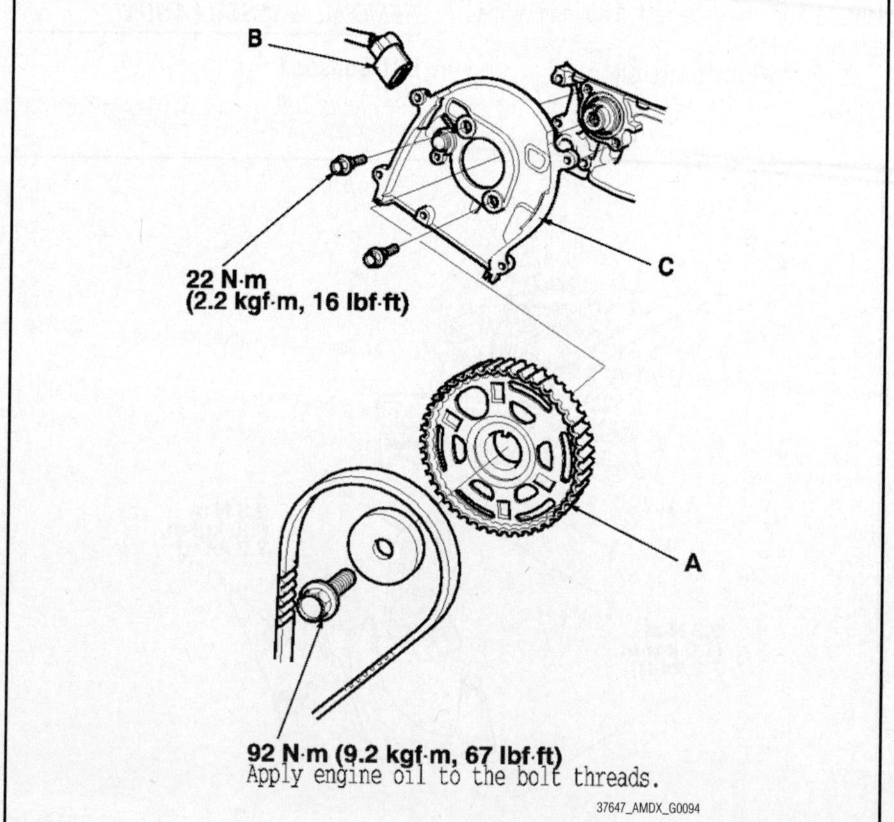
B
C
22 N·m
(2.2 kgf·m, 16 lbf·ft)
A
92 N·m (9.2 kgf·m, 67 lbf·ft)
Apply engine oil to the bolt threads.
37647_AMDX_G0094

Fig. 105 Disconnect the CMP sensor connector (B), and remove the back cover (C)

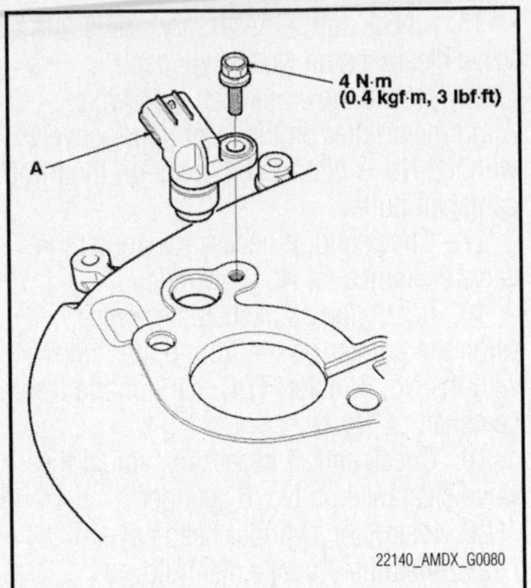
4 N·m
(0.4 kgf·m, 3 lbf·ft)
A
22140_AMDX_G0080

Fig. 106 Installing the camshaft position sensor into the back cover

CRANKSHAFT POSITION (CKP) SENSOR

LOCATION

The Crankshaft Position (CKP) sensor is located on the oil pump.

REMOVAL & INSTALLATION

See Figure 107.

1. Move the auto-tensioner to remove tension from the drive belt, then remove the accessory drive belt.
2. Remove the crankshaft pulley.
3. Remove the upper and lower front covers from the engine.
4. Remove the Crankshaft Position (CKP) sensor from the oil pump.
5. Installation is the reverse order of removal.
6. After installation, perform the CKP learn procedure as follows:
   a. Connect the HDS to the data link connector (DLC) located under the driver's side of the dashboard.
   b. Turn the ignition switch to ON (II).
   c. Make sure the HDS communicates with the PCM and other vehicle systems. If it doesn't, go to the DLC circuit troubleshooting.
   d. Select CRANK PATTERN in the ADJUSTMENT MENU with the HDS.
   e. Select CRANK PATTERN LEARNING with the HDS, and follow the screen prompts.

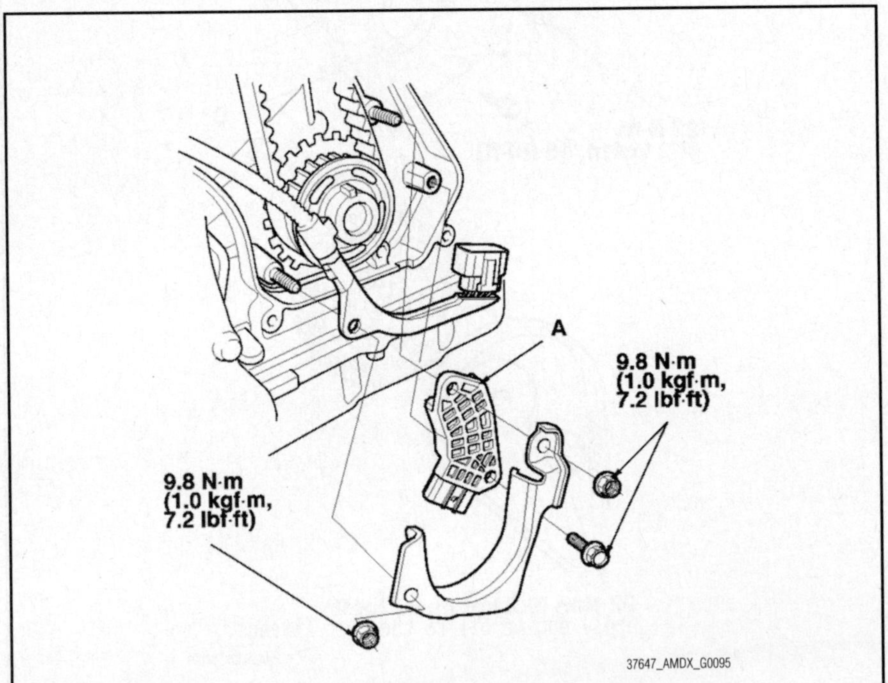
A
9.8 N·m
(1.0 kgf·m,
7.2 lbf·ft)
9.8 N·m
(1.0 kgf·m,
7.2 lbf·ft)
37647_AMDX_G0095

Fig. 107 Remove the Crankshaft Position (CKP) sensor (A) from the oil pump

ENGINE COOLANT TEMPERATURE (ECT) SENSOR

LOCATION

Sensor 1 is located in the engine compartment below the main underhood fuse box. Sensor 2 is located at the bottom of the radiator.

REMOVAL & INSTALLATION

ECT Sensor 1

See Figure 108.

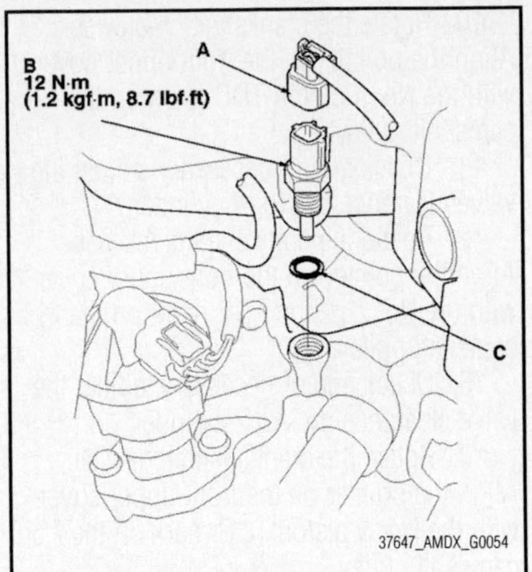
A
B
12 N·m
(1.2 kgf·m, 8.7 lbf·ft)
C
37647_AMDX_G0054

Fig. 108 Remove ECT sensor 1

1. Drain the engine cooling system.
2. Remove the engine appearance cover.
3. Remove the main under-hood fuse/relay box.
4. Disconnect the ECT sensor 1 connector.
5. Remove ECT sensor 1.
6. Install the parts in the reverse order of removal with a new O-ring, then refill the radiator with engine coolant.

ECT Sensor 2

See Figure 109.

1. Drain the engine cooling system.
2. Remove the splash shield.
3. Disconnect the ECT sensor 2 connector.
4. Remove ECT sensor 2.
5. Install the parts in the reverse order of removal with a new O-ring, then refill the radiator with engine coolant.

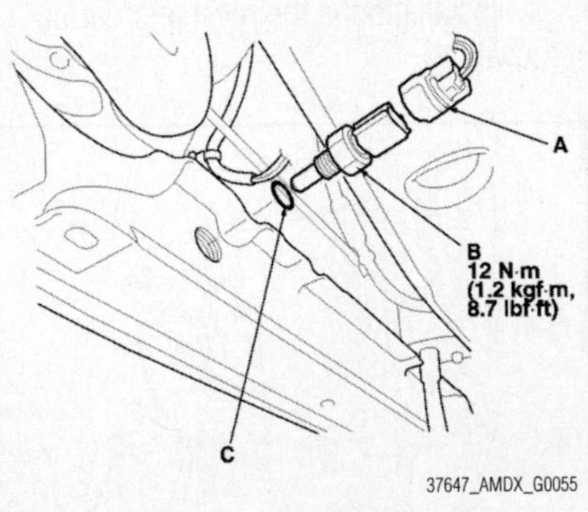
A
B
12 N·m
(1.2 kgf·m,
8.7 lbf·ft)
C
37647_AMDX_G0055

Fig. 109 Remove ECT sensor 2

EVAPORATIVE EMISSIONS (EVAP) CANISTER

LOCATION

The Evaporative Emissions (EVAP) canister is located under the vehicle body near the fuel tank.

REMOVAL & INSTALLATION

See Figures 110 through 113.

1. Remove the EVAP canister cover. Open the clamps.
2. Remove the hoses, the FTP sensor connector, and the EVAP canister vent shut valve connector.
3. Remove the bolts.
4. Remove the EVAP canister assembly.
5. Remove the EVAP canister bracket.
6. Pry the lock tabs outward, then remove the EVAP canister vent shut valve.

➡Be careful not to damage the lock tabs.

7. Install the EVAP canister vent shut valve in the new EVAP canister with a new O-ring.

➡Do not coat the O-ring with oil.

8. Install the parts in the reverse order of removal.

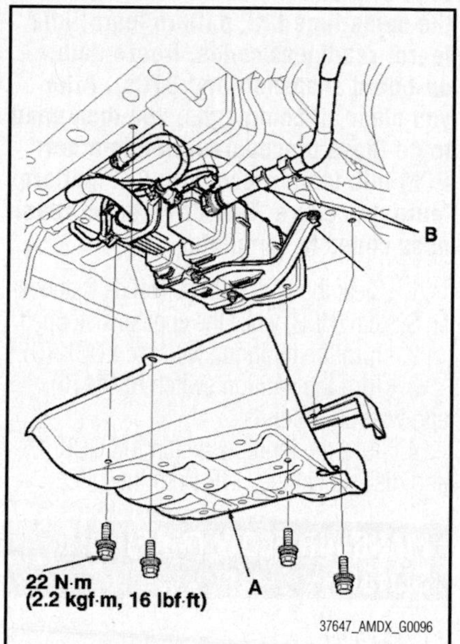
B
22 N·m
(2.2 kgf·m, 16 lbf·ft)
A
37647_AMDX_G0096
Fig. 110 Remove the EVAP canister cover (A). Open the clamps (B)

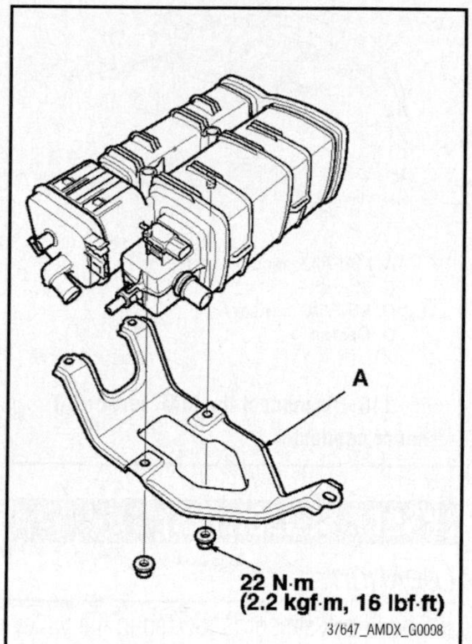
A
22 N·m
(2.2 kgf·m, 16 lbf·ft)
37647_AMDX_G0098
Fig. 112 Remove the EVAP canister bracket (A)

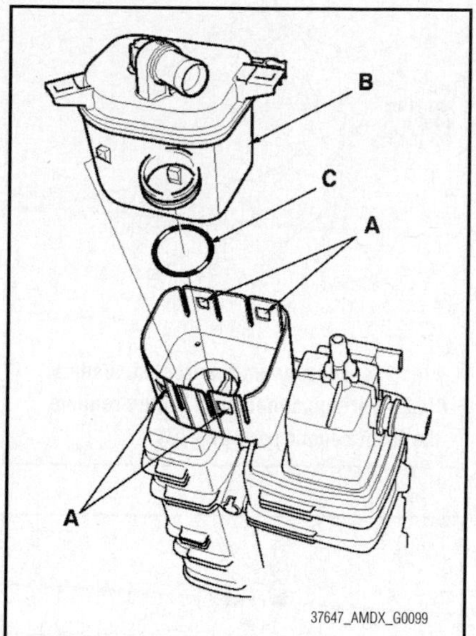
B
C
A
A
37647_AMDX_G0099
Fig. 113 Pry the lock tabs (A) outward, then remove the EVAP canister vent shut valve

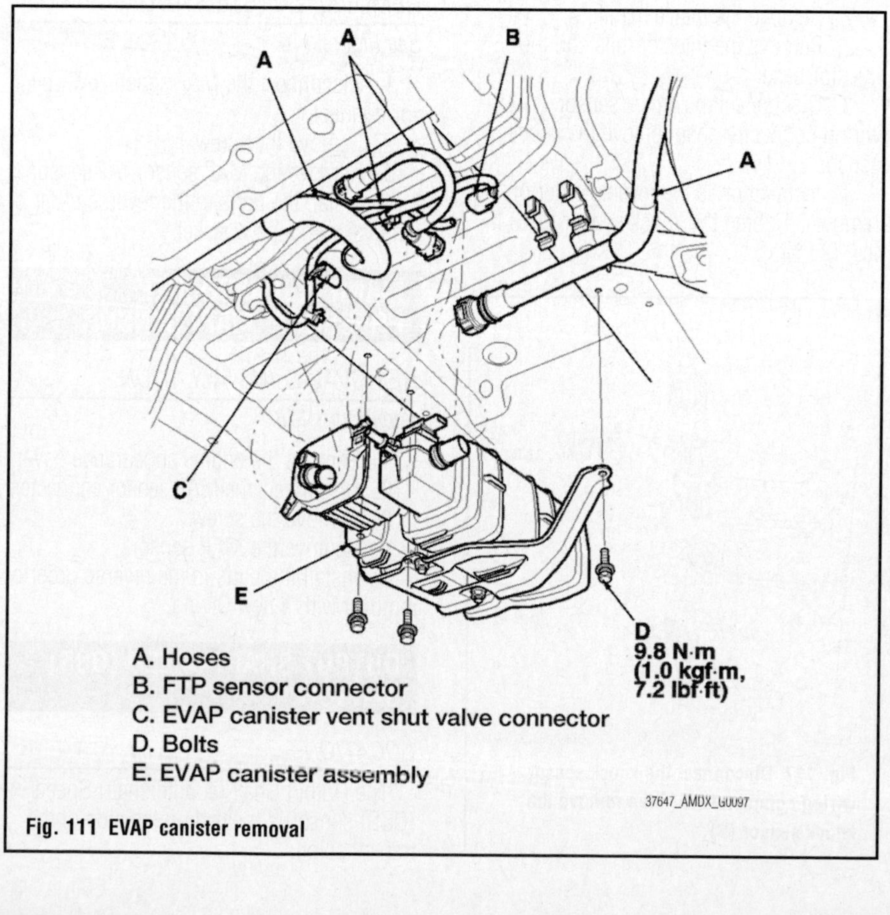
A
A
B
A
C
E
D
9.8 N·m
(1.0 kgf·m,
7.2 lbf·ft)
A. Hoses
B. FTP sensor connector
C. EVAP canister vent shut valve connector
D. Bolts
E. EVAP canister assembly
37647_AMDX_G0097
Fig. 111 EVAP canister removal

HEATED OXYGEN (HO2S) SENSOR

LOCATION

Refer to the graphics in the Removal and Installation section for locations.

REMOVAL & INSTALLATION

➡The following procedures require the use of an O2 sensor socket wrench.

Front Bank (Bank 2)

See Figure 114.

1. Disconnect the front secondary HO2S wiring connector, then remove the front secondary HO2S.
2. Install the parts in the reverse order of removal.

Rear Bank (Bank 1)

See Figure 115.

1. Disconnect the rear secondary HO2S wiring connector, then remove the rear secondary HO2S.
2. Install the parts in the reverse order of removal.

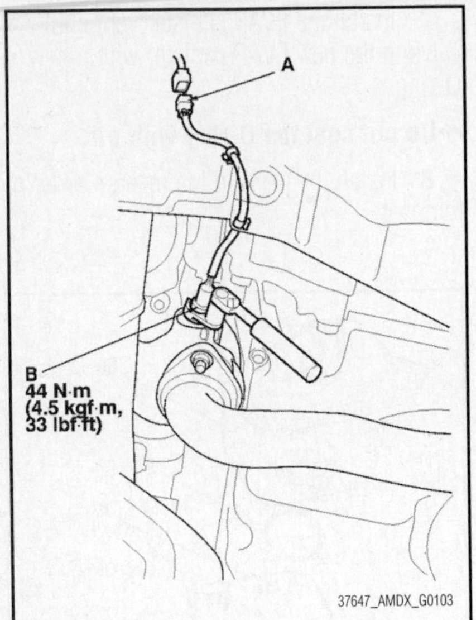
A
B
44 N·m
(4.5 kgf·m,
33 lbf·ft)
37647_AMDX_G0103

Fig. 114 Disconnect the front secondary HO2S wiring connector (A), then remove the front secondary HO2S (B)

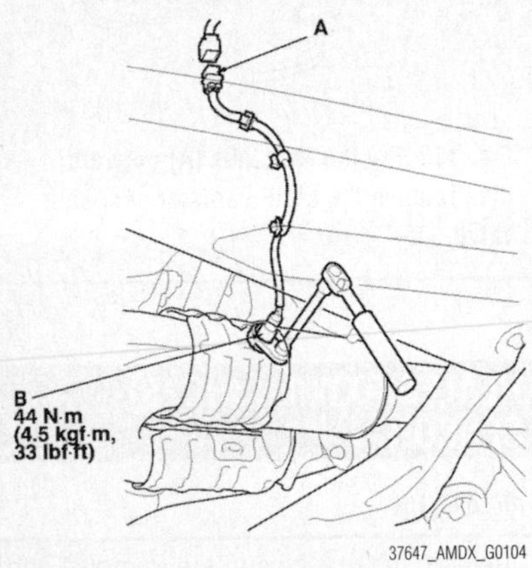
A
B
44 N·m
(4.5 kgf·m,
33 lbf·ft)
37647_AMDX_G0104

Fig. 115 Disconnect the rear secondary HO2S wiring connector (A), then remove the rear secondary HO2S (B)

INTAKE AIR TEMPERATURE (IAT) SENSOR

LOCATION

The MAF/IAT sensor is located in the air intake near the air cleaner.

REMOVAL & INSTALLATION

See Figure 116.

1. Disconnect the MAF sensor/IAT sensor connector.
2. Remove the screw.
3. Remove the MAF sensor/IAT sensor.
4. Install the parts in the reverse order of removal with a new gasket.

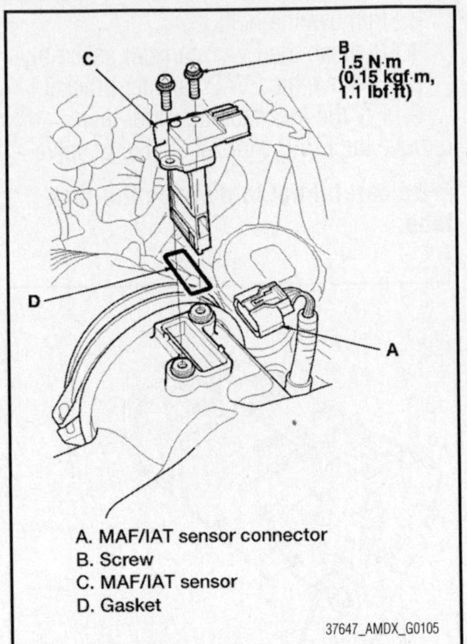
C
B
1.5 N·m
(0.15 kgf·m,
1.1 lbf·ft)
D
A
A. MAF/IAT sensor connector
B. Screw
C. MAF/IAT sensor
D. Gasket
37647_AMDX_G0105

Fig. 116 Disconnect the MAF sensor/IAT sensor connector

KNOCK SENSOR (KS)

LOCATION

The knock sensor is located in the valley of the engine block.

REMOVAL & INSTALLATION

See Figure 117.

1. Remove the intake manifold.
2. Remove the injector rails and the injector base.
3. Disconnect the knock sensor wiring connector, then remove the knock sensor.
4. Installation is the reverse order of removal. Tighten the knock sensor to 23 ft. lbs. (31 Nm).

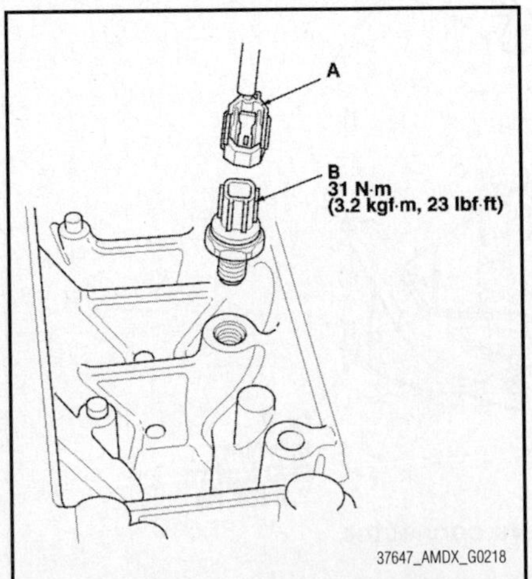
A
B
31 N·m
(3.2 kgf·m, 23 lbf·ft)
37647_AMDX_G0218

Fig. 117 Disconnect the knock sensor wiring connector (A), then remove the knock sensor (B)

MALFUNCTION INDICATOR LIGHT (MIL)

RESET PROCEDURE

➡If you are using a generic scan tool to clear commands, be aware that there is only one setting for clearing the PCM, and it clears all commands at the same time CKP pattern learn, idle learn, readiness codes, freeze data, on-board snapshot, and DTCs). After you clear all commands, you then need to do these procedures, in this order: PCM idle learn procedure; CKP pattern learn procedure; Test-drive to set readiness codes to complete.

1. Clear the DTC with the Honda Diagnostic System (HDS) while the engine is stopped.
2. Turn the ignition switch to LOCK (0).
3. Turn the ignition switch to ON (II), and wait 30 seconds.
4. Turn the ignition switch to LOCK (0), and disconnect the HDS from the DLC.

MASS AIR FLOW (MAF) SENSOR

LOCATION

The MAF/IAT sensor is located in the air intake near the air cleaner.

REMOVAL & INSTALLATION

See Figure 116.

1. Disconnect the MAF sensor/IAT sensor connector.
2. Remove the screw.
3. Remove the MAF sensor/IAT sensor.
4. Install the parts in the reverse order of removal with a new gasket.

MANIFOLD ABSOLUTE PRESSURE (MAP) SENSOR

REMOVAL & INSTALLATION

See Figure 118.

1. Remove the engine appearance cover.
2. Disconnect the MAP sensor connector.
3. Remove the screw.
4. Remove the MAP sensor.
5. Install the parts in the reverse order of removal with a new O-ring.

OUTPUT SHAFT SPEED (OSS) SENSOR

LOCATION

The Output Shaft (countershaft) Speed (OSS) sensor is located on the side of the transmission.

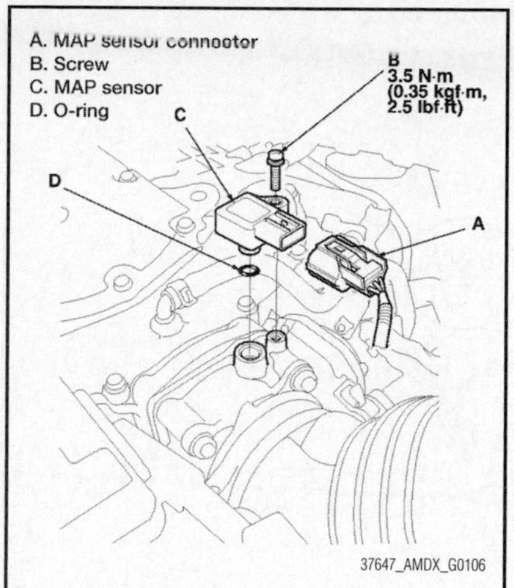
A. MAP sensor connector
B. Screw
C. MAP sensor
D. O-ring
B
3.5 N·m
(0.35 kgf·m,
2.5 lbf·ft)
C
D
A
37647_AMDX_G0106

Fig. 118 Disconnect the MAP sensor connector

REMOVAL & INSTALLATION

See Figures 119 through 121.

1. Raise the vehicle up on a lift, or apply the parking brake, block both rear wheels, and raise the front of the vehicle. Make sure it is securely supported.
2. Remove the transmission undercover and splash shield.
3. Remove the damper from the front subframe.
4. Disconnect the output shaft (countershaft) speed sensor connector.
5. Remove the output shaft (countershaft) speed sensor and sensor washer.

To install:

6. Install the new O-ring on the new output shaft (countershaft) speed sensor, then install the output shaft (countershaft) speed sensor and sensor washer.

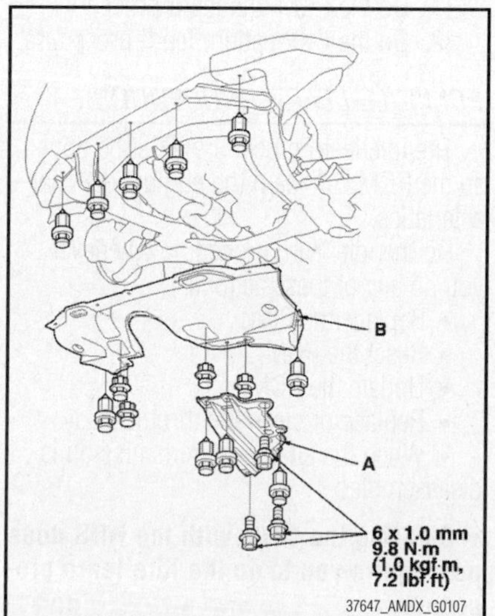
B
A
6 x 1.0 mm
9.8 N·m
(1.0 kgf·m,
7.2 lbf·ft)
37647_AMDX_G0107

Fig. 119 Remove the transmission undercover (A) and splash shield (B)

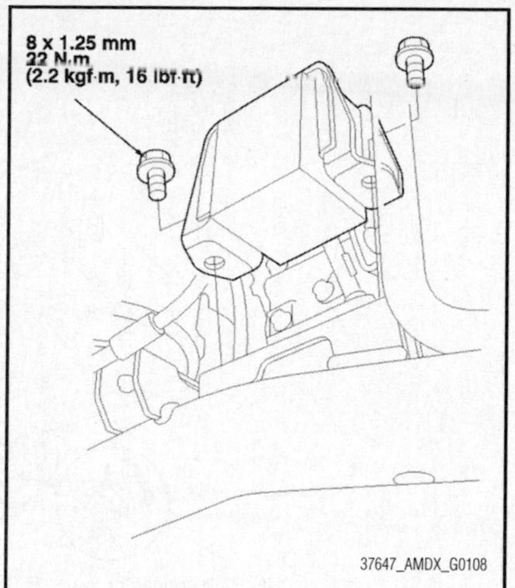
8 x 1.25 mm
22 N·m
(2.2 kgf·m, 16 lbf·ft)
37647_AMDX_G0108

Fig. 120 Remove the damper from the front subframe

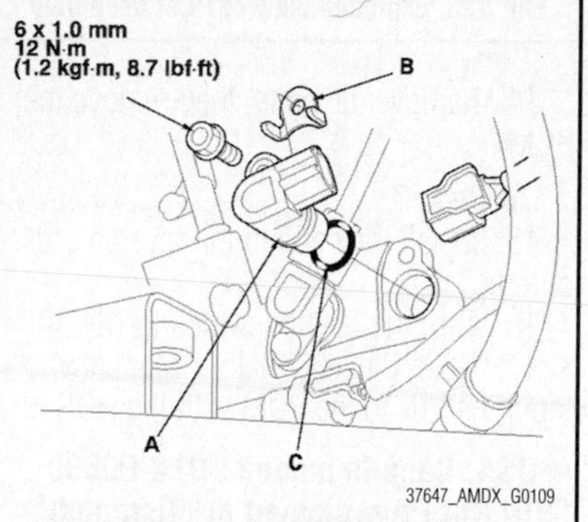
6 x 1.0 mm
12 N·m
(1.2 kgf·m, 8.7 lbf·ft)
B
A
C
37647_AMDX_G0109

Fig. 121 Remove the output shaft (countershaft) speed sensor (A) and sensor washer (B)

7. Check the connector for rust, dirt, or oil, then connect it securely.
8. Install the damper on the front subframe.
9. Install the splash shield and transmission undercover.

POSITIVE CRANKCASE VENTILATION (PCV) VALVE

LOCATION

The PCV valve is located on the intake manifold.

REMOVAL & INSTALLATION

See Figure 122.

1. Remove the engine cover.
2. Remove the bolt.

➡Take care not to spill oil on the hot exhaust manifold.

3. Remove the PCV valve.
4. Install the parts in the reverse order of removal.

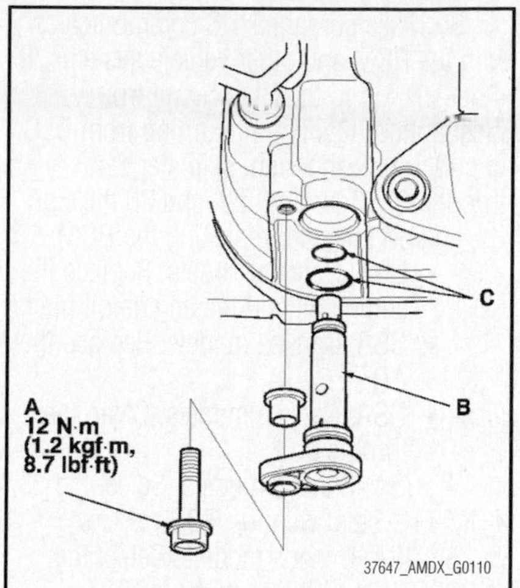
C
B
A
12 N·m
(1.2 kgf·m,
8.7 lbf·ft)
37647_AMDX_G0110

Fig. 122 Remove the bolt (A) and the PCV valve (B)

5. When installing a new PCV valve, use new O-rings and make sure they are in place.

POWERTRAIN CONTROL MODULE (PCM)

LOCATION

The Powertrain Control Module (PCM) is located in the engine compartment on the passenger side fender in front of the underhood fuse relay box.

REMOVAL & INSTALLATION

Special Tools Required

See Figure 123.

- Honda Diagnostic System (HDS) tablet tester
- Honda Interface Module (HIM) and an iN workstation with HDS and CM update software
- HDS pocket tester
- GNA600 and an iN workstation with HDS and CM update software

➡Use any one of these update tools.

- Make sure the HDS is loaded with the latest software version.
- If you are replacing the PCM after substituting a known-good PCM, reinstall the original PCM, then do this procedure.
- USA, Canada models: During the procedure, if any READ DATA, WRITE DATA, or other data checks fail, note the failure, then continue.

1. Connect the HDS to the Data Link Connector (DLC) located under the driver's side of the dashboard.
2. Turn the ignition switch to ON (II).

3. Make sure the HDS communicates with the PCM and other vehicle systems. If it doesn't, go to the DLC circuit troubleshooting. If you are returning from DLC circuit troubleshooting, skip steps 4 through 9, 18 through 23, and 26 through 28, and do this after replacing the PCM:

- USA, Canada models: Replace the engine oil and the engine oil filter.
- USA, Canada models: Replace the ATF.
- USA, Canada models: Clean the throttle body.

4. USA, Canada models: Select the PGM-FI system with the HDS.
5. USA, Canada models: Select the INSPECTION MENU with the HDS.
6. USA, Canada models: Select the ETCS TEST, then select the TP POSITION CHECK, and follow the screen prompts.

➡USA, Canada models: If the TP POSITION CHECK indicates FAILED, continue with this procedure.

7. USA, Canada models: Select the REPLACE PCM MENU, then select READ DATA and follow the screen prompts.

➡USA, Canada models: Doing this step copies (READS) the engine oil life data from the original PCM so you can later download (WRITES) it into the new PCM.

USA, Canada models: If READ DATA indicates FAILED, continue with this procedure.

8. USA, Canada models: Select the A/T system with the HDS.
9. USA, Canada models: Select the REPLACE TCM/PCM MENU, then select READ DATA and follow the screen prompts.

➡USA, Canada models: Doing this step copies (READS) the ATF life data from the original PCM so you can later download (WRITES) it into the new PCM.

USA, Canada models: If READ DATA indicates FAILED, continue with this procedure.

10. Turn the ignition switch to LOCK (0).
11. Jump the SCS line with the HDS.
12. Remove the bracket, then free the A/C discharge line from the clip and remove the A/C suction line mounting bracket bolt.
13. Remove the cover, then disconnect PCM connectors A, B, and C.

➡PCM connectors A, B, and C have symbols (A=□, B=△ , C=○) embossed on them for identification.

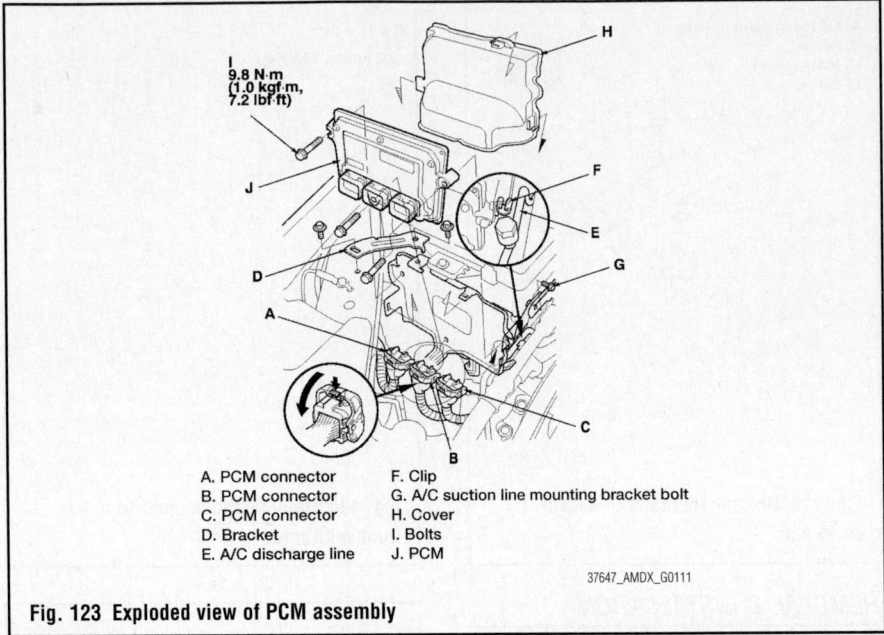

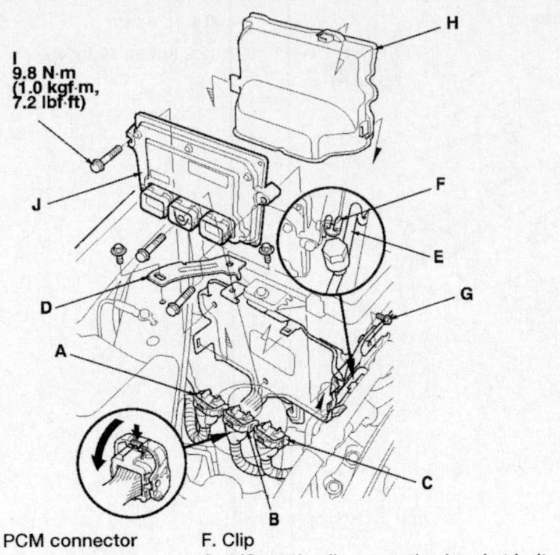
H
I
9.8 N·m
(1.0 kgf·m,
7.2 lbf·ft)
J
F
E
G
D
A
C
B
A. PCM connector
B. PCM connector
C. PCM connector
D. Bracket
E. A/C discharge line
F. Clip
G. A/C suction line mounting bracket bolt
H. Cover
I. Bolts
J. PCM
37647_AMDX_G0111

Fig. 123 Exploded view of PCM assembly

14. Remove the bolts, then remove the PCM.

To install:

15. Install the parts in the reverse order of removal.
16. Turn the ignition switch to ON (II).
17. USA, Canada models: Manually input the VIN to the PCM with the HDS.

➡USA, Canada models: DTC P0630 "VIN Not Programmed or Mismatch" may be stored because the VIN has not been programmed into the PCM; ignore it, and continue this procedure.

18. USA, Canada models: If the READ DATA (engine oil life) failed in step 7, go to step 21. Otherwise, go to step 19.
19. USA, Canada models: Select the PGM-FI system with the HDS.
20. USA, Canada models: Select the REPLACE PCM MENU, then select WRITE DATA and follow the screen prompts.

➡USA, Canada models: If the WRITE DATA indicates FAILED, continue with this procedure.

21. USA, Canada models: If the READ DATA (ATF life) failed in step 9, go to step 24. Otherwise go to step 22.
22. USA, Canada models: Select the A/T SYSTEM with the HDS.
23. USA, Canada models: Select the REPLACE TCM/PCM MENU, then select WRITE DATA and follow the screen prompts.

➡USA, Canada models: If the WRITE DATA indicates FAILED, continue with this procedure.

24. Select IMMOBI system with the HDS.
25. Enter the immobilizer code with the PCM replacement procedure in the HDS; it allows you to start the engine.
26. USA, Canada models: If the TP POSITION CHECK failed in step 6 clean the throttle body, then go to step 27.
27. USA, Canada models: If the READ DATA failed in step 7 or the WRITE DATA failed in step 20, replace the engine oil and engine oil filter, then go to step 28.
28. USA, Canada models: If the READ DATA failed in step 9 or the WRITE DATA failed in step 21, replace the ATF, then go to step 29.
29. Select PGM-FI system and reset the PCM with the HDS.
30. Reconnect all connectors, then update the PCM if it does not have the latest software.
31. Do the PCM idle learn procedure.
32. Do the CKP pattern learn procedure.

PCM IDLE LEARN PROCEDURE

The idle learn procedure must be done so the PCM can learn the engine idle characteristics.

Do the idle learn procedure whenever you do any of these actions:

- Replace the PCM.
- Reset the PCM.
- Update the PCM.
- Replace or clean the throttle body.
- When the engine or transmission is disassembled.

➡Clearing the DTCs with the HDS does not require you to do the idle learn procedure.

1. Make sure all electrical items (A/C, audio, lights, etc.) are off.

2. Reset the PCM with the HDS.
3. Turn the ignition switch to ON (II), and wait 2 seconds.
4. Start the engine. Hold the engine speed at 3,000 rpm without load (in Park or neutral) until the radiator fan comes on, or until the engine coolant temperature reaches 194°F (90°C).
5. Let the engine idle for about 5 minutes with the throttle fully closed.

➡If the radiator fan comes on, do not include its running time in the 5 minutes.

6. Verify on the HDS data list that the idle learn procedure is complete.

CKP PATTERN LEARN PROCEDURE

1. Connect the HDS to the data link connector (DLC) located under the driver's side of the dashboard.
2. Turn the ignition switch to ON (II). Make sure the HDS communicates with the PCM and other vehicle systems. If it doesn't, go to the DLC circuit troubleshooting.
3. Select CRANK PATTERN in the ADJUSTMENT MENU with the HDS.
4. Select CRANK PATTERN LEARNING with the HDS, and follow the screen prompts.

PCM UPDATE

- Use this procedure when you need to update the PCM during troubleshooting procedures.
- Make sure the HDS/HIM has the latest software version.
- Before you update the PCM, make sure the battery in the vehicle is fully charged and connect a jumper battery (not a battery charger) to maintain system voltage.
- Never turn the ignition switch to LOCK (0) or ACC (I) during the update. If there is a problem with the update, leave the ignition switch ON (II).
- To prevent PCM damage, do not operate anything electrical (headlights, audio system, brakes, A/C, power windows, door locks, etc.) during the update.
- To ensure the latest program is installed, do a PCM update whenever the PCM is substituted or replaced.
- You cannot update a PCM with a program it already has. It will only accept a new program.
- High temperature in the engine compartment might cause the PCM to become too hot to run the update. If the engine has been running before this procedure, open the hood and cool the engine compartment.
- If you need to diagnose the Honda Interface Module (HIM) because the HIM's red (#3) light came on or was flashing during the update, leave the ignition switch in ON (II) position when you disconnect the HIM from the Data Link Connector (DLC). This prevents damage to the PCM.

1. Turn the ignition switch to ON (II), but do not start the engine.
2. Connect the HDS to the data link connector (DLC) located under the driver's side of the dashboard.
3. Make sure the HDS communicates with the PCM and other vehicle systems. If it doesn't, go to the DLC circuit troubleshooting. If you are returning from the DLC circuit troubleshooting, skip steps 4 and 5 and clean the throttle body after updating the PCM.
4. USA, Canada models: Select the INSPECTION MENU with the HDS.
5. USA, Canada models: Select the ETCS TEST, then select the TP POSITION CHECK, and follow the HDS screen prompts.

➡USA, Canada models: If the TP POSITION CHECK indicates FAILED, continue this procedure.

6. Exit the HDS (USA, Canada models), then select the update mode, and follow the screen prompts to update the PCM.
7. If the software in the PCM is the latest, disconnect the HDS/HIM from the DLC, and go back to the procedure that you were doing. If the software in the PCM is not the latest, follow the instructions on the screen. If prompted to choose the PGM-FI system or the A/T system, make sure you update both.

➡If the PCM update system requires you to cool the PCM, follow the instructions on screen. If you run into a problem during the update procedure (programming takes over 15 minutes, status bar goes over 100 percent, D or immobilizer indicator flashes, HDS tablet freezes, etc.), follow these steps to minimize the chance of damaging the PCM:

- Leave the ignition switch in ON (II) position.
- Connect a jumper battery (do not connect a battery charger).
- Shut down the HDS.
- Disconnect the HDS from the DLC.
- Reboot the HDS.
- Reconnect the HDS to the DLC, and try the update procedure again.

8. USA, Canada models: If the TP POSITION CHECK failed in step 6, clean the throttle body.
9. Do the PCM idle learn procedure.
10. Do the CKP learn procedure.

THROTTLE POSITION SENSOR (TPS)

LOCATION

See Figure 124.

REMOVAL & INSTALLATION

➡The manufacturer does not provide a specific Removal and Installation procedure for this component. Refer to the graphic(s) when servicing this component.

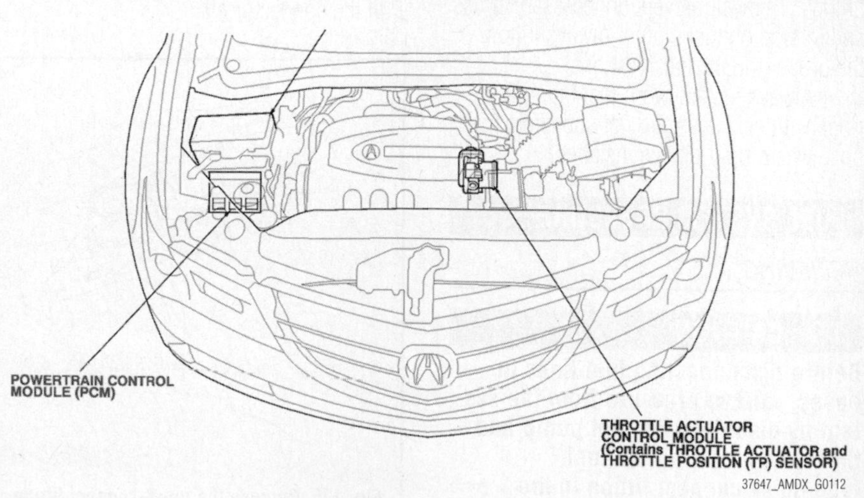

POWERTRAIN CONTROL MODULE (PCM)
THROTTLE ACTUATOR CONTROL MODULE (Contains THROTTLE ACTUATOR and THROTTLE POSITION (TP) SENSOR)
37647_AMDX_G0112

Fig. 124 Throttle actuator control module location

FUEL SYSTEM SERVICE PRECAUTIONS

Safety is the most important factor when performing not only fuel system maintenance, but any type of maintenance. Failure to conduct maintenance and repairs in a safe manner may result in serious personal injury or death. Work on a vehicle's fuel system components can be accomplished safely and effectively by adhering to the following rules and guidelines.

- To avoid the possibility of fire and personal injury, always disconnect the negative battery cable unless the repair or test procedure requires that battery voltage be applied.
- Always relieve the fuel system pressure prior to disconnecting any fuel system component (injector, fuel rail, pressure regulator, etc.) fitting or fuel line connection. Exercise extreme caution whenever relieving fuel system pressure to avoid exposing skin, face and eyes to fuel spray. Please be advised that fuel under pressure may penetrate the skin or any part of the body that it contacts.
- Always place a shop towel or cloth around the fitting or connection prior to loosening to absorb any excess fuel due to spillage. Ensure that all fuel spillage is quickly removed from engine surfaces. Ensure that all fuel-soaked cloths or towels are deposited into a flame-proof waste container with a lid.
- Always keep a dry chemical (Class B) fire extinguisher near the work area.
- Do not allow fuel spray or fuel vapors to come into contact with a spark or open flame.
- Always use a second wrench when loosening or tightening fuel line connection fittings. This will prevent unnecessary stress and torsion on fuel piping. Always follow the proper torque specifications.
- Always replace worn fuel fitting O-rings with new ones. Do not substitute fuel hose where rigid pipe is installed.

FUEL SYSTEM PRESSURE

RELIEVING

CAUTION

Before disconnecting fuel lines or hoses, relieve pressure from the system by disabling the fuel pump and then disconnecting the fuel tube/quick connect fitting in the engine compartment.

With the HDS

See Figure 125.

1. Turn the ignition switch to LOCK (0).
2. Connect the HDS to the data link connector (DLC) located under the driver's side of the dashboard.
3. Turn the ignition switch to ON (II).
4. Make sure the HDS communicates with the PCM. If it doesn't, go to the DLC circuit troubleshooting.
5. Turn the ignition switch to LOCK (0).
6. Remove the fuel fill cap to relieve the pressure in the fuel tank.
7. Turn the ignition switch to ON (II).
8. From the INSPECTION MENU of the HDS, select Fuel Pump OFF, then start the engine, and let it idle until it stalls.

➡Do not allow the engine to idle above 1,000 rpm or the PCM will continue to operate the fuel pump. A DTC or a Temporary DTC may be set during this procedure. Check for DTCs, and clear them as needed.

9. Turn the ignition switch to LOCK (0).
10. Do the battery terminal disconnection procedure.
11. Remove the quick-connect fitting cover.
12. Check the fuel quick-connect fitting for dirt, and clean it if needed.
13. Place a rag or shop towel over the quick-connect fitting.
14. Disconnect the quick-connect fitting: Hold the connector with one hand, and squeeze the retainer tabs with the other hand to release them from the locking tabs. Pull the connector off.

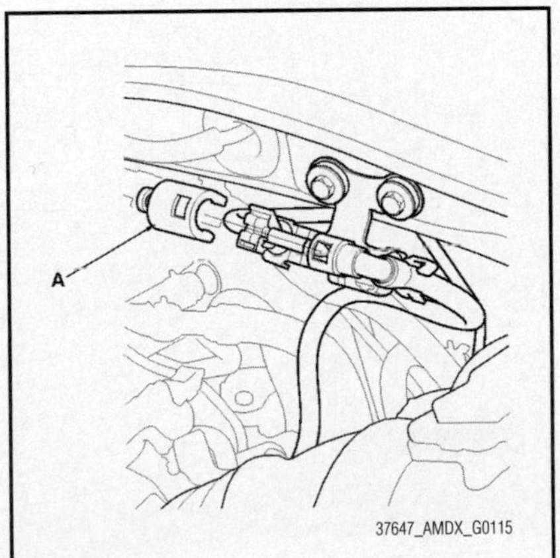
A
37647_AMDX_G0115

Fig. 125 Remove the quick-connect fitting cover (A)

➡Note the following

- Be careful not to damage the line or other parts.
- Do not use tools.
- If the connector does not move, keep the retainer tabs pressed down, and alternately pull and push the connector until it comes off easily.
- Do not remove the retainer from the line; once removed, the retainer must be replaced with a new one.

15. After disconnecting the quick-connect fitting, check it for dirt or damage.
16. Do the battery terminal reconnection procedure.

Without the HDS

See Figure 125.

1. Remove PGM-FI main relay 2 from the under-dash fuse/relay box.
2. Start the engine, and let it idle until it stalls.

➡If any DTCs are stored, clear and ignore them.

3. Turn the ignition switch to LOCK (0).
4. Remove the fuel fill cap to relieve the pressure in the fuel tank.
5. Do the battery terminal disconnection procedure.
6. Remove the quick-connect fitting cover.
7. Check the fuel quick-connect fitting for dirt, and clean it if needed.
8. Place a rag or shop towel over the quick-connect fitting.
9. Disconnect the quick-connect fitting: Hold the connector with one hand, and squeeze the retainer tabs with the other hand to release them from the locking tabs. Pull the connector off.
   - Be careful not to damage the line or other parts.
   - Do not use tools.
   - If the connector does not move, keep the retainer tabs pressed down, and alternately pull and push the connector until it comes off easily.
   - Do not remove the retainer from the line; once removed, the retainer must be replaced with a new one.
10. After disconnecting the quick-connect fitting, check it for dirt or damage.
11. Do the battery terminal reconnection procedure.

FUEL FILTER

REMOVAL & INSTALLATION

See Figure 126.

The fuel filter should be replaced whenever the fuel pressure drops below the specified value, after making sure that the fuel pump and the fuel pressure regulator are OK.

1. Remove the fuel tank unit.
2. Remove the fuel filter set.
3. Check these items before installing the fuel tank unit:
   - When connecting the wire harness, make sure the connection is secure and the connectors are firmly locked into place.
   - When installing the fuel gauge sending unit, make sure the connection is secure and the connector is firmly locked into place. Be careful not to bend or twist it excessively.
4. Install the parts in the reverse order of removal with new O-rings. When installing the fuel tank unit, align the marks on the unit and the fuel tank.

➡Coat the O-rings with clean engine oil.

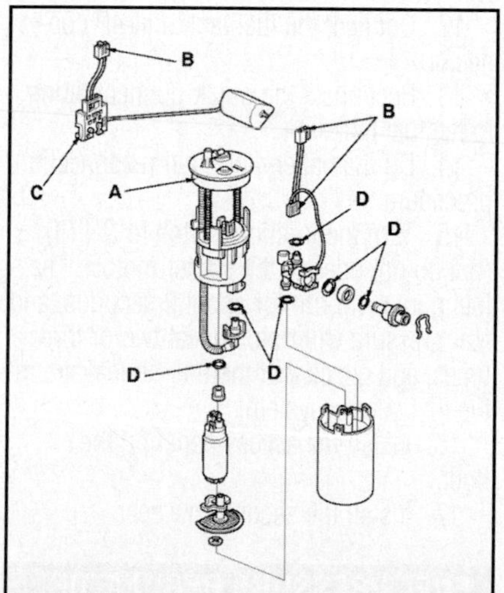
B
B
C
A
D
D
D
D
A. Fuel filter set
B. Connectors
C. Fuel gauge sending unit
D. O-rings
37647_AMDX_G0116

Fig. 126 Remove the fuel filter set

FUEL LEVEL SENDING UNIT

LOCATION

The fuel level sending unit is an integral part of the fuel tank unit.

REMOVAL & INSTALLATION

See Figures 127 through 129.

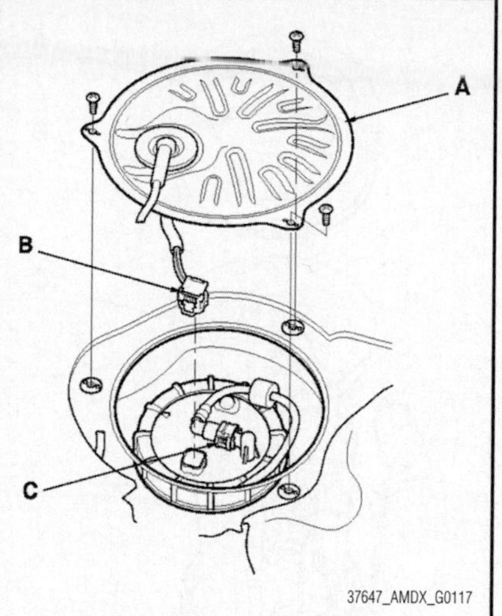
A
B
C
37647_AMDX_G0117

Fig. 127 Remove the access panel from the floor

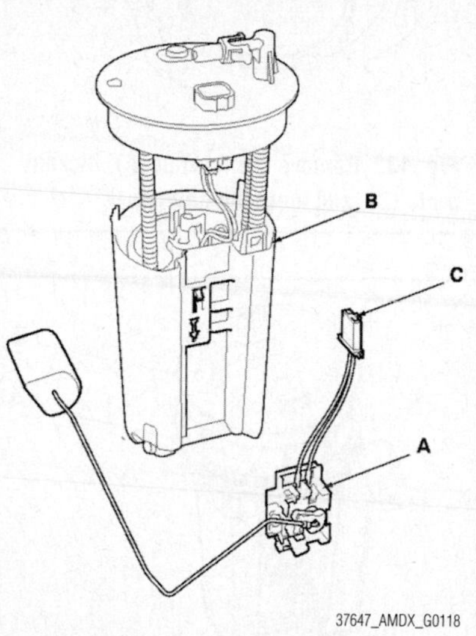
B
C
A
37647_AMDX_G0118

Fig. 128 Remove the fuel level sensor (fuel sending unit) (A) from the fuel tank unit (B)

1. Remove the fuel tank unit.
   a. Relieve the fuel pressure.
   b. Remove the second row seat.
   c. Remove the access panel from the floor.
   d. Disconnect the fuel tank unit 4P connector.
   e. Disconnect the quick-connect fitting from the fuel tank unit.
2. Remove the fuel level sensor (fuel sending unit) from the fuel tank unit.

To install:

3. Check these items before installing the fuel tank unit:
   a. When connecting the wire harness, make sure the connection is secure and the connector is firmly locked into place.

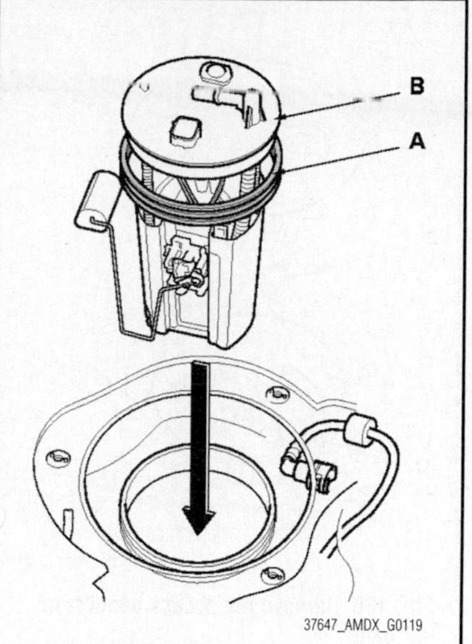
B
A
37647_AMDX_G0119

Fig. 129 Insert the fuel tank unit partially into the fuel tank

   b. When installing the fuel gauge sending unit, make sure the connection is secure and the connector is firmly locked into place. Be careful not to bend or twist it excessively.
4. Install the parts in the reverse order of removal.
5. When installing the fuel tank unit, align the marks on the unit and the fuel tank.
6. Temporarily attach a new base gasket to the fuel tank unit, then insert the fuel tank unit partially into the fuel tank.
   - Be careful not to damage the new base gasket.
   - Be careful not to bend the fuel gauge sending unit.
   - Do not coat the base gasket with oil.

FUEL TANK UNIT (FUEL PUMP)

REMOVAL & INSTALLATION

See Figures 130 through 135.

1. Relieve the fuel pressure.
2. Remove the second row seat.
3. Remove the access panel from the floor.
4. Disconnect the fuel tank unit 4P connector.
5. Disconnect the quick-connect fitting from the fuel tank unit.
6. Using the fuel sender wrench, loosen the locknut.
7. Remove the locknut, locknut plate, and fuel tank unit.

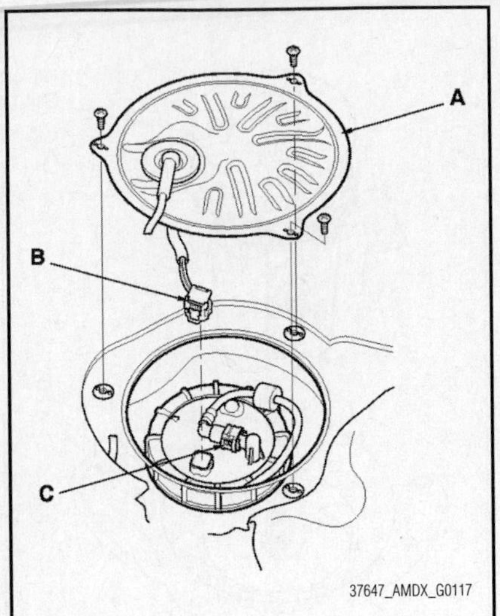
A
B
C
37647_AMDX_G0117

Fig. 130 Remove the access panel from the floor

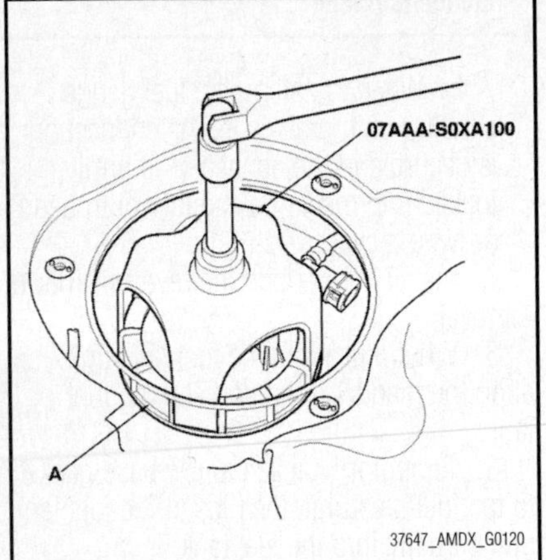
07AAA-S0XA100
A
37647_AMDX_G0120

Fig. 131 Using the fuel sender wrench, loosen the locknut

To install:

8. Temporarily attach a new base gasket to the fuel tank unit, then insert the fuel tank unit partially into the fuel tank.

➡:

- Be careful not to damage the new base gasket.
- Be careful not to bend the fuel gauge sending unit.
- Do not coat the base gasket with oil.

9. Transfer the base gasket from the fuel tank unit to the fuel tank.

10. Align the marks on the fuel tank and the fuel tank unit, then insert the fuel tank unit into the fuel tank until it sits on the base gasket.

➡To prevent a fuel leak, check the base gasket, visually or by hand, to make sure it is not pinched.

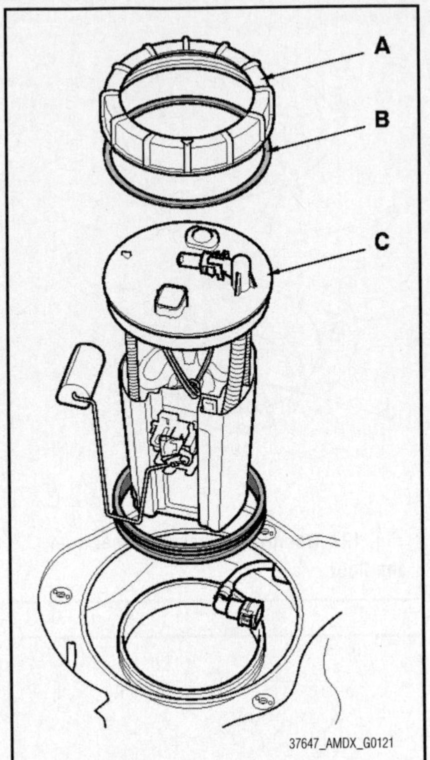
A
B
C
37647_AMDX_G0121

Fig. 132 Remove the locknut (A), locknut plate (B), and fuel tank unit (C)

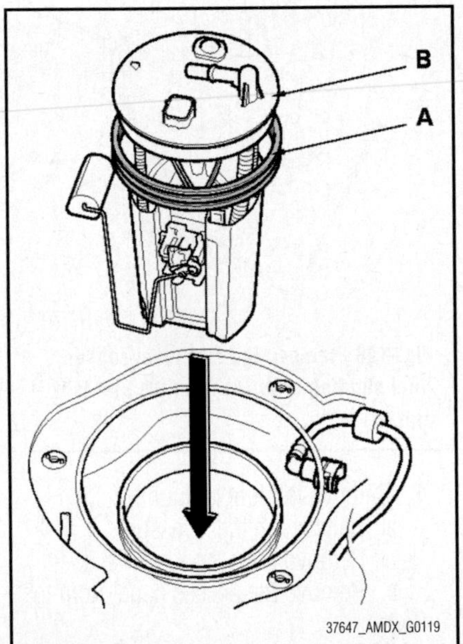
B
A
37647_AMDX_G0119

Fig. 133 Insert the fuel tank unit partially into the fuel tank

11. Using the tool, tighten the new locknut with a new locknut plate to the specified torque.

➡Before tightening, align the marks on the fuel tank and the locknut. After tightening, make sure the marks are still aligned. After installation, check the base gasket visually or by hand to be sure it is not pinched.

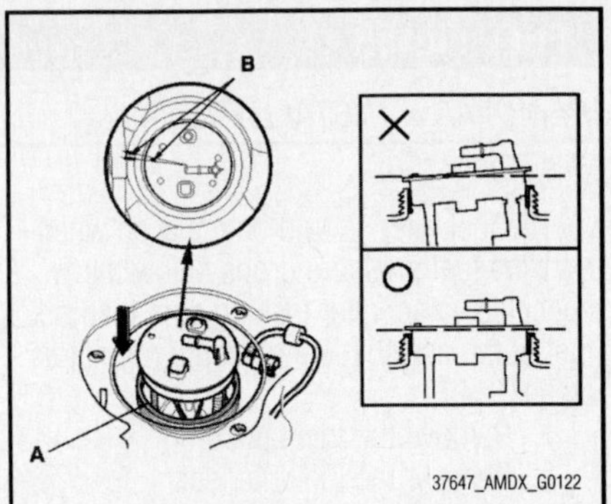
B
A
37647_AMDX_G0122

Fig. 134 Transfer the base gasket (A) from the fuel tank unit to the fuel tank

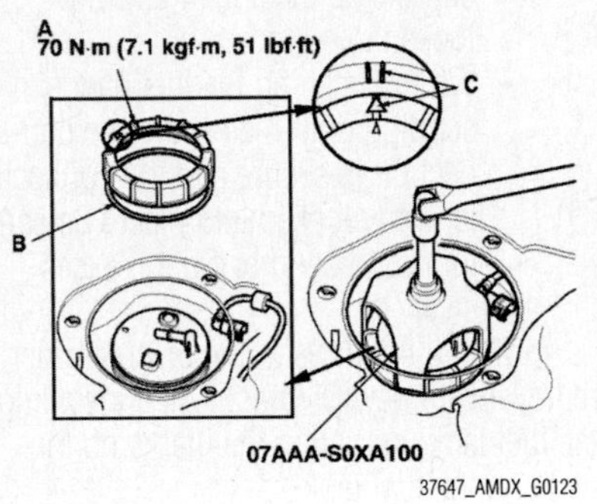
A
70 N·m (7.1 kgf·m, 51 lbf·ft)
C
B
07AAA-S0XA100
37647_AMDX_G0123

Fig. 135 Tighten the new locknut with a new locknut plate

12. Connect the fuel tank unit 4P connector.

13. Reconnect the quick-connect fitting to the fuel tank unit.

14. Do the battery terminal reconnection procedure.

15. Turn the ignition switch to ON (II) (but do not operate the starter motor). The fuel pump will run for about 2 seconds, and fuel pressure will rise. Repeat two or three times, and check that there is no leakage in the fuel supply system.

16. Install the access panel on the floor.

17. Install the second row seat.

FUEL PRESSURE REGULATOR

REMOVAL & INSTALLATION

See Figures 136 and 137.

1. Remove the fuel tank unit.
2. Remove the reservoir.
3. Remove the bracket.
4. Remove the clip.
5. Remove the fuel pressure regulator.

To install:

6. Install the parts in the reverse order of removal with new O-rings.

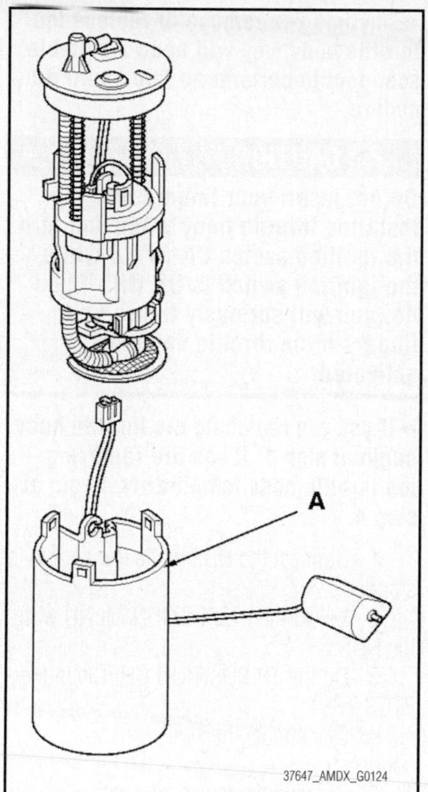
A
37647_AMDX_G0124
Fig. 136 Remove the reservoir (A)

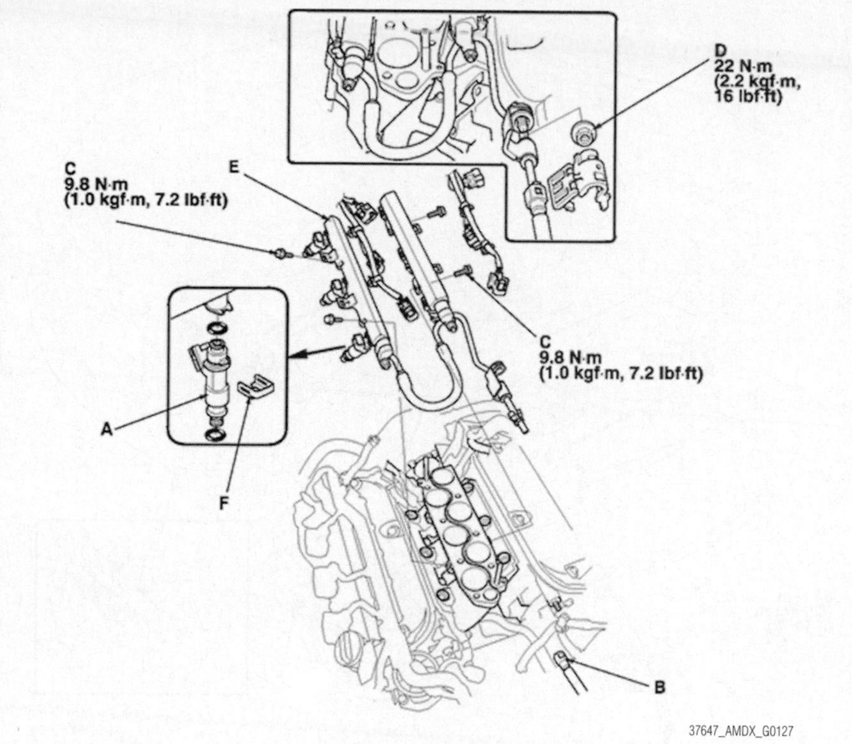
D
22 N·m
(2.2 kgf·m,
16 lbf·ft)
E
C
9.8 N·m
(1.0 kgf·m, 7.2 lbf·ft)
C
9.8 N·m
(1.0 kgf·m, 7.2 lbf·ft)
A
F
B
37647_AMDX_G0127
Fig. 138 Exploded view of fuel rail and injector system

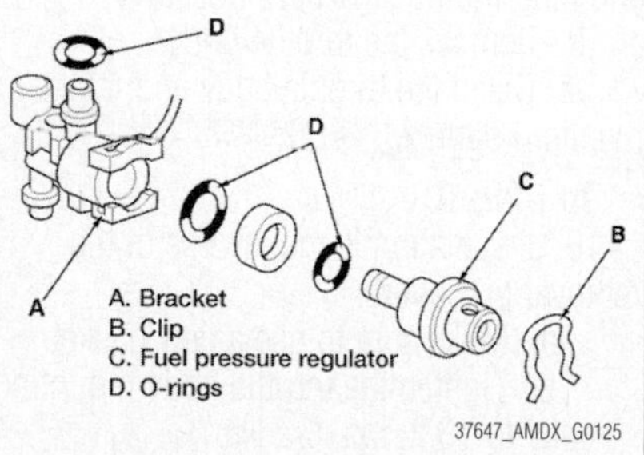
D
D
C
B
A
A. Bracket
B. Clip
C. Fuel pressure regulator
D. O-rings
37647_AMDX_G0125
Fig. 137 Exploded view of fuel pressure regulator assembly

➡Coat the O-rings with clean engine oil; do not use any other oils or fluids. Do not pinch the O-rings during installation.

FUEL RAIL AND INJECTOR

REMOVAL & INSTALLATION

See Figure 138.

1. Relieve the fuel pressure.
2. Remove the intake manifold.
3. Disconnect the connectors from the injectors.
4. Disconnect the quick-connect fitting.
5. Remove the fuel rail mounting bolts and nut from the fuel rails.
6. Remove the injector clips from the fuel rail.
7. Remove the injectors from the rails.

To install:

8. Coat the new O-ring with clean engine oil, and insert the injectors into the fuel rails.
9. Install the injector clips.
10. Coat the new injector O-ring with clean engine oil.
11. Install the injectors in the injector base.
12. Install the fuel rail mounting bolts and nut.
13. Install the injector connectors.
14. Connect the quick-connect fitting.
15. Turn the ignition switch to ON (II), but do not operate the starter. After the fuel pump runs for about 2 seconds, the fuel pressure in the fuel line rises. Repeat this two or three times, then check for fuel leaks.
16. Install the intake manifold.

FUEL TANK

DRAINING

1. Remove the fuel tank unit. Refer to Fuel Tank Unit Removal and Installation section.
2. Using a hand pump, a hose, and a container suitable for fuel, draw the fuel from the fuel tank.
3. Reinstall the fuel tank unit.

REMOVAL & INSTALLATION

See Figure 139.

1. Relieve the fuel pressure.
2. Drain the fuel tank.
3. Disconnect the fuel tank unit wiring harness and the quick-connect fittings from the fuel tank unit.
4. Raise the vehicle on a lift.
5. Remove the exhaust pipe.
6. Remove the driveshaft, and support it with jackstands.
7. Remove the fuel tank guard and fuel tank protector.
8. Loosen the clamp, and disconnect the tube. Slide back the clamps, then twist the hoses as you pull to avoid damaging them.
9. Open the clamps.
10. Disconnect the hoses.
11. Place a jack or other support under the fuel tank.
12. Remove the strap bolts and the straps.
13. Remove the fuel tank.
14. Install the parts in the reverse order of removal.

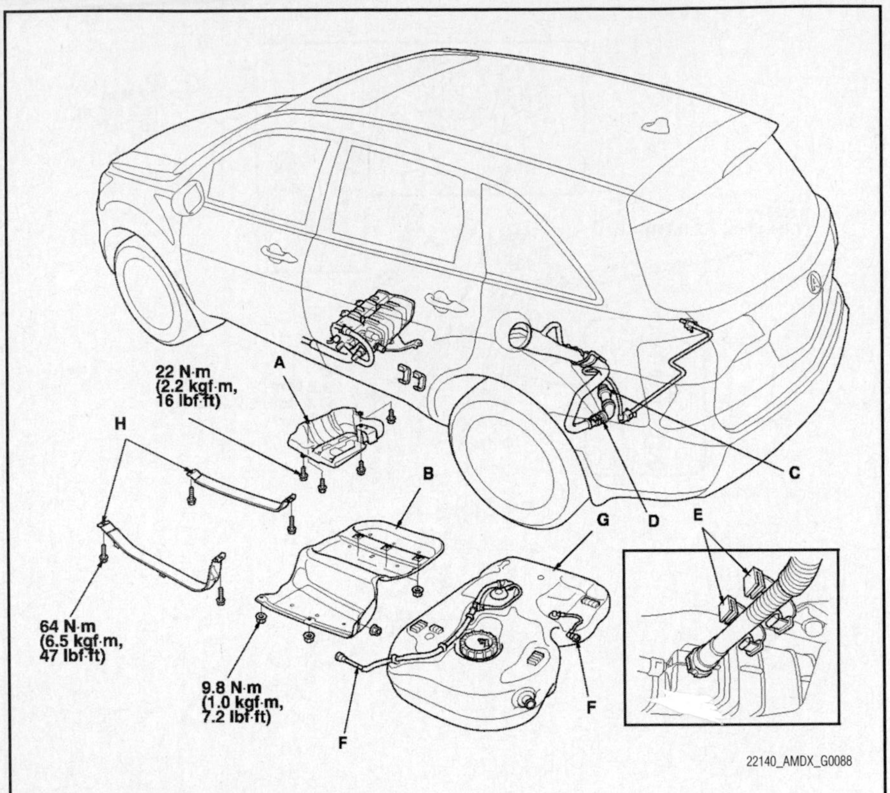
22 N·m
(2.2 kgf·m,
16 lbf·ft)
A
H
B
C
G
D
E
64 N·m
(6.5 kgf·m,
47 lbf·ft)
9.8 N·m
(1.0 kgf·m,
7.2 lbf·ft)
F
F
22140_AMDX_G0088

Fig. 139 Removing the fuel tank assembly

➡**Anytime you remove or replace the throttle body, you will need a suitable scan tool to perform an Idle Learn procedure.**

⁂ CAUTION

Do not insert your fingers into the installed throttle body when you turn the ignition switch ON (II) or while the ignition switch is ON (II). If you do, you will seriously injure your fingers if the throttle valve is activated.

➡**If you are replacing the throttle body, begin at step 1. If you are removing the throttle body temporarily, begin at step 4.**

1. Connect the HDS while the engine is stopped.
2. Select the INSPECTION MENU with the HDS.
3. Do the TP POSITION CHECK in the ETCS TEST.
4. Disconnect the MAP sensor connector.
5. Remove the intake air duct.
6. Disconnect the throttle body connector.
7. Disconnect the water bypass hoses, and plug the water bypass hoses.
8. Remove the throttle body.
9. Clean the throttle body and intake manifold surface.

To install:

10. Installation is the reverse of the removal procedure.
    a. Make sure to use a new gasket.
    b. Tighten the throttle body mounting bolts to 16 ft. lbs. (22 Nm).
    c. Refill the radiator with engine coolant.
11. Do the PCM idle learn procedure after the throttle body has been replaced.

IDLE SPEED

ADJUSTMENT

➡**Before checking the idle speed, check these items:**

- The Malfunction Indicator Lamp (MIL) has not been reported on, and there are no DTCs.
- Ignition timing
- Spark plugs
- Air cleaner
- PCV system

1. Apply the parking brake and make sure the headlights are off.
2. Disconnect the evaporative emission (EVAP) canister purge valve connector.
3. Connect the HDS to the data link connector (DLC) located under the driver's side of the dashboard.
4. Turn the ignition switch to ON (II).
5. Make sure the HDS communicates with the PCM. If it doesn't, go to the DLC circuit troubleshooting.
6. Start the engine. Hold the engine speed at 3,000 rpm without load (in Park or neutral) until the radiator fan comes on, then let it idle.
7. Check the idle speed without load conditions: headlights, blower fan, radiator fan, and air conditioner off. Idle speed should be: 710±50 rpm (in Park or neutral)
8. Let the engine idle for 1 minute with high electric load (A/C switch ON, temperature set to max cool, blower fan on high, and headlights on high beam). Idle speed should be: 710±50 rpm (in Park or neutral)

➡**If the idle speed is not within specification, do the PCM idle learn procedure. If the idle speed is still not within specification, go to symptom troubleshooting.**

9. Reconnect the EVAP canister purge valve connector.

THROTTLE BODY

REMOVAL & INSTALLATION

See Figure 140.

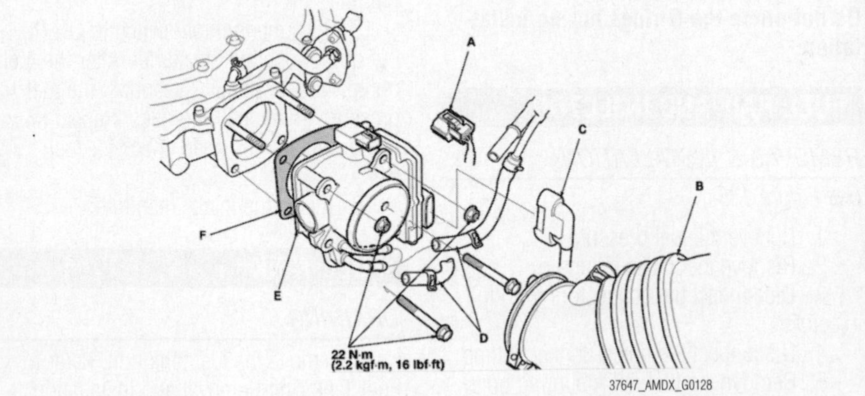
A
C
B
F
E
D
22 N·m
(2.2 kgf·m, 16 lbf·ft)
37647_AMDX_G0128

Fig. 140 Exploded view of throttle body assembly

HEATING, VENTILATION & AIR CONDITIONING SYSTEM

BLOWER MOTOR

REMOVAL & INSTALLATION

See Figures 141 and 142.

1. Remove the passenger's dashboard undercover.
2. Open the glove box.
3. Remove the cap, then disconnect the glove box damper from the pivot on the glove box.
4. Close the glove box.
5. Remove the mounting bolts.
6. While holding the glove box, release the glove box stop on each side from the dashboard by pushing them in, then remove the glove box.
7. Remove the harness clips, the bolts, and the glove box frame.

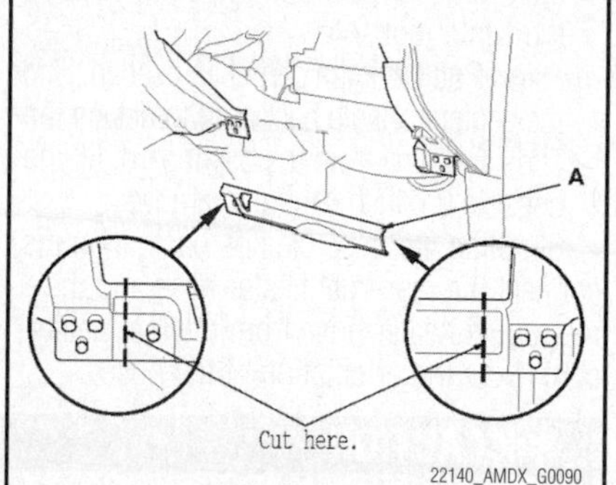
A
Cut here.
22140_AMDX_G0090

Fig. 141 Cut the plastic cross brace (A) in the glove box opening with diagonal cutters in the area shown, and discard it.

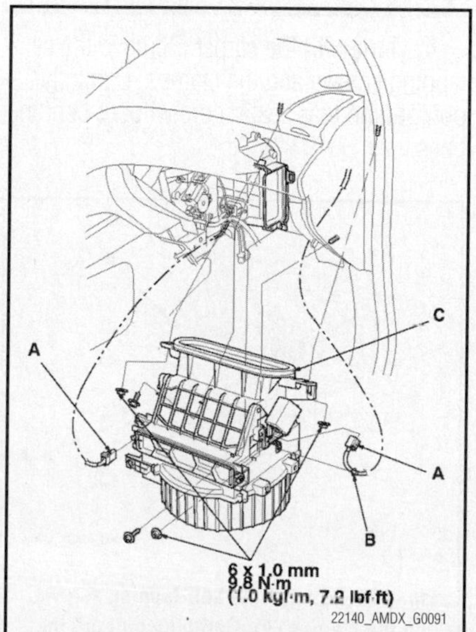
A
C
A
B
6 x 1.0 mm
9.8 N·m
(1.0 kgf·m, 7.2 lbf·ft)
22140_AMDX_G0091

Fig. 142 Removing the blower unit

8. Cut the plastic cross brace in the glove box opening with diagonal cutters in the area shown, and discard it.
9. Remove the wire harness clips, the self-tapping screws, and the passenger's heater duct.
10. Disconnect the connector from the front blower motor. Remove the wire harness clips.
11. Disconnect the connectors from the recirculation control motor and adaptive front lighting control unit, then remove the harness clip. Remove the self-tapping screws, the mounting nuts, and the blower unit.
12. Install the unit in the reverse order of removal. Make sure that there is no air leakage.

HEATER CORE

REMOVAL & INSTALLATION

See Figures 143 and 144.

1. Disconnect the negative battery cable.
2. Recover the refrigerant with a recovery/recycling/charging station, as outlined in this section.
3. Disconnect the front receiver line and front suction line from the front evaporator core.
4. From under the hood, open the cable clamp, then disconnect the heater valve cable from the heater valve arm. Turn the heater valve arm to the fully opened position.
5. Drain the engine cooling system.
6. Remove the clamp and slide the hose clamps back. Remove the nut and the water valve, then disconnect the inlet heater hose and the outlet heater hose from the heater unit.

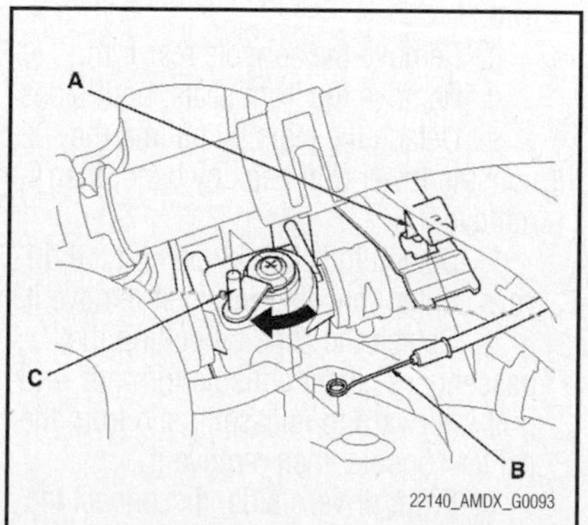
A
C
B
22140_AMDX_G0093

Fig. 143 Open the cable clamp (A), then disconnect the heater valve cable (B) from the heater valve arm (C). Turn the heater valve arm to the fully opened position.

⁂ CAUTION

Engine coolant will run out when the hoses are disconnected; drain it into a clean drip pan. Be sure not to let coolant spill on the electrical parts or the painted surfaces. If any coolant spills, rinse it off immediately.

7. Remove the mounting nuts from the heater unit. Take care not to damage or bend the fuel lines or brake lines, etc.
8. Remove the dashboard.
9. Disconnect the wiring harness from the front blower motor. Remove the wire harness clips.
10. Disconnect the wiring harness from the adaptive front lighting control unit, then remove the harness clip.
11. Disconnect the wiring harness from the front mode control motor, the passenger's air mix control motor, the recirculation control motor, and the front power transistor. Remove the wire harness clips.
12. Disconnect the wiring harness from the driver's air mix control motor and the front evaporator temperature sensor. Remove the wire harness clips and the wire harness.
13. Turn over the carpet. Remove the wire harness clips, the rear heater duct

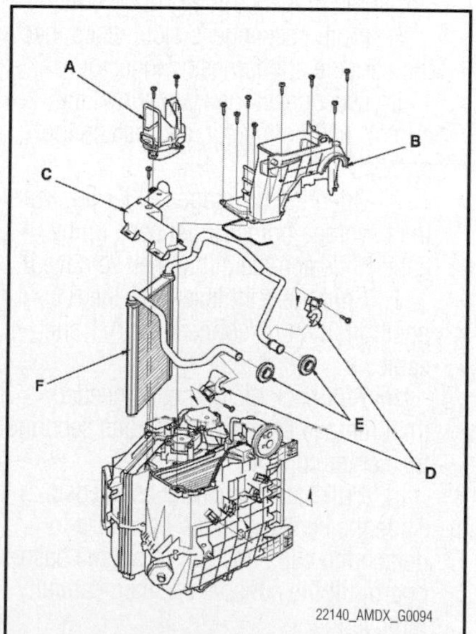
A
B
C
F
E
D
22140_AMDX_G0094

Fig. 144 Removing the heater core from the blower-heater unit

mounting clips, and the rear heater duct.

14. Remove the mounting nuts. Slide the blower-heater unit, then remove the drain hose and blower-heater unit.

15. Remove the self-tapping screws and the passenger's heater duct.

16. Remove the self-tapping screws and the expansion valve cover.

17. Remove the self-tapping screw and the front heater core cover.

18. Remove the self-tapping screws, the heater pipe brackets, the grommets, and carefully pull out the front heater core.

To install:

19. Installation is the reverse order of removal. Tighten the heater mounting nuts to 9 ft. lbs. (12 Nm).

20. Refill the cooling system to the correct level.

21. Adjust the heater valve cable if necessary.

AUXILIARY HEATING & AIR CONDITIONING SYSTEM

BLOWER MOTOR

REMOVAL & INSTALLATION

See Figure 145.

1. Disconnect the negative battery cable.
2. Remove the center console as follows:
   a. Remove the driver's inner dashboard trim.
   b. Remove the passenger's dashboard trim.
   c. Remove the console rear trim.
   d. Remove the front seats, both sides.
   e. Detach the clips by pulling the console lower trim back by hand, then remove it.
   f. Detach the clips by pulling the driver's center console trim, then remove it.
   g. Detach the clips by pulling the passenger's center console trim out, and pull it forward to release its pin from the center console, then remove it.
   h. From driver's side, disconnect the parking pin shift connector and shift lock solenoid connector. Remove the rear blower motor relay mounting bolt, second row seat heater relay mounting bolt (for some models), and ground bolt.
   i. From passenger's side, disconnect the console subharness connector.
   j. Disconnect the HandsFreeLink control unit connector, and detach the harness clips.
   k. Open the beverage holder lid. Pull the beverage holder inner case up by hand to detach the clips, then remove it.
   l. Shift the shift lever into the R position, then disconnect the A/T shift cable.
   m. Pull back the carpet as needed, then remove the mounting bolts securing the center console.
   n. Pull back the carpet as needed. Slide the center console backward to detach the clips and pins from the dashboard, lift the console up, then carefully remove it.
3. Disconnect the wiring harness from the rear blower motor, then remove the wire harness clips.

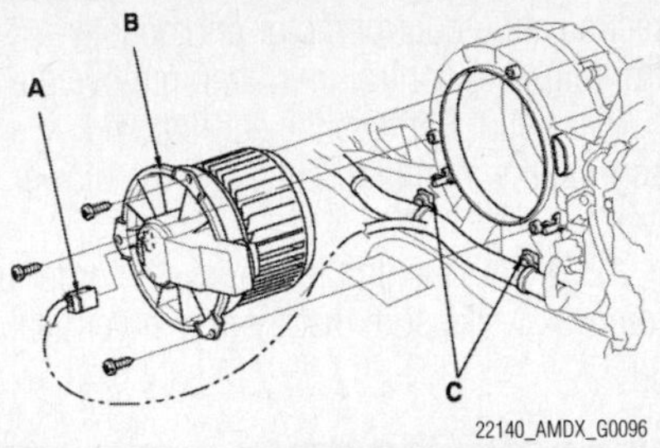
A
B
C
22140_AMDX_G0096

Fig. 145 Disconnect the 2P connector (A) from the rear blower motor (B), then remove the wire harness clips (C). Remove the self-tapping screws and the rear blower motor from the rear HVAC unit.

4. Remove the self-tapping screws and the rear blower motor from the rear HVAC unit.
5. Installation is the reverse order of removal.

HEATER CORE

REMOVAL & INSTALLATION

See Figure 146.

1. Drain the engine cooling system.
2. Remove the center console as follows:
   a. Remove the driver's inner dashboard trim.
   b. Remove the passenger's dashboard trim.
   c. Remove the console rear trim.
   d. Remove the front seats, both sides.
   e. Detach the clips by pulling the console lower trim back by hand, then remove it.
   f. Detach the clips by pulling the driver's center console trim, then remove it.
   g. Detach the clips by pulling the passenger's center console trim out, and pull it forward to release its pin from the center console, then remove it.
   h. From driver's side, disconnect the parking pin shift connector and shift lock solenoid connector. Remove the rear blower motor relay mounting bolt, second row seat heater relay mounting bolt (for some models), and ground bolt.
   i. From passenger's side, disconnect the console subharness connector.
   j. Disconnect the HandsFreeLink control unit connector, and detach the harness clips.
   k. Open the beverage holder lid. Pull the beverage holder inner case up by hand to detach the clips, then remove it.
   l. Shift the shift lever into the R position, then disconnect the A/T shift cable.
   m. Pull back the carpet as needed, then remove the mounting bolts securing the center console.
   n. Pull back the carpet as needed. Slide the center console backward to detach the clips and pins from the dashboard, lift the console up, then carefully remove it.
3. Slide the hose clamps back, then disconnect the rear inlet heater hose and the rear outlet heater hose from the rear heater core. Note the orientation of the hose.

✲✲ CAUTION

Engine coolant will run out when the hoses are disconnected; drain it into a clean drip pan. Be sure not to let coolant spill on the electrical parts or the painted surfaces. If any coolant spills, rinse it off immediately.

4. Turn over the carpet. Remove the self-tapping screws and the clamps. Carefully pull out the rear heater core without bending lines.

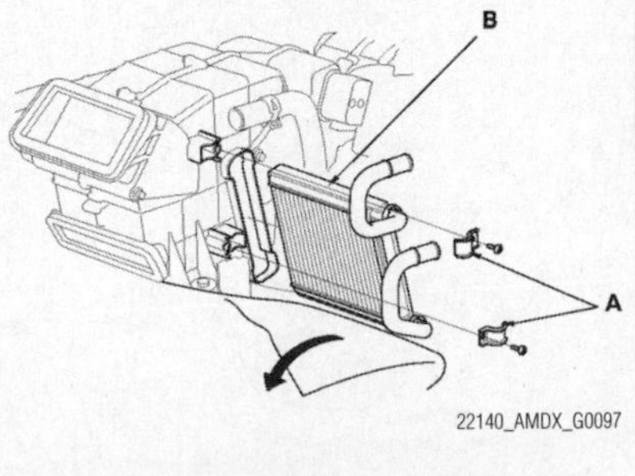
B
A
22140_AMDX_G0097

Fig. 146 Remove the self-tapping screws and the clamps (A). Carefully pull out the rear heater core (B) without bending lines

STEERING

POWER RACK & PINION STEERING GEAR

REMOVAL & INSTALLATION

See Figures 147 through 171.

1. Make sure you have the anti-theft code for the radio and the navigation system, then write down the frequencies for the radio's preset buttons.
2. Disconnect the negative cable from the battery first, then disconnect the positive cable. Wait at least 3 minutes before proceeding.
3. Drain the power steering fluid.
4. Raise and safely support the vehicle.
5. Remove the front wheels.
6. Remove the driver's dashboard undercover.
7. Remove the steering wheel.
8. Remove steering joint cover.
9. Remove the steering joint bolts, disconnect the steering joint by moving the steering joint toward the column.
10. Remove the center guide (if equipped) from the top of the pinion shaft, and discard it.

➡The center guide is for factory assembly use only.

11. Apply vinyl tape to the splines on the pinion shaft.
12. Remove the air cleaner housing.
13. Remove the service caps for the front damper flange nuts from the cowl cover. Position the engine hanger adapters (VSB02C000031) with the "FRONT" mark facing forward over the damper flange nuts.
14. Install the engine hanger balancer bar (VSB02C000019); attach the front arm to the front cylinder head with a spacer and 10 x 1.25 mm bolt, attach the rear arm to the rear cylinder head with 8 x 1.25 mm bolt.
15. Install the engine support hanger (AAR-T-12566) to the vehicle, and attach the hook to the slotted hole in the engine hanger balancer bar. Tighten the wing nut by hand to lift and support the engine/transmission.
16. Loosen the adjustable hose clamp, and disconnect the return hose.
17. Loosen the 16 mm flare nut, and disconnect the inlet line.
18. Remove the cotter pins from the tie-rod ball joints and remove the nuts.
19. Separate the tie-rod ball joints and knuckles using the ball joint remover.
20. Disconnect the lower arm ball joints from the knuckles.
21. Disconnect the stabilizer links from the lower arms.

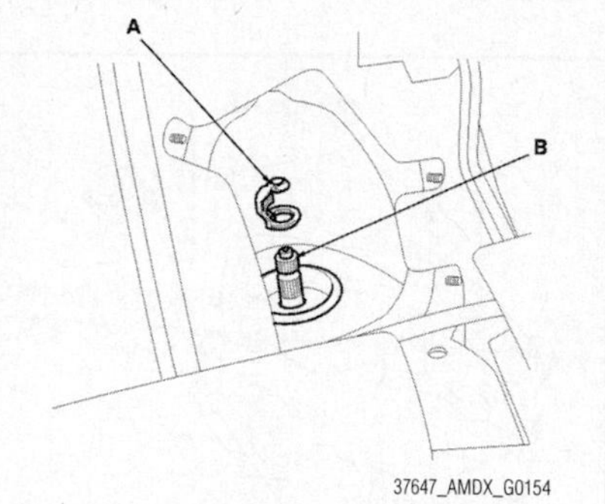
A
B
37647_AMDX_G0154

Fig. 148 Remove the center guide (A) (if equipped) from the top of the pinion shaft

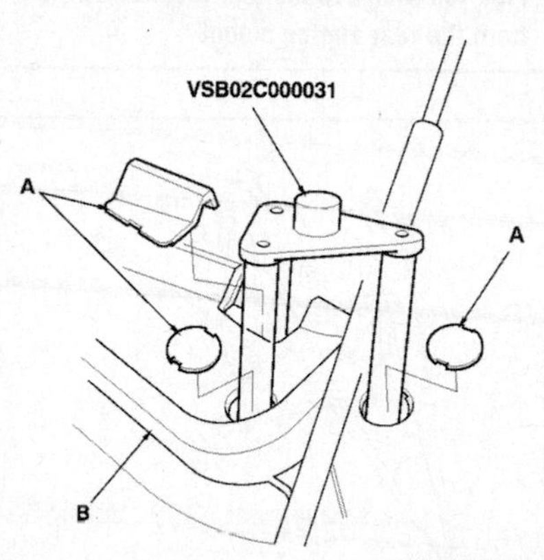
VSB02C000031
A
A
B
37647_AMDX_G0155

Fig. 149 Remove the service caps (A) for the front damper flange nuts from the cowl cover

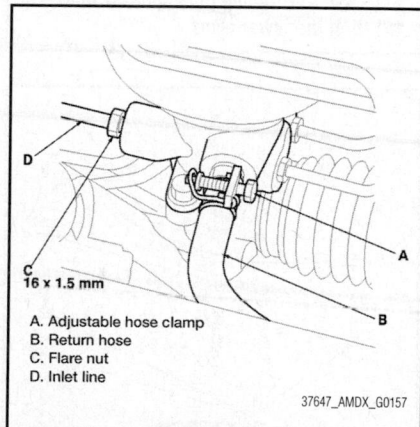

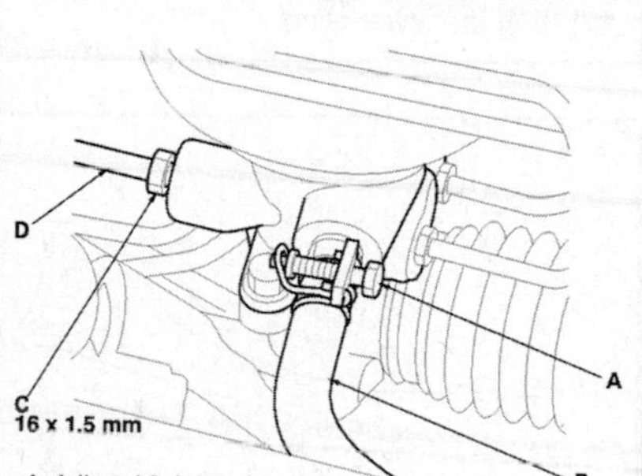
D
C
16 x 1.5 mm
A
B
A. Adjustable hose clamp
B. Return hose
C. Flare nut
D. Inlet line
37647_AMDX_G0157

Fig. 151 Loosen the adjustable hose clamp (A), and disconnect the return hose (B)

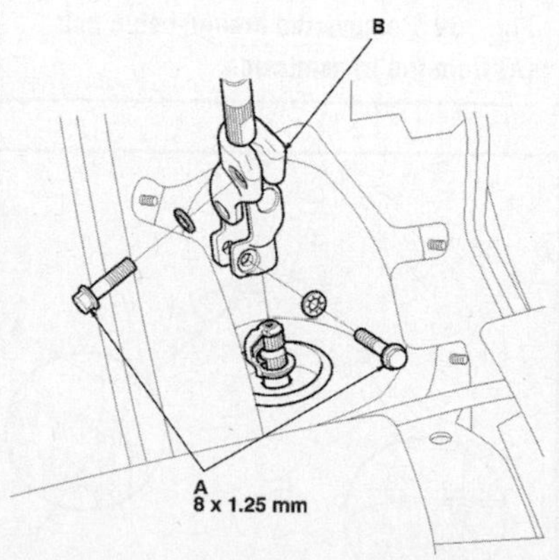
B
A
8 x 1.25 mm
37647_AMDX_G0153

Fig. 147 Remove the steering joint bolts (A), disconnect the steering joint by moving the steering joint (B) toward the column

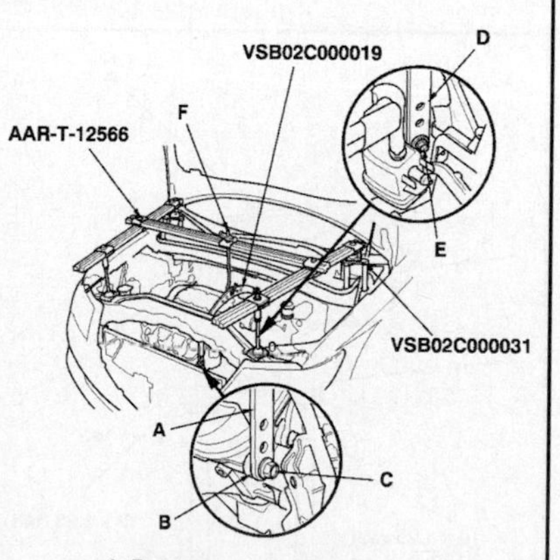
VSB02C000019
D
AAR-T-12566
F
E
VSB02C000031
A
C
B
A. Front arm
B. Spacer
C. Bolt
D. Rear arm
E. Bolt
F. Wing nut
37647_AMDX_G0156

Fig. 150 Install the engine hanger balancer bar (VSB02C000019)

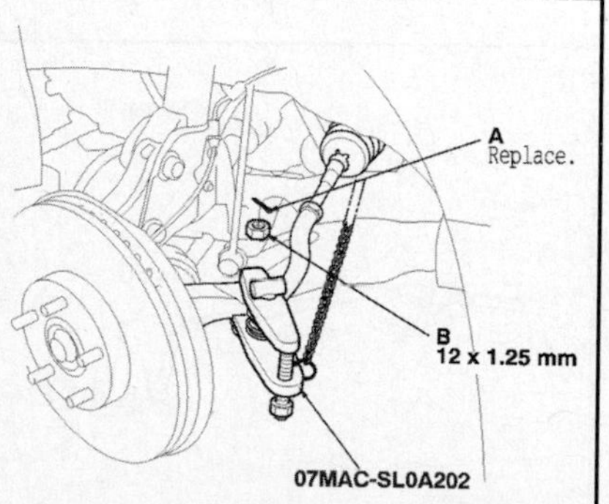
A
Replace.
B
12 x 1.25 mm
07MAC-SL0A202
37647_AMDX_G0158

Fig. 152 Remove the cotter pins (A) from the tie-rod ball joints and remove the nuts (B)

22. Remove the front undercover and the front splash shield.
23. Remove the front subframe stiffener plate.
24. Remove exhaust pipe A.
25. Remove the driveshaft and driveshaft protectors.
26. Remove the inlet line clamp bolts.
27. Remove the rear mount stop from the rear engine mount.
28. Remove the rear engine mount from the base bracket, and move it aside.

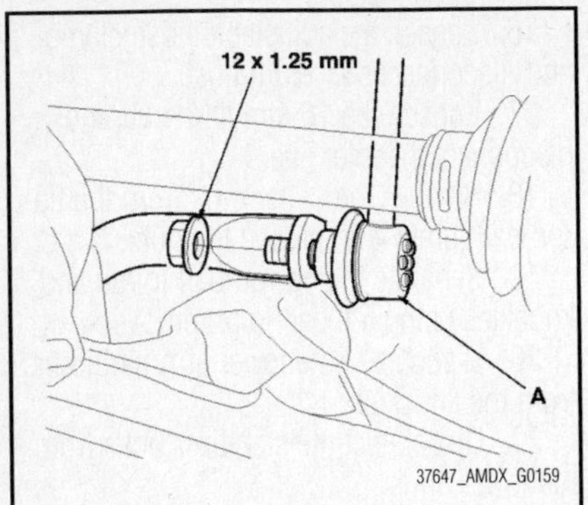
12 x 1.25 mm
A
37647_AMDX_G0159
Fig. 153 Disconnect the stabilizer links (A) from the lower arms

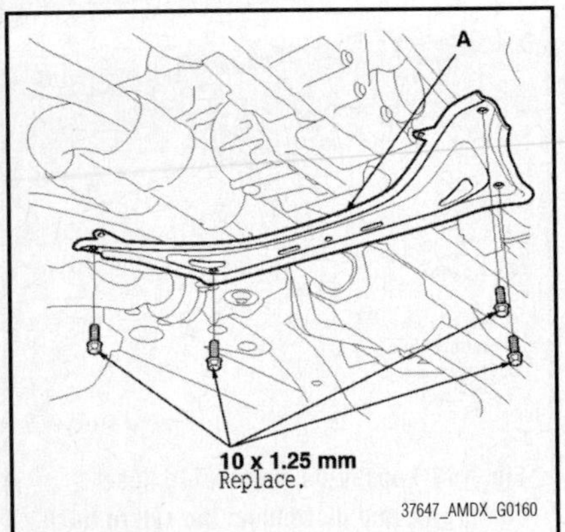
A
10 x 1.25 mm
Replace.
37647_AMDX_G0160
Fig. 154 Remove the front subframe stiffener plate (A)

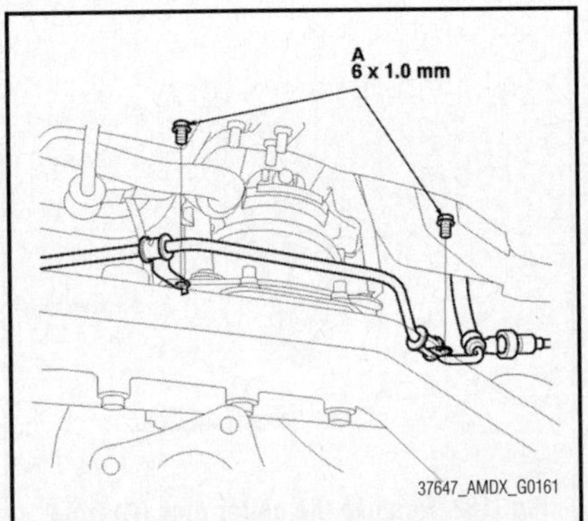
A
6 x 1.0 mm
37647_AMDX_G0161
Fig. 155 Remove the inlet line clamp bolts (A)

29. Remove the base bracket from the front subframe.
30. Remove the ground cable bolt from the transmission.
31. Make reference marks on the body across the marks on the edge of the front subframe.

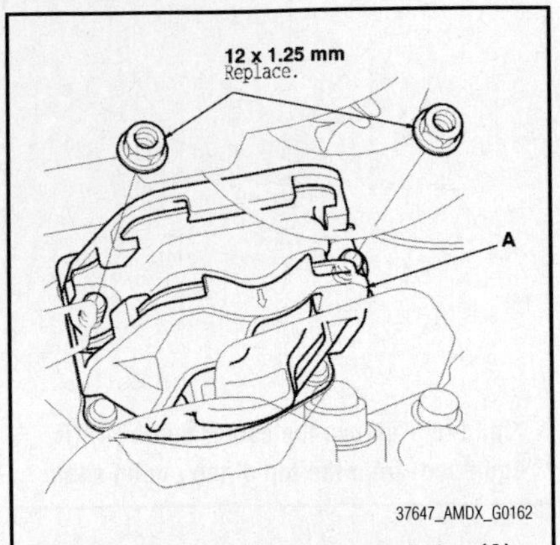
12 x 1.25 mm
Replace.
A
37647_AMDX_G0162
Fig. 156 Remove the rear mount stop (A) from the rear engine mount

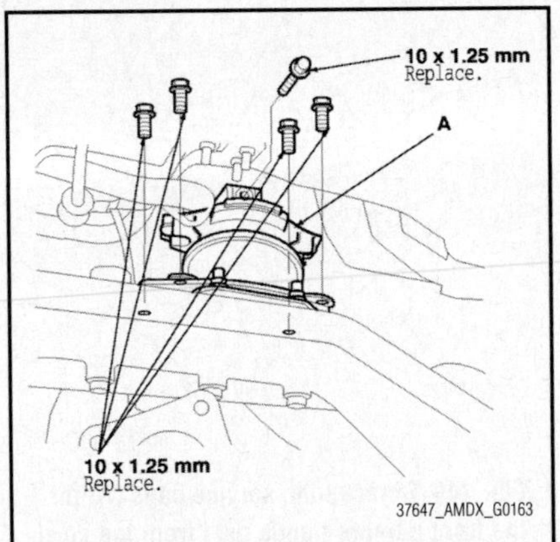
10 x 1.25 mm
Replace.
A
10 x 1.25 mm
Replace.
37647_AMDX_G0163
Fig. 157 Remove the rear engine mount (A) from the base bracket

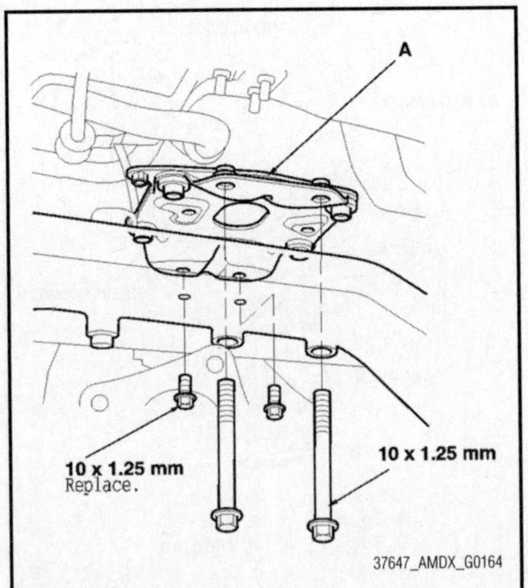
A
10 x 1.25 mm
Replace.
10 x 1.25 mm
37647_AMDX_G0164
Fig. 158 Remove the base bracket (A) from the front subframe

32. Loosen the four bolts holding the adjustable arms of the front subframe adapter to its center plate.
33. Line up the slots in the arms with the bolt holes on the corner of the jack base, then attach the front subframe adapter to the jack base with the bolts that came with the jack. Tighten all bolts securely.
34. Raise the jack to vehicle height, then attach the front subframe adapter to the front subframe using the subframe stiffener mounting bolts and bolt holes.
35. Support the front subframe securely by raising the transmission jack.
36. Remove the mounting bolts from the front subframe front brackets.
37. Loosen the special bolts on the front subframe front brackets so they are about 1⁹⁄₁₆ inches (40 mm) from the mounting surface. Do not loosen the special bolts more than necessary.
38. Remove the mounting bolts from the front subframe rear brackets.
39. Loosen the special bolts on the front subframe rear brackets so they are about 1⁹⁄₁₆ inches (40 mm) from the mounting surface. Do not loosen the special bolts more than necessary.

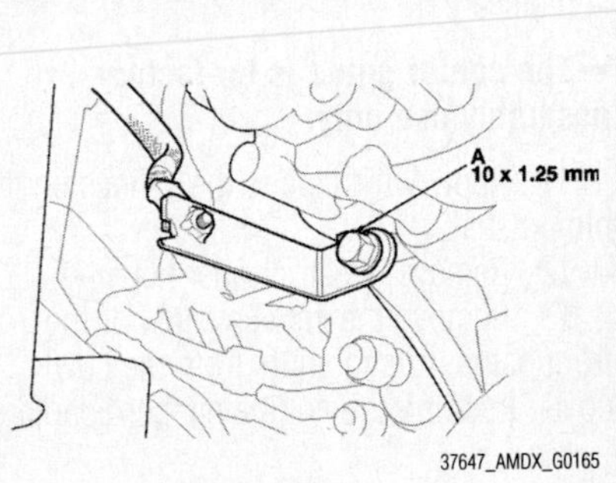
A
10 x 1.25 mm
37647_AMDX_G0165
Fig. 159 Remove the ground cable bolt (A) from the transmission

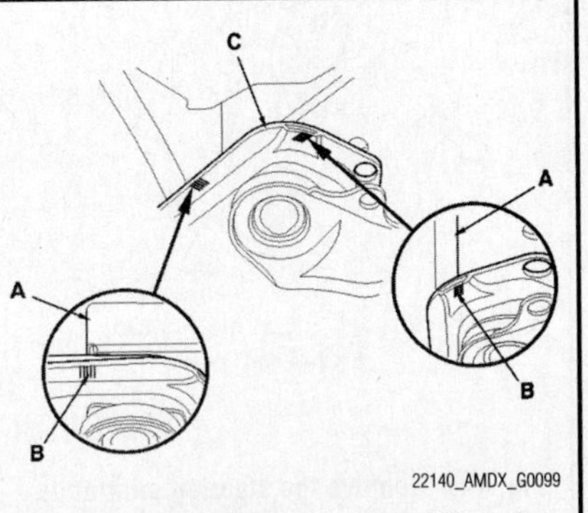
C
A
A
B
B
22140_AMDX_G0099
Fig. 160 Make reference marks (A) on the body across the marks (B) on the edge of the front subframe (C).

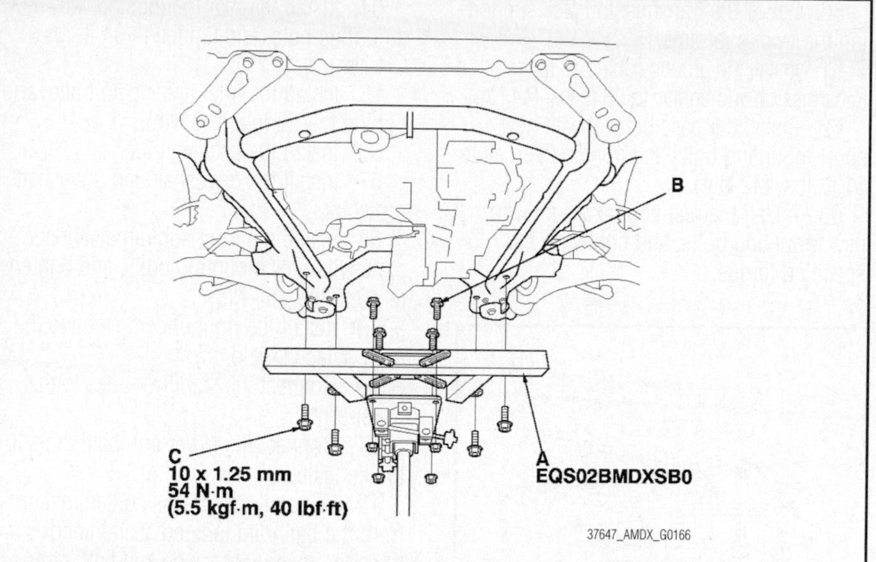
B
C
10 x 1.25 mm
54 N·m
(5.5 kgf·m, 40 lbf·ft)
A
EQS02BMDXSB0
37647_AMDX_G0166

Fig. 161 Line up the slots in the arms with the bolt holes on the corner of the jack base

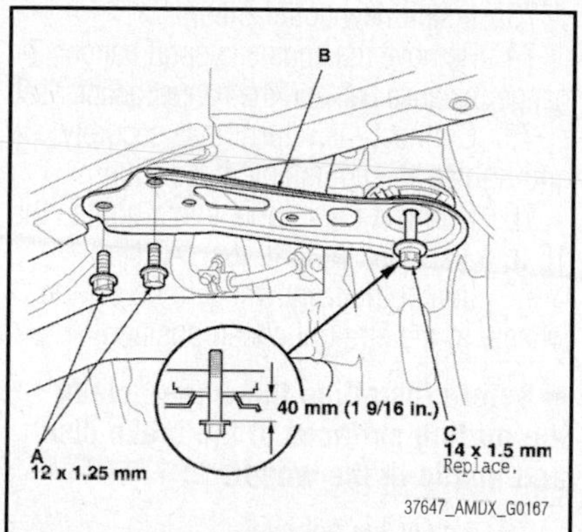
B
40 mm (1 9/16 in.)
C
14 x 1.5 mm
Replace.
A
12 x 1.25 mm
37647_AMDX_G0167

Fig. 162 Remove the mounting bolts (A) from the front subframe front brackets (B)

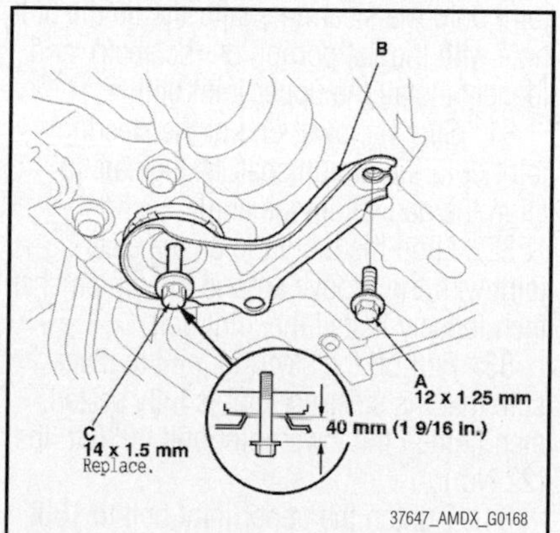
B
A
12 x 1.25 mm
40 mm (1 9/16 in.)
C
14 x 1.5 mm
Replace.
37647_AMDX_G0168

Fig. 163 Remove the mounting bolts (A) from the front subframe rear brackets (B)

40. Lower the transmission jack slowly until the front subframe has dropped about 1 9/16 inches (40 mm).

41. Remove the mounting bolts from the left side of the steering gearbox.

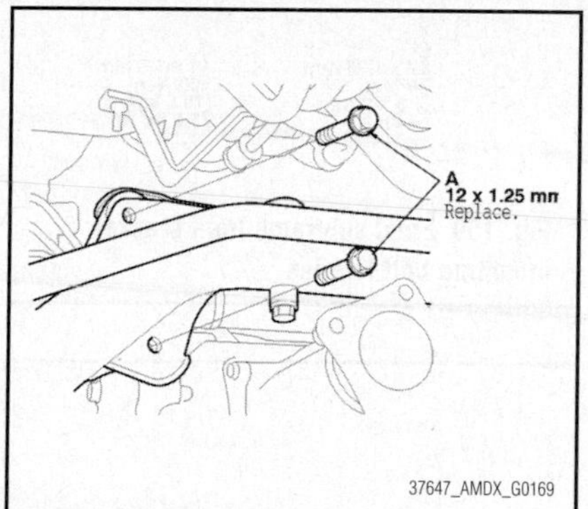
A
12 x 1.25 mm
Replace.
37647_AMDX_G0169

Fig. 164 Remove the mounting bolts (A) from the left side of the steering gear box

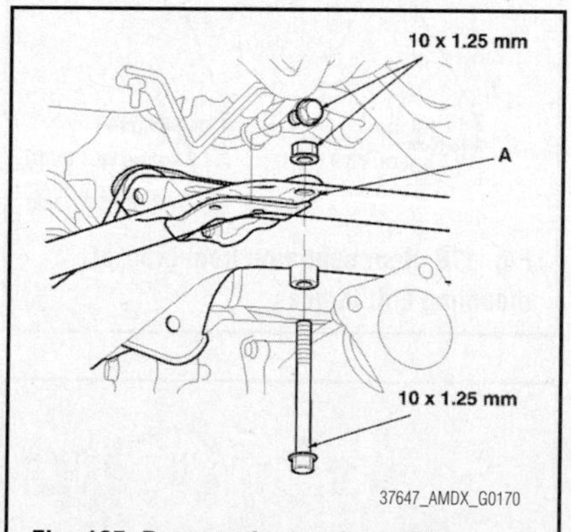
10 x 1.25 mm
A
10 x 1.25 mm
37647_AMDX_G0170

Fig. 165 Remove the gearbox stiffener bracket (A) from the left side of the front subframe

42. Remove the gearbox stiffener bracket from the left side of the front subframe.

43. Remove the mounting bolts from the right side of the steering gearbox, then remove the mounting bracket and cushion.

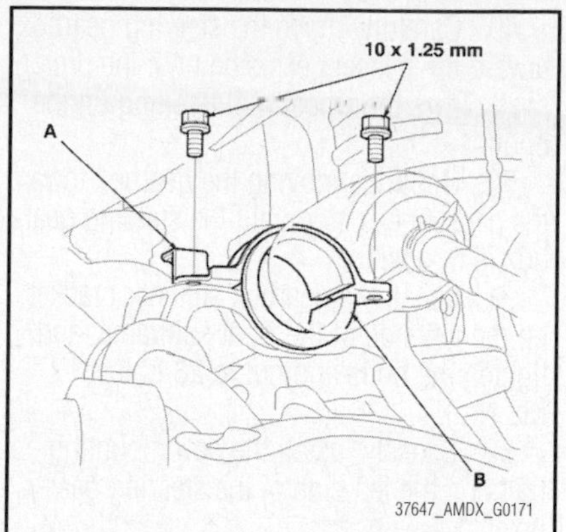
10 x 1.25 mm
A
B
37647_AMDX_G0171

Fig. 166 Remove the mounting bolts from the right side of the steering gearbox, then remove the mounting bracket (A) and cushion (B)

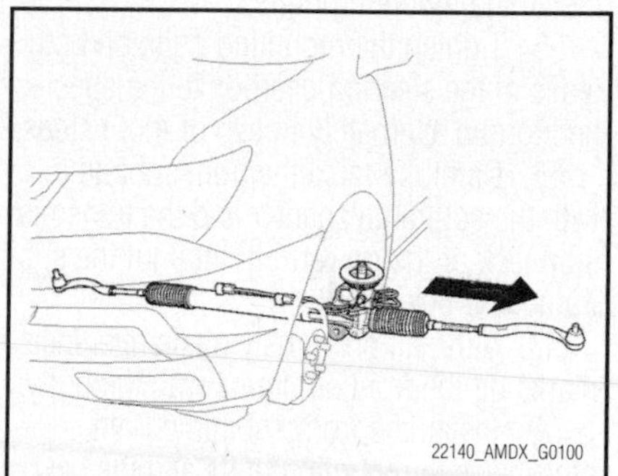
22140_AMDX_G0100

Fig. 167 Remove the steering gearbox through the driver's side wheel well opening

44. Carefully move the steering gearbox toward the driver's side until the pinion shaft clears the wheel well opening on the frame.

45. Remove the steering gearbox through the wheel well opening on the driver's side.

To install:

CAUTION

Before installing the steering gearbox, make sure that no power steering fluid is on the mating surface of the steering gearbox and front subframe. To prevent the gearbox mounting bolts from loosening after the installation, remove any power steering fluid from the mount cushions and bolt holes.

46. Apply vinyl tape to the splines on the pinion shaft.

47. Install the pinion shaft grommet on the top of the valve housing.

48. Slide the steering gearbox between the front subframe and body from the driver's side.

49. Carefully move the steering gearbox toward the passenger's side until the pinion shaft clears the wheel well opening on the body.

50. Continue moving the gearbox toward the passenger's side until the steering gearbox is in position.

51. Install the gearbox stiffener bracket on the left side of the front subframe, and tighten the bolts and nut to 28 ft. lbs. (38 Nm).

52. Loosely install the new mounting bolts on the left side of the steering gearbox.

53. Position the cutout on the mounting cushion as shown, and install it on the right side of the steering gearbox.

54. Install the gearbox mounting bracket over the mounting cushion, and loosely install the mounting bolts.

55. Tighten the mounting bolts on both sides of the steering gearbox to the specified torque alternately in two or more steps.

56. Carefully raise the front subframe with the subframe adapter and the transmission jack or the powertrain lift until the subframe is it position.

57. Align all previously made reference marks on the front subframe with the body.

58. Install the front subframe front brackets with mounting bolts and the new special bolts, and tighten to the specified torque.

59. Install the front subframe rear brackets with mounting bolts and the new special bolts, and tighten to the specified torque.

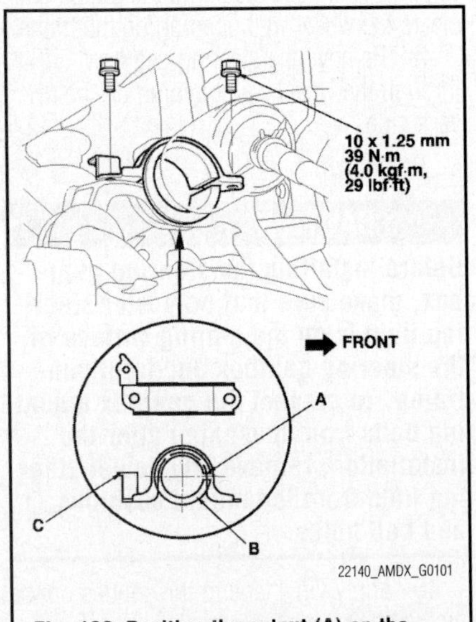
10 x 1.25 mm
39 N·m
(4.0 kgf·m,
29 lbf·ft)
FRONT
A
B
C
22140_AMDX_G0101

Fig. 168 Position the cutout (A) on the mounting cushion (B), and install it on the right side of the steering gearbox. Install the gearbox mounting bracket (C) over the mounting cushion, and loosely install the mounting bolts.

60. Lower the transmission jack supporting the front subframe.

61. Install the ground cable bolt to the transmission and tighten to 33 ft. lbs. (44 Nm).

62. Install the base bracket with new short mounting bolts and tighten them all to 31 ft. lbs. (42 Nm).

63. Install the rear engine mount with new mounting bolts, and tighten to the specified torque.

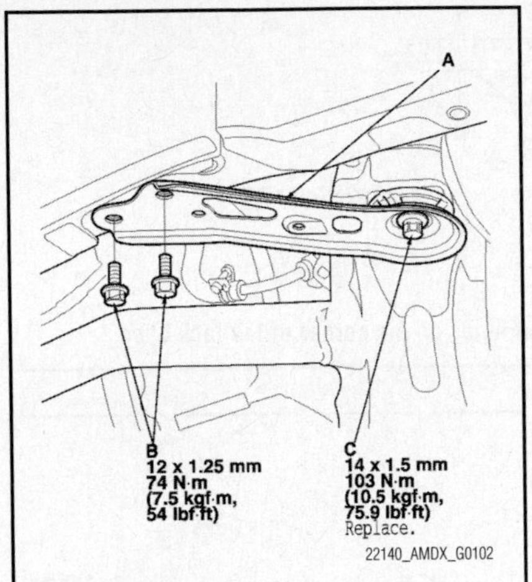
A
B
12 x 1.25 mm
74 N·m
(7.5 kgf·m,
54 lbf·ft)
C
14 x 1.5 mm
103 N·m
(10.5 kgf·m,
75.9 lbf·ft)
Replace.
22140_AMDX_G0102

Fig. 169 Front subframe front bracket mounting bolt torques

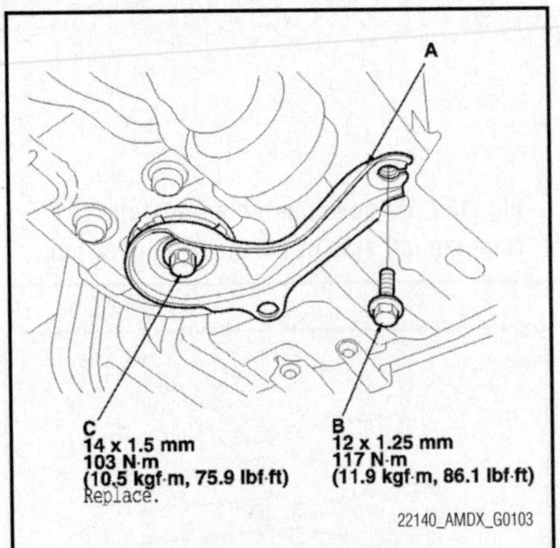
A
C
14 x 1.5 mm
103 N·m
(10.5 kgf·m, 75.9 lbf·ft)
Replace.
B
12 x 1.25 mm
117 N·m
(11.9 kgf·m, 86.1 lbf·ft)
22140_AMDX_G0103

Fig. 170 Rear subframe front bracket mounting bolt torques

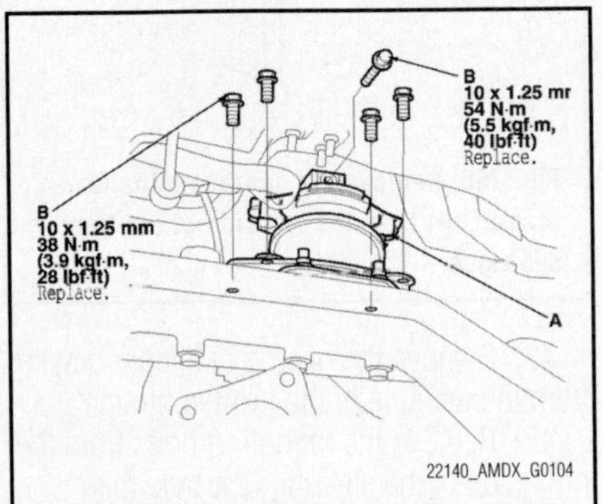
B
10 x 1.25 mm
54 N·m
(5.5 kgf·m,
40 lbf·ft)
Replace.
B
10 x 1.25 mm
38 N·m
(3.9 kgf·m,
28 lbf·ft)
Replace.
A
22140_AMDX_G0104

Fig. 171 Rear engine mount bolt torques

64. Install the rear mount stop with new mounting nuts, and tighten to 54 ft. lbs. (74 Nm).

65. Install the inlet line clamp bolts, and tighten to 7 ft. lbs. (9.8 Nm).

66. Install the exhaust pipe.

67. Install the driveshaft and driveshaft protectors.

68. Install the front subframe stiffener plate with new mounting bolts, and tighten to 40 ft. lbs. (54 Nm).

69. Install the front undercover and the front splash shield.

70. Connect the stabilizer links to the lower arms.

71. Connect the lower arm ball joints to the knuckles.

72. Wipe off any grease contamination from the ball joint tapered section and threads. Reconnect the tie-rod ball joints to the knuckles. Install the 12 mm nuts and tighten it to 40 ft. lbs. (54 Nm).

73. Install new cotter pins.

74. Remove the engine support hanger, the hanger balance bar, and the hanger adapter set.

75. Connect the return hose securely, and tighten the adjustable hose clamp.

76. Connect the inlet line and tighten the 16 mm flare nut to 31 ft. lbs. (42 Nm).

77. Install the front wheel, then set the wheels in the straight ahead position.

➡Before installing the wheel, clean the mating surfaces of the brake disc and inside of the wheel.

78. Lower the vehicle.

79. Center the steering rack within its stroke.

80. Insert the upper end of the steering joint onto the steering shaft (line up the bolt hole with the flat portion on the shaft), and loosely install the upper joint bolt.

81. Slip the lower end of the steering joint onto the pinion shaft taking care to align the gap within the angle.

82. Align the bolt hole on the steering joint with the groove around the pinion shaft then loosely install the joint bolt.

83. Pull on the steering joint to make sure that the steering joint is fully seated, then tighten the lower joint bolt to 16 ft. lbs. (22 Nm).

84. Tighten the upper joint bolt to 16 ft. lbs. (22 Nm).

85. Install the steering joint cover.

86. Install the steering wheel.

87. Connect the battery cables.

88. Turn the ignition switch to ON (II); the SRS indicator should come on for about 6 seconds and then go off.

89. Make sure the horn and turn signal switches work properly.

90. Make sure the steering wheel switches work properly.

91. Refill the power steering system to the correct level.

92. Bleed air from the system.

93. After installation, do these checks:

- Start the engine, allow it to idle, and turn the steering wheel from lock-to-lock several times to warm up the fluid. Check the gearbox for leaks.
- Check the steering wheel spoke angle. If steering spoke angles to the right and left are not equal (steering wheel and rack are not centered), correct the engagement of the joint/pinion shaft serrations, then adjust the front toe by turning the tie-rod ends, if necessary.
- Set the steering column to the center tilt position, then do the front toe inspection.

POWER STEERING PUMP

REMOVAL & INSTALLATION

See Figure 172.

1. Drain the power steering fluid from the reservoir.
2. Remove the engine appearance cover.
3. Remove the accessory drive belt from the pump pulley.
4. Cover the auto-tensioner, alternator, and A/C compressor with several shop towels to protect them from spilled power steering fluid. Disconnect the pump inlet hose and pump outlet hose from the pump, and plug them.

✽✽ WARNING

Take care not to spill the fluid on the body or parts. Wipe off any spilled fluid at once. Do not turn the steering wheel with the pump removed.

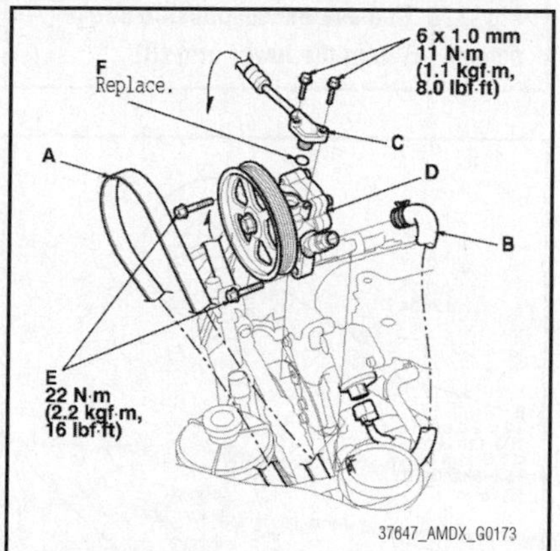
6 x 1.0 mm
11 N·m
(1.1 kgf·m,
8.0 lbf·ft)
F
Replace.
A
C
D
B
E
22 N·m
(2.2 kgf·m,
16 lbf·ft)
37647_AMDX_G0173

Fig. 172 Exploded view of power steering pump assembly

5. Remove the pump mounting bolts.
6. Cover the opening of the pump with a piece of tape to prevent foreign material from entering the pump.

To install:

7. Connect the pump inlet hose and pump outlet hose onto the new pump with the new O-ring.
8. Loosely install the pump in the pump bracket with the mounting bolts, then tighten the pump fittings securely.
9. Tighten the pump mounting bolts to 16 ft. lbs. (22 Nm).
10. Install the accessory drive belt.
11. Refill the power steering system to the correct level.

BLEEDING

1. Fill the reservoir to the upper level line.
2. Start the engine and run it at idle, then turn the steering from lock-to-lock several times to bleed air from the system.
3. Recheck the fluid level and add some if necessary. Do not fill the reservoir beyond the upper level line.
4. If the fluid is contaminated, dark, or discolored, repeat the procedure as necessary.

SUSPENSION

FRONT SUSPENSION

CONTROL LINKS

REMOVAL & INSTALLATION

Front Stabilizer Link

See Figure 173.

1. Raise the front of the vehicle, and support it with safety stands in the proper locations.

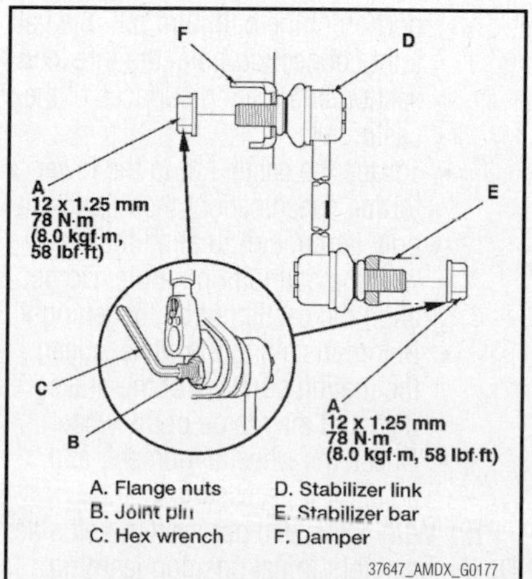
F
D
A
12 x 1.25 mm
78 N·m
(8.0 kgf·m,
58 lbf·ft)
E
C
B
A
12 x 1.25 mm
78 N·m
(8.0 kgf·m, 58 lbf·ft)
A. Flange nuts
B. Joint pin
C. Hex wrench
D. Stabilizer link
E. Stabilizer bar
F. Damper
37647_AMDX_G0177

Fig. 173 Remove the stabilizer link

2. Remove the front wheel.
3. Remove the flange nuts while holding the respective joint pin with a hex wrench, then remove the stabilizer link.

To install:

4. Install the stabilizer link on the stabilizer bar and the damper with the joint pins set at the center of their range of movement.
5. Install the flange nuts, and lightly tighten them.
6. Place a jack under the lower arm, and raise the suspension to load it with the vehicle's weight.
7. Tighten the flange nuts to the specified torque value while holding the respective joint pin with a hex wrench.
8. Clean the mating surfaces of the brake disc and the inside of the wheel, then install the front wheel.
9. Test-drive the vehicle.

LOWER BALL JOINT

REMOVAL & INSTALLATION

See Figures 174 and 175.

1. Remove the lower control arm. For additional information, refer to the following section, "Lower Control Arm, Removal & Installation."
2. Install a hex nut onto the threads of the ball joint. Make sure the nut is flush with the ball joint pin end to prevent damage to the threaded end of the ball joint pin.
3. Apply grease to the ball joint remover on the areas shown. This will ease installation of the tool and prevent damage to the pressure bolt threads.
4. Loosen the pressure bolt, and install the ball joint remover. Insert the jaws care-

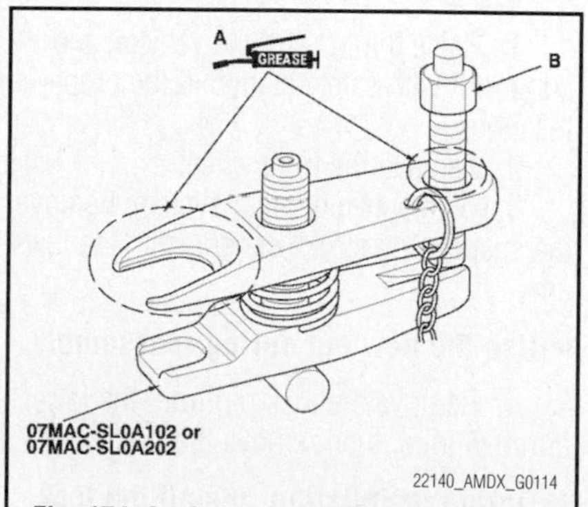
A
GREASE
B
07MAC-SL0A102 or
07MAC-SL0A202
22140_AMDX_G0114

Fig. 174 Apply grease to the ball joint remover on the areas shown (A)

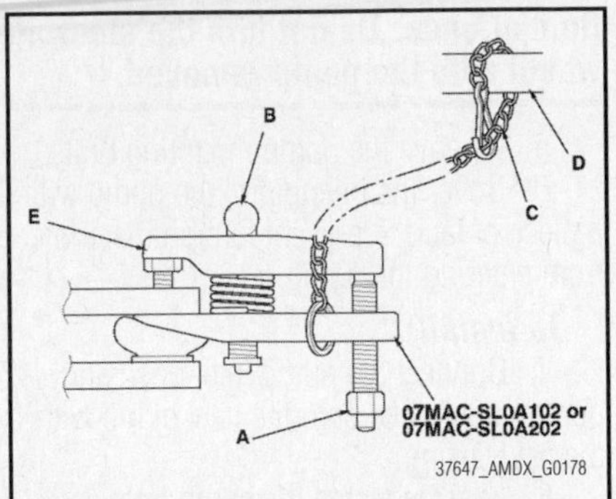
B
D
C
E
A
07MAC-SL0A102 or
07MAC-SL0A202
37647_AMDX_G0178

Fig. 175 Install the ball joint remover

fully, making sure not to damage the ball joint boot. Adjust the jaw spacing by turning the adjusting bolt.

CAUTION

Fasten the safety chain securely to a suspension arm or the subframe. Do not fasten it to a brake line or wire harness.

5. After adjusting the adjusting bolt, make sure the head of the adjusting bolt is in the position shown to allow the jaw to pivot.

6. With a wrench, tighten the pressure bolt until the ball joint pin pops loose from the ball joint connecting hole. If necessary, apply penetrating type lubricant to loosen the ball joint pin.

CAUTION

Do not use pneumatic or electric tools on the pressure bolt.

7. Remove the ball joint remover, then remove the nut from the end of the ball joint pin, and pull the ball joint out of the ball joint connecting hole. Inspect the ball joint boot, and replace it if damaged.

LOWER CONTROL ARM

REMOVAL & INSTALLATION

See Figures 176 through 179.

1. Raise the front of the vehicle, and support it with safety stands in the proper locations.

2. Remove the front wheel.

3. With active damper system: Remove the suspension stroke sensor from the lower arm.

➡Use the new nut during reassembly.

4. Remove the lock pin from the lower arm ball joint, then remove the castle nut.

➡During installation, install the lock pin as shown after tightening the new castle nut.

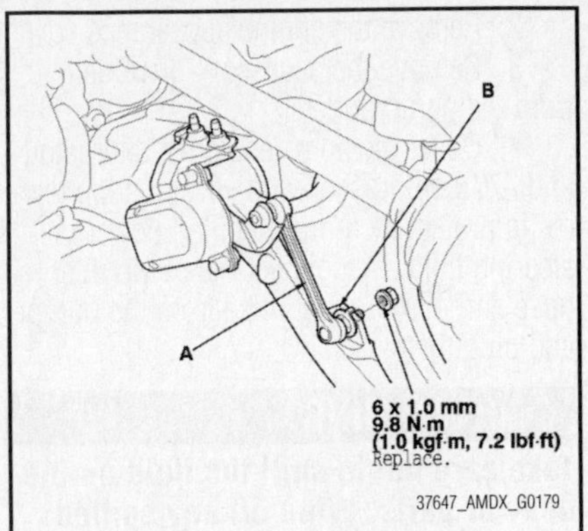
B
A
6 x 1.0 mm
9.8 N·m
(1.0 kgf·m, 7.2 lbf·ft)
Replace.
37647_AMDX_G0179

Fig. 176 Remove the suspension stroke sensor (A) from the lower arm (B)

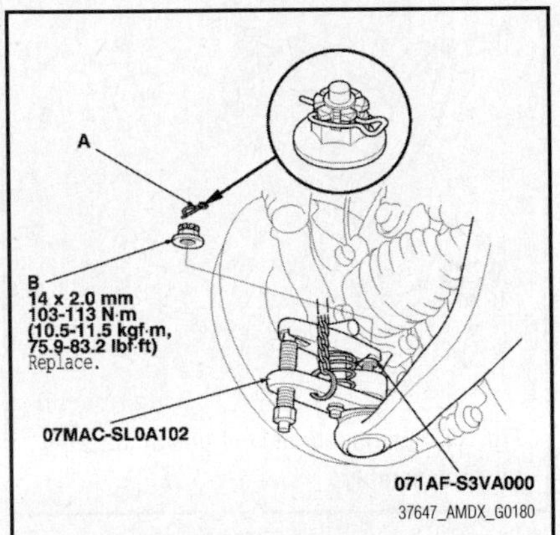
A
B
14 x 2.0 mm
103-113 N·m
(10.5-11.5 kgf·m,
75.9-83.2 lbf·ft)
Replace.
07MAC-SL0A102
071AF-S3VA000
37647_AMDX_G0180

Fig. 177 Remove the lock pin (A) from the lower arm ball joint, then remove the castle nut (B)

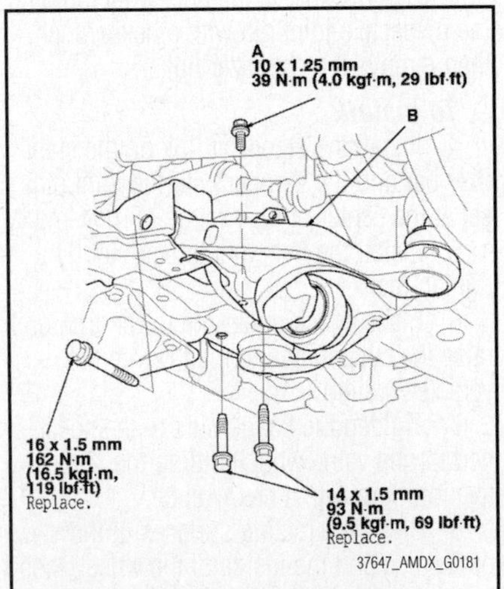
A
10 x 1.25 mm
39 N·m (4.0 kgf·m, 29 lbf·ft)
B
16 x 1.5 mm
162 N·m
(16.5 kgf·m,
119 lbf·ft)
Replace.
14 x 1.5 mm
93 N·m
(9.5 kgf·m, 69 lbf·ft)
Replace.
37647_AMDX_G0181

Fig. 178 Remove the stabilizer bar bushing holder mounting bolt (A)

5. Disconnect the lower arm ball joint from the knuckle using the ball joint thread protector and the ball joint remover.

6. Remove the stabilizer bar bushing holder mounting bolt.

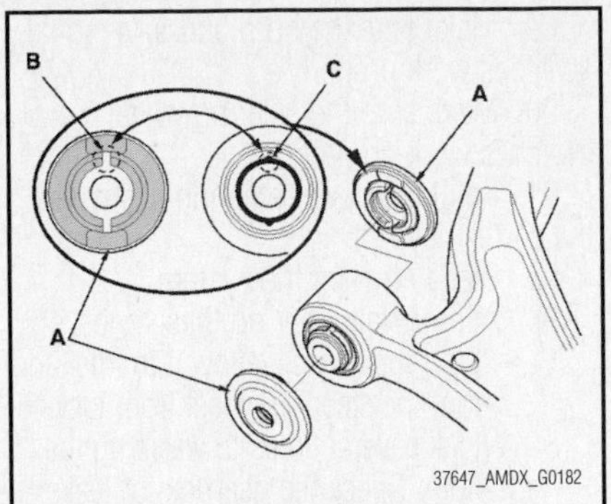
B
C
A
A
37647_AMDX_G0182

Fig. 179 Remove the lower arm stops (A)

➡During installation, install the mounting bolt after tightening the lower arm mounting 14 mm bolts to the specified torque value.

7. Remove the lower arm mounting 14 mm and 16 mm bolts, then remove the lower arm from the front suspension subframe.

➡Use the new mounting bolts during reassembly.

8. Remove the lower arm stops.

➡During installation, align the slot on the lower arm stop with the lug portion on the front side of lower arm bushing.

To install:

9. Install the lower arm in the reverse order of removal, and note these items:
   - First install all the components, and lightly tighten the bolts and nuts, then raise the suspension to load it with the vehicle's weight before fully tightening to the specified torque values.
   - Be careful not to damage the ball joint boot when connecting the knuckle.
   - Before connecting the lower arm ball joint to the knuckle, degrease the threaded section and tapered portion of the ball joint pin, the ball joint connecting hole, the threaded section and mating surfaces of the castle nut.
   - Torque the castle nut to the lower torque specification, then tighten it only far enough to align the slot with the ball joint pin hole. Do not align the castle nut by loosening it.
   - Before installing the wheel, clean the mating surfaces of the brake disc and the inside of the wheel.
   - Check the wheel alignment, and adjust it if necessary.

10. With active damper system/Left side: Do the headlight initial position learning procedure.

STEERING KNUCKLE

REMOVAL & INSTALLATION

See Figures 180 and 181.

1. Raise and safely support the vehicle.
2. Remove the front wheel.
3. Remove the brake hose mounting bolt from the damper.
4. Remove the brake caliper bracket mounting bolts, then remove the caliper assembly from the knuckle.

➡To prevent damage to the caliper assembly or brake hose, use a short piece of wire to hang the caliper assembly from the undercarriage. Do not twist the brake hose excessively.

5. Raise the stake, then remove the spindle nut.
6. Remove the brake rotor.
7. Remove the hub bearing unit and the splash guard from the knuckle.

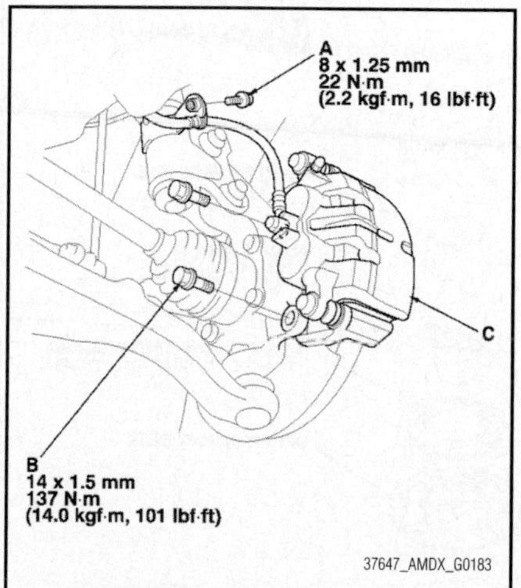
A
8 x 1.25 mm
22 N·m
(2.2 kgf·m, 16 lbf·ft)
C
B
14 x 1.5 mm
137 N·m
(14.0 kgf·m, 101 lbf·ft)
37647_AMDX_G0183

Fig. 180 Remove the brake hose mounting bolt (A) from the damper

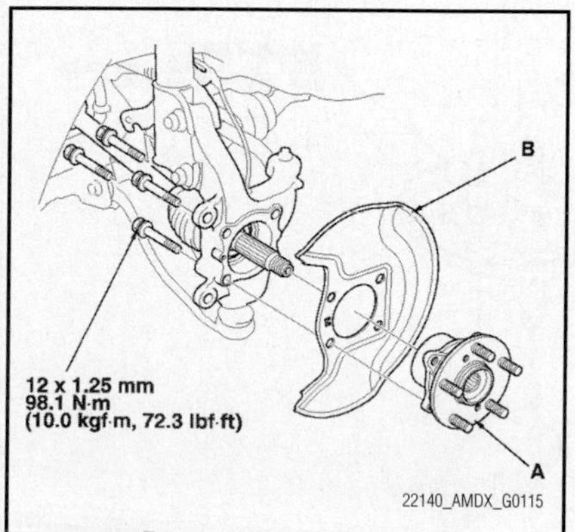
B
12 x 1.25 mm
98.1 N·m
(10.0 kgf·m, 72.3 lbf·ft)
A
22140_AMDX_G0115

Fig. 181 Remove the mounting bolts to remove the hub bearing unit (A) and splash guard (B).

To install:

8. Installation is the reverse order of removal. Note the following torque values.
   - Hub bearing unit: 72 ft. lbs. (98 Nm)
   - New spindle nut: 242 ft. lbs. (328 Nm)

STRUT

REMOVAL & INSTALLATION

See Figures 182 through 186.

➡Except where called out, the illustrations show without active damper system.

1. Raise the front of the vehicle, and support it with safety stands in the proper locations.
2. Remove the front wheel.
3. With active damper system: Disconnect the damper coil connector, and remove the flange bolt, and the harness clip.

➡Be careful not to damage or contaminate the damper coil connector.

4. Remove the wheel speed sensor

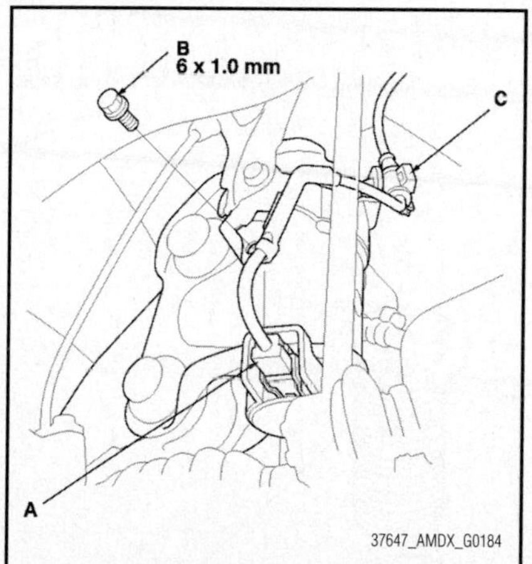
B
6 x 1.0 mm
C
A
37647_AMDX_G0184

Fig. 182 Disconnect the damper coil connector (A), and remove the flange bolt (B), and the harness clip (C)

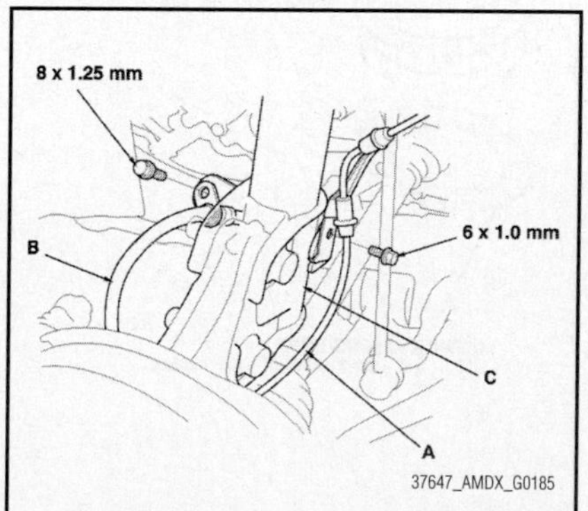
8 x 1.25 mm
B
6 x 1.0 mm
C
A
37647_AMDX_G0185

Fig. 183 Remove the wheel speed sensor harness (A) and the brake hose (B) from the damper (C)

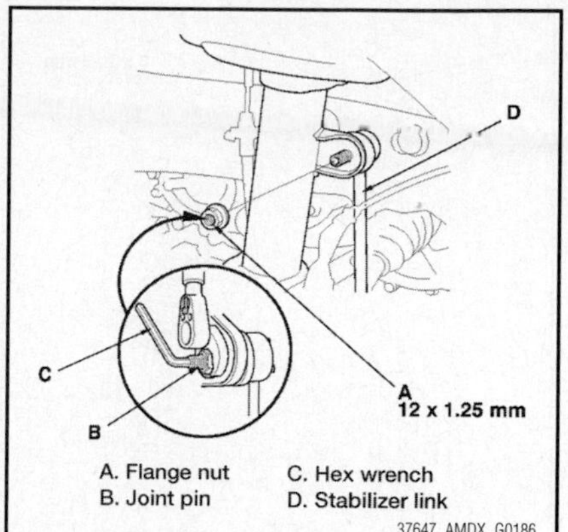
D
C
A
12 x 1.25 mm
B
A. Flange nut
B. Joint pin
C. Hex wrench
D. Stabilizer link
37647_AMDX_G0186

Fig. 184 Remove the flange nut, while holding the joint pin with a hex wrench, and disconnect the stabilizer link from the damper

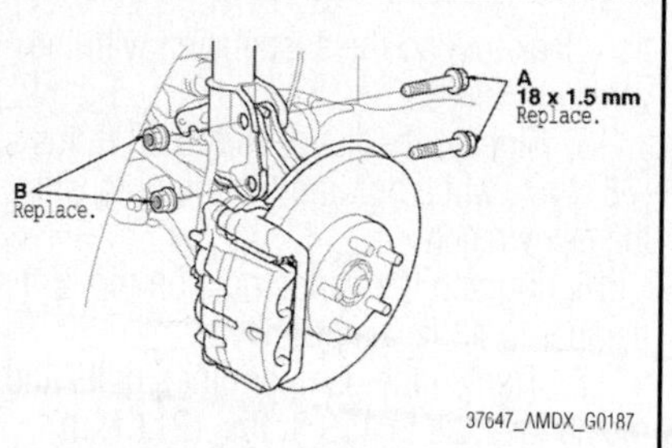
A
18 x 1.5 mm
Replace.
B
Replace.
37647_AMDX_G0187

Fig. 185 Remove the damper pinch bolts (A) and flange nuts (B) from the damper

harness and the brake hose from the damper. Do not disconnect the wheel speed sensor connector.

5. Remove the flange nut, while holding the joint pin with a hex wrench, and disconnect the stabilizer link from the damper.
6. Remove the damper pinch bolts and flange nuts from the damper.
7. Remove the cover and the service caps.
8. Remove the flange nuts from the top of the damper.
9. Remove the damper assembly.

➡The damper springs are different, left and right. Mark the springs L and R before you continue.

To install:

10. Install the damper assembly on to the frame. Note the direction of the damper mounting base.

➡Be careful not to damage the body.

11. Loosely install the new flange nuts to the top of the damper.
12. Loosely install new damper pinch bolts and new flange nuts to the damper.
13. Connect the stabilizer link to the damper, and loosely install a new flange nut.
14. Raise the front suspension with a

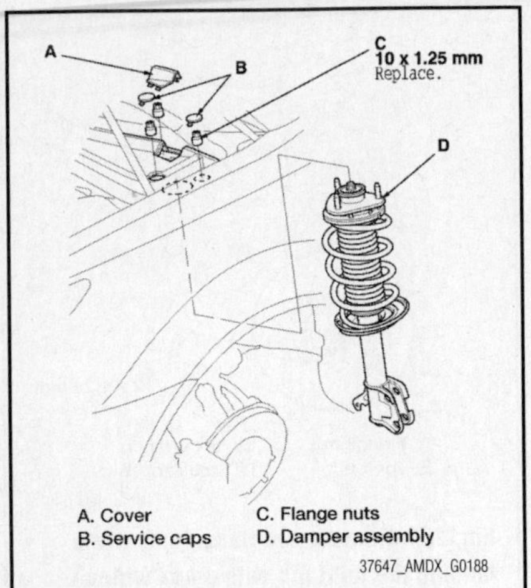
A
B
C
10 x 1.25 mm
Replace.
D
A. Cover
B. Service caps
C. Flange nuts
D. Damper assembly
37647_AMDX_G0188

Fig. 186 Remove the cover and service caps, the flange nuts and the damper assembly

floor jack to load the suspension with the vehicle's weight.

15. Tighten the flange nut to 58 ft. lbs. (78 N m), while holding the joint pin with the hex wrench

16. Tighten the flange nuts on top of the damper to 43 ft. lbs. (59 N m).

17. Tighten the damper pinch bolts and the flange nuts to 159 ft. lbs. (211 N m).

18. Install the cover and the service caps.

19. Install the wheel speed sensor harness and the brake hose to the damper.

20. With active damper system: Connect the damper coil connector, and install the flange bolt, and the harness clip.

➡Be careful not to damage or contaminate the damper coil connecter.

21. Clean the mating surfaces of the brake disc and the inside of the wheel, then install the front wheel.

22. With active damper system: Start the engine, then make sure there are no active damper system DTCs with the HDS.

23. With active damper system: Do the DAMPER FORCE OPERATION in the ACTIVE DAMPER SYSTEM INSPECTION MENU with the HDS, then make sure the all four damper units function normally.

24. Check the wheel alignment, and adjust it if necessary.

OVERHAUL

➡When compressing the damper spring, use a commercially available strut spring compressor (Branick MST-580A or Model 7200, or equivalent) according to the manufacturer's instructions.

➡Except where called out, the illustrations are shown without active damper system.

Disassembly

See Figures 187 through 189.

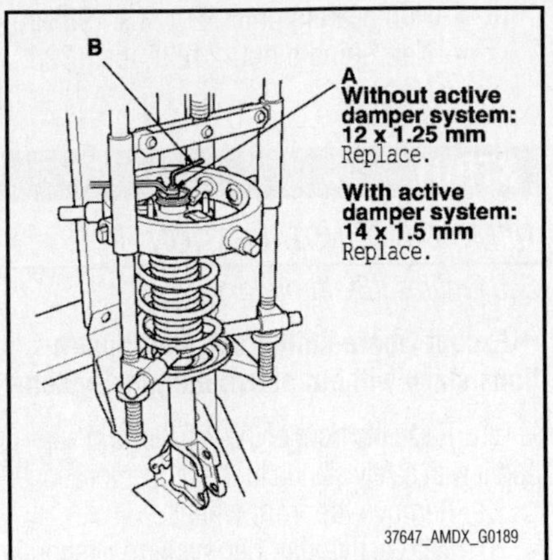
B
A
Without active damper system:
12 x 1.25 mm
Replace.
With active damper system:
14 x 1.5 mm
Replace.
37647_AMDX_G0189

Fig. 187 Compress the damper spring, then remove the self-locking nut (A) while holding the damper shaft with a hex wrench or TORX® wrench (B)

1. Compress the damper spring, then remove the self-locking nut while holding the damper shaft with a hex wrench or TORX® wrench. Do not compress the damper spring more than necessary to remove the nut.

2. Release the pressure from the strut spring compressor, then disassemble the damper as shown in the Exploded View.

Inspection

1. Reassemble the damper mounting base, the damper mounting washer, and the self-locking nut.

2. Compress the damper assembly by hand, and check for smooth operation through a full stroke, both compression and extension. The damper should extend smoothly and constantly when compression is released. If it does not, the gas is leaking and the damper should be replaced.

3. Check for oil leaks, abnormal noises, and binding during these tests.

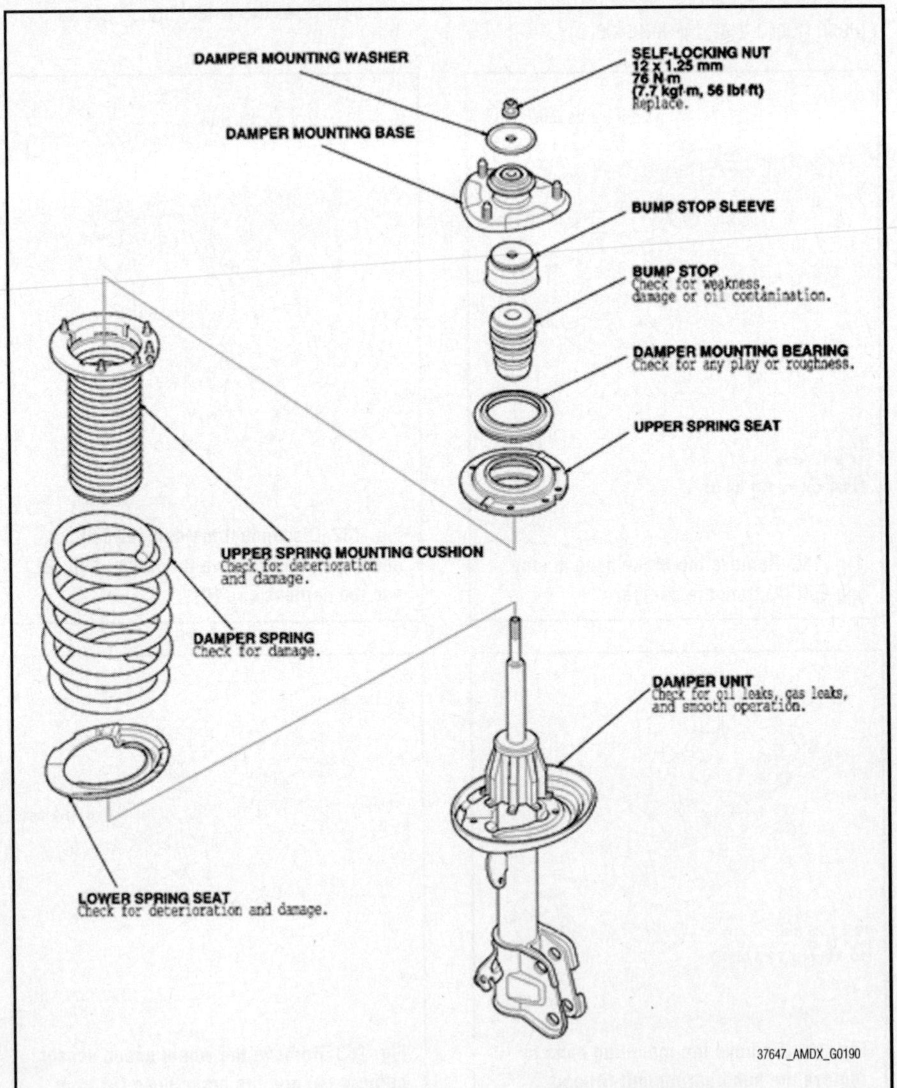
DAMPER MOUNTING WASHER
SELF-LOCKING NUT
12 x 1.25 mm
78 N m
(7.7 kgf m, 56 lbf ft)
Replace.
DAMPER MOUNTING BASE
BUMP STOP SLEEVE
BUMP STOP
Check for weakness, damage or oil contamination.
DAMPER MOUNTING BEARING
Check for any play or roughness.
UPPER SPRING SEAT
UPPER SPRING MOUNTING CUSHION
Check for deterioration and damage.
DAMPER SPRING
Check for damage.
DAMPER UNIT
Check for oil leaks, gas leaks, and smooth operation.
LOWER SPRING SEAT
Check for deterioration and damage.
37647_AMDX_G0190

Fig. 188 Exploded view of damper assembly without active damper system

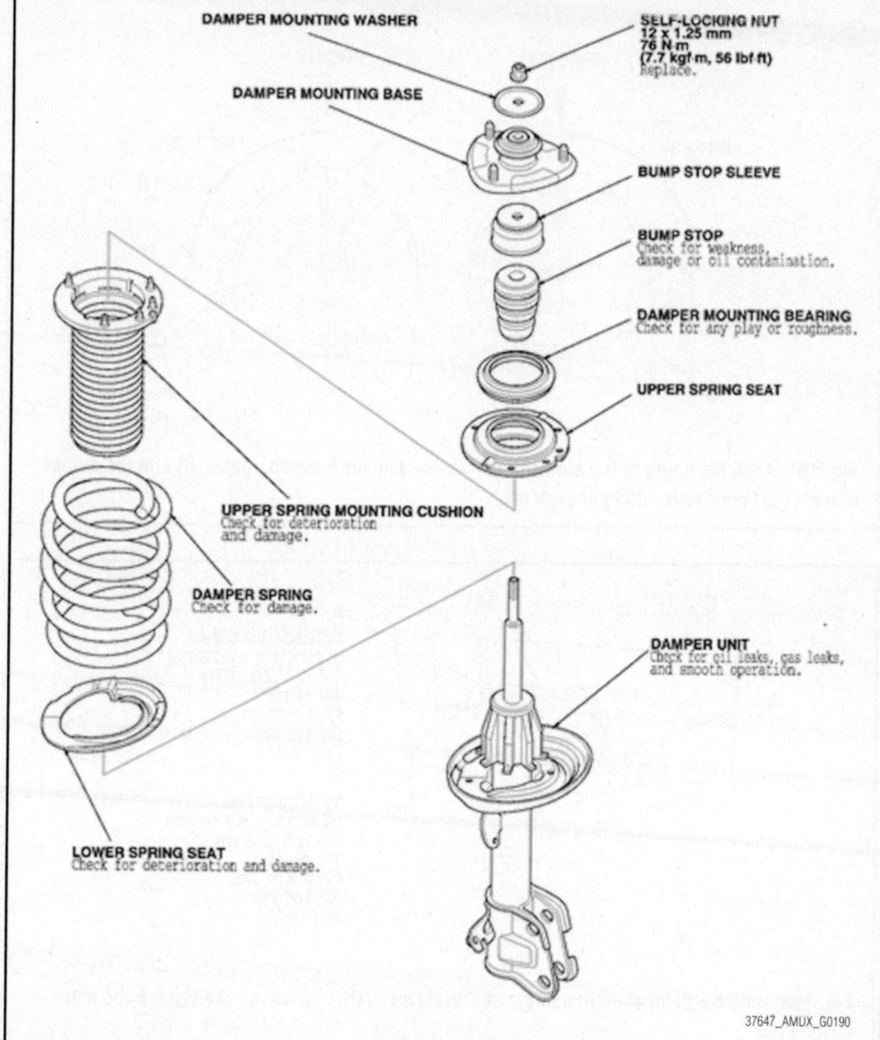
DAMPER MOUNTING WASHER
SELF-LOCKING NUT
12 x 1.25 mm
76 N·m
(7.7 kgf·m, 56 lbf·ft)
Replace.
DAMPER MOUNTING BASE
BUMP STOP SLEEVE
BUMP STOP
Check for weakness, damage or oil contamination.
DAMPER MOUNTING BEARING
Check for any play or roughness.
UPPER SPRING SEAT
UPPER SPRING MOUNTING CUSHION
Check for deterioration and damage.
DAMPER SPRING
Check for damage.
DAMPER UNIT
Check for oil leaks, gas leaks, and smooth operation.
LOWER SPRING SEAT
Check for deterioration and damage.
37647_AMDX_G0190

Fig. 189 Exploded view of damper assembly with active damper system

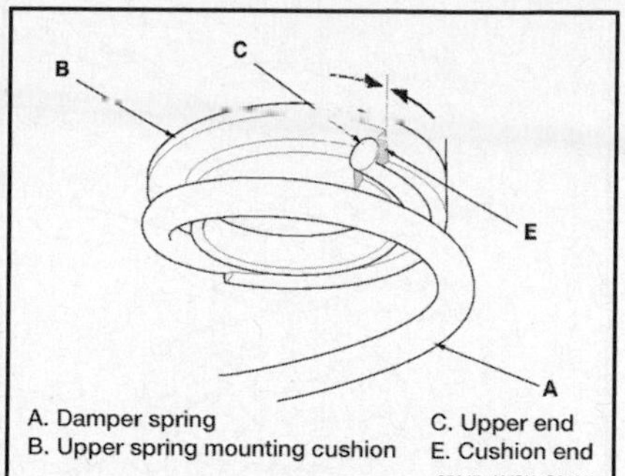
A. Damper spring
B. Upper spring mounting cushion
C. Upper end
E. Cushion end
37647_AMDX_G0193

Fig. 191 Install the damper spring on the upper spring mounting cushion, then align the upper end of the damper spring with the cushion end (With active damper system)

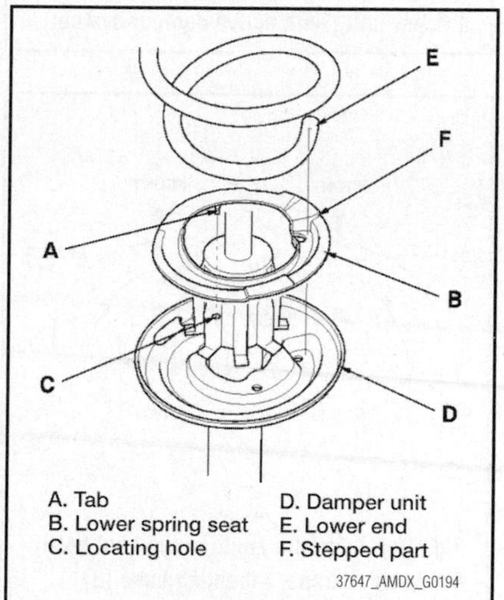
A. Tab
B. Lower spring seat
C. Locating hole
D. Damper unit
E. Lower end
F. Stepped part
37647_AMDX_G0194

Fig. 192 Install the tab on the lower spring seat in the locating hole on the damper unit (Without active damper system)

Reassembly

See Figures 190 through 196.

1. Install the damper spring on the upper spring mounting cushion, then align the upper end of the damper spring with the cushion stop or the cushion end.
2. Compress the damper spring.
3. Install the tab on the lower spring seat in the locating hole on the damper unit.
4. Align the lower end of the damper spring with the stepped part of the lower spring seat.
5. Install all the parts except the damper mounting washer and self-locking nut onto the damper unit by referring to the Exploded View.
6. Align the angle of the stud bolt on the damper mounting base and the damper bracket as shown.
7. Install the damper mounting washer and a new self-locking nut.

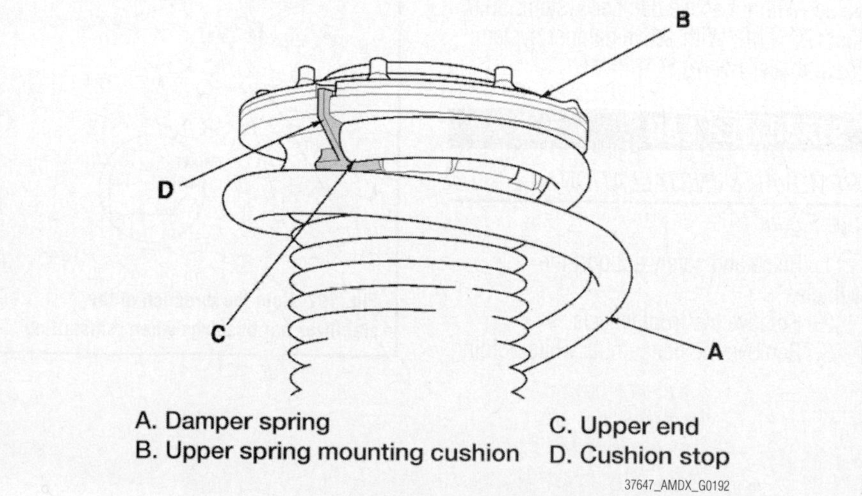
A. Damper spring
B. Upper spring mounting cushion
C. Upper end
D. Cushion stop
37647_AMDX_G0192

Fig. 190 Install the damper spring on the upper spring mounting cushion, then align the upper end of the damper spring with the cushion stop (Without active damper system)

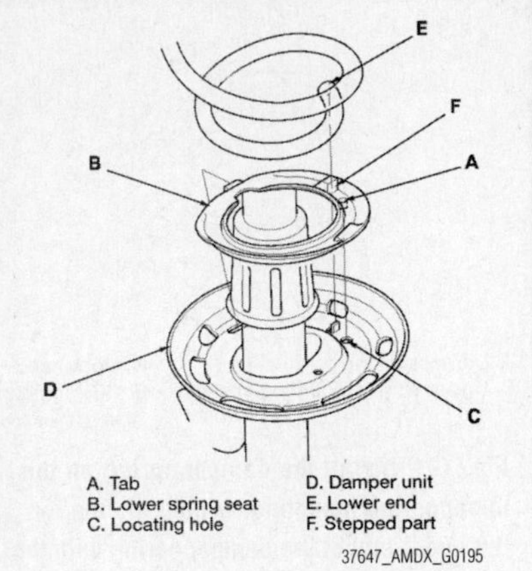

E
F
B
A
D
C
A. Tab
B. Lower spring seat
C. Locating hole
D. Damper unit
E. Lower end
F. Stepped part
37647_AMDX_G0195

Fig. 193 Install the tab on the lower spring seat in the locating hole on the damper unit (With active damper system)

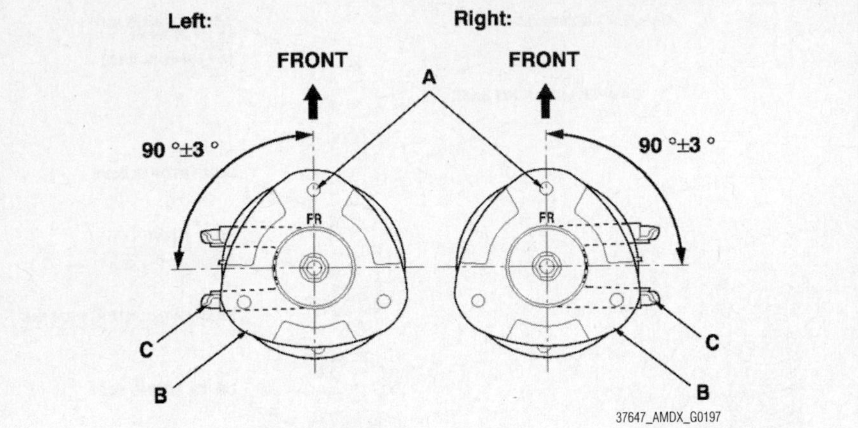

Left:
Right:
FRONT
FRONT
A
90 °±3 °
90 °±3 °
FR
FR
C
C
B
B
37647_AMDX_G0197

Fig. 195 Align the angle of the stud bolt (A) on the damper mounting base (B) and the damper bracket (C) (With active damper system)

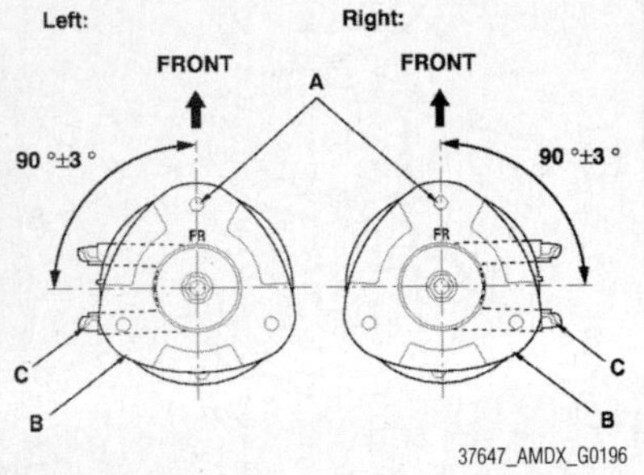

Left:
Right:
FRONT
FRONT
A
90 °±3 °
90 °±3 °
C
C
B
B
37647_AMDX_G0196

Fig. 194 Align the angle of the stud bolt (A) on the damper mounting base (B) and the damper bracket (C) (Without active damper system)

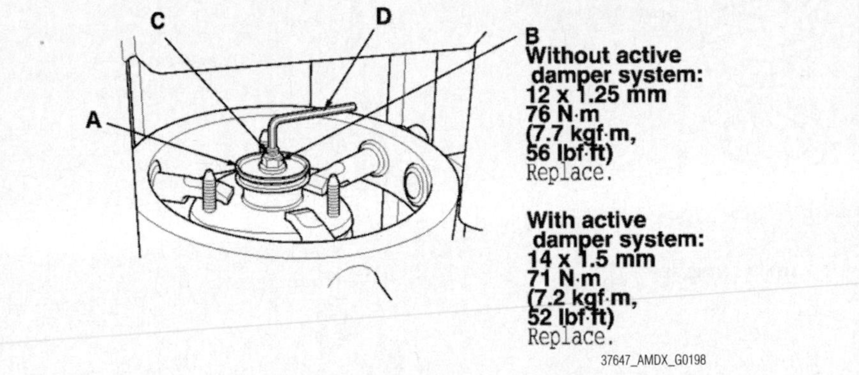

C
D
A
B
Without active damper system:
12 x 1.25 mm
76 N·m
(7.7 kgf·m, 56 lbf·ft)
Replace.
With active damper system:
14 x 1.5 mm
71 N·m
(7.2 kgf·m, 52 lbf·ft)
Replace.
37647_AMDX_G0198

Fig. 196 Hold the damper shaft using a hex wrench or TORX® wrench, and tighten the self-locking nut

8. Hold the damper shaft using a hex wrench or TORX;rM wrench, and tighten the self-locking nut to the specified torque value; Without active damper system: 56 ft. lbs. (76 N m); With active damper system: 52 ft. lbs. (71 N m).

STABILIZER BAR

REMOVAL & INSTALLATION

See Figure 197.

1. Raise and safely support the vehicle.
2. Remove the front wheels.
3. Remove the flange nuts while holding the respective joint pin with a hex wrench, and remove the stabilizer link.
4. Remove the flange bolts and the bushing holders, then remove the bushings and the stabilizer bar from the front suspension subframe.
5. Installation is the reverse order of removal. Tighten the bushing mounting bolts to 29 ft. lbs. (39 Nm).

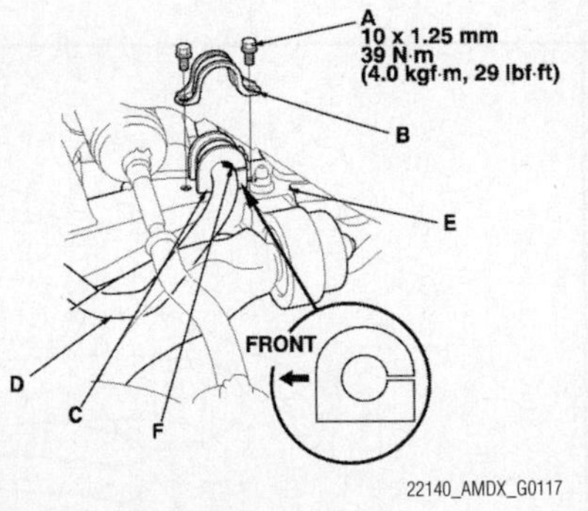

A
10 x 1.25 mm
39 N·m
(4.0 kgf·m, 29 lbf·ft)
B
E
FRONT
D
C
F
22140_AMDX_G0117

Fig. 197 Note the direction of the stabilizer bar bushings when reinstalling

WHEEL HUB & BEARING

REMOVAL & INSTALLATION

Refer to Steering knuckle Removal and Installation.

BALL JOINT

REMOVAL & INSTALLATION

See Figures 198 and 199.

1. Install a hex nut onto the threads of the ball joint. Make sure the nut is flush with the ball joint pin end to prevent damage to the threaded end of the ball joint pin.
2. Apply grease to the ball joint remover on the areas shown. This will ease installation of the tool and prevent damage to the pressure bolt threads.
3. Loosen the pressure bolt, and install the ball joint remover. Insert the jaws carefully, making sure not to damage the ball joint boot. Adjust the jaw spacing by turning the adjusting bolt.

**** CAUTION**

Fasten the safety chain securely to a suspension arm or the subframe. Do not fasten it to a brake line or wire harness.

4. After adjusting the adjusting bolt, make sure the head of the adjusting bolt is in the position shown to allow the jaw to pivot.
5. With a wrench, tighten the pressure bolt until the ball joint pin pops loose from the ball joint connecting hole. If necessary, apply penetrating type lubricant to loosen the ball joint pin.

**** CAUTION**

Do not use pneumatic or electric tools on the pressure bolt.

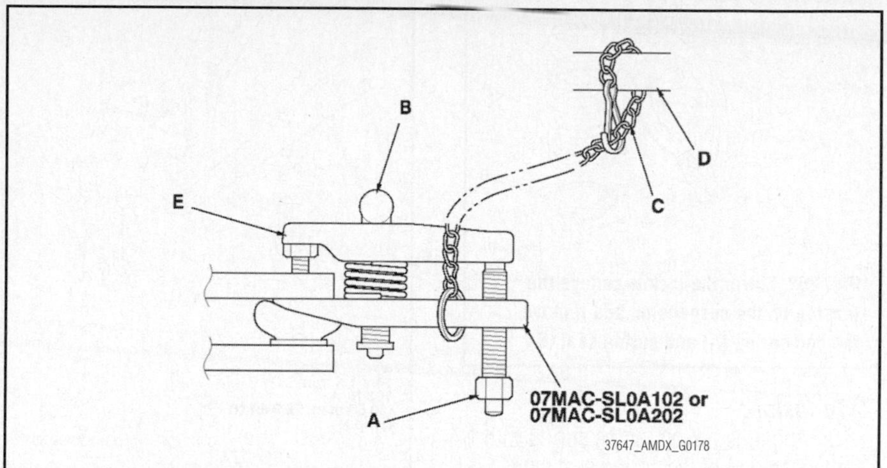
B
E
D
C
A
07MAC-SL0A102 or
07MAC-SL0A202
37647_AMDX_G0178

Fig. 199 Install the ball joint remover

6. Remove the ball joint remover, then remove the nut from the end of the ball joint pin, and pull the ball joint out of the ball joint connecting hole. Inspect the ball joint boot, and replace it if damaged.

COIL SPRING

REMOVAL & INSTALLATION

See Figures 200 through 202.

1. Raise and safely support the vehicle.
2. Remove the rear wheel.
3. Remove the muffler from the muffler hanger.
4. Remove the flange nut while holding the joint pin with a hex wrench, and disconnect the stabilizer link from the lower control arm B. Discard the flange nut.
5. Position a floor jack under the lower arm B and raise the floor jack until the suspension begins to compress.
6. Remove the flange bolt from the bottom of the damper.
7. Remove the flange bolt from the knuckle.
8. Lower the floor jack gradually and remove the spring and spring seat.

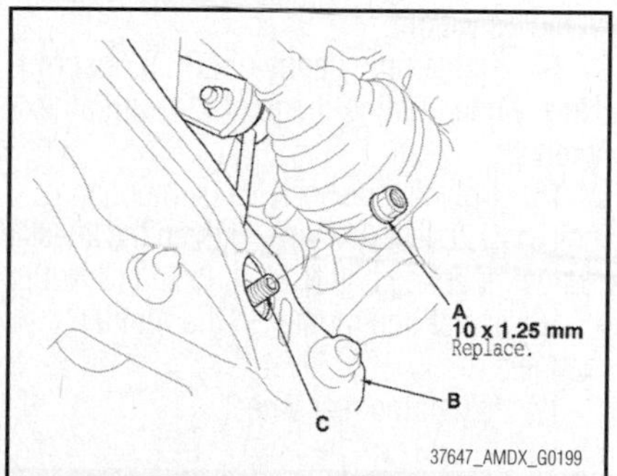
A
10 x 1.25 mm
Replace.
B
C
37647_AMDX_G0199

Fig. 200 Remove the flange nut (A) while holding the joint pin (C) with a hex wrench, and disconnect the stabilizer link from the lower arm B

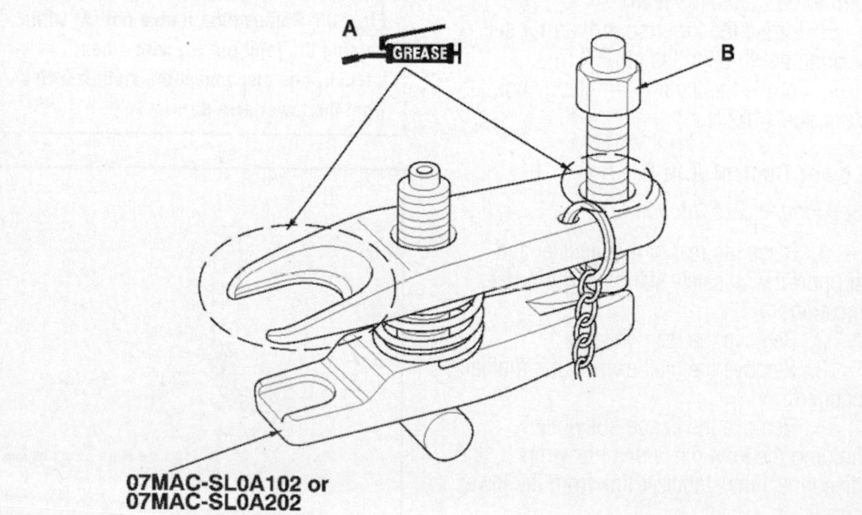
A
GREASE
B
07MAC-SL0A102 or
07MAC-SL0A202
22140_AMDX_G0114

Fig. 198 Apply grease to the ball joint remover on the areas shown (A)

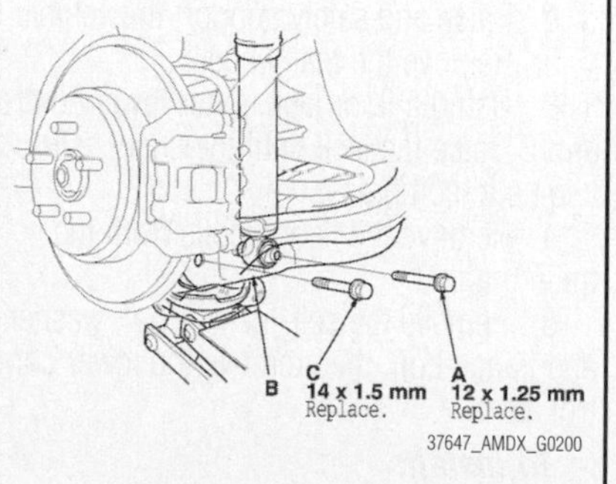
B
C
14 x 1.5 mm
Replace.
A
12 x 1.25 mm
Replace.
37647_AMDX_G0200

Fig. 201 Remove the flange bolt (A) from the bottom of the damper

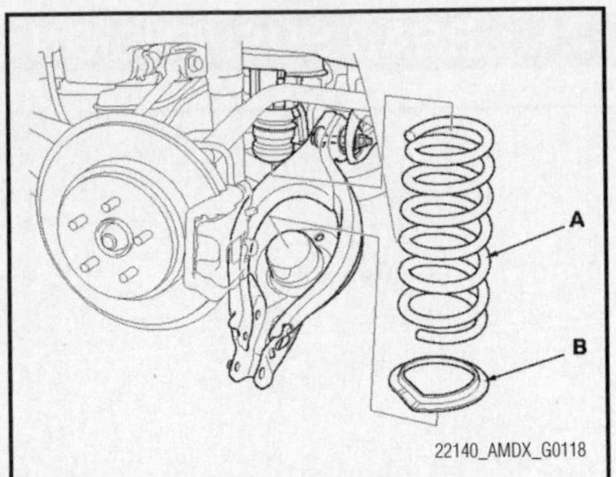
A
B
22140_AMDX_G0118

Fig. 202 Lower the jack to relieve the tension on the suspension and remove the coil spring (A) and spring seat (B)

To install:

9. Align the bottom of the spring with the stepped part of the spring seat, and install into the control arm.

10. With the jack under the control arm, raise the floor jack until the mounting hole in the lower control arm B aligns with the hole in the shock, then loosely install the new flange bolt to the bottom of the shock.

11. Loosely install the new flange bolt to the knuckle.

12. Connect the stabilizer link to the lower arm B, then loosely install the new flange nut.

13. Raise the rear suspension with the floor jack to load the suspension with the vehicle's weight.

14. Tighten the flange nut 36 ft. lbs. (49 Nm), while holding the joint pin with a hex wrench.

15. Tighten the shock flange mounting bolt to 47 ft. lbs. (64 Nm). Tighten the knuckle flange mounting bolt to 83 ft. lbs. (113 Nm).

16. Install the muffler to the muffler hanger.

17. Install the rear wheel.

CONTROL ARMS/LINKS

REMOVAL & INSTALLATION

Lower Control Arm A

See Figures 203 and 204.

1. Raise and safely support the vehicle.
2. Remove the rear wheel.
3. Using a floor jack under lower control arm B, raise the jack until the suspension begins to compress.
4. Remove the flange bolts from the trailing arm.
5. Remove the self-locking nut, washer, and flange bolt, then remove the lower control arm.

To install:

6. Installation is the reverse order of removal.

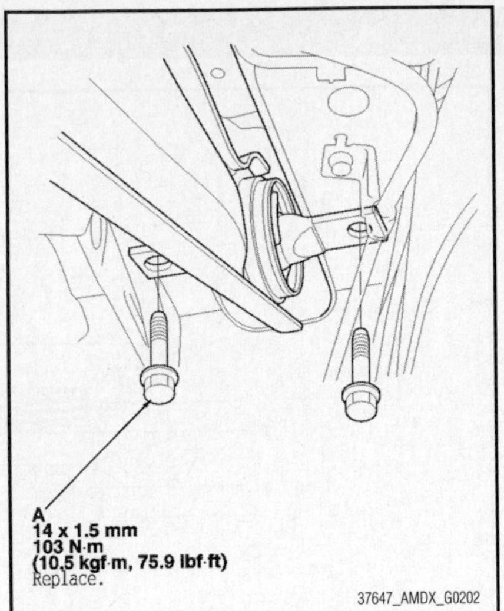
A
14 x 1.5 mm
103 N·m
(10.5 kgf·m, 75.9 lbf·ft)
Replace.
37647_AMDX_G0202

Fig. 203 Remove the flange bolts from the trailing arm

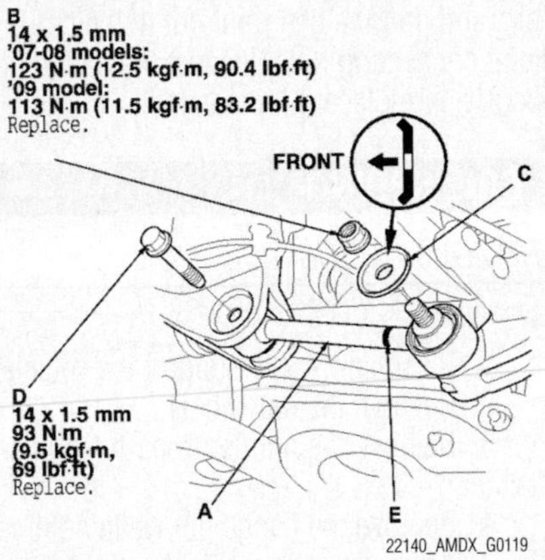
B
14 x 1.5 mm
'07-08 models:
123 N·m (12.5 kgf·m, 90.4 lbf·ft)
'09 model:
113 N·m (11.5 kgf·m, 83.2 lbf·ft)
Replace.
FRONT
C
D
14 x 1.5 mm
93 N·m
(9.5 kgf·m,
69 lbf·ft)
Replace.
A
E
22140_AMDX_G0119

Fig. 204 Lower control arm A mounting

7. Torque the lower control arm A flange bolt to 69 ft. lbs. (93 N m).

8. Torque the lower control arm A self-locking nut to 83 ft. lbs (113 N m).

9. Torque the trailing arm flange bolts to 76 ft. lbs. (103 N m).

Lower Control Arm B

See Figures 205 through 207.

1. Raise the rear of the vehicle, and support it with safety stands in the proper locations.

2. Remove the rear wheel.

3. Remove the muffler from the muffler hangers.

4. Remove the flange nut while holding the joint pin with a hex wrench, and disconnect the stabilizer link from the lower arm B.

5. Position a floor jack under the lower arm B. Raise the floor jack until the suspension begins to compress.

6. Remove the flange bolt from the bottom of the damper.

7. Remove the flange bolt from the knuckle.

8. Lower the floor jack gradually.

9. Remove the spring and the lower spring seat.

➡During installation, align the bottom of the spring with the stepped part of the lower spring seat and the lower arm B as shown.

10. Mark the cam positions of the adjusting bolt and adjusting cam plate, then remove the self-locking nut, adjusting cam plate, and adjusting bolt.

11. Remove the lower arm B.

To install:

12. Install the lower arm B in the reverse order of removal, and note these items:

- Replace all bolts and nuts with new during installation.
- First install all the suspension components, and lightly tighten the bolts and nuts, then raise the suspension to load it with the vehicle's

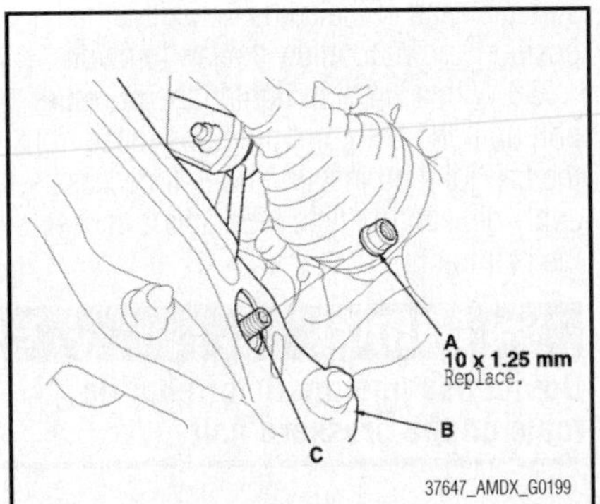
A
10 x 1.25 mm
Replace.
B
C
37647_AMDX_G0199

Fig. 205 Remove the flange nut (A) while holding the joint pin (C) with a hex wrench, and disconnect the stabilizer link from the lower arm B

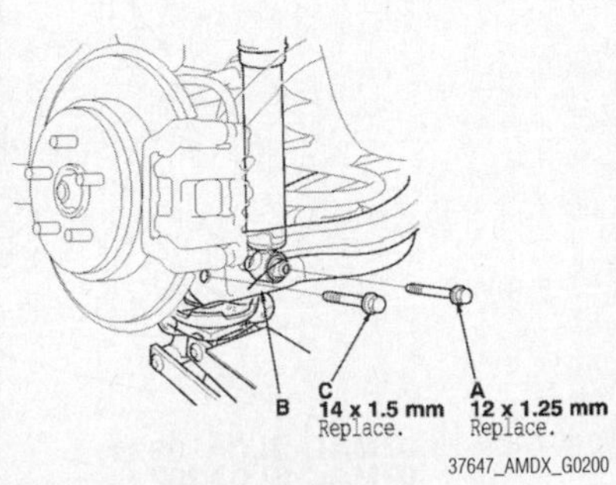
B
C
14 x 1.5 mm
Replace.
A
12 x 1.25 mm
Replace.
37647_AMDX_G0200

Fig. 206 Remove the flange bolt (A) from the bottom of the damper

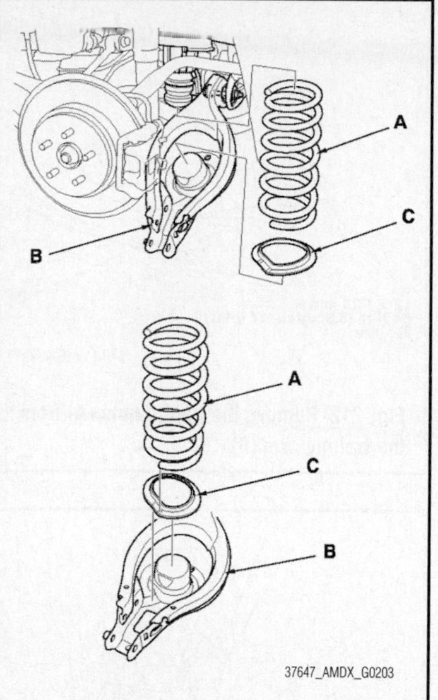
A
C
B
A
C
B
37647_AMDX_G0203

Fig. 207 Remove the spring (A) and the lower spring seat (C)

weight before fully tightening to the specified torque values.

- Align the cam positions of the adjusting bolt and the adjusting cam plate with the marked positions when tightening.
- Before installing the wheel, clean the mating surfaces of the brake disc and the inside of the wheel.
- Check the wheel alignment, and adjust it if necessary.

13. Torque the self-locking nut to 98 ft. lbs. (132 N m).
14. Torque the flange bolt to knuckle to 83 ft. lbs. (113 N m).
15. Torque the lower damper flange bolt to 47 ft. lbs. (64 N m).
16. Torque the stabilizer link flange nut to 36 ft. lbs. (49 N m).

Upper Control Arm

See Figures 208 and 209.

1. Raise the rear of the vehicle, and support it with safety stands in the proper locations.
2. Remove the rear wheel.
3. Position a floor jack under the lower arm B. Raise the floor jack until the suspension begins to compress.
4. With active damper system: Remove the suspension stroke sensor from the upper arm.
5. Remove the lock pin from the upper arm ball joint, then remove the castle nut.

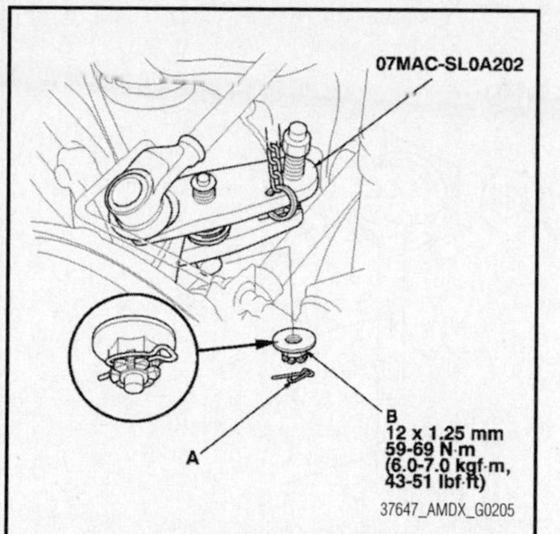
07MAC-SL0A202
B
12 x 1.25 mm
59-69 N·m
(6.0-7.0 kgf·m,
43-51 lbf·ft)
A
37647_AMDX_G0205

Fig. 208 Remove the lock pin (A) from the upper arm ball joint, then remove the castle nut (B)

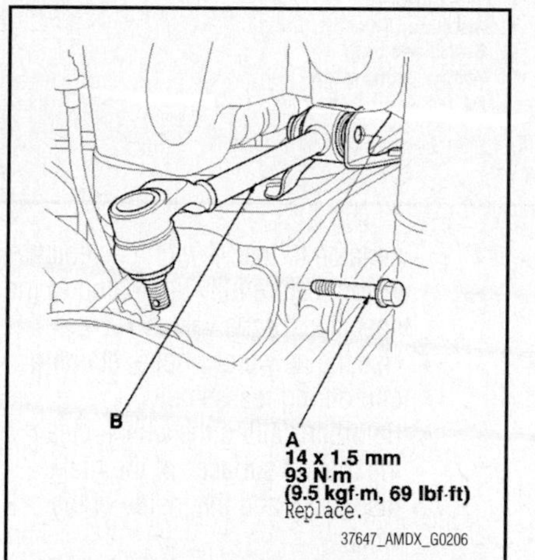
B
A
14 x 1.5 mm
93 N·m
(9.5 kgf·m, 69 lbf·ft)
Replace.
37647_AMDX_G0206

Fig. 209 Remove the flange bolt (A) from the vehicle, and remove the upper arm (B)

➡During installation, install the lock pin as shown after tightening the castle nut.

6. Disconnect the upper arm ball joint from the knuckle using the ball joint remover.
7. Remove the flange bolt from the vehicle, and remove the upper arm.

➡Use the new flange bolt during reassembly.

To install:

8. Install the upper arm in the reverse order of removal, and note these items:

- First install all the components, and lightly tighten the bolts and nuts, then raise the suspension to load it with the vehicle's weight before fully tightening to the specified torque values.
- Be careful not to damage the ball joint boot when connecting the knuckle.
- Before connecting the upper arm ball joint to the knuckle, degrease the threaded section and tapered portion of the ball joint pin, the ball joint connecting hole, the threaded section and mating surfaces of the castles nut.
- Torque the castle nut to the lower torque specification, then tighten it only far enough to align the slot with the ball joint pin hole. Do not align the castle nut by loosening it.
- Before installing the wheel, clean the mating surfaces of the brake disc/drum and the inside of the wheel.
- Check the rear wheel alignment, and adjust if necessary.

9. With active damper system: Do the memorizing rear suspension full rebound position.
10. With active damper system/Left side: Do the headlight initial position learning procedure.

Rear Stabilizer Link

See Figure 210.

1. Raise the rear of the vehicle, and support it with safety stands in the proper locations.
2. Remove the rear wheel.
3. Remove the self-locking nut and the flange nut while holding the respective joint pin with a hex wrench, then remove the stabilizer link.

To install:

4. Install the stabilizer link on the stabilizer bar and lower arm B with the joint pins set at the center of their range of movement.

➡The left stabilizer link has a yellow paint mark, while the right stabilizer link has a white mark. Align the paint mark to the outside.

5. Install the new self-locking nut and the new flange nut, and lightly tighten them.
6. Place a jack under the lower arm B, and raise the suspension to load it with the vehicle's weight.
7. Tighten the self-locking nut and flange nut to the specified torque values while holding the respective joint pin with a hex wrench.
8. Clean the mating surfaces of the brake disc and the inside of the wheel, then install the rear wheel.
9. Test-drive the vehicle.
10. After 5 minutes of driving, tighten the self-locking nut again to the specified torque value.

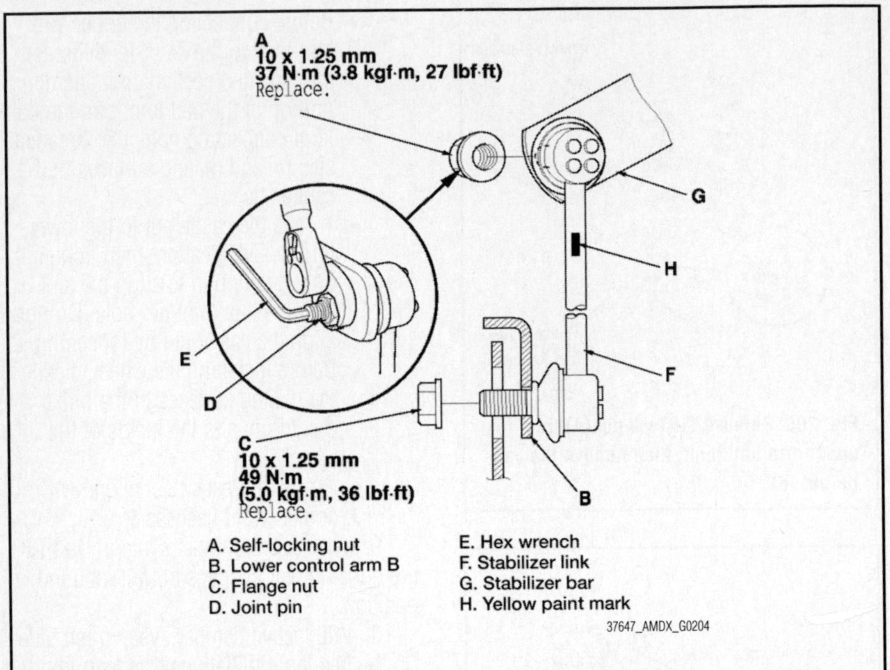
A
10 x 1.25 mm
37 N·m (3.8 kgf·m, 27 lbf·ft)
Replace.
G
H
F
E
D
C
10 x 1.25 mm
49 N·m
(5.0 kgf·m, 36 lbf·ft)
Replace.
B
A. Self-locking nut
B. Lower control arm B
C. Flange nut
D. Joint pin
E. Hex wrench
F. Stabilizer link
G. Stabilizer bar
H. Yellow paint mark
37647_AMDX_G0204

Fig. 210 Rear stabilizer link assembly

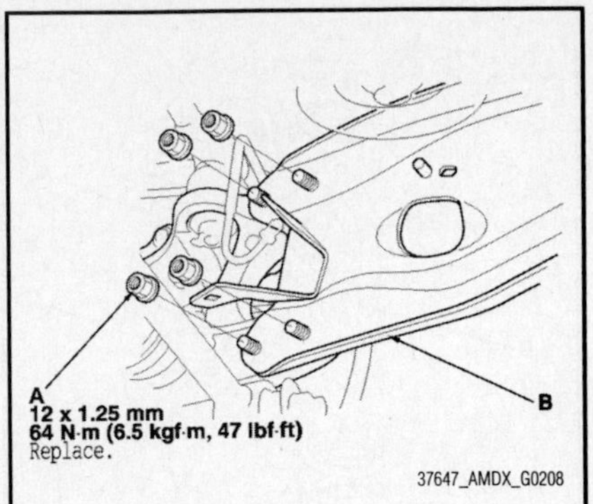
A
12 x 1.25 mm
64 N·m (6.5 kgf·m, 47 lbf·ft)
Replace.
B
37647_AMDX_G0208

Fig. 212 Remove the flange nuts (A) from the trailing arm (B)

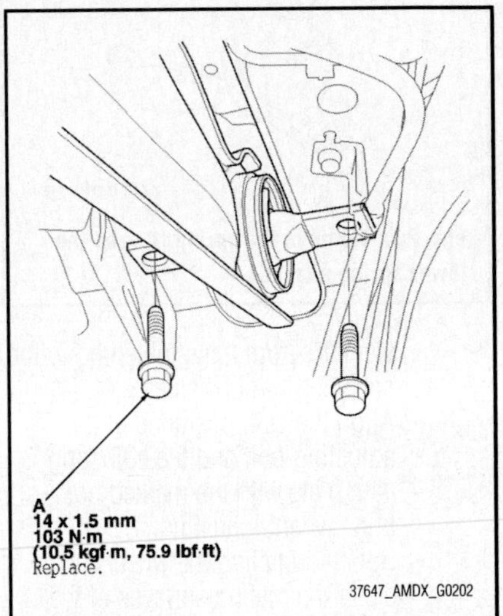
A
14 x 1.5 mm
103 N·m
(10.5 kgf·m, 75.9 lbf·ft)
Replace.
37647_AMDX_G0202

Fig. 213 Remove the flange bolts (A) from the trailing arm

Rear Trailing Arm

See Figures 211 through 213.

1. Raise the rear of the vehicle, and support it with safety stands in the proper locations.
2. Remove the rear wheel.
3. Remove the parking brake cable from the trailing arm.
4. Disconnect the brake line from the brake hose, then remove the brake hose retaining clip.

➡To prevent the brake fluid from flowing, plug and cover the hose ends and joints with a shop towel or equivalent material.

5. Disconnect the parking brake cable from the parking brake lever.
6. Remove the flange nuts from the trailing arm.

➡Use the new flange nuts during reassembly.

7. Remove the flange bolts from the trailing arm, then remove the trailing arm.

➡Use the new flange bolts during reassembly.

To install:

8. Install the trailing arm in the reverse order of removal, and note these items:
   - First install all the suspension components, and lightly tighten the bolts and nuts, then raise the suspension to load it with the vehicle's weight before fully tightening to the specified torque values.
   - Use the new brake hose retaining clip during reassembly.
   - Before installing the wheel, clean the mating surfaces of the brake disc/drum and the inside of the wheel.
   - Fill the master cylinder reservoir to the MAX (upper) level line, and bleed the brake system. Check for a

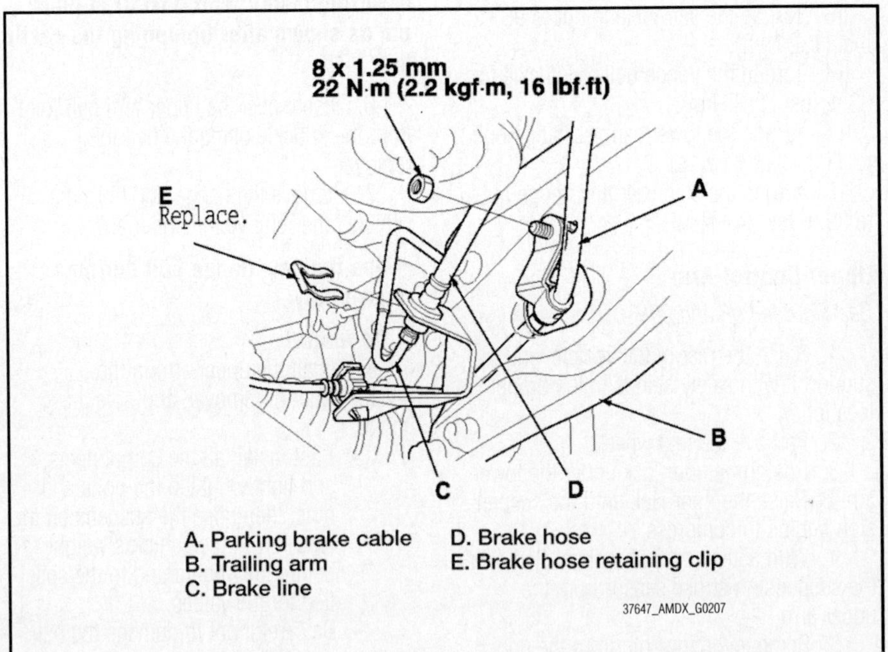
8 x 1.25 mm
22 N·m (2.2 kgf·m, 16 lbf·ft)
E
Replace.
A
B
C
D
A. Parking brake cable
B. Trailing arm
C. Brake line
D. Brake hose
E. Brake hose retaining clip
37647_AMDX_G0207

Fig. 211 Remove the parking brake cable from the trailing arm

leak at the brake hose/line joint, and retighten it if necessary.
- Check the wheel alignment, and adjust it if necessary.

SHOCK ABSORBER (DAMPER)

REMOVAL & INSTALLATION

See Figures 214 through 217.

1. Raise the rear of the vehicle, and support it with safety stands in the proper locations.
2. Remove the rear wheel.
3. With active damper system: Disconnect the damper coil connector.

➡Be careful not to damage or contaminate the damper coil connector.

4. Remove the flange nut while holding the joint pin with a hex wrench, and disconnect the stabilizer link from the lower arm B.
5. Position a floor jack under the lower arm B. Raise the floor jack until the suspension begins to compress.
6. Remove the flange bolt from the bottom of the damper.
7. Remove the flange bolt from the knuckle.
8. Remove the flange bolt from the top of the damper.
9. Lower the floor jack gradually, then remove the damper from the vehicle.

To install:

10. Position the damper between the body and the lower arm B.

➡With active damper system: Position the damper coil connector on the damper facing rearward.

11. Loosely install a new flange bolt to the top of the damper.
12. Raise the floor jack until the hole in the lower arm B aligns with the hole in the damper, then loosely install the new flange bolt to the bottom of the damper.
13. Loosely install the new flange bolt to the knuckle.
14. Connect the stabilizer link to the lower arm B, then loosely install the new flange nut.

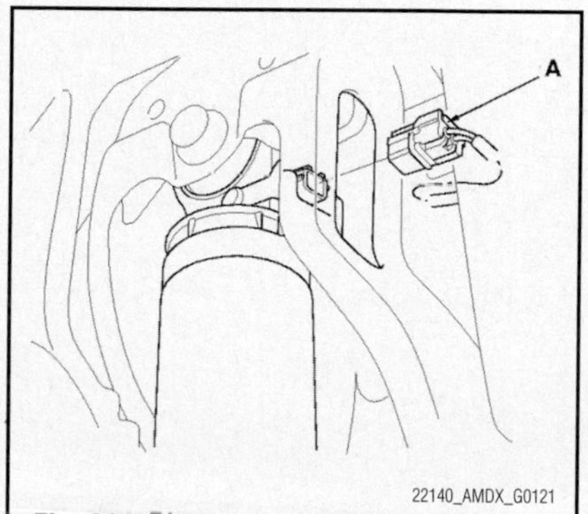
A
22140_AMDX_G0121

Fig. 214 Disconnect the wiring harness (A) of the active damper system, if equipped.

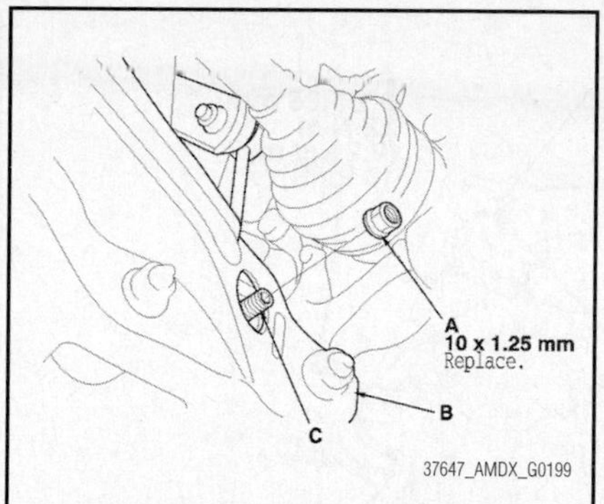
A
10 x 1.25 mm
Replace.
B
C
37647_AMDX_G0199

Fig. 215 Remove the flange nut (A) while holding the joint pin (C) with a hex wrench, and disconnect the stabilizer link from the lower arm B

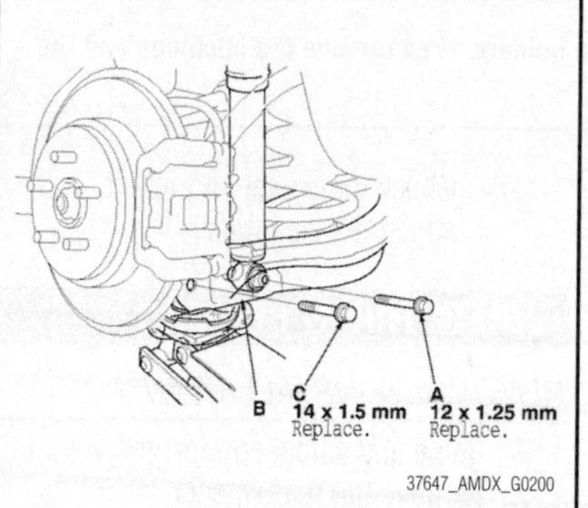
B
C
14 x 1.5 mm
Replace.
A
12 x 1.25 mm
Replace.
37647_AMDX_G0200

Fig. 216 Remove the flange bolt (A) from the bottom of the damper

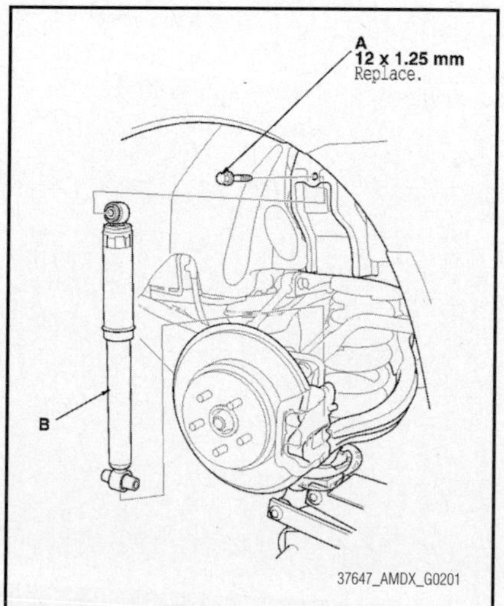
A
12 x 1.25 mm
Replace.
B
37647_AMDX_G0201

Fig. 217 Remove the flange bolt (A) from the top of the damper

15. Raise the rear suspension with the floor jack to load the suspension with the vehicle's weight.
16. Tighten the flange nut to 36 ft. lbs. (49 N m), while holding the joint pin with a hex wrench.
17. Tighten the flange bolts to the specified torque values
    - Upper and lower damper flange bolts: 47 ft. lbs. (64 N m)
    - Flange bolt to knuckle: 83 ft. lbs. (113 N m)

TESTING

1. Compress the damper assembly by hand, and check for smooth operation through a full stroke, both compression and extension. The damper should extend smoothly and constantly when compression is released. If it does not, the gas is leaking and the damper should be replaced.
2. Check for oil leaks, abnormal noises, and binding during these tests.

STABILIZER BAR

REMOVAL & INSTALLATION

See Figures 218 and 219.

1. Raise the rear of the vehicle, and support it with safety stands in the proper locations.
2. Remove the rear wheels.
3. Remove the spare tire from the vehicle.
4. Remove the spare tire support bracket.
5. Disconnect both stabilizer links from the stabilizer bar.
6. Remove the flange bolts and the bushing holders, then remove the bushings and the stabilizer bar from the rear suspension subframe.

To install:

7. Install the stabilizer bar in the reverse order of removal, and note these items:

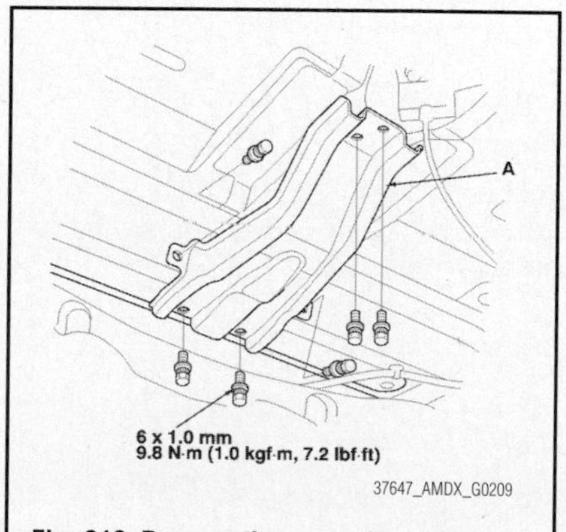
A
6 x 1.0 mm
9.8 N·m (1.0 kgf·m, 7.2 lbf·ft)
37647_AMDX_G0209

Fig. 218 Remove the spare tire support bracket (A)

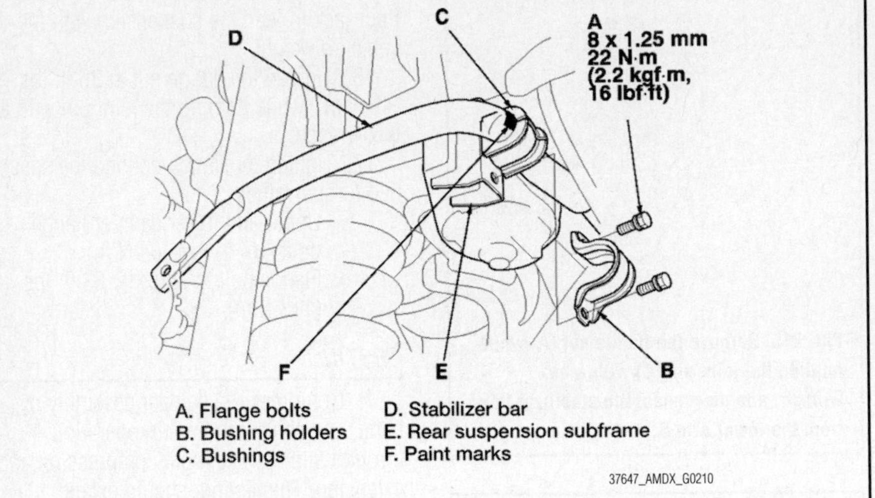
C
D
A
8 x 1.25 mm
22 N·m
(2.2 kgf·m,
16 lbf·ft)
F
E
B
A. Flange bolts
B. Bushing holders
C. Bushings
D. Stabilizer bar
E. Rear suspension subframe
F. Paint marks
37647_AMDX_G0210

Fig. 219 Remove the flange bolts and the bushing holders, then remove the bushings and the stabilizer bar from the rear suspension subframe

- Note the right and left direction of the stabilizer bar.
- Align the paint marks on the stabilizer bar with the sides of the bushings.
- Refer to stabilizer link removal/installation to connect the stabilizer bar to the links.
- Before installing the wheel, clean the mating surfaces of the brake disc/drum and the inside of the wheel.
- Check the wheel alignment, and adjust it if necessary.

WHEEL HUB & BEARING

REMOVAL & INSTALLATION

1. Raise and safely support the vehicle.
2. Remove the rear wheel.
3. Remove the brake caliper bracket mounting bolts, then remove the caliper assembly from the knuckle.

WARNING

To prevent damage to the caliper assembly or brake hose, use a short piece of wire to hang the caliper assembly from the undercarriage. Do not twist the brake hose excessively.

4. Remove the two mounting washers.
5. Raise the stake, then remove the spindle nut.
6. Release the parking brake, and remove the brake disc/drum.
7. Remove the mounting bolts, then remove the hub bearing unit.

To install:

8. Installation is the reverse order of removal. Note the following during installation:
   a. Tighten hub mounting bolts to 72 ft. lbs. (98 Nm).
   b. Using a new spindle nut, apply a small amount of clean engine oil to the seating surface of the nut.
   c. Tighten the nut to 181 ft. lbs. (245 Nm), then use a drift to stake the spindle nut shoulder against the driveshaft.

ADJUSTMENT

1. No adjustment is possible. If the hub bearing unit end play is outside of the service limits, the hub bearing unit must be replaced.

ACURA

RDX

2

BRAKES 2-9

ANTI-LOCK BRAKE SYSTEM (ABS) 2-10
Wheel Speed Sensor Rings (Toothed Rings) 2-11
Removal & Installation..........2-11
Wheel Speed Sensors 2-10
Removal & Installation..........2-10

BLEEDING THE BRAKE SYSTEM 2-9
Bleeding Procedure....................2-9
Bleeding Procedure 2-9

FRONT DISC BRAKES..........2-11
Brake Calipers..........................2-11
Removal & Installation..........2-11
Brake Pads 2-11
Removal & Installation..........2-11

INFORMATION AND PRECAUTIONS..................2-9
Anti-lock Systems..................2-9
Disc and Drum Systems 2-9

PARKING BRAKE 2-13
Parking Brake Cables 2-13
Adjustment 2-14
Removal & Installation..........2-13
Parking Brake Shoes.................2-14
Removal & Installation..........2-14

REAR DISC BRAKES 2-12
Brake Caliper............................2-12
Removal & Installation..........2-12
Brake Pads 2-12
Removal & Installation..........2-12

CHASSIS ELECTRICAL.......2-15

AIRBAG (Supplemental Restraint System) 2-15
General Information..................2-15
Arming the System 2-15
Clockspring Centering..........2-15
Disarming the System...........2-15
Service Precautions 2-15

DRIVE TRAIN..................2-16

Front Halfshaft..........................2-16
CV-boot Inspection..............2-18
Removal & Installation..........2-16
Rear Axle Housing....................2-18
Removal & Installation..........2-18
Rear Halfshaft...........................2-18
Removal & Installation..........2-18
Rear Side Case Seal 2-19
Removal & Installation..........2-19

ENGINE COOLING 2-20

Engine Fan 2-20
Removal & Installation..........2-20
Radiator...................................2-20
Removal & Installation..........2-20
Thermostat 2-21
Removal & Installation..........2-21
Water Pump 2-22
Removal & Installation..........2-22

ENGINE ELECTRICAL.........2-23

CHARGING SYSTEM 2-23
Alternator 2-23
Battery Terminal Disconnect/reconnect Procedure..........................2-23
Removal & Installation..........2-23
Voltage Regulator.....................2-23
Adjustment 2-23

IGNITION SYSTEM 2-24
Ignition Coil 2-24
Removal & Installation..........2-24
Ignition Timing.........................2-24
Adjustment 2-24
Inspection 2-24
Spark Plugs..............................2-24
Removal & Installation..........2-24

STARTING SYSTEM 2-25
Starter 2-25
Removal & Installation..........2-25

ENGINE MECHANICAL 2-25

ENGINE MECHANICAL COMPONENTS 2-25
Accessory Drive Belts 2-25
Adjustment 2-25
Drive Belt Routing................2-25
Inspection 2-25
Removal & Installation..........2-26
Camshaft & Valve Lifters...........2-26
Inspection 2-26
Removal & Installation..........2-27
Catalytic Converter....................2-29
Removal & Installation..........2-29
Crankshaft Damper...................2-31
Removal & Installation..........2-31
Crankshaft Front Seal...2-31
Removal & Installation..........2-31
Cylinder Head 2-32
Removal & Installation..........2-32
Exhaust Manifold 2-33
Removal & Installation..........2-33
Intake Manifold 2-33
Removal & Installation..........2-33
Oil Pan 2-36
Removal & Installation..........2-36
Oil Pump..................................2-37
Inspection 2-39
Removal & Installation..........2-37
Piston & Ring...........................2-39
Rear Main Seal.........................2-39
Removal & Installation..........2-39
Rocker Arms.............................2-40
Removal & Installation..........2-40
Timing Chain & Sprockets 2-45
Removal & Installation..........2-45
Timing Chain Front Cover 2-42
Removal & Installation..........2-42
Turbocharger 2-48
Removal & Installation..........2-48
Valve Lash................................2-49
Adjustment & Inspection 2-49

ENGINE PERFORMANCE & EMISSION CONTROLS.......2-51

Accelerator Pedal Position (APP) Sensor 2-51
Location..............................2-51
Removal & Installation..........2-51
Camshaft Position (CMP) Sensor 2-51
Location..............................2-51
Removal & Installation..........2-51
Crankshaft Position (CKP) sensor 2-51

CKP Pattern Clear/relearn Procedure....2-51
Location....2-51
Removal & Installation....2-51
Engine Coolant Temperature (ECT) Sensor....2-52
Location....2-52
Removal & Installation....2-52
Evaporative Emissions (EVAP) Canister....2-53
Location....2-53
Removal & Installation....2-53
Evaporative Emissions (EVAP) Canister Purge Valve....2-54
Location....2-54
Removal & Installation....2-54
Heated Oxygen Sensor (HO2S)....2-54
Location....2-54
Removal & Installation....2-55
Intake Air Temperature (IAT) Sensor....2-55
Location....2-55
Removal & Installation....2-55
Knock Sensor (KS)....2-55
Location....2-55
Removal & Installation....2-55
Malfunction Indicator Light....2-56
Reset Procedure....2-56
Manifold Absolute Pressure (MAP) Sensor....2-56
Location....2-56
Removal & Installation....2-56
Mass Air Flow/Intake Air Temperature (MAF/IAT) Sensor....2-56
Location....2-56
Removal & Installation....2-57
Output Shaft Speed (OSS) Sensor....2-56
Location....2-56
Removal & Installation....2-56
Positive Crankcase Ventilation (PCV) Valve....2-56
Location....2-56
Removal & Installation....2-56
Powertrain Control Module....2-57
Location....2-57
PCM Reset Procedure....2-57
Removal & Installation....2-57

FUEL....2-58

GASOLINE FUEL INJECTION SYSTEM....2-58
Fuel Filter....2-59
Removal & Installation....2-59
Fuel Level Sending Unit....2-59
Location....2-59
Removal & Installation....2-59
Fuel Pump....2-59
Removal & Installation....2-59
Fuel Rail & Injector....2-60
Removal & Installation....2-60
Fuel System Pressure....2-58
Relieving....2-58
Fuel System Service Precautions....2-58
Fuel Tank....2-61
Removal & Installation....2-61
Idle Speed....2-62
Adjustment....2-62

HEATING & AIR CONDITIONING SYSTEM....2-64

Blower Motor....2-64
Removal & Installation....2-64
Heater Core....2-64
Removal & Installation....2-64

SPECIFICATIONS AND MAINTENANCE CHARTS....2-03

Brake Specifications....2-07
Camshaft Specifications....2-05
Capacities....2-04
Crankshaft and Connecting Rod Specifications....2-05
Engine and Vehicle Identification Chart....2-03
Engine Tune-Up Specifications....2-03
Fluid Specifications....2-04
General Engine Specifications....2-03
Piston and Ring Specifications....2-05
Scheduled Maintenance Intervals....2-08
Tire, Wheel and Ball Joint Specifications....2-07
Torque Specifications....2-06
Valve Specifications....2-04
Wheel Alignment....2-07

STEERING....2-65

Power Rack & Pinion Steering Gear....2-65
Removal & Installation....2-65
Power Steering Pump....2-67
Bleeding....2-68
Fluid Fill Procedure....2-68
Removal & Installation....2-67

SUSPENSION....2-69

FRONT SUSPENSION....2-69
Lower Ball Joints....2-69
Removal & Installation....2-69
Lower Control Arms....2-69
Removal & Installation....2-69
MacPherson Struts....2-69
Overhaul....2-70
Removal & Installation....2-69
Stabilizer Bar....2-72
Removal & Installation....2-72
Steering Knuckle....2-72
Removal & Installation....2-72
Wheel Bearings....2-73
Adjustment....2-74
Removal & Installation....2-73

REAR SUSPENSION....2-75
Coil Springs....2-75
Removal & Installation....2-75
Control Arms & Links....2-75
Removal & Installation....2-75
Shock Absorbers....2-77
Removal & Installation....2-77
Stabilizer Bar....2-78
Removal & Installation....2-78
Steering Knuckle....2-77
Removal & Installation....2-77
Trailing Arm....2-79
Removal & Installation....2-79
Wheel Bearings....2-79
Adjustment....2-80
Removal & Installation....2-79

SPECIFICATIONS AND MAINTENANCE CHARTS

ENGINE AND VEHICLE IDENTIFICATION CHART

Engine Code							Model Year	
Code	Liters (cc)	Cu. In.	Cyl.	Fuel Sys.	Engine Type	Eng. Mfg.	Code ①	Year
K23A1	2.3 (2350)	140	4	SMFI	DOHC	Honda	9	2009
							A	2010

DOHC: Dual Overhead Cam

SMFI: Sequential Multi-port Fuel Injection

① 10th position of VIN

37647_ARDX_C0001

GENERAL ENGINE SPECIFICATIONS

Year	Model	Engine Displacement Liters (VIN)	Net Horsepower @ rpm	Net Torque @ rpm (ft. lbs.)	Bore x Stroke (in.)	Com-pression Ratio	Oil Pressure @ rpm
2009	RDX	2.3 (K23A1)	NA	NA	3.39x3.90	8.8:1	44@3000
2010	RDX	2.3 (K23A1)	NA	NA	3.39x3.90	8.8:1	44@3000

SMFI: Sequential Multi-port Fuel Injection

NA: Not Available

37647_ARDX_C0002

ENGINE TUNE-UP SPECIFICATIONS

Year	Engine Displacement Liters (VIN)	Spark Plug Gap (in.)	Ignition Timing (deg.)		Fuel Pump (psi)	Idle Speed (rpm)		Valve Clearance (in.)	
			MT	AT		MT	AT	In.	Ex.
2009	2.3 (K23A1)	0.028-0.031	—	12-14B	47-54	—	670-770	0.008-0.010	0.011-0.013
2010	2.3 (K23A1)	0.028-0.031	—	12-14B	47-54	—	670-770	0.008-0.010	0.011-0.013

NOTE: The Vehicle Emission Control Information label often reflects changes made during production and must be used if they differ from this chart.

NOTE: The fuel pressure readings are given with the vacuum hose disconnected

B: Before top dead center

37647_ARDX_C0003

CAPACITIES

Year	Model	Engine Displacement Liters (VIN)	Engine Oil with Filter (qts.)	Transmission (pts.) 5-Spd	Transmission (pts.) Auto.	Transfer Case (pts.)	Drive Axle Front (pts.)	Drive Axle Rear (pts.)	Fuel Tank (gal.)	Cooling System (qts.)
2009	RDX	2.3 (K23A1)	4.7	—	①	②	—	③	18.0	④
2010	RDX	2.3 (K23A1)	4.7	—	①	②	—	③	18.0	④

NOTE: All capacities are approximate. Add fluid gradually and check to be sure a proper fluid level is obtained.

① Fluid change: 3.5 quarts
Overhaul: 8.2 quarts

② Fluid change: .45 quart
Overhaul: .48 quart

③ Fluid change: 2.67 quarts
Overhaul: 2.93 quarts

④ Fluid change: 1.85 gallons
Overhaul (engine): 2.22 gallons

37647_ARDX_C0004

FLUID SPECIFICATIONS

Year	Model	Engine Displacement Liters (VIN)	Engine Oil	Auto. Trans.	Drive Axle	Power Steering Fluid	Brake Master Cylinder
2009	RDX	2.3 (K23A1)	①	②	③	④	DOT 3
2010	RDX	2.3 (K23A1)	①	②	③	④	DOT 3

DOT: Department Of Transportation

Note: If specification disagrees with specification in owners manual, use specification in owners manaual

① Mobil 1 (P/N 5w-30-MB1-000) or equivalent that meets Acure HTO-06 standard

② Acura ATF-Z1 fluid

③ Transfer case: API classified GL4 or GL5 only. SAE 90 or SAE 80W-90 viscosity.
Rear differential: Acura ATF-Z1 fluid

④ Acura power steering fluid

37647_ARDX_C0014

VALVE SPECIFICATIONS

Year	Engine Displacement Liters (VIN)	Seat Angle (deg.)	Face Angle (deg.)	Spring Test Pressure (lbs. @ in.)	Spring Installed Height (in.)	Stem-to-Guide Clearance (in.) Intake	Stem-to-Guide Clearance (in.) Exhaust	Stem Diameter (in.) Intake	Stem Diameter (in.) Exhaust
2009	2.3 (K23A1)	NA	NA	NA	①	0.0012-0.0022	0.0022-0.0031	0.2156-0.2159	0.2146-0.2150
2010	2.3 (K23A1)	NA	NA	NA	①	0.0012-0.0022	0.0022-0.0031	0.2156-0.2159	0.2146-0.2150

NA: Not Available

① Valve spring free length:
Intake: 1.874 in.
Exhaust: 1.954 in.

37647_ARDX_C0005

CAMSHAFT SPECIFICATIONS

All measurements in inches unless noted

Year	Model	Engine Displacement Liters (VIN)	Journal Dia.	Brg. Oil Clearance	Shaft End-play	Circle Runout	Lobe Height Intake	Lobe Height Exhaust
2009	RDX	2.3 (K23A1)	NA	①	0.0020-0.0080	NA	②	1.3422
2010	RDX	2.3 (K23A1)	NA	①	0.0020-0.0080	NA	②	1.3422

NA: Not Available

① No. 1 Journal: 0.001-0.003 inch
All others: 0.002-0.004 inch

② Intake primary: 1.3356 inch. Intake secondary: 1.1668 inch.
All others: 0.002-0.004 inch

37647_ARDX_C0006

CRANKSHAFT AND CONNECTING ROD SPECIFICATIONS

All measurements are given in inches

Year	Engine Displacement Liters (VIN)	Crankshaft Main Brg. Journal Dia.	Crankshaft Main Brg. Oil Clearance	Crankshaft Shaft End-play	Crankshaft Thrust on No.	Connecting Rod Journal Diameter	Connecting Rod Oil Clearance	Connecting Rod Side Clearance
2009	2.3 (K23A1)	①	②	0.0040-0.0140	NA	1.8888 1.8898	0.0013-0.0025	0.0060-0.0140
2010	2.3 (K23A1)	①	②	0.0040-0.0140	NA	1.8888 1.8898	0.0013-0.0025	0.0060-0.0140

NA: Not Available

① Except No. 3: 2.1648-2.1657 inches
No. 3: 2.1644-2.1654 inches

② Except No. 3: 0.0007-0.0016 inches
No. 3: 0.0010-0.0019 inches

37647_ARDX_C0007

PISTON AND RING SPECIFICATIONS

All measurements are given in inches

Year	Engine Displacement Liters (VIN)	Piston Clearance	Ring Gap Top Compression	Ring Gap Bottom Compression	Ring Gap Oil Control	Ring Side Clearance Top Compression	Ring Side Clearance Bottom Compression	Ring Side Clearance Oil Control
2009	2.3 (K23A1)	0.0008-0.0016	0.0008-0.0012	0.016-0.0200	0.004-0.0120	0.0018-0.0035	0.0016-0.0026	0.0008-0.002
2010	2.3 (K23A1)	0.0008-0.0016	0.0008-0.0012	0.016-0.0200	0.004-0.0120	0.0018-0.0035	0.0016-0.0026	0.0008-0.002

37647_ARDX_C0008

TORQUE SPECIFICATIONS

All readings in ft. lbs.

Year	Engine Displacement Liters (VIN)	Cylinder Head Bolts	Main Bearing Bolts	Rod Bearing Bolts	Crankshaft Damper Bolts	Flywheel Bolts	Manifold Intake	Manifold Exhaust	Spark Plugs	Oil Pan Drain Plug
2009	2.3 (K23A1)	①	②	③	④	NA	⑤	⑥	13	29
2010	2.3 (K23A1)	①	②	③	④	NA	⑤	⑥	13	29

NA: Not Available

① Step 1: 29 ft. lbs.
Step 2: Tighten all bolts an additional 90 degrees. If using a new bolt tighten an extra 90 degrees

② Step 1: 22 ft. lbs.
Step 2: additional 67 degrees

③ Step 1: 22 ft. lbs.
Step 2: 90 degrees

④ Used bolt: 36 ft. lbs.
New bolt: 130 ft. lbs., loosen and retighten to 36 ft. lbs.

⑤ Tighten bolts and nuts in a crisscross pattern beginning with the inner bolt, in three stages to 16 ft. lbs.

⑥ Tighten bolts and nuts in a crisscross pattern beginning with the inner bolt, in three stages to 33 ft. lbs.

37647_ARDX_C0009

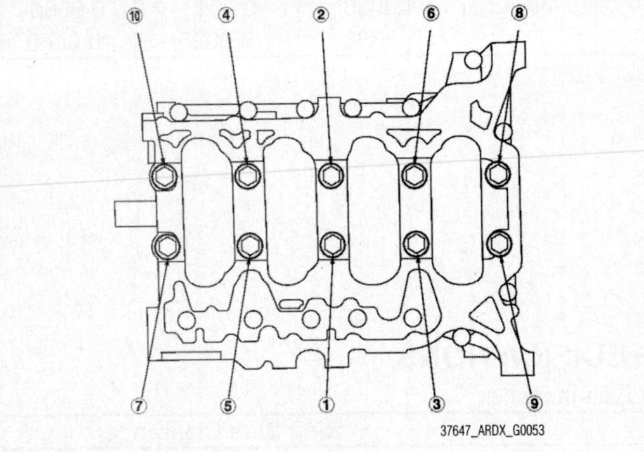
⑩
④
②
⑥
⑧
⑦
⑤
①
③
⑨
37647_ARDX_G0053

Fig. 1 Main bearing torque sequence

WHEEL ALIGNMENT

Year	Model		Caster Range (+/-Deg.)	Caster Preferred Setting (Deg.)	Camber Range (+/-Deg.)	Camber Preferred Setting (Deg.)	Toe-in (in.)
2009	RDX	F	1.00	1.57	0.45	0	0+/-0.08
		R	—	—	0.45	-1.00	0.10+/-0.08
2010	RDX	F	1.00	1.57	0.45	0	0+/-0.08
		R	—	—	0.45	-1.00	0.10+/-0.08

37647_ARDX_C0010

TIRE, WHEEL AND BALL JOINT SPECIFICATIONS

Year	Model	OEM Tires Standard	OEM Tires Optional	Tire Pressures (psi) Front	Tire Pressures (psi) Rear	Wheel Size	Ball Joint Inspection	Lug Nut (ft. lbs.)
2009	RDX	P235/65R18	①	32	32	R18 ②	NA	80
2010	RDX	P235/65R18	①	32	32	R18 ②	NA	80

OEM: Original Equipment Manufacturer

PSI: Pounds Per Square Inch

NA: Not Available

① Optional 18 inch tire sizes are: P225/60R18 and P245/55R18

② Optional 19 inch wheel available. Tire size is P245/55R19.

37647_ARDX_C0011

BRAKE SPECIFICATIONS

All measurements in inches unless noted

Year	Model		Brake Disc Original Thickness	Brake Disc Minimum Thickness	Brake Disc Maximum Runout	Brake Drum Diameter Original Inside Diameter	Brake Drum Diameter Max. Wear Limit	Brake Drum Diameter Maximum Machine Diameter	Minimum Lining Thickness Front	Minimum Lining Thickness Rear	Brake Caliper Bracket Bolts (ft. lbs.)	Brake Caliper Mounting Bolts (ft. lbs.)
2009	RDX	F	①	1.020	0.0016	—	—	—	0.060	—	101	37
		R	②	0.300	NA	—	—	—	—	0.060	80	17
2010	RDX	F	①	1.020	0.0016	—	—	—	0.060	—	101	37
		R	②	0.300	NA	—	—	—	—	0.060	80	17

NA: Not Available

F: Front

R: Rear

① 1.09-1.11 inch

② 0.35-0.36 inch

37647_ARDX_C0012

SCHEDULED MAINTENANCE INTERVALS

ACURA—RDX

TO BE SERVICED	OF SERVIC	VEHICLE MILEAGE INTERVAL (x1000)															
		7.5	15	22.5	30	37.5	45	52.5	60	67.5	75	82.5	90	97.5	105	112.5	120
Accessory drive belts	I & A				✓				✓				✓				✓
Air cleaner element	R				✓				✓				✓				✓
Brake fluid	R	Every 3 years															
Brake hoses & lines (incl. ABS)	I		✓		✓		✓		✓		✓		✓		✓		✓
Cooling system hoses & connections	I		✓		✓		✓		✓		✓		✓		✓		✓
Engine coolant ①	R						✓						✓				
Engine oil	R	✓	✓	✓	✓	✓	✓	✓	✓	✓	✓	✓	✓	✓	✓	✓	✓
Engine oil and coolant levels	I	Inspect at each fuel stop															
Engine oil filter	R		✓		✓		✓		✓		✓		✓		✓		✓
Exhaust system	I		✓		✓		✓		✓		✓		✓		✓		✓
Fluid levels and condition	I		✓		✓		✓		✓		✓		✓		✓		✓
Front and rear brakes	I		✓		✓		✓		✓		✓		✓		✓		✓
Fuel lines & connection	I		✓		✓		✓		✓		✓		✓		✓		✓
Halfshaft boots	I		✓		✓		✓		✓		✓		✓		✓		✓
Idle speed	I & A														✓		
Parking brake system	I & A		✓		✓		✓		✓		✓		✓		✓		✓
Rear differential fluid	R	✓			✓		✓		✓				✓				✓
Rotate and inspect tires	I	✓	✓	✓	✓	✓	✓	✓	✓	✓	✓	✓	✓	✓	✓	✓	✓
Spark plugs	R														✓		
Supplemental Restraint System	I	Inspect the SRS 10 years after production															
Suspension components	I		✓		✓		✓		✓		✓		✓		✓		✓
Tie rod ends, steering gear box & boots	I		✓		✓		✓		✓		✓		✓		✓		✓
Timing belt	R														✓		
Transmission fluid	R						✓				✓				✓		
Valve clearance	I	Adjust if valves are noisy															
Water pump	S/I														✓		

R: Replace I: Inspect A: Adjust

① Every 12,000 miles or 10 years, then every 60,000 miles or 5 years

FREQUENT OPERATION MAINTENANCE (SEVERE SERVICE)

If a vehicle is operated under any of the following conditions it is considered severe service:

- **Towing a trailer or using a camper or car-top carrier.**
- **Repeated short trips of less than 5 miles in temperatures below freezing, or trips of less than 10 miles in any temperature.**
- **Extensive idling or low-speed driving for long distances as in heavy commercial use, such as delivery, taxi or police cars.**
- **Operating on rough, muddy or salt-covered roads.**
- **Operating on unpaved or dusty roads.**
- **Driving in extremely hot (over 90°) conditions.**

Air cleaner element: replace every 15,000 miles

Engine oil and filter: replace every 3750 miles or 6 months, whichever occurs first.

Timing belt: replace every 60,000 miles if the vehicle is regularly driven in temperatures above 110°F or below -20°F, or if frequently towing a trailer.

Transmission fluid: replace every 30,000 miles.

Rear differential fluid: replace every 60,000 miles.

Front and rear brakes: inspect every 7500 miles or 6 months, whichever occurs first.

Locks and hinges: lubricate every 15,000 miles.

Tie rods, steering gear box, boots: inspect every 7500 miles or 6 months, whichever occurs first.

Suspension components: inspect every 7500 miles or 6 months, whichever occurs first.

Halfshaft boots: inspect every 7500 miles or 6 months, whichever occurs first.

37647_ARDX_C0013

BRAKES — INFORMATION AND PRECAUTIONS

ANTI-LOCK SYSTEMS

- Certain components within the ABS system are not intended to be serviced or repaired individually.
- Do not use rubber hoses or other parts not specifically specified for and ABS system. When using repair kits, replace all parts included in the kit. Partial or incorrect repair may lead to functional problems and require the replacement of components.
- Lubricate rubber parts with clean, fresh brake fluid to ease assembly. Do not use shop air to clean parts; damage to rubber components may result.
- Use only DOT 3 brake fluid from an unopened container.
- If any hydraulic component or line is removed or replaced, it may be necessary to bleed the entire system.
- A clean repair area is essential. Always clean the reservoir and cap thoroughly before removing the cap. The slightest amount of dirt in the fluid may plug an orifice and impair the system function. Perform repairs after components have been thoroughly cleaned; use only denatured alcohol to clean components. Do not allow ABS components to come into contact with any substance containing mineral oil; this includes used shop rags.
- The Anti-Lock control unit is a microprocessor similar to other computer units in the vehicle. Ensure that the ignition switch is **OFF** before removing or installing controller harnesses. Avoid static electricity discharge at or near the controller.
- If any arc welding is to be done on the vehicle, the control unit should be unplugged before welding operations begin.

DISC AND DRUM SYSTEMS

**** CAUTION**

Dust and dirt accumulating on brake parts during normal use may contain asbestos fibers from production or aftermarket brake linings. Breathing excessive concentrations of asbestos fibers can cause serious bodily harm. Exercise care when servicing brake parts. Do not sand or grind brake lining unless equipment used is designed to contain the dust residue. Do not clean brake parts with compressed air or by dry brushing. Cleaning should be done by dampening the brake components with a fine mist of water, then wiping the brake components clean with a dampened cloth. Dispose of cloth and all residue containing asbestos fibers in an impermeable container with the appropriate label. Follow practices prescribed by the Occupational Safety and Health Administration (OSHA) and the Environmental Protection Agency (EPA) for the handling, processing, and disposing of dust or debris that may contain asbestos fibers.

BRAKES — BLEEDING THE BRAKE SYSTEM

BLEEDING PROCEDURE

BLEEDING PROCEDURE

See Figure 2.

**** WARNING**

Do not reuse the drained fluid. Use only clean Honda DOT 3 Brake Fluid from an unopened container. Using a non-Honda brake fluid can cause corrosion and shorten the life of the system.

**** WARNING**

Make sure no dirt or other foreign matter is allowed to contaminate the brake fluid.

**** WARNING**

Do not spill brake fluid on the vehicle, it may damage the paint; if brake fluid does contact the paint, wash it off immediately with water.

1. The reservoir on the master cylinder must be at the MAX (upper) level mark at the start of the bleeding procedure and checked after bleeding each brake caliper. Add fluid as required.
2. Make sure the brake fluid level in the reservoir is at the MAX (upper) level line.
3. Slide a piece of clear plastic hose over the first bleed screw, and submerge the other end in a container of new brake fluid.
4. Have someone slowly pump the brake pedal several times, and then apply steady pressure.
5. Starting at the left-front, loosen the brake bleed screw to allow air to escape from the system. Then tighten the bleed screw securely.
6. Repeat the procedure for each wheel in the sequence shown following until air bubbles no longer appear in the fluid.
7. Refill the master cylinder reservoir to the MAX (upper) level line.

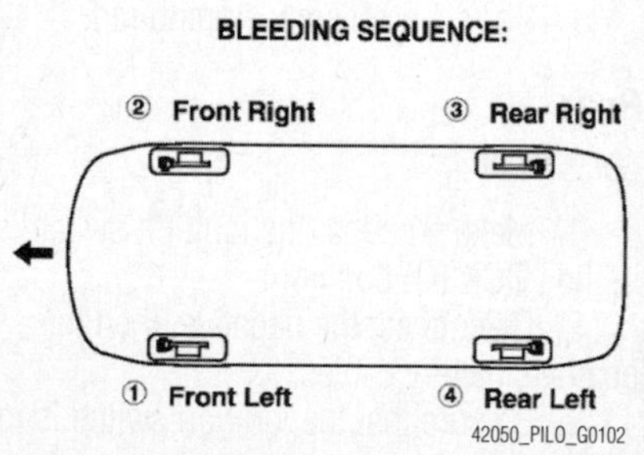
BLEEDING SEQUENCE:
② Front Right
③ Rear Right
① Front Left
④ Rear Left
42050_PILO_G0102

Fig. 2 Brake bleeding sequence

WHEEL SPEED SENSORS

REMOVAL & INSTALLATION

Front

See Figure 3.

1. Make sure that the ignition switch is in the LOCK (0) position.
2. Disconnect the negative, then the positive, battery cables.
3. Be sure that the ignition switch is in LOCK (0).
4. Raise and support the vehicle safely.
5. Release the connector holding clamps. Disconnect the wheel speed sensor connector.
6. Remove the clips, bolt and wheel speed sensor.

To install:

7. Installation is the reverse of the removal procedure.

➡Install the sensor carefully to avoid twisting the wires. If the sensor comes in contact with the wheel bearing, it is faulty and must be replaced. Do not drop the sensor.

8. Connect the battery cables.
9. Turn the ignition switch ON (II), the SRS indicator should come on for about six seconds, then go off.
10. Start the engine and check that the ABS and VSA indicators do not stay on. Test drive the vehicle to ensure that the indicator lights do not come on.
11. Correct problems, as required.

Rear

See Figure 4.

1. Make sure that the ignition switch is in the LOCK (0) position.
2. Disconnect the negative, then the positive, battery cables.
3. Be sure that the ignition switch is in LOCK (0).
4. Raise and support the vehicle safely.
5. Disconnect the wheel speed sensor connector.
6. Remove the clips, bolt and wheel speed sensor.

To install:

7. Installation is the reverse of the removal procedure.

➡Install the sensor carefully to avoid twisting the wires. If the sensor comes

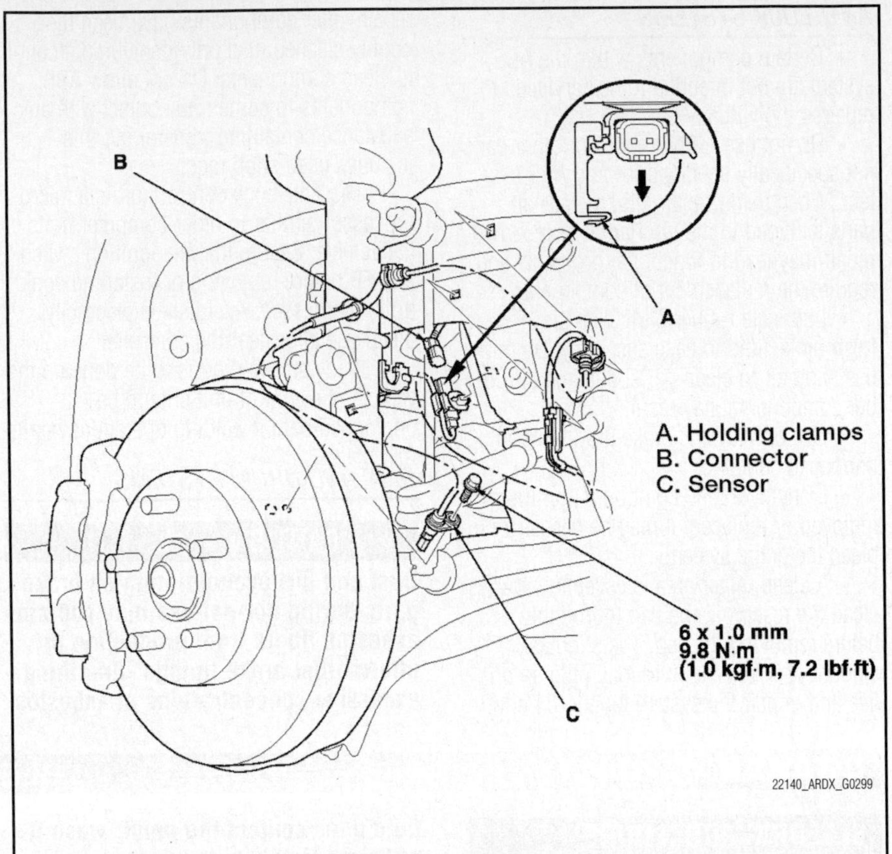
B
A
A. Holding clamps
B. Connector
C. Sensor
6 x 1.0 mm
9.8 N·m
(1.0 kgf·m, 7.2 lbf·ft)
C
22140_ARDX_G0299

Fig. 3 Front wheel speed sensor and related components

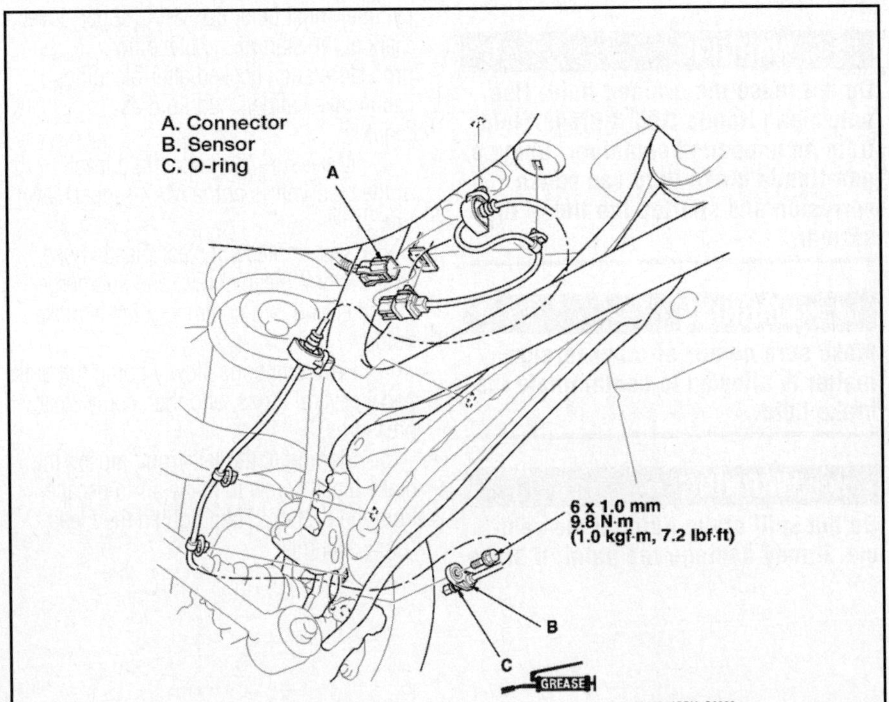
A. Connector
B. Sensor
C. O-ring
A
6 x 1.0 mm
9.8 N·m
(1.0 kgf·m, 7.2 lbf·ft)
B
C
GREASE
22140_ARDX_G0300

Fig. 4 Rear wheel speed sensor and related components

in contact with the wheel bearing, it is faulty and must be replaced. Do not drop the sensor.

8. Connect the battery cables.
9. Turn the ignition switch ON (II), the SRS indicator should come on for about six seconds, then go off.
10. Start the engine and check that the ABS and VSA indicators do not stay on. Test drive the vehicle to ensure that the indicator lights do not come on.
11. Correct problems, as required.

WHEEL SPEED SENSOR RINGS (TOOTHED RINGS)

REMOVAL & INSTALLATION

➡The Wheel Speed Sensor Rings (Magnetic Encoders) are replaced along with the Hub and Bearing.

BRAKES

FRONT DISC BRAKES

BRAKE CALIPERS

REMOVAL & INSTALLATION

1. Before servicing the vehicle, refer to the "PRECAUTIONS" section.
2. As required, remove a small amount of brake fluid from the reservoir using a suction pump.
3. Raise and safely support the vehicle.
4. Remove the tire and wheel assembly.
5. Remove the brake hose mounting bolt. Disconnect and plug the brake line hose at the caliper.
6. Remove the brake caliper mounting bolts.
7. Remove the caliper assembly from its mounting.

⁂ WARNING

Do not allow the caliper to hang by the brake line hose, as damage to the hose may result.

8. As required, remove the disc brake pads and shims from the caliper.

To install:

9. Clean the caliper thoroughly; remove any dirt or dust. Check the brake rotor for grooves or cracks and machine or replace, as necessary.
10. Installation is the reverse of the removal procedure.
11. Be sure to fill the brake system using the proper grade and type brake fluid.
12. Bleed the brake system.
13. Check for leaks and correct as required.

BRAKE PADS

REMOVAL & INSTALLATION

See Figure 5.

1. Before servicing the vehicle, refer to the "PRECAUTIONS" section.
2. As required, remove a small amount of brake fluid from the reservoir using a suction pump.
3. Raise and safely support the vehicle.
4. Remove the tire and wheel assembly.
5. Remove the flange bolt while holding the caliper pin with a wrench.

➡Be careful not to damage the pin boot, and pivot the caliper up and out of the way.

6. Remove the pad shims and pads. Remove the pad retainers.

To install:

7. Install the pad retainers.

➡Apply molybdenum brake grease to both surfaces of the shims and the back of the disc brake pads, prior to installation.

8. Install the pads and shims. See illustration for proper positioning.
9. Use a suitable tool to push caliper piston into its bore and enable the caliper to fit over the pads.
10. Lubricate the piston boot with silicon grease. Avoid twisting the boot.
11. Continue the installation in the reverse order of the removal procedure.
12. Add brake fluid to the master cylinder reservoir. Depress the brake pedal several times to seat the pads. Bleed the brakes if necessary.

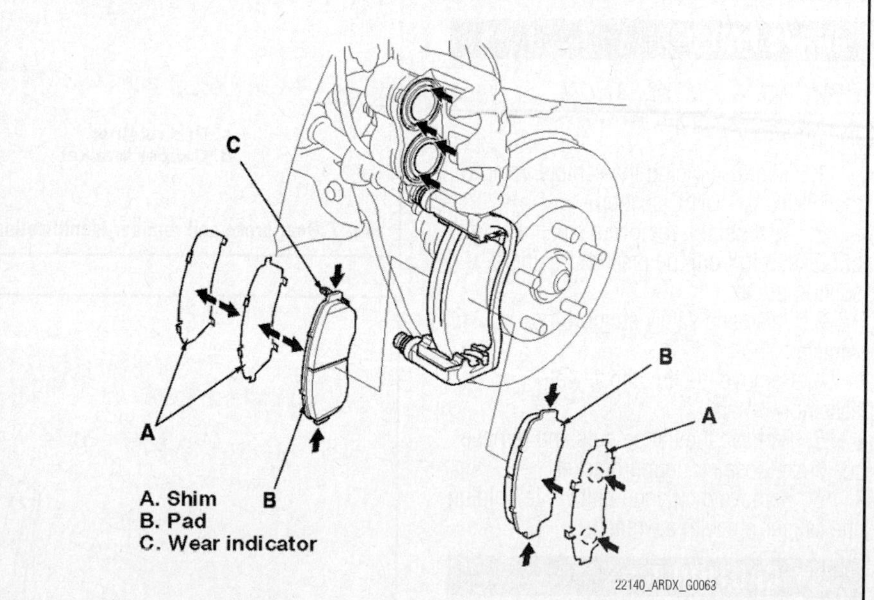
C
A
B
B
A
A. Shim
B. Pad
C. Wear indicator
22140_ARDX_G0063

Fig. 5 Front brake pad positioning

BRAKE CALIPER

REMOVAL & INSTALLATION

See Figure 6.

1. Before servicing the vehicle, refer to the "PRECAUTIONS" section.
2. As required, remove a small amount of brake fluid from the reservoir using a suction pump.
3. Raise and safely support the vehicle.
4. Remove the tire and wheel assembly.
5. Remove the caliper from its mounting.

WARNING

Position the unit aside, using mechanics wire. Avoid allowing the caliper to hang by the brake hose. Do not bend or twist the brake hose excessively.

To install:

6. Installation is the reverse of the removal procedure.
7. If the parking brake shoes were loosened, perform the parking brake adjustment (major).

BRAKE PADS

REMOVAL & INSTALLATION

See Figures 7 and 8.

1. Before servicing the vehicle, refer to the "PRECAUTIONS" section.
2. As required, remove a small amount of brake fluid from the reservoir using a suction pump.
3. Raise and safely support the vehicle.
4. Remove the tire and wheel assembly.
5. Remove the flange nuts and remove the brake hose mounting bracket.
6. Remove the flange bolt while holding the caliper pin with a wrench.

CAUTION

Be careful not to damage the pin boot, and pivot the caliper up and out of the way.

7. Remove the pad shims and pads.
8. Remove the pad retainers.

➡The upper and lower pad retainers are different. During installation, make

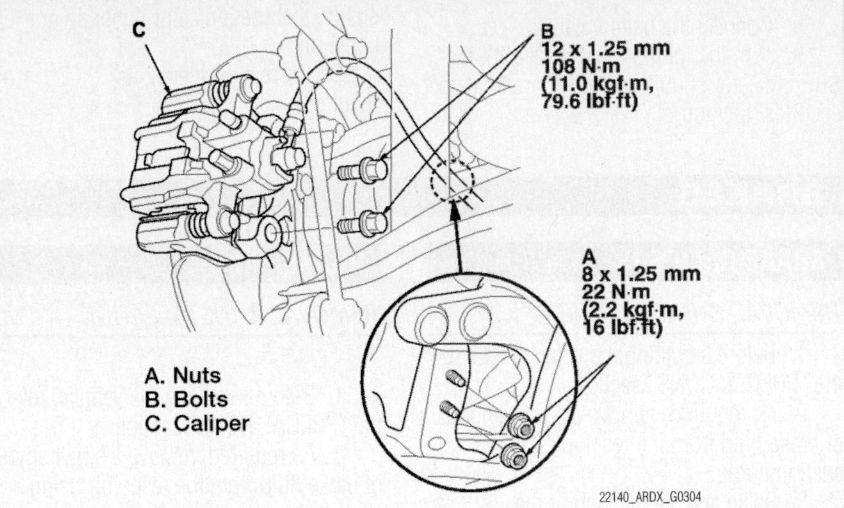
C
B
12 x 1.25 mm
108 N·m
(11.0 kgf·m,
79.6 lbf·ft)
A
8 x 1.25 mm
22 N·m
(2.2 kgf·m,
16 lbf·ft)
A. Nuts
B. Bolts
C. Caliper
22140_ARDX_G0304

Fig. 6 Rear caliper mounting bolt location

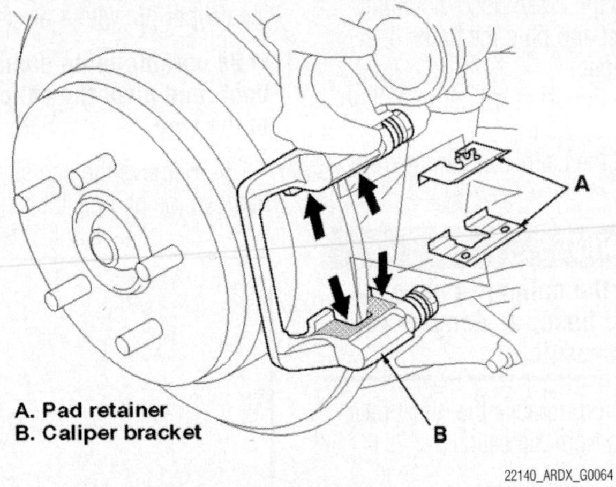
A
A. Pad retainer
B. Caliper bracket
B
22140_ARDX_G0064

Fig. 7 Rear brake pad retainer identification and positioning

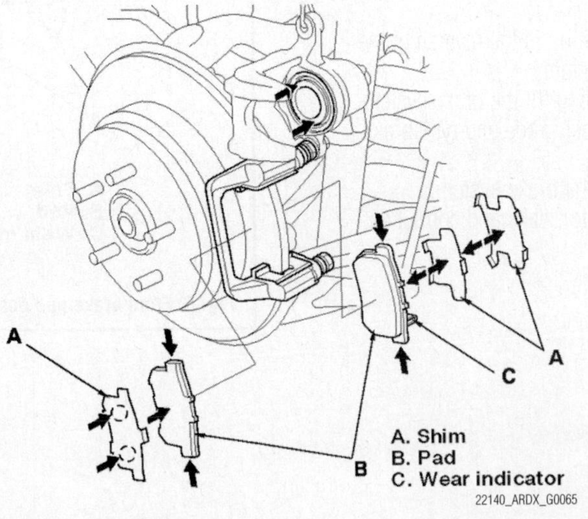
A
C
A
B
A. Shim
B. Pad
C. Wear indicator
22140_ARDX_G0065

Fig. 8 Rear brake pad positioning

sure the pad retainers are in the proper position.

To install:

9. Install the pad retainers.
10. Apply molybdenum brake grease to both surfaces of the shims and the back of the disc brake pads, prior to installation.
11. Install the pads and shims. See illustration for proper positioning.
12. Use a suitable tool to push caliper piston into its bore and enable the caliper to fit over the pads.
13. Lubricate the piston boot with silicon grease. Avoid twisting the boot.
14. Continue the installation in the reverse order of the removal procedure.
15. Add brake fluid to the master cylinder reservoir. Depress the brake pedal several times to seat the pads. Bleed the brakes if necessary.

BRAKES — PARKING BRAKE

PARKING BRAKE CABLES

REMOVAL & INSTALLATION

See Figure 9.

1. Before servicing the vehicle, refer to the "PRECAUTIONS" section.
2. Raise and safely support the vehicle.
3. Remove the tire and wheel assembly.
4. Loosen the parking brake cable adjusting nut, located underneath the dash at the parking brake.
5. Remove the parking brake shoes.
6. Disconnect the parking brake cable from the parking brake lever.
7. Remove the parking brake cable mounting bolts from the backing plate.
8. Pull the cable and remove it from the backing plate.

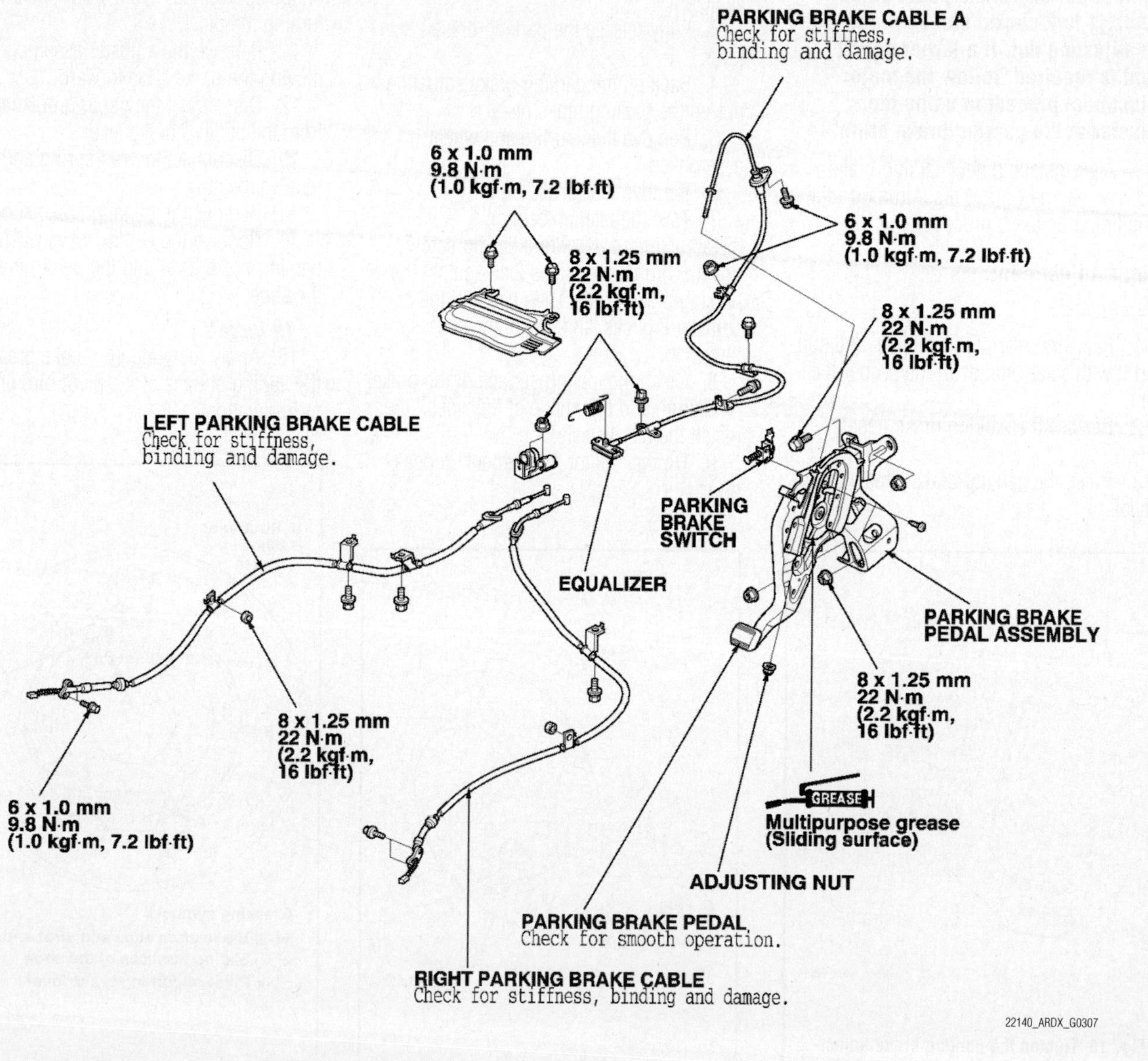

PARKING BRAKE CABLE A
Check for stiffness, binding and damage.
6 x 1.0 mm
9.8 N·m
(1.0 kgf·m, 7.2 lbf·ft)
8 x 1.25 mm
22 N·m
(2.2 kgf·m, 16 lbf·ft)
6 x 1.0 mm
9.8 N·m
(1.0 kgf·m, 7.2 lbf·ft)
8 x 1.25 mm
22 N·m
(2.2 kgf·m, 16 lbf·ft)
LEFT PARKING BRAKE CABLE
Check for stiffness, binding and damage.
PARKING BRAKE SWITCH
EQUALIZER
PARKING BRAKE PEDAL ASSEMBLY
8 x 1.25 mm
22 N·m
(2.2 kgf·m, 16 lbf·ft)
8 x 1.25 mm
22 N·m
(2.2 kgf·m, 16 lbf·ft)
6 x 1.0 mm
9.8 N·m
(1.0 kgf·m, 7.2 lbf·ft)
GREASE
Multipurpose grease (Sliding surface)
ADJUSTING NUT
PARKING BRAKE PEDAL
Check for smooth operation.
RIGHT PARKING BRAKE CABLE
Check for stiffness, binding and damage.
22140_ARDX_G0307

Fig. 9 Parking brake cables and related components

To install:

9. Installation is the reverse of the removal procedure.

10. Be careful not to bend or distort the cable.

11. Perform the parking brake adjustment procedure.

12. Apply the parking brake firmly ten times.

13. Check and readjust the parking brake.

ADJUSTMENT

1. Press the parking brake pedal with enough force to fully apply the parking brake. The parking brake pedal should be locked within 6–7 clicks.

2. Adjust the parking brake if the pedal clicks are not within the specification.

➡Minor parking brake pedal adjustments (1 to 2 clicks) can be made with the adjusting nut. If a larger adjustment is required, follow the major adjustment procedure using the adjuster at the parking brake drum.

3. After installing new parking brake shoes and/or new brake disc/drum, do the shoe lining break-in procedure.

Minor Adjustment

See Figure 10.

1. Raise the rear of the vehicle, and support it with safety stands in the proper locations.

2. Release the parking brake pedal fully.

3. Press the parking brake pedal 1 click.

4. Tighten the parking brake adjusting nut until the parking brakes drag slightly when the rear wheels are rotated.

5. Release the parking brake pedal fully, and check that the parking brakes do not drag when the rear wheels are rotated. Readjust if necessary.

6. Make sure the parking brakes are fully applied when the parking brake pedal is pressed all the way.

Major Adjustment

See Figure 11.

➡This adjustment is to be done when replacing parking brake shoes.

1. Before servicing the vehicle, refer to the "PRECAUTIONS" section.

2. Raise and safely support the vehicle.

3. Fully release the parking brake lever.

4. Back off the parking brake adjusting nut on the parking brake pedal.

5. Remove the rear tire and wheel assemblies.

6. Remove the access plug.

7. Turn the ratchet teeth on the adjuster nut (B) with a flat-tip screwdriver (C) until the shoes lock against the parking brake drum. Then back off the adjuster 8 clicks, and install the access plug.

8. Clean the mating surface of the brake disc/drum and the inside of the wheel, then install the rear wheels.

9. Do the "Minor Adjustment" procedure.

PARKING BRAKE SHOES

REMOVAL & INSTALLATION

See Figures 12 and 13.

1. Before servicing the vehicle, refer to the "PRECAUTIONS" section.

2. As required, remove a small amount of brake fluid from the reservoir using a suction pump.

3. Raise and safely support the vehicle.

4. Remove the tire and wheel assembly.

5. Release the parking brake.

6. Remove the rear brake caliper.

7. Remove the rear disc brake rotor.

8. Disconnect and remove the brake spring and the upper return spring.

9. Disconnect and remove the lower return spring.

10. Remove the tension pins, by pushing on the appropriate retainer springs and turning the pin.

11. Remove the adjuster assembly, by moving the brake shoe forward.

12. Disconnect the parking brake cable from the parking brake lever.

13. Disconnect the rod spring and remove the strut.

14. Remove the parking brake shoes.

15. Remove the U-clip, wave washer, parking brake lever and the pivot pin from the shoe.

To install:

16. Apply molybdenum brake grease to the sliding surface of the pivot pin, prior to installation.

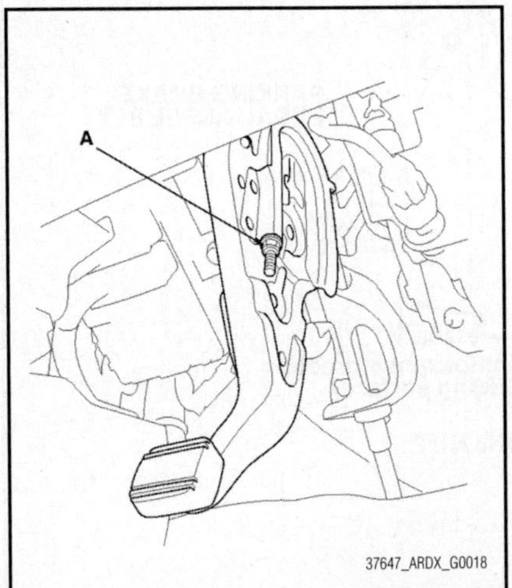
A
37647_ARDX_G0018

Fig. 10 Tighten the parking brake adjusting nut (A) until the parking brakes drag slightly when the rear wheels are rotated.

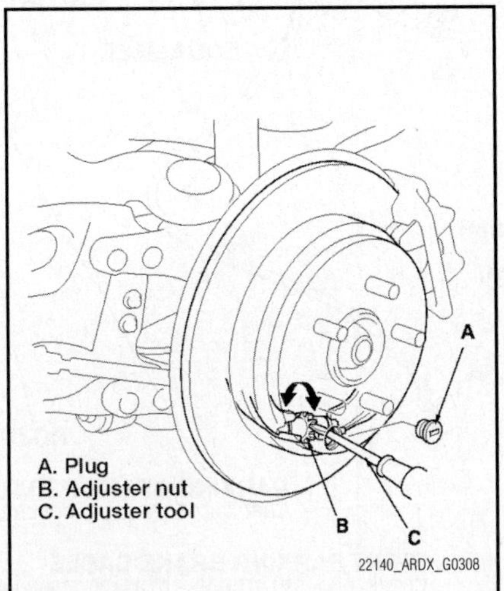
A
A. Plug
B. Adjuster nut
C. Adjuster tool
B
C
22140_ARDX_G0308

Fig. 11 Parking brake cable adjustment locating point

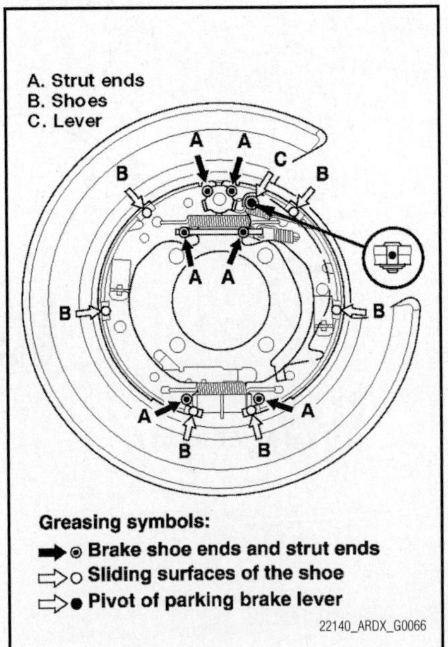
A. Strut ends
B. Shoes
C. Lever
Greasing symbols:
Brake shoe ends and strut ends
Sliding surfaces of the shoe
Pivot of parking brake lever
22140_ARDX_G0066

Fig. 12 Rear parking brake grease application points

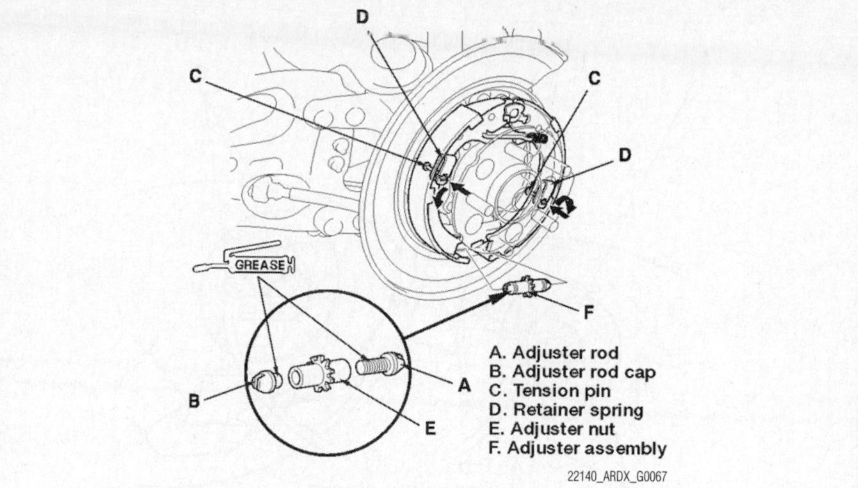
D
C
C
D
GREASE
F
B
A
E
A. Adjuster rod
B. Adjuster rod cap
C. Tension pin
D. Retainer spring
E. Adjuster nut
F. Adjuster assembly
22140_ARDX_G0067

Fig. 13 Rear parking brake adjuster pin positioning

17. Installation is the reverse of the removal procedure.
18. Install the wave washer with its convex side facing out.
19. Pinch the U-clip securely to prevent the parking brake lever from coming out of the brake shoe.
20. Apply molybdenum brake grease to surfaces indicated in the illustration.
21. Be sure that the adjuster pin is positioned correctly.
22. Fully release the parking brake pedal. Back off the pedal adjusting nut, on the parking brake pedal.
23. Remove the access plug.
24. Turn the ratchet teeth on the adjuster nut until the shoes lock against the parking brake drum.
25. Back off the adjuster eight clicks. Install the access plug.

CHASSIS ELECTRICAL

AIR BAG (SUPPLEMENTAL RESTRAINT SYSTEM)

GENERAL INFORMATION

CAUTION

These vehicles are equipped with an air bag system. The system must be disarmed before performing service on, or around, system components, the steering column, instrument panel components, wiring and sensors. Failure to follow the safety precautions and the disarming procedure could result in accidental air bag deployment, possible injury and unnecessary system repairs.

SERVICE PRECAUTIONS

CAUTION

Disconnect and isolate the battery negative cable before beginning any airbag system component diagnosis, testing, removal, or installation procedures. Wait at least 90 seconds after the ignition switch is turned off and the negative (-) terminal cable is disconnected from the battery before starting the operation. The SRS is equipped with a backup power source, so if work is started within 90 seconds after disconnecting the negative (-) terminal cable from the battery, the SRS may be deployed. Failure to disable the airbag system may result in accidental airbag deployment, personal injury, or death.

DISARMING THE SYSTEM

1. Before servicing the vehicle, refer to the “PRECAUTIONS” section.
2. Some system store data in memory is lost when the battery is disconnected. Do the following procedures before disconnecting the battery:
   a. Make sure you have the anti-theft code(s) for the audio and/or the navigation system (if equipped).
   b. If you are replacing the audio unit, write down the audio presets (AM and FM), and the XM audio presets (if equipped), because the audio unit does not retain the presets after the battery is disconnected.
3. Make sure that the ignition switch is in the LOCK (0) position.
4. Disconnect the negative, then the positive, battery cables.

➡Wait at least three minutes after disconnecting the battery cables before starting the repair procedure.

ARMING THE SYSTEM

➡Some system store data in memory is lost when the battery is disconnected. Do the following procedures to restore the system back to normal operation.

1. Clean the battery terminals.
2. Test the battery.
3. Reconnect the positive cable to the battery first, then reconnect the negative cable to the battery.
4. Apply multipurpose grease to the terminals to prevent corrosion.
5. Enter the anti-theft code(s) for the audio system and/or the navigation system (if equipped).
6. Enter the audio presets (if applicable), and enter the XM audio presets (if equipped).
7. Set the clock (for vehicles without navigation).

CLOCKSPRING CENTERING

See Figures 14 and 15.

1. Before servicing the vehicle, refer to the “PRECAUTIONS” section.
2. Make sure that the ignition switch is in the LOCK (0) position.
3. Disconnect the negative, then the positive, battery cables.

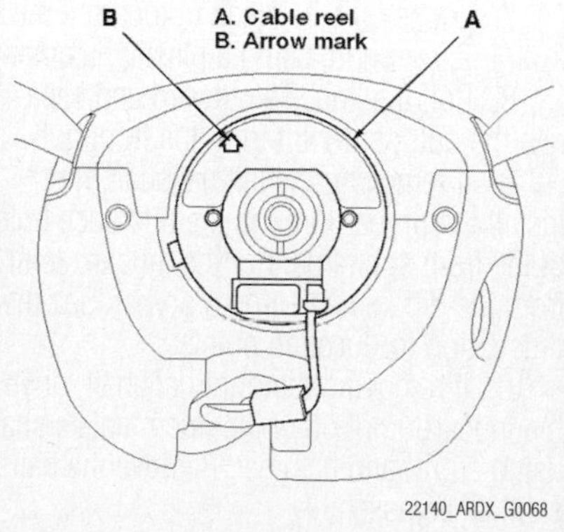
B
A. Cable reel
B. Arrow mark
A
22140_ARDX_G0068

Fig. 14 Cable reel arrow location and positioning: the arrow (mark B) on the cable reel label should point straight up.

4. Be sure that the front wheels are in the straight ahead position.
5. Remove the steering wheel.
6. To center the cable reel, first rotate the cable reel clockwise until it stops.
7. Rotate it counterclockwise about three full turns.
8. The arrow on the cable reel label should point straight up.
9. Position the two tabs of the turn signal canceling sleeve, as shown.
10. Install the steering wheel onto the steering column shaft, making sure that the steering wheel hub engages the pins of the cable reel and the tabs of the turn signal canceling sleeve.

CAUTION

Do not tap on the steering wheel of the steering column shaft when installing the steering wheel.

11. Install the steering wheel. Tighten the retaining bolt to 29 ft. lbs. (39 Nm).

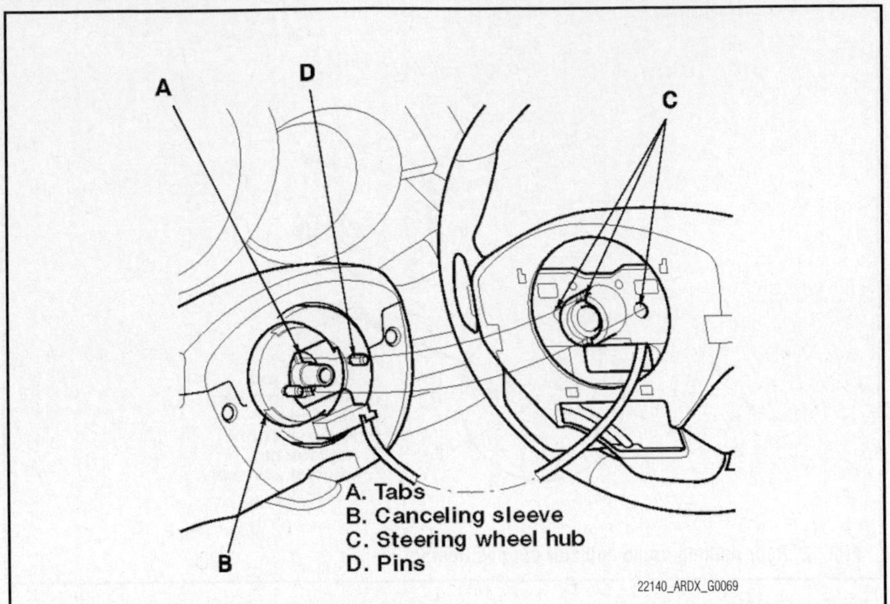
A
D
C
B
A. Tabs
B. Canceling sleeve
C. Steering wheel hub
D. Pins
22140_ARDX_G0069

Fig. 15 Turn signal canceling cam positioning

DRIVE TRAIN

FRONT HALFSHAFT

REMOVAL & INSTALLATION

See Figures 16 through 23.

1. Raise and support the vehicle safely.
2. Remove the front wheels.
3. Pry up the locking tab on the spindle nut, then remove the nut.
4. Remove the drain plug and drain the transmission fluid. Dispose of used fluid properly. Install the drain plug using a new sealing washer.
5. Pry up the locking tab on the spindle nut and remove the nut.
6. Remove the nuts and bolts and separate the lower arm, using a prybar.
7. Loosen the halfshaft outboard joint from the front hub, using a plastic hammer.
8. Pull the knuckle outward and separate the outboard joint from the front hub.
9. If removing the left halfshaft, first, install the prybar through the reference hole of the front subframe. Pry the inboard joint from the differential using a prybar. See the illustration for locating points.
10. If removing the right halfshaft, drive the inboard joint off of the intermediate shaft using a drift and hammer. Remove the halfshaft as an assembly.

➡Do not pull the halfshaft or the inboard joint may come apart.

11. Remove the set ring from the inboard ring.

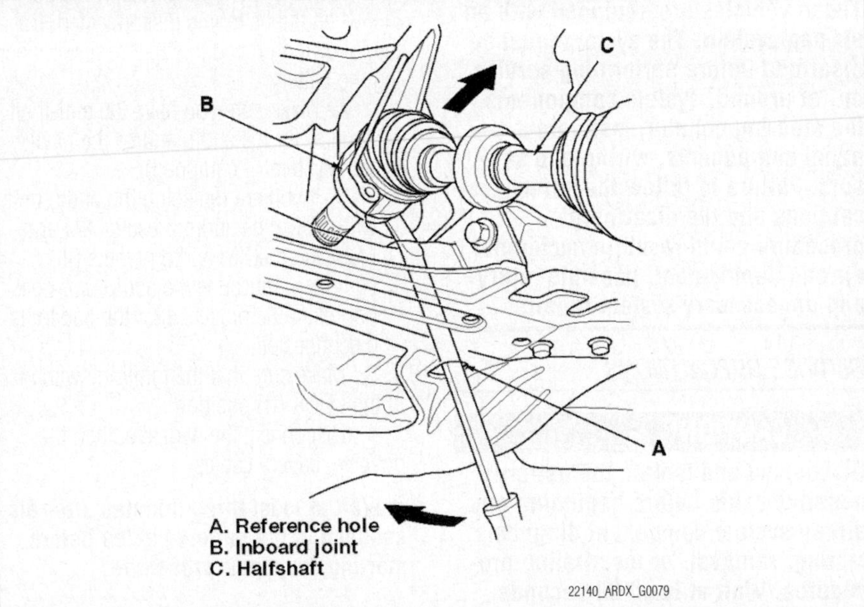
C
B
A
A. Reference hole
B. Inboard joint
C. Halfshaft
22140_ARDX_G0079

Fig. 16 If removing the left halfshaft, install the prybar through the reference hole (A) of the front subframe. Pry the inboard joint (B) from the differential using a prybar. See the illustration for locating points.

12. Remove the set ring from the intermediate shaft.

To install:

13. Installation is the reverse of the removal procedure.
14. Be sure that the mating surfaces of the joint and splined section are clean and free of debris.
15. Apply about 0.18 ounce of moly 60 paste (PN 08734-0001 or equivalent) to the contact area of the outboard joint and front wheel bearing.
16. Be sure to use new set rings on both shafts.
17. Apply 0.02–0.04 ounce of super high grease to the whole splined surface of the right halfshaft.

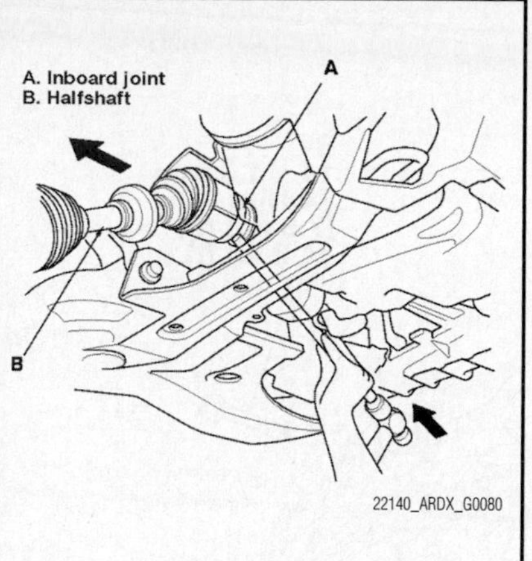
A. Inboard joint
B. Halfshaft
A
B
22140_ARDX_G0080

Fig. 17 Right halfshaft removal locating points

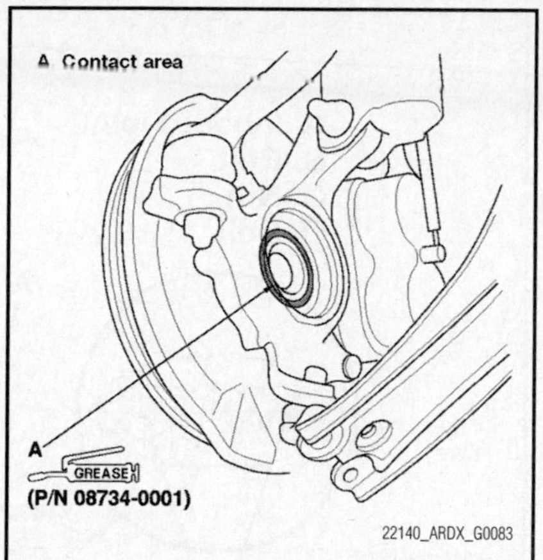
A. Contact area
A
GREASE
(P/N 08734-0001)
22140_ARDX_G0083

Fig. 20 Grease application—front halfshaft

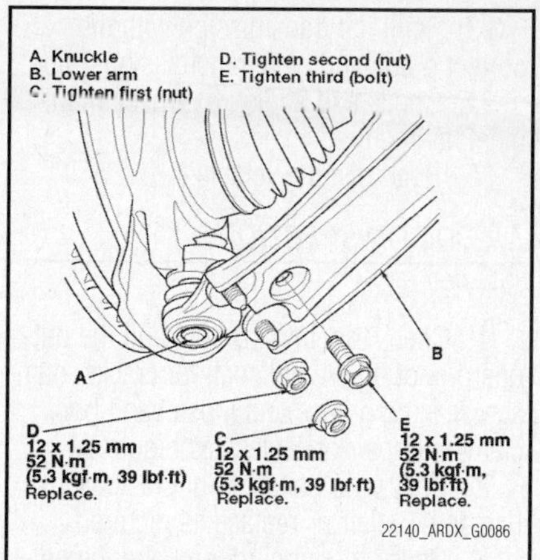
A. Knuckle
B. Lower arm
C. Tighten first (nut)
D. Tighten second (nut)
E. Tighten third (bolt)
A
B
D
12 x 1.25 mm
52 N·m
(5.3 kgf·m, 39 lbf·ft)
Replace.
C
12 x 1.25 mm
52 N·m
(5.3 kgf·m, 39 lbf·ft)
Replace.
E
12 x 1.25 mm
52 N·m
(5.3 kgf·m,
39 lbf·ft)
Replace.
22140_ARDX_G0086

Fig. 23 Lower arm to knuckle tightening sequence

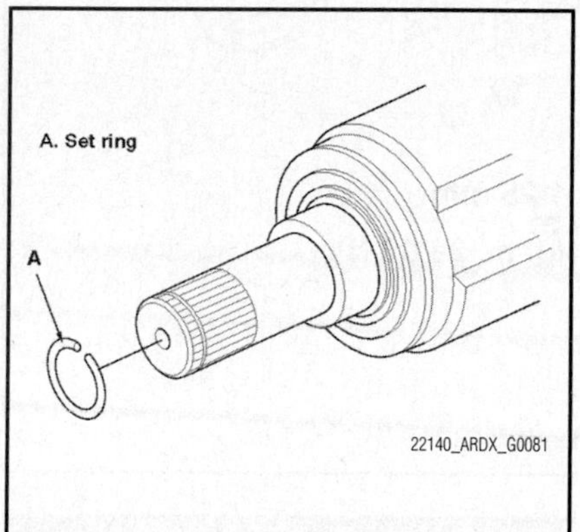
A. Set ring
A
22140_ARDX_G0081

Fig. 18 Inboard joint set ring

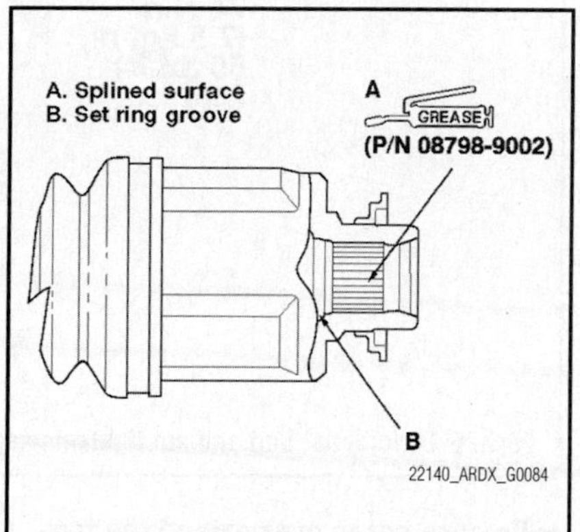
A. Splined surface
B. Set ring groove
A
GREASE
(P/N 08798-9002)
B
22140_ARDX_G0084

Fig. 21 Right halfshaft grease application

19. Insert the inboard end of the halfshaft into the differential or intermediate shaft until the set ring locks in the groove.

➡Insert the halfshaft horizontally to prevent damage to the oil seal.

20. Continue the installation in the reverse order of the removal procedure.

21. Tighten the lower arm to knuckle retaining bolt and nuts to 39 ft. lbs. (53 Nm). Be sure to use new bolt and nuts. Use the sequence in the illustration.

22. Use a new spindle nut and tighten to 242 ft. lbs. (328 Nm).

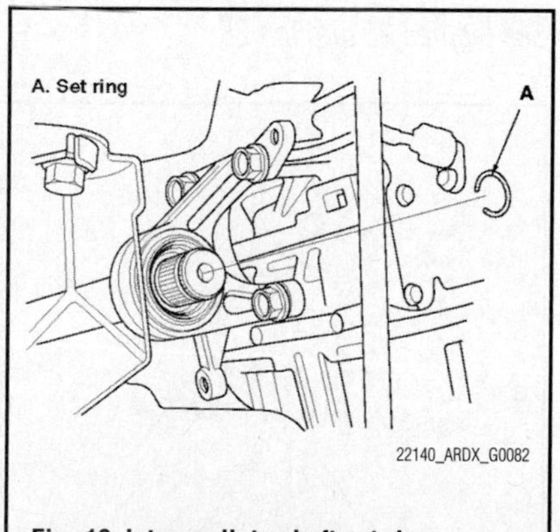
A. Set ring
A
22140_ARDX_G0082

Fig. 19 Intermediate shaft set ring

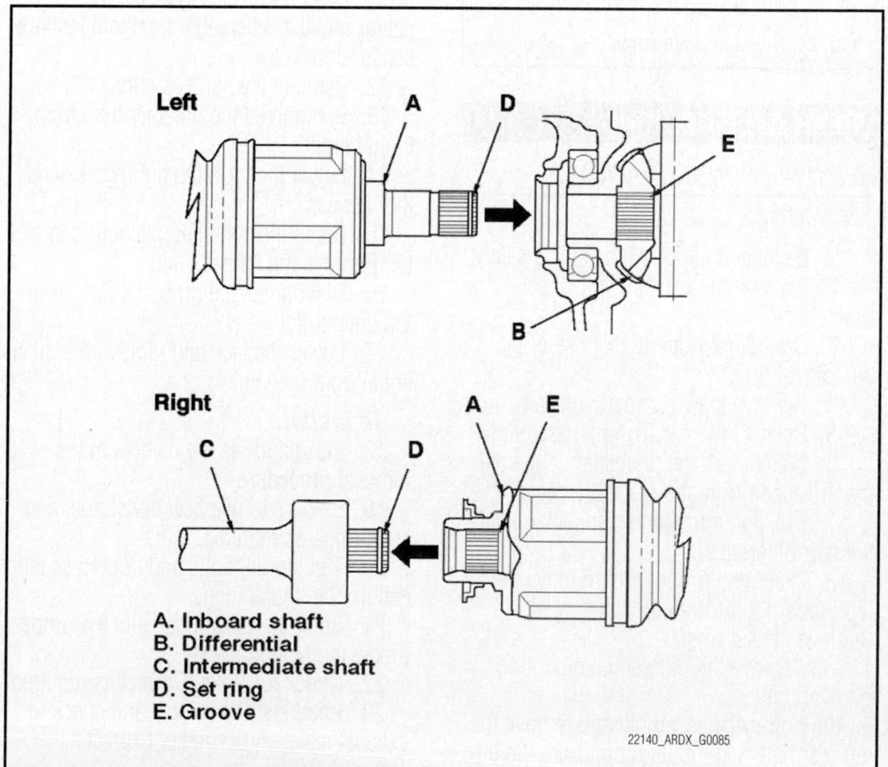
Left
A
D
E
B
Right
A
E
C
D
A. Inboard shaft
B. Differential
C. Intermediate shaft
D. Set ring
E. Groove
22140_ARDX_G0085

Fig. 22 Proper halfshaft installation

➡After applying the grease, remove the grease from the splined grooves at intervals of 2 to 3 splines and from the set ring groove so that air can bleed from the intermediate shaft.

18. Clean the areas where the halfshaft contacts the differential thoroughly with solvent and dry with compressed air.

➡Do not wash rubber parts with solvent.

23. Refill the transmission with the correct grade and type transmission fluid.
24. Check and adjust the wheel alignment, as required.
25. Road test the vehicle.

CV-BOOT INSPECTION

See Figure 24.

1. Check the inboard boot and the outboard boot on the halfshaft for cracks, damage, leaking grease and loose band boot clamps. Repair or replace as required.
2. Check the halfshaft for cracks and damage. Repair or replace as required.
3. Check the inboard joint and the outboard joint for cracks and damage. Replace the joint as an assembly.
4. Hold the inboard joint and turn the front wheel by hand, make sure that the joint is not excessively loose. If necessary, replace the joint as an assembly.

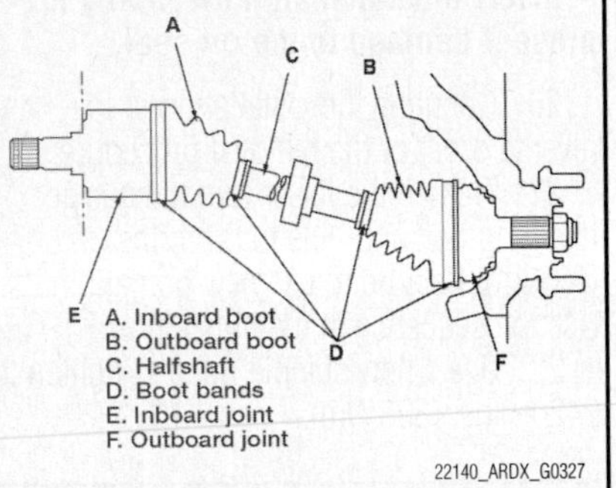
A
C
B
E
D
F
A. Inboard boot
B. Outboard boot
C. Halfshaft
D. Boot bands
E. Inboard joint
F. Outboard joint
22140_ARDX_G0327

Fig. 24 Halfshaft inspection

REAR AXLE HOUSING

REMOVAL & INSTALLATION

See Figure 25.

1. Raise and support the vehicle safely.
2. Remove the rear tire and wheel assemblies.
3. Drain and properly dispose of the differential fluid.
4. Remove the right rear halfshaft.
5. Remove the exhaust pipe assembly.
6. Matchmark the driveshaft. Separate the driveshaft from the rear differential.
7. Position a transmission jack under the rear differential.
8. Using the proper tool, remove the left inboard joint from the differential. Remove the set ring.
9. Remove the differential mounting bolts and nuts.
10. Lower the assembly and remove the left rear halfshaft inboard joint from the differential.

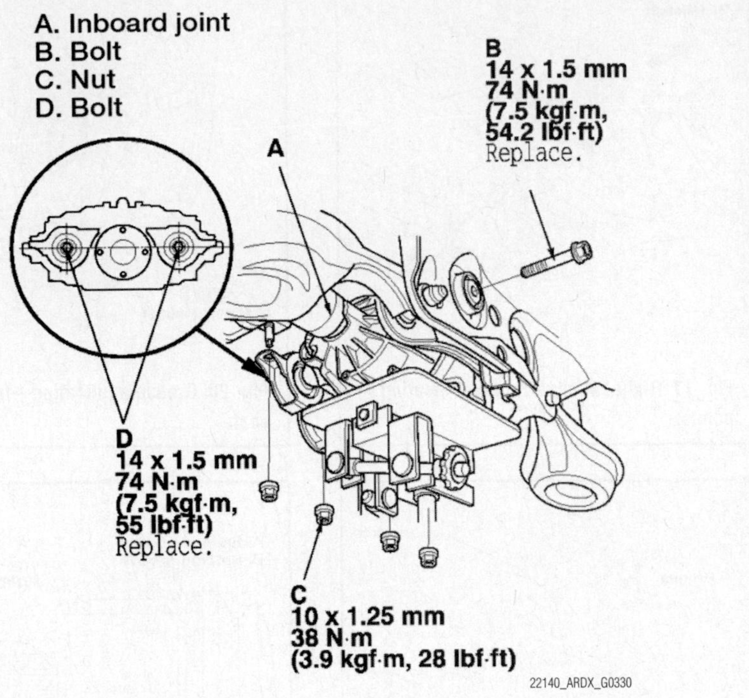
A. Inboard joint
B. Bolt
C. Nut
D. Bolt
B
14 x 1.5 mm
74 N·m
(7.5 kgf·m,
54.2 lbf·ft)
Replace.
A
D
14 x 1.5 mm
74 N·m
(7.5 kgf·m,
55 lbf·ft)
Replace.
C
10 x 1.25 mm
38 N·m
(3.9 kgf·m, 28 lbf·ft)
22140_ARDX_G0330

Fig. 25 Differential bolt and nut tightening specifications

➡Be sure not to over extend the harness and the tube.

11. Disconnect the right solenoid connector and the differential fluid temperature sensor connector.
12. Remove the harness clips.
13. Disconnect the breather tube from the differential.
14. Lower the jack. Do not over extend the harness.
15. Disconnect the left solenoid connector. Remove the harness clip.
16. Disconnect the ground cable, from the differential.
17. Lower the jack and remove the differential from the vehicle.

To install:

18. Installation is the reverse of the removal procedure.
19. Be sure to use new bolts, nuts and snap rings as required.
20. Tighten the bolts and nuts to specification. See illustration.
21. Refill the differential with the proper grade and type fluid.
22. Check for leaks. Correct as required.
23. Road test the vehicle and check for exhaust leaks and/or rattle. Correct as required.

REAR HALFSHAFT

REMOVAL & INSTALLATION

See Figures 26 through 28.

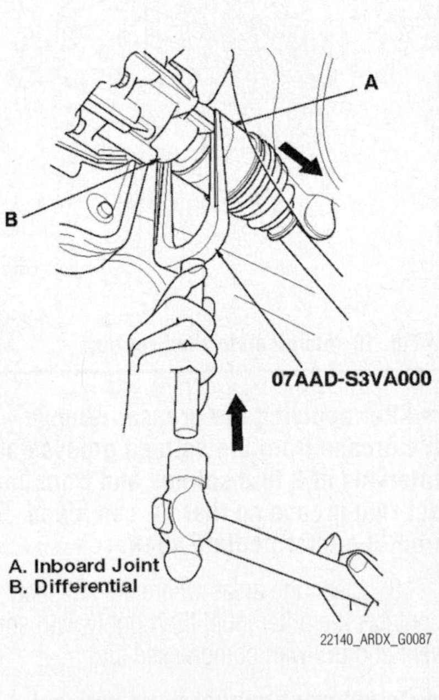
A
B
07AAD-S3VA000
A. Inboard Joint
B. Differential
22140_ARDX_G0087

Fig. 26 Rear halfshaft removal tool installation

1. Raise and support the vehicle safely.
2. Remove the rear tire and wheel assemblies.
3. Remove the locking tab on the spindle nut. Remove the nut.
4. Remove the lower arm.
5. Remove the flange bolt and separate the knuckle from the lower arm.
6. Loosen the rear halfshaft outboard joint from the rear hub housing, using a plastic hammer.
7. Pull the knuckle outward, and separate the rear halfshaft outboard joint.

➡When removing the outboard joint, make sure the knuckle is supported. Make sure not to overextend the brake line hose.

8. Wedge the halfshaft removal tool between the inboard joint and the differential.
9. Remove the rear halfshaft. Remove the set ring.

To install:

10. Installation is the reverse of the removal procedure.
11. Be sure that the mating surfaces of the joint and splined section are clean and free of debris.
12. Apply 0.02–0.04 ounce of super high grease to the whole splined surface of the right halfshaft.

➡After applying the grease, remove the grease from the splined grooves at intervals of 2 to 3 splines and from the set ring groove so that air can bleed from the intermediate shaft.

13. Be sure to use new set rings.
14. Clean the areas where the halfshaft contacts the differential thoroughly with solvent and dry with compressed air.
15. Insert the halfshaft into the differential, until the set ring locks in place.
16. Tighten the new flange bolt to 76 ft. lbs. (103 Nm).

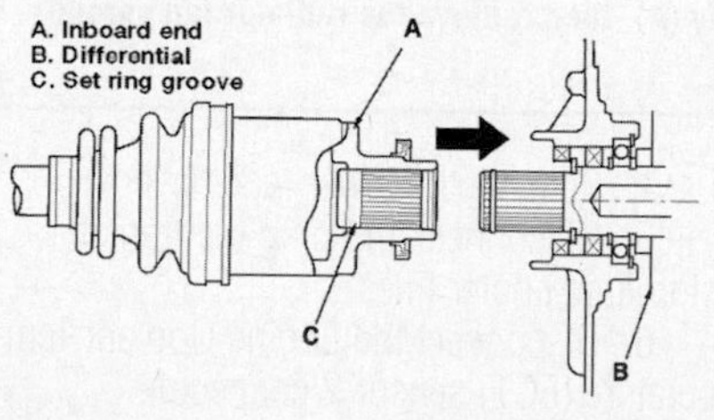
A. Inboard end
B. Differential
C. Set ring groove
A
C
B
22140_ARDX_G0089

Fig. 27 Rear halfshaft to differential alignment

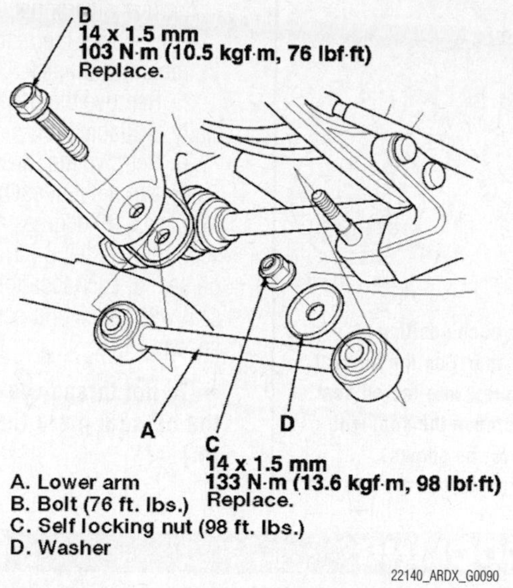
D
14 x 1.5 mm
103 N·m (10.5 kgf·m, 76 lbf·ft)
Replace.
A
D
C
14 x 1.5 mm
133 N·m (13.6 kgf·m, 98 lbf·ft)
Replace.
A. Lower arm
B. Bolt (76 ft. lbs.)
C. Self locking nut (98 ft. lbs.)
D. Washer
22140_ARDX_G0090

Fig. 28 Rear lower arm bolt location and torque data

17. Tighten the lower arm bolt and nuts to specification shown in the illustration. Be sure to use new a bolt and nut.
18. Tighten the spindle locknut to 181 ft. lbs. (245 Nm). Be sure to use a new nut.
19. Check and adjust the wheel alignment, as required.
20. Road test the vehicle.

REAR SIDE CASE SEAL

REMOVAL & INSTALLATION

See Figures 29 and 30.

1. Remove the rear differential as follows:
   a. Drain the differential fluid.
   b. Remove the right rear driveshaft.
   c. Remove the exhaust pipe.
   d. Make a reference marks across the propeller shaft and the rear differential companion flange. Separate the propeller shaft from the rear differential.
   e. Place the transmission jack under the rear differential.
   f. Using the driveshaft remover and a hammer, remove the left inboard joint from the rear differential, then remove the set ring from the end of the driveshaft connection point.
   g. Remove the rear differential mounting bolt and nuts.
   h. Lower the rear differential a little on the transmission jack, then remove the left rear driveshaft inboard joint from the rear differential.

⁂ CAUTION

Make sure not to over extend the harness and the tube.

   i. Disconnect the right solenoid 4P connector and the rear differential fluid temperature sensor 2P connector, then remove the harness clip.
   j. Disconnect the breather tube from the rear differential.
   k. Lower the rear differential a little on the transmission jack.

⁂ CAUTION

Make sure not to over extend the harness.

   l. Disconnect the left solenoid 4P connector, then remove the harness clip.

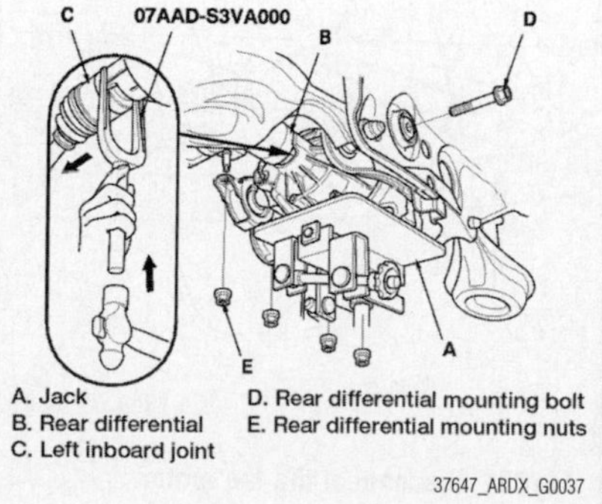
C
07AAD-S3VA000
B
D
E
A
A. Jack
B. Rear differential
C. Left inboard joint
D. Rear differential mounting bolt
E. Rear differential mounting nuts
37647_ARDX_G0037

Fig. 29 Lowering the rear differential

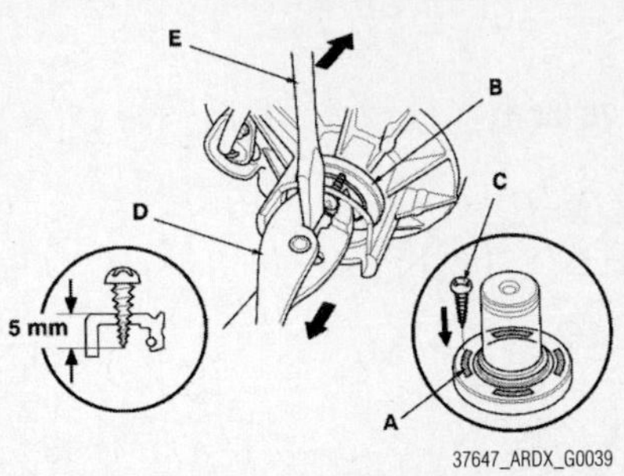
E
B
C
D
5 mm
A
37647_ARDX_G0039

Fig. 30 Cut a slit at each position (4 places, 90 degrees apart) on the oil seal, then install 6 mm screw into the oil seal at each location. Remove the seal with pliers and screwdriver as shown.

m. Disconnect the ground cable from the rear differential.

n. Lower the rear differential on the transmission jack.

2. Remove the dust seal with a commercially available tool.

3. Remove the thrust washer.

4. Cut a slit at each position (4 places, 90 degrees apart) on the oil seal, then install 6 mm screw into the oil seal at each location. Remove the seal with pliers and screwdriver as shown.

➡**Do not thread the 6 mm screws into the oil seal more than 5 mm (0.197 in.).**

To install:

5. Install a new oil seal into the rear differential side case with the fork seal driver weight and the driver.

6. Install the thrust washer, then install a new dust seal with the fork seal driver weight and oil seal driver.

7. Install the rear differential in reverse of the removal procedure, noting the following:

a. Align the reference marks on the propeller shaft flange.

b. Use a new differential rear mounting bolt and tighten to 54 ft. lbs. (74 Nm).

c. Tighten the differential rear mounting nuts to 28 ft. lbs. (38 Nm).

d. Tighten the differential front mounting bolts to 54 ft. lbs. (74 Nm).

ENGINE COOLING

ENGINE FAN

REMOVAL & INSTALLATION

See Figures 31 and 32.

1. Remove the splash shield.

2. Disconnect the fan motor connectors, and remove the harness clamps.

3. Remove the clip, then remove the coolant reservoir from the holder.

4. Remove the clips, then remove the support rod clamp bracket.

5. Remove the condenser fan shroud assembly, then remove the radiator fan shroud assembly from the condenser fan shroud side.

6. Installation is the reverse of the removal procedure.

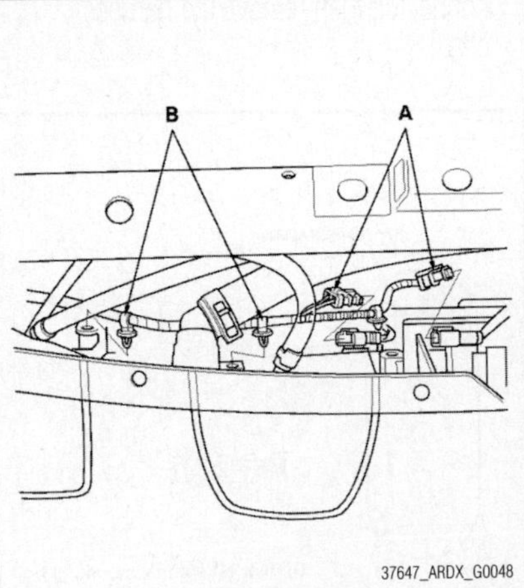
B
A
37647_ARDX_G0048

Fig. 31 Disconnect the fan motor connectors (A), and remove the harness clamps (B).

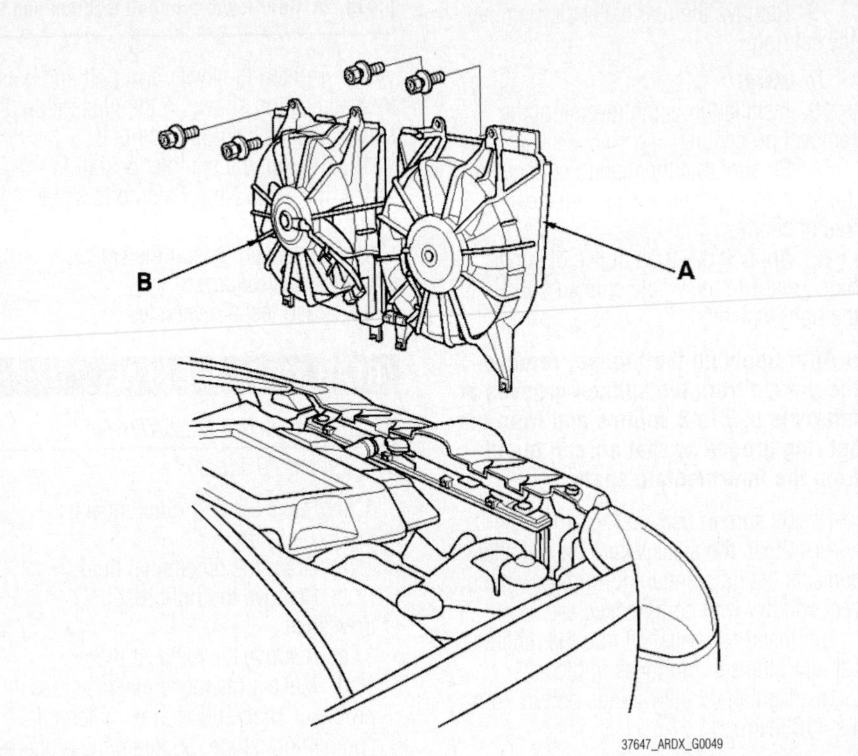
B
A
37647_ARDX_G0049

Fig. 32 Remove the condenser fan shroud assembly (A), then remove the radiator fan shroud assembly (B) from the condenser fan shroud side.

RADIATOR

REMOVAL & INSTALLATION

See Figures 33 and 34.

1. Drain the cooling system. Properly dispose of used coolant.

2. Remove the condenser and radiator fan shroud assembly.

3. Remove the upper radiator hose.

4. Remove the lower radiator hose.

5. Disconnect and plug the transmission fluid cooler lines.

6. Disconnect the Engine Coolant Temperature (ECT) sensor 2 connector.

7. Remove the air intake duct cover and the four clips. Remove the upper radiator brackets.

8. Pull the radiator up and out of its mounting.

To install:

9. Installation is the reverse of the removal procedure.
10. Be sure to use new O-rings.
11. When installing the radiator be sure the lower cushions are set securely.
12. Tighten the upper radiator bracket screws to 7.2 ft. lbs. (9.8 Nm).
13. Check the lower hose quick connector and set ring for cracks and damage. Replace as required.
14. Be sure that the set ring is in place inside the quick connector. If the set ring is not properly seated in the connector, replace the connector.
15. Replace the O-ring in the quick connector.
16. Check the lock. If it is damaged, replace it.
17. When installing a new lock to the connector, slide it straight down along the groove.
18. Install the lower hose on the quick connector. Install the clamp.
19. Push down the lock, then push the quick connector onto the radiator until you hear a click.
20. Continue the installation in the reverse order of the removal procedure.
21. Fill the cooling system with the proper grade and type coolant.

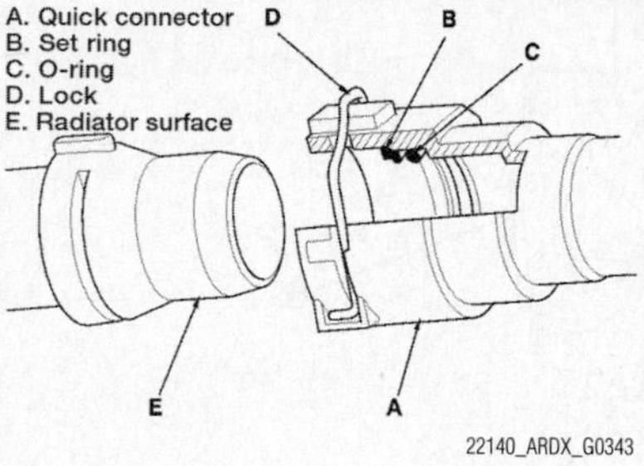
A. Quick connector
B. Set ring
C. O-ring
D. Lock
E. Radiator surface
D
B
C
E
A
22140_ARDX_G0343

Fig. 34 Quick connector alignment and installation

22. Bleed the air from the cooling system with the heater valve open.
23. Loosely install the radiator cap.
24. Start the engine and allow it to reach operating temperature until the fan comes on twice.
25. Turn the engine off. Check the coolant level. Correct as required.
26. Install the radiator cap. Run the engine and check for leaks.
27. Connect the Honda Diagnostic System (HDS) tool, or equivalent scan tool, to the DLC.

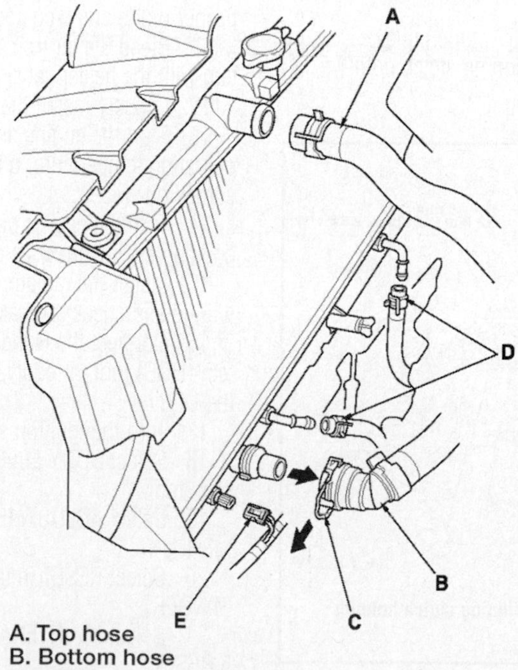
A
D
B
E
C
A. Top hose
B. Bottom hose
C. Lock
D. Transmission fluid hoses
E. ECT two connector
22140_ARDX_G0342

Fig. 33 Radiator removal disconnection points

28. Turn the ignition switch ON (II).
29. Select BODY ELECTRICAL on the scan tool.
30. Select ADJUSTMENT in the GAUGE MENU.
31. Select RESET in the MAINTENANCE MINDER.
32. Select MAINTENANCE SUB ITEM 5 RESET.

THERMOSTAT

REMOVAL & INSTALLATION

See Figures 35 and 36.

1. Drain the cooling system. Be sure to properly dispose of used coolant.
2. Remove the splash shield.
3. Disconnect the fan motor connectors and remove the harness clamp. Remove the coolant reservoir from the holder.
4. Remove the clips and the support rod clamp bracket.
5. Remove the condenser fan shroud assembly, then remove the radiator fan shroud assembly from the condenser fan shroud side.
6. Remove the lower hose.
7. Remove the thermostat retaining bolts. Remove the thermostat from its mounting.

To install:

8. Installation is the reverse of the removal procedure.
9. Be sure to use a new O-ring.
10. Tighten the retaining bolts to 8.8 ft. lbs. (11.9 Nm).
11. Fill the cooling system with the proper grade and type coolant.
12. Bleed the air from the cooling system with the heater valve open.
13. Loosely install the radiator cap.

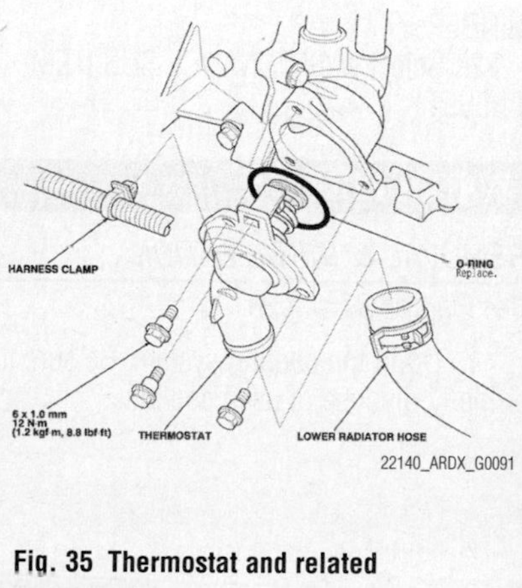
HARNESS CLAMP
O-RING
Replace.
6 x 1.0 mm
12 N·m
(1.2 kgf·m, 8.8 lbf·ft)
THERMOSTAT
LOWER RADIATOR HOSE
22140_ARDX_G0091

Fig. 35 Thermostat and related components

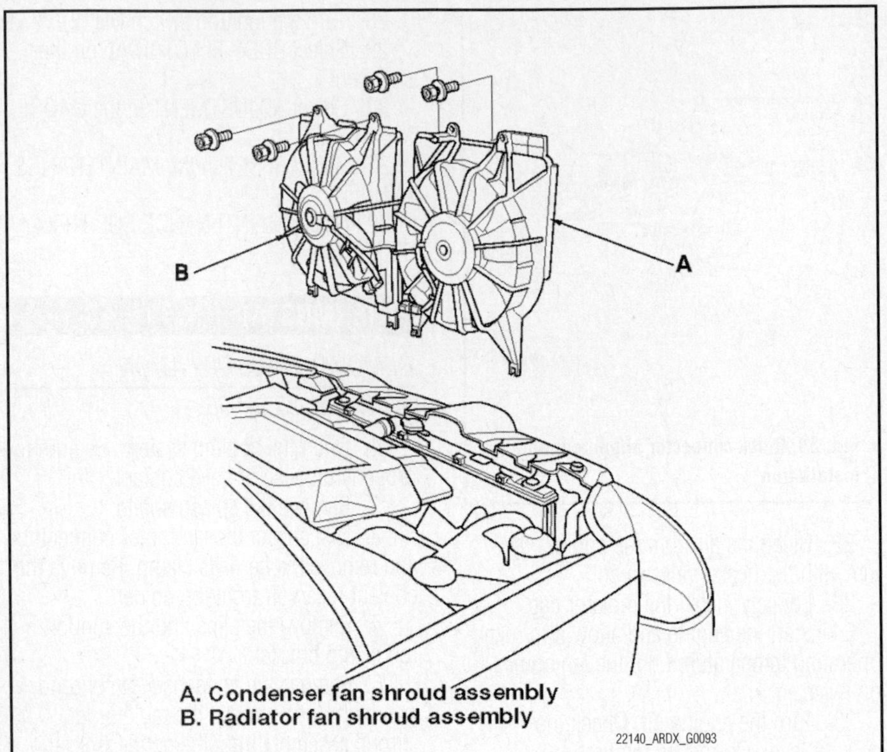
A
B
A. Condenser fan shroud assembly
B. Radiator fan shroud assembly
22140_ARDX_G0093

Fig. 36 Condenser fan shroud and related components

14. Start the engine and allow it to reach operating temperature, until the radiator fan comes on twice.

15. Turn the engine off. Check the coolant level. Correct as required.

16. Install the radiator cap. Run the engine and check for leaks.

17. Connect the Honda Diagnostic System (HDS) tool, or equivalent scan tool, to the DLC.

18. Turn the ignition switch ON (II).

19. Select BODY ELECTRICAL on the scan tool.

20. Select ADJUSTMENT in the GAUGE MENU.

21. Select RESET in the MAINTENANCE MINDER.

22. Select MAINTENANCE SUB ITEM 5 RESET.

WATER PUMP

REMOVAL & INSTALLATION

See Figures 37 and 38.

1. Drain the cooling system. Be sure to properly dispose of used coolant.

2. Remove the splash shield.

3. Move the drive belt auto tensioner using a belt tension release tool, to relieve the tension on the drive belt. Remove the drive belt.

4. Hold the water pump pulley using an air conditioning clutch holder tool.

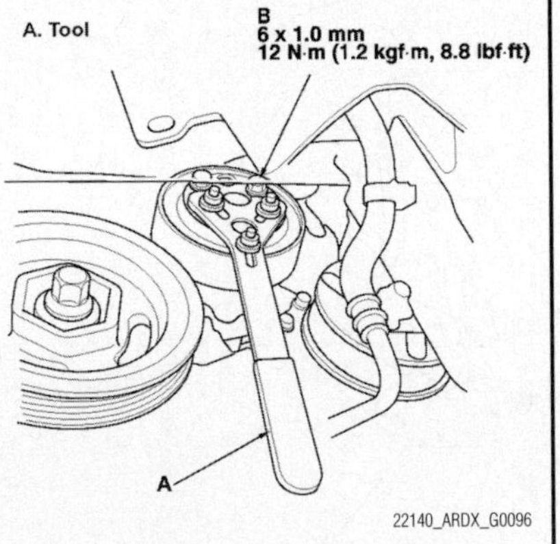
A. Tool
B
6 x 1.0 mm
12 N·m (1.2 kgf·m, 8.8 lbf·ft)
A
22140_ARDX_G0096

Fig. 37 Air conditioning clutch holding tool positioning

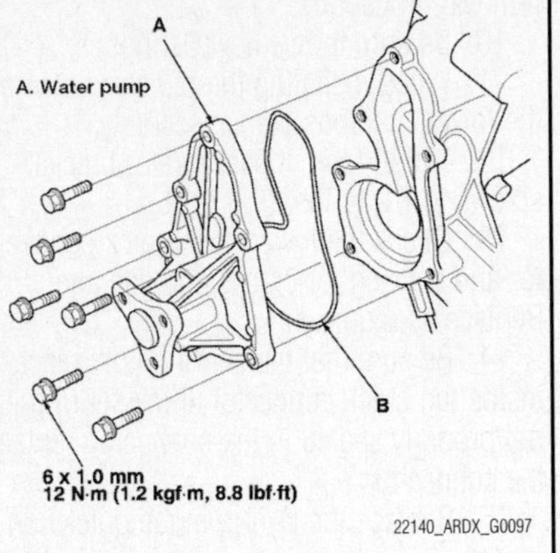
A
A. Water pump
B
6 x 1.0 mm
12 N·m (1.2 kgf·m, 8.8 lbf·ft)
22140_ARDX_G0097

Fig. 38 Water pump and related components

5. Remove the water pump pulley mounting bolts. Remove the water pump pulley.

6. Remove the water pump retaining bolts. Remove the water pump from its mounting.

To install:

7. Installation is the reverse of the removal procedure.

8. Be sure to use a new O-ring.

9. Tighten the retaining bolts to 8.8 ft. lbs. (11.9 Nm).

10. Fill the cooling system with the proper grade and type coolant.

11. Bleed the air from the cooling system with the heater valve open.

12. Loosely install the radiator cap.

13. Start the engine and allow it to reach operating temperature, until the radiator fan comes on twice.

14. Turn the engine off. Check the coolant level. Correct as required.

15. Install the radiator cap. Run the engine and check for leaks.

16. Connect the Honda Diagnostic System (HDS) tool, or equivalent scan tool, to the DLC.

17. Turn the ignition switch ON (II).

18. Select BODY ELECTRICAL on the scan tool.

19. Select ADJUSTMENT in the GAUGE MENU.

20. Select RESET in the MAINTENANCE MINDER.

21. Select MAINTENANCE SUB ITEM 5 RESET.

ALTERNATOR

REMOVAL & INSTALLATION

See Figures 39 and 40.

1. Make sure you have the anti-theft code for the audio system and the navigation system (if equipped).
2. Make sure the ignition switch to LOCK (0).
3. Disconnect the negative cable from the battery.
4. Move the drive belt auto tensioner using a belt tension release tool, to relieve the tension on the drive belt. Remove the drive belt.
5. Remove the clip and remove the coolant reservoir from its holder.
6. Remove the clips and the support rod clamp bracket.
7. Disconnect the condenser fan motor connector. Remove the condenser fan shroud assembly.
8. Disconnect the alternator electrical connectors.
9. Remove the harness clamp from the alternator.
10. Remove the alternator retaining bolts. Remove the component from its mounting.

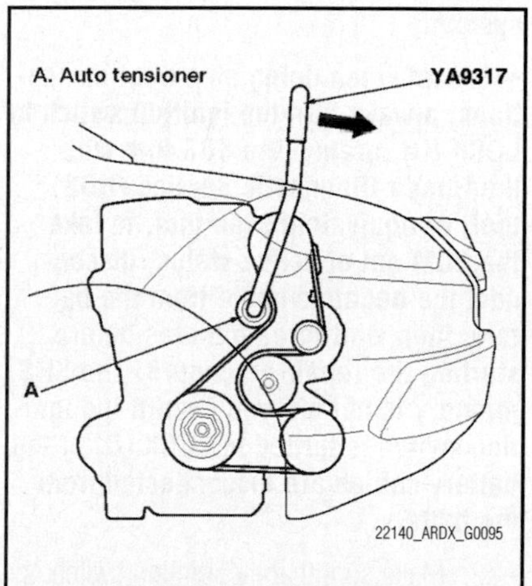
A. Auto tensioner
YA9317
A
22140_ARDX_G0095
Fig. 39 Drive belt tension tool and removal direction

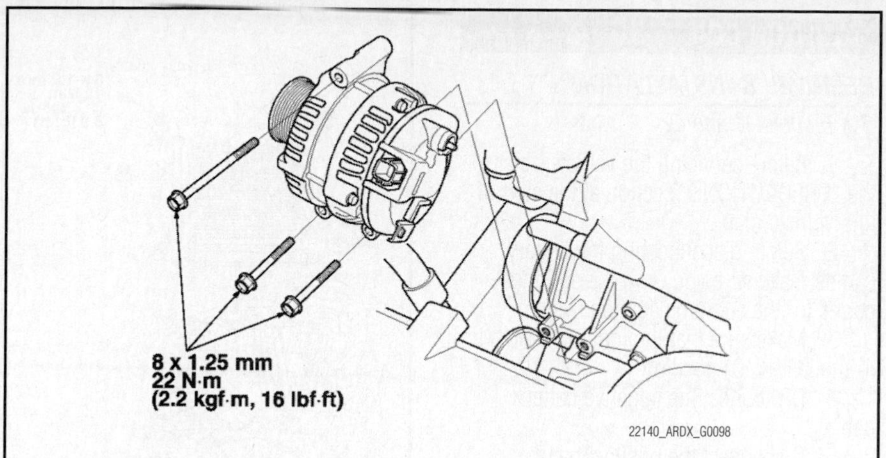
8 x 1.25 mm
22 N·m
(2.2 kgf·m, 16 lbf·ft)
22140_ARDX_G0098
Fig. 40 Alternator and related components

To install:

11. Installation is the reverse of the removal procedure.
12. Tighten the retaining bolts to 16 ft. lbs. (22 Nm).

BATTERY TERMINAL DISCONNECT/ RECONNECT PROCEDURE

Disconnect

➡Some system store data in memory is lost when the battery is disconnected. Do the following procedures before disconnecting the battery.

1. Make sure you have the anti-theft code(s) for the audio and/or the navigation system (if equipped).
2. If you are replacing the audio unit, write down the audio presets (AM and FM), and the XM audio presets (if equipped), because the audio unit does not retain the presets after the battery is disconnected.
3. Make sure the ignition switch is in LOCK (0).
4. Disconnect and isolate the negative cable from the battery.

➡Always disconnect the negative cable from the battery first.

5. Disconnect the positive cable from the battery.

Reconnect

1. Clean the battery terminals.
2. Test the battery.
3. Reconnect the positive cable to the battery first, then reconnect the negative cable to the battery.

➡Always connect the positive cable to the battery first.

4. Apply multipurpose grease to the terminals to prevent corrosion.
5. Enter the anti-theft code(s) for the audio system and/or the navigation system (if equipped).
6. Enter the audio presets (if applicable), and enter the XM audio presets (if equipped).
7. Set the clock (for vehicles without navigation).

VOLTAGE REGULATOR

ADJUSTMENT

➡The voltage regulator is an integral, electronic component of the alternator and no adjustment is needed.

IGNITION COIL

REMOVAL & INSTALLATION

See Figures 41 and 42.

1. Before servicing the vehicle, refer to the "PRECAUTIONS" section at the start of this vehicle menu.
2. Before disconnecting the battery cables make sure you have the anti theft codes for the audio/navigation system.
3. Make sure that the ignition switch is in the LOCK (0) position.
4. Disconnect the negative battery cable.
5. Disconnect the positive battery cable.

➡Wait at least three minutes after disconnecting the battery cables before starting the repair procedure.

6. Remove the air charge cooler cover.
7. Disconnect the turbocharger boost sensor connector. Remove the vacuum hoses and the air bypass outlet connecting tube.
8. Remove the air charge cooler retaining bolts. Remove the air charge cooler.
9. Remove the ignition coil cover. Disconnect the ignition coil connectors.
10. Remove the ignition coils.

To install:

11. Tighten the coil retaining bolt to 8.8 ft. lbs. (11.9 Nm).

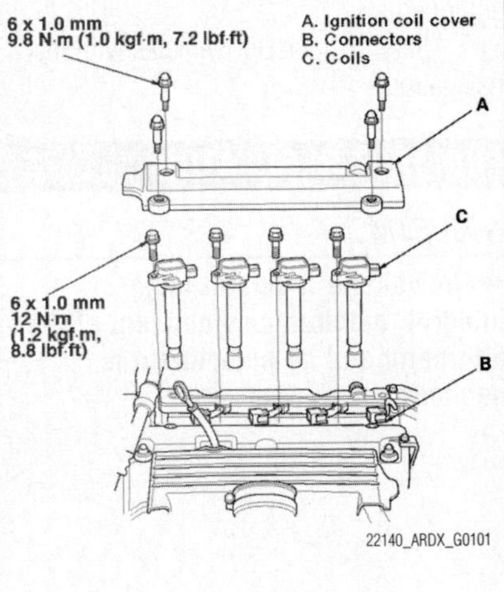
6 x 1.0 mm
9.8 N·m (1.0 kgf·m, 7.2 lbf·ft)
A. Ignition coil cover
B. Connectors
C. Coils
A
C
6 x 1.0 mm
12 N·m
(1.2 kgf·m,
8.8 lbf·ft)
B
22140_ARDX_G0101

Fig. 41 Ignition coils and related components

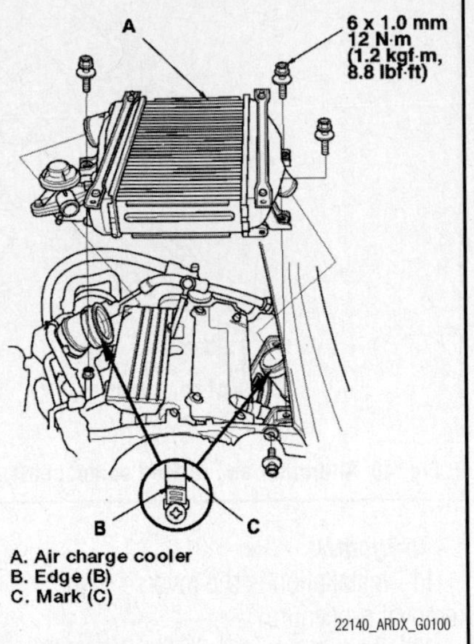
A
6 x 1.0 mm
12 N·m
(1.2 kgf·m,
8.8 lbf·ft)
B
C
A. Air charge cooler
B. Edge (B)
C. Mark (C)
22140_ARDX_G0100

Fig. 42 Air charge cooler band tightening

12. Install the air charge cooler and connect the intake air ducts.
13. Tighten the hose bands until edge (B) of the band aligns with the mark (C) on the band.

➡If the hose band edge exceeds the mark, replace the hose band.

14. Continue the installation in the reverse order of the removal procedure.

IGNITION TIMING

INSPECTION

1. Before servicing the vehicle, refer to the "PRECAUTIONS" section at the start of this vehicle menu.
2. Connect the Honda Diagnostic Service (HDS) tool, or equivalent scan tool, to the DLC.
3. Turn the ignition switch ON (II).
4. Make sure that the tool communicates with the PCM, if not troubleshoot the DLC circuit.
5. Start the engine. Hold the engine speed at 3,000 RPM, without a load in Park or Drive until the radiator fan comes on, then let the engine idle.
6. Check the idle speed.
7. Jump the SCS line using the HDS tool.
8. Connect the timing light in the service loop (white tape).
9. Check the timing specification under the no load (all accessories off) condition.

➡If not within specification, check cam timing. If cam timing is ok, update the PCM with the latest software package or substitute with known good PCM. If system works properly, and the PCM was substituted, replace the original PCM.

10. Disconnect the tool and the timing light.

ADJUSTMENT

The ignition timing is not adjustable. It is controlled by the PCM.

SPARK PLUGS

REMOVAL & INSTALLATION

1. Before servicing the vehicle, refer to the "PRECAUTIONS" section at the start of this vehicle menu.

➡Before disconnecting the battery cables make sure you have the anti theft codes for the audio/navigation system.

➡Except when doing electrical inspections, always turn the ignition switch to LOCK (0), ground the SCS line with the Honda Diagnostic Service (HDS) tool, or equivalent scan tool, to take the PCM out of active status, disconnect the negative cable from the battery, then wait three minutes before starting the repair procedure. The SRS memory is not cleared even if the ignition switch is turned to LOCK (0) or the battery cables are disconnected from the battery.

2. Make sure that the ignition switch is in the LOCK (0) position.
3. Disconnect the negative battery cable.
4. Disconnect the positive battery cable.
5. Remove the ignition coils.
6. Remove the spark plugs.
7. Installation is the reverse of the removal procedure.

STARTER

REMOVAL & INSTALLATION

See Figure 43.

1. Before servicing the vehicle, refer to the "PRECAUTIONS" section at the start of this vehicle menu.

➡Before disconnecting the battery cables make sure you have the anti theft codes for the audio/navigation system.

➡Except when doing electrical inspections, always turn the ignition switch to LOCK (0), ground the SCS line with the Honda Diagnostic Service (HDS) tool, or equivalent scan tool, to take the PCM out of active status, disconnect the negative cable from the battery, then wait three minutes before starting the repair procedure. The SRS memory is not cleared even if the ignition switch is turned to LOCK (0) or the battery cables are disconnected from the battery.

2. Make sure that the ignition switch is in the LOCK (0) position.
3. Disconnect the negative battery cable.
4. Disconnect the positive battery cable.
5. Remove the clip and remove the coolant reservoir from its holder.
6. Remove the clips and the support rod clamp bracket.
7. Disconnect the condenser fan motor connector. Remove the condenser fan shroud assembly.
8. Remove the harness clamp and connector. Remove the intake manifold bracket.
9. Disconnect the electrical connectors from the starter. Disconnect the harness clamps.
10. Remove the starter retaining bolts. Remove the starter from the vehicle.

To install:

11. Installation is the reverse of the removal procedure.
12. Tighten the starter retaining bolts as shown.

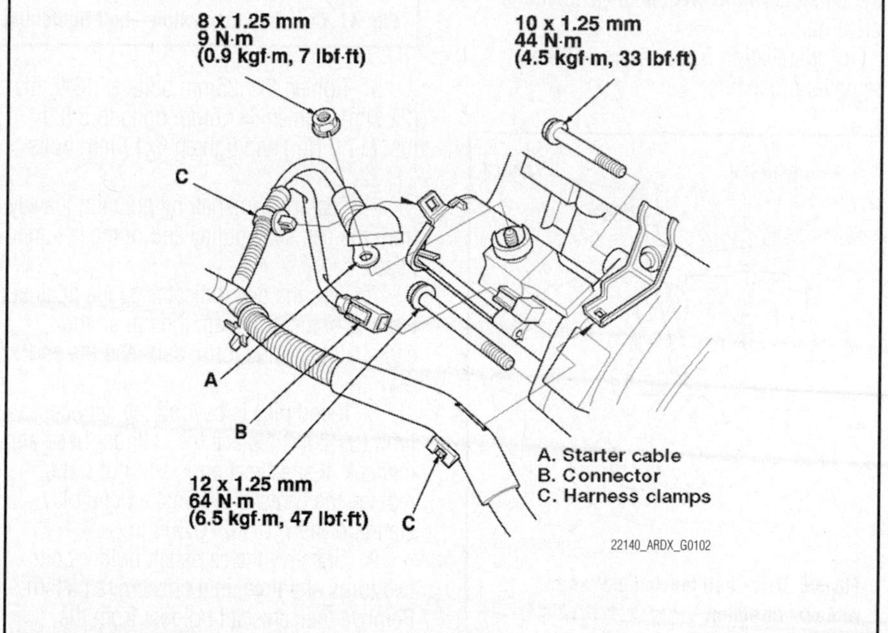
8 x 1.25 mm
9 N·m
(0.9 kgf·m, 7 lbf·ft)
10 x 1.25 mm
44 N·m
(4.5 kgf·m, 33 lbf·ft)
C
A
B
12 x 1.25 mm
64 N·m
(6.5 kgf·m, 47 lbf·ft)
C
A. Starter cable
B. Connector
C. Harness clamps
22140_ARDX_G0102

Fig. 43 Starter and related components

ACCESSORY DRIVE BELTS

DRIVE BELT ROUTING

See Figure 44.

Refer to the accompanying illustration.

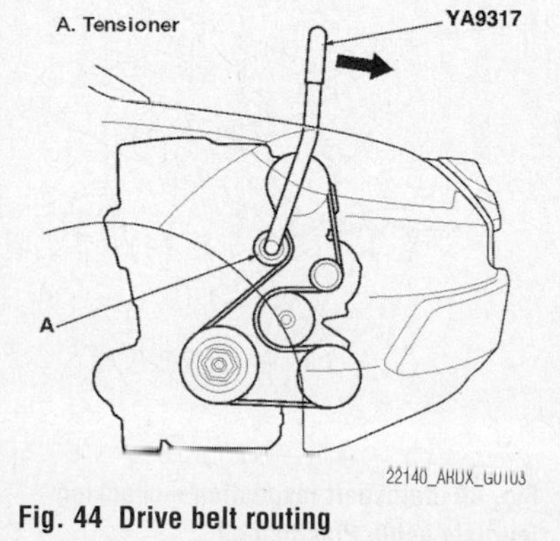
A. Tensioner
YA9317
A
22140_ARDX_G0103

Fig. 44 Drive belt routing

INSPECTION

See Figure 45.

1. Before servicing the vehicle, refer to the "PRECAUTIONS" section at the start of this vehicle menu.
2. Inspect the belt for cracks and/or damage.
3. If damage exists, replace the belt.
4. Check that the auto tensioner indicator is within range. If not replace the belt.

ADJUSTMENT

The drive belt is adjusted automatically. If out of specification, it must be replaced.

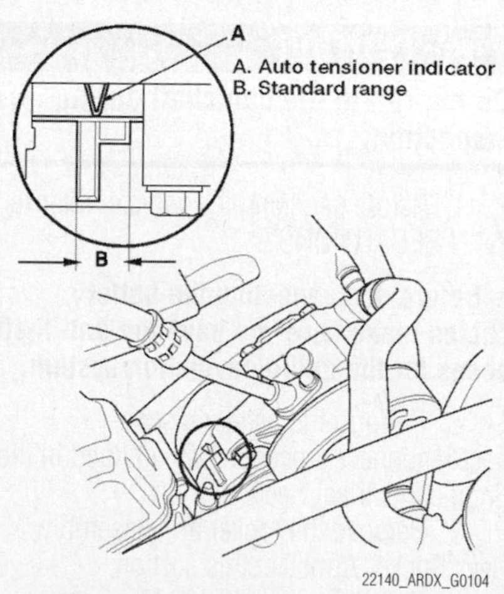
A
A. Auto tensioner indicator
B. Standard range
B
22140_ARDX_G0104

Fig. 45 Drive belt replacement specification

REMOVAL & INSTALLATION

See Figure 46.

1. Before servicing the vehicle, refer to the "PRECAUTIONS".

➡Before disconnecting the battery cables make sure you have the anti theft codes for the audio/navigation system.

2. Perform the battery disconnect/reconnect procedure, as outlined in the Engine Electrical Section.
3. Move the drive belt auto tensioner using a belt tension release tool, to relieve the tension on the drive belt. Remove the drive belt.
4. Installation is the reverse of the removal procedure.

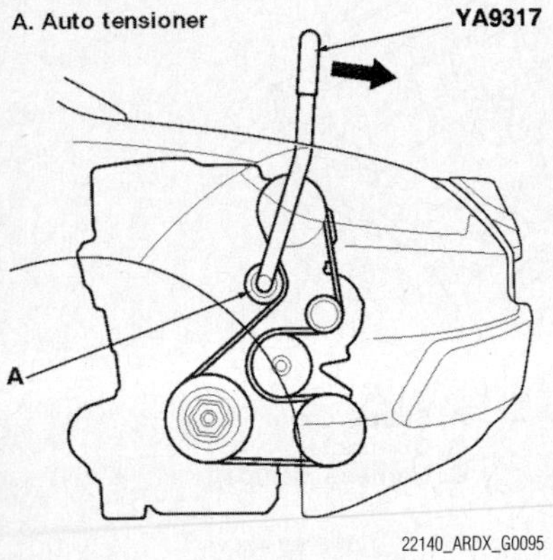
A. Auto tensioner
YA9317
A
22140_ARDX_G0095

Fig. 46 Drive belt tension tool and removal direction

CAMSHAFT & VALVE LIFTERS

INSPECTION

See Figures 47 through 51.

CAUTION

Do not rotate the camshaft during inspection.

1. Before servicing the vehicle, refer to the "PRECAUTIONS".

➡Before disconnecting the battery cables make sure you have the anti theft codes for the audio/navigation system.

2. Perform the battery disconnect/reconnect procedure, as outlined in the Engine Electrical Section.
3. Remove the rocker arm assembly. See "Rocker Arms" in this section.
4. Put the rocker shaft holders, the camshaft and the camshaft holders on the cylinder head. Tighten in sequence to specification.

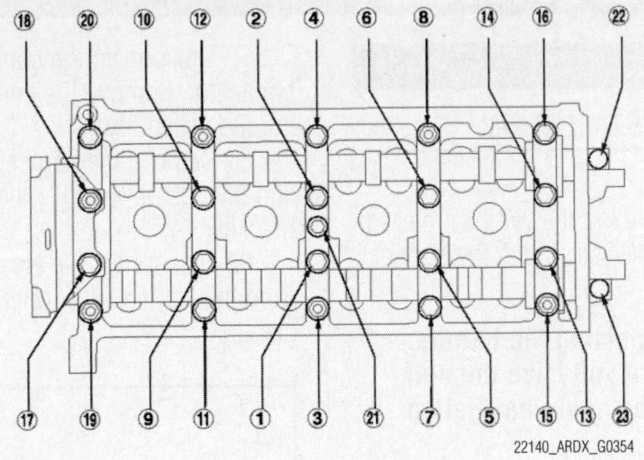
18 20 10 12 2 4 6 8 14 16 22
17 19 9 11 1 3 21 7 5 15 13 23
22140_ARDX_G0354

Fig. 47 Camshaft inspection—bolt tightening sequence

5. Tighten 8x1.25mm bolts to 16 ft. lbs. (22 Nm), tighten 6x1.0mm bolts to 8.8 ft. lbs. (11.9 Nm) and tighten 6x1.0mm bolts 21, 22, 23.
6. Seat the camshaft by pushing it away from the camshaft pulley end of the cylinder head.
7. Zero the dial indicator gauge against the end of the camshaft, then push the camshaft back and forth, and read the end play.
8. If end play is beyond the service limit (0.02 in.) replace the cylinder head and recheck. If specification is still not right, replace the camshaft. Camshaft end play standard (new) 0.002–0.008 in.
9. Unscrew the camshaft holder bolts two turns at a time, in a crisscross pattern. Remove the camshaft holders from the cylinder head.
10. Lift the camshafts out of the cylinder head, wipe them clean and inspect the lift ramps. Replace the camshafts, as required.

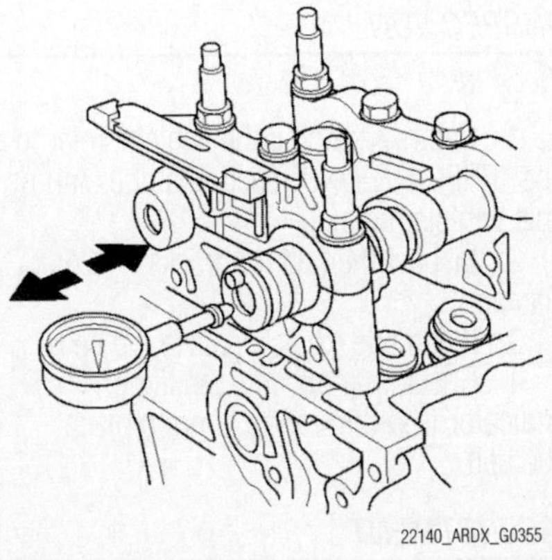
22140_ARDX_G0355

Fig. 48 Camshaft inspection—dial indicator gauge positioning

11. Install the camshaft holders, then tighten the bolts to specification and in sequence.
12. Remove the camshaft holders. Measure the widest portion of Plastigage®on each journal.
13. If the camshaft to holder oil clearance is within limits, measure the lobe height. Primary camshaft lobe height specification is 1.3356 in. (intake) and 1.3422 in. (exhaust). Secondary camshaft lobe height specification is 1.1668 in. (intake) and 1.3422 in. (exhaust).
14. If the camshaft to holder oil clearance is beyond the service limit (0.006 in.) and the camshaft has been replaced, replace the cylinder head. Camshaft to holder oil clearance standard (new) No. 1 journal 0.001–0.003 in., all other journals 0.002–0.004 in.
15. If the camshaft to holder oil clearance is beyond the service limit and the camshaft has not been replaced, check the total runout with the camshaft supported in V-blocks.
16. If total runout is within specification, replace the cylinder head. Specification is

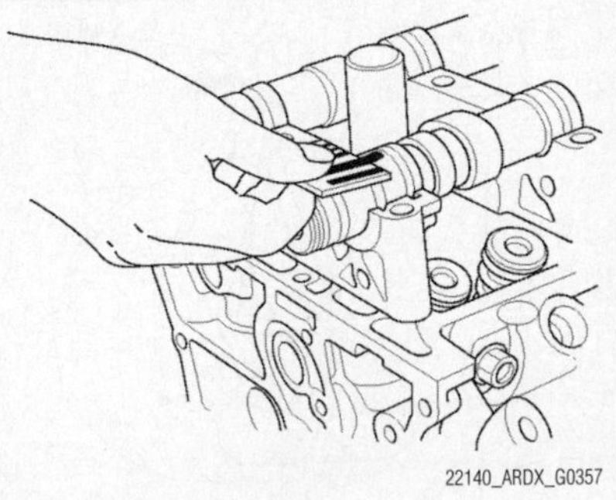
22140_ARDX_G0357

Fig. 49 Camshaft inspection—checking journals using Plastigage®

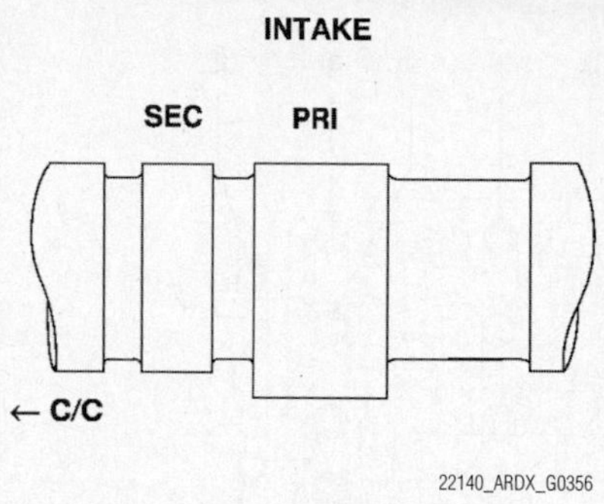
INTAKE
SEC
PRI
← C/C
22140_ARDX_G0356

Fig. 50 Camshaft inspection—primary (PRI) and secondary (SEC) lobe height identification

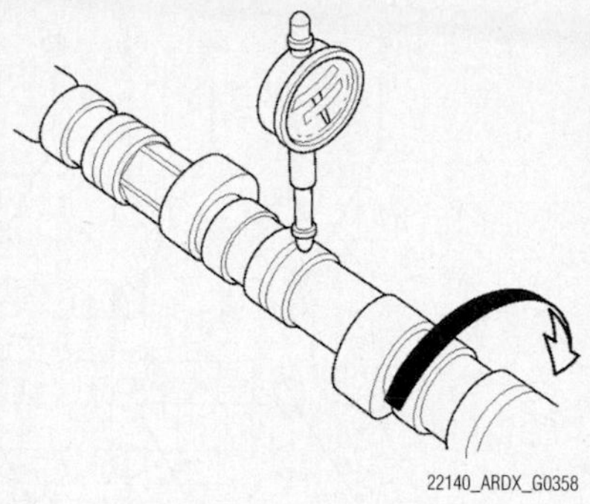
22140_ARDX_G0358

Fig. 51 Camshaft inspection—checking total runout

standard (new) 0.001 in. max, service limit 0.002 in.

17. If total runout is beyond service limit, replace the camshaft and recheck the camshaft to holder oil clearance. If oil clearance is still beyond service limit, replace the cylinder head.

REMOVAL & INSTALLATION

See Figures 52 through 57.

1. Before servicing the vehicle, refer to the "PRECAUTIONS".
2. Perform the battery disconnect/reconnect procedure, as outlined in the Engine Electrical Section.

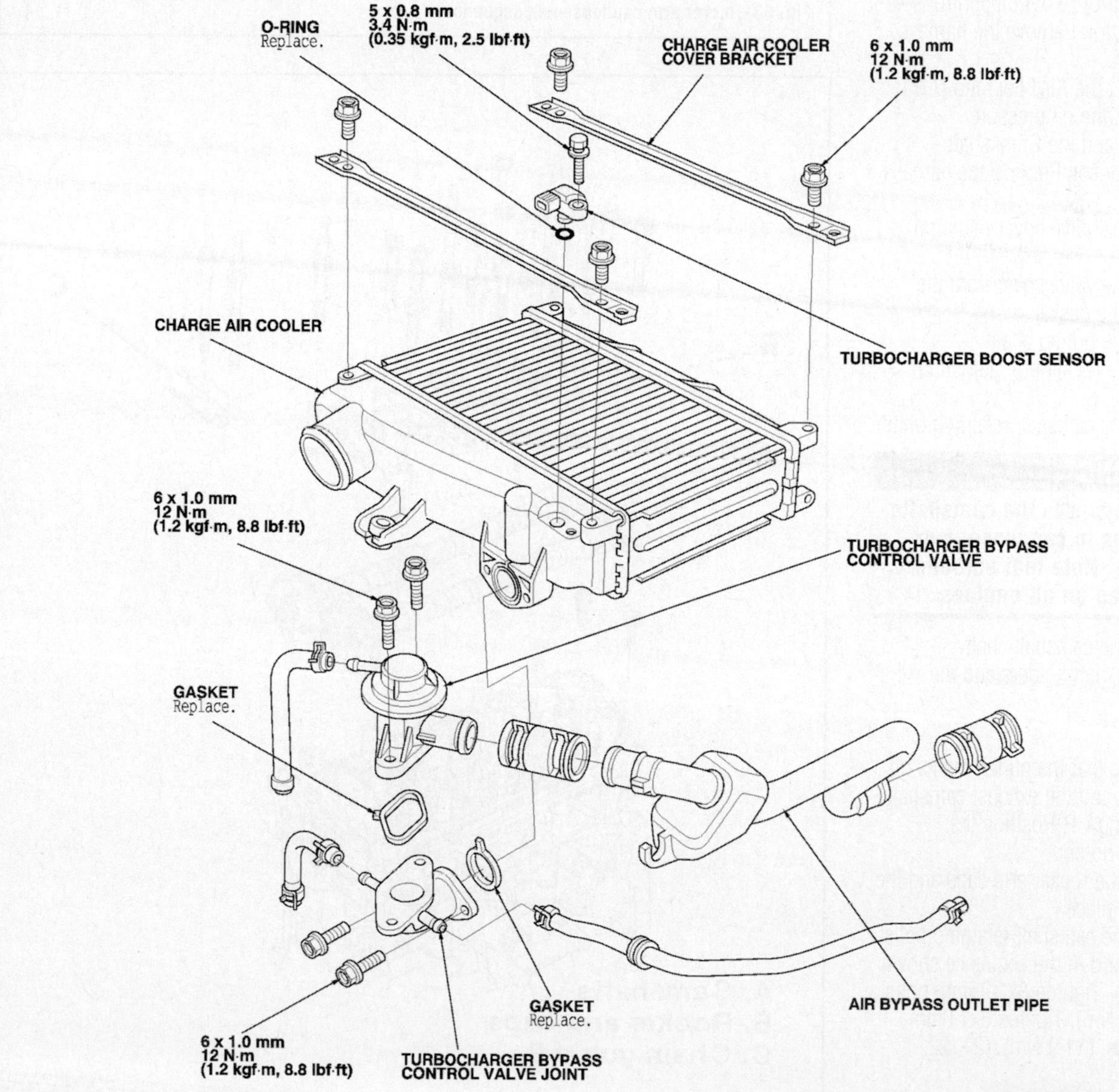
O-RING
Replace.
5 x 0.8 mm
3.4 N·m
(0.35 kgf·m, 2.5 lbf·ft)
CHARGE AIR COOLER
COVER BRACKET
6 x 1.0 mm
12 N·m
(1.2 kgf·m, 8.8 lbf·ft)
CHARGE AIR COOLER
TURBOCHARGER BOOST SENSOR
6 x 1.0 mm
12 N·m
(1.2 kgf·m, 8.8 lbf·ft)
TURBOCHARGER BYPASS
CONTROL VALVE
GASKET
Replace.
GASKET
Replace.
AIR BYPASS OUTLET PIPE
6 x 1.0 mm
12 N·m
(1.2 kgf·m, 8.8 lbf·ft)
TURBOCHARGER BYPASS
CONTROL VALVE JOINT
22140_ARDX_G0099

Fig. 52 Air charge cooler and related components

3. Remove the air charge cooler cover.
4. Disconnect the turbocharger boost sensor connector. Remove the vacuum hoses and the air bypass outlet connecting tube.
5. Remove the air charge cooler retaining bolts. Remove the air charge cooler.
6. Remove the fuel injector cover.
7. Remove the ignition coil cover. Disconnect the ignition coil connectors.
8. Remove the ignition coils.
9. Remove the power steering hose bracket, the breather hose and the dipstick.
10. Remove the bracket mounting bolts, from the valve cover.
11. Disconnect the rocker arm oil control solenoid connector, the rocker are oil pressure switch connector and the Variable Valve Timing Control (VTC) oil control solenoid valve connector. Remove the harness clamp.
12. Disconnect the Air/Fuel ratio (A/F) sensor connector, the oil pressure switch connector and the Crankshaft Position (CKP) sensor. Remove the harness clamps.
13. Remove the valve cover retaining bolts.
14. Remove the valve cover from the engine. Discard the gasket.
15. Remove the timing chain.
16. Loosen the rocker arm adjusting screws.
17. Remove the camshaft retaining bolts.

⁂ WARNING

To prevent damage to the camshafts, loosen the bolts in sequence, two turns at a time. Note that bolt one may not be used on all engines.

18. Remove the camshaft chain guide B, the camshaft holders and the camshafts.

To install:

19. Make sure that the punch marks on the VTC actuator and the exhaust camshaft sprocket are facing UP. Position the camshafts in the holder.
20. Position the rocker arm caps and the chain guide B in place.
21. Tighten the camshaft retaining bolts to specification and in the sequence shown in the illustration. Tighten 8x1.25mm bolts to 16 ft. lbs. (22 Nm). Tighten 6x1.0mm bolts to 8.8 ft. lbs. (11.9 Nm). (21, 22 and 23).
22. Continue the installation in the reverse order of the removal procedure.

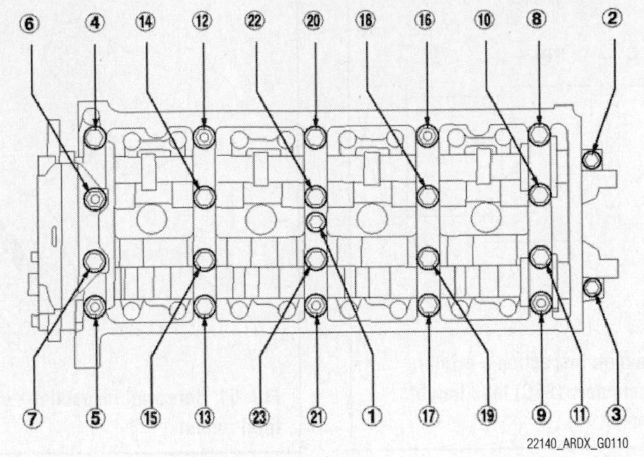
6
4
14
12
22
20
18
16
10
8
2
7
5
15
13
23
21
1
17
19
9
11
3
22140_ARDX_G0110

Fig. 53 Rocker arm cap loosening sequence

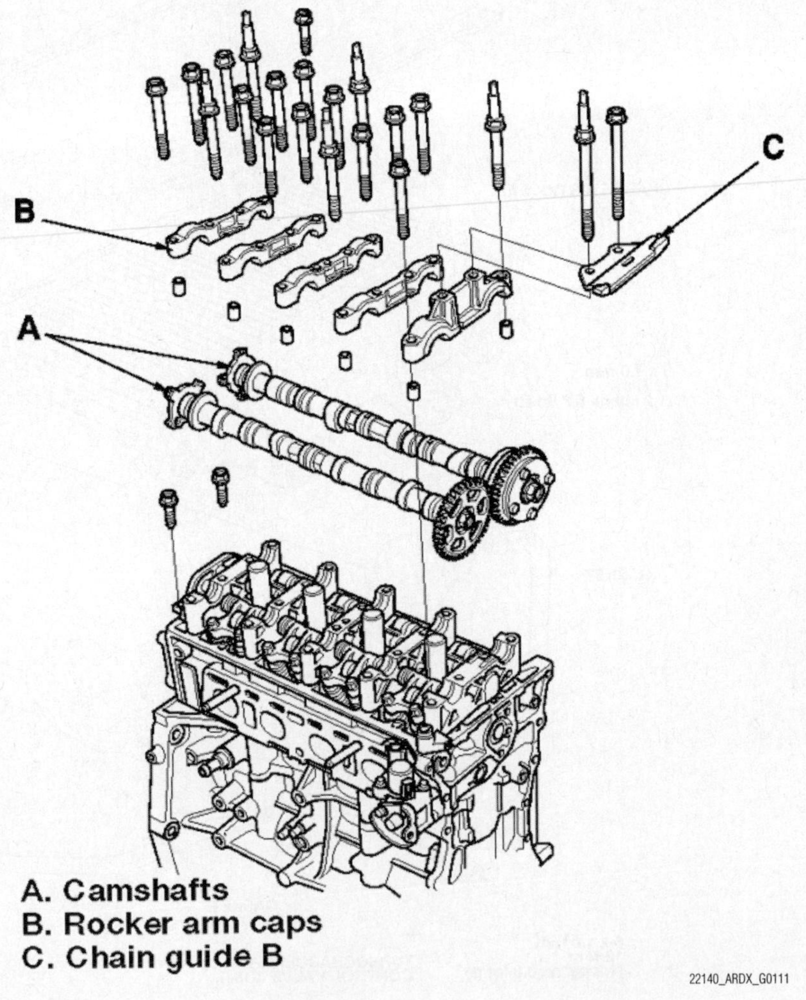
A
B
C
A. Camshafts
B. Rocker arm caps
C. Chain guide B
22140_ARDX_G0111

Fig. 54 Camshafts and related components

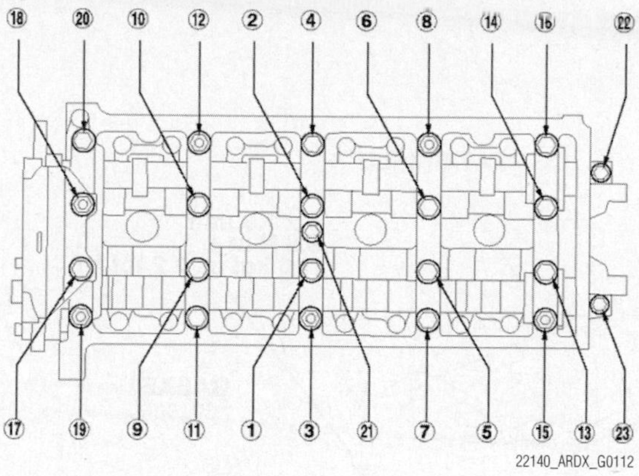
18
20
10
12
2
4
6
8
14
16
22
17
19
9
11
1
3
21
7
5
15
13
23
22140_ARDX_G0112
Fig. 55 Rocker arm cap tightening sequence

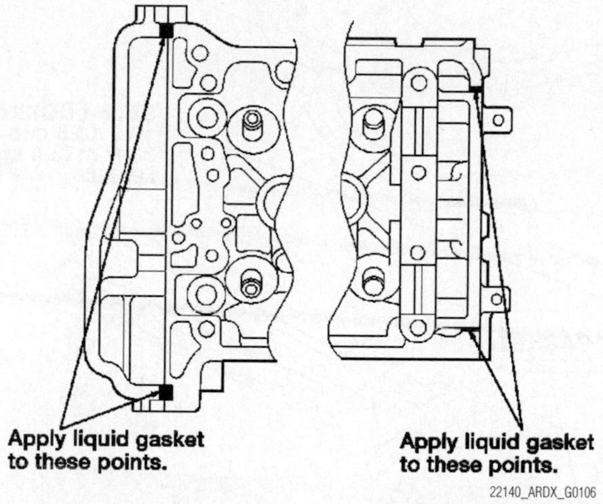
Apply liquid gasket to these points.
Apply liquid gasket to these points.
22140_ARDX_G0106
Fig. 56 Liquid gasket application points

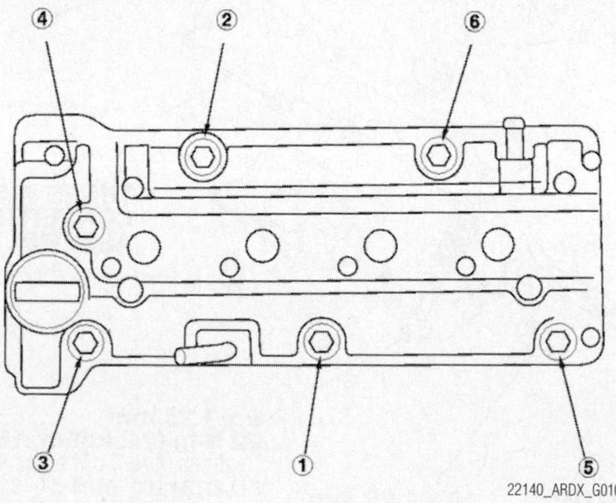
4
2
6
3
1
5
22140_ARDX_G0108
Fig. 57 Valve cover torque sequence

23. Before installing the valve cover, clean the mating surfaces of the cylinder head and valve cover.
24. Install a new gasket in the groove of the valve cover.
25. Be sure that the mating surfaces are clean and dry.
26. Apply liquid gasket part number 08717-0004 or equivalent on the chain case and the number five rocker shaft holder mating surface areas.

✻✻ CAUTION

Install the component within five minutes of applying the liquid gasket. Some liquid gasket materials require installation within four minutes. Be sure to read and follow the manufacturer's instruction. If too much time has passed before installing the component remove the liquid gasket and residue, then reapply new liquid gasket.

27. Position the spark plug seals on the spark plug tubes.
28. Position the valve cover on the cylinder head. Slide the cover back and forth to seat the gasket.
29. Inspect the valve cover washers, replace as required.
30. Tighten the retaining bolts in three steps to 8.8 ft. lbs. (11.9 Nm), using the sequence shown in the illustration.

➡Wait at least 30 minutes to allow the gasket to cure before filling the engine with oil.

✻✻ CAUTION

Do not run the engine for at least three hours after installing the valve cover.

31. Continue the installation in the reverse order of the removal procedure.
32. Install the air charge cooler and connect the intake air ducts.
33. Tighten the hose bands until edge (B) of the band aligns with the mark (C) on the band.

➡If the hose band edge exceeds the mark, replace the hose band.

34. Start the engine and check for leaks, correct as required.
35. Adjust the valves.
36. Turn the ignition switch ON (II), the SRS indicator should come on for about six seconds, and then go off.

CATALYTIC CONVERTER

REMOVAL & INSTALLATION

See Figure 58.

➡Manufacturer does not provide a specific removal and installation procedure for this component. Use the illustration as a guide when servicing this component.

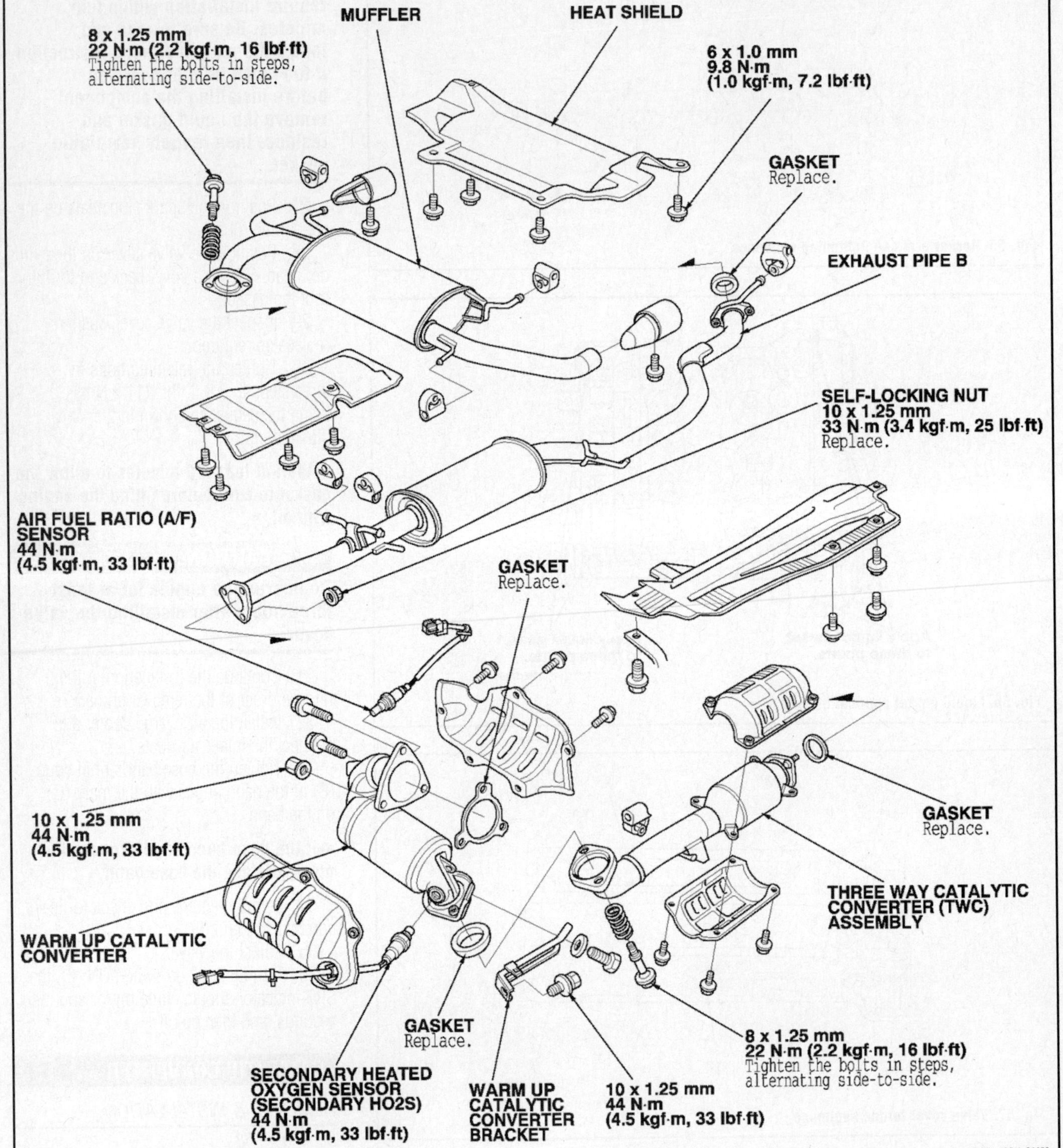
MUFFLER
HEAT SHIELD
8 x 1.25 mm
22 N·m (2.2 kgf·m, 16 lbf·ft)
Tighten the bolts in steps, alternating side-to-side.
6 x 1.0 mm
9.8 N·m
(1.0 kgf·m, 7.2 lbf·ft)
GASKET
Replace.
EXHAUST PIPE B
SELF-LOCKING NUT
10 x 1.25 mm
33 N·m (3.4 kgf·m, 25 lbf·ft)
Replace.
AIR FUEL RATIO (A/F) SENSOR
44 N·m
(4.5 kgf·m, 33 lbf·ft)
GASKET
Replace.
10 x 1.25 mm
44 N·m
(4.5 kgf·m, 33 lbf·ft)
GASKET
Replace.
THREE WAY CATALYTIC CONVERTER (TWC) ASSEMBLY
WARM UP CATALYTIC CONVERTER
GASKET
Replace.
8 x 1.25 mm
22 N·m (2.2 kgf·m, 16 lbf·ft)
Tighten the bolts in steps, alternating side-to-side.
SECONDARY HEATED OXYGEN SENSOR (SECONDARY HO2S)
44 N·m
(4.5 kgf·m, 33 lbf·ft)
WARM UP CATALYTIC CONVERTER BRACKET
10 x 1.25 mm
44 N·m
(4.5 kgf·m, 33 lbf·ft)
37647_ARDX_G0052

Fig. 58 Exploded view of the exhaust system components, showing the 3-way catalytic converter

CRANKSHAFT DAMPER

REMOVAL & INSTALLATION

See Figures 59 and 60.

1. Before servicing the vehicle, refer to the "PRECAUTIONS".
2. Perform the battery disconnect/reconnect procedure, as outlined in the Engine Electrical Section.
3. Raise and support the vehicle safely. Remove the front tire and wheel assemblies.
4. Remove the splash shield.
5. Remove the drive belt.
6. Hold the pulley with the holder handle and holder attachment. Remove the bolt with a socket and breaker bar.
7. Remove the crankshaft pulley.

To install:

8. Clean the crankshaft pulley, the crankshaft, the bolt and the washer. Lubricate components with clean engine oil as shown in illustration.
9. Install the pulley and retaining bolt. Tighten the bolt to 36 ft. lbs. (49 Nm). Do not use an impact wrench.
10. If using a new bolt, tighten to 130 ft. lbs. (176 Nm), loosen and retighten to 36 ft. lbs. (49 Nm). Do not use an impact wrench.
11. Tighten the bolt an additional 90 degrees.
12. Continue the installation in the reverse order of the removal procedure.

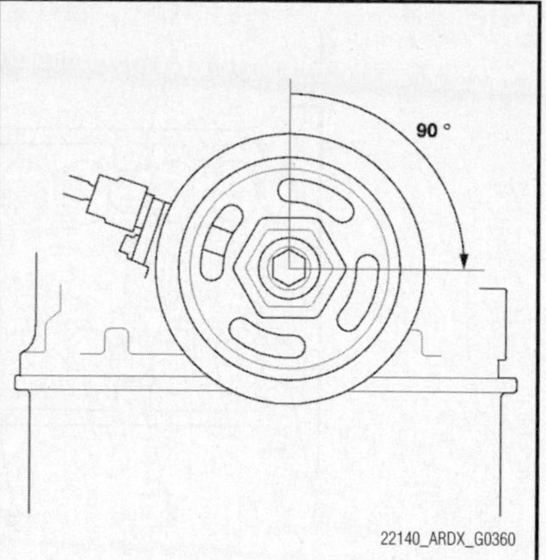
90 °
22140_ARDX_G0360

Fig. 60 Crankshaft pulley bolt tightening

13. Turn the ignition switch ON (II), the SRS indicator should come on for about six seconds, and then go off.

CRANKSHAFT FRONT SEAL

REMOVAL & INSTALLATION

See Figures 61 and 62.

1. Before servicing the vehicle, refer to the "PRECAUTIONS".
2. Perform the battery disconnect/reconnect procedure, as outlined in the Engine Electrical Section.
3. Remove the timing chain cover.
4. Carefully pry the old seal out of its mounting, using a suitable tool.

To install:

5. Position a new seal on the cover. Coat it with clean engine oil.
6. Use a seal driver tool to drive the new seal squarely into the chain case to the specified installed height.
7. Measure the distance between the chain case surface and the oil seal.
8. The oil seal installed height should be 1.30–1.33 in.

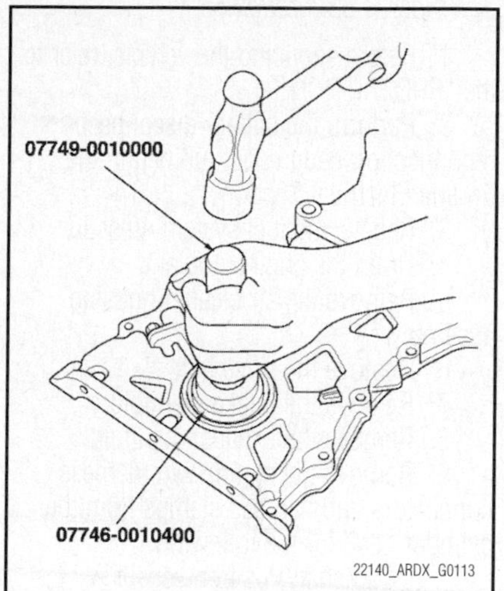
07749-0010000
07746-0010400
22140_ARDX_G0113

Fig. 61 Timing cover oil seal tool installation

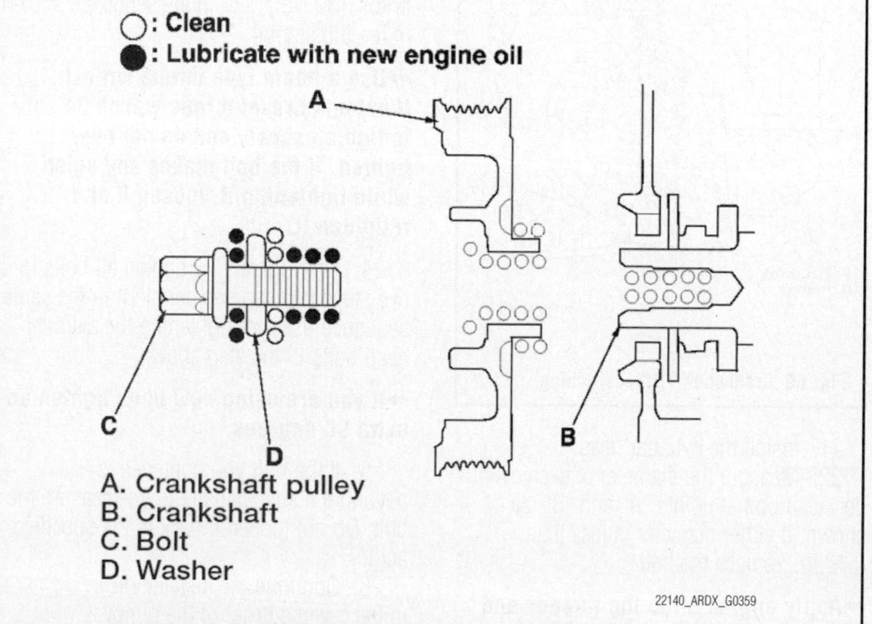
○: Clean
●: Lubricate with new engine oil
A
B
C
D
A. Crankshaft pulley
B. Crankshaft
C. Bolt
D. Washer
22140_ARDX_G0359

Fig. 59 Crankshaft pulley lubrication points

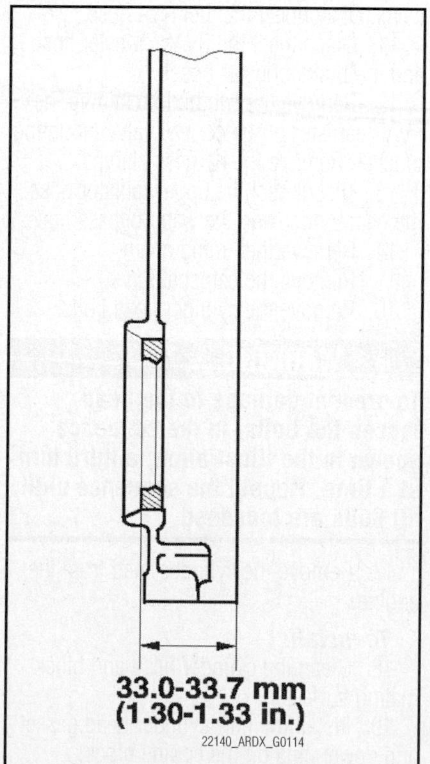
33.0-33.7 mm
(1.30-1.33 in.)
22140_ARDX_G0114

Fig. 62 Timing cover oil seal height specification

9. Continue the installation in the reverse order of the removal procedure.
10. Start the engine and check for leaks, correct as required.
11. Turn the ignition switch ON (II), the SRS indicator should come on for about six seconds, and then go off.

CYLINDER HEAD

REMOVAL & INSTALLATION

See Figures 63 through 68.

1. Before servicing the vehicle, refer to the "PRECAUTIONS".
2. Perform the battery disconnect/reconnect procedure, as outlined in the Engine Electrical Section.
3. Relieve the fuel system pressure.
4. Drain the engine coolant.
5. Remove the air cleaner housing assembly.
6. Remove the drive belt.
7. Remove the intake manifold.
8. Remove the exhaust manifold.
9. Remove the engine wire harness connectors and harness clamps from the cylinder head for the following:
   - Camshaft Position sensor A
   - Engine Coolant Temperature sensor
   - Camshaft Position sensor B
   - EVAP solenoid
10. Remove the quick connect fitting cover. Disconnect the fuel feed hose.
11. Disconnect the EVAP canister hose and the brake booster hoses.
12. Remove the two bolts retaining the EVAP canister purge control valve mounting bracket. Remove the harness clamps.
13. Disconnect the upper radiator hose, the heater hose and the water bypass hose.
14. Remove the timing chain.
15. Remove the camshafts.
16. Remove the cylinder head bolts.

⁂ WARNING

To prevent damage to the head loosen the bolts, in the sequence shown in the illustration, a third turn at a time. Repeat the sequence until all bolts are loosened.

17. Remove the cylinder head from the engine.

To install:

18. Clean the cylinder head and block mating surfaces.
19. Install the new cylinder head gasket and dowel pins on the engine block.
20. Position the crankshaft to TDC. Align the TDC mark on the crankshaft sprocket with the pointer on the engine block.

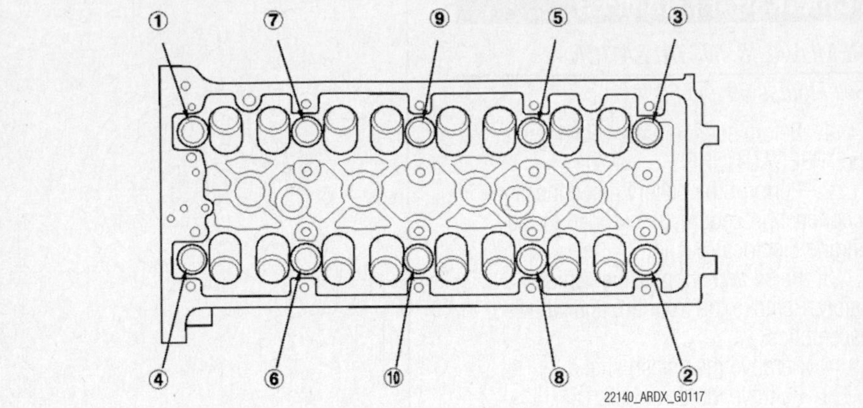
1
7
9
5
3
4
6
10
8
2
22140_ARDX_G0117

Fig. 63 Cylinder head bolt removal sequence

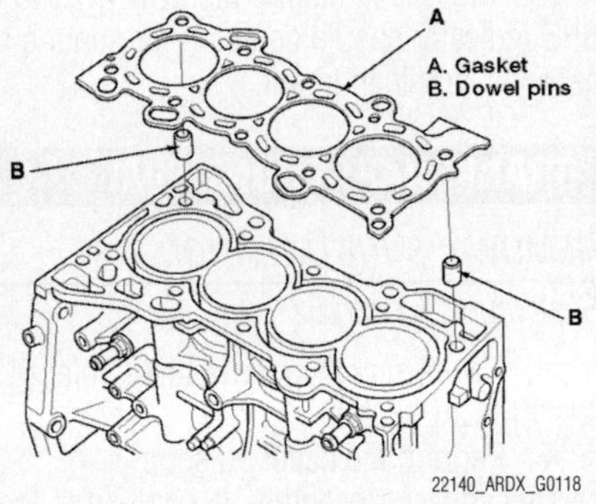
A
A. Gasket
B. Dowel pins
B
B
22140_ARDX_G0118

Fig. 64 Cylinder head gasket positioning

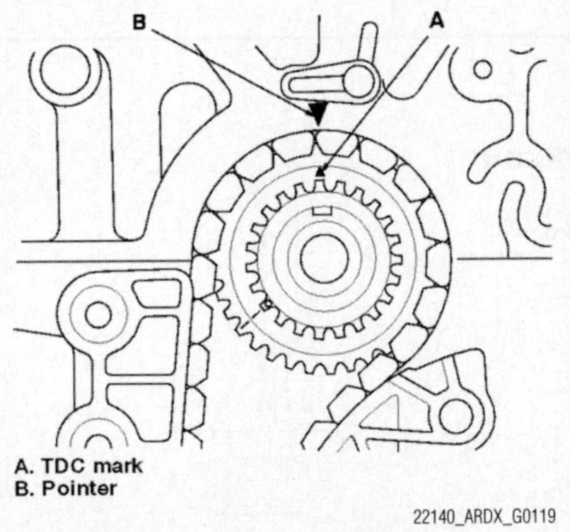
B
A
A. TDC mark
B. Pointer
22140_ARDX_G0119

Fig. 65 Crankshaft TDC positioning

21. Install the cylinder head.
22. Measure the diameter of each cylinder head bolt at points "A" and "B", as shown. If either diameter is less than 0.42 in., replace the bolt.

➡Apply engine oil to the threads and under the bolt heads, prior to installation.

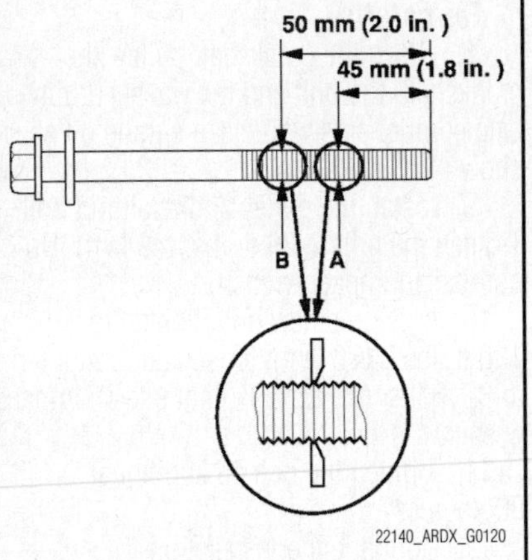
50 mm (2.0 in.)
45 mm (1.8 in.)
B
A
22140_ARDX_G0120

Fig. 66 Cylinder head bolt measurement

23. Tighten the cylinder head bolts to 29 ft. lbs. (39 Nm), and in the sequence shown in the illustration.

➡Use a beam type torque wrench. If using a preset torque wench be sure to tighten slowly and do not over tighten. If the bolt makes any noise while tightening it, loosen it and retighten it.

24. After tightening, tighten all bolts in two steps (90 degrees per step) in the same sequence used for tightening the cylinder head bolts in the step above.

➡If you are using new bolts tighten an extra 90 degrees.

25. If the bolt you tightened is beyond the specified angle, re-measure the bolt. Do not loosen it back to the specified angle.
26. Continue the installation in the reverse order of the removal procedure.

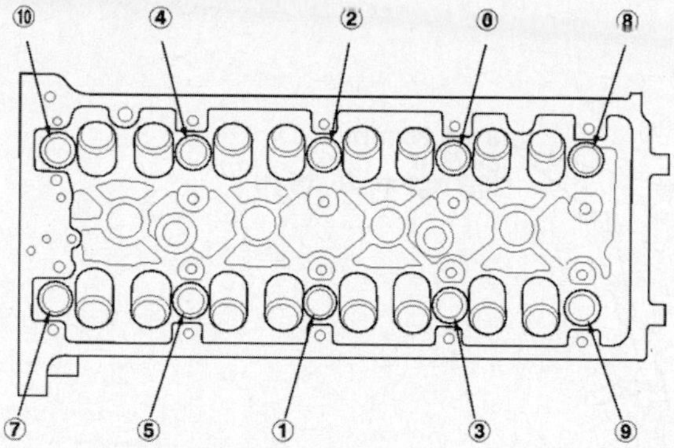
10
4
2
6
8
7
5
1
3
9
22140_ARDX_G0121

Fig. 67 Cylinder head bolt torquing sequence—part one

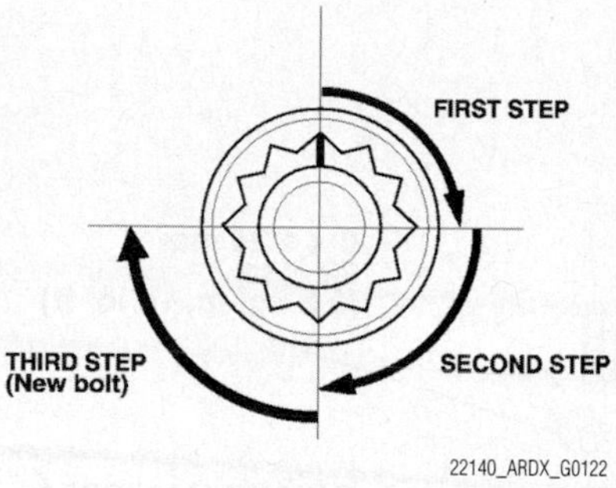
FIRST STEP
THIRD STEP
(New bolt)
SECOND STEP
22140_ARDX_G0122

Fig. 68 Cylinder head bolt torquing sequence—part two

27. Start the engine and check for leaks, correct as required.
28. Inspect the idle speed. Inspect the ignition timing.
29. Turn the ignition switch ON (II), the SRS indicator should come on for about six seconds, and then go off.

EXHAUST MANIFOLD

REMOVAL & INSTALLATION

See Figure 69.

1. Before servicing the vehicle, refer to the "PRECAUTIONS".
2. Perform the battery disconnect/reconnect procedure, as outlined in the Engine Electrical Section.
3. Remove the turbocharger. See "Turbocharger" in this section.
4. Remove the water bypass hoses.
5. Remove the exhaust manifold retaining bolts.
6. Remove the exhaust manifold from the engine.

To install:

7. Clean the mating surfaces of the engine and the exhaust manifold of dirt and debris.
8. Be sure to use a new gasket.
9. Tighten the retaining bolts and nuts to 33 ft. lbs. (45 Nm).
10. Continue the installation in the reverse order of the removal procedure.
11. Turn the ignition switch ON (II), the SRS indicator should come on for about six seconds, and then go off.
12. Road test the vehicle, correct problems as required.

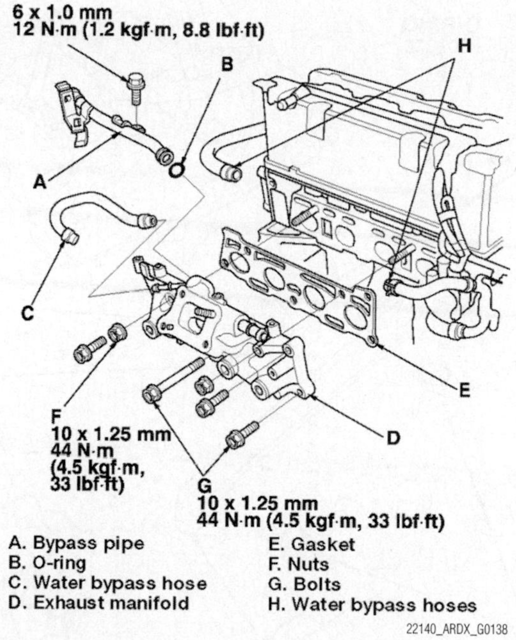
6 x 1.0 mm
12 N·m (1.2 kgf·m, 8.8 lbf·ft)
B
H
A
C
E
F
10 x 1.25 mm
44 N·m
(4.5 kgf·m,
33 lbf·ft)
D
G
10 x 1.25 mm
44 N·m (4.5 kgf·m, 33 lbf·ft)
A. Bypass pipe
B. O-ring
C. Water bypass hose
D. Exhaust manifold
E. Gasket
F. Nuts
G. Bolts
H. Water bypass hoses
22140_ARDX_G0138

Fig. 69 Exhaust manifold and related components

INTAKE MANIFOLD

REMOVAL & INSTALLATION

See Figures 70 through 72.

1. Before servicing the vehicle, refer to the "PRECAUTIONS".
2. Perform the battery disconnect/reconnect procedure, as outlined in the Engine Electrical Section.
3. Relieve the fuel system pressure. See "Relieving Fuel System Pressure" in "FUEL SYSTEM" section.
4. Remove the air charge cooler cover.
5. Disconnect the turbocharger boost sensor connector. Remove the vacuum hoses and the air bypass outlet connecting tube.
6. Remove the air charge cooler retaining bolts. Remove the air charge cooler.
7. Remove the fuel injector cover. Remove the injector electrical connectors.
8. Remove the throttle actuator connector. Remove the Manifold Absolute Pressure (MAP) sensor connector. Remove the Intake Air Temperature (IAT) sensor connector. Remove the turbocharger bypass control solenoid valve connector.
9. Remove the quick connect fitting cover. Disconnect the fuel feed hose.
10. Remove the ground cable, the vacuum hoses, the brake booster vacuum hoses and the turbocharger bypass control solenoid valve bracket mounting bolt.
11. Remove the EVAP canister hose and the water bypass line bracket mounting bolt.
12. Remove the water bypass hoses, plug the lines.

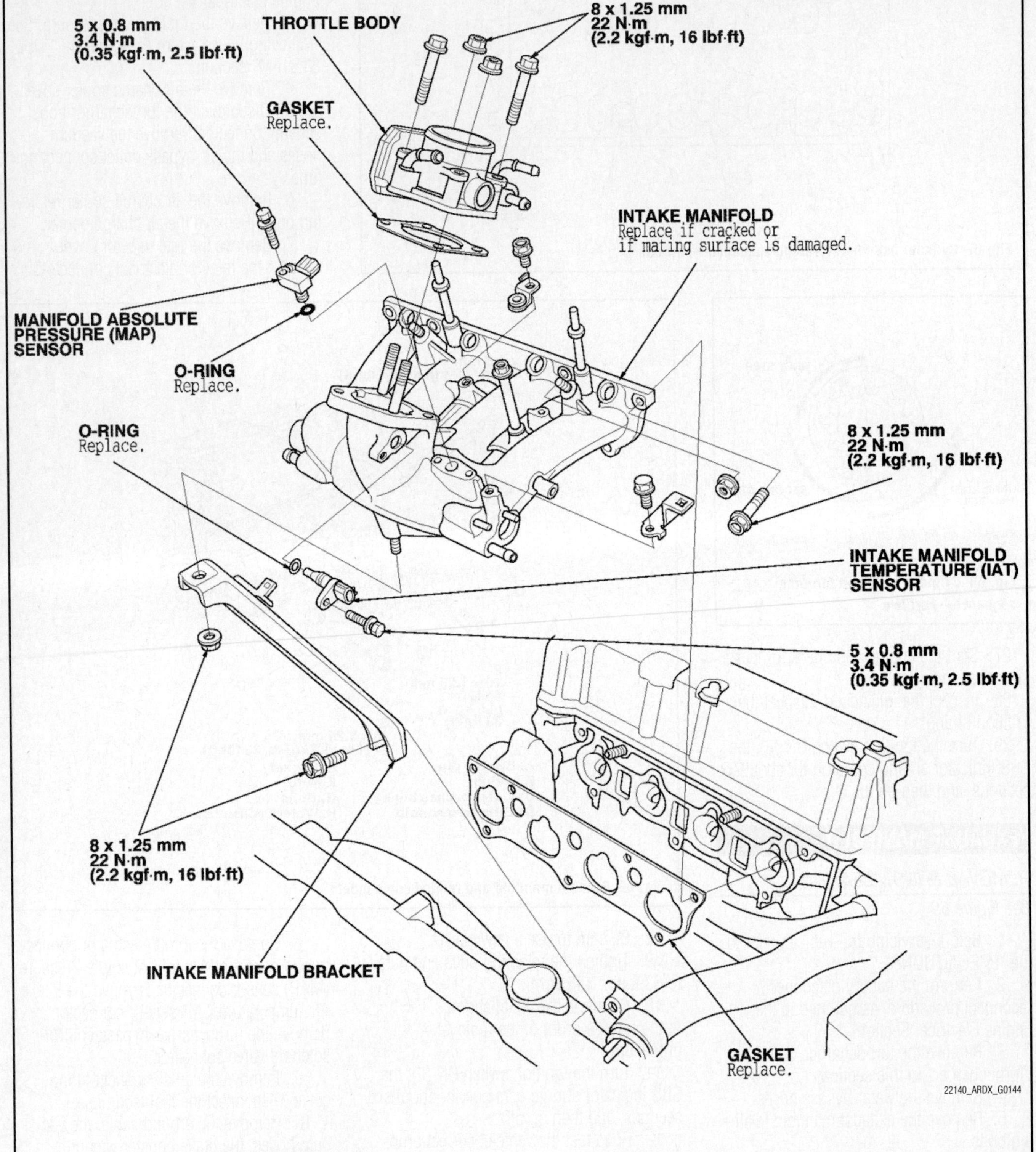
5 x 0.8 mm
3.4 N·m
(0.35 kgf·m, 2.5 lbf·ft)
THROTTLE BODY
8 x 1.25 mm
22 N·m
(2.2 kgf·m, 16 lbf·ft)
GASKET
Replace.
INTAKE MANIFOLD
Replace if cracked or
if mating surface is damaged.
MANIFOLD ABSOLUTE
PRESSURE (MAP)
SENSOR
O-RING
Replace.
O-RING
Replace.
8 x 1.25 mm
22 N·m
(2.2 kgf·m, 16 lbf·ft)
INTAKE MANIFOLD
TEMPERATURE (IAT)
SENSOR
5 x 0.8 mm
3.4 N·m
(0.35 kgf·m, 2.5 lbf·ft)
8 x 1.25 mm
22 N·m
(2.2 kgf·m, 16 lbf·ft)
INTAKE MANIFOLD BRACKET
GASKET
Replace.
22140_ARDX_G0144

Fig. 70 Intake manifold and related components

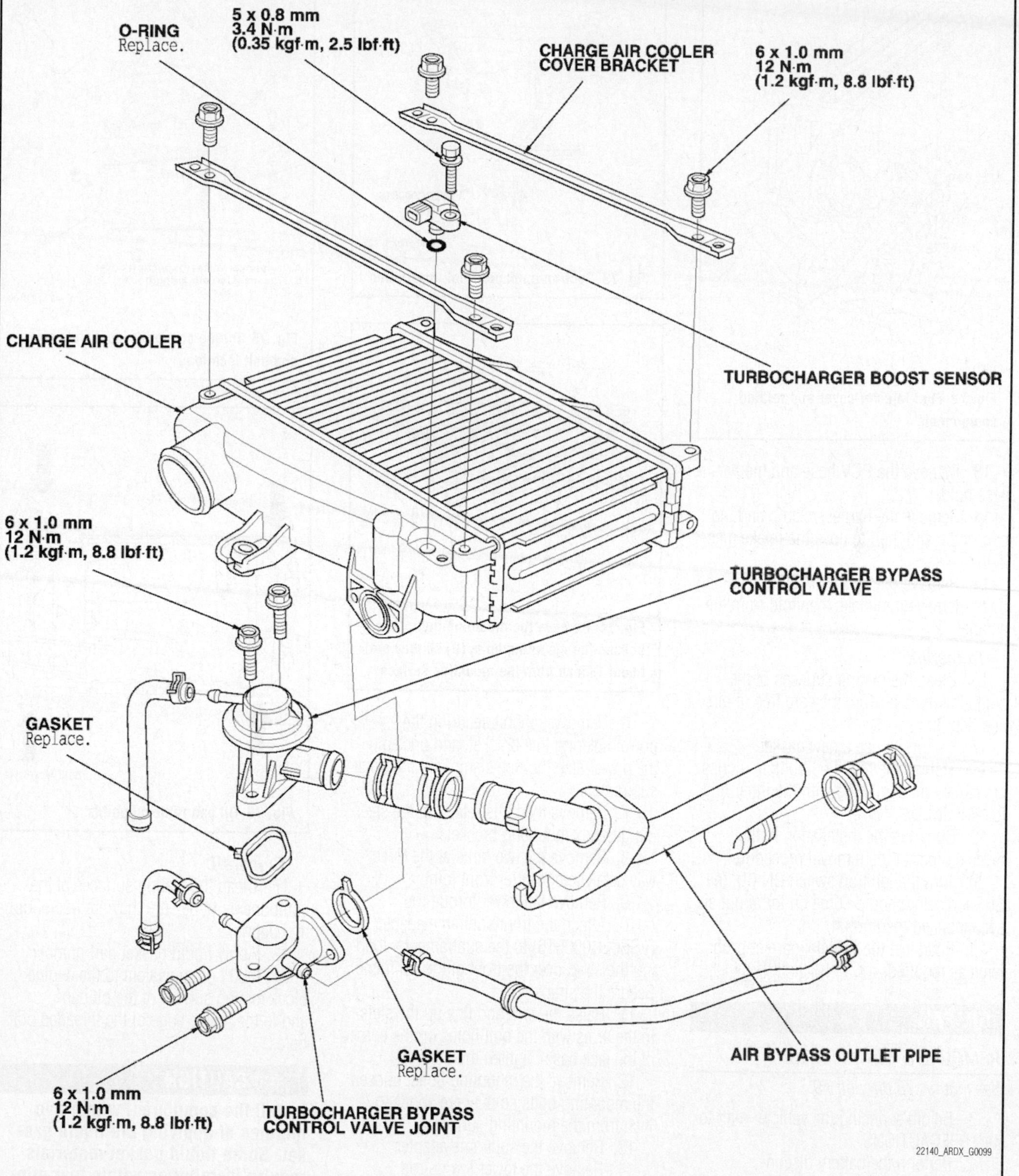
O-RING
Replace.
5 x 0.8 mm
3.4 N·m
(0.35 kgf·m, 2.5 lbf·ft)
CHARGE AIR COOLER
COVER BRACKET
6 x 1.0 mm
12 N·m
(1.2 kgf·m, 8.8 lbf·ft)
CHARGE AIR COOLER
TURBOCHARGER BOOST SENSOR
6 x 1.0 mm
12 N·m
(1.2 kgf·m, 8.8 lbf·ft)
TURBOCHARGER BYPASS
CONTROL VALVE
GASKET
Replace.
GASKET
Replace.
AIR BYPASS OUTLET PIPE
6 x 1.0 mm
12 N·m
(1.2 kgf·m, 8.8 lbf·ft)
TURBOCHARGER BYPASS
CONTROL VALVE JOINT
22140_ARDX_G0099

Fig. 71 Air charge cooler and related components

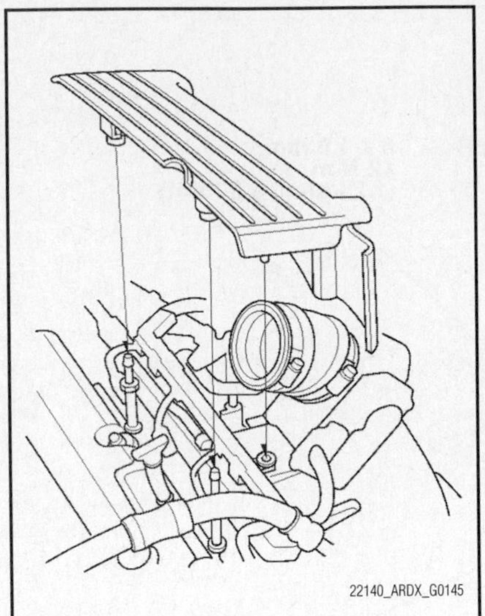
22140_ARDX_G0145

Fig. 72 Fuel injector cover and related components

13. Remove the PCV hose and the harness holder.

14. Remove the harness clamp and the connector, and then remove the intake manifold bracket.

15. Remove the intake manifold retaining bolts. Remove the intake manifold from the engine.

To install:

16. Clean the mating surfaces of the components to be sure they are free of dirt and debris.

17. Be sure to use a new gasket.

18. Tighten the retaining bolts in a crisscross pattern, beginning with the inner bolt to 16 ft. lbs. (22 Nm).

19. Continue the installation in the reverse order of the removal procedure.

20. Turn the ignition switch ON (II), the SRS indicator should come on for about six seconds, and then go off.

21. Road test the vehicle, correct problems as required.

OIL PAN

REMOVAL & INSTALLATION

See Figures 73 through 79.

1. Before servicing the vehicle, refer to the "PRECAUTIONS".

2. Perform the battery disconnect/reconnect procedure, as outlined in the Engine Electrical Section.

3. Raise and support the vehicle safely. Remove the front tire and wheel assemblies.

4. Drain the engine oil.

5. Remove the splash shield.

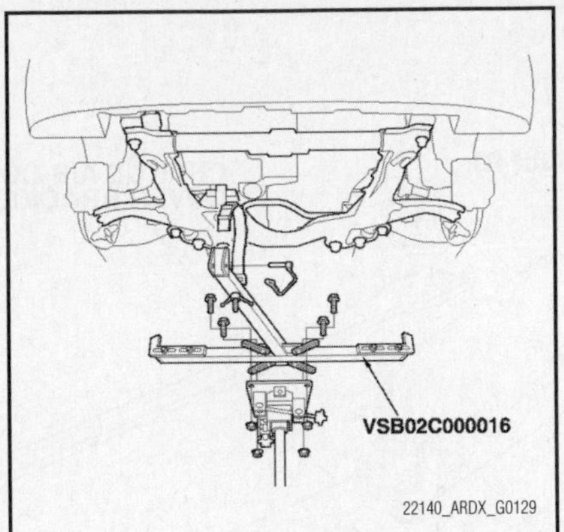
VSB02C000016
22140_ARDX_G0129

Fig. 73 Subframe adapter tool installation

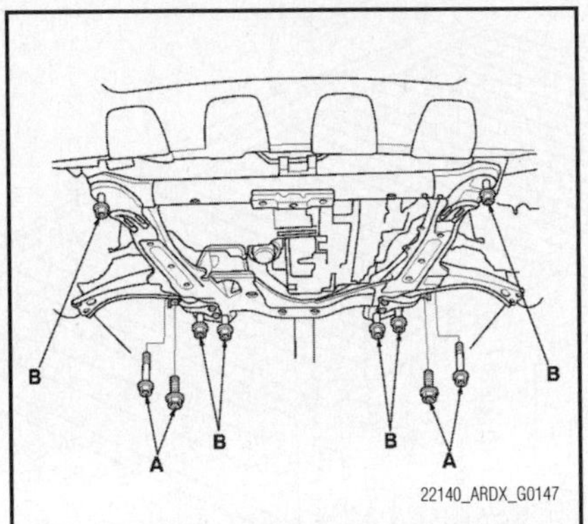
B
B
A
B
B
A
B
22140_ARDX_G0147

Fig. 74 Remove the mounting bolts (A). Loosen the mounting bolts (B) so they are about ½ inch from the mounting surface.

6. Remove the bolt securing the power steering line bracket, and unclamp the power steering line clamps on the front subframe.

7. Remove the bolts retaining the steering gearbox mounting brackets.

8. Remove the two bolts at the three way catalytic converter front joint.

9. Remove the lower torque rod.

10. Attach the front subframe adapter (VSB02C000016) to the subframe, by looping the strap over the front of the subframe. Secure the strap.

11. Raise the jack and line up the slots in the arms with the bolt holes on the corner of the jack base. Tighten the bolts.

12. Remove the mounting bolts. Loosen the mounting bolts so they are about ½ inch from the mounting surface.

13. Remove the subframe adapter.

14. Remove the lower torque rod bracket. Remove the torque converter cover inspection plate. Remove the transmission mounting bolts.

15. Remove the oil pan retaining bolts.

16. Using a suitable pry tool, separate the oil pan from the block.

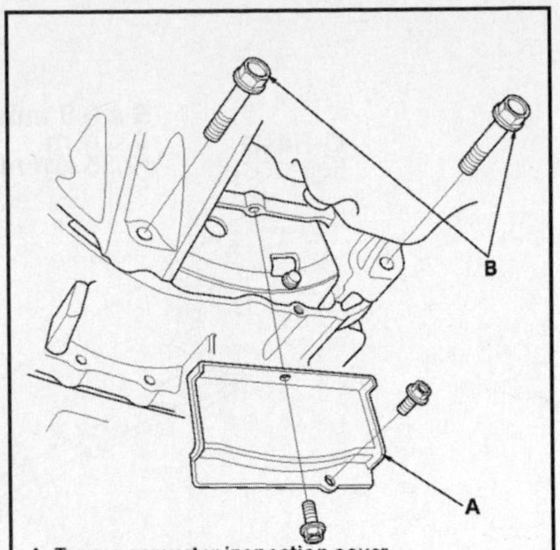
B
A
A. Torque converter inspection cover
B. Transmission mounting bolts
22140_ARDX_G0148

Fig. 75 Torque converter inspection cover and bolt locations

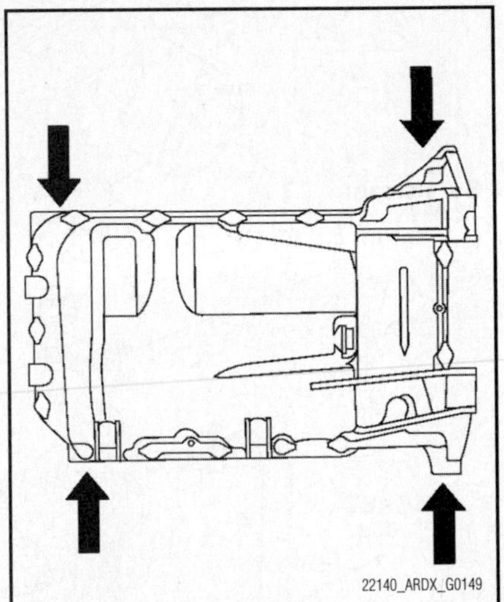
22140_ARDX_G0149

Fig. 76 Oil pan removal points

To install:

17. Clean the mating surfaces of the components to be sure they are free of dirt and debris.

18. Apply liquid gasket part number 08717-0004 or equivalent to the engine block mating surface of the oil pan and to the inside edge of the threaded bolt holes.

⁂ CAUTION

Install the component within five minutes of applying the liquid gasket. Some liquid gasket materials require installation within four minutes. Be sure to read and follow the manufacturer's instruction. If too much time has passed before installing the component remove the liquid gasket and residue, then reapply new liquid gasket.

19. Apply liquid gasket about 0.12 in. diameter bead along the broken line.

20. Install the oil pan.

21. Tighten the bolts in three steps. In the final step, tighten all bolts in the sequence shown in the illustration, to 8.8 ft. lbs. (11.9 Nm). Wipe off the excess liquid gasket on each side of the crankshaft pulley and drive plate.

➡**Wait at least 30 minutes to allow the gasket to cure before filling the engine with oil.**

⁂ CAUTION

Do not run the engine for at least three hours after installing the oil pan.

22. Continue the installation in the reverse order of the removal procedure.

23. When installing the subframe be sure to tighten the bolts as shown.

24. Fill the engine with the proper grade and type engine oil.

25. Turn the ignition switch ON (II), the SRS indicator should come on for about six seconds, and then go off.

26. Road test the vehicle, correct problems as required.

OIL PUMP

REMOVAL & INSTALLATION

See Figures 80 through 86.

1. Before servicing the vehicle, refer to the “PRECAUTIONS”.

2. Perform the battery disconnect/reconnect procedure, as outlined in the Engine Electrical Section.

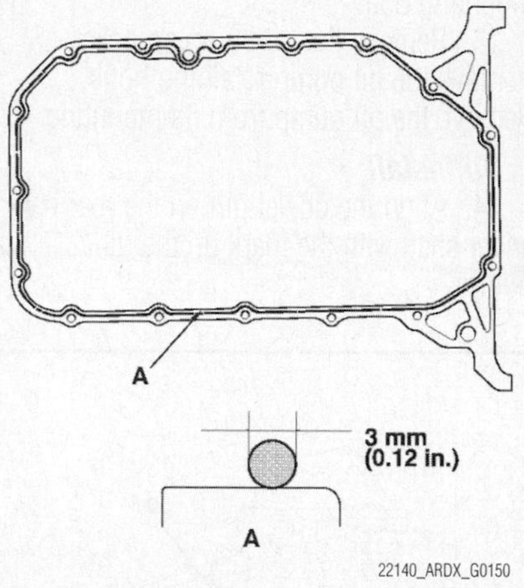
A
3 mm
(0.12 in.)
A
22140_ARDX_G0150

Fig. 77 Apply liquid gasket about 0.12 in. diameter bead along the broken line (A).

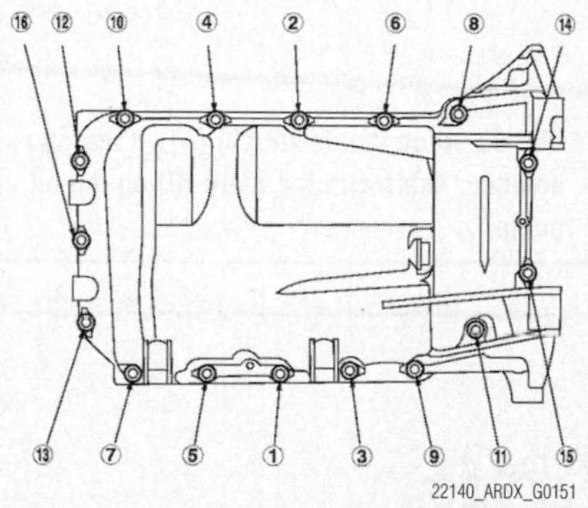
16 12 10 4 2 6 8 14
13 7 5 1 3 9 11 15
22140_ARDX_G0151

Fig. 78 Oil pan bolt tightening sequence

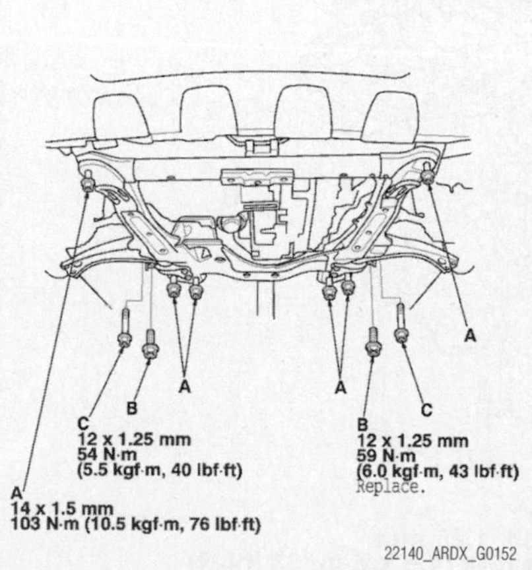
A
C B A
12 x 1.25 mm
54 N·m
(5.5 kgf·m, 40 lbf·ft)
A B C
12 x 1.25 mm
59 N·m
(6.0 kgf·m, 43 lbf·ft)
Replace.
A
14 x 1.5 mm
103 N·m (10.5 kgf·m, 76 lbf·ft)
22140_ARDX_G0152

Fig. 79 Subframe bolt tightening specifications

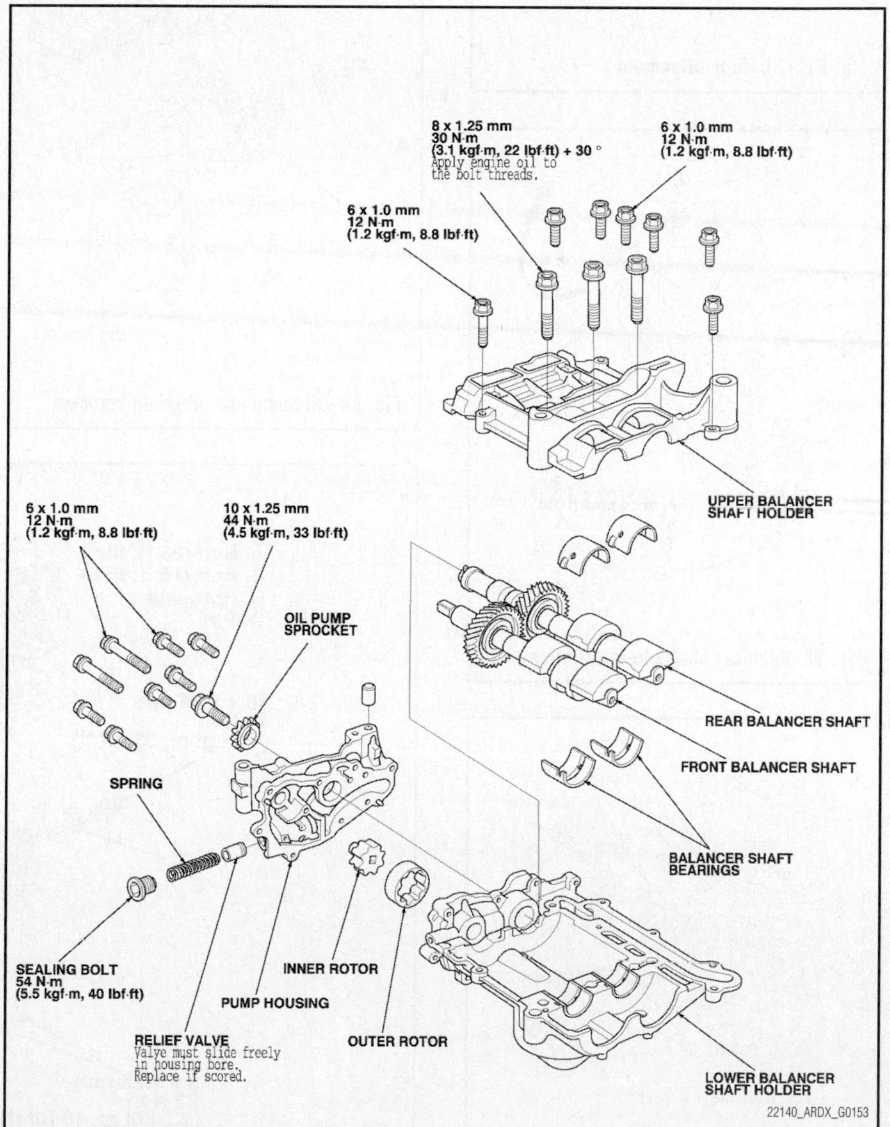
8 x 1.25 mm
30 N·m
(3.1 kgf·m, 22 lbf·ft) + 30 °
Apply engine oil to the bolt threads.
6 x 1.0 mm
12 N·m
(1.2 kgf·m, 8.8 lbf·ft)
6 x 1.0 mm
12 N·m
(1.2 kgf·m, 8.8 lbf·ft)
UPPER BALANCER SHAFT HOLDER
6 x 1.0 mm
12 N·m
(1.2 kgf·m, 8.8 lbf·ft)
10 x 1.25 mm
44 N·m
(4.5 kgf·m, 33 lbf·ft)
OIL PUMP SPROCKET
REAR BALANCER SHAFT
FRONT BALANCER SHAFT
SPRING
BALANCER SHAFT BEARINGS
SEALING BOLT
54 N·m
(5.5 kgf·m, 40 lbf·ft)
INNER ROTOR
PUMP HOUSING
RELIEF VALVE
Valve must slide freely in housing bore.
Replace if scored.
OUTER ROTOR
LOWER BALANCER SHAFT HOLDER
22140_ARDX_G0153

Fig. 80 Oil pump and related components

3. Turn the crankshaft pulley so that TDC mark (on the pulley) lines up with the pointer (on the engine).

4. Raise and support the vehicle safely. Remove the front tire and wheel assemblies.

5. Drain the engine oil.

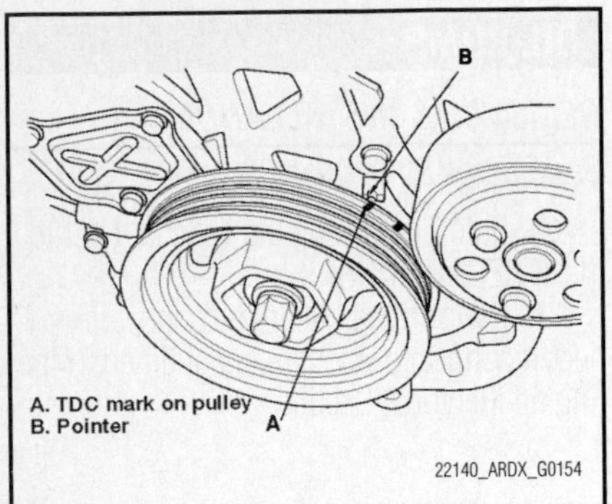
B
A. TDC mark on pulley
B. Pointer
A
22140_ARDX_G0154

Fig. 81 TDC mark alignment

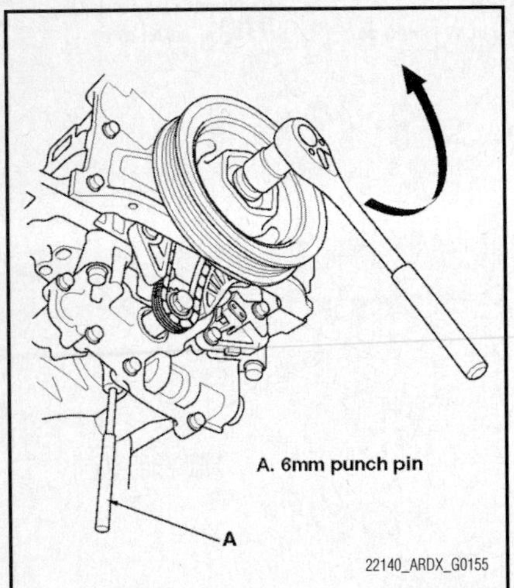
A. 6mm punch pin
A
22140_ARDX_G0155

Fig. 82 Balancer shaft pin installation

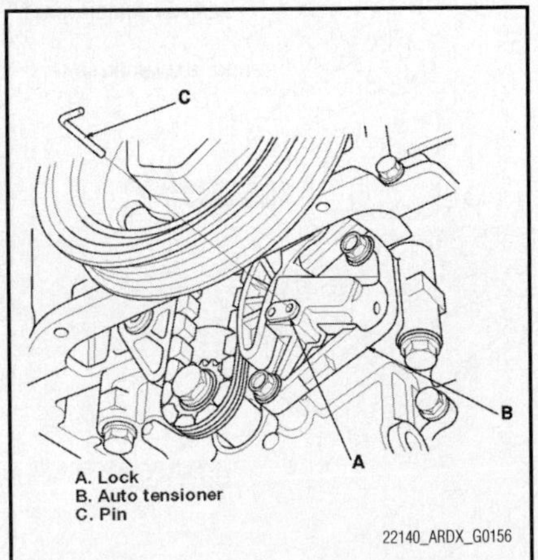
C
B
A
A. Lock
B. Auto tensioner
C. Pin
22140_ARDX_G0156

Fig. 83 Oil pump auto tensioner and related components

6. Remove the oil pan.

7. Loosely install the crankshaft pulley.

8. To hold the rear balancer shaft, insert a 6mm pin into the maintenance hole in the balancer shaft holder and through the rear balancer shaft.

9. Turn the crankshaft counterclockwise to compress the oil pump chain auto tensioner.

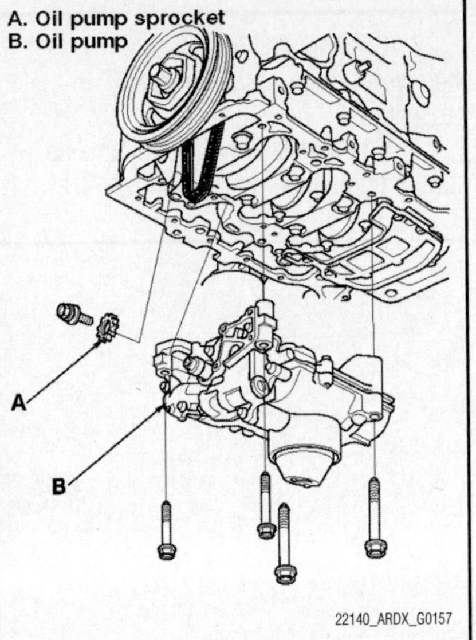
A. Oil pump sprocket
B. Oil pump
A
B
22140_ARDX_G0157

Fig. 84 Oil pump and oil pump sprocket

10. Align the holes on the lock and the oil pump chain auto tensioner. Insert a 0.06 in. diameter pin into the holes. Turn the crankshaft clockwise to secure the pin.

11. Remove the oil pump chain auto tensioner.

12. Loosen the oil pump sprocket mounting bolt.

13. Remove the oil pump sprocket. Remove the oil pump retaining bolts. Remove the oil pump from its mounting.

To install:

14. Align the dowel pin on the rear balancer shaft with the mark on the oil pump.

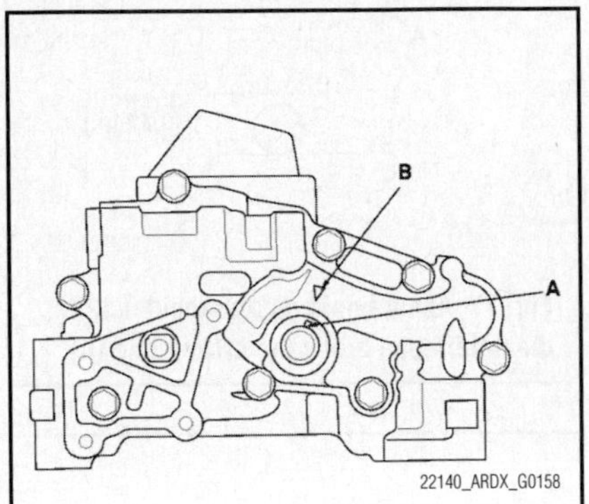
B
A
22140_ARDX_G0158

Fig. 85 Align the dowel pin (A) on the rear balancer shaft with the mark (B) on the oil pump.

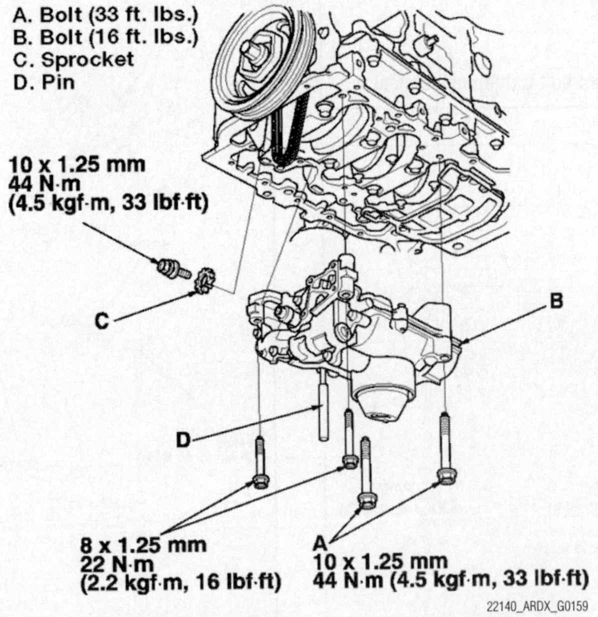
A. Bolt (33 ft. lbs.)
B. Bolt (16 ft. lbs.)
C. Sprocket
D. Pin
10 x 1.25 mm
44 N·m
(4.5 kgf·m, 33 lbf·ft)
C
B
D
8 x 1.25 mm
22 N·m
(2.2 kgf·m, 16 lbf·ft)
A
10 x 1.25 mm
44 N·m (4.5 kgf·m, 33 lbf·ft)
22140_ARDX_G0159

Fig. 86 Oil pump bolt identification and tightening specifications

15. To hold the balancer shaft, insert a 6mm pin into the maintenance hole in the lower balancer shaft holder and through the rear balancer shaft.

16. Apply clean engine oil to the bolt threads of the oil pump retaining bolts.

17. Loosely install the oil pump, then install the oil pump sprocket. Tighten the retaining bolt to 33 ft. lbs. (45 Nm).

18. Remove the pin driver.

19. Tighten the oil pump retaining bolts to specification. Tighten the 8x1.25mm bolts to 16 ft. lbs. (22 Nm). Tighten the 10x1.25mm bolts to 33 ft. lbs. (45 Nm).

20. Install the oil pump chain auto tensioner. Tighten the retaining bolts to 8.8 ft. lbs. (11.9 Nm). Remove the pin from the assembly.

21. Install the oil pan.

22. Continue the installation in the reverse order of the removal procedure.

23. Fill the engine with the proper grade and type engine oil.

24. Reconnect the battery. Perform the battery disconnect/reconnect procedure. See "BATTERY SYSTEM" section.

25. Road test the vehicle, correct problems as required.

INSPECTION

See Figures 87 through 89.

1. Remove the oil pump from the engine.

2. Remove the pump housing retaining bolts. Remove the pump housing.

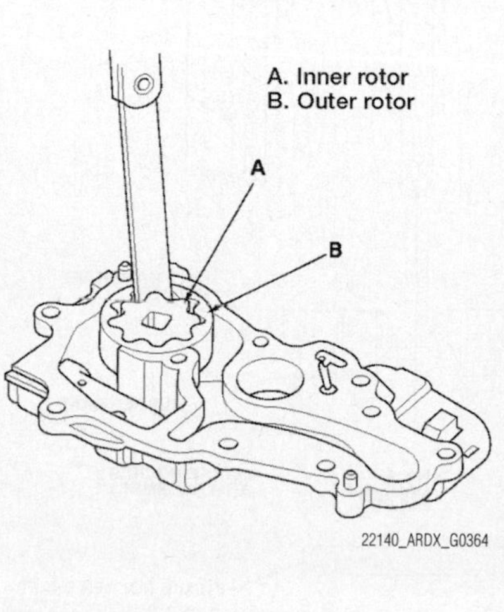
A. Inner rotor
B. Outer rotor
A
B
22140_ARDX_G0364

Fig. 87 Oil pump inner to outer rotor radial clearance check

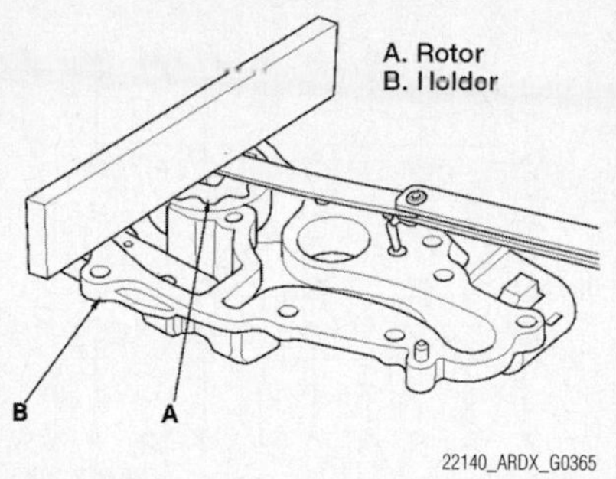
A. Rotor
B. Holder
B
A
22140_ARDX_G0365

Fig. 88 Oil pump housing to rotor axial clearance check

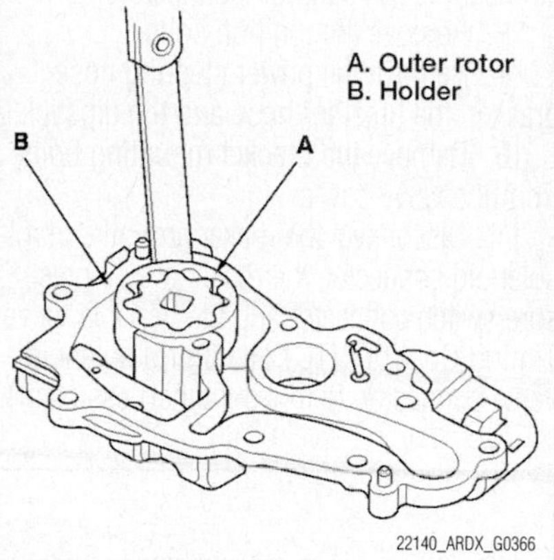
A. Outer rotor
B. Holder
B
A
22140_ARDX_G0366

Fig. 89 Oil pump housing to outer rotor clearance check

3. Check the inner to outer radial clearance between the inner rotor and the outer rotor. It specification exceeds the service limit, replace the oil pump. Standard (new) service specification is 0.004-0.007 in. Service limit specification is 0.008 in.

4. Check the housing to rotor axial clearance between the rotor and the lower balancer shaft holder. It specification exceeds the service limit, replace the oil pump. Standard (new) service specification is 0.0012-0.0030 in. Service limit specification is 0.005 in.

5. Check the housing to outer rotor clearance between the outer rotor and the lower balancer shaft holder. It specification exceeds the service limit, replace the oil pump. Standard (new) service specification is 0.0014-0.0035 in. Service limit specification is 0.009 in.

PISTON & RING

See Figure 90

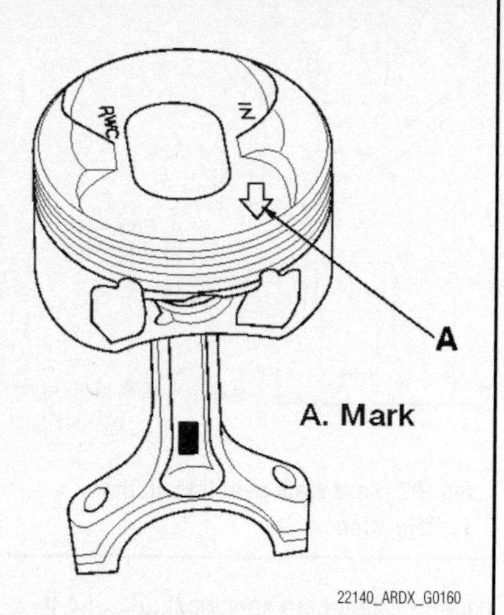
A
A. Mark
22140_ARDX_G0160

Fig. 90 Piston identification mark

REAR MAIN SEAL

REMOVAL & INSTALLATION

See Figures 91 and 92.

1. Remove the transmission.

2. Remove the drive plate.

3. Remove the seal from its mounting, using the proper seal removal tool.

To install:

4. Clean and dry the crankshaft oil seal housing.

5. Use the driver tool and the attachment to drive a new seal squarely into the block, to the specified installed height.

6. Measure the distance between the engine block and the oil seal. Specified height should be 0.001–0.047 in.

7. Install the drive plate. Tighten the bolts to specification and in a crisscross

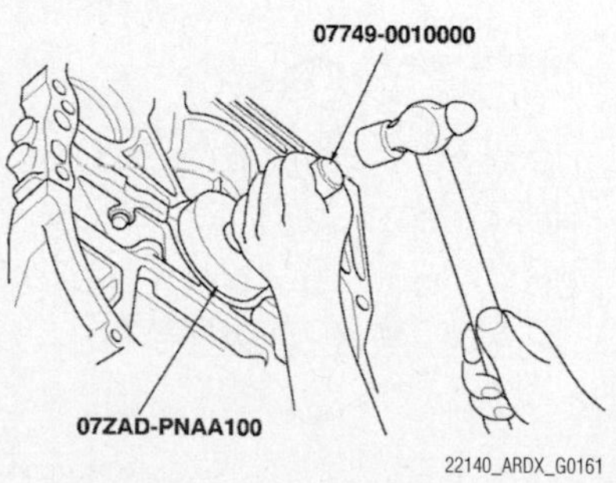
07749-0010000
07ZAD-PNAA100
22140_ARDX_G0161

Fig. 91 Rear main seal installation

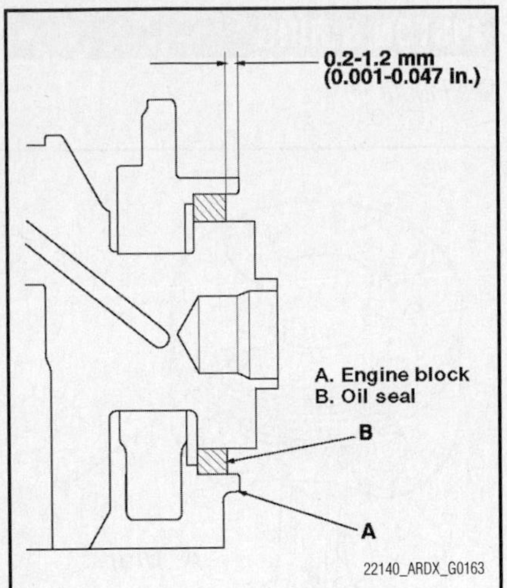
0.2-1.2 mm
(0.001-0.047 in.)
A. Engine block
B. Oil seal
B
A
22140_ARDX_G0163

Fig. 92 Rear main seal installation specification

pattern. Tightening specification is 54 ft. lbs. (73 Nm).

8. Continue the installation in the reverse order of the removal procedure.

9. Road test the vehicle, correct problems as required.

ROCKER ARMS

REMOVAL & INSTALLATION

See Figures 93 through 102.

1. Before servicing the vehicle, refer to the "PRECAUTIONS".

2. Perform the battery disconnect/reconnect procedure, as outlined in the Engine Electrical Section.

3. Remove the air charge cooler cover.

4. Disconnect the turbocharger boost sensor connector. Remove the vacuum hoses and the air bypass outlet connecting tube.

5. Remove the air charge cooler retaining bolts. Remove the air charge cooler.

6. Remove the fuel injector cover.

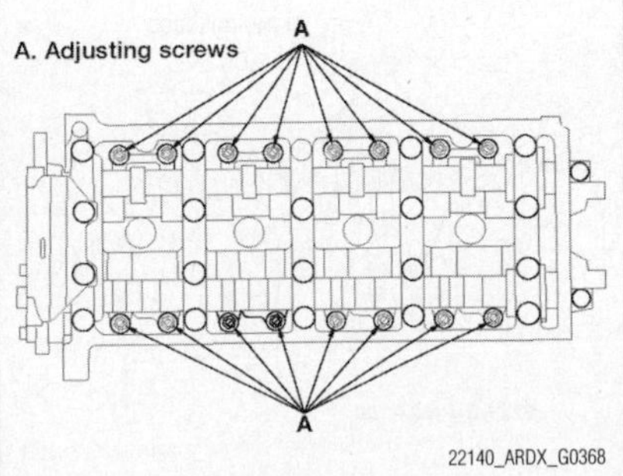
A. Adjusting screws
A
A
22140_ARDX_G0368

Fig. 93 Rocker arm adjusting screws

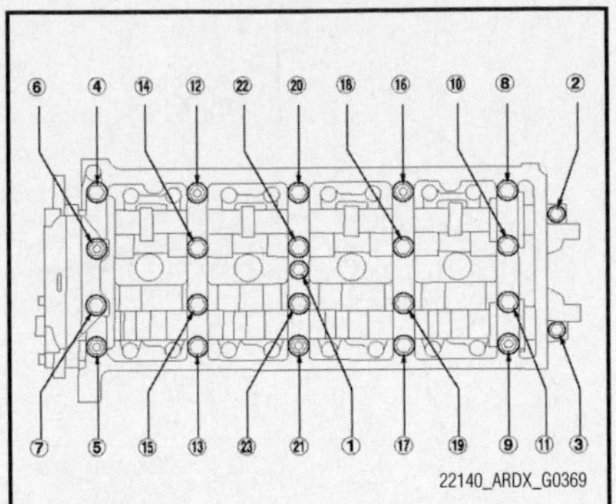
6 4 14 12 22 20 18 16 10 8 2
7 5 15 13 23 21 1 17 19 9 11 3
22140_ARDX_G0369

Fig. 94 Insert the bolts (A) into the rocker shaft holder. Remove the rocker arm shaft assembly.

7. Remove the ignition coil cover. Disconnect the ignition coil connectors.

8. Remove the ignition coils.

9. Remove the power steering hose bracket, the breather hose and the dipstick.

10. Remove the bracket mounting bolts from the valve cover.

11. Disconnect the rocker arm oil control solenoid connector, the rocker are oil pressure switch connector and the Variable Valve Timing Control (VTC) oil control solenoid valve connector. Remove the harness clamp.

12. Disconnect the Air/Fuel ratio (A/F) sensor connector, the oil pressure switch connector and the Crankshaft Position (CKP) sensor. Remove the harness clamps.

13. Remove the valve cover retaining bolts.

14. Remove the valve cover from the engine. Discard the gasket.

15. Remove the timing chain.

16. Loosen the rocker arm adjusting screws.

17. Remove the camshaft holder bolts.

CAUTION

To prevent damage to the camshafts, loosen all bolts in sequence two turns at a time. Bolt No. 1 is not used on all engines.

18. Remove the cam chain guide B, the camshaft holders and the camshafts.

19. Insert the bolts into the rocker shaft holder. Remove the rocker arm shaft assembly.

To install:

20. If disassembled, reassemble the rocker arm assembly.

21. Clean and dry the No. five rocker shaft holder mating surface.

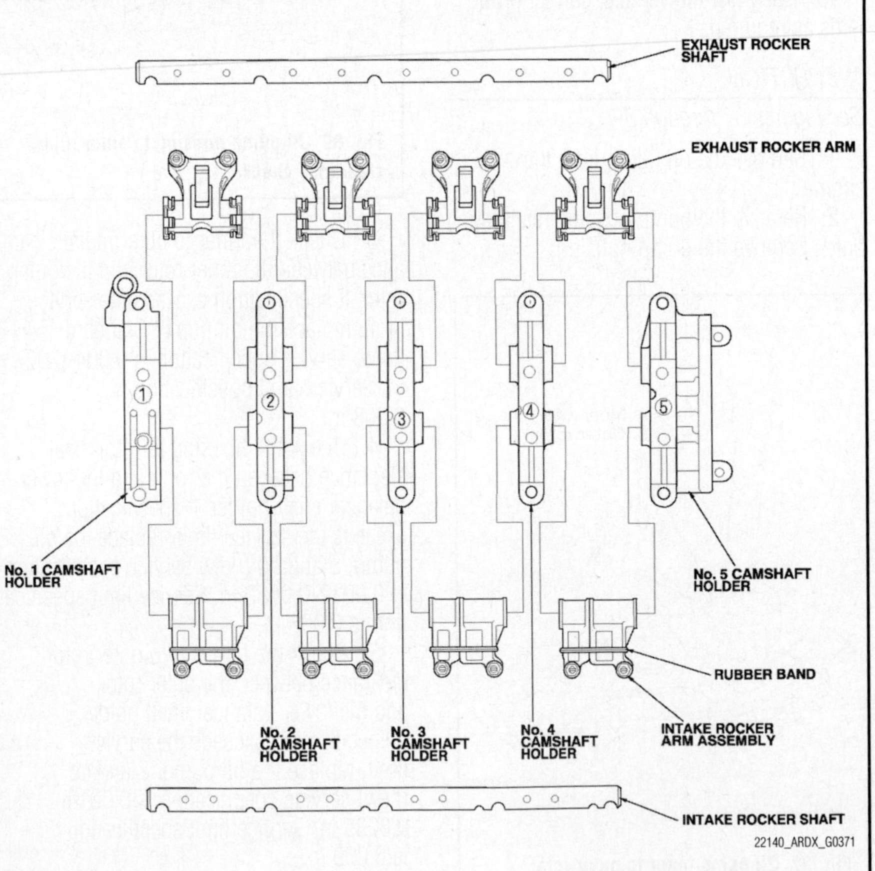
EXHAUST ROCKER SHAFT
EXHAUST ROCKER ARM
No. 1 CAMSHAFT HOLDER
No. 5 CAMSHAFT HOLDER
RUBBER BAND
INTAKE ROCKER ARM ASSEMBLY
No. 2 CAMSHAFT HOLDER
No. 3 CAMSHAFT HOLDER
No. 4 CAMSHAFT HOLDER
INTAKE ROCKER SHAFT
22140_ARDX_G0371

Fig. 95 Rocker arm assembly exploded view

22. Apply liquid gasket to the cylinder head mating surface of the No. five rocker shaft holder and to the inner threads of the bolt holes.

➡Install the component within five minutes of applying the liquid gasket. Some liquid gasket materials require installation within four minutes. Be sure to read and follow the manufacturer's instruction. If too much time has passed before installing the component remove the liquid gasket and residue, then reapply new liquid gasket.

23. Apply the liquid gasket about ½ inch diameter bead along the broken line indicated.
24. Install the assembly onto the cylinder head, using the bolts that you used to remove the assembly from the cylinder head during the removal procedure. Remove the bolts.
25. Make sure that the punch marks on the VTC actuator and the exhaust camshaft sprocket face up. Position the camshafts in the holder.
26. Position the camshaft holders and the cam chain guide into position.
27. Tighten the bolts to specification and in the proper sequence.
28. Tighten 8x1.25mm bolts to 16 ft. lbs. (22 Nm), tighten 6x1.0mm bolts to 8.8 ft. lbs. (11.9 Nm) and tighten 6x1.0mm bolts 21, 22, 23.
29. Install the timing chain.
30. Clean the mating surfaces of the cylinder head and valve cover.
31. Install a new gasket in the groove of the valve cover.
32. Be sure that the mating surfaces are clean and dry.

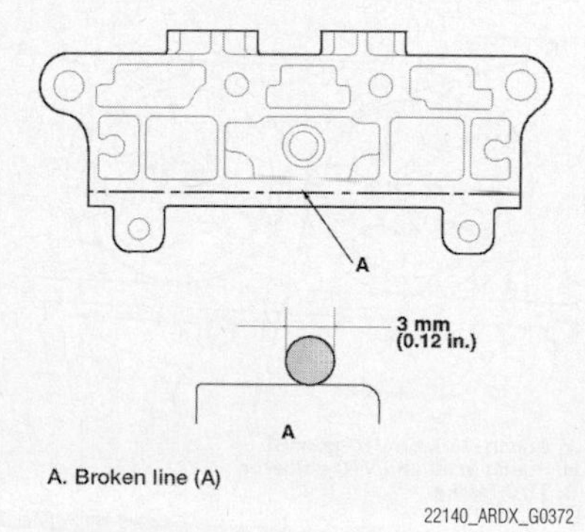
A
3 mm
(0.12 in.)
A
A. Broken line (A)
22140_ARDX_G0372

Fig. 96 Apply the liquid gasket about ½ inch diameter bead along the broken line indicated (A).

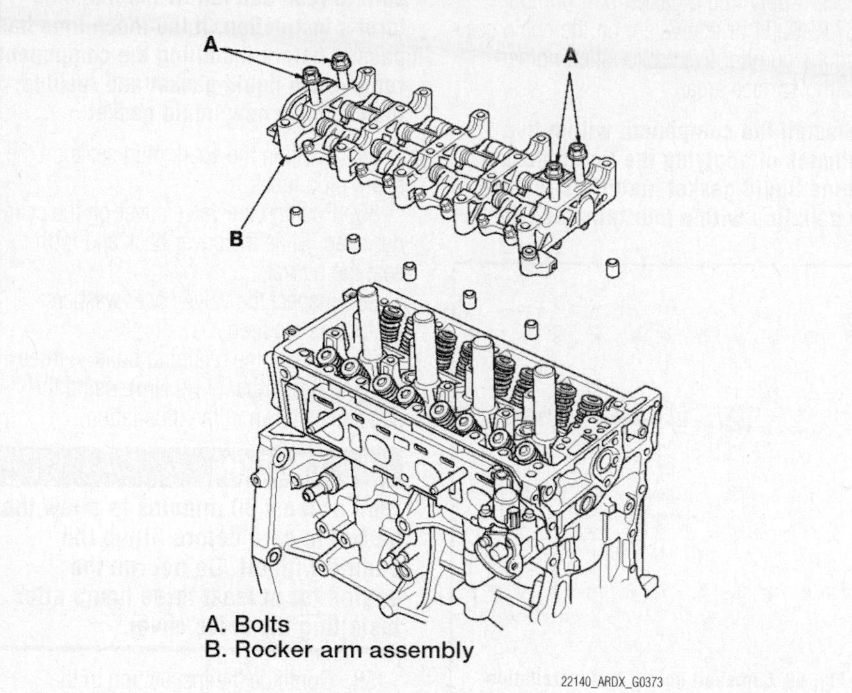
A
A
B
A. Bolts
B. Rocker arm assembly
22140_ARDX_G0373

Fig. 97 Install the assembly onto the cylinder head, using the bolts (A) that you used to remove the assembly from the cylinder head during the removal procedure. Remove the bolts (A).

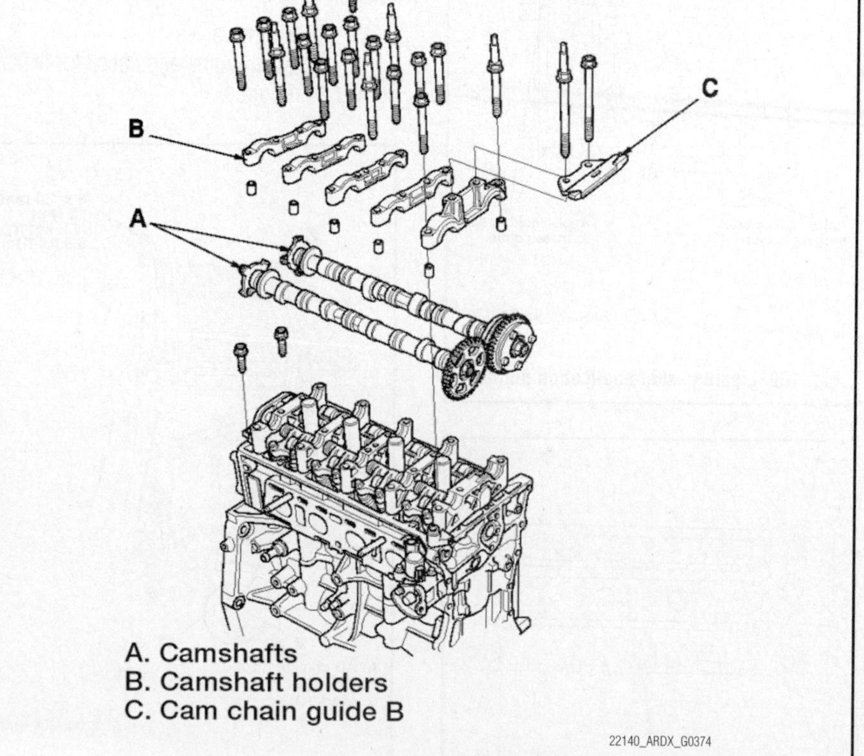
B
C
A
A. Camshafts
B. Camshaft holders
C. Cam chain guide B
22140_ARDX_G0374

Fig. 98 Position the camshaft holders and the cam chain guide (B) into position.

33. Apply liquid gasket part number 08717-0004 or equivalent on the chain case and the number five rocker shaft holder mating surface areas.

➡Install the component within five minutes of applying the liquid gasket. Some liquid gasket materials require installation within four minutes. Be sure to read and follow the manufacturer's instruction. If too much time has passed before installing the component remove the liquid gasket and residue, then reapply new liquid gasket.

34. Position the spark plug seals on the spark plug tubes.
35. Position the valve cover on the cylinder head. Slide the cover back and forth to seat the gasket.
36. Inspect the valve cover washers, replace as required.
37. Tighten the retaining bolts in three steps to 8.8 ft. lbs. (11.9 Nm), using the sequence shown in the illustration.

CAUTION

Wait at least 30 minutes to allow the gasket to cure before filling the engine with oil. Do not run the engine for at least three hours after installing the valve cover.

38. Continue the installation in the reverse order of the removal procedure.
39. Install the air charge cooler and connect the intake air ducts.
40. Tighten the hose bands until edge of the band aligns with the mark on the band.

➡If the hose band edge exceeds the mark, replace the hose band.

41. Adjust the valves.
42. Start the engine and check for leaks, correct as required.

43. Turn the ignition switch ON (II), the SRS indicator should come on for about six seconds, and then go off.

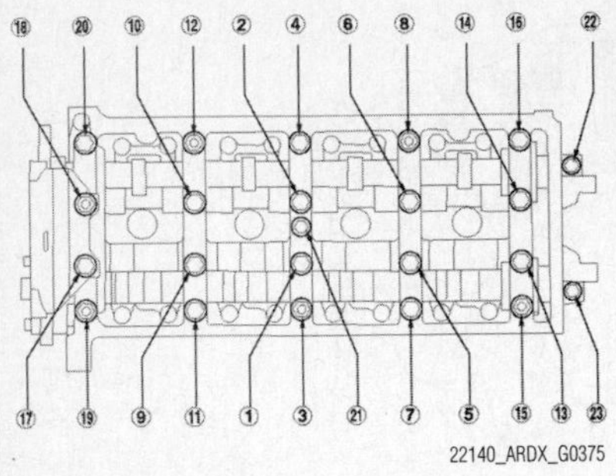
22140_ARDX_G0375

Fig. 99 Camshaft holder bolt installation sequence

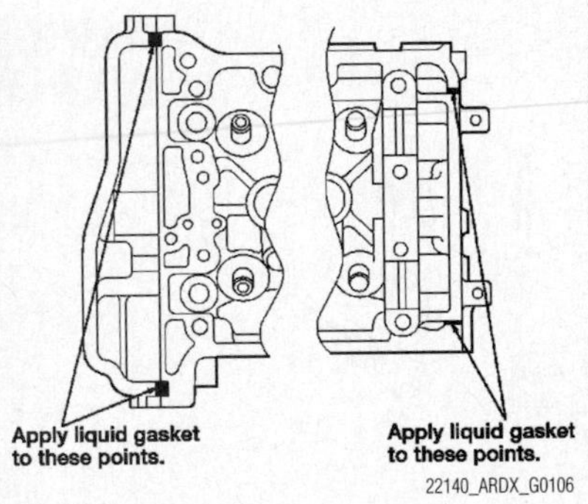
Apply liquid gasket to these points.
Apply liquid gasket to these points.
22140_ARDX_G0106

Fig. 100 Liquid gasket application points

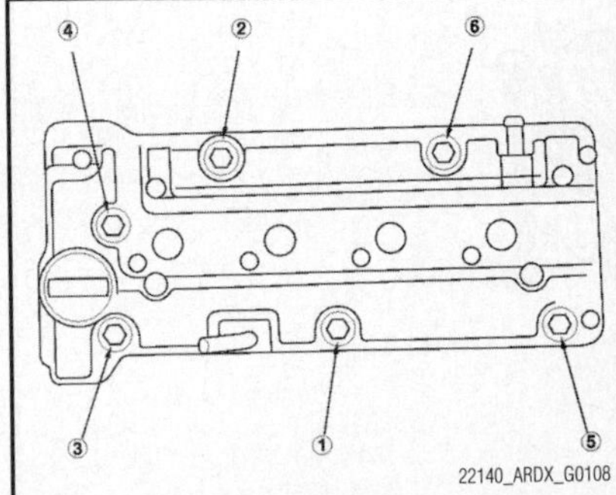
22140_ARDX_G0108

Fig. 101 Valve cover torque sequence

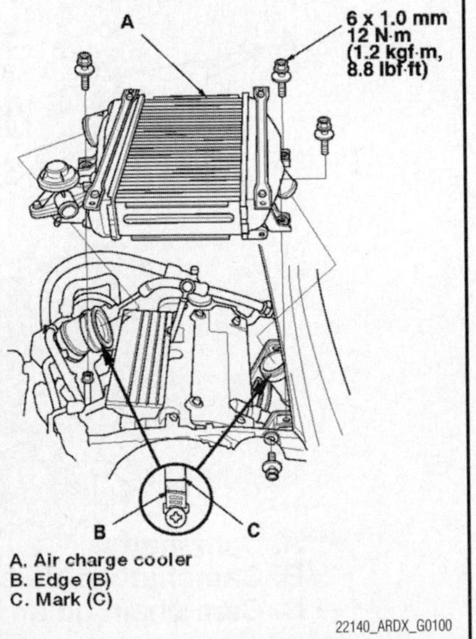
A
6 x 1.0 mm
12 N·m
(1.2 kgf·m,
8.8 lbf·ft)
B
C
A. Air charge cooler
B. Edge (B)
C. Mark (C)
22140_ARDX_G0100

Fig. 102 Tighten the hose bands until edge (B) of the band aligns with the mark (C) on the band.

TIMING CHAIN FRONT COVER

REMOVAL & INSTALLATION

See Figures 103 through 113.

1. Before servicing the vehicle, refer to the "PRECAUTIONS".
2. Perform the battery disconnect/ reconnect procedure, as outlined in the Engine Electrical Section.
3. Remove the front splash shield.
4. Remove the drive belt.
5. Remove the valve cover.
6. Position the number one piston to TDC. The punch mark on the VTC actuator and the punch mark on the exhaust camshaft sprocket should be at the top. Align the TDC marks on the VTC actuator and the exhaust camshaft sprocket.
7. Disconnect the VTC oil control solenoid valve 2P connector. Remove the bolt and the VTC oil control solenoid valve.

➡Check the VTC oil control solenoid valve strainer for clogging. If the strainer is clogged, replace the control solenoid.

8. Remove the crankshaft pulley retaining bolt. Remove the crankshaft pulley.
9. Properly support the engine with a suitable jack and block of wood under the oil pan.
10. Remove the torque rod stiffener and upper torque rod.
11. Remove the ground cable. Remove the side engine mount bracket.
12. Remove the water bypass mounting bolt. Remove the engine mount bracket.

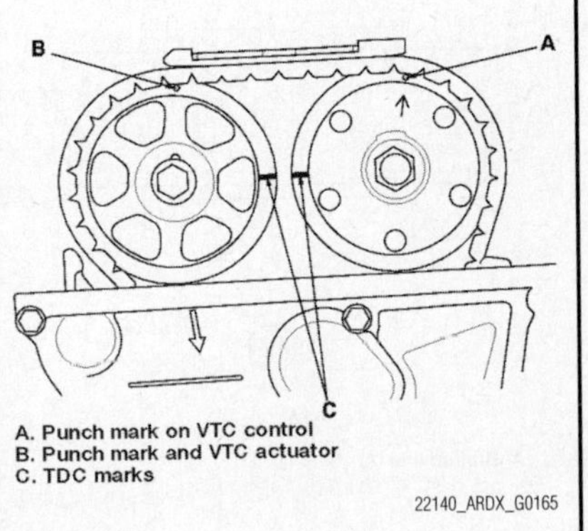
B
A
C
A. Punch mark on VTC control
B. Punch mark and VTC actuator
C. TDC marks
22140_ARDX_G0165

Fig. 103 Camshaft sprocket positioning at TDC

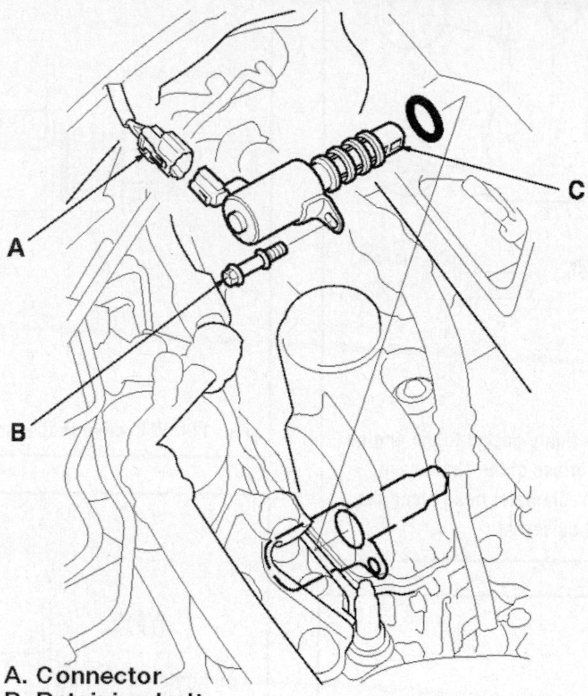
A
B
C
A. Connector
B. Retaining bolt
C. VTC valve
22140_ARDX_G0166

Fig. 104 VTC solenoid valve location

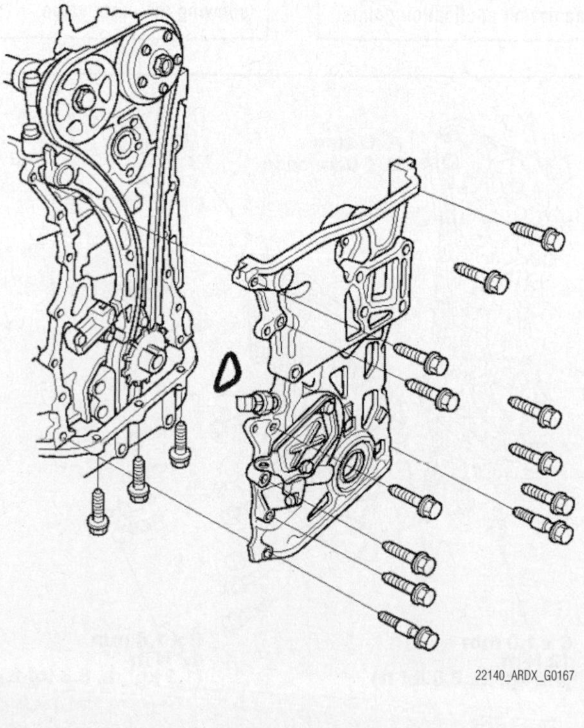
22140_ARDX_G0167

Fig. 105 Timing chain cover and retaining bolt locations

13. Remove the timing chain case cover retaining bolts. Remove the cover from its mounting on the engine

To install:

14. Installation is the reverse of the removal procedure.
15. Replace the timing cover oil seal. Be sure to use the proper seal installation tools.
16. Measure the distance between the chain case surface and the oil seal. Specification should be 1.30–1.33 in.
17. Remove all the old liquid gasket material from the chain case mating surfaces, the bolts and the bolts holes of all mating surfaces. Clean and dry the chain case mating surfaces.
18. Apply liquid gasket to the engine block mating surface of the chain case. Apply a 0.12 in. diameter bead along the broken line, as shown.

➡Install the component within five minutes of applying the liquid gasket. Some liquid gasket materials require installation within four minutes. Be sure to read and follow the manufacturer's instruction. If too much time has passed before installing the component remove the liquid gasket and residue, then reapply new liquid gasket.

19. Apply liquid gasket to the engine block upper surface contact areas on the chain case and on the lower block upper surface contact areas of the chain case. Apply a 0.43 in. diameter and about 0.12 in. thickness of gasket.
20. Apply liquid gasket to the oil pan mating surface of the chain case. Apply a 0.12 in. diameter bead along the broken line as shown.

⁂ CAUTION

Install the component within five minutes of applying the liquid gasket. Some liquid gasket materials require installation within four minutes. Be sure to read and follow the manufacturer's instruction. If too much time has passed before installing the component remove the liquid gasket and residue, then reapply new liquid gasket.

21. Install a new O-ring on the chain case. Set the edge of the chain case to the edge of the oil pan. Install the chain case on the engine block. Wipe any excess liquid gasket on the oil pan and chain case mating

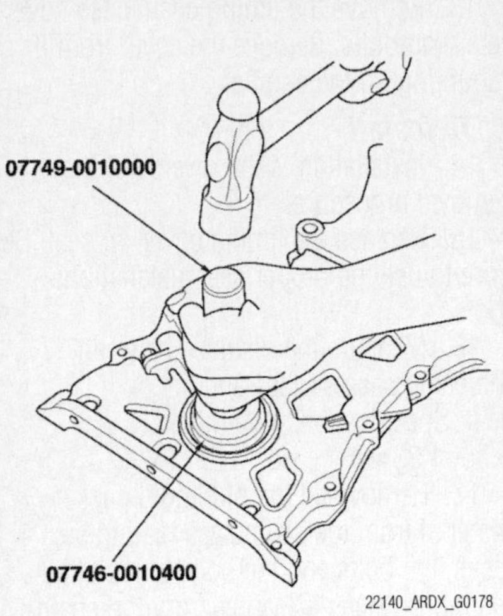
07749-0010000
07746-0010400
22140_ARDX_G0178

Fig. 106 Timing chain cover seal installation

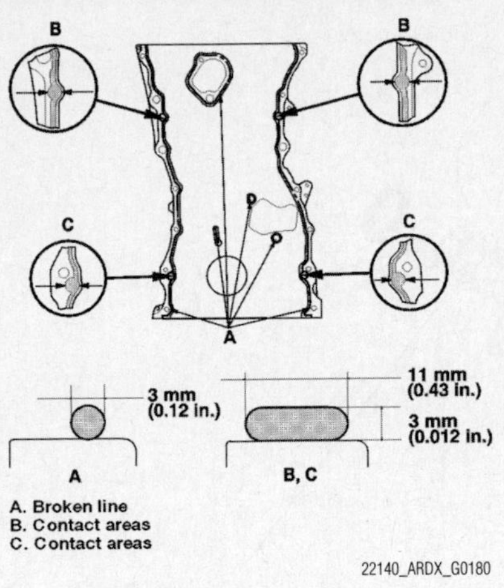
B
B
C
C
D
A
3 mm
(0.12 in.)
11 mm
(0.43 in.)
3 mm
(0.012 in.)
A
B, C
A. Broken line
B. Contact areas
C. Contact areas
22140_ARDX_G0180

Fig. 108 Apply liquid gasket to the engine block mating surface of the chain case. Apply a 0.12 in. diameter bead along the broken line (A) as shown.

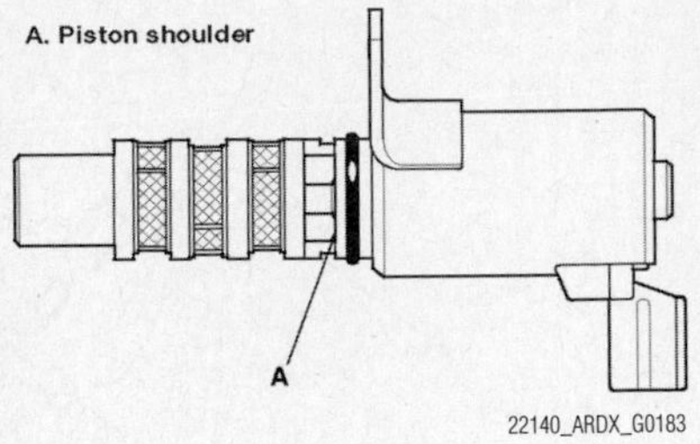
A. Piston shoulder
A
22140_ARDX_G0183

Fig. 111 VTC valve in the closed position

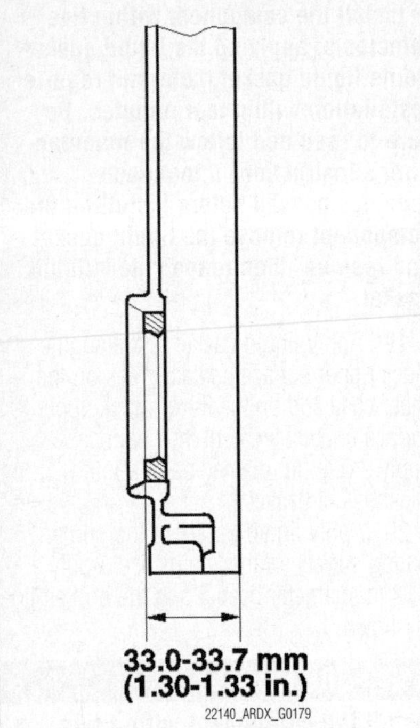
33.0-33.7 mm
(1.30-1.33 in.)
22140_ARDX_G0179

Fig. 107 Timing chain cover seal positioning

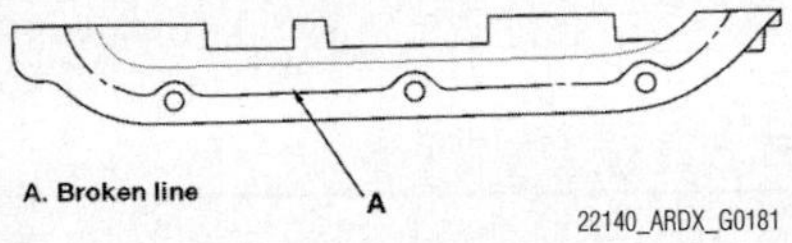
A. Broken line
A
22140_ARDX_G0181

Fig. 109 Oil pan gasket application points

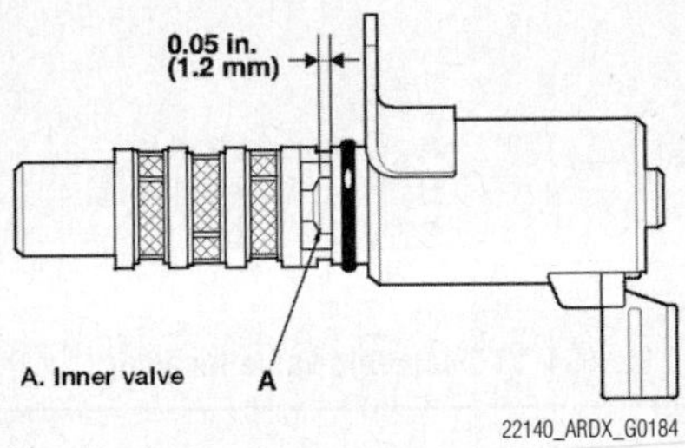
0.05 in.
(1.2 mm)
A. Inner valve
A
22140_ARDX_G0184

Fig. 112 VTC valve in the open position, showing the inner valve

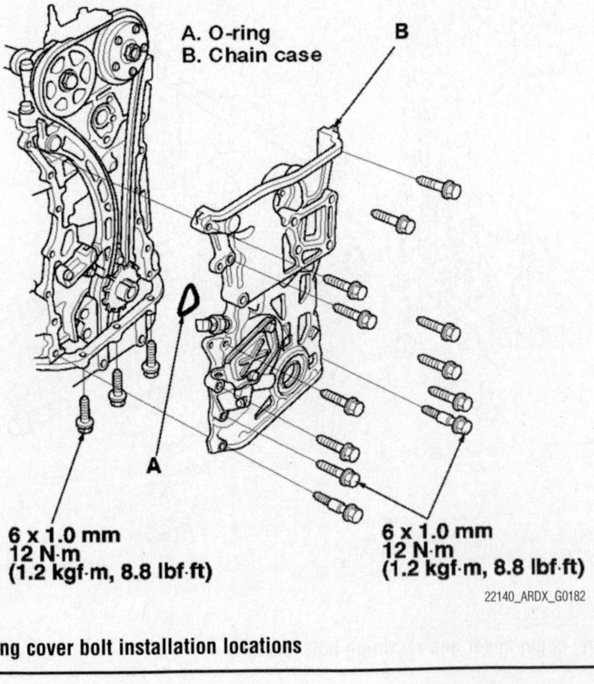
A. O-ring
B. Chain case
B
A
6 x 1.0 mm
12 N·m
(1.2 kgf·m, 8.8 lbf·ft)
6 x 1.0 mm
12 N·m
(1.2 kgf·m, 8.8 lbf·ft)
22140_ARDX_G0182

Fig. 110 Timing cover bolt installation locations

surfaces. When installing the chain case, do not slide the bottom surface onto the oil pan mounting surface.

➡Wait at least 30 minutes to allow the gasket to cure before filling the engine with oil.

⁂ CAUTION

Do not run the engine for at least three hours after installing the valve cover.

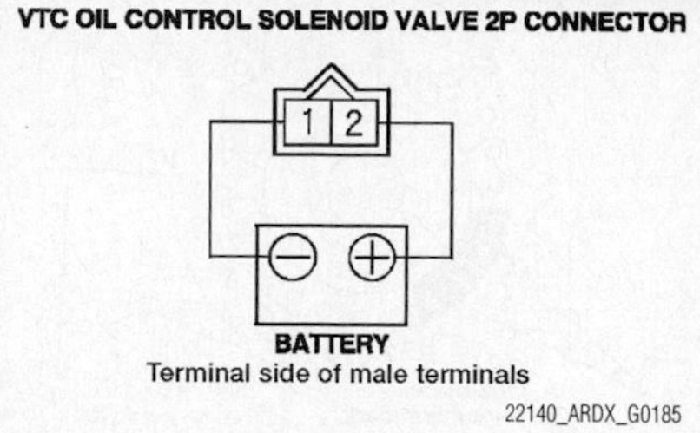
VTC OIL CONTROL SOLENOID VALVE 2P CONNECTOR
1 2
BATTERY
Terminal side of male terminals
22140_ARDX_G0185

Fig. 113 VTC terminal positioning and connection points

22. Tighten the cover retaining bolts to specification. Tightening specification is 8.8 ft. lbs. (11.9 Nm).
23. Continue the installation in the reverse order of the removal procedure.
24. When installing the VTC valve note the amount of valve opening by observing the position of the shoulder piston, thru the valve retard drain valve. If you see the shoulder of the piston, the valve is open and must be replaced.
25. Connect the battery positive terminal of the VTC terminal to the solenoid valve 2P connector terminal number two. Connect the battery negative terminal to the VTC solenoid valve 2P connector terminal number one.
26. Appearance of the inner valve in the port should be at least 0.05 in. If the inner valve does not open, replace it and repeat the procedure.
27. Perform the CKP pattern clear/relearn procedure as follows:
 a. Start the engine.
 b. Hold the engine speed at 3000 RPM, without a load in either PARK or NEUTRAL until the radiator fan comes on.
 c. Test drive the vehicle on a level road.
 d. Decelerate with the throttle fully closed from an engine speed of 2500 RPM down to 1000 RPM with the transmission in the S position 1st or 2nd gear.
 e. Repeat the above step several times.
 f. Turn the ignition switch to LOCK (0).
 g. Turn the ignition switch to ON (II) and wait thirty seconds.

TIMING CHAIN & SPROCKETS

REMOVAL & INSTALLATION

See Figures 114 through 128.

1. Before servicing the vehicle, refer to the "PRECAUTIONS".
2. Perform the battery disconnect/reconnect procedure, as outlined in the Engine Electrical Section.
3. Remove the front splash shield.
4. Remove the drive belt.
5. Remove the valve cover.
6. Position the number one piston to TDC. The punch mark on the VTC actuator and the punch mark on the exhaust camshaft sprocket should be at the top. Align the TDC marks on the VTC actuator and the exhaust camshaft sprocket.
7. Disconnect the VTC oil control solenoid valve 2P connector. Remove the bolt and the VTC oil control solenoid valve.

➡Check the VTC oil control solenoid valve strainer for clogging. If the strainer is clogged, replace the control solenoid.

8. Remove the crankshaft pulley retaining bolt. Remove the crankshaft pulley.
9. Properly support the engine with a suitable jack and block of wood under the oil pan.
10. Remove the torque rod stiffener and upper torque rod.
11. Remove the ground cable. Remove the side engine mount bracket.
12. Remove the water bypass mounting bolt. Remove the engine mount bracket.
13. Remove the timing chain case cover retaining bolts. Remove the cover from its mounting on the engine.

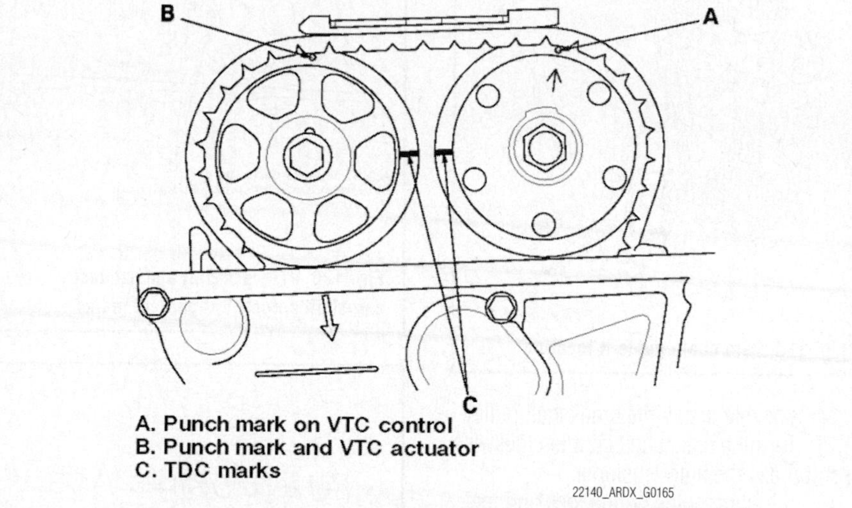
B
A
C
A. Punch mark on VTC control
B. Punch mark and VTC actuator
C. TDC marks
22140_ARDX_G0165

Fig. 114 Camshaft sprocket positioning at TDC

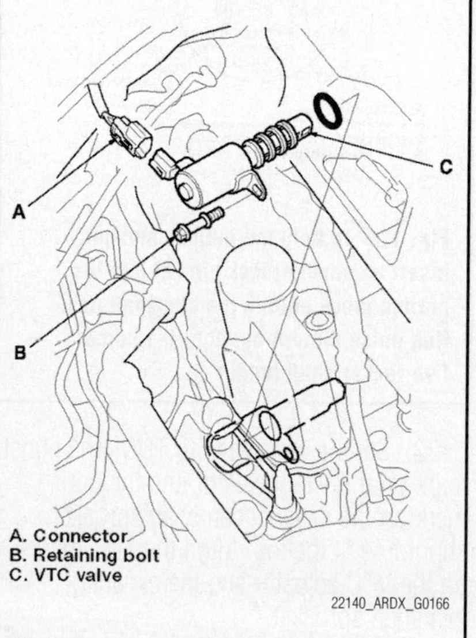
A
B
C
A. Connector
B. Retaining bolt
C. VTC valve
22140_ARDX_G0166

Fig. 115 VTC solenoid valve location

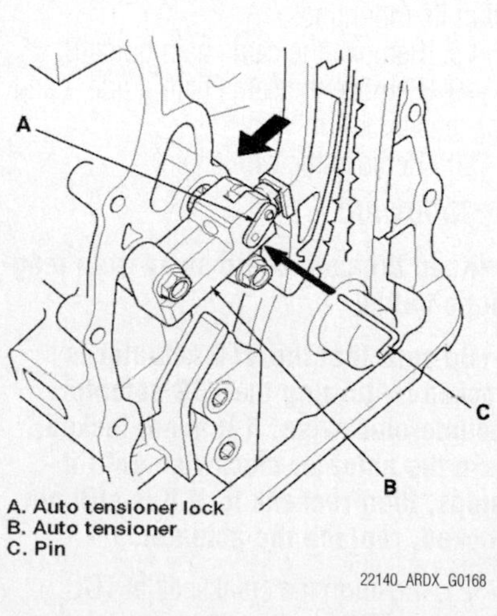
A
B
C
A. Auto tensioner lock
B. Auto tensioner
C. Pin
22140_ARDX_G0168

Fig. 116 Auto tensioner pin Installation

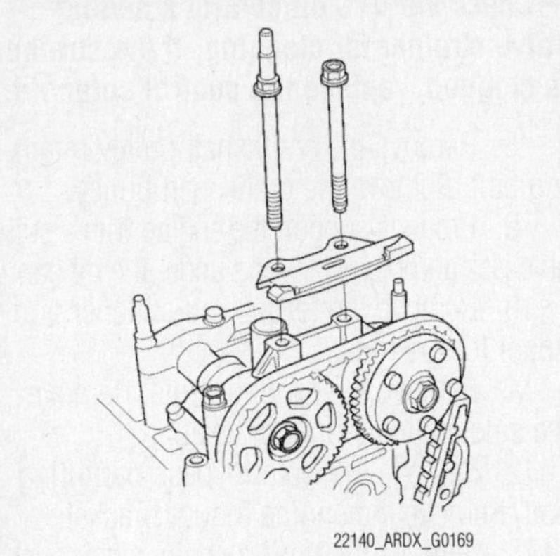
22140_ARDX_G0169

Fig. 117 Cam chain guide B location

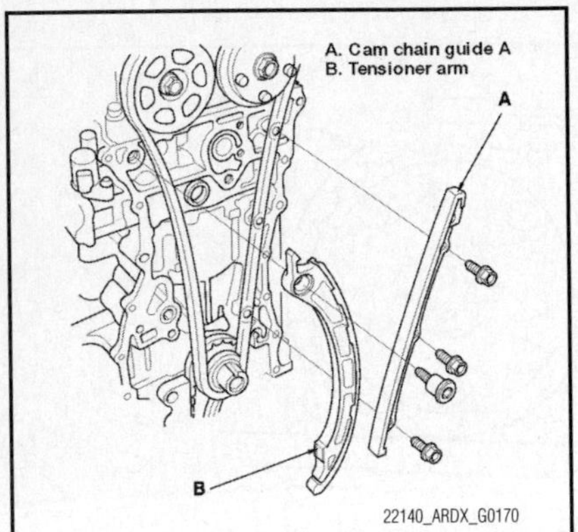
A. Cam chain guide A
B. Tensioner arm
A
B
22140_ARDX_G0170

Fig. 118 Cam chain guide A location

14. Loosely install the crankshaft pulley.

15. Turn the crankshaft counterclockwise to compress the auto tensioner.

16. Align the holes on the lock and the auto tensioner. Insert a 0.06 in. diameter pin into the holes. Turn the crankshaft clockwise to secure the pin.

17. Remove the auto tensioner assembly retaining bolts. Remove the auto tensioner from its mounting.

18. Remove the cam chain guide B.

19. Remove the cam chain guide A and the tensioner arm.

20. Remove the cam chain.

To install:

➡Keep the cam chain away from magnetic fields.

➡Be sure that the VTC actuator is locked by turning the VTC actuator counterclockwise. If it is not locked, turn the actuator clockwise until it stops, then recheck it. If it is still not locked, replace the actuator.

21. Position the crankshaft at TDC. Align the TDC mark on the crankshaft sprocket with the pointer on the engine block.

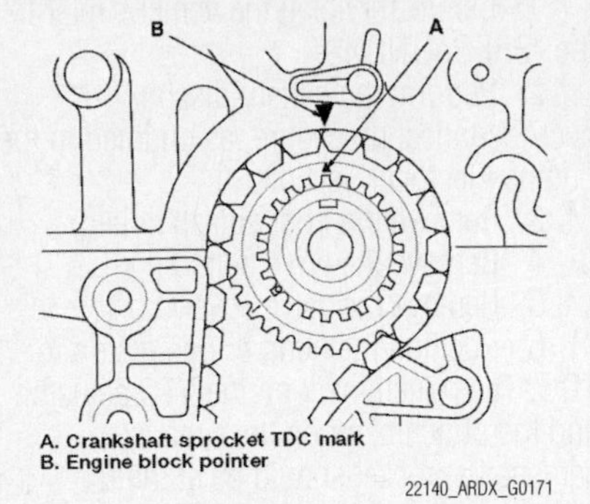
B
A
A. Crankshaft sprocket TDC mark
B. Engine block pointer
22140_ARDX_G0171

Fig. 119 Crankshaft sprocket positioning at TDC

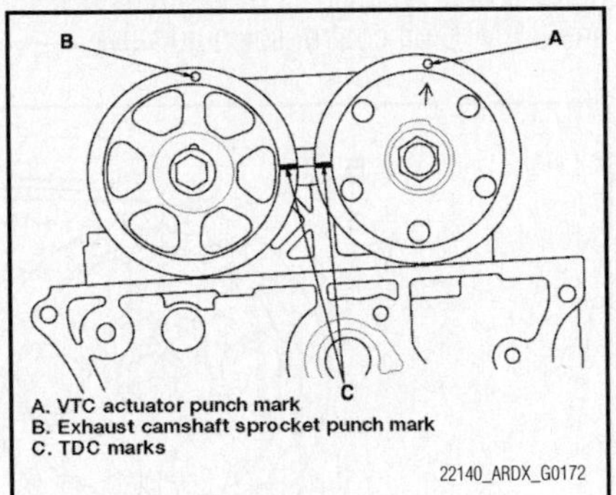
B
A
C
A. VTC actuator punch mark
B. Exhaust camshaft sprocket punch mark
C. TDC marks
22140_ARDX_G0172

Fig. 120 VTC camshaft and exhaust camshaft sprocket alignment marks

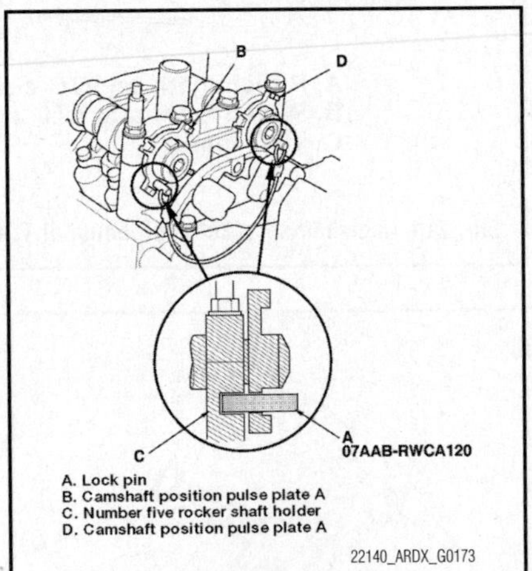
B
D
C
A
07AAB-RWCA120
A. Lock pin
B. Camshaft position pulse plate A
C. Number five rocker shaft holder
D. Camshaft position pulse plate A
22140_ARDX_G0173

Fig. 121 To hold the intake camshaft, insert a camshaft lock pin (A) into the maintenance hole in the camshaft position pulse plate A and thru the number five rocker shaft holder

22. Set the camshafts to TDC. The punch mark on the VTC actuator and the punch mark on the exhaust camshaft sprocket should be at the top. Align the TDC marks on the VTC actuator and the exhaust camshaft sprocket.

23. To hold the intake camshaft, insert a camshaft lock pin into the maintenance hole

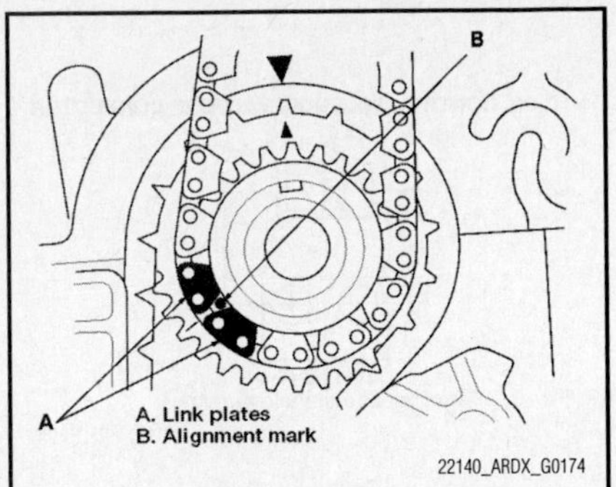
B
A
A. Link plates
B. Alignment mark
22140_ARDX_G0174

Fig. 122 Timing chain link plate location and identification at the crankshaft

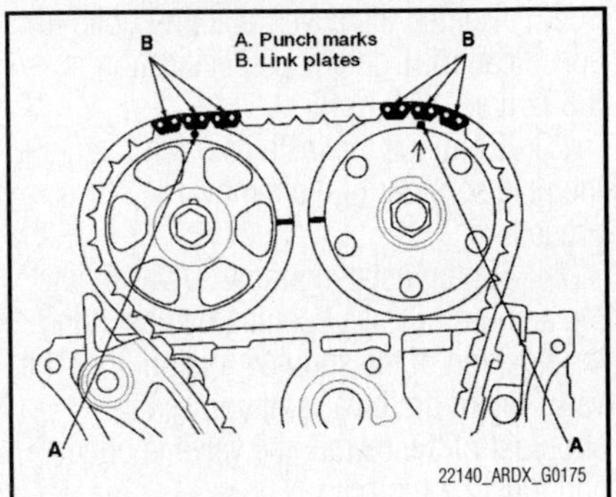
A. Punch marks
B. Link plates
B
B
A
A
22140_ARDX_G0175

Fig. 123 Timing chain link plate location and identification at the camshaft

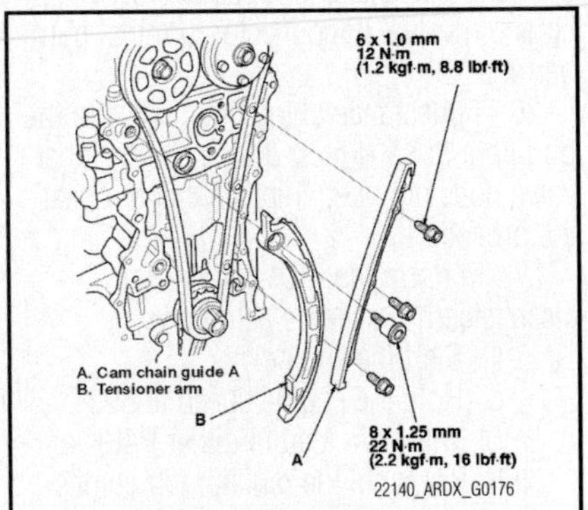
6 x 1.0 mm
12 N·m
(1.2 kgf·m, 8.8 lbf·ft)
A. Cam chain guide A
B. Tensioner arm
B
A
8 x 1.25 mm
22 N·m
(2.2 kgf·m, 16 lbf·ft)
22140_ARDX_G0176

Fig. 124 Timing chain guide tightening specifications

in the camshaft position pulse plate A and thru the number five rocker shaft holder.

24. Install the cam chain on the crankshaft sprocket with the center of the two colored link plates aligned with the mark on the crankshaft sprocket.

25. Install the chain on the VTC actuator and the exhaust camshaft sprocket with the punch marks aligned with the center of the three colored link plates.

26. Install the cam chain guide A and tensioner arm. Tighten the tensioner arm bolt to 16 ft. lbs (22 Nm) and the cam chain guide A bolts to 8.8 ft. lbs. (11.9 Nm).

27. Compress the auto tensioner when

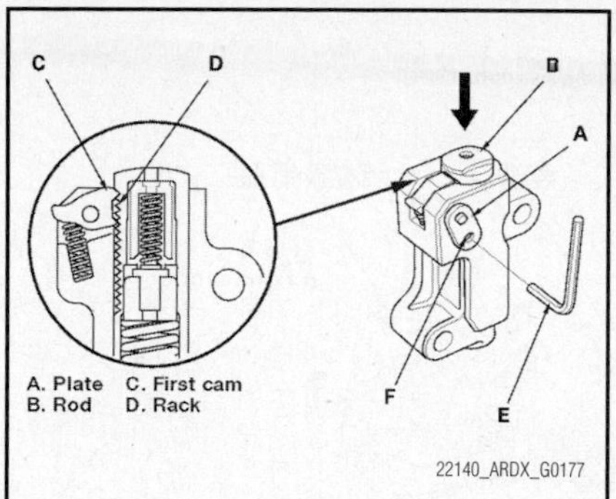
A. Plate C. First cam
B. Rod D. Rack
22140_ARDX_G0177

Fig. 125 Auto tensioner compression

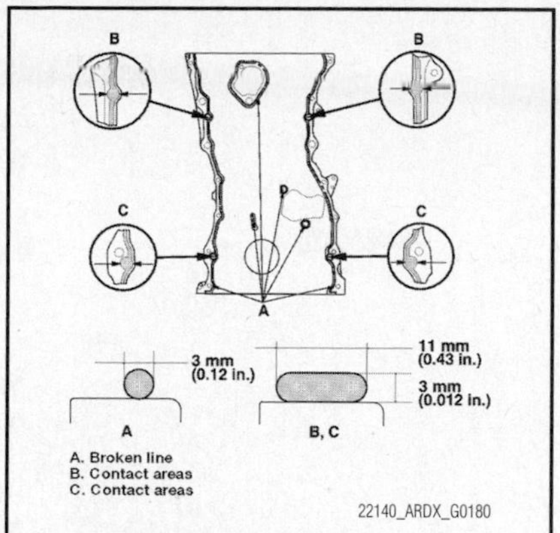
3 mm (0.12 in.)
11 mm (0.43 in.)
3 mm (0.012 in.)
A. Broken line
B. Contact areas
C. Contact areas
22140_ARDX_G0180

Fig. 126 Apply a 0.12 in. diameter bead along the broken line (A) as shown

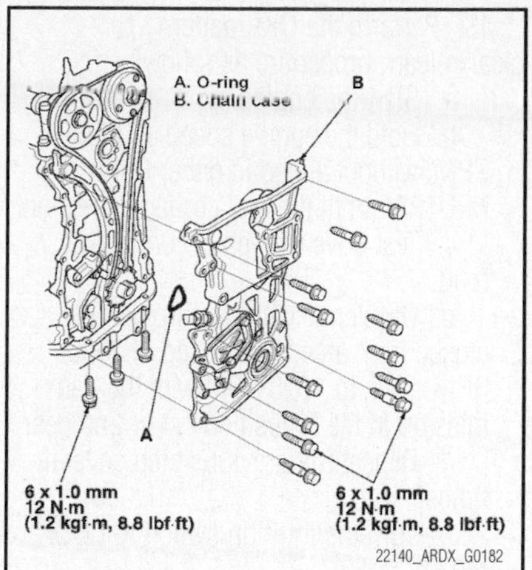
A. O-ring
B. Chain case
6 x 1.0 mm
12 N·m
(1.2 kgf·m, 8.8 lbf·ft)
6 x 1.0 mm
12 N·m
(1.2 kgf·m, 8.8 lbf·ft)
22140_ARDX_G0182

Fig. 128 Timing cover bolt installation locations

replacing the camshaft. Remove the pin from the auto tensioner. Turn the plate counterclockwise to release the lock. Press the rod and set the first cam to the edge of the rack. Insert a 0.05 in. diameter pin into the holes.

➡If the chain tensioner is not set up as described, the tensioner will be damaged.

28. Install the auto tensioner. Tighten the retaining bolts to 8.8 ft. lbs. (11.9 Nm).

29. Install the cam chain guide B. Tighten the retaining bolts to 16 ft. lbs. (22 Nm).

30. Remove the lock pin from the auto tensioner.

31. Remove the camshaft lock pin set (tool).

32. Replace the timing cover oil seal. Be sure to use the proper seal installation tools.

➡Measure the distance between the chain case surface and the oil seal. Specification should be 1.30–1.33 in.

33. Remove all the old liquid gasket material from the chain case mating surfaces, the bolts and the bolts holes of all mating surfaces. Clean and dry the chain case mating surfaces.

34. Apply liquid gasket to the engine block mating surface of the chain case. Apply a 0.12 in. diameter bead along the broken line as shown.

⁂ CAUTION

Install the component within five minutes of applying the liquid gasket. Some liquid gasket materials require installation within four minutes. Be sure to read and follow the manufacturer's instruction. If too much time has passed before installing the component remove the liquid gasket and residue, then reapply new liquid gasket.

35. Apply liquid gasket to the engine block upper surface contact areas on the chain case and on the lower block upper surface contact areas of the chain case. Apply a 0.43 in. diameter and about 0.12 in. thickness of gasket to areas shown.

36. Apply liquid gasket to the oil pan mating surface of the chain case. Apply a 0.12 in. diameter bead along the broken line, shown in the illustration.

Install the component within five minutes of applying the liquid gasket. Some liquid gasket materials require installation within four minutes. Be sure to read and follow the manufacturer's instruction. If too much time has passed before installing the component remove the liquid gasket and residue, then reapply new liquid gasket.

37. Install a new O-ring on the chain case. Set the edge of the chain case to the edge of the oil pan. Install the chain case on the engine block. Wipe any excess liquid gasket on the oil pan and chain case mating surfaces. When installing the chain case, do not slide the bottom surface onto the oil pan mounting surface.

➡Wait at least 30 minutes to allow the gasket to cure before filling the engine with oil.

⁂ CAUTION

Do not run the engine for at least three hours after installing the valve cover.

38. Tighten the cover retaining bolts to specification. Tightening specification is 8.8 ft. lbs. (11.9 Nm).

39. Continue the installation in the reverse order of the removal procedure.

40. When installing the VTC valve note the amount of valve opening by observing the position of the shoulder piston, thru the valve retard drain valve. If you see the shoulder of the piston, the valve is open and must be replaced.

41. Connect the battery positive terminal of the VTC terminal to the solenoid valve 2P connector terminal number two. Connect the battery negative terminal to the VTC solenoid valve 2P connector terminal number one. Refer to the accompanying illustration.

42. Appearance of the inner valve in the port should be at least 0.05 in. If the inner valve does not open, replace it and repeat the procedure.

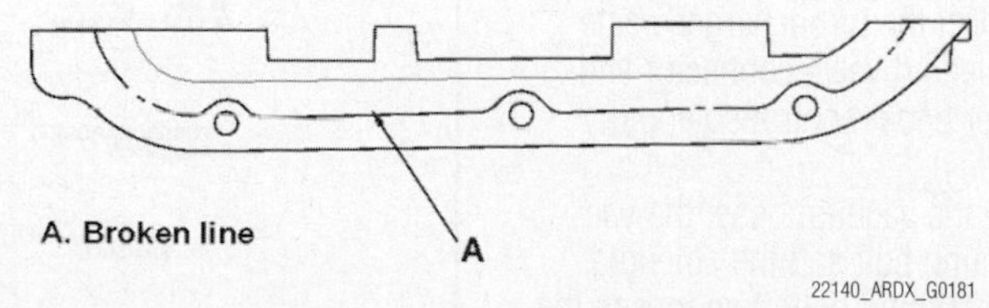
A. Broken line
A
22140_ARDX_G0181

Fig. 127 Apply liquid gasket part to the engine block upper surface contact areas (B) on the chain case and on the lower block upper surface contact areas (C) of the chain case, also, apply a 0.43 in. diameter and about 0.12 in. thickness of gasket to areas (B) and (C). Apply liquid gasket to the oil pan mating surface of the chain case. Apply a 0.12 in. diameter bead along the broken line (A) as shown.

43. Perform the CKP pattern clear/relearn procedure as follows:
   a. Start the engine.
   b. Hold the engine speed at 3000 RPM, without a load in either PARK or NEUTRAL until the radiator fan comes on.
   c. Test drive the vehicle on a level road.
   d. Decelerate with the throttle fully closed from an engine speed of 2500 RPM down to 1000 RPM with the transmission in the S position 1st or 2nd gear.
   e. Repeat the previous step several times.
   f. Turn the ignition switch to LOCK (0).
   g. Turn the ignition switch to ON (II) and wait thirty seconds.

TURBOCHARGER

REMOVAL & INSTALLATION

See Figures 129 and 130.

1. Before servicing the vehicle, refer to the "PRECAUTIONS".
2. Perform the battery disconnect/reconnect procedure, as outlined in the Engine Electrical Section.
3. Drain the coolant. Be sure to properly dispose of used coolant.
4. Remove the cowl panel and the under cowl panel, removing fasteners on the passenger and driver sides.
5. Remove the under hood fuse/relay box from the bracket.
6. Remove the master cylinder reservoir from the bracket. Remove the bracket.
7. Remove the heater hoses, then open the cable clamp and remove the heater valve cable.
8. Remove the heater valve mounting bolt. Remove the air bypass outlet pipe.
9. Remove the EVAP canister hose. Remove the intake air duct.
10. Remove the intake air ducts.
11. Remove the breather hose and the water bypass hose then remove the water bypass line mounting bolt.
12. Disconnect the turbocharger waste gate control solenoid valve connector and the turbocharger boost control solenoid valve connector.
13. Remove the vacuum hose, the vacuum line mounting bolt and the solenoid valve bracket mounting bolt, then loosen the solenoid valve bracket mounting bolt. Remove the solenoid valve bracket.
14. Remove the charge air cooler bracket. Remove the water bypass hose and the vacuum hoses. Remove the water bypass pipe from the turbocharger.

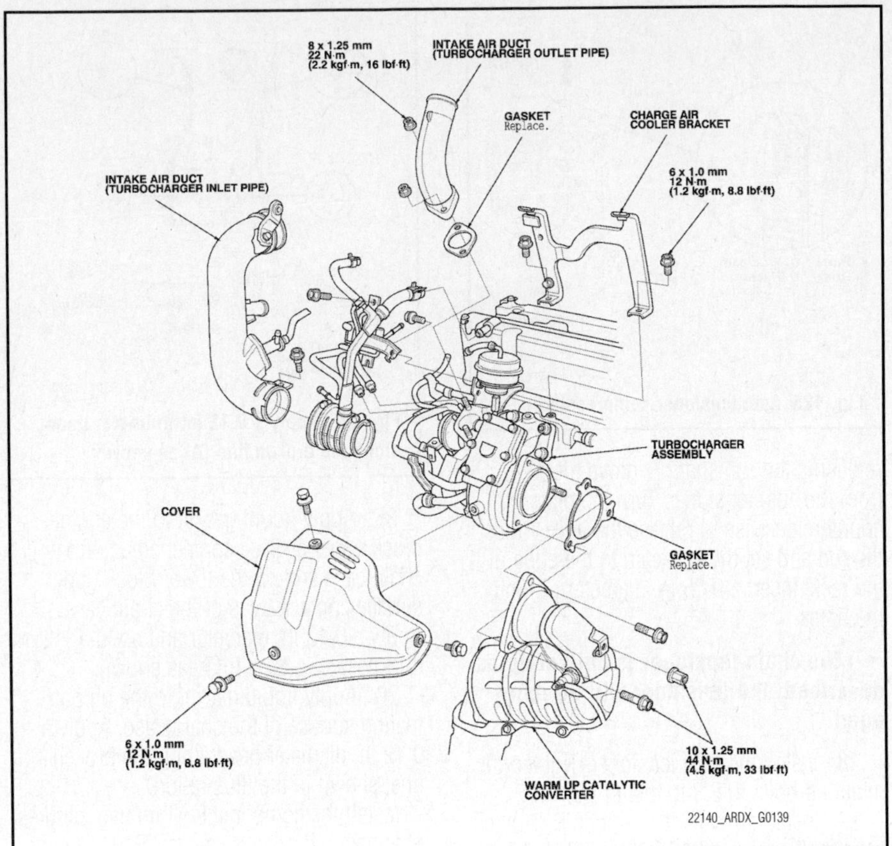
8 x 1.25 mm
22 N·m
(2.2 kgf·m, 16 lbf·ft)
INTAKE AIR DUCT
(TURBOCHARGER OUTLET PIPE)
GASKET
Replace.
CHARGE AIR
COOLER BRACKET
6 x 1.0 mm
12 N·m
(1.2 kgf·m, 8.8 lbf·ft)
INTAKE AIR DUCT
(TURBOCHARGER INLET PIPE)
TURBOCHARGER
ASSEMBLY
COVER
GASKET
Replace.
6 x 1.0 mm
12 N·m
(1.2 kgf·m, 8.8 lbf·ft)
10 x 1.25 mm
44 N·m
(4.5 kgf·m, 33 lbf·ft)
WARM UP CATALYTIC
CONVERTER
22140_ARDX_G0139

Fig. 129 Turbocharger and related components—part one

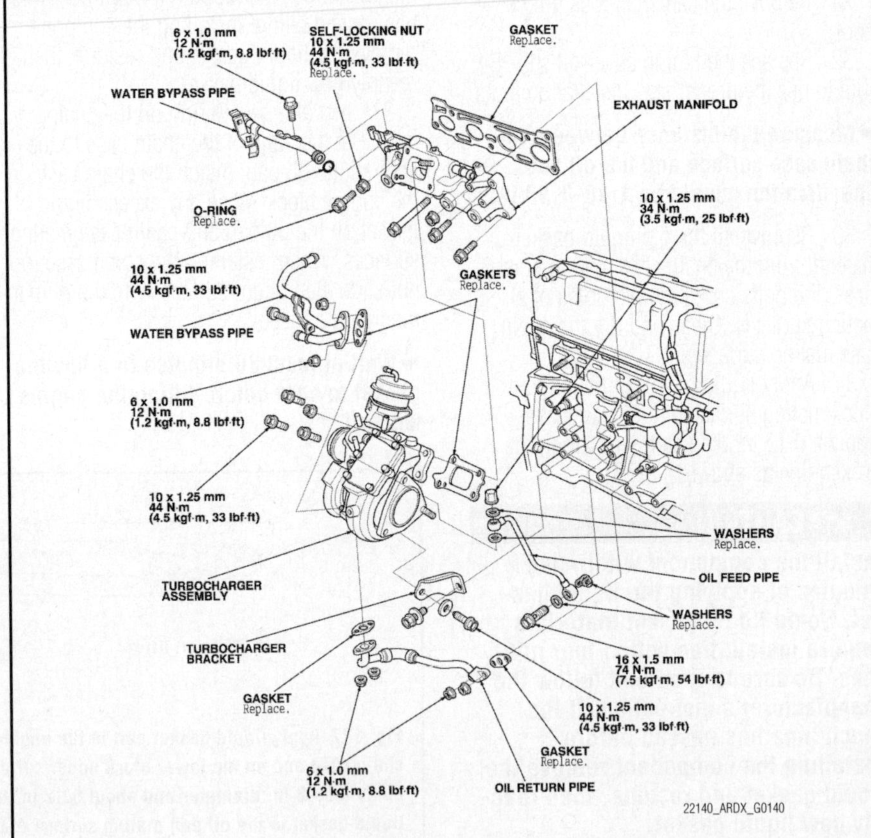
6 x 1.0 mm
12 N·m
(1.2 kgf·m, 8.8 lbf·ft)
SELF-LOCKING NUT
10 x 1.25 mm
44 N·m
(4.5 kgf·m, 33 lbf·ft)
Replace.
GASKET
Replace.
WATER BYPASS PIPE
EXHAUST MANIFOLD
O-RING
Replace.
10 x 1.25 mm
34 N·m
(3.5 kgf·m, 25 lbf·ft)
10 x 1.25 mm
44 N·m
(4.5 kgf·m, 33 lbf·ft)
GASKETS
Replace.
WATER BYPASS PIPE
6 x 1.0 mm
12 N·m
(1.2 kgf·m, 8.8 lbf·ft)
10 x 1.25 mm
44 N·m
(4.5 kgf·m, 33 lbf·ft)
WASHERS
Replace.
OIL FEED PIPE
TURBOCHARGER
ASSEMBLY
WASHERS
Replace.
TURBOCHARGER
BRACKET
16 x 1.5 mm
74 N·m
(7.5 kgf·m, 54 lbf·ft)
GASKET
Replace.
10 x 1.25 mm
44 N·m
(4.5 kgf·m, 33 lbf·ft)
GASKET
Replace.
6 x 1.0 mm
12 N·m
(1.2 kgf·m, 8.8 lbf·ft)
OIL RETURN PIPE
22140_ARDX_G0140

Fig. 130 Turbocharger and related components—part two

15. Remove the intake air duct.
16. Raise and safely support the vehicle.
17. Remove the oil return pipe, then remove the 16x1.5mm bolt at the oil feed pipe.
18. Remove the turbocharger bracket.
19. Lower the vehicle.
20. Remove the oil feed pipe.
21. Remove the turbocharger assembly retaining bolts. Remove the turbocharger from its mounting.

➡Be sure to seal the air inlet, the air outlet, the oil line hole, and the water line hole with tape.

➡Use care when handling the turbocharger assembly. Do not turn the actuator rod adjusting nut, or touch the impeller.

To install:

22. Clean the mating surfaces of the components to be sure they are free of dirt and debris.
23. Be sure to use a new gasket.
24. Tighten the turbocharger retaining bolts to 33 ft. lbs. (45 Nm).
25. Continue the installation in the reverse order of the removal procedure.
26. Turn the ignition switch ON (II), the SRS indicator should come on for about six seconds, and then go off.
27. Road test the vehicle, correct problems as required.

VALVE LASH

ADJUSTMENT & INSPECTION

See Figures 131 through 140.

1. Remove the air charge cooler cover.
2. Disconnect the turbocharger boost sensor connector. Remove the vacuum hoses and the air bypass outlet connecting tube.
3. Remove the air charge cooler retaining bolts. Remove the air charge cooler.
4. Remove the fuel injector cover.
5. Remove the ignition coil cover. Disconnect the ignition coil connectors.
6. Remove the ignition coils.
7. Remove the power steering hose bracket, the breather hose, and the dipstick.
8. Remove the bracket mounting bolts from the valve cover.
9. Disconnect the rocker arm oil control solenoid connector, the rocker are oil pressure switch connector and the Variable Valve Timing Control (VTC) oil control solenoid valve connector. Remove the harness clamp.

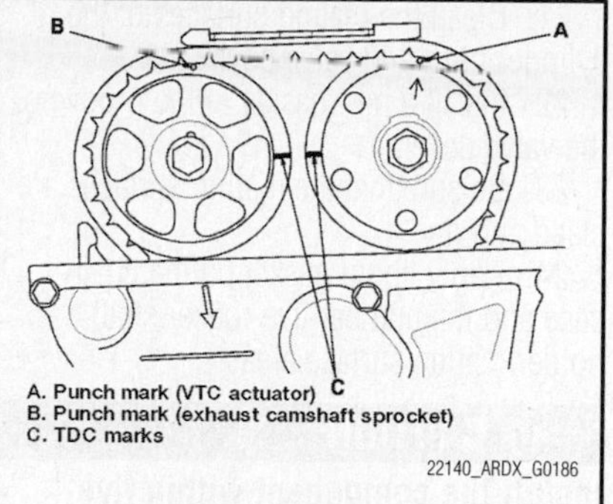
B
A
C
A. Punch mark (VTC actuator)
B. Punch mark (exhaust camshaft sprocket)
C. TDC marks
22140_ARDX_G0186

Fig. 131 VTC actuator and exhaust camshaft sprocket alignment marks

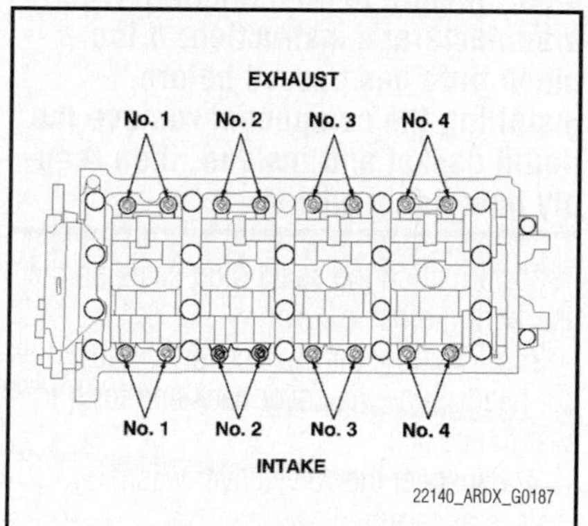
EXHAUST
No. 1
No. 2
No. 3
No. 4
No. 1
No. 2
No. 3
No. 4
INTAKE
22140_ARDX_G0187

Fig. 132 Intake/exhaust valve location

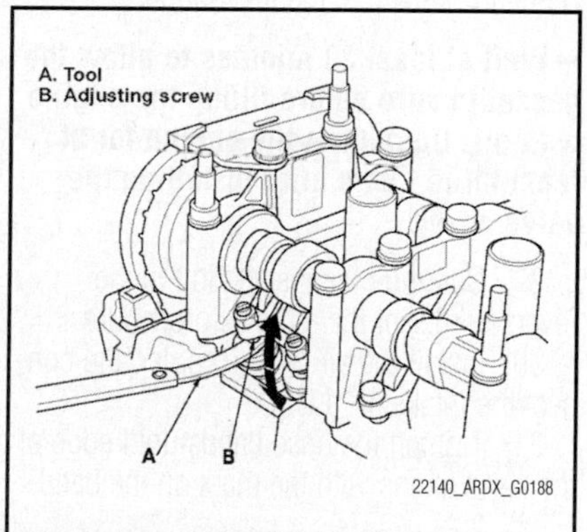
A. Tool
B. Adjusting screw
A
B
22140_ARDX_G0188

Fig. 133 Checking the valve adjustment

10. Disconnect the Air/Fuel ratio (A/F) sensor connector, the oil pressure switch connector and the Crankshaft Position (CKP) sensor. Remove the harness clamps.
11. Remove the valve cover retaining bolts.
12. Remove the valve cover from the engine. Discard the gasket.
13. Position the number one piston at TDC. The punch mark on the VTC actuator and the punch mark on the exhaust camshaft sprocket should be at the top. Align the TDC marks on the VTC actuator and exhaust camshaft sprocket.

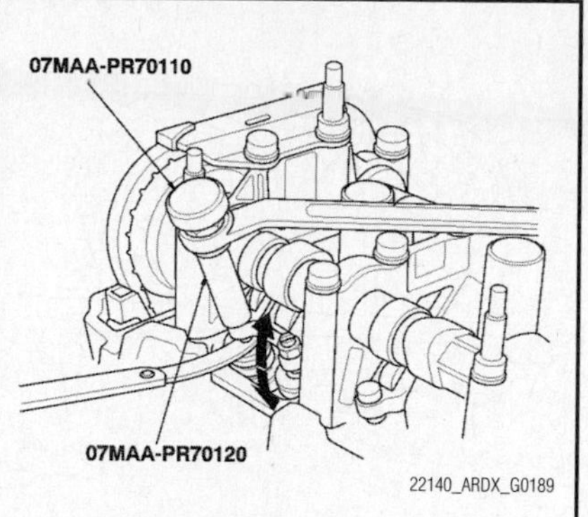
07MAA-PR70110
07MAA-PR70120
22140_ARDX_G0189

Fig. 134 Adjusting the valve adjustment

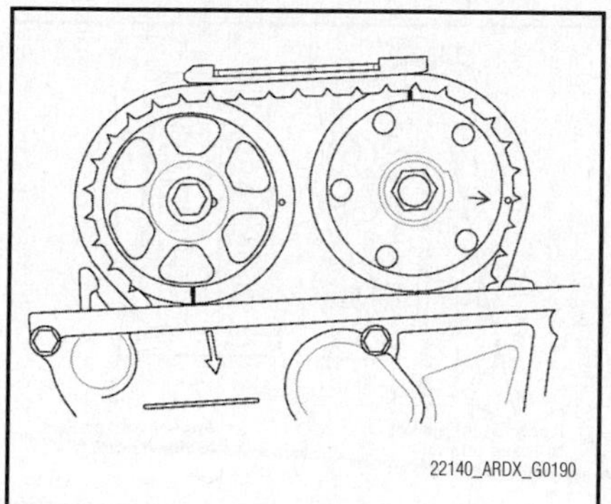
22140_ARDX_G0190

Fig. 135 Positioning number three cylinder

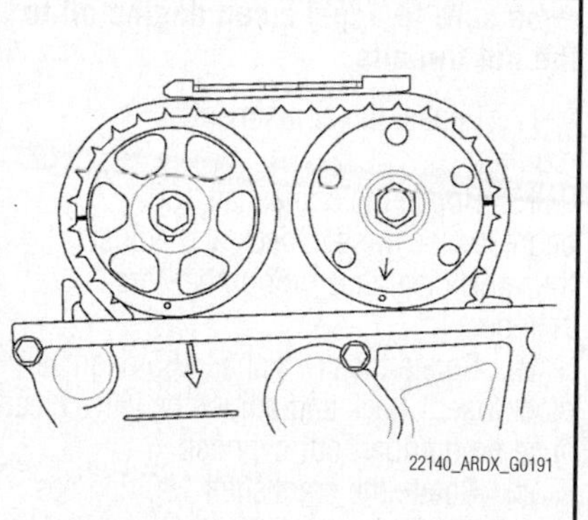
22140_ARDX_G0191

Fig. 136 Positioning number four cylinder

14. Check the valve clearance, using a gauge tool. Specification should be 0.008–0.010 in. for intake valves and 0.011–0.013 in. for exhaust valves.
15. Insert the feeler gauge between the adjusting screw and the end of the valve stem. Slide the gauge back and forth, you should feel a slight drag.
16. If you feel too much/little drag, loosen the locknut, turn the adjusting screw until the drag on the feeler gauge is correct. Tighten the locknut to specification:
   - Intake: 14 ft. lbs. (19 Nm)
   - Exhaust: 10 ft. lbs. (14 Nm)

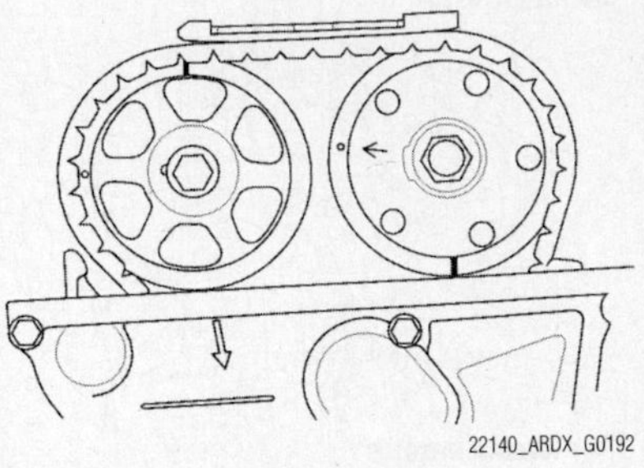
22140_ARDX_G0192

Fig. 137 Positioning number two cylinder

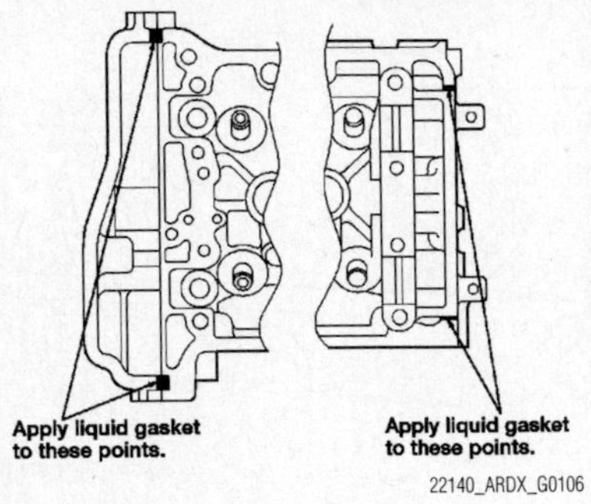
Apply liquid gasket to these points.
Apply liquid gasket to these points.
22140_ARDX_G0106

Fig. 138 Liquid gasket application points

➡**Be sure to apply clean engine oil to the nut threads.**

17. Repeat the adjustment as required.

18. Rotate the crankshaft 180 degrees clockwise. Check and adjust the valve clearance on number three cylinder.

19. Rotate the crankshaft 180 degrees clockwise. Check and adjust the valve clearance on number four cylinder.

20. Rotate the crankshaft 180 degrees clockwise. Check and adjust the valve clearance on number two cylinder.

21. Clean the mating surfaces of the cylinder head and valve cover.

22. Install a new gasket in the groove of the valve cover.

23. Be sure that the mating surfaces are clean and dry.

24. Apply liquid gasket on the chain case and the number five rocker shaft holder mating surface areas.

CAUTION

Install the component within five minutes of applying the liquid gasket. Some liquid gasket materials require installation within four minutes. Be sure to read and follow the manufacturer's instruction. If too much time has passed before installing the component remove the liquid gasket and residue, then reapply new liquid gasket.

25. Position the spark plug seals on the spark plug tubes.

26. Position the valve cover on the cylinder head. Slide the cover back and forth to seat the gasket.

27. Inspect the valve cover washers, replace as required.

28. Tighten the retaining bolts in three steps to 8.8 ft. lbs. (11.9 Nm), using the sequence shown in the illustration.

➡**Wait at least 30 minutes to allow the gasket to cure before filling the engine with oil. Do not run the engine for at least three hours after installing the valve cover.**

29. Continue the installation in the reverse order of the removal procedure.

30. Install the air charge cooler and connect the intake air ducts.

31. Tighten the hose bands until edge of the band aligns with the mark on the band.

➡**If the hose band edge exceeds the mark, replace the hose band.**

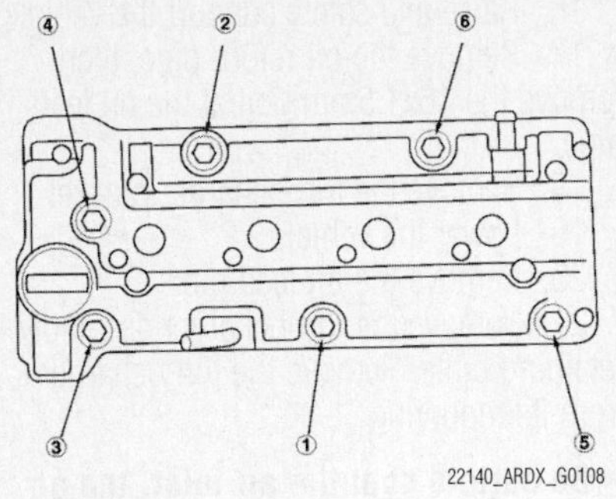
4
2
6
3
1
5
22140_ARDX_G0108

Fig. 139 Valve cover torque sequence

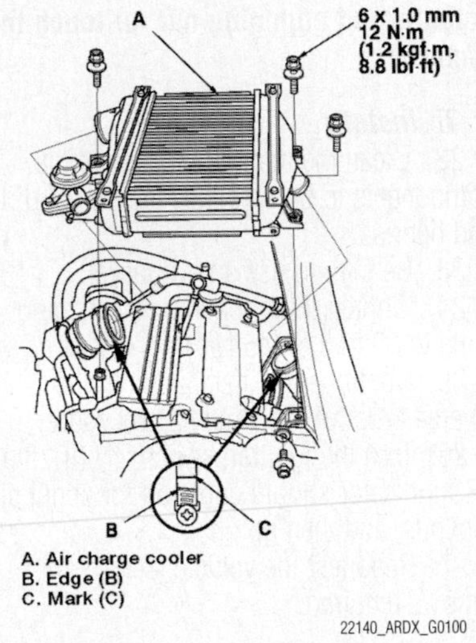
A
6 x 1.0 mm
12 N·m
(1.2 kgf·m,
8.8 lbf·ft)
B
C
A. Air charge cooler
B. Edge (B)
C. Mark (C)
22140_ARDX_G0100

Fig. 140 Tighten the hose bands until edge (B) of the band aligns with the mark (C) on the band.

32. Start the engine and check for leaks, correct as required.

33. Turn the ignition switch ON (II), the SRS indicator should come on for about six seconds, and then go off.

ENGINE PERFORMANCE & EMISSION CONTROLS

ACCELERATOR PEDAL POSITION (APP) SENSOR

LOCATION

The Accelerator Pedal Position (APP) sensor is located on the top of the accelerator pedal assembly.

REMOVAL & INSTALLATION

➡Manufacturer does not provide a specific removal and installation procedure for this component.

CAMSHAFT POSITION (CMP) SENSOR

LOCATION

This sensor is located on the top of the engine mid way between the passenger's side of the vehicle and the driver's side of the vehicle.

REMOVAL & INSTALLATION

See Figures 141 and 142.

1. Remove the air cleaner, if removing sensor A.
2. Remove the air charge cooler, if removing sensor B.
3. Remove the air bypass outlet pipe.
4. Disconnect the electrical connector.
5. Remove the component retaining bolt.
6. Remove the component from its mounting.
7. Discard the O-ring.

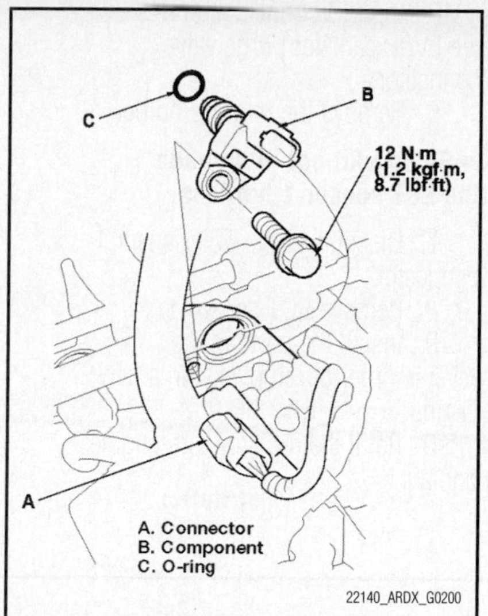
C
B
12 N·m
(1.2 kgf·m,
8.7 lbf·ft)
A
A. Connector
B. Component
C. O-ring
22140_ARDX_G0200

Fig. 142 Camshaft Position sensor B and related components

To install:

8. Be sure to use a new O-ring.
9. Installation is the reverse of the removal procedure.
10. Tighten the retaining bolt to 9 ft. lbs. (12 Nm).

CRANKSHAFT POSITION (CKP) SENSOR

LOCATION

This sensor is located on the passenger's side of the vehicle down by the crankshaft damper pulley.

REMOVAL & INSTALLATION

See Figure 143.

1. Disconnect the electrical connector.
2. Remove the component retaining bolt.
3. Remove the component from its mounting.
4. Discard the O-ring.

To install:

5. Be sure to use a new O-ring.
6. Installation is the reverse of the removal procedure.
7. Tighten the retaining bolt to 9 ft. lbs. (12 Nm).
8. Perform the "CKP Pattern Clear/Relearn Procedure".

CKP PATTERN CLEAR/RELEARN PROCEDURE

1. Start the engine.
2. Hold the engine speed at 3000 RPM, without a load in either PARK or NEUTRAL until the radiator fan comes on.
3. Test drive the vehicle on a level road.
4. Decelerate with the throttle fully closed from an engine speed of 2500 RPM down to 1000 RPM with the transmission in the S position 1st or 2nd gear.
5. Repeat the above step several times.
6. Turn the ignition switch to LOCK (0).
7. Turn the ignition switch to ON (II) and wait thirty seconds.

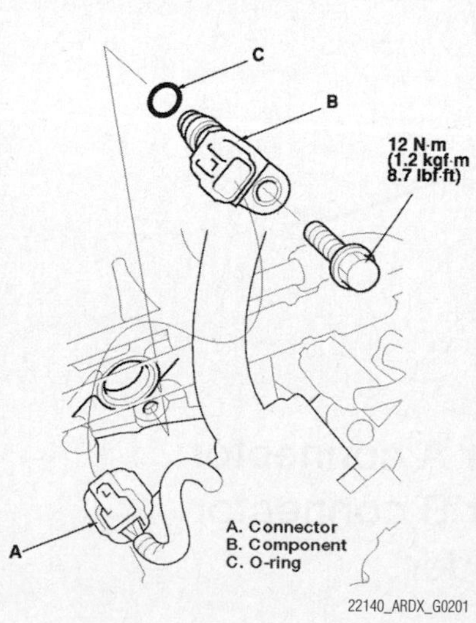
C
B
12 N·m
(1.2 kgf·m
8.7 lbf·ft)
A
A. Connector
B. Component
C. O-ring
22140_ARDX_G0201

Fig. 141 Camshaft Position sensor A and related components

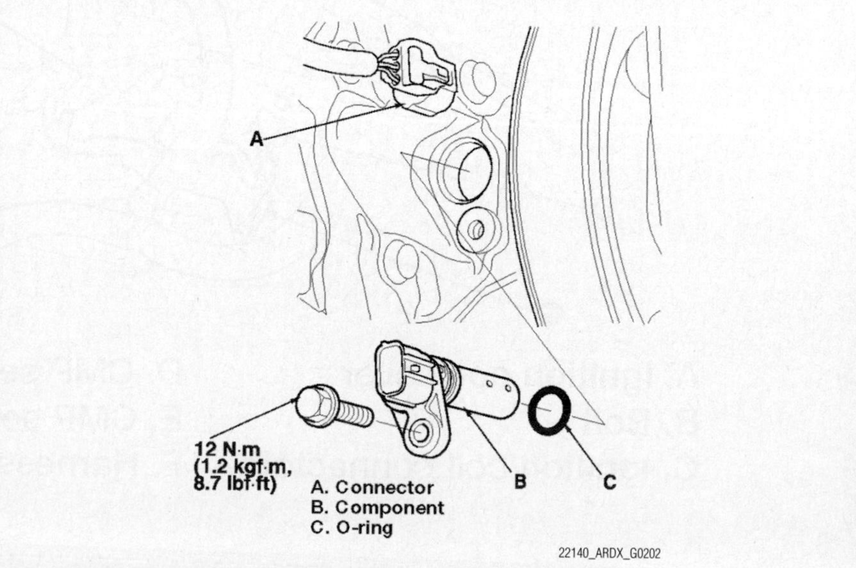
A
12 N·m
(1.2 kgf·m,
8.7 lbf·ft)
A. Connector
B. Component
C. O-ring
B
C
22140_ARDX_G0202

Fig. 143 Crankshaft Position (CKP) sensor and related components

ENGINE COOLANT TEMPERATURE (ECT) SENSOR

LOCATION

The Engine Coolant Temperature (ECT) Sensor 1 is located in the front of the engine block.

REMOVAL & INSTALLATION

ECT Sensor 1

See Figures 144 and 145.

1. Drain the engine coolant.
2. Remove the air cleaner.
3. Remove the charge air cooler.
4. Remove the ignition coil cover and bolt.
5. Disconnect the ignition coil connectors, CMP sensor A connector, CMP sensor B connector, and the EVAP canister purge valve connector.
6. Remove the harness holder.

➡**Be careful not to damage the ECT sensor 1 harness.**

7. Disconnect the ECT sensor 1 connector.
8. Remove ECT sensor 1.
9. Install the parts in the reverse order of removal with a new O-ring.
10. Refill the radiator with engine coolant.

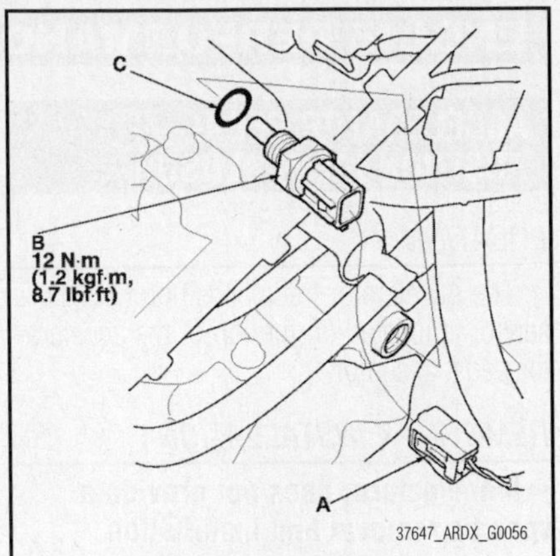
C
B
12 N·m
(1.2 kgf·m,
8.7 lbf·ft)
A
37647_ARDX_G0056

Fig. 145 Disconnect the ECT sensor 1 connector (A), then remove ECT sensor 1 (B). Discard the O-ring (C).

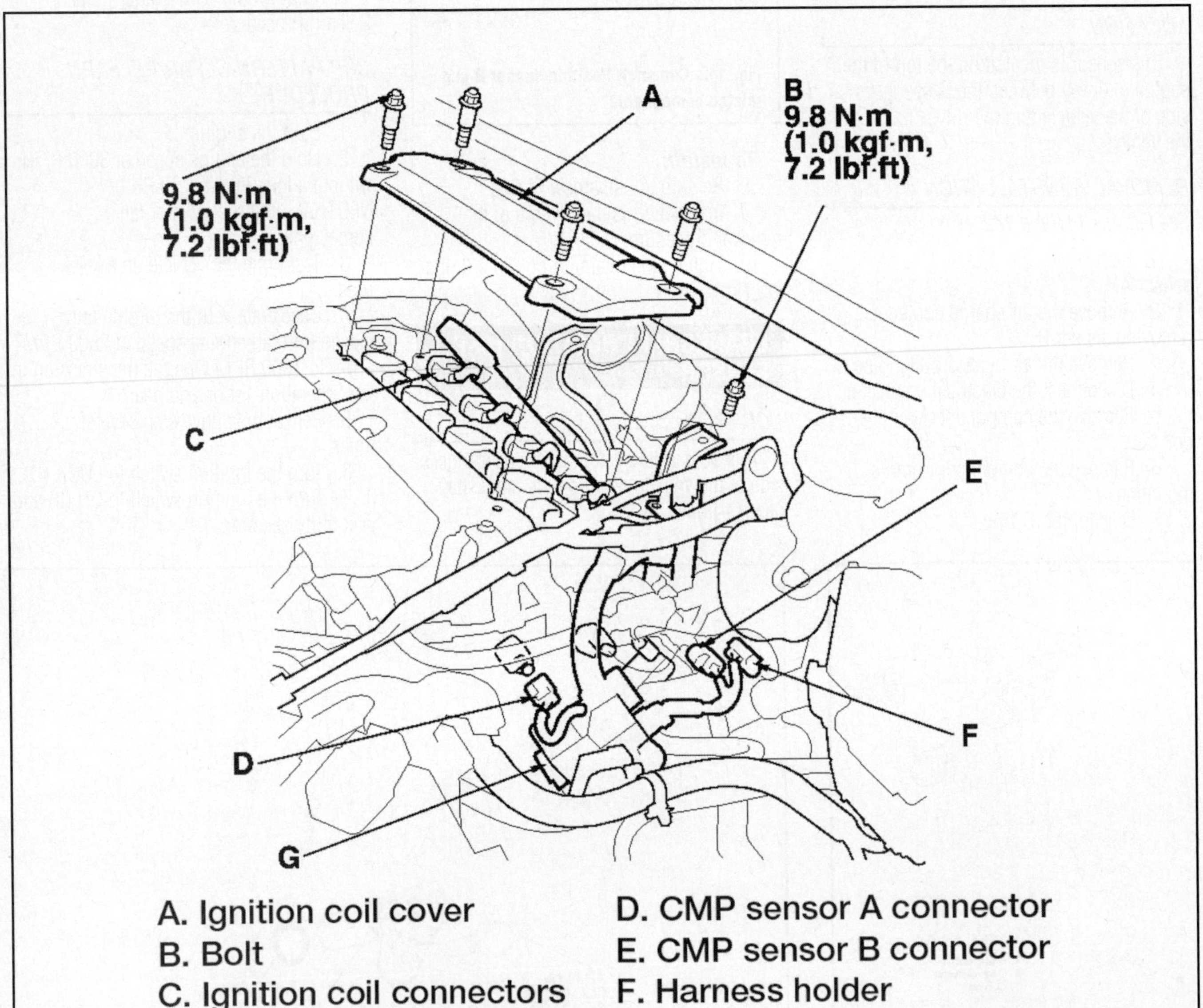
A
B
9.8 N·m
(1.0 kgf·m,
7.2 lbf·ft)
9.8 N·m
(1.0 kgf·m,
7.2 lbf·ft)
C
E
F
D
G
A. Ignition coil cover
B. Bolt
C. Ignition coil connectors
D. CMP sensor A connector
E. CMP sensor B connector
F. Harness holder
37647_ARDX_G0055

Fig. 144 Removing components for access to ECT sensor 1

ECT Sensor 2

See Figure 146.

1. Drain the engine coolant.
2. Remove the splash shield.
3. Disconnect the ECT sensor 2 connector, then remove ECT sensor 2.
4. Install ECT sensor 2 with a new O-ring (C).
5. Install the splash shield.
6. Refill the radiator with engine coolant.

EVAPORATIVE EMISSIONS (EVAP) CANISTER

LOCATION

See Figures 147 and 148.

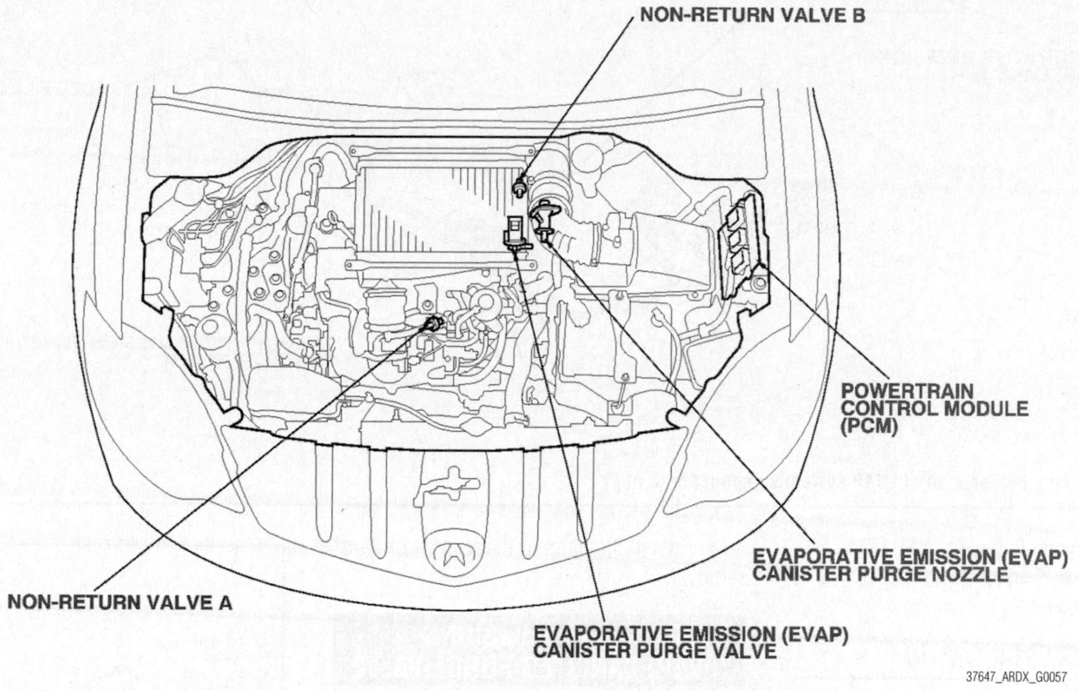
NON-RETURN VALVE B
POWERTRAIN CONTROL MODULE (PCM)
EVAPORATIVE EMISSION (EVAP) CANISTER PURGE NOZZLE
NON-RETURN VALVE A
EVAPORATIVE EMISSION (EVAP) CANISTER PURGE VALVE
37647_ARDX_G0057

Fig. 146 Disconnect the ECT sensor 2 connector (A), then remove ECT sensor 2 (B). Discard the O-ring (C).

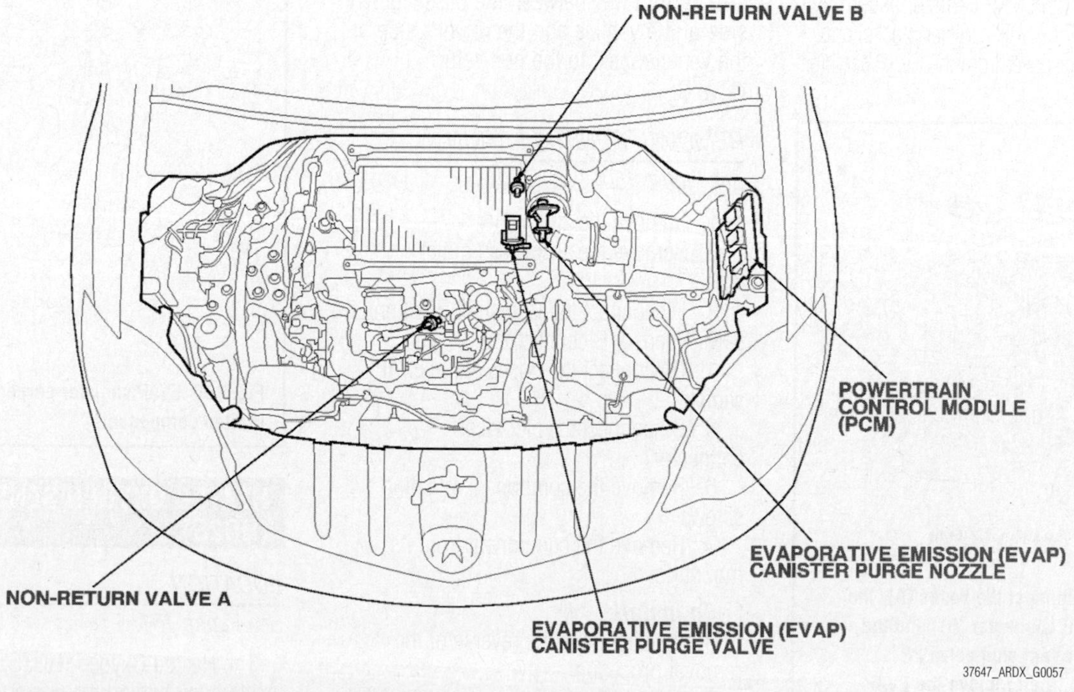
NON-RETURN VALVE B
POWERTRAIN CONTROL MODULE (PCM)
EVAPORATIVE EMISSION (EVAP) CANISTER PURGE NOZZLE
NON-RETURN VALVE A
EVAPORATIVE EMISSION (EVAP) CANISTER PURGE VALVE
37647_ARDX_G0057

Fig. 147 Showing the location of EVAP system components (1 of 2)

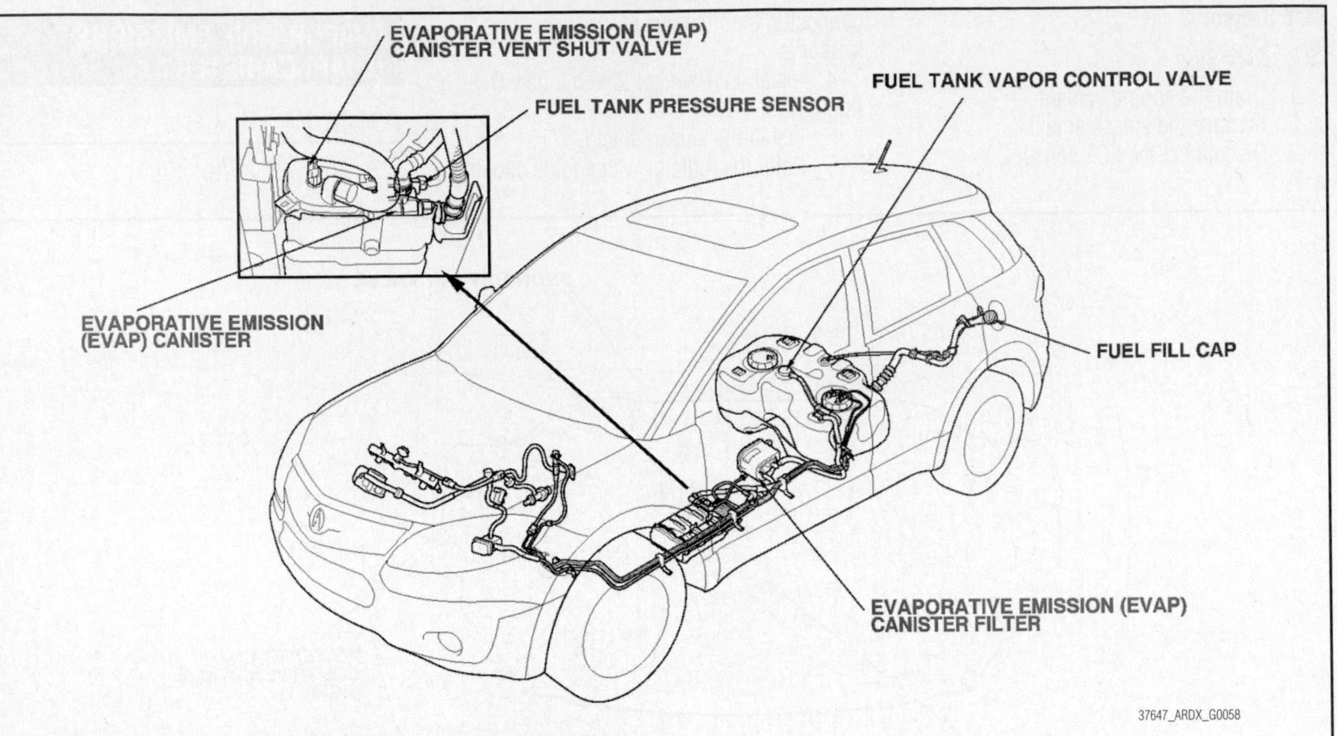
EVAPORATIVE EMISSION (EVAP)
CANISTER VENT SHUT VALVE
FUEL TANK PRESSURE SENSOR
FUEL TANK VAPOR CONTROL VALVE
EVAPORATIVE EMISSION
(EVAP) CANISTER
FUEL FILL CAP
EVAPORATIVE EMISSION (EVAP)
CANISTER FILTER
37647_ARDX_G0058

Fig. 148 Showing the location of EVAP system components (2 of 2)

The Evaporative Emissions (EVAP) canister is located under the center of the vehicle, as shown.

REMOVAL & INSTALLATION

See Figure 149.

1. Disconnect the hoses, the FTP sensor 3P connector, and the EVAP canister vent shut valve 2P connector.
2. Remove the EVAP canister assembly.
3. Remove the EVAP canister bracket and the EVAP canister cover from the EVAP canister.

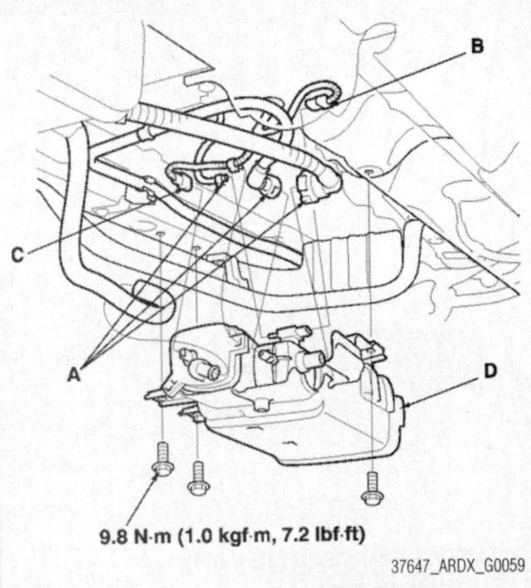
B
C
A
D
9.8 N·m (1.0 kgf·m, 7.2 lbf·ft)
37647_ARDX_G0059

Fig. 149 Disconnect the hoses (A), the FTP sensor 3P connector (B), and the EVAP canister vent shut valve 2P connector (C), then remove the EVAP canister assembly (D).

4. Install the parts in the reverse order of removal.

EVAPORATIVE EMISSIONS (EVAP) CANISTER PURGE VALVE

LOCATION

The Evaporative Emissions (EVAP) canister purge valve is located on the firewall mid way between the passenger's side of the vehicle and the driver's side of the vehicle, next to the non-return valve B.

REMOVAL & INSTALLATION

See Figure 150.

1. Remove the air cleaner.
2. Remove the air bypass outlet pipe.
3. Disconnect the Camshaft Position (CMP) sensor B connector.
4. Disconnect the hoses at the EVAP canister.
5. Disconnect the electrical connector.
6. Remove the component retaining screws.
7. Remove the component from its mounting.

To install:

8. Installation is the reverse of the removal procedure.
9. Tighten the retaining screws to 4.4 ft. lbs. (5.97 Nm).

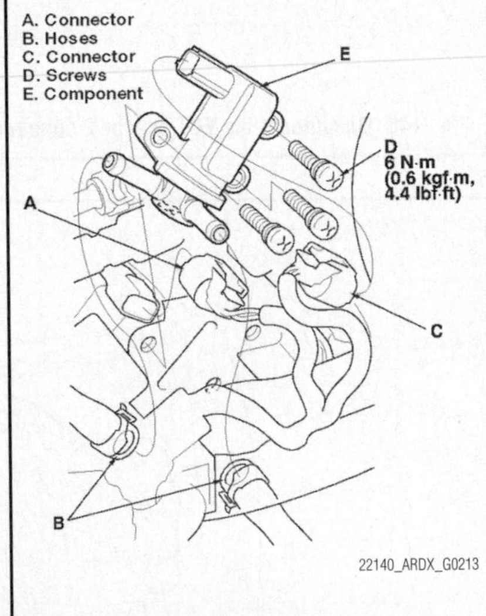
A. Connector
B. Hoses
C. Connector
D. Screws
E. Component
E
D
6 N·m
(0.6 kgf·m,
4.4 lbf·ft)
A
C
B
22140_ARDX_G0213

Fig. 150 EVAP canister purge valve and related components

HEATED OXYGEN SENSOR (HO2S)

LOCATION

See Figure 151.

The Heated Oxygen (HO2S) sensor is located just behind the warm-up three way catalytic converter.

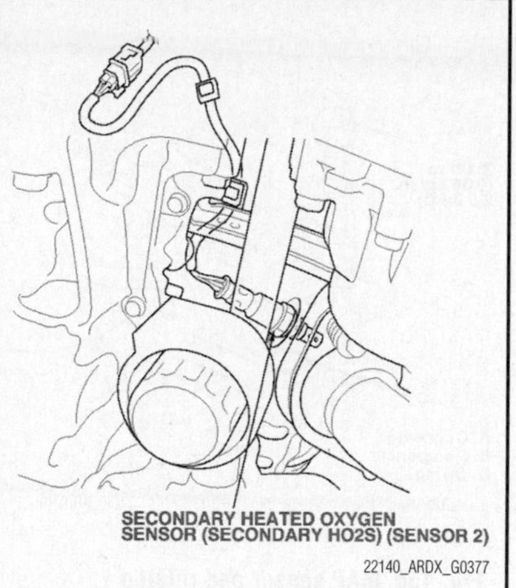
SECONDARY HEATED OXYGEN SENSOR (SECONDARY HO2S) (SENSOR 2)
22140_ARDX_G0377

Fig. 151 Heated Oxygen (HO2S) sensor location

REMOVAL & INSTALLATION

See Figure 152.

1. Remove the WU-TWC stay.
2. Disconnect the electrical connector.
3. Remove the component from its mounting, using a sensor removal tool.

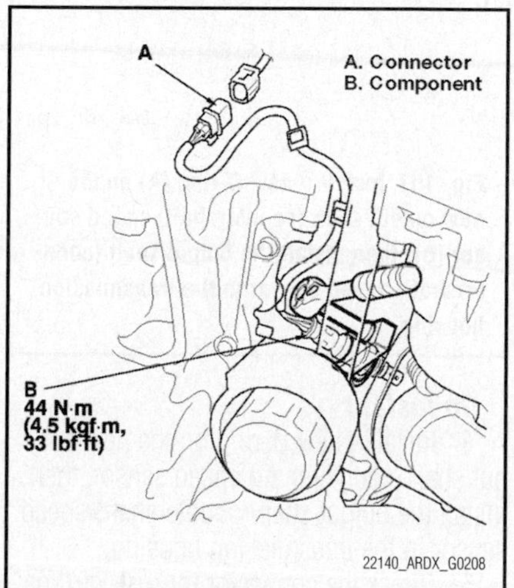
A
A. Connector
B. Component
B
44 N·m
(4.5 kgf·m,
33 lbf·ft)
22140_ARDX_G0208

Fig. 152 Heated Oxygen sensor (HO2S) and related components

To install:

4. Installation is the reverse of the removal procedure.
5. Tighten the sensor to 33 ft. lbs. (45 Nm).

INTAKE AIR TEMPERATURE (IAT) SENSOR

LOCATION

The Intake Air Temperature (IAT) sensor is located on the engine mid way between the passenger's side of the vehicle and the driver's side of the vehicle, next to the Manifold Absolute Pressure (MAP) sensor.

REMOVAL & INSTALLATION

See Figure 153.

1. Remove the air charge cooler cover.
2. Disconnect the electrical connector.
3. Remove the component retaining bolt.
4. Remove the component from its mounting.
5. Discard the O-ring.

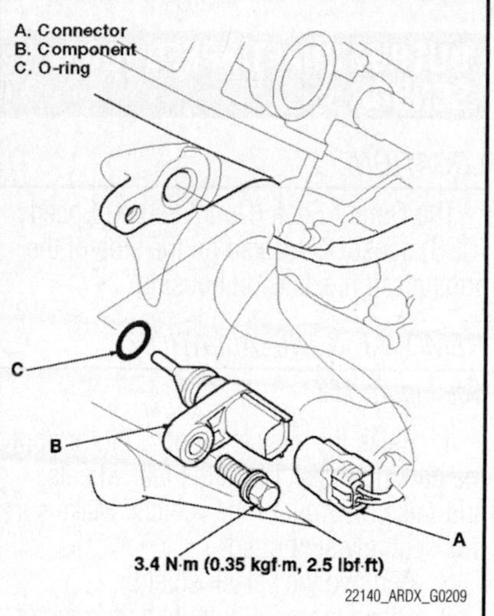
A. Connector
B. Component
C. O-ring
C
B
A
3.4 N·m (0.35 kgf·m, 2.5 lbf·ft)
22140_ARDX_G0209

Fig. 153 IAT sensor and related components

To install:

6. Be sure to use a new O-ring.
7. Installation is the reverse of the removal procedure.
8. Tighten the retaining bolt to 3 ft. lbs. (4 Nm).

KNOCK SENSOR (KS)

LOCATION

The Knock Sensor (KS) is located on the engine, in the middle, mid way between the passenger's side of the vehicle and the driver's side of the vehicle, next to the Intake Air Temperature (IAT) sensor and the Manifold Absolute Pressure (MAP) sensor.

REMOVAL & INSTALLATION

See Figure 154.

1. Perform the battery disconnect/reconnect procedure, as outlined in the Engine Electrical Section.
2. Disconnect the positive battery cable.
3. Disconnect the electrical connector.
4. Remove the component from its mounting.

To install:

5. Installation is the reverse of the removal procedure.
6. Tighten the retaining bolt to 23 ft. lbs. (31 Nm).

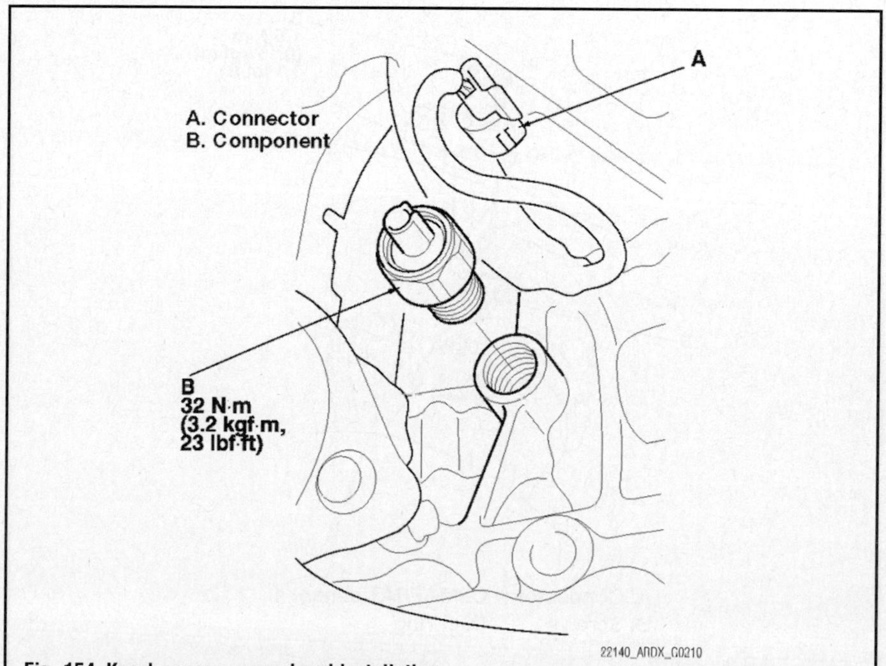
A
A. Connector
B. Component
B
32 N·m
(3.2 kgf·m,
23 lbf·ft)
22140_ARDX_G0210

Fig. 154 Knock sensor removal and installation

MALFUNCTION INDICATOR LIGHT

RESET PROCEDURE

1. Use a scan tool with the latest software and clear the DTC codes.

MASS AIR FLOW/INTAKE AIR TEMPERATURE (MAF/IAT) SENSOR

LOCATION

The Mass Air Flow (MAF) sensor is located on the air intake snorkel on the driver's side of the vehicle, near the PGI relay.

REMOVAL & INSTALLATION

See Figure 155.

1. Disconnect the MAF/IAT sensor 1 connector.
2. Remove the screws.
3. Remove the MAF/IAT sensor 1 and discard the O-ring.
4. Install the parts in the reverse order of removal with a new O-ring.

MANIFOLD ABSOLUTE PRESSURE (MAP) SENSOR

LOCATION

The Manifold Absolute Pressure (MAP) sensor is located on the engine mid way between the passenger's side of the vehicle and the driver's side of the vehicle, next to the Intake Air Temperature (IAT) sensor.

REMOVAL & INSTALLATION

See Figure 156.

1. Disconnect the electrical connector.
2. Remove the component retaining bolt.
3. Remove the component from its mounting.
4. Discard the O-ring.

To install:

5. Be sure to use a new O-ring.
6. Installation is the reverse of the removal procedure.
7. Tighten the retaining bolt to 3 ft. lbs. (4 Nm).

OUTPUT SHAFT SPEED (OSS) SENSOR

LOCATION

The Output Shaft (Countershaft) Speed (OSS) sensor is located on the side of the transmission extension housing.

REMOVAL & INSTALLATION

See Figure 157.

1. Raise the vehicle up on a lift, or apply the parking brake, block the rear wheels, and raise the front of the vehicle. Make sure it is securely supported.
2. Remove the splash shield.
3. Disconnect the output shaft (countershaft) speed sensor connector, and remove the output shaft (countershaft) speed sensor.

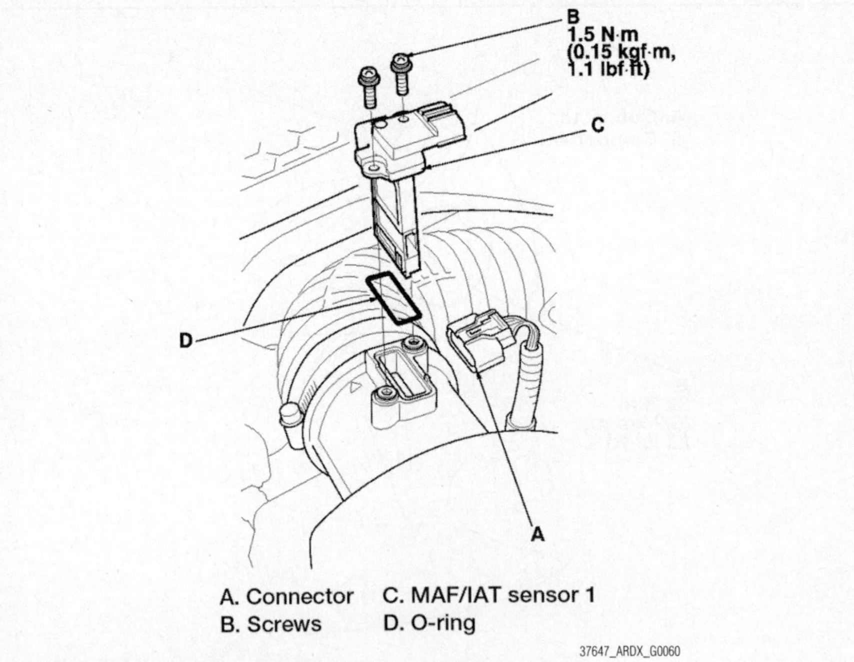

B
1.5 N·m
(0.15 kgf·m,
1.1 lbf·ft)
C
D
A
A. Connector
B. Screws
C. MAF/IAT sensor 1
D. O-ring
37647_ARDX_G0060

Fig. 155 Showing the MAF/IAT sensor

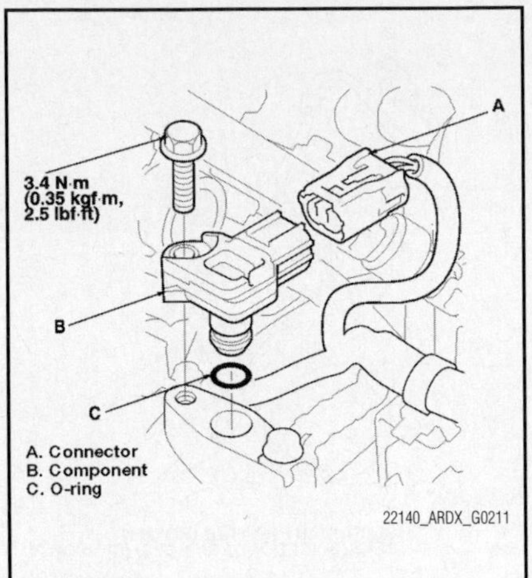

A
3.4 N·m
(0.35 kgf·m,
2.5 lbf·ft)
B
C
A. Connector
B. Component
C. O-ring
22140_ARDX_G0211

Fig. 156 MAP sensor and related components

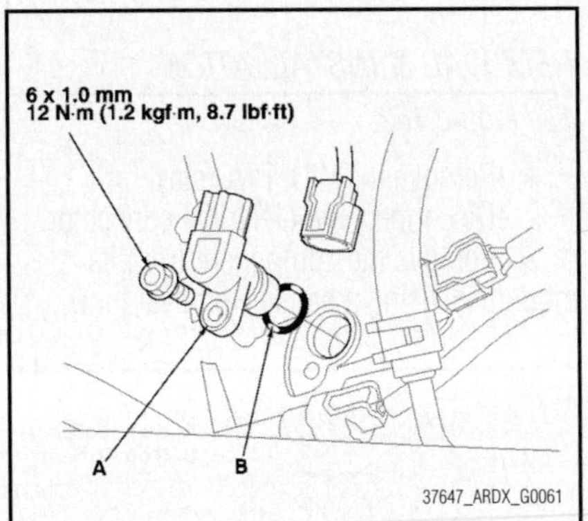

6 x 1.0 mm
12 N·m (1.2 kgf·m, 8.7 lbf·ft)
A
B
37647_ARDX_G0061

Fig. 157 Install a new O-ring (A) on the new output shaft (countershaft) speed sensor (B), then install the output shaft (countershaft) speed sensor in the transmission housing.

To install:

4. Install a new O-ring on the new output shaft (countershaft) speed sensor, then install the output shaft (countershaft) speed sensor in the transmission housing.
5. Check the connector for rust, dirt, or oil, then connect the connector securely.
6. Install the splash shield.

POSITIVE CRANKCASE VENTILATION (PCV) VALVE

LOCATION

The Positive Crankcase Ventilation (PCV) valve is located in the front right side of the engine.

REMOVAL & INSTALLATION

See Figure 158.

1. Disconnect the PCV hose.
2. Remove the PCV valve.

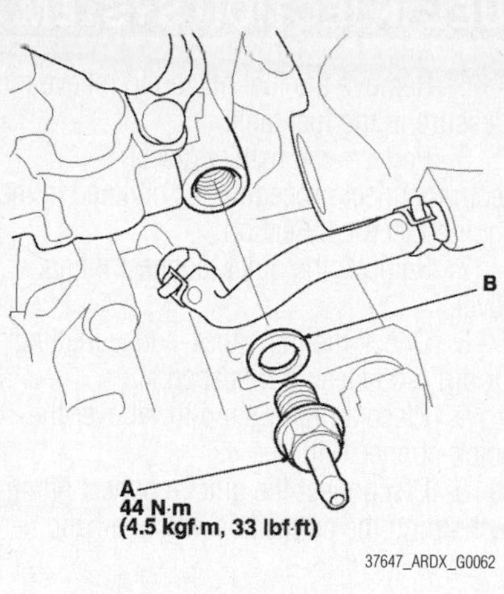
B
A
44 N·m
(4.5 kgf·m, 33 lbf·ft)
37647_ARDX_G0062

Fig. 158 Showing the PCV valve (A) and the washer (B)

3. Install the parts in the reverse order of removal with a new washer.

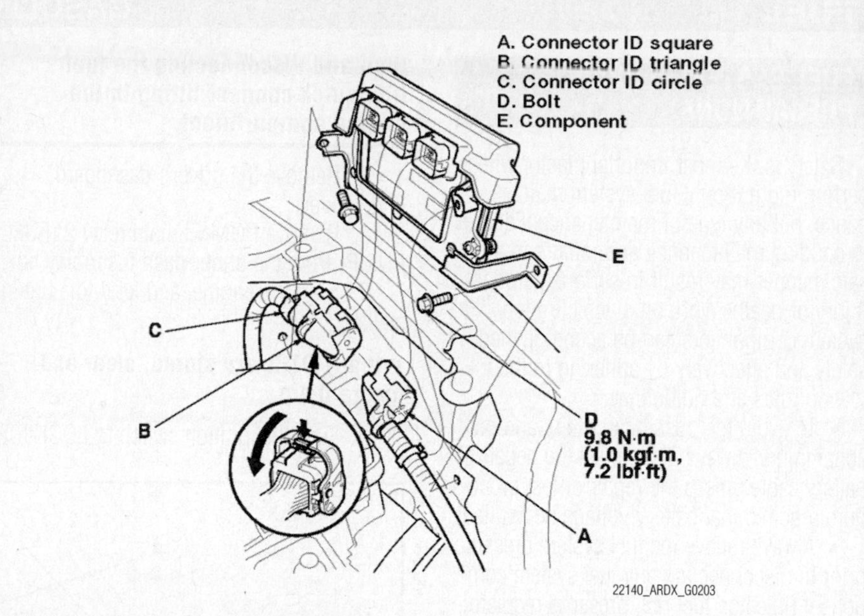
A. Connector ID square
B. Connector ID triangle
C. Connector ID circle
D. Bolt
E. Component
E
C
B
D
9.8 N·m
(1.0 kgf·m,
7.2 lbf·ft)
A
22140_ARDX_G0203

Fig. 159 PCM and related components

POWERTRAIN CONTROL MODULE

LOCATION

The Powertrain Control Module (PCM) is located on the driver's side of the vehicle, on the inner fender well.

REMOVAL & INSTALLATION

See Figure 159.

1. Perform the battery disconnect/reconnect procedure, as outlined in the Engine Electrical Section.
2. Refer to the scan tool instructions for the particular model you are repairing. Use those procedures to position the PCM for replacement. Diagnostics for engine oil replacement, ATF fluid replacement, and throttle body cleaning will be addressed on both US and Canadian vehicles.
3. Disconnect the electrical connectors.

➡These connectors have symbols embossed in them for identification, see illustration.

4. Remove the component retaining bolts.
5. Remove the component from its mounting.
6. Remove the cover and the bracket from the PCM

To install:

7. Installation is the reverse of the removal procedure.
8. Tighten the retaining bolt to 7 ft. lbs. (9 Nm).
9. Update the PCM if it does not have the latest software.

➡The PCM relearn procedure must be performed whenever you replace the PCM, Reset the PCM, update the PCM, remove the engine or replace/clean the throttle body. Erasing DTC's with the scan tool does not require you to perform the relearn procedure.

PCM RESET PROCEDURE

1. Reset the PCM with the Honda Diagnostic System (HDS) tool, or equivalent scan tool, as follows:
   a. Ensure the engine is stopped.
   b. Turn the ignition switch to LOCK (0).
   c. Turn the ignition switch to ON (II), and wait thirty seconds.
   d. Turn the ignition switch to LOCK (0) and disconnect the tool from the DLC.
   e. Make sure that all electrical items are turned off.
   f. Reset the PCM with the scan tool.
   g. Turn the ignition switch ON (II) and wait two seconds.
   h. Start the engine. Hold the engine speed at 3000 RPM, without a load in either PARK or NEUTRAL until the radiator fan comes on or until the coolant temperature reaches 194° F.
   i. Let the engine idle for about five minutes with the throttle fully closed

➡If the radiator fan comes on, do not include its running time in the five minutes.

   j. Perform the "CKP Pattern Clear/Relearn Procedure". See "Crankshaft Position (CKP) Sensor" in this section.

FUEL SYSTEM SERVICE PRECAUTIONS

Safety is the most important factor when performing not only fuel system maintenance, but any type of maintenance. Failure to conduct maintenance and repairs in a safe manner may result in serious personal injury or death. Work on a vehicle's fuel system components can be accomplished safely and effectively by adhering to the following rules and guidelines.

- To avoid the possibility of fire and personal injury, always disconnect the negative battery cable unless the repair or test procedure requires that battery voltage be applied.
- Always relieve the fuel system pressure prior to disconnecting any fuel system component (injector, fuel rail, pressure regulator, etc.) fitting or fuel line connection. Exercise extreme caution whenever relieving fuel system pressure to avoid exposing skin, face and eyes to fuel spray. Please be advised that fuel under pressure may penetrate the skin or any part of the body that it contacts.
- Always place a shop towel or cloth around the fitting or connection prior to loosening to absorb any excess fuel due to spillage. Ensure that all fuel spillage is quickly removed from engine surfaces. Ensure that all fuel-soaked cloths or towels are deposited into a flame-proof waste container with a lid.
- Always keep a dry chemical (Class B) fire extinguisher near the work area.
- Do not allow fuel spray or fuel vapors to come into contact with a spark or open flame.
- Always use a second wrench when loosening or tightening fuel line connection fittings. This will prevent unnecessary stress and torsion on fuel piping. Always follow the proper torque specifications.
- Always replace worn fuel fitting O-rings with new ones. Do not substitute fuel hose where rigid pipe is installed.

FUEL SYSTEM PRESSURE

RELIEVING

See Figures 160 and 161.

➡Before doing this procedure, make sure the engine is cold.

✲✲ CAUTION

Before disconnecting fuel lines or hoses, relieve pressure from the system by disabling the fuel pump, running the engine until it stalls, then and disconnecting the fuel line/quick connect fitting in the engine compartment.

1. Remove the driver's dashboard undercover.
2. Remove PGM-FI main relay 2 (FUEL PUMP) from the under-dash fuse/relay box.
3. Start the engine, and let it idle until it stalls.

➡If any DTCs are stored, clear and ignore them.

4. Turn the ignition switch to LOCK (0).
5. Remove the fuel fill cap to relieve the pressure in the fuel tank.
6. Perform the battery disconnect/reconnect procedure, as outlined in the Engine Electrical Section.
7. Remove the quick-connect fitting cover.
8. Check the fuel quick-connect fitting for dirt, and clean it if needed.
9. Place a rag or shop towel over the quick-connect fitting.
10. Disconnect the quick-connect fitting by holding the connector with one hand,

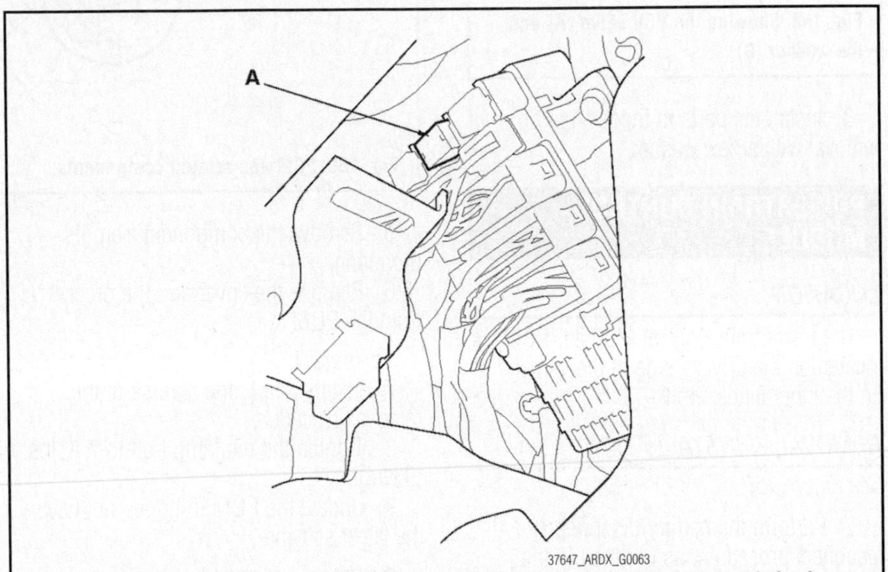
A
37647_ARDX_G0063

Fig. 160 Remove PGM-FI main relay 2 (FUEL PUMP) (A) from the under-dash fuse/relay box.

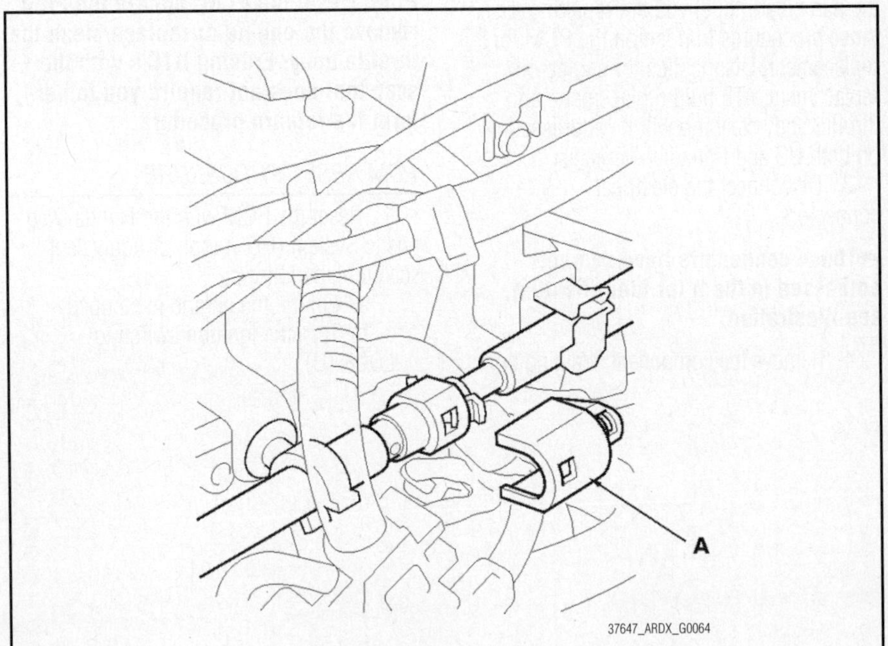
A
37647_ARDX_G0064

Fig. 161 Remove the quick-connect fitting cover (A).

and squeezing the retainer tabs with the other hand to release them from the locking tabs. Pull the connector off.

CAUTION

Be careful not to damage the line or other parts. Do not use tools.

11. If the connector does not move, keep the retainer tabs pressed down, and alternately pull and push the connector until it comes off easily.

➡Do not remove the retainer from the line; once removed, the retainer must be replaced with a new one.

12. After disconnecting the quick-connect fitting, check it for dirt or damage.
13. Perform the battery disconnect/reconnect procedure, as outlined in the Engine Electrical Section.

FUEL FILTER

REMOVAL & INSTALLATION

See Figure 162.

1. Properly relieve the fuel system pressure. See "Fuel System Pressure" in this section.
2. Remove the fuel tank.
3. Remove the fuel strainer set.

To install:

4. Installation is the reverse of the removal procedure.
5. Be sure to use new O-rings. Coat the O-rings with clean engine oil.

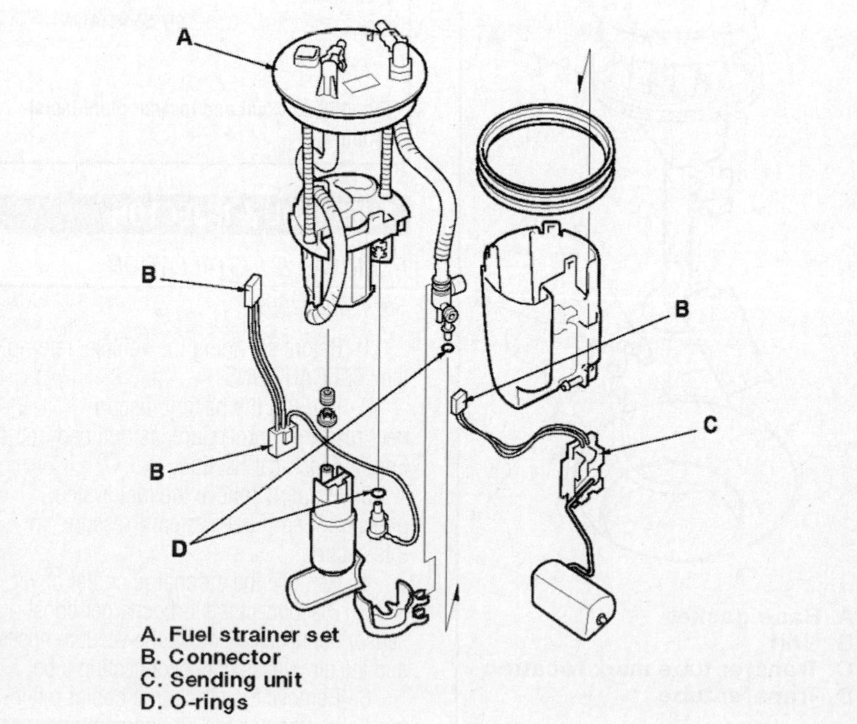
A
B
B
B
C
D
A. Fuel strainer set
B. Connector
C. Sending unit
D. O-rings
22140_ARDX_G0216

Fig. 162 Fuel filter and related components

FUEL LEVEL SENDING UNIT

LOCATION

The fuel level sending unit is an integral part of the fuel pump assembly on top of the fuel tank.

REMOVAL & INSTALLATION

The fuel level sending unit procedures are listed under "Fuel Pump". Please refer to the appropriate procedure.

FUEL PUMP

REMOVAL & INSTALLATION

Fuel Pump/Fuel Gauge Sending Unit

See Figure 163.

1. Before servicing the vehicle, refer to the "PRECAUTIONS".
2. Perform the battery disconnect/reconnect procedure, as outlined in the Engine Electrical Section.

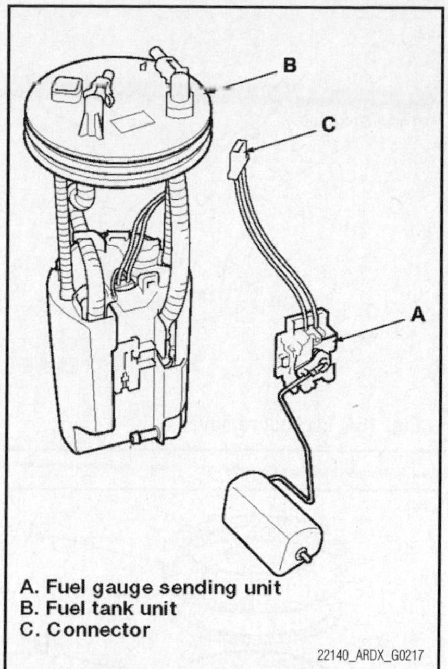
B
C
A
A. Fuel gauge sending unit
B. Fuel tank unit
C. Connector
22140_ARDX_G0217

Fig. 163 Fuel pump/fuel gauge sending unit related components

3. Properly relieve the fuel system pressure. See "Fuel System Pressure" in this section.
4. Remove the fuel tank.
5. Remove the fuel level sensor (fuel gauge sending unit) from the fuel tank unit.
6. Installation is the reverse of the removal procedure.

Fuel Tank Unit

See Figures 164 through 168.

1. Before servicing the vehicle, refer to the "PRECAUTIONS".
2. Perform the battery disconnect/reconnect procedure, as outlined in the Engine Electrical Section.
3. Properly relieve the fuel system pressure. See "Fuel System Pressure" in this section.
4. Remove the fuel cap.
5. Fold the rear seat down and remove the left rear door seal trim.
6. Pull back the carpet and expose the access panel.
7. Remove the access panel from the left side of the floor.
8. Disconnect the fuel tank unit connector.
9. Disconnect the quick connect fittings from the fuel tank.
10. Using tool 07AAA-SOXA100 or equivalent, loosen the locknut.
11. Remove the locknut. Lift up the fuel tank unit from the fuel tank. Disconnect the transfer tube.

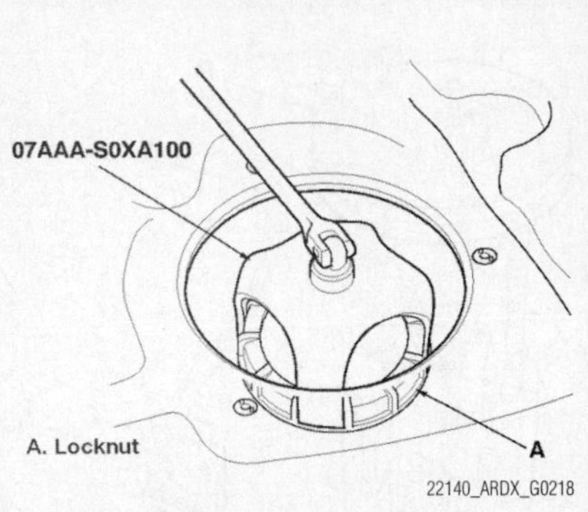
07AAA-S0XA100
A. Locknut
A
22140_ARDX_G0218

Fig. 164 Locknut removal

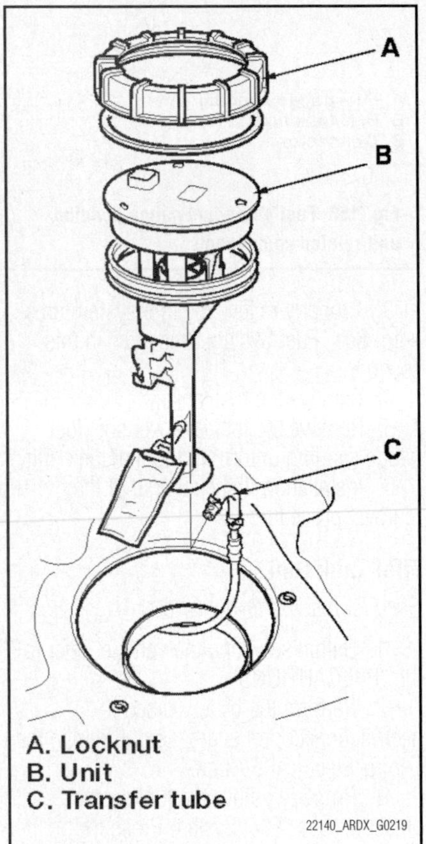
A
B
C
A. Locknut
B. Unit
C. Transfer tube
22140_ARDX_G0219

Fig. 165 Fuel tank unit

To install:

12. Temporarily attach a new base gasket to the assembly.
13. Position the mark on the transfer tube up.
14. Insert the assembly partially into the fuel tank.

➡**Be careful not to damage the new base gasket. Be careful not to bend the sending unit. Do not coat the base gasket with oil.**

15. Transfer the base gasket from the sending unit to the fuel tank.
16. Align the marks on the fuel tank and the fuel gauge sending unit. Insert the sending unit and allow it to contact the base gasket. See the illustration for proper alignment.

CAUTION

To avoid a fuel leak, check the base gasket visually or with your hand to be sure it is not pinched. Correct as required.

17. Using the tool, tighten the locknut with the new locknut plate to specification. Specification is 52 ft. lbs. (70 Nm).

➡**After tightening, be sure the marks are still aligned. To avoid a fuel leak, check the base gasket visually or with your hand to be sure it is not pinched. Correct as required.**

18. Connect the electrical connector.
19. Reconnect the negative battery cable and turn the ignition switch ON (II). The fuel pump will run for about two seconds and fuel pressure will rise.

➡**Do not operate the starter motor.**

20. Repeat the above Step two or three times and check that there is no leakage in the fuel supply system.
21. Continue the installation in the reverse order of the removal procedure.

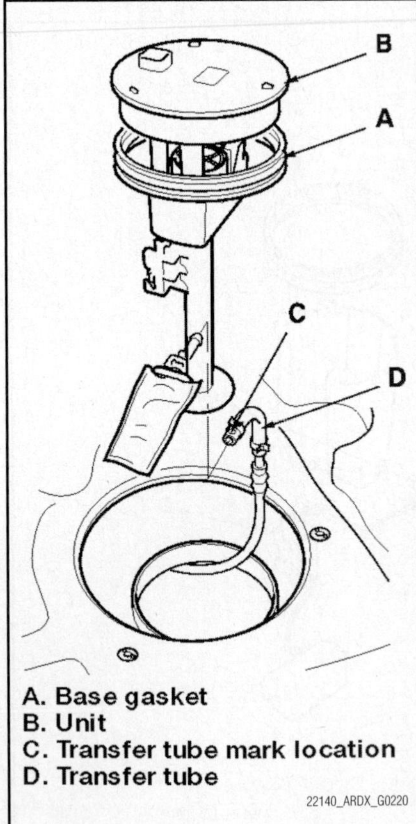
B
A
C
D
A. Base gasket
B. Unit
C. Transfer tube mark location
D. Transfer tube
22140_ARDX_G0220

Fig. 166 Transfer tube alignment

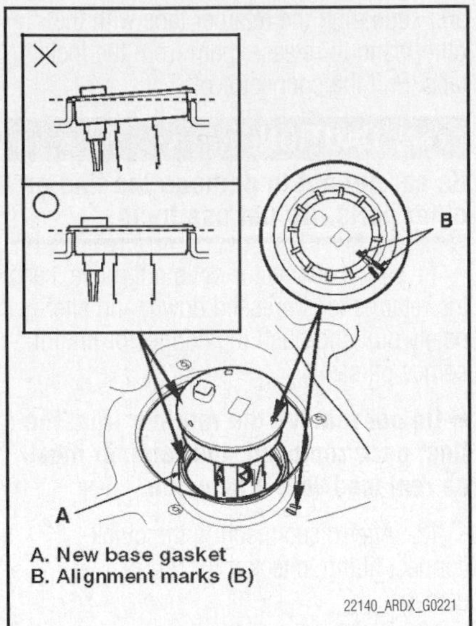
B
A
A. New base gasket
B. Alignment marks (B)
22140_ARDX_G0221

Fig. 167 Alignment marks

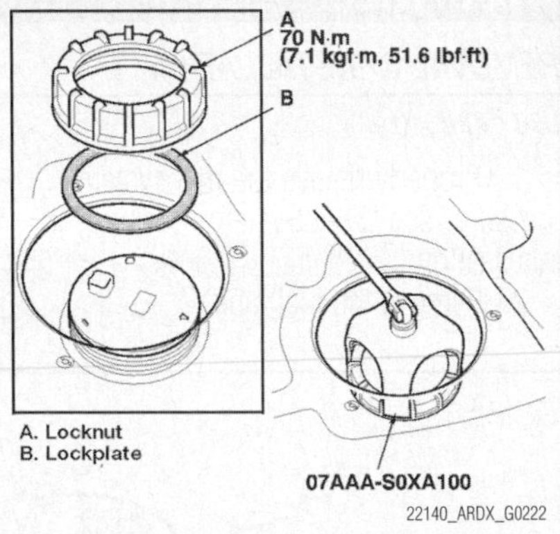
A
70 N·m
(7.1 kgf·m, 51.6 lbf·ft)
B
A. Locknut
B. Lockplate
07AAA-S0XA100
22140_ARDX_G0222

Fig. 168 Locknut and locknut plate installation

FUEL RAIL & INJECTOR

REMOVAL & INSTALLATION

See Figure 169.

1. Before servicing the vehicle, refer to the "PRECAUTIONS".
2. Perform the battery disconnect/reconnect procedure, as outlined in the Engine Electrical Section.
3. Properly relieve the fuel system pressure. See "Fuel System Pressure" in this section.
4. Remove the air charge cooler cover.
5. Disconnect the turbocharger boost sensor connector. Remove the vacuum hoses and the air bypass outlet connecting tube.
6. Remove the air charge cooler retaining bolts. Remove the air charge cooler.
7. Disconnect the throttle body connector. Remove the injector cover.

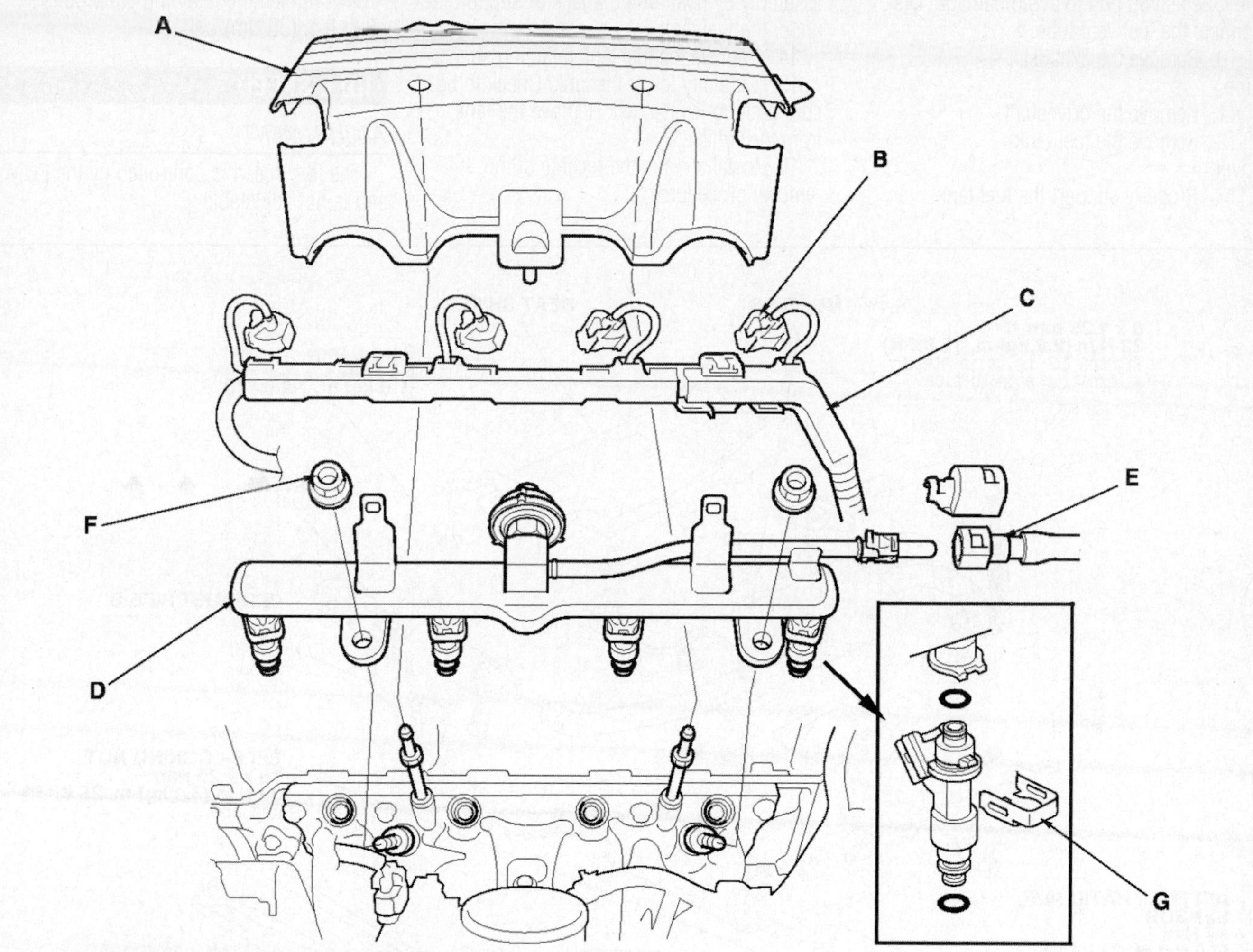
A
B
C
D
E
F
G
22140_ARDX_G0226

A. Cover
B. Connectors
C. Harness holder
D. Fuel rail
E. Quick connect fitting
F. Nuts
G. Injector clips

Fig. 169 Fuel injectors and related components

8. Disconnect the connectors from the injectors. Remove the harness holder from the fuel rail.
9. Disconnect the quick connectors.
10. Remove the fuel retaining nuts. Remove the fuel rail from its mounting.
11. Remove the injector clips from the injectors. Remove the injectors from the fuel rail.

To install:

12. Coat the new O-rings (black) with clean engine oil, and then insert the injectors into the fuel rail. Install the injector clips.
13. Coat the new O-rings (green) with clean engine oil, and then install the fuel rail and the injectors in the injector base.
14. Install the fuel rail retaining nuts. Tighten them to 16 ft. lbs. (22 Nm).
15. Connect the quick connect fitting.
16. Turn the ignition switch ON (II). The fuel pump will run for about two seconds and fuel pressure will rise.

➡Do not operate the starter motor.

17. Repeat the above step two or three times and check that there is no leakage in the fuel supply system.
18. Continue the installation in the reverse order of the removal procedure.
19. Clear any stored DTC codes, as required.

FUEL TANK

REMOVAL & INSTALLATION

See Figures 170 and 171.

1. Before servicing the vehicle, refer to the "PRECAUTIONS".
2. Perform the battery disconnect/ reconnect procedure, as outlined in the Engine Electrical Section.
3. Properly relieve the fuel system pressure. See "Fuel System Pressure" in this section.
4. Remove the fuel tank unit. Using a hand pump, hose and container drain the fuel from the tank.
5. Reinstall the fuel tank unit without connecting the quick connect fittings and the electrical connector.
6. Disconnect the secondary fuel gauge electrical connector.
7. Remove the fuel fill pipe cover. Disconnect the quick connect fitting.
8. Raise and support the vehicle safely.
9. Disconnect the fuel fill tube from the fuel tank. Slide back the clamp, then twist

the tube as you pull to avoid damage. Disconnect the fuel vent tube.

10. Remove the exhaust pipe.
11. Remove the driveshaft.
12. Remove the fuel tank cover.
13. Properly support the fuel tank assembly by positioning a jack or support under the fuel tank.
14. Remove the fuel tank retaining straps.
15. Carefully lower the tank. Check to be sure nothing is attached. Remove the tank from the vehicle.
16. Installation is the reverse of the removal procedure.
17. Tighten the retaining strap bolts to 28 ft. lbs. (38 Nm).

IDLE SPEED

ADJUSTMENT

The idle speed is controlled by the PCM and is not adjustable.

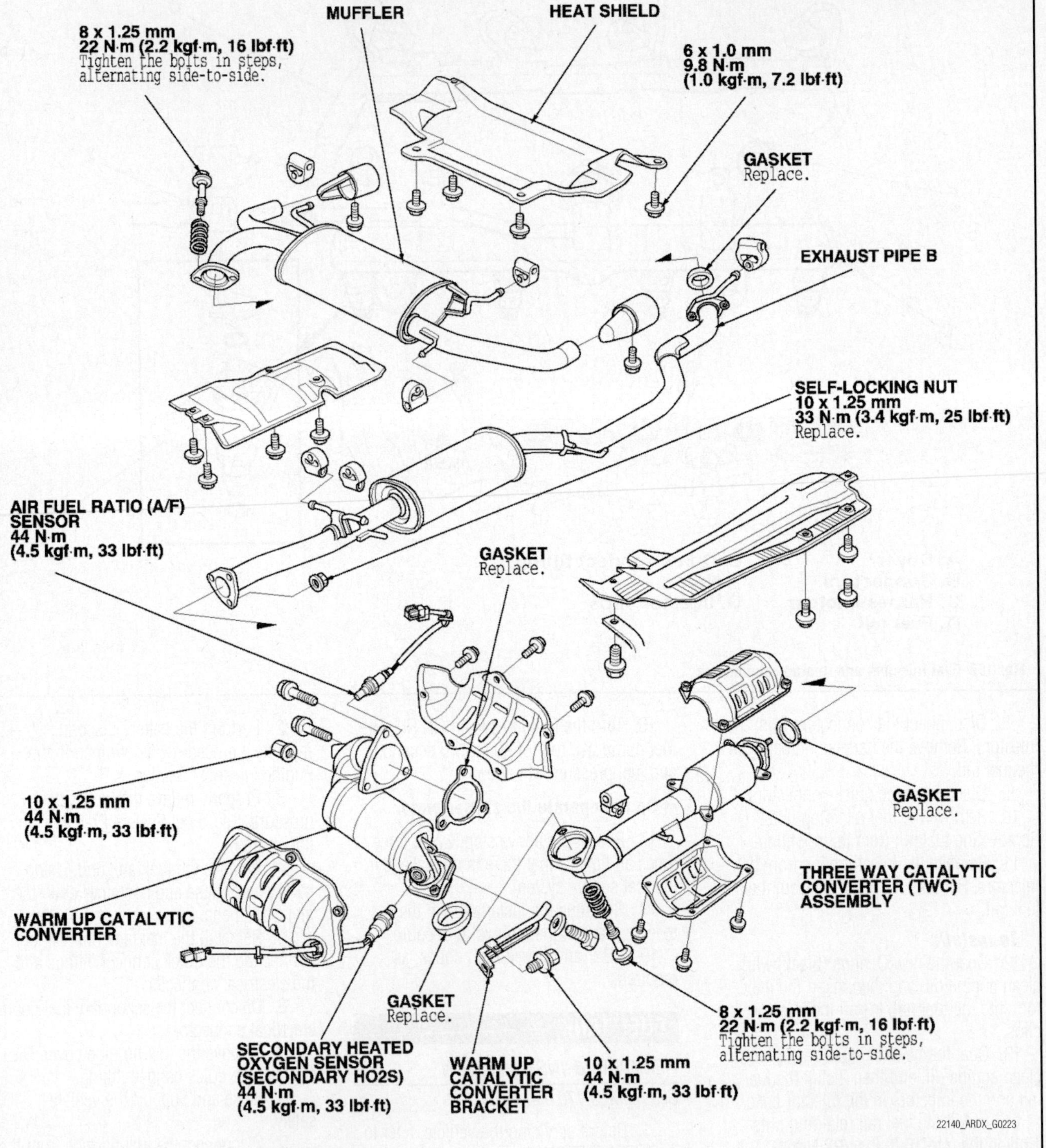

MUFFLER
HEAT SHIELD
8 x 1.25 mm
22 N·m (2.2 kgf·m, 16 lbf·ft)
Tighten the bolts in steps, alternating side-to-side.
6 x 1.0 mm
9.8 N·m
(1.0 kgf·m, 7.2 lbf·ft)
GASKET
Replace.
EXHAUST PIPE B
SELF-LOCKING NUT
10 x 1.25 mm
33 N·m (3.4 kgf·m, 25 lbf·ft)
Replace.
AIR FUEL RATIO (A/F) SENSOR
44 N·m
(4.5 kgf·m, 33 lbf·ft)
GASKET
Replace.
10 x 1.25 mm
44 N·m
(4.5 kgf·m, 33 lbf·ft)
GASKET
Replace.
THREE WAY CATALYTIC CONVERTER (TWC) ASSEMBLY
WARM UP CATALYTIC CONVERTER
GASKET
Replace.
8 x 1.25 mm
22 N·m (2.2 kgf·m, 16 lbf·ft)
Tighten the bolts in steps, alternating side-to-side.
SECONDARY HEATED OXYGEN SENSOR (SECONDARY HO2S)
44 N·m
(4.5 kgf·m, 33 lbf·ft)
WARM UP CATALYTIC CONVERTER BRACKET
10 x 1.25 mm
44 N·m
(4.5 kgf·m, 33 lbf·ft)
22140_ARDX_G0223

Fig. 170 Exhaust pipe and related components

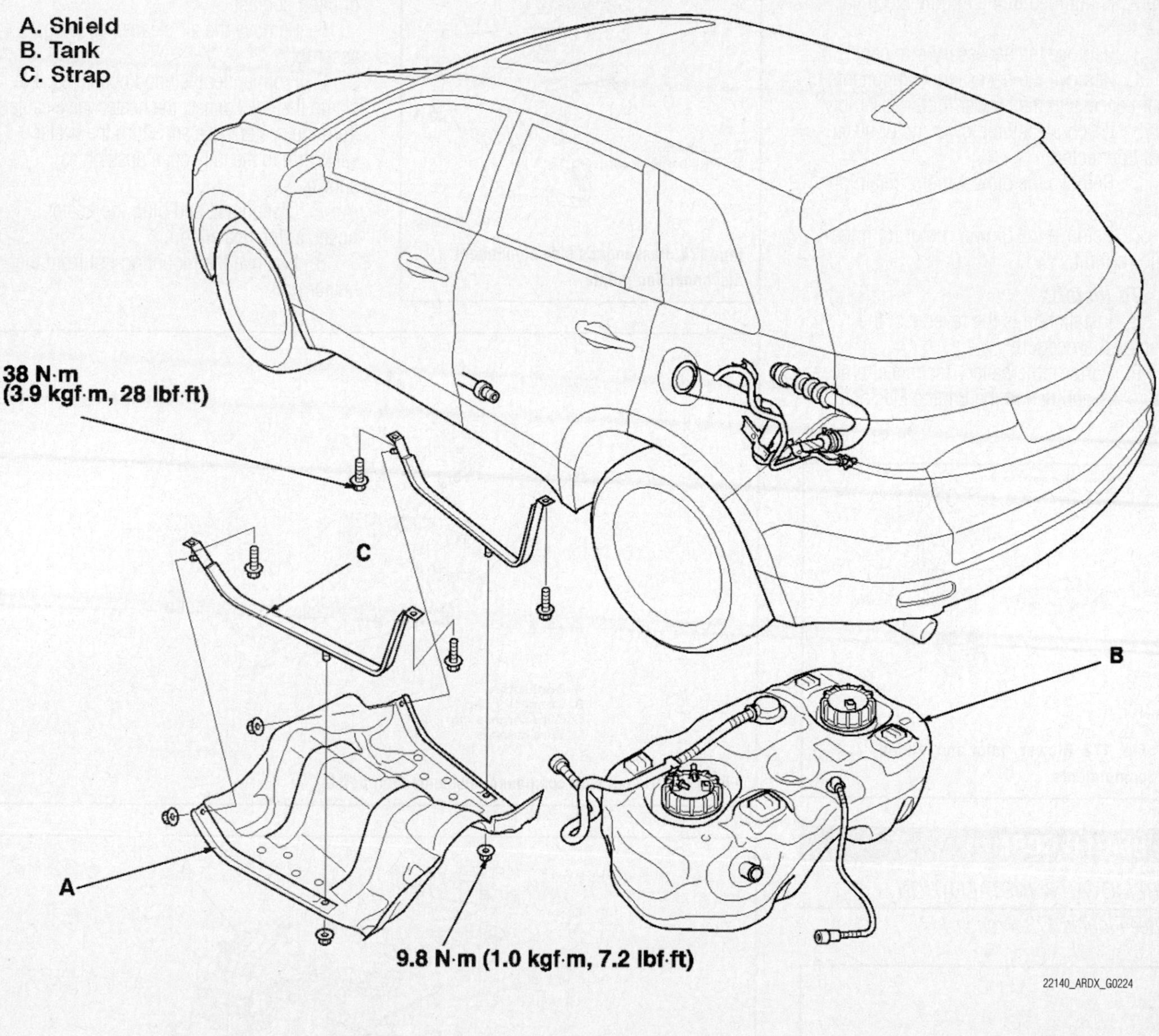
A. Shield
B. Tank
C. Strap
38 N·m
(3.9 kgf·m, 28 lbf·ft)
C
B
A
9.8 N·m (1.0 kgf·m, 7.2 lbf·ft)
22140_ARDX_G0224

Fig. 171 Fuel tank and related components

HEATING & AIR CONDITIONING SYSTEM

BLOWER MOTOR

REMOVAL & INSTALLATION

See Figure 172.

1. Before servicing the vehicle, refer to the "PRECAUTIONS" section at the start of this section.
2. Perform the battery disconnect procedure, as outlined in the Engine Electrical Section.
3. Remove the necessary trim panels.
4. Remove the necessary components to gain access to the blower motor assembly.
5. Disconnect the blower motor electrical connectors.
6. Remove the blower motor retaining screws.
7. Remove the blower motor from the blower unit.

To install:

8. Installation is the reverse of the removal procedure.
9. Perform the battery reconnect procedure, as outlined in the Engine Electrical Section.

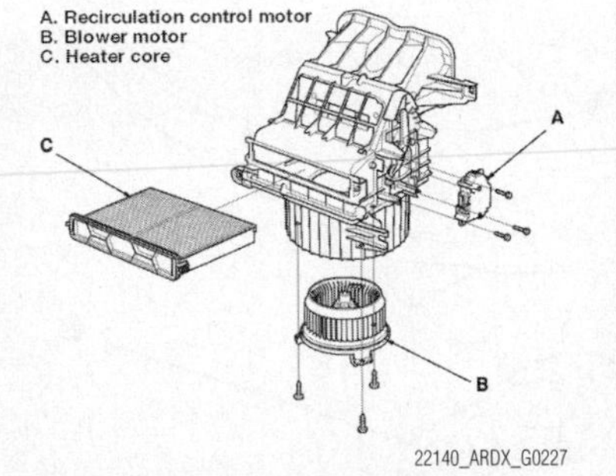
A. Recirculation control motor
B. Blower motor
C. Heater core
A
B
C
22140_ARDX_G0227

Fig. 172 Blower motor and related components

HEATER CORE

REMOVAL & INSTALLATION

See Figures 173 through 177.

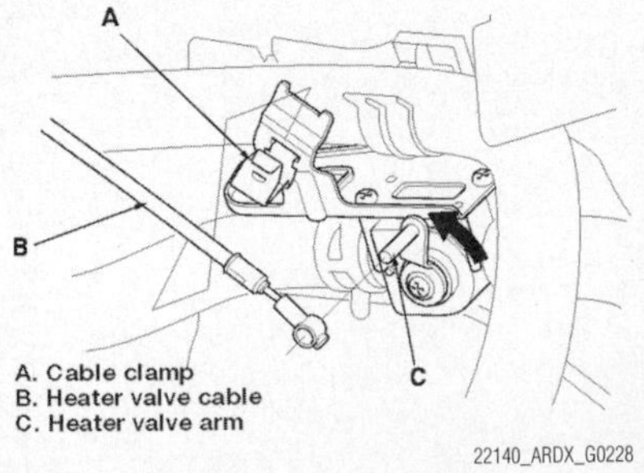
A
B
C
A. Cable clamp
B. Heater valve cable
C. Heater valve arm
22140_ARDX_G0228

Fig. 173 Heater valve arm positioning

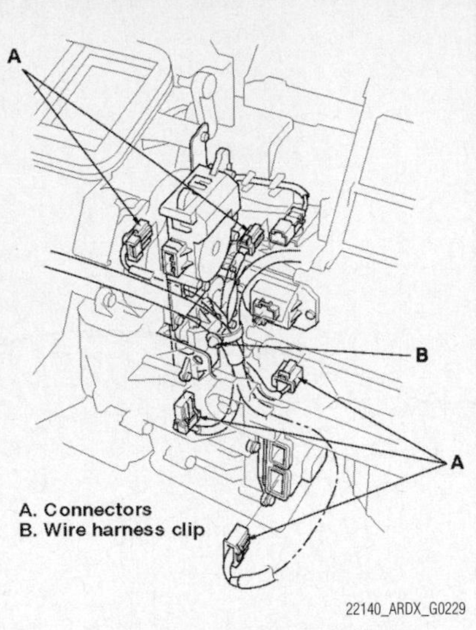
A
B
A
A. Connectors
B. Wire harness clip
22140_ARDX_G0229

Fig. 174 Passenger's side component disconnection points

1. Before servicing the vehicle, refer to the "PRECAUTIONS" section at the start of this section.
2. Perform the battery disconnect procedure, as outlined in the Engine Electrical Section.
3. Properly discharge the air conditioning system.
4. Drain the coolant. Properly dispose of used coolant.
5. Remove the air cleaner housing assembly.
6. From under the hood open the cable clamp then disconnect the heater valve cable from the heater valve arm. Turn the heater valve arm to the fully open position as shown.
7. Disconnect and plug the heater hoses at the heater core.
8. Remove the mounting nut from the heater unit.

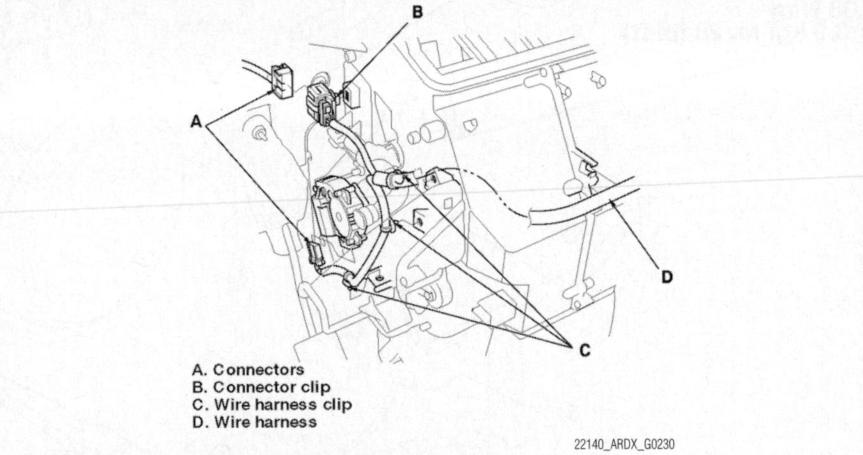
B
A
D
C
A. Connectors
B. Connector clip
C. Wire harness clip
D. Wire harness
22140_ARDX_G0230

Fig. 175 Driver's side component disconnection points

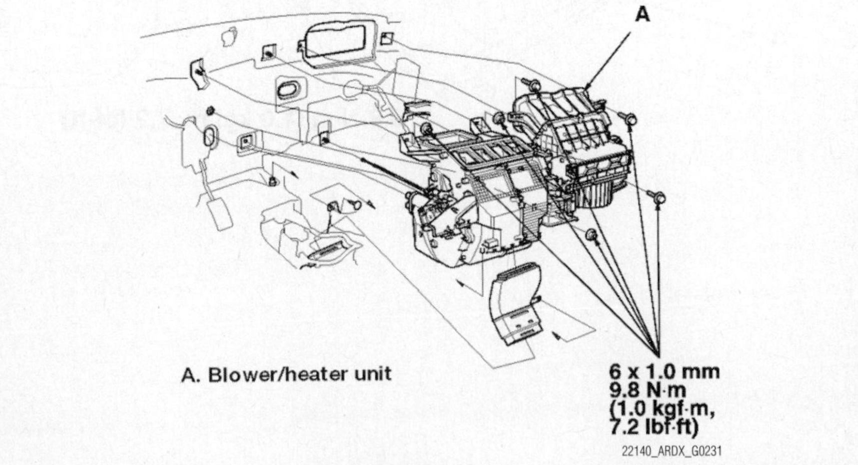
A
A. Blower/heater unit
6 x 1.0 mm
9.8 N·m
(1.0 kgf·m,
7.2 lbf·ft)
22140_ARDX_G0231

Fig. 176 Blower/heater unit and related components

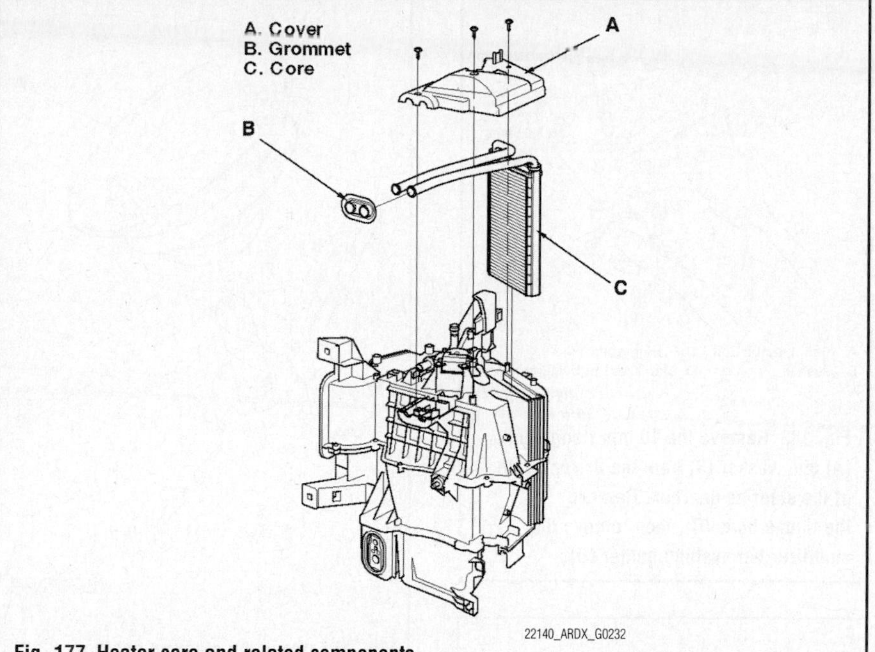
A. Cover
B. Grommet
C. Core
A
B
C
22140_ARDX_G0232

Fig. 177 Heater core and related components

9. Remove the lower instrument panel.
10. Disconnect the blower motor electrical connector. Remove the wire harness clip and ground terminal bolt.
11. Disconnect the connector from the recirculation control motor.
12. Disconnect the connectors from the climate control unit, mode control motor, passenger's air mix door motor, evaporator temperature sensor and the power transistor.
13. Remove the wire harness.
14. Disconnect the connectors from the driver's air mix door motor and air conditioning wirc harness.
15. Remove the connector clip, the wire harness clips and the wire harness.
16. Remove the mounting bolt, mounting nuts, and remove the blower/heater unit from the vehicle.
17. Remove the self-taping screws. Remove the heater core cover.
18. Remove the grommet and carefully pull the heater core out of its mounting.

To install:

19. Installation is the reverse of the removal procedure.
20. Do not interchange the heater inlet and outlet hoses.
21. Be sure to use new O-rings, as required.
22. Perform the battery reconnect procedure, as outlined in the Engine Electrical Section.
23. Evacuate, recharge and leak test the system.
24. Fill the engine with the proper grade and type engine coolant.
25. Start the engine and check for leaks. Correct as required.
26. Be sure that the air conditioning system is functioning properly.

STEERING

POWER RACK & PINION STEERING GEAR

REMOVAL & INSTALLATION

See Figures 178 through 188.

1. Drain the power steering fluid.
2. Perform the battery disconnect procedure, as outlined in the Engine Electrical Section.
3. Raise the front of vehicle, and support it with safety stands in the proper locations.
4. Remove the front wheels.
5. Remove the steering wheel.
6. Remove the driver's dashboard undercover.
7. Remove the steering joint cover at the floorboard.
8. Remove the steering joint bolt, and disconnect the steering joint by moving the steering joint toward the column.
9. Hold the lower slide shaft on the column with a piece of wire between the joint yoke on the lower slide shaft to the joint yoke on the upper shaft.
10. Remove the center guide (if equipped), and discard it.

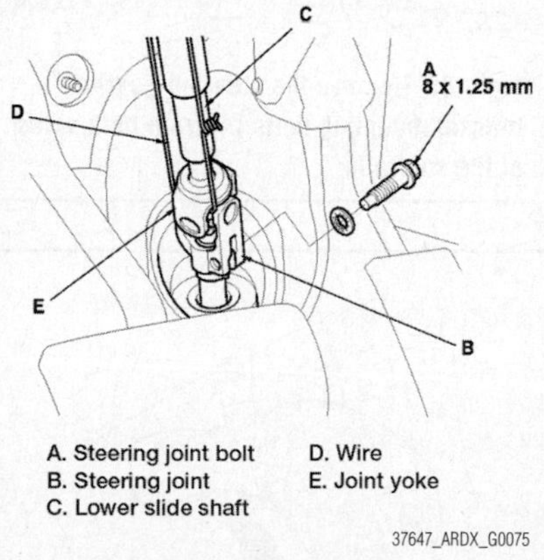
C
A
8 x 1.25 mm
D
E
B
A. Steering joint bolt
B. Steering joint
C. Lower slide shaft
D. Wire
E. Joint yoke
37647_ARDX_G0075

Fig. 178 Removing the steering joint from the lower steering shaft

➡The center guide is for factory assembly only.

11. Remove the tie rod ball joint nut cotter pin from and loosen the nut.
12. Separate the tie rod ball joint and knuckle using the ball joint remover.
13. Remove the air cleaner housing and the under-hood fuse/relay box.

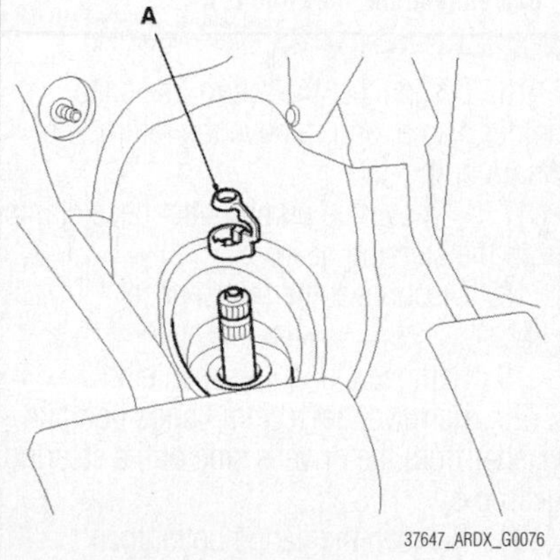
A
37647_ARDX_G0076

Fig. 179 Remove the center guide (A) (if equipped), and discard it.

14. Loosen the adjustable hose clamp, and disconnect the return hose. Remove the inlet line clamp bolt. Loosen the 18 mm flare nut, and disconnect the inlet line.
15. Remove the 10 mm flange bolt from the driver's side of the steering gearbox.

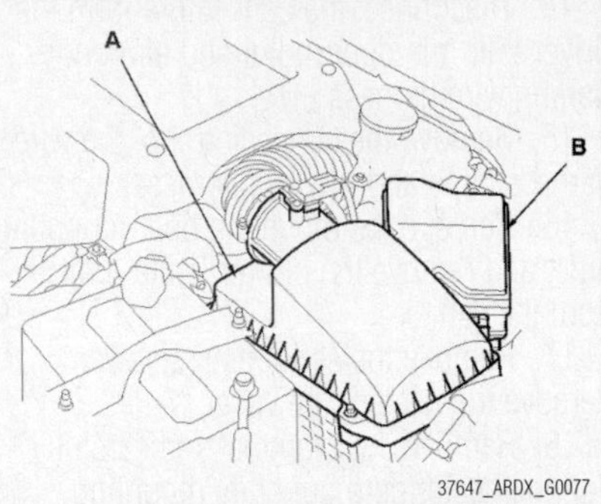
A
B
37647_ARDX_G0077

Fig. 180 Remove the air cleaner housing (A) and the under-hood fuse/relay box (B).

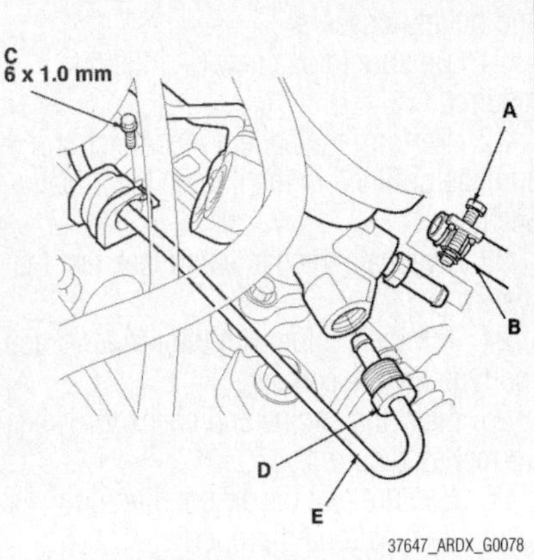
C
6 x 1.0 mm
A
B
D
E
37647_ARDX_G0078

Fig. 181 Loosen the adjustable hose clamp (A), and disconnect the return hose (B). Remove the inlet line clamp bolt (C). Loosen the 18 mm flare nut (D), and disconnect the inlet line (E).

16. Disconnect the return line snap holder clamp, and remove the return line clamp bolt.

17. Remove the pump outlet hose clamp from the steering gearbox.

18. Disconnect the return hose clips.

19. Remove the P/S heat shield.

20. Remove the 10 mm flange bolt and washer from the driver's side of the steering gearbox.

21. Remove the flange bolts, then remove the stabilizer bar bushing holder.

22. Repeat this process on the passenger's side of the steering gearbox.

23. Remove the steering gearbox bracket mounting bolts from both sides of the subframe.

24. Remove the gearbox mounting brackets.

25. Move the steering gearbox toward the front, and remove the pinion shaft grommet from the top of the valve housing. Apply vinyl tape to the splines on the pinion shaft.

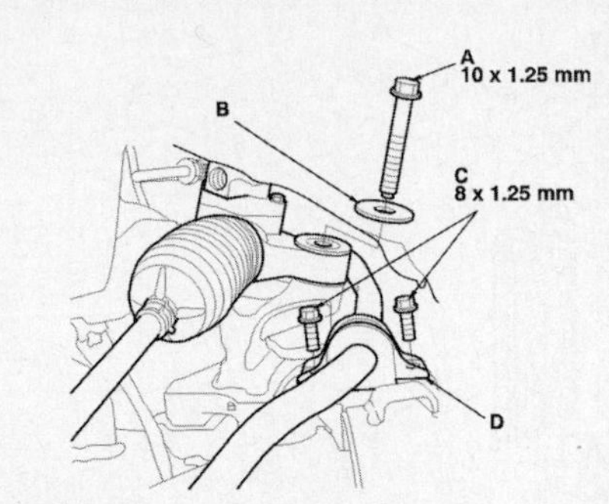
A
10 x 1.25 mm
B
C
8 x 1.25 mm
D

A. 10 mm flange bolt C. Flange bolts
B. Washer D. Stabilizer bar bushing holder
37647_ARDX_G0080

Fig. 182 Remove the 10 mm flange bolt (A) and washer (B) from the driver's side of the steering gearbox. Remove the flange bolts (C), then remove the stabilizer bar bushing holder (D).

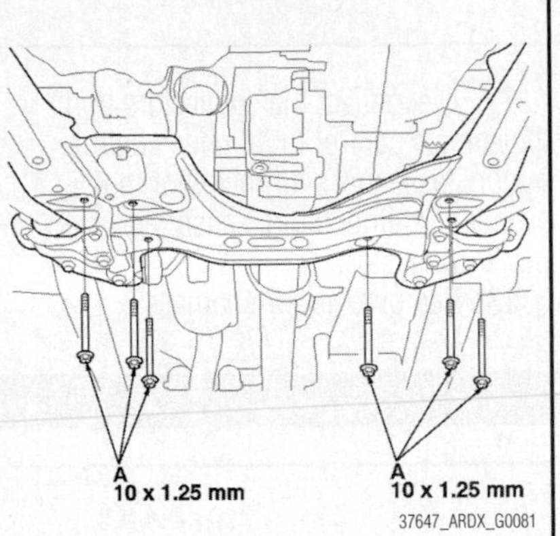
A
10 x 1.25 mm
A
10 x 1.25 mm
37647_ARDX_G0081

Fig. 183 Remove the steering gearbox bracket mounting bolts (A) from both sides of the subframe.

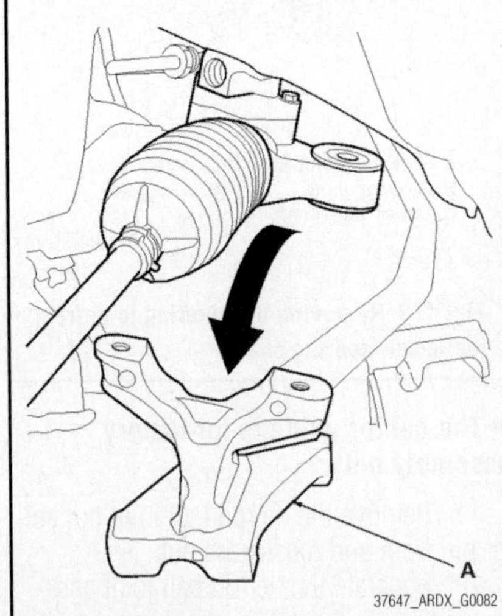
A
37647_ARDX_G0082

Fig. 184 Remove the gearbox mounting brackets (A).

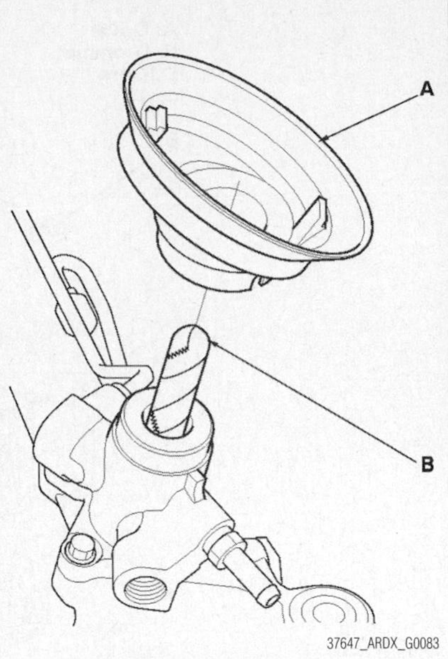
A
B
37647_ARDX_G0083

Fig. 185 Move the steering gearbox toward the front, and remove the pinion shaft grommet (A) from the top of the valve housing. Apply vinyl tape (B) to the splines on the pinion shaft.

26. Carefully move the steering gearbox and tie rods as an assembly toward the front side until the pinion shaft clears the wheel well opening on the frame.

27. Remove the steering gearbox through the wheel well opening on the driver's side.

✱✱ CAUTION

After removing the steering gearbox, make sure that no power steering fluid gets on the gearbox mount cushions, gearbox housing, or the surface of the front subframe. Wipe off any spilled fluid at once.

To install:

28. Using solvent and a brush, wash any oil and dirt off the valve body unit, its lines, and the end of the gearbox. Blow dry with compressed air.

29. Slide the steering gearbox between the front subframe and the body from the driver's side. Place the gearbox in position on the front subframe.

30. Rotate the steering gearbox so the pinion shaft points downward.

31. Continue moving the gearbox toward the passenger's side until the steering gearbox is in position.

32. Make sure the power steering return line and feed line are routed above the gearbox.

CAUTION

Be sure to remove the steering wheel before disconnecting the steering joint. Damage to the cable reel can occur.

33. Remove the vinyl tape from pinion shaft, and install the pinion shaft grommet. Align the slot in the pinion shaft grommet with the lug portion on the valve housing.
34. Install the gearbox mounting brackets.
35. Install and tighten the steering gearbox bracket mounting bolts (through the subframe) to 54 ft. lbs. (74 Nm).
36. Loosely install the passenger's side gearbox mounting bolts and washer. Install the passenger's side stabilizer bar bushing holder and tighten the holder bolts to 16 ft. lbs. (22 Nm).
37. Repeat the process on the driver's side.
38. Install the P/S heat shield.
39. Install the pump outlet hose clamp on the steering gearbox and connect the return hose clips
40. Connect the return line holder, and install the return line clamp bolt.
41. Loosely install the remaining 10 mm gearbox mounting bolt.
42. Tighten the gearbox mounting bolts, in an alternating pattern, to 43 ft. lbs. (59 Nm).
43. Connect the inlet line, and tighten the 18 mm flare nut to 35 ft. lbs. (47 Nm).
44. Connect the return hose securely, and tighten the adjustable hose clamp.
45. Install the inlet line clamp bolt
46. Install the under-hood fuse/relay box and the air cleaner housing.
47. Wipe off any grease contamination from the ball joint tapered section and threads. Reconnect the tie rod ends to the steering knuckles. Install the 12 mm nut, and tighten it to 40 ft. lbs. (54 Nm). Install a new cotter pin, and bend it as shown.
48. Center the steering rack within its stroke in steering joint connection.
49. With the rack in the straight ahead driving position, cut the wire and slip the lower end of the steering joint onto the pinion shaft in the range shown.
50. Align the bolt hole on the steering joint with the groove around the pinion shaft, then loosely install the joint bolt. Be sure that the joint bolt is securely in the groove in the pinion shaft. Pull on the steering joint to make sure that the steering joint is fully seated. Tighten the steering joint bolt to 21 ft. lbs. (28 Nm).
51. Install the steering joint cover at the floorboard.
52. Install the driver's dashboard undercover.
53. Install the front wheel, then set the wheels in the straight ahead position.
54. Install the steering wheel.
55. Perform the battery terminal reconnection procedure. See "Battery System" in "ENGINE ELECTRICAL" section.
56. Perform the following tasks:
a. Verify cruise control, audio remote, navigation and HFL voice controls, horn, and turn signal switch operation.
b. Make sure the steering wheel is centered.
c. Fill the system with power steering fluid, and bleed air from the system. See "Bleeding" under "Power Steering Pump" in this section.
57. After installation, do the following checks:
a. Start the engine, allow it to idle, and turn the steering wheel from lock-to-lock several times to warm up the fluid Check the gearbox for leaks..
b. Check the steering wheel spoke angle.
c. If steering spoke angles to the right and left are not equal (steering wheel and rack are not centered), correct the engagement of the joint/pinion shaft serrations.
d. Set the steering column to the center tilt position, and to the center telescopic position, then do the front toe inspection/adjustment.

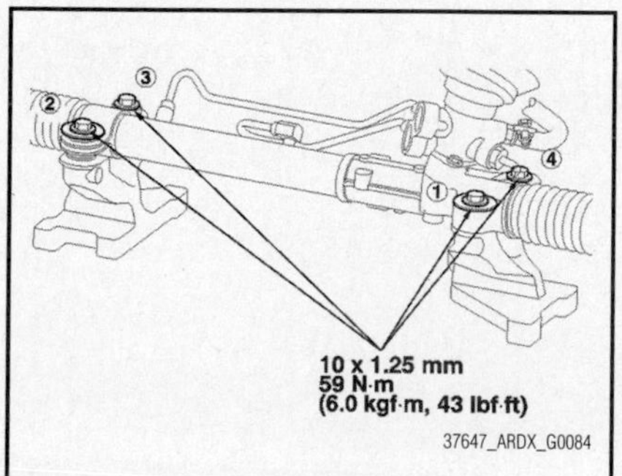
10 x 1.25 mm
59 N·m
(6.0 kgf·m, 43 lbf·ft)
37647_ARDX_G0084

Fig. 186 Tighten the gearbox mounting bolts, in an alternating pattern, to 43 ft. lbs. (59 Nm).

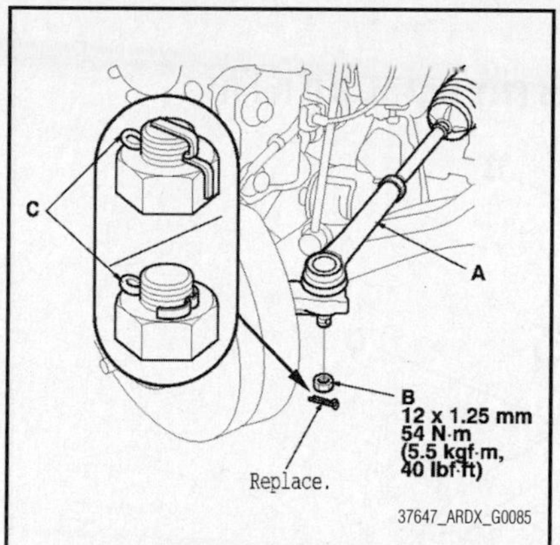
C
A
B
12 x 1.25 mm
54 N·m
(5.5 kgf·m,
40 lbf·ft)
Replace.
37647_ARDX_G0085

Fig. 187 Wipe off any grease contamination from the ball joint tapered section and threads. Reconnect the tie rod ends (A) to the steering knuckles. Install the 12 mm nut (B), and tighten it to 40 ft. lbs. (54 Nm). Install a new cotter pin (C), and bend it as shown.

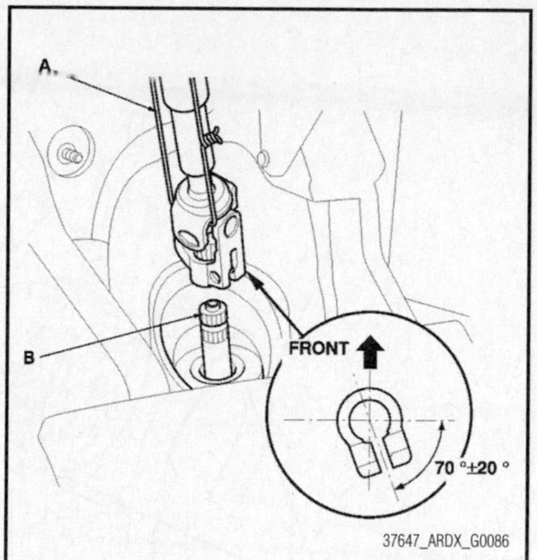
A
B
FRONT
70 °±20 °
37647_ARDX_G0086

Fig. 188 With the rack in the straight ahead driving position, cut the wire (A) and slip the lower end of the steering joint onto the pinion shaft (B) in the range shown.

POWER STEERING PUMP

REMOVAL & INSTALLATION

See Figure 189.

1. Drain the power steering reservoir. Be sure to properly dispose of used power steering fluid.
2. Remove the drive belt from the pump pulley.
3. Disconnect and plug the power steering pump fluid lines.

➡Be sure to cover the auto tensioner, alternator and compressor with a shop towel in order to prevent fluid from spilling on these components.

4. Remove the power steering pump retaining bolts. Remove the power steering pump from its mounting.

➡Cover the opening of the pump with a piece of tape to prevent dirt and debris from entering the pump.

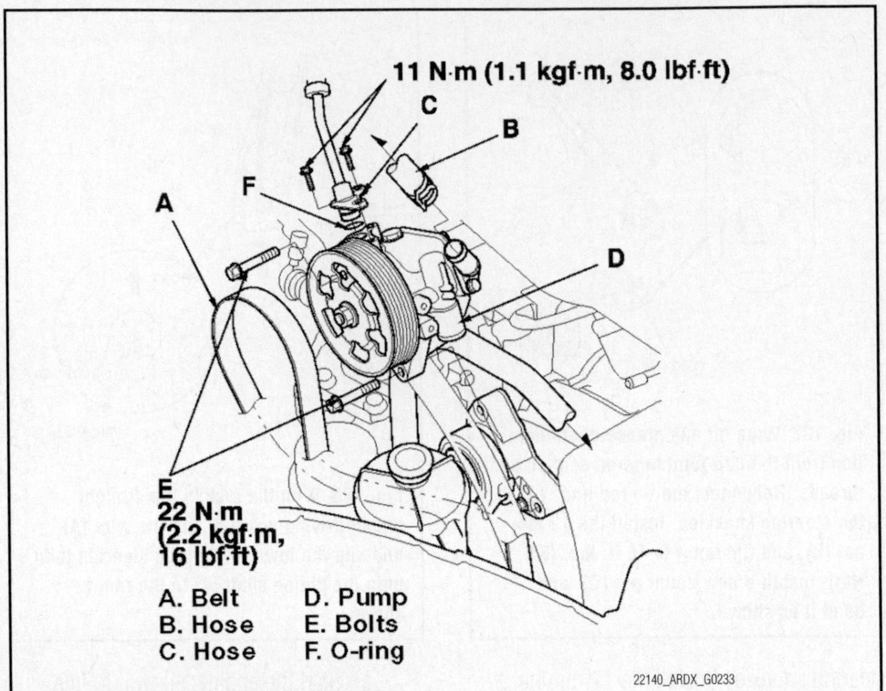
11 N·m (1.1 kgf·m, 8.0 lbf·ft)
C
B
F
A
D
E
22 N·m
(2.2 kgf·m,
16 lbf·ft)
A. Belt
B. Hose
C. Hose
D. Pump
E. Bolts
F. O-ring
22140_ARDX_G0233

Fig. 189 Power steering pump and related components

To install:

5. Connect the power steering pump hoes using new O-rings.
6. Loosely install the pump in the pump bracket with the mounting bolts. Tighten to 16 ft. lbs. (22 Nm).
7. Install the drive belt. Make sure that the belt is properly positioned on the pulleys.
8. Fill the reservoir with the proper grade and type power steering fluid.

BLEEDING

1. Before servicing the vehicle, refer to the "PRECAUTIONS" section at the start of this vehicle menu.
2. Fill the power steering reservoir with the proper grade and type power steering fluid.
3. Start the engine and run it at fast idle.
4. Turn the steering wheel from lock-to-lock several times to bleed the air from the system.
5. Recheck the fluid level, correct as required.
6. Do not fill the reservoir past the upper level line.

FLUID FILL PROCEDURE

1. Check the reservoir at regular intervals, and add the recommended fluid as necessary.

CAUTION

Always use Acura Power Steering Fluid. Using any other type of power steering fluid or automatic transmission fluid can cause increased wear and poor steering in cold weather.

➡If the fluid is contaminated, the screen in the reservoir may be partially blocked. Replace the reservoir if necessary.

➡Use the following procedure if fluid replacement of the power steering pump is required:

2. Remove the reservoir from its holder. Raise the reservoir, then disconnect the return hose to drain the reservoir. Take care not to spill the fluid on the body and parts. Wipe off any spilled fluid at once.

➡Inspect the reservoir screen for any debris. If the reservoir screen is clogged, replace the reservoir.

3. Connect a hose of suitable diameter to the disconnected return hose, and put the hose end in a suitable container.
4. Start the engine, let it run at idle, and turn the steering wheel from lock-to-lock several times. When fluid stops running out of the hose, shut off the engine. Discard the fluid.
5. Reinstall the return hose on the reservoir.
6. Fill the reservoir to the upper level line.
7. Start the engine and run it at fast idle, then turn the steering from lock-to-lock several times to bleed air from the system.
8. Recheck the fluid level and add some if necessary. Do not fill the reservoir beyond the upper level line.
9. If the fluid is contaminated, dark, or discolored, repeat the procedure as necessary.
10. Note the system capacity:
    - 0.96 U.S. qt. (0.9 L) at disassembly
    - Reservoir capacity: 0.32 U.S. qt. (0.3 L)

LOWER BALL JOINTS

REMOVAL & INSTALLATION

See Figures 190 and 191.

1. Disconnect the positive battery cable.
2. Raise and support the vehicle safely. Remove the tire and wheel assemblies.
3. Remove the flange bolt and flange nuts from the lower arm.
4. Disconnect the lower ball joint from the lower arm.
5. Remove the lock pin from the lower ball joint. Remove the castle nut.
6. Install the ball joint thread protector. Using the ball joint removal tool, remove the lower ball joint.

To install:

7. Installation is the reverse of the removal procedure.
8. Loosely install the new flange bolt and flange nuts. Tighten to 38 ft. lbs. (52 Nm) in the following order:
   - Front nut
   - Rear nut
   - Rear bolt
9. Tighten the castle nut to 43–54 ft. lbs. (58–73 Nm), and then tighten it only enough to install the cotter pin. Do not align the castle nut by loosening it.

➡First install all the components, and lightly tighten the bolts and nuts, then tighten the lower ball joint to the lower arm to specification. Raise the suspension to load it with the vehicles weight before fully tightening the lower ball joint to the knuckle to specification.

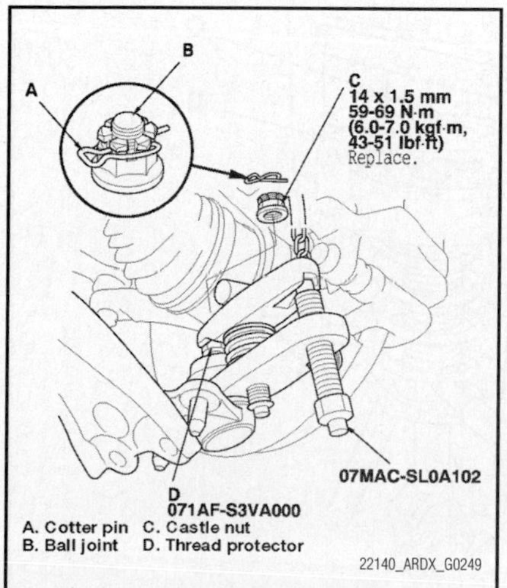

A
B
C
14 x 1.5 mm
59-69 N·m
(6.0-7.0 kgf·m,
43-51 lbf·ft)
Replace.
07MAC-SL0A102
D
071AF-S3VA000
A. Cotter pin C. Castle nut
B. Ball joint D. Thread protector
22140_ARDX_G0249

Fig. 190 Lower ball joint tool installation

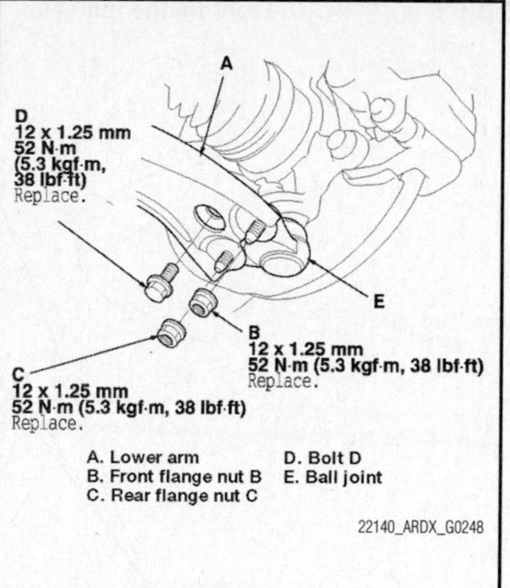

A
D
12 x 1.25 mm
52 N·m
(5.3 kgf·m,
38 lbf·ft)
Replace.
E
B
12 x 1.25 mm
52 N·m (5.3 kgf·m, 38 lbf·ft)
Replace.
C
12 x 1.25 mm
52 N·m (5.3 kgf·m, 38 lbf·ft)
Replace.
A. Lower arm D. Bolt D
B. Front flange nut B E. Ball joint
C. Rear flange nut C
22140_ARDX_G0248

Fig. 191 Lower ball joint and related components

10. Check and adjust the wheel alignment, as required.

LOWER CONTROL ARMS

REMOVAL & INSTALLATION

See Figures 191 and 192

1. Raise and support the vehicle safely. Remove the tire and wheel assemblies.
2. Remove the flange bolts and the bushing holder bolts from the front stabilizer bar.
3. Remove the flange bolt and flange nuts from the lower arm.
4. Disconnect the lower ball joint from the lower arm.
5. Remove the lower arm mounting bolt.
6. Remove the lower arm mounting bolts. Remove the lower arm from the front suspension subframe.

To install:

7. Installation is the reverse of the removal procedure.
8. Be sure to use new bolts when installing the lower control arm.
9. Tighten the stabilizer bushing bolts to 16 ft. lbs. (22 Nm).
10. Loosely install the new flange bolt and flange nuts. Tighten to 38 ft. lbs. (52 Nm) in the following order:
    - Front flange nut
    - Rear flange nut
    - Bolt
11. Tighten the castle nut to 43–54 ft. lbs. (58–73 Nm), and then tighten it only enough to install the cotter pin. Do not align the castle nut by loosening it.

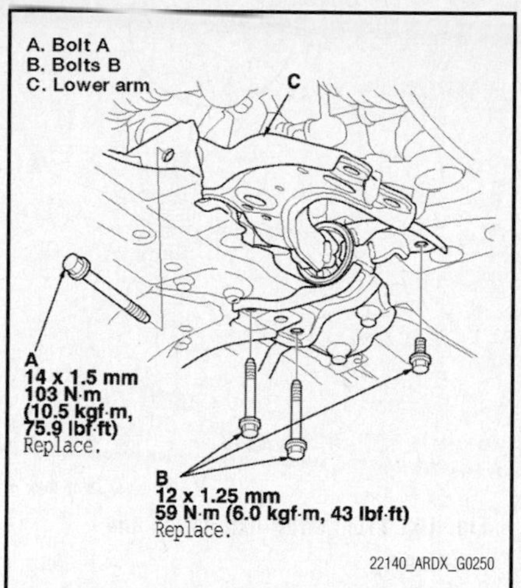

A. Bolt A
B. Bolts B
C. Lower arm
C
A
14 x 1.5 mm
103 N·m
(10.5 kgf·m,
75.9 lbf·ft)
Replace.
B
12 x 1.25 mm
59 N·m (6.0 kgf·m, 43 lbf·ft)
Replace.
22140_ARDX_G0250

Fig. 192 Lower control arm and related components

➡First install all the components, and lightly tighten the bolts and nuts, then tighten the lower ball joint to the lower arm to specification. Raise the suspension to load it with the vehicles weight before fully tightening the lower ball joint to the knuckle to specification.

12. Check and adjust the wheel alignment, as required.

MACPHERSON STRUTS

REMOVAL & INSTALLATION

See Figures 193 through 195.

1. Raise and support the vehicle safely. Remove the tire and wheel assemblies.
2. Remove the wheel sensor harness clip, the wire guide and the brake hose bracket from the strut. Do not disconnect the wheel sensor connector.
3. Disconnect the stabilizer link from the strut.
4. Remove the strut pinch bolts and self locking nuts from the strut.
5. Remove the lid from the cowl cover.
6. Remove the flange nuts from the top of the strut.

➡Strut springs are different, left and right. Mark the springs "L" and "R" before removing them.

7. Remove the strut from the vehicle.

To install:

8. Installation is the reverse of the removal procedure.

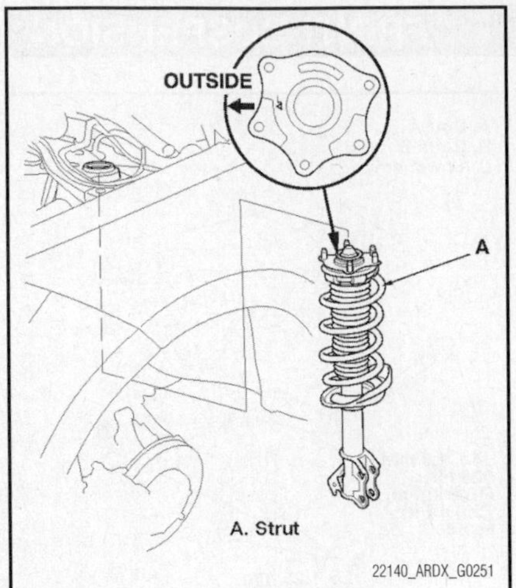
OUTSIDE
A
A. Strut
22140_ARDX_G0251

Fig. 193 Front strut installation and positioning

9. Be sure to use new bolts and nuts, as required.

10. Loosely install the new upper strut retaining flange nuts.

11. Loosely install the new front strut retaining bolts and new self locking nuts.

12. Connect the stabilizer link to the strut. Tighten the new nut to 58 ft. lbs. (78 Nm).

13. Install the wheel speed sensor harness clip, wire guide and brake hose bracket to the strut.

14. Raise the front suspension with a floor jack to load the suspension with the vehicles weight.

15. Tighten the strut bolts and self locking nuts to 122 ft. lbs. (165 Nm).

16. Tighten the upper strut nuts to specification and in the order shown in the illustration. Specification is 33 ft. lbs. (45 Nm). Install the cowl cover.

17. Check and adjust the wheel alignment, as required.

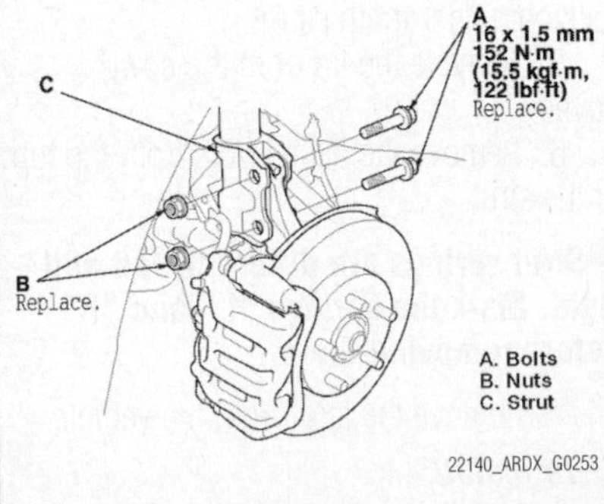
A
16 x 1.5 mm
152 N·m
(15.5 kgf·m,
122 lbf·ft)
Replace.
C
B
Replace.
A. Bolts
B. Nuts
C. Strut
22140_ARDX_G0253

Fig. 194 Front strut lower bolt installation

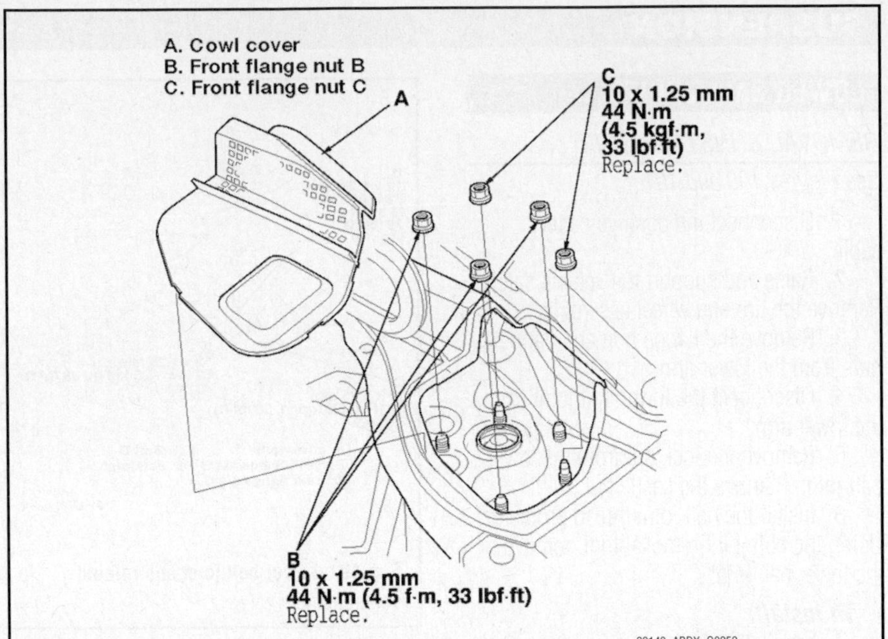
A. Cowl cover
B. Front flange nut B
C. Front flange nut C
A
C
10 x 1.25 mm
44 N·m
(4.5 kgf·m,
33 lbf·ft)
Replace.
B
10 x 1.25 mm
44 N·m (4.5 f·m, 33 lbf·ft)
Replace.
22140_ARDX_G0252

Fig. 195 Front strut flange nut torque sequence

OVERHAUL

See Figures 196 through 200.

1. Raise and support the vehicle safely. Remove the tire and wheel assemblies.

2. Remove the strut.

3. Position the strut in a suitable holding fixture.

4. Remove the cap from the top of the strut assembly.

5. Compress the strut spring and remove the self locking nut, using the strut nut adapter and a 17mm deep socket while holding the strut shaft with a wrench.

➡Do not compress the spring more than necessary to remove the nut.

6. Release the pressure from the strut spring compressor tool.

7. Disassemble the strut. Inspect and replace defective components, as required.

8. Install the strut spring on the upper spring mounting cushion, then align the upper end of the strut spring with the cushion stop.

9. Compress the strut spring.

10. Install all parts except the strut mounting washer and the self locking nut on to the strut unit.

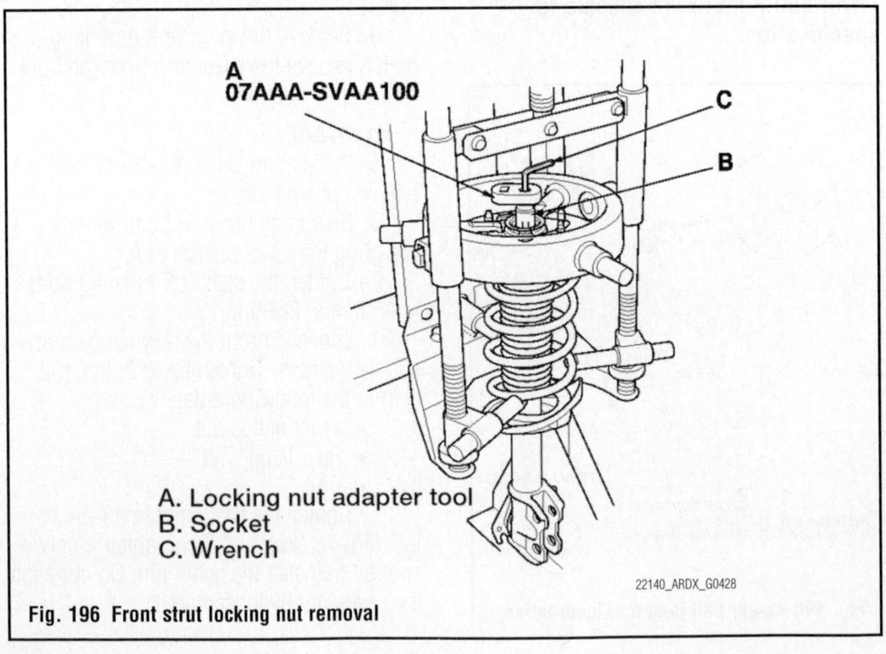
A
07AAA-SVAA100
C
B
A. Locking nut adapter tool
B. Socket
C. Wrench
22140_ARDX_G0428

Fig. 196 Front strut locking nut removal

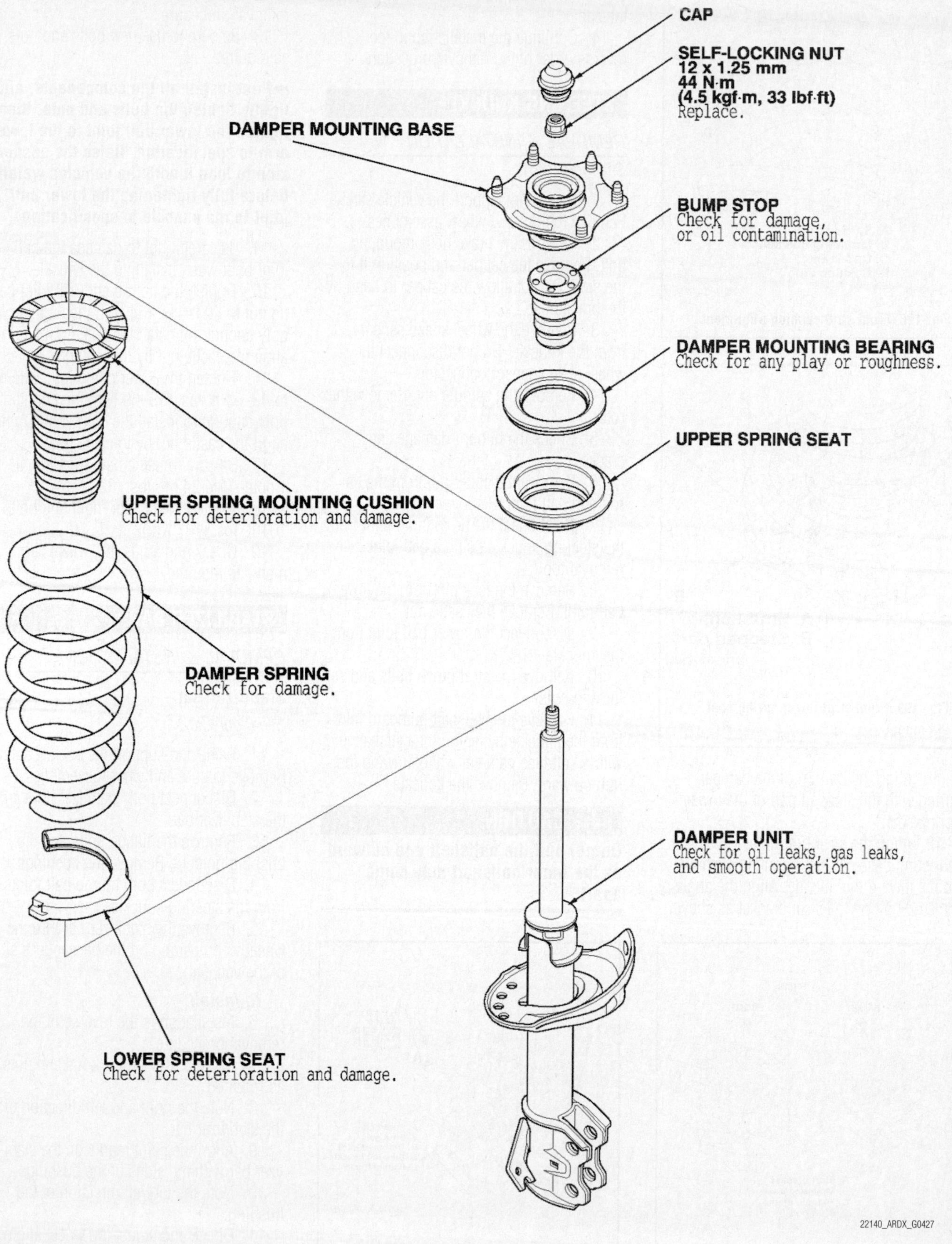
CAP
SELF-LOCKING NUT
12 x 1.25 mm
44 N·m
(4.5 kgf·m, 33 lbf·ft)
Replace.
DAMPER MOUNTING BASE
BUMP STOP
Check for damage, or oil contamination.
DAMPER MOUNTING BEARING
Check for any play or roughness.
UPPER SPRING SEAT
UPPER SPRING MOUNTING CUSHION
Check for deterioration and damage.
DAMPER SPRING
Check for damage.
DAMPER UNIT
Check for oil leaks, gas leaks, and smooth operation.
LOWER SPRING SEAT
Check for deterioration and damage.
22140_ARDX_G0427

Fig. 197 Front strut and related components

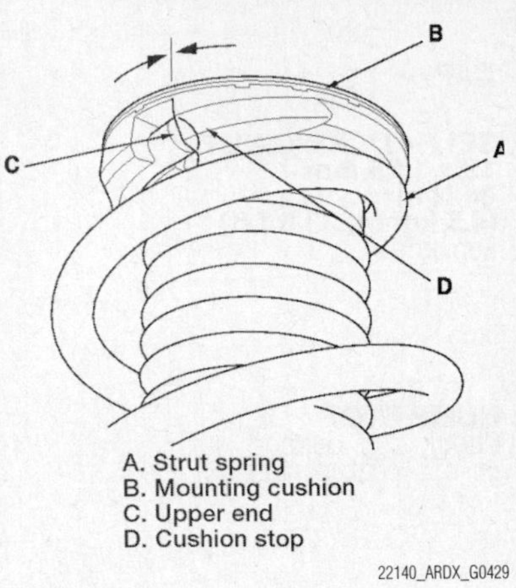
B
A
C
D
A. Strut spring
B. Mounting cushion
C. Upper end
D. Cushion stop
22140_ARDX_G0429

Fig. 198 Front strut cushion alignment

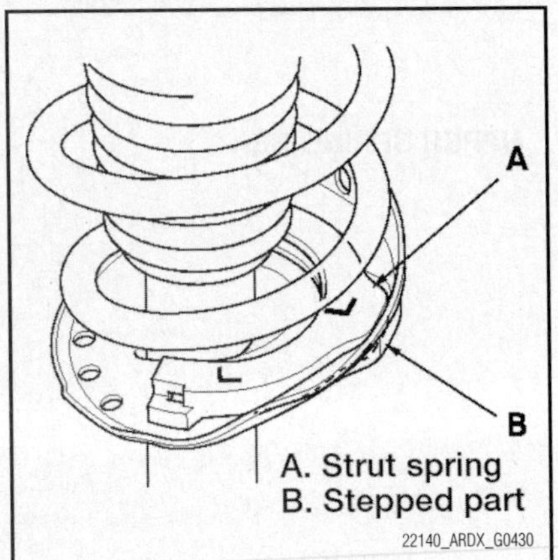
A
B
A. Strut spring
B. Stepped part
22140_ARDX_G0430

Fig. 199 Front strut lower spring seat alignment

11. Align the bottom of the damper spring with the stepped part of the lower spring seat.

12. Align the strut bracket and the strut mounting base so that the triangle stamp points toward the outside. Align the angle of the stud bolt on the strut bracket as shown.

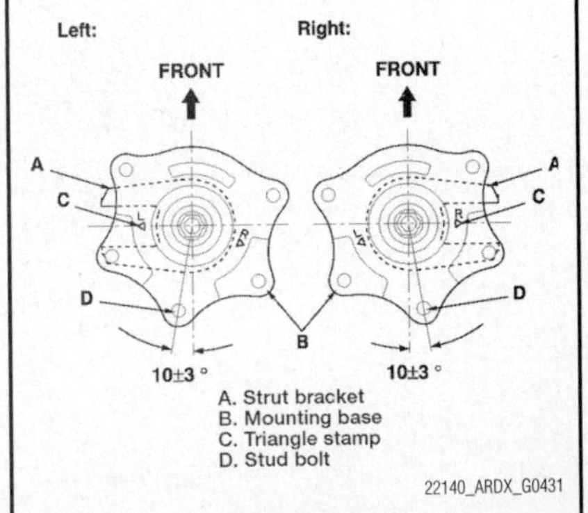
Left:
Right:
FRONT
FRONT
A
C
D
B
10±3 °
10±3 °
A. Strut bracket
B. Mounting base
C. Triangle stamp
D. Stud bolt
22140_ARDX_G0431

Fig. 200 Front strut alignment identification points

13. Install a new self locking nut. Tighten the nut to 33 ft. lbs. (45 Nm). Install the cap.

14. Continue the installation in the reverse order of the removal procedure.

STEERING KNUCKLE

REMOVAL & INSTALLATION

See Figure 201.

1. Raise and support the vehicle safely. Remove the tire and wheel assemblies.

2. Remove the brake hose mounting bolt. Remove the caliper and position it to the side. Do not allow the caliper to hang by the brake hose.

3. Remove the wheel speed sensor from the knuckle. Do not disconnect the wheel speed sensor connector.

4. Remove the spindle nut. Remove the rotor.

5. Check the hub for damage and cracks.

6. Remove the cotter pin from the tie rod end ball joint. Remove the nut.

7. Disconnect the tie rod ball joint from the steering knuckle using a ball joint removal tool.

8. Remove the lock pin from the lower ball joint. Remove the castle nut.

9. Disconnect the lower ball joint from the knuckle.

10. Remove the strut pinch bolts and self locking nuts.

11. Remove the halfshaft outboard joint from the knuckle by taping the halfshaft end with a soft face hammer while drawing the hub outward. Remove the knuckle.

✲✲ CAUTION

Do not pull the halfshaft end outward as the inner halfshaft may come apart.

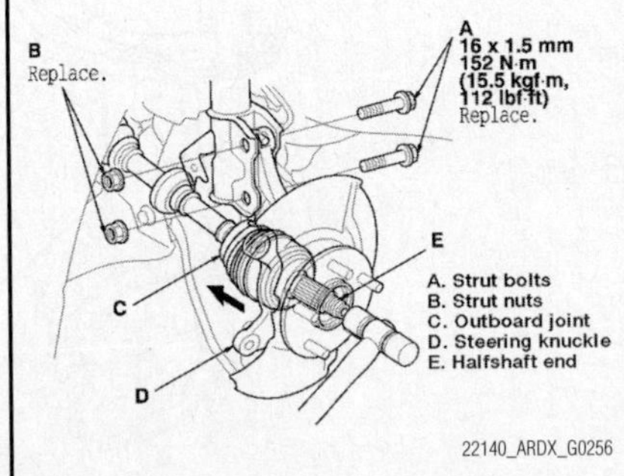
B
Replace.
A
16 x 1.5 mm
152 N·m
(15.5 kgf·m,
112 lbf·ft)
Replace.
E
C
D
A. Strut bolts
B. Strut nuts
C. Outboard joint
D. Steering knuckle
E. Halfshaft end
22140_ARDX_G0256

Fig. 201 Front steering knuckle and related components

To install:

12. Installation is the reverse of the removal procedure.

13. Be sure to use new bolts and nuts, as required.

➡First install all the components, and lightly tighten the bolts and nuts, then tighten the lower ball joint to the lower arm to specification. Raise the suspension to load it with the vehicles weight before fully tightening the lower ball joint to the knuckle to specification.

14. Be careful not to damage the ball joint boot when installing the knuckle.

15. Tighten the tie rod end ball joint castle nut to 40 ft. lbs. (54 Nm) then tighten it only enough to install the cotter pin. Do not align the castle nut by loosening it.

16. Tighten the lower ball joint castle nut to 43–51 ft. lbs. (58–69 Nm) then tighten it only enough to install the cotter pin. Do not align the castle nut by loosening it.

17. Be sure to use a new spindle nut. Tighten it to 242 ft. lbs. (328 Nm).

18. Tighten the caliper mounting bolts to 101 ft. lbs. (137 Nm).

19. Check and adjust the wheel alignment, as required.

STABILIZER BAR

REMOVAL & INSTALLATION

Stabilizer Bar

See Figure 202.

1. Raise and support the vehicle safely. Remove the tire and wheel assemblies.

2. Disconnect both stabilizer links from the stabilizer bar.

3. Remove the flange bolts and the bushing holders. Remove the bushings.

4. Disconnect both tie rod ball joints from the steering knuckles.

5. Remove the stabilizer bar thru the wheel well opening on the passenger's side of the vehicle.

To install:

6. Installation is the reverse of the removal procedure.

7. Be sure to use new bolts and nuts, as required.

8. Note the right and left direction of the stabilizer bar.

9. Align the paint marks on the stabilizer bar with the sides of the bushings.

10. Note the fore/aft direction of the bushing.

11. Check and adjust the wheel alignment, as required.

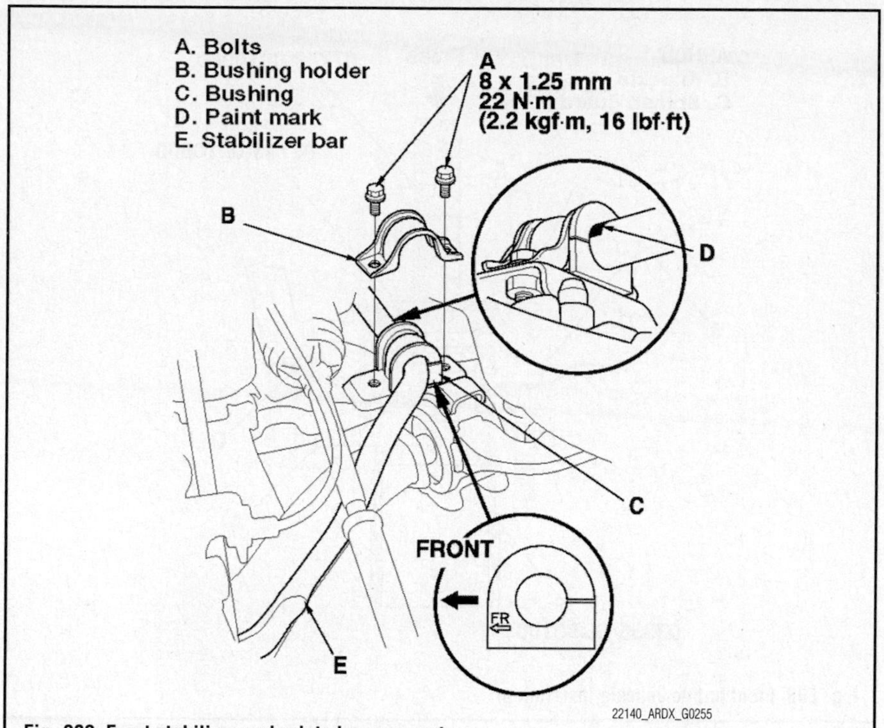
A. Bolts
B. Bushing holder
C. Bushing
D. Paint mark
E. Stabilizer bar
A
8 x 1.25 mm
22 N·m
(2.2 kgf·m, 16 lbf·ft)
B
D
C
FRONT
FR
E
22140_ARDX_G0255

Fig. 202 Front stabilizer and related components

Stabilizer Link

See Figure 203.

1. Raise and support the vehicle safely. Remove the tire and wheel assemblies.
2. Remove the flange nuts while holding the respective joint pin, with the hex wrench. Remove the stabilizer link.

To install:

3. Installation is the reverse of the removal procedure.
4. Be sure to use new bolts and nuts, as required.

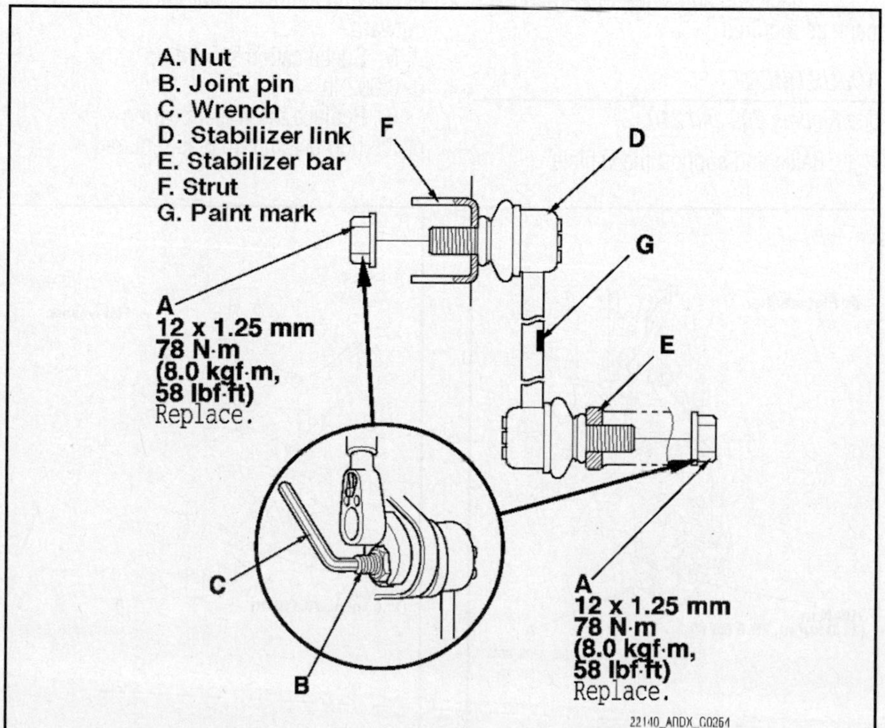
A. Nut
B. Joint pin
C. Wrench
D. Stabilizer link
E. Stabilizer bar
F. Strut
G. Paint mark
F
D
G
E
A
12 x 1.25 mm
78 N·m
(8.0 kgf·m,
58 lbf·ft)
Replace.
C
B
A
12 x 1.25 mm
78 N·m
(8.0 kgf·m,
58 lbf·ft)
Replace.
22140_ARDX_G0254

Fig. 203 Front stabilizer link

5. Install the stabilizer link on the stabilizer bar and the strut with the joint pins set at the center of their range movement.

➡The stabilizer link has a paint mark. Align the paint mark on the stabilizer link facing inward.

6. Install the new flange nuts and lightly tighten them. Tighten them to specification while holding the respective joint pin with the hex wrench. Specification is 58 ft. lbs. (79 Nm).
7. Check and adjust the wheel alignment, as required.

WHEEL BEARINGS

REMOVAL & INSTALLATION

See Figures 204 through 208.

1. Raise and support the vehicle safely. Remove the tire and wheel assemblies.
2. Remove the steering knuckle.
3. Separate the hub from the knuckle using a hub disassembly tool and a hydraulic press.

⁂ CAUTION

Be careful not to deform the splash shield. Hold the hub to prevent it from falling after it is pressed free from the tool.

4. Press the inner wheel bearing race off of the hub.
5. Remove the splash guard and the snapring from the knuckle.
6. Press the wheel bearing out of the knuckle, using the driver attachment and the press.

To install:

7. Press a new wheel bearing into the knuckle using the old bearing, a steel plate, the tool attachment, the support base and a press.

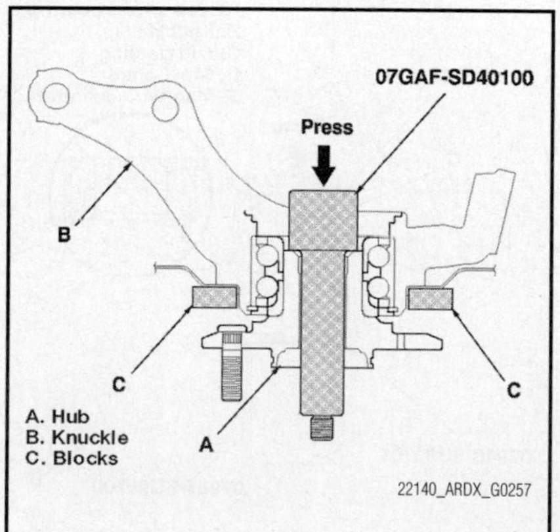
07GAF-SD40100
Press
B
C
C
A. Hub
B. Knuckle
C. Blocks
A
22140_ARDX_G0257

Fig. 204 Front hub removal—positioning in hydraulic press

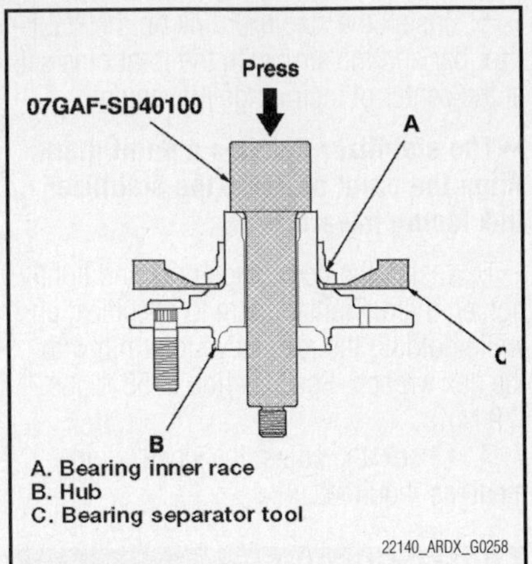
Press
07GAF-SD40100
A
C
B
A. Bearing inner race
B. Hub
C. Bearing separator tool
22140_ARDX_G0258

Fig. 205 Front hub removal—pressing off the inner race

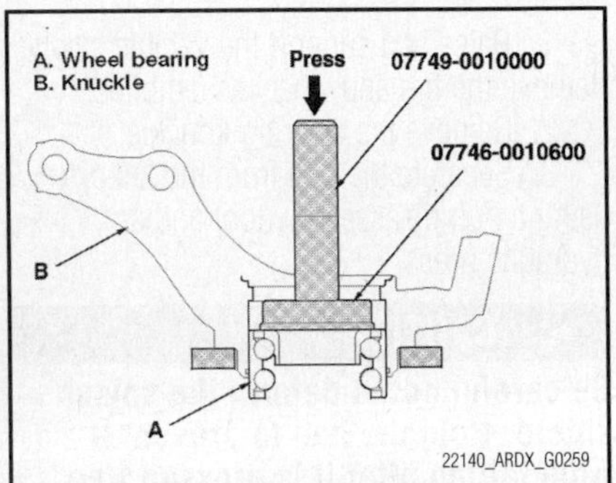
A. Wheel bearing
B. Knuckle
Press
07749-0010000
07746-0010600
B
A
22140_ARDX_G0259

Fig. 206 Front hub removal—pressing the wheel bearing out of the knuckle

8. Install the wheel bearing with the wheel speed sensor magnetic encoder (brown color) toward the inside of the knuckle.

9. Remove any oil, grease, dust, debris etc from the encoder surface. Keep all magnetic tools away from the encoder surface. Be careful not to damage the encoder surface when inserting the wheel bearing.

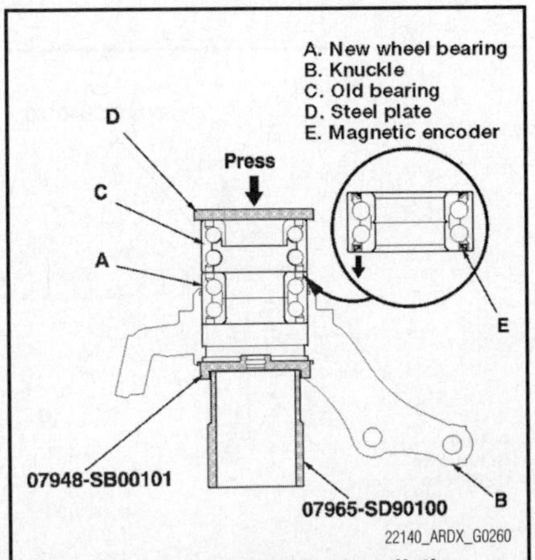
A. New wheel bearing
B. Knuckle
C. Old bearing
D. Steel plate
E. Magnetic encoder
D
Press
C
A
E
07948-SB00101
07965-SD90100
B
22140_ARDX_G0260

Fig. 207 Front hub bearing installation and encoder positioning

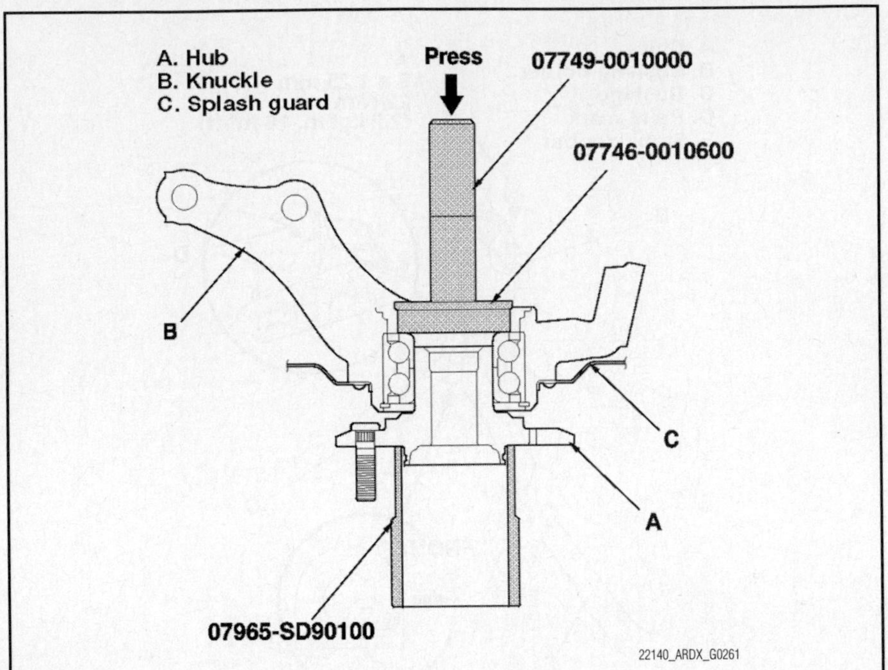
A. Hub
B. Knuckle
C. Splash guard
Press
07749-0010000
07746-0010600
B
C
A
07965-SD90100
22140_ARDX_G0261

Fig. 208 Front hub to knuckle installation

10. Install the snapring securely into the steering knuckle. Install the splash guard and tighten the retaining screws to 4 ft. lbs. (5 Nm).

11. Install the hub onto the knuckle as shown.

12. Continue the installation in the reverse order of the removal procedure.

13. Check and adjust the wheel alignment, as required.

ADJUSTMENT

See Figures 209 and 210.

1. Raise and support the vehicle safely. Remove the tire and wheel assemblies.

2. Install suitable flat washers and the wheel lug nuts. Tighten the lug nuts to specification.

3. Install a dial indicator gauge. Position the dial gauge against the hub flange.

4. Measure the bearing endplay by moving the disc inward and outward.

5. Specification should be 0–0.002 in.

6. Replace the wheel bearing or the hub bearing unit, as required.

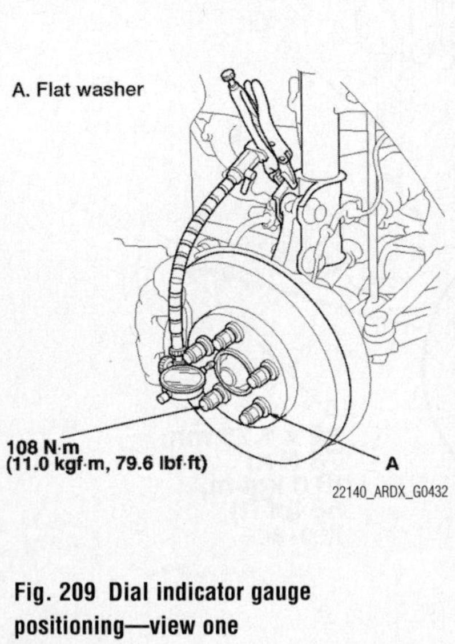
A. Flat washer
108 N·m
(11.0 kgf·m, 79.6 lbf·ft)
A
22140_ARDX_G0432

Fig. 209 Dial indicator gauge positioning—view one

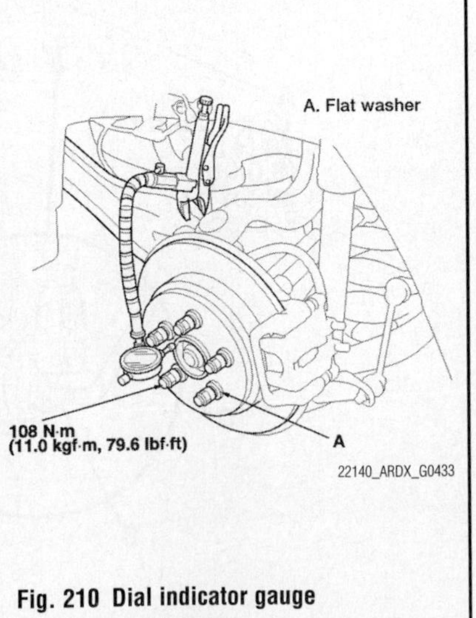
A. Flat washer
108 N·m
(11.0 kgf·m, 79.6 lbf·ft)
A
22140_ARDX_G0433

Fig. 210 Dial indicator gauge positioning—view two

COIL SPRINGS

REMOVAL & INSTALLATION

See Figures 211 through 213.

1. Raise and support the vehicle safely. Remove the tire and wheel assemblies.
2. Position the floor jack at the connecting point of the lower control arm and the lower strut mount.
3. Disconnect the stabilizer link from the lower arm.
4. Remove the flange bolt that connects the lower arm and the knuckle.
5. Remove the flange bolt that connects the lower arm and the bottom of the strut.
6. Lower the floor jack, slowly.
7. Remove the spring and the lower spring cushion.
8. Remove the flange bolt that connects to the body. Remove the spring mounting collar and the spring mounting cushion, as required.

To install:

9. Install the spring and the lower spring cushion. Align the bottom of the spring and the lower spring cushion, as shown.
10. Position the floor jack at the connecting point of the lower control arm.

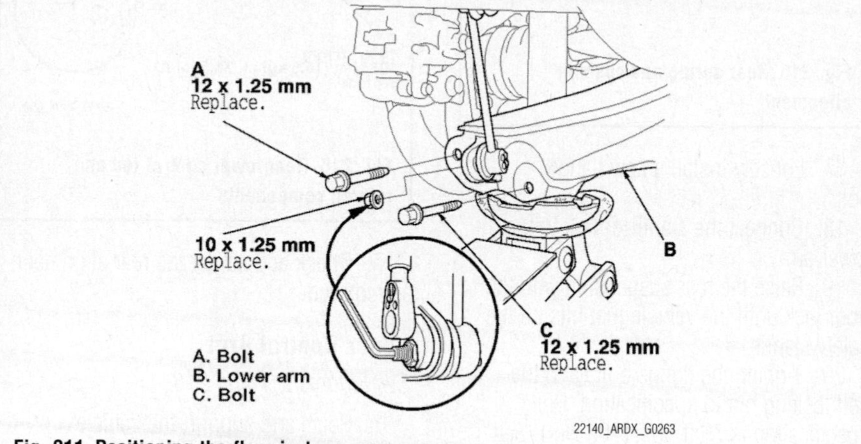
A
12 x 1.25 mm
Replace.
10 x 1.25 mm
Replace.
B
C
12 x 1.25 mm
Replace.
A. Bolt
B. Lower arm
C. Bolt
22140_ARDX_G0263

Fig. 211 Positioning the floor jack

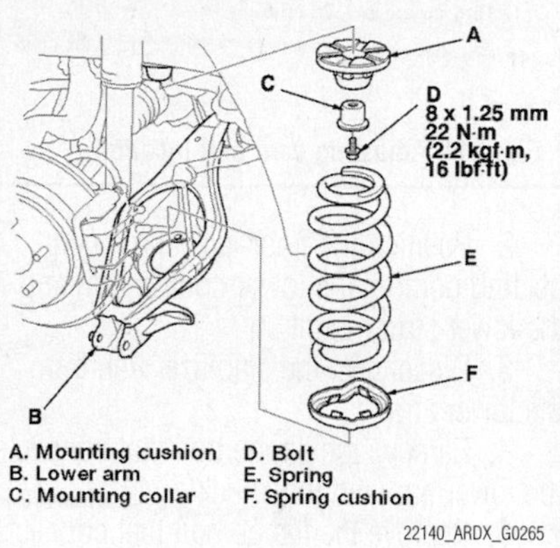
A
C
D
8 x 1.25 mm
22 N·m
(2.2 kgf·m,
16 lbf·ft)
E
F
B
A. Mounting cushion
B. Lower arm
C. Mounting collar
D. Bolt
E. Spring
F. Spring cushion
22140_ARDX_G0265

Fig. 213 Rear spring and cushion alignment

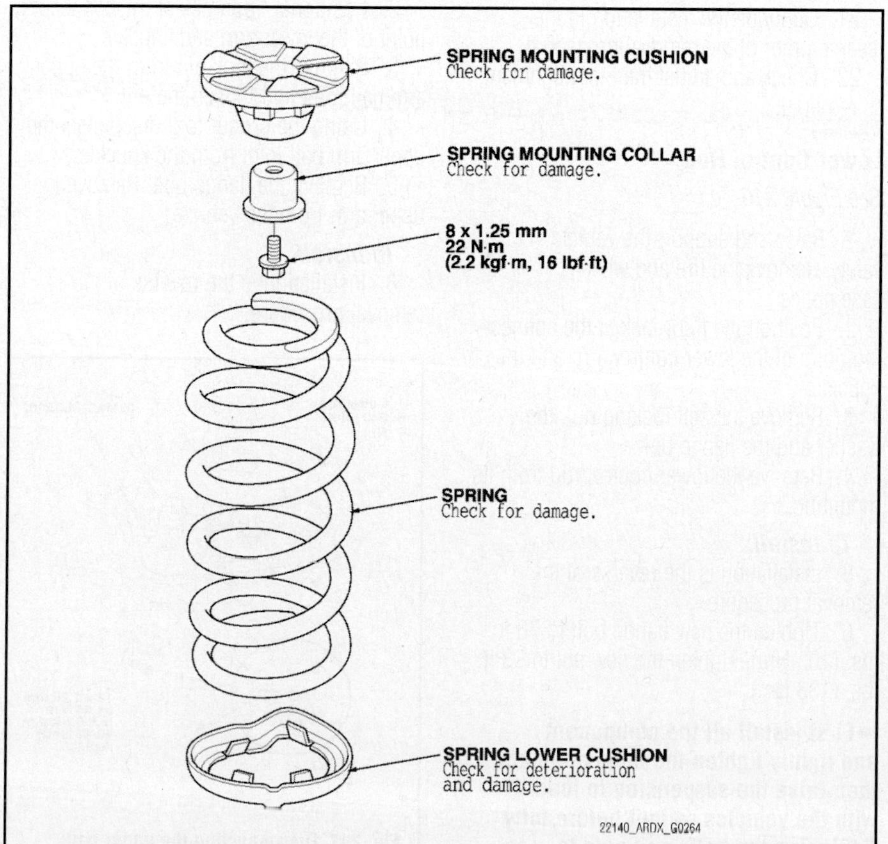
SPRING MOUNTING CUSHION
Check for damage.
SPRING MOUNTING COLLAR
Check for damage.
8 x 1.25 mm
22 N·m
(2.2 kgf·m, 16 lbf·ft)
SPRING
Check for damage.
SPRING LOWER CUSHION
Check for deterioration
and damage.
22140_ARDX_G0264

Fig. 212 Rear spring and related components

11. Slowly raise the jack until you align the bolt hole with the holes in the lower arm and the knuckle.
12. Loosely install a new flange bolt.
13. Compress the strut by hand until you can align the bolt hole with the holes in the lower arm and the strut.
14. Loosely install a new flange bolt.
15. Connect the stabilizer link to the lower arm.
16. Raise the rear suspension with the floor jack until the vehicle just lifts off the safety stands.
17. Tighten the flange bolts and the self locking nut to specification. Bolt specification is 76 ft. lbs. (103 Nm). Self locking nut specification is 29 ft. lbs. (39 Nm).
18. Continue the installation in the reverse order of the removal procedure.
19. Check and adjust the rear alignment, as required.

CONTROL ARMS & LINKS

REMOVAL & INSTALLATION

Lower Control Arm

See Figures 214 and 215.

1. Raise and support the vehicle safely. Remove the tire and wheel assemblies.

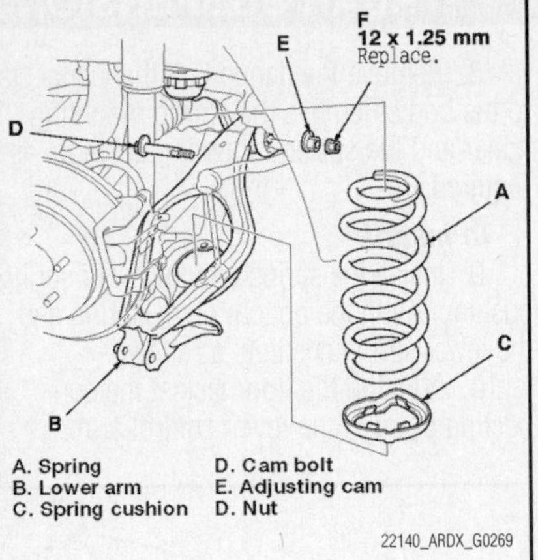
F
12 x 1.25 mm
Replace.
E
D
A
C
B
A. Spring
B. Lower arm
C. Spring cushion
D. Cam bolt
E. Adjusting cam
D. Nut
22140_ARDX_G0269

Fig. 214 Adjusting cam bolt location

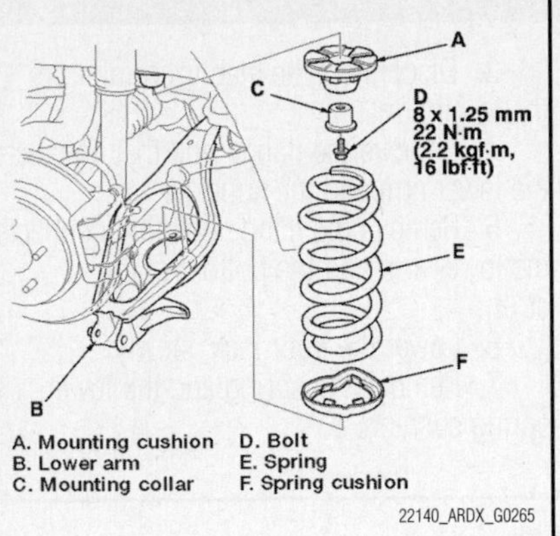
A
C
D
8 x 1.25 mm
22 N·m
(2.2 kgf·m,
16 lbf·ft)
E
F
B
A. Mounting cushion
B. Lower arm
C. Mounting collar
D. Bolt
E. Spring
F. Spring cushion
22140_ARDX_G0265

Fig. 215 Rear spring and cushion alignment

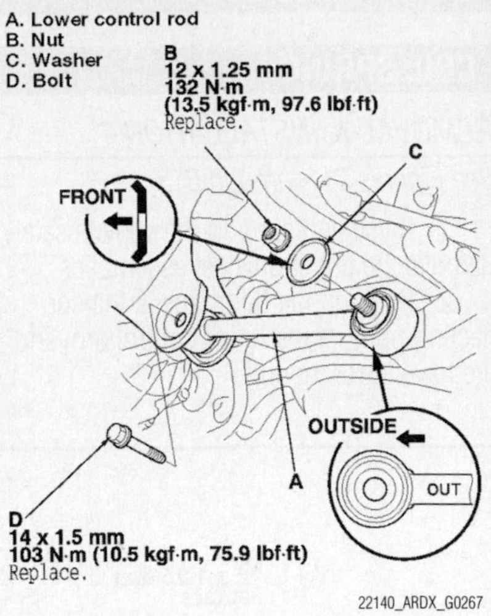
A. Lower control rod
B. Nut
C. Washer
D. Bolt
B
12 x 1.25 mm
132 N·m
(13.5 kgf·m, 97.6 lbf·ft)
Replace.
C
FRONT
OUTSIDE
A
OUT
D
14 x 1.5 mm
103 N·m (10.5 kgf·m, 75.9 lbf·ft)
Replace.
22140_ARDX_G0267

Fig. 216 Rear lower control rod and related components

2. Position the floor jack at the connecting point of the lower control arm and the lower strut mount.

3. Disconnect the stabilizer link from the lower arm.

4. Remove the flange bolt that connects the lower arm and the knuckle.

5. Remove the flange bolt that connects the lower arm and the bottom of the strut.

6. Lower the floor jack, slowly.

7. Remove the spring and the lower spring cushion.

8. Remove the flange bolt that connects to the body. Remove the spring mounting collar and the spring mounting cushion, as required.

9. Mark the cam positions of the adjusting bolt and the adjusting cam. Remove the nut, adjusting cam and adjusting bolt.

10. Set aside the nut and control arm mounting nut. Remove the lower arm.

To install:

11. Position the lower arm. Loosely install a new adjusting bolt, adjusting cam and self locking nut.

➡At final tightening the torque the nut to 43 ft. lbs. (58 Nm).

12. Install the spring and the lower spring cushion. Align the bottom of the spring and the lower spring cushion, as shown.

13. Position the floor jack at the connecting point of the lower control arm.

14. Slowly raise the jack until you align the bolt hole with the holes in the lower arm and the knuckle.

15. Loosely install a new flange bolt.

16. Compress the strut by hand until you can align the bolt hole with the holes in the lower arm and the strut.

17. Loosely install a new flange bolt.

18. Connect the stabilizer link to the lower arm.

19. Raise the rear suspension with the floor jack until the vehicle just lifts off the safety stands.

20. Tighten the flange bolts and the self locking nut to specification. Bolt specification is 76 ft. lbs. (103 Nm). Self locking nut specification is 29 ft. lbs. (39 Nm).

21. Continue the installation in the reverse order of the removal procedure.

22. Check and adjust the rear alignment, as required.

Lower Control Rod

See Figure 216.

1. Raise and support the vehicle safely. Remove the tire and wheel assemblies.

2. Position the floor jack at the connecting point of the lower control arm and the knuckle.

3. Remove the self locking nut, the washer and the flange bolt.

4. Remove the lower control rod from its mounting.

To install:

5. Installation is the reverse of the removal procedure.

6. Tighten the new flange bolt to 76 ft. lbs. (103 Nm). Tighten the new nut to 98 ft. lbs. (133 Nm).

➡First install all the components and lightly tighten the bolts and nuts, then raise the suspension to load it with the vehicles weight before fully tightening the bolts and nuts to specification.

7. Check and adjust the rear alignment, as required.

Upper Control Arm

See Figures 217 and 218.

1. Raise and support the vehicle safely. Remove the tire and wheel assemblies.

2. Position a floor jack at the connecting point of the lower arm and knuckle.

3. Remove the lock pin from the upper arm ball joint and loosen the nut.

4. Using the proper tool disconnect the upper arm ball joint from the knuckle.

5. Remove the flange bolt. Remove the upper arm from the vehicle.

To install:

6. Installation is the reverse of the removal procedure.

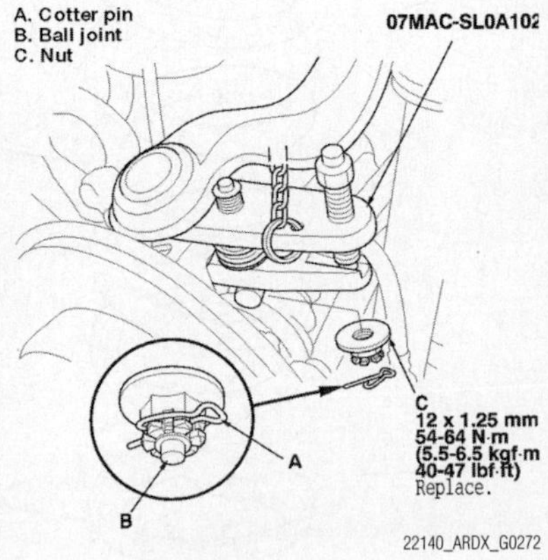
A. Cotter pin
B. Ball joint
C. Nut
07MAC-SL0A10
C
12 x 1.25 mm
54-64 N·m
(5.5-6.5 kgf·m
40-47 lbf·ft)
Replace.
A
B
22140_ARDX_G0272

Fig. 217 Disconnecting the upper ball joint

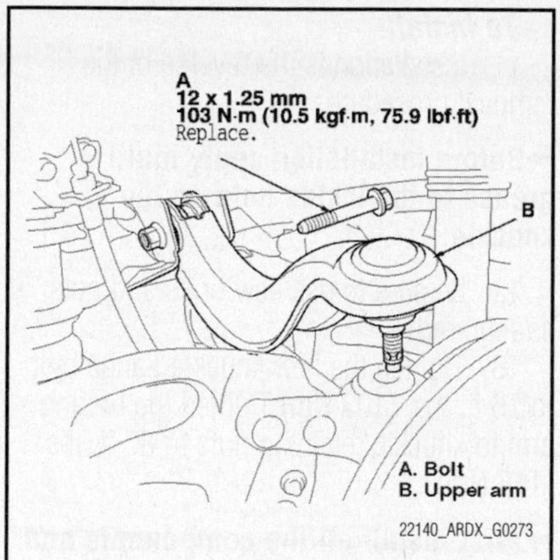
A
12 x 1.25 mm
103 N·m (10.5 kgf·m, 75.9 lbf·ft)
Replace.
B
A. Bolt
B. Upper arm
22140_ARDX_G0273

Fig. 218 Rear upper arm and related components

➡First install all components and lightly tighten the new flange bolt and the castle nut. Raise the suspension to load it with the vehicle's weight before fully tightening.

7. Tighten the castle nut to 43–47 ft. lbs. (58–64 Nm), and then tighten it only enough to install the cotter pin. Do not align the castle nut by loosening it.
8. Check and adjust the rear alignment, as required.

SHOCK ABSORBERS

REMOVAL & INSTALLATION

See Figure 219.

1. Raise and support the vehicle safely. Remove the tire and wheel assemblies.
2. Position a floor jack at the connecting point of the lower arm and the knuckle. Raise the floor jack until the suspension begins to compress.
3. Remove the flange bolt from the bottom of the strut.
4. Remove the lid from the cargo area side trim panel.
5. Remove the self locking nut, while holding the shock absorber shaft, using a hex wrench.
6. Remove the shock absorber washer and mounting cushion from the top of the shock absorber.
7. Compress the shock absorber and remove it from the vehicle.

To install:

8. Position the floor jack under the lower arm, to support the suspension.
9. Slowly raise the floor jack until you align the bolt hole with the holes in the lower arm and the shock absorber. Loosely install the new flange bolt.
10. First install the component and lightly tighten the new flange bolt, then raise the suspension to load it with the vehicles weight before fully tightening the bolt to specification.
11. Tighten the new flange bolt to 76 ft. lbs. (103 Nm).
12. Continue the installation in the reverse order of the removal procedure.
13. Tighten the new self locking nut to 22 ft. lbs. (30 Nm).
14. Check and adjust the rear alignment, as required.

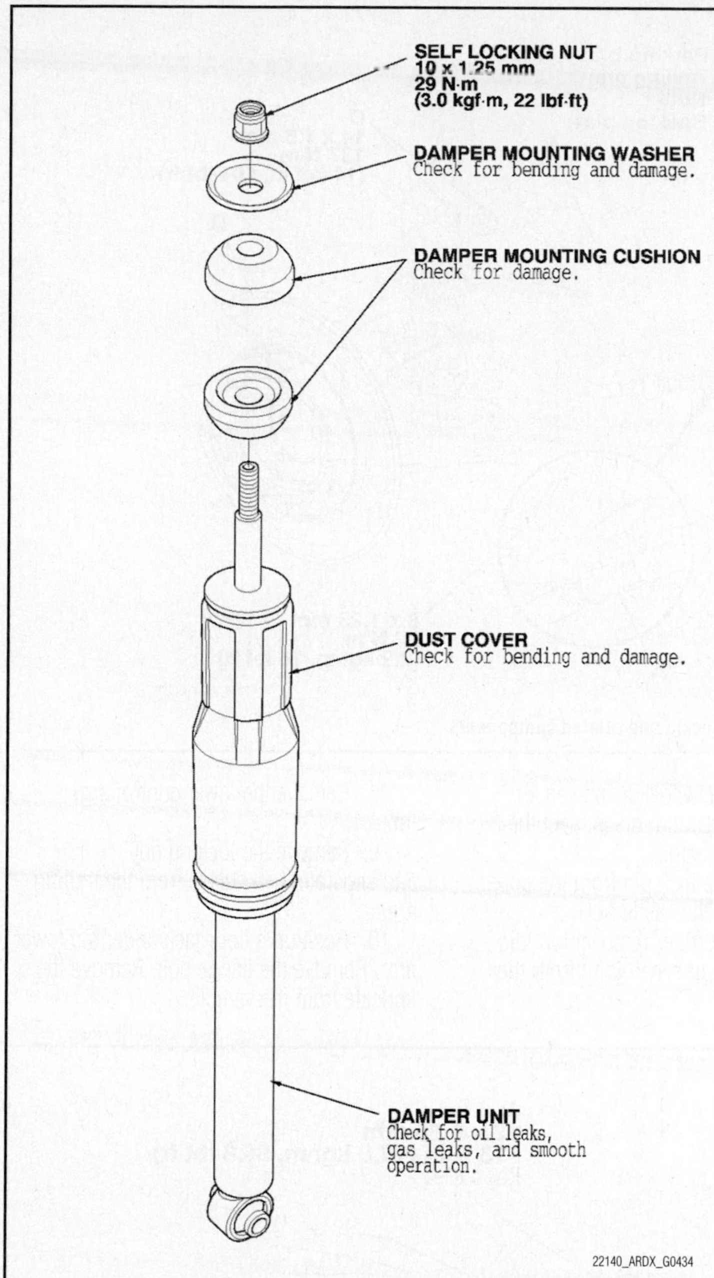
SELF LOCKING NUT
10 x 1.25 mm
29 N·m
(3.0 kgf·m, 22 lbf·ft)
DAMPER MOUNTING WASHER
Check for bending and damage.
DAMPER MOUNTING CUSHION
Check for damage.
DUST COVER
Check for bending and damage.
DAMPER UNIT
Check for oil leaks, gas leaks, and smooth operation.
22140_ARDX_G0434

Fig. 219 Rear shock absorber and related components

STEERING KNUCKLE

REMOVAL & INSTALLATION

See Figures 220 and 221.

1. Raise and support the vehicle safely. Remove the tire and wheel assemblies.
2. Remove the hub bearing unit.
3. Remove the parking brake cable from the trailing arm.
4. Remove the flange nuts and remove the backing plate.

➡Hang the backing plate, using mechanics wire, to prevent damage to the parking brake cable and the backing plate.

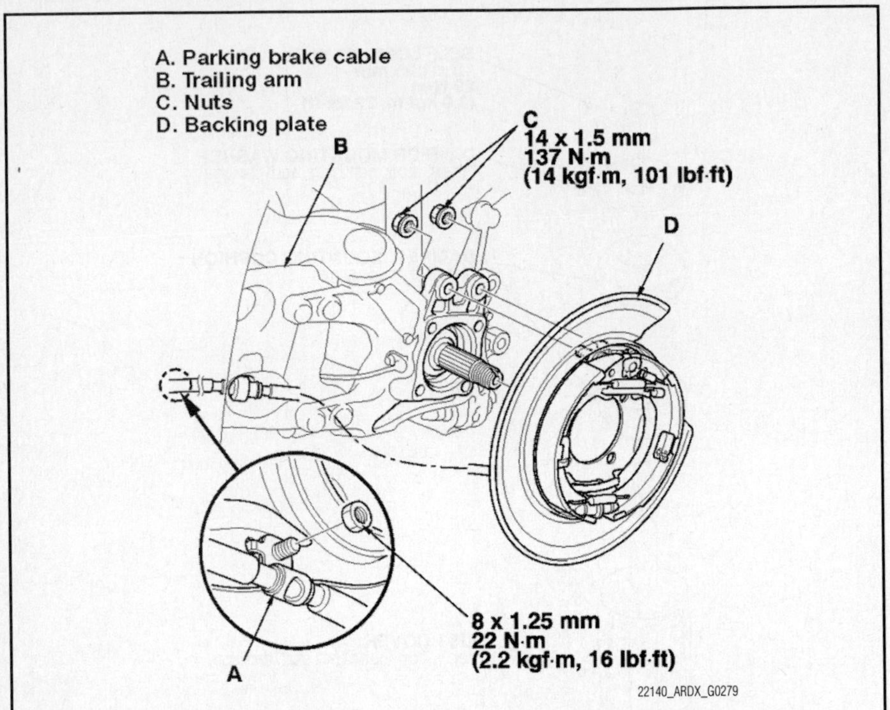
A. Parking brake cable
B. Trailing arm
C. Nuts
D. Backing plate
B
C
14 x 1.5 mm
137 N·m
(14 kgf·m, 101 lbf·ft)
D
8 x 1.25 mm
22 N·m
(2.2 kgf·m, 16 lbf·ft)
A
22140_ARDX_G0279

Fig. 220 Rear knuckle and related components

5. Remove the wheel speed sensor from the knuckle. Do not disconnect the sensor at the connector.

6. Remove the lock pin from the upper ball joint. Loosen the castle nut.

7. Using the proper removal tool disconnect the upper arm ball joint from the knuckle.

8. Remove the lower control arm link.

9. Remove the locking nut and separate the knuckle from the trailing arm.

10. Position a floor jack under the lower arm. Remove the flange bolt. Remove the knuckle from the vehicle.

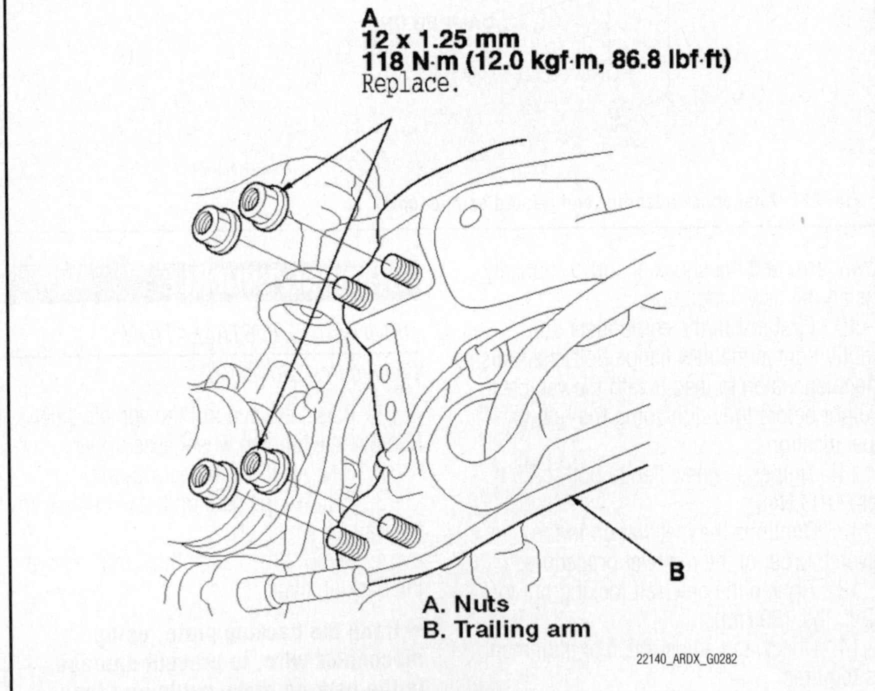
A
12 x 1.25 mm
118 N·m (12.0 kgf·m, 86.8 lbf·ft)
Replace.
B
A. Nuts
B. Trailing arm
22140_ARDX_G0282

Fig. 221 Trailing arm to knuckle retaining nut locations

To install:

11. Installation is the reverse of the removal procedure.

➡Before installation apply multi grease to the inside hole on the knuckle.

12. Be sure to use new bolts and nuts, as required.

13. Tighten the new knuckle flange bolt to 76 ft. lbs. (103 Nm).Tighten the trailing arm to knuckle retaining nuts to 87 ft. lbs. (118 Nm).

➡First install all the components and lightly tighten the bolts and nuts, then raise the suspension to load it with the vehicles weight before fully tightening the bolts and nuts to specification.

14. Check and adjust the rear alignment, as required.

STABILIZER BAR

REMOVAL & INSTALLATION

Stabilizer Bar

See Figure 222.

1. Raise and support the vehicle safely. Remove the tire and wheel assemblies.

2. Disconnect both stabilizer links from the stabilizer bar.

3. Remove the flange bolts and the bushing holders.

4. Remove the stabilizer bar from its mounting.

To install:

5. Installation is the reverse of the removal procedure.

6. Tighten the stabilizer bar bushing bolts to 29 ft. lbs. (30 Nm).

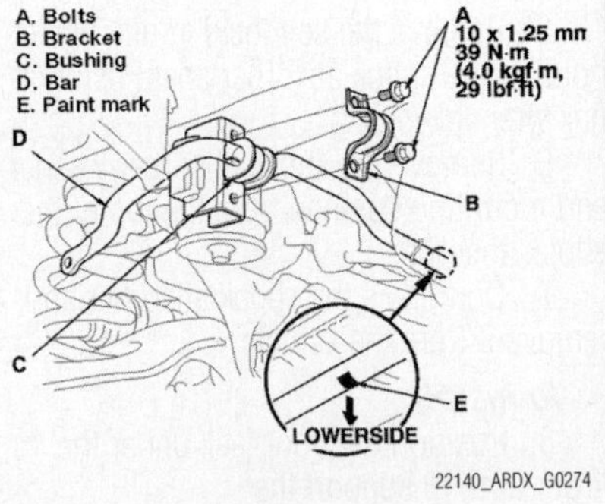
A. Bolts
B. Bracket
C. Bushing
D. Bar
E. Paint mark
A
10 x 1.25 mm
39 N·m
(4.0 kgf·m,
29 lbf·ft)
D
B
C
E
LOWERSIDE
22140_ARDX_G0274

Fig. 222 Rear stabilizer bar and related components

➡Note the right and left direction of the stabilizer bar. The bar has a paint mark on the center facing downward.

7. Check and adjust the rear alignment, as required.

Stabilizer Link

See Figure 223.

1. Raise and support the vehicle safely. Remove the tire and wheel assemblies.
2. Remove the self locking nut and the flange nut while holding the respective joint pin with the hex wrench.
3. Remove the stabilizer link from the vehicle.

To install:

4. Installation is the reverse of the removal procedure.

➡The stabilizer link has a paint mark. Align the paint mark facing inward.

5. Install a new self-locking nut and flange bolt. Lightly tighten them.
6. Position a floor jack under the trailing arm and raise the suspension to load it with the vehicle's weight.
7. Tighten the nuts to 30 ft. lbs. (41 Nm).
8. Test drive the vehicle. After five minutes of driving, tighten the self adjusting nut, again to specification.
9. Check and adjust the rear alignment, as required.

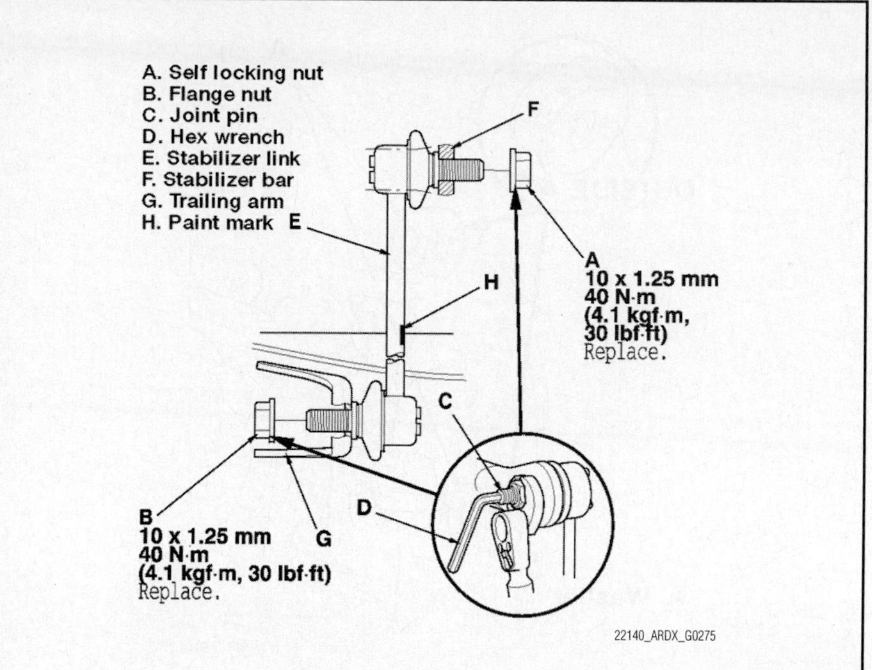
A. Self locking nut
B. Flange nut
C. Joint pin
D. Hex wrench
E. Stabilizer link
F. Stabilizer bar
G. Trailing arm
H. Paint mark
A
10 x 1.25 mm
40 N·m
(4.1 kgf·m, 30 lbf·ft)
Replace.
B
10 x 1.25 mm
40 N·m
(4.1 kgf·m, 30 lbf·ft)
Replace.
22140_ARDX_G0275

Fig. 223 Rear stabilizer bar link and related components

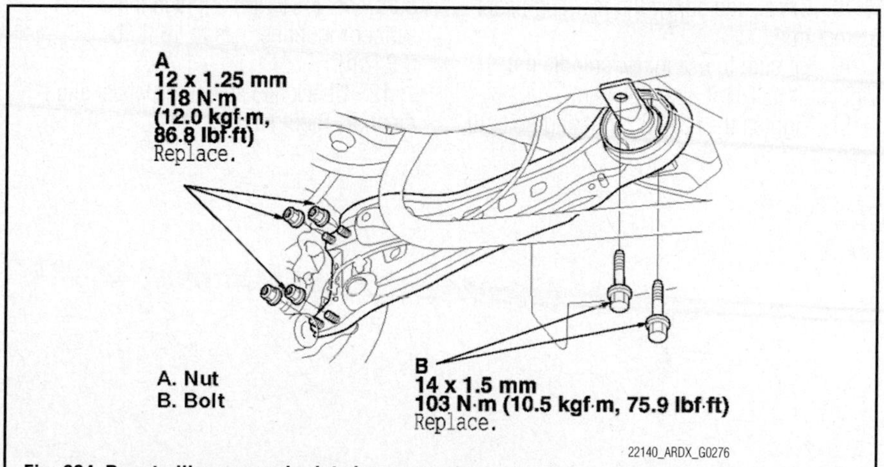
A
12 x 1.25 mm
118 N·m
(12.0 kgf·m, 86.8 lbf·ft)
Replace.
A. Nut
B. Bolt
B
14 x 1.5 mm
103 N·m (10.5 kgf·m, 75.9 lbf·ft)
Replace.
22140_ARDX_G0276

Fig. 224 Rear trailing arm and related components

TRAILING ARM

REMOVAL & INSTALLATION

See Figure 224.

1. Raise and support the vehicle safely. Remove the tire and wheel assemblies.
2. Remove the parking brake shoe. Remove the parking brake cable from the backing plate.
3. Remove the parking brake cable from the trailing arm.
4. Remove the brake hose mounting bracket from the trailing arm.
5. Position a floor jack at the connecting point of the lower arm and knuckle.
6. Remove the self locking nuts and flange bolts from the trailing arm.
7. Remove the trailing arm from its mounting.

To install:

8. Installation is the reverse of the removal procedure.
9. First install all the component and lightly tighten the new flange bolts and nuts, then raise the suspension to load it with the vehicles weight before fully tightening the bolts and nuts to specification.
10. Tighten the new flange bolt to 76 ft. lbs. (103 Nm). Tighten the new nuts to 87 ft. lbs. (118 Nm).
11. Check the brake hose for interference and twisting, correct as required.
12. Check and adjust the rear alignment, as required.

WHEEL BEARINGS

REMOVAL & INSTALLATION

See Figures 225 and 226.

1. Raise and support the vehicle safely. Remove the tire and wheel assemblies.
2. Remove the caliper and position it to the side. Do not allow the caliper to hang by the brake hose. Remove the two washers.
3. Remove the spindle nut.
4. Release the parking brake, if necessary. Remove the rotor.
5. Remove the hub bearing unit retaining bolts. Remove the component from the vehicle.

To install:

6. Installation is the reverse of the removal procedure.
7. Be sure to use new bolts and nuts, as required.
8. First install all the components and lightly tighten the bolts and nuts, then raise the suspension to load it with the vehicles weight before fully tightening the bolts and nuts to specification.

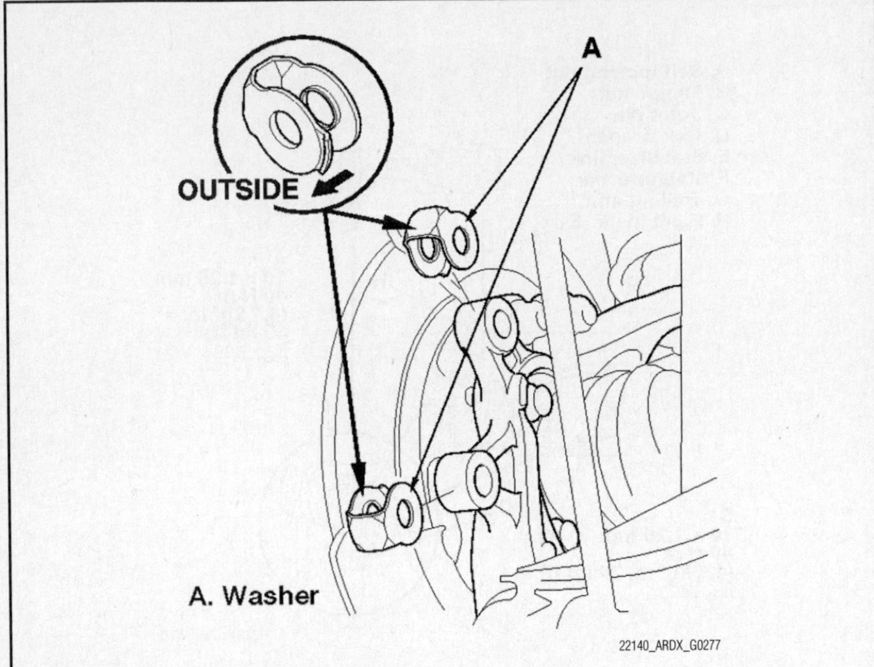
A
OUTSIDE
A. Washer
22140_ARDX_G0277

Fig. 225 Washer identification and location

9. Check and adjust the rear alignment, as required.

10. Be sure to use a new spindle nut. Tighten it to 181 ft. lbs. (245 Nm).

11. Tighten the caliper mounting bolts to 80 ft. lbs. (108 Nm). Tighten the caliper mounting nuts to 16 ft. lbs. (22 Nm).

12. Check and adjust the wheel alignment, as required.

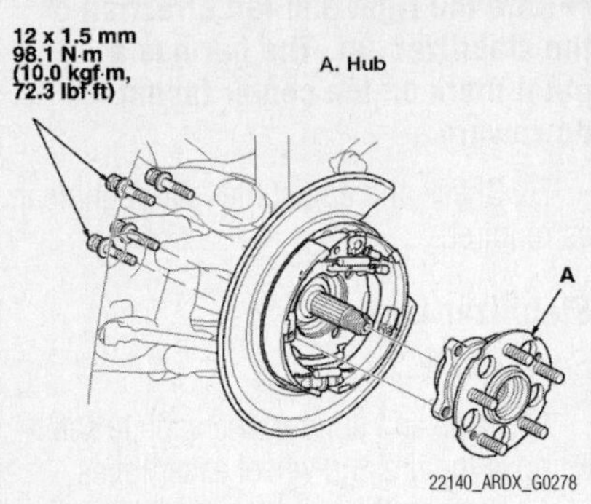
12 x 1.5 mm
98.1 N·m
(10.0 kgf·m,
72.3 lbf·ft)
A. Hub
A
22140_ARDX_G0278

Fig. 226 Rear hub and related components

ADJUSTMENT

See Figures 210 and 211.

1. Raise and support the vehicle safely. Remove the tire and wheel assemblies.
2. Install suitable flat washers and the wheel lug nuts. Tighten the lug nuts to specification.
3. Install a dial indicator gauge. Position the dial gauge against the hub flange.
4. Measure the bearing endplay by moving the disc inward and outward.
5. Specification should be 0–0.002 in.
6. Replace the wheel bearing or the hub bearing unit, as required.

ACURA

RL

3

BRAKES 3-9

ANTI-LOCK BRAKE SYSTEM (ABS) 3-10
Wheel Speed Sensors 3-10
Removal & Installation 3-10
BLEEDING THE BRAKE SYSTEM 3-9
Bleeding Procedure 3-9
Bleeding Procedure 3-9
Bleeding the ABS System 3-9
FRONT DISC BRAKES 3-12
Brake Caliper 3-12
Removal & Installation 3-12
Disc Brake Pads 3-12
Removal & Installation 3-12
INFORMATION AND PRECAUTIONS 3-9
Anti-lock Systems 3-9
Disc and Drum Systems 3-9
PARKING BRAKE 3-14
Parking Brake Adjustment 3-14
Adjustment 3-14
Parking Brake Shoes 3-14
Adjustment 3-15
Removal & Installation 3-14
REAR DISC BRAKES 3-13
Brake Caliper 3-13
Removal & Installation 3-13
Disc Brake Pads 3-13
Removal & Installation 3-13

CHASSIS ELECTRICAL 3-16

AIR BAG (SUPPLEMENTAL RESTRAINT SYSTEM) 3-16
General Information 3-16
Arming the System 3-16
Disarming the System 3-16
Service Precautions 3-16

DRIVE TRAIN 3-16

Front Driveshaft 3-16
CV-Joint Overhaul 3-18
Removal & Installation 3-16
Intermediate Shaft 3-19
Removal & Installation 3-19
Rear Differential 3-19
Removal & Installation 3-19
Rear Halfshaft 3-19
Removal & Installation 3-19
Transfer Assembly 3-20
Removal & Installation 3-20

ENGINE COOLING 3-21

Engine Fan & Shroud 3-21
Removal & Installation 3-21
Radiator 3-21
Removal & Installation 3-21
Thermostat 3-21
Removal & Installation 3-21
Water Pump 3-22
Removal & Installation 3-22

ENGINE ELECTRICAL 3-22

CHARGING SYSTEM 3-22
Alternator 3-22
Battery System 3-22
Battery Disconnect Procedure 3-22
Battery Reconnect Procedure 3-23
Removal & Installation 3-22
IGNITION SYSTEM 3-23
Firing Order 3-23
Ignition Coil 3-23
Removal & Installation 3-23
Ignition Timing 3-23
Adjustment 3-23
Inspection 3-23
Spark Plugs 3-23
Removal & Installation 3-23
STARTING SYSTEM 3-24
Starter 3-24
Removal & Installation 3-24
Solenoid or Relay Replacement 3-24

ENGINE MECHANICAL 3-24

Accessory Drive Belts 3-24
Accessory Belt Routing 3-24
Adjustment 3-24
Inspection 3-24
Removal & Installation 3-24
Camshaft and Valve Lifters 3-24
Inspection 3-24
Removal & Installation 3-25
Catalytic Converter 3-26
Removal & Installation 3-26
Crankshaft Damper (Pulley) 3-26
Removal & Installation 3-26
Crankshaft Front Seal 3-26
Removal & Installation 3-26
Cylinder Head 3-26
Removal & Installation 3-26
Exhaust Manifold 3-29
Removal & Installation 3-29
Flywheel (Drive Plate) 3-30
Removal & Installation 3-30
Intake Manifold 3-30
Removal & Installation 3-30
Oil Pan 3-30
Removal & Installation 3-30
Oil Pump 3-31
Inspection 3-31
Removal & Installation 3-31
Piston and Ring 3-31
Positioning 3-31
Rear Main Seal 3-31
Removal & Installation 3-31
Rocker Arms 3-32
Removal & Installation 3-32
Timing Belt & Sprockets 3-32
Removal & Installation 3-32
Valve Covers 3-34
Removal & Installation 3-34
Valve Lash 3-35
Adjustment 3-35

ENGINE PERFORMANCE & EMISSION CONTROLS 3-36

Accelerator Pedal Position (APP) Sensor 3-36
Location 3-36
Removal & Installation 3-36
Camshaft Position (CMP) Sensor 3-36
Location 3-36
Removal & Installation 3-36

Crankshaft Position (CKP) Sensor 3-36
CKP Pattern Clear/CKP Pattern Learn Procedure 3-37
Location 3-36
Removal & Installation 3-36
Engine Coolant Temperature (ECT) Sensor 3-37
Location 3-37
Operation 3-37
Removal & Installation 3-37
Evaporative Emissions (EVAP) Canister 3-37
Location 3-37
Removal & Installation 3-37
Evaporative Emissions (EVAP) Purge Control Valve 3-37
Location 3-37
Removal & Installation 3-37
Exhaust Gas Recirculation (EGR) Valve 3-39
Location 3-39
Removal & Installation 3-39
Heated Oxygen (HO2S) Sensor 3-39
Location 3-39
Removal & Installation 3-39
Knock Sensor (KS) 3-39
Location 3-39
Removal & Installation 3-39
Malfunction Indicator Light (MIL) 3-39
Reset Procedure 3-39
Manifold Absolute Pressure (MAP) Sensor 3-40
Location 3-40
Removal & Installation 3-40
Mass Air Flow/Intake Air Temperature (MAF/IAT) Sensor 3-40
Location 3-40
Removal & Installation 3-40
Output Shaft Speed (OSS) Sensor 3-40
Location 3-40
Removal & Installation 3-40
Positive Crankcase Ventilation (PCV) Valve 3-40
Location 3-40
Removal & Installation 3-40
Powertrain Control Module (PCM) 3-41
Location 3-41
PCM Idle Learn Procedure 3-42
Removal & Installation 3-41
Throttle Position Sensor (TPS) 3-42
Location 3-42
Removal & Installation 3-42

FUEL 3-42

GASOLINE FUEL INJECTION SYSTEM 3-42
Fuel Filter 3-43
Removal & Installation 3-43
Fuel Gauge Sending Unit 3-43
Location 3-43
Removal & Installation 3-43
Fuel Pump & Fuel Tank Unit 3-43
Removal & Installation 3-43
Fuel Rail & Injectors 3-44
Removal & Installation 3-44
Fuel System Pressure 3-42
Relieving 3-42
Fuel System Service Precautions 3-42
Fuel Tank 3-45
Draining 3-45
Removal & Installation 3-45
Idle Speed 3-45
Inspection 3-45
Throttle Body 3-45
Removal & Installation 3-45

HEATING & AIR CONDITIONING SYSTEM 3-46
Blower Motor 3-46
Removal & Installation 3-46
Heater Unit & Core 3-46
Removal & Installation 3-46

SPECIFICATIONS AND MAINTENANCE CHARTS 3-3
Brake Specifications 3-7
Camshaft and Bearing Specifications 3-5
Capacities 3-4
Crankshaft and Connecting Rod Specifications 3-5
Engine and Vehicle Identification 3-3
Gasoline Engine Tune-Up Specifications 3-3
Fluid Specifications 3-4
General Engine Specifications 3-3
Piston and Ring Specifications 3-5
Scheduled Maintenance Intervals 3-8
Tire, Wheel and Ball Joint Specifications 3-7
Torque Specifications 3-6
Valve Specifications 3-4
Wheel Alignment 3-7

STEERING 3-48
Power Rack & Pinion Steering Gear 3-48
Removal & Installation 3-48
Power Steering Pump 3-50
Bleeding 3-50
Fluid Fill Procedure 3-50
Removal & Installation 3-50

SUSPENSION 3-51

FRONT SUSPENSION 3-51
Lower Ball Joint 3-51
Removal & Installation 3-51
Lower Control Arm 3-51
Removal & Installation 3-51
Stabilizer Bar 3-51
Removal & Installation 3-51
Stabilizer Link 3-52
Removal & Installation 3-52
Strut (Front Damper & Spring) 3-52
Overhaul 3-53
Removal & Installation 3-52
Upper Ball Joint 3-53
Removal & Installation 3-53
Upper Control Arm 3-54
Removal & Installation 3-54
Wheel Hub & Bearing 3-54
Inspection 3-54
Removal & Installation 3-54

REAR SUSPENSION 3-55
Coil Spring 3-55
Removal & Installation 3-55
Control Arms 3-55
Removal & Installation 3-55
Shock Absorber (Damper) 3-57
Adaptive Front Lighting Control Unit Learning Procedure 3-57
Removal & Installation 3-57
Stabilizer Bar 3-57
Removal & Installation 3-57
Stabilizer Link 3-57
Removal & Installation 3-57
Wheel Bearings 3-58
Inspection 3-58
Removal & Installation 3-58

SPECIFICATIONS AND MAINTENANCE CHARTS

ENGINE AND VEHICLE IDENTIFICATION

Engine							Model Year	
Code	Liters (cc)	Cu. In.	Cyl.	Fuel Sys.	Engine Type	Eng. Mfg.	Code ①	Year
J37A2	3.7 (3700)	226	6	SMFI	SOHC	Honda	9	2009
							A	2010

SMFI: Sequential Multi-Port Fuel Injection

SOHC: Single Overhead Camshaft

① 10th digit of the Vehicle Identification Number (VIN)

37647_ACRL_C0001

GENERAL ENGINE SPECIFICATIONS

Year	Model	Engine Displacement Liters	Engine ID	Net Horsepower @ rpm	Net Torque @ rpm (ft. lbs.)	Bore x Stroke (in.)	Com-pression Ratio	Oil Pressure @ rpm
2009	RL	3.7	J37A2	300@6300	271@5000	3.42X3.89	10.5:1	44@3000
2010	RL	3.7	J37A2	300@6300	271@5000	3.42X3.89	10.5:1	44@3000

37647_ACRL_C0002

GASOLINE ENGINE TUNE-UP SPECIFICATIONS

Year	Engine Displacement Liters	Spark Plug Gap (in.)	Ignition Timing (deg.)		Fuel Pump (psi)	Idle Speed (rpm)		Valve Clearance	
			MT	AT		MT	AT	In.	Ex.
2009	3.7	0.039-0.043	NA	8-12 B	57-64 ①	NA	630-730	0.008-0.009	0.011-0.013
2010	3.7	0.039-0.043	NA	8-12 B	57-64 ①	NA	630-730	0.008-0.009	0.011-0.013

NOTE: The Vehicle Emission Control Information label reflects specification changes during production and must be used if they differ from this chart.

NA: Not Applicable

B: Before Top Dead Center

① At idle, pressure regulator vacuum hose disconnected

37647_ACRL_C0003

CAPACITIES

Year	Model	Engine Displacement Liters	Engine ID/VIN	Engine Oil with Filter (qts.)	Transmission (pts.) Service	Transmission (pts.) Overhaul	Drive Axle Front (pts.)	Drive Axle Rear (pts.)	Fuel Tank (gal.)	Cooling System (qts.)
2009	RL	3.7	J37A2	4.5	3.0	8.1	4.5 ①	0.77	19.4	6.4
2010	RL	3.7	J37A2	4.5	3.0	8.1	4.5 ①	0.77	19.4	6.4

NOTE: All capacities are approximate. Add fluid gradually and ensure a proper fluid level is obtained.

① Transfer case fluid capacity.

37647_ACRL_C0004

FLUID SPECIFICATIONS

Year	Model	Engine Displ. Liters	Engine Oil	Man. Trans.	Auto. Trans.	Drive Axle Front	Drive Axle Rear	Transfer Case	Power Steering Fluid	Brake Master Cylinder	Cooling System
2009	RL	3.7	5W-20 Honda	NA	Acura ATF-Z1	NA	Acura ATF-Z1	GL4, GL5 SAE 90	Acura PS Fluid	Acura DOT 3	①
2010	RL	3.7	5W-20 Honda	NA	Acura ATF-Z1	NA	Acura ATF-Z1	GL4, GL5 SAE 90	Acura PS Fluid	Acura DOT 3	①

DOT: Department Of Transportation

NA: Not Applicable

① Acura Long Life Antifreeze/Coolant-Type2

37647_ACRL_C0005

VALVE SPECIFICATIONS

Year	Engine Displacement Liters	Engine ID/VIN	Seat Angle (deg.)	Face Angle (deg.)	Spring Test Pressure (lbs. @ in.)	Spring Installed Height (in.)	Stem-to-Guide Clearance (in.) Intake	Stem-to-Guide Clearance (in.) Exhaust	Stem Diameter (in.) Intake	Stem Diameter (in.) Exhaust
2009	3.7	J37A2	45	45	NA	NA	0.0008-0.0009	0.0011-0.0013	0.2159-0.2163	0.2146-0.2150
2010	3.7	J37A2	45	45	NA	NA	0.0008-0.0009	0.0011-0.0013	0.2159-0.2163	0.2146-0.2150

NA: Not Available

37647_ACRL_C0006

CAMSHAFT AND BEARING SPECIFICATIONS

All measurements are given in inches.

Year	Engine Displacement Liters	Engine VIN	Journal Diameter	Brg. Oil Clearance	Shaft End-play	Runout	Journal Bore	Lobe Height Intake	Lobe Height Exhaust
2009	3.7	J37A2	NA	0.0020-0.0035	0.0020-0.0080	0.0010	NA	①	②
2010	3.7	J37A2	NA	0.0020-0.0035	0.0020-0.0080	0.0010	NA	①	②

NA: Not Available

① Primary: 1.3504 in.
Secondary: 1.4146 in.

② Primary: 1.4462 in.
Secondary: 1.4713 in.

37647_ACRL_C0009

CRANKSHAFT AND CONNECTING ROD SPECIFICATIONS

All measurements are given in inches.

Year	Engine Displacement Liters	Engine ID/VIN	Crankshaft Main Brg. Journal Dia.	Crankshaft Main Brg. Oil Clearance	Crankshaft Shaft End-play	Crankshaft Thrust on No.	Connecting Rod Journal Diameter	Connecting Rod Oil Clearance	Connecting Rod Side Clearance
2009	3.7	J37A2	2.8337-2.8346	0.0007-0.0018	0.004-0.014	NA	2.2431-2.2441	0.0008-0.0017	0.006-0.014
2010	3.7	J37A2	2.8337-2.8346	0.0007-0.0018	0.004-0.014	NA	2.2431-2.2441	0.0008-0.0017	0.006-0.014

NA: Not available.

37647_ACRL_C0007

PISTON AND RING SPECIFICATIONS

All measurements are given in inches

Year	Engine Displacement Liters	Engine ID/VIN	Piston Clearance	Ring Gap Top Compression	Ring Gap Bottom Compression	Ring Gap Oil Control	Ring Side Clearance Top Compression	Ring Side Clearance Bottom Compression	Ring Side Clearance Oil Control
2009	3.7	J37A2	0.0002-0.0013	0.012-0.016	0.016-0.022	0.0080-0.0280	0.0022-0.0033	0.0012-0.0024	NA
2010	3.7	J37A2	0.0002-0.0013	0.012-0.016	0.016-0.022	0.0080-0.0280	0.0022-0.0033	0.0012-0.0024	NA

NA: Not Applicable

37647_ACRL_C0008

TORQUE SPECIFICATIONS

All readings in ft. lbs.

Year	Engine Displacement Liters	Engine ID/VIN	Cylinder Head Bolts	Main Bearing Bolts	Rod Bearing Bolts	Crankshaft Damper Bolts	Flywheel Bolts	Manifold Intake	Manifold Exhaust	Spark Plugs	Oil Pan Drain Plug
2009	3.7	J37A2	①	②	③	④	54	16	23	13	29
2010	3.7	J37A2	①	②	③	④	54	16	23	13	29

① Step 1: 22 ft. lbs.
Step 2: Rotate 90 degrees
Step 3: Rotate an additional 90 degrees

② Main bearing cap bolts: 54 ft. lbs.
Bearing cap side bolts: 36 ft. lbs.

③ Step 1: 14 ft. lbs.
Step 2: Rotate 90 degrees

④ Step 1: 47 ft. lbs.
Step 2: Rotate 90 degrees

37647_ACRL_C0010

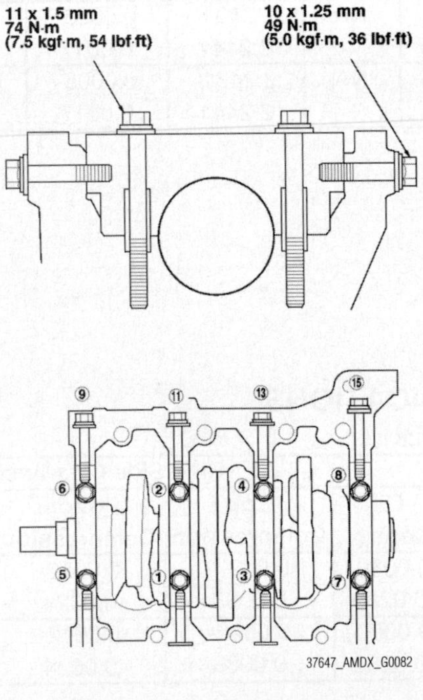
11 x 1.5 mm
74 N·m
(7.5 kgf·m, 54 lbf·ft)
10 x 1.25 mm
49 N·m
(5.0 kgf·m, 36 lbf·ft)
9
11
13
15
6
2
4
8
5
1
3
7
37647_AMDX_G0082

Fig. 1 Main bearing torque sequence—3.7L engine

WHEEL ALIGNMENT

Year	Model		Caster Range (+/-Deg.)	Caster Preferred Setting (Deg.)	Camber Range (+/-Deg.)	Camber Preferred Setting (Deg.)	Toe-in (in.)
2009	RL	F	0.30	+2.10	0.30	-0.08	0 +/- 0.08
		R	—	—	0.30	-1.15	0.08 +/- 0.08
2010	RL	F	0.30	+2.10	0.30	0.00	0 +/- 0.08
		R	—	—	0.30	0.08	0.08 +/- 0.08

37647_ACRL_C0011

TIRE, WHEEL AND BALL JOINT SPECIFICATIONS

Year	Model	OEM Tires Standard	OEM Tires Optional	Tire Pressures (psi) Front	Tire Pressures (psi) Rear	Wheel Size	Ball Joint Inspection	Lug Nut (ft. lbs.)
2009	RL	P245/45/R18	①	32	30 ②	18	NS	94
2010	RL	P245/45/R18	①	32	30 ②	18	NS	94

OEM: Original Equipment Manufacturer

PSI: Pounds Per Square Inch

NS: Not Specified by manufacturer

① Optional tire sizes available in 18 in. 19 in., and 20 in. wheel applications.

② Normal driving shown above; for high-speed driving: 36 front; 35 rear.

37647_ACRL_C0012

BRAKE SPECIFICATIONS

All measurements in inches unless noted

Year	Model		Brake Disc Original Thickness	Brake Disc Minimum Thickness	Brake Disc Maximum Runout	Minimum Lining Thickness Front	Minimum Lining Thickness Rear	Brake Caliper Bracket Bolts (ft. lbs.)	Brake Caliper Mounting Bolts (ft. lbs.)
2009	RL	F	1.10-1.11	1.020	0.002	0.06	—	—	58
		R	0.625-0.634	0.550	0.002	—	0.06	79	17
2010	RL	F	1.10-1.11	1.020	0.002	0.06	—	—	58
		R	0.625-0.634	0.550	0.002	—	0.06	79	17

NA: Not Available

F: Front

R: Rear

37647_ACRL_C0013

SCHEDULED MAINTENANCE INTERVALS

Acura RL

TO BE SERVICED	TYPE OF SERVICE	SERVICE INTERVALS						
		Symbol A	Symbol B	1	2	3	4	5
Air cleaner element	Replace				✓			
Brake fluid	Inspect		✓					
Brake hoses and lines	Inspect		✓					
Clutch fluid (if equipped)	Inspect		✓					
Drive belt	Inspect				✓			
Driveshaft boots	Inspect		✓					
Dust & pollen filter	Replace				✓			
Engine coolant	Inspect		✓					
Engine coolant	Replace							✓
Engine oil	Replace	✓						
Engine oil & filter	Replace		✓					
Exhaust system	Inspect		✓					
Front & rear brakes	Inspect		✓					
Fuel lines	Inspect		✓					
Parking brake adjustment	Inspect		✓					
Power steering fluid	Inspect		✓					
Spark plugs	Replace						✓	
Suspension components	Inspect		✓					
Tie-rod ends, steering gearbox and boots	Inspect		✓					
Timing belt	Replace						✓	
Tires	Rotate			✓				
Transmission fluid	Inspect		✓					
Transmission fluid	Replace					✓		
Valve clearance	Inspect						✓	
Windshield washer fluid	Inspect		✓					

NOTES:

Refer to the Owner's Manual for complete explanation of the service maintenance reminder system.

The maintenance minder is an important feature of the multi-information display. Based on engine operating conditions, the RL's onboard computer (PCM)

remaining engine oil life. The system also displays the remaining engine oil life along with the code(s) for other scheduled maintenance items needing service.

If the message "SERVICE DUE NOW" does not appear more than 12 months after the display is reset, change the engine oil every year.

Independent of the maintenance messages in the multi-information display, replace the brake fluid every 3 years.

Inspect idle speed every 160,000 miles (256,000 km).

Adjust the valves during services A, B, 1, 2 or 3 only if they are noisy.

37647_ACRL_C0014

BRAKES — INFORMATION AND PRECAUTIONS

ANTI-LOCK SYSTEMS

- Certain components within the ABS system are not intended to be serviced or repaired individually.
- Do not use rubber hoses or other parts not specifically specified for and ABS system. When using repair kits, replace all parts included in the kit. Partial or incorrect repair may lead to functional problems and require the replacement of components.
- Lubricate rubber parts with clean, fresh brake fluid to ease assembly. Do not use shop air to clean parts; damage to rubber components may result.
- Use only DOT 3 brake fluid from an unopened container.
- If any hydraulic component or line is removed or replaced, it may be necessary to bleed the entire system.
- A clean repair area is essential. Always clean the reservoir and cap thoroughly before removing the cap. The slightest amount of dirt in the fluid may plug an orifice and impair the system function. Perform repairs after components have been thoroughly cleaned; use only denatured alcohol to clean components. Do not allow ABS components to come into contact with any substance containing mineral oil; this includes used shop rags.
- The Anti-Lock control unit is a microprocessor similar to other computer units in the vehicle. Ensure that the ignition switch is **OFF** before removing or installing controller harnesses. Avoid static electricity discharge at or near the controller.
- If any arc welding is to be done on the vehicle, the control unit should be unplugged before welding operations begin.

DISC AND DRUM SYSTEMS

CAUTION

Dust and dirt accumulating on brake parts during normal use may contain asbestos fibers from production or aftermarket brake linings. Breathing excessive concentrations of asbestos fibers can cause serious bodily harm. Exercise care when servicing brake parts. Do not sand or grind brake lining unless equipment used is designed to contain the dust residue. Do not clean brake parts with compressed air or by dry brushing. Cleaning should be done by dampening the brake components with a fine mist of water, then wiping the brake components clean with a dampened cloth. Dispose of cloth and all residue containing asbestos fibers in an impermeable container with the appropriate label. Follow practices prescribed by the Occupational Safety and Health Administration (OSHA) and the Environmental Protection Agency (EPA) for the handling, processing, and disposing of dust or debris that may contain asbestos fibers.

BRAKES — BLEEDING THE BRAKE SYSTEM

BLEEDING PROCEDURE

BLEEDING PROCEDURE

➡See "Bleeding the ABS System" in this section.

BLEEDING THE ABS SYSTEM

See Figure 2.

WARNING

Read and observe the "PRECAUTIONS" in this section before starting work.

1. The reservoir on the master cylinder must be at the MAX (upper) level mark at the start of the bleeding procedure and checked after bleeding each brake caliper. Add fluid as required.
2. Make sure the brake fluid level in the reservoir is at the MAX (upper) level line.
3. Slide a piece of clear plastic hose over the first bleed screw, and submerge the other end in a container of new brake fluid.
4. Have someone slowly pump the brake pedal several times, and then apply steady pressure.
5. Starting at the left-front, loosen the brake bleed screw to allow air to escape from the system. Then tighten the bleed screw securely.
6. Repeat the procedure for each wheel in the sequence shown following until air bubbles no longer appear in the fluid.
7. Refill the master cylinder reservoir to the MAX (upper) level line.

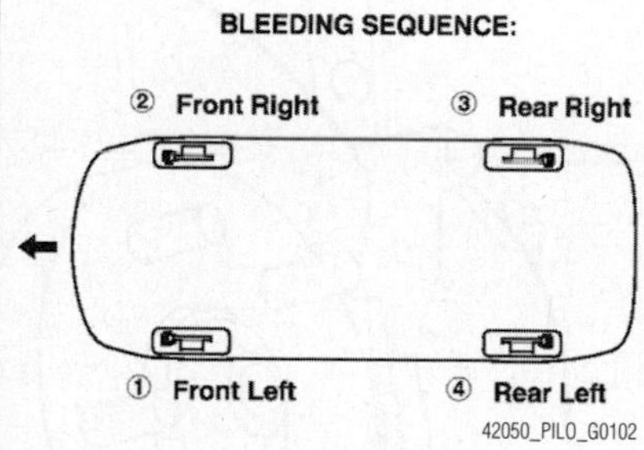
BLEEDING SEQUENCE:
② Front Right
③ Rear Right
① Front Left
④ Rear Left
42050_PILO_G0102

Fig. 2 Brake bleeding sequence

WHEEL SPEED SENSORS

REMOVAL & INSTALLATION

See Figures 3 and 4.

**** WARNING**

Read and observe the "PRECAUTIONS" in this section before starting work.

➡Install the sensor carefully to avoid twisting the wires.

1. Replace the clamps (A) when the wheel speed sensors are removed or replaced.
2. Use the figures for reference when removing and installing the wheel speed sensors.

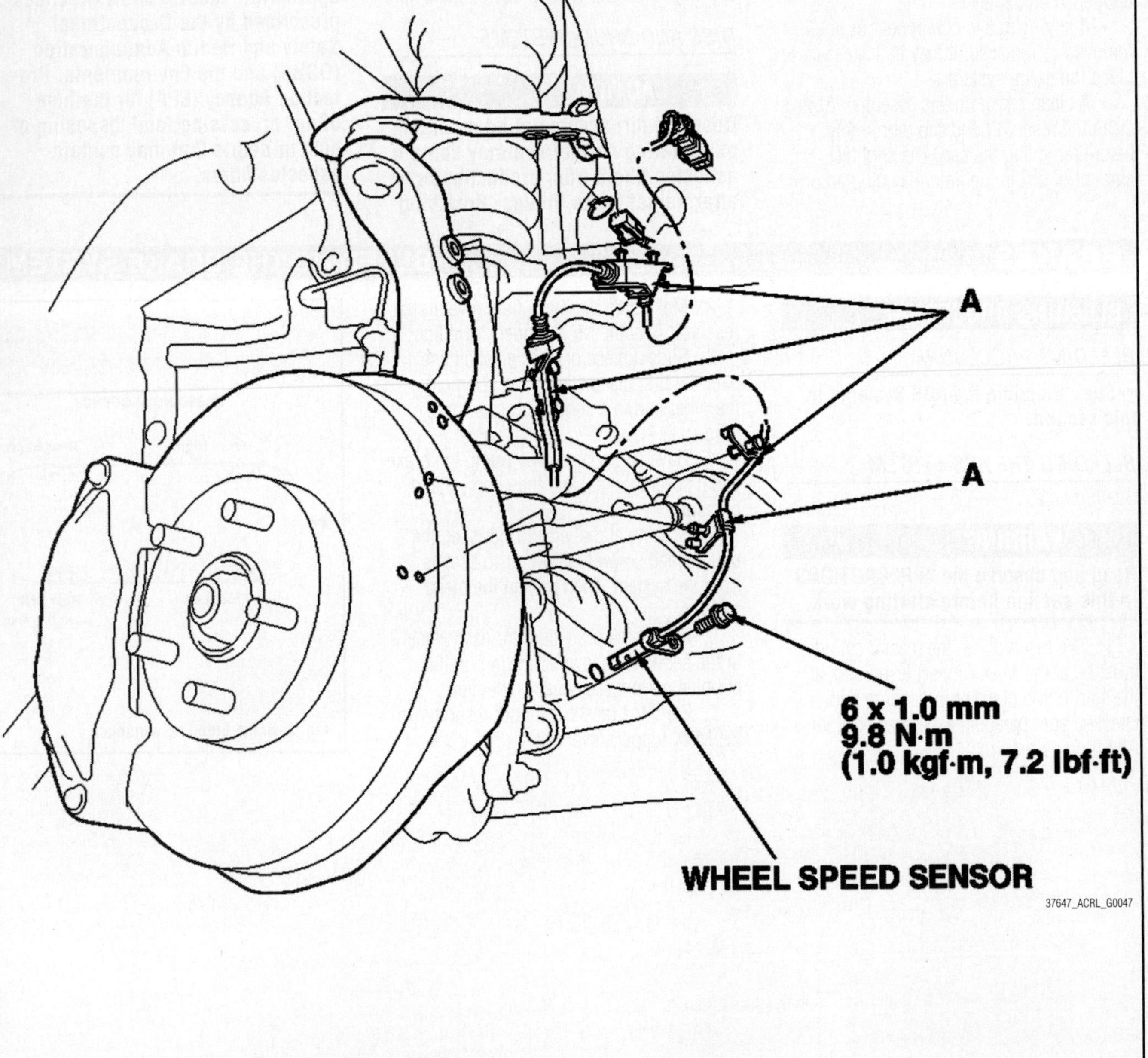
A
A
6 x 1.0 mm
9.8 N·m
(1.0 kgf·m, 7.2 lbf·ft)
WHEEL SPEED SENSOR
37647_ACRL_G0047

Fig. 3 Replace the clamps (A) when the wheel speed sensors are removed or replaced–front wheel sensors

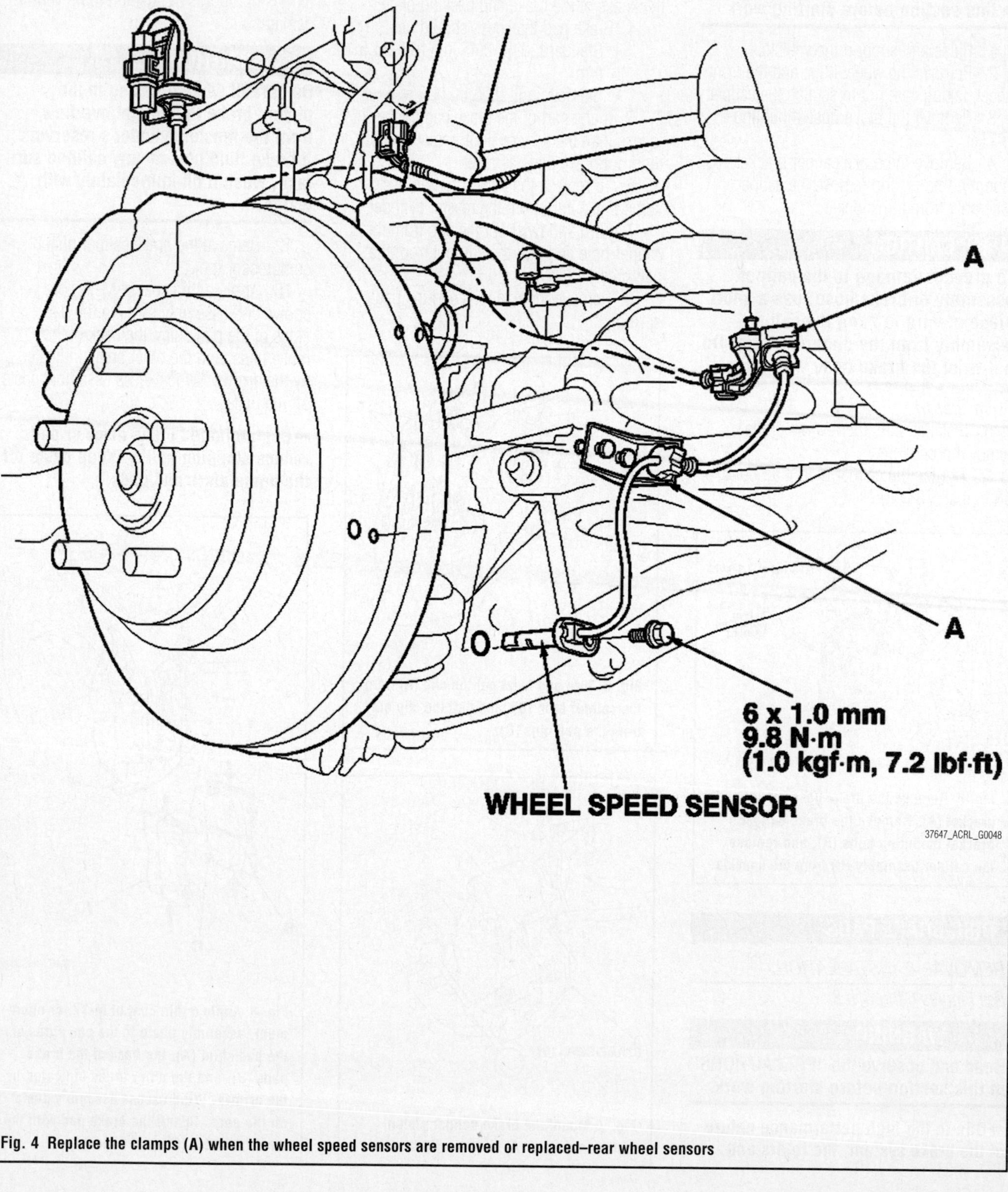
A
A
6 x 1.0 mm
9.8 N·m
(1.0 kgf·m, 7.2 lbf·ft)
WHEEL SPEED SENSOR
37647_ACRL_G0048

Fig. 4 Replace the clamps (A) when the wheel speed sensors are removed or replaced–rear wheel sensors

BRAKE CALIPER

REMOVAL & INSTALLATION

See Figure 5.

⁂ WARNING

Read and observe the "PRECAUTIONS" in this section before starting work.

1. Raise and support the vehicle.
2. Remove the wheel nuts, and the front wheel, taking care not to scratch the caliper.
3. Remove the brake hose mounting bracket.
4. Remove the brake caliper bracket mounting bolts, and remove the caliper assembly from the knuckle.

⁂ CAUTION

To prevent damage to the caliper assembly or brake hose, use a short piece of wire to hang the caliper assembly from the undercarriage. Do not twist the brake hose with force.

To install:

5. Installation is the reverse of the removal procedure.
6. Tighten the caliper mounting bolts to 58 ft. lbs. (78 Nm).

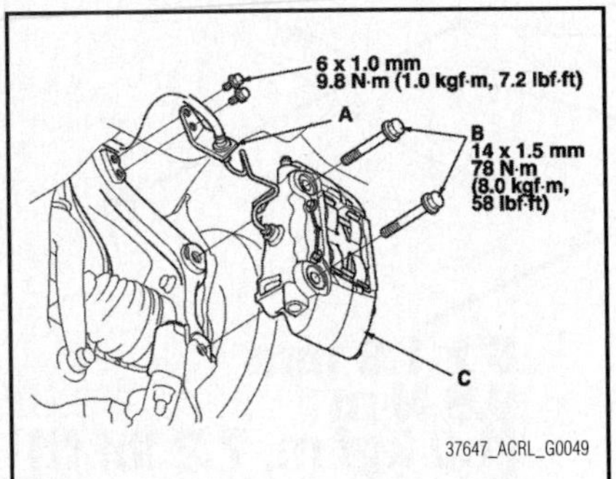
6 x 1.0 mm
9.8 N·m (1.0 kgf·m, 7.2 lbf·ft)
A
B
14 x 1.5 mm
78 N·m
(8.0 kgf·m,
58 lbf·ft)
C
37647_ACRL_G0049

Fig. 5 Remove the brake hose mounting bracket (A). Remove the brake caliper bracket mounting bolts (B), and remove the caliper assembly (C) from the knuckle.

DISC BRAKE PADS

REMOVAL & INSTALLATION

See Figures 6 through 8.

⁂ WARNING

Read and observe the "PRECAUTIONS" in this section before starting work.

➡Due to the high performance nature of the brake system, the rotors and pads may wear faster. Be sure to inspect the front rotor thickness anytime the front pads are replaced.

1. Raise and support the vehicle.
2. Remove the front wheels.
3. Check the thickness of the inner brake pad and outer pad. Do not include the thickness of the brake pad backing plate.
4. Brake pad thickness should be:
    - Standard: 0.41-0.45 in. (10.5-11.5 mm)
    - Service limit: 0.06 in. (1.6 mm)
5. If any part of the brake pad thickness is less than the service limit, replace all of the front brake pads as a set.
6. To replace the pads, first remove some brake fluid from the master cylinder.
7. Turn and twist out the clip from the caliper hole and pull the clip out from the pad pins.
8. Remove the pad pins and the pad spring.

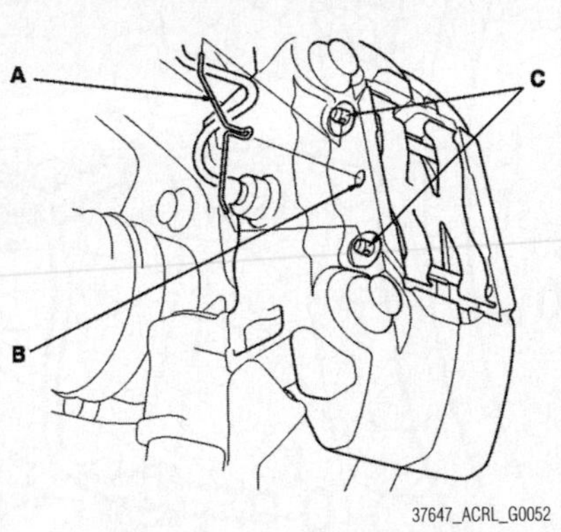
A
C
B
37647_ACRL_G0052

Fig. 6 Turn and twist out the clip (A) from the caliper hole (B), and pull the clip out from the pad pins (C).

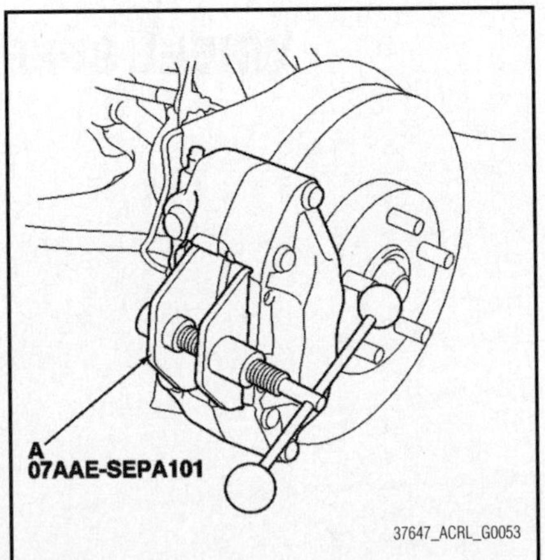
A
07AAE-SEPA101
37647_ACRL_G0053

Fig. 7 Mount the brake caliper piston compressor tool (A) on the caliper.

9. Remove the pads.

To install:

10. Mount the brake caliper piston compressor tool on the caliper.
11. Press in the piston with the brake caliper piston compressor tool so that the caliper will fit over the pads. Make sure the piston boot is in position to prevent damaging it.

⁂ CAUTION

Be careful when pressing in the piston; brake fluid might overflow from the master cylinder's reservoir. If brake fluid gets on any painted surface, wash it off immediately with water.

12. Remove the brake caliper piston compressor tool.
13. Apply a thin coat of M-77, or equivalent, assembly paste to the pad sides of the pad shim, the back of the brake pads, and the other areas indicated by the arrows. Wipe excess assembly paste off the pads.

➡Contaminated brake discs or pads reduce stopping ability. Keep paste off the brake discs and pads.

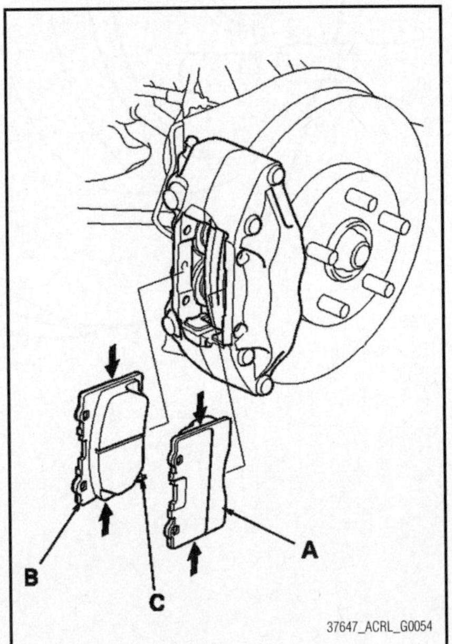
A
B
C
37647_ACRL_G0054

Fig. 8 Apply a thin coat of M-77, or equivalent, assembly paste to the pad sides of the pad shim (A), the back of the brake pads (B), and the other areas indicated by the arrows. Wipe excess assembly paste off the pads. Install the brake pad with the wear indicator (C) on the inside.

14. Install the brake pads correctly. Install the brake pad with the wear Indicator on the inside.

15. If reusing the brake pads, always reinstall the brake pads in their original positions to prevent a temporary loss of braking efficiency.

16. Install the pad spring. Hold the pad spring and install the pad pins into the caliper from the outside to the inside of vehicle.

17. First, insert the clip ends to the pad pins, and then twist the clip into the caliper hole to stabilize.

18. Clean the mating surface between the brake disc and the inside of the wheel, then install the front wheels.

19. Press the brake pedal several times to make sure the brakes work.

➡Engagement may require a greater pedal stroke immediately after the brake pads have been replaced as a set. Several applications of the brake pedal will restore the normal pedal stroke.

20. Add brake fluid as needed.

21. After installation, check for leaks at the brake hose and line joints or connections, and retighten if necessary.

22. Reinstall the front wheels, then test-drive the vehicle.

23. Check for leaks.

BRAKES — REAR DISC BRAKES

BRAKE CALIPER

REMOVAL & INSTALLATION

See Figure 9.

1. Raise and support the vehicle.
2. Remove the wheel nuts, and the rear wheel.
3. Remove the brake caliper bracket mounting bolts (A), and remove the caliper assembly (B) from the knuckle. To prevent damage to the caliper assembly or brake hose, use a short piece of wire to hang the caliper assembly from the undercarriage. Do not twist the brake hose excessively.
4. Installation is the reverse of the removal procedure.
5. Tighten the caliper mounting bolts to 80 ft. lbs. (108 Nm).

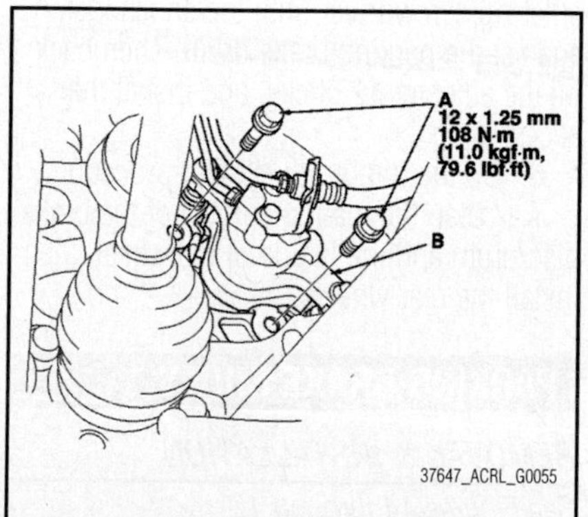
A
12 x 1.25 mm
108 N·m
(11.0 kgf·m,
79.6 lbf·ft)
B
37647_ACRL_G0055

Fig. 9 Remove the brake caliper bracket mounting bolts (A), and remove the caliper assembly (B) from the knuckle.

DISC BRAKE PADS

REMOVAL & INSTALLATION

See Figures 10 and 11.

⁂ WARNING

Read and observe the "PRECAUTIONS" in this section before starting work.

➡Due to the high performance nature of the brake system, the rotors and pads may wear faster. Be sure to inspect the front rotor thickness anytime the front pads are replaced.

1. Raise and support the vehicle.
2. Remove the rear wheels.
3. Check the thickness of the inner brake pad and outer pad. Do not include the thickness of the brake pad backing plate.
4. Brake pad thickness should be:
   - Standard: 0.37–0.41 in. (9.5–10.5 mm)
   - Service limit: 0.06 in. (1.6 mm)
5. If any part of the brake pad thickness is less than the service limit, replace all of the rear brake pads as a set.
6. To replace the rear pads, first remove some brake fluid from the master cylinder.
7. Release the parking brake.
8. Remove the caliper bracket mounting bolts (A) while holding the caliper pins (B) with a wrench being careful not to damage the pin boot, and remove the caliper (C). Check the hose and pin boots for damage and deterioration. Thoroughly clean the outside of the caliper to prevent dust and dirt from entering inside. Support the caliper with a piece of wire so it does not hang from the brake hose.

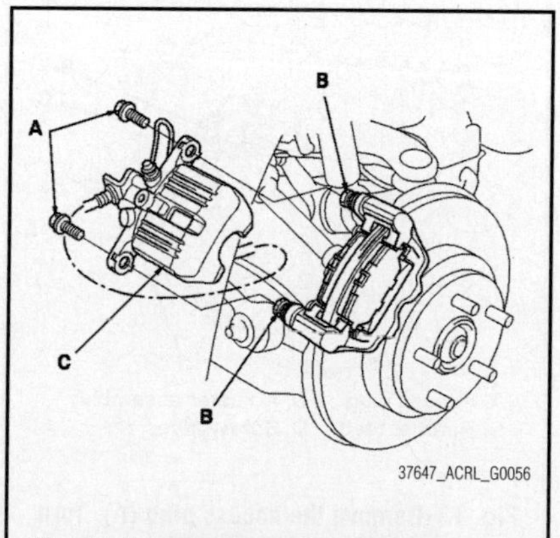
B
A
C
B
37647_ACRL_G0056

Fig. 10 Remove the caliper bracket mounting bolts (A) while holding the caliper pins (B) with a wrench being careful not to damage the pin boot, and remove the caliper (C).

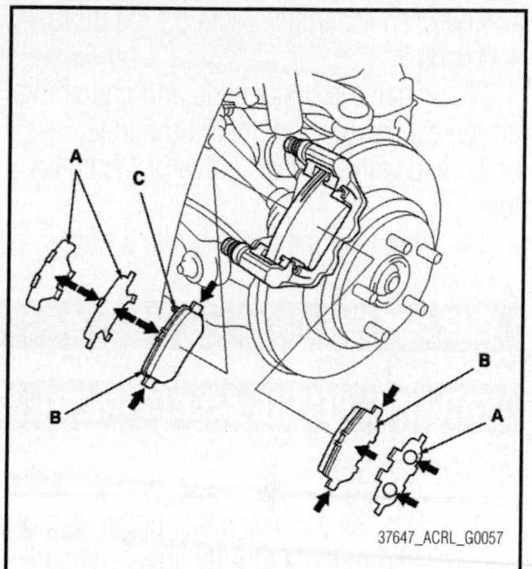
A
C
B
B
A
37647_ACRL_G0057

Fig. 11 Apply M-77 assembly paste to the pad side of the shims (A) and the back of the brake pads (B) and the other areas indicated by the arrows. Install the inner brake pad with its wear indicator (C) facing on top.

9. Remove the pad shims and brake pads.
10. Remove the pad retainers.

To install:

11. Clean the caliper bracket thoroughly; remove any rust, and check for grooves and cracks.
12. Verify that the caliper pins move in and out smoothly. Clean and lube if needed.
13. Inspect the brake disc for runout, thickness, parallelism, and check for damage and cracks.
    - Runout service limit: 0.0016 in. (0.04 mm)
    - Thickness max. refinishing limit: 0.55 in. (14.0 mm)
    - Parallelism: 0.0006 in. (0.015 mm) max.
14. Apply a thin coat of M-77 assembly paste to the retainers mating surfaces of the caliper bracket.
15. Install the pad retainers. Wipe excess assembly paste off the retainers.

⁂ CAUTION

Contaminated brake discs and pads reduce stopping ability. Keep assembly paste off the discs and pads.

16. Install the pad retainers.
17. Apply M-77 assembly paste to the pad side of the shims and the back of the brake pads and the other areas indicated by the arrows. Wipe excess assembly paste off the pad shims and brake pads. Contaminated brake discs or pads reduce stopping ability. Keep assembly paste off the discs and pads.
18. Install the brake pads and pad shims on the caliper bracket. Install the inner brake pad with its wear indicator facing on top.
19. If reusing the brake pads, always reinstall the brake pads in their original positions to prevent a temporary loss of braking efficiency.
20. Push in the piston so the caliper will fit over the brake pads. Make sure the piston boot is in position to prevent damaging it when installing the caliper.

⁂ CAUTION

Be careful when pushing in the caliper, brake fluid might overflow from the master cylinder's reservoir.

21. Apply a thin coat of M-77 assembly paste to the piston edges on their mating surfaces against the inner pad shim.
22. Install the brake caliper.
23. Install the caliper bolts, and torque them to 17 ft. lbs. (23 Nm) while holding the caliper pins with a wrench, being careful not to damage the pin boot.
24. Clean the mating surface between the brake disc/drum and the inside of the wheel, then install the rear wheels.
25. Press the brake pedal several times to make sure the brakes work, then road-test the vehicle.

➡Engagement may require a greater pedal stroke immediately after the brake pads have been replaced as a set. Several applications of the brake pedal will restore the normal pedal stroke.

26. Add brake fluid as needed.
27. After installation, check for leaks at the hose and line joints and connections, and retighten if necessary. Test-drive the vehicle, then recheck for leaks.

BRAKES — PARKING BRAKE

PARKING BRAKE ADJUSTMENT

ADJUSTMENT

1. Press the parking brake pedal with 66 lbs. (294 N) of force to fully apply the parking brake. The parking brake pedal should be locked within the 5–6 clicks.
2. Adjust the parking brake if the pedal clicks are not within the specification.

➡Minor parking brake pedal adjustments (1 to 2 clicks) can be made with the adjusting nut. If a larger adjustment is required, follow the major adjustment procedure using the adjuster at the parking brake drum.

3. After installing new parking brake shoes and/or new brake disc, make sure you drive the vehicle for "break-in".

Minor Adjustment

See Figure 12.

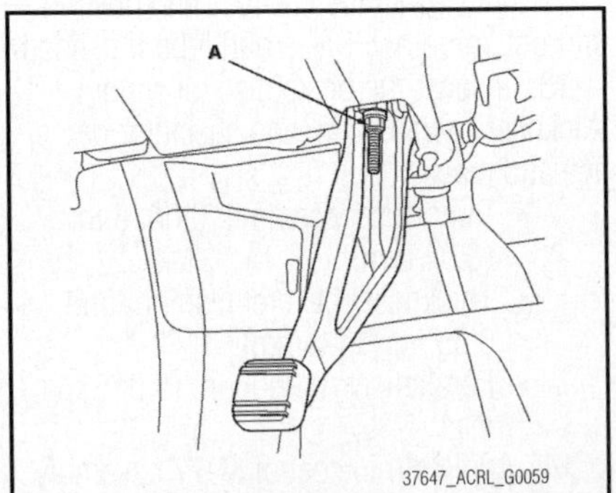
A
37647_ACRL_G0059

Fig. 12 Tighten the parking brake adjusting nut (A) until the parking brakes drag slightly when the rear wheels are rotated.

1. Raise and support the vehicle.
2. Release the parking brake pedal fully.
3. Press the parking brake pedal 1 click.
4. Tighten the parking brake adjusting nut until the parking brakes drag slightly when the rear wheels are rotated.
5. Release the parking brake pedal fully, and check that the parking brakes do not drag when the rear wheels are rotated. Readjust if necessary.
6. Make sure the parking brakes pedal is within 5 to 6 clicks.

Major Adjustment

See Figure 13.

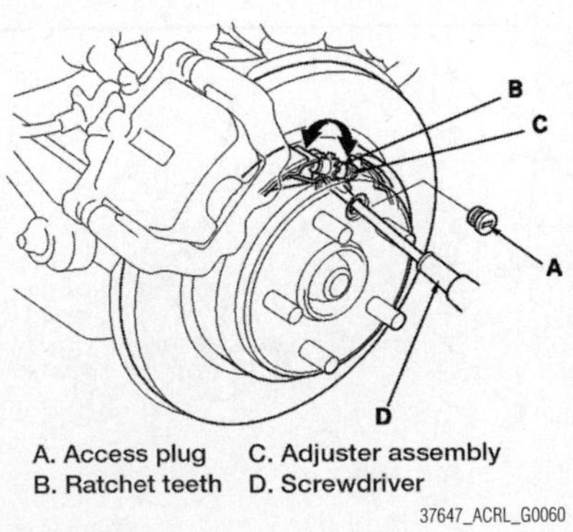
B
C
A
D
A. Access plug
B. Ratchet teeth
C. Adjuster assembly
D. Screwdriver
37647_ACRL_G0060

Fig. 13 Remove the access plug (A). Turn the ratchet teeth (B) on the adjuster assembly (C) with a flat-tip screwdriver (D) until the shoes lock against the parking brake drum. Then back off the adjuster 12 clicks, and install the access plug.

➡To be done when replacing parking brake shoes and after lining surface break-in.

1. Raise and support the vehicle.
2. Release the parking brake pedal fully.
3. Loosen the parking brake adjusting nut.
4. Remove the rear wheels.
5. Remove the access plug. Turn the ratchet teeth on the adjuster assembly with a flat-tip screwdriver until the shoes lock against the parking brake drum. Then back off the adjuster 12 clicks, and install the access plug.
6. Do the minor adjustment procedure.
7. Clean the mating surface of the brake disc/drum and the inside of the wheel, then install the rear wheels.

PARKING BRAKE SHOES

REMOVAL & INSTALLATION

See Figures 14 through 17.

⁂ WARNING

Read and observe the "PRECAUTIONS" in this section before starting work.

1. Raise and support the vehicle.
2. Remove the rear wheels.
3. Release the parking brake and remove the rear brake caliper and the brake disc/drum.
4. Disconnect and remove the upper brake spring.
5. Remove the adjuster assembly.
6. Disconnect and remove the lower brake spring.

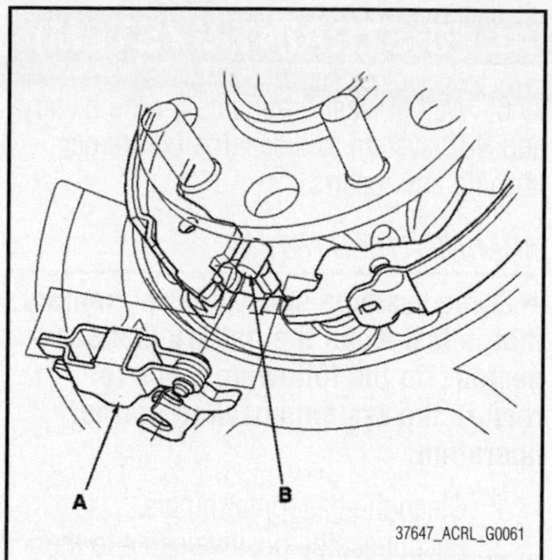
A
B
37647_ACRL_G0061

Fig. 14 Remove the parking brake lever assembly (A), and disconnect it from the parking brake cable end (B).

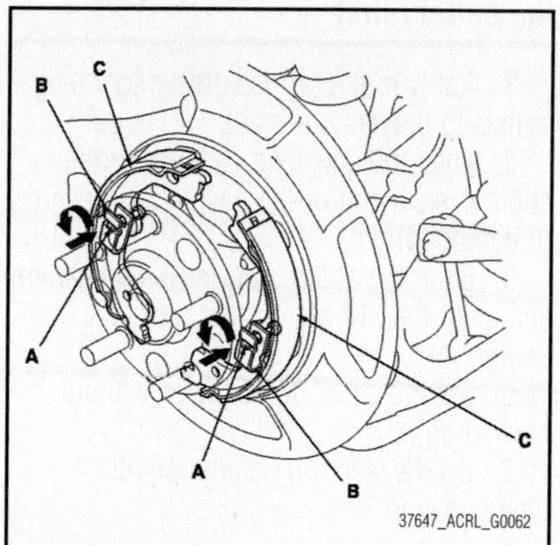
B
C
A
A
B
C
37647_ACRL_G0062

Fig. 15 Remove the tension pins (A) by pushing the retainer springs (B) and turning the pins. Remove the parking brake shoes (C).

7. Remove the parking brake lever assembly and disconnect it from the parking brake cable end.

8. Remove the tension pins by pushing the retainer springs and turning the pins.

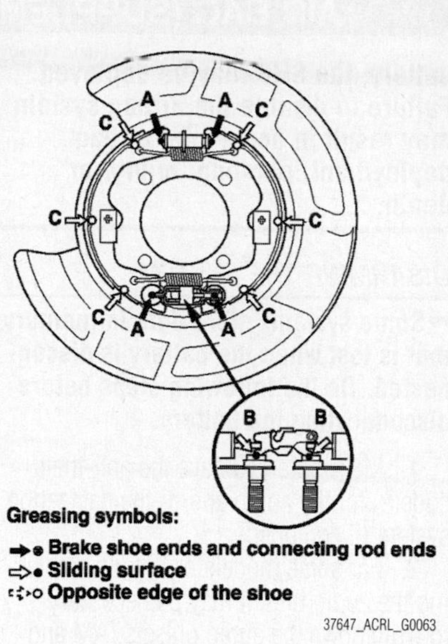
A
A
C
C
C
C
C
A
A
C
B
B
Greasing symbols:
Brake shoe ends and connecting rod ends
Sliding surface
Opposite edge of the shoe
37647_ACRL_G0063

Fig. 16 Apply Molykote 44MA grease, or equivalent, to the shoe ends and connecting rod ends (A), sliding surfaces (B), and opposite edges of the parking brake shoe (C) as shown. Wipe off any excess. Keep grease off the brake linings.

9. Remove the parking brake shoes.

To install:

10. Apply Molykote 44MA grease, or equivalent, to the shoe ends and connecting rod ends (A), sliding surfaces (B), and opposite edges of the parking brake shoe (C) as shown. Wipe off any excess. Keep grease off the brake linings.

11. Reinstall the tension pins and retainer springs.

12. Connect the parking brake cable end to the parking brake lever assembly, and install the parking brake lever assembly.

13. Reinstall the lower brake spring.

14. Clean the threaded portions of the longer clevis pin (A), and coat the threads of the longer clevis pin (A) with grease.

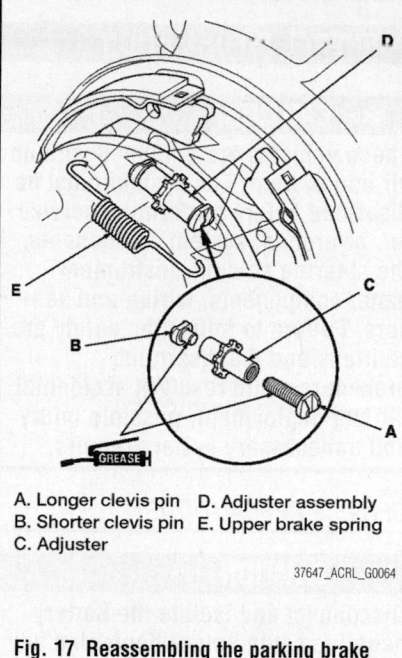

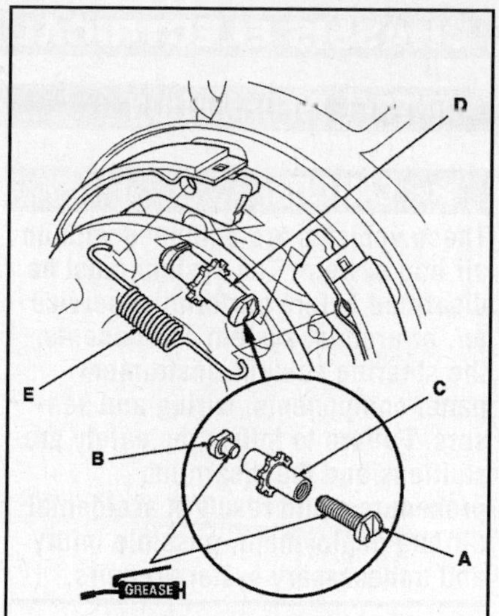
D
E
C
B
A
GREASE
A. Longer clevis pin
B. Shorter clevis pin
C. Adjuster
D. Adjuster assembly
E. Upper brake spring
37647_ACRL_G0064

Fig. 17 Reassembling the parking brake shoe adjuster

15. Clean the sliding surface of the shorter clevis pin (B), and coat the sliding surface of the shorter clevis pin (B) with grease.

16. Install the longer clevis pin (A) and pin (B) on the adjuster (C) and shorten the longer clevis pin (A) by turning the adjuster.

17. Position the brake shoe adjuster assembly (D) on the parking brake shoes.

18. Reinstall the upper brake spring (E).

19. Install the rear brake disc/drum and the rear brake caliper.

20. Do the "Major Parking Brake Adjustment".

ADJUSTMENT

➡See "Parking Brake Adjustment" in this section.

CHASSIS ELECTRICAL

AIR BAG (SUPPLEMENTAL RESTRAINT SYSTEM)

GENERAL INFORMATION

⁂ CAUTION

These vehicles are equipped with an air bag system. The system must be disarmed before performing service on, or around, system components, the steering column, instrument panel components, wiring and sensors. Failure to follow the safety precautions and the disarming procedure could result in accidental air bag deployment, possible injury and unnecessary system repairs.

SERVICE PRECAUTIONS

⁂ CAUTION

Disconnect and isolate the battery negative cable before beginning any airbag system component diagnosis, testing, removal, or installation procedures. Wait at least 90 seconds after the ignition switch is turned off and the negative (-) terminal cable is disconnected from the battery before starting the operation. The SRS is equipped with a backup power source, so if work is started within 90 seconds after disconnecting the negative (-) terminal cable from the battery, the SRS may be deployed. Failure to disable the airbag system may result in accidental airbag deployment, personal injury, or death.

DISARMING THE SYSTEM

➡Some systems store data in memory that is lost when the battery is disconnected. Do the following steps before disconnecting the battery.

1. Make sure you have the anti-theft code(s) for the audio and/or the navigation system (if equipped).
2. For some models or if you're replacing the audio unit, it may be necessary to write down the audio presets (AM and FM), and the XM radio presets (if equipped), because the audio unit does not retain the presets after the battery is disconnected.
3. Make sure the ignition switch is in LOCK (0).
4. Disconnect and isolate the negative cable from the battery.

⁂ CAUTION

Always disconnect the negative cable from the battery first.

5. Disconnect the positive cable from the battery.
6. Wait at least 3 minutes for the battery and SRS system to discharge before performing any repairs.

ARMING THE SYSTEM

➡Some systems store data in memory that is lost when the battery is disconnected. Do the following steps to restore the systems back to normal operation.

1. Clean the battery terminals.
2. Reconnect the positive cable to the battery first, then reconnect the negative cable to the battery.

⁂ CAUTION

Always connect the positive cable to the battery first.

3. Apply multipurpose grease to the terminals to prevent corrosion.
4. Enter the anti-theft code(s) for the audio system and/or the navigation system (if equipped).
5. Enter the audio presets (if applicable), and enter the XM radio presets (if equipped).
6. Set the clock (for vehicles without navigation).
7. Do the steering column position memorization.

DRIVE TRAIN

FRONT DRIVESHAFT

REMOVAL & INSTALLATION

See Figures 18 through 22.

1. Raise and support the vehicle.
2. Remove the front wheels.
3. Pry up the stake on the spindle nut, then remove and discard the nut.
4. Drain the transmission fluid, then reinstall the drain plug with a new sealing washer. Tighten the drain plug to 36 ft. lbs. (49 Nm).
5. Hold the stabilizer joint pin, using a hex wrench, and remove the flange nut. Separate the front stabilizer link from the lower arm.
6. Remove the damper fork mounting nut, the flange bolt, and the damper pinch bolt, then remove the damper fork.
7. Remove the steering knuckle holder bolt and nut.
8. Pull the knuckle outward, and separate the outboard joint from the front hub using a plastic hammer.

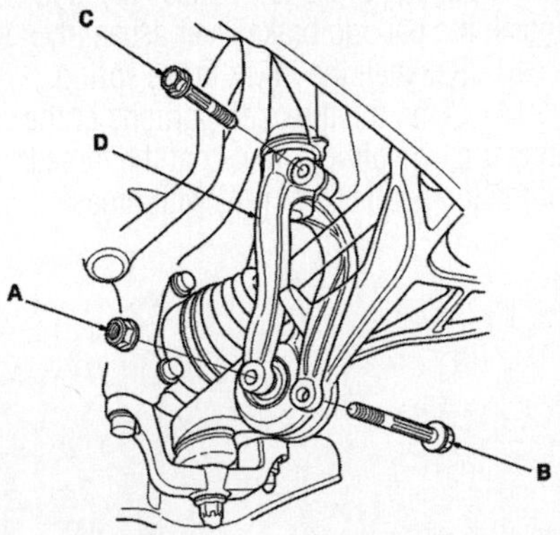
C
D
A
B

A. Damper fork nut C. Damper pinch bolt
B. Flange bolt D. Damper fork

37647_ACRL_G0115

Fig. 18 Remove the damper fork mounting nut, the flange bolt, and the damper pinch bolt, then remove the damper fork.

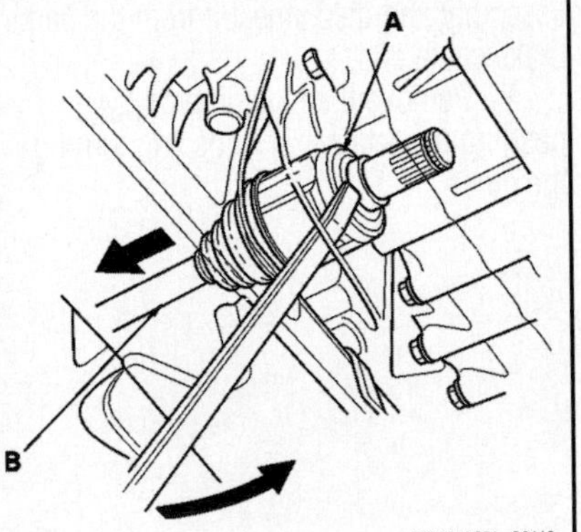
A
B

37647_ACRL_G0116

Fig. 19 Remove the left driveshaft by prying the inboard joint (A) from the differential using a prybar. Remove the driveshaft (B) as an assembly.

B
A
37647_ACRL_G0117

Fig. 20 Remove the right driveshaft by driving the inboard joint (A) off of the intermediate shaft, using a drift and a hammer. Remove the driveshaft (B) as an assembly.

9. Remove exhaust pipe A.

10. Remove the left driveshaft by prying the inboard joint from the differential using a prybar. Remove the driveshaft as an assembly.

➡Do not pull on the driveshaft or the inboard joint may come apart. Pull the inboard joint straight out to avoid damaging the oil seal. Be careful not to damage the oil seal or the end of the inboard joint with the prybar.

11. Remove the right driveshaft by driving the inboard joint off of the intermediate shaft, using a drift and a hammer. Remove the driveshaft as an assembly.

➡Do not pull on the driveshaft, or the inboard joint may come apart.

12. If necessary, remove the set ring from the left driveshaft inboard joint and the set ring from the intermediate shaft.

To install:

➡Before starting installation, make sure the mating surfaces of the joint and the splined section are clean.

13. Apply about 0.18 oz. of moly 60 paste (P/N 08734-0001) to the contact area of the outboard joint and front wheel bearing.

➡The paste helps prevent noise and vibration.

14. Install a new set ring into the set ring groove of the left driveshaft inboard joint.

15. Install a new set ring into the set ring groove of the intermediate shaft.

16. Apply 0.02-0.04 oz. of super high temp urea grease (P/N 08798-9002) to the whole splined surface of the right driveshaft.

17. After applying grease, remove the grease from the splined grooves at intervals of 2-3 splines and from the set ring groove so that air can bleed from the intermediate shaft.

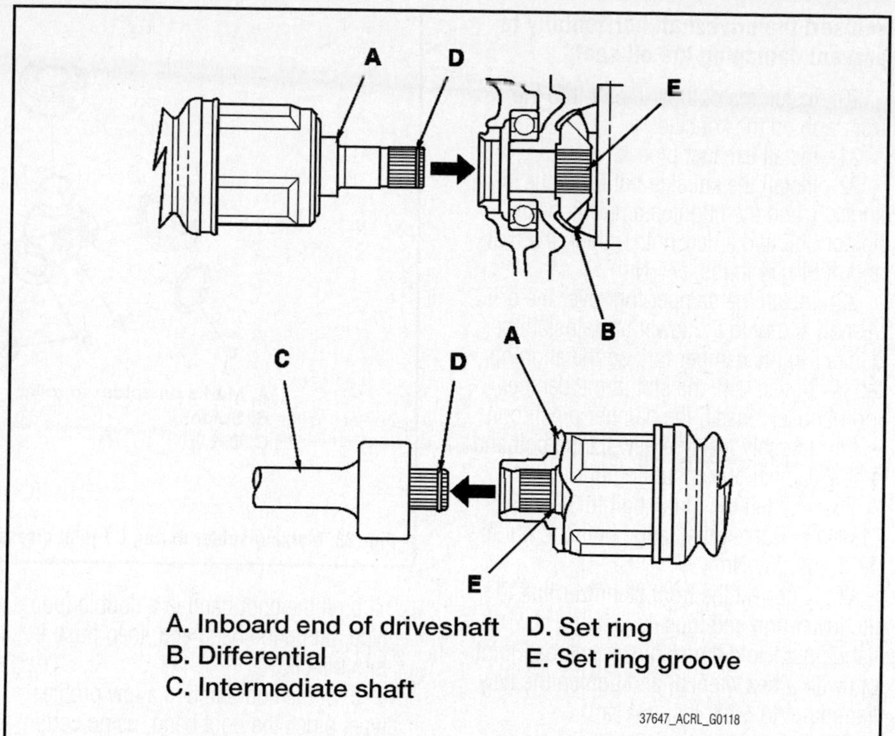
A
D
E
B
C
D
A
E
A. Inboard end of driveshaft
B. Differential
C. Intermediate shaft
D. Set ring
E. Set ring groove
37647_ACRL_G0118

Fig. 21 Insert the inboard end of the driveshaft into the differential or intermediate shaft until the set ring locks in the groove.

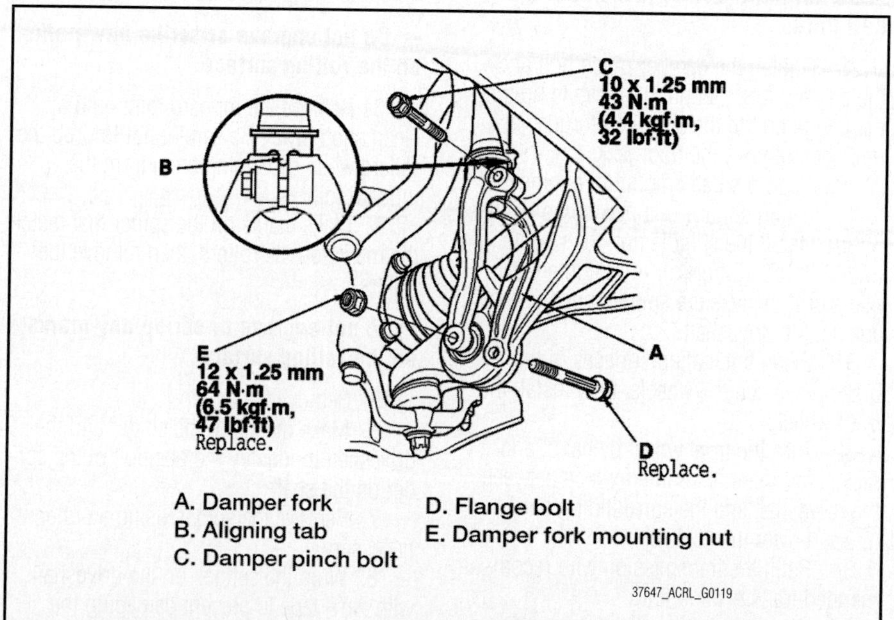
C
10 x 1.25 mm
43 N·m
(4.4 kgf·m,
32 lbf·ft)
B
E
12 x 1.25 mm
64 N·m
(6.5 kgf·m,
47 lbf·ft)
Replace.
A
D
Replace.
A. Damper fork
B. Aligning tab
C. Damper pinch bolt
D. Flange bolt
E. Damper fork mounting nut
37647_ACRL_G0119

Fig. 22 Install the damper fork over the driveshaft and onto the lower arm. Install the damper in the damper fork so the aligning tab is aligned with the slot in the damper fork. Loosely install the damper pinch bolt. Loosely install a new flange bolt and a new damper fork mounting nut.

18. Clean the areas where the driveshaft contacts the differential thoroughly with solvent and dry them with compressed air.

➡Do not wash the rubber parts with solvent.

19. Insert the inboard end (A) of the driveshaft into the differential (B) or intermediate shaft (C) until the set ring (D) locks in the groove (E).

➡Insert the driveshaft horizontally to prevent damaging the oil seal.

20. Install the outboard joint into the front hub on the knuckle.
21. Install exhaust pipe A.
22. Install the knuckle holder to the knuckle, and then tighten a new knuckle holder bolt and a new nut. Tighten the bolt and nut to 54 ft. lbs. (74 Nm).
23. Install the damper fork over the driveshaft and onto the lower arm. Install the damper in the damper fork so the aligning tab is aligned with the slot in the damper fork. Loosely install the damper pinch bolt.
24. Loosely install a new flange bolt and a new damper fork mounting nut.
25. Tighten the pinch bolt to 32 ft. lbs. (43 Nm). Tighten the flange bolt and nut to 47 ft. lbs. (64 Nm).
26. Connect the front stabilizer link to the lower arm and loosely install a new flange nut. Hold the stabilizer link ball joint pin with a hex wrench, and tighten the new flange nut to 40 ft. lbs. (54 Nm).
27. Place a floor jack under the lower arm, and raise the suspension to load it with the vehicle's weight.

➡Do not put the floor jack under the ball joint.

28. Tighten the damper pinch bolt and the damper fork mounting nut while holding the flange bolt to the specified torque values, then remove the floor jack.
29. Apply a small amount of engine oil to the seating surface of a new spindle nut.
30. Install the spindle nut, then tighten it to 242 ft. lbs. (329 Nm). After tightening, use a drift to stake the spindle nut shoulder against the driveshaft.
31. Clean the mating surfaces of the brake discs and the wheels, then install the front wheels.
32. Turn the front wheel by hand, and make sure there is no interference between the driveshaft and the surrounding parts.
33. Lower the vehicle.
34. Refill the transmission with recommended transmission fluid.
35. Check the wheel alignment, and adjust it if necessary.
36. Test-drive the vehicle.

CV-JOINT OVERHAUL

Inboard Joint Side

See Figure 23.

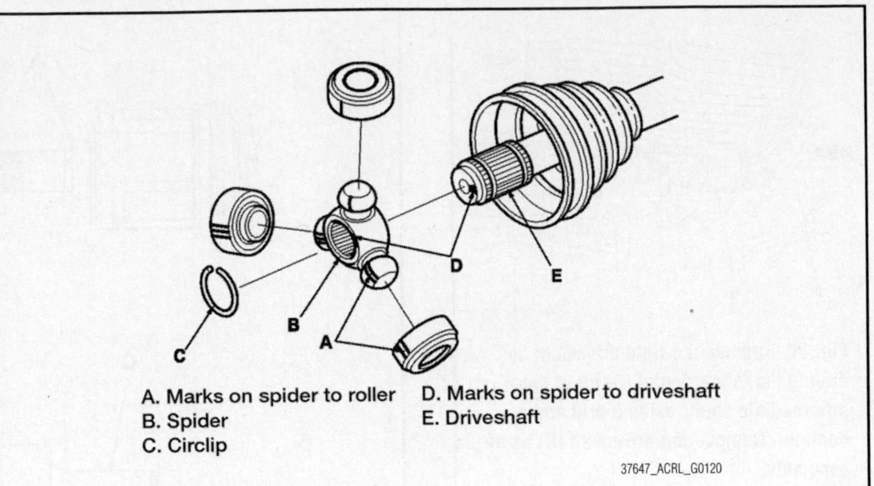
A
B
C
D
E
A. Marks on spider to roller
B. Spider
C. Circlip
D. Marks on spider to driveshaft
E. Driveshaft
37647_ACRL_G0120

Fig. 23 Marking spider during CV joint disassembly

1. Remove the boot bands. Be careful not to damage the boot.
   a. If the boot band is a welded type, cut the boot band.
   b. If the boot band is a double loop type, lift up the band end, then push it into the clip.
   c. If the boot band is a low profile type, pinch the boot band, using commercially available boot band pliers.
2. Make marks on each roller and the inboard joint to identify the locations of the rollers to the grooves on the inboard joint.

➡ Do not engrave or scribe any marks on the rolling surface.

3. Remove the inboard joint onto a clean shop towel. Be careful not to drop the rollers when separating them from the inboard joint.
4. Make marks on the spider that match the marks on the rollers, then remove the rollers.

➡Do not engrave or scribe any marks on the rolling surface.

5. Remove the circlip.
6. Make marks on the spider and the driveshaft to identify the position of the spider on the shaft.
7. Remove the spider, using a puller, if necessary.
8. Wrap the splines on the driveshaft with vinyl tape to prevent damaging the boot.
9. Remove the inboard boot. Be careful not to damage the boot.
10. Remove the vinyl tape.
11. Reassemble in reverse order, making note to align all reference marks.

Outboard Joint Side

See Figure 24.

1. Remove the boot bands. Lift up the three tabs, using a screwdriver, then release the band. Be careful not to damage the boot.

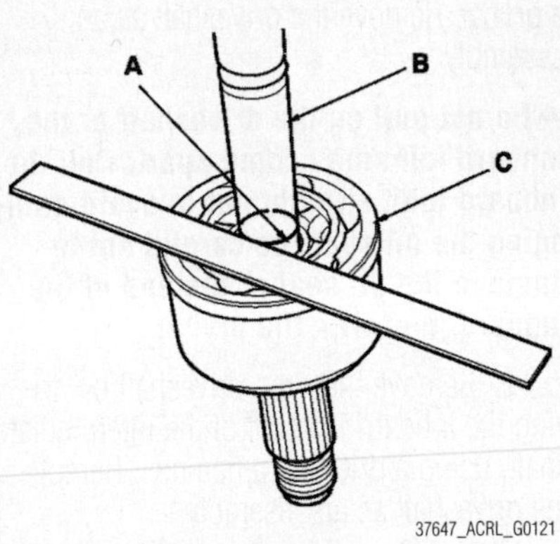
A
B
C
37647_ACRL_G0121

Fig. 24 Make a mark (A) on the driveshaft (B) at the same level as the outboard joint end (C).

2. Slide the outboard boot partially toward the inboard joint side. Be careful not to damage the boot.
3. Wipe off the grease to expose the driveshaft and the outboard joint inner race.
4. Make a mark on the driveshaft at the same level as the outboard joint end.
5. Securely clamp the driveshaft in a bench vise with a shop towel.
6. Remove the outboard joint, using the 26 x 1.5 mm threaded adapter and a commercially available 5/8" x 18 UNF slide hammer.
7. Remove the driveshaft from the bench vise.
8. Remove the stop ring from the driveshaft.
9. Wrap the splines on the driveshaft with vinyl tape to prevent damaging the boot.
10. Remove the outboard boot. Be careful not to damage the boot.

11. Remove the vinyl tape.
12. Reassemble in reverse order, making note to align all reference marks.

INTERMEDIATE SHAFT

REMOVAL & INSTALLATION

See Figures 25 and 26.

1. Drain the transmission fluid, then reinstall the drain plug with a new sealing washer.
2. Remove exhaust pipe A.
3. Remove the right front driveshaft. See "Axle Shaft" in this section.
4. Remove the rear warm up three way catalytic converter (WU-TWC) bracket.
5. Remove the flange bolt and the two dowel bolts.
6. Remove the intermediate shaft from the front differential. Hold the intermediate shaft horizontal until it is clear of the differential to prevent damaging the oil seal.

To install:

7. Clean the areas where the driveshaft contacts the differential thoroughly with solvent, and dry them with compressed air.

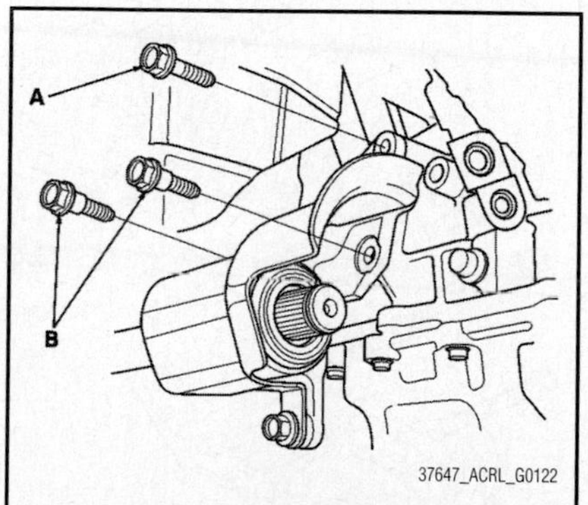
A
B
37647_ACRL_G0122
Fig. 25 Remove the flange bolt (A) and the two dowel bolts (B).

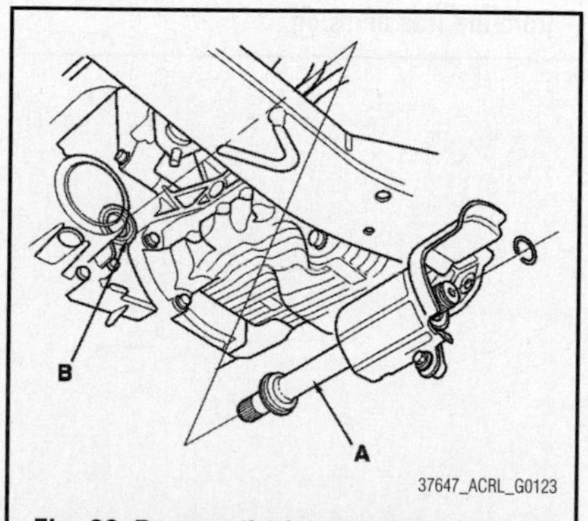
B
A
37647_ACRL_G0123
Fig. 26 Remove the intermediate shaft (A) from the front differential. Hold the intermediate shaft horizontal until it is clear of the differential to prevent damaging the oil seal (B).

➡Do not wash the rubber parts with solvent.

8. Install a new set ring into the set ring groove of the intermediate shaft.
9. Insert the intermediate shaft into the differential correctly.

➡Insert the intermediate shaft carefully to prevent damaging the oil seal.

10. Install the flange bolt (A) and the two dowel bolts. Torque to 29 ft. lbs. (39 Nm).
11. Install the rear warm up three way catalytic converter (WU-TWC) bracket. Torque the bolts to 16 ft. lbs. (22 Nm).
12. Install the right front driveshaft.
13. Install exhaust pipe A.
14. Refill the transmission with the recommended transmission fluid.
15. Check the wheel alignment, and adjust it if necessary.
16. Test-drive the vehicle.

REAR DIFFERENTIAL

REMOVAL & INSTALLATION

See Figures 27 and 28.

1. Raise the vehicle on a lift.
2. Drain the rear differential fluid.
3. Remove the propeller shaft.
4. Disconnect the rear differential harness connectors.
5. Disconnect the breather hose.
6. Disconnect the inboard driveshaft joints from the rear differential using a pair of screwdrivers.
7. Place the transmission jack under the rear differential.
8. Remove the two rear differential rear mounting bolts (horizontal through the crossmember).
9. Remove the four rear differential front mounting bolts (vertical at the front housing).
10. While lowering the rear differential (A) with the transmission jack, remove the rear driveshaft inboard joints (B) from the rear differential. Remove the set rings (C) from the rear differential.

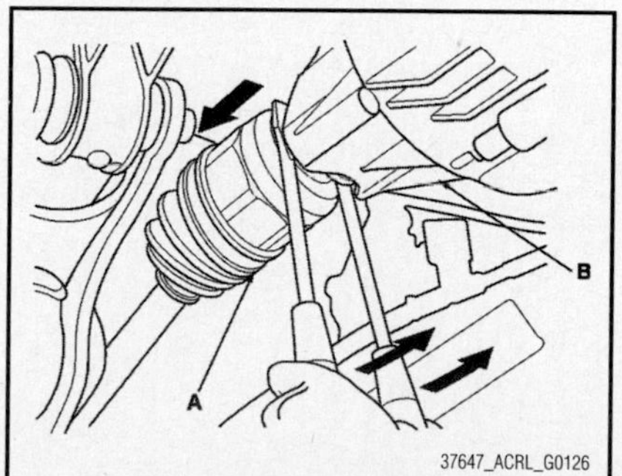
A
B
37647_ACRL_G0126
Fig. 27 Disconnect the inboard driveshaft joints (A) from the rear differential (D) using a pair of screwdrivers.

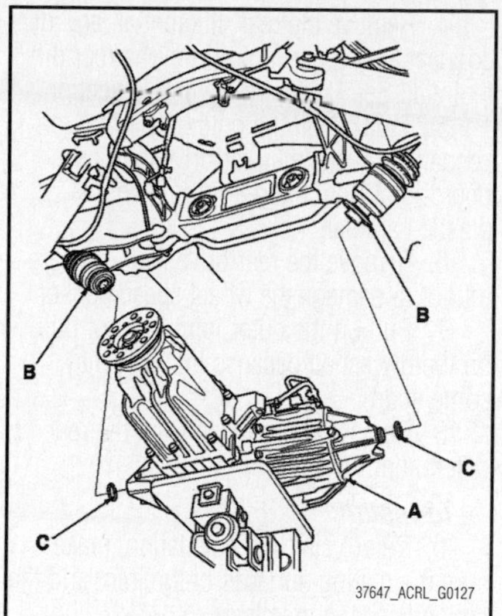
B
B
C
A
C
37647_ACRL_G0127
Fig. 28 While lowering the rear differential (A) with the transmission jack, remove the rear driveshaft inboard joints (B) from the rear differential. Remove the set rings (C) from the rear differential.

To install:

11. Place the rear differential on a transmission jack.
12. Install new set rings into the side case grooves of the rear differential (where the driveshafts will join).
13. Apply appropriate grease to the splines of the rear driveshaft inboard joints.
14. Raise the rear differential to the mounting level, then install the rear driveshaft inboard joints.
15. Install four new rear differential front mounting bolts. Torque to 22 ft. lbs. (30 Nm).
16. Install two new rear differential rear mounting bolts (through the crossmember). Torque to 54 ft. lbs. (74 Nm).
17. Connect the breather hose.
18. Connect the rear differential harness connectors and.
19. Install the propeller shaft.
20. Refill the rear differential fluid.
21. Test-drive the vehicle.

REAR HALFSHAFT

REMOVAL & INSTALLATION

See Figures 29 and 30.

1. Raise and support the vehicle.
2. Remove the rear wheels.
3. Pry up the stake on the spindle nut, then remove the nut.

4. Remove the rear differential, and disconnect the inboard joint from the rear differential. See "Differential" in this section.

5. Tapping on the outer end of the driveshaft, separate the rear driveshaft outboard joint from the rear hub, using a plastic hammer.

6. Remove the rear driveshaft. Be careful not to damage the wheel speed sensor.

7. Pull on the outer joint. Do not pull on the driveshaft because the joint may come apart.

8. Remove the set ring from the rear differential.

To install:

9. Before starting installation, make sure the mating surfaces of the joint and the splined section are clean.

10. Apply super-high temp urea grease to the whole splined surface end of the driveshaft.

11. After applying grease, remove the grease from the splined grooves at intervals of 2-3 splines and from the set ring groove so that air can bleed from the differential.

12. Install the outboard joint of the driveshaft into the rear hub. Be careful not to damage the wheel speed sensor.

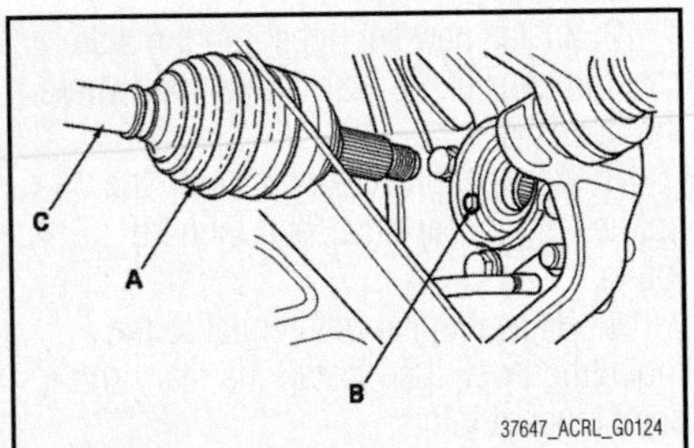
C
A
B
37647_ACRL_G0124
Fig. 29 Remove the rear driveshaft (A). Be careful not to damage the wheel speed sensor (B). Pull on the outer joint. Do not pull on the driveshaft (C) because the joint may come apart.

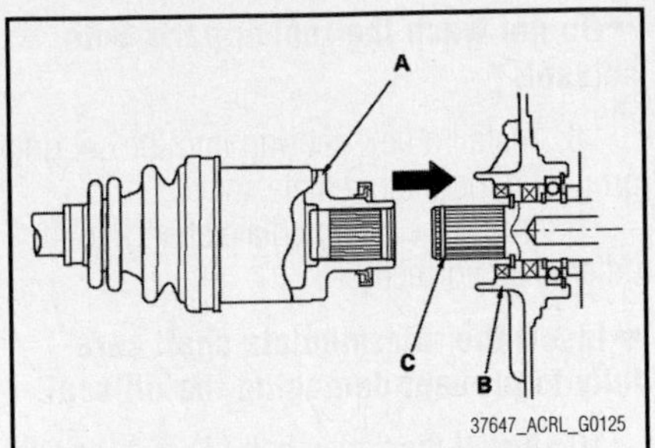
A
C
B
37647_ACRL_G0125
Fig. 30 Make sure the inboard joint (A) is installed all the way into the rear differential (B) and to ensure the set ring (C) is properly seated.

13. Install new set rings in the set ring groove of the rear differential.

14. Clean the areas where the driveshaft contacts the rear differential thoroughly with solvent, and dry with compressed air.

➡Do not wash the rubber parts with solvent.

15. Make sure the inboard joint is installed all the way into the rear differential and to ensure the set ring is properly seated.

16. Install the rear differential. See "Differential" in this section.

17. Apply a small amount of engine oil to the seating surface of a new spindle nut.

18. Install the spindle nut, then tighten it to 181 ft. lbs. (245 Nm). After tightening, use a drift to stake the spindle nut shoulder against the driveshaft.

19. Clean the mating surfaces of the brake discs and the wheels, then install the rear wheels.

20. Turn the rear wheel by hand, and make sure there is no interference between the driveshaft and the surrounding parts.

21. Lower the vehicle.

22. Check the wheel alignment, and adjust it if necessary.

23. Test-drive the vehicle.

TRANSFER ASSEMBLY

REMOVAL & INSTALLATION

See Figure 31.

1. Raise the vehicle on a lift, and make sure it is supported securely.
2. Shift the transmission into N.
3. Remove the drain plug and drain the automatic transmission fluid (ATF).
4. Reinstall the drain plug with a new sealing washer. Torque the plug to 36 ft. lbs. (49 Nm).
5. Remove exhaust pipe A.
6. Remove the bolt securing the transfer breather hose bracket and disconnect the breather hose from the breather pipe on the transfer assembly.
7. Make a reference mark across the propeller shaft and the transfer companion flange. Separate the propeller shaft from the transfer companion flange.
8. Remove the transfer assembly from the transmission.

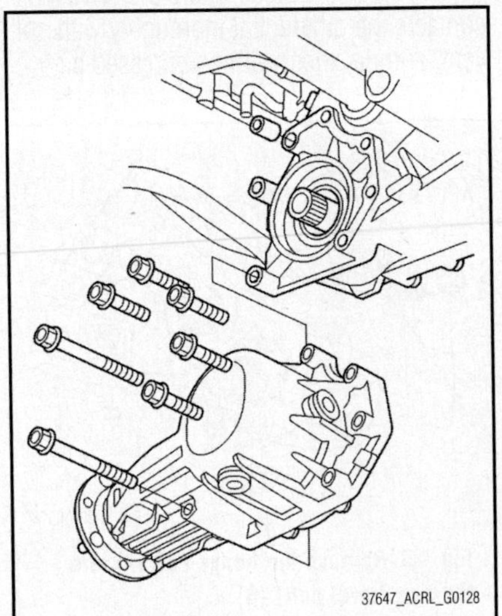
37647_ACRL_G0128
Fig. 31 Remove the transfer assembly from the transmission.

ENGINE COOLING

ENGINE FAN & SHROUD

REMOVAL & INSTALLATION

See Figures 32 through 34.

1. Remove the right side fender trim.
2. Raise the vehicle on the lift to full height.
3. Remove the splash shield.
4. Remove the harness connectors, the ground cable, and the harness clamps. Loosen the A/C condenser fan shroud mounting bolts.
5. Lower the vehicle on the lift.

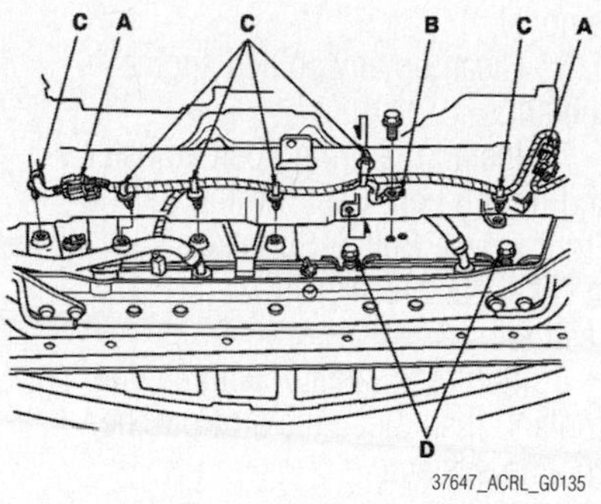
C A C B C A
D
37647_ACRL_G0135

Fig. 32 Remove the harness connectors (A), the ground cable (B), and the harness clamps (C). Loosen the A/C condenser fan shroud mounting bolts (D).

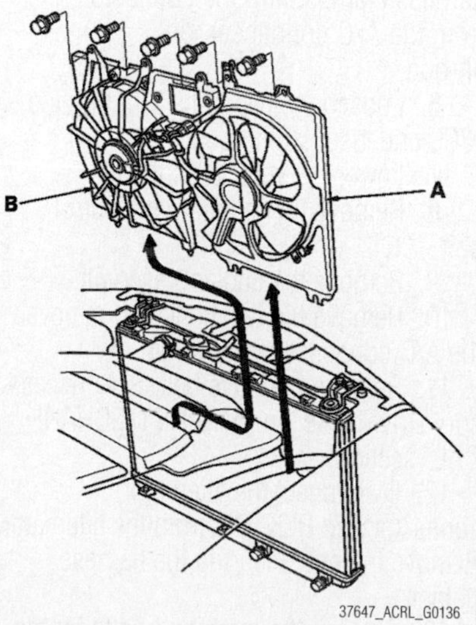
B
A
37647_ACRL_G0136

Fig. 33 Remove the condenser fan shroud assembly (A) and the radiator fan shroud assembly (B) from right side of the engine compartment.

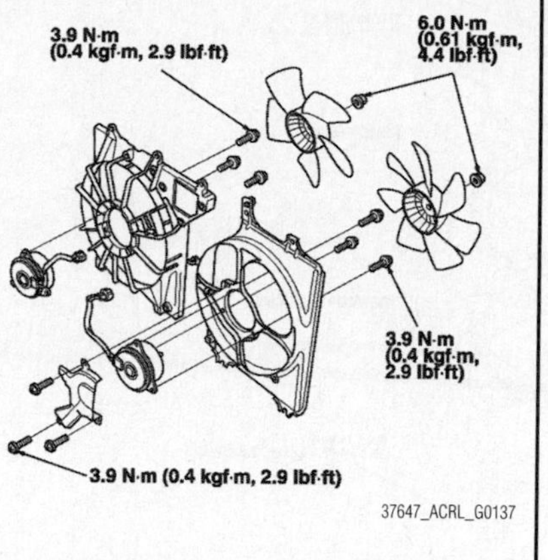
3.9 N·m (0.4 kgf·m, 2.9 lbf·ft)
6.0 N·m (0.61 kgf·m, 4.4 lbf·ft)
3.9 N·m (0.4 kgf·m, 2.9 lbf·ft)
3.9 N·m (0.4 kgf·m, 2.9 lbf·ft)
37647_ACRL_G0137

Fig. 34 Exploded view of the condenser and radiator fan and shroud assemblies

6. Remove the air intake duct cover and the coolant reservoir.
7. Disconnect the radiator fan control (RFC) connectors and the A/C condenser fan motor connector, then remove the harness clamps.
8. Remove the Radiator Fan Control (RFC) unit.
9. Remove the condenser fan shroud assembly and the radiator fan shroud assembly from right side of the engine compartment.
10. Installation is the reverse of the removal procedure.

RADIATOR

REMOVAL & INSTALLATION

See Figure 35.

1. Drain the engine coolant.
2. Remove the right side fender trim.
3. Raise the vehicle on the lift to full height.
4. Remove the splash shield.
5. Disconnect the lower radiator hose.
6. Disconnect the automatic transmission fluid (ATF) cooler hoses, then plug the hose and line.
7. Disconnect the engine coolant temperature (ECT) sensor 2 connector.
8. Lower the vehicle on the lift.
9. Disconnect the upper radiator hose and the coolant reservoir hose.
10. Remove the air intake duct cover, the coolant reservoir, and the radiator upper brackets.
11. Remove the A/C condenser fan shroud assembly and the radiator fan shroud assembly. See "Engine Fan & Shroud" in this section.

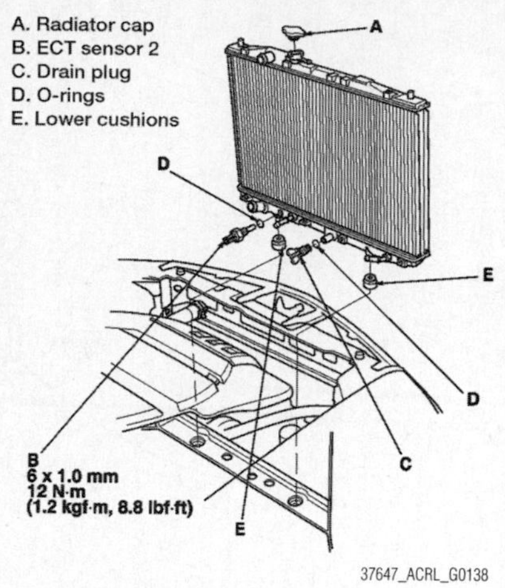
A. Radiator cap
B. ECT sensor 2
C. Drain plug
D. O-rings
E. Lower cushions
A
D
E
D
C
B
6 x 1.0 mm
12 N·m
(1.2 kgf·m, 8.8 lbf·ft)
E
37647_ACRL_G0138

Fig. 35 Removing/installing the radiator

12. Pull up the radiator, then remove the radiator cap (A), the ETC sensor 2 (B), and the drain plug (C).

To install:

13. Installation is the reverse of the removal procedure. Note the following:
    a. Reassemble the radiator with new O-rings (D).
    b. Install the radiator. Make sure the lower cushions (E) are set securely.
    c. Install the radiator fan shroud assembly and the A/C condenser fan shroud assembly. See "Engine Fan & Shroud" in this section.

THERMOSTAT

REMOVAL & INSTALLATION

See Figure 36.

1. Do the battery disconnect procedure. See "Battery System" in "ENGINE ELECTRICAL" section.
2. Drain the engine coolant.
3. Remove the thermostat cover, then remove the thermostat.
4. Install the thermostat with a new rubber seal and with the pin on the top side.
5. Do the battery reconnection procedure. See "Battery System" in "ENGINE ELECTRICAL" section.
6. Refill the radiator with engine coolant, then bleed the air from the cooling system.
7. Clean up any spilled engine coolant.

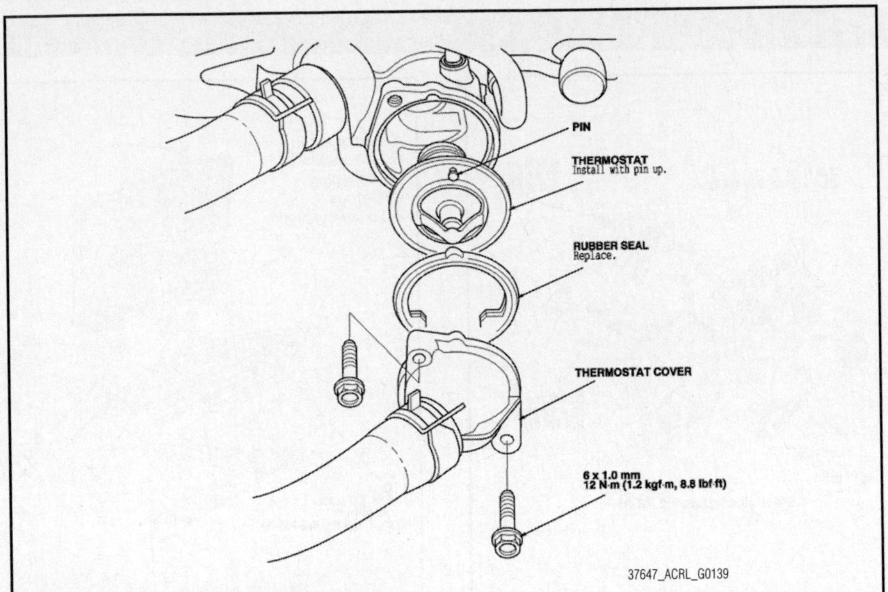
PIN
THERMOSTAT
Install with pin up.
RUBBER SEAL
Replace.
THERMOSTAT COVER
6 x 1.0 mm
12 N·m (1.2 kgf·m, 8.8 lbf·ft)
37647_ACRL_G0139

Fig. 36 Exploded view of the thermostat

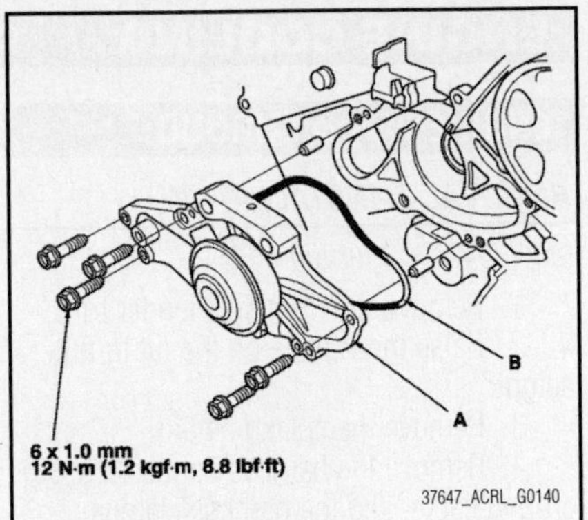
B
A
6 x 1.0 mm
12 N·m (1.2 kgf·m, 8.8 lbf·ft)
37647_ACRL_G0140

Fig. 37 Exploded view of the water pump (A), showing the O-ring (B) and related components

WATER PUMP

REMOVAL & INSTALLATION

See Figure 37.

1. Drain the engine coolant.
2. Remove the timing belt and belt adjuster. See "Timing Belt Front Cover, Belt & Sprockets" in "ENGINE MECHANICAL" section.
3. Remove the water pump by removing the five bolts.

To install:

4. Inspect and clean the O-ring groove and the mating surface of the engine block.
5. Install the water pump with a new O-ring in the reverse order of removal.
6. Clean up any spilled engine coolant.
7. Install the timing belt adjuster and timing belt. See "Timing Belt Front Cover, Belt & Sprockets" in "ENGINE MECHANICAL" section.
8. Refill the radiator with engine coolant, then bleed the air from the cooling system.

ENGINE ELECTRICAL

CHARGING SYSTEM

ALTERNATOR

REMOVAL & INSTALLATION

See Figures 38 and 39.

1. Do the battery terminal disconnect procedure, then wait at least 3 minutes before beginning work. See "Battery System" in "ENGINE ELECTRICAL" section.
2. Remove the upper grille cover.

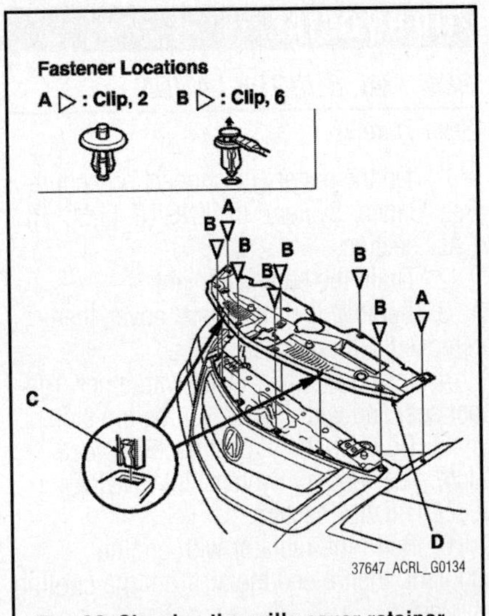
Fastener Locations
A ▷ : Clip, 2 B ▷ : Clip, 6
A
B
C
D
37647_ACRL_G0134

Fig. 38 Showing the grille cover retainer locations

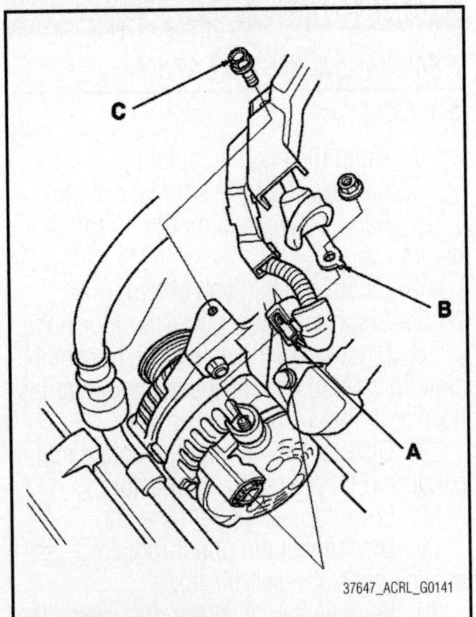
C
B
A
37647_ACRL_G0141

Fig. 39 Disconnect the alternator connector (A) and BLK wire (B) from the alternator. Remove the bolt (C) securing the harness holder.

3. Raise the vehicle on the lift to full height.
4. Remove the splash shield.
5. Remove the ground cable, the harness clamps, and the connector from the A/C condenser fan shroud.
6. Loosen the two bolts securing the A/C condenser fan shroud.
7. Lower the vehicle on the lift.
8. Remove the radiator fan control unit.
9. Remove the coolant reservoir.
10. Remove the two bolts and remove the A/C condenser fan shroud.
11. Remove the drive belt. See "Accessory Drive Belts" in "ENGINE MECHANICAL" section.
12. Disconnect the alternator connector and BLK wire from the alternator. Remove the bolt securing the harness holder.
13. Remove the mounting bolts for the alternator bracket and alternator, then remove the alternator.
14. Installation is the reverse of the removal procedure.

15. Torque the long alternator mounting bolt to 33 ft. lbs. (44 Nm) and the alternator bracket bolt to 16 ft. lbs. (22 Nm).

BATTERY DISCONNECT PROCEDURE

➡Some systems store data in memory that is lost when the battery is disconnected. Do the following steps before disconnecting the battery.

1. Make sure you have the anti-theft code(s) for the audio and/or the navigation system (if equipped).

➡For some models, or if replacing the audio unit, it may be necessary to write down the audio presets (AM and FM), and the XM radio presets (if equipped), because the audio unit does not retain the presets after the battery is disconnected.

2. Make sure the ignition switch is in LOCK (0).
3. Disconnect and isolate the negative cable from the battery.

➡Always disconnect the negative cable from the battery first.

4. Disconnect the positive cable from the battery.

BATTERY RECONNECT PROCEDURE

➡Some systems store data in memory that is lost when the battery is disconnected. Do the following steps to restore the systems back to normal operation.

1. Clean the battery terminals.
2. Test the battery.
3. Reconnect the positive cable to the battery first, then reconnect the negative cable to the battery.

➡Always connect the positive cable to the battery first.

4. Apply multipurpose grease to the terminals to prevent corrosion.
5. Enter the anti-theft code(s) for the audio system and/or the navigation system (if equipped).
6. Enter the audio presets (if applicable), and enter the XM radio presets (if equipped).
7. Set the clock (for vehicles without navigation).
8. Do the steering column position memorization.

ENGINE ELECTRICAL

IGNITION SYSTEM

FIRING ORDER

Firing order for this engine is: 1-4-2-5-3-6

GNITION COIL

REMOVAL & INSTALLATION

Front

1. Remove the engine cover.
2. Remove the upper grille cover.
3. Remove the coolant reservoir from the bracket.
4. Disconnect the ignition coil connectors, then remove the front ignition coils.
5. Install the front ignition coils in the reverse order of removal.

Rear

1. Disconnect the ignition coil connectors, then remove the rear ignition coils.
2. Install the rear ignition coils in the reverse order of removal.

GNITION TIMING

INSPECTION

See Figure 40.

1. Connect the Honda Diagnostic System (HDS), or equivalent scan tool, to the data link connector (DLC).
2. Turn the ignition switch to ON (II).
3. Make sure the scan tool communicates with the vehicle and the powertrain control module (PCM). If it doesn't communicate, troubleshoot the DLC circuit.
4. Check for DTCs. If a DTC is present, diagnose and repair the cause before continuing with this test.
5. Start the engine. Hold the engine speed at 3,000 RPM with no load (in N or P) until the radiator fan comes on, then let it idle.
6. Check the idle speed. See "Idle Speed" in "FUEL SYSTEMS" section.
7. Jump the SCS line with the scan tool.
8. Connect the timing light to the No. 1 ignition coil harness.
9. Aim the timing light toward the pointer on the timing belt cover and the timing mark on the crankshaft pulley. The other pointer is not used.
10. Check the ignition timing under a no load condition (headlights, blower fan, rear window defogger, and air conditioner are turned off).
11. Ignition timing should be 10° ± 2° BTDC (RED mark) at idle in N or P
   a. If the ignition timing differs from the specification, check the cam timing. If the cam timing is OK, update the PCM if it does not have the latest software, or substitute a known-good PCM, then recheck.

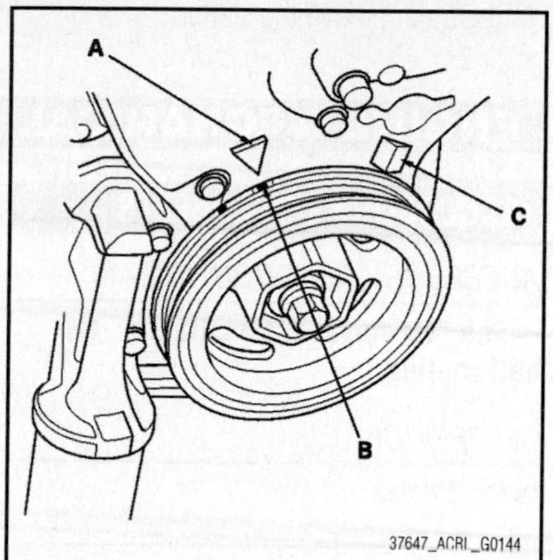
A
C
B
37647_ACRL_G0144

Fig. 40 Aim the timing light toward the pointer (A) on the timing belt cover and the timing mark (B) on the crankshaft pulley. The other pointer (C) is not used.

   b. If the system works properly, and the PCM was substituted, replace the original PCM.
12. Disconnect the HDS and the timing light.

ADJUSTMENT

➡No adjustment is possible.

SPARK PLUGS

REMOVAL & INSTALLATION

➡See "Ignition Coil" in this section.

ENGINE ELECTRICAL

STARTER

REMOVAL & INSTALLATION

See Figure 41.

1. Do the battery terminal disconnect procedure, then wait at least 3 minutes before beginning work. See "Battery System" in "ENGINE ELECTRICAL" section.
2. Remove the engine cover.
3. Remove the vacuum hose, transmission dipstick, and vacuum line mounting bolt (all located above the starter).
4. Remove the harness clamp. Disconnect the starter cable from the B terminal, then disconnect the BLK/WHT wire from the S terminal.
5. Remove the two bolts holding the starter, then remove the starter.

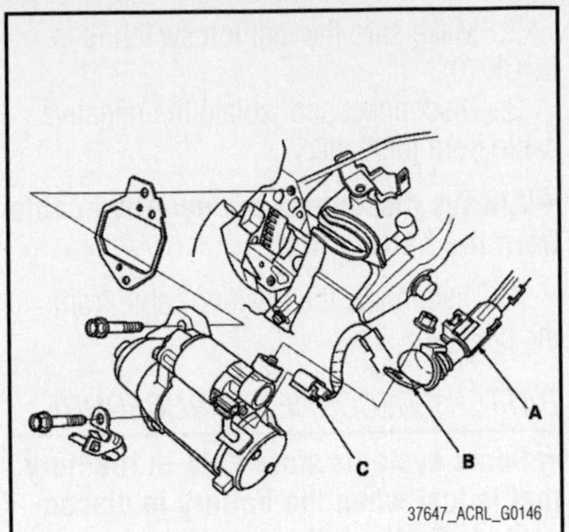
A
B
C
37647_ACRL_G0146

Fig. 41 Remove the harness clamp (A) and disconnect the starter cable (B) from the B terminal, then disconnect the BLK/WHT wire (C) from the S terminal.

To install:

6. Install the starter, using a new gasket, then install the harness clamp and connect the B terminal and BLK/WHT wire. Make sure the crimped side of the B terminal faces away from the starter when you connect it. Tighten the starter mounting bolts to 33 ft. lbs. (44 Nm).
7. Install the vacuum hose, transmission dipstick, and vacuum line mounting bolt.
8. Install the engine cover.
9. Do the battery terminal reconnect procedure. See "Battery System" in "ENGINE ELECTRICAL" section.
10. Start the engine to make sure the starter works properly.

SOLENOID OR RELAY REPLACEMENT

➡Manufacturer does not provide a specific removal and installation procedure for this component.

ENGINE MECHANICAL

ACCESSORY DRIVE BELTS

ACCESSORY BELT ROUTING

➡See "Removal & Installation" for belt routing.

INSPECTION

See Figure 42.

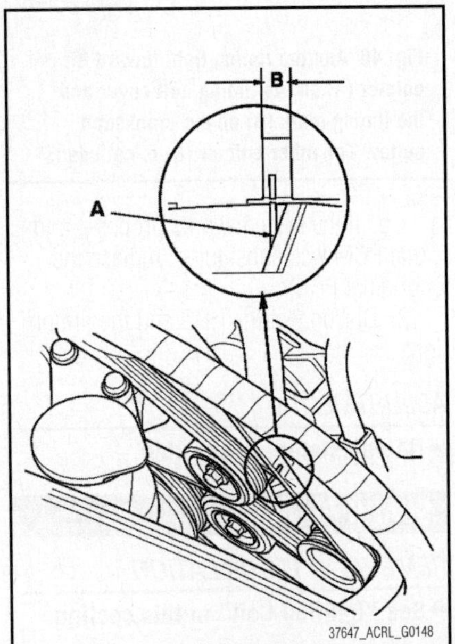
B
A
37647_ACRL_G0148

Fig. 42 Check that the auto-tensioner indicator (A) is within the standard range (B) as shown. If it is out of the standard range, replace the drive belt.

1. Remove the right upper fender trim.
2. Inspect the belt for cracks or damage. If the belt is cracked or damaged, replace it.
3. Check that the auto-tensioner indicator is within the standard range. If it is out of the standard range, replace the drive belt.

ADJUSTMENT

➡Accessory drive belt is automatically adjusted by the tensioner.

REMOVAL & INSTALLATION

See Figure 43.

1. Remove the right upper fender trim.
2. Remove the engine cover.

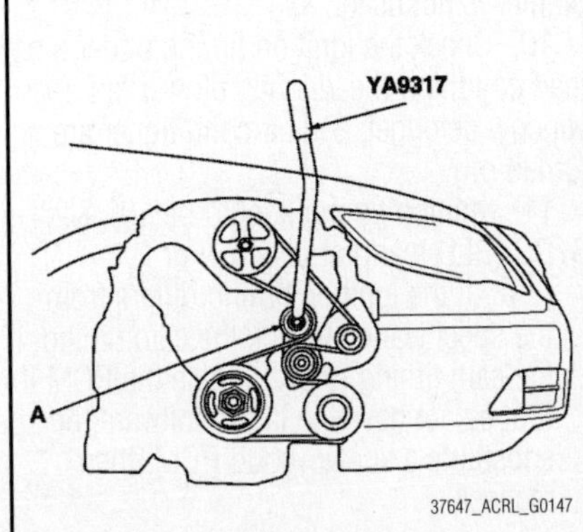
YA9317
A
37647_ACRL_G0147

Fig. 43 Move the auto-tensioner (A) using the belt tension release tool in the direction shown to relieve tension from the drive belt, then remove the drive belt.

3. Move the auto-tensioner (A) using the belt tension release tool in the direction shown to relieve tension from the drive belt, then remove the drive belt.
4. Install the new belt in the reverse order of removal.

CAMSHAFT AND VALVE LIFTERS

INSPECTION

See Figures 44 and 45.

1. Remove the cylinder head. See "Cylinder Head" in this section.
2. Remove the rocker arm assembly. See "Rocker Arms" in this section.
3. On the front camshaft, put the rocker shafts bridge and the rocker shaft holder on

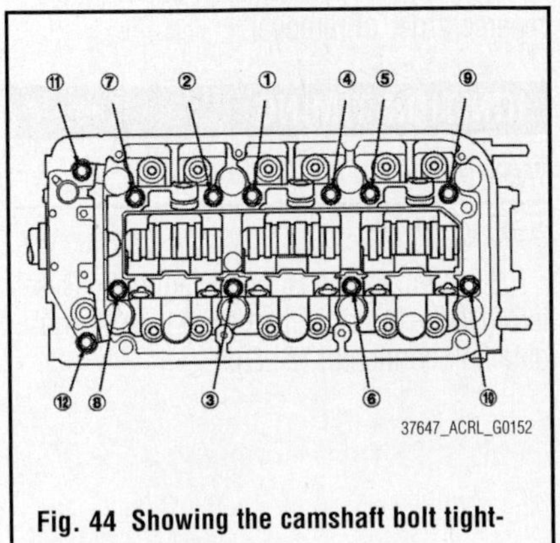
⑪ ⑦ ② ① ④ ⑤ ⑨
⑫ ⑧ ③ ⑥ ⑩
37647_ACRL_G0152

Fig. 44 Showing the camshaft bolt tightening sequence

the front cylinder head, then tighten the bolts to 16 ft. lbs. (22 Nm), in the sequence shown.

4. On the rear camshaft, put the rocker shaft bridge and the rocker shaft holder on the rear cylinder head, then tighten the bolts to 16 ft. lbs. (22 Nm), in the sequence shown.

5. Seat the camshaft by pushing it toward the rear of the cylinder head.

6. Zero a dial indicator against the end of the camshaft. Push the camshaft back and forth and read the end play.

a. If the end play is beyond the service limit, replace the thrust cover and recheck.

b. If it is still beyond the service limit, replace the camshaft.

7. Camshaft End Play:

- Standard (New): 0.002-0.008 in. (0.05-0.20 mm)
- Service Limit: 0.008 in. (0.20 mm)

8. Remove the camshaft thrust cover, then pull out the camshaft.

9. Wipe the camshaft clean, then inspect the lift ramps. Replace the camshaft if any lobes are pitted, scored, or excessively worn.

10. Measure the diameter of each camshaft journal.

11. Zero the gauge to the journal diameter.

12. Clean the camshaft bearing surfaces in the cylinder head. Measure the inside diameter of each camshaft bearing surface, and check for an out-of-round condition.

a. If the camshaft-to-holder clearance is within limits, skip the next step. If the camshaft-to-holder clearance is beyond the service limit and the camshaft has been replaced, replace the cylinder head.

b. If the camshaft-to-holder clearance is beyond the service limit and the camshaft has not been replaced, go to the next step.

13. Camshaft-to-Holder Oil Clearance:

- Standard (New): 0.0020-0.0035 in. (0.050-0.089 mm)

14. Install or connect the following:

- Service Limit: 0.006 in. (0.15 mm)

15. Check total runout with the camshaft supported on V-blocks.

a. If the total runout of the camshaft is within the service limit, replace the cylinder head.

b. If the total runout is beyond the service limit, replace the camshaft and recheck the oil clearance. If the oil clearance is still out of tolerance, replace the cylinder head.

16. Camshaft Total Runout:

- Standard (New): 0.001 in. (0.03 mm) max.

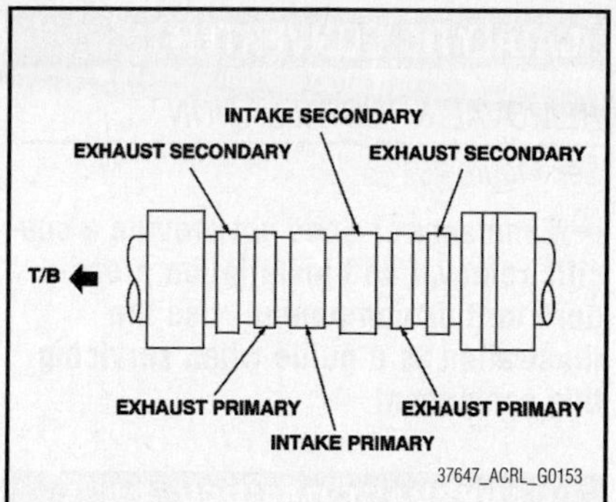
INTAKE SECONDARY
EXHAUST SECONDARY
EXHAUST SECONDARY
T/B
EXHAUST PRIMARY
EXHAUST PRIMARY
INTAKE PRIMARY
37647_ACRL_G0153

Fig. 45 Showing the camshaft cam lobes

- Service Limit: 0.002 in. (0.04 mm)

17. Measure the cam lobe height.

18. Cam Lobe Height Standard (New):

- Primary Intake: 1.3504 in. (34.299 mm)
- Primary Exhaust: 1.4462 in. (36.734 mm)
- Secondary Intake: 1.4146 in. (35.932 mm)
- Secondary Exhaust: 1.4713 in. (37.370 mm)

REMOVAL & INSTALLATION

Front

See Figure 46.

1. Do the battery terminal disconnect procedure, then wait at least 3 minutes before beginning work. See "Battery System" in "ENGINE ELECTRICAL" section.

2. Remove the battery and battery box.

3. Drain the engine cooling system.

4. Disconnect the upper radiator hose.

5. Remove the exhaust gas recirculation valve and the stud bolts.

6. Remove the timing belt. See "Timing Belt & Sprockets" in this section.

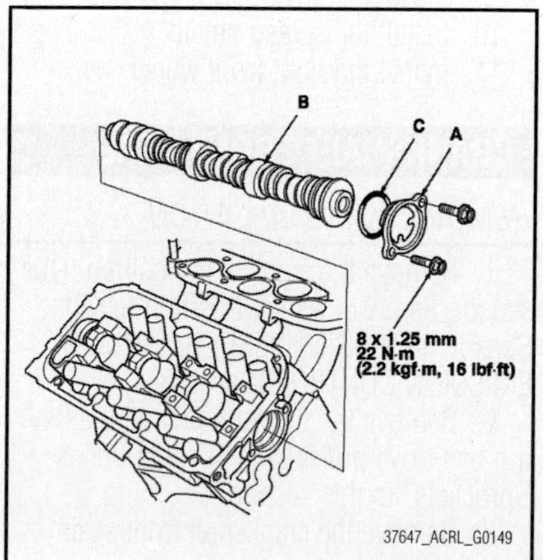
B
C
A
8 x 1.25 mm
22 N·m
(2.2 kgf·m, 16 lbf·ft)
37647_ACRL_G0149

Fig. 46 Remove the thrust cover (A), then remove the front camshaft (B).

7. Remove the rocker arm assembly. See "Rocker Arms" in this section.

8. Remove the front camshaft pulley.

9. Remove the thrust cover, then remove the front camshaft.

To install:

10. Install the front camshaft and thrust cover. Tighten the bolts in the sequence shown. See "Inspection".

➡Always use a new O-ring (C).

11. Apply new engine oil to the journals and the cam lobes.

12. Install the rocker arm assembly.

13. Install the timing belt.

14. Install the stud bolts, then install the EGR valve.

15. Install the upper radiator hose and the lower radiator hose.

16. Adjust the valve clearance.

17. Do the battery terminal reconnect procedure. See "Battery System" in "ENGINE ELECTRICAL" section.

18. Fill the radiator with engine coolant and bleed the air out.

19. Do the crankshaft position (CKP) pattern clear/CKP pattern learn procedure. See "Crankshaft Position (CKP) Sensor" in "ENGINE PERFORMANCE & EMISSION CONTROLS" section.

Rear

See Figures 47 and 48.

1. Do the battery terminal disconnect procedure, then wait at least 3 minutes before beginning work. See "Battery System" in "ENGINE ELECTRICAL" section.

2. Relieve the fuel pressure. See "Relieving Fuel System Pressure" in "FUEL SYSTEMS" section.

3. Remove the quick-connect fitting cover, then disconnect the fuel feed hose.

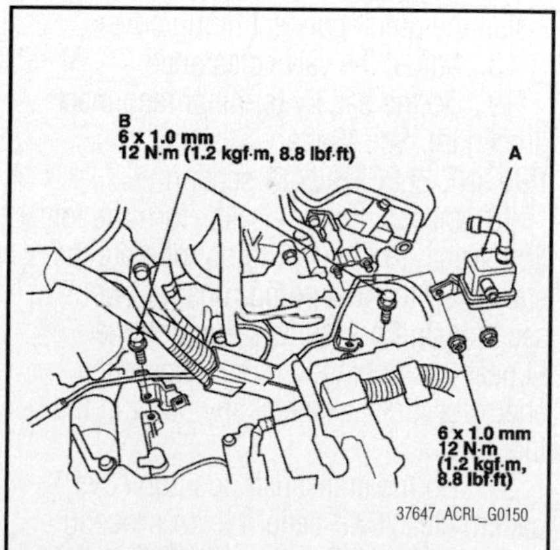
B
6 x 1.0 mm
12 N·m (1.2 kgf·m, 8.8 lbf·ft)
A
6 x 1.0 mm
12 N·m
(1.2 kgf·m,
8.8 lbf·ft)
37647_ACRL_G0150

Fig. 47 Remove the purge joint (A). Remove the two bolts (B) securing the vacuum line.

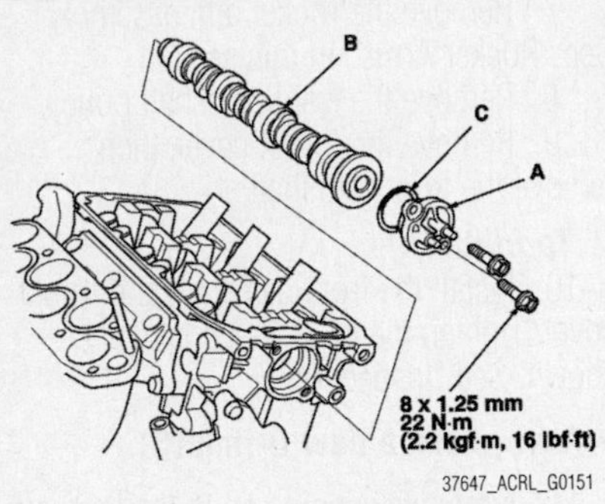
B
C
A
8 x 1.25 mm
22 N·m
(2.2 kgf·m, 16 lbf·ft)
37647_ACRL_G0151

Fig. 48 Remove the thrust cover (A), then remove the rear camshaft (B).

4. Remove the intake manifold. See "Intake Manifold" in this section.
5. Remove the purge joint.
6. Remove the two bolts securing the vacuum line.
7. Remove the timing belt. See "Timing Belt & Sprockets" in this section.
8. Remove the rocker arm assembly. See "Rocker Arms" in this section.
9. Remove the rear camshaft pulley.
10. Remove the thrust cover (A), then remove the rear camshaft (B).

To install:

11. Install the rear camshaft in the reverse order of removal. Tighten the bolts in the sequence shown. See "Inspection".

➡Always use a new O-ring (C).

12. Apply new engine oil to the journals and cam lobes.
13. Install the rocker arm assembly.
14. Install the timing belt. See "Timing Belt & Sprockets" in this section.
15. Install the vacuum line and the purge joint.
16. Install the intake manifold. See "Intake Manifold" in this section.
17. Connect the fuel feed hose, then install the quick-connect fitting cover.
18. Adjust the valve clearance.
19. Do the battery terminal reconnect procedure. See "Battery System" in "ENGINE ELECTRICAL" section.
20. Inspect for fuel leaks. Turn the ignition switch to ON (II) (do not operate the starter) so the fuel pump runs for about 2 seconds and pressurizes the fuel line. Repeat this operation three times, then check for fuel leakage at any point in the fuel line.
21. Do the crankshaft position (CKP) pattern clear/CKP pattern learn procedure. See "Crankshaft Position (CKP) Sensor" in "ENGINE PERFORMANCE & EMISSION CONTROLS" section.

CATALYTIC CONVERTER

REMOVAL & INSTALLATION

See Figure 49.

➡Manufacturer does not provide a specific removal and installation procedure for this component. Use the illustration as a guide when servicing this component.

CRANKSHAFT DAMPER (PULLEY)

REMOVAL & INSTALLATION

1. Raise the vehicle on the lift.
2. Remove the right front wheel.
3. Remove the splash shield.
4. Remove the drive belt.
5. Hold the pulley and crankshaft with a proper holder handle and attachment.
6. Remove the pulley bolt with a heavy duty 19 mm socket and a breaker bar, then remove the crankshaft pulley (damper) and components.

To install:

7. Remove any oil and clean the pulley components, the crankshaft, the bolt, and the washer. Lubricate with new engine oil at all contact surfaces.
8. Install the crankshaft pulley, and tighten the bolt. Do not use an impact wrench.
   a. Hold the pulley and crankshaft with the holder handle and the holder attachment.
   b. Tighten the bolt to 47 ft. lbs. (64 Nm) with a torque wrench and a 19 mm socket.
   c. Mark the bolt head and the crankshaft pulley, then tighten the bolt an additional 60°.
9. Install the drive belt.
10. Install the splash shield.
11. Install the right front wheel.

CRANKSHAFT FRONT SEAL

REMOVAL & INSTALLATION

1. Remove the crankshaft position (CKP) sensor. See "Crankshaft Position (CKP) Sensor" in "ENGINE PERFORMANCE & EMISSION CONTROLS" section.
2. Remove the timing belt, and the timing belt drive pulley. See "Timing Belt & Sprockets" in this section.
3. Remove the crankshaft front seal.

To install:

4. Clean and dry the crankshaft oil seal housing.
5. Apply a light coat of multipurpose grease to the crankshaft and to the lip of the seal.
6. Using a proper seal driver, squarely drive in the crankshaft oil seal until the driver bottoms against the oil pump.
7. When the seal is in place, clean any excess grease off the crankshaft, and check that the oil seal lip is not distorted.
8. Install the timing belt drive pulley, the timing belt, and the CKP sensor.

CYLINDER HEAD

REMOVAL & INSTALLATION

See Figures 50 through 55.

1. Properly relieve the fuel system pressure.
2. Make sure you have the anti-theft code for the radio and the navigation system, then write down the frequencies for the radio's preset buttons.
3. Disconnect the negative cable from the battery first, then disconnect the positive cable. Wait at least 3 minutes before proceeding.
4. Drain the engine cooling system.
5. Remove the accessory drive belt.
6. Remove the power steering pump and power steering hose clamp.
7. Remove the alternator.
8. Remove the timing belt.
9. Remove the intake manifold.
10. Remove the ignition coils.
11. Disconnect the upper and lower radiator hoses from the engine assembly.
12. Remove the quick-connect cover, then disconnect the fuel supply hose.
13. Remove the purge joint.
14. Remove the following engine wiring harnesses:
    - Six injector wiring harnesses
    - Engine coolant temperature (ECT) sensor 1 wiring harness
    - Crankshaft position (CKP) sensor wiring harness
    - Exhaust gas recirculation (EGR) valve wiring harness
    - Rocker arm oil control solenoid wiring harness
    - Rocker arm oil pressure switch wiring harness
    - Oil pressure switch wiring harness
    - Two air fuel ratio (A/F) sensor wiring harnesses
    - Two secondary heated oxygen sensor (secondary HO2S) wiring harnesses
15. Remove the front warm up three way catalytic converter (front WU-TWC) and rear

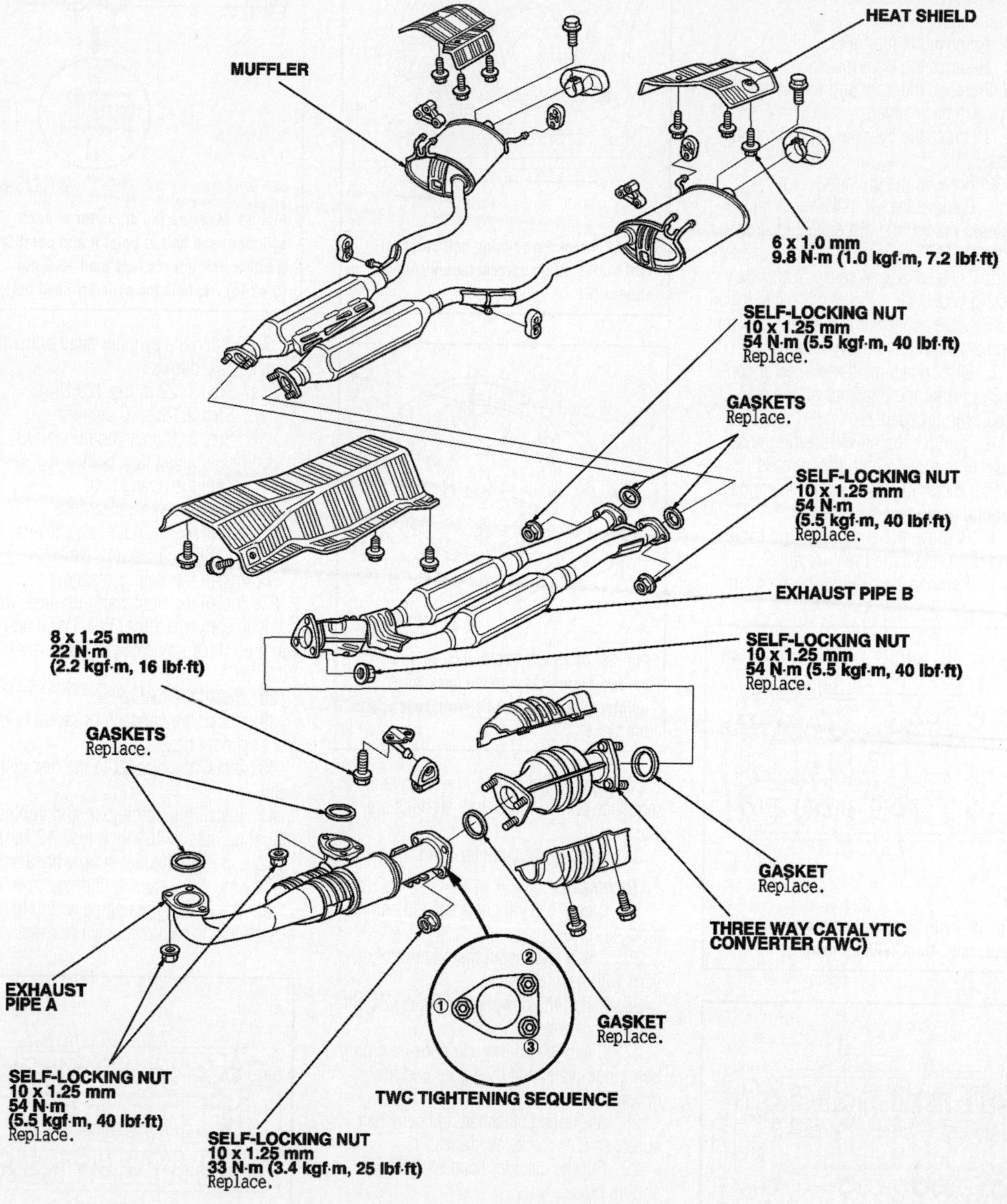
HEAT SHIELD
MUFFLER
6 x 1.0 mm
9.8 N·m (1.0 kgf·m, 7.2 lbf·ft)
SELF-LOCKING NUT
10 x 1.25 mm
54 N·m (5.5 kgf·m, 40 lbf·ft)
Replace.
GASKETS
Replace.
SELF-LOCKING NUT
10 x 1.25 mm
54 N·m
(5.5 kgf·m, 40 lbf·ft)
Replace.
EXHAUST PIPE B
8 x 1.25 mm
22 N·m
(2.2 kgf·m, 16 lbf·ft)
SELF-LOCKING NUT
10 x 1.25 mm
54 N·m (5.5 kgf·m, 40 lbf·ft)
Replace.
GASKETS
Replace.
GASKET
Replace.
THREE WAY CATALYTIC CONVERTER (TWC)
EXHAUST PIPE A
GASKET
Replace.
SELF-LOCKING NUT
10 x 1.25 mm
54 N·m
(5.5 kgf·m, 40 lbf·ft)
Replace.
TWC TIGHTENING SEQUENCE
SELF-LOCKING NUT
10 x 1.25 mm
33 N·m (3.4 kgf·m, 25 lbf·ft)
Replace.
37647_ACRL_G0155

Fig. 49 Exploded view of the exhaust system, including the catalytic converter components

warm up three way catalytic converter (rear WU-TWC).

16. Remove the connector bracket from the front cylinder head.
17. Remove the bracket from the rear cylinder head.
18. Remove the fuel rails.
19. Remove the water passage.
20. Remove the front and rear camshaft pulleys and back covers.
21. Remove the cylinder head cover as follows:
    a. Remove the dipstick.
    b. Remove the two bolts securing the harness holder, and disconnect the front air fuel ratio (A/F) sensor connector, front secondary heated oxygen sensor (secondary HO2S) connector, exhaust gas recirculation (EGR) valve connector and engine coolant temperature (ECT) sensor 1 connector.
    c. Disconnect the three injector connectors from the injectors on the rear side cylinder head.
    d. Remove the power steering hose bracket mounting bolt, the harness holder mounting bolts, and the engine ground cable bolt.
    e. Remove the engine ground cable and disconnect the breather hose.
    f. Remove the cylinder head covers.

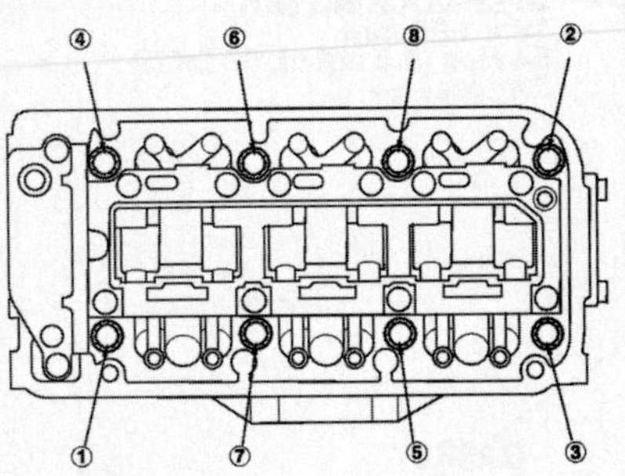
4
6
8
2
1
7
5
3

37647_ACRL_G0156

Fig. 50 Cylinder head bolt loosening sequence—front cylinder head

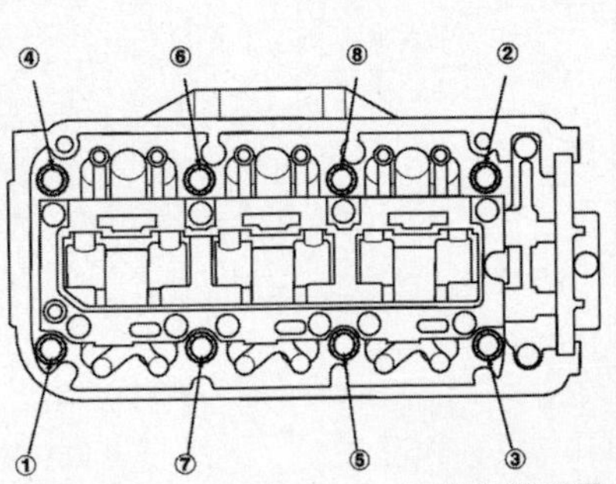
4
6
8
2
1
7
5
3

37647_ACRL_G0157

Fig. 51 Cylinder head bolt loosening sequence—rear cylinder head

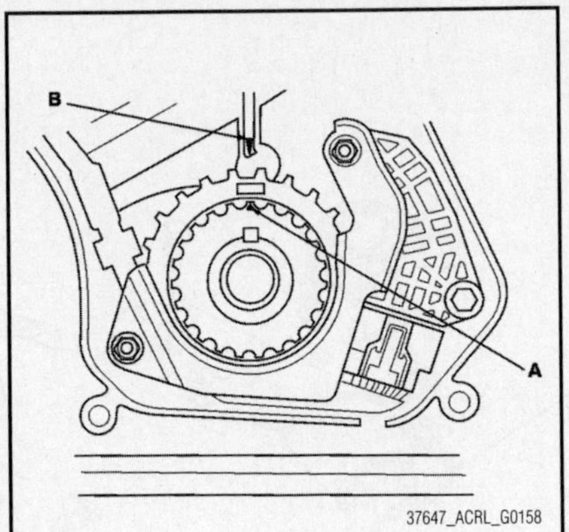
B
A

37647_ACRL_G0158

Fig. 52 Crankshaft timing belt sprocket TDC marks. Align sprocket mark (A) with pointer (B)

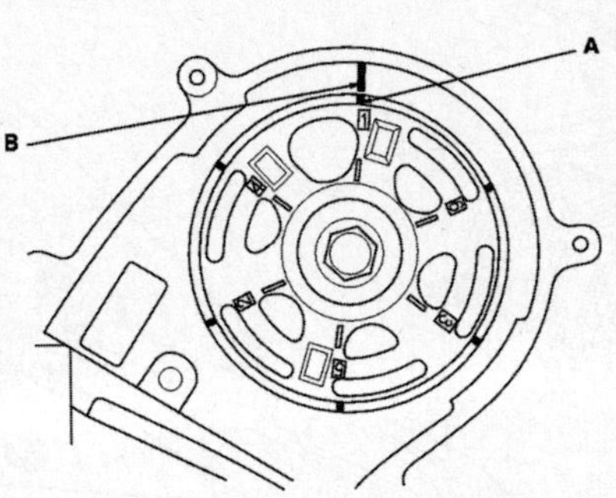
A
B

37647_ACRL_G0159

Fig. 53 Camshaft TDC marks. Align sprocket mark (A) with the back cover pointer (B)—front head shown; rear aligns the same

22. Loosen the cylinder head bolts in sequence and ⅓ turns until all bolts are loose.
23. Remove the cylinder head.

To install:

24. Clean the cylinder head and engine block surface.
25. Clean and install the oil control orifices with new O-rings.
26. Install the dowel pins and new cylinder head gaskets.
27. Clean the timing belt pulleys, timing belt guide plate, and the upper and lower covers.
28. Align the crankshaft and camshaft sprocket TDC marks as shown.
29. Put the cylinder head onto the engine block.
30. Measure the diameter of each cylinder head bolt at point A and point B. If either diameter is less than 10.6 mm (0.42 in), replace the cylinder head bolt.
31. Apply clean engine oil to the cylinder head bolt threads and flanges.

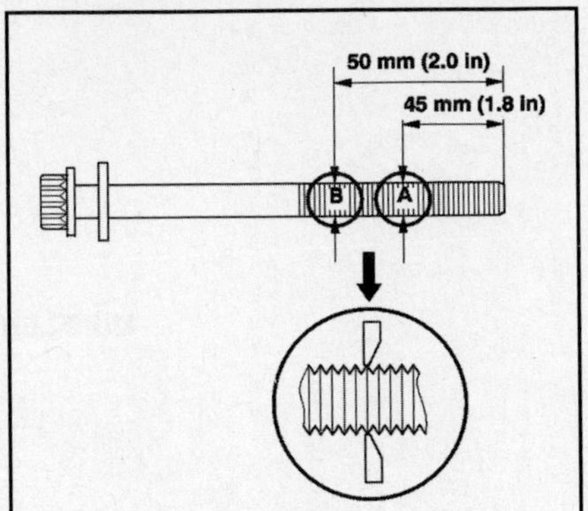
50 mm (2.0 in)
45 mm (1.8 in)
B
A

37647_ACRL_G0160

Fig. 54 Measure the diameter of each cylinder head bolt at point A and point B. If either diameter is less than 10.6 mm (0.42 in), replace the cylinder head bolt.

32. Tighten the cylinder head bolts in sequence as follows:
    a. Step 1: 22 ft. lbs. (29 Nm)
    b. Step 2: Plus 90 degrees
    c. Step 3: Plus an additional 90 degrees, if using new bolts
33. Install the timing belt.
34. Adjust the valve clearance.
35. Install the cylinder head covers and tighten to 108 inch lbs. (12 Nm).
36. Install the water passage.
37. Install the front warm up three way catalytic converter (front WU-TWC) and rear warm up three way catalytic converter (rear WU-TWC).
38. Install the fuel rails.
39. Install the connector bracket to the front cylinder head.
40. Install the bracket to the rear cylinder head.
41. Install the purge joint and tighten the mounting nuts to 109 inch lbs. (12 Nm).
42. Connect the fuel supply hose and replace the quick-connect fitting cover.
43. Reconnect the engine wiring harnesses that were previously removed.

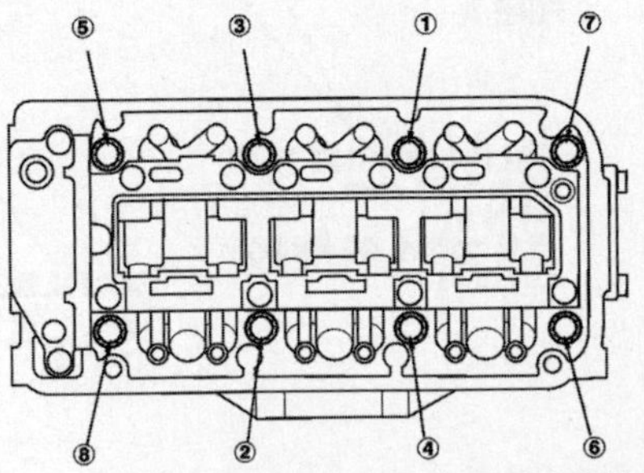
5
3
1
7
8
2
4
6

37647_ACRL_G0161

Fig. 55 Cylinder head bolt tightening sequence—front head shown; rear is the same sequence

44 Connect the upper and lower radiator hoses.

45. Install the intake manifold. See "Intake Manifold" in this section.

46. Install the alternator.

47. Install the power steering pump and hose bracket.

48. Install the accessory drive belt.

49. Reconnect the battery cables.

50. After installation, check that all tubes, hoses and connectors are installed correctly.

51. Refill the engine cooling system to the correct level.

52. Start the engine and check for leaks.

EXHAUST MANIFOLD

REMOVAL & INSTALLATION

See Figure 56.

➡Manufacturer does not provide a specific removal and installation procedure for this component. Use the

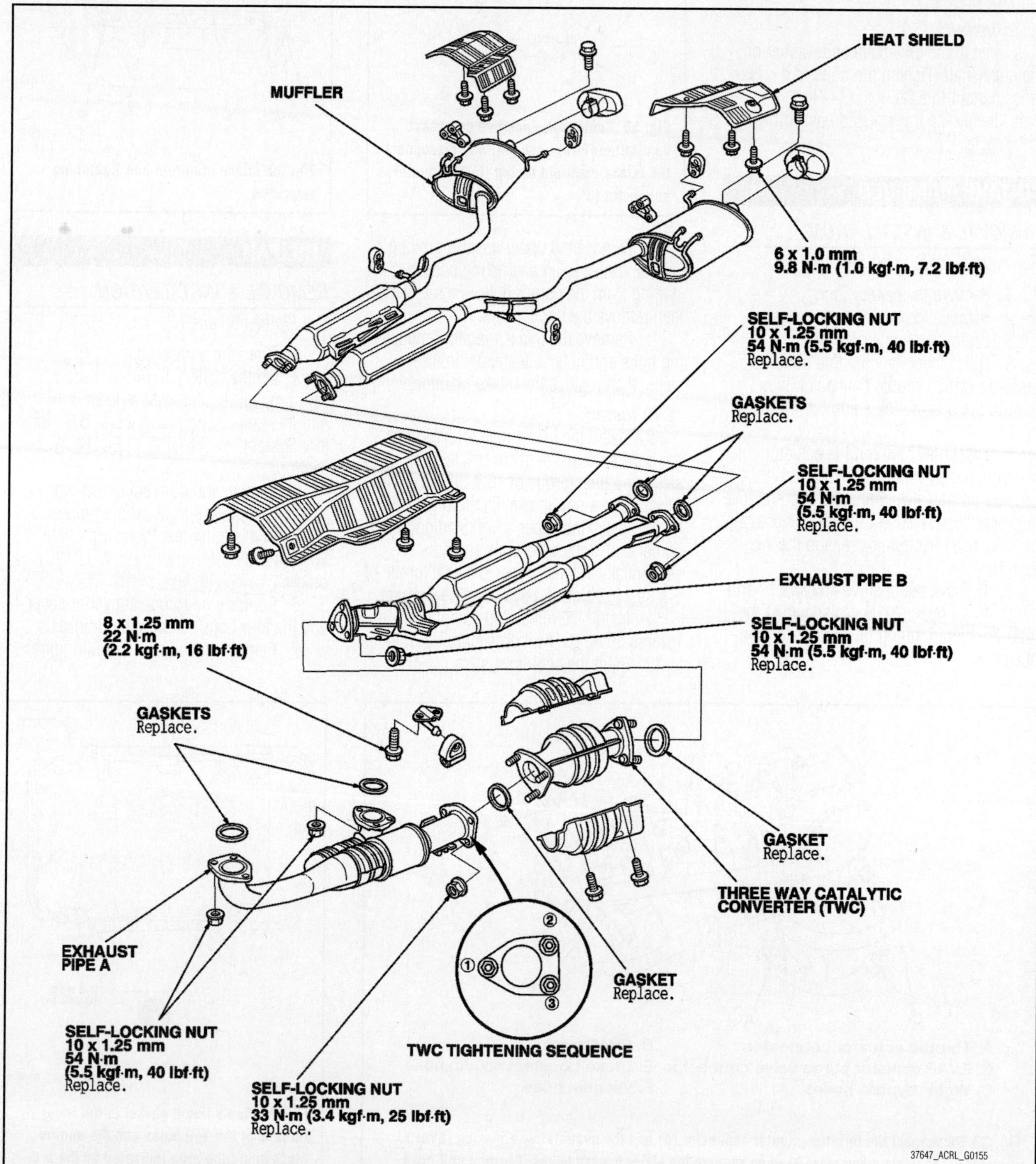

HEAT SHIELD
MUFFLER
6 x 1.0 mm
9.8 N·m (1.0 kgf·m, 7.2 lbf·ft)
SELF-LOCKING NUT
10 x 1.25 mm
54 N·m (5.5 kgf·m, 40 lbf·ft)
Replace.
GASKETS
Replace.
SELF-LOCKING NUT
10 x 1.25 mm
54 N·m
(5.5 kgf·m, 40 lbf·ft)
Replace.
EXHAUST PIPE B
SELF-LOCKING NUT
10 x 1.25 mm
54 N·m (5.5 kgf·m, 40 lbf·ft)
Replace.
8 x 1.25 mm
22 N·m
(2.2 kgf·m, 16 lbf·ft)
GASKETS
Replace.
GASKET
Replace.
THREE WAY CATALYTIC CONVERTER (TWC)
EXHAUST PIPE A
GASKET
Replace.
SELF-LOCKING NUT
10 x 1.25 mm
54 N·m
(5.5 kgf·m, 40 lbf·ft)
Replace.
TWC TIGHTENING SEQUENCE
SELF-LOCKING NUT
10 x 1.25 mm
33 N·m (3.4 kgf·m, 25 lbf·ft)
Replace.
37647_ACRL_G0155

Fig. 56 Exploded view of the exhaust system components

illustration as a guide when servicing this component.

FLYWHEEL (DRIVE PLATE)

REMOVAL & INSTALLATION

1. Remove the transmission assembly.
2. Remove the driveplate and washer from the engine.

To install:

3. Install the driveplate and washer on the crankshaft. Tighten the bolts in a criss-cross pattern to 54 ft. lbs. (74 Nm).
4. Install the transmission assembly into the vehicle.

INTAKE MANIFOLD

REMOVAL & INSTALLATION

See Figures 57 through 59.

1. Remove the battery trim.
2. Remove the engine cover.
3. Drain the engine coolant.
4. Disconnect the manifold absolute pressure (MAP) sensor connector, then remove the breather pipe and the intake air duct.
5. Disconnect the throttle actuator connector and the evaporative emission (EVAP) canister purge valve connector, then remove the water bypass hoses, the EVAP hose, the brake booster vacuum hose, and the vacuum hose.
6. Remove the positive crankcase ventilation (PCV) hose (A), and disconnect the intake manifold tuning (IMT) actuator connector (B).

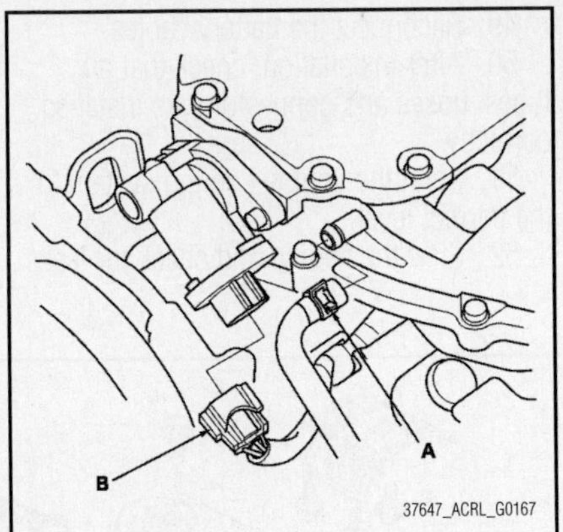
A
B
37647_ACRL_G0167

Fig. 58 Remove the positive crankcase ventilation (PCV) hose (A), and disconnect the intake manifold tuning (IMT) actuator connector (B).

7. Remove the upper cover mounting bolts and nuts in an alternating pattern, starting from the corner bolt, in three steps, then remove the upper cover.
8. Remove the intake manifold mounting bolts and nuts sequentially in three steps, then remove the intake manifold.

To install:

9. Install the intake manifold. Tighten the bolts and nuts sequentially in three steps to a final torque of 16 ft. lbs. (22 Nm). Always use a new intake manifold gasket.
10. Install the upper cover. Starting from a corner bolt, tighten the bolts and nuts sequentially in three steps, to a final torque of 9 ft. lbs. (12 Nm). Always use a new gasket.
11. Install the remaining components in reverse of the removal procedure.
12. Refill the cooling system.

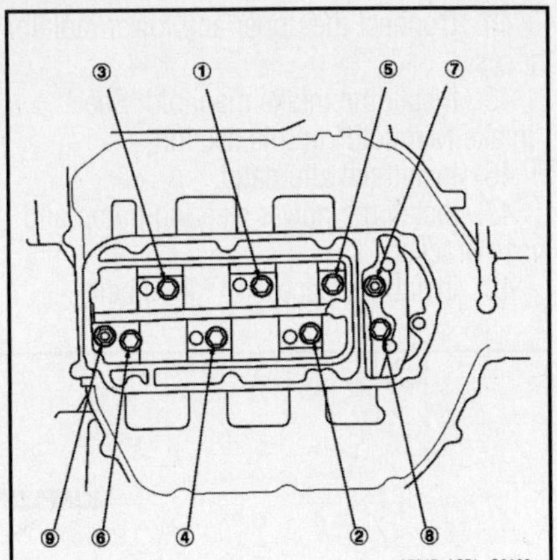
3
1
5
7
9
6
4
2
8
37647_ACRL_G0168

Fig. 59 Intake manifold bolt tightening sequence

OIL PAN

REMOVAL & INSTALLATION

See Figures 60 and 61.

1. Drain the engine oil.
2. Do the battery terminal disconnect procedure, then wait at least 3 minutes before beginning work. See "Battery System" in "ENGINE ELECTRICAL" section.
3. Remove the front splash shield.
4. Remove the front exhaust pipe.
5. Remove the rear Warm Up Three Way Catalytic Converter (rear WU-TWC) bracket.
6. Remove the torque converter cover and the four bolts securing the transaxle.
7. Remove the oil pan mounting bolts.

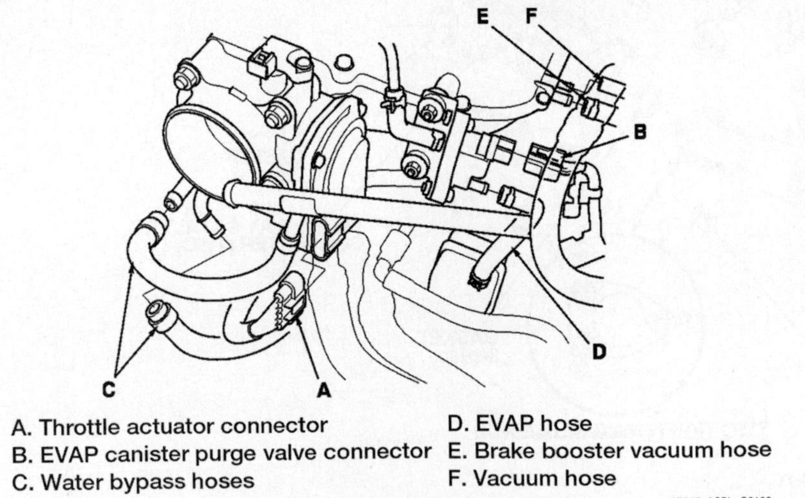
E
F
B
D
C
A
A. Throttle actuator connector
B. EVAP canister purge valve connector
C. Water bypass hoses
D. EVAP hose
E. Brake booster vacuum hose
F. Vacuum hose
37647_ACRL_G0166

Fig. 57 Disconnect the throttle actuator connector (A) and the evaporative emission (EVAP) canister purge valve connector (B), then remove the water bypass hoses (C), the EVAP hose (D), the brake booster vacuum hose (E), and the vacuum hose (F).

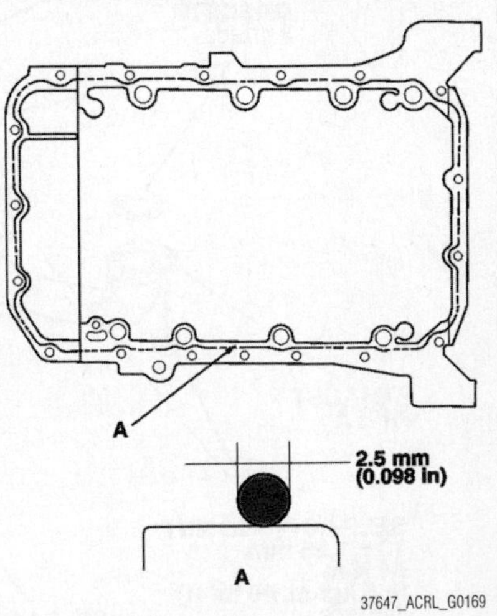
A
2.5 mm
(0.098 in)
A
37647_ACRL_G0169

Fig. 60 Apply liquid gasket to the inner threads of the bolt holes and the engine block along the area indicated by the broken line

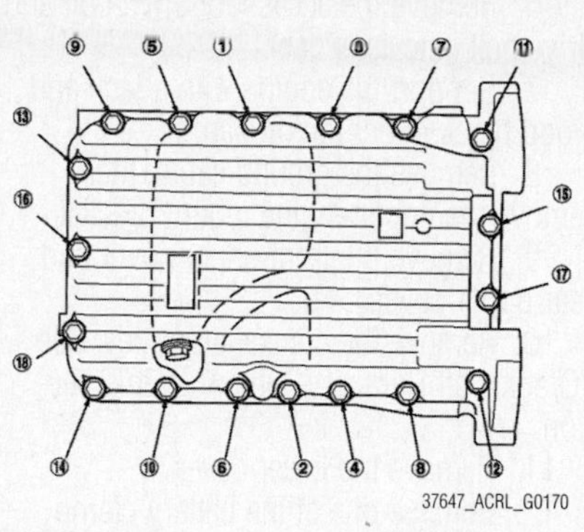
37647_ACRL_G0170
Fig. 61 Oil pan tightening sequence

8. Use a flat bladed screwdriver to separate the oil pan from the block.

To install:

9. Remove all of the old liquid gasket material.
10. Apply liquid gasket to the mating surfaces of the oil pan as shown.
11. Install the oil pan on the engine block. Tighten the mounting bolts in sequence to 9 ft. lbs. (12 Nm).
12. Tighten the four bolts securing the transmission. Torque the bolts to 54 ft. lbs. (74 N m).
13. Install the torque converter cover.
14. Install the rear Warm Up Three Way Catalytic Converter (rear WU-TWC) bracket.
15. Install the front exhaust pipe using new gaskets and new self-locking nuts.
16. Install the splash shield.
17. Do the battery terminal reconnect procedure. See "Battery System" in "ENGINE ELECTRICAL" section.

➡Wait at least 30 minutes before adding oil to the engine.

18. Refill the engine with oil to the correct level.

OIL PUMP

REMOVAL & INSTALLATION

See Figure 62.

➡Manufacturer does not provide a specific removal and installation procedure for this component. Use the illustration as a guide when servicing this component.

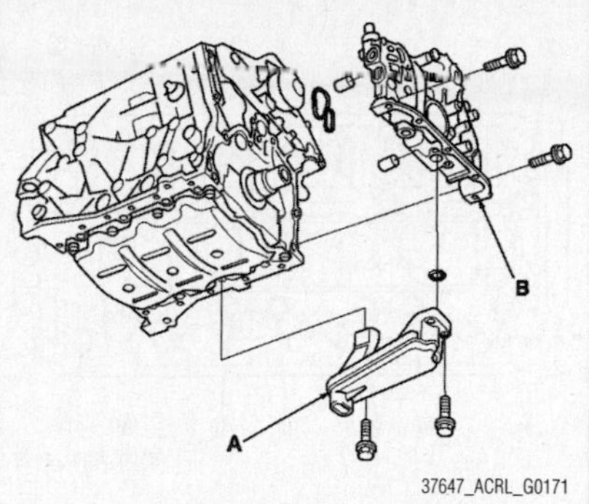
A
B
37647_ACRL_G0171
Fig. 62 Showing the location of the oil pump screen (A) and oil pump (B)

INSPECTION

See Figures 63 through 65.

1. Remove the oil pump assembly from the engine.
2. Remove the screws from the pump housing, then separate the housing and cover.
3. Check the inner-to-outer rotor radial clearance between the inner rotor and outer rotor. If the inner-to-outer rotor clearance exceeds 0.008 inches (0.20 mm), replace the oil pump assembly.
4. Check the housing-to-rotor axial clearance between the rotors and pump housing. If the housing-to-rotor axial clearance exceeds 0.005 inches (0.12 mm), replace the oil pump assembly.
5. Check the housing-to-outer rotor radial clearance between the outer rotor and pump housing. If the housing-to-outer rotor radial clearance exceeds 0.008 inches (0.20 mm), replace the oil pump assembly.
6. Inspect both rotors and pump housing for scoring or other damage. Replace the parts, if necessary.
7. Apply liquid thread lock to the pump housing screws, then install the oil pump cover.
8. Check that the oil pump turns freely.

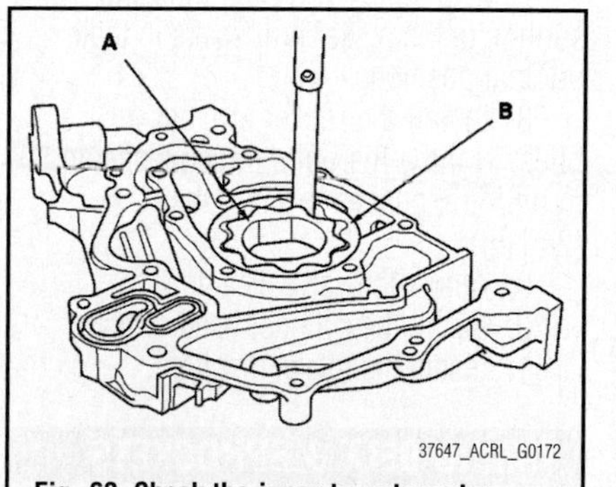
A
B
37647_ACRL_G0172
Fig. 63 Check the inner-to-outer rotor radial clearance between the inner rotor (A) and outer rotor (B)

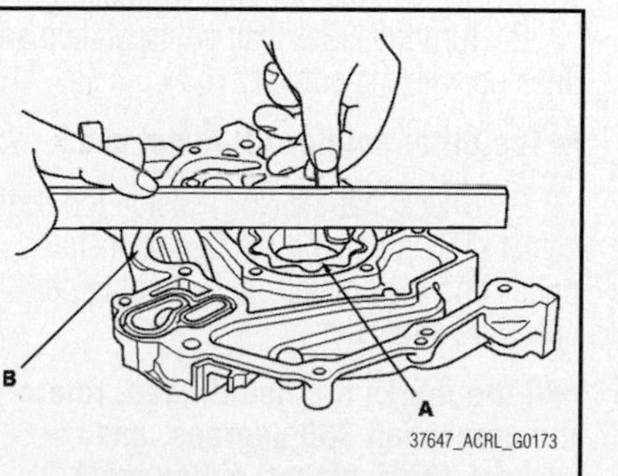
B
A
37647_ACRL_G0173
Fig. 64 Check the housing-to-rotor axial clearance between the rotors (A) and pump housing (B)

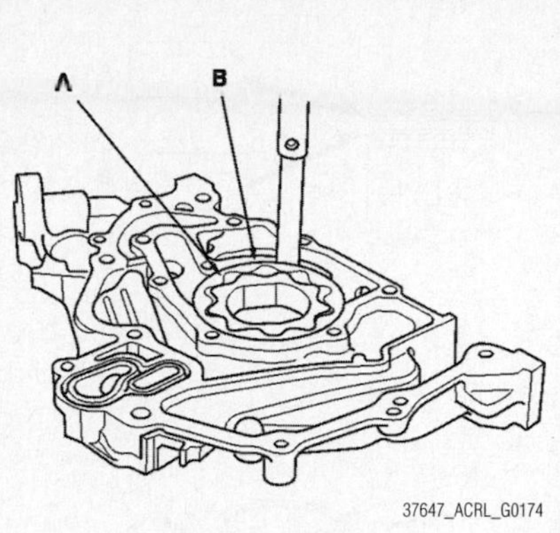
A
B
37647_ACRL_G0174
Fig. 65 Check the housing-to-outer rotor radial clearance between the outer rotor (A) and pump housing (B)

PISTON AND RING

POSITIONING

See Figures 66 and 67.

REAR MAIN SEAL

REMOVAL & INSTALLATION

1. Remove the transmission assembly.
2. Remove the driveplate.
3. Remove the rear main seal.

To install:

4. Clean and dry the crankshaft oil seal housing.
5. Apply a light coat of multipurpose grease to the crankshaft and to the lip of the seal.
6. Using a suitable seal driver tool, drive in the rear main seal until the driver attachment bottoms against the engine block end cover. Align the hole in the driver attachment with the pin on the crankshaft.
7. Clean any excess grease off the crankshaft, and check that the oil seal lip is not distorted.
8. Install the driveplate and transmission assembly.

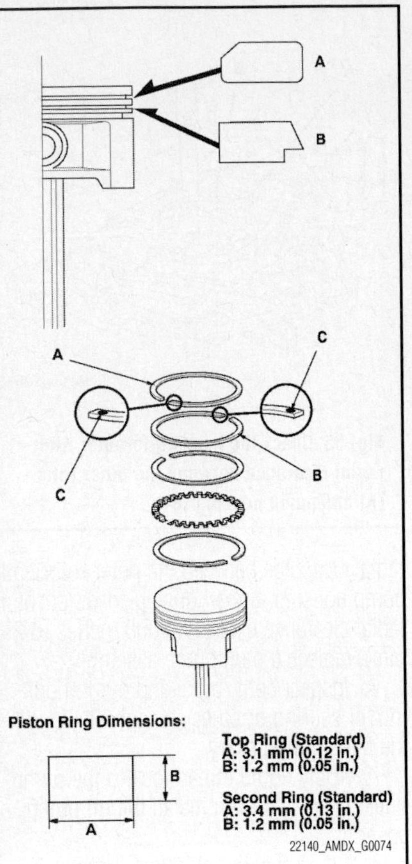
A
B
A
C
B
C
Piston Ring Dimensions:
Top Ring (Standard)
A: 3.1 mm (0.12 in.)
B: 1.2 mm (0.05 in.)
Second Ring (Standard)
A: 3.4 mm (0.13 in.)
B: 1.2 mm (0.05 in.)
22140_AMDX_G0074

Fig. 66 Piston ring installation—3.7L Engine

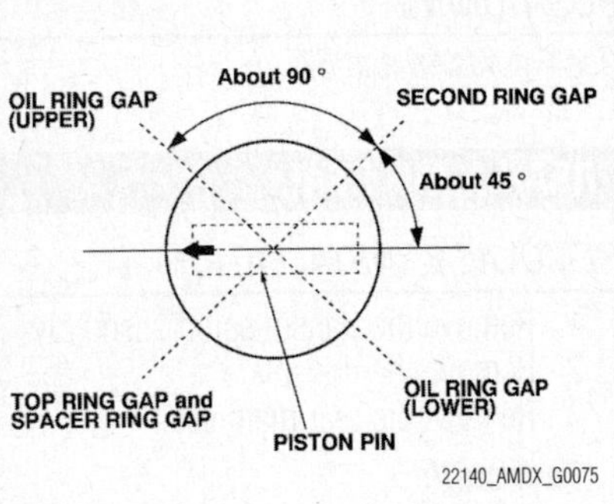
About 90 °
OIL RING GAP (UPPER)
SECOND RING GAP
About 45 °
TOP RING GAP and SPACER RING GAP
OIL RING GAP (LOWER)
PISTON PIN
22140_AMDX_G0075

Fig. 67 Ring end gap positioning—3.7 Engine

ROCKER ARMS

REMOVAL & INSTALLATION

See Figure 68.

1. Remove the intake manifold. See "Intake Manifold" in this section.
2. Remove the cylinder head cover.
3. Loosen the rocker arm locknuts and adjusting screws.
4. Remove the rocker arm assembly as follows:

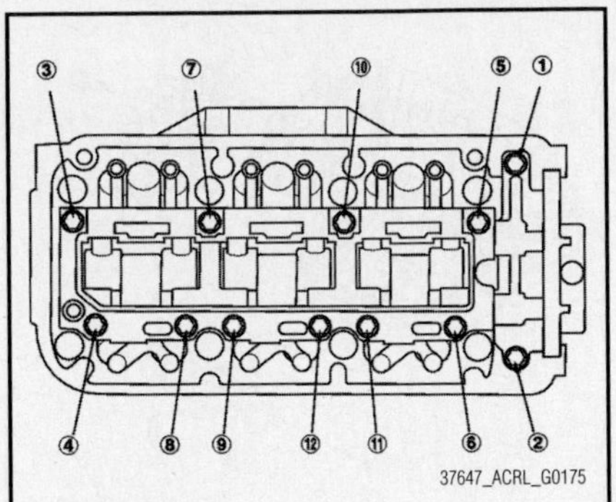
37647_ACRL_G0175

Fig. 68 Rocker arm shaft tightening sequence

a. Unscrew the rocker shaft bolts 2 turns at a time in a criss-cross pattern to avoid damaging the vales or rocker assembly.
b. Do not remove the rocker shaft bolts. These bolts keep the springs and rocker arms on the shafts.

5. Remove the rocker arms and shafts from the vehicle as an assembly.

➡Keep all valve train components in order for assembly.

6. Remove the rocker arms and springs from the rocker arm shafts.

To install:

7. Assemble the rocker arms and springs to the rocker arm shafts in their original positions.
8. Install the rocker arm assemblies. Tighten the bolts in sequence and in multiple passes to 17 ft. lbs. (24 Nm).
9. Adjust the valve clearance.
10. Install the cylinder head cover.
11. Install the intake manifold.

TIMING BELT & SPROCKETS

REMOVAL & INSTALLATION

See Figures 69 through 75.

1. Remove the right upper fender trim.
2. Turn the crankshaft so its white mark lines up with the pointer.

➡The other pointer (C) is not used.

3. Check that the No. 1 piston top dead center (TDC) mark on the front camshaft pulley and the pointer on the front upper cover are aligned.

➡If the marks are not aligned, rotate the crankshaft 360 degrees, and recheck the camshaft pulley mark.

4. Remove the right front wheel.
5. Remove the splash shield.
6. Remove the accessory drive belt and drive belt auto-tensioner.
7. Support the engine with a jack and wood block under the oil pan.
8. Remove the ground cable, then remove the upper engine mount bracket
9. Remove the front upper cover and rear upper cover.
10. Remove the crankshaft pulley. See "Crankshaft Damper (Pulley)" in this section.
11. Remove the lower cover.
12. Remove one of the battery clamp bolts from the battery tray, and grind the end to a slight bevel.
13. Thread in the battery clamp bolt as shown to hold the timing belt adjuster in its current position. Tighten it by hand; do not use a wrench.
14. Remove the side engine mount bracket.
15. Remove the idler pulley bolt and the idler pulley, then remove the timing belt. Discard the idler pulley bolt.

To install:

16. Clean the timing belt pulleys, the timing belt guide plate, and the upper and lower covers.
17. Set the timing belt drive pulley to top dead center (TDC) by aligning the TDC mark on the tooth of the timing belt drive pulley with the pointer on the oil pump.
18. Set the camshaft pulleys to TDC by aligning the TDC marks on the camshaft pulleys with the pointers on the back covers.
19. Remove the battery clamp bolt from the back cover.
20. Remove the auto-tensioner.
21. Align the holes on the rod and housing of the auto-tensioner.
22. Use a hydraulic press to slowly compress the auto-tensioner. Insert a 0.08 in. (2.0 mm) pin through the housing and the rod.

CAUTION

The compression pressure should not exceed 2200 lbs (9800 N).

23. Install the auto-tensioner. Make sure the pin stays in place.
24. With the battery clamp bolt in place to hold the timing belt adjuster, loosely install the idler pulley with a new idler pulley bolt so the pulley can move but does not come off.
25. Install the timing belt in a counterclockwise sequence starting with the drive pulley.
26. Tighten the idler pulley bolt to 33 ft. lbs. (44 Nm).

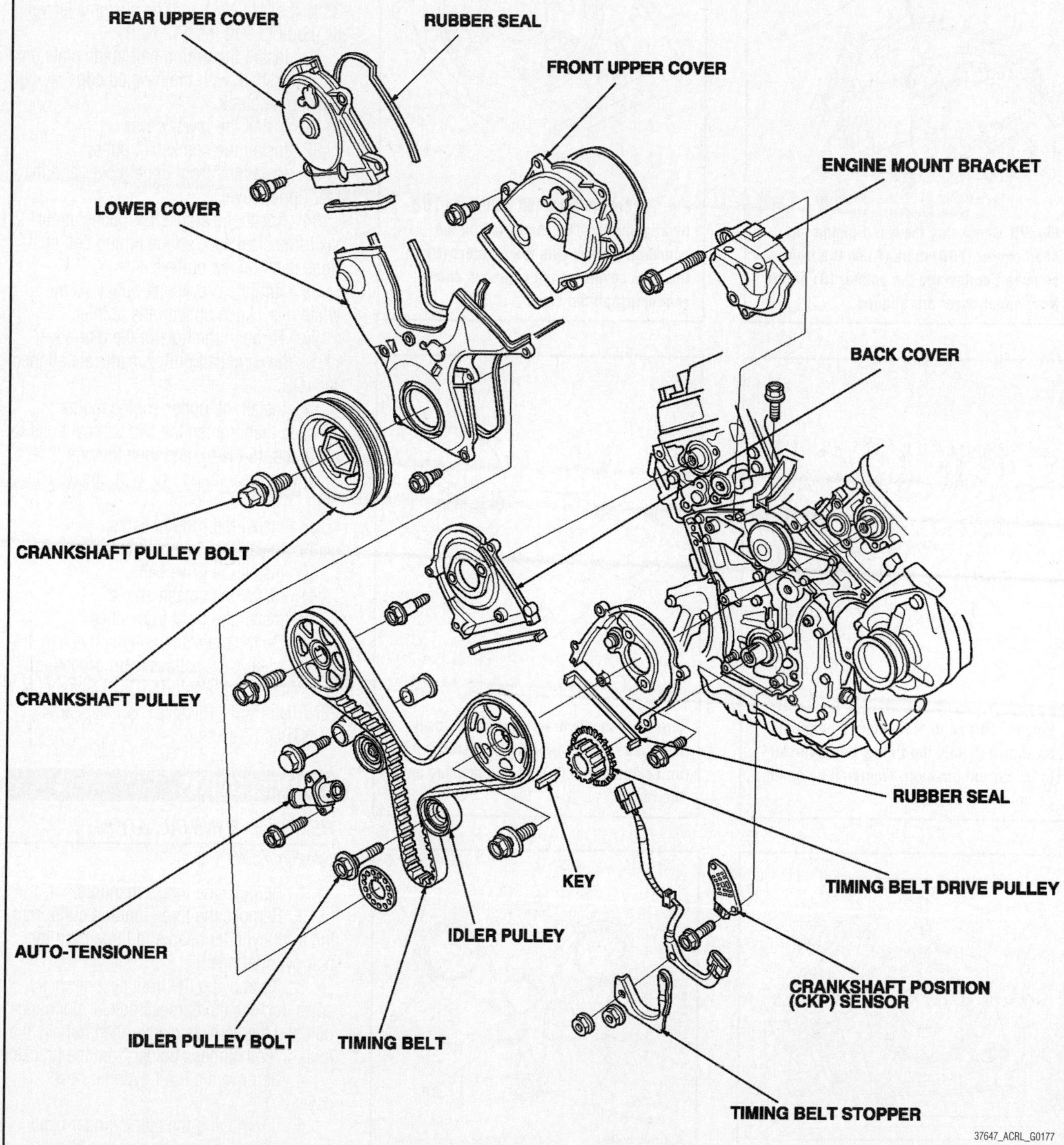
REAR UPPER COVER
RUBBER SEAL
FRONT UPPER COVER
ENGINE MOUNT BRACKET
LOWER COVER
BACK COVER
CRANKSHAFT PULLEY BOLT
CRANKSHAFT PULLEY
RUBBER SEAL
KEY
TIMING BELT DRIVE PULLEY
AUTO-TENSIONER
IDLER PULLEY
CRANKSHAFT POSITION (CKP) SENSOR
IDLER PULLEY BOLT
TIMING BELT
TIMING BELT STOPPER
37647_ACRL_G0177

Fig. 69 Exploded view of the front covers and timing belt components

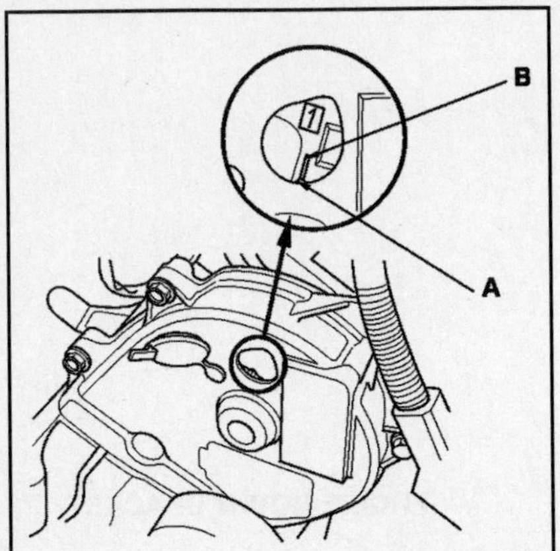
B
A
37647_ACRL_G0176

Fig. 70 Check that the No. 1 piston top dead center (TDC) mark (A) on the front camshaft pulley and the pointer (B) on the front upper cover are aligned.

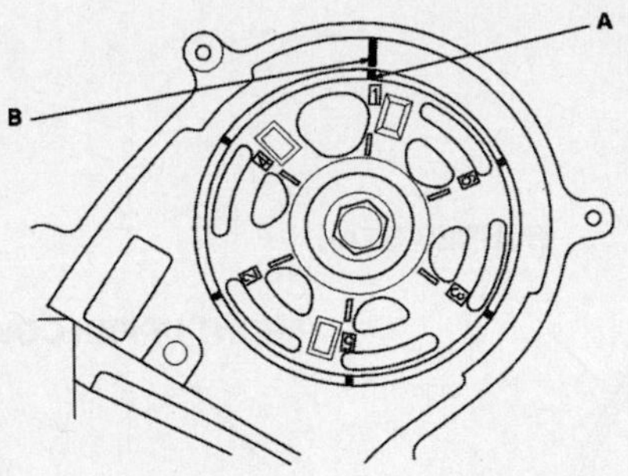
A
B
37647_ACRL_G0180

Fig. 73 Set the camshaft pulleys to TDC by aligning the TDC marks (A) on the camshaft pulleys with the pointers (B) on the back covers—front camshaft shown; rear camshaft the same

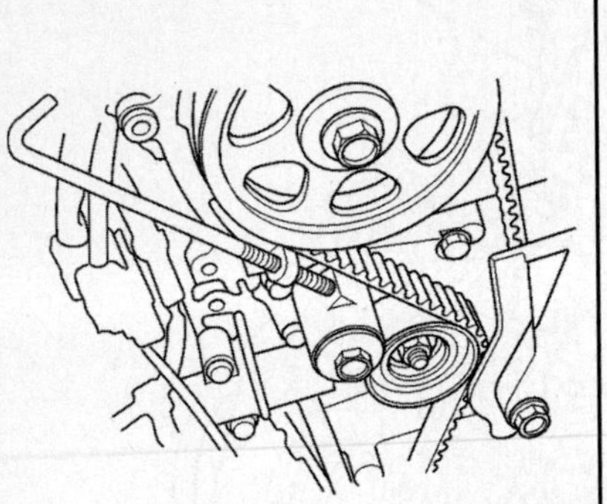
37647_ACRL_G0178

Fig. 71 Thread in the battery clamp bolt as shown to hold the timing belt adjuster in its current position. Tighten it by hand; do not use a wrench.

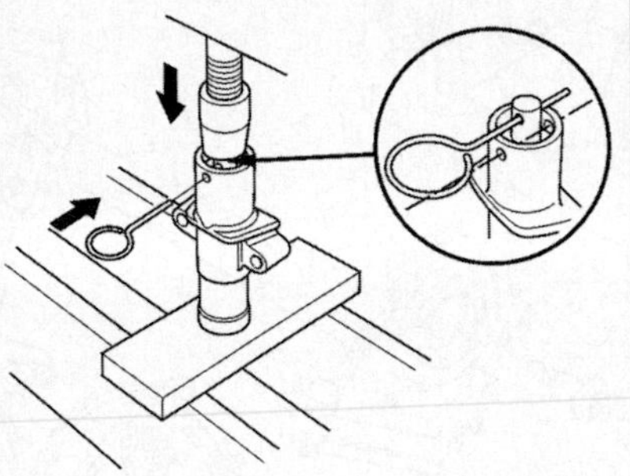
37647_ACRL_G0181

Fig. 74 Use a hydraulic press to slowly compress the auto-tensioner. Insert a 0.08 in. (2.0 mm) pin through the housing and the rod.

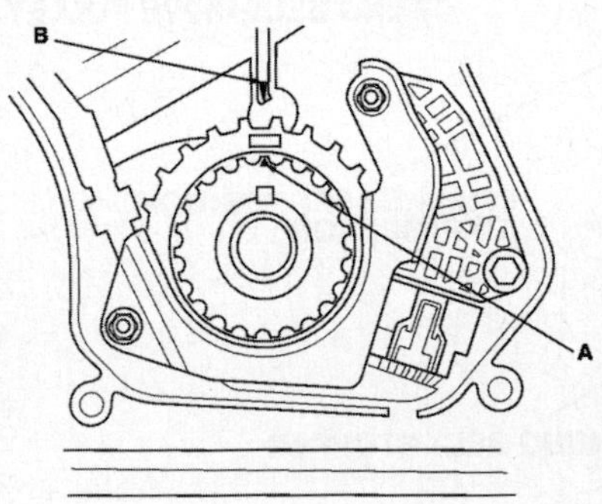
B
A
37647_ACRL_G0179

Fig. 72 Set the timing belt drive pulley to top dead center (TDC) by aligning the TDC mark (A) on the tooth of the timing belt drive pulley with the pointer (B) on the oil pump.

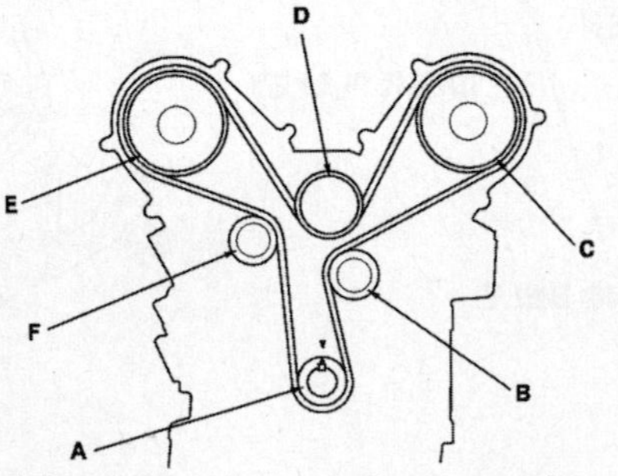
D
E
C
F
B
A
A. Drive pulley
B. Idler pulley
C. Front camshaft pulley
D. Water pump pulley
E. Rear camshaft pulley
F. Adjusting pulley
37647_ACRL_G0182

Fig. 75 Install the timing belt in a counterclockwise sequence starting with the drive pulley.

27. Remove the pin from the auto-tensioner.
28. Remove the battery clamp bolt from the back cover.
29. Install the side engine mount bracket. Tighten the three horizontal bolts to 33 ft. lbs. (44 Nm) and the one smaller vertical bolt to 9 ft. lbs. (12 Nm).
30. Install the timing belt guide plate on the crankshaft, with the beveled edge facing the engine block.
31. Install the lower cover.
32. Install the crankshaft pulley.
33. Install the front upper cover and the rear upper cover.
34. Rotate the crankshaft pulley about six turns clockwise so the timing belt positions itself on the pulleys.
35. Turn the crankshaft pulley so the white mark lines up with the pointer.
36. Through the hole in the rear cover, check the camshaft pulley marks are aligned for TDC.
37. Install the upper engine mount bracket, then tighten the two vertical bolts to 40 ft. lbs. (54 Nm), and then the one smaller horizontal bolt to 47 ft. lbs. (64 Nm).
38. Install the ground cable.
39. Install the drive belt auto-tensioner.
40. Install the drive belt.
41. Install the splash shield.
42. Install the right front wheel.
43. Do the crankshaft position (CKP) pattern clear/CKP pattern learn procedure. See "Crankshaft Position (CKP) Sensor" in "ENGINE PERFORMANCE & EMISSION CONTROLS" section.

VALVE COVERS

REMOVAL & INSTALLATION

See Figure 76.

1. Remove the intake manifold.
2. Remove the three ignition coils from the front cylinder head and three from the rear cylinder head.
3. If removing the front cylinder head cover, remove the harness holder mounting bolt and the harness clamp, then remove the ignition coil harness holder from the bracket.
4. Remove the front cylinder head cover.
5. If removing the rear cylinder head cover, remove the power steering hose clamp mounting bolt, the harness holder mounting bolts, the harness clamp, and the breather hose.
6. Remove the wire harness from the rear upper cover.
7. Remove the rear cylinder head cover.

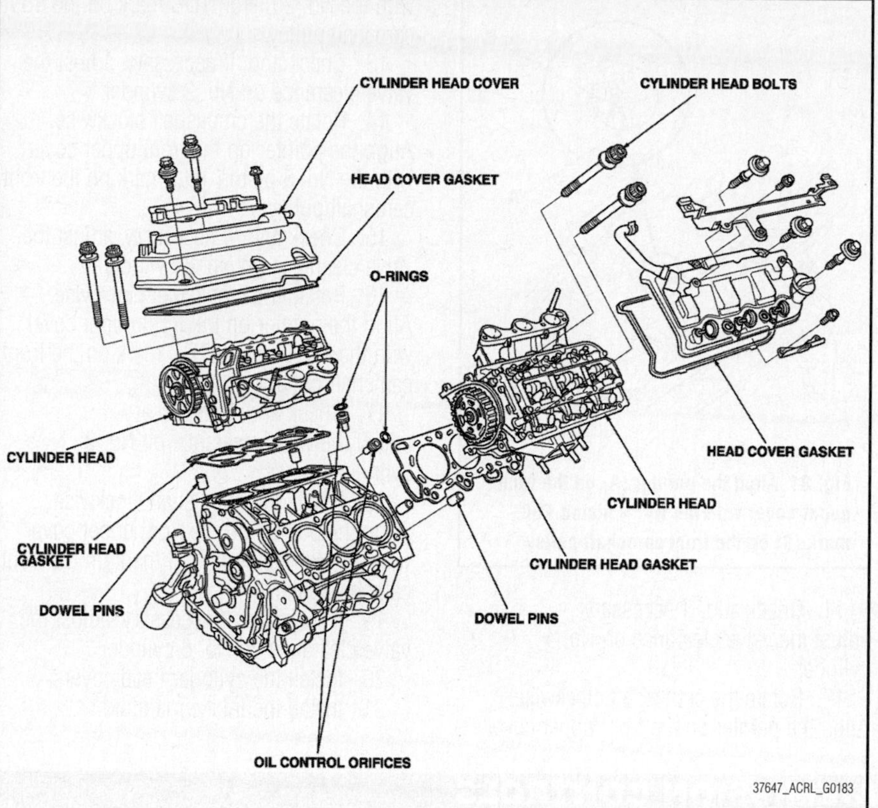
CYLINDER HEAD COVER
CYLINDER HEAD BOLTS
HEAD COVER GASKET
O-RINGS
CYLINDER HEAD
HEAD COVER GASKET
CYLINDER HEAD
CYLINDER HEAD GASKET
CYLINDER HEAD GASKET
DOWEL PINS
DOWEL PINS
OIL CONTROL ORIFICES
37647_ACRL_G0183

Fig. 76 Exploded view of the cylinder head components, showing the valve covers (cylinder head covers)

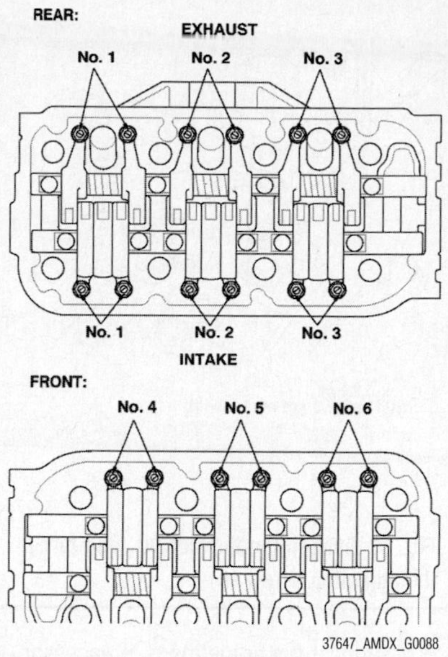
REAR:
EXHAUST
No. 1
No. 2
No. 3
No. 1
No. 2
No. 3
INTAKE
FRONT:
No. 4
No. 5
No. 6
37647_AMDX_G0088

Fig. 78 Intake and exhaust valve locations

To install:

8. Thoroughly clean the head cover gasket and the groove.
9. Install the head cover gasket in the groove of the cylinder head cover. Make sure the head cover gasket is seated securely.
10. Remove all of the old liquid gasket from the rocker shaft holder and the cylinder head.
11. Clean the head cover contacting surfaces with a shop towel.
12. Apply liquid gasket (P/N 08717-0004 or equivalent) evenly to the rocker shaft holder mating areas.
13. Install the component within 5 minutes of applying the liquid gasket.
14. Install all remaining components in reverse of the removal procedure.
15. Tighten the valve cover (cylinder head cover) bolts to 9 ft. lbs. (12 Nm).

VALVE LASH

ADJUSTMENT

See Figures 77 through 81.

➡Adjust the valves only when the cylinder head temperature is less than 100°F (38°C).

1. Remove the intake manifold.
2. Remove the cylinder head (valve) covers.
3. Set the No. 1 piston at top dead center (TDC).
4. Align the pointer on the front upper cover with the No. 1 piston TDC mark on the front camshaft pulley.

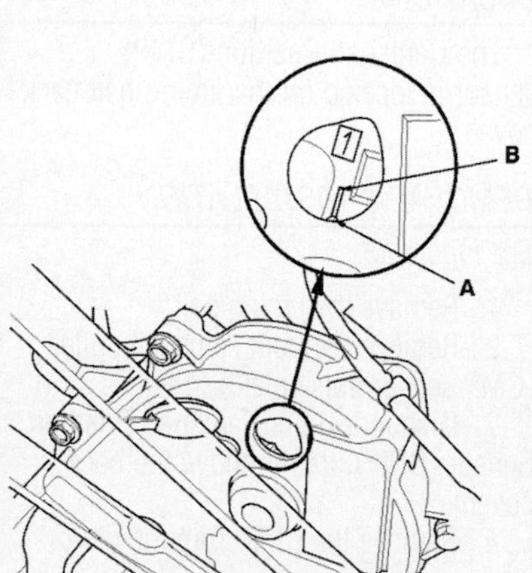
B
A
37647_AMDX_G0087

Fig. 77 Align the pointer (A) on the front upper cover with the No. 1 piston TDC mark (B) on the front camshaft pulley

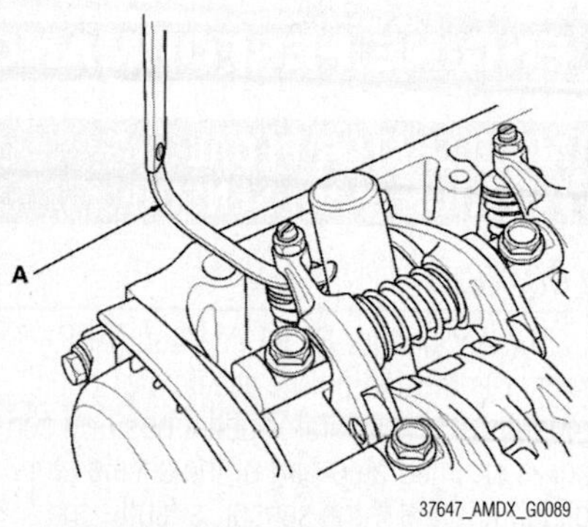
A
37647_AMDX_G0089

Fig. 79 Insert the feeler gauge (A) between the adjusting screw and the end of the valve stem on No. 1

5. Select the correct thickness feeler gauge for the valves you're going to check.
   a. Valve Clearance: Intake: 0.008–0.009 inches (0.20–0.24 mm); Exhaust: 0.011–0.013 inches (0.28–0.32 mm)
6. Insert the feeler gauge between the adjusting screw and the end of the valve stem on No. 1 cylinder and slide it back and forth; you should feel a slight amount of drag.
7. If you feel too much or too little drag, loosen the locknut, and turn the adjusting screw until the drag on the feeler gauge is correct.
8. Tighten the locknut and recheck the clearance.

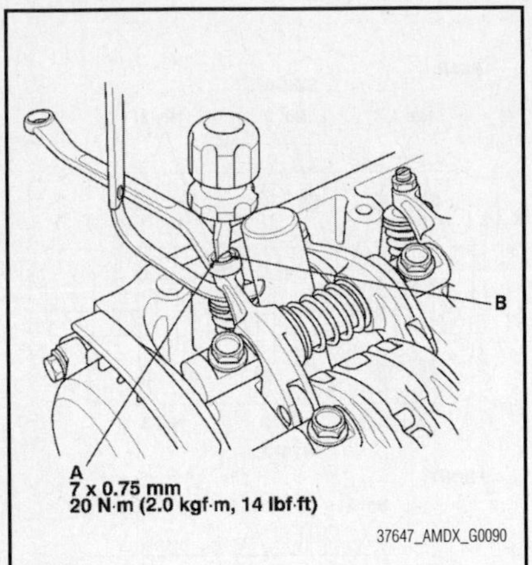
B
A
7 x 0.75 mm
20 N·m (2.0 kgf·m, 14 lbf·ft)
37647_AMDX_G0090

Fig. 80 Loosen the locknut (A), and turn the adjusting screw (B)

9. Repeat the adjustment, if necessary.
10. Rotate the crankshaft clockwise. Align the pointer on the front upper cover with the No. 4 piston TDC mark on the front camshaft pulley.

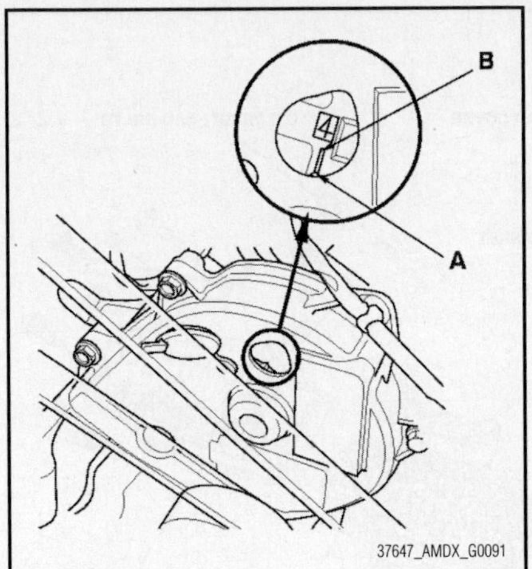
B
A
37647_AMDX_G0091

Fig. 81 Align the pointer (A) on the front upper cover with the No. 4 piston TDC mark (B) on the front camshaft pulley

11. Check and, if necessary, adjust the valve clearance on No. 4 cylinder.
12. Rotate the crankshaft clockwise. Align the pointer on the front upper cover with the No. 2 piston TDC mark on the front camshaft pulley.
13. Check and, if necessary, adjust the valve clearance on No. 2 cylinder.
14. Rotate the crankshaft clockwise. Align the pointer on the front upper cover with the No. 5 piston TDC mark on the front camshaft pulley.
15. Check and, if necessary, adjust the valve clearance on No. 5 cylinder.
16. Rotate the crankshaft clockwise. Align the pointer on the front upper cover with the No. 3 piston TDC mark on the front camshaft pulley.
17. Check and, if necessary, adjust the valve clearance on No. 3 cylinder.
18. Rotate the crankshaft clockwise. Align the pointer on the front upper cover with the No. 6 piston TDC mark on the front camshaft pulley.
19. Check and, if necessary, adjust the valve clearance on No. 6 cylinder.
20. Install the cylinder head covers.
21. Install the intake manifold.

ENGINE PERFORMANCE & EMISSION CONTROLS

ACCELERATOR PEDAL POSITION (APP) SENSOR

LOCATION

The Accelerator Pedal Position (APP) sensor is located on the accelerator pedal module. The accelerator pedal position sensor is an integrated part of the accelerator pedal module. If the sensor is faulty, the entire module must be replaced.

REMOVAL & INSTALLATION

See Figure 82.

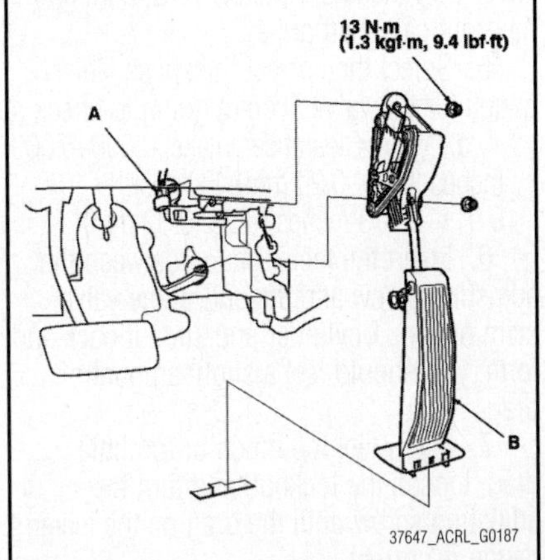
13 N·m
(1.3 kgf·m, 9.4 lbf·ft)
A
B
37647_ACRL_G0187

Fig. 82 Showing the accelerator pedal module

1. Disconnect the APP sensor connector (A).
2. Release the lock tab on the base of the accelerator pedal, then remove the accelerator pedal module (B).
3. Install the parts in the reverse order of removal.

CAMSHAFT POSITION (CMP) SENSOR

LOCATION

The Camshaft Position (CMP) sensor is located on the timing belt back cover.

REMOVAL & INSTALLATION

See Figure 83.

1. Remove the timing belt.
2. Remove the front camshaft pulley (CMP sensor pulse plate).
3. Disconnect the Camshaft Position Sensor (CMP), then remove the back cover.
4. Remove the CMP from the back cover.
5. Installation is the reverse order of removal.
6. Do the CKP pattern clear/CKP pattern learn procedure. See "Crankshaft Position (CKP) Sensor" in this section.

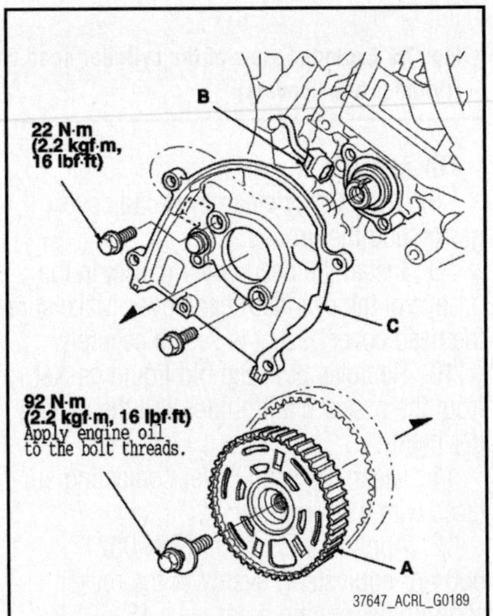
B
22 N·m
(2.2 kgf·m,
16 lbf·ft)
C
92 N·m
(2.2 kgf·m, 16 lbf·ft)
Apply engine oil
to the bolt threads.
A
37647_ACRL_G0189

Fig. 83 Disconnect the CMP sensor connector (B), and remove the back cover (C)

CRANKSHAFT POSITION (CKP) SENSOR

LOCATION

The Crankshaft Position (CKP) sensor is located on the oil pump.

REMOVAL & INSTALLATION

See Figure 84.

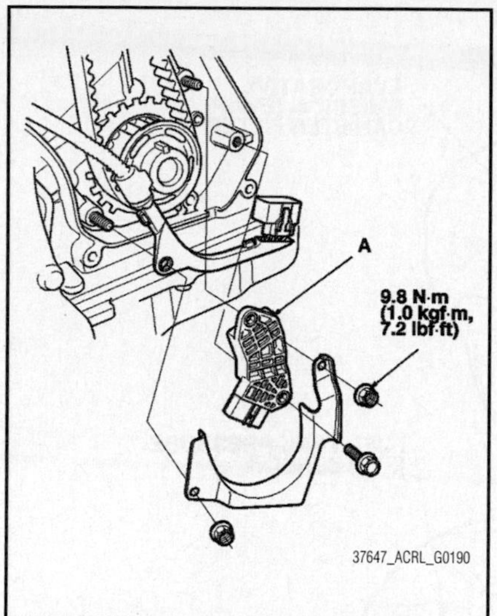
A
9.8 N·m
(1.0 kgf·m,
7.2 lbf·ft)
37647_ACRL_G0190

Fig. 84 Remove the Crankshaft Position (CKP) sensor (A) from the oil pump

1. Move the auto-tensioner to remove tension from the drive belt, then remove the accessory drive belt.
2. Remove the crankshaft pulley.
3. Remove the upper and lower front covers from the engine.
4. Remove the Crankshaft Position (CKP) sensor from the oil pump.
5. Installation is the reverse order of removal.
6. Do the CKP Pattern Clear/CKP Pattern Learn Procedure.

CKP PATTERN CLEAR/CKP PATTERN LEARN PROCEDURE

➡This procedure allows the clear and learn process without a scan tool.

1. Start the engine. Hold the engine speed at 3,000 RPM without load (in P or N) until the radiator fan comes on.
2. Test-drive the vehicle on a level road: Decelerate (with the throttle fully closed) from an engine speed of 2,500 RPM down to 1,000 RPM with the transmission in 2.
3. Test-drive the vehicle on a level road: Decelerate (with the throttle fully closed) from an engine speed of 5,000 RPM down to 3,000 RPM with the transmission in 2.
4. Repeat step 2 and 3 several times.
5. Turn the ignition switch to LOCK (0).
6. Turn the ignition switch to ON (II), and wait 30 seconds.

ENGINE COOLANT TEMPERATURE (ECT) SENSOR

LOCATION

Sensor 1 is located in the engine compartment below the main underhood fuse box. Sensor 2 is located at the bottom of the radiator.

OPERATION

ECT sensors 1 and 2 are temperature dependent resistors (thermistors). The resistance decreases as the engine coolant temperature increases.

REMOVAL & INSTALLATION

ECT Sensor 1

See Figure 85.

1. Drain the engine cooling system.
2. Remove the engine appearance cover.
3. Remove the main under-hood fuse/relay box.
4. Disconnect the ECT sensor 1 connector.
5. Remove ECT sensor 1.
6. Install the parts in the reverse order of removal with a new O-ring, then refill the radiator with engine coolant.

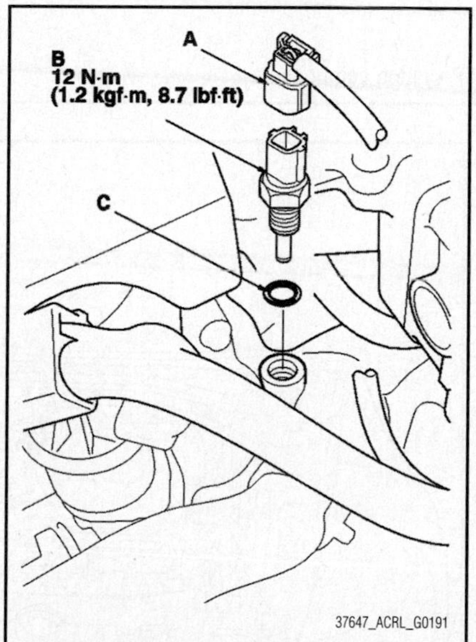
A
B
12 N·m
(1.2 kgf·m, 8.7 lbf·ft)
C
37647_ACRL_G0191

Fig. 85 Remove ECT sensor 1 (B)

ECT Sensor 2

See Figure 86.

1. Drain the engine cooling system.
2. Remove the splash shield.
3. Disconnect the ECT sensor 2 connector.
4. Remove ECT sensor 2.
5. Install the parts in the reverse order of removal with a new O-ring, then refill the radiator with engine coolant.

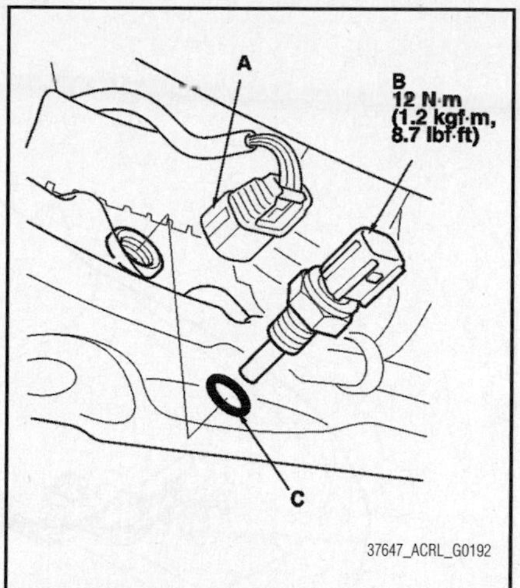
A
B
12 N·m
(1.2 kgf·m,
8.7 lbf·ft)
C
37647_ACRL_G0192

Fig. 86 Remove ECT sensor 2 (B)

EVAPORATIVE EMISSIONS (EVAP) CANISTER

LOCATION

See Figure 87.

The Evaporative Emissions (EVAP) canister is located under the vehicle body near the fuel tank.

REMOVAL & INSTALLATION

See Figure 88.

1. Remove the propeller shaft.
2. Remove the rear differential.
3. Disconnect the connectors (A) and the hoses (B).
4. If needed, remove the EVAP canister from its holding bracket.
5. Installation is the reverse of the removal procedure.

EVAPORATIVE EMISSIONS (EVAP) PURGE CONTROL VALVE

LOCATION

See Figure 89.

The EVAP Purge Control Valve is located in the engine compartment.

REMOVAL & INSTALLATION

See Figure 90.

1. Remove the engine cover.
2. Disconnect the hoses (A) and the EVAP canister purge valve 2P connector (B).
3. Remove the EVAP canister purge valve and the bracket (C) and remove the purge valve.
4. Installation is the reverse of the removal procedure.

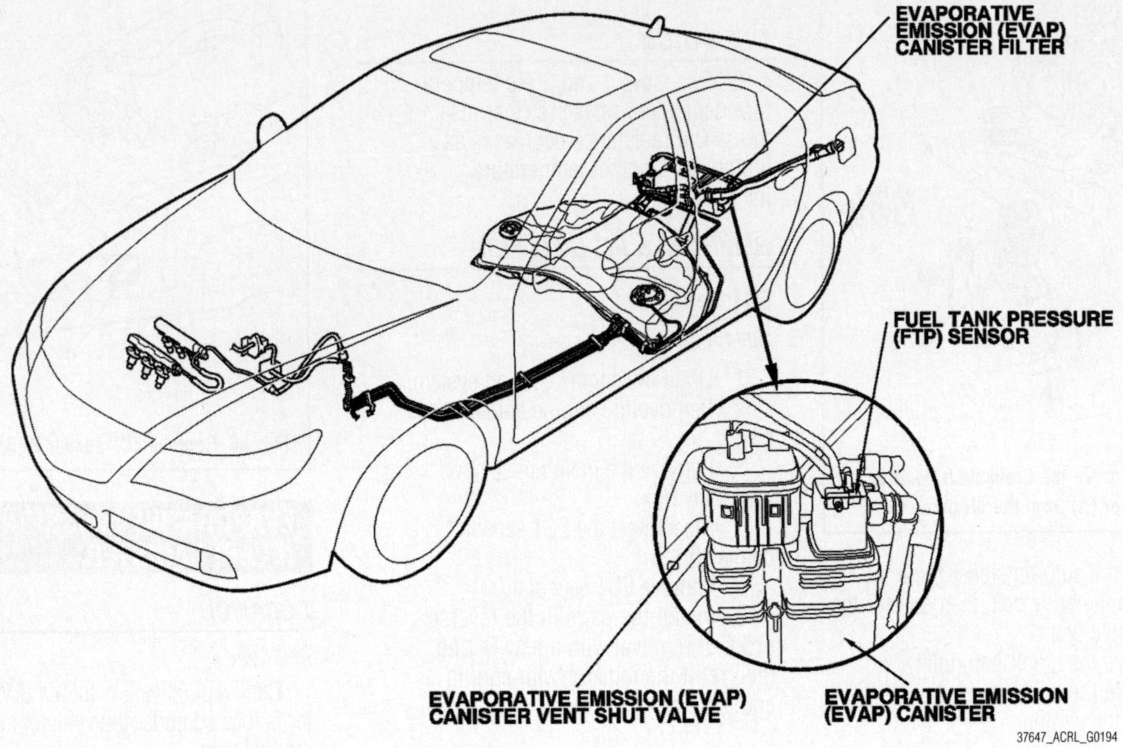
EVAPORATIVE EMISSION (EVAP) CANISTER FILTER
FUEL TANK PRESSURE (FTP) SENSOR
EVAPORATIVE EMISSION (EVAP) CANISTER VENT SHUT VALVE
EVAPORATIVE EMISSION (EVAP) CANISTER
37647_ACRL_G0194

Fig. 87 Showing the under car locations of the EVAP system components

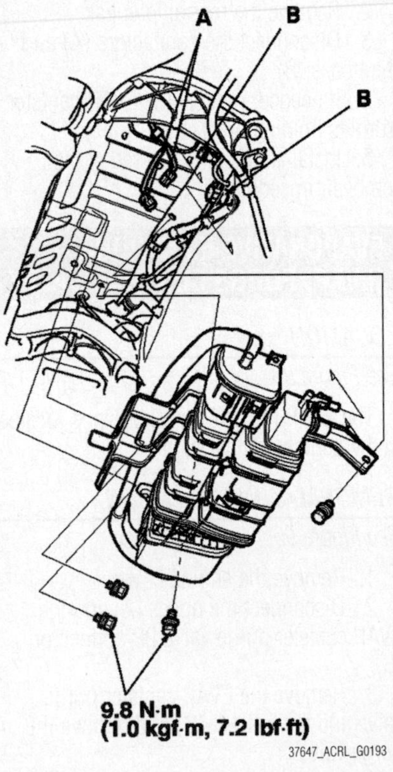
A
B
B
9.8 N·m
(1.0 kgf·m, 7.2 lbf·ft)
37647_ACRL_G0193

Fig. 88 Disconnect the connectors (A) and the hoses (B) and remove the EVAP canister.

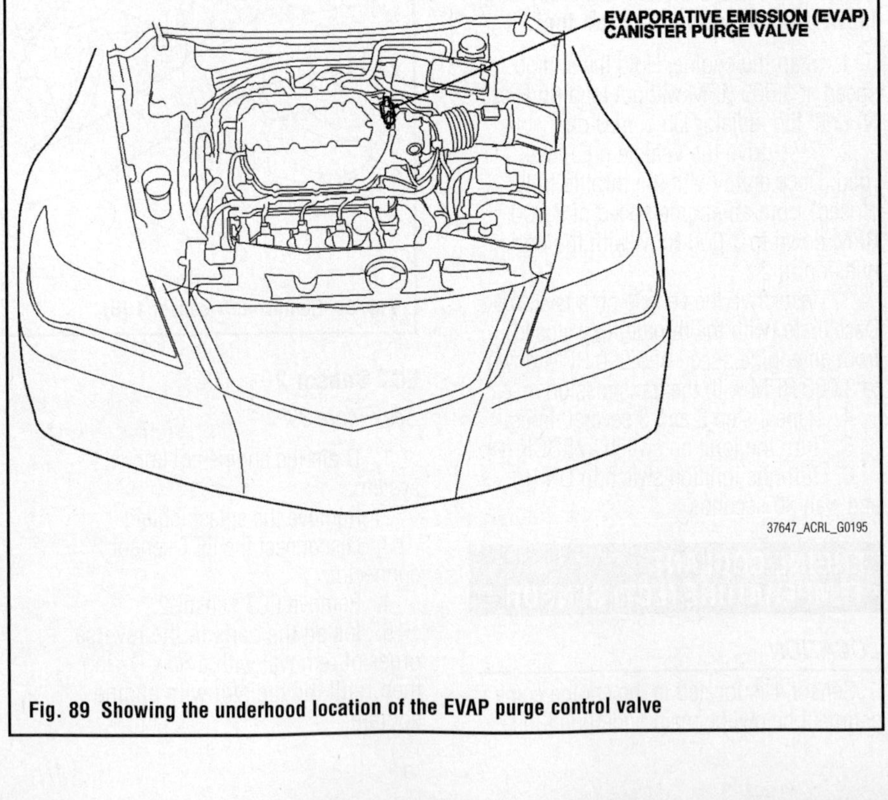
EVAPORATIVE EMISSION (EVAP) CANISTER PURGE VALVE
37647_ACRL_G0195

Fig. 89 Showing the underhood location of the EVAP purge control valve

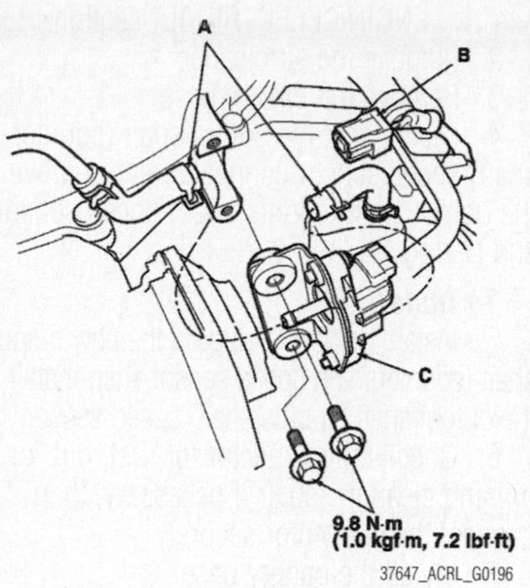
A
B
C
9.8 N·m
(1.0 kgf·m, 7.2 lbf·ft)
37647_ACRL_G0196

Fig. 90 Disconnect the hoses (A) and the EVAP canister purge valve 2P connector (B). Remove the EVAP canister purge valve and the bracket (C).

EXHAUST GAS RECIRCULATION (EGR) VALVE

LOCATION

The Exhaust Gas Recirculation (EGR) valve is located near the front left side of the engine.

REMOVAL & INSTALLATION

See Figure 91.

1. Remove the engine cover.
2. Disconnect the EGR valve 5P connector (A).
3. Remove the EGR valve (B) and gasket (C).
4. Install the parts in the reverse order of removal with a new gasket.

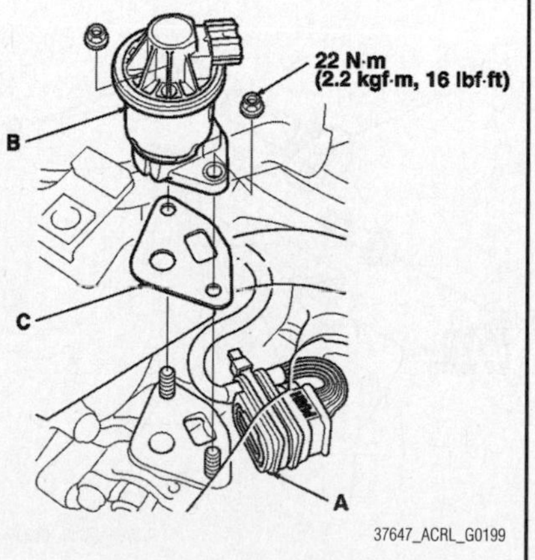
22 N·m
(2.2 kgf·m, 16 lbf·ft)
B
C
A
37647_ACRL_G0199

Fig. 91 Disconnect the EGR valve 5P connector (A), then remove the EGR valve (B) and gasket (C).

HEATED OXYGEN (HO2S) SENSOR

LOCATION

Refer to the graphics in the "Removal and Installation" section for locations.

REMOVAL & INSTALLATION

Front Bank (Bank 2)

See Figure 92.

1. Disconnect the front secondary HO2S wiring connector, then remove the front secondary HO2S.
2. Install the parts in the reverse order of removal.

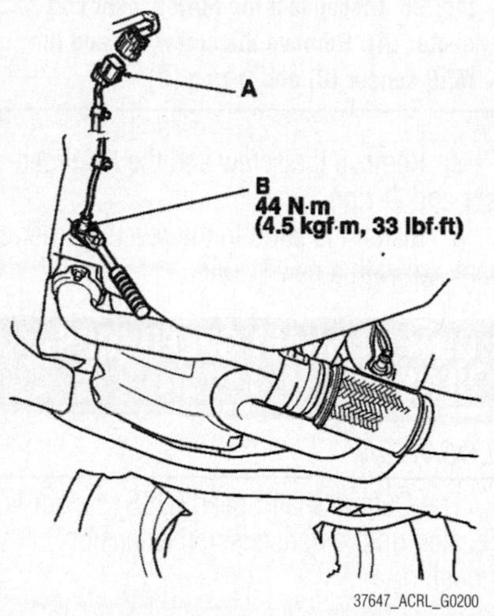
A
B
44 N·m
(4.5 kgf·m, 33 lbf·ft)
37647_ACRL_G0200

Fig. 92 Disconnect the front secondary HO2S wiring connector (A), then remove the front secondary HO2S (B)

Rear Bank (Bank 1)

See Figure 93.

1. Disconnect the rear secondary HO2S wiring connector, then remove the rear secondary HO2S.

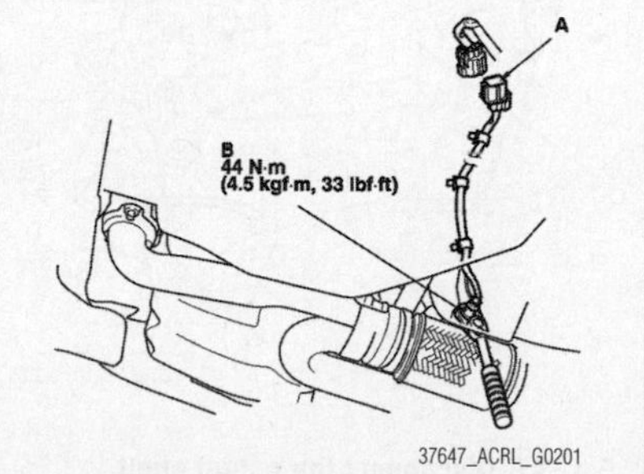
A
B
44 N·m
(4.5 kgf·m, 33 lbf·ft)
37647_ACRL_G0201

Fig. 93 Disconnect the rear secondary HO2S wiring connector (A), then remove the rear secondary HO2S (B)

2. Install the parts in the reverse order of removal.

KNOCK SENSOR (KS)

LOCATION

The knock sensor is located in the valley of the engine block.

REMOVAL & INSTALLATION

See Figure 94.

1. Remove the intake manifold.
2. Remove the injector rails and the injector base.
3. Disconnect the knock sensor wiring connector, then remove the knock sensor.
4. Installation is the reverse order of removal. Tighten the knock sensor to 23 ft. lbs. (31 Nm).

MALFUNCTION INDICATOR LIGHT (MIL)

RESET PROCEDURE

➡If you are using a generic scan tool to clear commands, be aware that there is only one setting for clearing the PCM, and it clears all commands at the same time CKP pattern learn, idle learn, readiness codes, freeze data, on-board snapshot, and DTCs). After you clear all commands, you then need to do these procedures, in this order: PCM idle learn procedure; CKP pattern learn procedure; Test-drive to set readiness codes to complete.

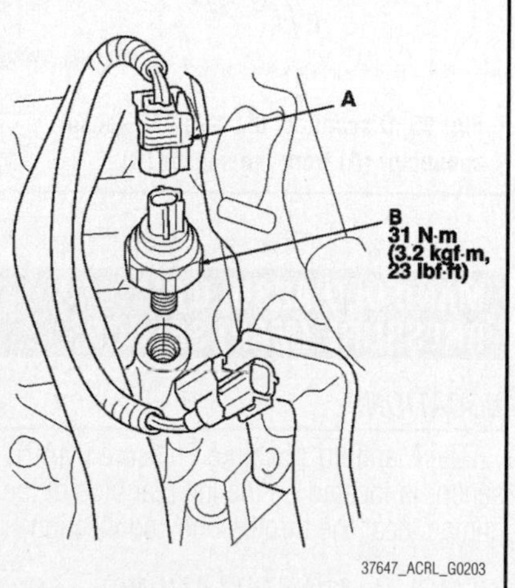
A
B
31 N·m
(3.2 kgf·m,
23 lbf·ft)
37647_ACRL_G0203

Fig. 94 Disconnect the knock sensor wiring connector (A), then remove the knock sensor (B)

1. Clear the DTC with the scan tool while the engine is stopped.
2. Turn the ignition switch to LOCK (0).
3. Turn the ignition switch to ON (II), and wait 30 seconds.
4. Turn the ignition switch to LOCK (0), and disconnect the scan tool from the DLC.

MASS AIR FLOW/INTAKE AIR TEMPERATURE (MAF/IAT) SENSOR

LOCATION

The MAF/IAT sensor is located in the air intake near the air cleaner.

REMOVAL & INSTALLATION

See Figure 95.

1. Disconnect the MAF/IAT sensor connector.
2. Remove the screw.
3. Remove the MAF/IAT sensor.
4. Install the parts in the reverse order of removal with a new gasket.

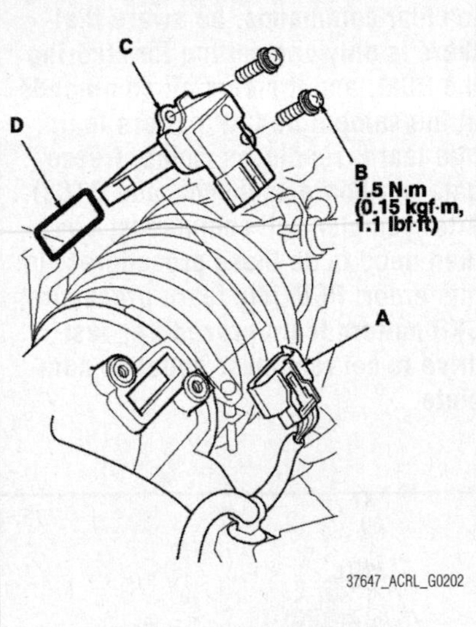
C
D
B
1.5 N·m
(0.15 kgf·m,
1.1 lbf·ft)
A
37647_ACRL_G0202

Fig. 95 Disconnect the MAF/IAT sensor connector (A) from the sensor (C)

MANIFOLD ABSOLUTE PRESSURE (MAP) SENSOR

LOCATION

The Manifold Absolute Pressure (MAP) sensor is located on the left rear side of the engine, near the throttle body connection.

REMOVAL & INSTALLATION

See Figure 96.

1. Remove the engine appearance cover.
2. Disconnect the MAP sensor connector.

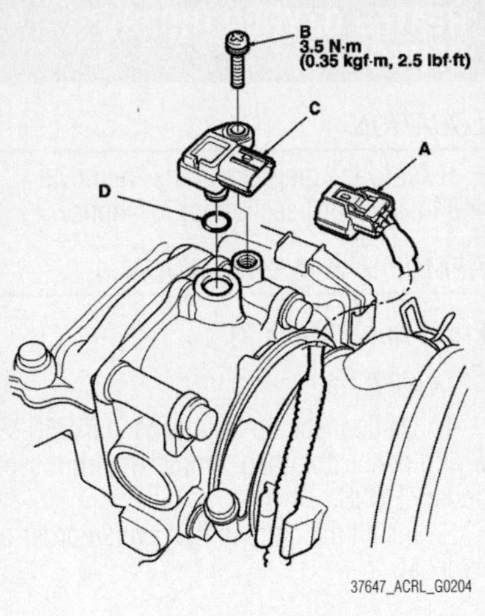
B
3.5 N·m
(0.35 kgf·m, 2.5 lbf·ft)
C
A
D
37647_ACRL_G0204

Fig. 96 Disconnect the MAP sensor connector (A). Remove the screw (B) and the MAP sensor (C) and O-ring (D).

3. Remove the screw and the MAP sensor and O-ring.
4. Install the parts in the reverse order of removal with a new O-ring.

OUTPUT SHAFT SPEED (OSS) SENSOR

LOCATION

The Output Shaft Speed (OSS) sensor is located on the transmission extension housing.

REMOVAL & INSTALLATION

See Figure 97.

1. Do the battery terminal disconnect procedure, then wait at least 3 minutes before beginning work. See "Battery System" in "ENGINE ELECTRICAL" section.
2. Remove the air cleaner.
3. Remove the battery base.
4. Disconnect the output shaft (countershaft) speed sensor connector, and remove the output shaft (countershaft) speed sensor and O-ring.

To install:

5. Install the new O-ring on the new output shaft (countershaft) speed sensor, then install the output shaft (countershaft) speed sensor.
6. Check the connector for rust, dirt, or oil, and clean or repair if necessary, then connect the connector securely.
7. Install the battery base.
8. Install the air cleaner.
9. Do the battery terminal reconnect procedure, then wait at least 3 minutes before beginning work. See "Battery System" in "ENGINE ELECTRICAL" section.

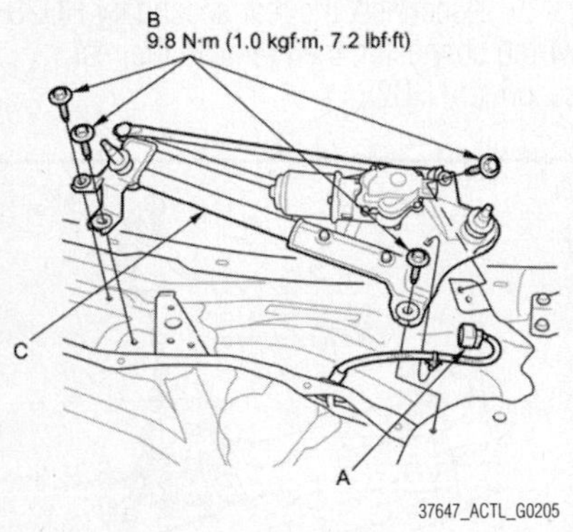
B
9.8 N·m (1.0 kgf·m, 7.2 lbf·ft)
C
A
37647_ACTL_G0205

Fig. 97 Disconnect the output shaft (countershaft) speed sensor connector, and remove the output shaft (countershaft) speed sensor (A) and O-ring (B).

POSITIVE CRANKCASE VENTILATION (PCV) VALVE

LOCATION

The PCV valve is located on the intake manifold.

REMOVAL & INSTALLATION

See Figure 98.

1. Remove the engine cover.
2. Remove the bolt.

➡Take care not to spill oil on the hot exhaust manifold.

3. Remove the PCV valve.
4. Install the parts in the reverse order of removal.

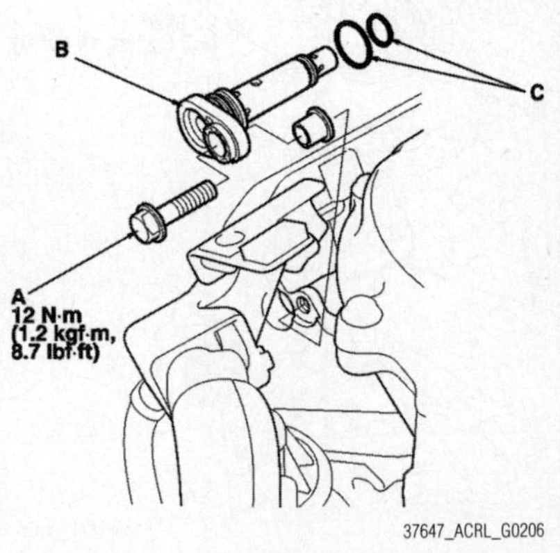
B
C
A
12 N·m
(1.2 kgf·m,
8.7 lbf·ft)
37647_ACRL_G0206

Fig. 98 Remove the bolt (A) and the PCV valve (B)

5. When installing a new PCV valve, use new O-rings and make sure they are in place.

POWERTRAIN CONTROL MODULE (PCM)

LOCATION

See Figure 99.

The Powertrain Control Module (PCM) is located under the center of the instrument panel, behind the center console.

REMOVAL & INSTALLATION

➡Make sure any scan tool being used has the latest software version.

➡If replacing the PCM after substituting a known-good PCM, reinstall the original PCM, then do this procedure.

➡During the procedure, if any READ DATA, WRITE DATA, or other data checks fail, note the failure, then continue.

1. Connect the HDS, or equivalent scan tool, to the data link connector (DLC) located under the driver's side of the dashboard.
2. Turn the ignition switch to ON (II).
3. Make sure the scan tool communicates with the PCM and other vehicle systems.
   a. If it doesn't, go to the DLC circuit troubleshooting.
   b. If you are returning from DLC circuit troubleshooting, skip steps 4 through 9, 17 through 19, 22, and 23, then do this after replacing the PCM:
4. Replace the engine oil and the engine oil filter.
5. Clean the throttle body.
6. Select the PGM-FI system with the scan tool.
7. Select the INSPECTION MENU with the scan tool.
8. Select the ETCS TEST, then select the TP POSITION CHECK, and follow the screen prompts.

➡If the TP POSITION CHECK indicates FAILED, continue with this procedure.

9. Select the REPLACE PCM MENU, then select READ DATA, and follow the screen prompts.

➡Doing this step copies (READS) the engine oil life data from the original PCM so you can later download (WRITE) it into the new PCM.

10. If the READ DATA indicates FAILED, continue with this procedure.
11. Turn the ignition switch to LOCK (0).
12. Do the battery terminal disconnect procedure, then wait at least 3 minutes before beginning work. See "Battery System" in "ENGINE ELECTRICAL" section.
13. Remove the front console covers, and pull back the carpet.
14. Remove the ducts.
15. Disconnect the PCM connectors.
16. Remove the bolts, then remove the PCM.

To install:

17. Install the parts in the reverse order of removal.
18. Do the battery terminal reconnect procedure. See "Battery System" in "ENGINE ELECTRICAL" section.
19. Turn the ignition switch to ON (II).
20. Manually input the VIN to the PCM with the scan tool.

➡DTC P0630 VIN Not Programmed or Mismatch may be stored because the VIN has not been programmed into the PCM, ignore it, and continue this procedure.

21. If the READ DATA (engine oil life) failed in step 7, go to step 21. Otherwise, go to step 19.

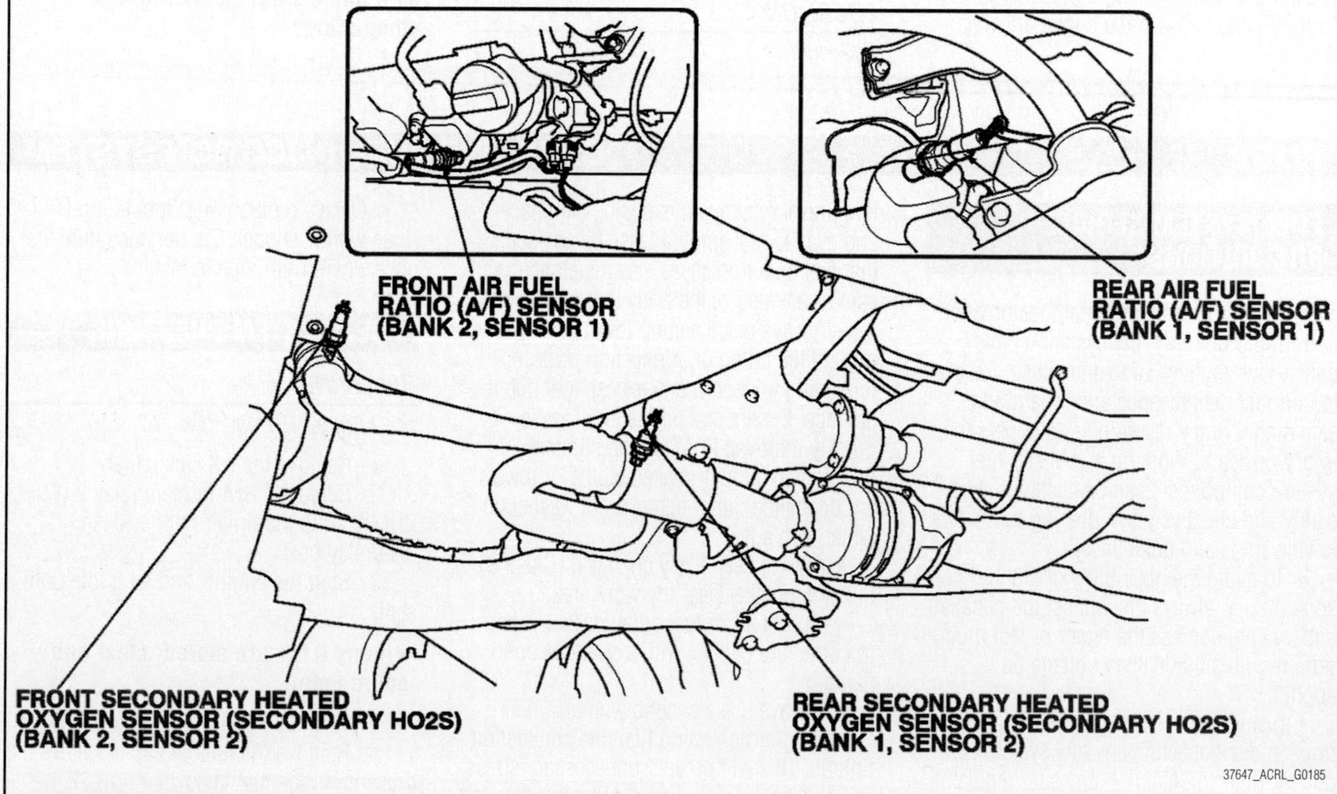
FRONT AIR FUEL RATIO (A/F) SENSOR (BANK 2, SENSOR 1)
REAR AIR FUEL RATIO (A/F) SENSOR (BANK 1, SENSOR 1)
FRONT SECONDARY HEATED OXYGEN SENSOR (SECONDARY HO2S) (BANK 2, SENSOR 2)
REAR SECONDARY HEATED OXYGEN SENSOR (SECONDARY HO2S) (BANK 1, SENSOR 2)
37647_ACRL_G0185

Fig. 99 Showing the in-car location of the PCM and related components

22. Select the PGM-FI system with the HDS.
23. Select the REPLACE PCM MENU, then select WRITE DATA, and follow the screen prompts.

➡If the WRITE DATA indicates FAILED, continue this procedure.

24. Select IMMOBI system with the scan tool.
25. Enter the immobilizer PCM code that you got from the software, and use the PCM replacement procedure in the scan tool; it allows you to start the engine.
26. If the TP POSITION CHECK failed in step 6, clean the throttle body, then go to step 24.
27. If the READ DATA failed in step 7 or the WRITE DATA failed in step 20, replace the engine oil and engine oil filter, then go to step 25.
28. Select PGM-FI system, and reset the PCM with the scan tool.
29. Update the PCM if it does not have the latest software.
30. Do the PCM idle learn procedure.
31. Do the CKP Clear/Learn procedure. See the Learn Procedure in "Crankshaft Position (CKP) Sensor" in this section.

PCM IDLE LEARN PROCEDURE

The idle learn procedure must be done so the PCM can learn the engine idle characteristics.

1. Do the idle learn procedure whenever you do any of these actions:
   - Replace the PCM.
   - Reset the PCM.
   - Update the PCM.
   - Replace or clean the throttle body.
   - Disassemble the engine or the transmission.

➡Clearing the DTCs with the scan tool does not require you to do the idle learn procedure.

Procedure

1. Make sure all electrical items (the A/C, the audio, the rear window defogger, the lights, etc.) are off.
2. Reset the PCM with the scan tool.
3. Turn the ignition switch to ON (II), and wait 2 seconds.
4. Start the engine. Hold the engine speed at 3,000 RPM without load (in P or N) until the radiator fan comes on, or until the engine coolant temperature reaches 194°F (90°C).
5. Let the engine idle for about 5 minutes with the throttle fully closed.
6. If the radiator fan comes on, do not include its running time in the 5 minutes.

THROTTLE POSITION SENSOR (TPS)

LOCATION

See Figure 100.

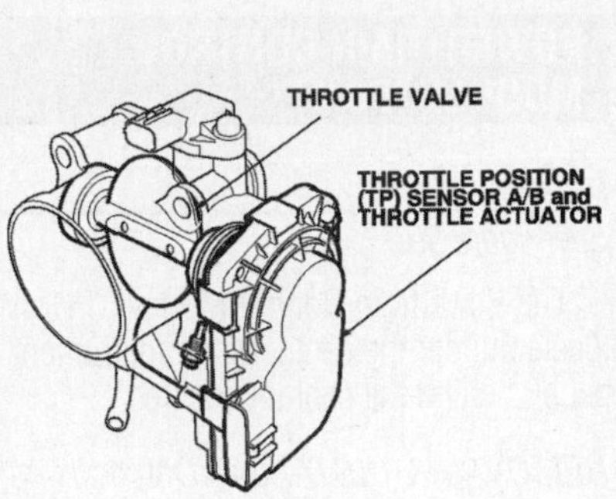
THROTTLE VALVE
THROTTLE POSITION (TP) SENSOR A/B and THROTTLE ACTUATOR
37647_ACRL_G0208

Fig. 100 The Throttle Position Sensor (TPS) is located on the throttle body housing

The Throttle Position Sensor (TPS) is located on the throttle body housing.

REMOVAL & INSTALLATION

➡Manufacturer does not provide a specific removal and installation procedure for this component. Use the location illustration as a guide when servicing this component.

FUEL — GASOLINE FUEL INJECTION SYSTEM

FUEL SYSTEM SERVICE PRECAUTIONS

Safety is the most important factor when performing not only fuel system maintenance, but any type of maintenance. Failure to conduct maintenance and repairs in a safe manner may result in serious personal injury or death. Work on a vehicle's fuel system components can be accomplished safely and effectively by adhering to the following rules and guidelines.

- To avoid the possibility of fire and personal injury, always disconnect the negative battery cable unless the repair or test procedure requires that battery voltage be applied.
- Always relieve the fuel system pressure prior to disconnecting any fuel system component (injector, fuel rail, pressure regulator, etc.) fitting or fuel line connection. Exercise extreme caution whenever relieving fuel system pressure to avoid exposing skin, face and eyes to fuel spray. Please be advised that fuel under pressure may penetrate the skin or any part of the body that it contacts.
- Always place a shop towel or cloth around the fitting or connection prior to loosening to absorb any excess fuel due to spillage. Ensure that all fuel spillage is quickly removed from engine surfaces. Ensure that all fuel-soaked cloths or towels are deposited into a flame-proof waste container with a lid.
- Always keep a dry chemical (Class B) fire extinguisher near the work area.
- Do not allow fuel spray or fuel vapors to come into contact with a spark or open flame.
- Always use a second wrench when loosening or tightening fuel line connection fittings. This will prevent unnecessary stress and torsion on fuel piping. Always follow the proper torque specifications.
- Always replace worn fuel fitting O-rings with new ones. Do not substitute fuel hose where rigid pipe is installed.

FUEL SYSTEM PRESSURE

RELIEVING

See Figures 101 and 102.

1. Remove the left kick panel.
2. Remove PGM-FI main relay 2 (FUEL PUMP) from the driver's under-dash fuse/relay box.
3. Start the engine, and let it idle until it stalls.

➡If any DTCs are stored, clear and ignore them.

4. Turn the ignition switch to LOCK (0).
5. Remove the fuel fill cap to relieve the pressure in the fuel tank.
6. Do the battery terminal disconnect procedure, then wait at least 3 minutes

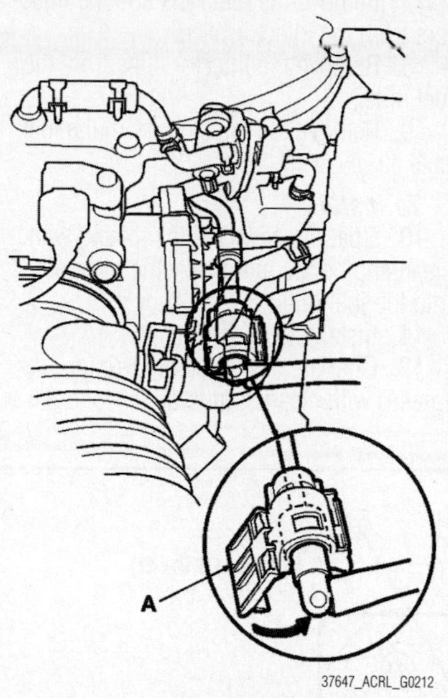
A
37647_ACRL_G0212

Fig. 101 Remove the quick-connect fitting cover (A).

before beginning work. See "Battery System" in "ENGINE ELECTRICAL" section.

7. Remove the quick-connect fitting cover.

8. Check the fuel quick-connect fitting for dirt, and clean it if needed.

9. Place a rag or a shop towel over the quick-connect fitting.

10. Disconnect the quick-connect fitting as follows:

a. Hold the connector with one hand, and squeeze the retainer tabs with the other hand to release them from the locking tabs.

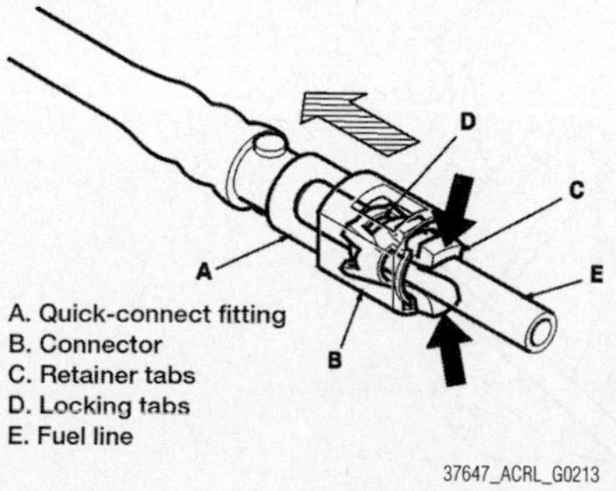
D
C
A
E
B
A. Quick-connect fitting
B. Connector
C. Retainer tabs
D. Locking tabs
E. Fuel line
37647_ACRL_G0213

Fig. 102 Disconnect the quick-connect fitting (A): Hold the connector (B) with one hand, and squeeze the retainer tabs (C) with the other hand to release them from the locking tabs (D). Pull the connector off. Be careful not to damage the line (E) or other parts.

b. Pull the connector off. Be careful not to damage the line or other parts.

➡Do not use tools.

11. If the connector does not move, keep the retainer tabs pressed down, and alternately pull and push the connector until it comes off easily.

12. Do not remove the retainer from the line; once removed, the retainer must be replaced with a new one. After disconnecting the quick-connect fitting, check it for dirt or damage.

FUEL FILTER

REMOVAL & INSTALLATION

See Figure 103.

The fuel filter should be replaced whenever the fuel pressure drops below the specified value, after making sure that the fuel pump and the fuel pressure regulator are OK.

1. If the fuel tank is full, drain the fuel.

2. Remove the fuel tank unit. See "Fuel Pump & Fuel Tank Unit" in this section.

3. Remove the fuel filter set.

4. Check these items before installing the fuel tank unit:

a. When connecting the wire harness, make sure the connection is secure and the connectors are firmly locked into place.

b. When installing the fuel gauge sending unit, make sure the connection is secure and the connector is firmly locked into place. Be careful not to bend or twist it excessively.

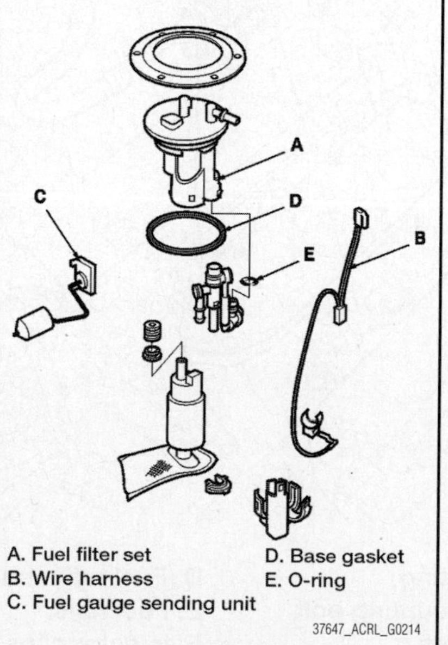
A
C
D
B
E
A. Fuel filter set
B. Wire harness
C. Fuel gauge sending unit
D. Base gasket
E. O-ring
37647_ACRL_G0214

Fig. 103 Showing the component of the fuel tank unit, with the fuel filter

To install:

5. Install the parts in the reverse order of removal with a new base gasket (D) and a new O-ring (E).

6. Coat the O-rings with clean engine oil; do not use any other oils or fluids.

7. Do not pinch the O-rings during installation.

8. Use all the new parts supplied in the fuel filter replacement kit.

FUEL GAUGE SENDING UNIT

LOCATION

The fuel gauge sent unit is a component of the fuel pump unit in the fuel tank.

REMOVAL & INSTALLATION

See Figure 104.

1. Remove the fuel tank unit. See "Fuel Pump & Fuel Tank Unit" in this section.

2. Remove the fuel level sensor (fuel sending unit) from the fuel tank unit.

3. Before installing the fuel tank unit. Make sure the connection is secure. Be careful not to bend or twist it excessively.

4. Install the parts in the reverse order of removal. When installing the fuel tank unit, align the marks on the unit and the fuel tank.

FUEL PUMP & FUEL TANK UNIT

REMOVAL & INSTALLATION

1. Relieve the fuel pressure. See "Fuel System Pressure" in this section.

2. Remove the fuel fill cap.

3. If the fuel tank is full, drain the fuel.

4. Remove the rear seat cushion.

5. Remove the access panel from the left side of the floor.

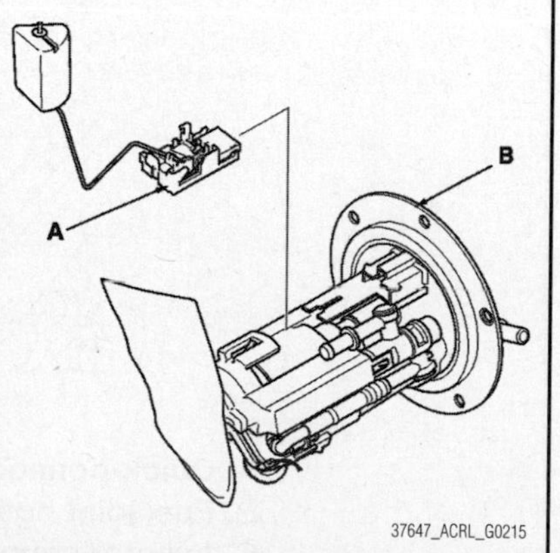
B
A
37647_ACRL_G0215

Fig. 104 Remove the fuel level sensor (fuel sending unit) (A) from the fuel tank unit (B).

6. Disconnect the fuel tank pump 5P connector.
7. Disconnect the quick-connect fitting from the fuel tank unit.
8. Remove the fuel tank unit and disconnect the transfer tube.
9. Installation is the reverse of the removal procedure.

FUEL RAIL & INJECTORS

REMOVAL & INSTALLATION

See Figure 105.

1. Relieve the fuel pressure. See "Fuel Pressure Relieving" in this section.
2. Remove the intake manifold. See "Intake Manifold" in "ENGINE MECHANICAL" section.
3. Disconnect the quick-connect fitting.
4. Remove the fuel joint hose mounting bolt.
5. Disconnect the connectors from the injectors.
6. Remove the fuel rail mounting bolts from the fuel rails.
7. Remove the fuel rails and the injectors from the injector base.
8. Remove the injector clips from the fuel rails.
9. Remove the injectors from the fuel rails.

To install:

10. Coat the new O-rings (black) with clean engine oil, and insert the injectors into the fuel rails.
11. Install the injector clips.
12. Coat the new injector O-rings (green) with clean engine oil.

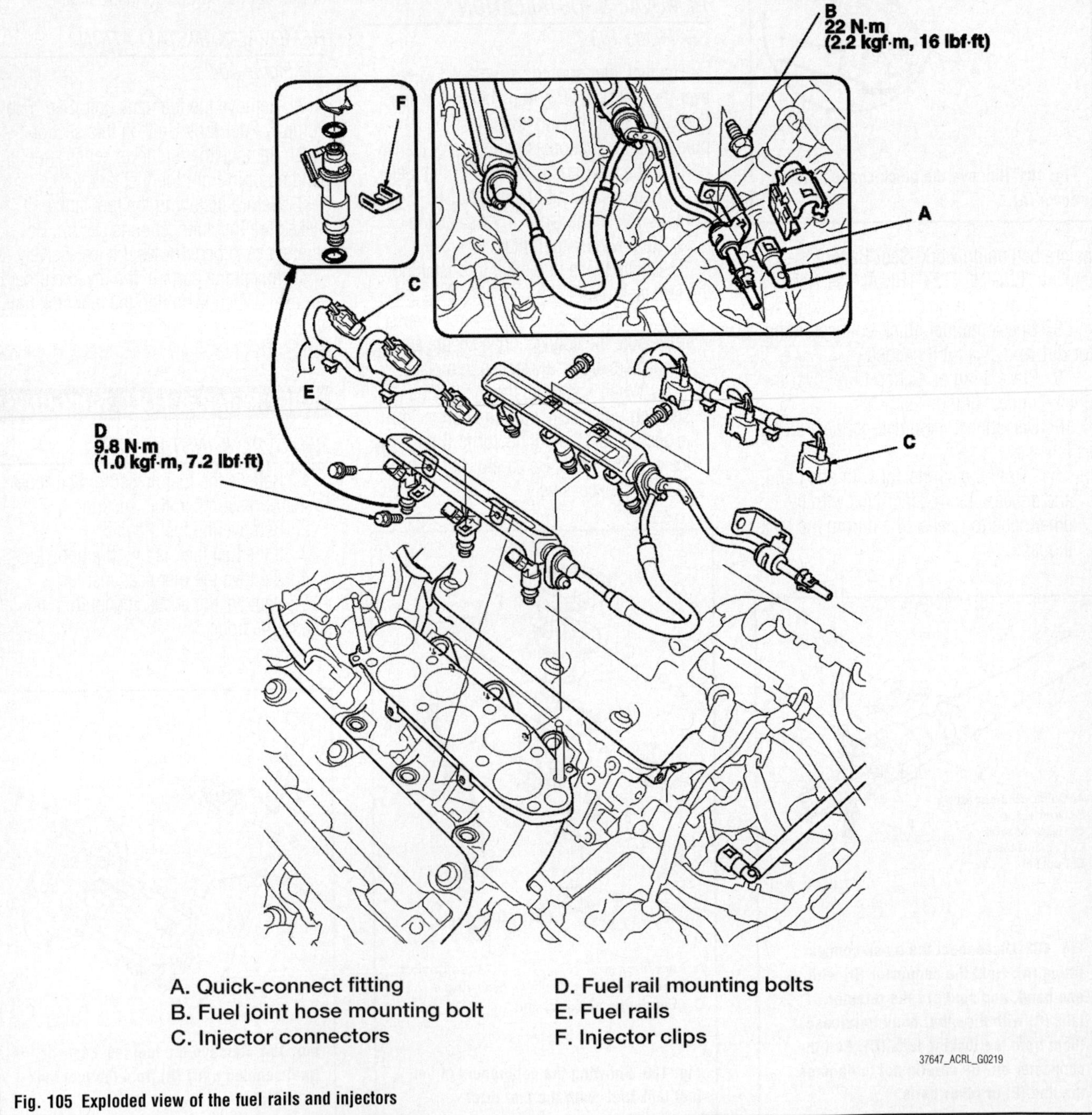
B
22 N·m
(2.2 kgf·m, 16 lbf·ft)
F
A
C
E
C
D
9.8 N·m
(1.0 kgf·m, 7.2 lbf·ft)
A. Quick-connect fitting
B. Fuel joint hose mounting bolt
C. Injector connectors
D. Fuel rail mounting bolts
E. Fuel rails
F. Injector clips
37647_ACRL_G0219

Fig. 105 Exploded view of the fuel rails and injectors

13. Install the fuel rails and the injectors in the injector base.

14. Install the fuel rail mounting bolts, and connect the injector connectors.

15. Install the fuel joint hose mounting bolt.

16. Connect the quick-connect fitting.

17. Turn the ignition switch to ON (II), but do not operate the starter. After the fuel pump runs for about 2 seconds, the fuel rail will be pressurized. Repeat this two or three times, then check for fuel leaks.

18. Install the intake manifold with a new gasket.

FUEL TANK

DRAINING

1. Remove the secondary fuel gauge sending unit.

2. Using a hand pump, a hose, and a container suitable for fuel, draw the fuel from the secondary fuel gauge sending unit side of the tank.

3. Remove the fuel tank unit. See "Fuel Pump & Fuel Tank Unit" in this section.

4. Using a hand pump, a hose, and a container suitable for fuel, draw the fuel from the fuel tank unit side of the tank.

REMOVAL & INSTALLATION

1. Drain the fuel tank.

2. Reinstall the fuel tank unit and the secondary fuel gauge sending unit without connecting the fuel tank unit 5P connector, the secondary fuel gauge sending unit 5P connector and the quick -connect fitting.

3. Remove the propeller shaft.

4. Remove the rear differential.

5. Raise the vehicle on a lift.

6. Remove the fuel tank covers and parking brake cable bolts.

7. Remove the clip from the EVAP canister bracket.

8. Disconnect the hoses. Slide back the clamps, then twist the hoses as you pull to avoid damaging them.

9. Place a jack or other support under the tank.

10. Remove the strap bolts and the straps.

11. Remove the fuel tank. If it sticks to the undercoat on its mount, carefully pry it off the mount.

To install:

Install the parts in the reverse order of removal.

➡The new fuel vent hose and the fuel vent return hose have ring pull at the connectors. When you connect the hoses and confirm that the connections are secure, remove the ring pulls by pulling them down.

12. Before connecting the fuel fill pipe and the quick-connect fitting, check for dirt, and clean it if needed, taking care not to damage the fuel fill pipe and other parts.

IDLE SPEED

INSPECTION

1. Before checking the idle speed, check these items:

- The malfunction indicator lamp (MIL) has not been reported on, and there are no DTCs.
- Ignition timing
- Spark plugs
- Air cleaner
- PCV system

2. Apply the parking brake, and make sure the headlights are off.

3. Disconnect the evaporative emission (EVAP) canister purge valve connector.

4. Connect the scan tool to the data link connector (DLC) (A) located under the driver's side of the dashboard.

5. Make sure the scan tool communicates with the PCM. If it does not, go to the DLC circuit troubleshooting.

6. Start the engine. Hold the engine speed at 3,000 RPM without load (in P or N) until the radiator fan comes on, then let it idle.

7. Check the idle speed without load conditions: headlights, blower fan, radiator fan, and air conditioner off. Idle speed should be 680±50 RPM (in Park or neutral)

8. Let the engine idle for 1 minute with high electric load (the A/C switch on, the temperature set to max cool, the blower fan on high, the rear window defogger on, and the headlights on high beam).

a. If the idle speed is not within specification, do the PCM idle learn procedure. See "Powertrain Control Module (PCM)" in "ENGINE PERFORMANCE & EMISSION CONTROLS" section.

b. If the idle speed is still not within specification, go to the symptom troubleshooting.

9. Reconnect the EVAP canister purge valve connector.

THROTTLE BODY

REMOVAL & INSTALLATION

See Figure 106.

⁂ WARNING

Do not insert your fingers into the installed throttle body when you turn the ignition switch to ON (II) or while the ignition switch is ON (II). If you do, you will seriously injure your fingers if the throttle valve is activated.

➡If you are replacing the throttle body, begin at step 1. If you are

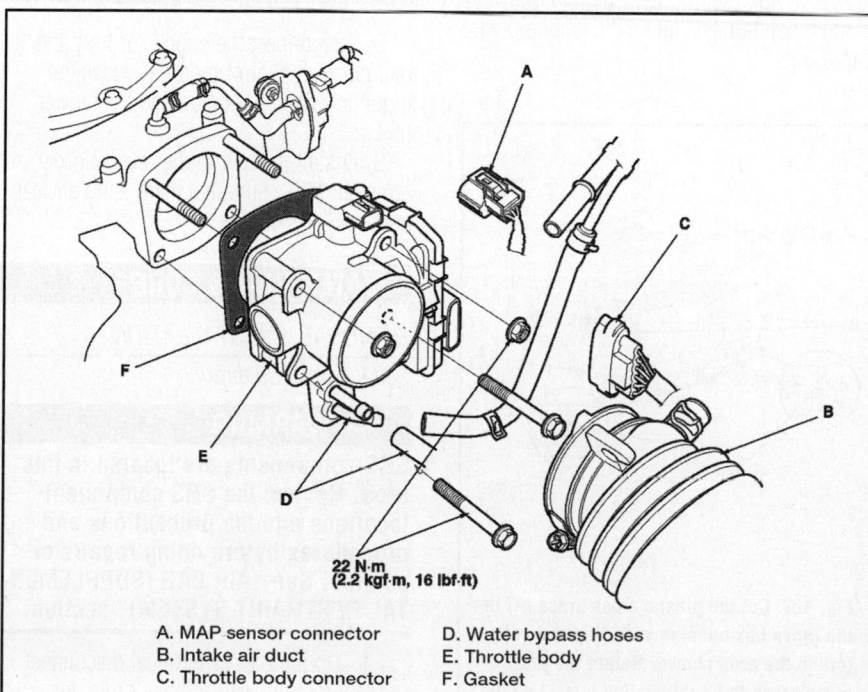

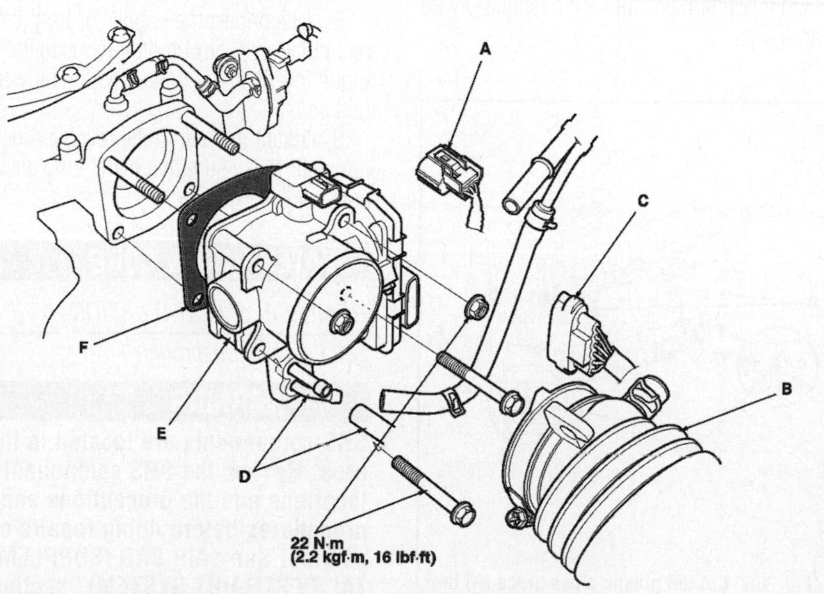
A
B
C
D
E
F
22 N·m
(2.2 kgf·m, 16 lbf·ft)

A. MAP sensor connector
B. Intake air duct
C. Throttle body connector
D. Water bypass hoses
E. Throttle body
F. Gasket

37647_ACRL_G0220

Fig. 106 Removing/installing the throttle body

removing the throttle body temporarily, begin at step 4.

1. Connect the scan tool while the engine is stopped.
2. Select the INSPECTION MENU with the scan tool.
3. Do the TP POSITION CHECK in the ETCS TEST.
4. Turn the ignition switch to ON (II).
5. Disconnect the MAP sensor connector (A).
6. Remove the intake air duct (B).
7. Disconnect the throttle body connector (C).
8. Disconnect and plug the water bypass hoses (D).
9. Remove the throttle body (E).
10. Install the parts in the reverse order of removal with a new gasket (F), then do this:
    a. Refill the radiator with engine coolant.
    b. Do the PCM idle learn procedure. See "Powertrain Control Module (PCM)" in this section.

HEATING & AIR CONDITIONING SYSTEM

BLOWER MOTOR

REMOVAL & INSTALLATION

See Figures 107 and 108.

1. Read the "Precautions" information in this section before beginning work.
2. Remove the glove box.
3. Remove the right kick panel.
4. Remove the USB adapter unit.
5. Cut the plastic cross brace in the glove box opening with diagonal cutters in the area shown. Retain the plastic cross brace to be reinstalled later.

➡Use the grommets and the self-tapping screws to reinstall the plastic cross brace.

6. Remove the wire harness clips and the connector clips and then remove the bolts and the glove box frame.
7. Disconnect the connectors from the blower motor, the throttle actuator control module subharness, and the dashboard wire harnesses.

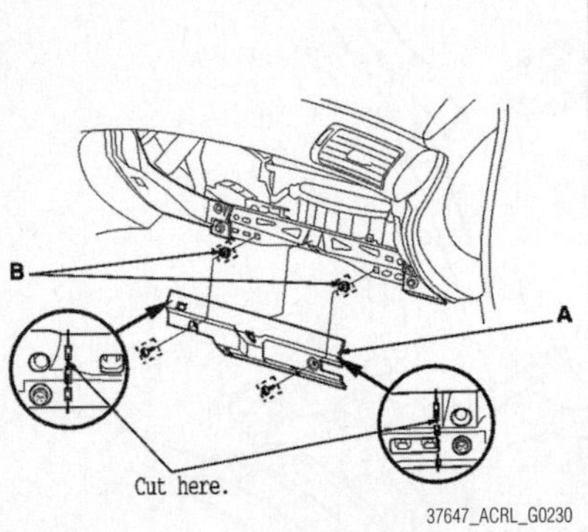
B
A
Cut here.
37647_ACRL_G0230

Fig. 107 Cut the plastic cross brace (A) in the glove box opening with diagonal cutters in the area shown. Retain the plastic cross brace to be reinstalled later. Use the grommets (B) and the self-tapping screws to reinstall the plastic cross brace.

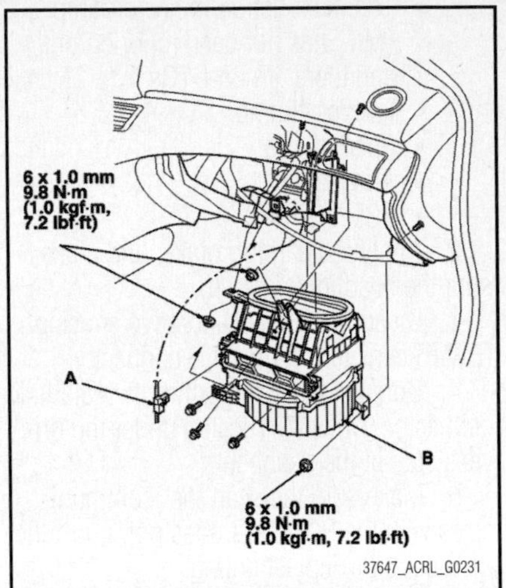
6 x 1.0 mm
9.8 N·m
(1.0 kgf·m,
7.2 lbf·ft)
A
B
6 x 1.0 mm
9.8 N·m
(1.0 kgf·m, 7.2 lbf·ft)
37647_ACRL_G0231

Fig. 108 Disconnect the connector (A) from the recirculation control motor. Remove the mounting nuts, the bolts, and the blower unit (B).

8. Disconnect the connector from the recirculation control motor. Remove the mounting nuts, the bolts, and the blower unit.
9. Install the unit in the reverse order of removal. Make sure that there is no air leakage.

HEATER UNIT & CORE

REMOVAL & INSTALLATION

See Figures 109 through 114.

⁑ WARNING

SRS components are located in this area. Review the SRS component locations and the precautions and procedures before doing repairs or service. See "AIR BAG (SUPPLEMENTAL RESTRAINT SYSTEM)" section.

1. Do the battery terminal disconnect procedure, then wait at least 3 minutes before beginning work. See "Battery System" in "ENGINE ELECTRICAL" section.

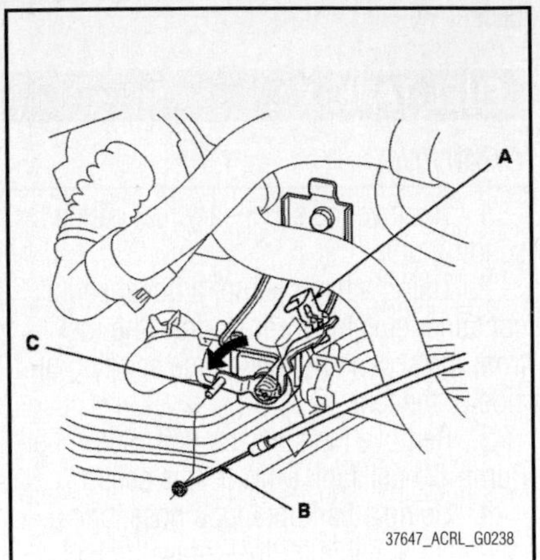
A
C
B
37647_ACRL_G0238

Fig. 109 From under the hood open the cable clamp (A), then disconnect the heater valve cable (B) from the heater valve arm (C). Turn the heater valve arm to the fully opened position as shown.

2. Disconnect the A/C lines from the evaporator core.
3. From under the hood open the cable clamp, then disconnect the heater valve cable from the heater valve arm. Turn the heater valve arm to the fully opened position as shown.
4. When the engine is cool, drain the engine coolant from the radiator.
5. Slide the hose clamps back and remove the bolt and the water valve bracket, then disconnect the inlet heater hose and the outlet heater hose from the heater unit at the firewall connections.

⁑ CAUTION

Engine coolant will run out when the hoses are disconnected; drain it into a clean drip pan. Be sure not to let coolant spill on the electrical parts or the painted surfaces. If any coolant spills, rinse it off immediately.

6. Remove the firewall mounting nut that retains the heater unit. Take care not to

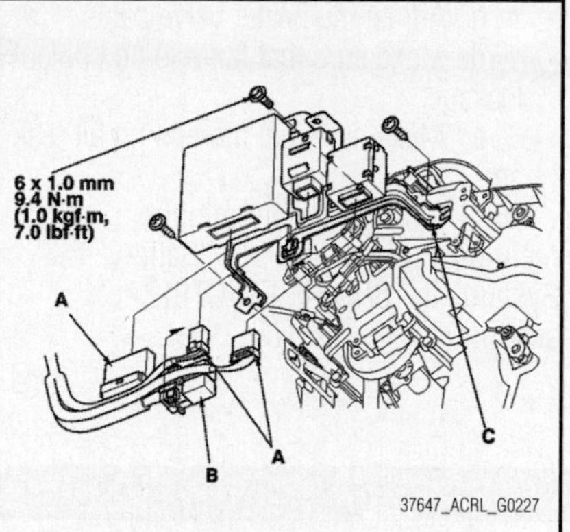
6 x 1.0 mm
9.4 N·m
(1.0 kgf·m,
7.0 lbf·ft)
A
B
A
C
37647_ACRL_G0227

Fig. 110 Disconnect the connectors (A) from the adaptive front lighting control unit, the electronically controlled power steering control unit, and the daytime running lights control unit. Remove the relay (B), the bolt, and the self-tapping screws, then remove the bracket (C).

damage or bend the fuel lines or the brake lines.

7. Remove the dashboard.

8. Disconnect the connectors from the adaptive front lighting control unit, the electronically controlled power steering control unit, and the daytime running lights control unit. Remove the relay, the bolt, and the self-tapping screws, then remove the bracket.

9. Disconnect the connectors from the driver's cool vent control motor, the driver's air mix control motor, the mode control motor, and the evaporator temperature

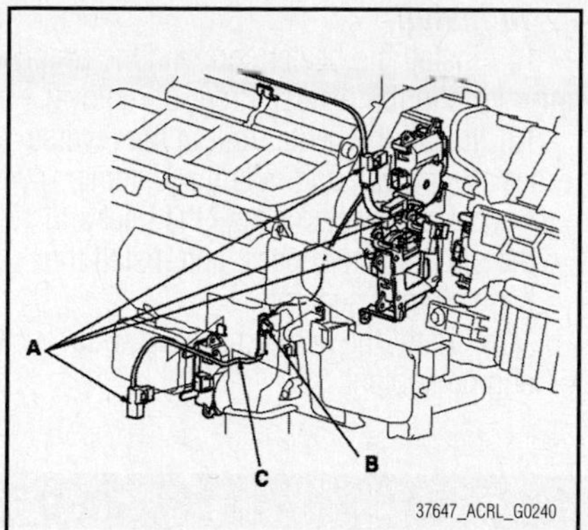
A
C
B
37647_ACRL_G0240

Fig. 112 Disconnect the connectors (A) from the recirculation control motor, the passenger's cool vent control motor, the passenger's air mix control motor, and the rear vent control motor, and then remove the wire harness clip (B) and the wire harness (C).

sensor, then remove the wire harness clips.

10. Disconnect the connectors from the blower motor, the power transistor, the con-

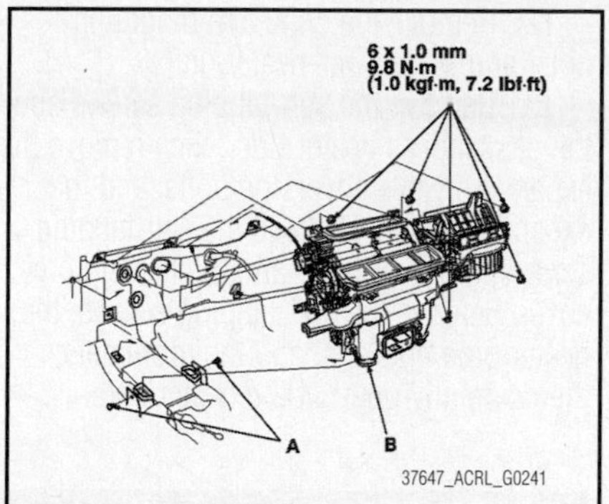
6 x 1.0 mm
9.8 N·m
(1.0 kgf·m, 7.2 lbf·ft)
A
B
37647_ACRL_G0241

Fig. 113 Remove the clips (A), the mounting nuts, and the blower-heater unit (B).

trol motor relay, the throttle actuator control module subharness, and the dashboard wire harnesses. Remove the self-tapping screw, the connector clip and the wire harness clips.

11. Disconnect the connectors from the recirculation control motor, the passenger's cool vent control motor, the passenger's air mix control motor, and the rear vent control motor, and then remove the wire harness clip and the wire harness.

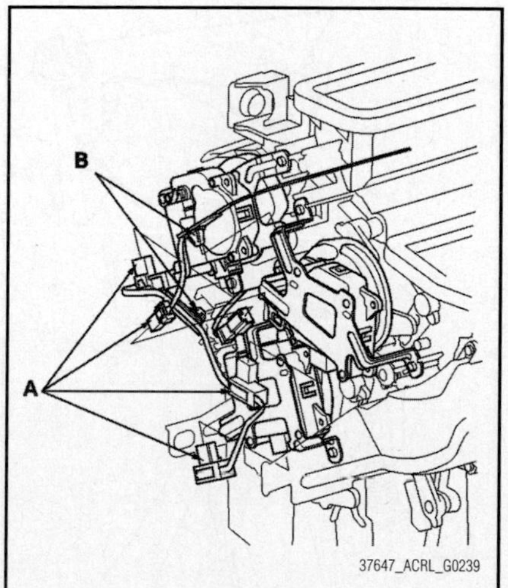
B
A
37647_ACRL_G0239

Fig. 111 Disconnect the connectors (A) from the driver's cool vent control motor, the driver's air mix control motor, the mode control motor, and the evaporator temperature sensor, then remove the wire harness clips (B).

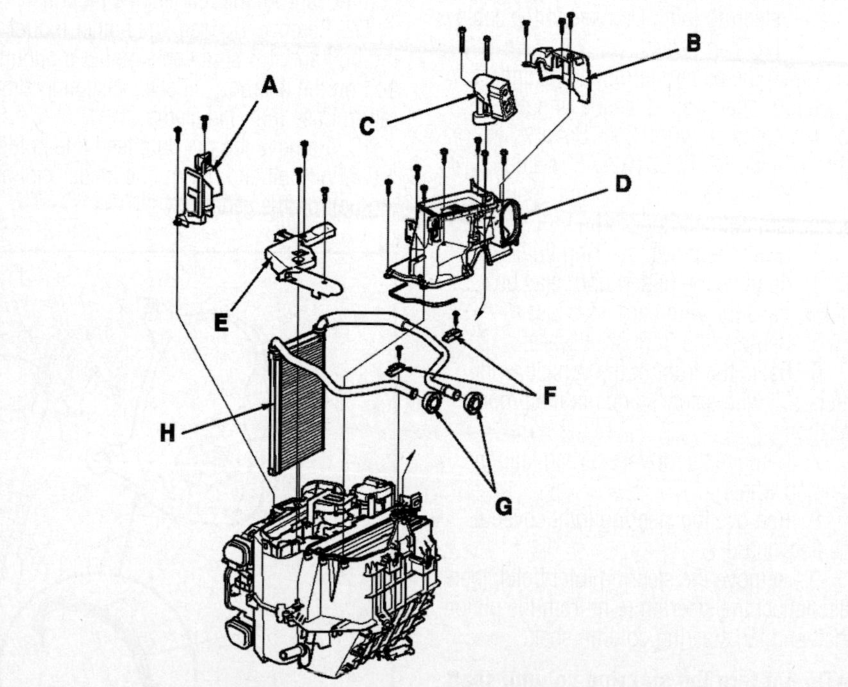
A
B
C
D
E
F
G
H
A. Heater air duct
B. Expansion valve cover
C. Expansion valve
D. Joint duct
E. Heater core cover
F. Heater pip brackets
G. Grommets
H. Heater core
37647_ACRL_G0242

Fig. 114 Remove the self-tapping screws and the passenger's heater duct (A), then remove the expansion valve cover (B), the bolts, and the expansion valve (C). Remove the self-tapping screws, the joint duct (D), and the heater core cover (E). Remove the self-tapping screws, the heater pipe brackets (F), and the grommets (G), then carefully pull out the heater core (H).

12. Remove the clips, the mounting nuts, and the blower-heater unit.

13. Remove the self-tapping screws and the passenger's heater duct, then remove the expansion valve cover, the bolts, and the expansion valve. Remove the self-tapping screws, the joint duct, and the heater core cover. Remove the self-tapping screws, the heater pipe brackets, and the grommets, then carefully pull out the heater core.

To install:

14. Install the heater core and the evaporator core in the reverse order of removal.

15. Install the heater unit in the reverse order of removal, and note these items:

a. Do not interchange the inlet and the outlet heater hoses, and install the hose clamps securely.

b. Refill the cooling system with engine coolant.

c. Adjust the heater valve cable.

d. Make sure that there is no coolant leakage.

e. Make sure that there is no air leakage.

16. Do the battery terminal reconnect procedure. See "Battery System" in "ENGINE ELECTRICAL" section.

STEERING

POWER RACK & PINION STEERING GEAR

REMOVAL & INSTALLATION

See Figures 115 through 121.

➡Note these items during removal:

- Using solvent and a brush, wash any oil and dirt off the valve body unit, its lines, and the end of the gearbox. Blow dry with compressed air.
- Make sure to remove the steering wheel before disconnecting the steering joint. Damage to the cable reel can occur.

1. Do the battery terminal disconnect procedure, then wait at least 3 minutes before beginning work. See "Battery System" in "ENGINE ELECTRICAL" section.

2. Remove the left and right upper fender trim and left upper trim.

3. Drain the power steering fluid.

4. Remove the underhood relay box, then release the wire harness clips.

5. Remove the front strut brace.

6. Raise the front of the vehicle, and support it with safety stands in the proper locations.

7. Remove the driver's airbag, and the steering wheel.

8. Remove the steering joint cover at the floorboard.

9. Remove the steering joint bolts, then disconnect the steering joint from the pinion shaft and the steering column shaft.

➡Do not turn the steering column shaft after disconnecting the steering joint. If the column shaft is turned, do the "VSA Sensor Neutral Position Memorization" after installing the steering gearbox.

10. Remove the steering shaft center guide. Remove steering joint cover. Be careful not to damage the mating surface on joint cover and the pinion shaft grommet. Replace the cover seal if necessary.

11. Remove the pinion shaft grommet from the top of the valve body unit. Apply vinyl tape to the splines on the pinion shaft.

12. Remove and discard the cotter pin from the tie-rod ball joint nut, and loosen the nut.

13. Separate the tie-rod ball joint from the steering knuckle.

14. Remove the nuts from the rear engine mount.

15. Remove the rear engine mount and the engine mount vacuum hose holder mounting bolt on the gearbox stiffener.

16. Disconnect the engine mount vacuum hose from the rear engine mount.

17. Remove the feed line holder mounting bolt and A/T shift cable holder mounting bolt on the gearbox stiffener. Disconnect the return hose from the clamp.

18. Remove the p/s fluid feed line holder mounting bolt and return line holder mounting bolt on the gearbox.

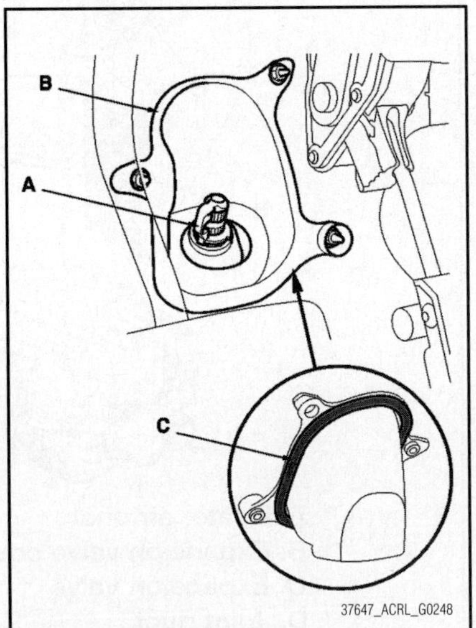
B
A
C
37647_ACRL_G0248

Fig. 115 Remove the center guide (A). Remove steering joint cover (B). Be careful not to damage the mating surface on joint cover and the pinion shaft grommet. Replace the cover seal (C) if necessary.

19. Disconnect the Electronically Controlled Power Steering (ECPS) valve motor connector.

20. Place several shop towels under the line connections, and cover the gearbox mounting cushion to protect it from the power steering fluid.

21. Loosen the adjustable hose clamp and disconnect the return hose from the steering gearbox.

22. Loosen the 14 mm flare nut and disconnect the feed line from the steering gearbox.

23. Remove the return line connector nipple from the steering gearbox.

24. Remove the P/S gearbox heat shield.

25. Remove the gearbox stiffener, the dynamic damper between the gearbox stiffener and the front suspension subframe.

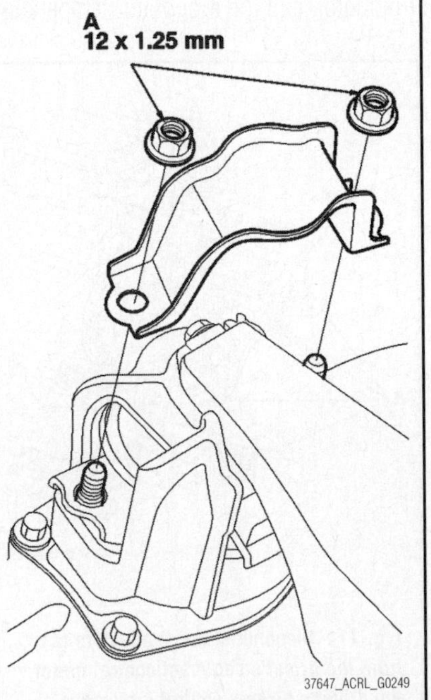
A
12 x 1.25 mm
37647_ACRL_G0249

Fig. 116 Remove the nuts (A) from the rear engine mount.

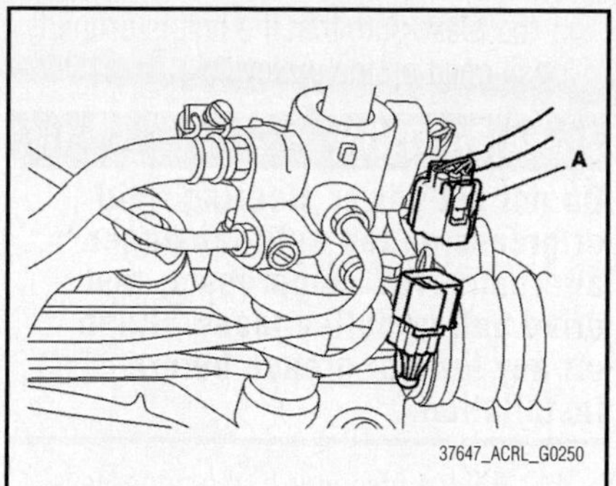
A
37647_ACRL_G0250

Fig. 117 Disconnect the ECPS valve motor connector (A).

26. Remove the steering gearbox mounting bolt and washer on the right gearbox mount. Repeat for the other side.

27. Remove the 4-way brake line joint bolts. Do not disconnect the lines. Then disconnect the brake pressure sensor connectors if equipped with Adaptive Cruise Control (ACC).

28. Move the steering gearbox to the passenger's side, and rotate it so the pinion shaft points toward the front of the vehicle.

29. Carefully move the steering gearbox as an assembly toward the left side of the vehicle until the pinion shaft clears the wheel well opening. Be careful not to damage the brake lines with the pinion shaft.

30. Lift the passenger's side of the gearbox, and remove the steering gearbox through the wheel well opening on the driver's side as shown by the arrow.

31. After removing the steering gearbox, make sure that no power steering fluid gets on the gearbox mount cushions, the gearbox housing, the surface of the front suspension the subframe, and the stiffener. Wipe off any spilled fluid at once.

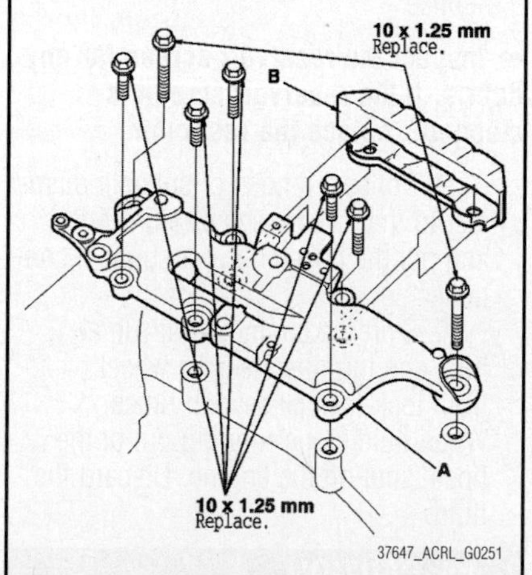
10 x 1.25 mm
Replace.
B
A
10 x 1.25 mm
Replace.
37647_ACRL_G0251

Fig. 118 Remove the gearbox stiffener (A) the dynamic damper (B) between the gearbox stiffener and the front suspension subframe.

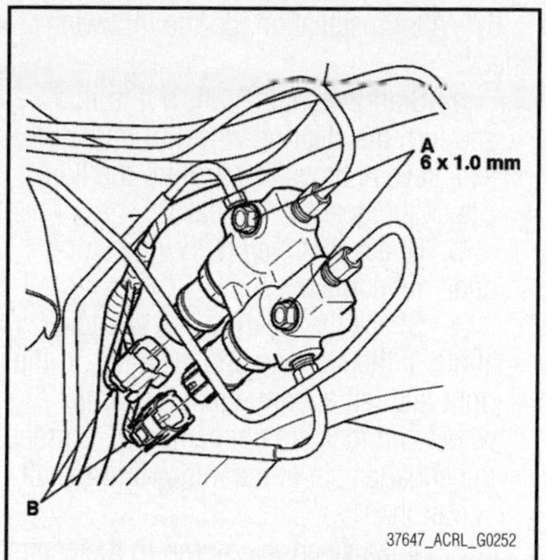
A
6 x 1.0 mm
B
37647_ACRL_G0252

Fig. 119 Remove the 4-way brake line joint bolts (A). Do not disconnect the lines. Then disconnect the brake pressure sensor connectors (B)—with ACC

To install:

➡Read information in "Power Steering Hoses, Lines & Switch" in this section during installation.

⁂ CAUTION

Do not turn the steering column shaft. If the column shaft is turned, do the "VSA Sensor Neutral Position Memorization" with the scan tool after installing the steering gearbox.

32. Before installing the steering gearbox, make sure that no power steering fluid is on the mating surface of the gearbox and front suspension subframe. To prevent the gearbox mounting bolts from loosening after the installation, remove any power steering fluid from the mount cushions and bolt holes.

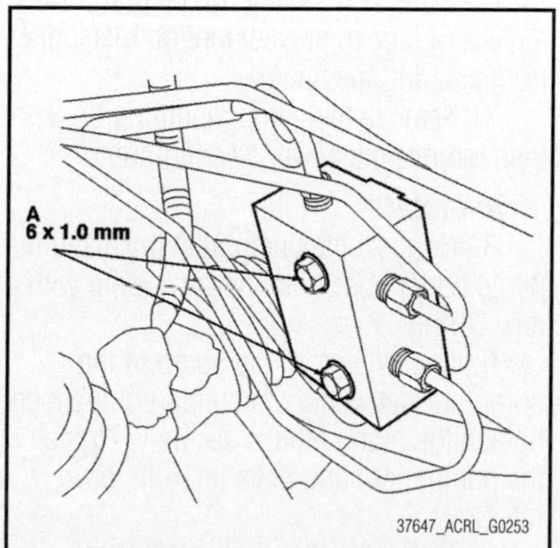
A
6 x 1.0 mm
37647_ACRL_G0253

Fig. 120 Remove the 4-way brake line joint bolts (A). Do not disconnect the lines—without ACC

33. Apply vinyl tape to the splines on the pinion shaft for protection.

34. Apply a mild soap and water solution to both sides of the mount cushion mating surfaces.

35. Lift and pass the passenger's side of the steering gearbox through the wheel well opening on the driver's side.

36. Carefully move the steering gearbox toward the passenger's side until the pinion shaft clears the wheel well opening on the body.

37. Rotate the steering gearbox so the pinion shaft points upward, then move the steering gearbox to the driver's side.

38. Install the 4-way brake line joint. If equipped with ACC, reconnect the brake pressure sensor connectors.

39. Install the new steering gearbox mounting bolt and washer on the left front gearbox mount, then tighten the steering gearbox mounting bolt to 58 ft. lbs. (78 Nm). Repeat on the right side.

40. Install the gearbox stiffener and the dynamic damper, then tighten the new mounting bolts to 25 ft. lbs. (34 Nm).

41. Install the new steering gearbox mounting bolts, then tighten the steering gearbox mounting bolts to 58 ft. lbs. (78 Nm).

42. Install the P/S heat shield.

43. Install the return line connector nipple in the steering gearbox, then tighten it to 21 ft. lbs. (28 Nm).

44. Connect the feed line and tighten the 14 mm flare nut to 27 ft. lbs. (37 Nm).

45. Connect the return hose securely, and tighten the adjustable hose clamp.

46. Connect the ECPS valve motor connector.

47. Install the feed line holder mounting bolt and return line holder mounting bolt on the gearbox.

48. Install the feed line holder mounting bolt and A/T shift cable holder mounting bolt on the gearbox stiffener.

49. Install the return hose in the clamp.

50. Connect the engine mount vacuum hose in the rear engine mount.

51. Install the rear engine mount, then tighten the new mounting bolts and the engine mount vacuum hose holder mounting bolt on the gearbox stiffener to 40 ft. lbs. (54 Nm).

52. Install the nuts on the rear engine mount, then tighten them to 47 ft. lbs. (64 Nm).

53. Wipe off any grease from the ball joint tapered section and threads, then reconnect the tie-rod end to the knuckle arms.

54. Install the tie-rod end ball joint nut, and tighten it to 43 ft. lbs. (59 Nm), then install the new cotter pin.

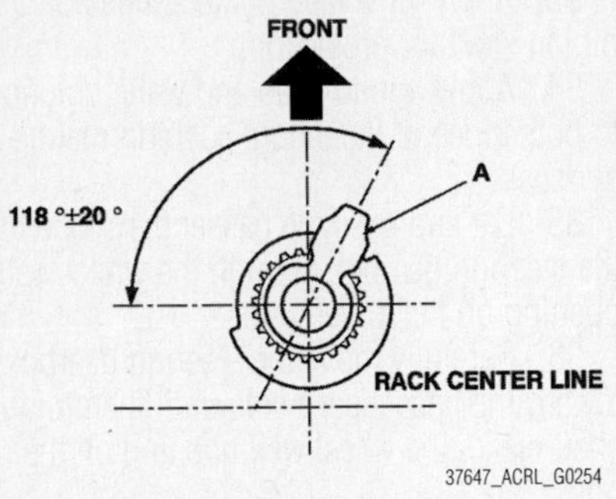
FRONT
A
118 °±20 °
RACK CENTER LINE
37647_ACRL_G0254

Fig. 121 Install, and position the center guide (A) as shown.

55. Install the front wheels, and set them in the straight ahead position.
56. Remove the vinyl tape from the pinion shaft, and install the pinion shaft grommet. Align the slot in the pinion shaft grommet with the lug portion on the valve housing.
57. Install the new cover seal all the way around in the steering joint cover. Make sure there are no wrinkles in the seal, then install the steering joint cover.
58. Install, and position the center guide as shown.
59. Install the steering joint, aligning the slit of the steering joint with the tab of the center guide.
60. Align the bolt hole on the steering joint with the groove around the pinion shaft, and loosely install the joint bolts. Be sure that the joint bolt is securely in the groove in the pinion shaft. Pull on the steering joint to make sure that the steering joint is fully seated. Tighten the steering joint bolts to 21 ft. lbs. (28 Nm).
61. Install the steering joint cover at the floorboard.
62. Install the relay box, then connect the wire harness clips.
63. Install the front strut brace.
64. Install the steering wheel and the driver's airbag.
65. Do the battery terminal reconnect procedure. See "Battery System" in "ENGINE ELECTRICAL" section.
66. Do these tasks:
  a. Turn the ignition switch to ON (II) and check that the SRS indicator comes on for about 6 seconds and then goes off.
  b. Make sure the horn and the turn signal switches work properly.
  c. Make sure the steering wheel switches work properly.
  d. Fill the system with power steering fluid, and bleed air from the system.
67. After installation, do the following checks.
  a. Start the engine, allow it to idle, and turn the steering wheel from lock-to-lock several times to warm up the fluid. Check the gearbox for leaks.
  b. Check that the ECPS indicator does not come on.
  c. Check the steering wheel spoke angle. If the steering spoke angles to the right and left are not equal (steering wheel and rack are not centered), correct the engagement of the joint/pinion shaft serrations.
68. Set the steering column to the center tilt position, then do the front toe inspection.
69. Install the left and right upper fender trim and left upper trim.

POWER STEERING PUMP

REMOVAL & INSTALLATION

1. Place a suitable container under the vehicle.
2. Remove the right upper fender trim.
3. Drain the power steering fluid from the reservoir.
4. Remove the drive belt from the pump pulley.
5. Cover the auto-tensioner, the alternator, and the A/C compressor with several shop towels to protect them from spilled power steering fluid. Disconnect the pump inlet hose and the pump outlet hose from the pump, and plug them.

⁂ CAUTION

Take care not to spill the fluid on the body or parts. Wipe off any spilled fluid at once. Do not turn the steering wheel with the pump removed.

6. Cover the opening of the pump with a piece of tape to prevent foreign material from entering the pump.
7. Remove the pump mounting bolts, then remove the power steering pump.

To install:

8. Connect the pump inlet hose and the pump outlet hose onto the new pump with a new O-ring.
9. Loosely install the pump in the pump bracket with the mounting bolts, then tighten the pump fittings securely. Tighten the pump mounting bolts to 16 ft. lbs. (22 Nm).
10. Install the drive belt. Note these items during belt installation:
  a. Inspect the belt for wear and cracks. Replace the belt if necessary.
  b. Make sure that the belt is properly positioned on the pulleys.

⁂ CAUTION

Do not get power steering fluid or grease on the auto-tensioner, alternator, A/C compressor, and drive belt or pulley faces. Clean off any fluid or grease before installation.

11. Fill the reservoir to the upper level line.
12. Start the engine, and check for fluid leaks.
13. Install the right upper fender trim.

BLEEDING

➡See "Fluid Fill Procedure" in this section.

FLUID FILL PROCEDURE

1. Remove the right upper fender trim.
2. Check the reservoir at regular intervals, and add the recommended fluid as necessary to fill the reservoir to the upper level line.

➡Manufacturers recommends to always use Acura-brand Power Steering Fluid. Using any other type of power steering fluid or automatic transmission fluid can cause increased wear, and poor steering in cold weather.

3. If drain and refill is required, proceed as follows:
  a. Raise the reservoir, then disconnect the return hose to drain the reservoir. Take care not to spill the fluid on the body and parts. Wipe off any spilled fluid at once.

➡ Inspect the reservoir screen for any debris. If the reservoir screen is clogged, replace the reservoir.

  b. Connect a hose of suitable diameter to the disconnected return hose, and put the hose end in a suitable container.
  c. Start the engine, let it run at idle, and turn the steering wheel from lock-to-lock several times. When fluid stops running out of the hose, shut off the engine. Discard the fluid.

⁂ CAUTION

Stop the motor immediately once the fluid stops running out of the hose to prevent pump damage.

d. Reinstall the return hose on the reservoir.

e. Fill the reservoir to the upper level line.

f. Start the engine and run it at fast idle, then turn the steering from lock-to-lock several times to bleed air from the system.

g. Recheck the fluid level and add some if necessary. Do not fill the reservoir beyond the upper level line.

h. If the fluid is contaminated, dark, or discolored, repeat the procedure as necessary until the system is clean.

4. Install the right upper fender trim.

SUSPENSION

FRONT SUSPENSION

LOWER BALL JOINT

REMOVAL & INSTALLATION

➡Always use a ball joint remover to disconnect a ball joint. Do not strike the housing or any other part of the ball joint connection to disconnect it.

1. Install a hex nut or the ball joint thread protector onto the threads of the ball joint.

➡Using a hex nut, make sure the nut is flush with the ball joint pin end to prevent damage to the threaded end of the ball joint pin.

2. Apply grease to the ball joint remover on both ends. This will ease the installation of the tool, and prevent damage to the pressure bolt threads.
3. Install the ball joint remover. Insert the jaws carefully, making sure not to damage the ball joint boot.
4. Adjust the jaw spacing as required.

➡Fasten a safety chain securely to a suspension arm or the subframe. Do not fasten it to a brake line or wire harness.

5. Remove the ball joint.
6. Remove the tool, then remove the nut from the end of the ball joint pin, and pull the ball joint out of the ball joint connecting hole. Inspect the ball joint boot, and replace it if damaged.
7. Installation is the reverse of the removal procedure.

LOWER CONTROL ARM

REMOVAL & INSTALLATION

1. Raise and support the vehicle.
2. Remove the wheel nuts, and the front wheels taking care not to scratch the caliper.
3. Remove the bolt and the nut and remove the holder from the steering knuckle.
4. Remove the cotter pin from the lower ball joint castle nut, then remove the nut.

⁂ CAUTION

To avoid damaging the ball joint, install a hex nut on the threads of the ball joint. Be careful not to damage the ball joint boot when installing the remover. Do not force or hammer on the lower arm, or pry between the lower arm and knuckle. You could damage the ball joint.

5. Disconnect the lower arm ball joint from the holder using the ball joint thread protector and the ball joint remover.
6. Remove the self-locking nut and the washer, then remove the damper fork mounting nut and the bolt.
7. Remove the flange bolts and the self-locking nut, then remove the lower control arm.

To install:

8. In case of a loose lower arm bracket, follow the steps to retighten its bolts:
   a. Lightly tighten the three new bolts, aligning the center of the cam to the edge.
   b. Tighten the adjusting bolt to 36 ft. lbs. (49 Nm).
   c. Tighten the lower control arm bracket mounting bolt to 36 ft. lbs. (49 Nm).
   d. Tighten the lower control arm bracket mounting bolt to 36 ft. lbs. (49 Nm).
9. Install the lower arm in the reverse order of removal, and note these items:
   a. First install all the components, and lightly tighten the bolts and the nuts, then raise the suspension to load it with the vehicle's weight before fully tightening it to the specified torque. Do not place the jack against the ball joint pin of the knuckle.
   b. Be careful not to damage the ball joint boot when installing the knuckle.
   c. Before connecting the lower ball joint to the holder, degrease the threaded section and the tapered portion of the ball joint pin, the holder connecting hole, the threaded section and the mating surface of the castle nut.
   d. Torque the castle nut to the lower torque specification, then tighten it only far enough to align the slot with the ball joint pin hole. Do not align the castle nut by loosening it.
   e. Install a new cotter pin on the castle nut after torquing.
   f. Before installing the wheel, clean the mating surfaces of the brake disc and the inside of the wheel.
   g. Note the tightening specifications:
      - Vertical flange bolt: 69 ft. lbs. (93 Nm)
      - Horizontal flange bolt: 65 ft. lbs. (88 Nm)
      - Damper fork bolt: 47 ft. lbs. (64 Nm)
      - Stabilizer link locking nut: 40 ft. lbs. (54 Nm)
10. Check the wheel alignment, and adjust it if necessary.

STABILIZER BAR

REMOVAL & INSTALLATION

1. Raise and support the vehicle.
2. Remove the wheel nuts, and the front wheels, taking care not to scratch the caliper.
3. Disconnect the stabilizer links from the stabilizer bar on the right and left sides.
4. Remove the front suspension subframe from the body.
5. Remove the flange bolts and the stabilizer bar bushing holders, then remove the bushings and the stabilizer bar.

To install:

6. Install the stabilizer bar in the reverse order of removal, and note these items:
   a. Note the right and left direction of the stabilizer bar.
   b. Align the ends of the paint marks on the stabilizer bar with each end of the bushings.
   c. Note the fore/aft direction of the bushing holders.
   d. See "Stabilizer Link" to connect the stabilizer bar to the links.
7. Clean the mating surface of the brake disc and the inside of wheel, then install the front wheel.
8. Check the wheel alignment, and adjust it if necessary.

STABILIZER LINK

REMOVAL & INSTALLATION

See Figures 122 and 123.

1. Raise and support the vehicle.
2. Remove the wheel nuts, and the front wheels, taking care not to scratch the caliper.
3. Remove the self-locking nuts while holding the respective joint pin with a hex wrench and remove the washer and the stabilizer link.
4. Place the floor jack under the lower arm, and raise the suspension.

➡Do not place the jack against the knuckle.

To install:

5. Install the stabilizer link on the stabilizer bar and the lower arm with the joint pins set at the center of their range of movement. The paint mark on the stabilizer link is near the upper end of the stabilizer link.
6. Install the washer and the new self-locking nuts and lightly tighten them.
7. Tighten the new self-locking nuts to the specified torque while holding the respective joint pin with a hex wrench:
   - Upper nut: 32 ft. lbs. (43 Nm)
   - Lower nut: 40 ft. lbs. (54 Nm)
8. Clean the mating surface of the brake disc and the inside of the wheel, then install the front wheel.
9. Test-drive the vehicle.
10. After 5 minutes of driving, tighten the self-locking nuts again to the specified torque.

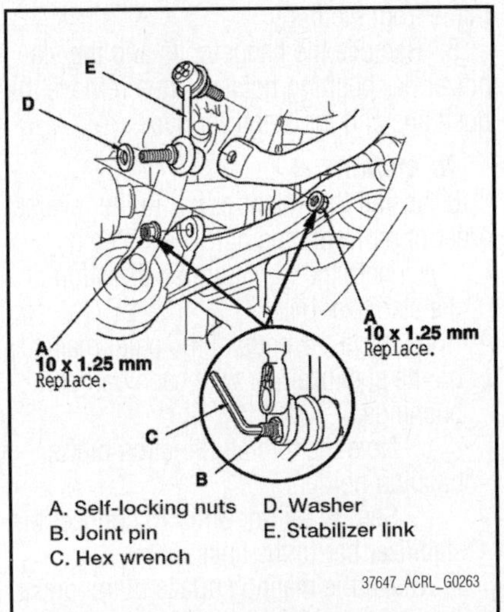
E
D
A
10 x 1.25 mm
Replace.
A
10 x 1.25 mm
Replace.
C
B
A. Self-locking nuts
B. Joint pin
C. Hex wrench
D. Washer
E. Stabilizer link
37647_ACRL_G0263

Fig. 122 Remove the self-locking nuts (A) while holding the respective joint pin (B) with a hex wrench (C), and remove the washer (D) and the stabilizer link (E).

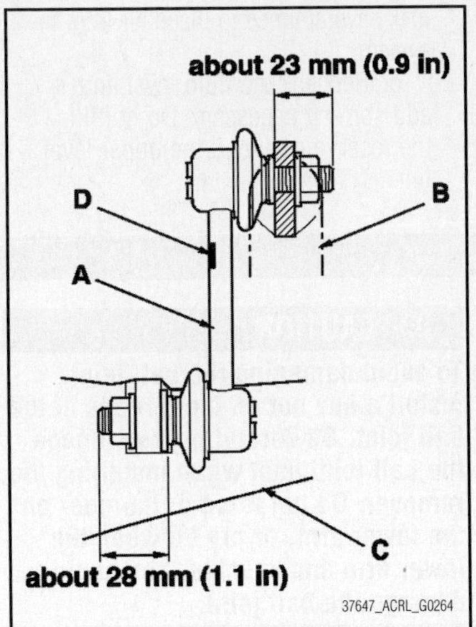
about 23 mm (0.9 in)
D
B
A
C
about 28 mm (1.1 in)
37647_ACRL_G0264

Fig. 123 Install the stabilizer link (A) on the stabilizer bar (B) and the lower arm (C) with the joint pins set at the center of their range of movement. The paint mark (D) on the stabilizer link is near the upper end of the stabilizer link.

STRUT (FRONT DAMPER & SPRING)

REMOVAL & INSTALLATION

See Figures 124 and 125.

1. Remove the upper fender trims and left upper trim.
2. Raise and support the vehicle.
3. Remove the front wheel, taking care not to scratch the caliper.
4. Remove the damper fork from the damper and the lower arm while pushing the lower arm down.
5. Disconnect the stabilizer link from the lower arm.
6. Remove the 8 mm flange nuts and the 10 mm flange nuts from the top of the damper.
7. Before removing the damper assembly, make a space to remove the damper assembly by rotating the steering wheel clockwise until it stops, then remove the front damper and spring assembly.

⁂ CAUTION

Be careful not to damage or scratch the aluminum parts. It may weaken the suspension.

To install:

8. Position the damper assembly in the body, with the aligning tab at the lower end of the strut facing inside.

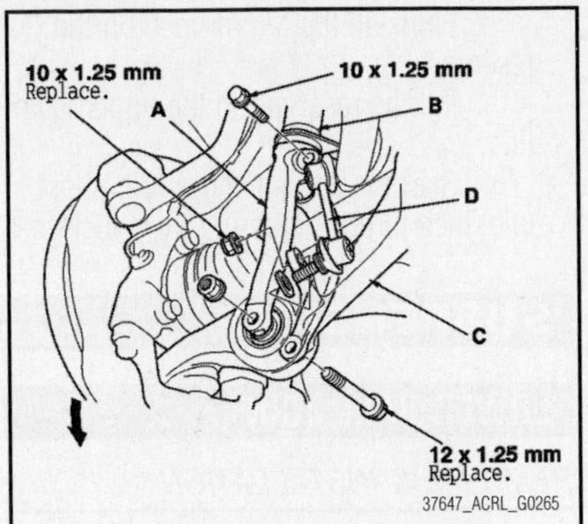
10 x 1.25 mm
Replace.
A
10 x 1.25 mm
B
D
C
12 x 1.25 mm
Replace.
37647_ACRL_G0265

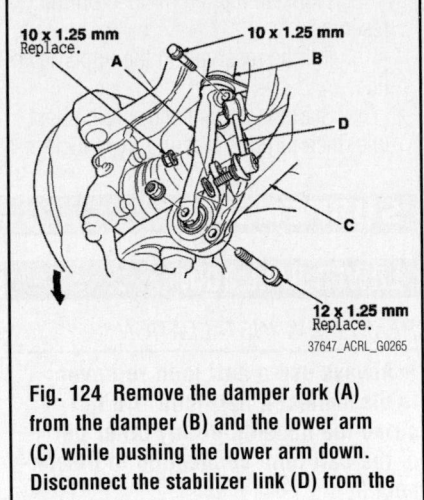

Fig. 124 Remove the damper fork (A) from the damper (B) and the lower arm (C) while pushing the lower arm down. Disconnect the stabilizer link (D) from the lower arm.

9. Loosely install the 8 mm flange nuts and the 10 mm new flange nuts on the top of the damper.
10. Install the damper fork over the driveshaft and onto the lower arm while pushing the lower arm down. Install the front damper in the damper fork so that the aligning tab is aligned with the slot in the damper fork.
11. Loosely install the damper pinch bolt into the damper fork.
12. Install the new flange bolt to the damper fork and lower arm, then lightly tighten the new damper fork mounting nut.
13. Install the stabilizer link on the lower arm with the washer and the new self-locking nut, and lightly tighten them.
14. Place the floor jack under the lower arm, and raise the suspension to load it with the vehicle's weight.
15. Tighten the flange nuts on the top of the damper to the specified torque.
16. Tighten the damper pinch bolt, the flange nut on the damper fork, and the self-locking nut to the specified torque.
17. Tighten torques are as follows:
    - Damper pinch bolt: 32 ft. lbs. (43 Nm)
    - Lower fork nut: 47 ft. lbs. (64 Nm)
    - Lower link nut: 40 ft. lbs. (54 Nm)
    - 8 mm upper strut nuts: 25 ft. lbs. (34 Nm)
    - 10 mm upper strut nuts: 39 ft. lbs. (55 Nm)
18. Clean the mating surfaces of the brake disc and the inside of the wheel, then install the front wheel.
19. Install the upper fender trims.

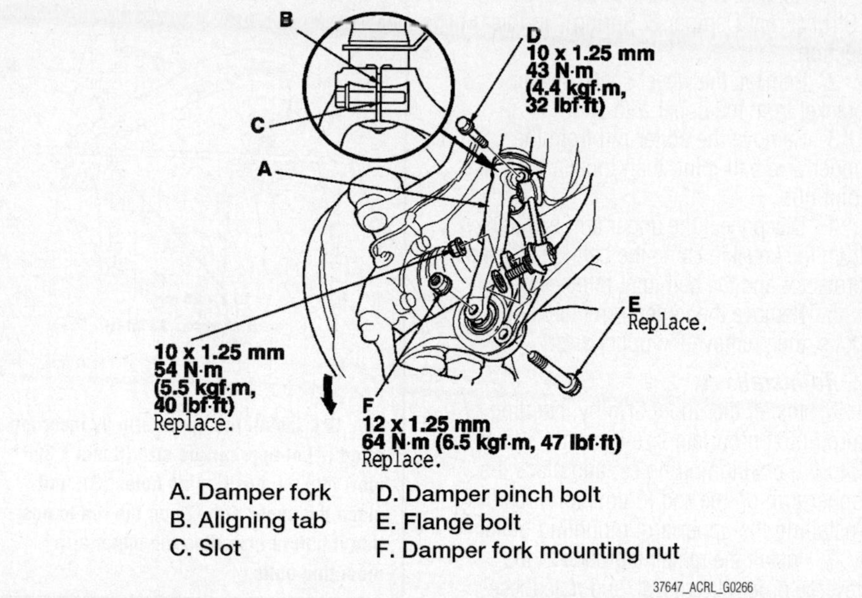
B
C
A
D
10 x 1.25 mm
43 N·m
(4.4 kgf·m,
32 lbf·ft)
E
Replace.
10 x 1.25 mm
54 N·m
(5.5 kgf·m,
40 lbf·ft)
Replace.
F
12 x 1.25 mm
64 N·m (6.5 kgf·m, 47 lbf·ft)
Replace.
A. Damper fork
B. Aligning tab
C. Slot
D. Damper pinch bolt
E. Flange bolt
F. Damper fork mounting nut
37647_ACRL_G0266

Fig. 125 Install the damper fork over the driveshaft and onto the lower arm while pushing the lower arm down. Install the front damper in the damper fork so that the aligning tab is aligned with the slot in the damper fork. Loosely install the damper pinch bolt into the damper fork. Install the new flange bolt to the damper fork and lower arm, then lightly tighten the new damper fork mounting nut.

Reassembly

See Figures 127 and 128.

1. Align the damper mounting base and spring mounting cushion, then reassemble the damper except for the washer and self-locking nut.
2. Install the damper assembly on the strut spring compressor and compress the spring lightly.
3. Align the bottom of the spring and the stepped part of the lower spring seat.
4. Position the damper mounting base so the stud bolt is aligned with the aligning tab in the damper unit.
5. Compress the damper spring. Do not compress the spring excessively.
6. Install the washer and a new 10 mm self-locking nut. Hold the damper shaft with a hex wrench and tighten the 10 mm self-locking nut to the specified torque of 22 ft. lbs. (29 Nm).
7. Remove the damper assembly from the strut spring compressor.

OVERHAUL

Disassembly & Inspection

See Figure 126.

➡When compressing the damper spring, use a commercially available strut spring compressor (Branick MST-580A or Model 7200, or equivalent).

1. Compress the damper spring, then remove the self-locking nut while holding the damper shaft with a hex wrench. Do not compress the damper spring more than is necessary to remove the self-locking nut.
2. Release the pressure from the strut spring compressor, then disassemble the damper as shown.
3. Reassemble all the parts, except for the spring mounting cushion and spring.
4. Compress the damper assembly by hand, and check for smooth operation through a full stroke, both compression and extension. The damper should extend smoothly and constantly when compression is released. If it does not, the gas is leaking and the damper should be replaced.
5. Check for oil leaks, abnormal noises, and binding during these tests.

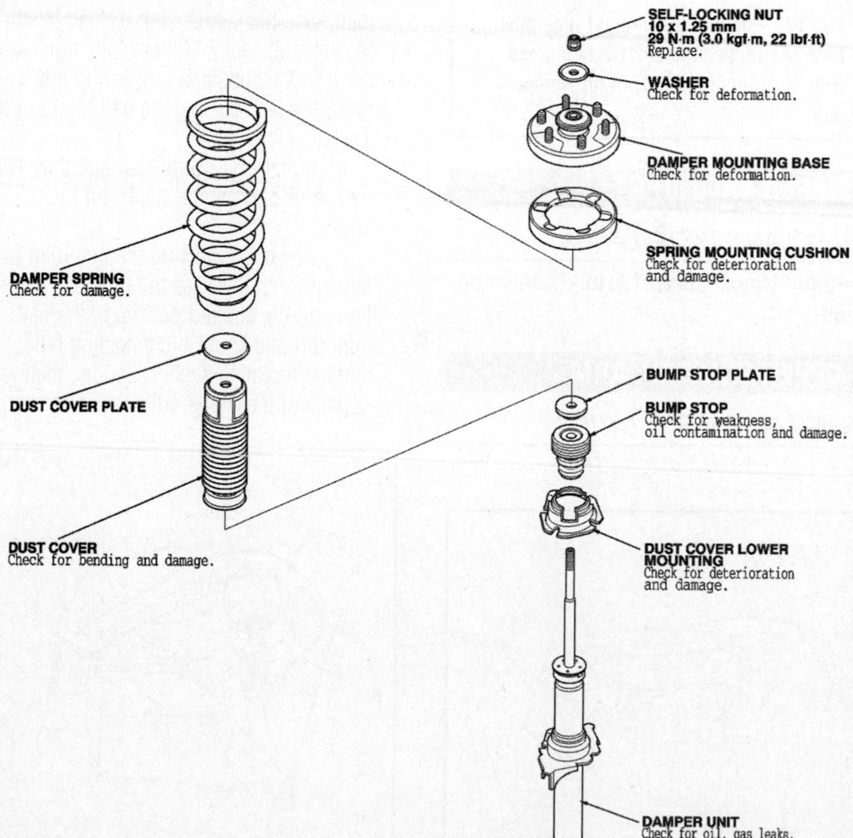
SELF-LOCKING NUT
10 x 1.25 mm
29 N·m (3.0 kgf·m, 22 lbf·ft)
Replace.
WASHER
Check for deformation.
DAMPER MOUNTING BASE
Check for deformation.
SPRING MOUNTING CUSHION
Check for deterioration and damage.
DAMPER SPRING
Check for damage.
DUST COVER PLATE
BUMP STOP PLATE
BUMP STOP
Check for weakness, oil contamination and damage.
DUST COVER
Check for bending and damage.
DUST COVER LOWER MOUNTING
Check for deterioration and damage.
DAMPER UNIT
Check for oil, gas leaks, and smooth operation.
37647_ACRL_G0267

Fig. 126 Exploded view of the damper (strut) and spring assembly

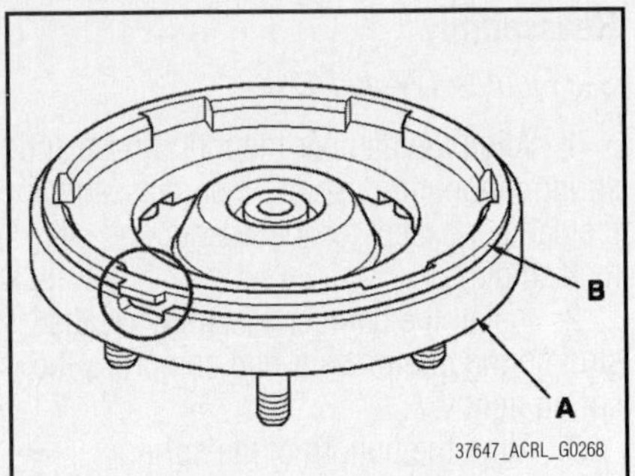
B
A
37647_ACRL_G0268

Fig. 127 Align the damper mounting base (A) and spring mounting cushion (B) as shown, then reassemble the damper except for the washer and self-locking nut.

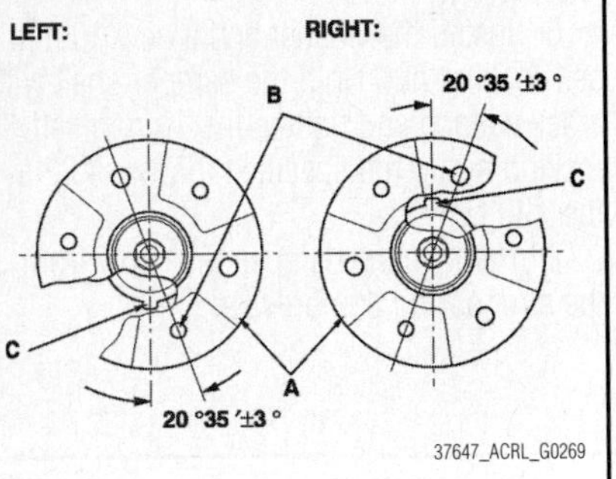
LEFT:
RIGHT:
B
20 °35 ′±3 °
C
C
A
20 °35 ′±3 °
37647_ACRL_G0269

Fig. 128 Position the damper mounting base (A) so the stud bolt (B) is aligned with the aligning tab (C) in the damper unit.

UPPER BALL JOINT

REMOVAL & INSTALLATION

➡See "Upper Control Arm" in this section.

UPPER CONTROL ARM

REMOVAL & INSTALLATION

See Figures 129 through 131.

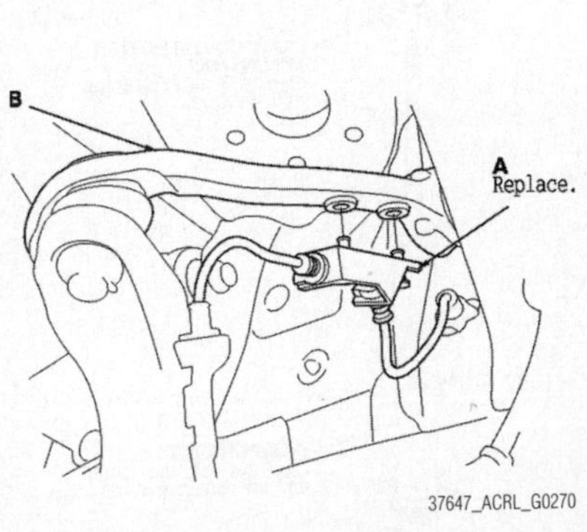
B
A
Replace.
37647_ACRL_G0270

Fig. 129 Remove the wheel speed sensor bracket (A) from the upper arm (B).

1. Remove the front damper. See "Strut (Front Damper & Spring)" in this section.
2. Remove the wheel speed sensor bracket from the upper arm.
3. Remove the cotter pin from the upper arm ball joint, then loosen the ball joint nut.
4. Disconnect the upper arm ball joint from the knuckle using the ball joint thread protector and the ball joint remover.
5. Remove the upper arm mounting bolts, then remove the upper arm.

To install:

6. Install the upper arm by inserting a rod of appropriate size (6 mm x 300 mm) into the positioning holes, and place the upper arm on the rod to position it before installing the upper arm mounting bolts.
7. Install the remaining parts in the reverse order of removal, and note these items:
   a. Torque specifications are:
      - Upper arm retaining bolts: 23 ft. lbs. (31 Nm)
      - Upper ball joint castle nut (new): 43–51 ft. lbs. (59–69 Nm)

   b. First install all the components, and lightly tighten the bolts and the nuts, then raise the suspension to load it with the vehicle's weight before fully tightening it to the specified torque. Do not place the jack against the ball joint pin of the knuckle.
   c. Be careful not to damage the ball joint boot when installing the knuckle.
   d. Before connecting the ball joint to the knuckle, degrease the threaded section and the tapered portion of the ball joint pin, the knuckle connecting hole, and the threaded section and the mating surface of the castle nut.

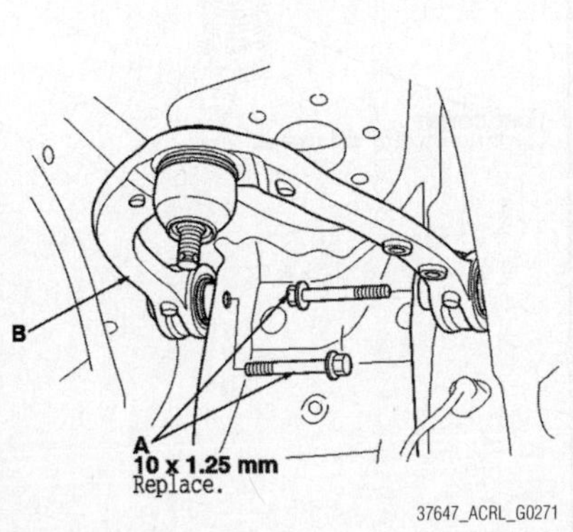
B
A
10 x 1.25 mm
Replace.
37647_ACRL_G0271

Fig. 130 Remove the upper arm mounting bolts (A), then remove the upper arm (B).

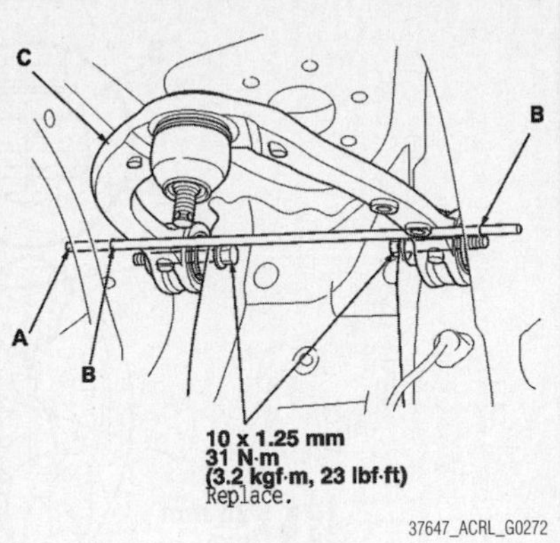
C
B
A
B
10 x 1.25 mm
31 N·m
(3.2 kgf·m, 23 lbf·ft)
Replace.
37647_ACRL_G0272

Fig. 131 Install the upper arm by inserting a rod (A) of appropriate size (6 mm x 300 mm) into the positioning holes (B), and place the upper arm (C) on the rod to position it before installing the upper arm mounting bolts.

   e. Torque the castle nut to the lower torque specification, then tighten it only far enough to align the slot with the ball joint pin hole. Do not align the castle nut by loosening it.

➡Use a new wheel speed sensor bracket on reassembly.

   f. Before installing the wheel, clean the mating surfaces on the brake disc and the inside of the wheel.
   g. Check the front wheel alignment, and adjust it if necessary.

WHEEL HUB & BEARING

REMOVAL & INSTALLATION

➡Manufacturer does not provide a specific removal and installation procedure for this component.

INSPECTION

1. Raise and support the vehicle.
2. Remove the wheel nuts, and the wheels, taking care not to scratch the caliper.
3. Install suitable flat washers (A) and the wheel nuts. Tighten the nuts to the specified torque to hold the brake disc securely against the hub.
4. Attach the dial gauge. Place the dial gauge against the hub flange or hub cap.
5. Measure the bearing end play while moving the brake disc or the brake disc/drum inward and outward.
6. Bearing end play should not exceed 0.00–0.002 in. (0.00–0.05 mm).
7. If the bearing end play measurement is more than the standard, replace the wheel bearing or the hub bearing unit.

COIL SPRING

REMOVAL & INSTALLATION

1. Raise and support the vehicle.
2. Remove the wheel nuts, and the rear wheel.
3. Remove the self-locking nut and the washer while holding the joint pin with a hex wrench, and disconnect the stabilizer link from the lower arm B.
4. Place a floor jack at the connecting point of the lower arm B and the stabilizer link.
5. Remove the cotter pin from the lower arm B ball joint, and loosen the castle nut.
6. Disconnect the lower arm ball joint from the knuckle using the ball joint thread protector and the ball joint remover.
7. Lower the floor jack gradually.
8. Remove the spring, the spring mounting cushion, and the lower spring seat.

To install:

9. Install the spring mounting cushion, the spring mount cushion, then align the bottom of the spring, the stepped part of the lower spring seat and lower arm B.
10. Place the floor jack at the connecting point of the lower arm and the stabilizer link.
11. Align the bottom of the spring and the stepped part of the lower spring seat, and the lower arm B.
12. Place the floor jack at the connecting point of lower arm and the stabilizer link.
13. Raise the jack slowly until you can align the bolt hole of the lower arm B and the knuckle ball joint pin, then loosely install a new castle nut.
14. Install the stabilizer link on the lower arm B with the washer and a new self-locking nut, and lightly tighten them.
15. Raise the rear suspension with a floor jack to load it with the vehicle's weight.
16. Tighten the castle nut and self-locking nut to the specified torque of 54–61 ft. lbs. (74–83 Nm).

➡Torque the castle nut to the lower torque specification, then tighten it only far enough to align the slot with the ball joint pin hole. Do not align the castle nut by loosening it.

17. Insert a new cotter pin into the ball joint pin from the rear to the front of the vehicle, and bend its end.
18. Clean the mating surface of the brake disc/drum and the inside of the wheel, then install the rear wheel.
19. Check the rear wheel alignment, and adjust it if necessary.

CONTROL ARMS

REMOVAL & INSTALLATION

Rear Control Arm

See Figure 132.

1. Raise and support the vehicle.
2. Remove the wheel nuts, and the rear wheel.
3. Remove the control arm mounting nut and the washer from the knuckle side.
4. Mark the cam positions of the adjusting bolt and the adjusting cam, then remove the self-locking nut, the adjusting cam, and the adjusting bolt.
5. Remove the control arm.

To install:

6. Install the control arm in the reverse order of removal, and note these items:

➡Use the figure to identify nuts when tightening to specification.

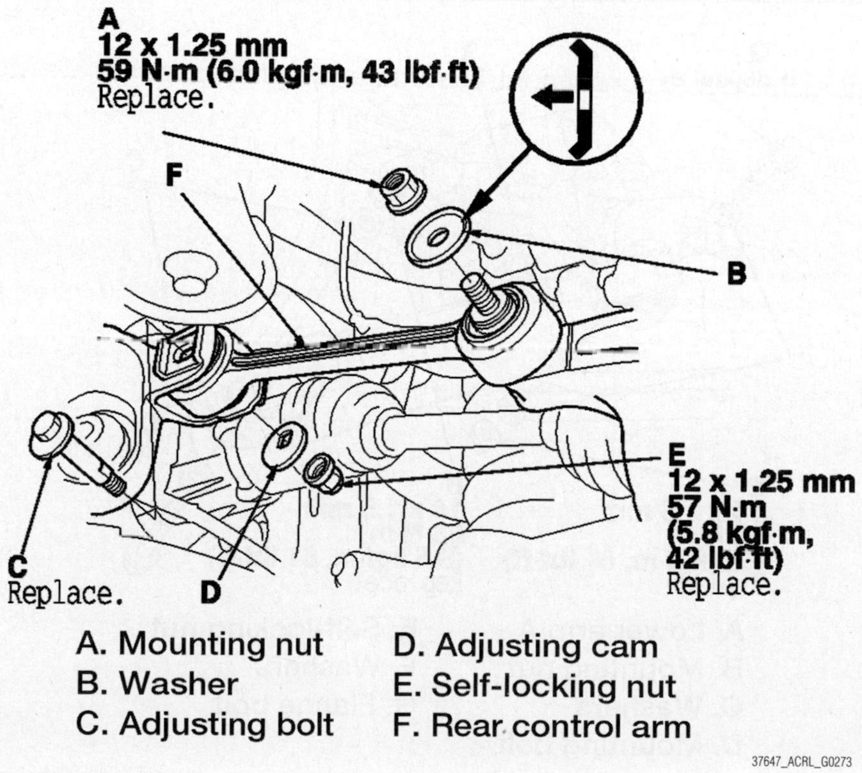
A
12 x 1.25 mm
59 N·m (6.0 kgf·m, 43 lbf·ft)
Replace.
F
B
E
12 x 1.25 mm
57 N·m
(5.8 kgf·m,
42 lbf·ft)
Replace.
C
Replace.
D
A. Mounting nut
B. Washer
C. Adjusting bolt
D. Adjusting cam
E. Self-locking nut
F. Rear control arm
37647_ACRL_G0273

Fig. 132 Removing/installing the rear control arm

a. First install all of the components, and lightly tighten the bolts and nuts, then raise the suspension to load it with the vehicle's weight before fully tightening to the specified torque.

b. Align the cam positions of the adjusting bolt and the adjusting cam with the marked positions when tightening.

c. Tighten all mounting hardware to the specified torque.

d. Before installing the wheel, clean the mating surfaces on the brake disc/drum and inside of the wheel.

7. Check the wheel alignment, and adjust it if necessary.

Lower Arm A

See Figure 133.

1. Raise and support the vehicle.
2. Remove the wheel nuts, and the rear wheel.
3. Remove the lower arm A mounting nut (B), the washers (C) and the mounting bolt (D) from the knuckle side.
4. Remove the self-locking nut (E), the washers (F), and the flange bolt (G), then remove the lower arm A.

To install:

5. Install the lower arm A in the reverse order of removal, and note these items:

➡Use the figure to identify nuts when tightening to specification.

a. First install all of the components, and lightly tighten the bolts and nuts, then raise the suspension to load it with the vehicle's weight before fully tightening to the specified torque.

b. Before installing the wheel, clean the mating surfaces on the brake disc/drum and the inside of the wheel.

6. Check the wheel alignment, and adjust it if necessary.

Lower Arm B

1. Raise and support the vehicle.
2. Remove the wheel nuts, and the rear wheel.
3. Remove the self-locking nut and the washer while holding the joint pin with a hex wrench, and disconnect the stabilizer link from lower arm B.
4. Place a floor jack at the connecting point of lower arm B and the stabilizer link.
5. Remove the cotter pin from the lower arm B ball joint, and loosen the castle nut.
6. Disconnect the lower arm ball joint from the knuckle using the ball joint thread protector and the ball joint remover.
7. Lower the floor jack gradually.
8. Remove the coil spring, the spring mounting cushion, and the lower spring seat.
9. Remove the self-locking nut and the flange bolt, then remove lower arm B.

To install:

10. Position lower arm B, install a new flange bolt, and loosely install a new self-locking nut.
11. Install the spring mounting cushion and the spring.
12. Align the bottom of the coil spring, the stepped part of the lower spring seat and lower arm B.
13. Place a floor jack at the connecting point of lower arm B and the stabilizer link.
14. Raise the jack slowly until you can align the bolt hole of lower arm B and the knuckle ball joint pin, then loosely install a new castle nut.
15. Install the stabilizer link on the lower arm B with the washer and a new self-locking nut, and lightly tighten them.
16. Raise the rear suspension with a floor jack to load it with the vehicle's weight.
17. Final torque specifications are:
    - Lower arm B ball joint castle nut: 54–61 ft. lbs. (74–83 Nm)

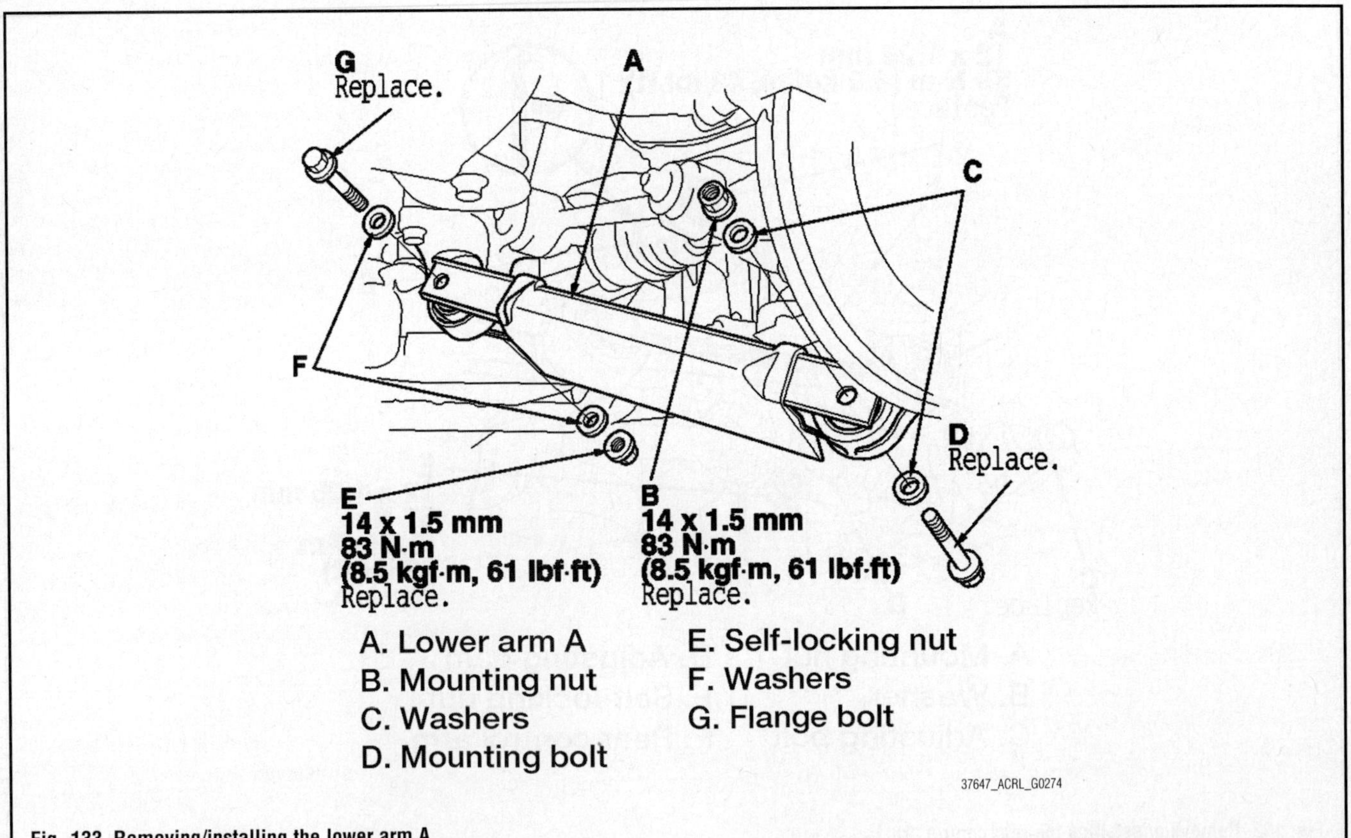
G
Replace.
A
C
F
D
Replace.
E
14 x 1.5 mm
83 N·m
(8.5 kgf·m, 61 lbf·ft)
Replace.
B
14 x 1.5 mm
83 N·m
(8.5 kgf·m, 61 lbf·ft)
Replace.
A. Lower arm A
B. Mounting nut
C. Washers
D. Mounting bolt
E. Self-locking nut
F. Washers
G. Flange bolt
37647_ACRL_G0274

Fig. 133 Removing/installing the lower arm A

- Lower arm B inner retaining bolt/nut: 64 ft. lbs. (86 Nm)
- Lower arm B self-locking outer nut: 40 ft. lbs. (54 Nm)

18. Tighten the castle nut and the self-locking nut to the specified torque.
19. Torque the castle nut to the lower torque specification, then tighten it only far enough to align the slot with the ball joint pin hole. Do not align the castle nut by loosening it.
20. Insert a new cotter pin into the ball joint pin from the rear to the front of the vehicle, and bend its end.
21. Clean the mating surface of the brake disc/drum the inside of the wheel, then install the rear wheel.
22. Check the rear wheel alignment, and adjust it if necessary.

SHOCK ABSORBER (DAMPER)

REMOVAL & INSTALLATION

See Figure 134.

1. Raise and support the vehicle.
2. Remove the wheel nuts, and the rear wheel.
3. Place a floor jack at the connecting point of lower arm B and the stabilizer link to support them.
4. Disconnect the headlight leveling sensor linkage from the damper.
5. Remove the rear shelf.
6. Remove the two flange nuts retaining the top end of the damper.
7. Remove the damper lower bolt from the knuckle.
8. Lower the rear suspension with the floor jack, then remove the damper from the vehicle.

To install:

9. Place a floor jack at the connecting point of lower arm B and the stabilizer link.
10. Compress the damper by hand, and move it into position.

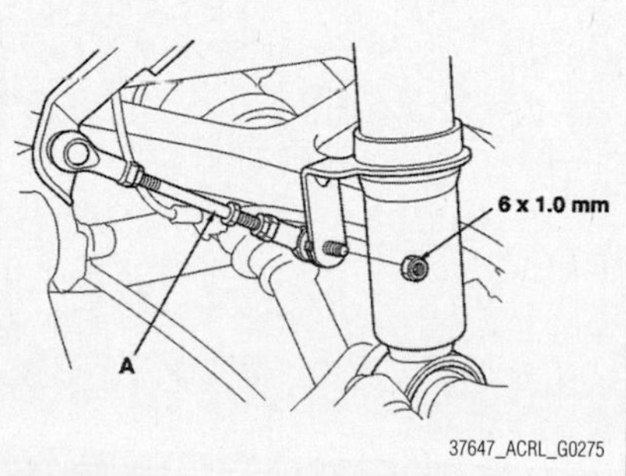
6 x 1.0 mm
A
37647_ACRL_G0275

Fig. 134 Disconnect the headlight leveling sensor linkage (A) from the damper.

11. Loosely tighten a new damper lower mounting bolt.
12. Loosely install new flange nuts.
13. Raise the rear suspension with a floor jack to load it with the vehicle's weight.
14. Tighten the flange nuts and the damper lower mounting bolt to the specified torque:
    - Upper flange nuts: 28 ft. lbs. (38 Nm)
    - Lower mounting bolt: 43 ft. lbs. (59 Nm)
15. Connect the headlight leveling sensor linkage to the damper with the self-locking nut.
16. Clean the mating surface of the brake disc/drum and the inside of the wheel, then install the rear wheel.
17. Check the rear wheel alignment, and adjust it if necessary.
18. Do the "Adaptive Front Lighting Control Unit Learning Procedure".

ADAPTIVE FRONT LIGHTING CONTROL UNIT LEARNING PROCEDURE

➡If the AFS indicator is blinking, check the DTCs.

1. Remove all baggage from the vehicle.
2. Turn the ignition switch to ON (II).
3. Make sure the AFS indicator does not turn off. (If the AFS indicator turns off, the adaptive front lighting control unit is not a new or previously learned unit.)
4. Turn the ignition switch to LOCK (0) and then back to ON (II).
5. Within 5 seconds, turn the headlight switch ON/OFF or to passing three times.
6. Make sure the AFS indicator blinks in a 1 second cycle. (If the indicator does not blink, return to Step 2.)
7. Turn the headlights ON, and make sure the headlights move while the AFS control unit runs the initialization function.
8. Turn the steering wheel to the left or to the right 90 degrees or more from the center position.
9. Test-drive, with only the driver in the vehicle, in a straight line for 98.4 feet (30 m) at least 16 mph (25 km/h).
10. Make sure the AFS indicator turns off. (If the indicator does not turn off, repeat the Steps 8 and 9.)
11. Adjust the headlights.

STABILIZER BAR

REMOVAL & INSTALLATION

1. Raise and support the vehicle.
2. Remove the wheel nuts, and the rear wheels.
3. Disconnect the stabilizer links from the stabilizer bar on the right and left sides.
4. Remove the flange bolts and the bushing holders, then remove the bushing and the stabilizer bar.

To install:

5. Install the stabilizer bar in the reverse order of removal, and note these items:
    a. Note the right and left direction of the stabilizer bar.
    b. Align the ends of the paint marks on the stabilizer bar with each end of the bushings.
6. Tighten the stabilizer bar mounting bolts to 36 ft. lbs. (49 Nm).
    a. Clean the mating surface of the brake disc/drum and the inside of the wheel, then install the rear wheel.
7. Check the wheel alignment, and adjust it if necessary.

STABILIZER LINK

REMOVAL & INSTALLATION

See Figure 135.

1. Raise and support the vehicle.
2. Remove the wheel nuts, and the rear wheel.
3. Remove the self-locking nuts (A) while holding the respective joint pin (B)

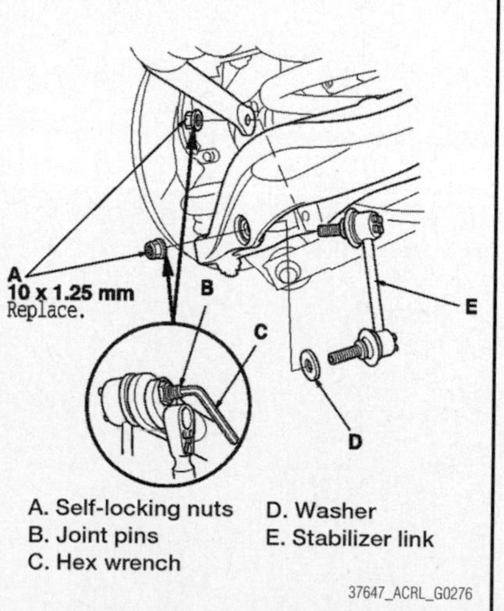
A
10 x 1.25 mm
Replace.
B
C
D
E
A. Self-locking nuts
B. Joint pins
C. Hex wrench
D. Washer
E. Stabilizer link
37647_ACRL_G0276

Fig. 135 Remove the self-locking nuts (A) while holding the respective joint pin (B) with a hex wrench (C), then remove the washer (D) and the stabilizer link (E).

with a hex wrench (C), then remove the washer (D) and the stabilizer link (E).

To install:

4. Place a floor jack at the connecting point of lower arm B and the stabilizer link, and raise the suspension.

5. Install the stabilizer link on the stabilizer bar and lower arm B with the joint pins set at the center of their range of the movement.

➡The stabilizer link has a paint mark in the middle of the link. The paint mark indicates the difference between the left and right stabilizer link.

6. Install the washer and new self-locking nuts, and lightly tighten them.

7. Tighten the self-locking nuts to the specified torque while holding the respective joint pin with a hex wrench.

8. Torque specifications are:
   - Lower link nut: 40 ft. lbs. (54 Nm)
   - Upper link nut: 32 ft. lbs. (43 Nm)

9. Clean the mating surface of the brake disc/drum and the inside of the wheel, then install the rear wheel.

10. Test-drive the vehicle.

11. After 5 minutes of driving, tighten the self-locking nuts again to the specified torque.

WHEEL BEARINGS

REMOVAL & INSTALLATION

➡Manufacturer does not provide a specific removal and installation procedure for this component.

INSPECTION

1. Raise and support the vehicle.
2. Remove the wheel nuts, and the wheels, taking care not to scratch the caliper.
3. Install suitable flat washers (A) and the wheel nuts. Tighten the nuts to the specified torque to hold the brake disc securely against the hub.
4. Attach the dial gauge. Place the dial gauge against the hub flange or hub cap.
5. Measure the bearing end play while moving the brake disc or the brake disc/drum inward and outward.
6. Bearing end play should not exceed 0.00–0.002 in. (0.00–0.05 mm).
7. If the bearing end play measurement is more than the standard, replace the wheel bearing or the hub bearing unit.

ACURA

TL

4

BRAKES 4-9

ANTI-LOCK BRAKE SYSTEM (ABS) 4-10
Wheel Speed Sensors 4-10
Removal & Installation..........4-10
BLEEDING THE BRAKE SYSTEM 4-9
Bleeding Procedure.....................4-9
Bleeding Procedure 4-9
Bleeding the ABS System 4-9
FRONT DISC BRAKES 4-12
Brake Caliper.............................4-12
Removal & Installation..........4-12
Disc Brake Pads.........................4-12
Removal & Installation..........4-12
INFORMATION AND PRECAUTIONS..................4-9
Anti-lock Systems...................4-9
Disc and Drum Systems.........4-9
PARKING BRAKE 4-14
Parking Brake Shoes................4-14
Adjustment 4-15
Removal & Installation..........4-14
REAR DISC BRAKES 4-13
Brake Caliper.............................4-13
Removal & Installation..........4-13
Disc Brake Pads.........................4-13
Removal & Installation..........4-13

CHASSIS ELECTRICAL.......4-16

AIR BAG (SUPPLEMENTAL RESTRAINT SYSTEM)........4-16
General Information..................4-16
Arming the System 4-16
Disarming the System...........4-16
Precautions..........................4-16

DRIVE TRAIN.................4-17

Clutch...4-17
Removal & Installation..........4-17
Clutch Driven Disc & Pressure Plate........................4-20
Removal & Installation..........4-20
Clutch Master Cylinder 4-20
Removal & Installation..........4-20
Clutch Slave Cylinder................4-23
Removal & Installation..........4-23
Front Halfshaft...........................4-24
Removal & Installation..........4-24
Hydraulic System Bleeding 4-24
Bleeding Procedure 4-24
Rear Halfshaft...........................4-26
Removal & Installation..........4-26
Transfer Case Assembly 4-27
Removal & Installation..........4-27

ENGINE COOLING 4-28

Engine Fan 4-28
Removal & Installation..........4-28
Radiator.....................................4-29
Removal & Installation..........4-29
Thermostat 4-29
Removal & Installation..........4-29
Water Pump 4-30
Removal & Installation..........4-30

ENGINE ELECTRICAL.........4-31

CHARGING SYSTEM 4-31
Alternator 4-31
Removal & Installation..........4-31
IGNITION SYSTEM 4-32
Firing Order..............................4-32
Ignition Coil 4-32
Removal & Installation..........4-32
Ignition Timing..........................4-33
Adjustment 4-33
Spark Plugs...............................4-33
Removal & Installation..........4-33
STARTING SYSTEM 4-33
Starter 4-33
Removal & Installation..........4-33
Solenoid or Relay Replacement.......................4-34

ENGINE MECHANICAL 4-34

Accessory Drive Belts 4-34
Accessory Belt Routing.........4-34
Adjustment 4-34
Inspection 4-34
Removal & Installation..........4-34
Camshaft and Valve Lifters........4-34
Inspection 4-34
Removal & Installation..........4-35
Catalytic Converter...................4-36
Removal & Installation..........4-36
Crankshaft Front Seal...............4-40
Removal & Installation..........4-40
Crankshaft Pulley 4-39
Removal & Installation..........4-39
Cylinder Head 4-40
Removal & Installation..........4-40
Flywheel/Drive Plate.................4-42
Removal & Installation..........4-42
Intake Manifold 4-42
Removal & Installation..........4-42
Oil Pan 4-43
Removal & Installation..........4-43
Oil Pump...................................4-45
Inspection 4-46
Removal & Installation..........4-45
Piston and Ring.........................4-46
Positioning 4-46
Rear Main Seal..........................4-47
Removal & Installation..........4-47
Rocker Arms.............................4-47
Removal & Installation..........4-47
Timing Belt & Sprockets 4-49
Removal & Installation..........4-49
Timing Belt Front Cover 4-48
Removal & Installation..........4-48
Valve Lash.................................4-52
Adjustment 4-52

ENGINE PERFORMANCE & EMISSION CONTROLS 4-54

Accelerator Pedal Position (APP) Sensor 4-54
Location................................4-54
Removal & Installation..........4-54
Camshaft Position (CMP) Sensor 4-54
Removal & Installation..........4-54
Crankshaft Position (CKP) Sensor 4-55
CKP Pattern Clear/CKP Pattern Learn 4-55
Removal & Installation..........4-55

Engine Control Module (ECM)...4-55
ECM/PCM Idle Learn Procedure....4-56
ECM/PCM Update....4-57
Location....4-55
Removal & Installation....4-55
Engine Coolant Temperature (ECT) Sensor....4-57
Location....4-57
Removal & Installation....4-57
Evaporative Emissions (EVAP) Canister....4-58
Location....4-58
Removal & Installation....4-58
Exhaust Gas Recirculation (EGR) Valve....4-61
Location....4-61
Removal & Installation....4-61
Heated Oxygen Sensor (HO2S)....4-62
Location....4-62
Removal & Installation....4-62
Intake Air Temperature (IAT) Sensor....4-63
Location....4-63
Removal & Installation....4-63
Knock Sensor (KS)....4-63
Location....4-63
Removal & Installation....4-63
Manifold Absolute Pressure (MAP) Sensor....4-63
Location....4-63
Removal & Installation....4-63
Mass Air Flow (MAF)/Intake Air Temperature (IAT) Sensor....4-63
Location....4-63
Removal & Installation....4-63
Output Shaft Speed (OSS) Sensor....4-64
Location....4-64
Removal & Installation....4-64
Positive Crankcase Ventilation (PCV) Valve....4-64
Location....4-64
Removal & Installation....4-64

FUEL....4-64

GASOLINE FUEL INJECTION SYSTEM....4-64
Fuel Filter....4-65
Removal & Installation....4-65
Fuel Level Sending Unit....4-66
Location....4-66
Removal & Installation....4-66
Fuel Pressure Regulator....4-67
Removal & Installation....4-67
Fuel Pump....4-67
Removal & Installation....4-67
Fuel Pump Control Module (with SH-AWD)....4-67
Removal & Installation....4-67
Fuel Rail and Injector....4-68
Removal & Installation....4-68
Fuel System Pressure....4-64
Relieving....4-64
Fuel System Service Precautions....4-64
Fuel Tank....4-69
Draining....4-69
Removal & Installation....4-72
Idle Speed....4-73
Adjustment....4-73
Throttle Body....4-74
Removal & Installation....4-74

HEATING & AIR CONDITIONING SYSTEM....4-76

Blower Motor....4-76
Removal & Installation....4-76
Heater Core....4-78
Removal & Installation....4-78

SPECIFICATIONS AND MAINTENANCE CHARTS....4-3

Brake Specifications....4-7
Camshaft and Bearing Specifications....4-5
Capacities....4-4
Crankshaft and Connecting Rod Specifications....4-5
Engine and Vehicle Identification....4-3
Fluid Specifications....4-4
Gasoline Engine Tune-Up Specifications....4-3
General Engine Specifications....4-3
Piston and Ring Specifications....4-5
Scheduled Maintenance Intervals....4-8
Tire, Wheel and Ball Joint Specifications....4-7
Torque Specifications....4-6
Valve Specifications....4-4
Wheel Alignment....4-7

STEERING....4-81

Electric Power Steering (EPS) Control Unit....4-81
Removal & Installation....4-81
Torque Sensor Neutral Position Memorization....4-81
Power Rack & Pinion Steering Gear....4-81
Removal & Installation....4-81
Torque Sensor Neutral Position Memorization....4-91
VSA Sensor Neutral Position Memorization Procedure....4-91

SUSPENSION....4-91

FRONT SUSPENSION....4-91
Control Links....4-91
Removal & Installation....4-91
Lower Ball Joint....4-92
Removal & Installation....4-92
Lower Control Arm....4-92
Removal & Installation....4-92
Stabilizer Bar....4-98
Removal & Installation....4-98
Steering Knuckle....4-94
Removal & Installation....4-94
Strut (Damper/Spring)....4-95
Overhaul....4-96
Removal & Installation....4-95
Upper Ball Joint....4-103
Removal & Installation....4-103
Upper Control Arm....4-103
Removal & Installation....4-103
Wheel Hub & Bearing....4-104
Removal & Installation....4-104
REAR SUSPENSION....4-105
Control Arms/Links....4-105
Removal & Installation....4-105
Stabilizer Bar....4-106
Removal & Installation....4-106
Strut (Damper/Spring)....4-106
Removal & Installation....4-106
Wheel Hub & Bearing....4-107
Removal & Installation....4-107

SPECIFICATIONS AND MAINTENANCE CHARTS

ENGINE AND VEHICLE IDENTIFICATION

Engine							Model Year	
Code	Liters (cc)	Cu. In.	Cyl.	Fuel Sys.	Engine Type	Eng. Mfg.	Code ①	Year
J35Z6	3.5 (3490)	213	6	PGM-FI	SOHC	Honda	9	2009
J37A4	3.7 (3704)	226	6	PGM-FI	SOHC	Honda	A	2010

PGM-FI: Programmed Fuel Injection

SOHC: Single Overhead Camshaft

① 10th digit of the Vehicle Identification Number (VIN)

37647_ACTL_C0001

GENERAL ENGINE SPECIFICATIONS

Year	Model	Engine Displacement Liters	Engine ID	Net Horsepower @ rpm	Net Torque @ rpm (ft. lbs.)	Bore x Stroke (in.)	Com-pression Ratio	Oil Pressure @ rpm
2009	TL	3.5	J35Z6	280@6200	254@5000	3.50x3.66	11.2:1	71@3000
	TL	3.7	J37A4	305@6300	273@5000	3.54x3.78	11.2:1	71@3000
2010	TL	3.5	J35Z6	280@6200	254@5000	3.50x3.66	11.2:1	71@3000
	TL	3.7	J37A4	305@6300	273@5000	3.54x3.78	11.2:1	71@3000

37647_ACTL_C0002

GASOLINE ENGINE TUNE-UP SPECIFICATIONS

Year	Engine Displacement Liters	Engine ID/VIN	Spark Plug Gap (in.)	Ignition Timing (deg.)		Fuel Pump (psi)	Idle Speed (rpm)		Valve Clearance	
				MT	AT		MT	AT	In.	Ex.
2009	3.5	J35Z6	0.039-0.043	NA	8-12 B	57-64 ①	NA	630-730	0.008-0.009	0.011-0.013
	3.7	J37A4	0.039-0.043	NA	8-12 B	57-64 ①	NA	630-730	0.008-0.009	0.011-0.013
2010	3.5	J35Z6	0.039-0.043	NA	8-12 B	57-64 ①	700-800	630-730	0.008-0.009	0.011-0.013
	3.7	J37A4	0.039-0.043	NA	8-12 B	57-64 ①	700-800	630-730	0.008-0.009	0.011-0.013

NOTE: The Vehicle Emission Control Information label reflects specification changes during production and must be used if they differ from this chart.

① Pressure with fuel pressure gauge connected

B: Before Top Dead Center

NA: Not Available

37647_ACTL_C0003

CAPACITIES

Year	Model	Engine Displacement Liters	Engine ID/VIN	Engine Oil with Filter (qts.)	Transmission (pts.) 6-Spd	Transmission (pts.) Auto.	Fuel Tank (gal.)	Cooling System (qts.)
2009	TL	3.5	J35Z6	4.5	—	①	18.5	6.6
		3.7	J37A4	4.5	—	①	18.5	6.6
2010	TL	3.5	J35Z6	4.5	4.4	①	18.5	6.6
		3.7	J37A4	4.5	4.4	①	18.5	6.6

NOTE: All capacities are approximate. Add fluid gradually and ensure a proper fluid level is obtained.

NOTE: Capacities given are service, not overhaul capacities

① Fluid change with SH-AWD: 6.0 pts.

Fluid change without SH-AWD: 7.0 pts.

37647_ACTL_C0004

FLUID SPECIFICATIONS

Year	Model	Engine Displacement Liters	Engine Oil	Man. Trans.	Auto. Trans.	Transfer Case	Power Steering Fluid	Brake Master Cylinder	Cooling System
2009	TL	3.5	5W-20 Acura	—	Acura ATF-Z1	SAE 90 GL4 /GL5	Acura PS Fluid	Acura DOT 3	①
	TL	3.7	5W-20 Acura	—	Acura ATF-Z1	SAE 90 GL4 /GL5	Acura PS Fluid	Acura DOT 3	①
2010	TL	3.5	5W-20 Acura	Acura MTF	Acura ATF-Z1	SAE 90 GL4 /GL5	Acura PS Fluid	Acura DOT 3	①
	TL	3.7	5W-20 Acura	Acura MTF	Acura ATF-Z1	SAE 90 GL4 /GL5	Acura PS Fluid	Acura DOT 3	①

DOT: Department of Transportation

① Acura Long Life Antifreeze/Coolant-Type2

37647_ACTL_C0005

VALVE SPECIFICATIONS

Year	Engine Displacement Liters	Engine ID/VIN	Seat Angle (deg.)	Face Angle (deg.)	Spring Test Pressure (lbs. @ in.)	Spring Installed Height (in.)	Stem-to-Guide Clearance (in.) Intake	Stem-to-Guide Clearance (in.) Exhaust	Stem Diameter (in.) Intake	Stem Diameter (in.) Exhaust
2009	3.5	J35Z6	45	45	NA	NA	0.0008-0.0018	0.0022-0.0031	0.2159-0.2163	0.2146-0.2150
	3.7	J37A4	45	45	NA	NA	0.0008-0.0018	0.0022-0.0031	0.2159-0.2163	0.2146-0.2150
2010	3.5	J35Z6	45	45	NA	NA	0.0008-0.0018	0.0022-0.0031	0.2159-0.2163	0.2146-0.2150
	3.7	J37A4	45	45	NA	NA	0.0008-0.0018	0.0022-0.0031	0.2159-0.2163	0.2146-0.2150

NA: Not Available

37647_ACTL_C0006

CAMSHAFT AND BEARING SPECIFICATIONS

All measurements are given in inches.

Year	Engine Displacement Liters	Engine VIN	Journal Diameter	Brg. Oil Clearance	Shaft End-play	Runout	Journal Bore	Lobe Height Intake	Lobe Height Exhaust
2009	3.5	J35Z6	NA	0.0020-0.0035	0.0020-0.0080	0.0010	NA	①	1.4472
	3.7	J37A4	NA	0.0020-0.0035	0.0020-0.0080	0.0010	NA	②	③
2010	3.5	J35Z6	NA	0.0020-0.0035	0.0020-0.0080	0.0010	NA	①	1.4472
	3.7	J37A4	NA	0.0020-0.0035	0.0020-0.0080	0.0010	NA	②	③

NA: Information not available

① Primary: 1.3504 inches
Secondary: 1.4024 inches

② Primary: 1.3504 innches
Secondary: 1.4146 inches

③ Primary: 1.4462 inches
Secondary: 1.4713 inches

37647_ACTL_C0009

CRANKSHAFT AND CONNECTING ROD SPECIFICATIONS

All measurements are given in inches.

Year	Engine Displacement Liters	Engine ID/VIN	Crankshaft Main Brg. Journal Dia.	Crankshaft Main Brg. Oil Clearance	Crankshaft Shaft End-play	Crankshaft Thrust on No.	Connecting Rod Journal Diameter	Connecting Rod Oil Clearance	Connecting Rod Side Clearance
2009	3.5	J35Z6	2.8337-2.8346	0.0007-0.0018	0.0040-0.0140	3	2.1644-2.1654	0.0008-0.0017	0.0060-0.0140
	3.7	J37A4	2.8337-2.8346	0.0007-0.0018	0.0040-0.0140	3	2.2431-2.2441	0.0008-0.0017	0.0060-0.0140
2010	3.5	J35Z6	2.8337-2.8346	0.0007-0.0018	0.0040-0.0140	3	2.1644-2.1654	0.0008-0.0017	0.0060-0.0140
	3.7	J37A4	2.8337-2.8346	0.0007-0.0018	0.0040-0.0140	3	2.2431-2.2441	0.0008-0.0017	0.0060-0.0140

37647_ACTL_C0007

PISTON AND RING SPECIFICATIONS

All measurements are given in inches

Year	Engine Displacement Liters	Engine ID/VIN	Piston Clearance	Ring Gap Top Compression	Ring Gap Bottom Compression	Ring Gap Oil Control	Ring Side Clearance Top Compression	Ring Side Clearance Bottom Compression	Ring Side Clearance Oil Control
2009	3.5	J35Z6	0.0006-0.0016	0.0080-0.0014	0.0160-0.0220	0.0080-0.0280	0.0022-0.0031	0.0012-0.0022	NA
	3.7	J37A4	0.0006-0.00160	0.0120-0.0160	0.0160-0.0220	0.0080-0.0140	0.0022-0.0033	0.0012-0.0024	NA
2010	3.5	J35Z6	0.0006-0.0016	0.0080-0.0014	0.0160-0.0220	0.0080-0.0280	0.0022-0.0031	0.0012-0.0022	NA
	3.7	J37A4	0.0006-0.00160	0.0120-0.0160	0.0160-0.0220	0.0080-0.0140	0.0022-0.0033	0.0012-0.0024	NA

NA: Not Applicable

37647_ACTL_C0008

TORQUE SPECIFICATIONS

All readings in ft. lbs.

Year	Engine Displacement Liters	Engine ID/VIN	Cylinder Head Bolts	Main Bearing Bolts	Rod Bearing Bolts	Crankshaft Damper Bolts	Flywheel Bolts	Manifold Intake	Manifold Exhaust	Spark Plugs	Oil Pan Drain Plug
2009	3.5	J35Z6	①	②	③	④	54	16	23	13	29
	3.7	J37A4	①	②	③	④	54	16	23	13	29
2010	3.5	J35Z6	①	②	③	④	54	16	23	13	29
	3.7	J37A4	①	②	③	④	54	16	23	13	29

① Step 1: 22 ft. lbs.
Step 2: Rotate 90 degrees
Step 3: Rotate an additional 90 degrees
Step 4 (new bolts only): additional 90 degrees

② Step 1: 54 ft. lbs.
Step 2: Repeat Step 1 to ensure 54 ft. lbs.
Step 3: Side bolts to 36 ft. lbs.
Step 4: Repeat Step 3 to ensure 36 ft. lbs.

③ Step 1: 14 ft. lbs.
Step 2: Additional 90 degrees

④ Step 1: 47 ft. lbs.
Step 2: Additional 60 degrees

37647_ACTL_C0010

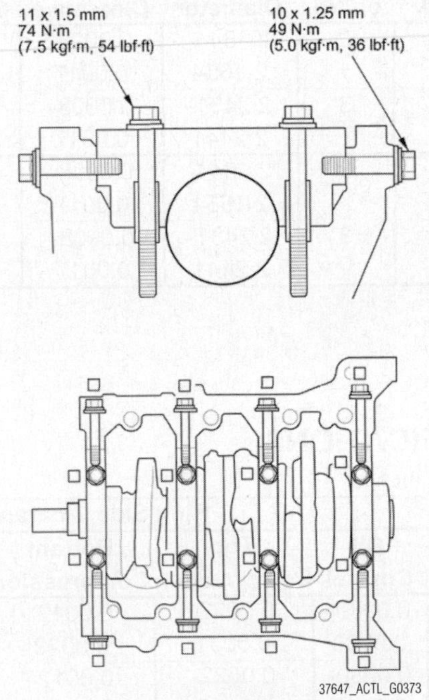
11 x 1.5 mm
74 N·m
(7.5 kgf·m, 54 lbf·ft)
10 x 1.25 mm
49 N·m
(5.0 kgf·m, 36 lbf·ft)
37647_ACTL_G0373

Fig. 1 Main bearing torque sequence

WHEEL ALIGNMENT

Year	Model		Caster Range (+/-Deg.)	Caster Preferred Setting (Deg.)	Camber Range (+/-Deg.)	Camber Preferred Setting (Deg.)	Toe-in (in.)
2009	TL	F	0.45	+3.28	+30, -45	-0.30	0 +/- 0.08
		R	—	—	+30, -45	-1.00	0 +/- 0.08
2010	TL	F	0.45	+3.28	+30, -45	-0.30	0 +/- 0.08
		R	—	—	+30, -45	-1.00	0 +/- 0.08

37647_ACTL_C0011

TIRE, WHEEL AND BALL JOINT SPECIFICATIONS

Year	Model	OEM Tires Standard	OEM Tires Optional	Tire Pressures (psi) Front	Tire Pressures (psi) Rear	Wheel Size	Ball Joint Inspection	Lug Nut (ft. lbs.)
2009	TL	P245/45VR18	None	32	32	18 x 8	NS	94
2010	TL	P245/45VR18	None	32	32	18 x 8	NS	94

OEM: Original Equipment Manufacturer

PSI: Pounds Per Square Inch

NS: Not Specified by manufacturer

37647_ACTL_C0012

BRAKE SPECIFICATIONS

All measurements in inches unless noted

Year	Model		Brake Disc Original Thickness	Brake Disc Minimum Thickness	Brake Disc Maximum Runout	Brake Drum Diameter Original Inside Diameter	Brake Drum Diameter Max. Wear Limit	Brake Drum Diameter Maximum Machine Diameter	Minimum Lining Thickness Front	Minimum Lining Thickness Rear	Brake Caliper Bracket Bolts (ft. lbs.)	Brake Caliper Mounting Bolts (ft. lbs.)
2009	TL	F	1.100-1.110	1.020	0.002	—	—	—	0.06	—	101	53
		R	0.430-0.440	0.350	0.002	—	—	—	—	0.06	65	16
2010	TL	F	1.100-1.110	1.020	0.002	—	—	—	0.06	—	101	53
		R	0.430-0.440	0.350	0.002	—	—	—	—	0.06	65	16

F: Front

R: Rear

37647_ACTL_C0013

SCHEDULED MAINTENANCE INTERVALS
Acura TL

TO BE SERVICED	TYPE OF SERVICE	SERVICE INTERVALS						
		Symbol A	Symbol B	1	2	3	4	5
Air cleaner element	Replace				✓			
Brake fluid	Inspect		✓					
Brake hoses and lines	Inspect		✓					
Clutch fluid (if equipped)	Inspect		✓					
Drive belt	Inspect				✓			
Driveshaft boots	Inspect		✓					
Dust & pollen filter	Replace				✓			
Engine coolant	Inspect		✓					
Engine coolant	Replace							✓
Engine oil	Replace	✓						
Engine oil & filter	Replace		✓					
Exhaust system	Inspect		✓					
Front & rear brakes	Inspect		✓					
Fuel lines	Inspect		✓					
Parking brake adjustment	Inspect		✓					
Power steering fluid	Inspect		✓					
Spark plugs	Replace						✓	
Suspension components	Inspect		✓					
Tie-rod ends, steering gearbox and boots	Inspect		✓					
Timing belt	Replace						✓	
Tires	Rotate			✓				
Transmission fluid	Inspect		✓					
Transmission fluid	Replace					✓		
Valve clearance	Inspect						✓	
Windshield washer fluid	Inspect		✓					

37647_ACTL_C0014

BRAKES — INFORMATION AND PRECAUTIONS

ANTI-LOCK SYSTEMS

- Certain components within the ABS system are not intended to be serviced or repaired individually.
- Do not use rubber hoses or other parts not specifically specified for and ABS system. When using repair kits, replace all parts included in the kit. Partial or incorrect repair may lead to functional problems and require the replacement of components.
- Lubricate rubber parts with clean, fresh brake fluid to ease assembly. Do not use shop air to clean parts; damage to rubber components may result.
- Use only DOT 3 brake fluid from an unopened container.
- If any hydraulic component or line is removed or replaced, it may be necessary to bleed the entire system.
- A clean repair area is essential. Always clean the reservoir and cap thoroughly before removing the cap. The slightest amount of dirt in the fluid may plug an orifice and impair the system function. Perform repairs after components have been thoroughly cleaned; use only denatured alcohol to clean components. Do not allow ABS components to come into contact with any substance containing mineral oil; this includes used shop rags.
- The Anti-Lock control unit is a microprocessor similar to other computer units in the vehicle. Ensure that the ignition switch is **OFF**before removing or installing controller harnesses. Avoid static electricity discharge at or near the controller.
- If any arc welding is to be done on the vehicle, the control unit should be unplugged before welding operations begin.

DISC AND DRUM SYSTEMS

CAUTION

Dust and dirt accumulating on brake parts during normal use may contain asbestos fibers from production or aftermarket brake linings. Breathing excessive concentrations of asbestos fibers can cause serious bodily harm. Exercise care when servicing brake parts. Do not sand or grind brake lining unless equipment used is designed to contain the dust residue. Do not clean brake parts with compressed air or by dry brushing. Cleaning should be done by dampening the brake components with a fine mist of water, then wiping the brake components clean with a dampened cloth. Dispose of cloth and all residue containing asbestos fibers in an impermeable container with the appropriate label. Follow practices prescribed by the Occupational Safety and Health Administration (OSHA) and the Environmental Protection Agency (EPA) for the handling, processing, and disposing of dust or debris that may contain asbestos fibers.

BRAKES — BLEEDING THE BRAKE SYSTEM

BLEEDING PROCEDURE

BLEEDING PROCEDURE

➡See "Bleeding the ABS System" in this section.

BLEEDING THE ABS SYSTEM

See Figure 2.

WARNING

Read and observe the "PRECAUTIONS" in this section before starting work.

1. The reservoir on the master cylinder must be at the MAX (upper) level mark at the start of the bleeding procedure and checked after bleeding each brake caliper. Add fluid as required.
2. Make sure the brake fluid level in the reservoir is at the MAX (upper) level line.
3. Slide a piece of clear plastic hose over the first bleed screw, and submerge the other end in a container of new brake fluid.
4. Have someone slowly pump the brake pedal several times, and then apply steady pressure.
5. Starting at the left-front, loosen the brake bleed screw to allow air to escape from the system. Then tighten the bleed screw securely.
6. Repeat the procedure for each wheel in the sequence shown following until air bubbles no longer appear in the fluid.
7. Refill the master cylinder reservoir to the MAX (upper) level line.

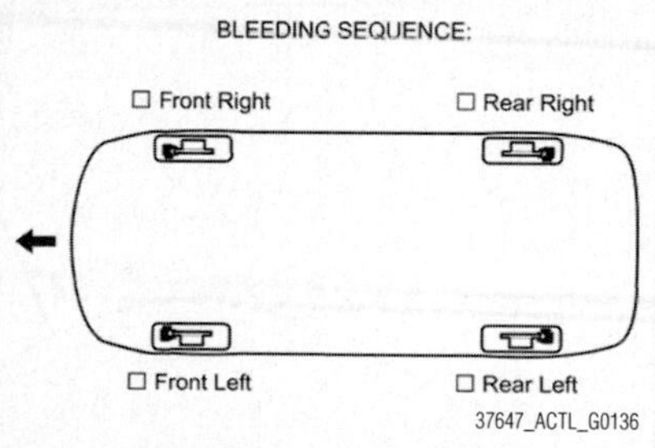
BLEEDING SEQUENCE:
Front Right
Rear Right
Front Left
Rear Left
37647_ACTL_G0136
Fig. 2 Brake bleeding sequence

WHEEL SPEED SENSORS

REMOVAL & INSTALLATION

Front

See Figure 3.

1. Turn the ignition switch to LOCK (0), or press the engine start/stop button to select the OFF mode.
2. Remove the front wheel.
3. Release the clamp, then disconnect the wheel speed sensor connector.
4. Remove the bolts and the wheel speed sensor.

To install:

5. Install the wheel speed sensor in the reverse order of removal, and note these items:
   - Do not twist the sensor wires.
   - If the wheel speed sensor comes in contact with the wheel bearing, it is faulty.
   - Make sure there is no debris in the sensor mounting hole.
6. Start the engine, and make sure the ABS and the VSA indicators go off.

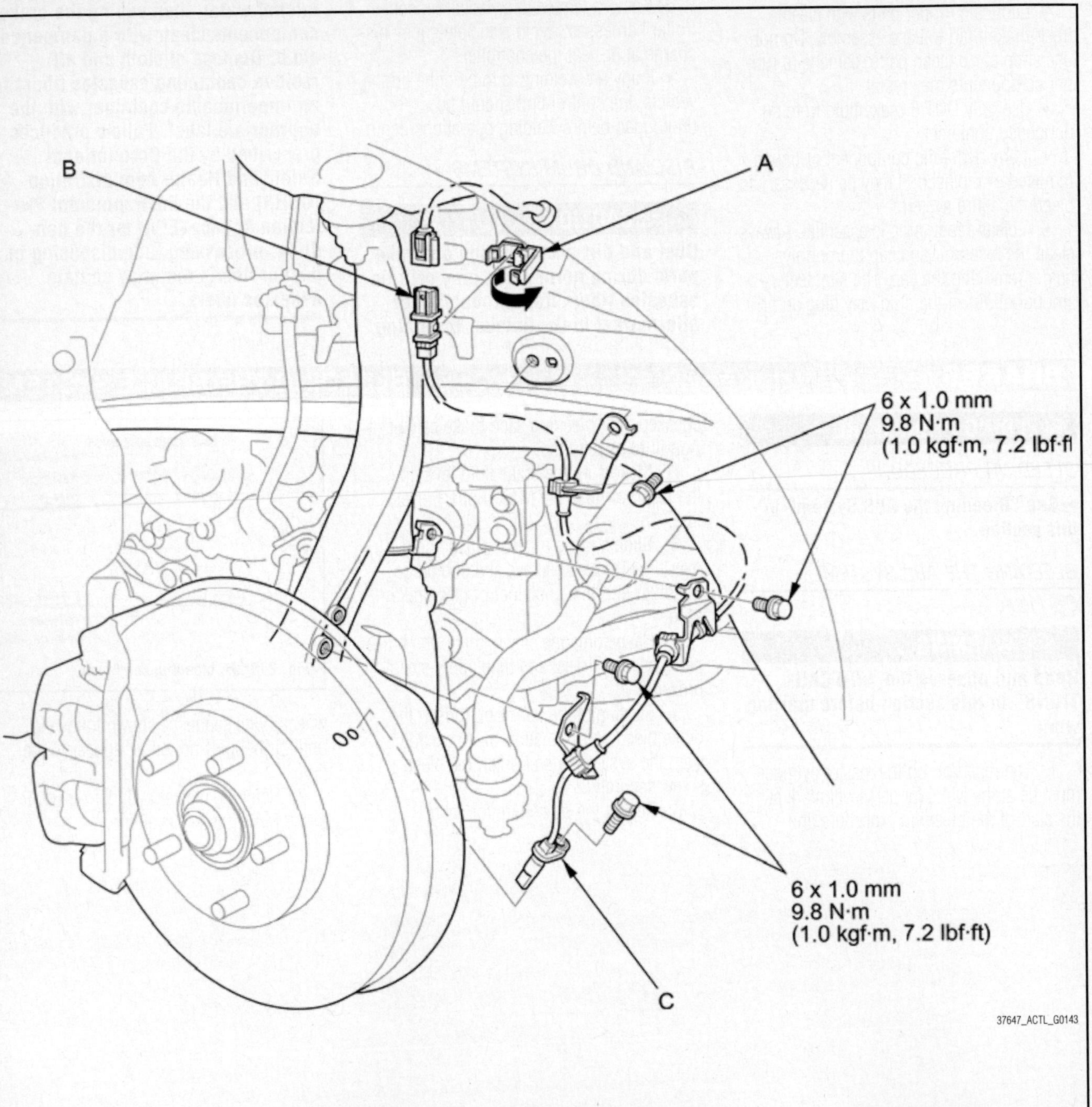
B
A
6 x 1.0 mm
9.8 N·m
(1.0 kgf·m, 7.2 lbf·ft)
6 x 1.0 mm
9.8 N·m
(1.0 kgf·m, 7.2 lbf·ft)
C
37647_ACTL_G0143

Fig. 3 Release the clamp (A), then disconnect the wheel speed sensor connector (B)

7. Test-drive the vehicle, and make sure the ABS and the VSA indicators do not come on.

Rear

See Figure 4.

1. Turn the ignition switch to LOCK (0), or press the engine start/stop button to select the OFF mode.
2. Remove the rear wheel.
3. Release the clamp, then disconnect the wheel speed sensor connector.
4. Remove the clamps, the bolt, and the wheel speed sensor.

To install:

5. Install the wheel speed sensor in the reverse order of removal, and note these items:
   - Do not twist the sensor wires.
   - If the wheel speed sensor comes in contact with the hub bearing unit, it is faulty.
   - Make sure there is no debris in the sensor mounting hole.
   - Lubricate the O-ring on the wheel speed sensor.
6. Start the engine, and make sure the ABS and the VSA indicators go off.
7. Test-drive the vehicle, and make sure the ABS and the VSA indicators do not come on.

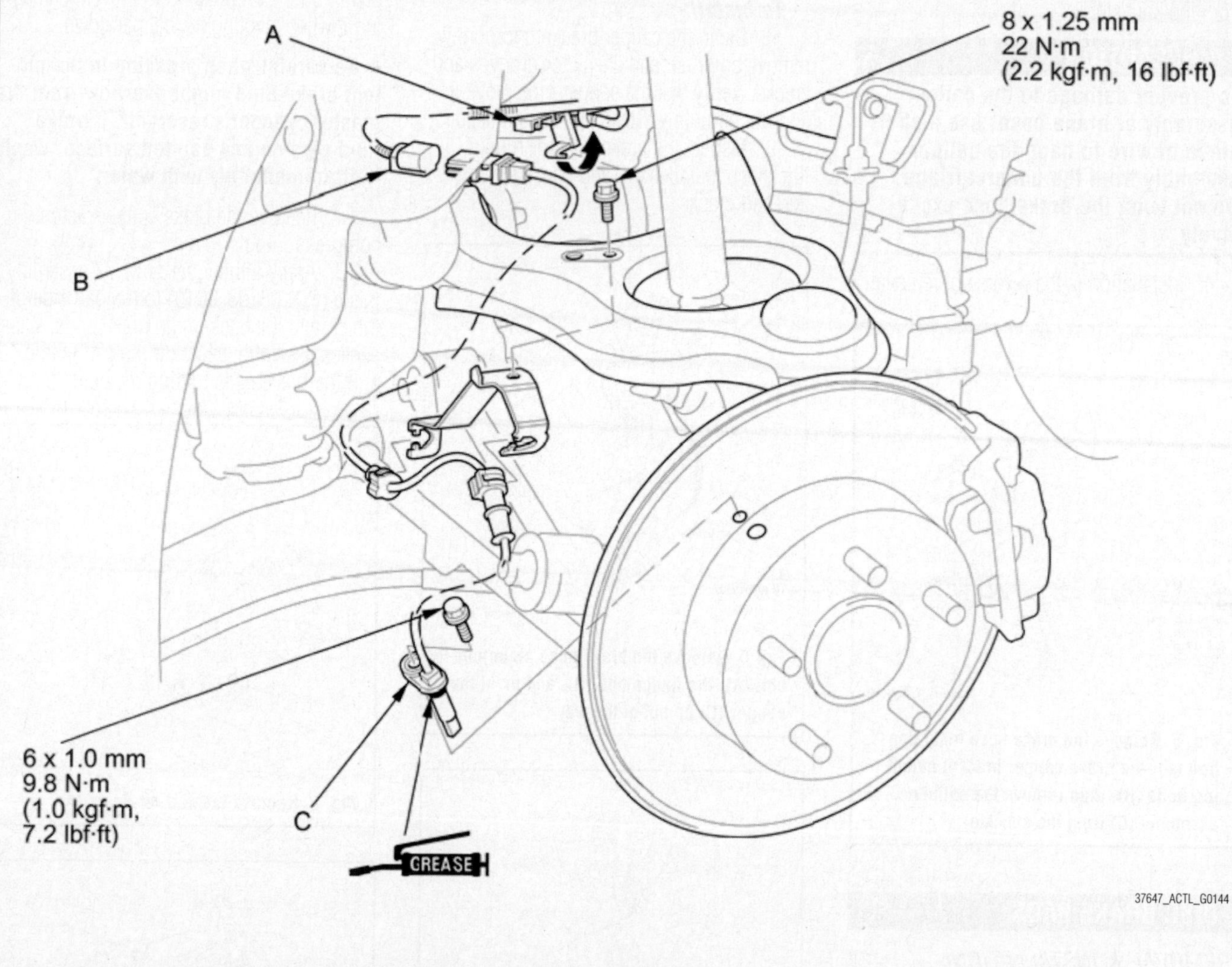
A
8 x 1.25 mm
22 N·m
(2.2 kgf·m, 16 lbf·ft)
B
6 x 1.0 mm
9.8 N·m
(1.0 kgf·m,
7.2 lbf·ft)
C
GREASE
37647_ACTL_G0144

Fig. 4 Release the clamp (A), then disconnect the wheel speed sensor connector (B)

BRAKE CALIPER

REMOVAL & INSTALLATION

See Figure 5.

➡**Keep any grease off the brake disc and the brake pads.**

1. Raise and support the vehicle.
2. Remove the front wheel.
3. Remove the brake hose mounting bolt, the brake caliper bracket mounting bolts, then remove the caliper assembly from the knuckle.

⁂ CAUTION

To prevent damage to the caliper assembly or brake hose, use a short piece of wire to hang the caliper assembly from the undercarriage. Do not twist the brake hose excessively.

4. Installation is the reverse of removal.

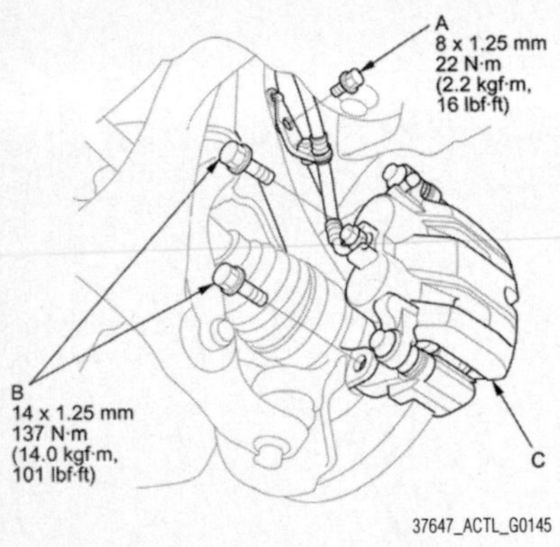
A
8 x 1.25 mm
22 N·m
(2.2 kgf·m,
16 lbf·ft)
B
14 x 1.25 mm
137 N·m
(14.0 kgf·m,
101 lbf·ft)
C
37647_ACTL_G0145

Fig. 5 Remove the brake hose mounting bolt (A), the brake caliper bracket mounting bolts (B), then remove the caliper assembly (C) from the knuckle

DISC BRAKE PADS

REMOVAL & INSTALLATION

See Figures 6 through 9.

⁂ CAUTION

Frequent inhalation of brake pad dust, regardless of material composition, could be hazardous to your health. Avoid breathing dust particles. Never use an air hose or brush to clean brake assemblies. Use an OSHA-approved vacuum cleaner.

1. Remove some brake fluid from the master cylinder.
2. Raise and support the vehicle.
3. Remove the front wheels.
4. Remove the brake hose mounting bolt, the flange bolt, and pivot the caliper up out of the way.
5. Check the hose and the pin boots for damage and deterioration.
6. Remove the pad shims and the brake pads.
7. Remove the pad retainers.

To install:

8. Clean the caliper bracket thoroughly; remove any rust, and check for grooves and cracks. Verify that the caliper pins move in and out smoothly. Clean and lube if needed.
9. Inspect the brake disc for runout, thickness, parallelism, and check for damage and cracks.

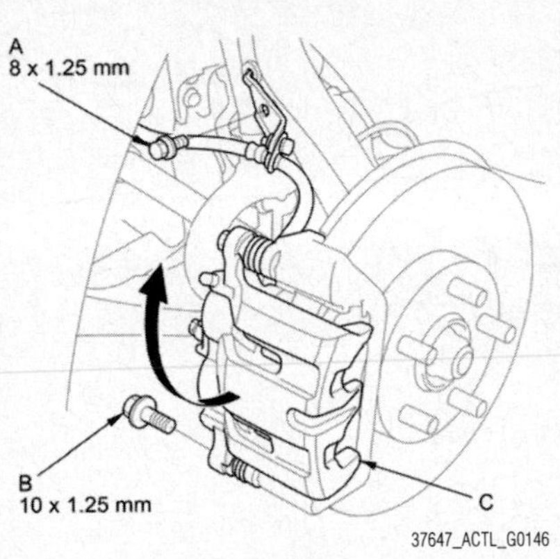
A
8 x 1.25 mm
B
10 x 1.25 mm
C
37647_ACTL_G0146

Fig. 6 Remove the brake hose mounting bolt (A), the flange bolt (B), and pivot the caliper (C) up out of the way

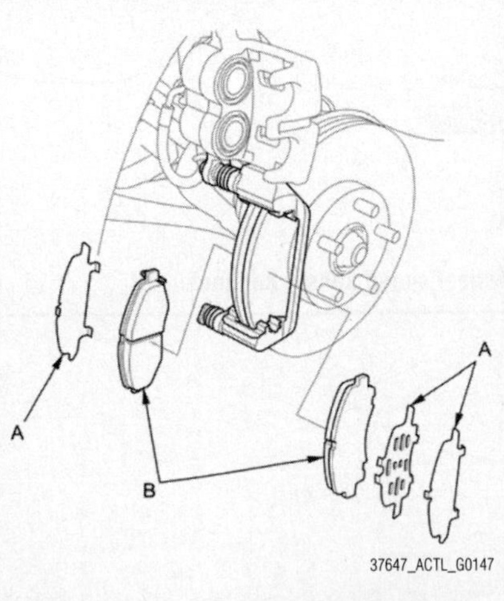
A
A
B
37647_ACTL_G0147

Fig. 7 Remove the pad shims (A) and the brake pads (B)

10. Apply a thin coat of M-77 assembly paste (P/N 08798-9010) to the retainer mating surface of the caliper bracket.
11. Install the pad retainers. Wipe excess assembly paste off the retainers. Keep any assembly paste off the brake disc and the brake pads.
12. Install the brake caliper piston compressor tool on the caliper body.
13. Press in the piston with the brake caliper piston compressor tool so the caliper will fit over the brake pads. Make sure the piston boot is in position to prevent damaging it when pivoting the caliper down.

➡**Be careful when pressing in the piston; brake fluid might overflow from the master cylinder's reservoir. If brake fluid gets on any painted surface, wash it off immediately with water.**

14. Remove the brake caliper piston compressor tool.
15. Apply a thin coat of M-77 assembly paste (P/N 08798-9010) to the pad side of the shims, the back of the brake pads. Wipe excess assembly paste off the pad shims and the brake pads friction material.

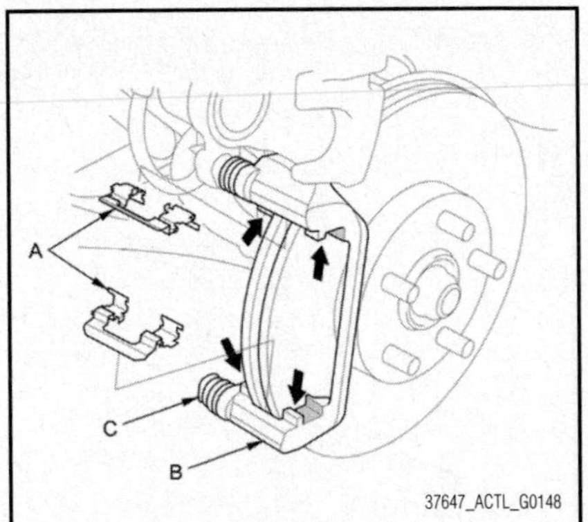
A
C
B
37647_ACTL_G0148

Fig. 8 Remove the pad retainers (A)

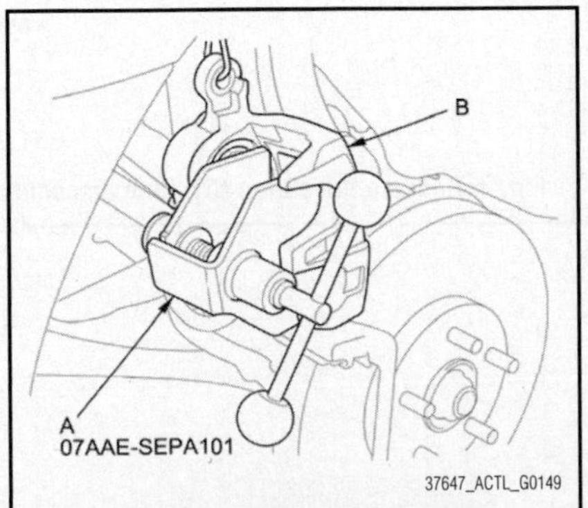
B
A
07AAE-SEPA101
37647_ACTL_G0149

Fig. 9 Install the brake caliper piston compressor tool (A) on the caliper body (B)

⁂ CAUTION

Keep grease and assembly paste off the brake disc and the brake pads. Contaminated brake disc or brake pads reduce stopping ability.

16. Install the brake pads and pad shims correctly. Install the brake pad with the wear indicator on the upper inside. If you are reusing the brake pads, always reinstall the brake pads in their original positions to prevent a temporary loss of braking efficiency.
17. Pivot the caliper down into position. Install the flange bolt, and tighten it to 53 ft. lbs. (72 Nm).
18. Install the brake hose mounting bolt.
19. Clean the mating surface between the brake disc and the inside of the wheel, then install the front wheels.
20. Press the brake pedal several times to make sure the brakes work.

➡Engagement may require a greater pedal stroke immediately after the brake pads have been replaced as a set. Several applications of the brake pedal will restore the normal pedal stroke.

21. Add brake fluid as needed.
22. After installation, check for leaks at hose and line joints or connections, and retighten if necessary.
23. Test-drive the vehicle, then check for leaks.

REAR DISC BRAKES

BRAKE CALIPER

REMOVAL & INSTALLATION

See Figure 10.

➡Keep any grease off the brake disc/drum and the brake pads.

1. Raise and support the vehicle.
2. Remove the rear wheel.
3. Release the parking brake lever fully.
4. Remove the brake hose mounting bolt and the brake caliper bracket mounting bolts, then remove the caliper assembly from the knuckle.

⁂ CAUTION

To prevent damage to the caliper assembly or brake hose, use a short piece of wire to hang the caliper assembly from the undercarriage. Do not twist the brake hose excessively.

5. Installation is the reverse of removal.
6. Tighten the caliper bracket mounting bolts to 65 ft. lbs. (88 Nm).

DISC BRAKE PADS

REMOVAL & INSTALLATION

See Figures 10 through 13.

⁂ CAUTION

Frequent inhalation of brake pad dust, regardless of material composition, could be hazardous to your health. Avoid breathing dust particles. Never use an air hose or brush to clean brake assemblies. Use an OSHA-approved vacuum cleaner.

1. Remove some brake fluid from the master cylinder.
2. Raise and support the vehicle.
3. Remove the rear wheels.
4. Remove the brake hose mounting bolt and the brake caliper bracket mounting bolts, then remove the caliper assembly from the knuckle.

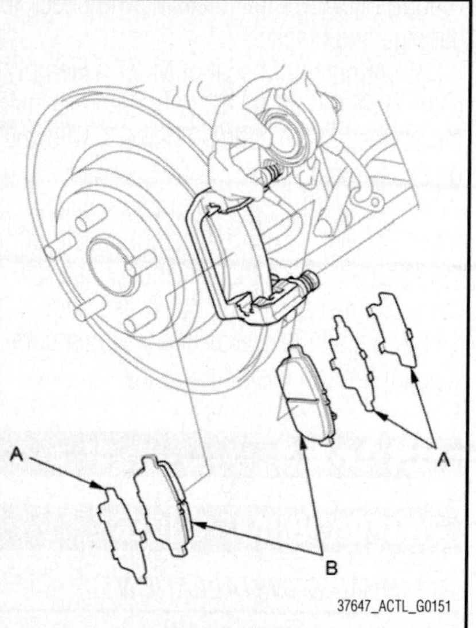
A
A
B
37647_ACTL_G0151

Fig. 11 Remove the pad shims (A) and the brake pads (B)

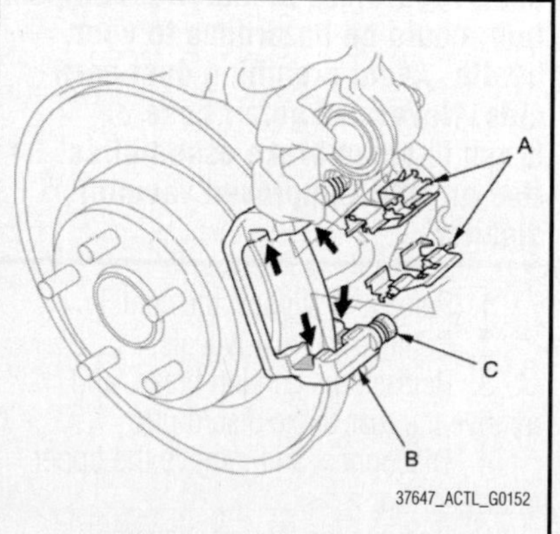
A
C
B
37647_ACTL_G0152

Fig. 12 Remove the upper and lower pad retainers (A)

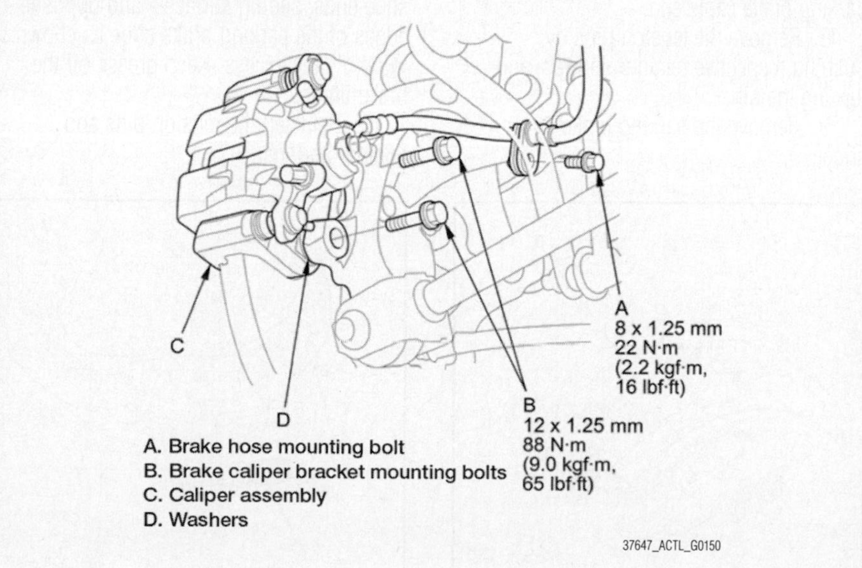
C
D
A
8 x 1.25 mm
22 N·m
(2.2 kgf·m,
16 lbf·ft)
B
12 x 1.25 mm
88 N·m
(9.0 kgf·m,
65 lbf·ft)
A. Brake hose mounting bolt
B. Brake caliper bracket mounting bolts
C. Caliper assembly
D. Washers
37647_ACTL_G0150

Fig. 10 Remove the brake hose mounting bolt and the brake caliper bracket mounting bolts, then remove the caliper assembly from the knuckle

CAUTION

To prevent damage to the caliper assembly or brake hose, use a short piece of wire to hang the caliper assembly from the undercarriage. Do not twist the brake hose excessively.

5. Remove the pad shims and the brake pads.
6. Remove the upper and lower pad retainers.

➡The upper and lower pad retainers are different. During installation, make sure the pad retainers are in their proper positions.

To install:

7. Clean the caliper bracket thoroughly; remove any rust, and check for grooves and cracks. Verify that the caliper pins move in and out smoothly. Clean and lube if needed.
8. Inspect the brake disc/drum for runout, thickness, parallelism, and check for damage and cracks.
9. Apply a thin coat of M-77 assembly paste (P/N 08798-9010) to the retainer mating surface of the caliper bracket (indicated by the arrows).
10. Install the upper and lower pad retainers. Wipe excess assembly paste off the retainers. Keep the assembly paste off the brake disc/drum and the brake pads.
11. Install the brake caliper piston compressor tool on the caliper body.

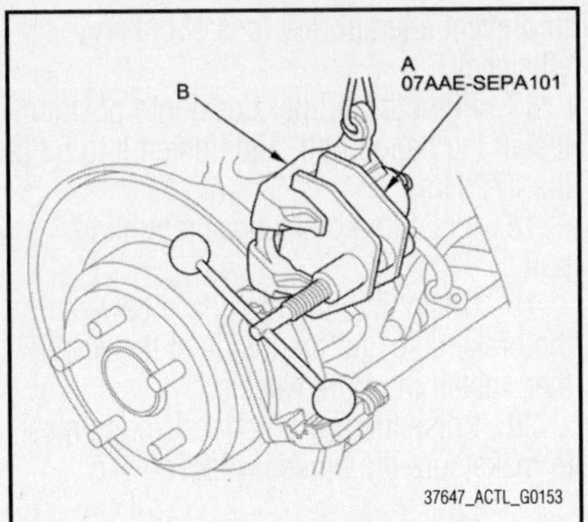
A
07AAE-SEPA101
B
37647_ACTL_G0153

Fig. 13 Install the brake caliper piston compressor tool (A) on the caliper body (B)

12. Press in the piston with the brake caliper piston compressor tool so the caliper will fit over the brake pads. Make sure the piston boot is in position to prevent damaging it when pivoting the caliper down.

➡Be careful when pressing in the piston; brake fluid might overflow from the master cylinder's reservoir. If brake fluid gets on any painted surface, wash it off immediately with water.

13. Remove the brake caliper piston compressor tool.
14. Apply a thin coat of M-77 assembly paste (P/N 08798-9010) to the pad side of the shims and the back of the brake pads. Wipe excess assembly paste off the pad shims and the brake pads friction material. Keep grease and assembly paste off the brake disc/drum and the brake pads. Contaminated brake disc/drum or brake pads reduce stopping ability.
15. Install the brake pads and pad shims correctly. Install the brake pad with the wear indicator on the bottom inside. If you are reusing the brake pads, always reinstall the brake pads in their original positions to prevent a temporary loss of braking efficiency.
16. Pivot the caliper down into position. Install the flange bolt, and tighten it to 16 ft. lbs. (22 Nm).
17. Install the brake hose mounting bolt.
18. Clean the mating surfaces between the brake disc/drum and the inside of the wheel, then install the rear wheels.
19. Press the brake pedal several times to make sure the brakes work.

➡Engagement may require a greater pedal stroke immediately after the brake pads have been replaced as a set. Several applications of the brake pedal will restore the normal pedal stroke.

20. Add brake fluid as needed.
21. After installation, check for leaks at hose and line joints or connections, and retighten if necessary.
22. Test-drive the vehicle, then recheck for leaks.

BRAKES

PARKING BRAKE

PARKING BRAKE SHOES

REMOVAL & INSTALLATION

See Figures 14 through 19.

CAUTION

Frequent inhalation of brake pad dust, regardless of material composition, could be hazardous to your health. Avoid breathing dust particles. Never use an air hose or brush to clean brake assemblies. Use an OSHA-approved vacuum cleaner.

1. Raise and support the vehicle.
2. Remove the rear wheels.
3. Release the parking brake, and remove the rear brake disc/drum.
4. Disconnect and remove the upper return spring.
5. Remove the adjuster assembly.
6. Disconnect and remove the lower return spring.
7. Remove the parking brake lever assembly, and disconnect it from the parking brake cable end.
8. Remove the tension pins by pushing respective retainer springs, and turning the pin.
9. Remove the parking brake shoes.

To install:

10. Apply Molykote 44MA grease to the shoe ends, sliding surfaces, and opposite edges of the parking brake shoe as shown. Wipe off any excess. Keep grease off the brake linings.
11. Reinstall the tension pins and retainer springs.

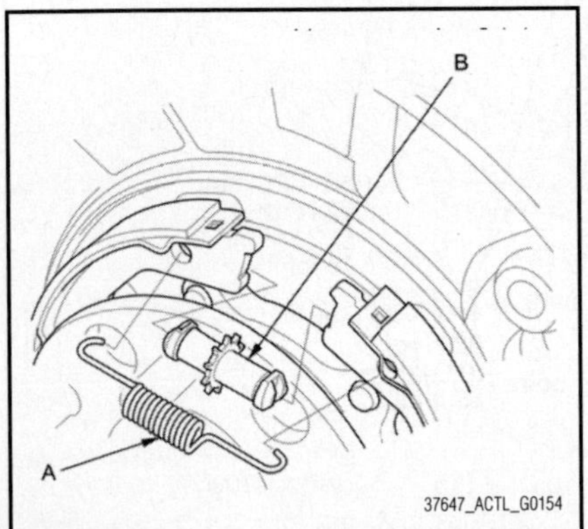
B
A
37647_ACTL_G0154

Fig. 14 Disconnect and remove the upper return spring (A)and adjuster assembly (B)

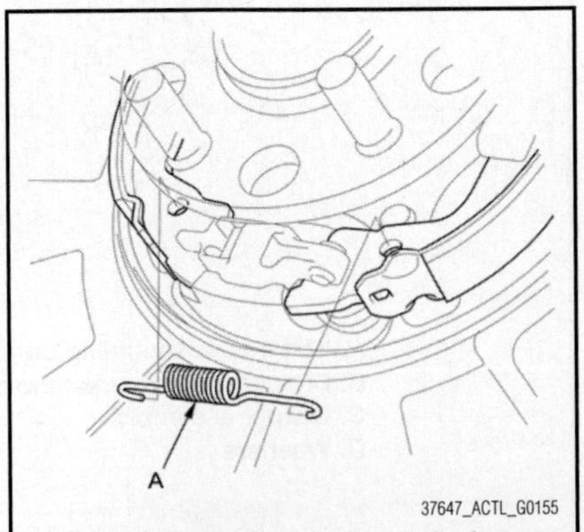
A
37647_ACTL_G0155

Fig. 15 Disconnect and remove the lower return spring (A)

12. Connect the parking brake cable end to the parking brake lever assembly, and install the parking brake lever assembly.
13. Reinstall the lower return spring.
14. Clean the threaded portions of the clevis pin A, and coat the threads of the clevis pin A with grease. Clean the sliding surface of the clevis pin B, and coat the sliding surface of the clevis pin B with grease. Install the clevis pin A and B on the adjuster and shorten the clevis pin A by turning the adjuster.
15. Position the brake shoe adjuster assembly on the parking brake shoes.
16. Reinstall the upper return spring.
17. Install the rear brake disc/drum and the rear brake caliper.
18. Do the major parking brake adjustment.
19. Clean the mating surfaces between the brake disc/drum and the inside of the wheel, then install the rear wheels.

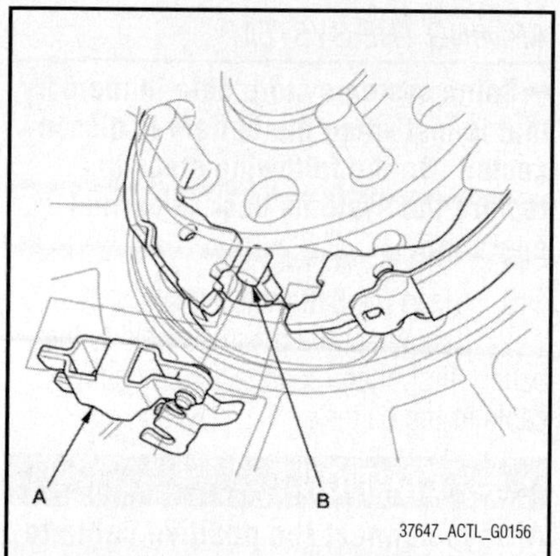
A
B
37647_ACTL_G0156

Fig. 16 Remove the parking brake lever assembly (A), and disconnect it from the parking brake cable end (B)

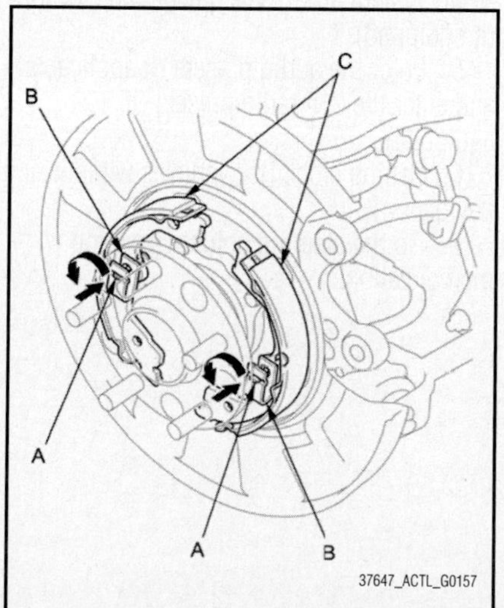
B
C
A
A
B
37647_ACTL_G0157

Fig. 17 Remove the tension pins (A) by pushing respective retainer springs (B), and turning the pin

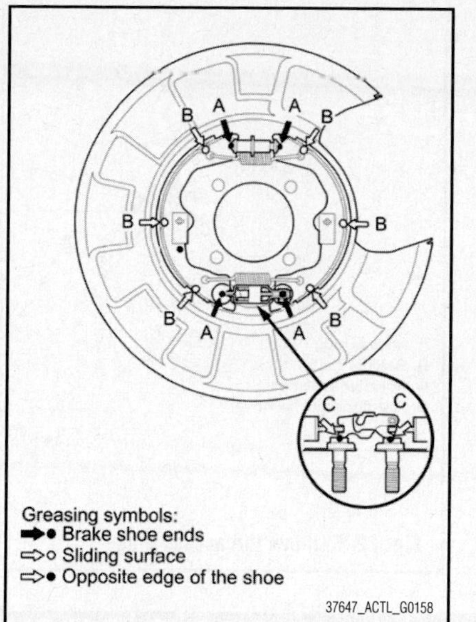
B A A B
B B
B A A B
C C
Greasing symbols:
Brake shoe ends
Sliding surface
Opposite edge of the shoe
37647_ACTL_G0158

Fig. 18 Apply Molykote 44MA grease to the shoe ends (A), sliding surfaces (B), and opposite edges of the parking brake shoe (C)

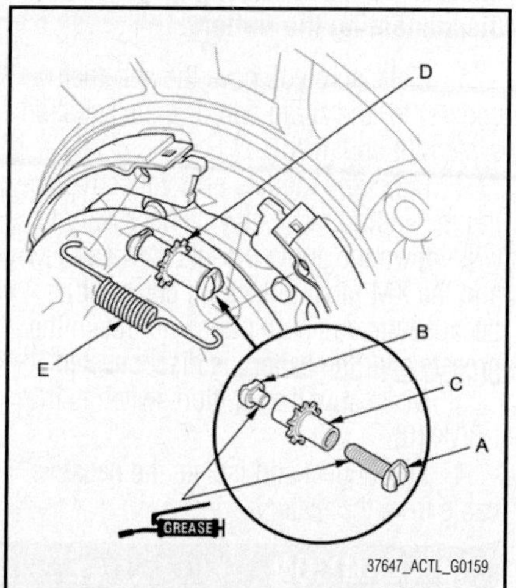
D
E
B
C
A
GREASE
37647_ACTL_G0159

Fig. 19 Installation of adjuster and upper return spring

ADJUSTMENT

1. Pull the parking brake lever with 44 lbs. (196 N) of force to fully apply the parking brake. The parking brake lever should be locked within the 6 to 8 clicks.
2. If the number of lever clicks is not as specified, adjust the parking brake.

➡**Minor parking brake lever adjustments (1 to 2 clicks) can be made with the adjusting nut (see parking brake minor adjustment). If a larger adjustment is required, follow the major adjustment procedure using the adjuster nut at the parking brake drum (see parking brake major adjustment).**

3. After installing new parking brake shoes and/or new brake disc/drum, make sure you drive the vehicle for "break-in".

Minor Adjustment

See Figure 20.

1. Raise and support the vehicle.
2. Release the parking brake lever fully.
3. Remove the console box mat, then remove the lid.

➡**It is easier to use packing tape to remove the console box mat.**

4. Pull the parking brake lever 1 click.
5. Tighten the parking brake adjusting nut until the parking brakes drag slightly when the rear wheels are turned.
6. Release the parking brake lever fully, and check that the parking brakes do not drag when the rear wheels are turned. Readjust if necessary.
7. Make sure the parking brake lever is within the specified number of clicks (6 to 8 clicks).
8. Install the lid, then install the console box mat.

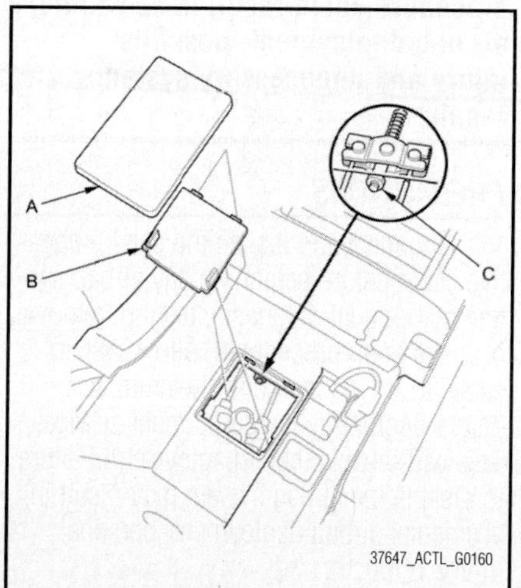
A
B
C
37647_ACTL_G0160

Fig. 20 Remove the console box mat (A), then remove the lid (B)

Major Adjustment (to be done when replacing parking brake shoes)

See Figures 21 and 22.

1. Raise and support the vehicle.
2. Release the parking brake lever fully.

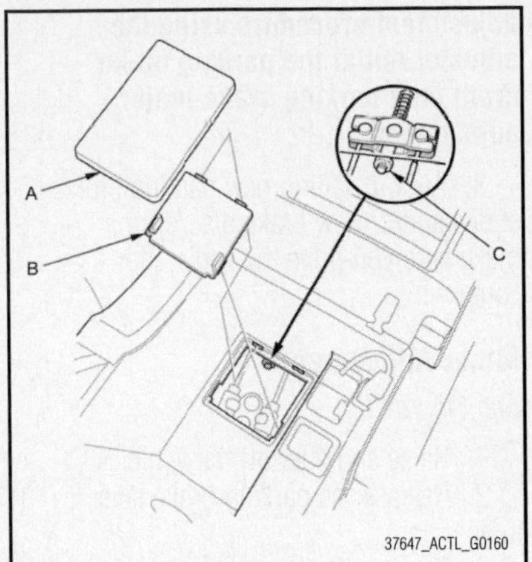
A
B
C
37647_ACTL_G0160

Fig. 21 Remove the console box mat (A), then remove the lid (B)

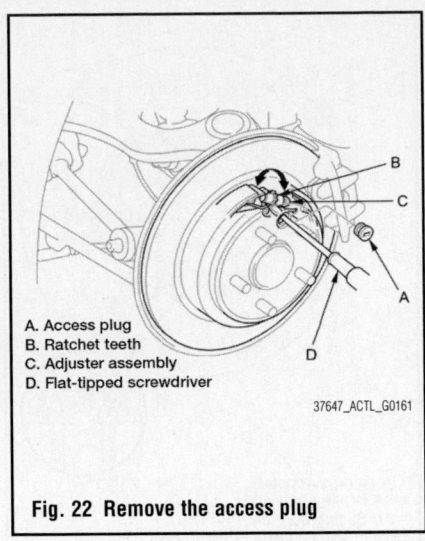

A. Access plug
B. Ratchet teeth
C. Adjuster assembly
D. Flat-tipped screwdriver
37647_ACTL_G0161

Fig. 22 Remove the access plug

3. Remove the console box mat, then remove the lid.

➡**It is easier to use packing tape to remove the console box mat.**

4. Loosen the parking brake adjusting nut.
5. Remove the rear wheels.
6. Remove the access plug.
7. Turn the ratchet teeth on the adjuster assembly with a flat-tip screwdriver until the shoes lock against the parking brake drum. Then back off the adjuster 8 clicks, and install the access plug.
8. Do the minor adjustment procedure.
9. Clean the mating surfaces between the brake disc/drum and the inside of the wheel, then install the rear wheels.

CHASSIS ELECTRICAL

AIR BAG (SUPPLEMENTAL RESTRAINT SYSTEM)

GENERAL INFORMATION

✲✲ CAUTION

These vehicles are equipped with an air bag system. The system must be disarmed before performing service on, or around, system components, the steering column, instrument panel components, wiring and sensors. Failure to follow the safety precautions and the disarming procedure could result in accidental air bag deployment, possible injury and unnecessary system repairs.

PRECAUTIONS

Disconnect and isolate the battery negative cable before beginning any airbag system component diagnosis, testing, removal, or installation procedures. Allow system capacitor to discharge for two minutes before beginning any component service. This will disable the airbag system. Failure to disable the airbag system may result in accidental airbag deployment, personal injury, or death.

DISARMING THE SYSTEM

➡**Some systems store data in memory that is lost when the battery is disconnected. Do the following steps before disconnecting the battery.**

1. Make sure you have the anti-theft code(s) for the audio and/or the navigation system (if equipped).
2. For some models or if you're replacing the audio unit, it may be necessary to write down the audio presets (AM and FM), and the XM radio presets (if equipped), because the audio unit does not retain the presets after the battery is disconnected.
3. Make sure the ignition switch is in LOCK (0).
4. Disconnect and isolate the negative cable from the battery.

✲✲ CAUTION

Always disconnect the negative cable from the battery first.

5. Disconnect the positive cable from the battery.
6. Wait at least 3 minutes for the battery and SRS system to discharge before performing any repairs.

ARMING THE SYSTEM

➡**Some systems store data in memory that is lost when the battery is disconnected. Do the following steps to restore the systems back to normal operation.**

1. Clean the battery terminals.
2. Reconnect the positive cable to the battery first, then reconnect the negative cable to the battery.

✲✲ CAUTION

Always connect the positive cable to the battery first.

3. Apply multipurpose grease to the terminals to prevent corrosion.
4. Enter the anti-theft code(s) for the audio system and/or the navigation system (if equipped).
5. Enter the audio presets (if applicable), and enter the XM radio presets (if equipped).
6. Set the clock (for vehicles without navigation).
7. Do the steering column position memorization.

DRIVE TRAIN

CLUTCH

REMOVAL & INSTALLATION

Special Tools Required:

- Clutch Alignment Disc 07JAF-PM7011A
- Ring Gear Holder 07LAB-PV00100
- Clutch Alignment Tool Set 07PAF-0020000
- Clutch Alignment Pilot, 20 mm 07PAF-0020360
- Clutch Alignment Shaft 07ZAF-PR8A100
- Attachment, 22 x 24 mm 07746-0010800
- Driver Handle, 15 x 135L 07749-0010000
- Remover Handle 07936-3710100
- Bearing Remover Shaft Set, 20 mm 07936-3710600
- Slide Hammer 07936-371020A

Engine Side

Pressure Plate Inspection and Removal

See Figures 23 through 25.

1. Remove the transmission.
2. Check the evenness of the height of the diaphragm spring fingers using the clutch alignment disc, the clutch alignment shaft, the remover handle, and a feeler gauge. If the height difference is more than the service limit—0.03inches (0.8 mm), replace the pressure plate.

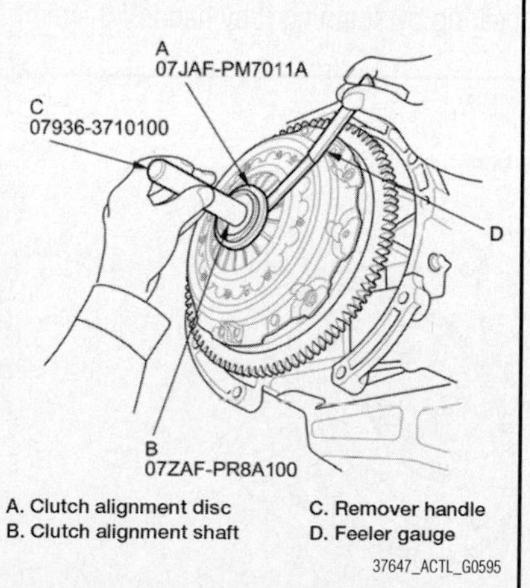
A
07JAF-PM7011A
C
07936-3710100
D
B
07ZAF-PR8A100
A. Clutch alignment disc
B. Clutch alignment shaft
C. Remover handle
D. Feeler gauge
37647_ACTL_G0595

Fig. 23 Check the evenness of the height of the diaphragm spring fingers using the clutch alignment disc, the clutch alignment shaft, the remover handle, and a feeler gauge

3. Install the ring gear holder, the clutch alignment tool set, and the 20 mm clutch alignment pilot.
4. To prevent warping, loosen the pressure plate mounting bolts in a crisscross pattern in several steps, then remove the pressure plate.
5. Inspect the fingers of the diaphragm spring for wear at the release bearing contact area.
6. Inspect the pressure plate surface for wear, cracks, and burning.
7. Inspect for warpage using a straight edge and a feeler gauge. Measure across the pressure plate. If the most measurement difference is more than the service limit—0.006 inches (0.15 mm), replace the pressure plate.

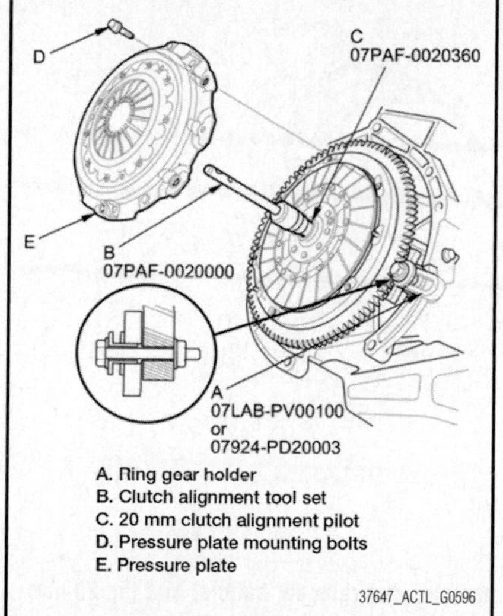
D
C
07PAF-0020360
E
B
07PAF-0020000
A
07LAB-PV00100
or
07924-PD20003
A. Ring gear holder
B. Clutch alignment tool set
C. 20 mm clutch alignment pilot
D. Pressure plate mounting bolts
E. Pressure plate
37647_ACTL_G0596

Fig. 24 Install the ring gear holder, the clutch alignment tool set, and the 20 mm clutch alignment pilot

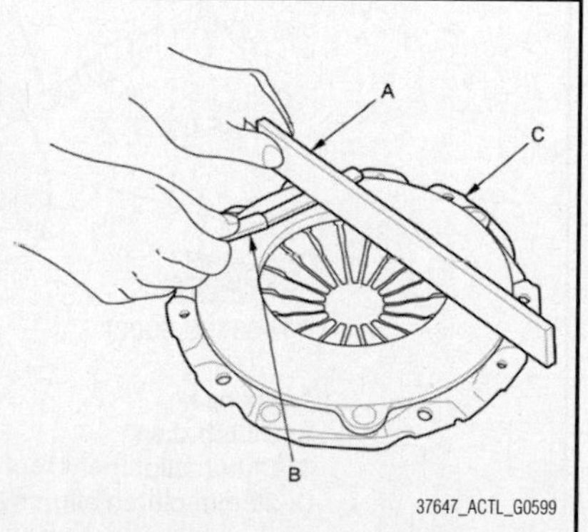
A
C
B
37647_ACTL_G0599

Fig. 25 Inspect for warpage using a straight edge (A) and a feeler gauge (B). Measure across the pressure plate (C)

Clutch Disc Inspection and Removal

1. Remove the clutch disc, the clutch alignment tool set, and the 20 mm clutch alignment pilot.
2. Inspect the lining of the clutch disc for signs of slippage or oil. If the clutch disc looks burnt or is oil soaked, replace the clutch disc. If the clutch disc is oil soaked, find and repair the source of the oil leak.
3. Measure the clutch disc thickness. If the measurement is less than the service limit—0.24 inches (6.0 mm), replace the clutch disc.
4. Measure the depths of the rivets from the clutch disc lining surface to the rivets on both sides. If the measurement is less than the service limit—0.03 inches (0.7 mm), replace the clutch disc.

Flywheel Inspection

1. Remove the ring gear holder.
2. Inspect the ring gear teeth for wear and damage.
3. Inspect the clutch disc mating surface on the flywheel for wear, cracks, and burning.
4. Measure the flywheel runout using a dial indicator. Through at least two full turns with pushing against the flywheel each time you turn it to take up the crankshaft thrust washer clearance. If the measurement is not within the standard (new)—0.01 inches (0.3 mm), replace the flywheel.

Flywheel Replacement

See Figure 26.

1. Install the ring gear holder.
2. Loosen the flywheel mounting bolts in a crisscross pattern in several steps. Remove the bolts, then remove the flywheel and the ring gear holder.

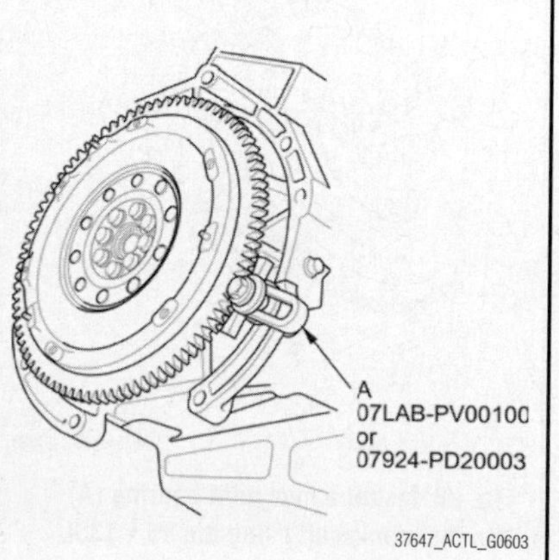
A
07LAB-PV00100
or
07924-PD20003
37647_ACTL_G0603

Fig. 26 Install the ring gear holder (A)

3. Install the flywheel on the crankshaft, and install the mounting bolts finger-tight.

4. Install the ring gear holder, then torque the flywheel mounting bolts in a crisscross pattern in several steps to 76 ft. lbs. (103 Nm).

Pilot Bearing Inspection

1. Turn the inner race of the pilot bearing with your finger. The pilot bearing should turn smoothly and quietly. Check that the pilot bearing outer race fits tightly in the flywheel. If the race does not turn smoothly, quietly, or fit tight in the flywheel, replace the pilot bearing.

Pilot Bearing Replacement

See Figures 27 and 28.

1. Remove the pilot bearing using the slide hammer, the remover handle, and the 20 mm bearing remover shaft set.

2. Install a new pilot bearing into the crankshaft using the 15 x 135L driver handle and the 22 x 24 mm attachment.

Clutch Disc and Pressure Plate Installation

See Figure 29.

1. Temporarily install the clutch disc onto the splines of the transmission mainshaft. Make sure the clutch disc slides freely on the mainshaft.

2. Apply super high temp urea grease (P/N 08798- 9002) to the splines of the clutch disc, then install the clutch disc using the clutch alignment tool set and the 20 mm clutch alignment pilot.

3. Align (±60 degrees) the paint marks across the pressure plate and the flywheel.

4. Install the pressure plate and the mounting bolts, finger-tight.

5. Torque the mounting bolts in a crisscross pattern. Tighten the bolts in several steps to prevent warping the diaphragm spring. Make sure that there is not clearance between the pressure plate and the flywheel. The final torque should be: 19 ft. lbs. (25 Nm)

6. Remove the ring gear holder, the clutch alignment tool set, and the 20 mm clutch alignment pilot.

7. Make sure the diaphragm spring fingers are all the same height.

8. Do the release bearing inspection, and replace it if necessary.

9. Install the transmission.

Transmission Side

Release Bearing Removal

See Figure 30.

1. Remove the transmission.

2. Remove the release fork boot from the clutch housing.

3. Remove the release fork from the clutch housing by squeezing the release fork set spring. Remove the release bearing.

Release Bearing Inspection

1. Check the play of the release bearing by spinning it by hand. If there

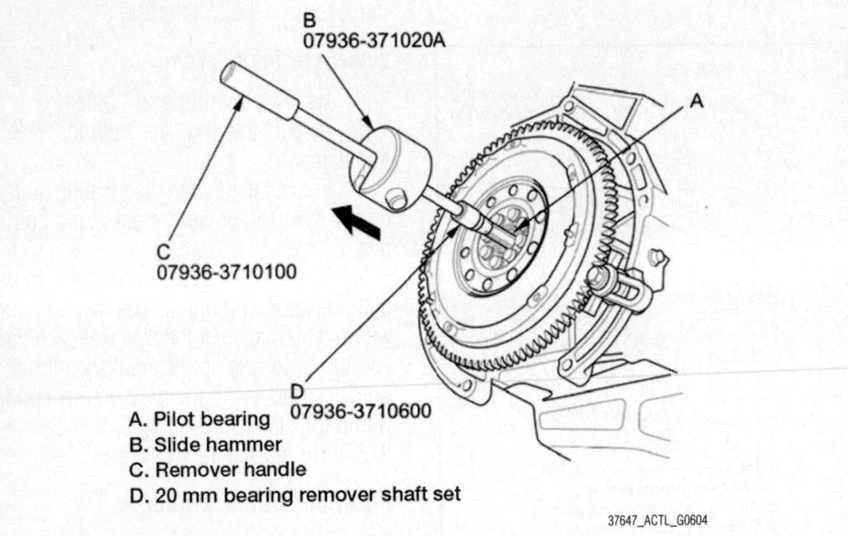
B
07936-371020A
A
C
07936-3710100
D
07936-3710600
A. Pilot bearing
B. Slide hammer
C. Remover handle
D. 20 mm bearing remover shaft set
37647_ACTL_G0604

Fig. 27 Remove the pilot bearing using the slide hammer, the remover handle, and the 20 mm bearing remover shaft set

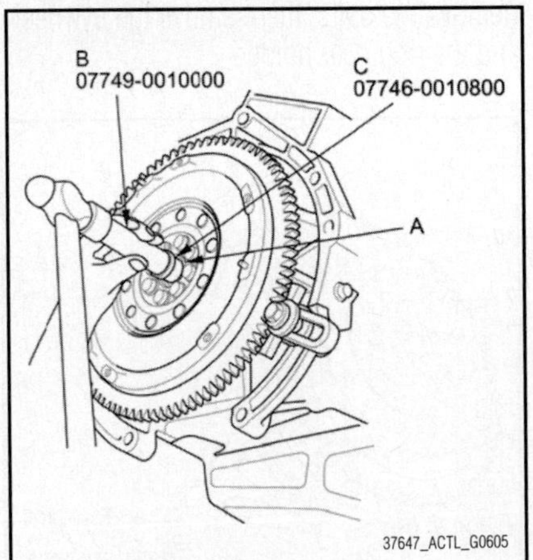
B
07749-0010000
C
07746-0010800
A
37647_ACTL_G0605

Fig. 28 Install a new pilot bearing (A) into the crankshaft using the 15 x 135L driver handle (B) and the 22 x 24 mm attachment (C)

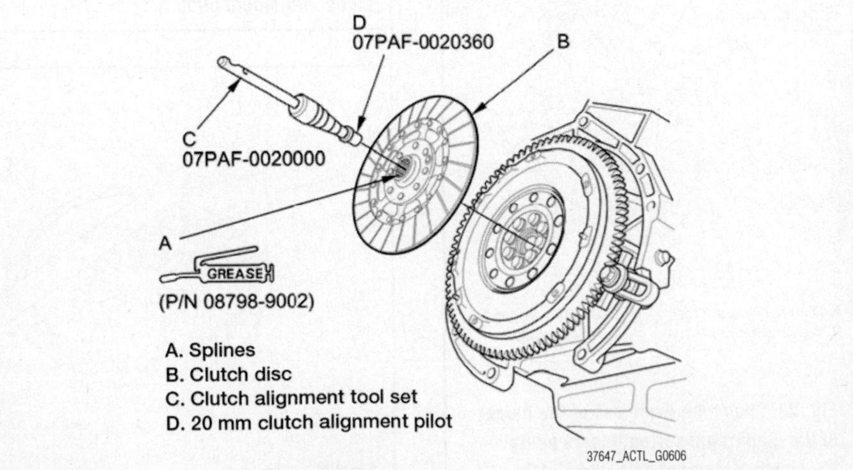
D
07PAF-0020360
B
C
07PAF-0020000
A
GREASE
(P/N 08798-9002)
A. Splines
B. Clutch disc
C. Clutch alignment tool set
D. 20 mm clutch alignment pilot
37647_ACTL_G0606

Fig. 29 Apply super high temp urea grease to the splines of the clutch disc, then install the clutch disc using the clutch alignment tool set and the 20 mm clutch alignment pilot

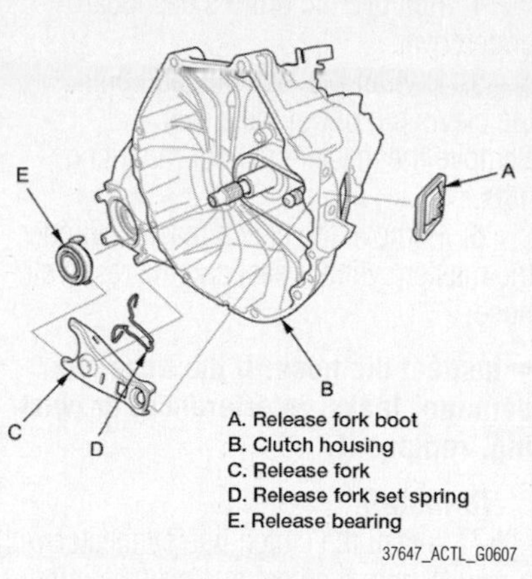
E
A
B
C
D
A. Release fork boot
B. Clutch housing
C. Release fork
D. Release fork set spring
E. Release bearing
37647_ACTL_G0607

Fig. 30 Remove the release fork boot from the clutch housing

is excessive play or noise, replace the release bearing.

➡The release bearing is packed with grease. Do not wash it in solvent.

Release Bearing Installation

See Figures 31 and 32.

1. Apply super high temp urea grease (P/N 08798-9002) to the release fork, the release fork bolt, the release bearing, and the release bearing guide in the shaded areas, then set the release fork set spring.

➡Replace the release fork bolt if necessary.

2. With the release fork slide between the release bearing pawls, install the release bearing on the mainshaft while inserting the release fork through the hole in the clutch housing.
3. Align the detent of the release fork with the release fork bolt, then press the detent of the release fork over the release fork bolt squarely.
4. Install the release fork boot. Make sure the boot seals around the release fork and the clutch housing.
5. Move the release fork right and left to make sure that it fits properly against the release bearing and that the release bearing slides smoothly. Wipe off any excess grease.
6. Install the transmission.

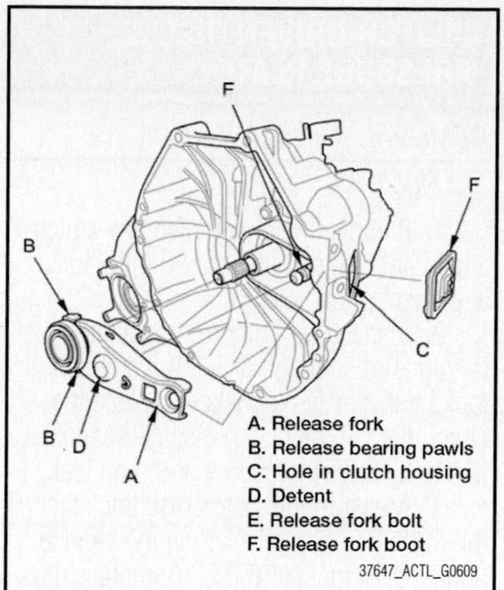
F
F
B
C
B D
A
A. Release fork
B. Release bearing pawls
C. Hole in clutch housing
D. Detent
E. Release fork bolt
F. Release fork boot
37647_ACTL_G0609

Fig. 32 Install the release bearing on the mainshaft while inserting the release fork through the hole in the clutch housing

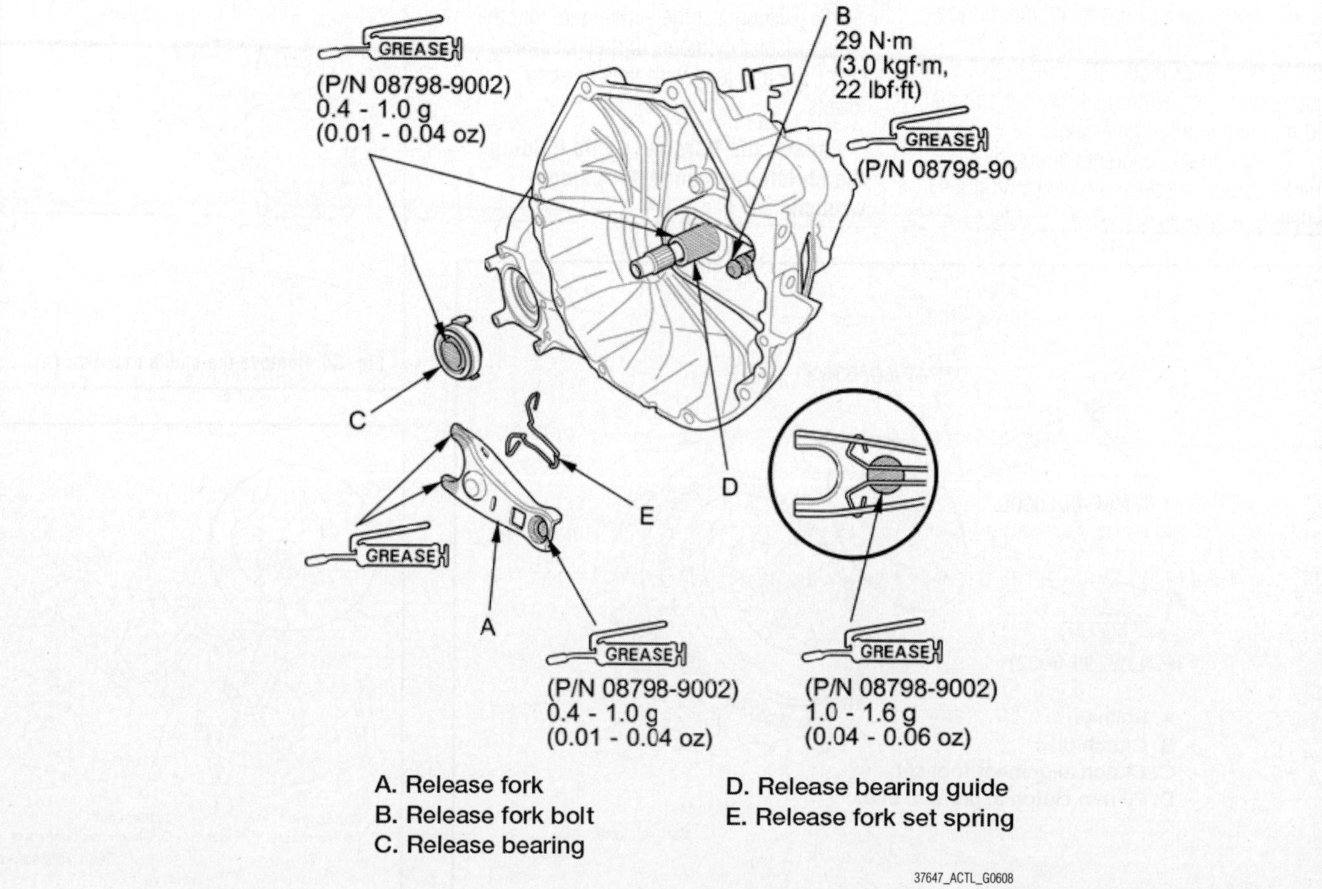
GREASE
(P/N 08798-9002)
0.4 - 1.0 g
(0.01 - 0.04 oz)
B
29 N·m
(3.0 kgf·m,
22 lbf·ft)
GREASE
(P/N 08798-90
C
D
E
GREASE
A
GREASE
(P/N 08798-9002)
0.4 - 1.0 g
(0.01 - 0.04 oz)
GREASE
(P/N 08798-9002)
1.0 - 1.6 g
(0.04 - 0.06 oz)
A. Release fork
B. Release fork bolt
C. Release bearing
D. Release bearing guide
E. Release fork set spring
37647_ACTL_G0608

Fig. 31 Apply super high temp urea grease to the release fork, the release fork bolt, the release bearing, and the release bearing guide in the shaded areas, then set the release fork set spring

CLUTCH DRIVEN DISC & PRESSURE PLATE

REMOVAL & INSTALLATION

See Figure 33.

1. Remove the clutch disc, the clutch alignment tool set, and the 20 mm clutch alignment pilot.
2. Inspect the lining of the clutch disc for signs of slippage or oil. If the clutch disc looks burnt or is oil soaked, replace the clutch disc. If the clutch disc is oil soaked, find and repair the source of the oil leak.
3. Measure the clutch disc thickness. If the measurement is less than the service limit—0.24 inches (6.0 mm), replace the clutch disc.
4. Measure the depths of the rivets from the clutch disc lining surface to the rivets on both sides. If the measurement is less than the service limit—0.03 inches (0.7 mm), replace the clutch disc.

To install:

5. Temporarily install the clutch disc onto the splines of the transmission mainshaft. Make sure the clutch disc slides freely on the mainshaft.
6. Apply super high temp urea grease (P/N 08798- 9002) to the splines of the clutch disc, then install the clutch disc using the clutch alignment tool set and the 20 mm clutch alignment pilot.
7. Align (±60 degrees) the paint marks across the pressure plate and the flywheel.
8. Install the pressure plate and the mounting bolts, finger-tight.
9. Torque the mounting bolts in a crisscross pattern. Tighten the bolts in several steps to prevent warping the diaphragm spring. Make sure that there is not clearance between the pressure plate and the flywheel. The final torque should be: 19 ft. lbs. (25 Nm)
10. Remove the ring gear holder, the clutch alignment tool set, and the 20 mm clutch alignment pilot.
11. Make sure the diaphragm spring fingers are all the same height.
12. Do the release bearing inspection, and replace it if necessary.
13. Install the transmission.

CLUTCH MASTER CYLINDER

REMOVAL & INSTALLATION

See Figures 34 through 44.

1. Remove and discard the brake fluid from the clutch master cylinder reservoir with a syringe or other suitable device.
2. Remove the clutch reservoir.
3. Disconnect the clutch line from the clutch master cylinder. To prevent spills, cover the line joint with rags or shop towels.

➡Loosen the flare nut while holding the clutch line connector using a wrench.

4. Remove the driver's dashboard undercover.
5. Pry out the lock pin, and pull the clevis pin out of the clevis. Remove the master cylinder mounting nuts.
6. Remove the clutch master cylinder, the master cylinder seal and the reservoir hose.

➡Inspect the hose. If the hose has damage, leaks, interference, or twisting, replace it.

To install:

7. Install the clutch line connector with a new O-ring, then set in a new retaining clip to the master cylinder.

➡Apply the silicone grease (P/N 08C30-B0234M) on the O-ring and the joint part of the clutch line connector. Make sure not to get any silicone grease on the terminal part of the connectors and switches, especially if you have silicone grease on your hands or gloves.

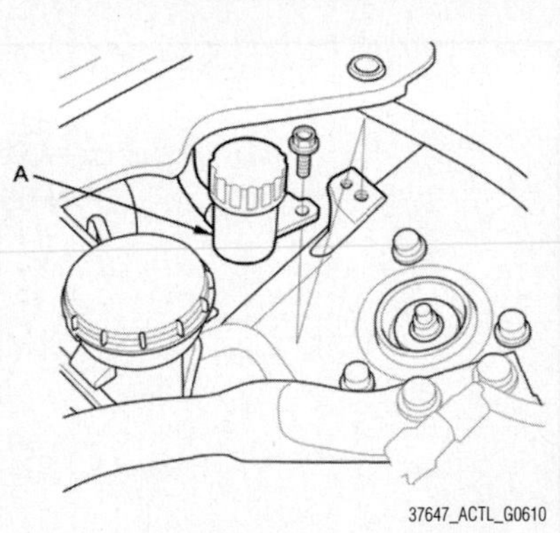
A
37647_ACTL_G0610

Fig. 34 Remove the clutch reservoir (A)

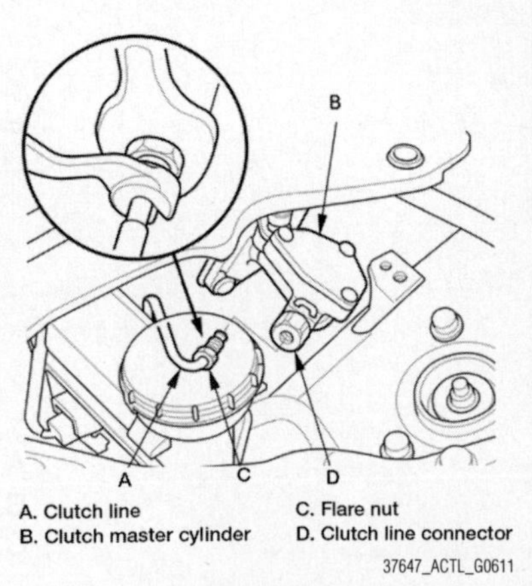
B
A
C
D
A. Clutch line
B. Clutch master cylinder
C. Flare nut
D. Clutch line connector
37647_ACTL_G0611

Fig. 35 Disconnect the clutch line from the clutch master cylinder

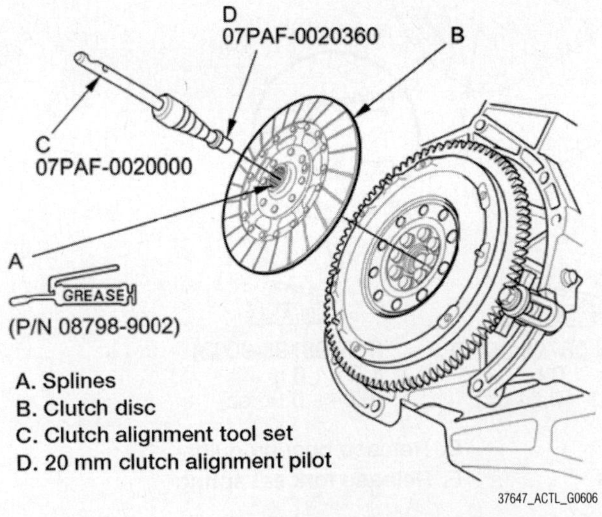
D
07PAF-0020360
B
C
07PAF-0020000
A
GREASE
(P/N 08798-9002)
A. Splines
B. Clutch disc
C. Clutch alignment tool set
D. 20 mm clutch alignment pilot
37647_ACTL_G0606

Fig. 33 Apply super high temp urea grease to the splines of the clutch disc, then install the clutch disc using the clutch alignment tool set and the 20 mm clutch alignment pilot

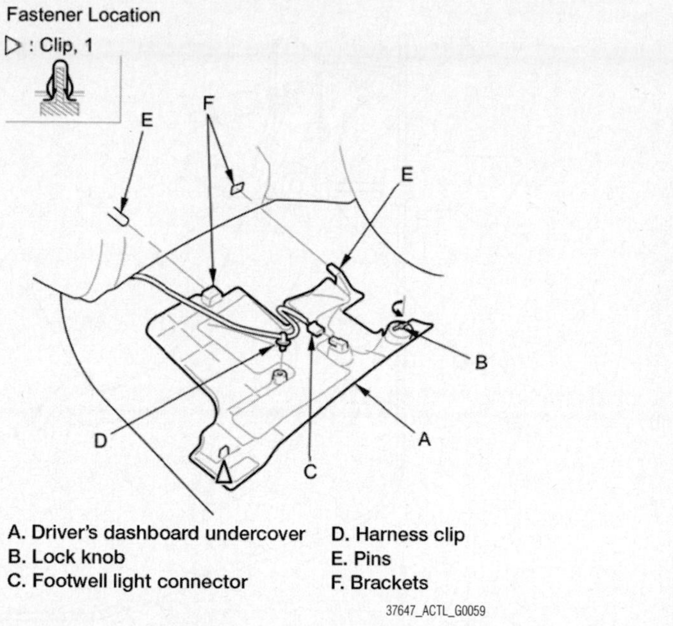
Fastener Location
▷ : Clip, 1
A. Driver's dashboard undercover
B. Lock knob
C. Footwell light connector
D. Harness clip
E. Pins
F. Brackets
37647_ACTL_G0059

Fig. 36 Remove the driver's dashboard undercover

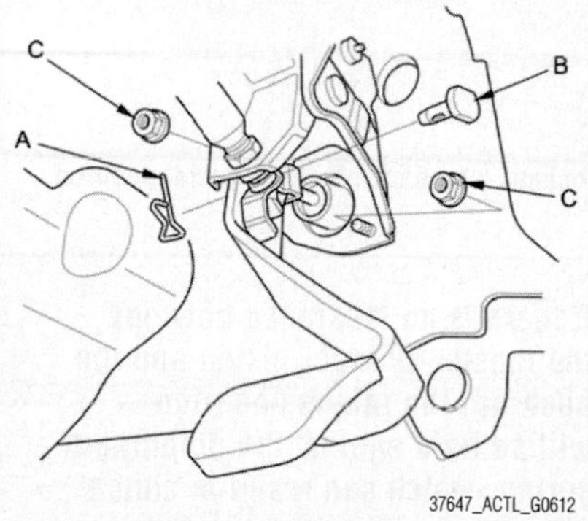
37647_ACTL_G0612

Fig. 37 Pry out the lock pin (A), and pull the clevis pin (B) out of the clevis. Remove the master cylinder mounting nuts (C).

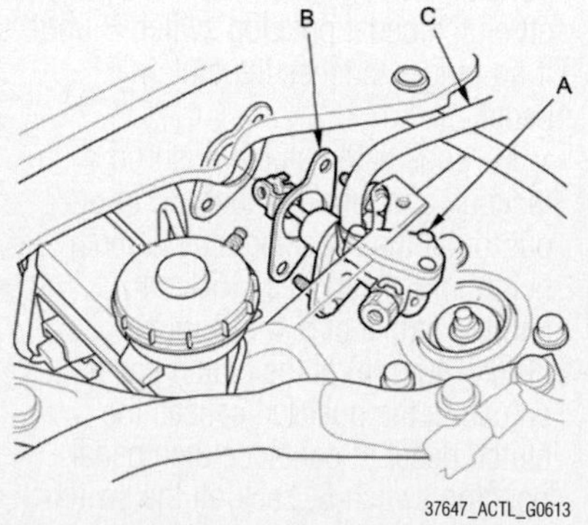
37647_ACTL_G0613

Fig. 38 Remove the clutch master cylinder (A), the master cylinder seal (B) and the reservoir hose (C)

8. To prevent the retaining clip from coming off, pry apart the tip of the clip with a screwdriver.

9. Install the reservoir hose, and adjust the hose clamp positions and its direction.

➡**When attaching the reservoir hose, align the yellow mark on the hose to the rib on reservoir connection area, and the blue mark on the hose to the rib on the clutch master cylinder connection area.**

10. Install a new master cylinder seal and the clutch master cylinder.

11. Install the master cylinder mounting nuts.

12. Apply multipurpose grease to the clevis pin, and the mating surfaces of the clevis and the pedal. Slide the clevis pin into the clevis, then install the lock pin.

13. Connect the clutch line to the clutch master cylinder.

➡**Tighten the flare nut while holding the clutch line connector using a wrench.**

14. Install the clutch reservoir.

15. Bleed the clutch hydraulic system.

16. Adjust the clutch pedal, clutch pedal position switch B, and clutch pedal position switch A.

➡**Remove the driver's floor mat before adjusting the clutch pedal.**

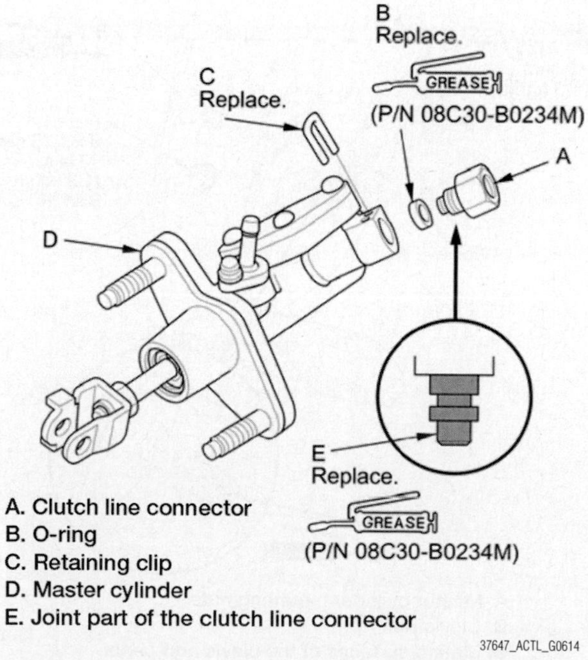
B
Replace.
GREASE
(P/N 08C30-B0234M)
C
Replace.
A
D
E
Replace.
GREASE
(P/N 08C30-B0234M)
A. Clutch line connector
B. O-ring
C. Retaining clip
D. Master cylinder
E. Joint part of the clutch line connector
37647_ACTL_G0614

Fig. 39 Install the clutch line connector with a new O-ring, then set in a new retaining clip to the master cylinder

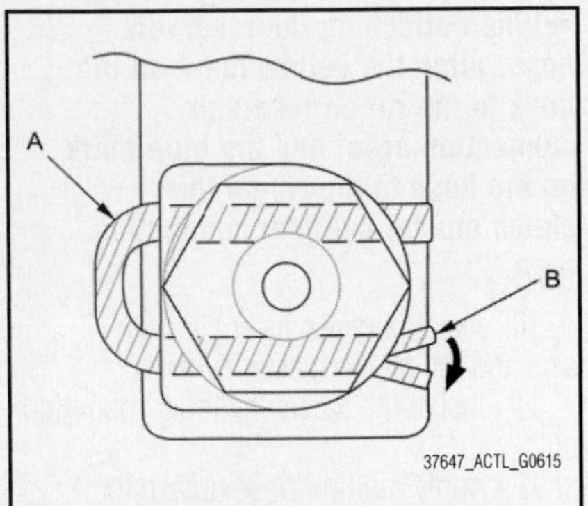
A
B
37647_ACTL_G0615

Fig. 40 To prevent the retaining clip (A) from coming off, pry apart the tip (B) of the clip with a screwdriver

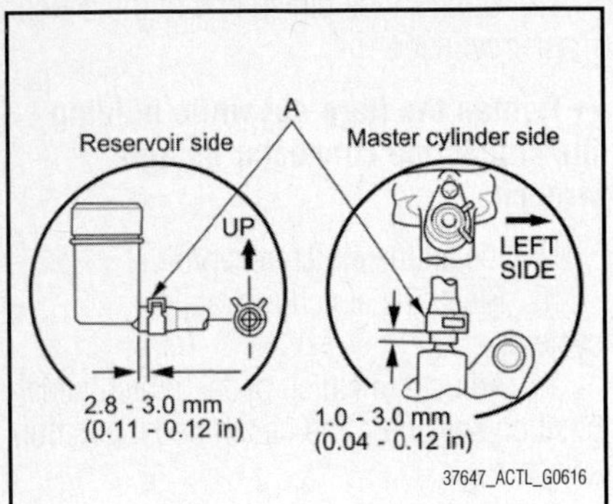
A
Reservoir side
Master cylinder side
UP
LEFT SIDE
2.8 - 3.0 mm
(0.11 - 0.12 in)
1.0 - 3.0 mm
(0.04 - 0.12 in)
37647_ACTL_G0616

Fig. 41 Install the reservoir hose, and adjust the hose clamp (A) positions and its direction

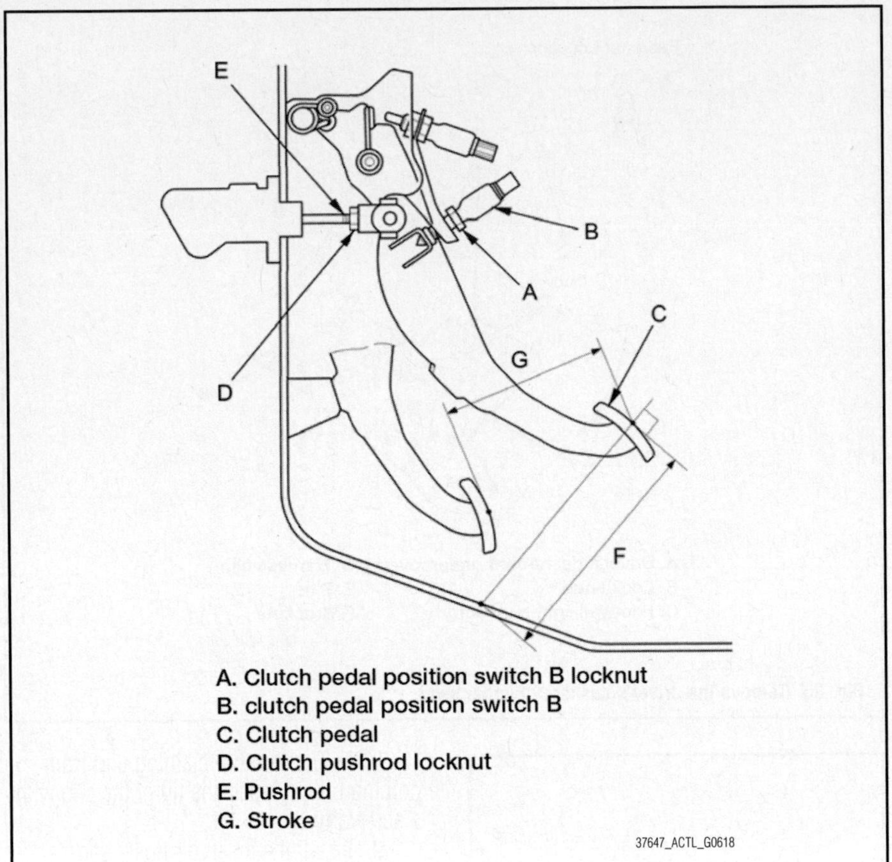
E
B
A
C
G
D
F
A. Clutch pedal position switch B locknut
B. clutch pedal position switch B
C. Clutch pedal
D. Clutch pushrod locknut
E. Pushrod
G. Stroke
37647_ACTL_G0618

Fig. 43 Loosen the clutch pedal position switch B locknut, and back off clutch pedal position switch B until it no longer touches the clutch pedal

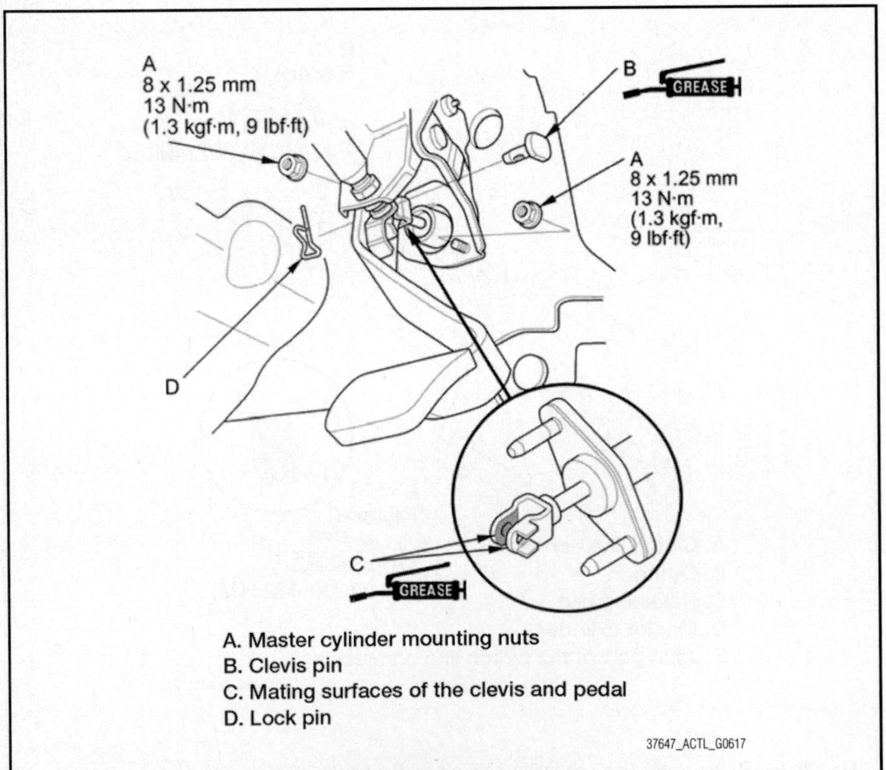
A
8 x 1.25 mm
13 N·m
(1.3 kgf·m, 9 lbf·ft)
B
GREASE
A
8 x 1.25 mm
13 N·m
(1.3 kgf·m,
9 lbf·ft)
D
C
GREASE
A. Master cylinder mounting nuts
B. Clevis pin
C. Mating surfaces of the clevis and pedal
D. Lock pin
37647_ACTL_G0617

Fig. 42 Install the master cylinder mounting nuts

If there is no clearance between the master cylinder piston and the pushrod, the release bearing will be held against the diaphragm spring, which can result in clutch slippage or other clutch problems.

a. Disconnect the clutch pedal position switch A connector and the clutch pedal position switch B connector.

b. Loosen the clutch pedal position switch B locknut, and back off clutch pedal position switch B until it no longer touches the clutch pedal.

c. Loosen the clutch pushrod locknut, and turn the pushrod in or out to get the specified clutch pedal height of 6.14 inches (156 mm), and the stroke of 5.1–5.5 inches (130–140 mm) at the clutch pedal. If adjusting the pushrod causes the clutch pedal to contact clutch pedal position switch B, back off the switch further.

d. Tighten the clutch pushrod locknut.

e. With the clutch pedal released, turn

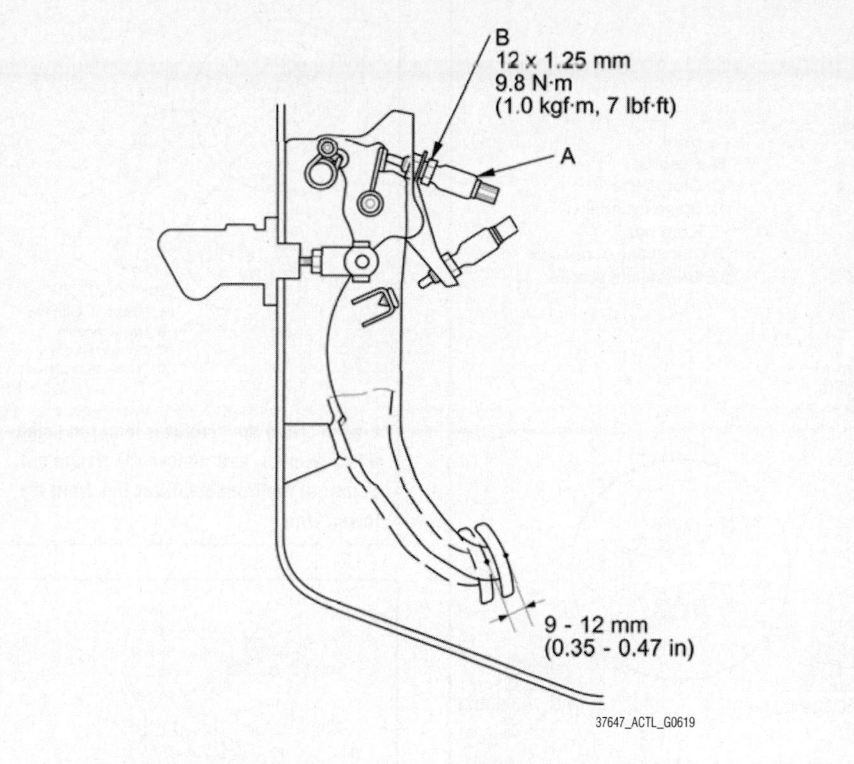
B
12 x 1.25 mm
9.8 N·m
(1.0 kgf·m, 7 lbf·ft)
A
9 - 12 mm
(0.35 - 0.47 in)
37647_ACTL_G0619

Fig. 44 Loosen the clutch pedal position switch A locknut (B)

in clutch pedal position switch B until it contacts the clutch pedal.

f. Turn in clutch pedal position switch B an additional 3/4 to 1 turn. Make sure the clutch pedal height did not change.

g. While holding clutch pedal position switch B, tighten the locknut.

h. Loosen the clutch pedal position switch A locknut.

i. Fully press the clutch pedal to the floor, then release the clutch pedal 0.35–0.47 inches (9–12 mm) and hold it there.

j. Adjust the position of clutch pedal position switch A so the engine starts with the clutch pedal in this position.

k. While holding clutch pedal position switch A, tighten the locknut.

l. Check the clutch operation.

m. Connect the clutch pedal position switch A connector and the clutch pedal position switch B connector, then check the cruise control and the clutch pedal position switch operation.

17. Check the clutch operation and check for leaks.
18. Reinstall the driver's dashboard undercover.
19. Test-drive the vehicle.

CLUTCH SLAVE CYLINDER

REMOVAL & INSTALLATION

See Figures 45 and 46.

1. Remove the clutch line bracket and the slave cylinder.
2. Disconnect the clutch line from the slave cylinder. To prevent spills, cover the line joint with rags or shop towels.

➡Loosen the flare nut while holding the clutch line connector using a wrench.

To install:

3. Pull back the boot, and apply silicone grease (P/N 08C30-B0234M) to the boot and slave cylinder.
4. Reinstall the boot.
5. Apply a light coat of super high temp urea grease (P/N 08798-9002) to the pushrod of the slave cylinder.
6. Connect the clutch line to the slave cylinder, and tighten the flare nut.

➡Tighten the flare nut while holding the clutch line connector using a wrench.

7. Install the slave cylinder and the clutch line bracket.
8. Bleed the clutch hydraulic system.
9. Check the clutch operation, and check for leaks.
10. Test-drive the vehicle.

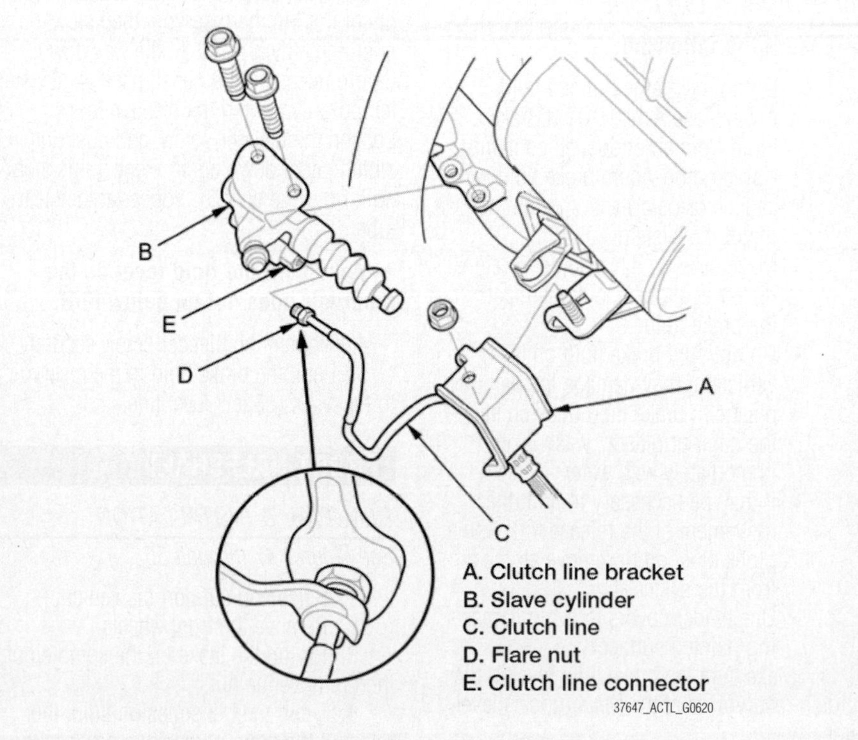
B
E
D
A
C
A. Clutch line bracket
B. Slave cylinder
C. Clutch line
D. Flare nut
E. Clutch line connector
37647_ACTL_G0620

Fig. 45 Exploded view of slave cylinder assembly

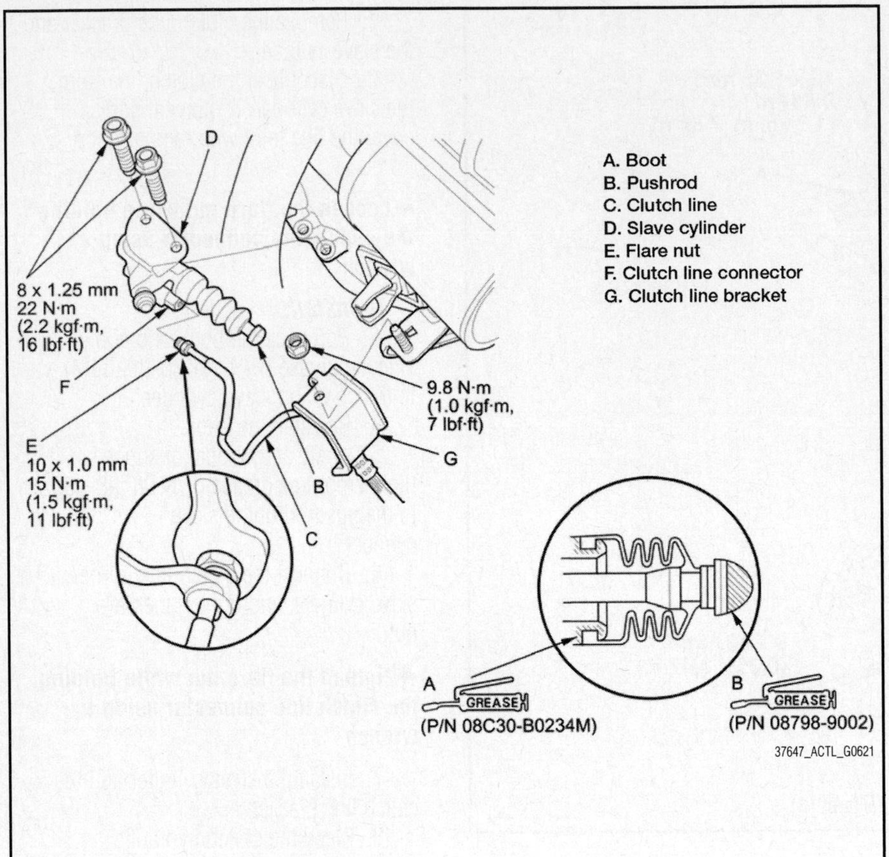
A. Boot
B. Pushrod
C. Clutch line
D. Slave cylinder
E. Flare nut
F. Clutch line connector
G. Clutch line bracket
8 x 1.25 mm
22 N·m
(2.2 kgf·m,
16 lbf·ft)
9.8 N·m
(1.0 kgf·m,
7 lbf·ft)
10 x 1.0 mm
15 N·m
(1.5 kgf·m,
11 lbf·ft)
GREASE
(P/N 08C30-B0234M)
GREASE
(P/N 08798-9002)
37647_ACTL_G0621

Fig. 46 Installing the slave cylinder

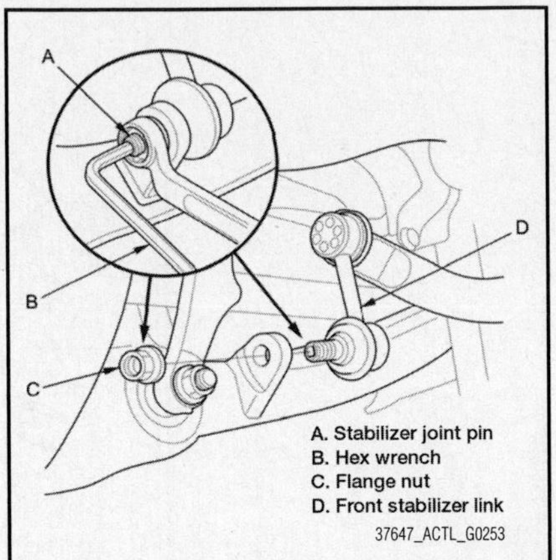
A. Stabilizer joint pin
B. Hex wrench
C. Flange nut
D. Front stabilizer link
37647_ACTL_G0253

Fig. 47 Hold the stabilizer joint pin using a hex wrench, and remove the flange nut. Separate the front stabilizer link from the lower arm.

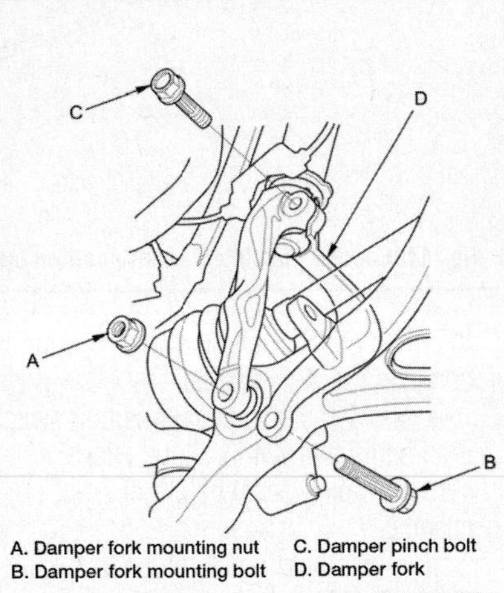
A. Damper fork mounting nut
B. Damper fork mounting bolt
C. Damper pinch bolt
D. Damper fork
37647_ACTL_G0254

Fig. 48 Remove the damper fork mounting nut, the damper fork mounting bolt, and the damper pinch bolt, then remove the damper fork

HYDRAULIC SYSTEM BLEEDING

BLEEDING PROCEDURE

➡Ensure the following:

- Do not reuse the drained fluid. Always use Acura DOT 3 Brake Fluid from an unopened container. Using a non-Acura brake fluid can cause corrosion and shorten the life of the system.
- Make sure no dirt or other foreign matter is allowed to contaminate the brake fluid.
- Do not spill brake fluid on the vehicle; it may damage the paint or plastic. If brake fluid does contact the paint or plastic, wash it off immediately with water.
- It may be necessary to limit the movement of the release fork with a block of wood to remove all the air from the system.
- Use fender covers to avoid damaging painted surfaces.

1. Make sure the brake fluid level in the clutch reservoir is at the MAX (upper) level line.
2. Attach one end of a clear tube to the bleeder screw, and put the other end into a container. Loosen the bleeder screw to allow air to escape from the system.
3. Make sure there is an adequate supply of fluid in the reservoir, then slowly push the clutch pedal all the way down. Before releasing the pedal, have an assistant temporarily tighten the bleeder screw. Loosen the bleeder screw, and push the clutch pedal down again. Repeat this step until no more bubbles appear at the clear tube.

➡Make sure the fluid level on the reservoir does not go below MIN.

4. Tighten the bleeder screw securely.
5. Refill the brake fluid in the reservoir to the MAX (upper) level line.

FRONT HALFSHAFT

REMOVAL & INSTALLATION

See Figures 47 through 56.

1. Raise and support the vehicle.
2. Remove the front wheels.
3. Pry up the stake on the spindle nut, then remove the nut.
4. Drain the transmission fluid, then reinstall the drain plug with a new sealing washer:
5. Hold the stabilizer joint pin using a hex wrench, and remove the flange nut. Separate the front stabilizer link from the lower arm.
6. Remove the damper fork mounting nut, the damper fork mounting bolt, and the damper pinch bolt, then remove the damper fork.
7. Remove the cotter pin from the knuckle ball joint, then remove the castle nut. Separate the ball joint from the lower arm using the ball joint remover.

➡Be careful not to damage the ball joint boot when installing the remover. Do not force or hammer on the lower arm, or pry between the lower arm and the knuckle. You could damage the ball joint.

8. Pull the knuckle outward, and separate the outboard joint from the front hub using a soft face hammer.

9. Remove exhaust pipe A.

10. Left driveshaft: Pry the inboard joint from the differential using a prybar.

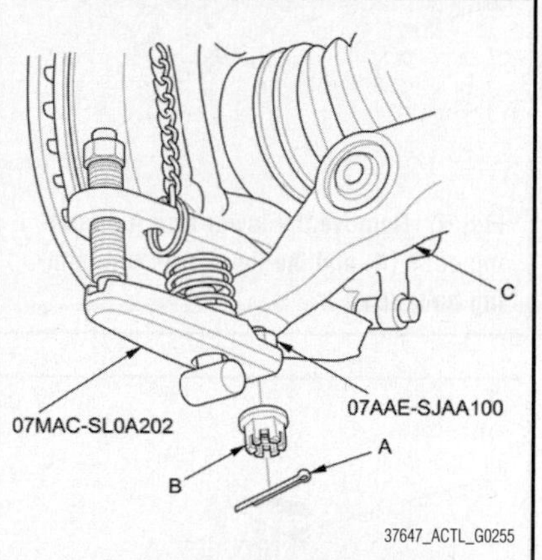
C
07MAC-SL0A202
07AAE-SJAA100
A
B
37647_ACTL_G0255

Fig. 49 Remove the cotter pin (A) from the knuckle ball joint, then remove the castle nut (B). Separate the ball joint from the lower arm (C) using the ball joint remover.

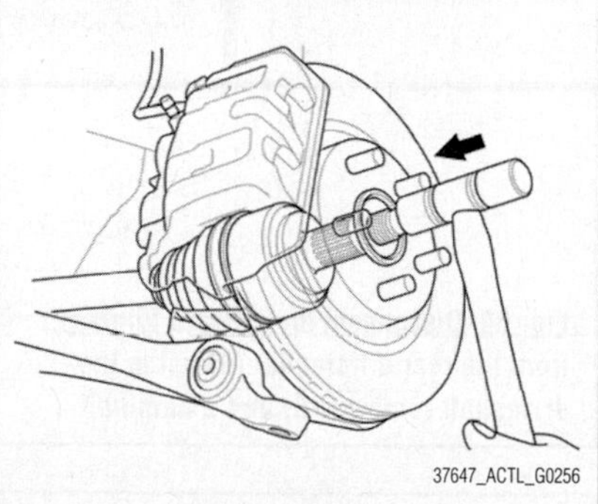
37647_ACTL_G0256

Fig. 50 Pull the knuckle outward, and separate the outboard joint from the front hub using a soft face hammer

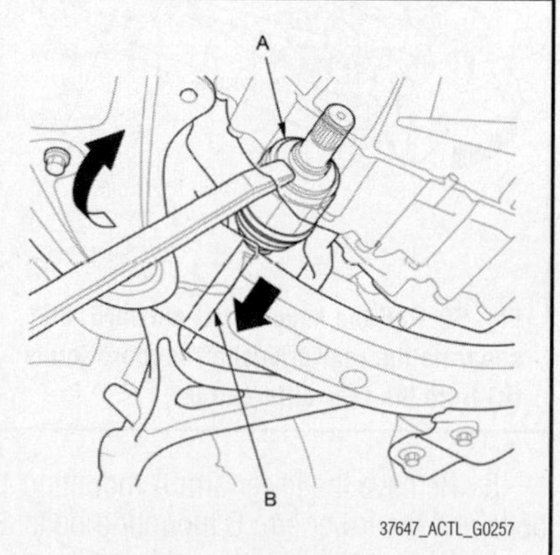
A
B
37647_ACTL_G0257

Fig. 51 Pry the inboard joint (A) from the differential using a prybar

11. Remove the driveshaft as an assembly.

➡**Do not pull on the driveshaft, or the inboard joint may come apart. Pull the inboard joint straight out to avoid damaging the oil seal. Be careful not to damage the oil seal or the end of the inboard joint using the prybar.**

12. Right driveshaft: Drive the inboard joint off of the intermediate shaft using a drift punch and a hammer.

13. Remove the driveshaft as an assembly.

➡**Do not pull on the driveshaft, or the inboard joint may come apart.**

To install:

➡**Before starting installation, make sure the mating surfaces of the joint and the splined section are clean.**

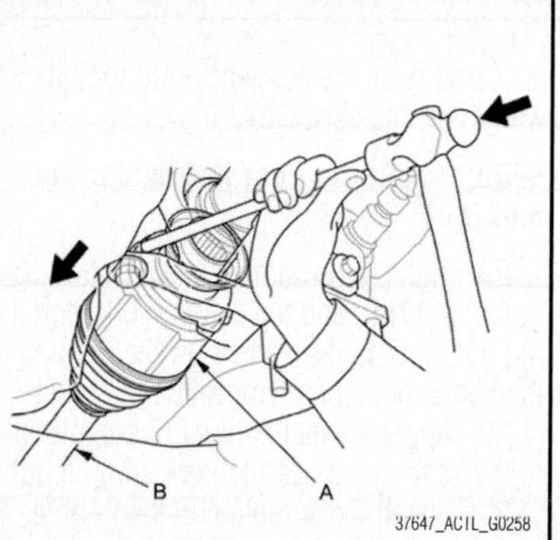
B
A
37647_ACTL_G0258

Fig. 52 Drive the inboard joint (A) off of the intermediate shaft using a drift punch and a hammer

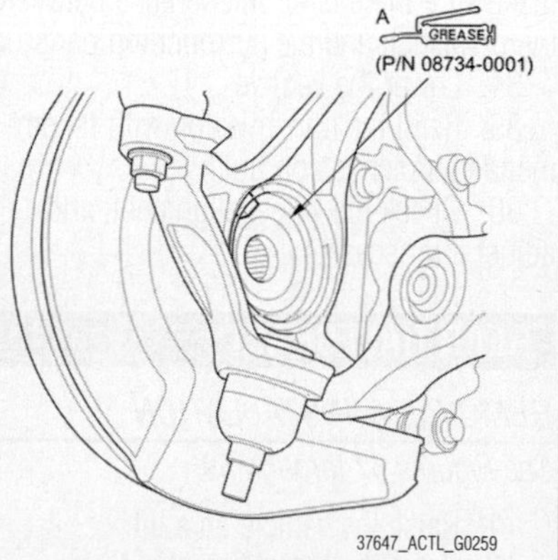
A
GREASE
(P/N 08734-0001)
37647_ACTL_G0259

Fig. 53 Apply Moly 60 paste to the contact area (A) of the outboard joint and the front wheel bearing

14. Apply Moly 60 paste (P/N 08734-0001) to the contact area of the outboard joint and the front wheel bearing.

➡**The paste helps prevent noise and vibration.**

15. Install a new set ring into the set ring groove of the left driveshaft inboard joint.

16. Install a new set ring into the set ring groove of the intermediate shaft.

17. Apply small quantity of super high temperature grease (P/N 08798-9002) to the whole splined surface of the right driveshaft.

18. After applying grease, remove the grease from the splined grooves at intervals of 2–3 splines and from the set ring groove so that air can bleed from the intermediate shaft.

19. Clean the areas where the driveshaft contacts the differential thoroughly with solvent, and dry them with compressed air.

➡**Do not wash the rubber parts with solvent.**

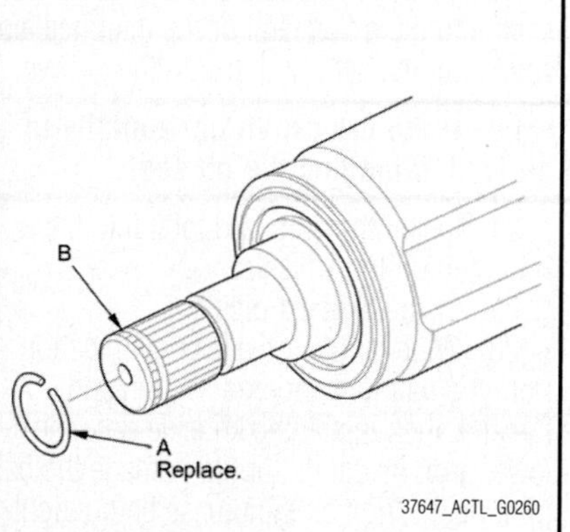
B
A
Replace.
37647_ACTL_G0260

Fig. 54 Install a new set ring (A) into the set ring groove (B) of the left driveshaft inboard joint

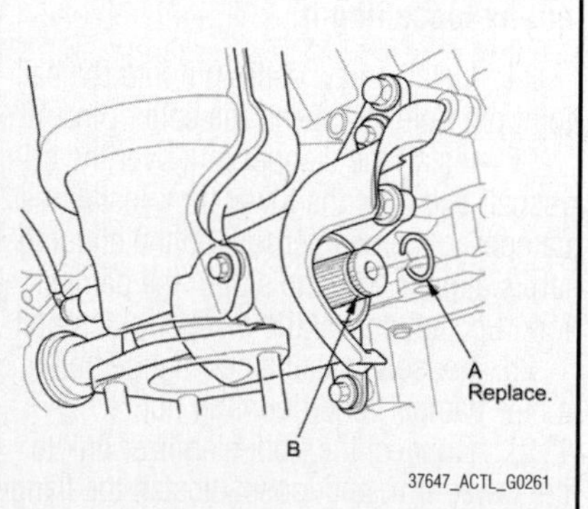
A
Replace.
B
37647_ACTL_G0261

Fig. 55 Install a new set ring (A) into the set ring groove (B) of the intermediate shaft

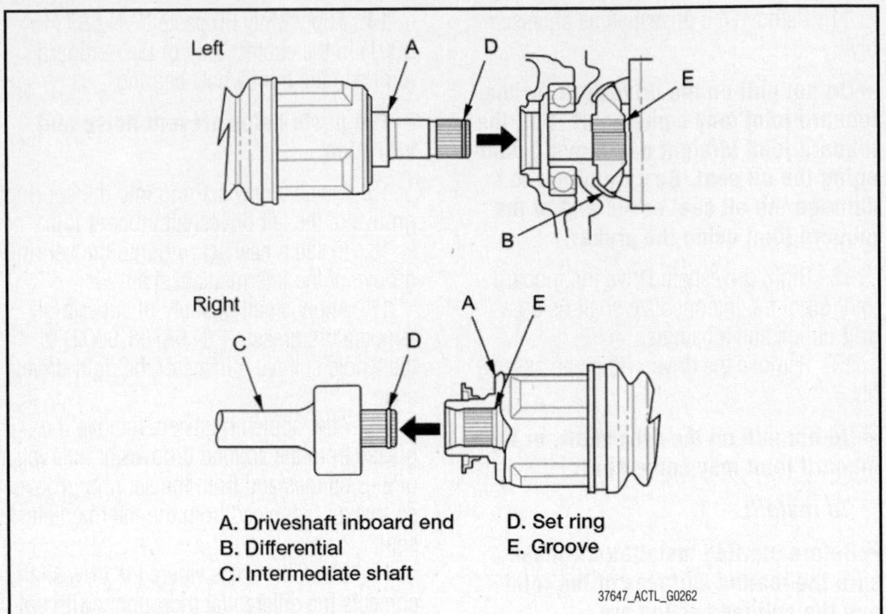
Left
A
D
E
B
Right
C
D
A
E
A. Driveshaft inboard end
B. Differential
C. Intermediate shaft
D. Set ring
E. Groove
37647_ACTL_G0262

Fig. 56 Insert the inboard end of the driveshaft into the differential or the intermediate shaft until the set ring locks in the groove

20. Insert the inboard end of the driveshaft into the differential or the intermediate shaft until the set ring locks in the groove.

➡Insert the driveshaft horizontally to prevent damaging the oil seal.

21. Install the outboard joint into the front hub on the knuckle.
22. Install exhaust pipe A.
23. Wipe off any grease contamination from the ball joint tapered section and threads, then install the knuckle onto the lower arm. Be careful not to damage the ball joint boot. Wipe off the grease before tightening the nut at the ball joint. Torque the castle nut to 69–76 ft. lbs. (93–103 N m), then tighten it only far enough to align the slot with the ball joint pin hole.

➡Make sure the ball joint boot is not damaged or cracked. Do not align the nut by loosening it.

24. Install a new cotter pin into the ball joint pin hole, and bend the cotter pin.
25. Install the damper fork over the driveshaft and onto the lower arm. Install the damper in the damper fork so the aligning tab is aligned with the slot in the damper fork. Loosely install the damper pinch bolt.
26. Loosely install a new flange bolt and a new damper fork mounting nut.
27. Connect the front stabilizer link to the lower arm, and loosely install the flange nut. Hold the stabilizer link ball joint pin using a hex wrench, and tighten the flange nut to 58 ft. lbs. (78 Nm).
28. Place a floor jack under the lower arm, and raise the suspension to load it with the vehicle's weight.

➡Do not put the floor jack under the ball joint.

29. Tighten the damper pinch bolt to 36 ft. lbs. (49 Nm) and the damper fork mounting nut to 47 ft. lbs. (64 Nm)while holding the flange bolt, then remove the floor jack.
30. Apply a small amount of engine oil to the seating surface of a new spindle nut.
31. Install the spindle nut, then tighten it to 242 ft. lbs. (329 Nm). After tightening, use a drift to stake the spindle nut shoulder against the driveshaft.
32. Clean the mating surfaces of the brake disc and the wheel, then install the front wheels.
33. Turn the front wheel by hand, and make sure there is no interference between the driveshaft and the surrounding parts.
34. Lower the vehicle .
35. Refill the transmission with recommended transmission fluid:
36. Check the wheel alignment, and adjust it if necessary.

REAR HALFSHAFT

REMOVAL & INSTALLATION

See Figures 57 through 59.

1. Raise the vehicle on a lift.
2. Drain the rear differential fluid.
3. Remove the rear wheels.
4. Remove exhaust pipe B and the muffler.

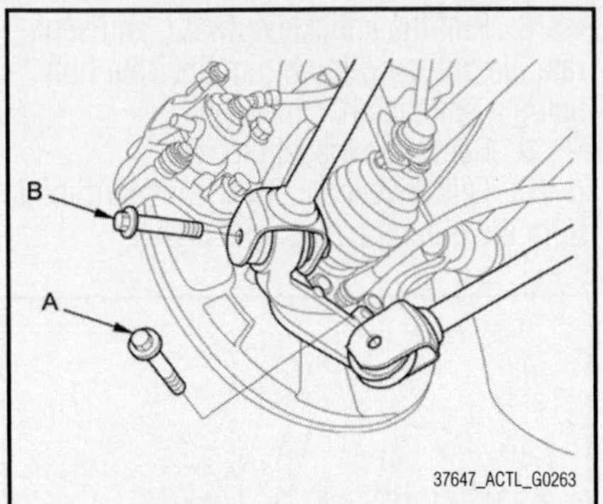
B
A
37647_ACTL_G0263

Fig. 57 Remove the lower arm A mounting bolts (A) and the lower arm B mounting bolts (B)

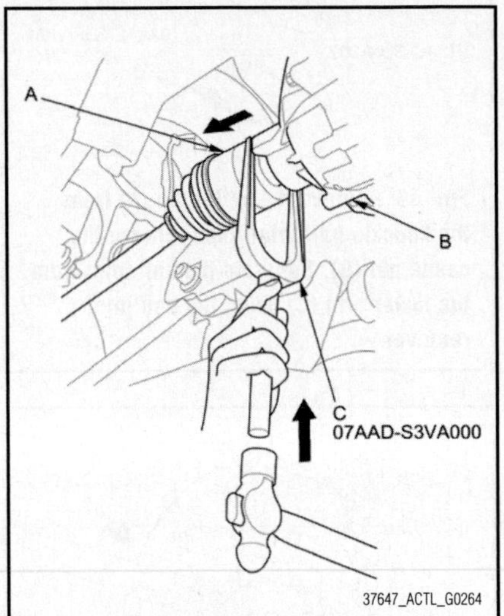
A
B
C
07AAD-S3VA000
37647_ACTL_G0264

Fig. 58 Disconnect the inboard joints (A) from the rear differential (B) using the driveshaft remover (C) and a hammer

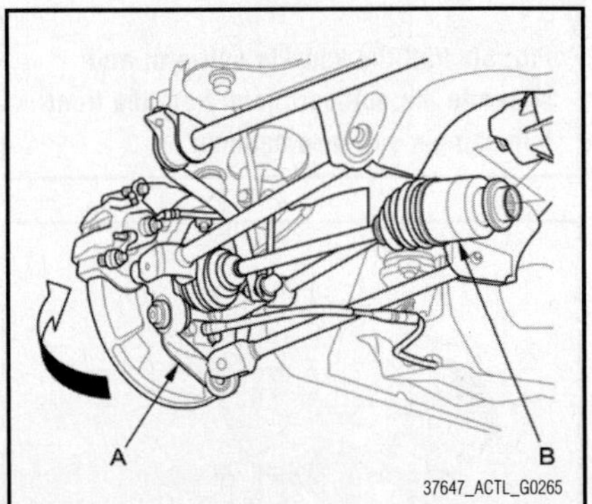
A
B
37647_ACTL_G0265

Fig. 59 Pull the knuckle (A) outward, and separate the rear driveshaft inboard joints (B) from the rear differential

5. Remove the lower arm A mounting bolts and the lower arm B mounting bolts.
6. Disconnect the inboard joints from the rear differential using the driveshaft remover and a hammer.

7. Pull the knuckle outward, and separate the rear driveshaft inboard joints from the rear differential.

To install:

8. Pull the knuckle outward, and insert the rear driveshaft inboard joints into the rear differential.

9. Loosely install new lower arm A mounting bolts and new lower arm B mounting bolts.

10. Place a floor jack under the lower arm, and raise the suspension to load it with the vehicle's weight.

11. Tighten the lower arm A mounting bolt and the lower arm B mounting bolt to 43 ft. lbs. (59 Nm).

12. Install exhaust pipe B and the muffler.

13. Install the rear wheels.

14. Refill the rear differential with the recommended fluid.

15. Test-drive the vehicle.

TRANSFER CASE ASSEMBLY

REMOVAL & INSTALLATION

See Figures 60 through 63.

1. Raise the vehicle on a lift, and make sure it is supported securely.

2. Move the shift lever to N.

3. Remove the drain plug, and drain the Automatic Transmission Fluid (ATF).

4. Reinstall the drain plug with a new sealing washer.

5. Remove the front subframe stiffener.

6. Remove the exhaust pipe A.

7. Remove the bolt securing the transfer breather hose bracket, and disconnect the breather hose from the breather pipe on the transfer assembly.

8. Make a reference mark across the No. 1 propeller shaft and the transfer companion flange.

9. Separate the No. 1 propeller shaft from the transfer companion flange.

10. Remove the transfer assembly and the dowel pin from the transmission.

To install:

11. Clean the areas where the transfer assembly contacts the transmission with solvent, and dry with compressed air. Then apply transmission fluid to the contact area.

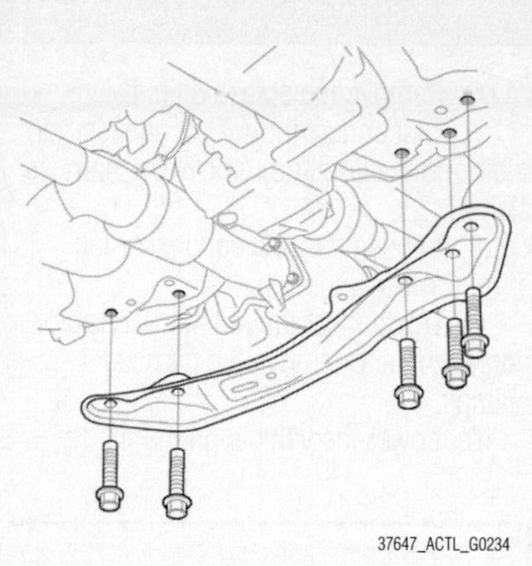
37647_ACTL_G0234

Fig. 60 Remove the front subframe stiffener

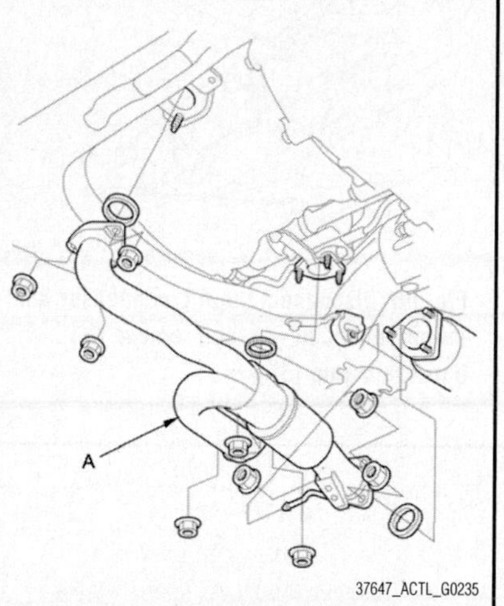
A
37647_ACTL_G0235

Fig. 61 Remove exhaust pipe A

12. Install the dowel pin in the transmission, and install the transfer assembly on the transmission. Tighten the transfer assembly mounting bolts to 38 ft. lbs. (51 Nm).

13. Install the No. 1 propeller shaft to the transfer companion flange by aligning the reference mark. Tighten the mounting nuts to 53 ft. lbs. (72 Nm).

14. Secure the transfer breather tube bracket on the transfer assembly with the bolt, and install the breather tube over the breather pipe.

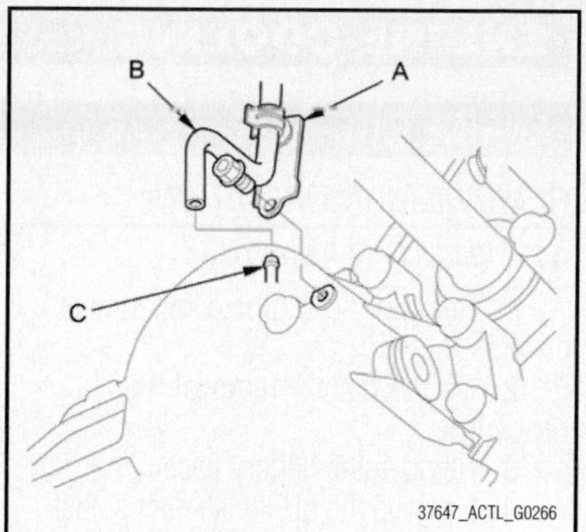
B
A
C
37647_ACTL_G0266

Fig. 62 Remove the bolt securing the transfer breather hose bracket (A), and disconnect the breather hose (B) from the breather pipe (C) on the transfer assembly

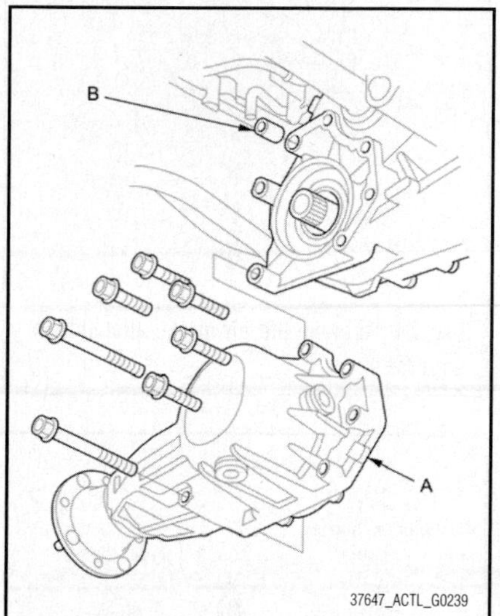
B
A
37647_ACTL_G0239

Fig. 63 With SH-AWD: Remove the transfer assembly (A) and dowel pin (B) from the transmission

15. Install the exhaust pipe A with the new self-locking nuts and new gaskets.

16. Install the front subframe stiffener with new mounting bolts.

17. Refill the transfer assembly with the transfer fluid (hypoid gear oil), if necessary.

18. Move the shift lever to N.

19. Refill the transmission with ATF.

ENGINE COOLING

ENGINE FAN

REMOVAL & INSTALLATION

See Figures 64 through 70.

1. Remove the engine compartment covers.
2. Do the battery removal procedure.
3. Remove the battery base.
4. Remove the air intake duct splash separator.
5. Raise the vehicle on the lift.
6. Remove the splash shield.
7. Disconnect the A/C condenser fan motor connector, and remove the harness clamp.
8. Loosen the A/C condenser fan shroud mounting bolts.
9. Disconnect the radiator fan motor connector, and remove the harness clamps.
10. Lower the vehicle on the lift.

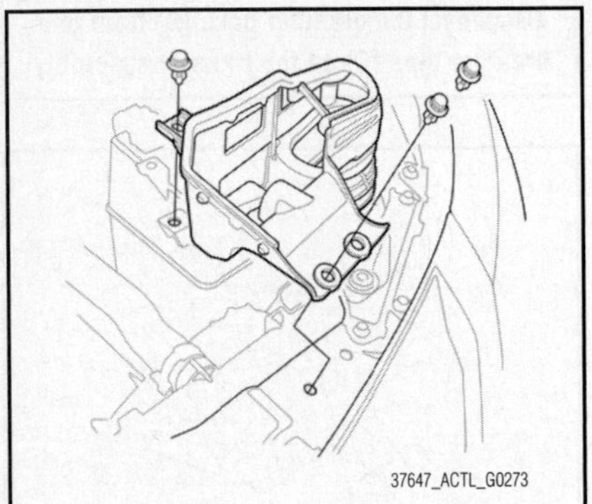
37647_ACTL_G0273

Fig. 64 Remove the air intake duct splash separator

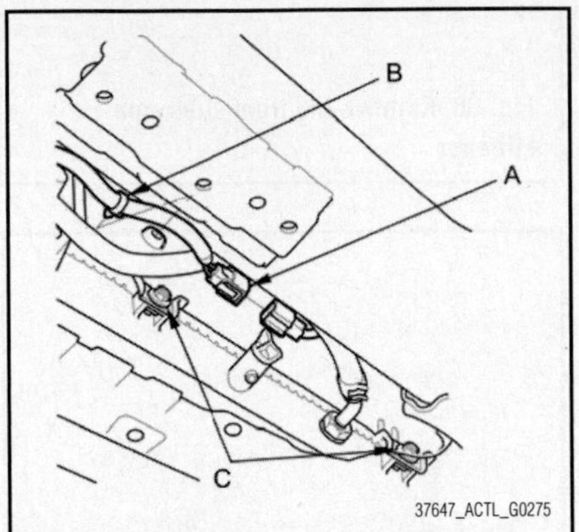
B
A
C
37647_ACTL_G0275

Fig. 66 Disconnect the A/C condenser fan motor connector (A), and remove the harness clamp (B)

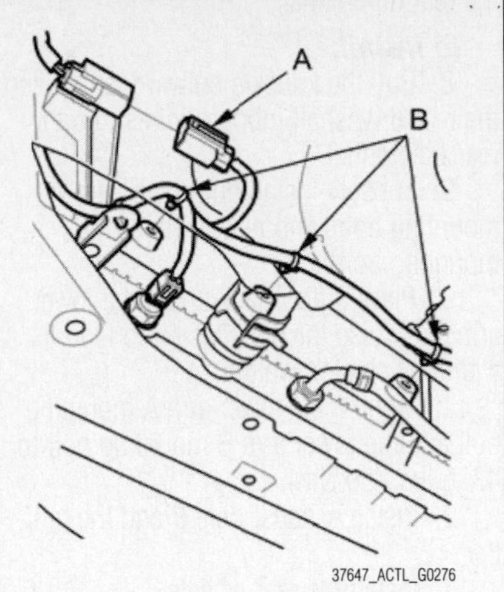
A
B
37647_ACTL_G0276

Fig. 67 Disconnect the radiator fan motor connector (A), and remove the harness clamps (B)

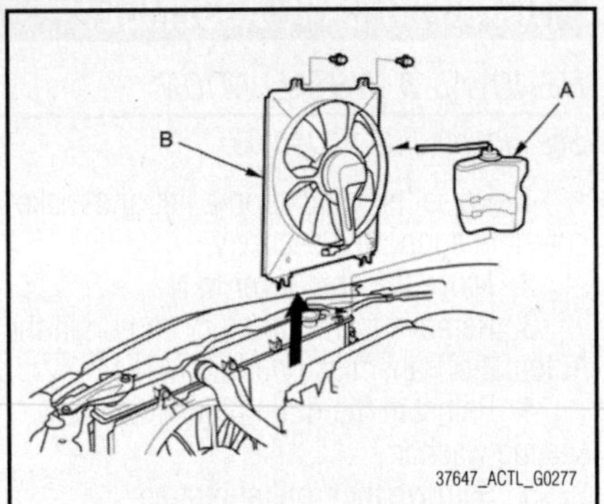
A
B
37647_ACTL_G0277

Fig. 68 Remove the coolant reservoir (A), then remove the A/C condenser fan shroud assembly (B)

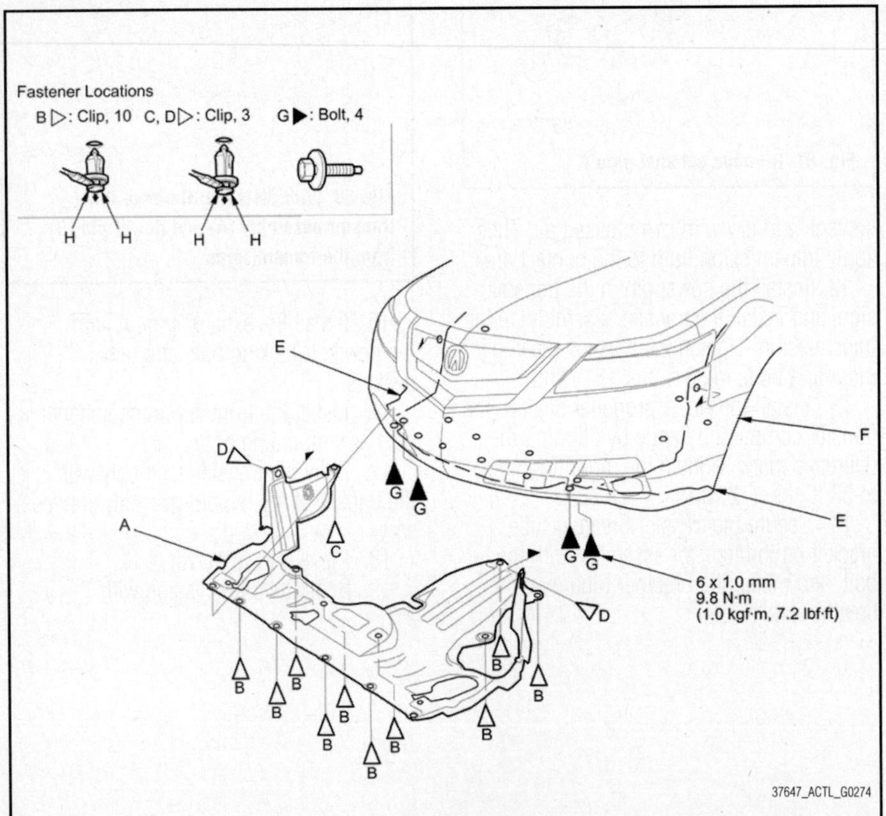
Fastener Locations
B ▷: Clip, 10 C, D ▷: Clip, 3 G ▶: Bolt, 4
H H H H
E
F
E
D
A
C
G G
G G
6 x 1.0 mm
9.8 N·m
(1.0 kgf·m, 7.2 lbf·ft)
D
B B B B B B B B B
37647_ACTL_G0274

Fig. 65 Remove the splash shield

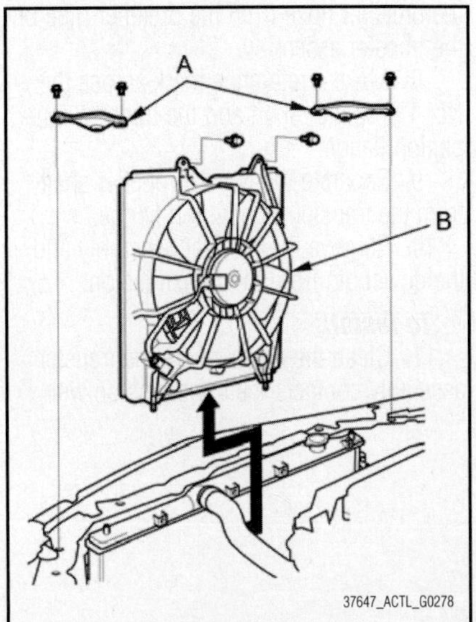
A
B
37647_ACTL_G0278

Fig. 69 Remove the upper brackets (A)

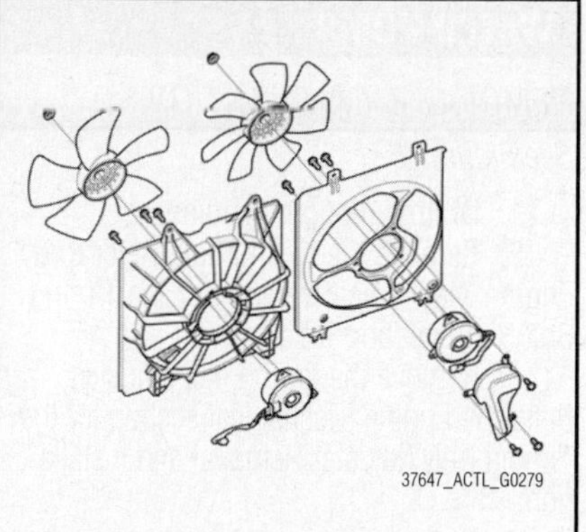
37647_ACTL_G0279

Fig. 70 Disassemble the fan shrouds

11. Remove the coolant reservoir, then remove the A/C condenser fan shroud assembly.
12. Remove the upper brackets.
13. Remove the radiator fan shroud assembly.

➡Pull-up the radiator, then move the radiator fan shroud assembly toward the right side of the vehicle to allow for enough space to lift it up and away from the A/C condenser fan shroud assembly.

14. Disassemble the fan shrouds.

To install:

15. Assemble the fan shrouds.
16. Install the left radiator fan shroud assembly.
17. Install the upper brackets.
18. Install the A/C condenser fan shroud assembly, then install the coolant reservoir.
19. Raise the vehicle on the lift.
20. Connector the radiator fan motor connector, and install the harness clamps.
21. Tighten the A/C condenser fan shroud mounting bolts.
22. Connect the A/C condenser fan motor connector, and install the harness clamp.
23. Install the splash shield.
24. Lower the vehicle on the lift.
25. Install the air intake duct splash separator.
26. Install the battery base.
27. Do the battery installation procedure.
28. Install the engine compartment covers.

RADIATOR

REMOVAL & INSTALLATION

See Figures 64 and 65, 67, 71 and 72.

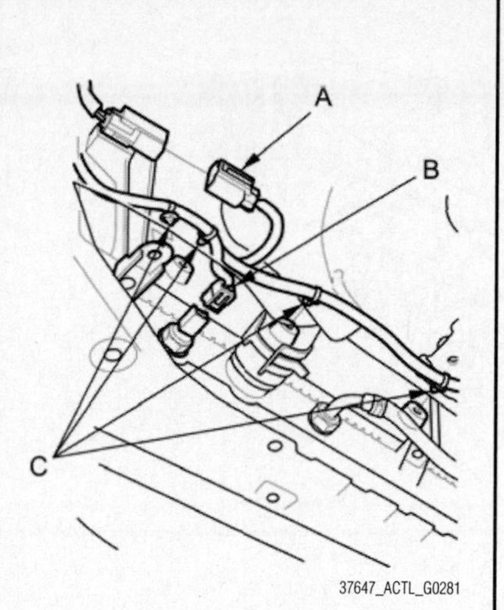
A
B
C
37647_ACTL_G0281

Fig. 71 Disconnect the radiator fan motor connector (A) and the ECT sensor 2 connector (B), then remove the harness clamps (C)

1. Remove the engine compartment covers.
2. Drain the engine coolant.
3. Remove the air intake duct splash separator.
4. Raise the vehicle on the lift.
5. Remove the splash shield.
6. Disconnect the A/C condenser fan motor connector, and remove the harness clamp.
7. Disconnect the radiator fan motor connector and the Engine Coolant Temperature (ECT) sensor 2 connector, then remove the harness clamps.
8. Disconnect the lower radiator hose.
9. A/T model: Disconnect the Automatic Transmission Fluid (ATF) cooler hoses from the radiator, then plug the cooler hoses and the lines.
10. Lower the vehicle on the lift.
11. Disconnect the upper radiator hose and remove the coolant reservoir.
12. Remove the upper brackets, then pull up the radiator with fan shrouds.
13. Remove the fan shrouds and other parts from the radiator.

To install:

14. Install radiator in the reverse order of removal. Make sure the upper and lower cushions are set securely.
15. A/T model: Refill the transmission with ATF.
16. Fill the radiator with engine coolant, and bleed the air from the cooling system.

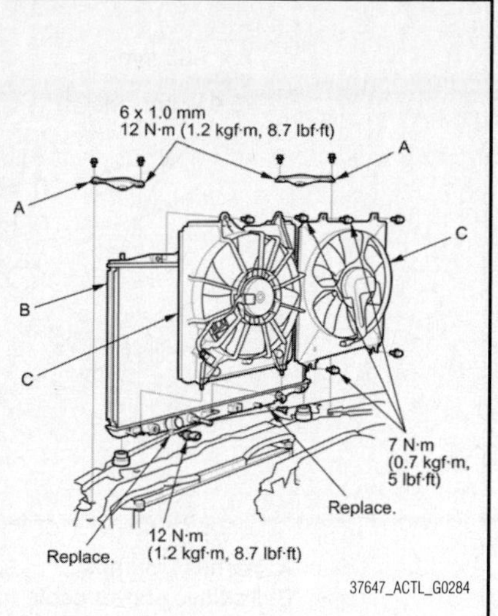
6 x 1.0 mm
12 N·m (1.2 kgf·m, 8.7 lbf·ft)
A
A
C
B
C
7 N·m
(0.7 kgf·m,
5 lbf·ft)
Replace.
12 N·m
(1.2 kgf·m, 8.7 lbf·ft)
Replace.
37647_ACTL_G0284

Fig. 72 Remove the upper brackets (A), then pull up the radiator (B) with fan shrouds (C)

17. Clean up any spilled engine coolant.
18. Install the engine compartment covers.

THERMOSTAT

REMOVAL & INSTALLATION

See Figures 73 and 74.

1. Remove the engine compartment covers.
2. Drain the engine coolant.
3. Do the battery disconnect procedure.
4. A/T model: Remove harness clamps and disconnect the positive starter cable.
5. A/T model: Disconnect the A/T clutch pressure control solenoid valve C connector, and remove the bolt and the harness holder.
6. Remove the thermostat cover, then remove the thermostat.

To install:

7. Install the new thermostat with a new rubber seal in the reverse order of the removal.
8. Do the battery installation procedure.
9. Refill the radiator with engine coolant, and bleed the air from the cooling system.
10. Clean up any spilled engine coolant.
11. Install the engine compartment covers.

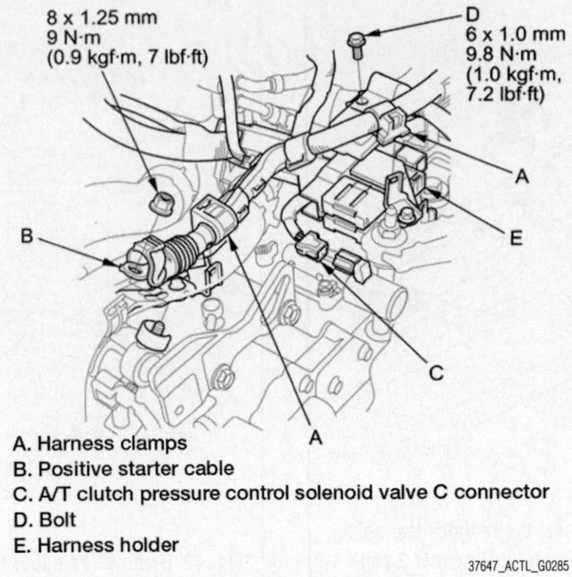
8 x 1.25 mm
9 N·m
(0.9 kgf·m, 7 lbf·ft)
D
6 x 1.0 mm
9.8 N·m
(1.0 kgf·m,
7.2 lbf·ft)
A
B
E
C
A
A. Harness clamps
B. Positive starter cable
C. A/T clutch pressure control solenoid valve C connector
D. Bolt
E. Harness holder
37647_ACTL_G0285

Fig. 73 Remove harness clamps and disconnect the positive starter cable

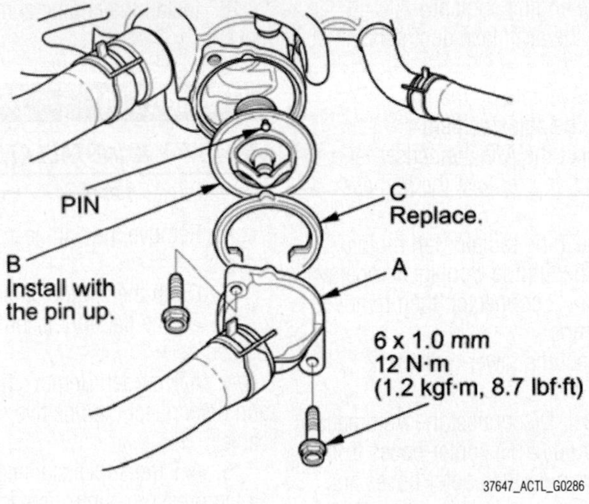
PIN
C
Replace.
B
Install with
the pin up.
A
6 x 1.0 mm
12 N·m
(1.2 kgf·m, 8.7 lbf·ft)
37647_ACTL_G0286

Fig. 74 Remove the thermostat cover (A), then remove the thermostat (B)

WATER PUMP

REMOVAL & INSTALLATION

See Figure 75.

1. Drain the engine coolant.
2. Remove the timing belt. Refer the Engine Mechanical section for the Timing Belt Removal and Installation.
3. Remove the timing belt adjuster. Refer the Engine Mechanical section for the Timing Belt Adjuster Removal and Installation.
4. Remove the five bolts securing the water pump, then remove the water pump and gasket.

To install:

5. Inspect and clean the O-ring groove and the mating surface of the engine block.
6. Install the water pump with a new gasket.
7. Clean up any spilled engine coolant.
8. Install the timing belt adjuster:
9. Install the timing belt:
10. Refill the radiator with engine coolant, and bleed the air from the cooling system.

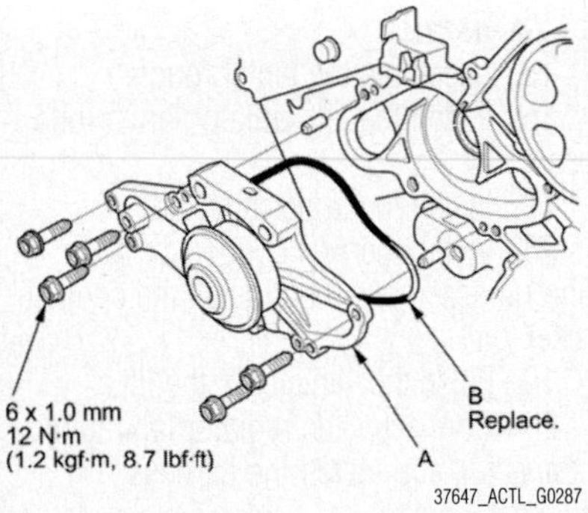
6 x 1.0 mm
12 N·m
(1.2 kgf·m, 8.7 lbf·ft)
B
Replace.
A
37647_ACTL_G0287

Fig. 75 Remove the five bolts securing the water pump (A), then remove the water pump and gasket (B)

ALTERNATOR

REMOVAL & INSTALLATION

See Figures 76 through 80.

1. Remove the engine compartment covers. Refer to Engine Mechanical section for Engine Compartment Covers Removal and Installation.
2. Do the battery terminal disconnection procedure.
3. Raise the vehicle on the lift.
4. Remove the splash shield.
5. Disconnect the A/C condenser fan motor connector and remove the harness clamp.
6. Loosen the A/C condenser fan shroud mounting bolts.
7. Lower the vehicle on the lift.
8. Remove the coolant reservoir, then remove the A/C condenser fan shroud assembly.
9. Remove the drive belt.
10. Disconnect the alternator connector and the positive alternator cable from the alternator.
11. Remove the harness clamp from the alternator and disconnect the A/C compressor clutch connector from the A/C compressor.

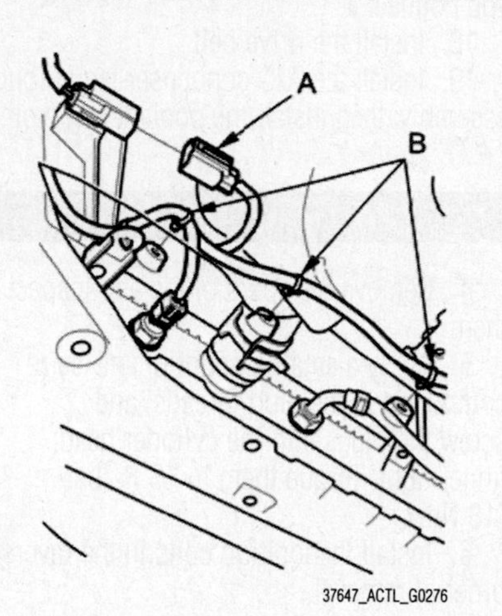
A
B
37647_ACTL_G0276

Fig. 77 Disconnect the radiator fan motor connector (A), and remove the harness clamps (B)

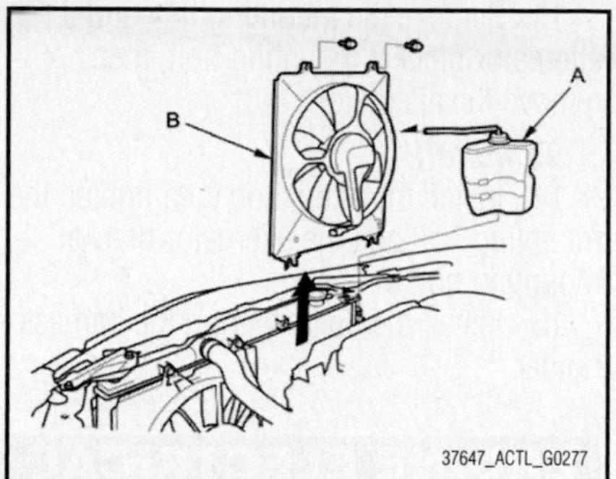
A
B
37647_ACTL_G0277

Fig. 78 Remove the coolant reservoir (A), then remove the A/C condenser fan shroud assembly (B)

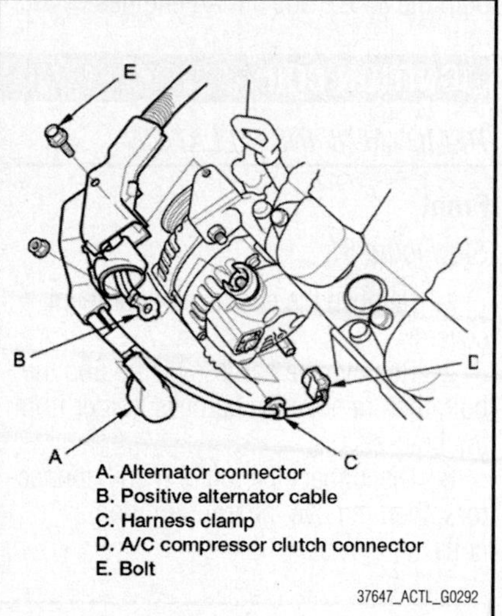
E
B
D
A
C
A. Alternator connector
B. Positive alternator cable
C. Harness clamp
D. A/C compressor clutch connector
E. Bolt
37647_ACTL_G0292

Fig. 79 Disconnect the alternator connector and the positive alternator cable from the alternator

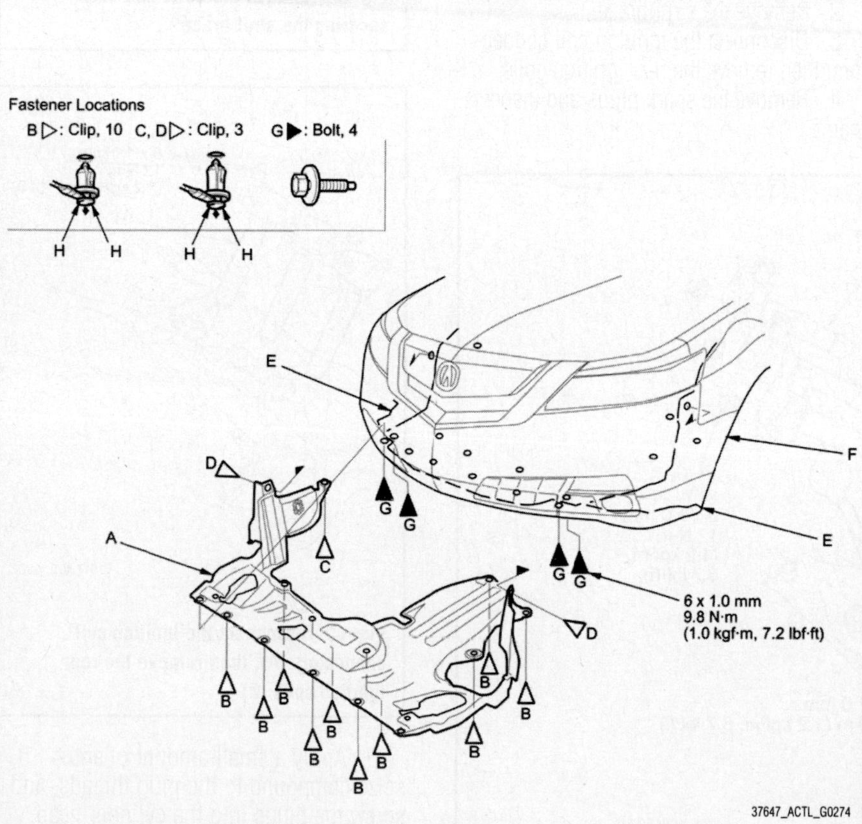
Fastener Locations
B ▷: Clip, 10 C, D ▷: Clip, 3 G ▶: Bolt, 4
H
H
H
H
E
F
D
A
C
G
G
E
G
G
6 x 1.0 mm
9.8 N·m
(1.0 kgf·m, 7.2 lbf·ft)
D
B
37647_ACTL_G0274

Fig. 76 Remove the splash shield

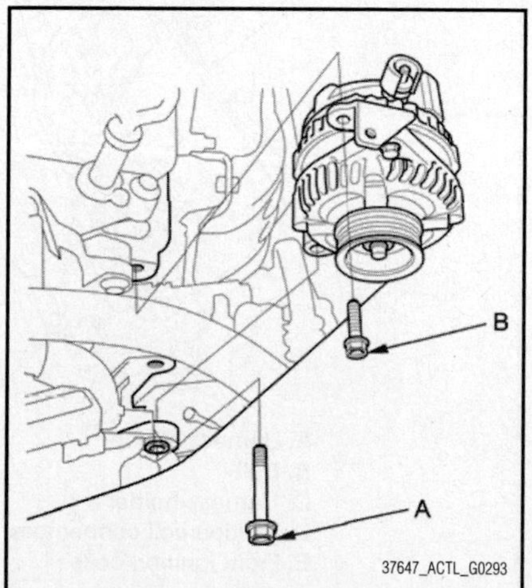
B
A
37647_ACTL_G0293

Fig. 80 Remove the mounting bolt (A) and the alternator bracket mounting bolt (B), then remove the alternator

12. Remove the bolt securing the harness holder.
13. Remove the mounting bolt and the alternator bracket mounting bolt, then remove the alternator.

To install:

14. Install the alternator, then tighten the mounting bolt and the alternator bracket mounting bolt.
15. Install the bolt securing the harness holder.
16. Install the harness clamp to the alternator and connect the A/C compressor clutch connector to the A/C compressor.
17. Connect the alternator connecter and the positive alternator cable to the alternator. Make sure the crimped side of the ring terminal faces away from the alternator when you connect it.
18. Install the drive belt.
19. Install the A/C condenser fan shroud assembly, then install the coolant reservoir.
20. Raise the vehicle on the lift.
21. Tighten the A/C condenser fan shroud mounting bolts.
22. Connect the A/C condenser fan motor connector and install the harness clamp.
23. Install the splash shield.
24. Lower the vehicle on the lift.
25. Do the battery terminal reconnection procedure.
26. Install the engine compartment covers.

ENGINE ELECTRICAL

IGNITION SYSTEM

FIRING ORDER

The firing order is 1–4–2–5–3–6 for both the J35Z6 and J37A4 engines.

IGNITION COIL

REMOVAL & INSTALLATION

Front

See Figure 81.

1. Remove the engine compartment covers.
2. Remove the harness clamp and the bolt, then remove the harness holder from the bracket.
3. Disconnect the ignition coil connectors, then remove the front ignition coils.
4. Remove the spark plugs and inspect them.
5. Apply a small amount of anti-seize compound to the plug threads, and screw the plugs into the cylinder head, finger tight. Torque them to 13 ft. lbs. (18 Nm).
6. Install the ignition coils in the reverse order of removal.

Rear

See Figures 82 and 83.

1. Remove the engine compartment covers.
2. Remove the strut brace.
3. Disconnect the ignition coil connectors, then remove the rear ignition coils.
4. Remove the spark plugs and inspect them.

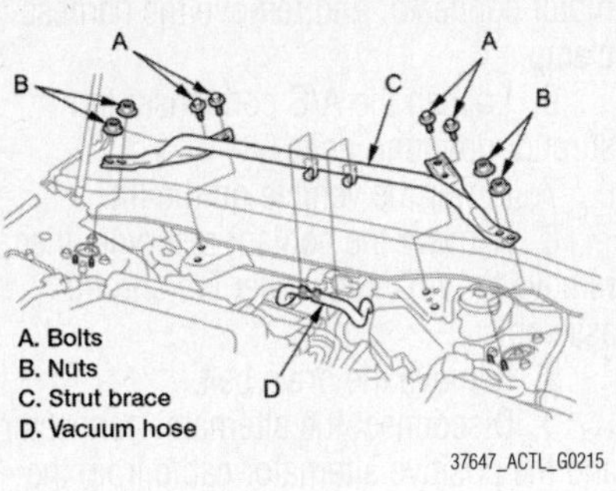
A
A
B
C
B
D
A. Bolts
B. Nuts
C. Strut brace
D. Vacuum hose
37647_ACTL_G0215

Fig. 82 Remove the bolts and nuts securing the strut brace

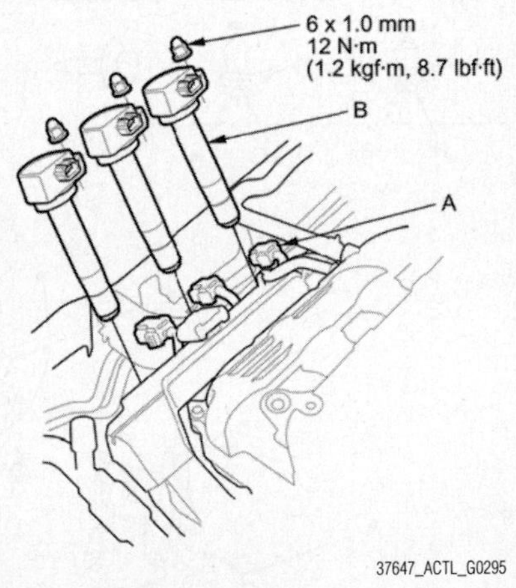
6 x 1.0 mm
12 N·m
(1.2 kgf·m, 8.7 lbf·ft)
B
A
37647_ACTL_G0295

Fig. 83 Disconnect the ignition coil connectors (A), then remove the rear ignition coils (B)

5. Apply a small amount of anti-seize compound to the plug threads, and screw the plugs into the cylinder head, finger tight. Torque them to 13 ft. lbs. (18 Nm).
6. Install the ignition coils in the reverse order of removal.

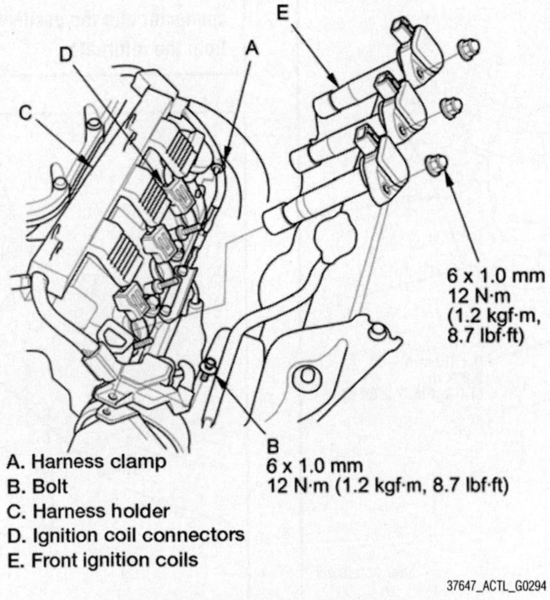
E
D
A
C
6 x 1.0 mm
12 N·m
(1.2 kgf·m,
8.7 lbf·ft)
B
6 x 1.0 mm
12 N·m (1.2 kgf·m, 8.7 lbf·ft)
A. Harness clamp
B. Bolt
C. Harness holder
D. Ignition coil connectors
E. Front ignition coils
37647_ACTL_G0294

Fig. 81 Remove the harness clamp and the bolt, then remove the harness holder from the bracket

IGNITION TIMING

ADJUSTMENT

1. Remove the engine compartment covers.
2. Connect the Honda Diagnostic System (HDS) to the Data Link Connector (DLC).
3. Turn the ignition switch to ON (II), or press the engine start/stop button to select the ON mode.
4. Make sure the HDS communicates with the vehicle and the Engine Control Module (ECM)/Powertrain Control Module (PCM). If it does not communicate, troubleshoot the DLC circuit.
5. Check for DTCs. If a DTC is present, diagnose and repair the cause before continuing with this test.
6. Start the engine. Hold the engine speed at 3,000 rpm with no load (in neutral (M/T model) or N or P (A/T model)) until the radiator fan comes on, then let it idle.
7. Check the idle speed.
8. Jump the SCS line with the HDS.
9. Connect the timing light to the No. 1 ignition coil harness.
10. Aim the light toward the pointer on the timing belt cover. Check the ignition timing under a no load condition (headlights, blower fan, rear window dofogger, and air conditioner are turned off).

➡The other pointer is not used.

Ignition timing specification: 10 °±2 ° BTDC (RED mark) at idle in N or P

11. If the ignition timing differs from the specification, check the cam timing. If the cam timing is OK, update the ECM/PCM if it does not have the latest software, or substitute a known-good ECM/PCM, then recheck. If the system works properly, and the ECM/PCM was substituted, replace the original ECM/PCM.
12. Disconnect the HDS and the timing light.
13. Install the engine compartment covers.

SPARK PLUGS

REMOVAL & INSTALLATION

Front

See Figure 81.

1. Remove the engine compartment covers.
2. Remove the harness clamp and the bolt, then remove the harness holder from the bracket.
3. Disconnect the ignition coil connectors, then remove the front ignition coils.
4. Remove the spark plugs and inspect them.
5. Apply a small amount of anti-seize compound to the plug threads, and screw the plugs into the cylinder head, finger tight. Torque them to 13 ft. lbs. (18 Nm).
6. Install the ignition coils in the reverse order of removal.

Rear

See Figures 82 and 83.

1. Remove the engine compartment covers.
2. Remove the strut brace.
3. Disconnect the ignition coil connectors, then remove the rear ignition coils.
4. Remove the spark plugs and inspect them.
5. Apply a small amount of anti-seize compound to the plug threads, and screw the plugs into the cylinder head, finger tight. Torque them to 13 ft. lbs. (18 Nm).
6. Install the ignition coils in the reverse order of removal.

ENGINE ELECTRICAL — STARTING SYSTEM

STARTER

REMOVAL & INSTALLATION

See Figures 84 and 85.

1. Remove the engine compartment covers.
2. Do the battery removal procedure.

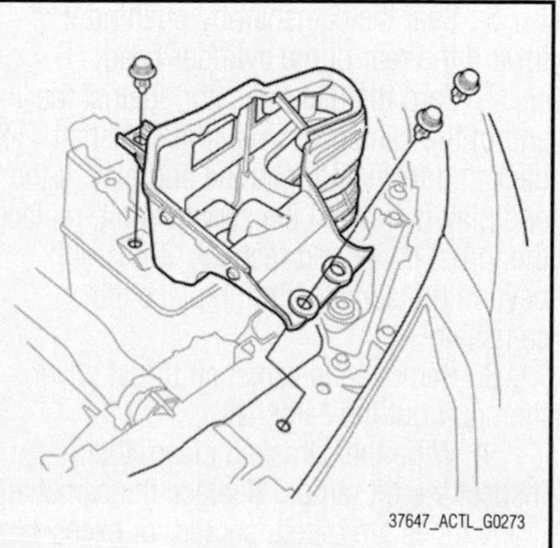
37647_ACTL_G0273

Fig. 84 Remove the air intake duct splash separator

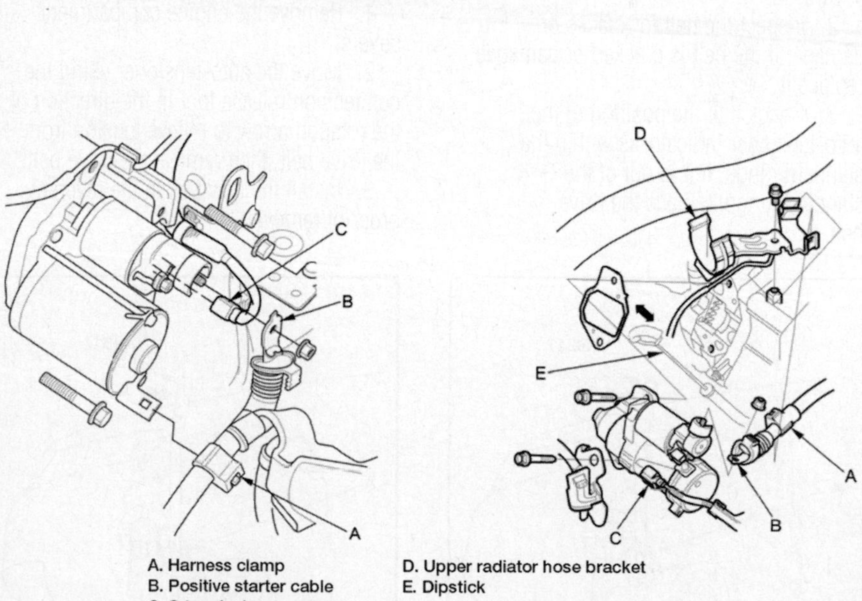
A
B
C
D
E
A. Harness clamp
B. Positive starter cable
C. S terminal connector
D. Upper radiator hose bracket
E. Dipstick
37647_ACTL_G0297

Fig. 85 Remove the harness clamp

3. Remove the battery base.
4. Remove the air intake duct splash separator.
5. Remove the harness clamp.
6. Disconnect the positive starter cable and the S terminal connector.
7. A/T model: Remove the upper radiator hose bracket and the dipstick.
8. Remove the two bolts holding the starter, then remove the starter.

To install:

9. Install the starter, then tighten the mounting bolts. Tighten to 54 ft. lbs. (74 Nm).

➡Always use a new gasket (A/T model).

10. A/T model: Install the upper radiator hose bracket and the dipstick.
11. Connect the positive starter cable and the S terminal connector. Make sure the crimped side of the ring terminal faces away from the starter when you connect it.
12. Install the harness clamp.
13. Install the air intake duct splash separator.
14. Install the battery base.
15. Do the battery installation procedure.
16. Start the engine to make sure the starter works properly.
17. Install the engine compartment covers.

SOLENOID OR RELAY REPLACEMENT

1. Remove the engine compartment covers.
2. Do the battery terminal disconnection procedure.
3. Remove the air intake duct splash separator.
4. Remove the harness clamps, then disconnect the positive starter cable, the motor wire, and the S terminal connector.
5. Check the hold-in coil for continuity between the S terminal and the armature housing (ground). There should be continuity.
   a. If there is continuity, go to step 6.
   b. If there is no continuity, replace the solenoid.
6. Check the pull-in coil for continuity between the S terminal and the M terminal. There should be continuity.
   a. If there is continuity, the solenoid is OK.
   b. If there is no continuity, replace the solenoid.
7. Install the wire and the connector in the reverse order of removal.
8. Do the battery terminal reconnection procedure.
9. Install the engine compartment covers.

ENGINE MECHANICAL

ACCESSORY DRIVE BELTS

ACCESSORY BELT ROUTING

See Figure 86.

Refer to the accompanying illustration for belt routing.

INSPECTION

1. Remove the engine compartment covers.
2. Inspect the belt for cracks or damage. If the belt is cracked or damaged, replace it.
3. Check that the position of the auto-tensioner indicator is within the standard range. If it is out of the standard range, replace the drive belt.

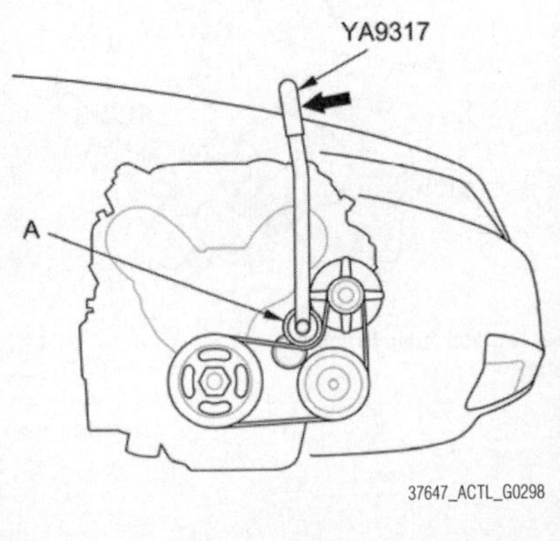

YA9317
A
37647_ACTL_G0298

Fig. 86 Accessory drive belt routing

4. Install the engine compartment covers.

ADJUSTMENT

No adjustment is necessary or possible. Proper tension is provided by the auto-tensioner.

REMOVAL & INSTALLATION

See Figure 87.

1. Remove the engine compartment covers.
2. Move the auto-tensioner using the belt tension release tool in the direction of the rotation arrow to relieve tension from the drive belt, then remove the drive belt.
3. Install the new belt in the reverse order of removal.

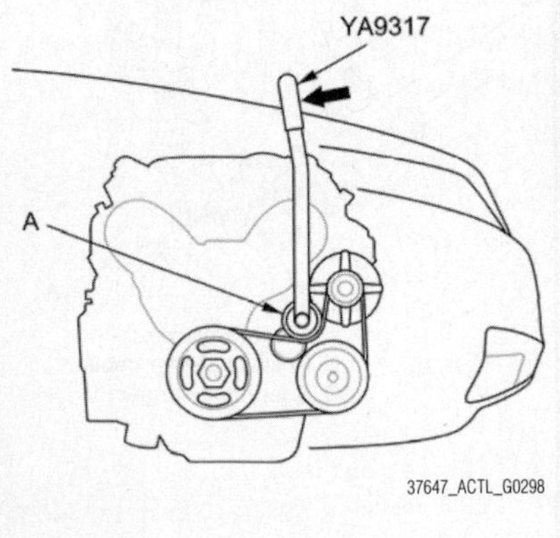

YA9317
A
37647_ACTL_G0298

Fig. 87 Remove the drive belt

CAMSHAFT AND VALVE LIFTERS

INSPECTION

J35Z6

See Figure 88.

1. Remove the cylinder head.
2. Remove the rocker arm assembly.
3. Disassemble the rocker arm assembly.
4. Front: Put the rocker shafts bridge and the rocker shaft holder on the front cylinder head, then tighten the bolts to 16 ft. lbs. (22 Nm).
5. Rear: Put the rocker shaft bridge and the rocker shaft holder on the rear cylinder head, then tighten the bolts to 16 ft. lbs. (22 Nm).
6. Seat the camshaft by pushing it toward the rear of the cylinder head.
7. Zero the dial indicator against the end of the camshaft. Push the camshaft back and forth and read the end play. If the end play is beyond the service limit, replace the thrust cover and recheck. If it is still beyond the service limit, replace the camshaft.
8. Remove the camshaft thrust cover, then pull out the camshaft.
9. Wipe the camshaft clean, then inspect the lift ramps. Replace the camshaft if any lobes are pitted, scored, or excessively worn.
10. Measure the diameter of each camshaft journal.

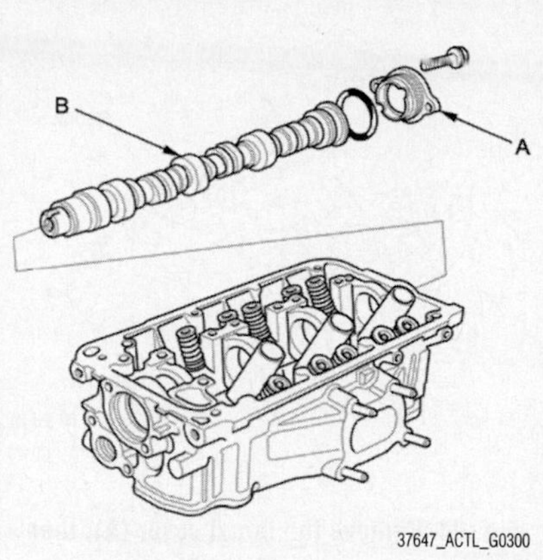
B
A
37647_ACTL_G0300
Fig. 88 Remove the camshaft thrust cover (A), then pull out the camshaft (B)

11. Zero the gauge to the journal diameter.
12. Clean the camshaft bearing surfaces in the cylinder head. Measure the inside diameter of each camshaft bearing surface, and check for an out-of-round condition.
    a. If the camshaft-to-holder clearance is within limits, go to step 14.
    b. If the camshaft-to-holder clearance is beyond the service limit and the camshaft has been replaced, replace the cylinder head.
    c. If the camshaft-to-holder clearance is beyond the service limit and the camshaft has not been replaced, go to step 13.

➡Camshaft-to-Holder Oil Clearance specifications:

- Standard (New): 0.0020–0.0035 inches (0.050–0.089 mm)
- Service Limit: 0.006 inches (0.15 mm)

13. Check total runout with the camshaft supported on V-blocks.
    a. If the total runout of the camshaft is within the service limit, replace the cylinder head.
    b. If the total runout is beyond the service limit, replace the camshaft and recheck the oil clearance. If the oil clearance is still out of tolerance, replace the cylinder head.

➡Camshaft Total Runout

- Standard (New): 0.001 inches (0.03 mm) max.
- Service Limit: 0.002 inches (0.04 mm)

14. Measure the cam lobe height.

➡Cam Lobe Height Standard (New):

- Intake Primary: 1.3504 inches
- Intake Secondary: 1.4024 inches
- Exhaust: 1.4472 inches

J37A4

See Figure 88.

1. Remove the cylinder head.
2. Remove the rocker arm assembly.
3. Disassemble the rocker arm assembly.
4. Front: Put the rocker shafts bridge and the rocker shaft holder on the front cylinder head, then tighten the bolts to 16 ft. lbs. (22 Nm).
5. Rear: Put the rocker shaft bridge and the rocker shaft holder on the rear cylinder head, then tighten the bolts to 16 ft. lbs. (22 Nm).
6. Seat the camshaft by pushing it toward the rear of the cylinder head.
7. Zero the dial indicator against the end of the camshaft. Push the camshaft back and forth and read the end play. If the end play is beyond the service limit, replace the thrust cover and recheck. If it is still beyond the service limit, replace the camshaft.
8. Remove the camshaft thrust cover, then pull out the camshaft.
9. Wipe the camshaft clean, then inspect the lift ramps. Replace the camshaft if any lobes are pitted, scored, or excessively worn.
10. Measure the diameter of each camshaft journal.
11. Zero the gauge to the journal diameter.
12. Clean the camshaft bearing surfaces in the cylinder head. Measure the inside diameter of each camshaft bearing surface, and check for an out-of-round condition.
    a. If the camshaft-to-holder clearance is within limits, go to step 14.
    b. If the camshaft-to-holder clearance is beyond the service limit and the camshaft has been replaced, replace the cylinder head.
    c. If the camshaft-to-holder clearance is beyond the service limit and the camshaft has not been replaced, go to step 13.

➡Camshaft-to-Holder Oil Clearance specifications:

- Standard (New): 0.0020–0.0035 inches (0.050–0.089 mm)
- Service Limit: 0.006 inches (0.15 mm)

13. Check total runout with the camshaft supported on V-blocks.
    a. If the total runout of the camshaft is within the service limit, replace the cylinder head.
    b. If the total runout is beyond the service limit, replace the camshaft and recheck the oil clearance. If the oil clearance is still out of tolerance, replace the cylinder head.

➡Camshaft Total Runout

- Standard (New): 0.001 inches (0.03 mm) max.
- Service Limit: 0.002 inches (0.04 mm)

14. Measure the cam lobe height.

➡Cam Lobe Height Standard (New):

- Intake Primary: 1.3504 inches
- Intake Secondary: 1.4146 inches
- Exhaust Primary: 1.4462 inches
- Exhaust Secondary: 1.4713 inches

REMOVAL & INSTALLATION

J35Z6 and J37A4 Engines

Front

See Figures 89 through 91.

1. Remove the engine compartment covers.

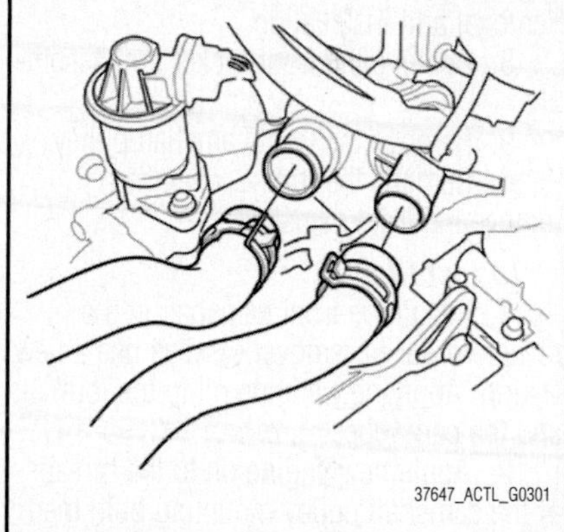
37647_ACTL_G0301
Fig. 89 Disconnect the radiator hoses

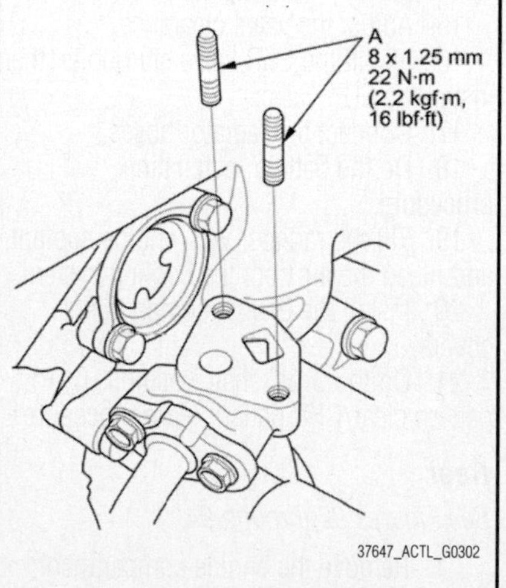
A
8 x 1.25 mm
22 N·m
(2.2 kgf·m,
16 lbf·ft)
37647_ACTL_G0302
Fig. 90 Remove the EGR valve stud bolts (A)

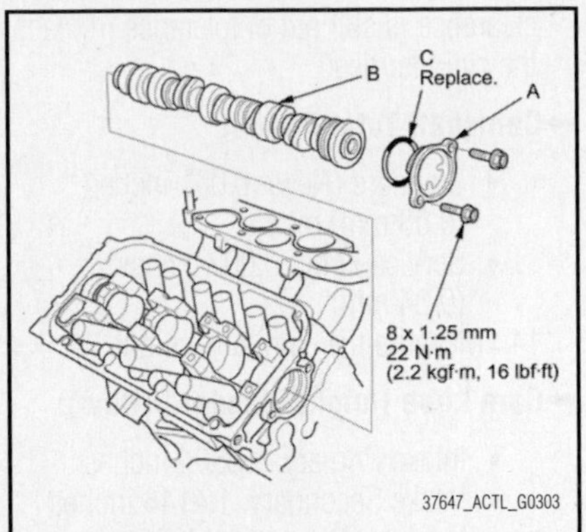
B
C
Replace.
A
8 x 1.25 mm
22 N·m
(2.2 kgf·m, 16 lbf·ft)
37647_ACTL_G0303

Fig. 91 Remove the thrust cover (A), then remove the front camshaft (B)

2. Do the battery disconnect procedure.
3. Drain the engine coolant.
4. Disconnect the radiator hoses.
5. Remove the exhaust gas recirculation (EGR) valve.
6. Remove the EGR valve stud bolts.
7. Remove the timing belt. Refer to Engine Mechanical section for Timing Belt Removal and Installation.
8. Remove the front rocker arm assembly.
9. Remove the front camshaft pulley.
10. Remove the thrust cover, then remove the front camshaft.

To install:

11. Install the front camshaft in the reverse order of removal. Always use a new O-ring. Apply new engine oil to the journals and the cam lobes.
12. Apply new engine oil to the threads of the camshaft pulley mounting bolt, then install the front camshaft pulley.
13. Install the front rocker arm assembly, then tighten the mounting bolts.
14. Install the timing belt.
15. Adjust the valve clearance.
16. Install the EGR valve stud bolts, then install the EGR valve.
17. Connect the radiator hoses.
18. Do the battery installation procedure.
19. Fill the radiator with engine coolant, and bleed the air from the cooling system.
20. Install the engine compartment covers.
21. Do the Crankshaft Position (CKP) pattern clear/CKP pattern learn procedure.

Rear

See Figures 92 through 94.

1. Remove the engine compartment covers.
2. Relieve the fuel pressure.
3. Do the battery disconnect procedure.
4. Drain the engine coolant.
5. Remove the under-hood fuse/relay box from the bracket.
6. Remove the air cleaner assembly.
7. Remove the quick-connect fitting cover, then disconnect the fuel feed hose.
8. Disconnect the heater hoses.
9. Disconnect the Evaporative Emission (EVAP) canister hose, then remove the EVAP canister joint with the bracket.
10. Remove the timing belt. Refer to Engine Mechanical section for Timing Belt Removal and Installation.
11. Remove the rear rocker arm assembly.
12. Remove the rear camshaft pulley.
13. Remove the thrust cover, then remove the rear camshaft.

To install:

14. Install the rear camshaft in the reverse order of removal. Always use a new O-ring. Apply new engine oil to the journals and cam lobes.
15. Apply new engine oil to the threads of the camshaft pulley mounting bolt, then install the rear camshaft pulley.

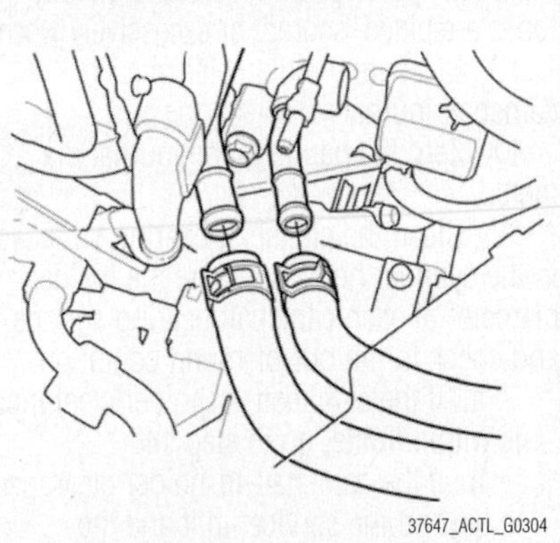
37647_ACTL_G0304

Fig. 92 Disconnect the heater hoses

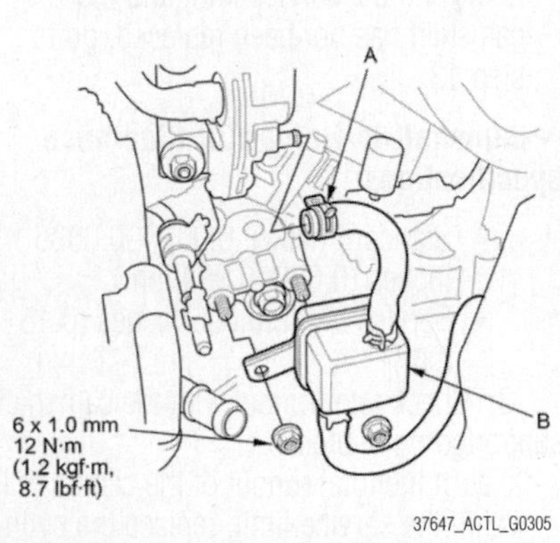
A
B
6 x 1.0 mm
12 N·m
(1.2 kgf·m,
8.7 lbf·ft)
37647_ACTL_G0305

Fig. 93 Disconnect the EVAP canister hose (A), then remove the EVAP canister joint (B) with the bracket

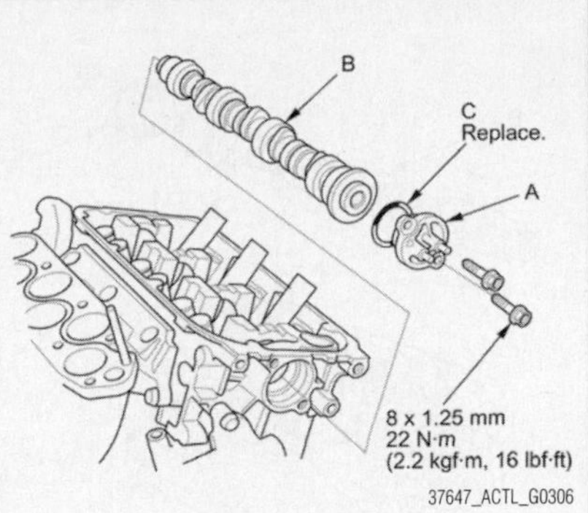
B
C
Replace.
A
8 x 1.25 mm
22 N·m
(2.2 kgf·m, 16 lbf·ft)
37647_ACTL_G0306

Fig. 94 Remove the thrust cover (A), then remove the rear camshaft (B)

16. Install the rear rocker arm assembly, then tighten the mounting bolts.
17. Install the timing belt.
18. Adjust the valve clearance.
19. Install the EVAP canister joint with the bracket, then connect the EVAP canister hose.
20. Connect the heater hoses.
21. Connect the fuel feed hose, then install the quick-connect fitting cover.
22. Install the air cleaner assembly.
23. Install the under-hood fuse/relay box to the bracket.
24. Do the battery installation procedure.
25. Inspect for fuel leaks. Turn the ignition switch to ON (II), or press the engine start/stop button to select the ON mode (do not operate the starter) so the fuel pump runs for about 2 seconds and pressurizes the fuel line. Repeat this operation three times, then check for fuel leakage at any point in the fuel line.
26. Fill the radiator with engine coolant, and bleed the air from the cooling system.
27. Install the engine compartment covers.
28. Do the Crankshaft Position (CKP) pattern clear/CKP pattern learn procedure.

CATALYTIC CONVERTER

REMOVAL & INSTALLATION

Under-Floor TWC

See Figures 95 and 96.

1. Raise the vehicle on a lift.
2. Remove the exhaust pipe hangers.
3. Remove the under-floor TWC.
4. Remove the converter cover.
5. Install the parts in the reverse order of removal with new gaskets and new self-locking nuts.

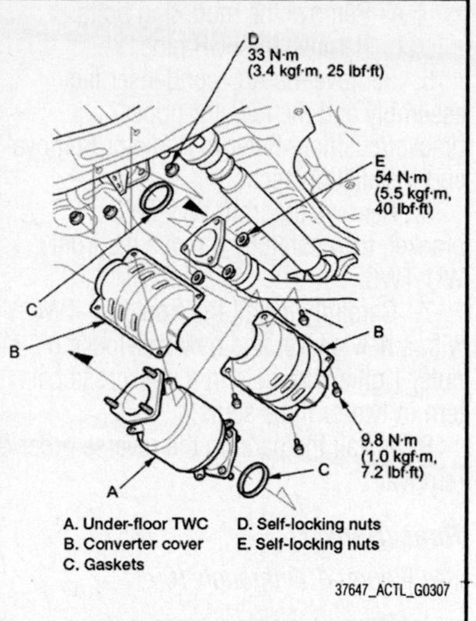
D
33 N·m
(3.4 kgf·m, 25 lbf·ft)
E
54 N·m
(5.5 kgf·m,
40 lbf·ft)
C
B
B
9.8 N·m
(1.0 kgf·m,
7.2 lbf·ft)
A
C
A. Under-floor TWC
B. Converter cover
C. Gaskets
D. Self-locking nuts
E. Self-locking nuts
37647_ACTL_G0307

Fig. 95 Remove the under-floor TWC with SH-AWD

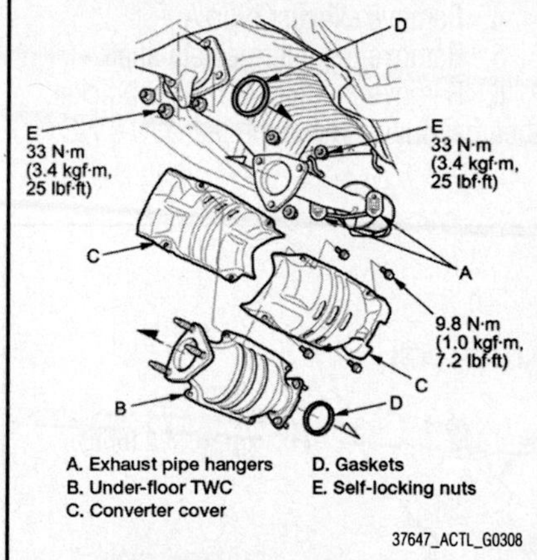
D
E
33 N·m
(3.4 kgf·m,
25 lbf·ft)
E
33 N·m
(3.4 kgf·m,
25 lbf·ft)
C
A
9.8 N·m
(1.0 kgf·m,
7.2 lbf·ft)
C
B
D
A. Exhaust pipe hangers
B. Under-floor TWC
C. Converter cover
D. Gaskets
E. Self-locking nuts
37647_ACTL_G0308

Fig. 96 Remove the under-floor TWC without SH-AWD

Warm Up TWC

Front (Bank 2)

See Figures 97 through 102.

1. Remove the No. 2 the ignition coil and the ignition coil heat insulator.
2. Remove the front A/F sensor (Sensor 1) and the front secondary HO2S (Sensor 2). Refer to Engine Performance and Emission Controls section for Heated Oxygen Sensor Removal and Installation.
3. Remove the exhaust pipe A mounting nuts (front WU-TWC side).
4. Remove the EGR pipe.
   a. Raise the vehicle on a lift.
   b. Remove the splash shield.
   c. Remove the mounting nuts.
   d. Lower the vehicle.

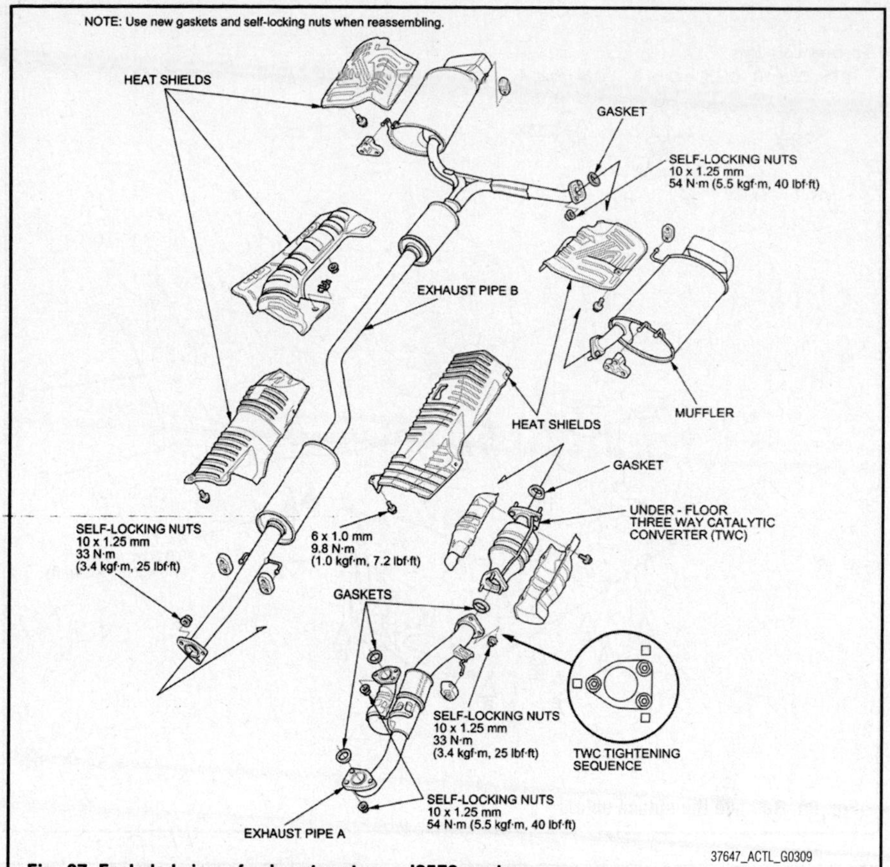
NOTE: Use new gaskets and self-locking nuts when reassembling.
HEAT SHIELDS
GASKET
SELF-LOCKING NUTS
10 x 1.25 mm
54 N·m (5.5 kgf·m, 40 lbf·ft)
EXHAUST PIPE B
MUFFLER
HEAT SHIELDS
GASKET
UNDER - FLOOR
THREE WAY CATALYTIC
CONVERTER (TWC)
SELF-LOCKING NUTS
10 x 1.25 mm
33 N·m
(3.4 kgf·m, 25 lbf·ft)
6 x 1.0 mm
9.8 N·m
(1.0 kgf·m, 7.2 lbf·ft)
GASKETS
SELF-LOCKING NUTS
10 x 1.25 mm
33 N·m
(3.4 kgf·m, 25 lbf·ft)
TWC TIGHTENING
SEQUENCE
SELF-LOCKING NUTS
10 x 1.25 mm
54 N·m (5.5 kgf·m, 40 lbf·ft)
EXHAUST PIPE A
37647_ACTL_G0309

Fig. 97 Exploded view of exhaust system—J35Z6 engine

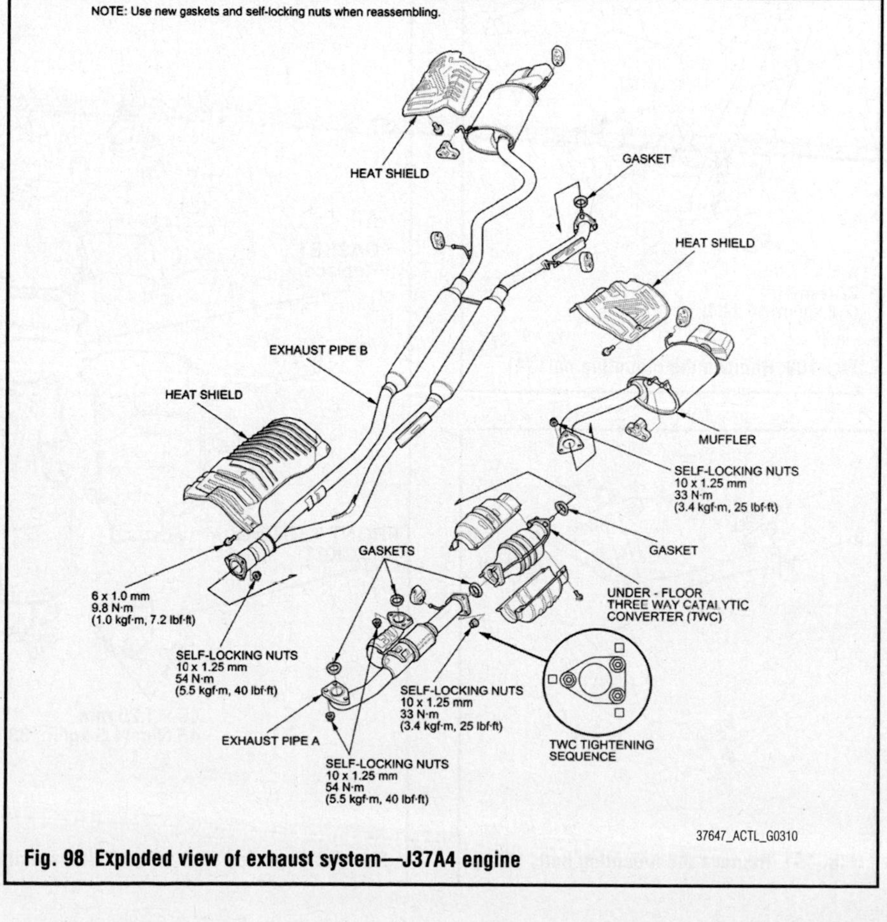
NOTE: Use new gaskets and self-locking nuts when reassembling.
HEAT SHIELD
GASKET
HEAT SHIELD
EXHAUST PIPE B
HEAT SHIELD
MUFFLER
SELF-LOCKING NUTS
10 x 1.25 mm
33 N·m
(3.4 kgf·m, 25 lbf·ft)
GASKETS
GASKET
6 x 1.0 mm
9.8 N·m
(1.0 kgf·m, 7.2 lbf·ft)
UNDER - FLOOR
THREE WAY CATALYTIC
CONVERTER (TWC)
SELF-LOCKING NUTS
10 x 1.25 mm
54 N·m
(5.5 kgf·m, 40 lbf·ft)
SELF-LOCKING NUTS
10 x 1.25 mm
33 N·m
(3.4 kgf·m, 25 lbf·ft)
EXHAUST PIPE A
TWC TIGHTENING
SEQUENCE
SELF-LOCKING NUTS
10 x 1.25 mm
54 N·m
(5.5 kgf·m, 40 lbf·ft)
37647_ACTL_G0310

Fig. 98 Exploded view of exhaust system—J37A4 engine

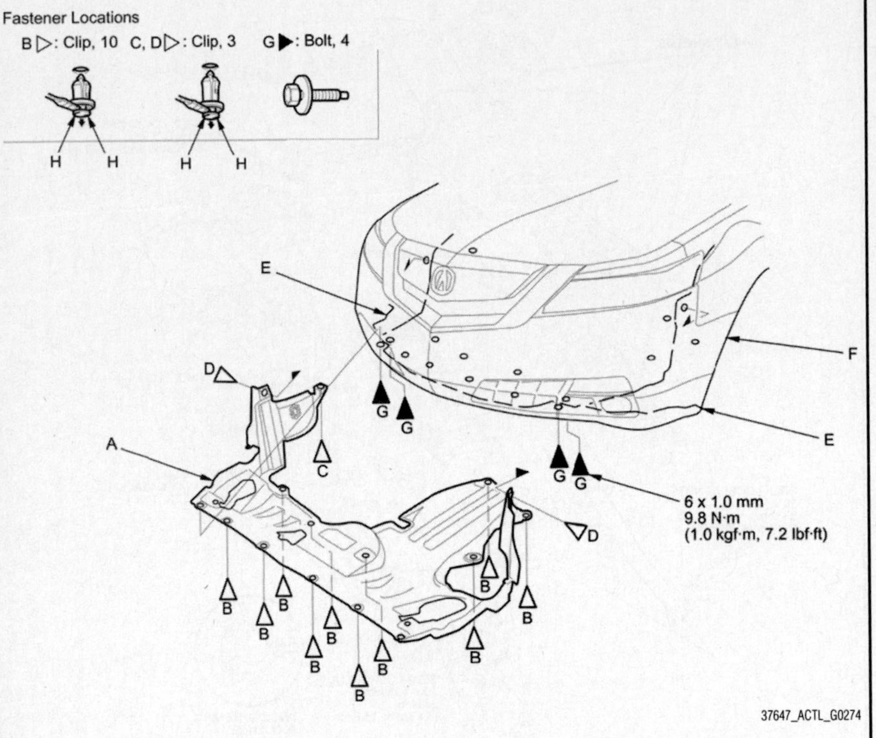
Fastener Locations
B ▷: Clip, 10 C, D ▷: Clip, 3 G ▶: Bolt, 4
H H H H
E
F
D
G
G
E
A
C
G G
6 x 1.0 mm
9.8 N·m
(1.0 kgf·m, 7.2 lbf·ft)
D
B B B B B B B B B B
37647_ACTL_G0274

Fig. 99 Remove the splash shield

e. Remove the mounting bolts.
f. Remove the EGR pipe.

5. Remove the A/C condenser fan assembly and the radiator upper bracket/cushion. Refer to Radiator Removal and Installation section.
6. Remove the front WU-TWC bracket, then carefully remove the front WU-TWC.
7. Carefully install the front WU-TWC with a new gasket and new self-locking nuts. Tighten the nuts in a crisscross pattern in two or three steps.
8. Install the parts in the reverse order of removal.

Rear (Bank 1)

See Figures 103 through 105.

1. Remove the strut brace.
2. Remove the rear A/F sensor (Sensor 1) and the rear secondary HO2S (Sensor 2).
3. Remove the P/S heat shield.
4. Remove exhaust pipe A.
5. Remove the intermediate shaft.
6. Remove the rear WU-TWC bracket, then carefully remove the rear WU-TWC.

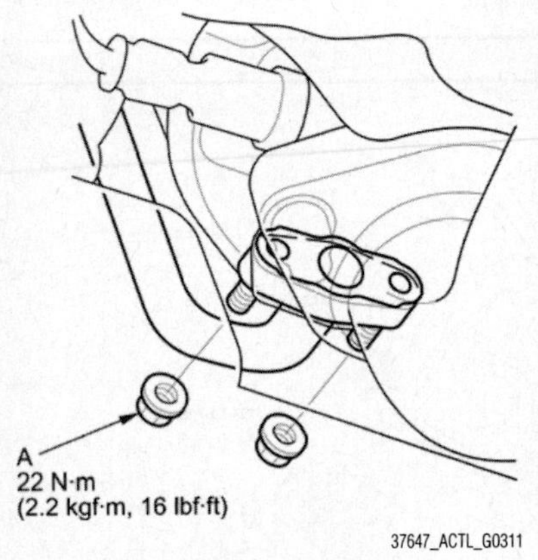
A
22 N·m
(2.2 kgf·m, 16 lbf·ft)
37647_ACTL_G0311

Fig. 100 Remove the mounting nuts (A)

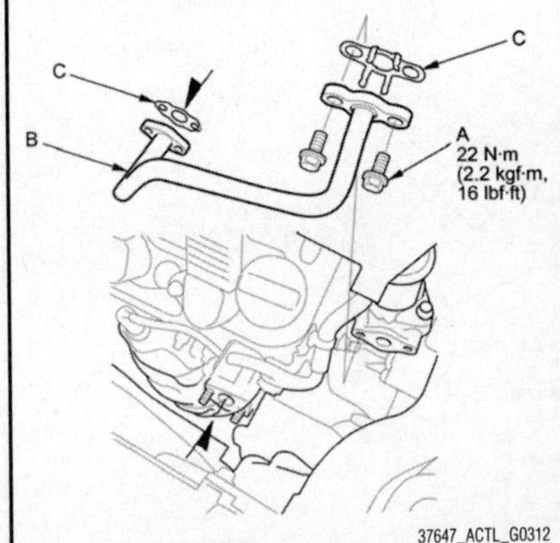
C
C
B
A
22 N·m
(2.2 kgf·m,
16 lbf·ft)
37647_ACTL_G0312

Fig. 101 Remove the mounting bolts (A)

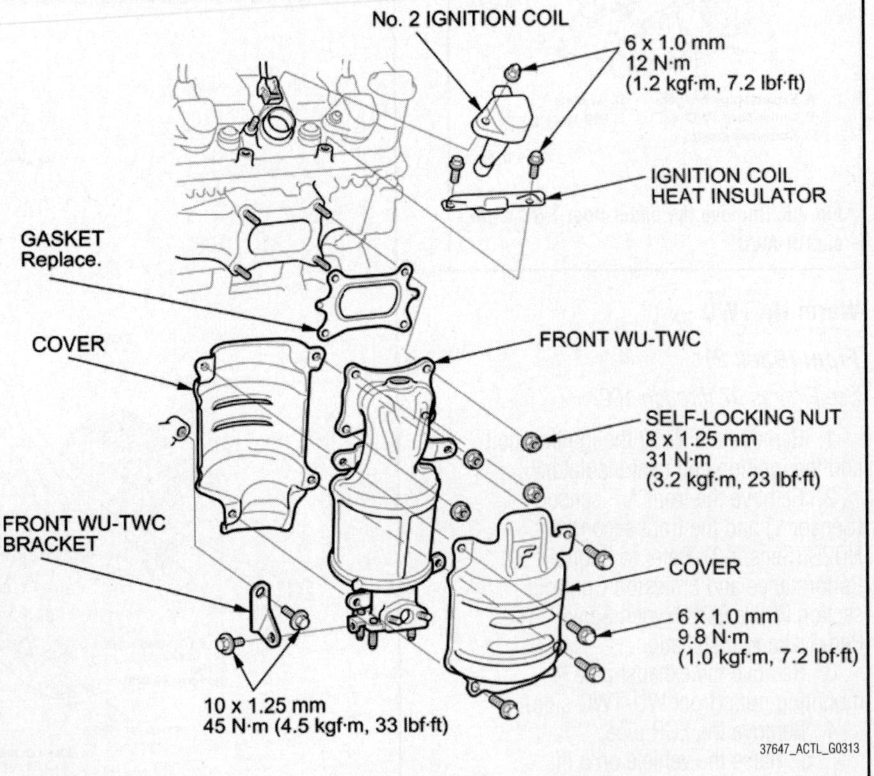
No. 2 IGNITION COIL
6 x 1.0 mm
12 N·m
(1.2 kgf·m, 7.2 lbf·ft)
IGNITION COIL
HEAT INSULATOR
GASKET
Replace.
COVER
FRONT WU-TWC
SELF-LOCKING NUT
8 x 1.25 mm
31 N·m
(3.2 kgf·m, 23 lbf·ft)
FRONT WU-TWC
BRACKET
COVER
6 x 1.0 mm
9.8 N·m
(1.0 kgf·m, 7.2 lbf·ft)
10 x 1.25 mm
45 N·m (4.5 kgf·m, 33 lbf·ft)
37647_ACTL_G0313

Fig. 102 Exploded view of WU-TWC assembly—Front (Bank 2)

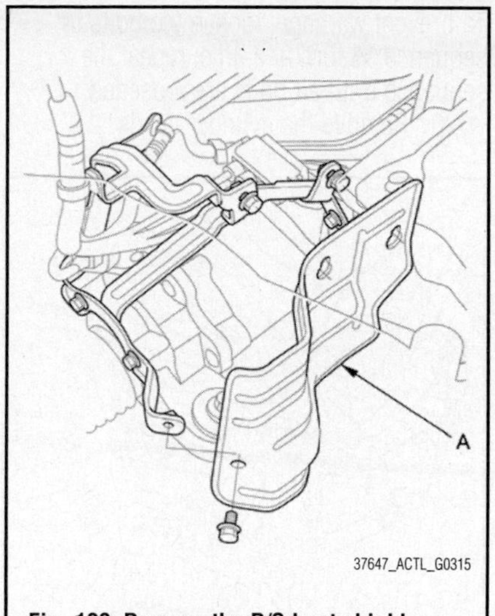
A
37647_ACTL_G0315

Fig. 103 Remove the P/S heat shield

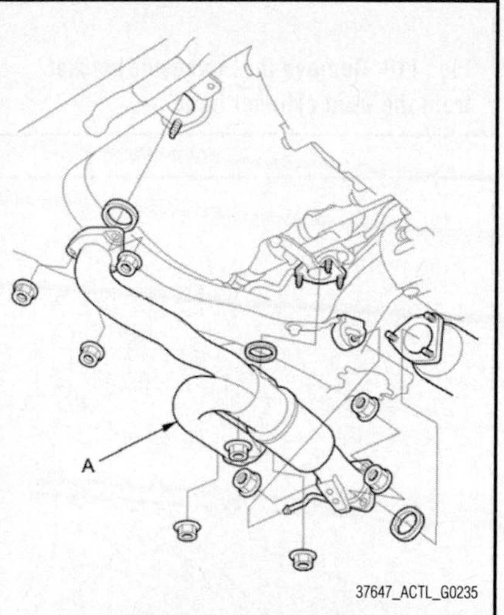
A
37647_ACTL_G0235

Fig. 104 Remove exhaust pipe A

7. Carefully install the rear WU-TWC with a new gasket and new self-locking nuts. Tighten the nuts in a crisscross pattern in two or three steps.
8. Install the parts in the reverse order of removal.

CRANKSHAFT PULLEY

REMOVAL & INSTALLATION

See Figures 99, 106 through 108.

1. Raise the vehicle on the lift.
2. Remove the right front wheel.
3. Remove the splash shield.
4. Remove the drive belt.
5. Hold the pulley with the handle, 6-25-660L and the holder attachment, 50 mm offset.
6. Remove the bolt with a heavy duly 19 mm socket and a breaker bar, then remove the crankshaft pulley.

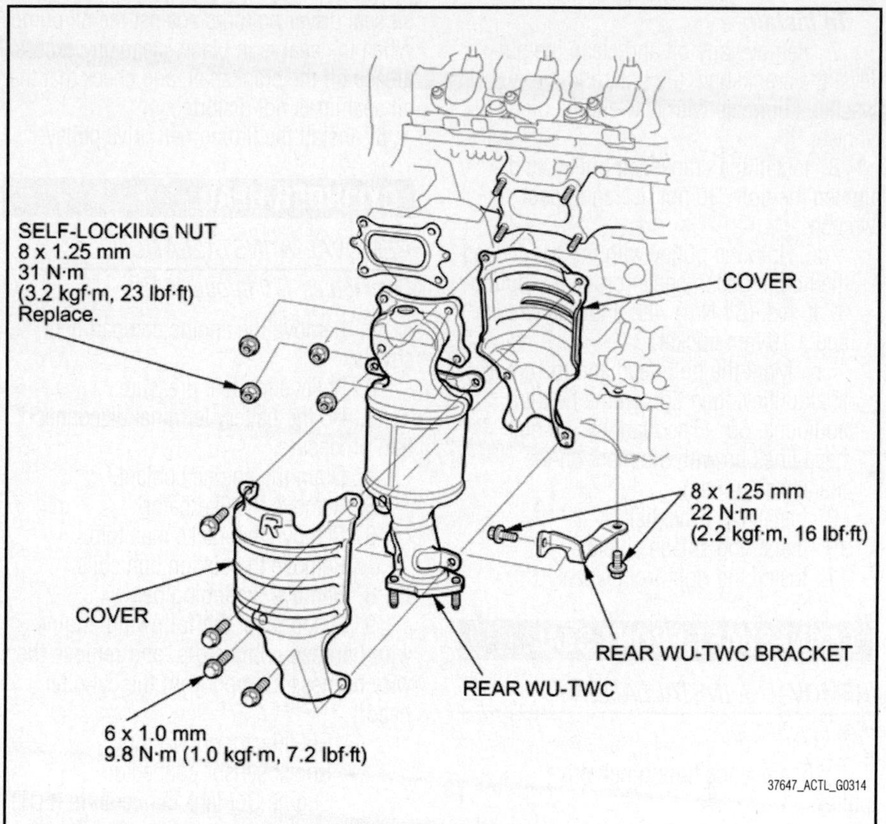
SELF-LOCKING NUT
8 x 1.25 mm
31 N·m
(3.2 kgf·m, 23 lbf·ft)
Replace.
COVER
8 x 1.25 mm
22 N·m
(2.2 kgf·m, 16 lbf·ft)
COVER
REAR WU-TWC BRACKET
REAR WU-TWC
6 x 1.0 mm
9.8 N·m (1.0 kgf·m, 7.2 lbf·ft)
37647_ACTL_G0314

Fig. 105 Exploded view of WU-TWC assembly—Rear (Bank 1)

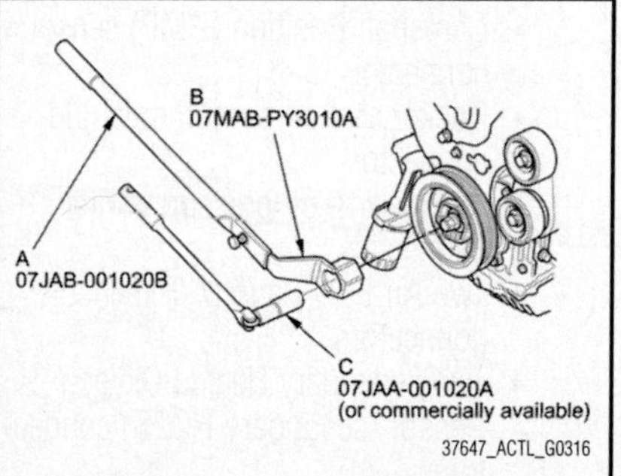
B
07MAB-PY3010A
A
07JAB-001020B
C
07JAA-001020A
(or commercially available)
37647_ACTL_G0316

Fig. 106 Removing the crankshaft pulley bolt

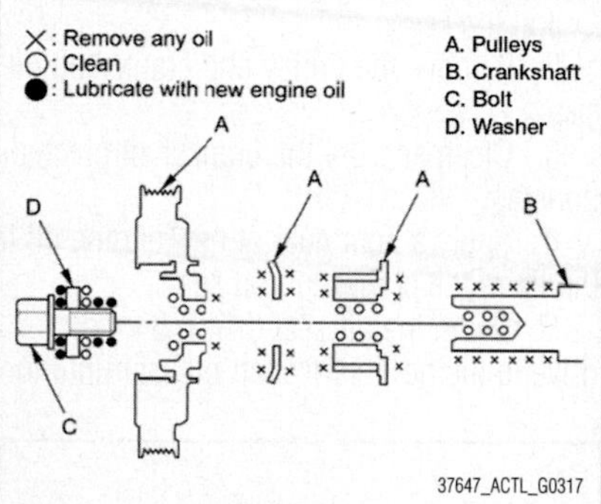
✕: Remove any oil
○: Clean
●: Lubricate with new engine oil
A. Pulleys
B. Crankshaft
C. Bolt
D. Washer
A
A
A
B
D
C
37647_ACTL_G0317

Fig. 107 Lubricate with new engine oil

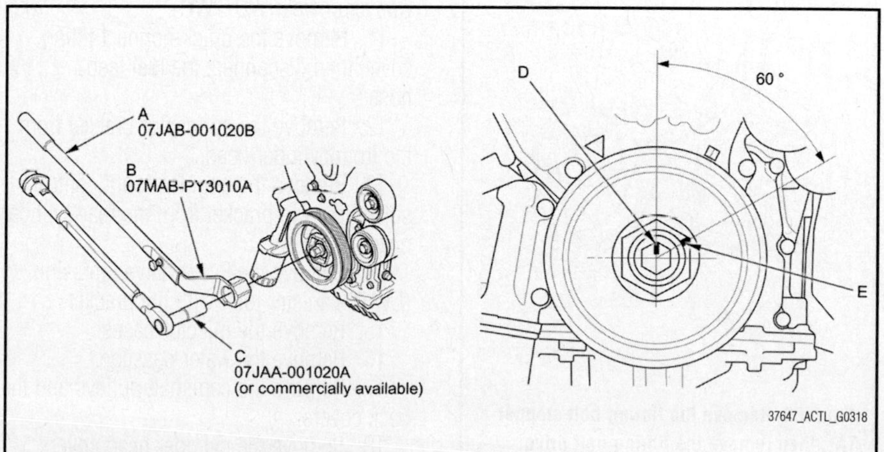
A
07JAB-001020B
B
07MAB-PY3010A
C
07JAA-001020A
(or commercially available)
D
60 °
E
37647_ACTL_G0318

Fig. 108 Install the crankshaft pulley, and tighten the bolt

To install:

7. Remove any oil and clean the pulleys, the crankshaft, the bolt, and the washer. Lubricate with new engine oil as shown.

8. Install the crankshaft pulley, and tighten the bolt. Do not use an impact wrench.

a. Hold the pulley with the handle and the holder attachment. Torque the bolt to 47 ft. lbs. (64 Nm) with a torque wrench and a 19 mm socket.

b. Mark the bolt head and the crankshaft pulley, then tighten the bolt an additional 60° (The mark on the bolt head lines up with the mark on the crankshaft pulley).

9. Install the drive belt.

10. Install the splash shield.

11. Install the right front wheel.

CRANKSHAFT FRONT SEAL

REMOVAL & INSTALLATION

See Figure 109.

1. Remove the timing belt drive pulley.

a. Remove the timing belt.

b. Remove the timing belt stopper, then remove the timing belt drive pulley and the key.

2. Remove the pulley end crankshaft oil seal.

3. Clean and dry the crankshaft oil seal housing.

4. Apply a light coat of new engine oil to the lip of the crankshaft oil seal.

5. Using the oil seal driver, 64 mm, drive in the new crankshaft oil seal until the oil seal driver bottoms against the oil pump. When the seal is in place, clean any excess grease off the crankshaft, and check that the oil seal lip is not distorted.

6. Install the timing belt drive pulley:

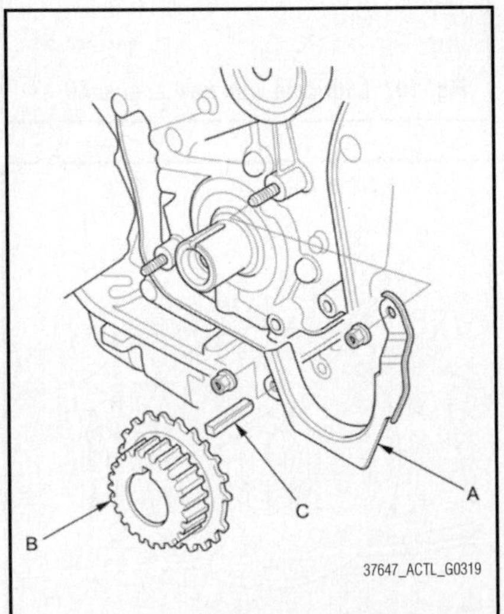
A
B
C
37647_ACTL_G0319

Fig. 109 Remove the timing belt stopper (A), then remove the timing belt drive pulley (B) and the key (C)

CYLINDER HEAD

REMOVAL & INSTALLATION

See Figures 110 through 118.

1. Remove the engine compartment covers.

2. Relieve the fuel pressure.

3. Do the battery terminal disconnection procedure.

4. Drain the engine coolant.

5. Remove the alternator.

6. Remove the intake manifold.

7. Remove the six ignition coils.

8. Remove the timing belt.

9. Disconnect the following engine wire harness connectors, and remove the wire harness clamps from the cylinder head:

- Six injector connectors
- Knock sensor connector
- Engine Coolant Temperature (ECT) sensor 1 connector
- Engine mount control solenoid valve connector
- Camshaft Position (CMP) sensor connector
- Rocker arm oil control solenoid connector
- Rocker arm oil pressure switch connector
- Two Air Fuel Ratio (A/F) sensor connectors
- Two secondary Heated Oxygen Sensor (secondary HO2S) connectors

10. Remove the front Warm Up Three Way Catalytic Converter (front WU-TWC) and the rear Warm Up Three Way Catalytic Converter (rear WU-TWC).

11. Remove the quick-connect fitting cover, then disconnect the fuel feed hose.

12. Remove the connector bracket from the front cylinder head.

13. Remove the engine mount control solenoid valve bracket from the rear cylinder head.

14. Remove the Evaporative Emission (EVAP) canister joint with the bracket.

15. Remove the injector bases.

16. Remove the water passage.

17. Remove the camshaft pulleys and the back covers.

18. Remove the cylinder head covers.

19. Remove the cylinder head bolts. To prevent warpage, loosen the bolts in sequence ⅓ turn at a time; repeat the sequence until all bolts are loosened.

20. Remove the cylinder heads.

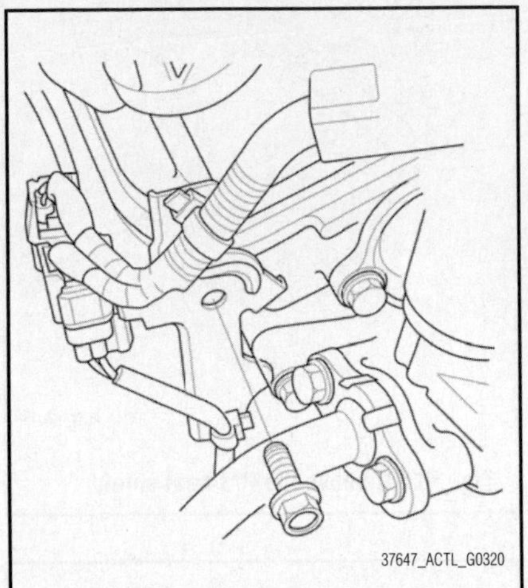
37647_ACTL_G0320

Fig. 110 Remove the connector bracket from the front cylinder head

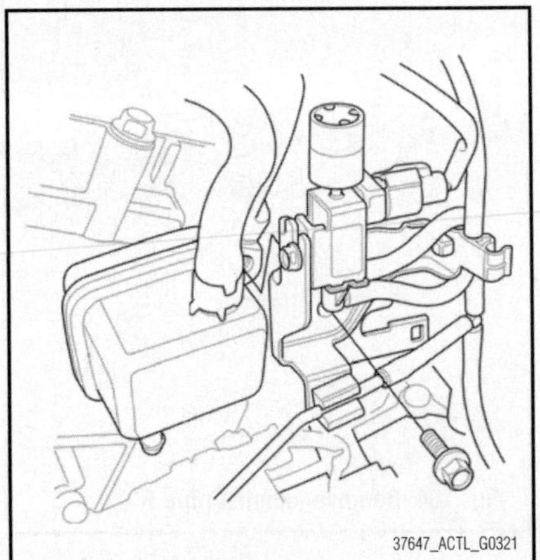
37647_ACTL_G0321

Fig. 111 Remove the engine mount control solenoid valve bracket from the rear cylinder head

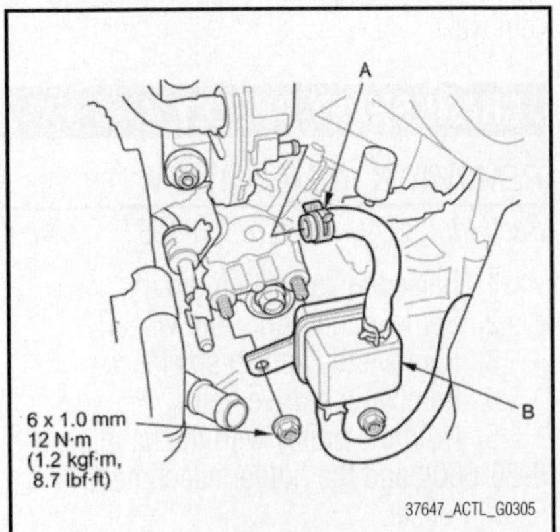
A
B
6 x 1.0 mm
12 N·m
(1.2 kgf·m,
8.7 lbf·ft)
37647_ACTL_G0305

Fig. 112 Remove the Evaporative Emission (EVAP) canister joint with the bracket

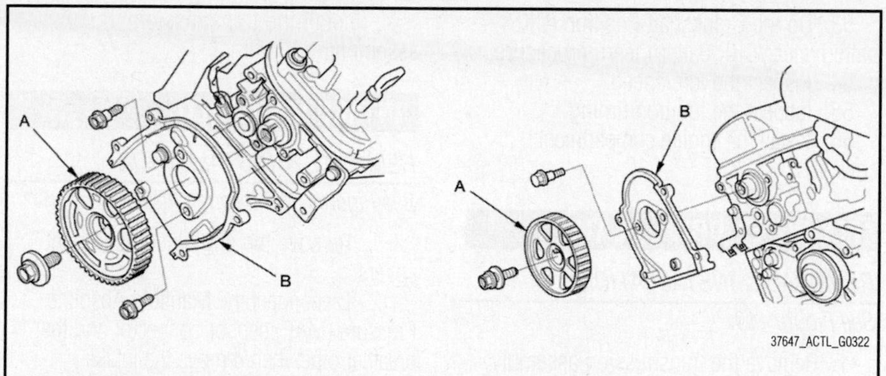
A
B
A
B
37647_ACTL_G0322

Fig. 113 Remove the camshaft pulleys (A) and the back covers (B)

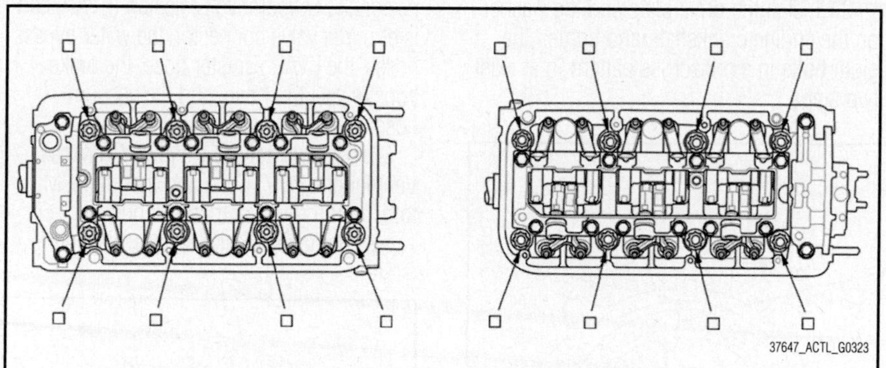
37647_ACTL_G0323

Fig. 114 Remove the cylinder head bolts

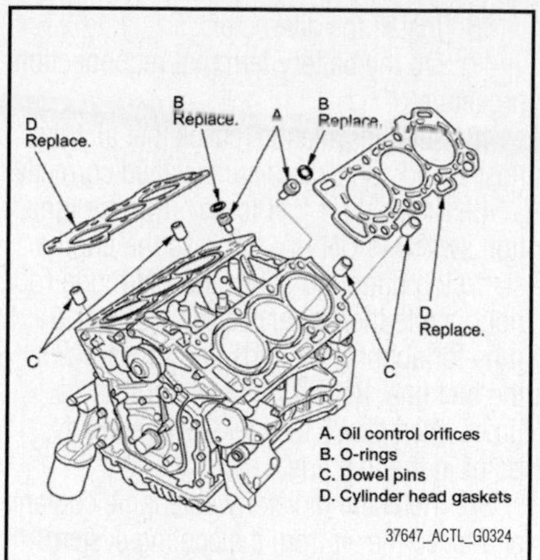
D
Replace.
B
Replace.
A
B
Replace.
D
Replace.
C
C
A. Oil control orifices
B. O-rings
C. Dowel pins
D. Cylinder head gaskets
37647_ACTL_G0324

Fig. 115 Clean and install the oil control orifices with new O-rings

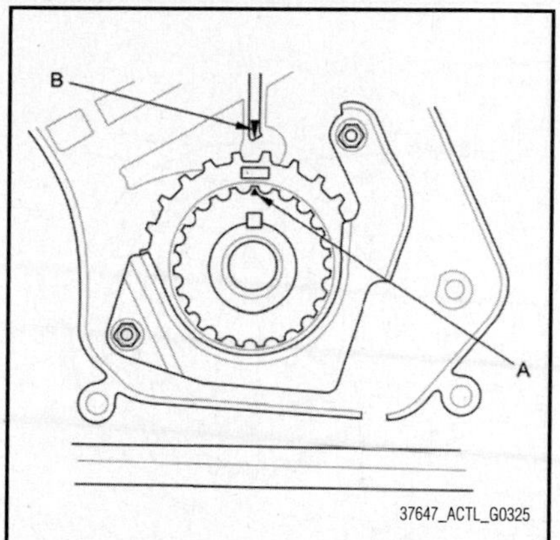
B
A
37647_ACTL_G0325

Fig. 116 Set the timing belt drive pulley to TDC by aligning the TDC mark (A) on the tooth of the timing belt drive pulley with the pointer (B) on the oil pump

To install:

21. Clean the cylinder head and the engine block surface.
22. Clean and install the oil control orifices with new O-rings.
23. Install the dowel pins and the new cylinder head gaskets.
24. Clean the timing belt pulleys, the timing belt guide plate, and the upper and lower covers.
25. Set the timing belt drive pulley to Top Dead Center (TDC) by aligning the TDC mark on the tooth of the timing belt drive pulley with the pointer on the oil pump.
26. Set the camshaft pulleys to TDC by aligning the TDC marks on the camshaft pulleys with the pointers on the back covers.
27. Install the cylinder heads on the engine block.
28. Measure the diameter of each cylinder head bolt at point A and point B.
29. If either diameter is less than 0.44 inches (11.3 mm), replace the cylinder head bolt.
30. Apply new engine oil to the threads and under the bolt heads of all cylinder head bolts.
31. Torque the cylinder head bolts in sequence to 22 ft. lbs. (29 Nm) using a beam-type torque wrench. When using a preset click-type torque wrench, be sure to tighten slowly and do not overtighten. If a bolt makes any noise while you are torquing it, loosen the bolt and retighten it from the first step.
32. After torquing, tighten all cylinder head bolts in two steps (90° per step) using the sequence shown in step 11. If you are using a new cylinder head bolt, tighten the bolt an extra 90°.

➡Remove the cylinder head bolt if you tightened it beyond the specified angle, and go back to step 8 of the procedure. Do not loosen it back to the specified angle.

33. Install the timing belt.
34. Adjust the valve clearance.
35. Install the cylinder head covers.
36. Install the water passage.
37. Install the injector bases.
38. Install the connector bracket to the front cylinder head.
39. Install the Evaporative Emission (EVAP) canister joint with the bracket.
40. Install the engine mount control solenoid valve bracket to the rear cylinder head.
41. Connect the fuel feed hose, then install the quick-connect fitting cover.
42. Install the front Warm Up Three Way Catalytic Converter (front WU-TWC) and the rear Warm Up Three Way Catalytic Converter (rear WU-TWC).
43. Connect the following engine wire harness connectors, and install the wire harness clamps to the cylinder head:
    - Six injector connectors
    - Knock sensor connector
    - Engine Coolant Temperature (ECT) sensor 1 connector
    - Engine mount control solenoid valve connector
    - Camshaft Position (CMP) sensor connector
    - Rocker arm oil control solenoid connector
    - Rocker arm oil pressure switch connector
    - Two Air Fuel Ratio (A/F) sensor connectors
    - Two secondary Heated Oxygen Sensor (secondary HO2S) connectors
44. Install the six ignition coils.
45. Install the intake manifold.

46. Install the alternator.

47. Do the battery terminal reconnection procedure.

48. After installation, check that all tubes, hoses, and connectors are installed correctly.

49. Inspect for fuel leaks. Turn the ignition switch to ON (II), or press the engine start/stop button to select the ON mode (do not operate the starter) so the fuel pump runs for about 2 seconds and pressurizes the fuel line. Repeat this operation three times, then check for fuel leakage at any point in the fuel line.

50. Refill the radiator with engine coolant, and bleed the air from the cooling system.

51. Check for fluid leaks.

52. Do the Powertrain Control Module (PCM) idle learn procedure.

53. Do the Crankshaft Position (CKP) pattern clear/CKP pattern learn procedure.

54. Inspect the idle speed.

55. Inspect the ignition timing.

56. Install the engine compartment covers.

FLYWHEEL/DRIVE PLATE

REMOVAL & INSTALLATION

See Figure 119.

1. Remove the transmission assembly.
2. Remove the drive plate and the washer from the engine crankshaft.
3. Install the drive plate and the washer on the engine crankshaft, and tighten the eight bolts in a crisscross pattern in at least two steps.
4. Install the transmission assembly.

INTAKE MANIFOLD

REMOVAL & INSTALLATION

See Figures 120 through 125.

1. Remove the engine compartment covers.
2. Disconnect the Manifold Absolute Pressure (MAP) sensor connector and the breather pipe, then remove the intake air duct.
3. Disconnect the throttle actuator connector, the Evaporative Emission (EVAP) canister purge valve connector, the water bypass hoses, the EVAP canister hose, the brake booster vacuum hose, and the vacuum hose.
4. Disconnect the Positive Crankcase Ventilation (PCV) hose and the Intake Manifold Tuning (IMT) actuator connector.
5. Remove the upper cover mounting

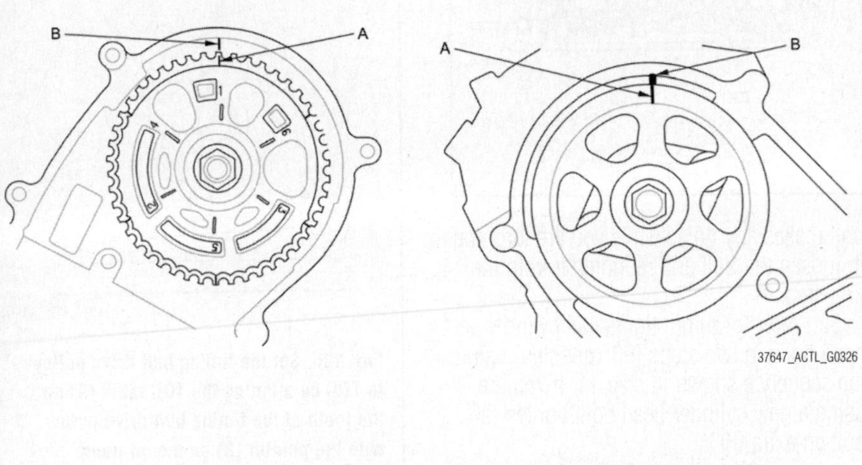
B
A
A
B
37647_ACTL_G0326

Fig. 117 Set the camshaft pulleys to TDC by aligning the TDC marks (A) on the camshaft pulleys with the pointers (B) on the back covers

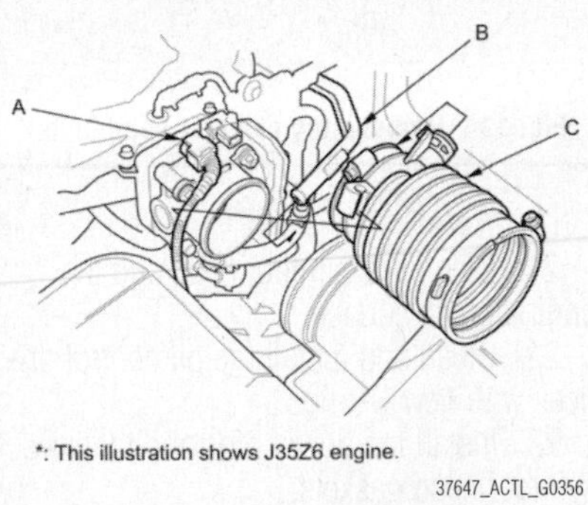
B
A
C
*: This illustration shows J35Z6 engine.
37647_ACTL_G0356

Fig. 120 Disconnect the MAP sensor connector (A) and the breather pipe (B), then remove the intake air duct (C)

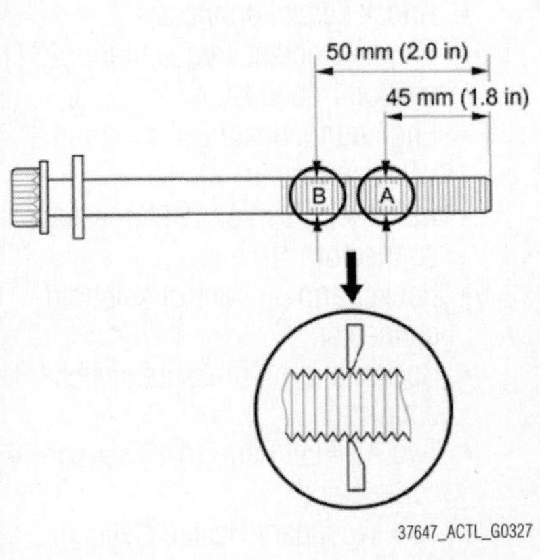
50 mm (2.0 in)
45 mm (1.8 in)
B
A
37647_ACTL_G0327

Fig. 118 Measure the diameter of each cylinder head bolt at point A and point B

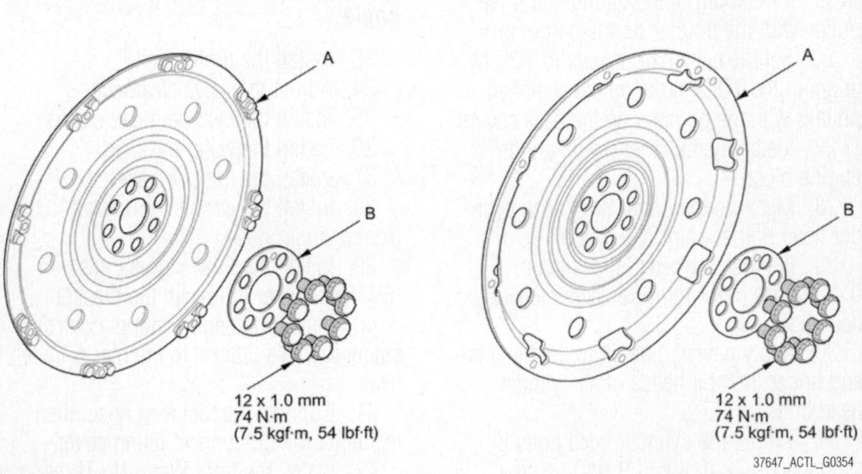
A
B
12 x 1.0 mm
74 N·m
(7.5 kgf·m, 54 lbf·ft)
A
B
12 x 1.0 mm
74 N·m
(7.5 kgf·m, 54 lbf·ft)
37647_ACTL_G0354

Fig. 119 Remove the drive plate (A) and the washer (B) from the engine crankshaft

bolts and nuts sequentially in three steps, then remove the upper cover.

6. Remove the intake manifold mounting bolts and nuts sequentially in three steps, then remove the intake manifold.

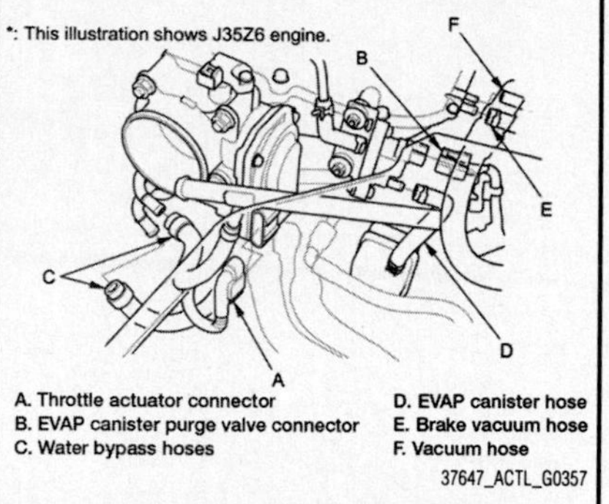
*: This illustration shows J35Z6 engine.
A. Throttle actuator connector
B. EVAP canister purge valve connector
C. Water bypass hoses
D. EVAP canister hose
E. Brake vacuum hose
F. Vacuum hose
37647_ACTL_G0357

Fig. 121 Disconnect the throttle actuator connector, the EVAP canister purge valve connector, the water bypass hoses, the EVAP canister hose, the brake booster vacuum hose, and the vacuum hose

To install:

7. Install the intake manifold. Tighten the mounting bolts and nuts sequentially in three steps to 16 ft. lbs. (22 Nm). Always use a new intake manifold gasket.

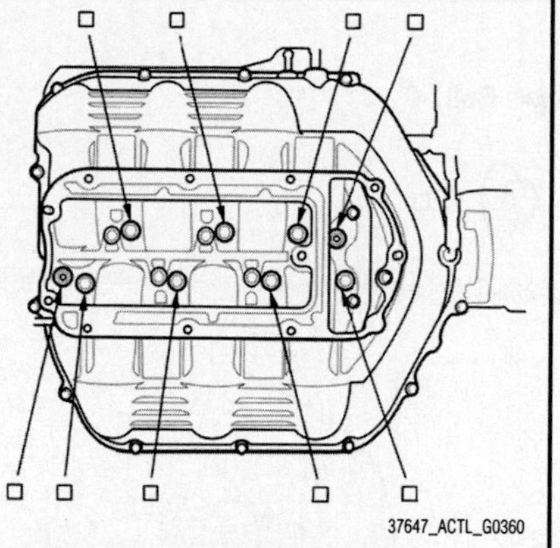
37647_ACTL_G0360

Fig. 124 Remove the intake manifold mounting bolts and nuts sequentially in three steps, then remove the intake manifold

8. Install the upper cover. Tighten the mounting bolts and nuts sequentially in three steps to 9 ft. lbs. (12 Nm). Always use a new gasket.

9. Connect the PCV hose and the IMT actuator connector.

10. Connect the throttle actuator connector, the EVAP canister purge valve connector, the water bypass hoses, the EVAP canister hose, the brake booster vacuum hose, and the vacuum hose.

11. Install the intake air duct, then connect the breather pipe and the MAP sensor connector.

12. Clean up any spilled engine coolant.

13. After installation, check that all tubes, hoses, and connectors are installed correctly.

14. Refill the radiator with engine coolant, and bleed the air from the cooling system..

15. Install the engine compartment covers.

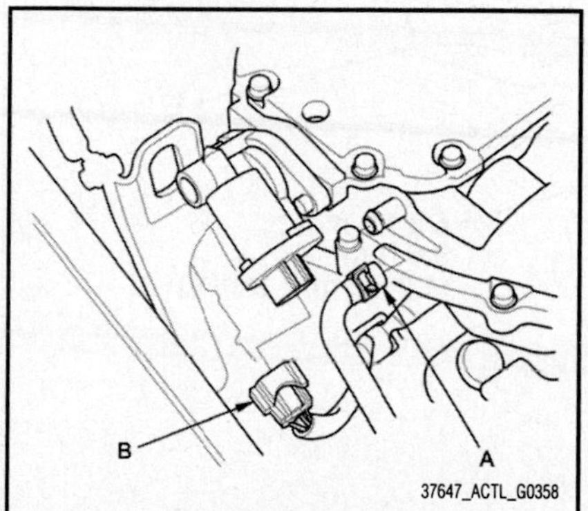
B
A
37647_ACTL_G0358

Fig. 122 Disconnect the PCV hose (A) and the IMT actuator connector (B)

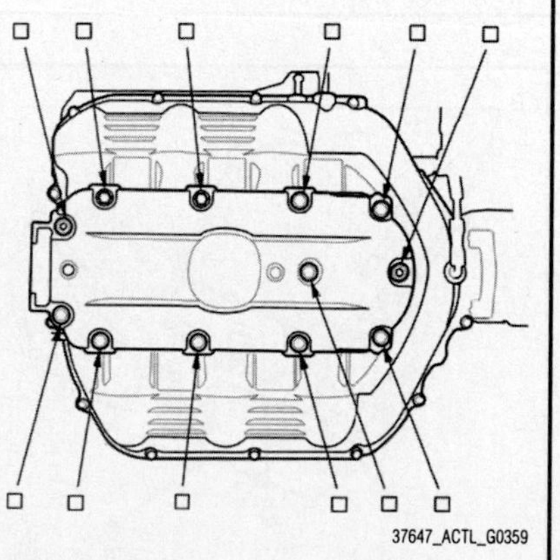
37647_ACTL_G0359

Fig. 123 Remove the upper cover mounting bolts and nuts sequentially in three steps

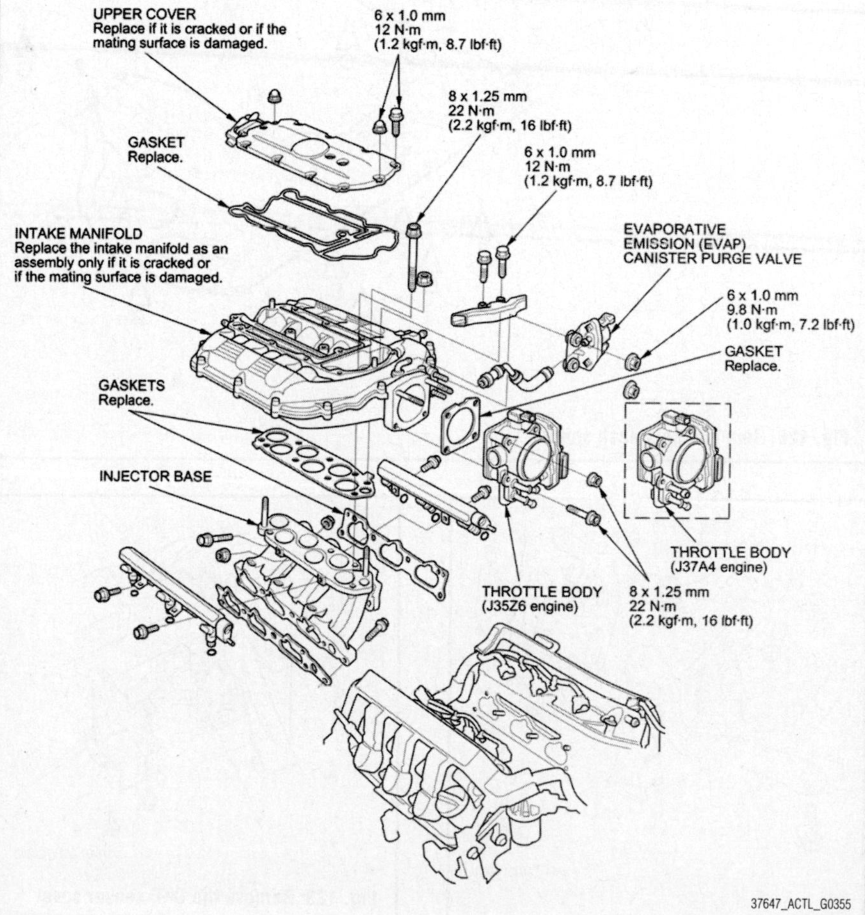
UPPER COVER
Replace if it is cracked or if the mating surface is damaged.
6 x 1.0 mm
12 N·m
(1.2 kgf·m, 8.7 lbf·ft)
8 x 1.25 mm
22 N·m
(2.2 kgf·m, 16 lbf·ft)
GASKET
Replace.
6 x 1.0 mm
12 N·m
(1.2 kgf·m, 8.7 lbf·ft)
INTAKE MANIFOLD
Replace the intake manifold as an assembly only if it is cracked or if the mating surface is damaged.
EVAPORATIVE EMISSION (EVAP) CANISTER PURGE VALVE
6 x 1.0 mm
9.8 N·m
(1.0 kgf·m, 7.2 lbf·ft)
GASKET
Replace.
GASKETS
Replace.
INJECTOR BASE
THROTTLE BODY
(J37A4 engine)
THROTTLE BODY
(J35Z6 engine)
8 x 1.25 mm
22 N·m
(2.2 kgf·m, 16 lbf·ft)
37647_ACTL_G0355

Fig. 125 Exploded view of the intake manifold assembly

OIL PAN

REMOVAL & INSTALLATION

See Figures 126 through 130.

➡If the engine is already out of the vehicle, go to step 6.

1. Raise the vehicle on the lift.
2. Drain the engine oil.
3. Remove the splash shield.
4. Remove exhaust pipe A.
5. Remove the rear Warm Up Three Way Catalytic Converter (rear WU-TWC) bracket.
6. Remove the Crankshaft Position (CKP) sensor cover and the bolt, then disconnect the CKP sensor connector.
7. Remove the clutch/torque converter cover and the four bolts securing the transmission.

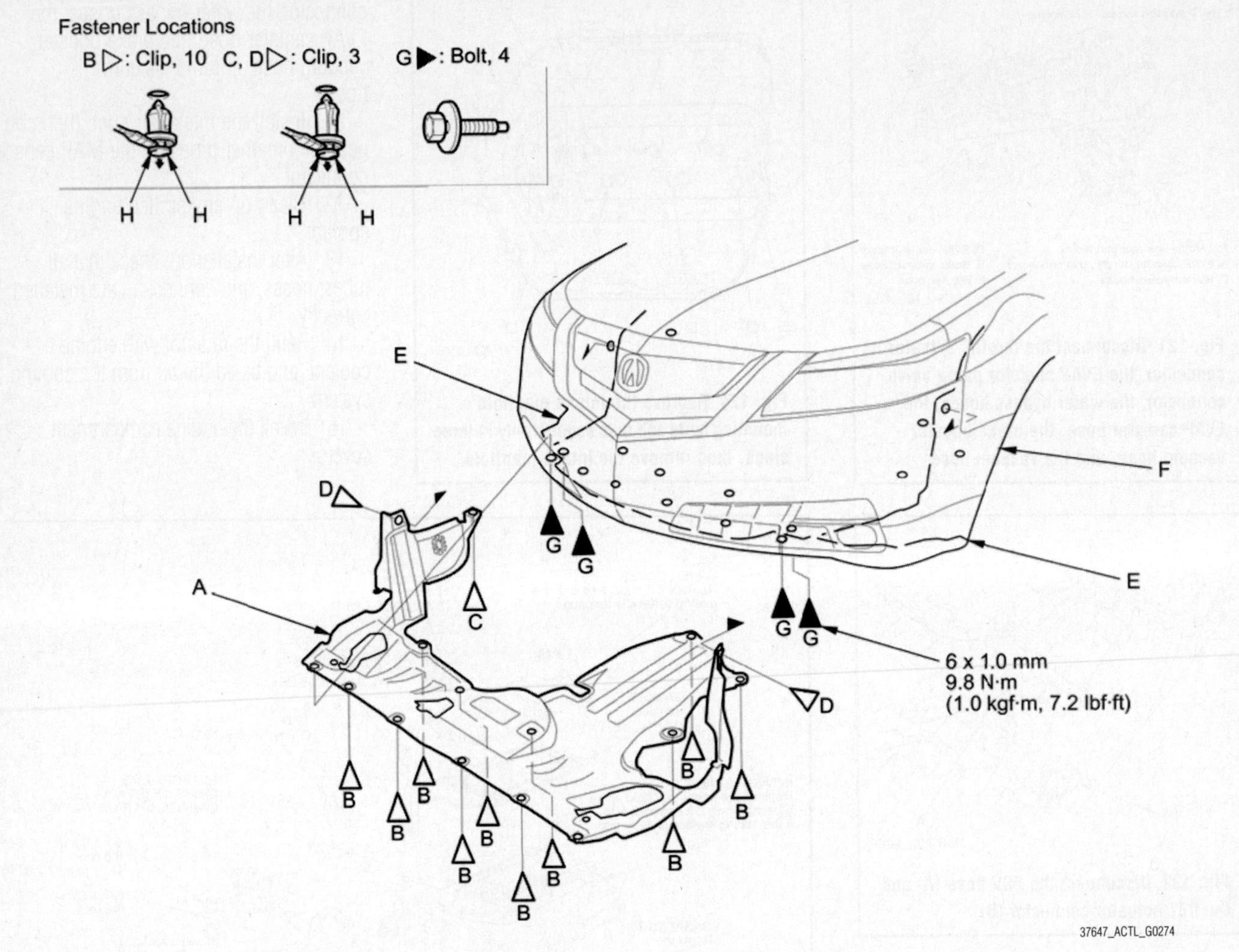
Fastener Locations
B ▷: Clip, 10 C, D ▷: Clip, 3 G ▶: Bolt, 4
H
H
H
H
A
B
C
D
E
F
G
6 x 1.0 mm
9.8 N·m
(1.0 kgf·m, 7.2 lbf·ft)
37647_ACTL_G0274

Fig. 126 Remove the splash shield

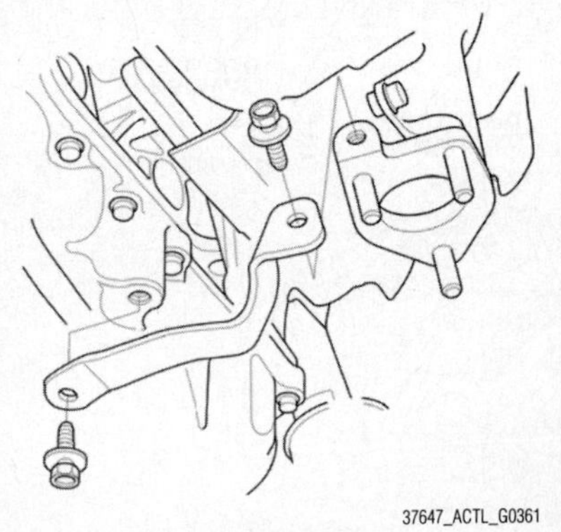
37647_ACTL_G0361

Fig. 127 Remove the rear WU-TWC bracket

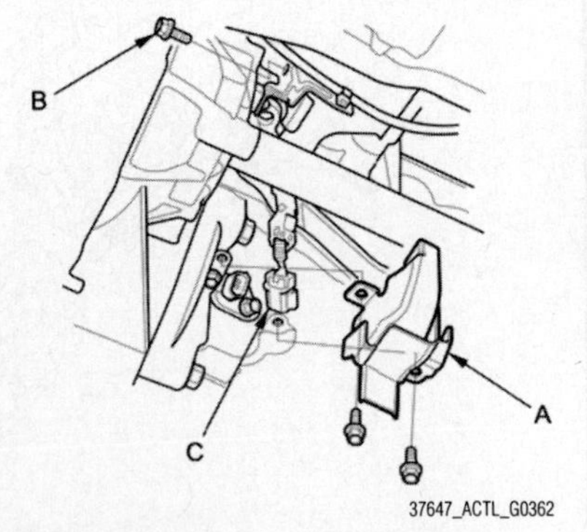
A
B
C
37647_ACTL_G0362

Fig. 128 Remove the CKP sensor cover (A) and the bolt (B), then disconnect the CKP sensor connector (C)

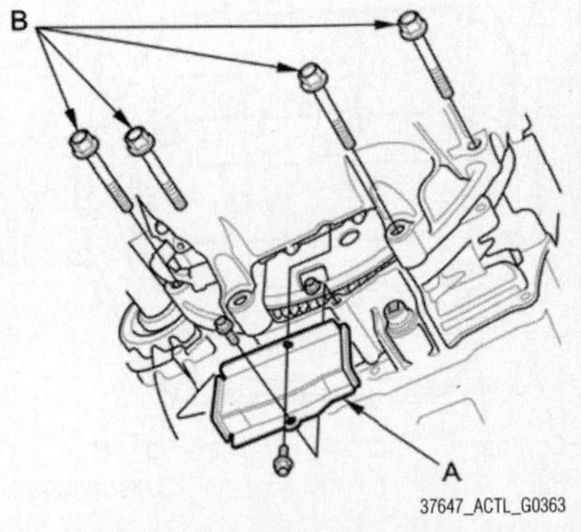
A
B
37647_ACTL_G0363

Fig. 129 Remove the clutch/torque converter cover (A) and the four bolts (B) securing the transmission

8. Remove the bolts securing the oil pan.

9. Using a flat blade screwdriver, separate the oil pan from the engine block in the places shown.

10. Remove the oil pan.

To install:

11. Remove all of the old liquid gasket from the oil pan mating surfaces, the bolts, and the bolt holes.

12. Clean and dry the oil pan mating surfaces.

13. Apply liquid gasket, P/N 08717-0004, 08718-0003, or 08718-0009 to the oil pan mating surface of the engine block and to the inside edge of the threaded bolt holes. Install the component within 5 minutes of applying the liquid gasket.

➡Apply a bead of liquid gasket along the mating edge of the oil pan. If you apply liquid gasket P/N 08718-0012, the component must be installed within 4 minutes. If too much time has passed after applying the liquid gasket, remove the old liquid gasket and residue, then reapply the new liquid gasket.

14. Install the oil pan on the engine block.

15. Tighten the bolts in three steps. In the final step, torque all bolts, in sequence, to 9 ft. lbs. (12 Nm).

➡Wait at least 30 minutes before filling the engine with oil. Do not run the engine for at least 3 hours after installing the oil pan.

16. Tighten the four bolts securing the transmission, then install the clutch/torque converter cover.

17. Connect the Crankshaft Position (CKP) sensor connector, then install the CKP sensor cover and the bolt.

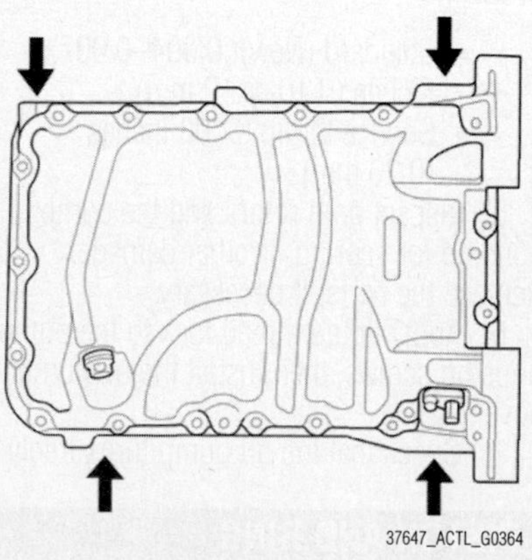
37647_ACTL_G0364

Fig. 130 Using a flat blade screwdriver, separate the oil pan from the engine block in the places shown

18. Install the rear Warm Up Three Way Catalytic Converter (rear WU-TWC) bracket.

19. If the engine is still in the vehicle, do the following steps.

20. Install exhaust pipe A using new gaskets and new self-locking nuts.

21. Install the splash shield.

22. Refill the engine with the recommended engine oil.

OIL PUMP

REMOVAL & INSTALLATION

See Figures 131 through 134.

1. Drain the engine oil.
2. Remove the timing belt:
3. Remove the timing belt drive pulley from the crankshaft:
4. Attach the chain hoist to the engine hook on the engine hanger bracket.
5. Remove the rocker arm oil control solenoid/oil filter assembly.
6. Remove the oil pan.
7. Remove the oil strainer.
8. Remove the mounting bolts, then remove the oil pump assembly.

To install:

9. Remove the old oil seal from the oil pump.

10. Clean and dry the crankshaft oil seal housing.

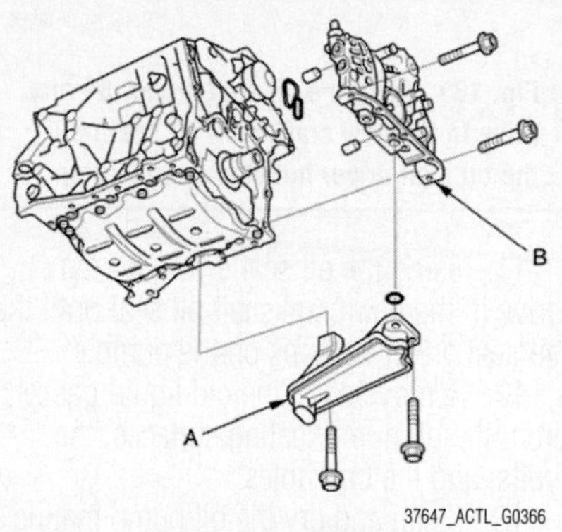
B
A
37647_ACTL_G0366

Fig. 131 Remove the oil strainer (A), then remove the oil pump assembly (B)

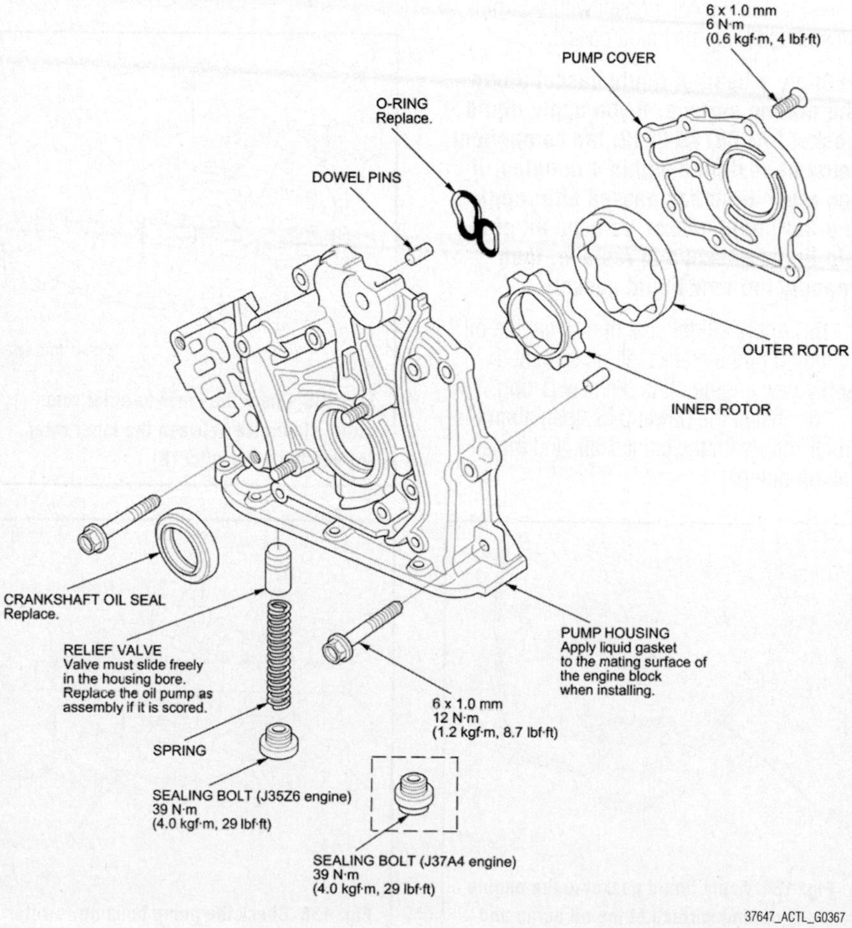
6 x 1.0 mm
6 N·m
(0.6 kgf·m, 4 lbf·ft)
PUMP COVER
O-RING
Replace.
DOWEL PINS
OUTER ROTOR
INNER ROTOR
CRANKSHAFT OIL SEAL
Replace.
RELIEF VALVE
Valve must slide freely in the housing bore.
Replace the oil pump as assembly if it is scored.
PUMP HOUSING
Apply liquid gasket to the mating surface of the engine block when installing.
6 x 1.0 mm
12 N·m
(1.2 kgf·m, 8.7 lbf·ft)
SPRING
SEALING BOLT (J35Z6 engine)
39 N·m
(4.0 kgf·m, 29 lbf·ft)
SEALING BOLT (J37A4 engine)
39 N·m
(4.0 kgf·m, 29 lbf·ft)
37647_ACTL_G0367

Fig. 132 Exploded view of oil pump assembly

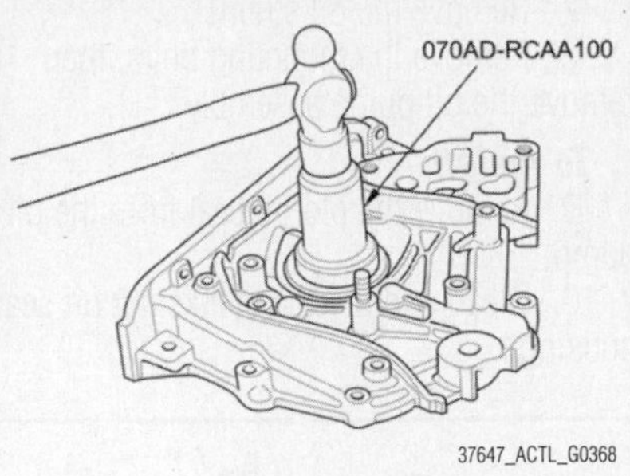
070AD-RCAA100
37647_ACTL_G0368

Fig. 133 Using the oil seal driver, 64 mm, drive in the new crankshaft oil seal until the oil seal driver bottoms on the pump

11. Using the oil seal driver, 64 mm, drive in the new crankshaft oil seal until the oil seal driver bottoms on the pump.
12. Remove all of the old liquid gasket from the oil pump mating surfaces, the bolts, and the bolt holes.
13. Clean and dry the oil pump mating surfaces.
14. Apply liquid gasket, P/N 08717-0004, 08718-0003, or 08718-0009, to the engine block mating surface of the oil pump and to the inside edge of the threaded bolt holes. Install the component within 5 minutes of applying the liquid gasket.

➡Apply a bead of liquid gasket along the mating surface. If you apply liquid gasket P/N 08718-0012, the component must be installed within 4 minutes. If too much time has passed after applying the liquid gasket, remove all of the old liquid gasket and residue, then reapply the new liquid gasket.

15. Apply a light coat of new engine oil to the lip of the crankshaft oil seal, and apply new engine oil to the new O-ring.
16. Install the dowel pins, then align the inner rotor with the crankshaft, and install the oil pump.

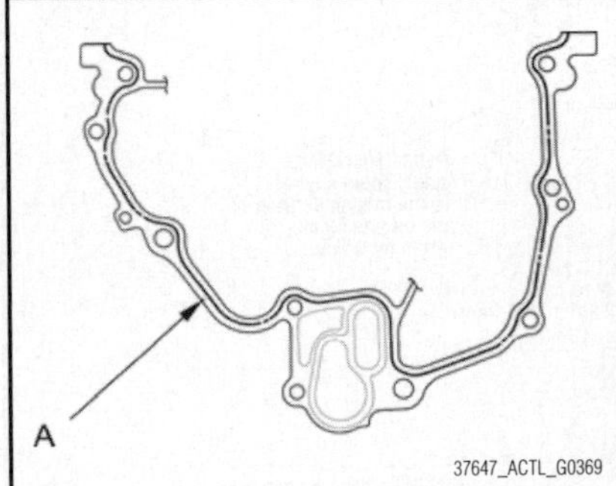
A
37647_ACTL_G0369

Fig. 134 Apply liquid gasket to the engine block mating surface of the oil pump and to the inside edge of the threaded bolt holes

➡Wait at least 30 minutes before filling the engine with oil. Do not run the engine for at least 3 hours after installing the oil pump.

17. Clean the excess grease off the crankshaft, and check the seal for distortion.
18. Install the oil strainer with a new O-ring.
19. Install the oil pan.
20. Install the rocker arm oil control solenoid/oil filter assembly with a new rocker arm oil control solenoid filter.
21. Install the timing belt drive pulley to the crankshaft:
22. Install the timing belt:
23. Remove the chain hoist.

INSPECTION

See Figures 135 through 137.

1. Remove the screws from the pump housing, then separate the pump housing and the cover.
2. Check the inner-to-outer rotor radial clearance between the inner rotor and the outer rotor. If the inner-to-outer rotor radial clearance exceeds the service limit, replace the oil pump assembly.

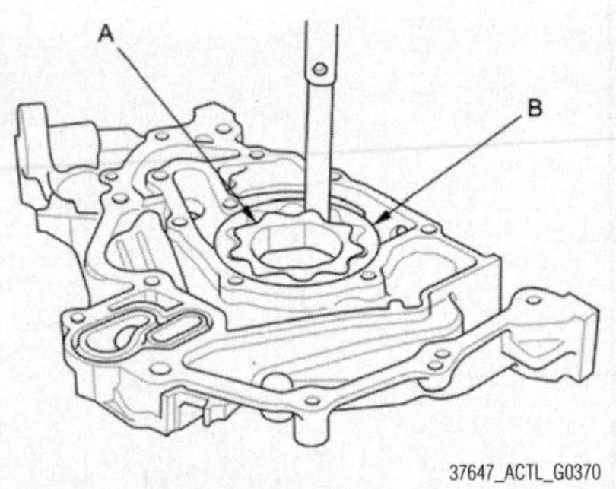
A
B
37647_ACTL_G0370

Fig. 135 Check the inner-to-outer rotor radial clearance between the inner rotor (A) and the outer rotor (B)

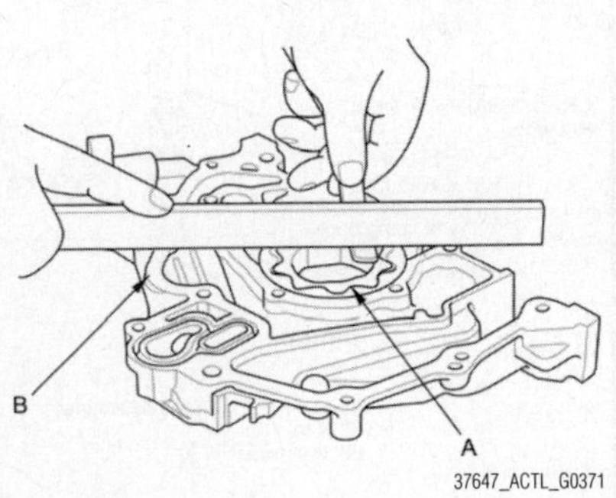
B
A
37647_ACTL_G0371

Fig. 136 Check the pump housing-to-rotor axial clearance between the rotors (A) and the pump housing (B)

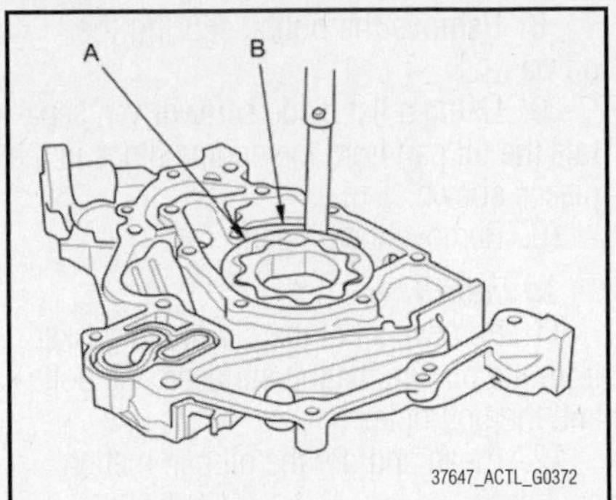
A
B
37647_ACTL_G0372

Fig. 137 Check the pump housing-to-outer rotor radial clearance between the outer rotor (A) and the pump housing (B)

➡Inner Rotor-to-Outer Rotor Radial Clearance

- Standard (New): 0.002–0.006 inches (0.04–0.16 mm)
- Service Limit: 0.008 inches (0.20 mm)

3. Check the pump housing-to-rotor axial clearance between the rotors and the pump housing. If the pump housing-to-rotor axial clearance exceeds the service limit, replace the oil pump assembly.

➡Pump Housing-to-Rotor Axial Clearance

- Standard (New): 0.001–0.003 inches (0.02–0.07 mm)
- Service Limit: 0.005 inches (0.12 mm)

4. Check the pump housing-to-outer rotor radial clearance between the outer rotor and the pump housing. If the pump housing-to-outer rotor radial clearance exceeds the service limit, replace the oil pump assembly.

➡Pump Housing-to-Outer Rotor Radial Clearance

- Standard (New): 0.004–0.007 inches (0.10–0.19 mm)
- Service Limit: 0.008 inches (0.20 mm)

5. Inspect both rotors and the pump housing for scoring or other damage. Replace the parts, if necessary.
6. Apply liquid thread lock to the pump housing screws, then install the oil pump cover.
7. Check that the oil pump turns freely.

PISTON AND RING

POSITIONING

See Figure 138.

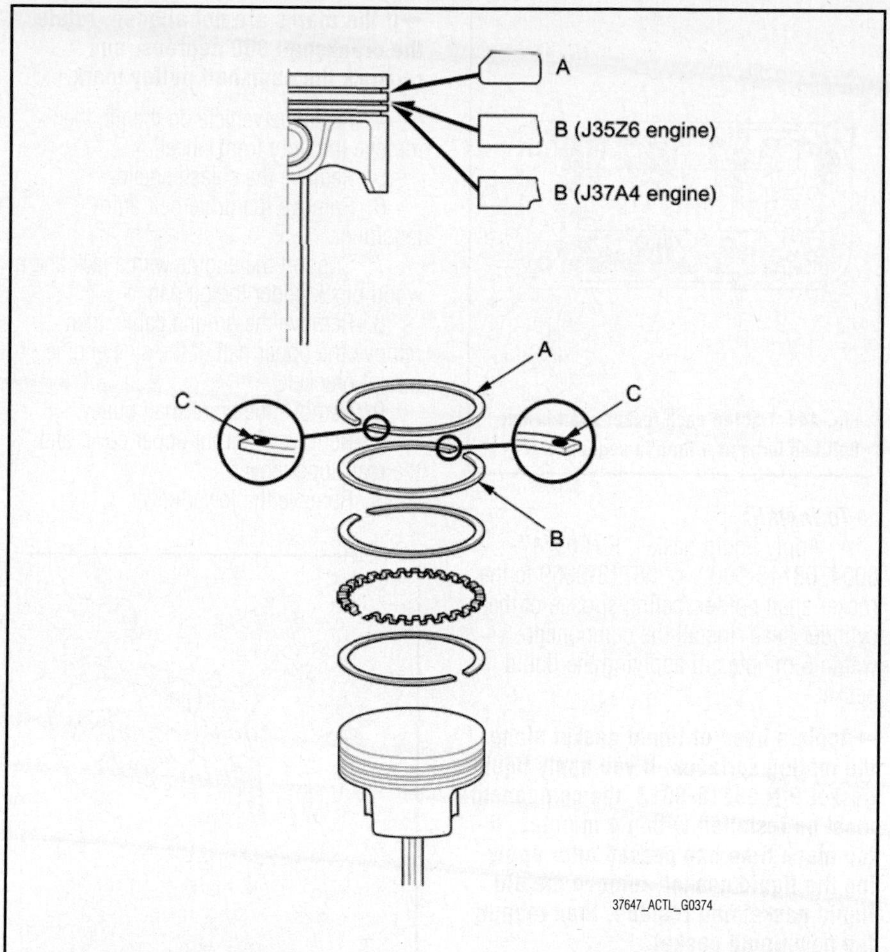
A
B (J35Z6 engine)
B (J37A4 engine)
A
C
C
B
37647_ACTL_G0374

Fig. 138 Piston ring positioning

REAR MAIN SEAL

REMOVAL & INSTALLATION

1. M/T model: Remove the transmission, and the flywheel.
2. A/T model: Remove the transmission, and the drive plate.
3. Remove the transmission end crankshaft oil seal.
4. Clean and dry the crankshaft oil seal housing.
5. Apply a light coat of new engine oil to the lip of the crankshaft oil seal.
6. Using the driver handle, 15 x 135 L and the oil seal driver attachment, 106 mm, drive in the new crankshaft oil seal until the oil seal driver attachment bottoms against the engine block end cover. Align the hole in the oil seal driver attachment with the pin on the crankshaft.
7. Clean any excess grease off the crankshaft, and check that the oil seal lip is not distorted.
8. M/T model: Install the flywheel, and the transmission.
9. A/T model: Install the drive plate, and the transmission.

ROCKER ARMS

REMOVAL & INSTALLATION

Front Cylinder Head

See Figures 139 through 141.

1. Remove the cylinder head cover.
2. Loosen the locknuts and the adjusting screws
3. Remove the rocker shaft bridge mounting bolts, the rocker shaft holder mounting bolts, and the rocker arm assembly.
   a. Loosen the rocker shaft bridge mounting bolts and the rocker shaft holder mounting bolts in sequence two turns at a time, to prevent damaging the valves or the rocker arm assembly.
   b. When removing the rocker arm assembly, do not remove the rocker shaft bridge mounting bolts and the rocker shaft holder mounting bolts. The bolts will keep the rocker arms on the shafts.

To install:

4. Apply liquid gasket, P/N 08717-0004, 08718-0003, or 08718-0009 to the rocker shaft holder mating surface of the cylinder head. Install the component within 5 minutes of applying the liquid gasket.

➡Apply a bead of liquid gasket along the mating surfaces. If you apply liquid gasket P/N 08718-0012, the component must be installed within 4 minutes. If too much time has passed after applying the liquid gasket, remove the old liquid gasket and residue, then reapply the new liquid gasket.

5. Set the rocker arm assembly in place, and loosely install the bolts. Make sure that

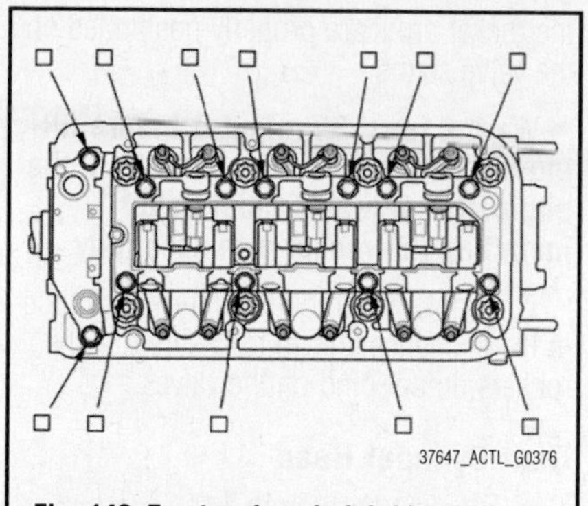
37647_ACTL_G0376

Fig. 140 Front rocker shaft bridge bolt loosening sequence

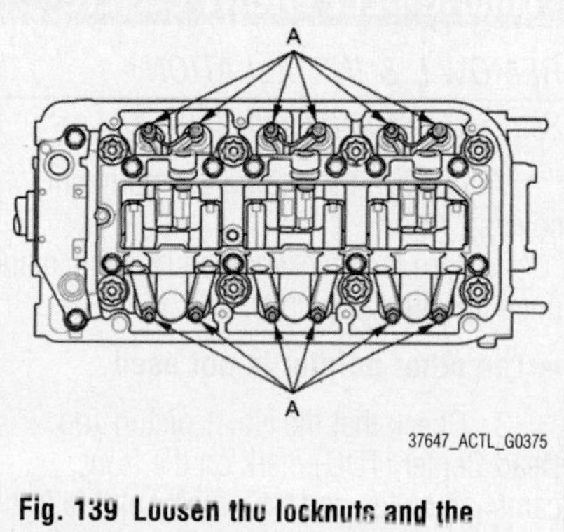
A
A
37647_ACTL_G0375

Fig. 139 Loosen the locknuts and the adjusting screws (A)

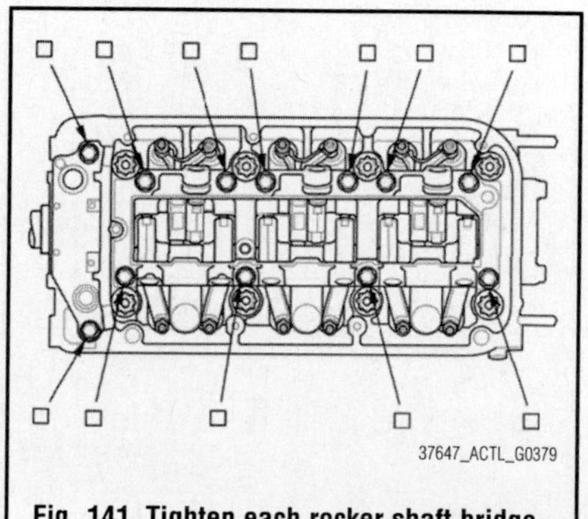
37647_ACTL_G0379

Fig. 141 Tighten each rocker shaft bridge bolt two turns at a time in sequence

the rocker arms are properly positioned on the valve stems.

➡**Wait at least 30 minutes before filling the engine with oil. Do not run the engine for at least 3 hours after installing the rocker arm assembly.**

6. Tighten each bolt two turns at a time in the sequence shown to ensure that the rockers do not bind on the valves.

Rear Cylinder Head

See Figures 142 through 144.

1. Remove the cylinder head cover.
2. Loosen the locknuts and the adjusting screws.
3. Remove the rocker shaft bridge mounting bolts, the rocker shaft holder mounting bolts, and the rocker arm assembly.
   a. Loosen the rocker shaft bridge mounting bolts and the rocker shaft holder mounting bolts in sequence two turns at a time, to prevent damaging the valves or the rocker arm assembly.
   b. When removing the rocker arm assembly, do not remove the rocker shaft bridge mounting bolts and the rocker shaft holder mounting bolts. The bolts will keep the rocker arms on the shafts.

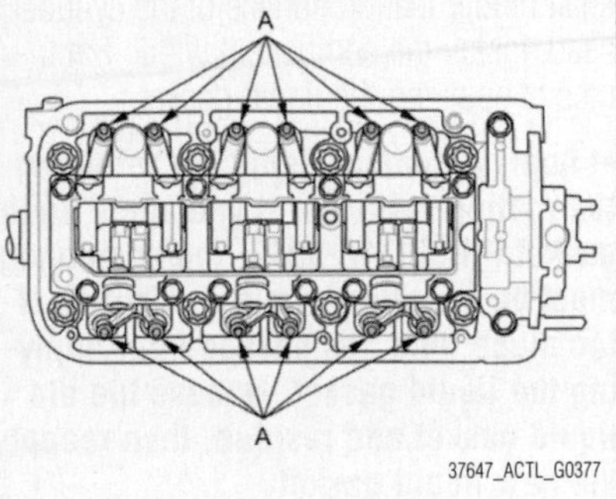
A
A
37647_ACTL_G0377

Fig. 142 Loosen the locknuts and the adjusting screws (A)

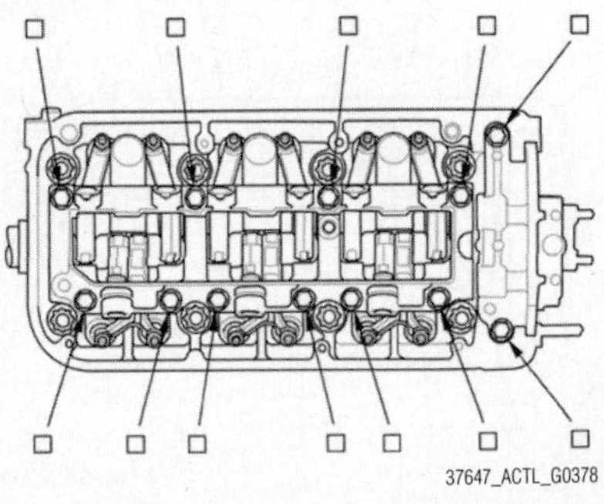
37647_ACTL_G0378

Fig. 143 Rear cylinder head bolt loosening sequence

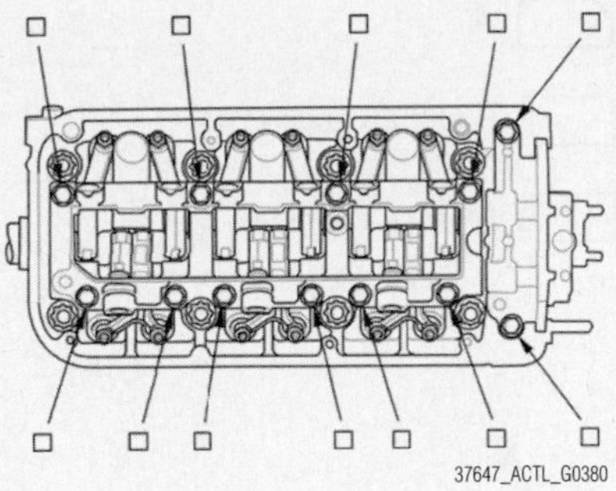
37647_ACTL_G0380

Fig. 144 Tighten each rocker shaft bridge bolt two turns at a time in sequence

To install:

4. Apply liquid gasket, P/N 08717-0004, 08718-0003, or 08718-0009 to the rocker shaft holder mating surface of the cylinder head. Install the component within 5 minutes of applying the liquid gasket.

➡**Apply a bead of liquid gasket along the mating surfaces. If you apply liquid gasket P/N 08718-0012, the component must be installed within 4 minutes. If too much time has passed after applying the liquid gasket, remove the old liquid gasket and residue, then reapply the new liquid gasket.**

5. Set the rocker arm assembly in place, and loosely install the bolts. Make sure that the rocker arms are properly positioned on the valve stems.

➡**Wait at least 30 minutes before filling the engine with oil. Do not run the engine for at least 3 hours after installing the rocker arm assembly.**

6. Tighten each bolt two turns at a time in the sequence shown to ensure that the rockers do not bind on the valves.

TIMING BELT FRONT COVER

REMOVAL & INSTALLATION

See Figures 145 through 150.

1. Remove the engine compartment covers.
2. Turn the crankshaft so its white mark lines up with the pointer.

➡**The other pointer is not used.**

3. Check that the No. 1 piston Top Dead Center (TDC) mark on the front camshaft pulley and the pointer on the front upper cover are aligned.

➡**If the marks are not aligned, rotate the crankshaft 360 degrees, and recheck the camshaft pulley mark.**

4. Raise the vehicle on the lift, then remove the right front wheel.
5. Remove the splash shield.
6. Remove the drive belt auto-tensioner.
7. Support the engine with a jack and a wood block under the oil pan.
8. Remove the ground cable, then remove the upper half of the side engine mount bracket.
9. Remove the crankshaft pulley.
10. Remove the front upper cover and the rear upper cover.
11. Remove the lower cover.

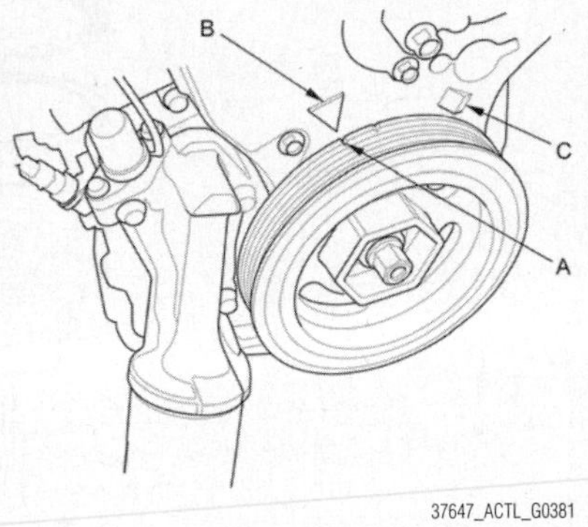
B
C
A
37647_ACTL_G0381

Fig. 145 Turn the crankshaft so its white mark (A) lines up with the pointer (B)

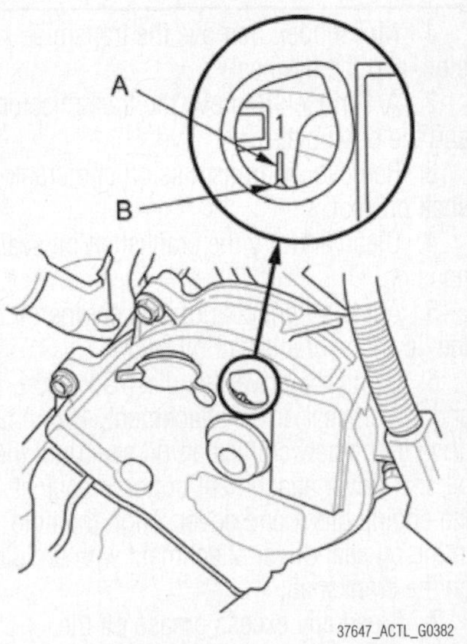
A
1
B
37647_ACTL_G0382

Fig. 146 Check that the No. 1 piston TDC mark (A) on the front camshaft pulley and the pointer (B) on the front upper cover are aligned

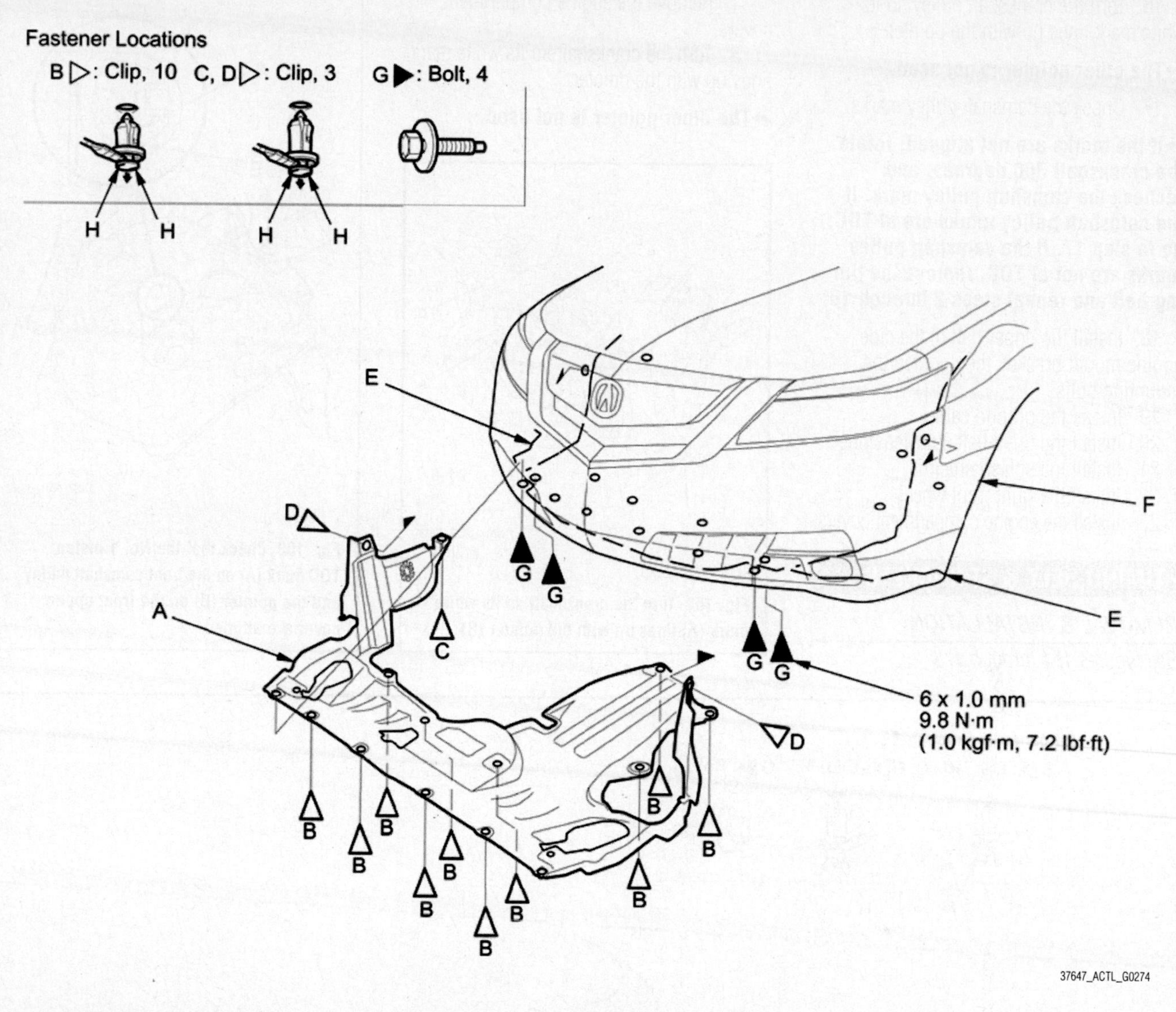
Fastener Locations
B ▷: Clip, 10 C, D ▷: Clip, 3 G ▶: Bolt, 4
H H H H
E
F
D
G G
A
C
G G
6 x 1.0 mm
9.8 N·m
(1.0 kgf·m, 7.2 lbf·ft)
D
B B B B B B B B B B
37647_ACTL_G0274

Fig. 147 Remove the splash shield

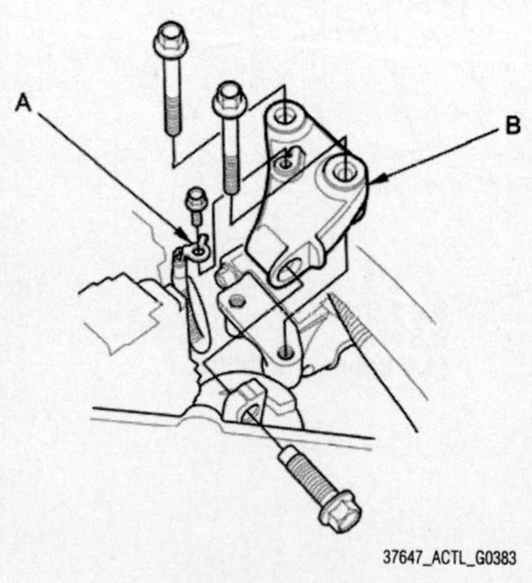
A
B
37647_ACTL_G0383

Fig. 148 Remove the ground cable (A), then remove the upper half of the side engine mount bracket (B)

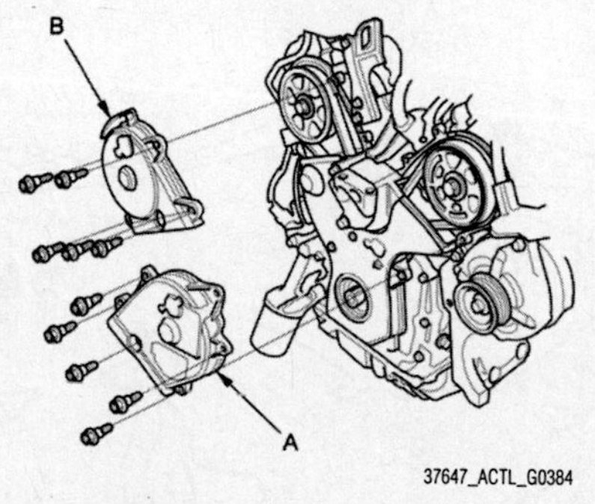
B
A
37647_ACTL_G0384

Fig. 149 Remove the front upper cover (A) and the rear upper cover (B)

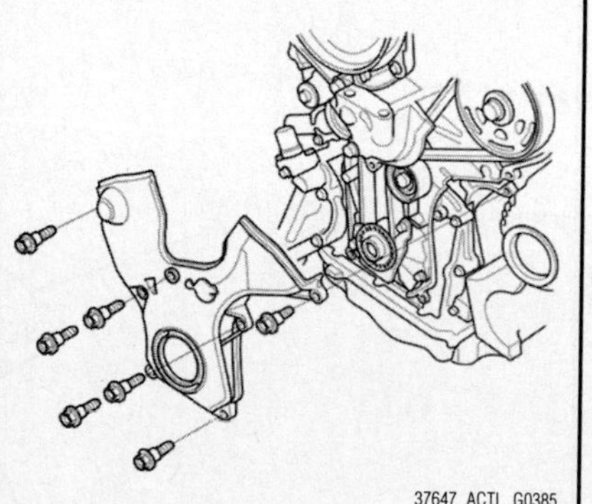
37647_ACTL_G0385

Fig. 150 Remove the lower cover

To install:

12. Install the lower cover.
13. Install the front upper cover and the rear upper cover.
14. Install the crankshaft pulley.
15. Rotate the crankshaft pulley about six turns clockwise so the timing belt positions itself on the pulleys.

16. Turn the crankshaft pulley so its white mark lines up with the pointer.

➡**The other pointer is not used.**

17. Check the camshaft pulley marks.

➡**If the marks are not aligned, rotate the crankshaft 360 degrees, and recheck the camshaft pulley mark. If the camshaft pulley marks are at TDC, go to step 17. If the camshaft pulley marks are not at TDC, remove the timing belt and repeat steps 2 through 16.**

18. Install the upper half of the side engine mount bracket, then tighten the mounting bolts.
19. Install the ground cable.
20. Install the drive belt auto-tensioner.
21. Install the splash shield.
22. Install the right front wheel.
23. Install the engine compartment covers.

TIMING BELT & SPROCKETS

REMOVAL & INSTALLATION

See Figures 151 through 173.

1. Remove the engine compartment covers.
2. Turn the crankshaft so its white mark lines up with the pointer.

➡**The other pointer is not used.**

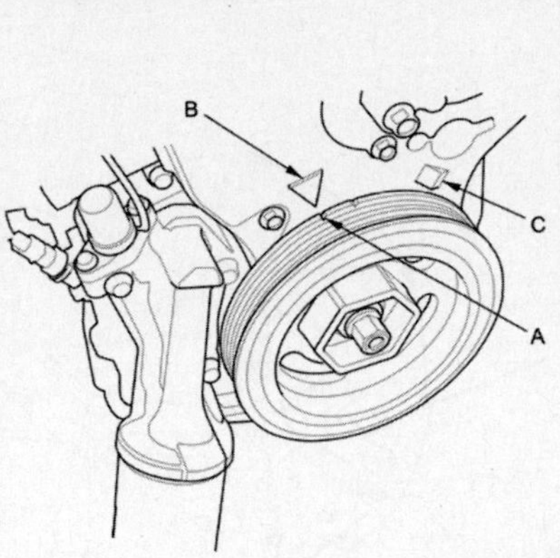
B
C
A

37647_ACTL_G0381

Fig. 152 Turn the crankshaft so its white mark (A) lines up with the pointer (B)

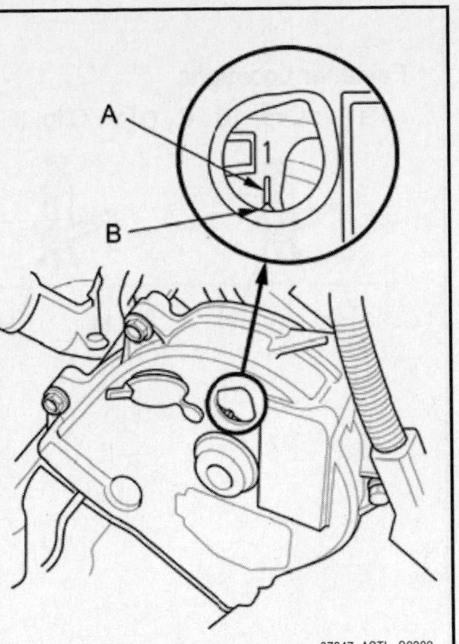
A
1
B

37647_ACTL_G0382

Fig. 153 Check that the No. 1 piston TDC mark (A) on the front camshaft pulley and the pointer (B) on the front upper cover are aligned

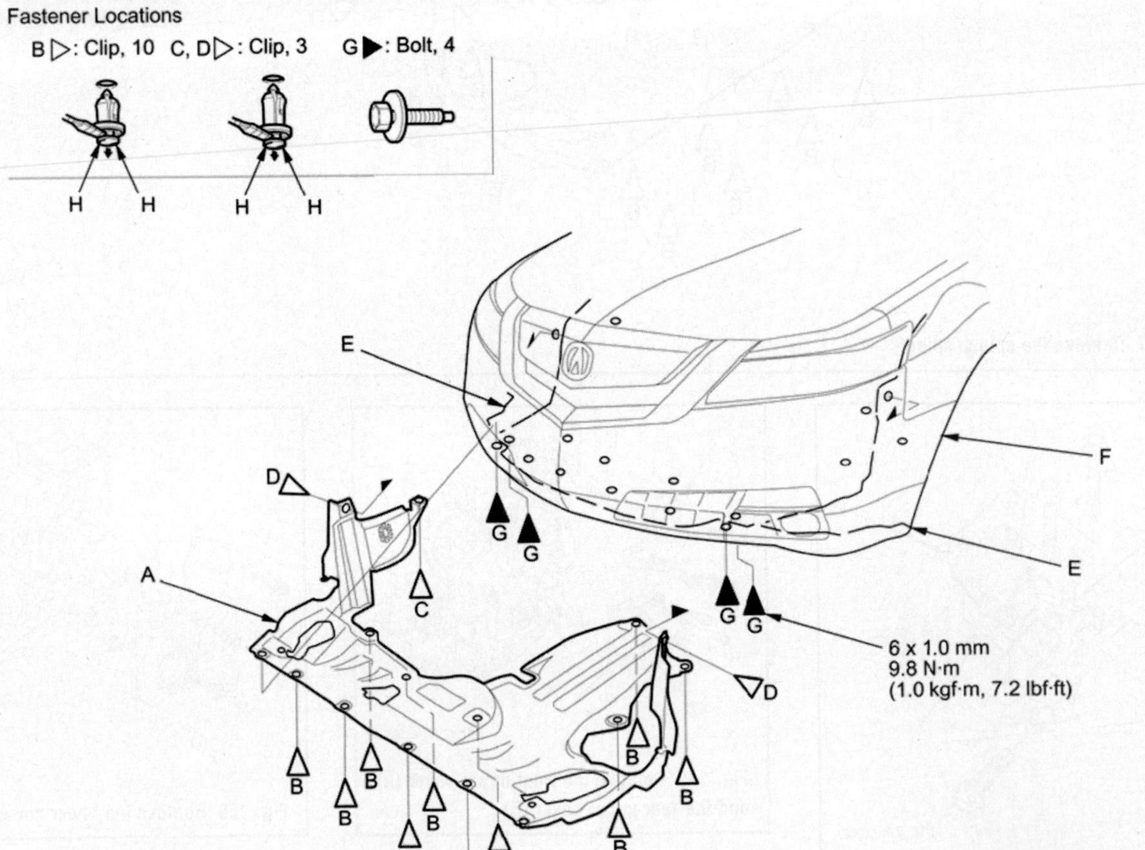
Fastener Locations
B ▷: Clip, 10 C, D ▷: Clip, 3 G ▶: Bolt, 4
H H H H
E
F
D
A
C
G G
G G
E
D
6 x 1.0 mm
9.8 N·m
(1.0 kgf·m, 7.2 lbf·ft)
B B B B B B B B B B

37647_ACTL_G0274

Fig. 151 Remove the splash shield

3. Check that the No. 1 piston Top Dead Center (TDC) mark on the front camshaft pulley and the pointer on the front upper cover are aligned.

➡If the marks are not aligned, rotate the crankshaft 360 degrees, and recheck the camshaft pulley mark.

4. Raise the vehicle on the lift, then remove the right front wheel.
5. Remove the splash shield.
6. Remove the drive belt auto-tensioner.
7. Support the engine with a jack and a wood block under the oil pan.
8. Remove the ground cable, then remove the upper half of the side engine mount bracket.
9. Remove the crankshaft pulley.
10. Remove the front upper cover and the rear upper cover.
11. Remove the lower cover.
12. Remove one of the battery clamp bolts from the battery tray, and grind the end of it.

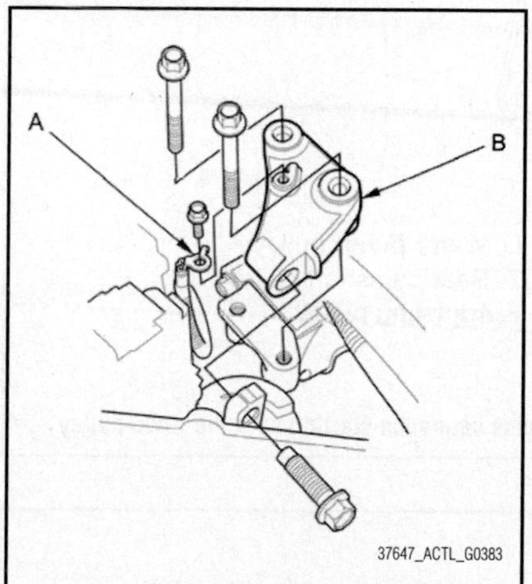
A
B
37647_ACTL_G0383

Fig. 154 Remove the ground cable (A), then remove the upper half of the side engine mount bracket (B)

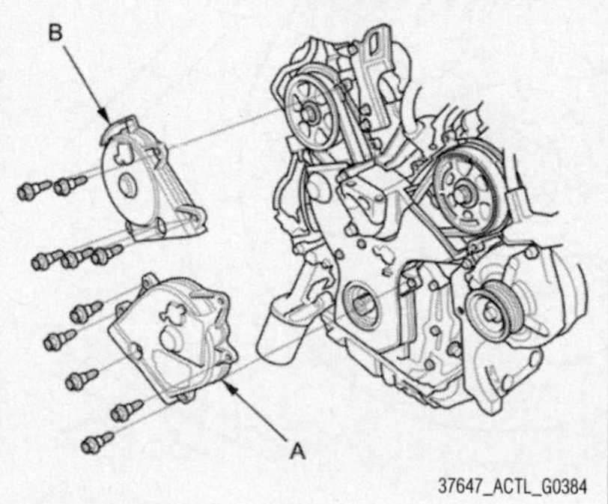
B
A
37647_ACTL_G0384

Fig. 155 Remove the front upper cover (A) and the rear upper cover (B)

13. Thread the battery clamp bolt into hold the timing belt adjuster in its current position. Tighten it by hand, do not use a wrench.
14. Remove the timing belt guide plate.
15. Remove the lower half of the side engine mount bracket.
16. Remove the idler pulley bolt and the idler pulley, then remove the timing belt. Discard the idler pulley bolt.

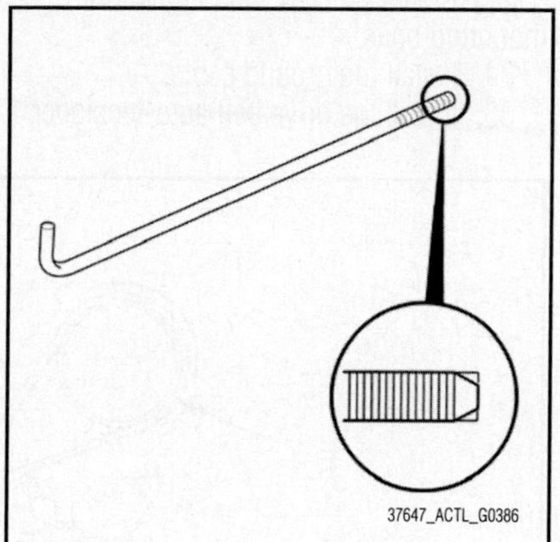
37647_ACTL_G0386

Fig. 156 Remove one of the battery clamp bolts from the battery tray, and grind the end of it

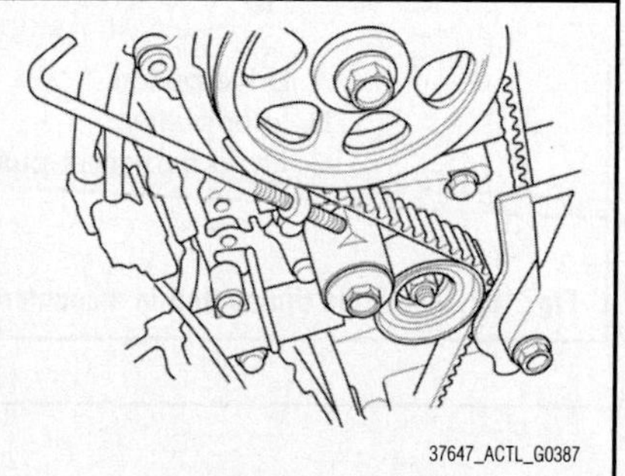
37647_ACTL_G0387

Fig. 157 Thread the battery clamp bolt into hold the timing belt adjuster in its current position

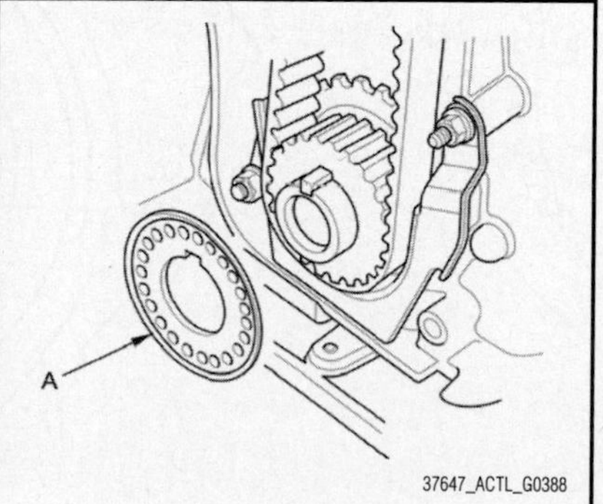
A
37647_ACTL_G0388

Fig. 158 Remove the timing belt guide plate (A)

To install:

17. Clean the timing belt pulleys, the timing belt guide plate, and the upper and lower covers.
18. Set the timing belt drive pulley to Top Dead Center (TDC) by aligning the TDC mark on the tooth of the timing belt drive pulley with the pointer on the oil pump.
19. Set the camshaft pulleys to TDC by aligning the TDC marks on the camshaft pulleys with the pointers on the back covers.
20. Loosely install the idler pulley with a new idler pulley bolt so the pulley can move but does not come off.
21. If the auto-tensioner has extended and the timing belt cannot be installed, do the timing belt replacement procedure.
22. Install the timing belt in a counterclockwise sequence starting with the drive pulley. Take care not to damage the timing belt during installation.
23. Tighten the idler pulley bolt to 33 ft. lbs. (44 Nm).

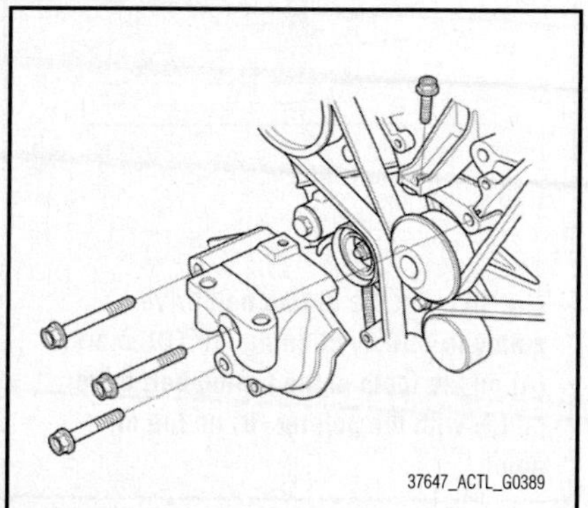
37647_ACTL_G0389

Fig. 159 Remove the lower half of the side engine mount bracket

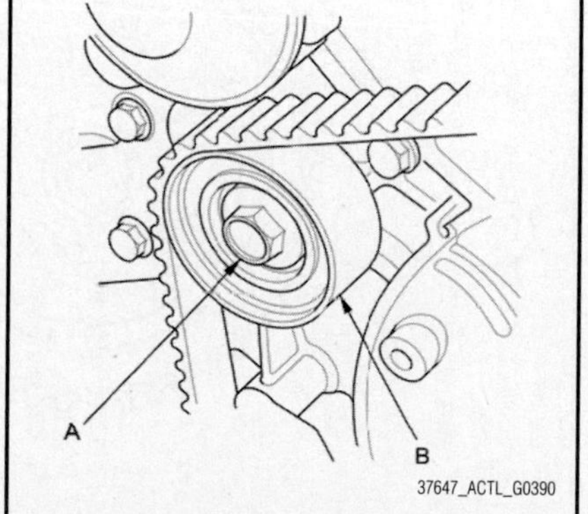
A
B
37647_ACTL_G0390

Fig. 160 Remove the idler pulley bolt (A) and the idler pulley (B), then remove the timing belt

24. Remove the battery clamp bolt from the back cover.
25. Install the lower half of the side engine mount bracket. Tighten the three long bolts to 33 ft. lbs. (44 Nm).
26. Install the timing belt guide plate.
27. Install the lower cover.
28. Install the front upper cover and the rear upper cover.
29. Install the crankshaft pulley.
30. Rotate the crankshaft pulley about six turns clockwise so the timing belt positions itself on the pulleys.
31. Turn the crankshaft pulley so its white mark lines up with the pointer.

➡The other pointer is not used.

32. Check the camshaft pulley marks.

➡If the marks are not aligned, rotate the crankshaft 360 degrees, and recheck the camshaft pulley mark. If the camshaft pulley marks are at TDC, go to step 17. If the camshaft pulley marks are not at TDC, remove the timing belt and repeat steps 2 through 16.

33. Install the upper half of the side engine mount bracket, then tighten the mounting bolts.
34. Install the ground cable.
35. Install the drive belt auto-tensioner.
36. Install the splash shield.
37. Install the right front wheel.
38. Install the engine compartment covers.

VALVE LASH

ADJUSTMENT

See Figures 164 and 165.

➡Connect the Honda Diagnostic System (HDS) to the data link connector (DLC), and monitor the Engine Coolant Temperature (ECT) sensor 1. Adjust the valve clearance only when the ECT sensor 1 is less than 100°F (38°C).

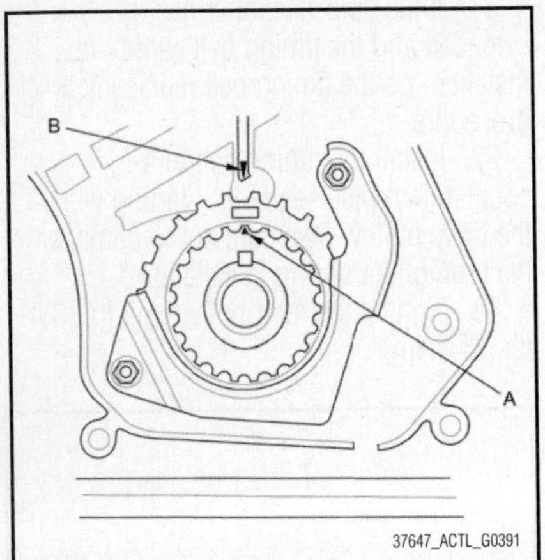
B
A
37647_ACTL_G0391

Fig. 161 Set the timing belt drive pulley to TDC by aligning the TDC mark (A) on the tooth of the timing belt drive pulley with the pointer (B) on the oil pump

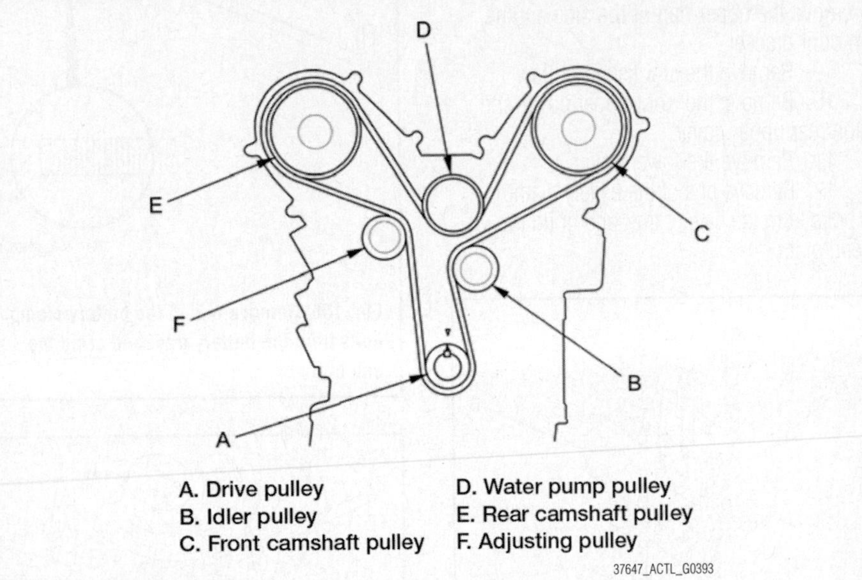
D
E
C
F
B
A
A. Drive pulley
B. Idler pulley
C. Front camshaft pulley
D. Water pump pulley
E. Rear camshaft pulley
F. Adjusting pulley
37647_ACTL_G0393

Fig. 163 Install the timing belt in a counterclockwise sequence starting with the drive pulley

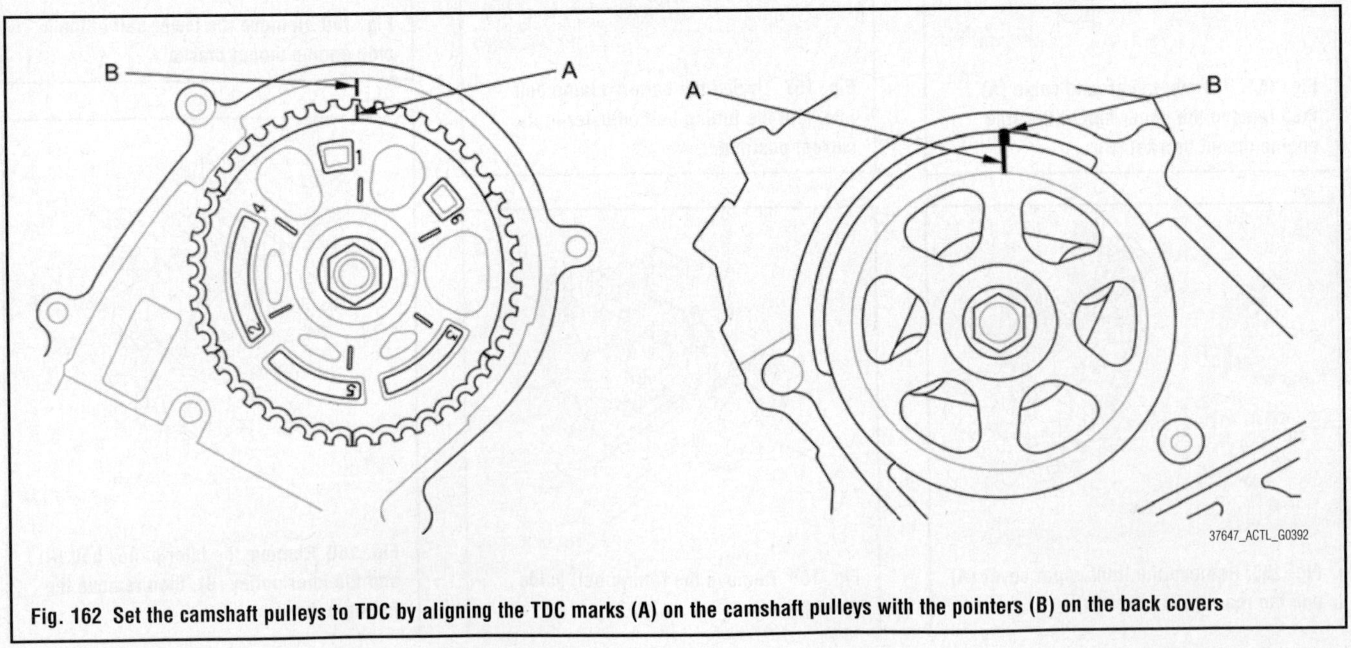
B
A
A
B
37647_ACTL_G0392

Fig. 162 Set the camshaft pulleys to TDC by aligning the TDC marks (A) on the camshaft pulleys with the pointers (B) on the back covers

1. Remove the cylinder head covers.
2. Set the No. 1 piston at Top Dead Center (TDC). Align the pointer on the front upper cover with the No. 1 piston TDC mark on the front camshaft pulley.
3. Select the correct feeler gauge for the valve clearance you are going to check.

➡Valve Clearance:

- Intake: 0.008–0.009 inches (0.20–0.24 mm)
- Exhaust: 0.011–0.013 inches (0.28–0.32 mm)

4. Insert the feeler gauge between the adjusting screw and the end of the valve stem on the No. 1 cylinder, and slide it back and forth; you should feel a slight amount of drag.
5. If you feel too much or too little drag, loosen the locknut, and turn the adjusting screw until the drag on the feeler gauge is correct.
6. While holding the adjusting screw with the screw driver, tighten the locknut, then recheck the clearance. Repeat the adjustment, if necessary.

➡Specified Torque:

- Intake: 14 ft. lbs. (20 Nm)
- Exhaust: 10 ft. lbs. (14 Nm)

➡Apply new engine oil to the nut threads.

7. Rotate the crankshaft clockwise. Align the pointer on the front upper cover with the No. 4 piston TDC mark on the front camshaft pulley.
8. Check and, if necessary, adjust the valve clearance on the No. 4 cylinder.

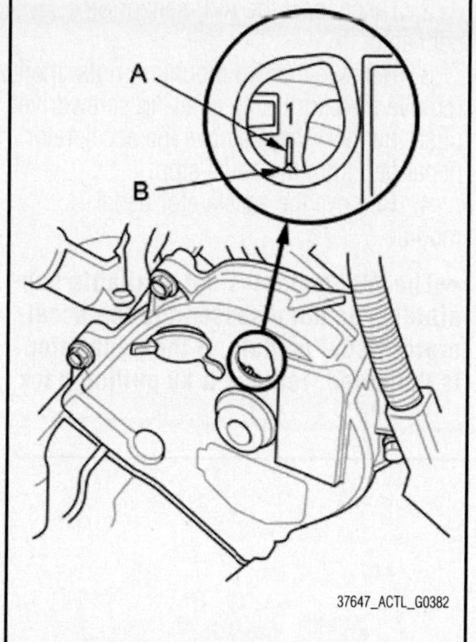
A
1
B
37647_ACTL_G0382
Fig. 164 Set the No. 1 piston at TDC. Align the pointer (A) on the front upper cover with the No. 1 piston TDC mark (B) on the front camshaft pulley

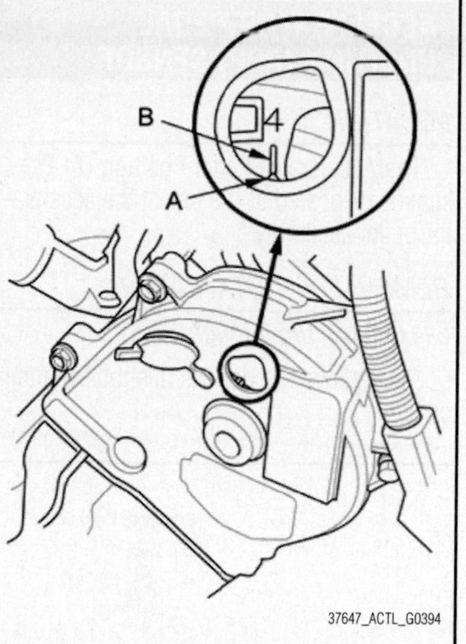
B
4
A
37647_ACTL_G0394
Fig. 165 Rotate the crankshaft clockwise. Align the pointer (A) on the front upper cover with the No. 4 piston TDC mark (B) on the front camshaft pulley

9. Rotate the crankshaft clockwise. Align the pointer on the front upper cover with the No. 2 piston TDC mark on the front camshaft pulley.

Check and, if necessary, adjust the valve clearance on the No. 2 cylinder.

10. Rotate the crankshaft clockwise. Align the pointer on the front upper cover with the No. 5 piston TDC mark on the front camshaft pulley.
11. Check and, if necessary, adjust the valve clearance on the No. 5 cylinder.
12. Rotate the crankshaft clockwise. Align the pointer on the front upper cover with the No. 3 piston TDC mark on the front camshaft pulley.
13. Check and, if necessary, adjust the valve clearance on the No. 3 cylinder.
14. Rotate the crankshaft clockwise. Align the pointer on the front upper cover with the No. 6 piston TDC mark on the front camshaft pulley.
15. Check and, if necessary, adjust the valve clearance on the No. 6 cylinder.
16. Install the cylinder head covers.

ENGINE PERFORMANCE & EMISSION CONTROLS

ACCELERATOR PEDAL POSITION (APP) SENSOR

LOCATION

The Accelerator Pedal Position (APP) sensor is located at the top of the accelerator pedal assembly.

REMOVAL & INSTALLATION

See Figures 166 through 168.

1. Remove the driver's dashboard undercover.
2. Disconnect the APP sensor 6P connector.
3. Remove the two mounting nuts, then remove the clip. Using a flat-tip screwdriver, push the lock, then remove the accelerator pedal pad from the pedal stop.
4. Remove the accelerator pedal module.

➡The APP sensor is not available separately. Do not disassemble the accelerator pedal module. If the pedal stop is damaged, replace it by pulling back the carpet, and removing the nuts, the bolt, and the pedal stop.

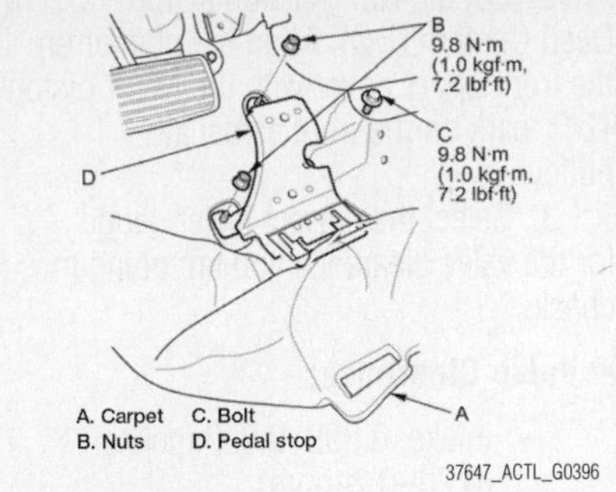
B
9.8 N·m
(1.0 kgf·m,
7.2 lbf·ft)
C
9.8 N·m
(1.0 kgf·m,
7.2 lbf·ft)
D
A
A. Carpet C. Bolt
B. Nuts D. Pedal stop
37647_ACTL_G0396

Fig. 168 Replace it by pulling back the carpet, and removing the nuts, the bolt, and the pedal stop

To install:

5. Install the accelerator pedal pad to the pedal stop. Slide the pad until it locks with a clicking sound.
6. Install a new clip.

➡The clip must be replaced whenever the accelerator pedal module is installed. Make sure that the accelerator pedal module and the clip are secure.

7. Install the nuts, and connect the APP sensor 6P connector.

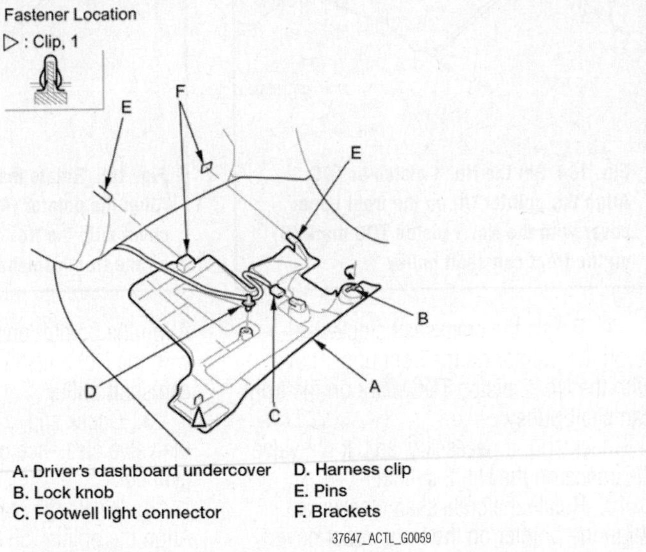
Fastener Location
▷ : Clip, 1
E
F
E
B
A
D
C
A. Driver's dashboard undercover
B. Lock knob
C. Footwell light connector
D. Harness clip
E. Pins
F. Brackets
37647_ACTL_G0059

Fig. 166 Remove the driver's dashboard undercover

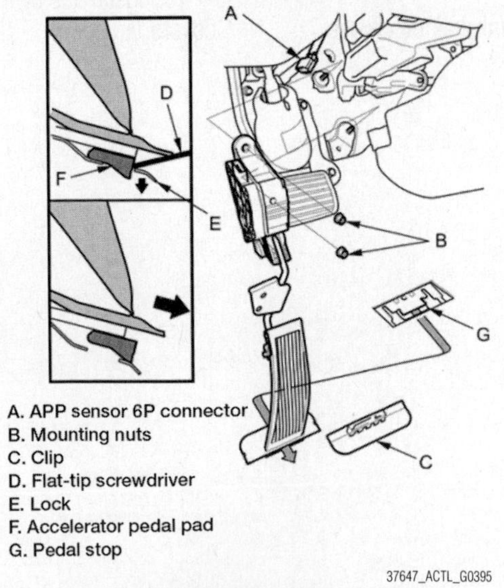
A
D
F
E
B
G
C
A. APP sensor 6P connector
B. Mounting nuts
C. Clip
D. Flat-tip screwdriver
E. Lock
F. Accelerator pedal pad
G. Pedal stop
37647_ACTL_G0395

Fig. 167 Disconnect the APP sensor 6P connector

CAMSHAFT POSITION (CMP) SENSOR

REMOVAL & INSTALLATION

See Figures 169 and 170.

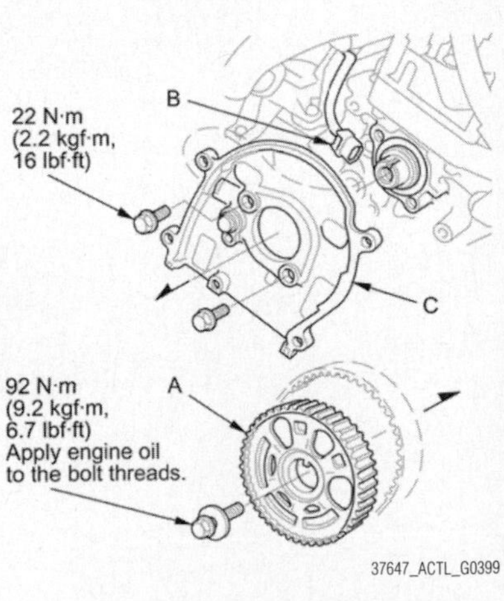
22 N·m
(2.2 kgf·m,
16 lbf·ft)
B
C
92 N·m
(9.2 kgf·m,
6.7 lbf·ft)
Apply engine oil
to the bolt threads.
A
37647_ACTL_G0399

Fig. 169 Remove the front camshaft pulley (CMP sensor pulse plate) (A)

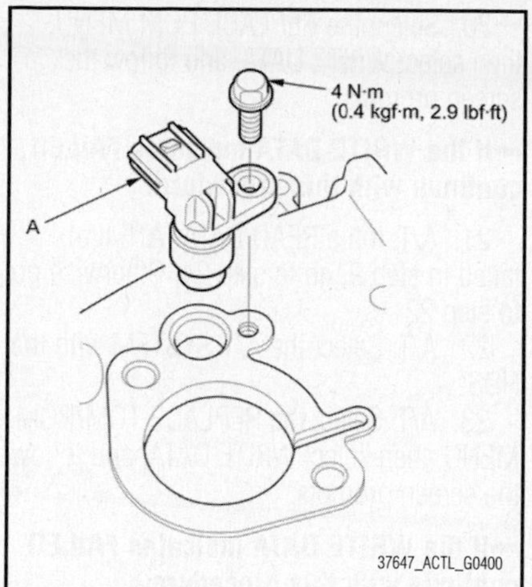
4 N·m
(0.4 kgf·m, 2.9 lbf·ft)
A
37647_ACTL_G0400

Fig. 170 Remove the CMP sensor (A) from the back cover

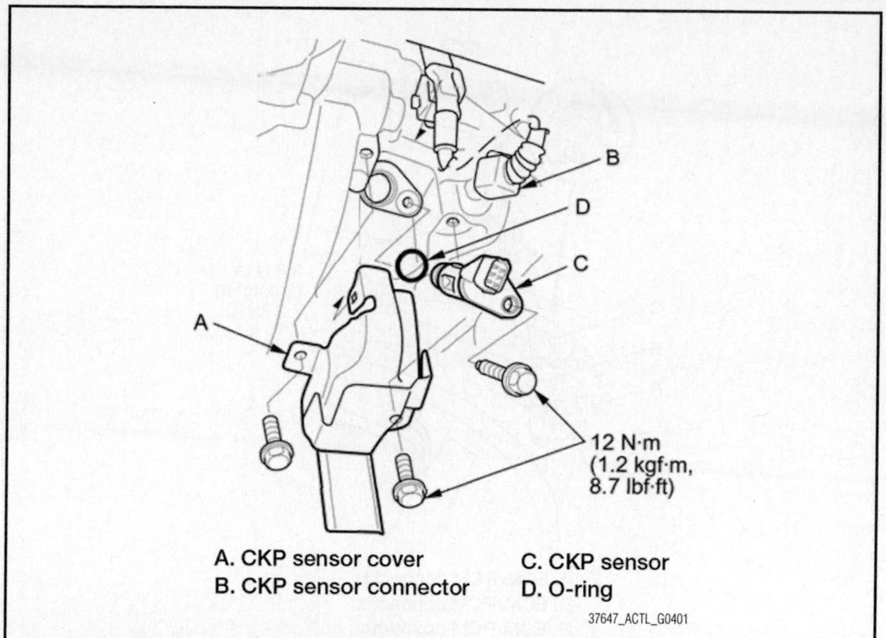
B
D
C
A
12 N·m
(1.2 kgf·m,
8.7 lbf·ft)
A. CKP sensor cover
B. CKP sensor connector
C. CKP sensor
D. O-ring
37647_ACTL_G0401

Fig. 171 Exploded view of CKP sensor assembly

1. Remove the timing belt.
2. Remove the front camshaft pulley (CMP sensor pulse plate).
3. Disconnect the CMP sensor connector, then remove the back cover.
4. Remove the CMP sensor from the back cover.
5. Install the parts in the reverse order of removal. Install the timing belt.
6. Do the CKP pattern clear/CKP pattern learn procedure.

CRANKSHAFT POSITION (CKP) SENSOR

REMOVAL & INSTALLATION

See Figure 171.

1. Raise the vehicle on a lift.

➡Make the vehicle is level, because engine oil will drip out when you remove the sensor.

2. Remove the CKP sensor cover.
3. Disconnect the CKP sensor connector.
4. Remove the CKP sensor.
5. Install the parts in the reverse order of removal with a new O-ring.
6. Do the CKP pattern clear/CKP pattern learn procedure.
7. Check for engine oil level, and add more oil if needed.

CKP PATTERN CLEAR/CKP PATTERN LEARN

Clear/Learn Procedure (with the HDS)

1. Connect the HDS to the Data Link Connector (DLC) located under the driver's side of the dashboard.
2. Turn the ignition switch to ON (II), or press the engine start/stop button to select the ON mode.
3. Make sure the HDS communicates with the ECM/PCM and other vehicle systems. If it doesn't, go to the DLC circuit troubleshooting.
4. Select CRANK PATTERN in the ADJUSTMENT MENU with the HDS.
5. Select CRANK PATTERN CLEAR, and clear the CKP pattern.
6. Select CRANK PATTERN LEARNING with the HDS, and follow the screen prompts.

Learn Procedure (without the HDS)

1. Start the engine. Hold the engine speed at 3,000 rpm without load (A/T in P or N, M/T in neutral) until the radiator fan comes on.
2. Test-drive the vehicle on a level road: Decelerate (with the throttle fully closed) from an engine speed of 2,500 rpm down to 1,000 rpm with the Transmission in 2nd.
3. Repeat step 2 several times.
4. Turn the ignition switch to LOCK (0), or press the engine start/stop button to select the OFF mode.
5. Turn the ignition switch to ON (II), or press the engine start/stop button to select the ON mode, and wait 30 seconds.

ENGINE CONTROL MODULE (ECM)

LOCATION

The Engine control Module (ECM)/ Powertrain Control Module (PCM) is located in the front of the engine compartment on the passenger's side.

REMOVAL & INSTALLATION

See Figures 172 and 173.

1. Connect HDS to the Data Link Connector (DLC) located under the driver's side of the dashboard.
2. Turn the ignition switch to ON (II), or press the engine start/stop button to select the ON mode.
3. Make sure the HDS communicates with the ECM/PCM and other vehicle systems. If it doesn't, go to the DLC circuit troubleshooting. If you are returning from the DLC circuit troubleshooting, skip steps 4 through 9, 18 through 23, and 26 through 28, and do this after replacing the ECM/PCM:
   - Replace the engine oil and the engine oil filter.
   - Replace the ATF (A/T).
   - Clean the throttle body.
4. Select the PGM-FI system with the HDS.
5. Select the INSPECTION MENU with the HDS.
6. Select the ETCS TEST, then select the TP POSITION CHECK, and follow the screen prompts.

➡If the TP POSITION CHECK indicates FAILED, continue with this procedure.

7. Select the REPLACE PCM MENU, then select READ DATA, and follow the screen prompts.

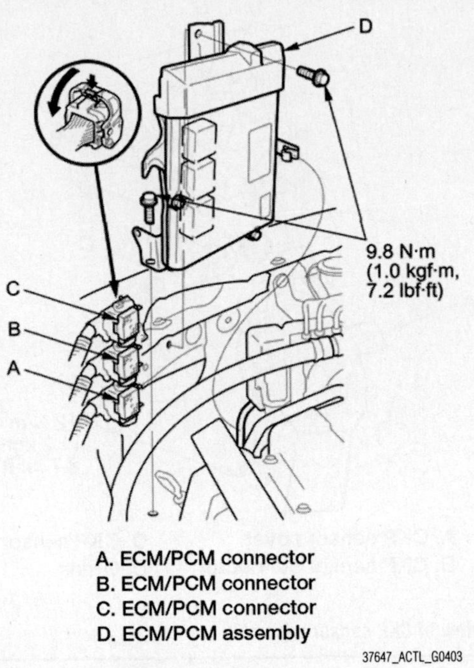
D
9.8 N·m
(1.0 kgf·m,
7.2 lbf·ft)
C
B
A
A. ECM/PCM connector
B. ECM/PCM connector
C. ECM/PCM connector
D. ECM/PCM assembly
37647_ACTL_G0403

Fig. 172 Disconnect ECM/PCM connectors, then remove the ECM/PCM assembly

➡Doing this step copies (READS) the engine oil life data from the original ECM/PCM so you can later download (WRITES) it into the new ECM/PCM. If READ DATA indicates FAILED, continue with this procedure.

8. A/T: Select the A/T system with the HDS.
9. A/T: Select the REPLACE TCM/PCM MENU, then select READ DATA, and follow the screen prompts.

➡Doing this step copies (READS) the ATF life data from the original PCM so you can later download (WRITES) it into the new PCM. If READ DATA indicates FAILED, continue with this procedure.

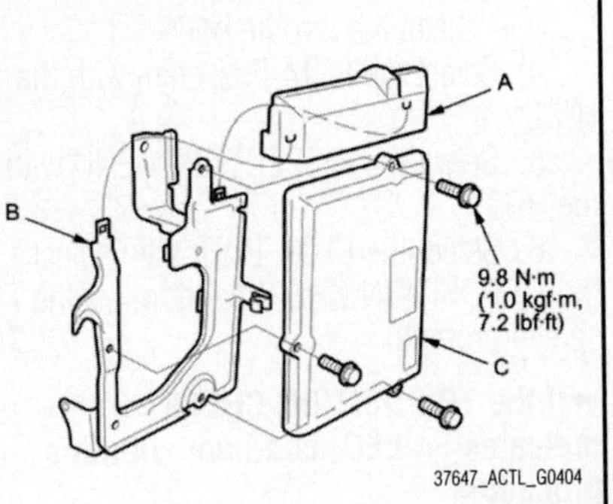
A
B
9.8 N·m
(1.0 kgf·m,
7.2 lbf·ft)
C
37647_ACTL_G0404

Fig. 173 Remove the cover (A) and the bracket (B) from the ECM/PCM

10. Turn the ignition switch to LOCK (0), or press the engine start/stop button to select the OFF mode.
11. Jump the SCS line with the HDS.
12. Canada model: If equipped, remove the ECM/PCM connector guard.
13. Disconnect ECM/PCM connectors, then remove the ECM/PCM assembly.

➡ECM/PCM connectors A, B, and C have symbols (A=□, B=□, C=○) embossed on them for identification.

14. Remove the cover and the bracket from the ECM/PCM.

To install:

15. Install the ECM/PCM in the reverse order of removal.
16. Turn the ignition switch to ON (II), or press the engine start/stop button to select the ON mode.
17. Manually input the VIN to the ECM/PCM with the HDS.

➡DTC P0630 VIN Not Programmed or Mismatch may be stored because the VIN has not been programmed into the ECM/PCM; ignore it, and continue this procedure.

18. If the READ DATA (engine oil life) failed in step 7, go to step 21 (A/T) or step 24 (M/T). Otherwise, go to step 19.
19. Select the PGM-FI system with the HDS.
20. Select the REPLACE PCM MENU, then select WRITE DATA, and follow the screen prompts.

➡If the WRITE DATA indicates FAILED, continue with this procedure.

21. A/T: If the READ DATA (ATF life) failed in step 9, go to step 24. Otherwise go to step 22.
22. A/T: Select the A/T SYSTEM with the HDS.
23. A/T: Select the REPLACE TCM/PCM MENU, then select WRITE DATA, and follow the screen prompts.

➡If the WRITE DATA indicates FAILED, continue with this procedure.

24. Select IMMOBI system (without keyless access) or ONE PUSH START system (with keyless access) with the HDS.
25. Enter the immobilizer ECM/PCM code that you got from the iN, and use the ECM/PCM replacement procedure in the IMMOBI Menu (without keyless access) or ONE PUSH START Menu (with keyless access) of the HDS; it allows you to start the engine.
26. If the TP POSITION CHECK failed in step 6 clean the throttle body, then go to step 27.
27. If the READ DATA failed in step 7 or the WRITE DATA failed in step 20, replace the engine oil and engine oil filter, then go to step 28 (A/T) or step 29 (M/T).
28. If the READ DATA failed in step 9 or the WRITE DATA failed in step 23, replace the ATF, then go to step 29.
29. Select PGM-FI system, and reset the ECM/PCM with the HDS.
30. Update the ECM/PCM if it does not have the latest software.
31. Do the ECM/PCM idle learn procedure.
32. Do the CKP pattern clear/CKP pattern learn procedure.

ECM/PCM IDLE LEARN PROCEDURE

The idle learn procedure must be done so the ECM/PCM can learn the engine idle characteristics.

Do the idle learn procedure whenever you do any of these actions:

- Replace the ECM/PCM.
- Reset the ECM/PCM.
- Update the ECM/PCM.
- Replace or clean the throttle body.
- Disassemble the engine or transmission.

➡Erasing DTCs with the HDS does not require you to do the idle learn procedure.

1. Make sure all electrical items (the A/C, the audio, the lights, etc.) are off.
2. Reset the ECM/PCM with the HDS.
3. Turn the ignition switch to ON (II), or press the engine start/stop button to select the ON mode, and wait 2 seconds.
4. Start the engine. Hold the engine speed at 3,000 rpm without load (A/T in P or N, M/T in neutral) until the radiator fan comes on, or until the engine coolant temperature reaches 194°F (90°C).
5. Let the engine idle for about 5 minutes with the throttle fully closed.

➡If the radiator fan comes on, do not include its running time in the 5 minutes.

ECM/PCM UPDATE

➡Ensure the following prior to ECM/PCM update:

- Make sure the HDS/iN workstation has the latest HDS software version.
- Before you update the ECM/PCM, make sure the battery in the vehicle is fully charged, and connect a jumper battery (not a battery charger) to maintain system voltage.
- Never turn the ignition switch to ACC (I) or LOCK (0) or never press the engine start/stop button to select the OFF or ACC mode during the update. If there is a problem with the update, leave the ignition switch ON (II) or vehicle in the ON mode.
- To prevent ECM/PCM damage, do not operate anything electrical (headlights, audio system, brakes, A/C, power windows, moonroof, door locks, etc.) during the update.
- To ensure the latest program is installed, do a ECM/PCM update whenever the ECM/PCM is substituted or replaced.
- You cannot update a ECM/PCM with a program it already has. It will only accept a new program.
- High temperature in the engine compartment might cause the ECM/PCM to become too hot to run the update. If the engine has been running before this procedure, open the hood and cool the engine compartment.
- If you need to diagnose the Honda Interface Module (HIM) because the HIM's red (_3) light came on or was flashing during the update, leave the ignition switch in ON (II) or vehicle in the ON mode when you disconnect the HIM from the Data Link Connector (DLC). This will prevent ECM/PCM damage.

1. Turn the ignition switch to ON (II), or press the engine start/stop button to select the ON mode, but do not start the engine.
2. Connect the HDS to the data link connector (DLC) located under the driver's side of the dashboard.
3. Make sure the HDS communicates with the ECM/PCM and other vehicle systems. If it doesn't, go to the DLC circuit troubleshooting. If you are returning from the DLC circuit troubleshooting, skip steps 4 and 5, and clean the throttle body after updating the ECM/PCM.
4. Select the INSPECTION MENU with the HDS.
5. Select the ETCS TEST, then select the TP POSITION CHECK, and follow the HDS screen prompts.

➡If the TP POSITION CHECK indicates FAILED, continue this procedure.

6. Exit the HDS diagnostic system, then select the update mode, and follow the screen prompts to update the ECM/PCM.
7. If the software in the ECM/PCM is the latest, disconnect the HDS/HIM/GNA600 from the DLC, and go back to the procedure that you were doing. If the software in the ECM/PCM is not the latest, follow the instructions on the screen. If prompted to choose the PGM-FI system or the A/T system, make sure you update both.

➡If the ECM/PCM update system requires you to cool the ECM/PCM, follow the instructions on screen. If you run into a problem during the update procedure (programming takes over 15 minutes, status bar goes over 100 %, D or immobilizer indicator flashes, HDS tablet freezes, etc.), follow these steps to minimize the chance of damaging the ECM/PCM:

- Leave the ignition switch in ON (II) or vehicle in the ON mode.
- Connect a jumper battery (do not connect a battery charger).
- Shut down the HDS.
- Disconnect the HDS from the DLC.
- Reboot the HDS.
- Reconnect the HDS to the DLC, and try the update procedure again.

8. If the TP POSITION CHECK failed in step 5, clean the throttle body.
9. Do the ECM/PCM idle learn procedure.
10. Do the CKP pattern clear/CKP pattern learn procedure.

ENGINE COOLANT TEMPERATURE (ECT) SENSOR

LOCATION

Refer to graphics in the Removal and Installation section to locate the two Engine Coolant Temperature (ECT) sensors.

REMOVAL & INSTALLATION

ECT Sensor 1

See Figure 174.

1. Drain the engine coolant.
2. Disconnect the ECT sensor 1 2P connector.
3. Remove ECT sensor 1.
4. Install the parts in the reverse order of removal with a new O-ring, then refill the radiator with engine coolant.

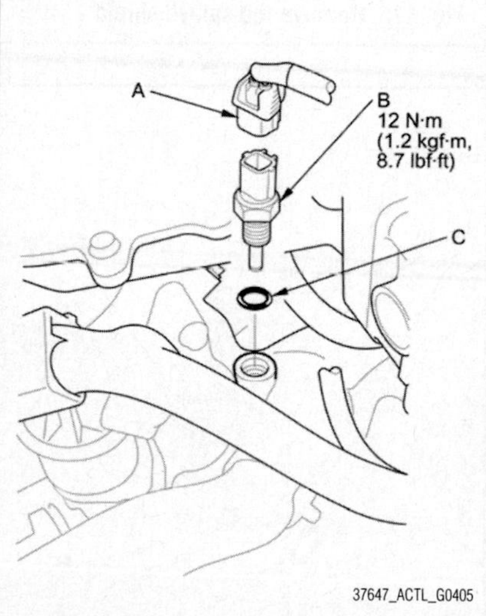
A
B
12 N·m
(1.2 kgf·m,
8.7 lbf·ft)
C
37647_ACTL_G0405

Fig. 174 Disconnect the ECT sensor 1 2P connector

ECT Sensor 2

See Figures 175 and 176.

1. Remove the splash shield.
2. Drain the engine coolant.
3. Disconnect the ECT sensor 2 2P connector.
4. Remove ECT sensor 2.
5. Install the parts in the reverse order of removal with a new O-ring, then refill the radiator with engine coolant.

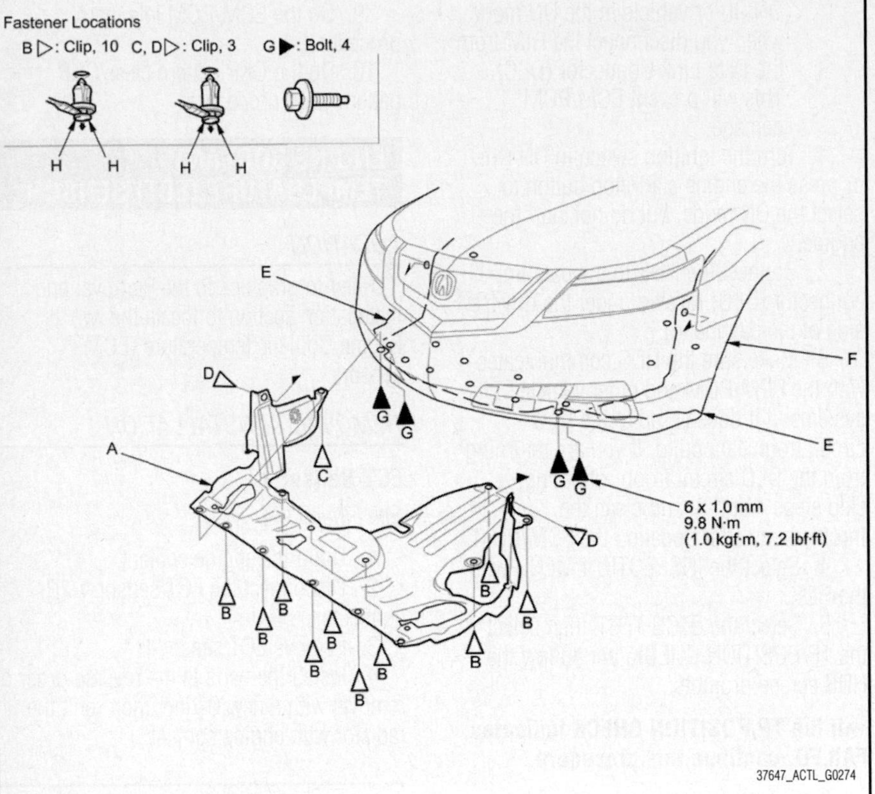
Fastener Locations
B ▷: Clip, 10 C, D▷: Clip, 3 G▶: Bolt, 4
H H H H
E
F
D
A
G G
C
E
G G
6 x 1.0 mm
9.8 N·m
(1.0 kgf·m, 7.2 lbf·ft)
D
B B B B B B B B
37647_ACTL_G0274

Fig. 175 Remove the splash shield

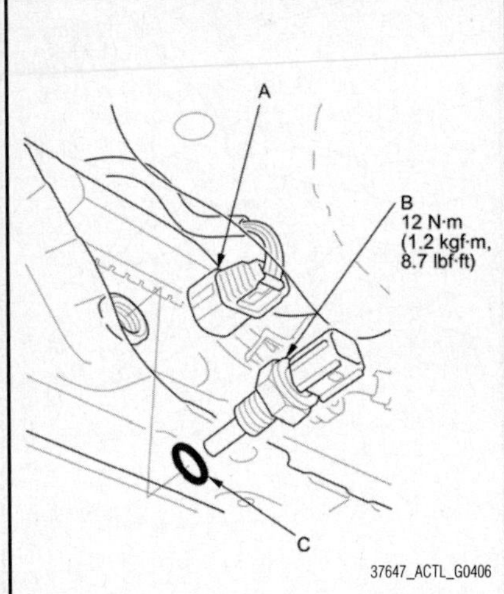
A
B
12 N·m
(1.2 kgf·m,
8.7 lbf·ft)
C
37647_ACTL_G0406

Fig. 176 Disconnect the ECT sensor 2 2P connector

EVAPORATIVE EMISSIONS (EVAP) CANISTER

LOCATION

See Figure 177.

Refer to the accompanying illustration.

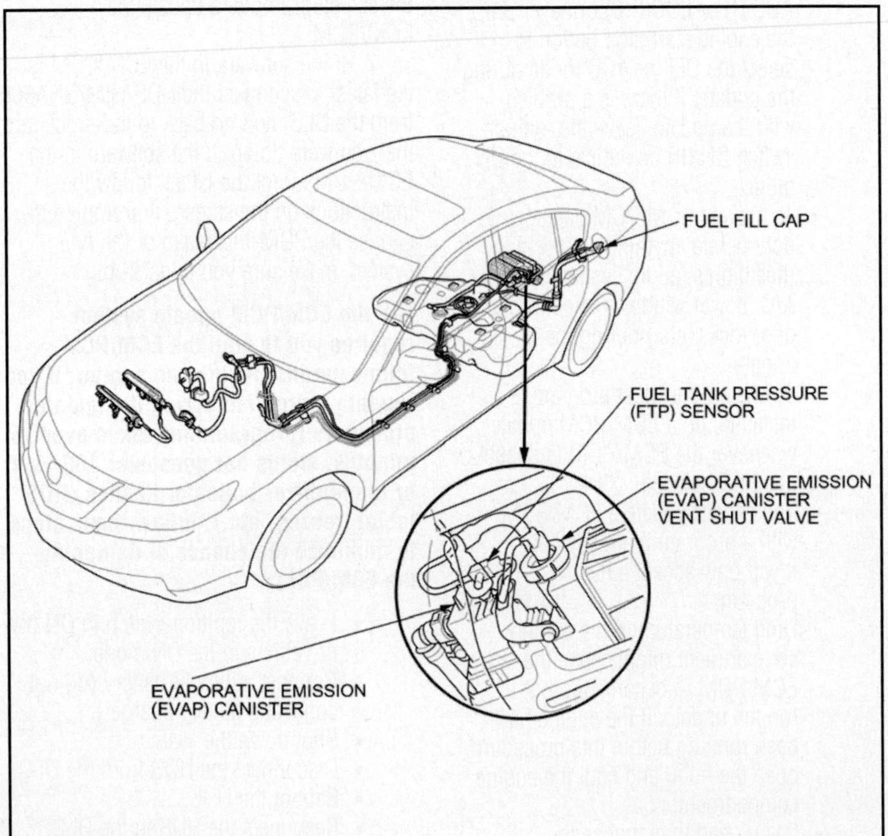
FUEL FILL CAP
FUEL TANK PRESSURE
(FTP) SENSOR
EVAPORATIVE EMISSION
(EVAP) CANISTER
VENT SHUT VALVE
EVAPORATIVE EMISSION
(EVAP) CANISTER
37647_ACTL_G0407

Fig. 177 EVAP canister system component locations

REMOVAL & INSTALLATION

With SH-AWD

See Figures 178 through 181.

1. Remove the rear differential.
2. Remove the fuel tank.
3. Remove the EVAP canister filter.

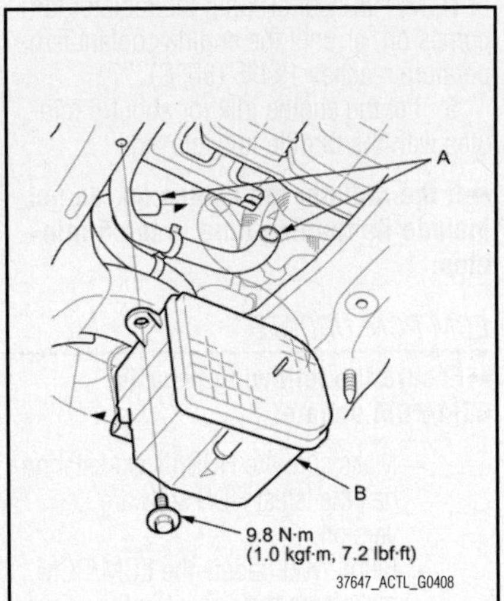
A
B
9.8 N·m
(1.0 kgf·m, 7.2 lbf·ft)
37647_ACTL_G0408

Fig. 178 Remove the EVAP canister filter

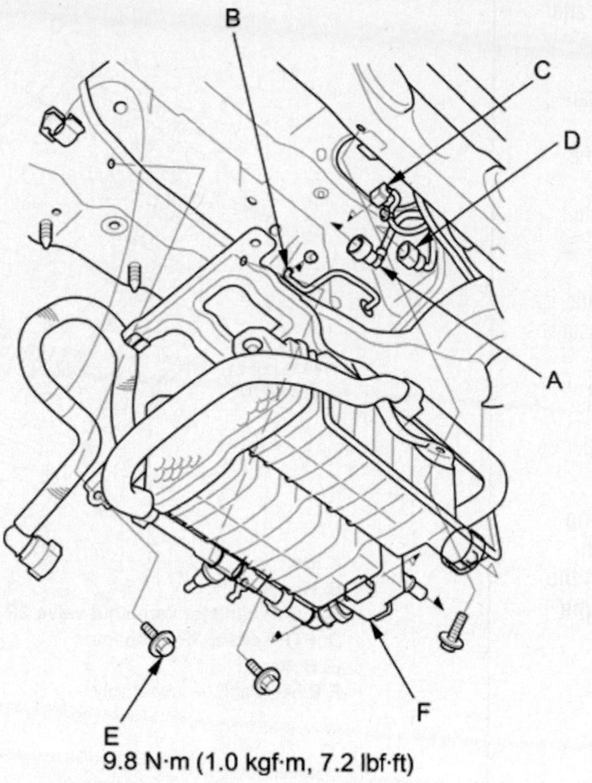
B
C
D
A
E
F
9.8 N·m (1.0 kgf·m, 7.2 lbf·ft)
A. Quick-connect fitting
B. Hose
C. FTP sensor 3P connector
D. EVAP canister vent shut valve 2P connector
E. Bolts
F. EVAP canister assembly
37647_ACTL_G0409

Fig. 179 Disconnect the quick-connect fitting, the hose, the FTP sensor 3P connector, and the EVAP canister vent shut valve 2P connector

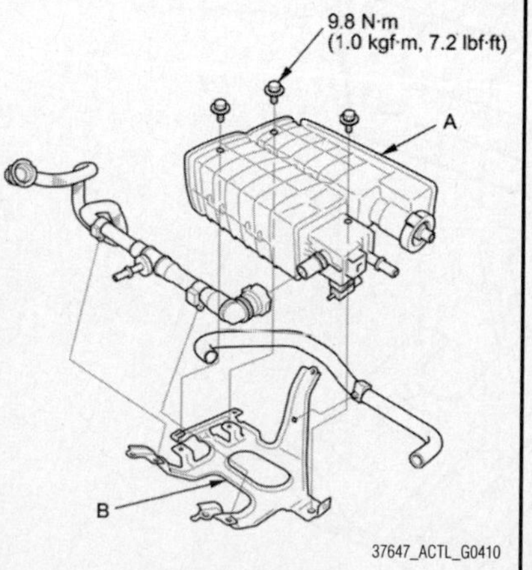
9.8 N·m
(1.0 kgf·m, 7.2 lbf·ft)
A
B
37647_ACTL_G0410

Fig. 180 Remove the EVAP canister (A) from the bracket (B)

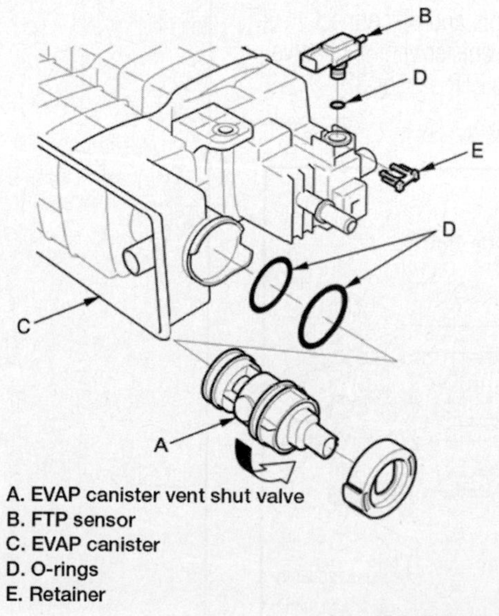
B
D
E
D
C
A
A. EVAP canister vent shut valve
B. FTP sensor
C. EVAP canister
D. O-rings
E. Retainer
37647_ACTL_G0411

Fig. 181 Remove the EVAP canister vent shut valve and the FTP sensor from the EVAP canister

4. Disconnect the quick-connect fitting, the hose, the FTP sensor 3P connector, and the EVAP canister vent shut valve 2P connector.

5. Remove the bolts and the EVAP canister assembly.

6. Remove the EVAP canister from the bracket.

7. Remove the EVAP canister vent shut valve and the FTP sensor from the EVAP canister.

8. Reassemble the EVAP canister with new O-rings and a new retainer, then install the EVAP canister bracket.

➡Do not coat the O-rings with oil.

9. Install the parts in the reverse order of removal.

➡The new fuel vapor hose has a ring pull at the connector. When you connect the connector and confirm that the connection is secure, remove the ring pull by pulling it down.

Without SH-AWD

See Figures 182 through 186.

1. Raise the vehicle on a lift.

2. Remove the wheel sensor harness clamps.

3. Support the rear subframe with a transmission jack and a wooden block.

4. Remove the rear subframe mounting bolts.

5. Lower the transmission jack and the rear subframe about 50 mm.

➡Be careful not to damage the connecting parts.

6. Remove the bolt, and disconnect the hoses, the EVAP canister vent shut valve 2P connector, and the FTP sensor 3P connector.

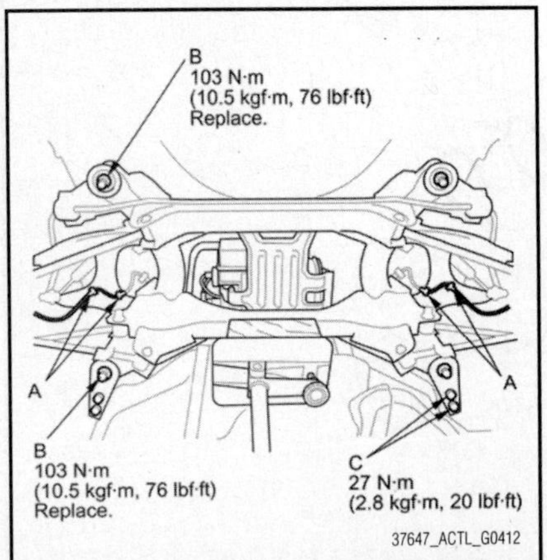
B
103 N·m
(10.5 kgf·m, 76 lbf·ft)
Replace.
A
A
B
103 N·m
(10.5 kgf·m, 76 lbf·ft)
Replace.
C
27 N·m
(2.8 kgf·m, 20 lbf·ft)
37647_ACTL_G0412

Fig. 182 Support the rear subframe with a transmission jack and a wooden block

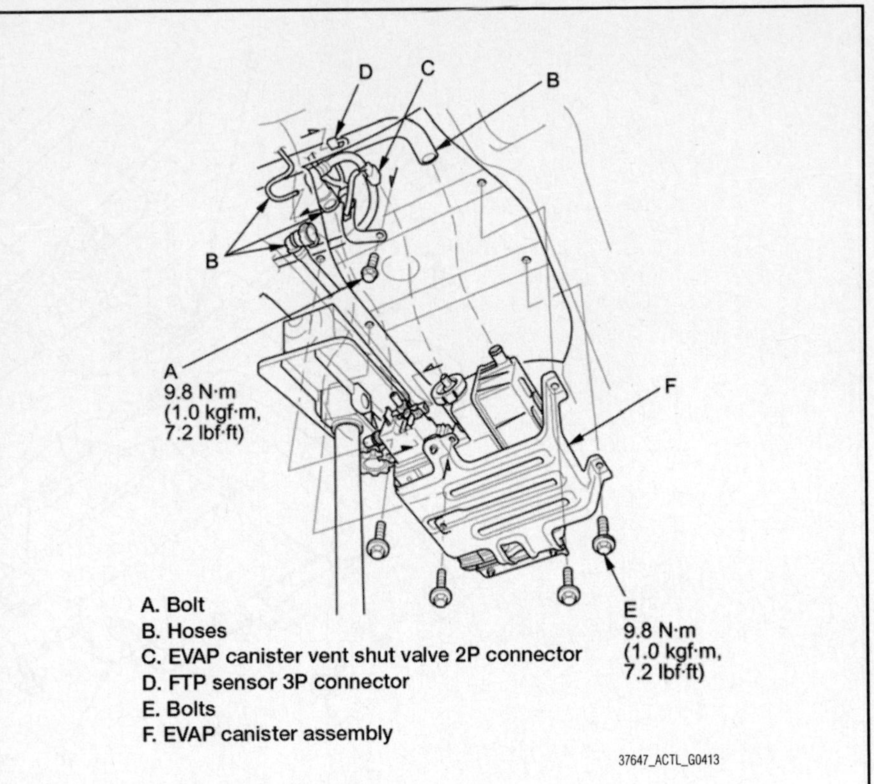
D
C
B
B
A
9.8 N·m
(1.0 kgf·m,
7.2 lbf·ft)
F
E
9.8 N·m
(1.0 kgf·m,
7.2 lbf·ft)
A. Bolt
B. Hoses
C. EVAP canister vent shut valve 2P connector
D. FTP sensor 3P connector
E. Bolts
F. EVAP canister assembly
37647_ACTL_G0413

Fig. 183 Remove the bolt, and disconnect the hoses, the EVAP canister vent shut valve 2P connector, and the FTP sensor 3P connector

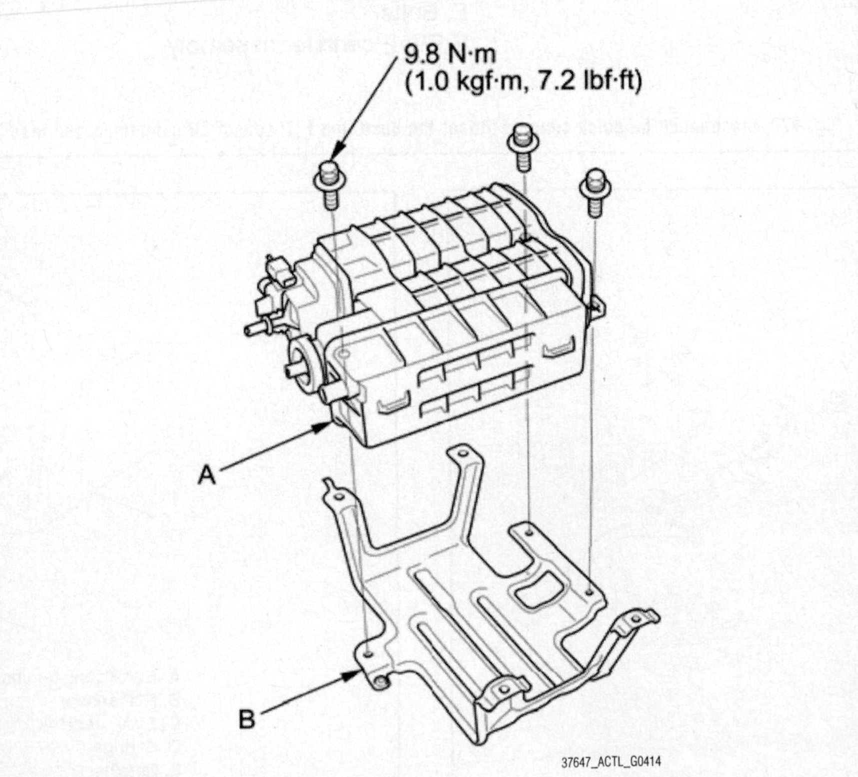
9.8 N·m
(1.0 kgf·m, 7.2 lbf·ft)
A
B
37647_ACTL_G0414

Fig. 184 Remove the EVAP canister (A) from the EVAP canister bracket (B)

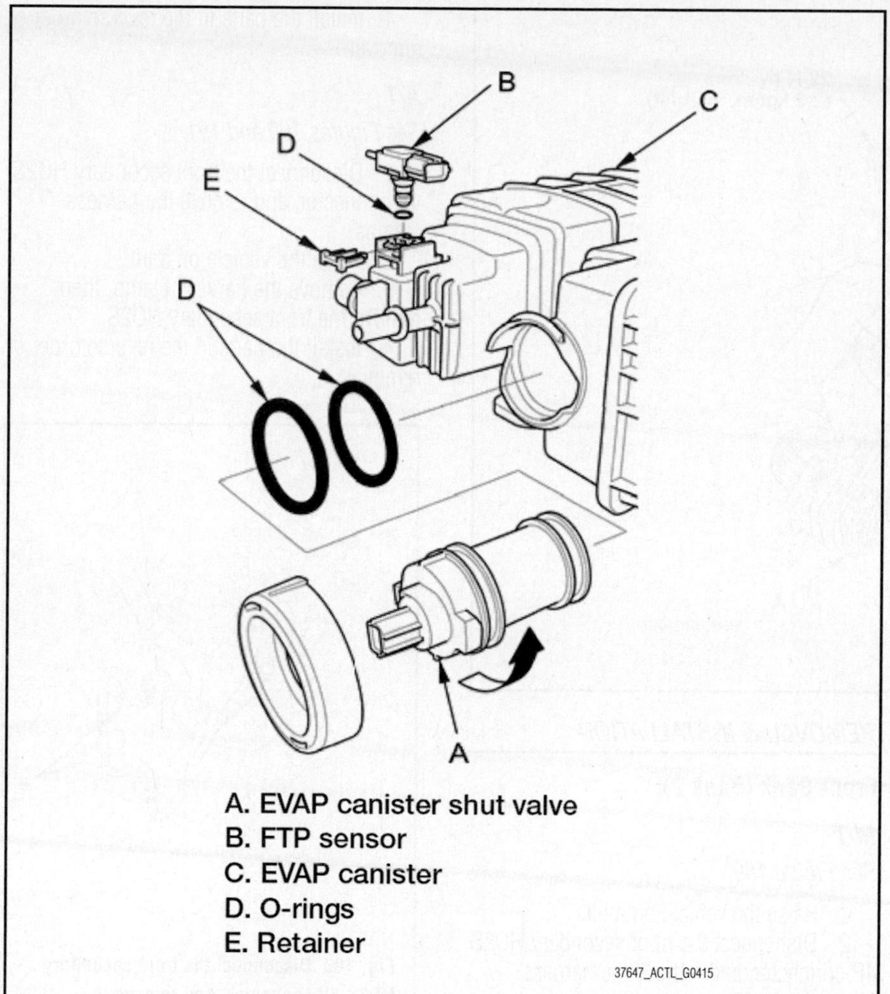
A
B
C
D
D
E
A. EVAP canister shut valve
B. FTP sensor
C. EVAP canister
D. O-rings
E. Retainer
37647_ACTL_G0415

Fig. 185 Remove the EVAP canister vent shut valve and the FTP sensor from the EVAP canister

7. Remove the bolts, then remove the EVAP canister assembly.

8. Remove the EVAP canister from the EVAP canister bracket.

9. Remove the EVAP canister vent shut valve and the FTP sensor from the EVAP canister.

10. Reassemble the EVAP canister with new O-rings and a new retainer, then install the EVAP canister bracket.

➡**Do not coat the O-rings with oil.**

11. Install the EVAP canister assembly to the body.

➡**Attach the bracket arm to the body.**

12. Install the parts in the reverse order of removal. Use new bolts when you install the rear subframe.

13. Check the wheel alignment.

EXHAUST GAS RECIRCULATION (EGR) VALVE

LOCATION

See Figure 187.

Refer to the accompanying illustration.

REMOVAL & INSTALLATION

See Figure 188.

1. Disconnect the EGR valve 5P connector.

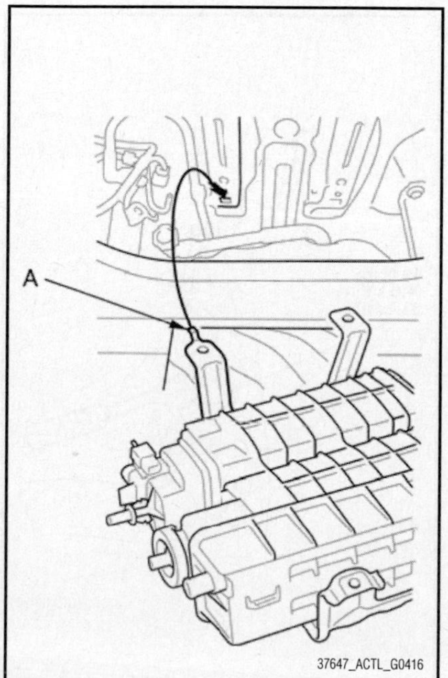
A
37647_ACTL_G0416

Fig. 186 Attach the bracket arm (A) to the body

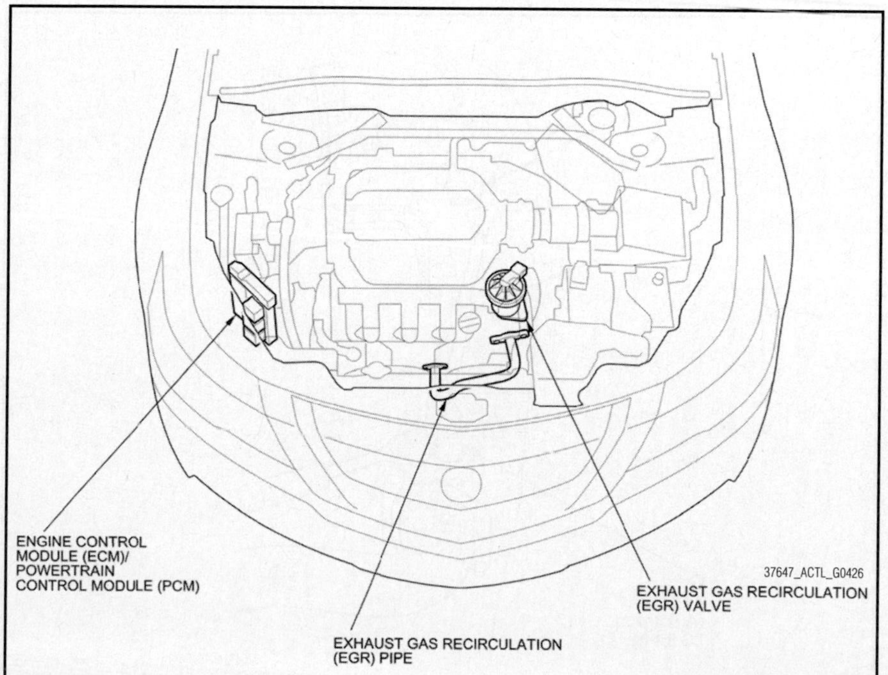
ENGINE CONTROL MODULE (ECM)/ POWERTRAIN CONTROL MODULE (PCM)
37647_ACTL_G0426
EXHAUST GAS RECIRCULATION (EGR) VALVE
EXHAUST GAS RECIRCULATION (EGR) PIPE

Fig. 187 EGR valve location

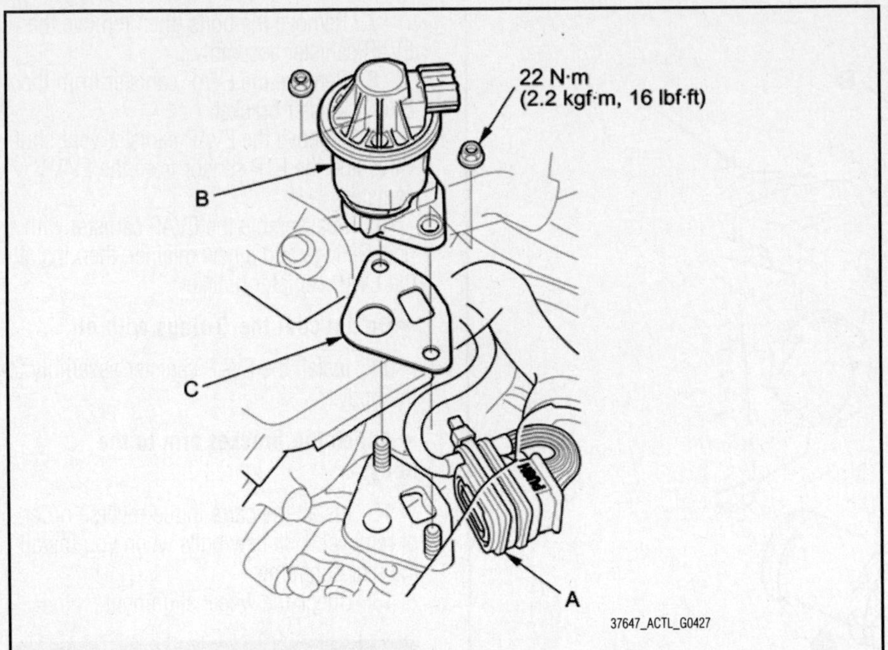
22 N·m
(2.2 kgf·m, 16 lbf·ft)
B
C
A
37647_ACTL_G0427

Fig. 188 Disconnect the EGR valve 5P connector

2. Remove the EGR valve.
3. Install the parts in the reverse order of removal with a new gasket.

HEATED OXYGEN SENSOR (HO2S)

LOCATION

The Heated Oxygen Sensors (HO2S) are located on the exhaust manifolds of each cylinder head.

REMOVAL & INSTALLATION

Front Bank (Bank 2)

M/T

See Figure 189.

1. Raise the vehicle on a lift.
2. Disconnect the front secondary HO2S 4P connector, and remove the harness clamp.
3. Remove the front secondary HO2S.
4. Install the parts in the reverse order of removal.

A/T

See Figures 190 and 191.

1. Disconnect the front secondary HO2S 4P connector, and remove the harness clamps.
2. Raise the vehicle on a lift.
3. Remove the harness clamp, then remove the front secondary HO2S.
4. Install the parts in the reverse order of removal.

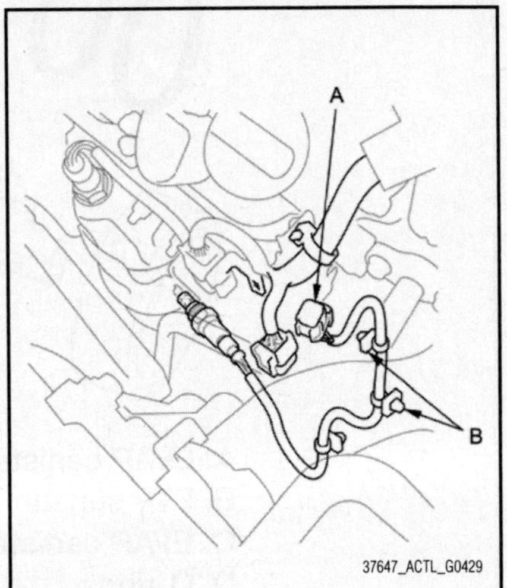
A
B
37647_ACTL_G0429

Fig. 190 Disconnect the front secondary HO2S 4P connector, and remove the harness clamps

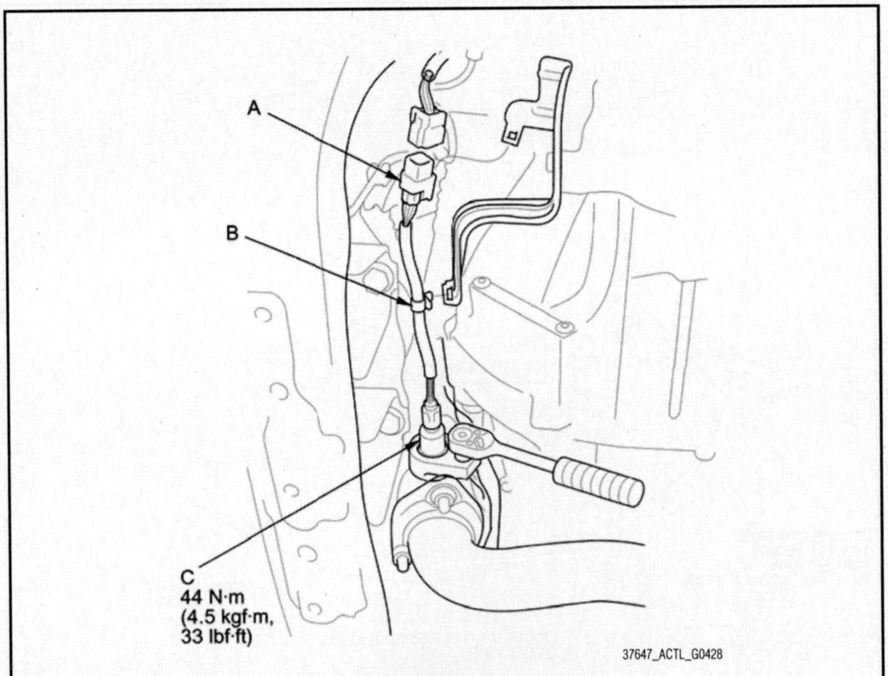
A
B
C
44 N·m
(4.5 kgf·m,
33 lbf·ft)
37647_ACTL_G0428

Fig. 189 Disconnect the front secondary HO2S 4P connector, and remove the harness clamp

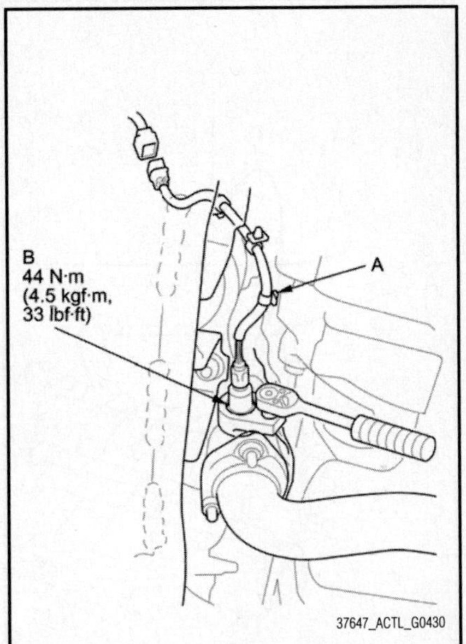
B
44 N·m
(4.5 kgf·m,
33 lbf·ft)
A
37647_ACTL_G0430

Fig. 191 Remove the harness clamp, then remove the front secondary HO2S

Rear Bank (Bank 1)

See Figure 192.

1. Raise the vehicle on a lift.
2. Disconnect the rear secondary HO2S 4P connector, and remove the harness clamps.
3. Remove the rear secondary HO2S.
4. Install the parts in the reverse order of removal.

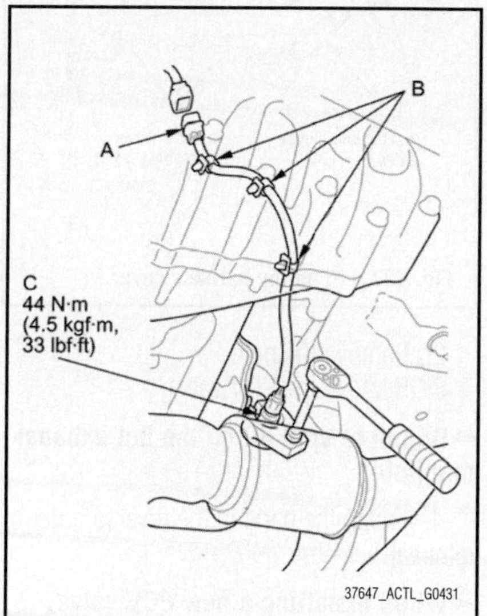
A
B
C
44 N·m
(4.5 kgf·m,
33 lbf·ft)
37647_ACTL_G0431

Fig. 192 Disconnect the rear secondary HO2S 4P connector, and remove the harness clamps

INTAKE AIR TEMPERATURE (IAT) SENSOR

LOCATION

The Intake Air Temperature (IAT) sensor is an integral part of the Mass Air Flow (MAF) sensor assembly. Refer to Mass Air Flow section for information regarding servicing the IAT.

REMOVAL & INSTALLATION

Refer to Mass Air Flow (MAF) in this section for Removal and Installation procedures.

KNOCK SENSOR (KS)

LOCATION

The Knock Sensor (KS) is located in the valley of the engine block between the two cylinder heads.

REMOVAL & INSTALLATION

See Figure 193.

1. Remove the intake manifold.
2. Remove the injector base.

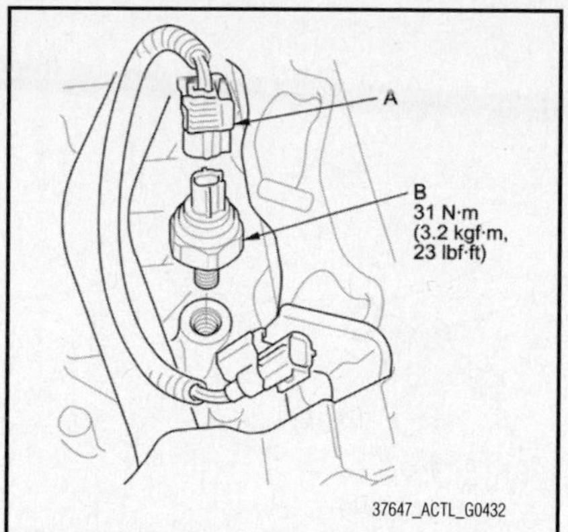
A
B
31 N·m
(3.2 kgf·m,
23 lbf·ft)
37647_ACTL_G0432

Fig. 193 Disconnect the knock sensor connector

3. Disconnect the knock sensor connector.
4. Remove the knock sensor.
5. Install the parts in the reverse order of removal.

MASS AIR FLOW (MAF)/INTAKE AIR TEMPERATURE (IAT) SENSOR

LOCATION

The MAF/IAT sensor is located on the engine air intake.

REMOVAL & INSTALLATION

See Figure 194.

1. Disconnect the MAF/IAT sensor 5P connector.

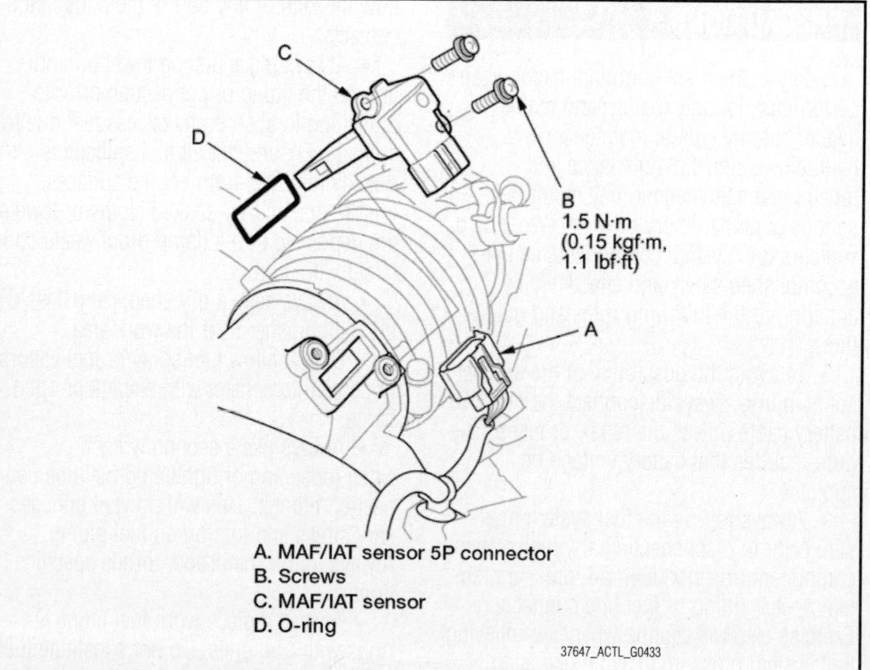
C
D
B
1.5 N·m
(0.15 kgf·m,
1.1 lbf·ft)
A
A. MAF/IAT sensor 5P connector
B. Screws
C. MAF/IAT sensor
D. O-ring
37647_ACTL_G0433

Fig. 194 Disconnect the MAF/IAT sensor 5P connector

2. Remove the screws.
3. Remove the MAF/IAT sensor.
4. Install the parts in the reverse order of removal with a new O-ring.

MANIFOLD ABSOLUTE PRESSURE (MAP) SENSOR

LOCATION

Refer to graphic in the Removal and Installation section for location.

REMOVAL & INSTALLATION

See Figure 195.

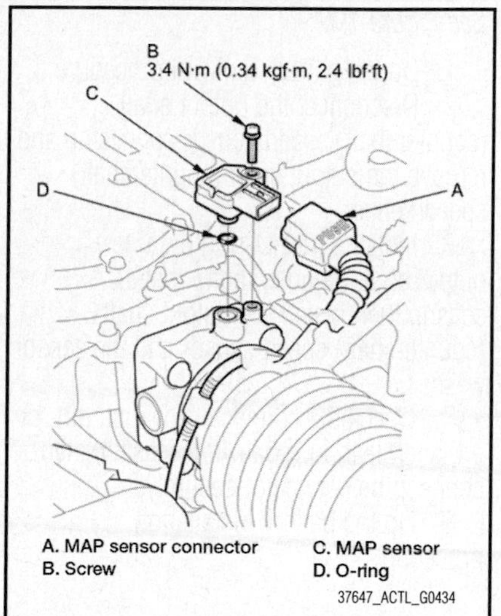
B
3.4 N·m (0.34 kgf·m, 2.4 lbf·ft)
C
D
A
A. MAP sensor connector
B. Screw
C. MAP sensor
D. O-ring
37647_ACTL_G0434

Fig. 195 Disconnect the MAP sensor connector

1. Disconnect the MAP sensor connector.
2. Remove the screw.
3. Remove the MAP sensor.
4. Install the parts in the reverse order of removal with a new O-ring.

OUTPUT SHAFT SPEED (OSS) SENSOR

LOCATION

The Output Shaft Speed (OSS) sensor is located on the side of the A/T transaxle.

REMOVAL & INSTALLATION

See Figure 196.

1. Do the battery removal procedure.
2. Disconnect the output shaft (countershaft) speed sensor connector, and remove the output shaft (countershaft) speed sensor.
3. Install a new O-ring on a new output shaft (countershaft) speed sensor, then install the output shaft (countershaft) speed sensor and the sensor washer.
4. Check the connector for rust, dirt, or oil, and clean or repair if necessary, then connect the connector securely.
5. Do the battery installation procedure.

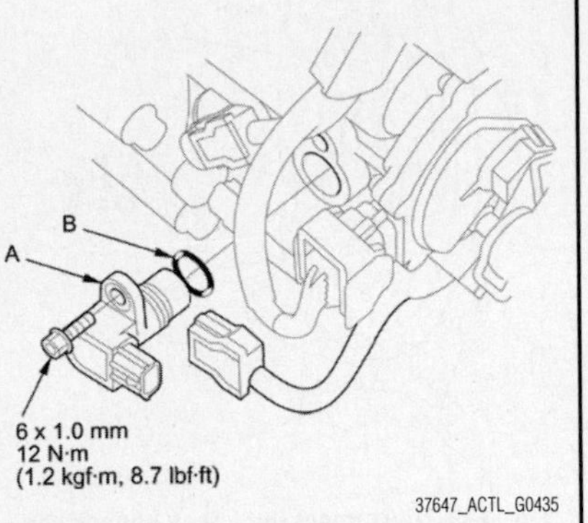
B
A
6 x 1.0 mm
12 N·m
(1.2 kgf·m, 8.7 lbf·ft)
37647_ACTL_G0435

Fig. 196 Disconnect the output shaft (countershaft) speed sensor connector, and remove the output shaft (countershaft) speed sensor

POSITIVE CRANKCASE VENTILATION (PCV) VALVE

LOCATION

The Positive Crankcase Ventilation (PCV) valve is located on the front right area of the engine below the harness cover.

REMOVAL & INSTALLATION

See Figure 197.

1. Lift up the harness cover.

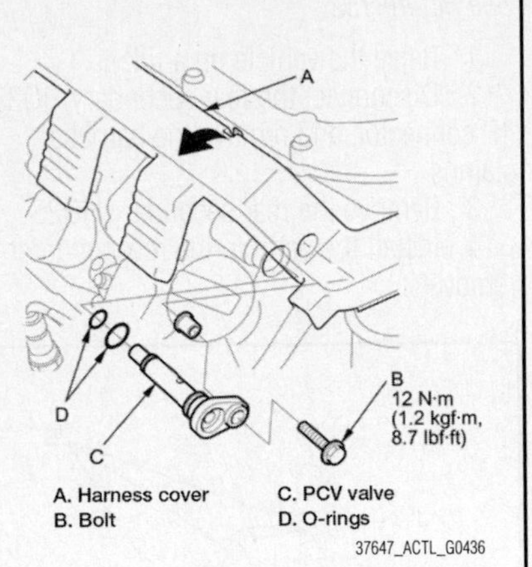
A
B
12 N·m
(1.2 kgf·m,
8.7 lbf·ft)
D
C
A. Harness cover
B. Bolt
C. PCV valve
D. O-rings
37647_ACTL_G0436

Fig. 197 Lift up the harness cover

2. Remove the bolt.
3. Remove the PCV valve.

➡Do not to spill oil on the hot exhaust manifold.

4. Install the parts in the reverse order of removal.

➡When installing a new PCV valve, make sure the O-rings are in place. When installing a used PCV valve, use new O-rings.

FUEL — GASOLINE FUEL INJECTION SYSTEM

FUEL SYSTEM SERVICE PRECAUTIONS

Safety is the most important factor when performing not only fuel system maintenance, but any type of maintenance. Failure to conduct maintenance and repairs in a safe manner may result in serious personal injury or death. Work on a vehicle's fuel system components can be accomplished safely and effectively by adhering to the following rules and guidelines.

- To avoid the possibility of fire and personal injury, always disconnect the negative battery cable unless the repair or test procedure requires that battery voltage be applied.
- Always relieve the fuel system pressure prior to disconnecting any fuel system component (injector, fuel rail, pressure regulator, etc.) fitting or fuel line connection. Exercise extreme caution whenever relieving fuel system pressure to avoid exposing skin, face and eyes to fuel spray. Please be advised that fuel under pressure may penetrate the skin or any part of the body that it contacts.
- Always place a shop towel or cloth around the fitting or connection prior to loosening to absorb any excess fuel due to spillage. Ensure that all fuel spillage is quickly removed from engine surfaces. Ensure that all fuel-soaked cloths or towels are deposited into a flame-proof waste container with a lid.
- Always keep a dry chemical (Class B) fire extinguisher near the work area.
- Do not allow fuel spray or fuel vapors to come into contact with a spark or open flame.
- Always use a second wrench when loosening or tightening fuel line connection fittings. This will prevent unnecessary stress and torsion on fuel piping. Always follow the proper torque specifications.
- Always replace worn fuel fitting O-rings with new ones. Do not substitute fuel hose where rigid pipe is installed.

FUEL SYSTEM PRESSURE

RELIEVING

CAUTION

Before disconnecting fuel lines or hoses, relieve pressure from the system by disabling the fuel pump and disconnecting the fuel line/quick connect fitting in the engine compartment.

With the HDS

1. Connect the HDS to the Data Link Connector (DLC) located under the driver's side of the dashboard.
2. Turn the ignition switch to ON (II), or press the engine start/stop button to select the ON mode.
3. Make sure the HDS communicates with the ECM/PCM. If it doesn't, go to the DLC circuit troubleshooting.
4. Turn the ignition switch to LOCK (0), or press the engine start/stop button to select the OFF mode.

5. Remove the fuel fill cap to relieve the pressure in the fuel tank.

6. Turn the ignition switch to ON (II), or press the engine start/stop button to select the ON mode.

7. From the INSPECTION MENU of the HDS, select Fuel Pump OFF, then start the engine, and let it idle until it stalls.

➡**Do not allow the engine to idle above 1,000 rpm or the ECM/PCM will continue to operate the fuel pump. Pending or Confirmed DTC may be set during this procedure. Check for DTCs, and clear them as needed.**

8. Turn the ignition switch to LOCK (0), or press the engine start/stop button to select the OFF mode.

9. Do the battery terminal disconnection procedure.

10. Remove the cover and the quick-connect fitting cover.

11. Check the quick-connect fitting for dirt, and clean it if needed.

12. Place a rag or a shop towel over the quick-connect fitting.

13. Disconnect the quick-connect fitting.

a. Hold the connector with one hand, and squeeze the retainer tabs with the other hand to release them from the locking tabs.

b. Pull the connector off.

➡**Be careful not to damage the line or other parts. Do not use tools. If the connector does not move, keep the retainer tabs pressed down, and alternately pull and push the connector until it comes off easily. Do not remove the retainer from the line; once removed, the retainer must be replaced with a new one.**

14. After disconnecting the quick-connect fitting, check it for dirt or damage; without SH-AWD system, with SH-AWD system.

15. Do the battery terminal reconnection procedure.

Without the HDS

See Figure 198.

1. Remove the driver's dashboard undercover.

2. Remove PGM-FI main relay 2 from the driver's under-dash fuse/relay box.

3. Start the engine, and let it idle until it stalls.

➡**If any DTCs are stored, clear and ignore them.**

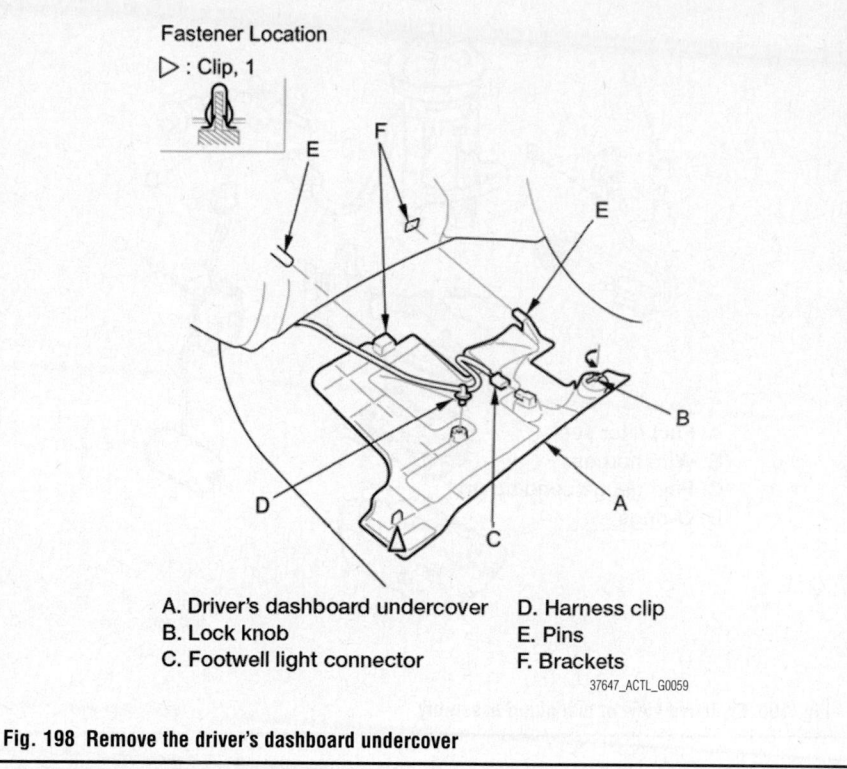

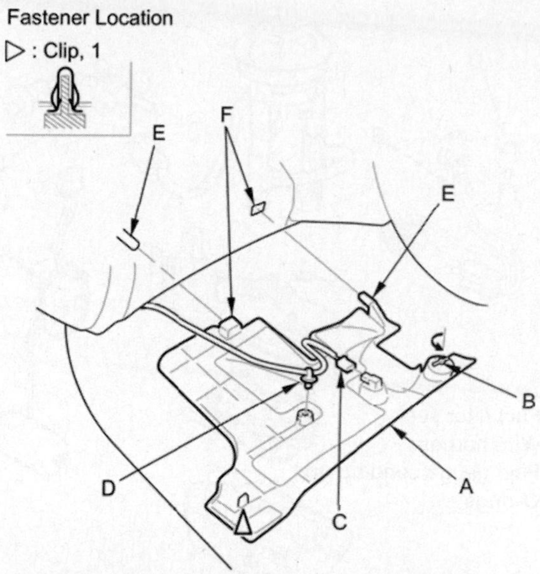
Fastener Location
▷ : Clip, 1
A
B
C
D
E
E
F
A. Driver's dashboard undercover
B. Lock knob
C. Footwell light connector
D. Harness clip
E. Pins
F. Brackets
37647_ACTL_G0059

Fig. 198 Remove the driver's dashboard undercover

4. Turn the ignition switch to LOCK (0), or press the engine start/stop button to select the OFF mode.

5. Remove the fuel fill cap to relieve the pressure in the fuel tank.

6. Do the battery terminal disconnection procedure.

7. Remove the cover and the quick-connect fitting cover.

8. Check the quick-connect fitting for dirt, and clean it if needed.

9. Place a rag or a shop towel over the quick-connect fitting. Disconnect the quick-connect fitting.

a. Hold the connector with one hand, and squeeze the retainer tabs with the other hand to release them from the locking tabs.

b. Pull the connector off.

➡**Be careful not to damage the line or other parts. Do not use tools. If the connector does not move, keep the retainer tabs pressed down, and alternately pull and push the connector until it comes off easily. Do not remove the retainer from the line; once removed, the retainer must be replaced with a new one.**

10. After disconnecting the quick-connect fitting, check it for dirt or damage; without SH-AWD system, with SH-AWD system.

11. Do the battery terminal reconnection procedure.

FUEL FILTER

REMOVAL & INSTALLATION

With SH-AWD

See Figure 199.

➡**The fuel filter should be replaced whenever the fuel pressure drops below the specified value, after making sure that the fuel pump and fuel pressure regulator are OK.**

1. Remove the fuel tank unit.

2. Remove the fuel filter set.

3. Check these items before installing the fuel tank unit:

- When connecting the wire harness, make sure the connection is secure and the connectors are firmly locked into place.
- When installing the fuel gauge sending unit, make sure the connection is secure and the connector is firmly locked into place. Be careful not to bend or twist it excessively.

4. Install the parts in the reverse order of removal with new O-rings. When installing the fuel tank unit, align the marks on the unit and the fuel tank.

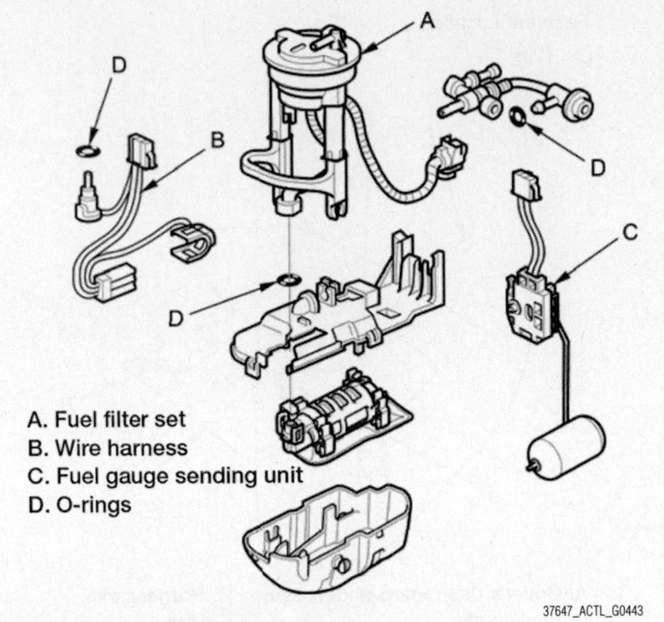
A
D
B
D
C
D
A. Fuel filter set
B. Wire harness
C. Fuel gauge sending unit
D. O-rings
37647_ACTL_G0443

Fig. 199 Exploded view of fuel pump assembly

➡Coat the O-rings with clean engine oil; do not use any other oil or fluid. Do not pinch the O-rings during installation. Use all the new parts supplied in the fuel filter replacement kit.

Without SH-AWD

See Figure 200.

➡The fuel filter should be replaced whenever the fuel pressure drops below the specified value, after making sure that the fuel pump and the fuel pressure regulator are OK.

1. Remove the fuel tank unit.
2. Remove the fuel filter set.
3. Check these items before installing the fuel tank unit:
   - When connecting the wire harness, make sure the connection is secure and the connectors are firmly locked into place.
   - When installing the fuel gauge sending unit, make sure the connection is secure and the connector is firmly locked into place. Be careful not to bend or twist it excessively.
4. Install the parts in the reverse order of removal with new O-rings and a new bracket. When installing the fuel tank unit, align the marks on the unit and the fuel tank.

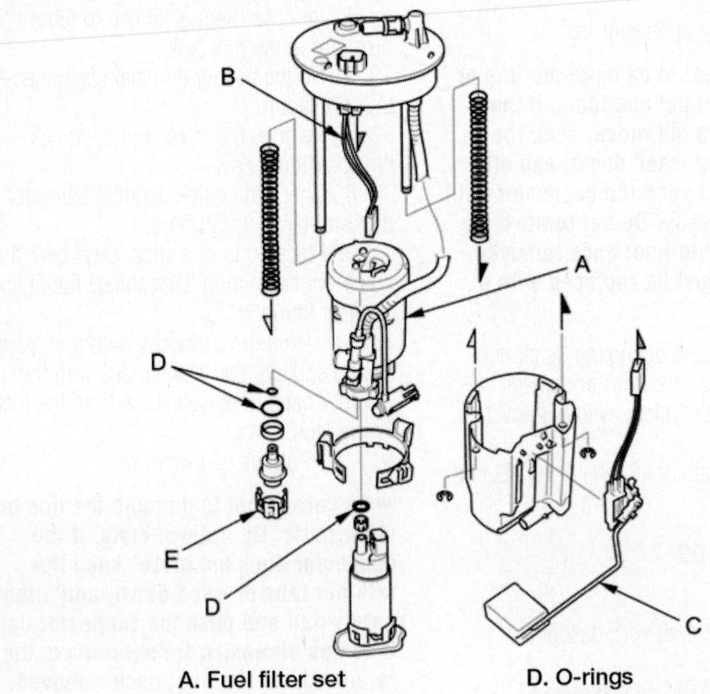
B
A
D
E
D
C
A. Fuel filter set
B. Wire harness
C. Fuel gauge sending unit
D. O-rings
E. Bracket
37647_ACTL_G0444

Fig. 200 Exploded view of fuel pump assembly

➡Coat the O-rings with clean engine oil; do not use any other oil or fluid. Do not pinch the O-rings during installation. Use all the new parts supplied in the fuel filter replacement kit.

FUEL LEVEL SENDING UNIT

LOCATION

The fuel level sending unit is an integral part of the fuel pump assembly which is located in the top of the fuel tank.

REMOVAL & INSTALLATION

See Figure 201.

1. Remove the fuel tank unit.
2. Remove the fuel level sensor (fuel gauge sending unit) from the fuel tank unit.
3. Check these items before installing the fuel tank unit:
   - When connecting the wire harness, make sure the connection is secure and the connector is firmly locked into place.

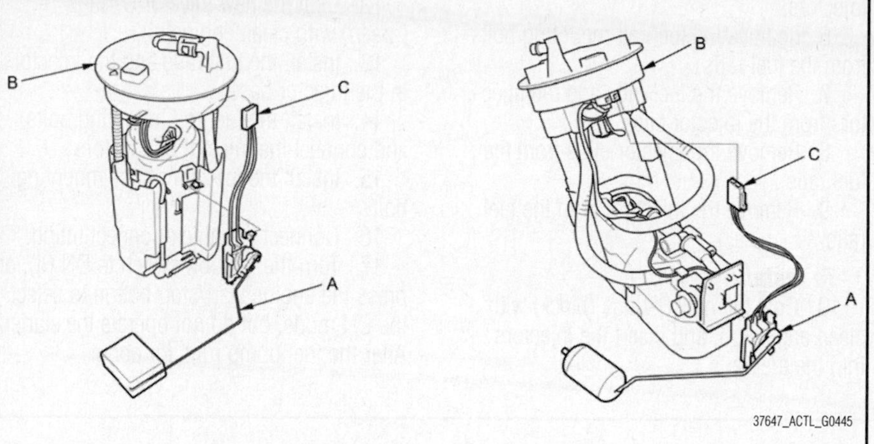
B
C
A
B
C
A
37647_ACTL_G0445

Fig. 201 Remove the fuel level sensor (fuel gauge sending unit) (A) from the fuel tank unit (B)

- When installing the fuel gauge sending unit, make sure the connection is secure. Be careful not to bend or twist it excessively.

4. Install the parts in the reverse order of removal. When installing the fuel tank unit, align the marks on the unit and the fuel tank.

FUEL PUMP

REMOVAL & INSTALLATION

1. Remove the fuel tank unit.
2. Remove the fuel pump from the fuel tank unit.
3. Check these items before installing the fuel tank unit:
   - When connecting the wire harness, make sure the connection is secure and the connector is firmly locked into place.
   - When installing the fuel gauge sending unit, make sure the connection is secure. Be careful not to bend or twist it excessively.
4. Install the parts in the reverse order of removal. When installing the fuel tank unit, align the marks on the unit and the fuel tank.

FUEL PUMP CONTROL MODULE (WITH SH-AWD)

REMOVAL & INSTALLATION

See Figure 202.

1. Remove the rear seat cushion and the seat-back.
2. Disconnect the fuel pump control module connector.
3. Remove the fuel pump control module.

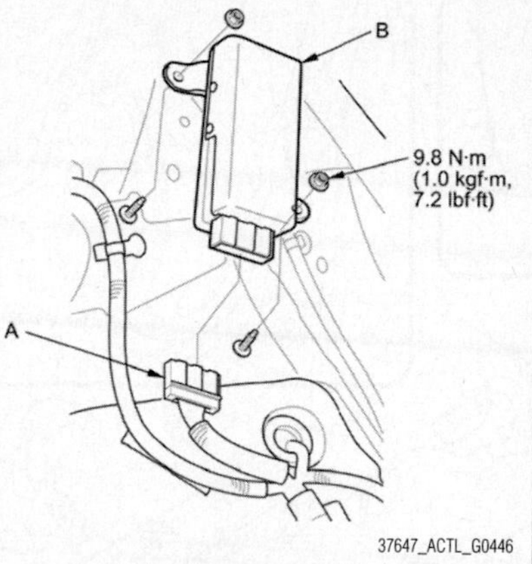
B
9.8 N·m
(1.0 kgf·m,
7.2 lbf·ft)
A
37647_ACTL_G0446

Fig. 202 Disconnect the fuel pump control module connector

4. Install the parts in the reverse order of removal.

FUEL PRESSURE REGULATOR

REMOVAL & INSTALLATION

With SH-AWD

See Figure 203.

1. Remove the fuel tank unit.
2. Remove the bracket.
3. Remove the pressure regulator.
4. Install the parts in the reverse order of removal with new O-ring. When installing the fuel tank unit, align the marks on the fuel tank unit and the fuel tank.

➡Coat the O-rings with clean engine oil; do not use any other oil or fluid. Do not pinch the O-rings during installation. Use all the new parts supplied in the pressure regulator replacement kit.

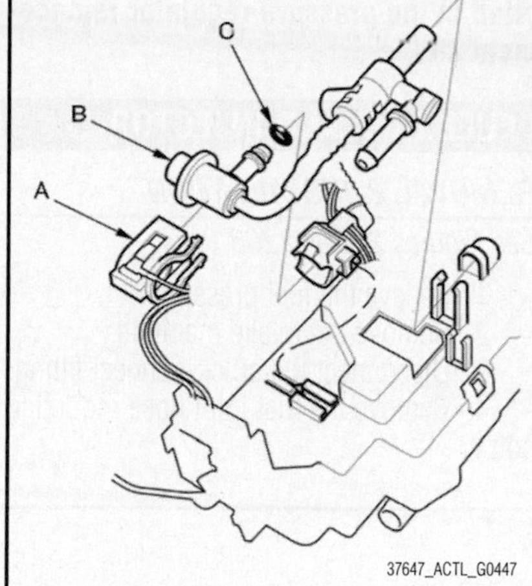
C
B
A
37647_ACTL_G0447

Fig. 203 Exploded view of fuel pressure regulator

Without SH-AWD

See Figure 204.

1. Remove the fuel tank unit.
2. Remove the reservoir.
3. Remove the bracket.
4. Remove the fuel pressure regulator and the backup ring.
5. Install the parts in the reverse order of removal with new O-rings and a new bracket. When installing the fuel tank unit, align the marks on the fuel tank unit and the fuel tank.

➡Coat the O-rings with clean engine oil; do not use any other type oils or

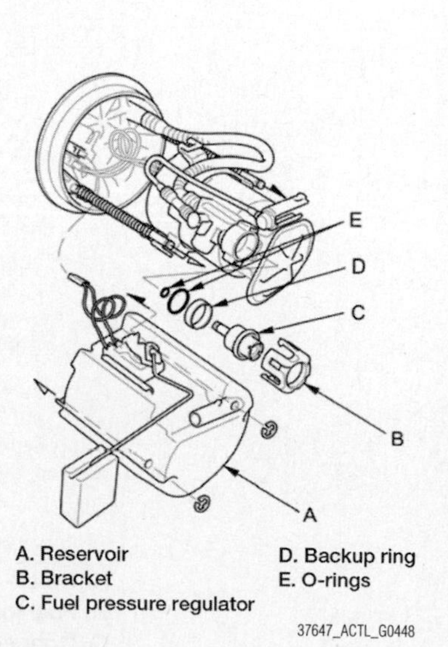
E
D
C
B
A
A. Reservoir
B. Bracket
C. Fuel pressure regulator
D. Backup ring
E. O-rings
37647_ACTL_G0448

Fig. 204 Exploded view of fuel pressure regulator

fluids. Do not pinch the O-rings during installation. Use all the new parts supplied in the pressure regulator replacement kit.

FUEL RAIL AND INJECTOR

REMOVAL & INSTALLATION

See Figures 205 and 206.

1. Relieve the fuel pressure.
2. Remove the intake manifold.
3. Disconnect the quick-connect fitting.
4. Remove the fuel joint hose mounting bolt.
5. Disconnect the connectors from the injectors.
6. Remove the fuel rail mounting bolts from the fuel rails.
7. Remove the fuel rails and the injectors from the injector base.
8. Remove the injector clips from the fuel rails.
9. Remove the injectors from the fuel rails.

To install:

10. Coat the new O-rings (black) with clean engine oil, and insert the injectors into the fuel rails.
11. Install the injector clips.
12. Coat the new injector O-rings (green) with clean engine oil.
13. Install the fuel rails and the injectors in the injector base.
14. Install the fuel rail mounting bolts, and connect the injector connectors.
15. Install the fuel joint hose mounting bolt.
16. Connect the quick-connect fitting.
17. Turn the ignition switch to ON (II), or press the engine start/stop button to select the ON mode, but do not operate the starter. After the fuel pump runs for about

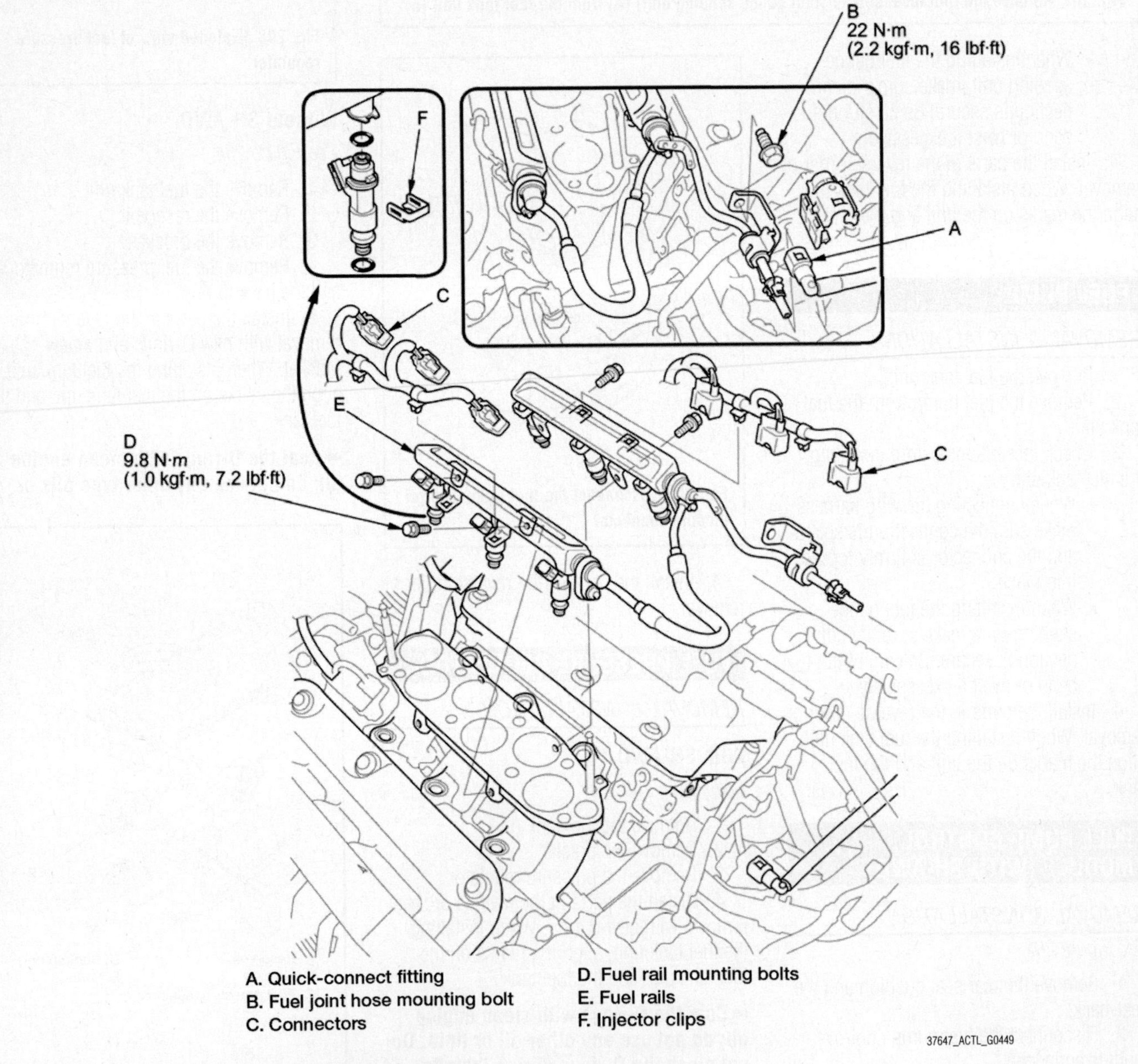
B
22 N·m
(2.2 kgf·m, 16 lbf·ft)
F
A
C
E
D
9.8 N·m
(1.0 kgf·m, 7.2 lbf·ft)
C
A. Quick-connect fitting
B. Fuel joint hose mounting bolt
C. Connectors
D. Fuel rail mounting bolts
E. Fuel rails
F. Injector clips
37647_ACTL_G0449

Fig. 205 Exploded view of fuel rail and injector assembly

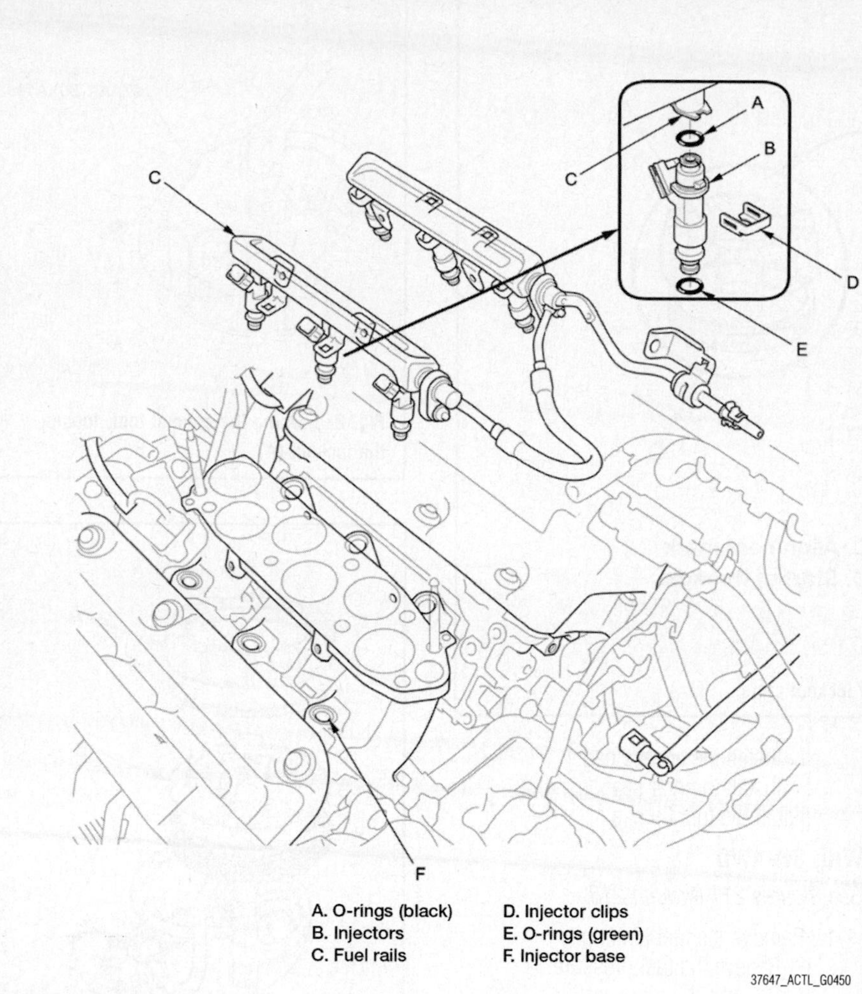
A
B
C
C
D
E
F
A. O-rings (black)
B. Injectors
C. Fuel rails
D. Injector clips
E. O-rings (green)
F. Injector base
37647_ACTL_G0450

Fig. 206 Installation of fuel rail and injectors

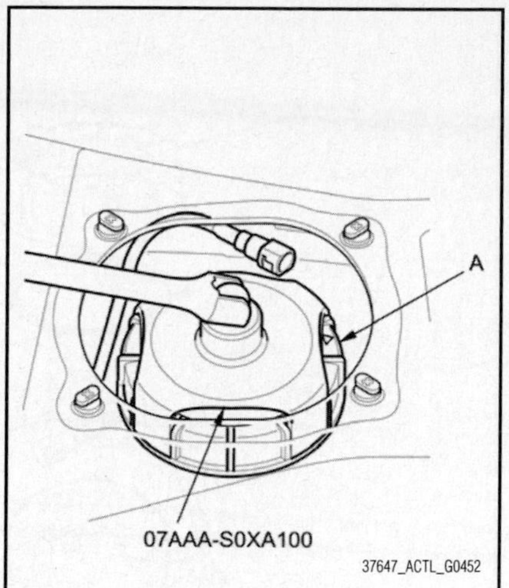
A
07AAA-S0XA100
37647_ACTL_G0452

Fig. 208 Using the special tool, loosen the locknut (A)

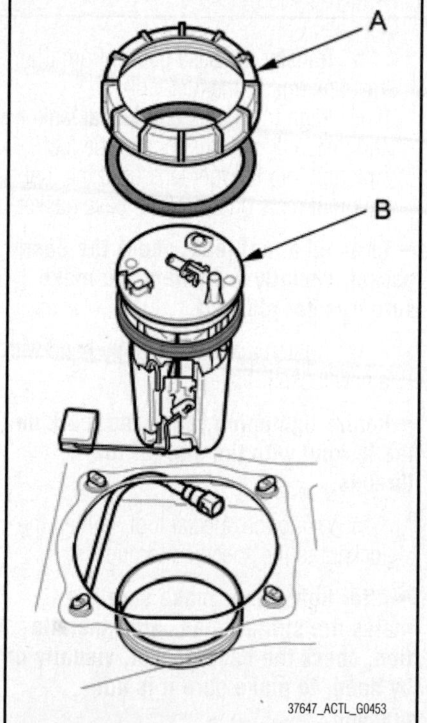
A
B
37647_ACTL_G0453

Fig. 209 Remove the locknut (A) and the fuel tank unit (B)

2 seconds, the fuel rail will be pressurized. Repeat this two or three times, then make sure there are no fuel leaks.

18. Install the intake manifold with a new gasket.

FUEL TANK

DRAINING

Without SH-AWD

See Figures 207 through 210.

1. Remove the fuel tank unit.
   a. Relieve the fuel pressure.
   b. Remove the rear seat cushion.
   c. Remove the access panel from the floor.
   d. Disconnect the fuel tank unit 4P connector.
   e. Disconnect the quick-connect fitting from the fuel tank unit.
   f. Using the special tool, loosen the locknut.

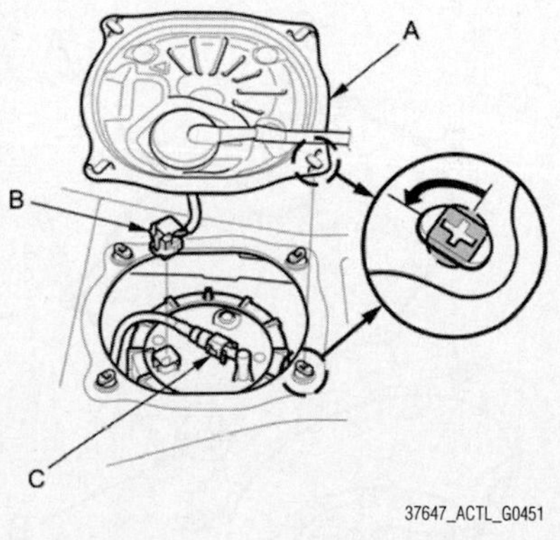
A
B
C
37647_ACTL_G0451

Fig. 207 Remove the access panel (A) from the floor

   g. Remove the locknut and the fuel tank unit.
2. Using a hand pump, draw the fuel from the fuel tank.

To install:

3. Install the fuel tank unit
   a. Temporarily attach a new base gasket to the fuel tank unit, then insert the fuel tank unit partially into the fuel tank.

➡**Be careful not to damage the new base gasket. Be careful not to bend the fuel gauge sending unit. Do not coat the base gasket with oil.**

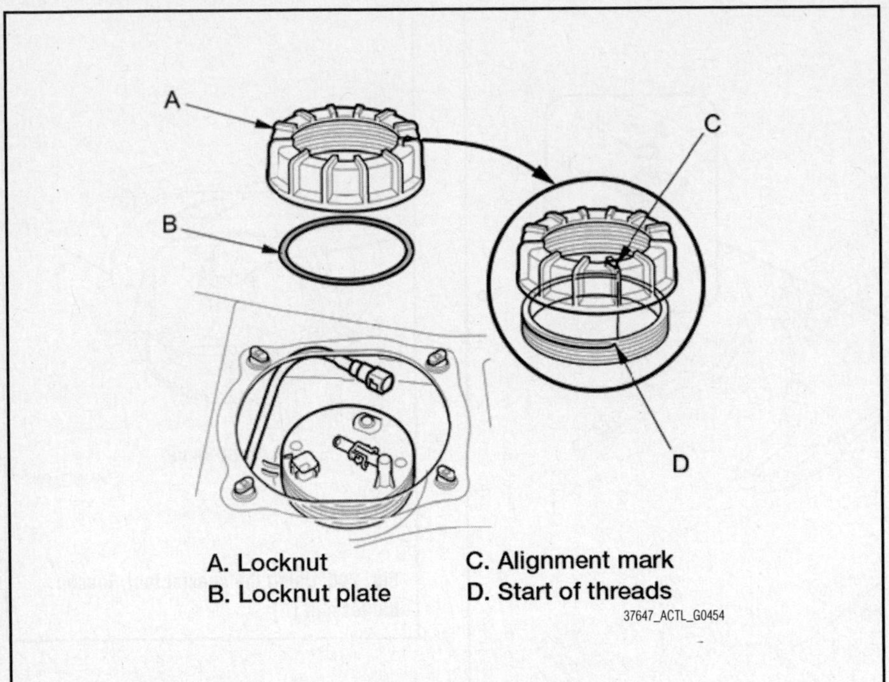
A
B
C
D
A. Locknut
B. Locknut plate
C. Alignment mark
D. Start of threads
37647_ACTL_G0454

Fig. 210 Tighten a new locknut by hand with a new locknut plate

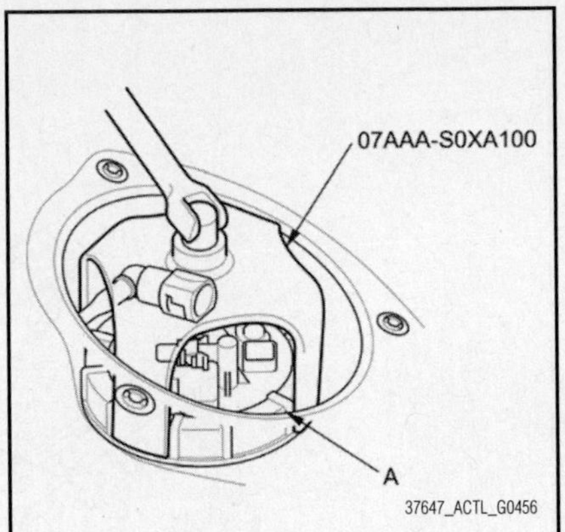
07AAA-S0XA100
A
37647_ACTL_G0456

Fig. 212 Using the special tool, loosen the locknut (A)

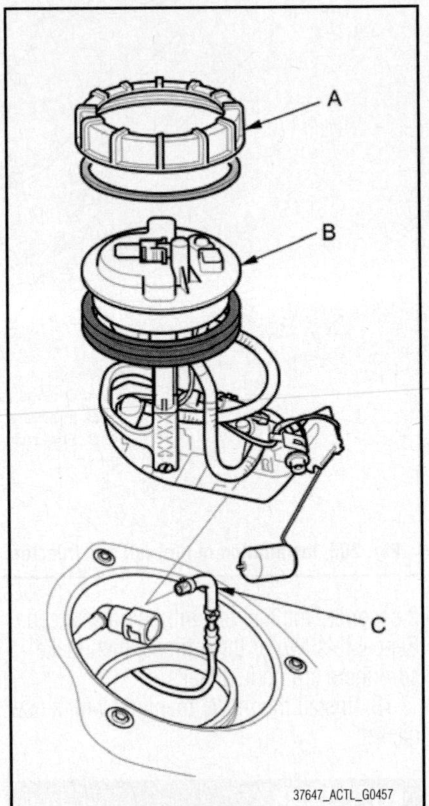
A
B
C
37647_ACTL_G0457

Fig. 213 Remove the locknut (A), and lift up the fuel tank unit (B)

b. Transfer the base gasket from the fuel tank unit to the fuel tank.

c. Align the marks on the fuel tank and fuel tank unit, then insert the fuel tank unit into the fuel tank until the fuel tank unit rests on top of the base gasket.

➡To avoid a fuel leak, check the base gasket, visually or by hand, to make sure it is not pinched.

d. Tighten a new locknut by hand with a new locknut plate.

➡Before tightening, align the mark on the locknut with the start of the threads.

e. Using the special tool, tighten the locknut to the specified torque.

➡After tightening, make sure the marks are still aligned. After installation, check the base gasket, visually or by hand, to make sure it is not pinched.

f. Connect the fuel tank unit 4P connector, then connect the quick-connect fitting.

g. Reconnect the negative cable to the battery, and turn the ignition switch to ON (II) or press the engine start /stop button to select the ON mode (but do not operate the starter motor). The fuel pump runs for about 2 seconds, and fuel pressure rises. Repeat this two or three times, and make sure there are no fuel leaks.

h. Install the access panel.

i. Install the rear seat cushion.

Install the fuel fill cap.

With SH-AWD

See Figures 211 through 218.

1. Remove the fuel tank unit.
   a. Relieve the fuel pressure.
   b. Remove the rear seat cushion.
   c. Remove the access panel from the left side of the floor.

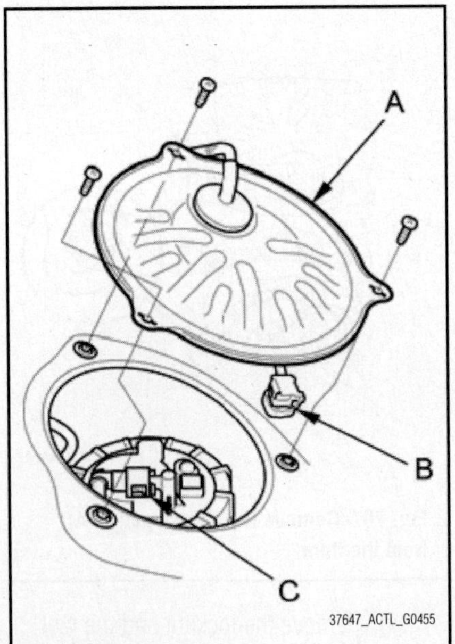
A
B
C
37647_ACTL_G0455

Fig. 211 Remove the access panel (A) from the left side of the floor

   d. Disconnect the fuel tank unit 4P connector.
   e. Disconnect the quick-connect fitting from the fuel tank unit.
   f. Using the special tool, loosen the locknut.
   g. Remove the locknut, and lift up the fuel tank unit.
   h. Disconnect the transfer tube, and remove the fuel tank unit.
2. Remove the secondary fuel gauge sending unit.

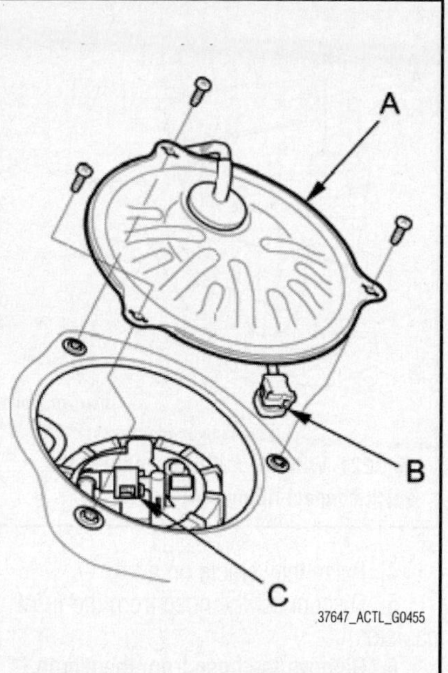
A
B
C
37647_ACTL_G0455
Fig. 214 Remove the access panel (A) from the left side of the floor—Left side shown, right side similar

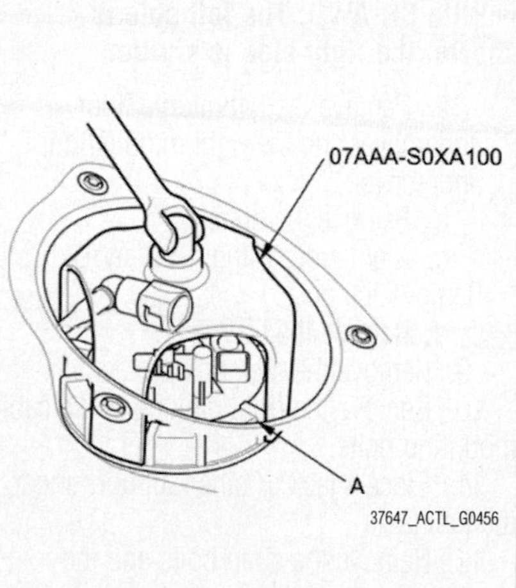
07AAA-S0XA100
A
37647_ACTL_G0456
Fig. 215 Using the special tool, loosen the locknut (A) —Left side shown, right side similar

a. Remove the access panel from the right side of the floor.

b. Disconnect the secondary fuel gauge sending unit 4P connector.

c. Using the special tool, loosen the locknut.

d. Remove the locknut, and lift up the secondary fuel gauge sending unit from the fuel tank.

e. Disconnect the transfer tube, and remove the secondary fuel gauge sending unit.

3. Using a hand pump, a hose, and a container suitable for fuel, draw the fuel from the fuel tank.

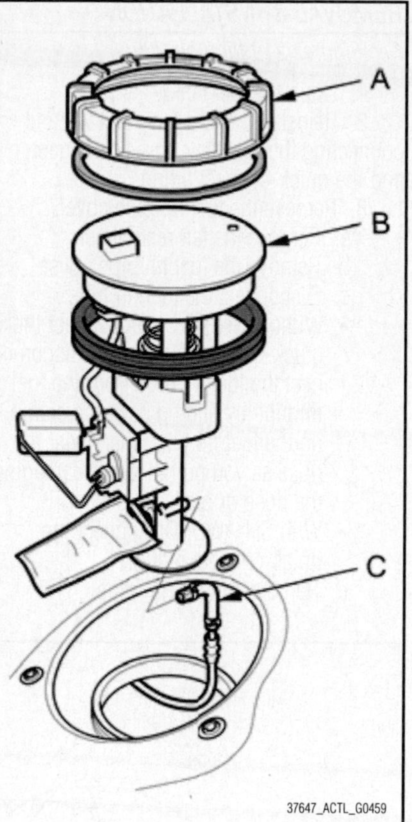
A
B
C
37647_ACTL_G0459
Fig. 216 Remove the locknut (A), and lift up the secondary fuel gauge sending unit (B) from the fuel tank

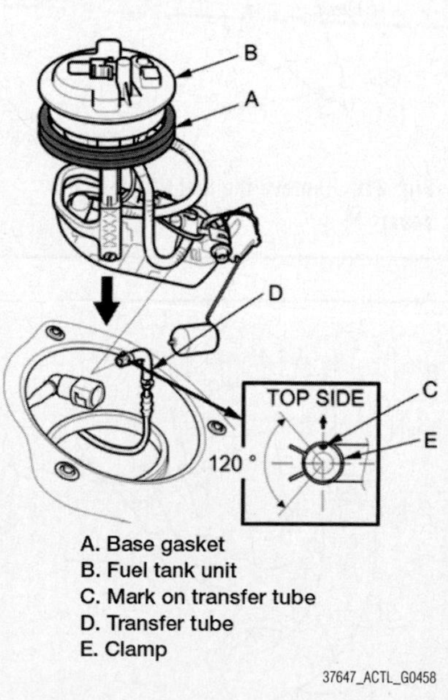
B
A
D
TOP SIDE
C
E
120 °
A. Base gasket
B. Fuel tank unit
C. Mark on transfer tube
D. Transfer tube
E. Clamp
37647_ACTL_G0458
Fig. 217 Temporarily attach a new base gasket to the fuel tank unit, position the mark of the transfer tube up, connect the transfer tube

To install:

4. Install the fuel tank unit.

a. Temporarily attach a new base gasket to the fuel tank unit, position the mark of the transfer tube up, connect the transfer tube, then insert the fuel tank unit partially into the fuel tank.

➡Be careful not to damage the new base gasket. Make sure the clamp is positioned as shown. Be careful not to bend the fuel gauge sending unit. Do not coat the base gasket with oil.

b. Transfer the base gasket from the fuel tank unit to the fuel tank.

c. Align the marks on the fuel tank and the fuel tank unit, then insert it straight into the fuel tank. Make sure it evenly contacts the base gasket.

➡To prevent a fuel leak, check the base gasket, visually or by hand, to make sure it is not pinched.

d. Tighten a new locknut by hand with a new locknut plate.

➡Before tightening, align the mark on the locknut with the start of the threads.

e. Using the special tool, tighten the locknut to the specified torque.

➡After tightening, make sure the marks are still aligned. After installation, check the base gasket, visually or by hand, to be sure it is not pinched.

f. Connect the fuel tank unit 4P connector, then connect the quick-connect fitting.

g. Reconnect the negative cable to the battery, and turn the ignition switch to ON (II) or press the engine start/stop button to select the ON mode (but do not operate the starter motor). The fuel pump runs for about 2 seconds, and fuel pressure rises. Repeat two or three times, and make sure there are no fuel leaks.

h. Install the access panel.

i. Install the rear seat cushion.

j. Install the fuel fill cap.

5. Install the secondary fuel gauge sending unit.

a. Temporarily attach a new b``ase gasket to the secondary fuel gauge sending unit, position the mark on the transfer tube up, connect the transfer tube, then insert the secondary fuel gauge sending unit partially into the fuel tank.

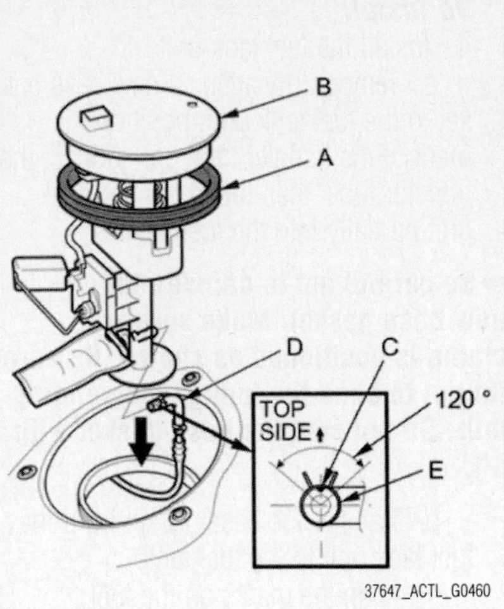
B
A
D
C
TOP
SIDE
120 °
E
37647_ACTL_G0460

Fig. 218 Temporarily attach a new base gasket to the secondary fuel gauge sending unit, position the mark on the transfer tube up, connect the transfer tube

➡Be careful not to damage the new base gasket. Make sure the clamp is positioned. Be careful not to bend the secondary fuel gauge sending unit. Do not coat the base gasket with oil.

b. Transfer the base gasket from the secondary fuel gauge sending unit to the fuel tank.

c. Align the marks on the fuel tank and the secondary fuel gauge sending unit, then insert the secondary fuel gauge sending unit, straight into the fuel tank. Make sure it evenly contacts the base gasket.

➡To avoid a fuel leak, check the base gasket, visually or by hand, to make sure it is not pinched.

d. Tighten a new locknut by hand with a new locknut plate.

➡Before tightening, align the mark on the locknut with the start of the threads.

e. Using the tool, tighten the locknut to the specified torque.

➡After tightening, make sure the marks are still aligned. After installation, check the base gasket, visually or by hand, to be sure it is not pinched.

f. Connect the secondary fuel gauge sending unit 4P connector.

g. Install the parts in the reverse order of removal.

REMOVAL & INSTALLATION

See Figures 219 through 225.

1. Drain the fuel tank.
2. Reinstall the fuel tank unit without connecting the fuel tank unit 4P connector and the quick-connect fitting.
3. Remove the fuel fill pipe cover.
   a. Remove the left rear wheel.
   b. Remove the fuel fill pipe cover.
   c. Disconnect the fuel fill tube:
   - Without SH-AWD: Disconnect the quick-connect fitting, and disconnect the fuel fill tube from the fuel fill pipe by sliding back the clamp on the fuel fill tube, then twist the hose as you pull to avoid damaging the hose or pipe.
   - With SH-AWD: Disconnect the quick-connect fittings.

   d. Remove the fuel fill pipe.

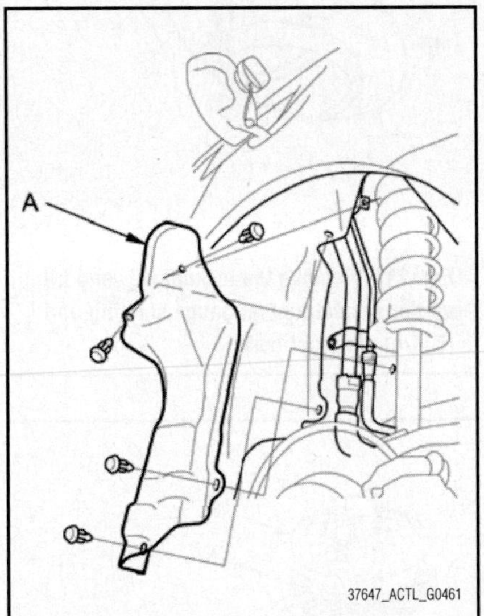
A
37647_ACTL_G0461

Fig. 219 Remove the fuel fill pipe cover (A)

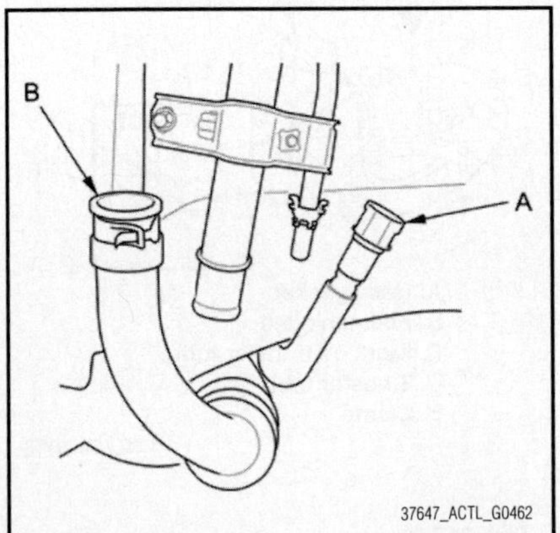
B
A
37647_ACTL_G0462

Fig. 220 Without SH-AWD: Disconnect the quick-connect fitting (A), and disconnect the fuel fill tube (B) from the fuel fill pipe

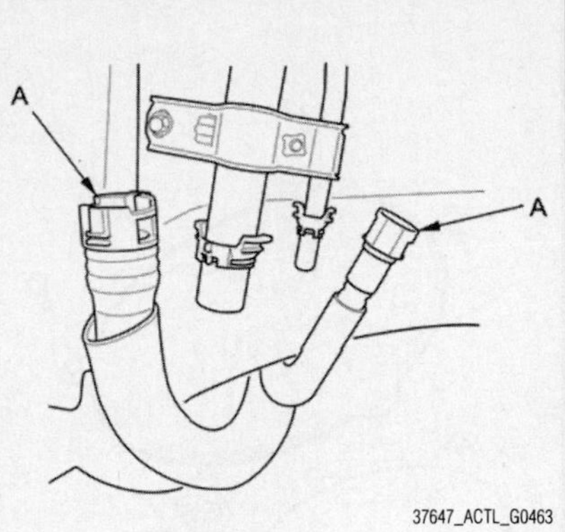
A
A
37647_ACTL_G0463

Fig. 221 With SH-AWD: Disconnect the quick-connect fittings (A)

4. Raise the vehicle on a lift.
5. Disconnect the hose from the EVAP canister.
6. Remove the hose from the clamp.

➡Be careful not to damage the hose.

7. Remove the exhaust pipe.
8. Remove the middle floor undercover.

➡With SH-AWD: The left side is shown; the right side is similar.

a. Remove the left middle floor undercover and the right middle floor undercover.
b. Remove the bolts.
c. Right side (without SH-AWD): Remove the nut.
d. Release the hook(s).

9. Remove the fuel tank protector.
10. Remove the right parking brake cable mounting bolts.
11. Place a jack or other support under the fuel tank.
12. Remove the strap bolts and the straps.

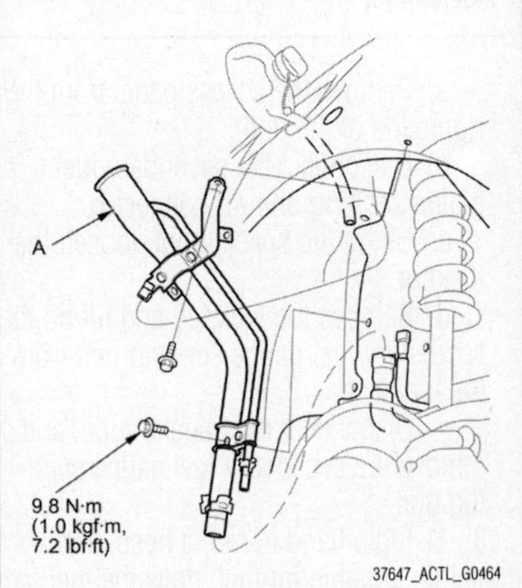
A
9.8 N·m
(1.0 kgf·m,
7.2 lbf·ft)
37647_ACTL_G0464

Fig. 222 Remove the fuel fill pipe (A)

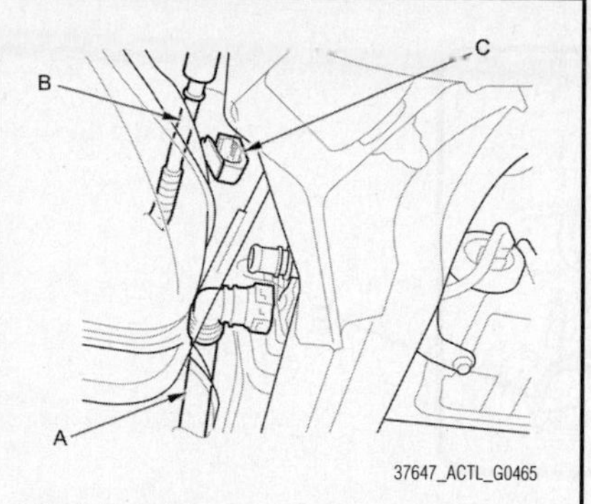
B
C
A
37647_ACTL_G0465

Fig. 223 Disconnect the hose (A) from the EVAP canister

13. Remove the fuel tank.

To install:

14. Install the parts in the reverse order of removal.

a. New fuel tanks have ring pull at the fuel vapor hose connector. When you connect the hose and confirm that the connection is secure, remove the ring pull by pulling it down.

b. Before connecting the fuel fill pipe and the quick-connect fitting, check for dirt, and clean it if needed, taking care not to damage the fuel fill pipe and other parts.

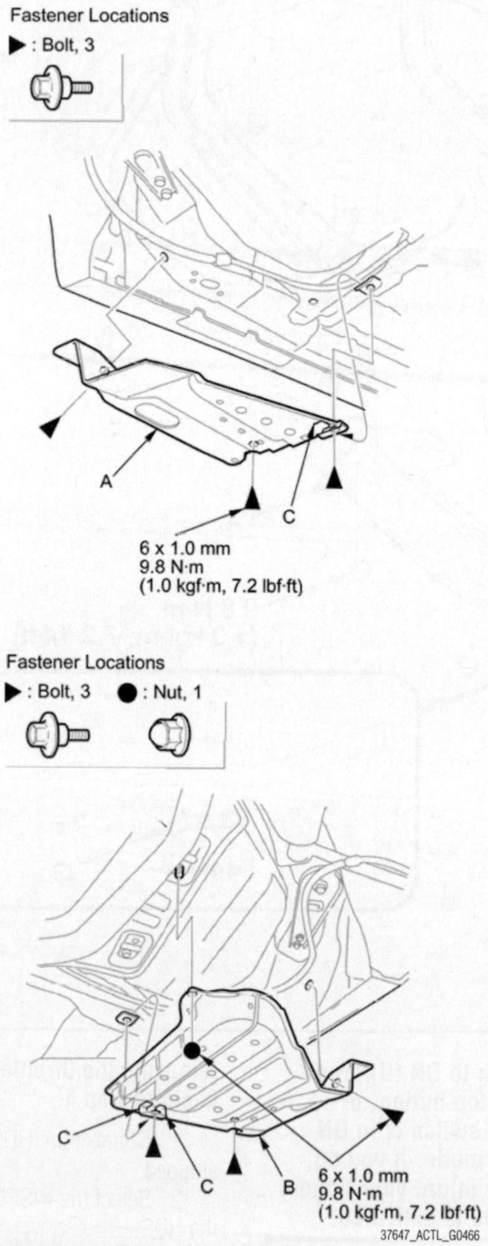
Fastener Locations
▶ : Bolt, 3
A
C
6 x 1.0 mm
9.8 N·m
(1.0 kgf·m, 7.2 lbf·ft)
Fastener Locations
▶ : Bolt, 3
● : Nut, 1
C
C
B
6 x 1.0 mm
9.8 N·m
(1.0 kgf·m, 7.2 lbf·ft)
37647_ACTL_G0466

Fig. 224 Remove the middle floor undercover

c. When installing the fuel tank protector, make sure to insert it into the clip in the direction shown.

IDLE SPEED

ADJUSTMENT

➡Before checking the idle speed, check these items:

- The Malfunction Indicator Lamp (MIL) has not been reported on, and there are no DTCs.
- Ignition timing
- Spark plugs
- Air cleaner
- PCV system
- Apply the parking brake, and make sure the headlights are off.

1. Disconnect the Evaporative Emission (EVAP) canister purge valve connector.
2. Connect the HDS to the Data Link Connector (DLC) located under the driver's side of the dashboard.
3. Make sure the HDS communicates with the ECM/PCM. If it doesn't, go to the DLC circuit troubleshooting.
4. Start the engine. Hold the engine speed at 3,000 rpm without load (A/T in P or N, M/T in neutral) until the radiator fan comes on, then let it idle.
5. Check the idle speed without load conditions: headlights, blower fan, radiator fan, and air conditioner off.

➡Idle speed should be:

- J35Z6 engine: 680 ± 50 rpm (in P or N)
- J37A4 engine (M/T): 750 ± 50 rpm (in neutral)
- J37A4 engine (A/T): 700 ± 50 rpm (in P or N)

6. Let the engine idle for 1 minute with high electric load (the A/C on, the temperature set to max cool, the blower fan on high, the headlights on high beam).

➡Idle speed should be:

- J35Z6 engine: 680 ± 50 rpm (in P or N)
- J37A4 engine (M/T): 750 ± 50 rpm (in neutral)
- J37A4 engine (A/T): 700 ± 50 rpm (in P or N)

If the idle speed is not within specification, do the ECM/PCM idle learn procedure. If the idle speed is still not within specification, go to the symptom troubleshooting.

7. Reconnect the EVAP canister purge valve connector.

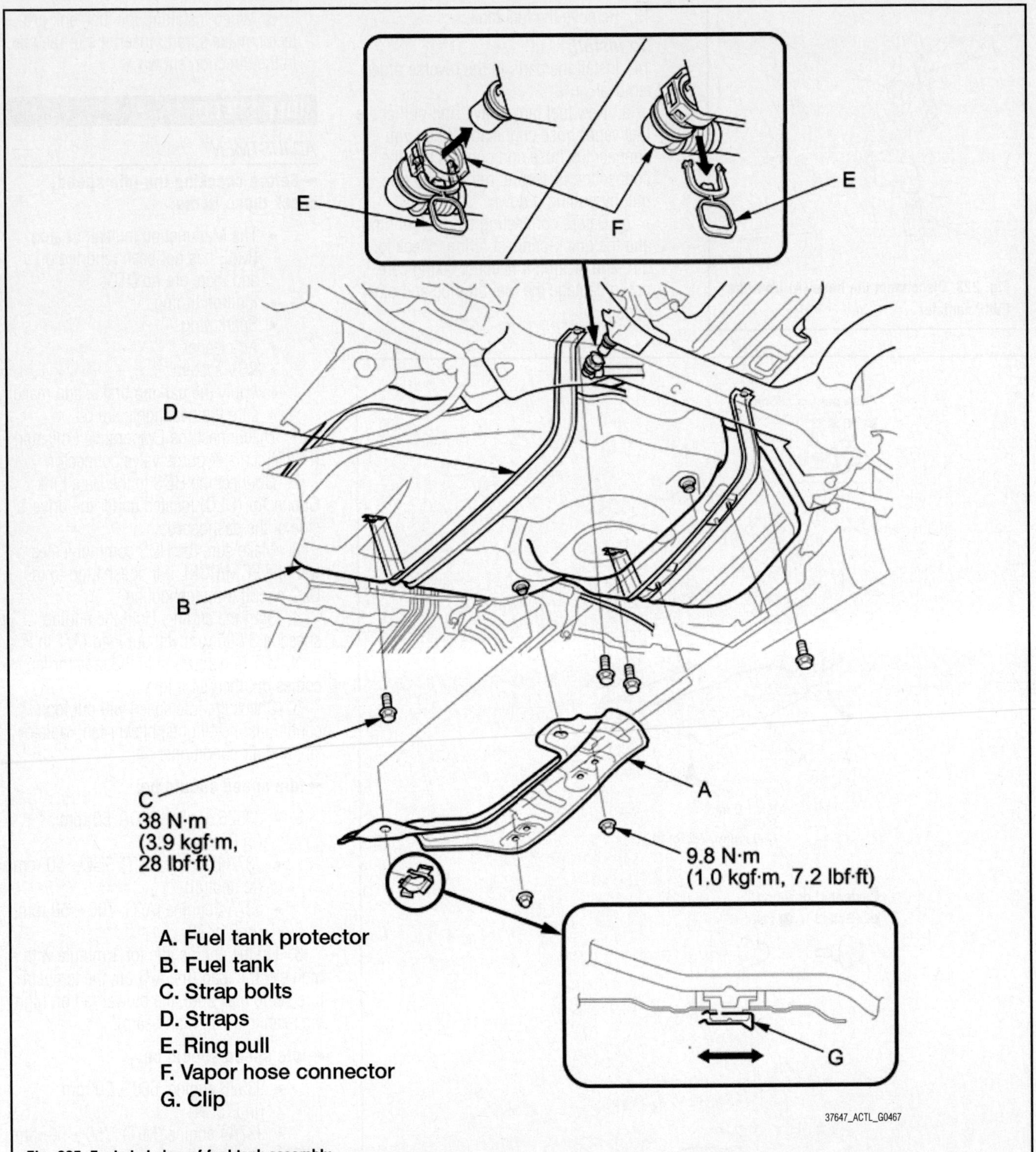
E
F
E
D
B
C
38 N·m
(3.9 kgf·m,
28 lbf·ft)
A
9.8 N·m
(1.0 kgf·m, 7.2 lbf·ft)
G
A. Fuel tank protector
B. Fuel tank
C. Strap bolts
D. Straps
E. Ring pull
F. Vapor hose connector
G. Clip
37647_ACTL_G0467

Fig. 225 Exploded view of fuel tank assembly

THROTTLE BODY

REMOVAL & INSTALLATION

See Figure 226.

CAUTION

Do not insert your fingers into the installed throttle body when you turn the ignition switch to ON (II)/press the engine start/stop button, or while the ignition switch is in ON (II)/vehicle in ON mode. If you do, you will seriously injure your fingers if the throttle valve is activated.

➡If you are replacing or cleaning the throttle body, begin at step 1. If you are removing the throttle body temporarily, begin at step 4.

1. Connect the HDS while the engine is stopped.
2. Select the INSPECTION MENU with the HDS.
3. Do the TP POSITION CHECK in the ETCS TEST.

4. Turn the ignition switch to LOCK (0), or press the engine start/stop button to select the OFF mode.

5. Disconnect the MAP sensor connector.

6. Remove the intake air duct.

7. Disconnect the throttle body connector.

8. Disconnect and plug the water bypass hoses, then remove the throttle body.

To install:

9. Install the parts in the reverse order of removal with a new gasket.

➡When tightening the screw of the hose band, align the edge of the hose band with the mark painted on the hose band.

10. After the installation, refill the radiator with engine coolant.

11. Do the ECM/PCM idle learn procedure.

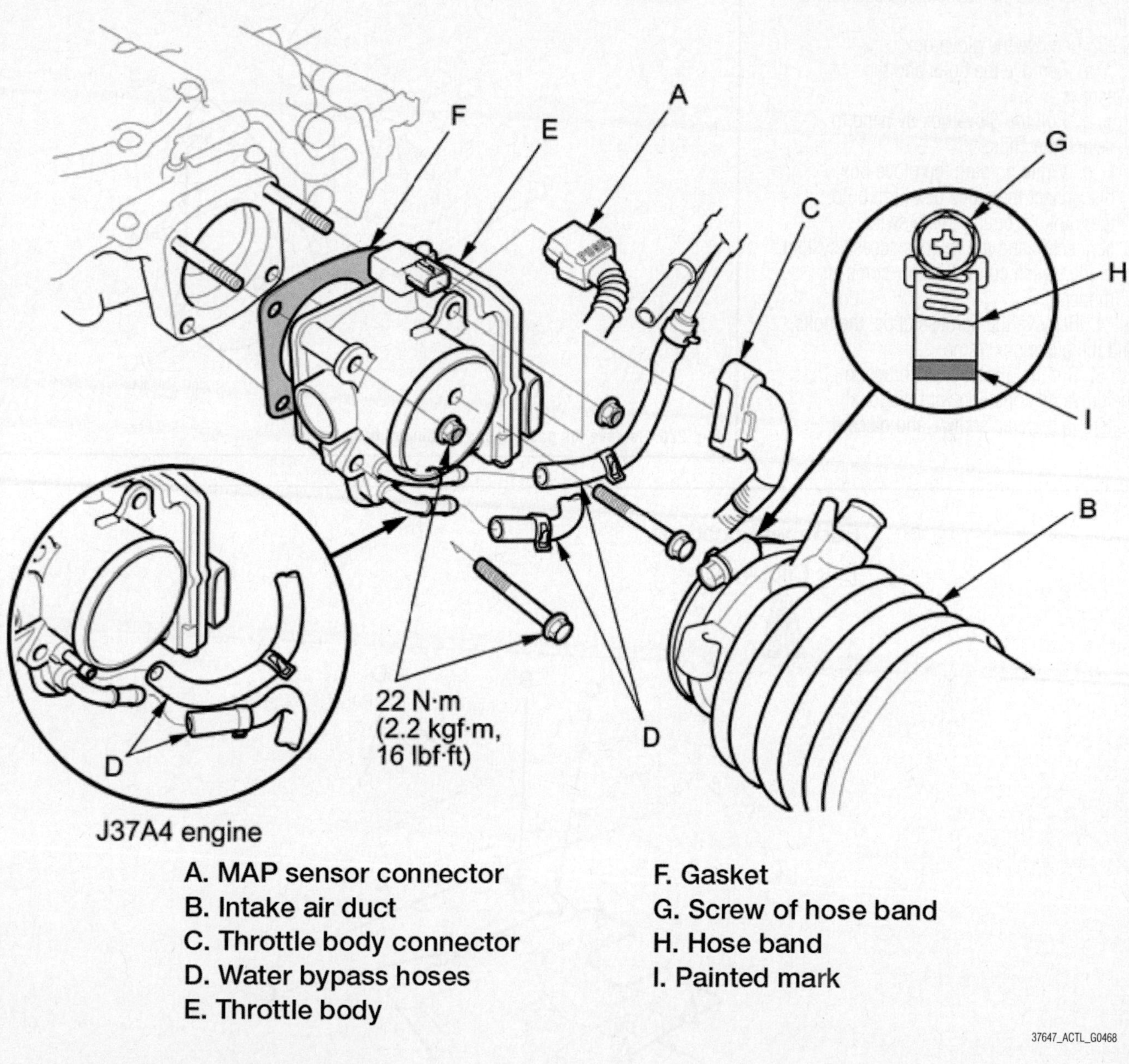
A
F
E
G
C
H
I
B
22 N·m
(2.2 kgf·m,
16 lbf·ft)
D
D
J37A4 engine
A. MAP sensor connector
B. Intake air duct
C. Throttle body connector
D. Water bypass hoses
E. Throttle body
F. Gasket
G. Screw of hose band
H. Hose band
I. Painted mark
37647_ACTL_G0468

Fig. 226 Exploded view of throttle body assembly

HEATING & AIR CONDITIONING SYSTEM

BLOWER MOTOR

REMOVAL & INSTALLATION

See Figures 227 through 237.

1. Remove the passenger's dashboard undercover.
2. Remove the passenger's dashboard trim.
3. Remove the glove box.
   a. Remove the bolts and the screws.
   b. Pull the glove box by hand to detach the clips.
   c. While holding the glove box, disconnect the glove box light bulb, the trunk lid opener main switch connector, and the keyless access system mode switch connector (for some models).
4. Remove the harness clips, the bolts, and the glove box frame.
5. Cut the plastic cross brace in the glove box opening with diagonal cutters in the area shown, and discard it.

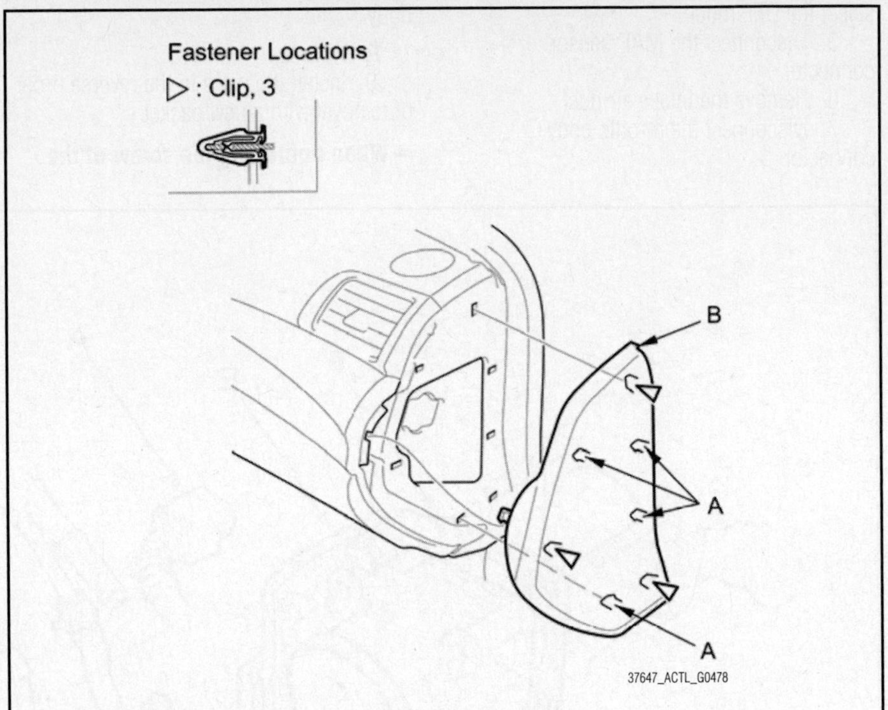
Fastener Locations
▷ : Clip, 3
B
A
A
37647_ACTL_G0478

Fig. 228 Remove the passenger's dashboard trim

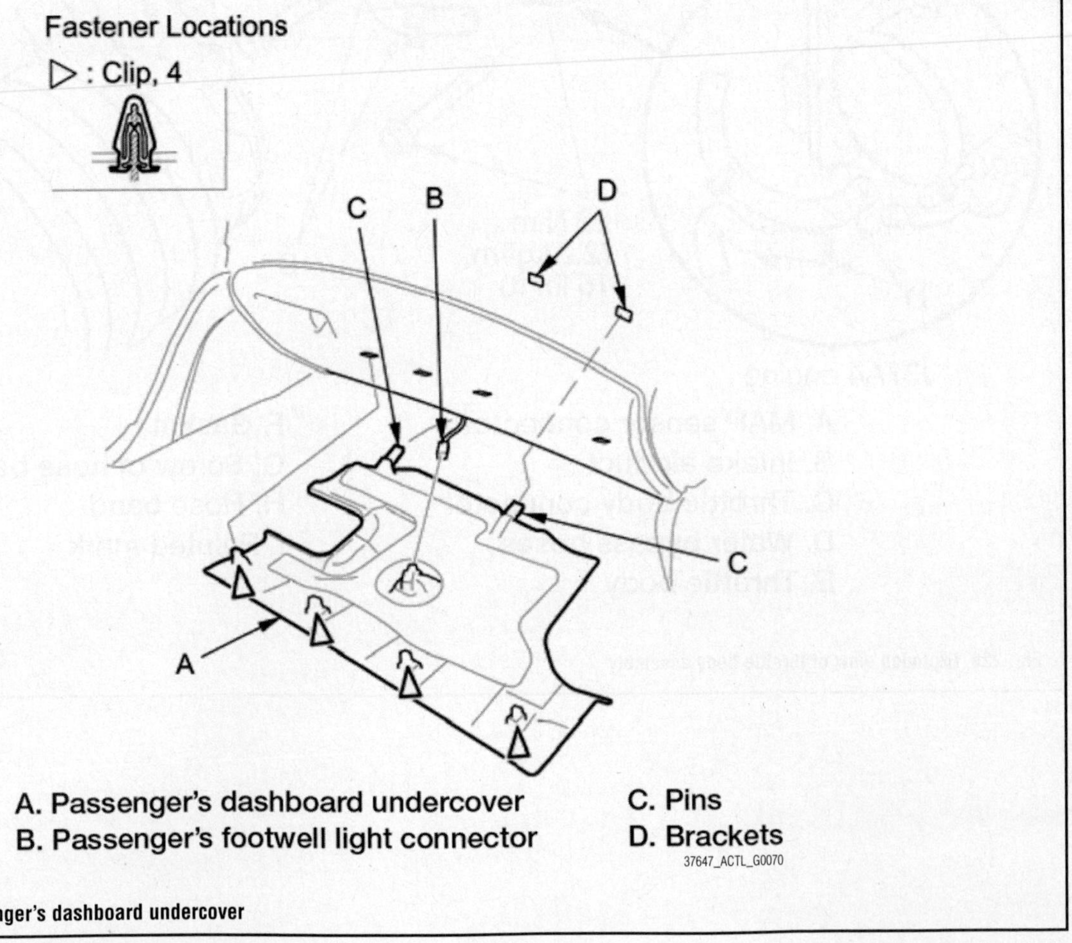
Fastener Locations
▷ : Clip, 4
C
B
D
C
A
A. Passenger's dashboard undercover
B. Passenger's footwell light connector
C. Pins
D. Brackets
37647_ACTL_G0070

Fig. 227 Remove the passenger's dashboard undercover

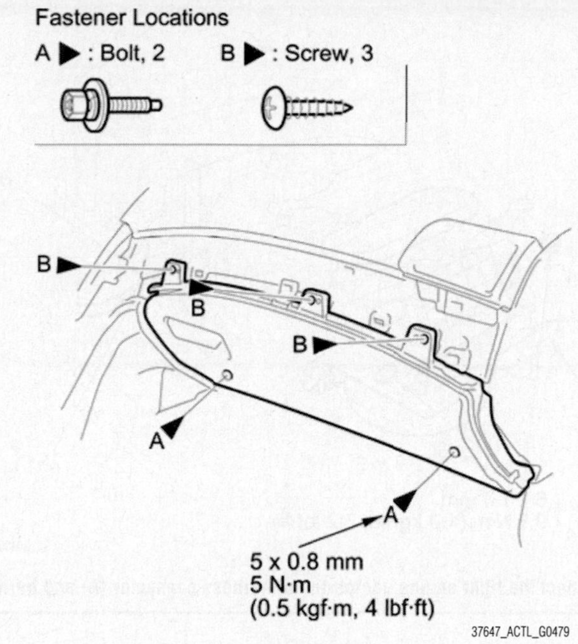
Fastener Locations
A ▶ : Bolt, 2
B ▶ : Screw, 3
B
B
B
A
A
5 x 0.8 mm
5 N·m
(0.5 kgf·m, 4 lbf·ft)
37647_ACTL_G0479

Fig. 229 Remove the bolts (A) and the screws (B)

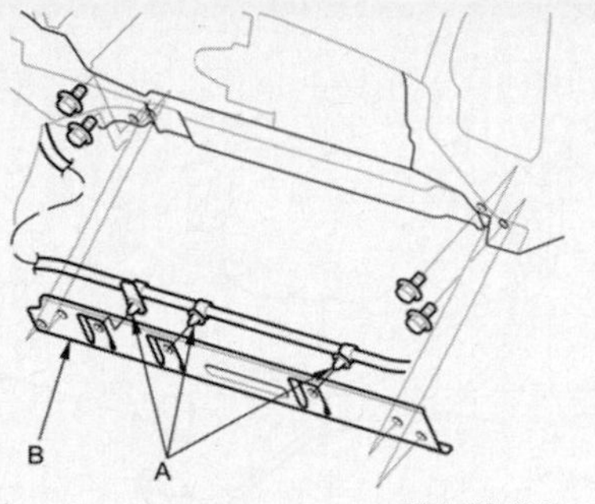
B
A
37647_ACTL_G0481

Fig. 231 Remove the harness clips (A), the bolts, and the glove box frame (B)

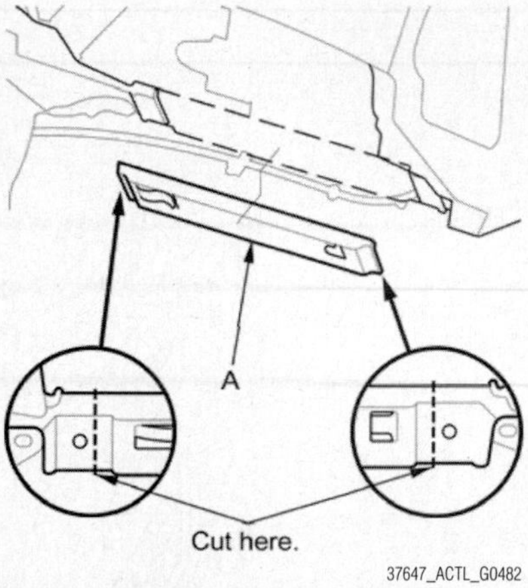
A
Cut here.
37647_ACTL_G0482

Fig. 232 Cut the plastic cross brace (A) in the glove box opening

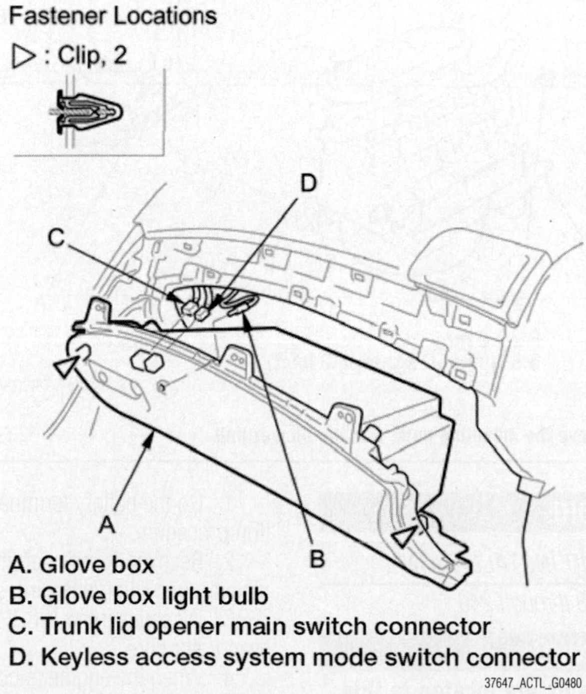
Fastener Locations
▷ : Clip, 2
D
C
A
B
A. Glove box
B. Glove box light bulb
C. Trunk lid opener main switch connector
D. Keyless access system mode switch connector
37647_ACTL_G0480

Fig. 230 Remove the glove box

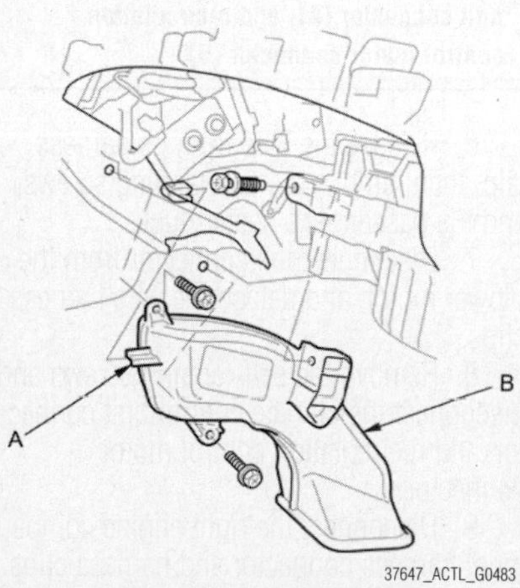
A
B
37647_ACTL_G0483

Fig. 233 Release the wire from its harness clip (A), remove the self-tapping screws, and the passenger's heater duct (B)

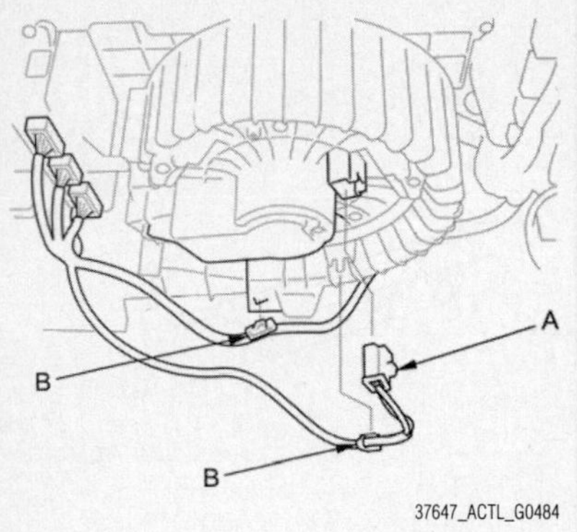
A
B
B
37647_ACTL_G0484

Fig. 234 Disconnect the connector (A) from the blower motor, and detach the wire harness clips (B)

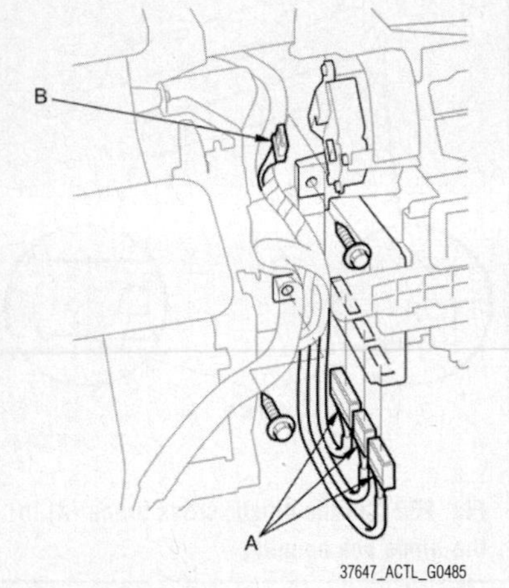
B
A
37647_ACTL_G0485

Fig. 235 Disconnect the climate control unit connector (A), and recirculation control motor connector (B)

6. Release the wire from its harness clip, then remove the self-tapping screws, and the passenger's heater duct.
7. Disconnect the connector from the blower motor, and detach the wire harness clips.
8. Remove the self-tapping screws and disconnect the climate control unit connector, and recirculation control motor connector.
9. Disconnect the right engine compartment harness connector and harness clips.
10. Remove the mounting nuts, and the blower unit.
11. Install the unit in the reverse order of removal. Make sure that there are no air leaks.

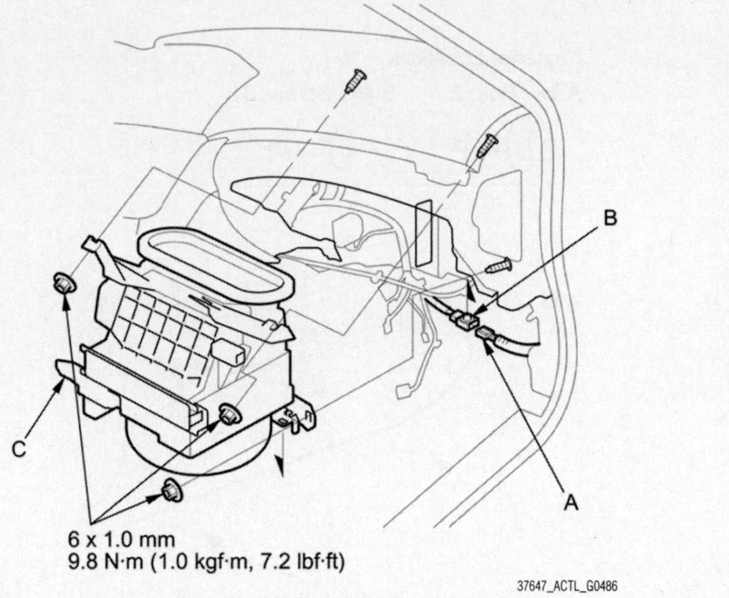
B
C
A
6 x 1.0 mm
9.8 N·m (1.0 kgf·m, 7.2 lbf·ft)
37647_ACTL_G0486

Fig. 236 Disconnect the right engine compartment harness connector (A) and harness clips (B)

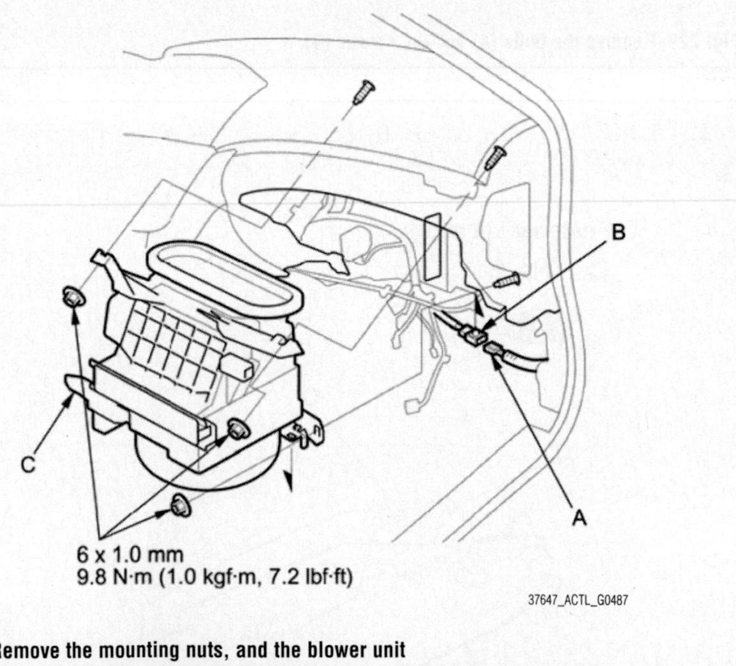
B
C
A
6 x 1.0 mm
9.8 N·m (1.0 kgf·m, 7.2 lbf·ft)
37647_ACTL_G0487

Fig. 237 Remove the mounting nuts, and the blower unit

HEATER CORE

REMOVAL AND INSTALLATION

See Figures 238 through 246.

⁂ WARNING

SRS components are located in this area. Review the SRS component locations and the precautions and procedures before doing repairs or service.

1. Do the battery terminal disconnection procedure.
2. Recover the refrigerant with a recovery/recycling/charging station.
3. Disconnect the A/C line from the evaporator core.
4. When the engine is cool, drain the engine coolant from the radiator.
5. From under the hood, slide the heater hose clamps back.
6. Disconnect the inlet heater hose and

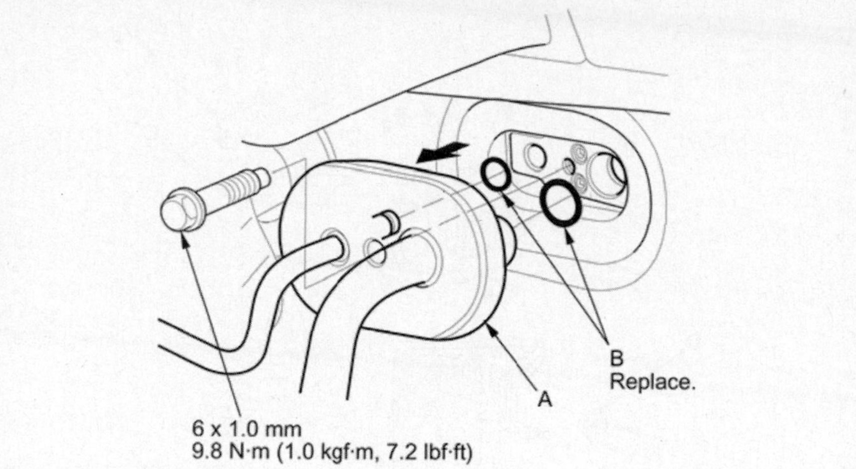

B
Replace.
A
6 x 1.0 mm
9.8 N·m (1.0 kgf·m, 7.2 lbf·ft)
37647_ACTL_G0501

Fig. 238 Disconnect the A/C line from the evaporator core

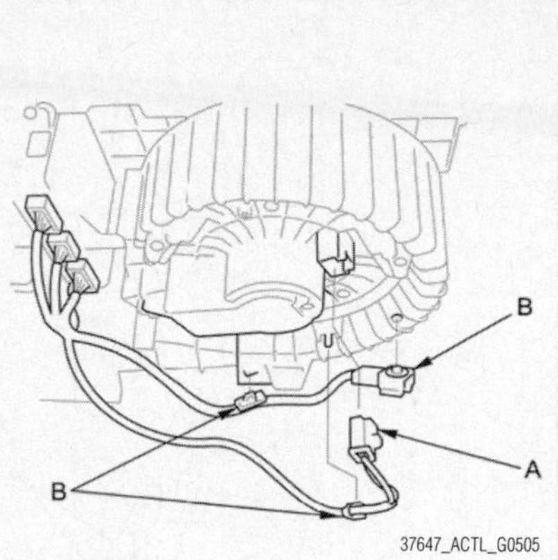

B
A
B
37647_ACTL_G0505

Fig. 242 Disconnect the connector (A) from the blower motor. Remove the wire harness clips (B).

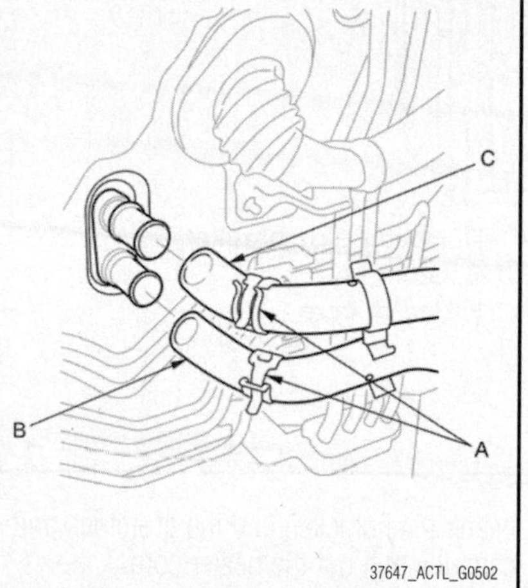

C
B
A
37647_ACTL_G0502

Fig. 239 Disconnect the inlet heater hose (B) and the outlet heater hose (C) from the heater unit

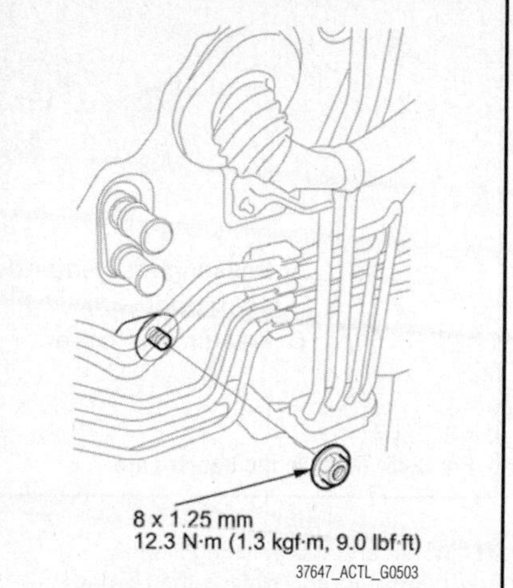

8 x 1.25 mm
12.3 N·m (1.3 kgf·m, 9.0 lbf·ft)
37647_ACTL_G0503

Fig. 240 Remove the mounting nut from the heater unit

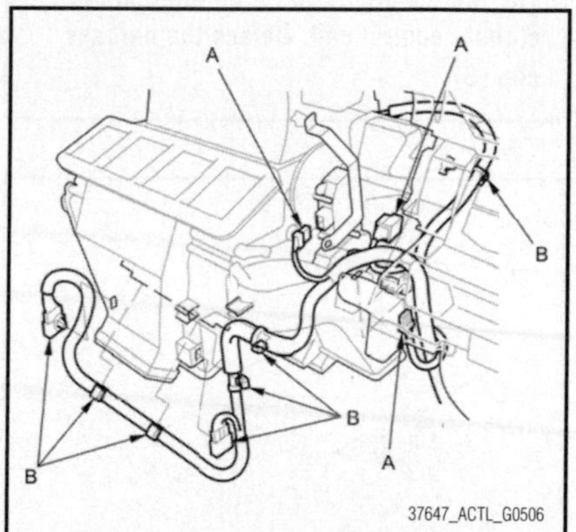

A
A
B
B
B
A
37647_ACTL_G0506

Fig. 243 Disconnect these connectors (A): The passenger's mode control motor, the power transistor, and the passenger's air mix control motor. Detach the harness clips (B)

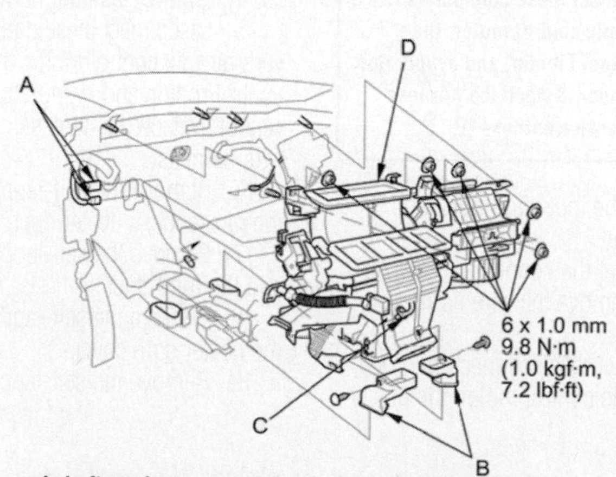

D
A
6 x 1.0 mm
9.8 N·m
(1.0 kgf·m,
7.2 lbf·ft)
C
B

A. Left engine compartment wire harness connectors
B. Ducts
C. Drain hose
D. Blower-heater unit

37647_ACTL_G0504

Fig. 241 Disconnect the left engine compartment wire harness connectors

the outlet heater hose from the heater unit. Note and/or mark the layout of the hoses.

➡Engine coolant will run out when the hoses are disconnected; drain it into a clean drip pan. Be sure not to let coolant spill on the electrical parts or the painted surfaces. If any coolant spills, rinse it off immediately.

7. Remove the mounting nut from the heater unit. Take care not to damage or bend the fuel lines or the brake lines.
8. Remove the dashboard.
9. Disconnect the left engine compartment wire harness connectors.
10. Remove the ducts, and the drain hose.

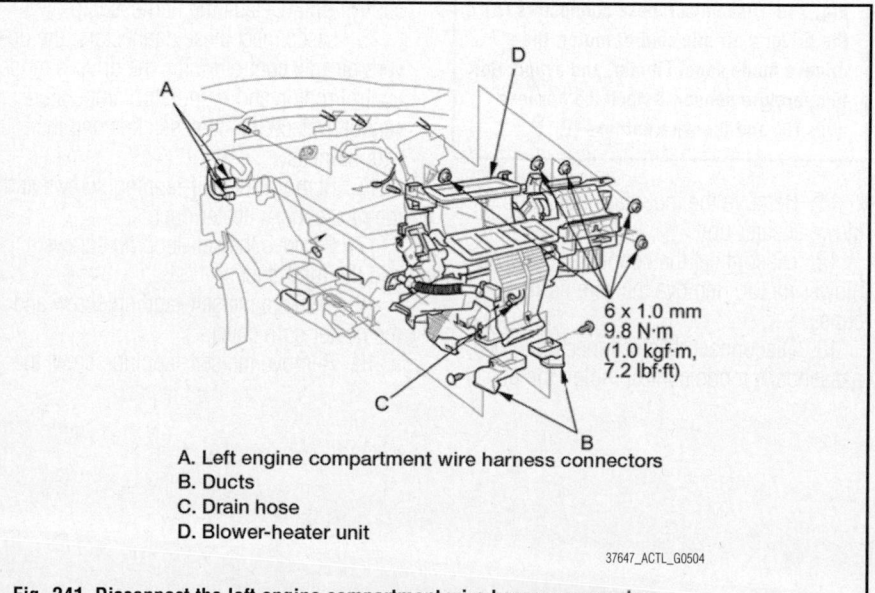

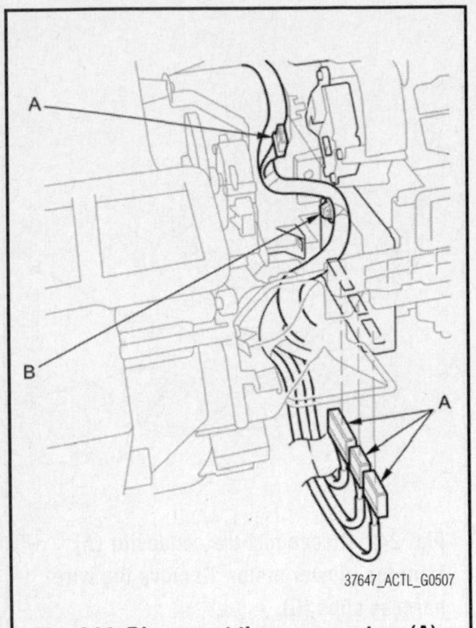
A
B
A
37647_ACTL_G0507

Fig. 244 Disconnect these connectors (A): The recirculation control motor, and the climate control unit. Detach the harness clip (B).

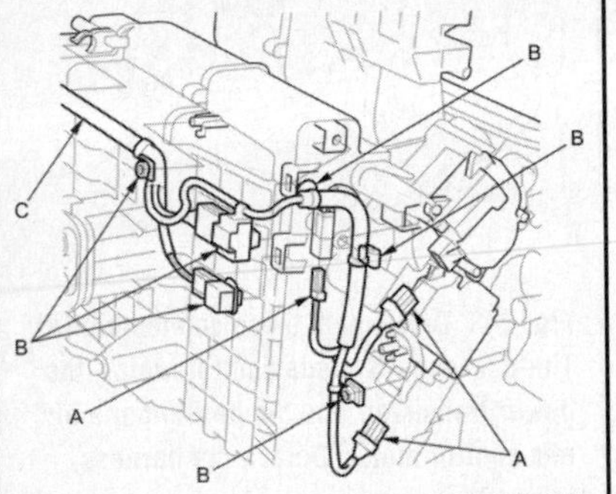
B
B
C
B
A
B
A
37647_ACTL_G0508

Fig. 245 Disconnect these connectors (A): the driver's air mix control motor, the driver's mode control motor, and evaporator temperature sensor. Detach the harness clips (B) and the wire harness (C).

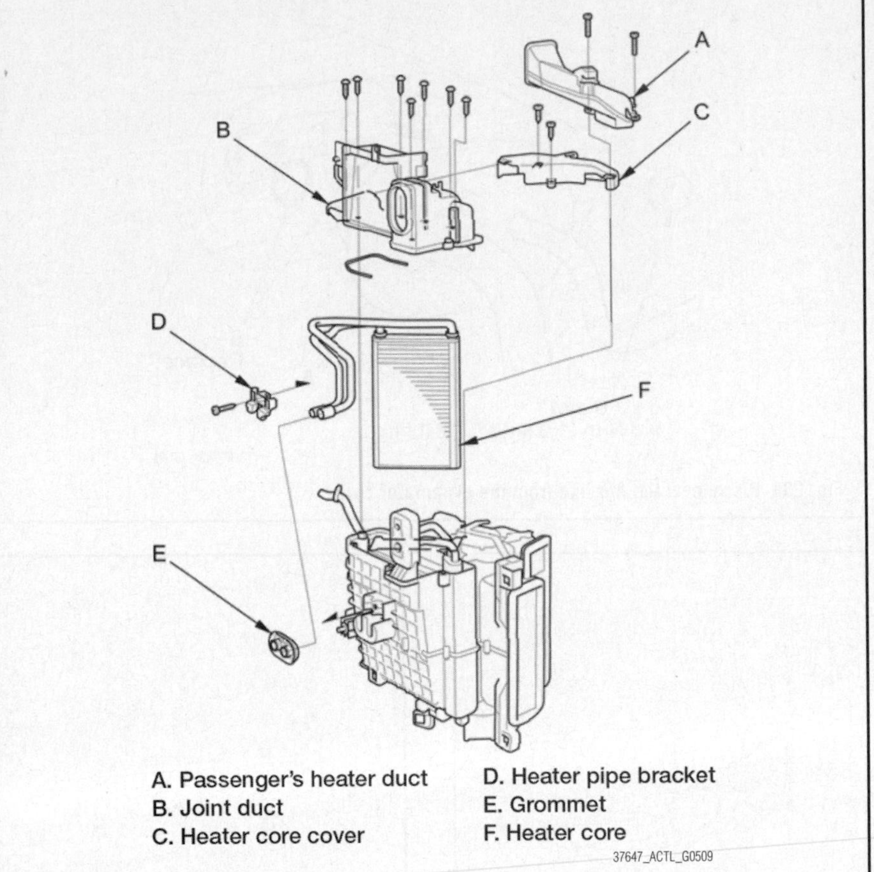
A
B
C
D
F
E
A. Passenger's heater duct
B. Joint duct
C. Heater core cover
D. Heater pipe bracket
E. Grommet
F. Heater core
37647_ACTL_G0509

Fig. 246 Remove the heater core

11. Remove the mounting nuts, and the blower-heater unit.
12. Disconnect the connector from the blower motor. Remove the wire harness clips.
13. Disconnect these connectors: The passenger's mode control motor, the power transistor, and the passenger's air mix control motor. Detach the harness clips.
14. Disconnect these connectors: The recirculation control motor, and the climate control unit. Detach the harness clip.
15. Disconnect these connectors: the driver's air mix control motor, the driver's mode control motor, and evaporator temperature sensor. Detach the harness clips and the wire harness.
16. Remove the self-tapping screws and the passenger's heater duct.
17. Remove the self-tapping screws and the joint duct.
18. Remove the self-tapping screw and the heater core cover.
19. Remove the self-tapping screw, the heater pipe bracket, and the grommet, and carefully pull out the heater core.

To install:

20. Install the heater core in the reverse order of removal.
21. Install the heater unit in the reverse order of removal, and note these items:
    - Do not interchange the inlet and outlet heater hoses, and install the hose clamps securely.
    - Refill the cooling system with engine coolant.
    - Make sure that there is no coolant leakage.
    - Make sure that there is no air leakage.
    - Charge the system.
22. Do the battery terminal reconnection procedure.

STEERING

ELECTRIC POWER STEERING (EPS) CONTROL UNIT

REMOVAL & INSTALLATION

See Figure 247.

1. Do the battery terminal disconnection procedure.
2. Remove these items:
   - Passenger's dashboard undercover
   - Passenger's front door sill trim
   - Passenger's kick panel
   - Passenger's under-dash fuse/relay box
3. Disconnect EPS control unit connector A (2P), connector B (2P), connector C (2P), and connector D (28P).
4. Remove the bolts from the EPS control unit.
5. Remove the EPS control unit.

To install:

6. Install the EPS control unit in the reverse order of removal.
7. Do the battery terminal reconnection procedure.
8. Do the torque sensor neutral position memorization.
9. After installation, start the engine, allow it to idle, and turn the steering wheel from lock to lock several times. Check that the EPS indicator does not come on.

TORQUE SENSOR NEUTRAL POSITION MEMORIZATION

The torque sensor neutral position must be memorized whenever the steering gearbox or the EPS control unit is replaced. Note that the torque sensor neutral position is not affected when erasing the DTC.

➡The torque sensor is temperature sensitive. When memorizing the torque sensor neutral position, the ambient temperature must be above 68°F (20°C).

1. With the ignition switch in LOCK (0) or with the vehicle ignition in the OFF mode, connect the HDS to the Data Link Connector (DLC) located under the driver's side of the dashboard.
2. Turn the ignition switch to ON (II), or press the engine start/stop button to select the ON mode.
3. Make sure the HDS communicates with the vehicle and the EPS control unit. If it doesn't, troubleshoot the DLC circuit.
4. From the EPS MENU, select MISCELLANEOUS TEST, then TORQUE SENSOR LEARN, and follow the screen prompts on the HDS.

➡See the HDS Help menu for specific instructions.

5. Turn the ignition switch to LOCK (0), or press the engine start/stop button to select the OFF mode.

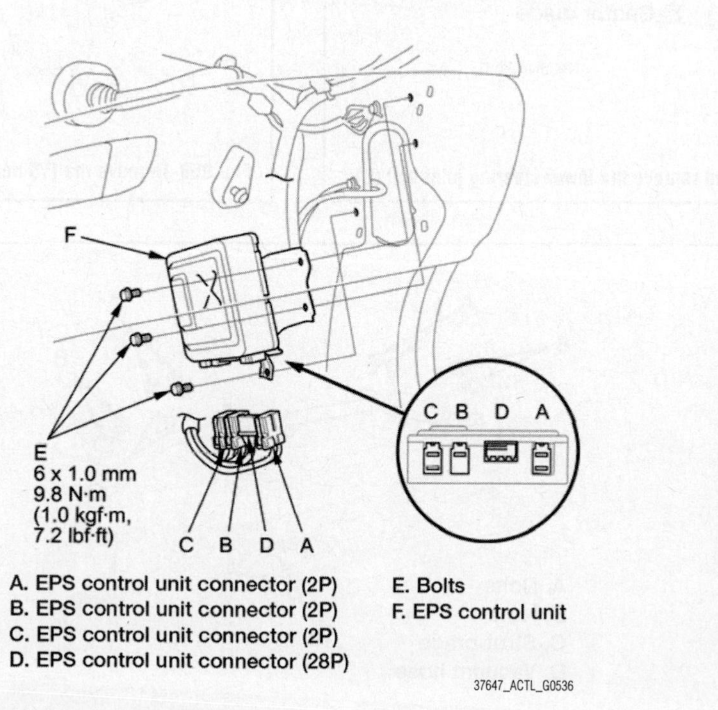
F
E
6 x 1.0 mm
9.8 N·m
(1.0 kgf·m,
7.2 lbf·ft)
C B D A
C B D A
A. EPS control unit connector (2P)
B. EPS control unit connector (2P)
C. EPS control unit connector (2P)
D. EPS control unit connector (28P)
E. Bolts
F. EPS control unit
37647_ACTL_G0536

Fig. 247 Disconnect EPS control unit connector A , connector B, connector C, and connector D

POWER RACK & PINION STEERING GEAR

REMOVAL & INSTALLATION

See Figures 248 through 277.

WARNING

SRS components are located in this area. Review the SRS component locations and the precautions and procedures before doing repairs or service.

➡Note these items during removal:

- Use solvent and a brush, wash any oil and dirt off the end of the steering gearbox. Avoid any electrical parts. Blow dry with compressed air.
- Be sure to remove the steering wheel before disconnecting the steering joint, or damage to the cable reel can occur.

1. Do the battery terminal disconnection procedure.
2. Raise and support the vehicle.
3. Remove the front wheels.
4. Remove the driver's airbag, and the steering wheel.
5. Remove the steering joint cover.
6. Loosen the upper steering joint bolt, and remove the lower steering joint bolt.
7. Disconnect the steering joint by sliding the steering joint into the column shaft. Tighten the upper steering joint bolt to hold the steering joint in place.

➡Do not disconnect the steering joint from the column shaft.

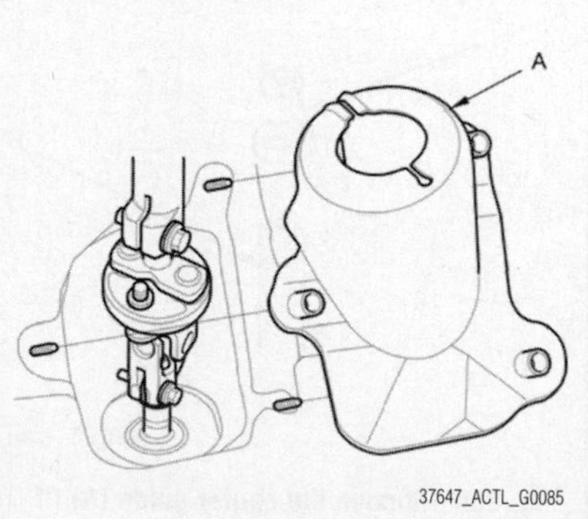
A
37647_ACTL_G0085

Fig. 248 Remove the steering joint cover

8. Remove the center guide (if equipped) from the top of the pinion shaft, and discard it.

9. Apply vinyl tape to the splines on the pinion shaft.

10. Disconnect the hood support struts from both sides of the pivot ball (bolted to the hood). Secure the hood in a vertical position. Remove the right side pivot ball and install it into the lower threaded hole, then reattach the support strut.

➡Do not attempt to close the hood with the support strut in the vertical position, as it will damage the support strut and hood.

11. Remove the left and right engine compartment covers, the strut brace cover, and the front bulkhead cover, then remove the cowl cover.

12. Remove the strut brace.

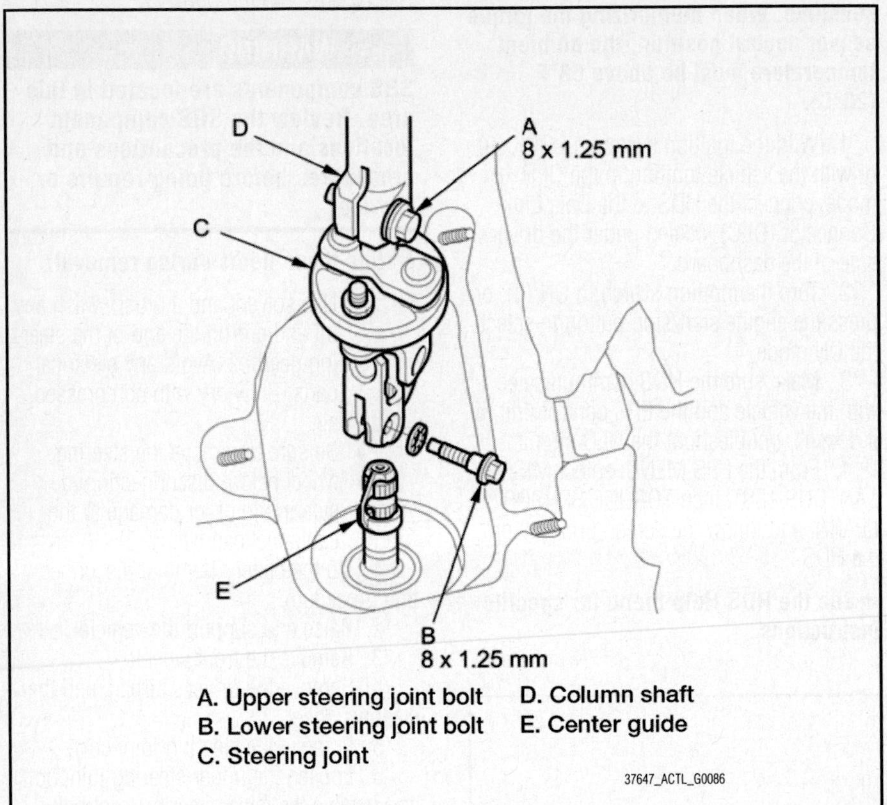
A
8 x 1.25 mm
D
C
E
B
8 x 1.25 mm
A. Upper steering joint bolt
B. Lower steering joint bolt
C. Steering joint
D. Column shaft
E. Center guide
37647_ACTL_G0086

Fig. 249 Loosen the upper steering joint bolt (A), and remove the lower steering joint bolt (B)

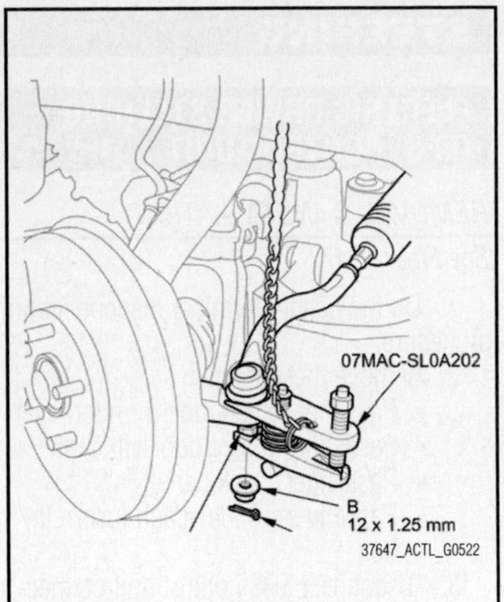
07MAC-SL0A202
A
B
12 x 1.25 mm
37647_ACTL_G0522

Fig. 252 Remove the cotter pin (A) from the tie-rod end ball joint, then remove the nut (B)

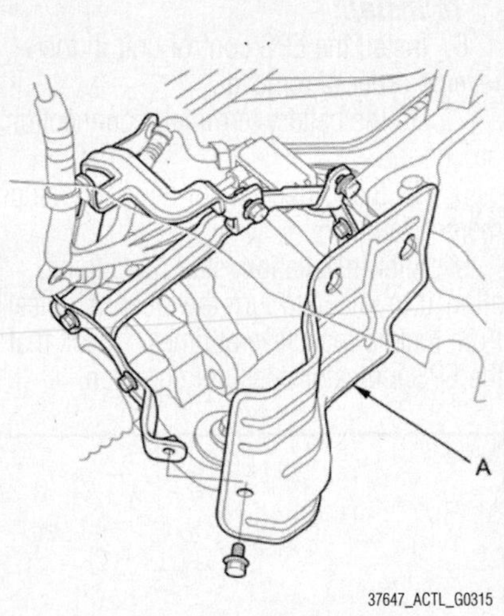
A
37647_ACTL_G0315

Fig. 253 Remove the P/S heat shield

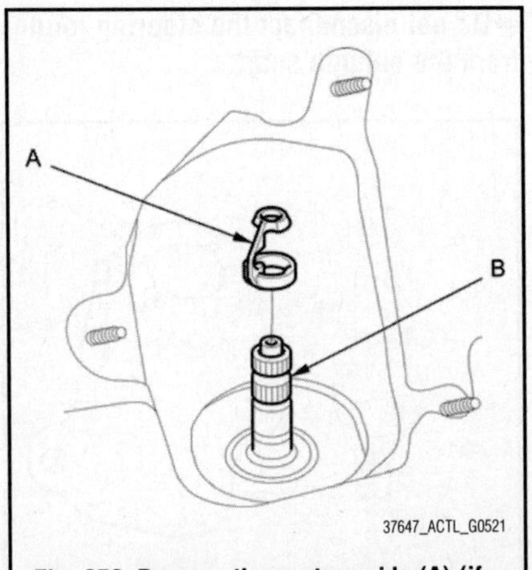
A
B
37647_ACTL_G0521

Fig. 250 Remove the center guide (A) (if equipped) from the top of the pinion shaft (B), and discard it

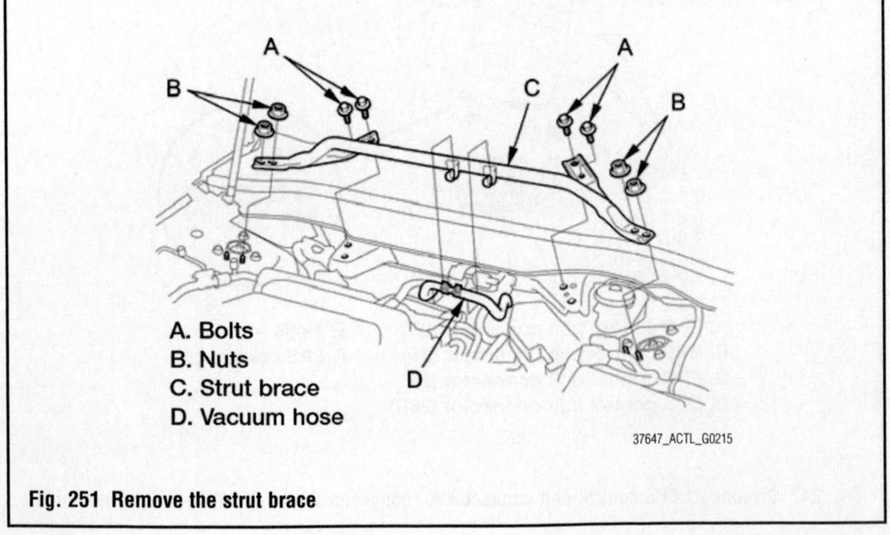
A
B
A
C
B
D
A. Bolts
B. Nuts
C. Strut brace
D. Vacuum hose
37647_ACTL_G0215

Fig. 251 Remove the strut brace

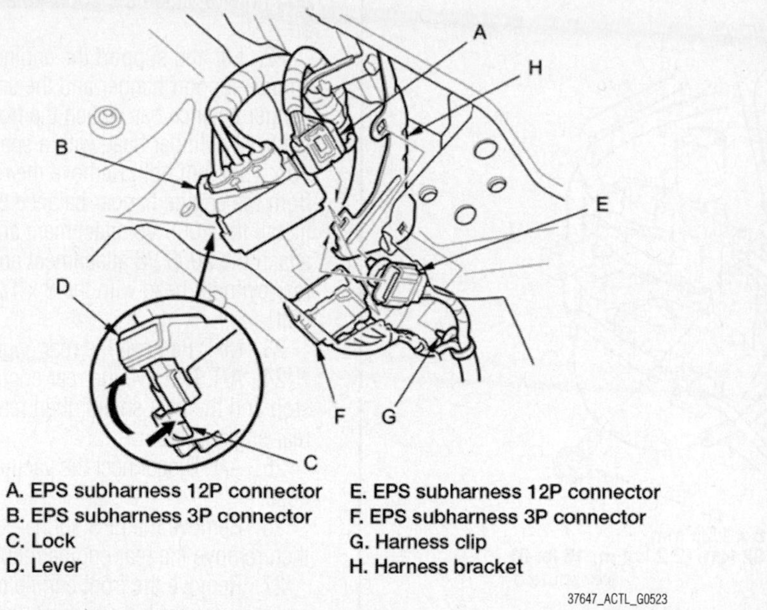
A
H
B
E
D
F G
C
A. EPS subharness 12P connector
B. EPS subharness 3P connector
C. Lock
D. Lever
E. EPS subharness 12P connector
F. EPS subharness 3P connector
G. Harness clip
H. Harness bracket
37647_ACTL_G0523

Fig. 254 Disconnect the EPS subharness 12P connector and the EPS subharness 3P connector by pushing the lock and pulling up the lever

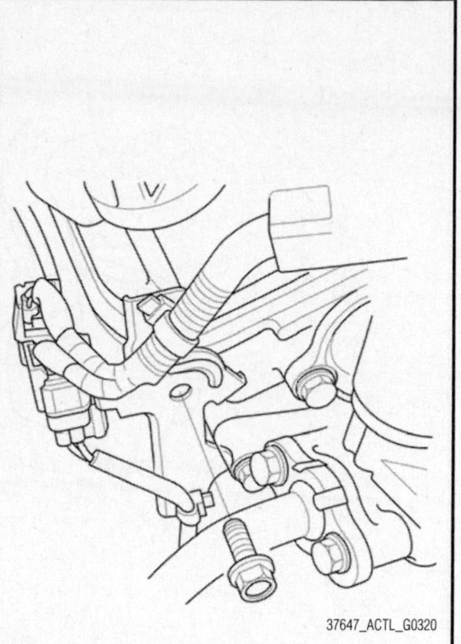
37647_ACTL_G0320

Fig. 257 Remove the connector bracket (A) from the front cylinder head

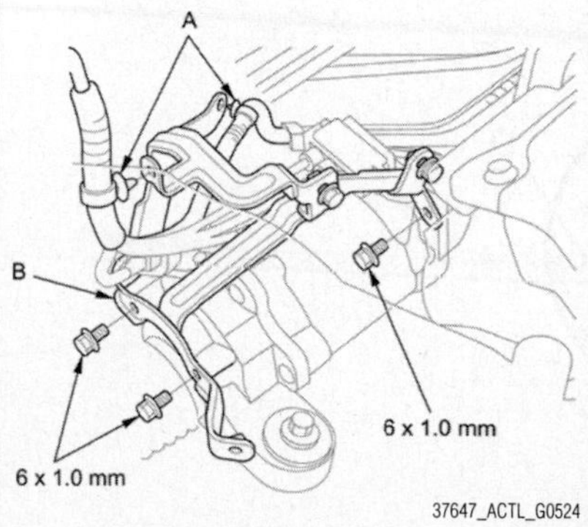
A
B
6 x 1.0 mm
6 x 1.0 mm
37647_ACTL_G0524

Fig. 255 Remove the harness clips (A), then remove the heat shield bracket (B) from the steering gearbox

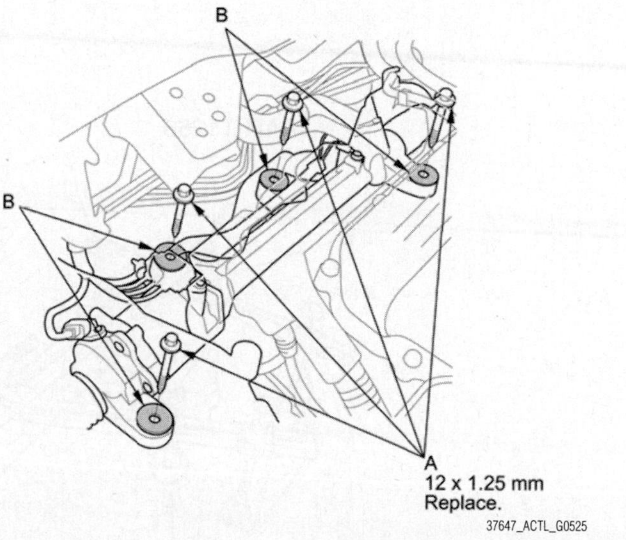
B
B
A
12 x 1.25 mm
Replace.
37647_ACTL_G0525

Fig. 256 Remove the steering gearbox mounting bolts (A) and washers (B) from the steering gearbox

13. Remove the cotter pin from the tie-rod end ball joint, then remove the nut on both sides.

14. Disconnect the tie-rod end ball joint from the knuckle using the ball joint thread protector and the ball joint remover.

➡Be careful not to damage the ball joint boot when installing the remover.

15. Remove the P/S heat shield.

16. Disconnect the EPS subharness 12P connector and the EPS subharness 3P connector by pushing the lock and pulling up the lever. Wrap the connectors with vinyl tape to avoid contamination from grease or water.

17. Remove the EPS subharness 12P connector, the EPS subharness 3P connector, and the harness clip from the harness bracket.

18. Remove the harness clips, then remove the heat shield bracket from the steering gearbox.

19. Remove the steering gearbox mounting bolts and washers from the steering gearbox.

20. Remove the connector bracket from the front cylinder head; use the bracket bolt hole to attach the engine hanger balance bar front arm.

21. Disconnect the vacuum hose, and remove the harness clamp, then remove the engine mount control solenoid valve bracket

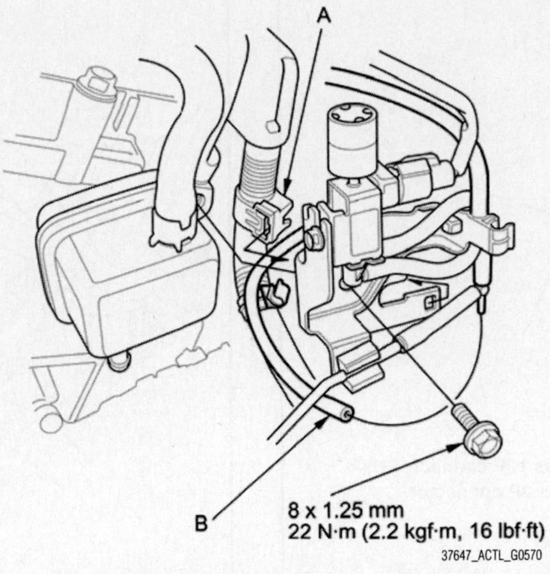
A
B
8 x 1.25 mm
22 N·m (2.2 kgf·m, 16 lbf·ft)
37647_ACTL_G0570

Fig. 258 Disconnect the vacuum hose, and remove the harness clamp, then remove the engine mount control solenoid valve bracket from the rear cylinder head

from the rear cylinder head; use the bracket bolt hole to attach the 2008 V6 attachment arm.

22. Lift and support the engine with the engine support hanger and the engine hanger balance bar. Attach the front arm to the front cylinder head with a spacer and a 10 x 1.25 mm bolt. Remove the rear arm from the engine hanger balance bar, and install the 2008 V6 attachment arm, then attach the 2008 V6 attachment arm to the rear cylinder head with the 8 x 1.25 mm bolt.

23. M/T: Remove the rear engine mount.

24. A/T: Remove the rear engine mount stop and the heat shield, then remove the rear engine mount.

25. A/T: Disconnect the vacuum hose.

26. Remove the EPS subharness holder, then remove the rear engine mount base.

27. Remove the front engine mount stop and the vacuum hose bracket, then remove the front engine mount bolt.

28. M/T: Remove the front engine mount mounting bolts, then remove the front

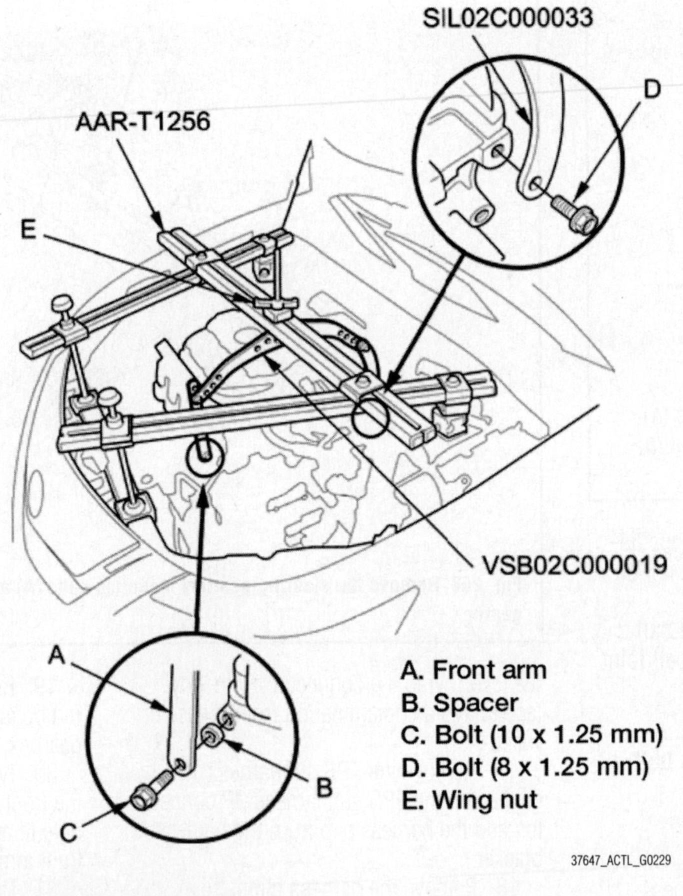
SIL02C000033
D
AAR-T1256
E
VSB02C000019
A
B
C
A. Front arm
B. Spacer
C. Bolt (10 x 1.25 mm)
D. Bolt (8 x 1.25 mm)
E. Wing nut
37647_ACTL_G0229

Fig. 259 Lift and support the engine with the engine support hanger and the engine hanger balance bar

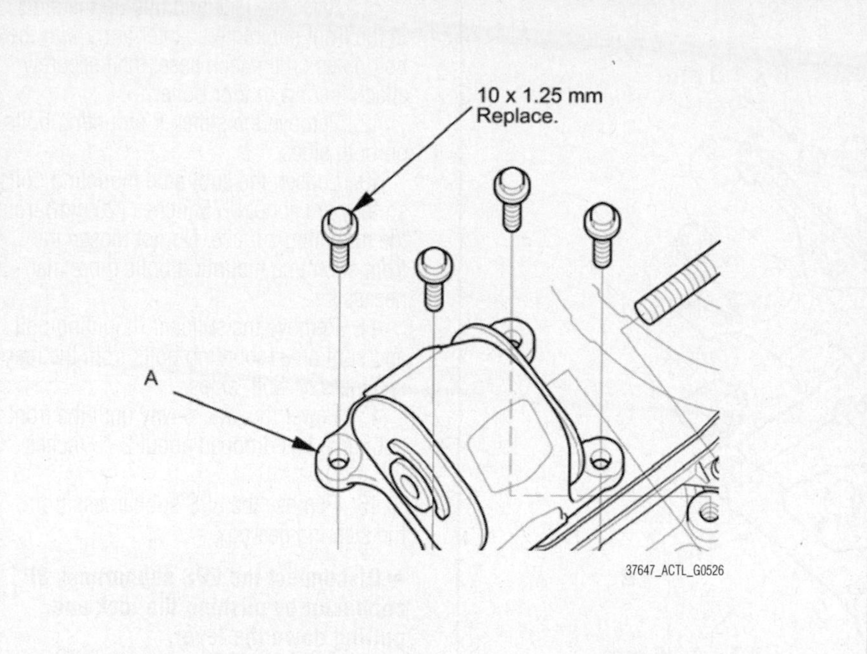
10 x 1.25 mm
Replace.
A
37647_ACTL_G0526

Fig. 260 M/T: Remove the rear engine mount (A)

engine mount, and disconnect the vacuum hose.

29. Raise the vehicle on a lift.
30. Remove the front splash shield.
31. Remove the front subframe stiffener.
32. Remove exhaust pipe A.
33. With SH-AWD: Remove the propeller shaft.
34. M/T: Remove the transfer assembly.
35. Remove the cotter pin from the knuckle ball joint, then remove the castle nut.
36. Disconnect the knuckle ball joint from the lower arm using the ball joint thread protector and the ball joint remover.

➡Be careful not to damage the ball joint boot when installing the remover. Do not force or hammer on the lower arm, or pry between the lower arm and the knuckle. You could damage the ball joint.

37. Remove the damper fork mounting nut while holding the mounting bolt, then remove the mounting bolt.

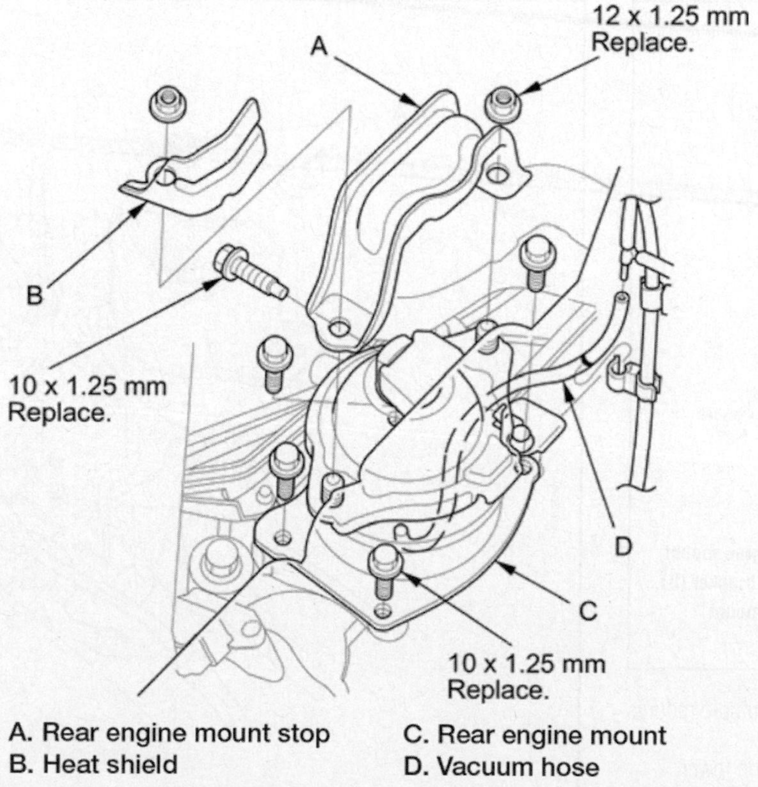
12 x 1.25 mm
Replace.
A
B
10 x 1.25 mm
Replace.
D
C
10 x 1.25 mm
Replace.
A. Rear engine mount stop
B. Heat shield
C. Rear engine mount
D. Vacuum hose
37647_ACTL_G0527

Fig. 261 A/T: Remove the rear engine mount stop and the heat shield, then remove the rear engine mount

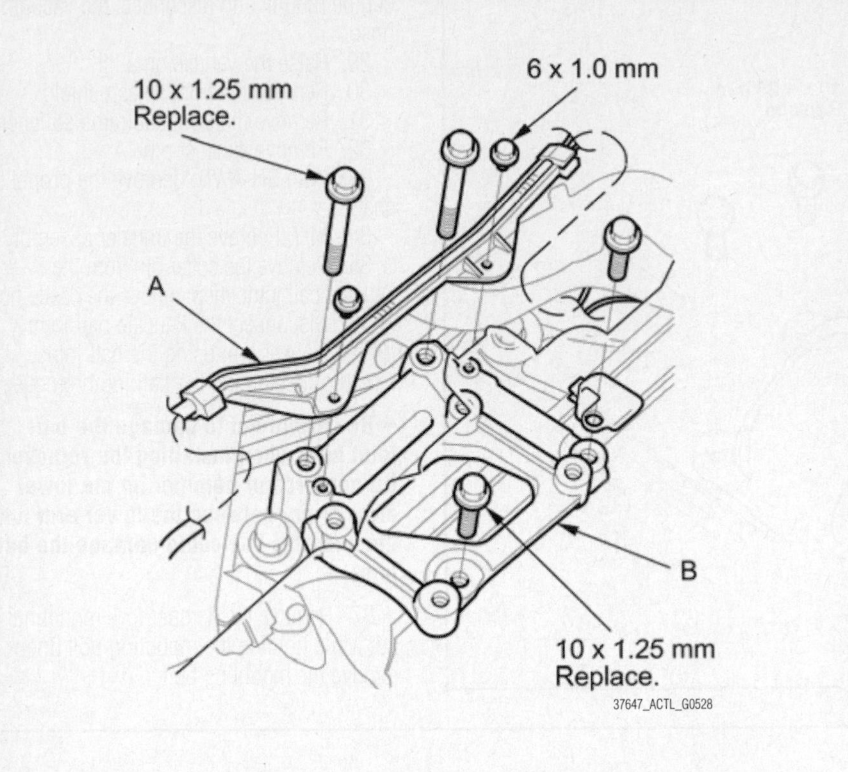
10 x 1.25 mm
Replace.
6 x 1.0 mm
A
B
10 x 1.25 mm
Replace.
37647_ACTL_G0528

Fig. 262 Remove the EPS subharness holder (A), then remove the rear engine mount base (B)

41. Raise the jack and line up the slots in the front subframe adapter arms with the bolt holes on the jack base, then securely attach them with four bolts.

42. Remove the stiffener mounting bolts on both sides.

43. Loosen the subframe mounting bolts so they are about 0.75 inches (20 mm) from the mounting surface. Do not loosen the front subframe mounting bolts more than necessary.

44. Remove the stiffener mounting bolt and subframe mounting bolts from the rear stiffeners on both sides.

45. Lower the jack slowly until the front subframe has dropped about 2.75 inches (70 mm).

46. Remove the EPS subharness from the steering gearbox.

➡Disconnect the EPS subharness 3P connector by pushing the lock and pulling down the lever.

47. Carefully move the steering gearbox toward the driver's side until the pinion shaft clears the fenderwell opening on the body.

48. Remove the steering gearbox

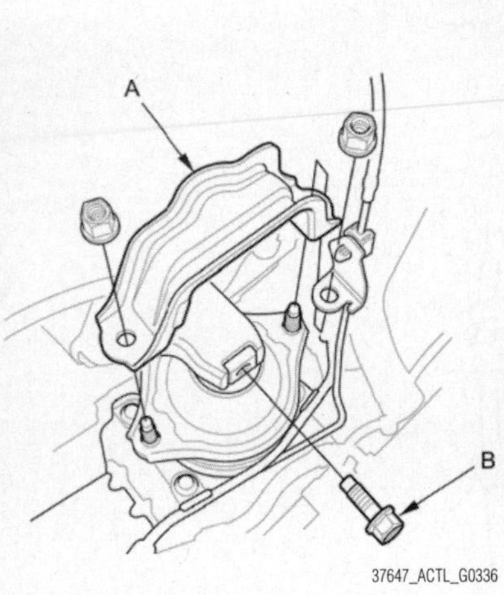
A
B
37647_ACTL_G0336

Fig. 263 Remove the front engine mount stop (A) and the vacuum hose bracket (B), then remove the front engine mount bolt (C)

38. Remove the subframe middle mounts on both sides.

39. Remove the transmission lower mount nuts.

40. Attach the front subframe adapter to the subframe by hanging the belt over the front of the subframe, then secure the belt with its stop.

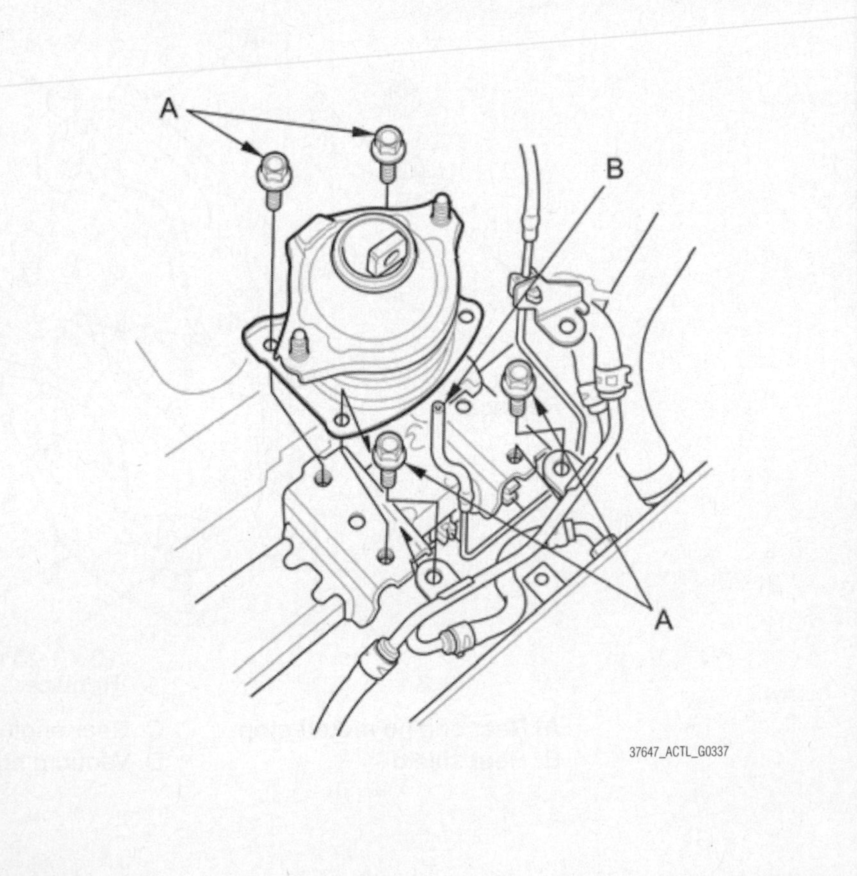
A
B
A
37647_ACTL_G0337

Fig. 264 M/T: Remove the front engine mount mounting bolts (A), then remove the front engine mount (B), and disconnect the vacuum hose (C)

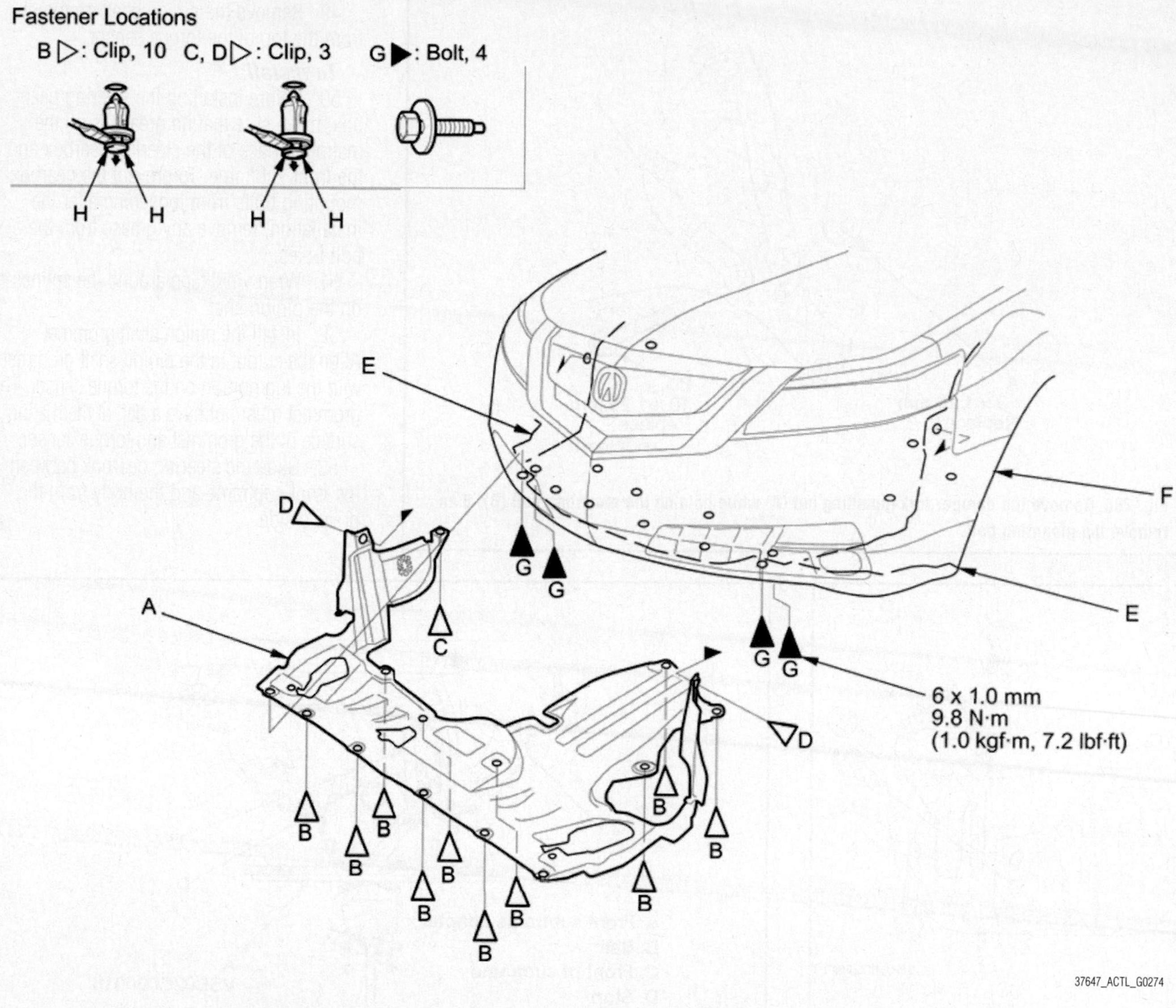
Fastener Locations
B ▷: Clip, 10 C, D ▷: Clip, 3 G ▶: Bolt, 4
H
H
H
H
E
F
D
A
C
G
G
E
G
G
6 x 1.0 mm
9.8 N·m
(1.0 kgf·m, 7.2 lbf·ft)
D
B
37647_ACTL_G0274

Fig. 265 Remove the front splash shield

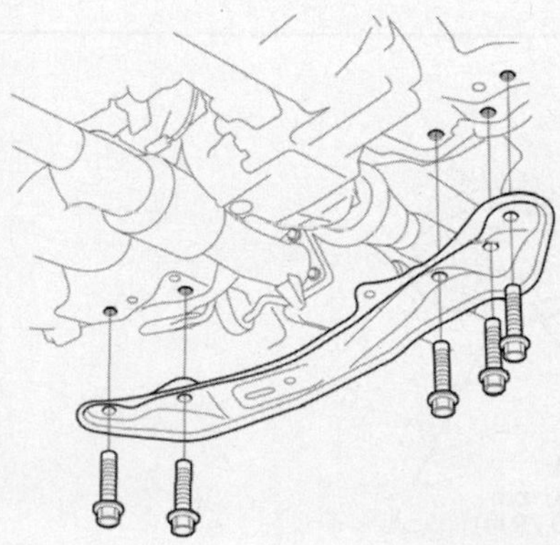
37647_ACTL_G0234

Fig. 266 Remove the front subframe stiffener

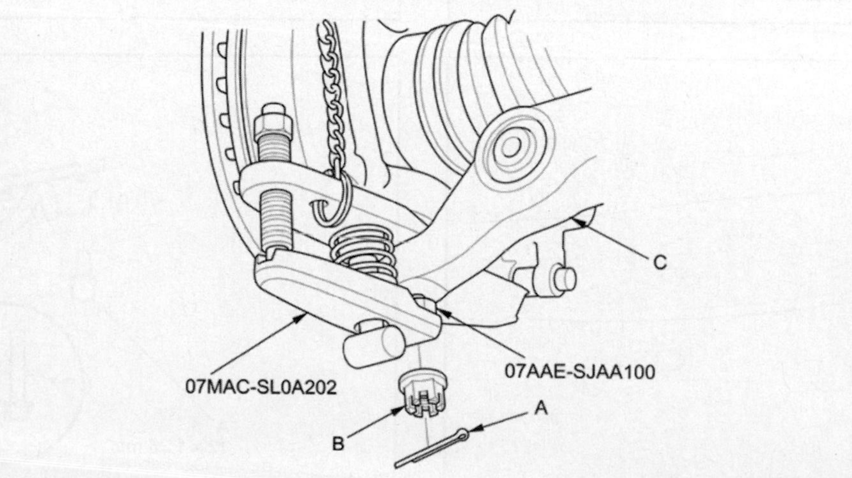
C
07MAC-SL0A202
07AAE-SJAA100
A
B
37647_ACTL_G0255

Fig. 267 Remove the cotter pin (A) from the knuckle ball joint, then remove the castle nut (B)

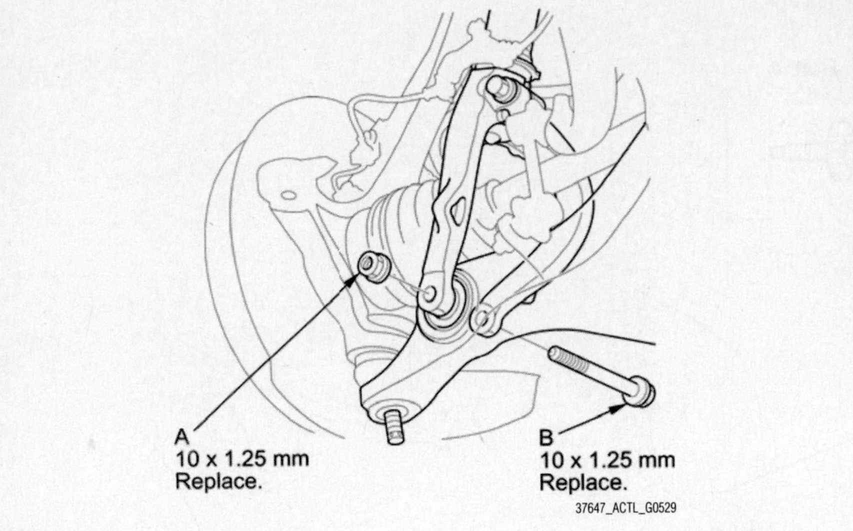
A
10 x 1.25 mm
Replace.
B
10 x 1.25 mm
Replace.
37647_ACTL_G0529

Fig. 268 Remove the damper fork mounting nut (A) while holding the mounting bolt (B), then remove the mounting bolt

through the fenderwell opening on the driver's side.

49. Remove the pinion shaft grommet from the top of the torque sensor.

To install:

50. Before installing the steering gearbox, make sure that no grease is on the mating surface of the steering gearbox and the front subframe. To prevent the gearbox mounting bolts from loosening after the installation, remove any grease from the bolt holes.

51. Wrap vinyl tape around the splines on the pinion shaft.

52. Install the pinion shaft grommet. Align the cutout in the pinion shaft grommet with the lug portion on the torque sensor. The grommet must not have a gap at the mating surface of the grommet and torque sensor.

53. Slide the steering gearbox between the front subframe and the body from the driver's side.

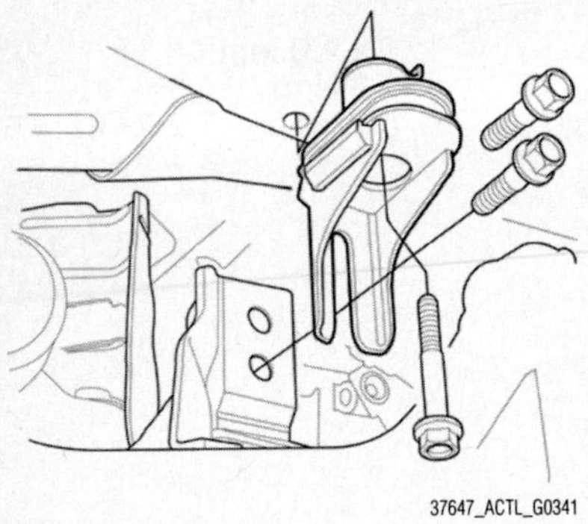
37647_ACTL_G0341

Fig. 269 Remove the subframe middle mounts

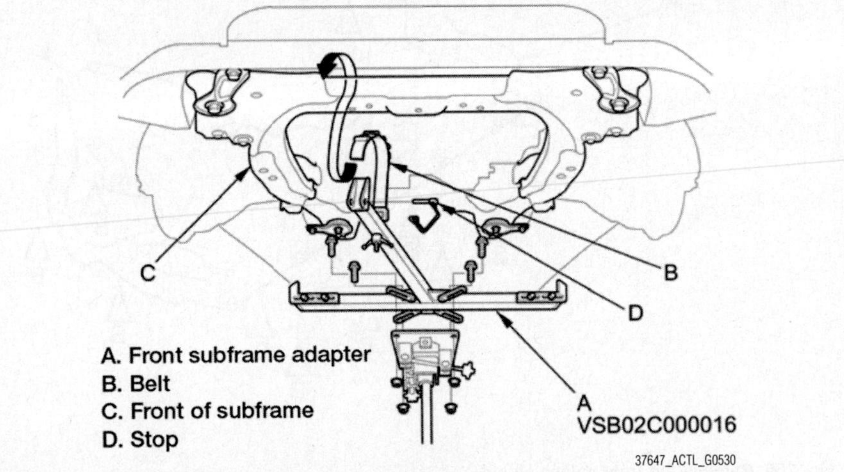
C
B
D
A. Front subframe adapter
B. Belt
C. Front of subframe
D. Stop
A
VSB02C000016
37647_ACTL_G0530

Fig. 271 Attach the front subframe adapter to the subframe by hanging the belt over the front of the subframe, then secure the belt with its stop

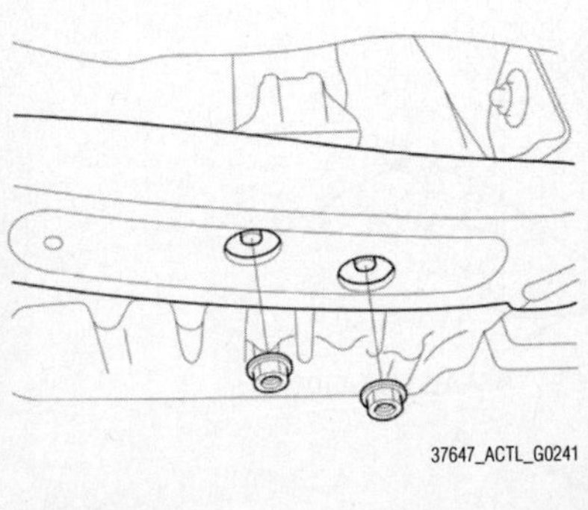
37647_ACTL_G0241

Fig. 270 Remove the transmission lower mount nuts

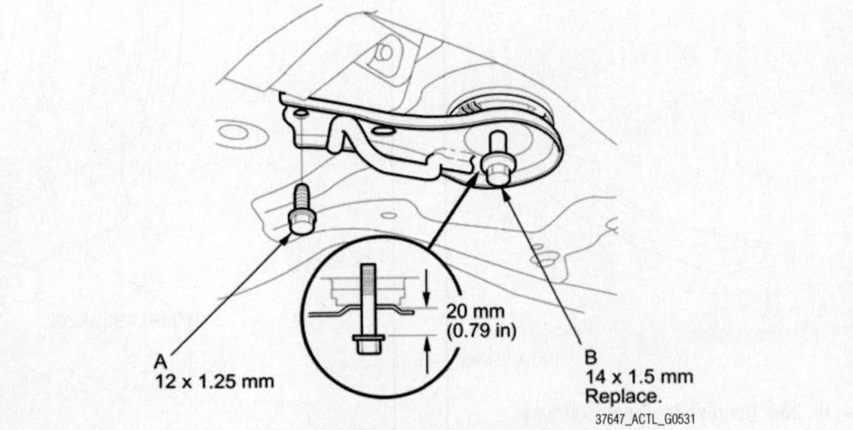
20 mm
(0.79 in)
A
12 x 1.25 mm
B
14 x 1.5 mm
Replace.
37647_ACTL_G0531

Fig. 272 Remove the stiffener mounting bolts (A) on both sides

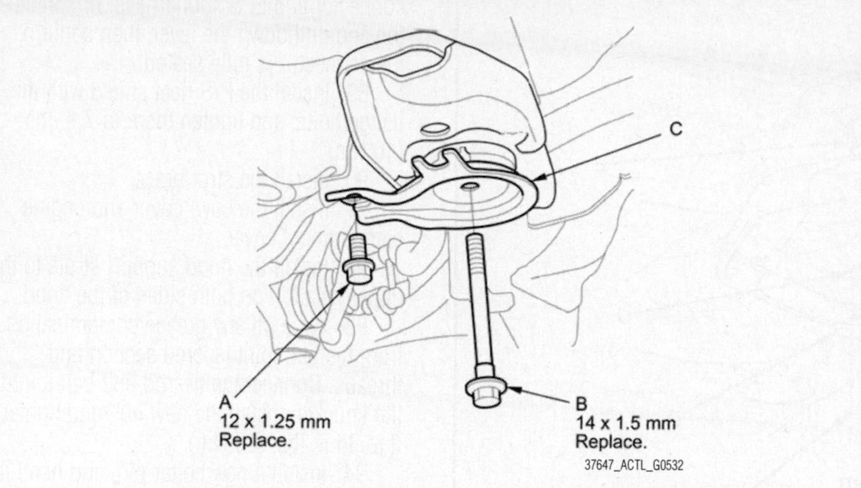
C
A
12 x 1.25 mm
Replace.
B
14 x 1.5 mm
Replace.
37647_ACTL_G0532

Fig. 273 Remove the stiffener mounting bolt (A) and subframe mounting bolts (B) from the rear stiffeners (C)

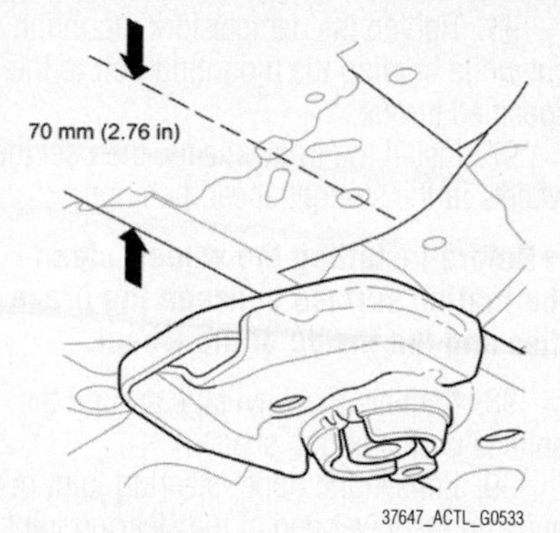
70 mm (2.76 in)
37647_ACTL_G0533

Fig. 274 Lower the jack slowly until the front subframe has dropped about 2.75 inches (70 mm)

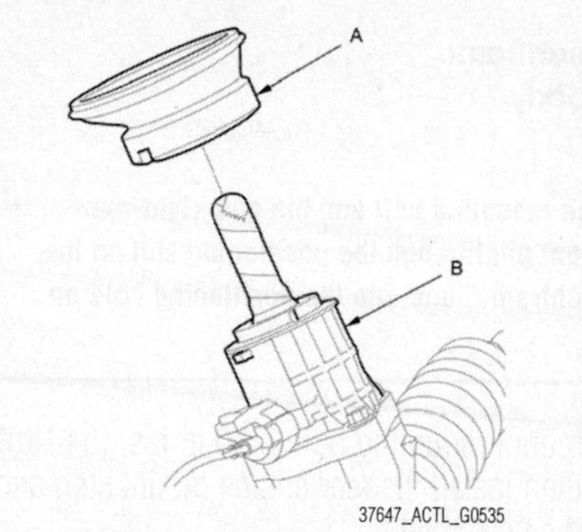
A
B
37647_ACTL_G0535

Fig. 276 Remove the pinion shaft grommet (A) from the top of the torque sensor (B)

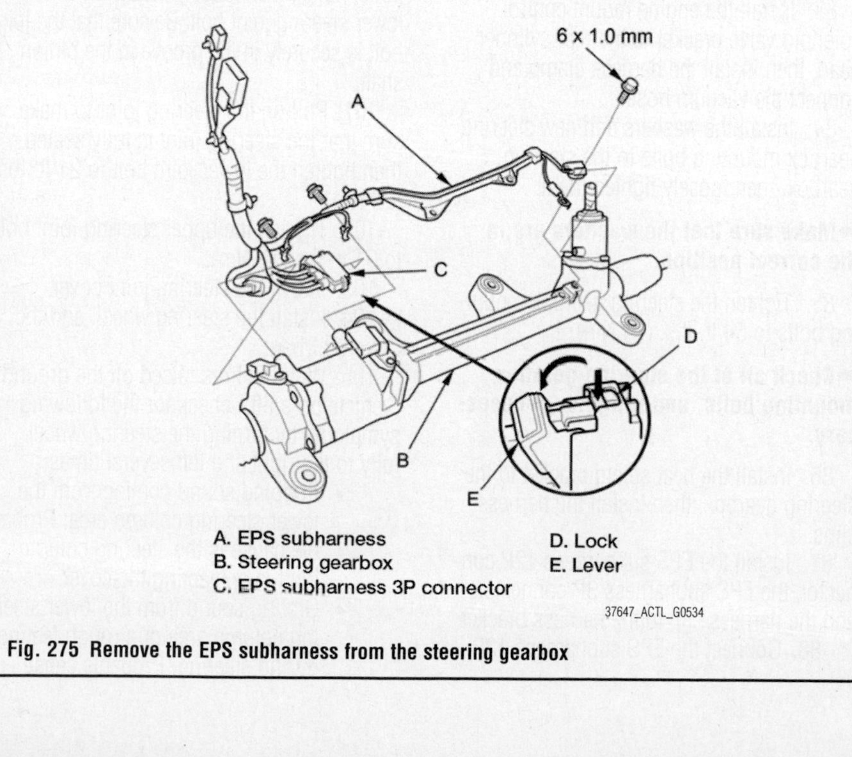
6 x 1.0 mm
A
C
D
B
E
A. EPS subharness
B. Steering gearbox
C. EPS subharness 3P connector
D. Lock
E. Lever
37647_ACTL_G0534

Fig. 275 Remove the EPS subharness from the steering gearbox

54. Carefully move the steering gearbox toward the passenger's side until the pinion shaft clears the wheel well opening on the body.

55. Continue moving the steering gearbox toward the passenger's side until the steering gearbox is in position.

56. Install the EPS subharness to the steering gearbox.

➡Pull up the lever of the EPS motor 3P connector, then confirm the connector is fully seated.

57. Loosely tighten the new right-rear subframe mounting bolt and the new stiffener mounting bolt; insert the 15.7 mm side of the subframe alignment pin through the positioning slot on the rear stiffener, through the positioning hole on the subframe, and into the positioning hole on the body.

58. Loosely tighten the new left-rear subframe mounting bolt and the new stiffener mounting bolt with the same procedure as the right rear using the subframe alignment pin.

59. Loosely tighten the new front side subframe mounting bolt and the stiffener mounting bolt on both sides.

60. Tighten the right-rear subframe mounting bolt to 76 ft. lbs. (103 Nm).

61. Tighten the left-rear subframe mounting bolt to 76 ft. lbs. (103 Nm).

62. Tighten the front subframe mounting bolts to 76 ft. lbs. (103 Nm).

➡Check all of the subframe mounting bolts, and retighten if necessary.

63. Tighten the stiffener mounting bolts to 69 ft. lbs. (93 Nm).

➡Before tightening the stiffener mounting bolts, check that the positioning holes and slot are aligned using the subframe alignment pin.

64. Install the transmission lower mount nuts), and tighten them to 33 ft. lbs. (44 Nm).

65. Install the subframe middle mounts on both sides with new mounting bolts, and tighten the long bolt to 33 ft. lbs. (44 Nm). Tighten the two short bolts to 36 ft. lbs. (49 Nm).

66. Remove the transmission jack supporting the front subframe.

67. Install the new damper fork mounting bolt and new mounting nut, and loosely tighten the nut while holding the mounting bolt.

68. Tighten the new castle nut to 69–76 ft. lbs. (93–103 Nm), then tighten it only far enough to align the slot with the ball

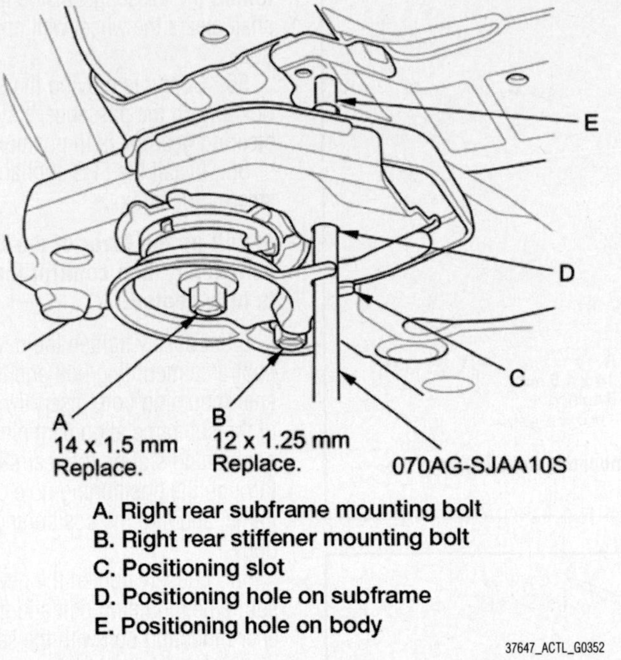
E
D
C
A
14 x 1.5 mm
Replace.
B
12 x 1.25 mm
Replace.
070AG-SJAA10S
A. Right rear subframe mounting bolt
B. Right rear stiffener mounting bolt
C. Positioning slot
D. Positioning hole on subframe
E. Positioning hole on body
37647_ACTL_G0352

Fig. 277 Loosely install the new right-rear subframe mounting bolt and the new right-rear stiffener mounting bolt; insert the subframe alignment pin through the positioning slot on the rear stiffener, through the positioning hole on the subframe, and into the positioning hole on the body

joint pin hole. Do not align the castle nut by loosening it.

➡**Insert the new cotter pin into the ball joint pin hole from the front to the rear of the vehicle, and bend its end. Check the ball joint pin hole direction before connecting the ball joint.**

69. M/T: Install the transfer assembly.
70. With SH-AWD: Install the propeller shaft.
71. Install exhaust pipe A.
72. Install the front subframe stiffener. Tighten the bolts to 40 ft. lbs. (54 Nm).
73. Install the front splash shield.
74. Lower the vehicle.
75. M/T: Connect the vacuum hose, then align the front engine mount with its mounting holes, and install new front engine mount mounting bolts.
76. Tighten the new front engine mount bolt to 40 ft. lbs. (54 Nm), then install the front engine mount stop and the vacuum hose bracket tightening the nuts to 47 ft. lbs. (64 Nm).
77. Install the rear engine mount base, then install the EPS subharness.
78. M/T: Tighten the new rear engine mount mounting bolts to 40 ft. lbs. (54 Nm) in a crisscross sequence.
79. A/T: Tighten the new rear engine mount mounting bolt to 40 ft. lbs. (54 Nm), then install the rear engine mount stop and the heat shield.
80. A/T: Connect the vacuum hose.
81. Remove the engine support hanger and the engine hanger balance bar.
82. Install the connector bracket to the front cylinder head.
83. Install the engine mount control solenoid valve bracket to the rear cylinder head, then install the harness clamp and connect the vacuum hose.
84. Install the washers and new steering gearbox mounting bolts to the steering gearbox, then loosely tighten them.

➡**Make sure that the washers are in the correct position.**

85. Tighten the steering gearbox mounting bolts to 54 ft. lbs. (74 Nm).

➡**Check all of the steering gearbox mounting bolts, and retighten if necessary.**

86. Install the heat shield bracket to the steering gearbox, then install the harness clips.
87. Install the EPS subharness 12P connector, the EPS subharness 3P connector, and the harness clip to the harness bracket.
88. Connect the EPS subharness 12P connector and EPS subharness 3P connector, and pull down the lever, then confirm the connector is fully seated.
89. Install the P/S heat shield with the flange bolts, and tighten them to 7 ft. lbs. (10 Nm).
90. Install the strut brace.
91. Install the cowl cover and engine compartment cover.
92. Install the hood support struts to the upper location on both sides of the hood.
93. Wipe off any grease contamination from the ball joint tapered section and threads. Connect the tie-rod end ball joint to the knuckle. Install the new nut, and tighten it to 44 ft. lbs. (60 Nm).
94. Install a new cotter pin, and bend it.
95. Place a floor jack under the lower arm, and raise the suspension to load it with the vehicle's weight. Do not place the jack against the ball joint pin of the knuckle.
96. Tighten the damper fork mounting nut while holding the mounting bolt to the specified torque.
97. Install the front wheels, then set the wheels in the straight ahead position.

➡**Before installing the wheel, clean the mating surface between the brake disc and the inside of the wheel.**

98. Remove the vinyl tape around the splines on the pinion shaft.
99. Loosen the upper steering joint bolt, and slip the lower end of the steering joint onto the pinion shaft taking care to align the gap within the angle.
100. Align the bolt hole on the steering joint with the groove around the pinion shaft, then loosely install the lower steering joint bolt. Be sure that the joint bolt is securely in the groove in the pinion shaft.
101. Pull on the steering joint to make sure that the steering joint is fully seated, then tighten the lower joint bolt to 21 ft. lbs (28 Nm).
102. Tighten the upper steering joint bolt to 21 ft. lbs (28 Nm).
103. Install the steering joint cover.
104. Install the steering wheel, and the driver's airbag.
105. With the tires raised off the ground (vehicle on a lift), check for the following symptoms by turning the steering wheel fully to the right and left several times:

- Rubbing sound coming from the lower steering column area: Probable cause is the steering column joint is contacting the cover.
- Grating sound from the lower steering column area or a rough feeling during steering: Probable cause is

poor engagement of the pinion shaft splines.

- Noise from around the steering wheel during steering: Probable cause is poor engagement of the SRS cable reel with the steering wheel or a damaged cable reel.

106. Do the battery terminal reconnection procedure, and check these items:

- Turn the ignition switch to ON (II), or press the engine start/stop button to select the ON mode, and check that the SRS indicator comes on for about 6 seconds, and then goes off.
- Make sure the horn and turn signal switches work properly.
- Make sure the steering wheel switches work properly.

107. After installation, check these items:

- Start the engine, allow it to idle, and turn the steering wheel from lock to lock several times.
- Check that the EPS indicator does not come on.
- Check the steering wheel spoke angle. If steering spoke angles to the right and left are not equal (steering wheel and rack are not centered), correct the engagement of the joint/pinion shaft splines.
- Do the torque sensor neutral position memorization.
- With SH-AWD: Do the VSA sensor neutral position memorization.
- Check the wheel alignment, and adjust it if necessary.

VSA SENSOR NEUTRAL POSITION MEMORIZATION PROCEDURE

➡Do not press the brake pedal during this procedure.

1. Park the vehicle on a flat and level surface, with the steering wheel in the straight ahead position.
2. With the ignition switch in LOCK (0), or press the engine start/stop button to select the OFF mode, connect the HDS to the Data Link Connector (DLC) under the driver's side of the dashboard.
3. Turn the ignition switch to ON (II), or press the engine start/stop button to select the ON mode.
4. Make sure the HDS communicates with the vehicle and the VSA modulator-control unit. If it doesn't, troubleshoot the DLC circuit.
5. Select VSA ADJUSTMENT with the HDS, and follow the screen prompts.

➡See the HDS Help menu for specific instructions.

6. Turn the ignition switch to LOCK (0), or press the engine start/stop button to select the OFF mode.

TORQUE SENSOR NEUTRAL POSITION MEMORIZATION

The torque sensor neutral position must be memorized whenever the steering gearbox or the EPS control unit is replaced. Note that the torque sensor neutral position is not affected when erasing the DTC.

➡The torque sensor is temperature sensitive. When memorizing the torque sensor neutral position, the ambient temperature must be above 68°F (20°C).

1. With the ignition switch in LOCK (0) or with the vehicle ignition in the OFF mode, connect the HDS to the Data Link Connector (DLC) located under the driver's side of the dashboard.
2. Turn the ignition switch to ON (II), or press the engine start/stop button to select the ON mode.
3. Make sure the HDS communicates with the vehicle and the EPS control unit. If it doesn't, troubleshoot the DLC circuit.
4. From the EPS MENU, select MISCELLANEOUS TEST, then TORQUE SENSOR LEARN, and follow the screen prompts on the HDS.

➡See the HDS Help menu for specific instructions.

5. Turn the ignition switch to LOCK (0), or press the engine start/stop button to select the OFF mode.

SUSPENSION

FRONT SUSPENSION

CONTROL LINKS

REMOVAL & INSTALLATION

Front Stabilizer Link

See Figure 278.

1. Raise and support the vehicle.
2. Remove the front wheel.
3. Remove the flange nuts while holding the respective joint pin with a hex wrench, then remove the stabilizer link.
4. Install the stabilizer link on the stabilizer bar and the lower arm with the joint pins set at the center of their range of movement.

➡The stabilizer link has a paint mark. The left stabilizer link is marked with yellow paint, and the right stabilizer link is marked with white paint.

5. Install the new flange nuts, and tighten them to 58 ft. lbs. (78 Nm) while holding the respective joint pin with a hex wrench.

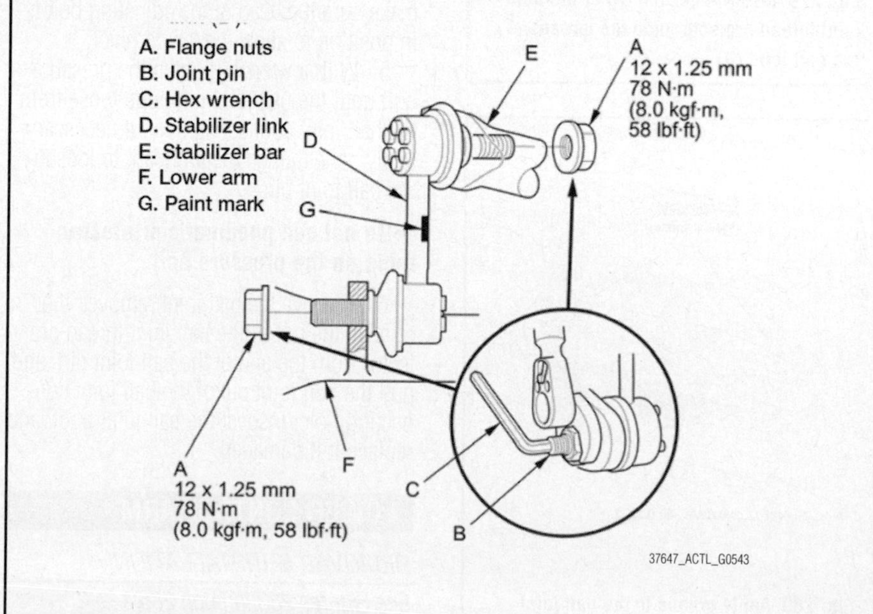
A. Flange nuts
B. Joint pin
C. Hex wrench
D. Stabilizer link
E. Stabilizer bar
F. Lower arm
G. Paint mark
E
A
12 x 1.25 mm
78 N·m
(8.0 kgf·m,
58 lbf·ft)
D
G
A
12 x 1.25 mm
78 N·m
(8.0 kgf·m, 58 lbf·ft)
F
C
B
37647_ACTL_G0543

Fig. 278 Remove the front stabilizer link

6. Clean the mating surfaces of the brake disc and the inside of the wheel, then install the front wheel.

LOWER BALL JOINT

REMOVAL & INSTALLATION

See Figures 279 through 281.

➡Always use a ball joint remover to disconnect a ball joint. Do not strike the housing or any other part of the ball joint connection to disconnect it.

1. Install a hex nut or the ball joint thread protector onto the threads of the ball joint.

➡Using a hex nut, make sure the nut is flush with the ball joint pin end to prevent damage to the threaded end of the ball joint pin.

2. Apply grease to the ball joint remover on the areas shown. This will ease the installation of the tool, and prevent damage to the pressure bolt threads.

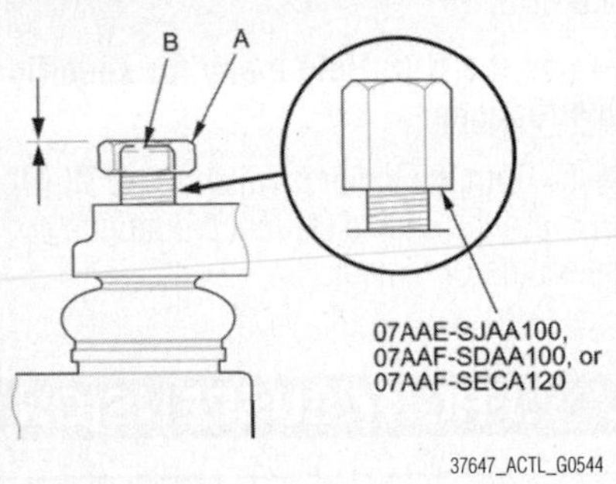
B
A
07AAE-SJAA100,
07AAF-SDAA100, or
07AAF-SECA120
37647_ACTL_G0544

Fig. 279 Install a hex nut (A) or the ball joint thread protector onto the threads of the ball joint (B)

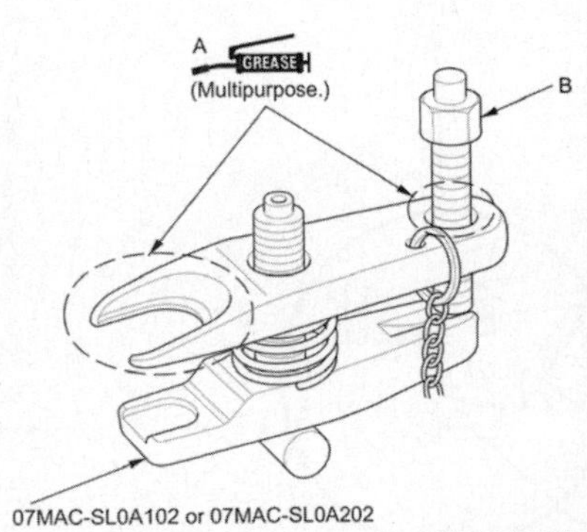
A
GREASE
(Multipurpose.)
B
07MAC-SL0A102 or 07MAC-SL0A202
37647_ACTL_G0545

Fig. 280 Apply grease to the ball joint remover on the areas shown (A)

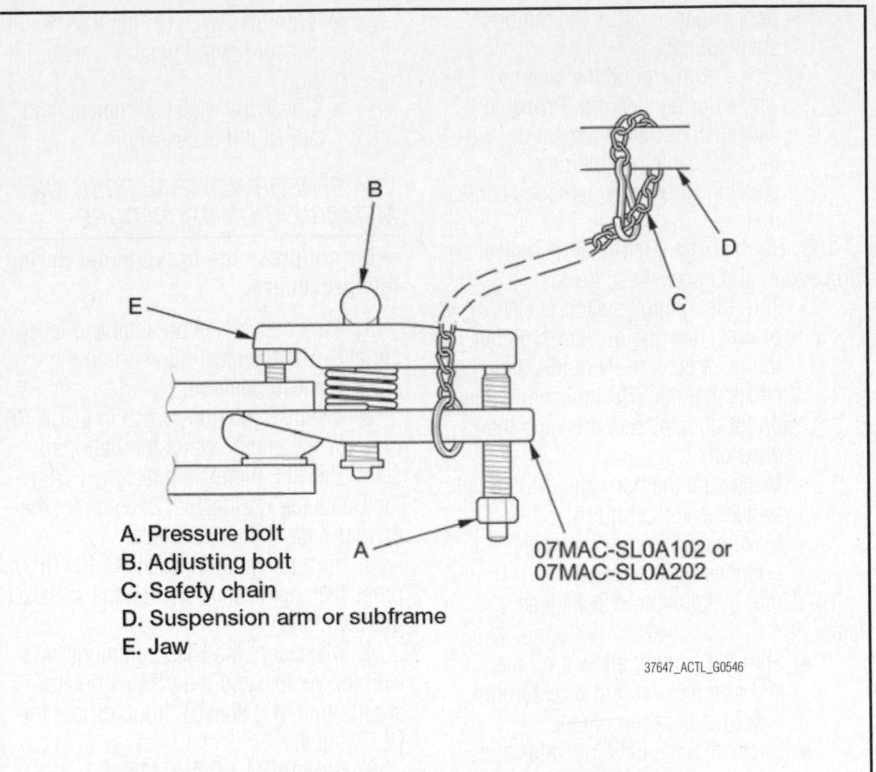
B
D
E
C
A
A. Pressure bolt
B. Adjusting bolt
C. Safety chain
D. Suspension arm or subframe
E. Jaw
07MAC-SL0A102 or
07MAC-SL0A202
37647_ACTL_G0546

Fig. 281 Loosen the pressure bolt, and install the ball joint remover as shown

3. Loosen the pressure bolt, and install the ball joint remover as shown. Insert the jaws carefully, making sure not to damage the ball joint boot. Adjust the jaw spacing by turning the adjusting bolt.

➡Fasten the safety chain securely to a suspension arm or the subframe. Do not fasten it to a brake line or wire harness.

4. After adjusting the adjusting bolt, make sure the head of the adjusting bolt is in position to allow the jaw to pivot.
5. With a wrench, tighten the pressure bolt until the ball joint pin pops loose from the ball joint connecting hole. If necessary, apply penetrating type lubricant to loosen the ball joint pin.

➡Do not use pneumatic or electric tools on the pressure bolt.

6. Remove the ball joint remover, then remove the nut or the ball joint thread protector from the end of the ball joint pin, and pull the ball joint out of the ball joint connecting hole. Inspect the ball joint boot, and replace it if damaged.

LOWER CONTROL ARM

REMOVAL & INSTALLATION

See Figures 282 through 285.

1. Raise and support the vehicle.
2. Remove the front wheel.
3. Remove the spindle nut.
4. Remove the front splash shield.
5. Remove the damper fork mounting nut while holding the mounting bolt, then remove the mounting bolt.

➡Use the new damper fork mounting bolt and the new mounting nut, and torque the nut while holding the bolt during reassembly.

6. Disconnect the stabilizer link from the lower arm.
7. Remove the cotter pin from the knuckle ball joint, then remove the castle nut.

➡Use the new castle nut during reassembly. During installation, insert the new cotter pin into the ball joint pin hole from the front to the rear of the vehicle, and bend it. Check the ball joint pin hole direction before connecting the ball joint.

8. Disconnect the knuckle ball joint from the lower arm using the ball joint thread protector and the ball joint remover.

➡Be careful not to damage the ball joint boot when installing the remover. Do not force or hammer on the lower arm, or pry between the lower arm and the knuckle. You could damage the ball joint.

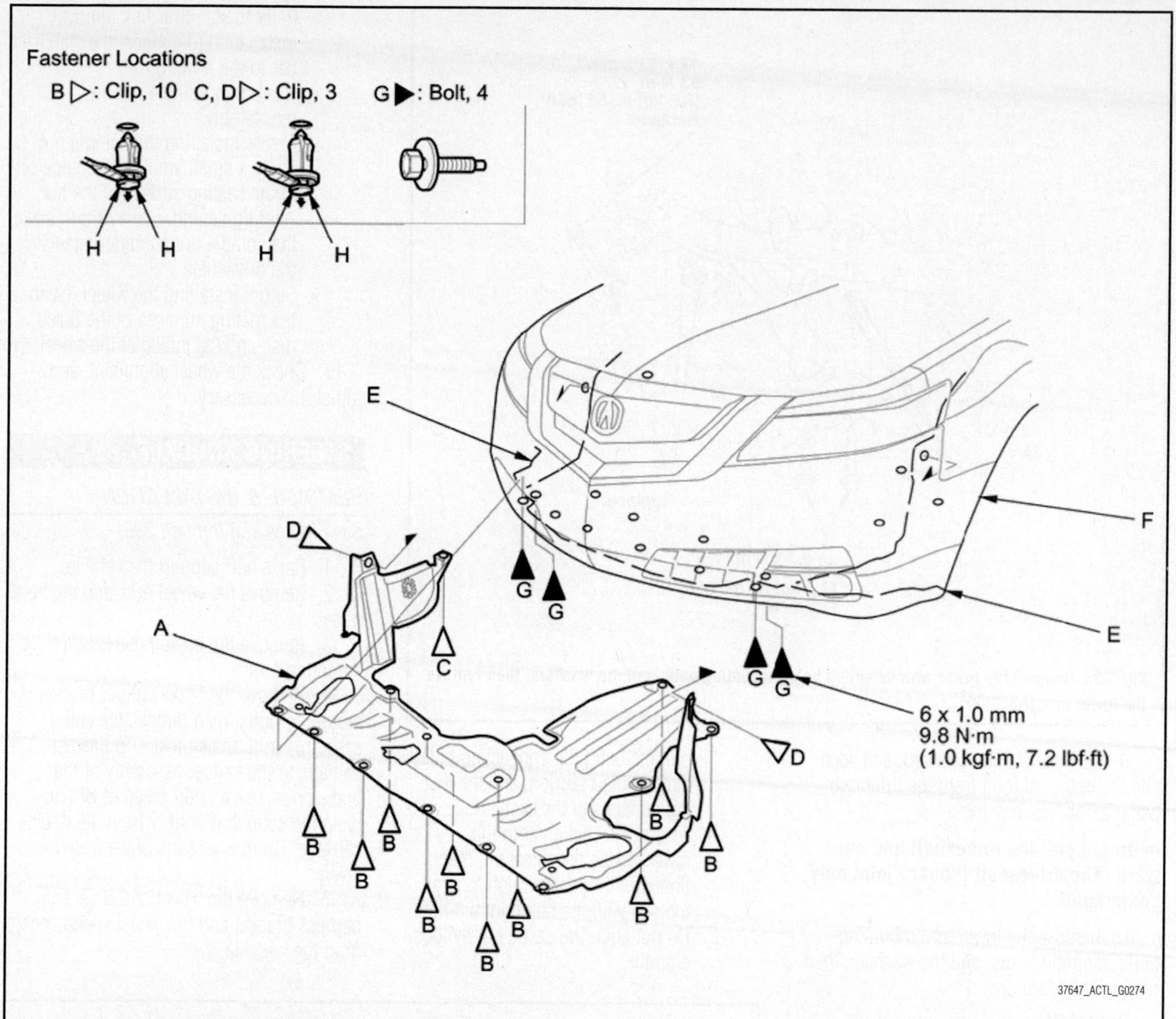
Fastener Locations
B ▷: Clip, 10 C, D ▷: Clip, 3 G ▶: Bolt, 4
H
H
H
H
E
F
D
G
G
A
C
E
G G
6 x 1.0 mm
9.8 N·m
(1.0 kgf·m, 7.2 lbf·ft)
D
B
37647_ACTL_G0274

Fig. 282 Remove the front splash shield

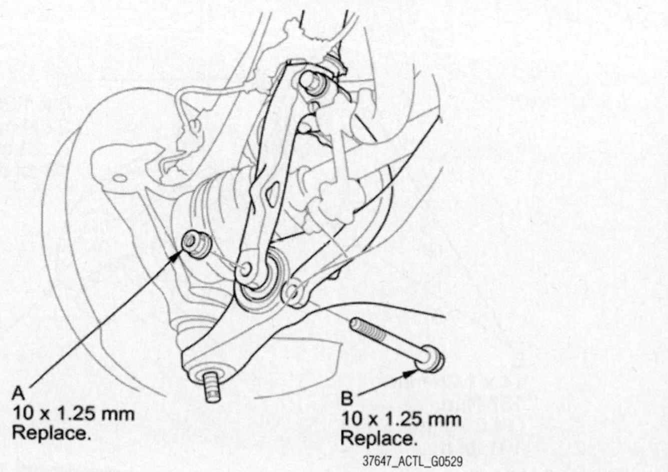
A
10 x 1.25 mm
Replace.
B
10 x 1.25 mm
Replace.
37647_ACTL_G0529

Fig. 283 Remove the damper fork mounting nut (A) while holding the mounting bolt (B), then remove the mounting bolt

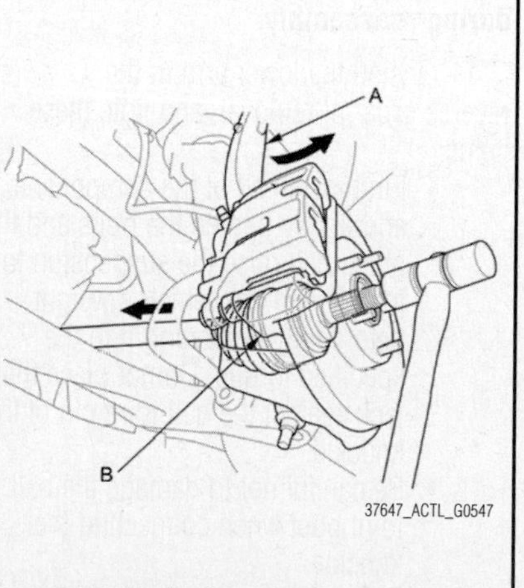
A
B
37647_ACTL_G0547

Fig. 284 Pull the knuckle (A) outward, and separate the outboard joint (B) from the front hub using a soft face hammer

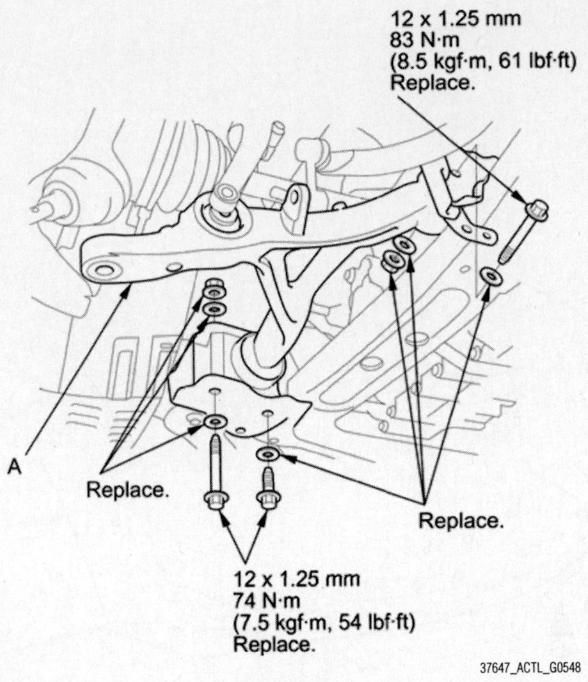
12 x 1.25 mm
83 N·m
(8.5 kgf·m, 61 lbf·ft)
Replace.
A
Replace.
Replace.
12 x 1.25 mm
74 N·m
(7.5 kgf·m, 54 lbf·ft)
Replace.
37647_ACTL_G0548

Fig. 285 Remove the lower arm mounting bolts, mounting nuts, and the washers, then remove the lower arm (A)

9. Pull the knuckle outward, and separate the outboard joint from the front hub using a soft face hammer.

➡Do not pull the driveshaft end outward. The driveshaft inboard joint may come apart.

10. Remove the lower arm mounting bolts, mounting nuts, and the washers, then remove the lower arm.

To install:

➡Use new lower arm mounting bolts, new mounting nuts, and new washers during reassembly.

11. Install the lower arm in the reverse order of removal, and note these items:

- First install all of the components, and lightly tighten the bolts and the nuts, then raise the suspension to load it with the vehicle's weight before fully tightening it to the specified torque. Do not place the jack against the ball joint pin of the knuckle.
- Be careful not to damage the ball joint boot when connecting the knuckle.
- Before connecting the ball joint, degrease the threaded section and the tapered portion of the ball joint pin, the ball joint connecting hole, the threaded section, and the mating surfaces of the castle nut.
- Torque the castle nut to the lower torque specification, then tighten it only far enough to align the slot with the ball joint pin hole. Do not align the castle nut by loosening it.
- Refer to stabilizer link removal/installation to connect the stabilizer bar to the lower arm.
- Use a new spindle nut on reassembly.
- Before installing the spindle nut, apply a small amount of engine oil to the seating surface of the nut. After tightening, use a drift to stake the spindle nut shoulder against the driveshaft.
- Before installing the wheel, clean the mating surfaces of the brake disc and the inside of the wheel.

12. Check the wheel alignment, and adjust it if necessary.

STEERING KNUCKLE

REMOVAL & INSTALLATION

See Figures 286 through 288.

1. Raise and support the vehicle.
2. Remove the wheel nuts and the front wheel.
3. Remove the brake hose bracket mounting bolt.
4. Remove the brake caliper bracket mounting bolts, then remove the caliper assembly from the knuckle. To prevent damage to the caliper assembly or the brake hose, use a short piece of wire to hang the caliper assembly from the undercarriage. Do not twist the brake hose excessively.
5. Remove the wheel speed sensor harness bracket and the wheel speed sensor from the knuckle.

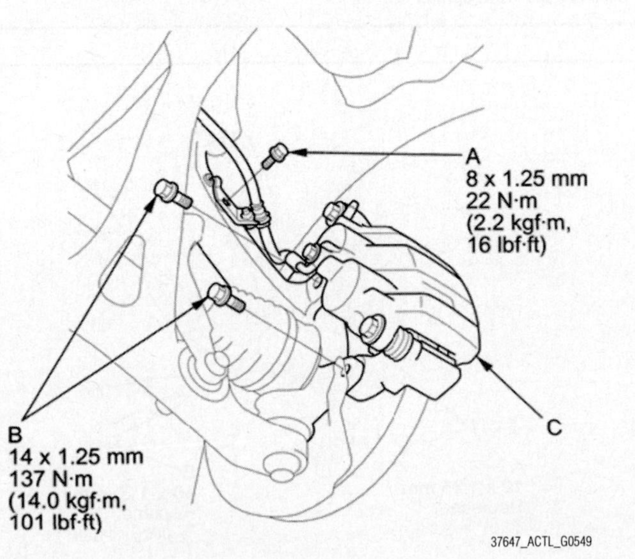
A
8 x 1.25 mm
22 N·m
(2.2 kgf·m,
16 lbf·ft)
C
B
14 x 1.25 mm
137 N·m
(14.0 kgf·m,
101 lbf·ft)
37647_ACTL_G0549

Fig. 286 Remove the brake hose bracket mounting bolt (A)

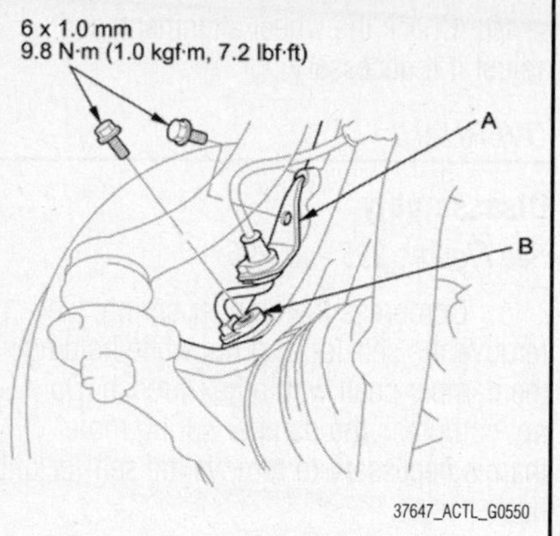
6 x 1.0 mm
9.8 N·m (1.0 kgf·m, 7.2 lbf·ft)
A
B
37647_ACTL_G0550

Fig. 287 Remove the wheel speed sensor harness bracket (A) and the wheel speed sensor (B) from the knuckle

➡Do not disconnect the wheel speed sensor connector.

6. Pry up the stake on the spindle nut, then remove the nut.
7. Remove the front brake disc.
8. Check the front hub for damage and cracks.
9. Remove the cotter pin from the tie-rod end ball joint, then remove the nut.
10. Disconnect the tie-rod end ball joint from the knuckle using the ball joint thread protector and the ball joint remover.
11. Remove the cotter pin from the knuckle ball joint, then remove the castle nut.
12. Disconnect the knuckle ball joint from the lower arm using the ball joint thread protector and the ball joint remover.

➡Be careful not to damage the ball joint boot when installing the remover. Do not force or hammer on the lower arm, or pry between the lower arm and the knuckle. You could damage the ball joint.

13. Remove the cotter pin from the upper arm ball joint, then remove the castle nut.
14. Disconnect the upper arm ball joint from the knuckle using the ball joint thread protector and the ball joint remover.
15. Pull the knuckle outward, and separate the outboard joint from the front hub using a soft face hammer, then remove the knuckle/hub.

➡Do not pull the driveshaft end outward. The driveshaft inboard joint may come apart. During installation, apply grease to the mating surfaces of the wheel bearing and the driveshaft outboard joint.

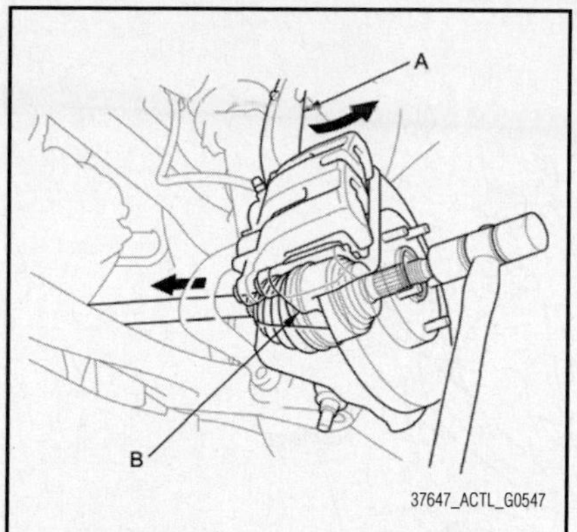
A
B
37647_ACTL_G0547

Fig. 288 Pull the knuckle (A) outward, and separate the outboard joint (B) from the front hub using a soft face hammer

To install:

16. Install the knuckle/hub in the reverse order of removal, and note these items:
   - First install all of the components, and lightly tighten the bolts and the nuts, then raise the suspension to load it with the vehicle's weight before fully tightening to the specified torque. Do not place the jack against the ball joint pin of the knuckle.
   - Be careful not to damage the ball joint boot when connecting the knuckle.
   - Before connecting the ball joint, degrease the threaded section and the tapered portion of the ball joint pin, the ball joint connecting hole, the threaded section, and the mating surfaces of the castle nut.
   - Torque the castle nut to the lower torque specification, then tighten it only far enough to align the slot with the ball joint pin hole. Do not align the castle nut by loosening it.
   - Use a new spindle nut on reassembly.
   - Before installing the spindle nut, apply a small amount of engine oil to the seating surface of the nut. After tightening, use a drift to stake the spindle nut shoulder against the driveshaft.
   - Before installing the brake disc, clean the mating surfaces of the front hub and the inside of the brake disc.
   - Before installing the wheel, clean the mating surfaces of the brake disc and the inside of the wheel.
17. Check the wheel alignment, and adjust it if necessary.

STRUT (DAMPER/SPRING)

REMOVAL & INSTALLATION

See Figures 289 through 294.

1. Raise and support the vehicle.
2. Remove the front wheel.
3. Remove the wheel speed sensor harness bracket mounting bolt.
4. Remove the damper pinch bolt and the damper fork mounting nut while holding the mounting bolt, then remove the damper fork from the damper and the lower arm.
5. Remove the front strut brace mounting nuts.
6. Remove the damper mounting nuts from the top of the damper. Do not let the damper/spring drop down under its own weight.
7. Remove the damper/spring.

To install:

8. Position the damper/spring in the body with the aligning tab facing inside.
9. Loosely install the new damper mounting nuts to the top of the damper.
10. Loosely install the front strut brace mounting nuts.
11. Install the damper fork over the driveshaft and onto the lower arm. Install the aligning tab on the damper unit into the slot of the damper fork.
12. Loosely install the damper pinch bolt into the damper fork.
13. Connect the damper fork and the lower arm with the new damper fork mounting bolt, then lightly tighten the new mounting nut.
14. Place a floor jack under the lower arm, and raise the suspension to load it with the vehicle's weight.
15. Tighten the damper pinch bolt to 36 ft. lbs. (49 Nm) and the damper fork mounting nut while holding the mounting bolt to 47 ft. lbs. (64 Nm).

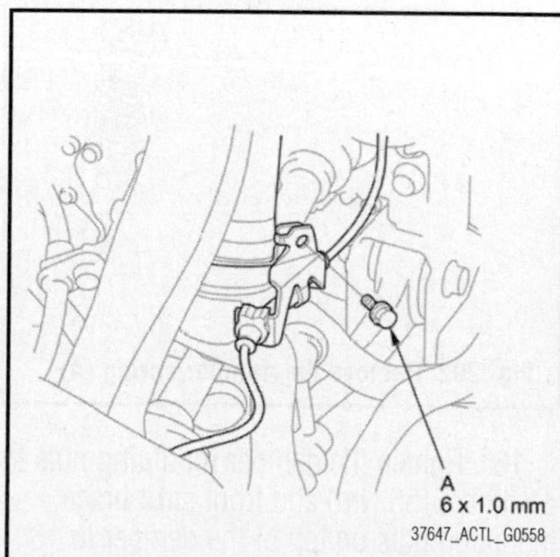
A
6 x 1.0 mm
37647_ACTL_G0558

Fig. 289 Remove the wheel speed sensor harness bracket mounting bolt (A)

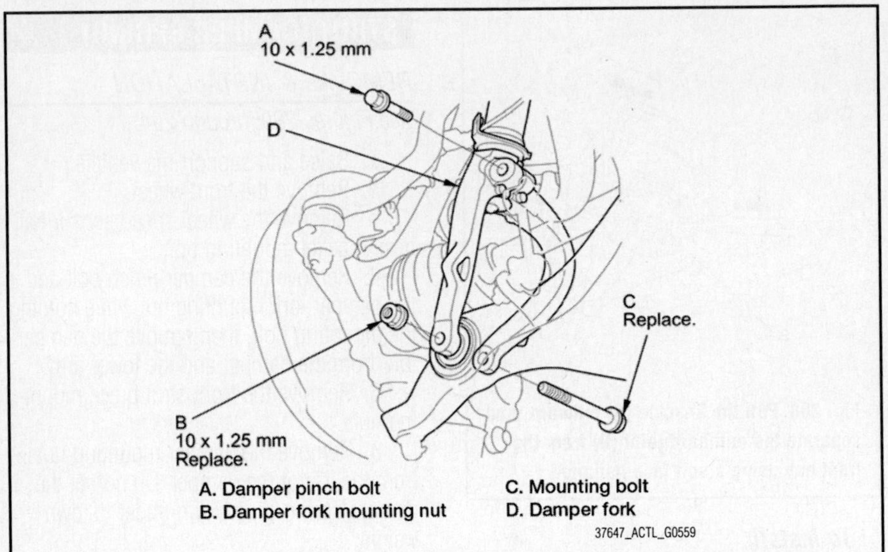
A
10 x 1.25 mm
D
C
Replace.
B
10 x 1.25 mm
Replace.
A. Damper pinch bolt
B. Damper fork mounting nut
C. Mounting bolt
D. Damper fork
37647_ACTL_G0559

Fig. 290 Remove the damper pinch bolt and the damper fork mounting nut while holding the mounting bolt, then remove the damper fork from the damper and the lower arm

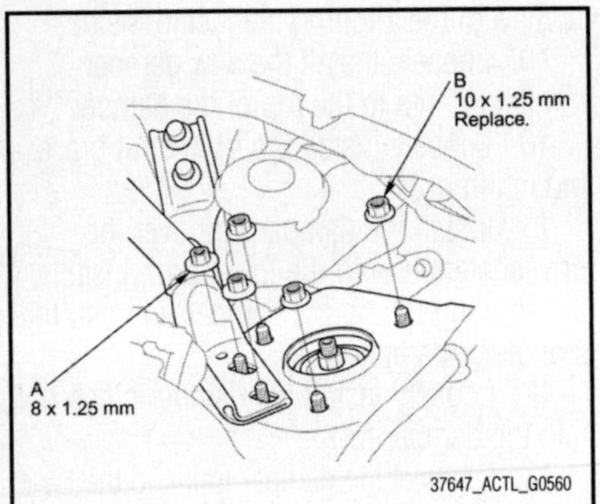
B
10 x 1.25 mm
Replace.
A
8 x 1.25 mm
37647_ACTL_G0560

Fig. 291 Remove the front strut brace mounting nuts (A) and the damper mounting nuts (B) from the top of the damper

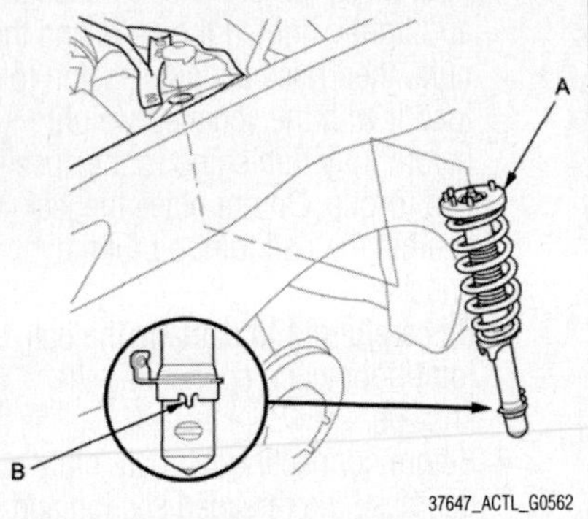
A
B
37647_ACTL_G0562

Fig. 293 Position the damper/spring (A) in the body with the aligning tab (B) facing inside

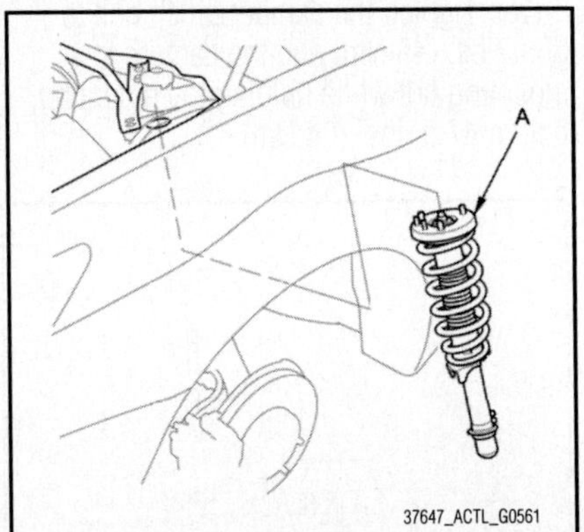
A
37647_ACTL_G0561

Fig. 292 Remove the damper/spring (A)

16. Tighten the damper mounting nuts to 41 ft. lbs. (55 Nm) and front strut brace mounting nuts on top of the damper to 16 ft. lbs. (22 Nm).

17. Install the wheel speed sensor harness bracket.

18. Clean the mating surfaces of the brake disc and the inside of the wheel, then install the front wheel.

19. Check the wheel alignment, and adjust it if necessary.

OVERHAUL

Disassembly

See Figures 295 and 296.

1. Compress the damper spring, then remove the self-locking nut while holding the damper shaft with a hex wrench. Do not compress the damper spring more than is necessary to remove the self-locking nut.

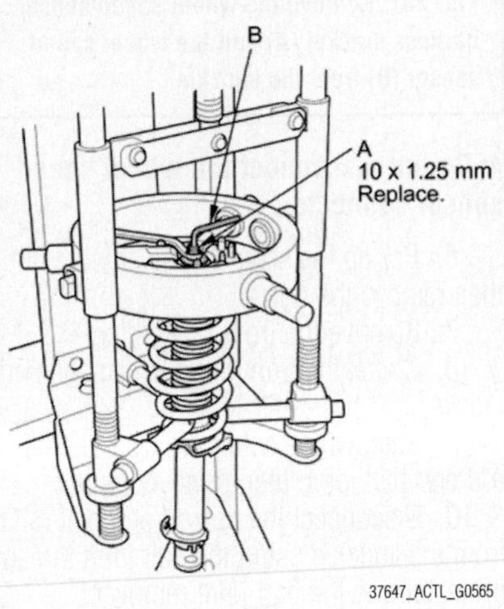
B
A
10 x 1.25 mm
Replace.
37647_ACTL_G0565

Fig. 295 Compress the damper spring, then remove the self-locking nut (A) while holding the damper shaft with a hex wrench (B)

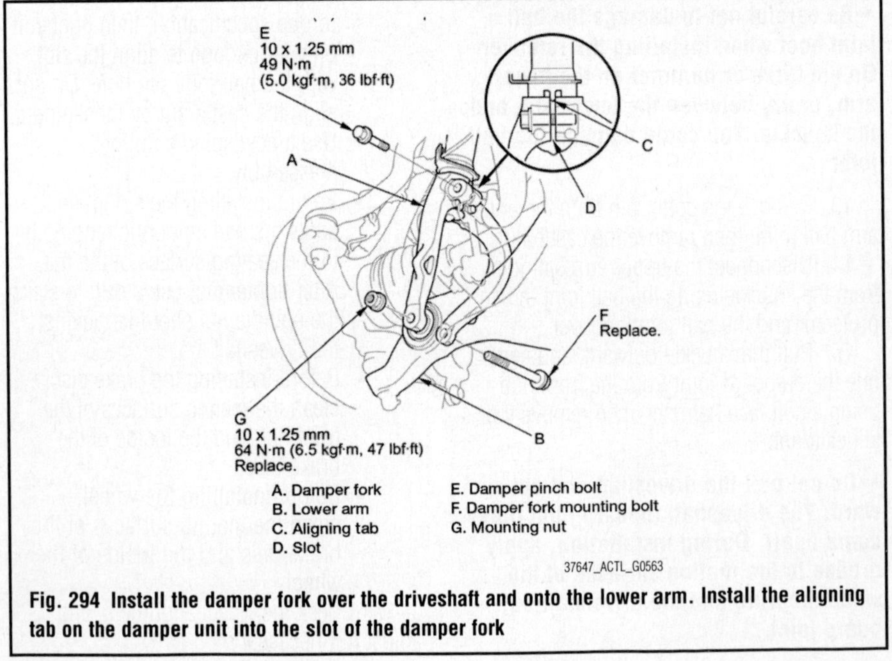
E
10 x 1.25 mm
49 N·m
(5.0 kgf·m, 36 lbf·ft)
A
C
D
F
Replace.
G
10 x 1.25 mm
64 N·m (6.5 kgf·m, 47 lbf·ft)
Replace.
B
A. Damper fork
B. Lower arm
C. Aligning tab
D. Slot
E. Damper pinch bolt
F. Damper fork mounting bolt
G. Mounting nut
37647_ACTL_G0563

Fig. 294 Install the damper fork over the driveshaft and onto the lower arm. Install the aligning tab on the damper unit into the slot of the damper fork

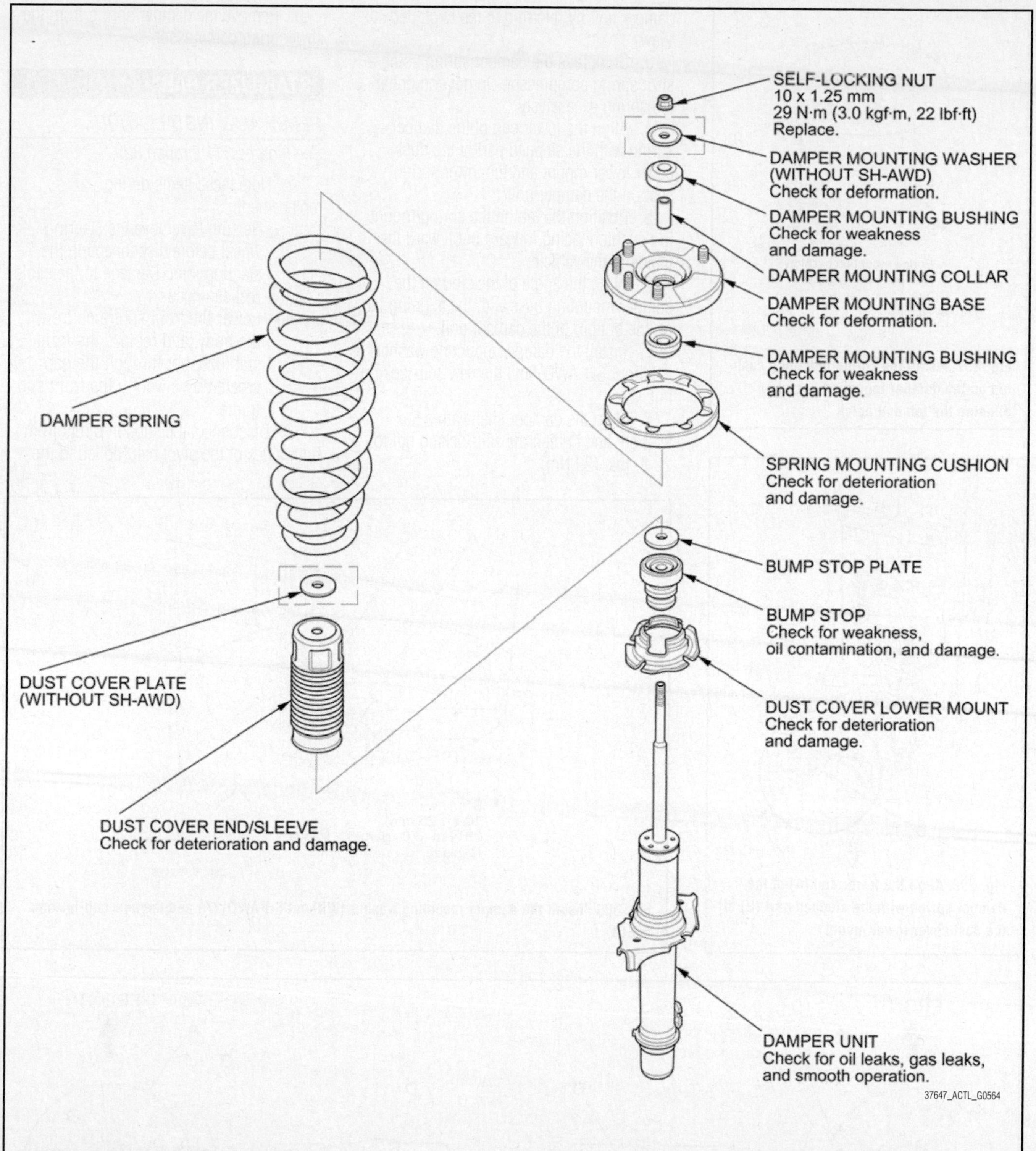
SELF-LOCKING NUT
10 x 1.25 mm
29 N·m (3.0 kgf·m, 22 lbf·ft)
Replace.
DAMPER MOUNTING WASHER
(WITHOUT SH-AWD)
Check for deformation.
DAMPER MOUNTING BUSHING
Check for weakness
and damage.
DAMPER MOUNTING COLLAR
DAMPER MOUNTING BASE
Check for deformation.
DAMPER MOUNTING BUSHING
Check for weakness
and damage.
DAMPER SPRING
SPRING MOUNTING CUSHION
Check for deterioration
and damage.
BUMP STOP PLATE
BUMP STOP
Check for weakness,
oil contamination, and damage.
DUST COVER PLATE
(WITHOUT SH-AWD)
DUST COVER LOWER MOUNT
Check for deterioration
and damage.
DUST COVER END/SLEEVE
Check for deterioration and damage.
DAMPER UNIT
Check for oil leaks, gas leaks,
and smooth operation.
37647_ACTL_G0564

Fig. 296 Exploded view of damper/spring assembly

2. Release the pressure from the strut spring compressor, then disassemble the damper as shown in the Exploded View.

Inspection

1. Reassemble all the parts, except for the damper spring.
2. Compress the damper assembly by hand, and check for smooth operation through a full stroke, both compression and extension. The damper should extend smoothly and constantly when compression is released. If it does not, the gas is leaking and the damper should be replaced.
3. Check for oil leaks, abnormal noises, and binding during these tests.

Reassembly

See Figures 297 through 300.

1. Install the spring mounting cushion on the damper mounting base by aligning the tab and notch.
2. Install all the parts except the damper mounting washer (Without SH-AWD) and the self-locking nut onto the

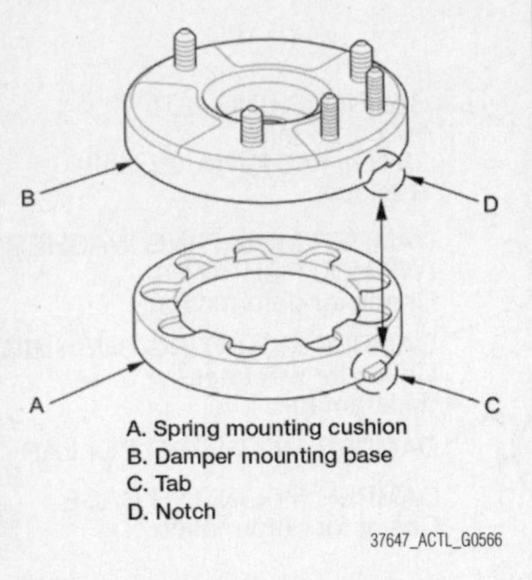
B
D
A
C
A. Spring mounting cushion
B. Damper mounting base
C. Tab
D. Notch
37647_ACTL_G0566

Fig. 297 Install the spring mounting cushion on the damper mounting base by aligning the tab and notch

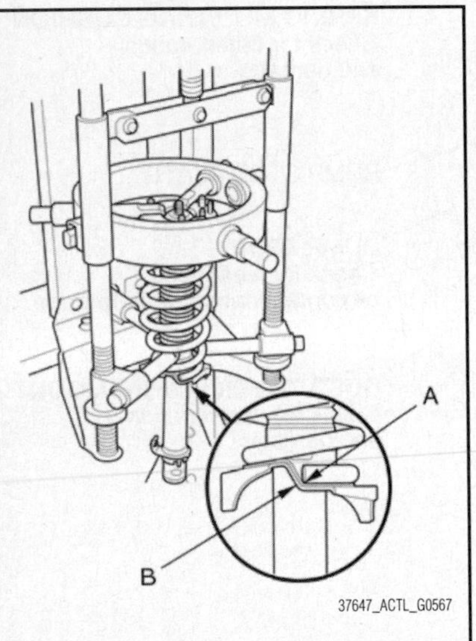
A
B
37647_ACTL_G0567

Fig. 298 Align the lower end (A) of the damper spring with the stepped part (B) of the dust cover lower mount

damper unit by referring to the Exploded View.

3. Compress the damper spring using a strut spring compressor. Do not compress the spring excessively.

4. Align the lower end of the damper spring with the stepped part of the dust cover lower mount and the lower spring seat on the damper unit.

5. Position the tab on the spring mounting cushion facing forward but toward the inside of the vehicle.

6. Align the angle of the stud on the damper mounting base with the aligning tab on the bottom of the damper unit.

7. Install the damper mounting washer (Without SH-AWD) and the new self-locking nut.

8. Hold the damper shaft with a hex wrench, and tighten the self-locking nut to 22 ft. lbs. (29 Nm).

9. Remove the damper/spring from the strut spring compressor.

STABILIZER BAR

REMOVAL & INSTALLATION

See Figures 311 through 329.

1. Note these items during replacement:
   - Be sure to remove the steering wheel before disconnecting the steering joint. Damage to the cable reel can occur.
   - Lower the front subframe from the body, and replace the front stabilizer bar through the gap created by lowering the front subframe.

2. Disconnect the support struts from both sides of the pivot ball (bolted to the

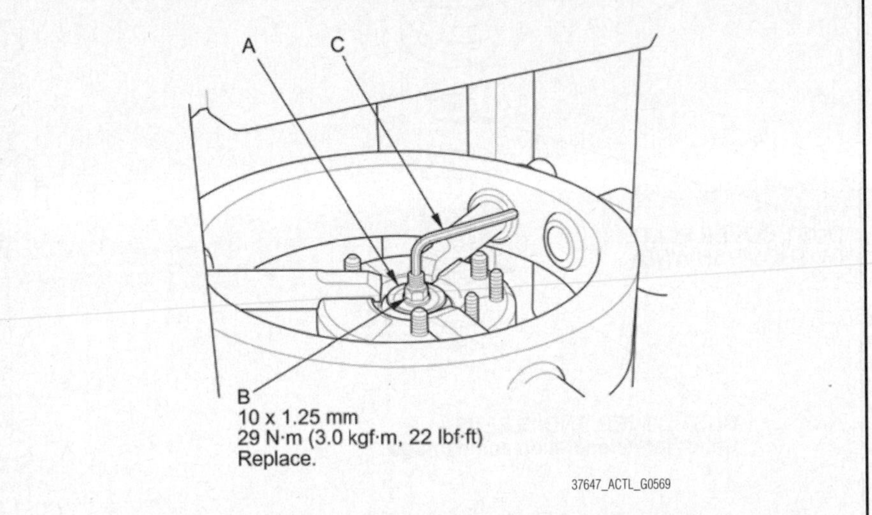
A
C
B
10 x 1.25 mm
29 N·m (3.0 kgf·m, 22 lbf·ft)
Replace.
37647_ACTL_G0569

Fig. 300 Install the damper mounting washer (Without SH-AWD) (A) and the new self-locking nut (B)

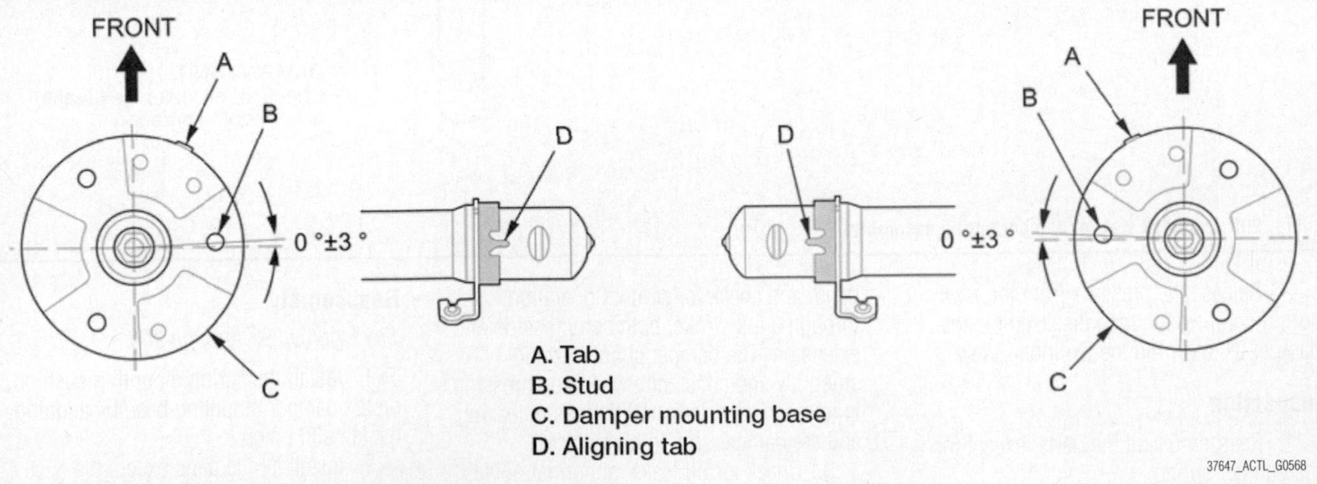
FRONT
A
B
0 °±3 °
C
D
D
0 °±3 °
FRONT
A
B
C
A. Tab
B. Stud
C. Damper mounting base
D. Aligning tab
37647_ACTL_G0568

Fig. 299 Position the tab on the spring mounting cushion facing forward but toward the inside of the vehicle

hood). Secure the hood in a vertical position. Install the right side pivot ball into the lower threaded hole.

➡Do not attempt to close the hood with the support strut in the vertical position, as it will damage the support strut and hood.

3. Remove the engine compartment covers.
4. Do the battery terminal disconnection procedure.
5. Remove the front strut brace.
6. Raise and support the vehicle.
7. Remove the front wheels.
8. Disconnect both sides of the stabilizer link from the stabilizer bar.
9. Disconnect both sides of the tie-rod end ball joint from the knuckle.
10. Remove the driver's airbag and the steering wheel.
11. Remove the steering joint cover.
12. Loosen the upper steering joint bolt, and remove the lower steering joint bolt. Disconnect the steering joint by sliding the steering joint into the column shaft. Tighten the steering joint upper bolt to hold the column shaft.

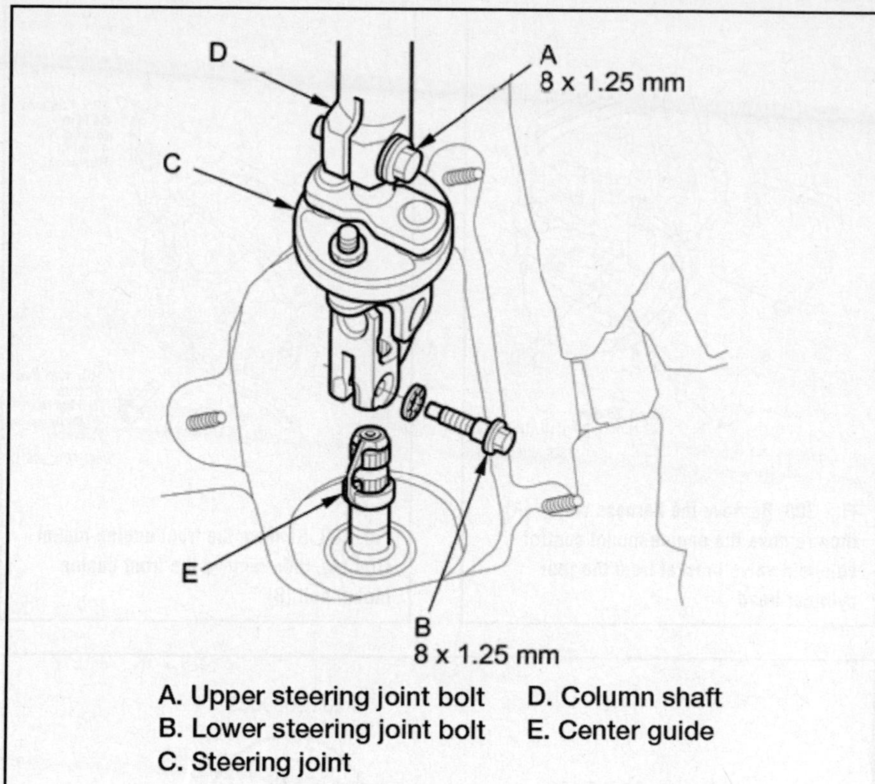
D
A
8 x 1.25 mm
C
E
B
8 x 1.25 mm
A. Upper steering joint bolt
B. Lower steering joint bolt
C. Steering joint
D. Column shaft
E. Center guide
37647_ACTL_G0086

Fig. 303 Loosen the upper steering joint bolt (A), and remove the lower steering joint bolt (B)

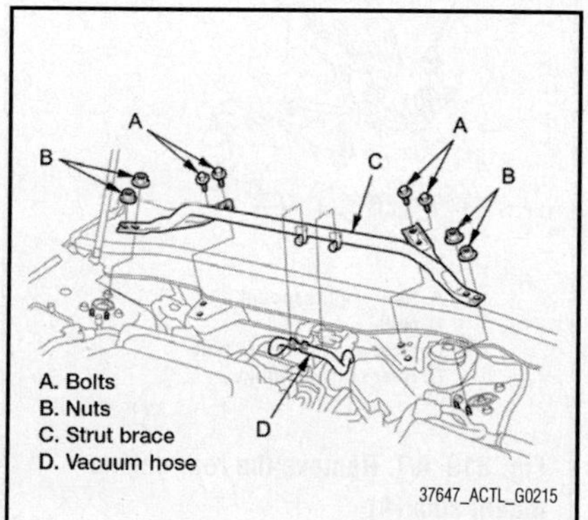
A
B
A
C
A
B
A. Bolts
B. Nuts
C. Strut brace
D. Vacuum hose
D
37647_ACTL_G0215

Fig. 301 Remove the front strut brace

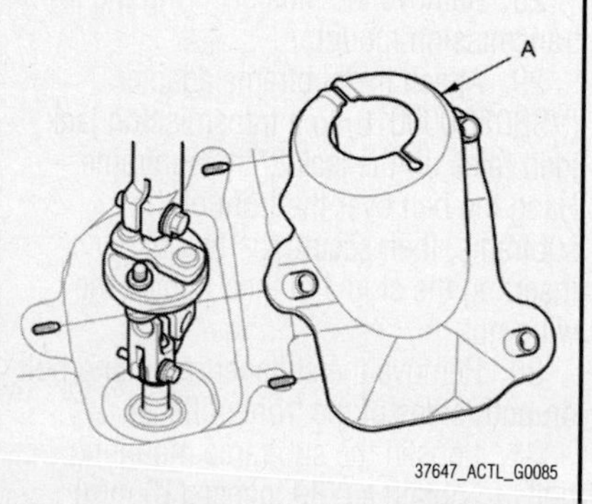
A
37647_ACTL_G0085

Fig. 302 Remove the steering joint cover

➡If the center guide is in place and has not moved, leave it in place. If the center guide has moved or been removed, discard it.

13. M/T: Remove the P/S heat shield.
14. Remove the connector bracket from the front cylinder head; use the bracket bolt hole to attach the engine hanger balance bar front arm.
15. Remove the harness clamp, then remove the engine mount control solenoid valve bracket from the rear cylinder head; use the bracket bolt hole to attach the 2008 V6 attachment arm.
16. A/T: Disconnect the vacuum hose.
17. Lift and support the engine with the engine support hanger (AAR-T1256) and the engine hanger balance bar (VSB02C000019). Attach the front arm to the front cylinder head with a spacer and a 10 x 1.25 mm bolt. Remove the rear arm from the engine hanger balance bar, and install the 2008 V6 attachment arm (SIL02C000033), then attach the 2008 V6 attachment arm to the rear cylinder head with a 8 x 1.25 mm bolt.

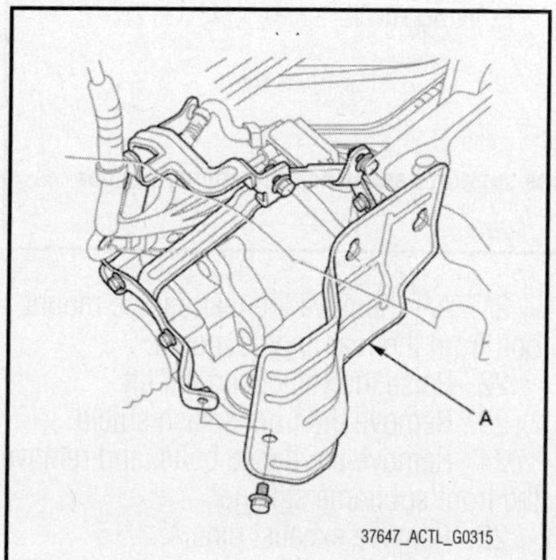
A
37647_ACTL_G0315

Fig. 304 M/T: Remove the P/S heat shield (A)

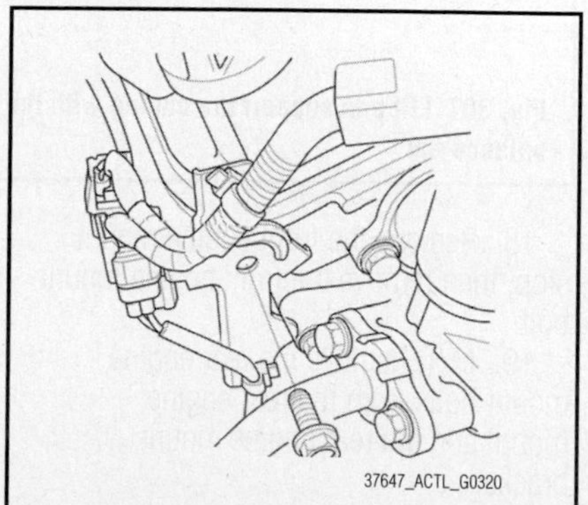
37647_ACTL_G0320

Fig. 305 Remove the connector bracket from the front cylinder head

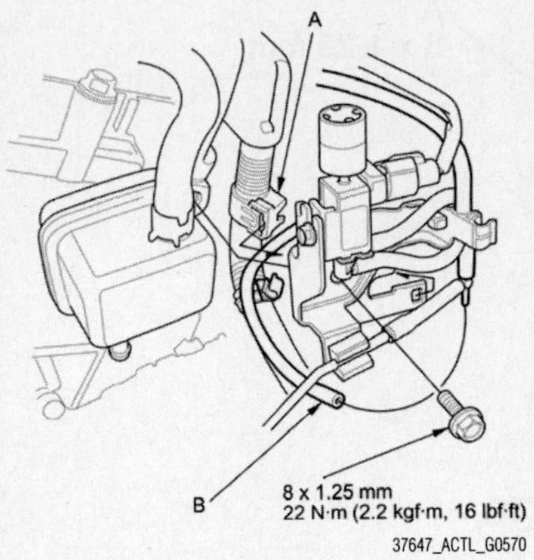
A
B
8 x 1.25 mm
22 N·m (2.2 kgf·m, 16 lbf·ft)
37647_ACTL_G0570

Fig. 306 Remove the harness clamp (A), then remove the engine mount control solenoid valve bracket from the rear cylinder head

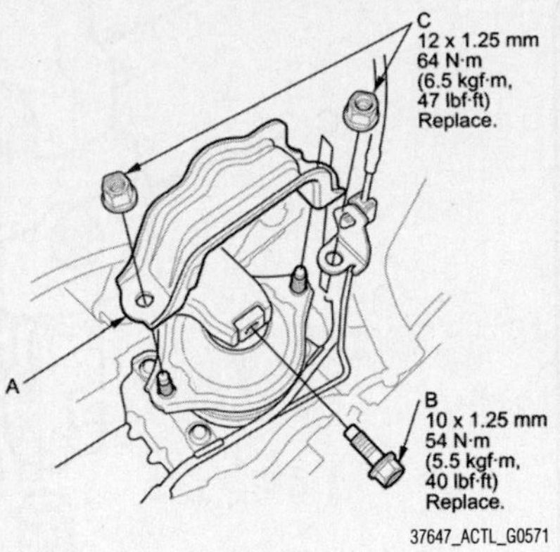
C
12 x 1.25 mm
64 N·m
(6.5 kgf·m,
47 lbf·ft)
Replace.
A
B
10 x 1.25 mm
54 N·m
(5.5 kgf·m,
40 lbf·ft)
Replace.
37647_ACTL_G0571

Fig. 308 Remove the front engine mount stop (A), then remove the front engine mount bolt (B)

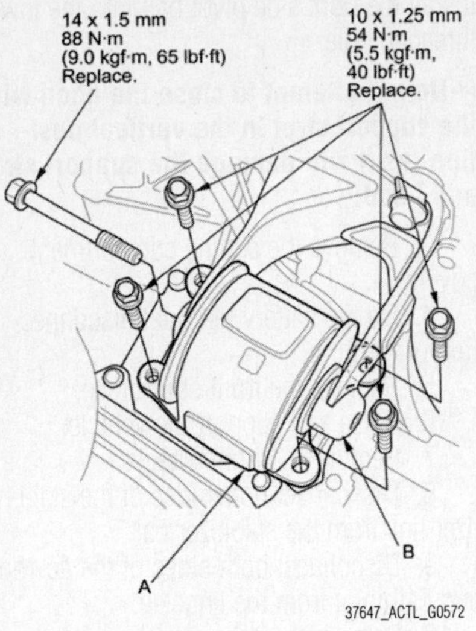
14 x 1.5 mm
88 N·m
(9.0 kgf·m, 65 lbf·ft)
Replace.
10 x 1.25 mm
54 N·m
(5.5 kgf·m,
40 lbf·ft)
Replace.
B
A
37647_ACTL_G0572

Fig. 309 M/T: Remove the rear engine mount bolts from the rear engine mount (A) and the rear engine mount bracket (B)

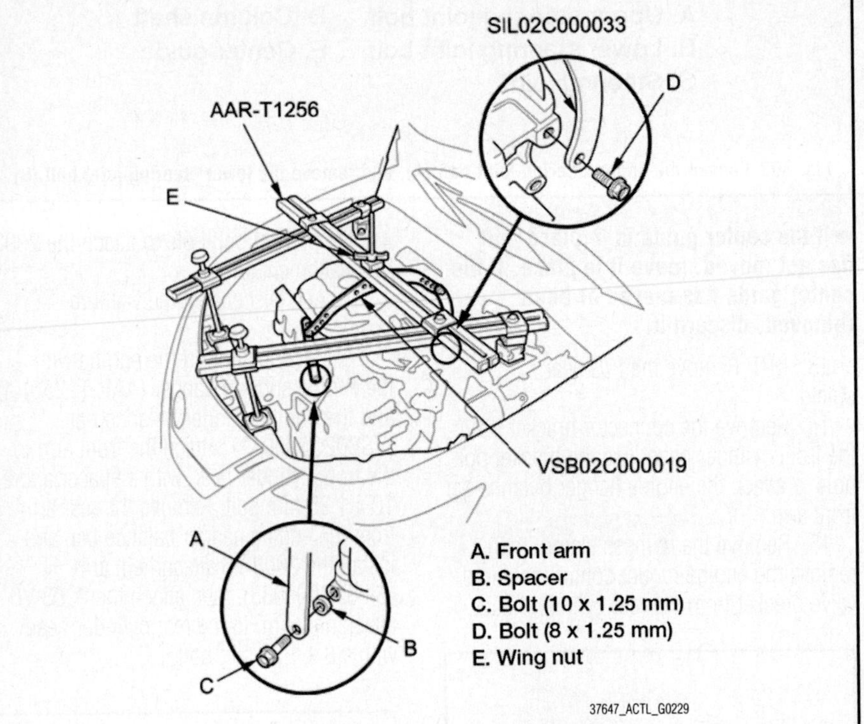
SIL02C000033
D
AAR-T1256
E
VSB02C000019
A
B
C
A. Front arm
B. Spacer
C. Bolt (10 x 1.25 mm)
D. Bolt (8 x 1.25 mm)
E. Wing nut
37647_ACTL_G0229

Fig. 307 Lift and support the engine with the engine support hanger and the engine hanger balance bar

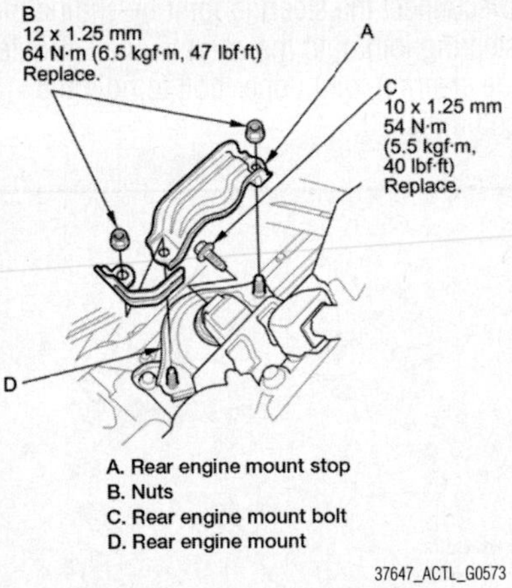
B
12 x 1.25 mm
64 N·m (6.5 kgf·m, 47 lbf·ft)
Replace.
A
C
10 x 1.25 mm
54 N·m
(5.5 kgf·m,
40 lbf·ft)
Replace.
D
A. Rear engine mount stop
B. Nuts
C. Rear engine mount bolt
D. Rear engine mount
37647_ACTL_G0573

Fig. 310 A/T: Remove the rear engine mount stop (A)

18. Remove the front engine mount stop, then remove the front engine mount bolt.
19. M/T: Remove the rear engine mount bolts from the rear engine mount and the rear engine mount bracket.
20. A/T: Remove the rear engine mount stop.
21. A/T: Remove the rear engine mount bolt from the rear engine mount.
22. Raise the vehicle on the lift.
23. Remove the front splash shield.
24. Remove the flange bolts, and remove the front subframe stiffener.
25. Remove exhaust pipe A:
26. With SH-AWD: Remove the propeller shaft.
27. Remove the front subframe mounting bolts, and the front subframe middle mounts.
28. Remove the nuts securing the lower transmission mount.
29. Attach the subframe adapter (VSB02C000016) to a transmission jack, then raise up the jack to the subframe. Hang the belt over the front of the subframe, then secure the belt by inserting the stop and then tighten the wing nut.
30. Remove the stiffener mounting bolt on both sides of the front stiffener.
31. Loosen the subframe mounting bolts to obtain a 0.39 inches (10 mm) distance between the bolt seat and the

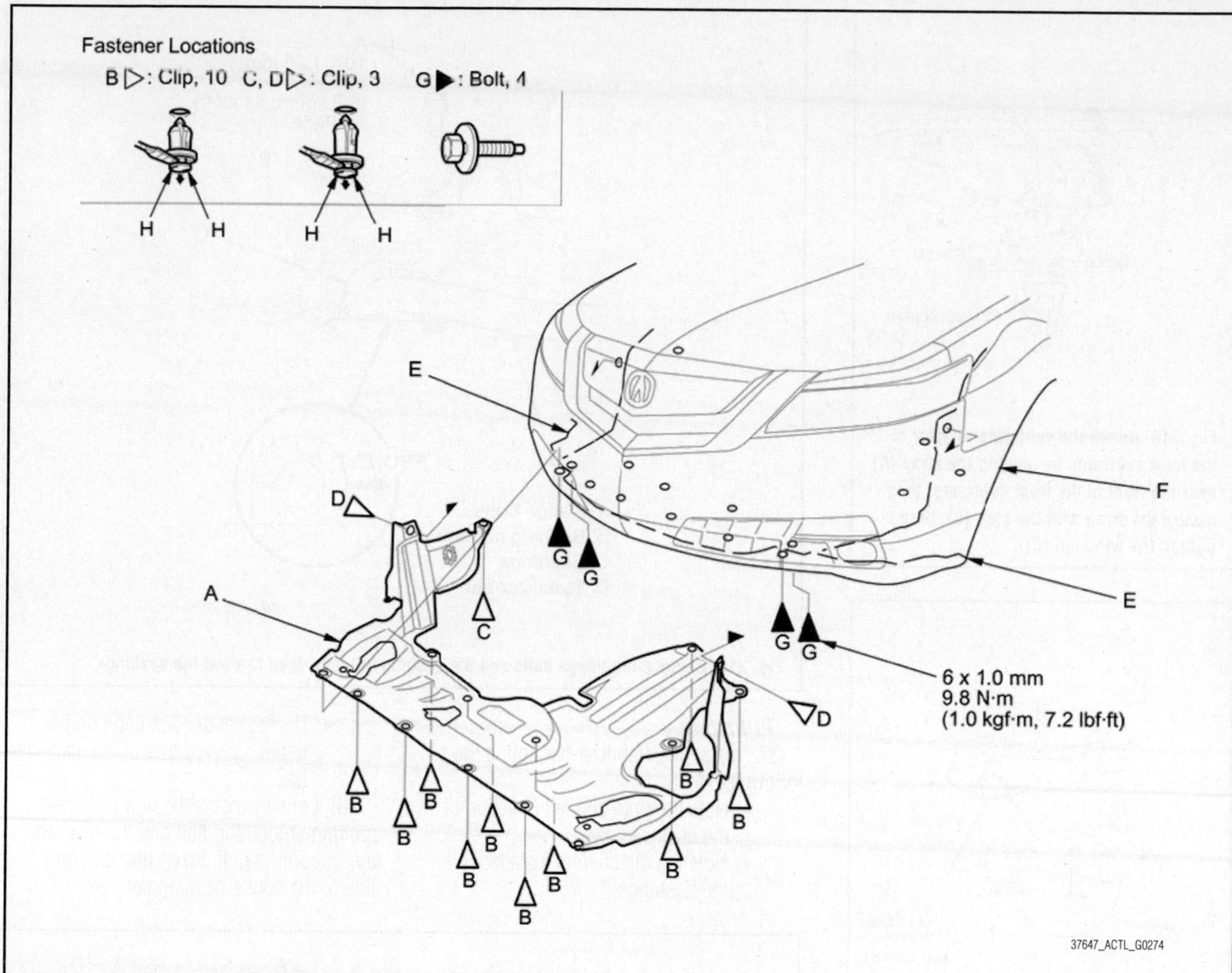
Fastener Locations
B ▷: Clip, 10 C, D ▷: Clip, 3 G ▶: Bolt, 4
H
H
H
H
E
F
D
A
C
G
G
G
G
E
6 x 1.0 mm
9.8 N·m
(1.0 kgf·m, 7.2 lbf·ft)
D
B
B
B
B
B
B
B
B
B
B
37647_ACTL_G0274

Fig. 311 Remove the front splash shield

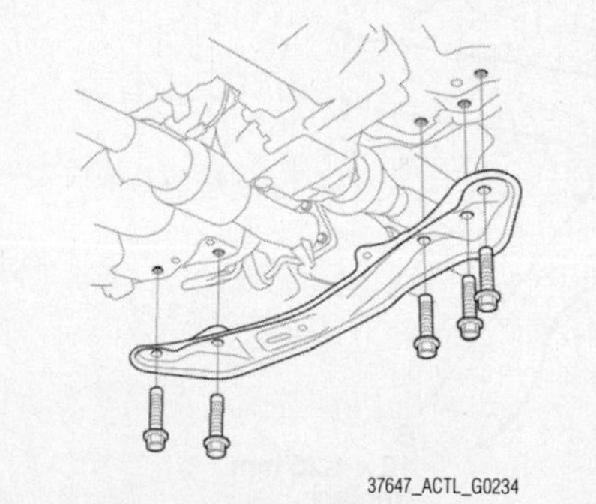
37647_ACTL_G0234

Fig. 312 Remove the front subframe stiffener

37647_ACTL_G0341

Fig. 313 Remove the subframe middle mounts

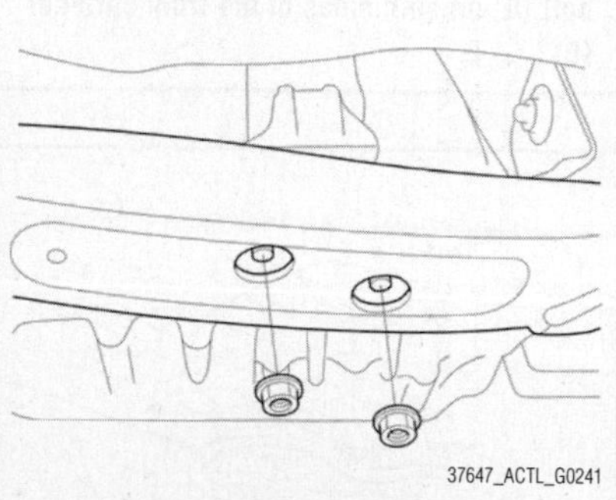
37647_ACTL_G0241

Fig. 314 Remove the nuts securing the lower transmission mount

mounting surface. Do not loosen the mounting bolts more than necessary.

32. Remove the stiffener mounting bolt on both sides of the rear stiffener.

33. Loosen the subframe mounting bolts to obtain a 1.18 inches (30 mm) distance between the bolt seat and the mounting surface. Do not loosen the mounting bolts more than necessary.

34. Lower the transmission jack slowly with the front subframe adapter until the front subframe has dropped about 1.18 inches (30 mm).

➡Do not lower the front subframe beyond the loosened subframe mounting bolts clearance.

35. Remove the flange bolts and the bushing holders, then remove the bushings.

36. Move the stabilizer bar toward the passenger's side, and remove the stabilizer bar.

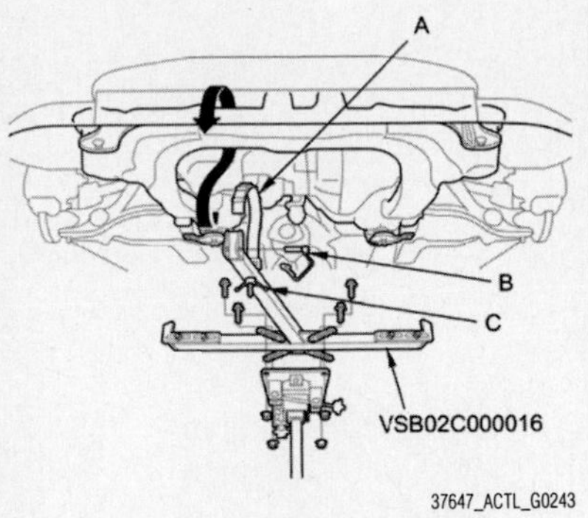
A
B
C
VSB02C000016
37647_ACTL_G0243

Fig. 315 Attach the subframe adapter to the front subframe by looping the strap (A) over the front of the front subframe, then secure the strap with the stop (B), then tighten the wing nut (C)

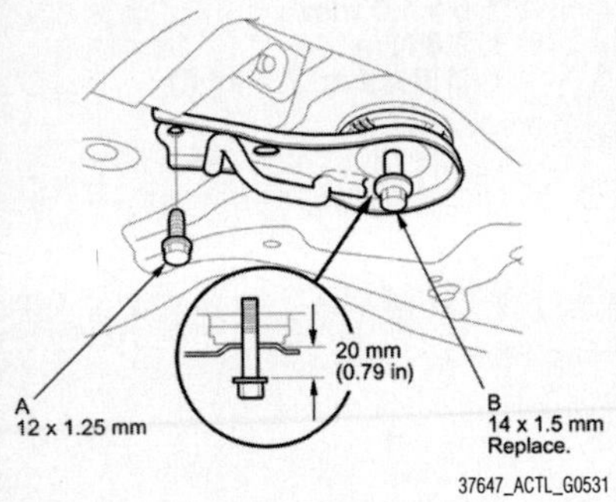
20 mm
(0.79 in)
A
12 x 1.25 mm
B
14 x 1.5 mm
Replace.
37647_ACTL_G0531

Fig. 316 Remove the stiffener mounting bolt (A) on both sides of the front stiffener (B)

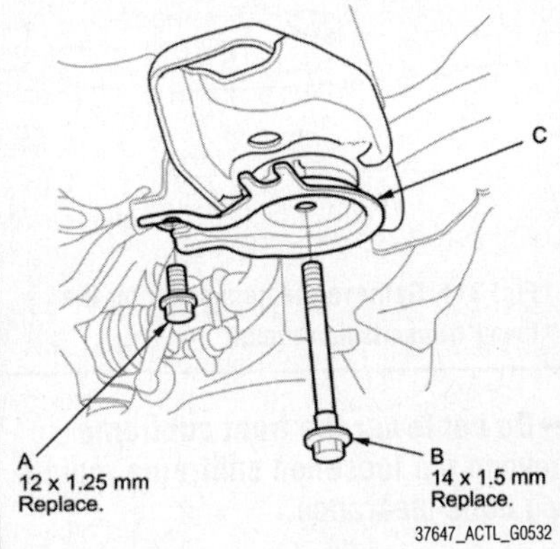
C
A
12 x 1.25 mm
Replace.
B
14 x 1.5 mm
Replace.
37647_ACTL_G0532

Fig. 317 Remove the stiffener mounting bolt on both sides of the rear stiffener

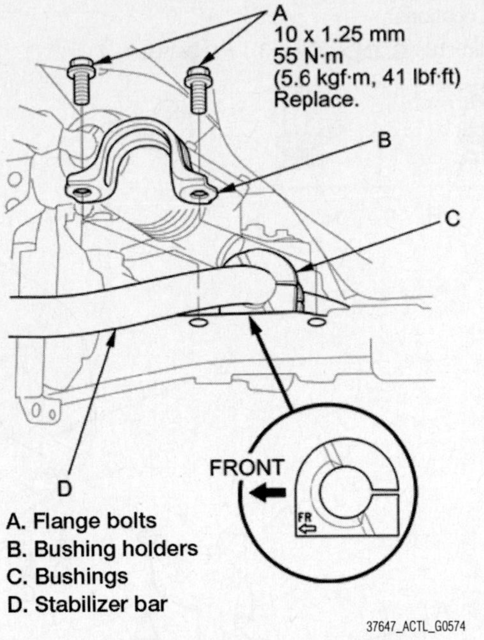
A
10 x 1.25 mm
55 N·m
(5.6 kgf·m, 41 lbf·ft)
Replace.
B
C
D
FRONT
FR
A. Flange bolts
B. Bushing holders
C. Bushings
D. Stabilizer bar
37647_ACTL_G0574

Fig. 318 Remove the flange bolts and the bushing holders, then remove the bushings

To install:

37. Install the stabilizer bar noting the following:

- Note the right and left direction of the stabilizer bar.
- Note the direction of installation for the bushings.
- Do not set the bushings on the bent or curved part of the stabilizer bar.

38. Loosely install the new right rear subframe mounting bolt and the new stiffener mounting bolt; insert the 15.7 mm side of the subframe alignment pin

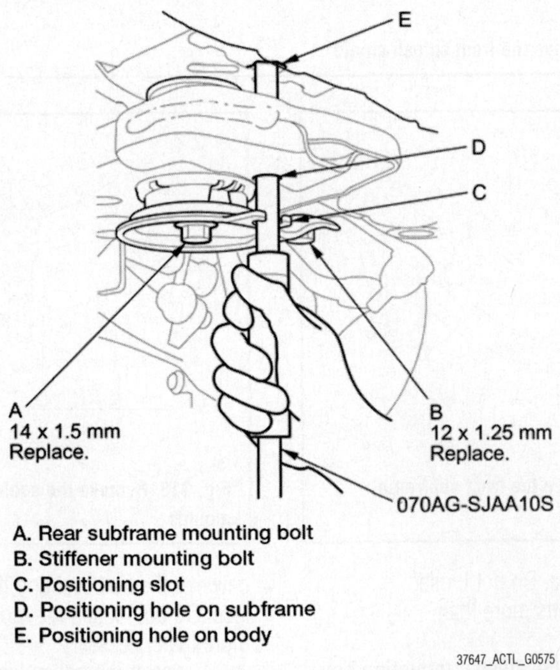
E
D
C
A
14 x 1.5 mm
Replace.
B
12 x 1.25 mm
Replace.
070AG-SJAA10S
A. Rear subframe mounting bolt
B. Stiffener mounting bolt
C. Positioning slot
D. Positioning hole on subframe
E. Positioning hole on body
37647_ACTL_G0575

Fig. 319 Insert the 15.7 mm side of the subframe alignment pin through the positioning slot on the rear stiffener, through the positioning hole on the front subframe, and into the positioning hole on the body

(070AG-SJAA10S) through the positioning slot on the rear stiffener, through the positioning hole on the front subframe, and into the positioning hole on the body, then tighten the right rear subframe mounting bolt.

39. Loosely install the new left rear subframe mounting bolt and the new stiffener mounting bolt with the same procedure as the right rear using the subframe alignment pin.

40. Loosely install the new subframe mounting bolt and the stiffener mounting bolt to both front stiffeners.

41. Tighten the right rear subframe mounting bolt to 76 ft. lbs. (103 Nm).

42. Tighten the left rear subframe mounting bolt to 76 ft. lbs. (103 Nm).

43. Tighten the front side subframe mounting bolts to 76 ft. lbs. (103 Nm).

➡Check all of the subframe mounting bolts, and retighten if necessary.

44. Tighten the stiffener mounting bolts to 69 ft. lbs. (93 Nm).

➡Before tightening the stiffener mounting bolts, check that the positioning holes and slot are aligned using the subframe alignment pin.

45. Install all of the removed parts in the reverse order of removal, and note these items:

- Refer to stabilizer link removal/installation to connect the stabilizer bar to the links.
- If the center guide is in place, use it to determine the steering joint installation angle.
- If the center guide is gone, check the steering joint installation angle.
- Check the steering wheel installation.
- When connecting the front and rear engine mount, first lightly tighten the mounting bolts, supporting the engine/transmission with a floor jack, then remove the jack, and tighten the bolts to the specified torque.
- Before installing the wheel, clean the mating surfaces of the brake disc and inside of the wheel.

46. Do the battery terminal reconnection procedure, then turn the ignition switch to ON (II) and check that the SRS indicator should come on for about 6 seconds and then go off.

47. Check the wheel alignment, and adjust it if necessary.

UPPER BALL JOINT

REMOVAL & INSTALLATION

See Figures 320 through 322.

➡Always use a ball joint remover to disconnect a ball joint. Do not strike the housing or any other part of the ball joint connection to disconnect it.

1. Install a hex nut or the ball joint thread protector onto the threads of the ball joint.

➡Using a hex nut, make sure the nut is flush with the ball joint pin end to prevent damage to the threaded end of the ball joint pin.

2. Apply grease to the ball joint remover on the areas shown. This will ease the installation of the tool, and prevent damage to the pressure bolt threads.

3. Loosen the pressure bolt, and install the ball joint remover as shown. Insert the jaws carefully, making sure not to damage the ball joint boot. Adjust the jaw spacing by turning the adjusting bolt.

➡Fasten the safety chain securely to a suspension arm or the subframe. Do not fasten it to a brake line or wire harness.

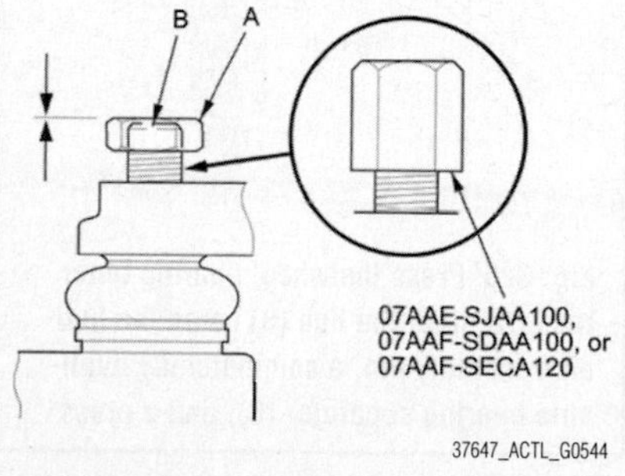
B
A
07AAE-SJAA100,
07AAF-SDAA100, or
07AAF-SECA120
37647_ACTL_G0544

Fig. 320 Install a hex nut (A) or the ball joint thread protector onto the threads of the ball joint (B)

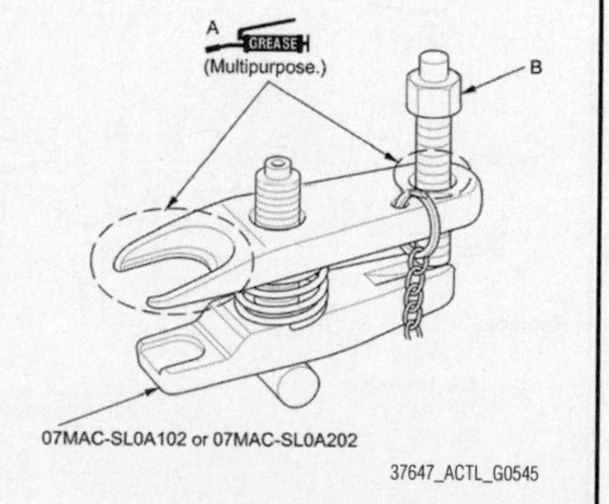
A
GREASE
(Multipurpose.)
B
07MAC-SL0A102 or 07MAC-SL0A202
37647_ACTL_G0545

Fig. 321 Apply grease to the ball joint remover on the areas shown (A)

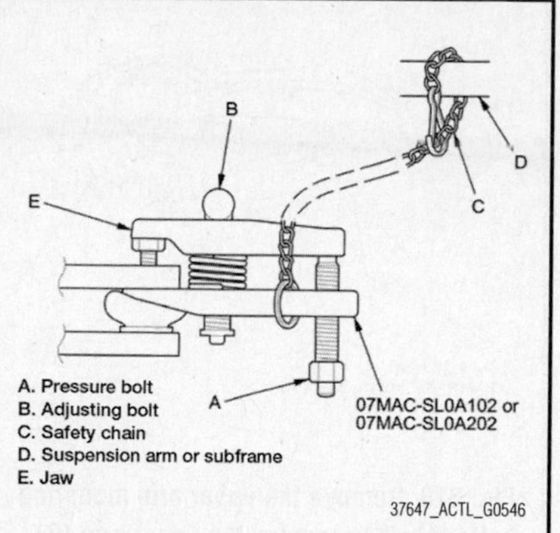
B
D
E
C
A
07MAC-SL0A102 or
07MAC-SL0A202
A. Pressure bolt
B. Adjusting bolt
C. Safety chain
D. Suspension arm or subframe
E. Jaw
37647_ACTL_G0546

Fig. 322 Loosen the pressure bolt, and install the ball joint remover as shown

4. After adjusting the adjusting bolt, make sure the head of the adjusting bolt is in the position to allow the jaw to pivot.

5. With a wrench, tighten the pressure bolt until the ball joint pin pops loose from the ball joint connecting hole. If necessary, apply penetrating type lubricant to loosen the ball joint pin.

➡Do not use pneumatic or electric tools on the pressure bolt.

6. Remove the ball joint remover, then remove the nut or the ball joint thread protector from the end of the ball joint pin, and pull the ball joint out of the ball joint connecting hole. Inspect the ball joint boot, and replace it if damaged.

UPPER CONTROL ARM

REMOVAL & INSTALLATION

See Figure 323.

1. Raise and support the vehicle.
2. Remove the front wheel.
3. Remove the front damper/spring.
4. Remove the cotter pin from the upper arm ball joint, then remove the castle nut.
5. Disconnect the upper arm ball joint from the knuckle using the ball joint thread protector and the ball joint remover.
6. Remove the upper arm mounting bolts, then remove the upper arm.
7. Install the upper arm in the reverse order of removal, and note these items:

- First install all of the components, and lightly tighten the bolts and the nuts, then raise the suspension to load it with the vehicle's weight before fully tightening it to the specified torque. Tighten the upper arm mounting bolts to 23 ft. lbs. (31 Nm).

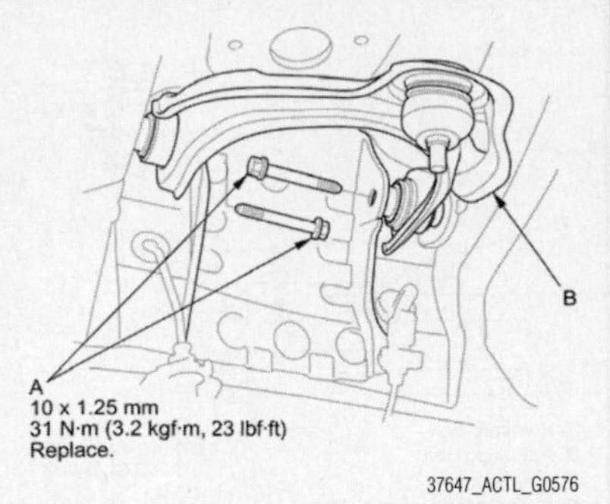
B
A
10 x 1.25 mm
31 N·m (3.2 kgf·m, 23 lbf·ft)
Replace.
37647_ACTL_G0576

Fig. 323 Remove the upper arm mounting bolts (A), then remove the upper arm (B)

✲✲ CAUTION

Do not place the jack against the ball joint pin of the knuckle.

- Be careful not to damage the ball joint boot when connecting the knuckle.
- Before connecting the ball joint, degrease the threaded section and the tapered portion of the ball joint pin, the ball joint connecting hole, the threaded section, and the mating surfaces of the castle nut.
- Torque the castle nut to 33–38 ft. lbs. (44–52 Nm), then tighten it only far enough to align the slot with the ball joint pin hole. Do not align the castle nut by loosening it.
- Before installing the wheel, clean the mating surfaces of the brake disc and the inside of the wheel.

8. Check the wheel alignment, and adjust it if necessary.

WHEEL HUB & BEARING

REMOVAL & INSTALLATION

See Figures 324 through 330.

1. Remove the knuckle/hub.
2. Separate the hub from the knuckle using the hub dis/assembly pin and a hydraulic press. Hold the knuckle with the attachment of the hydraulic press or equivalent tool. Be careful not to damage or deform the splash guard. Hold onto the hub to keep it from falling when pressed clear.
3. Press the wheel bearing inner race off of the hub using the hub dis/assembly pin, a commercially available bearing separator, and a press.
4. Remove the snap ring and the splash guard from the knuckle.
5. Press the wheel bearing out of the knuckle using the attachment, the driver handle, and a press.

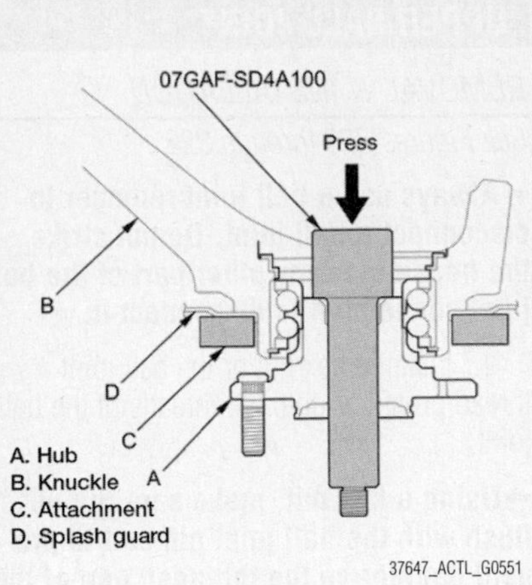
07GAF-SD4A100
Press
B
D
C
A
A. Hub
B. Knuckle
C. Attachment
D. Splash guard
37647_ACTL_G0551

Fig. 324 Separate the hub from the knuckle using the hub dis/assembly pin and a hydraulic press

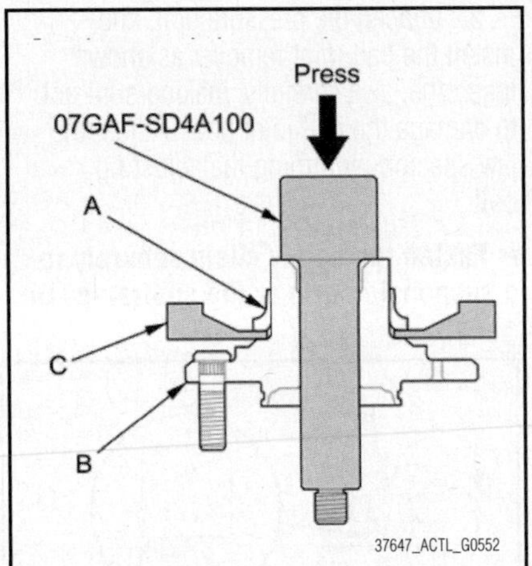
Press
07GAF-SD4A100
A
C
B
37647_ACTL_G0552

Fig. 325 Press the wheel bearing inner race (A) off of the hub (B) using the hub dis/assembly pin, a commercially available bearing separator (C), and a press

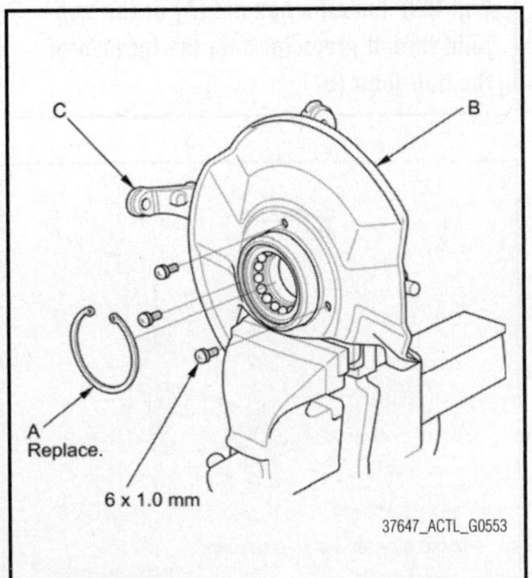
C
B
A
Replace.
6 x 1.0 mm
37647_ACTL_G0553

Fig. 326 Remove the snap ring (A) and the splash guard (B) from the knuckle (C)

6. Wash the knuckle and the hub thoroughly in high flash point solvent before reassembly.
7. Press a new wheel bearing into the knuckle using the old bearing, a steel plate, the attachment, the support base, and a press.

➡Install the wheel bearing with the wheel speed sensor magnetic encoder (brown color), toward the inside of the knuckle. Remove any oil, grease, dust, metal debris, and other foreign material from the magnetic encoder surface.

Keep any magnetic tools away from the magnetic encoder surface. Be careful not to damage the magnetic encoder surface when you insert the wheel bearing.

8. Check the front knuckle ring for damage or deformation, and replace it if necessary.

➡When installing the new front knuckle ring, position the knuckle ring notch portion toward the cut out near the ball joint in the knuckle, and align

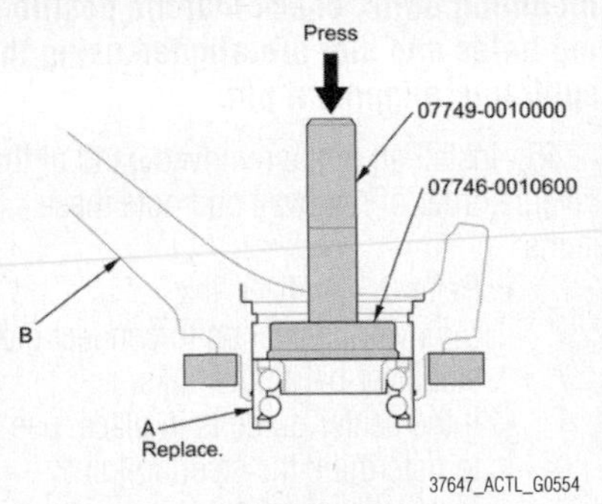
Press
07749-0010000
07746-0010600
B
A
Replace.
37647_ACTL_G0554

Fig. 327 Press the wheel bearing (A) out of the knuckle (B) using the attachment, the driver handle, and a press

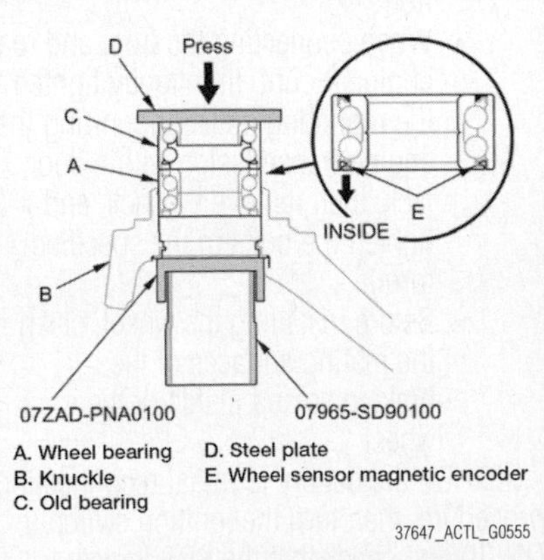
D
Press
C
A
INSIDE
E
B
07ZAD-PNA0100
07965-SD90100
A. Wheel bearing
B. Knuckle
C. Old bearing
D. Steel plate
E. Wheel sensor magnetic encoder
37647_ACTL_G0555

Fig. 328 Press a new wheel bearing into the knuckle using the old bearing, a steel plate, the attachment, the support base, and a press

the center of the knuckle ring ledge portion with the center of the wheel speed sensor hole on the knuckle.

9. Install the new snap ring securely in the knuckle.
10. Install the splash guard, and tighten the screws to 7 ft. lbs. (10 Nm).
11. Install the hub onto the knuckle using the attachment, the driver handle, the support base, and a hydraulic press. Be careful not to damage the splash guard.
12. Install the knuckle/hub.

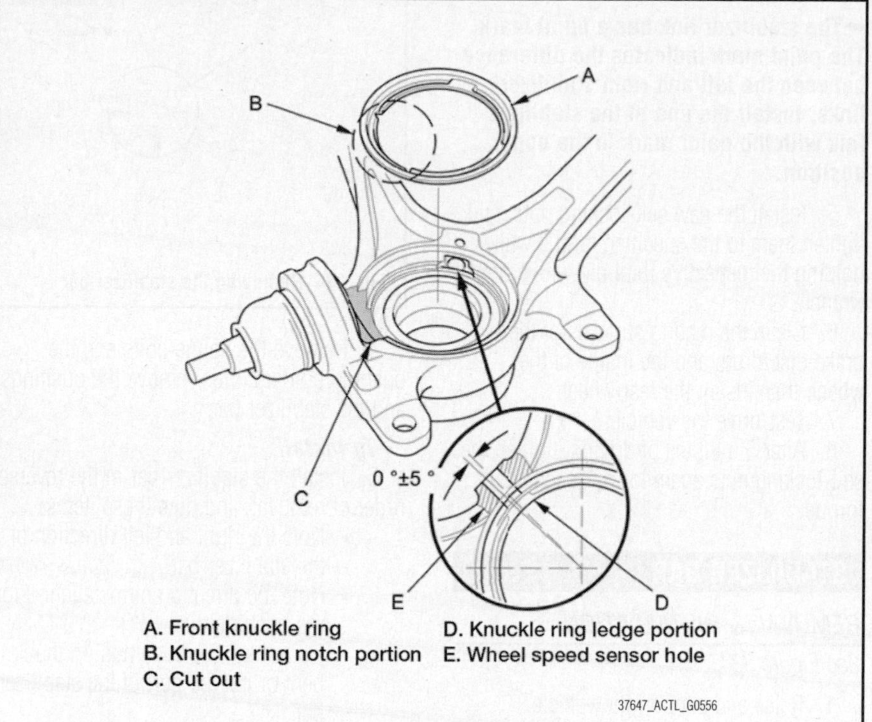

A
B
C
0 °±5 °
E
D
A. Front knuckle ring
B. Knuckle ring notch portion
C. Cut out
D. Knuckle ring ledge portion
E. Wheel speed sensor hole
37647_ACTL_G0556

Fig. 329 Check the front knuckle ring for damage or deformation, and replace it if necessary

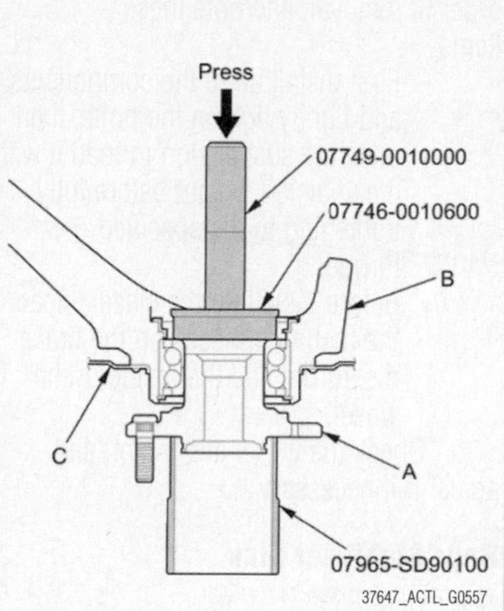

Press
07749-0010000
07746-0010600
B
C
A
07965-SD90100
37647_ACTL_G0557

Fig. 330 Install the hub (A) onto the knuckle (B) using the attachment, the driver handle, the support base, and a hydraulic press

SUSPENSION — REAR SUSPENSION

CONTROL ARMS/LINKS

REMOVAL & INSTALLATION

Rear Lower Arm A

See Figure 331.

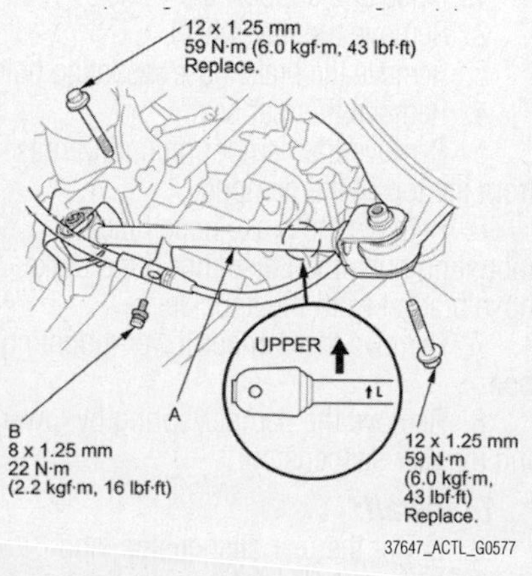

12 x 1.25 mm
59 N·m (6.0 kgf·m, 43 lbf·ft)
Replace.
UPPER
A
B
8 x 1.25 mm
22 N·m
(2.2 kgf·m, 16 lbf·ft)
12 x 1.25 mm
59 N·m
(6.0 kgf·m,
43 lbf·ft)
Replace.
37647_ACTL_G0577

Fig. 331 Remove the parking brake cable mounting bolt (B)

1. Raise and support the vehicle.
2. Remove the rear wheel.
3. Remove the parking brake cable mounting bolt.
4. Remove the lower arm A mounting bolts, then remove lower arm A.

To install:

➡Use new mounting bolts during reassembly.

5. Install lower arm A in the reverse order of removal, and note these items:
   - First install all of the components, and lightly tighten the bolts, then raise the suspension to load it with the vehicle's weight before fully tightening to the specified torque.
   - Before installing the wheel, clean the mating surfaces on the brake disc/drum and the inside of the wheel.
6. Check the wheel alignment, and adjust it if necessary.

Rear Lower Arm B

See Figure 332.

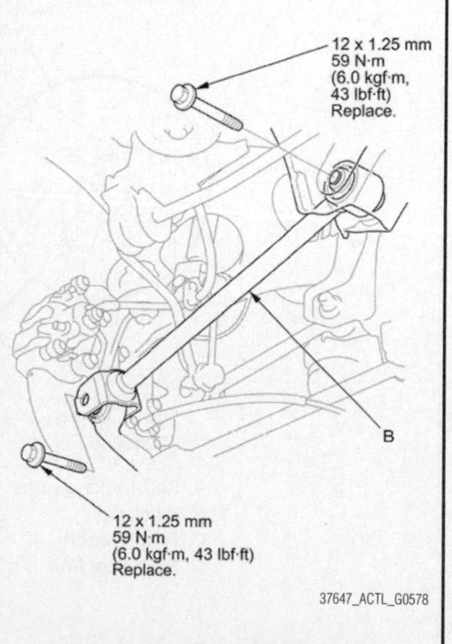

12 x 1.25 mm
59 N·m
(6.0 kgf·m,
43 lbf·ft)
Replace.
B
12 x 1.25 mm
59 N·m
(6.0 kgf·m, 43 lbf·ft)
Replace.
37647_ACTL_G0578

Fig. 332 Remove the lower arm B mounting bolts, then remove lower arm B

1. Raise and support the vehicle.
2. Remove the rear wheel.
3. Remove the lower arm B mounting bolts, then remove lower arm B.

To install:

➡Use new mounting bolts during reassembly.

4. Install lower arm B in the reverse order of removal, and note these items:
   - First install all of the components, and lightly tighten the bolts, then raise the suspension to load it with the vehicle's weight before fully tightening to the specified torque.
   - Before installing the wheel, clean the mating surfaces on the brake disc/drum and the inside of the wheel.
5. Check the wheel alignment, and adjust it if necessary.

Rear Stabilizer Link

See Figure 333.

1. Raise and support the vehicle.
2. Remove the rear wheel.
3. Remove the self-locking nuts while holding the respective joint pin with a hex wrench, then remove the stabilizer link.

To install:

4. Install the stabilizer link on the stabilizer bar and the brake hose bracket with the joint pins set at the center of their range of movement.

➡The stabilizer link has a paint mark. The paint mark indicates the difference between the left and right stabilizer links. Install the end of the stabilizer link with the paint mark in the upper position.

5. Install the new self-locking nuts, and tighten them to the specified torque while holding the respective joint pin with a hex wrench.
6. Clean the mating surfaces of the brake disc/drum and the inside of the wheel, then install the rear wheel.
7. Test-drive the vehicle.
8. After 5 minutes of driving, tighten the self-locking nuts again to the specified torque.

STABILIZER BAR

REMOVAL & INSTALLATION

See Figure 334.

1. Raise and support the vehicle.
2. Remove the rear wheels.
3. Disconnect both stabilizer links from the stabilizer bar.

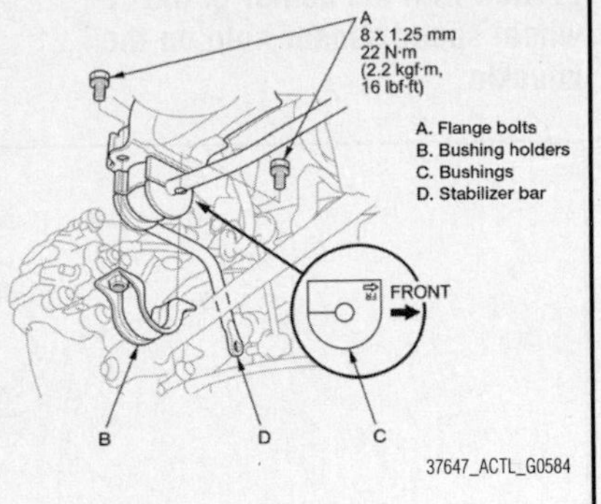
A
8 x 1.25 mm
22 N·m
(2.2 kgf·m,
16 lbf·ft)
A. Flange bolts
B. Bushing holders
C. Bushings
D. Stabilizer bar
FRONT
B
D
C
37647_ACTL_G0584

Fig. 334 Removing the stabilizer bar

4. Remove the flange bolts and the bushing holders, then remove the bushings and the stabilizer bar.

To install:

5. Install the stabilizer bar in the reverse order of removal, and note these items:
   - Note the right and left direction of the stabilizer bar.
   - Note the direction of installation for the bushing.
   - Do not set the bushings on the bent or curved part of the stabilizer bar.
   - Refer to the stabilizer link removal/installation to connect the stabilizer bar to the links.
   - Before installing the wheel, clean the mating surfaces of the brake disc/drum and the inside of the wheel.

STRUT (DAMPER/SPRING)

REMOVAL & INSTALLATION

See Figures 335 through 338.

1. Raise and support the vehicle.
2. Remove the rear wheel.
3. Remove the brake hose mounting bolt.
4. Remove the rear seat.
5. Remove the damper mounting nuts from the top of the damper.
6. Remove the brake hose bracket mounting nuts, then disconnect the brake hose bracket from the knuckle.
7. Remove the damper lower mounting bolt.
8. Remove the damper/spring by lowering the rear suspension.

To install:

9. Lower the rear suspension, and position the damper/spring in the body with the welded nut on the bottom of the damper facing forward.

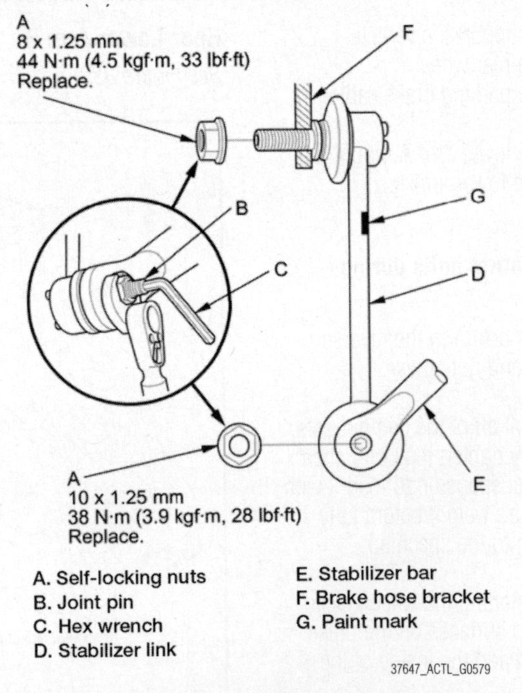
A
8 x 1.25 mm
44 N·m (4.5 kgf·m, 33 lbf·ft)
Replace.
F
B
G
C
D
A
10 x 1.25 mm
38 N·m (3.9 kgf·m, 28 lbf·ft)
Replace.
E
A. Self-locking nuts
B. Joint pin
C. Hex wrench
D. Stabilizer link
E. Stabilizer bar
F. Brake hose bracket
G. Paint mark
37647_ACTL_G0579

Fig. 333 Remove the self-locking nuts while holding the respective joint pin with a hex wrench, then remove the stabilizer link

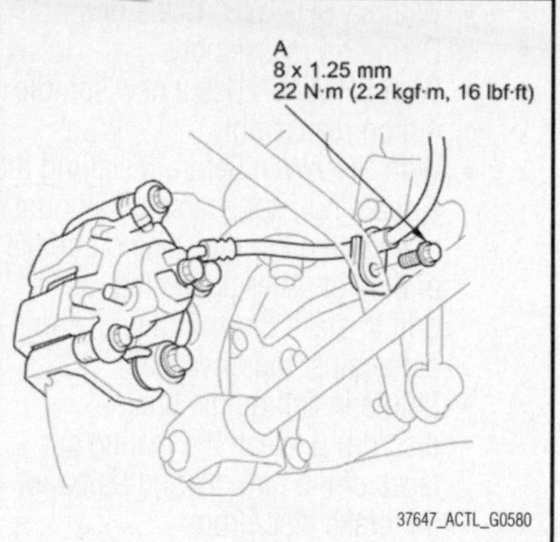
A
8 x 1.25 mm
22 N·m (2.2 kgf·m, 16 lbf·ft)
37647_ACTL_G0580

Fig. 335 Remove the brake hose mounting bolt (A)

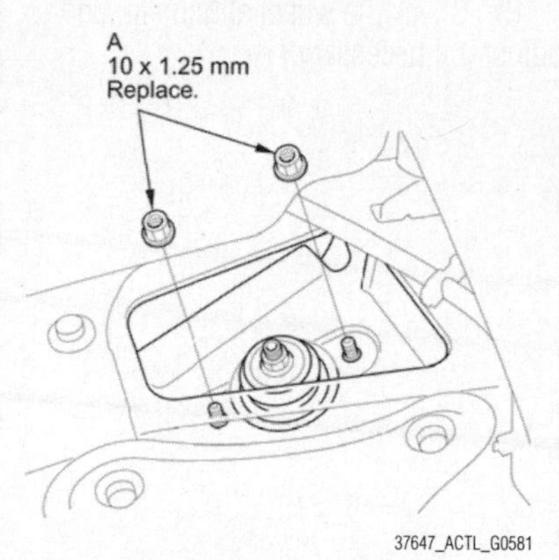
A
10 x 1.25 mm
Replace.
37647_ACTL_G0581

Fig. 336 Remove the damper mounting nuts (A) from the top of the damper

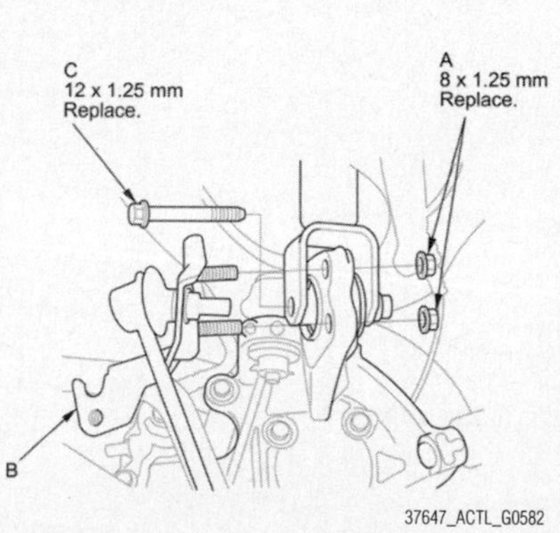
C
12 x 1.25 mm
Replace.
A
8 x 1.25 mm
Replace.
B
37647_ACTL_G0582

Fig. 337 Remove the brake hose bracket mounting nuts (A), then disconnect the brake hose bracket (B) from the knuckle

➡Make sure the damper is installed in the correct direction.

10. Loosely install the new damper mounting nuts to the top of the damper.

11. Loosely install the new damper lower mounting bolt on the bottom of the damper. Connect the brake hose bracket to the knuckle, and loosely install the new brake hose bracket mounting nuts.

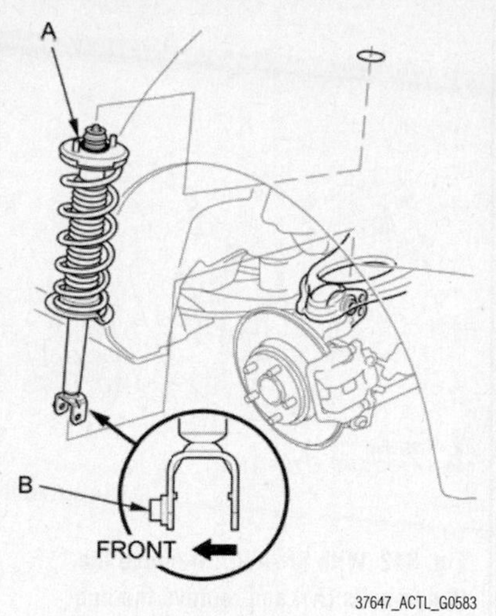
A
B
FRONT
37647_ACTL_G0583

Fig. 338 Make sure the damper is installed in the correct direction

12. Place a floor jack under the connecting point of the knuckle and lower arm A, and raise the suspension to load it with the vehicle's weight.

13. Tighten the damper lower mounting bolt to 47 ft. lbs. (64 Nm) and the brake hose bracket mounting nuts to 13 ft. lbs. (18 Nm).

14. Tighten the damper mounting nuts on top of the damper to 41 ft. lbs. (55 Nm).

15. Install the brake hose mounting bolt, and tighten the mounting bolt to 16 ft. lbs. (22 Nm).

16. Install the rear seat.

17. Clean the mating surfaces of the brake disc/drum and the inside of the wheel, then install the rear wheel.

18. Check the wheel alignment, and adjust it if necessary.

WHEEL HUB & BEARING

REMOVAL & INSTALLATION

See Figures 339 through 342.

1. Raise and support the vehicle.

2. Remove the wheel nuts, and the rear wheel.

3. Release the parking brake lever fully.

4. Remove the brake hose mounting bolt.

5. Remove the brake caliper bracket mounting bolts, then remove the caliper assembly from the knuckle. To prevent damage to the caliper assembly or the brake hose, use a short piece of wire to hang the caliper assembly from the undercarriage. Do not twist the brake hose excessively.

6. Remove the two washers.

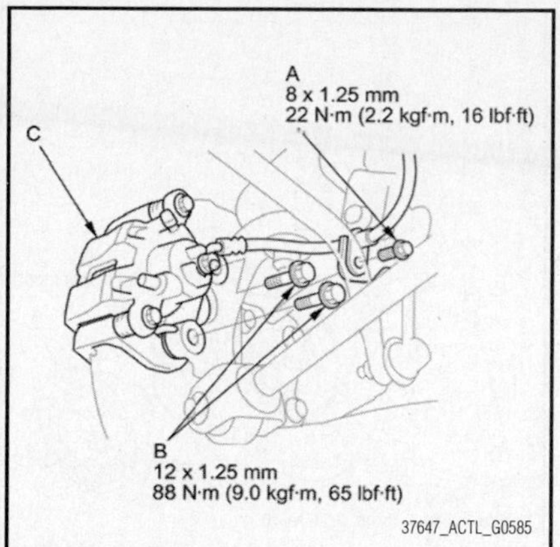
A
8 x 1.25 mm
22 N·m (2.2 kgf·m, 16 lbf·ft)
C
B
12 x 1.25 mm
88 N·m (9.0 kgf·m, 65 lbf·ft)
37647_ACTL_G0585

Fig. 339 Remove the brake hose mounting bolt (A) and the brake caliper bracket mounting bolts (B), then remove the caliper assembly (C)

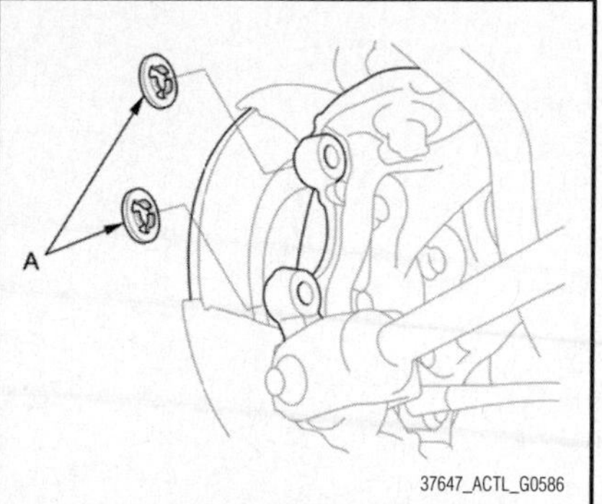
A
37647_ACTL_G0586

Fig. 340 Remove the two washers (A)

➡During installation, make sure the washers are installed between the brake caliper bracket and the knuckle.

7. With SH-AWD: Pry up the stake on the spindle nut, then remove the spindle nut.

8. Remove the rear brake disc/drum.

9. Without SH-AWD: Remove the hub bearing unit and the O-ring.

10. With SH-AWD: Remove the flange bolts, and remove the hub bearing unit by tapping the driveshaft end with a soft face hammer while drawing the hub bearing unit outward.

➡Do not pull the driveshaft end outward. The driveshaft inboard joint may come apart.

11. Check the hub bearing unit for damage and cracks.

To install:

12. Install the hub bearing unit in the reverse order of removal, and note these items:

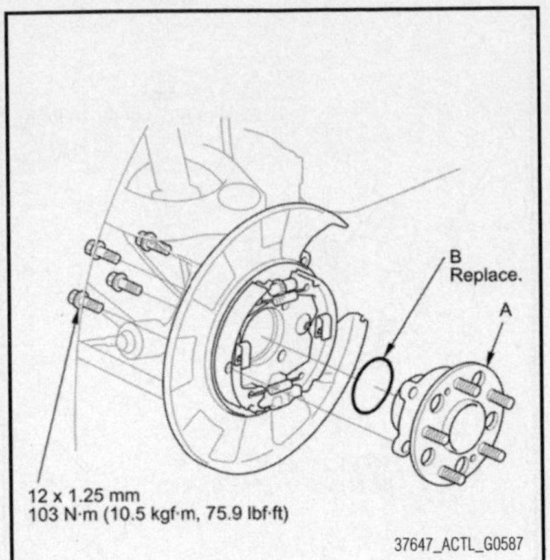
B
Replace.
A
12 x 1.25 mm
103 N·m (10.5 kgf·m, 75.9 lbf·ft)
37647_ACTL_G0587

Fig. 341 Without SH-AWD: Remove the hub bearing unit (A) and the O-ring (B)

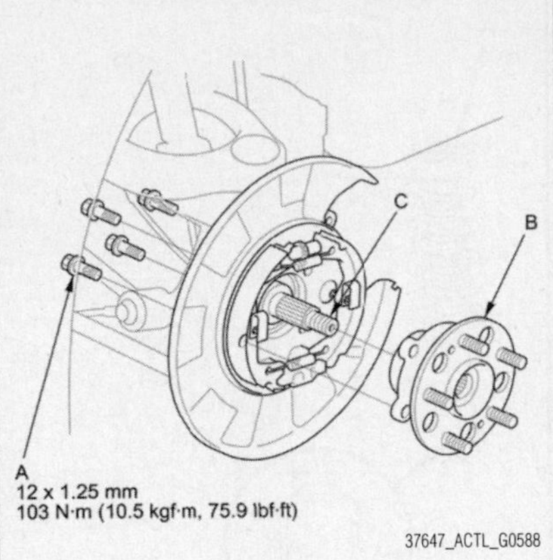
C
B
A
12 x 1.25 mm
103 N·m (10.5 kgf·m, 75.9 lbf·ft)
37647_ACTL_G0588

Fig. 342 With SH-AWD: Remove the flange bolts (A), and remove the hub bearing unit (B) by tapping the driveshaft end (C) with a soft face hammer

- Without SH-AWD: Use a new O-ring on reassembly.
- With SH-AWD: Use a new spindle nut on reassembly.
- With SH-AWD: Before installing the spindle nut, apply a small amount of engine oil to the seating surface of the nut. After tightening, use a drift to stake the spindle nut shoulder against the driveshaft.
- Before installing the brake disc/drum, clean the mating surfaces of the hub bearing unit and the brake disc/drum.
- Before installing the wheel, clean the mating surfaces of the brake disc/drum and the inside of the wheel.

13. Check the wheel alignment, and adjust it if necessary.

ACURA

TSX

5

BRAKES........................5-10

ANTI-LOCK BRAKE SYSTEM (ABS)................5-11
Wheel Speed Sensors...............5-11
Removal & Installation..........5-11
BLEEDING THE BRAKE SYSTEM........................5-10
Bleeding Procedure..................5-10
Bleeding Procedure.............5-10
Bleeding the ABS System.....5-10
FRONT DISC BRAKES.........5-12
Brake Caliper..........................5-12
Removal & Installation..........5-12
Disc Brake Pads......................5-12
Removal & Installation..........5-12
INFORMATION AND PRECAUTIONS...............5-10
Anti-lock Systems................5-10
Disc and Drum Systems.............................5-10
PARKING BRAKE...............5-14
Parking Brake Cables...............5-14
Adjustment..........................5-14
REAR DISC BRAKES..........5-13
Brake Caliper...........................5-13
Removal & Installation..........5-13
Disc Brake Pads......................5-13
Removal & Installation..........5-13

CHASSIS ELECTRICAL.......5-15

AIR BAG (SUPPLEMENTAL RESTRAINT SYSTEM)........5-15
General Information..................5-15
Arming the System..............5-15
Disarming the System...........5-15
Precautions..........................5-15

DRIVE TRAIN.................5-15

Clutch......................................5-15
Removal & Installation..........5-15
Front Halfshafts........................5-18
Removal & Installation..........5-18
Hydraulic System Bleeding.......5-18
Bleeding Procedure.............5-18
Fluid Fill Procedure.............5-18
Intermediate Shaft.....................5-22
Removal & Installation..........5-22

ENGINE COOLING...........5-23

Engine Fan...............................5-23
Removal & Installation..........5-23
Radiator...................................5-25
Removal & Installation..........5-25
Thermostat...............................5-28
Removal & Installation..........5-28
Water Pump.............................5-28
Battery Reconnect/Relearn Procedure..........................5-29
Removal & Installation..........5-28

ENGINE ELECTRICAL.........5-29

CHARGING SYSTEM...........5-29
Alternator.................................5-29
Removal & Installation..........5-29
IGNITION SYSTEM.............5-31
Firing Order.............................5-31
Ignition Coil.............................5-31
Removal & Installation..........5-31
Ignition Timing.........................5-32
Adjustment..........................5-32
Spark Plugs.............................5-32
Removal & Installation..........5-32
STARTING SYSTEM...........5-33
Starter......................................5-33
Removal & Installation..........5-33

ENGINE MECHANICAL.......5-34

Accessory Drive Belts..............5-34
Accessory Belt Routing.............................5-34
Adjustment..........................5-34
Inspection...........................5-34
Removal & Installation..........5-34
Cam Chain Front Cover............5-60
Removal & Installation..........5-60
Camshaft..................................5-34
Inspection...........................5-34
Removal & Installation..........5-36
Catalytic Converter...................5-38
Removal & Installation..........5-38
Crankshaft Front Seal...............5-41
Removal & Installation..........5-41
Crankshaft Pulley.....................5-40
Removal & Installation..........5-40
Cylinder Head..........................5-41
Removal & Installation..........5-41
Drive Plate...............................5-46
Removal & Installation..........5-46
Intake Manifold........................5-46
Removal & Installation..........5-46
Oil Pan.....................................5-48
Removal & Installation..........5-48
Oil Pump..................................5-50
Inspection...........................5-53
Removal & Installation..........5-50
Piston & Ring...........................5-54
Positioning..........................5-54
Rear Main Seal.........................5-54
Removal & Installation..........5-54
Rocker Arms.............................5-55
Removal & Installation..........5-55
Timing (Cam) Chain & Sprockets..............................5-61
Removal & Installation..........5-61
Timing Belt & Sprockets...........5-58
Removal & Installation..........5-58
Timing Belt Front Cover............5-57
Removal & Installation..........5-57
Valve Lash................................5-64
Adjustment..........................5-64

ENGINE PERFORMANCE & EMISSION CONTROLS.......5-66

Accelerator Pedal Position (APP) Sensor.........................5-66
Location..............................5-66
Removal & Installation..........5-66
Camshaft Position (CMP) Sensor....................................5-66
Location..............................5-66
Removal & Installation..........5-66
Crankshaft Position (CKP) Sensor....................................5-67
CKP Pattern Clear/CKP Pattern Learn.....................5-67
Location..............................5-67
Removal & Installation..........5-67

Engine Control Module (ECM)/Powertrain Control Module (PCM)....5-67
- ECM/PCM IDLE Learn Procedure....5-70
- ECM/PCM Update....5-70
- Location....5-67
- Removal & Installation....5-68

Engine Coolant Temperature (ECT) Sensor....5-70
- Location....5-70
- Removal & Installation....5-70

Evaporative Emissions (EVAP) Canister....5-71
- Location....5-71
- Removal & Installation....5-71

Exhaust Gas Recirculation (EGR) Valve....5-72
- Location....5-72
- Removal & Installation....5-73

Heated Oxygen (HO2S) Sensor....5-73
- Removal & Installation....5-73

Intake Air Temperature (IAT) Sensor....5-73
- Location....5-73
- Removal & Installation....5-74

Knock Sensor (KS)....5-74
- Removal & Installation....5-74

Manifold Absolute Pressure (MAP) Sensor....5-74
- Location....5-74
- Removal & Installation....5-74

Mass Air Flow (MAF)/Intake Air Temperature (IAT) Sensor (Hot Wire)....5-74
- Location....5-74
- Removal & Installation....5-74

Output Shaft Speed (OSS) Sensor....5-75
- Location....5-75
- Removal & Installation....5-75

Positive Crankcase Ventilation (PCV) Valve....5-76
- Removal & Installation....5-76

Variable Timing Camshaft (VTC) Oil Control Solenoid Valve....5-76
- Location....5-76
- Removal & Installation....5-76

FUEL....5-77

GASOLINE FUEL INJECTION SYSTEM....5-77

Fuel Filter....5-78
- Removal & Installation....5-78

Fuel Pump/Fuel Gauge Sending Unit....5-78
- Removal & Installation....5-78

Fuel Rail & Injector....5-79
- Removal & Installation....5-79

Fuel System Pressure....5-77
- Relieving....5-77

Fuel System Service Precautions....5-77

Fuel Tank....5-82
- Draining....5-82
- Removal & Installation....5-82

Idle Speed....5-83
- Adjustment....5-83

Throttle Body....5-86
- Removal & Installation....5-86

HEATING & AIR CONDITIONING SYSTEM....5-86

Blower Motor....5-86
- Removal & Installation....5-86

Heater Core....5-86
- Removal & Installation....5-86

Heater Unit....5-88
- Removal & Installation....5-88

SPECIFICATIONS AND MAINTENANCE CHARTS....5-3

Brake Specifications....5-8
Camshaft Specifications....5-5
Capacities....5-4
Crankshaft and Connecting Rod....5-5
Engine and Vehicle Identification....5-3
Gasoline Engine Tune-Up Specifications....5-3
Fluid Specifications....5-4
General Engine Specifications....5-3
Piston and Ring Specifications....5-6
Scheduled Maintenance Intervals....5-9
Tire, Wheel and Ball Joint Specifications....5-8
Torque Specifications....5-6
Valve Specifications....5-4
Wheel Alignment....5-8

STEERING....5-91

Electric Power Steering (EPS) Control Unit....5-91
- Removal & Installation....5-91
- Torque Sensor Neutral Position Memorization....5-91

EPS Steering Gearbox....5-92
- Removal & Installation....5-92
- Torque Sensor Neutral Position Memorization....5-97

SUSPENSION....5-103

FRONT SUSPENSION....5-103

Control Links....5-103
- Removal & Installation....5-103

Lower Ball Joint....5-103
- Removal & Installation....5-103

Lower Control Arm....5-104
- Removal & Installation....5-104

Stabilizer Bar....5-105
- Removal & Installation....5-105

Steering Knuckle....5-105
- Removal & Installation....5-105

Strut (Damper/Spring)....5-106
- Overhaul....5-108
- Removal & Installation....5-106

Upper Ball Joint....5-109
- Removal & Installation....5-109

Upper Control Arm....5-110
- Removal & Installation....5-110

Wheel Hub & Bearing....5-111
- Removal & Installation....5-111

REAR SUSPENSION....5-112

Control Arms/Links....5-112
- Removal & Installation....5-112

Stabilizer Bar....5-115
- Removal & Installation....5-115

Strut (Damper/Spring)....5-115
- Removal & Installation....5-115

SPECIFICATIONS AND MAINTENANCE CHARTS

ENGINE AND VEHICLE IDENTIFICATION

Engine							Model Year	
Code	Liters (cc)	Cu. In.	Cyl.	Fuel Sys.	Engine Type	Eng. Mfg.	Code ①	Year
K24Z3	2.4 (2354)	144	4	MPFI	DOHC	Honda	9	2009
J35Z6	3.5 (3490)	213	6	PGM-FI	SOHC	Honda	A	2010

MPFI: Multi-Port Fuel Injection

PGM-FI: Programmed fuel injection

DOHC: Double Overhead Camshaft

SOHC: Single Overhead Camshaft

① 10th digit of the Vehicle Identification Number (VIN)

37647_ATSX_C0001

GENERAL ENGINE SPECIFICATIONS

Year	Model	Engine Displacement Liters	Engine ID	Net Horsepower @ rpm	Net Torque @ rpm (ft. lbs.)	Bore x Stroke (in.)	Com-pression Ratio	Oil Pressure @ rpm
2009	TSX	2.4	K24Z3	201@7000	172@3400	3.43X3.90	11.0:1	44@3000
2010	TSX	2.4	K24Z3	201@7000	172@3400	3.43X3.90	11.0:1	44@3000
		3.5	J35Z6	280@6200	254@5000	3.50X3.66	11.2:1	71@3000

37647_ATSX_C0002

GASOLINE ENGINE TUNE-UP SPECIFICATIONS

Year	Engine Displacement Liters	Engine ID/VIN	Spark Plug Gap (in.)	Ignition Timing (deg.)		Fuel Pump (psi)	Idle Speed (rpm)		Valve Clearance	
				MT	AT		MT	AT	In.	Ex.
2009	2.4	K24Z3	0.039-0.043	6-10 B	6-10 B	48-55	700-800	750-850	0.008-0.010	0.010-0.011
2010	2.4	K24Z3	0.039-0.043	6-10 B	6-10 B	48-55	700-800	750-850	0.008-0.010	0.010-0.011
	3.5	J35Z6	0.039-0.043	8-12 B	8-12 B	57-64	—	630-730	0.008-0.009	0.011-0.013

NOTE: The Vehicle Emission Control Information label reflects specification changes during production and must be used if they differ from this chart.

B: Before Top Dead Center

37647_ATSX_C0003

CAPACITIES

Year	Model	Engine Displacement Liters	Engine ID/VIN	Engine Oil with Filter (qts.)	Transmission (pts.) 6-Spd	Transmission (pts.) Auto.	Fuel Tank (gal.)	Cooling System (qts.)
2009	TSX	2.4	K24Z3	4.2	4.2	5.2	18.5	6.5
2010	TSX	2.4	K24Z3	4.2	4.2	5.2	18.5	6.5
		3.5	J35Z6	4.5	—	6.0	18.5	7.0

NOTE: All capacities are approximate. Add fluid gradually and ensure a proper fluid level is obtained.

NOTE: Capacities given are service, not overhaul capacities

37647_ATSX_C0004

FLUID SPECIFICATIONS

Year	Model	Engine Displacement Liters	Engine Oil	Man. Trans.	Auto. Trans.	Power Steering Fluid	Brake Master Cylinder	Cooling System
2009	TSX	2.4	5W-20 Acura	Acura MTF	Acura ATF-Z1	Acura PS Fluid	Acura DOT 3	①
2010	TSX	2.4	5W-20 Acura	Acura MTF	Acura ATF-Z1	Acura PS Fluid	Acura DOT 3	①
		3.5	5W-20 Honda	—	Acura ATF-Z1	Acura PS Fluid	Acura DOT 3	①

DOT: Department Of Transportation

① Acura Long Life Antifreeze/Coolant-Type2

37647_ATSX_C0005

VALVE SPECIFICATIONS

Year	Engine Displacement Liters	Engine ID/VIN	Seat Angle (deg.)	Face Angle (deg.)	Spring Test Pressure (lbs. @ in.)	Spring Installed Height (in.)	Stem-to-Guide Clearance (in.) Intake	Stem-to-Guide Clearance (in.) Exhaust	Stem Diameter (in.) Intake	Stem Diameter (in.) Exhaust
2009	2.4	K24Z3	45	45	NA	NA	0.0012-0.0022	0.0022-0.0031	0.2156-0.2159	0.2146-0.2150
2010	2.4	K24Z3	45	45	NA	NA	0.0012-0.0022	0.0022-0.0031	0.2156-0.2159	0.2146-0.2150
	3.5	J35Z6	45	45	NA	NA	0.0008-0.0018	0.0022-0.0031	0.2159-0.2163	0.2146-0.2150

NA: Not Available

37647_ATSX_C0006

CAMSHAFT AND BEARING SPECIFICATIONS

All measurements are given in inches.

Year	Engine Displacement Liters	Engine VIN	Journal Diameter	Brg. Oil Clearance	Shaft End-play	Runout	Journal Bore	Lobe Height Intake	Lobe Height Exhaust
2009	2.4	K24Z3	NA	①	0.0020-0.0080	0.0010	NA	②	1.3500
2010	2.4	K24Z3	NA	①	0.0020-0.0080	0.0010	NA	②	1.3500
	3.5	J35Z6	NA	0.0020-0.0035	0.0020-0.0080	0.0010	NA	③	1.4472

NA: Information not available

① No. 1 Journal: 0.0010-0.0030 inches

Other Journals: 0.0020-0.0040 inches

② Primary: 1.3285 inches

Mid: 1.359 inches

Secondary: 1.3285 inches

③ Primary: 1.3504 inches

Secondary: 1.4024 inches

37647_ATSX_C0009

CRANKSHAFT AND CONNECTING ROD SPECIFICATIONS

All measurements are given in inches.

Year	Engine Displacement Liters	Engine ID/VIN	Crankshaft Main Brg. Journal Dia.	Crankshaft Main Brg. Oil Clearance	Crankshaft Shaft End-play	Crankshaft Thrust on No.	Connecting Rod Journal Diameter	Connecting Rod Oil Clearance	Connecting Rod Side Clearance
2009	2.4	K24Z3	①	②	0.0040-0.0140	4	1.8888-1.8898	0.0013-0.0026	0.0060-0.0140
2010	2.4	K24Z3	①	②	0.0040-0.0140	4	1.8888-1.8898	0.0013-0.0026	0.0060-0.0140
	3.5	J35Z6	2.8337-2.8346	0.0007-0.0017	0.0040-0.0140	3	2.1644-2.1654	0.0008-0.0017	0.0060-0.0140

① Nos. 1, 2, 4 and 5: 2.1647-2.1657 inches

No. 3: 2.1644-2.1654 inches

② Nos. 1, 2, 4 and 5: 0.0007-0.0016 inches

No. 3: 0.0010-0.0019 inches

37647_ATSX_C0007

PISTON AND RING SPECIFICATIONS

All measurements are given in inches

Year	Engine Displacement Liters	Engine ID/VIN	Piston Clearance	Ring Gap Top Compression	Ring Gap Bottom Compression	Ring Gap Oil Control	Ring Side Clearance Top Compression	Ring Side Clearance Bottom Compression	Ring Side Clearance Oil Control
2009	2.4	K24Z3	0.0008-0.0016	0.0080-0.0014	0.0200-0.0260	0.0080-0.0280	0.0024-0.0033	0.0016-0.0026	NA
2010	2.4	K24Z3	0.0008-0.0016	0.0080-0.0014	0.0200-0.0260	0.0080-0.0280	0.0024-0.0033	0.0016-0.0026	NA
	3.5	J35Z6	0.0006-0.0016	0.0080-0.0140	0.0160-0.0220	0.0080-0.0280	0.0022-0.0031	0.0012-0.0022	NA

NA; Not Available

37647_ATSX_C0008

TORQUE SPECIFICATIONS

All readings in ft. lbs.

Year	Engine Displacement Liters	Engine ID/VIN	Cylinder Head Bolts	Main Bearing Bolts	Rod Bearing Bolts	Crankshaft Damper Bolts	Flywheel Bolts	Manifold Intake	Manifold Exhaust	Spark Plugs	Oil Pan Drain Plug
2009	2.4	K24Z3	①	②	③	④	54	16	33	13	30
2010	2.4	K24Z3	①	②	③	④	54	16	33	13	30
	3.5	J35Z6	⑤	⑥	⑦	⑧	54	16	23	13	29

① Step 1: 29 ft. lbs.
Step 2: Rotate 90 degrees
Step 3: Rotate an additional 90 degrees
Step 4 (new bolts only): additional 90 degrees

② Step 1: 22 ft. lbs.
Step 2: Rotate 48 degrees

③ Step 1: 30 ft. lbs.
Step 2: Rotate 120 degrees

④ Step 1: 36 ft. lbs.
Step 2: Reotate 90 degrees

⑤ Step 1: 22 ft. lbs.
Step 2: Rotate 90 degrees
Step 3: Rotate an additional 90 degrees
Step 4 (new bolts only): additional 90 degrees

⑥ Step 1: 54 ft. lbs.
Step 2: Repeat Step 1 to ensure 54 ft. lbs.
Step 3: Side bolts to 36 ft. lbs.
Step 4: Repeat Step 3 to ensure 36 ft. lbs.

⑦ Step 1: 14 ft. lbs.Cap bolts: 29 ft. lbs.
Step 2: Rotate 90 degrees

⑧ Step 1: 47 ft. lbs.
Step 2: Rotate 60 degrees

37647_ATSX_C0010

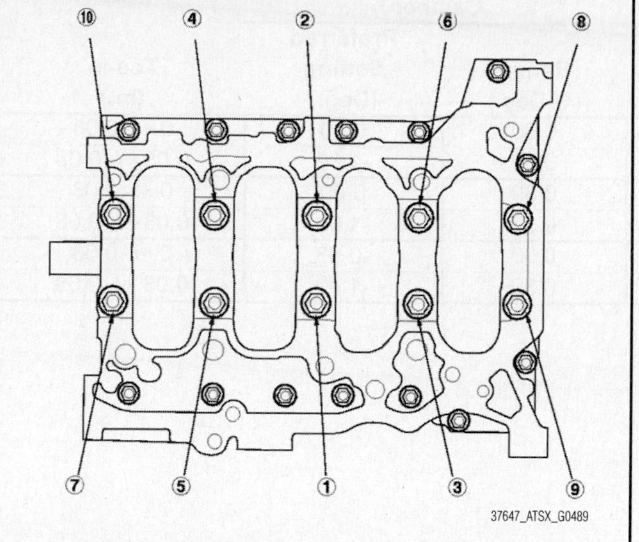
10
4
2
6
8
7
5
1
3
9
37647_ATSX_G0489

Fig. 1 Main bearing torque sequence—4 cylinder engine, 1 of 2

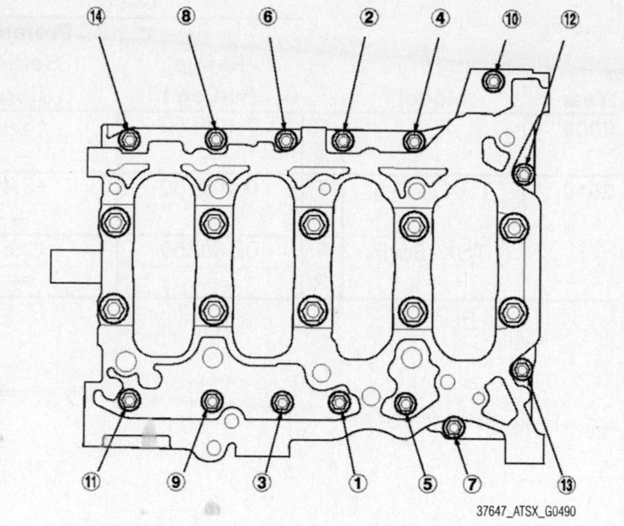
14
8
6
2
4
10
12
11
9
3
1
5
7
13
37647_ATSX_G0490

Fig. 2 Main bearing torque sequence—4 cylinder engine, 2 of 2

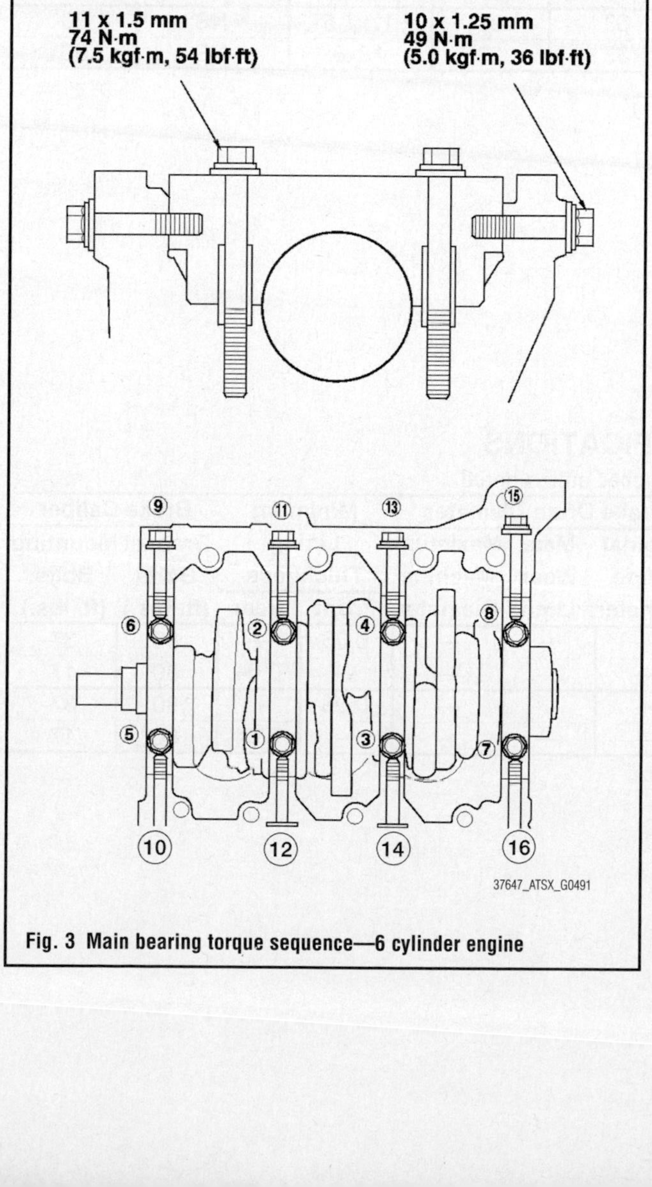
11 x 1.5 mm
74 N·m
(7.5 kgf·m, 54 lbf·ft)
10 x 1.25 mm
49 N·m
(5.0 kgf·m, 36 lbf·ft)
9
11
13
15
6
2
4
8
5
1
3
7
10
12
14
16
37647_ATSX_G0491

Fig. 3 Main bearing torque sequence—6 cylinder engine

WHEEL ALIGNMENT

Year	Model		Caster Range (+/-Deg.)	Caster Preferred Setting (Deg.)	Camber Range (+/-Deg.)	Camber Preferred Setting (Deg.)	Toe-in (in.)
2009	TSX	F	0.10/0.50	+3.47	0.30	0.00	0 +/- 0.08
		R	—	—	0.30	-1.00	0.08 +/- 0.08
2010	TSX (4 cyl)	F	0.10/0.50	+3.47	0.30	0.00	0 +/- 0.08
		R	—	—	0.30	-1.00	0.08 +/- 0.08
	TSX (6cyl)	F	0.10/0.50	+3.52	0.30	-0.03	0 +/- 0.08
		R	—	—	0.30	-1.12	0.08 +/- 0.08

37647_ATSX_C0011

TIRE, WHEEL AND BALL JOINT SPECIFICATIONS

Year	Model	OEM Tires Standard	OEM Tires Optional	Tire Pressures (psi) Front	Tire Pressures (psi) Rear	Wheel Size	Ball Joint Inspection	Lug Nut (ft. lbs.)
2009	TSX	P225/50R17	None	33	33	17x7.5	NS	80
2010	TSX (4 cyl)	P225/50R17	None	33	33	17x7.5	NS	80
	TSX (6cyl)	P235/45R18	None	33	33	18x8	NS	80

OEM: Original Equipment Manufacturer

PSI: Pounds Per Square Inch

NS: Not Specified by manufacturer

37647_ATSX_C0012

BRAKE SPECIFICATIONS

All measurements in inches unless noted

Year	Model		Brake Disc Original Thickness	Brake Disc Minimum Thickness	Brake Disc Maximum Runout	Brake Drum Diameter Original Inside Diameter	Brake Drum Diameter Max. Wear Limit	Brake Drum Diameter Maximum Machine Diameter	Minimum Lining Thickness Front	Minimum Lining Thickness Rear	Brake Caliper Bracket Bolts (ft. lbs.)	Brake Caliper Mounting Bolts (ft. lbs.)
2009	TSX	F	1.100-1.100	1.020	0.002	—	—	—	0.06	—	80	37
		R	0.350-0.360	0.310	0.002	—	—	—	—	0.04	80	17
2010	TSX	F	1.100-1.100	1.020	0.002	—	—	—	0.06	—	80	37
		R	0.350-0.360	0.310	0.002	—	—	—	—	0.04	80	17

NA: Not Available

F: Front

R: Rear

37647_ATSX_C0013

SCHEDULED MAINTENANCE INTERVALS

Acura TSX

TO BE SERVICED	TYPE OF SERVICE	SERVICE INTERVALS: Symbol A	Symbol B	1	2	3	4	5
Air cleaner element	Replace				✓			
Brake fluid	Inspect		✓					
Brake hoses and lines	Inspect		✓					
Clutch fluid (if equipped)	Inspect		✓					
Drive belt	Inspect				✓			
Driveshaft boots	Inspect		✓					
Dust & pollen filter	Replace				✓			
Engine coolant	Inspect		✓					
Engine Coolant	Replace							✓
Engine oil	Replace	✓						
Engine oil & filter	Replace		✓					
Exhaust system	Inspect		✓					
Front & Rear brakes	Inspect		✓					
Fuel lines	Inspect		✓					
Parking brake adjustment	Inspect		✓					
Power steering fluid	Inspect		✓					
Spark plugs	Replace						✓	
Suspension components	Inspect		✓					
Tie-rod ends, steering gearbox and boots	Inspect		✓					
Timing belt	Replace						✓	
Tires	Rotate			✓				
Transmission fluid	Inspect		✓					
Transmission fluid	Replace					✓		
Valve clearance	Inspect						✓	
Windshield washer fluid	Inspect		✓					

37647_ATSX_C0014

BRAKES — INFORMATION AND PRECAUTIONS

ANTI-LOCK SYSTEMS

- Certain components within the ABS system are not intended to be serviced or repaired individually.
- Do not use rubber hoses or other parts not specifically specified for and ABS system. When using repair kits, replace all parts included in the kit. Partial or incorrect repair may lead to functional problems and require the replacement of components.
- Lubricate rubber parts with clean, fresh brake fluid to ease assembly. Do not use shop air to clean parts; damage to rubber components may result.
- Use only DOT 3 brake fluid from an unopened container.
- If any hydraulic component or line is removed or replaced, it may be necessary to bleed the entire system.
- A clean repair area is essential. Always clean the reservoir and cap thoroughly before removing the cap. The slightest amount of dirt in the fluid may plug an orifice and impair the system function. Perform repairs after components have been thoroughly cleaned; use only denatured alcohol to clean components. Do not allow ABS components to come into contact with any substance containing mineral oil; this includes used shop rags.
- The Anti-Lock control unit is a microprocessor similar to other computer units in the vehicle. Ensure that the ignition switch is **OFF** before removing or installing controller harnesses. Avoid static electricity discharge at or near the controller.
- If any arc welding is to be done on the vehicle, the control unit should be unplugged before welding operations begin.

DISC AND DRUM SYSTEMS

CAUTION

Dust and dirt accumulating on brake parts during normal use may contain asbestos fibers from production or aftermarket brake linings. Breathing excessive concentrations of asbestos fibers can cause serious bodily harm. Exercise care when servicing brake parts. Do not sand or grind brake lining unless equipment used is designed to contain the dust residue. Do not clean brake parts with compressed air or by dry brushing. Cleaning should be done by dampening the brake components with a fine mist of water, then wiping the brake components clean with a dampened cloth. Dispose of cloth and all residue containing asbestos fibers in an impermeable container with the appropriate label. Follow practices prescribed by the Occupational Safety and Health Administration (OSHA) and the Environmental Protection Agency (EPA) for the handling, processing, and disposing of dust or debris that may contain asbestos fibers.

BRAKES — BLEEDING THE BRAKE SYSTEM

BLEEDING PROCEDURE

BLEEDING PROCEDURE

See Figure 4.

1. Make sure the brake fluid level in the reservoir is at the MAX (upper) level line.
2. Have someone slowly pump the brake pedal several times, then apply steady pressure.
3. Start the bleeding at the driver's side of the front brake system.

➡Bleed the calipers in the sequence shown.

4. Attach a length of clear drain tube to the bleed screw, then loosen the bleed screw to allow air to escape from the system. Then tighten the bleed screw securely.
5. Refill the master cylinder reservoir to the MAX (upper) level line.
6. Repeat the procedure for each brake circuit until there are no air bubbles in the fluid.

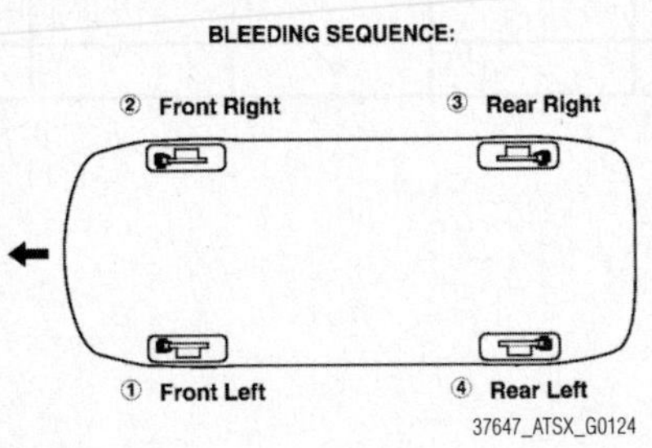
BLEEDING SEQUENCE:
② Front Right
③ Rear Right
① Front Left
④ Rear Left
37647_ATSX_G0124

Fig. 4 Proper caliper bleeding sequence

BLEEDING THE ABS SYSTEM

The ABS brake system is bled in the usual fashion with no special procedures required. Refer to the bleeding procedure located in this section. Make certain the master cylinder reservoir is filled before the bleeding is begun and check the level frequently.

WHEEL SPEED SENSORS

REMOVAL & INSTALLATION

Front

See Figure 5.

1. Turn the ignition switch to LOCK (0).
2. Release the clamp, then disconnect the wheel speed sensor connector.
3. Remove the bolts and the wheel speed sensor.
4. Install the wheel speed sensor in the reverse order of removal, and note these items:
   - Do not twist the sensor wires.
   - If the wheel speed sensor comes in contact with the wheel bearing, it is faulty.
   - Make sure there is no debris in the sensor mounting hole.
5. Start the engine, and test-drive the vehicle. Make sure that the ABS and VSA indicators turn off during the test-drive and do not come back on.

Rear

See Figure 6.

1. Turn the ignition switch to LOCK (0).
2. Release the clamp, then disconnect the wheel speed sensor connector.
3. Remove the clamps, the bolt, and the wheel speed sensor.
4. Install the wheel speed sensor in the reverse order of removal, and note these items:
   - Do not twist the sensor wires.
   - If the wheel speed sensor comes in contact with the hub bearing unit, it is faulty.
   - Make sure there is no debris in the sensor mounting hole.
   - Lubricate the O-ring on the wheel speed sensor.
5. Start the engine, and test-drive the vehicle. Make sure that the ABS and VSA indicators turn off during the test-drive and do not come back on.

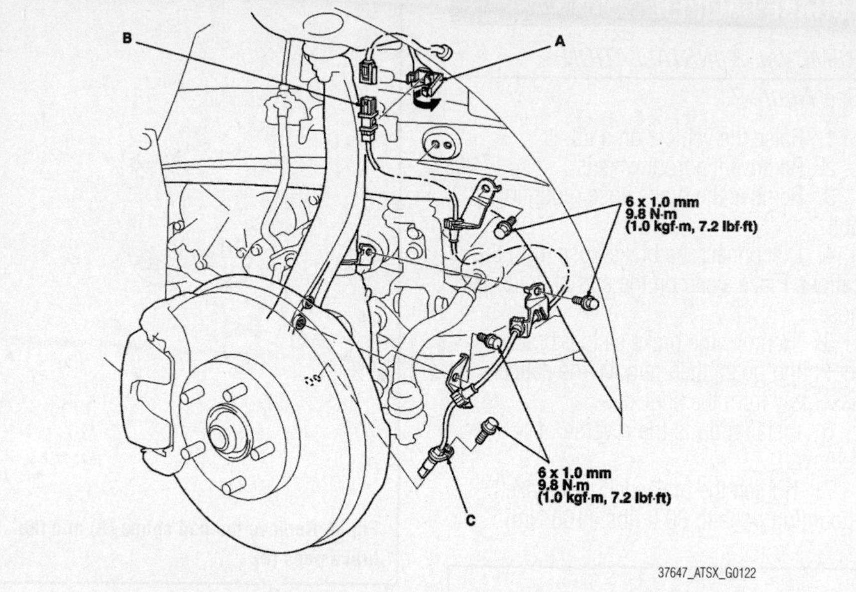
B
A
6 x 1.0 mm
9.8 N·m
(1.0 kgf·m, 7.2 lbf·ft)
6 x 1.0 mm
9.8 N·m
(1.0 kgf·m, 7.2 lbf·ft)
C
37647_ATSX_G0122

Fig. 5 Release the clamp (A), disconnect the wheel speed sensor connector (B) and remove the wheel speed sensor (C)

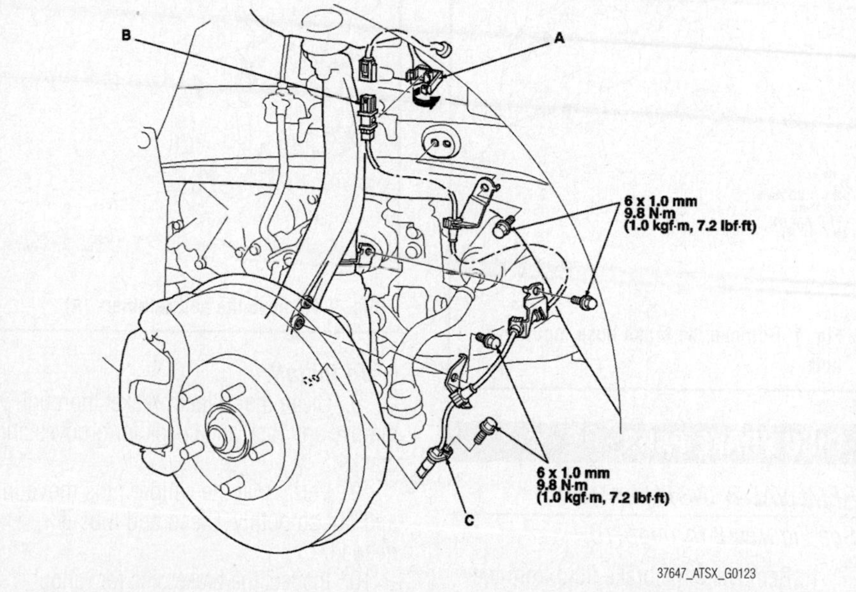
B
A
6 x 1.0 mm
9.8 N·m
(1.0 kgf·m, 7.2 lbf·ft)
6 x 1.0 mm
9.8 N·m
(1.0 kgf·m, 7.2 lbf·ft)
C
37647_ATSX_G0123

Fig. 6 Release the clamp (A), disconnect the wheel speed sensor connector (B) and remove the wheel speed sensor (C)

BRAKE CALIPER

REMOVAL & INSTALLATION

See Figure 7.

1. Raise the vehicle on a lift.
2. Remove the front wheels.
3. Remove the brake hose mounting bolt.
4. Disconnect the brake hose from the caliper. Place a cap on the end of the brake hose.
5. Remove the brake caliper bracket mounting bolts, then remove the caliper assembly from the knuckle.
6. Installation is the reverse of removal.
7. Tighten the brake caliper bracket mounting bolts to 80 ft. lbs. (108 Nm).

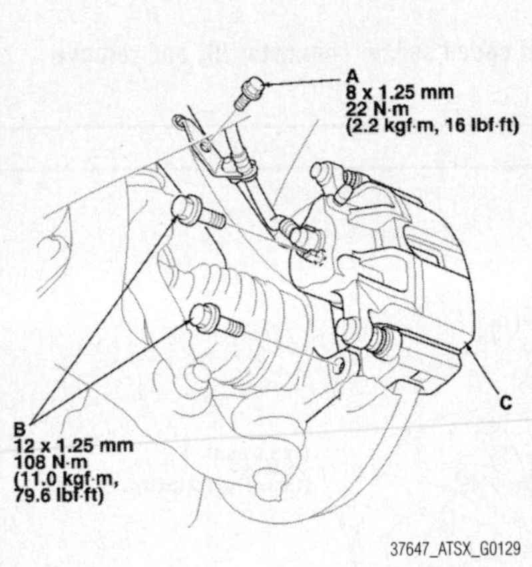
A
8 x 1.25 mm
22 N·m
(2.2 kgf·m, 16 lbf·ft)
C
B
12 x 1.25 mm
108 N·m
(11.0 kgf·m,
79.6 lbf·ft)
37647_ATSX_G0129

Fig. 7 Remove the brake hose mounting bolt

DISC BRAKE PADS

REMOVAL & INSTALLATION

See Figures 8 through 10.

1. Remove some brake fluid from the master cylinder.
2. Raise the vehicle on a lift.
3. Remove the front wheels.
4. Remove the brake hose mounting bolt.
5. Remove the flange bolt while holding the caliper pin with a wrench. Be careful not to damage the pin boot, and pivot the caliper up out of the way. Check the hose and the pin boots for damage and deterioration.
6. Remove the pad shims and the brake pads.
7. Remove the pad retainers.

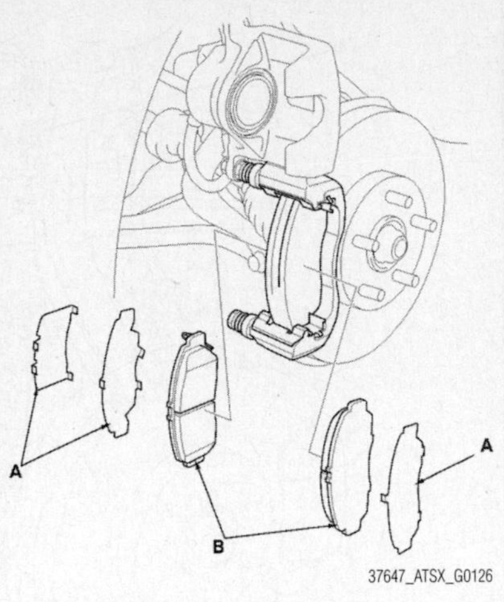
A
A
B
37647_ATSX_G0126

Fig. 8 Remove the pad shims (A) and the brake pads (B)

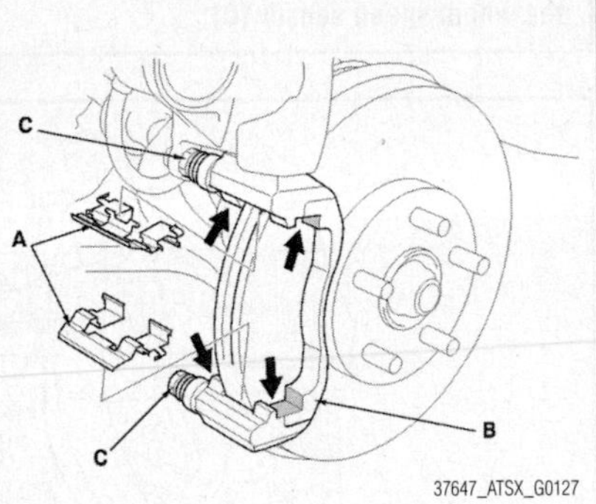
C
A
C
B
37647_ATSX_G0127

Fig. 9 Remove the pad retainers (A)

To install:

8. Clean the caliper bracket thoroughly; remove any rust, and check for grooves and cracks.
9. Verify that the caliper pins move in and out smoothly. Clean and lube if needed.
10. Inspect the brake disc for runout, thickness, parallelism, and check for damage and cracks.
11. Apply a thin coat of M-77 assembly paste (P/N 08798-9010) to the retainer mating surface of the caliper bracket (indicated by the arrows).
12. Install the pad retainers. Wipe excess assembly paste off the retainers. Keep the assembly paste off the brake disc and the brake pads.
13. Install the brake caliper piston compressor tool on the caliper body.

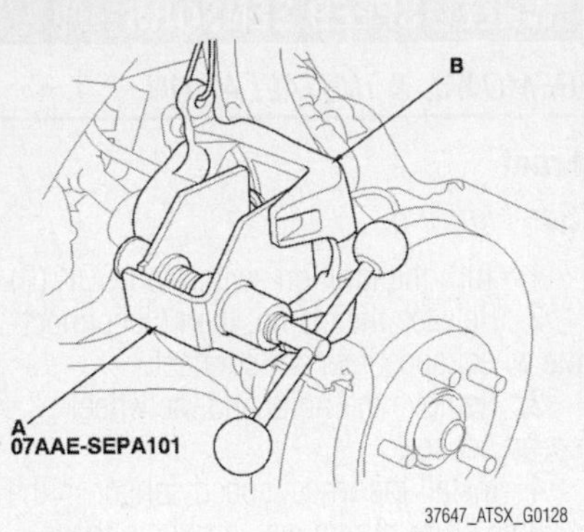
B
A
07AAE-SEPA101
37647_ATSX_G0128

Fig. 10 Install the brake caliper piston compressor tool (A) on the caliper body (B)

14. Press in the piston with the brake caliper piston compressor tool so the caliper will fit over the brake pads. Make sure the piston boot is in position to prevent damaging it when pivoting the caliper down.

➡Be careful when pressing in the piston; brake fluid might overflow from the master cylinder's reservoir. If brake fluid gets on any painted surface, wash it off immediately with water.

15. Remove the brake caliper piston compressor tool.
16. Apply a thin coat of M-77 assembly paste (P/N 08798-9010) to the pad side of the shims, the back of the brake pads and the other areas indicated by the arrows. Wipe excess assembly paste off the pad shims and the brake pads friction material. Keep grease and assembly paste off the brake disc and the brake pads. Contaminated brake disc or brake pads reduce stopping ability.
17. Install the brake pads and the pad shims correctly. Install the brake pad with the wear indicator on the upper inside. If you are reusing the brake pads, always reinstall the brake pads in their original positions to prevent a temporary loss of braking efficiency.
18. Pivot the caliper down into position. Install the flange bolt, and tighten it to 37 ft. lbs. (50 Nm) while holding the caliper pin with a wrench being careful not to damage the pin boot.
19. Install the brake hose mounting bolt. Tighten the bolt to 16 ft. lbs. (22 Nm).
20. Clean the mating surfaces between the brake disc and the inside of the wheel, then install the front wheels.

21. Press the brake pedal several times to make sure the brakes work.

➡**Engagement may require a greater pedal stroke immediately after the brake pads have been replaced as a set. Several applications of the brake pedal will restore the normal pedal stroke.**

22. Add brake fluid as needed.
23. After installation, check for leaks at hose and line joints or connections, and retighten if necessary.
24. Test-drive the vehicle, then recheck for leaks.

BRAKES

REAR DISC BRAKES

BRAKE CALIPER

REMOVAL & INSTALLATION

See Figures 11 through 13.

1. Raise the vehicle on a lift.
2. Remove the rear wheel.
3. Release the parking brake lever fully.
4. Loosen the parking brake cable adjusting nut.
5. Remove the flange bolt from the arm.
6. Disconnect the parking brake cable from the lever.
7. Remove the brake hose mounting bolt.

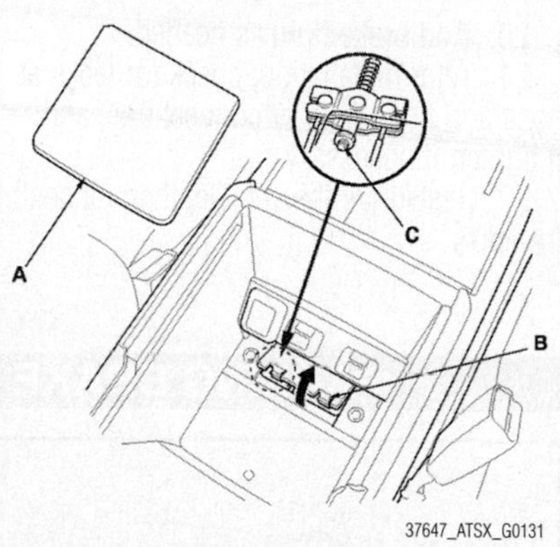
A
B
C
37647_ATSX_G0131

Fig. 11 Remove the console box mat (A), open the lid (B) then loosen the parking brake cable adjusting nut (C)

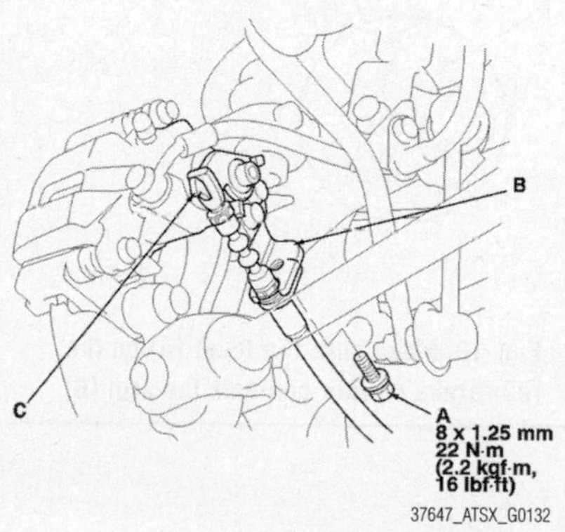
A
8 x 1.25 mm
22 N·m
(2.2 kgf·m, 16 lbf·ft)
B
C
37647_ATSX_G0132

Fig. 12 Remove the flange bolt (A) from the arm (B)and disconnect the parking brake cable from the lever (C)

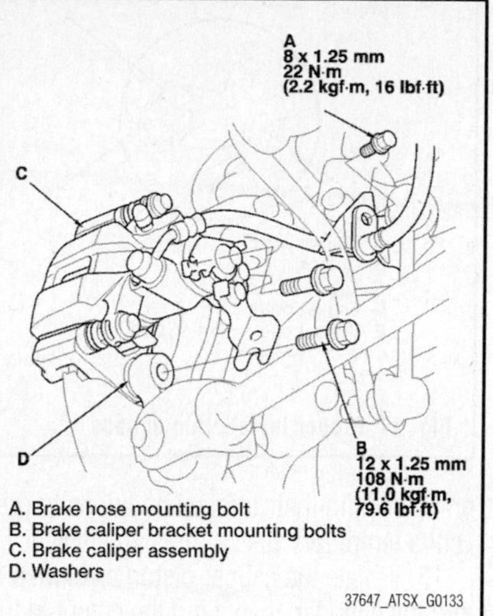
A
8 x 1.25 mm
22 N·m
(2.2 kgf·m, 16 lbf·ft)
B
12 x 1.25 mm
108 N·m
(11.0 kgf·m, 79.6 lbf·ft)
C
D
A. Brake hose mounting bolt
B. Brake caliper bracket mounting bolts
C. Brake caliper assembly
D. Washers
37647_ATSX_G0133

Fig. 13 Remove the brake caliper assembly

8. Disconnect the brake hose from the caliper. Place a cap on the end of the brake hose.
9. Remove the brake caliper bracket mounting bolts, and remove the caliper assembly from the knuckle.

➡**Make sure the washers are in position on reassembly, if they are removed.**

10. Installation is the reverse of removal.

DISC BRAKE PADS

REMOVAL & INSTALLATION

See Figures 14 through 17.

1. Remove some brake fluid from the master cylinder.
2. Raise the vehicle on a lift.
3. Remove the rear wheels.
4. Remove the brake hose mounting bolt.
5. Remove the flange bolts while holding respective caliper pin with a wrench. Be careful not to damage the pin boot, and remove the caliper. Check the hose, the pin boots, and the parking brake cable boots for damage and deterioration.

➡**Do not twist the brake hose and the parking brake cable to prevent damage.**

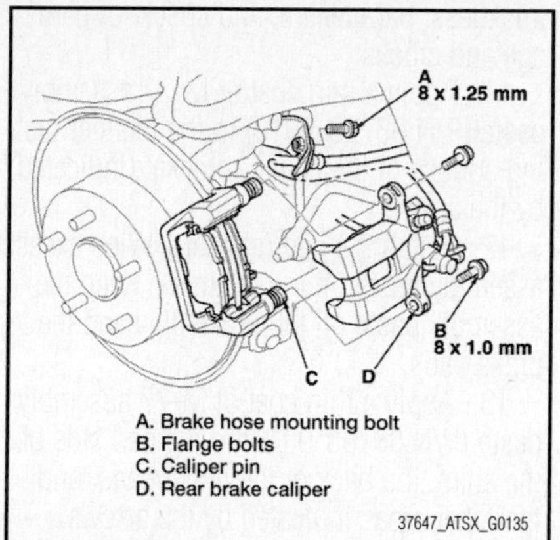
A
8 x 1.25 mm
B
8 x 1.0 mm
C
D
A. Brake hose mounting bolt
B. Flange bolts
C. Caliper pin
D. Rear brake caliper
37647_ATSX_G0135

Fig. 14 Rear brake caliper assembly

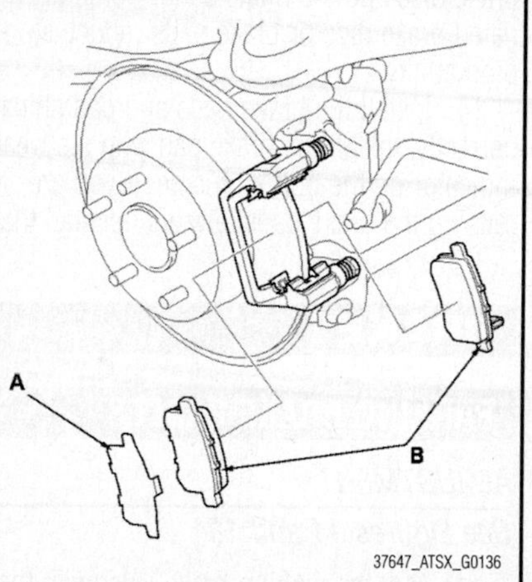
A
B
37647_ATSX_G0136

Fig. 15 Remove the pad shim (A) and the brake pads (B)

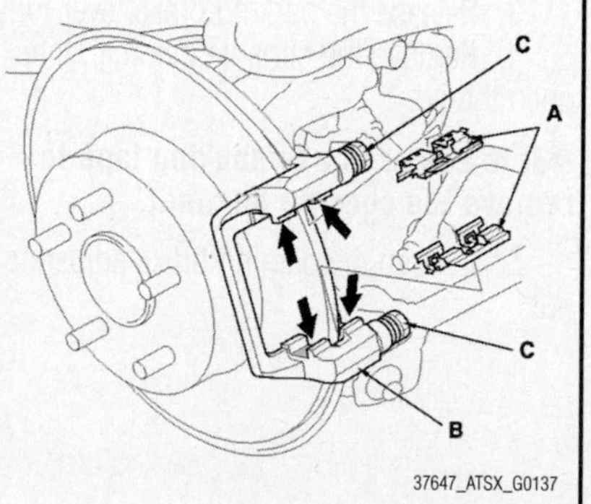
A
B
C
C
37647_ATSX_G0137

Fig. 16 Remove the pad retainers (A)

6. Remove the pad shim and the brake pads.
7. Remove the pad retainers (A).
8. Clean the caliper bracket thoroughly; remove any rust, and check for grooves and cracks.
9. Verify that the caliper pins move in and out smoothly. Clean and lube if needed.
10. Inspect the brake disc for runout, thickness, parallelism, and check for damage and cracks.
11. Apply a thin coat of M-77 assembly paste (P/N 08798-9010) to the retainer mating surface of the caliper bracket (indicated by the arrows).
12. Install the pad retainers. Wipe excess assembly paste off the retainers. Keep the assembly paste off the brake disc and the brake pads.
13. Apply a thin coat of M-77 assembly paste (P/N 08798-9010) to the pad side of the shim, the back of the brake pads, and the other areas indicated by the arrows. Wipe excess assembly paste off the pad shim and the brake pads friction material. Keep grease and assembly paste off the brake disc and the brake pads. Contaminated brake disc or brake pads reduce stopping ability.
14. Install the brake pads and pad shim correctly. Install the brake pad with the wear indicator on the bottom inside. If you are reusing the brake pads, always reinstall the brake pads in their original positions to prevent a temporary loss of braking efficiency.

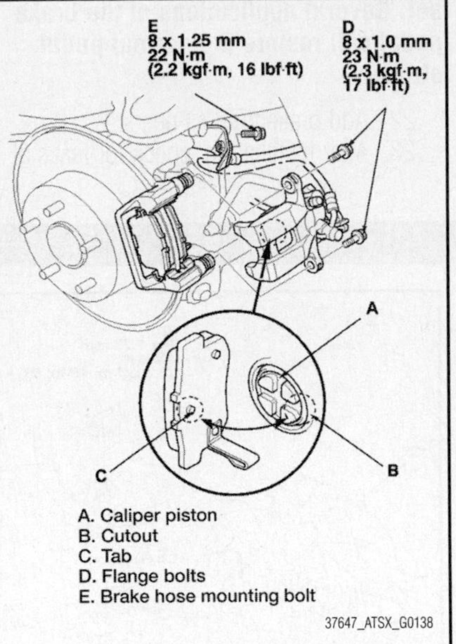
E
8 x 1.25 mm
22 N·m
(2.2 kgf·m, 16 lbf·ft)
D
8 x 1.0 mm
23 N·m
(2.3 kgf·m, 17 lbf·ft)
A
B
C
A. Caliper piston
B. Cutout
C. Tab
D. Flange bolts
E. Brake hose mounting bolt
37647_ATSX_G0138

Fig. 17 Proper installation of pads

15. Rotate the caliper piston clockwise into the cylinder, then align the cutout in the piston with the tab on the inner pad by turning the piston back. Lubricate the boot with rubber grease to avoid twisting the piston boot. If the piston boot is twisted, back it out so it is positioned properly.

➡Be careful when moving the piston back in the caliper; brake fluid might overflow from the master cylinder's reservoir. If brake fluid gets on any painted surface, wash it off immediately with water.

16. Install the caliper. Install the flange bolts, and tighten it to the specified torque while holding the respective caliper pins with a wrench being careful not to damage the pin boots and parking brake cable boots.
17. Install the brake hose mounting bolt.
18. Clean the mating surfaces between the brake disc and the inside of the wheel, then install the rear wheels.
19. Press the brake pedal several times to make sure the brakes work.

➡Engagement may require a greater pedal stroke immediately after the brake pads have been replaced as a set. Several applications of the brake pedal will restore the normal pedal stroke.

20. Add brake fluid as needed.
21. After installation, check for leaks at hose and line joints or connections, and retighten if necessary.
22. Test-drive the vehicle, then recheck for leaks.

BRAKES

PARKING BRAKE

PARKING BRAKE CABLES

ADJUSTMENT

See Figures 11 and 18

1. Pull the parking brake lever with the force to fully apply the parking brake. The parking brake lever should be locked within 6 to 8 clicks.
2. If the number of lever clicks is not as specified, adjust the parking brake.
3. Release the parking brake lever fully.
4. Remove the console box mat, then open the lid.

➡It is easier to use packing tape to remove the console box mat.

5. Loosen the parking brake adjusting nut.
6. Make sure the lever on the rear brake caliper contacts the arm.

➡The lever will contact the arm when the parking brake adjusting nut is loosened.

7. Pull the parking brake lever 1 click.
8. Tighten the parking brake adjusting nut until the parking brakes drag slightly when the rear wheels are turned.
9. Release the parking brake lever fully, and check that the parking brakes do not drag when the rear wheels are turned. Readjust if necessary.
10. Make sure the parking brake lever is within the specified number of clicks (6 to 8 clicks).
11. Close the lid, then install the console mat.

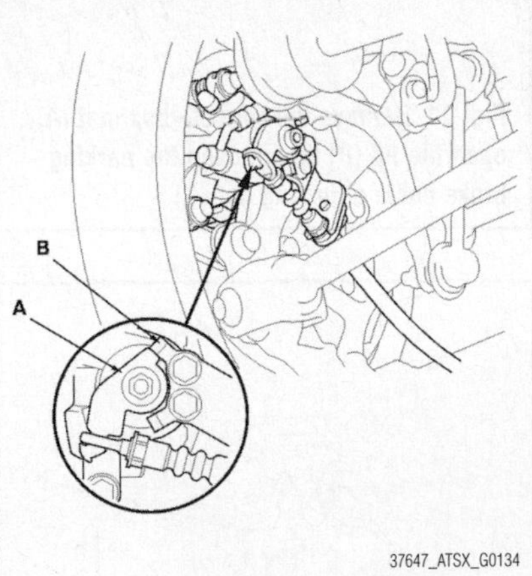
B
A
37647_ATSX_G0134

Fig. 18 Make sure the lever (A) on the rear brake caliper contacts the arm (B)

CHASSIS ELECTRICAL

AIR BAG (SUPPLEMENTAL RESTRAINT SYSTEM)

GENERAL INFORMATION

CAUTION

These vehicles are equipped with an air bag system. The system must be disarmed before performing service on, or around, system components, the steering column, instrument panel components, wiring and sensors. Failure to follow the safety precautions and the disarming procedure could result in accidental air bag deployment, possible injury and unnecessary system repairs.

PRECAUTIONS

Disconnect and isolate the battery negative cable before beginning any airbag system component diagnosis, testing, removal, or installation procedures. Allow system capacitor to discharge for two minutes before beginning any component service. This will disable the airbag system. Failure to disable the airbag system may result in accidental airbag deployment, personal injury, or death.

DISARMING THE SYSTEM

➡Some systems store data in memory that is lost when the battery is disconnected. Do the following steps before disconnecting the battery.

1. Make sure you have the anti-theft code(s) for the audio and/or the navigation system (if equipped).
2. For some models or if you're replacing the audio unit, it may be necessary to write down the audio presets (AM and FM), and the XM radio presets (if equipped), because the audio unit does not retain the presets after the battery is disconnected.
3. Make sure the ignition switch is in LOCK (0).
4. Disconnect and isolate the negative cable from the battery.

CAUTION

Always disconnect the negative cable from the battery first.

5. Disconnect the positive cable from the battery.
6. Wait at least 3 minutes for the battery and SRS system to discharge before performing any repairs.

ARMING THE SYSTEM

➡Some systems store data in memory that is lost when the battery is disconnected. Do the following steps to restore the systems back to normal operation.

1. Clean the battery terminals.
2. Reconnect the positive cable to the battery first, then reconnect the negative cable to the battery.

CAUTION

Always connect the positive cable to the battery first.

3. Apply multipurpose grease to the terminals to prevent corrosion.
4. Enter the anti-theft code(s) for the audio system and/or the navigation system (if equipped).
5. Enter the audio presets (if applicable), and enter the XM radio presets (if equipped).
6. Set the clock (for vehicles without navigation).
7. Do the steering column position memorization.

DRIVE TRAIN

CLUTCH

REMOVAL & INSTALLATION

Special Tools Required

Special Tools Required:

- Clutch Alignment Disc 07JAF-PM7011A
- Ring Gear Holder 07LAB-PV00100
- Clutch Alignment Tool Set 07PAF-0020000
- Clutch Alignment Pilot, 21 mm 07PAF-0020370
- Clutch Alignment Shaft 07ZAF-PR8A100
- Bearing Driver Attachment, 22 x 24 mm 07746-0010800
- Driver Handle, 15 x 135L 07749-0010000
- Remover Handle 07936-3710100
- Bearing Remover Shaft Set, 20 mm 07936-3710600
- Slide Hammer 07936-371020A

Engine Side

Pressure Plate Inspection and Removal

See Figures 19 through 21.

1. Remove the transmission.
2. Check the evenness of the height of the diaphragm spring fingers using the clutch alignment disc, the clutch alignment shaft, the remover handle, and a feeler gauge. If the height difference is more than the service limit—0.03inches (0.8 mm), replace the pressure plate.
3. Install the ring gear holder, the clutch alignment tool set, and the 20 mm clutch alignment pilot.
4. To prevent warping, loosen the pressure plate mounting bolts in a crisscross pattern in several steps, then remove the pressure plate.
5. Inspect the fingers of the diaphragm spring for wear at the release bearing contact area.
6. Inspect the pressure plate surface for wear, cracks, and burning.
7. Inspect for warpage using a straight edge and a feeler gauge. Measure across the pressure plate. If the most measurement difference is more than the service limit—0.006 inches (0.15 mm), replace the pressure plate.

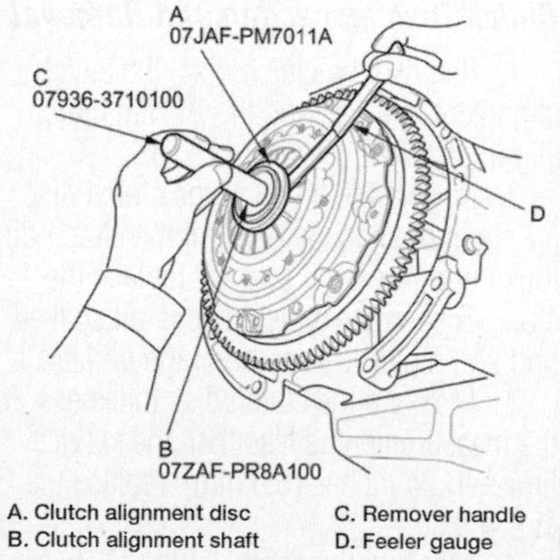
A
07JAF-PM7011A
C
07936-3710100
D
B
07ZAF-PR8A100
A. Clutch alignment disc
B. Clutch alignment shaft
C. Remover handle
D. Feeler gauge
37647_ACTL_G0595

Fig. 19 Check the evenness of the height of the diaphragm spring fingers using the clutch alignment disc, the clutch alignment shaft, the remover handle, and a feeler gauge

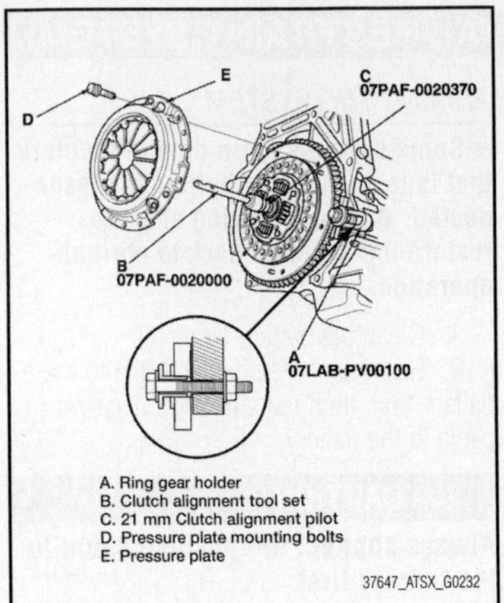
E
C
07PAF-0020370
D
B
07PAF-0020000
A
07LAB-PV00100
A. Ring gear holder
B. Clutch alignment tool set
C. 21 mm Clutch alignment pilot
D. Pressure plate mounting bolts
E. Pressure plate
37647_ATSX_G0232

Fig. 20 Install the ring gear holder, the clutch alignment tool set, and the 21 mm clutch alignment pilot

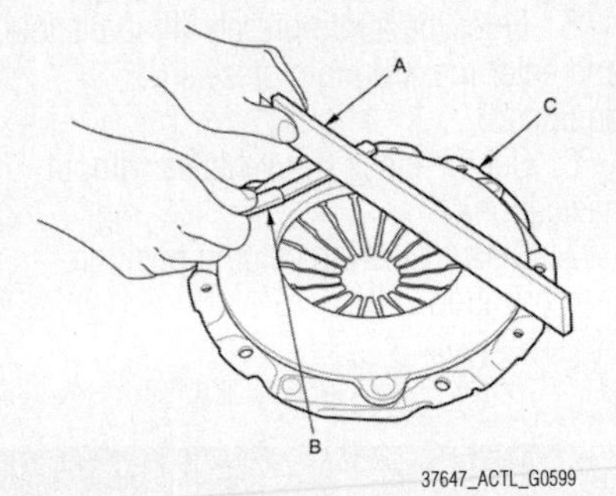
A
C
B
37647_ACTL_G0599

Fig. 21 Inspect for warpage using a straight edge (A) and a feeler gauge (B). Measure across the pressure plate (C)

Clutch Disc Inspection and Removal

1. Remove the clutch disc, the clutch alignment tool set, and the 21 mm clutch alignment pilot.
2. Inspect the lining of the clutch disc for signs of slippage or oil. If the clutch disc looks burnt or is oil soaked, replace the clutch disc. If the clutch disc is oil soaked, find and repair the source of the oil leak.
3. Measure the clutch disc thickness. If the measurement is less than the service limit—0.24 inches (6.0 mm), replace the clutch disc.
4. Measure the depths of the rivets from the clutch disc lining surface to the rivets on both sides. If the measurement is less than the service limit—0.03 inches (0.7 mm), replace the clutch disc.

Flywheel Inspection

1. Remove the ring gear holder.
2. Inspect the ring gear teeth for wear and damage.
3. Inspect the clutch disc mating surface on the flywheel for wear, cracks, and burning.
4. Measure the flywheel runout using a dial indicator. Through at least two full turns with pushing against the flywheel each time you turn it to take up the crankshaft thrust washer clearance. If the measurement is not within the standard (new)—0.002 inches (0.05 mm), replace the flywheel.

Flywheel Replacement

See Figure 22.

1. Install the ring gear holder.
2. Loosen the flywheel mounting bolts in a crisscross pattern in several steps. Remove the bolts, then remove the flywheel and the ring gear holder.
3. Install the flywheel on the crankshaft, and install the mounting bolts finger-tight.
4. Install the ring gear holder, then torque the flywheel mounting bolts in a crisscross pattern in several steps to 90 ft. lbs. (123 Nm).

Pilot Bearing Inspection

1. Turn the inner race of the pilot bearing with your finger. The pilot bearing should turn smoothly and quietly. Check that the pilot bearing outer race fits tightly in the flywheel. If the race does not turn smoothly, quietly, or fit tight in the flywheel, replace the pilot bearing.

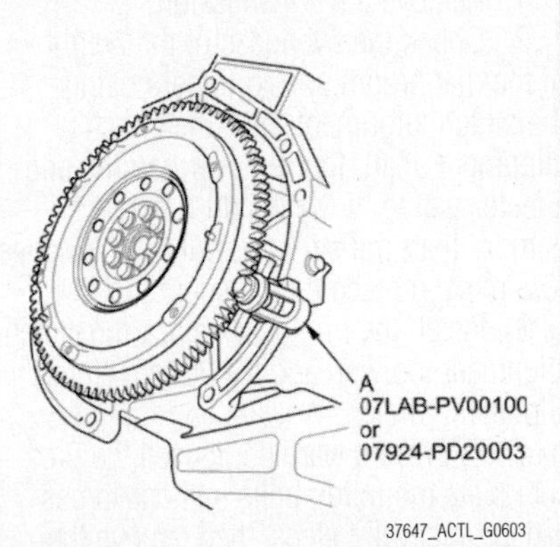
A
07LAB-PV00100
or
07924-PD20003
37647_ACTL_G0603

Fig. 22 Install the ring gear holder (A)

Pilot Bearing Replacement

See Figures 23 and 24.

1. Remove the pilot bearing using the slide hammer, the remover handle, and the bearing remover shaft set.
2. Install a new pilot bushing into the crankshaft using the 15 x 135L driver handle and the 22 x 24 mm attachment.

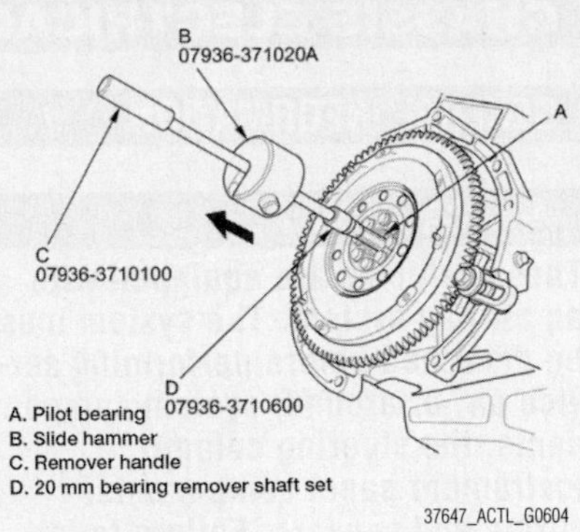
B
07936-371020A
A
C
07936-3710100
D
07936-3710600
A. Pilot bearing
B. Slide hammer
C. Remover handle
D. 20 mm bearing remover shaft set
37647_ACTL_G0604

Fig. 23 Remove the pilot bearing using the slide hammer, the remover handle, and the bearing remover shaft set

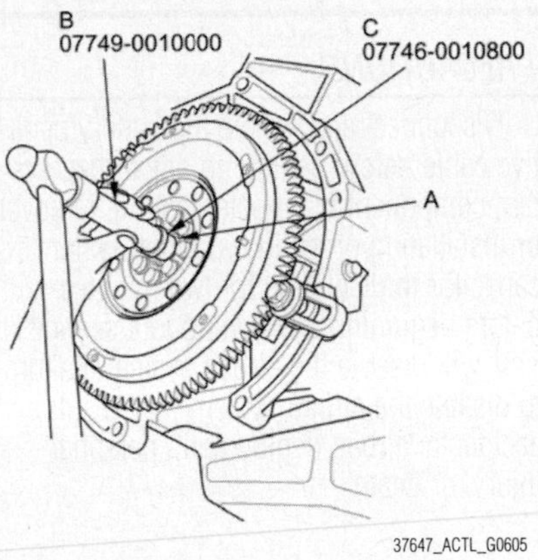
B
07749-0010000
C
07746-0010800
A
37647_ACTL_G0605

Fig. 24 Install a new pilot bushing (A) into the crankshaft using the 15 x 135L driver handle (B) and the 22 x 24 mm attachment (C)

Clutch Disc and Pressure Plate Installation

See Figures 25 and 26.

1. Temporarily install the clutch disc onto the splines of the transmission mainshaft. Make sure the clutch disc slides freely on the mainshaft.
2. Apply super high temp urea grease (P/N 08798- 9002) to the splines of the clutch disc, then install the clutch disc using the clutch alignment tool set and the clutch alignment pilot.
3. Install the pressure plate and the mounting bolts, finger-tight.
4. Torque the mounting bolts in a crisscross pattern. Tighten the bolts in several steps to prevent warping the diaphragm spring. Make sure that there is not clearance between the pressure plate and the flywheel. The final torque should be: 19 ft. lbs. (25 Nm)
5. Remove the ring gear holder, the clutch alignment tool set, and the clutch alignment pilot.

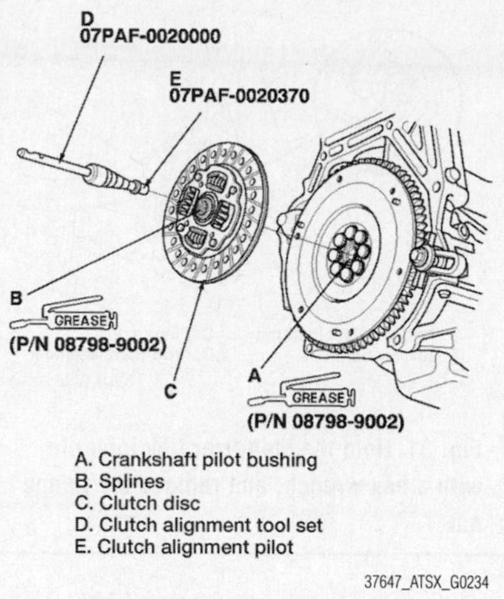
D
07PAF-0020000
E
07PAF-0020370
B
GREASE
(P/N 08798-9002)
C
A
GREASE
(P/N 08798-9002)
A. Crankshaft pilot bushing
B. Splines
C. Clutch disc
D. Clutch alignment tool set
E. Clutch alignment pilot
37647_ATSX_G0234

Fig. 25 Apply super high temp urea grease to the splines of the clutch disc, then install the clutch disc using the clutch alignment tool set and the clutch alignment pilot

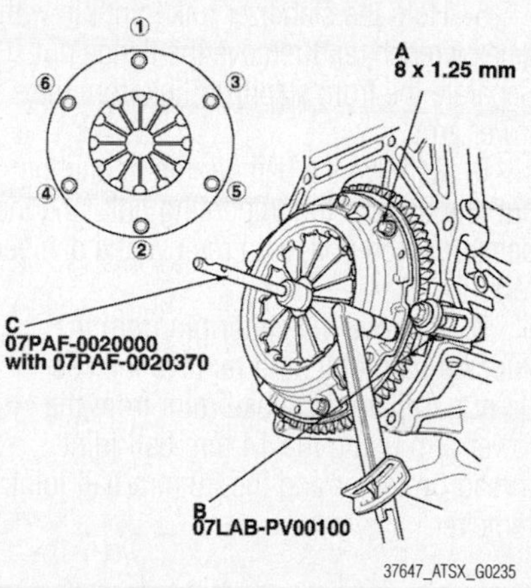
A
8 x 1.25 mm
①
②
③
④
⑤
⑥
C
07PAF-0020000
with 07PAF-0020370
B
07LAB-PV00100
37647_ATSX_G0235

Fig. 26 Torque the mounting bolts in a crisscross pattern

6. Make sure the diaphragm spring fingers are all the same height.
7. Do the release bearing inspection, and replace it if necessary.
8. Install the transmission.

Transmission Side

Release Bearing Removal

See Figure 27.

1. Remove the transmission.
2. Remove the release fork boot from the clutch housing.
3. Remove the release fork from the clutch housing by squeezing the release fork set spring. Remove the release bearing.

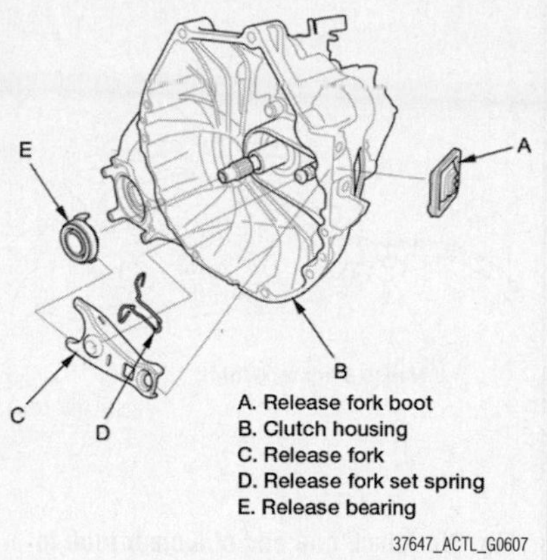
E
A
B
C
D
A. Release fork boot
B. Clutch housing
C. Release fork
D. Release fork set spring
E. Release bearing
37647_ACTL_G0607

Fig. 27 Remove the release fork boot from the clutch housing

Release Bearing Inspection

1. Check the play of the release bearing by spinning it by hand. If there is excessive play or noise, replace the release bearing.

➡**The release bearing is packed with grease. Do not wash it in solvent.**

Release Bearing Installation

See Figures 28 and 29.

1. Apply super high temp urea grease (P/N 08798-9002) to the release fork, the release fork bolt, the release bearing, and the release bearing guide in the shaded areas, then set the release fork set spring.

➡**Replace the release fork bolt if necessary.**

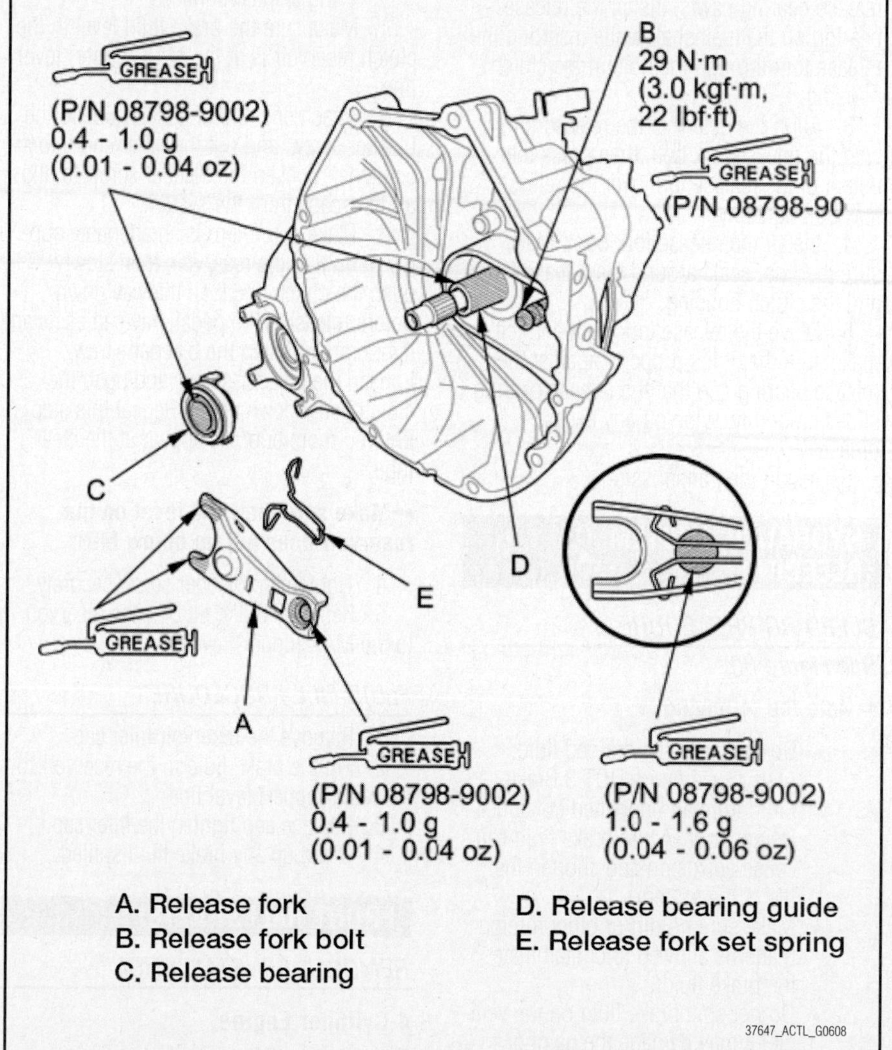
GREASE
(P/N 08798-9002)
0.4 - 1.0 g
(0.01 - 0.04 oz)
B
29 N·m
(3.0 kgf·m,
22 lbf·ft)
GREASE
(P/N 08798-90
C
D
E
GREASE
A
GREASE
(P/N 08798-9002)
0.4 - 1.0 g
(0.01 - 0.04 oz)
GREASE
(P/N 08798-9002)
1.0 - 1.6 g
(0.04 - 0.06 oz)
A. Release fork
B. Release fork bolt
C. Release bearing
D. Release bearing guide
E. Release fork set spring
37647_ACTL_G0608

Fig. 28 Apply super high temp urea grease to the release fork, the release fork bolt, the release bearing, and the release bearing guide in the shaded areas, then set the release fork set spring

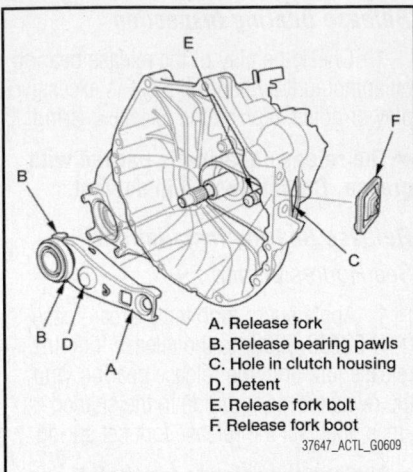

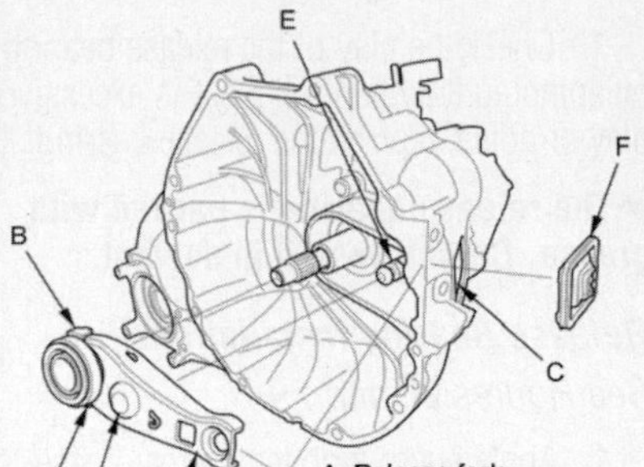
E
F
B
C
B D
A
A. Release fork
B. Release bearing pawls
C. Hole in clutch housing
D. Detent
E. Release fork bolt
F. Release fork boot
37647_ACTL_G0609

Fig. 29 Install the release bearing on the mainshaft while inserting the release fork through the hole in the clutch housing

2. With the release fork slid between the release bearing pawls, install the release bearing on the mainshaft while inserting the release fork through the hole in the clutch housing.

3. Align the detent of the release fork with the release fork bolt, then press the detent of the release fork over the release fork bolt squarely.

4. Install the release fork boot. Make sure the boot seals around the release fork and the clutch housing.

5. Move the release fork right and left to make sure that it fits properly against the release bearing and that the release bearing slides smoothly. Wipe off any excess grease.

6. Install the transmission.

HYDRAULIC SYSTEM BLEEDING

BLEEDING PROCEDURE

See Figure 30.

➡Note the following:

- Do not reuse the drained fluid. Always use Acura DOT 3 Brake Fluid from an unopened container. Using a non-Acura brake fluid can cause corrosion and shorten the life of the system.
- Make sure no dirt or other foreign matter is allowed to contaminate the brake fluid.
- Do not spill brake fluid on the vehicle; it may damage the paint or plastic. If brake fluid does contact the paint or plastic, wash it off immediately with water.
- If may be necessary to limit the movement of the release fork with a block of wood to remove all the air from the system.
- Use fender covers to avoid damaging painted surfaces.

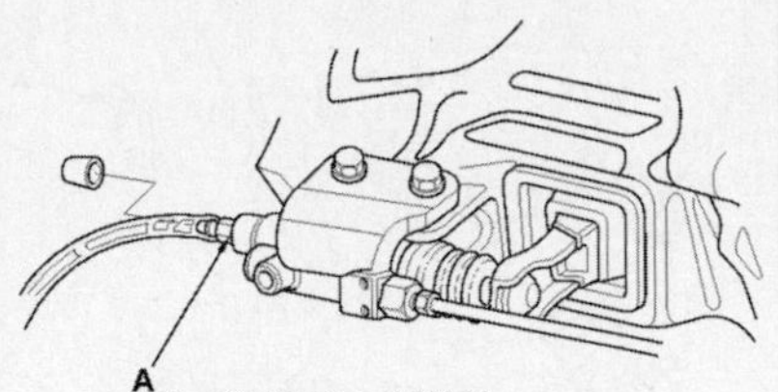
A
8 N·m (0.8 kgf·m, 6 lbf·ft)
37647_ATSX_G0244

Fig. 30 Attach one end of a clear tube to the bleeder screw (A)

1. Make sure the brake fluid level in the clutch reservoir is at the MAX (upper) level line.

2. Attach one end of a clear tube to the bleeder screw, and put the other end into a container. Loosen the bleeder screw to allow air to escape from the system.

3. Make sure there is an adequate supply of fluid in the reservoir, then slowly push the clutch pedal all the way down. Before releasing the pedal, have an assistant temporarily tighten the bleeder screw. Loosen the bleeder screw, and push the clutch pedal down again. Repeat this step until no more bubbles appear at the clear tube.

➡Make sure the fluid level on the reservoir does not go below MIN.

4. Tighten the bleeder screw securely.

5. Refill the brake fluid in the reservoir to the MAX (upper) level line.

FLUID FILL PROCEDURE

1. Remove the reservoir filler cap.
2. Fill the brake fluid in the reservoir to the MAX (upper) level line.
3. Replace and tighten the filler cap.
4. Clean up any brake fluid spilled.

FRONT HALFSHAFTS

REMOVAL & INSTALLATION

4-Cylinder Engine

See Figures 31 through 41.

Special Tools Required:

- Ball joint threaded protector, 14 mm 07AAE-SJAA100
- Ball joint remover, 28 mm 07MAC-SL0A202

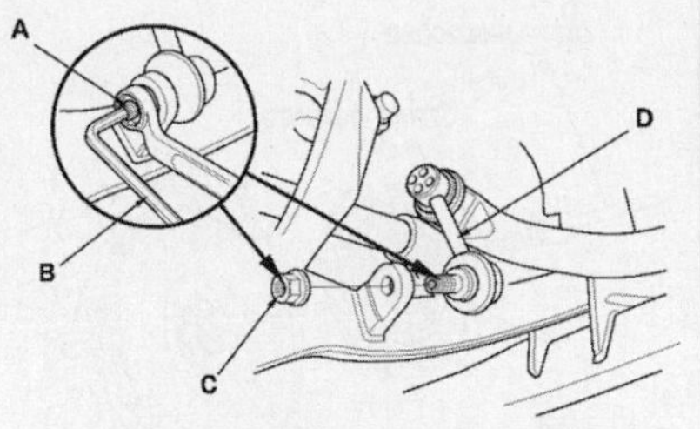
A
D
B
C
A. Stabilizer link joint pin
B. Hex wrench
C. Flange nut
D. Front stabilizer link
37647_ATSX_G0248

Fig. 31 Hold the stabilizer link joint pin with a hex wrench, and remove the flange nut

1. Raise the vehicle on a lift.
2. Remove the front wheels.
3. Pry up the stake on the spindle nut, then remove the nut.
4. Drain the transmission fluid, then reinstall the drain plug with a new sealing washer.
5. Hold the stabilizer link joint pin with a hex wrench, and remove the flange nut. Separate the front stabilizer link from the lower arm.
6. Remove the damper fork mounting nut, the damper fork mounting bolt, and the damper pinch bolt, then remove the damper fork.
7. Remove the cotter pin from the knuckle ball joint, and remove the castle nut. Separate the ball joint from the lower arm using the 14 mm ball joint thread protector and the 28 mm ball joint remover.

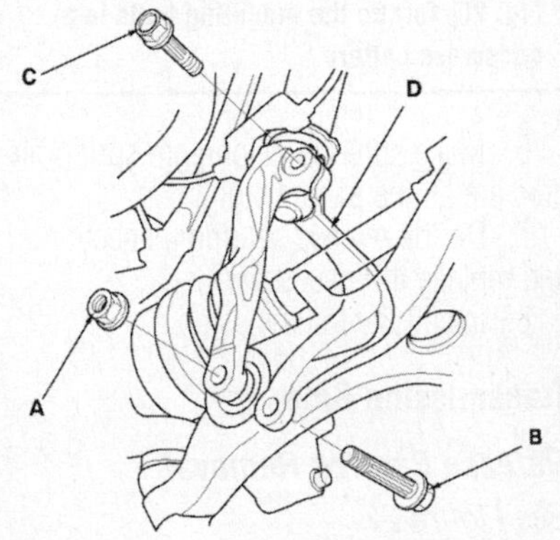
C
D
A
B
A. Damper fork mounting nut
B. Damper fork mounting bolt
C. Damper pinch bolt
D. Damper fork
37647_ATSX_G0249

Fig. 32 Remove the damper fork mounting nut, the damper fork mounting bolt, and the damper pinch bolt, then remove the damper fork

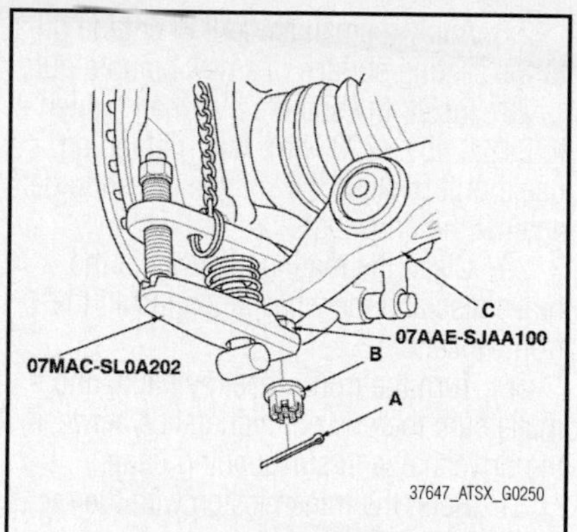
C
07AAE-SJAA100
07MAC-SL0A202
B
A
37647_ATSX_G0250

Fig. 33 Remove the cotter pin (A) from the knuckle ball joint, remove the castle nut (B) and separate the ball joint from the lower arm (C)

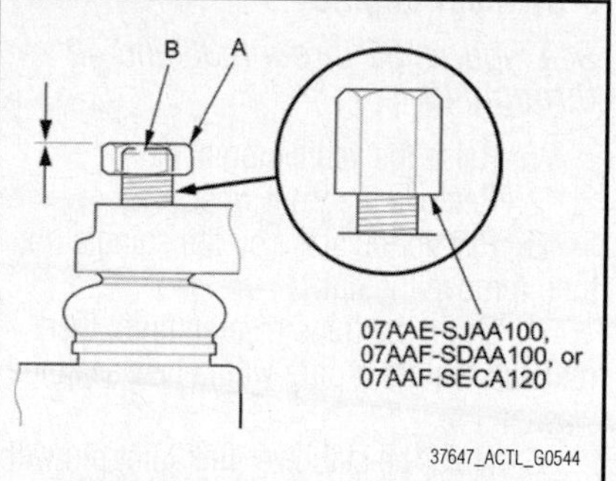
B
A
07AAE-SJAA100,
07AAF-SDAA100, or
07AAF-SECA120
37647_ACTL_G0544

Fig. 34 Install a hex nut (A) or the ball joint thread protector onto the threads of the ball joint (B)

➡Always use a ball joint remover to disconnect a ball joint. Do not strike the housing or any other part of the ball joint connection to disconnect it.

a. Install a hex nut or the ball joint thread protector onto the threads of the ball joint.

➡Using a hex nut, make sure the nut is flush with the ball joint pin end to prevent damage to the threaded end of the ball joint pin.

b. Apply grease to the ball joint remover on the areas shown. This will ease the installation of the tool, and prevent damage to the pressure bolt threads.

c. Loosen the pressure bolt, and install the ball joint remover as shown. Insert the jaws carefully, making sure not to damage the ball joint boot. Adjust the jaw spacing by turning the adjusting bolt.

➡Fasten the safety chain securely to a suspension arm or the subframe. Do not fasten it to a brake line or wire harness.

d. After adjusting the adjusting bolt, make sure the head of the adjusting bolt

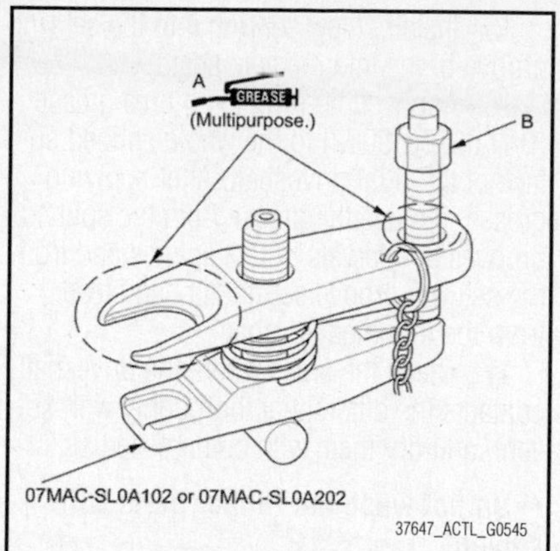
A
GREASE
(Multipurpose.)
B
07MAC-SL0A102 or 07MAC-SL0A202
37647_ACTL_G0545

Fig. 35 Apply grease to the ball joint remover on the areas shown (A)

is in position to allow the jaw to pivot.

e. With a wrench, tighten the pressure bolt until the ball joint pin pops loose from the ball joint connecting hole. If necessary, apply penetrating type lubricant to loosen the ball joint pin.

➡Do not use pneumatic or electric tools on the pressure bolt.

f. Remove the ball joint remover, then remove the nut or the ball joint thread protector from the end of the ball joint pin, and pull the ball joint out of the ball joint connecting hole. Inspect the ball joint boot, and replace it if damaged.

➡Be careful not to damage the ball joint boot when installing the remover. Do not force or hammer on the lower arm, or pry between the lower arm and the knuckle. You could damage the ball joint.

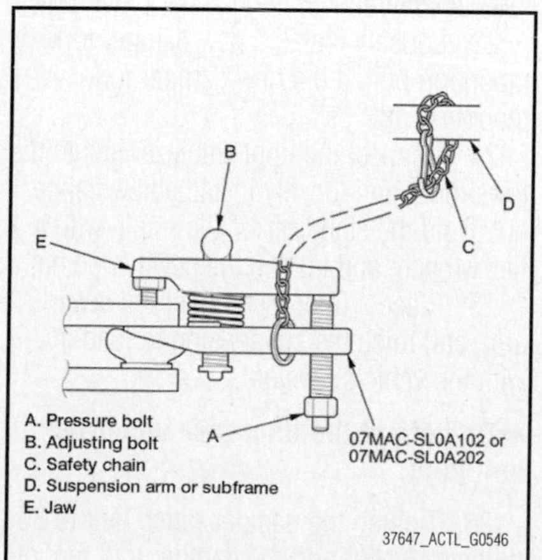
B
D
E
C
A
07MAC-SL0A102 or
07MAC-SL0A202
A. Pressure bolt
B. Adjusting bolt
C. Safety chain
D. Suspension arm or subframe
E. Jaw
37647_ACTL_G0546

Fig. 36 Loosen the pressure bolt, and install the ball joint remover as shown

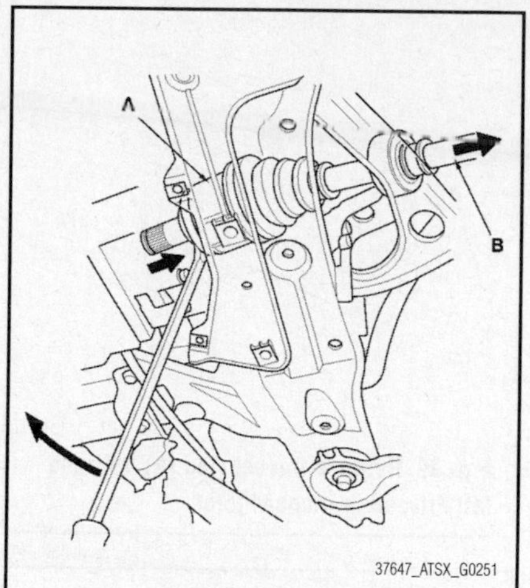
A
B
37647_ATSX_G0251

Fig. 37 Pry the inboard joint (A) from the differential using a prybar

8. Pull the knuckle outward, and separate the outboard joint from the front hub using a plastic hammer.

9. Left driveshaft: Pry the inboard joint from the differential using a prybar. Remove the driveshaft as an assembly.

➡Do not pull on the driveshaft, or the inboard joint may come apart. Pull the inboard joint straight out to avoid damaging the oil seal. Be careful not to damage the oil seal or the end of the inboard joint with the prybar.

10. Right driveshaft: Drive the inboard joint off of the intermediate shaft using a drift and a hammer. Remove the driveshaft as an assembly.

➡Do not pull on the driveshaft, or the inboard joint may come apart. Be careful not to damage the end of the inboard joint with the drift.

11. Remove the set ring from the left driveshaft inboard joint.

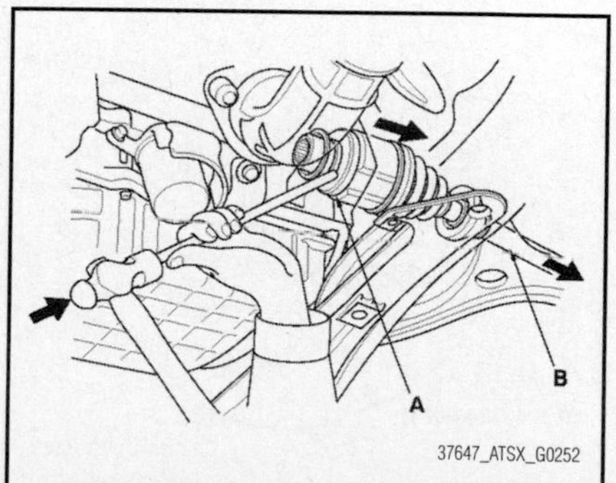
A
B
37647_ATSX_G0252

Fig. 38 Drive the inboard joint (A) off of the intermediate shaft using a drift and a hammer

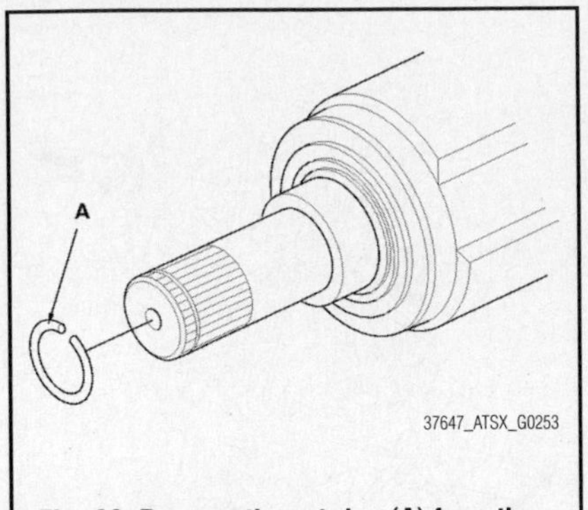
A
37647_ATSX_G0253

Fig. 39 Remove the set ring (A) from the left driveshaft inboard joint

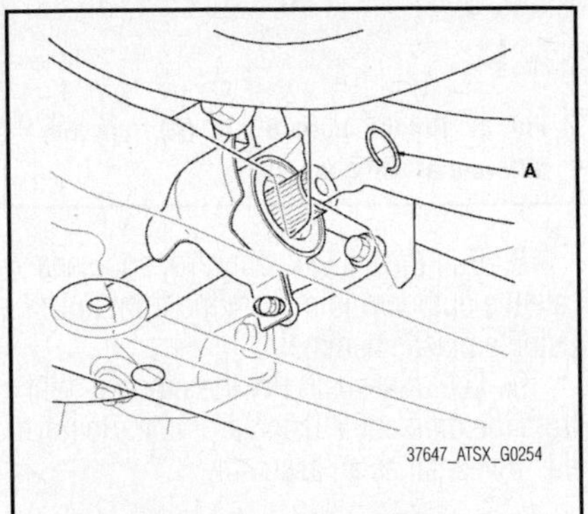
A
37647_ATSX_G0254

Fig. 40 Remove the set ring (A) from the intermediate shaft

12. Remove the set ring from the intermediate shaft.

To install:

➡**Before starting installation, make sure the mating surfaces of the joint and the splined section are clean.**

13. Apply Moly 60 paste (P/N 08734-0001) to the contact area of the outboard joint and the front wheel bearing.

➡**The paste helps prevent noise and vibration.**

14. Install a new set ring into the set ring groove of the left driveshaft inboard joint.

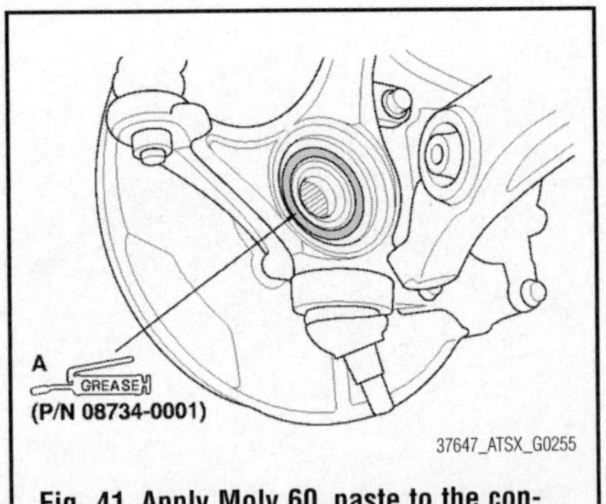
A
GREASE
(P/N 08734-0001)
37647_ATSX_G0255

Fig. 41 Apply Moly 60 paste to the contact area (A) of the outboard joint and the front wheel bearing

15. Install a new set ring into the set ring groove of the intermediate shaft.

16. Apply super high temp urea grease (P/N 08798-9002) to the whole splined surface of the right driveshaft. After applying grease, remove the grease from the splined grooves at intervals of 2-3 splines and from the set ring groove so that air can bleed from the intermediate shaft.

17. Clean the areas where the driveshaft contacts the differential thoroughly with solvent, and dry them with compressed air.

➡**Do not wash the rubber parts with solvent.**

18. Insert the inboard end of the driveshaft into the differential or the intermediate shaft until the set ring locks in the groove.

➡**Insert the driveshaft horizontally to prevent damaging the oil seal.**

19. Install the outboard joint into the front hub on the knuckle.

20. Wipe off any grease contamination from the ball joint tapered section and threads, then install the knuckle onto the lower arm. Be careful not to damage the ball joint boot. Wipe off the grease before tightening the nut at the ball joint. Torque the castle nut to the lower torque specification, then tighten it only far enough to align the slot with the ball joint pin hole. Castle nut torque specifications: 58–65 ft.lbs. (78–88 Nm).

➡**Make sure the ball joint boot is not damaged or cracked. Do not align the nut by loosening it.**

21. Install a new cotter pin into the ball joint pin hole, and bend the cotter pin.

22. Install the damper fork over the driveshaft and onto the lower arm. Install the damper in the damper fork so the aligning tab is aligned with the slot in the damper fork. Loosely install the damper pinch bolt.

23. Loosely install a new damper fork mounting bolt and a new damper fork mounting nut.

24. Connect the front stabilizer link to the lower arm, and loosely install a new flange nut. Hold the stabilizer link joint pin with a hex wrench, and tighten the new flange nut.

25. Place a floor jack under the lower arm, and raise the suspension to load it with the vehicle's weight.

➡**Do not put the floor jack under the ball joint.**

26. Tighten the damper pinch bolt to 36 ft. lbs. (49 Nm) and the damper fork mounting nut while holding the damper fork mounting bolt to 47 ft. lbs. (64 Nm), then remove the floor jack.

27. Apply a small amount of engine oil to the seating surface of a new spindle nut.

28. Install the spindle nut, then tighten it to 242 ft. lbs. (328 Nm). After tightening, use a drift to stake the spindle nut shoulder against the driveshaft.

29. Clean the mating surfaces of the brake disc and the wheel, then install the front wheels.

30. Turn the front wheel by hand, and make sure there is no interference between the driveshaft and surrounding parts.

31. Refill the transmission with the recommended transmission fluid:

32. Lower the vehicle on the lift.

33. Check the wheel alignment, and adjust it if necessary.

34. Test-drive the vehicle.

6-Cylinder Engine

See Figures 31 through 36 and 42 through 46.

1. Raise the vehicle on a lift.
2. Remove the front wheels.
3. Pry up the stake on the spindle nut, then remove the nut.
4. Drain the transmission fluid, then reinstall the drain plug with a new sealing washer.
5. Hold the stabilizer link joint pin with a hex wrench, and remove the flange nut. Separate the front stabilizer link from the lower arm.
6. Remove the damper fork mounting nut, the damper fork mounting bolt, and the damper pinch bolt, then remove the damper fork.
7. Remove the cotter pin from the knuckle ball joint, and remove the castle nut. Separate the ball joint from the lower arm using the 14 mm ball joint thread protector and the 28 mm ball joint remover.

➡**Always use a ball joint remover to disconnect a ball joint. Do not strike the housing or any other part of the ball joint connection to disconnect it.**

a. Install a hex nut or the ball joint thread protector onto the threads of the ball joint.

➡**Using a hex nut, make sure the nut is flush with the ball joint pin end to prevent damage to the threaded end of the ball joint pin.**

b. Apply grease to the ball joint remover on the areas shown. This will ease the installation of the tool, and prevent damage to the pressure bolt threads.

c. Loosen the pressure bolt, and install the ball joint remover as shown.

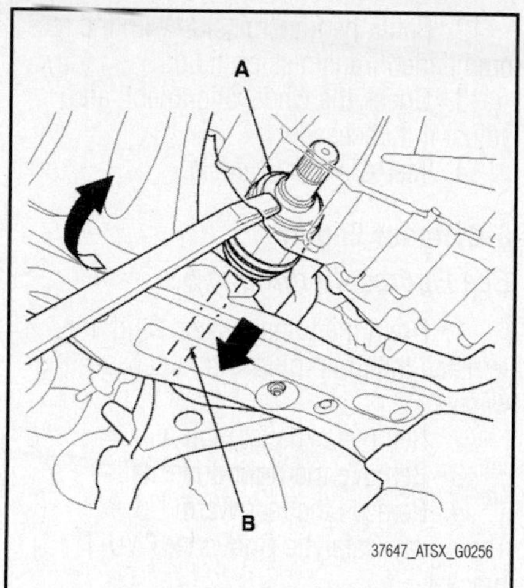
A
B
37647_ATSX_G0256

Fig. 42 Pry the inboard joint (A) from the differential using a prybar

Insert the jaws carefully, making sure not to damage the ball joint boot. Adjust the jaw spacing by turning the adjusting bolt.

➡Fasten the safety chain securely to a suspension arm or the subframe. Do not fasten it to a brake line or wire harness.

d. After adjusting the adjusting bolt, make sure the head of the adjusting bolt is in position to allow the jaw to pivot.

e. With a wrench, tighten the pressure bolt until the ball joint pin pops loose from the ball joint connecting hole. If necessary, apply penetrating type lubricant to loosen the ball joint pin.

➡Do not use pneumatic or electric tools on the pressure bolt.

f. Remove the ball joint remover, then remove the nut or the ball joint thread protector from the end of the ball joint pin, and pull the ball joint out of the ball joint connecting hole. Inspect the ball joint boot, and replace it if damaged.

➡Be careful not to damage the ball joint boot when installing the remover. Do not force or hammer on the lower arm, or pry between the lower arm and the knuckle. You could damage the ball joint.

8. Pull the knuckle outward, and separate the outboard joint from the front hub using a plastic hammer.
9. Remove exhaust pipe A.
10. Left driveshaft: Pry the inboard joint from the differential using a prybar. Remove the driveshaft as an assembly.

➡Do not pull on the driveshaft, or the inboard joint may come apart. Pull the inboard joint straight out to avoid damaging the oil seal. Be careful not to damage the oil seal or the end of the inboard joint with the prybar.

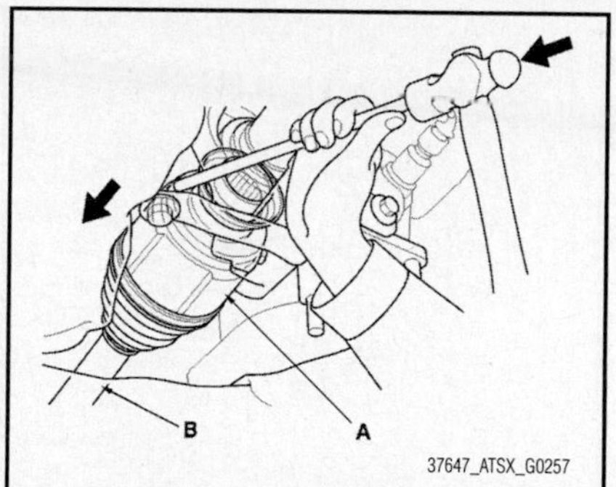
B
A
37647_ATSX_G0257

Fig. 43 Drive the inboard joint (A) off of the intermediate shaft using a drift and a hammer

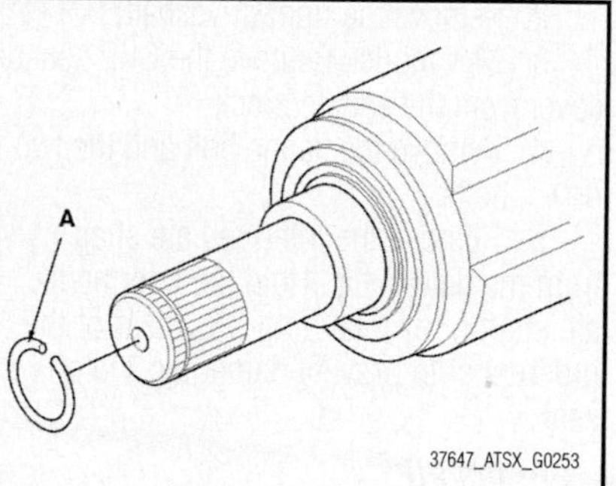
A
37647_ATSX_G0253

Fig. 44 Remove the set ring (A) from the left driveshaft inboard joint

11. Right driveshaft: Drive the inboard joint off of the intermediate shaft using a drift and a hammer. Remove the driveshaft as an assembly.

➡Do not pull on the driveshaft, or the inboard joint may come apart. Be careful not to damage the end of the inboard joint with the drift.

12. Remove the set ring from the left driveshaft inboard joint.

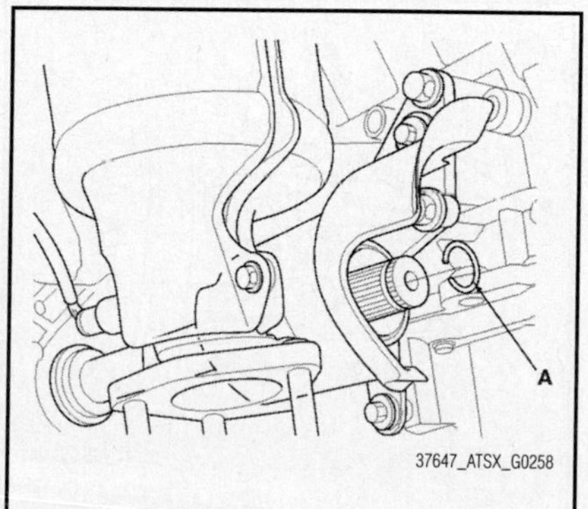
A
37647_ATSX_G0258

Fig. 45 Remove the set ring (A) from the intermediate shaft

13. Remove the set ring from the intermediate shaft.

To install:

➡Before starting installation, make sure the mating surfaces of the joint and the splined section are clean.

14. Apply Moly 60 paste (P/N 08734-0001) to the contact area of the outboard joint and the front wheel bearing.

➡The paste helps prevent noise and vibration.

15. Install a new set ring into the set ring groove of the left driveshaft inboard joint.
16. Install a new set ring into the set ring groove of the intermediate shaft.
17. Apply super high temp urea grease (P/N 08798-9002) to the whole splined surface of the right driveshaft. After applying grease, remove the grease from the splined grooves at intervals of 2-3 splines and from the set ring groove so that air can bleed from the intermediate shaft.
18. Clean the areas where the driveshaft contacts the differential thoroughly with solvent, and dry them with compressed air.

➡Do not wash the rubber parts with solvent.

19. Insert the inboard end of the driveshaft into the differential or the intermediate shaft until the set ring locks in the groove.

➡Insert the driveshaft horizontally to prevent damaging the oil seal.

20. Install the outboard joint into the front hub on the knuckle.
21. Install exhaust pipe A.
22. Wipe off any grease contamination from the ball joint tapered section and threads, then install the knuckle onto the lower arm. Be careful not to damage the ball joint boot. Wipe off the grease before tightening the nut at the ball joint. Torque the castle nut to the lower torque specification,

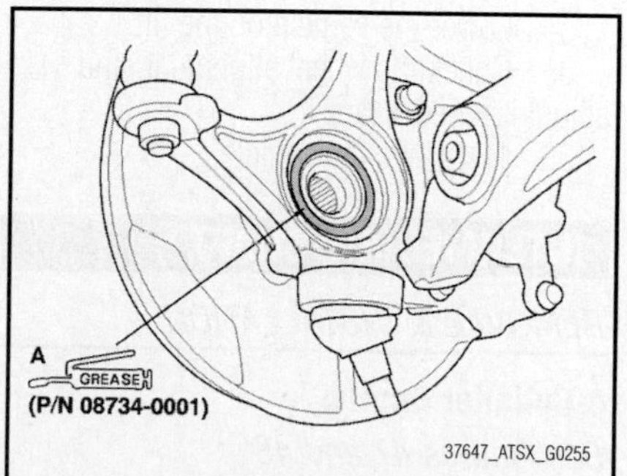
A
GREASE
(P/N 08734-0001)
37647_ATSX_G0255

Fig. 46 Apply Moly 60 paste to the contact area (A) of the outboard joint and the front wheel bearing

then tighten it only far enough to align the slot with the ball joint pin hole. Castle nut torque specifications: 58–65 ft. lbs. (78–88 Nm).

➡Make sure the ball joint boot is not damaged or cracked. Do not align the nut by loosening it.

23. Install a new cotter pin into the ball joint pin hole, and bend the cotter pin.
24. Install the damper fork over the driveshaft and onto the lower arm. Install the damper in the damper fork so the aligning tab is aligned with the slot in the damper fork. Loosely install the damper pinch bolt.
25. Loosely install a new damper fork mounting bolt and a new damper fork mounting nut.
26. Connect the front stabilizer link to the lower arm, and loosely install a new flange nut. Hold the stabilizer link joint pin with a hex wrench, and tighten the new flange nut.
27. Place a floor jack under the lower arm, and raise the suspension to load it with the vehicle's weight.

➡Do not put the floor jack under the ball joint.

28. Tighten the damper pinch bolt to 36 ft. lbs. (49 Nm) and the damper fork mounting nut while holding the damper fork mounting bolt to 47 ft. lbs. (64 Nm), then remove the floor jack.
29. Apply a small amount of engine oil to the seating surface of a new spindle nut.
30. Install the spindle nut, then tighten it to 242 ft. lbs. (328 Nm). After tightening, use a drift to stake the spindle nut shoulder against the driveshaft.
31. Clean the mating surfaces of the brake disc and the wheel, then install the front wheels.
32. Turn the front wheel by hand, and make sure there is no interference between the driveshaft and surrounding parts.
33. Refill the transmission with the recommended transmission fluid:
34. Lower the vehicle on the lift.
35. Check the wheel alignment, and adjust it if necessary.
36. Test-drive the vehicle.

INTERMEDIATE SHAFT

REMOVAL & INSTALLATION

4-Cylinder Engine

See Figures 47 and 48.

1. Drain the transmission fluid, then reinstall the drain plug with a new sealing washer:

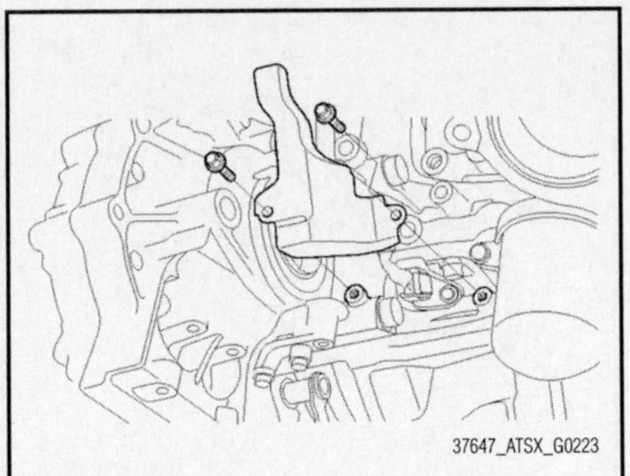
37647_ATSX_G0223

Fig. 47 M/T model: Remove the CKP sensor cover (A) from the engine block

2. Remove the right driveshaft.
3. M/T model: Remove the CKP sensor cover from the engine block.
4. Remove the flange bolt and the two dowel bolts.
5. Remove the intermediate shaft from the differential. Hold the intermediate shaft horizontal until it is clear of the differential to prevent damaging the oil seal.

To install:

6. Clean the areas where the intermediate shaft contacts the differential thoroughly with solvent, and dry them with compressed air.

➡Do not wash the rubber parts with solvent.

7. Install a new set ring onto the set ring groove of the intermediate shaft.
8. Insert the intermediate shaft into the differential correctly.

➡Insert the intermediate shaft carefully to prevent damaging the oil seal.

9. Install the flange bolt and the two dowel bolts.
10. Install the right driveshaft.
11. M/T model: Install the CKP sensor cover to the engine block.

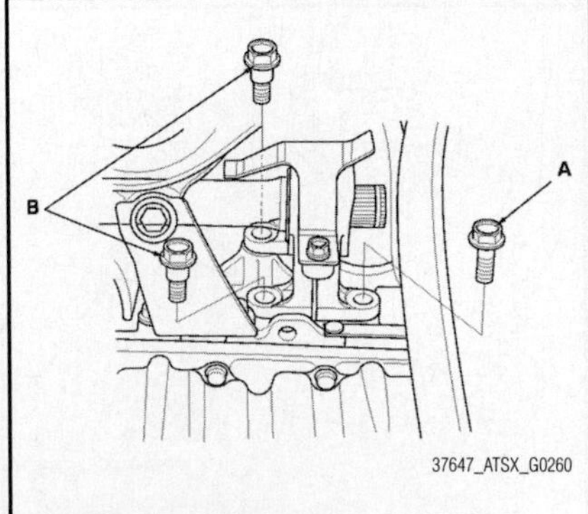
A
B
37647_ATSX_G0260

Fig. 48 Remove the flange bolt (A) and the two dowel bolts (B)

12. Refill the transmission with the recommended transmission fluid:
13. Check the wheel alignment, and adjust it if necessary.
14. Test-drive the vehicle.

6-Cylinder Engine

See Figures 49 through 53.

1. Drain the transmission fluid, then reinstall the drain plug with a new sealing washer.
2. Remove exhaust pipe A.
3. Remove the right driveshaft.
4. Remove the rear Warm Up Three Way Catalytic Converter (WU-TWC) bracket.
5. Remove the Crankshaft Position (CKP) sensor cover.
6. Disconnect the CKP sensor connector, and remove the sensor harness clamp bracket bolt.
7. Swing the sensor harness out of the way to prevent damaging it.
8. Remove the flange bolt and the two dowel bolts.

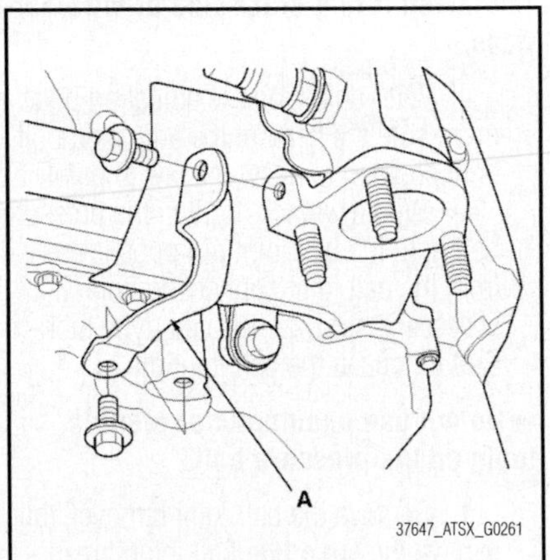
A
37647_ATSX_G0261

Fig. 49 Remove the rear WU-TWC bracket (A)

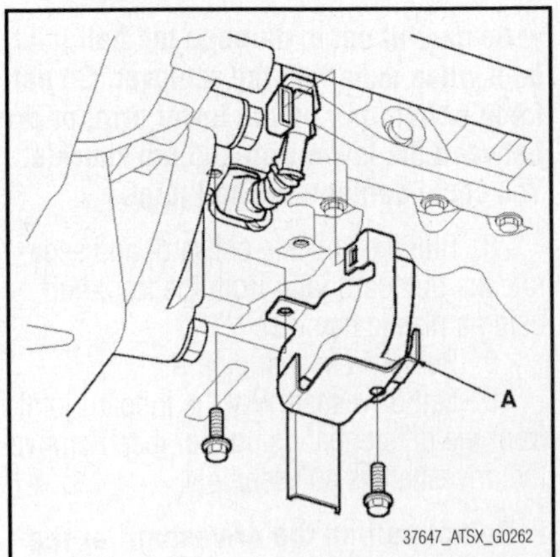
A
37647_ATSX_G0262

Fig. 50 Remove the CKP sensor cover (A)

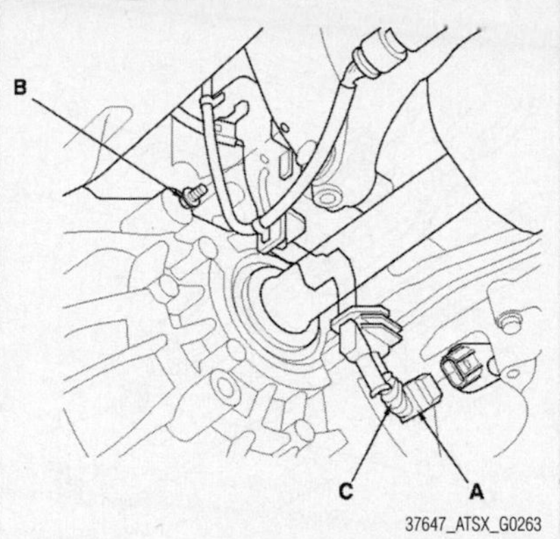
B
C
A
37647_ATSX_G0263

Fig. 51 Disconnect the CKP sensor connector (A), and remove the sensor harness clamp bracket bolt (B)

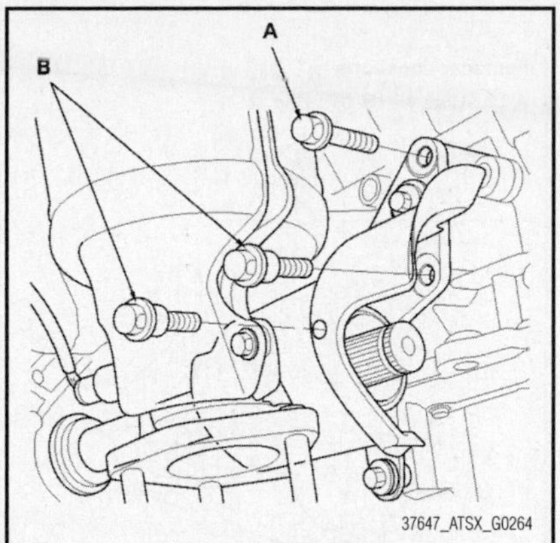
A
B
37647_ATSX_G0264

Fig. 52 Remove the flange bolt (A) and the two dowel bolts (B)

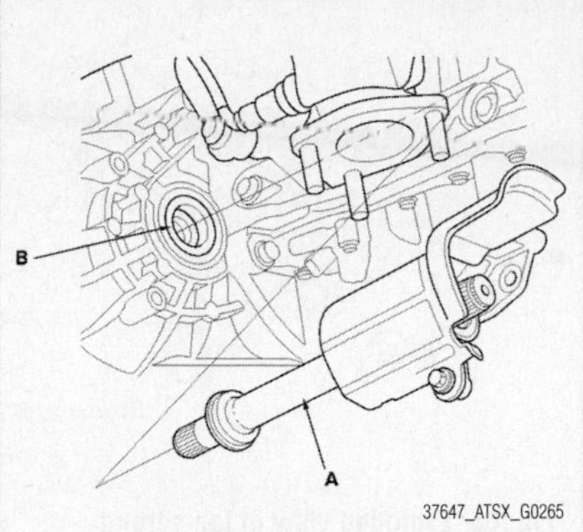
B
A
37647_ATSX_G0265

Fig. 53 Remove the intermediate shaft (A) from the differential; prevent damaging the oil seal (B)

9. Remove the intermediate shaft from the differential. Hold the intermediate shaft horizontal until it is clear of the differential to prevent damaging the oil seal.

To install:

10. Clean the areas where the intermediate shaft contacts the differential thoroughly with solvent, and dry them with compressed air.

➡Do not wash the rubber parts with solvent.

11. Install a new set ring into the set ring groove of the intermediate shaft.
12. Insert the intermediate shaft into the differential correctly.

➡Insert the intermediate shaft carefully to prevent damaging the oil seal.

13. Install the flange bolt and the two dowel bolts.
14. Install the rear WU-TWC bracket.
15. Install the right driveshaft.
16. Connect the CKP sensor connector, and install the sensor harness clamp bracket.
17. Install the CKP sensor cover.
18. Install exhaust pipe A.
19. Refill the transmission with the recommended transmission fluid.
20. Check the wheel alignment, and adjust it if necessary.
21. Test-drive the vehicle.

ENGINE COOLING

ENGINE FAN

REMOVAL & INSTALLATION

4-Cylinder Engine

See Figures 54 and 55.

1. Disconnect the fan motor connectors, and remove the harness clamp.
2. Remove the A/C condenser fan shroud assembly.
3. Remove the radiator fan shroud assembly.

➡Move the radiator fan shroud assembly toward the right side of the vehicle to allow enough space to lift it up and away from the A/C condenser fan shroud assembly.

4. Disassemble the fan shrouds.

To install:

5. Assemble the fan shrouds.
6. Install the radiator fan shroud assembly.
7. Install the A/C condenser fan shroud assembly.
8. Connect the fan motor connectors, and install the harness clamp.

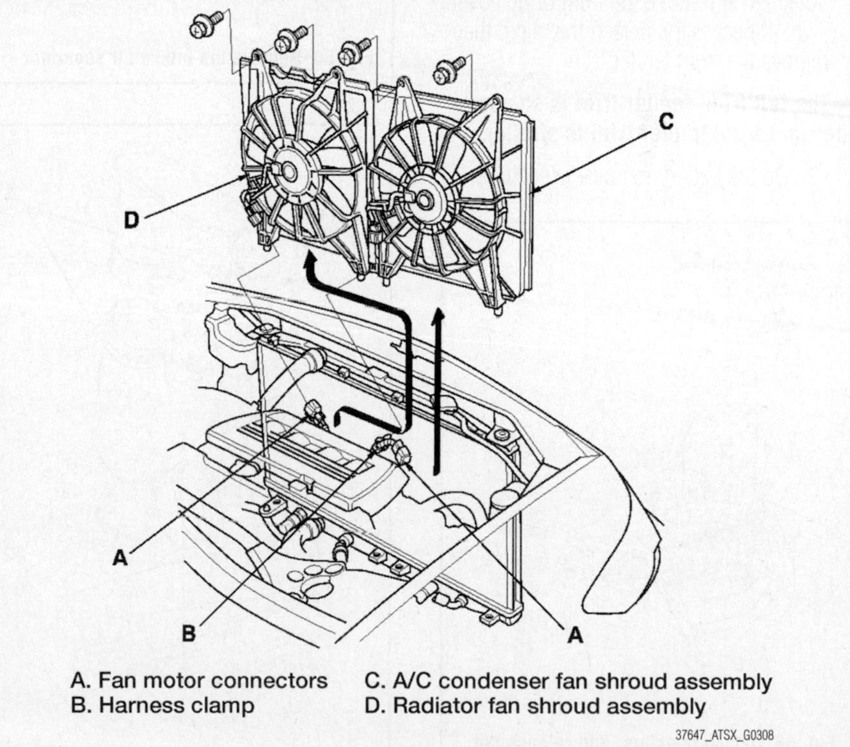
C
D
A
B
A
A. Fan motor connectors
B. Harness clamp
C. A/C condenser fan shroud assembly
D. Radiator fan shroud assembly
37647_ATSX_G0308

Fig. 54 Remove the radiator fan shroud assembly

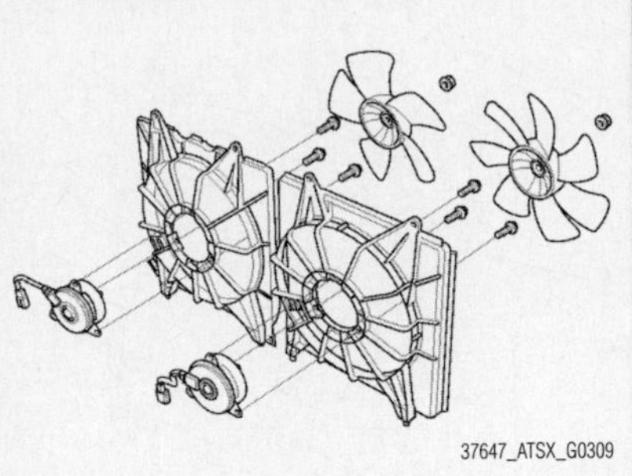
37647_ATSX_G0309

Fig. 55 Exploded view of fan shroud assemblies

6-Cylinder Engine

See Figures 56 through 64.

1. Remove the engine compartment covers.
   a. Remove the left engine compartment cover.
   b. Detach the clip, and release the hooks.
   c. Pull up the left engine compartment cover to release the pins.
   d. Remove the right engine compartment cover.
   e. Detach the clip, and release the hooks.
   f. Pull up the right engine compartment cover to release the pins.
   g. Detach the clips, and release the hooks, then remove the front grille cover.
   h. If necessary, detach the clips, then remove the front fender trim.

➡The left front fender trim is shown; the right front fender trim is similar.

2. Do the battery removal procedure.

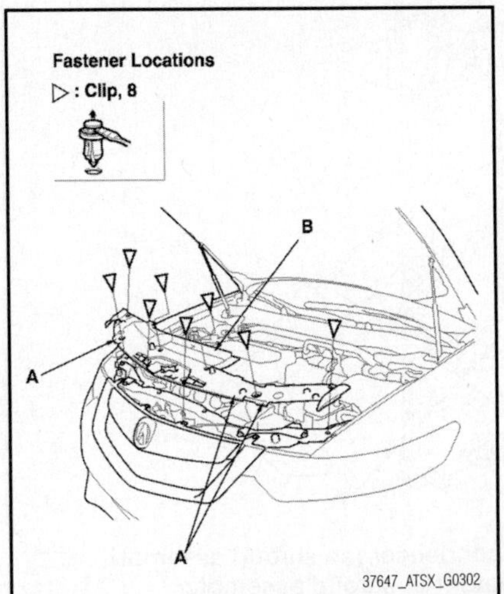
Fastener Locations
▷ : Clip, 8
B
A
A
37647_ATSX_G0302

Fig. 56 Detach the clips, and release the hooks (A), then remove the front grille cover (B)

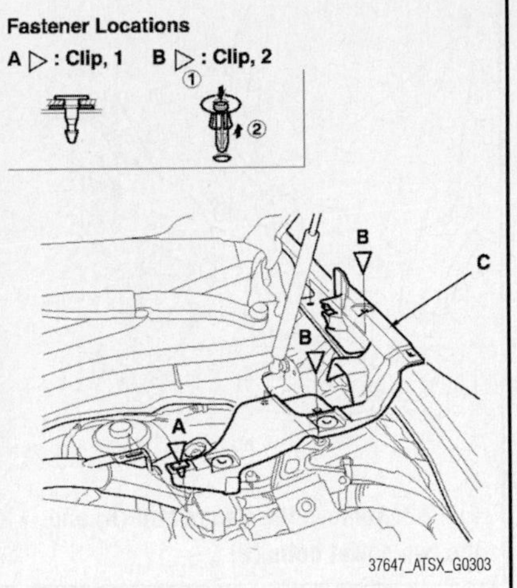
Fastener Locations
A ▷ : Clip, 1
B ▷ : Clip, 2
①
②
B
C
B
A
37647_ATSX_G0303

Fig. 57 Detach the clips (A, B), then remove the front fender trim (C)

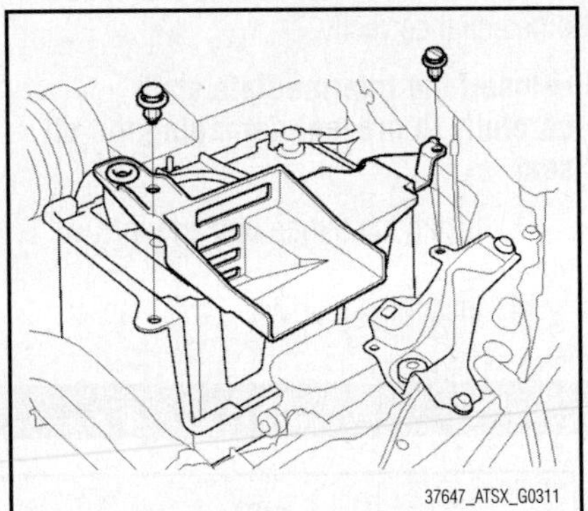
37647_ATSX_G0311

Fig. 58 Remove the intake air separator

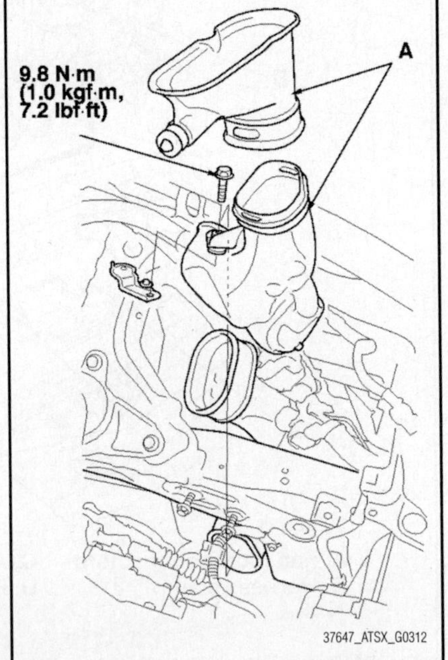
9.8 N·m
(1.0 kgf·m,
7.2 lbf·ft)
A
37647_ATSX_G0312

Fig. 59 Remove the intake air ducts (A)

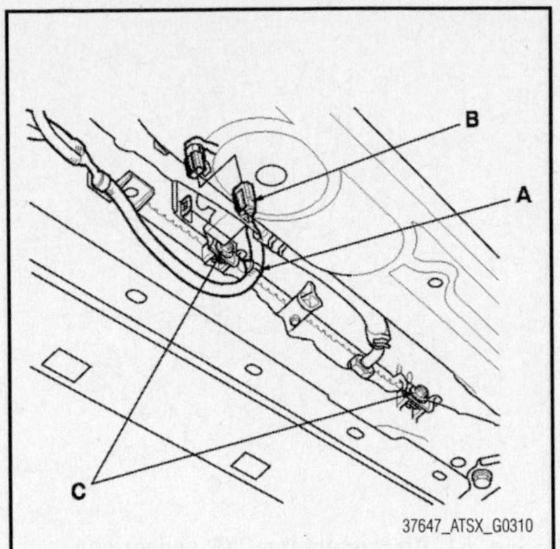
B
A
C
37647_ATSX_G0310

Fig. 60 Disconnect the A/C condenser fan motor connector (A), remove the harness clamp (B) and loosen the A/C condenser fan shroud mounting bolts (C)

3. Remove the battery base.
4. Remove the intake air separator.
5. Remove the intake air ducts.
7. Remove the splash shield.
8. Disconnect the A/C condenser fan motor connector, and remove the harness clamp.
9. Loosen the A/C condenser fan shroud mounting bolts.
10. Disconnect the radiator fan motor connector.
11. Lower the vehicle on the lift.
12. Remove the A/C condenser fan shroud assembly.
13. Remove the upper brackets.
14. Remove the radiator fan shroud assembly.

➡Pull-up the radiator, then move the radiator fan shroud assembly toward the right side of the vehicle to allow for

A
37647_ATSX_G0313

Fig. 61 Disconnect the radiator fan motor connector (A)

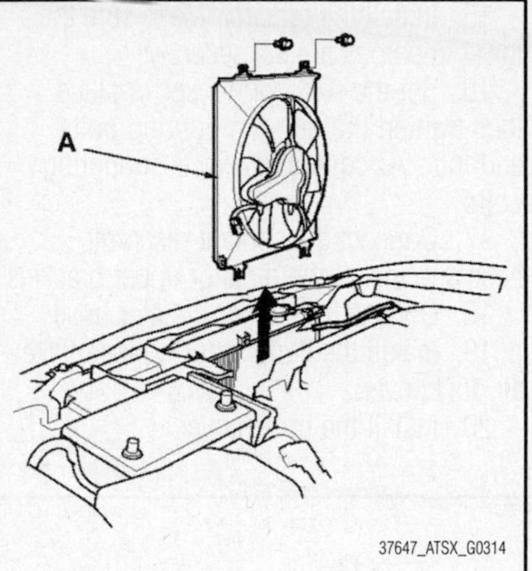
A
37647_ATSX_G0314

Fig. 62 Remove the A/C condenser fan shroud assembly (A)

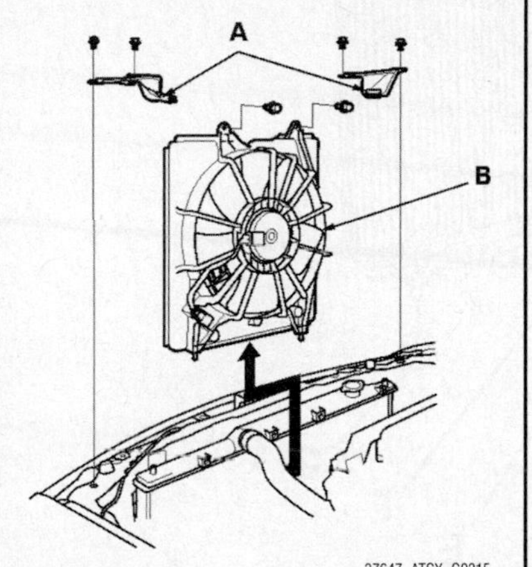
A
B
37647_ATSX_G0315

Fig. 63 Remove the upper brackets (A) and the radiator fan shroud assembly (B)

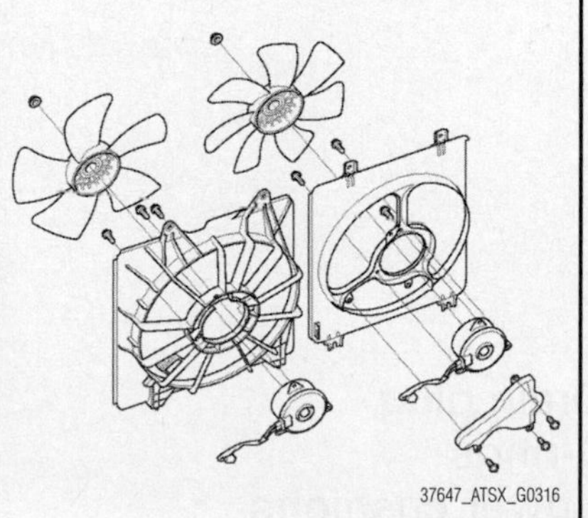
37647_ATSX_G0316

Fig. 64 Exploded view of fan shroud assemblies

enough space to lift it up and away from the A/C condenser fan shroud assembly.

15. Disassemble the fan shrouds.

To install:

16. Assemble the fan shrouds.
17. Install the radiator fan shroud assembly.
18. Install the upper brackets.
19. Install the A/C condenser fan shroud assembly.
20. Raise the vehicle on the lift.
21. Connector the radiator fan motor connector.
22. Tighten the A/C condenser fan shroud mounting bolts.
23. Connect the A/C condenser fan motor connector, and install the harness clamp.
24. Install the splash shield.
25. Lower the vehicle on the lift.
26. Install the intake air duct.
27. Install the intake air separator.
28. Install the battery base.
29. Do the battery installation procedure.
30. Install the engine compartment covers.

RADIATOR

REMOVAL & INSTALLATION

4-Cylinder Engine

See Figures 56, 65 through 68.

1. Drain the engine coolant.
2. Raise the vehicle on the lift to full height.

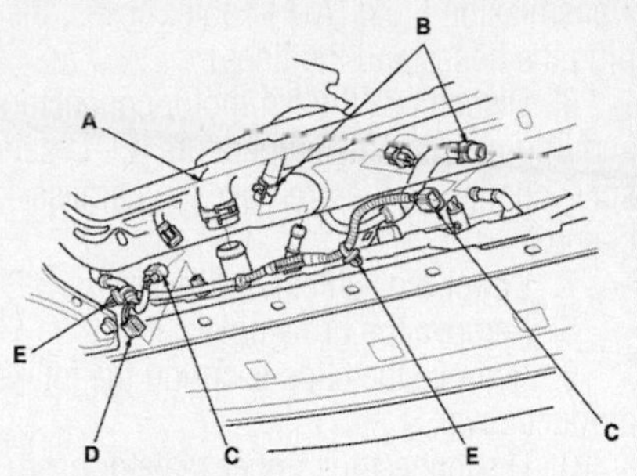
A
B
C
D
E

A. Lower radiator hose
B. ATF cooler hoses
C. Fan motor connectors
D. ECT sensor 2 connector
E. Harness clamps

37647_ATSX_G0320

Fig. 65 Disconnect the lower radiator hose

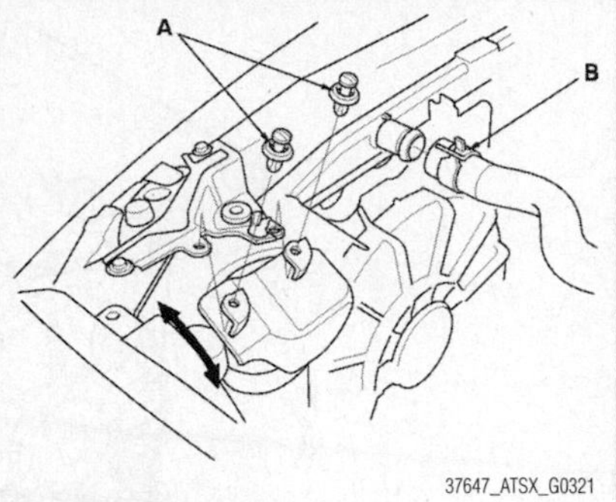
A
B
37647_ATSX_G0321

Fig. 66 Remove the clips (A) securing the intake air duct cover; disconnect the upper radiator hose (B)

3. Remove the splash shield.
4. Disconnect the lower radiator hose.
5. A/T model: Disconnect the Automatic

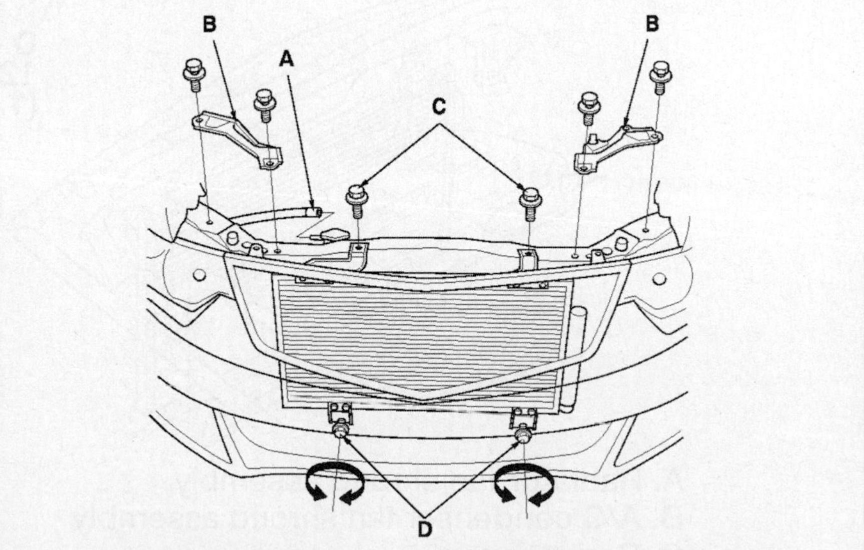
A
B
C
D

A. Coolant reservoir hose
B. Radiator upper brackets
C. A/C condenser upper mounting bolts
D. A/C condenser lower mounting bolts

37647_ATSX_G0322

Fig. 67 Disconnect the coolant reservoir hose and remove the radiator upper brackets

Transmission Fluid (ATF) cooler hoses, then plug the hoses and the lines.

6. Disconnect the fan motor connectors and Engine Coolant Temperature (ECT) sensor 2 connector, then remove the harness clamps.
7. Lower the vehicle on the lift.
8. Remove the grille cover.
9. Remove the clips securing the intake air duct cover.
10. Disconnect the upper radiator hose.
11. Disconnect the coolant reservoir hose and remove the radiator upper brackets.
12. Remove the A/C condenser upper mounting bolts and loosen the A/C condenser lower mounting bolts.
13. Pull up the radiator, then remove the radiator fan shroud assembly, the A/C condenser fan shroud assembly, the radiator cap, ECT sensor 2, and the drain plug.

To install:

14. Reassemble the radiator with new O-rings.
15. Install the radiator. Make sure the lower cushions are set securely.
16. Set the A/C condenser in place, then tighten the upper mounting bolts and the A/C condenser lower mounting bolts.
17. Connect the coolant reservoir hose and install the radiator upper brackets.
18. Connect the upper radiator hose.
19. Install the clips securing the intake air duct cover.
20. Install the grille cover.

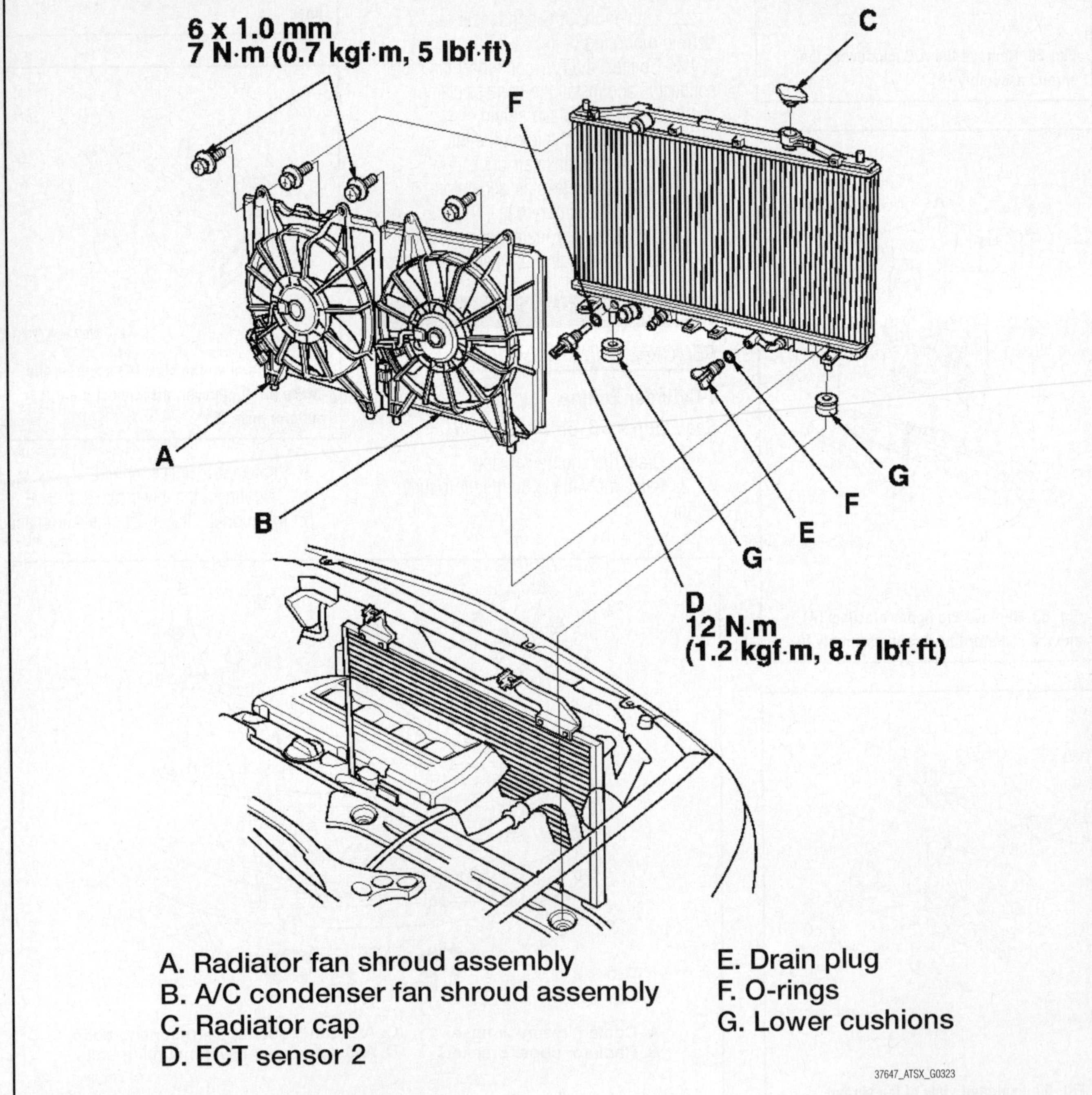
6 x 1.0 mm
7 N·m (0.7 kgf·m, 5 lbf·ft)
C
F
A
B
G
F
E
G
D
12 N·m
(1.2 kgf·m, 8.7 lbf·ft)
A. Radiator fan shroud assembly
B. A/C condenser fan shroud assembly
C. Radiator cap
D. ECT sensor 2
E. Drain plug
F. O-rings
G. Lower cushions
37647_ATSX_G0323

Fig. 68 Remove the radiator

21. Raise the vehicle on the lift to full height.
22. Connect the fan motor connectors and ECT sensor 2 connector, then install the harness clamps.
23. Connect the lower radiator hose.
24. A/T model: Connect the Automatic Transmission Fluid (ATF) cooler hoses.
25. Install the splash shield.
26. Lower the vehicle on the lift.
27. Refill the radiator with engine coolant, and bleed the air from the cooling system with the heater valve open.
28. Clean up any spilled engine coolant.

6-Cylinder Engine

See Figures 56 through 58, 60 and 69 through 73.

1. Remove the engine compartment covers.
   a. Remove the left engine compartment cover.
   b. Detach the clip, and release the hooks.
   c. Pull up the left engine compartment cover to release the pins.
   d. Remove the right engine compartment cover.
   e. Detach the clip, and release the hooks.
   f. Pull up the right engine compartment cover to release the pins.
   g. Detach the clips, and release the hooks, then remove the front grille cover.
   h. If necessary, detach the clips, then remove the front fender trim.

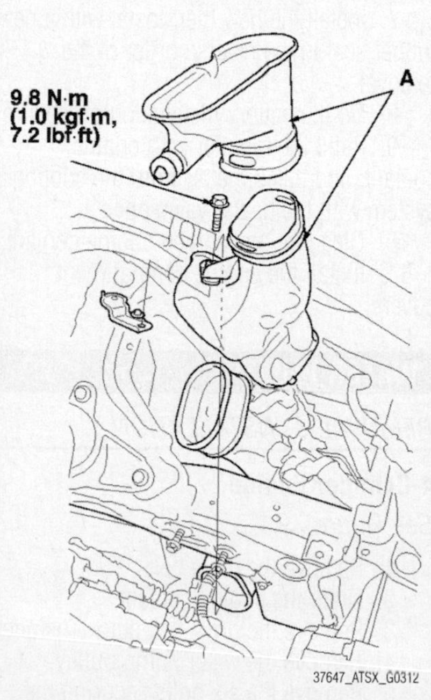
9.8 N·m
(1.0 kgf·m,
7.2 lbf·ft)
A
37647_ATSX_G0312
Fig. 69 Remove the intake air ducts

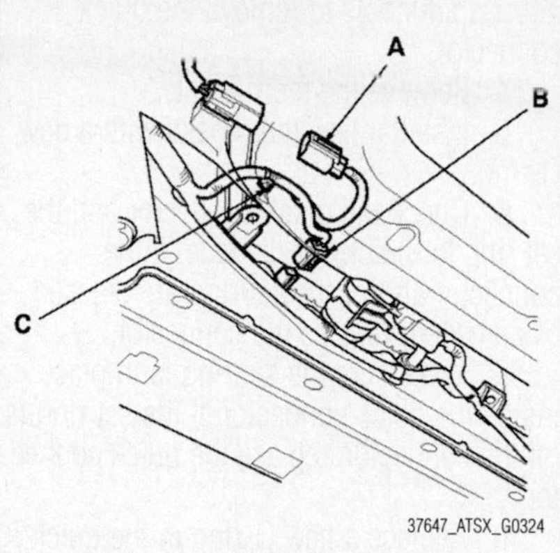
A
B
C
37647_ATSX_G0324
Fig. 70 Disconnect the radiator fan motor connector (A) and the ECT sensor 2 connector (B), then remove the harness clamp (C)

➡The left front fender trim is shown; the right front fender trim is similar.

2. Drain the engine coolant.
3. Remove the intake air separator.
4. Remove the intake air ducts.
5. Raise the vehicle on the lift.
6. Remove the splash shield.
7. Disconnect the A/C condenser fan motor connector, and remove the harness clamp.
8. Loosen the A/C condenser fan shroud mounting bolts.
9. Disconnect the radiator fan motor connector and the Engine Coolant Temperature (ECT) sensor 2 connector, then remove the harness clamp.
10. Lower the vehicle on the lift.
11. Disconnect the upper radiator hose, then remove the radiator fan shroud assembly and the A/C condenser fan shroud assembly.

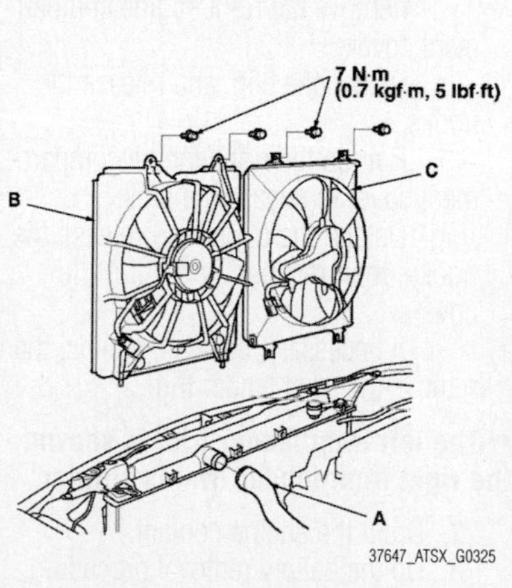
7 N·m
(0.7 kgf·m, 5 lbf·ft)
B
C
A
37647_ATSX_G0325
Fig. 71 Disconnect the upper radiator hose (A), then remove the radiator fan shroud assembly (B) and the A/C condenser fan shroud assembly (C)

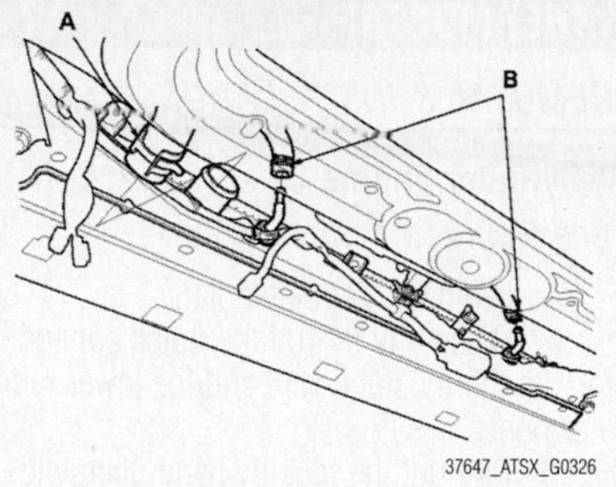
A
B
37647_ATSX_G0326
Fig. 72 Disconnect the lower radiator hose (A) and the ATF cooler hoses (B)

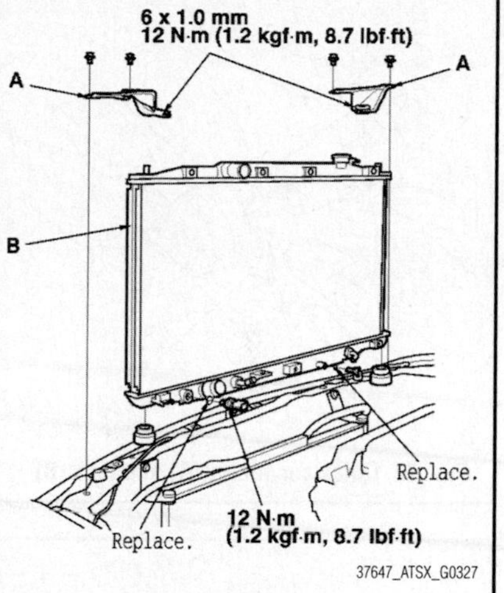
6 x 1.0 mm
12 N·m (1.2 kgf·m, 8.7 lbf·ft)
A
A
B
Replace.
Replace.
12 N·m
(1.2 kgf·m, 8.7 lbf·ft)
37647_ATSX_G0327
Fig. 73 Remove the upper brackets (A), then pull up the radiator (B)

12. Raise the vehicle on the lift.
13. Disconnect the lower radiator hose.
14. Disconnect the Automatic Transmission Fluid (ATF) cooler hoses from the radiator, then plug the cooler hoses and the lines.
15. Lower the vehicle on the lift.
16. Remove the upper brackets, then pull up the radiator.
17. Remove the other parts from the radiator.

To install:

18. Install radiator in the reverse order of removal.

➡Make sure the upper and lower cushions are set securely.

19. Fill the radiator with engine coolant, and bleed the air from the cooling system with the heater valve open.
20. Clean up any spilled engine coolant.
21. Install the engine compartment covers.

THERMOSTAT

REMOVAL & INSTALLATION

4-Cylinder Engine

See Figures 74 through 76.

1. Drain the engine coolant.
2. Clean any dirt off the quick connector, the thermostat cover, and the lower radiator hose.
3. Pull out the lock by hand, then wiggle the quick connector loose, and remove it from the thermostat cover. Do not use any tools to remove the quick connector.
4. Remove the thermostat.
5. Install a new thermostat with a new O-ring.
6. Check the quick connector and the set ring for cracks or damage. If the connector and/or the set ring are cracked or damaged, replace the connector.
7. Make sure the set ring is in place inside the quick connector. If the set ring is off the connector, replace the quick connector.
8. Replace a new O-ring in the quick connector.
9. Check the lock. If the lock is damaged or deformed, replace it. When installing the new lock to the connector, push it straight down along the groove.
10. Clean the connecting surface of the thermostat cover, then apply clean engine coolant around the connecting surface.
11. Push the lock down, then push the quick connector onto the thermostat cover until you hear it click.
12. Refill the radiator with engine coolant, and bleed the air from the cooling system with the heater valve open.

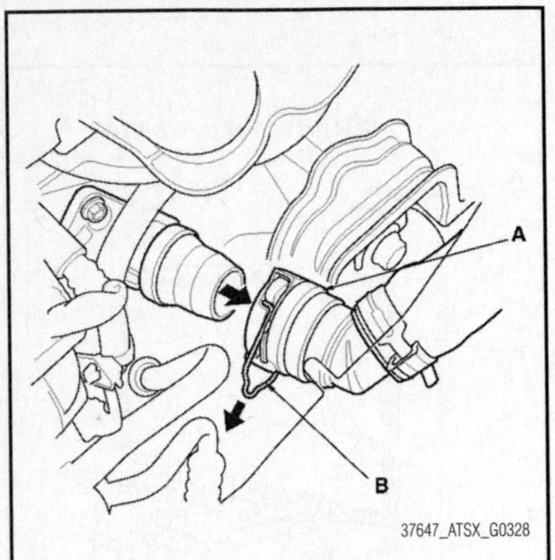
A
B
37647_ATSX_G0328
Fig. 74 Quick connector (A) and lock (B)

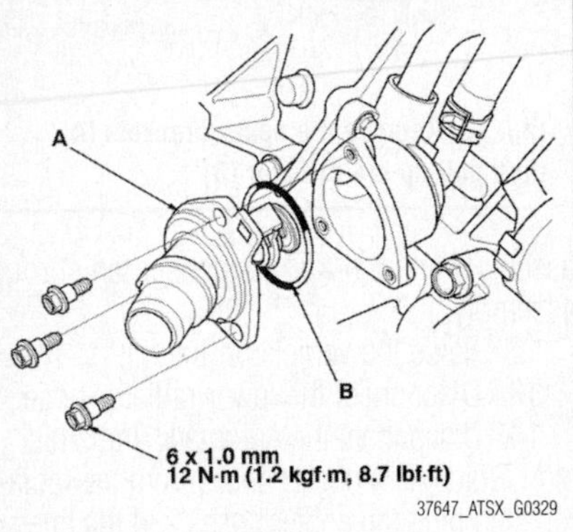
A
B
6 x 1.0 mm
12 N·m (1.2 kgf·m, 8.7 lbf·ft)
37647_ATSX_G0329
Fig. 75 Thermostat (A) and O-ring (B)

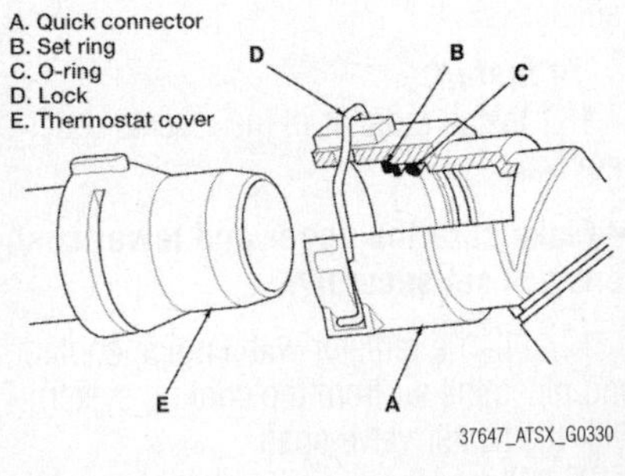
A. Quick connector
B. Set ring
C. O-ring
D. Lock
E. Thermostat cover
D
B
C
E
A
37647_ATSX_G0330
Fig. 76 Quick connector assembly

6-Cylinder Engine

See Figures 56 and 57, 77 and 78.

1. Remove the engine compartment covers.
   a. Remove the left engine compartment cover.
   b. Detach the clip, and release the hooks.
   c. Pull up the left engine compartment cover to release the pins.
   d. Remove the right engine compartment cover.
   e. Detach the clip, and release the hooks.
   f. Pull up the right engine compartment cover to release the pins.
   g. Detach the clips, and release the hooks, then remove the front grille cover.
   h. If necessary, detach the clips, then remove the front fender trim.

➡The left front fender trim is shown; the right front fender trim is similar.

2. Drain the engine coolant.
3. Do the battery removal procedure.
4. Remove harness clamps and disconnect the positive starter cable.
5. Disconnect the A/T clutch pressure control solenoid valve C connector, and remove the bolt and the harness holder.

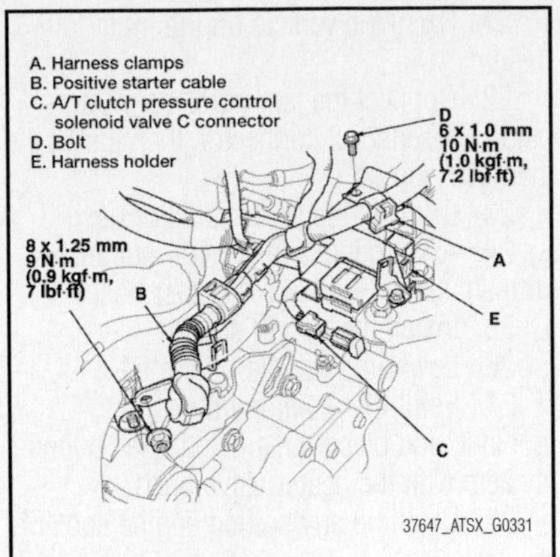
A. Harness clamps
B. Positive starter cable
C. A/T clutch pressure control solenoid valve C connector
D. Bolt
E. Harness holder
D
6 x 1.0 mm
10 N·m
(1.0 kgf·m,
7.2 lbf·ft)
8 x 1.25 mm
9 N·m
(0.9 kgf·m,
7 lbf·ft)
B
A
E
C
37647_ATSX_G0331
Fig. 77 Remove harness clamps and disconnect the positive starter cable

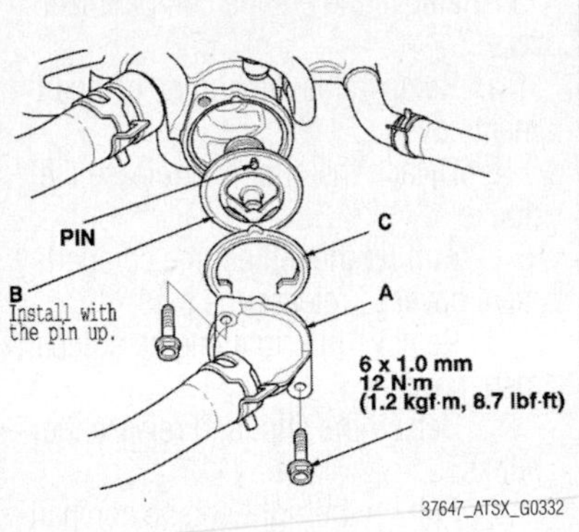
PIN
C
B
Install with the pin up.
A
6 x 1.0 mm
12 N·m
(1.2 kgf·m, 8.7 lbf·ft)
37647_ATSX_G0332
Fig. 78 Thermostat cover (A), thermostat (B), rubber seal (C)

6. Remove the thermostat cover, then remove the thermostat.
7. Install the new thermostat with a new rubber seal in the reverse order of the removal.
8. Do the battery installation procedure.
9. Refill the radiator with engine coolant, and bleed the air from the cooling system with the heater valve open.
10. Clean up any spilled engine coolant.
11. Install the engine compartment covers.

WATER PUMP

REMOVAL & INSTALLATION

4-Cylinder Engine

See Figure 79.

1. Remove the drive belt.
2. Drain the engine coolant.
3. Remove the drive belt auto-tensioner.
4. Remove the water pump pulley.
5. Remove the six bolts securing the water pump, then remove the water pump.

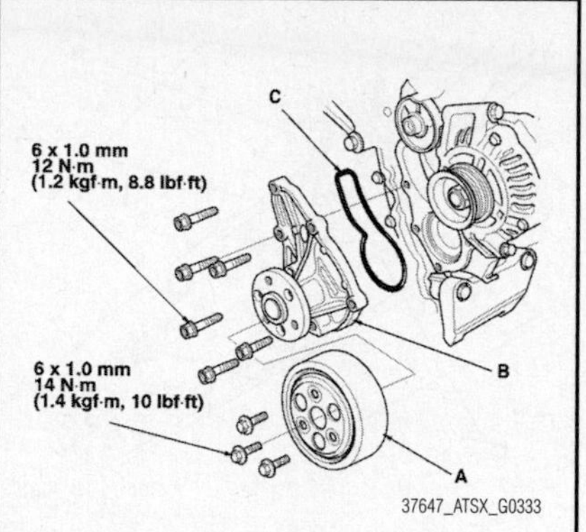
C
6 x 1.0 mm
12 N·m
(1.2 kgf·m, 8.8 lbf·ft)
B
6 x 1.0 mm
14 N·m
(1.4 kgf·m, 10 lbf·ft)
A
37647_ATSX_G0333

Fig. 79 Exploded view of water pump assembly

To install:

6. Inspect and clean the O-ring groove and the mating surface of the water passage.
7. Install the water pump with a new O-ring and the water pump pulley.
8. Clean up any spilled engine coolant.
9. Install the drive belt auto-tensioner.
10. Install the drive belt.
11. Refill the radiator with engine coolant, and bleed the air from the cooling system with the heater valve open.

6-Cylinder Engine

See Figure 80.

1. Drain the engine coolant.
2. Remove the timing belt. Refer the Engine Mechanical section for the Timing Belt Removal and Installation.
3. Remove the timing belt adjuster. Refer the Engine Mechanical section for the Timing Belt Adjuster Removal and Installation.
4. Remove the five bolts securing the water pump, then remove the water pump and gasket.

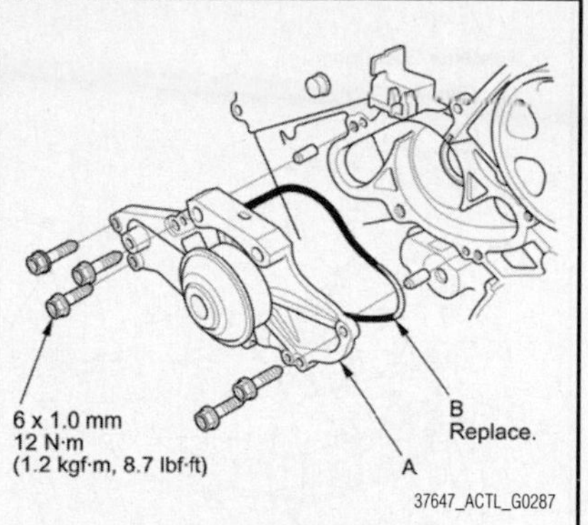
6 x 1.0 mm
12 N·m
(1.2 kgf·m, 8.7 lbf·ft)
B
Replace.
A
37647_ACTL_G0287

Fig. 80 Remove the five bolts securing the water pump (A), then remove the water pump and gasket (B)

To install:

5. Inspect and clean the O-ring groove and the mating surface of the engine block.
6. Install the water pump with a new gasket.
7. Clean up any spilled engine coolant.
8. Install the timing belt adjuster:
9. Install the timing belt:
10. Refill the radiator with engine coolant, and bleed the air from the cooling system.

BATTERY RECONNECT/RELEARN PROCEDURE

Disconnection

➡Some systems store data in memory (including seat position, mirror position, etc) that is lost when the battery is disconnected. Do the following procedures before disconnecting the battery.

1. Make sure you have the anti-theft code(s) for the audio and/or the navigation system (if equipped).
2. If you are replacing the audio unit, write down the audio presets (AM and FM), and the XM audio presets (if equipped), because the audio unit does not retain the presets after the battery is disconnected.
3. Make sure the ignition switch is in LOCK (0).
4. Disconnect and isolate the negative cable from the battery.

➡Always disconnect the negative cable from the battery first.

5. Disconnect the positive cable from the battery.

Reconnection

➡Some systems store data in memory (including seat position, mirror position, etc) that is lost when the battery is disconnected. Do the following procedures to restore the systems back to normal operation.

1. Clean the battery terminals.
2. Test the battery.
3. Reconnect the positive cable to the battery first, then reconnect the negative cable to the battery.

➡Always connect the positive cable to the battery first.

4. Apply multipurpose grease to the terminals to prevent corrosion.
5. Enter the anti-theft code(s) for the audio system and/or the navigation system (if equipped).
6. Enter the audio presets (if applicable), and enter the XM audio presets (if equipped).
7. Set the clock (for vehicles without navigation).

ENGINE ELECTRICAL — CHARGING SYSTEM

ALTERNATOR

REMOVAL & INSTALLATION

4-Cylinder Engine

See Figures 81 and 82.

1. Do the battery terminal disconnection procedure.
2. Remove the accessory drive belt.
3. Remove the two bolts securing the alternator.
4. Disconnect the alternator connector and the positive alternator cable, and remove the harness clamp, then remove the alternator.

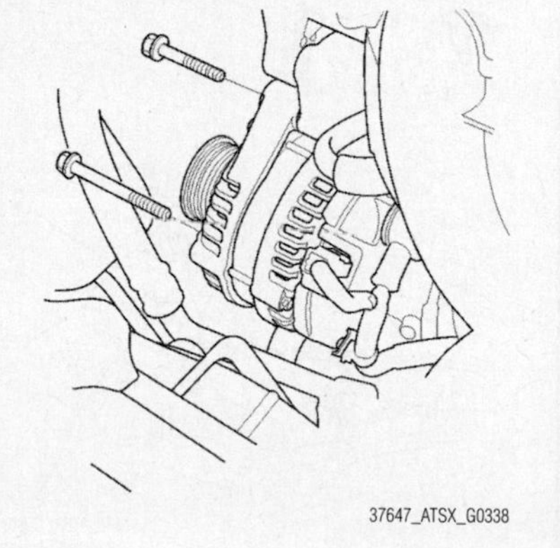
37647_ATSX_G0338

Fig. 81 Remove the two bolts securing the alternator

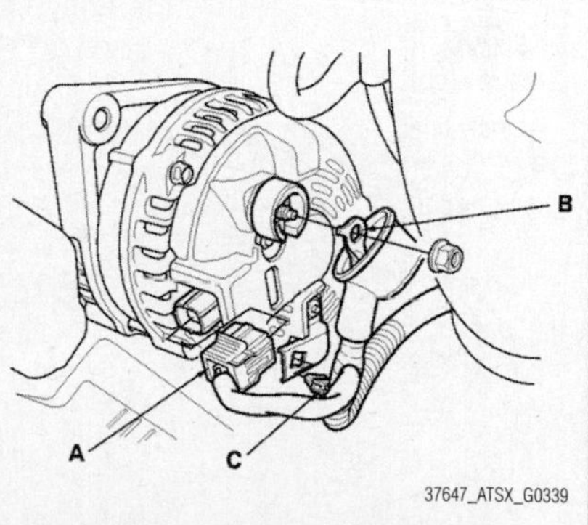
B
A
C
37647_ATSX_G0339

Fig. 82 Disconnect the alternator connector (A) and the positive alternator cable (B), and remove the harness clamp (C)

To install:

5. Install the alternator, then connect the alternator connector and the positive alternator cable, and install the harness clamp. Make sure the crimped side of the ring terminal faces away from the alternator when you connect it.

6. Tighten the two bolts securing the alternator. Tighten the short bolt to 16 ft. lbs. (22 Nm). Tighten the long bolt to 33 ft. lbs. 44 Nm).

7. Install the drive belt.

8. Do the battery terminal reconnection procedure.

6-Cylinder Engine

See Figures 83 through 90.

1. Remove the engine compartment covers.
   a. Remove the left engine compartment cover.
   b. Detach the clip, and release the hooks.

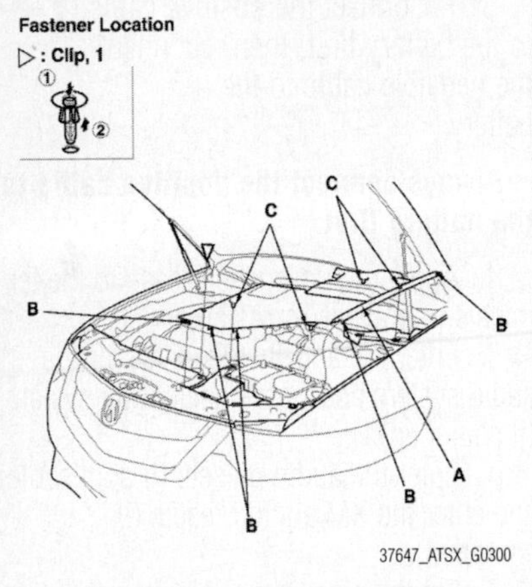
Fastener Location
▷ : Clip, 1
C
B
A
37647_ATSX_G0300

Fig. 83 Remove the left engine compartment cover (A)

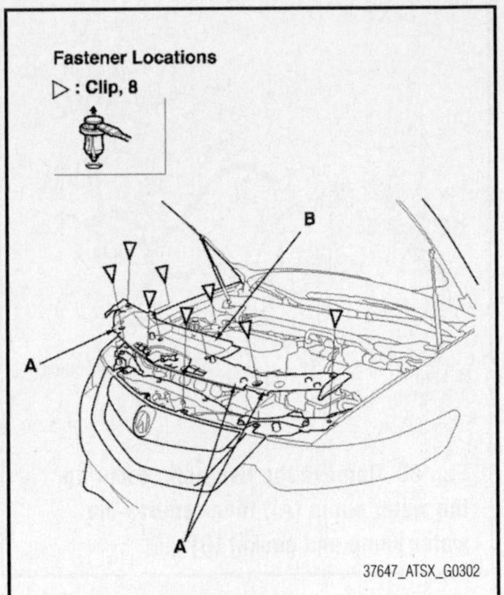
Fastener Locations
▷ : Clip, 8
B
A
37647_ATSX_G0302

Fig. 85 Detach the clips, and release the hooks (A), then remove the front grille cover (B)

   c. Pull up the left engine compartment cover to release the pins.
   d. Remove the right engine compartment cover.
   e. Detach the clip, and release the hooks.
   f. Pull up the right engine compartment cover to release the pins.
   g. Detach the clips, and release the hooks, then remove the front grille cover.
   h. If necessary, detach the clips, then remove the front fender trim.

➡The left front fender trim is shown; the right front fender trim is similar.

2. Do the battery terminal disconnection procedure.
3. Raise the vehicle on the lift.
4. Remove the splash shield.
5. Disconnect the A/C condenser fan

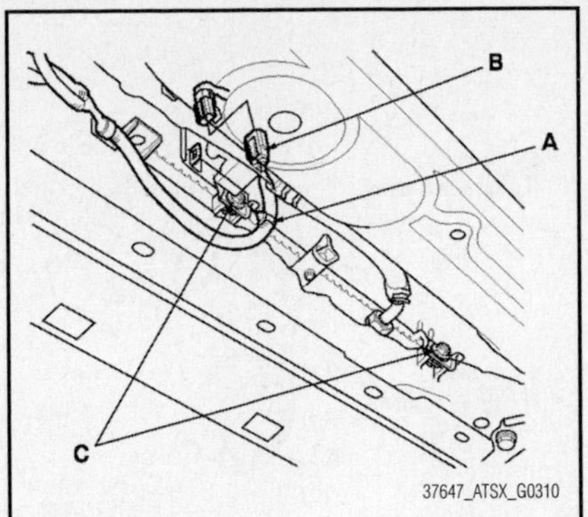
B
A
C
37647_ATSX_G0310

Fig. 87 Disconnect the A/C condenser fan motor connector (A), remove the harness clamp (B) and loosen the A/C condenser fan shroud mounting bolts (C)

motor connector and remove the harness clamp.

6. Loosen the A/C condenser fan shroud mounting bolts.
7. Lower the vehicle on the lift.
8. Remove the A/C condenser fan shroud assembly.
9. Remove the drive belt.
10. Disconnect the alternator connector and the positive alternator cable from the alternator.
11. Remove the harness clamp from the alternator and disconnect the A/C compressor clutch connector from the A/C compressor.
12. Remove the bolt securing the harness holder.
13. Remove the mounting bolt and the alternator bracket mounting bolt, then remove the alternator.

To install:

14. Install the alternator, then tighten the mounting bolt to 33 ft. lbs. (44 Nm) and the

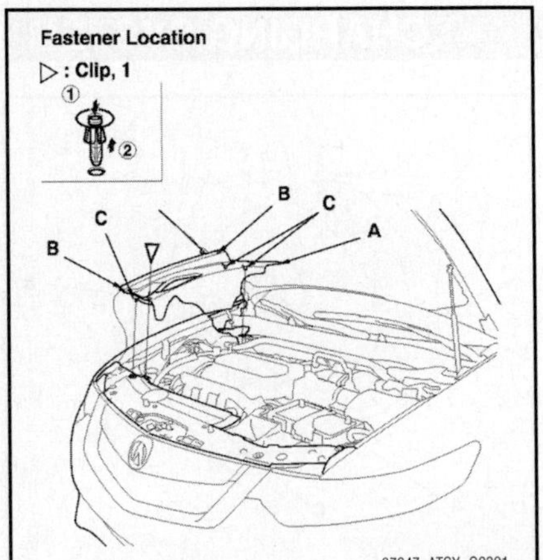
Fastener Location
▷ : Clip, 1
B
C
A
37647_ATSX_G0301

Fig. 84 Remove the right engine compartment cover (A)

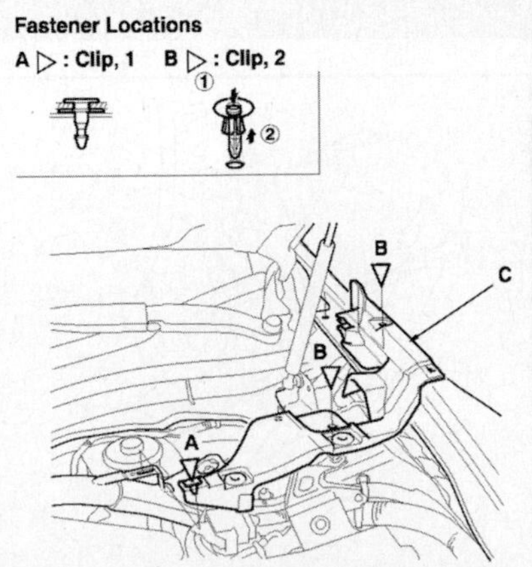
Fastener Locations
A ▷ : Clip, 1
B ▷ : Clip, 2
B
C
A
37647_ATSX_G0303

Fig. 86 Detach the clips (A, B), then remove the front fender trim (C)

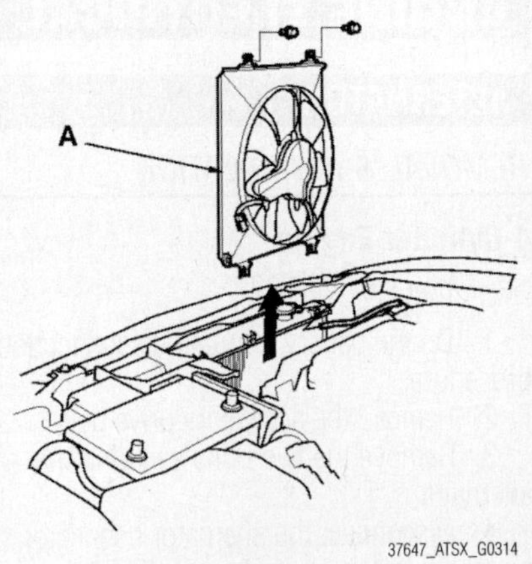
A
37647_ATSX_G0314

Fig. 88 Remove the A/C condenser fan shroud assembly (A)

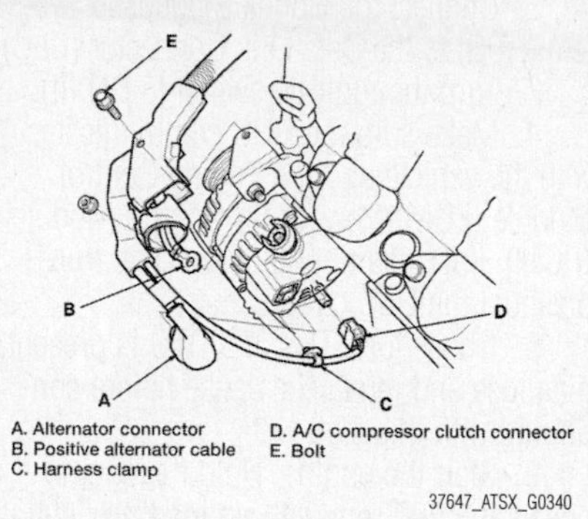
E
B
D
A
C
A. Alternator connector
B. Positive alternator cable
C. Harness clamp
D. A/C compressor clutch connector
E. Bolt
37647_ATSX_G0340

Fig. 89 Disconnect the alternator connector and the positive alternator cable from the alternator

alternator bracket mounting bolt to 16 ft. lbs. (22 Nm).

15. Install the bolt securing the harness holder.

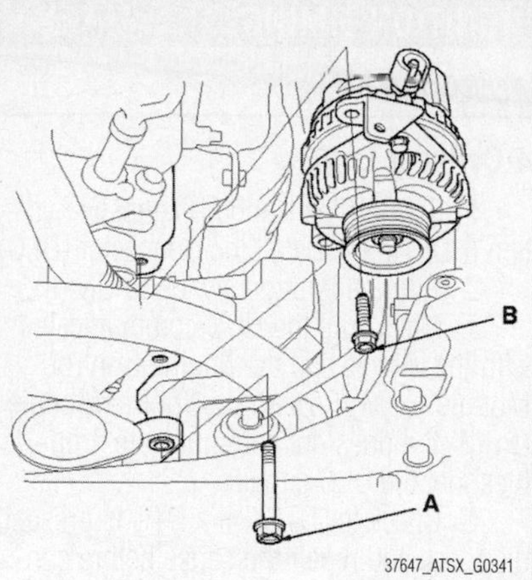
B
A
37647_ATSX_G0341

Fig. 90 Remove the mounting bolt (A) and the alternator bracket mounting bolt (B), then remove the alternator

16. Install the harness clamp to the alternator and connect the A/C compressor clutch connector to the A/C compressor.
17. Connect the alternator connecter and the positive alternator cable to the alternator. Make sure the crimped side of the ring terminal faces away from the alternator when you connect it.
18. Install the drive belt.
19. Install the A/C condenser fan shroud assembly.
20. Raise the vehicle on the lift.
21. Tighten the A/C condenser fan shroud mounting bolts.
22. Connect the A/C condenser fan motor connector and install the harness clamp.
23. Install the splash shield.
24. Lower the vehicle on the lift.
25. Do the battery terminal reconnection procedure.
26. Install the engine compartment covers.

ENGINE ELECTRICAL

IGNITION SYSTEM

FIRING ORDER

4-Cylinder Engine

The firing order is 1–3–4–2.

6-Cylinder Engine

The firing order is 1–4–2–5–3–6.

IGNITION COIL

REMOVAL & INSTALLATION

4-Cylinder Engine

See Figure 91.

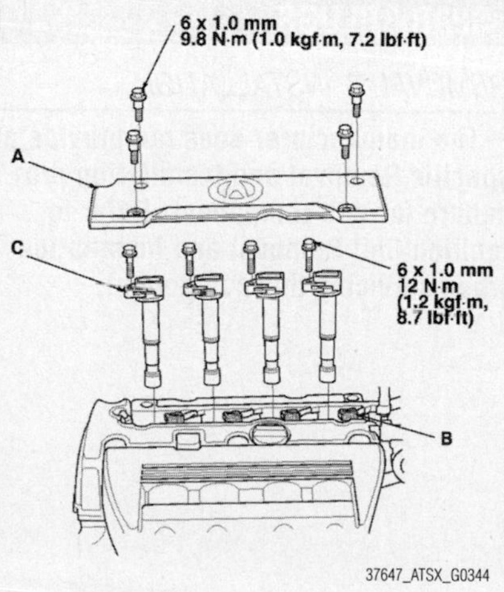
6 x 1.0 mm
9.8 N·m (1.0 kgf·m, 7.2 lbf·ft)
A
C
6 x 1.0 mm
12 N·m
(1.2 kgf·m,
8.7 lbf·ft)
B
37647_ATSX_G0344

Fig. 91 Remove the ignition coil cover (A), disconnect the ignition coil connectors (B), then remove the ignition coils (C)

1. Remove the ignition coil cover.
2. Disconnect the ignition coil connectors, then remove the ignition coils.
3. Install the ignition coils in the reverse order of removal.

6-Cylinder Engine

Front

See Figure 92.

1. Remove the engine compartment covers.
2. Remove the harness clamp and the bolt, then remove the harness holder from the bracket.
3. Disconnect the ignition coil connectors, then remove the front ignition coils.
4. Install the front ignition coils in the reverse order of removal.

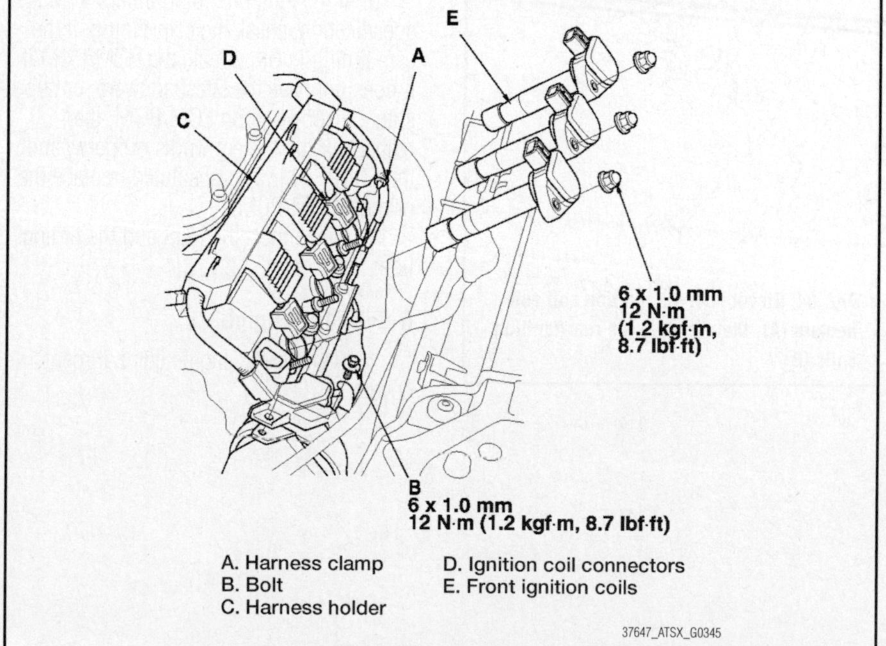
E
D
A
C
6 x 1.0 mm
12 N·m
(1.2 kgf·m,
8.7 lbf·ft)
B
6 x 1.0 mm
12 N·m (1.2 kgf·m, 8.7 lbf·ft)
A. Harness clamp
B. Bolt
C. Harness holder
D. Ignition coil connectors
E. Front ignition coils
37647_ATSX_G0345

Fig. 92 Remove the harness clamp and the bolt, then remove the harness holder from the bracket

Rear

See Figures 93 and 94.

1. Remove the engine compartment covers.
2. Remove the strut brace.
3. Disconnect the ignition coil connectors, then remove the rear ignition coils.
4. Install the rear ignition coils in the reverse order of removal.

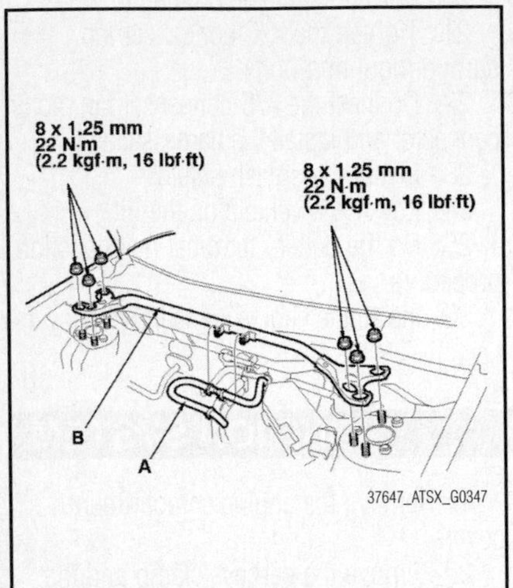
8 x 1.25 mm
22 N·m
(2.2 kgf·m, 16 lbf·ft)
8 x 1.25 mm
22 N·m
(2.2 kgf·m, 16 lbf·ft)
B
A
37647_ATSX_G0347

Fig. 93 Remove the strut brace

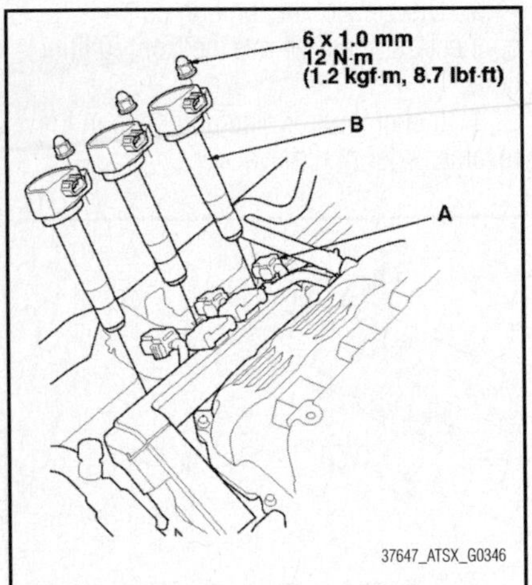
6 x 1.0 mm
12 N·m
(1.2 kgf·m, 8.7 lbf·ft)
B
A
37647_ATSX_G0346

Fig. 94 Disconnect the ignition coil connectors (A), then remove the rear ignition coils (B)

IGNITION TIMING

ADJUSTMENT

4-Cylinder Engine

1. Connect the Honda Diagnostic System (HDS) to the Data Link Connector (DLC).
2. Turn the ignition switch to ON (II).
3. Make sure the HDS communicates with the vehicle and the Engine Control Module (ECM)/Powertrain Control Module (PCM) If it does not communicate, troubleshoot the DLC circuit..
4. Check for DTCs. If a DTC is present, diagnose and repair the cause before continuing with this test.
5. Start the engine. Hold the engine speed at 3,000 rpm with no load (in N or P (A/T model) or neutral (M/T model)) until the radiator fan comes on, then let it idle.
6. Check the idle speed.
7. Jump the SCS line with the HDS.
8. Connect the timing light to the service loop (white tape).
9. Aim the light toward the pointer on the cam chain case. Check the ignition timing under a no load condition (headlights, blower fan, rear window defogger, and air conditioner are turned off).

➡The other pointer is not used.

Ignition Timing specifications:

- M/T model: 8°±2° BTDC (RED mark) at idle in neutral
- A/T model: 8°±2° BTDC (RED mark) at idle in N or P

10. If the ignition timing differs from the specification, check the cam timing. If the cam timing is OK, update the ECM/PCM if it does not have the latest software, or substitute a known-good ECM/PCM, then recheck. If the system works properly, and the ECM/PCM was substituted, replace the original ECM/PCM.
11. Disconnect the HDS and the timing light.

6-Cylinder Engine

1. Remove the engine compartment covers.
2. Connect the Honda Diagnostic System (HDS) to the Data Link Connector (DLC).
3. Turn the ignition switch to ON (II).
4. Make sure the HDS communicates with the vehicle and the Engine Control Module (ECM)/Powertrain Control Module (PCM). If it does not communicate, troubleshoot the DLC circuit.
5. Check for DTCs. If a DTC is present, diagnose and repair the cause before continuing with this test.
6. Start the engine. Hold the engine speed at 3,000 rpm with no load (in neutral (M/T model) or N or P (A/T model)) until the radiator fan comes on, then let it idle.
7. Check the idle speed.
8. Jump the SCS line with the HDS.
9. Connect the timing light to the No. 1 ignition coil harness.
10. Aim the light toward the pointer on the timing belt cover. Check the ignition timing under a no load condition (headlights, blower fan, rear window defogger, and air conditioner are turned off).

➡The other pointer is not used.

Ignition timing specification: 10 °±2 ° BTDC (RED mark) at idle in N or P

11. If the ignition timing differs from the specification, check the cam timing. If the cam timing is OK, update the ECM/PCM if it does not have the latest software, or substitute a known-good ECM/PCM, then recheck. If the system works properly, and the ECM/PCM was substituted, replace the original ECM/PCM.
12. Disconnect the HDS and the timing light.
13. Install the engine compartment covers.

SPARK PLUGS

REMOVAL & INSTALLATION

➡The manufacturer does not provide a specific Removal and Installation procedure for this component. Refer to Ignition Coil Removal and Installation when servicing this component.

STARTER

REMOVAL & INSTALLATION

4-Cylinder Engine

See Figures 95 and 96.

1. Do the battery terminal disconnection procedure.
2. Remove the intake manifold.
3. Disconnect the positive starter cable from the B terminal, and the S terminal connector.
4. Remove the harness clamp.
5. Remove the two bolts securing the starter, then remove the starter.

To install:

6. Install the starter, then tighten the two bolts. Tighten the 12 x 1.25 mm bolt to 47 ft. lbs. (64 Nm). Tighten the 10 x 1.25 mm bolt to 33 ft. lbs. (44 Nm).
7. Install the harness clamp.
8. Connect the positive starter cable to the B terminal, and the S terminal connector. Make sure the crimped side of the ring terminal faces away from the starter when you connect it.
9. Install the intake manifold.
10. Do the battery terminal reconnection procedure.
11. Start the engine to make sure the starter works properly.

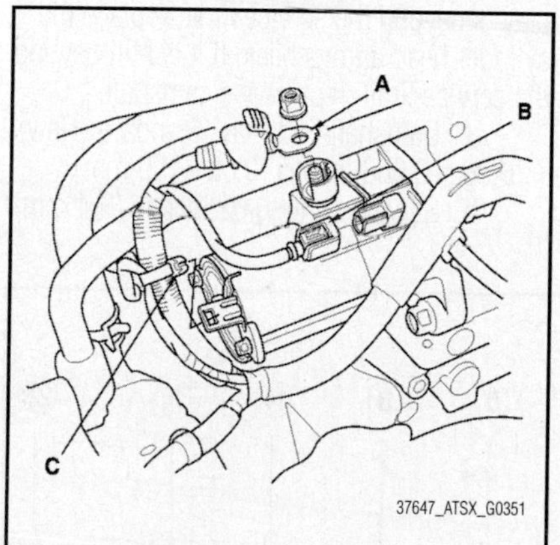
A
B
C
37647_ATSX_G0351

Fig. 95 Disconnect the positive starter cable (A) from the B terminal, and the S terminal connector (B)

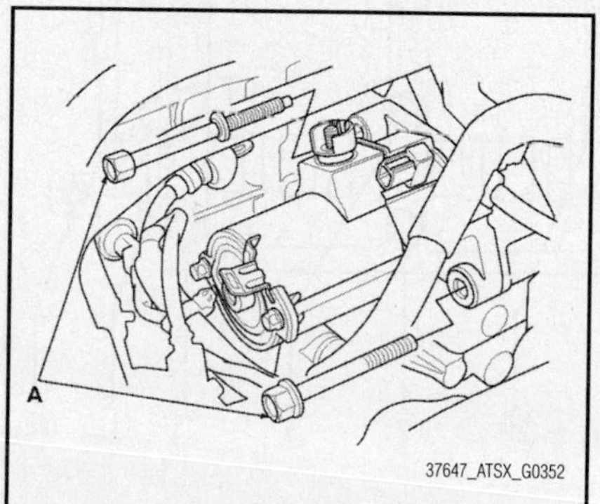
A
37647_ATSX_G0352

Fig. 96 Remove the two bolts (A) securing the starter, then remove the starter

6-Cylinder Engine

See Figures 97 and 98.

1. Remove the engine compartment covers.
2. Remove the intake air separator.
3. Do the battery removal procedure.
4. Remove the battery base.
5. Remove the harness clamp.
6. Disconnect the positive starter cable and the S terminal connector.
7. Remove the upper radiator hose bracket and the dipstick, then disconnect the vacuum hose.
8. Remove the two bolts holding the starter, then remove the starter.

To install:

9. Install the starter, then tighten the mounting bolts to 33 ft. lbs. (44 Nm).

➡Always use a new gasket.

10. Connect the vacuum hose, then install the upper radiator hose bracket and the dipstick.
11. Connect the positive starter cable and the S terminal connector. Make sure the crimped side of the ring terminal faces away from the starter when you connect it.
12. Install the harness clamp.
13. Install the battery base.
14. Do the battery installation procedure.
15. Start the engine to make sure the starter works properly.
16. Install the intake air separator.
17. Install the engine compartment covers.

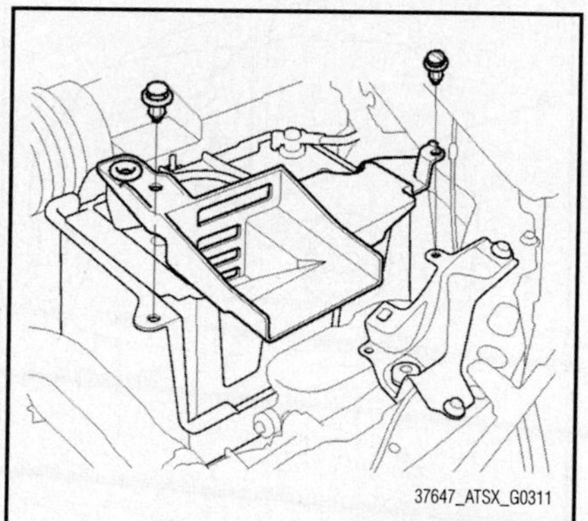
37647_ATSX_G0311

Fig. 97 Remove the intake air separator

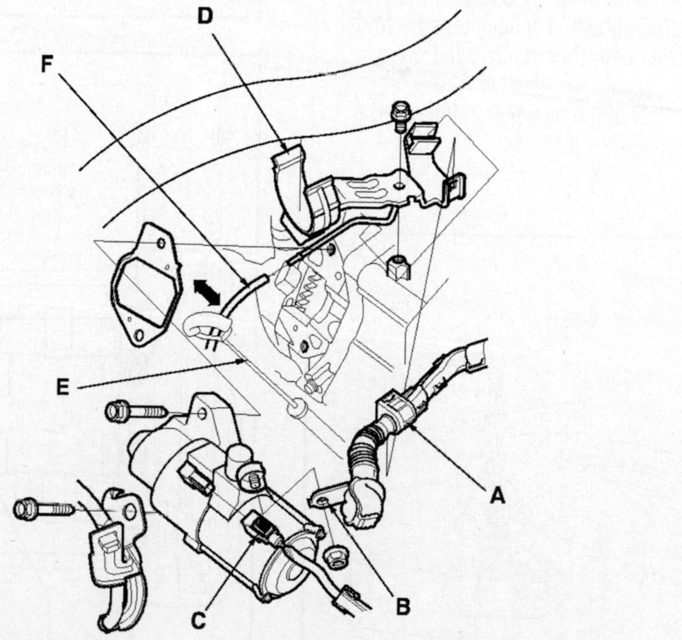
D
F
E
A
B
C
A. Harness clamp
B. Positive starter cable
C. S terminal connector
D. Upper radiator hose bracket
E. Dipstick
F. Vacuum hose
37647_ATSX_G0353

Fig. 98 Exploded view of the starter assembly

ENGINE MECHANICAL

ACCESSORY DRIVE BELTS

ACCESSORY BELT ROUTING

Refer to the graphic in the Removal and Installation procedure for the proper routing diagram.

INSPECTION

1. Inspect the belt for cracks or damage. If the belt is cracked or damaged, replace it.
2. Check that the position of the auto-tensioner indicator is within the standard range. If it is out of the standard range, replace the drive belt.

ADJUSTMENT

No adjustment is necessary or possible. Proper tension is provided by the auto-tensioner.

REMOVAL & INSTALLATION

4-Cylinder Engine

See Figure 99.

Special Tools Required: Belt tension release tool Snap-on® YA9317 or equivalent, commercially available

1. Move the auto-tensioner with the belt tension release tool in the direction shown to relieve tension from the drive belt, then remove the drive belt.
2. Install the new belt in the reverse order of removal.

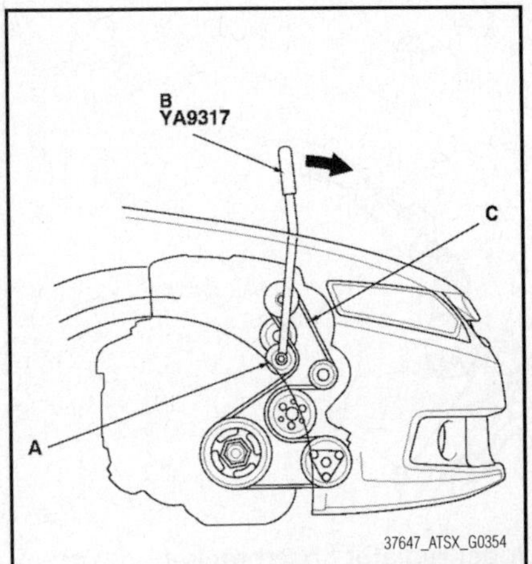
B
YA9317
C
A
37647_ATSX_G0354

Fig. 99 Move the auto-tensioner (A) with the belt tension release tool (B) to relieve tension from the drive belt (C), then remove the drive belt

6-Cylinder Engine

See Figure 100.

1. Move the auto-tensioner with the belt tension release tool in the direction shown to relieve tension from the drive belt, then remove the drive belt.
2. Install the new belt in the reverse order of removal.

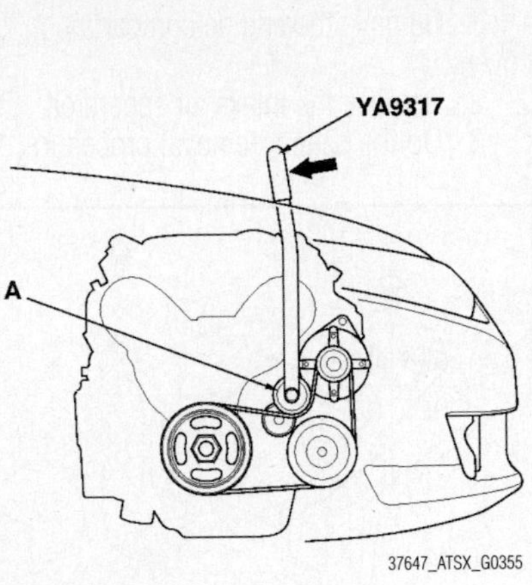
YA9317
A
37647_ATSX_G0355

Fig. 100 Move the auto-tensioner (A) with the belt tension release tool to relieve tension from the drive belt, then remove the drive belt

CAMSHAFT

INSPECTION

4-Cylinder Engine

See Figures 101 through 105.

➡**Do not rotate the camshaft during inspection.**

1. Remove the rocker arm assembly.
2. Put the rocker shaft holders, the camshaft, and the camshaft holders on the cylinder head, then tighten the bolts, in sequence, to 16 ft. lbs. (22 Nm) except for bolts 21, 22, and 23. Tighten these three bolts to 9 ft. lbs. (12 Nm).

➡**If the engine does not have bolt 21, skip it and continue the torque sequence.**

3. Seat the camshaft by pushing it away from the camshaft pulley end of the cylinder head.
4. Zero the dial indicator against the end of the camshaft, then push the camshaft back and forth, and read the end play. If the end play is beyond the service limit, replace the cylinder head and recheck. If it is still beyond the service limit, replace the camshaft.
    a. Camshaft End Play: Standard (New): 0.002–0.008 inches (0.05–0.20 mm)
    b. Service Limit: 0.02 inches (0.4 mm)

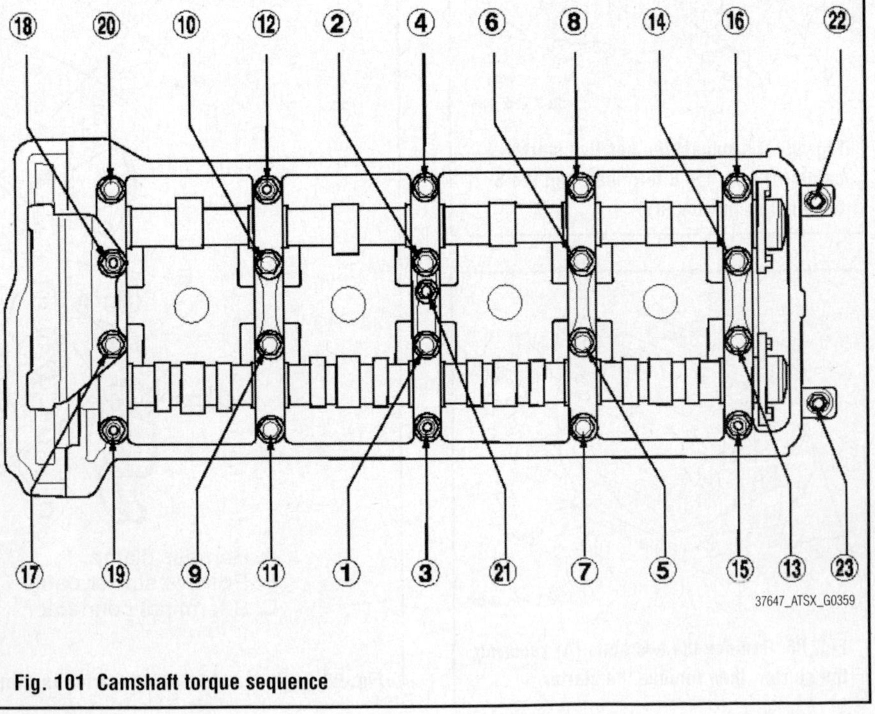
18
20
10
12
2
4
6
8
14
16
22
17
19
9
11
1
3
21
7
5
15
13
23
37647_ATSX_G0359

Fig. 101 Camshaft torque sequence

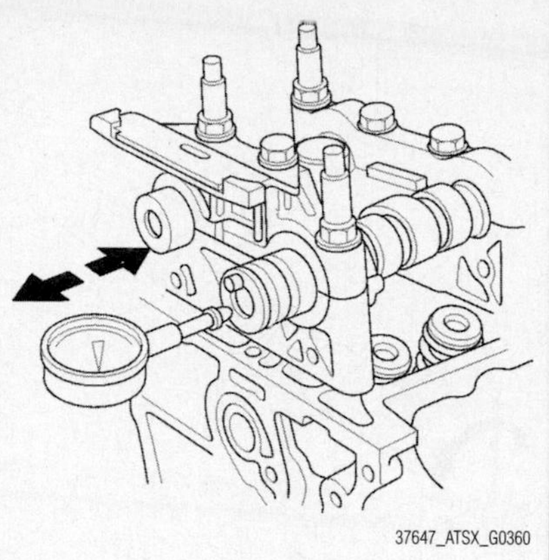
37647_ATSX_G0360

Fig. 102 Checking camshaft end play

5. Loosen the camshaft holder bolts two turns at a time, in a crisscross pattern. Then remove the camshaft holders from the cylinder head.

6. Lift the camshafts out of the cylinder head, wipe them clean, then inspect the lift ramps. Replace the camshaft if any lobes are pitted, scored, or excessively worn.

7. Clean the camshaft journal surfaces in the cylinder head, then set the camshafts back in place. Place a Plastigage strip across each journal.

8. Install the camshaft holders, then tighten the bolts to the specified torque as shown in step 2.

9. Remove the camshaft holders. Measure the widest portion of Plastigage on each journal.

a. If the camshaft-to-holder clearance is within limits, go to step 11.

b. If the camshaft-to-holder clearance is beyond the service limit, and the camshaft has been replaced, replace the cylinder head.

c. If the camshaft-to-holder clearance is beyond the service limit, and the camshaft has not been replaced, go to step 10.

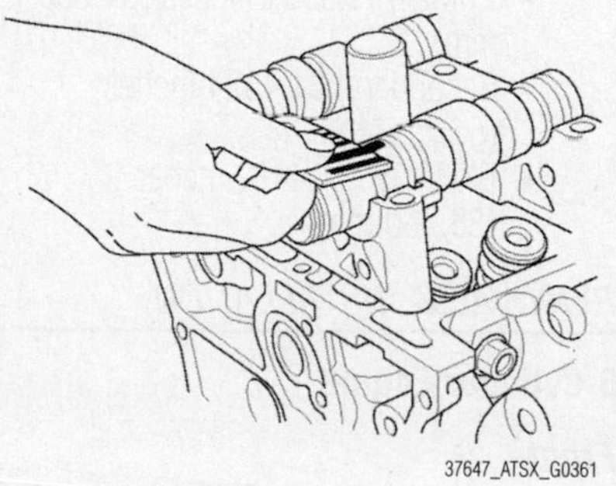
37647_ATSX_G0361

Fig. 103 Measure the widest portion of Plastigage on each journal

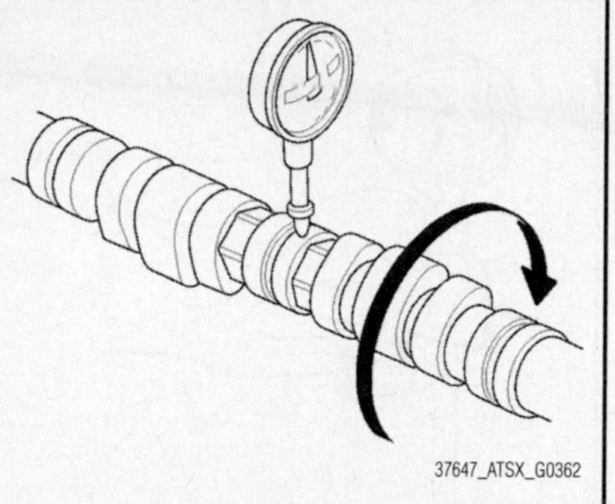
37647_ATSX_G0362

Fig. 104 Check the total runout with the camshaft supported on V-blocks

➡Camshaft-to-Holder Oil Clearance: Standard (New):

- No. 1 Journal: 0.001–0.003 inches (0.030–0.069 mm)
- No. 2, 3, 4, 5 Journals: 0.002–0.004 inches (0.060–0.099 mm)
- Service Limit: 0.006 inches (0.15 mm)

10. Check the total runout with the camshaft supported on V-blocks.

a. If the total runout of the camshaft is within the service limit, replace the cylinder head.

b. If the total runout is beyond the service limit, replace the camshaft and recheck the camshaft-to-holder oil clearance. If the oil clearance is still beyond the service limit, replace the cylinder head.

c. Camshaft Total Runout: Standard (New): 0.001 inches (0.03 mm) max.

d. Service Limit: 0.002 inches (0.04 mm)

11. Measure cam lobe height.

➡Cam Lobe Height Standard (New):

- INTAKE
- Primary: 1.3285 inches (33.744 mm)

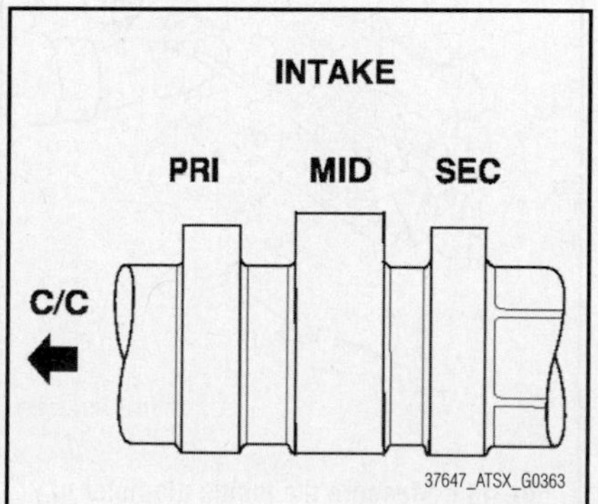
INTAKE
PRI
MID
SEC
C/C
37647_ATSX_G0363

Fig. 105 Measure cam lobe height

- Mid: 1.3959 inches (35.456 mm)
- Secondary: 1.3285 inches (33.744 mm)
- EXHAUST
- 1.3500 inches (34.291 mm)

6-Cylinder Engine

See Figures 106 through 113.

1. Remove the cylinder head.
2. Remove the rocker arm assembly.
3. Front: Put the rocker shafts bridge and the rocker shaft holder on the front cylinder head, then tighten the bolts to 16 ft. lbs. (22 Nm).
4. Rear: Put the rocker shaft bridge and the rocker shaft holder on the rear cylinder head, then tighten the bolts to 16 ft. lbs. (22 Nm).
5. Seat the camshaft by pushing it toward the rear of the cylinder head.
6. Zero the dial indicator against the end of the camshaft. Push the camshaft back and forth and read the end play. If the end play is beyond the service limit, replace the thrust cover and recheck. If it is still beyond the service limit, replace the camshaft.

a. Camshaft End Play: Standard (New): 0.002–0.008 inches (0.05-0.20 mm)

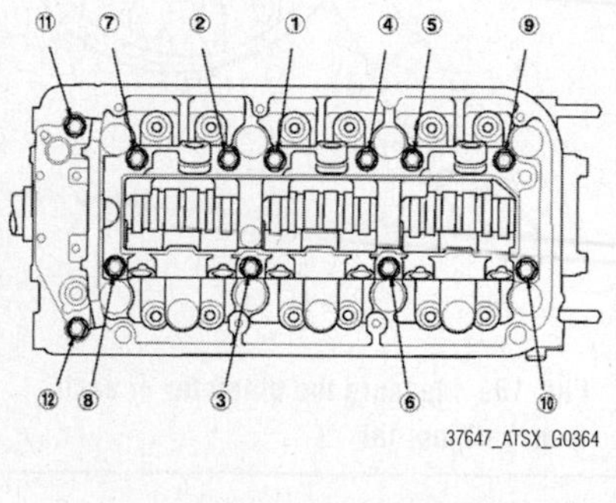
37647_ATSX_G0364

Fig. 106 Front rocker shaft bridge tightening sequence

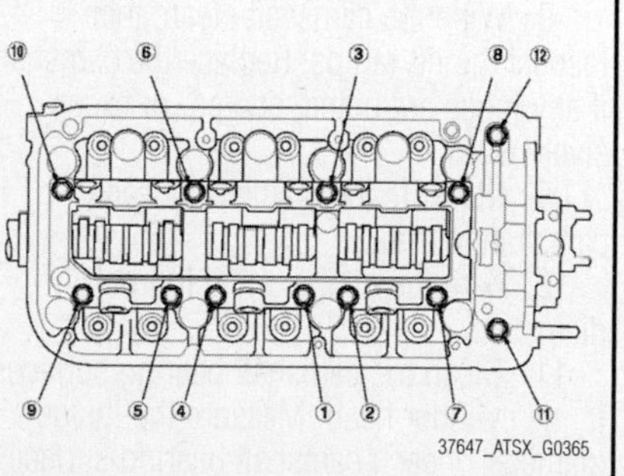
37647_ATSX_G0365

Fig. 107 Rear rocker shaft bridge tightening sequence

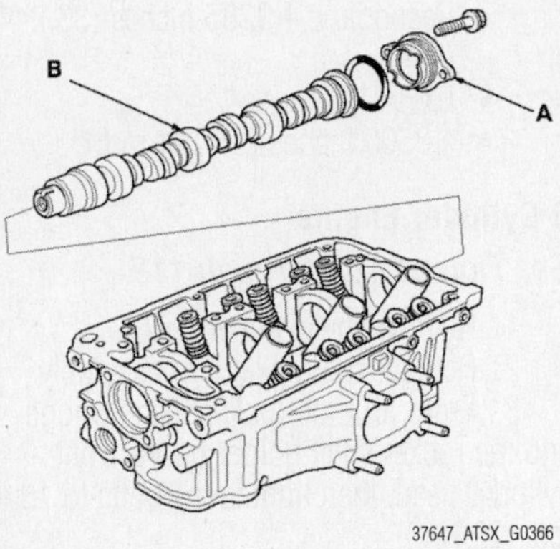
B
A
37647_ATSX_G0366

Fig. 108 **Remove the camshaft thrust cover (A), then pull out the camshaft (B)**

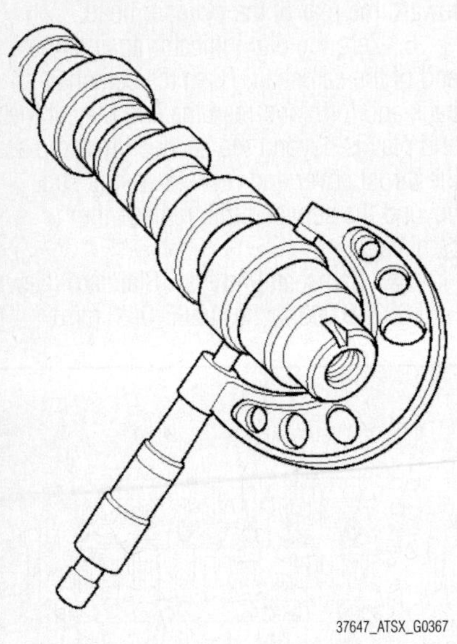
37647_ATSX_G0367

Fig. 109 **Measure the diameter of each camshaft journal**

b. Service Limit: 0.008 inches (0.20 mm)

7. Remove the camshaft thrust cover, then pull out the camshaft.

8. Wipe the camshaft clean, then inspect the lift ramps. Replace the camshaft if any lobes are pitted, scored, or excessively worn.

9. Measure the diameter of each camshaft journal.

10. Zero the gauge to the journal diameter.

11. Clean the camshaft bearing surfaces in the cylinder head. Measure the inside diameter of each camshaft bearing surface, and check for an out-of-round condition.

a. If the camshaft-to-holder clearance is within limits, go to step 13.

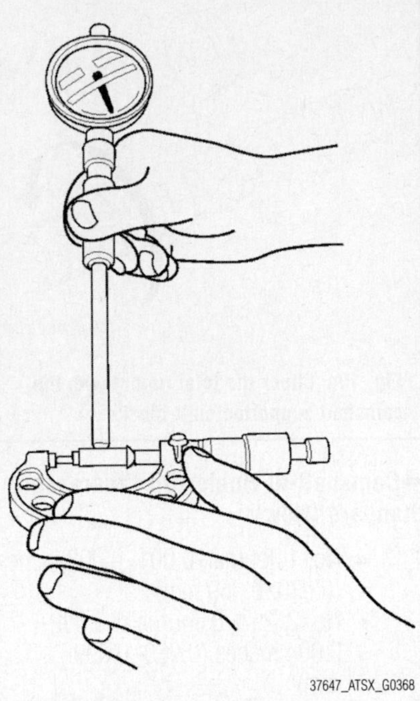
37647_ATSX_G0368

Fig. 110 **Zero the gauge to the journal diameter**

b. If the camshaft-to-holder clearance is beyond the service limit and the camshaft has been replaced, replace the cylinder head.

Camshaft-to-Holder Oil Clearance: Standard (New): 0.0020–0.0035 inches (0.050–0.089 mm)

Service Limit: 0.0060 inches (0.15 mm)

a. If the camshaft-to-holder clearance is beyond the service limit and the camshaft has not been replaced, go to step 12.

12. Check total runout with the camshaft supported on V-blocks.

a. If the total runout of the camshaft is within the service limit, replace the cylinder head.

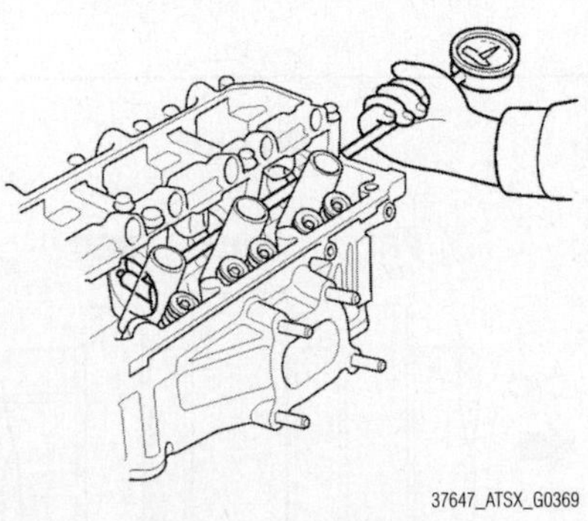
37647_ATSX_G0369

Fig. 111 **Measure the inside diameter of each camshaft bearing surface, and check for an out-of-round condition**

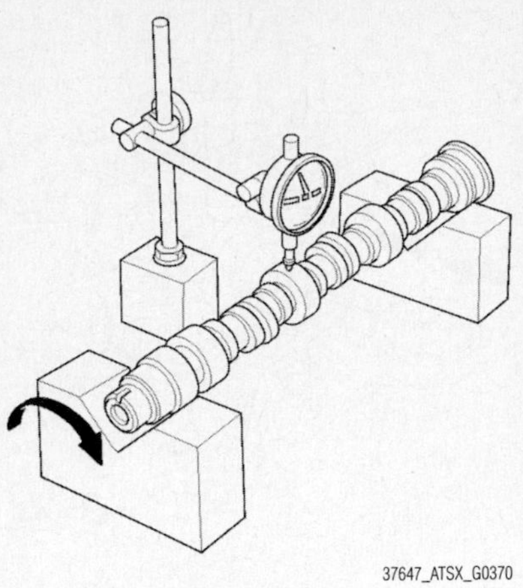
37647_ATSX_G0370

Fig. 112 **Check total runout with the camshaft supported on V-blocks**

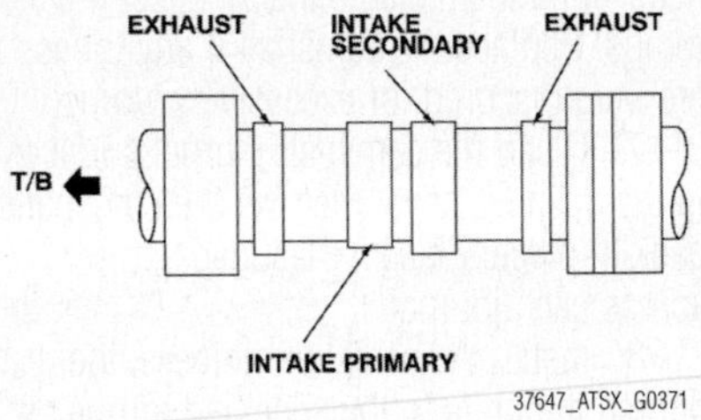
EXHAUST
INTAKE SECONDARY
EXHAUST
T/B
INTAKE PRIMARY
37647_ATSX_G0371

Fig. 113 **Measure the cam lobe height**

b. If the total runout is beyond the service limit, replace the camshaft and recheck the oil clearance. If the oil clearance is still out of tolerance, replace the cylinder head.

Camshaft Total Runout: Standard (New): 0.0012 inches (0.03 mm) max.

Service Limit: 0.002 inches (0.04 mm)

13. Measure the cam lobe height.

➡Cam Lobe Height Standard (New):

- INTAKE
- Primary: 1.3504 inches (34.299 mm)
- Secondary: 31.4024 inches (5.621 mm)
- EXHAUST: 1.4472 inches (36.760 mm)

REMOVAL & INSTALLATION

6-Cylinder Engine

Front

See Figures 114 and 115.

1. Remove the engine compartment covers.

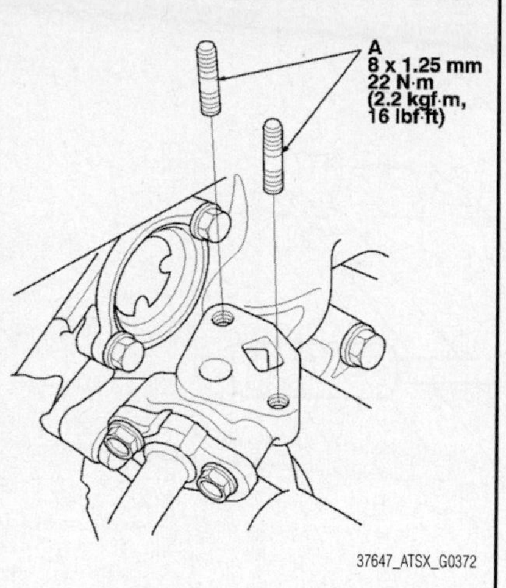
A
8 x 1.25 mm
22 N·m
(2.2 kgf·m,
16 lbf·ft)
37647_ATSX_G0372

Fig. 114 Remove the EGR valve stud bolts (A)

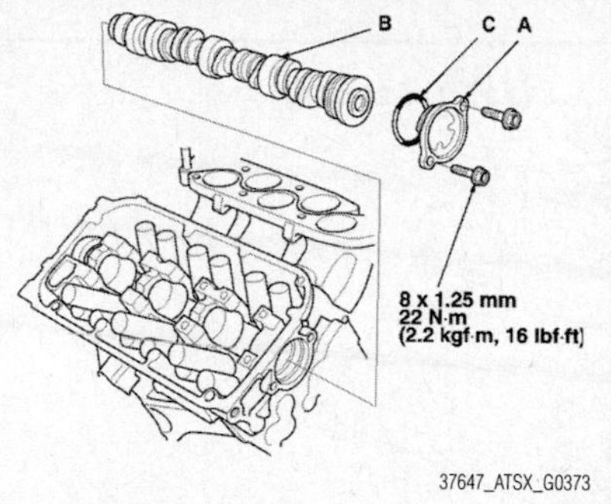
B
C
A
8 x 1.25 mm
22 N·m
(2.2 kgf·m, 16 lbf·ft)
37647_ATSX_G0373

Fig. 115 Remove the thrust cover (A), then remove the front camshaft (B)

2. Do the battery removal procedure.
3. Drain the engine coolant.
4. Disconnect the radiator hoses.
5. Remove the exhaust gas recirculation (EGR) valve.
6. Remove the EGR valve stud bolts.
7. Remove the timing belts.
8. Remove the front rocker arm assembly.
9. Remove the front camshaft pulley.
10. Remove the thrust cover, then remove the front camshaft.

To install:

11. Install the front camshaft in the reverse order of removal. Always use a new O-ring. Apply new engine oil to the journals and the cam lobes.
12. Apply new engine oil to the threads of the camshaft pulley mounting bolt, then install the front camshaft pulley.
13. Install the front rocker arm assembly, then tighten the mounting bolts.
14. Install the timing belt.
15. Adjust the valve clearance.
16. Install the EGR valve stud bolts, then install the EGR valve.
17. Connect the radiator hoses.
18. Do the battery installation procedure.
19. Fill the radiator with engine coolant, and bleed the air from the cooling system with the heater valve open.
20. Install the engine compartment covers.
21. Do the Crankshaft Position (CKP) pattern clear/CKP pattern learn procedure.

Rear

See Figures 116 through 118.

1. Remove the engine compartment covers.
2. Relieve the fuel pressure.
3. Do the battery removal procedure.
4. Drain the engine coolant.
5. Remove the under-hood fuse/relay box from the bracket.
6. Remove the air cleaner assembly.
7. Remove the quick-connect fitting cover, then disconnect the fuel feed hose.

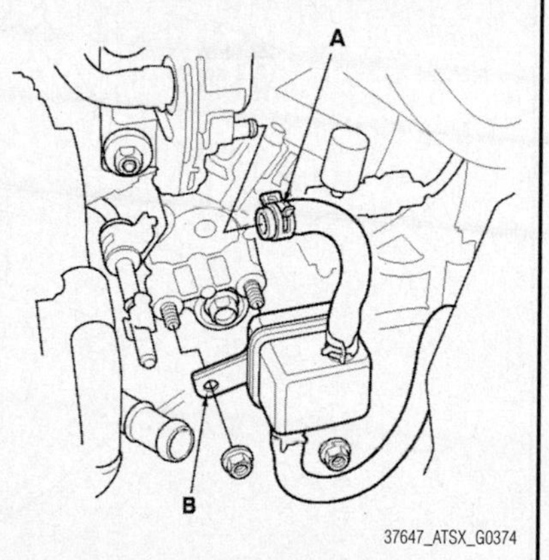
A
B
37647_ATSX_G0374

Fig. 116 Disconnect the EVAP canister hose (A), then remove the EVAP canister joint (B)

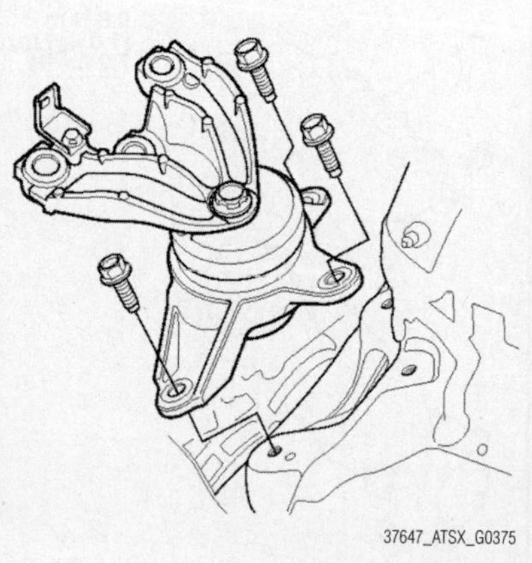
37647_ATSX_G0375

Fig. 117 Remove the upper transmission mount

8. Disconnect the heater hoses.
9. Disconnect the Evaporative Emission (EVAP) canister hose, then remove the EVAP canister joint.
10. Remove the upper transmission mount.
11. Remove the timing belt.
12. Remove the rear rocker arm assembly.
13. Remove the rear camshaft pulley.
14. Remove the thrust cover, then remove the rear camshaft.

To install:

15. Install the rear camshaft in the reverse order of removal. Always use a new O-ring. Apply new engine oil to the journals and cam lobes.
16. Apply new engine oil to the threads of the camshaft pulley mounting bolt, then install the rear camshaft pulley.
17. Install the rear rocker arm assembly, then tighten the mounting bolts.
18. Install the timing belt.
19. Install the upper transmission mount.
20. Adjust the valve clearance.
21. Install the EVAP canister joint, then connect the EVAP canister hose.
22. Connect the heater hoses.
23. Connect the fuel feed hose, then install the quick-connect fitting cover.
24. Install the air cleaner assembly.
25. Install the under-hood fuse/relay box to the bracket.
26. Do the battery installation procedure.
27. Inspect for fuel leaks. Turn the ignition switch to ON (II) (do not operate the starter) so the fuel pump runs for about 2 seconds and pressurizes the fuel line. Repeat this operation three times, then check for fuel leakage at any point in the fuel line.
28. Fill the radiator with engine coolant, and bleed the air from the cooling system with the heater valve open.

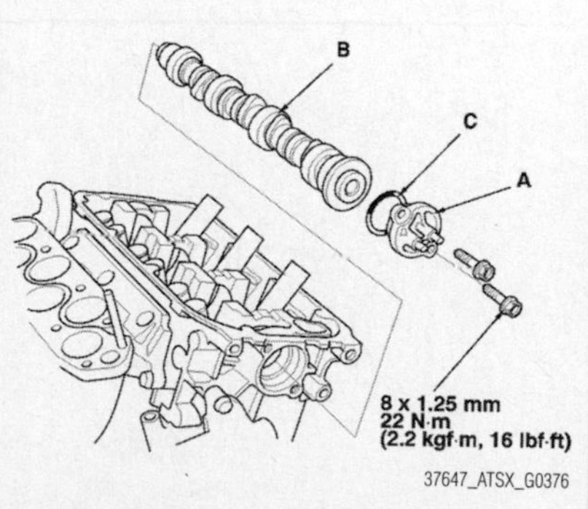
B
C
A
8 x 1.25 mm
22 N·m
(2.2 kgf·m, 16 lbf·ft)
37647_ATSX_G0376

Fig. 118 Remove the thrust cover (A), then remove the rear camshaft (B)

29. Install the engine compartment covers.
30. Do the Crankshaft Position (CKP) pattern clear/CKP pattern learn procedure.

CATALYTIC CONVERTER

REMOVAL & INSTALLATION

4-Cylinder Engine

Warm up TWC

See Figures 119 through 123.

➡Special Tools Required: O2 sensor wrench, Snap-on® YA8875, or SWR2, or equivalent, commercially available

1. Raise the vehicle on a lift.
2. Remove the secondary HO2S (Sensor 2).
   a. Remove the front splash shield.
   b. Disconnect the secondary HO2S 4P connector, then remove secondary HO2S.

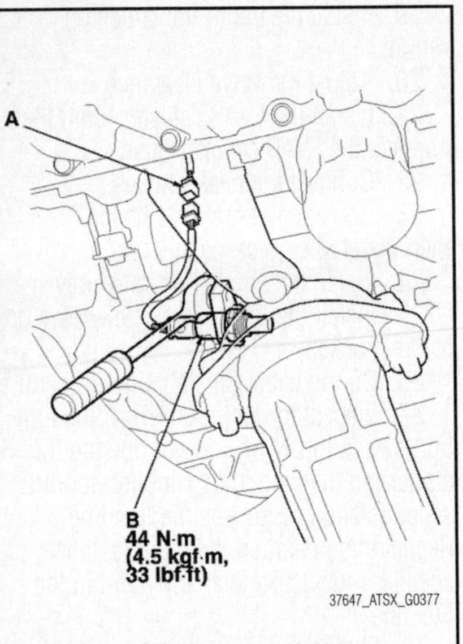
A
B
44 N·m
(4.5 kgf·m,
33 lbf·ft)
37647_ATSX_G0377

Fig. 119 Disconnect the secondary HO2S 4P connector (A), then remove secondary HO2S (B)

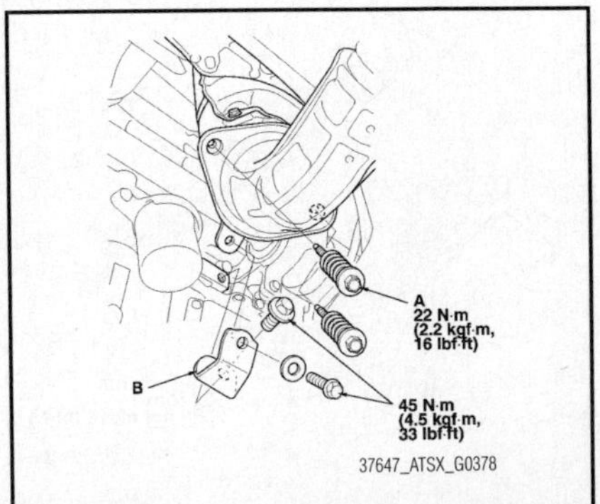
A
22 N·m
(2.2 kgf·m,
16 lbf·ft)
B
45 N·m
(4.5 kgf·m,
33 lbf·ft)
37647_ATSX_G0378

Fig. 120 Remove the bolts (A); remove the WU-TWC bracket (B)

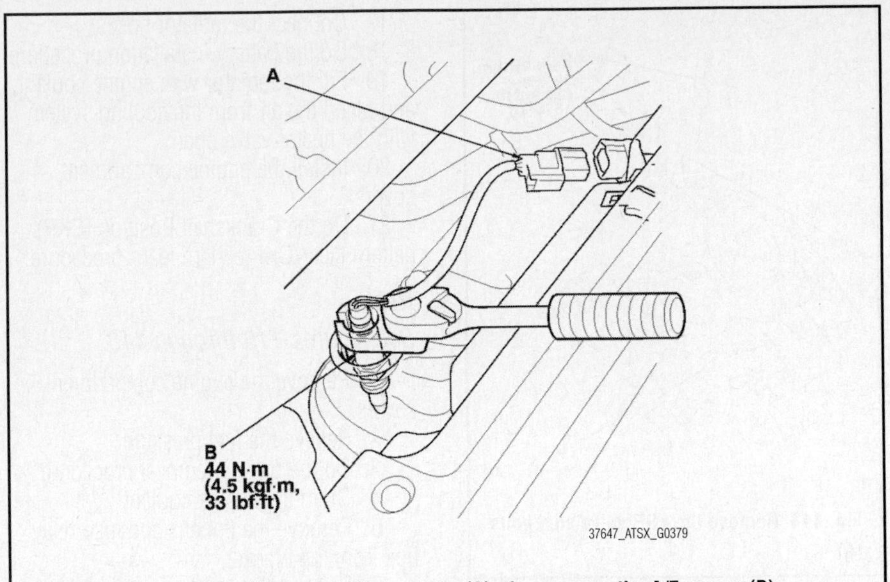
A
B
44 N·m
(4.5 kgf·m,
33 lbf·ft)
37647_ATSX_G0379

Fig. 121 Disconnect the A/F sensor 4P connector (A), then remove the A/F sensor (B)

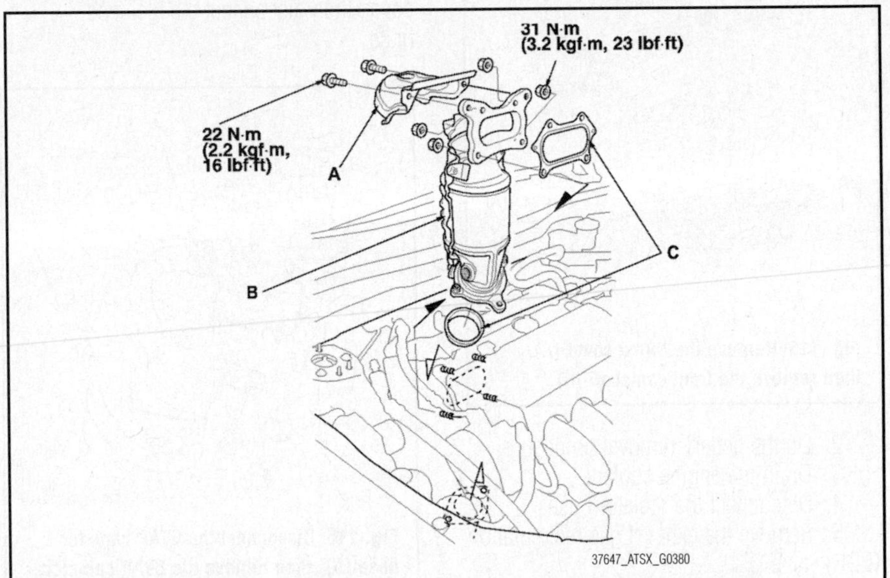
31 N·m
(3.2 kgf·m, 23 lbf·ft)
22 N·m
(2.2 kgf·m,
16 lbf·ft)
A
B
C
37647_ATSX_G0380

Fig. 122 Remove the upper converter cover (A), the WU-TWC (B) and the gaskets (C)

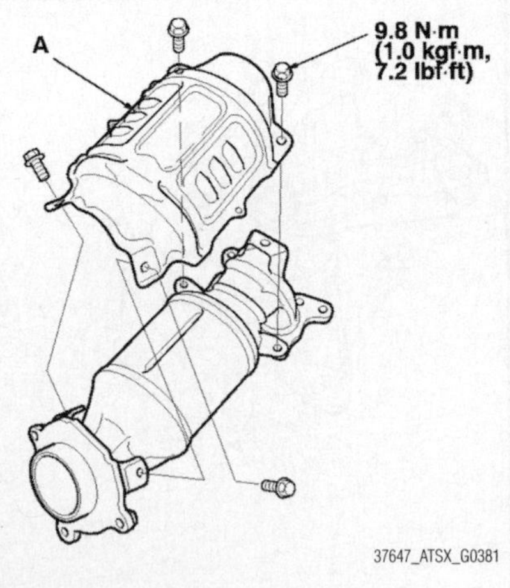
A
9.8 N·m
(1.0 kgf·m,
7.2 lbf·ft)
37647_ATSX_G0381

Fig. 123 Remove the converter cover (A)

3. Remove the bolts.
4. Remove the WU-TWC bracket.
5. Lower the vehicle.
6. Remove the frame brace.
7. Remove the A/F sensor (Sensor 1).
   a. Disconnect the A/F sensor 4P connector, then remove the A/F sensor.
8. Remove the upper converter cover.
9. Remove the WU-TWC and the gaskets.
10. Remove the converter cover.
11. Install the parts in the reverse order of removal with new gaskets.

Under-Floor TWC

See Figure 124.

1. Raise the vehicle on a lift.
2. Remove the exhaust pipe hangers.

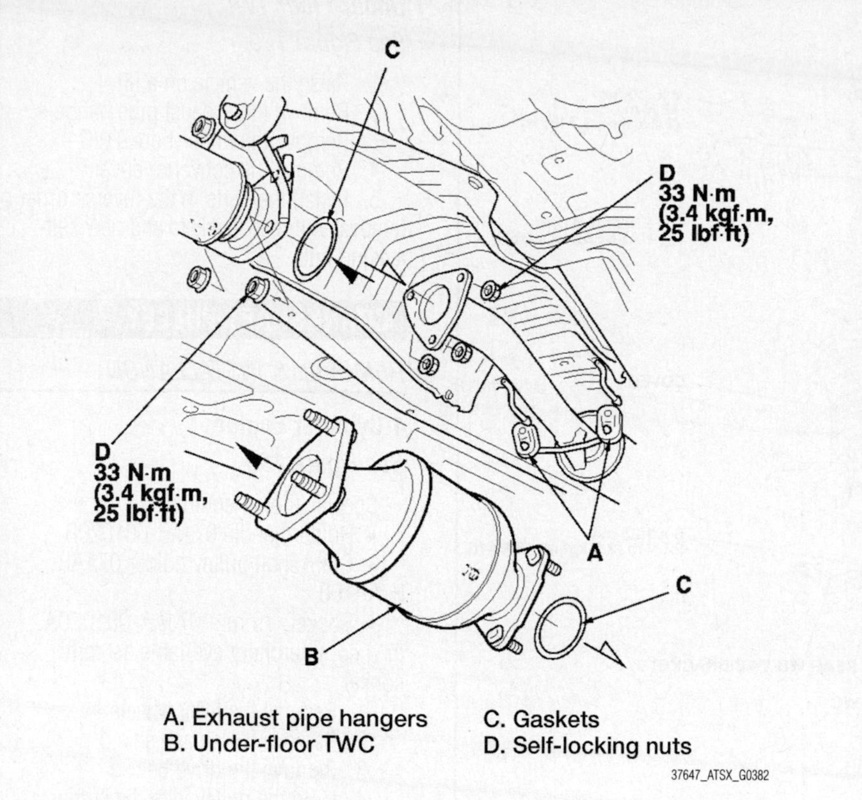
C
D
33 N·m
(3.4 kgf·m,
25 lbf·ft)
D
33 N·m
(3.4 kgf·m,
25 lbf·ft)
A
C
B
A. Exhaust pipe hangers
B. Under-floor TWC
C. Gaskets
D. Self-locking nuts
37647_ATSX_G0382

Fig. 124 Exploded view of under-floor TWC assembly

3. Remove the under-floor TWC.
4. Install the parts in the reverse order of removal with new gaskets and new self-locking nuts.

6-Cylinder Engine

Warm Up TWC Front (Bank 2)

See Figure 125.

1. Remove the No. 2 the ignition coil and the ignition coil heat insulator.
2. Remove the front A/F sensor (Sensor 1) and the front secondary HO2S (Sensor 2).
3. Remove the exhaust pipe A mounting nuts (front WU-TWC side).
4. Remove the EGR pipe.
5. Remove the A/C condenser fan assembly and the radiator upper bracket/cushion.
6. Remove the front WU-TWC bracket, then carefully remove the front WU-TWC.
7. Carefully install the front WU-TWC with a new gasket and new self-locking nuts. Tighten the nuts in a crisscross pattern in two or three steps.
8. Install the parts in the reverse order of removal.

Warm Up TWC Front (Bank 1)

See Figures 126 through 128.

1. Remove the strut brace.
2. Remove the cowl cover.
3. Remove the engine wire harness holder mounting bolts.
4. Remove the No. 4 ignition coil and the ignition coil heat insulator.
5. Remove the P/S heat shield.
6. Remove the rear A/F sensor (Sensor 1) and the rear secondary HO2S (Sensor 2).

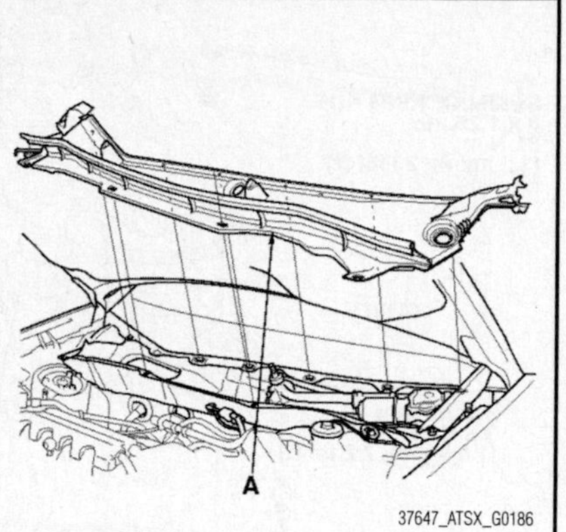
A
37647_ATSX_G0186

Fig. 126 Remove the cowl cover

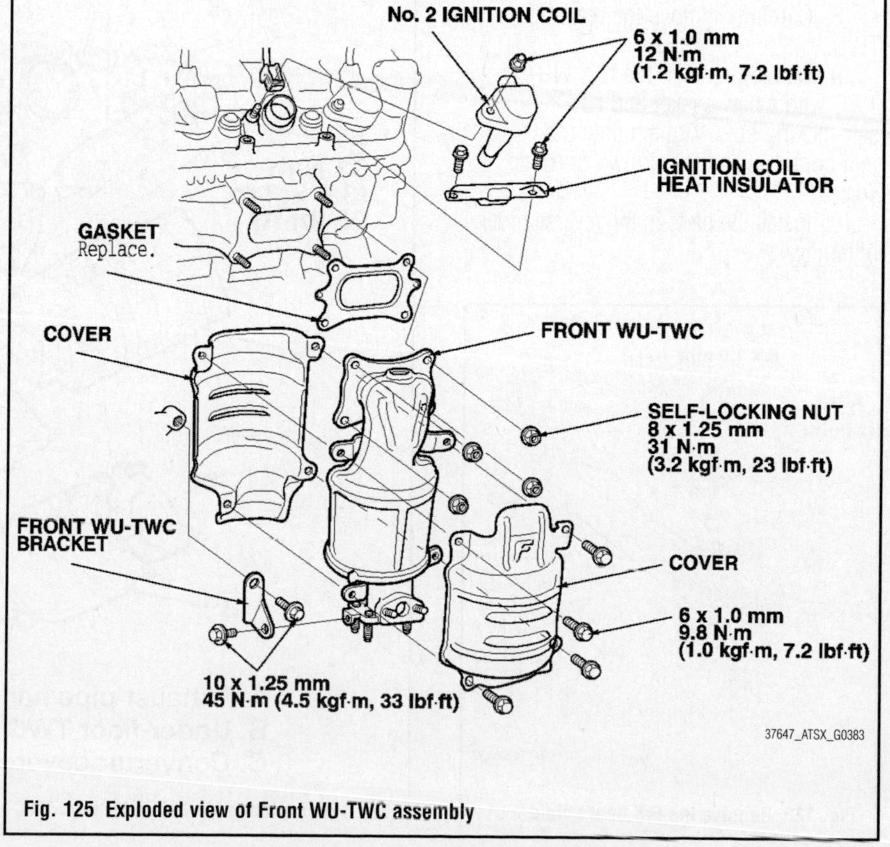
No. 2 IGNITION COIL
6 x 1.0 mm
12 N·m
(1.2 kgf·m, 7.2 lbf·ft)
IGNITION COIL
HEAT INSULATOR
GASKET
Replace.
COVER
FRONT WU-TWC
SELF-LOCKING NUT
8 x 1.25 mm
31 N·m
(3.2 kgf·m, 23 lbf·ft)
FRONT WU-TWC
BRACKET
COVER
6 x 1.0 mm
9.8 N·m
(1.0 kgf·m, 7.2 lbf·ft)
10 x 1.25 mm
45 N·m (4.5 kgf·m, 33 lbf·ft)
37647_ATSX_G0383

Fig. 125 Exploded view of Front WU-TWC assembly

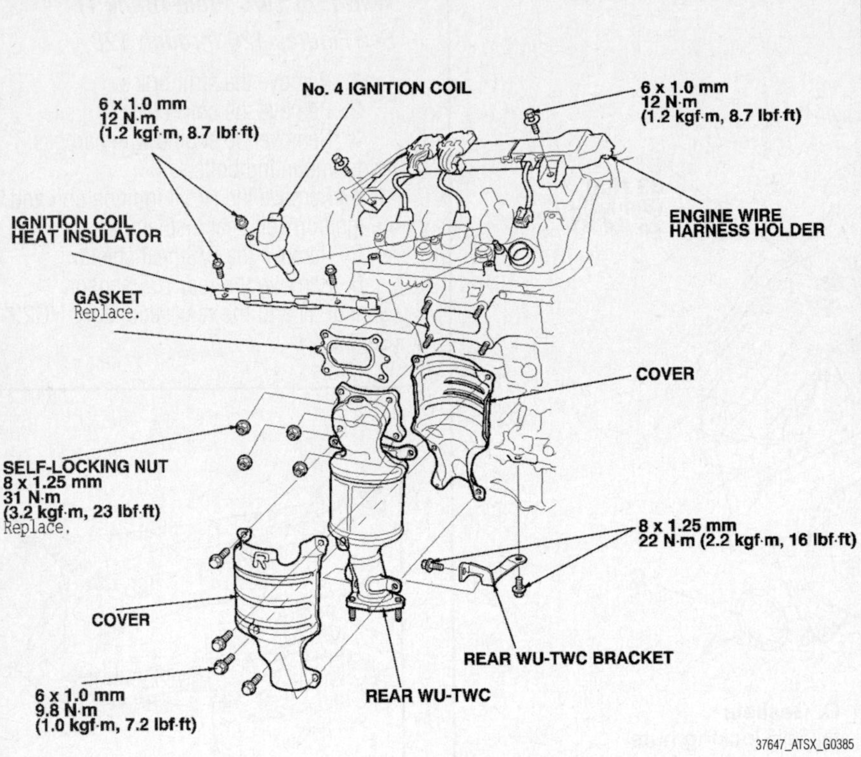
No. 4 IGNITION COIL
6 x 1.0 mm
12 N·m
(1.2 kgf·m, 8.7 lbf·ft)
6 x 1.0 mm
12 N·m
(1.2 kgf·m, 8.7 lbf·ft)
IGNITION COIL
HEAT INSULATOR
ENGINE WIRE
HARNESS HOLDER
GASKET
Replace.
COVER
SELF-LOCKING NUT
8 x 1.25 mm
31 N·m
(3.2 kgf·m, 23 lbf·ft)
Replace.
8 x 1.25 mm
22 N·m (2.2 kgf·m, 16 lbf·ft)
COVER
REAR WU-TWC BRACKET
6 x 1.0 mm
9.8 N·m
(1.0 kgf·m, 7.2 lbf·ft)
REAR WU-TWC
37647_ATSX_G0385

Fig. 127 Exploded view of rear WU-TWC assembly

7. Remove the exhaust pipe A mounting nuts (rear WU-TWC side).
8. Carefully remove the rear WU-TWC.
9. Carefully install the rear WU-TWC with a new gasket and new self-locking nuts. Tighten the nuts in a crisscross pattern in two or three steps.
10. Install the parts in the reverse order of removal.

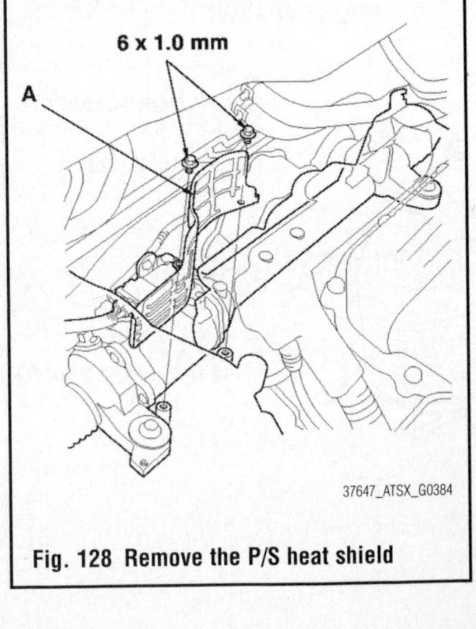
6 x 1.0 mm
A
37647_ATSX_G0384

Fig. 128 Remove the P/S heat shield

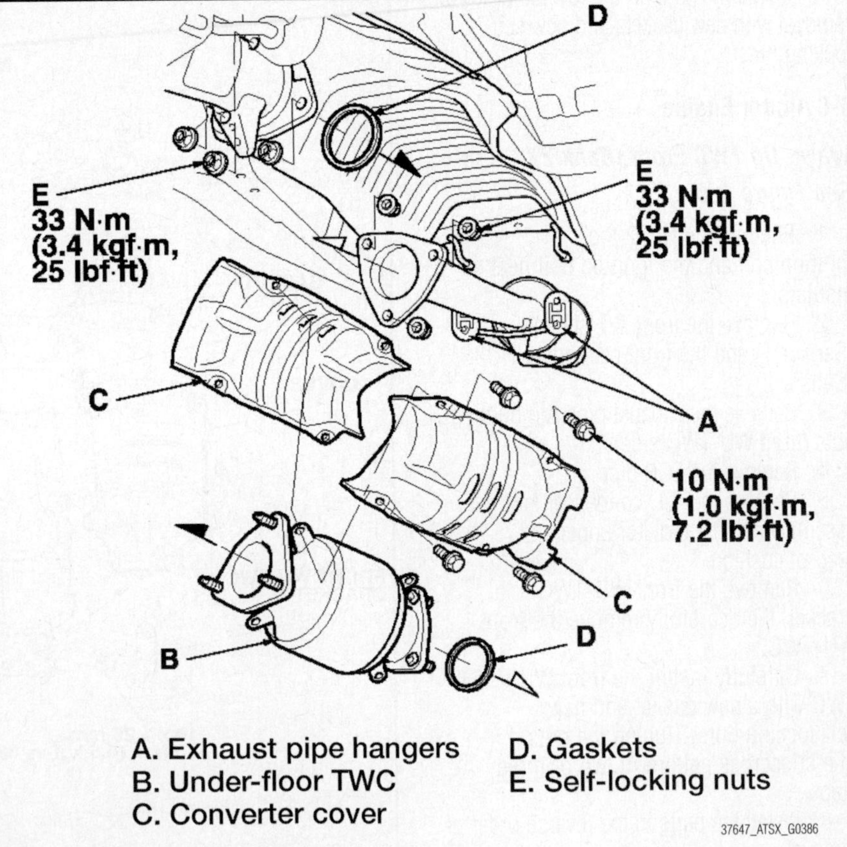
D
E
33 N·m
(3.4 kgf·m,
25 lbf·ft)
E
33 N·m
(3.4 kgf·m,
25 lbf·ft)
C
A
10 N·m
(1.0 kgf·m,
7.2 lbf·ft)
C
D
B
A. Exhaust pipe hangers
B. Under-floor TWC
C. Converter cover
D. Gaskets
E. Self-locking nuts
37647_ATSX_G0386

Fig. 129 Exploded view of under-floor TWC assembly

Under-Floor TWC

See Figure 129.

1. Raise the vehicle on a lift.
2. Remove the exhaust pipe hangers.
3. Remove the under-floor TWC.
4. Remove the converter cover.
5. Install the parts in the reverse order of removal with new gaskets and new self-locking nuts.

CRANKSHAFT PULLEY

REMOVAL & INSTALLATION

4-Cylinder Engine

See Figure 130.

Special Tools Required:

- Holder handle 07JAB-001020B
- Crankshaft pulley holder 07AAB-RJAA100
- Socket, 19 mm 07JAA-001020A or a commercially available 19 mm socket

1. Remove the front wheels.
2. Remove the splash shield.
3. Remove the drive belt.
4. Hold the pulley with the holder handle and the holder attachment.
5. Remove the bolt with a socket, 19 mm

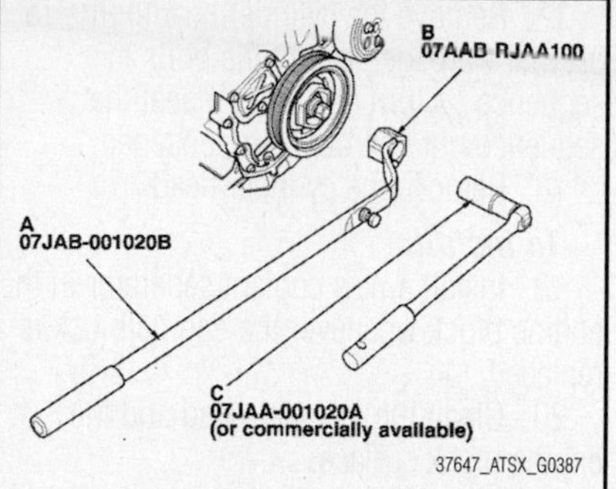
B
07AAB RJAA100
A
07JAB-001020B
C
07JAA-001020A
(or commercially available)
37647_ATSX_G0387

Fig. 130 Hold the pulley with the holder handle (A) and the holder attachment (B)

and a breaker bar, then remove the crankshaft pulley.

To install:

6. Install the crankshaft pulley, and hold the pulley with the holder handle and the holder attachment.
7. Torque the bolt to 36 ft. lbs. (49 Nm) with a torque wrench and a socket, 19 mm. Do not use an impact wrench. If the pulley bolt or crankshaft are new, torque the bolt to 130 ft. lbs. (177 Nm), then remove the bolt and torque it to 36 ft. lbs. (49 Nm).
8. Tighten the pulley bolt an additional 90°.
9. Install the drive belt.
10. Install the splash shield.
11. Install the front wheels.

6-Cylinder Engine

See Figures 131 and 132.

Special Tools Required:

- Holder handle 07JAB-001020B
- Holder attachment, 50 mm, offset 07MAB-PY3010A
- Socket, 19 mm 07JAA-001020A, or a commercially available 19 mm socket

1. Raise the vehicle on the lift.
2. Remove the right front wheel.
3. Remove the splash shield.

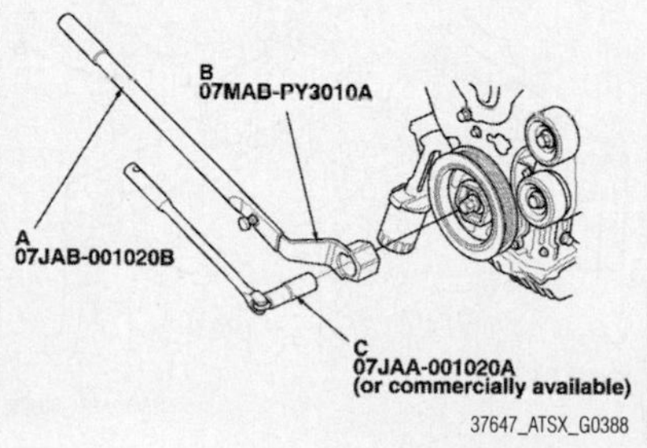
B
07MAB-PY3010A
A
07JAB-001020B
C
07JAA-001020A
(or commercially available)
37647_ATSX_G0388

Fig. 131 Hold the pulley with the holder handle (A) and the holder attachment (B)

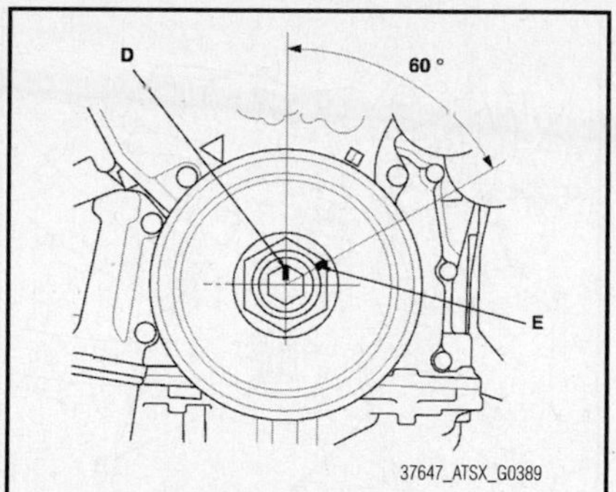
D
60°
E
37647_ATSX_G0389

Fig. 132 Mark the bolt head (D) and the crankshaft pulley (E)

4. Remove the drive belt.
5. Hold the pulley with the holder handle and the holder attachment.
6. Remove the bolt with a heavy duty 19 mm socket and a breaker bar, then remove the crankshaft pulley.

To install:

7. Install the crankshaft pulley, and tighten the bolt. Do not use an impact wrench.
8. Hold the pulley with the holder handle and the holder attachment. Torque the bolt to 47 ft. lbs. (64 Nm) with a torque wrench and a 19 mm socket.
9. Mark the bolt head and the crankshaft pulley, then tighten the bolt an additional 60° (The mark on the bolt head lines up with the mark on the crankshaft pulley).
10. Install the drive belt.
11. Install the splash shield.
12. Install the right front wheel.

CRANKSHAFT FRONT SEAL

REMOVAL & INSTALLATION

6-Cylinder Engine (In Car)

Special Tools Required: Oil seal driver, 64 mm 070AD-RCA0100

1. Remove the timing belt, the timing belt stopper, and the timing belt drive pulley.
2. Remove the pulley end crankshaft oil seal.
3. Clean and dry the crankshaft oil seal housing.
4. Apply a light coat of new engine oil to the lip of the crankshaft oil seal.
5. Using the oil seal driver, 64 mm, drive in the new crankshaft oil seal until the oil seal driver bottoms against the oil pump. When the seal is in place, clean any excess grease off the crankshaft, and check that the oil seal lip is not distorted.
6. Install the timing belt drive pulley, the timing belt stopper, and the timing belt.

CYLINDER HEAD

REMOVAL & INSTALLATION

4-Cylinder Engine

See Figures 133 through 145.

➡Prior to starting the removal of the cylinder head, note the following:

- Use fender covers to avoid damaging painted surfaces.
- To avoid damage, unplug the wiring connectors carefully while holding the connector portion.
- Connect the Honda Diagnostic System (HDS) to the Data Link Connector (DLC), and monitor the Engine Coolant Temperature (ECT) sensor 1. To avoid damaging the cylinder head, wait until the ECT sensor 1 temperature drops below 100°F (38°C) before loosening the cylinder head bolts.
- Mark all wiring and hoses to avoid misconnection. Also, be sure that they do not contact other wiring or hoses, or interfere with other parts.

1. Remove the strut brace.
2. Relieve the fuel pressure.
3. Drain the engine coolant.

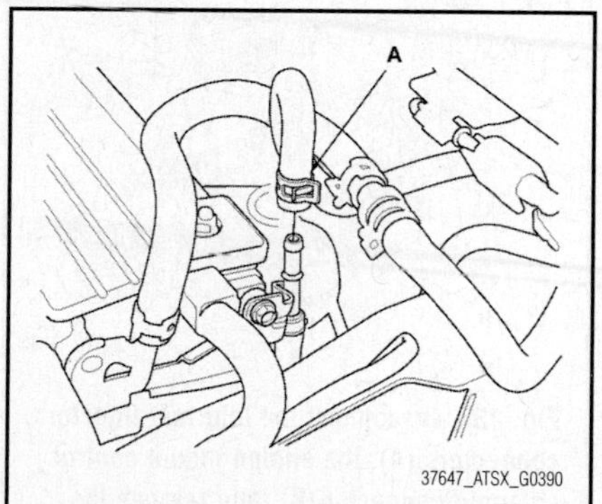
A
37647_ATSX_G0390

Fig. 133 Remove the EVAP canister hose (A)

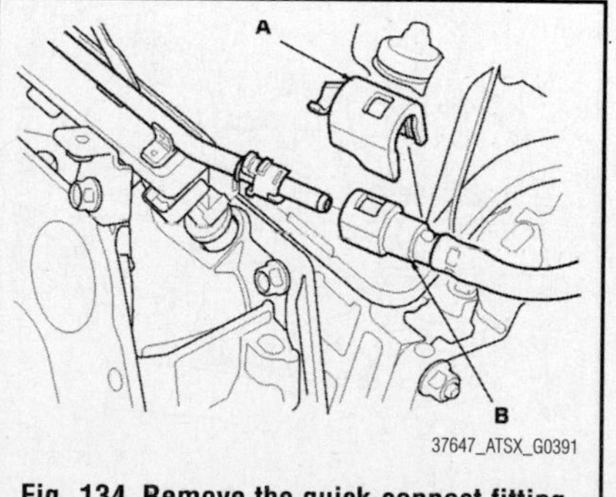
A
B
37647_ATSX_G0391

Fig. 134 Remove the quick-connect fitting cover (A), then disconnect the fuel feed hose (B)

4. Remove the drive belt.
5. Remove the intake manifold.
6. Remove the catalytic converter.
7. Remove the Evaporative Emission (EVAP) canister hose.
8. Remove the quick-connect fitting cover, then disconnect the fuel feed hose.
9. Disconnect the four fuel injector connectors, the engine mount control solenoid connector, and remove the ground cables.
10. Remove the four bolts securing the EVAP canister purge valve bracket.
11. Disconnect the upper radiator hose, the heater hoses, and the water bypass hose.
12. Remove the two bolts securing the connecting pipe.
13. Disconnect the water bypass hose.
14. Disconnect the following engine wire harness connectors, and remove the wire harness clamps from the cylinder head:
  - Engine Coolant Temperature (ECT) sensor 1 connector
  - Camshaft Position (CMP) sensor A (Intake) connector
  - Camshaft Position (CMP) sensor B (Exhaust) connector

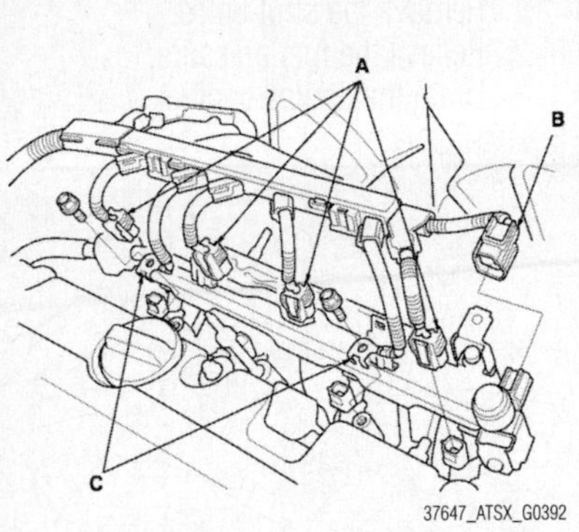
A
B
C
37647_ATSX_G0392

Fig. 135 Disconnect the four fuel injector connectors (A), the engine mount control solenoid connector (B), and remove the ground cables (C)

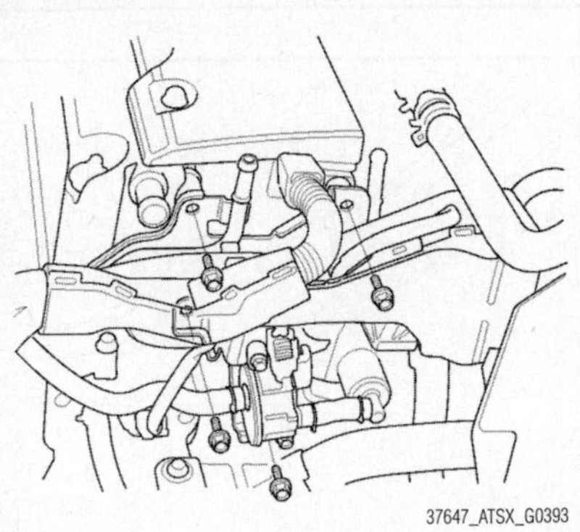
37647_ATSX_G0393

Fig. 136 Remove the four bolts securing the EVAP canister purge valve bracket

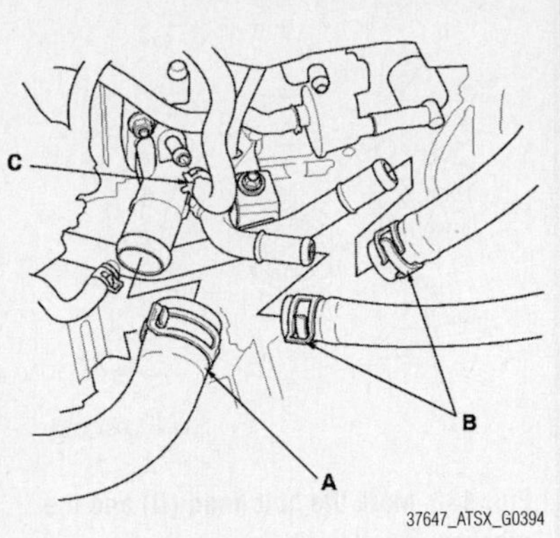
C
B
A
37647_ATSX_G0394

Fig. 137 Disconnect the upper radiator hose (A), the heater hoses (B), and the water bypass hose (C)

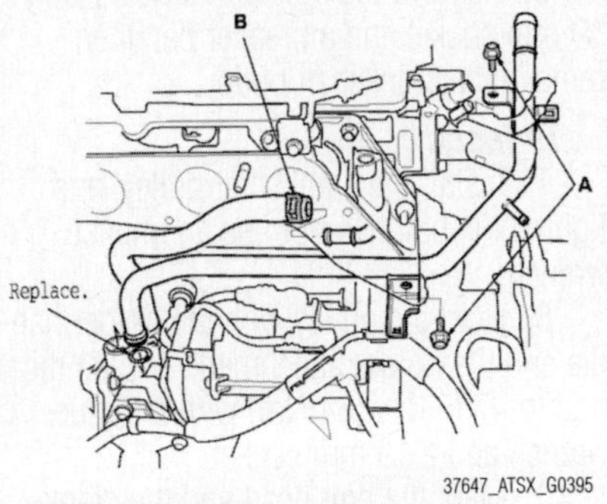
B
A
Replace.
37647_ATSX_G0395

Fig. 138 Remove the two bolts (A) securing the connecting pipe and disconnect the water bypass hose (B)

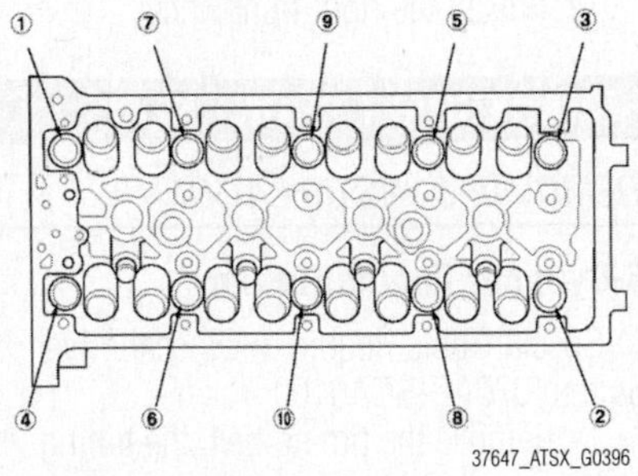
①
⑦
⑨
⑤
③
④
⑥
⑩
⑧
②
37647_ATSX_G0396

Fig. 139 Remove the cylinder head bolts in sequence

  - Rocker arm oil control solenoid connector
  - Rocker arm oil pressure switch connector
  - EVAP canister purge valve connector
  - Variable Valve Timing Control (VTC) oil control solenoid valve connector
  - Engine oil pressure switch connector

15. Remove the cam chain.
16. Remove the rocker arm assembly.
17. Remove the cylinder head bolts. To prevent warpage, loosen the bolts in sequence ⅓ turn at a time; repeat the sequence until all bolts are loosened.
18. Remove the cylinder head.

To install:

19. Install a new coolant separator in the engine block whenever the engine block is replaced.
20. Clean the cylinder head and the engine block surface.
21. Install the new cylinder head gasket and the dowel pins on the engine block. Always use a new cylinder head gasket.
22. Set the crankshaft to Top Dead Center (TDC). Align the TDC mark on the crankshaft sprocket with the pointer on the engine block.

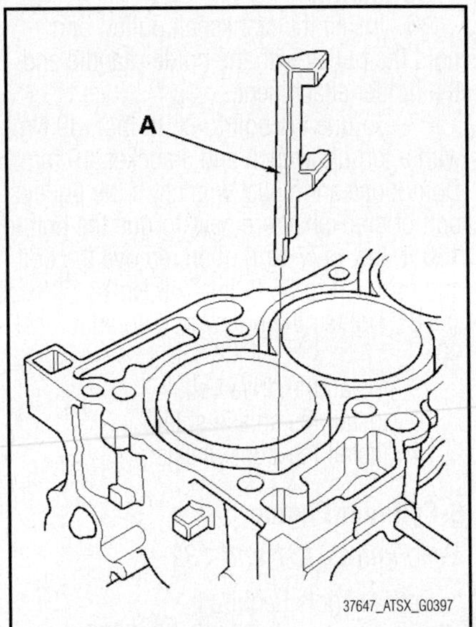
A
37647_ATSX_G0397

Fig. 140 Install a new coolant separator (A) in the engine block

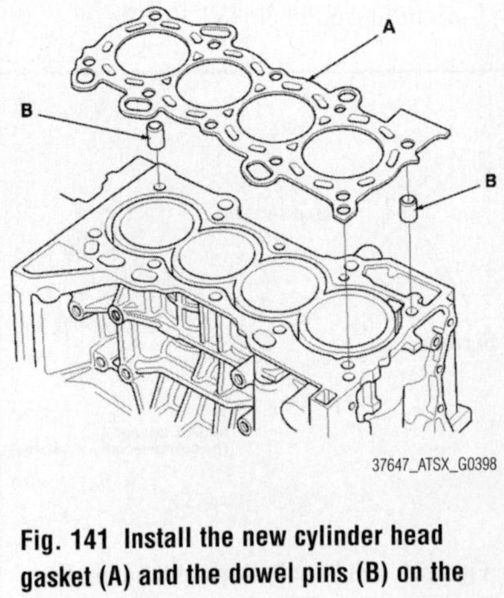
A
B
B
37647_ATSX_G0398

Fig. 141 Install the new cylinder head gasket (A) and the dowel pins (B) on the engine block

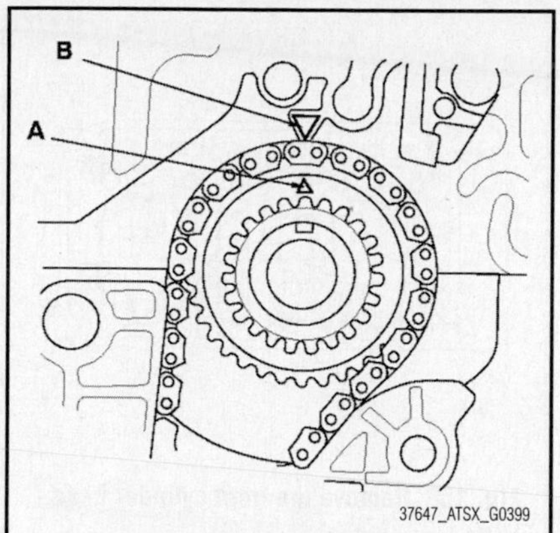
B
A
37647_ATSX_G0399
Fig. 142 Align the TDC mark (A) on the crankshaft sprocket with the pointer (B) on the engine block

23. Install the cylinder head on the engine block.

24. Measure the diameter of each cylinder head bolt at point A and point B.

25. If either diameter is less than 0.42 inches (10.6 mm), replace the cylinder head bolt.

26. Apply new engine oil to the threads and under the bolt heads of all cylinder head bolts.

27. Torque the cylinder head bolts in sequence to 29 ft. lbs. (39 Nm). Use a beam-type torque wrench. When using a preset click-type torque wrench, be sure to tighten slowly and do not overtighten. If a bolt makes any noise while you are torquing it, loosen the bolt and retighten it from the first step.

28. After torquing, tighten all cylinder head bolts in two steps (90° per step) using the sequence shown in step 9. If you are using a new cylinder head bolt, tighten the bolt an extra 90°.

➡Remove the cylinder head bolt if you tightened it beyond the specified angle,

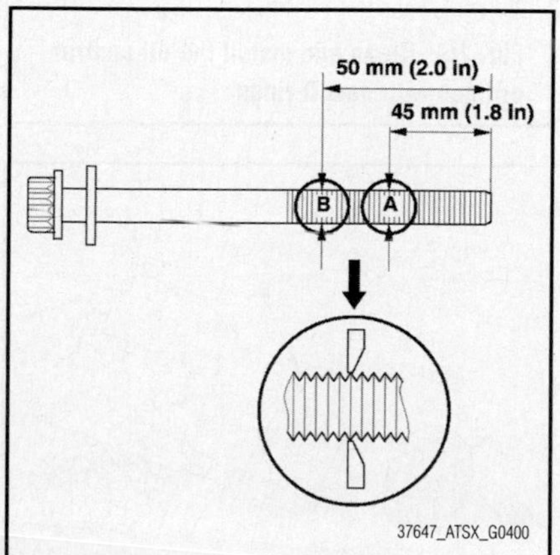
50 mm (2.0 in)
45 mm (1.8 in)
B
A
37647_ATSX_G0400
Fig. 143 Measure the diameter of each cylinder head bolt at point A and point B

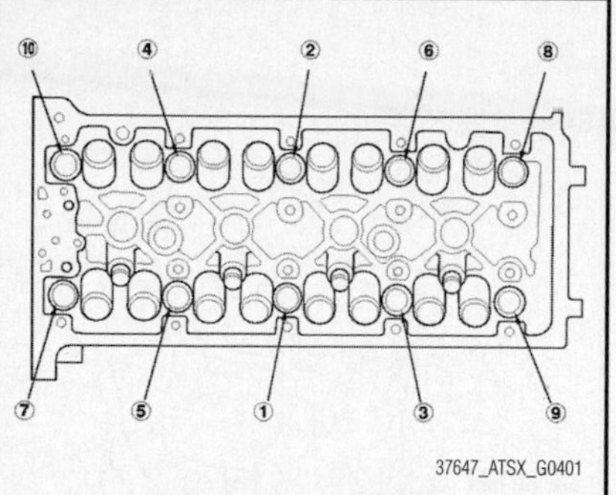
10
4
2
6
8
7
5
1
3
9
37647_ATSX_G0401
Fig. 144 Torque the cylinder head bolts in sequence

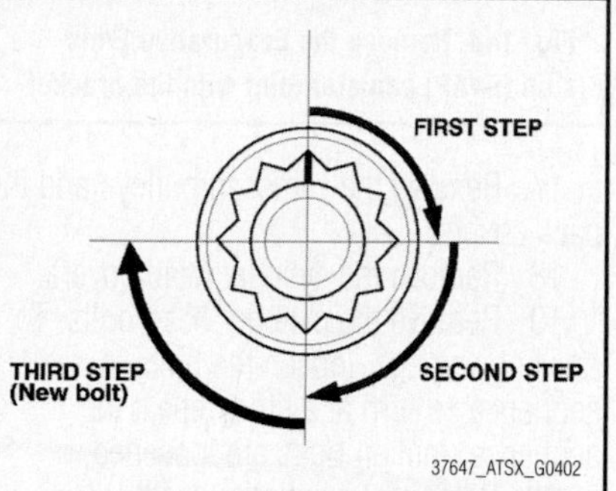
FIRST STEP
THIRD STEP
(New bolt)
SECOND STEP
37647_ATSX_G0402
Fig. 145 After torquing, tighten all cylinder head bolts in two steps (90° per step); tighten new bolts an extra 90°

and go back to step 6 of the procedure. Do not loosen it back to the specified angle.

29. Install the rocker arm assembly.

30. Install the cam chain.

31. Connect the following engine wire harness connectors, and install the wire harness clamps to the cylinder head:

- Engine Coolant Temperature (ECT) sensor 1 connector
- Camshaft Position (CMP) sensor A (Intake) connector
- Camshaft Position (CMP) sensor B (Exhaust) connector
- Rocker arm oil control solenoid connector
- Rocker arm oil pressure switch connector
- Evaporative Emission (EVAP) canister purge valve connector
- Variable Valve Timing Control (VTC) oil control solenoid valve connector
- Engine oil pressure switch connector

32. Install the two bolts securing the connecting pipe.

33. Install the water bypass hose.

34. Connect the upper radiator hose, the heater hoses, and the water bypass hose.

35. Install the four bolts securing the EVAP canister purge valve bracket.

36. Connect the four fuel injector connectors, the engine mount control solenoid connector, and install the ground cables.

37. Connect the fuel feed hose, then install the quick-connect fitting cover.

38. Connect the EVAP canister hose.

39. Install the catalytic converter.

40. Install the intake manifold.

41. Install the drive belt.

42. Install the strut brace.

43. After installation, check that all tubes, hoses, and connectors are installed correctly.

44. Inspect for fuel leaks. Turn the ignition switch to ON (II) (do not operate the starter) so the fuel pump runs for about 2 seconds and pressurizes the fuel line. Repeat this operation three times, then check for fuel leakage at any point in the fuel line.

45. Refill the radiator with engine coolant, and bleed the air from the cooling system with the heater valve open.

46. Check for fluid leaks.

47. Do the Engine Control Module (ECM)/Powertrain Control Module (PCM) idle lean procedure.

48. Do the Crankshaft Position (CKP) pattern clear/CKP pattern lean procedure.

49. Inspect the idle speed.

50. Inspect the ignition timing.

6-Cylinder Engine

See Figures 146 through 158.

1. Remove the engine compartment covers.

2. Relieve the fuel pressure.

3. Do the battery terminal disconnection procedure.

4. Drain the engine coolant.

5. Remove the alternator.

6. Remove the intake manifold.

7. Remove the six ignition coils.

8. Remove the timing belt.

9. Disconnect the following engine wire harness connectors, and remove the wire harness clamps from the cylinder head:

- Six injector connectors
- Knock sensor connector
- Engine Coolant Temperature (ECT) sensor 1 connector
- Engine mount control solenoid valve connector
- Camshaft Position (CMP) sensor connector
- Rocker arm oil control solenoid connector

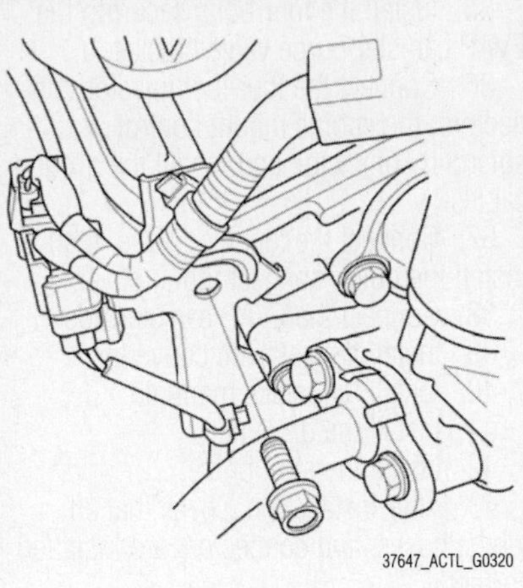
37647_ACTL_G0320

Fig. 146 Remove the connector bracket from the front cylinder head

- Rocker arm oil pressure switch connector
- Two Air Fuel Ratio (A/F) sensor connectors
- Two secondary Heated Oxygen Sensor (secondary HO2S) connectors

10. Remove the front Warm Up Three Way Catalytic Converter (front WU-TWC) and the rear Warm Up Three Way Catalytic Converter (rear WU-TWC).
11. Remove the quick-connect fitting cover, then disconnect the fuel feed hose.
12. Remove the connector bracket from the front cylinder head.
13. Remove the engine mount control solenoid valve bracket from the rear cylinder head.
14. Remove the Evaporative Emission (EVAP) canister joint with the bracket.
15. Remove the injector bases.
16. Remove the water passage.

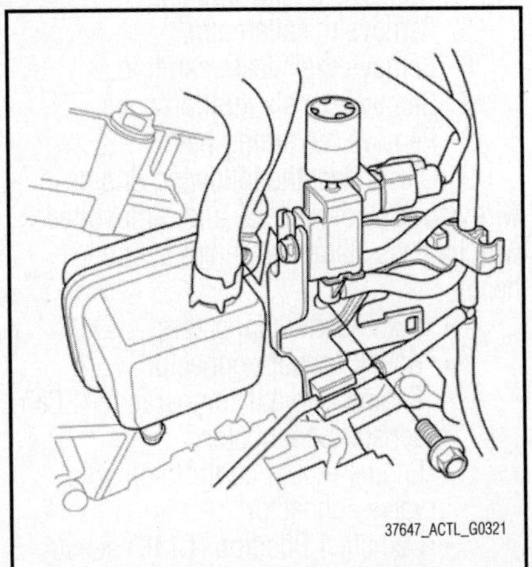
37647_ACTL_G0321

Fig. 147 Remove the engine mount control solenoid valve bracket from the rear cylinder head

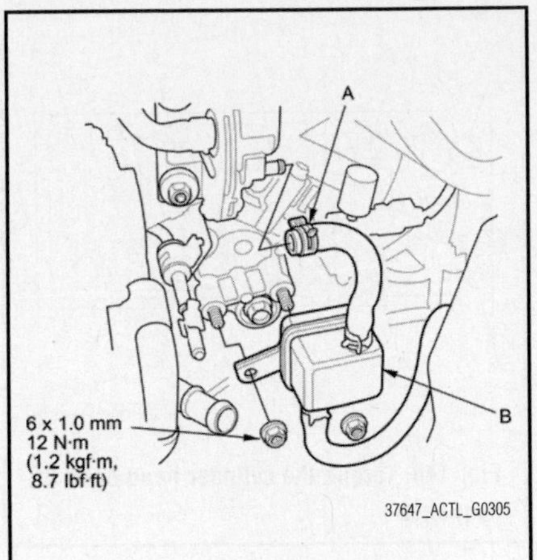
A
B
6 x 1.0 mm
12 N·m
(1.2 kgf·m,
8.7 lbf·ft)
37647_ACTL_G0305

Fig. 148 Remove the Evaporative Emission (EVAP) canister joint with the bracket

17. Remove the camshaft pulleys and the back covers.
18. Remove the cylinder head covers.
19. Remove the cylinder head bolts. To prevent warpage, loosen the bolts in sequence ⅓ turn at a time; repeat the sequence until all bolts are loosened.
20. Remove the cylinder heads.

To install:

21. Clean the cylinder head and the engine block surface.
22. Clean and install the oil control orifices with new O-rings.
23. Install the dowel pins and the new cylinder head gaskets.
24. Clean the timing belt pulleys, the timing belt guide plate, and the upper and lower covers.
25. Set the timing belt drive pulley to Top Dead Center (TDC) by aligning the TDC mark on the tooth of the timing belt drive pulley with the pointer on the oil pump.
26. Set the camshaft pulleys to TDC by aligning the TDC marks on the camshaft pulleys with the pointers on the back covers.
27. Install the cylinder heads on the engine block.

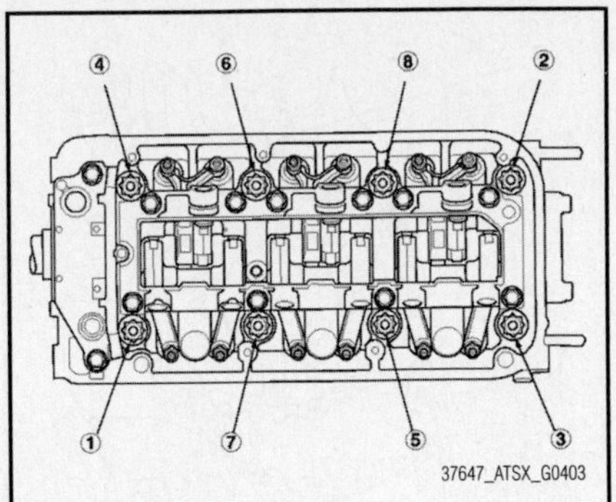
37647_ATSX_G0403

Fig. 150 Remove the front cylinder head bolts

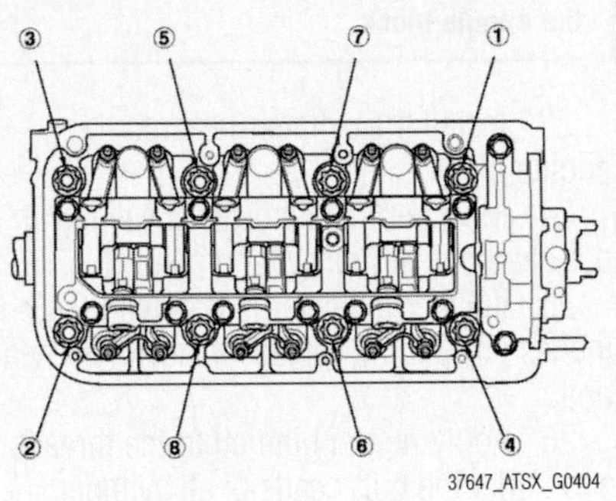
37647_ATSX_G0404

Fig. 151 Remove the rear cylinder head bolts

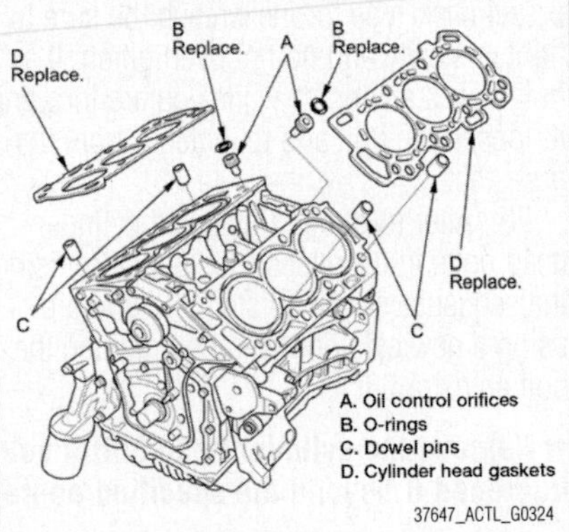
D
Replace.
B
Replace.
A
B
Replace.
D
Replace.
C
C
A. Oil control orifices
B. O-rings
C. Dowel pins
D. Cylinder head gaskets
37647_ACTL_G0324

Fig. 152 Clean and install the oil control orifices with new O-rings

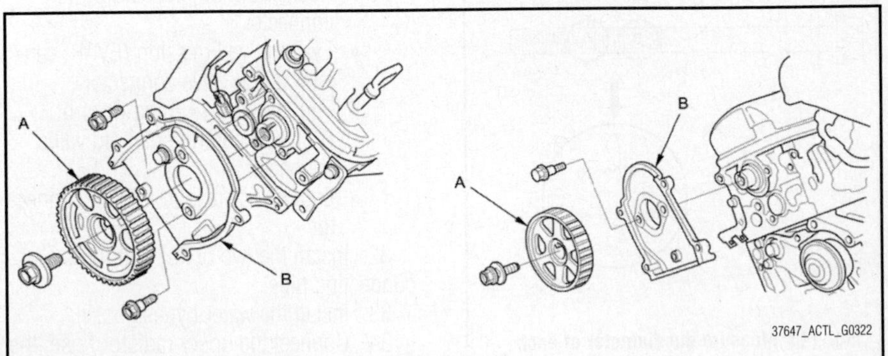
A
B
A
B
37647_ACTL_G0322

Fig. 149 Remove the camshaft pulleys (A) and the back covers (B)

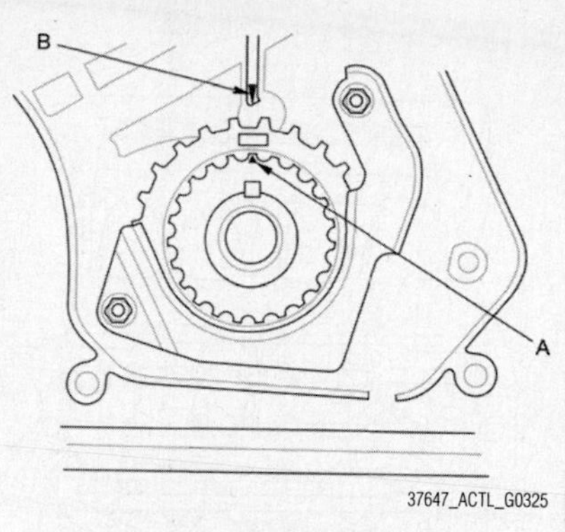
B
A
37647_ACTL_G0325

Fig. 153 Set the timing belt drive pulley to TDC by aligning the TDC mark (A) on the tooth of the timing belt drive pulley with the pointer (B) on the oil pump

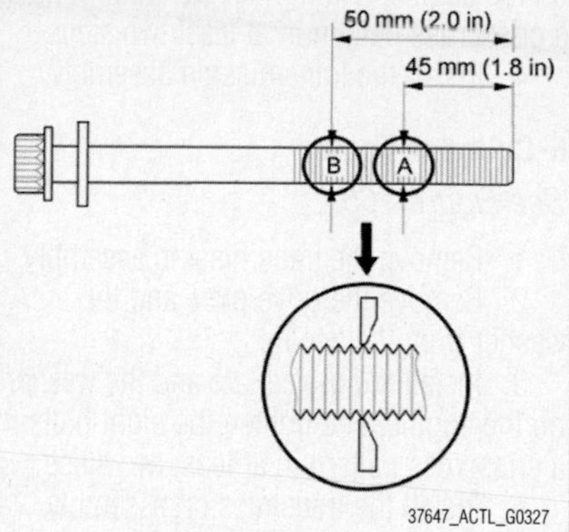
50 mm (2.0 in)
45 mm (1.8 in)
B
A
37647_ACTL_G0327

Fig. 155 Measure the diameter of each cylinder head bolt at point A and point B

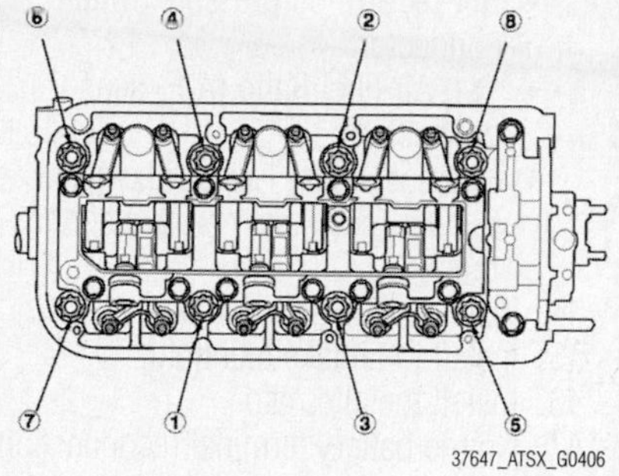
6
4
2
8
7
1
3
5
37647_ATSX_G0406

Fig. 157 Torque the rear cylinder head bolts in sequence

28. Measure the diameter of each cylinder head bolt at point A and point B.

29. If either diameter is less than 0.44 inches (11.3 mm), replace the cylinder head bolt.

30. Apply new engine oil to the threads and under the bolt heads of all cylinder head bolts.

31. Torque the cylinder head bolts in sequence to 22 ft. lbs. (29 Nm) using a beam-type torque wrench. When using a preset click-type torque wrench, be sure to tighten slowly and do not overtighten. If a bolt makes any noise while you are torquing it, loosen the bolt and retighten it from the first step.

32. After torquing, tighten all cylinder head bolts in two steps (90° per step) using the sequence shown in step 11. If you are using a new cylinder head bolt, tighten the bolt an extra 90°.

➡Remove the cylinder head bolt if you tightened it beyond the specified angle, and go back to step 8 of the procedure. Do not loosen it back to the specified angle.

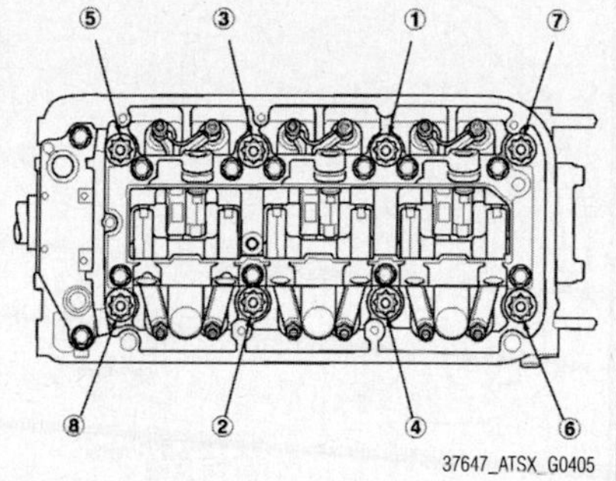
5
3
1
7
8
2
4
6
37647_ATSX_G0405

Fig. 156 Torque the front cylinder head bolts in sequence

33. Install the timing belt.

34. Adjust the valve clearance.

35. Install the cylinder head covers.

36. Install the water passage.

37. Install the injector bases.

38. Install the connector bracket to the front cylinder head.

39. Install the Evaporative Emission (EVAP) canister joint with the bracket.

40. Install the engine mount control solenoid valve bracket to the rear cylinder head.

41. Connect the fuel feed hose, then install the quick-connect fitting cover.

42. Install the front Warm Up Three Way Catalytic Converter (front WU-TWC) and the rear Warm Up Three Way Catalytic Converter (rear WU-TWC).

43. Connect the following engine wire harness connectors, and install the wire harness clamps to the cylinder head:

- Six injector connectors
- Knock sensor connector
- Engine Coolant Temperature (ECT) sensor 1 connector
- Engine mount control solenoid valve connector
- Camshaft Position (CMP) sensor connector

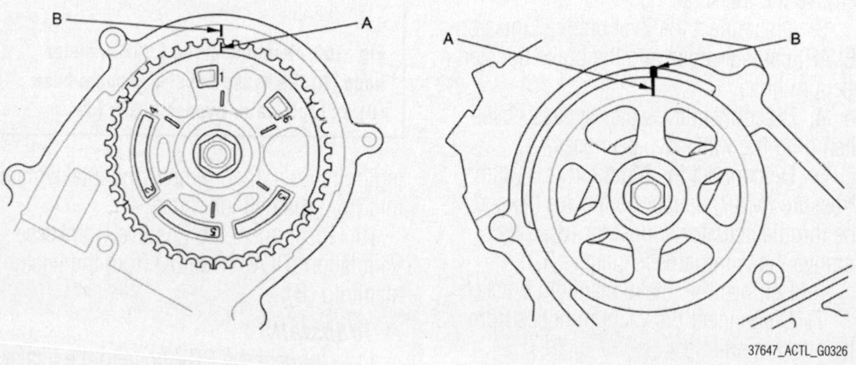
B
A
A
B
37647_ACTL_G0326

Fig. 154 Set the camshaft pulleys to TDC by aligning the TDC marks (A) on the camshaft pulleys with the pointers (B) on the back covers

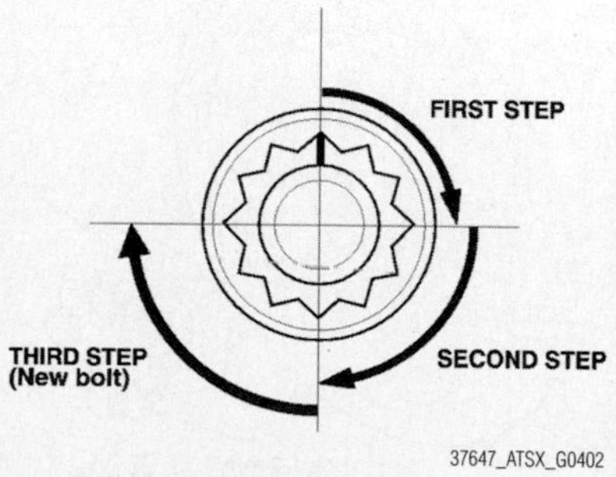
FIRST STEP
THIRD STEP
(New bolt)
SECOND STEP
37647_ATSX_G0402

Fig. 158 After torquing, tighten all cylinder head bolts in two steps (90° per step); tighten new bolts an extra 90°

- Rocker arm oil control solenoid connector
- Rocker arm oil pressure switch connector
- Two Air Fuel Ratio (A/F) sensor connectors
- Two secondary Heated Oxygen Sensor (secondary HO2S) connectors

44. Install the six ignition coils.
45. Install the intake manifold.
46. Install the alternator.
47. Do the battery terminal reconnection procedure.
48. After installation, check that all tubes, hoses, and connectors are installed correctly.
49. Inspect for fuel leaks. Turn the ignition switch to ON (II), (do not operate the starter) so the fuel pump runs for about 2 seconds and pressurizes the fuel line. Repeat this operation three times, then check for fuel leakage at any point in the fuel line.
50. Refill the radiator with engine coolant, and bleed the air from the cooling system.
51. Check for fluid leaks.
52. Do the Powertrain Control Module (PCM) idle learn procedure.
53. Do the Crankshaft Position (CKP) pattern clear/CKP pattern learn procedure.
54. Inspect the idle speed.
55. Inspect the ignition timing.
56. Install the engine compartment covers.

DRIVE PLATE

REMOVAL & INSTALLATION

4-Cylinder Engine

See Figure 159.

1. Remove the transmission assembly.
2. Remove the drive plate and the washer from the engine.

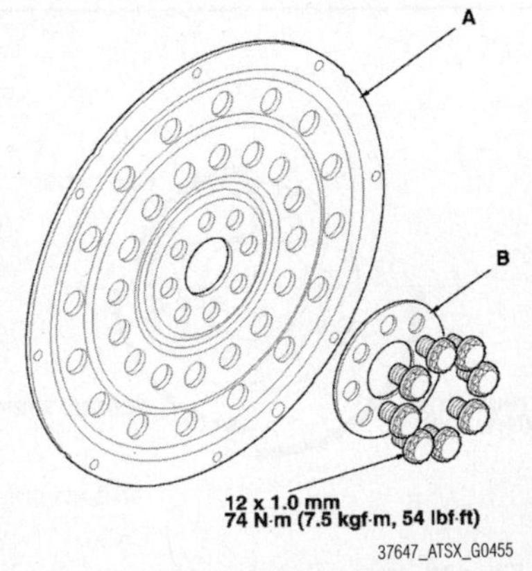
A
B
12 x 1.0 mm
74 N·m (7.5 kgf·m, 54 lbf·ft)
37647_ATSX_G0455

Fig. 159 Remove the drive plate (A) and the washer (B) from the engine

3. Install the drive plate and the washer on the engine, and tighten the eight bolts in a crisscross pattern in at least two steps.
4. Install the transmission assembly.

6-Cylinder Engine

See Figure 160.

1. Remove the transmission assembly.
2. Remove the drive plate and the washer from the engine.
3. Install the drive plate and the washer on the engine, and tighten the eight bolts in a crisscross pattern in at least two steps.
4. Install the transmission assembly.

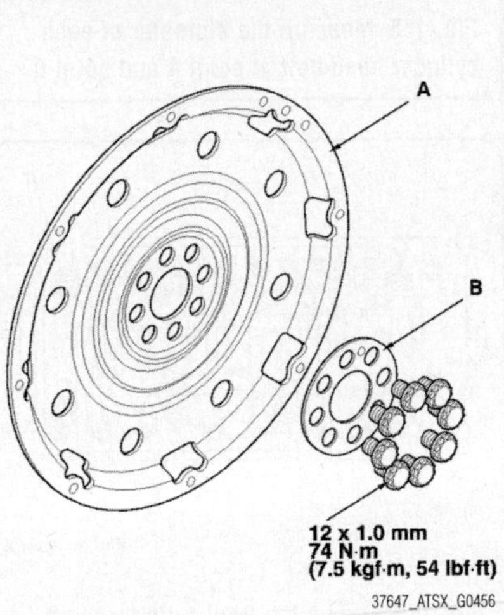
A
B
12 x 1.0 mm
74 N·m
(7.5 kgf·m, 54 lbf·ft)
37647_ATSX_G0456

Fig. 160 Remove the drive plate (A) and the washer (B) from the engine

INTAKE MANIFOLD

REMOVAL & INSTALLATION

4-Cylinder Engine

See Figures 161 through 168.

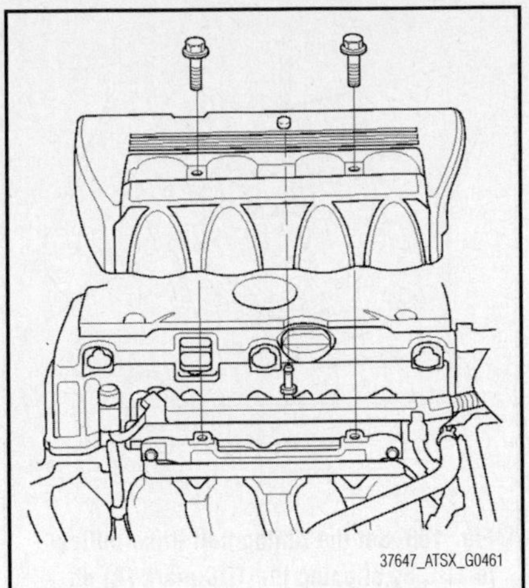
37647_ATSX_G0461

Fig. 161 Remove the intake manifold cover

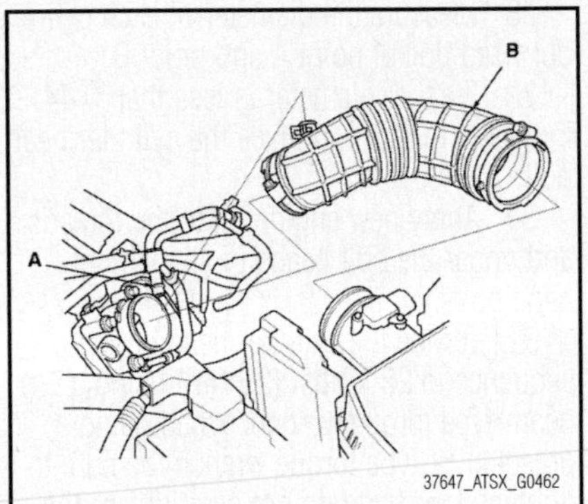
B
A
37647_ATSX_G0462

Fig. 162 Disconnect the breather pipe (A), then remove the intake air duct (B)

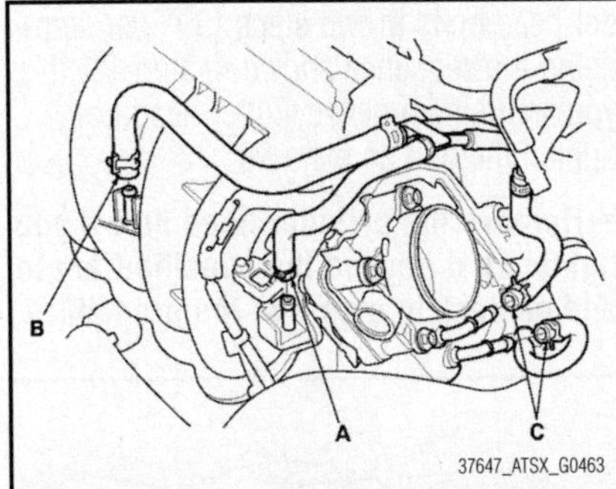
B
A
C
37647_ATSX_G0463

Fig. 163 Disconnect the EVAP canister hose (A) the brake booster vacuum hose (B) and the water bypass hoses (C)

1. Remove the intake manifold cover.
2. Disconnect the breather pipe, then remove the intake air duct.
3. Disconnect the Evaporative Emission (EVAP) canister hose and the brake booster vacuum hose.
4. Disconnect the water bypass hoses, then plug the water bypass hoses.
5. Disconnect the Manifold Absolute Pressure (MAP) sensor connector (A) and the throttle actuator connector (B), then remove the wire harness clamps (C).
6. Remove the intake manifold bracket.
7. Disconnect the vacuum hoses from the intake manifold.
8. Remove the vacuum hoses from the clamps.
9. Remove the intake manifold mounting bolts/nuts, then remove the intake manifold from the cylinder head.
10. Disconnect the Positive Crankcase Ventilation (PCV) hose (A) from the intake manifold (B).

To install:

11. Connect the PCV hose to the intake manifold.
12. Install the intake manifold with new gaskets, and tighten the bolts and nuts in a

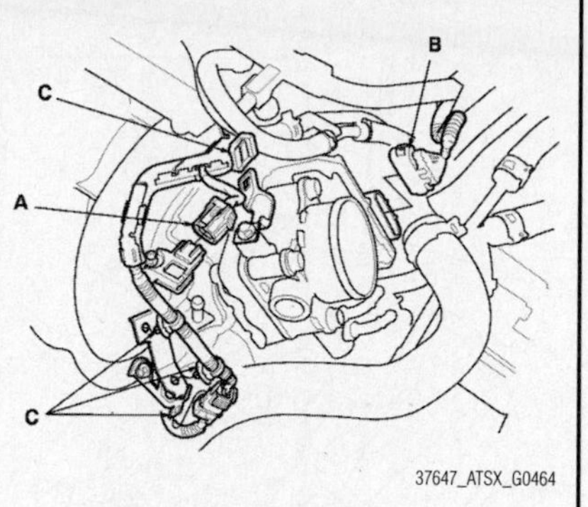
B
C
A
C
37647_ATSX_G0464

Fig. 164 Disconnect the MAP sensor connector (A) and the throttle actuator connector (B), then remove the wire harness clamps (C)

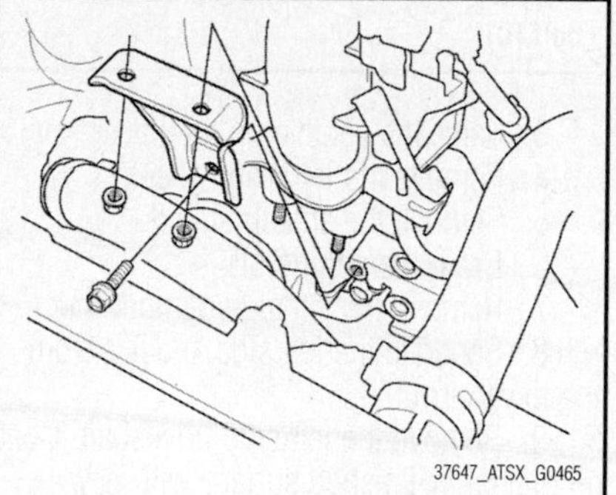
37647_ATSX_G0465

Fig. 165 Remove the intake manifold bracket

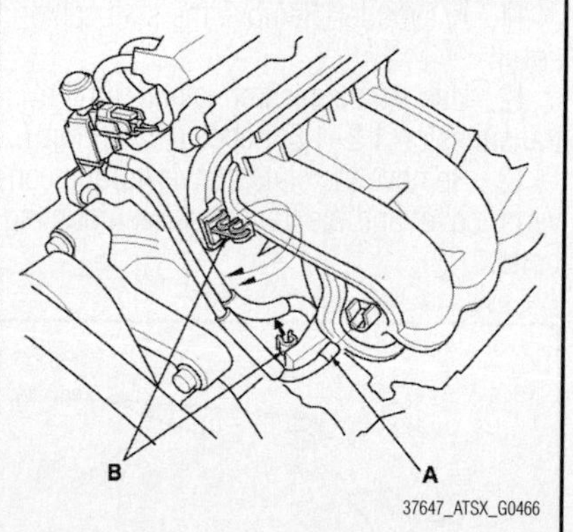
B
A
37647_ATSX_G0466

Fig. 166 Disconnect the vacuum hoses (A) and remove from the clamps (B)

crisscross pattern in three steps, beginning with the inner bolt. Tighten the bolts and nuts to 16 ft. lbs. (22 Nm).

13. Connect the vacuum hoses to the intake manifold.
14. Install the vacuum hoses to the clamps.
15. Install the intake manifold bracket. Tighten the bolt and nuts to 16 ft. lbs. (22 Nm).

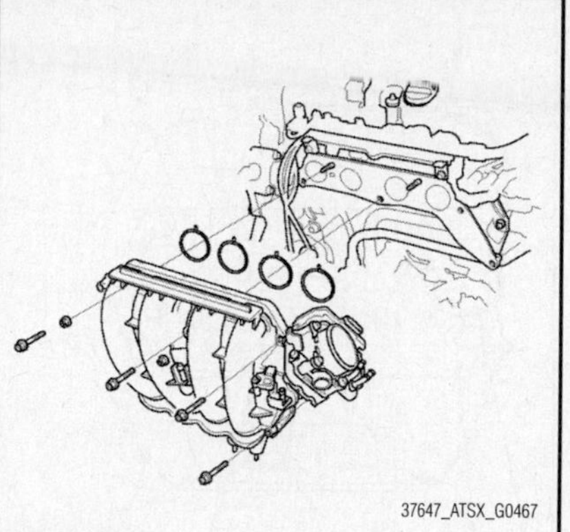
37647_ATSX_G0467

Fig. 167 Remove the intake manifold mounting bolts/nuts, then remove the intake manifold from the cylinder head

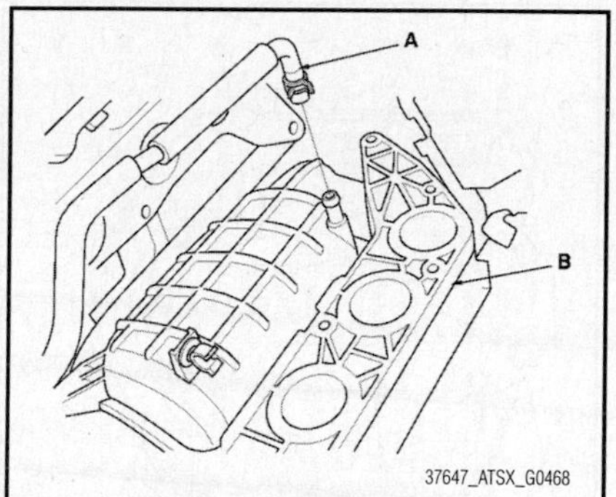
A
B
37647_ATSX_G0468

Fig. 168 Disconnect the PCV hose (A) from the intake manifold (B)

16. Connect the MAP sensor connector and the throttle actuator connector, then install the wire harness clamps.
17. Connect the water bypass hoses.
18. Connect the EVAP canister hose and the brake booster vacuum hose.
19. Install the intake air duct, then connect the breather pipe.

➡When tightening the screw of the hose band, align the edge of the hose band with the mark painted on the hose band. If you tighten the screw over the mark, replace the hose band.

20. Install the intake manifold cover.
21. Clean up any spilled engine coolant.
22. After installation, check that all tubes, hoses, and connectors are installed correctly.
23. Refill the radiator with engine coolant, and bleed the air from the cooling system with the heater valve open.

6-Cylinder Engine

See Figures 169 through 175.

1.Remove the engine compartment covers.
2. Disconnect the Manifold Absolute Pressure (MAP) sensor connector and the breather pipe, then remove the intake air duct.
3. Disconnect the throttle actuator connector, the Evaporative Emission (EVAP) canister purge valve connector, the water bypass hoses, the EVAP canister hose, the brake booster vacuum hose, and the vacuum hose.

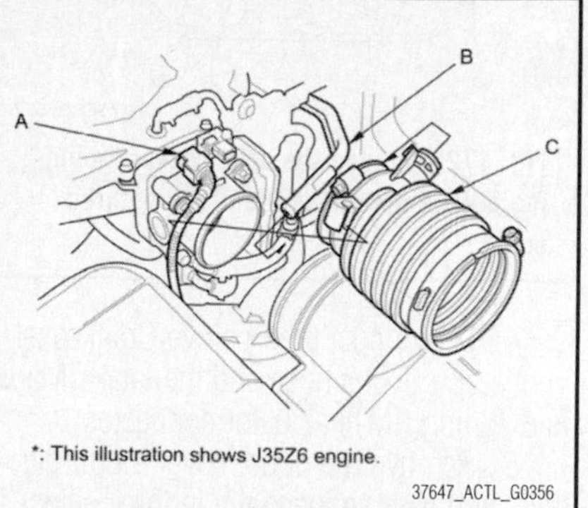
B
A
C
*: This illustration shows J35Z6 engine.
37647_ACTL_G0356

Fig. 169 Disconnect the MAP sensor connector (A) and the breather pipe (B), then remove the intake air duct (C)

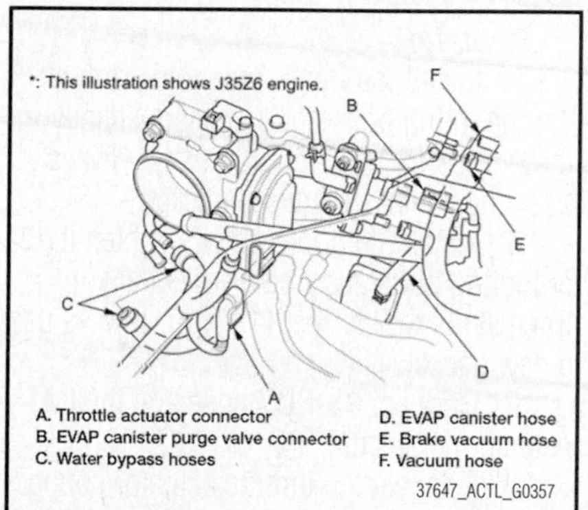
*: This illustration shows J35Z6 engine.
F
B
E
C
A
D
A. Throttle actuator connector
B. EVAP canister purge valve connector
C. Water bypass hoses
D. EVAP canister hose
E. Brake vacuum hose
F. Vacuum hose
37647_ACTL_G0357

Fig. 170 Disconnect the throttle actuator connector, the EVAP canister purge valve connector, the water bypass hoses, the EVAP canister hose, the brake booster vacuum hose, and the vacuum hose

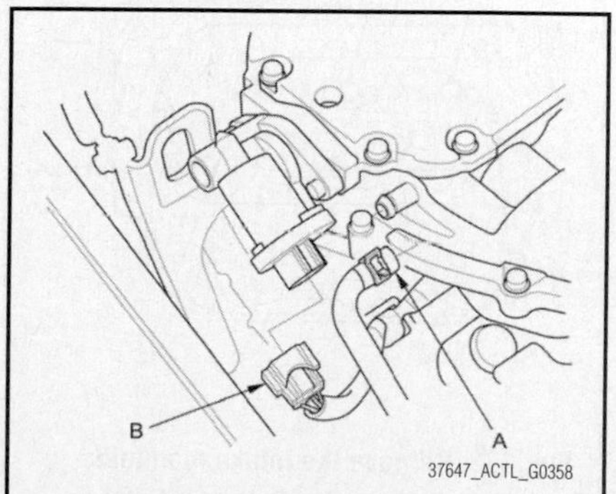
B
A
37647_ACTL_G0358

Fig. 171 Disconnect the PCV hose (A) and the IMT actuator connector (B)

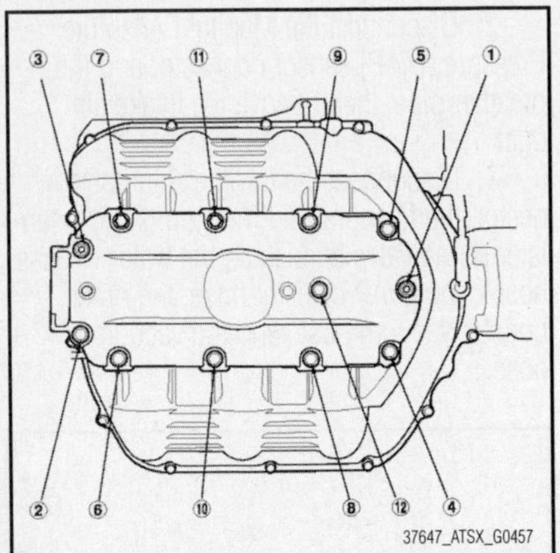
37647_ATSX_G0457
Fig. 172 Remove the upper cover mounting bolts and nuts sequentially in three steps

4. Disconnect the Positive Crankcase Ventilation (PCV) hose and the Intake Manifold Tuning (IMT) actuator connector.

5. Remove the upper cover mounting bolts and nuts sequentially in three steps, then remove the upper cover.

6. Remove the intake manifold mounting bolts and nuts sequentially in three steps, then remove the intake manifold.

To install:

7. Install the intake manifold. Tighten the mounting bolts and nuts sequentially in three steps to 16 ft. lbs. (22 Nm). Always use a new intake manifold gasket.

8. Install the upper cover. Tighten the mounting bolts and nuts sequentially in three steps to 9 ft. lbs. (12 Nm). Always use a new gasket.

9. Connect the PCV hose and the IMT actuator connector.

10. Connect the throttle actuator connector, the EVAP canister purge valve connector, the water bypass hoses, the EVAP canister hose, the brake booster vacuum hose, and the vacuum hose.

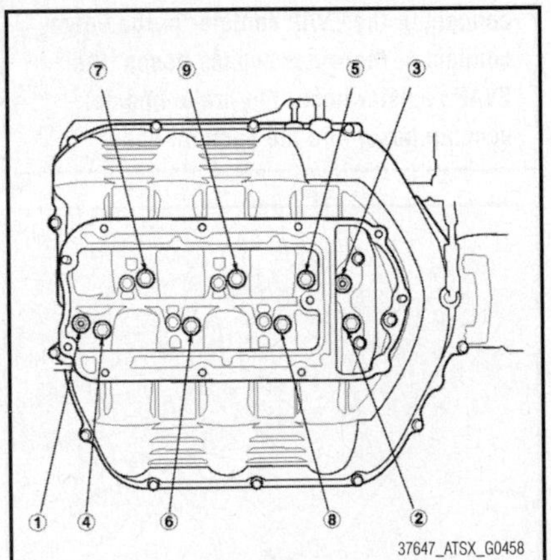
37647_ATSX_G0458
Fig. 173 Remove the intake manifold mounting bolts and nuts sequentially in three steps, then remove the intake manifold

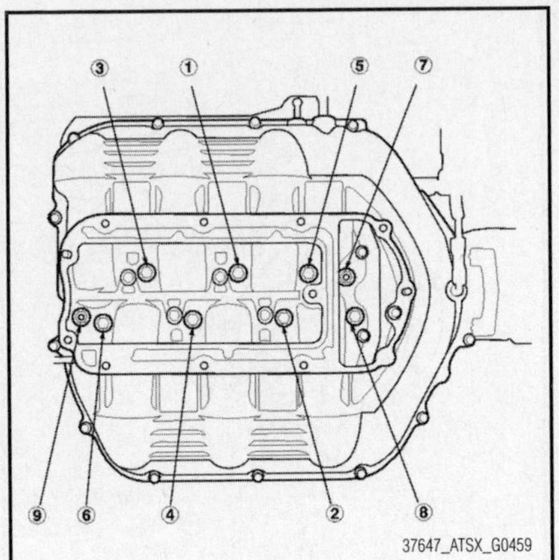
37647_ATSX_G0459
Fig. 174 Install the intake manifold

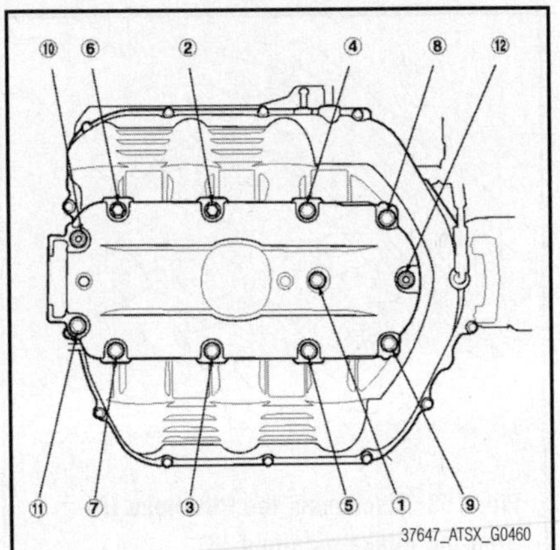
37647_ATSX_G0460
Fig. 175 Install the upper cover

11. Install the intake air duct, then connect the breather pipe and the MAP sensor connector.

12. Clean up any spilled engine coolant.

13. After installation, check that all tubes, hoses, and connectors are installed correctly.

14. Refill the radiator with engine coolant, and bleed the air from the cooling system..

15. Install the engine compartment covers.

OIL PAN

REMOVAL & INSTALLATION

4-Cylinder Engine

See Figures 176 through 181.

1. If the engine is already out of the vehicle, go to step 13.

2. Disconnect the vacuum hose and remove the, front engine mount stop, then remove the front engine mount bolt.

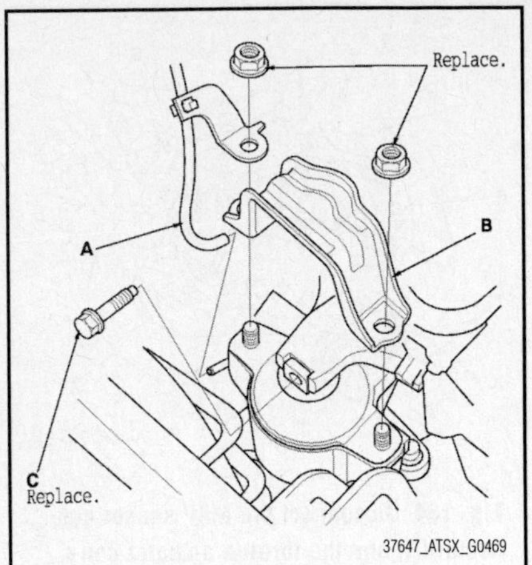
Replace.
A
B
C
Replace.
37647_ATSX_G0469
Fig. 176 Disconnect the vacuum hose (A) and remove the, front engine mount stop (B), then remove the front engine mount bolt (C)

3. Raise the vehicle on the lift.

4. Remove the left front wheel.

5. Remove the splash shield.

6. Drain the engine oil.

7. Remove the left side damper fork.

8. Separate the left side knuckle from the lower arm.

9. Remove the left side driveshaft. Coat all precision-finished surface with new engine oil. Tie a plastic bag over the driveshaft end.

10. Remove the lower transmission mount bracket mounting bolts and nuts.

11. A/T model: Remove the shift cable cover.

12. Use a transmission jack to lift the transmission 1.2–1.6 inches (30–40 mm).

13. Remove the clutch cover/torque converter cover and the transmission mounting bolts.

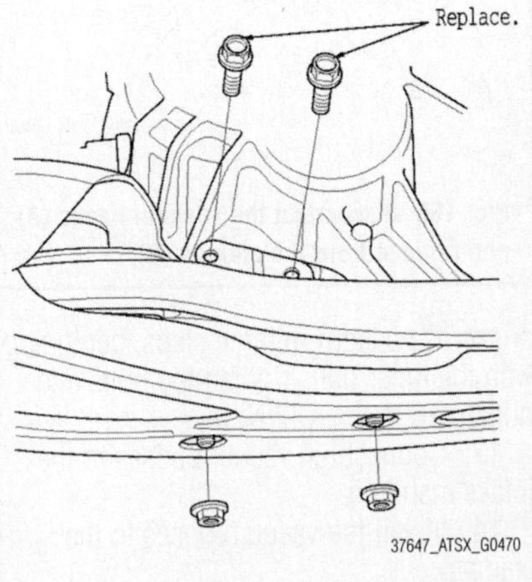
Replace.
37647_ATSX_G0470
Fig. 177 Remove the lower transmission mount bracket mounting bolts and nuts

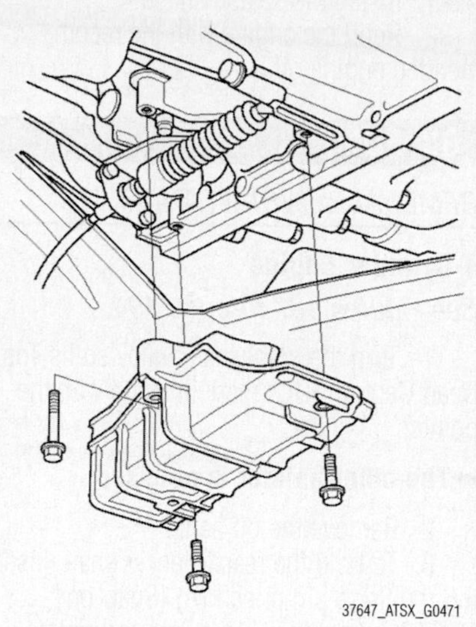

Fig. 178 A/T model: Remove the shift cable cover

14. Remove the bolts securing the oil pan.
15. Using a flat blade screwdriver, separate the oil pan from the engine block in the places shown.
16. Remove the oil pan.

To install:

17. Remove all of the old liquid gasket from the oil pan mating surfaces, the bolts, and the bolt holes.
18. Clean and dry the oil pan mating surfaces.
19. Apply liquid gasket (P/N 08717-0004, 08718-0003, or 08718-0009) to the engine block mating surface of the oil pan

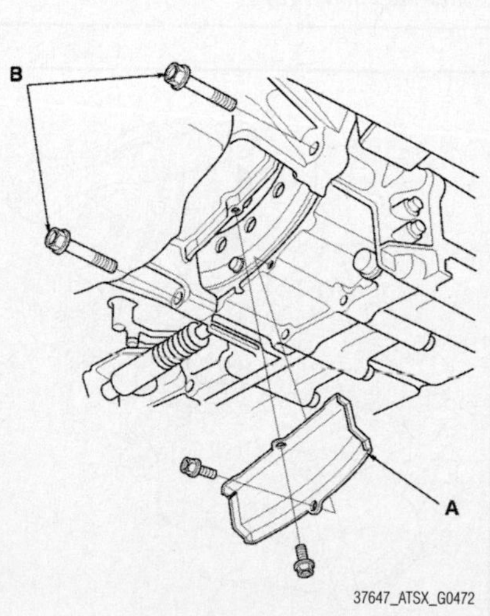
B
A
37647_ATSX_G0472

Fig. 179 Remove the clutch cover/torque converter cover (A) and the transmission mounting bolts (B)

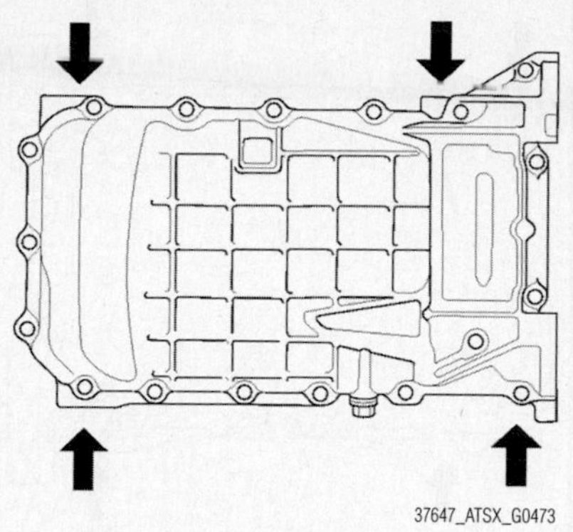

Fig. 180 Separate the oil pan from the engine block in the places shown

and to the inside edge of the threaded bolt holes. Install the component within 5 minutes of applying the liquid gasket.

➡If too much time has passed after applying the liquid gasket, remove the old liquid gasket and residue, then reapply new liquid gasket.

20. Install the oil pan.
21. Tighten the bolts in three steps. In the final step, torque all bolts, in sequence, to 9 ft. lbs. (12 Nm). Wipe off the excess liquid gasket on the each side of crankshaft pulley and the flywheel/drive plate.

➡Wait at least 30 minutes before filling the engine with oil. Do not run the engine for at least 3 hours after installing the oil pan.

22. Install the clutch cover/torque converter cover and the transmission mounting bolts.
23. If the engine is still in the vehicle, do steps 8 through 18.
24. Lower the transmission jack from the transmission.
25. A/T model: Install the shift cable cover.

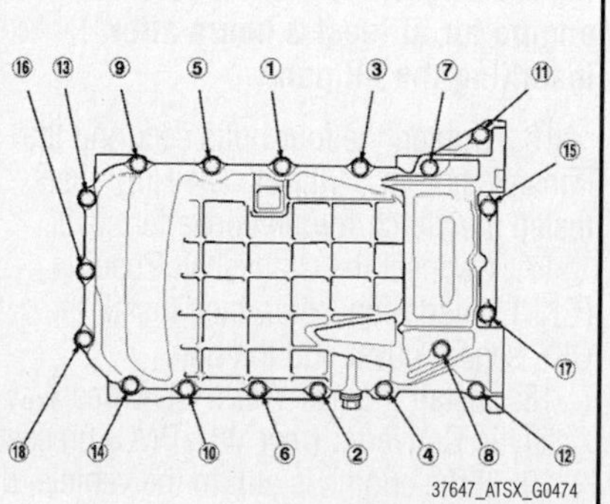
16 13 9 5 1 3 7 11
15
17
18 14 10 6 2 4 8 12
37647_ATSX_G0474

Fig. 181 Oil pan bolt installation sequence

26. Tighten the lower transmission mount bracket mounting bolts and nuts. Tighten the bolts to 40 ft. lbs. (54 Nm). Tighten the nuts to 33 ft. lbs. (44 Nm).
27. Install a new set ring on the end of driveshaft, then install the driveshaft. Make sure the ring "clicks" into place in the differential.
28. Connect the lower arm to the left side knuckle.
29. Install the left side damper fork.
30. Install the splash shield.
31. Install the left front wheel.
32. Lower the vehicle on the lift.
33. Tighten the front engine mount bolt to 40 ft. lbs. (54 Nm), then install the front engine mount stop and connect the vacuum hose.
34. Refill the engine with recommended engine oil.

6-Cylinder Engine

See Figures 182 through 186.

➡If the engine is already out of the vehicle, go to step 6.

1. Raise the vehicle on the lift.
2. Drain the engine oil.
3. Remove the splash shield.
4. Remove exhaust pipe A.
5. Remove the rear Warm Up Three Way Catalytic Converter (rear WU-TWC) bracket.
6. Remove the Crankshaft Position (CKP) sensor cover and the bolt, then disconnect the CKP sensor connector.
7. Remove the clutch/torque converter cover and the four bolts securing the transmission.
8. Remove the bolts securing the oil pan.

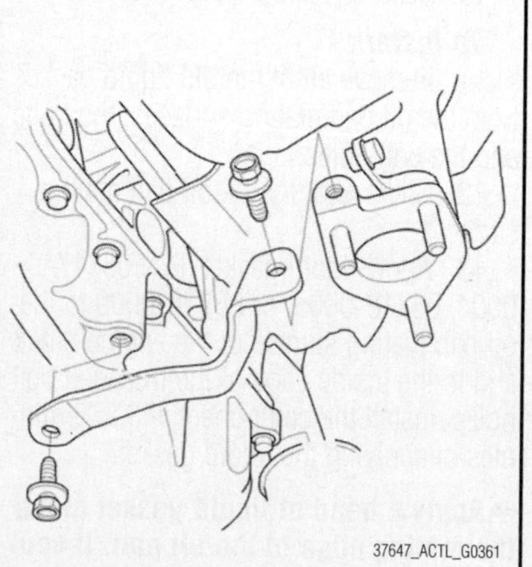

Fig. 182 Remove the rear WU-TWC bracket

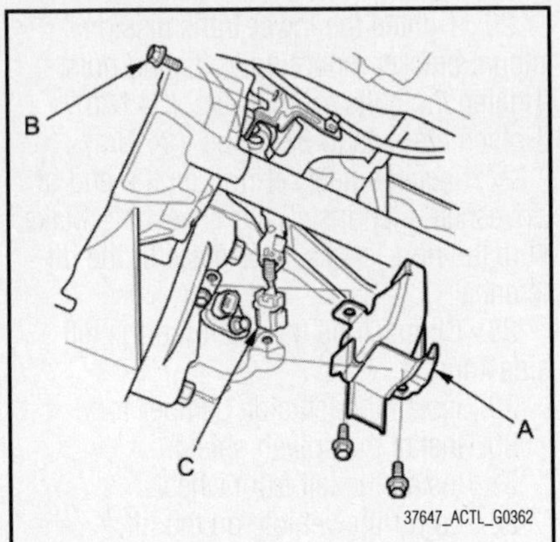
B
A
C
37647_ACTL_G0362

Fig. 183 Remove the CKP sensor cover (A) and the bolt (B), then disconnect the CKP sensor connector (C)

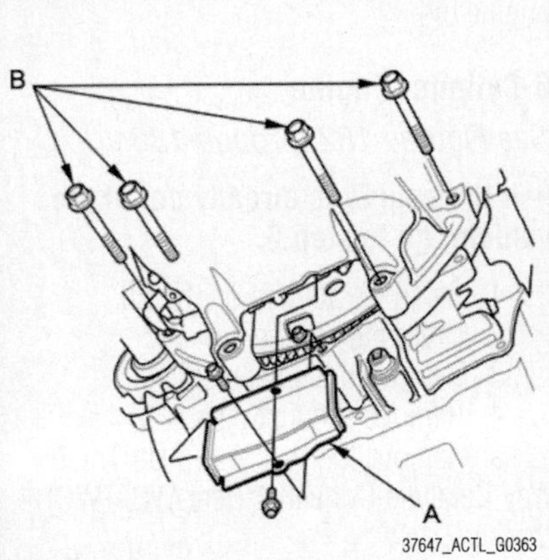
B
A
37647_ACTL_G0363

Fig. 184 Remove the clutch/torque converter cover (A) and the four bolts (B) securing the transmission

9. Using a flat blade screwdriver, separate the oil pan from the engine block in the places shown.
10. Remove the oil pan.

To install:

11. Remove all of the old liquid gasket from the oil pan mating surfaces, the bolts, and the bolt holes.
12. Clean and dry the oil pan mating surfaces.
13. Apply liquid gasket, P/N 08717-0004, 08718-0003, or 08718-0009 to the oil pan mating surface of the engine block and to the inside edge of the threaded bolt holes. Install the component within 5 minutes of applying the liquid gasket.

➡Apply a bead of liquid gasket along the mating edge of the oil pan. If you apply liquid gasket P/N 08718-0012, the component must be installed within 4 minutes. If too much time has passed after applying the liquid gasket, remove the old liquid gasket and

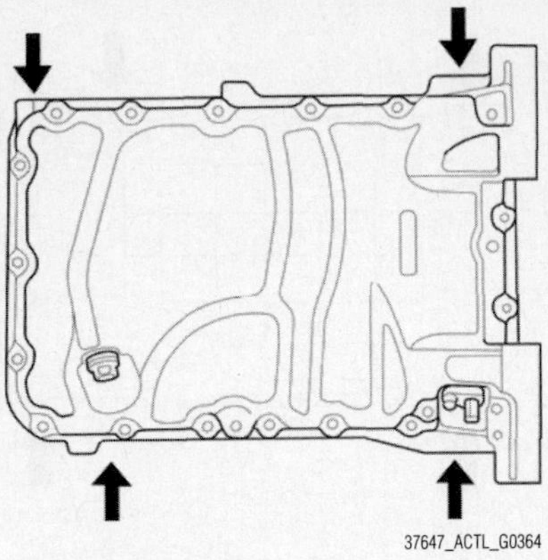
37647_ACTL_G0364

Fig. 185 Using a flat blade screwdriver, separate the oil pan from the engine block in the places shown

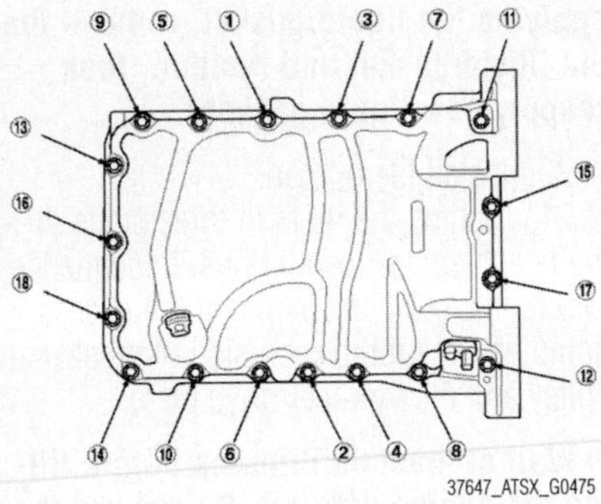
37647_ATSX_G0475

Fig. 186 Oil pan bolt installation sequence

residue, then reapply the new liquid gasket.

14. Install the oil pan on the engine block.
15. Tighten the bolts in three steps. In the final step, torque all bolts, in sequence, to 9 ft. lbs. (12 Nm).

➡Wait at least 30 minutes before filling the engine with oil. Do not run the engine for at least 3 hours after installing the oil pan.

16. Tighten the four bolts securing the transmission to 54 ft. lbs. (74 Nm), then install the clutch/torque converter cover.
17. Connect the Crankshaft Position (CKP) sensor connector, then install the CKP sensor cover and the bolt.
18. Install the rear Warm Up Three Way Catalytic Converter (rear WU-TWC) bracket.
19. If the engine is still in the vehicle, do the following steps.
20. Install exhaust pipe A using new gaskets and new self-locking nuts.
21. Install the splash shield.
22. Refill the engine with the recommended engine oil.

OIL PUMP

REMOVAL & INSTALLATION

4-Cylinder Engine

See Figures 187 through 194.

1. Turn the crankshaft pulley so its Top Dead Center (TDC) mark lines up with the pointer.

➡The other pointer is not used.

2. Remove the oil pan.
3. To hold the rear balancer shaft, insert a 6 mm long pin punch (A) (Snap-on® PPC108LA or equivalent) into the maintenance hole in the balancer shaft holder and through the rear balancer shaft.
4. Turn the crankshaft counterclockwise to compress the oil pump chain auto-tensioner.

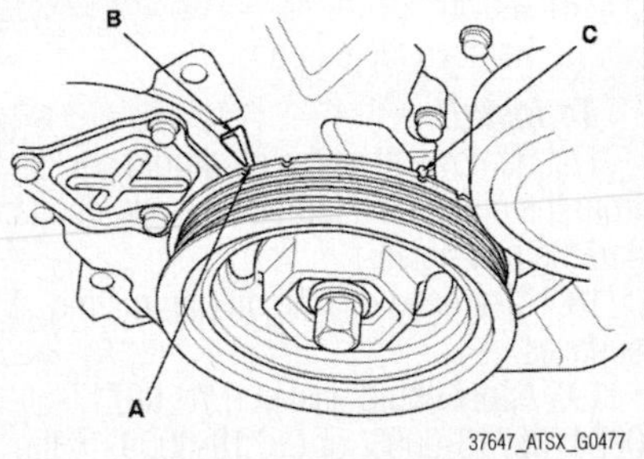
B
C
A
37647_ATSX_G0477

Fig. 187 Turn the crankshaft pulley so its Top Dead Center (TDC) mark (A) lines up with the pointer (B)

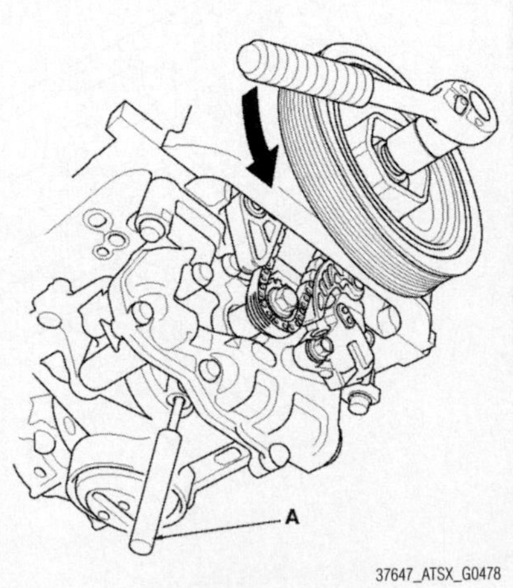
A
37647_ATSX_G0478

Fig. 188 To hold the rear balancer shaft, insert a 6 mm long pin punch (A)

5. Align the holes on the lock and the oil pump chain auto-tensioner, then insert a 0.12 inches (3.0 mm) diameter pin into the holes. Turn the crankshaft clockwise to secure the pin.

6. Remove the oil pump chain auto-tensioner.

7. Loosen the oil pump sprocket mounting bolt.

8. Remove the oil pump sprocket, then remove the oil pump.

To install:

9. Make sure the No. 1 piston Top Dead Center (TDC) mark lines up with the pointer.

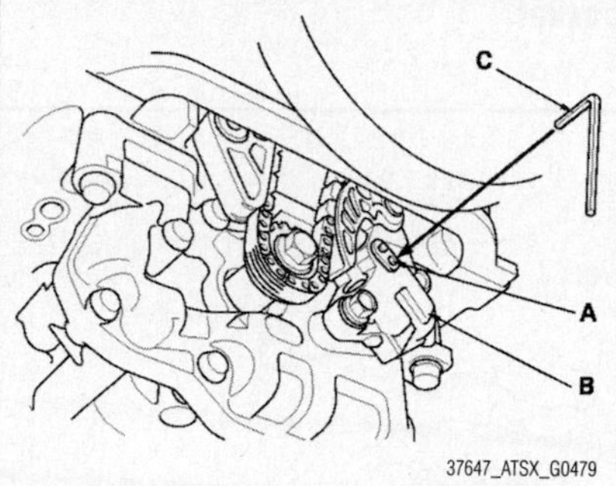
C
A
B
37647_ATSX_G0479

Fig. 189 Align the holes on the lock (A) and the oil pump chain auto-tensioner (B), then insert a 0.12 inches (3.0 mm) diameter pin (C) into the holes

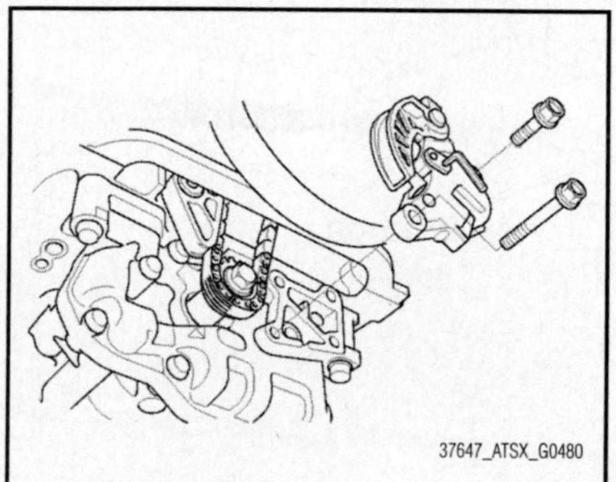
37647_ATSX_G0480

Fig. 190 Remove the oil pump chain auto-tensioner

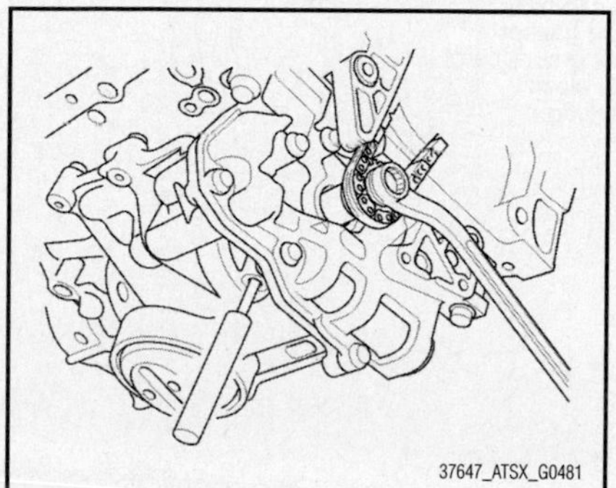
37647_ATSX_G0481

Fig. 191 Loosen the oil pump sprocket mounting bolt.

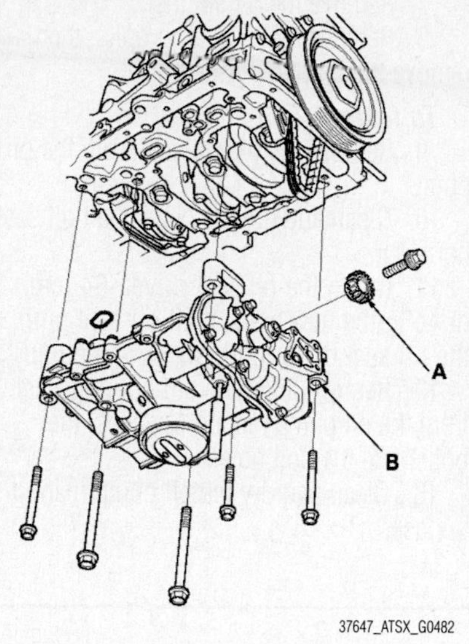
A
B
37647_ATSX_G0482

Fig. 192 Remove the oil pump sprocket (A), then remove the oil pump (B).

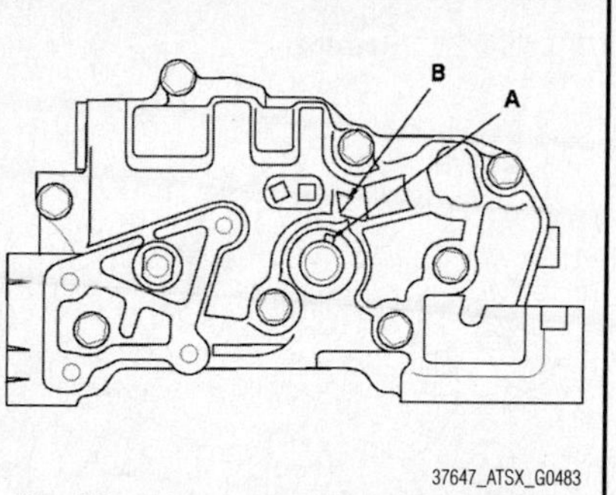
B
A
37647_ATSX_G0483

Fig. 193 Align the dowel pin (A) on the rear balancer shaft with the mark (B) on the oil pump.

10. Align the dowel pin on the rear balancer shaft with the mark on the oil pump.

11. To hold the rear balancer shaft, insert a 6 mm long pin punch (Snap-on® PPC108LA or equivalent) into the maintenance hole in the balancer shaft holder and through the rear balancer shaft.

12. Apply new engine oil to the threads of the oil pump mounting bolts and oil pump sprocket mounting bolt, then loosely install the oil pump with a new O-ring, then install the oil pump sprocket.

13. Tighten the oil pump mounting bolts and the oil pump sprocket mounting bolt.

14. Remove the 6 mm pin punch.

15. Install the oil pump chain auto-tensioner.

➡If the 1.2 inches (3.0 mm) diameter pin come off the auto-tensioner, compress the auto-tensioner.

16. Remove the pin from the oil pump chain auto-tensioner.

17. Install the oil pan.

6-Cylinder Engine

See Figures 195 through 198.

1. Drain the engine oil.
2. Remove the timing belt:
3. Remove the timing belt drive pulley from the crankshaft:
4. Attach the chain hoist to the engine hook on the engine hanger bracket.
5. Remove the rocker arm oil control solenoid/oil filter assembly.
6. Remove the oil pan.

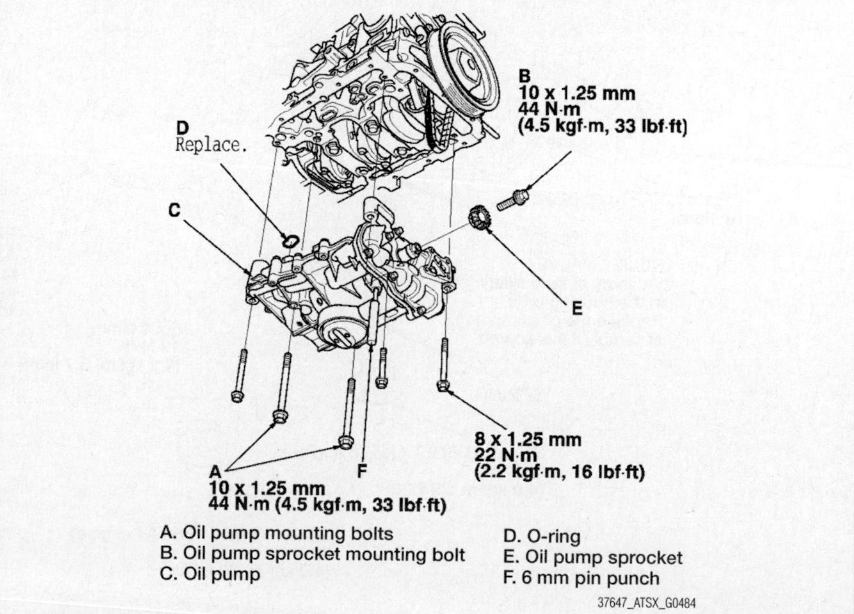
B
10 x 1.25 mm
44 N·m
(4.5 kgf·m, 33 lbf·ft)
D
Replace.
C
E
A
F
10 x 1.25 mm
44 N·m (4.5 kgf·m, 33 lbf·ft)
8 x 1.25 mm
22 N·m
(2.2 kgf·m, 16 lbf·ft)
A. Oil pump mounting bolts
B. Oil pump sprocket mounting bolt
C. Oil pump
D. O-ring
E. Oil pump sprocket
F. 6 mm pin punch
37647_ATSX_G0484

Fig. 194 Install the oil pump with a new O-ring, then install the oil pump sprocket

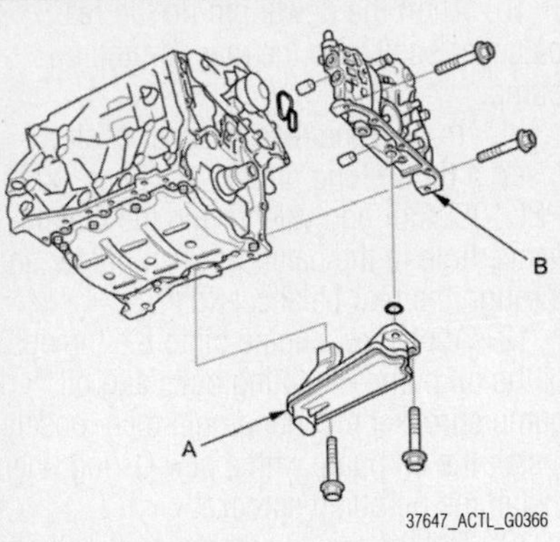
B
A
37647_ACTL_G0366

Fig. 195 Remove the oil strainer (A), then remove the oil pump assembly (B)

7. Remove the oil strainer.

8. Remove the mounting bolts, then remove the oil pump assembly.

To install:

9. Remove the old oil seal from the oil pump.

10. Clean and dry the crankshaft oil seal housing.

11. Using the oil seal driver, 64 mm, drive in the new crankshaft oil seal until the oil seal driver bottoms on the pump.

12. Remove all of the old liquid gasket from the oil pump mating surfaces, the bolts, and the bolt holes.

13. Clean and dry the oil pump mating surfaces.

14. Apply liquid gasket, P/N 08717-0004, 08718-0003, or 08718-0009, to the engine block mating surface of the oil pump and to the inside edge of the threaded bolt holes. Install the component within 5 minutes of applying the liquid gasket.

➡Apply a bead of liquid gasket along the mating surface. If you apply liquid gasket P/N 08718-0012, the component must be installed within 5 minutes. If too much time has passed after applying the liquid gasket, remove all of the old liquid gasket and residue, then reapply the new liquid gasket.

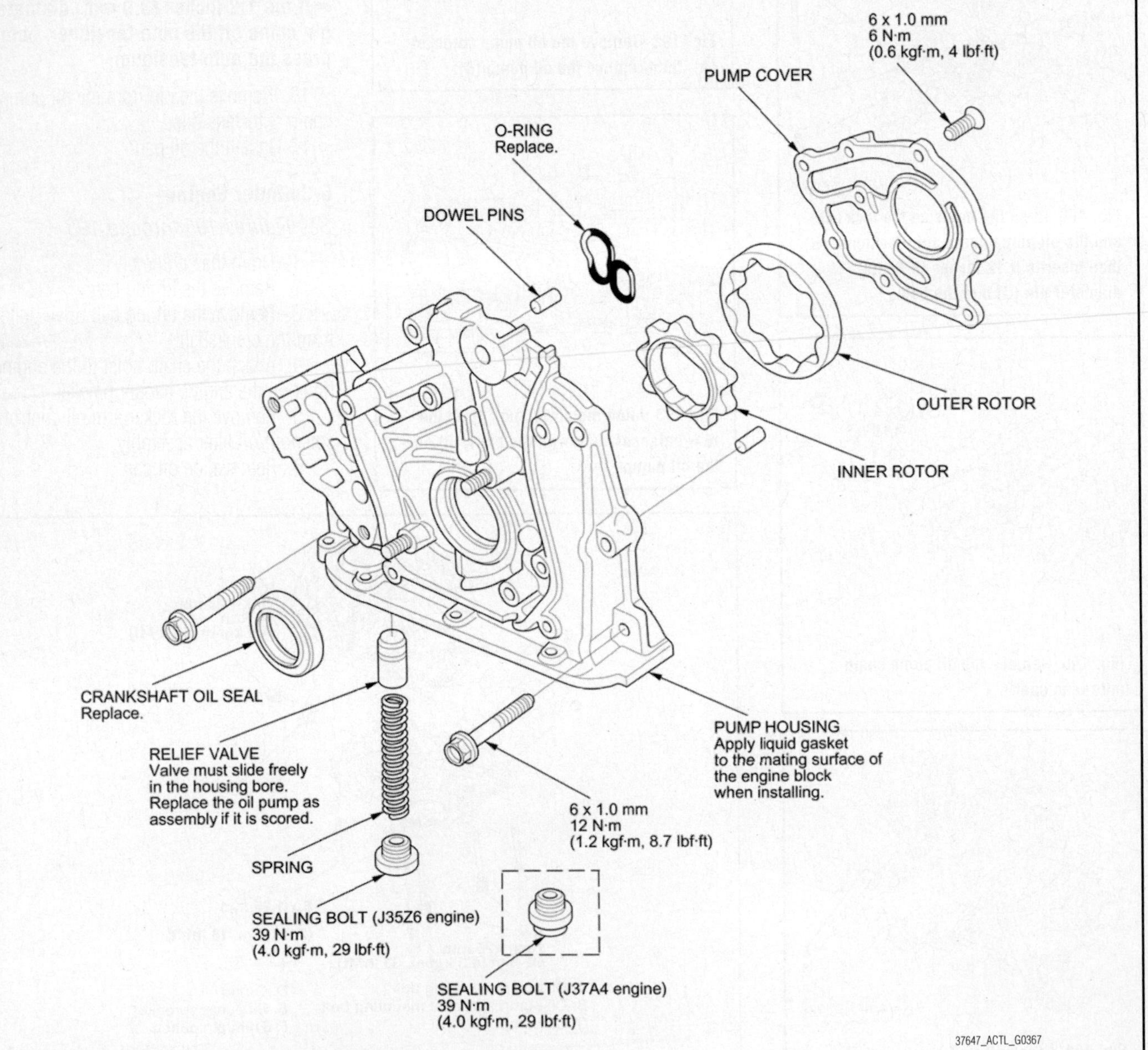
6 x 1.0 mm
6 N·m
(0.6 kgf·m, 4 lbf·ft)
PUMP COVER
O-RING
Replace.
DOWEL PINS
OUTER ROTOR
INNER ROTOR
CRANKSHAFT OIL SEAL
Replace.
RELIEF VALVE
Valve must slide freely in the housing bore.
Replace the oil pump as assembly if it is scored.
PUMP HOUSING
Apply liquid gasket to the mating surface of the engine block when installing.
6 x 1.0 mm
12 N·m
(1.2 kgf·m, 8.7 lbf·ft)
SPRING
SEALING BOLT (J35Z6 engine)
39 N·m
(4.0 kgf·m, 29 lbf·ft)
SEALING BOLT (J37A4 engine)
39 N·m
(4.0 kgf·m, 29 lbf·ft)
37647_ACTL_G0367

Fig. 196 Exploded view of oil pump assembly

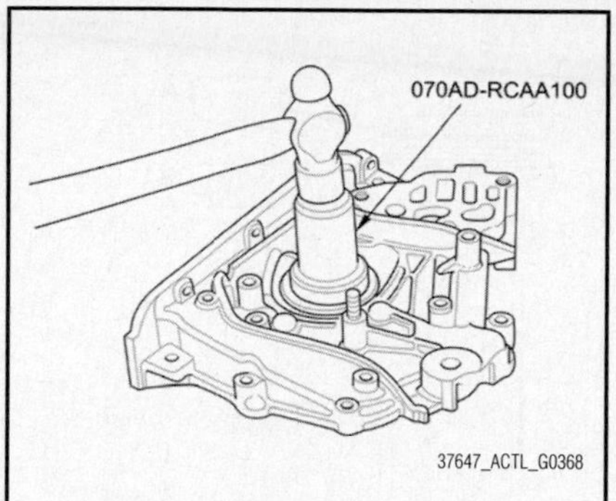
070AD-RCAA100
37647_ACTL_G0368

Fig. 197 Using the oil seal driver, 64 mm, drive in the new crankshaft oil seal until the oil seal driver bottoms on the pump

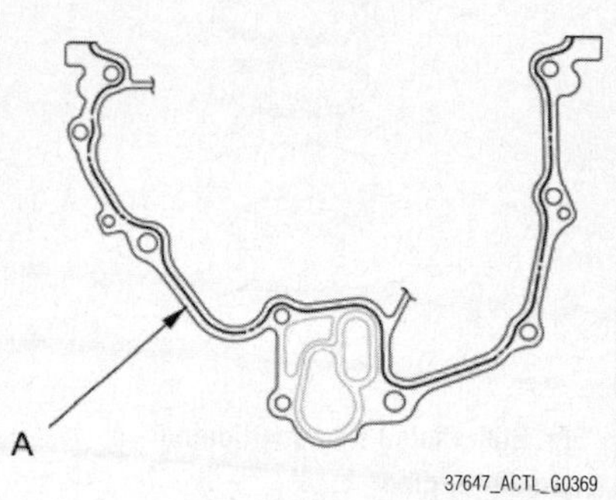
A
37647_ACTL_G0369

Fig. 198 Apply liquid gasket to the engine block mating surface of the oil pump and to the inside edge of the threaded bolt holes

15. Apply a light coat of new engine oil to the lip of the crankshaft oil seal, and apply new engine oil to the new O-ring.
16. Install the dowel pins, then align the inner rotor with the crankshaft, and install the oil pump.

➡Wait at least 30 minutes before filling the engine with oil. Do not run the engine for at least 3 hours after installing the oil pump.

17. Clean the excess grease off the crankshaft, and check the seal for distortion.
18. Install the oil strainer with a new O-ring.
19. Install the oil pan.
20. Install the rocker arm oil control solenoid/oil filter assembly with a new rocker arm oil control solenoid filter.
21. Install the timing belt drive pulley to the crankshaft:
22. Install the timing belt:
23. Remove the chain hoist.

INSPECTION

4-Cylinder Engine

See Figures 199 through 202.

1. Remove the pump housing.
2. Align the tip of the inner rotor tooth with the mark on the outer rotor, then check the inner-to-outer rotor radial clearance between the inner rotor and the outer rotor. If the inner-to-outer rotor radial clearance exceeds the service limit, replace the oil pump.

➡Inner Rotor-to-Outer Rotor Radial Clearance

- Standard (New): 0.002–0.006 inches (0.05–0.15 mm)
- Service Limit: 0.007 inches (0.19 mm)

3. Check the pump housing-to-rotor axial clearance between the rotor and the pump housing. If the pump housing-to-rotor axial clearance exceeds the service limit, replace the oil pump.

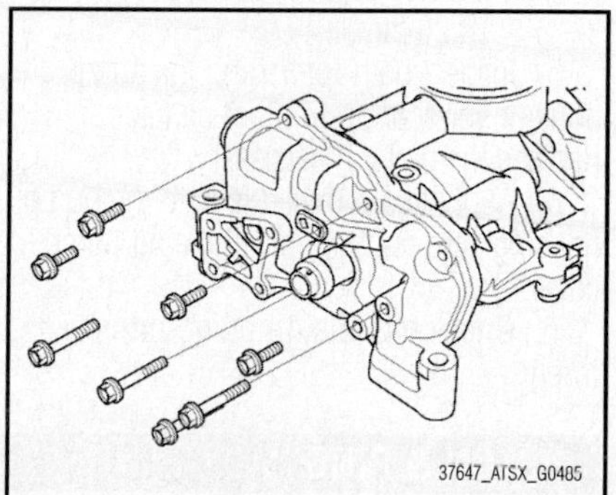
37647_ATSX_G0485

Fig. 199 Remove the pump housing

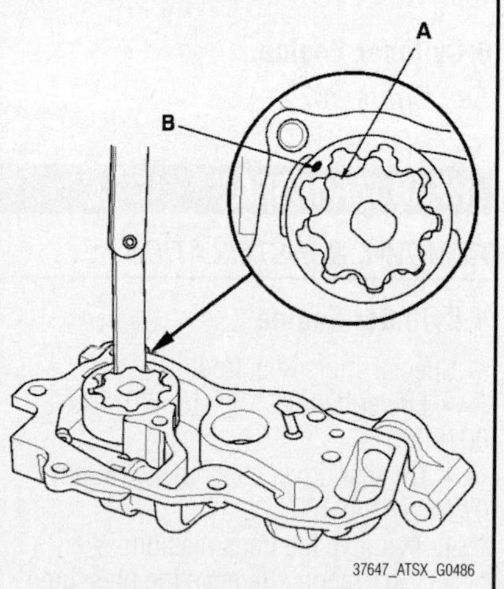
A
B
37647_ATSX_G0486

Fig. 200 Align the tip of the inner rotor tooth (A) with the mark (B) on the outer rotor

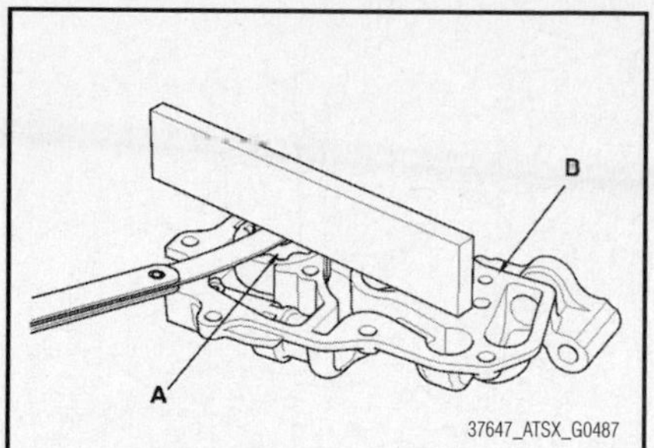
D
A
37647_ATSX_G0487

Fig. 201 Check the pump housing-to-rotor axial clearance between the rotor (A) and the pump housing (B)

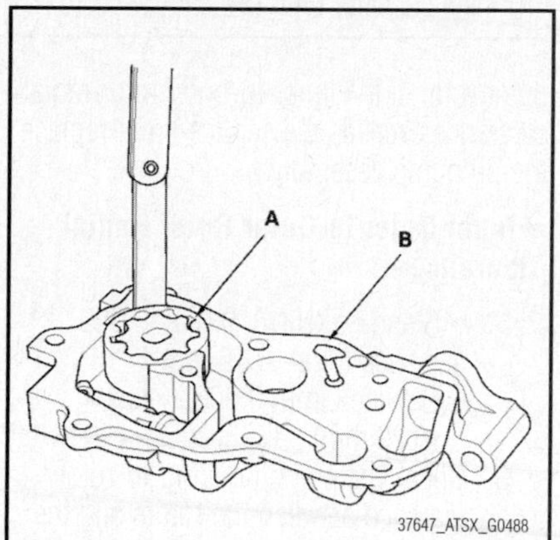
A
B
37647_ATSX_G0488

Fig. 202 Check the pump housing-to-outer rotor radial clearance between the outer rotor (A) and the pump housing (B)

➡Pump Housing-to-Rotor Axial Clearance

- Standard (New): 0.0014–0.0028 inches (0.035–0.070 mm)
- Service Limit: 0.005 inches (0.12 mm)

4. Check the pump housing-to-outer rotor radial clearance between the outer rotor and the pump housing. If the pump housing-to-outer rotor radial clearance exceeds the service limit, replace the oil pump.

➡Pump Housing-to-Outer Rotor Radial Clearance

- Standard (New): 0.0059–0.0083 inches (0.15–0.21 mm)
- Service Limit: 0.009 inches (0.23 mm)

6-Cylinder Engine

See Figures 203 through 205.

1. Remove the screws from the pump housing, then separate the pump housing and the cover.
2. Check the inner-to-outer rotor radial clearance between the inner rotor and the

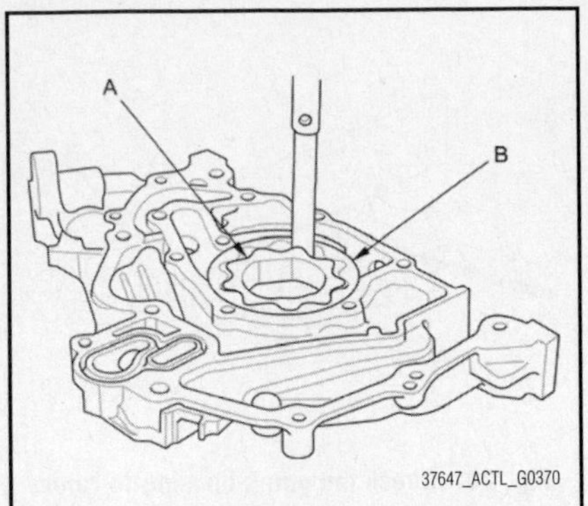
A
B
37647_ACTL_G0370

Fig. 203 Check the inner-to-outer rotor radial clearance between the inner rotor (A) and the outer rotor (B)

outer rotor. If the inner-to-outer rotor radial clearance exceeds the service limit, replace the oil pump assembly.

➡Inner Rotor-to-Outer Rotor Radial Clearance

- Standard (New): 0.002–0.006 inches (0.04–0.16 mm)
- Service Limit: 0.008 inches (0.20 mm)

3. Check the pump housing-to-rotor axial clearance between the rotors and the pump housing. If the pump housing-to-rotor axial clearance exceeds the service limit, replace the oil pump assembly.

➡Pump Housing-to-Rotor Axial Clearance

- Standard (New): 0.001–0.003 inches (0.02–0.07 mm)
- Service Limit: 0.005 inches (0.12 mm)

4. Check the pump housing-to-outer rotor radial clearance between the outer rotor and the pump housing. If the pump housing-to-outer rotor radial clearance

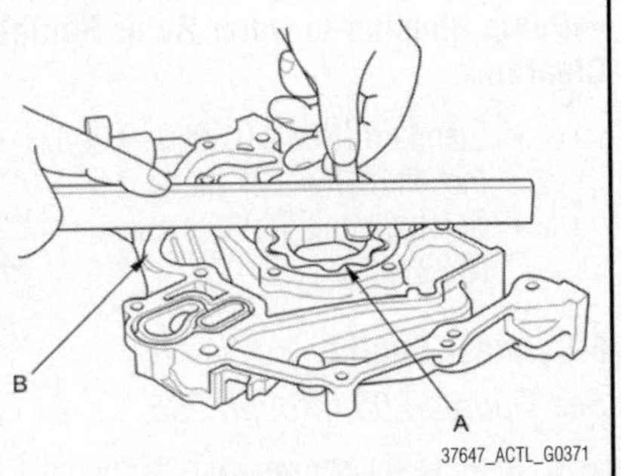
B
A
37647_ACTL_G0371

Fig. 204 Check the pump housing-to-rotor axial clearance between the rotors (A) and the pump housing (B)

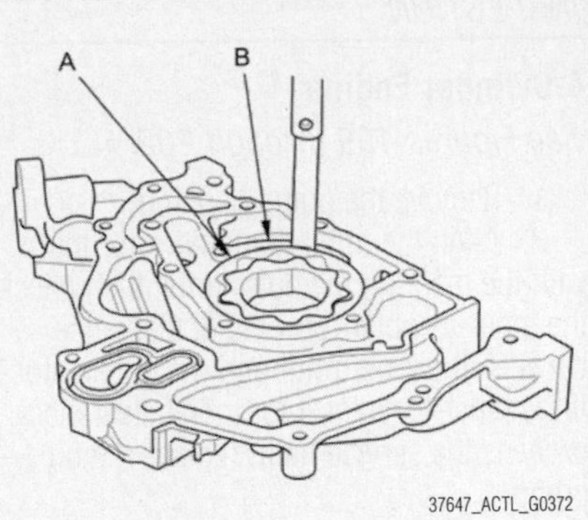
A
B
37647_ACTL_G0372

Fig. 205 Check the pump housing-to-outer rotor radial clearance between the outer rotor (A) and the pump housing (B)

exceeds the service limit, replace the oil pump assembly.

➡Pump Housing-to-Outer Rotor Radial Clearance

- Standard (New): 0.004–0.007 inches (0.10–0.19 mm)
- Service Limit: 0.008 inches (0.20 mm)

5. Inspect both rotors and the pump housing for scoring or other damage. Replace the parts, if necessary.
6. Apply liquid thread lock to the pump housing screws, then install the oil pump cover.
7. Check that the oil pump turns freely.

PISTON & RING

POSITIONING

4-Cylinder Engine

See Figure 206.

6-Cylinder Engine

See Figure 207.

REAR MAIN SEAL

REMOVAL & INSTALLATION

4-Cylinder Engine

Special Tools Required:

- Driver handle, 15 x 135L 07749-0010000
- Oil seal driver attachment 96 mm 07ZAD-PNAA100

1. Remove the transmission:
2. M/T model: Remove the pressure plate, the clutch disk, and the flywheel.
3. A/T model: Remove the drive plate.
4. Clean and dry the crankshaft oil seal housing.

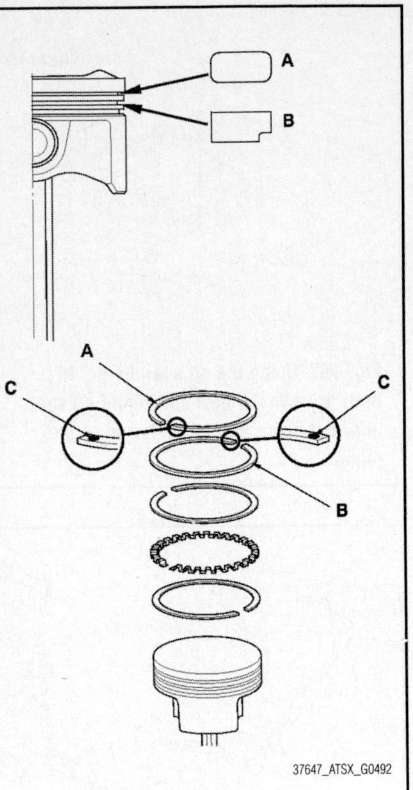
A
B
A
C
C
B
37647_ATSX_G0492

Fig. 206 Piston ring positioning—4 cylinder engine

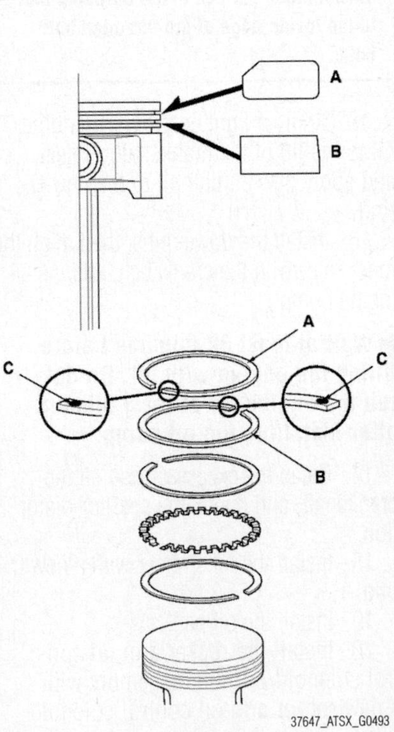
A
B
A
C
C
B
37647_ATSX_G0493

Fig. 207 Piston ring positioning—6 cylinder engine

5. Use the driver handle, 15 x 135L and the oil seal driver attachment 90 mm to drive a new crankshaft oil seal squarely into the engine block to the specified installed height.

6. M/T model: Install the flywheel, the clutch disk, and the pressure plate.

7. A/T model: Install the drive plate.

8. Install the transmission:

6-Cylinder Engine

Special Tools Required:

• Driver handle, 15 x 135L 07749-0010000

• Oil seal driver attachment 106 mm 070AD-RCA0200

1. M/T model: Remove the transmission, and the flywheel.

2. A/T model: Remove the transmission, and the drive plate.

3. Remove the transmission end crankshaft oil seal.

4. Clean and dry the crankshaft oil seal housing.

5. Apply a light coat of new engine oil to the lip of the crankshaft oil seal.

6. Using the driver handle, 15 x 135 L and the oil seal driver attachment, 106 mm, drive in the new crankshaft oil seal until the oil seal driver attachment bottoms against the engine block end cover. Align the hole in the oil seal driver attachment with the pin on the crankshaft.

7. Clean any excess grease off the crankshaft, and check that the oil seal lip is not distorted.

8. M/T model: Install the flywheel, and the transmission.

9. A/T model: Install the drive plate, and the transmission.

ROCKER ARMS

REMOVAL & INSTALLATION

4-Cylinder Engine

See Figures 208 through 212.

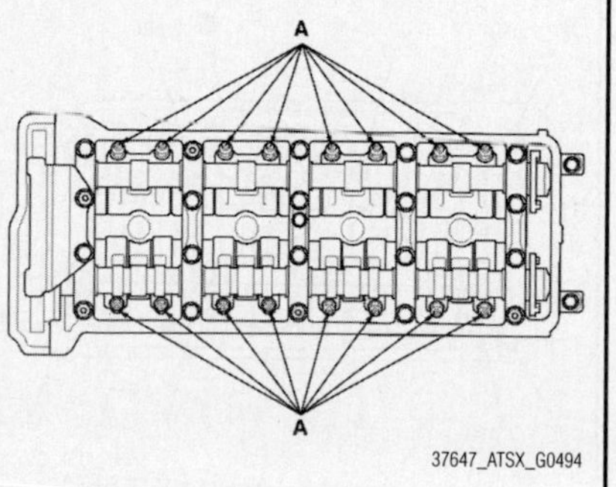
A
A
37647_ATSX_G0494

Fig. 208 Loosen the rocker arm adjusting screws (A)

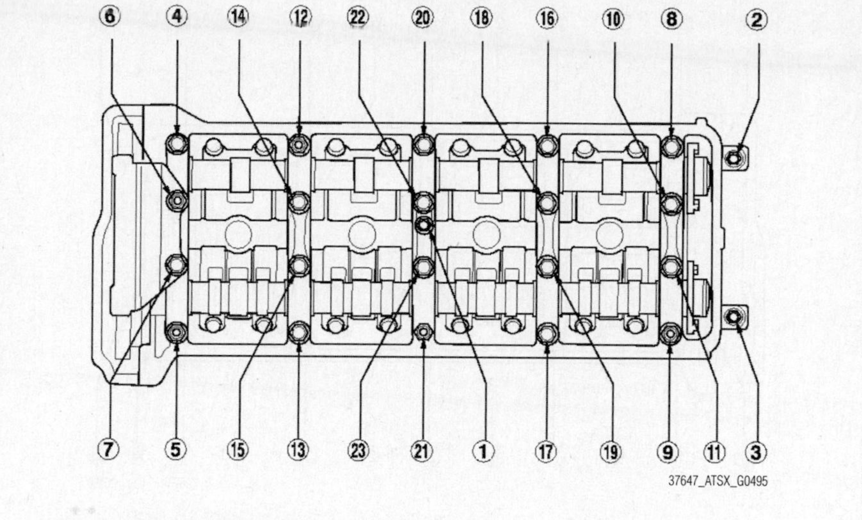
6 4 14 12 22 20 18 16 10 8 2
7 5 15 13 23 21 1 17 19 9 11 3
37647_ATSX_G0495

Fig. 209 Remove the camshaft holder bolts in sequence

1. Remove the cam chain.

2. Loosen the rocker arm adjusting screws.

3. Remove the camshaft holder bolts. To prevent damaging the camshafts, loosen the bolts, in sequence, two turns at a time.

➡Bolt #1 is not on all engines.

4. Remove cam chain guide B, the camshaft holders, and the camshafts.

5. Insert the bolts into the rocker shaft holder, then remove the rocker arm assembly.

To install:

6. Reassemble the rocker arm assembly.

7. Clean and dry the No. 5 rocker shaft holder mating surface.

8. Apply liquid gasket (P/N 08717-0004, 08718-0003, or 08718-0009) to the cylinder head mating surface of the No. 5 rocker shaft holder and to the inside edge of the threaded bolt holes. Install the component within 5 minutes of applying the liquid gasket.

➡ If too much time has passed after applying the liquid gasket, remove the old liquid gasket and residue, then reapply new liquid gasket.

9. Insert the bolts into the rocker shaft holder, then install the rocker arm assembly on the cylinder head.

10. Remove the bolts from the rocker shaft holder.

11. Make sure the punch marks on the Variable Valve Timing Control (VTC) actuator and the exhaust camshaft sprocket are facing up, then set the camshafts in the holder.

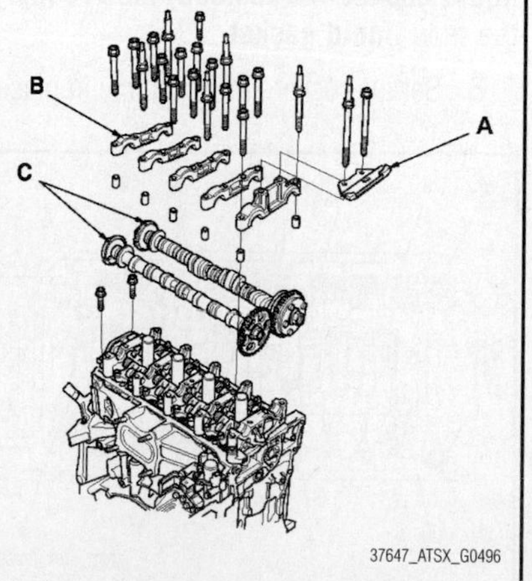
B
A
C
37647_ATSX_G0496

Fig. 210 Remove cam chain guide B (A), the camshaft holders (B), and the camshafts (C)

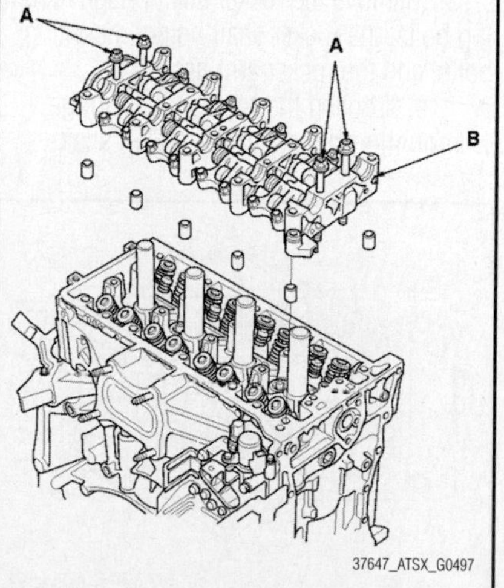
A
A
B
37647_ATSX_G0497

Fig. 211 Insert the bolts (A) into the rocker shaft holder, then remove the rocker arm assembly (B)

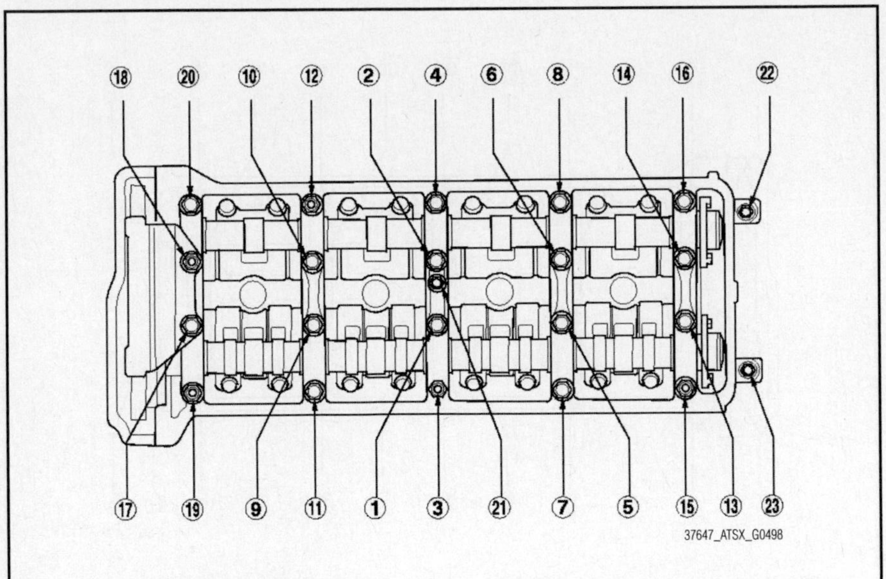
18
20
10
12
2
4
6
8
14
16
22
17
19
9
11
1
3
21
7
5
15
13
23
37647_ATSX_G0498

Fig. 212 Camshaft holder bolt sequence

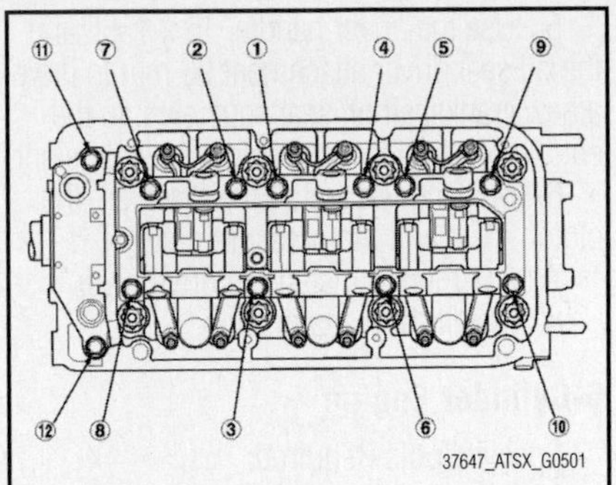
11
7
2
1
4
5
9
12
8
3
6
10
37647_ATSX_G0501

Fig. 215 Tighten each rocker shaft bridge bolt two turns at a time in sequence

12. Set the camshaft holders and cam chain guide B in place

13. Tighten the camshaft holder bolts to 16 ft. lbs. (22 Nm) except for bolts 21, 22, and 23. Tighten these bolts to 9 ft. lbs. (12 Nm).

➡**If the engine does not have bolt 21, skip it and continue the torque sequence.**

14. Install the cam chain, then adjust the valve clearance.

6-Cylinder Engine

Front Cylinder Head

See Figures 213 through 215.

1. Remove the cylinder head cover.
2. Loosen the locknuts and the adjusting screws
3. Remove the rocker shaft bridge mounting bolts, the rocker shaft holder mounting bolts, and the rocker arm assembly.
    a. Loosen the rocker shaft bridge mounting bolts and the rocker shaft holder mounting bolts in sequence two turns at a time, to prevent damaging the valves or the rocker arm assembly.
    b. When removing the rocker arm assembly, do not remove the rocker shaft bridge mounting bolts and the rocker shaft holder mounting bolts. The bolts will keep the rocker arms on the shafts.

To install:

4. Apply liquid gasket, P/N 08717-0004, 08718-0003, or 08718-0009 to the rocker shaft holder mating surface of the cylinder head. Install the component within 5 minutes of applying the liquid gasket.

➡**Apply a bead of liquid gasket along the mating surfaces. If you apply liquid gasket P/N 08718-0012, the component must be installed within 4 minutes. If too much time has passed after applying the liquid gasket, remove the old liquid gasket and residue, then reapply the new liquid gasket.**

5. Set the rocker arm assembly in place, and loosely install the bolts. Make sure that the rocker arms are properly positioned on the valve stems.

➡**Wait at least 30 minutes before filling the engine with oil. Do not run the engine for at least 3 hours after installing the rocker arm assembly.**

6. Tighten each bolt two turns at a time in the sequence shown to ensure that the rockers do not bind on the valves.

Rear Cylinder Head

See Figures 216 through 218.

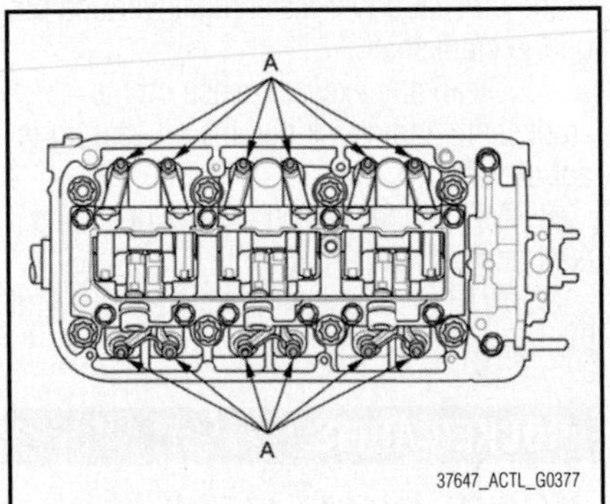
A
A
37647_ACTL_G0377

Fig. 216 Loosen the locknuts and the adjusting screws (A)

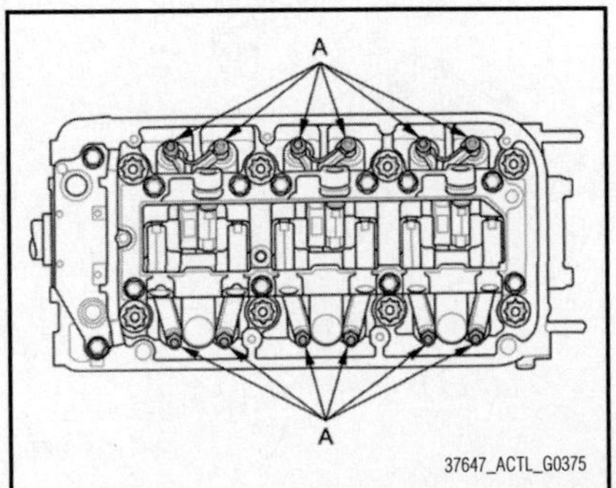
A
A
37647_ACTL_G0375

Fig. 213 Loosen the locknuts and the adjusting screws (A)

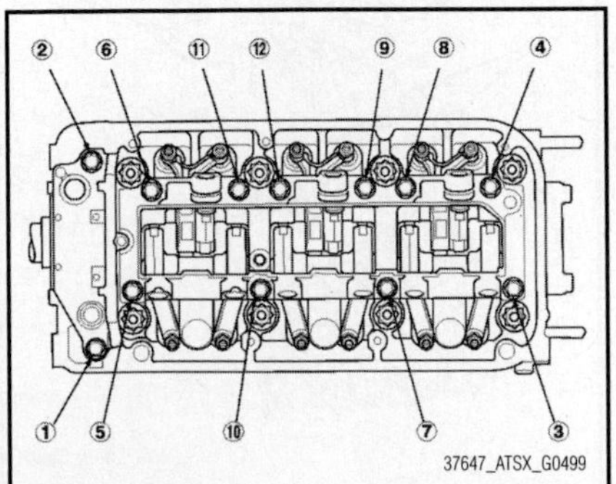
2
6
11
12
9
8
4
1
5
10
7
3
37647_ATSX_G0499

Fig. 214 Front rocker shaft bridge bolt loosening sequence

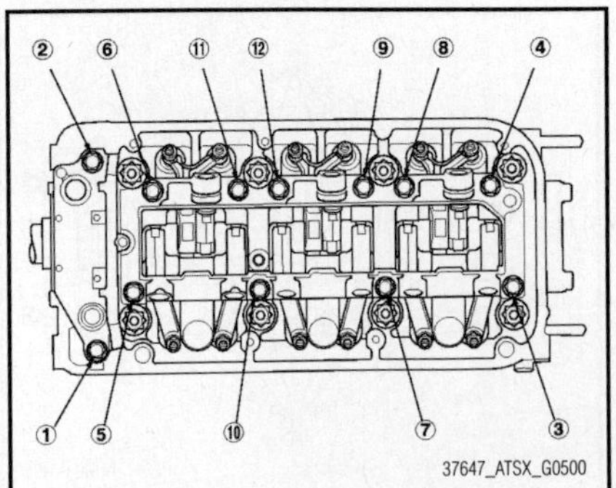
2
6
11
12
9
8
4
1
5
10
7
3
37647_ATSX_G0500

Fig. 217 Rear cylinder head bolt loosening sequence

1. Remove the cylinder head cover.
2. Loosen the locknuts and the adjusting screws.
3. Remove the rocker shaft bridge mounting bolts, the rocker shaft holder mounting bolts, and the rocker arm assembly.
   a. Loosen the rocker shaft bridge mounting bolts and the rocker shaft holder mounting bolts in sequence two turns at a time, to prevent damaging the valves or the rocker arm assembly.
   b. When removing the rocker arm assembly, do not remove the rocker shaft bridge mounting bolts and the rocker shaft holder mounting bolts. The bolts will keep the rocker arms on the shafts.

To install:

4. Apply liquid gasket, P/N 08717-0004, 08718-0003, or 08718-0009 to the rocker shaft holder mating surface of the cylinder head. Install the component within 5 minutes of applying the liquid gasket.

➡Apply a bead of liquid gasket along the mating surfaces. If you apply liquid gasket P/N 08718-0012, the component must be installed within 4 minutes. If too much time has passed after applying the liquid gasket, remove the old liquid gasket and residue, then reapply the new liquid gasket.

5. Set the rocker arm assembly in place, and loosely install the bolts. Make sure that the rocker arms are properly positioned on the valve stems.

➡Wait at least 30 minutes before filling the engine with oil. Do not run the engine for at least 3 hours after installing the rocker arm assembly.

6. Tighten each bolt two turns at a time in the sequence shown to ensure that the rockers do not bind on the valves.

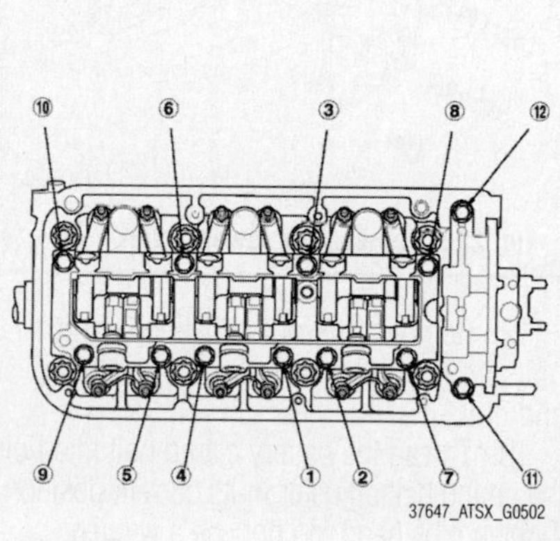
⑩
⑥
③
⑧
⑫
⑨
⑤
④
①
②
⑦
⑪
37647_ATSX_G0502

Fig. 218 Tighten each rocker shaft bridge bolt two turns at a time in sequence

TIMING BELT FRONT COVER

REMOVAL & INSTALLATION

6-Cylinder Engine

See Figures 219 through 223.

1. Remove the engine compartment covers.
2. Turn the crankshaft so its white mark lines up with the pointer.

➡The other pointer is not used.

3. Check that the No. 1 piston Top Dead Center (TDC) mark on the front camshaft pulley and the pointer on the front upper cover are aligned.

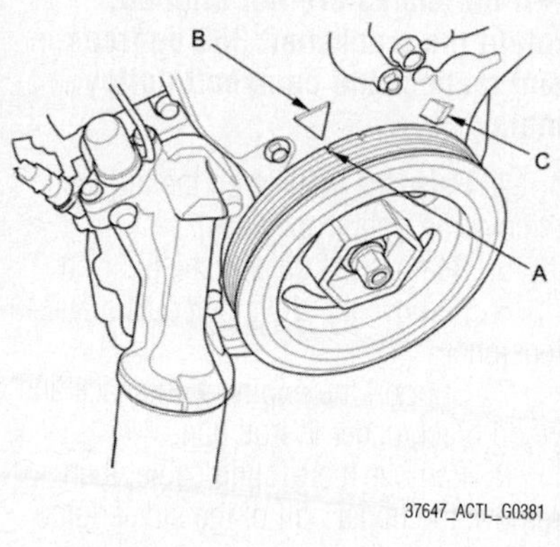
B
C
A
37647_ACTL_G0381

Fig. 219 Turn the crankshaft so its white mark (A) lines up with the pointer (B)

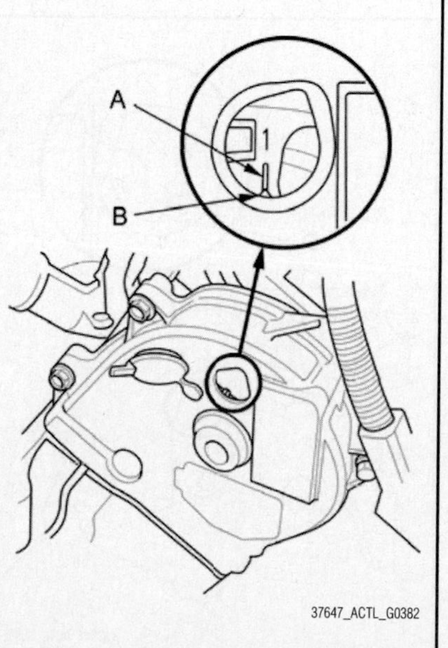
A
1
B
37647_ACTL_G0382

Fig. 220 Check that the No. 1 piston TDC mark (A) on the front camshaft pulley and the pointer (B) on the front upper cover are aligned

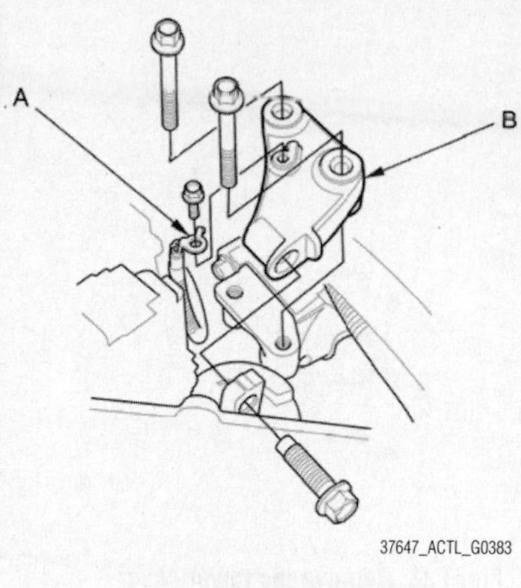
A
B
37647_ACTL_G0383

Fig. 221 Remove the ground cable (A), then remove the upper half of the side engine mount bracket (B)

➡If the marks are not aligned, rotate the crankshaft 360 degrees, and recheck the camshaft pulley mark.

4. Raise the vehicle on the lift, then remove the right front wheel.
5. Remove the splash shield.
6. Remove the drive belt auto-tensioner.
7. Support the engine with a jack and a wood block under the oil pan.
8. Remove the ground cable, then remove the upper half of the side engine mount bracket.
9. Remove the crankshaft pulley.
10. Remove the front upper cover and the rear upper cover.
11. Remove the lower cover.

To install:

12. Install the lower cover.
13. Install the front upper cover and the rear upper cover.
14. Install the crankshaft pulley.

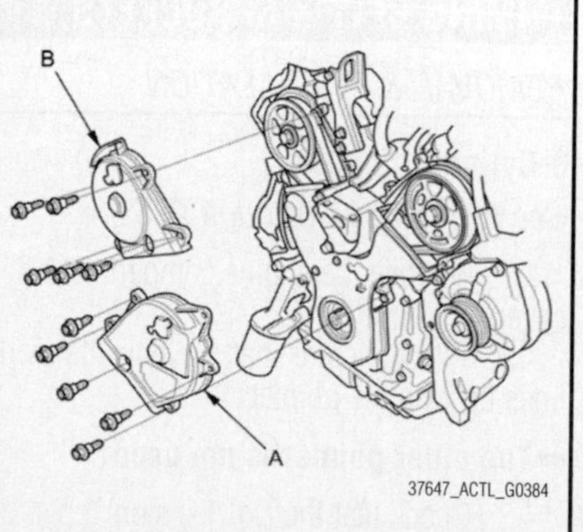
B
A
37647_ACTL_G0384

Fig. 222 Remove the front upper cover (A) and the rear upper cover (B)

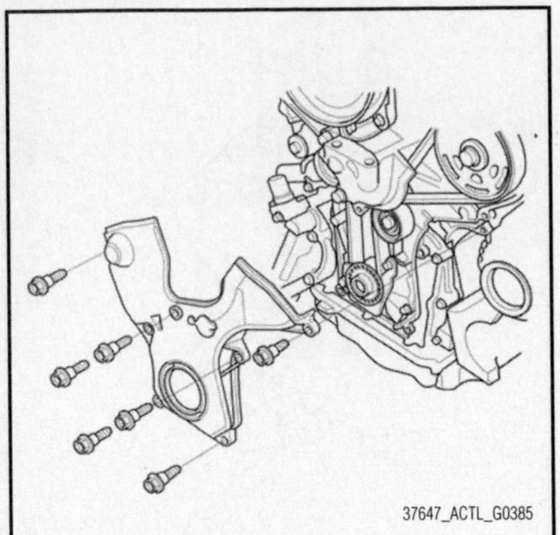
37647_ACTL_G0385

Fig. 223 Remove the lower cover

15. Rotate the crankshaft pulley about six turns clockwise so the timing belt positions itself on the pulleys.

16. Turn the crankshaft pulley so its white mark lines up with the pointer.

➡The other pointer is not used.

17. Check the camshaft pulley marks.

➡If the marks are not aligned, rotate the crankshaft 360 degrees, and recheck the camshaft pulley mark. If the camshaft pulley marks are at TDC, go to step 17. If the camshaft pulley marks are not at TDC, remove the timing belt and repeat steps 2 through 16.

18. Install the upper half of the side engine mount bracket, then tighten the mounting bolts.

19. Install the ground cable.

20. Install the drive belt auto-tensioner.

21. Install the splash shield.

22. Install the right front wheel.

23. Install the engine compartment covers.

TIMING BELT & SPROCKETS

REMOVAL & INSTALLATION

6-Cylinder Engine

See Figures 224 through 236.

1. Remove the engine compartment covers.

2. Turn the crankshaft so its white mark lines up with the pointer.

➡The other pointer is not used.

3. Check that the No. 1 piston Top Dead Center (TDC) mark on the front camshaft pulley and the pointer on the front upper cover are aligned.

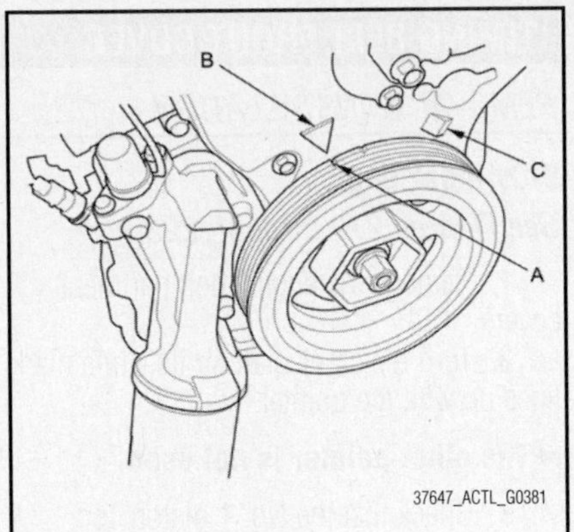
B
C
A
37647_ACTL_G0381

Fig. 224 Turn the crankshaft so its white mark (A) lines up with the pointer (B)

➡If the marks are not aligned, rotate the crankshaft 360 degrees, and recheck the camshaft pulley mark.

4. Raise the vehicle on the lift, then remove the right front wheel.

5. Remove the splash shield.

6. Remove the drive belt auto-tensioner.

7. Support the engine with a jack and a wood block under the oil pan.

8. Remove the ground cable, then remove the upper half of the side engine mount bracket.

9. Remove the crankshaft pulley.

10. Remove the front upper cover and the rear upper cover.

11. Remove the lower cover.

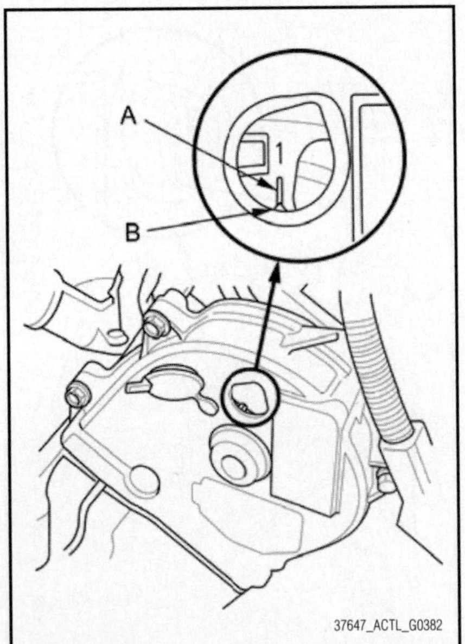
A
1
B
37647_ACTL_G0382

Fig. 225 Check that the No. 1 piston TDC mark (A) on the front camshaft pulley and the pointer (B) on the front upper cover are aligned

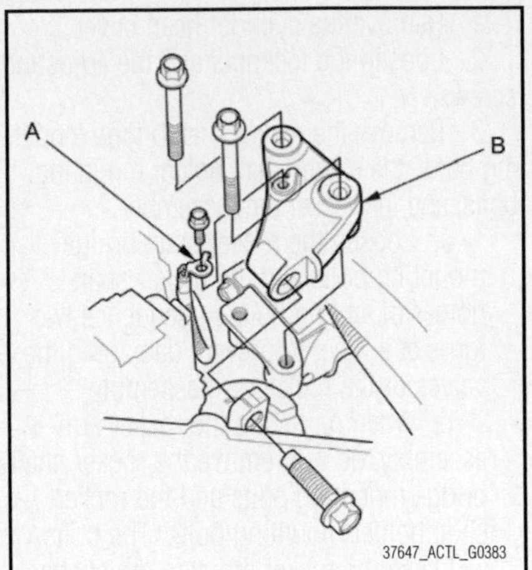
A
B
37647_ACTL_G0383

Fig. 226 Remove the ground cable (A), then remove the upper half of the side engine mount bracket (B)

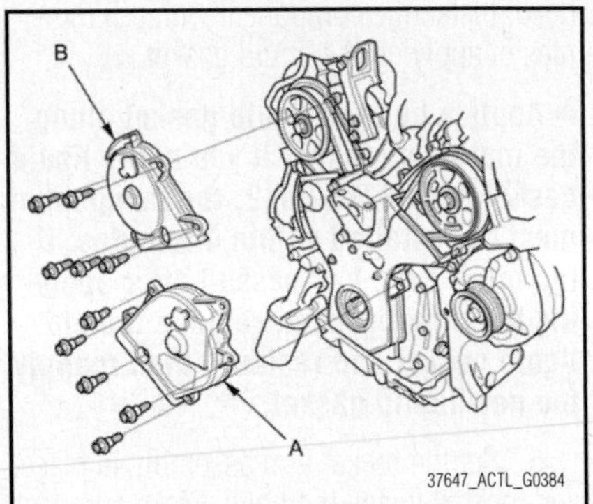
B
A
37647_ACTL_G0384

Fig. 227 Remove the front upper cover (A) and the rear upper cover (B)

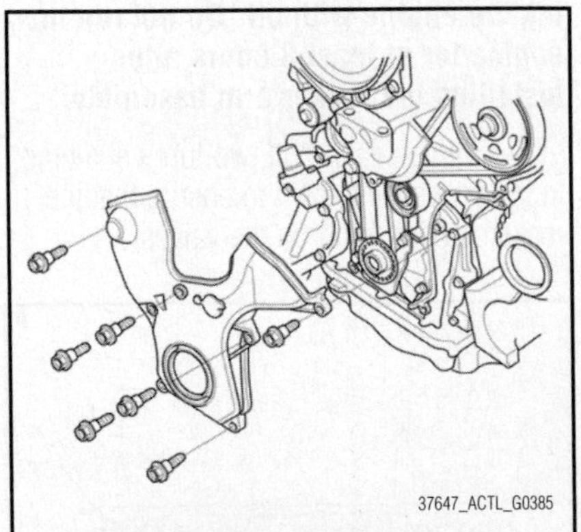
37647_ACTL_G0385

Fig. 228 Remove the lower cover

12. Remove one of the battery clamp bolts from the battery tray, and grind the end of it.

13. Thread the battery clamp bolt into hold the timing belt adjuster in its current position. Tighten it by hand, do not use a wrench.

14. Remove the timing belt guide plate.

15. Remove the lower half of the side engine mount bracket.

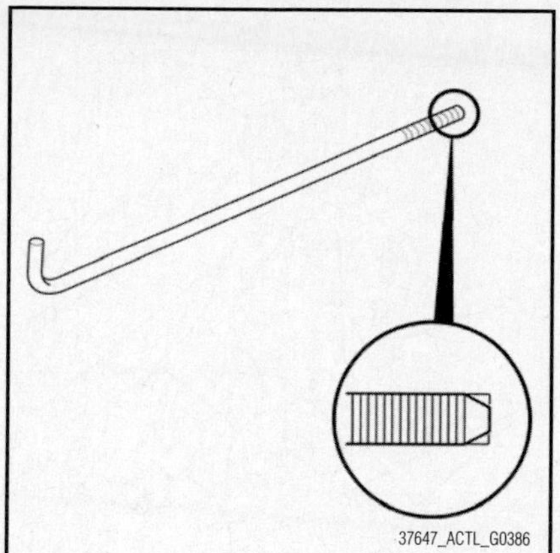
37647_ACTL_G0386

Fig. 229 Remove one of the battery clamp bolts from the battery tray, and grind the end of it

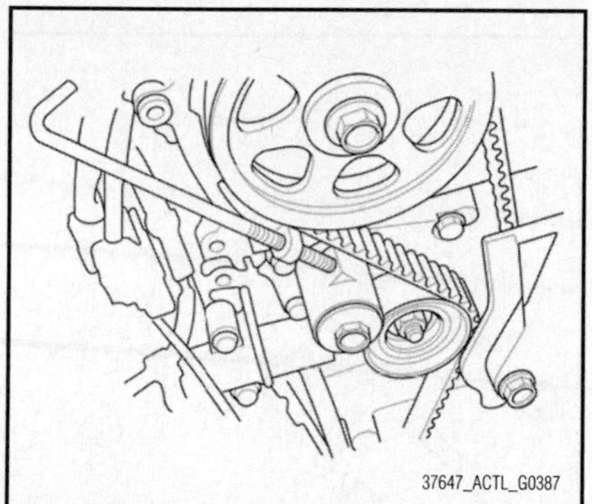
37647_ACTL_G0387

Fig. 230 Thread the battery clamp bolt into hold the timing belt adjuster in its current position

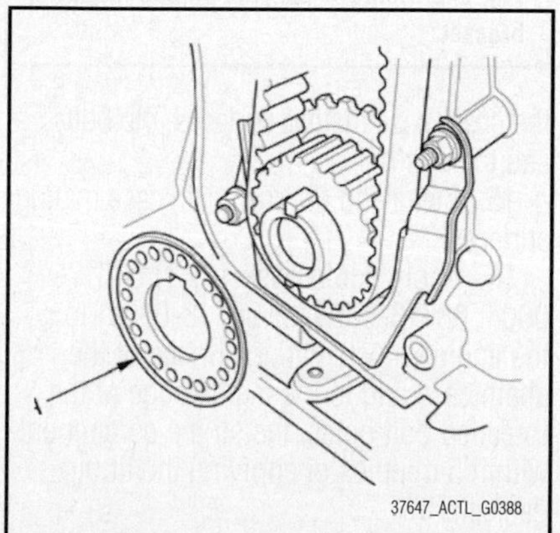
37647_ACTL_G0388

Fig. 231 Remove the timing belt guide plate (A)

16. Remove the idler pulley bolt and the idler pulley, then remove the timing belt. Discard the idler pulley bolt.

To install:

17. Clean the timing belt pulleys, the timing belt guide plate, and the upper and lower covers.

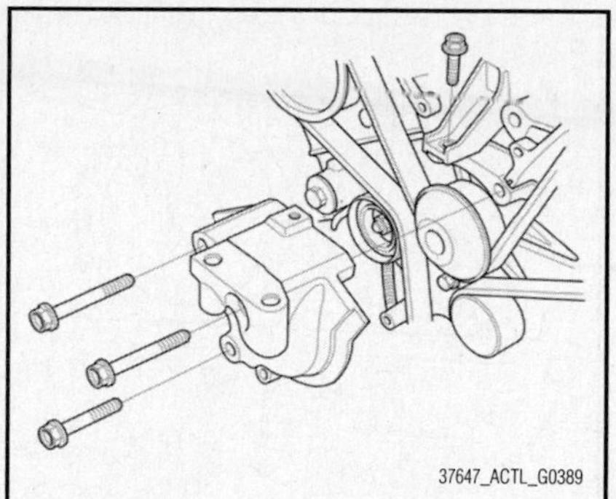
37647_ACTL_G0389

Fig. 232 Remove the lower half of the side engine mount bracket

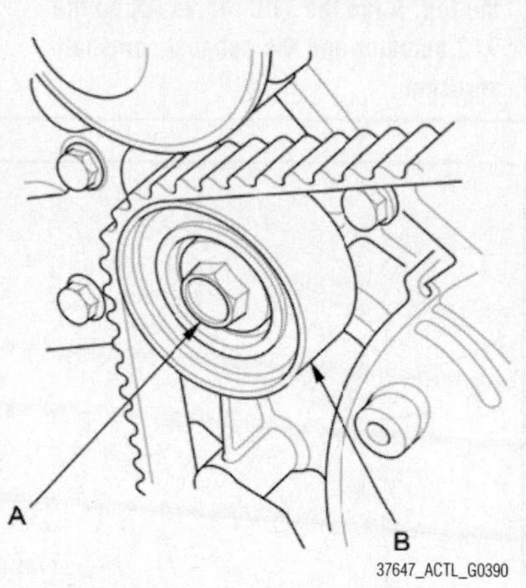
A
B
37647_ACTL_G0390

Fig. 233 Remove the idler pulley bolt (A) and the idler pulley (B), then remove the timing belt

18. Set the timing belt drive pulley to Top Dead Center (TDC) by aligning the TDC mark on the tooth of the timing belt drive pulley with the pointer on the oil pump.

19. Set the camshaft pulleys to TDC by aligning the TDC marks on the camshaft pulleys with the pointers on the back covers.

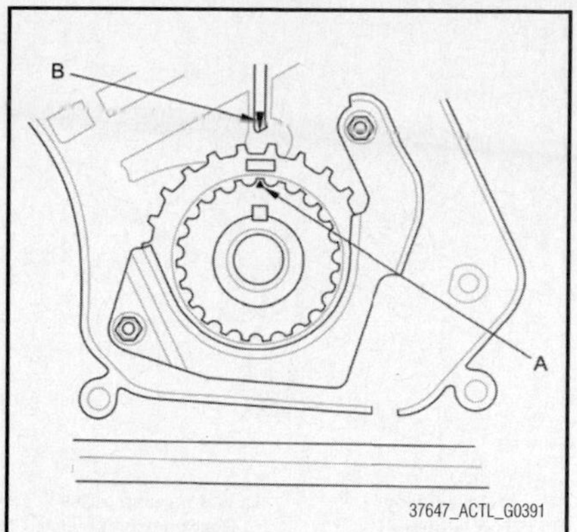
B
A
37647_ACTL_G0391

Fig. 234 Set the timing belt drive pulley to TDC by aligning the TDC mark (A) on the tooth of the timing belt drive pulley with the pointer (B) on the oil pump

20. Loosely install the idler pulley with a new idler pulley bolt so the pulley can move but does not come off.

21. If the auto-tensioner has extended and the timing belt cannot be installed, do the timing belt replacement procedure.

22. Install the timing belt in a counterclockwise sequence starting with the drive pulley. Take care not to damage the timing belt during installation.

23. Tighten the idler pulley bolt to 33 ft. lbs. (44 Nm).

24. Remove the battery clamp bolt from the back cover.

25. Install the lower half of the side engine mount bracket. Tighten the three long bolts to 33 ft. lbs. (44 Nm).

26. Install the timing belt guide plate.

27. Install the lower cover.

28. Install the front upper cover and the rear upper cover.

29. Install the crankshaft pulley.

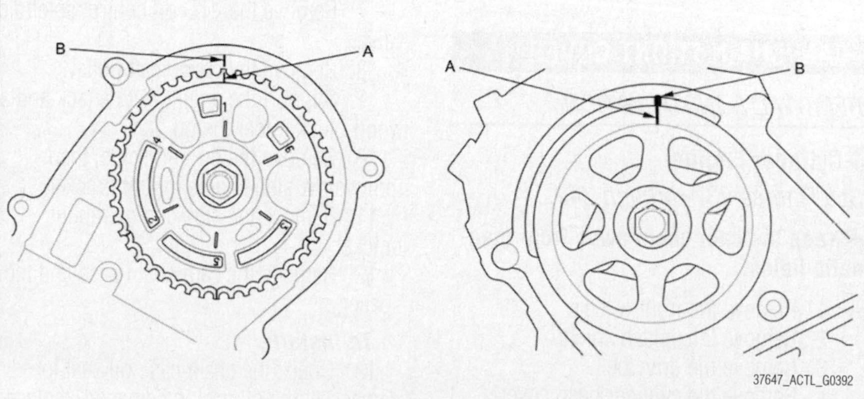
B
A
A
B
37647_ACTL_G0392

Fig. 235 Set the camshaft pulleys to TDC by aligning the TDC marks (A) on the camshaft pulleys with the pointers (B) on the back covers

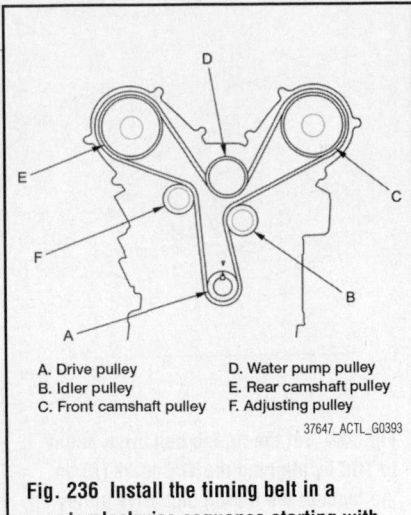

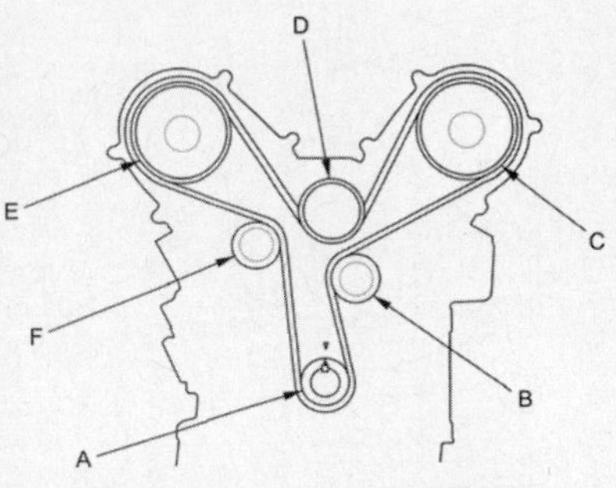
D
E
C
F
B
A
A. Drive pulley
B. Idler pulley
C. Front camshaft pulley
D. Water pump pulley
E. Rear camshaft pulley
F. Adjusting pulley
37647_ACTL_G0393

Fig. 236 Install the timing belt in a counterclockwise sequence starting with the drive pulley

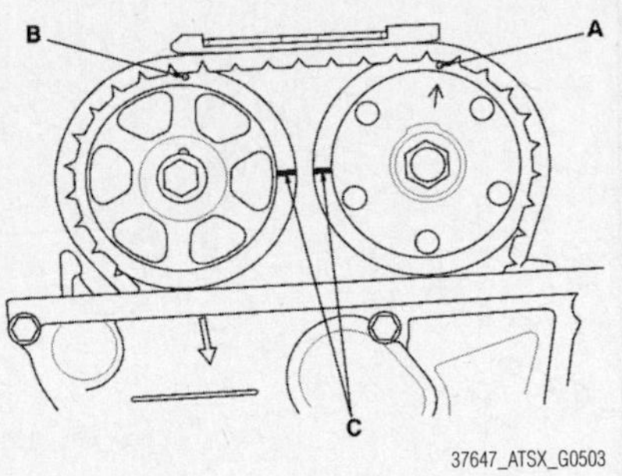
B
A
C
37647_ATSX_G0503

Fig. 237 The punch mark (A) on the VTC actuator and the punch mark (B) on the exhaust camshaft sprocket should be at the top. Align the TDC marks (C) on the VTC actuator and the exhaust camshaft sprocket

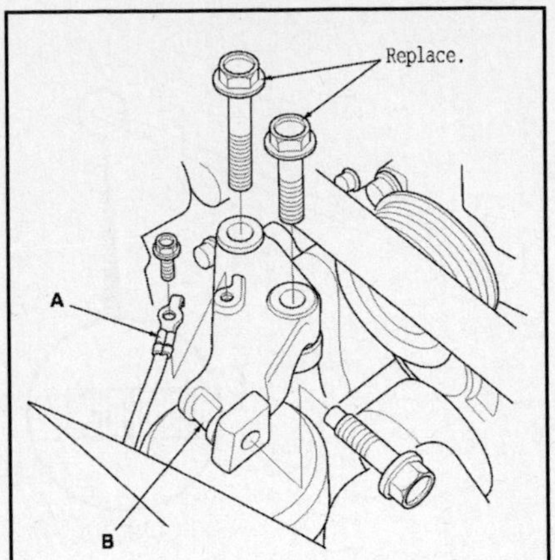
Replace.
A
B
37647_ATSX_G0419

Fig. 239 Remove the ground cable (A), then remove the side engine mount bracket (B)

30. Rotate the crankshaft pulley about six turns clockwise so the timing belt positions itself on the pulleys.
31. Turn the crankshaft pulley so its white mark lines up with the pointer.

➡The other pointer is not used.

32. Check the camshaft pulley marks.

➡If the marks are not aligned, rotate the crankshaft 360 degrees, and recheck the camshaft pulley mark. If the camshaft pulley marks are at TDC, go to step 17. If the camshaft pulley marks are not at TDC, remove the timing belt and repeat steps 2 through 16.

33. Install the upper half of the side engine mount bracket, then tighten the mounting bolts.
34. Install the ground cable.
35. Install the drive belt auto-tensioner.
36. Install the splash shield.
37. Install the right front wheel.
38. Install the engine compartment covers.

CAM CHAIN FRONT COVER

REMOVAL & INSTALLATION

4-Cylinder Engine

See Figures 237 through 242.

➡Keep the cam chain away from magnetic fields.

1. Remove the front wheels.
2. Remove the splash shield.
3. Remove the drive belt.
4. Remove the cylinder head cover.
5. Set the No. 1 piston at Top Dead Center (TDC). The punch mark on the Variable Valve Timing Control (VTC) actuator and the punch mark on the exhaust camshaft sprocket should be at the top. Align the TDC marks on the VTC actuator and the exhaust camshaft sprocket.

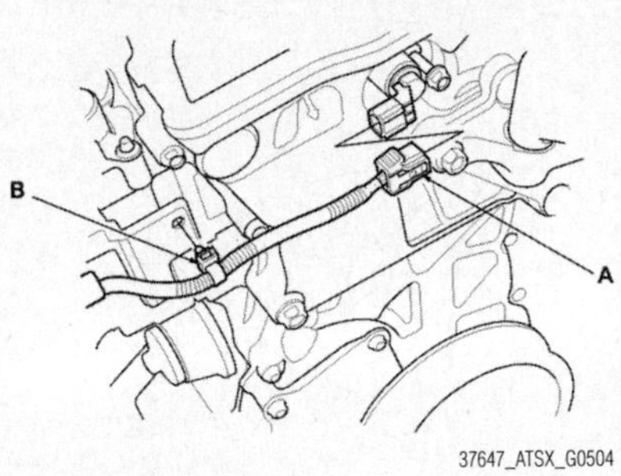
B
A
37647_ATSX_G0504

Fig. 238 Disconnect the VTC oil control solenoid valve connector (A) and remove the harness clamp (B)

6. Disconnect the VTC oil control solenoid valve connector and remove the harness clamp.
7. Remove the VTC oil control solenoid valve.
8. Remove the crankshaft pulley.
9. Support the engine with a jack and a wood block under the oil pan.
10. Remove the ground cable, then remove the side engine mount bracket.
11. Remove the side engine mount bracket.
12. Remove the cam chain case and the spacer.

To install:

13. Check the chain case oil seal for damage. If the oil seal is damaged, replace the chain case oil seal.
14. Remove the old liquid gasket from the chain case mating surfaces, the bolts, and the bolt holes.

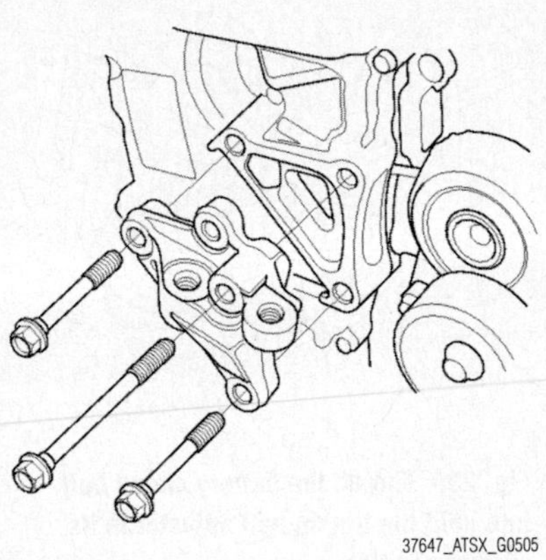
37647_ATSX_G0505

Fig. 240 Remove the side engine mount bracket

15. Clean and dry the chain case mating surfaces.
16. Apply liquid gasket (P/N 08717-0004, 08718-0003, or 08718-0009) to the engine block mating surface of the chain case and to the inside edge of the threaded bolt holes. Install the component within 5 minutes of applying the liquid gasket.

➡If too much time has passed after applying the liquid gasket, remove the old liquid gasket and residue, then reapply new liquid gasket.

17. Apply liquid gasket to the engine block upper surface contact areas and the lower block upper surface contact areas on the chain case.

➡Apply about 0.43 inches (11 mm) diameter and about 0.12 inches (3 mm)

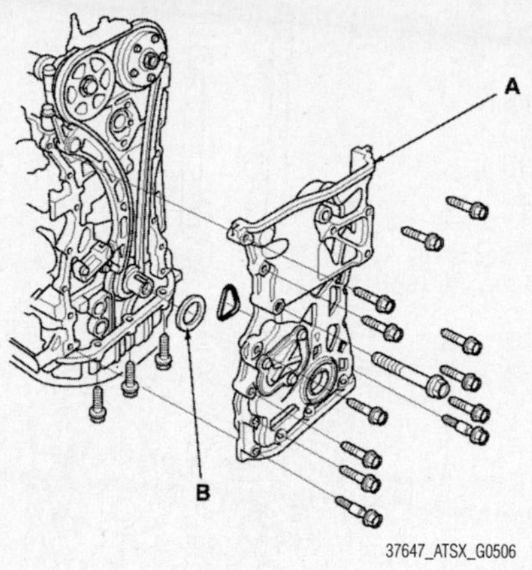
A
B
37647_ATSX_G0506

Fig. 241 Remove the cam chain case (A) and the spacer (B)

thickness of liquid gasket to the areas B and C.

18. Apply liquid gasket (P/N 08717-0004, 08718-0003, or 08718-0009) to the oil pan mating surface of the oil pump. Install the component within 5 minutes of applying the liquid gasket.

19. Install the spacer, then install the new O-ring on the chain case. Set the edge of the chain case to the edge of the oil pan, then install the chain case on the engine block. Wipe off the excess liquid gasket on the oil pan and chain case mating area.

➡When installing the chain case, do not slide the bottom surface onto the oil pan mounting surface. Wait at least 30 minutes before filling the engine with oil. Do not run the engine for at least 3 hours after installing the chain case.

20. Install the side engine mount bracket, then tighten the side engine mount bracket mounting bolts to 33 ft. lbs. (44 Nm).
21. Install the side engine mount bracket, and tighten the top two new bolts to 40 ft. lbs. (54 Nm), then tighten the lower bolt to 47 ft. lbs. (64 Nm).
22. Install the ground cable.
23. Remove the jack and the wood block.
24. Install the crankshaft pulley.
25. Install the VTC oil control solenoid valve.
26. Connect the VTC oil control solenoid valve connector and install the harness clamp.
27. Install the cylinder head cover.
28. Install the drive belt.
29. Install the splash shield.
30. Install the front wheels.
31. Do the crankshaft position (CKP) pattern clear/CKP pattern learn procedure.

TIMING (CAM) CHAIN & SPROCKETS

REMOVAL & INSTALLATION

4-Cylinder Engine

See Figures 243 through 258.

➡Keep the cam chain away from magnetic fields.

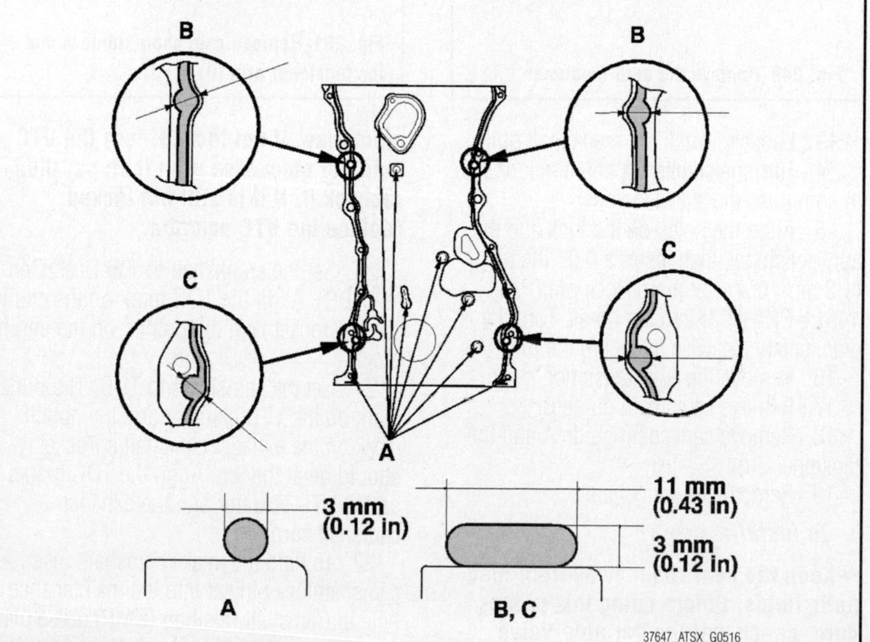
B
B
C
C
A
3 mm
(0.12 in)
11 mm
(0.43 in)
3 mm
(0.12 in)
A
B, C
37647_ATSX_G0516

Fig. 242 Apply about 0.43 inches (11 mm) diameter and about 0.12 inches (3 mm) thickness of liquid gasket to the areas (B) and (C)

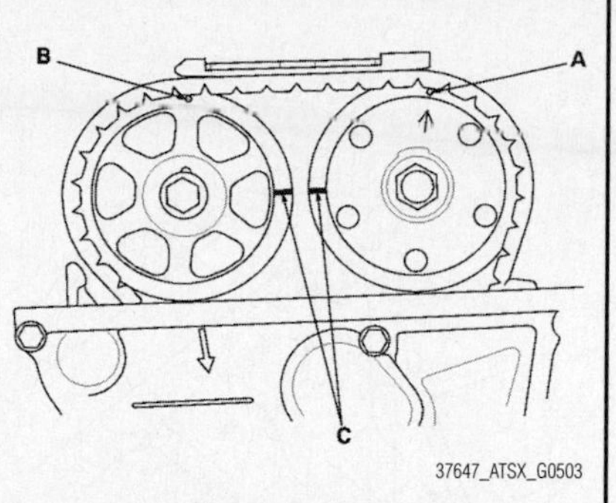
B
A
C
37647_ATSX_G0503

Fig. 243 The punch mark (A) on the VTC actuator and the punch mark (B) on the exhaust camshaft sprocket should be at the top. Align the TDC marks (C) on the VTC actuator and the exhaust camshaft sprocket

1. Remove the front wheels.
2. Remove the splash shield.
3. Remove the drive belt.
4. Remove the cylinder head cover.
5. Set the No. 1 piston at Top Dead Center (TDC). The punch mark on the Variable Valve Timing Control (VTC) actuator and the punch mark on the exhaust camshaft sprocket should be at the top. Align the TDC marks on the VTC actuator and the exhaust camshaft sprocket.
6. Disconnect the VTC oil control solenoid valve connector and remove the harness clamp.
7. Remove the VTC oil control solenoid valve.
8. Remove the crankshaft pulley.
9. Support the engine with a jack and a wood block under the oil pan.
10. Remove the ground cable, then remove the side engine mount bracket.
11. Remove the side engine mount bracket.
12. Remove the cam chain case and the spacer.

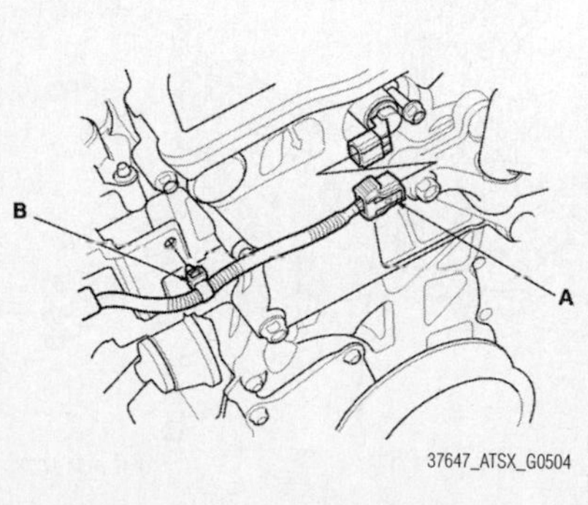
B
A
37647_ATSX_G0504

Fig. 244 Disconnect the VTC oil control solenoid valve connector (A) and remove the harness clamp (B)

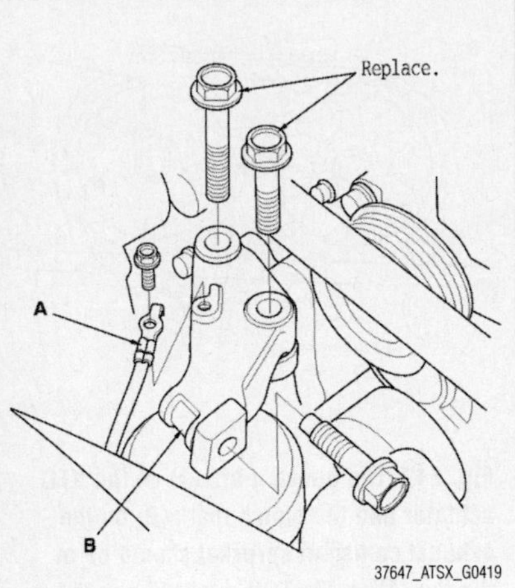
Replace.
A
B
37647_ATSX_G0419

Fig. 245 Remove the ground cable (A), then remove the side engine mount bracket (B)

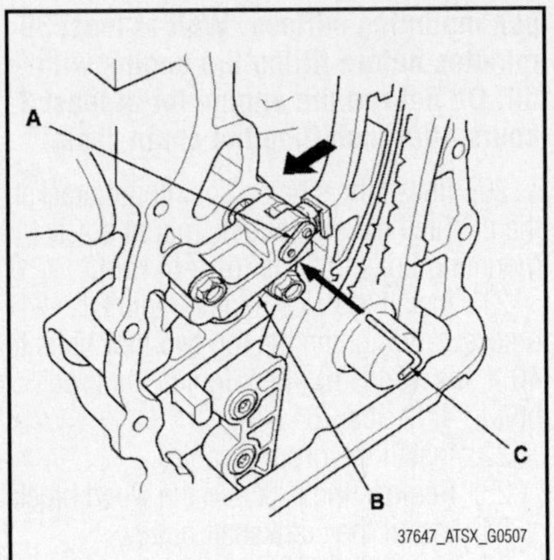
A
B
C
37647_ATSX_G0507

Fig. 248 Align the holes on the lock (A) and the auto-tensioner (B), then insert a 0.05 inches (1.2 mm) diameter pin (C) into the holes

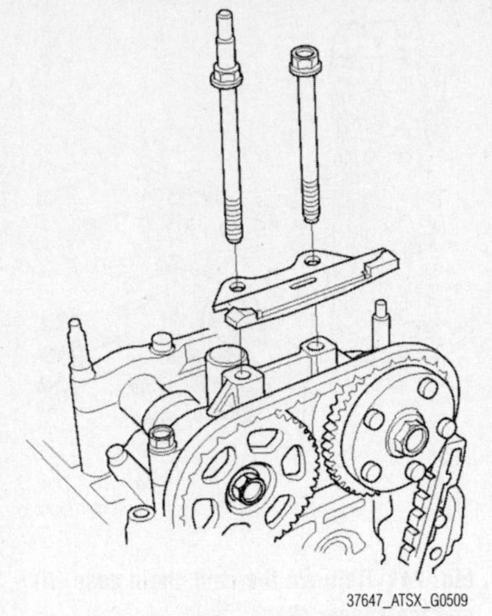
37647_ATSX_G0509

Fig. 250 Remove cam chain guide B

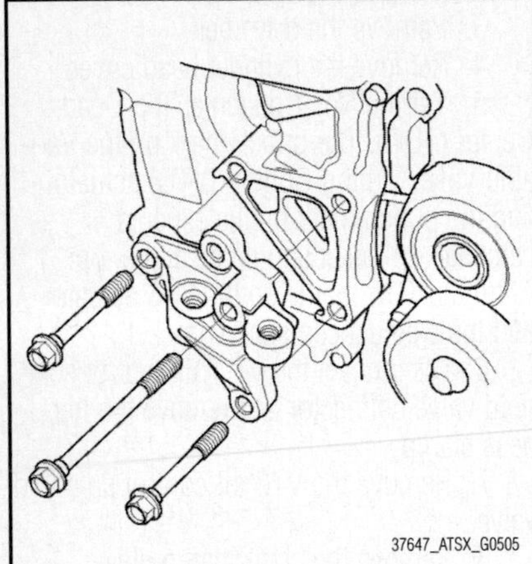
37647_ATSX_G0505

Fig. 246 Remove the side engine mount bracket

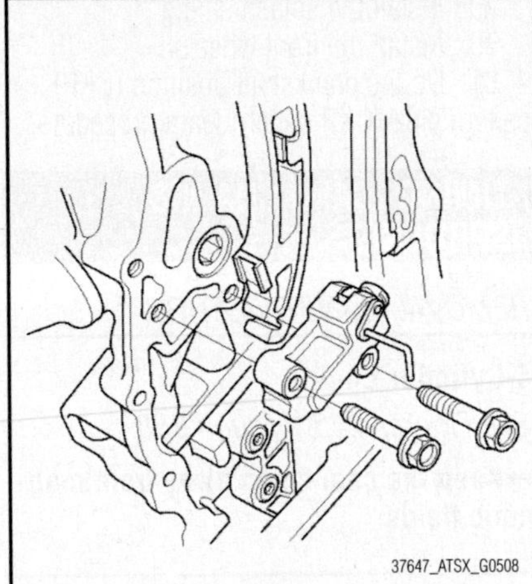
37647_ATSX_G0508

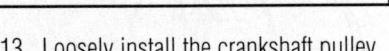

Fig. 249 Remove the auto-tensioner

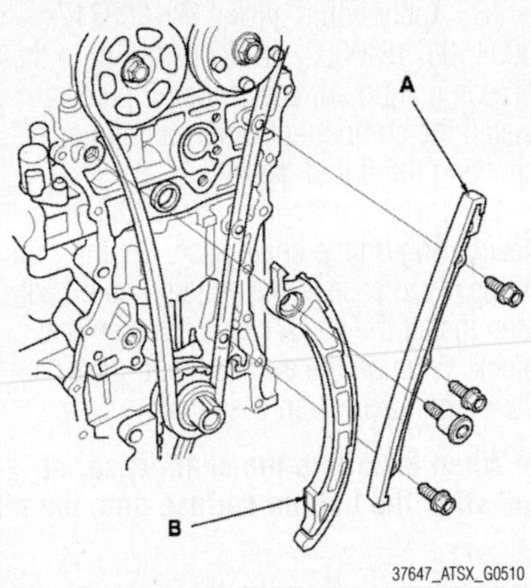
A
B
37647_ATSX_G0510

Fig. 251 Remove cam chain guide A and the tensioner arm (B)

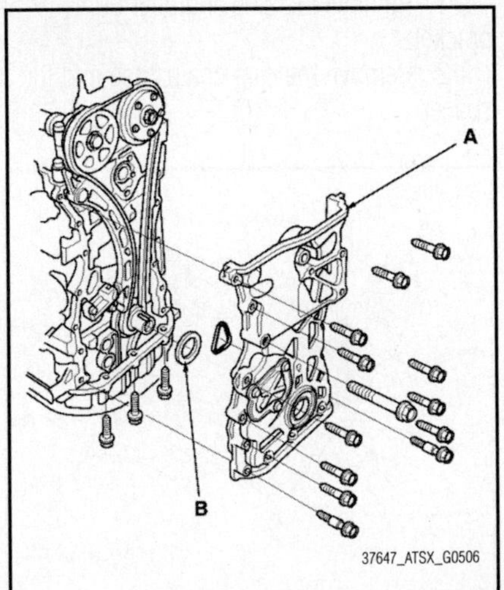
A
B
37647_ATSX_G0506

Fig. 247 Remove the cam chain case (A) and the spacer (B)

13. Loosely install the crankshaft pulley.

14. Turn the crankshaft counterclockwise to compress the auto-tensioner.

15. Align the holes on the lock and the auto-tensioner, then insert a 0.05 inches (1.2 mm) diameter pin or lock pin (P/N 14511-PNA-003) into the holes. Turn the crankshaft clockwise to secure the pin.

16. Remove the auto-tensioner.

17. Remove cam chain guide B.

18. Remove cam chain guide A and the tensioner arm.

19. Remove the cam chain.

To install:

➡Keep the cam chain away from magnetic fields. Before doing this procedure, check that the Variable Valve Timing Control (VTC) actuator is locked by turning the VTC actuator counterclockwise. If not locked, turn the VTC actuator clockwise until it stops, then recheck it. If it is still not locked, replace the VTC actuator.

20. Set the crankshaft to Top Dead Center (TDC). Align the TDC mark on the crankshaft sprocket with the pointer on the engine block.

21. Set the camshafts to TDC. The punch mark on the VTC actuator and the punch mark on the exhaust camshaft sprocket should be at the top. Align the TDC marks on the VTC actuator and the exhaust camshaft sprocket.

22. To hold the intake camshaft, insert a camshaft lock pin set into the maintenance hole in Camshaft Position (CMP) pulse plate A and through the No. 5 rocker shaft holder.

23. To hold the exhaust camshaft, insert a camshaft lock pin set into the

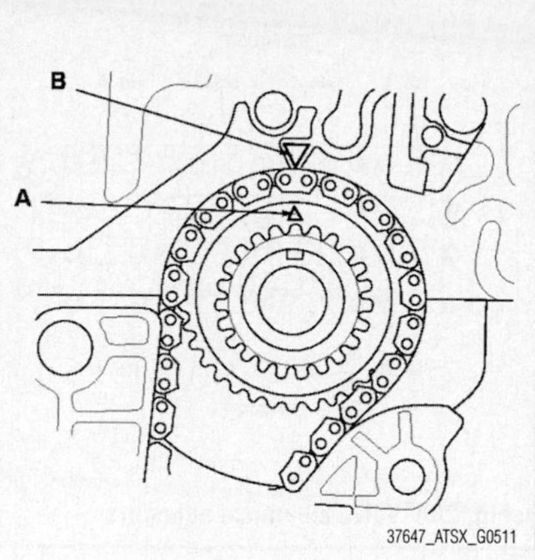
B
A
37647_ATSX_G0511

Fig. 252 Align the TDC mark (A) on the crankshaft sprocket with the pointer (B) on the engine block

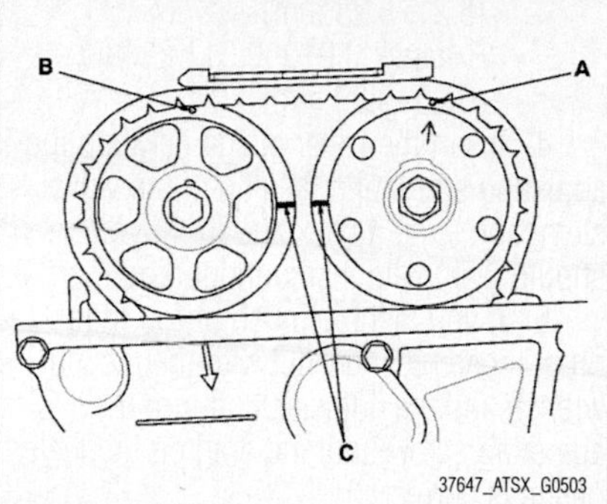
B
A
C
37647_ATSX_G0503

Fig. 253 The punch mark (A) on the VTC actuator and the punch mark (B) on the exhaust camshaft sprocket should be at the top. Align the TDC marks (C) on the VTC actuator and the exhaust camshaft sprocket

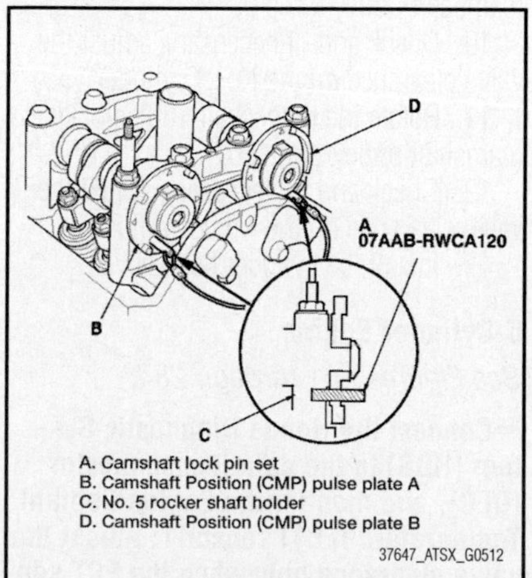
D
A
07AAB-RWCA120
B
C
A. Camshaft lock pin set
B. Camshaft Position (CMP) pulse plate A
C. No. 5 rocker shaft holder
D. Camshaft Position (CMP) pulse plate B
37647_ATSX_G0512

Fig. 254 To hold the intake camshaft, insert a camshaft lock pin set into the maintenance hole in CMP pulse plate A and through the No. 5 rocker shaft holder

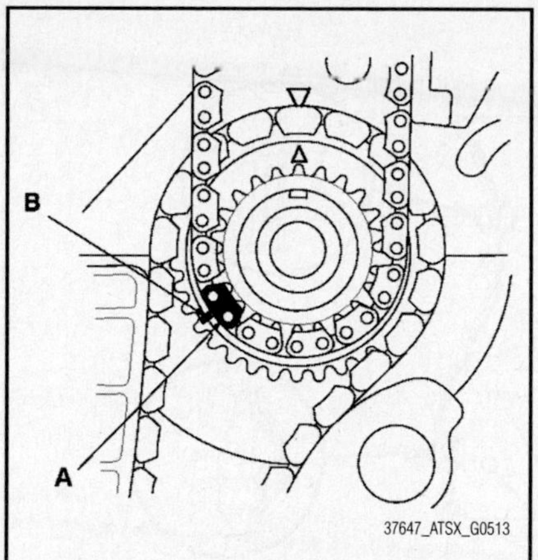
B
A
37647_ATSX_G0513

Fig. 255 Install the cam chain on the crankshaft sprocket with the colored link plate (A) aligned with the mark (B) on the crankshaft sprocket

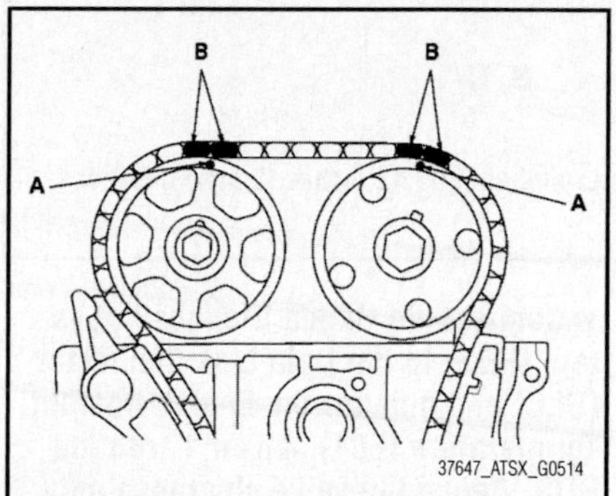
B
B
A
A
37647_ATSX_G0514

Fig. 256 Install the cam chain on the VTC actuator and the exhaust camshaft sprocket with the punch marks (A) aligned with the center of the two colored link plates (B)

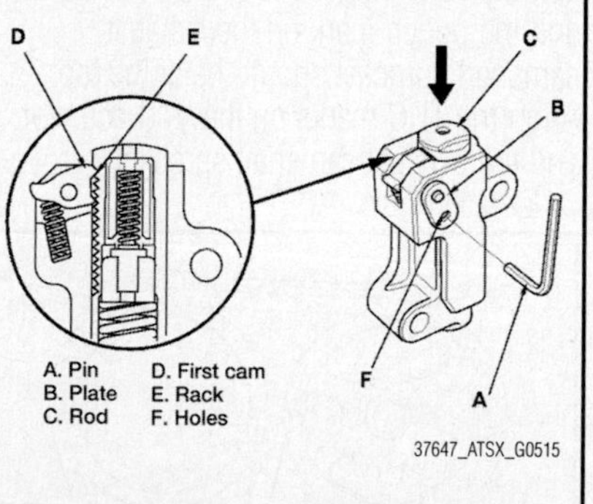
D
E
C
B
F
A
A. Pin
B. Plate
C. Rod
D. First cam
E. Rack
F. Holes
37647_ATSX_G0515

Fig. 257 Compress the auto-tensioner when replacing the cam chain

maintenance hole in CMP pulse plate B and through the No. 5 rocker shaft holder.

24. Install the cam chain on the crankshaft sprocket with the colored link plate aligned with the mark on the crankshaft sprocket.

25. Install the cam chain on the VTC actuator and the exhaust camshaft sprocket with the punch marks aligned with the center of the two colored link plates.

26. Install cam chain guide A and the tensioner arm.

27. Compress the auto-tensioner when replacing the cam chain. Remove the pin (P/N 14511-PNA-003) from the auto-tensioner that was installed during removal. Turn the plate counterclockwise, to release the lock, then press the rod, and set the first cam to the edge of the rack. Insert the 0.05 inches (1.2 mm) diameter pin or lock pin into the holes.

➡If the chain tensioner is not set up as described, the tensioner will become damaged.

28. Install the auto-tensioner.

29. Install cam chain guide B.

30. Remove the pin or lock pin from the auto-tensioner.

31. Remove the camshaft lock pin set.

32. Check the chain case oil seal for damage. If the oil seal is damaged, replace the chain case oil seal.

33. Remove the old liquid gasket from the chain case mating surfaces, the bolts, and the bolt holes.

34. Clean and dry the chain case mating surfaces.

35. Apply liquid gasket (P/N 08717-0004, 08718-0003, or 08718-0009) to the engine block mating surface of the chain case and to the inside edge of the threaded bolt holes. Install the component within 5 minutes of applying the liquid gasket.

➡If too much time has passed after applying the liquid gasket, remove the old liquid gasket and residue, then reapply new liquid gasket.

36. Apply liquid gasket to the engine block upper surface contact areas and the lower block upper surface contact areas on the chain case.

➡Apply about 0.43 inches (11 mm) diameter and about 0.12 inches (3 mm) thickness of liquid gasket to the areas B and C.

37. Apply liquid gasket (P/N 08717-0004, 08718-0003, or 08718-0009) to the oil pan mating surface of the oil pump. Install the component within 5 minutes of applying the liquid gasket.

38. Install the spacer, then install the new O-ring on the chain case. Set the edge of the chain case to the edge of the oil pan, then install the chain case on the engine block. Wipe off the excess liquid gasket on the oil pan and chain case mating area.

➡When installing the chain case, do not slide the bottom surface onto the

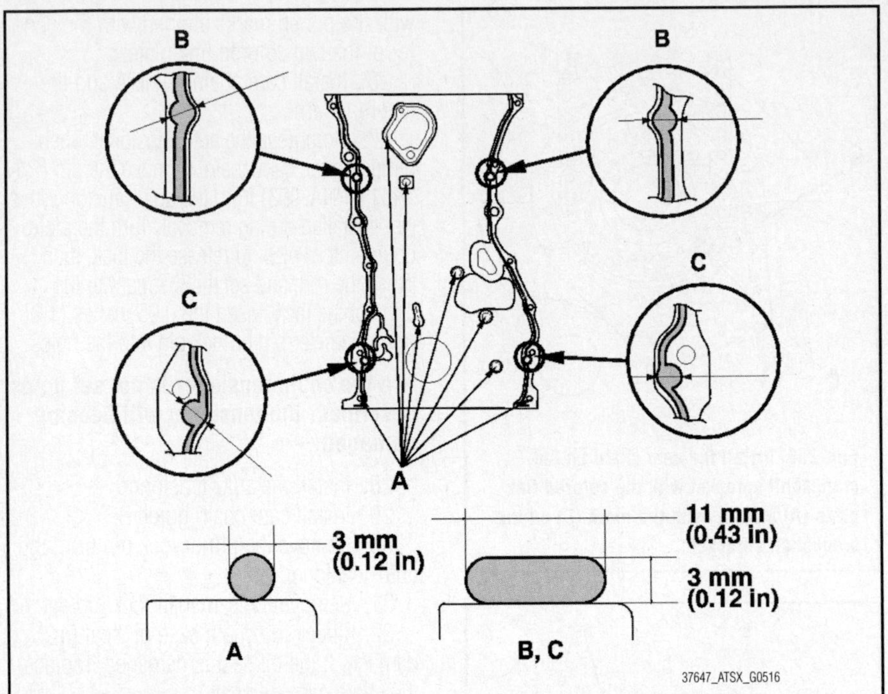
B
B
C
C
A
3 mm (0.12 in)
11 mm (0.43 in)
3 mm (0.12 in)
A
B, C
37647_ATSX_G0516

Fig. 258 Apply about 0.43 inches (11 mm) diameter and about 0.12 inches (3 mm) thickness of liquid gasket to the areas (B) and (C)

oil pan mounting surface. Wait at least 30 minutes before filling the engine with oil. Do not run the engine for at least 3 hours after installing the chain case.

39. Install the side engine mount bracket, then tighten the side engine mount bracket mounting bolts to 33 ft. lbs. (44 Nm).
40. Install the side engine mount bracket, and tighten the top two new bolts to 40 ft. lbs. (54 Nm), then tighten the lower bolt to 47 ft. lbs. (64 Nm).
41. Install the ground cable.
42. Remove the jack and the wood block.
43. Install the crankshaft pulley.
44. Install the VTC oil control solenoid valve.
45. Connect the VTC oil control solenoid valve connector and install the harness clamp.
46. Install the cylinder head cover.
47. Install the drive belt.
48. Install the splash shield.
49. Install the front wheels.
50. Do the crankshaft position (CKP) pattern clear/CKP pattern learn procedure.

VALVE LASH

ADJUSTMENT

4-Cylinder Engine

See Figures 259 and 260.

Special Tools Required:

- Adjuster 07MAA-PR70110
- Locknut wrench 07MAA-PR70120

➡Connect the Honda Diagnostic System (HDS) to the Data Link Connector (DLC) and monitor the Engine Coolant Temperature (ECT) sensor 1 with the HDS. Adjust the valve clearance only when the ECT sensor 1 temperature is less than 100°F (38°C).

1. Remove the cylinder head cover.
2. Set the No. 1 piston at Top Dead Center (TDC). The punch mark on the Variable Valve Timing Control (VTC) actuator and the punch mark on the exhaust camshaft sprocket should be at the top. Align the TDC marks on the VTC actuator and the exhaust camshaft sprocket.

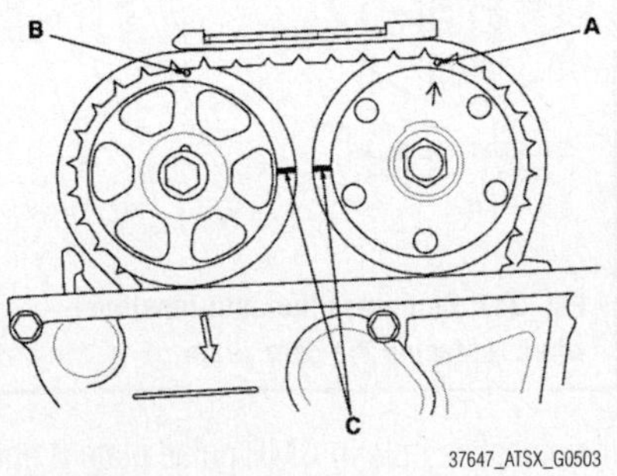
B
A
C
37647_ATSX_G0503

Fig. 259 The punch mark (A) on the VTC actuator and the punch mark (B) on the exhaust camshaft sprocket should be at the top. Align the TDC marks (C) on the VTC actuator and the exhaust camshaft sprocket

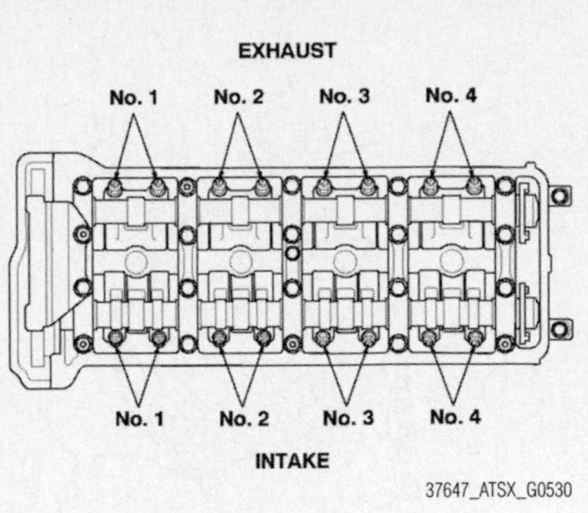
EXHAUST
No. 1
No. 2
No. 3
No. 4
No. 1
No. 2
No. 3
No. 4
INTAKE
37647_ATSX_G0530

Fig. 260 Valve clearance adjusters

3. Select the correct feeler gauge for the valve clearance you are going to check.

➡Valve Clearance:

- Intake: 0.008–0.010 inches (0.21–0.25 mm)
- Exhaust: 0.010–0.011 inches (0.25–0.29 mm)

4. Insert the feeler gauge between the adjusting screw and the end of the valve stem, and slide it back and forth; you should feel a slight amount of drag.
5. If you feel too much or too little drag, loosen the locknut with the locknut wrench and the adjuster, and turn the adjusting screw until the drag on the feeler gauge is correct.
6. Tighten the locknut to 10 ft. lbs. (14 Nm), and recheck the clearance. Repeat the adjustment if necessary.
7. Rotate the crankshaft 180° clockwise (camshaft pulley turns 90°).
8. Check and, if necessary, adjust the valve clearance on the No. 3 cylinder.
9. Rotate the crankshaft 180° clockwise (camshaft pulley turns 90°).
10. Check and, if necessary, adjust the valve clearance on the No. 4 cylinder.
11. Rotate the crankshaft 180° clockwise (camshaft pulley turns 90°).
12. Check and, if necessary, adjust the valve clearance on the No. 2 cylinder.
13. Install the cylinder head cover.

6-Cylinder Engine

See Figures 261 through 264.

➡Connect the Honda Diagnostic System (HDS) to the data link connector (DLC), and monitor the Engine Coolant Temperature (ECT) sensor 1. Adjust the valve clearance only when the ECT sensor 1 is less than 100°F (38°C).

1. Remove the cylinder head covers.
2. Set the No. 1 piston at Top Dead Center (TDC). Align the pointer on the front

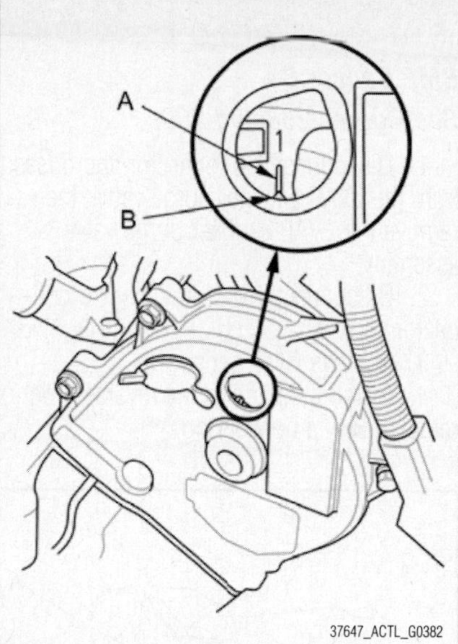
A
B
37647_ACTL_G0382

Fig. 261 Set the No. 1 piston at TDC. Align the pointer (A) on the front upper cover with the No. 1 piston TDC mark (B) on the front camshaft pulley

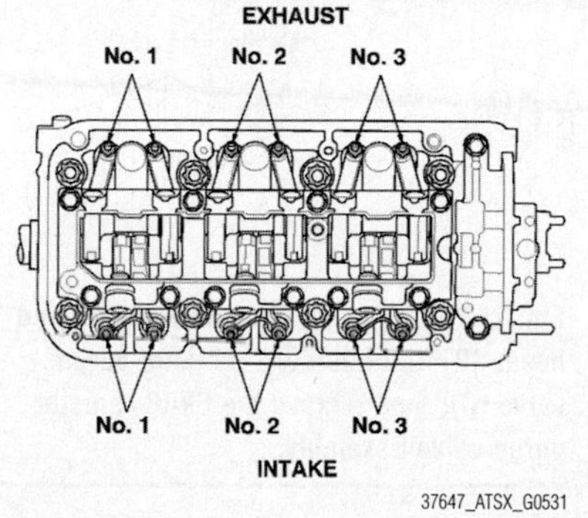
EXHAUST
No. 1
No. 2
No. 3
No. 1
No. 2
No. 3
INTAKE
37647_ATSX_G0531

Fig. 262 Valve clearance adjusters (Rear)

upper cover with the No. 1 piston TDC mark on the front camshaft pulley.

3. Select the correct feeler gauge for the valve clearance you are going to check.

➡Valve Clearance:

- Intake: 0.008–0.009 inches (0.20–0.24 mm)

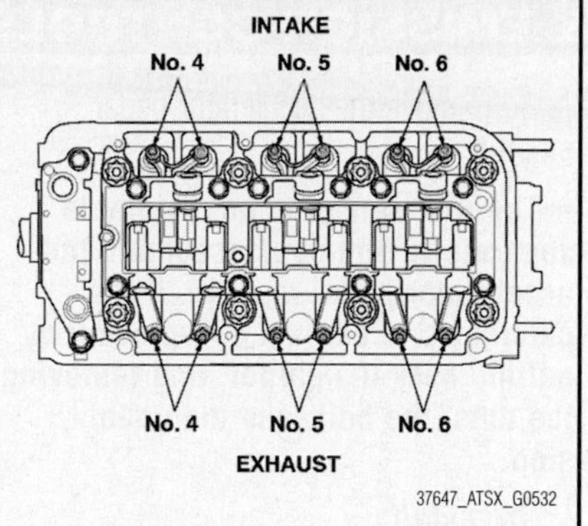
INTAKE
No. 4
No. 5
No. 6
No. 4
No. 5
No. 6
EXHAUST
37647_ATSX_G0532

Fig. 263 Valve clearance adjusters (Front)

- Exhaust: 0.011–0.013 inches (0.28–0.32 mm)

➡Apply new engine oil to the nut threads.

4. Insert the feeler gauge between the adjusting screw and the end of the valve stem on the No. 1 cylinder, and slide it back and forth; you should feel a slight amount of drag.

5. If you feel too much or too little drag, loosen the locknut, and turn the adjusting screw until the drag on the feeler gauge is correct.

6. While holding the adjusting screw with the screw driver, tighten the locknut, then recheck the clearance. Repeat the adjustment, if necessary.

➡Specified Torque:

- Intake: 14 ft. lbs. (20 Nm)
- Exhaust: 10 ft. lbs. (14 Nm)

➡Apply new engine oil to the nut threads.

7. Rotate the crankshaft clockwise. Align the pointer on the front upper cover with the No. 4 piston TDC mark on the front camshaft pulley.

8. Check and, if necessary, adjust the valve clearance on the No. 4 cylinder.

9. Rotate the crankshaft clockwise. Align the pointer on the front upper cover with the No. 2 piston TDC mark on the front camshaft pulley.

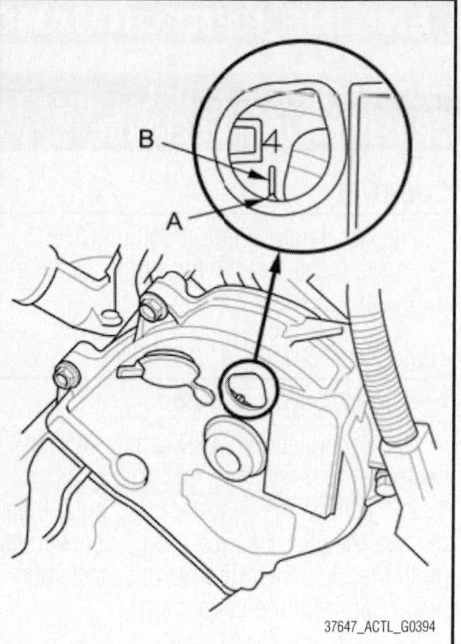
B
A
37647_ACTL_G0394

Fig. 264 Rotate the crankshaft clockwise. Align the pointer (A) on the front upper cover with the No. 4 piston TDC mark (B) on the front camshaft pulley

10. Check and, if necessary, adjust the valve clearance on the No. 2 cylinder.

11. Rotate the crankshaft clockwise. Align the pointer on the front upper cover with the No. 5 piston TDC mark on the front camshaft pulley.

12. Check and, if necessary, adjust the valve clearance on the No. 5 cylinder.

13. Rotate the crankshaft clockwise. Align the pointer on the front upper cover with the No. 3 piston TDC mark on the front camshaft pulley.

14. Check and, if necessary, adjust the valve clearance on the No. 3 cylinder.

15. Rotate the crankshaft clockwise. Align the pointer on the front upper cover with the No. 6 piston TDC mark on the front camshaft pulley.

16. Check and, if necessary, adjust the valve clearance on the No. 6 cylinder.

17. Install the cylinder head covers.

ENGINE PERFORMANCE & EMISSION CONTROLS

ACCELERATOR PEDAL POSITION (APP) SENSOR

LOCATION

The Accelerator Pedal Position (APP) sensor is located at the top of the accelerator pedal assembly.

REMOVAL & INSTALLATION

See Figures 265 and 266.

1. Disconnect the APP sensor 6P connector.
2. Remove the two mounting nuts, then remove the clip. Using a flat-tip screwdriver, push the lock, then remove the accelerator pedal pad from the pedal stop.

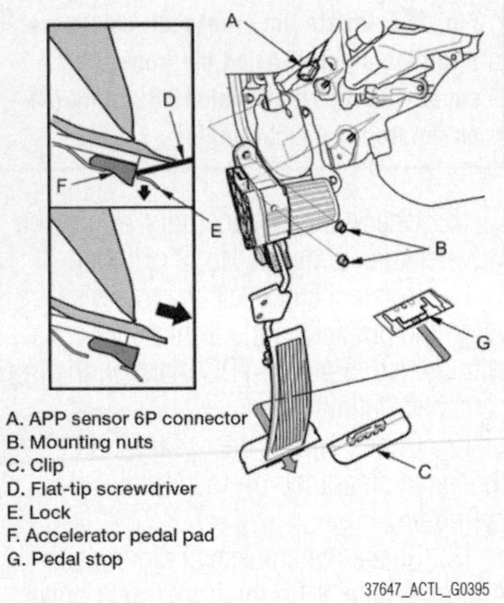
A
D
F
E
B
G
C
A. APP sensor 6P connector
B. Mounting nuts
C. Clip
D. Flat-tip screwdriver
E. Lock
F. Accelerator pedal pad
G. Pedal stop
37647_ACTL_G0395

Fig. 265 Disconnect the APP sensor 6P connector

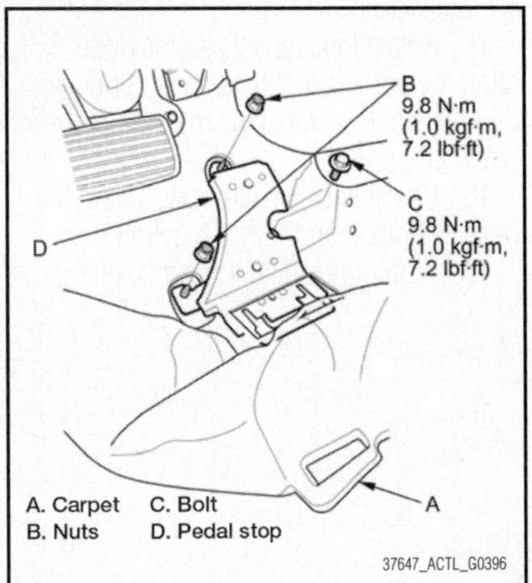
B
9.8 N·m
(1.0 kgf·m,
7.2 lbf·ft)
C
9.8 N·m
(1.0 kgf·m,
7.2 lbf·ft)
D
A
A. Carpet C. Bolt
B. Nuts D. Pedal stop
37647_ACTL_G0396

Fig. 266 Replace it by pulling back the carpet, and removing the nuts, the bolt, and the pedal stop

3. Remove the accelerator pedal module.

➡**The APP sensor is not available separately. Do not disassemble the accelerator pedal module. If the pedal stop is damaged, replace it by pulling back the carpet, and removing the nuts, the bolt, and the pedal stop.**

To install:

4. Install the accelerator pedal pad to the pedal stop. Slide the pad until it locks with a clicking sound.
5. Install a new clip.

➡**The clip must be replaced whenever the accelerator pedal module is installed. Make sure that the accelerator pedal module and the clip are secure.**

6. Install the nuts, and connect the APP sensor 6P connector.

CAMSHAFT POSITION (CMP) SENSOR

LOCATION

Refer to graphics in the Removal and Installation section for location.

REMOVAL & INSTALLATION

4-Cylinder Engine

CMP Sensor A

See Figure 267.

1. Disconnect the CMP sensor A connector.
2. Remove CMP sensor A from the intake camshaft side of the cylinder head.
3. Install the parts in the reverse order of removal with a new O-ring.

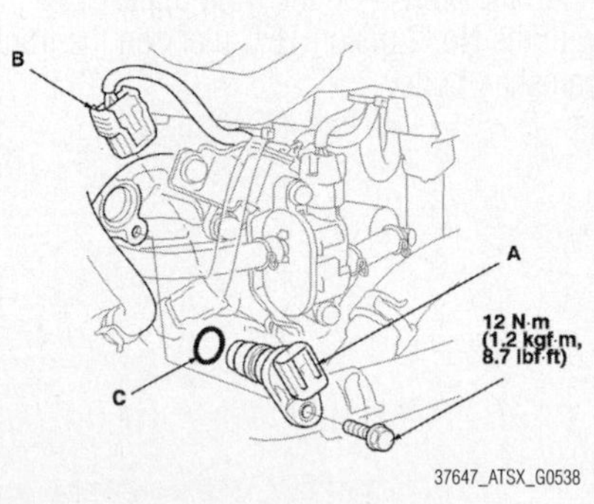
B
A
12 N·m
(1.2 kgf·m,
8.7 lbf·ft)
C
37647_ATSX_G0538

Fig. 267 Removing CMP sensor A

CMP Sensor B

See Figures 268 and 269.

1. Disconnect the connector and hoses from the EVAP canister purge valve, then remove the EVAP canister purge valve assembly.
2. Disconnect the CMP sensor B connector.
3. Remove CMP sensor B.
4. Install the parts in the reverse order of removal with a new O-ring.

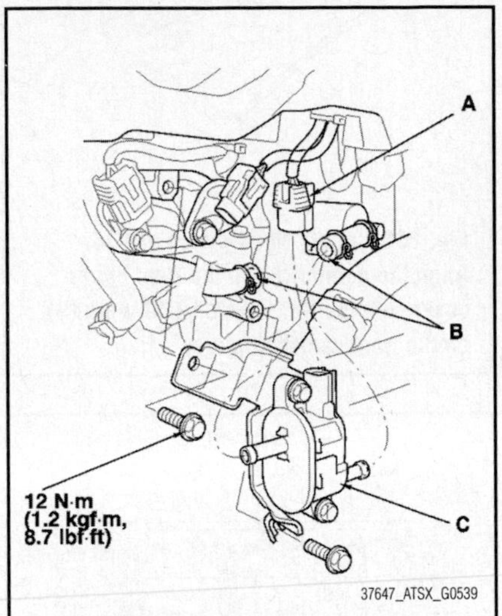
A
B
12 N·m
(1.2 kgf·m,
8.7 lbf·ft)
C
37647_ATSX_G0539

Fig. 268 Disconnect the connector (A) and hoses (B) from the EVAP canister purge valve (C), then remove the EVAP canister purge valve assembly.

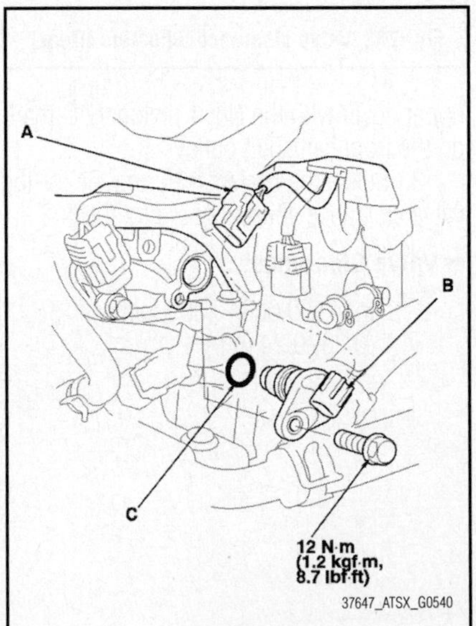
A
B
C
12 N·m
(1.2 kgf·m,
8.7 lbf·ft)
37647_ATSX_G0540

Fig. 269 Removing CMP sensor B

6-Cylinder Engine

See Figures 270 and 271.

1. Remove the timing belt.
2. Remove the front camshaft pulley (CMP sensor pulse plate).
3. Disconnect the CMP sensor connector, then remove the back cover.
4. Remove the CMP sensor from the back cover.
5. Install the parts in the reverse order of removal.
6. Install the timing belt.
7. Do the CKP pattern clear/CKP pattern learn procedure.

CRANKSHAFT POSITION (CKP) SENSOR

LOCATION

Refer to graphics in the Removal and Installation section for location.

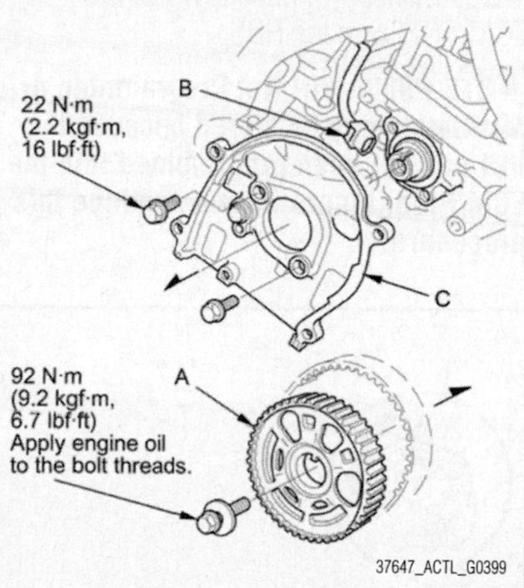
22 N·m
(2.2 kgf·m,
16 lbf·ft)
B
C
92 N·m
(9.2 kgf·m,
6.7 lbf·ft)
Apply engine oil
to the bolt threads.
A
37647_ACTL_G0399

Fig. 270 Remove the front camshaft pulley (CMP sensor pulse plate) (A)

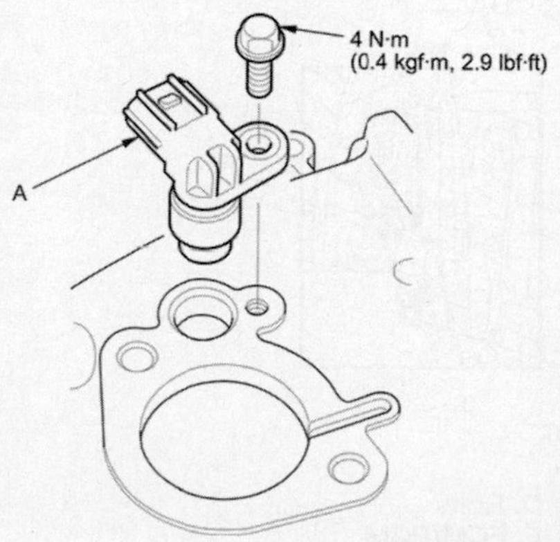
4 N·m
(0.4 kgf·m, 2.9 lbf·ft)
A
37647_ACTL_G0400

Fig. 271 Remove the CMP sensor (A) from the back cover

REMOVAL & INSTALLATION

4-Cylinder Engine

See Figure 272.

1. Raise the vehicle on a lift.

➡Make sure the vehicle is level, because engine oil will drip out when you remove the sensor.

2. Remove the CKP sensor cover.
3. Disconnect the CKP sensor connector.
4. Remove the CKP sensor.
5. Install the parts in the reverse order of removal with a new O-ring.
6. Do the CKP pattern clear/CKP pattern learn procedure.
7. Check the engine oil level, and add more oil if needed.

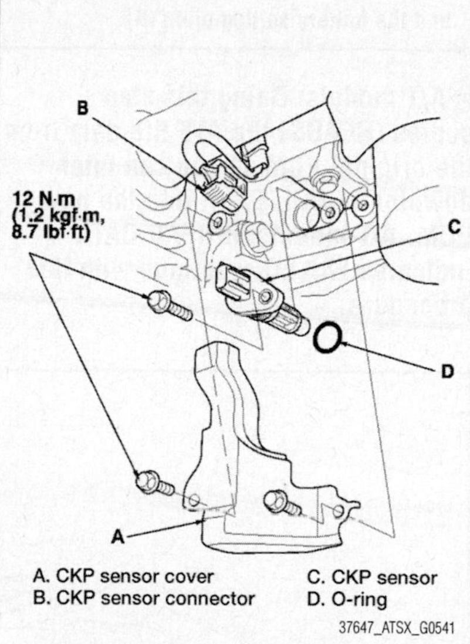
B
12 N·m
(1.2 kgf·m,
8.7 lbf·ft)
C
D
A
A. CKP sensor cover
B. CKP sensor connector
C. CKP sensor
D. O-ring
37647_ATSX_G0541

Fig. 272 Removing CKP sensor

6-Cylinder Engine

See Figure 273.

1. Raise the vehicle on a lift.

➡Make sure the vehicle is level, because engine oil will drip out when you remove the sensor.

2. Remove the CKP sensor cover.
3. Disconnect the CKP sensor connector.
4. Remove the CKP sensor.
5. Install the parts in the reverse order of removal with a new O-ring.
6. Do the CKP pattern clear/CKP pattern learn procedure.
7. Check for engine oil level, and add more oil if needed.

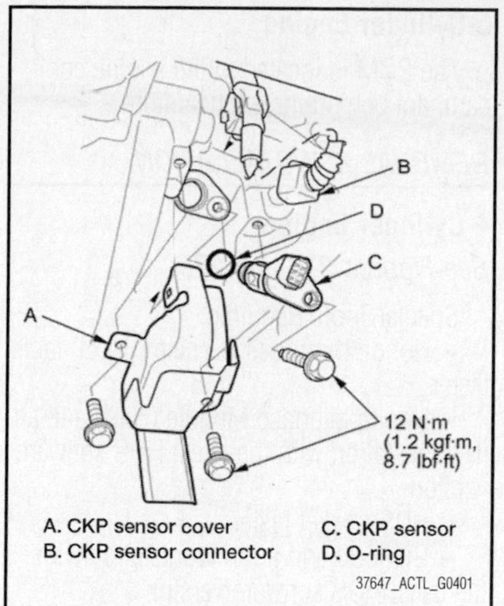
B
D
C
A
12 N·m
(1.2 kgf·m,
8.7 lbf·ft)
A. CKP sensor cover
B. CKP sensor connector
C. CKP sensor
D. O-ring
37647_ACTL_G0401

Fig. 273 Exploded view of CKP sensor assembly

CKP PATTERN CLEAR/CKP PATTERN LEARN

Clear/Learn Procedure (with the HDS)

1. Connect the HDS to the Data Link Connector (DLC) located under the driver's side of the dashboard.
2. Turn the ignition switch to ON (II).
3. Make sure the HDS communicates with the ECM/PCM and other vehicle systems. If it doesn't, go to the DLC circuit troubleshooting.
4. Select CRANK PATTERN in the ADJUSTMENT MENU with the HDS.
5. Select CRANK PATTERN CLEAR, and clear the CKP pattern.
6. Select CRANK PATTERN LEARNING with the HDS, and follow the screen prompts.

Learn Procedure (without the HDS)

1. Start the engine. Hold the engine speed at 3,000 rpm without load (A/T in P or N, M/T in neutral) until the radiator fan comes on.
2. Test-drive the vehicle on a level road: Decelerate (with the throttle fully closed) from an engine speed of 2,500 rpm down to 1,000 rpm with the Transmission in 2nd.
3. Repeat step 2 several times.
4. Turn the ignition switch to LOCK (0).
5. Turn the ignition switch to ON (II), and wait 30 seconds.

ENGINE CONTROL MODULE (ECM)/POWERTRAIN CONTROL MODULE (PCM)

LOCATION

4-Cylinder Engine

The ECM/PCM is located in the engine compartment, driver's side, near the battery.

6-Cylinder Engine

The PCM is located in the engine compartment behind the left headlight.

REMOVAL & INSTALLATION

4-Cylinder Engine

See Figures 274 and 275.

Special Tools Required:

- Honda Diagnostic System (HDS) tablet tester
- Honda Interface Module (HIM) and an iN workstation with the latest HDS software version
- HDS pocket tester
- GNA600 and an iN workstation with the latest HDS software version

➡Any one of the above updating tools can be used.

1. Connect the HDS to the Data Link Connector (DLC) located under the driver's side of the dashboard.
2. Turn the ignition switch to ON (II).
3. Make sure the HDS communicates with the ECM/PCM and other vehicle systems. If it doesn't, go to the DLC circuit troubleshooting. If you are returning from the DLC circuit troubleshooting, skip steps 4 through 9, 19 through 24, and 27 through 29, and do these procedures after replacing the ECM/PCM:
   a. Replace the engine oil and the engine oil filter.
   b. Replace the ATF (A/T).
   c. Clean the throttle body.
4. Select the PGM-FI system with the HDS.
5. Select the INSPECTIONMENU with the HDS.
6. Select the ETCS TEST, then select the TP POSITION CHECK, and follow the screen prompts.

➡If the TP POSITION CHECK indicates FAILED, continue with this procedure.

7. Select the REPLACE ECM/PCM MENU, then READ DATA, and follow the screen prompts.

➡Doing this step copies (READS) the engine oil life data from the original ECM/PCM so you can later download (WRITES) it into the new ECM/PCM. If READ DATA indicates FAILED, continue with this procedure.

8. A/T models: Select the A/T system with the HDS.
9. A/T models: Select the REPLACE TCM/PCM MENU, then READ DATA, and follow the screen prompts.

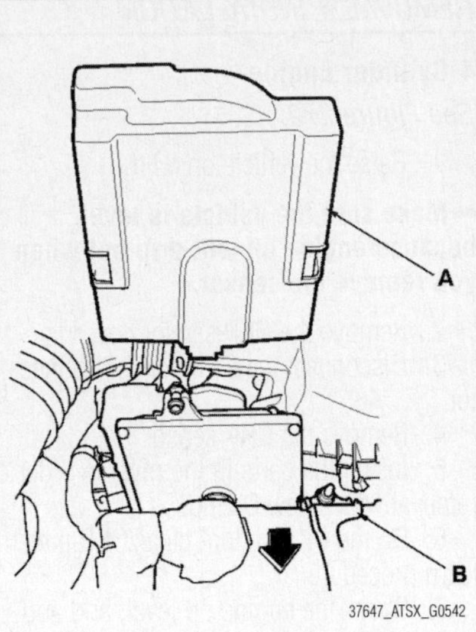
A
B
37647_ATSX_G0542

Fig. 274 Remove the ECM/PCM cover (A) and the battery setting plate (B)

➡A/T models: Doing this step copies (READS) the ATF life data from the original PCM so you can later download (WRITES) it into the new PCM. A/T models: If READ DATA indicates FAILED, continue with this procedure.

10. Turn the ignition switch to LOCK (0).
11. Jump the SCS line with the HDS.
12. Remove the ECM/PCM cover.
13. Remove the battery setting plate, and reposition the battery away from the ECM/PCM.

➡Do not disconnect the battery terminals.

14. Remove the bolts.
15. Disconnect ECM/PCM connectors A, B, and C, then remove the ECM/PCM.

➡ECM/PCM connectors A, B, and C have symbols (A = □, B = ◎, C = ○) embossed on them for identification. Canada model: When disconnecting the ECM/PCM connectors, remove the ECM/PCM sub-bracket.

16. Install the parts in the reverse order of removal.
17. Turn the ignition switch to ON (II).
18. Manually input the VIN to the ECM/PCM with the HDS.

➡DTC P0630 VIN Not Programmed or Mismatch may be stored because the VIN has not been programmed into the ECM/PCM; ignore it, and continue this procedure.

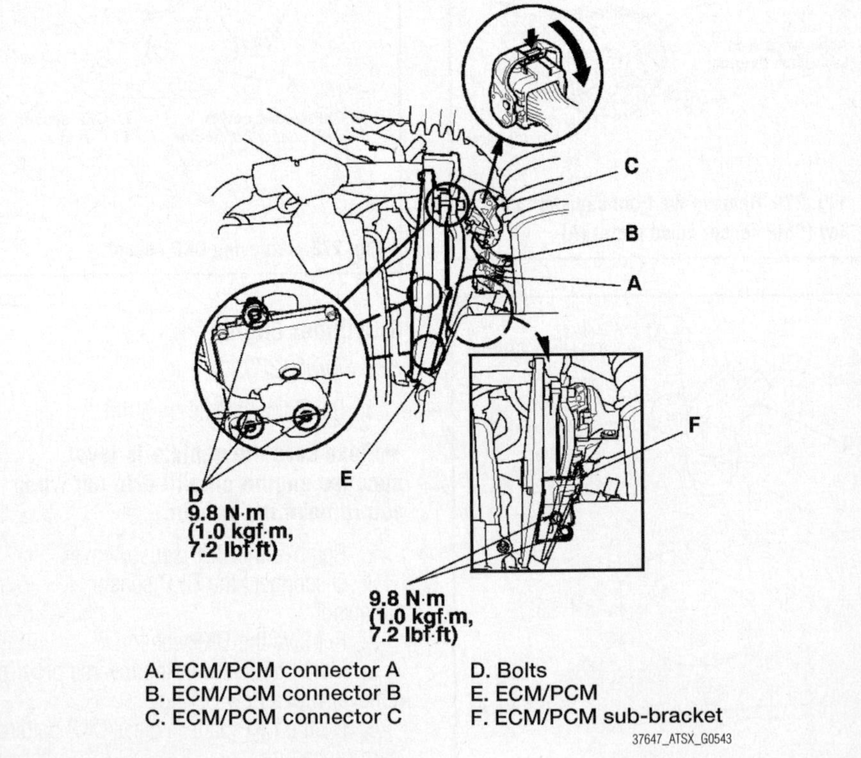
C
B
A
F
E
D
9.8 N·m
(1.0 kgf·m,
7.2 lbf·ft)
9.8 N·m
(1.0 kgf·m,
7.2 lbf·ft)
A. ECM/PCM connector A
B. ECM/PCM connector B
C. ECM/PCM connector C
D. Bolts
E. ECM/PCM
F. ECM/PCM sub-bracket
37647_ATSX_G0543

Fig. 275 ECM/PCM disconnect and removal

19. If the READ DATA (engine oil life) failed in step 7, go to step 22 (A/T) or step 25 (M/T). Otherwise, go to step 20.

20. Select the PGM-FI system with the HDS.

21. Select the REPLACE ECM/PCM MENU, then WRITE DATA, and follow the screen prompts.

➡If the WRITE DATA indicates FAILED, continue with this procedure.

22. A/T: If the READ DATA (ATF life) failed in step 8, go to step 25. Otherwise go to step 23.

23. A/T: Select the A/T SYSTEM with the HDS.

24. A/T: Select the REPLACE TCM/PCM MENU, then WRITE DATA, and follow the screen prompts.

➡A/T: If the WRITE DATA indicates FAILED, continue with this procedure.

25. Select IMMOBI system with the HDS.

26. Enter the immobilizer ECM/PCM code that you got from the iN, and use the ECM/PCM replacement procedure in the IMMOBI MENU of the HDS; it allows you to start the engine.

27. If the TP POSITION CHECK failed in step 6, clean the throttle body, then go to step 28.

28. If the READ DATA failed in step 7 or the WRITE DATA failed in step 21, replace the engine oil and engine oil filter, then go to step 29 (A/T) or step 30 (M/T).

29. A/T: If the READ DATA failed in step 9 or the WRITE DATA failed in step 24, replace the ATF, then go to step 30.

30. Select PGM-FI system, and reset the ECM/PCM with the HDS.

31. Update the ECM/PCM if it does not have the latest software.

32. Do the ECM/PCM idle learn procedure.

33. Do the CKP pattern clear/CKP pattern learn procedure.

6-Cylinder Engine

See Figure 276.

Special Tools Required:

- Honda Diagnostic System (HDS) tablet tester
- Honda Interface Module (HIM) and an iN workstation with the latest HDS software version
- HDS pocket tester
- GNA600 and an iN workstation with the latest HDS software version

➡Any one of the above updating tools can be used.

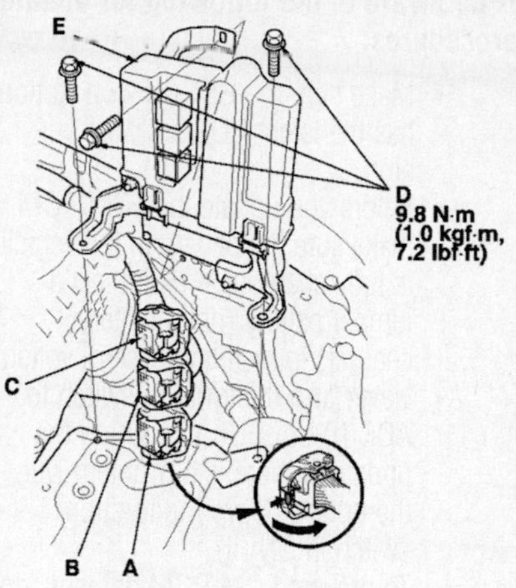
E
D
9.8 N·m
(1.0 kgf·m,
7.2 lbf·ft)
C
B
A

A. PCM connector A
B. PCM connector B
C. PCM connector C
D. Bolts
E. PCM assembly

37647_ATSX_G0544

Fig. 276 Disconnecting and removing the PCM

1. Connect HDS to the Data Link Connector (DLC) located under the driver's side of the dashboard.

2. Turn the ignition switch to ON (II).

3. Make sure the HDS communicates with the PCM and other vehicle systems. If it doesn't, go to the DLC circuit troubleshooting. If you are returning from the DLC circuit troubleshooting, skip steps 4 through 9, 20 through 25, and 28 through 30, and do this after replacing the PCM:

- Replace the engine oil and the engine oil filter.
- Replace the ATF.
- Clean the throttle body.

4. Select the PGM-FI system with the HDS.

5. Select the INSPECTIONMENU with the HDS.

6. Select the ETCS TEST, then select the TP POSITION CHECK, and follow the screen prompts.

➡If the TP POSITION CHECK indicates FAILED, continue with this procedure.

7. Select the REPLACE PCM MENU, then select READ DATA, and follow the screen prompts.

➡Doing this step copies (READS) the engine oil life data from the original PCM so you can later download (WRITES) it into the new PCM. If READ DATA indicates FAILED, continue with this procedure.

8. Select the A/T system with the HDS.

9. Select the REPLACE TCM/PCM MENU, then select READ DATA, and follow the screen prompts.

➡Doing this step copies (READS) the ATF life data from the original PCM so you can later download (WRITES) it into the new PCM. If READ DATA indicates FAILED, continue with this procedure.

10. Turn the ignition switch to LOCK (0).

11. Jump the SCS line with the HDS.

12. Do the battery removal procedure.

13. Remove the bolts.

14. Disconnect PCM connectors A, B, and C, then remove the PCM assembly.

➡PCM connectors A, B, and C have symbols (A = □, B = ◎, C = ○) embossed on them for identification.

15. Disconnect the cover and the bracket from the PCM.

16. Install a known-good PCM in the reverse order of removal.

17. Do the battery installation procedure.

18. Turn the ignition switch to ON (II).

➡DTC P0630 VIN Not Programmed or Mismatch may be stored because the VIN has not been programmed into the PCM; ignore it, and continue this procedure.

19. Manually input the VIN to the PCM with the HDS.

20. If the READ DATA (engine oil life) failed in step 7, go to step 23. Otherwise, go to step 21.

21. Select the PGM-FI system with the HDS.

22. Select the REPLACE PCM MENU, then select WRITE DATA, and follow the screen prompts.

➡If the WRITE DATA indicates FAILED, continue with this procedure.

23. If the READ DATA (ATF life) failed in step 9, go to step 26. Otherwise go to step 24.

24. Select the A/T SYSTEM with the HDS.

25. Select the REPLACE TCM/PCM MENU, then select WRITE DATA, and follow the screen prompts.

➡If the WRITE DATA indicates FAILED, continue with this procedure.

26. Select IMMOBI system with the HDS.

27. Enter the immobilizer PCM code that you got from the iN, and use the PCM replacement procedure in the IMMOBI MENU of the HDS; it allows you to start the engine.

28. If the TP POSITION CHECK failed in step 6 clean the throttle body, then go to step 26.

29. If the READ DATA failed in step 7 or the WRITE DATA failed in step 22, replace the engine oil and engine oil filter, then go to step 30.

30. If the READ DATA failed in step 9 or the WRITE DATA failed in step 25, replace the ATF, then go to step 31.

31. Select PGM-FI system, and reset the PCM with the HDS.

32. Update the PCM if it does not have the latest software.

33. Do the PCM idle learn procedure.

34. Do the CKP pattern clear/CKP pattern learn procedure.

ECM/PCM IDLE LEARN PROCEDURE

The idle learn procedure must be done so the ECM/PCM can learn the engine idle characteristics.

Do the idle learn procedure whenever you do any of these actions:

- Replace the ECM/PCM.
- Reset the ECM/PCM.
- Update the ECM/PCM.
- Replace or clean the throttle body.
- Disassemble the engine or the transmission.

➡Clearing the DTCs with the HDS does not require you to do the idle learn procedure.

1. Make sure all electrical items (the A/C, the audio system, the lights, etc.) are off.
2. Reset the ECM/PCM with the HDS.
3. Turn the ignition switch to ON (II), and wait 2 seconds.
4. Start the engine. Hold the engine speed at 3,000 rpm without load (A/T in P or N, M/T in neutral) until the radiator fan comes on, or until the engine coolant temperature reaches 194°F (90°C).
5. Let the engine idle for about 5 minutes with the throttle fully closed.

➡If the radiator fan comes on, do not include its running time in the 5 minutes.

ECM/PCM UPDATE

Special Tools Required:

- Honda diagnostic system (HDS) tablet tester
- Honda interface module (HIM) and an iN workstation with the latest HDS software version
- HDS pocket tester
- GNA600 and an iN workstation with the latest HDS software version

➡Any one of the above updating tools can be used.

➡Be aware of the following for update procedures:

- Make sure the HDS/iN workstation has the latest HDS software version.
- Before you update the ECM/PCM, make sure the battery in the vehicle is fully charged, and connect a jumper battery (not a battery charger) to maintain system voltage.
- Never turn the ignition switch to ACC (I) or LOCK (0) during the update. If there is a problem with the update, leave the ignition switch in ON (II).
- To prevent ECM/PCM damage, do not operate anything electrical (headlights, audio system, brakes, A/C, power windows, moonroof (if equipped), door locks, etc.) during the update.
- To ensure the latest program is installed, do an ECM/PCM update whenever the ECM/PCM is substituted or replaced.
- You cannot update an ECM/PCM with a program it already has. It will only accept a new program.
- High temperature in the engine compartment might cause the ECM/PCM to become too hot to run the update. If the engine has been running before this procedure, open the hood and cool the engine compartment.
- If you need to diagnose the Honda Interface Module (HIM) because the HIM's red (#3) light came on or was flashing during the update, leave the ignition switch in ON (II) when you disconnect the HIM from the Data Link Connector (DLC). This will prevent ECM/PCM damage.

1. Turn the ignition switch to ON (II), but do not start the engine.
2. Connect the HDS to the Data Link Connector (DLC) located under the driver's side of the dashboard.
3. Make sure the HDS communicates with the ECM/PCM and other vehicle systems. If it doesn't, go to the DLC circuit troubleshooting. If you are returning from the DLC circuit troubleshooting, skip steps 4 and 5, and clean the throttle body after updating the ECM/PCM.
4. Select the INSPECTIONMENU with the HDS.
5. Select the ETCS TEST, then select the TP POSITION CHECK, and follow the HDS screen prompts.

➡If the TP POSITION CHECK indicates FAILED, continue this procedure.

6. Exit the HDS diagnostic system, then select the update mode, and follow the screen prompts to update the ECM/PCM.
7. If the software in the ECM/PCM is the latest, disconnect the HDS/HIM/GNA600 from the DLC, and go back to the procedure that you were doing. If the software in the ECM/PCM is not the latest, follow the instructions on the screen. If prompted to choose the PGM-FI system or the A/T system (A/T), make sure you update both.

➡If the ECM/PCM update system requires you to cool the ECM/PCM, follow the instructions on the screen. If you run into a problem during the update procedure (programming takes over 15 minutes, status bar goes over 100 %, D (A/T) or immobilizer indicator flashes, HDS tablet freezes, etc.), follow these steps to minimize the chance of damaging the ECM/PCM:

- Leave the ignition switch in ON (II).
- Connect a jumper battery (do not connect a battery charger).
- Shut down the HDS.
- Disconnect the HDS from the DLC.
- Reboot the HDS.
- Reconnect the HDS to the DLC, and do the update again.

8. If the TP POSITION CHECK failed in step 5, clean the throttle body.
9. Do the ECM/PCM idle learn procedure.
10. Do the CKP pattern clear/CKP pattern learn procedure.

ENGINE COOLANT TEMPERATURE (ECT) SENSOR

LOCATION

Refer to graphics in the Removal and Installation section to locate the two Engine Coolant Temperature (ECT) sensors for each engine.

REMOVAL & INSTALLATION

4-Cylinder Engine

ECT sensor 1

See Figure 277.

1. Drain the engine coolant.
2. Disconnect the ECT sensor 1 connector.
3. Remove ECT sensor 1.
4. Install the parts in the reverse order of removal with a new O-ring, then refill the radiator with engine coolant.

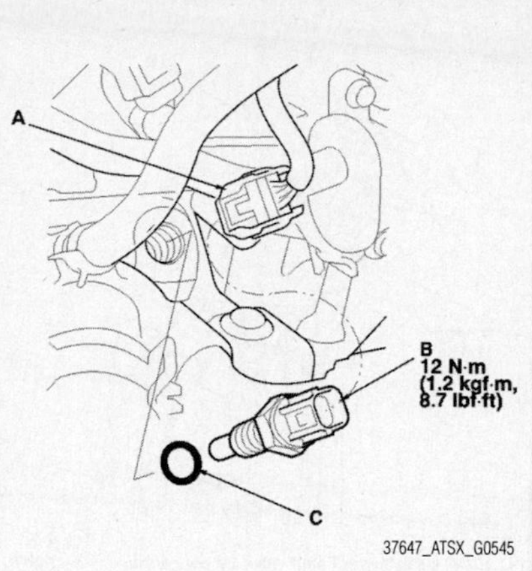
A
B
12 N·m
(1.2 kgf·m,
8.7 lbf·ft)
C
37647_ATSX_G0545

Fig. 277 ECT sensor 1 (B)

ECT sensor 2

See Figure 278.

1. Remove the front splash shield.
2. Drain the engine coolant.
3. Disconnect the ECT sensor 2 connector, then remove ECT sensor 2.
4. Install ECT sensor 2 with a new O-ring.
5. Install the front splash shield.
6. Refill the radiator with engine coolant.

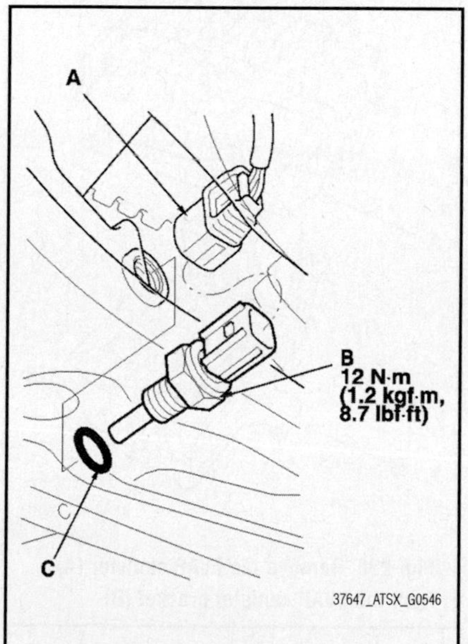
A
B
12 N·m
(1.2 kgf·m,
8.7 lbf·ft)
C
37647_ATSX_G0546

Fig. 278 ECT sensor 2 (B)

6-Cylinder Engine

ECT sensor 1

See Figure 279.

1. Drain the engine coolant.
2. Disconnect the ECT sensor 1 2P connector.
3. Remove ECT sensor 1.

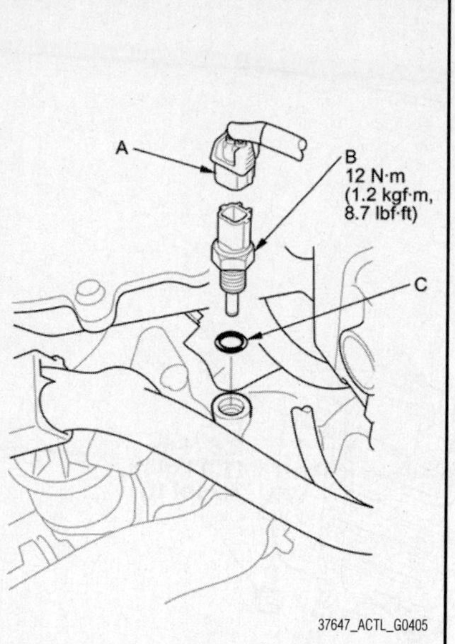
A
B
12 N·m
(1.2 kgf·m,
8.7 lbf·ft)
C
37647_ACTL_G0405

Fig. 279 Disconnect the ECT sensor 1 2P connector (A)

4. Install the parts in the reverse order of removal with a new O-ring, then refill the radiator with engine coolant.

ECT sensor 2

See Figure 280.

1. Remove the splash shield.
2. Drain the engine coolant.
3. Disconnect the ECT sensor 2 2P connector.
4. Remove ECT sensor 2.
5. Install the parts in the reverse order of removal with a new O-ring, then refill the radiator with engine coolant.

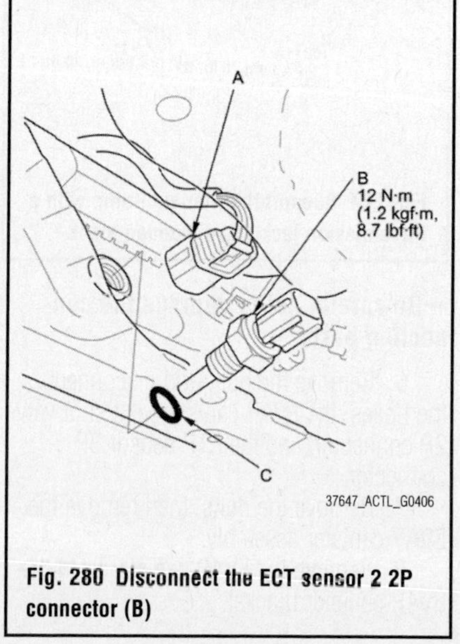
A
B
12 N·m
(1.2 kgf·m,
8.7 lbf·ft)
C
37647_ACTL_G0406

Fig. 280 Disconnect the ECT sensor 2 2P connector (B)

EVAPORATIVE EMISSIONS (EVAP) CANISTER

LOCATION

The EVAP canister is located under the rear of the vehicle near the fuel tank.

REMOVAL & INSTALLATION

4-Cylinder Engine

See Figures 281 through 283.

1. Raise the vehicle on a lift.
2. Remove the wheel sensor harness clamps.
3. Support the rear subframe with a transmission jack and a wooden block as shown.
4. Remove the rear subframe mounting bolts.
5. Lower the transmission jack and the rear subframe about 50 mm.

➡Be careful not to damage the connecting parts.

6. Remove the bolt, and disconnect the hoses, the EVAP canister vent shut valve 2P connector, and the FTP sensor 3P connector.
7. Remove the bolts, then remove the EVAP canister assembly.
8. Remove the EVAP canister from the EVAP canister bracket.

To install:

9. Install the EVAP canister assembly to the body.

➡Attach the bracket arm to the body.

10. Install the parts in the reverse order of removal. Use new bolts when you install the rear subframe.
11. Check the wheel alignment.

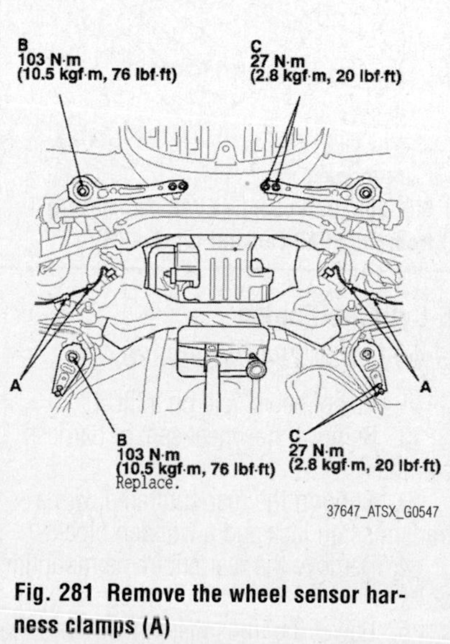
B
103 N·m
(10.5 kgf·m, 76 lbf·ft)
C
27 N·m
(2.8 kgf·m, 20 lbf·ft)
A
A
B
103 N·m
(10.5 kgf·m, 76 lbf·ft)
Replace.
C
27 N·m
(2.8 kgf·m, 20 lbf·ft)
37647_ATSX_G0547

Fig. 281 Remove the wheel sensor harness clamps (A)

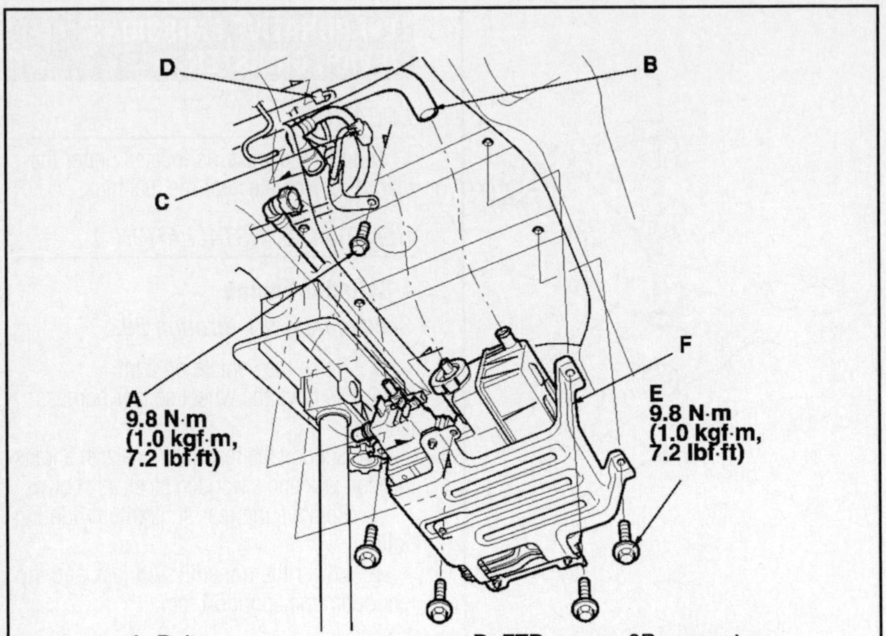
D
B
C
A
9.8 N·m
(1.0 kgf·m,
7.2 lbf·ft)
F
E
9.8 N·m
(1.0 kgf·m,
7.2 lbf·ft)

A. Bolt
B. Hoses
C. EVAP canister vent shut valve 2P connector
D. FTP sensor 3Pconnector
E. Bolts
F. EVAP canister assembly

37647_ATSX_G0548

Fig. 282 Remove the EVAP canister assembly

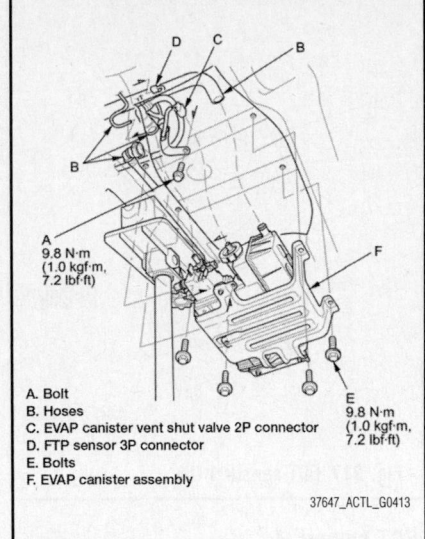

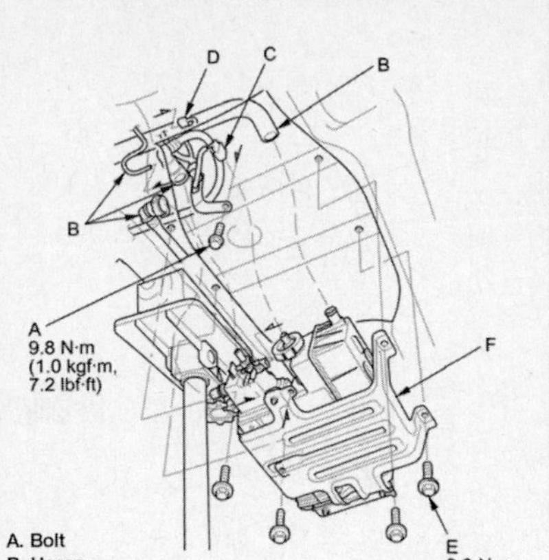
D
C
B
B
A
9.8 N·m
(1.0 kgf·m,
7.2 lbf·ft)
F
E
9.8 N·m
(1.0 kgf·m,
7.2 lbf·ft)

A. Bolt
B. Hoses
C. EVAP canister vent shut valve 2P connector
D. FTP sensor 3P connector
E. Bolts
F. EVAP canister assembly

37647_ACTL_G0413

Fig. 285 Remove the bolt, and disconnect the hoses, the EVAP canister vent shut valve 2P connector, and the FTP sensor 3P connector

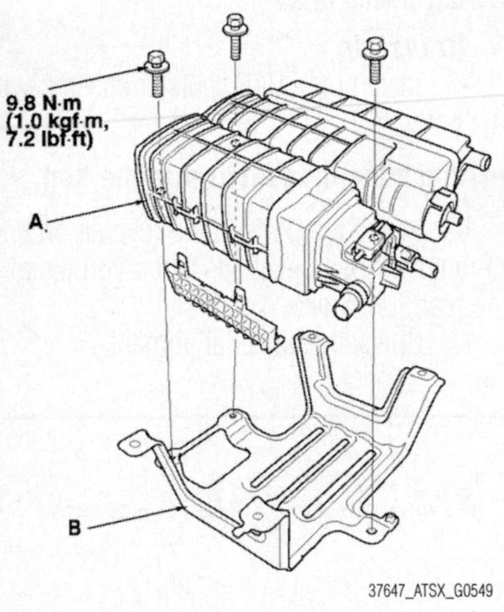
9.8 N·m
(1.0 kgf·m,
7.2 lbf·ft)
A
B

37647_ATSX_G0549

Fig. 283 Remove the EVAP canister (A) from the EVAP canister bracket (B)

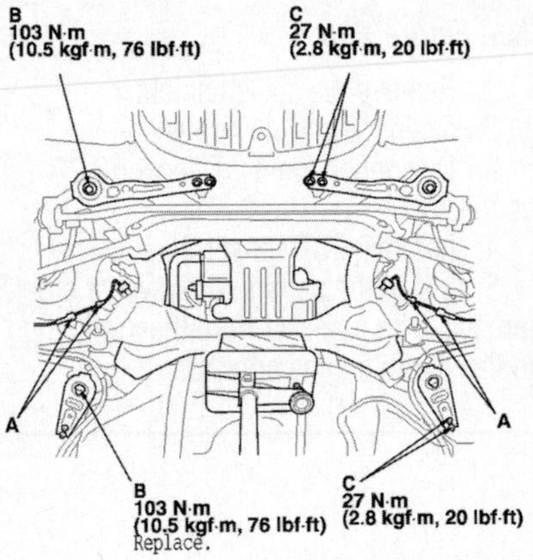
B
103 N·m
(10.5 kgf·m, 76 lbf·ft)
C
27 N·m
(2.8 kgf·m, 20 lbf·ft)
A
A
B
103 N·m
(10.5 kgf·m, 76 lbf·ft)
Replace.
C
27 N·m
(2.8 kgf·m, 20 lbf·ft)

37647_ATSX_G0550

Fig. 284 Support the rear subframe with a transmission jack and a wooden block

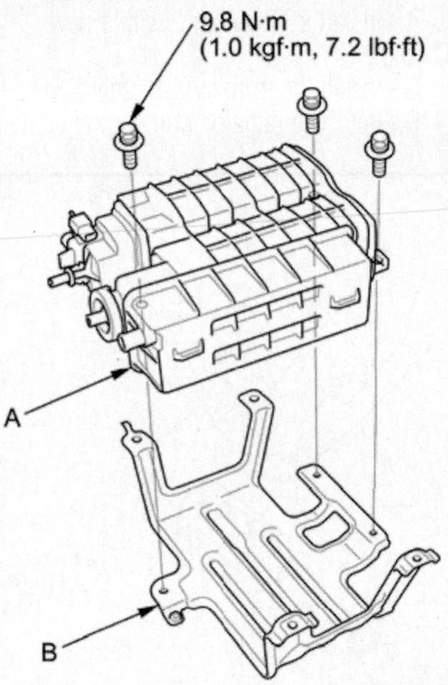
9.8 N·m
(1.0 kgf·m, 7.2 lbf·ft)
A
B

37647_ACTL_G0414

Fig. 286 Remove the EVAP canister (A) from the EVAP canister bracket (B)

6-Cylinder Engine

See Figures 284 through 287.

1. Raise the vehicle on a lift.
2. Remove the wheel sensor harness clamps.
3. Support the rear subframe with a transmission jack and a wooden block.
4. Remove the rear subframe mounting bolts.
5. Lower the transmission jack and the rear subframe about 50 mm.

➡Be careful not to damage the connecting parts.

6. Remove the bolt, and disconnect the hoses, the EVAP canister vent shut valve 2P connector, and the FTP sensor 3P connector.
7. Remove the bolts, then remove the EVAP canister assembly.
8. Remove the EVAP canister from the EVAP canister bracket.

To install:

9. Install the EVAP canister assembly to the body.

➡Attach the bracket arm to the body.

10. Install the parts in the reverse order of removal. Use new bolts when you install the rear subframe.
11. Check the wheel alignment.

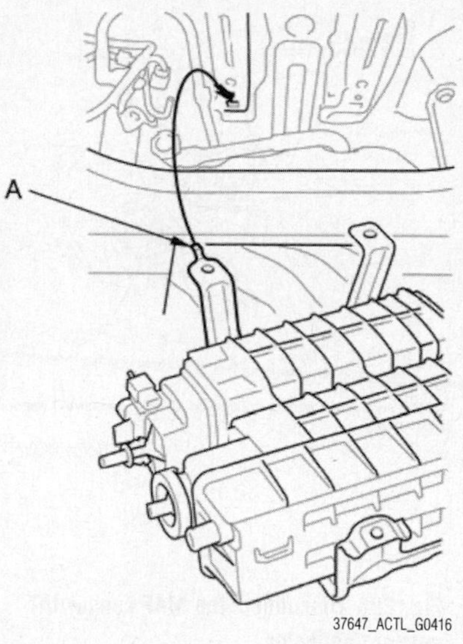
A
37647_ACTL_G0416

Fig. 287 Attach the bracket arm (A) to the body

EXHAUST GAS RECIRCULATION (EGR) VALVE

LOCATION

The EGR valve is located at the front left of the engine.

REMOVAL & INSTALLATION

6-Cylinder Engine

See Figure 288.

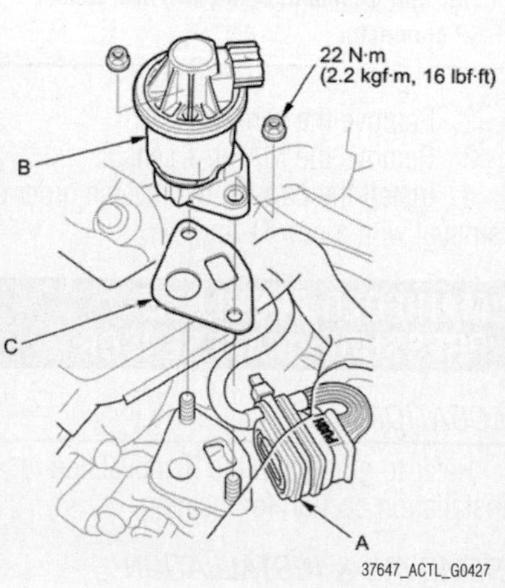
22 N·m
(2.2 kgf·m, 16 lbf·ft)
B
C
A
37647_ACTL_G0427

Fig. 288 Disconnect the EGR valve 5P connector

1. Disconnect the EGR valve 5P connector.
2. Remove the EGR valve.
3. Install the parts in the reverse order of removal with a new gasket.

HEATED OXYGEN (HO2S) SENSOR

REMOVAL & INSTALLATION

4-Cylinder Engine

See Figure 289.

Special Tools Required: O2 sensor wrench, Snap-on® YA8875, SWR2, or equivalent, commercially available

1. Remove the front splash shield.
2. Disconnect the secondary HO2S 4P connector, then remove secondary HO2S.
3. Install the parts in the reverse order of removal.

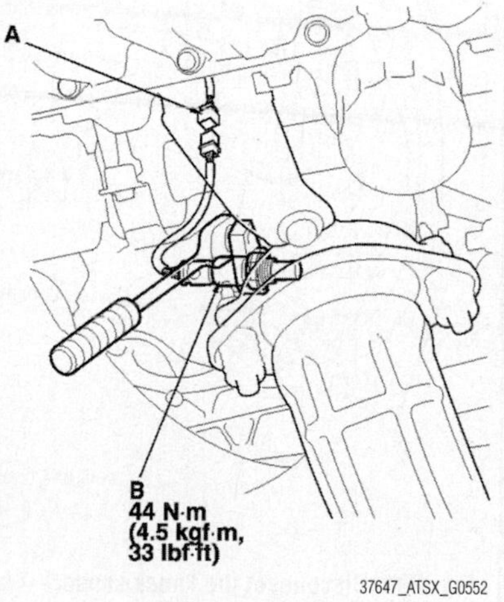
A
B
44 N·m
(4.5 kgf·m,
33 lbf·ft)
37647_ATSX_G0552

Fig. 289 Disconnect the secondary HO2S 4P connector (A), then remove secondary HO2S (B)

6-Cylinder Engine

Special Tools Required: O2 sensor wrench, Snap-on® YA8875 or SWR2, or equivalent, commercially available.

Front Bank (Bank 2)

See Figures 290 and 291.

1. Disconnect the front secondary HO2S 4P connector, and remove the harness clamps.
2. Raise the vehicle on a lift.
3. Remove the harness clamp, then remove the front secondary HO2S.

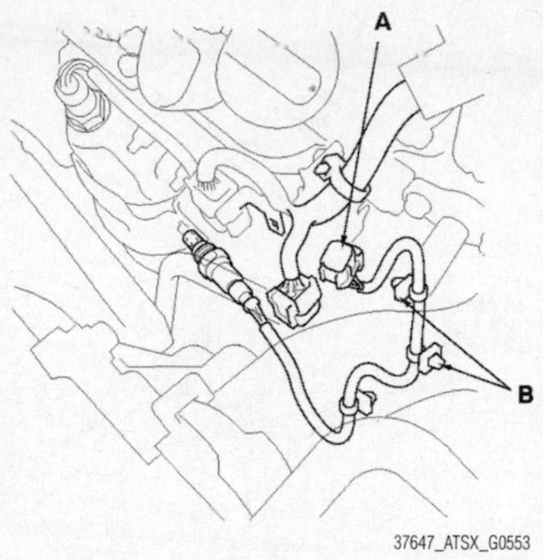
A
B
37647_ATSX_G0553

Fig. 290 Disconnect the front secondary HO2S 4P connector (A), and remove the harness clamps (B)

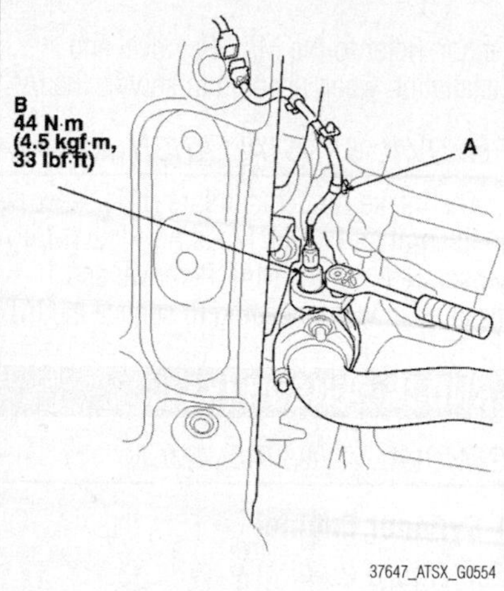
B
44 N·m
(4.5 kgf·m,
33 lbf·ft)
A
37647_ATSX_G0554

Fig. 291 Remove the harness clamp (A), then remove the front secondary HO2S (B)

4. Install the parts in the reverse order of removal.

Rear Bank (Bank 1)

See Figure 292.

1. Raise the vehicle on a lift.
2. Disconnect the rear secondary HO2S 4P connector, and remove the harness clamps.
3. Remove the rear secondary HO2S.
4. Install the parts in the reverse order of removal.

INTAKE AIR TEMPERATURE (IAT) SENSOR

LOCATION

The Intake Air Temperature (IAT) sensor is an integral part of the Mass Air Flow (MAF)

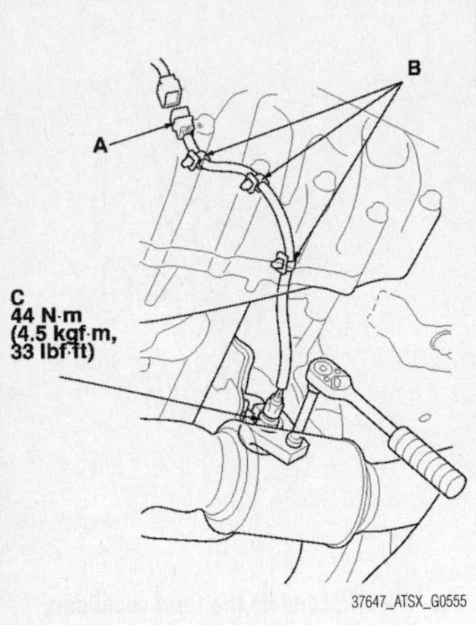
B
A
C
44 N·m
(4.5 kgf·m,
33 lbf·ft)
37647_ATSX_G0555

Fig. 292 Disconnect the rear secondary HO2S 4P connector (A), and remove the harness clamps (B)

sensor. Refer to the MAF Removal and Installation when needing to service the IAT.

REMOVAL & INSTALLATION

The Intake Air Temperature (IAT) sensor is an integral part of the Mass Air Flow (MAF) sensor. Refer to the MAF Removal and Installation when needing to service the IAT.

KNOCK SENSOR (KS)

REMOVAL & INSTALLATION

4-Cylinder Engine

See Figure 293.

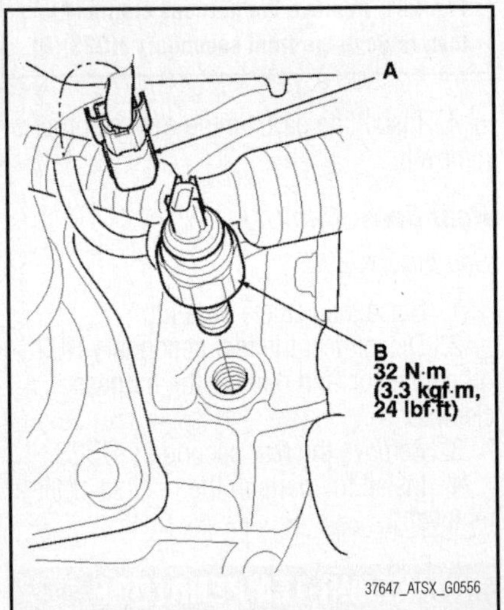
A
B
32 N·m
(3.3 kgf·m,
24 lbf·ft)
37647_ATSX_G0556

Fig. 293 Disconnect the knock sensor connector (A) and remove the knock sensor (B)

1. Remove the intake manifold.
2. Disconnect the knock sensor connector.
3. Remove the knock sensor.
4. Install the parts in the reverse order of removal.

6-Cylinder Engine

See Figure 294.

1. Remove the intake manifold.
2. Remove the injector base.
3. Disconnect the knock sensor connector.
4. Remove the knock sensor.
5. Install the parts in the reverse order of removal.

MASS AIR FLOW (MAF)/INTAKE AIR TEMPERATURE (IAT) SENSOR (HOT WIRE)

LOCATION

The MAF/IAT sensor is located on the engine air intake.

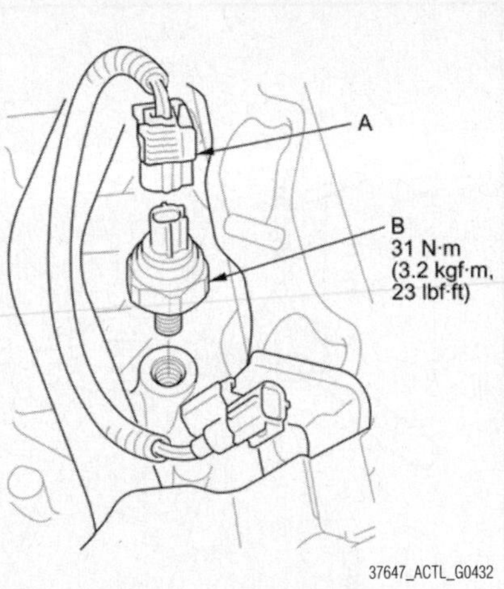
A
B
31 N·m
(3.2 kgf·m,
23 lbf·ft)
37647_ACTL_G0432

Fig. 294 Disconnect the knock sensor connector

REMOVAL & INSTALLATION

4-Cylinder Engine

See Figure 295.

1. Disconnect the MAF sensor/IAT sensor connector.
2. Remove the screws.
3. Remove the MAF sensor/IAT sensor.
4. Install the parts in the reverse order of removal with a new O-ring.

6-Cylinder Engine

See Figure 296.

1. Disconnect the MAF/IAT sensor 5P connector.

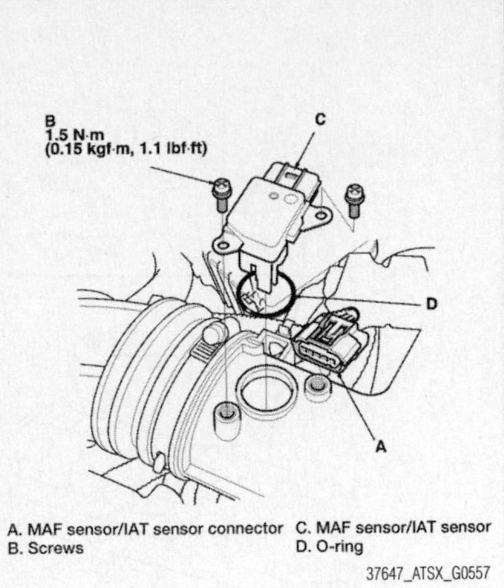
B
1.5 N·m
(0.15 kgf·m, 1.1 lbf·ft)
C
D
A
A. MAF sensor/IAT sensor connector
B. Screws
C. MAF sensor/IAT sensor
D. O-ring
37647_ATSX_G0557

Fig. 295 Disconnect the MAF sensor/IAT sensor connector

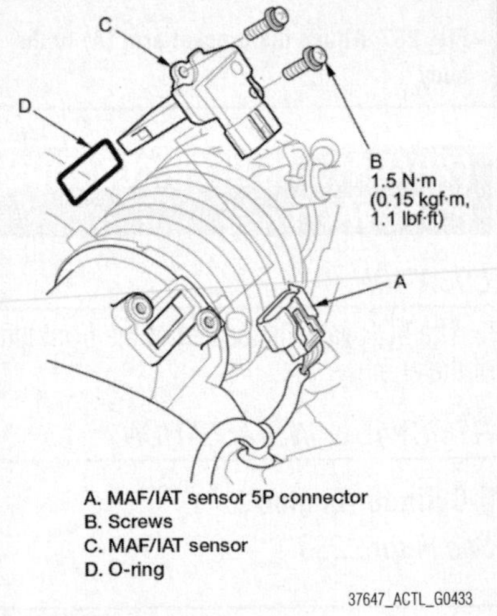
C
D
B
1.5 N·m
(0.15 kgf·m,
1.1 lbf·ft)
A
A. MAF/IAT sensor 5P connector
B. Screws
C. MAF/IAT sensor
D. O-ring
37647_ACTL_G0433

Fig. 296 Disconnect the MAF/IAT sensor 5P connector

2. Remove the screws.
3. Remove the MAF/IAT sensor.
4. Install the parts in the reverse order of removal with a new O-ring.

MANIFOLD ABSOLUTE PRESSURE (MAP) SENSOR

LOCATION

Refer to graphic in the Removal and Installation section for location.

REMOVAL & INSTALLATION

4-Cylinder Engine

See Figure 297.

1. Disconnect the MAP sensor connector.

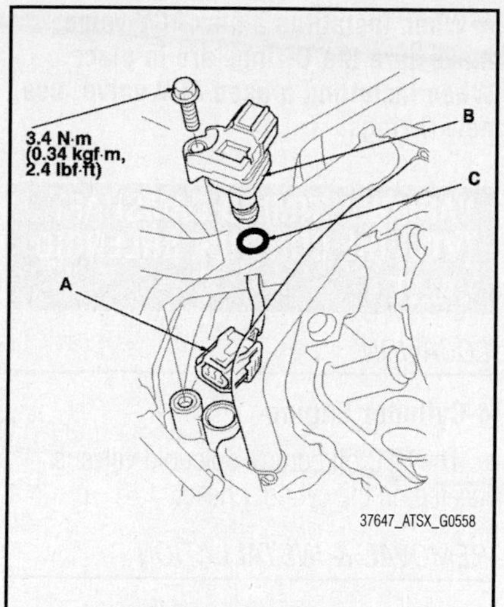
3.4 N·m
(0.34 kgf·m,
2.4 lbf·ft)
B
C
A
37647_ATSX_G0558

Fig. 297 Disconnect the MAP sensor connector (A) and remove the MAP sensor (B)

2. Remove the MAP sensor.
3. Install the parts in the reverse order of removal with a new O-ring.

6-Cylinder Engine

See Figure 298.

1. Disconnect the MAP sensor connector.
2. Remove the screw.
3. Remove the MAP sensor.
4. Install the parts in the reverse order of removal with a new O-ring.

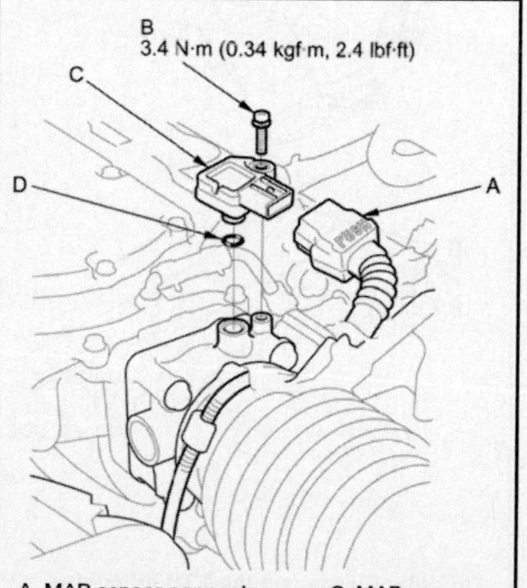
B
3.4 N·m (0.34 kgf·m, 2.4 lbf·ft)
C
D
A

A. MAP sensor connector
B. Screw
C. MAP sensor
D. O-ring

37647_ACTL_G0434

Fig. 298 Disconnect the MAP sensor connector

OUTPUT SHAFT SPEED (OSS) SENSOR

LOCATION

The Output Shaft Speed (OSS) sensor is located on the A/T transaxle.

REMOVAL & INSTALLATION

4-Cylinder Engine

See Figures 299 through 301.

1. Remove the intake air duct.
2. Remove the air cleaner housing.
3. Disconnect the output shaft (countershaft) speed sensor connector, and remove the output shaft (countershaft) speed sensor and the speed sensor washer.

To install:

4. Install a new O-ring on a new output shaft (countershaft) speed sensor, then install the output shaft (countershaft) speed sensor with the speed sensor washer in the transmission housing.
5. Check the connector for rust, dirt, or oil, and clean or repair if necessary. Then connect the connector securely.
6. Install the air cleaner housing.
7. Install the intake air duct.

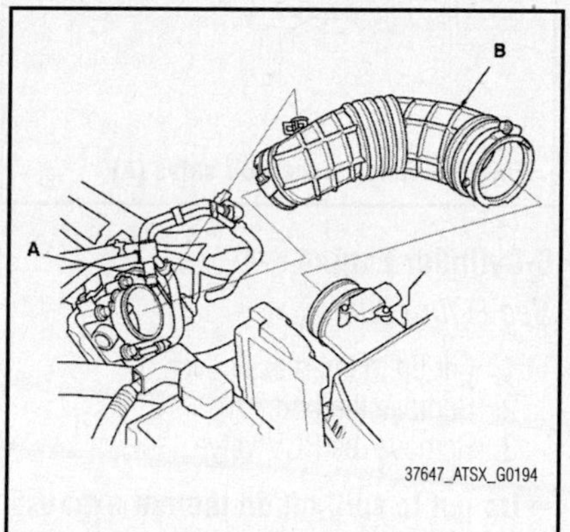
B
A
37647_ATSX_G0194

Fig. 299 Remove the intake air duct

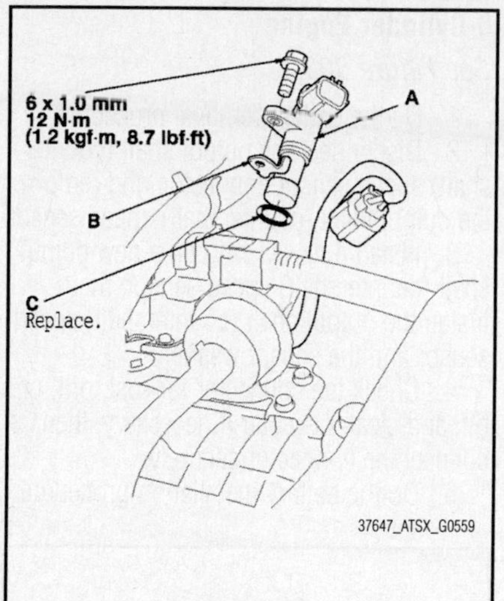
6 x 1.0 mm
12 N·m
(1.2 kgf·m, 8.7 lbf·ft)
A
B
C
Replace.
37647_ATSX_G0559

Fig. 301 Disconnect the output shaft (countershaft) speed sensor connector, and remove the output shaft (countershaft) speed sensor (A) and the speed sensor washer (B)

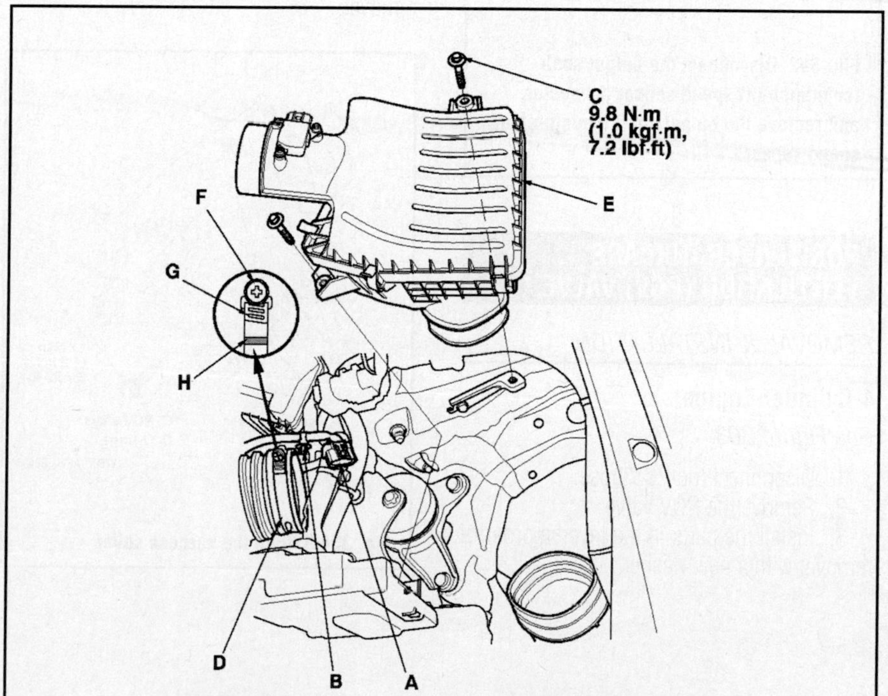
C
9.8 N·m
(1.0 kgf·m,
7.2 lbf·ft)
E
F
G
H
D
B
A

A. MAF/IAT sensor connector
B. Harness clamp
C. Bolts
D. Band
E. Air cleaner housing
F. Hose band
G. Edge of hose band
H. Painted mark

37647_ATSX_G0195

Fig. 300 Remove the air cleaner housing

6-Cylinder Engine

See Figure 302.

1. Do the battery removal procedure.
2. Disconnect the output shaft (countershaft) speed sensor connector, and remove the output shaft (countershaft) speed sensor.
3. Install a new O-ring on a new output shaft (countershaft) speed sensor, then install the output shaft (countershaft) speed sensor and the sensor washer.
4. Check the connector for rust, dirt, or oil, and clean or repair if necessary, then connect the connector securely.
5. Do the battery installation procedure.

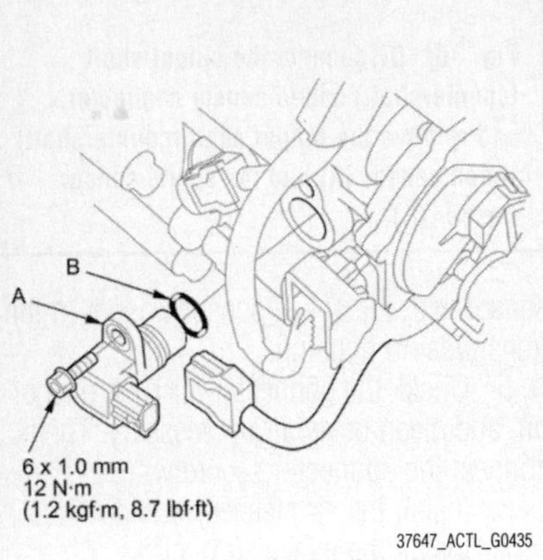
B
A
6 x 1.0 mm
12 N·m
(1.2 kgf·m, 8.7 lbf·ft)
37647_ACTL_G0435

Fig. 302 Disconnect the output shaft (countershaft) speed sensor connector, and remove the output shaft (countershaft) speed sensor

POSITIVE CRANKCASE VENTILATION (PCV) VALVE

REMOVAL & INSTALLATION

4-Cylinder Engine

See Figure 303.

1. Disconnect the PCV hose.
2. Remove the PCV valve.
3. Install the parts in the reverse order of removal with a new washer.

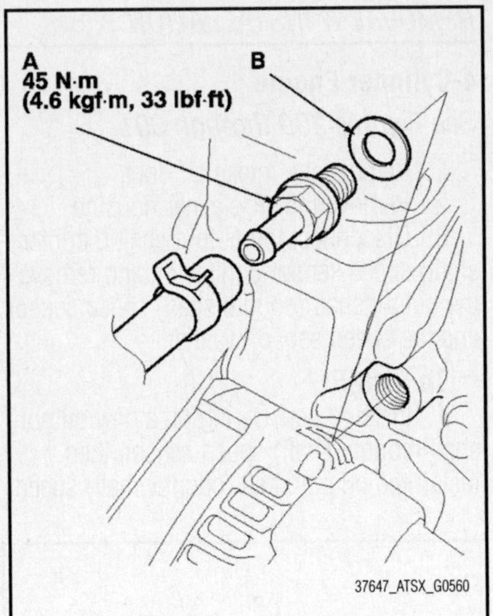
A
45 N·m
(4.6 kgf·m, 33 lbf·ft)
B
37647_ATSX_G0560

Fig. 303 Remove the PCV valve (A)

6-Cylinder Engine

See Figure 304.

1. Lift up the harness cover.
2. Remove the bolt.
3. Remove the PCV valve.

➡Do not to spill oil on the hot exhaust manifold.

4. Install the parts in the reverse order of removal.

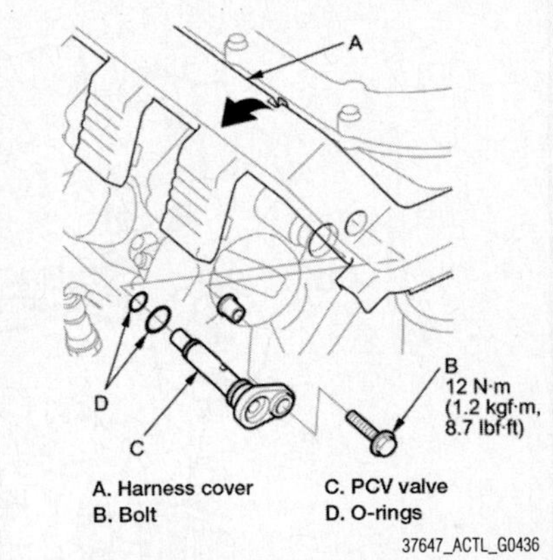
A
B
12 N·m
(1.2 kgf·m,
8.7 lbf·ft)
D
C
A. Harness cover
B. Bolt
C. PCV valve
D. O-rings
37647_ACTL_G0436

Fig. 304 Lift up the harness cover

➡When installing a new PCV valve, make sure the O-rings are in place. When installing a used PCV valve, use new O-rings.

VARIABLE TIMING CAMSHAFT (VTC) OIL CONTROL SOLENOID VALVE

LOCATION

4-Cylinder Engine

The VTC oil control solenoid valve is located on the cylinder head.

REMOVAL & INSTALLATION

4-Cylinder Engine

See Figure 305.

1. Disconnect the VTC oil control solenoid valve 2P connector.
2. Remove the bolt and the VTC oil control solenoid valve.
3. Installation is the reverse of removal.

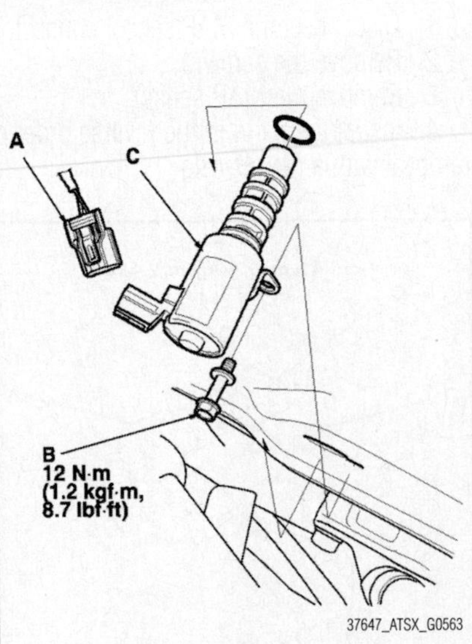
A
C
B
12 N·m
(1.2 kgf·m,
8.7 lbf·ft)
37647_ATSX_G0563

Fig. 305 Disconnect the VTC oil control solenoid valve 2P connector

FUEL SYSTEM SERVICE PRECAUTIONS

Safety is the most important factor when performing not only fuel system maintenance, but any type of maintenance. Failure to conduct maintenance and repairs in a safe manner may result in serious personal injury or death. Work on a vehicle's fuel system components can be accomplished safely and effectively by adhering to the following rules and guidelines.

- To avoid the possibility of fire and personal injury, always disconnect the negative battery cable unless the repair or test procedure requires that battery voltage be applied.
- Always relieve the fuel system pressure prior to disconnecting any fuel system component (injector, fuel rail, pressure regulator, etc.) fitting or fuel line connection. Exercise extreme caution whenever relieving fuel system pressure to avoid exposing skin, face and eyes to fuel spray. Please be advised that fuel under pressure may penetrate the skin or any part of the body that it contacts.
- Always place a shop towel or cloth around the fitting or connection prior to loosening to absorb any excess fuel due to spillage. Ensure that all fuel spillage is quickly removed from engine surfaces. Ensure that all fuel-soaked cloths or towels are deposited into a flame-proof waste container with a lid.
- Always keep a dry chemical (Class B) fire extinguisher near the work area.
- Do not allow fuel spray or fuel vapors to come into contact with a spark or open flame.
- Always use a second wrench when loosening or tightening fuel line connection fittings. This will prevent unnecessary stress and torsion on fuel piping. Always follow the proper torque specifications.
- Always replace worn fuel fitting O-rings with new ones. Do not substitute fuel hose where rigid pipe is installed.

FUEL SYSTEM PRESSURE

RELIEVING

CAUTION

Before disconnecting fuel lines or hoses, relieve pressure from the system by disabling the fuel pump and disconnecting the fuel line/quick connect fitting in the engine compartment.

With the HDS

1. Connect the HDS to the Data Link Connector (DLC) located under the driver's side of the dashboard.
2. Turn the ignition switch to ON (II).
3. Make sure the HDS communicates with the ECM/PCM. If it doesn't, go to the DLC circuit troubleshooting.
4. Turn the ignition switch to LOCK (0).
5. Remove the fuel fill cap to relieve the pressure in the fuel tank.
6. Turn the ignition switch to ON (II).
7. From the INSPECTIONMENU of the HDS, select Fuel Pump OFF, then start the engine, and let it idle until it stalls.

➡Do not allow the engine to idle above 1,000 rpm or the ECM/PCM will continue to operate the fuel pump. Pending or Confirmed DTC may be set during this procedure. Check for DTCs, and clear them as needed.

8. Turn the ignition switch to LOCK (0).
9. Do the battery terminal disconnection procedure.
10. Remove the cover and the quick-connect fitting cover.
11. Check the quick-connect fitting for dirt, and clean it if needed.
12. Place a rag or a shop towel over the quick-connect fitting.
13. Disconnect the quick-connect fitting.
    a. Hold the connector with one hand, and squeeze the retainer tabs with the other hand to release them from the locking tabs.
    b. Pull the connector off.

➡Be careful not to damage the line or other parts. Do not use tools. If the connector does not move, keep the retainer tabs pressed down, and alternately pull and push the connector until it comes off easily. Do not remove the retainer from the line; once removed, the retainer must be replaced with a new one.

14. After disconnecting the quick-connect fitting, check it for dirt or damage.
15. Do the battery terminal reconnection procedure.

Without the HDS

See Figure 306.

1. Remove the driver's dashboard undercover.
2. Remove PGM-FI main relay 2 from the driver's under-dash fuse/relay box.
3. Start the engine, and let it idle until it stalls.

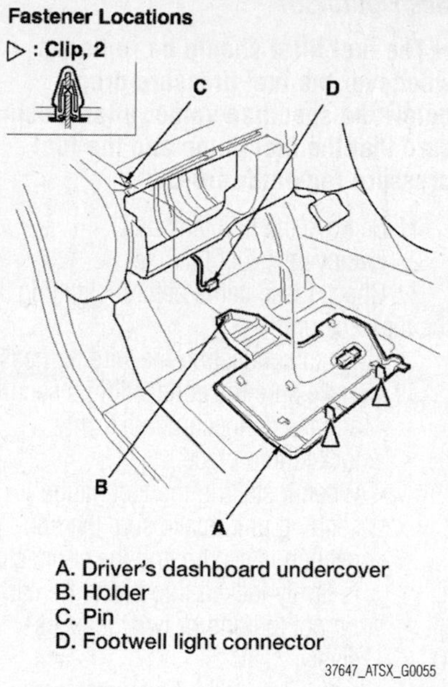
Fastener Locations
▷ : Clip, 2
C
D
B
A
A. Driver's dashboard undercover
B. Holder
C. Pin
D. Footwell light connector
37647_ATSX_G0055

Fig. 306 Remove driver's dashboard undercover

➡If any DTCs are stored, clear and ignore them.

4. Turn the ignition switch to LOCK (0).
5. Remove the fuel fill cap to relieve the pressure in the fuel tank.
6. Do the battery terminal disconnection procedure.
7. Remove the cover and the quick-connect fitting cover.
8. Check the quick-connect fitting for dirt, and clean it if needed.
9. Place a rag or a shop towel over the quick-connect fitting.
10. Disconnect the quick-connect fitting.
    a. Hold the connector with one hand, and squeeze the retainer tabs with the other hand to release them from the locking tabs.
    b. Pull the connector off.

➡Be careful not to damage the line or other parts. Do not use tools. If the connector does not move, keep the retainer tabs pressed down, and alternately pull and push the connector until it comes off easily. Do not remove the retainer from the line; once removed, the retainer must be replaced with a new one.

11. After disconnecting the quick-connect fitting, check it for dirt or damage.
12. Do the battery terminal reconnection procedure.

FUEL FILTER

REMOVAL & INSTALLATION

See Figure 307.

➡The fuel filter should be replaced whenever the fuel pressure drops below the specified value, after making sure that the fuel pump and the fuel pressure regulator are OK.

1. Remove the fuel tank unit.
2. Remove the fuel filter set.
3. Check these items before installing the fuel tank unit:
   - When connecting the wire harness, make sure the connection is secure and the connectors are firmly locked into place.
   - When installing the fuel gauge sending unit, make sure the connection is secure and the connector is firmly locked into place. Be careful not to bend or twist it excessively.
4. Install the parts in the reverse order of removal with new O-rings and a new bracket. When installing the fuel tank unit, align the marks on the unit and the fuel tank.

➡Coat the O-rings with clean engine oil; do not use any other oil or fluid. Do not pinch the O-rings during installation. Use all the new parts supplied in the fuel filter replacement kit.

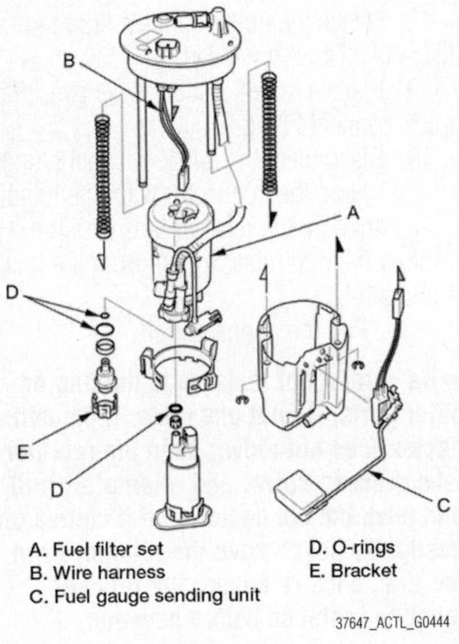
B
A
D
E
D
C
A. Fuel filter set
B. Wire harness
C. Fuel gauge sending unit
D. O-rings
E. Bracket
37647_ACTL_G0444

Fig. 307 Exploded view of fuel pump assembly

FUEL PUMP/FUEL GAUGE SENDING UNIT

REMOVAL & INSTALLATION

See Figures 308 through 311.

1. Remove the fuel tank unit.

Special Tools Required: Fuel sender wrench 07AAA-S0XA100

   a. Relieve the fuel pressure.
   b. Remove the rear seat cushion cover.
   c. Remove the access panel from the floor.
   d. Disconnect the fuel tank unit 4P connector.
   e. Disconnect the quick-connect fitting from the fuel tank unit.
   f. Using the special tool, loosen the locknut.
   g. Remove the locknut and the fuel tank unit.
2. Remove the fuel level sensor (fuel gauge sending unit) from the fuel tank unit.
3. Check these items before installing the fuel tank unit:

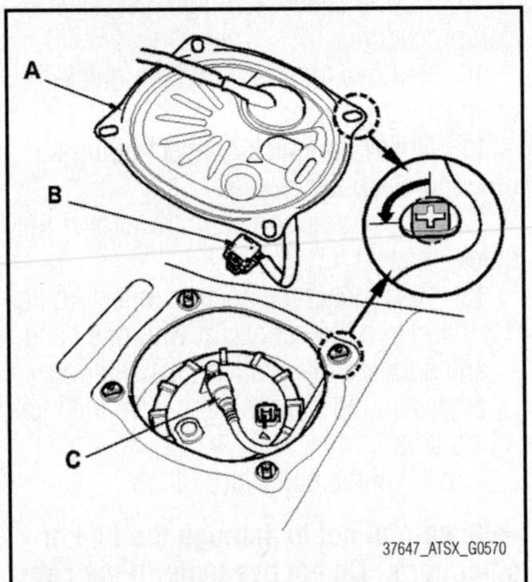
A
B
C
37647_ATSX_G0570

Fig. 308 Remove the access panel (A) and disconnect the fuel tank unit 4P connector (B) and the quick-connect fitting (C)

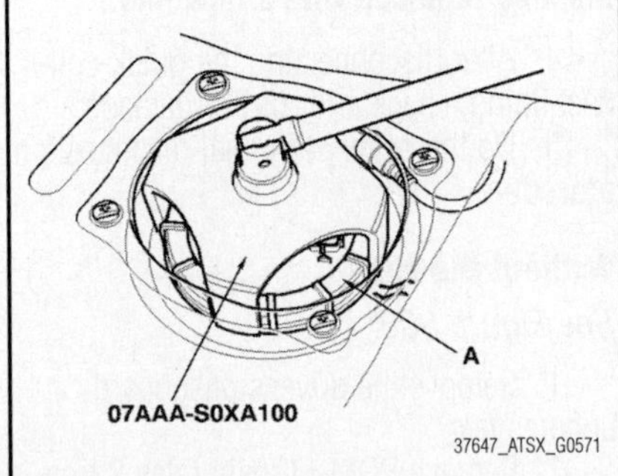
A
07AAA-S0XA100
37647_ATSX_G0571

Fig. 309 Using the special tool, loosen the locknut (A)

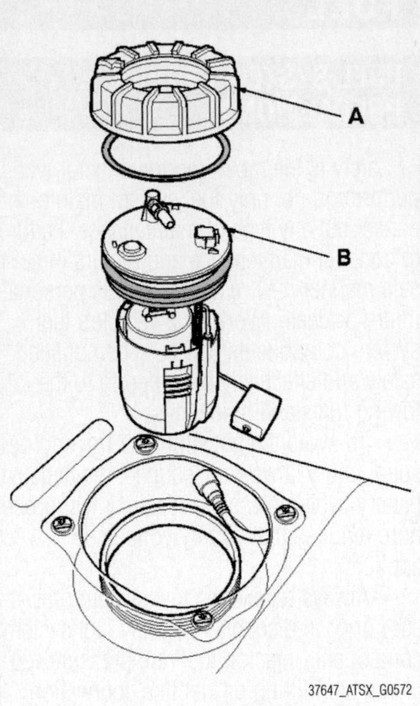
A
B
37647_ATSX_G0572

Fig. 310 Remove the locknut (A) and the fuel tank unit (B)

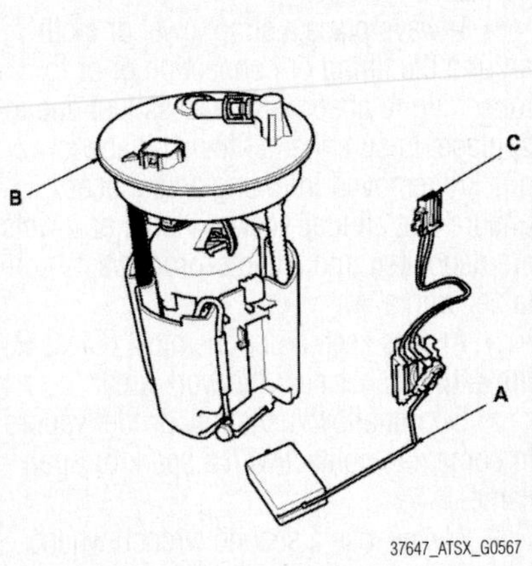
C
B
A
37647_ATSX_G0567

Fig. 311 Remove the fuel level sensor (fuel gauge sending unit) (A) from the fuel tank unit (B)

   a. When connecting the wire harness, make sure the connection is secure and the connector is firmly locked into place.
   b. When installing the fuel gauge sending unit, make sure the connection is secure. Be careful not to bend or twist it excessively.
4. Install the parts in the reverse order of removal. When installing the fuel tank unit, align the marks on the unit and the fuel tank.

FUEL RAIL & INJECTOR

REMOVAL & INSTALLATION

4-Cylinder Engine

See Figures 312 and 313.

1. Relieve the fuel pressure.
2. Remove the engine cover.
3. Remove the quick-connect fitting cover, then disconnect the quick-connect fitting.
4. Disconnect the injector connectors and the engine mount control solenoid valve connector.
5. Remove the ground cable bolts.
6. Remove the fuel rail mounting nuts from the fuel rail.
7. Remove the fuel rail and the injectors from the injector base.
8. Remove the injector clips from the fuel rail.
9. Remove the injectors from the fuel rail.

To install:

10. Coat the new O-rings (black) with clean engine oil, and insert the injectors into the fuel rail.
11. Install the injectors clips.
12. Coat the new injector O-rings (brown) with clean engine oil.
13. Install the fuel rail and the injectors in the injector base.
14. Install the fuel rail mounting nuts and the ground cable bolts.
15. Connect the injector connectors and the engine mount control solenoid valve connector.
16. Connect the quick-connect fitting, and quick-connect fitting cover.
17. Turn the ignition switch to ON (II), but do not operate the starter. After the fuel pump runs for about 2 seconds, the fuel rail will be pressurized. Repeat this two or three times, then make sure there are no fuel leaks.
18. Reinstall the engine cover.

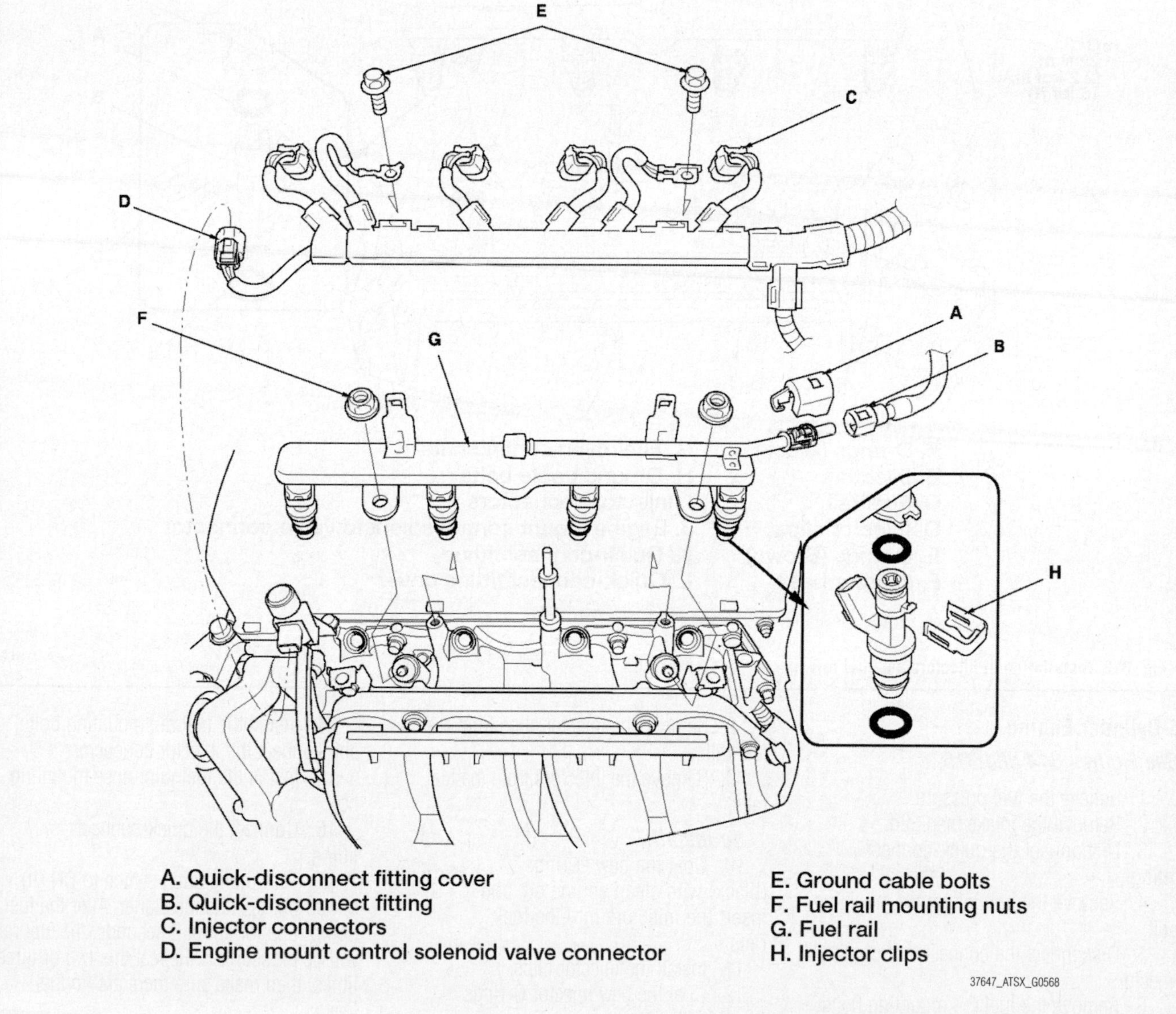
E
C
D
A
F
G
B
H
A. Quick-disconnect fitting cover
B. Quick-disconnect fitting
C. Injector connectors
D. Engine mount control solenoid valve connector
E. Ground cable bolts
F. Fuel rail mounting nuts
G. Fuel rail
H. Injector clips
37647_ATSX_G0568

Fig. 312 Fuel rail assembly

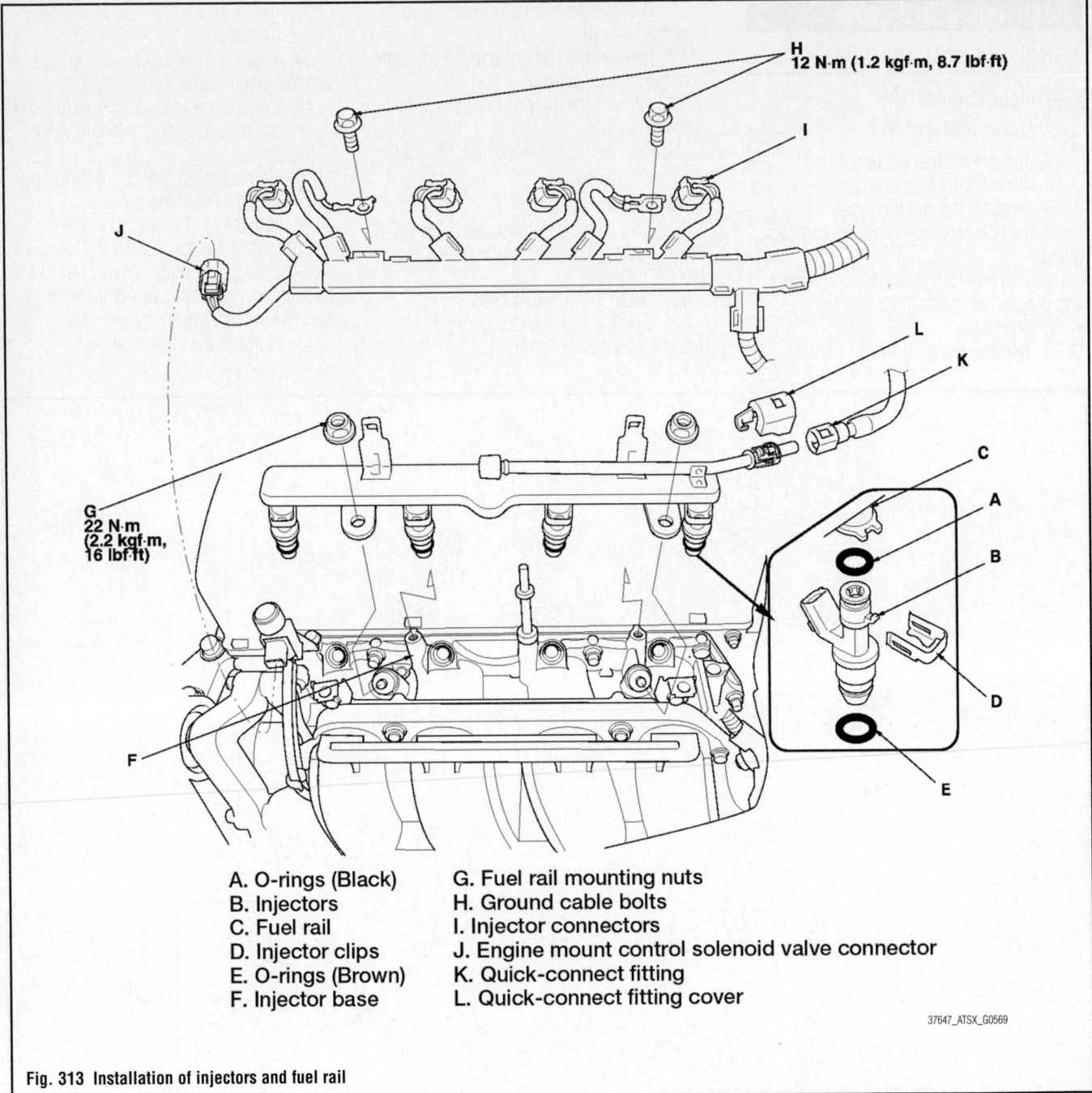

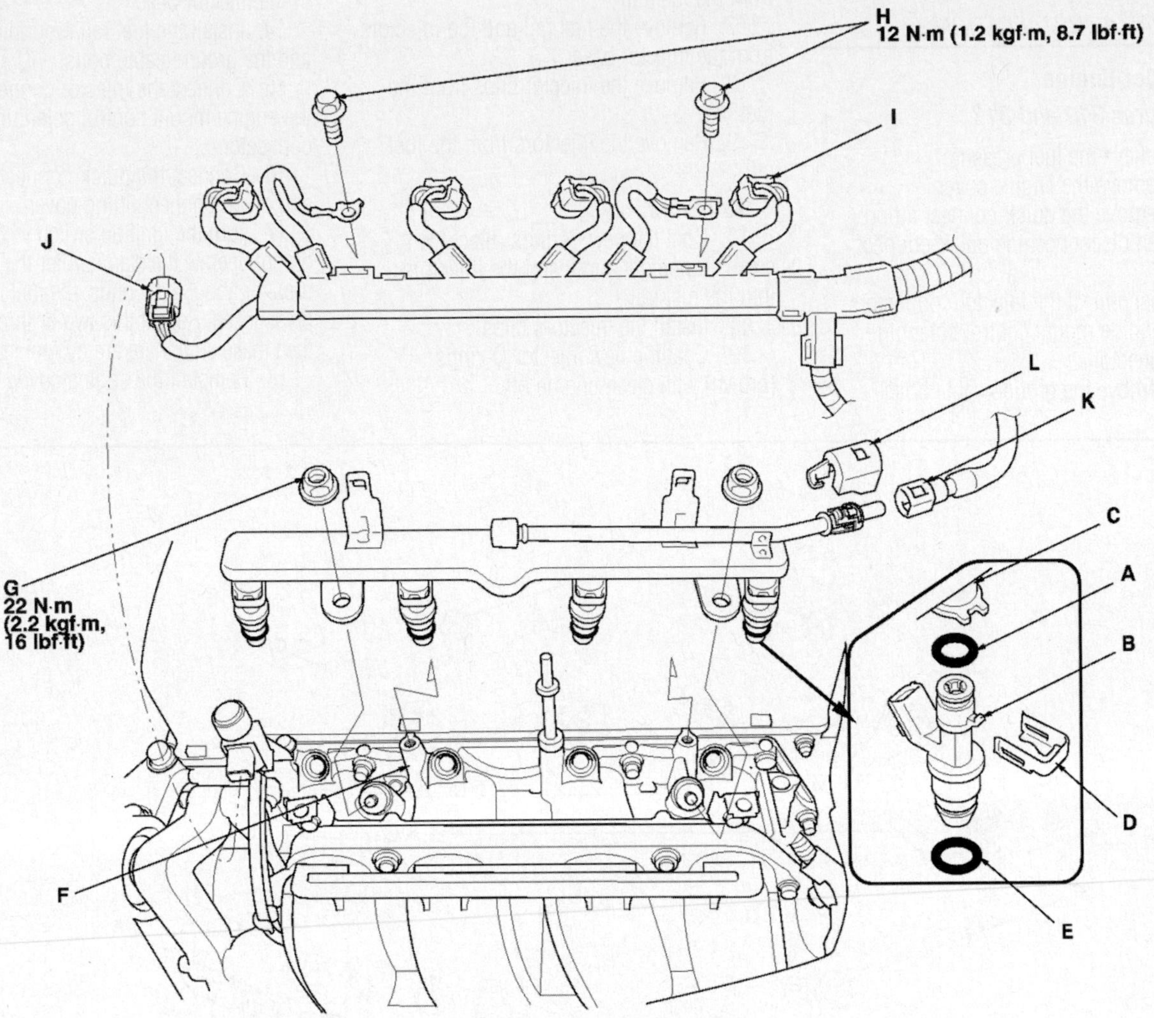
H
12 N·m (1.2 kgf·m, 8.7 lbf·ft)
I
J
L
K
C
A
B
D
E
G
22 N·m
(2.2 kgf·m,
16 lbf·ft)
F
A. O-rings (Black)
B. Injectors
C. Fuel rail
D. Injector clips
E. O-rings (Brown)
F. Injector base
G. Fuel rail mounting nuts
H. Ground cable bolts
I. Injector connectors
J. Engine mount control solenoid valve connector
K. Quick-connect fitting
L. Quick-connect fitting cover
37647_ATSX_G0569

Fig. 313 Installation of injectors and fuel rail

6-Cylinder Engine

See Figures 314 and 315.

1. Relieve the fuel pressure.
2. Remove the intake manifold.
3. Disconnect the quick-connect fitting.
4. Remove the fuel joint hose mounting bolt.
5. Disconnect the connectors from the injectors.
6. Remove the fuel rail mounting bolts from the fuel rails.
7. Remove the fuel rails and the injectors from the injector base.
8. Remove the injector clips from the fuel rails.
9. Remove the injectors from the fuel rails.

To install:

10. Coat the new O-rings (black) with clean engine oil, and insert the injectors into the fuel rails.
11. Install the injector clips.
12. Coat the new injector O-rings (green) with clean engine oil.
13. Install the fuel rails and the injectors in the injector base.
14. Install the fuel rail mounting bolts, and connect the injector connectors.
15. Install the fuel joint hose mounting bolt.
16. Connect the quick-connect fitting.
17. Turn the ignition switch to ON (II), but do not operate the starter. After the fuel pump runs for about 2 seconds, the fuel rail will be pressurized. Repeat this two or three times, then make sure there are no fuel leaks.
18. Install the intake manifold with a new gasket.

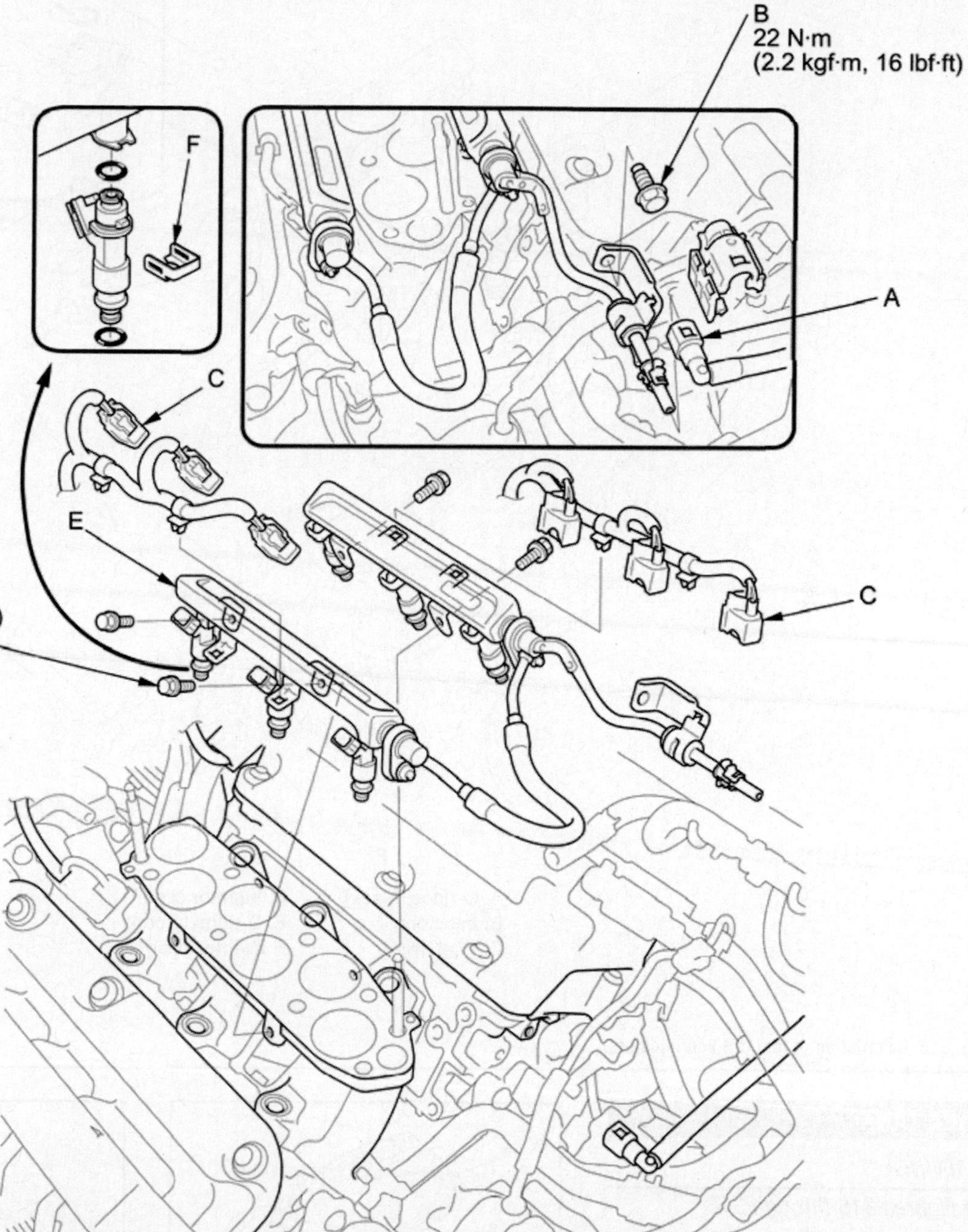
B
22 N·m
(2.2 kgf·m, 16 lbf·ft)
F
A
C
E
D
9.8 N·m
(1.0 kgf·m, 7.2 lbf·ft)
C

A. Quick-connect fitting
B. Fuel joint hose mounting bolt
C. Connectors
D. Fuel rail mounting bolts
E. Fuel rails
F. Injector clips

37647_ACTL_G0449

Fig. 314 Exploded view of fuel rail and injector assembly

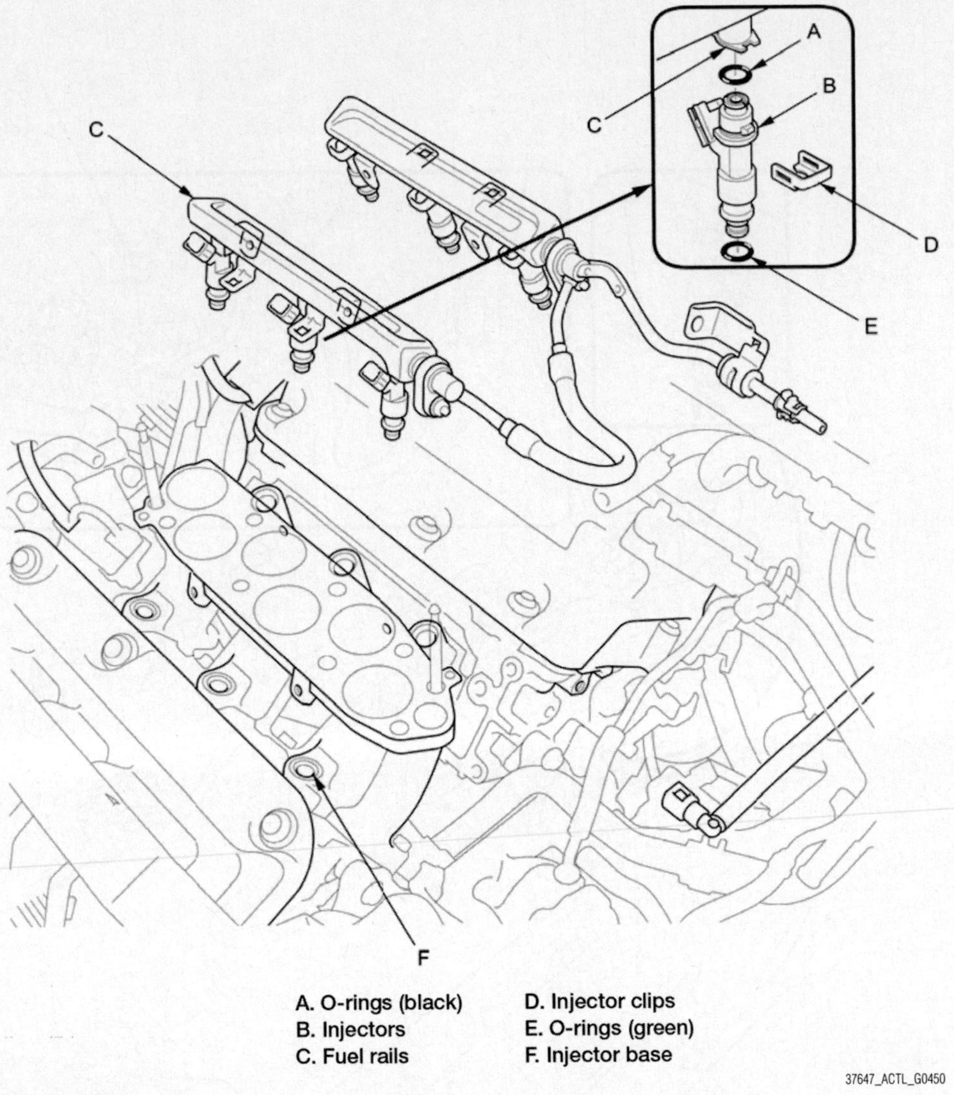
A
B
C
C
D
E
F
A. O-rings (black)
B. Injectors
C. Fuel rails
D. Injector clips
E. O-rings (green)
F. Injector base
37647_ACTL_G0450

Fig. 315 Installation of fuel rail and injectors

FUEL TANK

DRAINING

See Figures 316 through 318.

1. Remove the fuel tank unit.

Special Tools Required: Fuel sender wrench 07AAA-S0XA100

   a. Relieve the fuel pressure.
   b. Remove the rear seat cushion cover.
   c. Remove the access panel from the floor.
   d. Disconnect the fuel tank unit 4P connector.
   e. Disconnect the quick-connect fitting from the fuel tank unit.

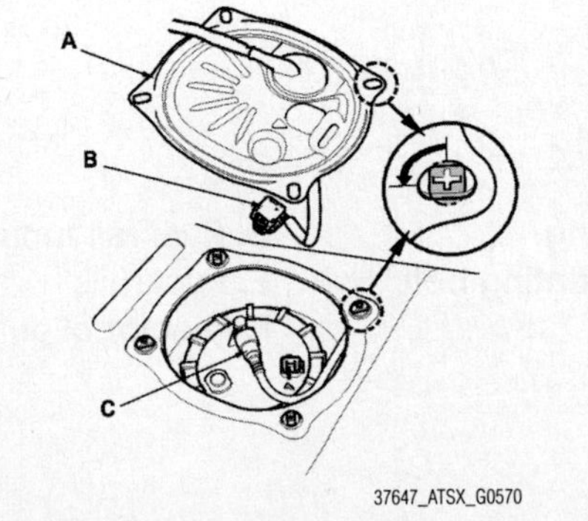
A
B
C
37647_ATSX_G0570

Fig. 316 Remove the access panel (A) and disconnect the fuel tank unit 4P connector (B) and the quick-connect fitting (C)

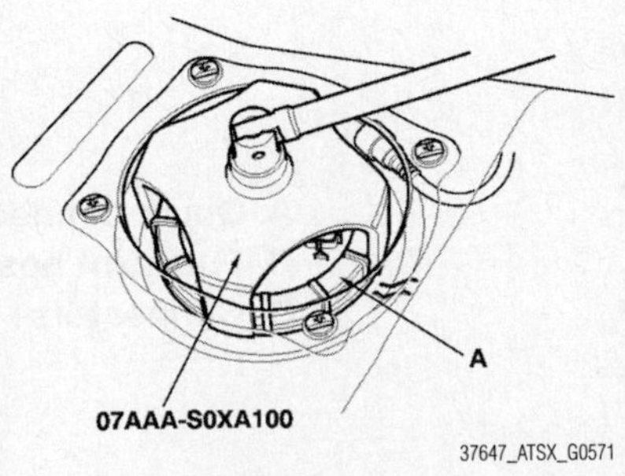
07AAA-S0XA100
A
37647_ATSX_G0571

Fig. 317 Using the special tool, loosen the locknut (A)

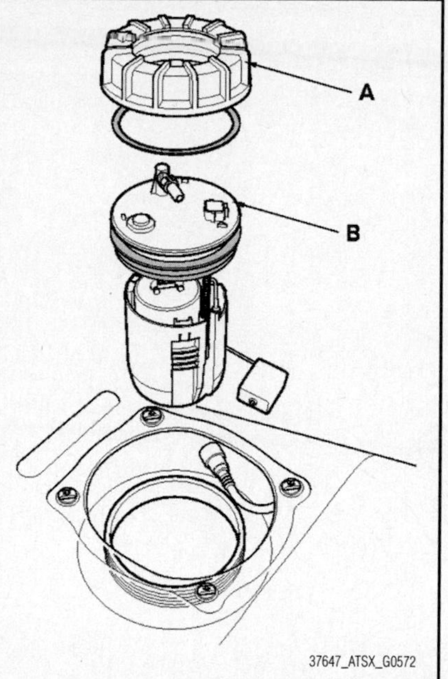
A
B
37647_ATSX_G0572

Fig. 318 Remove the locknut (A) and the fuel tank unit (B)

f. Using the special tool, loosen the locknut.

g. Remove the locknut and the fuel tank unit.

2. Using a hand pump, a hose, and a container suitable for fuel, draw the fuel from the fuel tank.
3. Reinstall the fuel tank unit.

REMOVAL & INSTALLATION

See Figures 319 through 323.

1. Drain the fuel tank.
2. Reinstall the fuel tank unit without connecting the fuel tank unit

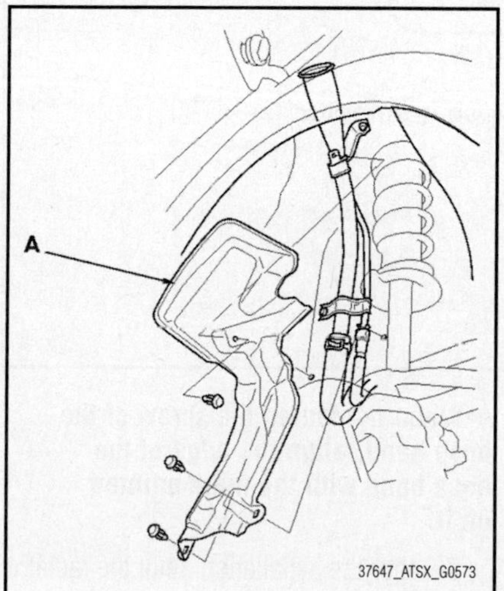
A
37647_ATSX_G0573

Fig. 319 Remove the fuel fill pipe cover (A)

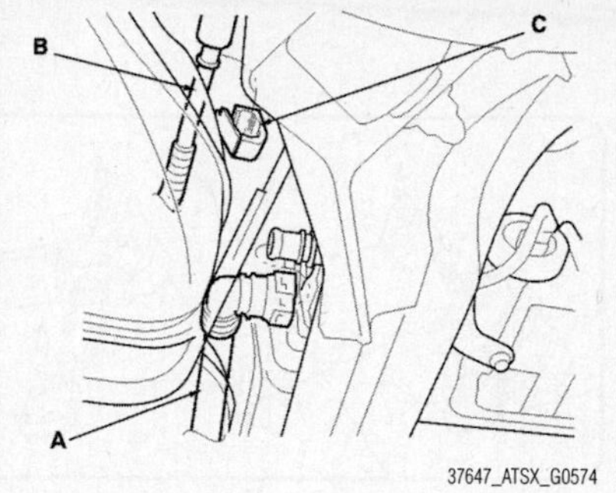
B
C
A
37647_ATSX_G0574

Fig. 320 Disconnect the hose (A) from the EVAP canister

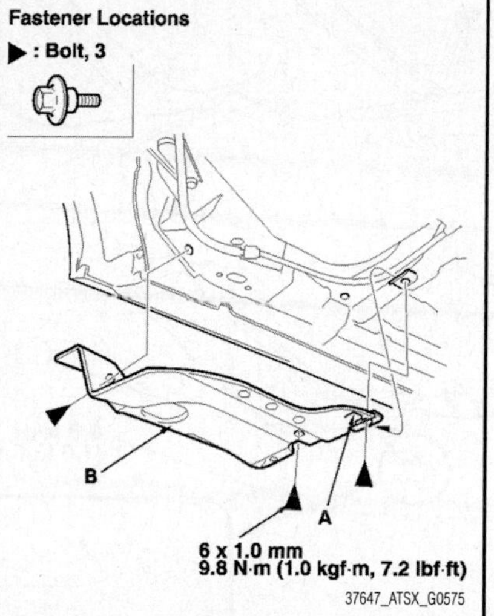
Fastener Locations
▶ : Bolt, 3
B
A
6 x 1.0 mm
9.8 N·m (1.0 kgf·m, 7.2 lbf·ft)
37647_ATSX_G0575

Fig. 321 Remove the middle floor under-cover (Left)

4P connector and the quick-connect fitting.

3. Remove the fuel fill pipe cover.
4. Disconnect the quick-connect fittings and the fuel fill tube from the fuel fill pipe. Slide back the clamps, then twist the hose as you pull to avoid damaging them.
5. Raise the vehicle on a lift.
6. Disconnect the hose from the EVAP canister.
7. Remove the hose from the clamp.

➡Be careful not to damage the hose.

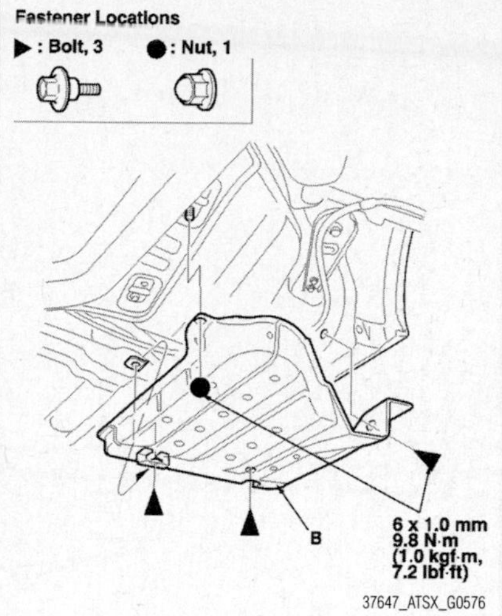
Fastener Locations
▶ : Bolt, 3
● : Nut, 1
B
6 x 1.0 mm
9.8 N·m
(1.0 kgf·m,
7.2 lbf·ft)
37647_ATSX_G0576

Fig. 322 Remove the middle floor under-cover (Right)

8. Remove the exhaust pipe.
9. Remove the middle floor undercover.
10. Remove the fuel tank protector.
11. Place a jack or other support under the fuel tank.
12. Remove the strap bolts and the straps.
13. Remove the fuel tank.
14. Install the parts in the reverse order of removal.

➡New fuel tanks have a ring pull at the fuel vapor hose connector. When you connect the hose and confirm that the connection is secure, remove the ring pull by pulling it down.

Before connecting the fuel fill pipe and the quick-connect fitting, check for dirt, and clean them if needed, taking care not to damage the fuel fill pipe or other parts.

When installing the fuel tank protector, make sure to insert it into the clip in the direction shown.

IDLE SPEED

ADJUSTMENT

The idle speed is controlled by the ECM/PCM. No adjustment is possible or necessary.

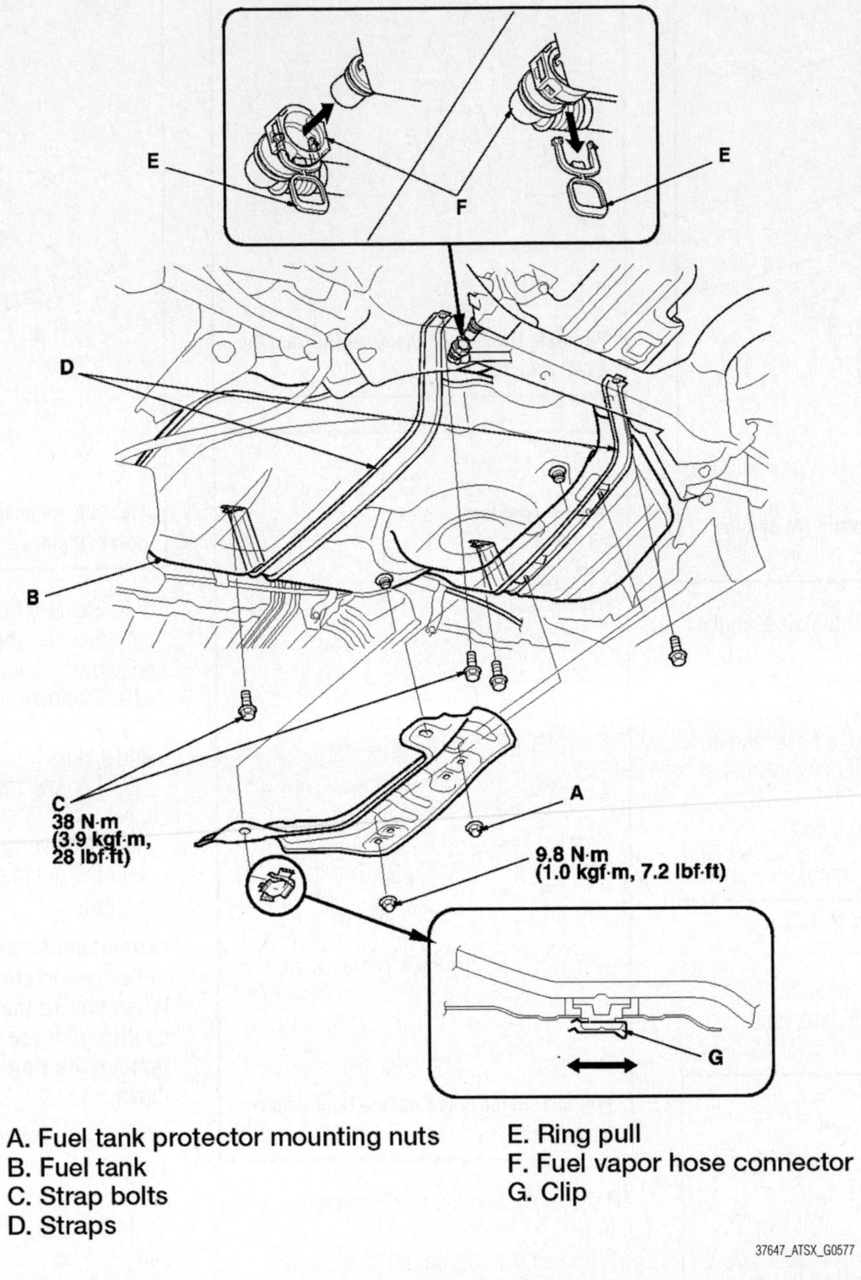
E
E
F
D
B
C
38 N·m
(3.9 kgf·m,
28 lbf·ft)
A
9.8 N·m
(1.0 kgf·m, 7.2 lbf·ft)
G
A. Fuel tank protector mounting nuts
B. Fuel tank
C. Strap bolts
D. Straps
E. Ring pull
F. Fuel vapor hose connector
G. Clip
37647_ATSX_G0577

Fig. 323 Removing the fuel tank

THROTTLE BODY

REMOVAL & INSTALLATION

4-Cylinder Engine

See Figure 324.

1. Turn the ignition switch to LOCK (0).
2. Remove the intake air duct.
3. Disconnect the throttle body connector.
4. Disconnect and plug the water bypass hoses.
5. Remove the throttle body.
6. Install the parts in the reverse order of removal with a new gasket.

➡When tightening the screw of the hose band, align the edge of the hose band with the mark painted on it.

7. After the installation, refill the radiator with engine coolant.
8. Do the ECM/PCM idle learn procedure.

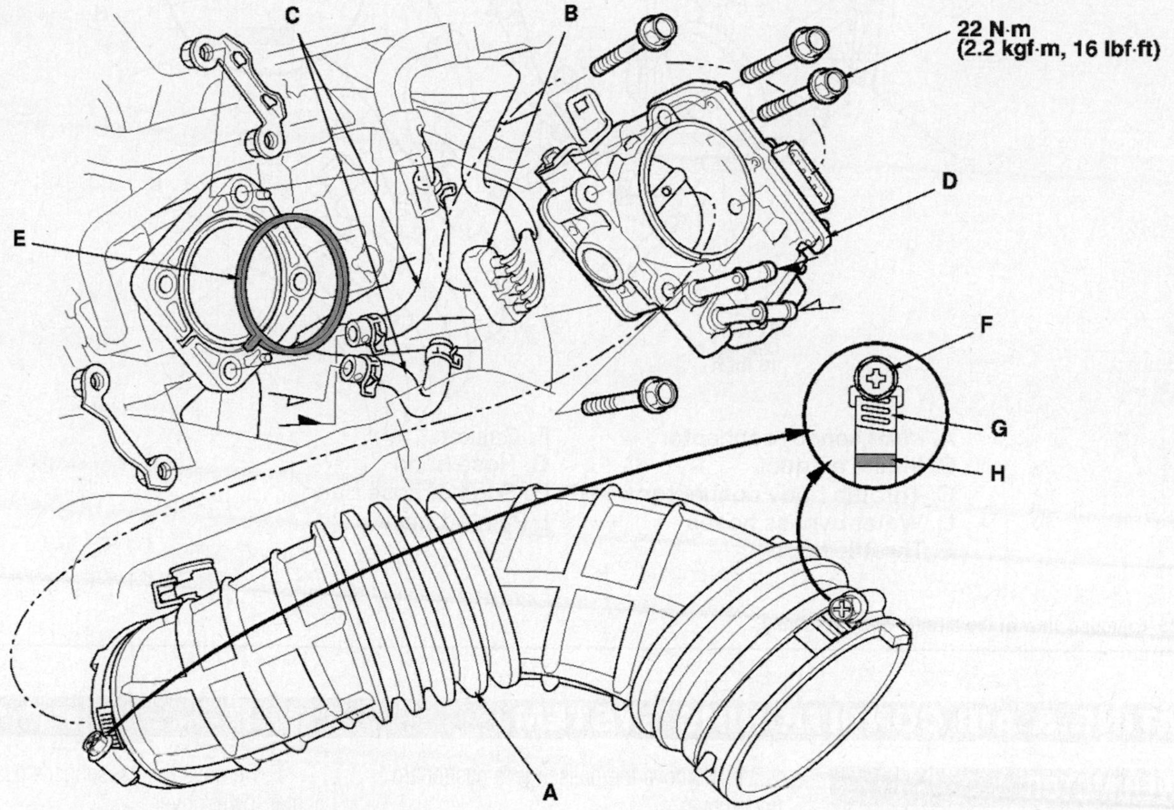
C
B
22 N·m
(2.2 kgf·m, 16 lbf·ft)
D
E
F
G
H
A

A. Intake air duct
B. Throttle body connector
C. Water bypass hoses
D. Throttle body
E. Gasket
F. Hose band
G. Edge of hose band
H. Painted mark

37647_ATSX_G0578

Fig. 324 Exploded view of throttle body removal

6-Cylinder Engine

See Figure 325.

1. Turn the ignition switch to LOCK (0).
2. Disconnect the MAP sensor connector.
3. Remove the intake air duct.
4. Disconnect the throttle body connector.
5. Disconnect and plug the water bypass hoses, then remove the throttle body.
6. Install the parts in the reverse order of removal with a new gasket.

➡When tightening the screw of the hose band, align the edge of the hose band with the mark painted on it.

7. After the installation, refill the radiator with engine coolant.
8. Do the PCM idle learn procedure.

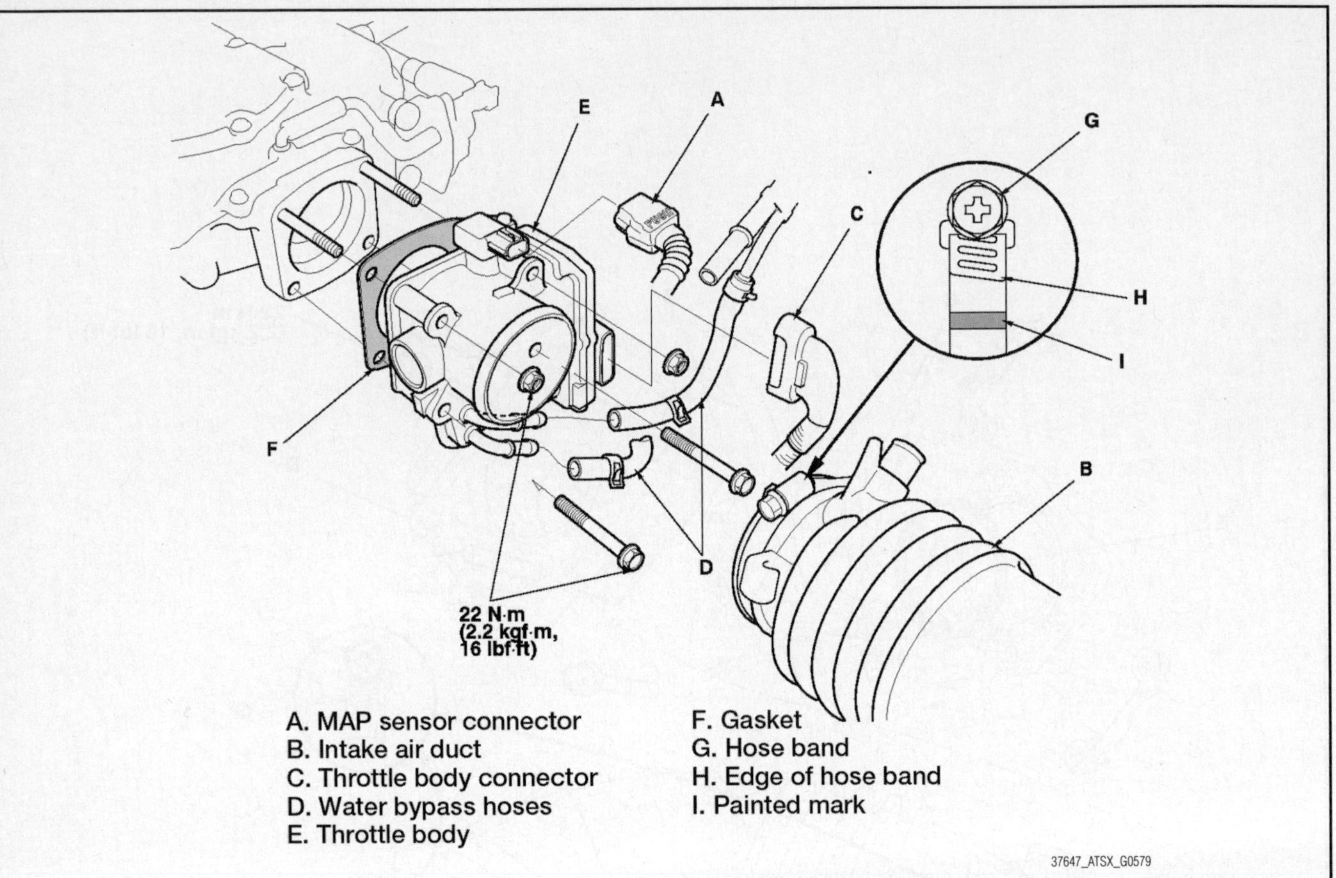
E
A
G
C
H
I
F
B
D
22 N·m
(2.2 kgf·m,
16 lbf·ft)
A. MAP sensor connector
B. Intake air duct
C. Throttle body connector
D. Water bypass hoses
E. Throttle body
F. Gasket
G. Hose band
H. Edge of hose band
I. Painted mark
37647_ATSX_G0579

Fig. 325 Exploded view of the throttle body assembly

HEATING & AIR CONDITIONING SYSTEM

BLOWER MOTOR

REMOVAL & INSTALLATION

See Figures 326 through 331.

1. Remove the passenger's dashboard undercover.
2. Remove the passenger's dashboard trim.
3. Remove the glove box.
4. Remove the passenger's dashboard center lower cover.
5. Remove the bolts and the connector clips. Then cut the plastic cross brace in the glove box opening with

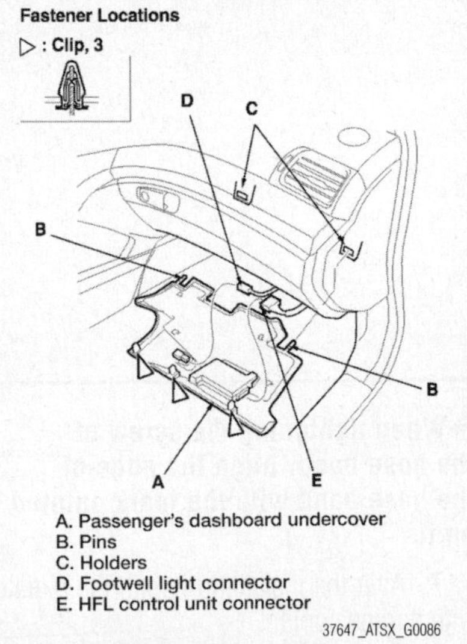
Fastener Locations
▷ : Clip, 3
D
C
B
B
A
E
A. Passenger's dashboard undercover
B. Pins
C. Holders
D. Footwell light connector
E. HFL control unit connector
37647_ATSX_G0086

Fig. 326 Remove the passenger's dashboard undercover

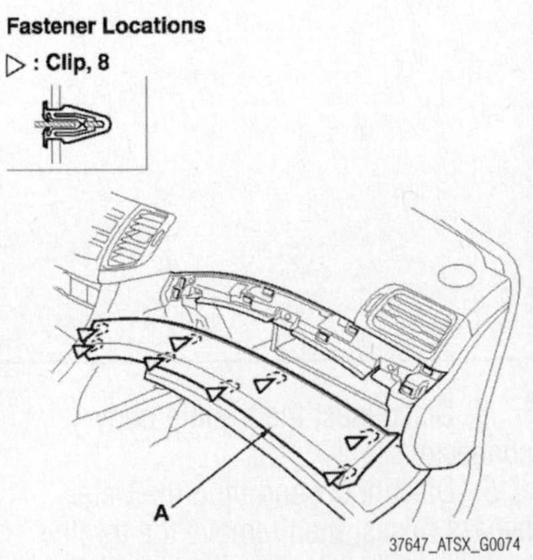
Fastener Locations
▷ : Clip, 8
A
37647_ATSX_G0074

Fig. 327 Remove the passenger's dashboard trim (A)

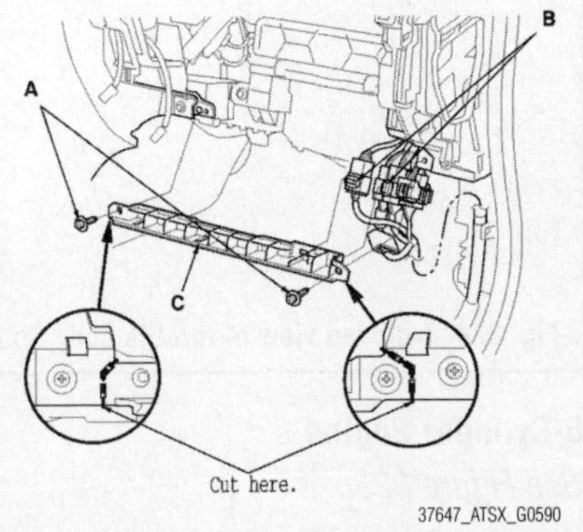
B
A
C
Cut here.
37647_ATSX_G0590

Fig. 328 Remove the bolts (A) and the connector clips (B), then cut the plastic cross brace (C)

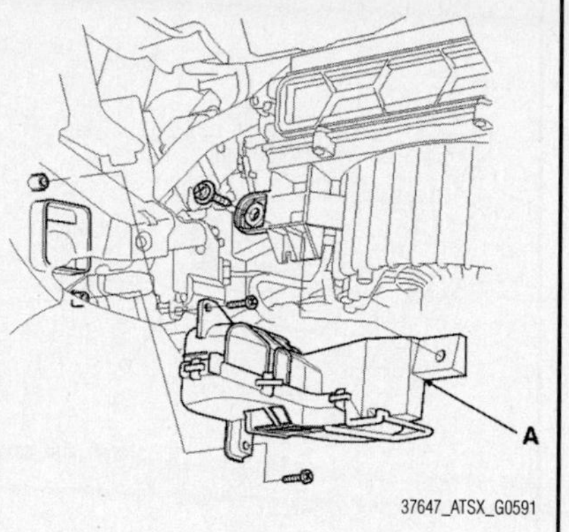
A
37647_ATSX_G0591

Fig. 329 Remove the self-tapping screws, and the passenger's heater duct (A)

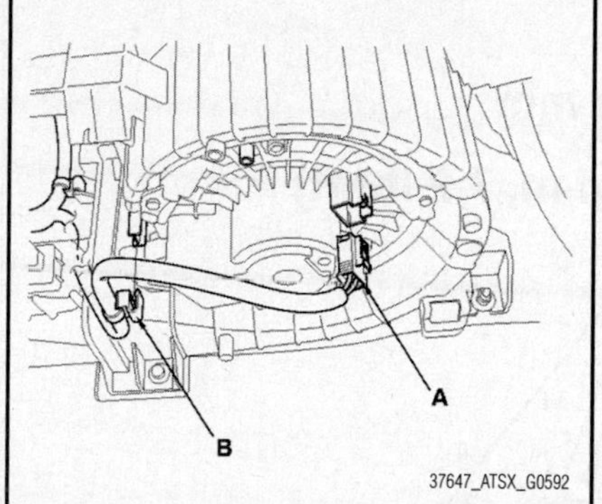
A
B
37647_ATSX_G0592

Fig. 330 Disconnect the connector (A) from the blower motor and the wire harness clip (B)

diagonal cutters in the area shown. Retain the plastic cross brace it will be reinstalled.

6. Remove the self-tapping screws, and the passenger's heater duct.

7. Disconnect the connector from the blower motor and the wire harness clip.

8. Disconnect the left engine compartment subharness connector from the bracket and harness clips.

9. Disconnect the connector from the recirculation control motor.

10. Remove the self-tapping screw, the mounting nuts, and the blower unit.

11. Install the unit in the reverse order of removal. Make sure that there are no air leaks.

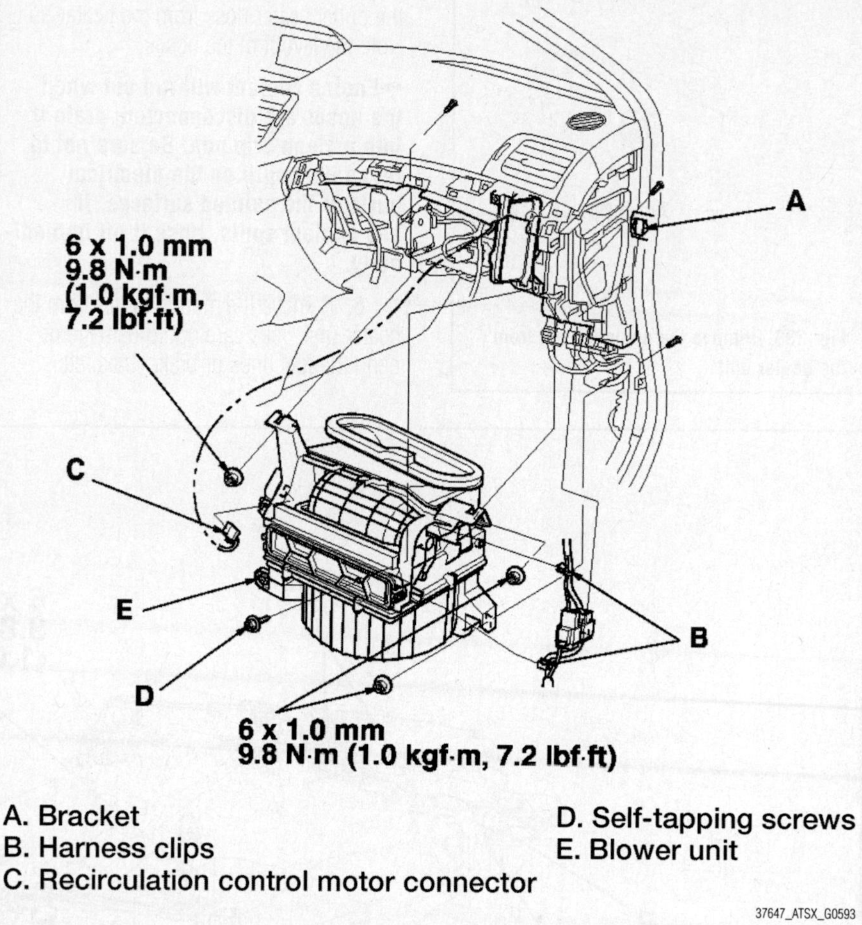
6 x 1.0 mm
9.8 N·m
(1.0 kgf·m,
7.2 lbf·ft)
A
C
E
B
D
6 x 1.0 mm
9.8 N·m (1.0 kgf·m, 7.2 lbf·ft)
A. Bracket
B. Harness clips
C. Recirculation control motor connector
D. Self-tapping screws
E. Blower unit
37647_ATSX_G0593

Fig. 331 Disconnect and remove the blower unit

HEATER CORE

REMOVAL & INSTALLATION

See Figures 332 through 338.

⁂ WARNING

SRS components are located in this area. Review the SRS component locations, and the precautions and procedures before doing repairs or service.

1. Do the battery terminal disconnection procedure.

2. Recover the refrigerant with a recovery/recycling/charging station.

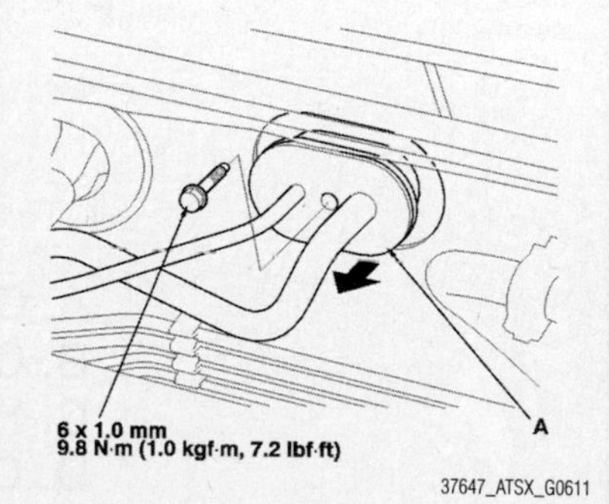
6 x 1.0 mm
9.8 N·m (1.0 kgf·m, 7.2 lbf·ft)
A
37647_ATSX_G0611

Fig. 332 Disconnect the A/C line from the evaporator core

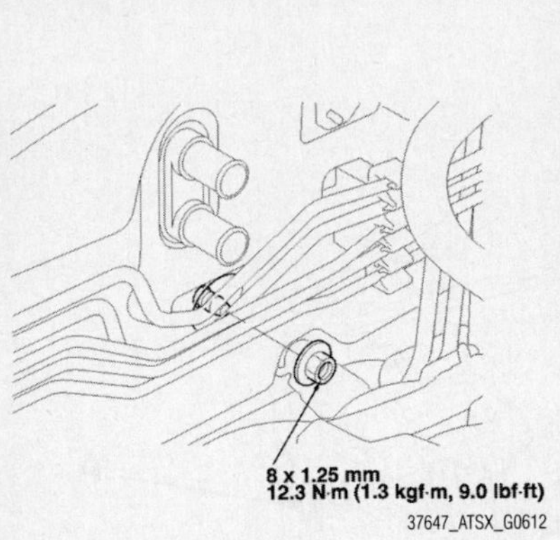
8 x 1.25 mm
12.3 N·m (1.3 kgf·m, 9.0 lbf·ft)
37647_ATSX_G0612

Fig. 333 Remove the mounting nut from the heater unit

3. Disconnect the A/C line from the evaporator core.
4. When the engine is cool, drain the engine coolant from the radiator.
5. Disconnect the inlet heater hose and the outlet heater hose from the heater unit. Note the layout of the hoses.

➡Engine coolant will run out when the hoses are disconnected; drain it into a clean drip pan. Be sure not to let coolant spill on the electrical parts or the painted surfaces. If any coolant spills, rinse it off immediately.

6. Remove the mounting nut from the heater unit. Take care not to damage or bend the fuel lines or brake lines, etc..

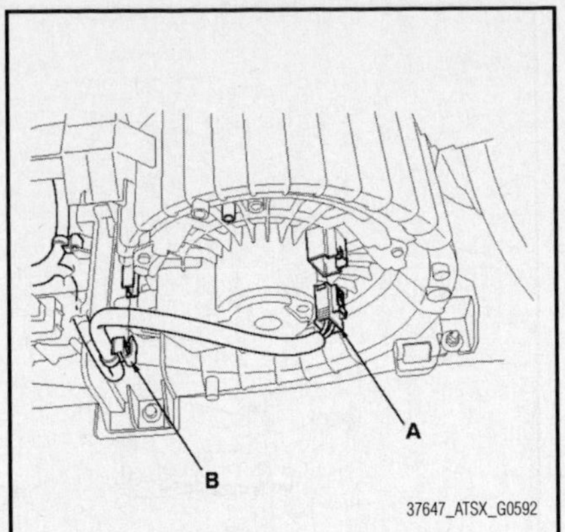
A
B
37647_ATSX_G0592

Fig. 335 Disconnect the connector (A) from the blower motor and the wire harness clip (B)

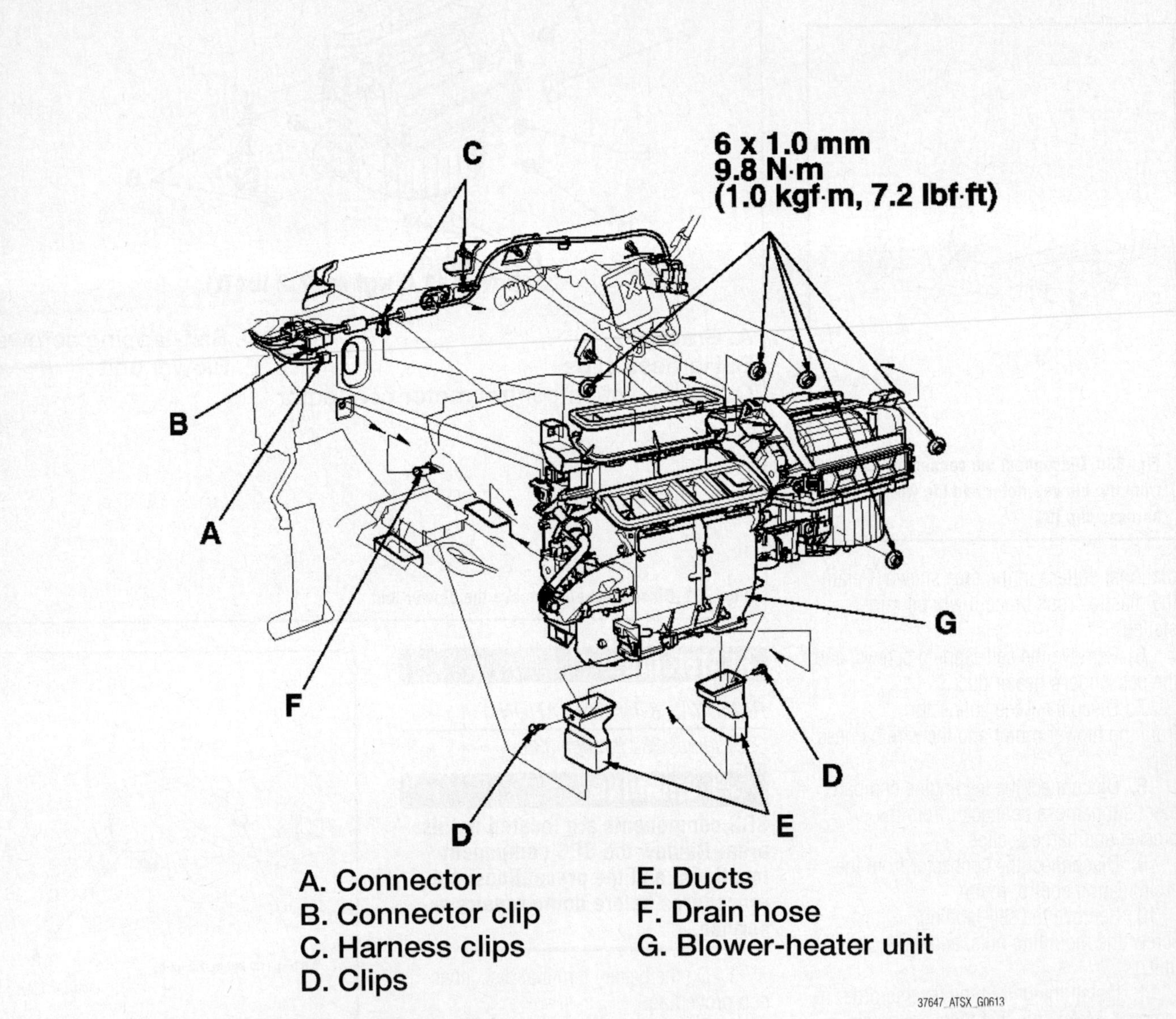
6 x 1.0 mm
9.8 N·m
(1.0 kgf·m, 7.2 lbf·ft)
C
B
A
F
G
D
E
D
A. Connector
B. Connector clip
C. Harness clips
D. Clips
E. Ducts
F. Drain hose
G. Blower-heater unit
37647_ATSX_G0613

Fig. 334 Remove the blower-heater unit

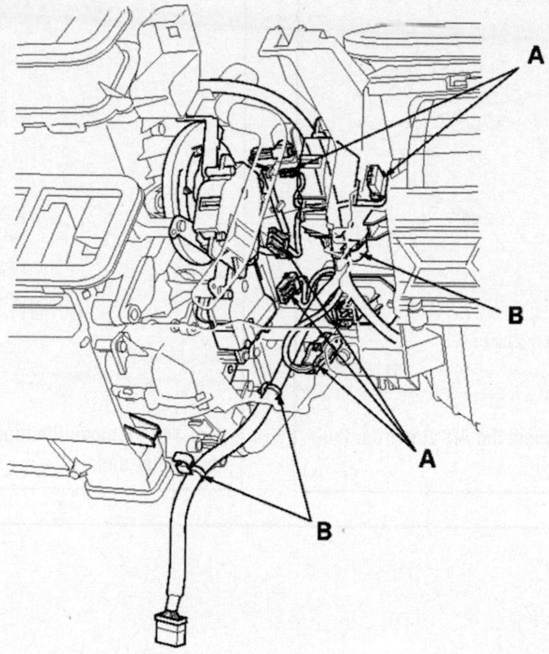
A
B
A
B
37647_ATSX_G0614

Fig. 336 Disconnect the connectors (A) and remove the wire harness clips (B)

7. Remove the dashboard.

8. Disconnect the connector. Remove the connector clip, the harness clips, the clips, ducts, and the drain hose. Then remove the mounting nuts, and the blower-heater unit.

9. Disconnect the connector (A) from the blower motor. Remove the wire harness clip (B).

10. Disconnect the connectors: The mode control motor, the power transistor, the evaporator temperature sensor, the passenger's air mix control motor, and the recirculation control motor. Remove the wire harness clips.

11. Disconnect the connector from the driver's air mix control motor. Remove the wire harness clips, the connector clip, and the wire harness.

12. Remove the self-tapping screws and the passenger's heater duct. Remove the self-tapping screws and the joint duct. Remove the self-tapping screw and the heater core cover. Remove the self-tapping screw, the heater pipe bracket, and the grommet, and carefully pull out the heater core.

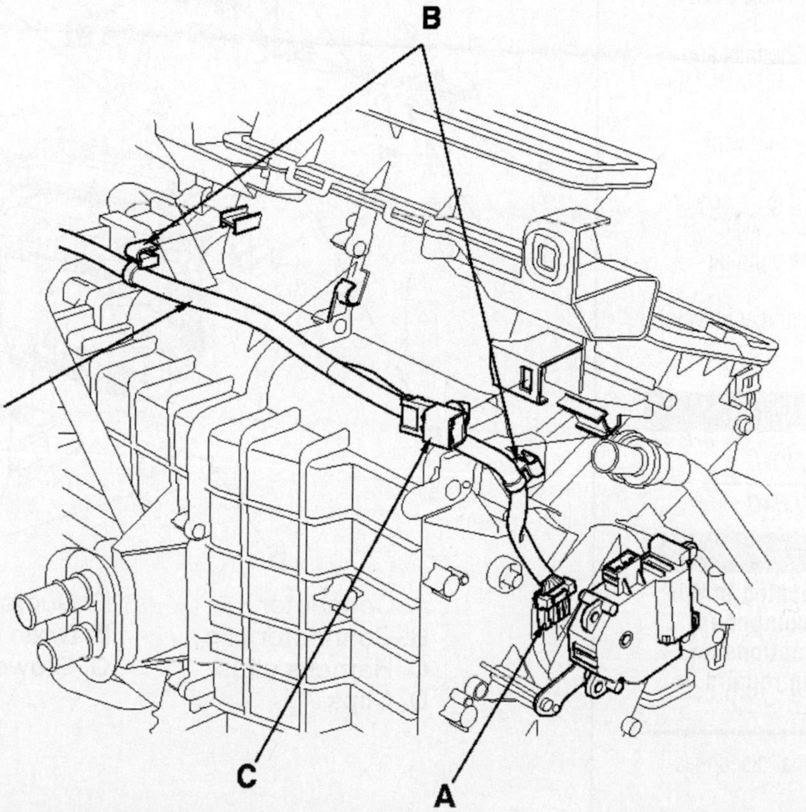
B
D
C
A
A. Driver's air mix control motor connector
B. Wire harness clips
C. Connector clip
D. Wire harness
37647_ATSX_G0615

Fig. 337 Disconnect the connector from the driver's air mix control motor

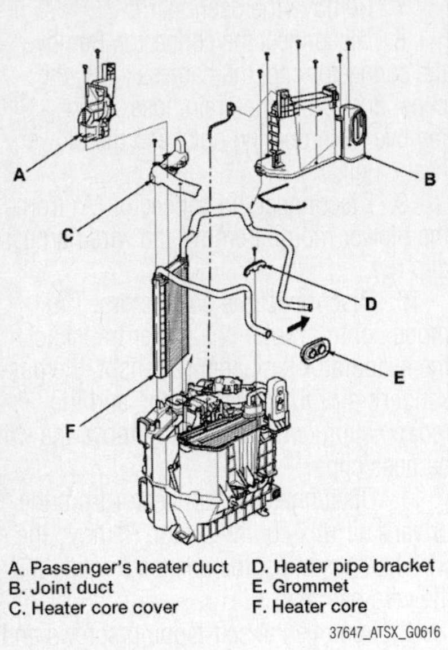
A
B
C
D
E
F
A. Passenger's heater duct
B. Joint duct
C. Heater core cover
D. Heater pipe bracket
E. Grommet
F. Heater core
37647_ATSX_G0616

Fig. 338 Remove the heater core

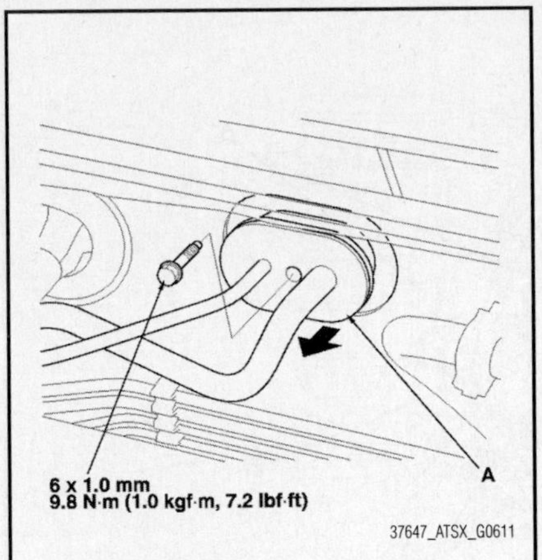
6 x 1.0 mm
9.8 N·m (1.0 kgf·m, 7.2 lbf·ft)
A
37647_ATSX_G0611

Fig. 339 Disconnect the A/C line from the evaporator core

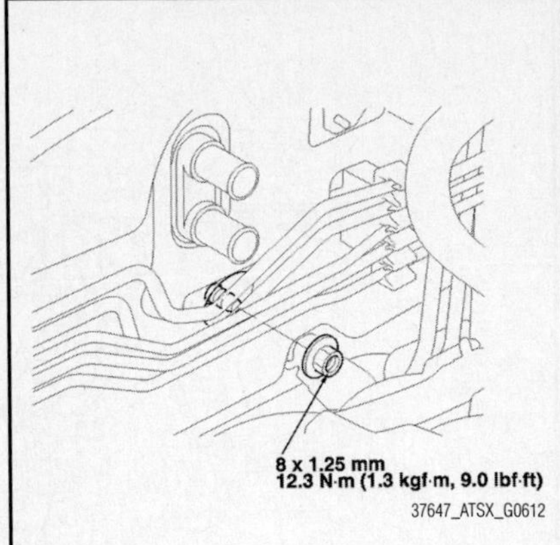
8 x 1.25 mm
12.3 N·m (1.3 kgf·m, 9.0 lbf·ft)
37647_ATSX_G0612

Fig. 340 Remove the mounting nut from the heater unit

To install:

13. Install the heater core, and the evaporator core in the reverse order of removal.
14. Install the heater unit in the reverse order of removal, and note these items:
    - Do not interchange the inlet and outlet heater hoses, and install the hose clamps securely.
    - Refill the cooling system with engine coolant.
    - Make sure that there is no coolant leakage.
    - Make sure that there is no air leakage.
15. Do the battery terminal reconnection procedure.

HEATER UNIT

REMOVAL & INSTALLATION

See Figures 339 through 340.

WARNING

SRS components are located in this area. Review the SRS component locations, and the precautions and procedures before doing repairs or service.

1. Do the battery terminal disconnection procedure.
2. Recover the refrigerant with a recovery/recycling/charging station.
3. Disconnect the A/C line from the evaporator core.

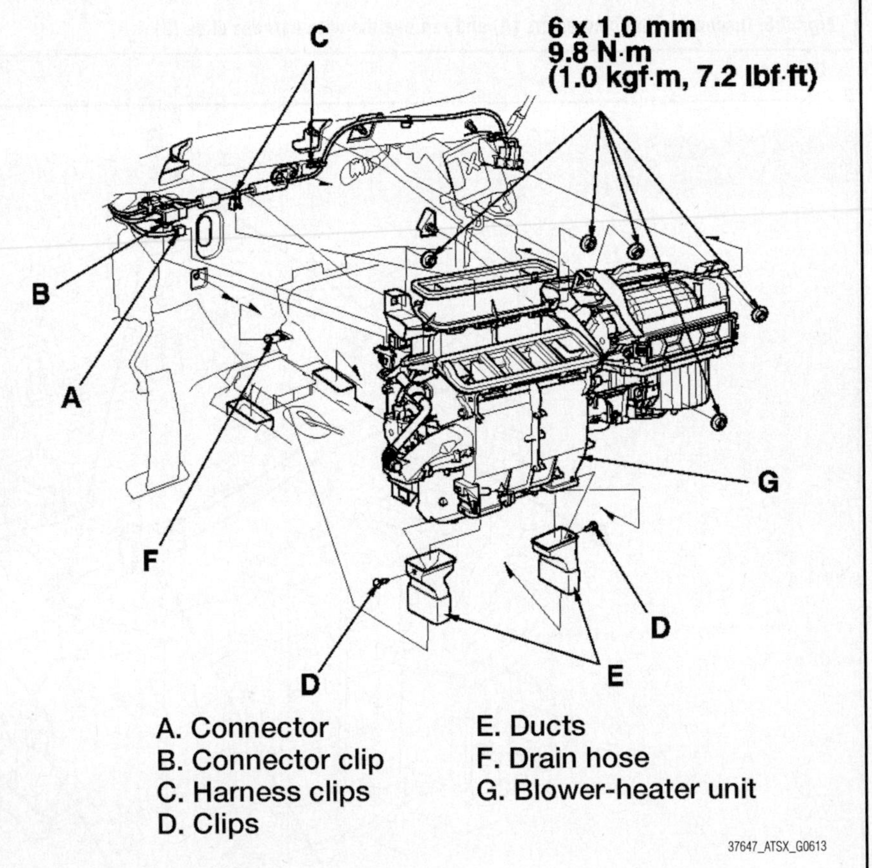
C
6 x 1.0 mm
9.8 N·m
(1.0 kgf·m, 7.2 lbf·ft)
B
A
F
G
D
D
E
A. Connector
B. Connector clip
C. Harness clips
D. Clips
E. Ducts
F. Drain hose
G. Blower-heater unit
37647_ATSX_G0613

Fig. 341 Remove the blower-heater unit

4. When the engine is cool, drain the engine coolant from the radiator.

5. Disconnect the inlet heater hose and the outlet heater hose from the heater unit. Note the layout of the hoses.

➡Engine coolant will run out when the hoses are disconnected; drain it into a clean drip pan. Be sure not to let coolant spill on the electrical parts or the painted surfaces. If any coolant spills, rinse it off immediately.

6. Remove the mounting nut from the heater unit. Take care not to damage or bend the fuel lines or brake lines, etc..

7. Remove the dashboard.

8. Disconnect the connector. Remove the connector clip, the harness clips, the clips, ducts, and the drain hose. Then remove the mounting nuts, and the blower-heater unit.

To install:

9. Install the heater unit in the reverse order of removal, and note these items:

- Do not interchange the inlet and outlet heater hoses, and install the hose clamps securely.
- Refill the cooling system with engine coolant.
- Make sure that there is no coolant leakage.
- Make sure that there is no air leakage.

10. Do the battery terminal reconnection procedure.

STEERING

ELECTRIC POWER STEERING (EPS) CONTROL UNIT

REMOVAL & INSTALLATION

See Figures 342 through 346.

1. Do the battery terminal disconnection procedure.

2. Remove these items:

- Passenger's dashboard undercover.
- Passenger's front door sill trim.
- Passenger's kick panel.
- Passenger's under-dash fuse/relay box.
- Harness clip from the EPS control unit bracket.

3. Disconnect EPS control unit connector A (2P), connector B (2P), connector C (2P), and connector D (28P).

4. Remove the bolts from the EPS control unit.

5. Remove the EPS control unit.

To install:

6. Install the EPS control unit in the reverse order of removal.

7. Do the battery terminal reconnection procedure.

8. Do the torque sensor neutral position memorization.

9. After installation, start the engine, allow it to idle, and turn the steering wheel from lock to lock several times. Check that the EPS indicator does not come on.

TORQUE SENSOR NEUTRAL POSITION MEMORIZATION

The torque sensor neutral position must be memorized whenever the steering gearbox, the torque sensor, or the EPS control unit is replaced. Note that the torque sensor neutral position is not affected when erasing the DTC.

➡The torque sensor is temperature sensitive. When memorizing the torque

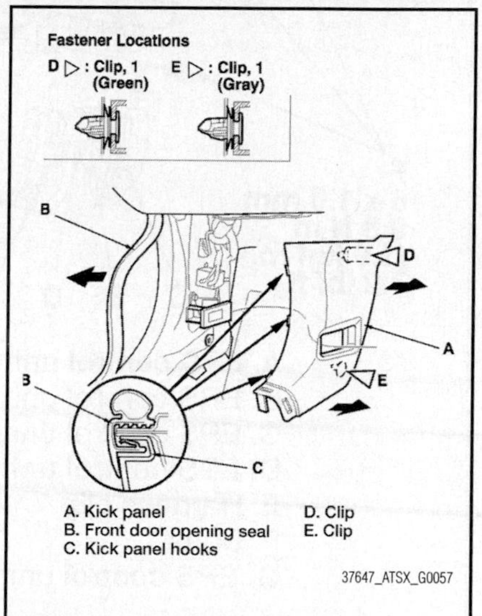
Fastener Locations
D ▷ : Clip, 1 (Green)
E ▷ : Clip, 1 (Gray)
A
B
C
D
E
A. Kick panel
B. Front door opening seal
C. Kick panel hooks
D. Clip
E. Clip
37647_ATSX_G0057

Fig. 344 Remove driver's kick panel (passenger side similar)

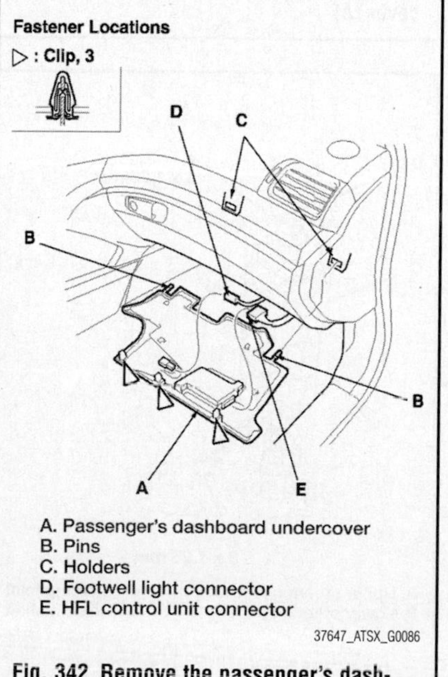
Fastener Locations
▷ : Clip, 3
A
B
C
D
E
A. Passenger's dashboard undercover
B. Pins
C. Holders
D. Footwell light connector
E. HFL control unit connector
37647_ATSX_G0086

Fig. 342 Remove the passenger's dashboard undercover

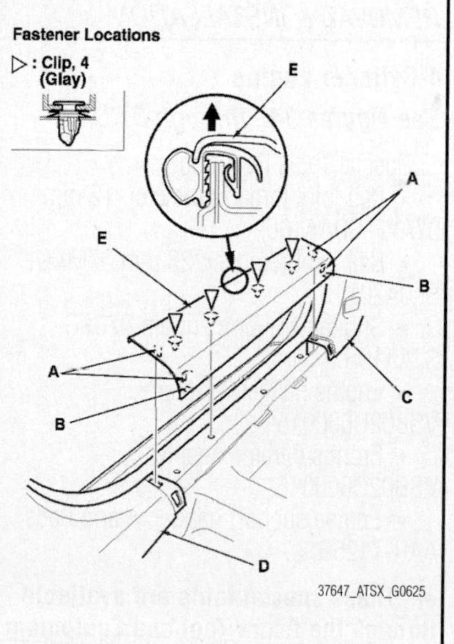
Fastener Locations
▷ : Clip, 4 (Glay)
A
B
C
D
E
37647_ATSX_G0625

Fig. 343 Passenger's front door sill trim

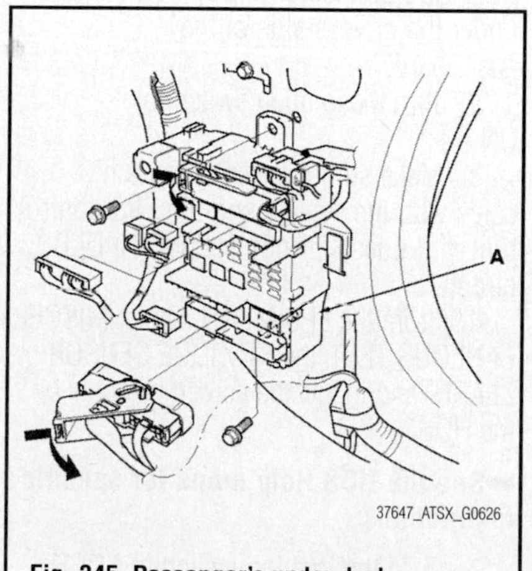
A
37647_ATSX_G0626

Fig. 345 Passenger's under-dash fuse/relay box

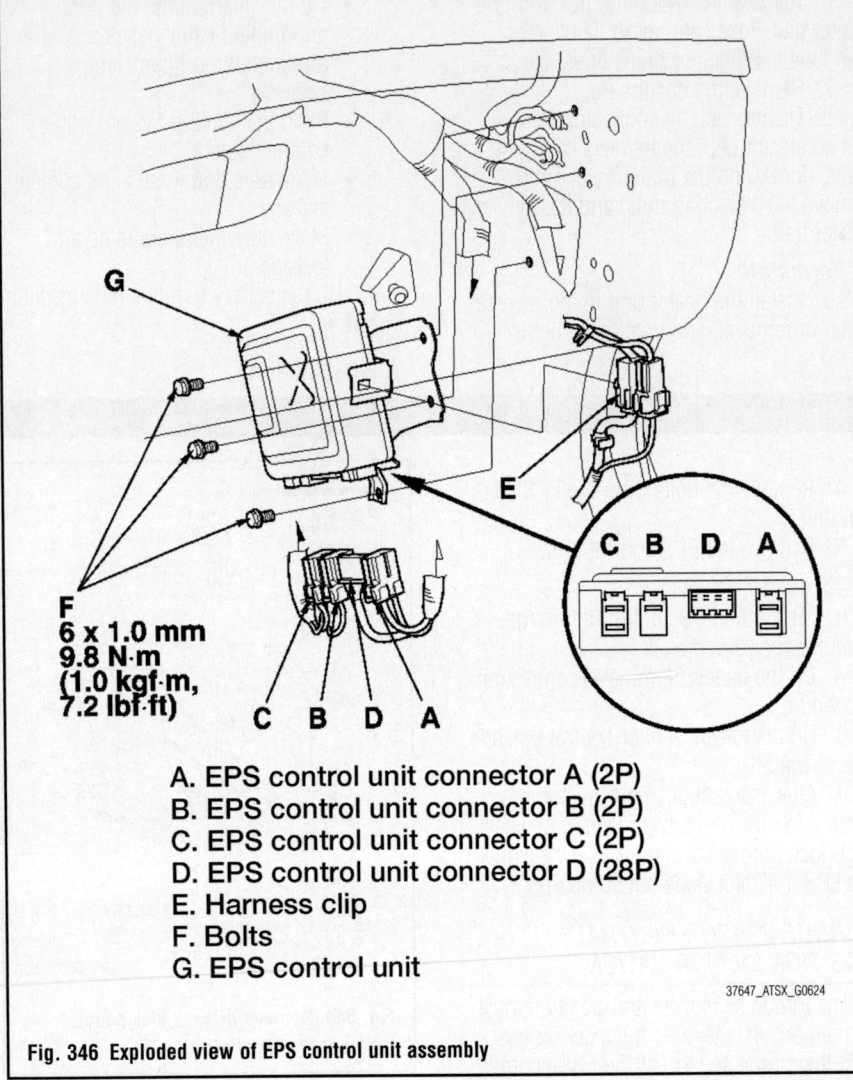

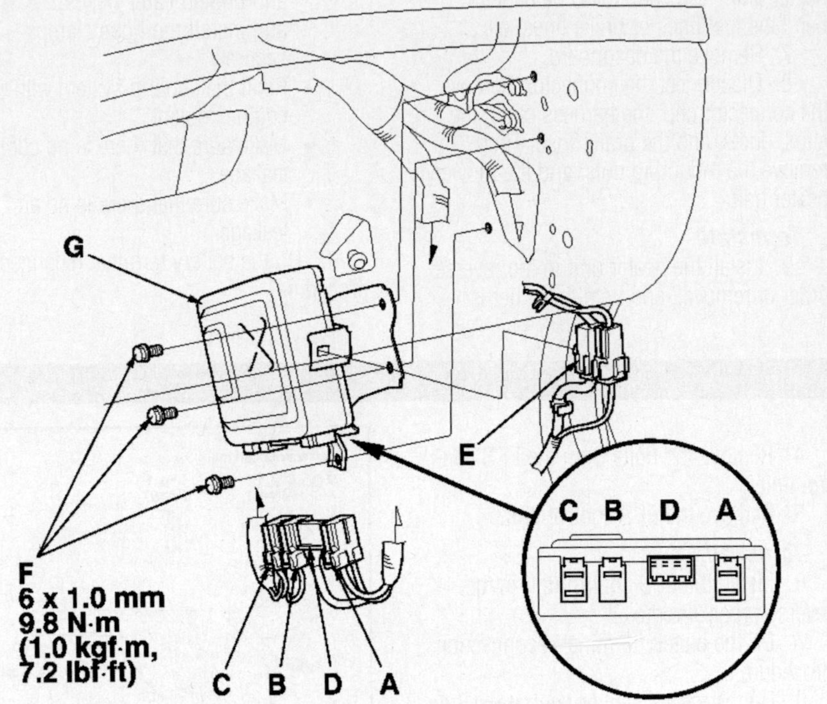
G
E
C B D A
F
6 x 1.0 mm
9.8 N·m
(1.0 kgf·m,
7.2 lbf·ft)
C B D A

A. EPS control unit connector A (2P)
B. EPS control unit connector B (2P)
C. EPS control unit connector C (2P)
D. EPS control unit connector D (28P)
E. Harness clip
F. Bolts
G. EPS control unit

37647_ATSX_G0624

Fig. 346 Exploded view of EPS control unit assembly

sensor neutral position, the ambient temperature must be above 68°F (20°C).

1. With the ignition switch in LOCK (0), connect the HDS to the Data Link Connector (DLC) located under the driver's side of the dashboard.
2. Turn the ignition switch to ON (II).
3. Make sure the HDS communicates with the vehicle and the EPS control unit. If it doesn't, troubleshoot the DLC circuit.
4. From the EPS MENU, select MISCELLANEOUS TEST, then TORQUE SENSOR LEARN, and follow the screen prompts on the HDS.

➡See the HDS Help menu for specific instructions.

5. Turn the ignition switch to LOCK (0).

EPS STEERING GEARBOX

REMOVAL & INSTALLATION

4-Cylinder Engine

See Figures 347 through 373.

Special Tools Required:

- Ball joint thread protector, 12 mm 07AAF-SDAA100
- Ball joint remover, 28 mm 07MAC-SL0A202
- Subframe alignment pin 070AG-SJAA10S
- Engine hanger adapter VSB02C000015*
- Front subframe adapter VSB02C000016*
- Engine support hanger, A and Reds AAR-T1256*

➡*These special tools are available through the Acura Tool and Equipment Program, 888-424-6857.

➡Note these items during removal:

- Use solvent and a brush, wash any oil and dirt off the end of the steering gearbox. Avoid any electrical parts. Blow dry with compressed air.
- Be sure to remove the steering wheel before disconnecting the steering joint, or damage to the cable reel can occur.

WARNING

SRS components are located in this area. Review the SRS components locations and the precautions and procedures before doing repairs or service.

1. Do the battery terminal disconnection procedure.
2. Raise the vehicle on a lift.

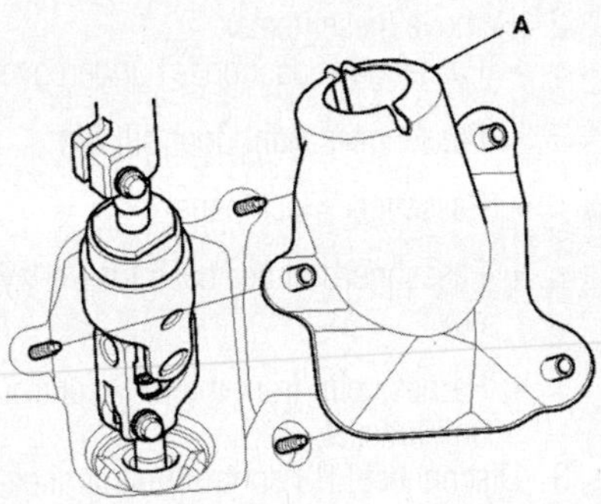
A

37647_ATSX_G0627

Fig. 347 Remove the steering joint cover (A)

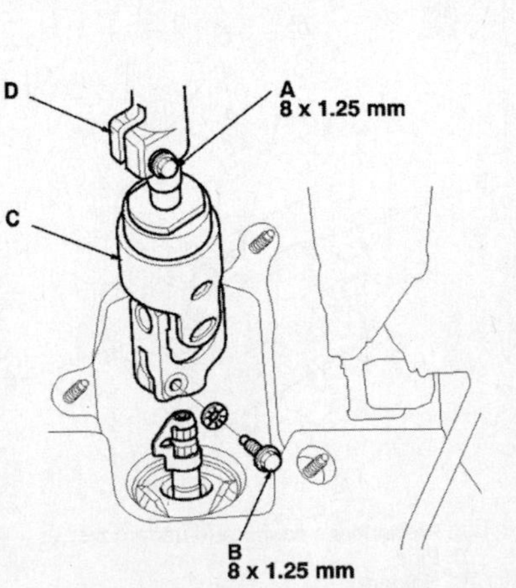
D
A
8 x 1.25 mm
C
B
8 x 1.25 mm

A. Upper steering joint bolt
B. Lower steering joint bolt
C. Steering joint
D. Column shaft

37647_ATSX_G0628

Fig. 348 Steering joint

3. Remove the front wheels.
4. Remove the drivers' airbag and the steering wheel.
5. Remove the steering joint cover.
6. Loosen the upper steering joint bolt, and remove the lower steering joint bolt.
7. Disconnect the steering joint by sliding the steering joint into the column shaft. Tighten the upper steering joint bolt to hold the column shaft.

➡Do not disconnect the steering joint from the column shaft.

8. Remove the center guide (if equipped) from the top of the pinion shaft, and discard it. The center guide is for factory assembly use only.
9. Apply vinyl tape to the splines on the pinion shaft.
10. Secure the hood in the wide open position (support rod in the lower hole).
11. Remove the strut brace.
12. Remove the front grille cover.

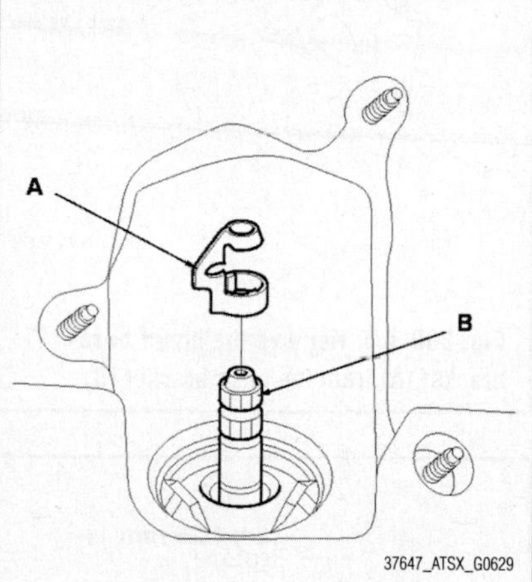
A
B
37647_ATSX_G0629

Fig. 349 Remove the center guide (A) (if equipped) from the top of the pinion shaft (B)

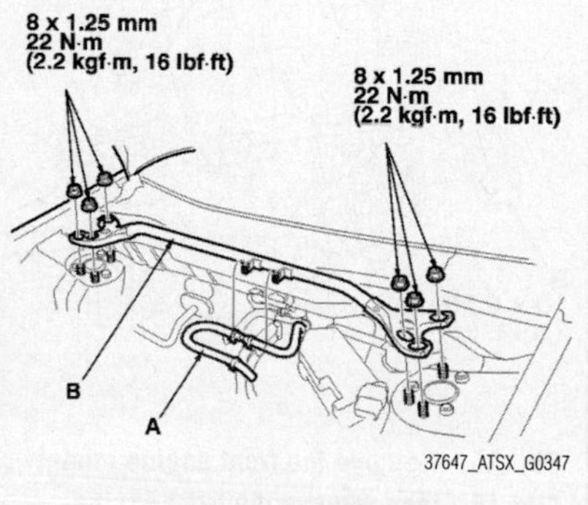
8 x 1.25 mm
22 N·m
(2.2 kgf·m, 16 lbf·ft)
8 x 1.25 mm
22 N·m
(2.2 kgf·m, 16 lbf·ft)
B
A
37647_ATSX_G0347

Fig. 350 Remove the strut brace

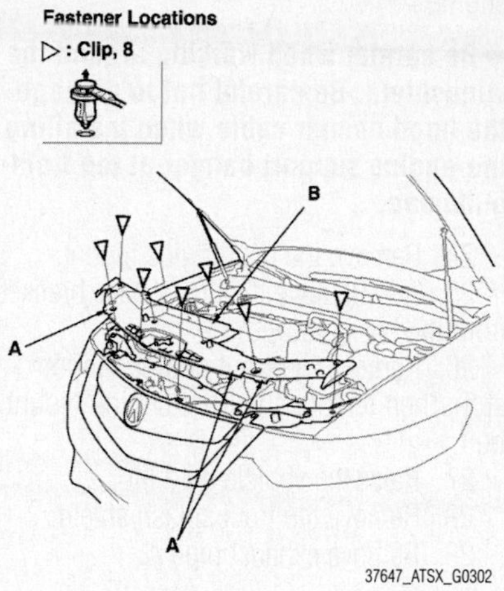
Fastener Locations
▷: Clip, 8
B
A
A
37647_ATSX_G0302

Fig. 351 Detach the clips, and release the hooks (A), then remove the front grille cover (B)

13. Remove the cotter pin from the tie-rod end ball joint, then remove the nut on both sides.
14. Disconnect the tie-rod end ball joint from the knuckle using the ball joint thread protector and the ball joint remover on both sides.

➡Be careful not to damage the ball joint boot when installing the remover.

15. Remove the P/S heat shield.
16. Disconnect the EPS motor angle sensor 6P connector.
17. Disconnect the EPS motor 3P connector by pushing the lock and pulling down the lever. Wrap the connectors with vinyl tape to avoid contamination from grease or water.

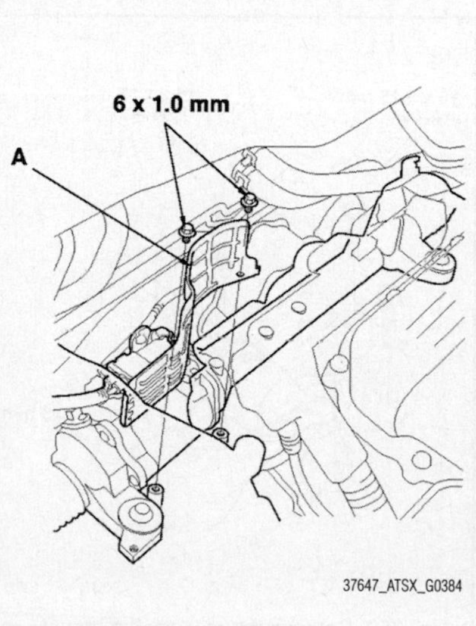
6 x 1.0 mm
A
37647_ATSX_G0384

Fig. 352 Remove the P/S heat shield (A)

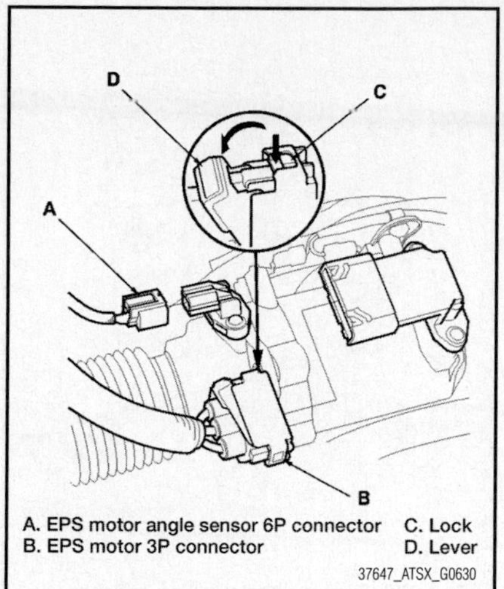
D
C
A
B
A. EPS motor angle sensor 6P connector
B. EPS motor 3P connector
C. Lock
D. Lever
37647_ATSX_G0630

Fig. 353 Disconnect the EPS motor angle sensor 6P connector

18. Disconnect the EPS subharness 6P connector. Wrap the connectors with vinyl tape to avoid contamination from grease or water.
19. Remove the steering gearbox mounting bolts and washers from the steering gearbox.
20. Remove the ignition coil cover.
21. Remove the harness cover bracket bolts.
22. Attach the engine hanger adapter (VSB02C000015) to the threaded hole located on the rear side of the cylinder head.
23. Install the engine support hanger (AAR-T1256) to the vehicle, and attach the hook to the slotted hole in the engine hanger adapter (VSB02C000015). Tighten

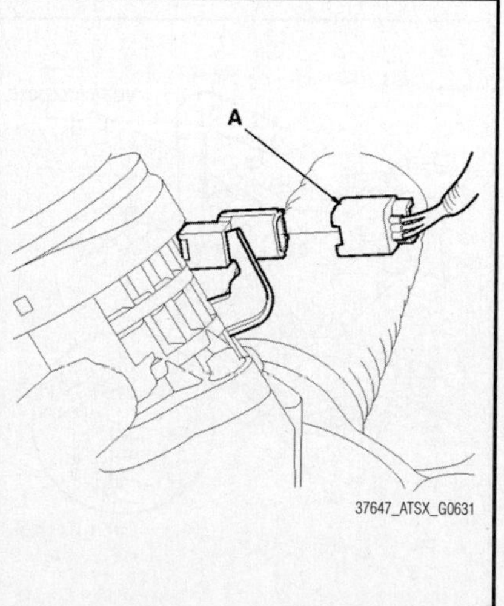
A
37647_ATSX_G0631

Fig. 354 Disconnect the EPS subharness 6P connector (A)

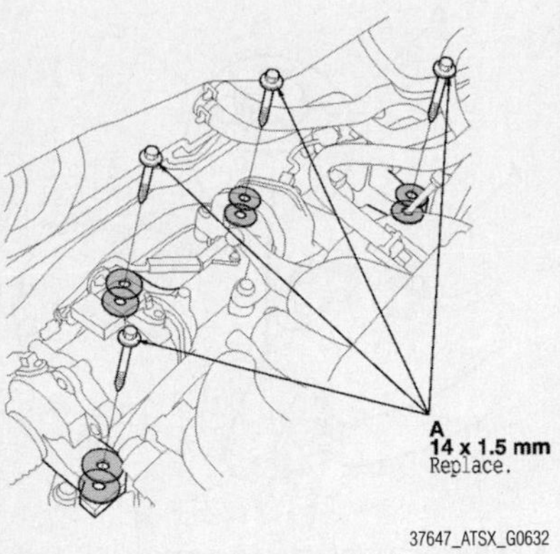
A
14 x 1.5 mm
Replace.
37647_ATSX_G0632

Fig. 355 Remove the steering gearbox mounting bolts (A) and washers (B) from the steering gearbox

the wing nut by hand to lift and support the engine.

➡Be careful when working around the windshield. Be careful not to damage the hood opener cable when installing the engine support hanger at the front bulkhead.

24. Remove the rear engine mount.
25. A/T: Remove the upper base bracket from the base bracket.
26. Remove the front engine mount stop, then remove the front engine mount bolt.
27. Raise the vehicle on a lift.
28. Remove the front splash shield.
29. Remove exhaust pipe A.
30. A/T: Disconnect the shift cable from the selector control lever.
31. '09 model: Remove the ground cable from the front subframe.
32. Attach the front subframe adapter (VSB02C000016) to the subframe by hanging the belt over the front of the subframe, then secure the belt with its stop.
33. Raise the jack and line up the slots in the front subframe adapter arms with the bolt holes on the jack base, then securely attach them with four bolts.
34. Remove the transmission lower mount bolts.
35. Remove the subframe middle mount on both sides.

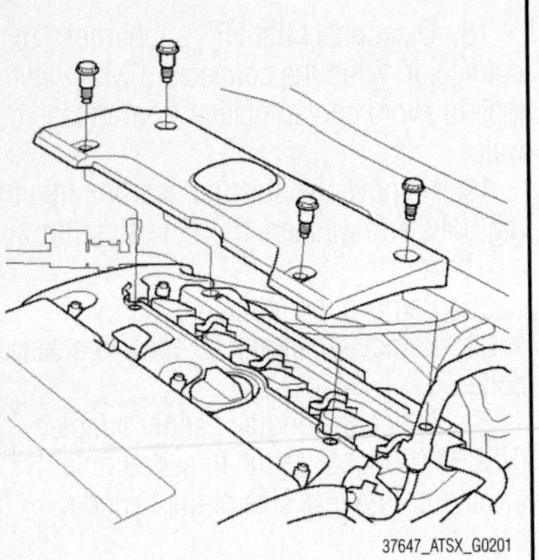
37647_ATSX_G0201

Fig. 356 Remove the ignition coil cover (A)

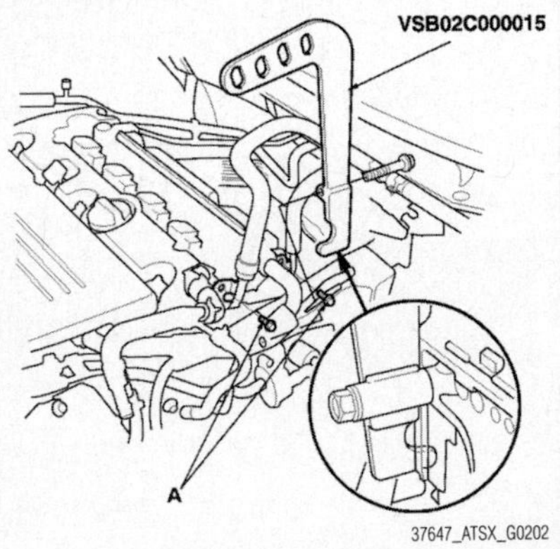
VSB02C000015
A
37647_ATSX_G0202

Fig. 357 Remove the harness cover bracket bolts (A)

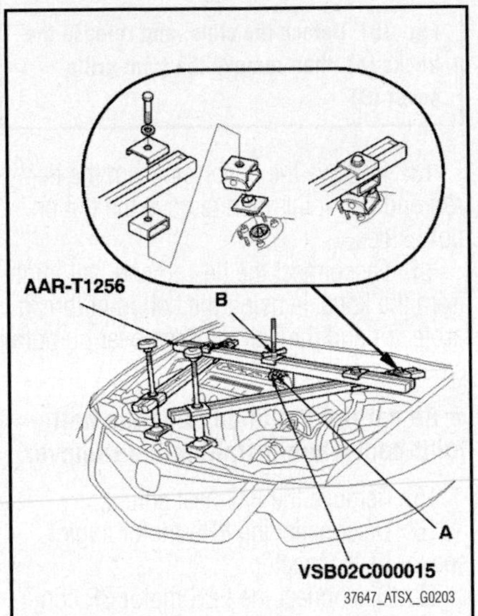
AAR-T1256
B
A
VSB02C000015
37647_ATSX_G0203

Fig. 358 Install the engine support hanger (AAR-T1256) to the vehicle

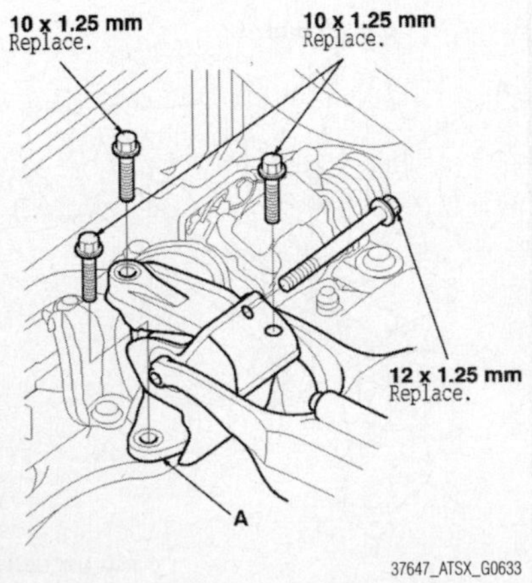
10 x 1.25 mm
Replace.
10 x 1.25 mm
Replace.
12 x 1.25 mm
Replace.
A
37647_ATSX_G0633

Fig. 359 Remove the rear engine mount (A)

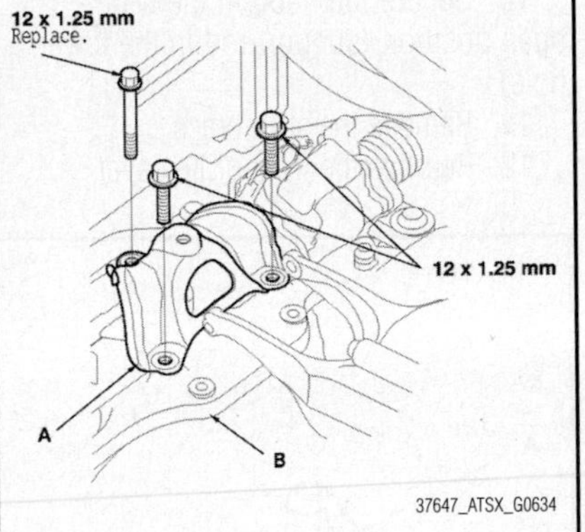
12 x 1.25 mm
Replace.
12 x 1.25 mm
A
B
37647_ATSX_G0634

Fig. 360 A/T: Remove the upper base bracket (A) from the base bracket (B)

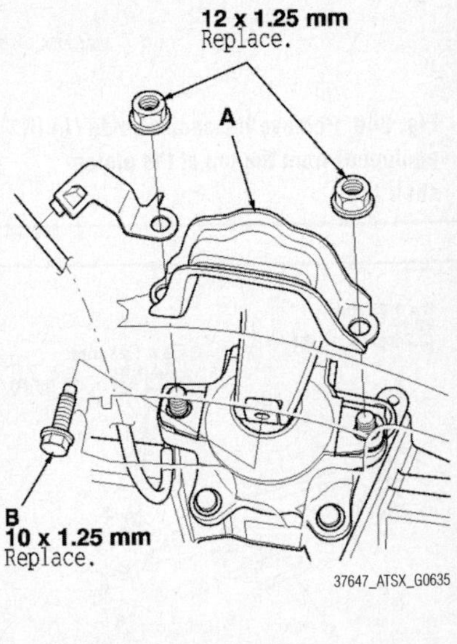
12 x 1.25 mm
Replace.
A
B
10 x 1.25 mm
Replace.
37647_ATSX_G0635

Fig. 361 Remove the front engine mount stop (A), then remove the front engine mount bolt (B)

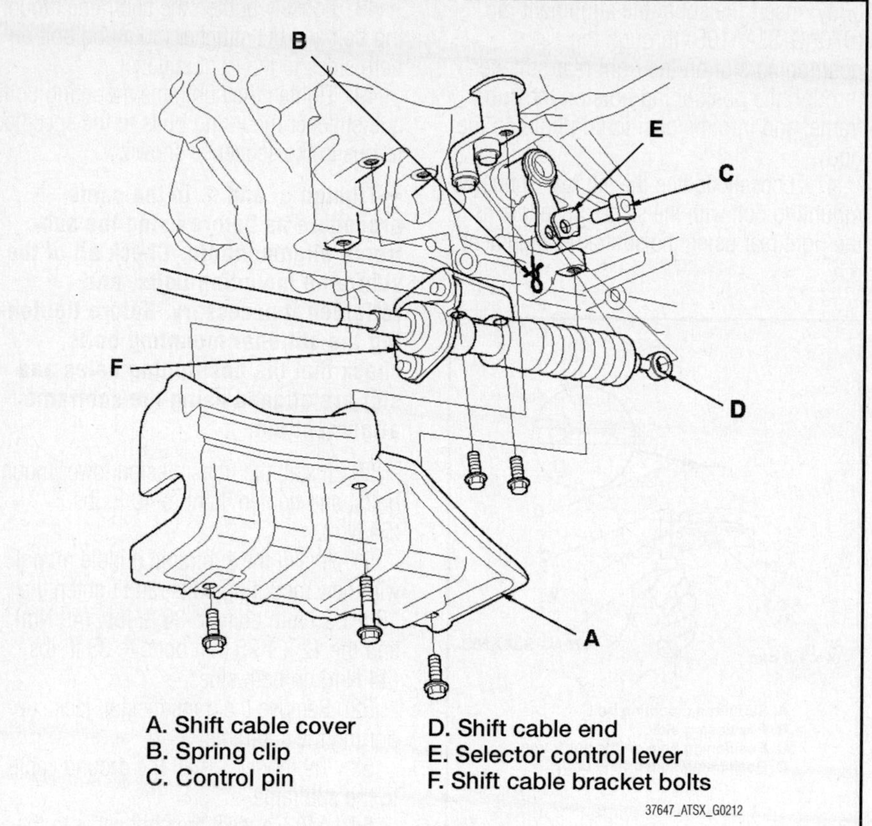
B
E
C
F
D
A
A. Shift cable cover
B. Spring clip
C. Control pin
D. Shift cable end
E. Selector control lever
F. Shift cable bracket bolts
37647_ATSX_G0212

Fig. 362 A/T: Disconnect the shift cable from the selector control lever

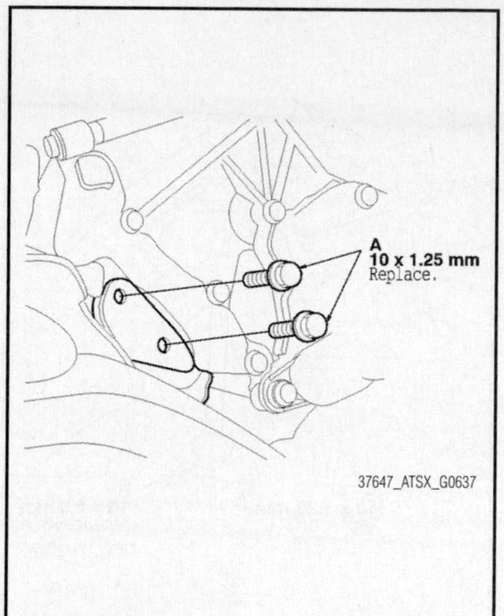
A
10 x 1.25 mm
Replace.
37647_ATSX_G0637

Fig. 365 Remove the transmission lower mount bolts (A)

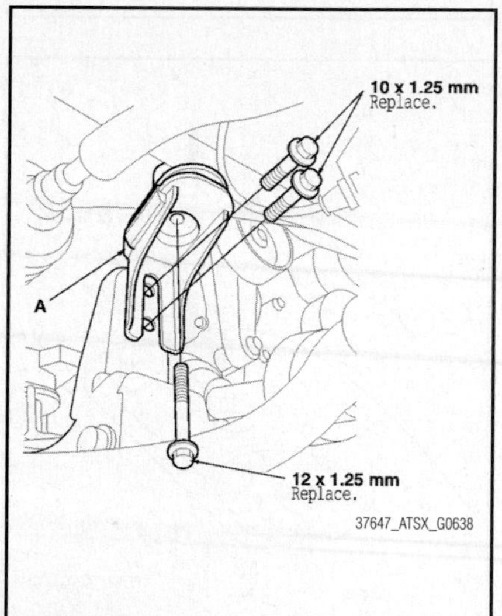
10 x 1.25 mm
Replace.
A
12 x 1.25 mm
Replace.
37647_ATSX_G0638

Fig. 366 Remove the subframe middle mount (A) on both sides

36. Remove the stiffeners mounting bolt on both sides.

37. Loosen the subframe mounting bolts so they are about 13⁄16 inches (20 mm) from the mounting surface. Do not loosen the subframe mounting bolts more than necessary.

38. Remove the stiffener mounting bolts and subframe mounting bolts from the rear stiffeners.

39. Lower the jack slowly until the subframe has dropped about 2 ¾ inches (70 mm).

40. Carefully move the steering gearbox toward the driver's side until the pinion shaft clears the fenderwell opening on the body.

41. Remove the steering gearbox through the fenderwell opening on the driver's side.

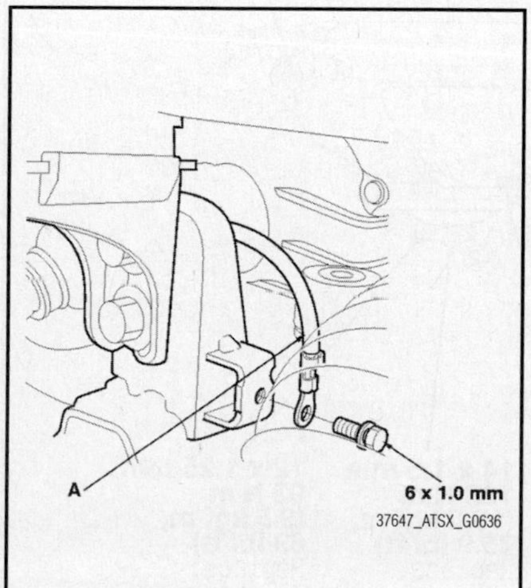
A
6 x 1.0 mm
37647_ATSX_G0636

Fig. 363 '09 model: Remove the ground cable (A) from the front subframe

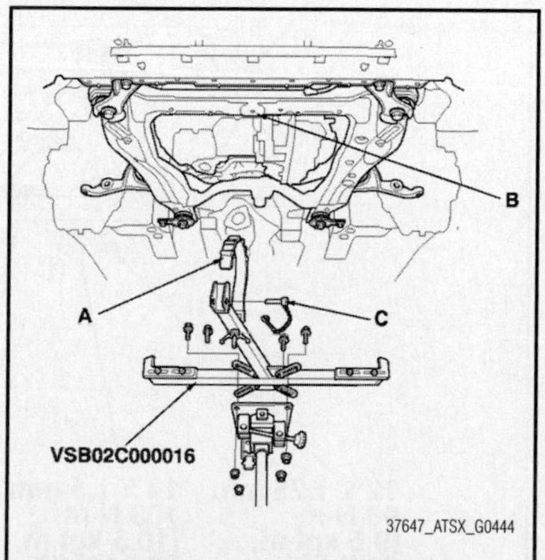
B
A
C
VSB02C000016
37647_ATSX_G0444

Fig. 364 Attach the front subframe adapter to the subframe by hanging the belt (A) over the front of the subframe (B), then secure the belt with its stop (C)

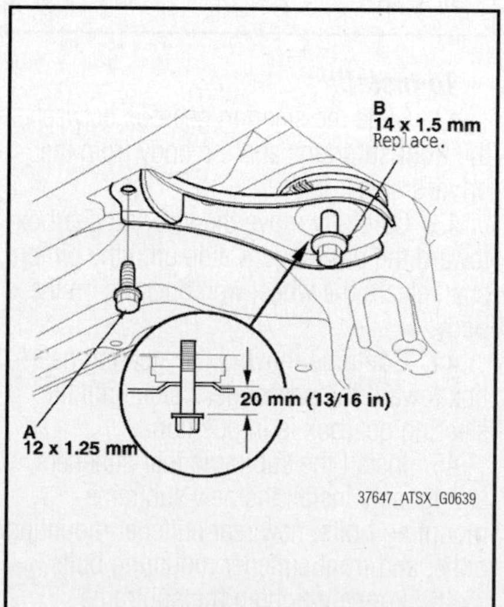
B
14 x 1.5 mm
Replace.
20 mm (13/16 in)
A
12 x 1.25 mm
37647_ATSX_G0639

Fig. 367 Remove the stiffeners mounting bolt (A) on both sides

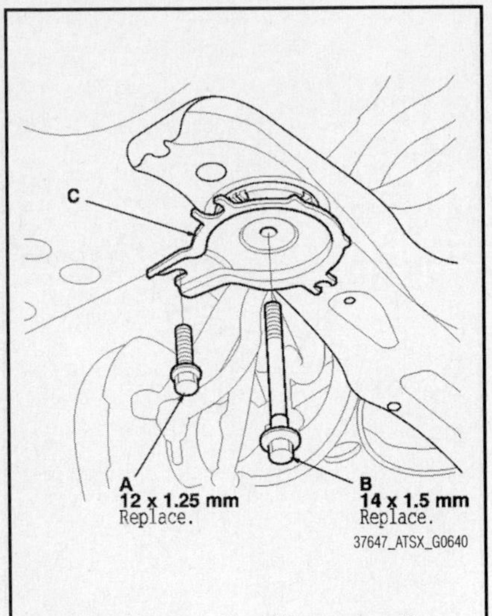

C
A
12 x 1.25 mm
Replace.
B
14 x 1.5 mm
Replace.
37647_ATSX_G0640

Fig. 368 Remove the stiffener mounting bolts (A) and subframe mounting bolts (B) from the rear stiffeners (C)

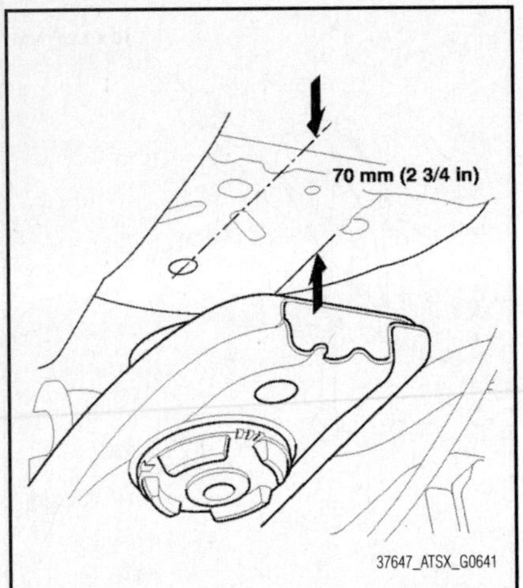

70 mm (2 3/4 in)
37647_ATSX_G0641

Fig. 369 Lower the jack slowly until the subframe has dropped about 2 ¾ inches (70 mm)

To install:

42. Slide the steering gearbox between the front subframe and the body from the driver's side.

43. Carefully move the steering gearbox toward the passenger's side until the pinion shaft clears the wheel well opening on the body.

44. Continue moving the steering gearbox toward the passenger's side until the steering gearbox is in position.

45. Install the subframe rear stiffeners, and loosely install the new subframe mounting bolts, new rear stiffener mounting bolts, and front stiffener mounting bolts.

46. Loosely tighten the subframe mounting bolt in the right rear stiffener until the subframe insulator contacts the body; insert the subframe alignment pin (070AG-SJAA10S) through the positioning slot on the right rear stiffener, through the positioning hole on the subframe, and into the positioning hole on the body.

47. Loosely tighten the left rear subframe mounting bolt with the same procedure as the right rear using the subframe alignment pin.

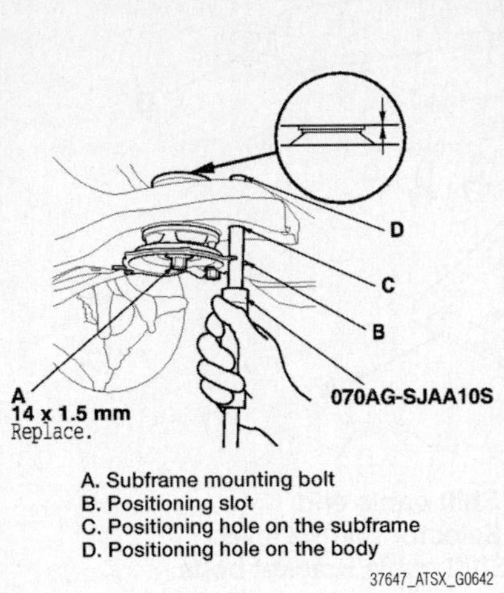

D
C
B
A
14 x 1.5 mm
Replace.
070AG-SJAA10S
A. Subframe mounting bolt
B. Positioning slot
C. Positioning hole on the subframe
D. Positioning hole on the body
37647_ATSX_G0642

Fig. 370 Insert the subframe alignment pin through the positioning slot on the right rear stiffener, through the positioning hole on the subframe, and into the positioning hole on the body

48. Loosely tighten the subframe mounting bolt and the stiffener mounting bolt on both sides to the front stiffener.

49. Tighten the subframe mounting bolts and stiffener mounting bolts to the specified torque in the sequence shown.

➡Tighten ① and ② in the same procedure as before using the subframe alignment pin. Check all of the subframe mounting bolts, and retighten if necessary. Before tightening the stiffener mounting bolts, check that the positioning holes and slot are aligned using the subframe alignment pin.

50. Install the transmission lower mount bolts, and tighten them to 40 ft. lbs. (54 Nm).

51. Install the subframe middle mount with new mounting bolts, and tighten the 10 x 1.25 mm bolts to 36 ft. lbs. (49 Nm) and the 12 x 1.25 mm bolts to 33 ft. lbs. (44 Nm) on both sides.

52. Remove the transmission jack supporting the subframe.

53. '09 model: Install the ground cable to the subframe.

54. A/T: Connect the shift cable to the selector control lever.

55. Install the exhaust pipe A.

56. Install the front splash shield.

57. Lower the vehicle.

58. Tighten the new front engine mount bolt to 40 ft. lbs. (54 Nm), then install the

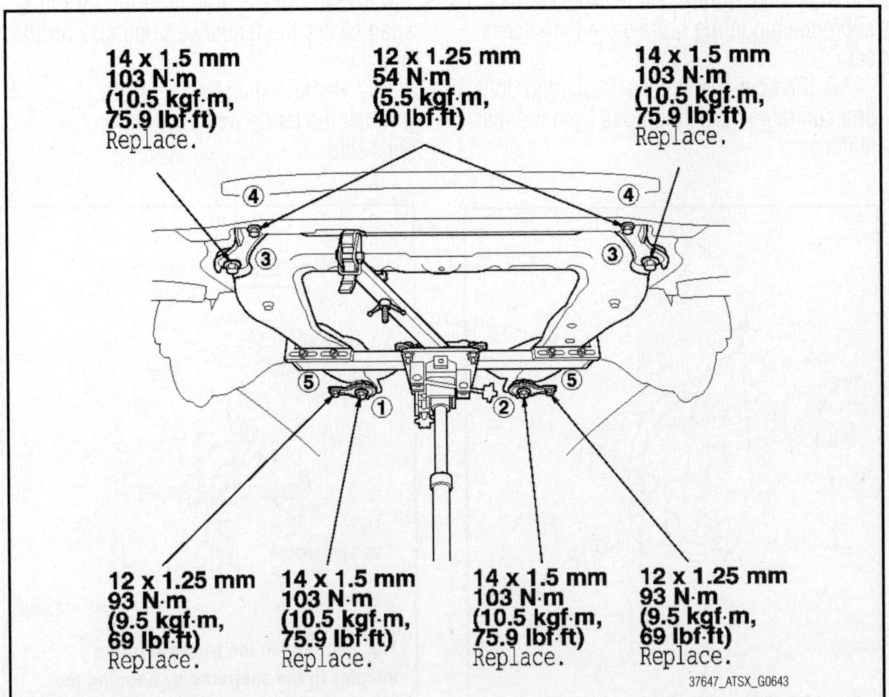

14 x 1.5 mm
103 N·m
(10.5 kgf·m,
75.9 lbf·ft)
Replace.
12 x 1.25 mm
54 N·m
(5.5 kgf·m,
40 lbf·ft)
14 x 1.5 mm
103 N·m
(10.5 kgf·m,
75.9 lbf·ft)
Replace.
12 x 1.25 mm
93 N·m
(9.5 kgf·m,
69 lbf·ft)
Replace.
14 x 1.5 mm
103 N·m
(10.5 kgf·m,
75.9 lbf·ft)
Replace.
14 x 1.5 mm
103 N·m
(10.5 kgf·m,
75.9 lbf·ft)
Replace.
12 x 1.25 mm
93 N·m
(9.5 kgf·m,
69 lbf·ft)
Replace.
37647_ATSX_G0643

Fig. 371 Tighten the subframe mounting bolts and stiffener mounting bolts in the sequence

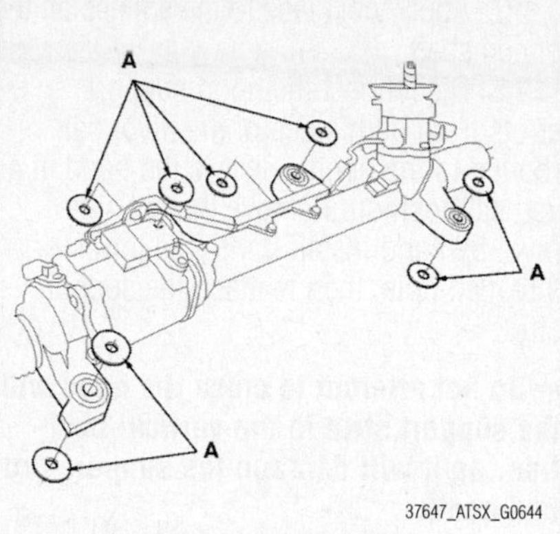
A
A
A
37647_ATSX_G0644

Fig. 372 Install the washers (A) and new steering gearbox mounting bolts to the steering gearbox

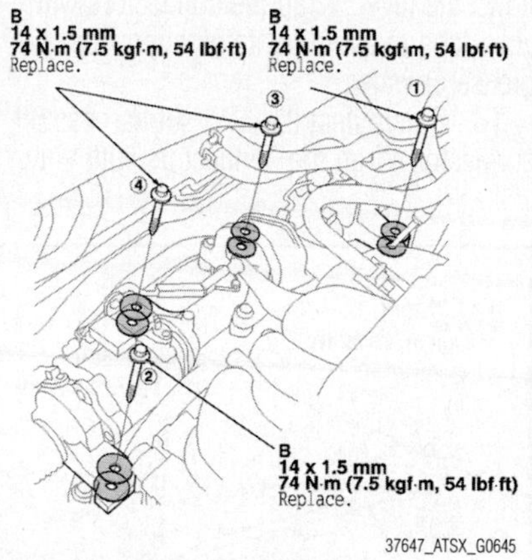
B
14 x 1.5 mm
74 N·m (7.5 kgf·m, 54 lbf·ft)
Replace.
B
14 x 1.5 mm
74 N·m (7.5 kgf·m, 54 lbf·ft)
Replace.
B
14 x 1.5 mm
74 N·m (7.5 kgf·m, 54 lbf·ft)
Replace.
37647_ATSX_G0645

Fig. 373 Tighten the steering gearbox mounting bolts (B) in sequence

front engine mount stop with new mount nuts. Tighten the nuts to 58 ft. lbs (78 Nm).

59. A/T: Install the upper base bracket to the base bracket with a new mounting bolt and mounting bolts, then tighten them to 40 ft. lbs. (54 Nm).
60. Install the rear engine mount with new mounting bolts, and tighten the 10 x 1.25 mm bolts to 40 ft. lbs. (54 Nm) and the 12 x 1.25 mm bolt to 58 ft. lbs. (78 Nm).
61. Remove the engine support hanger.
62. Remove the engine hanger adapter.
63. Install the harness cover bracket.
64. Install the ignition coil cover.
65. Install the washers and new steering gearbox mounting bolts to the steering gearbox, then loosely tighten them.

➡Make sure that the washers are in the correct position.

66. Tighten the steering gearbox mounting bolts to 54 ft. lbs. (74 Nm) in the sequence shown.

➡Check all of the steering gearbox mounting bolts, and retighten if necessary.

67. Connect the EPS subharness 6P connector to the steering gearbox.
68. Remove the vinyl tape, then connect the EPS motor angle sensor 6P connector to the steering gearbox.
69. Connect the EPS motor 3P connector, then pull up the lever, and confirm the connector is fully seated.
70. Install the P/S heat shield with the flange bolts.
71. Install the strut brace.
72. Install the front grille cover.
73. Reinstall the hood support rod to the upper location of the hood.
74. Wipe off any grease contamination from the ball joint tapered section and threads. Reconnect the tie-rod end ball joint to the knuckle. Install the nut, and tighten it to 40 ft, lbs. (54 Nm).
75. Install a new cotter pin, and bend it.
76. Install the front wheels, then set the wheels in the straight ahead position.

➡Before installing the wheel, clean the mating surface between the brake disc and the inside of the wheel.

77. Remove the vinyl tape around the splines on the pinion shaft.
78. Loosen the upper steering joint bolt and slip the lower end of the steering joint onto the pinion shaft taking care to align the gap within the angle of 35 degrees +/- 20 degrees.

➡Pick up the tabs of the pinion shaft grommet, and turn up the lip of the pinion shaft grommet securely in place. Make sure that light does not enter from the space between the pinion shaft grommet and the body.

79. Align the bolt hole on the steering joint with the groove around the pinion shaft, then loosely install the lower steering joint bolt. Be sure that the joint bolt is securely in the groove in the pinion shaft.
80. Pull on the steering joint to make sure that the steering joint is fully seated, then tighten the lower joint bolt to the specified torque.
81. Tighten the upper steering joint bolt to 21 ft. lbs. (28 Nm).
82. Install the steering joint cover.
83. Install the steering wheel, and the driver's airbag.
84. With the tires raised off the ground (vehicle on a lift), check for rubbing or grating noises by turning the steering wheel fully to the right and left several times.
85. Do the battery terminal reconnection procedure, and check these items:
    - Turn the ignition switch to ON (II), and check that the SRS indicator comes on for about 6 seconds, and then goes off.
    - Make sure the horn and turn signal switches work properly.
    - Make sure the steering wheel switches work properly.
86. After installation, check these items:
    - Start the engine, allow it to idle, and turn the steering wheel from lock to lock several times.
    - Check that the EPS indicator does not come on.
    - Check the steering wheel spoke angle. If steering spoke angles to the right and left are not equal (steering wheel and rack are not centered), correct the engagement of the joint/pinion shaft serrations.
87. Do the torque sensor neutral position memorization.
88. Check the wheel alignment, and adjust it if necessary.

TORQUE SENSOR NEUTRAL POSITION MEMORIZATION

The torque sensor neutral position must be memorized whenever the steering gearbox, the torque sensor, or the EPS control unit is replaced. Note that the torque sensor neutral position is not affected when erasing the DTC.

➡The torque sensor is temperature sensitive. When memorizing the torque sensor neutral position, the ambient temperature must be above 68°F (20°C).

1. With the ignition switch in LOCK (0), connect the HDS to the Data Link Connector (DLC) located under the driver's side of the dashboard.
2. Turn the ignition switch to ON (II).
3. Make sure the HDS communicates with the vehicle and the EPS control unit. If it doesn't, troubleshoot the DLC circuit.
4. From the EPS MENU, select MISCELLANEOUS TEST, then TORQUE SENSOR LEARN, and follow the screen prompts on the HDS.

➡See the HDS Help menu for specific instructions.

5. Turn the ignition switch to LOCK (0).

6-Cylinder Engine

See Figures 374 through 397.

Special Tools Required:

- Ball joint thread protector, 14 mm 07AAE-SJAA100
- Ball joint thread protector, 12 mm 07AAF-SDAA100
- Ball joint remover, 28 mm 07MAC-SL0A202
- Subframe alignment pin 070AG-SJAA10S
- Engine support hanger, A and Reds AAR-T1256*
- Engine hanger balance bar VSB02C000019*
- Front subframe adapter VSB02C000016*

➡*Available through the Acura Tool and Equipment Program, 888-424-6857.

⁂ WARNING

SRS components are located in this area Review the SRS component locations. and the precautions and procedures before doing repairs or service.

➡Note these items during removal:

- Use solvent and a brush, wash any oil and dirt off the end of the steering gearbox. Avoid any electrical parts. Blow dry with compressed air.
- Be sure to remove the steering wheel before disconnecting the steering joint, or damage to the cable reel can occur.

1. Remove the engine compartment covers.
2. Do the battery terminal disconnection procedure.

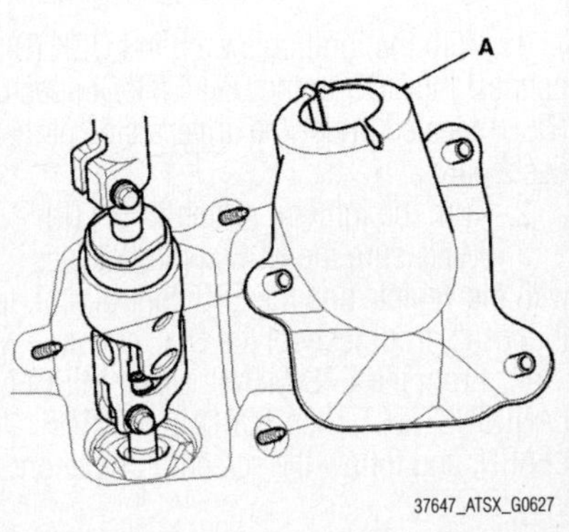
A
37647_ATSX_G0627

Fig. 374 Remove the steering joint cover (A)

3. Raise and support the vehicle.
4. Remove the front wheels.
5. Remove the driver's airbag, and the steering wheel.
6. Remove the steering joint cover.
7. Loosen the upper steering joint bolt, and remove the lower steering joint bolt.
8. Disconnect the steering joint by sliding the steering joint into the column shaft. Tighten the upper steering joint bolt to hold the column shaft.

➡Do not disconnect the steering joint from the column shaft.

9. Remove the center guide (if equipped) from the top of the pinion shaft, and discard it. The center guide is for factory assembly use only.

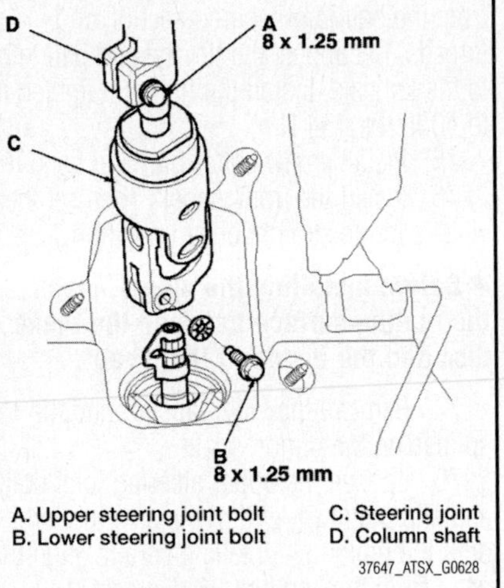
D
A
8 x 1.25 mm
C
B
8 x 1.25 mm
A. Upper steering joint bolt
B. Lower steering joint bolt
C. Steering joint
D. Column shaft
37647_ATSX_G0628

Fig. 375 Steering joint

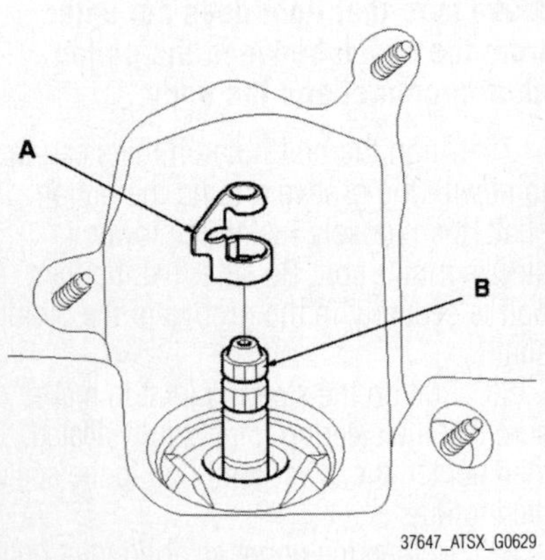
A
B
37647_ATSX_G0629

Fig. 376 Remove the center guide (A) (if equipped) from the top of the pinion shaft (B)

10. Apply vinyl tape to the splines on the pinion shaft.
11. Disconnect the hood support struts from both sides of the pivot ball (bolted to the hood). Secure the hood in a vertical position. Remove the left side pivot ball and install it into the lower threaded hole, then reattach the support strut.

➡Do not attempt to close the hood with the support strut in the vertical position, as it will damage the support strut and hood.

12. Remove the strut brace.
13. Remove the P/S heat shield.
14. Disconnect the EPS motor angle sensor 6P connector.
15. Disconnect the EPS motor 3P connector by pushing the lock and pulling down the lever. Wrap the connectors with vinyl tape to avoid contamination from grease or water.
16. Disconnect the EPS subharness 6P connector. Wrap the connectors with vinyl

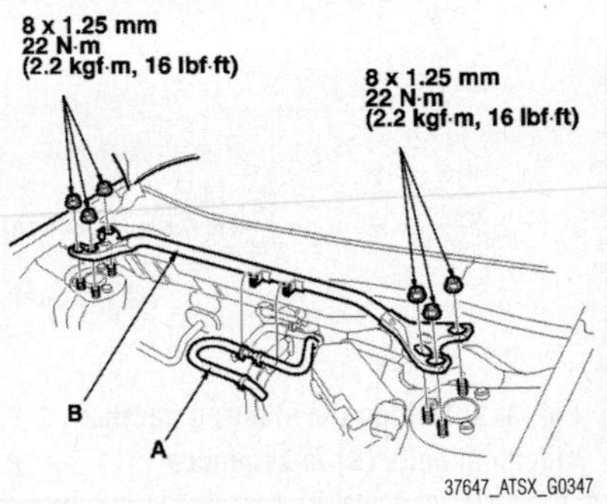
8 x 1.25 mm
22 N·m
(2.2 kgf·m, 16 lbf·ft)
8 x 1.25 mm
22 N·m
(2.2 kgf·m, 16 lbf·ft)
B
A
37647_ATSX_G0347

Fig. 377 Remove the strut brace

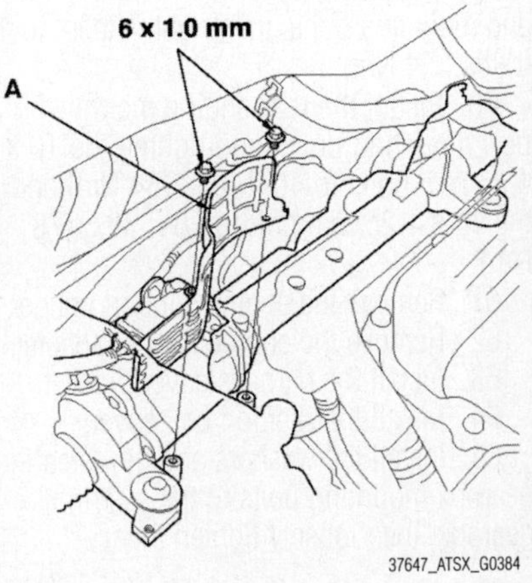
6 x 1.0 mm
A
37647_ATSX_G0384

Fig. 378 Remove the P/S heat shield (A)

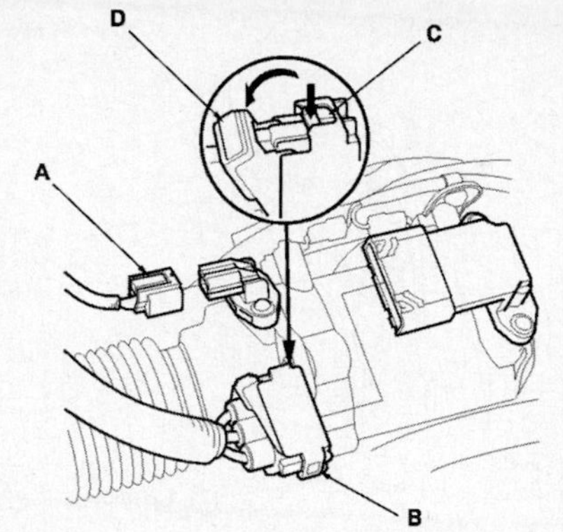
D
C
A
B

A. EPS motor angle sensor 6P connector C. Lock
B. EPS motor 3P connector D. Lever

37647_ATSX_G0630

Fig. 379 Disconnect the EPS motor angle sensor 6P connector

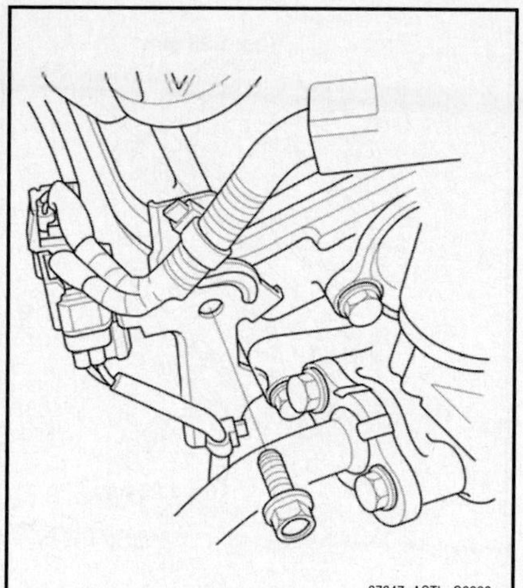

37647_ACTL_G0320

Fig. 382 Remove the connector bracket from the front cylinder head

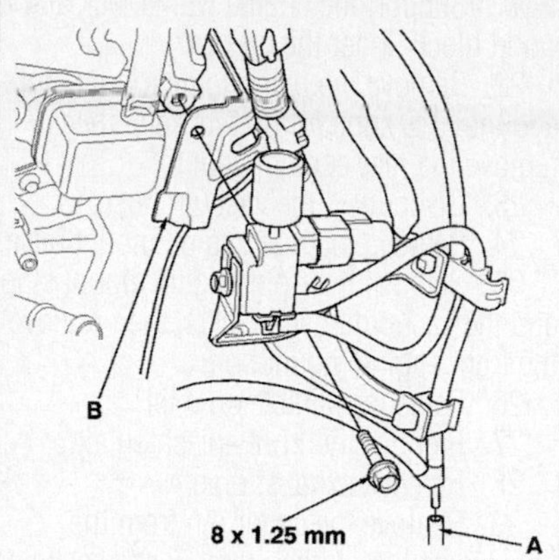
B
8 x 1.25 mm
A

37647_ATSX_G0646

Fig. 383 Disconnect the vacuum hose (A), then remove the engine mount control solenoid valve bracket (B) from the rear cylinder head

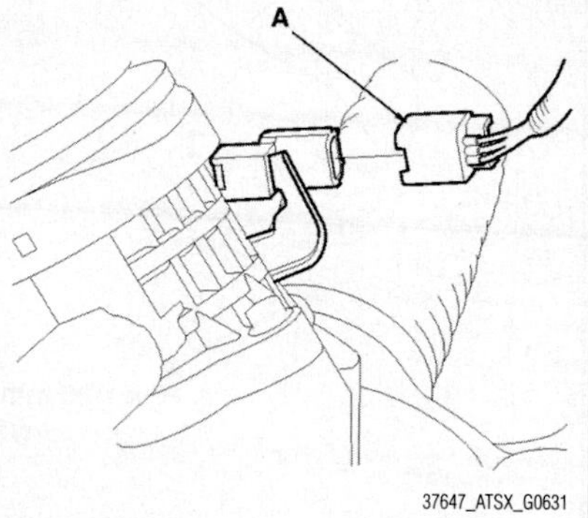
A

37647_ATSX_G0631

Fig. 380 Disconnect the EPS subharness 6P connector (A)

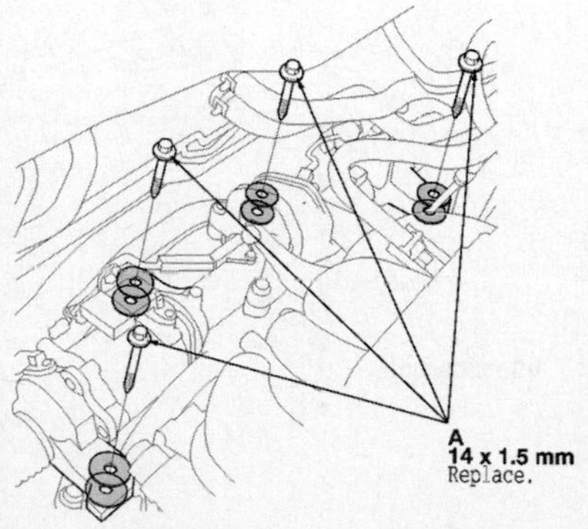
A
14 x 1.5 mm
Replace.

37647_ATSX_G0632

Fig. 381 Remove the steering gearbox mounting bolts (A) and washers (B) from the steering gearbox

tape to avoid contamination from grease or water.

17. Remove the steering gearbox mounting bolts and washers from the steering gearbox.

18. Remove the connector bracket from the front cylinder head; use the bracket bolt hole to attach the engine hanger balance bar front arm.

19. Disconnect the vacuum hose, then remove the engine mount control solenoid valve bracket from the rear cylinder head; use the bracket bolt hole to attach the engine hanger balance bar rear.

20. Lift and support the engine with the engine support hanger (AAR-T1256) and the engine hanger balance bar (VSB02C000019). Attach the front arm to the front cylinder head with a spacer and a 10 x 1.25 mm bolt. Attach the rear arm to the rear cylinder head with a spacer and a 8 x 1.25 mm bolt.

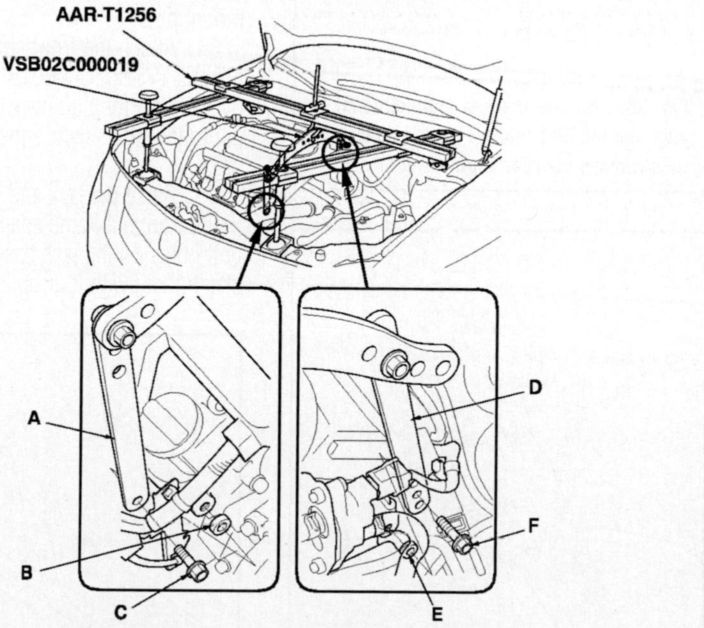
AAR-T1256
VSB02C000019
A
B
C
D
E
F

A. Front arm D. Rear arm
B. Spacer E. Spacer
C. 10 x 1.25 mm bolt F. 8 x 1.25 mm bolt

37647_ATSX_G0439

Fig. 384 Lift and support the engine with the engine support hanger and the engine hanger balance bar

21. Support the engine with a jack and a wood block under the oil pan.

22. Remove the rear engine mount stop and the EPS subharness brackets, then remove the rear engine mount.

23. Disconnect the vacuum hose.

24. Remove the rear engine mount base.

25. Remove the front engine mount stop and the vacuum hose bracket, then remove the front engine mount bolt.

26. Raise the vehicle on a lift.

27. Remove the front splash shield.

28. Remove exhaust pipe A.

29. Remove the cotter pin from the knuckle ball joint, then remove the castle nut.

30. Disconnect the knuckle ball joint from the lower arm using the ball joint thread protector and the ball joint remover.

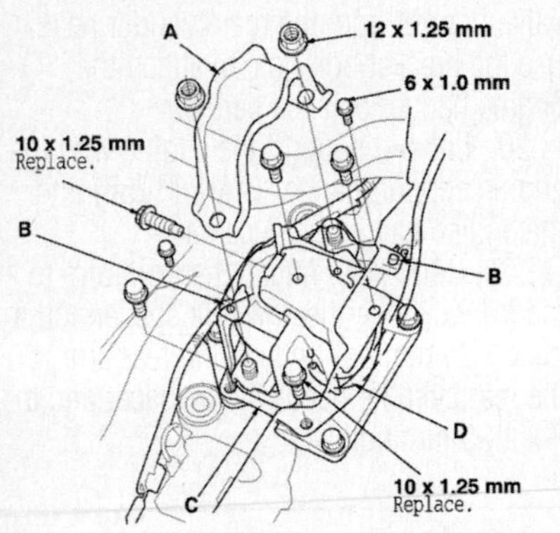
A
12 x 1.25 mm
6 x 1.0 mm
10 x 1.25 mm
Replace.
B
B
D
C
10 x 1.25 mm
Replace.
A. Rear engine mount stop
B. EPS subharness brackets
C. Rear engine mount
D. Vacuum hose
37647_ATSX_G0647

Fig. 385 Remove the rear engine mount stop and the EPS subharness brackets, then remove the rear engine mount

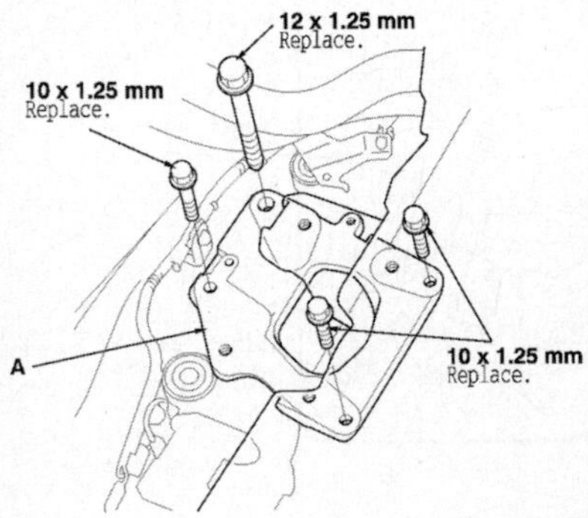
12 x 1.25 mm
Replace.
10 x 1.25 mm
Replace.
A
10 x 1.25 mm
Replace.
37647_ATSX_G0648

Fig. 386 Remove the rear engine mount base (A)

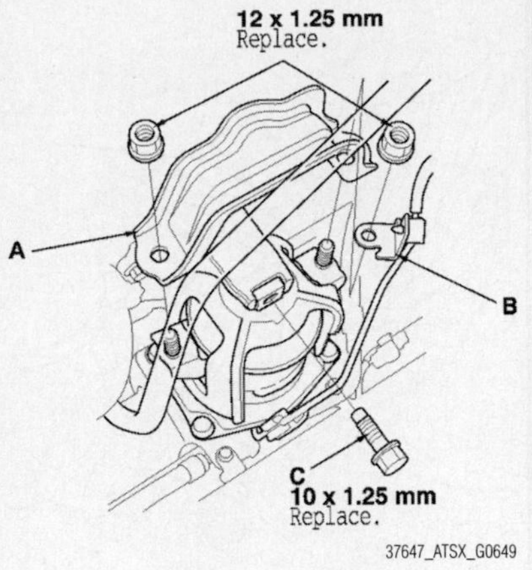
12 x 1.25 mm
Replace.
A
B
C
10 x 1.25 mm
Replace.
37647_ATSX_G0649

Fig. 387 Remove the front engine mount stop (A) and the vacuum hose bracket (B), then remove the front engine mount bolt (C)

➡Be careful not to damage the ball joint boot when installing the remover. Do not force or hammer on the lower arm, or pry between the lower arm and the knuckle. You could damage the ball joint.

31. Remove the damper fork mounting nut while holding the mounting bolt, then remove the mounting bolt.

32. Remove the subframe middle mount on both sides.

33. Remove the lower transmission mount nuts.

34. Attach the front subframe adapter (VSB02C000016) to the subframe by hanging the belt over the front of the subframe, then secure the belt with its stop.

35. Raise the jack and line up the slots in the front subframe adapter arms with the bolt holes on the jack base, then securely attach them with four bolts.

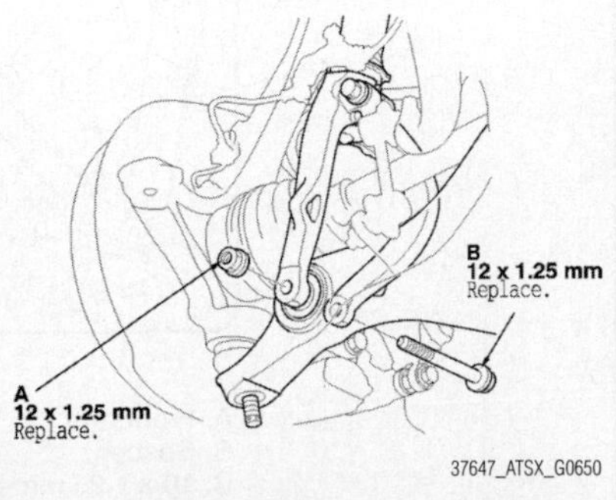
B
12 x 1.25 mm
Replace.
A
12 x 1.25 mm
Replace.
37647_ATSX_G0650

Fig. 388 Remove the damper fork mounting nut (A) while holding the mounting bolt (B), then remove the mounting bolt

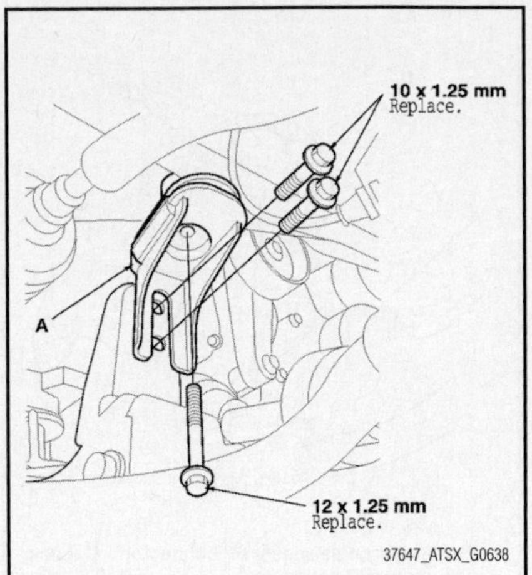
10 x 1.25 mm
Replace.
A
12 x 1.25 mm
Replace.
37647_ATSX_G0638

Fig. 389 Remove the subframe middle mount (A) on both sides

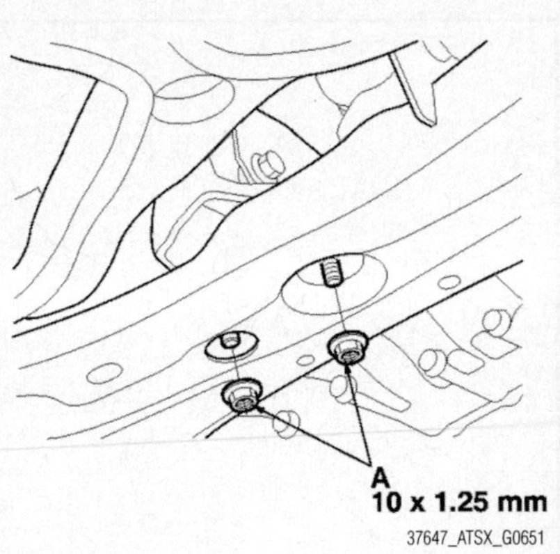
A
10 x 1.25 mm
37647_ATSX_G0651

Fig. 390 Remove the lower transmission mount nuts (A)

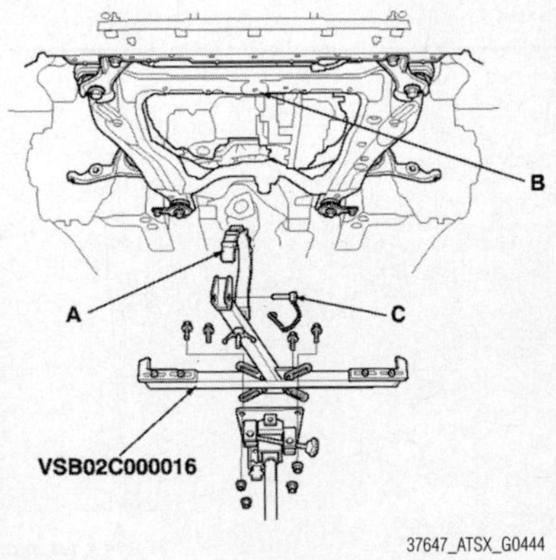
B
A
C
VSB02C000016
37647_ATSX_G0444

Fig. 391 Attach the front subframe adapter to the subframe by hanging the belt (A) over the front of the subframe (B), then secure the belt with its stop (C)

36. Remove the stiffener mounting bolt on both sides.

37. Loosen the subframe mounting bolts so they are about ¾ inches (20 mm) from the mounting surface. Do not loosen the subframe mounting bolts more than necessary.

38. Remove the stiffener mounting bolts and subframe mounting bolts from the rear stiffeners.

39. Lower the jack slowly until the subframe has dropped about 2 ¾ inches (70 mm).

40. Carefully move the steering gearbox toward the driver's side until the pinion shaft clears the fenderwell opening on the body.

41. Remove the steering gearbox through the fenderwell opening on the driver's side.

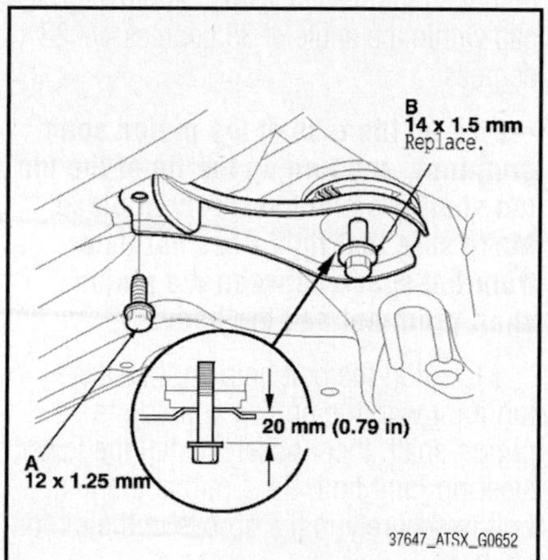
B
14 x 1.5 mm
Replace.
20 mm (0.79 in)
A
12 x 1.25 mm
37647_ATSX_G0652

Fig. 392 Remove the stiffener mounting bolt (A) on both sides

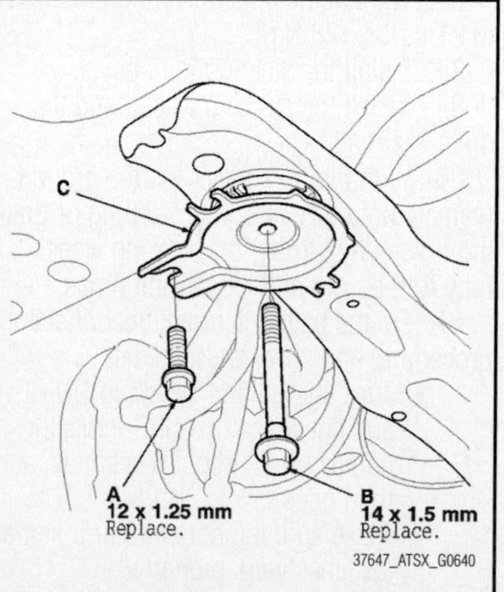
C
A
12 x 1.25 mm
Replace.
B
14 x 1.5 mm
Replace.
37647_ATSX_G0640

Fig. 393 Remove the stiffener mounting bolts (A) and subframe mounting bolts (B) from the rear stiffeners (C)

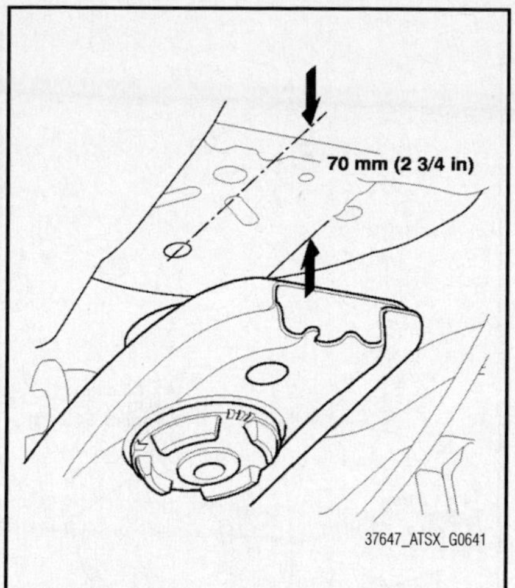
70 mm (2 3/4 in)
37647_ATSX_G0641

Fig. 394 Lower the jack slowly until the subframe has dropped about 2 ¾ inches (70 mm)

To install:

42. Slide the steering gearbox between the front subframe and the body from the driver's side.

43. Carefully move the steering gearbox toward the passenger's side until the pinion shaft clears the wheel well opening on the body.

44. Continue moving the steering gearbox toward the passenger's side until the steering gearbox is in position.

45. Install the subframe rear stiffeners, and loosely install the new subframe mounting bolts, new rear stiffener mounting bolts, and front stiffener mounting bolts.

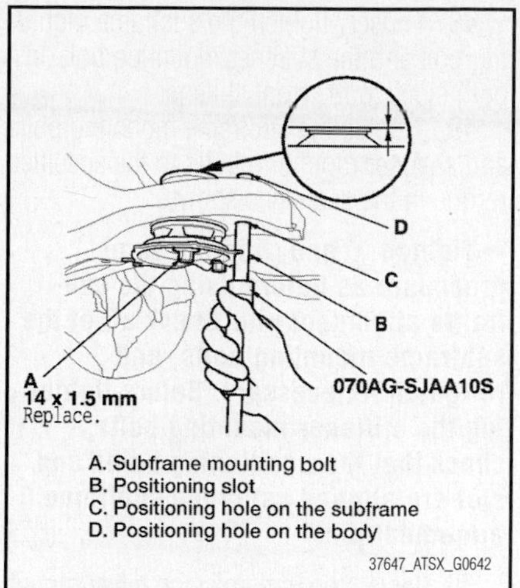
D
C
B
A
14 x 1.5 mm
Replace.
070AG-SJAA10S
A. Subframe mounting bolt
B. Positioning slot
C. Positioning hole on the subframe
D. Positioning hole on the body
37647_ATSX_G0642

Fig. 395 Insert the subframe alignment pin through the positioning slot on the right rear stiffener, through the positioning hole on the subframe, and into the positioning hole on the body

46. Loosely tighten the subframe mounting bolt in the right rear stiffener until the subframe insulator contacts the body; insert the subframe alignment pin (070AG-SJAA10S) through the positioning slot on the right rear stiffener, through the positioning hole on the subframe, and into the positioning hole on the body.

47. Loosely tighten the left rear subframe mounting bolt with the same procedure as the right rear using the subframe alignment pin.

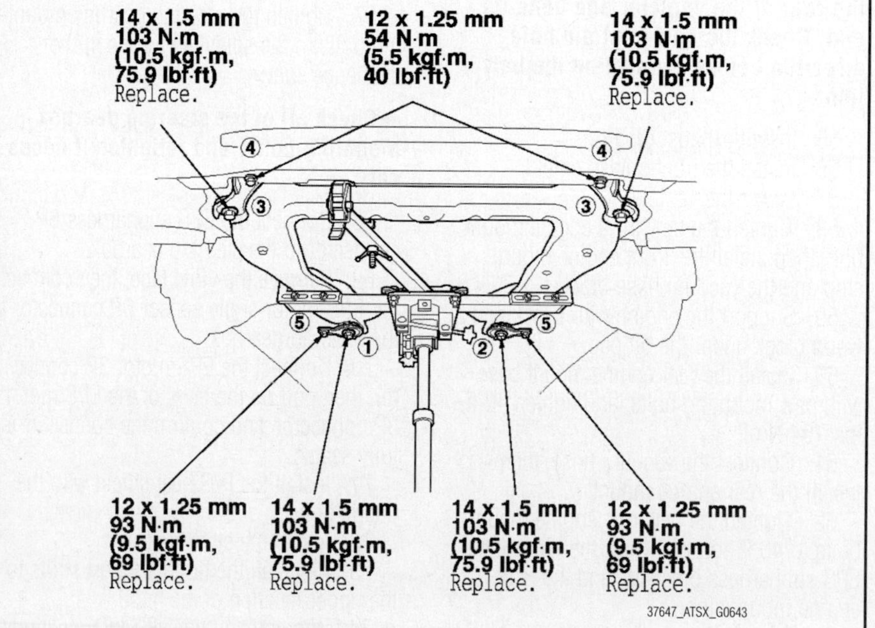
14 x 1.5 mm
103 N·m
(10.5 kgf·m,
75.9 lbf·ft)
Replace.
12 x 1.25 mm
54 N·m
(5.5 kgf·m,
40 lbf·ft)
14 x 1.5 mm
103 N·m
(10.5 kgf·m,
75.9 lbf·ft)
Replace.
12 x 1.25 mm
93 N·m
(9.5 kgf·m,
69 lbf·ft)
Replace.
14 x 1.5 mm
103 N·m
(10.5 kgf·m,
75.9 lbf·ft)
Replace.
14 x 1.5 mm
103 N·m
(10.5 kgf·m,
75.9 lbf·ft)
Replace.
12 x 1.25 mm
93 N·m
(9.5 kgf·m,
69 lbf·ft)
Replace.
37647_ATSX_G0643

Fig. 396 Tighten the subframe mounting bolts and stiffener mounting bolts in the sequence

48. Loosely tighten the subframe mounting bolt and the stiffener mounting bolt on both sides to the front stiffener.

49. Tighten the subframe mounting bolts and stiffener mounting bolts to the specified torque in the sequence shown.

➡Tighten ① and ② in the same procedure as before using the subframe alignment pin. Check all of the subframe mounting bolts, and retighten if necessary. Before tightening the stiffener mounting bolts, check that the positioning holes and slot are aligned using the subframe alignment pin.

50. Install the transmission lower mount nuts, and tighten them to 33 ft. lbs. (44 Nm).

51. Install the subframe middle mount with new mounting bolts, and tighten the 10 x 1.25 mm bolts to 36 ft. lbs. (49 Nm) and the 12 x 1.25 mm bolts to 33 ft. lbs. (44 Nm) on both sides.

52. Remove the transmission jack supporting the subframe.

53. Install the new damper fork mounting bolt and new mounting nut, and loosely tighten the nut while holding the mounting bolt.

54. Tighten the new castle nut to the lower torque specification, 69–76 ft. lbs. (93–103 Nm), then tighten it only far enough to align the slot with the ball joint pin hole. Do not align the castle nut by loosening it.

➡Insert the new cotter pin into the ball joint pin hole from the front to the rear of the vehicle, and bend its end. Check the ball joint pin hole direction before connecting the ball joint.

55. Install exhaust pipe A.

56. Install the front splash shield.

57. Lower the vehicle.

58. Tighten the new front engine mount bolt, then install the front engine mount stop and the vacuum hose bracket.

59. Support the engine with a jack and a wood block under the oil pan.

60. Install the rear engine mount base with new mounting bolts, and tighten 40 ft. lbs. (54 Nm).

61. Connect the vacuum hose, then install the rear engine mount.

62. Tighten the new rear engine mount bolts to 40 ft. lbs. (54 Nm), then install the EPS subharness brackets, and the rear engine mount stop.

63. Remove the engine support hanger and the engine hanger balance bar.

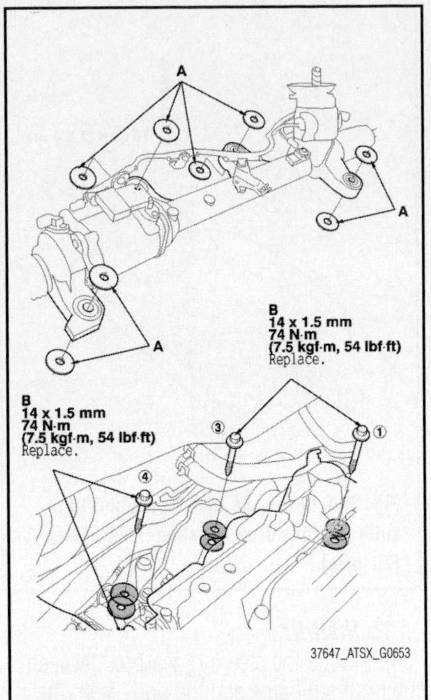
A
A
A
B
14 x 1.5 mm
74 N·m
(7.5 kgf·m, 54 lbf·ft)
Replace.
B
14 x 1.5 mm
74 N·m
(7.5 kgf·m, 54 lbf·ft)
Replace.
①
③
④
37647_ATSX_G0653

Fig. 397 Install the washers (A) and new steering gearbox mounting bolts (B) to the steering gearbox

64. Install the engine mount control solenoid valve bracket to the rear cylinder head, and connect the vacuum hose.

65. Install the connector bracket to the front cylinder head.

66. Install the washers and new steering gearbox mounting bolts to the steering gearbox, then loosely tighten them.

➡Make sure that the washers are in the correct position.

67. Tighten the steering gearbox mounting bolts to the specified torque in the sequence shown.

➡Check all of the steering gearbox mounting bolts, and retighten if necessary.

68. Connect the EPS subharness 6P connector to the steering gearbox.

69. Remove the vinyl tape, then connect the EPS motor angle sensor 6P connector to the steering gearbox.

70. Connect the EPS motor 3P connector, then pull up the lever of the EPS motor 3P connector, and confirm the connector is fully seated.

71. Install the P/S heat shield with the flange bolts.

72. Install the strut brace.

73. Reinstall the hood support struts to the upper location of the hood.

74. Wipe off any grease contamination from the ball joint tapered section and threads. Connect the tie-rod end ball joint to the knuckle. Install the nut, and tighten it to 44 ft. lbs. (60 Nm).

75. Install a new cotter pin, and bend it.

76. Place a floor jack under the lower arm, and raise the suspension to load it with the vehicle's weight. Do not place the jack against the ball joint pin of the knuckle.

77. Tighten the damper fork mounting nut while holding the mounting bolt to 47 ft. lbs. (64 Nm).

78. Install the front wheels, then set the wheels in the straight ahead position.

➡Before installing the wheel, clean the mating surface between the brake disc and the inside of the wheel.

79. Remove the vinyl tape around the splines on the pinion shaft.

80. Loosen the upper steering joint bolt and slip the lower end of the steering joint onto the pinion shaft taking care to align the gap within the angle of 35 degrees +/- 20 degrees.

➡Pick up the tabs of the pinion shaft grommet, and turn up the lip of the pinion shaft grommet securely in place. Make sure that light does not enter from the space between the pinion shaft grommet and the body.

81. Align the bolt hole on the steering joint with the groove around the pinion shaft, then loosely install the lower steering joint bolt. Be sure that the joint bolt is securely in the groove in the pinion shaft.

82. Pull on the steering joint to make sure that the steering joint is fully seated, then tighten the lower joint bolt to the specified torque.

83. Tighten the upper steering joint bolt to 21 ft. lbs. (28 Nm).

84. Install the steering joint cover.

85. Install the steering wheel, and the driver's airbag.

86. With the tires raised off the ground (vehicle on a lift), check for rubbing or grating noises by turning the steering wheel fully to the right and left several times.

87. Do the battery terminal reconnection procedure, and check these items:

- Turn the ignition switch to ON (II), and check that the SRS indicator comes on for about 6 seconds, and then goes off.
- Make sure the horn and turn signal switches work properly.
- Make sure the steering wheel switches work properly.

88. After installation, check these items:

- Start the engine, allow it to idle,

and turn the steering wheel from lock to lock several times.

- Check that the EPS indicator does not come on.
- Check the steering wheel spoke angle. If steering spoke angles to the right and left are not equal (steering wheel and rack are not centered), correct the engagement of the joint/pinion shaft serrations.

89. Do the torque sensor neutral position memorization.

90. Check the wheel alignment, and adjust it if necessary.

TORQUE SENSOR NEUTRAL POSITION MEMORIZATION

The torque sensor neutral position must be memorized whenever the steering gearbox, the torque sensor, or the EPS control unit is replaced. Note that the torque sensor neutral position is not affected when erasing the DTC.

➡The torque sensor is temperature sensitive. When memorizing the torque sensor neutral position, the ambient temperature must be above 68°F (20°C).

1. With the ignition switch in LOCK (0), connect the HDS to the Data Link Connector (DLC) located under the driver's side of the dashboard.
2. Turn the ignition switch to ON (II).
3. Make sure the HDS communicates with the vehicle and the EPS control unit. If it doesn't, troubleshoot the DLC circuit.
4. From the EPS MENU, select MISCELLANEOUS TEST, then TORQUE SENSOR LEARN, and follow the screen prompts on the HDS.

➡See the HDS Help menu for specific instructions.

5. Turn the ignition switch to LOCK (0).

SUSPENSION — FRONT SUSPENSION

CONTROL LINKS

REMOVAL & INSTALLATION

Front Stabilizer Link

See Figure 398.

1. Raise the vehicle on a lift.
2. Remove the front wheel.
3. Remove the self-locking nut and the flange nut while holding the respective joint pin with a hex wrench, then remove the stabilizer link.

To install:

4. Install the stabilizer link on the stabilizer bar and the lower arm with the joint pins set at the center of their range of movement.

➡The stabilizer link has a paint mark. The left stabilizer link is marked with yellow paint, and the right stabilizer link is marked with white paint.

5. Install the new self-locking nut and the new flange nut, and tighten them to the specified torque while holding the respective joint pin with a hex wrench.
6. Clean the mating surfaces of the brake disc and the inside of the wheel, then install the front wheel.
7. Test-drive the vehicle.
8. After 5 minutes of driving, tighten the self-locking nut again to the specified torque.

LOWER BALL JOINT

REMOVAL & INSTALLATION

See Figures 399 through 401.

Special Tools Required:

- Ball joint thread protector, 14 mm 07AAE-SJAA100
- Ball joint thread protector, 12 mm 07AAF-SDAA100
- Ball joint thread protector, 10 mm 07AAF-SECA120
- Ball joint remover, 32 mm 07MAC-SL0A102
- Ball joint remover, 28 mm 07MAC-SL0A202

➡Always use a ball joint remover to disconnect a ball joint. Do not strike the housing or any other part of the ball joint connection to disconnect it.

1. Install a hex nut or the ball joint thread protector onto the threads of the ball joint.

➡Using a hex nut, make sure the nut is flush with the ball joint pin end to prevent damage to the threaded end of the ball joint pin.

2. Apply grease to the ball joint remover on the areas shown. This will ease the installation of the tool, and prevent damage to the pressure bolt threads.

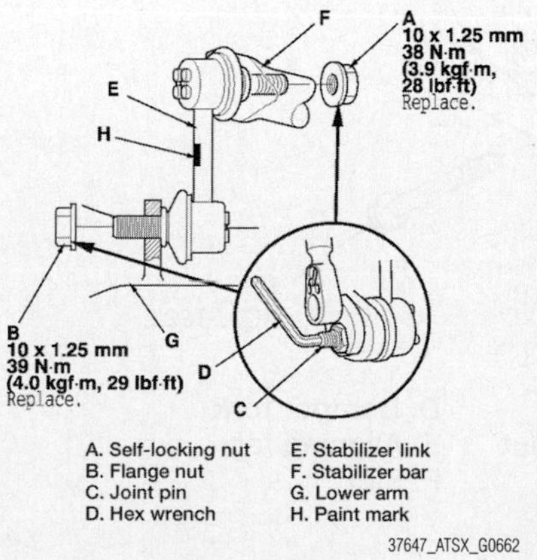
F
A
10 x 1.25 mm
38 N·m
(3.9 kgf·m, 28 lbf·ft)
Replace.
E
H
B
10 x 1.25 mm
39 N·m
(4.0 kgf·m, 29 lbf·ft)
Replace.
G
D
C
A. Self-locking nut
B. Flange nut
C. Joint pin
D. Hex wrench
E. Stabilizer link
F. Stabilizer bar
G. Lower arm
H. Paint mark
37647_ATSX_G0662

Fig. 398 Front stabilizer link

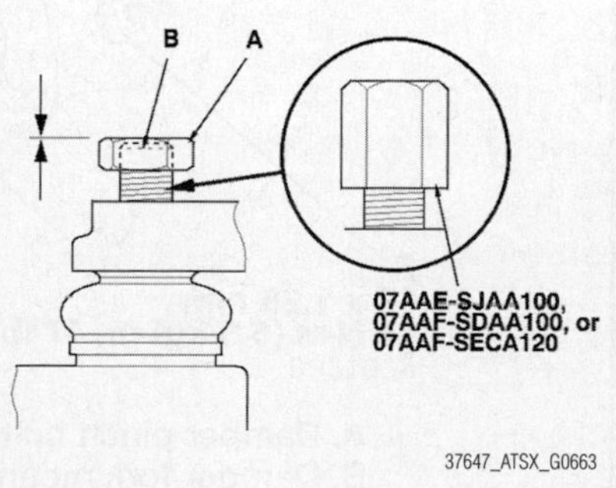
B
A
07AAE-SJAA100,
07AAF-SDAA100, or
07AAF-SECA120
37647_ATSX_G0663

Fig. 399 Install a hex nut (A) or the ball joint thread protector onto the threads of the ball joint (B)

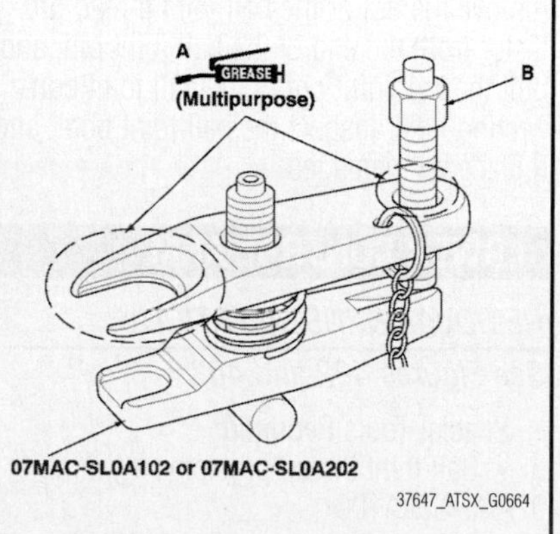
A
GREASE
(Multipurpose)
B
07MAC-SL0A102 or 07MAC-SL0A202
37647_ATSX_G0664

Fig. 400 Apply grease to the ball joint remover on the areas shown (A)

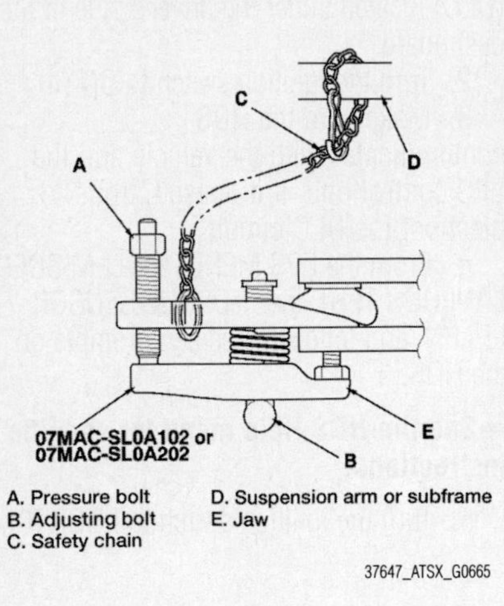
A
B
C
D
E
07MAC-SL0A102 or
07MAC-SL0A202
A. Pressure bolt
B. Adjusting bolt
C. Safety chain
D. Suspension arm or subframe
E. Jaw
37647_ATSX_G0665

Fig. 401 Loosen the pressure bolt, and install the ball joint remover as shown

3. Loosen the pressure bolt, and install the ball joint remover as shown. Insert the jaws carefully, making sure not to damage the ball joint boot. Adjust the jaw spacing by turning the adjusting bolt.

➡Fasten the safety chain securely to a suspension arm or the subframe. Do not fasten it to a brake line or wire harness.

4. After adjusting the adjusting bolt, make sure the head of the adjusting bolt is in the position shown to allow the jaw to pivot.

5. With a wrench, tighten the pressure bolt until the ball joint pin pops loose from the ball joint connecting hole. If necessary, apply penetrating type lubricant to loosen the ball joint pin.

➡Do not use pneumatic or electric tools on the pressure bolt.

6. Remove the ball joint remover, then remove the nut or the ball joint thread protector from the end of the ball joint pin, and pull the ball joint out of the ball joint connecting hole. Inspect the ball joint boot, and replace it if damaged.

LOWER CONTROL ARM

REMOVAL & INSTALLATION

See Figures 402 and 403.

Special Tools Required:

- Ball joint thread protector, 14 mm 07AAE-SJAA100
- Ball joint remover, 28 mm 07MAC-SL0A202
- Bushing driver 070AF-TA0A100
- Bushing receiver set 070AF-TA0A220

1. Raise the vehicle on a lift.
2. Remove the front wheel.
3. Remove the front splash shield.
4. Remove the damper pinch bolt and the damper fork mounting nut while holding the mounting bolt, then remove the damper fork from the damper and the lower arm.

➡During installation, insert the aligning tab on the damper unit into the slot of the damper fork. Use the new damper fork mounting bolt and the new mounting nut, and torque the nut while holding the bolt during reassembly.

5. Disconnect the stabilizer link from the lower arm.

6. Remove the cotter pin from the knuckle ball joint, then remove the castle nut.

➡During installation, insert the new cotter pin into the ball joint pin hole from the front to the rear of the vehicle, and bend it. Check the ball joint pin hole direction before connecting the ball joint.

7. Disconnect the knuckle ball joint from the lower arm using the ball joint thread protector and the ball joint remover.

➡Be careful not to damage the ball joint boot when installing the remover. Do not force or hammer on the lower arm, or pry between the lower arm and the knuckle. You could damage the ball joint.

8. Remove the lower arm mounting bolts, and remove the lower arm.

➡Use new lower arm mounting bolts during reassembly.

To install:

9. Install the lower arm in the reverse order of removal, and note these items:
   - First install all of the components, and lightly tighten the bolts and the nuts, then raise the suspension to load it with the vehicle's weight before fully tightening it to the specified torque. Do not place the jack against the ball joint pin of the knuckle.
   - Be careful not to damage the ball joint boot when connecting the knuckle.
   - Before connecting the ball joint, degrease the threaded section and the tapered portion of the ball

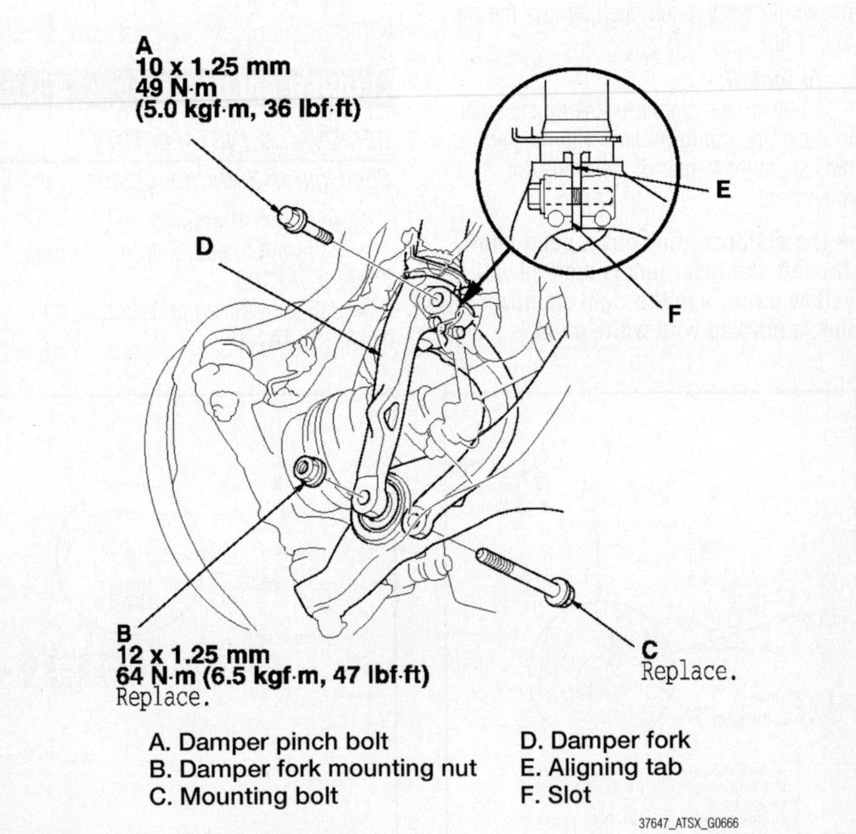
A
10 x 1.25 mm
49 N·m
(5.0 kgf·m, 36 lbf·ft)
D
E
F
B
12 x 1.25 mm
64 N·m (6.5 kgf·m, 47 lbf·ft)
Replace.
C
Replace.
A. Damper pinch bolt
B. Damper fork mounting nut
C. Mounting bolt
D. Damper fork
E. Aligning tab
F. Slot
37647_ATSX_G0666

Fig. 402 Remove the damper pinch bolt and the damper fork mounting nut while holding the mounting bolt

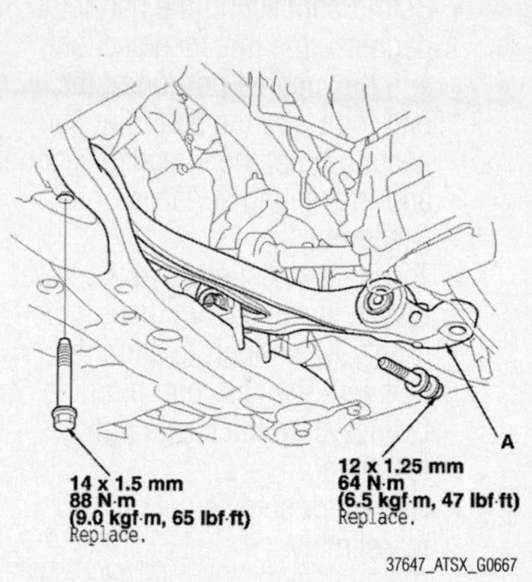
A
14 x 1.5 mm
88 N·m
(9.0 kgf·m, 65 lbf·ft)
Replace.
12 x 1.25 mm
64 N·m
(6.5 kgf·m, 47 lbf·ft)
Replace.
37647_ATSX_G0667

Fig. 403 Remove the lower arm mounting bolts, and remove the lower arm (A)

joint pin, the ball joint connecting hole, the threaded section, and the mating surfaces of the castle nut.

- Torque the castle nut to the lower torque specification, 58–65 ft. lbs. (78–88 Nm), then tighten it only far enough to align the slot with the ball joint pin hole. Do not align the castle nut by loosening it.
- Before installing the wheel, clean the mating surfaces of the brake disc and the inside of the wheel.

10. Check the wheel alignment, and adjust it if necessary.

STABILIZER BAR

REMOVAL & INSTALLATION

See Figure 404.

1. Raise the vehicle on a lift.
2. Remove the front wheels.
3. Disconnect both tie-rod ball joints from the knuckle.
4. Disconnect both stabilizer links from the stabilizer bar.
5. Remove the flange bolts and the bushing holders, then remove the bushings and the stabilizer bar.

To install:

➡During installation, align the paint marks on the stabilizer bar with the side of the bushings.

6. Install the stabilizer bar in the reverse order of removal, and note these items:
   - Note the right and left direction of the stabilizer bar.
   - Note the direction of installation for the bushings.

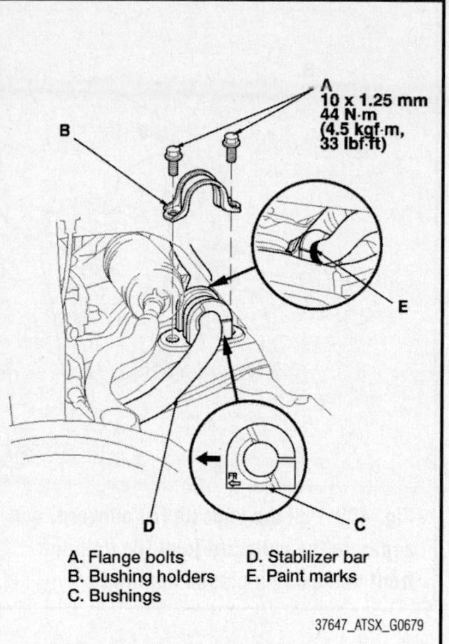
A
10 x 1.25 mm
44 N·m
(4.5 kgf·m,
33 lbf·ft)
B
E
D
C
A. Flange bolts
B. Bushing holders
C. Bushings
D. Stabilizer bar
E. Paint marks
37647_ATSX_G0679

Fig. 404 Remove the flange bolts and the bushing holders, then remove the bushings and the stabilizer bar

- Refer to stabilizer link removal/installation to connect the stabilizer bar to the links.
- Before installing the wheel, clean the mating surfaces of the brake disc and the inside of the wheel.

7. Check the wheel alignment, and adjust it if necessary.

STEERING KNUCKLE

REMOVAL & INSTALLATION

See Figures 405 through 409.

Special Tools Required:

- Ball joint thread protector, 14 mm 07AAE-SJAA100
- Ball joint thread protector, 12 mm 07AAF-SDAA100
- Ball joint thread protector, 10 mm 07AAF-SECA120
- Ball joint remover, 28 mm 07MAC-SL0A202
- Hub dis/assembly tool 07GAF-SD4A100
- Attachment, 72 x 75 mm 07746-0010600
- Driver handle 07749-0010000
- Attachment, 80 x 96 mm 07ZAD-PNA0100
- Support base 07965-SD90100

1. Raise the vehicle on a lift.
2. Remove the wheel nuts and the front wheel.
3. Remove the brake hose bracket mounting bolt.

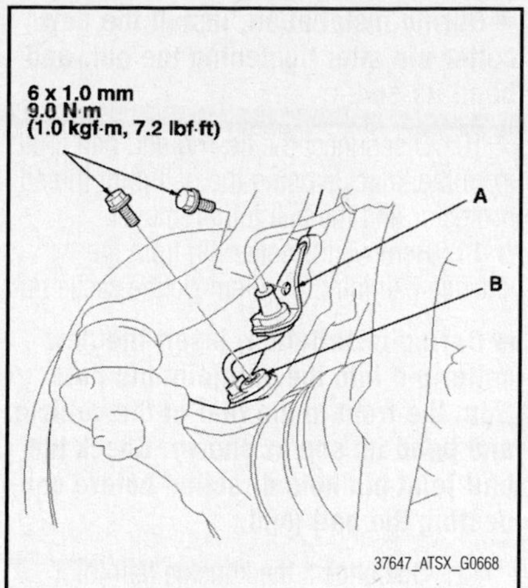
6 x 1.0 mm
9.0 N·m
(1.0 kgf·m, 7.2 lbf·ft)
A
B
37647_ATSX_G0668

Fig. 405 Remove the wheel speed sensor harness bracket (A) and the wheel speed sensor (B)

4. Remove the brake caliper bracket mounting bolts, then remove the caliper assembly from the knuckle. To prevent damage to the caliper assembly or the brake hose, use a short piece of wire to hang the caliper assembly from the undercarriage. Do not twist the brake hose excessively.
5. Remove the wheel speed sensor harness bracket and the wheel speed sensor from the knuckle. Do not disconnect the wheel speed sensor connector.
6. Pry up the stake on the spindle nut, then remove the nut.
7. Remove the front brake disc.
8. Check the front hub for damage and cracks.
9. Remove the cotter pin from the tie-rod end ball joint, then remove the nut.

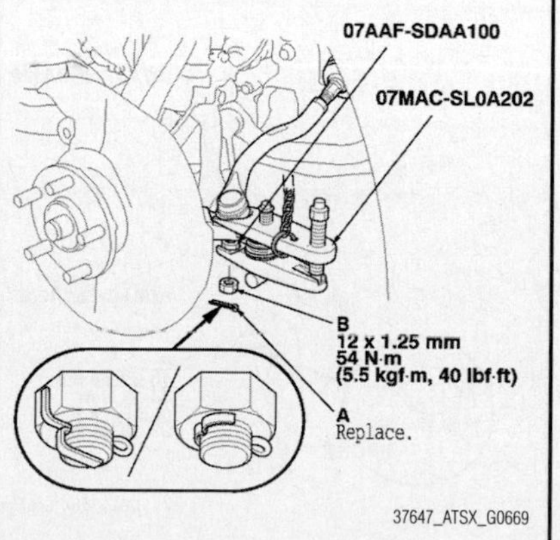
07AAF-SDAA100
07MAC-SL0A202
B
12 x 1.25 mm
54 N·m
(5.5 kgf·m, 40 lbf·ft)
A
Replace.
37647_ATSX_G0669

Fig. 406 Remove the cotter pin (A) from the tie-rod end ball joint, then remove the nut (B)

➡During installation, install the new cotter pin after tightening the nut, and bend its end.

10. Disconnect the tie-rod end ball joint from the knuckle using the ball joint thread protector and the ball joint remover.

11. Remove the cotter pin from the knuckle ball joint, then remove the castle nut.

➡During installation, insert the new cotter pin into the ball joint pin hole from the front to the rear of the vehicle, and bend its end as shown. Check the ball joint pin hole direction before connecting the ball joint.

12. Disconnect the knuckle ball joint from the lower arm using the ball joint thread protector and the ball joint remover.

➡Be careful not to damage the ball joint boot when installing the remover.

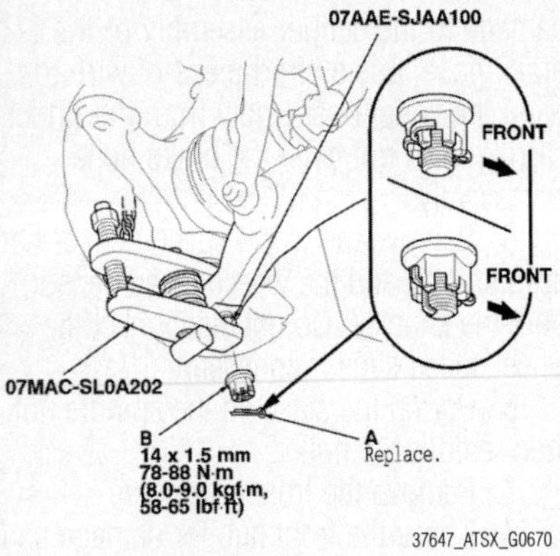
07AAE-SJAA100
FRONT
FRONT
07MAC-SL0A202
B
14 x 1.5 mm
78-88 N·m
(8.0-9.0 kgf·m,
58-65 lbf·ft)
A
Replace.
37647_ATSX_G0670

Fig. 407 Remove the cotter pin (A) from the knuckle ball joint, then remove the castle nut (B)

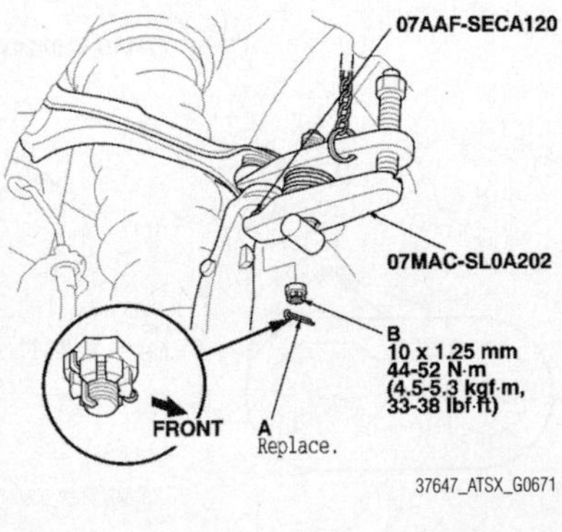
07AAF-SECA120
07MAC-SL0A202
B
10 x 1.25 mm
44-52 N·m
(4.5-5.3 kgf·m,
33-38 lbf·ft)
FRONT
A
Replace.
37647_ATSX_G0671

Fig. 408 Remove the cotter pin (A) from the upper arm ball joint, then remove the castle nut (B)

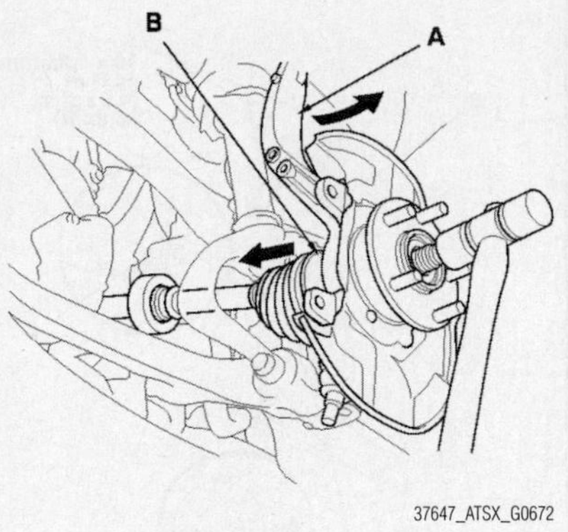
B
A
37647_ATSX_G0672

Fig. 409 Pull the knuckle (A) outward, and separate the outboard joint (B) from the front hub using a plastic hammer

Do not force or hammer on the lower arm, or pry between the lower arm and the knuckle. You could damage the ball joint.

13. Remove the cotter pin from the upper arm ball joint, then remove the castle nut.

➡During installation, insert the new cotter pin into the ball joint pin hole from the front to the rear of the vehicle, and bend its end as shown. Check the ball joint pin hole direction before connecting the ball joint.

14. Disconnect the upper arm ball joint from the knuckle using the ball joint thread protector and the ball joint remover.

15. Pull the knuckle outward, and separate the outboard joint from the front hub using a plastic hammer.

➡Do not pull the driveshaft end outward. The driveshaft inboard joint may come apart. During installation, apply grease to the mating surfaces of the wheel bearing and the driveshaft outboard joint.

To install:

16. Install the knuckle/hub in the reverse order of removal, and note these items:

- First install all of the components, and lightly tighten the bolts and the nuts, then raise the suspension to load it with the vehicle's weight before fully tightening to the specified torque. Do not place the jack against the ball joint pin of the knuckle.
- Be careful not to damage the ball joint boot when connecting the knuckle.
- Before connecting the ball joint, degrease the threaded section and the tapered portion of the ball joint pin, the ball joint connecting hole, the threaded section, and the mating surfaces of the castle nut.
- Torque the castle nut to the lower torque specification, then tighten it only far enough to align the slot with the ball joint pin hole. Do not align the castle nut by loosening it.
- Use a new spindle nut on reassembly.
- Before installing the spindle nut, apply a small amount of engine oil to the seating surface of the nut. After tightening, use a drift to stake the spindle nut shoulder against the driveshaft.
- Before installing the brake disc, clean the mating surfaces of the front hub and the inside of the brake disc.
- Before installing the wheel, clean the mating surfaces of the brake disc and the inside of the wheel.

17. Check the wheel alignment, and adjust it if necessary.

STRUT (DAMPER/SPRING)

REMOVAL & INSTALLATION

See Figures 410 through 415.

1. Raise the vehicle on a lift.
2. Remove the front wheel.
3. Remove the wheel speed sensor harness bracket mounting bolt.
4. Remove the damper pinch bolt and the damper fork mounting nut while holding

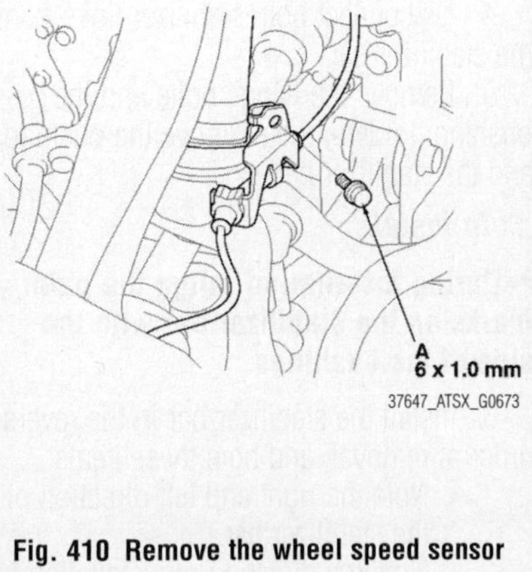
A
6 x 1.0 mm
37647_ATSX_G0673

Fig. 410 Remove the wheel speed sensor harness bracket mounting bolt (A)

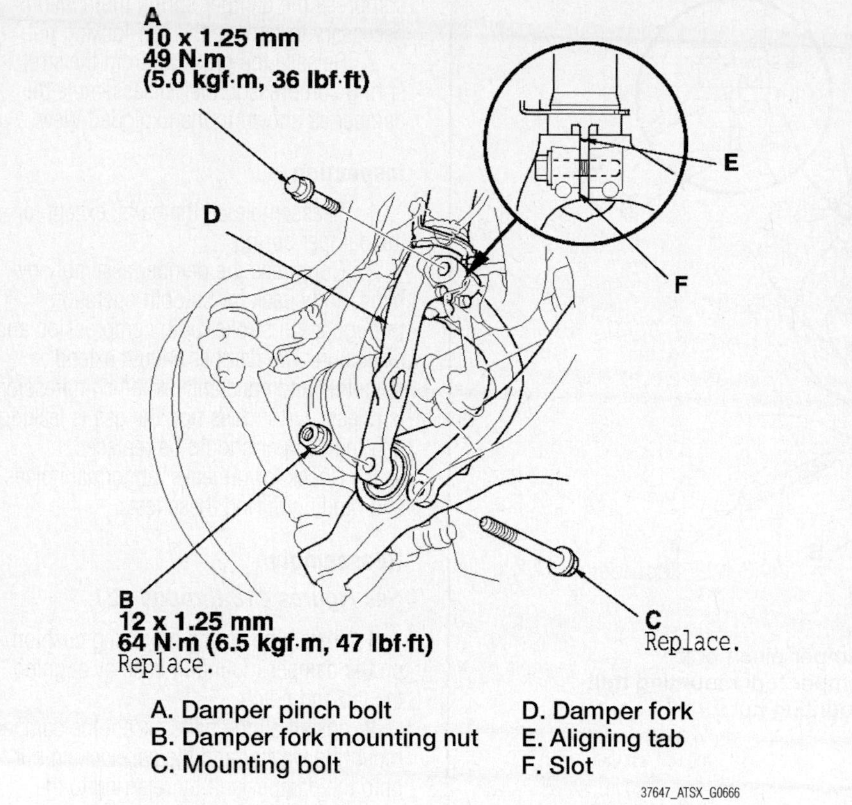
A
10 x 1.25 mm
49 N·m
(5.0 kgf·m, 36 lbf·ft)
D
E
F
B
12 x 1.25 mm
64 N·m (6.5 kgf·m, 47 lbf·ft)
Replace.
C
Replace.
A. Damper pinch bolt
B. Damper fork mounting nut
C. Mounting bolt
D. Damper fork
E. Aligning tab
F. Slot
37647_ATSX_G0666

Fig. 411 Remove the damper pinch bolt and the damper fork mounting nut while holding the mounting bolt

the mounting bolt, then remove the damper fork from the damper and the lower arm.

5. Remove the front strut brace mounting nuts.

6. Remove the damper mounting nuts from the top of the damper. Do not let the damper/spring drop down under its own weight.

7. Remove the damper/spring.

➡**Be careful not to damage the body.**

To install:

8. Position the damper/spring in the body with the aligning tab facing inside.

9. Loosely install the new damper mounting nuts to the top of the damper.

10. Loosely install the front strut brace mounting nuts.

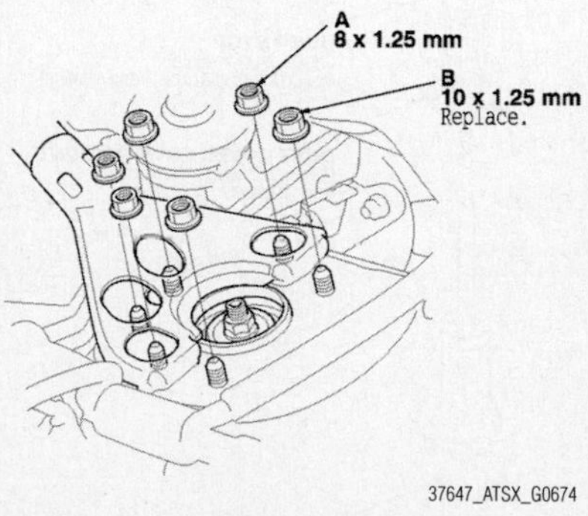
A
8 x 1.25 mm
B
10 x 1.25 mm
Replace.
37647_ATSX_G0674

Fig. 412 Remove the front strut brace mounting nuts (A) and the damper mounting nuts (B)

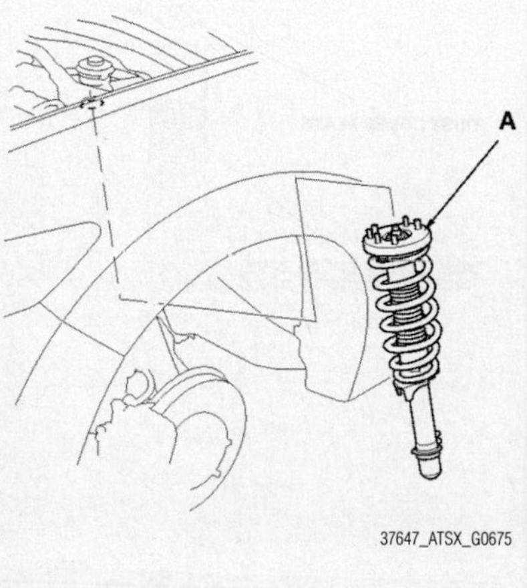
A
37647_ATSX_G0675

Fig. 413 Remove the damper/spring (A)

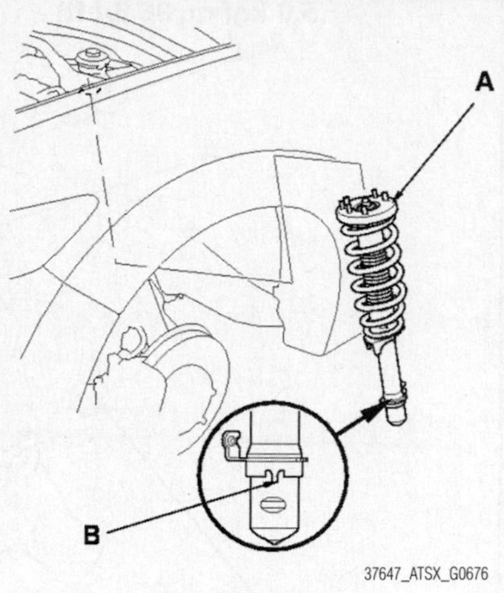
A
B
37647_ATSX_G0676

Fig. 414 Position the damper/spring (A) in the body with the aligning tab (B) facing inside

11. Install the damper fork over the driveshaft and onto the lower arm. Install the aligning tab on the damper unit into the slot of the damper fork.

12. Loosely install the damper pinch bolt into the damper fork.

13. Connect the damper fork and the lower arm with the new damper fork mounting bolt, then lightly tighten the new mounting nut.

14. Place a floor jack under the lower arm, and raise the suspension to load it with the vehicle's weight.

15. Tighten the damper pinch bolt and the damper fork mounting nut while holding the mounting bolt to the specified torque. Damper pinch bolt: tighten to 36 ft. lbs. (49 Nm); damper fork mounting nut: tighten to 47 ft. lbs. (64 Nm).

16. Tighten the damper mounting nuts and front strut brace mounting nuts on top of the damper to the specified torque. Damper mounting nuts: tighten to 41 ft. lbs. (55 Nm); strut brace mounting nuts: tighten to 16 ft. lbs. (22 Nm).

17. Install the wheel speed sensor harness bracket.

18. Clean the mating surfaces of the brake disc and the inside of the wheel, then install the front wheel.

19. Check the wheel alignment, and adjust it if necessary.

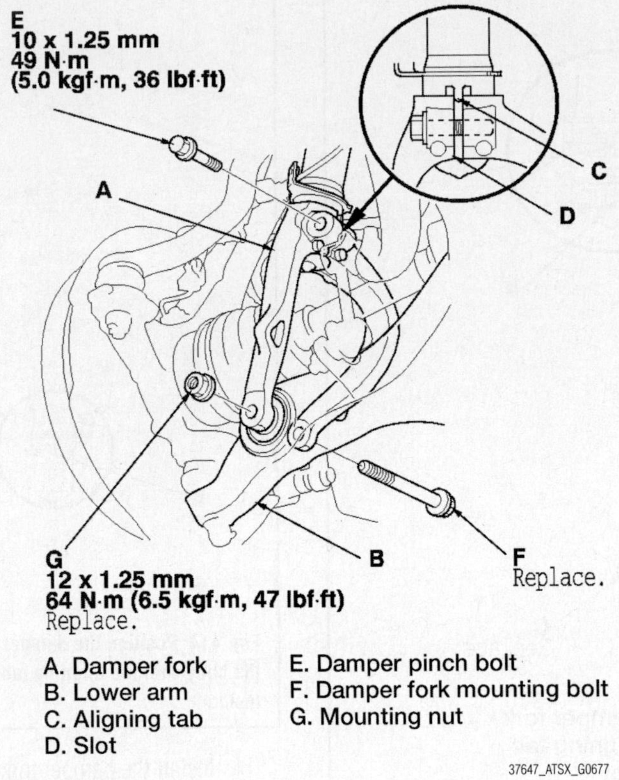
E
10 x 1.25 mm
49 N·m
(5.0 kgf·m, 36 lbf·ft)
A
C
D
G
12 x 1.25 mm
64 N·m (6.5 kgf·m, 47 lbf·ft)
Replace.
B
F
Replace.
A. Damper fork
B. Lower arm
C. Aligning tab
D. Slot
E. Damper pinch bolt
F. Damper fork mounting bolt
G. Mounting nut
37647_ATSX_G0677

Fig. 415 Installing damper fork

OVERHAUL

Disassembly

See Figures 415 and 416.

1. Compress the damper spring, then remove the self-locking nut while holding the damper shaft with a hex wrench. Do not compress the damper spring more than is necessary to remove the self-locking nut.
2. Release the pressure from the strut spring compressor, then disassemble the damper as shown in the Exploded View.

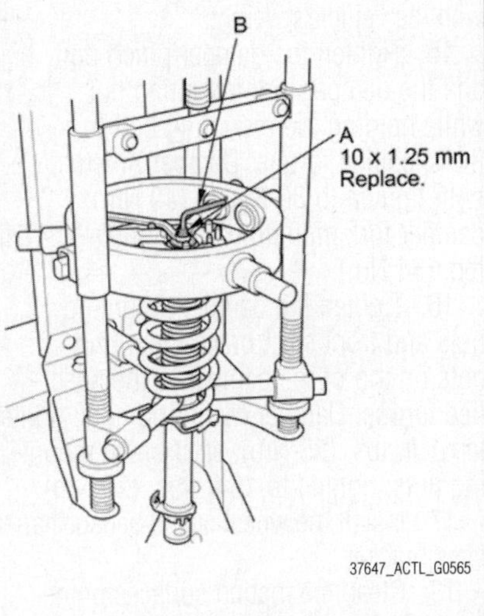
B
A
10 x 1.25 mm
Replace.
37647_ACTL_G0565

Fig. 416 Compress the damper spring, then remove the self-locking nut (A) while holding the damper shaft with a hex wrench (B)

Inspection

1. Reassemble all the parts, except for the damper spring.
2. Compress the damper assembly by hand, and check for smooth operation through a full stroke, both compression and extension. The damper should extend smoothly and constantly when compression is released. If it does not, the gas is leaking and the damper should be replaced.
3. Check for oil leaks, abnormal noises, and binding during these tests.

Reassembly

See Figures 418 through 421.

1. Install the spring mounting cushion on the damper mounting base by aligning the tab and notch.
2. Install all the parts except the damper mounting washer and the self-locking nut onto the damper unit by referring to the Exploded View.

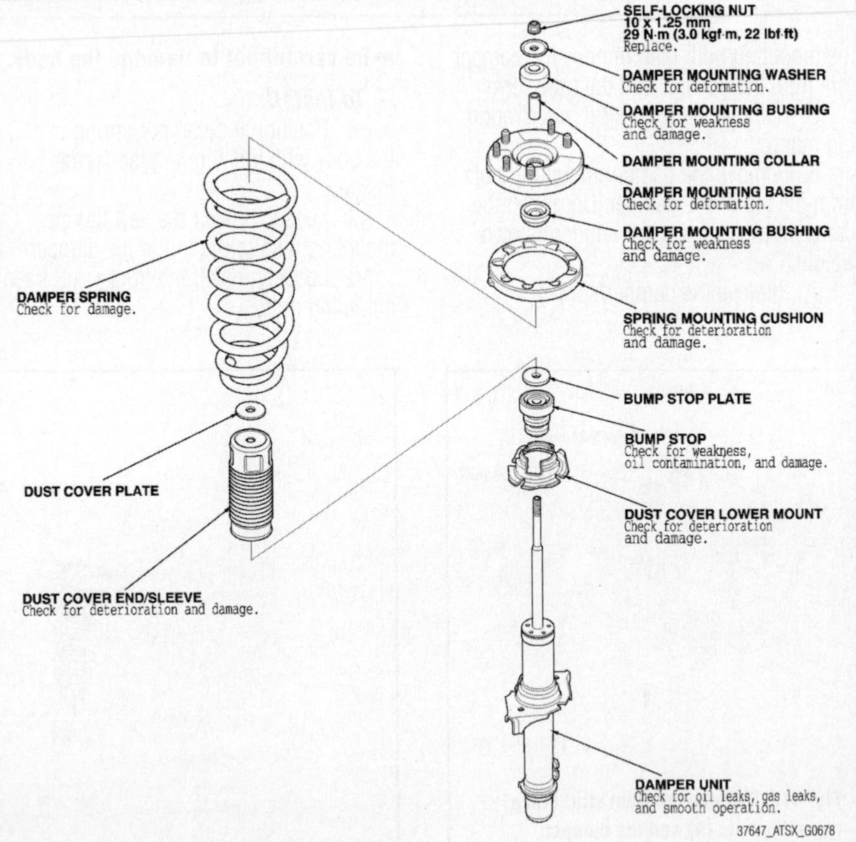
SELF-LOCKING NUT
10 x 1.25 mm
29 N·m (3.0 kgf·m, 22 lbf·ft)
Replace.
DAMPER MOUNTING WASHER
Check for deformation.
DAMPER MOUNTING BUSHING
Check for weakness and damage.
DAMPER MOUNTING COLLAR
DAMPER MOUNTING BASE
Check for deformation.
DAMPER MOUNTING BUSHING
Check for weakness and damage.
DAMPER SPRING
Check for damage.
SPRING MOUNTING CUSHION
Check for deterioration and damage.
BUMP STOP PLATE
BUMP STOP
Check for weakness, oil contamination, and damage.
DUST COVER PLATE
DUST COVER LOWER MOUNT
Check for deterioration and damage.
DUST COVER END/SLEEVE
Check for deterioration and damage.
DAMPER UNIT
Check for oil leaks, gas leaks, and smooth operation.
37647_ATSX_G0678

Fig. 417 Exploded view of damper/spring assembly

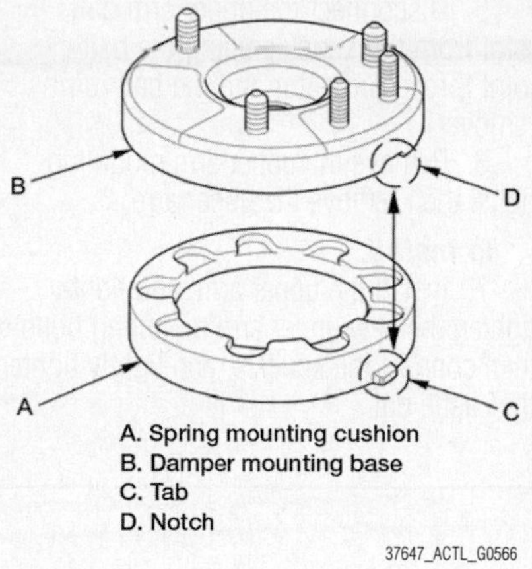
B
D
A
C
A. Spring mounting cushion
B. Damper mounting base
C. Tab
D. Notch
37647_ACTL_G0566

Fig. 418 Install the spring mounting cushion on the damper mounting base by aligning the tab and notch

3. Compress the damper spring using a strut spring compressor. Do not compress the spring excessively.

4. Align the lower end of the damper spring with the stepped part of the dust cover lower mount and the lower spring seat on the damper unit.

5. Position the tab on the spring mounting cushion facing forward but toward the inside of the vehicle.

6. Align the angle of the stud on the damper mounting base with the aligning tab on the bottom of the damper unit.

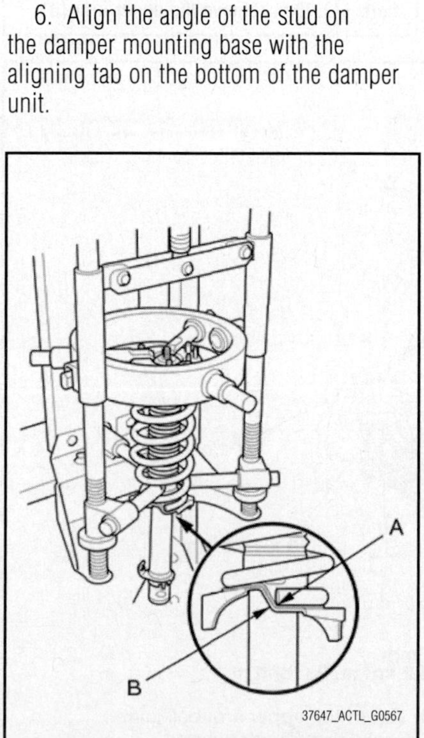
A
B
37647_ACTL_G0567

Fig. 419 Align the lower end (A) of the damper spring with the stepped part (B) of the dust cover lower mount

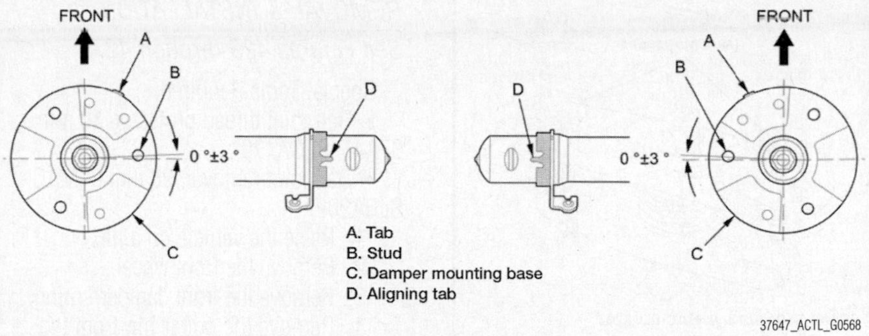
FRONT
A
B
D
0 °±3 °
C
A. Tab
B. Stud
C. Damper mounting base
D. Aligning tab
FRONT
A
B
D
0 °±3 °
C
37647_ACTL_G0568

Fig. 420 Position the tab on the spring mounting cushion facing forward but toward the inside of the vehicle

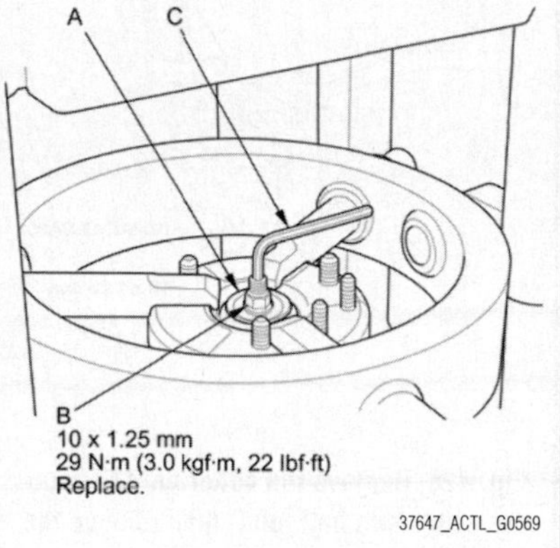
A
C
B
10 x 1.25 mm
29 N·m (3.0 kgf·m, 22 lbf·ft)
Replace.
37647_ACTL_G0569

Fig. 421 Install the damper mounting washer (A) and the new self-locking nut (B)

7. Install the damper mounting washer and the new self-locking nut.

8. Hold the damper shaft with a hex wrench, and tighten the self-locking nut to 22 ft. lbs. (29 Nm).

9. Remove the damper/spring from the strut spring compressor.

UPPER BALL JOINT

REMOVAL & INSTALLATION

See Figures 422 through 424.

Special Tools Required:

- Ball joint thread protector, 14 mm 07AAE-SJAA100
- Ball joint thread protector, 12 mm 07AAF-SDAA100
- Ball joint thread protector, 10 mm 07AAF-SECA120
- Ball joint remover, 32 mm 07MAC-SL0A102
- Ball joint remover, 28 mm 07MAC-SL0A202

➡Always use a ball joint remover to disconnect a ball joint. Do not strike the housing or any other part of the ball joint connection to disconnect it.

1. Install a hex nut or the ball joint thread protector onto the threads of the ball joint.

➡Using a hex nut, make sure the nut is flush with the ball joint pin end to prevent damage to the threaded end of the ball joint pin.

2. Apply grease to the ball joint remover on the areas shown. This will ease the installation of the tool, and prevent damage to the pressure bolt threads.

3. Loosen the pressure bolt, and install the ball joint remover as shown. Insert the jaws carefully, making sure not to damage the ball joint boot. Adjust the jaw spacing by turning the adjusting bolt.

➡Fasten the safety chain securely to a suspension arm or the subframe. Do not fasten it to a brake line or wire harness.

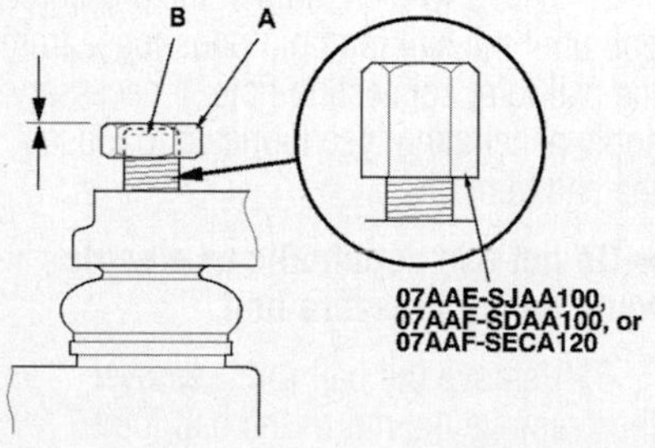
B
A
07AAE-SJAA100,
07AAF-SDAA100, or
07AAF-SECA120
37647_ATSX_G0663

Fig. 422 Install a hex nut (A) or the ball joint thread protector onto the threads of the ball joint (B)

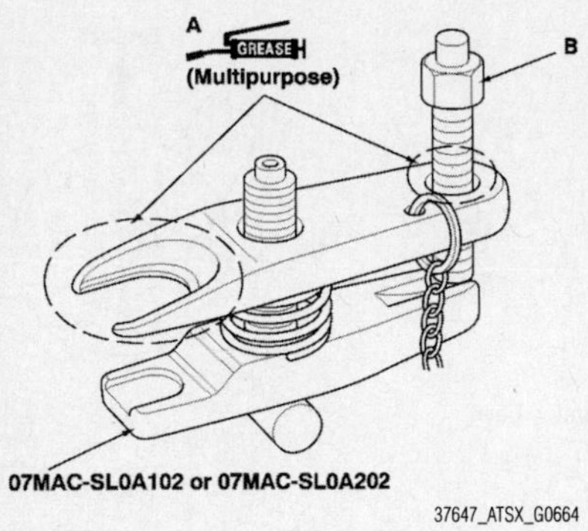
A
GREASE
(Multipurpose)
B
07MAC-SL0A102 or 07MAC-SL0A202
37647_ATSX_G0664

Fig. 423 Apply grease to the ball joint remover on the areas shown (A)

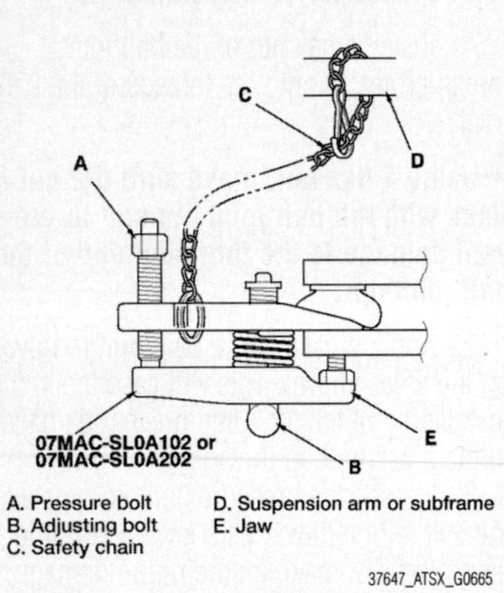
C
A
D
07MAC-SL0A102 or
07MAC-SL0A202
E
B
A. Pressure bolt
B. Adjusting bolt
C. Safety chain
D. Suspension arm or subframe
E. Jaw
37647_ATSX_G0665

Fig. 424 Loosen the pressure bolt, and install the ball joint remover as shown

4. After adjusting the adjusting bolt, make sure the head of the adjusting bolt is in the position shown to allow the jaw to pivot.

5. With a wrench, tighten the pressure bolt until the ball joint pin pops loose from the ball joint connecting hole. If necessary, apply penetrating type lubricant to loosen the ball joint pin.

➡Do not use pneumatic or electric tools on the pressure bolt.

6. Remove the ball joint remover, then remove the nut or the ball joint thread protector from the end of the ball joint pin, and pull the ball joint out of the ball joint connecting hole. Inspect the ball joint boot, and replace it if damaged.

UPPER CONTROL ARM

REMOVAL & INSTALLATION

See Figures 425 through 427.

Special Tools Required:

- Ball joint thread protector, 10 mm 07AAF-SECA120
- Ball joint remover, 28 mm 07MAC-SL0A202

1. Raise the vehicle on a lift.
2. Remove the front wheel.
3. Remove the front damper/spring.
4. Remove the cotter pin from the upper arm ball joint, then remove the castle nut.
5. Disconnect the upper arm ball joint from the knuckle using the ball joint thread protector and the ball joint remover.
6. Remove the upper arm mounting bolts, then remove the upper arm.

To install:

7. Install the upper arm, and lightly tighten the new upper arm mounting bolts, then connect the knuckle, and lightly tighten the castle nut.

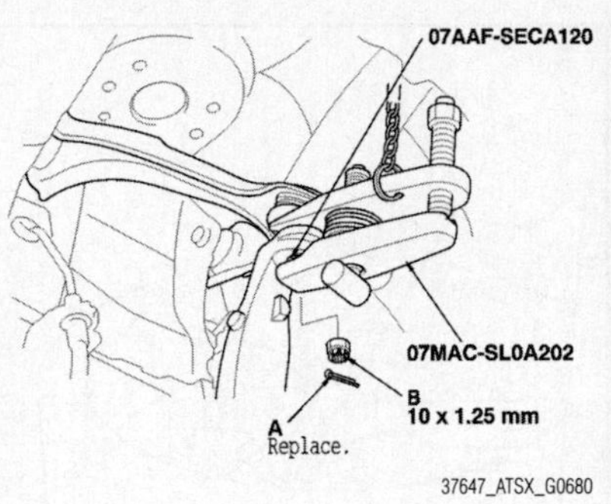
07AAF-SECA120
07MAC-SL0A202
A
Replace.
B
10 x 1.25 mm
37647_ATSX_G0680

Fig. 425 Remove the cotter pin (A) from the upper arm ball joint, then remove the castle nut (B)

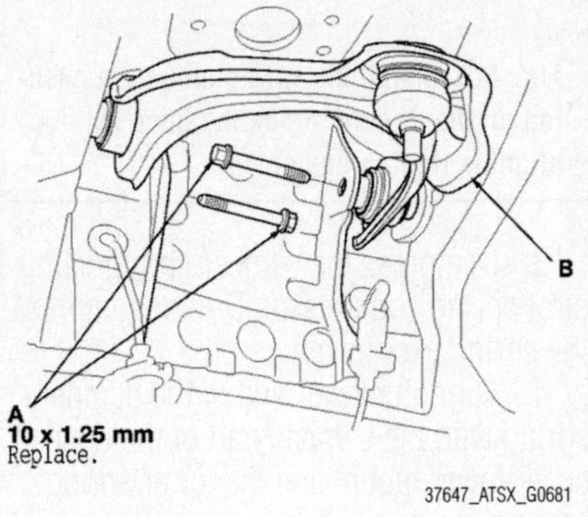
B
A
10 x 1.25 mm
Replace.
37647_ATSX_G0681

Fig. 426 Remove the upper arm mounting bolts (A), then remove the upper arm (B)

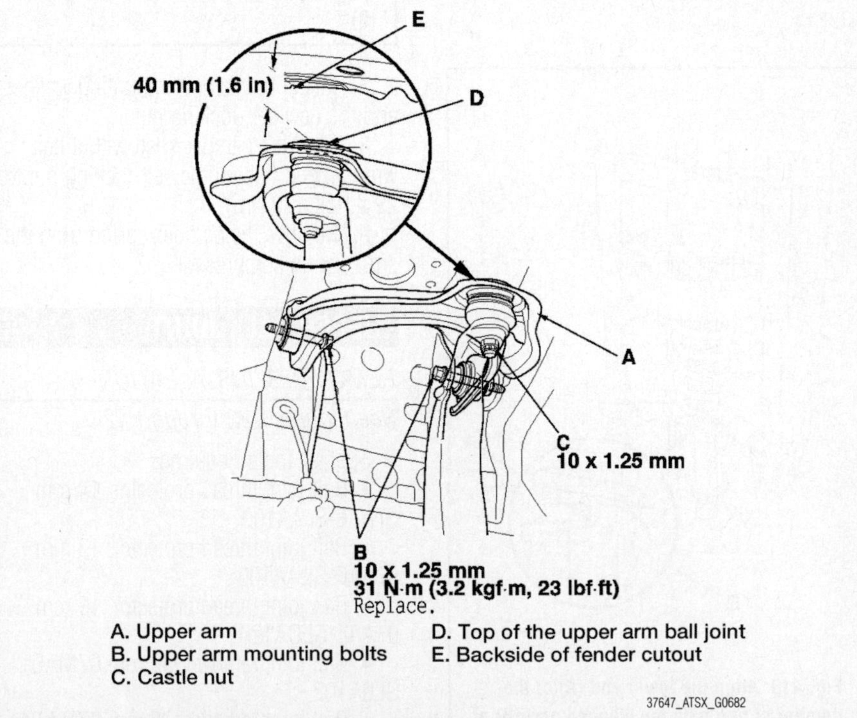
E
40 mm (1.6 in)
D
A
C
10 x 1.25 mm
B
10 x 1.25 mm
31 N·m (3.2 kgf·m, 23 lbf·ft)
Replace.
A. Upper arm
B. Upper arm mounting bolts
C. Castle nut
D. Top of the upper arm ball joint
E. Backside of fender cutout
37647_ATSX_G0682

Fig. 427 Install the upper arm, and lightly tighten the new upper arm mounting bolts

➡Be careful not to damage the ball joint boot when connecting the knuckle. Before connecting the ball joint, degrease the threaded section and the tapered portion of the ball joint pin, the ball joint connecting hole, the threaded section and the mating surfaces of the castle nut.

8. Place a floor jack under the lower arm, and raise the suspension until the clearance between the top of the upper arm ball joint and the backside of the fender cut out point is 1.6 inches (40 mm), then tighten the upper arm mounting bolts to 23 ft. lbs. (31 Nm).

➡To measure the specified clearance, temporarily remove the front inner fender.

9. Lower the floor jack
10. Install the front damper/spring.
11. Place the floor jack under the lower arm, and raise the suspension to load it with the vehicle's weight.
12. Tighten the castle nut on the upper arm ball joint to 33–38 ft. lbs. (44–52 Nm).

➡Torque the castle nut to the lower torque specification, then tighten it only far enough to align the slot with the ball joint pin hole. Do not align the castle nut by loosening it. Insert the new cotter pin into the ball joint pin hole from the front to the rear of the vehicle, and bend its end. Check the ball joint pin hole direction before connecting the ball joint.

13. Clean the mating surfaces of the brake disc and the inside of the wheel, then install the front wheel.
14. Check the wheel alignment, and adjust it if necessary.

WHEEL HUB & BEARING

REMOVAL & INSTALLATION

See Figures 428 through 434.

1. Remove the knuckle/hub.
2. Separate the hub from the knuckle using the hub dis/assembly pin and a hydraulic press. Hold the knuckle with the attachment of the hydraulic press or equivalent tool. Be careful not to damage or deform the splash guard. Hold onto the hub to keep it from falling when pressed clear.
3. Press the wheel bearing inner race off of the hub using the hub dis/assembly pin, a commercially available bearing separator, and a press.
4. Remove the snap ring and the splash guard from the knuckle.

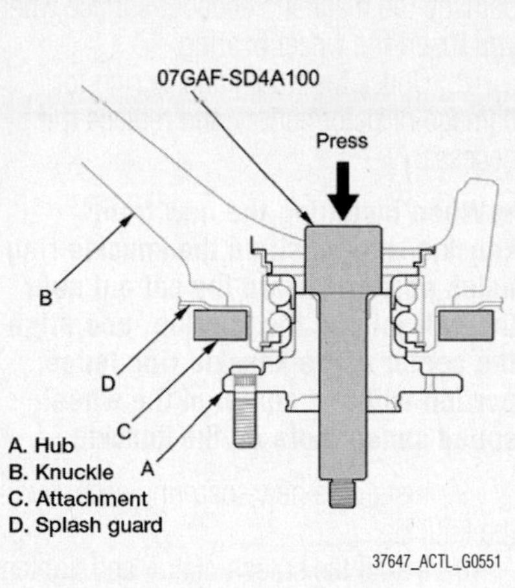
07GAF-SD4A100
Press
B
D
C
A
A. Hub
B. Knuckle
C. Attachment
D. Splash guard
37647_ACTL_G0551

Fig. 428 Separate the hub from the knuckle using the hub dis/assembly pin and a hydraulic press

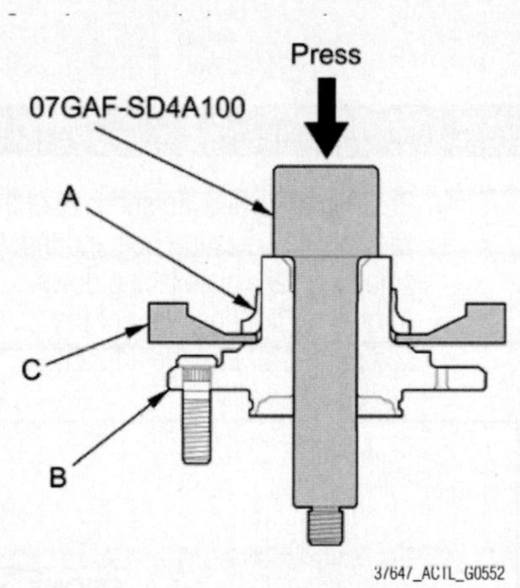
Press
07GAF-SD4A100
A
C
B
37647_ACTL_G0552

Fig. 429 Press the wheel bearing inner race (A) off of the hub (B) using the hub dis/assembly pin, a commercially available bearing separator (C), and a press

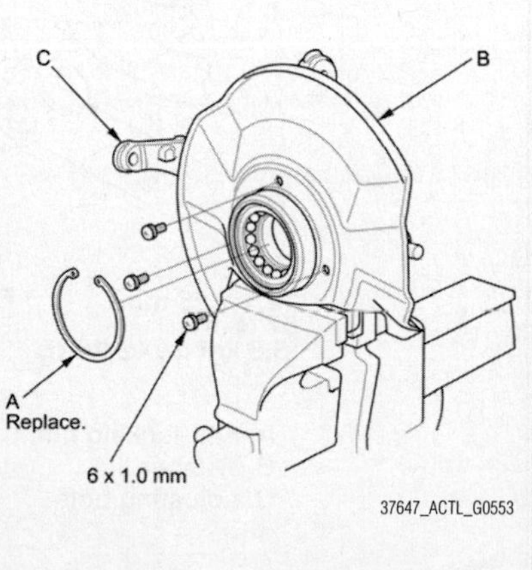
C
B
A
Replace.
6 x 1.0 mm
37647_ACTL_G0553

Fig. 430 Remove the snap ring (A) and the splash guard (B) from the knuckle (C)

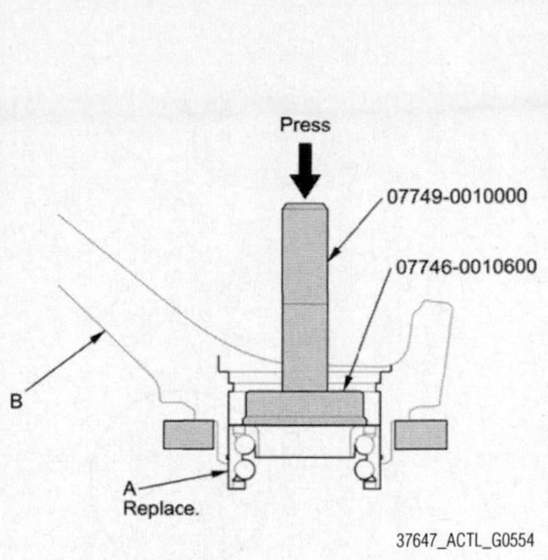
Press
07749-0010000
07746-0010600
B
A
Replace.
37647_ACTL_G0554

Fig. 431 Press the wheel bearing (A) out of the knuckle (B) using the attachment, the driver handle, and a press

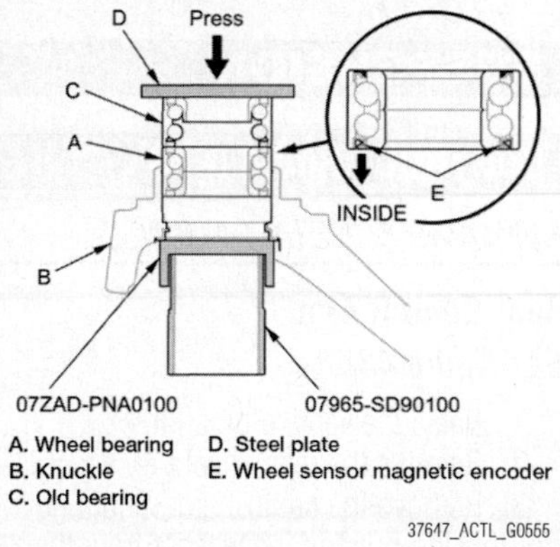
D
Press
C
A
E
INSIDE
B
07ZAD-PNA0100
07965-SD90100
A. Wheel bearing
B. Knuckle
C. Old bearing
D. Steel plate
E. Wheel sensor magnetic encoder
37647_ACTL_G0555

Fig. 432 Press a new wheel bearing into the knuckle using the old bearing, a steel plate, the attachment, the support base, and a press

5. Press the wheel bearing out of the knuckle using the attachment, the driver handle, and a press.
6. Wash the knuckle and the hub thoroughly in high flash point solvent before reassembly.
7. Press a new wheel bearing into the knuckle using the old bearing, a steel plate, the attachment, the support base, and a press.

➡Install the wheel bearing with the wheel speed sensor magnetic encoder (brown color), toward the inside of the knuckle. Remove any oil, grease, dust, metal debris, and other foreign material from the magnetic encoder surface.

Keep any magnetic tools away from the magnetic encoder surface. Be careful not to

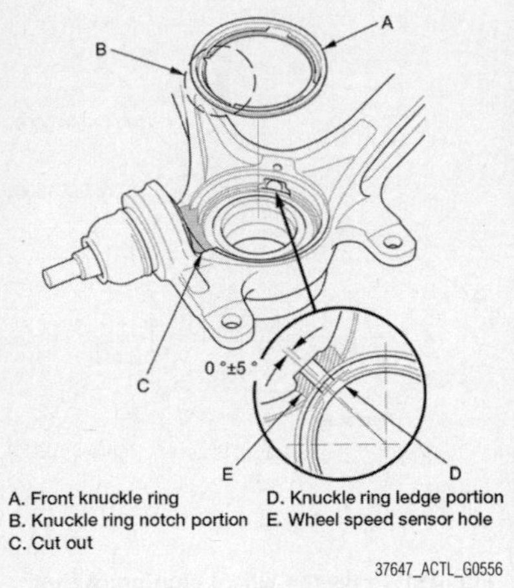
A
B
C
D
E
0 °±5 °
A. Front knuckle ring
B. Knuckle ring notch portion
C. Cut out
D. Knuckle ring ledge portion
E. Wheel speed sensor hole
37647_ACTL_G0556

Fig. 433 Check the front knuckle ring for damage or deformation, and replace it if necessary

damage the magnetic encoder surface when you insert the wheel bearing.

8. Check the front knuckle ring for damage or deformation, and replace it if necessary.

➡**When installing the new front knuckle ring, position the knuckle ring notch portion toward the cut out near the ball joint in the knuckle, and align the center of the knuckle ring ledge portion with the center of the wheel speed sensor hole on the knuckle.**

9. Install the new snap ring securely in the knuckle.

10. Install the splash guard, and tighten the screws to 7 ft. lbs. (10 Nm).

11. Install the hub onto the knuckle using the attachment, the driver handle, the support base, and a hydraulic press. Be careful not to damage the splash guard.

12. Install the knuckle/hub.

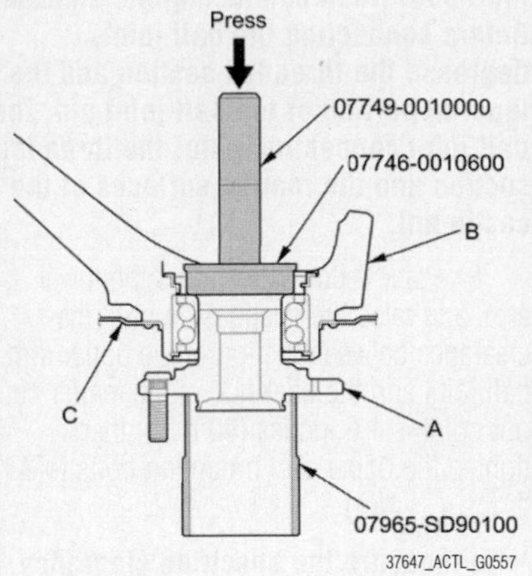
Press
07749-0010000
07746-0010600
B
C
A
07965-SD90100
37647_ACTL_G0557

Fig. 434 Install the hub (A) onto the knuckle (B) using the attachment, the driver handle, the support base, and a hydraulic press

SUSPENSION — REAR SUSPENSION

CONTROL ARMS/LINKS

REMOVAL & INSTALLATION

Rear Control Arm

See Figure 435.

1. Raise the vehicle on a lift.
2. Remove the rear wheel.
3. Remove the control arm mounting self-locking nut and the washer from the knuckle side.

➡**Use a new self-locking nut during reassembly.**

4. Mark the cam positions of the adjusting bolt and the adjusting cam plate with the frame.
5. Remove the self-locking nut while holding the adjusting bolt, then remove the adjusting cam plate, the adjusting bolt, and the control arm.

To install:

➡**Use a new adjusting bolt and a new self-locking nut during reassembly.**

6. Install the control arm in the reverse order of removal, and note these items:
   - First install all of the components, and lightly tighten the bolts and the nuts, then raise the suspension to load it with the vehicle's weight before fully tightening to the specified torque.
   - Position the extended surfaces of the cam on the adjusting bolt and the adjusting cam plate facing down.
   - Align the cam positions of the adjusting bolt and the adjusting cam plate with the marked positions on the frame when tightening the self-locking nut.
   - Before installing the wheel, clean the mating surfaces on the

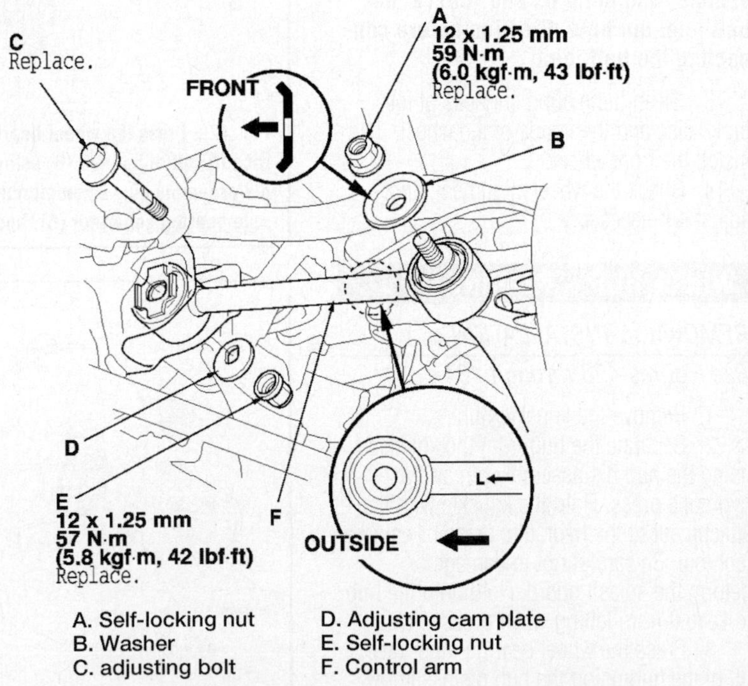
C
Replace.
FRONT
A
12 x 1.25 mm
59 N·m
(6.0 kgf·m, 43 lbf·ft)
Replace.
B
D
E
12 x 1.25 mm
57 N·m
(5.8 kgf·m, 42 lbf·ft)
Replace.
F
OUTSIDE
L
A. Self-locking nut
B. Washer
C. adjusting bolt
D. Adjusting cam plate
E. Self-locking nut
F. Control arm
37647_ATSX_G0691

Fig. 435 Remove the self-locking nut while holding the adjusting bolt, then remove the adjusting cam plate, the adjusting bolt, and the control arm

brake disc and the inside of the wheel.

7. Check the wheel alignment, and adjust it if necessary.

Rear Lower Arm A

See Figure 436.

1. Raise the vehicle on a lift.
2. Remove the rear wheel.
3. Remove the parking brake cable mounting bolt.
4. Remove the lower arm A mounting bolts, then remove lower arm A.

To install:

➡Use new mounting bolts during reassembly.

5. Install lower arm A in the reverse order of removal, and note these items:
   - First install all of the components, and lightly tighten the bolts, then raise the suspension to load it with the vehicle's weight before fully tightening to the specified torque.
   - Before installing the wheel, clean the mating surfaces on the brake disc and the inside of the wheel.
6. Check the wheel alignment, and adjust it if necessary.

Rear Lower Arm B

See Figure 436.

1. Raise the vehicle on a lift.
2. Remove the rear wheel.

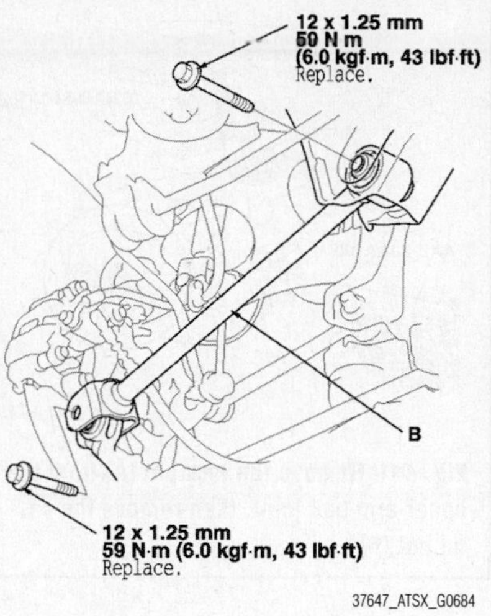
12 x 1.25 mm
59 N·m
(6.0 kgf·m, 43 lbf·ft)
Replace.
B
12 x 1.25 mm
59 N·m (6.0 kgf·m, 43 lbf·ft)
Replace.
37647_ATSX_G0684

Fig. 437 Remove the lower arm B mounting bolts, then remove lower arm B

3. Remove the lower arm B mounting bolts, then remove lower arm B.

To install:

➡Use new mounting bolts during reassembly.

4. Install lower arm B in the reverse order of removal, and note these items:
   - First install all of the components, and lightly tighten the bolts, then raise the suspension to load it with the vehicle's weight before fully tightening to the specified torque.
   - Make sure the clearance between lower arm B and the parking brake cable is more than 0.20 inches (5 mm).
   - Before installing the wheel, clean the mating surfaces on the brake disc and the inside of the wheel.
5. Check the wheel alignment, and adjust it if necessary.

Rear Stabilizer Link

See Figure 437.

1. Raise the vehicle on a lift.
2. Remove the rear wheel.
3. Remove the flange nut and the self-locking nut while holding the respective joint pin with a hex wrench, then remove the stabilizer link.

To install:

4. Install the stabilizer link on the stabilizer bar and the knuckle adding in the brake hose bracket with the joint pins set at the center of their range of movement.

➡The stabilizer link has a paint mark. The paint mark indicates the difference between the left and right stabilizer links. Install the end of the stabilizer link with the paint mark in the upper position.

5. Install the new flange nut and the new self-locking nut, and tighten them to the

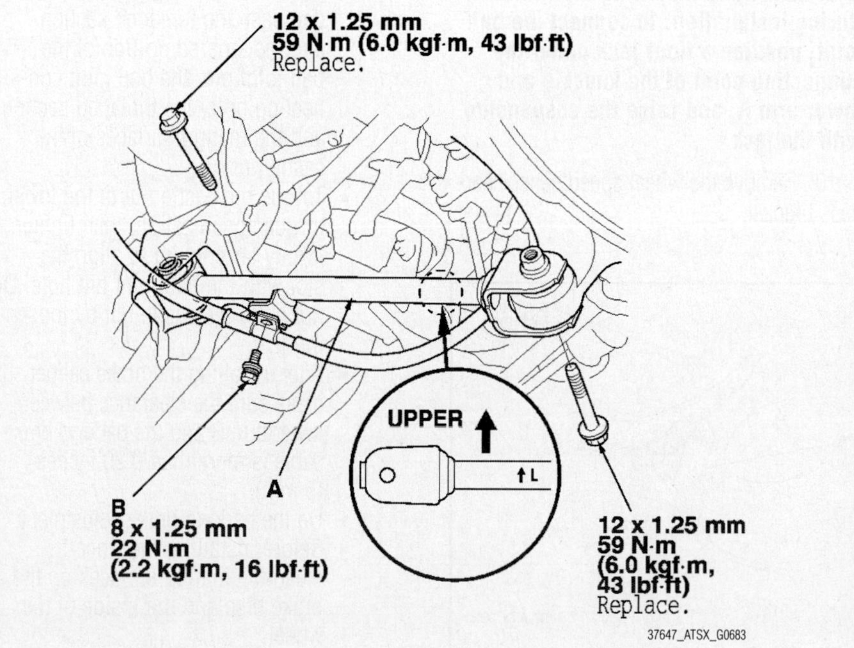
12 x 1.25 mm
59 N·m (6.0 kgf·m, 43 lbf·ft)
Replace.
UPPER
L
A
B
8 x 1.25 mm
22 N·m
(2.2 kgf·m, 16 lbf·ft)
12 x 1.25 mm
59 N·m
(6.0 kgf·m,
43 lbf·ft)
Replace.
37647_ATSX_G0683

Fig. 436 Remove the lower arm A mounting bolts, then remove lower arm A

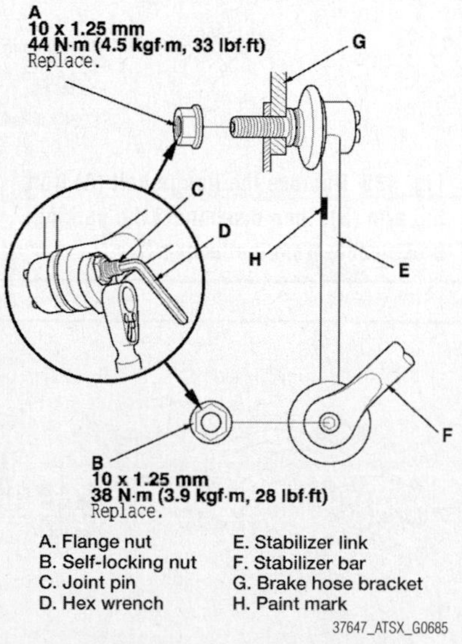
A
10 x 1.25 mm
44 N·m (4.5 kgf·m, 33 lbf·ft)
Replace.
G
C
D
H
E
F
B
10 x 1.25 mm
38 N·m (3.9 kgf·m, 28 lbf·ft)
Replace.
A. Flange nut
B. Self-locking nut
C. Joint pin
D. Hex wrench
E. Stabilizer link
F. Stabilizer bar
G. Brake hose bracket
H. Paint mark
37647_ATSX_G0685

Fig. 438 Remove the flange nut and the self-locking nut while holding the respective joint pin with a hex wrench, then remove the stabilizer link

specified torque while holding the respective joint pin with a hex wrench.

6. Clean the mating surfaces of the brake disc and the inside of the wheel, then install the rear wheel.
7. Test-drive the vehicle.
8. After 5 minutes of driving, tighten the self-locking nuts again to the specified torque.

Rear Upper Arm

See Figures 439 through 443.

Special Tools Required:

- Ball joint thread protector, 14 mm 07AAE-SJAA100
- Ball joint remover, 32 mm 07MAC-SL0A102

1. Raise the vehicle on a lift.
2. Remove the rear wheel.
3. Remove the rear damper/spring.
4. Release the parking brake lever fully.
5. Loosen the parking brake cable adjusting nut.
6. Remove the flange bolt from the arm. Then disconnect the parking brake cable from the lever.
7. Remove the brake caliper bracket mounting bolts, then remove the caliper assembly from the knuckle. To prevent damage to the caliper assembly or the brake hose, use a short piece of wire to hang the caliper assembly from the undercarriage. Do not twist the brake hose excessively.

➡Make sure the washers are reinstalled between the brake caliper bracket and the knuckle, if they are removed.

8. Remove the lock pin from the upper arm ball joint, then remove the castle nut.

➡During installation, install the lock pin as shown after tightening the new castle nut.

9. Disconnect the upper arm ball joint from the knuckle using the ball joint thread protector and the ball joint remover.

➡Be careful not to damage the ball joint boot when installing the remover. During installation, to connect the ball joint, position a floor jack under the connecting point of the knuckle and lower arm A, and raise the suspension with the jack.

10. Remove the wheel speed sensor harness bracket.

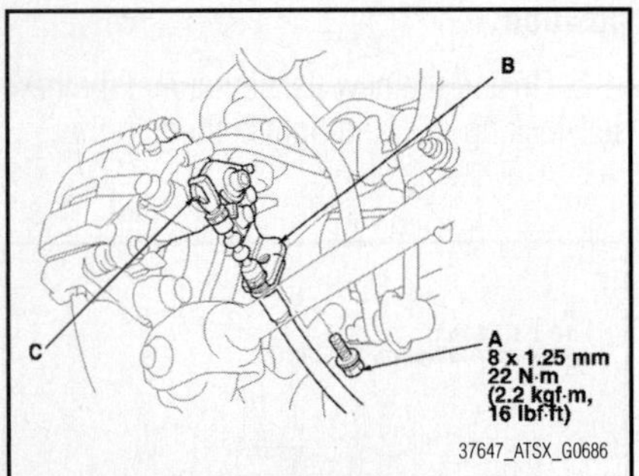
B
C
A
8 x 1.25 mm
22 N·m
(2.2 kgf·m,
16 lbf·ft)
37647_ATSX_G0686

Fig. 439 Remove the flange bolt (A) from the arm (B), then disconnect the parking brake cable from the lever (C)

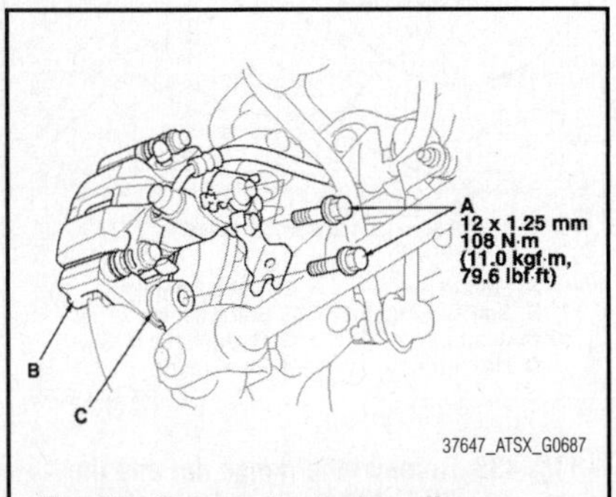
A
12 x 1.25 mm
108 N·m
(11.0 kgf·m,
79.6 lbf·ft)
B
C
37647_ATSX_G0687

Fig. 440 Remove the brake caliper bracket mounting bolts (A), then remove the caliper assembly (B) from the knuckle

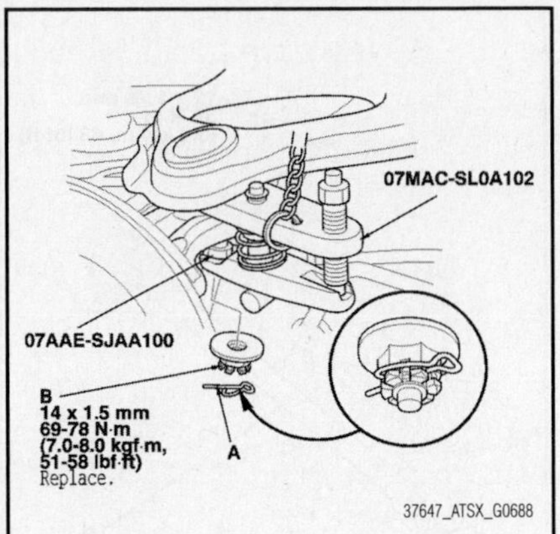
07MAC-SL0A102
07AAE-SJAA100
B
14 x 1.5 mm
69-78 N·m
(7.0-8.0 kgf·m,
51-58 lbf·ft)
Replace.
A
37647_ATSX_G0688

Fig. 441 Remove the lock pin (A) from the upper arm ball joint, then remove the castle nut (B)

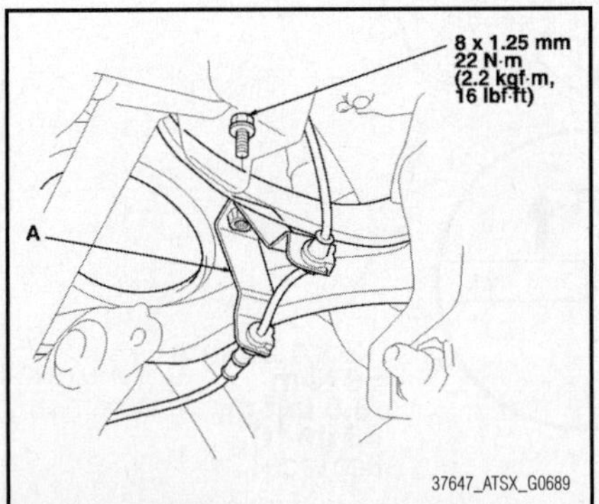
8 x 1.25 mm
22 N·m
(2.2 kgf·m,
16 lbf·ft)
A
37647_ATSX_G0689

Fig. 442 Remove the wheel speed sensor harness bracket (A)

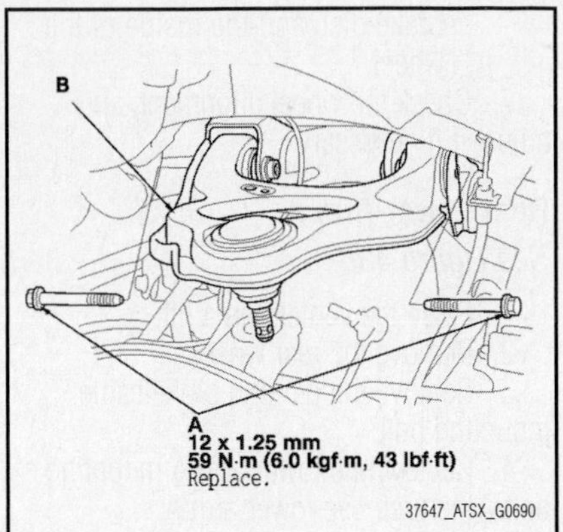
B
A
12 x 1.25 mm
59 N·m (6.0 kgf·m, 43 lbf·ft)
Replace.
37647_ATSX_G0690

Fig. 443 Remove the upper arm mounting bolts (A), then remove the upper arm (B)

11. Remove the upper arm mounting bolts, then remove the upper arm.

To install:

➡Use new mounting bolts during reassembly.

12. Install the upper arm in the reverse order of removal, and note these items:
    - First install all of the components, and lightly tighten the bolts and the nuts, then raise the suspension to load it with the vehicle's weight before fully tightening to the specified torque.
    - Be careful not to damage the ball joint boot when connecting the knuckle.
    - Before connecting the ball joint, degrease the threaded section and the tapered portion of the ball joint pin, the ball joint connecting hole, the threaded section, and the mating surfaces of the castle nut.
    - Torque the castle nut to the lower torque specification, then tighten it only far enough to align the slot with the ball joint pin hole. Do not align the castle nut by loosening it.
    - After installing the brake caliper, make sure the clearance between lower arm B and the parking brake cable is more than 0.20 inches (5 mm).
    - Do the parking brake adjustment.
    - Before installing the wheel, clean the mating surfaces on the brake disc and the inside of the wheel.
13. Check the wheel alignment, and adjust it if necessary.

STABILIZER BAR

REMOVAL & INSTALLATION

See Figure 444.

1. Raise the vehicle on a lift.
2. Remove the rear wheels.
3. Disconnect both stabilizer links from the stabilizer bar.
4. Remove the flange bolts and the bushing holders, then remove the bushings and the stabilizer bar.

To install:

➡During installation, align the paint marks on the stabilizer bar with the side of the bushings.

5. Install the stabilizer bar in the reverse order of removal, and note these items:
   - Note the right and left direction of the stabilizer bar.
   - Note the direction of installation for the bushing.
   - Refer to the stabilizer link removal/installation to connect the stabilizer bar to the links.
   - Before installing the wheel, clean the mating surfaces of the brake disc and the inside of the wheel.

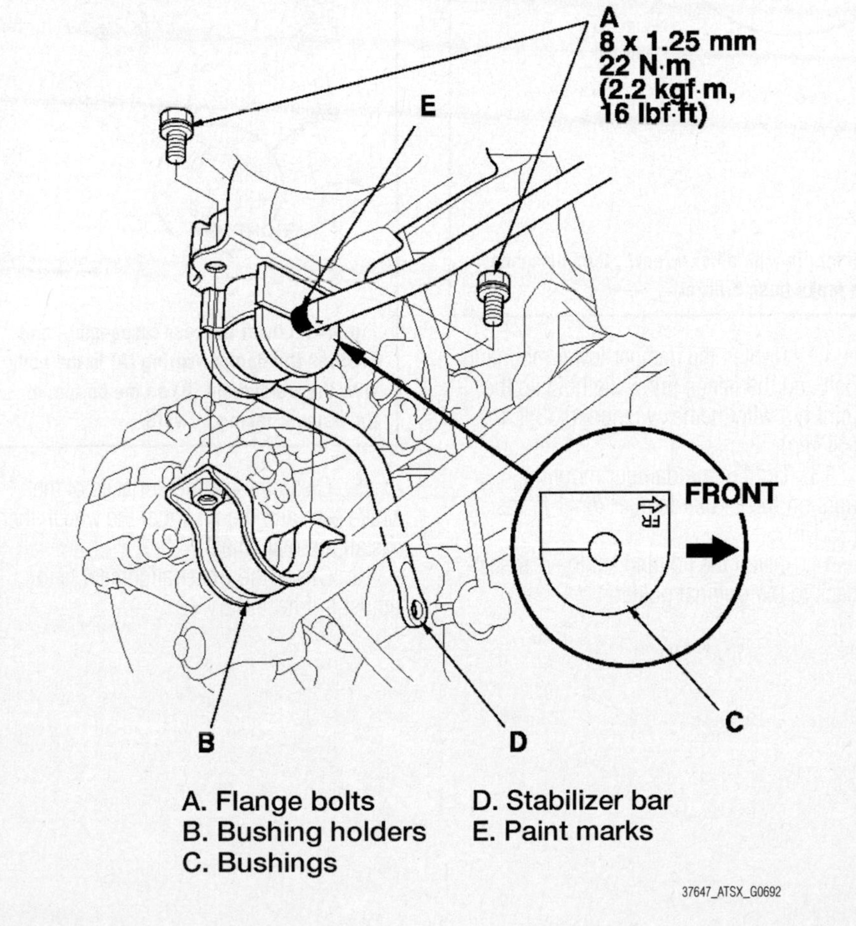
A
8 x 1.25 mm
22 N·m
(2.2 kgf·m,
16 lbf·ft)
E
FRONT
FR
B
D
C
A. Flange bolts
B. Bushing holders
C. Bushings
D. Stabilizer bar
E. Paint marks
37647_ATSX_G0692

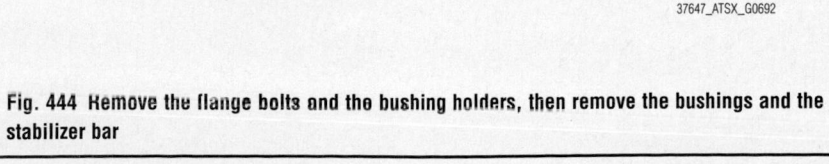

Fig. 444 Remove the flange bolts and the bushing holders, then remove the bushings and the stabilizer bar

STRUT (DAMPER/SPRING)

REMOVAL & INSTALLATION

See Figures 445 through 449.

1. Raise the vehicle on a lift.
2. Remove the rear wheel.
3. Fold down the rear seat-back, then remove the lid.

➡Lift up the tab inside underneath the lid first using a flat-tipped screwdriver, then release the hooks.

4. Remove the damper mounting nuts from the top of the damper.

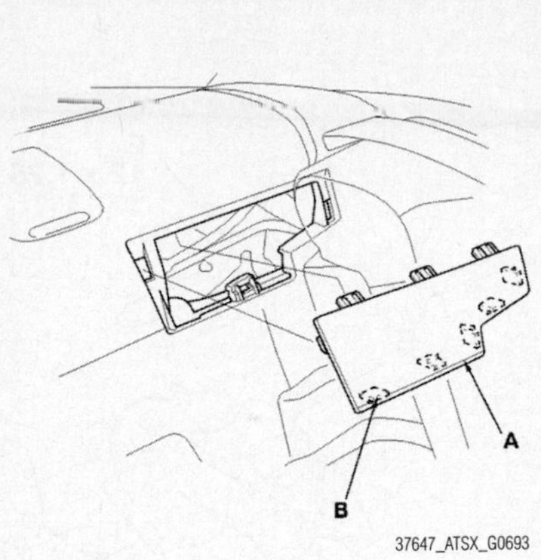
A
B
37647_ATSX_G0693

Fig. 445 Remove the lid (A) and lift up the tab (B)

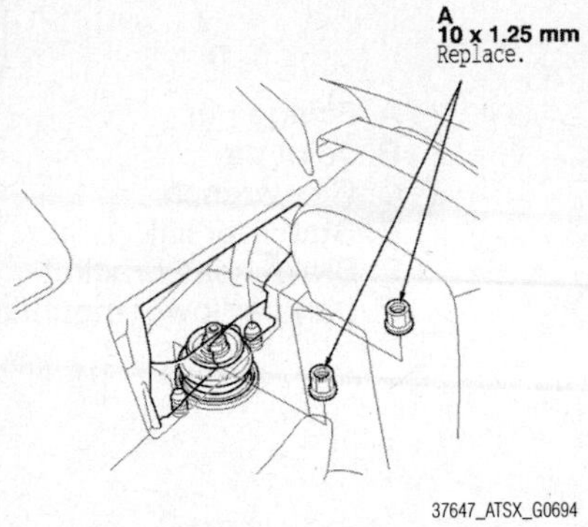
A
10 x 1.25 mm
Replace.
37647_ATSX_G0694

Fig. 446 Remove the damper mounting nuts (A) from the top of the damper

5. Remove the flange nut while holding the joint pin with a hex wrench, then disconnect the stabilizer link from the knuckle, and remove the brake hose bracket.
6. Remove the damper lower mounting bolt.
7. Remove the damper/spring by lowering the rear suspension.

➡Be careful not to damage the body.

To install:

8. Lower the rear suspension, and position the damper/spring in the body with the welded nut on the bottom of the damper facing forward.

➡Be careful not to damage the body. Make sure the damper is installed in the correct direction.

9. Loosely install the new damper mounting nuts to the top of the damper.

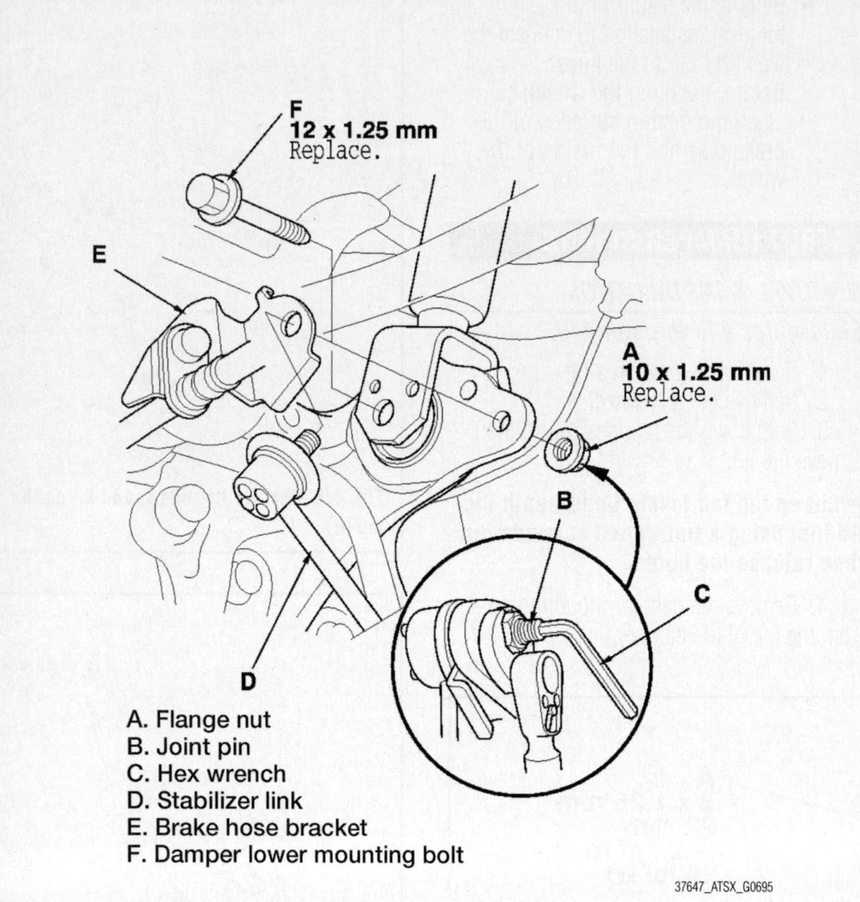
F
12 x 1.25 mm
Replace.
E
A
10 x 1.25 mm
Replace.
B
C
D
A. Flange nut
B. Joint pin
C. Hex wrench
D. Stabilizer link
E. Brake hose bracket
F. Damper lower mounting bolt
37647_ATSX_G0695

Fig. 447 Remove the flange nut while holding the joint pin with a hex wrench, then disconnect the stabilizer link from the knuckle, and remove the brake hose bracket

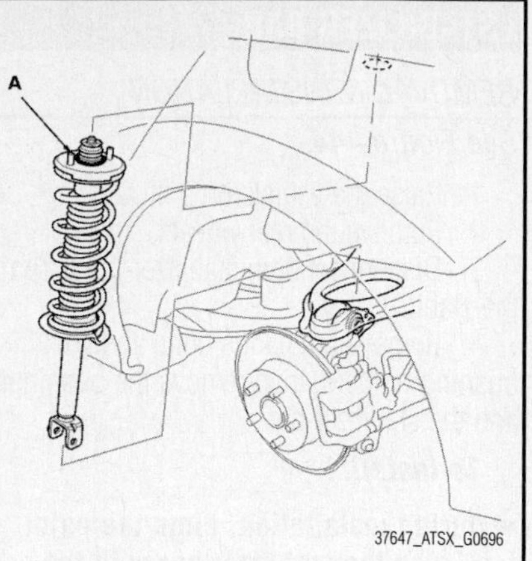
A
37647_ATSX_G0696

Fig. 448 Remove the damper/spring (A) by lowering the rear suspension

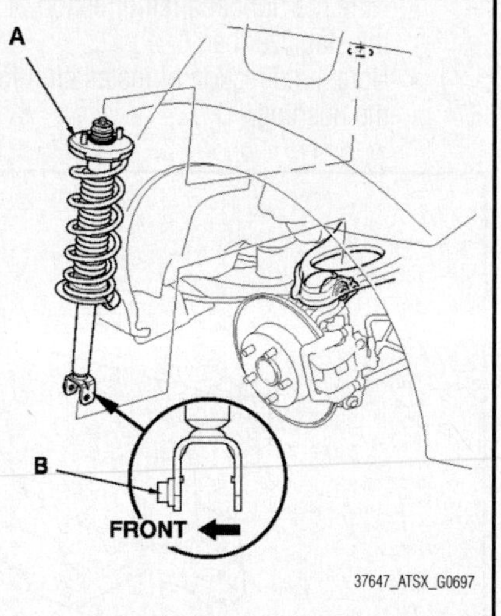
A
B
FRONT
37647_ATSX_G0697

Fig. 449 Lower the rear suspension, and position the damper/spring (A) in the body with the welded nut (B) on the bottom of the damper facing forward

10. Loosely install the new damper lower mounting bolt on the bottom of the damper. Connect the stabilizer link to the brake hose bracket to the knuckle, and loosely install the new flange nut.

11. Place a floor jack under the connecting point of the knuckle and lower arm A, and raise the suspension to load with the vehicle's weight.

12. Tighten the damper lower mounting bolt and the flange nut while holding the joint pin with the hex wrench to 33 ft. lbs. (44 Nm).

13. Tighten the damper mounting nuts on top of the damper to 41 ft. lbs. (55 Nm).

14. Install the lid, and set the rear seat-back to the original position.

15. Clean the mating surfaces of the brake disc and the inside of the wheel, then install the rear wheel.

16. Check the wheel alignment, and adjust it if necessary.

ACURA

6

ZDX

BRAKES 6-9

ANTI-LOCK BRAKE SYSTEM (ABS) 6-10
Wheel Speed Sensors 6-10
Removal & Installation..........6-10
BLEEDING THE BRAKE SYSTEM 6-9
Bleeding Procedure....................6-9
Bleeding Procedure 6-9
Bleeding the ABS System 6-10
FRONT DISC BRAKES 6-12
Brake Caliper...........................6-12
Removal & Installation..........6-12
Disc Brake Pads 6-12
Removal & Installation..........6-12
INFORMATION AND PRECAUTIONS.................6-9
Anti-lock Systems..................6-9
Disc and Drum Systems..............................6-9
PARKING BRAKE 6-15
Parking Brake Cable.................6-15
Removal & Installation..........6-15
Parking Brake Pedal 6-15
Adjustment 6-15
Parking Brake Shoes................6-17
Removal & Installation..........6-17
REAR DISC BRAKES 6-13
Brake Caliper...........................6-13
Removal & Installation..........6-13
Disc Brake Pads 6-14
Removal & Installation..........6-14

CHASSIS ELECTRICAL.......6-19

AIR BAG (SUPPLEMENTAL RESTRAINT SYSTEM)........6-19
General Information..................6-19
Arming the System 6-19
Clockspring Centering..........6-19
Disarming the System...........6-19
Precautions..........................6-19

DRIVE TRAIN.................6-19

Front Driveshaft........................6-19
Removal & Installation..........6-19
Rear Driveshaft.........................6-21
Removal & Installation..........6-21
Transfer Assembly....................6-22
Removal & Installation..........6-22

ENGINE COOLING 6-23

Engine Cooling Fans................6-23
Removal & Installation..........6-23
Radiator...................................6-24
Removal & Installation..........6-24
Thermostat 6-24
Removal & Installation..........6-24
Water Pump 6-24
Removal & Installation..........6-24

ENGINE ELECTRICAL.........6-25

CHARGING SYSTEM 6-25
Alternator 6-25
Battery Disconnect/ Reconnect Procedure..........6-25
Removal & Installation..........6-25
IGNITION SYSTEM 6-26
Firing Order.............................6-26
Ignition Coil 6-26
Removal & Installation..........6-26
Ignition Timing.........................6-26
Adjustment 6-26
Inspection 6-26
Spark Plugs..............................6-26
Removal & Installation..........6-26
STARTING SYSTEM...........6-26
Starter 6-26
Removal & Installation..........6-26

ENGINE MECHANICAL 6-27

Accessory Drive Belts 6-27
Accessory Belt Routing..........................6-27
Adjustment 6-27
Inspection 6-27
Removal & Installation..........6-27
Camshaft & Valve Lifters...........6-27
Removal & Installation..........6-27
Camshaft Pulley 6-28
Removal & Installation..........6-28
Catalytic Converter...................6-30
Removal & Installation..........6-30
Crankshaft Damper (Pulley) 6-31
Removal & Installation..........6-31
Crankshaft Front Seal...............6-32
Removal & Installation..........6-32
Cylinder Head 6-32
Removal & Installation..........6-32
Flexplate (Drive Plate)..............6-34
Removal & Installation..........6-34
Intake Manifold 6-34
Removal & Installation..........6-34
Oil Pan 6-36
Removal & Installation..........6-36
Oil Pump..................................6-36
Inspection 6-38
Removal & Installation..........6-36
Piston and Ring........................6-38
Positioning 6-38
Rear Main Seal.........................6-38
Removal & Installation..........6-38
Rocker Arms............................6-39
Removal & Installation..........6-39
Timing Belt & Sprockets 6-42
Removal & Installation..........6-42
Timing Belt Front Cover 6-40
Removal & Installation..........6-40
Valve Covers (Cylinder Head Covers) 6-44
Removal & Installation..........6-44
Valve Lash...............................6-45
Adjustment 6-45

ENGINE PERFORMANCE & EMISSION CONTROLS.......6-46

Accelerator Pedal Position (APP) Sensor 6-46
Location..............................6-46
Removal & Installation..........6-46
Camshaft Position (CMP) Sensor 6-46
Location..............................6-46
Removal & Installation..........6-46
Crankshaft Position (CKP) Sensor 6-46

CKP Pattern Clear/CKP Pattern Learn 6-46
Location 6-46
Removal & Installation 6-46
Engine Coolant Temperature (ECT) Sensor 6-47
Location 6-47
Removal & Installation 6-47
Evaporative Emissions (EVAP) Canister 6-47
Location 6-47
Removal & Installation 6-47
Exhaust Gas Recirculation (EGR) Valve 6-48
Location 6-48
Removal & Installation 6-48
Heated Oxygen Sensor (HO2S) 6-48
Location 6-48
Removal & Installation 6-49
Knock Sensor (KS) 6-49
Location 6-49
Removal & Installation 6-49
Manifold Absolute Pressure (MAP) Sensor 6-49
Location 6-49
Removal & Installation 6-49
Mass Air Flow/Intake Air Temperature (MAF/IAT) Sensor 6-49
Location 6-49
Removal & Installation 6-49
Positive Crankcase Ventilation (PCV) Valve 6-49
Location 6-49
Removal & Installation 6-50
Powertrain Control Module (PCM) 6-50
Location 6-50
PCM Idle Learn Procedure 6-50
Removal & Installation 6-50
Testing 6-51

FUEL 6-51

GASOLINE FUEL INJECTION SYSTEM 6-51
Fuel Filter 6-52
Removal & Installation 6-52
Fuel Level Sending Unit 6-52
Location 6-52
Removal & Installation 6-52
Fuel Pump Module 6-52
Removal & Installation 6-52
Fuel Rail and Injector 6-53
Removal & Installation 6-53
Fuel System Pressure 6-51
Relieving 6-51
Fuel System Service Precautions 6-51
Fuel Tank 6-54
Draining 6-54
Removal & Installation 6-54
Idle Speed 6-54
Adjustment 6-54
Throttle Body 6-55
Removal & Installation 6-55

HEATING & AIR CONDITIONING SYSTEM 6-56

Blower Motor 6-56
Removal & Installation 6-56
Heater Unit & Heater Core 6-56
Removal & Installation 6-56

SPECIFICATIONS AND MAINTENANCE CHARTS 6-3

Brake Specifications 6-7
Camshaft and Bearing Specifications 6-5
Capacities 6-4
Crankshaft and Connecting Rod Specifications 6-5
Engine and Vehicle Identification 6-3
Engine Tune-Up Specifications 6-3
Fluid Specifications 6-4
General Engine Specifications 6-3
Piston and Ring Specifications 6-5
Maintenance Minder Schedule 6-8
Tire, Wheel and Ball Joint Specifications 6-7
Torque Specifications 6-6
Valve Specifications 6-4
Wheel Alignment 6-7

STEERING 6-58

Electronically Controlled Power Steering (ECPS) Control Unit 6-58
Removal & Installation 6-58
Power Rack & Pinion Steering Gear 6-58
Removal & Installation 6-58
VSA Sensor Neutral Position Memorization 6-62
Power Steering Pump 6-62
Removal & Installation 6-62

SUSPENSION 6-63

FRONT SUSPENSION 6-63
Lower Ball Joint 6-63
Removal & Installation 6-63
Lower Control Arm 6-63
Removal & Installation 6-63
MacPherson Strut (Damper) 6-64
Overhaul 6-65
Removal & Installation 6-64
Stabilizer Bar 6-68
Removal & Installation 6-68
Stabilizer Link 6-70
Removal & Installation 6-70
Steering Knuckle/Hub Bearing Unit 6-67
Removal & Installation 6-67
Wheel Bearings 6-70
End Play Inspection 6-70
Removal & Installation 6-70

REAR SUSPENSION 6-71
Knuckle/Hub Bearing Unit 6-71
Removal & Installation 6-71
Lower Control Arm 6-72
Removal & Installation 6-72
McPherson Struts (Damper) 6-73
Overhaul 6-73
Removal & Installation 6-73
Stabilizer Bar 6-75
Removal & Installation 6-75
Stabilizer Link 6-75
Removal & Installation 6-75
Trailing Arm 6-75
Removal & Installation 6-75
Upper Control Arm 6-76
Removal & Installation 6-76
Wheel Bearings 6-76
End Play Inspection 6-76
Removal & Installation 6-76

SPECIFICATIONS AND MAINTENANCE CHARTS

ENGINE AND VEHICLE IDENTIFICATION CHART

Engine Code							Model Year	
Code	Liters (cc)	Cu. In.	Cyl.	Fuel Sys.	Engine Type	Eng. Mfg.	Code ①	Year
J37A5	3.7 (3664)	226	6	PGM-FI ③	SOHC	Honda	A	2010

SOHC: Single Overhead Cam

SMFI: Sequential Multi-port Fuel Injection

① 10th position of VIN

② VTEC: variable timing electronically controlled

③ PGM-FI: multi-point fuel injection

37647_AZDX_C0001

GENERAL ENGINE SPECIFICATIONS

Year	Model	Engine Displacement Liters	Engine ID	Net Horsepower @ rpm	Net Torque @ rpm (ft. lbs.)	Bore x Stroke (in.)	Com-pression Ratio	Oil Pressure @ rpm
2010	ZDX	3.7	J37A5	300@6300	270@4500	3.54x3.78	11.2:1	44@3000

37647_AZDX_C0002

ENGINE TUNE-UP SPECIFICATIONS

Year	Engine Displacement Liters	Engine ID	Spark Plug Gap (in.)	Ignition Timing (deg.)		Fuel Pump (psi)	Idle Speed (rpm)		Valve Clearance (in.)	
				MT	AT		MT	AT	In.	Ex.
2010	3.7	J37A5	0.039-0.043	—	8-12B	55-63	—	660-760	0.008-0.009	0.011-0.013

NOTE: The Vehicle Emission Control Information label often reflects changes made during production and must be used if they differ from this chart.

NOTE: The fuel pressure readings are given with the vacuum hose connected to the regulator and the engine running

B: Before top dead center

37647_AZDX_C0003

CAPACITIES

Year	Model	Engine Displacement Liters	Engine ID	Engine Oil with Filter (qts.)	Transmission (qts.) 5-Spd	Transmission (qts.) Auto.	Fuel Tank (gal.)	Cooling System (qts.)
2010	ZDX	3.7	J37A5	4.5	—	3.3 ①	21.0	7.1 ②

NOTE: All capacities are approximate. Add fluid gradually and check to be sure a proper fluid level is obtained.

① 3.3 qts for service; 8.5 qts for overhaul.

② 7.1 qts for service; 9.1 qts for engine overhaul.

37647_AZDX_C0004

FLUID SPECIFICATIONS

Year	Model	Engine Displ. Liters	Engine Oil	Man. Trans.	Auto. Trans.	Power Steering Fluid	Brake Master Cylinder	Cooling System
2010	ZDX	3.7	5W-20 Acura	N/A	Acura ATF-Z1	Acura PS Fluid	Acura DOT 3	①

DOT: Department Of Transpotation

N/A: Not Applicable

① Acura Long Life Antifreeze/Coolant-Type2

37647_AZDX_C0005

VALVE SPECIFICATIONS

Year	Engine Displacement Liters	Engine ID	Seat Angle (deg.)	Face Angle (deg.)	Spring Test Pressure (lbs. @ in.)	Spring Installed Height (in.)	Stem-to-Guide Clearance (in.) Intake	Stem-to-Guide Clearance (in.) Exhaust	Stem Diameter (in.) Intake	Stem Diameter (in.) Exhaust
2010	3.7	J37A5	NA	NA	NA	NA	0.00079-0.0018	0.00217-0.0032	0.2159-0.2163	0.2146-0.2150

NA: Not Available

37647_AZDX_C0007

CAMSHAFT AND BEARING SPECIFICATIONS

All measurements are given in inches.

Year	Engine Displacement Liters	Engine VIN	Journal Diameter	Brg. Oil Clearance	Shaft End-play	Runout	Journal Bore	Lobe Height Intake	Lobe Height Exhaust
2010	3.7	J37A5	NA	0.0020-0.0035	0.0020-0.0080	0.0012	NA	①	1.4472

NA: Information not available

① Primary: 1.35035 in.
Secondary: 1.41464 in.

37647_AZDX_C0006

CRANKSHAFT AND CONNECTING ROD SPECIFICATIONS

All measurements are given in inches

Year	Engine Displacement Liters	Engine ID	Crankshaft Main Brg. Journal Dia.	Crankshaft Main Brg. Oil Clearance	Crankshaft Shaft End-play	Crankshaft Thrust on No.	Connecting Rod Journal Diameter	Connecting Rod Oil Clearance	Connecting Rod Side Clearance
2010	3.7	J37A5	2.8337-2.8346	0.00075-0.0018	0.0039-0.0138	3	2.24315-2.24409	0.0002-0.00055	0.0059-0.0138

37647_AZDX_C0008

PISTON AND RING SPECIFICATIONS

All measurements are given in inches

Year	Engine Disp. Liters	Engine ID	Piston Clearance	Ring Gap Top Compression	Ring Gap Bottom Compression	Ring Gap Oil Control	Ring Side Clearance Top Compression	Ring Side Clearance Bottom Compression	Ring Side Clearance Oil Control
2010	3.7	J37A5	0.00016-0.00126	0.0118-0.0157	0.0157-0.0217	0.0079-0.0138	0.00217-0.00335	0.00118-0.0024	NA

NA: Not Applicable

37647_AZDX_C0009

TORQUE SPECIFICATIONS

All readings in ft. lbs.

Year	Engine Displacement Liters	Engine ID	Cylinder Head Bolts	Main Bearing Bolts	Rod Bearing Bolts	Crankshaft Damper Bolts	Flywheel Bolts	Manifold Intake	Manifold Exhaust	Spark Plugs	Oil Pan Drain Plug
2010	3.7	J37A5	①	②	③	④	55	16	40 ⑤	18	30

NOTE: Dip main bearing bolts and crankshaft damper bolt in clean engine oil prior to tightening.

① 12-point head bolts
Step 1: Torque all bolts to 22 ft. lbs.
Step 2: Torque all bolts an additional 90 deg.
Step 3: Torque all bolts an additional 90 deg.
New Bolt Only: an additional 90 deg.

② Cap bolts: 54 ft. lbs.
Side bolts: 36 ft. lbs.

③ Step 1: 14 ft. lbs.
Step 2: 90 degrees

④ Step 1: 48 ft. lbs.
Step 2: 65 degrees

⑤ Exhaust pipe-to-engine: 40 ft. lbs.; no exhaust manifold.

37647_AZDX_C0010

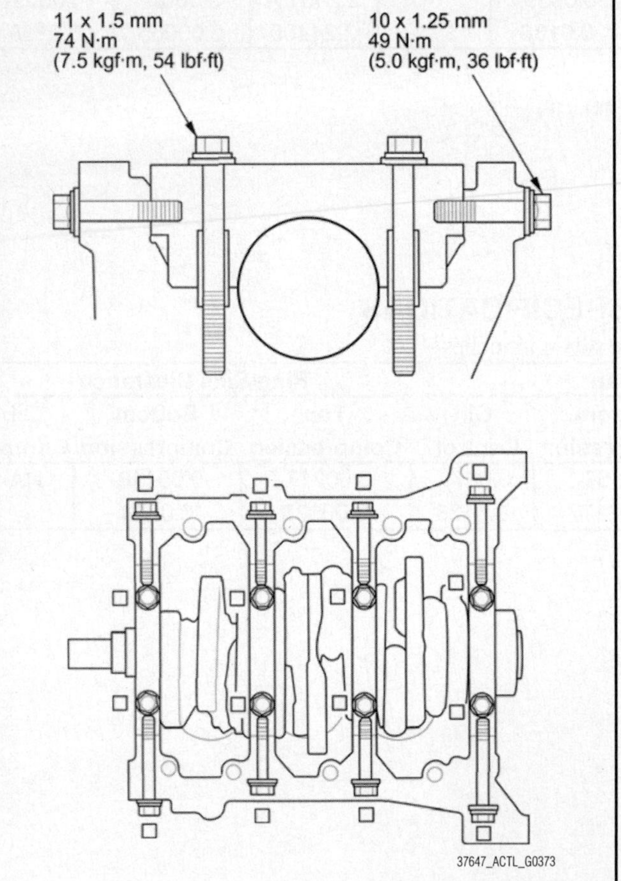
11 x 1.5 mm
74 N·m
(7.5 kgf·m, 54 lbf·ft)
10 x 1.25 mm
49 N·m
(5.0 kgf·m, 36 lbf·ft)
37647_ACTL_G0373

Fig. 1 Main bearing torque sequence

WHEEL ALIGNMENT

Year	Model		Caster Range (+/-Deg.)	Caster Preferred Setting (Deg.)	Camber Range (+/-Deg.)	Camber Preferred Setting (Deg.)	Toe-in (in.)
2010	ZDX	Front	0.60	+4.63	0.75	-0.50	0+/-0.08
		Rear	—	—	0.75	-0.50	0+/-0.08

37647_AZDX_C0011

TIRE, WHEEL AND BALL JOINT SPECIFICATIONS

Year	Model	OEM Tires Standard	OEM Tires Optional	Tire Pressures (psi) Front	Tire Pressures (psi) Rear	Wheel Size	Ball Joint Inspection	Lug Nut Torque (ft. lbs.)
2010	ZDX	P255-50R19	None	①	①	8.5	NS	94

OEM: Original Equipment Manufacturer

PSI: Pounds Per Square Inch

NS: Not specified by manufacturer

① See placard on vehicle

37647_AZDX_C0012

BRAKE SPECIFICATIONS

All measurements in inches unless noted

Year	Model		Brake Disc Original Thickness	Brake Disc Minimum Thickness	Brake Disc Maximum Runout	Minimum Lining Thickness Front	Minimum Lining Thickness Rear	Brake Caliper Bracket Bolts (ft. lbs.)	Brake Caliper Mounting Bolts (ft. lbs.)
2010	ZDX	F	1.102	1.024	0.002	0.063	—	101	53
		R	0.433	0.354	0.002	—	0.030	65	27

F: Front

R: Rear

37647_AZDX_C0013

MAINTENANCE MINDER SCHEDULE

Acura ZDX

The ZDX displays engine oil life and maintenance service items in the information display to indicate when to perform maintenance service. If the engine oil life is 15% or less, based on the onboard computer's caluculations, you will see SERVICE DUE SOON in the information display every time the ignition key is turned to ON. The maintenance minder indicator will also come on and the maintenance code(s) for other scheduled maintenance items needing service will be displayed below the message.

Symbol	Item	Service
A	Engine oil ①	Change
B	Engine oil and filter	Change
	Brakes	Inspect
	Parking brake adjustment	Check
	Steering gear and linkage	Inspect
	Suspension components	Inspect
	Driveshaft boots	Inspect
	Brake hoses and lines, including VSA lines	Inspect
	All fluid levels and condition	Inspect
	Exhaust system	Inspect
	Fuel lines and connections	Inspect
1	Tires	Rotate
2	Engine air filter ②	Replace
	Dust and pollen filter ③	Replace
	Accessory drive belt	Inspect
3	Transmission fluid	Replace
	Transfer case fluid	Replace
4	Spark plugs	Replace
	Timing belt ④	Replace
	Water pump	Inspect
	Valve clearance ⑤	Inspect
5	Engine coolant	Replace
6	Rear differential fluid ⑥	Replace

① If the message SERVICE DUE NOW does not appear more than 12 months after the display is reset, change every year.

② If driven in dusty conditions, replace every 15,000 miles.

③ If driven in urban areas that have a high concentration of soot and dust, replace every 15,000 miles

④ If driven regularly in temperatures over 110 deg.F or below -20 deg.F, or towing a trailer, replace every 60,000 miles.

⑤ Adjust if necessary.

⑥ Driving in mountainous areas at very low vehicle speeds or trailer towing results in higher level of mechanical (shear) stress to fluid. This requires differential fluid changes more frequently than recommended by the maintenance minder. If the vehicle is regularly driven in these conditions, have the differential fluid changed at 7,500 miles, thereafter every 15,000 miles.

Additionally, replace the brake fluid every 3 years, and inspect the idle speed every 160,000 miles.

To reset the Engine Oil Life Display:

- The vehicle must be stopped to reset the display.
- If a required service is done and the display is not reset, or if the maintenance display is reset without doing the service, the system will not show the proper maintenance timing. This can lead to serious mechanical problems because there will be no accurate record of when the required maintenance is needed.
- The engine oil life and the maintenance item(s) can be independently reset with the HDS, or equivalent scan tool.

1. Turn the ignition switch to ON (II), or press the engine start/stop button to select the ON mode.
2. If system message(s) are displayed, press the INFO button to cancel the display.
3. Push the SEL/RESET button repeatedly until the engine oil life indicator is displayed.
4. Press and hold the SEL/RESET button for about 10 seconds, the "OIL LIFE RESET" mode display appears.

37647_AZDX_C0014

BRAKES — INFORMATION AND PRECAUTIONS

ANTI-LOCK SYSTEMS

- Certain components within the ABS system are not intended to be serviced or repaired individually.
- Do not use rubber hoses or other parts not specifically specified for and ABS system. When using repair kits, replace all parts included in the kit. Partial or incorrect repair may lead to functional problems and require the replacement of components.
- Lubricate rubber parts with clean, fresh brake fluid to ease assembly. Do not use shop air to clean parts; damage to rubber components may result.
- Use only DOT 3 brake fluid from an unopened container.
- If any hydraulic component or line is removed or replaced, it may be necessary to bleed the entire system.
- A clean repair area is essential. Always clean the reservoir and cap thoroughly before removing the cap. The slightest amount of dirt in the fluid may plug an orifice and impair the system function. Perform repairs after components have been thoroughly cleaned; use only denatured alcohol to clean components. Do not allow ABS components to come into contact with any substance containing mineral oil; this includes used shop rags.
- The Anti-Lock control unit is a microprocessor similar to other computer units in the vehicle. Ensure that the ignition switch is **OFF** before removing or installing controller harnesses. Avoid static electricity discharge at or near the controller.
- If any arc welding is to be done on the vehicle, the control unit should be unplugged before welding operations begin.

DISC AND DRUM SYSTEMS

CAUTION

Dust and dirt accumulating on brake parts during normal use may contain asbestos fibers from production or aftermarket brake linings. Breathing excessive concentrations of asbestos fibers can cause serious bodily harm. Exercise care when servicing brake parts. Do not sand or grind brake lining unless equipment used is designed to contain the dust residue. Do not clean brake parts with compressed air or by dry brushing. Cleaning should be done by dampening the brake components with a fine mist of water, then wiping the brake components clean with a dampened cloth. Dispose of cloth and all residue containing asbestos fibers in an impermeable container with the appropriate label. Follow practices prescribed by the Occupational Safety and Health Administration (OSHA) and the Environmental Protection Agency (EPA) for the handling, processing, and disposing of dust or debris that may contain asbestos fibers.

BRAKES — BLEEDING THE BRAKE SYSTEM

BLEEDING PROCEDURE

BLEEDING PROCEDURE

See Figures 2 through 4.

➡Do not reuse the drained fluid. Use only clean Honda DOT 3 Brake Fluid from an unopened container. Using a non-Honda brake fluid can cause corrosion and shorten the life of the system.

Do not mix different brands of brake fluid; they may not be compatible.

Make sure no dirt or other foreign matter is allowed to contaminate the brake fluid.

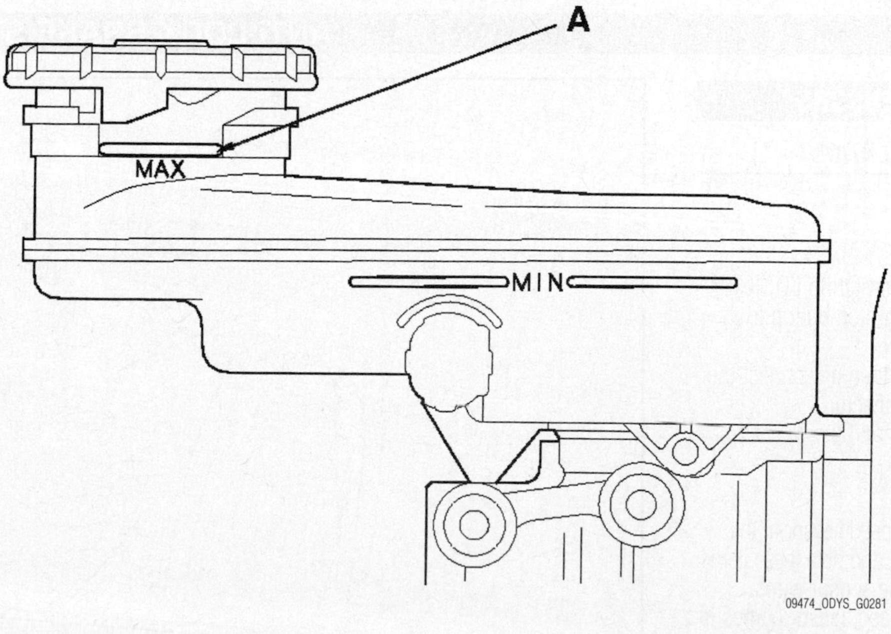
A
MAX
MIN
09474_ODYS_G0281

Fig. 2 The reservoir on the master cylinder must be at the MAX (A) level mark at the start of the bleeding procedure

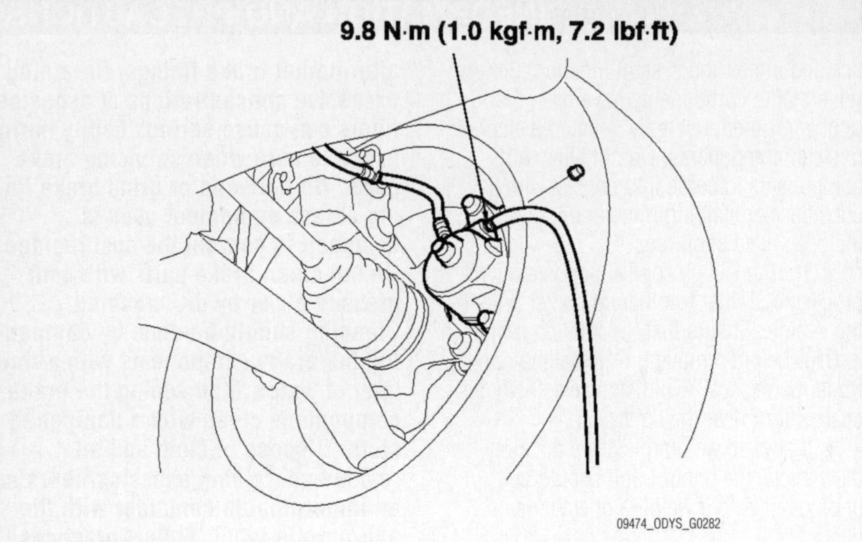
9.8 N·m (1.0 kgf·m, 7.2 lbf·ft)
09474_ODYS_G0282

Fig. 3 Bleed screw

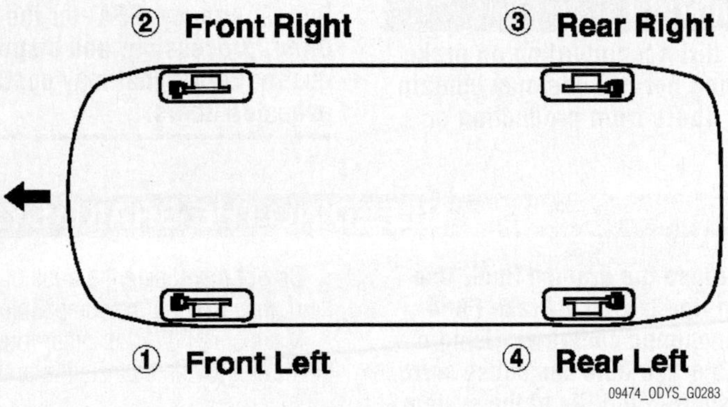
② Front Right
③ Rear Right
① Front Left
④ Rear Left
09474_ODYS_G0283

Fig. 4 Bleeding sequence

⁂ WARNING

Do not spill brake fluid on the vehicle, it may damage the paint; if brake fluid does contact the paint, wash it off immediately with water.

The reservoir on the master cylinder must be at the MAX (upper) level mark at the start of the bleeding procedure and checked after bleeding each brake caliper. Add fluid as required.

1. Make sure the brake fluid level in the reservoir is at the MAX (upper) level line.
2. Attach a length of clear drain tube to the bleed screw.
3. Have someone slowly pump the brake pedal several times, then apply steady pressure.
4. Starting at the left-front, loosen the brake bleed screw to allow air to escape from the system. Then tighten the bleed screw securely.
5. Repeat the procedure for each caliper until no air bubbles are in the fluid. Bleed the calipers in the sequence shown.
6. Refill the master cylinder reservoir to the MAX (upper) level line.

BLEEDING THE ABS SYSTEM

The bleeding procedure for the ABS System is the same as the Conventional Bleeding Procedure. See "Bleeding the Brake System" in this section.

BRAKES — ANTI-LOCK BRAKE SYSTEM (ABS)

WHEEL SPEED SENSORS

REMOVAL & INSTALLATION

Front

See Figure 5.

1. Turn the ignition switch to LOCK (0), or press the engine start/stop button to select the OFF mode.
2. Remove the clip, then disconnect the wheel speed sensor connector.
3. Remove the bolts and the wheel speed sensor.

To install:

4. Install the wheel speed sensor in the reverse order of removal, and note these items:
   a. Do not twist the sensor wires.
   b. If the wheel speed sensor comes in contact with the hub bearing unit, it is faulty.

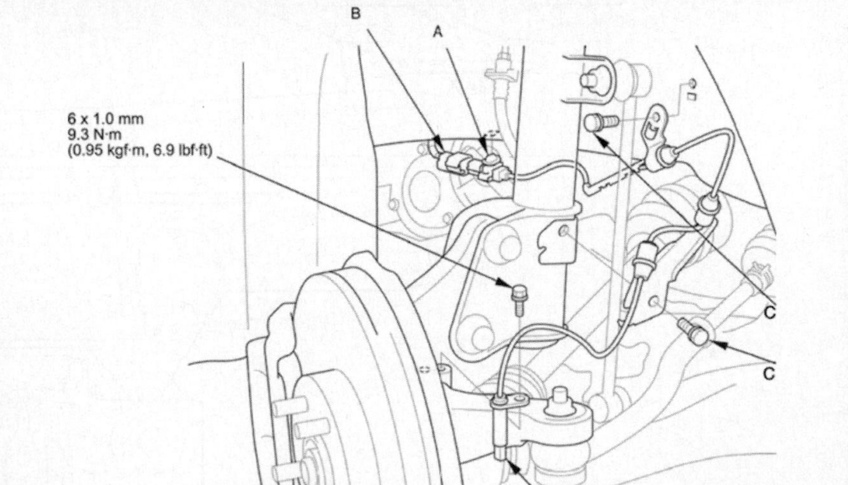
B
A
6 x 1.0 mm
9.3 N·m
(0.95 kgf·m, 6.9 lbf·ft)
C
C
C
37647_AZDX_G0068

Fig. 5 Remove the clip (A), then disconnect the wheel speed sensor connector (B). Remove the bolts and the wheel speed sensor (C).

c. Make sure there is no debris in the sensor mounting hole.

5. Start the engine, and make sure the ABS and the VSA indicators go off.

6. Test-drive the vehicle, and make sure the ABS and the VSA indicators do not come on.

Rear

See Figure 6.

1. Turn the ignition switch to LOCK (0), or press the engine start/stop button to select the OFF mode.

2. Remove the clip, then disconnect the wheel speed sensor connector.

3. Remove the clip (left side), the bolts, and the wheel speed sensor.

To install:

4. Install the wheel speed sensor in the reverse order of removal, and note these items:

a. Do not twist the sensor wires.

b. If the wheel speed sensor comes in contact with the hub bearing unit, it is faulty.

c. Make sure there is no debris in the sensor mounting hole.

5. Start the engine, and make sure the ABS and the VSA indicators go off.

6. Test-drive the vehicle, and make sure the ABS and the VSA indicators

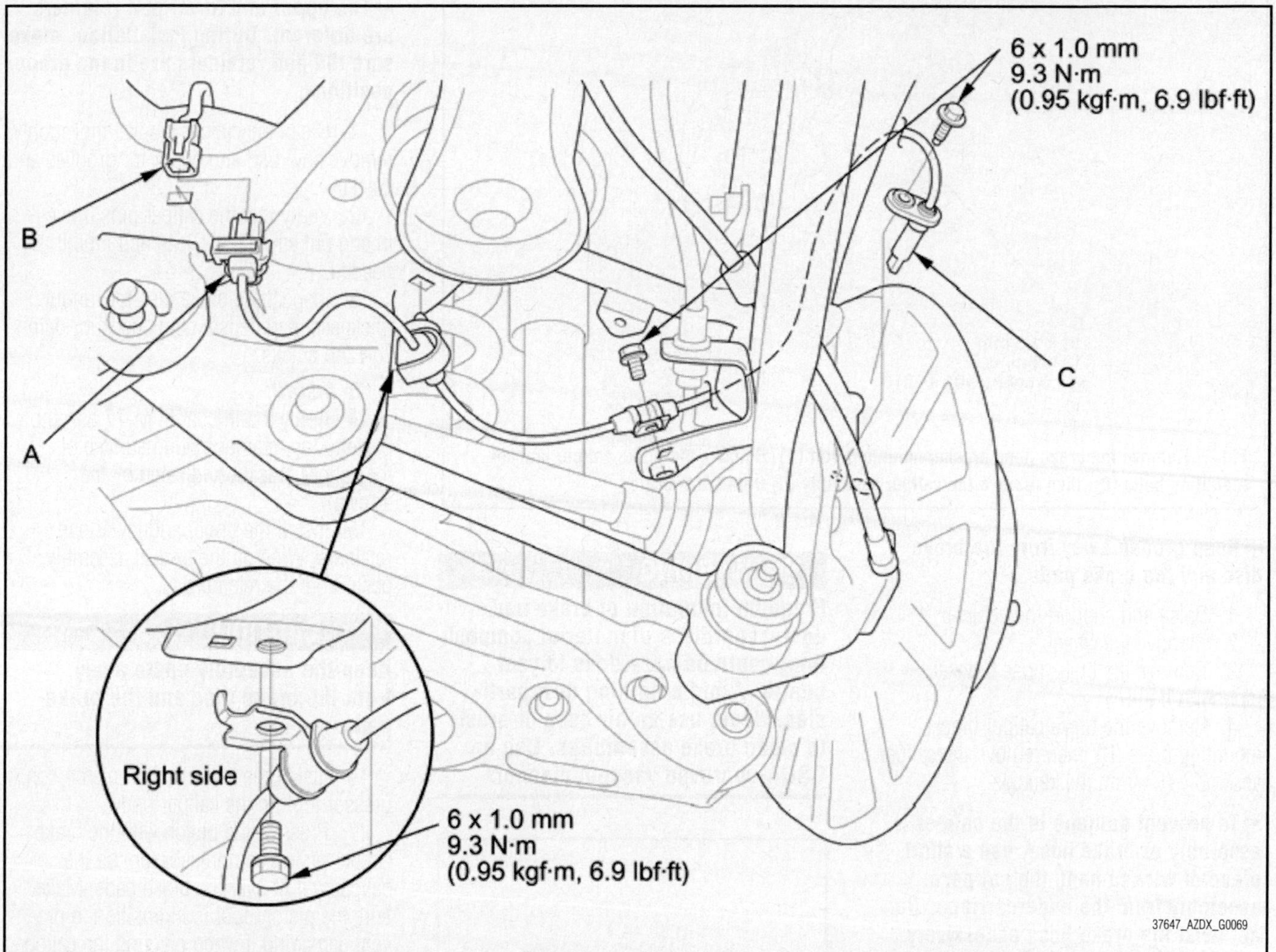
6 x 1.0 mm
9.3 N·m
(0.95 kgf·m, 6.9 lbf·ft)
B
C
A
Right side
6 x 1.0 mm
9.3 N·m
(0.95 kgf·m, 6.9 lbf·ft)
37647_AZDX_G0069

Fig. 6 Remove the clip (A), then disconnect the wheel speed sensor connector (B). Remove the bolts and the wheel speed sensor (C).

BRAKE CALIPER

REMOVAL & INSTALLATION

See Figure 7.

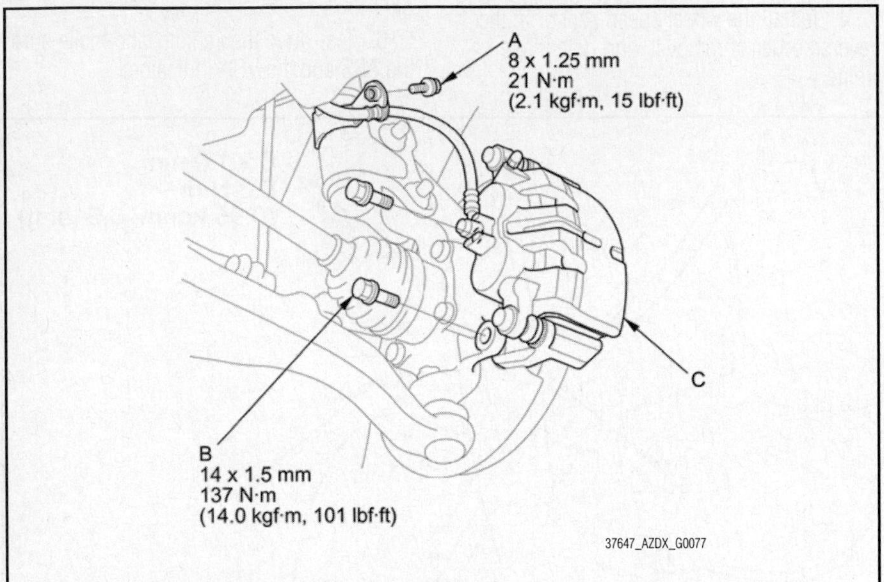
A
8 x 1.25 mm
21 N·m
(2.1 kgf·m, 15 lbf·ft)
C
B
14 x 1.5 mm
137 N·m
(14.0 kgf·m, 101 lbf·ft)
37647_AZDX_G0077

Fig. 7 Remove the brake hose bracket mounting bolt (A). Remove the brake caliper bracket mounting bolts (B), then remove the caliper assembly (C) from the knuckle.

➡Keep grease away from the brake disc and the brake pads.

1. Raise and support the vehicle.
2. Remove the wheel.
3. Remove the brake hose bracket mounting bolt (A).
4. Remove the brake caliper bracket mounting bolts (B), then remove the caliper assembly (C) from the knuckle.

➡To prevent damage to the caliper assembly or brake hose, use a short piece of wire to hang the caliper assembly from the undercarriage. Do not twist the brake hose excessively.

5. Installation is the reverse of the removal procedure.

DISC BRAKE PADS

REMOVAL & INSTALLATION

See Figures 8 through 10.

1. Raise and support the vehicle.
2. Remove the front wheels.
3. Check the thickness of the inner pad and the outer pad. Do not include the thickness of the backing plate.
4. If any part of the brake pad thickness is less than the service limit of 0.63 in. (1.6 mm), replace the front brake pads as a set.
5. Clean the mating surfaces between the brake disc and the inside of the wheel, then install the front wheels.
6. Remove some brake fluid from the master cylinder.

⁂ WARNING

Frequent inhalation of brake pad dust, regardless of material composition, could be hazardous to your health. Avoid breathing dust particles. Never use an air hose or brush to clean brake assemblies. Use an OSHA-approved vacuum cleaner.

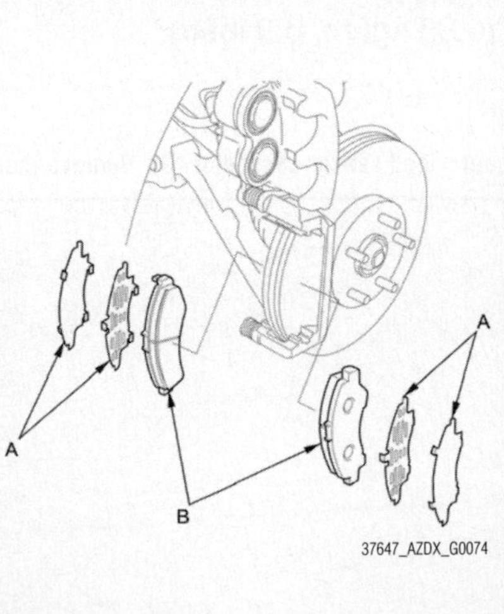
A
A
B
37647_AZDX_G0074

Fig. 8 Remove the pad shims (A) and the brake pads (B).

7. Remove the front wheels.
8. Remove the lower caliper flange bolt and pivot the caliper up out of the way. Check the hose and the pin boots for damage and deterioration.
9. Remove the pad shims and the brake pads.
10. Remove the upper and lower pad retainers.

➡The upper and lower pad retainers are different. During installation, make sure the pad retainers are in the proper positions.

11. Clean the caliper bracket thoroughly; remove any rust, and check for grooves and cracks.
12. Verify that the caliper pins move in and out smoothly. Clean and lube if needed.
13. Inspect the brake disc for runout, thickness, parallelism, and check for damage and cracks.

To install:

14. Apply a thin coat of M-77 assembly paste to the retainer mating surface of the caliper bracket (indicated by the arrows).
15. Install the upper and lower pad retainers. Wipe off the excess assembly paste from the retainers.

⁂ CAUTION

Keep the assembly paste away from the brake disc and the brake pads.

16. Install the brake caliper piston compressor tool on the caliper body.
17. Press in the piston with the brake caliper piston compressor tool so the caliper will fit over the brake pads. Make sure the piston boot is in position to prevent damaging it when pivoting the caliper down.

➡Be careful when pressing in the piston; brake fluid might overflow from the master cylinder's reservoir. If brake fluid gets on any painted surface, wash it off immediately with water.

18. Remove the brake caliper piston compressor tool.
19. Apply a thin coat of M-77 assembly paste to the pad side of the shims, the back of the brake pads (B), and the other areas indicated by the arrows. Wipe off the excess assembly paste from

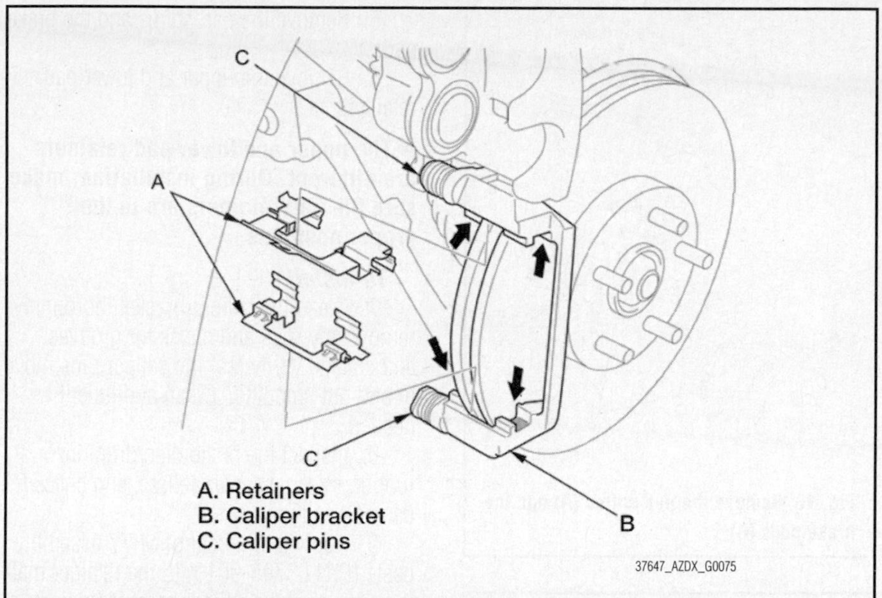
C
A
C
A. Retainers
B. Caliper bracket
C. Caliper pins
B
37647_AZDX_G0075

Fig. 9 Removing the upper and lower pad retainers and checking caliper bracket and pins

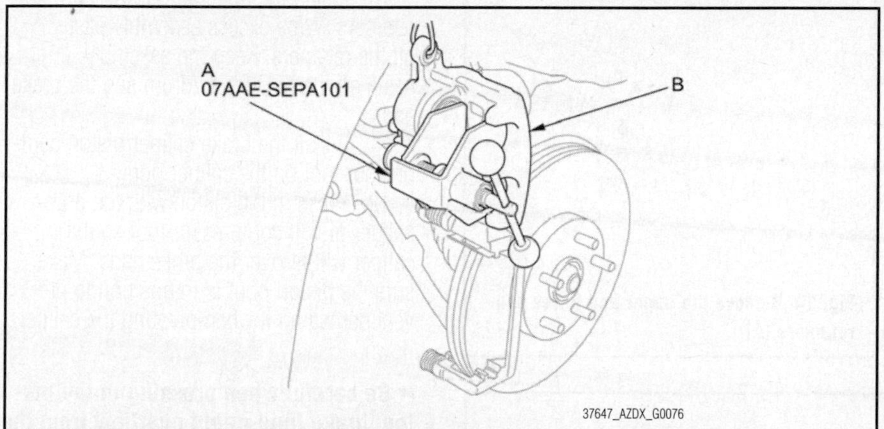
A
07AAE-SEPA101
B
37647_AZDX_G0076

Fig. 10 Install the brake caliper piston compressor tool (A) on the caliper body (B).

the pad shims and brake pads friction material.

⁂ CAUTION

Keep grease and assembly paste away from the brake disc and brake pads. Contaminated brake disc or brake pads reduce stopping ability.

20. Install the brake pads and pad shims correctly. Install the brake pad with the wear indicator on the upper inside. If you are reusing the brake pads, always reinstall the brake pads in their original positions to prevent a temporary loss of braking efficiency.
21. Pivot the caliper down into position. Install the flange bolt and tighten it to 53 ft. lbs. (72 Nm).
22. Clean the mating surfaces between the brake disc and the inside of the wheel, then install the front wheels.
23. Press the brake pedal several times to make sure the brakes work.

➡Engagement may require a greater pedal stroke immediately after the brake pads have been replaced as a set.

24. Several applications of the brake pedal will restore the normal pedal stroke.
25. Add brake fluid as needed.
26. After installation, check for leaks at hose and line joints or connections, and retighten if necessary. Test-drive the vehicle, then recheck for leaks.

BRAKES — REAR DISC BRAKES

BRAKE CALIPER

REMOVAL & INSTALLATION

See Figure 11.

1. Raise and support the vehicle.
2. Remove the rear wheel.
3. Release the parking brake.
4. Remove the brake caliper bracket mounting bolts, then remove the caliper assembly from the knuckle.

⁂ CAUTION

To prevent damage to the caliper assembly or brake hose, use a short piece of wire to hang the caliper assembly from the undercarriage. Do not twist the brake hose excessively.

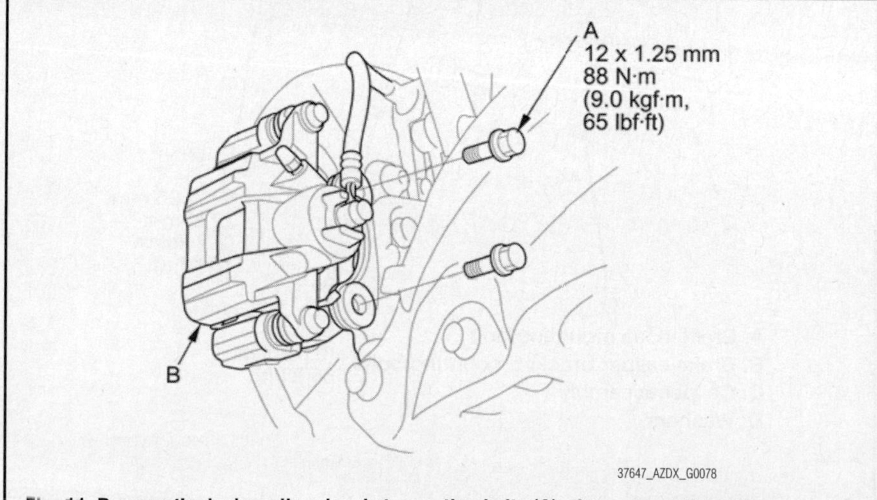
A
12 x 1.25 mm
88 N·m
(9.0 kgf·m,
65 lbf·ft)
B
37647_AZDX_G0078

Fig. 11 Remove the brake caliper bracket mounting bolts (A), then remove the caliper assembly (B) from the knuckle.

5. Installation is the reverse of the removal procedure.

6. Torque the caliper bracket mounting bolts to 65 ft. lbs. (88 Nm).

DISC BRAKE PADS

REMOVAL & INSTALLATION

See Figures 12 through 15.

CAUTION

Frequent inhalation of brake pad dust, regardless of material composition, could be hazardous to your health. Avoid breathing dust particles. Never use an air hose or brush to clean brake assemblies. Use an OSHA-approved vacuum cleaner.

1. Remove some brake fluid from the master cylinder.
2. Raise and support the vehicle.
3. Remove the rear wheels.
4. Remove the brake hose mounting bolt and the brake caliper bracket mounting bolts, then remove the caliper assembly from the knuckle.

CAUTION

To prevent damage to the caliper assembly or brake hose, use a short piece of wire to hang the caliper assembly from the undercarriage. Do not twist the brake hose excessively.

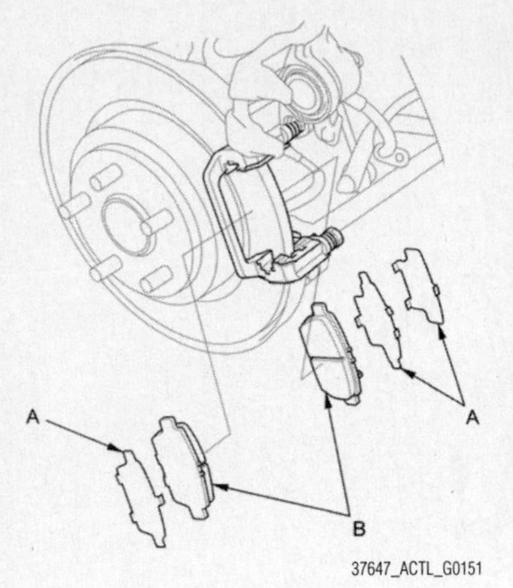
A
B
A
37647_ACTL_G0151

Fig. 13 Remove the pad shims (A) and the brake pads (B)

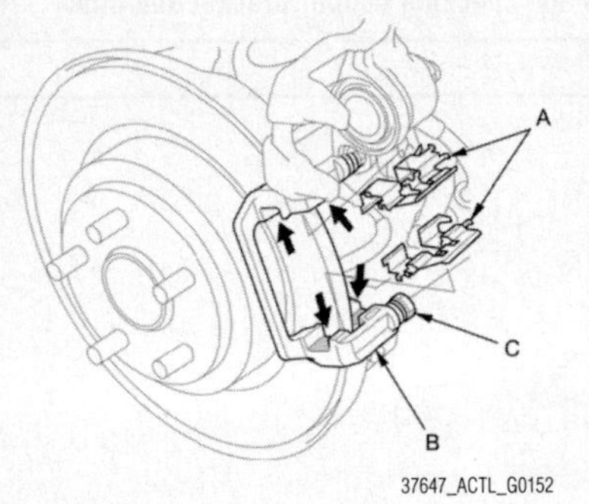
A
C
B
37647_ACTL_G0152

Fig. 14 Remove the upper and lower pad retainers (A)

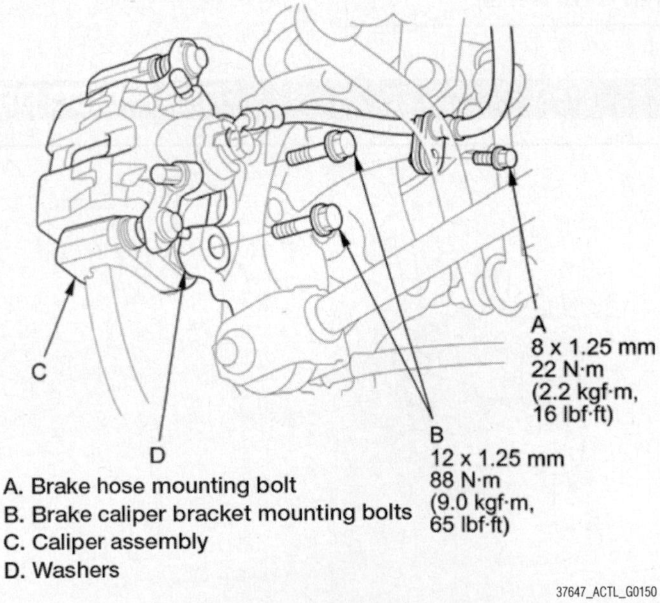
A
8 x 1.25 mm
22 N·m
(2.2 kgf·m,
16 lbf·ft)
C
D
B
12 x 1.25 mm
88 N·m
(9.0 kgf·m,
65 lbf·ft)
A. Brake hose mounting bolt
B. Brake caliper bracket mounting bolts
C. Caliper assembly
D. Washers
37647_ACTL_G0150

Fig. 12 Remove the brake hose mounting bolt and the brake caliper bracket mounting bolts, then remove the caliper assembly from the knuckle

5. Remove the pad shims and the brake pads.

6. Remove the upper and lower pad retainers.

➡The upper and lower pad retainers are different. During installation, make sure the pad retainers are in their proper positions.

To install:

7. Clean the caliper bracket thoroughly; remove any rust, and check for grooves and cracks. Verify that the caliper pins move in and out smoothly. Clean and lube if needed.

8. Inspect the brake disc/drum for runout, thickness, parallelism, and check for damage and cracks.

9. Apply a thin coat of M-77 assembly paste (P/N 08798-9010) to the retainer mating surface of the caliper bracket (indicated by the arrows).

10. Install the upper and lower pad retainers. Wipe excess assembly paste off the retainers. Keep the assembly paste off the brake disc/drum and the brake pads.

11. Install the brake caliper piston compressor tool on the caliper body.

12. Press in the piston with the brake caliper piston compressor tool so the caliper will fit over the brake pads. Make sure the piston boot is in position to prevent damaging it when pivoting the caliper down.

➡Be careful when pressing in the piston; brake fluid might overflow from the master cylinder's reservoir. If brake fluid gets on any painted surface, wash it off immediately with water.

13. Remove the brake caliper piston compressor tool.

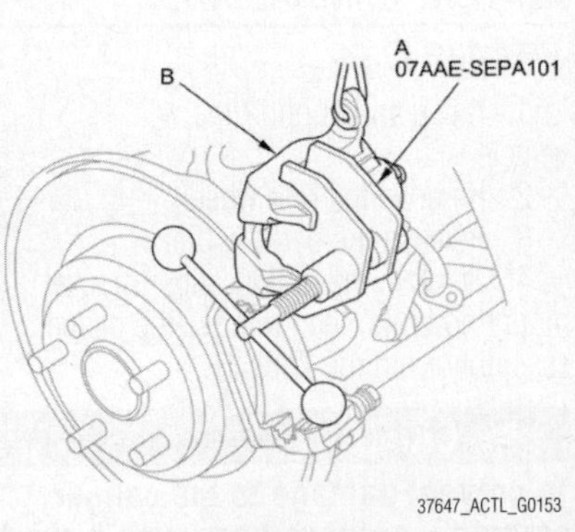
A
07AAE-SEPA101
B
37647_ACTL_G0153

Fig. 15 Install the brake caliper piston compressor tool (A) on the caliper body (B)

14. Apply a thin coat of M-77 assembly paste (P/N 08798-9010) to the pad side of the shims and the back of the brake pads. Wipe excess assembly paste off the pad shims and the brake pads friction material. Keep grease and assembly paste off the brake disc/drum and the brake pads. Contaminated brake disc/drum or brake pads reduce stopping ability.

15. Install the brake pads and pad shims correctly. Install the brake pad with the wear indicator on the bottom inside. If you are reusing the brake pads, always reinstall the brake pads in their original positions to prevent a temporary loss of braking efficiency.

16. Pivot the caliper down into position. Install the flange bolt, and tighten it to 16 ft. lbs. (22 Nm).

17. Install the brake hose mounting bolt.

18. Clean the mating surfaces between the brake disc/drum and the inside of the wheel, then install the rear wheels.

19. Press the brake pedal several times to make sure the brakes work.

➡Engagement may require a greater pedal stroke immediately after the brake pads have been replaced as a set. Several applications of the brake pedal will restore the normal pedal stroke.

20. Add brake fluid as needed.

21. After installation, check for leaks at hose and line joints or connections, and retighten if necessary.

22. Test-drive the vehicle, then recheck for leaks.

BRAKES — PARKING BRAKE

PARKING BRAKE PEDAL

ADJUSTMENT

1. Press the parking brake pedal fully apply the parking brake. The parking brake pedal should be locked within 8–10 clicks.
2. Adjust the parking brake if the pedal clicks are not within the specification.

➡Minor parking brake pedal adjustments (1 to 2 clicks) can be made with the adjusting nut. See "Minor Adjustment". If a larger adjustment is required, follow the "Major Adjustment" procedure using the adjuster nut at the parking brake drum.

➡After installing new parking brake shoes and/or new brake disc/drum, make sure you drive the vehicle for "break-in".

Minor Adjustment

See Figure 16.

1. Raise and support the vehicle.
2. Release the parking brake pedal fully.
3. Press the parking brake pedal 1 click.
4. Tighten the parking brake adjusting nut until the parking brakes drag slightly when the rear wheels are turned.
5. Release the parking brake pedal fully, and check that the parking brakes do not drag when the rear wheels are turned. Readjust if necessary.
6. Make sure the parking brake pedal is within 8–10 clicks.

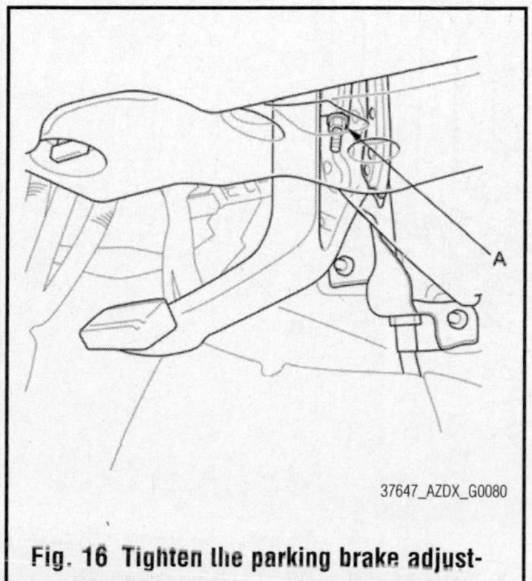
A
37647_AZDX_G0080

Fig. 16 Tighten the parking brake adjusting nut (A) until the parking brakes drag slightly when the rear wheels are turned.

Major Adjustment

➡Perform this adjustment after replacing parking brake shoes and after lining surface break-in.

1. Raise and support the vehicle.
2. Release the parking brake pedal fully.
3. Loosen the parking brake adjusting nut.
4. Remove the rear wheels.
5. Remove the access plug in the outer surface of the drum.
6. Turn the ratchet teeth on the adjuster nut with a brake adjuster tool or a flat-tip screwdriver until the shoes lock against the parking brake drum. Then back off the adjuster 10 clicks, and install the access plug.
7. Clean the mating surfaces between the brake disc/drum and the inside of the wheel, then install the rear wheels.
8. Do the "Minor Adjustment" procedure.

PARKING BRAKE CABLE

REMOVAL & INSTALLATION

➡The parking brake cables must not be bent or distorted. This will lead to stiff operation and premature cable failure.

Front Parking Brake Cable Replacement

See Figures 17 and 18.

1. Release the parking brake pedal.
2. Remove the driver's side front door sill trim and the kick panel.
3. Remove the footrest.
4. Remove the center console.
5. Pull back the carpet from the driver's footwell to access the parking brake cables.
6. Disconnect and remove the return spring.
7. Remove the wire guide base, then disconnect the rear parking brake cables from the equalizer.
8. Remove the parking brake cable from the parking brake pedal assembly.
9. Remove the parking brake cable mounting hardware, then remove the cable.
10. Install the parking brake cable in the reverse order of removal.
11. Adjust the parking brake.
12. Do the "Major Adjustment" first, then do the "Minor Adjustment". Apply the parking brake firmly 10 times then adjust it again.

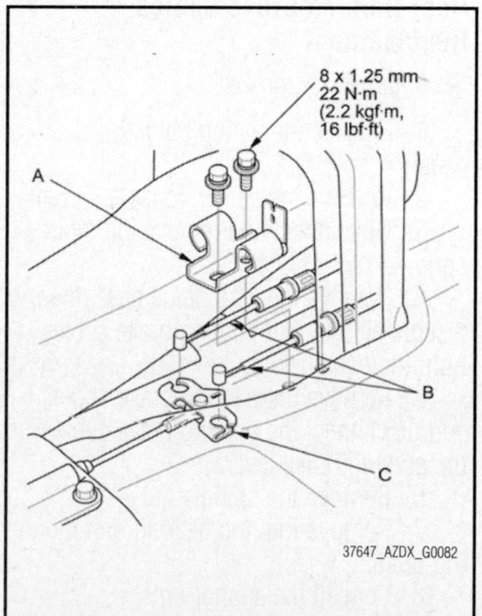
8 x 1.25 mm
22 N·m
(2.2 kgf·m,
16 lbf·ft)
A
B
C
37647_AZDX_G0082

Fig. 17 Remove the wire guide base (A), then disconnect the rear parking brake cables (B) from the equalizer (C).

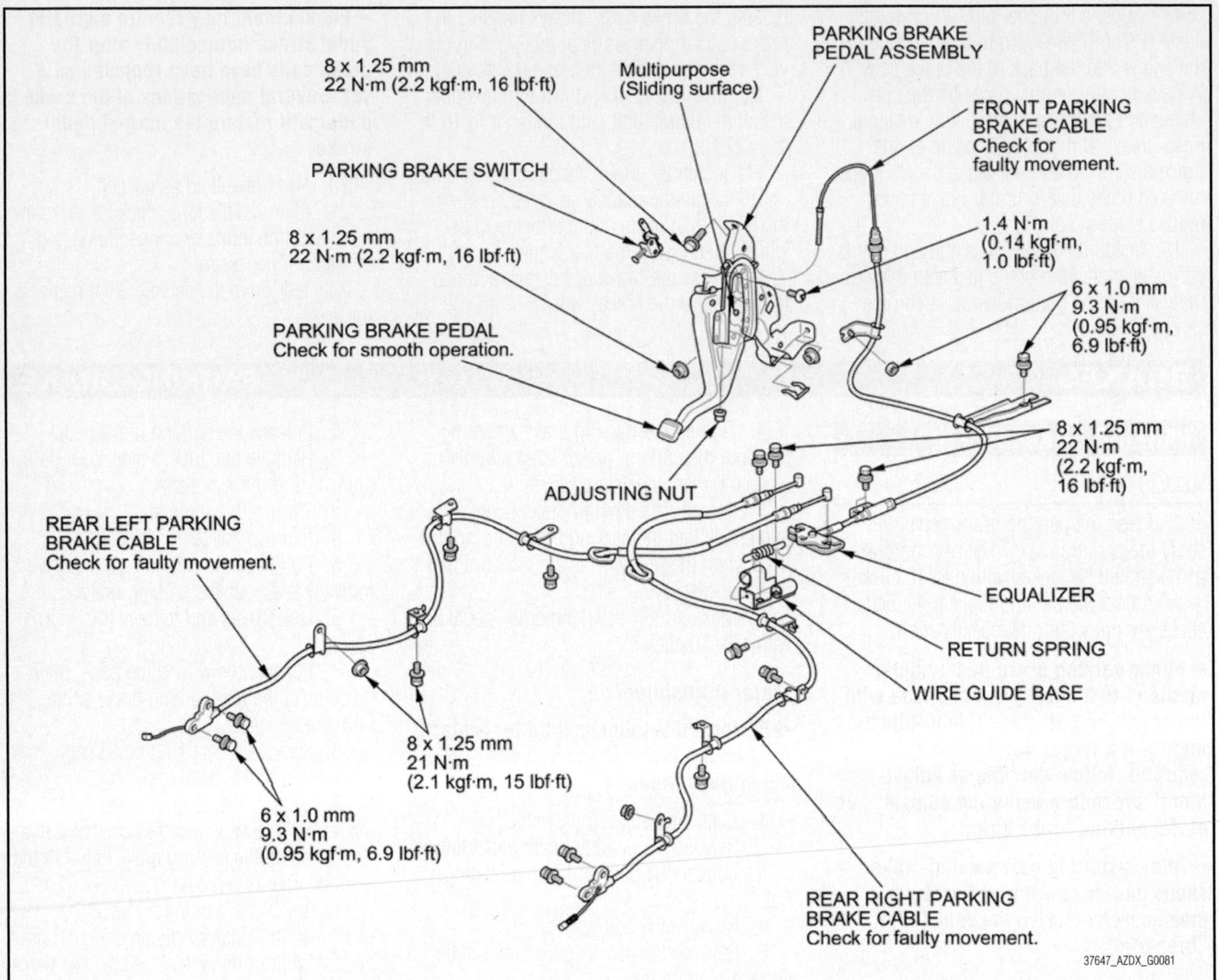
PARKING BRAKE
PEDAL ASSEMBLY
8 x 1.25 mm
22 N·m (2.2 kgf·m, 16 lbf·ft)
Multipurpose
(Sliding surface)
FRONT PARKING
BRAKE CABLE
Check for
faulty movement.
PARKING BRAKE SWITCH
1.4 N·m
(0.14 kgf·m,
1.0 lbf·ft)
8 x 1.25 mm
22 N·m (2.2 kgf·m, 16 lbf·ft)
6 x 1.0 mm
9.3 N·m
(0.95 kgf·m,
6.9 lbf·ft)
PARKING BRAKE PEDAL
Check for smooth operation.
8 x 1.25 mm
22 N·m
(2.2 kgf·m,
16 lbf·ft)
ADJUSTING NUT
REAR LEFT PARKING
BRAKE CABLE
Check for faulty movement.
EQUALIZER
RETURN SPRING
WIRE GUIDE BASE
8 x 1.25 mm
21 N·m
(2.1 kgf·m, 15 lbf·ft)
6 x 1.0 mm
9.3 N·m
(0.95 kgf·m, 6.9 lbf·ft)
REAR RIGHT PARKING
BRAKE CABLE
Check for faulty movement.
37647_AZDX_G0081

Fig. 18 View of the parking brake cable arrangement

Rear Parking Brake Cables Replacement

See Figures 19 and 20.

1. Release the parking brake pedal.
2. Remove the center console.
3. Disconnect and remove the return spring.
4. Remove the wire guide base, then disconnect the rear parking brake cables from the equalizer.
5. Remove the parking brake shoes, and disconnect the parking brake cable from the parking brake lever.
6. Remove the spindle nut.
7. Remove the hub bearing unit mounting bolts.
8. Loosen the anchor nuts.
9. Pull back the hub bearing unit and backing plate until the parking brake cable clears the insertion part of the backing plate.

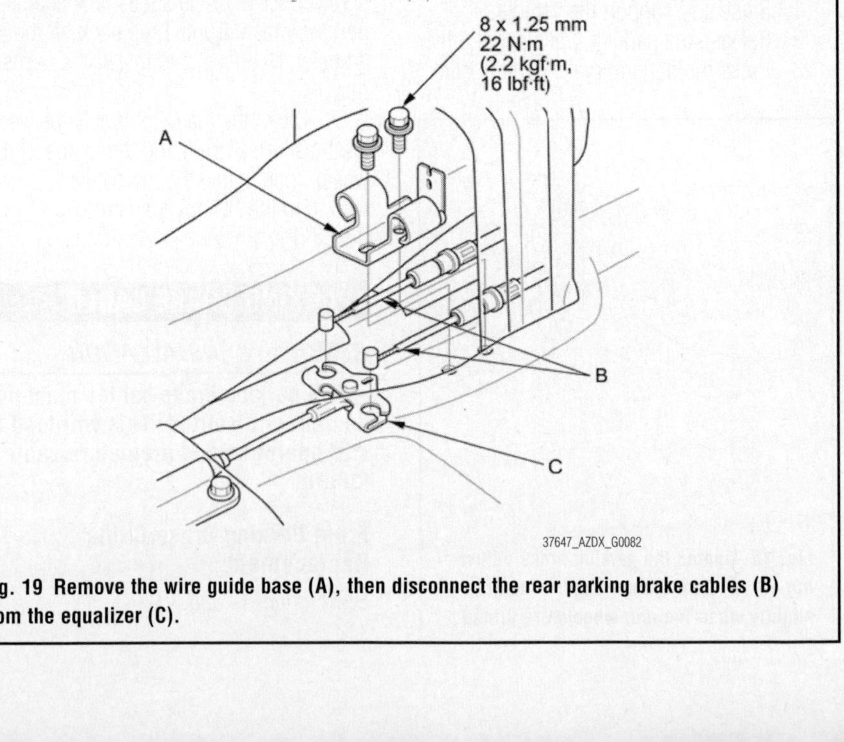
8 x 1.25 mm
22 N·m
(2.2 kgf·m,
16 lbf·ft)
A
B
C
37647_AZDX_G0082

Fig. 19 Remove the wire guide base (A), then disconnect the rear parking brake cables (B) from the equalizer (C).

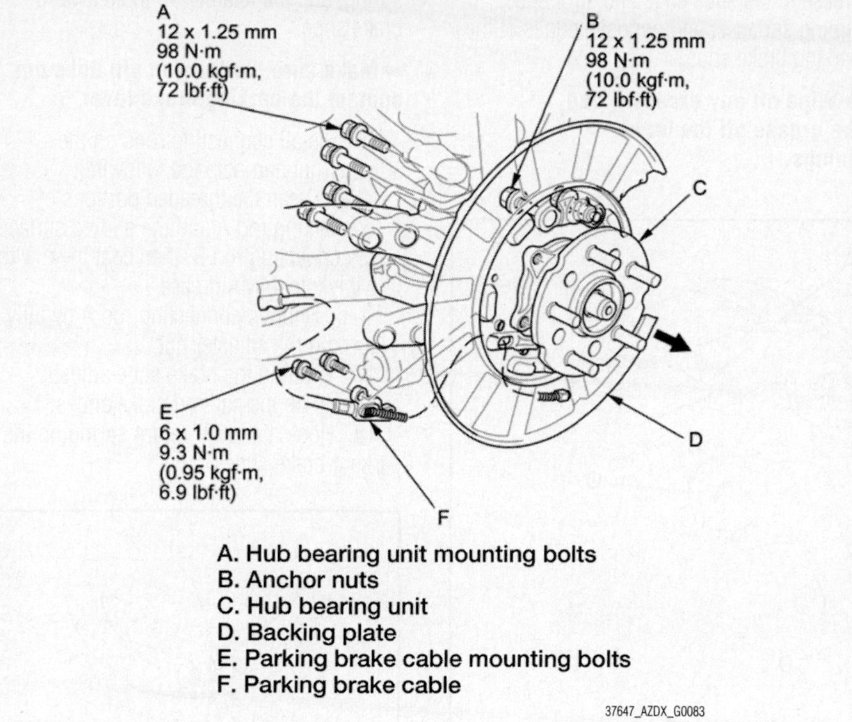
A
12 x 1.25 mm
98 N·m
(10.0 kgf·m,
72 lbf·ft)
B
12 x 1.25 mm
98 N·m
(10.0 kgf·m,
72 lbf·ft)
C
D
E
6 x 1.0 mm
9.3 N·m
(0.95 kgf·m,
6.9 lbf·ft)
F
A. Hub bearing unit mounting bolts
B. Anchor nuts
C. Hub bearing unit
D. Backing plate
E. Parking brake cable mounting bolts
F. Parking brake cable
37647_AZDX_G0083

Fig. 20 Showing the parking brake drum assembly

10. Remove the parking brake cable mounting bolts from the backing plate.
11. Pull the parking brake cable and remove it from the backing plate.
12. Remove the parking brake cable grommets and pull out the parking brake cable from the body floor.
13. Remove the parking brake cable mounting hardware, then remove the cable.
14. Install the parking brake cables in the reverse order of removal.

➡**Be careful not to bend or distort the cable.**

15. Adjust the parking brake. Do the "Major Adjustment" first, apply the parking brake firmly 10 times, then do the "Minor Adjustment".

PARKING BRAKE SHOES

REMOVAL & INSTALLATION

See Figures 21 through 29.

1. Raise and support the vehicle.
2. Remove the rear wheels.
3. Release the parking brake and remove the rear brake disc/drum.

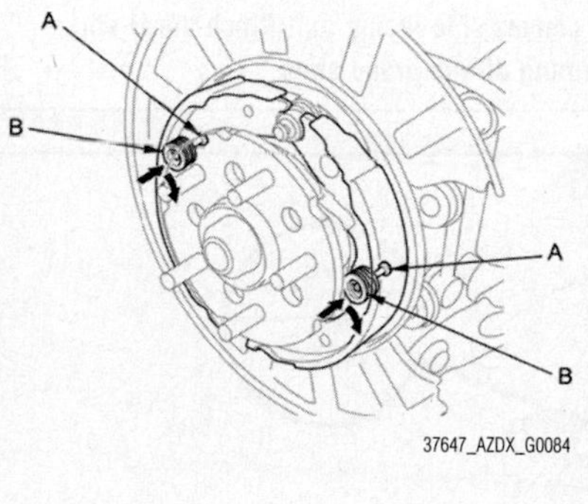
A
B
A
B
37647_AZDX_G0084

Fig. 21 Remove the tension pins (A) by pushing the respective retainers (B) and turning the pin.

4. Disconnect and remove the upper return springs, then remove the shoe guide plate.
5. Remove the tension pins by pushing the respective retainers and turning the pin.
6. Remove the strut and the rod spring.
7. Lower the parking brake shoe assembly.
8. Remove the forward brake shoe and the adjuster assembly by removing the lower return spring.

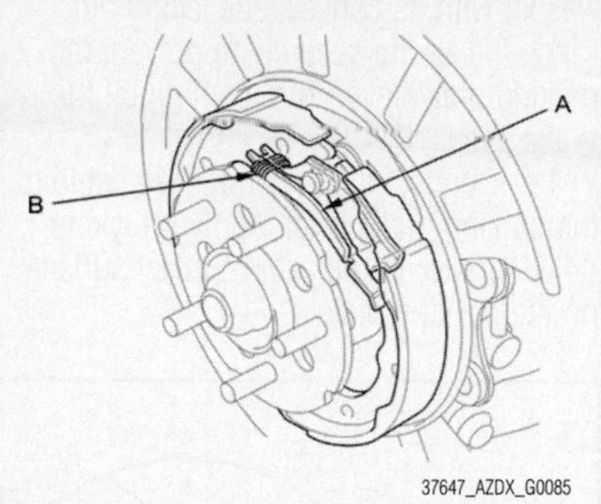
A
B
37647_AZDX_G0085

Fig. 22 Remove the strut (A) and the rod spring (B).

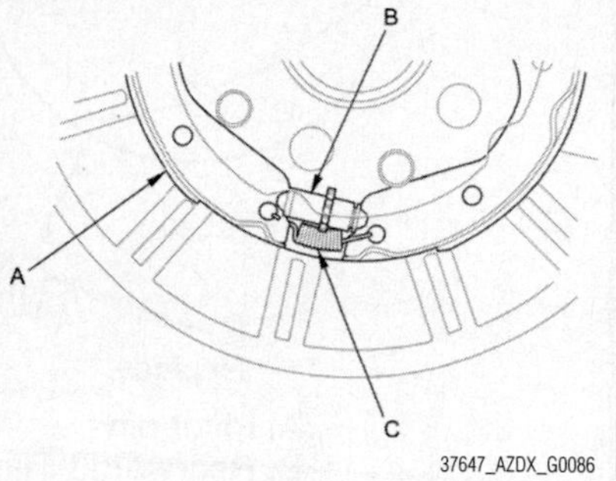
B
A
C
37647_AZDX_G0086

Fig. 23 Remove the forward brake shoe (A) and the adjuster assembly (B) by removing the lower return spring (C).

9. Remove the rearward brake shoe by disconnecting the parking brake cable from the parking brake lever.
10. Remove the U-clip, the wave washer, and the parking brake lever from the brake shoe.

To install:

11. Apply Molykote 44MA grease to the sliding surface of the pivot pin of the rearward brake shoe.
12. Install the parking brake lever and the wave washer on the pivot pin, and secure with a new U-clip. Install the wave

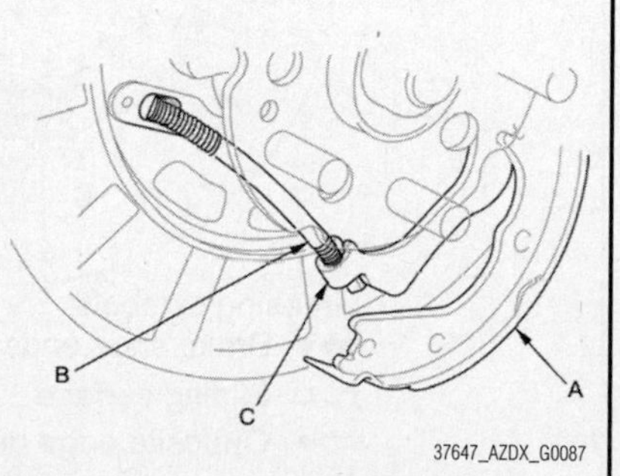
B
C
A
37647_AZDX_G0087

Fig. 24 Remove the rearward brake shoe (A) by disconnecting the parking brake cable (B) from the parking brake lever (C).

washer with its convex side facing out. Pinch the U-clip securely to prevent the parking brake lever from coming off the brake shoe.

13. Connect the parking brake cable to the parking brake lever. Apply Molykote 44MA grease to the cable contact surface on the backing plate.

14. Apply a thin coat of Molykote 44MA grease to the shoe ends and strut ends, sliding surfaces, and opposite edges of the parking brake shoe.

➡**Wipe off any excess. Keep the grease off the brake linings.**

15. Install the tension pin, the retainer spring, and the retainer on the rearward brake shoe.

➡**Make sure the tension pin does not contact the parking brake lever.**

16. Install connecting rods on the adjuster nut and note the following:

a. Clean the threaded portions of connecting rod A and the sliding surface of connecting rod B, then coat them with Molykote 44MA grease.

b. Shorten connecting rod A by fully turning the adjuster nut.

17. Position the brake shoe adjuster assembly on the parking brake shoes.

18. Hook the lower return spring on the parking brake shoes.

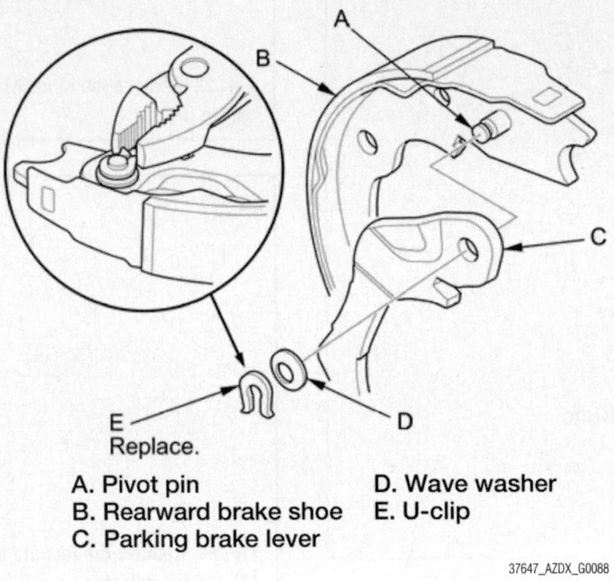
A
B
C
D
E
Replace.
A. Pivot pin
B. Rearward brake shoe
C. Parking brake lever
D. Wave washer
E. U-clip
37647_AZDX_G0088

Fig. 25 Apply Molykote 44MA grease to the sliding surface of the pivot pin of the rearward brake shoe. Install the parking brake lever and the wave washer on the pivot pin, and secure with a new U-clip. Install the wave washer with its convex side facing out. Pinch the U-clip securely to prevent the parking brake lever from coming off the brake shoe.

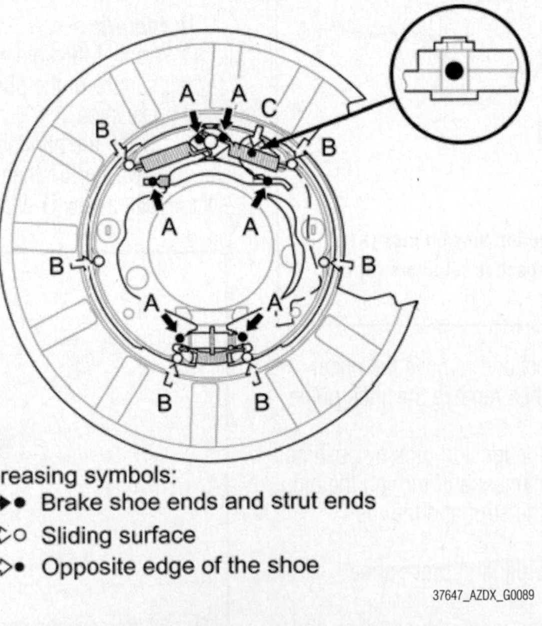
A
C
B
Greasing symbols:
Brake shoe ends and strut ends
Sliding surface
Opposite edge of the shoe
37647_AZDX_G0089

Fig. 26 Apply a thin coat of Molykote 44MA grease to the shoe ends and strut ends (A), sliding surfaces (B), and opposite edges of the parking brake shoe (C) as shown.

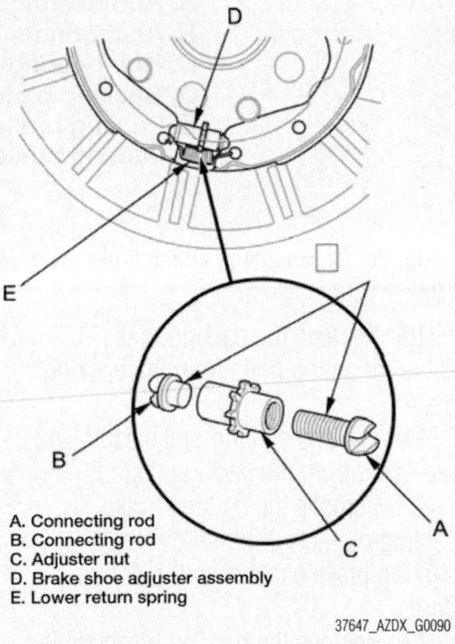
D
E
B
C
A
A. Connecting rod
B. Connecting rod
C. Adjuster nut
D. Brake shoe adjuster assembly
E. Lower return spring
37647_AZDX_G0090

Fig. 27 Installing the connecting rods on the adjuster

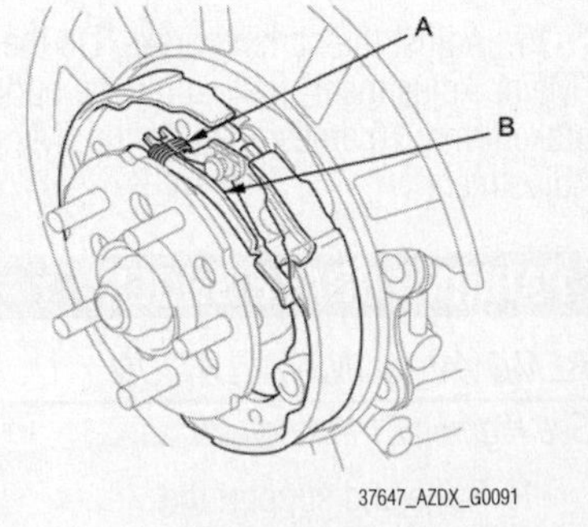
A
B
37647_AZDX_G0091

Fig. 28 Install the rod spring (A) to the strut (B) first. Then install the strut on the parking brake shoes.

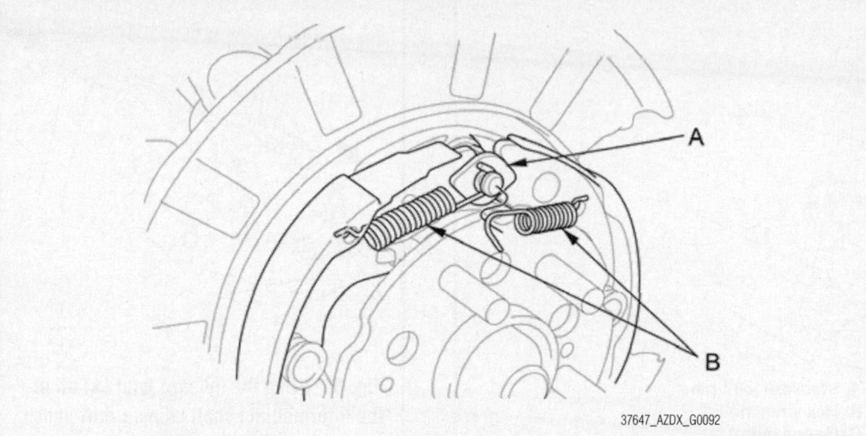
A
B
37647_AZDX_G0092
Fig. 29 Install the shoe guide plate (A), then install the upper return springs (B) as shown.

19. Install the rod spring to the strut first. Then install the strut on the parking brake shoes.
20. Install the tension pin, the retainer spring, and the retainer on the forward brake shoe.
21. Install the shoe guide plate, then install the upper return springs.
22. Install the rear brake disc/drum and the rear brake caliper bracket.
23. Do the parking brake "Major Adjustment".
24. Clean the mating surfaces between the brake/drum and the inside of the wheel, then install the rear wheels.

CHASSIS ELECTRICAL

AIR BAG (SUPPLEMENTAL RESTRAINT SYSTEM)

GENERAL INFORMATION

CAUTION

These vehicles are equipped with an air bag system. The system must be disarmed before performing service on, or around, system components, the steering column, instrument panel components, wiring and sensors. Failure to follow the safety precautions and the disarming procedure could result in accidental air bag deployment, possible injury and unnecessary system repairs.

PRECAUTIONS

Disconnect and isolate the battery negative cable before beginning any air bag system component diagnosis, testing, removal, or installation procedures. Allow system capacitor to discharge for two minutes before beginning any component service. This will disable the air bag system. Failure to disable the air bag system may result in accidental air bag deployment, personal injury, or death.

DISARMING THE SYSTEM

Disconnect and isolate the negative battery cable. Wait 3 minutes for the system capacitor to discharge before performing any service.

ARMING THE SYSTEM

1. After performing service, connect the negative battery cable to re-arm the SRS system.

CLOCKSPRING CENTERING

1. First rotate the cable reel clockwise until it stops.
2. Then rotate it counterclockwise (three full turns) until the arrow mark on the cable reel label points straight up.
3. Position the two tabs of the turn signal canceling sleeve as shown, and install the steering wheel on to the steering column shaft, making sure the steering wheel hub engages the pins of the cable reel and tabs of the turn signal canceling sleeve. Do not tap on the steering wheel or steering column shaft when installing the steering wheel.

DRIVE TRAIN

FRONT DRIVESHAFT

REMOVAL & INSTALLATION

See Figures 30 through 39.

1. Raise and support the vehicle.
2. Remove the front wheels.
3. Pry up the stake on the spindle nut, then remove the nut.
4. Drain the transmission fluid, then reinstall the drain plug with a new sealing washer:
5. Hold the stabilizer joint pin using a hex wrench, and remove the flange nut. Separate the front stabilizer link from the lower arm.
6. Remove the damper fork mounting nut, the damper fork mounting bolt, and the damper pinch bolt, then remove the damper fork.
7. Remove the cotter pin from the knuckle ball joint, then remove the castle nut. Separate the ball joint from the lower arm using the ball joint remover.

➡Be careful not to damage the ball joint boot when installing the remover. Do not force or hammer on the lower arm, or pry between the lower arm and the knuckle. You could damage the ball joint.

8. Pull the knuckle outward, and separate the outboard joint from the front hub using a soft face hammer.
9. Remove exhaust pipe A.
10. Left driveshaft: Pry the inboard joint from the differential using a prybar.
11. Remove the driveshaft as an assembly.

➡Do not pull on the driveshaft, or the inboard joint may come apart. Pull the inboard joint straight out to avoid damaging the oil seal. Be careful not to damage the oil seal or the end of the inboard joint using the prybar.

12. Right driveshaft: Drive the inboard joint off of the intermediate shaft using a drift punch and a hammer.
13. Remove the driveshaft as an assembly.

➡Do not pull on the driveshaft, or the inboard joint may come apart.

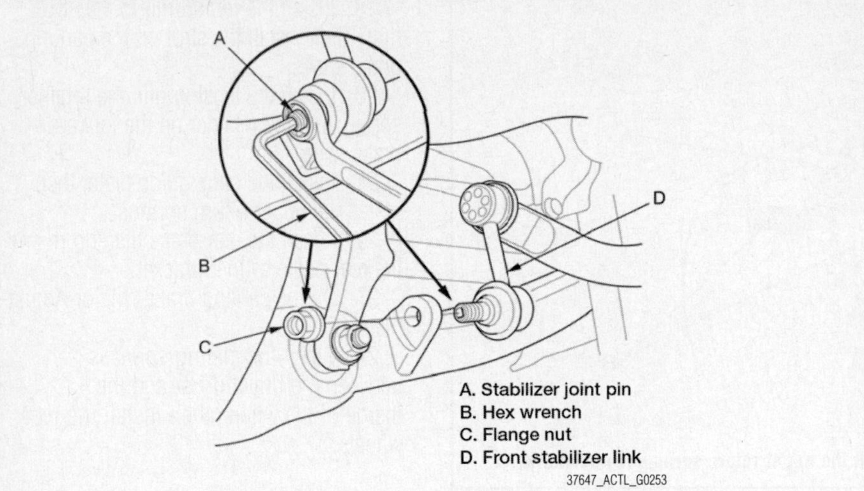
A
B
C
D
A. Stabilizer joint pin
B. Hex wrench
C. Flange nut
D. Front stabilizer link
37647_ACTL_G0253

Fig. 30 Hold the stabilizer joint pin using a hex wrench, and remove the flange nut. Separate the front stabilizer link from the lower arm.

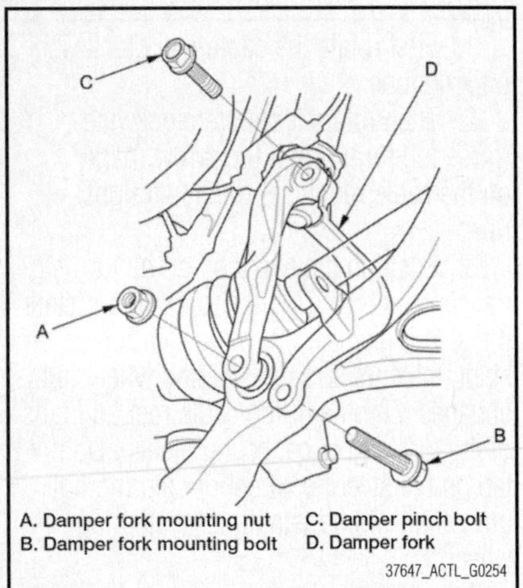
C
D
A
B
A. Damper fork mounting nut
B. Damper fork mounting bolt
C. Damper pinch bolt
D. Damper fork
37647_ACTL_G0254

Fig. 31 Remove the damper fork mounting nut, the damper fork mounting bolt, and the damper pinch bolt, then remove the damper fork

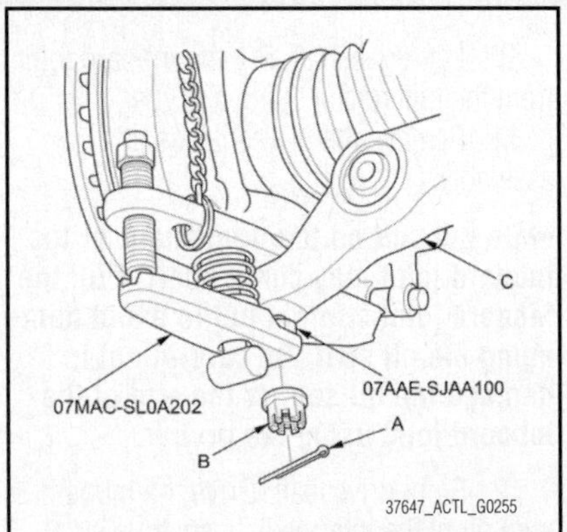
07MAC-SL0A202
07AAE-SJAA100
A
B
C
37647_ACTL_G0255

Fig. 32 Remove the cotter pin (A) from the knuckle ball joint, then remove the castle nut (B). Separate the ball joint from the lower arm (C) using the ball joint remover.

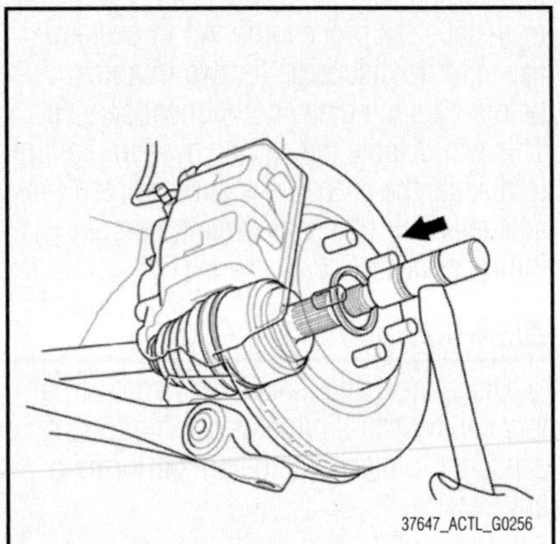
37647_ACTL_G0256

Fig. 33 Pull the knuckle outward, and separate the outboard joint from the front hub using a soft face hammer

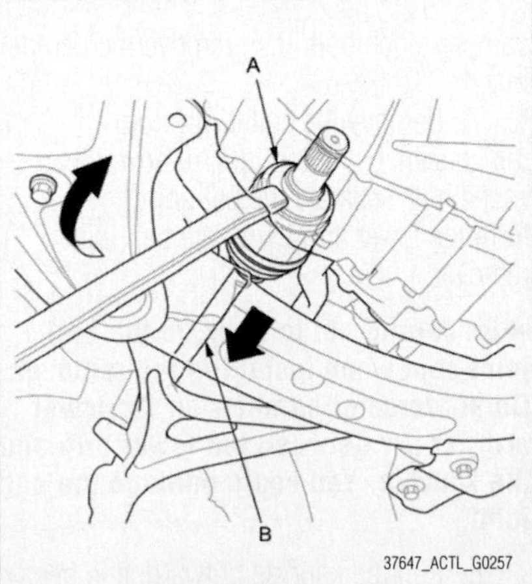
A
B
37647_ACTL_G0257

Fig. 34 Pry the inboard joint (A) from the differential using a prybar

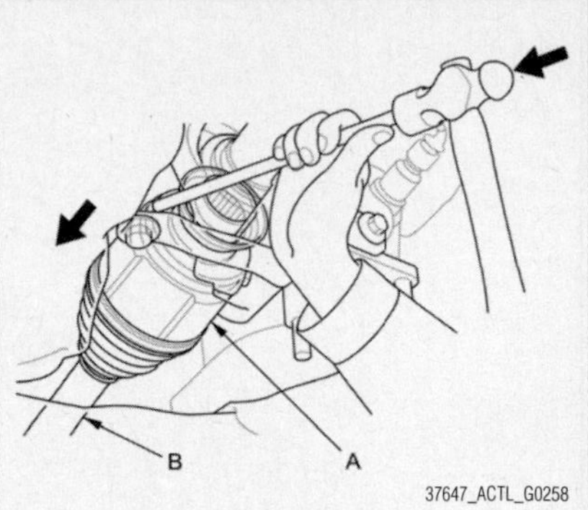
B
A
37647_ACTL_G0258

Fig. 35 Drive the inboard joint (A) off of the intermediate shaft using a drift punch and a hammer

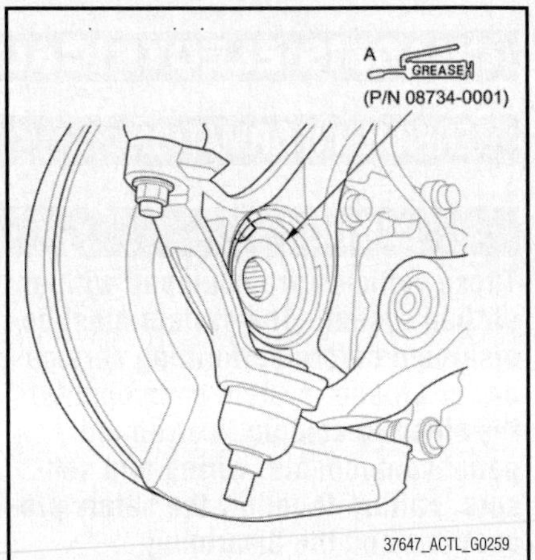
A
GREASE
(P/N 08734-0001)
37647_ACTL_G0259

Fig. 36 Apply Moly 60 paste to the contact area (A) of the outboard joint and the front wheel bearing

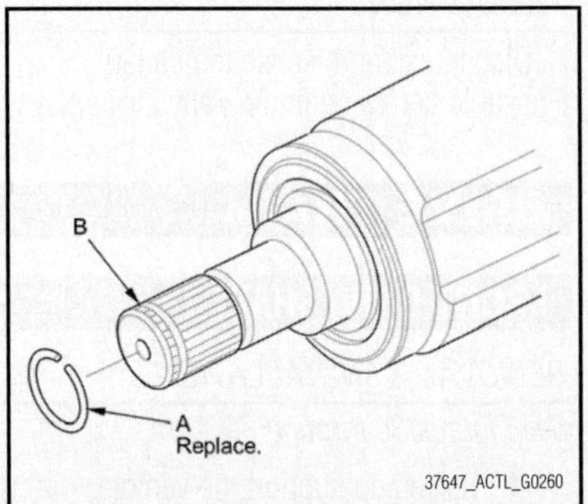
B
A
Replace.
37647_ACTL_G0260

Fig. 37 Install a new set ring (A) into the set ring groove (B) of the left driveshaft inboard joint

To install:

➡Before starting installation, make sure the mating surfaces of the joint and the splined section are clean.

14. Apply Moly 60 paste (P/N 08734-0001) to the contact area of the outboard joint and the front wheel bearing.

➡**The paste helps prevent noise and vibration.**

15. Install a new set ring into the set ring groove of the left driveshaft inboard joint.
16. Install a new set ring into the set ring groove of the intermediate shaft.
17. Apply small quantity of super high temperature grease (P/N 08798-9002) to the whole splined surface of the right driveshaft.
18. After applying grease, remove the grease from the splined grooves at intervals of 2–3 splines and from the set ring groove so that air can bleed from the intermediate shaft.

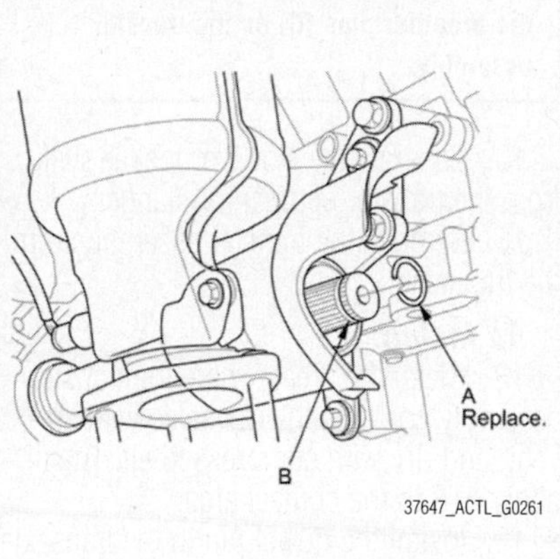
A
Replace.
B
37647_ACTL_G0261

Fig. 38 Install a new set ring (A) into the set ring groove (B) of the intermediate shaft

19. Clean the areas where the driveshaft contacts the differential thoroughly with solvent, and dry them with compressed air.

➡**Do not wash the rubber parts with solvent.**

20. Insert the inboard end of the driveshaft into the differential or the intermediate shaft until the set ring locks in the groove.

➡**Insert the driveshaft horizontally to prevent damaging the oil seal.**

21. Install the outboard joint into the front hub on the knuckle.
22. Install exhaust pipe A.
23. Wipe off any grease contamination from the ball joint tapered section and threads, then install the knuckle onto the lower arm. Be careful not to damage the ball joint boot. Wipe off the grease before tightening the nut at the ball joint. Torque the castle nut to 69–76 ft. lbs. (93–103 Nm), then tighten it only far enough to align the slot with the ball joint pin hole.

➡**Make sure the ball joint boot is not damaged or cracked. Do not align the nut by loosening it.**

24. Install a new cotter pin into the ball joint pin hole, and bend the cotter pin.
25. Install the damper fork over the driveshaft and onto the lower arm. Install the damper in the damper fork so the aligning tab is aligned with the slot in the damper fork. Loosely install the damper pinch bolt.
26. Loosely install a new flange bolt and a new damper fork mounting nut.
27. Connect the front stabilizer link to the lower arm, and loosely install the flange nut. Hold the stabilizer link ball joint pin using a hex wrench, and tighten the flange nut to 58 ft. lbs. (78 Nm).
28. Place a floor jack under the lower arm, and raise the suspension to load it with the vehicle's weight.

➡**Do not put the floor jack under the ball joint.**

29. Tighten the damper pinch bolt to 36 ft. lbs. (49 Nm) and the damper fork mounting nut to 47 ft. lbs. (64 Nm)while holding the flange bolt, then remove the floor jack.
30. Apply a small amount of engine oil to the seating surface of a new spindle nut.
31. Install the spindle nut, then tighten it to 242 ft. lbs. (329 Nm). After tightening, use a drift to stake the spindle nut shoulder against the driveshaft.
32. Clean the mating surfaces of the brake disc and the wheel, then install the front wheels.
33. Turn the front wheel by hand, and make sure there is no interference between the driveshaft and the surrounding parts.
34. Lower the vehicle .
35. Refill the transmission with recommended transmission fluid:
36. Check the wheel alignment, and adjust it if necessary.

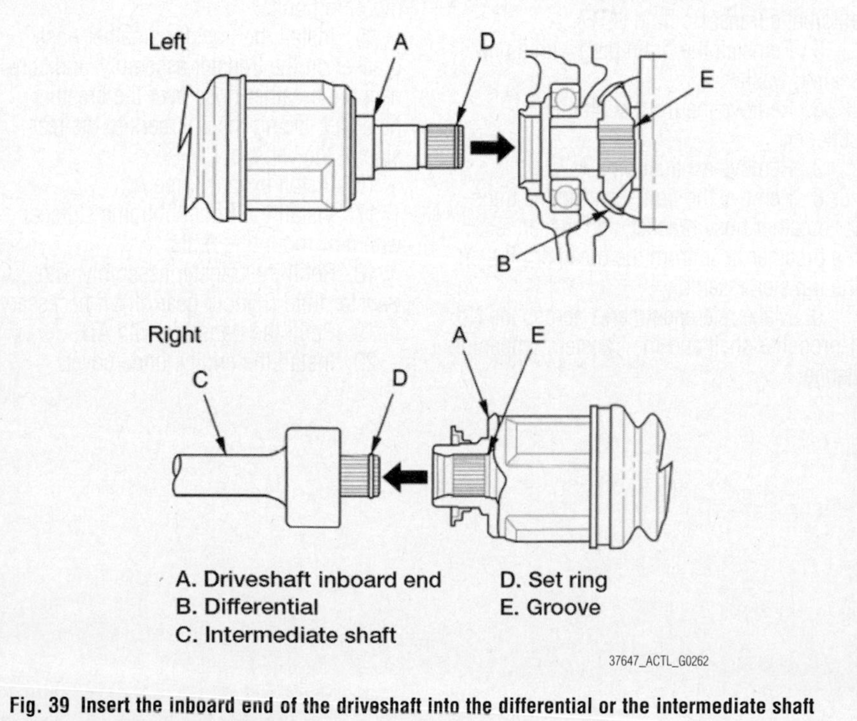
Left
A
D
E
B
Right
A
E
C
D
A. Driveshaft inboard end
B. Differential
C. Intermediate shaft
D. Set ring
E. Groove
37647_ACTL_G0262

Fig. 39 Insert the inboard end of the driveshaft into the differential or the intermediate shaft until the set ring locks in the groove

REAR DRIVESHAFT

REMOVAL & INSTALLATION

See Figures 40 through 42.

1. Raise the vehicle on a lift.
2. Drain the rear differential fluid.

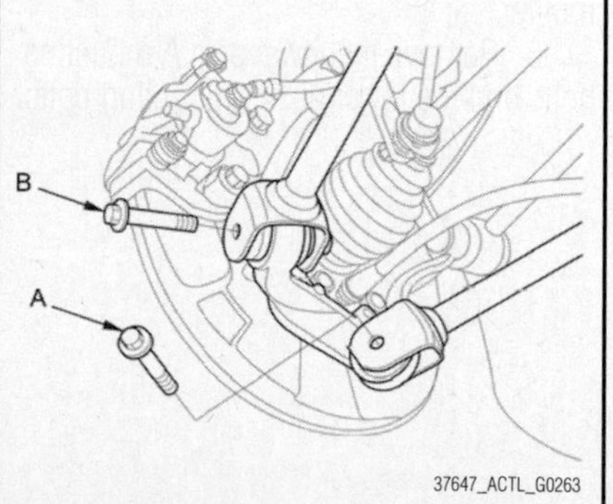
B
A
37647_ACTL_G0263

Fig. 40 Remove the lower arm A mounting bolts (A) and the lower arm B mounting bolts (B)

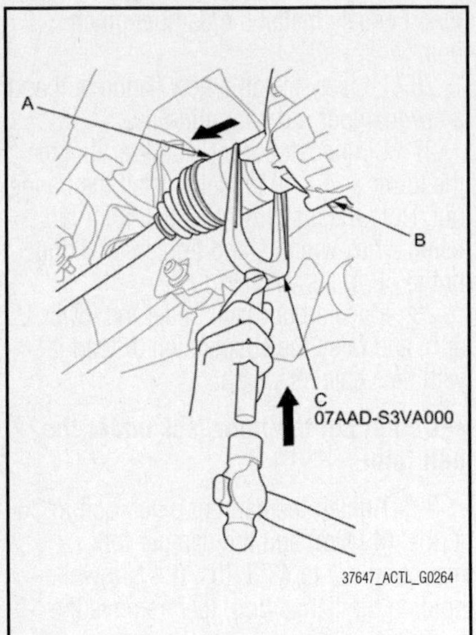
A
B
C
07AAD-S3VA000
37647_ACTL_G0264

Fig. 41 Disconnect the inboard joints (A) from the rear differential (B) using the driveshaft remover (C) and a hammer

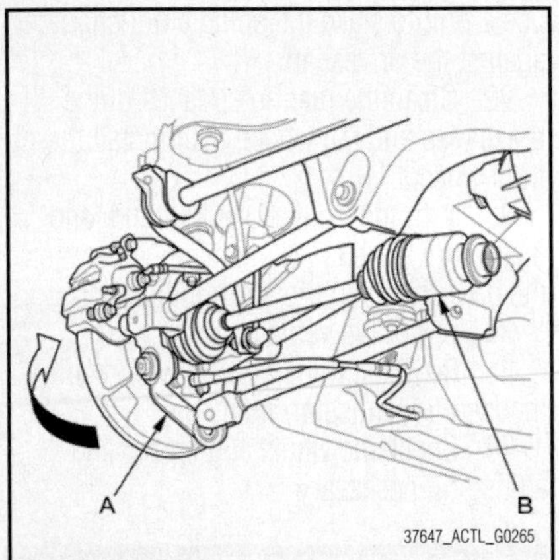
A
B
37647_ACTL_G0265

Fig. 42 Pull the knuckle (A) outward, and separate the rear driveshaft inboard joints (B) from the rear differential

3. Remove the rear wheels.
4. Remove exhaust pipe B and the muffler.
5. Remove the lower arm A mounting bolts and the lower arm B mounting bolts.
6. Disconnect the inboard joints from the rear differential using the driveshaft remover and a hammer.
7. Pull the knuckle outward, and separate the rear driveshaft inboard joints from the rear differential.

To install:

8. Pull the knuckle outward, and insert the rear driveshaft inboard joints into the rear differential.
9. Loosely install new lower arm A mounting bolts and new lower arm B mounting bolts.
10. Place a floor jack under the lower arm, and raise the suspension to load it with the vehicle's weight.
11. Tighten the lower arm A mounting bolt and the lower arm B mounting bolt to 43 ft. lbs. (59 Nm).
12. Install exhaust pipe B and the muffler.
13. Install the rear wheels.
14. Refill the rear differential with the recommended fluid.
15. Test-drive the vehicle.

TRANSFER ASSEMBLY

REMOVAL & INSTALLATION

See Figure 43.

1. Raise the vehicle on a lift, and make sure it is supported securely.
2. Shift the transaxle into N.
3. Remove the engine undercover.
4. Remove the drain plug and drain the automatic transaxle fluid (ATF).
5. Reinstall the drain plug with a new sealing washer.
6. Remove the front subframe stiffener.
7. Remove exhaust pipe A.
8. Remove the bolt securing the transfer breather hose bracket, and disconnect the breather hose from the breather pipe on the transfer assembly.
9. Make reference marks across the No. 1 propeller shaft and the transfer companion flange.

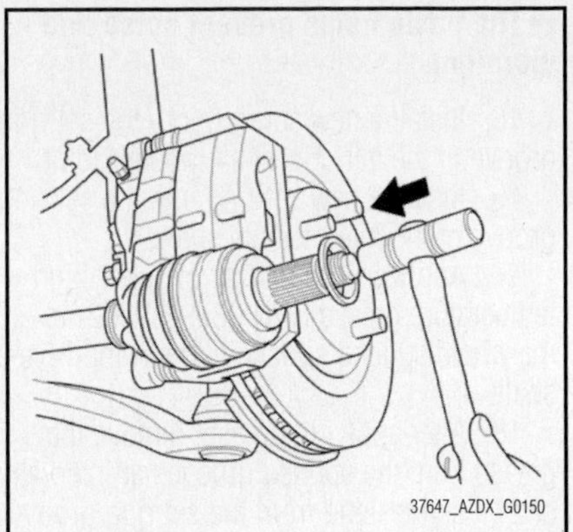
37647_AZDX_G0150

Fig. 43 Remove the bolt securing the transfer breather hose bracket (A), and disconnect the breather hose (B) from the breather pipe (C) on the transfer assembly.

10. Separate the No. 1 propeller shaft from the transfer companion flange.
11. Remove the transfer assembly from the transaxle.

To install:

12. Clean the areas where the transfer assembly contacts the transaxle with solvent, and dry with compressed air. Then apply ATF to the contact area.
13. Install the dowel pin in the transaxle, and install the transfer assembly on the transaxle.
14. Install the No. 1 propeller shaft to the transfer companion flange by aligning the reference marks. Make sure you use new mounting bolts.
15. Install the transfer breather hose bracket on the transfer assembly, and connect the breather hose over the breather pipe with facing the dot mark to the rear side of the vehicle.
16. Install exhaust pipe A.
17. Install the front subframe stiffener with new mounting bolts.
18. Refill the transfer assembly with transfer fluid (hypoid gear oil), if necessary.
19. Refill the transaxle with ATF.
20. Install the engine undercover.

ENGINE COOLING

ENGINE COOLING FANS

REMOVAL & INSTALLATION

A/C Condenser Fan Assembly

See Figures 44 and 45.

1. Disconnect the A/C condenser fan motor connector and remove the harness clamps.
2. Remove the coolant reservoir.
3. Unclamp the ATF cooler hose clamps.
4. Loosen the two bolts, then remove the A/C condenser fan shroud assembly.
5. Disassemble the A/C condenser fan shroud assembly.

To install:

6. Reassemble the A/C condenser fan shroud assembly.

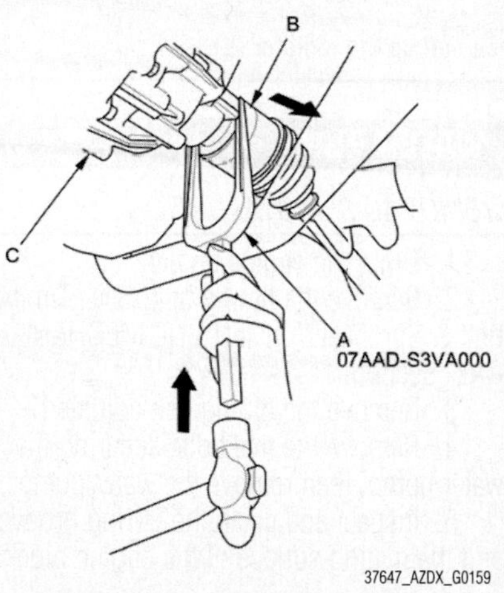
B
C
A
07AAD-S3VA000
37647_AZDX_G0159

Fig. 44 Disconnect the A/C condenser fan motor connector (A), and remove the harness clamps (B). Remove the coolant reservoir (C).

7. Install the A/C condenser fan shroud assembly, then tighten the bolts.
8. Clamp the ATF cooler hose clamps.
9. Install the coolant reservoir.
10. Connect the A/C condenser fan motor connector, and install the harness clamps.

Radiator Fan Assembly

See Figures 46 through 48.

1. Drain the engine coolant.
2. Remove the bulkhead cover.
3. Remove the battery.
4. Remove the battery base.
5. Disconnect the radiator fan motor connector, the engine coolant temperature (ECT) sensor 2 sub harness connector, and the ECT sensor 2 connector, then remove the harness clamp.
6. Disconnect the upper radiator hose.
7. Remove the radiator fan shroud assembly.
8. Remove the harness clamp and the wire harness.
9. Disassemble the radiator fan shroud assembly.

To install:

10. Reassemble the radiator fan shroud assembly.
11. Install the wire harness and the harness clamp.
12. Install the radiator fan shroud assembly, then tighten the bolts.
13. Connect the upper radiator hose.
14. Install the harness clamp, then connect the ECT sensor 2 connector, the ECT sensor 2 sub harness connector, and the radiator fan motor connector.
15. Install the battery base.
16. Do the battery installation procedure.

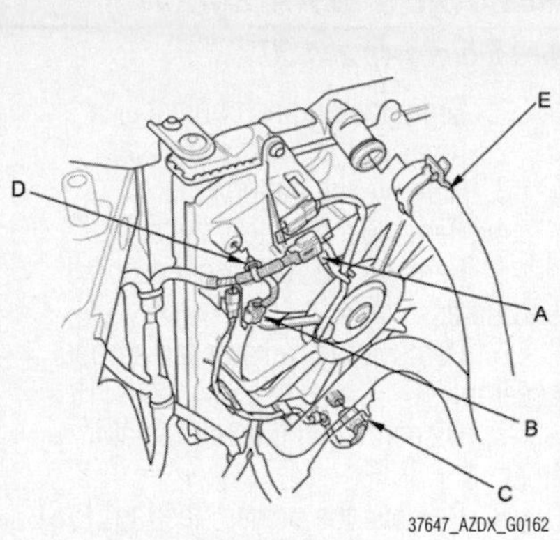
E
D
A
B
C
37647_AZDX_G0162

Fig. 46 Detaching connectors, clamps and radiator hose

17. Install the bulkhead cover.
18. Fill the radiator with engine coolant, and bleed the air from the cooling system.
19. Clean up any spilled engine coolant.

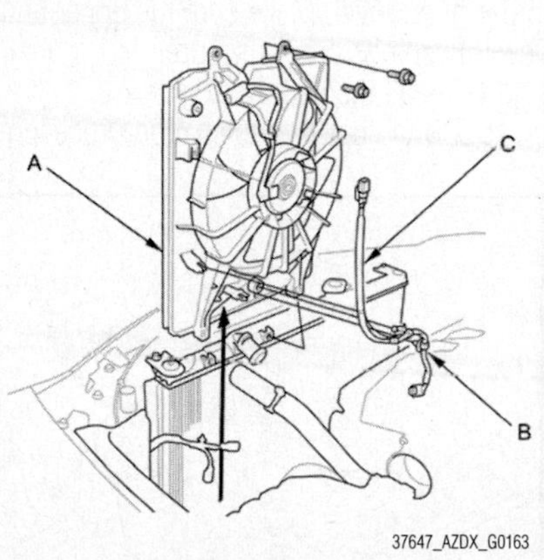
A
C
B
37647_AZDX_G0163

Fig. 47 Remove the radiator fan shroud assembly (A). Remove the harness clamp (B) and the wire harness (C).

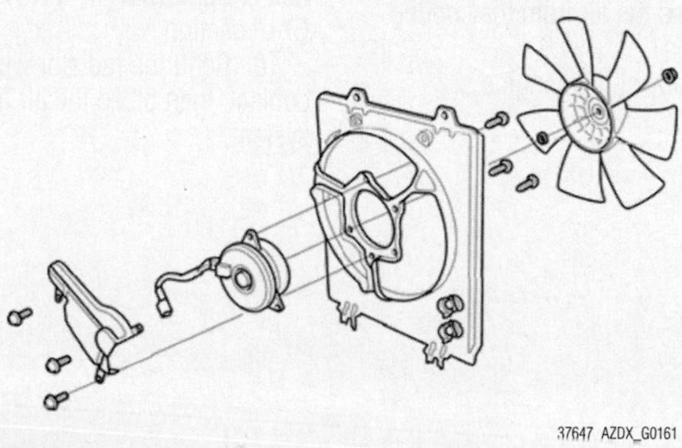
37647_AZDX_G0161

Fig. 45 Disassemble the A/C condenser fan shroud assembly.

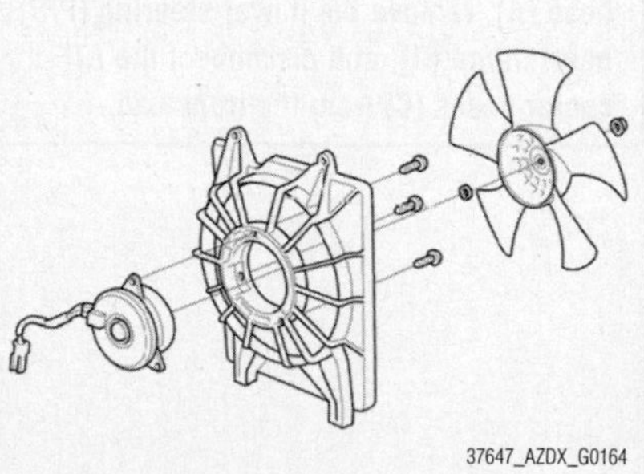
37647_AZDX_G0164

Fig. 48 Disassemble the radiator fan shroud assembly.

RADIATOR

REMOVAL & INSTALLATION

See Figures 49 and 50.

1. Drain the engine coolant.
2. Remove the bulkhead cover.
3. Remove the battery.
4. Remove the battery base.
5. Remove the A/C condenser fan shroud assembly..
6. Remove the radiator fan shroud assembly.
7. Disconnect the lower radiator hose.
8. Remove the power steering (P/S) hose clamp.
9. Disconnect the ATF cooler hoses from the transaxle, then plug the ATF cooler hoses and the lines.
10. Disconnect the ATF cooler hoses, then plug the ATF cooler hoses and the lines.
11. Remove the radiator upper brackets, then pull up the radiator.
12. Remove the other parts from the radiator.
13. Install the radiator in the reverse order of removal. Make sure the upper and lower cushions are set securely.
14. Install the battery base.
15. Do the battery installation procedure.

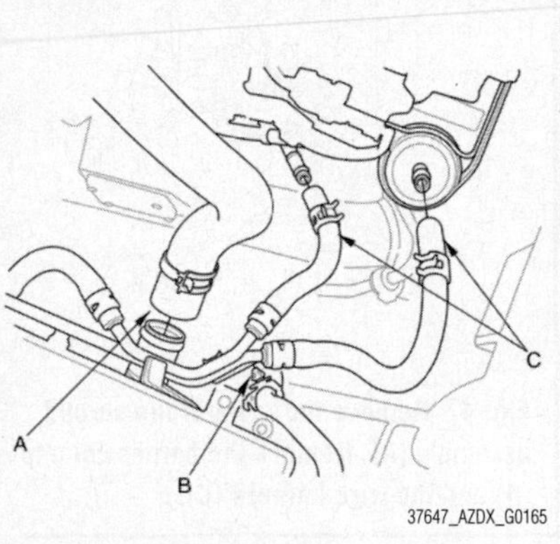
A
B
C
37647_AZDX_G0165

Fig. 49 Disconnect the lower radiator hose (A), remove the power steering (P/S) hose clamp (B), and disconnect the ATF cooler hoses (C) from the transaxle.

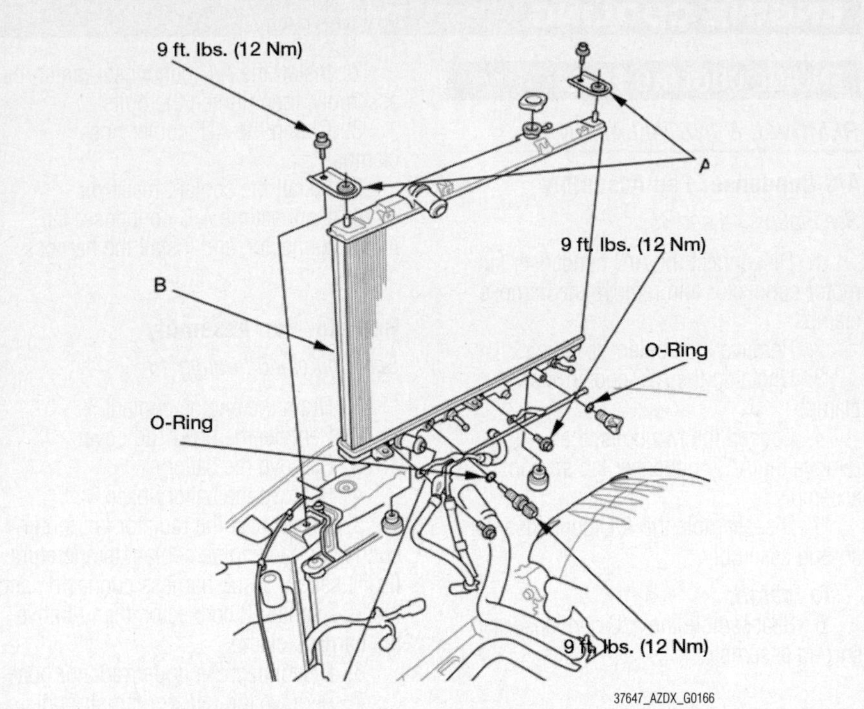
9 ft. lbs. (12 Nm)
A
9 ft. lbs. (12 Nm)
B
O-Ring
O-Ring
9 ft. lbs. (12 Nm)
37647_AZDX_G0166

Fig. 50 Remove the radiator upper brackets (A), then pull up the radiator (B).

16. Refill the transaxle with ATF.
17. Fill the radiator with engine coolant, and bleed the air from the cooling system.
18. Clean up any spilled engine coolant.

THERMOSTAT

REMOVAL & INSTALLATION

1. Drain the engine coolant.
2. Remove the intake air duct.
3. Remove the thermostat cover, then remove the thermostat.
4. Install the new thermostat with a new rubber seal, then install the thermostat cover.
5. Install the intake air duct.
6. Refill the radiator with engine coolant, and bleed the air from the cooling system.
7. Clean up any spilled engine coolant.

WATER PUMP

REMOVAL & INSTALLATION

1. Drain the engine coolant.
2. Remove the timing belt. See "Timing Belt & Sprockets" in "ENGINE MECHANICAL" section.
3. Remove the timing belt adjuster.
4. Remove the five bolts securing the water pump, then remove the water pump.
5. Inspect and clean the O-ring groove and the mating surface of the engine block.

To install:

6. Install the water pump with a new O-ring.
7. Clean up any spilled engine coolant.
8. Install the timing belt adjuster.
9. Install the timing belt. See "Timing Belt & Sprockets" in "ENGINE MECHANICAL" section.
10. Refill the radiator with engine coolant, then bleed the air from the cooling system.

ALTERNATOR

REMOVAL & INSTALLATION

See Figure 51.

1. Disconnect the battery terminals.
2. Remove the A/C condenser fan shroud assembly. See "COOLING SYSTEM" section.
3. Remove the drive belt.
4. Disconnect the alternator connector and the positive alternator cable from the alternator.
5. Disconnect the A/C compressor clutch connector from the A/C compressor.
6. Remove the bolt securing the harness holder.
7. Remove the mounting bolt and the alternator bracket mounting bolt, then remove the alternator.

To install:

8. Install the alternator, then tighten the mounting bolt to 35 ft. lbs. (45 Nm) and the alternator bracket mounting bolt to 16 ft. lbs. (22 Nm).
9. Install the bolt securing the harness holder.
10. Connect the A/C compressor clutch connector to the A/C compressor.
11. Connect the alternator connector and the positive alternator cable to the alternator. Make sure the crimped side of the ring terminal faces away from the alternator when you connect it.
12. Install the drive belt.
13. Install the A/C condenser fan shroud assembly.
14. Reconnect the battery terminals.

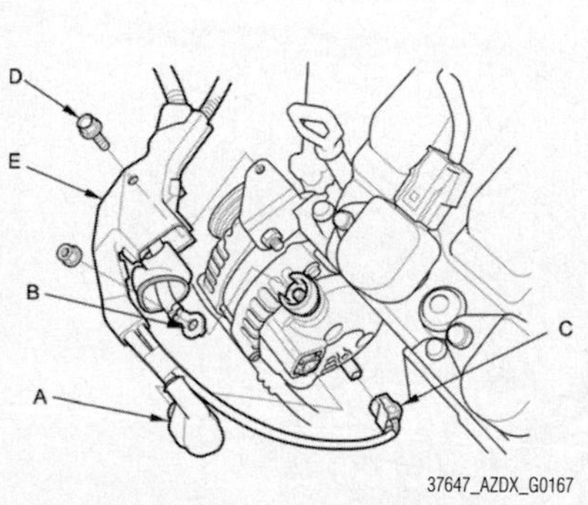
D
E
B
C
A
37647_AZDX_G0167

Fig. 51 Disconnecting wiring from the alternator

BATTERY DISCONNECT/RECONNECT PROCEDURE

Disconnection

➡Some systems store data in memory (including seat position, mirror position, etc) that is lost when the battery is disconnected. Do the following procedures before disconnecting the battery. Make sure you have the anti-theft code(s) for the audio system and/or the audio-navigation system (if equipped). If you are replacing the audio unit and/or the audio-navigation unit, write down the audio presets (AM and FM), and the XM audio presets (if equipped).

1. Make sure the ignition switch is in LOCK (0), or the vehicle ignition in the OFF mode.
2. Disconnect and isolate the negative cable from the battery.

➡Always disconnect the negative cable from the battery first.

3. Disconnect the positive cable from the battery.

Reconnection

➡Some systems store data in memory (including seat position, mirror position, etc) that is lost when the battery is disconnected. Do the following procedures to restore the systems back to normal operation.

1. Clean the battery terminals.
2. Test the battery.
3. Reconnect the positive cable to the battery first, then reconnect the negative cable to the battery.

➡Always connect the positive cable to the battery first.

4. Apply multipurpose grease to the terminals to prevent corrosion.
5. Do the "Steering Column Position Memorization".
6. Enter the anti-theft code(s) for the audio system and/or the audio-navigation system (if equipped).
7. Enter the audio presets (if applicable), and enter the XM audio presets (if equipped).
8. Set the clock (for vehicles without navigation).

REMOVAL & INSTALLATION

➡Proper battery terminal disconnection and reconnection must be done before and after doing this procedure. Some systems store data in the memory that is lost when the battery is disconnected.

1. Disconnect the negative battery cable, then the positive cable.
2. Remove the two nuts securing the battery setting plate, then remove the battery setting plate and the battery.

To install:

3. Install the battery, then install the battery setting plate.
4. Tighten the two nuts equally until the battery is stable.

➡Do not deform the battery setting plate by over-tightening the nuts.

5. Reconnect the battery cables.

➡Make sure the battery is correctly installed, and that the positive terminal and the negative terminal are not connected in reverse.

ENGINE ELECTRICAL — IGNITION SYSTEM

FIRING ORDER

Engine firing order is 1-4-2-5-3-6.

IGNITION COIL

REMOVAL & INSTALLATION

Front or Rear Cylinder Head

1. Remove the engine cover.
2. Disconnect the ignition coil connectors, then remove the front ignition coils.
3. Remove the spark plugs and inspect them.

To install:

4. Apply a small amount of anti-seize compound to the plug threads, and screw the plugs into the cylinder head, finger tight.
5. Torque them to 16 ft. lbs. (22 Nm).
6. Install the front ignition coils in the reverse order of removal.

IGNITION TIMING

INSPECTION

1. Connect a Honda Diagnostic System (HDS), or equivalent OBD-II compliant scan tool, to the data link connector (DLC) and check for DTCs. If a DTC is present, diagnose and repair the cause before inspecting the ignition timing.
2. Start the engine. Hold the engine at 3,000 rpm with no load (in Neutral) until the radiator fan comes on, then let it idle.
3. Check the idle speed, as outlined in the Fuel System Section.
4. Select "SCS" mode using the HDS.
5. Connect the timing light to the No. 1 ignition coil harness.
6. Aim the light toward the pointer on the timing belt cover. Check the ignition timing under a no load condition (headlights, blower fan, rear window defogger, and air conditioner are turned off). The ignition timing should be 8–12°BTDC (red mark) at idle in Park or Neutral.
7. If the ignition timing differs from the specification, check the cam timing. If the cam timing is OK, update the powertrain control module (PCM) if it does not have the latest software, or substitute a known-good PCM, then recheck. If the system works properly, and the PCM was substituted, replace the original PCM.
8. Disconnect the HDS and the timing light.

ADJUSTMENT

The ignition timing is controlled by the Powertrain Control Module (PCM). No adjustment is necessary or possible.

SPARK PLUGS

REMOVAL & INSTALLATION

➡**See "Ignition Coil" in this section.**

ENGINE ELECTRICAL — STARTING SYSTEM

STARTER

REMOVAL & INSTALLATION

See Figure 52.

1. Remove the battery and the battery base.
2. Disconnect the positive starter cable and the S terminal connector.
3. Remove the lower radiator hose bracket from the starter and remove the dipstick.
4. Remove the two bolts holding the starter, then remove the starter.

To install:

5. Install the starter, then tighten the mounting bolts to 33 ft. lbs. (45 Nm).

➡**Always use a new gasket.**

6. Install the lower radiator hose bracket to the starter and install the dipstick.
7. Connect the positive starter cable and the S terminal connector. Make sure the crimped side of the ring terminal faces away from the starter when you connect it.
8. Install the battery base and install the battery.
9. Start the engine to make sure the starter works properly.

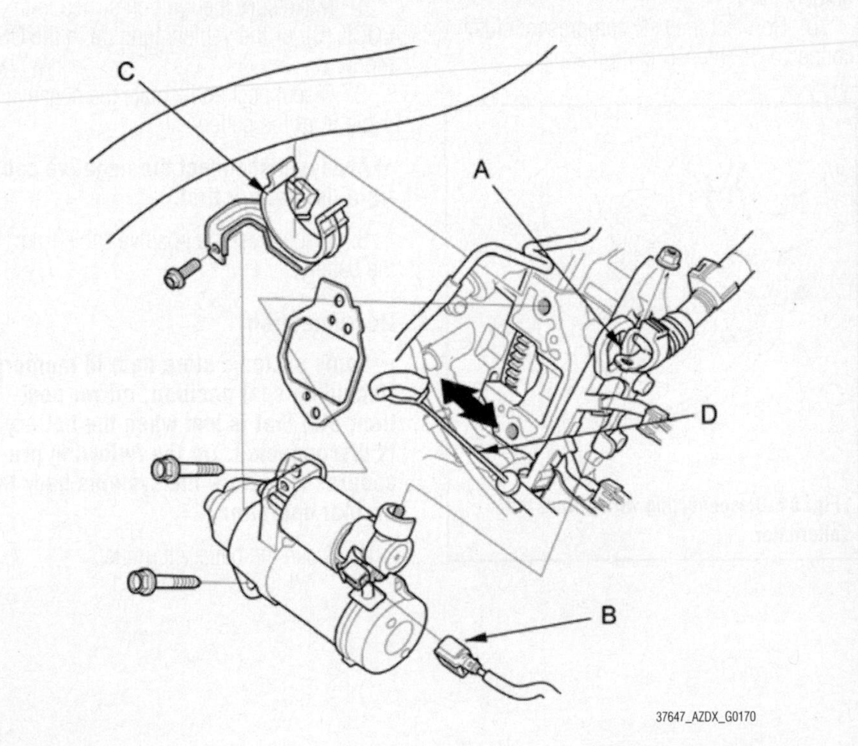
C
A
D
B
37647_AZDX_G0170

Fig. 52 Disconnect the positive starter cable and the S terminal connector. Remove the lower radiator hose bracket from the starter and remove the dipstick.

ENGINE MECHANICAL

ACCESSORY DRIVE BELTS

ACCESSORY BELT ROUTING

See Figure 53.

Refer to the accompanying illustration for belt routing.

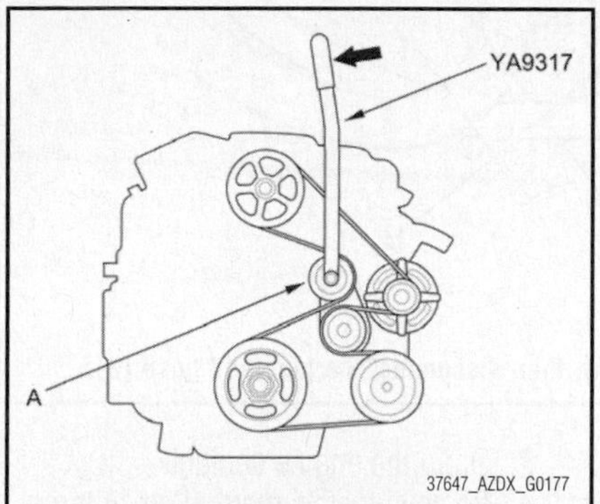
YA9317
A
37647_AZDX_G0177

Fig. 53 Showing the accessory belt routing, the location of the auto tensioner (A) and using a lever to remove tension during belt removal

INSPECTION

1. Remove the engine compartment covers. See "Valve Covers" in this section.
2. Inspect the belt for cracks or damage. If the belt is cracked or damaged, replace it.
3. Check that the position of the auto-tensioner indicator is within the standard range. If it is out of the standard range, replace the drive belt.
4. Install the engine compartment covers.

ADJUSTMENT

No adjustment is necessary or possible. Proper tension is provided by the auto-tensioner.

REMOVAL & INSTALLATION

➡See illustration under "Accessory Belt Routing".

CAMSHAFT & VALVE LIFTERS

REMOVAL & INSTALLATION

Front

See Figure 54.

1. Do the battery removal procedure.
2. Remove the battery base.
3. Drain the engine coolant.
4. Disconnect the radiator hoses.
5. Remove the EGR valve.
6. Remove the EGR valve stud bolts.

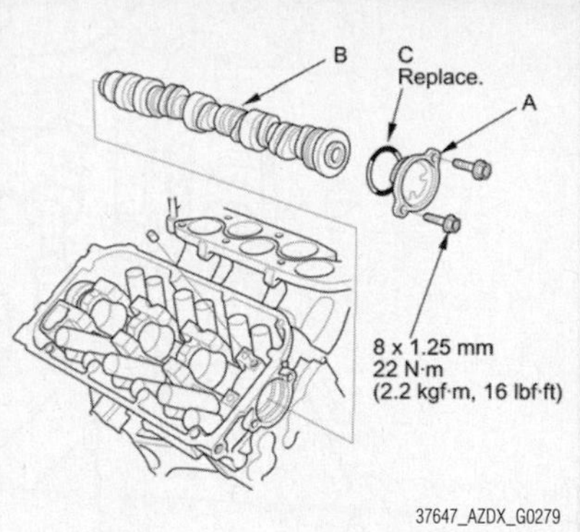
B
C
Replace.
A
8 x 1.25 mm
22 N·m
(2.2 kgf·m, 16 lbf·ft)
37647_AZDX_G0279

Fig. 54 Removing/installing the thrust cover (A), front camshaft (B) and O-ring (C)

7. Remove the timing belt. See "Timing Belt & Sprockets" in this section.
8. Remove the front rocker arm assembly. See "Rocker Arms" in this section.
9. Remove the front camshaft pulley.
10. Remove the thrust cover (A), then remove the front camshaft (B) and O-ring (C).

To install:

11. Install the front camshaft in the reverse order of removal.
    a. Always use a new O-ring.
    b. Apply new engine oil to the journals and the cam lobes.
    c. Apply new engine oil to the threads of the camshaft pulley mounting bolt, then install the front camshaft pulley.
12. Install the front rocker arm assembly, then tighten the mounting bolts, in an alternating sequence, to 16 ft. lbs. (22 Nm).
13. Install the timing belt. See "Timing Belt & Sprockets" in this section.
14. Adjust the valve clearance. See "Valve Clearance" in this section.
15. Install the EGR valve stud bolts, then install the EGR valve.
16. Connect the radiator hoses.
17. Install the battery base.
18. Install the battery.
19. Fill the radiator with engine coolant, and bleed the air from the cooling system.
20. Do the "Crankshaft Position (CKP) Pattern Clear/CKP Pattern Learn Procedure". See "Crankshaft Position (CKP) Sensor" in "ENGINE PERFORMANCE & EMISSION CONTROLS" section.

Rear

See Figure 55.

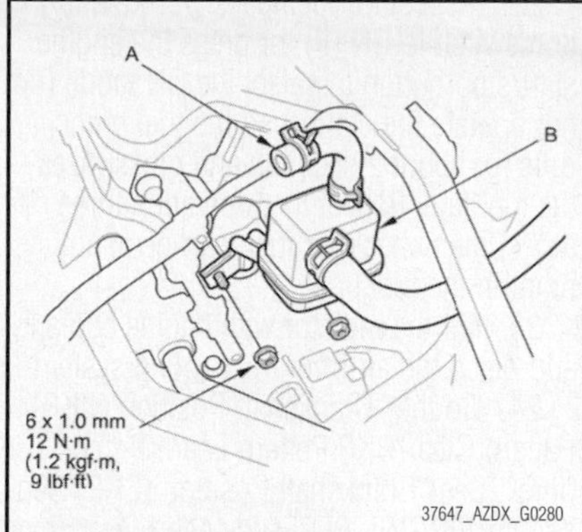
A
B
6 x 1.0 mm
12 N·m
(1.2 kgf·m,
9 lbf·ft)
37647_AZDX_G0280

Fig. 55 Disconnect the EVAP canister hose (A), then remove the EVAP canister purge joint (B) with the bracket.

1. Relieve the fuel pressure. See "GASOLINE FUEL INJECTION SYSTEM" section.
2. Remove the battery.
3. Drain the engine coolant.
4. Remove the air cleaner.
5. Remove the quick-connect fitting cover, then disconnect the fuel feed hose.
6. Disconnect the heater hoses.
7. Disconnect the EVAP canister hose, then remove the EVAP canister purge joint with the bracket.
8. Remove the timing belt. See "Timing Belt & Sprockets" in this section.
9. Remove the rear rocker arm assembly. See "Rocker Arms" in this section.
10. Remove the rear camshaft pulley. See "Camshaft Pulley" in this section.
11. Remove the thrust cover, then remove the rear camshaft.

To install:

12. Install the rear camshaft in the reverse order of removal.
    a. Always use a new O-ring.
    b. Apply new engine oil to the journals and cam lobes.
13. Apply new engine oil to the threads of the camshaft pulley mounting bolt, then install the rear camshaft pulley. See "Camshaft Pulley" in this section.
14. Install the rear rocker arm assembly, then tighten the mounting bolts, in an alternating sequence, to 16 ft. lbs. (22 Nm).
15. Install the timing belt. See "Timing Belt & Sprockets" in this section.
16. Adjust the valve clearance. See "Valve Clearance" in this section.
17. Install the EVAP canister purge joint with the bracket, then connect the EVAP canister hose.
18. Connect the heater hoses.
19. Connect the fuel feed hose, then install the quick-connect fitting cover.

20. Install the air cleaner.
21. Install the battery.
22. Inspect for fuel leaks. Turn the ignition switch to ON (II), or press the engine start/stop button to select the ON mode (do not operate the starter) so the fuel pump runs for about 2 seconds and pressurizes the fuel line. Repeat this operation three times, then check for fuel leakage at any point in the fuel line.
23. Fill the radiator with engine coolant, and bleed the air from the cooling system.
24. Do the "Crankshaft Position (CKP) Pattern Clear/CKP Pattern Learn Procedure". See "Crankshaft Position (CKP) Sensor" in "ENGINE PERFORMANCE & EMISSION CONTROLS" section.

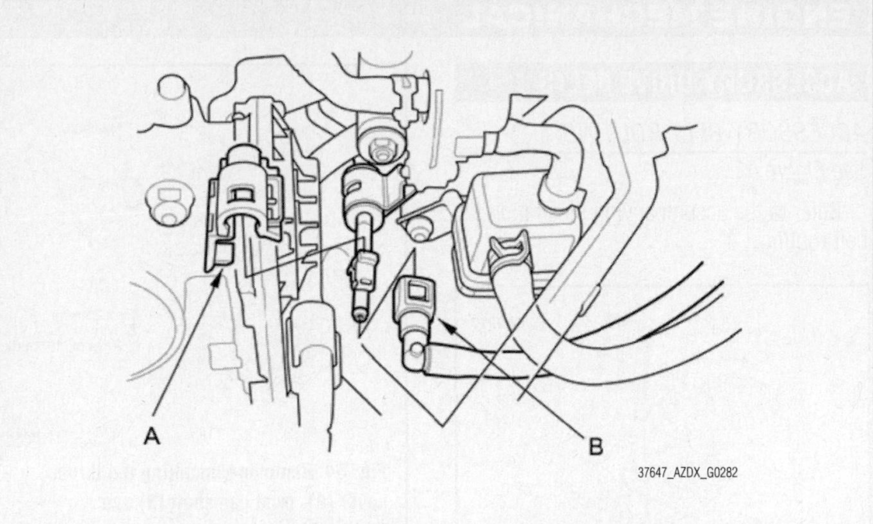
A
B
37647_AZDX_G0282
Fig. 57 Remove the quick-connect fitting cover (A), then disconnect the fuel feed hose (B).

CAMSHAFT PULLEY

REMOVAL & INSTALLATION

Front & Rear

See Figures 56 through 61.

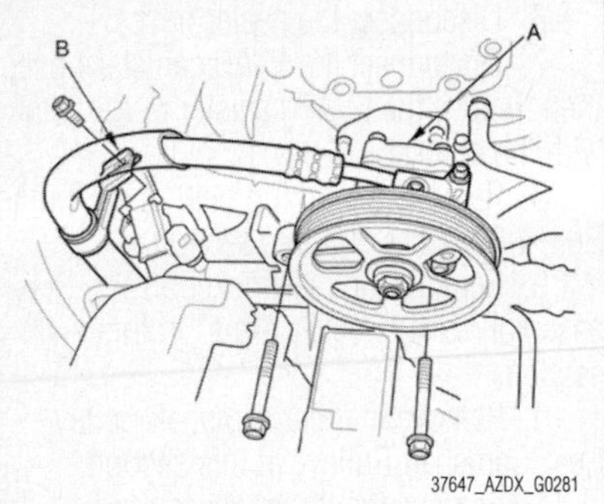
B
A
37647_AZDX_G0281
Fig. 56 Remove the power steering pump (A) and the P/S hose bracket (B) with its hoses connected.

**** CAUTION**

Monitor the engine coolant temperature (ECT) sensor 1. To avoid damaging the cylinder head, wait until the ECT drops below 100°F (38°C) before loosening the cylinder head bolts.

➡Mark all wiring and hoses to avoid misconnection. Also, be sure that they do not contact any other wiring or hoses, or interfere with any other parts.

1. Relieve the fuel pressure. See "Relieving Fuel System Pressure" in "GASOLINE FUEL INJECTION SYSTEM" section.
2. Do the battery terminal disconnection procedure. See "ENGINE ELECTRICAL" section. Wait at least 3 minutes before proceeding.
3. Drain the engine coolant.
4. Remove the alternator. See "Alternator" in "ENGINE ELECTRICAL" section.
5. Remove the power steering pump and the P/S hose bracket with its hoses connected.
6. Remove the intake manifold. See "Intake Manifold" in this section.
7. Remove the six ignition coils.
8. Remove the timing belt. See "Timing Belt & Sprockets" in this section.
9. Disconnect the following engine wire harness connectors, and remove the wire harness clamps from the cylinder head:
   - Six injector connectors
   - Knock sensor connector
   - Engine coolant temperature (ECT) sensor 1 connector
   - Engine mount control solenoid valve connector

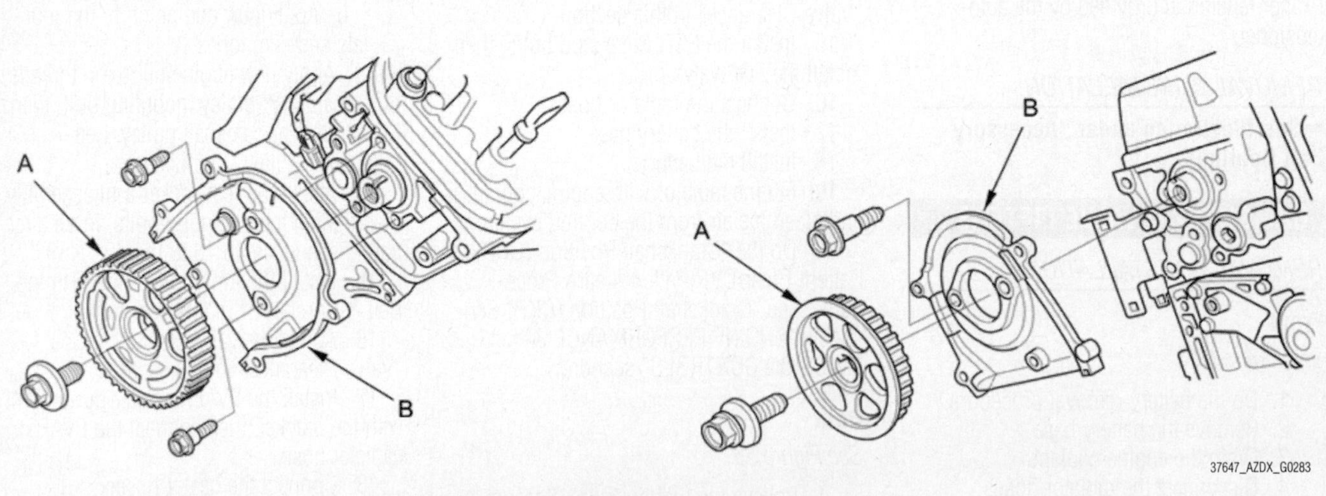
A
B
A
B
37647_AZDX_G0283
Fig. 58 Remove the camshaft pulleys (A) and the back covers (B).

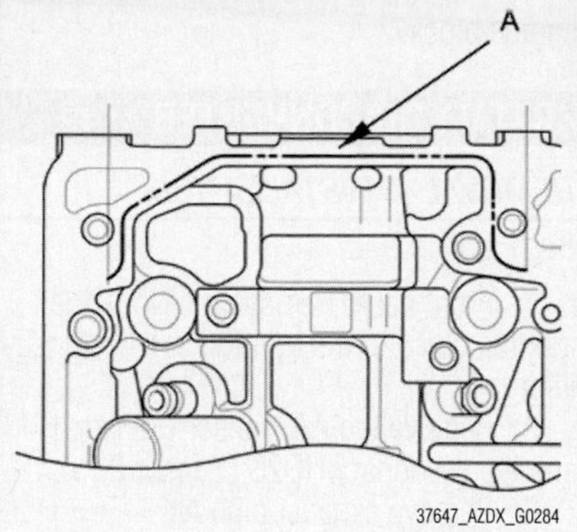
A
37647_AZDX_G0284

Fig. 59 Apply a 0.098 in. (2.5 mm) diameter bead of liquid gasket along the broken line (A).

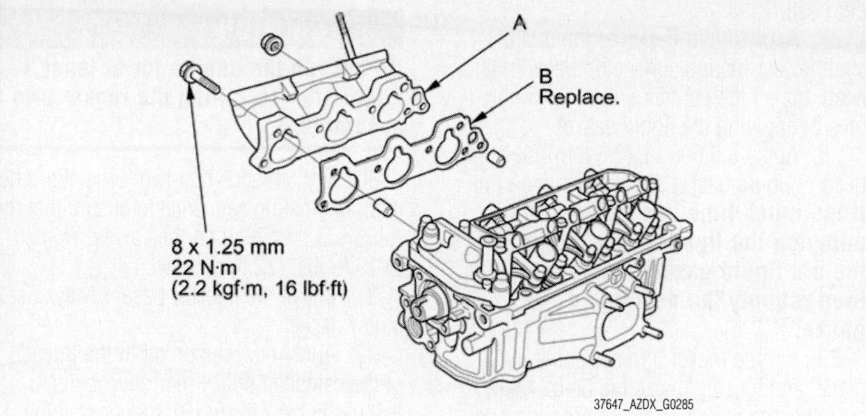
A
B
Replace.
8 x 1.25 mm
22 N·m
(2.2 kgf·m, 16 lbf·ft)
37647_AZDX_G0285

Fig. 60 Install the injector base (A). Always use a new gasket (B).

- Exhaust gas recirculation (EGR) valve connector
- Camshaft position (CMP) sensor connector
- Rocker arm oil control solenoid connector
- Rocker arm oil pressure switch connector
- Two air fuel ratio (A/F) sensor connectors
- Two secondary heated oxygen sensor (secondary HO2S) connectors

10. Remove the front warm up three way catalytic converter (front WU-TWC) and the rear warm up three way catalytic converter (rear WU-TWC). See "Catalytic Converter" in this section.
11. Remove the quick-connect fitting cover (A), then disconnect the fuel feed hose (B).
12. Remove the evaporative emission (EVAP) canister purge joint with the bracket.
13. Remove the connector bracket from the front cylinder head.
14. Remove the engine mount control solenoid valve bracket from the rear cylinder head.
15. Remove the injector bases.
16. Remove the water passage.
17. Remove the camshaft pulleys (A) and the back covers (B).

To install:

18. Clean and dry the camshaft oil seal housing.
19. Apply a light coat of new engine oil to the lip of the camshaft oil seal.
20. Gently tap the new camshaft oil seal into the cylinder head.
    a. Tap the camshaft oil seal in squarely.
    b. Install the oil seal about 0.020–0.059 in. (0.5–1.5 mm) below the surface of the cylinder head.
21. Insert the camshaft into the cylinder head, then install the camshaft thrust cover.

➡Always use a new O-ring.

22. Apply new engine oil to the journals and the cam lobes.
23. Clean the excess grease off the camshaft, and check that the oil seal lip is not distorted.
24. Install the dowel pin.
25. If the rocker arm assembly is disassembled, reassemble the rocker arm assembly.

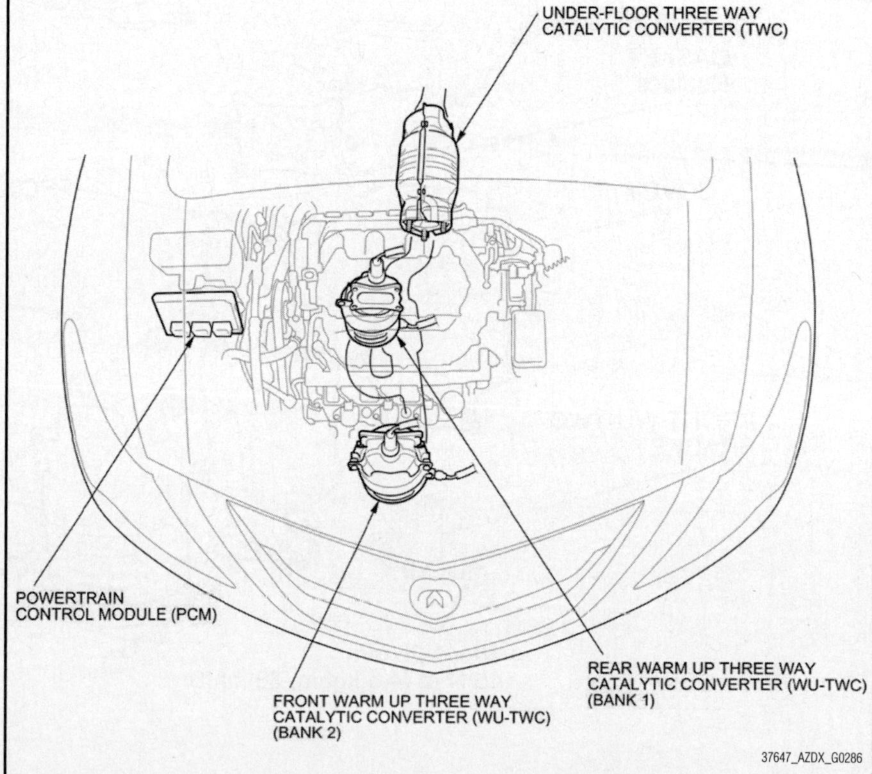
UNDER-FLOOR THREE WAY
CATALYTIC CONVERTER (TWC)
POWERTRAIN
CONTROL MODULE (PCM)
REAR WARM UP THREE WAY
CATALYTIC CONVERTER (WU-TWC)
(BANK 1)
FRONT WARM UP THREE WAY
CATALYTIC CONVERTER (WU-TWC)
(BANK 2)
37647_AZDX_G0286

Fig. 61 Apply new engine oil to the threads of the camshaft pulley mounting bolt (A). Install the back cover (B), then install the camshaft pulley (C). Torque the bolts to 16 ft. lbs. (22 Nm).

26. Remove all of the old liquid gasket from the rocker shaft holder and the cylinder head.

27. Apply liquid gasket to the rocker shaft holder mating surface of the cylinder head. Install the component within 5 minutes of applying the liquid gasket.

28. Apply a 0.098 in. (2.5 mm) diameter bead of liquid gasket along the broken line. **If too much time has passed after applying the liquid gasket, remove the old liquid gasket and residue, then reapply the new liquid gasket.**

29. Set the rocker arm assembly in place, and loosely install the bolts. Make sure that the rocker arms are properly positioned on the valve stems.

➡Wait at least 30 minutes before filling the engine with oil.

⁂ CAUTION

Do not run the engine for at least 3 hours after installing the rocker arm assembly.

30. Tighten each bolt two turns at a time in an alternating sequence to ensure that the rockers do not bind on the valves. Tighten to 16 ft. lbs. (22 Nm).

31. Install the injector base. Always use a new gasket.

32. Apply new engine oil to the threads of the camshaft pulley mounting bolt (A). Install the back cover (B), then install the camshaft pulley (C). Torque the bolts to 16 ft. lbs. (22 Nm).

33. Set the camshaft pulleys to top dead center (TDC) before bolting them onto the engine block.

CATALYTIC CONVERTER

REMOVAL & INSTALLATION

See Figure 62.

1. Front bank: Remove the No. 2 the ignition coil and the ignition coil heat insulator.
2. Remove the A/F sensor (Sensor 1) and the secondary HO2S (Sensor 2).
3. Remove exhaust pipe A.
4. Front bank: Remove the EGR pipe.
5. Rear bank: Remove the intermediate shaft.

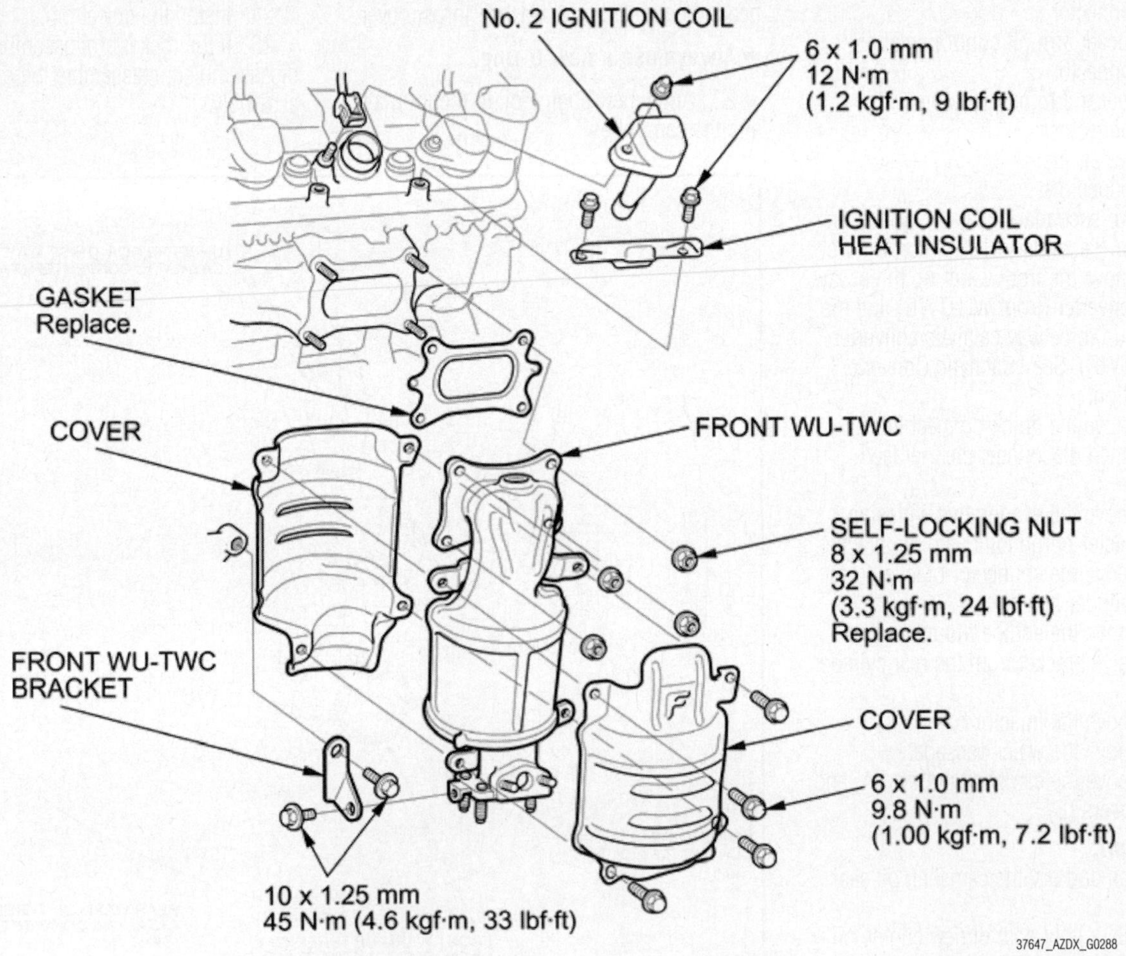
No. 2 IGNITION COIL
6 x 1.0 mm
12 N·m
(1.2 kgf·m, 9 lbf·ft)
IGNITION COIL
HEAT INSULATOR
GASKET
Replace.
COVER
FRONT WU-TWC
SELF-LOCKING NUT
8 x 1.25 mm
32 N·m
(3.3 kgf·m, 24 lbf·ft)
Replace.
FRONT WU-TWC
BRACKET
COVER
6 x 1.0 mm
9.8 N·m
(1.00 kgf·m, 7.2 lbf·ft)
10 x 1.25 mm
45 N·m (4.6 kgf·m, 33 lbf·ft)
37647_AZDX_G0288

Fig. 62 Carefully install the front WU-TWC with a new gasket and new self-locking nuts. Tighten the nuts in a crisscross pattern in two or three steps to 24 ft. lbs. (32 Nm).

6. Remove the A/C condenser fan assembly and the radiator upper bracket/cushion.

7. Remove the front WU-TWC bracket, then carefully remove the front WU-TWC.

To install:

8. Carefully install the front WU-TWC with a new gasket and new self-locking nuts. Tighten the nuts in a crisscross pattern in two or three steps to 24 ft. lbs. (32 Nm).

9. Install the parts in the reverse order of removal.

CRANKSHAFT DAMPER (PULLEY)

REMOVAL & INSTALLATION

See Figures 63 through 66.

1. Raise the vehicle on the lift.
2. Remove the right front wheel.
3. Remove the splash shield.
4. Remove the drive belt.
5. Hold the pulley with the handle, 6-25-660L and the holder attachment, 50 mm offset.

6. Remove the bolt with a heavy duty 19 mm socket and a breaker bar, then remove the crankshaft pulley.

To install:

7. Remove any oil and clean the pulleys, the crankshaft, the bolt, and the washer. Lubricate with new engine oil as shown.

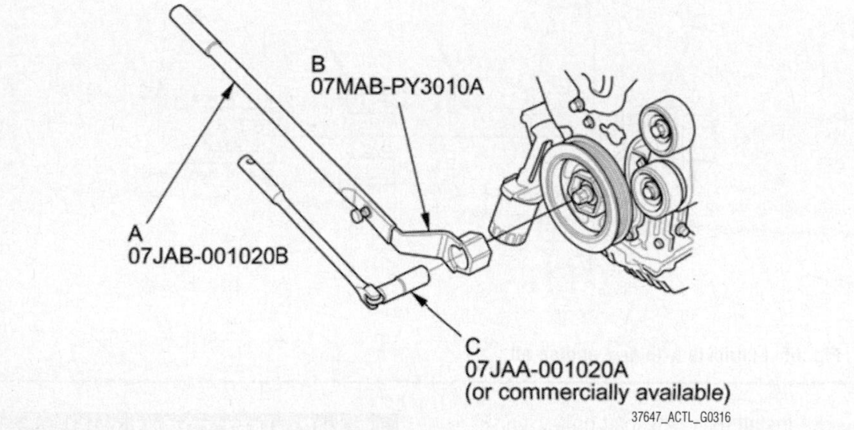

B
07MAB-PY3010A
A
07JAB-001020B
C
07JAA-001020A
(or commercially available)
37647_ACTL_G0316

Fig. 64 Removing the crankshaft pulley bolt

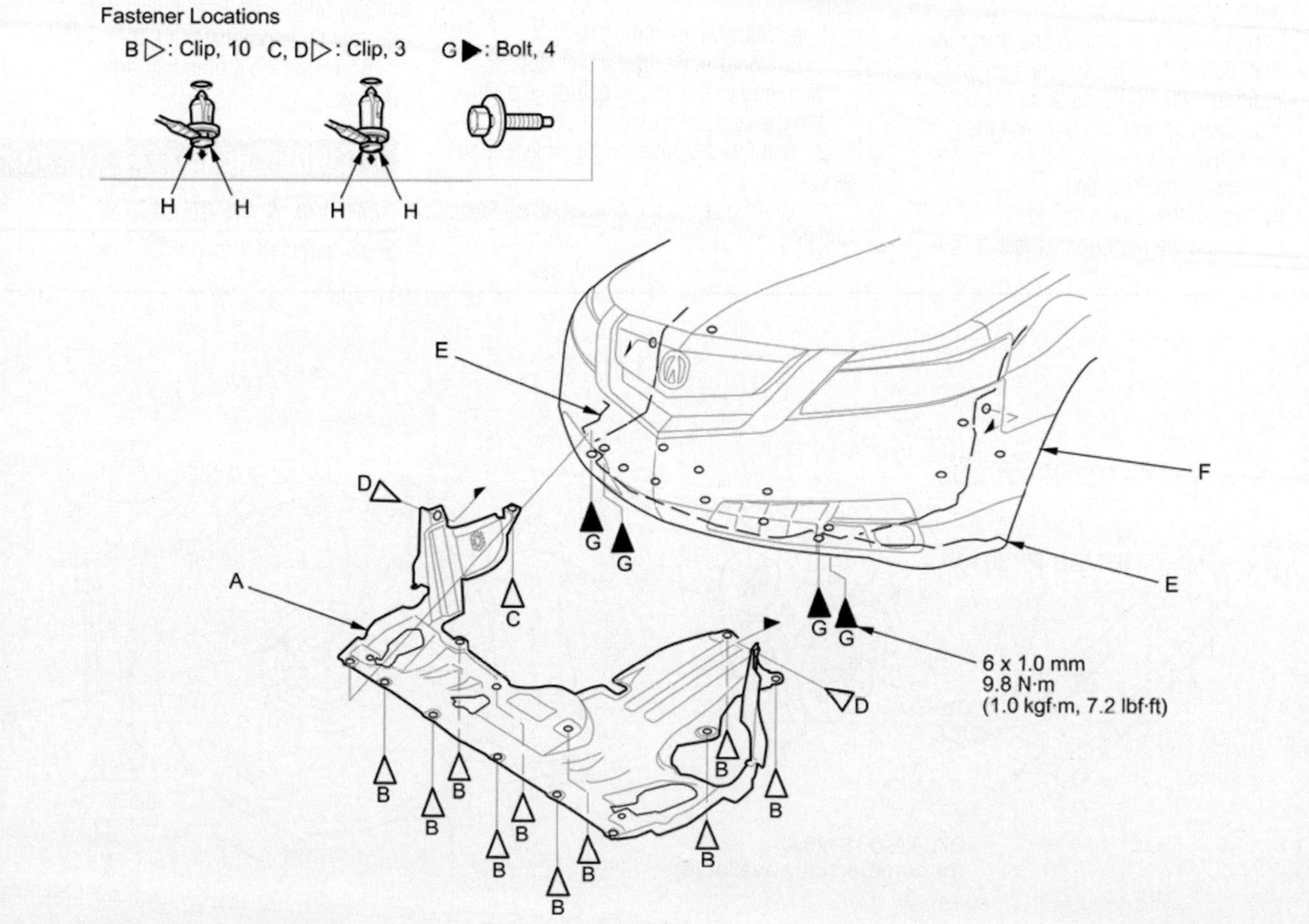

Fastener Locations
B ▷: Clip, 10 C, D▷: Clip, 3 G▶: Bolt, 4
H
H
H
H
E
D
A
C
G
G
F
E
G
G
6 x 1.0 mm
9.8 N·m
(1.0 kgf·m, 7.2 lbf·ft)
D
B
37647_ACTL_G0274

Fig. 63 Remove the splash shield

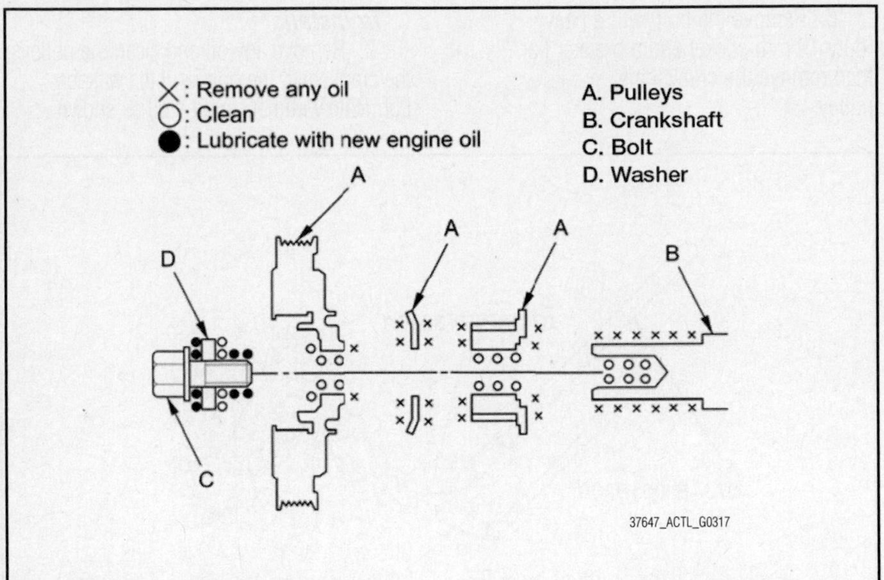

X: Remove any oil
O: Clean
●: Lubricate with new engine oil
A. Pulleys
B. Crankshaft
C. Bolt
D. Washer
A
A
A
B
D
C
37647_ACTL_G0317

Fig. 65 Lubricate with new engine oil

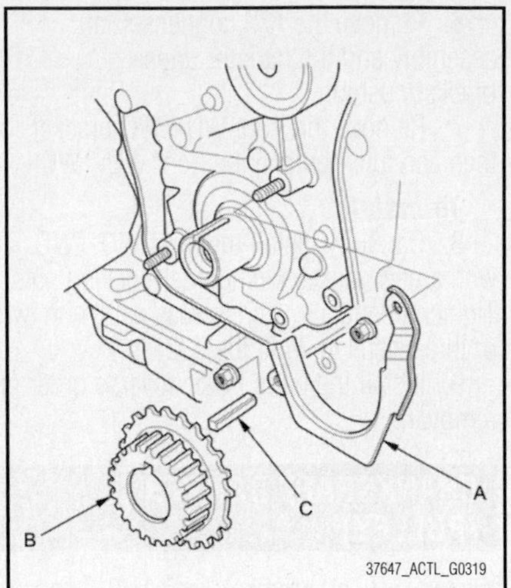

A
C
B
37647_ACTL_G0319

Fig. 67 Remove the timing belt stopper (A), then remove the timing belt drive pulley (B) and the key (C)

8. Install the crankshaft pulley, and tighten the bolt. Do not use an impact wrench.
   a. Hold the pulley with the handle and the holder attachment. Torque the bolt to 47 ft. lbs. (64 Nm) with a torque wrench and a 19 mm socket.
   b. Mark the bolt head and the crankshaft pulley, then tighten the bolt an additional 60 ° (The mark on the bolt head lines up with the mark on the crankshaft pulley).
9. Install the drive belt.
10. Install the splash shield.
11. Install the right front wheel.

CRANKSHAFT FRONT SEAL

REMOVAL & INSTALLATION

See Figure 67.

1. Remove the timing belt drive pulley.
   a. Remove the timing belt.
   b. Remove the timing belt stopper, then remove the timing belt drive pulley and the key.
2. Remove the pulley end crankshaft oil seal.
3. Clean and dry the crankshaft oil seal housing.
4. Apply a light coat of new engine oil to the lip of the crankshaft oil seal.
5. Using the oil seal driver, 64 mm, drive in the new crankshaft oil seal until the oil seal driver bottoms against the oil pump. When the seal is in place, clean any excess grease off the crankshaft, and check that the oil seal lip is not distorted.
6. Install the timing belt drive pulley:

CYLINDER HEAD

REMOVAL & INSTALLATION

See Figures 68 through 72.

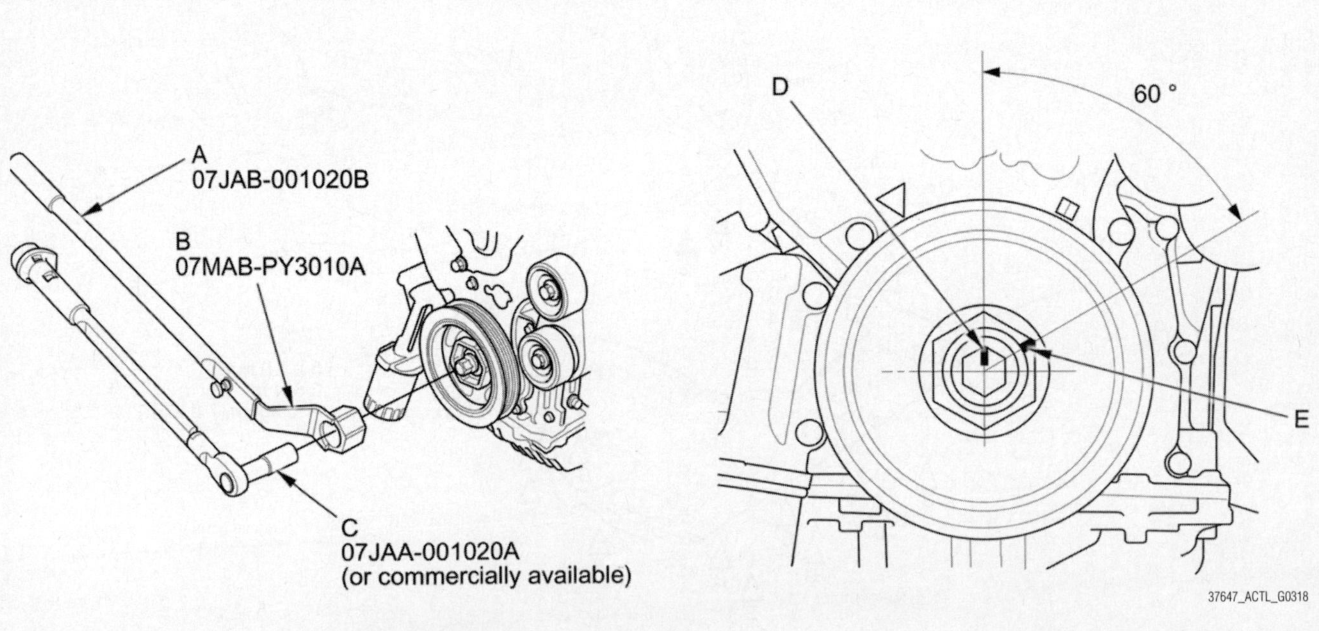

A
07JAB-001020B
B
07MAB-PY3010A
C
07JAA-001020A
(or commercially available)
D
60 °
E
37647_ACTL_G0318

Fig. 66 Install the crankshaft pulley, and tighten the bolt

CAUTION

Monitor the engine coolant temperature (ECT) sensor 1. To avoid damaging the cylinder head, wait until the ECT drops below 100°F (38°C) before loosening the cylinder head bolts.

➡Mark all wiring and hoses to avoid misconnection. Also, be sure that they do not contact any other wiring or hoses, or interfere with any other parts.

1. Relieve the fuel pressure. See "Relieving Fuel System Pressure" in "GASOLINE FUEL INJECTION SYSTEM" section.
2. Do the battery terminal disconnection procedure, as outlined in the "ENGINE ELECTRICAL" section.
3. Drain the engine coolant.
4. Remove the alternator. See "Alternator" in "ENGINE ELECTRICAL" section.
5. Remove the power steering pump and the P/S hose bracket with its hoses connected.
6. Remove the intake manifold. See "Intake Manifold" in this section.
7. Remove the six ignition coils.
8. Remove the timing belt. See "Timing Belt & Sprockets" in this section.
9. Disconnect the following engine wire harness connectors, and remove the wire harness clamps from the cylinder head:
    - Six injector connectors
    - Knock sensor connector
    - Engine coolant temperature (ECT) sensor 1 connector
    - Engine mount control solenoid valve connector
    - Exhaust gas recirculation (EGR) valve connector
    - Camshaft position (CMP) sensor connector
    - Rocker arm oil control solenoid connector
    - Rocker arm oil pressure switch connector
    - Two air fuel ratio (A/F) sensor connectors
    - Two secondary heated oxygen sensor (secondary HO2S) connectors
10. Remove the front warm up three way catalytic converter (front WU-TWC) and the rear warm up three way catalytic converter (rear WU-TWC). See "Catalytic Converter" in this section.
11. Remove the quick-connect fitting cover (A), then disconnect the fuel feed hose (B).
12. Remove the evaporative emission (EVAP) canister purge joint with the bracket.
13. Remove the connector bracket from the front cylinder head.
14. Remove the engine mount control solenoid valve bracket from the rear cylinder head.

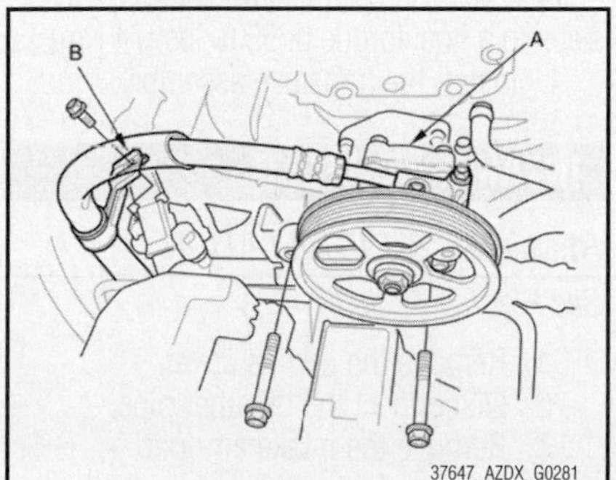
B
A
37647_AZDX_G0281

Fig. 68 Remove the power steering pump (A) and the P/S hose bracket (B) with its hoses connected.

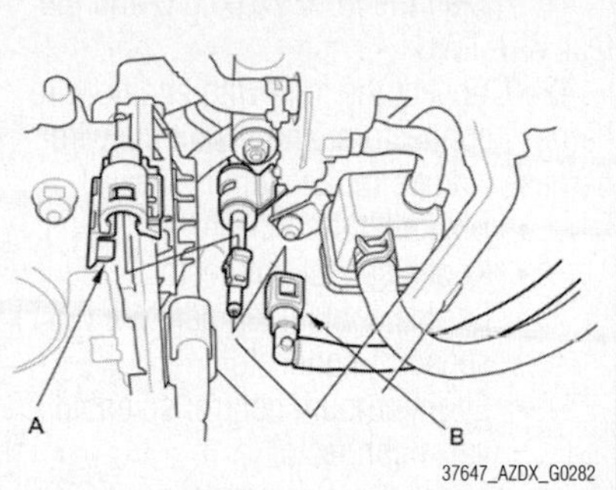
A
B
37647_AZDX_G0282

Fig. 69 Remove the quick-connect fitting cover (A), then disconnect the fuel feed hose (B).

15. Remove the injector bases.
16. Remove the water passage.
17. Remove the camshaft pulleys (A) and the back covers (B).
18. Remove the cylinder head covers.
19. Remove the cylinder head bolts.

CAUTION

To prevent warpage, loosen the bolts in sequence 1/3 turn at a time; repeat the sequence until all bolts are loosened.

20. Remove the cylinder heads.

To install:

21. Clean the cylinder head and the engine block surface.
22. Clean and install the oil control orifices with new O-rings. Install the dowel pins and the new cylinder head gaskets.
23. Clean the timing belt pulleys, the timing belt guide plate, and the upper and lower covers.
24. Set the timing belt drive pulley to top dead center (TDC) by aligning the TDC mark on the tooth of the timing belt drive pulley with the pointer on the oil pump.
25. Set the camshaft pulleys to TDC by aligning the TDC marks on the camshaft pulleys with the pointers on the back covers.
26. Install the cylinder heads on the engine block.
27. Measure the diameter of each cylinder head bolt at point A and point B.

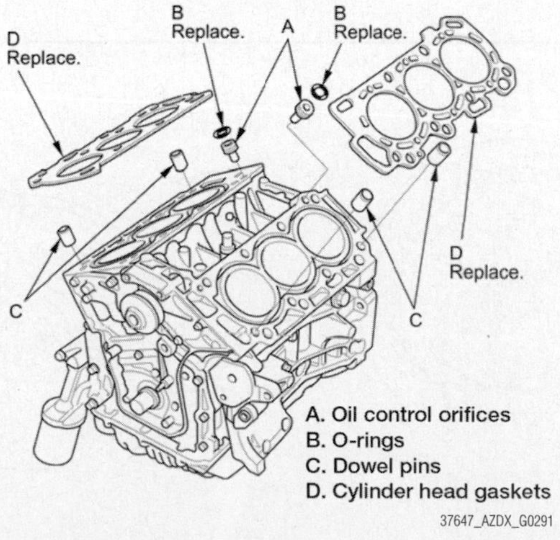
D
Replace.
B
Replace.
A
B
Replace.
D
Replace.
C
C
A. Oil control orifices
B. O-rings
C. Dowel pins
D. Cylinder head gaskets
37647_AZDX_G0291

Fig. 70 Clean and install the oil control orifices with new O-rings. Install the dowel pins and the new cylinder head gaskets.

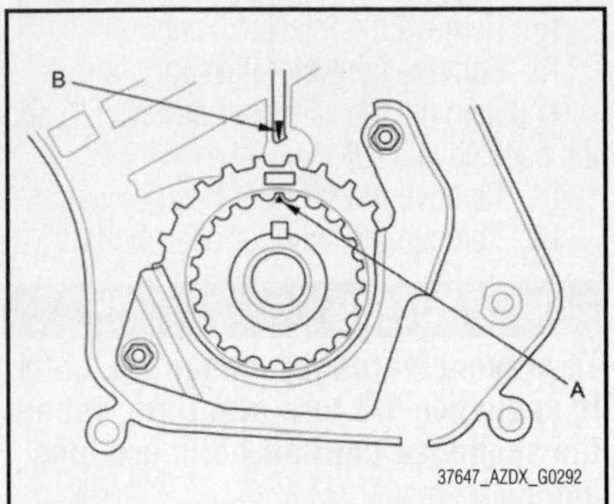
B
A
37647_AZDX_G0292

Fig. 71 Set the timing belt drive pulley to top dead center (TDC) by aligning the TDC mark (A) on the tooth of the timing belt drive pulley with the pointer (B) on the oil pump.

28. If either diameter is less than 11.3 mm (0.445 in), replace the cylinder head bolt.
29. Apply new engine oil to the threads and under the bolt heads of all cylinder head bolts.
30. Torque the cylinder head bolts in an alternating sequence to 22 ft. lbs. (30 Nm), using a beam-type torque wrench.

➡When using a preset click-type torque wrench, be sure to tighten slowly and do not overtighten. If a bolt makes any noise while you are torquing it, loosen the bolt and retighten it from the first step.

31. After torquing, tighten all cylinder head bolts in two steps (90 degrees per step) using the same alternating sequence as previous.

➡If using a new cylinder head bolt, tighten the bolt an extra 90 degrees.

✻✻ CAUTION

Remove the cylinder head bolt if tightened beyond the specified angle, and go back and repeat the tightening procedure. Do not loosen it back to the specified angle.

32. Install the timing belt. See "Timing Belt & Sprockets" in this section.
33. Adjust the valve clearance.
34. Install the cylinder head covers.
35. Install the water passage.
36. Install the injector bases.
37. Install the connector bracket to the front cylinder head.
38. Install the engine mount control solenoid valve bracket to the rear cylinder head.
39. Install the EVAP canister purge joint with the bracket.
40. Connect the fuel feed hose, then install the quick-connect fitting cover.
41. Install the front WU-TWC and the rear WU-TWC.
42. Connect the following engine wire harness connectors, and install the wire harness clamps to the cylinder head:
- Six injector connectors
- Knock sensor connector
- Engine coolant temperature (ECT) sensor 1 connector
- Engine mount control solenoid valve connector
- Exhaust gas recirculation (EGR) valve connector
- Camshaft position (CMP) sensor connector
- Rocker arm oil control solenoid connector
- Rocker arm oil pressure switch connector
- Two air fuel ratio (A/F) sensor connectors
- Two secondary heated oxygen sensor (secondary HO2S) connectors

43. Install the six ignition coils.
44. Install the intake manifold. See "Intake Manifold" in this section.
45. Install the power steering pump and the P/S hose bracket.
46. Install the alternator.
47. Reconnect the battery.
48. After installation, check that all tubes, hoses, and connectors are installed correctly.
49. Inspect for fuel leaks. Turn the ignition switch to ON (II), or press the engine start/stop button to select the ON mode (do not operate the starter) so the fuel pump runs for about 2 seconds and pressurizes the fuel line. Repeat this operation three times, then check for fuel leakage at any point in the fuel line.
50. Refill the radiator with engine coolant, and bleed the air from the cooling system.
51. Check for fluid leaks.
52. Do the "Idle Learn Procedure". See "Powertrain Control Module (PCM)" in "ENGINE PERFORMANCE & EMISSION CONTROL" section.
53. Do the "CKP Pattern Clear/CKP Pattern Learn Procedure". See "Crankshaft Position (CKP) Sensor" in "ENGINE PERFORMANCE & EMISSION CONTROL" section.
54. Inspect the idle speed.
55. Inspect the ignition timing.

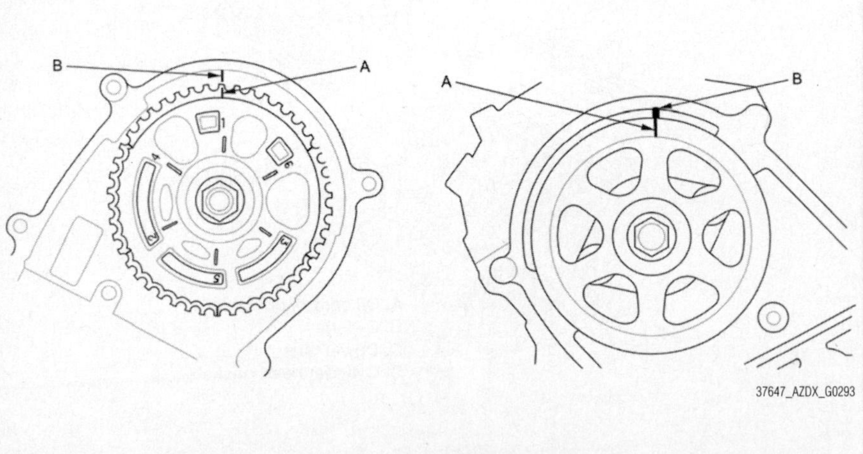
B
A
A
B
37647_AZDX_G0293

Fig. 72 Set the camshaft pulleys to TDC by aligning the TDC marks (A) on the camshaft pulleys with the pointers (B) on the back covers.

FLEXPLATE (DRIVE PLATE)

REMOVAL & INSTALLATION

1. Remove the transaxle assembly.
2. Remove the drive plate and the washer from the engine.
3. Install the drive plate and the washer on the engine crankshaft, and tighten the eight bolts in a crisscross pattern in at least two steps to a final torque of 55 ft. lbs (74 Nm).
4. Install the transaxle assembly.

INTAKE MANIFOLD

REMOVAL & INSTALLATION

See Figures 73 and 74.

1. Remove the engine cover.
2. Disconnect the breather pipe.
3. Remove the intake air duct.
4. Disconnect the PCV hose, the brake booster vacuum hose, the vacuum hose, and the IMT actuator connector from the intake manifold.

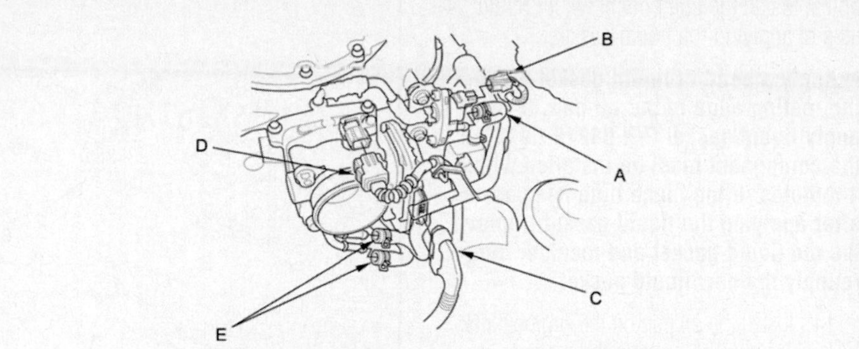
B
D
A
C
E

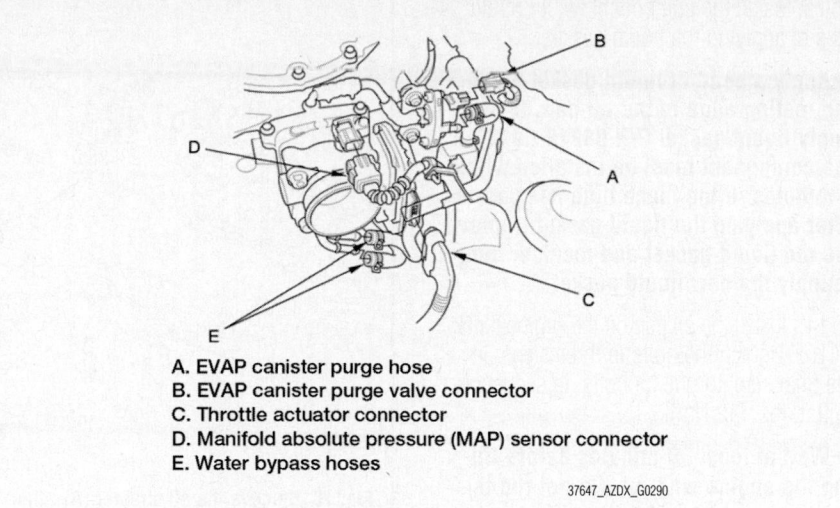

A. EVAP canister purge hose
B. EVAP canister purge valve connector
C. Throttle actuator connector
D. Manifold absolute pressure (MAP) sensor connector
E. Water bypass hoses

37647_AZDX_G0290

Fig. 73 Disconnect the EVAP canister purge hose, the EVAP canister purge valve connector, the throttle actuator connector, and the manifold absolute pressure (MAP) sensor connector. Disconnect and plug the water bypass hoses.

5. Disconnect the EVAP canister purge hose (A), the EVAP canister purge valve connector (B), the throttle actuator connector (C), and the manifold absolute pressure (MAP) sensor connector (D). Disconnect and plug the water bypass hoses (E).

6. Remove the upper cover mounting bolts and nuts, using an alternating sequence, in three steps, then remove the upper cover.

7. Remove the intake manifold mounting bolts and nuts, using an alternating sequence, in three steps.

8. Remove the intake manifold.

To install:

9. Install the intake manifold. Tighten the bolts and nuts in an alternating sequence, in three steps. Always use a new intake manifold gasket. Tighten the bolts and nuts to 16 ft. lbs. (22 Nm).

10. Install the upper cover. Tighten the

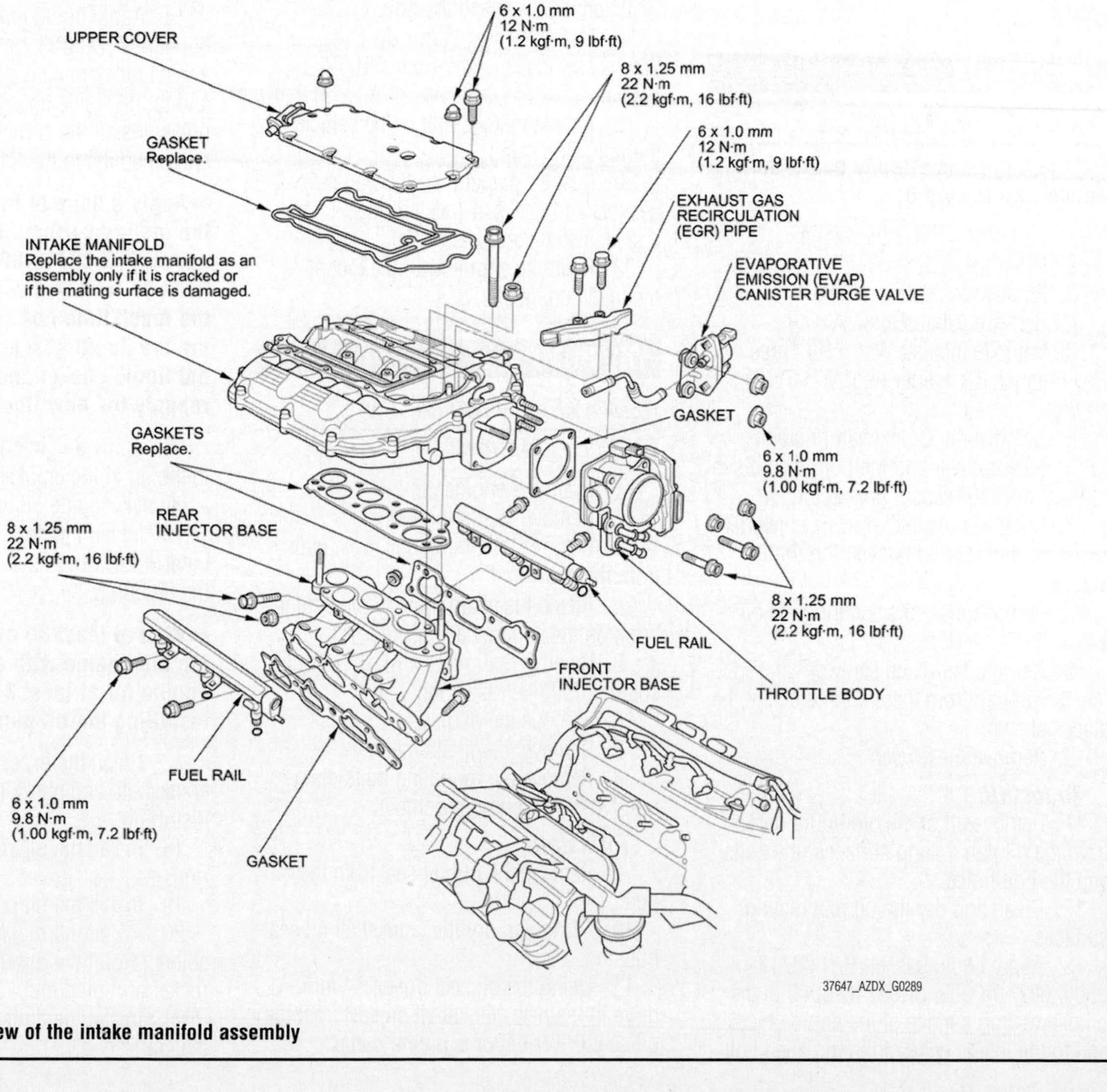
6 x 1.0 mm
12 N·m
(1.2 kgf·m, 9 lbf·ft)
UPPER COVER
8 x 1.25 mm
22 N·m
(2.2 kgf·m, 16 lbf·ft)
6 x 1.0 mm
12 N·m
(1.2 kgf·m, 9 lbf·ft)
GASKET
Replace.
EXHAUST GAS
RECIRCULATION
(EGR) PIPE
INTAKE MANIFOLD
Replace the intake manifold as an assembly only if it is cracked or if the mating surface is damaged.
EVAPORATIVE
EMISSION (EVAP)
CANISTER PURGE VALVE
GASKET
GASKETS
Replace.
6 x 1.0 mm
9.8 N·m
(1.00 kgf·m, 7.2 lbf·ft)
8 x 1.25 mm
22 N·m
(2.2 kgf·m, 16 lbf·ft)
REAR
INJECTOR BASE
8 x 1.25 mm
22 N·m
(2.2 kgf·m, 16 lbf·ft)
FUEL RAIL
FRONT
INJECTOR BASE
THROTTLE BODY
FUEL RAIL
6 x 1.0 mm
9.8 N·m
(1.00 kgf·m, 7.2 lbf·ft)
GASKET

37647_AZDX_G0289

Fig. 74 Exploded view of the intake manifold assembly

bolts and nuts sequentially in three steps to 9 ft. lbs. (12 Nm). Always use a new gasket.

11. Connect the water bypass hoses. Connect the MAP sensor connector, the throttle actuator connector, the EVAP canister purge valve connector, and the EVAP canister purge hose.

12. Connect the IMT actuator connector, the vacuum hose, the brake booster vacuum hose, and the PCV hose.

13. Install the intake air duct.

➡When tightening the screw of the hose band, align the edge of the hose band with the mark painted on the hose band. If you tighten the screw over the mark, replace the hose band.

14. Connect the breather pipe.
15. Install the engine cover.
16. Clean up any spilled engine coolant.
17. After installation, check that all tubes, hoses, and connectors are installed correctly.
18. Refill the radiator with engine coolant, and bleed the air from the cooling system.

OIL PAN

REMOVAL & INSTALLATION

➡If the engine is already out of the vehicle, go to step 6.

1. Raise the vehicle on the lift.
2. Drain the engine oil.
3. Remove the splash shield.
4. Remove exhaust pipe A.
5. Remove the rear Warm Up Three Way Catalytic Converter (rear WU-TWC) bracket.
6. Remove the Crankshaft Position (CKP) sensor cover and the bolt, then disconnect the CKP sensor connector.
7. Remove the clutch/torque converter cover and the four bolts securing the transaxle.
8. Remove the bolts securing the oil pan.
9. Using a flat blade screwdriver, separate the oil pan from the engine block in the places shown.
10. Remove the oil pan.

To install:

11. Remove all of the old liquid gasket from the oil pan mating surfaces, the bolts, and the bolt holes.
12. Clean and dry the oil pan mating surfaces.
13. Apply liquid gasket, P/N 08717-0004, 08718-0003, or 08718-0009 to the oil pan mating surface of the engine block and to the inside edge of the threaded bolt holes. Install the component within 5 minutes of applying the liquid gasket.

➡Apply a bead of liquid gasket along the mating edge of the oil pan. If you apply liquid gasket P/N 08718-0012, the component must be installed within 4 minutes. If too much time has passed after applying the liquid gasket, remove the old liquid gasket and residue, then reapply the new liquid gasket.

14. Install the oil pan on the engine block.
15. Tighten the bolts in three steps. In the final step, torque all bolts, in sequence, to 9 ft. lbs. (12 Nm).

➡Wait at least 30 minutes before filling the engine with oil. Do not run the engine for at least 3 hours after installing the oil pan.

16. Tighten the four bolts securing the transaxle, then install the clutch/torque converter cover.
17. Connect the Crankshaft Position (CKP) sensor connector, then install the CKP sensor cover and the bolt.
18. Install the rear Warm Up Three Way Catalytic Converter (rear WU-TWC) bracket.
19. If the engine is still in the vehicle, do the following steps.
20. Install exhaust pipe A using new gaskets and new self-locking nuts.
21. Install the splash shield.
22. Refill the engine with the recommended engine oil.

OIL PUMP

REMOVAL & INSTALLATION

See Figures 75 through 78.

1. Drain the engine oil.
2. Remove the timing belt:
3. Remove the timing belt drive pulley from the crankshaft:
4. Attach the chain hoist to the engine hook on the engine hanger bracket.
5. Remove the rocker arm oil control solenoid/oil filter assembly.
6. Remove the oil pan.
7. Remove the oil strainer.
8. Remove the mounting bolts, then remove the oil pump assembly.

To install:

9. Remove the old oil seal from the oil pump.
10. Clean and dry the crankshaft oil seal housing.
11. Using the oil seal driver, 64 mm, drive in the new crankshaft oil seal until the oil seal driver bottoms on the pump.

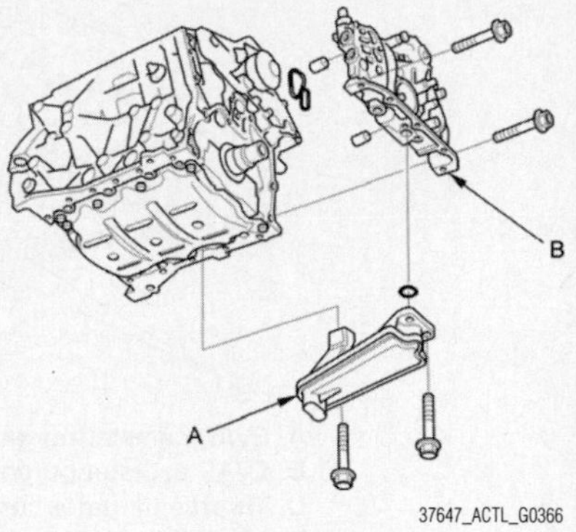
B
A
37647_ACTL_G0366

Fig. 75 Remove the oil strainer (A) , then remove the oil pump assembly (B)

12. Remove all of the old liquid gasket from the oil pump mating surfaces, the bolts, and the bolt holes.
13. Clean and dry the oil pump mating surfaces.
14. Apply liquid gasket, P/N 08717-0004, 08718-0003, or 08718-0009, to the engine block mating surface of the oil pump and to the inside edge of the threaded bolt holes. Install the component within 5 minutes of applying the liquid gasket.

➡Apply a bead of liquid gasket along the mating surface. If you apply liquid gasket P/N 08718-0012, the component must be installed within 4 minutes. If too much time has passed after applying the liquid gasket, remove all of the old liquid gasket and residue, then reapply the new liquid gasket.

15. Apply a light coat of new engine oil to the lip of the crankshaft oil seal, and apply new engine oil to the new O-ring.
16. Install the dowel pins, then align the inner rotor with the crankshaft, and install the oil pump.

➡Wait at least 30 minutes before filling the engine with oil. Do not run the engine for at least 3 hours after installing the oil pump.

17. Clean the excess grease off the crankshaft, and check the seal for distortion.
18. Install the oil strainer with a new O-ring.
19. Install the oil pan.
20. Install the rocker arm oil control solenoid/oil filter assembly with a new rocker arm oil control solenoid filter.
21. Install the timing belt drive pulley to the crankshaft:

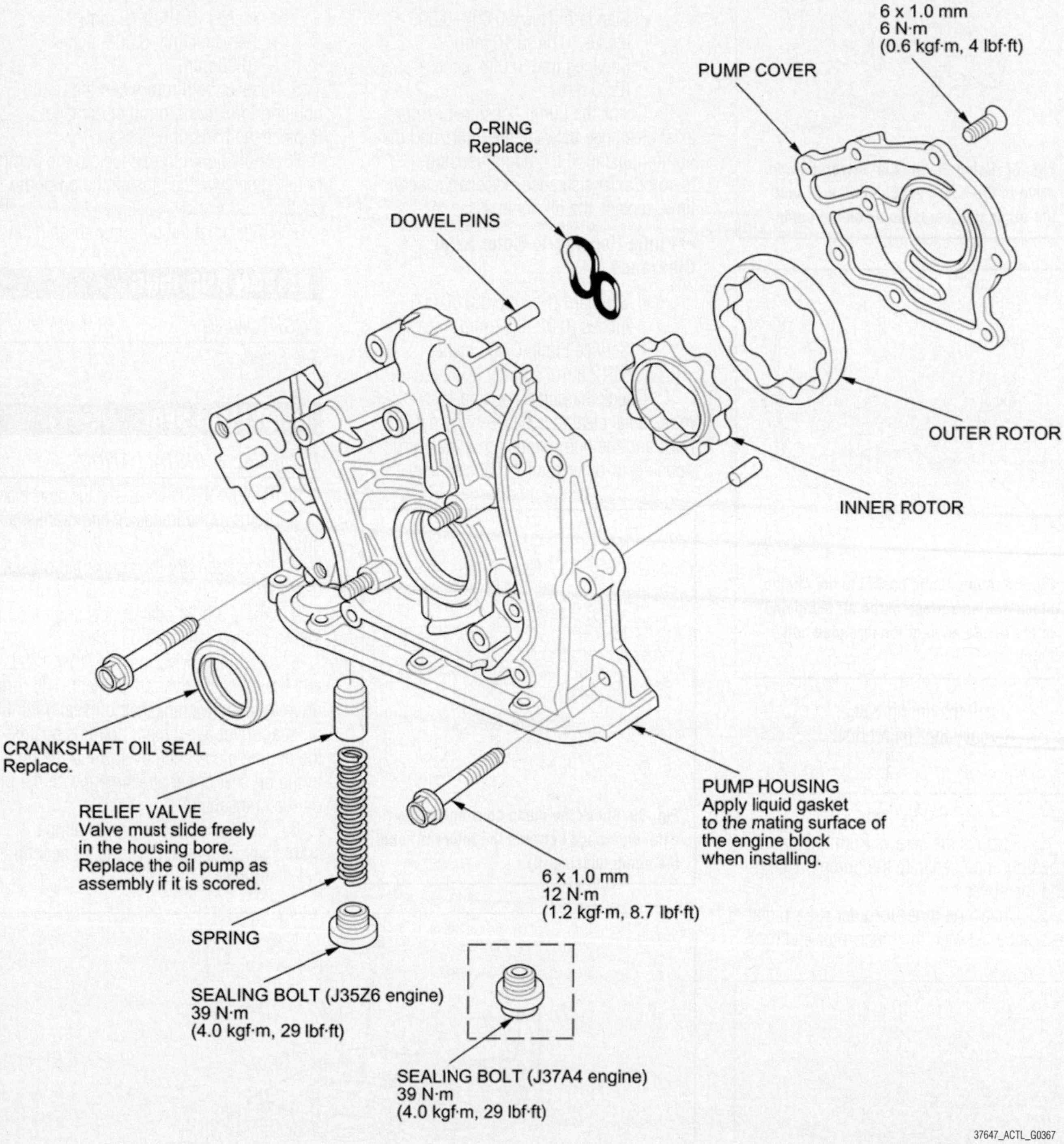
6 x 1.0 mm
6 N·m
(0.6 kgf·m, 4 lbf·ft)
PUMP COVER
O-RING
Replace.
DOWEL PINS
OUTER ROTOR
INNER ROTOR
CRANKSHAFT OIL SEAL
Replace.
RELIEF VALVE
Valve must slide freely
in the housing bore.
Replace the oil pump as
assembly if it is scored.
PUMP HOUSING
Apply liquid gasket
to the mating surface of
the engine block
when installing.
6 x 1.0 mm
12 N·m
(1.2 kgf·m, 8.7 lbf·ft)
SPRING
SEALING BOLT (J35Z6 engine)
39 N·m
(4.0 kgf·m, 29 lbf·ft)
SEALING BOLT (J37A4 engine)
39 N·m
(4.0 kgf·m, 29 lbf·ft)
37647_ACTL_G0367

Fig. 76 Exploded view of oil pump assembly

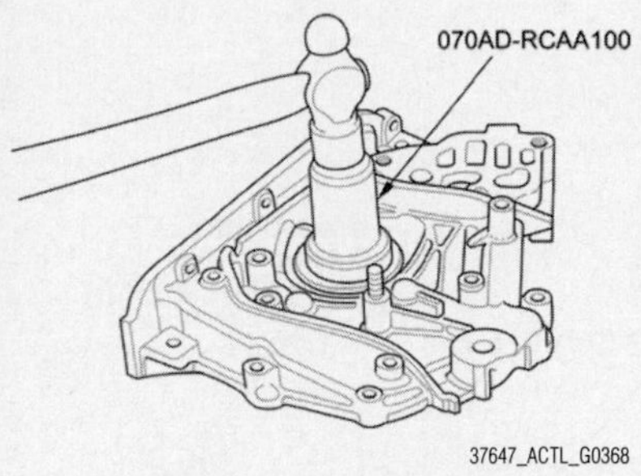
070AD-RCAA100
37647_ACTL_G0368

Fig. 77 Using the oil seal driver, 64 mm, drive in the new crankshaft oil seal until the oil seal driver bottoms on the pump

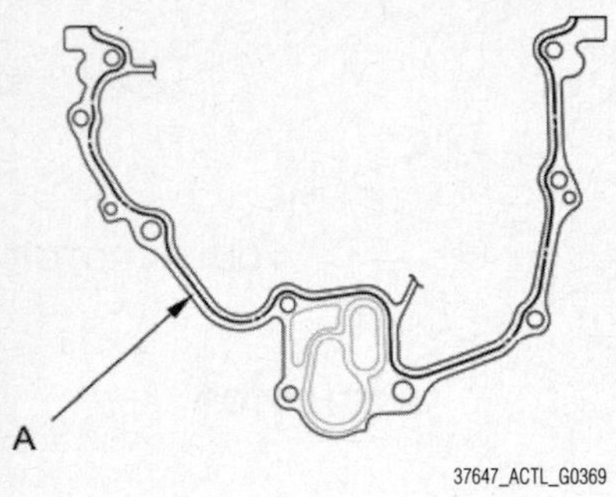
A
37647_ACTL_G0369

Fig. 78 Apply liquid gasket to the engine block mating surface of the oil pump and to the inside edge of the threaded bolt holes

22. Install the timing belt:
23. Remove the chain hoist.

INSPECTION

See Figures 79 through 81.

1. Remove the screws from the pump housing, then separate the pump housing and the cover.
2. Check the inner-to-outer rotor radial clearance between the inner rotor and the outer rotor. If the inner-to-outer rotor radial clearance exceeds the service limit, replace the oil pump assembly.

➡Inner Rotor-to-Outer Rotor Radial Clearance

- Standard (New): 0.002–0.006 inches (0.04–0.16 mm)
- Service Limit: 0.008 inches (0.20 mm)

3. Check the pump housing-to-rotor axial clearance between the rotors and the pump housing. If the pump housing-to-rotor axial clearance exceeds the service limit, replace the oil pump assembly.

➡Pump Housing-to-Rotor Axial Clearance

- Standard (New): 0.001–0.003 inches (0.02–0.07 mm)
- Service Limit: 0.005 inches (0.12 mm)

4. Check the pump housing-to-outer rotor radial clearance between the outer rotor and the pump housing. If the pump housing-to-outer rotor radial clearance exceeds the service limit, replace the oil pump assembly.

➡Pump Housing-to-Outer Rotor Radial Clearance

- Standard (New): 0.004–0.007 inches (0.10–0.19 mm)
- Service Limit: 0.008 inches (0.20 mm)

5. Inspect both rotors and the pump housing for scoring or other damage. Replace the parts, if necessary.
6. Apply liquid thread lock to the pump housing screws, then install the oil pump cover.
7. Check that the oil pump turns freely.

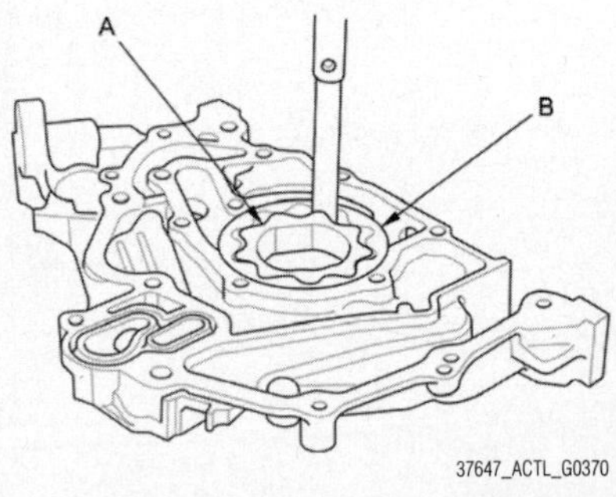
A
B
37647_ACTL_G0370

Fig. 79 Check the inner-to-outer rotor radial clearance between the inner rotor (A) and the outer rotor (B)

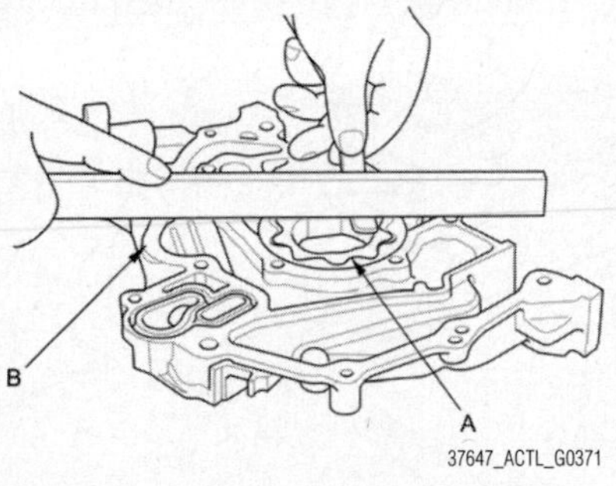
B
A
37647_ACTL_G0371

Fig. 80 Check the pump housing-to-rotor axial clearance between the rotors (A) and the pump housing (B)

PISTON AND RING

POSITIONING

See Figure 82.

REAR MAIN SEAL

REMOVAL & INSTALLATION

1. Remove the transaxle and the drive plate.
2. Remove the transaxle end crankshaft oil seal.
3. Clean and dry the crankshaft oil seal housing.
4. Apply a light coat of new engine oil to the lip of the crankshaft oil seal.
5. Using the driver handle, 15 x 135 L and the oil seal driver attachment, 106 mm, drive in the new crankshaft oil seal until the oil seal driver attachment bottoms against the engine block end cover. Align the hole in the oil seal driver attachment with the pin on the crankshaft.
6. Clean any excess grease off the crankshaft, and check that the oil seal lip is not distorted.

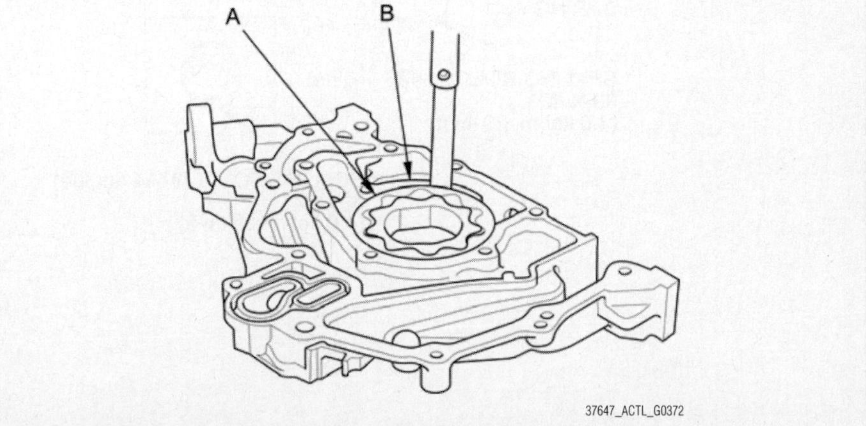
A
B
37647_ACTL_G0372

Fig. 81 Check the pump housing-to-outer rotor radial clearance between the outer rotor (A) and the pump housing (B)

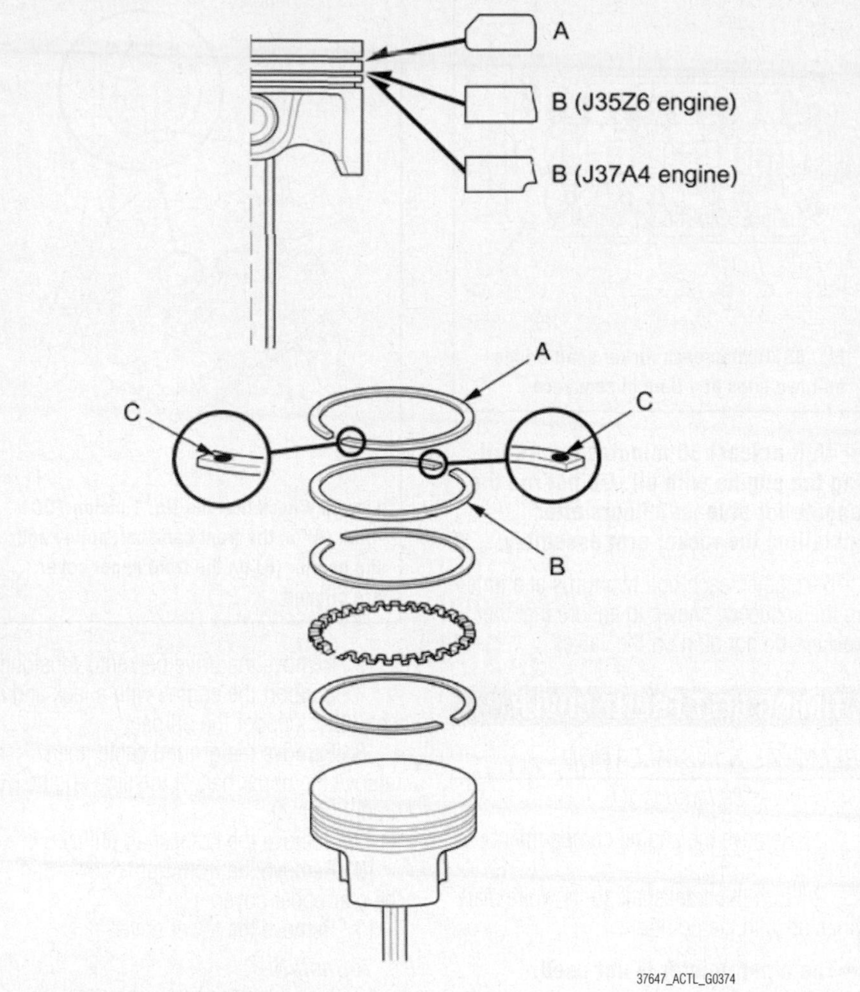
A
B (J35Z6 engine)
B (J37A4 engine)
A
C
C
B
37647_ACTL_G0374

Fig. 82 Piston ring positioning

7. Install the drive plate and the transaxle.

ROCKER ARMS

REMOVAL & INSTALLATION

Front Cylinder Head

See Figures 83 through 85.

1. Remove the cylinder head cover.
2. Loosen the locknuts and the adjusting screws
3. Remove the rocker shaft bridge mounting bolts, the rocker shaft holder mounting bolts, and the rocker arm assembly.
   a. Loosen the rocker shaft bridge mounting bolts and the rocker shaft holder mounting bolts in sequence two turns at a time, to prevent damaging the valves or the rocker arm assembly.
   b. When removing the rocker arm assembly, do not remove the rocker shaft bridge mounting bolts and the rocker shaft holder mounting bolts. The bolts will keep the rocker arms on the shafts.

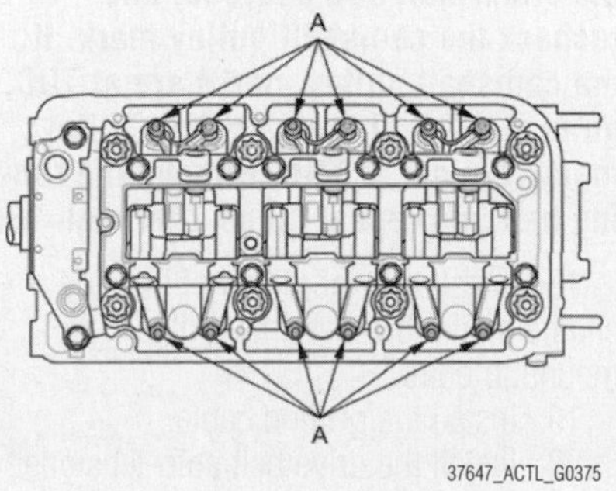
A
A
37647_ACTL_G0375

Fig. 83 Loosen the locknuts and the adjusting screws (A)

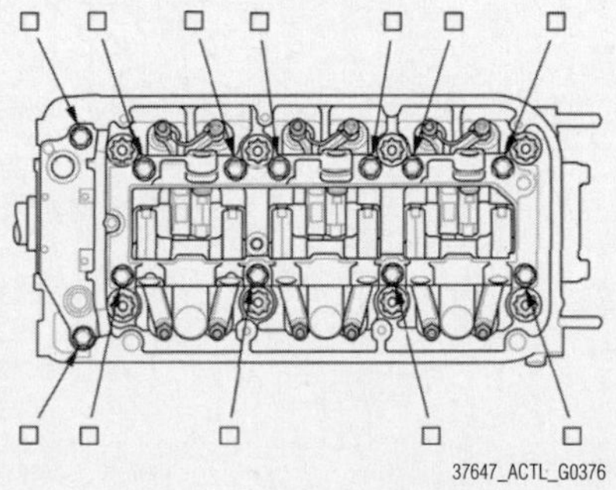
37647_ACTL_G0376

Fig. 84 Front rocker shaft bridge bolt loosening sequence

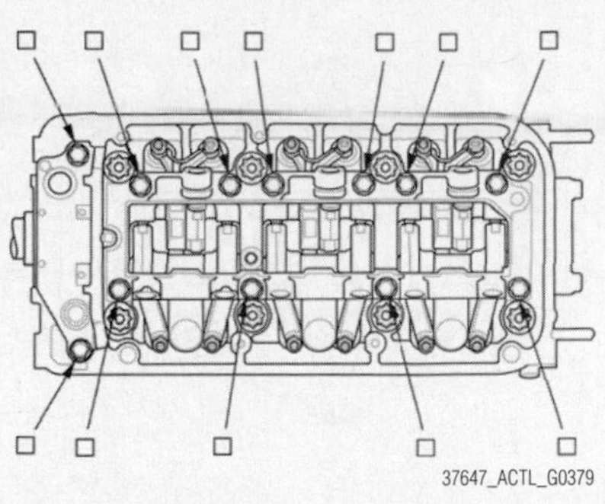
37647_ACTL_G0379

Fig. 85 Tighten each rocker shaft bridge bolt two turns at a time in sequence

To install:

4. Apply liquid gasket, P/N 08717-0004, 08718-0003, or 08718-0009 to the rocker shaft holder mating surface of the cylinder head. Install the component within 5 minutes of applying the liquid gasket.

➡Apply a bead of liquid gasket along the mating surfaces. If you apply liquid gasket P/N 08718-0012, the component must be installed within 4 minutes. If too much time has passed after applying the liquid gasket, remove the old liquid gasket and residue, then reapply the new liquid gasket.

5. Set the rocker arm assembly in place, and loosely install the bolts. Make sure that the rocker arms are properly positioned on the valve stems.

➡Wait at least 30 minutes before filling the engine with oil. Do not run the engine for at least 3 hours after installing the rocker arm assembly.

6. Tighten each bolt two turns at a time in the sequence shown to ensure that the rockers do not bind on the valves.

Rear Cylinder Head

See Figures 86 through 88.

1. Remove the cylinder head cover.
2. Loosen the locknuts and the adjusting screws.
3. Remove the rocker shaft bridge mounting bolts, the rocker shaft holder mounting bolts, and the rocker arm assembly.

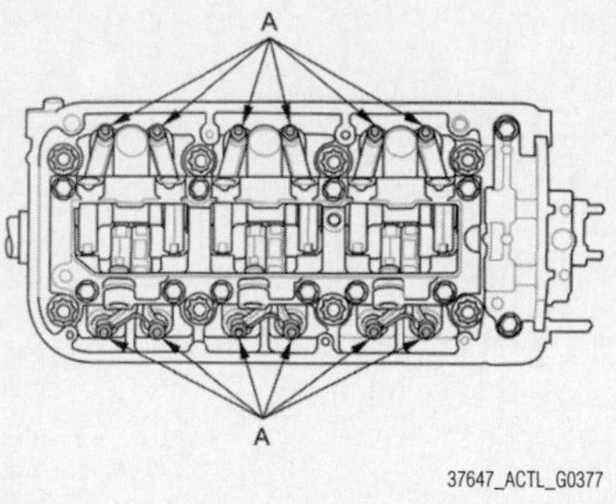
A
A
37647_ACTL_G0377

Fig. 86 Loosen the locknuts and the adjusting screws (A)

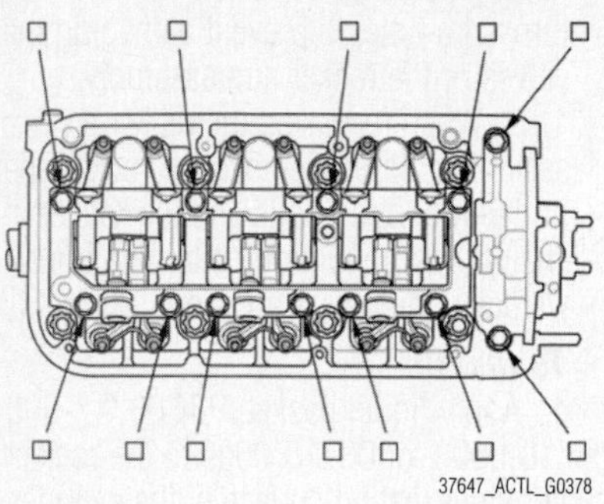
37647_ACTL_G0378

Fig. 87 Rear cylinder head bolt loosening sequence

a. Loosen the rocker shaft bridge mounting bolts and the rocker shaft holder mounting bolts in sequence two turns at a time, to prevent damaging the valves or the rocker arm assembly.

b. When removing the rocker arm assembly, do not remove the rocker shaft bridge mounting bolts and the rocker shaft holder mounting bolts. The bolts will keep the rocker arms on the shafts.

To install:

4. Apply liquid gasket, P/N 08717-0004, 08718-0003, or 08718-0009 to the rocker shaft holder mating surface of the cylinder head. Install the component within 5 minutes of applying the liquid gasket.

➡Apply a bead of liquid gasket along the mating surfaces. If you apply liquid gasket P/N 08718-0012, the component must be installed within 4 minutes. If too much time has passed after applying the liquid gasket, remove the old liquid gasket and residue, then reapply the new liquid gasket.

5. Set the rocker arm assembly in place, and loosely install the bolts. Make sure that the rocker arms are properly positioned on the valve stems.

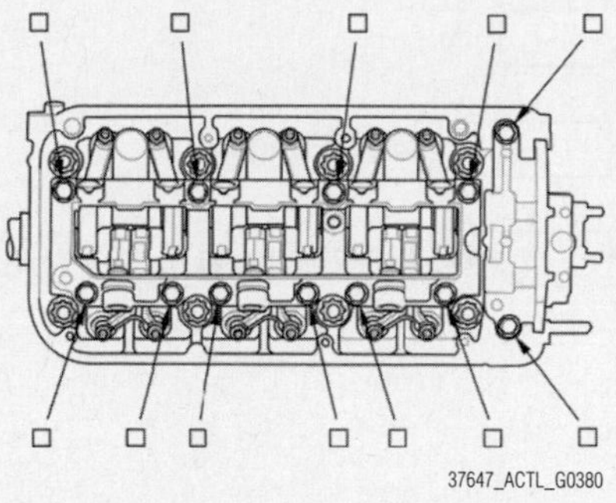
37647_ACTL_G0380

Fig. 88 Tighten each rocker shaft bridge bolt two turns at a time in sequence

➡Wait at least 30 minutes before filling the engine with oil. Do not run the engine for at least 3 hours after installing the rocker arm assembly.

6. Tighten each bolt two turns at a time in the sequence shown to ensure that the rockers do not bind on the valves.

TIMING BELT FRONT COVER

REMOVAL & INSTALLATION

See Figures 89 through 94.

1. Remove the engine compartment covers.
2. Turn the crankshaft so its white mark lines up with the pointer.

➡The other pointer is not used.

3. Check that the No. 1 piston Top Dead Center (TDC) mark on the front camshaft pulley and the pointer on the front upper cover are aligned.

➡If the marks are not aligned, rotate the crankshaft 360 degrees, and recheck the camshaft pulley mark.

4. Raise the vehicle on the lift, then remove the right front wheel.
5. Remove the splash shield.

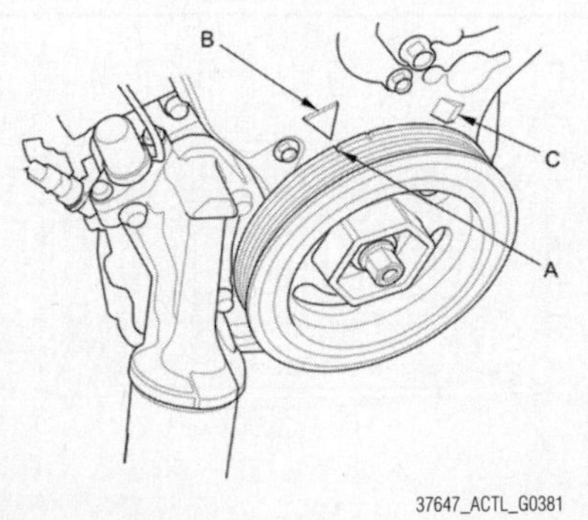
B
C
A
37647_ACTL_G0381

Fig. 89 Turn the crankshaft so its white mark (A) lines up with the pointer (B)

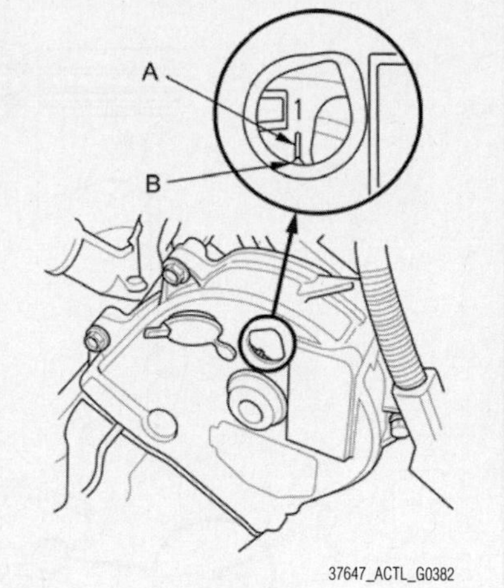
A
B
37647_ACTL_G0382

Fig. 90 Check that the No. 1 piston TDC mark (A) on the front camshaft pulley and the pointer (B) on the front upper cover are aligned

6. Remove the drive belt auto-tensioner.
7. Support the engine with a jack and a wood block under the oil pan.
8. Remove the ground cable, then remove the upper half of the side engine mount bracket.
9. Remove the crankshaft pulley.
10. Remove the front upper cover and the rear upper cover.
11. Remove the lower cover.

To install:

12. Install the lower cover.
13. Install the front upper cover and the rear upper cover.
14. Install the crankshaft pulley.
15. Rotate the crankshaft pulley about six turns clockwise so the timing belt positions itself on the pulleys.
16. Turn the crankshaft pulley so its white mark lines up with the pointer.

➡The other pointer is not used.

17. Check the camshaft pulley marks.

➡If the marks are not aligned, rotate the crankshaft 360 degrees, and recheck the camshaft pulley mark. If the camshaft pulley marks are at TDC, go to step 17. If the camshaft pulley marks are not at TDC, remove the timing belt and repeat steps 2 through 16.

18. Install the upper half of the side engine mount bracket, then tighten the mounting bolts.
19. Install the ground cable.
20. Install the drive belt auto-tensioner.
21. Install the splash shield.
22. Install the right front wheel.
23. Install the engine compartment covers.

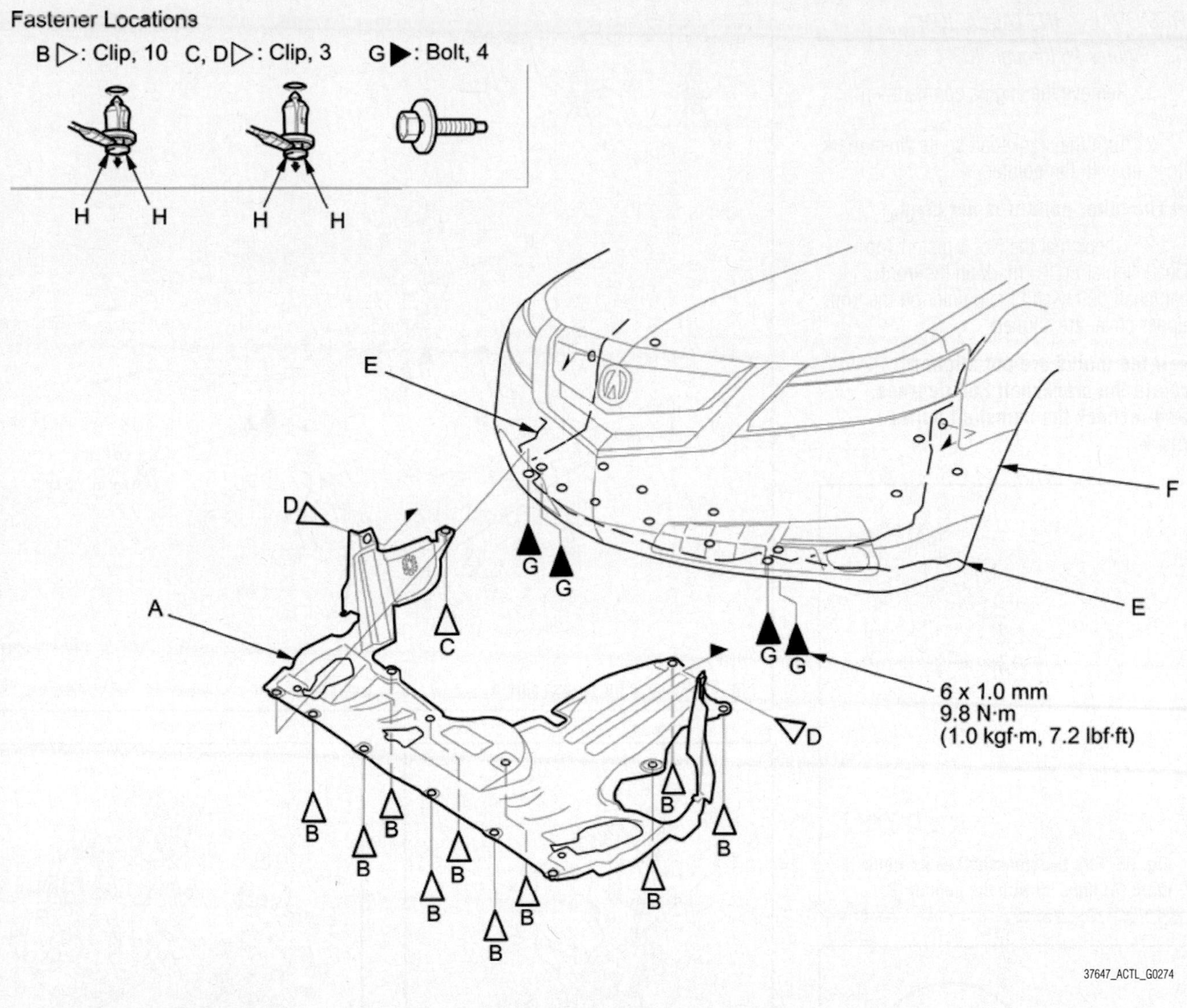

Fastener Locations
B ▷: Clip, 10 C, D ▷: Clip, 3 G ▶: Bolt, 4
H
H
H
H
E
F
D
A
C
G
G
G
G
E
6 x 1.0 mm
9.8 N·m
(1.0 kgf·m, 7.2 lbf·ft)
D
B
B
B
B
B
B
B
B
B
B
37647_ACTL_G0274

Fig. 91 Remove the splash shield

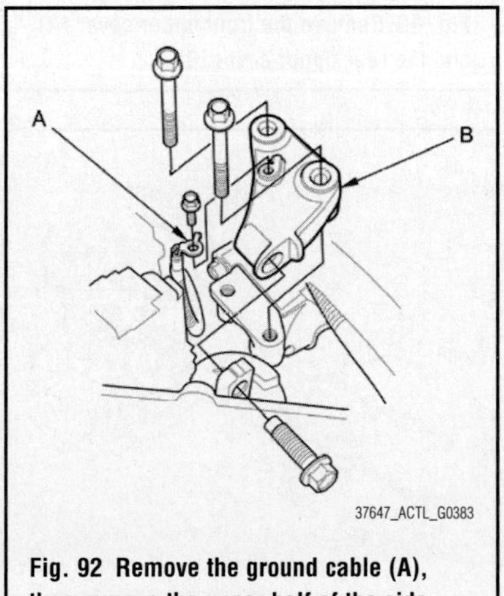

A
B
37647_ACTL_G0383

Fig. 92 Remove the ground cable (A), then remove the upper half of the side engine mount bracket (B)

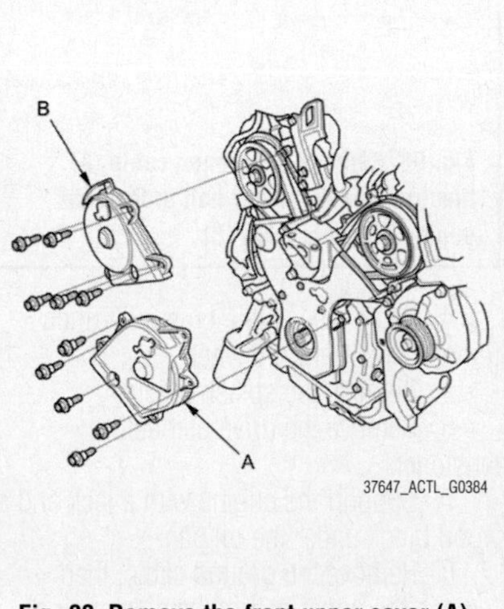

B
A
37647_ACTL_G0384

Fig. 93 Remove the front upper cover (A) and the rear upper cover (B)

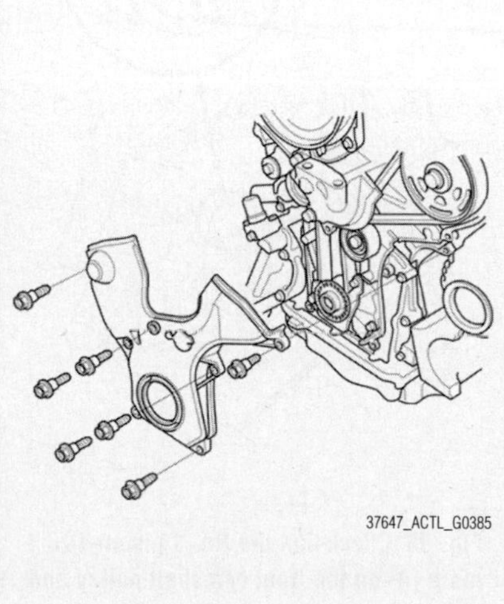

37647_ACTL_G0385

Fig. 94 Remove the lower cover

TIMING BELT & SPROCKETS

REMOVAL & INSTALLATION

See Figures 95 through 108.

1. Remove the engine compartment covers.
2. Turn the crankshaft so its white mark lines up with the pointer.

➡**The other pointer is not used.**

3. Check that the No. 1 piston Top Dead Center (TDC) mark on the front camshaft pulley and the pointer on the front upper cover are aligned.

➡**If the marks are not aligned, rotate the crankshaft 360 degrees, and recheck the camshaft pulley mark.**

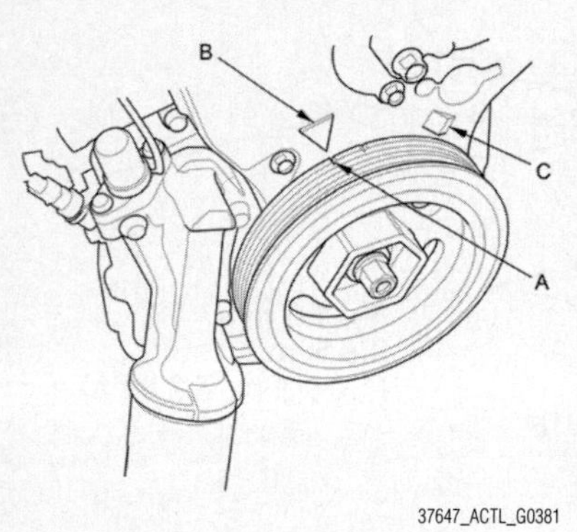
B
C
A
37647_ACTL_G0381

Fig. 95 Turn the crankshaft so its white mark (A) lines up with the pointer (B)

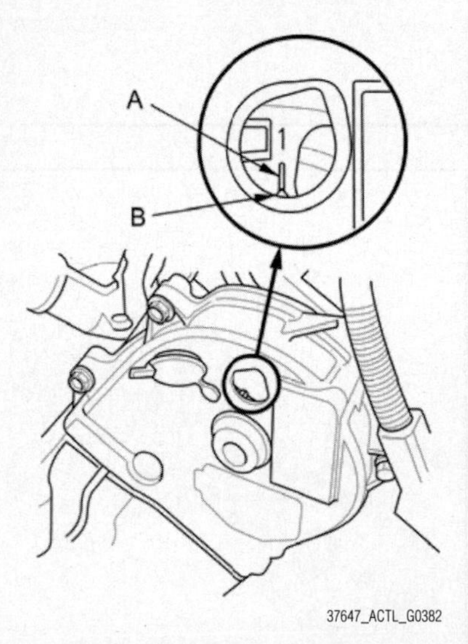
A
1
B
37647_ACTL_G0382

Fig. 96 Check that the No. 1 piston TDC mark (A) on the front camshaft pulley and the pointer (B) on the front upper cover are aligned

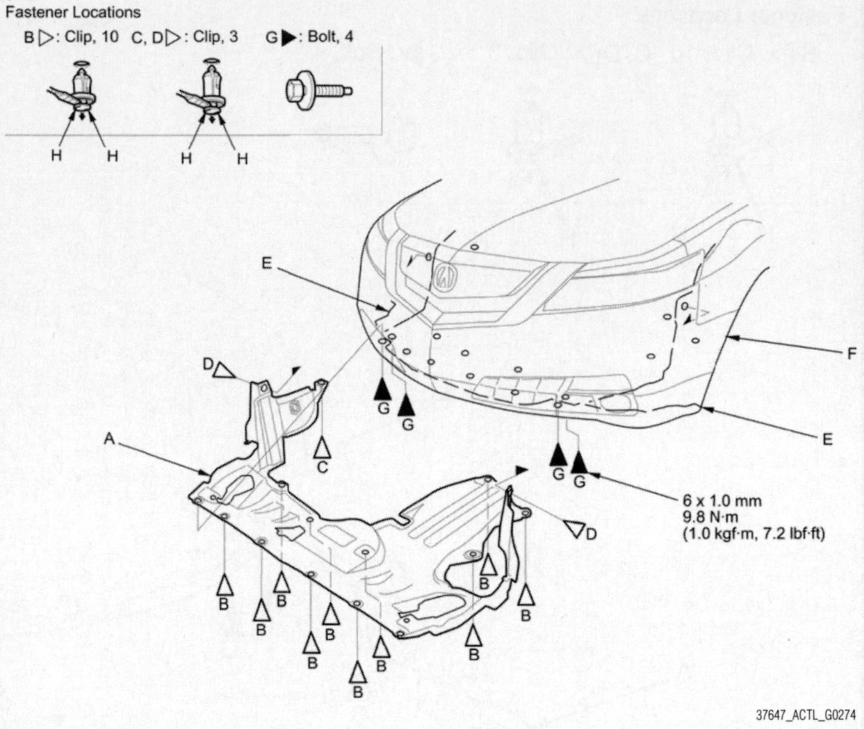
Fastener Locations
B ▷: Clip, 10 C, D▷: Clip, 3 G▶: Bolt, 4
H H H H
E
F
D
G G
A
C
E
G G
6 x 1.0 mm
9.8 N·m
(1.0 kgf·m, 7.2 lbf·ft)
D
B B B B B B B B B B
37647_ACTL_G0274

Fig. 97 Remove the splash shield

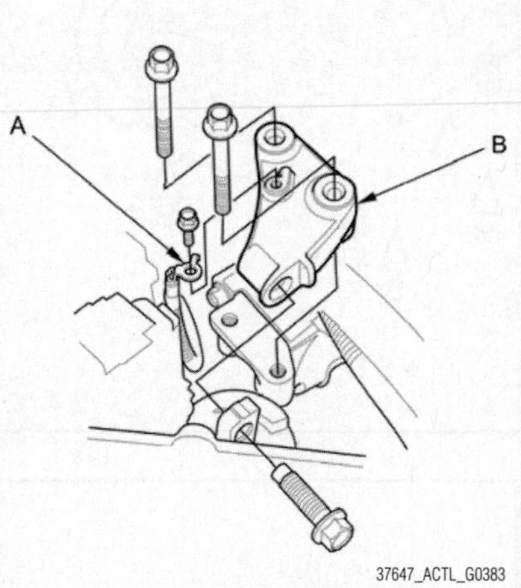
A
B
37647_ACTL_G0383

Fig. 98 Remove the ground cable (A), then remove the upper half of the side engine mount bracket (B)

4. Raise the vehicle on the lift, then remove the right front wheel.
5. Remove the splash shield.
6. Remove the drive belt auto-tensioner.
7. Support the engine with a jack and a wood block under the oil pan.
8. Remove the ground cable, then remove the upper half of the side engine mount bracket.

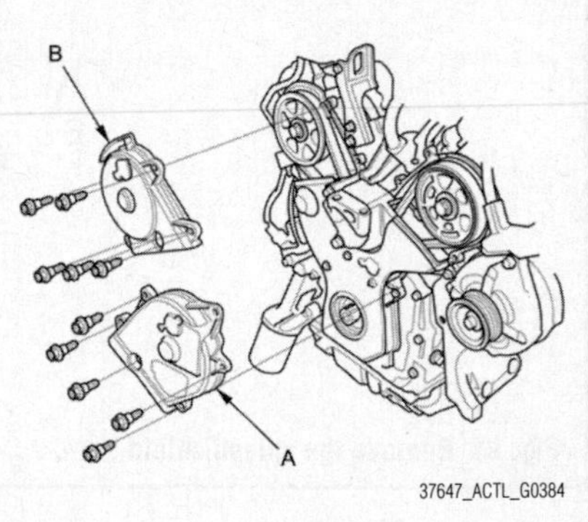
B
A
37647_ACTL_G0384

Fig. 99 Remove the front upper cover (A) and the rear upper cover (B)

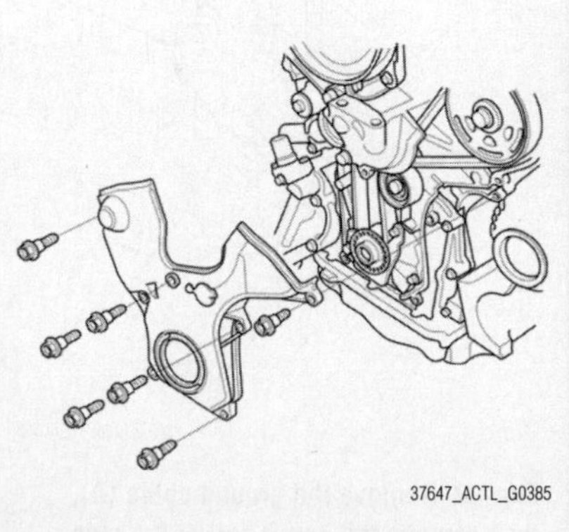
37647_ACTL_G0385

Fig. 100 Remove the lower cover

9. Remove the crankshaft pulley.
10. Remove the front upper cover and the rear upper cover.
11. Remove the lower cover.
12. Remove one of the battery clamp bolts from the battery tray, and grind the end of it.
13. Thread the battery clamp bolt into hold the timing belt adjuster in its current position. Tighten it by hand, do not use a wrench.
14. Remove the timing belt guide plate.
15. Remove the lower half of the side engine mount bracket.
16. Remove the idler pulley bolt and the idler pulley, then remove the timing belt. Discard the idler pulley bolt.

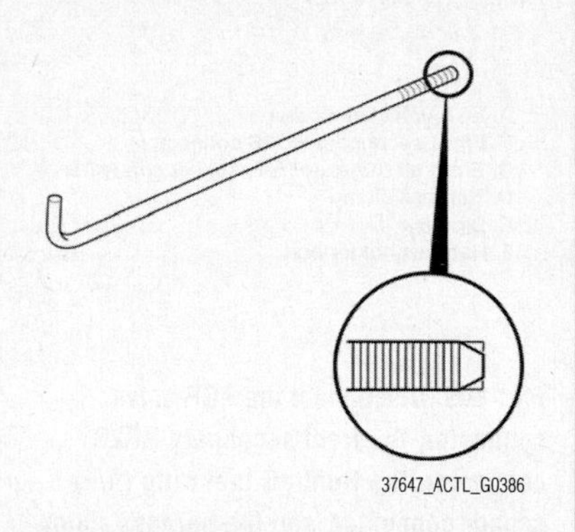
37647_ACTL_G0386

Fig. 101 Remove one of the battery clamp bolts from the battery tray, and grind the end of it

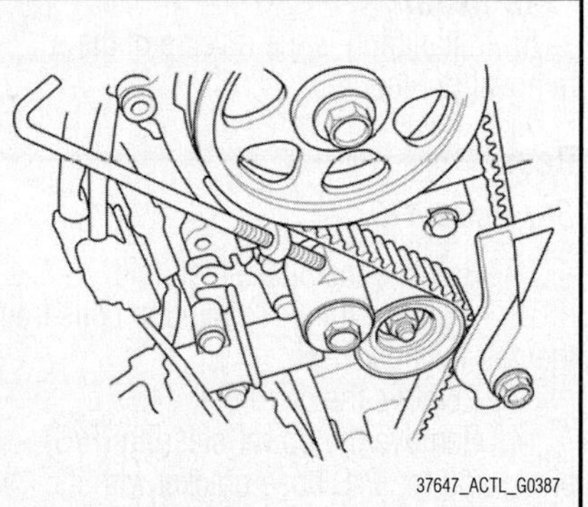
37647_ACTL_G0387

Fig. 102 Thread the battery clamp bolt into hold the timing belt adjuster in its current position

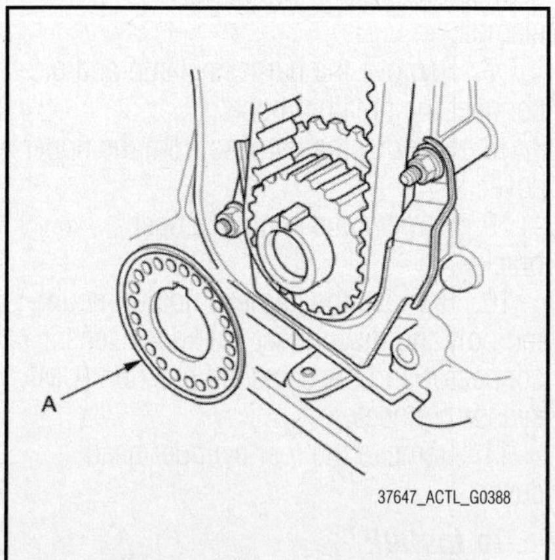
A
37647_ACTL_G0388

Fig. 103 Remove the timing belt guide plate (A)

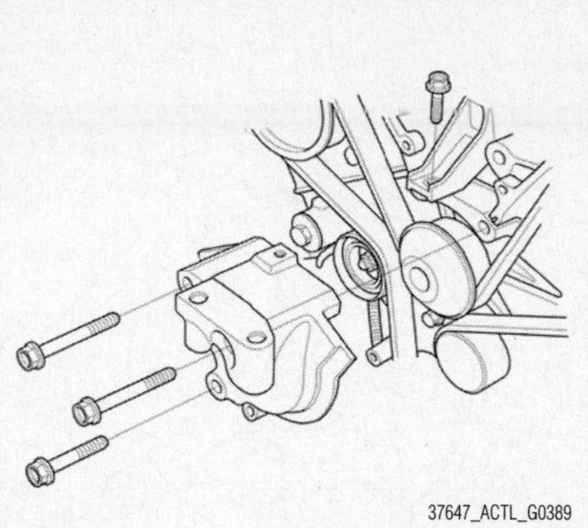
37647_ACTL_G0389

Fig. 104 Remove the lower half of the side engine mount bracket

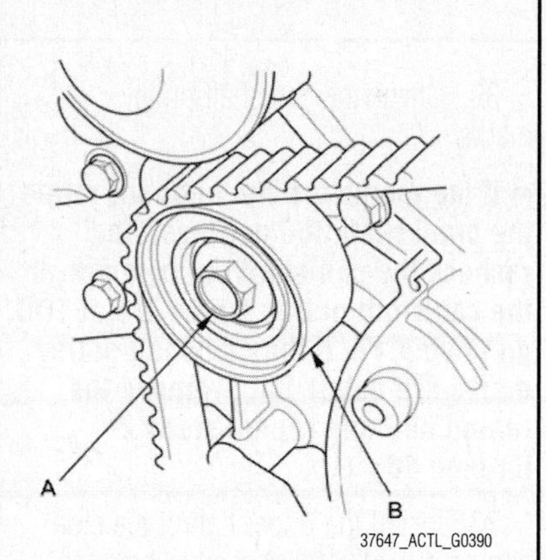
A
B
37647_ACTL_G0390

Fig. 105 Remove the idler pulley bolt (A) and the idler pulley (B), then remove the timing belt

To install:

17. Clean the timing belt pulleys, the timing belt guide plate, and the upper and lower covers.
18. Set the timing belt drive pulley to Top Dead Center (TDC) by aligning the TDC mark on the tooth of the timing belt drive pulley with the pointer on the oil pump.
19. Set the camshaft pulleys to TDC by aligning the TDC marks on the camshaft pulleys with the pointers on the back covers.
20. Loosely install the idler pulley with a new idler pulley bolt so the pulley can move but does not come off.
21. If the auto-tensioner has extended and the timing belt cannot be installed, do the timing belt replacement procedure.
22. Install the timing belt in a counter-clockwise sequence starting with the drive pulley. Take care not to damage the timing belt during installation.
23. Tighten the idler pulley bolt to 33 ft. lbs. (44 Nm).

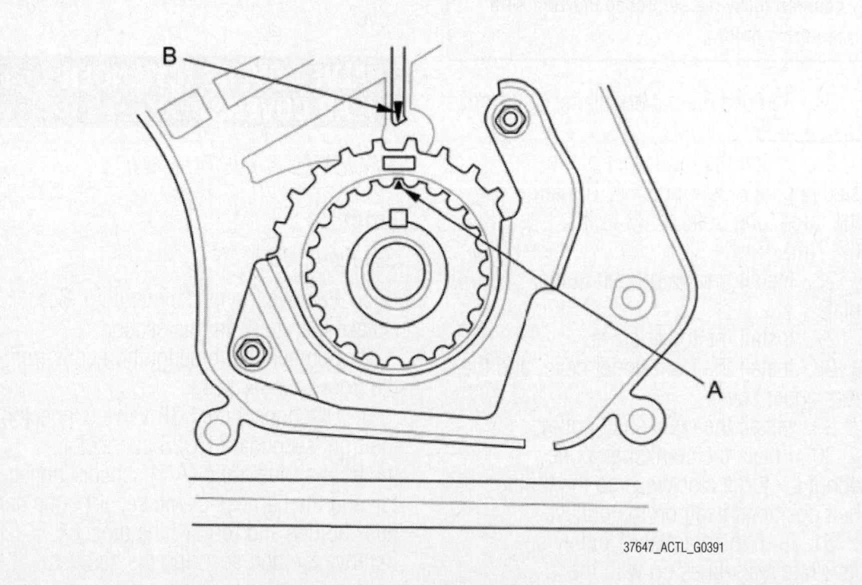
B
A
37647_ACTL_G0391

Fig. 106 Set the timing belt drive pulley to TDC by aligning the TDC mark (A) on the tooth of the timing belt drive pulley with the pointer (B) on the oil pump

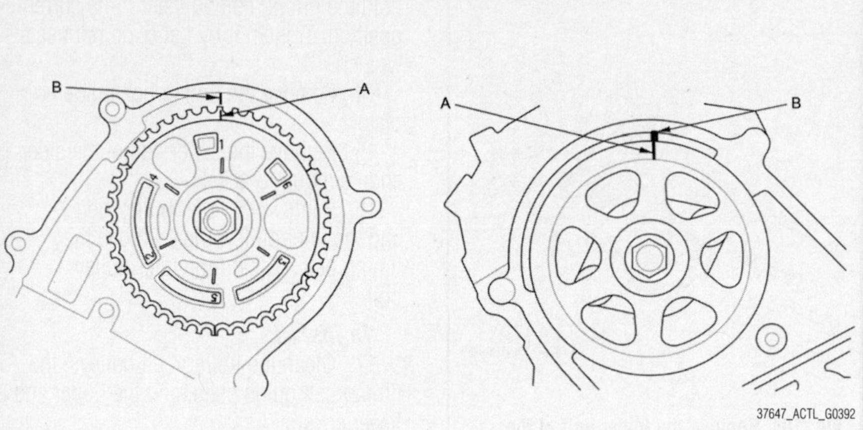
B
A
A
B
37647_ACTL_G0392

Fig. 107 Set the camshaft pulleys to TDC by aligning the TDC marks (A) on the camshaft pulleys with the pointers (B) on the back covers

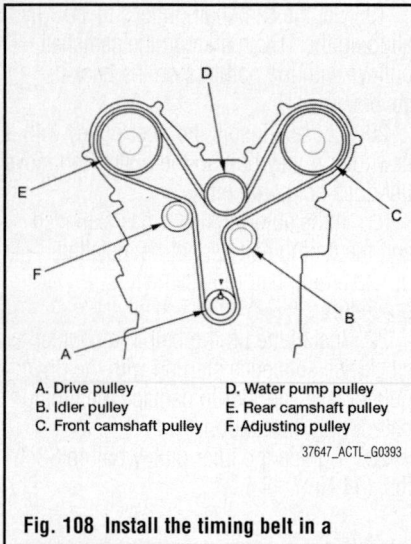

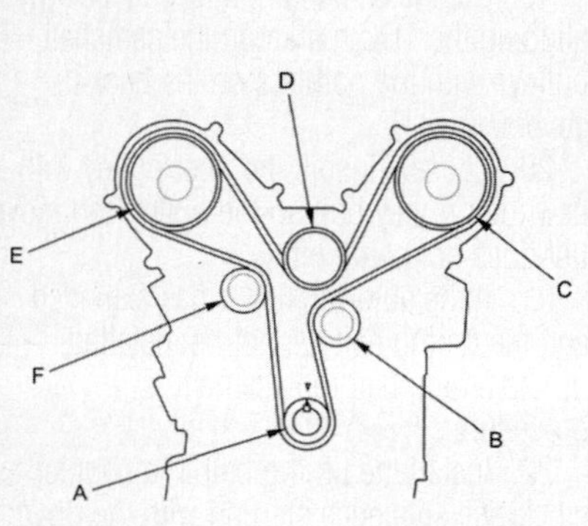
D
E
C
F
B
A
A. Drive pulley
B. Idler pulley
C. Front camshaft pulley
D. Water pump pulley
E. Rear camshaft pulley
F. Adjusting pulley
37647_ACTL_G0393

Fig. 108 Install the timing belt in a counterclockwise sequence starting with the drive pulley

24. Remove the battery clamp bolt from the back cover.
25. Install the lower half of the side engine mount bracket. Tighten the three long bolts to 33 ft. lbs. (44 Nm).
26. Install the timing belt guide plate.
27. Install the lower cover.
28. Install the front upper cover and the rear upper cover.
29. Install the crankshaft pulley.
30. Rotate the crankshaft pulley about six turns clockwise so the timing belt positions itself on the pulleys.
31. Turn the crankshaft pulley so its white mark lines up with the pointer.

➡The other pointer is not used.

32. Check the camshaft pulley marks.

➡If the marks are not aligned, rotate the crankshaft 360 degrees, and recheck the camshaft pulley mark. If the camshaft pulley marks are at TDC, go to step 17. If the camshaft pulley marks are not at TDC, remove the timing belt and repeat steps 2 through 16.

33. Install the upper half of the side engine mount bracket, then tighten the mounting bolts.
34. Install the ground cable.
35. Install the drive belt auto-tensioner.
36. Install the splash shield.
37. Install the right front wheel.
38. Install the engine compartment covers.

VALVE COVERS (CYLINDER HEAD COVERS)

REMOVAL & INSTALLATION

Front

See Figure 109.

1. Remove the intake manifold. See "Intake Manifold" in this section.
2. Remove the three ignition coils from the front cylinder head.
3. Disconnect the EGR valve connector, the front secondary HO2S connector, the front air fuel ratio (A/F) sensor connector, and the harness clamp securing the harness holder, and remove the dipstick. Remove the bolt securing the harness holder.
4. Remove the front cylinder head cover.

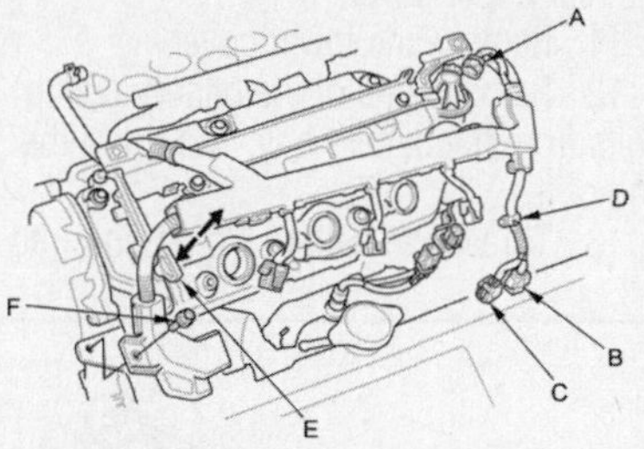
A
D
F
B
C
E
A. EGR valve connector
B. Front secondary HO2S connector
C. Front air fuel ratio (A/F) sensor connector
D. Harness clamp
E. Dipstick
F. Harness holder bolt
37647_AZDX_G0313

Fig. 109 Disconnect the EGR valve connector, the front secondary HO2S connector, the front air fuel ratio (A/F) sensor connector, and the harness clamp securing the harness holder, and remove the dipstick. Remove the bolt securing the harness holder.

To install:

5. Installation is the reverse of the removal procedure.

Rear

See Figure 110.

1. Remove the intake manifold.
2. Remove the three ignition coils from the rear cylinder head.
3. Remove the drive belt.
4. Remove the power steering (P/S) pump and the P/S hose bracket with its hoses connected.
5. Remove the harness holder mounting bolts and the engine ground cable.
6. Disconnect the three injector connectors and the two harness clamps.
7. Remove the harness clamp and disconnect the breather hose.
8. Remove the harness from the upper cover.
9. Remove the engine hanger bracket.
10. Remove the harness holder mounting bolt and disconnect the knock sensor connector and the camshaft position (CMP) sensor connector.
11. Remove the rear cylinder head cover.

To install:

12. Installation is the reverse of the removal procedure.

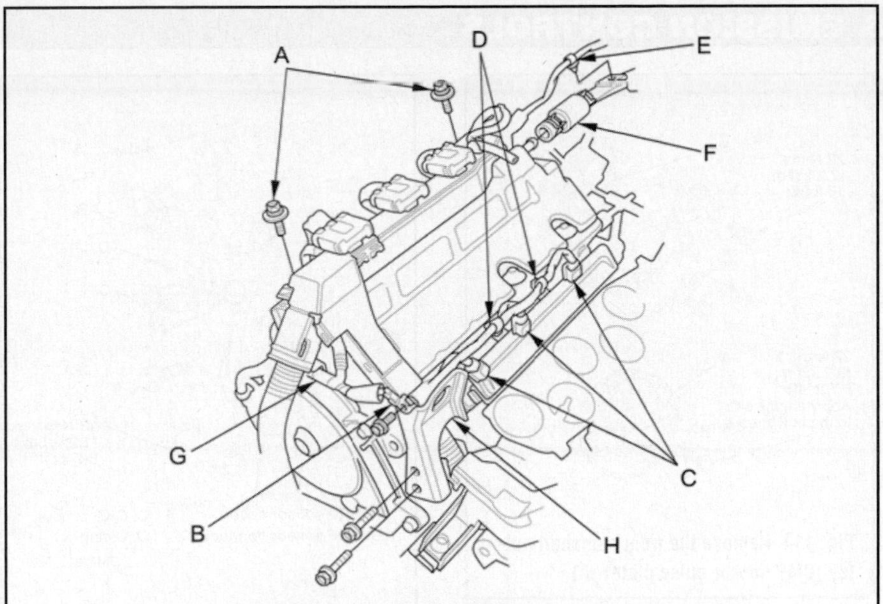
A
D
E
F
G
C
B
H

A. Harness holder mounting bolts
B. Engine ground cable
C. Injector connectors
D. Harness clamps
E. Harness clamp
F. Breather hose
G. Harness on the upper cover
H. Engine hanger bracket

37647_AZDX_G0314

Fig. 110 Disconnecting and removing component to access the rear cylinder head cover

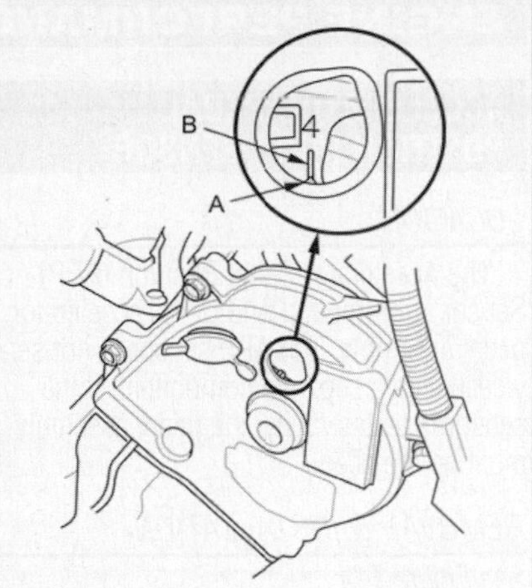
B
4
A

37647_ACTL_G0394

Fig. 112 Rotate the crankshaft clockwise. Align the pointer (A) on the front upper cover with the No. 4 piston TDC mark (B) on the front camshaft pulley

VALVE LASH

ADJUSTMENT

See Figures 111 and 112.

➡Connect the Honda Diagnostic System (HDS) to the data link connector (DLC), and monitor the Engine Coolant Temperature (ECT) sensor 1. Adjust the valve clearance only when the ECT sensor 1 is less than 100°F (38°C).

1. Remove the cylinder head covers.
2. Set the No. 1 piston at Top Dead Center (TDC). Align the pointer on the front upper cover with the No. 1 piston TDC mark on the front camshaft pulley.
3. Select the correct feeler gauge for the valve clearance you are going to check.

➡Valve Clearance:

- Intake: 0.008–0.009 inches (0.20–0.24 mm)
- Exhaust: 0.011–0.013 inches (0.28–0.32 mm)

4. Insert the feeler gauge between the adjusting screw and the end of the valve stem on the No. 1 cylinder, and slide it back and forth; you should feel a slight amount of drag.

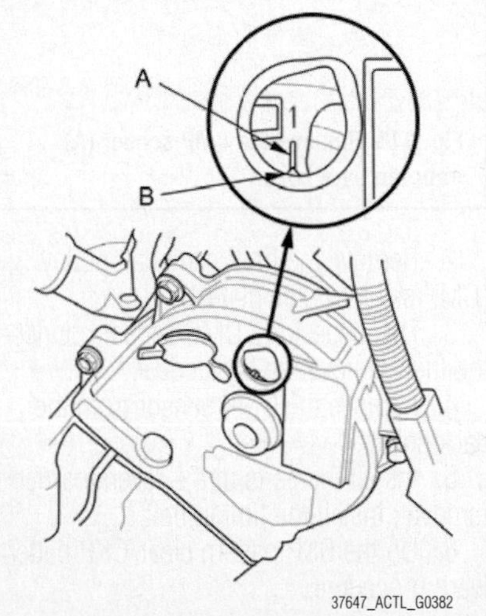
A
1
B

37647_ACTL_G0382

Fig. 111 Set the No. 1 piston at TDC. Align the polnter (A) on the front upper cover with the No. 1 piston TDC mark (B) on the front camshaft pulley

5. If you feel too much or too little drag, loosen the locknut, and turn the adjusting screw until the drag on the feeler gauge is correct.
6. While holding the adjusting screw with the screw driver, tighten the locknut, then recheck the clearance. Repeat the adjustment, if necessary.

➡Specified Torque:

- Intake: 14 ft. lbs. (20 Nm)
- Exhaust: 10 ft. lbs. (14 Nm)

➡Apply new engine oil to the nut threads.

7. Rotate the crankshaft clockwise. Align the pointer on the front upper cover with the No. 4 piston TDC mark on the front camshaft pulley.
8. Check and, if necessary, adjust the valve clearance on the No. 4 cylinder.
9. Rotate the crankshaft clockwise. Align the pointer on the front upper cover with the No. 2 piston TDC mark on the front camshaft pulley. Check and, if necessary, adjust the valve clearance on the No. 2 cylinder.
10. Rotate the crankshaft clockwise. Align the pointer on the front upper cover with the No. 5 piston TDC mark on the front camshaft pulley.
11. Check and, if necessary, adjust the valve clearance on the No. 5 cylinder.
12. Rotate the crankshaft clockwise. Align the pointer on the front upper cover with the No. 3 piston TDC mark on the front camshaft pulley.
13. Check and, if necessary, adjust the valve clearance on the No. 3 cylinder.
14. Rotate the crankshaft clockwise. Align the pointer on the front upper cover with the No. 6 piston TDC mark on the front camshaft pulley.
15. Check and, if necessary, adjust the valve clearance on the No. 6 cylinder.
16. Install the cylinder head covers.

ACCELERATOR PEDAL POSITION (APP) SENSOR

LOCATION

The Accelerator Pedal Position (APP) Sensor is on the top end of the accelerator pedal assembly. The APP sensor is not serviceable as a separate component. If the sensor is tested faulty, the pedal assembly must be replaced.

REMOVAL & INSTALLATION

See Figure 113.

1. Disconnect the accelerator pedal position (APP) sensor 6P connector.
2. Remove the accelerator pedal module.

➡The APP sensor is not available separately. Do not disassemble the accelerator pedal module.

3. Install the parts in the reverse order of removal.

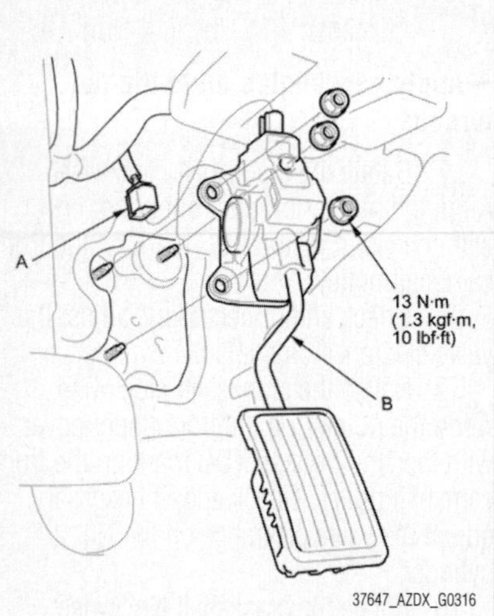
A
13 N·m
(1.3 kgf·m,
10 lbf·ft)
B
37647_AZDX_G0316

Fig. 113 Disconnect the accelerator pedal position (APP) sensor 6P connector (A). Remove the accelerator pedal module (B).

CAMSHAFT POSITION (CMP) SENSOR

LOCATION

The Camshaft Position (CMP) sensor is located in the front of the cylinder head, behind the upper camshaft pulley plate.

REMOVAL & INSTALLATION

See Figures 114 and 115.

1. Remove the timing belt.

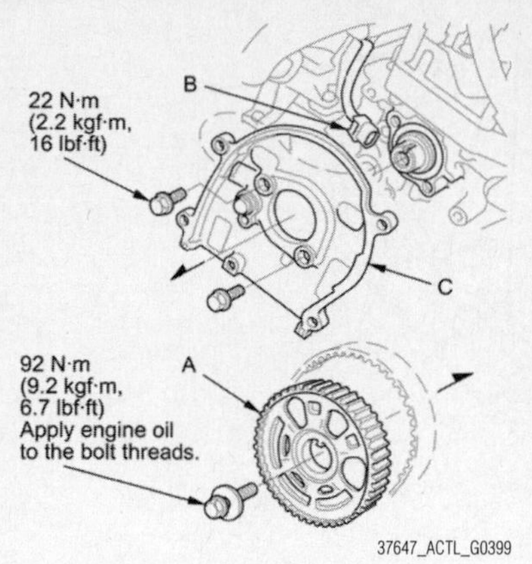
22 N·m
(2.2 kgf·m,
16 lbf·ft)
B
C
92 N·m
(9.2 kgf·m,
6.7 lbf·ft)
Apply engine oil
to the bolt threads.
A
37647_ACTL_G0399

Fig. 114 Remove the front camshaft pulley (CMP sensor pulse plate) (A)

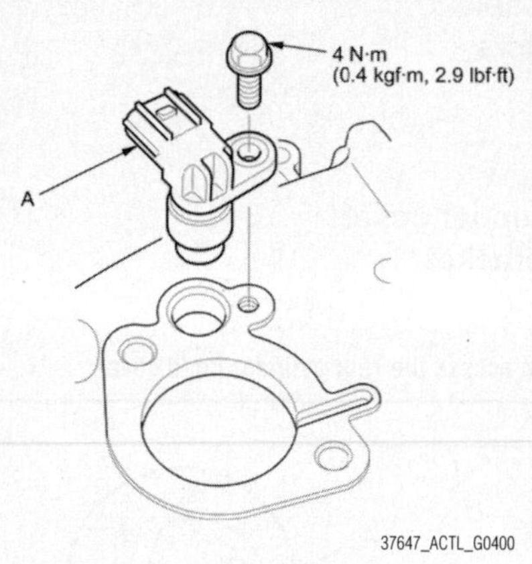
4 N·m
(0.4 kgf·m, 2.9 lbf·ft)
A
37647_ACTL_G0400

Fig. 115 Remove the CMP sensor (A) from the back cover

2. Remove the front camshaft pulley (CMP sensor pulse plate).
3. Disconnect the CMP sensor connector, then remove the back cover.
4. Remove the CMP sensor from the back cover.
5. Install the parts in the reverse order of removal. Install the timing belt.
6. Do the CKP pattern clear/CKP pattern learn procedure.

CRANKSHAFT POSITION (CKP) SENSOR

LOCATION

The Crankshaft Position (CKP) Sensor is located on the forward side of the engine block, near the crank pulley.

REMOVAL & INSTALLATION

See Figure 116.

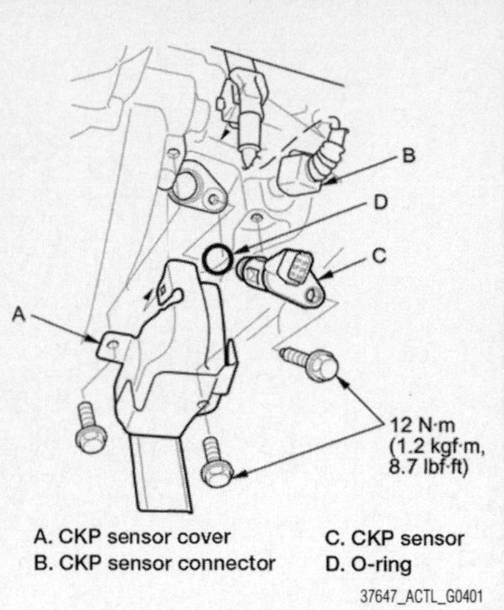
B
D
C
A
12 N·m
(1.2 kgf·m,
8.7 lbf·ft)
A. CKP sensor cover
B. CKP sensor connector
C. CKP sensor
D. O-ring
37647_ACTL_G0401

Fig. 116 Exploded view of CKP sensor assembly

1. Raise the vehicle on a lift.

➡Make the vehicle is level, because engine oil will drip out when you remove the sensor.

2. Remove the CKP sensor cover.
3. Disconnect the CKP sensor connector.
4. Remove the CKP sensor.
5. Install the parts in the reverse order of removal with a new O-ring.
6. Do the CKP pattern clear/CKP pattern learn procedure.
7. Check for engine oil level, and add more oil if needed.

CKP PATTERN CLEAR/CKP PATTERN LEARN

Clear/Learn Procedure (with the HDS)

1. Connect the HDS to the Data Link Connector (DLC) located under the driver's side of the dashboard.
2. Turn the ignition switch to ON (II), or press the engine start/stop button to select the ON mode.
3. Make sure the HDS communicates with the ECM/PCM and other vehicle systems. If it doesn't, go to the DLC circuit troubleshooting.
4. Select CRANK PATTERN in the ADJUSTMENT MENU with the HDS.
5. Select CRANK PATTERN CLEAR, and clear the CKP pattern.
6. Select CRANK PATTERN LEARNING with the HDS, and follow the screen prompts.

Learn Procedure (without the HDS)

1. Start the engine. Hold the engine speed at 3,000 rpm without load (A/T in P or N, M/T in neutral) until the radiator fan comes on.
2. Test-drive the vehicle on a level road: Decelerate (with the throttle fully closed) from an engine speed of 2,500 rpm down to 1,000 rpm with the Transmission in 2nd.
3. Repeat step 2 several times.
4. Turn the ignition switch to LOCK (0), or press the engine start/stop button to select the OFF mode.
5. Turn the ignition switch to ON (II), or press the engine start/stop button to select the ON mode, and wait 30 seconds.

ENGINE COOLANT TEMPERATURE (ECT) SENSOR

LOCATION

There are two Engine Coolant Temperature (ECT) sensors used on this engine. ECT 1 is located on top of the engine, while ECT 2 is located on the lower front edge of the engine.

REMOVAL & INSTALLATION

ECT Sensor 1

See Figure 117.

1. Drain the engine coolant.
2. Remove the engine cover.
3. Disconnect the ECT sensor 1 connector.
4. Remove ECT sensor 1 and O-ring.
5. Install the parts in the reverse order of removal with a new O-ring, then refill the radiator with engine coolant.

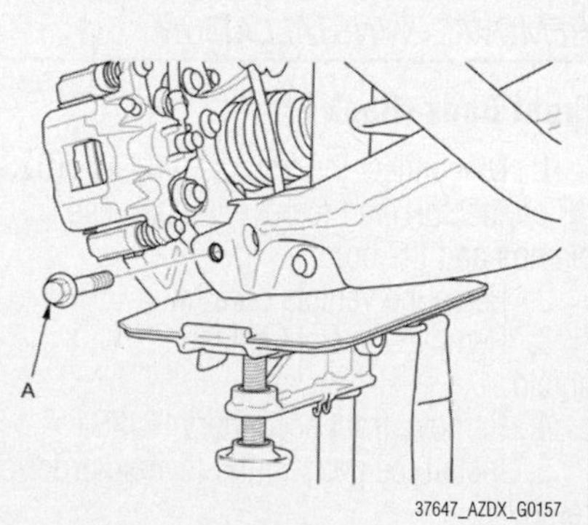
A
37647_AZDX_G0157

Fig. 117 Disconnect the ECT sensor 1 connector (A). Remove ECT sensor 1 (B) and O-ring (C).

ECT Sensor 2

See Figure 118.

1. Drain the engine coolant.
2. Remove the front splash shield.
3. Disconnect the ECT sensor 2 connector.
4. Remove ECT sensor 2 and O-ring.
5. Install the parts in the reverse order of removal with a new O-ring, then refill the radiator with engine coolant.

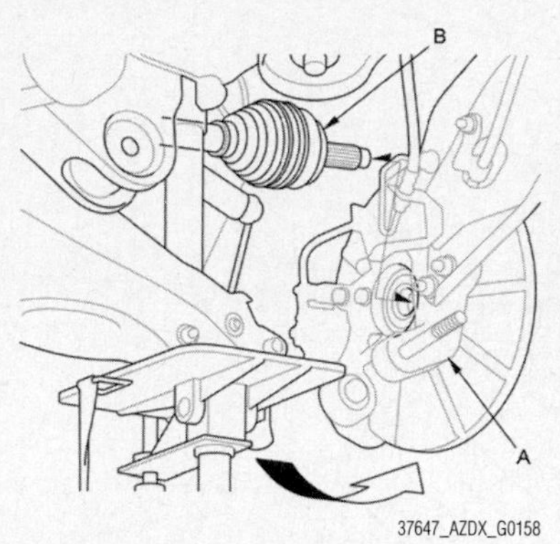
B
A
37647_AZDX_G0158

Fig. 118 Disconnect the ECT sensor 2 connector (A). Remove ECT sensor 2 (B) and O-ring (C).

EVAPORATIVE EMISSIONS (EVAP) CANISTER

LOCATION

See Figure 119.

The Evaporative Emissions (EVAP) canister is located under the vehicle, just in front of the fuel tank.

REMOVAL & INSTALLATION

See Figures 120 and 121.

1. Remove the EVAP canister cover.
2. Disconnect the quick-connect fitting, the hoses, the EVAP canister vent shut valve 2P connector, and the FTP sensor 3P connector, and remove the harness clamps.
3. Remove the bolts.
4. Remove the EVAP canister assembly.
5. Remove the EVAP canister from the bracket.

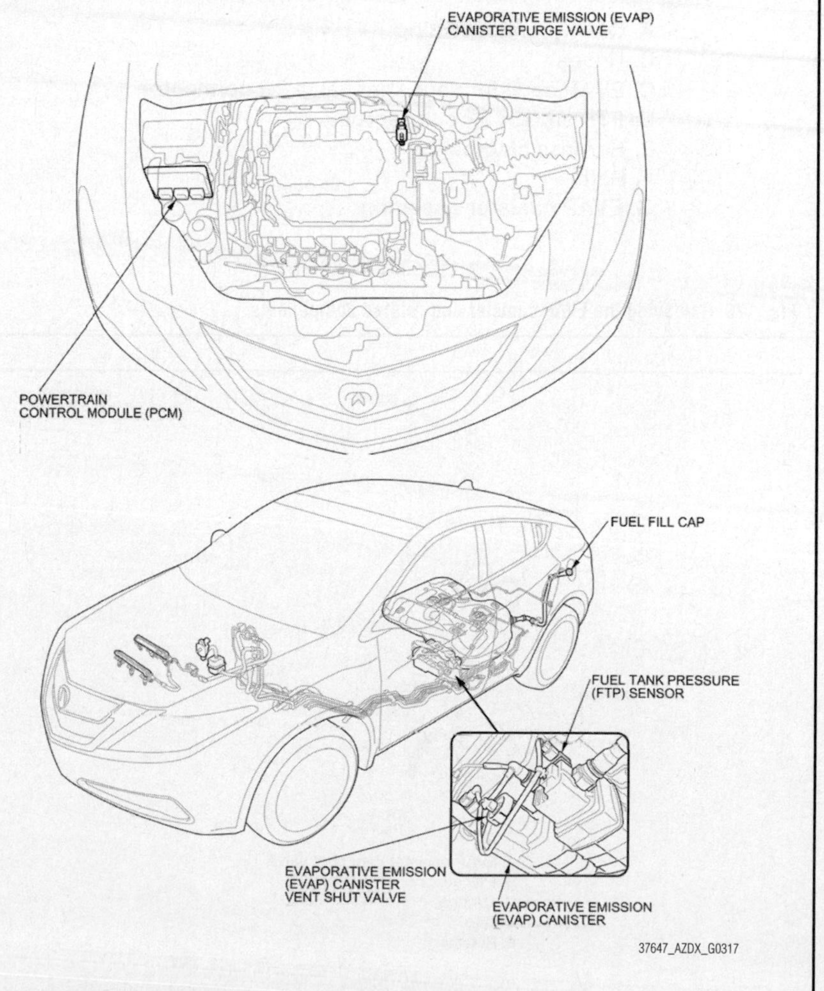
EVAPORATIVE EMISSION (EVAP) CANISTER PURGE VALVE
POWERTRAIN CONTROL MODULE (PCM)
FUEL FILL CAP
FUEL TANK PRESSURE (FTP) SENSOR
EVAPORATIVE EMISSION (EVAP) CANISTER VENT SHUT VALVE
EVAPORATIVE EMISSION (EVAP) CANISTER
37647_AZDX_G0317

Fig. 119 Showing the location of the EVAP system components

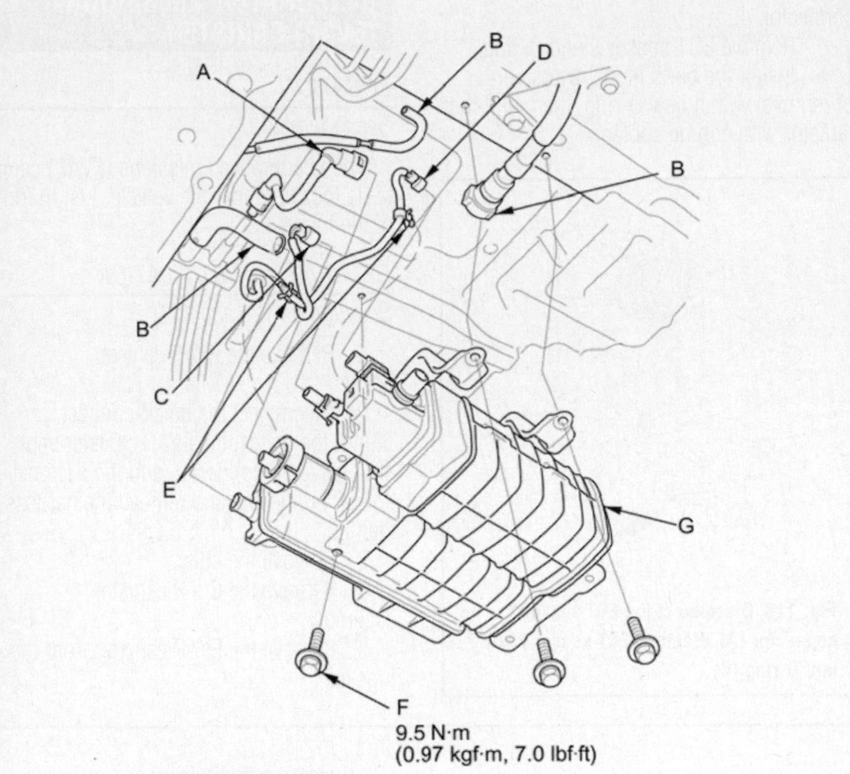
A
B
D
B
B
C
E
G
F
9.5 N·m
(0.97 kgf·m, 7.0 lbf·ft)
A. Quick-connect fitting
B. Hoses
C. EVAP canister vent shut valve 2P connector
D. FTP sensor 3P connector
E. Harness clamps
F. Bolts
G. EVAP canister assembly
37647_AZDX_G0318

Fig. 120 Removing the EVAP canister and related components

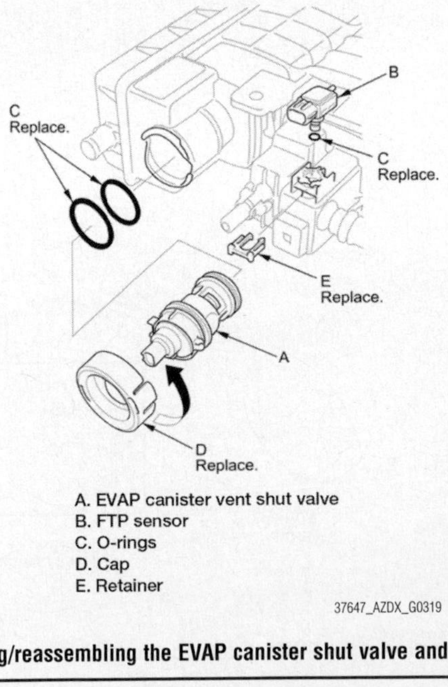
B
C
Replace.
C
Replace.
E
Replace.
A
D
Replace.
A. EVAP canister vent shut valve
B. FTP sensor
C. O-rings
D. Cap
E. Retainer
37647_AZDX_G0319

Fig. 121 Disassembling/reassembling the EVAP canister shut valve and FTP sensor

6. Remove the EVAP canister vent shut valve and the FTP sensor from the EVAP canister.

To install:

7. Reassemble the EVAP canister with new O-rings, a new cap, and a new retainer, then install the EVAP canister bracket.

➡Do not coat the O-rings with oil.

8. Install the parts in the reverse order of removal.

EXHAUST GAS RECIRCULATION (EGR) VALVE

LOCATION

See Figure 122.

Refer to the accompanying illustration.

REMOVAL & INSTALLATION

See Figure 123.

1. Disconnect the EGR valve 5P connector.
2. Remove the EGR valve.
3. Install the parts in the reverse order of removal with a new gasket.

HEATED OXYGEN SENSOR (HO2S)

LOCATION

The secondary Heated Oxygen Sensor (HO2S) is located in the lower end of the catalytic converter.

REMOVAL & INSTALLATION

Front Bank (Bank 2)

1. Disconnect the front secondary HO2S 4P connector, then remove the harness clamps and the bolt.
2. Raise the vehicle on a lift.
3. Remove the front splash shield.
4. Remove front secondary HO2S.
5. Install the parts in the reverse order of removal.

Rear Bank (Bank 1)

1. Raise the vehicle on a lift.
2. Disconnect the rear secondary HO2S 4P connector, and remove the harness clamps.
3. Remove rear secondary HO2S.
4. Install the parts in the reverse order of removal.

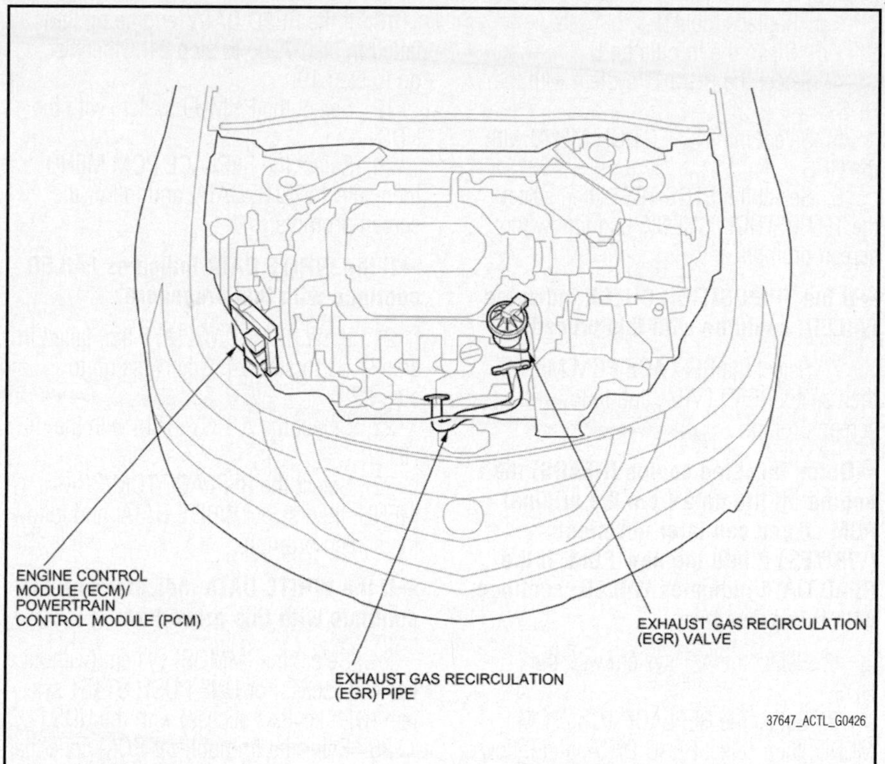
ENGINE CONTROL MODULE (ECM)/ POWERTRAIN CONTROL MODULE (PCM)
EXHAUST GAS RECIRCULATION (EGR) VALVE
EXHAUST GAS RECIRCULATION (EGR) PIPE
37647_ACTL_G0426

Fig. 122 EGR valve location

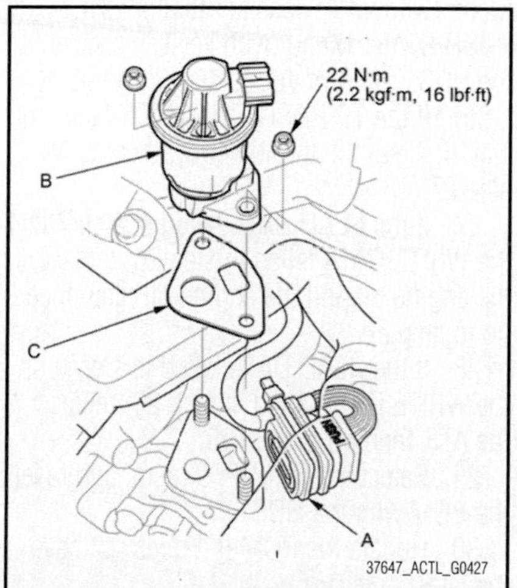
22 N·m (2.2 kgf·m, 16 lbf·ft)
B
C
A
37647_ACTL_G0427

Fig. 123 Disconnect the EGR valve 5P connector (A)

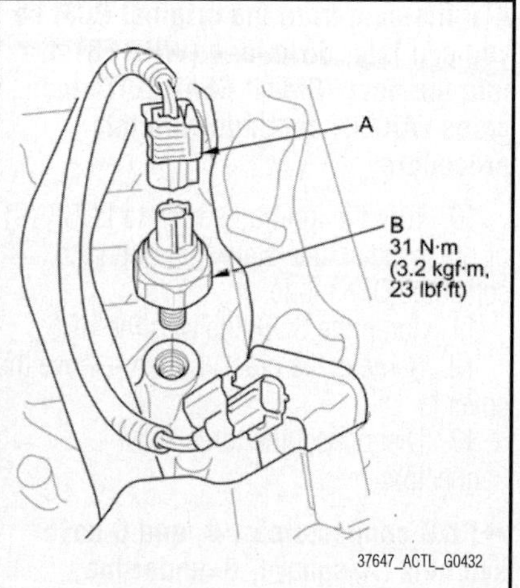
A
B
31 N·m (3.2 kgf·m, 23 lbf·ft)
37647_ACTL_G0432

Fig. 124 Disconnect the knock sensor connector (A)

KNOCK SENSOR (KS)

LOCATION

The knock sensor is located in the valley of the engine block between the two cylinder heads.

REMOVAL & INSTALLATION

See Figure 124.

1. Remove the intake manifold.
2. Remove the injector rail and base.
3. Disconnect the knock sensor connector.
4. Remove the knock sensor.
5. Install the parts in the reverse order of removal.

MASS AIR FLOW/INTAKE AIR TEMPERATURE (MAF/IAT) SENSOR

LOCATION

The MAF/IAT sensor is located on the engine air intake duct, near the throttle body end.

REMOVAL & INSTALLATION

See Figure 125.

1. Disconnect the MAF/IAT sensor 5P connector.
2. Remove the screws.
3. Remove the MAF/IAT sensor.
4. Install the parts in the reverse order of removal with a new O-ring.

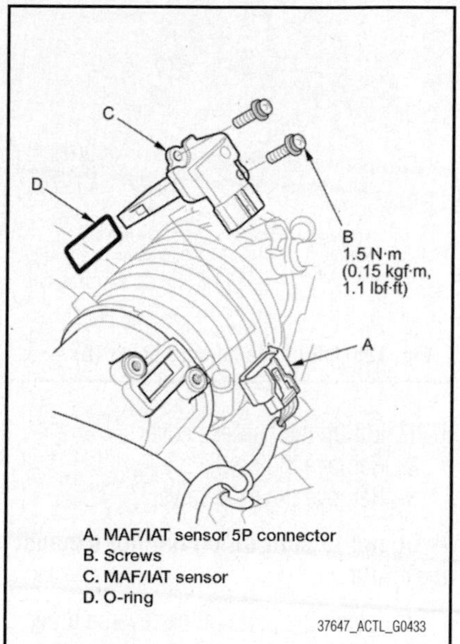
C
D
B
1.5 N·m (0.15 kgf·m, 1.1 lbf·ft)
A
A. MAF/IAT sensor 5P connector
B. Screws
C. MAF/IAT sensor
D. O-ring
37647_ACTL_G0433

Fig. 125 Disconnect the MAF/IAT sensor 5P connector

MANIFOLD ABSOLUTE PRESSURE (MAP) SENSOR

LOCATION

The Manifold Absolute Pressure (MAP) sensor is located near the throttle body.

REMOVAL & INSTALLATION

1. Remove the engine cover.
2. Disconnect the MAP sensor connector.
3. Remove the screw.
4. Remove the MAP sensor.
5. Install the parts in the reverse order of removal with a new O-ring.

POSITIVE CRANKCASE VENTILATION (PCV) VALVE

LOCATION

The Positive Crankcase Ventilation (PCV) valve is located on the front right area of the engine below the harness cover.

REMOVAL & INSTALLATION

See Figure 126.

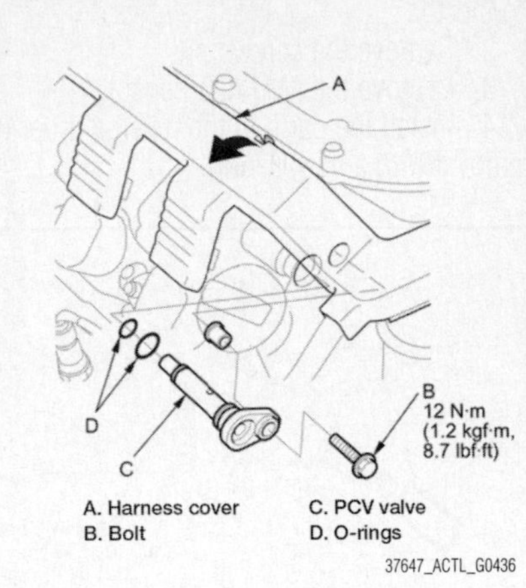
A
B
12 N·m
(1.2 kgf·m,
8.7 lbf·ft)
D
C
A. Harness cover
B. Bolt
C. PCV valve
D. O-rings
37647_ACTL_G0436

Fig. 126 Lift up the harness cover (A)

1. Lift up the harness cover.
2. Remove the bolt.
3. Remove the PCV valve.

➡Do not to spill oil on the hot exhaust manifold.

4. Install the parts in the reverse order of removal.

➡When installing a new PCV valve, make sure the O-rings are in place. When installing a used PCV valve, use new O-rings.

POWERTRAIN CONTROL MODULE (PCM)

LOCATION

The Powertrain Control Module (PCM) is located on the left side of the engine compartment.

REMOVAL & INSTALLATION

1. Connect HDS to the data link connector (DLC) located under the driver's side of the dashboard.
2. Turn the ignition switch to ON (II), or press the engine start/stop button to select the ON mode.
3. Make sure the HDS communicates with the PCM and all other vehicle systems. If it doesn't, go to the DLC circuit troubleshooting. If you are returning from the DLC circuit troubleshooting, skip steps 5 through 9, 18 through 23, and 26 through 28, and do this after replacing the PCM:
   a. Replace the engine oil and the engine oil filter.
   b. Replace the ATF.
   c. Clean the throttle body.
4. Select the PGM-FI system with the HDS.
5. Select the INSPECTION MENU with the HDS.
6. Select the ETCS TEST, then select the TP POSITION CHECK, and follow the screen prompts.

➡If the TP POSITION CHECK indicates FAILED, continue with this procedure.

7. Select the REPLACE PCM MENU, then select READ DATA, and follow the screen prompts.

➡Doing this step copies (READS) the engine oil life data from the original PCM so you can later download (WRITES) it into the new PCM. If the READ DATA indicates FAILED, continue with this procedure.

8. Select the A/T system with the HDS.
9. Select the REPLACE TCM/PCM MENU, then select READ DATA, and follow the screen prompts.

➡Doing this step copies (READS) the ATF life data from the original PCM so you can later download (WRITES) it into the new PCM. If READ DATA indicates FAILED, continue with this procedure.

10. Turn the ignition switch to LOCK (0), or press the engine start/stop button to select the OFF mode.
11. Jump the SCS line with the HDS.
12. Remove the bracket, then remove the cover.
13. Disconnect the three PCM connectors.

➡PCM connectors A, B, and C have symbols (A=square, B=underline, C=circle) embossed on them for identification.

14. Loosen or remove the bolts and remove the PCM.
15. Install the PCM in the reverse order of removal.
16. Turn the ignition switch to ON (II), or press the engine start/stop button to select the ON mode.
17. Manually input the VIN to the PCM with the HDS.

➡DTC P0630 VIN Not Programmed or Mismatch may be stored because the VIN has not been programmed into the PCM; ignore it, and continue this procedure.

18. If the READ DATA (engine oil life) failed in step 7, go to step 21. Otherwise, go to step 19.
19. Select the PGM-FI system with the HDS.
20. Select the REPLACE PCM MENU, then select WRITE DATA, and follow the screen prompts.

➡If the WRITE DATA indicates FAILED, continue with this procedure.

21. If the READ DATA (ATF life) failed in step 9, go to step 24. Otherwise go to step 22.
22. Select the A/T SYSTEM with the HDS.
23. Select the REPLACE TCM/PCM MENU, then select WRITE DATA, and follow the screen prompts.

➡If the WRITE DATA indicates FAILED, continue with this procedure.

24. Select the IMMOBI system (without keyless access) or ONE PUSH START system (with keyless access) with the HDS.
25. Enter the immobilizer PCM code that you got from the iN, and use the PCM replacement procedure in the IMMOBI Menu (without keyless access) or ONE PUSH START Menu (with keyless access) of the HDS; it allows you to start the engine.
26. If the TP POSITION CHECK failed in step 6, clean the throttle body, then go to step 27.
27. If the READ DATA failed in step 7 or the WRITE DATA failed in step 20, replace the engine oil and the engine oil filter, then go to step 28.
28. If the READ DATA failed in step 9 or the WRITE DATA failed in step 23, replace the ATF, then go to step 29.
29. Select the PGM-FI system, and reset the PCM with the HDS.
30. Update the PCM if it does not have the latest software.
31. Do the "PCM Idle Learn Procedure".
32. Do the "CKP Pattern Clear/CKP Pattern Learn Procedure". See "Crankshaft Position (CKP) Sensor" in this section.

PCM IDLE LEARN PROCEDURE

General Information

The idle learn procedure must be done so the PCM can learn the engine idle characteristics.

Do the idle learn procedure whenever you do any of these actions:

- Replace the PCM.
- Reset the PCM.
- Update the PCM.
- Replace or clean the throttle body.

• Disassemble the engine or the transmission.

➡**Clearing the DTCs with the HDS, or equivalent scan tool, does not require you to do the idle learn procedure.**

Procedure

1. Make sure all electrical items (the A/C, the audio, the lights, etc.) are off.
2. Reset the PCM with the HDS, or equivalent scan tool.
3. Turn the ignition switch to ON (II), or press the engine start/stop button to select the ON mode, and wait 2 seconds.
4. Start the engine. Hold the engine speed at 3,000 rpm without load (in P or N) until the radiator fan comes on, or until the engine coolant temperature reaches 194°F (90°C).
5. Let the engine idle for about 5 minutes with the throttle fully closed.

➡**If the radiator fan comes on, do not include its running time in the 5 minutes.**

TESTING

➡**See "DIAGNOSTICS" section for specific tests of the PCM as listed by DTC.**

FUEL — GASOLINE FUEL INJECTION SYSTEM

FUEL SYSTEM SERVICE PRECAUTIONS

Safety is the most important factor when performing not only fuel system maintenance, but any type of maintenance. Failure to conduct maintenance and repairs in a safe manner may result in serious personal injury or death. Work on a vehicle's fuel system components can be accomplished safely and effectively by adhering to the following rules and guidelines.

• To avoid the possibility of fire and personal injury, always disconnect the negative battery cable unless the repair or test procedure requires that battery voltage be applied.

• Always relieve the fuel system pressure prior to disconnecting any fuel system component (injector, fuel rail, pressure regulator, etc.) fitting or fuel line connection. Exercise extreme caution whenever relieving fuel system pressure to avoid exposing skin, face and eyes to fuel spray. Please be advised that fuel under pressure may penetrate the skin or any part of the body that it contacts.

• Always place a shop towel or cloth around the fitting or connection prior to loosening to absorb any excess fuel due to spillage. Ensure that all fuel spillage is quickly removed from engine surfaces. Ensure that all fuel-soaked cloths or towels are deposited into a flame-proof waste container with a lid.

• Always keep a dry chemical (Class B) fire extinguisher near the work area.

• Do not allow fuel spray or fuel vapors to come into contact with a spark or open flame.

• Always use a second wrench when loosening or tightening fuel line connection fittings. This will prevent unnecessary stress and torsion on fuel piping. Always follow the proper torque specifications.

• Always replace worn fuel fitting O-rings with new ones. Do not substitute fuel hose where rigid pipe is installed.

FUEL SYSTEM PRESSURE

RELIEVING

CAUTION

Before disconnecting fuel lines or hoses, relieve pressure from the system by disabling the fuel pump and then disconnecting the fuel line/quick-connect fitting in the engine compartment.

With the HDS

1. Turn the ignition switch to LOCK (0), or press the engine start/stop button to select the OFF mode.
2. Connect the HDS, or equivalent scan tool, to the data link connector (DLC) located under the driver's side of the dashboard.
3. Turn the ignition switch to ON (II), or press the engine start/stop button to select the ON mode.
4. Make sure the HDS communicates with the PCM. If it doesn't, go to the DLC circuit troubleshooting.
5. Turn the ignition switch to LOCK (0), or press the engine start/stop button to select the OFF mode.
6. Remove the fuel fill cap to relieve the pressure in the fuel tank.
7. Turn the ignition switch to ON (II), or press the engine start/stop button to select the ON mode.
8. From the INSPECTION MENU of the HDS, select Fuel Pump OFF, then start the engine, and let it idle until it stalls.

CAUTION

Do not allow the engine to idle above 1,000 rpm or the PCM will continue to operate the fuel pump.

➡**A Pending or Confirmed DTC may be set during this procedure. Check for DTCs, and clear them as needed.**

9. Turn the ignition switch to LOCK (0), or press the engine start/stop button to select the OFF mode.
10. Do the battery terminal disconnection procedure. See "ENGINE ELECTRICAL" section. Wait at least 3 minutes before proceeding.
11. Remove the quick-connect fitting cover.
12. Check the fuel quick-connect fitting for dirt, and clean it if needed.
13. Place a rag or shop towel over the quick-connect fitting.
14. Disconnect the quick-connect fitting as follows:
    a. Hold the connector with one hand, and squeeze the retainer tabs with the other hand to release them from the locking tabs.
    b. Pull the connector off.

CAUTION

Be careful not to damage the line (E) or other parts. Do not use tools.

    c. If the connector does not move, keep the retainer tabs pressed down, and alternately pull and push the connector until it comes off easily.
    d. Do not remove the retainer from the line; once removed, the retainer must be replaced with a new one.
15. After disconnecting the quick-connect fitting, check it for dirt or damage.
16. Do the battery terminal reconnection procedure.

Without the HDS

1. Remove PGM-FI main relay 2 from the driver's under-dash fuse/relay box.
2. Start the engine, and let it idle until it stalls.

➡**If any DTCs are stored, clear and ignore them.**

3. Turn the ignition switch to LOCK (0), or press the engine start/stop button to select the OFF mode.
4. Remove the fuel fill cap to relieve the pressure in the fuel tank

5. Do the battery terminal disconnection procedure. See "ENGINE ELECTRICAL" section. Wait at least 3 minutes before proceeding.

6. Remove the quick-connect fitting cover.

7. Check the fuel quick-connect fitting for dirt, and clean it if needed.

8. Place a rag or shop towel over the quick-connect fitting.

9. Disconnect the quick-connect fitting as follows:

a. Hold the connector with one hand, and squeeze the retainer tabs with the other hand to release them from the locking tabs.

b. Pull the connector off.

⁂ CAUTION

Be careful not to damage the line (E) or other parts. Do not use tools.

c. If the connector does not move, keep the retainer tabs pressed down, and alternately pull and push the connector until it comes off easily.

d. Do not remove the retainer from the line; once removed, the retainer must be replaced with a new one.

10. After disconnecting the quick-connect fitting, check it for dirt or damage.

11. Do the battery terminal reconnection procedure.

FUEL FILTER

REMOVAL & INSTALLATION

See Figure 127.

➡**The fuel filter should be replaced whenever the fuel pressure drops below the specified value, after making sure that the fuel pump and the fuel pressure regulator are OK.**

1. Remove the fuel tank unit.
2. Remove the fuel filter set.
3. Check these items before installing the fuel tank unit:
   - When connecting the wire harness, make sure the connection is secure and the connectors are firmly locked into place.
   - When installing the fuel gauge sending unit, make sure the connection is secure and the connector is firmly locked into place. Be careful not to bend or twist it excessively.
4. Install the parts in the reverse order of removal with new O-rings and a new bracket. When installing the fuel tank unit, align the marks on the unit and the fuel tank.

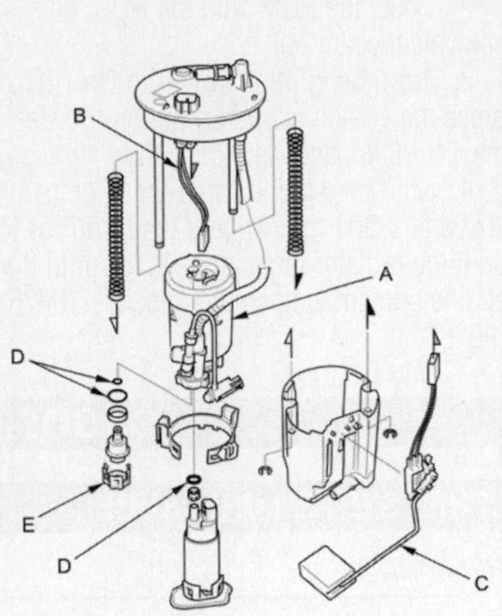
B
A
D
E
D
C
A. Fuel filter set
B. Wire harness
C. Fuel gauge sending unit
D. O-rings
E. Bracket
37647_ACTL_G0444

Fig. 127 Exploded view of fuel pump assembly

➡**Coat the O-rings with clean engine oil; do not use any other oil or fluid. Do not pinch the O-rings during installation. Use all the new parts supplied in the fuel filter replacement kit.**

FUEL LEVEL SENDING UNIT

LOCATION

The fuel level sending unit is an integral part of the fuel pump assembly which is located in the top of the fuel tank.

REMOVAL & INSTALLATION

See Figure 128.

1. Remove the fuel tank unit.
2. Remove the fuel level sensor (fuel gauge sending unit) from the fuel tank unit.
3. Check these items before installing the fuel tank unit:
   - When connecting the wire harness, make sure the connection is secure and the connector is firmly locked into place.
   - When installing the fuel gauge sending unit, make sure the connection is secure. Be careful not to bend or twist it excessively.
4. Install the parts in the reverse order of removal. When installing the fuel tank unit, align the marks on the unit and the fuel tank.

FUEL PUMP MODULE

REMOVAL & INSTALLATION

1. Remove the fuel tank unit.
2. Remove the fuel pump from the fuel tank unit.
3. Check these items before installing the fuel tank unit:
   - When connecting the wire harness, make sure the connection is secure and the connector is firmly locked into place.
   - When installing the fuel gauge sending unit, make sure the connection is secure. Be careful not to bend or twist it excessively.
4. Install the parts in the reverse order of removal. When installing the fuel tank unit, align the marks on the unit and the fuel tank.

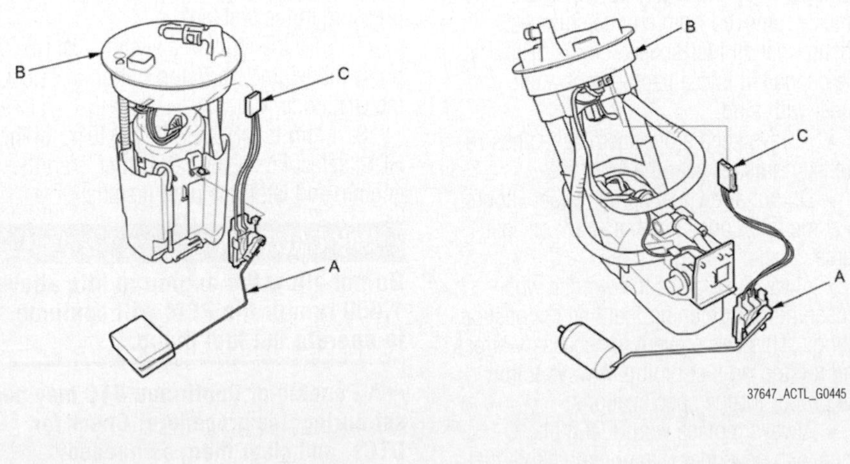
B
C
A
B
C
A
37647_ACTL_G0445

Fig. 128 Remove the fuel level sensor (fuel gauge sending unit) (A) from the fuel tank unit (B)

FUEL RAIL AND INJECTOR

REMOVAL & INSTALLATION

See Figure 129.

1. Relieve the fuel pressure. See "Fuel System Pressure" in this section.
2. Remove the intake manifold. See "Intake Manifold" in "ENGINE MECHANICAL" section.
3. Disconnect the quick-connect fitting.
4. Remove the fuel joint hose mounting bolt.
5. Disconnect the connectors from the injectors.
6. Remove the fuel rail mounting bolts from the fuel rails.
7. Remove the fuel rails and the injectors from the injector base.
8. Remove the injector clips from the fuel rails.
9. Remove the injectors from the fuel rails.

To install:

10. Coat the new O-rings (black) with clean engine oil, and insert the injectors into the fuel rails.
11. Install the injector clips.
12. Coat the new injector O-rings (green) with clean engine oil.
13. Install the fuel rails and the injectors in the injector base.
14. Reinstall the fuel rail mounting bolts, and connect the injector connectors.

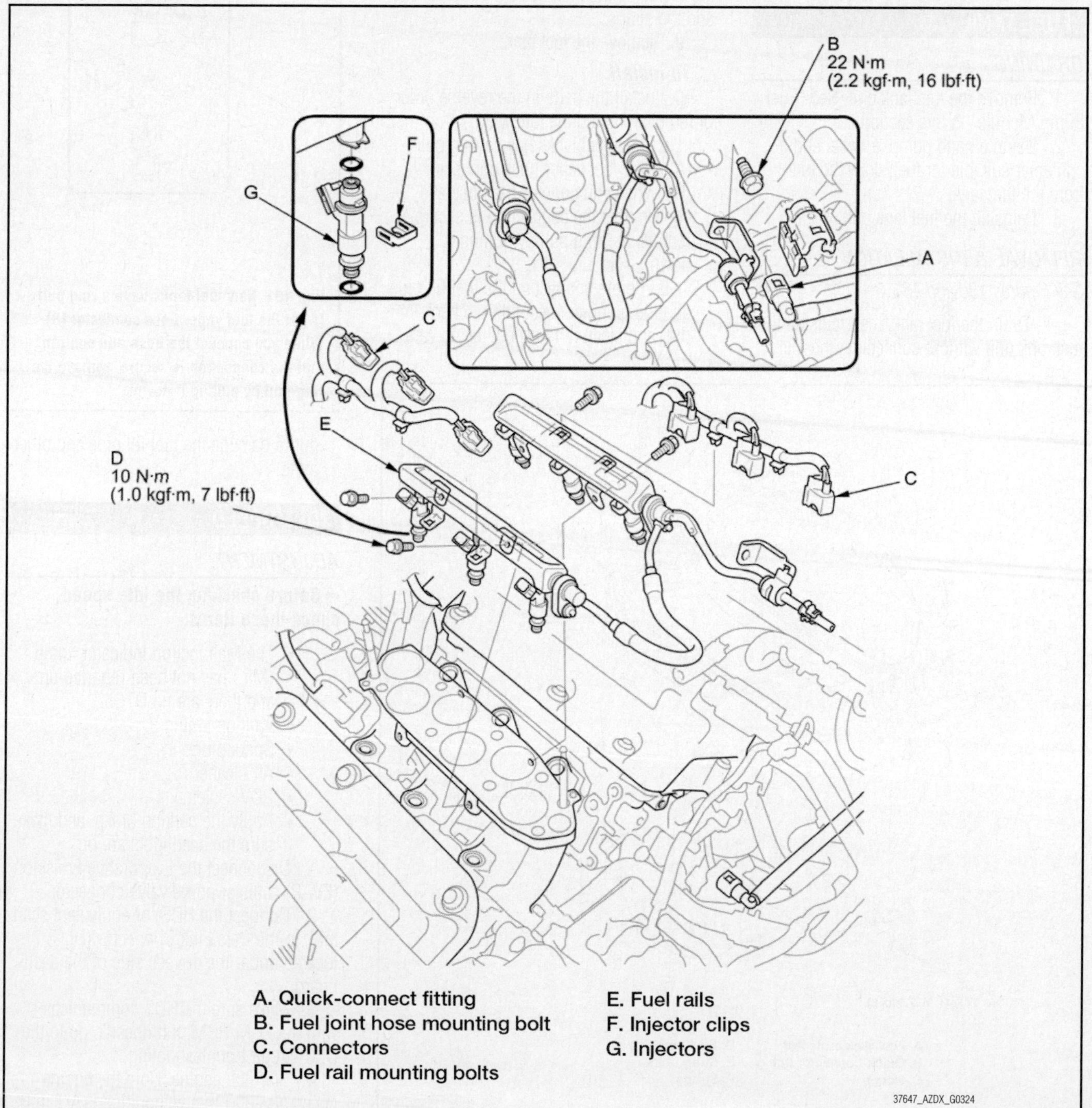
B
22 N·m
(2.2 kgf·m, 16 lbf·ft)
F
G
A
C
C
E
D
10 N·m
(1.0 kgf·m, 7 lbf·ft)
A. Quick-connect fitting
B. Fuel joint hose mounting bolt
C. Connectors
D. Fuel rail mounting bolts
E. Fuel rails
F. Injector clips
G. Injectors
37647_AZDX_G0324

Fig. 129 Exploded view of the fuel injector and rail components

15. Reinstall the fuel joint hose mounting bolt.

16. Connect the quick-connect fitting.

17. Turn the ignition switch to ON (II), or press the engine start/stop button to select the ON mode, but do not operate the starter. After the fuel pump runs for about 2 seconds, the fuel rail will be pressurized. 18. Repeat this two or three times, then make sure there are no fuel leaks.

19. Reinstall the intake manifold with a new gasket.

FUEL TANK

DRAINING

1. Remove the fuel tank unit. See "Fuel Pump Module" in this section.
2. Using a hand pump, a hose, and a container suitable for fuel, draw the fuel from the fuel tank.
3. Reinstall the fuel tank unit.

REMOVAL & INSTALLATION

See Figures 130 and 131.

1. Drain the fuel tank, then reinstall the fuel tank unit without connecting the fuel tank unit 4P connector and the fuel tank unit quick-connect fitting.
2. Raise the vehicle on a lift.
3. Remove the muffler.
4. Remove the propeller shaft.
5. Remove the EVAP canister cover.
6. Remove the fuel tank protector.
7. Disconnect the quick-connector fitting and the hoses. Slide back the clamps, then twist the hoses as you pull to avoid damaging them.
8. Place a jack or other support under the fuel tank, then remove the strap bolts and the straps.
9. Remove the fuel tank.

To install:

10. Install the parts in the reverse order of removal, noting the following:
    a. New fuel tanks have a ring pull (A) at the fuel vapor hose connector (B). When you connect the hose and confirm that the connection is secure, remove the ring pull by pulling it down.
    b. Before connecting the fuel fill pipe and the quick-connect fitting, check for dirt, and clean it if needed, taking care not to damage the fuel fill pipe and other parts.

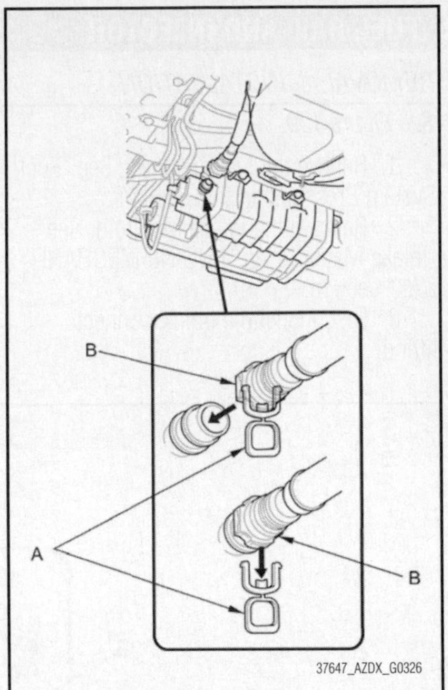
B
A
B
37647_AZDX_G0326

Fig. 131 New fuel tanks have a ring pull (A) at the fuel vapor hose connector (B). When you connect the hose and confirm that the connection is secure, remove the ring pull by pulling it down.

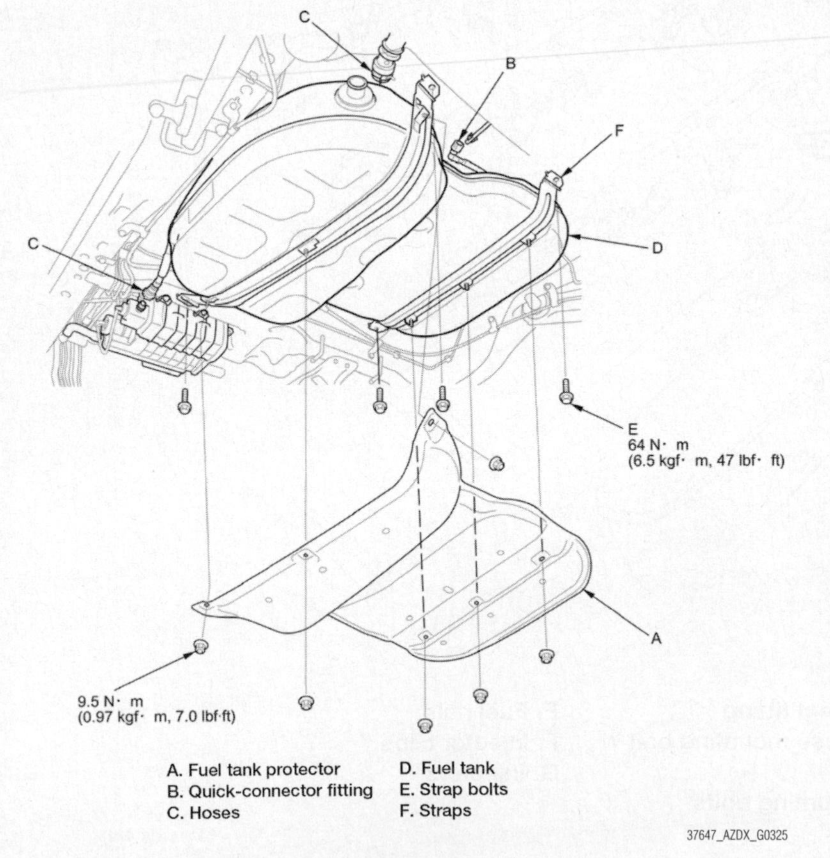
C
B
F
C
D
E
64 N· m
(6.5 kgf· m, 47 lbf· ft)
A
9.5 N· m
(0.97 kgf· m, 7.0 lbf·ft)
A. Fuel tank protector
B. Quick-connector fitting
C. Hoses
D. Fuel tank
E. Strap bolts
F. Straps
37647_AZDX_G0325

Fig. 130 Exploded view of the fuel tank and connections

IDLE SPEED

ADJUSTMENT

➡Before checking the idle speed, check these items:

- The Malfunction Indicator Lamp (MIL) has not been reported on, and there are no DTCs.
- Ignition timing
- Spark plugs
- Air cleaner
- PCV system
- Apply the parking brake, and make sure the headlights are off.

1. Disconnect the Evaporative Emission (EVAP) canister purge valve connector.
2. Connect the HDS, or equivalent scan tool, to the Data Link Connector (DLC) located under the driver's side of the dashboard.
3. Make sure the HDS communicates with the ECM/PCM. If it doesn't, go to the DLC circuit troubleshooting.
4. Start the engine. Hold the engine speed at 3,000 rpm without load (A/T in P or N, M/T in neutral) until the radiator fan comes on, then let it idle.

5. Check the idle speed without load conditions: headlights, blower fan, radiator fan, and air conditioner off.

➡**Idle speed should be 710 ± 50 rpm (in P or N)**

6. Let the engine idle for 1 minute with high electric load (the A/C on, the temperature set to max cool, the blower fan on high, the headlights on high beam).

➡**Idle speed should be 710 ± 50 rpm (in P or N)**

If the idle speed is not within specification, do the ECM/PCM idle learn procedure. If the idle speed is still not within specification, go to the symptom troubleshooting.

7. Reconnect the EVAP canister purge valve connector.

THROTTLE BODY

REMOVAL & INSTALLATION

See Figure 132.

⁂ CAUTION

Read the "Precautions" at the beginning of this section before starting work.

➡**If you are replacing the throttle body, begin at step 1. If you are removing the throttle body temporarily, begin at step 4.**

1. Connect the HDS, or equivalent scan tool, while the engine is stopped.
2. Select the INSPECTION MENU with the HDS.
3. Do the TP POSITION CHECK in the ETCS TEST.
4. Turn the ignition switch to LOCK (0), or press the engine start/stop button to select the OFF mode.

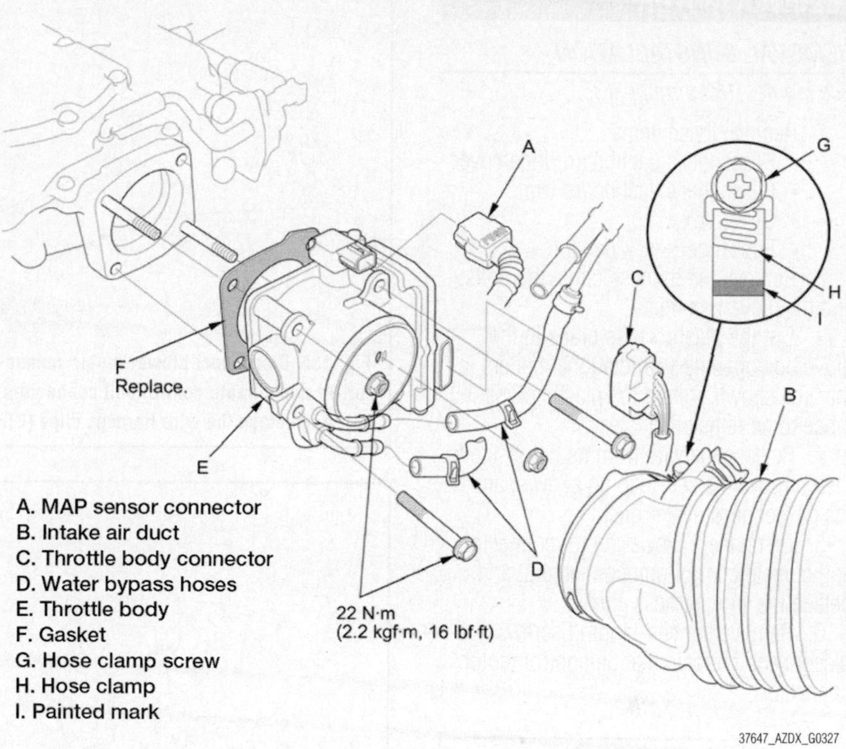
A
G
C
H
I
B
F
Replace.
E
D
A. MAP sensor connector
B. Intake air duct
C. Throttle body connector
D. Water bypass hoses
E. Throttle body
F. Gasket
G. Hose clamp screw
H. Hose clamp
I. Painted mark
22 N·m
(2.2 kgf·m, 16 lbf·ft)
37647_AZDX_G0327

Fig. 132 Exploded view of the throttle body assembly and related components

5. Disconnect the MAP sensor connector.
6. Remove the intake air duct.
7. Disconnect the throttle body connector.
8. Disconnect and plug the water bypass hoses, then remove the throttle body and gasket.

To install:

9. Install the parts in the reverse order of removal with a new gasket.
10. When tightening the screw of the hose clamp, align the edge of the hose clamp with the painted mark.
11. After installation, refill the radiator with engine coolant.
12. Do the "PCM Idle Learn Procedure". See "Powertrain Control Module (PCM)" in "ENGINE PERFORMANCE & EMISSION CONTROLS" section.

HEATING & AIR CONDITIONING SYSTEM

BLOWER MOTOR

REMOVAL & INSTALLATION

See Figures 133 through 136.

1. Remove these items:
   - Passenger's dashboard undercover
   - Passenger's dashboard trim
   - Glove box
   - Passenger's kick panel
2. Remove the harness clips, the bolts, and the glove box frame.
3. Cut the plastic cross brace in the glove box opening with diagonal cutters in the area shown. Retain the plastic cross brace to be reinstalled later.
4. Release the wire from its harness clip, then remove the self-tapping screws, and the passenger's heater duct.
5. Disconnect blower motor connector and climate control unit connectors, and detach the wire harness clips.
6. Remove the self-tapping screws and disconnect the recirculation control motor connector.

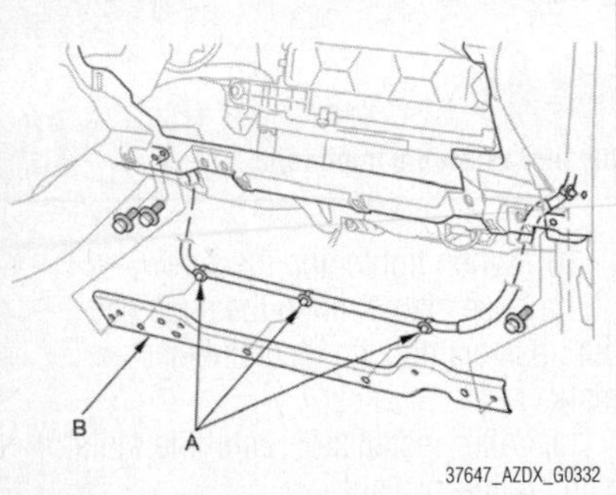
B
A
37647_AZDX_G0332

Fig. 133 Remove the harness clips (A), the bolts, and the glove box frame (B).

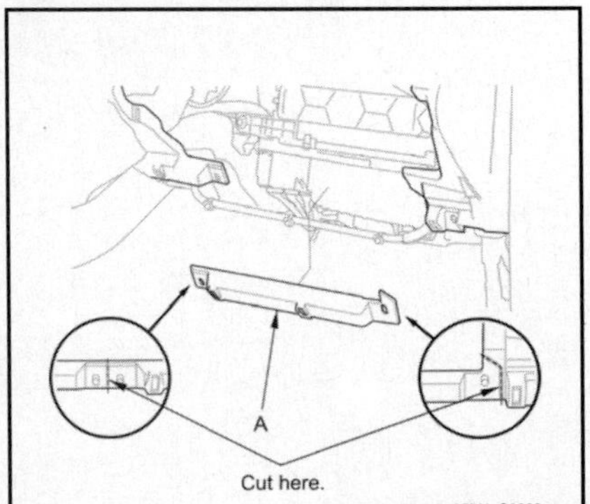
A
Cut here.
37647_AZDX_G0333

Fig. 134 Cut the plastic cross brace (A) in the glove box opening with diagonal cutters in the area shown. Retain the plastic cross brace to be reinstalled later.

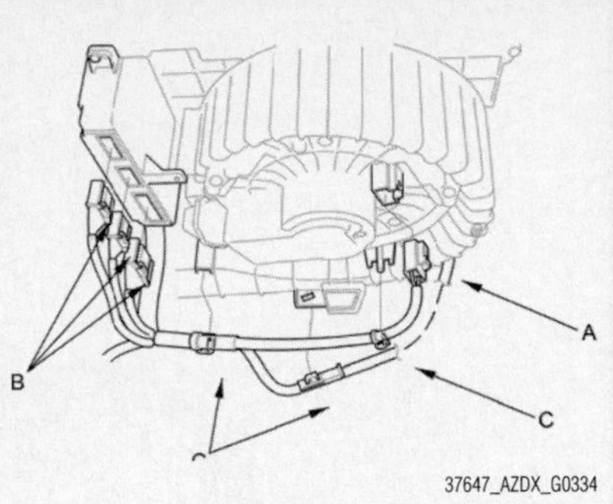
A
B
C
37647_AZDX_G0334

Fig. 135 Disconnect blower motor connector (A) and climate control unit connectors (B), and detach the wire harness clips (C).

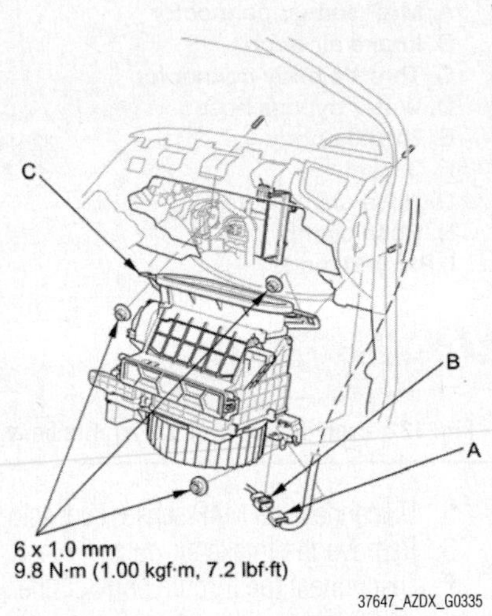
C
B
A
6 x 1.0 mm
9.8 N·m (1.00 kgf·m, 7.2 lbf·ft)
37647_AZDX_G0335

Fig. 136 Disconnect the right engine compartment harness connector (C205) (A) and harness clips (B). Remove the mounting nuts, and the blower unit (C).

7. Disconnect the right engine compartment harness connector (C205) and harness clips.
8. Remove the mounting nuts, and the blower unit.
9. Install the unit in the reverse order of removal. Make sure that there are no air leaks.

HEATER UNIT & HEATER CORE

REMOVAL & INSTALLATION

See Figures 137 through 145.

⁂ WARNING

SRS components are located in this area. Review the SRS component locations and the precautions and procedures before doing repairs or service.

1. Do the battery terminal disconnection procedure.
2. Recover the refrigerant with a recovery/recycling/charging station.
3. Disconnect the A/C line from the evaporator core.
4. When the engine is cool, drain the engine coolant from the radiator.
5. From under the hood, slide the heater hose clamps back.
6. Disconnect the inlet heater hose and the outlet heater hose from the heater unit. Note and/or mark the layout of the hoses.

➡Engine coolant will run out when the hoses are disconnected; drain it into a clean drip pan. Be sure not to let coolant spill on the electrical parts or the painted surfaces. If any coolant spills, rinse it off immediately.

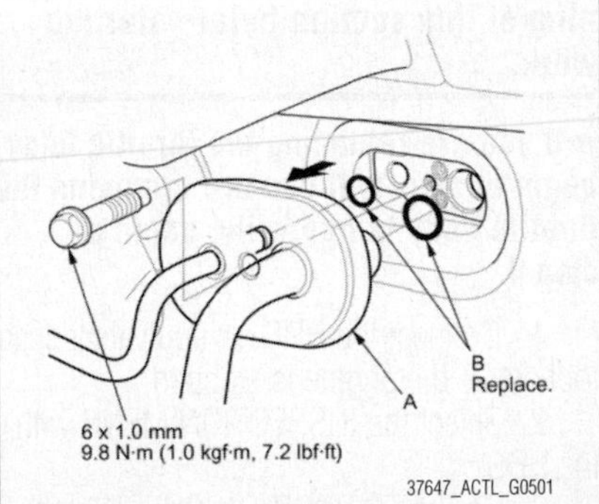
B
Replace.
A
6 x 1.0 mm
9.8 N·m (1.0 kgf·m, 7.2 lbf·ft)
37647_ACTL_G0501

Fig. 137 Disconnect the A/C line from the evaporator core

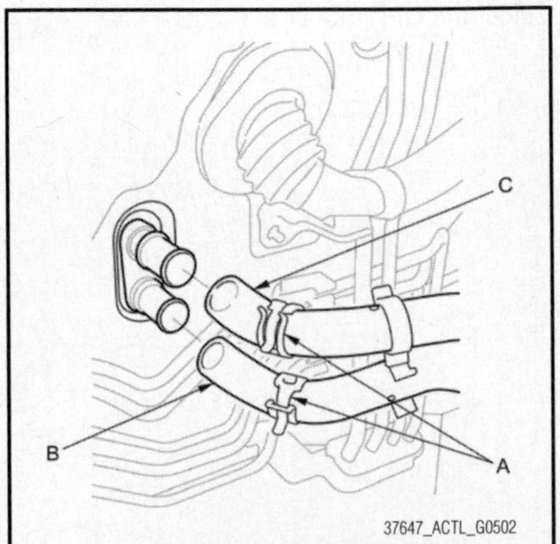
C
B
A
37647_ACTL_G0502

Fig. 138 Disconnect the inlet heater hose (B) and the outlet heater hose (C) from the heater unit

7. Remove the mounting nut from the heater unit. Take care not to damage or bend the fuel lines or the brake lines.

8. Remove the dashboard.

9. Disconnect the left engine compartment wire harness connectors.

10. Remove the ducts, and the drain hose.

11. Remove the mounting nuts, and the blower-heater unit.

12. Disconnect the connector from the blower motor. Remove the wire harness clips.

13. Disconnect these connectors: The passenger's mode control motor, the power transistor, and the passenger's air mix control motor. Detach the harness clips.

14. Disconnect these connectors: The recirculation control motor, and the climate control unit. Detach the harness clip.

15. Disconnect these connectors: the driver's air mix control motor, the driver's mode control motor, and evaporator temperature sensor. Detach the harness clips and the wire harness.

16. Remove the self-tapping screws and the passenger's heater duct.

17. Remove the self-tapping screws and the joint duct.

18. Remove the self-tapping screw and the heater core cover.

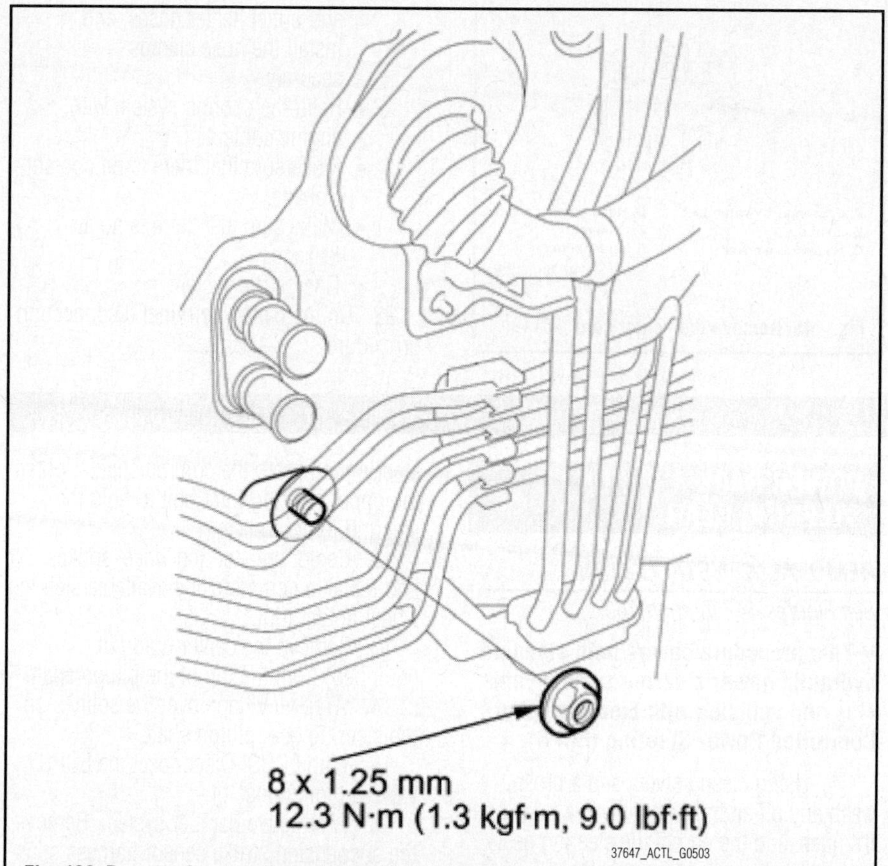
8 x 1.25 mm
12.3 N·m (1.3 kgf·m, 9.0 lbf·ft)
37647_ACTL_G0503

Fig. 139 Remove the mounting nut from the heater unit

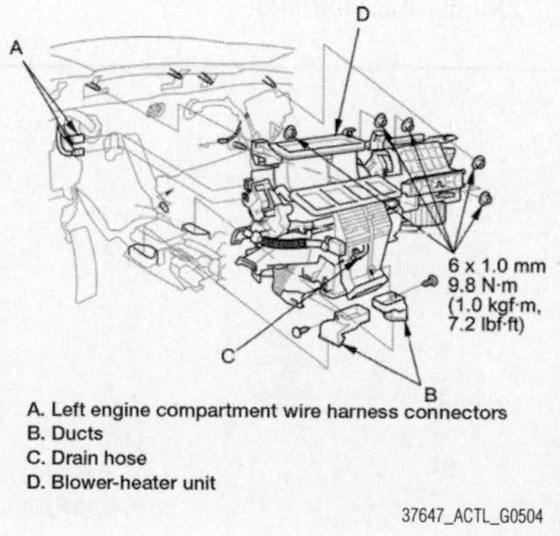
A
D
6 x 1.0 mm
9.8 N·m
(1.0 kgf·m,
7.2 lbf·ft)
C
B
A. Left engine compartment wire harness connectors
B. Ducts
C. Drain hose
D. Blower-heater unit
37647_ACTL_G0504

Fig. 140 Disconnect the left engine compartment wire harness connectors

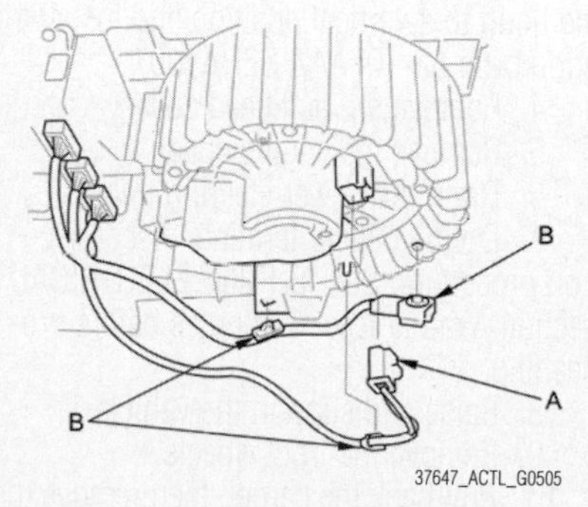
B
A
B
37647_ACTL_G0505

Fig. 141 Disconnect the connector (A) from the blower motor. Remove the wire harness clips (B).

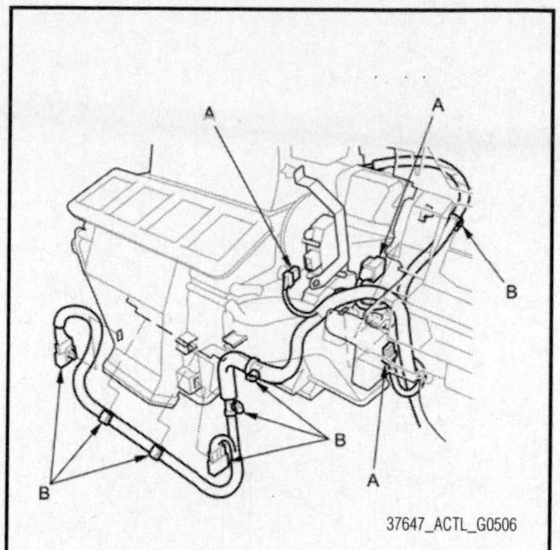
A
A
B
B
B
A
37647_ACTL_G0506

Fig. 142 Disconnect these connectors (A): The passenger's mode control motor, the power transistor, and the passenger's air mix control motor. Detach the harness clips (B)

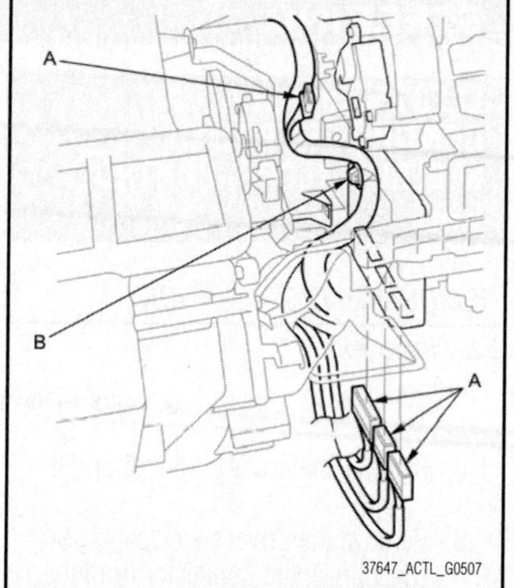
A
B
A
37647_ACTL_G0507

Fig. 143 Disconnect these connectors (A): The recirculation control motor, and the climate control unit. Detach the harness clip (B).

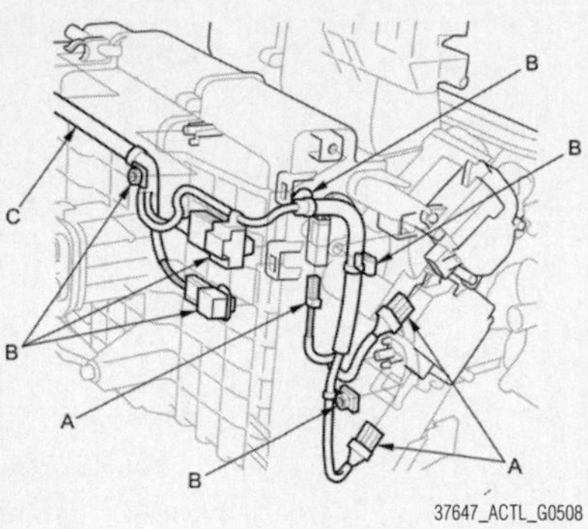
A
B
C
37647_ACTL_G0508

Fig. 144 Disconnect these connectors (A): the driver's air mix control motor, the driver's mode control motor, and evaporator temperature sensor. Detach the harness clips (B) and the wire harness (C).

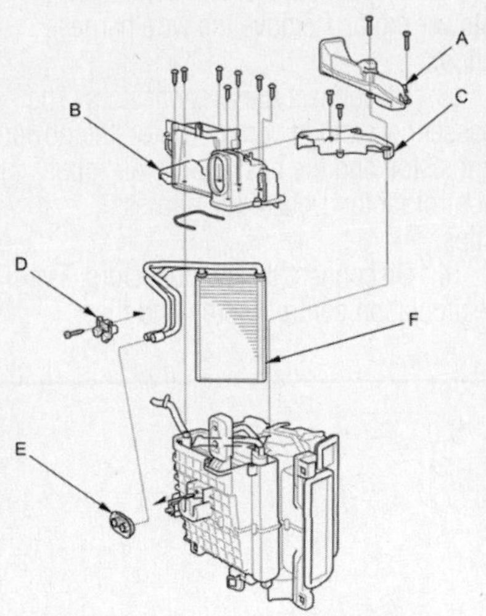
A
B
C
D
E
F
37647_ACTL_G0509

A. Passenger's heater duct
B. Joint duct
C. Heater core cover
D. Heater pipe bracket
E. Grommet
F. Heater core

Fig. 145 Remove the heater core

19. Remove the self-tapping screw, the heater pipe bracket, and the grommet, and carefully pull out the heater core.

To install:

20. Install the heater core in the reverse order of removal.
21. Install the heater unit in the reverse order of removal, and note these items:
    - Do not interchange the inlet and outlet heater hoses, and install the hose clamps securely.
    - Refill the cooling system with engine coolant.
    - Make sure that there is no coolant leakage.
    - Make sure that there is no air leakage.
    - Charge the system.
22. Do the battery terminal reconnection procedure.

STEERING

ELECTRONICALLY CONTROLLED POWER STEERING (ECPS) CONTROL UNIT

REMOVAL & INSTALLATION

See Figure 146.

1. Remove the driver's dashboard undercover.
2. Remove the driver's front door sill trim.
3. Remove the driver's kick panel.
4. Disconnect the connector from the ECPS control unit, then remove the control unit.
5. Install the ECPS control unit in the reverse order of removal.

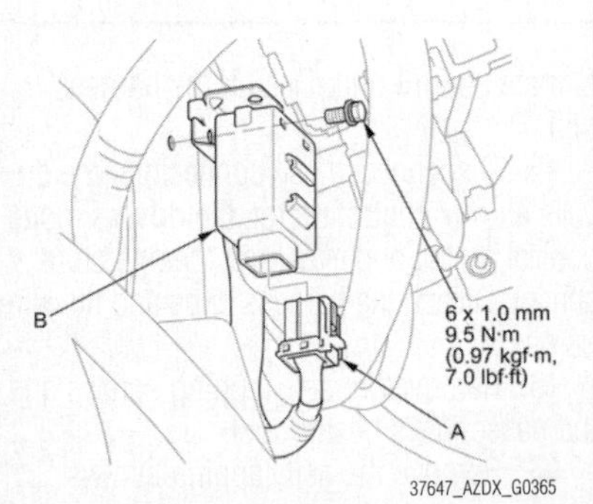
6 x 1.0 mm
9.5 N·m
(0.97 kgf·m,
7.0 lbf·ft)
A
B
37647_AZDX_G0365

Fig. 146 Showing the Electronically Controlled Power Steering (ECPS) Control Unit

POWER RACK & PINION STEERING GEAR

REMOVAL & INSTALLATION

See Figures 147 through 166.

➡This procedure covers both standard hydraulic power steering system vehicles and vehicles with Electronically Controlled Power Steering (ECPS).

1. Using clean solvent and a brush, wash any oil and dirt off the valve body unit, it's lines, and the end of the steering gearbox. Blow dry with compressed air.
2. Remove the passenger's side front fender trim.
3. Pry the hood support strut clip, and release the hood support strut, then move the hood to a vertical position and install a hood prop bar (07AAZ-SZNA100).
4. Remove the bulkhead cover.
5. Remove the engine cover.
6. Drain the power steering fluid.
7. Do the battery terminal disconnection procedure. See "ENGINE ELECTRICAL" section. Wait at least 3 minutes before proceeding.
8. Raise and support the vehicle.
9. Remove the front wheels.
10. Pull back the carpet, then remove the steering joint cover.
11. Loosen the upper steering joint bolt, and remove the lower steering joint bolt. Disconnect the steering joint by sliding the steering joint into the column shaft. Tighten the upper steering joint bolt to hold the steering joint in place.
12. Center the steering wheel spokes, and install a commercially available steering wheel holder tool.
13. Remove the center guide (if equipped) from the top of the pinion shaft.
14. Wrap vinyl tape over the splines on the steering gear pinion shaft.
15. With ECPS: Disconnect the ECPS valve motor connector.
16. With active damper system: Remove the suspension stroke sensor harness clips on both sides.
17. Loosen the adjustable hose clamp and disconnect the return hose.
18. Loosen the 16 mm flare nut and disconnect the inlet line.

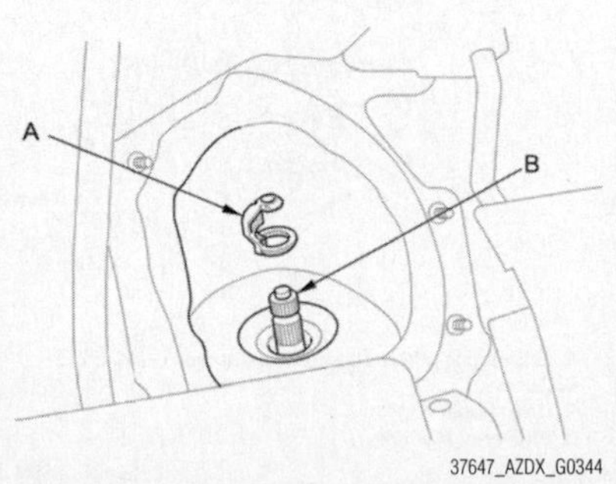
A
B
37647_AZDX_G0344

Fig. 147 Remove the center guide (A) (if equipped) from the top of the pinion shaft (B).

19. Remove the cotter pin from the tie-rod end ball joint, then remove the nut.

20. Disconnect the tie-rod end ball joint from the knuckle using the ball joint thread protector and the ball joint remover on both sides.

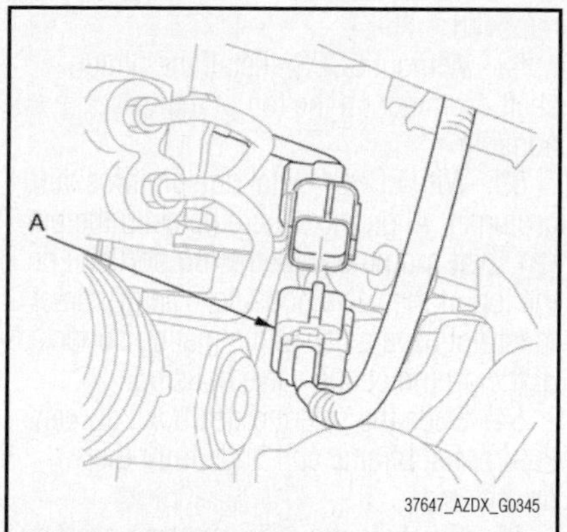
A
37647_AZDX_G0345

Fig. 148 With ECPS: Disconnect the ECPS valve motor connector (A).

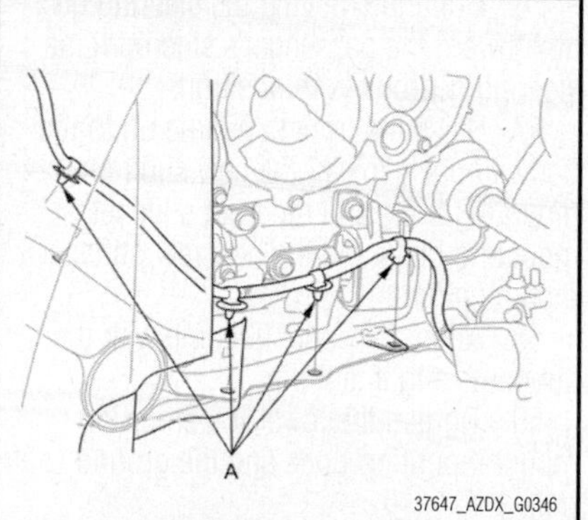
A
A
37647_AZDX_G0346

Fig. 149 With active damper system: Remove the suspension stroke sensor harness clips (A) on both sides.

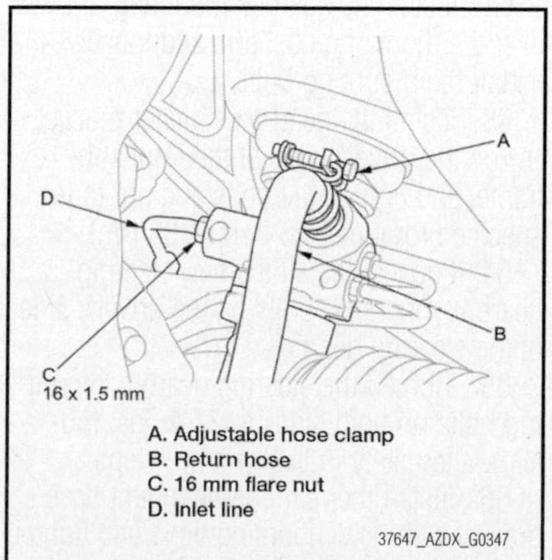
A
D
B
C
16 x 1.5 mm
A. Adjustable hose clamp
B. Return hose
C. 16 mm flare nut
D. Inlet line
37647_AZDX_G0347

Fig. 150 Loosen the adjustable hose clamp and disconnect the return hose. Loosen the 16 mm flare nut and disconnect the inlet line.

21. Disconnect the stabilizer link from the stabilizer bar on both sides.

22. Install an appropriate engine support assembly and apply tension to support the engine.

23. Remove the front engine mount stop and the front engine mount control solenoid valve line bracket, then remove the front engine mount mounting bolt.

24. Raise the vehicle.

25. Remove the engine undercover and the front splash shield.

26. Disconnect the front engine mount control solenoid valve tube and remove the clip. Remove the front engine mount mounting bolts.

27. Remove the front subframe stiffener plate.

28. Remove exhaust pipe A.

29. Remove the propeller shaft from the transfer shaft flange, after making a reference mark across the connection.

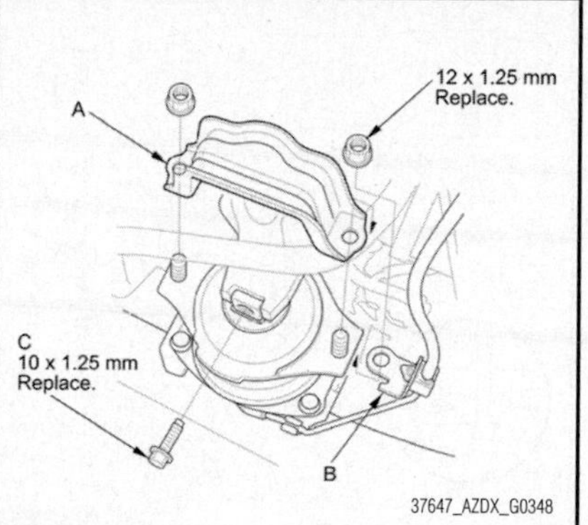
12 x 1.25 mm
Replace.
A
C
10 x 1.25 mm
Replace.
B
37647_AZDX_G0348

Fig. 151 Remove the front engine mount stop (A) and the front engine mount control solenoid valve line bracket (B), then remove the front engine mount mounting bolt (C).

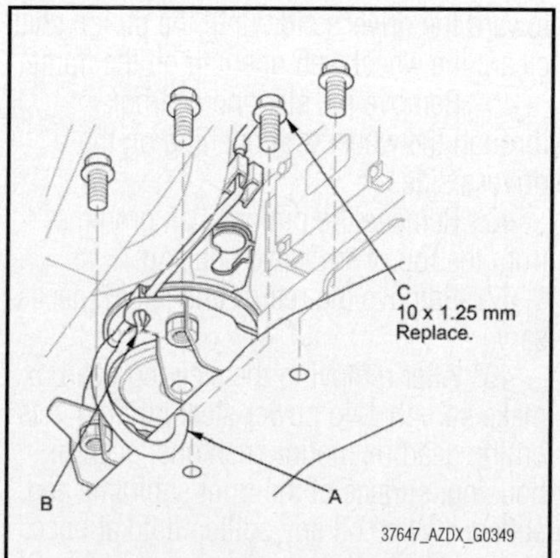
C
10 x 1.25 mm
Replace.
B
A
37647_AZDX_G0349

Fig. 152 Disconnect the front engine mount control solenoid valve tube (A), and remove the clip (B). Remove the front engine mount mounting bolts (C).

30. Remove the inlet line clamp bolts.

31. Remove the rear mount stop, heat shield, and the rear engine mount mounting bolts, then move the rear engine mount aside.

32. Remove the rear engine mount base bracket.

33. Remove the steering gearbox mounting bolts from the driver's side of the steering gearbox.

34. Remove the gearbox stiffener bracket from the driver's side of the front subframe.

35. Remove the steering gearbox mounting bolts from the passenger's side of the steering gearbox, then remove the mounting bracket and cushion.

36. Attach a subframe adapter (VSB02C000016 or equivalent) to the subframe by looping a strap over the front of the subframe, then secure the strap.

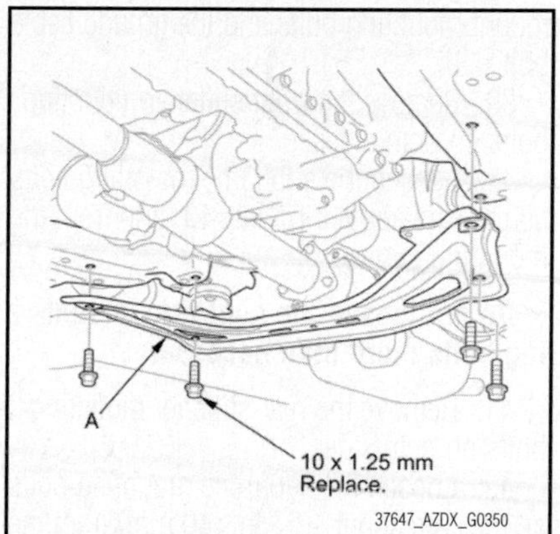
A
10 x 1.25 mm
Replace.
37647_AZDX_G0350

Fig. 153 Remove the front subframe stiffener plate (A).

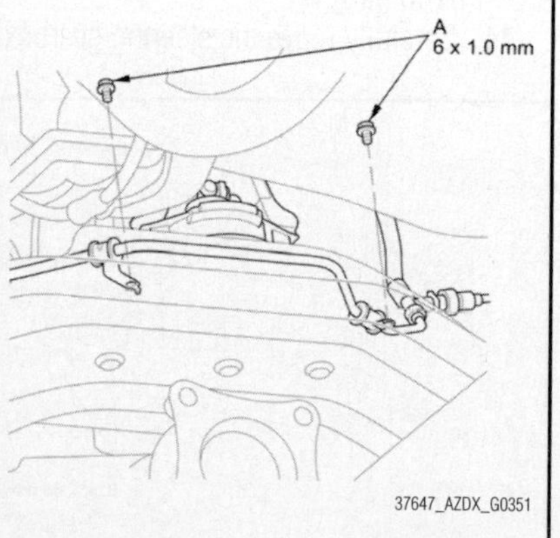
A
6 x 1.0 mm
37647_AZDX_G0351

Fig. 154 Remove the inlet line clamp bolts (A).

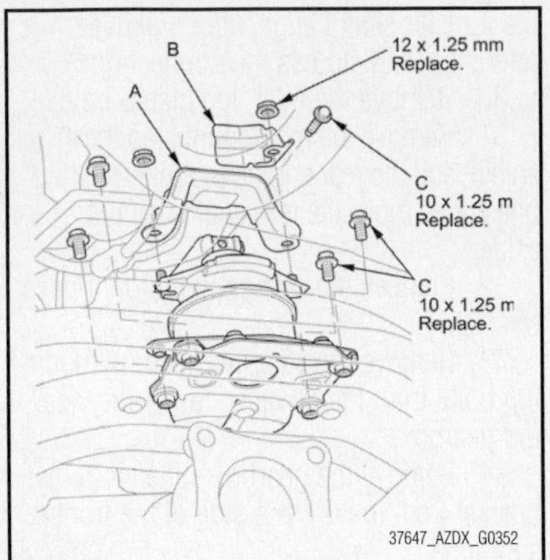
B
12 x 1.25 mm
Replace.
A
C
10 x 1.25 m
Replace.
C
10 x 1.25 m
Replace.
37647_AZDX_G0352

Fig. 155 Remove the rear mount stop (A), heat shield (B), and the rear engine mount mounting bolts (C), then move the rear engine mount aside.

37. Raise a jack and secure it with the bolt holes on the jack base, then securely attach them with four bolts.

38. Remove the lower transmission mount mounting bolts and the ground cable bolt.

39. Remove the front stiffener mounting bolts on both sides.

40. Loosen the subframe mounting bolts so they are about 1.57 in. (40 mm) from the mounting surface.

➡Do not loosen the subframe mounting bolts more than necessary.

41. Remove the rear stiffener mounting bolts on both sides.

42. Loosen the subframe mounting bolts so they are about 1.57 in. (40 mm) from the mounting surface.

➡Do not loosen the subframe mounting bolts more than necessary.

43. Lower the transmission jack slowly until the front subframe has dropped about 1.57 in. (40 mm).

44. Carefully move the steering gearbox

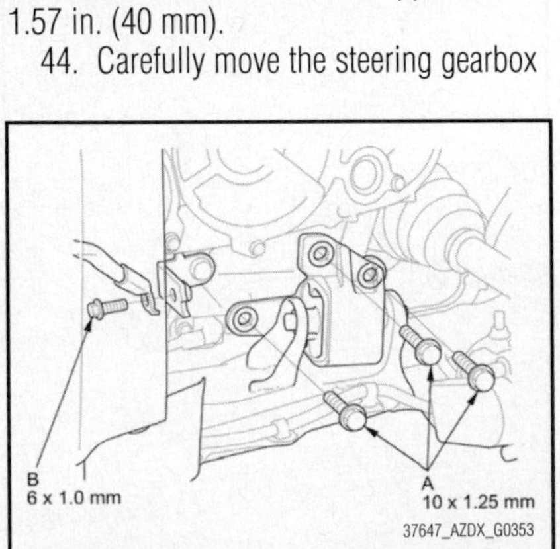
B
6 x 1.0 mm
A
10 x 1.25 mm
37647_AZDX_G0353

Fig. 156 Remove the lower transmission mount mounting bolts (A) and the ground cable bolt (B).

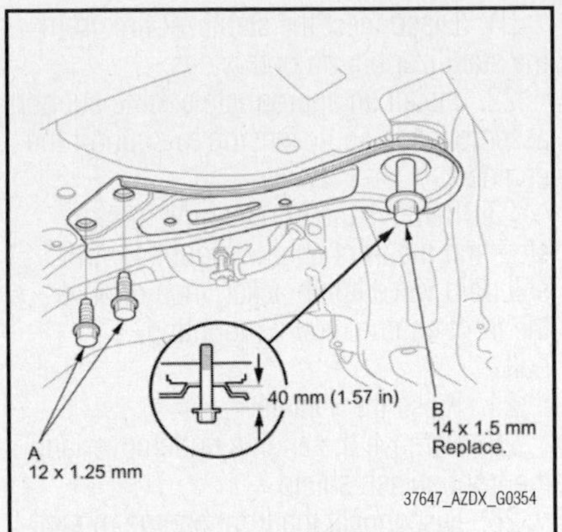
40 mm (1.57 in)
B
14 x 1.5 mm
Replace.
A
12 x 1.25 mm
37647_AZDX_G0354

Fig. 157 Remove the front stiffener mounting bolts (A) on both sides. Loosen the subframe mounting bolts (B) so they are about 1.57 in. (40 mm) from the mounting surface.

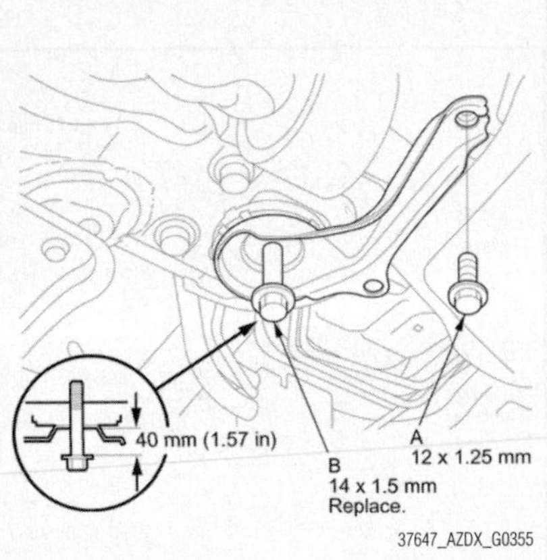
40 mm (1.57 in)
A
12 x 1.25 mm
B
14 x 1.5 mm
Replace.
37647_AZDX_G0355

Fig. 158 Remove the rear stiffener mounting bolts (A) on both sides. Loosen the subframe mounting bolts (B) so they are about 1.57 in. (40 mm) from the mounting surface.

toward the driver's side until the pinion shaft clears the wheel well opening on the frame.

45. Remove the steering gearbox through the wheel well opening on the driver's side.

46. Remove the pinion shaft grommet from the top of the valve housing.

47. Remove the return line joint if necessary.

48. After removing the steering gearbox, make sure that no power steering fluid gets on the gearbox mount cushions, gearbox housing, surface of the front subframe and stiffener. Wipe off any spilled fluid at once.

To install:

49. Before installing the steering gearbox, make sure that no power steering fluid is on the mating surface of the steering gearbox and front subframe. To prevent the gearbox mounting bolts from loosening after the installation, remove any power steering fluid from the mount cushions and bolt holes.

50. Wrap vinyl tape over the splines on the steering gear pinion shaft.

51. Install the return line joint if removed.

52. Without ECPS: Install the pinion shaft grommet on the top of the valve housing.

53. With ECPS: Install the pinion shaft grommet. Align the square holes in the pinion shaft grommet with the lug portions on the top of the valve housing. The grommet must not have a gap at the mating surface of the grommet and valve housing.

54. Slide the steering gearbox between the front subframe and body from the driver's side.

55. Carefully move the steering gearbox toward the passenger's side until the pinion shaft clears the wheel well opening on the body.

56. Continue moving the steering gearbox toward the passenger's side until the steering gearbox is in position.

57. Raise the front subframe up to the body. Loosely install the new subframe mounting bolts and the front stiffener mounting bolts and the new rear stiffener mounting bolts.

58. Align the front subframe with the subframe alignment pin.

59. Tighten the lower transmission mount mounting bolts and the ground cable bolt.

60. Lower the transmission jack supporting the subframe.

61. Position the cutout on the mounting cushion, as shown, and install it on the passenger's side of the steering gearbox.

62. Install the gearbox mounting bracket over the mounting cushion, and loosely install the mounting bolts.

63. Install the gearbox stiffener bracket (A) on the driver's side of the front subframe, and tighten the bolts and nut to the specified torque of 28 ft. lbs. (38 Nm).

64. Loosely install the new steering gearbox mounting bolts on the driver's side of the steering gearbox.

65. Tighten the steering gearbox mounting bolts on both sides to 37 ft. lbs. (50 Nm), alternately in two or more steps.

66. Install the rear engine mount base bracket with new mounting bolts and tighten to 31 ft. lbs. (42 Nm).

67. Install the rear engine mount with new mounting bolts and tighten to the respective specified torques shown in the figure.

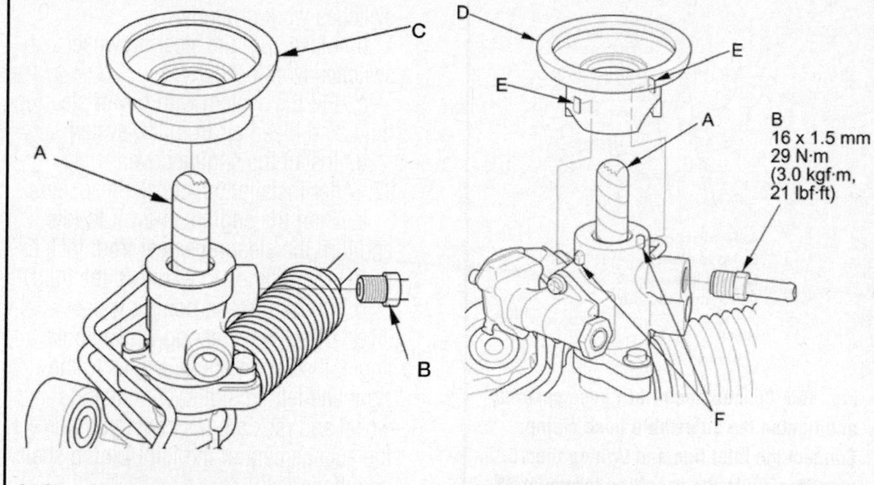
C
D
E
E
A
A
B
16 x 1.5 mm
29 N·m
(3.0 kgf·m,
21 lbf·ft)
B
F
37647_AZDX_G0356

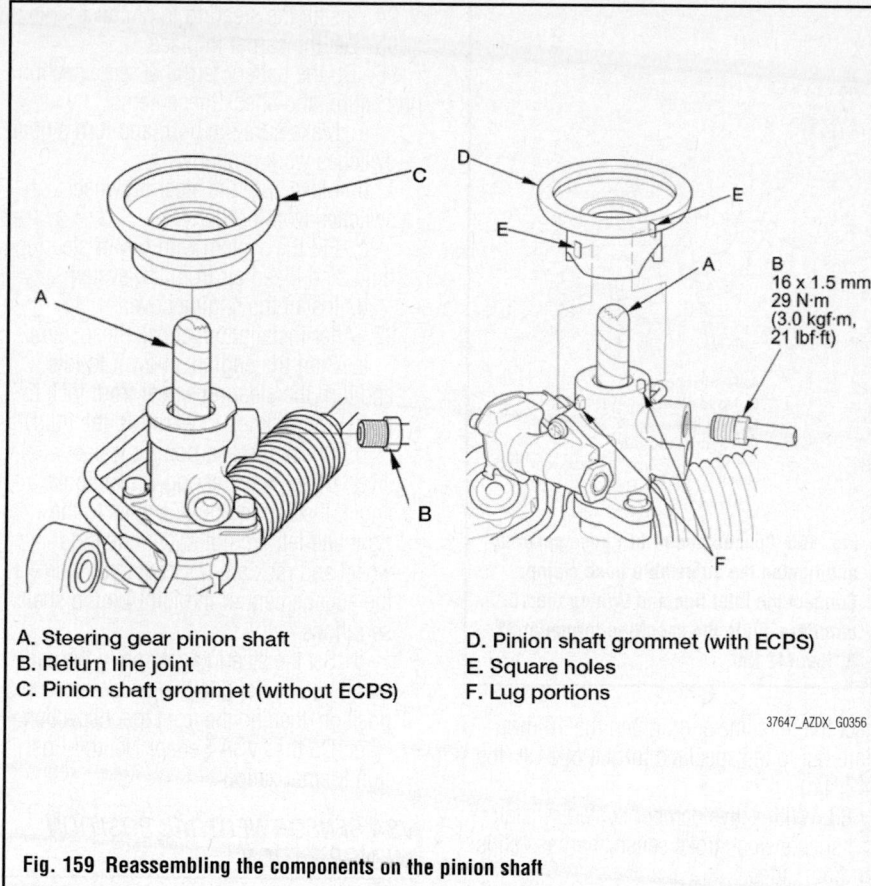

A. Steering gear pinion shaft
B. Return line joint
C. Pinion shaft grommet (without ECPS)
D. Pinion shaft grommet (with ECPS)
E. Square holes
F. Lug portions

Fig. 159 Reassembling the components on the pinion shaft

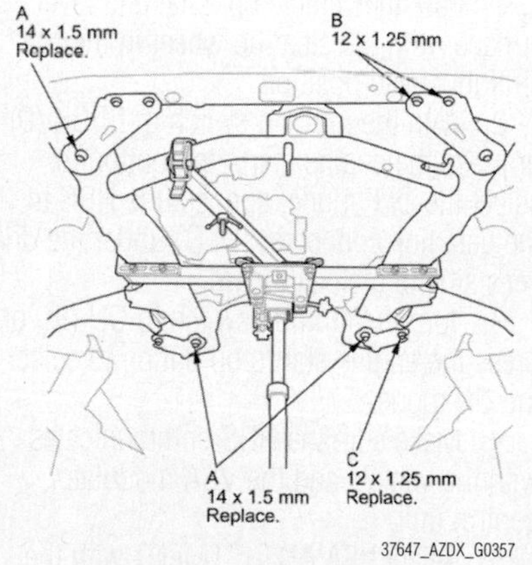
A
14 x 1.5 mm
Replace.
B
12 x 1.25 mm
A
14 x 1.5 mm
Replace.
C
12 x 1.25 mm
Replace.
37647_AZDX_G0357

Fig. 160 Raise the front subframe up to the body. Loosely install the new subframe mounting bolts (A) and the front stiffener mounting bolts (B), and the new rear stiffener mounting bolts (C).

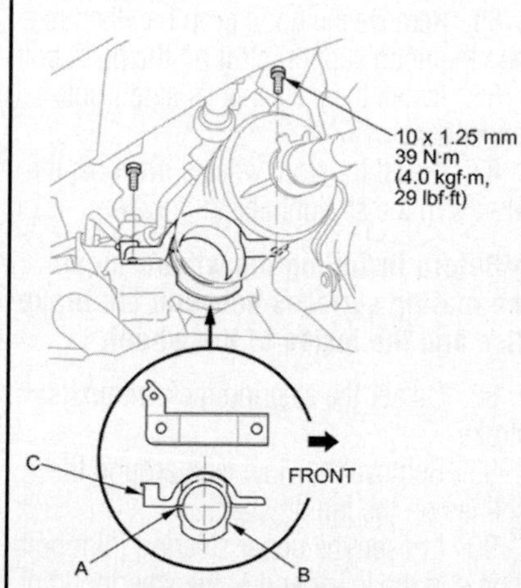
10 x 1.25 mm
39 N·m
(4.0 kgf·m,
29 lbf·ft)
FRONT
C
A
B

A. Rear engine mount
B. Mounting bolts
C. Rear mount stop
D. Heat shield
E. Mounting nuts

37647_AZDX_G0358

Fig. 161 Position the cutout (A) on the mounting cushion (B) as shown, and install it on the passenger's side of the steering gearbox. Install the gearbox mounting bracket (C) over the mounting cushion, and loosely install the mounting bolts.

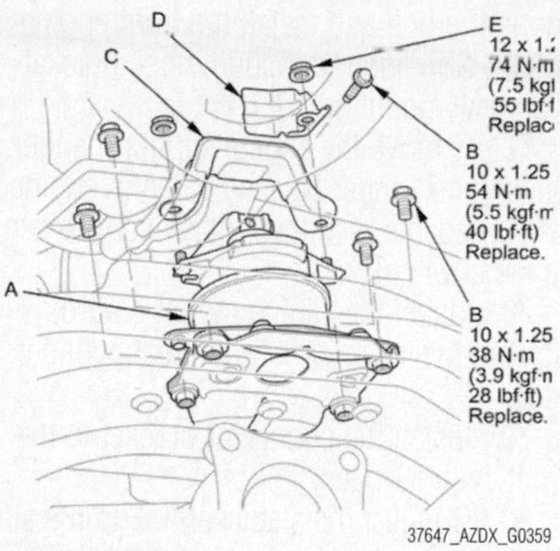
D
C
E
B
10 x 1.25
54 N·m
(5.5 kgf·m
40 lbf·ft)
Replace.
A
B
10 x 1.25
38 N·m
(3.9 kgf·m
28 lbf·ft)
Replace.
37647_AZDX_G0359

Fig. 162 Install the rear engine mount with new mounting bolts and tighten to the respective specified torques shown in the figure. Install the rear mount stop and the heat shield with new mounting nuts and tighten to the specified torque as shown.

68. Install the rear mount stop and the heat shield with new mounting nuts and tighten to the specified torque in the figure.

69. Install the power steering inlet line clamp bolts, and tighten to 7 ft. lbs. (10 Nm).

70. Install exhaust pipe A and tighten the new flange-to-converter nuts to 24 ft. lbs. (33 Nm), and the new flange-to-engine nuts to 40 ft. lbs. (54 Nm).

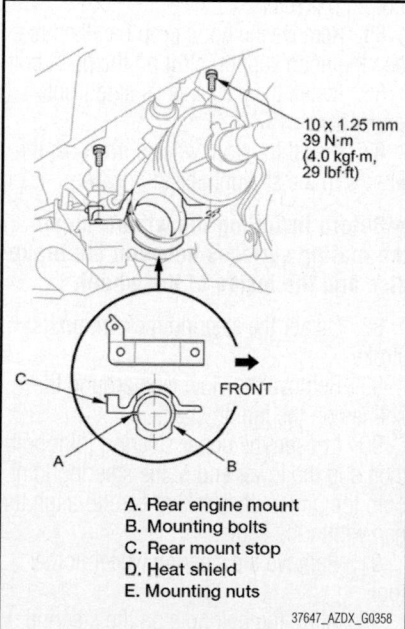

71. Install the propeller shaft onto the transfer shaft flange and tighten the new nuts to 53 ft. lbs. (72 Nm).

72. Install the front subframe stiffener plate with new mounting bolts and tighten to the specified torque of 44 ft. lbs. (59 Nm).

73. Install the engine undercover and the front splash shield.

74. Install the new front engine mount mounting bolts and tighten to 28 ft. lbs. (38 Nm), then connect the front engine mount control solenoid valve tube.

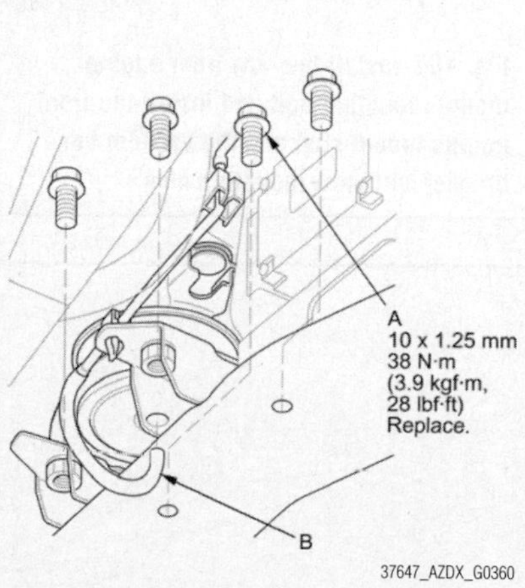
A
10 x 1.25 mm
38 N·m
(3.9 kgf·m,
28 lbf·ft)
Replace.
B
37647_AZDX_G0360

Fig. 163 Install the new front engine mount mounting bolts (A) and tighten to 28 ft. lbs. (38 Nm), then connect the front engine mount control solenoid valve tube (B).

75. Lower the vehicle.

76. Install the new front engine mount mounting bolt and install the front engine mount stop and the vacuum hose bracket with new mounting nuts.

77. Remove the engine support hanger, the engine hanger balance bar, and engine hanger adapter set, then install service caps to the cowl cover.

78. Install the engine mount control solenoid valve bracket to the rear cylinder head.

79. Install the connector bracket to the front cylinder head.

80. Connect the stabilizer link to the stabilizer bar on both sides. Tighten the bolts to 28 ft. lbs. (38 Nm).

81. On both sides, wipe off any grease contamination from the ball joint tapered section and threads. Reconnect the tie-rod end ball joint to the knuckle. Install the new nut and tighten it to the specified torque of 44 ft. lbs. (60 Nm). Install a new cotter pin.

82. Connect the return hose securely and tighten the adjustable hose clamp. Connect the inlet line and tighten the 16 mm flare nut to the specified torque of 31 ft. lbs. (42 Nm).

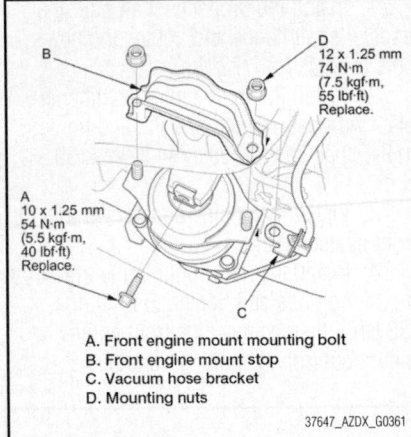

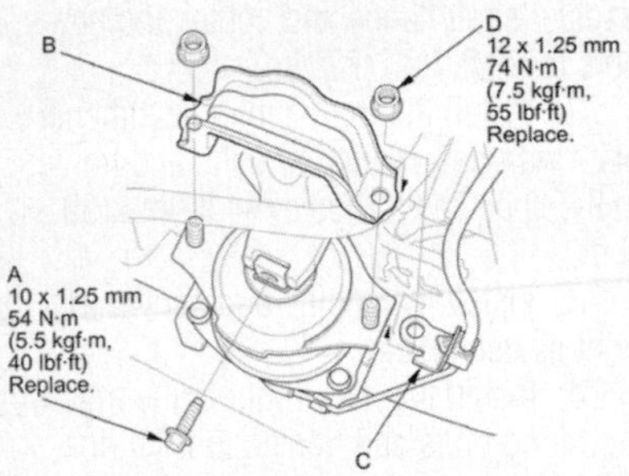
B
D
12 x 1.25 mm
74 N·m
(7.5 kgf·m,
55 lbf·ft)
Replace.
A
10 x 1.25 mm
54 N·m
(5.5 kgf·m,
40 lbf·ft)
Replace.
C
A. Front engine mount mounting bolt
B. Front engine mount stop
C. Vacuum hose bracket
D. Mounting nuts
37647_AZDX_G0361

Fig. 164 Install the new front engine mount mounting bolt and install the front engine mount stop and the vacuum hose bracket with new mounting nuts.

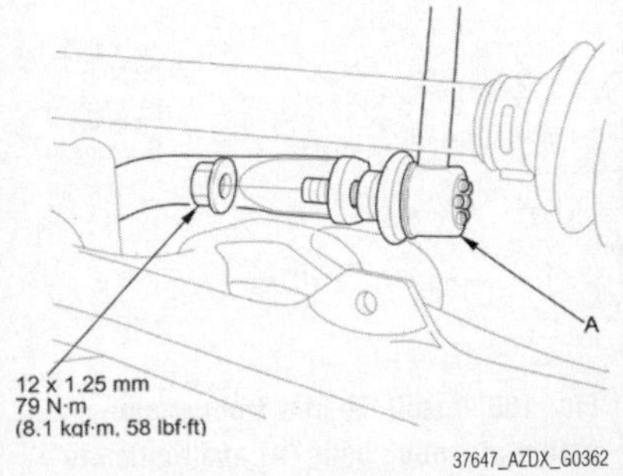
A
12 x 1.25 mm
79 N·m
(8.1 kgf·m, 58 lbf·ft)
37647_AZDX_G0362

Fig. 165 Connect the stabilizer link to the stabilizer bar on both sides.

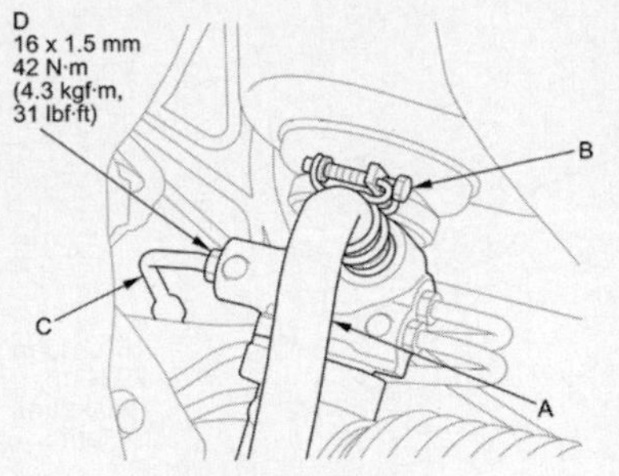
D
16 x 1.5 mm
42 N·m
(4.3 kgf·m,
31 lbf·ft)
B
C
A
A. Return hose
B. Hose clamp
C. Inlet line
D. 16 mm flare nut
37647_AZDX_G0363

Fig. 166 Connect the return hose securely and tighten the adjustable hose clamp. Connect the inlet line and tighten the 16 mm flare nut to the specified torque of 31 ft. lbs. (42 Nm).

83. With active damper system: Install the suspension stroke sensor harness clips on both sides.

84. With ECPS: Connect the ECPS valve motor connector.

85. Remove the hood prop bar, then reattach the hood support strut on the pivot bolt.

86. Install the passenger's side front fender trim.

87. Install the front wheels, then set the wheels in the straight ahead position.

➡**Before installing the wheel, clean the mating surfaces between the brake disc and the inside of the wheel.**

88. Center the steering rack within its stroke.

89. Remove the vinyl tape around the splines on the pinion shaft.

90. Loosen the upper steering joint bolt, then slip the lower end of the steering joint onto the pinion shaft taking care to align the gap within the angle.

91. Remove the steering wheel holder tool.

92. Align the bolt hole on the steering joint with the groove around the pinion shaft then loosely install the lower steering joint bolt.

93. Pull on the steering joint to make sure that the steering joint is fully seated, then tighten the lower joint bolt, then the upper steering joint bolt, to the specified torque of 16 ft. lbs. (22 Nm).

94. Install the steering joint cover.

95. Set the carpet in place.

96. Do the battery terminal reconnection procedure, and check these items:

a. Make sure the horn and turn signal switches work properly.

b. Make sure the steering wheel switches work properly.

c. Fill the system with power steering fluid, and bleed air from the system.

d. Install the engine cover.

97. After installation, check these items:

a. Start the engine, allow it to idle, and turn the steering wheel from lock to lock several times to warm up the fluid.

b. Check the gearbox for leaks.

c. Check the steering wheel spoke angle. If steering spoke angles to the right and left are not equal (steering wheel and rack are not centered), correct the engagement of the joint/pinion shaft serrations.

d. Set the steering column to the center tilt position and to the center telescopic position, then do the front toe inspection.

e. Do the "VSA Sensor Neutral Position Memorization".

VSA SENSOR NEUTRAL POSITION MEMORIZATION

➡**Do not press the brake pedal during this procedure.**

1. Park the vehicle on a flat and level surface, with the steering wheel in the straight ahead position.

2. With the ignition switch in LOCK (0), or press the engine start/stop button to select the OFF mode, connect the HDS to the data link connector (DLC) under the driver's side of the dashboard.

3. Turn the ignition switch to ON (II), or press the engine start/stop button to select the ON mode.

4. Make sure the HDS communicates with the vehicle and the VSA modulator-control unit.

5. Select VSA ADJUSTMENT with the HDS, and follow the screen prompts.

➡**See the HDS Help menu for specific instructions.**

6. Turn the ignition switch in LOCK (0), or press the engine start/stop button to select the OFF mode.

POWER STEERING PUMP

REMOVAL & INSTALLATION

1. Remove the engine cover.

2. Place a suitable container under the vehicle to catch any spilled fluid.

3. Drain the power steering fluid from the reservoir.
4. Remove the drive belt from the pump pulley.
5. Cover the auto-tensioner, the alternator, and A/C compressor with several shop towels to protect them from spilled power steering fluid. Disconnect the pump inlet hose and pump outlet hose from the pump, and plug them.

⁂ CAUTION

Take care not to spill the fluid on the vehicle. Wipe off any spilled fluid at once. Do not turn the steering wheel with the pump removed.

6. Remove the pump mounting bolts, then remove the pump.
7. Cover the opening of the pump to prevent foreign material from entering the pump.

To install:

8. Transfer the pump inlet hose and the pump outlet hose from the original pump onto the new pump with the new O-ring.
9. Loosely install the pump in the pump bracket with the mounting bolts, then tighten the pump fittings to the specified torque of 8 ft. lbs. (11 Nm).
10. Tighten the pump mounting bolts to the specified torque of 16 ft. lbs. (22 Nm).
11. Install the drive belt.
12. Note these items during drive belt installation:
    a. Inspect the belt for wear and cracks. Replace the belt if necessary.
    b. Make sure that the belt is properly positioned on the pulleys.
    c. Do not get power steering fluid or grease on the auto-tensioner, alternator, A/C compressor, and drive belt or pulley faces. Clean off any fluid or grease before installation.
13. Fill the reservoir to the upper level line.
14. Install the engine cover.
15. Start the engine, and check for leaks.

SUSPENSION

FRONT SUSPENSION

LOWER BALL JOINT

REMOVAL & INSTALLATION

See Figures 167 through 169.

➡Always use a ball joint remover to disconnect a ball joint. Do not strike the housing or any other part of the ball joint connection to disconnect it.

1. Install a hex nut or the ball joint thread protector onto the threads of the ball joint.

➡Using a hex nut, make sure the nut is flush with the ball joint pin end to prevent damage to the threaded end of the ball joint pin.

2. Apply grease to the ball joint remover on the areas shown. This will ease the installation of the tool, and prevent damage to the pressure bolt threads.
3. Loosen the pressure bolt, and install the ball joint remover as shown. Insert the jaws carefully, making sure not to damage the ball joint boot. Adjust the jaw spacing by turning the adjusting bolt.

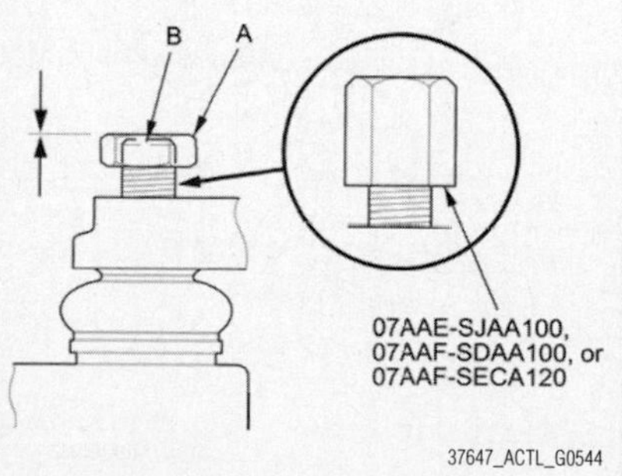

B
A
07AAE-SJAA100,
07AAF-SDAA100, or
07AAF-SECA120
37647_ACTL_G0544

Fig. 167 Install a hex nut (A) or the ball joint thread protector onto the threads of the ball joint (B)

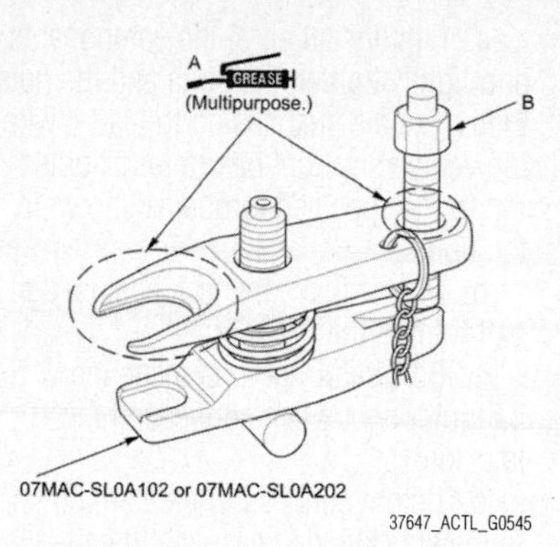

A
GREASE
(Multipurpose.)
B
07MAC-SL0A102 or 07MAC-SL0A202
37647_ACTL_G0545

Fig. 168 Apply grease to the ball joint remover on the areas shown (A)

➡Fasten the safety chain securely to a suspension arm or the subframe. Do not fasten it to a brake line or wire harness.

4. After adjusting the adjusting bolt, make sure the head of the adjusting bolt is in position to allow the jaw to pivot.
5. With a wrench, tighten the pressure bolt until the ball joint pin pops loose from the ball joint connecting hole. If necessary, apply penetrating type lubricant to loosen the ball joint pin.

➡Do not use pneumatic or electric tools on the pressure bolt.

6. Remove the ball joint remover, then remove the nut or the ball joint thread protector from the end of the ball joint pin, and pull the ball joint out of the ball joint connecting hole. Inspect the ball joint boot, and replace it if damaged.

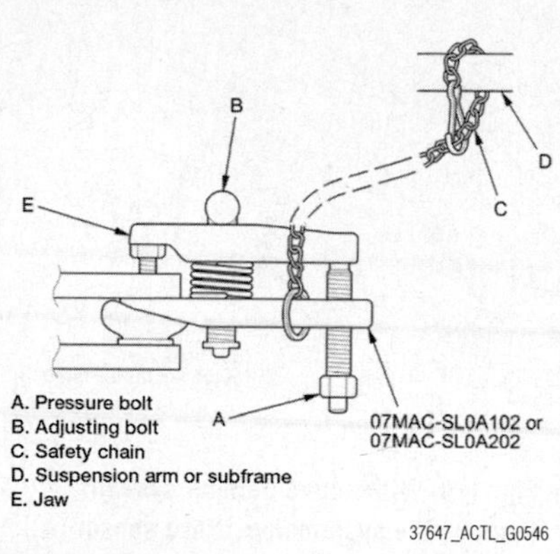

B
D
C
E
A
A. Pressure bolt
B. Adjusting bolt
C. Safety chain
D. Suspension arm or subframe
E. Jaw
07MAC-SL0A102 or
07MAC-SL0A202
37647_ACTL_G0546

Fig. 169 Loosen the pressure bolt, and install the ball joint remover as shown

LOWER CONTROL ARM

REMOVAL & INSTALLATION

See Figures 170 through 172.

1. Raise and support the vehicle.
2. Remove the front wheel.
3. With active damper system: Remove the suspension stroke sensor from the lower arm and detach the harness clip.
4. Remove the lock pin from the lower arm ball joint, then remove the castle nut.
5. Disconnect the lower arm ball joint from the knuckle using the ball joint thread protector and the ball joint remover.

⁂ CAUTION

Do not force or hammer on the lower arm, or pry between the lower arm and the knuckle. You could damage the ball joint.

6. Remove the mounting bolt of the rear side on the stabilizer bar bushing holder.

7. Remove the lower arm mounting 14 mm and 16 mm bolts, then remove the lower arm.

8. Remove the lower arm stops.

➡During installation, align the slot on the lower arm stop with the lug portion on the front side of lower arm bushing.

To install:

9. Install the lower arm in the reverse order of removal, and note these items:

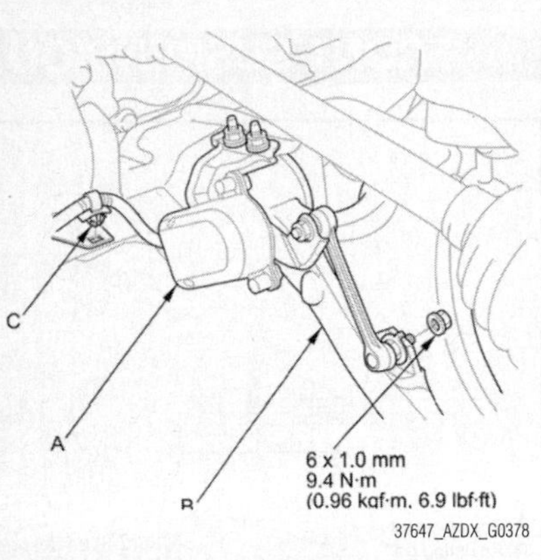
C
A
B
6 x 1.0 mm
9.4 N·m
(0.96 kgf·m, 6.9 lbf·ft)
37647_AZDX_G0378

Fig. 170 With active damper system: Remove the suspension stroke sensor (A) from the lower arm (B) and detach the harness clip (C).

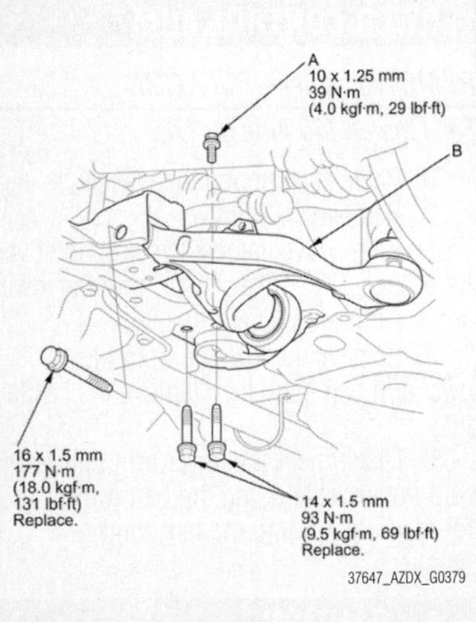
A
10 x 1.25 mm
39 N·m
(4.0 kgf·m, 29 lbf·ft)
B
16 x 1.5 mm
177 N·m
(18.0 kgf·m,
131 lbf·ft)
Replace.
14 x 1.5 mm
93 N·m
(9.5 kgf·m, 69 lbf·ft)
Replace.
37647_AZDX_G0379

Fig. 171 Remove the mounting bolt (A) of the rear side on the stabilizer bar bushing holder. Remove the lower arm mounting 14 mm and 16 mm bolts, then remove the lower arm (B).

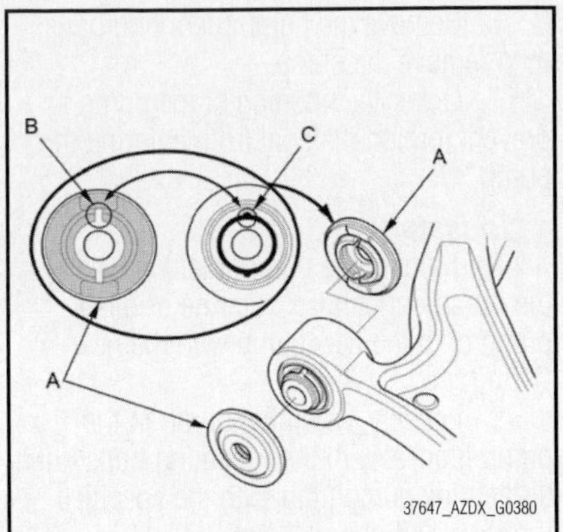
B
C
A
A
37647_AZDX_G0380

Fig. 172 Remove the lower arm stops (A). During installation, align the slot (B) on the lower arm stop with the lug portion (C) on the front side of lower arm bushing.

a. First install all of the components, and lightly tighten the bolts and the nuts, then raise the suspension to load it with the vehicle's weight before fully tightening to the specified torque as shown in the figure.

b. Do not place the jack against the ball joint of the lower arm.

c. Be careful not to damage the ball joint boot when connecting the knuckle.

d. Before connecting the ball joint to the knuckle, degrease the threaded section and the tapered portion of the ball joint pin, the ball joint connecting hole, the threaded section and the mating surfaces of the castle nut.

e. Torque the castle nut to the lower torque specification, then tighten it only far enough to align the slot with the ball joint pin hole. Do not align the castle nut by loosening it.

f. Before installing the wheel, clean the mating surfaces of the brake disc and the inside of the wheel.

10. Check the wheel alignment, and adjust it if necessary.

Bushing Replacement

See Figure 173.

1. Remove the self-locking nut, then remove the rear side lower arm bushing with bracket.

2. Align the angle of the lower arm center line and the under edge line of the bushing bracket as shown.

3. Install a new self-locking nut, then tighten the nut to the specified torque of 119 ft. lbs.(162 Nm).

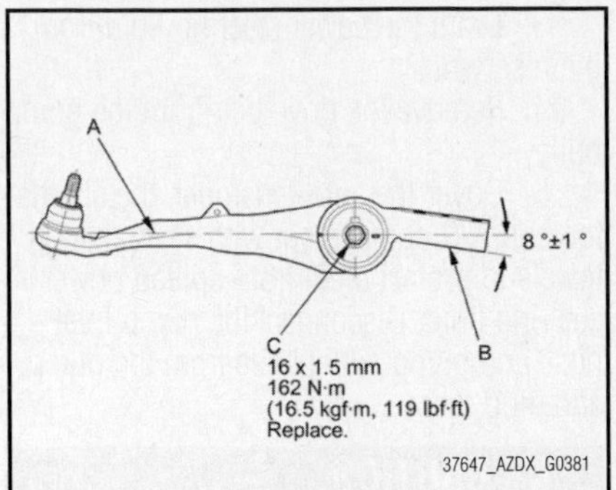
A
8 °±1 °
C
16 x 1.5 mm
162 N·m
(16.5 kgf·m, 119 lbf·ft)
Replace.
B
37647_AZDX_G0381

Fig. 173 Align the angle of the lower arm center line (A) and the under edge line (B) of the bushing bracket as shown. Install a new self-locking nut (C), then tighten the nut to the specified torque of 119 ft. lbs.(162 Nm).

MACPHERSON STRUT (DAMPER)

REMOVAL & INSTALLATION

See Figures 174 through 177.

1. Raise and support the vehicle.

2. Remove the front wheel.

3. With active damper system: Disconnect the damper coil 2P connector, and remove the flange bolt, and detach the harness clip.

➡Be careful not to damage or contaminate the connector.

4. Remove the wheel speed sensor harness and the brake hose from the damper. Do not disconnect the wheel speed sensor connector.

5. Remove the flange nut, while holding the joint pin with a hex wrench, and disconnect the stabilizer link from the damper.

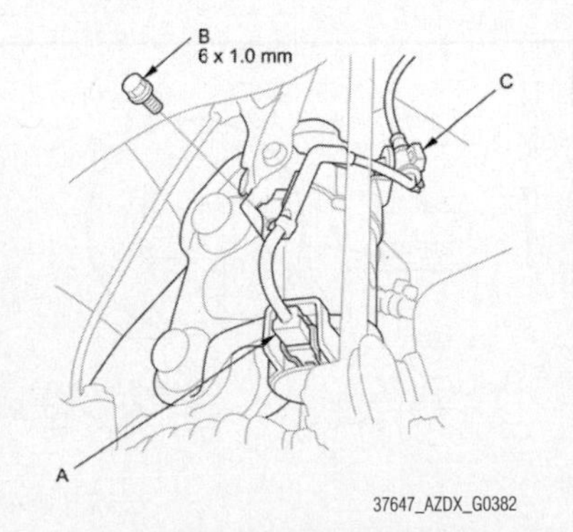
B
6 x 1.0 mm
C
A
37647_AZDX_G0382

Fig. 174 With active damper system: Disconnect the damper coil 2P connector (A), and remove the flange bolt (B), and detach the harness clip (C).

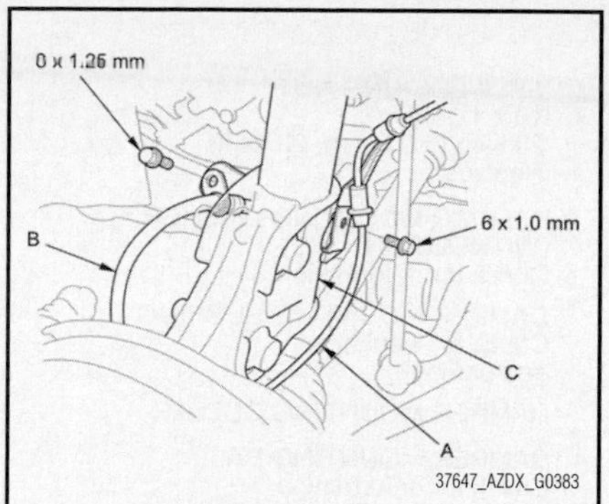
0 x 1.25 mm
6 x 1.0 mm
B
C
A
37647_AZDX_G0383
Fig. 175 Remove the wheel speed sensor harness (A) and the brake hose (B) from the damper (C). Do not disconnect the wheel speed sensor connector.

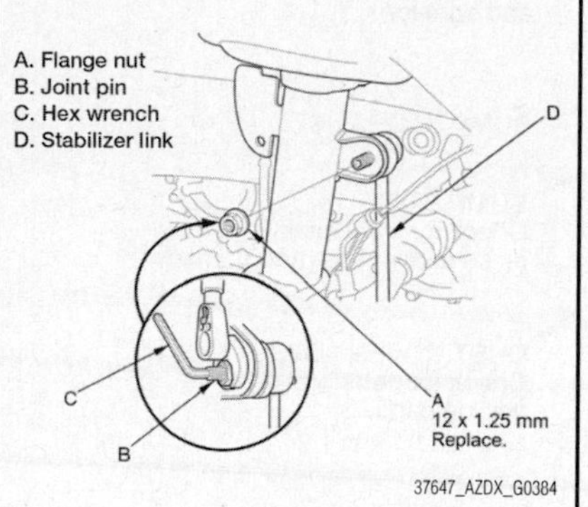
A. Flange nut
B. Joint pin
C. Hex wrench
D. Stabilizer link
D
C
A
12 x 1.25 mm
Replace.
B
37647_AZDX_G0384
Fig. 176 Remove the flange nut, while holding the joint pin with a hex wrench, and disconnect the stabilizer link from the damper.

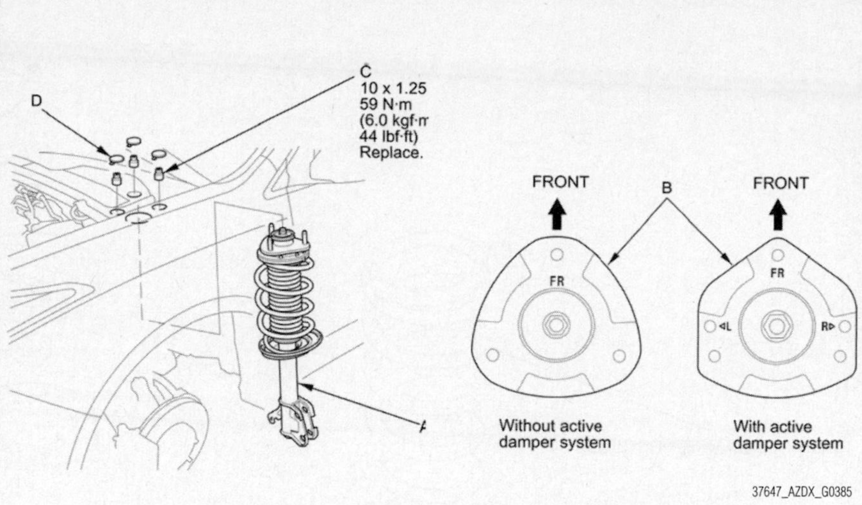
C
10 x 1.25
59 N·m
(6.0 kgf·m
44 lbf·ft)
Replace.
D
FRONT
B
FRONT
FR
FR
Without active damper system
With active damper system
37647_AZDX_G0385
Fig. 177 Install the damper/spring assembly (A) on to the frame. Note the direction of the damper mounting base (B) as shown.

6. Remove the damper pinch bolts and flange nuts from the damper.

CAUTION

Do not allow the knuckle to rotate too far outward. This may allow the driveshaft inboard joint to come apart.

7. Remove the service caps from the shock tower and remove the flange nuts from the top of the damper.

8. Remove the damper/spring assembly.

➡The left and right damper springs are different. Mark the springs L and R before you continue.

To install:

➡Do not fully tighten nuts or bolts until instructed to do so.

9. Install the damper/spring on to the frame. Note the direction of the damper mounting base as shown.

10. Loosely install the new flange nuts to the top of the damper.

11. Loosely install new damper pinch bolts and new flange nuts to the lower end of the damper.

12. Connect the stabilizer link to the damper, and loosely install a new flange nut.

13. Tighten the lower flange nut to 58 ft. lbs. (79 Nm), while holding the joint pin with the hex wrench.

14. Tighten the flange nuts on top of the damper to 44 ft. lbs. (59 Nm).

15. Tighten the damper pinch bolts while holding the flange nuts to 156 ft. lbs. (211 Nm).

16. Install the service caps.

17. Install the wheel speed sensor harness and the brake hose to the damper.

18. With active damper system: Connect the damper 2P coil connector and install the flange bolt and attach the harness clip.

CAUTION

Be careful not to damage or contaminate the connecter.

19. Clean the mating surfaces of the brake disc and the inside of the wheel, then install the front wheel.

20. With active damper system: Start the engine, then make sure there are no active damper system DTCs shown, using the HDS, or equivalent scan tool.

21. With active damper system: Do the DAMPER FORCE OPERATION TEST in the FUNCTIONAL TESTS MENU with the HDS, then make sure all four damper units function normally.

22. Check the wheel alignment, and adjust it if necessary.

OVERHAUL

Disassembly

See Figures 178 and 179.

1. Compress the damper spring, then remove the self-locking nut while holding the damper shaft with a hex wrench. Do not compress the damper spring more than is necessary to remove the self-locking nut.

2. Release the pressure from the strut spring compressor, then disassemble the damper as shown in the Exploded View.

Inspection

1. Reassemble all the parts, except for the damper spring.

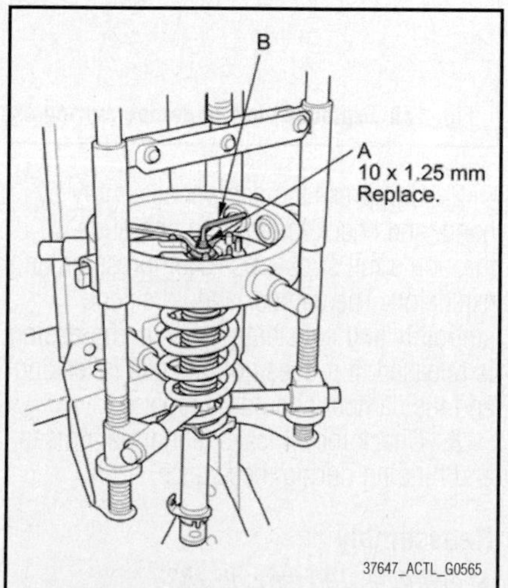
B
A
10 x 1.25 mm
Replace.
37647_ACTL_G0565
Fig. 178 Compress the damper spring, then remove the self-locking nut (A) while holding the damper shaft with a hex wrench (B)

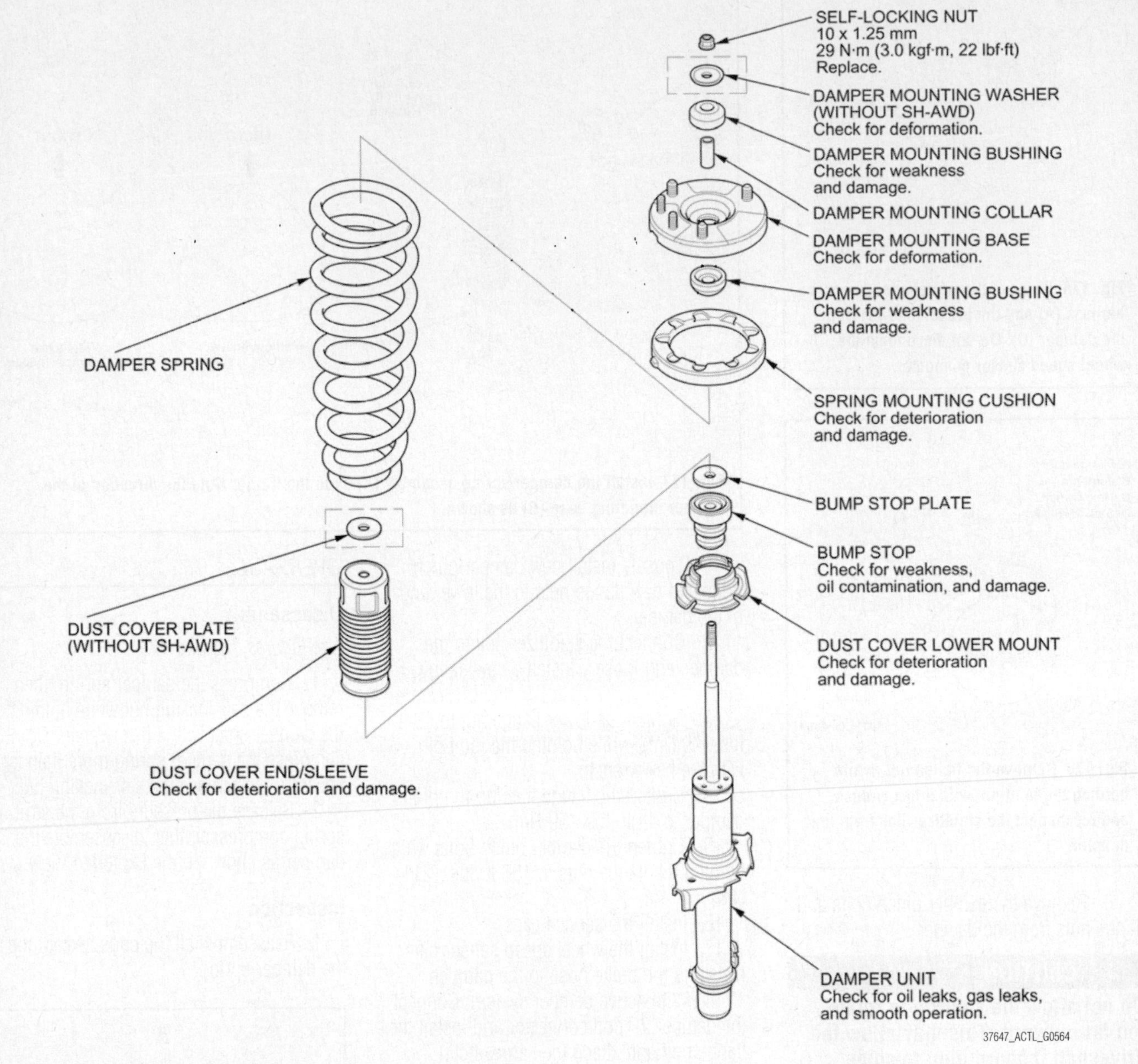
SELF-LOCKING NUT
10 x 1.25 mm
29 N·m (3.0 kgf·m, 22 lbf·ft)
Replace.
DAMPER MOUNTING WASHER
(WITHOUT SH-AWD)
Check for deformation.
DAMPER MOUNTING BUSHING
Check for weakness
and damage.
DAMPER MOUNTING COLLAR
DAMPER MOUNTING BASE
Check for deformation.
DAMPER MOUNTING BUSHING
Check for weakness
and damage.
SPRING MOUNTING CUSHION
Check for deterioration
and damage.
DAMPER SPRING
BUMP STOP PLATE
BUMP STOP
Check for weakness,
oil contamination, and damage.
DUST COVER PLATE
(WITHOUT SH-AWD)
DUST COVER LOWER MOUNT
Check for deterioration
and damage.
DUST COVER END/SLEEVE
Check for deterioration and damage.
DAMPER UNIT
Check for oil leaks, gas leaks,
and smooth operation.
37647_ACTL_G0564

Fig. 179 Exploded view of damper/spring assembly

2. Compress the damper assembly by hand, and check for smooth operation through a full stroke, both compression and extension. The damper should extend smoothly and constantly when compression is released. If it does not, the gas is leaking and the damper should be replaced.

3. Check for oil leaks, abnormal noises, and binding during these tests.

Reassembly

See Figures 180 through 183.

1. Install the spring mounting cushion on the damper mounting base by aligning the tab and notch.

2. Install all the parts except the damper mounting washer (Without SH-AWD) and the self-locking nut onto the damper unit by referring to the Exploded View.

3. Compress the damper spring using a strut spring compressor. Do not compress the spring excessively.

4. Align the lower end of the damper spring with the stepped part of the dust cover lower mount and the lower spring seat on the damper unit.

5. Position the tab on the spring mounting cushion facing forward but toward the inside of the vehicle.

6. Align the angle of the stud on the damper mounting base with the aligning tab on the bottom of the damper unit.

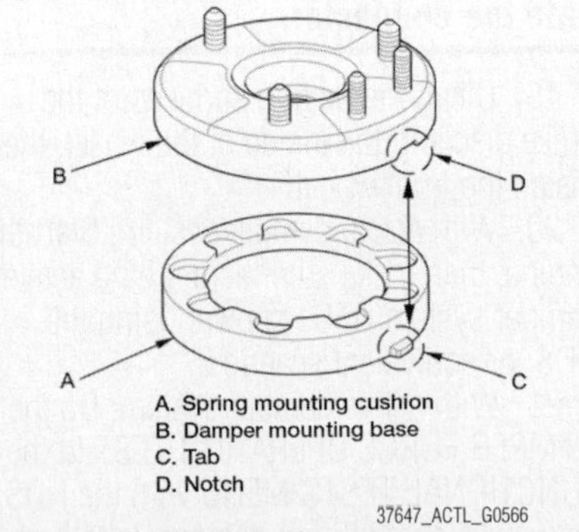
B
D
A
C
A. Spring mounting cushion
B. Damper mounting base
C. Tab
D. Notch
37647_ACTL_G0566

Fig. 180 Install the spring mounting cushion on the damper mounting base by aligning the tab and notch

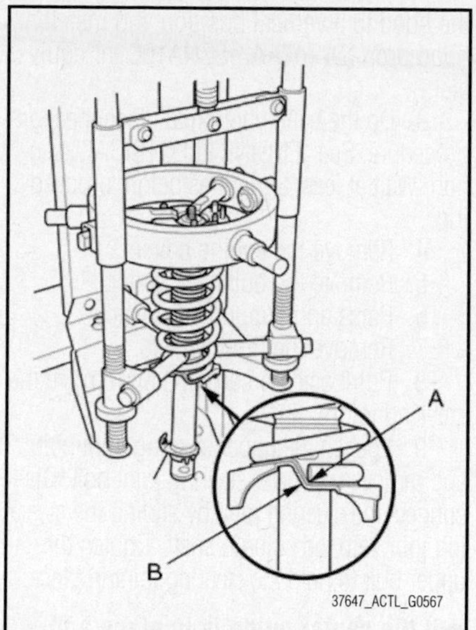
A
B
37647_ACTL_G0567

Fig. 181 Align the lower end (A) of the damper spring with the stepped part (B) of the dust cover lower mount

7. Install the damper mounting washer (Without SH-AWD) and the new self-locking nut.

8. Hold the damper shaft with a hex wrench, and tighten the self-locking nut to 22 ft. lbs. (29 Nm).

9. Remove the damper/spring from the strut spring compressor.

STEERING KNUCKLE/HUB BEARING UNIT

REMOVAL & INSTALLATION

Hub Bearing Unit

See Figure 184.

➡Refer to the exploded view figure for reference.

1. Raise and support the vehicle.

2. Remove the wheel nuts and the front wheel.

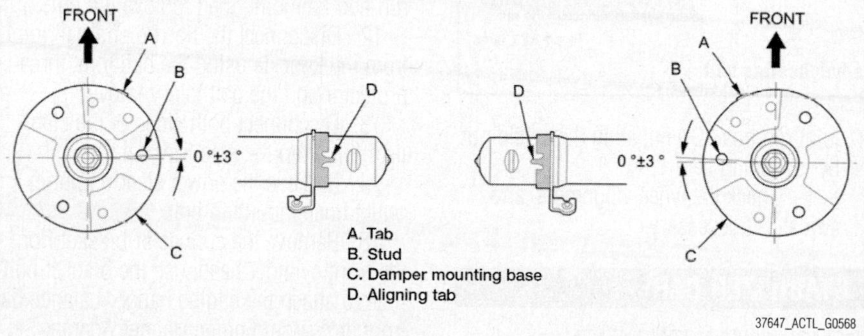
FRONT
A
B
D
0 °±3 °
C
A. Tab
B. Stud
C. Damper mounting base
D. Aligning tab
FRONT
A
B
D
0 °±3 °
C
37647_ACTL_G0568

Fig. 182 Position the tab on the spring mounting cushion facing forward but toward the inside of the vehicle

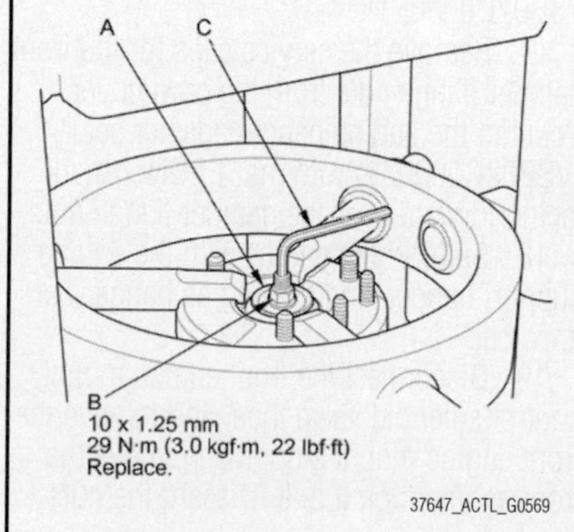
A
C
B
10 x 1.25 mm
29 N·m (3.0 kgf·m, 22 lbf·ft)
Replace.
37647_ACTL_G0569

Fig. 183 Install the damper mounting washer (Without SH-AWD) (A) and the new self-locking nut (B)

3. Remove the brake hose mounting bolt from the damper.

4. Remove the brake caliper bracket mounting bolts, then remove the caliper assembly from the knuckle.

⁂ CAUTION

To prevent damage to the caliper assembly or brake hose, use a short piece of wire to hang the caliper assembly from the undercarriage. Do not twist or stretch the brake hose excessively.

5. Pry up the stake on the spindle nut, then remove the nut.

6. Remove the front brake disc.

7. Remove the flange bolts, and remove the hub bearing unit by tapping the driveshaft end with a soft face hammer while drawing the hub bearing unit outward, then remove the splash guard.

⁂ CAUTION

Do not pull the driveshaft end outward. The driveshaft inboard joint may come apart.

8. Check the hub bearing unit for damage and cracks.

To install:

➡Torque settings are shown in the exploded view figure.

a. During installation, apply grease to the mating surfaces of the hub bearing unit and driveshaft outboard joint.

b. Install the hub bearing unit in the reverse order of removal, and note these items:

c. Use the new spindle nut during reassembly.

d. Before installing the spindle nut, apply a small amount of engine oil to the seating surface of the nut.

e. After tightening, use a drift to stake the spindle nut shoulder against the driveshaft.

f. Before installing the brake disc, clean the mating surfaces of the hub bearing unit and the inside of the brake disc.

g. Before installing the wheel, clean the mating surfaces of the brake disc and the inside of the wheel.

h. Check the wheel alignment, and adjust it if necessary.

Knuckle Replacement

➡Refer to the exploded view figure for reference.

1. Remove the hub bearing unit as above.

2. Remove the wheel speed sensor from the knuckle, but do not disconnect the wheel speed sensor connector.

3. Remove the cotter pin from the tie-rod end ball joint, then remove the nut.

4. Disconnect the tie-rod end ball joint from the knuckle using the ball joint thread protector and the ball joint remover.

5. Remove the lock pin from the lower arm ball joint, then remove the castle nut.

6. Disconnect the lower arm ball joint from the knuckle using the ball joint thread protector and the ball joint remover, using care not to damage the ball joint boot.

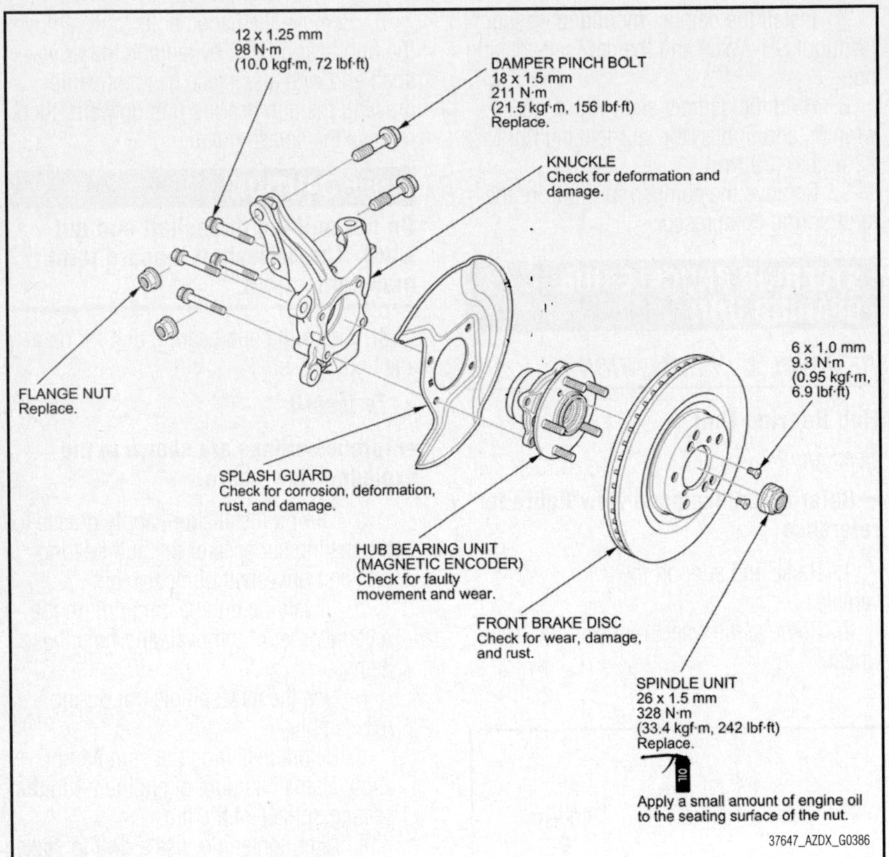

12 x 1.25 mm
98 N·m
(10.0 kgf·m, 72 lbf·ft)
DAMPER PINCH BOLT
18 x 1.5 mm
211 N·m
(21.5 kgf·m, 156 lbf·ft)
Replace.
KNUCKLE
Check for deformation and damage.
FLANGE NUT
Replace.
6 x 1.0 mm
9.3 N·m
(0.95 kgf·m, 6.9 lbf·ft)
SPLASH GUARD
Check for corrosion, deformation, rust, and damage.
HUB BEARING UNIT
(MAGNETIC ENCODER)
Check for faulty movement and wear.
FRONT BRAKE DISC
Check for wear, damage, and rust.
SPINDLE UNIT
26 x 1.5 mm
328 N·m
(33.4 kgf·m, 242 lbf·ft)
Replace.
Apply a small amount of engine oil to the seating surface of the nut.
37647_AZDX_G0386

Fig. 184 Exploded view of the front steering knuckle/hub bearing unit

✲✲ CAUTION

Do not force or hammer on the lower arm, or pry between the lower arm and the knuckle. You could damage the ball joint.

7. Remove the damper pinch bolts and flange nuts from the damper, then remove the knuckle.

To install:

8. Install the knuckle in the reverse order of removal, and note these items:
   a. During installation, install new damper pinch bolts and new flange nuts.
   b. Be careful not to damage the ball joint boot when connecting the knuckle.
   c. During installation, install the new cotter pin after tightening the new nut, and properly bend its end.
   d. Before connecting the ball joint to the knuckle, degrease the threaded section and the tapered portion of the ball joint pin, the ball joint connecting hole, the threaded section and the mating surfaces of the castle nut.
   e. Torque the castle nut to the lower torque specification, then tighten it only far enough to align the slot with the ball joint pin hole. Do not align the castle nut by loosening it.
   f. Check the wheel alignment, and adjust it if necessary.

STABILIZER BAR

REMOVAL & INSTALLATION

See Figures 185 through 190.

➡These special tools, or equivalent, are required during this procedure:

- Hood Prop Bar 07AAZ-SZNA100
- Ball Joint Thread Protector, 12 mm 07AAF-SDAA100
- Ball Joint Remover, 28 mm 07MAC-SL0A202
- Engine Hanger Adapter Set VSB02C000031
- Subframe Adapter VSB02C000016

➡When you replace the stabilizer bar, lower the front subframe from the body, and replace the stabilizer bar through the gap created by lowering the front subframe.

1. Remove the passenger's side front fender trim.
2. Pry the hood support strut clip, and release the hood support strut, then move the hood to a vertical position and install a hood prop bar (o7AAZ-SZNA100, or equivalent).
3. Do the battery terminal disconnection procedure. See "ENGINE ELECTRICAL" section. Wait at least 3 minutes before proceeding.
4. Remove the engine cover.
5. Remove the bulkhead cover.
6. Raise and support the vehicle.
7. Remove the front wheels.
8. Pull back the carpet, then remove the steering joint cover.
9. Loosen the upper steering joint bolt, and remove the lower steering joint bolt. Disconnect the steering joint by sliding the steering joint into the column shaft. Tighten the upper bolt to hold the steering joint in place.

➡If the center guide is in place and has not moved, leave it in place. If the center guide has moved or been removed, discard it.

10. Center the steering wheel spokes, and install a commercially available steering wheel holder tool.
11. Remove the cotter pin from the tie-rod end ball joint, then remove the nut.
12. Disconnect the tie-rod end ball joint from the knuckle using the ball joint thread protector and the ball joint remover.
13. Disconnect both sides of the stabilizer link from the stabilizer bar.
14. Remove the power steering pump outlet hose mounting bolt.
15. Remove the connector bracket from the front cylinder head; use the bracket bolt hole to attach the engine hanger balance bar front arm (from Engine Hanger Adapter Set VSB02C000031).
16. Remove the engine mount control solenoid valve bracket from the rear cylinder head; use the bracket bolt hole to attach the engine hanger balance bar rear arm (from Engine Hanger Adapter Set VSB02C000031).
17. Remove the service caps for the front damper flange nuts from the cowl cover. Position the engine hanger adapter set (VSB02C000031) with the "FRONT" mark facing forward over the damper flange nuts.
18. Support the engine with the engine support hanger, and the engine hanger balance bar.
19. Disconnect the front engine mount control solenoid valve tube, and remove the front engine mount stop, then remove the front engine mount bolt. Discard the nuts and bolt.
20. Raise the vehicle on a lift.
21. Remove the engine undercover.
22. Remove the front splash shield.

23. Remove the front subframe stiffener plate.

24. Remove exhaust pipe A.

25. Remove the P/S line bracket, then remove the rear engine mount base mounting bolts.

26. Remove the ground cable bolt and remove the lower transmission mount mounting bolts.

27. Attach the subframe adapter (VSB02C000016) to a transmission jack, then raise up the jack to the subframe. Loop the belt over the front of the subframe, then secure the belt by inserting the stop and then tightening the wing nut.

28. Raise the jack and line up the slots in the subframe adapter arms with the bolt holes on the jack base, then securely attach them with four bolts.

29. Remove the stiffener mounting bolts on both sides of the front stiffener.

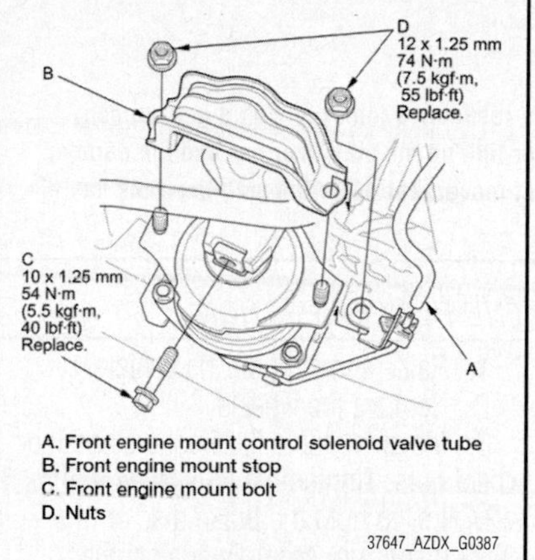
D
12 x 1.25 mm
74 N·m
(7.5 kgf·m,
55 lbf·ft)
Replace.
B
C
10 x 1.25 mm
54 N·m
(5.5 kgf·m,
40 lbf·ft)
Replace.
A
A. Front engine mount control solenoid valve tube
B. Front engine mount stop
C. Front engine mount bolt
D. Nuts
37647_AZDX_G0387

Fig. 185 Disconnect the front engine mount control solenoid valve tube, and remove the front engine mount stop, then remove the front engine mount bolt. Discard the nuts and bolt.

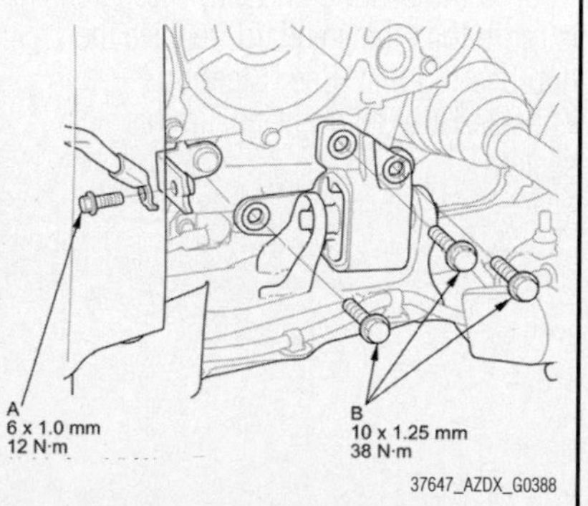
A
6 x 1.0 mm
12 N·m
B
10 x 1.25 mm
38 N·m
37647_AZDX_G0388

Fig. 186 Remove the ground cable bolt (A) and remove the lower transmission mount mounting bolts (B).

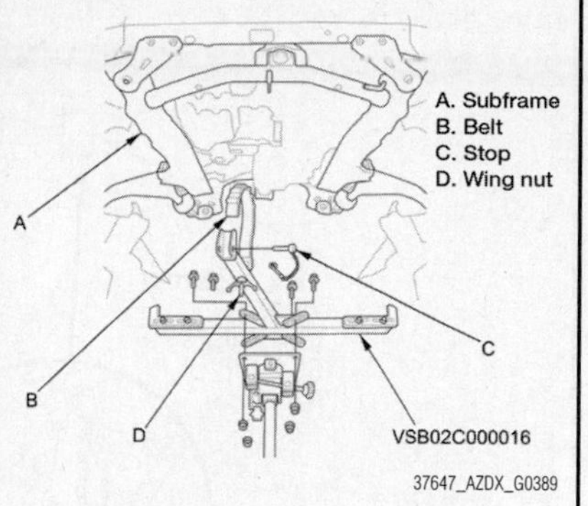
A. Subframe
B. Belt
C. Stop
D. Wing nut
A
B
C
D
VSB02C000016
37647_AZDX_G0389

Fig. 187 Attach the subframe adapter (VSB02C000016) to a transmission jack, then raise up the jack to the subframe. Loop the belt over the front of the subframe, then secure the belt by inserting the stop and then tightening the wing nut.

30. Loosen the subframe mounting bolts to obtain a 10 mm (0.39 in) distance between the bolt seat and the mounting surface. Do not loosen the mounting bolts more than necessary.

31. Remove the stiffener mounting bolts on both sides of the rear stiffener.

32. Loosen the subframe mounting bolts to obtain a 30 mm (1.18 in) distance between the bolt seat and the mounting surface. Do not loosen the mounting bolts more than necessary.

33. Lower the transmission jack slowly with the front subframe adapter until the front subframe has dropped about 30 mm (1.18 in). Do not lower the front subframe beyond the loosened subframe mounting bolts clearance.

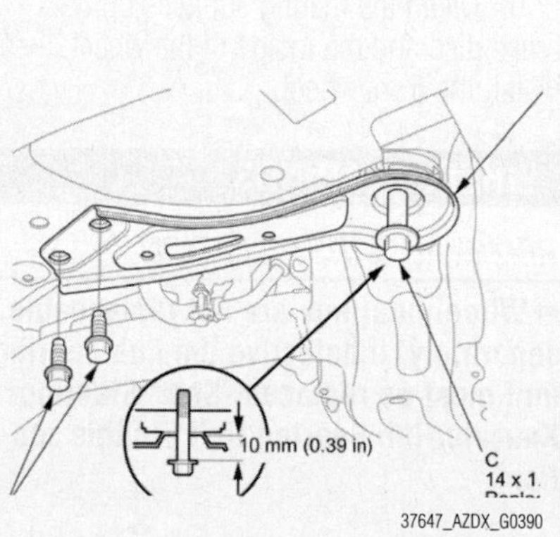
10 mm (0.39 in)
C
37647_AZDX_G0390

Fig. 188 Remove the stiffener mounting bolts (A) on both sides of the front stiffener (B). Loosen the subframe mounting bolts (C) to obtain a 10 mm (0.39 in) distance between the bolt seat and the mounting surface. Do not loosen the mounting bolts more than necessary.

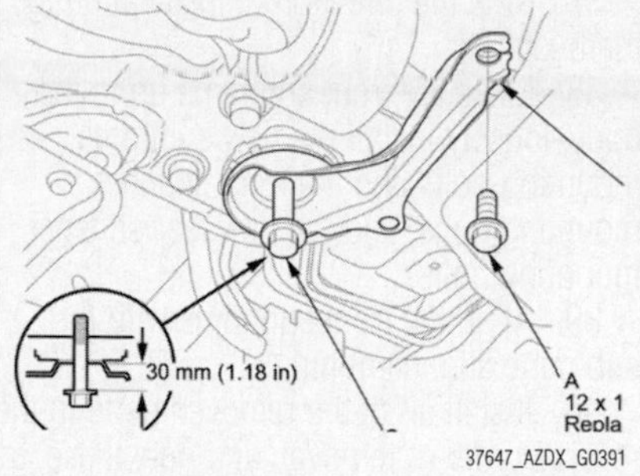
30 mm (1.18 in)
A
37647_AZDX_G0391

Fig. 189 Remove the stiffener mounting bolts (A) on both sides of the rear stiffener (B). Loosen the subframe mounting bolts (C) to obtain a 30 mm (1.18 in) distance between the bolt seat and the mounting surface. Do not loosen the mounting bolts more than necessary.

34. Remove the flange bolts and the bushing holders, then remove the bushings. Note the paint marks on the stabilizer bar which will be aligned with the side of the bushings during installation.

35. Move the stabilizer bar toward the passenger's side, and remove the stabilizer bar.

To install:

➡Refer to the appropriate figure in the removal steps for specified torque settings.

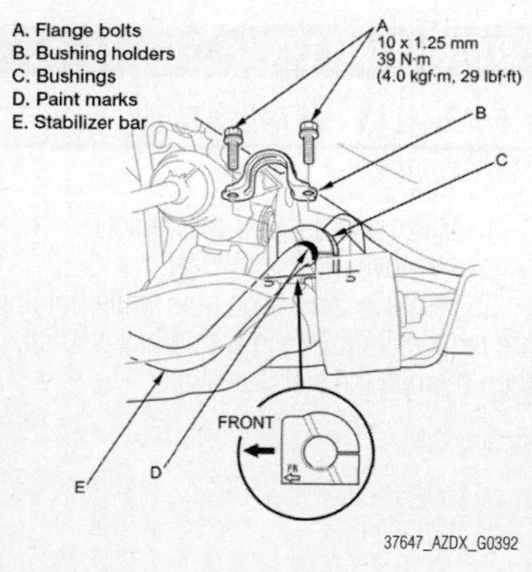
A. Flange bolts
B. Bushing holders
C. Bushings
D. Paint marks
E. Stabilizer bar
A
10 x 1.25 mm
39 N·m
(4.0 kgf·m, 29 lbf·ft)
B
C
FRONT
D
E
37647_AZDX_G0392

Fig. 190 Remove the flange bolts and the bushing holders, then remove the bushings. Note the paint marks on the stabilizer bar which will be aligned with the side of the bushings during installation.

36. Install the stabilizer bar.
37. Note the direction of installation for the bushings.
38. Raise the front subframe up to the body, loosely install the new subframe mounting bolts and the front stiffener mounting bolts, and the new rear stiffener mounting bolts.
39. Align the front subframe with the subframe alignment pin.
40. Install all of the removed parts in the reverse order of removal, and note these items:
a. Refer to stabilizer link removal/installation to connect the stabilizer bar to the links.
b. During installation, align the paint marks on the stabilizer bar with the side of the bushings.
c. Use new mounting bolts for the bolt of the front side during reassembly.
d. Use new front subframe stiffener flange bolts during reassembly.
e. Use new nuts and a front engine mount bolt during reassembly.
f. If the center guide is in place, use it to determine the steering joint installation angle.
g. If the center guide is gone, check the steering joint installation angle.
h. During installation, install the new cotter pin after tightening the new nut, and properly bend its end.
i. Before installing the wheel, clean the mating surfaces on the brake disc and the inside of the wheel.
41. Do the battery terminal reconnection procedure.
42. Check the wheel alignment, and adjust it if necessary.

STABILIZER LINK

REMOVAL & INSTALLATION

See Figure 191.

1. Raise and support the vehicle.
2. Remove the front wheel.
3. Remove the flange nuts while holding the respective joint pin with a hex wrench, then remove the stabilizer link.

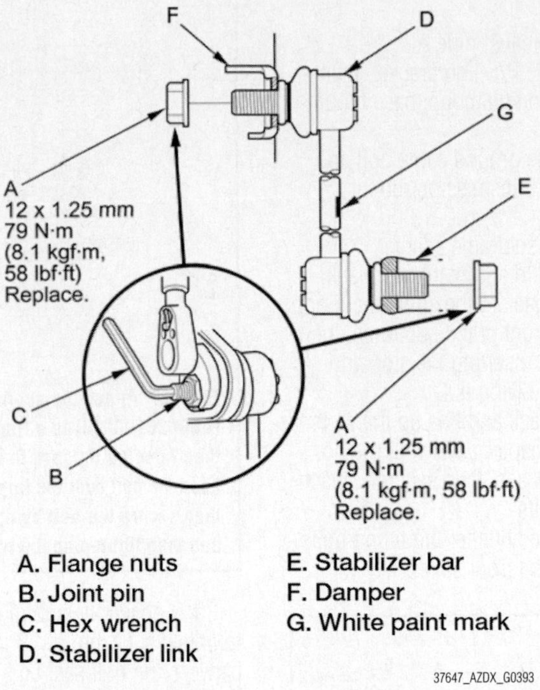
F
D
G
E
A
12 x 1.25 mm
79 N·m
(8.1 kgf·m,
58 lbf·ft)
Replace.
C
B
A
12 x 1.25 mm
79 N·m
(8.1 kgf·m, 58 lbf·ft)
Replace.
A. Flange nuts
B. Joint pin
C. Hex wrench
D. Stabilizer link
E. Stabilizer bar
F. Damper
G. White paint mark
37647_AZDX_G0393

Fig. 191 Remove the flange nuts while holding the respective joint pin with a hex wrench, then remove the stabilizer link. Install the stabilizer link on the stabilizer bar and the damper with the joint pins set at the center of their range of movement. Note the stabilizer link has a white paint mark.

To install:

4. Install the stabilizer link on the stabilizer bar and the damper with the joint pins set at the center of their range of movement. Note the stabilizer link has a white paint mark.
5. Install the flange nuts, and tighten them to the specified torque while holding the respective joint pin with a hex wrench.
6. Clean the mating surfaces of the brake disc and the inside of the wheel, then install the front wheel.

WHEEL BEARINGS

REMOVAL & INSTALLATION

➡Wheel bearings are not replaceable separately. If defective the hub bearing unit must be replaced. See "Steering Knuckle/Hub Bearing Unit" in this section.

END PLAY INSPECTION

1. Raise and support the vehicle.
2. Remove the wheels.
3. Install suitable flat washers and the wheel nuts. Tighten the nuts to 94 ft. lbs. (127 Nm) to hold the brake disc or the brake disc/drum securely against the hub.
4. Attach a dial gauge and place the dial gauge against the hub flange.
5. Measure the bearing end play while moving the brake disc or the brake disc/drum inward and outward. End play should be 0–0.0020 in. (0–0.05 mm).
6. If the bearing end play measurement is more than the standard, replace the hub bearing unit. See "Steering Knuckle/Hub Bearing Unit" in this section.

KNUCKLE/HUB BEARING UNIT

REMOVAL & INSTALLATION

Hub Bearing Unit Replacement

See Figure 192.

➡Refer to the exploded view illustration for reference during this procedure.

1. Raise and support the vehicle.
2. Remove the wheel nuts and rear wheel.
3. Remove the brake caliper bracket mounting bolts, then remove the caliper assembly from the knuckle.

CAUTION

To prevent damage to the caliper assembly or the brake hose, use a short piece of wire to hang the caliper assembly from the undercarriage. Do not twist the brake hose excessively.

4. Remove the two washers from the caliper-knuckle mounting tangs.
5. Pry up the stake on the spindle nut, then remove the nut.
6. Release the parking brake lever fully, and remove the rear brake disc/drum.
7. Remove the hub bearing unit.

CAUTION

Do not pull the driveshaft end outward. The driveshaft inboard joint may come apart.

8. Check the hub bearing unit for damage and cracks.

To install:

9. Install the hub bearing unit in the reverse order of removal, and note these items:
   a. Use a new spindle nut during reassembly.
   b. Before installing the spindle nut, apply a small amount of engine oil to the seating surface of the nut.
   c. After tightening, use a drift to stake the spindle nut shoulder against the driveshaft.
   d. Before installing the brake disc/drum, clean the mating surfaces of the hub bearing unit and the inside of the brake disc/drum.
   e. During installation, make sure the washers are installed between the brake caliper bracket and the knuckle.
   f. Before installing the wheel, clean the mating surfaces of the brake disc/drum and the inside of the wheel.
10. Check the wheel alignment, and adjust it if necessary.

Knuckle Replacement

➡Refer to the exploded view illustration for reference during this procedure.

1. Remove the hub bearing unit.
2. Remove the brake hose bracket mounting bolt from the knuckle.
3. Remove the flange nuts, then remove the backing plate with the parking brake shoe assembly.

CAUTION

To prevent damage to the backing plate with the parking brake shoes assembly and cable, use a short piece of wire to hang the backing plate from the undercarriage. Do not twist the parking brake cable excessively.

4. Remove the wheel speed sensor from the knuckle. Do not disconnect the wheel speed sensor connector.
5. Position a floor jack under lower arm B. Raise the floor jack until the suspension begins to compress.
6. Remove the lock pin from the upper arm ball joint, then remove the castle nut.
7. Disconnect the upper arm ball joint from the knuckle using the ball joint thread protector and the ball joint remover.

➡Be careful not to damage the ball joint boot when installing the remover.

8. Remove the self-locking nut, the washer, and the flange bolt, then remove lower arm A.
9. Remove the flange nuts from the trailing arm.
10. Remove the flange bolt, then remove the knuckle.

To install:

11. Install the knuckle in the reverse order of removal, and note these items:

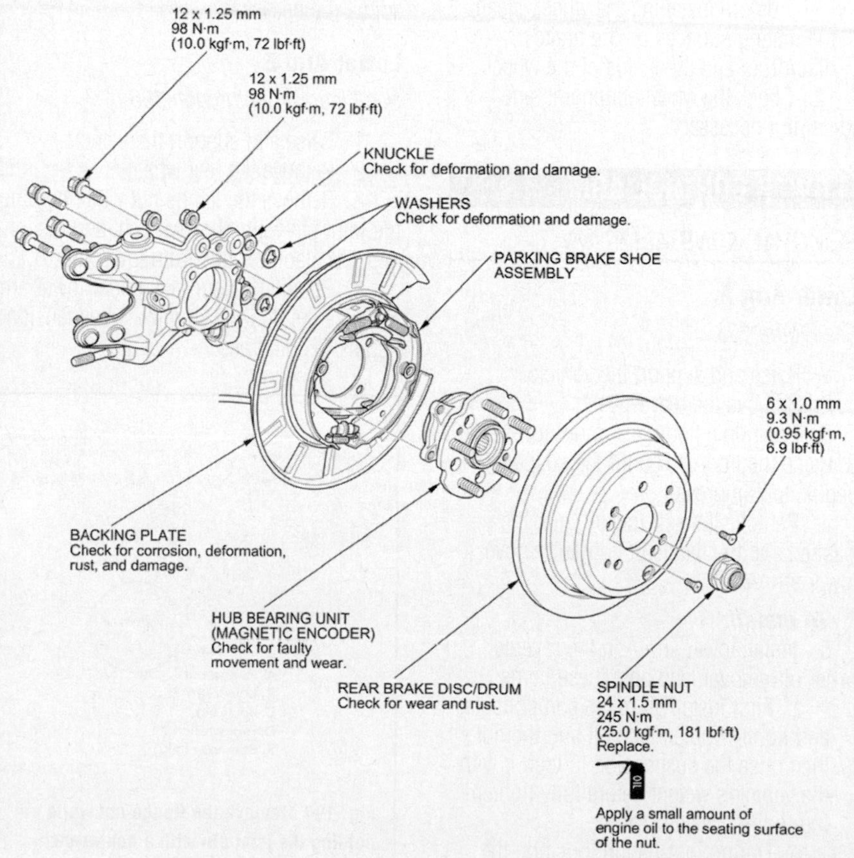
12 x 1.25 mm
98 N·m
(10.0 kgf·m, 72 lbf·ft)
12 x 1.25 mm
98 N·m
(10.0 kgf·m, 72 lbf·ft)
KNUCKLE
Check for deformation and damage.
WASHERS
Check for deformation and damage.
PARKING BRAKE SHOE ASSEMBLY
6 x 1.0 mm
9.3 N·m
(0.95 kgf·m, 6.9 lbf·ft)
BACKING PLATE
Check for corrosion, deformation, rust, and damage.
HUB BEARING UNIT
(MAGNETIC ENCODER)
Check for faulty movement and wear.
REAR BRAKE DISC/DRUM
Check for wear and rust.
SPINDLE NUT
24 x 1.5 mm
245 N·m
(25.0 kgf·m, 181 lbf·ft)
Replace.
Apply a small amount of engine oil to the seating surface of the nut.
37647_AZDX_G0404

Fig. 192 Exploded view of the rear knuckle/hub bearing unit

a. First install all of the components, and lightly tighten the bolts and the nuts, then raise the suspension to load it with the vehicle's weight before fully tightening to the specified torque.

b. Use new flange nuts during reassembly.

c. During installation, position the paint mark on lower arm A toward the outside of the vehicle.

d. Use a new self-locking nut and the new flange bolt, during reassembly.

e. Be careful not to damage the ball joint boot when connecting the knuckle.

f. Before connecting the ball joint, degrease the threaded section and the tapered portion of the ball joint pin, the ball joint connecting hole, and the threaded section and the mating surfaces of the castle nut.

g. During installation, install the lock pin by properly bending it over the edge of the castle nut after tightening the castle nut.

h. Torque the castle nut to the lower torque specification, then tighten it only far enough to align the slot with the ball joint pin hole. Do not align the castle nut by loosening it.

i. Before installing the wheel, clean the mating surfaces on the brake disc/drum and the inside of the wheel.

12. Check the wheel alignment, and adjust it if necessary.

LOWER CONTROL ARM

REMOVAL & INSTALLATION

Lower Arm A

See Figure 193.

1. Raise and support the vehicle.
2. Remove the rear wheel.
3. Position a floor jack under lower arm B. Raise the floor jack until the suspension begins to compress.
4. Remove the self-locking nut, the washer, and the flange bolt, then remove lower arm A.

To install:

5. Install lower arm A in the reverse order of removal, and note these items:

a. First install all of the components, and lightly tighten the bolt and the nut, then raise the suspension to load it with the vehicle's weight before fully tightening to the specified torque.

b. During installation, position the paint mark on lower arm A toward the outside of the vehicle.

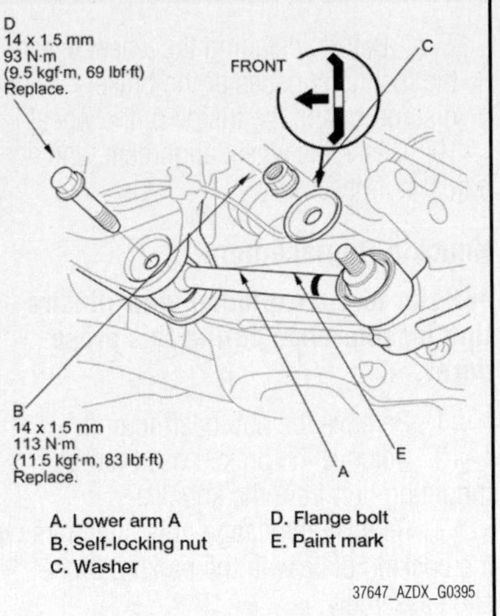
D
14 x 1.5 mm
93 N·m
(9.5 kgf·m, 69 lbf·ft)
Replace.
FRONT
C
B
14 x 1.5 mm
113 N·m
(11.5 kgf·m, 83 lbf·ft)
Replace.
A
E
A. Lower arm A
B. Self-locking nut
C. Washer
D. Flange bolt
E. Paint mark
37647_AZDX_G0395

Fig. 193 Remove the self-locking nut, the washer, and the flange bolt, then remove lower arm A.

c. Use a new self-locking nut and the new flange bolt, during assembly.

d. Before installing the wheel, clean the mating surfaces of the brake disc/drum and the inside of the wheel.

6. Check the wheel alignment, and adjust it if necessary.

Lower Arm B

See Figures 194 through 196.

1. Raise and support the vehicle.
2. Remove the rear wheel.
3. Remove the flange nut while holding the joint pin with a hex wrench, and disconnect the stabilizer link from lower arm B.
4. Position a floor jack under lower arm B. Raise the floor jack until the suspension begins to compress.

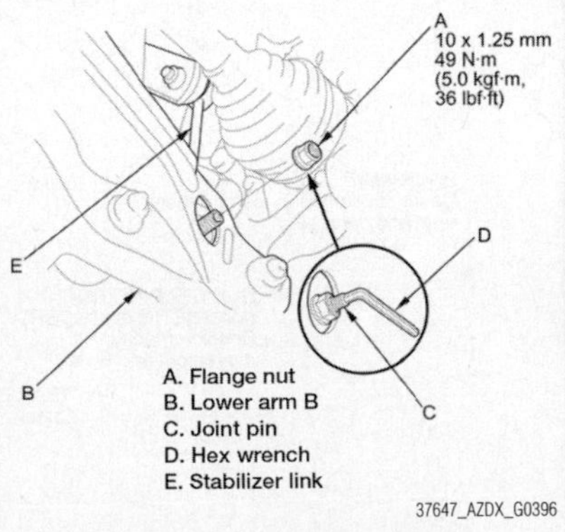
A
10 x 1.25 mm
49 N·m
(5.0 kgf·m,
36 lbf·ft)
D
E
B
C
A. Flange nut
B. Lower arm B
C. Joint pin
D. Hex wrench
E. Stabilizer link
37647_AZDX_G0396

Fig. 194 Remove the flange nut while holding the joint pin with a hex wrench, and disconnect the stabilizer link from lower arm B.

5. Remove the flange bolt from the bottom of the damper.
6. Remove the flange bolt from the knuckle.
7. Lower the floor jack gradually.
8. Remove the spring and the lower spring seat from the lower arm B.
9. Mark the cam positions of the adjusting bolt and the adjusting cam plate with the frame. Remove the self-locking nut while holding the adjusting bolt, then remove the adjusting cam plate, the adjusting bolt, and lower arm B.

To install:

10. Install lower arm B in the reverse order of removal, and note these items:

a. First install all of the suspension components, and lightly tighten the bolts and the nuts, then raise the suspension to load it with the vehicle's weight before fully tightening to the specified torques:

- Self-locking nut on cam adjusting bolt: 80 ft. lbs. (108 Nm)
- Knuckle flange bolt: 83 ft. lbs. (113 Nm)
- Lower damper flange bolt: 47 ft. lbs. (64 Nm)
- Stabilizer link flange nut: 36 ft. lbs. (49 Nm)

b. Use a new adjusting bolt and a new self-locking nut during reassembly.

c. Use the new flange bolts during reassembly.

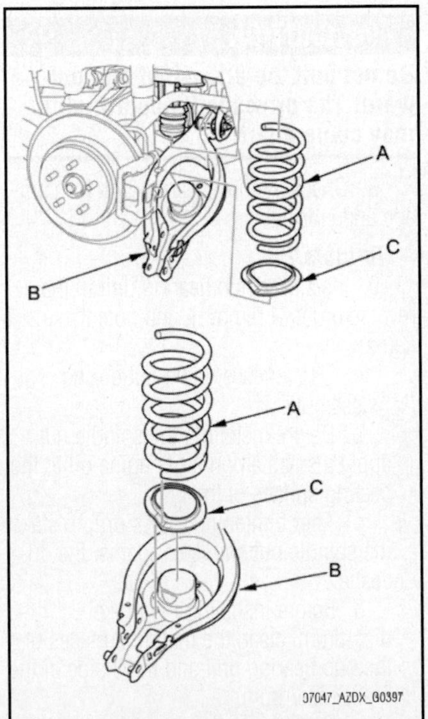
A
C
B
A
C
B
37047_AZDX_G0397

Fig. 195 Remove the spring (A) and the lower spring seat (C) from the lower arm B.

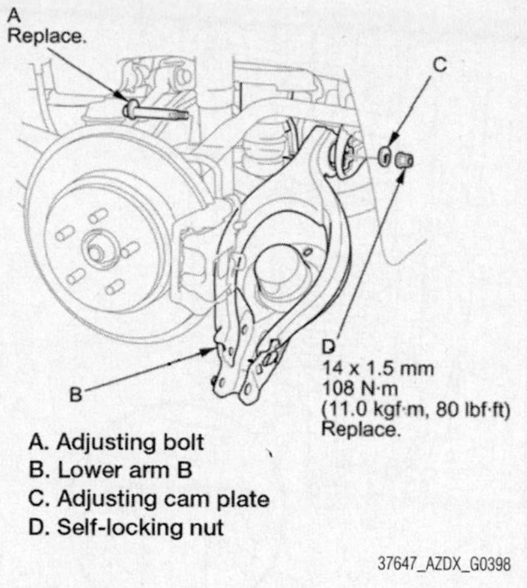
A
Replace.
C
D
14 x 1.5 mm
108 N·m
(11.0 kgf·m, 80 lbf·ft)
Replace.
B
A. Adjusting bolt
B. Lower arm B
C. Adjusting cam plate
D. Self-locking nut
37647_AZDX_G0398

Fig. 196 Mark the cam positions of the adjusting bolt and the adjusting cam plate with the frame. Remove the self-locking nut while holding the adjusting bolt, then remove the adjusting cam plate, the adjusting bolt, and lower arm B.

d. Align the bottom of the spring with the stepped part of the lower spring seat and lower arm B as shown in the figure under removal steps.

e. Align the cam positions of the adjusting bolt and the adjusting cam plate with the marked positions on the frame when tightening the self-locking nut.

f. Before installing the wheel, clean the mating surfaces of the brake disc/drum and the inside of the wheel.

11. Check the wheel alignment, and adjust it if necessary.

MCPHERSON STRUTS (DAMPER)

REMOVAL & INSTALLATION

See Figure 197.

1. Raise and support the vehicle.
2. Remove the rear wheel.
3. With active damper system: Disconnect the damper coil 2P connector, and detach the harness clip. Use care not to damage or contaminate the connector.
4. Remove the flange nut while holding the joint pin with a hex wrench, and disconnect the stabilizer link from lower arm B.
5. Position a floor jack under lower arm B. Raise the floor jack until the suspension begins to compress.
6. Remove the flange bolt from the bottom of the damper.
7. Remove the flange bolt from the knuckle.

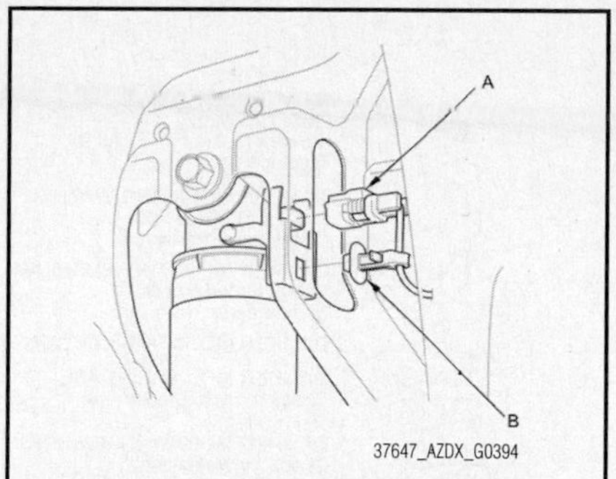
A
B
37647_AZDX_G0394

Fig. 197 With active damper system: Disconnect the damper coil 2P connector (A), and detach the harness clip (B). Use care not to damage or contaminate the connector.

8. Remove the flange bolt from the top of the damper.
9. Lower the floor jack gradually, then remove the damper from the vehicle.

Inspection:

10. Compress the damper assembly by hand, and check for smooth operation through a full stroke, both compression and extension. The damper should extend smoothly and constantly when compression is released. If it does not, the gas is leaking and the damper should be replaced.
11. Check for oil leaks, abnormal noises, and binding during these tests.

To install:

12. Position the damper between the body and lower arm B.
13. With active damper system: Position the damper coil 2P connector on the damper facing rearward.
14. Loosely install a new flange bolt to the top of the damper.
15. Raise the floor jack until the hole in lower arm B aligns with the hole in the damper, then loosely install the new flange bolt to the bottom of the damper.
16. Loosely install the new flange bolt to the knuckle.
17. Connect the stabilizer link to lower arm B, and install the new flange nut, and tighten the flange nut to 36 ft. lbs. (49 Nm), while holding the joint pin with the hex wrench.
18. Raise the rear suspension with the floor jack to load the suspension with the vehicle's weight.
19. Tighten the flange bolts on the top of the damper to 47 ft. lbs. (64 Nm) and bottom of the damper to 47 ft. lbs. (64 Nm).
20. Tighten the flange bolt to the knuckle to 83 ft. lbs. (113 Nm).
21. With active damper system: Connect the damper coil 2P connector, and attach the harness clip. Use care not to damage or contaminate the connector.
22. Clean the mating surfaces of the brake disc/drum and the inside of the wheel, then install the rear wheel.
23. With active damper system: Do the memorizing rear suspension full rebound position.
24. With active damper system: Start the engine, then make sure there are no active damper system DTCs with the HDS.
25. With active damper system: Do the DAMPER FORCE OPERATION TEST in the FUNCTIONAL TESTS MENU with the HDS, or equivalent scan tool, then make sure all four damper units function normally.
26. Check the wheel alignment, and adjust it if necessary.

OVERHAUL

Disassembly

See Figures 198 and 199.

1. Compress the damper spring, then remove the self-locking nut while holding the damper shaft with a hex wrench. Do not compress the damper spring more than is necessary to remove the self-locking nut.
2. Release the pressure from the strut spring compressor, then disassemble the damper as shown in the Exploded View.

Inspection

1. Reassemble all the parts, except for the damper spring.
2. Compress the damper assembly by hand, and check for smooth operation through a full stroke, both compression and extension. The damper should extend

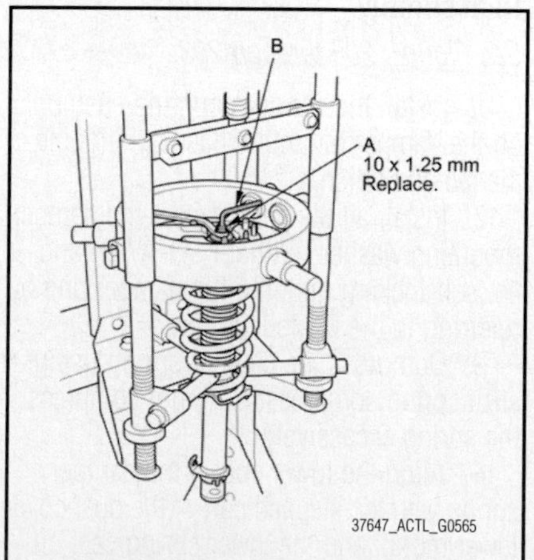
B
A
10 x 1.25 mm
Replace.
37647_ACTL_G0565

Fig. 198 Compress the damper spring, then remove the self-locking nut (A) while holding the damper shaft with a hex wrench (B)

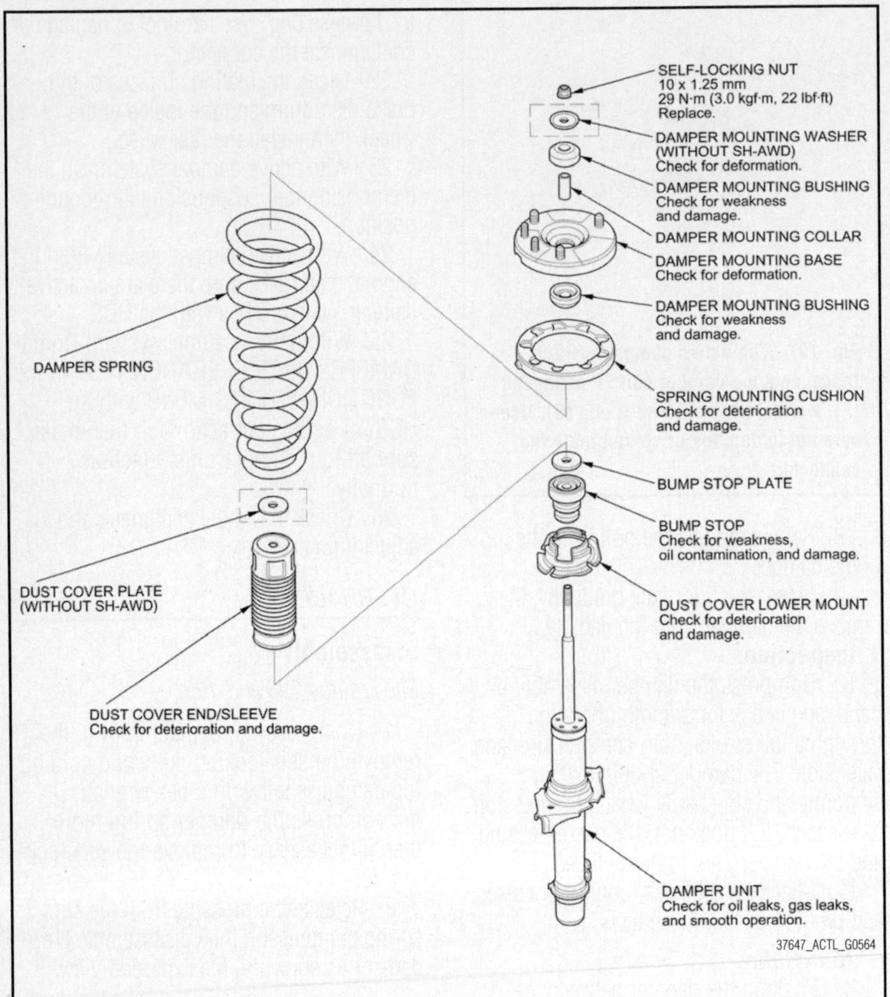
SELF-LOCKING NUT
10 x 1.25 mm
29 N·m (3.0 kgf·m, 22 lbf·ft)
Replace.
DAMPER MOUNTING WASHER
(WITHOUT SH-AWD)
Check for deformation.
DAMPER MOUNTING BUSHING
Check for weakness
and damage.
DAMPER MOUNTING COLLAR
DAMPER MOUNTING BASE
Check for deformation.
DAMPER MOUNTING BUSHING
Check for weakness
and damage.
DAMPER SPRING
SPRING MOUNTING CUSHION
Check for deterioration
and damage.
BUMP STOP PLATE
BUMP STOP
Check for weakness,
oil contamination, and damage.
DUST COVER PLATE
(WITHOUT SH-AWD)
DUST COVER LOWER MOUNT
Check for deterioration
and damage.
DUST COVER END/SLEEVE
Check for deterioration and damage.
DAMPER UNIT
Check for oil leaks, gas leaks,
and smooth operation.
37647_ACTL_G0564

Fig. 199 Exploded view of damper/spring assembly

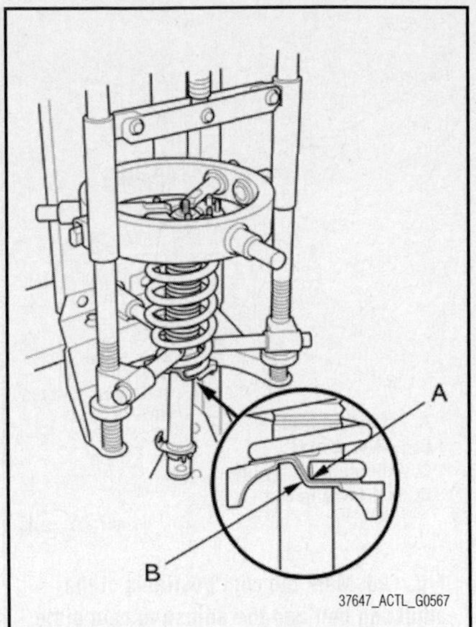
A
B
37647_ACTL_G0567

Fig. 201 Align the lower end (A) of the damper spring with the stepped part (B) of the dust cover lower mount

smoothly and constantly when compression is released. If it does not, the gas is leaking and the damper should be replaced.

3. Check for oil leaks, abnormal noises, and binding during these tests.

Reassembly

See Figures 200 through 203.

1. Install the spring mounting cushion on the damper mounting base by aligning the tab and notch.
2. Install all the parts except the damper mounting washer (Without SH-AWD) and the self-locking nut onto the damper unit by referring to the Exploded View.
3. Compress the damper spring using a strut spring compressor. Do not compress the spring excessively.
4. Align the lower end of the damper spring with the stepped part of the dust cover lower mount and the lower spring seat on the damper unit.
5. Position the tab on the spring mounting cushion facing forward but toward the inside of the vehicle.
6. Align the angle of the stud on the damper mounting base with the aligning tab on the bottom of the damper unit.
7. Install the damper mounting washer (Without SH-AWD) and the new self-locking nut.
8. Hold the damper shaft with a hex wrench, and tighten the self-locking nut to 22 ft. lbs. (29 Nm).
9. Remove the damper/spring from the strut spring compressor.

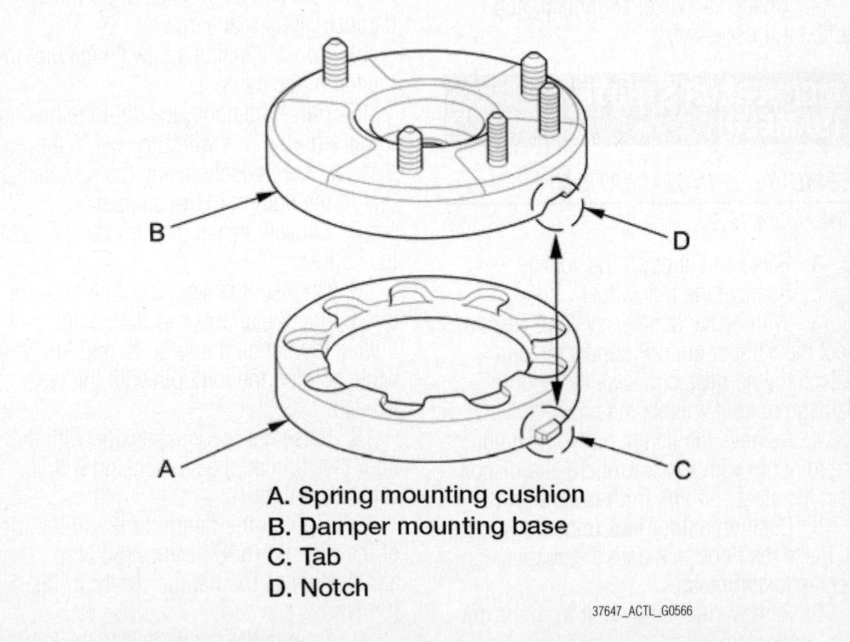
B
D
A
C
A. Spring mounting cushion
B. Damper mounting base
C. Tab
D. Notch
37647_ACTL_G0566

Fig. 200 Install the spring mounting cushion on the damper mounting base by aligning the tab and notch

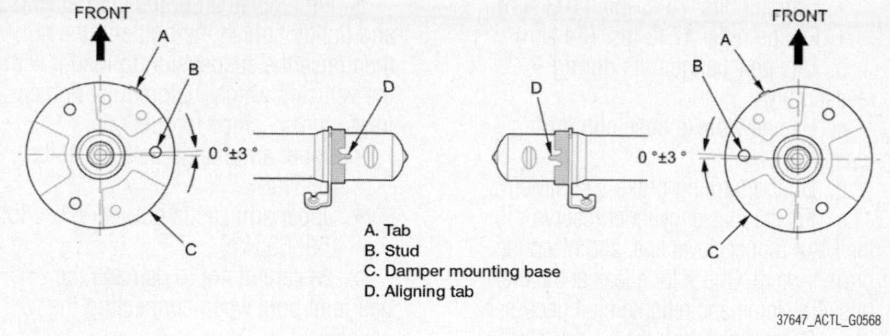
FRONT
A
B
C
D
0 °±3 °
A. Tab
B. Stud
C. Damper mounting base
D. Aligning tab
37647_ACTL_G0568

Fig. 202 Position the tab on the spring mounting cushion facing forward but toward the inside of the vehicle

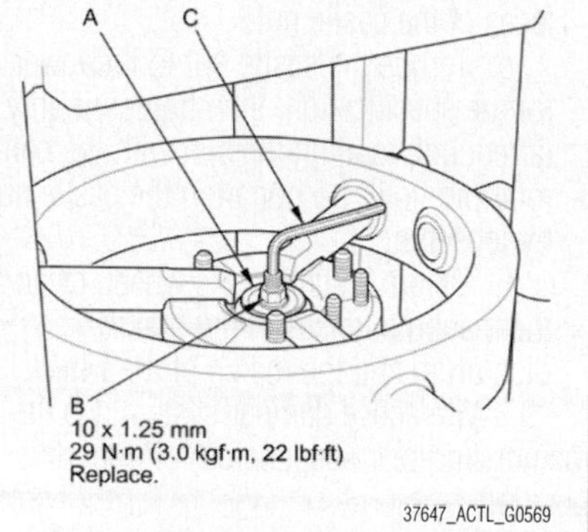
A
C
B
10 x 1.25 mm
29 N·m (3.0 kgf·m, 22 lbf·ft)
Replace.
37647_ACTL_G0569

Fig. 203 Install the damper mounting washer (Without SH-AWD) (A) and the new self-locking nut (B)

STABILIZER BAR

REMOVAL & INSTALLATION

See Figure 204.

1. Raise and support the vehicle.
2. Remove the rear wheels.
3. Remove the spare tire from the vehicle.
4. Remove the spare tire support bracket.
5. Disconnect both stabilizer links from the stabilizer bar.
6. Remove the flange bolts and the bushing holders, then remove the bushings and the stabilizer bar. Note the paint marks on the stabilizer bar which will be aligned with the side of the bushings during installation.

To install:

7. Install the stabilizer bar in the reverse order of removal, and note these items:
   a. Note the right and left direction of the stabilizer bar.
   b. Align the paint marks on the stabilizer bar with the side of the bushings.

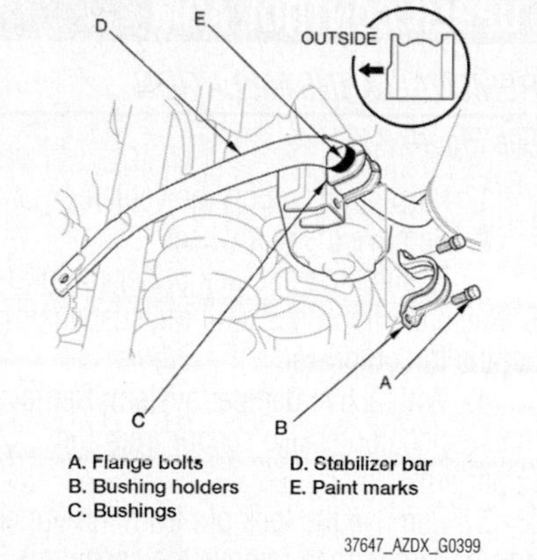
D
E
OUTSIDE
C
B
A
A. Flange bolts
B. Bushing holders
C. Bushings
D. Stabilizer bar
E. Paint marks
37647_AZDX_G0399

Fig. 204 Remove the flange bolts and the bushing holders, then remove the bushings and the stabilizer bar. Note the paint marks on the stabilizer bar which will be aligned with the side of the bushings during installation.

   c. See "Stabilizer Link" to connect the stabilizer bar to the links.
   d. Before installing the wheel, clean the mating surfaces of the brake disc/drum and the inside of the wheel.

STABILIZER LINK

REMOVAL & INSTALLATION

See Figure 205.

1. Raise and support the vehicle.
2. Remove the rear wheel.
3. Remove the self-locking nut and the flange nut while holding the respective joint pin with a hex wrench, then remove the stabilizer link.

To install:

4. Install the stabilizer link on the stabilizer bar and lower arm B with the joint pins set at the center of their range of movement.

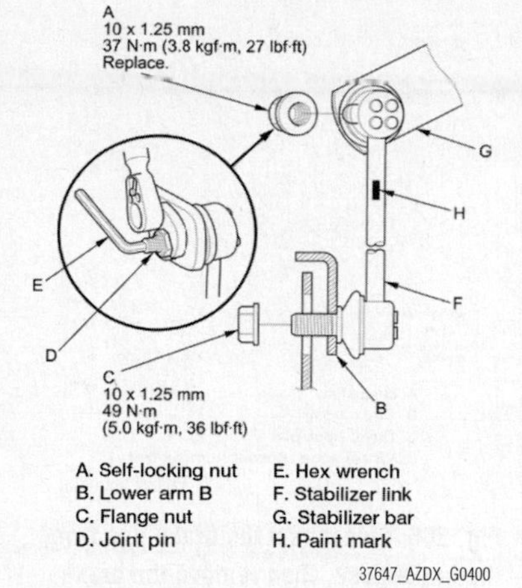
A
10 x 1.25 mm
37 N·m (3.8 kgf·m, 27 lbf·ft)
Replace.
G
H
E
F
D
C
10 x 1.25 mm
49 N·m
(5.0 kgf·m, 36 lbf·ft)
B
A. Self-locking nut
B. Lower arm B
C. Flange nut
D. Joint pin
E. Hex wrench
F. Stabilizer link
G. Stabilizer bar
H. Paint mark
37647_AZDX_G0400

Fig. 205 Remove the self-locking nut and the flange nut while holding the respective joint pin with a hex wrench, then remove the stabilizer link.

➡The stabilizer link has a paint mark. The paint mark indicates the difference between the left and right stabilizer links.

5. Install the end of the stabilizer link with the paint mark in the upper position.
6. Install a new self-locking nut and the flange nut, and tighten the self-locking upper nut to 27 ft. lbs. (37 Nm). Tighten the lower flange nut to 36 ft. lbs. (49 Nm) while holding the respective joint pin with a hex wrench.
7. Clean the mating surfaces of the brake disc/drum and the inside of the wheel, then install the rear wheel.
8. Test-drive the vehicle.
9. After 5 minutes of driving, tighten the self-locking nut again to 27 ft. lbs. (37 Nm).

TRAILING ARM

REMOVAL & INSTALLATION

See Figures 206 through 208.

1. Raise and support the vehicle.
2. Remove the rear wheel.
3. Disconnect the parking brake cable from the parking brake lever.
4. Remove the parking brake cable mounting bolts from the backing plate.
5. Remove the parking brake cable nut from the trailing arm.
6. Disconnect the brake line from the brake hose, then remove the brake hose clip.

➡Plug the end of a hose and joint to prevent spilling brake fluid.

7. Remove the wheel speed sensor harness bolt.
8. Remove the flange nuts from the trailing arm.

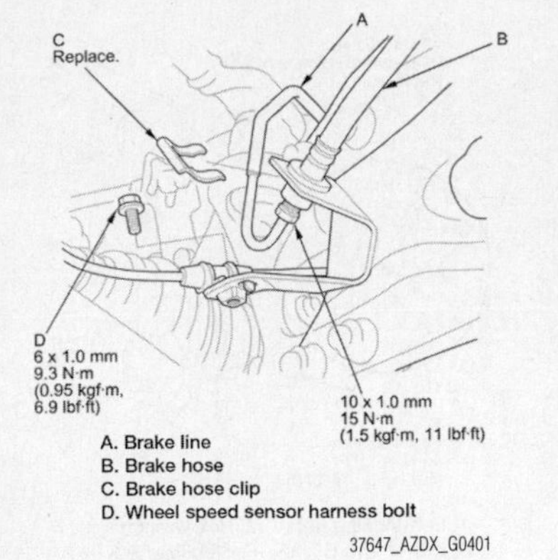
C
Replace.
A
B
D
6 x 1.0 mm
9.3 N·m
(0.95 kgf·m,
6.9 lbf·ft)
10 x 1.0 mm
15 N·m
(1.5 kgf·m, 11 lbf·ft)
A. Brake line
B. Brake hose
C. Brake hose clip
D. Wheel speed sensor harness bolt
37647_AZDX_G0401

Fig. 206 Disconnect the brake line from the brake hose, then remove the brake hose clip. Remove the wheel speed sensor harness bolt.

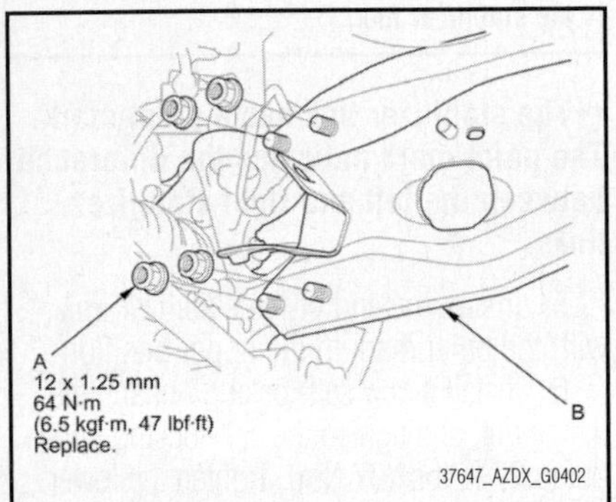
A
12 x 1.25 mm
64 N·m
(6.5 kgf·m, 47 lbf·ft)
Replace.
B
37647_AZDX_G0402

Fig. 207 Remove the flange nuts (A) from the trailing arm (B).

9. Remove the flange bolts from the trailing arm, then remove the trailing arm.

To install:

10. Install the trailing arm in the reverse order of removal, and note these items:

a. First install all of the components, and lightly tighten the bolts and the nuts, then raise the suspension to load it with the vehicle's weight before fully tightening to the specified torque:

- Flange bolts: 76 ft. lbs. (103 Nm)
- Flange nuts: 47 ft. lbs. (64 Nm)

b. Use new flange nuts during reassembly.

c. Use new brake hose clip during reassembly.

d. Do the parking brake adjustment.

e. Fill the master cylinder reservoir to the MAX (upper) level line, and bleed the brake system. Check for a leak at the brake hose/line joint, and retighten it if necessary.

f. Before installing the wheel, clean the mating surfaces on the brake disc/drum and the inside of the wheel.

11. Check the wheel alignment, and adjust it if necessary.

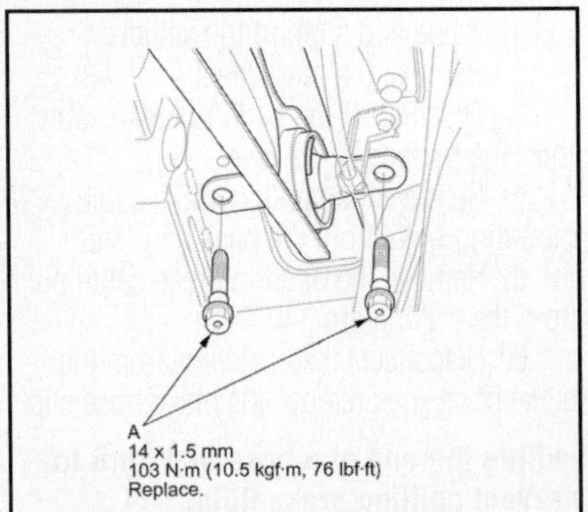
A
14 x 1.5 mm
103 N·m (10.5 kgf·m, 76 lbf·ft)
Replace.
37647_AZDX_G0403

Fig. 208 Remove the flange bolts (A) from the trailing arm (B), then remove the trailing arm.

UPPER CONTROL ARM

REMOVAL & INSTALLATION

See Figure 209.

1. Raise and support the vehicle.
2. Remove the rear wheel.
3. Position a floor jack under lower arm B. Raise the floor jack until the suspension begins to compress.
4. With active damper system: Remove the suspension stroke sensor from the upper arm.
5. Remove the lock pin from the upper arm ball joint, then remove the castle nut.
6. Disconnect the upper arm ball joint from the knuckle using the ball joint thread protector and the ball joint remover. Use care not to damage the ball joint boot when installing the remover.
7. Remove the flange bolt and remove the upper arm.

To install:

8. Install the upper arm in the reverse order of removal, and note these items:

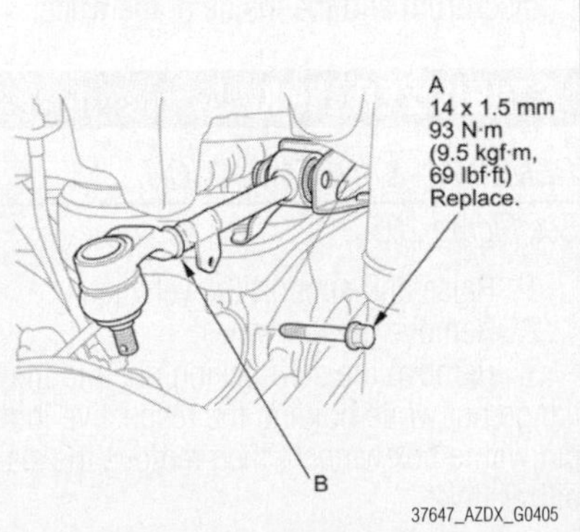
A
14 x 1.5 mm
93 N·m
(9.5 kgf·m,
69 lbf·ft)
Replace.
B
37647_AZDX_G0405

Fig. 209 Remove the flange bolt (A) and remove the upper arm (B).

a. Use the new flange bolt during reassembly.

b. First install all of the components, and lightly tighten the bolt and the nut, then raise the suspension to load it with the vehicle's weight before fully tightening to the specified torque:

- Upper arm flange bolt: 69 ft. lbs. (93 Nm)
- Upper arm castle nut: 44–51 ft. lbs. (59–69 Nm)

c. Be careful not to damage the ball joint boot when connecting the knuckle.

d. Before connecting the ball joint to the knuckle, degrease the threaded section and the tapered portion of the ball joint pin, the ball joint connecting hole, the threaded section and the mating surfaces of the castle nut.

e. Torque the castle nut to the lower torque specification, then tighten it only far enough to align the slot with the ball joint pin hole. Do not align the castle nut by loosening it.

f. Before installing the wheel, clean the mating surfaces of the brake disc/drum and the inside of the wheel.

9. With active damper system: Do the memorizing rear suspension full rebound position.

10. Check the wheel alignment, and adjust it if necessary.

WHEEL BEARINGS

REMOVAL & INSTALLATION

➡Wheel bearings are an integral part of the hub bearing unit. If replacement is needed, see "Knuckle/Hub Bearing Unit" in this section.

END PLAY INSPECTION

1. Raise and support the vehicle.
2. Remove the wheels.
3. Install suitable flat washers and the wheel nuts. Tighten the nuts to 94 ft. lbs. (127 Nm) to hold the brake disc or the brake disc/drum securely against the hub.
4. Attach a dial gauge and place the dial gauge against the hub flange.
5. Measure the bearing end play while moving the brake disc or the brake disc/drum inward and outward. End play should be 0–0.0020 in. (0–0.05 mm).
6. If the bearing end play measurement is more than the standard, replace the hub bearing unit. See "Steering Knuckle/Hub Bearing Unit" in this section.

ACURA

Diagnostic Trouble Codes

7

DIAGNOSTIC TROUBLE CODES 7-1
Gas Engine OBD II Trouble Code List (P0XXX Codes) 7-2
Gas Engine OBD II Trouble Code List (P1XXX Codes) 7-48
Gas Engine OBD II Trouble Code List (P2XXX Codes) 7-57
Gas Engine OBD II Trouble Code List (U0XXX Codes) 7-68
Gas Engine OBD II Trouble Code List (U1XXX Codes) 7-74
OBD II Vehicle Applications 7-1
Acura 7-1

DIAGNOSTIC TROUBLE CODES

OBD II VEHICLE APPLICATIONS

ACURA

MDX
2009–2010
- 3.7L V6 .... J37A1

RDX
2009–2010
- 2.3L I4 .... K23A1

RL
2009–2010
- 3.7L V6 .... J37A2

TL
2009–2010
- 3.5L V6 .... J35Z6
- 3.7L V6 .... J37A4

TSX
2009–2010
- 2.4L I4 .... K24A3
- 3.5L V6 .... J35Z6

ZDX
2010
- 3.7L V6 .... J37A5

Gas Engine OBD II Trouble Code List (P0XXX Codes)

DTC	Trouble Code Title, Conditions & Possible Causes
DTC: P0010 **1T PCM, MIL: Yes** **Year:** 2009, 2010 **Model:** CSX, RDX **Engine:** 2.0L L4, 2.3L L4 **Transmission:** All	**Variable Valve Timing Control (VTC) Oil Control Solenoid Valve Malfunction :** With the engine running one of the following conditions must be met: Condition 1: Output duty is 40% or more and the VTC current is 200 mA or less for at least 5 seconds. Condition 2: Output duty is 5% or less and the VTC current is 800 mA or more for at least 5 seconds. **NOTE: Before you troubleshoot, record all freeze data and any on-board snapshot, and review the general troubleshooting information.** **Possible causes:** • Poor connections or loose terminals at the VTC oil control solenoid valve and the ECM/PCM • "Open" circuit between the VTC oil control solenoid valve and ground • "Open" or "Short" circuit between the ECM/PCM and the VTC oil control solenoid valve • Faulty VTC oil control solenoid valve
DTC: P0011 **2T PCM, MIL: Yes** **Year:** 2009, 2010 **Model:** RDX, TSX **Engine:** 2.3L L4, 2.4L L4, 3.5L V6 **Transmission:** All	**Variable Valve Timing Control (VTC) System Malfunction :** With the engine running and at operating temperature. The difference between the timing control command and the actual timing of the camshaft is +/− 50 degrees or more for at least 15 seconds. **NOTE: Before you troubleshoot, record all freeze data and any on-board snapshot, and review the general troubleshooting information.** **Possible causes:** • Engine oil level low, If the level is OK, check the engine oil pressure • Poor connections or loose terminals at the VTC oil control solenoid valve and the ECM/PCM • Faulty VTC oil control solenoid valve or clogged VTC strainer • Faulty VTC actuator • Perform the ECM/PCM idle learn procedure and the CKP pattern clear/CKP pattern learn procedure
DTC: P0034 **1T PCM, MIL: Yes** **Year:** 2009, 2010 **Model:** RDX **Engine:** 2.3L L4 **Transmission:** All	**Turbocharger Bypass Control Solenoid Valve Circuit Low Voltage:** With the engine running, battery voltage a minimum of 10.5 V and DTC P0035 is not active. The return signal is OPEN and this condition continues at least 5 seconds, though the close signal is output to the turbocharger bypass control solenoid valve. **NOTE: Before you troubleshoot, record all freeze data and any on-board snapshot, and review the general troubleshooting information.** **Possible causes:** • Poor connections or loose terminals at the turbocharger bypass control solenoid valve and the PCM • Blown fuse • "Open" or "Short" circuit between the PCM and turbocharger bypass control solenoid valve • Faulty turbocharger bypass control solenoid valve • PCM may need to be updated with the latest software • Faulty PCM
DTC: P0035 **1T PCM, MIL: Yes** **Year:** 2009, 2010 **Model:** RDX **Engine:** 2.3L L4 **Transmission:** All	**Turbocharger Bypass Control Solenoid Valve Circuit High Voltage:** With the engine running, battery voltage a minimum of 10.5V and DTC P0034 is not active. The return signal is closed and this condition continues at least 5 seconds, though the open signal is output to the turbocharger bypass control solenoid valve. **NOTE: Before you troubleshoot, record all freeze data and any on-board snapshot, and review the general troubleshooting information.** **Possible causes:** • Poor connections or loose terminals at the turbocharger bypass control valve and the PCM • PCM may need to be updated with the latest software • Faulty PCM
DTC: P0045 **1T PCM, MIL: Yes** **Year:** 2009, 2010 **Model:** RDX **Engine:** 2.3L L4 **Transmission:** All	**Turbocharger Boost Control Solenoid Valve Circuit Malfunction:** Engine coolant temperature (ECT SENSOR 1) above 158 °F (70 °C), Transmission in D, Engine speed above 3,800 rpm, REL TP SENSOR above 59 deg, Drive at least 2 seconds. The return signal does not change as set by the output of the turbocharger boost control solenoid valve, and this condition continues for at least 2 seconds. When duty is between 98-2% and there is no return signal for 2 seconds or more continuously. **NOTE: Before you troubleshoot, record all freeze data and any on-board snapshot, and review the general troubleshooting information.** **Possible causes:** • Poor connections or loose terminals at the turbocharger boost control solenoid valve and the PCM • Blown fuse • "Open" or "Short" circuit between the PCM and turbocharger boost control solenoid valve • Faulty turbocharger boost control solenoid valve • PCM may need to be updated with the latest software • Faulty PCM

DTC	Trouble Code Title, Conditions & Possible Causes
DTC: P0096 **2T PCM, MIL: Yes** **Year:** 2009, 2010 **Model:** RDX **Engine:** 2.3L L4 **Transmission:** All	**Intake Air Temperature (IAT) Sensor 2 Circuit Range/Performance Problem:** With the engine OFF for a minimum of 6 hours, and DTCs P0112, P0113, P0116, P0117, P0118, P0125, P1116, P2183, P2184, P2185 and P2610 are not active. A malfunction is detected if these following 3 conditions are present after the engine and ignition switch have been turned OFF for at least 6 hours before restarting the engine. 1). When the temperature difference between the IAT2 and the ECT1 is 48 °F (27 °C) or more. 2). When the temperature difference between the IAT2 and the ECT2 is 52 °F (29 °C) or more. 1). When the temperature difference between the ECT2 and the ECT1 is 75 °F (42 °C) or more. **Possible causes:** • Poor connections or loose terminals at ECT sensors 1/2, IAT sensor 2 and the PCM • Faulty IAT sensor 2
DTC: P0097 **1T PCM, MIL: Yes** **Year:** 2009, 2010 **Model:** RDX **Engine:** 2.3L L4 **Transmission:** All	**Intake Air Temperature (IAT) Sensor 2 Circuit Low Voltage :** With the ignition ON the IAT sensor 2 voltage is 0.07 V or less for at least 5 seconds. **NOTE: Before you troubleshoot, record all freeze data and any on-board snapshot, and review the general troubleshooting information.** **Possible causes:** • Poor connections or loose terminals at IAT sensor 2 and the PCM • "Short" between IAT sensor 2 and the PCM • Faulty IAT sensor 2 • PCM needs to be updated with the latest software • Faulty PCM
DTC: P0098 **1T PCM, MIL: Yes** **Year:** 2009, 2010 **Model:** RDX **Engine:** 2.3L L4 **Transmission:** All	**Intake Air Temperature (IAT) Sensor 2 Circuit High Voltage:** With the ignition ON the IAT sensor output voltage is 4.93 V or more for at least 5 seconds. **NOTE: Before you troubleshoot, record all freeze data and any on-board snapshot, and review the general troubleshooting information.** **Possible causes:** • Poor connections or loose terminals at IAT sensor 2 and the PCM • "Open" circuit between the PCM and IAT sensor 2 • Faulty IAT sensor 2 • PCM needs to be updated with the latest software • Faulty PCM
DTC: P0101 **2T PCM, MIL: Yes** **Year:** 2009, 2010 **Model:** MDX, RDX, RL, TL, TSX, ZDX **Engine:** 2.3L L4, 2.4L L4, 3.5L V6, 3.7L V6 **Transmission:** All	**Mass Airflow (MAF) Sensor Circuit Range/Performance Problem:** Elapsed time after engine start 5 seconds, engine coolant temperature 156 °F (69 °C), engine speed 650-2,100 rpm. No active DTCs P0102, P0103, P0107, P0108, P0112, P0113, P0117, P0118, P0171, P0172, P0174, P0175, P0300, P0301, P0302, P0303, P0304, P0305, P0306, P0335, P0339, P0340, P0341, P0344, P0401, P0404, P0443, P0496, P0497, P0506, P0507, P1128, P1129, P145C, P2413, P2646, P2647, P2648, P2649. The difference between the amount of intake air measured by the MAF sensor and the amount of intake air calculated from the MAP sensor output is out of the normal area for at least 10 seconds. **Possible causes:** • Dirty air cleaner element • Faulty PCV valve or hose • Faulty EVAP canister purge valve • Vacuum leaks at the Throttle body, Intake manifold, Brake booster • Cracked or loose Air Intake Duct • Poor connections or loose terminals at the MAF sensor/IAT sensor and the PCM • Faulty MAF sensor/IAT sensor • Faulty PCM
DTC: P0102 **1T PCM, MIL: Yes** **Year:** 2009, 2010 **Model:** MDX, RDX, RL, TL, TSX, ZDX **Engine:** 2.3L L4, 2.4L L4, 3.5L V6, 3.7L V6 **Transmission:** All	**Mass Air Flow (MAF) Sensor Circuit Low Voltage:** The lower limit of the MAF sensor output is specified. If the output is below that limit, the PCM detects a malfunction and stores a DTC. Execution is continuous and the duration time is 2 seconds or more. The MAF sensor input voltage is 0.1 volt or less for at least 2 seconds. **NOTE: Before you troubleshoot, record all freeze data and any on-board snapshot.** **Possible causes:** • Poor connections or loose terminals at the MAF sensor/IAT sensor and the PCM • Blown fuse • "Open" or "Short" in the wire between the MAF sensor and the fuse • "Open" or "Short" in the wire between the PCM and the MAF sensor • Faulty MAF sensor/IAT sensor • PCM may need to be updated with the latest software • Faulty PCM

DTC	Trouble Code Title, Conditions & Possible Causes
DTC: P0103 **1T PCM, MIL: Yes** **Year:** 2009, 2010 **Model:** MDX, RDX, RL, TL, TSX, ZDX **Engine:** 2.3L L4, 2.4L L4, 3.5L V6, 3.7L V6 **Transmission:** All	**Mass Airflow (MAF) Sensor Circuit High Voltage:** The upper limit of the MAF sensor output is specified. If the output is above that limit, the PCM detects a malfunction and stores a DTC. Execution is continuous and the duration time is 2 seconds or more. The MAF sensor input voltage is 4.89 V or more for at least 2 seconds. P0102 is not active. **NOTE: Before you troubleshoot, record all freeze data and any on-board snapshot.** **Possible causes:** • Poor connections or loose terminals at the MAF sensor/IAT sensor and the PCM • "Short" in the wire between the PCM and the MAF sensor • Faulty MAF sensor/IAT sensor • Faulty PCM
DTC: P0107 **1T PCM, MIL: Yes** **Year:** 2009, 2010 **Model:** MDX, RL, ZDX **Engine:** 3.7L V6 **Transmission:** All	**Manifold Absolute Pressure (MAP) Sensor Circuit Low Voltage (PGM-FI System):** The MAP sensor outputs low signal voltage at high vacuum (throttle valve closed) and high signal voltage at low vacuum (throttle valve wide open). If a signal voltage from the MAP sensor is a set value or less, the Powertrain Control Module (PCM) detects a malfunction and a DTC is stored. The execution time is continuous and the duration time is 2 seconds or more. The MAP sensor output voltage is 0.23 V or less for at least 2 seconds. DTC P0108 is not active. **NOTE: Before you troubleshoot, record all freeze data and any on-board snapshot.** **Possible causes:** • Poor connections or loose terminals at the MAP sensor and the PCM • "Short" in the wire between the PCM and the MAP sensor • "Open" in the wire between the PCM and the MAP sensor • Faulty MAP sensor • Faulty PCM
DTC: P0107 **1T PCM, MIL: Yes** **Year:** 2009, 2010 **Model:** RDX, TSX **Engine:** 2.3L L4, 2.4L L4, 3.5L V6 **Transmission:** All	**Manifold Absolute Pressure (MAP) Sensor Circuit Low Voltage:** With the engine running the Map sensor output voltage is 0.24 volts or less for at least 2 seconds. **Possible causes:** • Poor connections or loose wires at the MAP sensor and at the PCM • "Open" or "Short circuit between the Map sensor and PCM • Faulty MAP sensor • PCM needs to be updated with the latest software • Faulty PCM
DTC: P0108 **1T PCM, MIL: Yes** **Year:** 2009, 2010 **Model:** RDX, TSX **Engine:** 2.3L L4, 2.4L L4, 3.5L V6 **Transmission:** All	**Manifold Absolute Pressure (MAP) Sensor Circuit High Voltage:** With the engine running the Map sensor output voltage value is 4.49 V for at least 2 seconds. **NOTE: Before you troubleshoot, record all freeze data and any on-board snapshot, and review the general troubleshooting information.** **Possible causes:** • Poor connections or loose wires at the MAP sensor and at the PCM • "Open" circuit between the MAP sensor and the PCM • Faulty MAP sensor • PCM needs to be updated with the latest software • Faulty PCM
DTC: P0108 **1T PCM, MIL: Yes** **Year:** 2009, 2010 **Model:** MDX, RL, TL, ZDX **Engine:** 3.5L V6, 3.7L V6 **Transmission:** All	**MAP Sensor Circuit High Voltage (A/T/System) (With Navigation):** The MAP sensor outputs low signal voltage at high vacuum (throttle valve closed) and high signal voltage at low vacuum (throttle valve wide open). If a signal voltage from the MAP sensor is a set value or more, the Powertrain Control Module (PCM) detects a malfunction and a DTC is stored. The execution time is continuous and the duration time is 2 seconds or more. DTC P0107 is not active. The MAP sensor output voltage is 4.49 V or more for at least 2 seconds. **NOTE: Before you troubleshoot, record all freeze data and any on-board snapshot.** **Possible causes:** • Poor connections or loose terminals at the MAP sensor and the PCM • "Open" in the wire between the PCM and the MAP sensor • "Open" in the wire between the PCM and the MAP sensor • PCM may need to be updated with the latest software • Faulty PCM

DTC	Trouble Code Title, Conditions & Possible Causes
DTC: P0111 **2T PCM, MIL: Yes** **Year:** 2009, 2010 **Model:** MDX, RL, TL, TSX, ZDX **Engine:** 2.4L L4, 3.5L V6, 3.7L V6 **Transmission:** All	**Intake Air Temperature (IAT) Sensor Circuit Range/Performance:** The execution is once per driving cycle and the duration time is 10 seconds or more. Engine OFF time is 6 hours. DTCs P0112, P0113, P0116, P0117, P0118, P0125, P1116, P2183, P2184, P2185, P2610 are not active. A malfunction is detected if the following three conditions are present after the engine has stopped and the ignition switch has been turned to LOCK (0) for at least 6 hours before restarting the engine: * When the temperature difference between the IAT and ECT1 is 57 °F (32 °C) or more. * When the temperature difference between the IAT and ECT2 is 30 °F (17 °C) or more. * When the temperature difference between the ECT2 and ECT1 is 73 °F (41 °C) or more. **NOTE: Before you troubleshoot, record all freeze data and any on-board snapshot.** **Possible causes:** • Poor connections or loose terminals at ECT sensor 1 and 2 and the MAF sensor/IAT sensor • Poor connections or loose terminals at the IAT sensor and the PCM • Faulty MAF sensor/IAT sensor • Faulty PCM
DTC: P0112 **1T PCM, MIL: Yes** **Year:** 2009, 2010 **Model:** RDX **Engine:** 2.3L L4 **Transmission:** All	**Intake Air Temperature (IAT) Sensor 1 Circuit Low Voltage:** With the ignition ON the IAT sensor 1 output voltage is 0.07 V or less for at least 5 seconds. **NOTE: Before you troubleshoot, record all freeze data and any on-board snapshot, and review the general troubleshooting information.** **Possible causes:** • Poor connections or loose wires at the IAT sensor and at the PCM • "Short" circuit between the IAT sensor and the PCM • Faulty IAT sensor • PCM needs to be updated with the latest software • Faulty PCM
DTC: P0112 **1T PCM, MIL: Yes** **Year:** 2009, 2010 **Model:** MDX, RL, TL, ZDX **Engine:** 3.5L V6, 3.7L V6 **Transmission:** All	**Intake Air Temperature (IAT) Sensor Circuit Low Voltage:** If the IAT sensor output voltage is excessively low, the Powertrain Control Module (PCM) detects a malfunction and a DTC is stored. The execution time is continuous and the duration time is 2 seconds or more. The IAT sensor output voltage is 0.08 V or less for at least 2 seconds. **NOTE: Before you troubleshoot, record all freeze data and any on-board snapshot.** **Possible causes:** • Poor connections or loose terminals at the MAF sensor/IAT sensor and the PCM • "Short" in the wire between the MAF sensor/IAT sensor and the PCM • Faulty MAF sensor/IAT sensor • PCM may need to be updated with the latest software • Faulty PCM
DTC: P0112 **1T PCM, MIL: Yes** **Year:** 2009, 2010 **Model:** TSX **Engine:** 2.4L L4, 3.5L V6 **Transmission:** All	**Intake Air Temperature (IAT) Sensor Circuit Low Voltage:** With the ignition ON the IAT sensor output voltage is 0.08 V or less for at least 2 seconds. **Possible causes:** • Poor connections or loose wires at the IAT sensor and at the PCM • "Short" circuit between the IAT sensor and the PCM • Faulty IAT sensor • Faulty PCM
DTC: P0113 **1T PCM, MIL: Yes** **Year:** 2009, 2010 **Model:** TSX **Engine:** 2.4L L4, 3.5L V6 **Transmission:** All	**Intake Air Temperature (IAT) Sensor Circuit High Voltage:** With ignition ON, the IAT sensor output voltage is 4.92 V or more for at least 2 seconds. **Possible causes:** • Poor connections or loose wires at the IAT sensor and at the PCM • "Open" circuit between the PCM and the IAT sensor • Faulty IAT sensor • Faulty PCM
DTC: P0113 **1T PCM, MIL: Yes** **Year:** 2009, 2010 **Model:** RDX **Engine:** 2.3L L4 **Transmission:** All	**Intake Air Temperature (IAT) Sensor 1 Circuit High Voltage:** With the ignition ON the IAT sensor 1 output voltage is 4.93 V or less for at least 5 seconds. **NOTE: Before you troubleshoot, record all freeze data and any on-board snapshot, and review the general troubleshooting information.** **Possible causes:** • Poor connections or loose wires at the IAT sensor and at the PCM • "Open" circuit between the IAT sensor 1 and the PCM • Faulty IAT sensor 1 • PCM needs to be updated with the latest software • Faulty PCM

DTC	Trouble Code Title, Conditions & Possible Causes
DTC: P0113 **1T PCM, MIL: Yes** **Year:** 2009, 2010 **Model:** MDX, TL, ZDX **Engine:** 3.5L V6, 3.7L V6 **Transmission:** All	**Intake Air Temperature (IAT) Sensor Circuit High Voltage:** The IAT sensor resistance varies depending on temperature. The output voltage and the sensor resistance increase as the intake air temperature decreases. Conversely, the output voltage and the sensor resistance decrease as the intake air temperature increases. If the IAT sensor output voltage is excessively high, the Powertrain Control Module (PCM) detects a malfunction and a DTC is stored. The execution time is continuous and the duration time is 2 seconds or more. P0112 is not active. The IAT sensor output voltage is 4.92 V or more for at least 2 seconds. **NOTE: Before you troubleshoot, record all freeze data and any on-board snapshot.** **Possible causes:** • Poor connections or loose terminals at the MAF sensor/IAT sensor and the PCM • "Open" in the wire between the PCM and the MAF sensor/IAT sensor • "Open" in the wire between the PCM and the MAF sensor/IAT sensor • Faulty MAF sensor/IAT sensor • PCM may need to be updated with the latest software • Faulty PCM
DTC: P0116 **2T PCM, MIL: Yes** **Year:** 2009, 2010 **Model:** RDX **Engine:** 2.3L L4 **Transmission:** All	**Engine Coolant Temperature (ECT) Sensor 1 Circuit Range/Performance Problem:** A malfunction is detected if the following three conditions are present after the engine has stopped and the ignition switch has been turned to LOCK (0) for at least 6 hours before restarting the engine: (1) When the temperature difference between the IAT and ECT1 is 57 °F (32 °C) or more. (2) When the temperature difference between the IAT and ECT2 is 30 °F (17 °C) or more. (3) When the temperature difference between the ECT2 and ECT1 is 73 °F (41 °C) or more. **NOTE: If DTC P0111 is stored at the same time as DTC P1116, troubleshoot DTC P0111 first, then recheck for DTC P1116.** **Possible causes:** • Poor connections or loose terminals at ECT sensor 1 and ECT sensor 2 and the PCM • Faulty ECT sensor 1 • Faulty ECT sensor 2
DTC: P0116 **2T PCM, MIL: Yes** **Year:** 2009, 2010 **Model:** TL **Engine:** 3.5L V6, 3.7L V6 **Transmission:** All	**Engine Coolant Temperature (ECT) Sensor Range/Performance Problem :** Malfunction 1 (Slow Response): The engine coolant temperature does not reach (15°C) 86°F within 1,200 seconds. Malfunction 2 (Stuck): The ECT sensor output voltage does not vary by 60 mV or more within 1,200 seconds. **NOTE: If DTC P0117 and/or P0118 are stored at the same time as DTC P0116, troubleshoot those DTCs first, then recheck for DTC P0116.** **Possible causes:** • Faulty thermostat (Stuck Open) • Faulty ECT sensor
DTC: P0116 **2T PCM, MIL: Yes** **Year:** 2009, 2010 **Model:** MDX, RL, TSX, ZDX **Engine:** 2.4L L4, 3.5L V6, 3.7L V6 **Transmission:** All	**Engine Coolant Temperature (ECT) Sensor 1 Circuit Range/Performance Problem :** The execution time is once per driving cycle and the duration time is 10 seconds or more. The following DTCs are not active P0101, P0102, P0103, P0107, P0108, P0117, P0118, P0134, P0135, P0154, P0155, P0171, P0172, P0174, P0175, P0300, P0301, P0302, P0303, P0304, P0305, P0306, P0335, P0339, P0340, P0344, P0401, P0404, P0443, P0496, P0497, P0506, P0507, P0627, P1077, P1078, P1128, P1129, P1172, P1174, P145C, P2195, P2197, P2227, P2228, P2229, P2237, P2238, P2240, P2241, P2243, P2245, P2247, P2249, P2251, P2252, P2254, P2255, P2413, P2610, P2646, P2647, P2648, P2649. Malfunction determination 1: With a completely cooled engine (one that has been off for at least 6 hours): When the change in coolant temperature after 42 minutes or more of running time or drive 5 miles (7 km) or more of driving time is 50 °F (10 °C) or less, a malfunction is detected. Malfunction determination 2: With a partially cooled engine (one that has been off for less than 6 hours): When the difference between the coolant temperature after 42 minutes or more of running time or drive 5 miles (7 km) or more of driving time the coolant temperature after the engine has been off for 150 minutes and then run for 10 seconds is 50°F (10 °C) or less, a malfunction is detected. **NOTE: Before you troubleshoot, record all freeze data and any on-board snapshot.** **Possible causes:** • Poor connections or loose terminals at ECT sensor 1 and the PCM • Faulty Replace ECT sensor (1) • Faulty PCM

DTC	Trouble Code Title, Conditions & Possible Causes
DTC: P0117 **1T PCM, MIL: Yes** **Year:** 2009, 2010 **Model:** MDX, RDX, RL, TL, TSX, ZDX **Engine:** 2.3L L4, 2.4L L4, 3.5L V6, 3.7L V6 **Transmission:** All	**Engine Coolant Temperature (ECT) Sensor 1 Circuit Low Voltage:** If the ECT sensor 1 output voltage is less than a set value when the engine coolant temperature is high, the PCM detects a malfunction and a DTC is stored. The execution time is continuous and the duration time is 2 seconds or more. P0118 is not active. The ECT sensor 1 output voltage is 0.08 V or less for at least 2 seconds. **NOTE: Before you troubleshoot, record all freeze data and any on-board snapshot.** **Possible causes:** • Poor connections or loose terminals at ECT sensor 1 and the PCM • "Short" in the wire between ECT sensor 1 and the PCM • Faulty ECT sensor (1) • PCM may need to be updated with the latest software • Faulty PCM
DTC: P0118 **1T PCM, MIL: Yes** **Year:** 2009, 2010 **Model:** MDX, RDX, RL, TL, TSX, ZDX **Engine:** 2.3L L4, 2.4L L4, 3.5L V6, 3.7L V6 **Transmission:** All	**Engine Coolant Temperature (ECT) Sensor 1 Circuit High Voltage:** If the ECT sensor 1 output voltage is less than the set value when the engine coolant temperature is low, the PCM detects a malfunction and a DTC is stored. The execution time is continuous and the duration time is 2 seconds or more. DTC P0117 is not active. The ECT sensor 1 output voltage is 4.92 V or more for at least 2 seconds. **NOTE: Before you troubleshoot, record all freeze data and any on-board snapshot.** **Possible causes:** • Poor connections or loose terminals at ECT sensor 1 and the PCM • "Open" in the wire between the PCM and ECT sensor 1 • "Open" in the wire between the PCM and ECT sensor 1 • Update the PCM if it does not have the latest software, • Faulty ECT sensor (1) • Faulty PCM
DTC: P0122 **1T PCM, MIL: Yes** **Year:** 2009, 2010 **Model:** MDX, ZDX **Engine:** 3.7L V6 **Transmission:** All	**Throttle Position (TP) Sensor A Circuit Low Voltage:** If the signal from TP sensor A is less than a fixed value for a set time, the PCM detects a TP sensor A malfunction and stores a DTC. The execution time is continuous and the duration time is 200 milliseconds or more. With the engine running and DTCs P0123, P2101, P2118, P2135, P2176 are not active. **NOTE: Before you troubleshoot, record all freeze data and any on-board snapshot.** **Possible causes:** • Poor connections or loose terminals at the throttle body and the PCM • "Short" in the wire between the throttle body and the PCM • "Open" in the wire between the throttle body and the PCM • Faulty throttle body • PCM may need to be updated with the latest software • Faulty PCM
DTC: P0122 **1T PCM, MIL: Yes** **Year:** 2009, 2010 **Model:** RDX, TL, TSX **Engine:** 2.3L L4, 2.4L L4, 3.5L V6, 3.7L V6 **Transmission:** All	**Throttle Position (TP) Sensor A Circuit Low Voltage :** With the ignition ON the TP sensor 1 output voltage value is 0.20 V or less for at least 0.1 second. **NOTE: Before you troubleshoot, record all freeze data and any on-board snapshot, and review the general troubleshooting information.** **Possible causes:** • Poor connections or loose terminals at the throttle body and the PCM • "Short" in the wire between the throttle body and the PCM • "Open" in the wire between the throttle body and the PCM • Faulty throttle body • PCM may need to be updated with the latest software • Faulty PCM
DTC: P0123 **1T PCM, MIL: Yes** **Year:** 2009, 2010 **Model:** MDX, RDX, TL, TSX, ZDX **Engine:** 2.3L L4, 2.4L L4, 3.5L V6, 3.7L V6 **Transmission:** All	**Throttle Position (TP) Sensor A Circuit High Voltage:** If the signal from TP sensor A is more than a fixed value for a set time, the PCM detects a TP sensor A malfunction and stores a DTC. The execution time is continuous and the duration time is 200 milliseconds or more. Ignition is running and DTCs P0122, P2101, P2118, P2135, P2176 are not active. The TP sensor A output voltage is 4.8 V or more for at least 200 milliseconds. **NOTE: Before you troubleshoot, record all freeze data and any on-board snapshot.** **Possible causes:** • Poor connections or loose terminals at the throttle body and the PCM • "Open" in the wire between the throttle body and the PCM • "Open" in the wire between the throttle body and the PCM • Faulty throttle body • PCM may need to be updated with the latest software • Faulty PCM

DTC	Trouble Code Title, Conditions & Possible Causes
DTC: P0125 **2T PCM, MIL: Yes** **Year:** 2009, 2010 **Model:** MDX, RDX, RL, TL, TSX, ZDX **Engine:** 2.3L L4, 2.4L L4, 3.5L V6, 3.7L V6 **Transmission:** All	**Engine Coolant Temperature (ECT) Sensor 1 Malfunction/Slow Response:** The execution time is once per driving cycle and the duration time is 20 minutes or less. The following DTCs are not active P0101, P0102, P0103, P0107, P0108, P0111, P0112, P0113, P0117, P0118, P0134, P0135, P0154, P0155, P0171, P0172, P0174, P0175, P0300, P0301, P0302, P0303, P0304, P0305, P0306, P0335, P0339, P0340, P0344, P0401, P0404, P0443, P0496, P0497, P0506, P0507, P0627, P1077, P1078, P1128, P1129, P1172, P1174, P145C, P2195, P2197, P2227, P2228, P2229, P2237, P2238, P2240, P2241, P2243, P2245, P2247, P2249, P2251, P2252, P2254, P2255, P2413, P2646, P2647, P2648, P2649. As the engine coolant warms, the ECT sensor 1 resistance decreases, and the PCM detects a low signal voltage. If the ECT sensor 1 output voltage does not reach a specified temperature at which closed-loop control for stoichiometric air/fuel ratio starts within a set time, depending on the initial coolant temperature after starting the engine, the PCM detects a malfunction. **NOTE: Before you troubleshoot, record all freeze data and any on-board snapshot.** **Possible causes:** • Poor connections or loose terminals at ECT sensor 1, ECT sensor 2, and the PCM • Low coolant level • Faulty thermostat • Faulty ECT sensor (1) • Faulty PCM
DTC: P0128 **2T PCM, MIL: Yes** **Year:** 2009, 2010 **Model:** MDX, RL, ZDX **Engine:** 3.7L V6 **Transmission:** All	**Cooling System Malfunction:** The execution time is once per drive cycle and the duration time is dependent on driving conditions. Malfunction determination 1: If the difference between the current measured coolant temperature at the radiator (ECT 2) and the initial coolant temperature at the radiator (ECT 2) is at least 21 °F (12 °C) when the calculated coolant temperature at the engine (ECT 1) reaches 159 °F (71 °C), a malfunction is detected (thermostat stuck open); or if the coolant temperature at the radiator (ECT 2) only reaches 68 °F (20 °C), a malfunction is detected (thermostat malfunction). Malfunction determination 2: When the calculated engine coolant temperature (ECT 1) reaches 158 °F (70 °C) before the measured engine coolant temperature (ECT 1) reaches 158 °F (70 °C), a malfunction is detected. **NOTE: Before you troubleshoot, record all freeze data and any on-board snapshot.** **Possible causes:** • Faulty thermostat • Faulty fan relay or circuit • Poor connections or loose terminals at ECT sensor 1, ECT sensor 2, and the PCM • PCM may need to be updated with the latest software • Faulty PCM
DTC: P0128 **2T PCM, MIL: Yes** **Year:** 2009, 2010 **Model:** RDX, TL, TSX **Engine:** 2.3L L4, 2.4L L4, 3.5L V6, 3.7L V6 **Transmission:** All	**Cooling System Malfunction:** With engine running the ECT sensor output (70C) 158F or less, an estimated engine coolant temperature is (75C) 168F or more. The difference between the estimated engine coolant temperature and the ECT sensor output is (15C) 27F or more. **NOTE: If the DTCs listed below are stored at the same time as DTC P0128, troubleshoot those DTCs first:** P0107, P0108, P1128, P1129, P1106, P1107, P1108, P1259, P0401, P0116, P0117, P0118, P0112, P0113, P0335, P0336, P0300-P0306, P0505, P1519, then recheck for P0128. **Possible causes:** • Low coolant level • Faulty thermostat (Stuck Open) • Radiator fan runs constantly • PCM may need to be updated with the latest software • Faulty PCM
DTC: P0133 **2T O2S1, MIL: Yes** **Year:** 2009, 2010 **Model:** MDX, RL, TL, TSX, ZDX **Engine:** 3.5L V6, 3.7L V6 **Transmission:** All	**Rear Air/Fuel Ratio (A/F) Sensor (Bank 1, Sensor 1) Malfunction/Slow Response:** The execution time is once per drive cycle and the duration time is 6.8 seconds or more. Engine coolant temperature 156 °F (69 °C), Intake air temperature, Fuel trim, 0.73-1.47, Engine speed 1,250-2,200 rpm, Vehicle speed 33 mph or more. The total rear A/F sensor output value is 11 or less in 6.5 seconds. **NOTE: Before you troubleshoot, record all freeze data and any on-board snapshot.** Conditions for illuminating the MIL When a malfunction is detected during the first drive cycle, a Temporary DTC is stored in the PCM memory. If the malfunction recurs during the next (second) drive cycle, the MIL comes on and the DTC and the freeze frame data are stored. **Possible causes:** • Poor connections or loose terminals at the A/F sensor (Sensor 1) and the PCM • Faulty A/F sensor (Sensor 1)

DTC	Trouble Code Title, Conditions & Possible Causes
DTC: P0133 **2T PCM, MIL: Yes** **Year:** 2009, 2010 **Model:** RDX, TSX **Engine:** 2.3L L4, 2.4L L4 **Transmission:** All	**Air/Fuel Ratio (A/F) Sensor (Sensor 1) Malfunction/Slow Response:** With the engine running and in closed loop, and the vehicle speed at least 33 mph. The total A/F sensor (SENSOR 1) output value for A/T models is 20 and for M/T models 38 or less in 0.8 seconds. **NOTE: Before you troubleshoot, record all freeze data and any on-board snapshot, and review the general troubleshooting information. If DTC P0139 is stored at the same time as DTC P0133, troubleshoot DTC P0139 first, then recheck for DTC P0133.** **Possible causes:** • Poor connections or loose terminals at the A/F sensor (Sensor 1) and the ECM/PCM • Faulty A/F sensor (Sensor 1)
DTC: P0134 **2T PCM, MIL: Yes** **Year:** 2009, 2010 **Model:** RDX, TSX **Engine:** 2.3L L4, 2.4L L4 **Transmission:** All	**Air/Fuel Ratio (A/F) Sensor (Sensor 1) Heater System Malfunction :** With the engine running for 40 seconds or more and battery voltage above 10.5 V. Malfunction 1: From start up the A/F sensor (SENSOR 1) internal resistance value is 90 ohms or more for at least 40 seconds right after the engine starts. Malfunction 2: With the engine hot and at operating temperature the A/F sensor (SENSOR 1) internal resistance value is 250 ohms or more for at least 1.0 second. **NOTE: Before you troubleshoot, record all freeze data and any on-board snapshot, and review the general troubleshooting information. If DTC P0135 is stored at the same time as DTC P0134, troubleshoot DTC P0135 first, then recheck for DTC P0134.** **Possible causes:** • Poor connections or loose terminals at the A/F sensor (Sensor 1), the A/F sensor relay, and the ECM/PCM • "Open" or "Short" between the A/F sensor (Sensor 1), the A/F sensor relay, and/or the ECM/PCM • Faulty A/F sensor relay • Faulty A/F sensor (Sensor 1)
DTC: P0134 **2T PCM, MIL: Yes** **Year:** 2009, 2010 **Model:** MDX, RL, TL, TSX **Engine:** 3.5L V6, 3.7L V6 **Transmission:** All	**Rear Air/Fuel Ratio (A/F) Sensor (Bank 1, Sensor 1) Heater System Malfunction:** The execution time is continuous and the duration time is 40 seconds or more. Battery voltage minimum 10.5 volts. Malfunction determination 1: The rear A/F sensor (bank 1, sensor 1) internal resistance value is 110 ohms or more for at least 40 seconds right after the engine starts. Malfunction determination 2: The rear A/F sensor (bank 1, sensor 1) internal resistance value is 110 ohms or more for at least 15 seconds. **NOTE: Before you troubleshoot, record all freeze data and any on-board snapshot.** The rear A/F sensor (bank 1, sensor 1) internal resistance value is 200 ohms or more for at least 1 second. **Possible causes:** • Poor connections or loose terminals at the A/F sensor (Sensor 1), the relay and the PCM • "Open" circuit between the PCM and the A/F sensor • Faulty A/F Sensor (Bank 1, Sensor 1) • PCM needs to be updated with the latest software • Faulty PCM
DTC: P0135 **1T PCM, MIL: Yes** **Year:** 2009, 2010 **Model:** MDX, RL, TL, TSX, ZDX **Engine:** 3.5L V6, 3.7L V6 **Transmission:** All	**Rear Air/Fuel Ratio (A/F) Sensor (Bank 1, Sensor 1) Heater Circuit Malfunction:** The execution time is continuous and the duration time is 2 seconds or more. Engine running and battery at 10 volts or more. No return signal HIGH is detected when the PCM output duty is less than 2%. Return signal does not change when the PCM output duty is more than 20% and less than 80%. No return signal LOW is detected when the PCM output duty is more than 8%. **NOTE: Before you troubleshoot, record all freeze data and any on-board snapshot.** **Possible causes:** • Poor connections or loose terminals at the A/F sensor (Sensor 1), the relay and the PCM • "Open" or Short" circuit between the A/F sensors, the relay box or the PCM • Faulty A/F sensor (Sensor 1) • Faulty PCM
DTC: P0135 **2T PCM, MIL: Yes** **Year:** 2009, 2010 **Model:** RDX, TSX **Engine:** 2.3L L4, 2.4L L4 **Transmission:** All	**Air/Fuel Ratio (A/F) Sensor (Sensor 1) Heater Circuit Malfunction :** With the engine running and the engine temperature at least 69 degrees the following malfunctions are detected. Malfunction 1: The heater current is 0.8 A or less for at least 4 seconds while the heater is active, and the heater current is 0.8 A or more for at least 4 seconds while the heater is not active. Malfunction 2: The heater current is 15.2 A or more for at least 0.6 second. **NOTE: Before you troubleshoot, record all freeze data and any on-board snapshot, and review the general troubleshooting information.** **Possible causes:** • Blown fuse • Poor connections or loose wires at the A/F sensor (Sensor 1), the relay, and the ECM/PCM • "Open" or "Short" circuit between the A/F sensor (Sensor 1), A/F sensor relay and/or the ECM/PCM • Faulty A/F sensor relay • Faulty A/F sensor (Sensor 1) • Faulty ECM/PCM

DTC	Trouble Code Title, Conditions & Possible Causes
DTC: P0137 **1T PCM, MIL: Yes** **Year:** 2009, 2010 **Model:** RDX, TSX **Engine:** 2.3L L4, 2.4L L4 **Transmission:** All	**Secondary HO2S (Sensor 2) Circuit Low Voltage :** With engine running the secondary HO2S (Sensor 2) output voltage remains in Zone 1 (0.29V or below) for at least 72 seconds. **Possible causes:** • Poor connections or loose wires at the secondary HO2S (Sensor 2) and the PCM • "Short" circuit between the secondary HO2S (Sensor 2) and the PCM • Faulty secondary HO2S (Sensor 2) • Faulty PCM
DTC: P0137 **1T PCM, MIL: Yes** **Year:** 2009, 2010 **Model:** MDX, RL, TL, TSX, ZDX **Engine:** 3.5L V6, 3.7L V6 **Transmission:** All	**Rear Secondary Heated Oxygen Sensor (Secondary HO2S (Bank 1, Sensor 2) Circuit Low Voltage:** The execution time is continuous and the duration time is 40 seconds or more. The system is in closed loop with the engine coolant temperature at 156 °F (69 °C) and the intake air temperature at -13 °F (-25 °C). The rear secondary HO2S output voltage is 0.05 V or less for at least 40 seconds. After current is applied to the rear secondary HO2S heater, if the rear secondary HO2S output continues low (lean) during feedback control, a malfunction is detected and a DTC is stored. **NOTE: Before you troubleshoot, record all freeze data and any on-board snapshot.** **Possible causes:** • Poor connections or loose terminals at the secondary HO2S (Sensor 2) and the PCM • "Open" or "Short circuit between the PCM and the secondary HO2S (Sensor 2) • Faulty secondary HO2S (Sensor 2) • Faulty PCM
DTC: P0138 **2T PCM, MIL: Yes** **Year:** 2009, 2010 **Model:** TSX **Engine:** 2.4L L4 **Transmission:** All	**Secondary HO2S (Sensor 2) Circuit High Voltage :** With engine running the Secondary HO2S (Sensor 2) remains in Zone 4 (0.80V or above) for at least 72 seconds. **Possible causes:** • Poor connections or loose wires at the secondary HO2S (Sensor 2) and at the PCM • "Open" circuit between the PCM and the secondary HO2S (Sensor 2) • Faulty secondary HO2S (Sensor 2) • Faulty PCM
DTC: P0138 **1T PCM, MIL: Yes** **Year:** 2009, 2010 **Model:** MDX, RL, TL, TSX, ZDX **Engine:** 3.5L V6, 3.7L V6 **Transmission:** All	**Rear Secondary Heated Oxygen Sensor (Secondary HO2S (Bank 1, Sensor 2) Circuit High Voltage:** The execution time is continuous and the duration time is 2 seconds or more. The rear secondary HO2S output voltage is 1.270V or more for at least 5 seconds. After current is applied to the rear secondary HO2S heater, if the rear secondary HO2S output continues to exceed the upper limit used during feedback control, a malfunction is detected and a DTC is stored. **NOTE: Before you troubleshoot, record all freeze data and any on-board snapshot.** **Possible causes:** • Poor connections or loose terminals at the secondary HO2S (Sensor 2) and the PCM • "Open" circuit between the PCM and the secondary HO2S (Sensor 2) • Faulty (Secondary HO2S (Bank 1, Sensor 2) • Faulty PCM
DTC: P0139 **2T PCM, MIL: Yes** **Year:** 2009, 2010 **Model:** MDX, RL, TSX, ZDX **Engine:** 3.5L V6, 3.7L V6 **Transmission:** All	**Rear Secondary Heated Oxygen Sensor (Secondary HO2S (Bank 1, Sensor 2) Circuit Slow Response:** If the response time of the rear secondary HO2S becomes longer than the specified time after current to the secondary HO2S heater is applied, a malfunction is detected and a DTC is stored. The execution time is once per driving cycle and the duration time is 20.7 seconds or less. When a malfunction is detected during the first drive cycle, a Temporary DTC is stored in the PCM memory. If the malfunction recurs during the next (second) drive cycle, the MIL comes on and the DTC and the freeze frame data are stored. **NOTE: Before you troubleshoot, record all freeze data and any on-board snapshot.** **Possible causes:** • Poor connections or loose terminals at the secondary HO2S (Sensor 2) and the PCM • Faulty (Secondary HO2S (Bank 1, Sensor 2) • Faulty PCM
DTC: P0139 **2T PCM, MIL: Yes** **Year:** 2009, 2010 **Model:** RDX, TSX **Engine:** 2.3L L4, 2.4L L4 **Transmission:** All	**Secondary HO2S (Sensor 2) Slow Response :** Engine started, vehicle driven in closed loop in 4th or 6th gear at over 55 mph at steady speed, and the PCM detected the HO2S response time to switch between 300-600 mv was too slow, or that the rich to lean or lean to rich switch time was too slow. **Possible causes:** • Poor connections or loose wires at the secondary HO2S (Sensor 2) and at the PCM • Faulty secondary HO2S (Sensor 2)

DTC	Trouble Code Title, Conditions & Possible Causes
DTC: P0141 **1T PCM, MIL: Yes** **Year:** 2009, 2010 **Model:** MDX, RL, TL, TSX, ZDX **Engine:** 3.5L V6, 3.7L V6 **Transmission:** All	**Rear Secondary Heated Oxygen Sensor (Bank 1 Sensor 2) Heater Circuit Malfunction:** The rear secondary HO2S heater output is 0.38 A or less, or 3.33 A or more, for at least 5 seconds when the heater is on. Engine running the battery voltage (IGP terminal of PCM) is 10.5-16 V, DTCs P0117, P0118 are not active. If the rear secondary HO2S heater draws more or less than a specified amperage, the PCM detects a malfunction and a DTC is stored. **NOTE: Before you troubleshoot, record all freeze data and any on-board snapshot.** **Possible causes:** • Poor connections or loose terminals at the secondary HO2S (Sensor 2), the relay and the PCM • "Open" or "Short circuit between the PCM and the secondary HO2S (Sensor 2) • "Open or "Short circuit between the A/F sensors, the relay • Faulty (Secondary HO2S (Bank Sensor 2)
DTC: P0141 **1T PCM, MIL: Yes** **Year:** 2009, 2010 **Model:** RDX, TSX **Engine:** 2.3L L4, 2.4L L4 **Transmission:** All	**Secondary HO2S (Sensor 2) Heater Circuit Malfunction :** With engine running the current is 0.5 A or less, or 3.6 A or more for at least 5 seconds when the heater is on. **Possible causes:** • Poor connections or loose wires at the primary HO2S (Sensor 1) and the PCM • "Open" or "Short" circuit between the primary HO2S (Sensor 1) and the PCM • Faulty primary HO2S (Sensor 1) • Faulty PCM
DTC: P0153 **2T PCM, MIL: Yes** **Year:** 2009, 2010 **Model:** MDX, RL, TL, TSX, ZDX **Engine:** 3.5L V6, 3.7L V6 **Transmission:** All	**Front Air/Fuel Ratio (A/F) Sensor (Bank 2, Sensor 1) Malfunction/Slow Response:** The execution time is once per drive cycle and the duration time is 6.8 seconds or more. Engine coolant temperature 156 °F (69 °C), Intake air temperature, Fuel trim, 0.73-1.47, Engine speed 1,250-2,200 rpm, Vehicle speed 33 mph or more. The total front A/F sensor (bank 2, sensor 1) output value is 34 or less in 6.8 seconds. **Possible causes:** • Poor connections or loose terminals at the A/F sensor (Sensor 1) and the PCM • Faulty A/F sensor (Sensor 1)
DTC: P0154 **2T PCM, MIL: Yes** **Year:** 2009, 2010 **Model:** MDX, TL, TSX, ZDX **Engine:** 3.5L V6, 3.7L V6 **Transmission:** All	**Front Air Fuel Ratio (A/F) Sensor (Bank 2, Sensor 1) Heater System Malfunction :** The execution time is continuous and the duration time is 40 seconds or more. Battery voltage minimum 10.5 volts. Malfunction determination 1: The front A/F sensor (bank 2, sensor 1) internal resistance value is 200 or more for at least 40 seconds right after the engine starts. Malfunction determination 2: The front A/F sensor (bank 2, sensor 1) internal resistance value is 270 or more for at least 1.0 second. **NOTE: Before you troubleshoot, record all freeze data and any on-board snapshot.** **Possible causes:** • Poor connections or loose terminals at the A/F sensor (bank 2, Sensor 1), the under-hood fuse/relay box (PGM-FI sub relay) and the PCM • "Open" circuit between the PCM and the A/F sensor • Faulty A/F Sensor (Bank 2, Sensor 1) • PCM needs to be updated with the latest software • Faulty PCM
DTC: P0155 **2T PCM, MIL: Yes** **Year:** 2009, 2010 **Model:** MDX, RL, TSX, ZDX **Engine:** 3.5L V6, 3.7L V6 **Transmission:** All	**Front Air/Fuel Ratio (A/F) Sensor (Bank 2, Sensor 1) Heater Circuit Malfunction:** The execution time is continuous and the duration time is 2 seconds or more. Engine running and battery voltage at 10 volts or more. One of these 2 conditions must be met for at least 2 seconds: (1) No return signal HIGH is detected when the PCM output duty is less than 20%. (2) Return signal does not change when the PCM output duty is more than 20% and less than 80%. **NOTE: Before you troubleshoot, record all freeze data and any on-board snapshot.** **Possible causes:** • Poor connections or loose terminals at the A/F sensor (Sensor 1), the relay and the PCM • "Open" or "Short" Circuit between the A/F sensors, the relay or PCM • Faulty (A/F) Sensor (Bank 2, Sensor 1) • Faulty PCM
DTC: P0157 **1T O2S HTR2, MIL: Yes** **Year:** 2009, 2010 **Model:** MDX, RL, TL, TSX, ZDX **Engine:** 3.5L V6, 3.7L V6 **Transmission:** All	**Front Secondary HO2S (Bank 2, Sensor 2) Circuit Low Voltage :** The execution time is continuous and the duration time is 40 seconds or more. The system is in closed loop with the engine coolant temperature at 156 °F (69 °C) and the intake air temperature at -13 °F (-25 °C). The rear secondary HO2S output voltage is 0.05 V or less for at least 40 seconds. After current is applied to the rear secondary HO2S heater, if the rear secondary HO2S output continues low (lean) during feedback control, a malfunction is detected and a DTC is stored. **NOTE: Before you troubleshoot, record all freeze data and any on-board snapshot.** **Possible causes:** • Poor connections or loose terminals at the secondary HO2S (Sensor 2) and the PCM • "Open" or "Short circuit between the PCM and the secondary HO2S (Sensor 2) • Faulty secondary HO2S (Sensor 2) • Faulty PCM

DTC	Trouble Code Title, Conditions & Possible Causes
DTC: P0158 **2T PCM, MIL: Yes** **Year:** 2009, 2010 **Model:** MDX, RL, TL, TSX, ZDX **Engine:** 3.5L V6, 3.7L V6 **Transmission:** All	**Front Secondary HO2S (Bank 2, Sensor 2) Circuit High Voltage :** With the vehicle at operating temperature and driven between 1,000-3,000 rpm for at least 1 minute, the rear secondary HO2S output voltage is 1.270 V or more for at least 5 seconds. After current is applied to the rear secondary HO2S heater, if the rear secondary HO2S output continues to exceed the upper limit used during feedback control, a malfunction is detected and a DTC is stored. **NOTE: Before you troubleshoot, record all freeze data and any on-board snapshot.** **Possible causes:** • Poor connections or loose terminals at the secondary HO2S (Sensor 2) and the PCM • "Open" circuit between the PCM and the secondary HO2S (Sensor 2) • Faulty (Secondary HO2S (Bank 1, Sensor 2) • Faulty PCM
DTC: P0159 **2T PCM, MIL: Yes** **Year:** 2009, 2010 **Model:** MDX, RL, TL, TSX, ZDX **Engine:** 3.5L V6, 3.7L V6 **Transmission:** All	**Front Secondary HO2S (Bank 2, Sensor 2) Slow Response :** The execution time is once per driving cycle and the duration time is 20.7 seconds or less. Engine coolant temperature is 156 °F (69 °C), intake air temperature -13 °F (-25 °C), engine speed 1,180-2,000 rpms and vehicle speed is 30 mph. When the front secondary HO2S output drops to the response deterioration judgment threshold value and the response characteristics measurement is finished. (0.77-4.58 seconds) **Possible causes:** • Poor connections or loose terminals at the secondary HO2S (Sensor 2) and the PCM • Faulty Front Secondary HO2S (Bank 2, Sensor 2) • Faulty PCM
DTC: P0161 **1T PCM, MIL: Yes** **Year:** 2009, 2010 **Model:** MDX, RL, TL, TSX, ZDX **Engine:** 3.5L V6, 3.7L V6 **Transmission:** All	**Front Secondary HO2S (Bank 2, Sensor 2) Heater Circuit Malfunction :** The front secondary HO2S heater output is 0.38 A or less, or 3.33 A or more, for at least 5 seconds when the heater is on. Engine running the battery voltage (IGP terminal of PCM) is 10.5-16 V, DTCs P0117, P0118 are not active. If the front secondary HO2S heater draws more or less than a specified amperage, the PCM detects a malfunction and a DTC is stored. **NOTE: Before you troubleshoot, record all freeze data and any on-board snapshot.** **Possible causes:** • Poor connections or loose terminals at the front secondary HO2S (Sensor 2), the relay and the PCM • "Open" or "Short circuit between the PCM and the front secondary HO2S (Sensor 2) • Faulty PCM
DTC: P0171 **2T PCM, MIL: Yes** **Year:** 2009, 2010 **Model:** RDX, TSX **Engine:** 2.3L L4, 2.4L L4 **Transmission:** All	**Fuel System Too Lean:** With the engine at operating temperature in closed loop, and the engine speed between 550-650 rpm. The Long term fuel trim is higher than 1.33 (+33%). If any related DTCs are listed first troubleshoot those DTCs first, then recheck for P0171. **NOTE: Before you troubleshoot, record all freeze data and any on-board snapshot, and review the general troubleshooting information.** **Possible causes:** • Vacuum leaks • Improper valve clearances • Faulty injectors • Clogged fuel filter • Faulty fuel pump or regulator • Faulty EVAP canister purge valve • Faulty A/F sensor (Sensor 1)
DTC: P0171 **2T PCM, MIL: Yes** **Year:** 2009, 2010 **Model:** MDX, RL, TL, TSX, ZDX **Engine:** 3.5L V6, 3.7L V6 **Transmission:** All	**Rear Bank (Bank 1) Fuel System Too Lean:** If long term fuel trim is higher than normal (too lean), a malfunction in the fuel metering components is detected and a DTC is stored. Long term fuel trim is higher than 1.188 (+18.8 %). Engine is at running temperature, vehicle speed at 550-4,000 rpms and the system is in closed loop. **NOTE: Before you troubleshoot, record all freeze data and any on-board snapshot.** **Possible causes:** • Vacuum leaks • Clogged fuel filter • Faulty fuel pump or regulator • Faulty MAF sensor/IAT sensor • Faulty EVAP canister purge valve • Faulty Throttle body • Faulty fuel injectors

DTC	Trouble Code Title, Conditions & Possible Causes
DTC: P0172 **2T PCM, MIL: Yes** **Year:** 2009, 2010 **Model:** MDX, RL, TL, TSX, ZDX **Engine:** 3.5L V6, 3.7L V6 **Transmission:** All	**Rear Bank (Bank 1) Fuel System Too Rich:** The execution time is once per driving cycle and the duration time is 13 seconds or more. If long term fuel trim is lower than normal (too rich), a malfunction in the fuel metering components is detected and a DTC is stored. Engine is at running temperature, vehicle speed at 550-4,000 rpm and the system is in closed loop the long term fuel trim is lower than 0.820 (-18.0 %). **NOTE: Before you troubleshoot, record all freeze data and any on-board snapshot.** **Possible causes:** • Faulty fuel pump or regulator • Engine valve clearance • Faulty coolant temp sensor • Faulty MAF sensor/IAT sensor • Faulty EVAP canister purge valve • Faulty Throttle body • Faulty fuel injectors
DTC: P0172 **2T PCM, MIL: Yes** **Year:** 2009, 2010 **Model:** RDX, TSX **Engine:** 2.3L L4, 2.4L L4 **Transmission:** All	**Fuel System Too Rich:** With the vehicle running, engine rpm between 550-650 and in closed loop. The long term fuel trim is lower than 0.785 (-215%). If any related DTCs are present, troubleshoot those DTCs first, then recheck P0172. **NOTE: Before you troubleshoot, record all freeze data and any on-board snapshot, and review the general troubleshooting information.** **Possible causes:** • Improper fuel pressure • Improper valve clearances • Leaking injectors • MAF sensor/IAT sensor • Faulty EVAP canister purge valve • Faulty A/F sensor (Sensor 1)
DTC: P0174 **1T PCM, MIL: Yes** **Year:** 2009, 2010 **Model:** MDX, RL, TL, TSX, ZDX **Engine:** 3.5L V6, 3.7L V6 **Transmission:** All	**Front Bank (Bank 2) Fuel System Too Lean :** If long term fuel trim is higher than normal (too lean), a malfunction in the fuel metering components is detected and a DTC is stored. Long term fuel trim is higher than 1.188 (+18.8 %). Engine is at running temperature, vehicle speed at 550-4,000 rpm and the system is in closed loop. **NOTE: Before you troubleshoot, record all freeze data and any on-board snapshot** **Possible causes:** • Vacuum leaks • Clogged fuel filter • Faulty fuel pump or regulator • Faulty MAF sensor/IAT sensor • Faulty EVAP canister purge valve • Faulty Throttle body • Faulty fuel injectors
DTC: P0175 **2T PCM, MIL: Yes** **Year:** 2009, 2010 **Model:** MDX, RL, TL, TSX, ZDX **Engine:** 3.5L V6, 3.7L V6 **Transmission:** All	**Front Bank (Bank 2) Fuel System Too Rich :** The execution time is once per driving cycle and the duration time is 13 seconds or more. If long term fuel trim is lower than normal (too rich), a malfunction in the fuel metering components is detected and a DTC is stored. Engine is at running temperature, vehicle speed at 550-4,000 rpm and the system is in closed loop the long term fuel trim is lower than 0.820 (-18.0 %). **NOTE: Before you troubleshoot, record all freeze data and any on-board snapshot.** **Possible causes:** • Faulty fuel pump or regulator • Engine valve clearance • Faulty coolant temp sensor • Faulty MAF sensor/IAT sensor • Faulty EVAP canister purge valve • Faulty Throttle body • Faulty fuel injectors
DTC: P0201 **1T CCM, MIL: Yes** **Year:** 2009, 2010 **Model:** MDX, ZDX **Engine:** 3.7L V6 **Transmission:** All	**Fuel Injector 1 Circuit Malfunction:** Engine started, system voltage over 9v, and PCM detected the injector voltage for Cylinder 1 did not equal the system voltage with the injector commanded "off", or that the injector voltage did not equal zero (0) volts with the injector commanded "on". **Possible causes:** • Fuel injector control circuit is open or shorted to ground • Fuel injector power circuit is open between injector and relay • Fuel Injector has failed • PCM has failed (injector driver circuit may be open or shorted)

DTC	Trouble Code Title, Conditions & Possible Causes
DTC: P0202 **1T CCM, MIL: Yes** **Year:** 2009, 2010 **Model:** MDX, ZDX **Engine:** 3.7L V6 **Transmission:** All	**Fuel Injector 2 Circuit Malfunction:** Engine started, system voltage over 9v, and PCM detected the injector voltage for Cylinder 2 did not equal the system voltage with the injector commanded "off", or that the injector voltage did not equal zero (0) volts with the injector commanded "on". **Possible causes:** • Fuel injector control circuit is open or shorted to ground • Fuel injector power circuit open between injector and ECM fuse • Fuel Injector has failed • PCM has failed (injector driver circuit may be open or shorted)
DTC: P0203 **1T CCM, MIL: Yes** **Year:** 2009, 2010 **Model:** MDX, ZDX **Engine:** 3.7L V6 **Transmission:** All	**Fuel Injector 3 Circuit Malfunction:** Engine started, system voltage over 9v, and PCM detected the injector voltage for Cylinder 3 did not equal the system voltage with the injector commanded "off", or that the injector voltage did not equal zero (0) volts with the injector commanded "on". **Possible causes:** • Fuel injector control circuit is open or shorted to ground • Fuel injector power circuit open between injector and ECM fuse • Fuel Injector has failed • PCM has failed (injector driver circuit may be open or shorted)
DTC: P0204 **1T CCM, MIL: Yes** **Year:** 2009, 2010 **Model:** MDX, ZDX **Engine:** 3.7L V6 **Transmission:** All	**Fuel Injector 4 Circuit Malfunction:** Engine started, system voltage over 9v, and PCM detected the injector voltage for Cylinder 4 did not equal the system voltage with the injector commanded "off", or that the injector voltage did not equal zero (0) volts with the injector commanded "on". **Possible causes:** • Fuel injector control circuit is open or shorted to ground • Fuel injector power circuit open between injector and ECM fuse • Fuel Injector has failed • PCM has failed (injector driver circuit may be open or shorted)
DTC: P0205 **1T CCM, MIL: Yes** **Year:** 2009, 2010 **Model:** MDX, ZDX **Engine:** 3.7L V6 **Transmission:** All	**Fuel Injector 5 Circuit Malfunction:** Engine started, system voltage over 9v, and PCM detected the injector voltage for Cylinder 5 did not equal the system voltage with the injector commanded "off", or that the injector voltage did not equal zero (0) volts with the injector commanded "on". **Possible causes:** • Fuel injector control circuit is open or shorted to ground • Fuel injector power circuit open between injector and ECM fuse • Fuel Injector has failed • PCM has failed (injector driver circuit may be open or shorted)
DTC: P0206 **1T CCM, MIL: Yes** **Year:** 2009, 2010 **Model:** MDX, ZDX **Engine:** 3.7L V6 **Transmission:** All	**Fuel Injector 6 Circuit Malfunction:** Engine started, system voltage over 9v, and PCM detected the injector voltage for Cylinder 6 did not equal the system voltage with the injector commanded "off", or that the injector voltage did not equal zero (0) volts with the injector commanded "on". **Possible causes:** • Fuel injector control circuit is open or shorted to ground • Fuel injector power circuit open between injector and ECM fuse • Fuel Injector has failed • PCM has failed (injector driver circuit may be open or shorted)
DTC: P0222 **1T PCM, MIL: Yes** **Year:** 2009, 2010 **Model:** MDX, RDX, RL, TL, TSX, ZDX **Engine:** 2.3L L4, 2.4L L4, 3.5L V6, 3.7L V6 **Transmission:** All	**Throttle Position (TP) Sensor B Circuit Low Voltage:** The execution time is continuous and the duration time is 200 milliseconds or more. Ignition on and DTCs P0223, P2101, P2118, P2135, P2176 are not active. The TP sensor B output voltage is 0.3 V or less for at least 200 milliseconds. If the signal from TP sensor B is less than a fixed value for a set time, the PCM detects a TP sensor B malfunction and stores a DTC. **NOTE: Before you troubleshoot, record all freeze data and any on-board snapshot.** **Possible causes:** • Poor connections or loose terminals at the throttle body and the PCM • "Open" or "Short" circuit between the throttle body and the PCM • Faulty throttle body • PCM may need to be updated with the latest software • Faulty PCM

DTC	Trouble Code Title, Conditions & Possible Causes
DTC. P0223 **1T PCM, MIL: Yes** **Year:** 2009, 2010 **Model:** MDX, RDX, RL, TL, TSX, ZDX **Engine:** 2.3L L4, 2.4L L4, 3.5L V6, 3.7L V6 **Transmission:** All	**Throttle Position (TP) Sensor B Circuit High Voltage :** The execution time is continuous and the duration time is 200 milliseconds or more. Ignition ON and DTCs P0222, P2101, P2118, P2135, P2176 are not active. The TP sensor B output voltage is 4.8 V or more for at least 200 milliseconds. If the signal from TP sensor B is more than a fixed value for a set time, the PCM detects a TP sensor B malfunction and stores a DTC. **NOTE: Before you troubleshoot, record all freeze data and any on-board snapshot.** **Possible causes:** • Poor connections or loose terminals at the throttle body and the PCM • "Open" or "Short" between the throttle body and the PCM • PCM may need to be updated with the latest software • Faulty throttle body • Faulty PCM
DTC: P0234 **2T PCM, MIL: Yes** **Year:** 2009, 2010 **Model:** RDX **Engine:** 2.3L L4 **Transmission:** All	**Turbocharger Overboost Problem:** With the engine running at a speed 3,800 rpm or more without a load and at operating temperature. The boost pressure is at a specified pressure or more, and the fuel cut for engine protection operates for at least 2 seconds. **NOTE: If any of the DTCs listed below are indicated at the same time as DTC P0234, troubleshoot those DTCs first, then recheck for P0234. P0107, P0108, P1128, P1129: Manifold Absolute Pressure (MAP) sensor. P0236, P0237, P0238: Turbocharger boost sensor.** **Possible causes:** • Faulty turbocharger wastegate control solenoid valve • Poor connections, blockage, or damage to these following parts: • 1). Hose between the turbocharger wastegate control solenoid valve and the turbocharger wastegate control actuator • 2). Hoses and pipe between the turbocharger wastegate control solenoid valve and turbocharger • 3). Turbocharger port
DTC: P0236 **2T PCM, MIL: Yes** **Year:** 2009, 2010 **Model:** RDX **Engine:** 2.3L L4 **Transmission:** All	**Turbocharger Boost Sensor Circuit Range/Performance Problem:** With the engine running at a speed of 1,750 or more, and the throttle position 22° or more for at least 5 seconds. When the fluctuation value of the turbocharger boost sensor measured at each detection of negative pressure area determination and high throttle area determination, a STUCK failure is detected. **NOTE: Before you troubleshoot, record all freeze data and any on-board snapshot, and review the general troubleshooting information.** **Possible causes:** • Poor connections or loose terminals at the turbocharger boost pressure sensor and the PCM • Faulty turbocharger boost sensor
DTC: P0237 **1T PCM, MIL: Yes** **Year:** 2009, 2010 **Model:** RDX **Engine:** 2.3L L4 **Transmission:** All	**Turbocharger Boost Sensor Circuit Low Voltage:** With the engine running the turbocharger boost sensor output voltage value is 0.23V or less for at least 2 seconds. **NOTE: Before you troubleshoot, record all freeze data and any on-board snapshot, and review the general troubleshooting information.** **Possible causes:** • Poor connections or loose terminals at the turbocharger boost sensor and the PCM • "Open" or "Short" circuit between the PCM and the turbocharger boost sensor • Faulty turbocharger boost sensor • PCM may need to be updated with the latest software • Faulty PCM
DTC: P0238 **1T PCM, MIL: Yes** **Year:** 2009, 2010 **Model:** RDX **Engine:** 2.3L L4 **Transmission:** All	**Turbocharger Boost Sensor Circuit High Voltage:** With the engine running, and DTC P0237 not active the turbocharger boost sensor output voltage value is 4.49V or more for at least 2 seconds. **NOTE: Before you troubleshoot, record all freeze data and any on-board snapshot, and review the general troubleshooting information.** **Possible causes:** • Poor connections or loose terminals at the turbocharger boost sensor and the PCM • "Open" circuit between the PCM and the turbocharger boost sensor • Faulty turbocharger boost sensor • PCM may need to be updated with the latest software • Faulty PCM

DTC	Trouble Code Title, Conditions & Possible Causes
DTC: P0243 **1T PCM, MIL: Yes** **Year:** 2009, 2010 **Model:** RDX **Engine:** 2.3L L4 **Transmission:** All	**Turbocharger Wastegate Control Solenoid Valve Circuit Malfunction:** With the engine running, battery voltage a minimum of 10.5V, waste gate solenoid valve output duty between 2-98%. The return signal does not change as set by the output of the turbocharger wastegate control solenoid valve and this condition continues at least 2 seconds. **NOTE: Before you troubleshoot, record all freeze data and any on-board snapshot, and review the general troubleshooting information.** **Possible causes:** • Poor connections or loose terminals at the turbocharger wastegate control solenoid valve and the PCM • Blown fuse • "Open" or "Short" circuit between the PCM and turbocharger wastegate control solenoid valve • Faulty turbocharger wastegate control solenoid valve • PCM may need to be updated with the latest software • Faulty PCM
DTC: P0299 **2T PCM, MIL: Yes** **Year:** 2009, 2010 **Model:** RDX **Engine:** 2.3L L4 **Transmission:** All	**Turbocharger Underboost Problem:** With the engine speed at 3,800 rpm without a load and at operating temperature, the difference between the upper limit boost pressure output and the actual boost pressure output is 26 kPa (7.8 in.HG, 200 mmHg) or more for at least 4 seconds. **NOTE: Before you troubleshoot, record all freeze data and any on-board snapshot, and review the general troubleshooting information.** **Possible causes:** • Clogged air cleaner element • Leaks or damage to intake air duct • Faulty turbocharger bypass control solenoid valve • Poor connections, blockage, or damage between the turbocharger bypass control solenoid valve and the turbocharger bypass control valve • Poor connections, blockage, or damage between the turbocharger bypass control solenoid valve and the turbocharger bypass control valve base • Poor connections, blockage, or damage between the turbocharger bypass control valve base port • Clogged exhaust • Faulty turbocharger • Faulty turbocharger wastegate control actuator
DTC: P0300 **2T PCM, MIL: Yes** **Year:** 2009, 2010 **Model:** CSX, MDX, RDX, RL, TL, TSX, ZDX **Engine:** 2.0L L4, 2.3L L4, 2.4L L4, 3.5L V6, 3.7L V6 **Transmission:** All	**Random Misfire Detected:** The execution time is once per driving cycle and the duration time is 200 revolutions for Type 1 and 1,000 revolutions for Type 2. Conditions for illuminating the MIL Type 1 and Type 2. * Misfire Type 1 (Severe) Per 200 revolutions 27-90 times. If a type 1 misfire (catalyst damaging) occurs once, the MIL blinks once per second, a Temporary DTC is stored, and the high rpm fuel injection stop system activates. The fuel injection stops, at high rpm only, on the cylinder that has the highest misfire counts. the MIL continues to blink, and the fuel injection stays off at high rpm, until the drive is completed. If a type 1 misfire occurs during a second drive cycle, the MIL and fuel injection behave the same and a DTC is stored. After a type 1 misfire has been detected during two drive cycles, the MIL comes on and stays on beginning with the third drive cycle, unless the Temporary DTC has been cleared by the PCM. Even if the MIL is on, it will start blinking if a type 1 misfire occurs. If the malfunction recurs during the next (second) drive cycle, the MIL comes on and the DTC and the freeze frame data are stored. * Misfire Type 2 (Light) Per 1,000 revolutions 55 times. If a type 2 misfire (emission-related but not severe enough to immediately damage the catalyst) occurs, a Temporary DTC is stored, but the MIL does come on or blink. If a type 2 misfire occurs during a second drive cycle, the MIL comes on and stays on unless the Temporary DTC has been cleared by the PCM. **NOTE: Before you troubleshoot, record all freeze data and any on-board snapshot.** **Possible causes:** • Poor fuel quality • Clogged fuel filter • Faulty spark plugs • Faulty Fuel pump and/or regulator • Poor connections or loose terminals at the ignition coil, the injector, and the PCM • Check the CKP pattern learn procedure

DTC	Trouble Code Title, Conditions & Possible Causes
DTC: P0301 **2T PCM, MIL: Yes** **Year:** 2009, 2010 **Model:** CSX, MDX, RDX, RL, TL, TSX, ZDX **Engine:** 2.0L L4, 2.3L L4, 2.4L L4, 3.5L V6, 3.7L V6 **Transmission:** All	**Cylinder 1 Misfire Detected :** The execution time is once per driving cycle and the duration time is 200 revolutions for Type 1 and 1,000 revolutions for Type 2. Conditions for illuminating the MIL Type 1 and Type 2. * Misfire Type 1 (Severe) Per 200 revolutions 27-90 times. If a type 1 misfire (catalyst damaging) occurs once, the MIL blinks once per second, a Temporary DTC is stored, and the high rpm fuel injection stop system activates. The fuel injection stops, at high rpm only, on the cylinder that has the highest misfire counts. the MIL continues to blink, and the fuel injection stays off at high rpm, until the drive is completed. If a type 1 misfire occurs during a second drive cycle, the MIL and fuel injection behave the same and a DTC is stored. After a type 1 misfire has been detected during two drive cycles, the MIL comes on and stays on beginning with the third drive cycle, unless the Temporary DTC has been cleared by the PCM. Even if the MIL is on, it will start blinking if a type 1 misfire occurs. If the malfunction recurs during the next (second) drive cycle, the MIL comes on and the DTC and the freeze frame data are stored. * Misfire Type 2 (Light) Per 1,000 revolutions 55 times. If a type 2 misfire (emission-related but not severe enough to immediately damage the catalyst) occurs, a Temporary DTC is stored, but the MIL does come on or blink. If a type 2 misfire occurs during a second drive cycle, the MIL comes on and stays on unless the Temporary DTC has been cleared by the PCM. **NOTE: Before you troubleshoot, record all freeze data and any on-board snapshot.** **Possible causes:** • Faulty spark plug • Faulty ignition coil • Faulty fuel injector • Faulty fuel pump and/or regulator • Poor connections or loose terminals at the ignition coil, the injector, and the PCM • "Open" or "Short" circuit between the ignition coil and the under-hood fuse/relay box • "Open" or "Short" circuit between the PCM and the ignition coil • Internal engine problem • Incorrect PCM idle learn procedure • Incorrect CKP pattern learn procedure • MAF sensor
DTC: P0302 **2T PCM, MIL: Yes** **Year:** 2009, 2010 **Model:** CSX, MDX, RDX, RL, TL, TSX, ZDX **Engine:** 2.0L L4, 2.3L L4, 2.4L L4, 3.5L V6, 3.7L V6 **Transmission:** All	**Cylinder 2 Misfire Detected :** The execution time is once per driving cycle and the duration time is 200 revolutions for Type 1 and 1,000 revolutions for Type 2. Conditions for illuminating the MIL Type 1 and Type 2. * Misfire Type 1 (Severe) Per 200 revolutions 27-90 times. If a type 1 misfire (catalyst damaging) occurs once, the MIL blinks once per second, a Temporary DTC is stored, and the high rpm fuel injection stop system activates. The fuel injection stops, at high rpm only, on the cylinder that has the highest misfire counts. the MIL continues to blink, and the fuel injection stays off at high rpm, until the drive is completed. If a type 1 misfire occurs during a second drive cycle, the MIL and fuel injection behave the same and a DTC is stored. After a type 1 misfire has been detected during two drive cycles, the MIL comes on and stays on beginning with the third drive cycle, unless the Temporary DTC has been cleared by the PCM. Even if the MIL is on, it will start blinking if a type 1 misfire occurs. If the malfunction recurs during the next (second) drive cycle, the MIL comes on and the DTC and the freeze frame data are stored. * Misfire Type 2 (Light) Per 1,000 revolutions 55 times. If a type 2 misfire (emission-related but not severe enough to immediately damage the catalyst) occurs, a Temporary DTC is stored, but the MIL does come on or blink. If a type 2 misfire occurs during a second drive cycle, the MIL comes on and stays on unless the Temporary DTC has been cleared by the PCM. **NOTE: Before you troubleshoot, record all freeze data and any on-board snapshot.** **Possible causes:** • Faulty spark plug • Faulty ignition coil • Faulty fuel injector • Faulty fuel pump and/or regulator • Poor connections or loose terminals at the ignition coil, the injector, and the PCM • "Open" or "Short" circuit between the ignition coil and the under-hood fuse/relay box • "Open" or "Short" circuit between the PCM and the ignition coil • Internal engine problem • Incorrect PCM idle learn procedure • Incorrect CKP pattern learn procedure • MAF sensor

DTC	Trouble Code Title, Conditions & Possible Causes
DTC: P0303 **2T PCM, MIL: Yes** **Year:** 2009, 2010 **Model:** CSX, MDX, RDX, RL, TL, TSX, ZDX **Engine:** 2.0L L4, 2.3L L4, 2.4L L4, 3.5L V6, 3.7L V6 **Transmission:** All	**Cylinder 3 Misfire Detected :** The execution time is once per driving cycle and the duration time is 200 revolutions for Type 1 and 1,000 revolutions for Type 2. Conditions for illuminating the MIL Type 1 and Type 2. * Misfire Type 1 (Severe) Per 200 revolutions 27-90 times. If a type 1 misfire (catalyst damaging) occurs once, the MIL blinks once per second, a Temporary DTC is stored, and the high rpm fuel injection stop system activates. The fuel injection stops, at high rpm only, on the cylinder that has the highest misfire counts. the MIL continues to blink, and the fuel injection stays off at high rpm, until the drive is completed. If a type 1 misfire occurs during a second drive cycle, the MIL and fuel injection behave the same and a DTC is stored. After a type 1 misfire has been detected during two drive cycles, the MIL comes on and stays on beginning with the third drive cycle, unless the Temporary DTC has been cleared by the PCM. Even if the MIL is on, it will start blinking if a type 1 misfire occurs. If the malfunction recurs during the next (second) drive cycle, the MIL comes on and the DTC and the freeze frame data are stored. * Misfire Type 2 (Light) Per 1,000 revolutions 55 times. If a type 2 misfire (emission-related but not severe enough to immediately damage the catalyst) occurs, a Temporary DTC is stored, but the MIL does come on or blink. If a type 2 misfire occurs during a second drive cycle, the MIL comes on and stays on unless the Temporary DTC has been cleared by the PCM. **NOTE: Before you troubleshoot, record all freeze data and any on-board snapshot.** **Possible causes:** • Faulty spark plug • Faulty ignition coil • Faulty fuel injector • Faulty fuel pump and/or regulator • Poor connections or loose terminals at the ignition coil, the injector, and the PCM • "Open" or "Short" circuit between the ignition coil and the under-hood fuse/relay box • "Open" or "Short" circuit between the PCM and the ignition coil • Internal engine problem • Incorrect PCM idle learn procedure • Incorrect CKP pattern learn procedure • MAF sensor
DTC: P0304 **2T PCM, MIL: Yes** **Year:** 2009, 2010 **Model:** CSX, MDX, RDX, RL, TL, TSX, ZDX **Engine:** 2.0L L4, 2.3L L4, 2.4L L4, 3.5L V6, 3.7L V6 **Transmission:** All	**Cylinder 4 Misfire Detected :** The execution time is once per driving cycle and the duration time is 200 revolutions for Type 1 and 1,000 revolutions for Type 2. Conditions for illuminating the MIL Type 1 and Type 2. * Misfire Type 1 (Severe) Per 200 revolutions 27-90 times. If a type 1 misfire (catalyst damaging) occurs once, the MIL blinks once per second, a Temporary DTC is stored, and the high rpm fuel injection stop system activates. The fuel injection stops, at high rpm only, on the cylinder that has the highest misfire counts. the MIL continues to blink, and the fuel injection stays off at high rpm, until the drive is completed. If a type 1 misfire occurs during a second drive cycle, the MIL and fuel injection behave the same and a DTC is stored. After a type 1 misfire has been detected during two drive cycles, the MIL comes on and stays on beginning with the third drive cycle, unless the Temporary DTC has been cleared by the PCM. Even if the MIL is on, it will start blinking if a type 1 misfire occurs. If the malfunction recurs during the next (second) drive cycle, the MIL comes on and the DTC and the freeze frame data are stored. * Misfire Type 2 (Light) Per 1,000 revolutions 55 times. If a type 2 misfire (emission-related but not severe enough to immediately damage the catalyst) occurs, a Temporary DTC is stored, but the MIL does come on or blink. If a type 2 misfire occurs during a second drive cycle, the MIL comes on and stays on unless the Temporary DTC has been cleared by the PCM. **NOTE: Before you troubleshoot, record all freeze data and any on-board snapshot.** **Possible causes:** • Faulty spark plug • Faulty ignition coil • Faulty fuel injector • Faulty fuel pump and/or regulator • Poor connections or loose terminals at the ignition coil, the injector, and the PCM • "Open" or "Short" circuit between the ignition coil and the under-hood fuse/relay box • "Open" or "Short" circuit between the PCM and the ignition coil • Internal engine problem • Incorrect PCM idle learn procedure • Incorrect CKP pattern learn procedure • MAF sensor

DTC	Trouble Code Title, Conditions & Possible Causes
DTC: P0305 **2T PCM, MIL: Yes** **Year:** 2009, 2010 **Model:** MDX, RL, ZDX **Engine:** 3.7L V6 **Transmission:** All	**Cylinder 5 Misfire Detected :** The execution time is once per driving cycle and the duration time is 200 revolutions for Type 1 and 1,000 revolutions for Type 2. Conditions for illuminating the MIL Type 1 and Type 2. * Misfire Type 1 (Severe) Per 200 revolutions 27-90 times. If a type 1 misfire (catalyst damaging) occurs once, the MIL blinks once per second, a Temporary DTC is stored, and the high rpm fuel injection stop system activates. The fuel injection stops, at high rpm only, on the cylinder that has the highest misfire counts. the MIL continues to blink, and the fuel injection stays off at high rpm, until the drive is completed. If a type 1 misfire occurs during a second drive cycle, the MIL and fuel injection behave the same and a DTC is stored. After a type 1 misfire has been detected during two drive cycles, the MIL comes on and stays on beginning with the third drive cycle, unless the Temporary DTC has been cleared by the PCM. Even if the MIL is on, it will start blinking if a type 1 misfire occurs. If the malfunction recurs during the next (second) drive cycle, the MIL comes on and the DTC and the freeze frame data are stored. * Misfire Type 2 (Light) Per 1,000 revolutions 55 times. If a type 2 misfire (emission-related but not severe enough to immediately damage the catalyst) occurs, a Temporary DTC is stored, but the MIL does come on or blink. If a type 2 misfire occurs during a second drive cycle, the MIL comes on and stays on unless the Temporary DTC has been cleared by the PCM. **NOTE: Before you troubleshoot, record all freeze data and any on-board snapshot.** **Possible causes:** • Faulty spark plug • Faulty ignition coil • Faulty fuel injector • Faulty fuel pump and/or regulator • Poor connections or loose terminals at the ignition coil, the injector, and the PCM • "Open" or "Short" circuit between the ignition coil and the under-hood fuse/relay box • "Open" or "Short" circuit between the PCM and the ignition coil • Internal engine problem • Incorrect PCM idle learn procedure • Incorrect CKP pattern learn procedure • MAF sensor
DTC: P0306 **1T PCM, MIL: Yes** **Year:** 2009, 2010 **Model:** MDX, RL, TL, TSX, ZDX **Engine:** 3.5L V6, 3.7L V6 **Transmission:** All	**Cylinder 6 Misfire Detected:** The execution time is once per driving cycle and the duration time is 200 revolutions for Type 1 and 1,000 revolutions for Type 2. Conditions for illuminating the MIL Type 1 and Type 2. * Misfire Type 1 (Severe) Per 200 revolutions 27-90 times. If a type 1 misfire (catalyst damaging) occurs once, the MIL blinks once per second, a Temporary DTC is stored, and the high rpm fuel injection stop system activates. The fuel injection stops, at high rpm only, on the cylinder that has the highest misfire counts. the MIL continues to blink, and the fuel injection stays off at high rpm, until the drive is completed. If a type 1 misfire occurs during a second drive cycle, the MIL and fuel injection behave the same and a DTC is stored. After a type 1 misfire has been detected during two drive cycles, the MIL comes on and stays on beginning with the third drive cycle, unless the Temporary DTC has been cleared by the PCM. Even if the MIL is on, it will start blinking if a type 1 misfire occurs. If the malfunction recurs during the next (second) drive cycle, the MIL comes on and the DTC and the freeze frame data are stored. * Misfire Type 2 (Light) Per 1,000 revolutions 55 times. If a type 2 misfire (emission-related but not severe enough to immediately damage the catalyst) occurs, a Temporary DTC is stored, but the MIL does come on or blink. If a type 2 misfire occurs during a second drive cycle, the MIL comes on and stays on unless the Temporary DTC has been cleared by the PCM. **NOTE: Before you troubleshoot, record all freeze data and any on-board snapshot** **Possible causes:** • Faulty spark plug • Faulty ignition coil • Faulty fuel injector • Faulty fuel pump and/or regulator • Poor connections or loose terminals at the ignition coil, the injector, and the PCM • "Open" or "Short" circuit between the ignition coil and the under-hood fuse/relay box • "Open" or "Short" circuit between the PCM and the ignition coil • Internal engine problem • Incorrect PCM idle learn procedure • Incorrect CKP pattern learn procedure • MAF sensor

DTC	Trouble Code Title, Conditions & Possible Causes
DTC: P0325 **1T PCM, MIL: Yes** **Year:** 2009, 2010 **Model:** MDX, RDX, RL, TSX, ZDX **Engine:** 2.3L L4, 2.4L L4, 3.5L V6, 3.7L V6 **Transmission:** All	**Knock Sensor Circuit Malfunction :** The execution time is continuous and the duration time is 5 seconds or more. Engine running, speed 2,000 rpm or more, coolant temperature at 140 °F (60 °C) or more. No signals from the knock sensor are detected for at least 5 seconds. If the signals from the knock sensor do not vary for a set time, the PCM detects a malfunction and stores a DTC. **NOTE: Before you troubleshoot, record all freeze data and any on-board snapshot.** **Possible causes:** • Poor connections or loose terminals at the knock sensor and the PCM • "Open" or "Short" circuit between the PCM and the knock sensor sub-harness • "Open" or "Short" circuit in the knock sensor subharness • Poor connections or loose terminals at the ignition coil, the injector, and the PCM • Faulty knock sensor • PCM may need to be updated with the latest software • Faulty PCM
DTC: P0335 **1T PCM, MIL: Yes** **Year:** 2009, 2010 **Model:** RDX, TSX **Engine:** 2.3L L4, 2.4L L4, 3.5L V6 **Transmission:** All	**Crankshaft Position (CKP) Sensor No Signal:** With the engine running the PCM has detected no CKP pulses over 75 times. **Possible causes:** • Poor connections or loose wires at the CKP sensor and at the PCM • "Open" or "Short" circuit between the PCM and the CKP sensor • Faulty CKP sensor • Faulty PCM
DTC: P0335 **1T PCM, TCIL: Yes** **Year:** 2009, 2010 **Model:** MDX, RL, ZDX **Engine:** 3.7L V6 **Transmission:** All	**Crankshaft Position (CKP) Sensor A No Signal :** The execution is continuous and the duration is for 2 seconds or more (when the engine speed is 750 rpm). MIL OFF, D indicator blinks. No signals from the CKP sensor are input at least 63 times. If no pulsing signals from the CKP sensor are detected, a malfunction is detected and a DTC is stored. **Possible causes:** • Poor connections or loose terminals at CKP sensor A and the PCM • "Open" or "Short" in the wire between the PCM and CKP sensor A • Faulty CKP sensor A • PCM may need to be updated with the latest software • Faulty PCM
DTC: P0339 **1T PCM, MIL: Yes** **Year:** 2009, 2010 **Model:** RDX, TSX **Engine:** 2.3L L4, 2.4L L4, 3.5L V6 **Transmission:** All	**Crankshaft Position (CKP) Sensor Circuit Intermittent Interruption :** With the engine running other than 58 pulses are detected during intervals between reference pulses for each crank revolution. This condition has been detected 30 times. **NOTE: Before you troubleshoot, record all freeze data and any on-board snapshot, and review the general troubleshooting information.** **Possible causes:** • Poor connections at the ECM/PCM, CKP sensor, engine ground, body ground • Damaged CKP sensor pulse plate • Faulty CKP sensor
DTC: P0339 **1T PCM, MIL: Yes** **Year:** 2009, 2010 **Model:** MDX, RL, ZDX **Engine:** 3.7L V6 **Transmission:** All	**Crankshaft Position (CKP) Sensor A Circuit Intermittent Interruption :** Other than 22 pulses are detected during intervals between reference pulses for each crankshaft revolution. This condition has been detected at least 30 times. An abnormal amount of pulsing signals are detected from CKP sensor A. **NOTE: Before you troubleshoot, record all freeze data and any on-board snapshot.** **Possible causes:** • Poor connections or loose terminals at CKP sensor A/B and the PCM • Poor grounds at engine, body, CKP sensor A/B or PCM • Damaged CKP sensor A/B pulse plate/timing belt drive pulley • Faulty CKP sensor A/B • PCM may need to be updated with the latest software • Faulty PCM

DTC	Trouble Code Title, Conditions & Possible Causes
DTC: P0340 **1T PCM, MIL: Yes** **Year:** 2009, 2010 **Model:** RDX **Engine:** 2.3L L4 **Transmission:** All	**Camshaft Position (CMP) Sensor No Signal :** With the engine running no signals from the CMP sensor are detected while signals from the CKP sensor are detected 352 times in succession. DTCs P0335, P0339 and P0344 are not active. **NOTE: Before you troubleshoot, record all freeze data and any on-board snapshot, and review the general troubleshooting information.** **Possible causes:** • Poor connections or loose terminals at the CMP sensor and the ECM/PCM • "Open" circuit between the CMP sensor and PGM-FI main relay 1 (FI MAIN) • "Open" circuit between the ECM/PCM and the CMP sensor • "Open" in the wire between the CMP sensor and ground • Faulty CMP sensor
DTC: P0340 **1T PCM, MIL: Yes** **Year:** 2009, 2010 **Model:** RL **Engine:** 3.7L V6 **Transmission:** All	**Camshaft Position (CMP) Sensor No Signal:** With the engine running no signals from CMP sensor A are detected while signals from the CKP sensor are detected 352 times in succession. The execution time is continuous and DTCs P0335, P0339, P0344 are not active. **NOTE: Before you troubleshoot, record all freeze data and any on-board snapshot.** **Possible causes:** • Poor connections or loose terminals at the CMP sensor and the PCM • "Open" circuit between the CMP sensor and PGM-FI main relay • "Open" or "Short" circuit between the PCM and the CMP sensor • "Open" in the wire between the CMP sensor and ground • Faulty CMP sensor • PCM needs to be updated with the latest software • Faulty PCM
DTC: P0341 **2T PCM, MIL: Yes** **Year:** 2009, 2010 **Model:** RDX, TSX **Engine:** 2.3L L4, 2.4L L4 **Transmission:** All	**Camshaft Position (CMP) Sensor A and CKP Sensor Incorrect Phase Detected :** With the engine running and at operating temperature, and the vehicle speed between 19-38 mph (30-60 km/h) for at least 5 seconds. Malfunctions are as follows: Malfunction with VTC OFF: The gap between the CMP sensor A pulse and the middle of the CMP sensor A assembly is 10 degrees or more for at least 5 seconds. Malfunction with VTC ACTIVE: The timing of the camshaft is out of specified range (Other than when BTDC is between 10-100 degrees) for at least 5 seconds. **NOTE: Before you troubleshoot, record all freeze data and any on-board snapshot, and review the general troubleshooting information.** **Possible causes:** • The camshaft timing needs to be reset • Damage or stretched cam chain • Faulty auto-tensioner • Poor connections or loose terminals at the VTC oil control solenoid valve and the ECM/PCM • Faulty VTC actuator • Faulty VTC oil control solenoid valve
DTC: P0344 **2T PCM, MIL: Yes** **Year:** 2009, 2010 **Model:** RDX, TSX **Engine:** 2.3L L4, 2.4L L4 **Transmission:** All	**Camshaft Position (CMP) Sensor A Circuit Intermittent Interruption:** With the engine running, more or less than two CMP sensor A pulses are detected during intervals between the CKP standard pulses. This condition occurs at least 30 times. **NOTE: Before you troubleshoot, record all freeze data and any on-board snapshot, and review the general troubleshooting information.** **Possible causes:** • Poor connections or loose terminals at the engine and/or body ground • Poor connections or loose terminals at CMP sensor A and the ECM/PCM • Damaged CMP pulse plate A • Faulty CMP sensor A
DTC: P0344 **1T PCM, MIL: Yes** **Year:** 2009, 2010 **Model:** RL **Engine:** 3.7L V6 **Transmission:** All	**Camshaft Position (CMP) Sensor Intermittent Interruption:** With the engine running more or less than two CMP sensor pulses are detected during intervals between the CKP standard pulses. This condition occurs at least 30 times. The execution time is continuous and DTCs P0335, P0339, P0340 are not active. **NOTE: Before you troubleshoot, record all freeze data and any on-board snapshot.** **Possible causes:** • Damage to the CMP sensor pulse projection on the front camshaft pulley • Poor or loose connections at CMP sensor, PCM, engine ground, body ground • Faulty CMP sensor

DTC	Trouble Code Title, Conditions & Possible Causes
DTC: P0351 **1T PCM, MIL: Yes** **Year:** 2009, 2010 **Model:** MDX, RL, TL, TSX, ZDX **Engine:** 2.4L L4, 3.5L V6, 3.7L V6 **Transmission:** All	**Cylinder 1 Ignition Coil Circuit Malfunction:** With the engine running, the execution time is continuous and the duration time is 5 seconds or more. (when the engine speed is 700 rpm) The return signal does not change for at least 5 seconds when the ignition coil is triggered. **NOTE: Before you troubleshoot, record all freeze data and any on-board snapshot.** **Possible causes:** • Blown fuse • Poor connections or loose terminals at the ignition coil and the PCM • "Open" or "Short" circuit between the ignition coils and the ignition coil relay • "Open" or "Short" circuit between the ignition coils and the PCM • "Open" in the wire between the ignition coil and ground • Faulty ignition coil relay • Faulty ignition coil
DTC: P0352 **1T PCM** **Year:** 2009, 2010 **Model:** MDX, RL, TL, TSX, ZDX **Engine:** 2.4L L4, 3.5L V6, 3.7L V6 **Transmission:** All	**Cylinder 2 Ignition Coil Circuit Malfunction:** With the engine running, the execution time is continuous and the duration time is 5 seconds or more. (when the engine speed is 700 rpm) The return signal does not change for at least 5 seconds when the ignition coil is triggered. **Possible causes:** • Blown fuse • Poor connections or loose terminals at the ignition coil and the PCM • "Open" or "Short" circuit between the ignition coils and the ignition coil relay • "Open" or "Short" circuit between the ignition coils and the PCM • "Open" in the wire between the ignition coil and ground • Faulty ignition coil relay • Faulty ignition coil
DTC: P0353 **1T PCM, MIL: Yes** **Year:** 2009, 2010 **Model:** MDX, RL, TL, TSX, ZDX **Engine:** 2.4L L4, 3.5L V6, 3.7L V6 **Transmission:** All	**Cylinder 3 Ignition Coil Circuit Malfunction:** With the engine running, the execution time is continuous and the duration time is 5 seconds or more. (when the engine speed is 700 rpm) The return signal does not change for at least 5 seconds when the ignition coil is triggered. **NOTE: Before you troubleshoot, record all freeze data and any on-board snapshot** **Possible causes:** • Blown fuse • Poor connections or loose terminals at the ignition coil and the PCM • "Open" or "Short" circuit between the ignition coils and the ignition coil relay • "Open" or "Short" circuit between the ignition coils and the PCM • "Open" in the wire between the ignition coil and ground • Faulty ignition coil relay • Faulty ignition coil
DTC: P0354 **1T PCM, MIL: Yes** **Year:** 2009, 2010 **Model:** MDX, RL, TL, TSX, ZDX **Engine:** 2.4L L4, 3.5L V6, 3.7L V6 **Transmission:** All	**Cylinder 4 Ignition Coil Circuit Malfunction:** With the engine running, the execution time is continuous and the duration time is 5 seconds or more. (when the engine speed is 700 rpm) The return signal does not change for at least 5 seconds when the ignition coil is triggered. **NOTE: Before you troubleshoot, record all freeze data and any on-board snapshot** **Possible causes:** • Blown fuse • Poor connections or loose terminals at the ignition coil and the PCM • "Open" or "Short" circuit between the ignition coils and the ignition coil relay • "Open" or "Short" circuit between the ignition coils and the PCM • "Open" in the wire between the ignition coil and ground • Faulty ignition coil relay • Faulty ignition coil
DTC: P0355 **1T PCM, MIL: Yes** **Year:** 2010 **Model:** MDX, RL, TL, TSX, ZDX **Engine:** 3.5L V6, 3.7L V6 **Transmission:** All	**Cylinder 5 Ignition Coil Circuit Malfunction:** With the engine running, the execution time is continuous and the duration time is 5 seconds or more. (when the engine speed is 700 rpm) The return signal does not change for at least 5 seconds when the ignition coil is triggered. **NOTE: Before you troubleshoot, record all freeze data and any on-board snapshot** **Possible causes:** • Blown fuse • Poor connections or loose terminals at the ignition coil and the PCM • "Open" or "Short" circuit between the ignition coils and the ignition coil relay • "Open" or "Short" circuit between the ignition coils and the PCM • "Open" in the wire between the ignition coil and ground • Faulty ignition coil relay • Faulty ignition coil

DTC	Trouble Code Title, Conditions & Possible Causes
DTC: P0356 **1T PCM, MIL: Yes** **Year:** 2010 **Model:** MDX, RL, TL, TSX, ZDX **Engine:** 3.5L V6, 3.7L V6 **Transmission:** All	**Cylinder 6 Ignition Coil Circuit Malfunction:** With the engine running, the execution time is continuous and the duration time is 5 seconds or more. (whon the engine speed is 700 rpm) The return signal does not change for at least 5 seconds when the ignition coil is triggered. **NOTE: Before you troubleshoot, record all freeze data and any on-board snapshot** **Possible causes:** • Blown fuse • Poor connections or loose terminals at the ignition coil and the PCM • "Open" or "Short" circuit between the ignition coils and the ignition coil relay • "Open" or "Short" circuit between the ignition coils and the PCM • "Open" in the wire between the ignition coil and ground • Faulty ignition coil relay • Faulty ignition coil
DTC: P0365 **1T PCM, MIL: Yes** **Year:** 2009, 2010 **Model:** RDX, TSX **Engine:** 2.3L L4, 2.4L L4 **Transmission:** All	**Camshaft Position (CMP) Sensor B No Signal:** With the engine running, NO CMP sensor B pulsing signals are detected for at least 50 times in succession. **Possible causes:** • Poor connections or loose terminals at CMP sensor B and the ECM/PCM • "Open" circuit between CMP sensor B and PGM-FI main relay 1 • "Open" or "Short" circuit between the ECM/PCM and CMP sensor B • "Open" circuit between CMP sensor B and ground • Faulty CMP sensor B • PCM may need to be updated with the latest software • Faulty PCM
DTC: P0365 **1T PCM, MIL: Yes** **Year:** 2010 **Model:** TSX, ZDX **Engine:** 3.5L V6, 3.7L V6 **Transmission:** All	**Camshaft Position (CMP) Sensor Circuit No Signal:** With the engine running at 750 rpm, DTCs P0335, P0339 and P0369 are not active. No input signals from the CMP sensor are detected while the signals from the CKP sensor are detected at least 300 times. **NOTE: Before you troubleshoot, record all freeze data and any on-board snapshot** **Possible causes:** • Poor connections or loose terminals at the CMP sensor and the PCM • "Open" circuit between the CMP sensor and PGM-FI main relay 1 • "Open" or "Short" circuit between the CMP and the PCM • "Open" in the wire between the CMP sensor and ground • Faulty CMP sensor • PCM needs to be updated with the latest software • Faulty PCM
DTC: P0369 **1T PCM, MIL: Yes** **Year:** 2009, 2010 **Model:** RDX, TSX **Engine:** 2.3L L4, 2.4L L4 **Transmission:** All	**Camshaft Position (CMP) Sensor B Intermittent Interruption:** With the engine running, more or less than four CMP B pulsing signals are detected during intervals between standard pulses at least 30 times in succession. **NOTE: Before you troubleshoot, record all freeze data and any on-board snapshot, and review the general troubleshooting information.** **Possible causes:** • Poor connections or loose terminals at the engine and/or body ground • Poor connections or loose terminals at CMP sensor B and the ECM/PCM • Damaged CMP pulse plate B • Faulty CMP sensor B
DTC: P0369 **1T PCM, MIL: Yes** **Year:** 2010 **Model:** TSX, ZDX **Engine:** 3.5L V6, 3.7L V6 **Transmission:** All	**Camshaft Position (CMP) Sensor Circuit Intermittent Interruption:** Other than 30 times of CMP pulses are detected during two cycles of CKP. This condition has been detected at least 10 times. **NOTE: Before you troubleshoot, record all freeze data and any on-board snapshot.** **Possible causes:** • Poor connections or loose terminals at the CMP and the PCM • Poor ground connections • Damaged front camshaft pulley (CMP pulse plate) • Faulty CMP sensor

DTC	Trouble Code Title, Conditions & Possible Causes
DTC: P0385 **1T PCM, MIL: Yes** **Year:** 2009, 2010 **Model:** MDX, RL **Engine:** 3.7L V6 **Transmission:** All	**Crankshaft Position (CKP) Sensor B No Signal :** With the engine running the execution is continuous and P0385, P0389 are not active. No signals from CKP sensor A are detected while signals from CKP sensor B are detected 127 times in succession. If no pulsing signals are detected from CKP sensor A, a malfunction is detected and a DTC is stored. **NOTE: Before you troubleshoot, record all freeze data and any on-board** snapshot. **Possible causes:** • Poor connections or loose terminals at CKP sensor A/B and the PCM • "Open" or "Short" in the wire between CKP sensor A/B and the under-hood fuse/relay box (PGM-FI main relay 1) • "Open" or "Short" in the wire between the PCM and CKP sensor A/B, • Faulty CKP sensor A/B • PCM may need to be updated with the latest software • Faulty PCM
DTC: P0389 **1T PCM, MIL: Yes** **Year:** 2009, 2010 **Model:** MDX, RL **Engine:** 3.7L V6 **Transmission:** All	**Crankshaft Position (CKP) Sensor B Intermittent Interruption :** Engine running, other than 22 pulses are detected during intervals between reference pulses for each crankshaft revolution. This condition has been detected at least 30 times. An abnormal amount of pulsing signals are detected from CKP sensor B. **NOTE: Before you troubleshoot, record all freeze data and any on-board snapshot.** **Possible causes:** • Poor connections or loose terminals at CKP sensor A/B and the PCM • Poor grounds at engine, body, CKP sensor A/B or PCM • Damaged CKP sensor A/B pulse plate/timing belt drive pulley • Faulty CKP sensor A/B • PCM may need to be updated with the latest software • Faulty PCM
DTC: P0400 **2T PCM, MIL: Yes** **Year:** 2009, 2010 **Model:** RL, TSX, ZDX **Engine:** 3.5L V6, 3.7L V6 **Transmission:** All	**Exhaust Gas Recirculation (EGR) System Leak Detected:** With engine running, rpm 1,500-3,500 and system in closed loop the A/F sensor response parameter according to the opening and closing of the EGR is 14,746 (LT80A) or less. **NOTE: Before you troubleshoot, record all freeze data and any on-board snapshot.** **Possible causes:** • Loose or damaged EGR pipe • Exhaust gas leakage between the EGR pipe and the EGR valve • Faulty EGR valve
DTC: P0401 **2T PCM, MIL: Yes** **Year:** 2009, 2010 **Model:** MDX, RL, TSX, ZDX **Engine:** 3.5L V6, 3.7L V6 **Transmission:** All	**Exhaust Gas Recirculation (EGR) Insufficient Flow:** Engine temperature is at a minimum of 156 °F (69 °C), engine speed is between 1,100-2,400 rpm with the map value of 14 kPa (4.0 in.Hg, 100 mmHg), throttle fully closed and battery 10.5 volt minimum. The execution time is once per driving cycle and the duration time is 3 seconds or more. The ratio of the current EGR flow to the normal EGR flow is 15% or less for at least 3 seconds. **NOTE: Before you troubleshoot, record all freeze data and any on-board snapshot.** **Possible causes:** • Clogged intake manifold, and/or EGR ports • Poor connections or loose terminals at the EGR valve and the PCM • Faulty EGR valve • PCM may need to be updated with the latest software • Faulty PCM
DTC: P0404 **2T PCM, MIL: Yes** **Year:** 2009, 2010 **Model:** MDX, RL, TSX, ZDX **Engine:** 3.5L V6, 3.7L V6 **Transmission:** All	**Exhaust Gas Recirculation (EGR) Control Circuit Range/Performance Problem :** Vehicle at a speed between 15-75 mph, actual valve lift command at 0.012 in. (0.3 mm). The execution time is once per driving cycle and the duration time is 5 seconds or more. The difference between the command value of the amount of EGR valve lift in the PCM and the actual amount of valve lift is 0.041 in. (1.020 mm) or more for at least 5 seconds. **NOTE: Before you troubleshoot, record all freeze data and any on-board snapshot.** **Possible causes:** • Carbon build-up on the EGR valve • Faulty EGR valve • Poor connections or loose terminals at the EGR valve and the PCM • "Short" or "Short" between the PCM and the EGR valve • "Open" in the wire between the EGR valve and ground • PCM may need to be updated with the latest software • Faulty PCM

DTC	Trouble Code Title, Conditions & Possible Causes
DTC: P0406 **1T PCM, MIL: Yes** **Year:** 2009, 2010 **Model:** MDX, RL, TSX, ZDX **Engine:** 3.5L V6, 3.7L V6 **Transmission:** All	**Exhaust Gas Recirculation (EGR) Valve Position Sensor Circuit High Voltage:** With the engine running the execution time is continuous and the duration time is 2 seconds or more. The EGR valve position sensor output voltage is 4.88 V or more for at least 2 seconds. **NOTE: Before you troubleshoot, record all freeze data and any on-board snapshot.** **Possible causes:** • Poor connections or loose terminals at the EGR valve and the PCM • "Open" circuit between the EGR valve and the PCM • PCM may need to be updated with the latest software • Faulty EGR valve • Faulty PCM
DTC: P0420 **2T PCM, MIL: Yes** **Year:** 2009, 2010 **Model:** RDX, TSX **Engine:** 2.3L L4, 2.4L L4 **Transmission:** All	**Catalytic System Efficiency Below Threshold :** With the vehicle driven to a speed of 16-75 mph at less than 3500 rpm in closed loop for 2-5 minutes, ECT sensor more than 140°F, MAF sensor from 8-50 g/sec, engine load less than 99% (±8%), predicted Catalyst temperature over 750°F, and the PCM detected the catalyst oxygen storage capacity was below an acceptable threshold during the test. **NOTE: If some of the DTCs listed below are stored at the same time as DTC P0420, troubleshoot those DTCs first, then recheck for DTC P0420.** **Possible causes:** • Air leaks in at the exhaust manifold or exhaust pipes • Poor fuel quality • Catalytic converter damaged or has failed (deteriorated) • Front HO2S is more aged than the rear HO2S (HO2S is lazy) • Faulty Secondary HO2S (Sensor 2) • PCM has failed
DTC: P0420 **2T PCM, MIL: Yes** **Year:** 2009, 2010 **Model:** MDX, RL, TSX, ZDX **Engine:** 3.5L V6, 3.7L V6 **Transmission:** All	**Rear Bank Catalyst System Efficiency Below Threshold (Bank 1):** The execution time is once per driving cycle and the duration time is 102 seconds or more. The number of detections is 752 (CTAGLT67) or more. **NOTE: Before you troubleshoot, record all freeze data and any on-board snapshot.** **Possible causes:** • Poor connections or loose terminals at the secondary HO2S (sensor 2) and the PCM • Faulty (Bank 1) WU-TWC
DTC: P0430 **2T PCM, MIL: Yes** **Year:** 2009, 2010 **Model:** MDX, TSX, ZDX **Engine:** 3.5L V6, 3.7L V6 **Transmission:** All	**Front Bank Catalyst System Efficiency Below Threshold (Bank 2) :** The execution time is once per driving cycle and the duration time is 102 seconds or more. The number of detections is 720 (CTAGLT68) or more. **NOTE: Before you troubleshoot, record all freeze data and any on-board snapshot.** **Possible causes:** • Poor connections or loose terminals at the secondary HO2S (sensor 2) and the PCM • Faulty (Bank 2) WU-TWC
DTC: P0443 **1T PCM, MIL: Yes** **Year:** 2009, 2010 **Model:** MDX, RDX, RL, TSX, ZDX **Engine:** 2.3L L4, 2.4L L4, 3.5L V6, 3.7L V6 **Transmission:** All	**Evaporative Emission (EVAP) Canister Purge Valve Circuit Malfunction:** With the engine running, EVAP canister purge valve output duty at 2% to 98% the execution time is continuous and the duration time is 5 seconds or more. When the return signal does not change according to the EVAP canister purge valve duty cycle for a set time, the PCM detects a malfunction. The return signal does not change according to the EVAP canister purge valve output for at least 5 seconds. **NOTE: Before you troubleshoot, record all freeze data and any on-board snapshot.** **Possible causes:** • Poor connections or loose terminals at the EVAP canister purge valve and the PCM • "Open" or "short" between the EVAP canister purge valve and the PCM • "Open" or "short" between the EVAP canister purge valve and the under-dash fuse/relay box • EVAP canister purge valve • PCM may need to be updated with the latest software • Faulty PCM
DTC: P0451 **2T PCM, MIL: Yes** **Year:** 2009, 2010 **Model:** MDX, RDX, RL, TSX, ZDX **Engine:** 2.3L L4, 2.4L L4, 3.5L V6, 3.7L V6 **Transmission:** All	**Fuel Tank Pressure (FTP) Sensor Circuit Range/Performance Problem:** Elapsed time after starting the engine is 2 seconds with the throttle fully closed. The execution time is once per driving cycle and the duration time is 20 seconds or more. The FTP sensor output fluctuates by 0.3 kPa (0.1 in.Hg, 2 mmHg) or more at least five times within 3 seconds. **NOTE: Before you troubleshoot, record all freeze data and any on-board snapshot.** **Possible causes:** • Poor connections or loose terminals at the FTP sensor and the PCM • Faulty FTP sensor

DTC	Trouble Code Title, Conditions & Possible Causes
DTC: P0452 **2T PCM, MIL: Yes** **Year:** 2009, 2010 **Model:** MDX, RDX, RL, TSX, ZDX **Engine:** 2.3L L4, 2.4L L4, 3.5L V6, 3.7L V6 **Transmission:** All	**Fuel Tank Pressure (FTP) Sensor Circuit Low Voltage:** Elapsed time after starting the engine is 2 seconds at idle. The execution time is once per driving cycle and the duration time is 3 seconds or more. The output from the fuel tank pressure sensor is less than -7 kPa (-2.1 in.Hg, -55 mmHg) for at least 3 seconds. **NOTE: Before you troubleshoot, record all freeze data and any on-board snapshot.** **Possible causes:** • Poor connections or loose terminals at the FTP sensor and the PCM • "Open" or "Short" in the wire(s) between the PCM and the FTP sensor • Faulty FTP sensor
DTC: P0453 **2T PCM, MIL: Yes** **Year:** 2009, 2010 **Model:** MDX, RDX, RL, TSX, ZDX **Engine:** 2.3L L4, 2.4L L4, 3.5L V6, 3.7L V6 **Transmission:** All	**Fuel Tank Pressure (FTP) Sensor Circuit High Voltage:** Elapsed time after starting the engine is 2 seconds at idle. The execution time is once per driving cycle and the duration time is 3 seconds or more. If the FTP sensor output voltage is higher than a target value within a set time after starting the engine in a cold condition. The output from the fuel tank pressure sensor is more than 8 kPa (2.2 in.Hg, 55 mmHg) for at least 3 seconds. **NOTE: Before you troubleshoot, record all freeze data and any on-board snapshot.** **Possible causes:** • Poor connections or loose terminals at the FTP sensor and the PCM • "Open" in the wire between the PCM and the FTP sensor • Faulty FTP sensor • PCM may need to be updated with the latest software • Faulty PCM
DTC: P0455 **2T PCM, MIL: Yes** **Year:** 2009, 2010 **Model:** MDX, RDX, TSX, ZDX **Engine:** 2.3L L4, 2.4L L4, 3.5L V6, 3.7L V6 **Transmission:** All	**Evaporative Emission (EVAP) System Large Leak Detected:** The execution time is once per driving cycle and the duration time is 36 minutes and 37 seconds, or less. Here is an overview of the malfunction detection for the EONV method: 1: Judgment of detection of 0.09 inch leak as normal operation 2: Judgment of detection of 0.02 inch leak as normal operation 3: Detection of 0.02 inch leak 4: Detection of atmospheric pressure failure 5: Flickering of the FTP sensor The execution time is once per driving cycle and the duration time is 36 minutes and 37 seconds, or less. The variation of pressure inside the fuel tank is 0.03 kPa (0.009 in.Hg, 0.24 mmHg) or more. **NOTE: Before you troubleshoot, record all freeze data and any on-board snapshot.** **Possible causes:** • Faulty or loose fuel fill cap • Poor connection or damage at the fuel tank vapor control valve hose • Poor connections or loose terminals at the FTP sensor, the EVAP canister • purge valve, or the EVAP canister vent shut valve, and the PCM • Faulty EVAP canister vent shut valve • Faulty FTP sensor O-ring • Faulty fuel tank vapor control valve hose • Faulty EVAP canister vent shut valve case and O-ring • Faulty EVAP canister • Faulty fuel tank unit base gasket, and/or fuel tank

DTC	Trouble Code Title, Conditions & Possible Causes
DTC: P0456 **2T PCM, MIL: Yes** **Year:** 2009, 2010 **Model:** MDX, RDX, RL, TSX **Engine:** 2.3L L4, 2.4L L4, 3.5L V6, 3.7L V6 **Transmission:** All	**EVAP System Very Small Leak Detected :** The execution time is once per driving cycle and the duration time is at least 16 minutes and 37 seconds, but not more than 36 minutes and 37 seconds. Here is an overview of the malfunction detection for the EONV method: 1: Judgment of detection of 0.09 inch leak as normal operation 2: Judgment of detection of 0.02 inch leak as normal operation 3: Detection of 0.02 inch leak 4: Detection of atmospheric pressure failure 5: Flickering of the FTP sensor The execution time is once per driving cycle and the duration time is 36 minutes and 37 seconds, or less. The variation of pressure inside the fuel tank is 0.03 kPa (0.009 in.Hg, 0.24 mmHg) or more. **NOTE: Before you troubleshoot, record all freeze data and any on-board snapshot.** **Possible causes:** • Faulty or loose fuel fill cap • Poor connection or damage at the fuel tank vapor control valve hose • Poor connections or loose terminals at the FTP sensor, the EVAP canister • purge valve, or the EVAP canister vent shut valve, and the PCM • Faulty EVAP canister vent shut valve • Faulty FTP sensor O-ring • Faulty fuel tank vapor control valve hose • Faulty EVAP canister vent shut valve case and O-ring • Faulty EVAP canister • Faulty fuel tank unit base gasket, and/or fuel tank
DTC: P0457 **3T PCM, MIL: Yes** **Year:** 2009, 2010 **Model:** MDX, RDX, TSX, ZDX **Engine:** 2.3L L4, 2.4L L4, 3.5L V6, 3.7L V6 **Transmission:** All	**Evaporative Emission (EVAP) System Leak Detected/Fuel Fill Cap Loose or Missing:** Engine coolant temperature before EVAP purge control starts 140 °F (60 °C), 2 mph (2 km/h), Barometric pressure, 76 kPa (22.5 in.Hg, 569 mmHg), Battery voltage 10.5 volts or more, fuel trim 0.73-1.47, system is in close loop. The execution time is continuous and the duration time is 12 seconds or more. The output from the fuel cap monitor is 0.053 or less for at least 12 seconds (when there is no NG judgment history in this drive cycle). P0455 or P0456 are judged as NG. **NOTE: Before you troubleshoot, record all freeze data and any on-board snapshot.** **Possible causes:** • Faulty fuel fill cap seal missing or damaged, fuel fill pipe damaged • Poor connections or loose terminals at the FTP sensor, the EVAP canister vent shut valve, and the PCM • Faulty routing of the EVAP canister vent tube • Faulty EVAP canister vent shut valve
DTC: P0461 **1T PCM** **Year:** 2009, 2010 **Model:** MDX, RDX, TSX **Engine:** 2.3L L4, 2.4L L4, 3.5L V6, 3.7L V6 **Transmission:** All	**Fuel Level Sensor (Fuel Gauge Sending Unit) Range/Performance Problem :** The execution time is every 125 miles (200 km), DTCs P0462, P0463, U0028, U0155 are not active. If the powertrain control module (PCM) receives no change in the fuel level sensor output after driving for a specified number of miles, it detects a malfunction. The change in the fuel level sensor output is 3.5 % or less. **Possible causes:** • Poor connections or loose terminals at the fuel gauge sending unit and the gauge control module • Faulty fuel gauge sending unit
DTC: P0461 **1T PCM** **Year:** 2009, 2010 **Model:** RL, ZDX **Engine:** 3.7L V6 **Transmission:** All	**Fuel Level Sensor (Fuel Gauge Sending Unit) Circuit Range/Performance Problem :** The execution time is every 125 miles (200 km).The change in the fuel level sensor output is 3.5 % or less. **NOTE: Because it requires 162 miles (260 km) of driving without refueling to complete this diagnosis, DTC P0461 cannot be duplicated during this troubleshooting.** **Possible causes:** • Poor connections or loose terminals at the fuel gauge sending unit and the gauge control module • Faulty fuel gauge sending unit
DTC: P0462 **1T PCM** **Year:** 2009, 2010 **Model:** MDX, RDX, RL, TSX, ZDX **Engine:** 2.3L L4, 2.4L L4, 3.5L V6, 3.7L V6 **Transmission:** All	**Fuel Level Sensor (Fuel Gauge Sending Unit) Circuit Low Voltage:** The fuel level sensor (fuel gauge sending unit) output voltage is 0.10 V or less for at least 5 seconds. The execution time is continuous DTCs P0463, U0155 are not active. **NOTE: Before you troubleshoot, record all freeze data and any on-board snapshot.** **Possible causes:** • Poor connections or loose terminals at the gauge control module, the fuel gauge sending unit, and the secondary fuel gauge sending unit • "Short" circuit between the gauge control module (signal line) and the fuel gauge sending unit • Faulty fuel gauge sending unit • Faulty gauge control module • PCM may need to be updated with the latest software • Faulty PCM

DTC	Trouble Code Title, Conditions & Possible Causes
DTC: P0463 **1T PCM** **Year:** 2009, 2010 **Model:** MDX, RDX, RL, TSX, ZDX **Engine:** 2.3L L4, 2.4L L4, 3.5L V6, 3.7L V6 **Transmission:** All	**Fuel Level Sensor (Fuel Gauge Sending Unit) Circuit High Voltage :** The execution time is every 125 miles (200 km), DTCs P0462, P0463, U0028, U0155 are not active. If the Powertrain Control Module (PCM) receives no change in the fuel level sensor output after driving for a specified number of miles, it detects a malfunction. The fuel level sensor output voltage is 4.92 V or more for at least 5 seconds. **Possible causes:** • Poor connections or loose terminals at the gauge control module and the fuel gauge sending unit • "Open" in the wire between the gauge control module (GND line) and the fuel gauge sending unit • "Open" in the wire between the gauge control module (signal line) and the fuel gauge sending unit • Faulty fuel gauge sending unit • Faulty gauge control module • PCM may need to be updated with the latest software • Faulty PCM
DTC: P0480 **1T PCM** **Year:** 2009, 2010 **Model:** RL **Engine:** 3.7L V6 **Transmission:** All	**Radiator Fan Control (RFC) System Malfunction:** With ignition switch ON (II) the RFC terminal voltage is low for at least 20 seconds, and the same condition continues even when the power reset is repeated 5 times. (Power is supplied for about 5 seconds after the power reset.) **NOTE: Before you troubleshoot, record all freeze data and any on-board snapshot.** **Possible causes:** • Blown fuse • Radiator fan interference, or debris in the fan • "Open" circuit between the PCM and the RFC unit • "Short" circuit between the RFC unit relay and the fuse • Faulty fan motor • Faulty RFC unit • PCM may need to be updated with the latest software • Faulty PCM
DTC: P0496 **2T PCM, MIL: Yes** **Year:** 2009, 2010 **Model:** MDX, RDX, RL, TSX, ZDX **Engine:** 2.3L L4, 2.4L L4, 3.5L V6, 3.7L V6 **Transmission:** All	**Evaporative Emission (EVAP) System High Purge Flow Detected:** Monitor execution is once per driving cycle. The output from the EVAP canister purge valve is 0.2 kPa (0.07 in.Hg, 2 mmHg) or more for at least 10 seconds. **NOTE: Before you troubleshoot, record all freeze data and any on-board snapshot.** **Possible causes:** • Poor connections or loose terminals at the FTP sensor, the EVAP canister purge valve, the EVAP canister vent shut valve, and the PCM • Faulty EVAP canister purge valve
DTC: P0497 **2T PCM, MIL: Yes** **Year:** 2009, 2010 **Model:** MDX, RDX, RL, TSX, ZDX **Engine:** 2.3L L4, 2.4L L4, 3.5L V6, 3.7L V6 **Transmission:** All	**Evaporative Emission (EVAP) System Low Purge Flow Detected:** The execution time is once per driving cycle and P145C is judged as NG. Enable conditions are low load duration time 10 seconds, Wait for 10 seconds after the ignition switch is turned to LOCK (0) and Engine coolant temperature before EVAP purge control starts is 140 °F (60 °C). The output from the fuel tank pressure sensor is 0.2 kPa (0.07 in.Hg, 2 mmHg) or less for at least 10 seconds. **NOTE: Before you troubleshoot, record all freeze data and any on-board snapshot.** **Possible causes:** • Faulty or Loose fuel fill cap • Poor connections or loose terminals at the FTP sensor, the EVAP canister purge valve, the EVAP canister vent shut valve, and the PCM • Blockage in the vacuum hose between the EVAP canister purge valve and the EVAP canister • Faulty EVAP canister purge valve
DTC: P0498 **1T PCM, MIL: Yes** **Year:** 2009, 2010 **Model:** RDX, TSX **Engine:** 2.3L L4, 2.4L L4, 3.5L V6 **Transmission:** All	**Evaporative Emission (EVAP) Canister Vent Shut Valve Control Circuit Low Voltage:** The execution time is continuous and the duration time is 5 seconds or more. If the return signal is OFF (Low) when the Powertrain Control Module (PCM) outputs the ON signal to the EVAP canister vent shut valve, the PCM detects a malfunction. The return signal is Low for at least 5 seconds when the PCM outputs the ON signal to the EVAP canister vent shut valve. **Possible causes:** • Poor connections or loose terminals at the EVAP canister vent shut valve and the PCM • "Open" or "Short" in the wire between the EVAP canister vent shut valve and the PCM • "Open" in the wire between the EVAP canister vent shut valve and the A/F relay • Faulty EVAP canister vent shut valve

DTC	Trouble Code Title, Conditions & Possible Causes
DTC: P0498 **1T PCM, MIL: Yes** **Year:** 2009, 2010 **Model:** MDX, RL, ZDX **Engine:** 3.7L V6 **Transmission:** All	**Evaporative Emission (EVAP) Canister Vent Shut Valve Control Circuit Low Voltage:** The execution time is continuous and the duration time is 5 seconds or more. If the return signal is OFF when the powertrain control module (PCM) outputs the ON signal to the EVAP canister vent shut valve, the PCM detects a malfunction. DTC P0499 is not active. The return signal is Low for at least 5 seconds when the PCM outputs the ON signal to the EVAP canister vent shut valve. **Possible causes:** • Poor connections or loose terminals at the EVAP canister vent shut valve and the PCM • "Open" or "Short" in the wire between the EVAP canister vent shut valve and the PCM • "Open" in the wire between the EVAP canister vent shut valve and the under-hood fuse/relay box (PGM-FI sub-relay) • Faulty EVAP canister vent shut valve
DTC: P0499 **1T PCM, MIL: Yes** **Year:** 2009, 2010 **Model:** MDX, RDX, RL, TSX, ZDX **Engine:** 2.3L L4, 2.4L L4, 3.5L V6, 3.7L V6 **Transmission:** All	**Evaporative Emission (EVAP) Canister Vent Shut Valve Control Circuit High Voltage:** If the return signal is ON when the Powertrain Control Module (PCM) outputs the OFF signal to the EVAP canister vent shut valve, the PCM detects a malfunction. The execution time is continuous and the duration time is 5 seconds or more. DTC P0498 is not active. The return signal is ON for at least 5 seconds when the PCM outputs the Low signal to the EVAP canister vent shut valve. **NOTE: Before you troubleshoot, record all freeze data and any on-board snapshot.** **Possible causes:** • Poor connections or loose terminals at the EVAP canister vent shut valve and the PCM • PCM may need to be updated with the latest software • Faulty PCM • Faulty EVAP canister vent shut valve
DTC: P0506 **2T PCM, MIL: Yes** **Year:** 2009, 2010 **Model:** MDX, RDX, TSX, ZDX **Engine:** 2.3L L4, 2.4L L4, 3.5L V6, 3.7L V6 **Transmission:** All	**Idle Control System RPM Lower Than Expected:** The execution time is once per driving cycle and the duration time is 20 seconds or more. Throttle fully closed, fuel trim 0.75-1.47, intake air temperature 19 °F (-7 °C) and battery voltage 10.5 or more. If the actual idle speed varies beyond a specified value from the target speed over a certain period of time, the PCM detects a malfunction. The actual idle speed is at least 100 rpm less than the target idle speed for at least 20 seconds. **NOTE: Before you troubleshoot, record all freeze data and any on-board snapshot.** **Possible causes:** • Dirt, carbon, or damage in the throttle bore • Damaged air cleaner element • Incorrect DATA LIST parameter conditions • Poor connections or loose terminals at the throttle body and the PCM • Faulty throttle body • PCM may need to be updated with the latest software • Faulty PCM
DTC: P0507 **2T PCM, MIL: Yes** **Year:** 2009, 2010 **Model:** MDX, RDX, RL, TSX, ZDX **Engine:** 2.3L L4, 2.4L L4, 3.5L V6, 3.7L V6 **Transmission:** All	**Idle Control System RPM Higher Than Expected:** Enable conditions are as follows: coolant temperature 156 °F (69 °C) minimum, intake air temperature 19 °F (-7 °C), fuel trim 0.73-1.47, throttle closed and battery voltage at least 10.5 volts. The execution time is once per driving cycle and the duration time is 20 seconds or more. If the actual idle speed varies beyond a specified value from the target speed over a certain period of time, the PCM detects a malfunction. The actual idle speed is at least 200 rpm greater than the target idle speed for at least 20 seconds. **NOTE: Before you troubleshoot, record all freeze data and any on-board snapshot.** **Possible causes:** • Vacuum leaks • Faulty PCV valve • Dirty throttle bore • Faulty EVAP canister purge valve • Poor connections or loose terminals at the throttle body and the PCM • Improper PCM idle learn procedure • PCM may need to be updated with the latest software • Faulty PCM

DTC	Trouble Code Title, Conditions & Possible Causes
DTC: P050A **2T PCM, MIL: Yes** **Year:** 2009, 2010 **Model:** MDX, RDX, RL, TSX, ZDX **Engine:** 2.3L L4, 2.4L L4, 3.5L V6, 3.7L V6 **Transmission:** All	**Cold Start Idle Air Control System Performance Problem:** The execution time is once per driving cycle and the duration time is 10 seconds or more. When the actual amount of air is less than the target amount, a malfunction is detected. The total airflow is decreased by a factor of 0.693 for at least 10 seconds. **NOTE: Before you troubleshoot, record all freeze data and any on-board snapshot.** **Possible causes:** • Dirty air cleaner element • Damage air cleaner element or housing • Dirty or damaged throttle bore • Poor connections or loose terminals at the throttle body, the MAF sensor/IAT sensor • Faulty throttle body • Faulty MAF sensor/IAT sensor
DTC: P050B **2T PCM, MIL: Yes** **Year:** 2009, 2010 **Model:** MDX, RL, TSX, ZDX **Engine:** 2.4L L4, 3.5L V6, 3.7L V6 **Transmission:** All	**Cold Start Ignition Timing Performance Problem:** The execution time is once per driving cycle and the duration time is 3.5 seconds or more and the engine speed is 2,100 rpm or more for at least 3.5 seconds. When the actual engine speed is a specified value or more, and it continues for a specified time, a malfunction is detected. **NOTE: Before you troubleshoot, record all freeze data and any on-board snapshot.** **Possible causes:** • Poor connections or blockage at the intake air duct • Damage at the air cleaner housing • Dirt or debris in the air cleaner element • Incorrect ignition timing • Faulty Throttle body • Faulty MAF sensor/IAT sensor • Poor connections or loose terminals at the CKP sensor, the throttle body, the MAF sensor/IAT sensor, ECT sensor 1, ECT sensor 2, and the PCM • Low engine coolant • Faulty ECT sensor 1, and/or ECT sensor 2
DTC: P050B **2T PCM, MIL: Yes** **Year:** 2009, 2010 **Model:** RDX **Engine:** 2.3L L4 **Transmission:** All	**Cold Start Ignition Timing Control System Performance Problem:** With the vehicle at idle and throttle position fully closed, the engine speed is 2,150 rpm or more for at least 3.5 seconds. **NOTE: Before you troubleshoot, record all freeze data and any on-board snapshot, and review the general troubleshooting information.** **Possible causes:** • Poor connections or blockage at the air intake duct • Damaged air cleaner housing or dirty air cleaner • Damaged CKP sensor and/or the CKP sensor pulser plate • Faulty throttle body • Dirty or faulty MAF sensor/IAT sensor 1 • Faulty ECT SENSOR 1 and/or ECT SENSOR 2 • Check and repair any problems with the following items, Engine compression, VTEC system, Engine oil, A/C system, Power steering system • PCM may need to be updated with the latest software • Faulty PCM
DTC: P0532 **2T PCM** **Year:** 2009, 2010 **Model:** RDX, TSX **Engine:** 2.3L L4, 2.4L L4, 3.5L V6 **Transmission:** All	**A/C Pressure Sensor Circuit Low Voltage:** Ignition switch ON (II), DTC P0533 is not active. The A/C pressure sensor output voltage is 0.24 V for at least 10 seconds. **NOTE: Before you troubleshoot, record all freeze data and any on-board snapshot, and review the general troubleshooting information.** **Possible causes:** • Poor connections or loose terminals at the A/C pressure sensor and the ECM/PCM • "Open" or "Short between the ECM/PCM and the A/C pressure sensor • Faulty A/C pressure sensor • PCM may need to be updated with the latest software • Faulty PCM

DTC	Trouble Code Title, Conditions & Possible Causes
DTC: P0533 **1T PCM** **Year:** 2009, 2010 **Model:** RDX **Engine:** 2.3L L4 **Transmission:** All	**A/C Pressure Sensor Circuit High Voltage:** With the ignition switch on, the A/C pressure sensor output voltage is 4.74 V or more for at least 10 seconds. DTC P0532 is not active. **NOTE: Before you troubleshoot, record all freeze data and any on-board snapshot, and review the general troubleshooting information.** **Possible causes:** • Poor connections or loose terminals at the A/C pressure sensor and the PCM • "Open" circuit between the PCM and the A/C pressure sensor • Faulty A/C pressure sensor • PCM may need to be updated with the latest software • Faulty PCM
DTC: P0562 **1T PCM** **Year:** 2009, 2010 **Model:** MDX, RDX, RL, TSX, ZDX **Engine:** 2.3L L4, 2.4L L4, 3.5L V6, 3.7L V6 **Transmission:** All	**Charging System Low Voltage:** The execution time is continuous and the duration time is 60 seconds or more, engine speed 550 rpm. When the IGP (power source) terminal voltage is a set value or less for a set time, the PCM detects a malfunction. The IGP terminal voltage is 11.0 V or less for at least 60 seconds. **NOTE: Before you troubleshoot, record all freeze data and any on-board snapshot. If any high current load accessories are installed, this DTC can be set.** **Possible causes:** • Faulty battery, or connections • Faulty alternator • Poor connections or loose terminals at the alternator and the main under-hood fuse box
DTC: P0563 **1T PCM, MIL: Yes** **Year:** 2009, 2010 **Model:** MDX, RL, ZDX **Engine:** 3.7L V6 **Transmission:** All	**Powertrain Control Module (PCM) Power Source Circuit Unexpected Voltage:** The PCM operates for at least 5 seconds after the ignition switch is turned to LOCK (0). Battery voltage 10.1 volts. (IGP terminal of PCM) **NOTE: Before you troubleshoot, record all freeze data and any on-board snapshot.** **Possible causes:** • Faulty PGM-FI main relay 1 • Poor connections or loose terminals under-hood fuse/relay box (PGM-FI main relay 1) and the fuse in the under-hood fuse/relay box and the PCM • "Short" to power in the wire between the PCM and under-hood fuse/relay box (PGM-FI main relay 1) • PCM may need to be updated with the latest software • Faulty PCM
DTC: P0563 **1T PCM, MIL: Yes** **Year:** 2009, 2010 **Model:** RDX, TSX **Engine:** 2.3L L4, 2.4L L4, 3.5L V6 **Transmission:** All	**Engine Control Module (ECM) Powertrain Control Module (PCM) Power Source Circuit Unexpected Voltage:** The ECM/PCM operates for at least 5 seconds after the ignition switch is turned to LOCK (0). Battery voltage 10.1 volts. (IGP terminal of PCM) **NOTE: Before you troubleshoot, record all freeze data and any on-board snapshot.** **Possible causes:** • Faulty PGM-FI main relay 1 • Poor connections or loose terminals under-hood fuse/relay box (PGM-FI main relay 1) and the fuse • "Short" to power in the wire between the PCM (PGM-FI main relay 1) • ECM/PCM may need to be updated with the latest software • Faulty ECM/PCM
DTC: P0602 **1T PCM, MIL: Yes** **Year:** 2009, 2010 **Model:** MDX, RDX, RL, TSX, ZDX **Engine:** 2.3L L4, 2.4L L4, 3.5L V6, 3.7L V6 **Transmission:** All	**ECM/PCM Programming Error:** With ignition on the execution time is continuous and the duration time is 1 second or less. The ECM/PCM program update stops 1 second before it is finished. **NOTE: This DTC is indicated when a PCM update is not completed.** WARNING: Do not turn the ignition switch to LOCK (0) or ACC (I) while updating the PCM. If you turn the ignition switch to LOCK (0) before completion, the ECM/PCM can be damaged. **Possible causes:** • ECM/PCM needs to be updated with the latest software • Faulty ECM/PCM
DTC: P0606 **1T PCM, MIL: Yes** **Year:** 2009, 2010 **Model:** RDX, TSX **Engine:** 2.3L L4, 2.4L L4, 3.5L V6 **Transmission:** All	**ECM/PCM Processor Malfunction:** After 30 seconds have elapsed since start-up, or after the engine speed exceeds 1,000 rpm once. No signal from the DKS CPU is detected or is abnormal for at least 5 seconds. **NOTE: Before you troubleshoot, record all freeze data and any on-board snapshot, and review the general troubleshooting information.** **Possible causes:** • ECM/PCM may need to be updated with the latest software • Faulty ECM/PCM

DTC	Trouble Code Title, Conditions & Possible Causes
DTC: P060A **1T PCM, MIL: Yes** **Year:** 2009, 2010 **Model:** RDX, ZDX **Engine:** 2.3L L4, 3.7L V6 **Transmission:** All	**Powertrain Control Module (PCM) Internal Control Module Malfunction:** With the ignition ON and battery voltage at a minimum of 10. 0 volts. The execution time is continuous and the duration time is 200 milliseconds or more. With keyless access system 110 milliseconds or more. One of these 2 symptoms occurs: (1) Symptom 1: The internal communication between the FI CPU and the A/T CPU is abnormal or the internal communication is interrupted for at least 200 milliseconds. (2) Symptom 2: The watchdog timer that monitors the A/T CPU detects an abnormality, and the FI CPU receives the A/T CPU check signal for at least 110 milliseconds. **Possible causes:** • PCM needs to be updated with the latest software • Faulty PCM
DTC: P060A **1T PCM, MIL: Yes** **Year:** 2009, 2010 **Model:** MDX, TSX **Engine:** 2.4L L4, 3.5L V6, 3.7L V6 **Transmission:** All	**Powertrain Control Module (PCM) (A/T System) Internal Control Module Malfunction:** (Symptom 1) The execution time is continuous and the duration time 500 milliseconds or more The serial communication between the FI CPU and A/T CPU is abnormal or the serial communication is interrupted for at least 500 milliseconds. (Symptom 2) The execution time is continuous and the duration time , 30 milliseconds or more The watchdog timer that monitors the A/T CPU detects an abnormality, and the FI CPU receives the A/T CPU check signal for at least 30 milliseconds. **NOTE: Before you troubleshoot, record all freeze data and any on-board snapshot.** **Possible causes:** • PCM needs to be updated with the latest software • Faulty PCM
DTC: P060F **1T PCM, MIL: Yes** **Year:** 2009, 2010 **Model:** TSX **Engine:** 2.4L L4, 3.5L V6 **Transmission:** All	**Powertrain Control Module (PCM) Internal Control Module Keep Alive Memory (KAM) Error :** With the ignition switch ON a malfunction is detected whenever the keep alive data retrieval and writing process is not completed normally. **NOTE: Before you troubleshoot, record all freeze data and any on-board snapshot with the HDS, and review General Troubleshooting Information.** **Possible causes:** • Poor connections or loose terminals at the PCM • PCM may need to be updated with the latest software • Faulty PCM
DTC: P0615 **1T PCM** **Year:** 2009, 2010 **Model:** MDX, ZDX **Engine:** 3.7L V6 **Transmission:** All	**Starter Cut Relay STRLD Circuit Malfunction:** With the ignition switch ON (II). The diagnosis line (STRLD) input voltage is between 2.4 V to 2.6 V for at least 1 second. The execution time is continuous and the duration time is 1 seconds or more. **NOTE: Before you troubleshoot, record all freeze data and any on-board snapshot.** **Possible causes:** • Blown fuse • Poor connections or loose terminals at starter cut relay 1, starter cut relay 2, and the PCM • "Open" circuit between the PCM and starter cut relay 1 • PCM needs to be updated with the latest software • Faulty PCM
DTC: P0627 **1T PCM, MIL: Yes** **Year:** 2009, 2010 **Model:** MDX, RDX, RL, ZDX **Engine:** 2.3L L4, 3.7L V6 **Transmission:** All	**Fuel Pump Control Module System Malfunction:** When the diagnosis signal from the fuel pump control module system is low for a set time or more, the PCM detects a malfunction. The execution time is continuous, diagnosis signal is low for at least 2 seconds. **NOTE: Before you troubleshoot, record all freeze data and any on-board snapshot.** **Possible causes:** • Blown Fuse • Faulty PGM-FI main relay 2 • Poor connections or loose terminals at the fuel pump control module, PGM-FI main relay 2, the fuel tank unit, and the PCM • "Open" in the wire between the under-hood fuse/relay box and the under-dash fuse/relay box • Faulty relay control module (under-hood fuse/relay box)

DTC	Trouble Code Title, Conditions & Possible Causes
DTC: P062F **1T PCM, TCIL: Yes** **Year:** 2009, 2010 **Model:** MDX, RDX, RL, ZDX **Engine:** 2.3L L4, 3.7L V6 **Transmission:** All	**ECM/PCM Internal Control Module Keep Alive Memory (KAM) Error :** A malfunction is detected whenever the keep alive data retrieval and writing process is not complotod normally **NOTE: Before you troubleshoot, record all freeze data and any on-board snapshot.** **Possible causes:** • ECM/PCM needs to be updated with the latest software • Faulty ECM/PCM
DTC: P0630 **1T PCM, MIL: Yes** **Year:** 2009, 2010 **Model:** MDX, RDX, RL, TSX, ZDX **Engine:** 2.3L L4, 2.4L L4, 3.5L V6, 3.7L V6 **Transmission:** All	**VIN Not Programmed or Mismatch:** The VIN is not registered in the keep-alive memory in the PCM. **NOTE: Before you troubleshoot, record all freeze data and any on-board snapshot.** **Possible causes:** • ECM/PCM needs to be updated with the latest software • Faulty ECM/PCM
DTC: P0641 **1T PCM, MIL: Yes** **Year:** 2010 **Model:** ZDX **Engine:** 3.7L V6 **Transmission:** All	**Sensor Reference Voltage A Malfunction:** With the ignition ON the execution time is continuous and the sensor power voltage is 5.2 V or more, or 4.5 V or less, for at least 2.0 seconds. **NOTE: Before you troubleshoot, record all freeze data and any on-board snapshot. It may be possible to locate the fault by disconnecting one component at a time from the 5-volt reference circuit while viewing the 5-Volt Reference circuit parameter on the scan tool. The scan tool parameter would change from Fault to OK when the source of the fault is disconnected. If all 5-volt reference components have been disconnected and a Fault is still indicated, the fault may exist in the wiring harness.** **Possible causes:** • Intermittent condition • "Open" or "Short" circuit in the following 5-volt reference circuits, APP sensor, Throttle body, Input shaft (mainshaft) speed sensor • PCM needs to be updated with the latest software • Faulty PCM
DTC: P0651 **1T VCM, MIL: Yes** **Year:** 2010 **Model:** ZDX **Engine:** 3.7L V6 **Transmission:** All	**Sensor Reference Voltage B Malfunction:** With the ignition ON the execution time is continuous and the sensor power voltage is 5.2 V or more, or 4.5 V or less, for at least 2.0 seconds. **NOTE: Before you troubleshoot, record all freeze data and any on-board snapshot. It may be possible to locate the fault by disconnecting one component at a time from the 5-volt reference circuit while viewing the 5-Volt Reference circuit parameter on the scan tool. The scan tool parameter would change from Fault to OK when the source of the fault is disconnected. If all 5-volt reference components have been disconnected and a Fault is still indicated, the fault may exist in the wiring harness.** **Possible causes:** • Intermittent condition • "Open" or "Short" circuit in the following 5-volt reference circuits, APP sensor, FTP sensor, IMT actuator, MAP sensor, EGR valve, Output shaft (countershaft) speed sensor • PCM needs to be updated with the latest software • Faulty PCM
DTC: P0685 **2T PCM, MIL: Yes** **Year:** 2009, 2010 **Model:** TSX **Engine:** 2.4L L4, 3.5L V6 **Transmission:** All	**ECM/PCM Power Control Circuit Malfunction:** When the voltage to the ECM/PCM is turned off and the ECM/PCM shuts down without the normal shut down procedure, a malfunction in the PGM-FI main relay 1 control circuit is detected. **NOTE: Before you troubleshoot, record all freeze data and any on-board snapshot.** **Possible causes:** • ECM/PCM needs to be updated with the latest software • Faulty ECM/PCM
DTC: P0685 **2T PCM, MIL: Yes** **Year:** 2009, 2010 **Model:** MDX, RDX, RL, ZDX **Engine:** 2.3L L4, 3.7L V6 **Transmission:** All	**Powertrain Control Module (PCM) Power Control Circuit/Internal Circuit Malfunction:** When the voltage to the PCM is turned off and the PCM shuts down without the normal shut down procedure, a malfunction in the PGM-FI main relay 1 control circuit is detected. **NOTE: Before you troubleshoot, record all freeze data and any on-board snapshot.** **Possible causes:** • Loose terminals at the IGP line connectors • PCM needs to be updated with the latest software • Faulty PCM

DTC	Trouble Code Title, Conditions & Possible Causes
DTC: P0705 **1T PCM, MIL: Yes, TCIL: Yes** **Year:** 2009, 2010 **Model:** RDX **Engine:** 2.3L L4 **Transmission:** All	**Transmission Range Switch Circuit (Multiple Shift-position Input):** Malfunction 1: The PCM detects the selected range switch input and another range switch (except L2 switch) input simultaneously for at least 1 second. Malfunction 2: The PCM detects the P,R or N range switch input and the L2 switch input simultaneously for at least 1 second. **NOTE: Record all freeze data and review General Troubleshooting Information before you troubleshoot. This code is caused by an electrical circuit problem and cannot be caused by a mechanical problem in the transmission.** **Possible causes:** • Intermittent "Short" in the wires between the transmission range switch and PCM • "Short" to ground in the wire between PCM connector terminal and the transmission range switch • "Open" in the wire between PCM connector terminals and ground • Faulty transmission range switch • PCM needs to be updated with the latest software • Faulty PCM
DTC: P0705 **1T PCM, MIL: Yes, TCIL: Yes** **Year:** 2009, 2010 **Model:** MDX, RL, TSX, ZDX **Engine:** 2.4L L4, 3.5L V6, 3.7L V6 **Transmission:** All	**Short in Transmission Range Switch Circuit (Multiple Shift-position Input):** One of 3 conditions occurs: (1) The PCM detects the selected range switch input and another range switch input simultaneously for at least 1 seconds. (2) The PCM detects the P, R, or N range switch input and the FWD switch input simultaneously for at least 1 seconds (3) The PCM detects the D or D3 range switch input and the RVS switch input simultaneously for at least 1 second. **NOTE: Before you troubleshoot, record all freeze data and any on-board** snapshot. This code is caused by an electrical circuit problem and cannot be caused by a mechanical problem in the transmission. **Possible causes:** • Intermittent "Short" in the wire between the transmission range switch and the PCM • Faulty transmission range switch • PCM needs to be updated with the latest software • Faulty PCM
DTC: P0706 **2T PCM, MIL: Yes** **Year:** 2009, 2010 **Model:** MDX, RDX, RL, ZDX **Engine:** 2.3L L4, 3.7L V6 **Transmission:** All	**Open in Transmission Range Switch Circuit:** No FWD position signal is detected when the vehicle speed changes from 6 mph (10 km/h) 25 mph (40 km/h) 6 mph (10 km/h) in D or D3. DTCs P0705, P0721, P0722 are not active. **NOTE: This code is caused by an electrical circuit problem and cannot be caused by a mechanical problem in the transmission.** **Possible causes:** • Poor connections or loose terminals at the transmission range switch and the PCM • "Open" in the wire between the transmission range switch and ground • "Open" in the wire between the transmission range switch and PCM • Faulty transmission range switch • PCM needs to be updated with the latest software • Faulty PCM
DTC: P0711 **1T PCM, TCIL: Yes** **Year:** 2009, 2010 **Model:** MDX, RDX, RL, TSX, ZDX **Engine:** 2.3L L4, 2.4L L4, 3.5L V6, 3.7L V6 **Transmission:** All	**Problem in ATF Temperature Sensor Circuit:** The ATF temperature sensor signal does not change. Stuck at low temperature or stuck at high temperature is detected. **NOTE: This code is caused by an electrical circuit problem and cannot be caused by a mechanical problem in the transmission.** **Possible causes:** • Faulty ATF temperature sensor or temperature sensor/shift solenoid harness • Poor connections or loose terminals between the ATF temperature sensor and the PCM • PCM needs to be updated with the latest software • Faulty PCM
DTC: P0712 **1T PCM, TCIL: Yes** **Year:** 2009, 2010 **Model:** MDX, RDX, RL, TSX, ZDX **Engine:** 2.3L L4, 2.4L L4, 3.5L V6, 3.7L V6 **Transmission:** All	**Short in ATF Temperature Sensor Circuit:** When the ATF temperature sensor signal voltage to the PCM is under the specification, indicating that the temperature is above the specification (a short to ground), a malfunction is detected. The ATF temperature sensor output voltage is less than 0.07 V for at least 10 seconds. **NOTE: This code is caused by an electrical circuit problem and cannot be caused by a mechanical problem in the transmission.** **Possible causes:** • "Short" between the ATF temperature sensor and the PCM • Faulty ATF temperature sensor or temperature sensor/shift solenoid harness • PCM needs to be updated with the latest software • Faulty PCM

DTC	Trouble Code Title, Conditions & Possible Causes
DTC: P0713 **1T PCM, TCIL: Yes** **Year:** 2009, 2010 **Model:** MDX, RDX, RL, TSX, ZDX **Engine:** 2.3L L4, 2.4L L4, 3.5L V6, 3.7L V6 **Transmission:** All	**Open in ATF Temperature Sensor Circuit:** When the ATF temperature sensor signal voltage to the PCM is above the specification, indicating that the temperature is under the specification (open), a malfunction is detected. The ATF temperature sensor output voltage is 4.93 V or more for at least 10 seconds. **NOTE: This code is caused by an electrical circuit problem and cannot be caused by a mechanical problem in the transmission.** **Possible causes:** • Poor connections or loose terminals at the ATF temperature sensor and the PCM • "Open" circuit between PCM connector terminal and the ATF temperature sensor • Faulty ATF temperature sensor or temperature sensor/shift solenoid harness • PCM needs to be updated with the latest software • Faulty PCM
DTC: P0716 **1T PCM, MIL: Yes, TCIL: Yes** **Year:** 2009, 2010 **Model:** MDX, RDX, RL, TSX, ZDX **Engine:** 2.3L L4, 2.4L L4, 3.5L V6, 3.7L V6 **Transmission:** All	**Problem in Input Shaft (Mainshaft) Speed Sensor Circuit:** If no pulses occur with the input shaft (mainshaft) rotating, the PCM detects a malfunction that may be caused by an open, a temporary open, or a short to ground. The vehicle speed measured by the input shaft (mainshaft) speed sensor/(divided by) the vehicle speed measured by the output shaft (countershaft) speed sensor is less than 0.156 for at least 10 seconds. **NOTE: This code is caused by an electrical circuit problem and cannot be caused by a mechanical problem in the transmission.** **Possible causes:** • Loose or poor connections at the PCM and input shaft (mainshaft) speed sensor connectors • Poor grounds • Faulty or improperly installed Input Shaft (Mainshaft) Speed Sensor • PCM needs to be updated with the latest software • Faulty PCM
DTC: P0717 **1T PCM, MIL: Yes, TCIL: Yes** **Year:** 2009, 2010 **Model:** MDX, RDX, RL, TSX **Engine:** 2.3L L4, 2.4L L4, 3.5L V6, 3.7L V6 **Transmission:** All	**Problem in Input Shaft (Mainshaft) Speed Sensor Circuit (No Signal Input):** If no pulses occur with the input shaft (mainshaft) rotating, the PCM detects a malfunction that may be caused by an open, a temporary open, or a short to ground. When the vehicle speed measured by the output shaft (countershaft) speed sensor is 13 mph (20 km/h) or more, the vehicle speed measured by the input shaft (mainshaft) speed sensor is 1 mph (2 km/h) or less for at least 10 seconds. **NOTE: This code is caused by an electrical circuit problem and cannot be caused by a mechanical problem in the transmission.** **Possible causes:** • Loose or poor connections at the PCM and input shaft (mainshaft) speed sensor connectors • "Open" in the wires between PCM connector terminals and ground (G101), or repair poor ground • Faulty or improperly installed Input Shaft (Mainshaft) Speed Sensor • PCM needs to be updated with the latest software • Faulty PCM
DTC: P0718 **2T PCM, MIL: Yes, TCIL: Yes** **Year:** 2009, 2010 **Model:** MDX, RDX, RL, TSX, ZDX **Engine:** 2.3L L4, 2.4L L4, 3.5L V6, 3.7L V6 **Transmission:** All	**Input Shaft (Mainshaft) Speed Sensor Intermittent Failure:** If no pulses occur with the input shaft (mainshaft) rotating, the PCM detects a malfunction that may be caused by an open, a temporary open, or a short to ground. The fluctuation of the vehicle speed measured by the input shaft (mainshaft) speed sensor in 10 milliseconds is 4 mph (6km/h) or more, and it fluctuates at least six times within 500 milliseconds. **NOTE: This code is caused by an electrical circuit problem and cannot be caused by a mechanical problem in the transmission.** **Possible causes:** • Poor connections or loose terminals at the input shaft (mainshaft) speed sensor and the PCM • "Open" or "Short" in the wire between PCM connector terminal and the input shaft (mainshaft) speed sensor connector • "Open" in the wires between PCM connector terminals and ground • Faulty input shaft (mainshaft) speed sensor • PCM needs to be updated with the latest software • Faulty PCM

DTC	Trouble Code Title, Conditions & Possible Causes
DTC: P0720 **2T PCM, MIL: Yes** **Year:** 2009, 2010 **Model:** RDX, TSX **Engine:** 2.3L L4, 2.4L L4 **Transmission:** All	**Output Shaft (Countershaft) Speed Sensor Circuit Malfunction :** With the engine speed at a minimum of 4,000 rpm and during fuel cut-off operation for deceleration. NO signal from the output shaft (Countershaft) speed sensor is detected for at least 5 seconds. **NOTE: Before you troubleshoot, record all freeze data and any on-board snapshot, and review the general troubleshooting information.** **Possible causes:** • Poor connections or loose terminals at the output shaft (countershaft) speed sensor and the ECM • "Open" or "Short" circuit between the ECM and the output shaft (countershaft) speed sensor • Faulty output shaft (countershaft) speed sensor • ECM may need to be updated with the latest software • Faulty ECM
DTC: P0721 **1T PCM, MIL: Yes, TCIL: Yes** **Year:** 2009, 2010 **Model:** MDX, RDX, RL, TSX, ZDX **Engine:** 2.3L L4, 2.4L L4, 3.7L V6 **Transmission:** All	**Problem in Output Shaft (Countershaft) Speed Sensor Circuit:** If pulse dropouts occur with the output shaft (countershaft) rotating, the PCM detects a malfunction that may be caused by an open, a temporary open, or a short to ground. The vehicle speed measured by the input shaft (mainshaft) speed sensor/(divided by) the vehicle speed measured by the output shaft (countershaft) speed sensor is greater than 6.0 for at least 10 seconds. **NOTE: This code is caused by an electrical circuit problem and cannot be caused by a mechanical problem in the transmission.** **Possible causes:** • Faulty or improperly installed output shaft (countershaft) speed sensor • "Open" in the wires between PCM connector terminals and ground • "Open" or "Short" in the wire between PCM connector and the output shaft (countershaft) speed sensor connector • PCM needs to be updated with the latest software • Faulty PCM
DTC: P0722 **1T PCM, MIL: Yes, TCIL: Yes** **Year:** 2009, 2010 **Model:** MDX, RDX, RL, TSX, ZDX **Engine:** 2.3L L4, 2.4L L4, 3.5L V6, 3.7L V6 **Transmission:** All	**Problem in Output Shaft (Countershaft) Speed Sensor Circuit (No Signal Input):** If pulse dropouts occur with the output shaft (countershaft) rotating, the PCM detects a malfunction that may be caused by an open, a temporary open, or a short to ground. When the vehicle speed measured by the input shaft (mainshaft) speed sensor is 13 mph (20 km/h) or more, the vehicle speed measured by the output shaft (countershaft) speed sensor is 1 mph (2 km/h) or less for at least 10 seconds. **NOTE: This code is caused by an electrical circuit problem and cannot be caused by a mechanical problem in the transmission.** **Possible causes:** • Faulty or improperly installed output shaft (countershaft) speed sensor • Loose or poor connections at the PCM and output shaft (countershaft) speed sensor connectors • "Open" in the wires between PCM connector terminals and ground or poor ground • "Open" or "Short" in the wire between PCM connector terminal and the output shaft (countershaft) speed sensor connector • PCM needs to be updated with the latest software • Faulty PCM
DTC: P0723 **2T PCM, MIL: Yes, TCIL: Yes** **Year:** 2009, 2010 **Model:** MDX, RDX, RL, TSX, ZDX **Engine:** 2.3L L4, 2.4L L4, 3.5L V6, 3.7L V6 **Transmission:** All	**Output Shaft (Countershaft) Speed Sensor Intermittent Failure:** If pulse dropouts occur with the output shaft (countershaft) rotating, the PCM detects a malfunction that may be caused by an open, a temporary open, or a short to ground. Based on the fluctuation of the vehicle speed measured by the output shaft (countershaft) speed sensor, a malfunction is detected. The fluctuation of the vehicle speed measured by the output shaft (countershaft) speed sensor in 10 milliseconds is 4 mph (6km/h) or more, and it fluctuates at least six times within 500 milliseconds. **NOTE: This code is caused by an electrical circuit problem and cannot be caused by a mechanical problem in the transmission.** **Possible causes:** • Faulty or improperly installed output shaft (countershaft) speed sensor • Poor connections and loose terminals at the output shaft (countershaft) speed sensor and the PCM • "Open" in the wires between PCM connector terminals and ground or poor ground • "Open" or "Short" in the wire between PCM connector terminal and the output shaft (countershaft) speed sensor connector • PCM needs to be updated with the latest software • Faulty PCM

DTC	Trouble Code Title, Conditions & Possible Causes
DTC: P0729 **2T PCM, TCIL: Yes** **Year:** 2010 **Model:** ZDX **Engine:** 3.7L V6 **Transmission:** All	**Problem in 6th Clutch and 6th Clutch Hydraulic Circuit:** With the vehicle running in drive and engine speed at a minimum of 7 mph, battery voltage 11 volts. The actual gear ratio must match one of these conditions for at least 12 seconds with the 6th gear shift command: • Actual gear ratio is greater than the 6th gear ratio by a factor of 1.25. • Actual gear ratio is less than the 6th gear ratio by a factor of 0.8. **NOTE: Before you troubleshoot, record all freeze data and any on-board snapshots.** **Possible causes:** • Low or dirty transmission fluid • Faulty ATF pump or regulator valve • Faulty 6th clutch, replace the 3rd/6th clutch or the transmission
DTC: P0731 **2T PCM, TCIL: Yes** **Year:** 2009, 2010 **Model:** MDX, RDX, RL, TL, TSX, ZDX **Engine:** 2.3L L4, 2.4L L4, 3.5L V6, 3.7L V6 **Transmission:** All	**Problem in 1st Clutch and 1st Clutch Hydraulic Circuit (1st gear incorrect ratio):** The Powertrain Control Module (PCM) computes the ratio of the input shaft (mainshaft) speed to the output shaft (countershaft) speed. When the ratio is not the 1st gear ratio, it is detected as a malfunction of the hydraulic circuit or the 1st clutch. (Symptom 1) The actual gear ratio must match one of these conditions for at least 12 seconds with the 1st gear command: * Actual gear ratio is greater than the 1st gear ratio by a factor of 1.2 * Actual gear ratio is less than the 1st gear ratio by a factor of 0.75 (Symptom 2) The actual gear position is neutral for at least 3 seconds and then the gear up-shifted from 2nd to 3rd, even though 1st gear shift is commanded. **NOTE: Before you troubleshoot, record all freeze data and any on-board. snapshot,** **Possible causes:** • Low or dirty transmission fluid • Faulty shift valves B and C • Faulty ATF pump and the regulator valve • Inspect the strainer for metal debris or excessive clutch material, if present • replace the transmission
DTC: P0732 **2T PCM, TCIL: Yes** **Year:** 2009, 2010 **Model:** MDX, RDX, RL, TL, TSX, ZDX **Engine:** 2.3L L4, 2.4L L4, 3.5L V6, 3.7L V6 **Transmission:** All	**Problem in 2nd Clutch and 2nd Clutch Hydraulic Circuit (2nd gear incorrect ratio):** The Powertrain Control Module (PCM) computes the ratio of the input shaft (mainshaft) speed to the output shaft (countershaft) speed. When the ratio is not the 2nd gear ratio, it is detected as a malfunction of the hydraulic circuit or the 2nd clutch. The actual gear ratio must match one of these conditions for at least 12 seconds with the 2nd gear command. * Actual gear ratio is greater than the 2nd gear ratio by a factor of 1.2. * Actual gear ratio is less than the 2nd gear ratio by a factor of 0.75. **NOTE: Before you troubleshoot, record all freeze data and any on-board snapshot.** **Possible causes:** • Low or dirty transmission fluid • Faulty shift valves A, B and C • Faulty ATF pump and the regulator valve • Inspect the strainer for metal debris or excessive clutch material, if present • replace the transmission
DTC: P0733 **2T PCM, TCIL: Yes** **Year:** 2009, 2010 **Model:** MDX, RDX, RL, TL, TSX, ZDX **Engine:** 2.3L L4, 2.4L L4, 3.5L V6, 3.7L V6 **Transmission:** All	**Problem in 3rd Clutch and 3rd Clutch Hydraulic Circuit (3rd gear incorrect ratio):** The powertrain control module (PCM) computes the ratio of the input shaft (mainshaft) speed to the output shaft (countershaft) speed. When the ratio is not the 3rd gear ratio, it is detected as a malfunction of the hydraulic circuit or the 3rd clutch. The actual gear ratio must match one of these conditions for at least 12 seconds with the 3rd gear command. * Actual gear ratio is greater than the 3rd gear ratio by a factor of 1.2. * Actual gear ratio is less than the 3rd gear ratio by a factor of 0.75. **NOTE: Before you troubleshoot, record all freeze data and any on-board snapshot.** **Possible causes:** • Low or dirty transmission fluid • Faulty shift valves A, B, and C are stuck • Faulty ATF pump and the regulator valve • Inspect the strainer for metal debris or excessive clutch material, if present • replace the transmission

DTC	Trouble Code Title, Conditions & Possible Causes
DTC: P0734 **2T PCM, TCIL: Yes** **Year:** 2009, 2010 **Model:** MDX, RDX, RL, TL, TSX, ZDX **Engine:** 2.3L L4, 2.4L L4, 3.5L V6, 3.7L V6 **Transmission:** All	**Problem in 4th Clutch and 4th Clutch Hydraulic Circuit (4th gear incorrect ratio):** The Powertrain Control Module (PCM) computes the ratio of the input shaft (mainshaft) speed to the output shaft (countershaft) speed. When the ratio is not the 4th gear ratio, it is detected as a malfunction of the hydraulic circuit or the 4th clutch. The actual gear ratio must match one of these conditions for at least 12 seconds with the 4th gear command. * Actual gear ratio is greater than the 4th gear ratio by a factor of 1.2. * Actual gear ratio is less than the 4th gear ratio by a factor of 0.75. **NOTE: Before you troubleshoot, record all freeze data and any on-board snapshot.** **Possible causes:** • Low or dirty transmission fluid • Faulty shift valves A, B, C, and D are stuck • Faulty ATF pump and the regulator valve • Inspect the strainer for metal debris or excessive clutch material, if present • replace the transmission
DTC: P0735 **2T PCM, TCIL: Yes** **Year:** 2009, 2010 **Model:** MDX, RDX, RL, TL, TSX, ZDX **Engine:** 2.3L L4, 2.4L L4, 3.5L V6, 3.7L V6 **Transmission:** All	**Problem in 5th Clutch and 5th Clutch Hydraulic Circuit (5th gear incorrect ratio):** The Powertrain Control Module (PCM) computes the ratio of the input shaft (mainshaft) speed to the output shaft (countershaft) speed. When the ratio is not the 5th gear ratio, it is detected as a malfunction of the hydraulic circuit or the 5th clutch. The actual gear ratio must match one of these conditions for at least 12 seconds with the 5th gear command. * Actual gear ratio is greater than the 5th gear ratio by a factor of 1.2 * Actual gear ratio is less than the 5th gear ratio by a factor of 0.75 **NOTE: Before you troubleshoot, record all freeze data and any on-board snapshot** **Possible causes:** • Low or dirty transmission fluid • Faulty shift valves A, B, C, and D are stuck • Faulty ATF pump and the regulator valve • Inspect the strainer for metal debris or excessive clutch material, if present • replace the transmission
DTC: P0741 **2T PCM, MIL: Yes** **Year:** 2009, 2010 **Model:** MDX, RDX, RL, TL, TSX, ZDX **Engine:** 2.3L L4, 2.4L L4, 3.5L V6, 3.7L V6 **Transmission:** All	**Torque Converter Clutch Hydraulic Circuit Stuck OFF:** If the ratio of engine speed and input shaft (mainshaft) speed is not about 1:1 while the PCM is issuing the command to turn shift solenoid valve D and A/T clutch pressure control solenoid valve C ON, the PCM detects a faulty lock-up control system. The ratio of the engine revolutions to the transmission input pulses does not reach about 100 % for at least 22 seconds. **NOTE: Before you troubleshoot, record all freeze data and any on-board snapshot.** **Possible causes:** • Low or dirty transmission fluid • Inspect the strainer for metal debris or excessive clutch material, if present replace the transmission • Faulty shift solenoid valve D • Faulty torque converter clutch mechanism, torque converter clutch hydraulic circuit, lock-up shift valve, lock-up control valve, or replace the transmission
DTC: P0746 **2T PCM, MIL: Yes, TCIL: Yes** **Year:** 2009, 2010 **Model:** MDX, RDX, RL, TL, TSX, ZDX **Engine:** 2.3L L4, 3.5L V6, 3.7L V6 **Transmission:** All	**A/T Clutch Pressure Control Solenoid Valve A Stuck OFF:** When an improper gear ratio is output compared to the predetermined gear ratio, an A/T clutch pressure control solenoid valve A OFF failure is detected. One of these symptoms occur: * Transmission is held in 1st gear. * The engine speed flares when upshifting to 2nd-3rd. * The engine speed flares when upshifting to 3rd-4th or 5th if applicable. **NOTE: Before you troubleshoot, record all freeze data and any on-board snapshot.** **Possible causes:** • Faulty hydraulic system related with shift valve A • Low or dirty transmission fluid • Inspect the strainer for metal debris or excessive clutch material, if present • replace the transmission

DTC	Trouble Code Title, Conditions & Possible Causes
DTC: P0747 **2T PCM, MIL: Yes, TCIL: Yes** **Year:** 2009, 2010 **Model:** MDX, RDX, RL, TL, TSX, ZDX **Engine:** 2.3L L4, 2.4L L4, 3.5L V6, 3.7L V6 **Transmission:** All	**A/T Clutch Pressure Control Solenoid Valve A Stuck ON:** When an improper gear ratio is output compared to the predetermined gear ratio, an A/T clutch pressure control solenoid valve A ON failure is detected. The execution time is continuous and the duration time is 20 seconds. One of these conditions occur: * The transmission is held in 2nd gear against the 2nd-3rd gear upshift command as long as 20 seconds though there is no record of being neutral when the 1st gear shift is commanded. * The transmission is held in 4th gear against the 4th-5th gear upshift command as long as 20 seconds though there is no record of being neutral when the 1st gear shift is commanded. **NOTE: Before you troubleshoot, record all freeze data and any on-board snapshot.** **Possible causes:** • Low or dirty transmission fluid • Inspect the strainer for metal debris or excessive clutch material, if present • replace the transmission • Faulty hydraulic system related with shift valve A
DTC: P0751 **2T PCM, MIL: Yes, TCIL: Yes** **Year:** 2009, 2010 **Model:** MDX, RDX, RL, TL, TSX, ZDX **Engine:** 2.3L L4, 2.4L L4, 3.5L V6, 3.7L V6 **Transmission:** All	**Shift Solenoid Valve A Stuck OFF:** When an improper gear ratio is output compared to the predetermined gear change mode, a shift solenoid valve A OFF failure is detected and a DTC is stored. The execution time is continuous and the duration time is 2 seconds or more. The transmission is held in 5th gear against the 3rd gear command for at least 2 seconds. **NOTE: Before you troubleshoot, record all freeze data and any on-board snapshot.** **Possible causes:** • Low or dirty transmission fluid • Faulty shift solenoid valve A • Inspect the strainer for metal debris or excessive clutch material, if present • replace the transmission
DTC: P0752 **2T PCM, MIL: Yes, TCIL: Yes** **Year:** 2009, 2010 **Model:** MDX, RDX, TL, TSX, ZDX **Engine:** 2.3L L4, 2.4L L4, 3.5L V6, 3.7L V6 **Transmission:** All	**Shift Solenoid Valve A Stuck ON:** When the wrong transmission fluid switch is turned on for a given speed change mode, a shift solenoid valve turn-on malfunction is detected. The execution time is continuous depending on the driving pattern. The 3rd clutch transmission fluid switch is ON against the 4th-5th gear upshift command for at least 11 seconds. **NOTE: Before you troubleshoot, record all freeze data and any on-board snapshot.** **Possible causes:** • Low or dirty transmission fluid • Faulty shift solenoid valve A • Inspect the strainer for metal debris or excessive clutch material, if present • replace the transmission
DTC: P0756 **2T PCM, MIL: Yes, TCIL: Yes** **Year:** 2009, 2010 **Model:** MDX, RDX, RL, TL, TSX, ZDX **Engine:** 2.3L L4, 2.4L L4, 3.5L V6, 3.7L V6 **Transmission:** All	**Shift Solenoid Valve B Stuck OFF:** When an improper gear ratio is output compared to the predetermined gear change mode, a shift solenoid valve B OFF failure is detected. The transmission is held in 4th gear against the 2nd gear command for at least 2 seconds. The execution time is continuous and the duration time is 2 seconds or more. **NOTE: Before you troubleshoot, record all freeze data and any on-board snapshot.** **Possible causes:** • Low or dirty transmission fluid • Faulty shift valve B • Inspect the strainer for metal debris or excessive clutch material, if present • replace the transmission
DTC: P0757 **2T PCM, MIL: Yes, TCIL: Yes** **Year:** 2009, 2010 **Model:** MDX, RDX, RL, TL, TSX, ZDX **Engine:** 2.3L L4, 2.4L L4, 3.5L V6, 3.7L V6 **Transmission:** All	**Shift Solenoid Valve B Stuck ON:** When the wrong gear ratio is output for a given speed change mode, or when the wrong transmission fluid pressure switch is turned-on, a shift solenoid valve turn-on malfunction is detected. The execution time is continuous depending on the driving pattern. One of these conditions occur: * The 2nd clutch transmission fluid switch is ON against the 3rd-4th gear upshift command for at least 11 seconds. * After the 3rd-4th gear upshift command is output, it is neutral for at least 2 seconds when the 5th gear shift command is output, though there is no history of being neutral. **NOTE: Before you troubleshoot, record all freeze data and any on-board snapshot.** **Possible causes:** • Low or dirty transmission fluid • Inspect the strainer for metal debris or excessive clutch material, if present • replace the transmission • Faulty shift solenoid valve B

DTC	Trouble Code Title, Conditions & Possible Causes
DTC: P0761 **2T PCM, MIL: Yes, TCIL: Yes** **Year:** 2009, 2010 **Model:** MDX, RDX, RL, TL, TSX, ZDX **Engine:** 2.3L L4, 2.4L L4, 3.5L V6, 3.7L V6 **Transmission:** All	**Shift Solenoid Valve C Stuck OFF:** The execution time is continuous and the duration time is 20 seconds. When an improper gear ratio is output compared to the predetermined gear change mode, a shift solenoid valve C OFF failure is detected. One of these symptoms occurred when the actual gear position was neutral when the 1st gear shift is commanded: * The transmission is held in 2nd gear against the 2nd-3rd gear upshift command for at least 17 seconds. * The transmission is held in 4th gear against the 4th-5th gear upshift command for at least 17 seconds. **NOTE: Before you troubleshoot, record all freeze data and any on-board snapshot.** **Possible causes:** • Low or dirty transmission fluid • Faulty shift solenoid valve C • Inspect the strainer for metal debris or excessive clutch material, if present • replace the transmission
DTC: P0762 **2T PCM, MIL: Yes, TCIL: Yes** **Year:** 2009, 2010 **Model:** MDX, RDX, RL, TL, ZDX **Engine:** 2.3L L4, 3.5L V6, 3.7L V6 **Transmission:** All	**Shift Solenoid Valve C Stuck ON:** When an improper gear ratio is output compared to the predetermined gear change mode, a shift solenoid valve C ON failure is detected. The execution time is continuous and the duration time is 20 seconds. The transmission is held in 3rd gear against the 3rd-4th gear upshift command for as long as 20 seconds, without records that the gear change time was short when the 2nd-3rd gear upshift were commanded. **NOTE: Before you troubleshoot, record all freeze data and any on-board snapshot.** **Possible causes:** • Low or dirty transmission fluid • Faulty shift solenoid valve C • Inspect the strainer for metal debris or excessive clutch material, if present replace the transmission
DTC: P0766 **2T PCM, MIL: Yes, TCIL: Yes** **Year:** 2009, 2010 **Model:** MDX, RDX, RL, TSX **Engine:** 2.3L L4, 3.5L V6, 3.7L V6 **Transmission:** All	**Shift Solenoid Valve D Stuck OFF:** When an improper gear ratio is output compared to the predetermined gear change mode, a shift solenoid valve D OFF failure is detected. **NOTE: Before you troubleshoot, record all freeze data and any on-board snapshot.** **Possible causes:** • Low or dirty transmission fluid • Faulty shift solenoid valve D • Inspect the strainer for metal debris or excessive clutch material, if present • replace the transmission
DTC: P0767 **2T PCM, MIL: Yes, TCIL: Yes** **Year:** 2009, 2010 **Model:** MDX, RDX, RL, TSX **Engine:** 2.3L L4, 3.5L V6, 3.7L V6 **Transmission:** All	**Shift Solenoid Valve D Stuck ON:** When an improper gear ratio is output compared to the predetermined gear change mode, a shift solenoid valve D ON failure is detected. One of these conditions occur: * The actual gear position is neutral for at least 3 seconds when 1st gear in-gear is commanded, though there is no history of being neutral when reverse gear in-gear is commanded. * The actual gear position is neutral for at least 3 seconds, though 1st gear in-gear is commanded and reverse drive occurred during this driving cycle. **NOTE: Before you troubleshoot, record all freeze data and any on-board snapshot.** **Possible causes:** • Low or dirty transmission fluid • Faulty shift solenoid valve D • Faulty ATF pump and the regulator valve • Inspect the strainer for metal debris or excessive clutch material, if present • replace the transmission
DTC: P0771 **2T PCM, MIL: Yes, TCIL: Yes** **Year:** 2009, 2010 **Model:** TSX **Engine:** 2.4L L4 **Transmission:** All	**Shift Solenoid Valve E Stuck OFF:** While driving the vehicle with the torque converter lock-up ON a malfunction is detected when shifting from 3rd gear into 4th gear. **NOTE: Before you troubleshoot, record all freeze data and any on-board snapshot with the HDS, and review General Troubleshooting Information.** **Possible causes:** • Low or dirty transmission fluid • Strainer has metal debris or excessive clutch material, replace the transmission • Faulty shift solenoid valve E • Faulty transmission.

DTC	Trouble Code Title, Conditions & Possible Causes
DTC: P0776 **2T PCM, MIL: Yes, TCIL: Yes** **Year:** 2009, 2010 **Model:** MDX, RDX, RL, TSX, ZDX **Engine:** 2.3L L4, 2.4L L4, 3.5L V6, 3.7L V6 **Transmission:** All	**A/T Clutch Pressure Control Solenoid Valve B Stuck OFF:** When an improper gear ratio is output compared to the predetermined gear change mode, an A/T clutch pressure control solenoid valve B OFF failure is detected. The transmission is held in 3rd gear against the 3rd-4th gear upshift command for as long as 20 seconds, with records that the gear change time was short when the 2nd-3rd gear upshift was commanded. **NOTE: Before you troubleshoot, record all freeze data and any on-board snapshot.** **Possible causes:** • Low or dirty transmission fluid • Faulty A/T clutch pressure control solenoid valve B • Inspect the strainer for metal debris or excessive clutch material, if present • replace the transmission
DTC: P0777 **2T PCM, MIL: Yes, TCIL: Yes** **Year:** 2009, 2010 **Model:** MDX, RDX, RL, TSX, ZDX **Engine:** 2.3L L4, 2.4L L4, 3.5L V6, 3.7L V6 **Transmission:** All	**A/T Clutch Pressure Control Solenoid Valve B Stuck ON:** When an improper gear ratio is output compared to the predetermined gear change mode, an A/T clutch pressure control solenoid valve B ON failure is detected. The engine speed flares during 2nd-3rd and 3rd-4th upshifts for at least 1 second. The execution time is continuous depending on the driving pattern. **NOTE: Before you troubleshoot, record all freeze data and any on-board snapshot.** **Possible causes:** • Low or dirty transmission fluid • Faulty A/T clutch pressure control solenoid valve B • Inspect the strainer for metal debris or excessive clutch material, if present • replace the transmission
DTC: P0780 **2T PCM, MIL: Yes, TCIL: Yes** **Year:** 2009, 2010 **Model:** TSX **Engine:** 2.4L L4, 3.5L V6 **Transmission:** All	**Shift Control System:** This code is stored whenever DTCs P1730, P1731, P1732, P1733, and P1734 are detected. Refer to specific DTC information. Before you troubleshoot, record all freeze data and any on-board snapshot with the HDS, and review General Troubleshooting Information. **Possible causes:** • Refer to specific DTC information.
DTC: P0796 **2T PCM, MIL: Yes, TCIL: Yes** **Year:** 2009, 2010 **Model:** MDX, RDX, RL, TSX, ZDX **Engine:** 2.3L L4, 2.4L L4, 3.5L V6, 3.7L V6 **Transmission:** All	**A/T Clutch Pressure Control Solenoid Valve C Stuck OFF:** When an improper gear ratio is output compared to the predetermined gear ratio, an A/T clutch pressure control solenoid valve C OFF failure is detected. The execution time is continuous depending on the driving pattern. **NOTE: Before you troubleshoot, record all freeze data and any on-board snapshot.** **Possible causes:** • Low or dirty transmission fluid • Faulty A/T clutch pressure control solenoid valve C • Inspect the strainer for metal debris or excessive clutch material, if present • replace the transmission
DTC: P0797 **2T PCM, MIL: Yes, TCIL: Yes** **Year:** 2009, 2010 **Model:** MDX, RDX, RL, TSX, ZDX **Engine:** 2.3L L4, 2.4L L4, 3.5L V6, 3.7L V6 **Transmission:** All	**A/T Clutch Pressure Control Solenoid Valve C Stuck ON:** When the wrong transmission fluid pressure switch is turned on, an A/T clutch pressure control solenoid valve C turn-on malfunction is detected. The 2nd clutch transmission fluid switch is ON against the 3rd-4th gear upshift command for at least 11 seconds. The execution time is continuous and the duration time is 20 seconds. **NOTE: Before you troubleshoot, record all freeze data and any on-board snapshot,.** **Possible causes:** • Low or dirty transmission fluid • Faulty hydraulic system related with shift valve C • Faulty A/T clutch pressure control solenoid valve C • Inspect the strainer for metal debris or excessive clutch material, if present • replace the transmission
DTC: P0812 **2T PCM, MIL: Yes, TCIL: Yes** **Year:** 2009, 2010 **Model:** MDX, RDX, RL, TSX, ZDX **Engine:** 2.3L L4, 2.4L L4, 3.5L V6, 3.7L V6 **Transmission:** All	**Open in Transmission Range Switch ATP RVS Switch Circuit:** If the R switch is OPEN with the shift lever in R, the PCM detects a switch OPEN failure. The RVS signal is detected but the R switch signal is not detected for at least 2 seconds. The execution time is continuous depending on the driving pattern. **NOTE: This code is caused by an electrical circuit problem and cannot be caused by a mechanical problem in the transmission.** **Possible causes:** • Faulty transmission range switch • Poor connections and loose terminals at the transmission range switch and the PCM • "Open" in the wire between PCM and the transmission range switch • "Open" in the wire between transmission range switch and ground, or poor ground • PCM needs to be updated with the latest software • Faulty PCM

DTC	Trouble Code Title, Conditions & Possible Causes
DTC: P0815 **1T PCM, MIL: Yes, TCIL: Yes** **Year:** 2009, 2010 **Model:** MDX **Engine:** 3.7L V6 **Transmission:** All	**Short in Transmission Gear Selection Switch Upshift Switch Circuit, or Transmission Gear Selection Switch Upshift Switch Stuck ON :** The S-UP switch signal is turned on in P, R, N, and D3 for at least 10 seconds. The execution time is continuous and the duration time is 10 seconds or more. **NOTE: This code is caused by an electrical circuit problem and cannot be caused by a mechanical problem in the transmission.** **Possible causes:** • Faulty transmission gear selection switch • "Open" or "Short" in the wire between PCM and the transmission gear selection switch • "Short" to ground between the transmission gear selection switch and the PCM • PCM needs to be updated with the latest software • Faulty PCM
DTC: P0816 **1T PCM, MIL: Yes, TCIL: Yes** **Year:** 2009, 2010 **Model:** MDX **Engine:** 3.7L V6 **Transmission:** All	**Short in Transmission Gear Selection Switch Downshift Switch Circuit, or Transmission Gear Selection Switch Downshift Switch Stuck ON :** The S-DN switch signal is turned on in P, R, N, and D3 for at least 10 seconds. DTCs P0705, P0706, P0815, P0957, P0958 are not active. The execution time is continuous and the duration time is 10 seconds or more. This code is caused by an electrical circuit problem and cannot be caused by a mechanical problem in the transmission. **Possible causes:** • Faulty transmission gear selection switch • Poor connections and loose terminals at the transmission gear selection switch and the PCM • Check the SDN (GRN) wire for an intermittent short to ground between the transmission gear selection switch and the PCM • "Short" to ground in the wire between PCM connector terminal and the transmission gear selection switch • PCM may need to be updated with the latest software • Faulty PCM
DTC: P0842 **1T PCM, MIL: Yes, TCIL: Yes** **Year:** 2009, 2010 **Model:** MDX, RDX, RL, TSX, ZDX **Engine:** 2.3L L4, 2.4L L4, 3.5L V6, 3.7L V6 **Transmission:** All	**Short in 2nd Clutch Transmission Fluid Pressure Switch Circuit, or 2nd Clutch Transmission Fluid Pressure Switch Stuck ON:** The input signal from the 2nd clutch transmission fluid pressure switch to the PCM is low when driving in 1st gear, 3rd gear, or 5th gear. The execution time is continuous and the duration time is 2 seconds or more. **NOTE: This code is caused by an electrical circuit problem and cannot be caused by a mechanical problem in the transmission.** **Possible causes:** • Faulty 2nd clutch transmission fluid pressure switch • Poor connections and loose terminals at the 2nd clutch transmission fluid pressure switch and the PCM • OP2SW wire for an intermittent short to ground between the 2nd clutch transmission fluid pressure switch and the PCM • "Short" in the wire between PCM connector terminal and the 2nd clutch transmission fluid pressure switch • PCM may need to be updated with the latest software • Faulty PCM
DTC: P0843 **1T PCM, MIL: Yes, TCIL: Yes** **Year:** 2009, 2010 **Model:** MDX, RDX, RL, TSX, ZDX **Engine:** 2.3L L4, 2.4L L4, 3.5L V6, 3.7L V6 **Transmission:** All	**Open in 2nd Clutch Transmission Fluid Pressure Switch Circuit, or 2nd Clutch Transmission Fluid Pressure Switch Stuck OFF :** The input signal from the 2nd clutch transmission fluid pressure switch to the PCM is high when driving in 2nd gear. The execution time is continuous and the duration time is 2 seconds or more. **NOTE: Before you troubleshoot, record all freeze data and any on-board snapshot.** **Possible causes:** • Faulty 2nd clutch transmission fluid pressure switch • Poor connections and loose terminals at the 2nd clutch transmission fluid pressure switch and the PCM. • "Open" circuit between PCM connector terminal and the 2nd clutch transmission fluid pressure switch • PCM needs to be updated with the latest software • Faulty PCM

DTC	Trouble Code Title, Conditions & Possible Causes
DTC: P0847 **1T PCM, MIL: Yes, TCIL: Yes** **Year:** 2009, 2010 **Model:** MDX, RDX, RL, TSX, ZDX **Engine:** 2.3L L4, 2.4L L4, 3.5L V6, 3.7L V6 **Transmission:** All	**Short In 3rd Clutch Transmission Fluid Pressure Switch Circuit, or 3rd Clutch Transmission Fluid Pressure Switch Stuck ON:** The input signal from the 3rd clutch transmission fluid pressure switch to the PCM is low when driving in 1st gear, 2nd gear, 4th gear, or 5th gear. The execution time is continuous and the duration time is 2 seconds or more. **NOTE: This code is caused by an electrical circuit problem and cannot be caused by a mechanical problem in the transmission.** **Possible causes:** • Faulty 3rd clutch transmission fluid pressure switch • Poor connections and loose terminals at the 3rd clutch transmission fluid pressure switch and the PCM • "Short" circuit between PCM connector terminal and the 3rd clutch transmission fluid pressure switch • PCM may need to be updated with the latest software • Faulty PCM
DTC: P0848 **1T PCM, MIL: Yes, TCIL: Yes** **Year:** 2009, 2010 **Model:** MDX, RDX, RL, TSX, ZDX **Engine:** 2.3L L4, 2.4L L4, 3.5L V6, 3.7L V6 **Transmission:** All	**Open in 3rd Clutch Transmission Fluid Pressure Switch Circuit, or 3rd Clutch Transmission Fluid Pressure Switch Stuck OFF :** The input signal from the 3rd clutch transmission fluid pressure switch to the PCM is high when driving in 3rd gear. The execution time is continuous and the duration time is 2 seconds or more. **NOTE: Before you troubleshoot, record all freeze data and any on-board snapshot.** **Possible causes:** • Faulty 3rd clutch transmission fluid pressure switch • Poor connections or loose terminals at the 3rd clutch transmission fluid pressure switch and the PCM • "Open" circuit between PCM connector terminal and the 3rd clutch transmission fluid pressure switch • PCM may need to be updated with the latest software • Faulty PCM
DTC: P084C **2T PCM, MIL: Yes, TCIL: Yes** **Year:** 2010 **Model:** ZDX **Engine:** 3.7L V6 **Transmission:** All	**Short in Line Pressure Switch Circuit, or Line Pressure Switch Stuck ON:** One of these 2 symptoms occurs: (1) Symptom 1: The input signal from the line pressure switch to the PCM is low during the self shut down mode after the signal is low when the normal line pressure mode was commanded. (2) Symptom 2: The input signal from the line pressure switch to the PCM is low. **NOTE: Before you troubleshoot, record all freeze data and any on-board snapshot** **Possible causes:** • "Short" to body ground in the wire between PCM and the line pressure switch • Faulty line pressure switch • Faulty PCM
DTC: P0872 **1T PCM, MIL: Yes, TCIL: Yes** **Year:** 2009, 2010 **Model:** MDX, RDX, RL, TSX, ZDX **Engine:** 2.3L L4, 2.4L L4, 3.5L V6, 3.7L V6 **Transmission:** All	**Short in 4th Clutch Transmission Fluid Pressure Switch Circuit, or 4th Clutch Transmission Fluid Pressure Switch Stuck ON :** The input signal from the 4th clutch transmission fluid pressure switch to the PCM is low when driving in 5th gear. The execution time is continuous and the duration time is 2 seconds or more. **NOTE: This code is caused by an electrical circuit problem and cannot be caused by a mechanical problem in the transmission.** **Possible causes:** • Faulty 4th clutch transmission fluid pressure switch • Check the OP4SW wire for an intermittent short to ground between the 4th clutch transmission fluid pressure switch and the PCM
DTC: P0873 **1T PCM, MIL: Yes, TCIL: Yes** **Year:** 2009, 2010 **Model:** MDX, RDX, RL, TSX, ZDX **Engine:** 2.3L L4, 2.4L L4, 3.5L V6, 3.7L V6 **Transmission:** All	**Open in 4th Clutch Transmission Fluid Pressure Switch Circuit, or 4th Clutch Transmission Fluid Pressure Switch Stuck OFF:** The input signal from the 4th clutch transmission fluid pressure switch to the PCM is high when driving in 4th gear. The execution time is continuous and the duration time is 2 seconds or more. **NOTE: Before you troubleshoot, record all freeze data and any on-board snapshot.** **Possible causes:** • Faulty 4th clutch transmission fluid pressure switch • Poor connections and loose terminals at the 4th clutch transmission fluid pressure switch and the PCM • "Open" in the wire between PCM connector terminal and the 4th clutch transmission fluid pressure switch • PCM may need to be updated with the latest software • Faulty PCM

DTC	Trouble Code Title, Conditions & Possible Causes
DTC: P0877 **2T PCM, MIL: Yes, TCIL: Yes** **Year:** 2010 **Model:** ZDX **Engine:** 3.7L V6 **Transmission:** All	**Short in Transmission Fluid Pressure Switch D (5th Clutch) Circuit, or Transmission Fluid Pressure Switch D (5th Clutch) Stuck ON:** One of these 3 conditions occurs: (1) The input signal from transmission fluid pressure switch D (5th clutch) to the PCM is low for at least 5 seconds when the shift lever is in P, R, or N. (2) The input signal from transmission fluid pressure switch D (5th clutch) to the PCM is low for at least 5 seconds when driving in 1st or 2nd gear. (3) The input signal from transmission fluid pressure switch D (5th clutch) to the PCM is low for at least 5 seconds when the 2nd gear shift is commanded or when the shift lever is in P, R, or N, after the signal is low for at least 3 seconds when driving in 3rd, 4th, or 6th gear. **NOTE: Before you troubleshoot, record all freeze data and any on-board snapshot. This code is caused by an electrical circuit problem and cannot be caused by a mechanical problem in the transmission.** **Possible causes:** • Poor connections or ground • "Short" to body ground in the wire between PCM and transmission fluid pressure switch D (5th clutch) • Faulty transmission fluid pressure switch D (5th clutch) • Faulty PCM
DTC: P0878 **2T PCM, MIL: Yes, TCIL: Yes** **Year:** 2010 **Model:** ZDX **Engine:** 3.7L V6 **Transmission:** All	**Open in Transmission Fluid Pressure Switch D (5th Clutch) Circuit, or Transmission Fluid Pressure Switch D (5th Clutch) Stuck OFF:** The input signal from transmission fluid pressure switch D (5th clutch) to the PCM is high when driving in 5th gear. The execution time is continuous and the duration time is 4 seconds or more. **NOTE: Before you troubleshoot, record all freeze data and any on-board snapshot. If DTC P0878 is stored in the PCM, the transmission does not shift to any gear other than 2nd through 5th gear because of the fail-safe function** **Possible causes:** • Poor connections and loose terminals between transmission fluid pressure switch D (5th clutch) and the PCM • "Open" circuit between PCM and transmission fluid pressure switch D (5th clutch) • Faulty transmission fluid pressure switch D (5th clutch) • Faulty PCM
DTC: P0957 **1T PCM, MIL: Yes, TCIL: Yes** **Year:** 2009, 2010 **Model:** MDX **Engine:** 3.7L V6 **Transmission:** All	**Short in Transmission Gear Selection Switch Circuit, or Transmission Gear Selection Switch Stuck ON :** The S-MODE switch signal is turned on in P, R, N, and D3 for at least 10 seconds. The execution time is continuous and the duration time is 10 seconds or more. DTCs P0705, P0706, P0815, P0816, P0958 are not active. **NOTE: This code is caused by an electrical circuit problem and cannot be caused by a mechanical problem in the transmission.** **Possible causes:** • Faulty transmission gear selection switch • Poor connections and loose terminals at the transmission gear selection switch and the PCM • "Short" circuit between PCM connector terminal and the transmission gear selection switch • PCM may need to be updated with the latest software • Faulty PCM
DTC: P0958 **1T PCM, MIL: Yes, TCIL: Yes** **Year:** 2009, 2010 **Model:** MDX **Engine:** 3.7L V6 **Transmission:** All	**Open in Transmission Gear Selection Switch Circuit, or Transmission Gear Selection Switch Stuck OFF:** The S-UP switch inputs or S-DN switch inputs are counted at least two times when the S-MODE switch signal is turned off in D. The execution time is continuous depending on the driving pattern. **NOTE: This code is caused by an electrical circuit problem and cannot be caused by a mechanical problem in the transmission.** **Possible causes:** • Poor connections and loose terminals at the transmission gear selection switch and the PCM • "Open" circuit between transmission gear selection switch/park pin switch/A/T gear position indicator panel light connector and ground, or poor ground • Faulty transmission gear selection switch • PCM may need to be updated with the latest software • Faulty PCM

DTC	Trouble Code Title, Conditions & Possible Causes
DTC: P0962 **1T PCM, MIL: Yes, TCIL: Yes** **Year:** 2009, 2010 **Model:** MDX, RDX, RL, TSX, ZDX **Engine:** 2.3L L4, 2.4L L4, 3.5L V6, 3.7L V6 **Transmission:** All	**Problem in A/T Clutch Pressure Control Solenoid Valve A Circuit:** If the measured current for the PCM output duty cycle is not within a specified range (open or short), a malfunction is detected. The execution time is continuous and the duration time is 2 seconds or more. The measured current for the PCM command value is Duty 57-89 %, current less than 0.2, low input. **NOTE: This code is caused by an electrical circuit problem and cannot be caused by a mechanical problem in the transmission.** **Possible causes:** • Poor connections and loose terminals at A/T clutch pressure control solenoid valve A and the PCM • Faulty A/T clutch pressure control solenoid valve A • "Open" circuit between A/T clutch pressure control solenoid valve A and ground, or poor ground • PCM may need to be updated with the latest software • Faulty PCM
DTC: P0963 **1T PCM, MIL: Yes, TCIL: Yes** **Year:** 2009, 2010 **Model:** MDX, RDX, RL, TSX, ZDX **Engine:** 2.3L L4, 2.4L L4, 3.5L V6, 3.7L V6 **Transmission:** All	**Problem in A/T Clutch Pressure Control Solenoid Valve A:** Duty cycle is less than 13-27 %, current (A) 0.6-0.9, high input failure. The execution time is continuous and the duration time is 2 seconds or more. DTCs P0962, P0966, P0967, P0970, P0971 are not active. **NOTE: This code is caused by an electrical circuit problem and cannot be caused by a mechanical problem in the transmission.** **Possible causes:** • Poor connections and loose terminals at A/T clutch pressure control solenoid valve A and the PCM • "Open" circuit between A/T clutch pressure control solenoid valve A and ground, or poor ground • Faulty A/T clutch pressure control solenoid valve A • PCM may need to be updated with the latest software • Faulty PCM
DTC: P0966 **1T PCM, MIL: Yes, TCIL: Yes** **Year:** 2009, 2010 **Model:** MDX, RDX, RL, TSX, ZDX **Engine:** 2.3L L4, 2.4L L4, 3.5L V6, 3.7L V6 **Transmission:** All	**Problem in A/T Clutch Pressure Control Solenoid Valve B Circuit:** The measured current is less than 0.2, duty cycle 57-89 %, low input. The execution time is continuous and the duration time is 1 seconds or more. **NOTE: This code is caused by an electrical circuit problem and cannot be caused by a mechanical problem in the transmission.** **Possible causes:** • Poor connections and loose terminals at A/T clutch pressure control solenoid valve B and the PCM • "Open" or "Short" circuit between PCM connector terminal and A/T clutch pressure control solenoid valve B • "Open" circuit between A/T clutch pressure control solenoid valve B and ground, or poor ground • Faulty A/T clutch pressure control solenoid valve B • PCM may need to be updated with the latest software • Faulty PCM
DTC: P0967 **1T PCM, MIL: Yes, TCIL: Yes** **Year:** 2009, 2010 **Model:** MDX, RDX, RL, TSX, ZDX **Engine:** 2.3L L4, 2.4L L4, 3.5L V6, 3.7L V6 **Transmission:** All	**Problem in A/T Clutch Pressure Control Solenoid Valve B:** The measured current for the PCMs command is 0.6-0.9, duty cycle 13-27%, high input failure. Engine running, DTCs P0962, P0963, P0966, P0970, P0971 are not active. The execution time is continuous and the duration time is 1 seconds or more. **NOTE: This code is caused by an electrical circuit problem and cannot be caused by a mechanical problem in the transmission.** **Possible causes:** • Poor connections and loose terminals at A/T clutch pressure control solenoid valve B and the PCM • "Open" circuit between A/T clutch pressure control solenoid valve C and ground, or poor ground • Faulty A/T clutch pressure control solenoid valve B • PCM may need to be updated with the latest software • Faulty PCM
DTC: P0970 **1T PCM, MIL: Yes, TCIL: Yes** **Year:** 2009, 2010 **Model:** MDX, RDX, RL, TSX, ZDX **Engine:** 2.3L L4, 2.4L L4, 3.5L V6, 3.7L V6 **Transmission:** All	**Problem in A/T Clutch Pressure Control Solenoid Valve C Circuit:** The measured current for the PCMs command value is 0.2-0.4, duty cycle 57-89%, low input failure. Engine is running with battery voltage at 11 volts, DTCs P0962, P0963, P0966, P0967, P0971 are not active. The execution time is continuous and the duration time is 1 seconds or more. **NOTE: This code is caused by an electrical circuit problem and cannot be caused by a mechanical problem in the transmission.** **Possible causes:** • Poor connections and loose terminals at A/T clutch pressure control solenoid valve C and the PCM • "Open" or "Short" circuit between PCM connector terminal and A/T clutch pressure control solenoid valve C • "Open" circuit between A/T clutch pressure control solenoid valve C and ground, or poor ground • Faulty A/T clutch pressure control solenoid valve C • PCM needs to be updated with the latest software • Faulty PCM

DTC	Trouble Code Title, Conditions & Possible Causes
DTC: P0971 **1T PCM, MIL: Yes, TCIL: Yes** **Year:** 2009, 2010 **Model:** MDX, RDX, RL, TSX, ZDX **Engine:** 2.3L L4, 2.4L L4, 3.5L V6, 3.7L V6 **Transmission:** All	**Problem in A/T Clutch Pressure Control Solenoid Valve C:** The measured current for the PCMs command value is 0.6-0.9, duty cycle 13-27, high input failure. The execution time is continuous and the duration time is 1 seconds or more. **NOTE: This code is caused by an electrical circuit problem and cannot be caused by a mechanical problem in the transmission.** **Possible causes:** • Poor connections and loose terminals at A/T clutch pressure control solenoid valve C and the PCM • "Open" in the wire between A/T clutch pressure control solenoid valve C and ground,, or poor ground • Faulty A/T clutch pressure control solenoid valve C • PCM needs to be updated with the latest software • Faulty PCM
DTC: P0973 **1T PCM, MIL: Yes, TCIL: Yes** **Year:** 2009, 2010 **Model:** MDX, RDX, RL, TSX, ZDX **Engine:** 2.3L L4, 2.4L L4, 3.5L V6, 3.7L V6 **Transmission:** All	**Short in Shift Solenoid Valve A Circuit:** The return signal does not match the command to turn ON shift solenoid valve A for at least 1 second. The execution time is continuous and the duration time is 2 seconds or more. DTCs P0974, P0982, P0983 are not active. **NOTE: This code is caused by an electrical circuit problem and cannot be caused by a mechanical problem in the transmission.** **Possible causes:** • Blown fuse • Poor connections and loose terminals at shift solenoid valve A and the PCM • SHA wire for an intermittent "Short" to ground between shift solenoid valve A and the PCM • "Short" circuit between PCM connector terminal and the shift solenoid harness connector • "Open" circuit between PCM connector terminals and ground, or poor ground • Faulty shift solenoid valve A or the shift solenoid harness • PCM needs to be updated with the latest software • Faulty PCM
DTC: P0974 **1T PCM, MIL: Yes, TCIL: Yes** **Year:** 2009, 2010 **Model:** MDX, RDX, RL, TSX, ZDX **Engine:** 2.3L L4, 2.4L L4, 3.5L V6, 3.7L V6 **Transmission:** All	**Open in Shift Solenoid Valve A Circuit:** The return signal does not match the command to turn OFF shift solenoid valve A for at least 1 second. The execution time is continuous and the duration time is 1 seconds or more. DTCs P0973, P0982, P0983 are not active. **NOTE: This code is caused by an electrical circuit problem and cannot be caused by a mechanical problem in the transmission.** **Possible causes:** • Poor connections or loose terminals at shift solenoid valve A and the PCM • "Open" in the wire between PCM connector terminal and the shift solenoid harness connector • Faulty shift solenoid valve A or shift solenoid harness • PCM may need to be updated with the latest software • Faulty PCM
DTC: P0976 **1T PCM, MIL: Yes, TCIL: Yes** **Year:** 2009, 2010 **Model:** MDX, RDX, RL, TSX, ZDX **Engine:** 2.3L L4, 2.4L L4, 3.5L V6, 3.7L V6 **Transmission:** All	**Short in Shift Solenoid Valve B Circuit:** The return signal does not match the command to turn ON shift solenoid valve B for at least 1 second. The execution time is continuous and the duration time is 2 seconds or more. DTCs P0977, P0979, P0980 are not active. **NOTE: This code is caused by an electrical circuit problem and cannot be caused by a mechanical problem in the transmission.** **Possible causes:** • Blown fuse • Poor connections and loose terminals at shift solenoid valve B and the PCM • "Open" in the wires between PCM connector terminals and ground, or poor ground • SHB wire for an intermittent short to ground between shift solenoid valve B and the PCM • "Short" in the wire between PCM connector terminal and the shift solenoid harness connector • Faulty shift solenoid valve B or the shift solenoid harness • PCM may need to be updated with the latest software • Faulty PCM
DTC: P0977 **1T PCM, MIL: Yes, TCIL: Yes** **Year:** 2009, 2010 **Model:** MDX, RDX, RL, TSX, ZDX **Engine:** 2.3L L4, 2.4L L4, 3.5L V6, 3.7L V6 **Transmission:** All	**Open in Shift Solenoid Valve B Circuit:** The return signal does not match the command to turn OFF shift solenoid valve B for at least 1 second. The execution time is continuous and the duration time is 1 seconds or more. Battery voltage minimum 11 volts, and DTCs P0976, P0979, P0980 are not active. **NOTE: This code is caused by an electrical circuit problem and cannot be caused by a mechanical problem in the transmission.** **Possible causes:** • Poor connections or loose terminals at shift solenoid valve B and the PCM • "Open" circuit between PCM connector terminal and the shift solenoid harness connector • Faulty shift solenoid valve B or shift solenoid harness • PCM may need to be updated with the latest software • Faulty PCM

DTC	Trouble Code Title, Conditions & Possible Causes
DTC: P0979 **1T PCM, MIL: Yes, TCIL: Yes** **Year:** 2009, 2010 **Model:** MDX, RDX, RL, TSX, ZDX **Engine:** 2.3L L4, 2.4L L4, 3.5L V6, 3.7L V6 **Transmission:** All	**Short in Shift Solenoid Valve C Circuit:** The return signal does not match the command to turn ON shift solenoid valve C for at least 1 second. The execution time is continuous and the duration time is 1 seconds or more. Battery voltage is 11 volts and DTCs P0976, P0977, P0980 are not active. **NOTE: This code is caused by an electrical circuit problem and cannot be caused by a mechanical problem in the transmission.** **Possible causes:** • Blown fuse • Poor connections and loose terminals at shift solenoid valve C and the PCM • SHC wire for an intermittent short to ground between shift solenoid valve C and the PCM • "Open circuit between PCM connector terminal and the under-dash fuse/relay box via the main relay • "Open" circuit between PCM connector terminals and ground, or poor ground • "Short" circuit between PCM connector terminal and the shift solenoid harness connector • Faulty shift solenoid valve C or the shift solenoid harness • PCM needs to be updated with the latest software • Faulty PCM
DTC: P0980 **1T PCM, MIL: Yes, TCIL: Yes** **Year:** 2009, 2010 **Model:** MDX, RDX, RL, TSX, ZDX **Engine:** 2.3L L4, 2.4L L4, 3.5L V6, 3.7L V6 **Transmission:** All	**Open in Shift Solenoid Valve C Circuit:** The return signal does not match the command to turn OFF shift solenoid valve C for at least 1 second. The execution time is continuous and the duration time is 1 seconds or more. Battery voltage is 11 V and DTCs P0976, P0977, P0979 are not active. **NOTE: This code is caused by an electrical circuit problem and cannot be caused by a mechanical problem in the transmission.** **Possible causes:** • Poor connections or loose terminals at shift solenoid valve C and the PCM • "Open" circuit between PCM connector terminal and the shift solenoid harness connector • Faulty shift solenoid valve C or shift solenoid harness • PCM needs to be updated with the latest software • Faulty PCM
DTC: P0982 **1T PCM, MIL: Yes, TCIL: Yes** **Year:** 2009, 2010 **Model:** MDX, RDX, RL, TSX **Engine:** 2.3L L4, 2.4L L4, 3.5L V6, 3.7L V6 **Transmission:** All	**Short in Shift Solenoid Valve D Circuit:** The return signal does not match the command to turn ON shift solenoid valve D for at least 1 second. The execution time is continuous and the duration time is 1 seconds or more. Battery voltage is 11v, and DTCs P0973, P0974, P0983 are not active. **NOTE: This code is caused by an electrical circuit problem and cannot be caused by a mechanical problem in the transmission.** **Possible causes:** • Blown Fuse • Poor connections and loose terminals at shift solenoid valve D and the PCM. If the PCM • SHD wire for an intermittent short to ground between shift solenoid valve D and the PCM • "Open" in the wire between PCM connector terminal and the under-dash fuse/relay box • "Short" in the wire between PCM connector terminal and the shift solenoid harness connector • Faulty shift solenoid valve D or the shift solenoid harness • PCM needs to be updated with the latest software • Faulty PCM
DTC: P0983 **1T PCM, MIL: Yes, TCIL: Yes** **Year:** 2009, 2010 **Model:** MDX, RDX, RL, TSX **Engine:** 2.3L L4, 2.4L L4, 3.5L V6, 3.7L V6 **Transmission:** All	**Open in Shift Solenoid Valve D Circuit:** The return signal does not match the command to turn OFF shift solenoid valve D for at least 1 second. The execution time is continuous and the duration time is seconds or more. Battery voltage is 11v and DTCs P0973, P0974, P0982 are not active. **NOTE: This code is caused by an electrical circuit problem and cannot be caused by a mechanical problem in the transmission.** **Possible causes:** • Poor connections or loose terminals at shift solenoid valve D and the PCM • "Open" circuit between PCM connector terminal C9 and the shift solenoid harness connector • Faulty shift solenoid valve D or shift solenoid harness • PCM may need to be updated with the latest software • Faulty PCM
DTC: P0985 **1T PCM, MIL: Yes, TCIL: Yes** **Year:** 2009, 2010 **Model:** TSX **Engine:** 2.4L L4, 3.5L V6 **Transmission:** All	**Short in Shift Solenoid Valve E Circuit:** With the engine started and in park the return signal does not match the command to turn ON shift solenoid valve E for at least 1 second. **NOTE: Before you troubleshoot, record all freeze data and any on-board snapshot with the HDS, and review General Troubleshooting Information. This code is caused by an electrical circuit problem and cannot be caused by a mechanical problem in the transmission.** **Possible causes:** • "Short" to body ground between PCM and the shift solenoid wire harness connector • "Short" between PCM and the shift solenoid wire harness connector • Faulty shift solenoid valve E or the shift solenoid wire harness

DTC	Trouble Code Title, Conditions & Possible Causes
DTC: P0986 **1T PCM, MIL: Yes, TCIL: Yes** **Year:** 2009, 2010 **Model:** TSX **Engine:** 2.4L L4, 3.5L V6 **Transmission:** All	**Open in Shift Solenoid Valve E Circuit:** With the engine running and driven in 2nd gear in the D position, the return signal does not match the command to turn OFF shift solenoid E for at least 1 second. **NOTE: Before you troubleshoot, record all freeze data and any on-board snapshot with the HDS, and review General Troubleshooting Information. This code is caused by an electrical circuit problem and cannot be caused by a mechanical problem in the transmission.** **Possible causes:** • Poor connections or loose terminals between shift solenoid valve E and the PCM • "Open" circuit between PCM connector and the shift solenoid wire harness connector • Faulty shift solenoid valve E, or shift solenoid wire harness • PCM may need to be updated with the latest software • Faulty PCM
DTC: P0989 **2T PCM, MIL: Yes, TCIL: Yes** **Year:** 2010 **Model:** ZDX **Engine:** 3.7L V6 **Transmission:** All	**Short in Transmission Fluid Pressure Switch E (6th Clutch) Circuit, or Transmission Fluid Pressure Switch E (6th Clutch) Stuck ON:** One of these 3 conditions occurs: (1) The input signal from transmission fluid pressure switch E (6th clutch) to the PCM is low for at least 4 seconds when the shift lever is in P, R, or N. (2) The input signal from transmission fluid pressure switch E (6th clutch) to the PCM is low for at least 4 seconds when driving in 1st, 2nd, or 3rd gear. (3) The input signal from transmission fluid pressure switch E (6th clutch) to the PCM is low for at least 4 seconds when the 3rd gear shift is commanded or when the shift lever is in P, R, or N, after the signal is low for at least 2 seconds when driving in 4th or 5th gear. **NOTE: Before you troubleshoot, record all freeze data and any on-board snapshot. This code is caused by an electrical circuit problem and cannot be caused by a mechanical problem in the transmission.** **Possible causes:** • Poor connections, loose terminals or faulty ground • "Short" to body ground in the wire between PCM and transmission fluid pressure switch E (6th clutch) • Faulty transmission fluid pressure switch E (6th clutch) • Faulty PCM
DTC: P0990 **2T PCM, MIL: Yes, TCIL: Yes** **Year:** 2010 **Model:** ZDX **Engine:** 3.7L V6 **Transmission:** All	**Open in Transmission Fluid Pressure Switch E (6th Clutch) Circuit, or Transmission Fluid Pressure Switch E (6th Clutch) Stuck OFF:** The input signal from transmission fluid pressure switch E (6th clutch) to the PCM is high when driving in 6th gear. **NOTE: Before you troubleshoot, record all freeze data and any on-board snapshot. If DTC P0990 is stored in the PCM, the transmission does not shift to any gear other than 3rd or 6th gear because of the fail-safe function.** **Possible causes:** • Poor connections and loose terminals between transmission fluid pressure switch E (6th clutch) and the PCM • "Open" circuit between PCM connector and transmission fluid pressure switch E (6th clutch) • Faulty transmission fluid pressure switch E (6th clutch) • Faulty PCM

Gas Engine OBD II Trouble Code List (P1XXX Codes)

DTC	Trouble Code Title, Conditions & Possible Causes
DTC: P1009 **1T PCM, MIL: Yes** **Year:** 2009, 2010 **Model:** RDX, TSX **Engine:** 2.3L L4, 2.4L L4, 3.5L V6 **Transmission:** All	**Variable Valve Timing Control (VTC) Advance Malfunction :** With the engine idling at operating temperature, the camshaft phase value is not 24.0 degrees or less within the monitored area (camshaft phase control directed value plus failure judgment value) after 3 seconds have passed, or when the camshaft phase value is 5.0 degrees for M/T and 20.0 degrees for A/T or less and continues for 0.5 seconds or more, even when the VTC does not actuate. **NOTE: Before you troubleshoot, record all freeze data and any on-board snapshot, and review the general troubleshooting information. If DTC P0341 is stored at the same time as DTC P1009, troubleshoot DTC P1009 first, then recheck for DTC P0341.** **Possible causes:** • Clogged VTC strainer (A) • Clogged oil passages at the VTC system • Faulty VTC oil control solenoid valve • Faulty VTC actuator

DTC	Trouble Code Title, Conditions & Possible Causes
DTC: P1077 **2T PCM, MIL: Yes** **Year:** 2009, 2010 **Model:** MDX **Engine:** 3.7L V6 **Transmission:** All	**Intake Manifold Tuning (IMT) Valve Stuck in High RPM Position:** When the PCM sends a close (long runner) command, no long runner return signal is received for at least 7 seconds. Execution is once per drive cycle, engine speed, 3,600 rpm and intake air temperature a minimum of 5 °F (-15 °C). **NOTE: Before you troubleshoot, record all freeze data and any on-board snapshot.** **Possible causes:** • Poor connections or loose terminals at the IMT actuator and the PCM • "Open" circuit between the PCM and the IMT actuator • Faulty IMT actuator • PCM may need to be updated with the latest software • Faulty PCM
DTC: P1077 **2T PCM, MIL: Yes** **Year:** 2009, 2010 **Model:** MDX, RL, TSX, ZDX **Engine:** 3.5L V6, 3.7L V6 **Transmission:** All	**Intake Manifold Tuning (IMT) Valve Stuck in High RPM Position:** When the PCM sends a close (long runner) command, no long runner return signal is received for at least 7 seconds. The execution time is once per driving cycle. **NOTE: Before you troubleshoot, record all freeze data and any on-board snapshot.** **Possible causes:** • Poor connections or loose terminals at the IMT actuator and the PCM • "Open" or "Short" circuit between the PCM and the IMT actuator • Faulty IMT actuator • Stuck valve, replace the intake manifold if necessary • PCM needs to be updated with the latest software • Faulty PCM
DTC: P1078 **2T PCM, MIL: Yes** **Year:** 2009, 2010 **Model:** MDX **Engine:** 3.7L V6 **Transmission:** All	**Intake Manifold Tuning (IMT) Valve Stuck in Low RPM Position:** When the PCM sends an open (short runner) command, no short runner return signal is received for at least 3 seconds. Intake air temperature 5 °F (-15 °C), engine speed is 3,800 rpm and battery voltage minimum of 10.5v. DTCs P0107, P0108, P0112, P0113, P0117, P0118, P0563, P1128, P1129, P2227, P2228, P2229 are not active. **NOTE:** **NOTE: Before you troubleshoot, record all freeze data and any on-board snapshot.** **Possible causes:** • Poor connections or loose terminals at the IMT actuator and the PCM • "Open" or "Short" circuit between the PCM and the IMT actuator • Faulty IMT actuator • PCM needs to be updated with the latest software • Faulty PCM
DTC: P1078 **2T PCM, MIL: Yes** **Year:** 2009, 2010 **Model:** RL, TSX, ZDX **Engine:** 3.5L V6, 3.7L V6 **Transmission:** All	**Intake Manifold Tuning (IMT) Valve Stuck in Low RPM Position:** When the PCM sends an open (short runner) command, no short runner return signal is received for at least 3 seconds. The execution time is once per driving cycle. **NOTE: Before you troubleshoot, record all freeze data and any on-board snapshot.** **Possible causes:** • Poor connections or loose terminals at the IMT actuator and the PCM • "Open" or "Short" circuit between the PCM and the IMT actuator • Faulty IMT actuator • Stuck valve, replace the intake manifold if necessary • PCM needs to be updated with the latest software • Faulty PCM
DTC: P1109 **1T PCM, MIL: Yes** **Year:** 2009, 2010 **Model:** MDX, RDX, RL, TSX, ZDX **Engine:** 2.3L L4, 2.4L L4, 3.5L V6, 3.7L V6 **Transmission:** All	**Barometric Pressure (BARO) Sensor Circuit Out of Range High:** The BARO sensor output voltage is between 3.59 V to 4.49 V for at least 2 seconds. The execution time is continuous and DTCs P2228, P2229 are not active. **NOTE: Before you troubleshoot, record all freeze data and any on-board snapshot.** **Possible causes:** • PCM needs to be updated with the latest software • Faulty PCM

DTC	Trouble Code Title, Conditions & Possible Causes
DTC: P1116 **2T PCM, MIL: Yes** **Year:** 2009, 2010 **Model:** MDX, RDX, RL, TSX, ZDX **Engine:** 2.3L L4, 2.4L L4, 3.5L V6, 3.7L V6 **Transmission:** All	**Engine Coolant Temperature (ECT) Sensor 1 Circuit Range/Performance Problem:** A malfunction is detected if the following 3 conditions are present after the engine has stopped and the ignition switch has been turned to LOCK (0) for at least 6 hours before restarting the engine: (1) When the temperature difference between the IAT and ECT1 is 57 °F (32 °C) or more. (2) When the temperature difference between the IAT and ECT2 is 30 °F (17 °C) or more. (3) When the temperature difference between the ECT2 and ECT1 is 73 °F (41 °C) or more. **NOTE: If DTC P0111 is stored at the same time as DTC P1116, troubleshoot DTC P0111 first, then recheck for DTC P1116.** **Possible causes:** • Poor connections or loose terminals at ECT sensor 1 and ECT sensor 2 • Faulty ECT sensor 1 • Faulty ECT sensor 2
DTC: P1128 **2T PCM, MIL: Yes** **Year:** 2009, 2010 **Model:** MDX, RDX, RL, TSX, ZDX **Engine:** 2.3L L4, 2.4L L4, 3.5L V6, 3.7L V6 **Transmission:** All	**Manifold Absolute Pressure (MAP) Sensor Signal Lower Than Expected:** The MAP sensor output is 33 kPa (9.8 in.Hg, 249 mmHg) or less for at least 2 seconds when atmospheric pressure is 52 kPa (15.4 in.Hg, 392 mmHg). The execution time is once per driving cycle. **NOTE: Before you troubleshoot, record all freeze data and any on-board snapshot.** **Possible causes:** • Dirty air cleaner element • Poor connections or loose terminals at the MAP sensor and the PCM • Faulty MAP sensor
DTC: P1129 **2T PCM, MIL: Yes** **Year:** 2009, 2010 **Model:** RDX **Engine:** 2.3L L4 **Transmission:** All	**MAP Sensor Signal Higher Than Expected:** With the vehicle speed 1,750 or more and the throttle position is at 22° or more. The difference between the barometric pressure sensor measured during boost conditions and negative pressure (NO boost) is a specified output or less, an upward shift in the MAP sensor signal is detected. **NOTE: Before you troubleshoot, record all freeze data and any on-board snapshot, and review the general troubleshooting information.** **Possible causes:** • Vacuum leaks • Poor connections or loose terminals at the MAP sensor and the PCM • Faulty MAP sensor
DTC: P1129 **2T PCM, MIL: Yes** **Year:** 2009, 2010 **Model:** MDX, RL, TSX, ZDX **Engine:** 2.4L L4, 3.5L V6, 3.7L V6 **Transmission:** All	**Manifold Absolute Pressure (MAP) Sensor Signal Higher Than Expected:** The MAP sensor output is 36 kPa (10.9 in.Hg, 277 mmHg) or more for at least 2 seconds. The execution time is once per driving cycle. **NOTE: Before you troubleshoot, record all freeze data and any on-board snapshot.** **Possible causes:** • Vacuum leaks • Poor connections or loose terminals at the MAP sensor and the PCM • Faulty MAP sensor
DTC: P1157 **1T PCM, MIL: Yes** **Year:** 2009, 2010 **Model:** RDX, TSX **Engine:** 2.3L L4, 2.4L L4, 3.5L V6 **Transmission:** All	**Air Fuel Ratio (A/F) Sensor (Sensor 1) AFS Line High Voltage:** With the engine running and battery voltage between 10.5-16.0 V, the A/F sensor (SENSOR 1) heater element resistance is 250 ohms or more for at least 5 seconds. **NOTE: Before you troubleshoot, record all freeze data and any on-board snapshot, and review the general troubleshooting information.** **Possible causes:** • Poor connections or loose terminals at the A/F sensor (Sensor 1) and the ECM/PCM • "Open" circuit between the ECM/PCM and the A/F sensor (Sensor 1) • Faulty A/F sensor (Sensor 1) • PCM may need to be updated with the latest software • Faulty PCM
DTC: P1172 **2T PCM, MIL: Yes** **Year:** 2009, 2010 **Model:** MDX, RL, TSX, ZDX **Engine:** 3.5L V6, 3.7L V6 **Transmission:** All	**Rear Air/Fuel Ratio (A/F) Sensor (Bank 1, Sensor 1) Circuit Out of Range High:** A malfunction is detected when the rear A/F sensor (bank 1, sensor 1) output voltage is 4.7 V or more. The execution time is continuous and the duration time is 7 seconds or more. **NOTE: Before you troubleshoot, record all freeze data and any on-board snapshot.** **Possible causes:** • Poor connections or loose terminals at the A/F sensor (Sensor 1) and the PCM • Faulty A/F sensor (Sensor 1)

DTC	Trouble Code Title, Conditions & Possible Causes
DTC: P1172 **1T PCM, MIL: Yes** **Year:** 2009, 2010 **Model:** RDX, TSX **Engine:** 2.3L L4, 2.4L L4 **Transmission:** All	**Air/Fuel Ratio (A/F) Sensor (Sensor 1) Circuit Out of Range High:** A malfunction is detected when the rear A/F sensor (Sensor 1) output voltage is 4.9 V or more. The execution time is continuous and the duration time is 7 seconds or more. **NOTE: Before you troubleshoot, record all freeze data and any on-board snapshot.** **Possible causes:** • Poor connections or loose terminals at the A/F sensor (Sensor 1) and the PCM • Faulty A/F sensor (Sensor 1)
DTC: P1174 **2T PCM, MIL: Yes** **Year:** 2009, 2010 **Model:** MDX, RL, ZDX **Engine:** 3.7L V6 **Transmission:** All	**Front Air/Fuel Ratio (A/F) Sensor (Bank 2, Sensor 1) Circuit Out of Range High:** A malfunction is detected when the front A/F sensor (bank 2, sensor 1) output voltage is 4.7 V or more. The execution time is continuous and the duration time is 7 seconds or more. **NOTE: Before you troubleshoot, record all freeze data and any on-board snapshot.** **Possible causes:** • Poor connections or loose terminals at the A/F sensor (Sensor 1) and the PCM • Faulty A/F sensor (Sensor 1)
DTC: P1233 **2T PCM, MIL: Yes** **Year:** 2009, 2010 **Model:** RDX **Engine:** 2.3L L4 **Transmission:** All	**Turbocharger Boost Sensor/BARO Sensor Incorrect Correlation:** Three seconds after start-up with the engine speed at a minimum of 550 rpm at idle. The difference between the BARO sensor output and the turbocharger boost sensor output is 26 kPa (7.8 in.Hg, 200 mmHG) or more for at least 5 seconds. **NOTE: Before you troubleshoot, record all freeze data and any on-board snapshot, and review the general troubleshooting information.** **Possible causes:** • If DTC P1233 is indicated alone, do the troubleshooting for DTC P0236 and P2227 using freeze data from P1233. • If any of the DTCs listed below are indicated at the same time as DTC P1233, troubleshoot those DTCs first, then recheck for P1233. • (P0236: Turbocharger boost sensor) • (P1128, P1129: Manifold Absolute Pressure (MAP) sensor) • (P2227: Barometric Pressure (BARO) sensor)
DTC: P1234 **2T PCM, MIL: Yes** **Year:** 2009, 2010 **Model:** RDX **Engine:** 2.3L L4 **Transmission:** All	**Turbocharger Boost Sensor/MAP Sensor Incorrect Correlation:** Engine speed at a minimum of 1,750 rpm, throttle position 8-22 °. The Map sensor output: LOW, or the turbocharger boost sensor: HIGH. The Map sensor output: HIGH, or the turbocharger boost sensor: LOW. **NOTE: Before you troubleshoot, record all freeze data and any on-board snapshot, and review the general troubleshooting information.** **Possible causes:** • If DTC P1234 is indicated alone, do the troubleshooting for DTC P0236, P1128, and P1129 using freeze data from P1234 • If any of the DTCs listed below are indicated at the same time as DTC P1234, troubleshoot those DTCs first, then recheck for P1234. • (P0236: Turbocharger boost sensor) • (P1128, P1129: Manifold Absolute Pressure (MAP) sensor) • (P2227: Barometric Pressure (BARO) sensor)
DTC: P1243 **T BCM** **Year:** 2009, 2010 **Model:** TL **Engine:** 3.5L V6, 3.7L V6 **Transmission:** All	**Problem in the Passenger's Mode Control Linkage, Doors, or Motor Circuit:** Passenger's Mode Control Malfunction detected. **Possible causes:** • Perform the Self-diagnostic Function with the HDS, or with the climate control unit • Loose wires or poor connections on the passenger's mode control motor circuit • "Open" or "Short" between the climate control unit and the passenger's mode control motor • Faulty ground • Faulty passenger's mode control motor, or repair the passenger's mode control linkage or doors • Faulty climate control unit
DTC: P1297 **1T PCM** **Year:** 2009, 2010 **Model:** MDX, RDX, RL, TSX, ZDX **Engine:** 2.3L L4, 2.4L L4, 3.5L V6, 3.7L V6 **Transmission:** All	**Electrical Load Detector (ELD) Circuit Low Voltage:** The ELD output voltage is 0.27 V or less for at least 5 seconds. The execution time is continuous and DTC P1298 is not active. **NOTE: Before you troubleshoot, record all freeze data and any on-board snapshot.** **Possible causes:** • Poor connections or loose terminals at the ELD and the PCM • "Short" circuit between the PCM and the ELD • Faulty left side engine compartment wire harness • PCM may need to be updated with the latest software • Faulty PCM

DTC	Trouble Code Title, Conditions & Possible Causes
DTC: P1298 **1T PCM** **Year:** 2009, 2010 **Model:** MDX, RDX, RL, TSX, ZDX **Engine:** 2.3L L4, 2.4L L4, 3.5L V6, 3.7L V6 **Transmission:** All	**Electrical Load Detector (ELD) Circuit High Voltage:** With the ignition switch ON (II) the ELD output voltage is 4.57 V or more for at least 5 seconds. The execution time is continuous and DTC P1297 is not active. **NOTE: Before you troubleshoot, record all freeze data and any on-board snapshot.** **Possible causes:** • Blown fuse • Poor connections or loose terminals at the ELD and the PCM • "Open" circuit between the fuse in the under-dash fuse/relay box and the ELD • "Open" in the wire between the ELD and ground • Faulty left side engine compartment wire harness • PCM may need to be updated with the latest software • Faulty PCM
DTC: P1454 **2T PCM, MIL: Yes** **Year:** 2009, 2010 **Model:** MDX, RDX, RL, TSX, ZDX **Engine:** 2.3L L4, 2.4L L4, 3.5L V6, 3.7L V6 **Transmission:** All	**Fuel Tank Pressure (FTP) Sensor Circuit Range/Performance Problem:** One of these 2 conditions is met. (1) The FTP sensor output fluctuates by 0.6 kPa (0.1 in.Hg, 5 mmHg) or more, or -0.6 kPa (-0.1 in.Hg, -5 mmHg) or less for at least 3 seconds. (2) The FTP sensor output value is -1.3 kPa (-0.3 in.Hg, -10 mmHg) or less for at least 3 seconds. DTCs P0452, P0453 are judged as OK. **NOTE: Before you troubleshoot, record all freeze data and any on-board snapshot.** **Possible causes:** • Poor connections or loose terminals at the FTP sensor, the EVAP canister vent shut valve, and the PCM • Blockage in the EVAP canister, canister filter, vent hoses, and drain joint, • Blockage in the FTP sensor air tube or vent • Faulty FTP sensor • Faulty EVAP canister vent shut valve
DTC: P145A **2T PCM, MIL: Yes** **Year:** 2009, 2010 **Model:** RDX **Engine:** 2.3L L4 **Transmission:** All	**EVAP System Incorrect Purge Flow Detected:** Engine coolant temperature 140 °F (60 °C) or more. Map value 20 kPa (6.0 in.Hg, 150 mm). Charging pressure 14 kPa (4.0 in.Hg, 100 mmHg). Battery voltage a minimum of 10.5 V. Evap Canister purge valve duty 15-90%. Fuel trim 0.69-1.47. Vehicle in closed loop. When the P145D is determined as failure and when P0496 is determined as normal, abnormal boost area purge flow is judged. **NOTE: Perform the EVAP function test in the inspection menu with the HDS.** **Possible causes:** • Poor connections or loose terminals at the FTP sensor and the PCM • Poor connection, blockage, or damage at the EVAP canister purge line between the EVAP canister purge nozzle and EVAP canister purge valve • Poor connection, blockage, or damage at the hose between EVAP canister purge nozzle and turbocharger bypass valve base • Blockage in the vacuum hose between the EVAP canister purge valve and the EVAP canister • Faulty vacuum hose between the turbocharger bypass control valve base and the EVAP canister purge nozzle • Faulty non-return valve
DTC: P145B **2T PCM, MIL: Yes** **Year:** 2009, 2010 **Model:** RDX **Engine:** 2.3L L4 **Transmission:** All	**EVAP System Purge Line Non-return Valve A Stuck Open:** Engine coolant temperature 140 °F (60 °C) or more. Map value 20 kPa (6.0 in.Hg, 150 mm). Charging pressure 14 kPa (4.0 in.Hg, 100 mmHg). Battery voltage a minimum of 10.5 V. Evap Canister purge valve duty 15-90%. Fuel trim 0.69-1.47. Vehicle in closed loop. The fuel tank pressure pulse phase misalignment duration is 20 milliseconds or more for at least 13 seconds. **NOTE: Perform the EVAP function test in the inspection menu with the HDS.** **Possible causes:** • Faulty non-return valve A • Faulty EVAP canister purge line
DTC: P145C **2T PCM, MIL: Yes** **Year:** 2009, 2010 **Model:** MDX, RDX, RL, TL, TSX, ZDX **Engine:** 2.3L L4, 2.4L L4, 3.5L V6, 3.7L V6 **Transmission:** All	**Evaporative Emission (EVAP) System Purge Flow Malfunction:** The pulses detected by the fuel tank pressure sensor are 1.0 % of the duty cycle or less for at least 31 seconds. The execution time is continuous. **NOTE: This DTC is representative of an EVAP system purge flow problem. If DTC P145C is indicated alone, troubleshoot P0496 and P0497 using the freeze data for P145C.** If any of the DTCs listed below are indicated at the same time as DTC P145C, troubleshoot those DTC first, then recheck for P145C: * P0496, P0497: EVAP system purge flow. **Possible causes:** • Troubleshoot appropriate DTCs

DTC	Trouble Code Title, Conditions & Possible Causes
DTC: P1549 **1T PCM** **Year:** 2009, 2010 **Model:** MDX, RDX, RL, TL, TSX, ZDX **Engine:** 2.3L L4, 2.4L L4, 3.5L V6, 3.7L V6 **Transmission:** All	**Charging System High Voltage:** The IGP terminal voltage is 16.0 volts or more for at least 60 seconds. The execution time is continuous. **NOTE: If a high voltage battery (24 V, etc.) is connected to the vehicle, this DTC can be stored.** **Possible causes:** • Poor connections or loose terminals at the alternator and the main under-hood fuse box • Faulty alternator
DTC: P1648 **1T BCM** **Year:** 2010 **Model:** ZDX **Engine:** 3.7L V6 **Transmission:** All	**Passenger's Door LF Antenna Circuit Open or Short:** Passenger's Door LF Antenna Circuit Open or Short condition present. **Possible causes:** • Loose or poor connections • "Open" or "Short condition in the wires • Faulty keyless access control unit • Faulty passenger's door LF antenna, replace the front passenger's door outer handle
DTC: P1658 **1T PCM, MIL: Yes** **Year:** 2009, 2010 **Model:** MDX, RDX, ZDX **Engine:** 2.3L L4, 3.7L V6 **Transmission:** All	**Electronic Throttle Control System (ETCS) Control Relay ON Malfunction:** The communication signal is input from the throttle actuator control module for at least 2 seconds, after the throttle actuator control module relay is turned off. The execution time is once per driving cycle. DTCs P0122, P0123, P0222, P0223, P1659, P2101, P2118, P2122, P2123, P2127, P2128, P2135, P2138, P2176 are not active. **NOTE: Before you troubleshoot, record all freeze data and any on-board snapshot.** **Possible causes:** • Poor connections or loose terminals at the under-hood fuse/relay box (ETCS control relay) and the PCM • "Short" in the wire between the PCM and the under-hood fuse/relay box (ETCS control relay) • "Short" to power in the wire between the PCM and the under-hood fuse/relay box (ETCS control relay) • Faulty relay control module (under-hood fuse/relay box) • PCM needs to be updated with the latest software • Faulty PCM
DTC: P1659 **1T PCM, MIL: Yes** **Year:** 2009, 2010 **Model:** MDX, RDX, ZDX **Engine:** 2.3L L4, 3.7L V6 **Transmission:** All	**Electronic Throttle Control System (ETCS) Control Relay OFF Malfunction:** Ignition switch ON (II), battery voltage 10 volts minimum and DTCs P0122, P0123, P0222, P0223, P2101, P2118, P2122, P2123, P2127, P2128, P2135, P2138, P2176 are not active. The voltage is not applied from the throttle actuator controller relay for at least 200 milliseconds. The execution time is once per driving cycle. **NOTE: Before you troubleshoot, record all freeze data and any on-board snapshot.** **Possible causes:** • Blown fuse • Poor connections or loose terminals at the under-hood fuse/relay box (ETCS control relay) and the PCM • "Open" or "Short" in the wire between the PCM and the under-hood fuse/relay box • Faulty right side engine compartment wire harness • Faulty relay control module (under-hood fuse/relay box) • PCM needs to be updated with the latest software • Faulty PCM
DTC: P1683 **1T PCM, MIL: Yes** **Year:** 2009, 2010 **Model:** MDX, RDX, RL, TL, TSX, ZDX **Engine:** 2.3L L4, 2.4L L4, 3.5L V6, 3.7L V6 **Transmission:** All	**Throttle Valve Default Position Spring Performance Problem:** Ignition switch LOCK (0), battery voltage a minimum of 6 volts, engine coolant minimum 158 °F (70 °C) and DTCs P0117, P0118, P0122, P0123, P0222, P0223, P2101, P2118, P2122, P2123, P2127, P2128, P2135, P2138, P2176 not active. The throttle valve position is more than +5 ° from fully closed, or less than +3 ° from fully closed for at least 3 seconds. **NOTE: Before you troubleshoot, record all freeze data and any on-board snapshot.** **Possible causes:** • Poor connections or loose terminals at the throttle body and the PCM • Faulty throttle body

DTC	Trouble Code Title, Conditions & Possible Causes
DTC: P1684 **1T PCM, MIL: Yes** **Year:** 2009, 2010 **Model:** MDX, RDX, RL, TL, TSX, ZDX **Engine:** 2.3L L4, 2.4L L4, 3.5L V6, 3.7L V6 **Transmission:** All	**Throttle Valve Return Spring Performance Problem:** Ignition switch LOCK (0), coolant temperature minimum 158 °F (70 °C), battery voltage minimum 6 volts and DTCs P0117, P0118, P0122, P0123, P0222, P0223, P2101, P2118, P2122, P2123, P2127, P2128, P2135, P2138, P2176 not active. The throttle valve opening angle is 17 ° or more, or 11 ° or less, for at least 3 seconds. Before you troubleshoot, record all freeze data and any on-board snapshot. **Possible causes:** • Poor connections or loose terminals at the throttle body and the PCM • Faulty throttle body
DTC: P16BB **1T PCM** **Year:** 2009, 2010 **Model:** MDX, RDX, RL, TL, TSX, ZDX **Engine:** 2.3L L4, 2.4L L4, 3.5L V6, 3.7L V6 **Transmission:** All	**Alternator B Terminal Circuit Low Voltage:** Engine speed 500-3,000 rpm, alternator control mode 14.5 volts. The IGP terminal voltage is 12.0 V or less, and the alternator power generation amount is within 1.0 % to 50.0 %, for at least 60 seconds. The execution time is continuous. **NOTE: Before you troubleshoot, record all freeze data and any on-board snapshot.** **Possible causes:** • Faulty battery • Poor connections or loose terminals at the alternator and the main under-hood fuse box • "Open" circuit between the alternator and the main under-hood fuse box • Faulty alternator
DTC: P16BC **1T PCM** **Year:** 2009, 2010 **Model:** MDX, RDX, RL, TL, TSX, ZDX **Engine:** 2.3L L4, 2.4L L4, 3.5L V6, 3.7L V6 **Transmission:** All	**Alternator FR Terminal Circuit/IGP Circuit Low Voltage:** Engine speed 500-3,000 rpm, alternator control mode 14.5 volts minimum. The IGP terminal voltage is 12.0 V or less, and the alternator power generation amount is 0.5 % or less, for at least 60 seconds. The execution time is continuous. **NOTE: Before you troubleshoot, record all freeze data and any on-board snapshot.** **Possible causes:** • Blown fuse • Poor connections or loose terminals at the alternator connector • Poor alternator ground • open in the wire between the alternator and the fuse in the under-dash fuse/relay box • Faulty alternator • PCM may need to be updated with the latest software • Faulty PCM
DTC: P16BD **1T PCM** **Year:** 2009, 2010 **Model:** MDX, RL, TL, TSX, ZDX **Engine:** 3.5L V6, 3.7L V6 **Transmission:** All	**Starter Cut Relay 2 Malfunction:** Engine running, starter switch OFF, DTC P16BE not active. The terminal voltage of the STRLD drops to 2.2 V for at least 5 seconds when the starter cut relay output is turned off. **NOTE: Before you troubleshoot, record all freeze data and any on-board snapshot.** **Possible causes:** • Poor connections or loose terminals at starter cut relay 2 and the PCM • Faulty starter relay 2
DTC: P16BE **1T PCM** **Year:** 2009, 2010 **Model:** MDX, RL, TL, TSX, ZDX **Engine:** 3.5L V6, 3.7L V6 **Transmission:** All	**Starter Cut Relay 1 Malfunction:** Engine running, Starter switch OFF cut relay turn on switching failure, ON STRLD open circuit failure. One of these conditions occurs. (1) The terminal voltage of the STRLD exceeds 3.0 volts for at least 5 seconds when the starter cut relay output is turned off. (2) The terminal voltage of the STRLD is 2.4-2.6 volts for at least 0.8 second when the starter cut relay output is turned on. **NOTE: Before you troubleshoot, record all freeze data and any on-board snapshot.** **Possible causes:** • Poor connections or loose terminals at starter cut relay 1 and the PCM • Faulty starter relay 1
DTC: P16BF **1T PCM** **Year:** 2009, 2010 **Model:** MDX, RL, TL, TSX, ZDX **Engine:** 3.5L V6, 3.7L V6 **Transmission:** All	**Starter Cut Relay STRLY Circuit Malfunction:** With ignition switch ON (II) the signal of the on command from the PCM to the starter and the return signal from the starter cut relay do not coincide for at least 5 seconds. The execution time is continuous. **NOTE: Before you troubleshoot, record all freeze data and any on-board snapshot.** **Possible causes:** • Poor connections or loose terminals at starter cut relay 1, starter cut relay 2, and the PCM • "Short" in the wire between the PCM and starter cut relay 1 or starter cut relay 2 • PCM needs to be updated with the latest software • Faulty PCM

DTC	Trouble Code Title, Conditions & Possible Causes
DTC: P16C0 **1T PCM, MIL: Yes** **Year:** 2009, 2010 **Model:** MDX, RDX, TSX, ZDX **Engine:** 2.3L L4, 2.4L L4, 3.5L V6, 3.7L V6 **Transmission:** All	**PCM A/T Control System Incomplete Update:** The program rewriting process does not finish normally due to an irregular process, such as turning off the PCM power during the A/T CPU rewriting process, and then turning the ignition switch to ON (II) again. The execution time is continuous and the duration time is 1 seconds or more. **NOTE: This code is indicated when PCM updating is incomplete.** **Possible causes:** • PCM may need to be updated with the latest software • Faulty PCM
DTC: P1730 **2T PCM, MIL: Yes, TCIL: Yes** **Year:** 2009, 2010 **Model:** TSX **Engine:** 2.4L L4 **Transmission:** All	**Problem in Shift Control System:** With the engine running and in Drive position allow trans mission to shift to 5th gear. Shift solenoid A or D stuck OFF, Shift solenoid B stuck ON, Shift Valves A, B or D stuck. **Possible causes:** • Low or dirty transmission fluid. • Repair the hydraulic system related to shift valves A, B, and D, or replace the transmission
DTC: P1731 **2T PCM, MIL: Yes, TCIL: Yes** **Year:** 2009, 2010 **Model:** TSX **Engine:** 2.4L L4 **Transmission:** All	**Problem in Shift Control System:** With the engine running and in Drive position allow transmission to shift to 5th gear. Shift solenoid E stuck ON, Shift solenoid E stuck, A/T Clutch pressure control solenoid A stuck OFF. **NOTE: Before you troubleshoot, record all freeze data and any on-board snapshot with the HDS, and review General Troubleshooting Information.** **Possible causes:** • Low or dirty transmission fluid • Faulty A/T clutch pressure control solenoid valve A • Repair the hydraulic system related to shift valve E, or replace the transmission
DTC: P1732 **2T PCM, MIL: Yes, TCIL: Yes** **Year:** 2009, 2010 **Model:** TSX **Engine:** 2.4L L4 **Transmission:** All	**Problem in Shift Control System: Shift Solenoid B or C Stuck ON:** With the engine running and in Drive position allow transmission to shift to 5th gear. Shift solenoid B or C stuck ON, Shift solenoid B or C Stuck. **NOTE: Before you troubleshoot, record all freeze data and any on-board snapshot with the HDS, and review General Troubleshooting Information.** **Possible causes:** • Low or dirty transmission fluid • Faulty shift solenoid valve B or C Stuck ON • Faulty shift solenoid valve B or C Stuck • Faulty transmission
DTC: P1733 **2T PCM, MIL: Yes, TCIL: Yes** **Year:** 2009, 2010 **Model:** TSX **Engine:** 2.4L L4 **Transmission:** All	**Problem in Shift Control System:** With the engine running and in Drive position allow transmission to shift to 5th gear. Shift solenoid D stuck ON, Shift solenoid D Stuck. A/T clutch pressure control switch valve C Stuck OFF. **NOTE: Before you troubleshoot, record all freeze data and any on-board snapshot with the HDS, and review General Troubleshooting Information.** **Possible causes:** • Low or dirty transmission fluid • Faulty shift solenoid valve D • Faulty A/T clutch pressure control switch valve C • Faulty transmission
DTC: P1734 **2T PCM, MIL: Yes, TCIL: Yes** **Year:** 2009, 2010 **Model:** TSX **Engine:** 2.4L L4 **Transmission:** All	**Problem in Shift Control System: :** With the engine running and in Drive position allow transmission to shift to 5th gear. Shift solenoid B or C stuck OFF, Shift solenoid B or C Stuck. **NOTE: Before you troubleshoot, record all freeze data and any on-board snapshot with the HDS, and review General Troubleshooting Information.** **Possible causes:** • Low or dirty transmission fluid • Faulty shift solenoid valve B or C Stuck OFF • Faulty shift solenoid valve B or C Stuck • Faulty transmission

DTC	Trouble Code Title, Conditions & Possible Causes
DTC: P1743 **2T PCM, TCIL: Yes** **Year:** 2009, 2010 **Model:** MDX, RDX, RL, TL **Engine:** 2.3L L4, 3.5L V6, 3.7L V6 **Transmission:** All	**Problem in Shift Control System; Shift Valve E Stuck OFF:** Vehicle speed is 5 mph (9 km/h), torque converter slip ratio 96-110 %. battery minimum of 11 volts. The transmission is neutral against the 5th gear shift command for at least 2 seconds, without records that the engine speed flares when upshifting to 3rd-4th. The execution time is continuous depending on the driving pattern. **NOTE: Before you troubleshoot, record all freeze data and any on-board snapshot,.** **Possible causes:** • Low transmission fluid • Dirty transmission fluid • If the strainer has metal debris or excessive clutch material, replace the transmission • Faulty shift valve E in the main valve body, replace the main valve body • Faulty transmission
DTC: P1744 **2T PCM, TCIL: Yes** **Year:** 2009, 2010 **Model:** MDX, RDX, RL, TL **Engine:** 2.3L L4, 3.5L V6, 3.7L V6 **Transmission:** All	**Problem in Shift Control System; Shift Valve E Stuck ON:** The transmission is held in 5th gear against the 1st gear shift command for at least 2 seconds. After driving the vehicle in 3rd gear in D for at least 2 seconds. The execution time is continuous depending on the driving pattern. **NOTE: Before you troubleshoot, record all freeze data and any on-board snapshot.** **Possible causes:** • Low or dirty transmission fluid • If the strainer has metal debris or excessive clutch material, replace the transmission • Faulty shift valve E in the main valve body, replace the main valve body • Faulty transmission
DTC: P1745 **2T PCM, TCIL: Yes** **Year:** 2009, 2010 **Model:** MDX, RDX, RL, TSX **Engine:** 2.3L L4, 3.5L V6, 3.7L V6 **Transmission:** All	**Problem in Shift Control System; Servo Control Valve Stuck OFF or Servo Valve Stuck OFF:** The engine speed flares when upshifting to 2nd-3rd, the engine speed flares when upshifting to 3rd-4th, driving in 5th gear neutral condition. The execution time is continuous and the duration time is 1-2 seconds or more. **NOTE: Before you troubleshoot, record all freeze data and any on-board snapshot.** **Possible causes:** • Low or dirty transmission fluid • If the strainer has metal debris or excessive clutch material, replace the transmission • Faulty servo control valve in the main valve body, servo valve in the regulator valve body, or replace the main valve body, regulator valve body • Faulty transmission
DTC: P177A **2T PCM, TCIL: Yes** **Year:** 2010 **Model:** ZDX **Engine:** 3.7L V6 **Transmission:** All	**Line Pressure Solenoid Valve A Stuck OFF, or Line Pressure Switch Stuck OFF:** The input signal from the line pressure switch to the PCM is high when the low line pressure mode is commanded. The execution time is continuous and the duration time is 4 seconds or more. **Possible causes:** • Blown fuse • Low or dirty transmission fluid • Inspect the strainer for metal debris or excessive clutch material, if present replace the transmission • "Short" to body ground in the wire between PCM and the driver's under-dash fuse/relay box • Poor connections and loose terminals between the line pressure switch and the PCM • "Open" circuit between line pressure switch and PCM • Faulty line pressure solenoid valve A • Faulty PCM
DTC: P1780 **2T PCM, MIL: Yes, TCIL: Yes** **Year:** 2009, 2010 **Model:** MDX, RDX, RL, TL, TSX **Engine:** 2.3L L4, 3.5L V6, 3.7L V6 **Transmission:** All	**Problem in Shift Control System (Transmission is in Default Mode):** The A/T control switches to the default mode due to a mechanical malfunction. The execution time is continuous depending on the driving pattern. **NOTE: DTC P1780 means there is one or more A/T DTCs about the shift control system. Before you troubleshoot, record all freeze data and any on-board snapshot.** **Possible causes:** • Poor connections and loose terminals at the PCM • PCM needs to be updated with the latest software • Faulty PCM

Gas Engine OBD II Trouble Code List (P2XXX Codes)

DTC	Trouble Code Title, Conditions & Possible Causes
DTC: P2101 **1T PCM, MIL: Yes** **Year:** 2009, 2010 **Model:** MDX, RDX, RL, TL, TSX, ZDX **Engine:** 2.3L L4, 2.4L L4, 3.5L V6, 3.7L V6 **Transmission:** All	**Electronic Throttle Control System (ETCS) Malfunction (Includes Hybrid Models):** Ignition switch ON (II), battery voltage minimum of 6 volts and DTCs P0122, P0123, P0222, P0223, P2118, P2122, P2123, P2127, P2128,P2135, P2138, P2176 are not active. Difference between throttle valve target position and actual throttle valve position, 4-6 ° or more. The execution time is continuous and the duration time is 250-500 milliseconds or more. **NOTE: Before you troubleshoot, record all freeze data and any on-board snapshot.** **Possible causes:** • Dirty throttle body • Poor connections or loose terminals at the throttle body and the PCM • "Open" circuit between the throttle body and the PCM • Faulty throttle body • PCM needs to be updated with the latest software • Faulty PCM
DTC: P2108 **1T PCM, MIL: Yes** **Year:** 2009, 2010 **Model:** RL **Engine:** 3.7L V6 **Transmission:** All	**Throttle Actuator Control Module Problem:** With the ignition ON (II) one of these 4 conditions must be met for at least 200 milliseconds. (1) Data read from the ROM is abnormal. (2) Data read from the RAM is abnormal. (3) The A/D converter standard voltage is out of specified value. (4) The serial signals between the PCM and the throttle actuator control module do not agree. **NOTE: Before you troubleshoot, record all freeze data and any on-board snapshot.** **Possible causes:** • Poor connections or loose terminals at the throttle body, the throttle actuator control module and the PCM • Faulty throttle actuator control module
DTC: P2118 **1T PCM, MIL: Yes** **Year:** 2009, 2010 **Model:** MDX, RDX, RL, TL, TSX, ZDX **Engine:** 2.3L L4, 2.4L L4, 3.5L V6, 3.7L V6 **Transmission:** All	**Throttle Actuator Current Range/Performance Problem (Includes Hybrid Models):** With ignition ON (II), battery voltage minimum of 6 volts and DTCs P0122, P0123, P0222, P0223, P2101, P2122, P2123, P2127, P2128, P2135, P2138, P2176 are not active. The current flow to the throttle actuator is 11 A or more for at least 200 milliseconds. The execution time is continuous. **NOTE: Before you troubleshoot, record all freeze data and any on-board snapshot.** **Possible causes:** • Poor connections or loose terminals at the throttle body and the PCM • "Short" circuit between the PCM (ETCSM-line) and (ETCSM+line) • Faulty throttle body • Faulty throttle actuator control module • PCM may need to be updated with the latest software • Faulty PCM
DTC: P2122 **1T PCM, MIL: Yes** **Year:** 2009, 2010 **Model:** MDX, RDX, RL, TL, TSX, ZDX **Engine:** 2.3L L4, 2.4L L4, 3.5L V6, 3.7L V6 **Transmission:** All	**APP Sensor A or 1 (TP Sensor D) Circuit Low Voltage (Includes Hybrid Models):** With the ignition switch ON (II) the APP sensor A output voltage is 0.2 V or less for at least 200 milliseconds. The execution time is continuous. DTC P2123 is not active. **NOTE: Before you troubleshoot, record all freeze data and any on-board snapshot.** **Possible causes:** • Poor connections or loose terminals at APP sensor A and the PCM • "Open" or "Short" circuit between the PCM and APP sensor A • Faulty APP sensor • Faulty accelerator pedal module • PCM needs to be updated with the latest software • Faulty PCM
DTC: P2123 **1T PCM, MIL: Yes** **Year:** 2009, 2010 **Model:** MDX, RDX, RL, TL, TSX, ZDX **Engine:** 2.3L L4, 2.4L L4, 3.5L V6, 3.7L V6 **Transmission:** All	**APP Sensor A or 1 (TP Sensor D) Circuit High Voltage (Includes Hybrid Models):** With engine running the APP sensor A output voltage is 4.9 V or more for at least 200 milliseconds. The execution time is continuous. DTC P2122 is not active. **Possible causes:** • Poor connections or loose terminals at APP sensor A and the PCM • "Open" circuit between the PCM and APP sensor A • Faulty APP sensor • Faulty accelerator pedal module • PCM needs to be updated with the latest software • Faulty PCM

DTC	Trouble Code Title, Conditions & Possible Causes
DTC: P2127 **1T PCM, MIL: Yes** **Year:** 2009, 2010 **Model:** MDX, RDX, RL, TL, TSX, ZDX **Engine:** 2.3L L4, 2.4L L4, 3.5L V6, 3.7L V6 **Transmission:** All	**APP Sensor B or 2 (Throttle Position (TP) Sensor E) Circuit Low Voltage (Includes Hybrid Models):** With the ignition ON (II) the APP sensor B output voltage is 0.2 V or less for at least 200 milliseconds. The execution time is continuous. DTC P2128 is not active. **NOTE: Before you troubleshoot, record all freeze data and any on-board snapshot.** **Possible causes:** • Poor connections or loose terminals at APP sensor B and the PCM • "Open" or "Short" circuit between the PCM and APP sensor B • Faulty accelerator pedal module • Faulty APP sensor • PCM needs to be updated with the latest software • Faulty PCM
DTC: P2128 **1T PCM, MIL: Yes** **Year:** 2009, 2010 **Model:** MDX, RDX, RL, TL, TSX, ZDX **Engine:** 2.3L L4, 2.4L L4, 3.5L V6, 3.7L V6 **Transmission:** All	**APP Sensor B or 2 (Throttle Position (TP) Sensor E) Circuit High Voltage (Includes Hybrid Models):** With engine running the APP sensor B output voltage is 4 V or more for at least 200 milliseconds. The execution time is continuous. DTC P2127 is not active. **NOTE: Before you troubleshoot, record all freeze data and any on-board snapshot.** **Possible causes:** • Poor connections or loose terminals APP sensor B and the PCM • "Open" circuit between the PCM and APP sensor B • Faulty accelerator pedal module • Faulty APP sensor • PCM may need to be updated with the latest software • Faulty PCM
DTC: P2135 **1T PCM, MIL: Yes** **Year:** 2009, 2010 **Model:** MDX, RDX, RL, TL, TSX, ZDX **Engine:** 2.3L L4, 2.4L L4, 3.5L V6, 3.7L V6 **Transmission:** All	**Throttle Position (TP) Sensor A/B or 1/2 Incorrect Voltage Correlation (Includes Hybrid Models):** The difference between the throttle valve positions indicated by TP sensor A and TP sensor B exceeds the value as follows, for at least 200 milliseconds. Difference between TP sensor A and TP sensor B 1.8-14.7° or more, or 5° or less. **NOTE: Before you troubleshoot, record all freeze data and any on-board snapshot.** **Possible causes:** • Poor connections or loose terminals at the throttle body and the PCM • "Short" circuit between the PCM (TPSA line) and the (TPSB line) • Faulty throttle body • PCM may need to be updated with the latest software • Faulty PCM
DTC: P2138 **1T PCM, MIL: Yes** **Year:** 2009, 2010 **Model:** MDX, RDX, RL, TL, TSX, ZDX **Engine:** 2.3L L4, 2.4L L4, 3.5L V6, 3.7L V6 **Transmission:** All	**APP Sensor A/B or 1/2 (Throttle Position (TP) Sensor D/E) Incorrect Voltage Correlation (Includes Hybrid Models):** With the ignition ON (II) one of these conditions must be met for at least 300 milliseconds: (1) APP sensor B voltage exceeds the range from 0 V or less to 0.37 V or more when the APP sensor A voltage is 0.37 V. (2) APP sensor B voltage exceeds the range from 2.31 V or less to 2.69 V or more when the APP sensor A voltage is 5 V. The execution time is continuous and DTCs P2122, P2123, P2127, P2128 are not active. **NOTE: Before you troubleshoot, record all freeze data and any on-board snapshot.** **Possible causes:** • Poor connections or loose terminals at the APP sensor and the PCM • "Short" circuit between PCM (APSA line) and (APSB line) • Faulty accelerator pedal module • PCM needs to be updated with the latest software • Faulty PCM
DTC: P2176 **1T PCM, MIL: Yes** **Year:** 2009, 2010 **Model:** MDX, RDX, RL, TL, TSX, ZDX **Engine:** 2.3L L4, 2.4L L4, 3.5L V6, 3.7L V6 **Transmission:** All	**Throttle Actuator Control System Idle Position Not Learned (Includes Hybrid Models):** One of these condition must be met for at least 0.7 seconds. (1) The registration of the throttle valve fully closed position is not completed within a predetermined time after the ignition switch is turned ON. (2) The registered value of the throttle valve fully closed position is 0.74 volts TP1, 1.61 volts TP2 or more, or 0.49 volts TP1, 1.37 volts TP2 or less. The execution time is once per driving cycle and DTCs P0122, P0123, P0222, P0223, P2101, P2118, P2135 are not active. **NOTE: If DTC P2135 is stored at the same time as DTC P2176, troubleshoot DTC P2135 first, then recheck for DTC P2176.** **Possible causes:** • Dirty throttle body • Poor connections or loose terminals at the throttle body and the PCM • "Open" circuit between the throttle body and the PCM • Faulty throttle body • PCM may need to be updated with the latest software • Faulty PCM

DTC	Trouble Code Title, Conditions & Possible Causes
DTC: P2183 **2T PCM, MIL: Yes** **Year:** 2009, 2010 **Model:** MDX, RDX, RL, TL, TSX, ZDX **Engine:** 2.3L L4, 2.4L L4, 3.5L V6, 3.7L V6 **Transmission:** All	**Engine Coolant Temperature (ECT) Sensor 2 Circuit Range/Performance Problem (Includes Hybrid Models):** A malfunction is detected if the following three conditions are present after the engine has stopped and the ignition switch has been turned to LOCK (0) for at least 6 hours before restarting the engine: (1) When the temperature difference between the IAT and ECT1 is 57 °F (32 °C) or more. (2) When the temperature difference between the IAT and ECT2 is 30 °F (17 °C) or more. (3) When the temperature difference between the ECT2 and ECT1 is 73 °F (41 °C) or more. The execution time is once per driving cycle and the duration time is 10 seconds or more. **NOTE: If DTC P0111 is stored at the same time as DTC P2183, troubleshoot DTC P0111 first, then recheck for DTC P2183.** **Possible causes:** • Poor connections or loose terminals at ECT sensor 1, ECT sensor 2, and the PCM • Faulty ECT sensor 1 • Faulty ECT sensor 2
DTC: P2184 **1T PCM, MIL: Yes** **Year:** 2009, 2010 **Model:** MDX, RDX, RL, TL, TSX, ZDX **Engine:** 2.3L L4, 2.4L L4, 3.5L V6, 3.7L V6 **Transmission:** All	**Engine Coolant Temperature (ECT) Sensor 2 Circuit Low Voltage (Includes Hybrid Models):** With ignition switch ON (II) the ECT sensor 2 output voltage is 0.08 V or less for at least 2 seconds. The execution time is continuous. DTC P2185 is not active. **NOTE: Before you troubleshoot, record all freeze data and any on-board snapshot.** **Possible causes:** • Poor connections or loose terminals at ECT sensor 2 and the PCM • "Short" in the wire between ECT sensor 2 and the PCM • Faulty ECT sensor 2 • PCM may need to be updated with the latest software • Faulty PCM
DTC: P2185 **1T PCM, MIL: Yes** **Year:** 2009, 2010 **Model:** MDX, RDX, RL, TL, TSX, ZDX **Engine:** 2.3L L4, 2.4L L4, 3.5L V6, 3.7L V6 **Transmission:** All	**Engine Coolant Temperature (ECT) Sensor 2 Circuit High Voltage (Includes Hybrid Models):** With ignition ON (II) the ECT sensor 2 output voltage is 4.92 volts or more for at least 2 seconds. The execution time is continuous and DTC P2184 is not active. **NOTE: Before you troubleshoot, record all freeze data and any on-board snapshot.** **Possible causes:** • Poor connections or loose terminals at ECT sensor 2 and the PCM • "Open" in the wire between the PCM and ECT sensor 2 • Faulty ECT sensor 2 • PCM may need to be updated with the latest software • Faulty PCM
DTC: P2195 **1T PCM, MIL: Yes** **Year:** 2009, 2010 **Model:** RDX, TSX **Engine:** 2.3L L4, 2.4L L4, 3.5L V6 **Transmission:** All	**A/F Sensor (Sensor 1) Signal Stuck Lean (Includes Hybrid Models):** The A/F sensor (Sensor 1) output voltage is 2.5 V or more for at least 7 seconds. The execution time is once per driving cycle and DTCs P0134, P0135, P0171, P0300, P0301, P0302, P0303, P0304, P0305, P0306, P0627, P1172, P2237, P2238, P2243, P2245, P2251, P2252 are not active. **NOTE: If DTC P2101, P2118, P2135, P2138, P2176, or a combination of P2122 and P2127, P2122 and P2138 or P2127 and P2138 is stored at the same time as DTC P2195, troubleshoot those DTCs first, then recheck for DTC P2195.** **Possible causes:** • Dirty or Faulty MAF sensor/IAT sensor (If equipped) • Loose A/F sensor • Poor connections or loose terminals at the A/F sensor (Sensor 1) and the PCM • Faulty A/F sensor (Sensor 1) • PCM may need to be updated with the latest software • Faulty PCM
DTC: P2195 **2T PCM, MIL: Yes** **Year:** 2009, 2010 **Model:** MDX, RL, TL, TSX, ZDX **Engine:** 3.5L V6, 3.7L V6 **Transmission:** All	**Rear Air/Fuel Ratio (A/F) Sensor (Bank 1, Sensor 1) Signal Stuck Lean:** The rear A/F sensor (bank 1, sensor 1) output voltage is 3.48 V or more for at least 7 seconds. The execution time is once per driving cycle and DTCs P0134, P0135, P0171, P0300, P0301, P0302, P0303, P0304, P0305, P0306, P0627, P1172, P2237, P2238, P2243, P2245, P2251, P2252 are not active. **NOTE: If DTC P2101, P2118, P2135, P2138, P2176, or a combination of P2122 and P2127, P2122 and P2138 or P2127 and P2138 is stored at the same time as DTC P2195, troubleshoot those DTCs first, then recheck for DTC P2195.** **Possible causes:** • Dirty or Faulty MAF sensor/IAT sensor (If equipped) • Loose A/F sensor • Poor connections or loose terminals at the A/F sensor (Sensor 1) and the PCM • Faulty A/F sensor (Sensor 1) • PCM may need to be updated with the latest software • Faulty PCM

DTC	Trouble Code Title, Conditions & Possible Causes
DTC: P2197 **2T PCM, MIL: Yes** **Year:** 2009, 2010 **Model:** MDX, RL, TL, ZDX **Engine:** 3.5L V6, 3.7L V6 **Transmission:** All	**Front Air/Fuel Ratio (A/F) Sensor (Bank 2, Sensor 1) Signal Stuck Lean:** The front A/F sensor (bank 2, sensor 1) output voltage is 3.48 V or more for at least 7 seconds. The execution time is once per driving cycle and DTCs P0154, P0155, P0174, P0300, P0301, P0302, P0303, P0304, P0305, P0306, P0627, P1174, P2240, P2241, P2247, P2249, P2254, P2255 are not active. **NOTE: If DTC P2101, P2118, P2135, P2138, P2176, or a combination of P2122 and P2127, P2122 and P2138 or P2127 and P2138 is stored at the same time as DTC P2195, P2197, troubleshoot those DTCs first, then recheck for DTC P2197.** **Possible causes:** • Loose A/F sensor • Poor connections or loose terminals at A/F sensor (Sensor 1) and the PCM • Faulty A/F sensor (Sensor 1) • PCM may need to be updated with the latest software • Faulty PCM
DTC: P2199 **2T PCM, MIL: Yes** **Year:** 2009, 2010 **Model:** RDX **Engine:** 2.3L L4 **Transmission:** All	**Intake Air Temperature (IAT) Sensor 1, 2 Correlation:** With the ignition ON for two seconds or OFF for 8 hours the voltage difference between IAT sensor 1 and IAT sensor 2 indicates a difference in intake air temperature of 77°F (25°C) or more. **NOTE: Before you troubleshoot, record all freeze data and any on-board snapshot, and review the general troubleshooting information.** **Possible causes:** • Poor connections or loose terminals at IAT sensor 1, IAT sensor 2 and the PCM • Replace the sensor that has the largest difference from room temperature
DTC: P2227 **2T PCM, MIL: Yes** **Year:** 2009, 2010 **Model:** MDX, RDX, RL, TL, TSX, ZDX **Engine:** 2.3L L4, 2.4L L4, 3.5L V6, 3.7L V6 **Transmission:** All	**Barometric Pressure (BARO) Sensor Circuit Range/Performance Problem:** Throttle position 17.0-33.0 degrees at 1,100-3,000 rpm. The difference between the BARO sensor output and the MAP sensor output is 26 kPa (7.5 in.Hg, 190 mmHg) or more for at least 2.5 seconds. The execution time is once per driving cycle and the duration time is 2.5 seconds or more. **NOTE: If DTC P0107, P0108, P1128, and/or P1129 are stored at the same time as DTC P2227, troubleshoot those DTCs first, then recheck for DTC P2227.** **Possible causes:** • Dirty air cleaner element • Faulty BARO sensor • PCM needs to be updated with the latest software • Faulty PCM
DTC: P2228 **1T PCM, MIL: Yes** **Year:** 2009, 2010 **Model:** MDX, RDX, RL, TL, TSX, ZDX **Engine:** 2.3L L4, 2.4L L4, 3.5L V6, 3.7L V6 **Transmission:** All	**Barometric Pressure (BARO) Sensor Circuit Low Voltage:** With ignition ON (II) the BARO sensor output voltage is 1.31 V or less for at least 2 seconds. The execution time is continuous and DTCs P1109, P2229 are not active. **NOTE: Before you troubleshoot, record all freeze data and any on-board snapshot.** **Possible causes:** • Poor connections or loose terminals at the PCM • Faulty BARO sensor • PCM needs to be updated with the latest software • Faulty PCM
DTC: P2229 **1T PCM, MIL: Yes** **Year:** 2009, 2010 **Model:** MDX, RDX, RL, TL, TSX, ZDX **Engine:** 2.3L L4, 2.4L L4, 3.5L V6, 3.7L V6 **Transmission:** All	**Barometric Pressure (BARO) Sensor Circuit High Voltage:** With ignition switch ON (II) the BARO sensor output voltage is 4.49 V or more for at least 2 seconds. The execution time is continuous and DTCs P1109, P2228 are not active. **NOTE: Before you troubleshoot, record all freeze data and any on-board snapshot.** **Possible causes:** • Faulty BARO sensor • PCM may need to be updated with the latest software • Faulty PCM
DTC: P2237 **1T PCM, MIL: Yes** **Year:** 2009, 2010 **Model:** MDX, RL, TL, ZDX **Engine:** 3.5L V6, 3.7L V6 **Transmission:** All	**Rear Air/Fuel Ratio (A/F) Sensor (Bank 1, Sensor 1) IP Circuit High Voltage:** With the engine running the IPB1 terminal voltage is 2 volts or less, or 5.6 volts or more, for at least 15 seconds. The execution time is continuous and DTCs P0135, P2195, P2238, P2243, P2245, P2251, P2252, P2627, P2628 are not active. **NOTE: Before you troubleshoot, record all freeze data and any on-board snapshot.** **Possible causes:** • Poor connections or loose terminals at the A/F sensor (Sensor 1) and the PCM • "Open" circuit between the PCM and the A/F sensor (Sensor 1) • Faulty A/F sensor (Sensor 1) • PCM may need to be updated with the latest software • Faulty PCM

DTC	Trouble Code Title, Conditions & Possible Causes
DTC: P2238 **1T PCM, MIL: Yes** **Year:** 2009, 2010 **Model:** RDX **Engine:** 2.3L L4 **Transmission:** All	**Air/Fuel Ratio (A/F) Sensor (Sensor 1) AFS+ Circuit Low Voltage:** With the engine running and battery voltage at a minimum of 10.5 V, the AFS+ voltage is 0.4 V or less for at least 4 seconds. **NOTE: Before you troubleshoot, record all freeze data and any on-board snapshot, and review the general troubleshooting information.** **Possible causes:** • Poor connections or loose terminals at the A/F sensor (Sensor 1) and the PCM • "Short" circuit between the PCM and the A/F sensor (Sensor 1) • Faulty A/F sensor (Sensor 1) • PCM may need to be updated with the latest software • Faulty PCM
DTC: P2238 **1T PCM, MIL: Yes** **Year:** 2009, 2010 **Model:** MDX, RL, TL, TSX, ZDX **Engine:** 3.5L V6, 3.7L V6 **Transmission:** All	**Rear Air/Fuel Ratio (A/F) Sensor (Bank 1, Sensor 1) IP Circuit Low Voltage:** With engine running the IPB1 input terminal voltage is 1 volts or less for at least 5 seconds after the sensor becomes active and 85 seconds before the sensor becomes active. The execution time is continuous, DTCs P0135, P2195, P2237, P2243, P2245, P2251, P2252, P2627, P2628 are not active. **NOTE: Before you troubleshoot, record all freeze data and any on-board snapshot.** **Possible causes:** • Poor connections or loose terminals at the A/F sensor (Sensor 1) and the PCM • "Short" circuit between the PCM and the A/F sensor (Sensor 1) • Faulty A/F sensor (Sensor 1) • PCM needs to be updated with the latest software • Faulty PCM
DTC: P2240 **1T PCM, MIL: Yes** **Year:** 2009, 2010 **Model:** MDX, RL, TL, TSX, ZDX **Engine:** 3.5L V6, 3.7L V6 **Transmission:** All	**Front Air/Fuel Ratio (A/F) Sensor (Bank 2, Sensor 1) IP Circuit High Voltage:** With the engine running the IPB2 terminal voltage is 5.6 volts or more, or 2 volts or less, for at least 15 seconds. The execution time is continuous and DTCs P0155, P2197, P2240, P2241, P2247, P2249, P2254, P2255, P2630, P2631 are not active. **NOTE: Before you troubleshoot, record all freeze data and any on-board snapshot.** **Possible causes:** • Poor connections or loose terminals at the A/F sensor (Sensor 1) • "Open" circuit between the PCM and the A/F sensor (Sensor 1) • Faulty A/F sensor (Sensor 1) • PCM needs to be updated with the latest software • Faulty PCM
DTC: P2241 **1T PCM, MIL: Yes** **Year:** 2009, 2010 **Model:** MDX, RL, TL, ZDX **Engine:** 3.5L V6, 3.7L V6 **Transmission:** All	**Front Air/Fuel Ratio (A/F) Sensor (Bank 2, Sensor 1) IP Circuit Low Voltage:** With engine running the IPB1 input terminal voltage is 1 volts or less for at least 5 seconds after the sensor becomes active and 85 seconds before the sensor becomes active. The execution time is continuous DTCs P0135, P2195, P2237, P2243, P2245, P2251, P2252, P2627, P2628 are not active. **NOTE: Before you troubleshoot, record all freeze data and any on-board snapshot.** **Possible causes:** • Poor connections or loose terminals at the A/F sensor (Sensor 1) and the PCM • "Short" circuit between the PCM and the A/F sensor (Sensor 1) • Faulty A/F sensor (Sensor 1) • PCM needs to be updated with the latest software • Faulty PCM
DTC: P2243 **1T PCM, MIL: Yes** **Year:** 2009, 2010 **Model:** MDX, RL, TL, TSX, ZDX **Engine:** 3.5L V6, 3.7L V6 **Transmission:** All	**Rear Air/Fuel Ratio (A/F) Sensor (Bank 1, Sensor 1) VCENT Circuit High Voltage:** With engine running the VSB1 terminal voltage repeatedly fluctuates from a value above 4.8 volts to a value below 3.4 volts, at least 150 times. The execution time is continuous and DTCs P0135, P2195, P2237, P2238, P2245, P2251, P2252, P2627, P2628 are not active. **NOTE: Before you troubleshoot, record all freeze data and any on-board snapshot.** **Possible causes:** • Poor connections or loose terminals at the A/F sensor (Sensor 1) and the PCM • "Open" circuit between the PCM and the A/F sensor (Sensor 1) • Faulty A/F sensor (Sensor 1) • PCM needs to be updated with the latest software • Faulty PCM

DTC	Trouble Code Title, Conditions & Possible Causes
DTC: P2245 **1T PCM, MIL: Yes** **Year:** 2009, 2010 **Model:** MDX, RL, TL, TSX, ZDX **Engine:** 3.5L V6, 3.7L V6 **Transmission:** All	**Rear Air/Fuel Ratio (A/F) Sensor (Bank 1, Sensor 1) VCENT Circuit Low Voltage:** With engine running the IPB1 input terminal voltage is 1 volts or less for at least 5 seconds after the sensor becomes active, 85 seconds before the sensor becomes active. The execution time is continuous and DTCs P0135, P2195, P2237, P2238, P2243, P2251, P2252, P2627, P2628 are not active. **Possible causes:** • Poor connections or loose terminals at the A/F sensor (Sensor 1) and the PCM • "Short" circuit between the PCM and the A/F sensor (Sensor 1) • Faulty A/F sensor (Sensor 1) • PCM needs to be updated with the latest software • Faulty PCM
DTC: P2246 **1T PCM, MIL: Yes** **Year:** 2009, 2010 **Model:** MDX **Engine:** 3.7L V6 **Transmission:** All	**Rocker Arm Oil Pressure Switch Circuit Low Voltage:** When the engine rpm is 4,300 (High lift cam operation) the rocker arm oil control solenoid is ON, the rocker arm oil pressure switch remains ON. **NOTE: Before you troubleshoot, record all freeze data and any on-board snapshot.** **Possible causes:** • Low engine oil level • Low oil pressure • Poor connections or loose terminals at the rocker arm oil pressure switch, the rocker arm oil control solenoid, and the PCM • "Short" circuit between the PCM and the rocker arm oil pressure switch • Faulty rocker arm oil pressure switch • Faulty rocker arm oil control solenoid • Rocker arm malfunction
DTC: P2247 **1T PCM, MIL: Yes** **Year:** 2009, 2010 **Model:** MDX, RL, TL, TSX, ZDX **Engine:** 3.5L V6, 3.7L V6 **Transmission:** All	**Front A/F Sensor (Bank 2, Sensor 1) VCENT Circuit High Voltage :** With engine running the VSB1 terminal voltage repeatedly fluctuates from a value above 4.8 volts to a value below 3.4 volts, at least 150 times. The execution time is continuous and DTCs P0135, P2195, P2237, P2238, P2245, P2251, P2252, P2627, P2628 are not active. **NOTE: Before you troubleshoot, record all freeze data and any on-board snapshot.** **Possible causes:** • Poor connections or loose terminals at the A/F sensor (Sensor 1) and the PCM • "Open" circuit between the PCM and the A/F sensor (Sensor 1) • Faulty A/F sensor (Sensor 1) • PCM needs to be updated with the latest software • Faulty PCM
DTC: P2249 **1T PCM, MIL: Yes** **Year:** 2009, 2010 **Model:** MDX, RL, TL, TSX, ZDX **Engine:** 3.5L V6, 3.7L V6 **Transmission:** All	**Front A/F Sensor (Bank 2, Sensor 1) VCENT Circuit Low Voltage :** With the engine running the VSB2 terminal voltage between reads 0.3-1.5 volts. The execution time is continuous and DTCs P0155, P2197, P2240, P2241, P2247, P2254, P2255 are not active. The IPB2 input terminal voltage is 1.0 volts or less for at least 5 seconds, after the sensor becomes active, 85 seconds before the sensor becomes active. **NOTE: Before you troubleshoot, record all freeze data and any on-board snapshot.** **Possible causes:** • Poor connections or loose terminals at the A/F sensor (Sensor 1) and the PCM • "Short" circuit between the PCM and the A/F sensor (Sensor 1) • Faulty A/F sensor (Sensor 1) • PCM needs to be updated with the latest software • Faulty PCM
DTC: P2251 **1T PCM, MIL: Yes** **Year:** 2009, 2010 **Model:** MDX, RL, TL, TSX, ZDX **Engine:** 3.5L V6, 3.7L V6 **Transmission:** All	**Rear Air/Fuel Ratio (A/F) Sensor (Bank 1, Sensor 1) VS Circuit High Voltage:** With engine running the VSB1 terminal voltage is 6 V or more, and the PCM internal signal voltage is 4.6 V or more, for at least 5 seconds. The execution time is continuous and DTCs P0135, P2195, P2237, P2238, P2243, P2245, P2252, P2627, P2628 are not active. **NOTE: If DTC P2251 is stored at the same time as DTC P0134, troubleshoot DTC P2251 first, then recheck for P0134.** **Possible causes:** • Poor connections or loose terminals at the A/F sensor (Sensor 1) and the PCM • "Open" circuit between the PCM and the A/F sensor (Sensor 1) • Faulty A/F sensor (Sensor 1) • PCM may need to be updated with the latest software • Faulty PCM

DTC	Trouble Code Title, Conditions & Possible Causes
DTC: P2252 **1T PCM, MIL: Yes** **Year:** 2009, 2010 **Model:** MDX, RL, TL, TSX, ZDX **Engine:** 3.5L V6, 3.7L V6 **Transmission:** All	**Rear Air/Fuel Ratio (A/F) Sensor (Bank 1, Sensor 1) VS Circuit Low Voltage:** With engine running the VSB1 terminal voltage is 0.3 V or less, and the IPD1 terminal voltage is 1 V or more, for at least 5 seconds after the sensor becomes active, 85 seconds before the sensor becomes active. **Possible causes:** • Poor connections or loose terminals at the A/F sensor (Sensor 1) and the PCM • "Short" circuit between the PCM and the A/F sensor (Sensor 1) • Faulty A/F sensor (Sensor 1) • PCM may need to be updated with the latest software • Faulty PCM
DTC: P2252 **1T PCM, MIL: Yes** **Year:** 2009, 2010 **Model:** RDX, TSX **Engine:** 2.3L L4, 2.4L L4 **Transmission:** All	**Air/Fuel Ratio (A/F) Sensor (Sensor 1) AFS- Circuit Low Voltage :** With the engine running the AFS-terminal voltage is 0.4 V or less for at least 5 seconds. **NOTE: Before you troubleshoot, record all freeze data and any on-board snapshot, and review the general troubleshooting information.** **Possible causes:** • Poor connections or loose terminals at the A/F sensor (Sensor 1) and the ECM/PCM • "Short" in the wire between the ECM/PCM and the A/F sensor (Sensor 1) • Faulty A/F sensor (Sensor 1) • PCM may need to be updated with the latest software • Faulty ECM/PCM
DTC: P2254 **1T PCM, MIL: Yes** **Year:** 2009, 2010 **Model:** MDX, RL, TL, TSX, ZDX **Engine:** 3.5L V6, 3.7L V6 **Transmission:** All	**Front A/F Sensor (Bank 2, Sensor 1) VS Circuit High Voltage:** With engine running the VSB2 terminal voltage is 6 V or more, and the PCM internal signal voltage is 4.6 V or more, for at least 5 seconds. The execution time is continuous and DTCs P0155, P2197, P2240, P2241, P2247, P2249, P2255, P2630, P2631 are not active. **NOTE: If DTC P2254 is stored at the same time as DTC P0154, troubleshoot DTC P2254 first, then recheck for P0154.** **Possible causes:** • Poor connections or loose terminals at the A/F sensor (Sensor 1) and the PCM • "Open" circuit between the PCM and the A/F sensor (Sensor 1) • Faulty A/F sensor (Sensor 1) • PCM may need to be updated with the latest software • Faulty PCM
DTC: P2255 **1T PCM, MIL: Yes** **Year:** 2009, 2010 **Model:** MDX, RL, TL, TSX, ZDX **Engine:** 3.5L V6, 3.7L V6 **Transmission:** All	**Front A/F Sensor (Bank 2, Sensor 1) VS Circuit Low Voltage :** With engine running the VSB2 terminal voltage is 0.3 V or less, and the IPB2 terminal voltage is 1 V or more, for at least 5 seconds after the sensor is active, 85 seconds before the sensor becomes active. **NOTE: Before you troubleshoot, record all freeze data and any on-board snapshot.** **Possible causes:** • Poor connections or loose terminals at the A/F sensor (Sensor 1) and the PCM • "Short" circuit between the PCM and the A/F sensor (Sensor 1) • Faulty A/F sensor (Sensor 1) • PCM may need to be updated with the latest software • Faulty PCM
DTC: P2261 **2T PCM, MIL: Yes** **Year:** 2009, 2010 **Model:** RDX **Engine:** 2.3L L4 **Transmission:** All	**Turbocharger Bypass Valve Stuck Closed:** When the throttle from full open supercharging condition returns to fully closed the pulse method of releasing pressure accumulation from the turbocharger bypass control solenoid valve is 350 or more for at least 2.5 seconds. **NOTE: Before you troubleshoot, record all freeze data and any on-board snapshot, and review the general troubleshooting information.** **Possible causes:** • Clogged air bypass outlet pipe • Faulty turbocharger bypass control solenoid valve and/or connecting hose

DTC	Trouble Code Title, Conditions & Possible Causes
DTC: P2263 **2T PCM, MIL: Yes** **Year:** 2009, 2010 **Model:** RDX **Engine:** 2.3L L4 **Transmission:** All	**Turbocharger Boost System Performance Problem:** With the engine at operating temperature and throttle not fully open, rpm 1,500 or more the degree of boost pressure rise delay is 0.1 or less for at least 2.0 seconds. **NOTE: If DTC P0229 is stored at the same time as DTC P2263, troubleshoot DTC P0229 first, then recheck for DTC P2263.** **Possible causes:** • Dirty air cleaner element • Leaks or damage at the intake air duct components • Faulty turbocharger boost sensor • Faulty turbocharger bypass control solenoid valve • Faulty turbocharger bypass control valve • Poor connections, blockage, or damage between the turbocharger bypass control solenoid valve and the turbocharger bypass control valve • Faulty turbocharger wastegate control solenoid valve • Blockage or clogging between the turbocharger wastegate control solenoid valve and intake air duct (turbocharger inlet connecting tube) • Faulty turbocharger and/or the turbocharger wastegate control actuator • Clogged exhaust • Faulty turbocharger boost control solenoid valve
DTC: P2270 **2T PCM, MIL: Yes** **Year:** 2009, 2010 **Model:** RDX **Engine:** 2.3L L4 **Transmission:** All	**Secondary HO2S (Sensor 2) Circuit Signal Stuck Lean:** With the engine running, and in closed loop with a speed of 30 mph or more, the secondary HO2S output voltage is 0.650 V or less. **NOTE: Before you troubleshoot, record all freeze data and any on-board snapshot, and review the general troubleshooting information.** **Possible causes:** • Poor connections or loose terminals at the secondary HO2S (Sensor 2) and the PCM • Faulty secondary HO2S (Sensor 2)
DTC: P2270 **2T PCM, MIL: Yes** **Year:** 2009, 2010 **Model:** MDX, RL, ZDX **Engine:** 3.7L V6 **Transmission:** All	**Rear Secondary Heated Oxygen Sensor (Secondary HO2S (Bank 1, Sensor 2) Circuit Signal Stuck Lean:** The rear secondary HO2S (bank 1, sensor 2) output voltage is 0.650 V or less. The execution time is once per driving cycle and the duration time is 36.3 seconds or less. **NOTE: Before you troubleshoot, record all freeze data and any on-board snapshot.** **Possible causes:** • Poor connections or loose terminals at the secondary HO2S (Sensor 2) and the PCM • Faulty secondary HO2S (Sensor 2)
DTC: P2271 **2T PCM, MIL: Yes** **Year:** 2009, 2010 **Model:** MDX, RL **Engine:** 3.7L V6 **Transmission:** All	**Rear Secondary HO2S (Bank 1, Sensor 2) Circuit Signal Stuck Rich :** If, after current is applied to the rear secondary HO2S heater, the rear secondary HO2S does not fluctuate and the output is stuck within the specified area, a malfunction is detected. The rear secondary HO2S (bank 1, sensor 2) output voltage is 0.293 V or more. The execution time is once per driving cycle and the duration time is 36.3 seconds or more. **NOTE: Before you troubleshoot, record all freeze data and any on-board snapshot.** **Possible causes:** • Poor connections or loose terminals at the secondary HO2S (Sensor 2) and the PCM • Faulty secondary HO2S (Sensor 2)
DTC: P2271 **2T PCM, MIL: Yes** **Year:** 2009, 2010 **Model:** RDX **Engine:** 2.3L L4 **Transmission:** All	**Secondary HO2S (Sensor 2) Circuit Signal Stuck Rich :** The secondary HO2S (Sensor 2) output voltage is 0.293 V or more. The execution time is once per driving cycle and the duration time is 24.2 seconds or more. **NOTE: Before you troubleshoot, record all freeze data and any on-board snapshot.** **Possible causes:** • Poor connections or loose terminals at the secondary HO2S (Sensor 2) and the PCM • Faulty secondary HO2S (Sensor 2)
DTC: P2272 **2T PCM, MIL: Yes** **Year:** 2009, 2010 **Model:** MDX, RL, ZDX **Engine:** 3.7L V6 **Transmission:** All	**Front Secondary HO2S (Bank 2, Sensor 2) Circuit Signal Stuck Lean :** If, after current is applied to the front secondary HO2S heater, the front secondary HO2S does not fluctuate and the output is stuck within the specified area, a malfunction is detected. The front secondary HO2S (bank 2, sensor 2) output voltage is 0.650 V or less. **Possible causes:** • Poor connections or loose terminals at the secondary HO2S (Sensor 2) and the PCM • Faulty secondary HO2S (Sensor 2)
DTC: P2273 **2T PCM, MIL: Yes** **Year:** 2009, 2010 **Model:** MDX, RL **Engine:** 3.7L V6 **Transmission:** All	**Front Secondary HO2S (Bank 2, Sensor 2) Circuit Signal Stuck Rich :** If, after current is applied to the front secondary HO2S heater, the front secondary HO2S does not fluctuate and the output is stuck within the specified area, a malfunction is detected. The front secondary HO2S (bank 2, sensor 2) output voltage is 0.293 V or more. **NOTE: Before you troubleshoot, record all freeze data and any on-board snapshot.** **Possible causes:** • Poor connections or loose terminals at the secondary HO2S (Sensor 2) and the PCM • Faulty secondary HO2S (Sensor 2)

DTC	Trouble Code Title, Conditions & Possible Causes
DTC: P2279 **2T PCM, MIL: Yes** **Year:** 2009, 2010 **Model:** MDX, RL, TL, TSX, ZDX **Engine:** 2.4L L4, 3.5L V6, 3.7L V6 **Transmission:** All	**Intake Air System Leak:** With the engine running in closed loop for a minimum of 15 seconds. Either of these 2 conditions is met: (1) The estimated volume of intake air is 310 l/min (327.6 US qt/min, 272.8 Imp qt/min) or more when the MAP value is 35 kPa (10.3 in. Hg, 260 mmHg). (2) The estimated volume of intake air is 302 l/min (319.2 US qt/min, 265.8 Imp qt/min) or more when the MAP value is 62 kPa (18.2 in. Hg, 460 mmHg). The execution time is once per driving cycle and the duration time is 22 seconds or more. **NOTE: If DTC P0443 is stored at the same time as DTC P2279, troubleshoot DTC P0443 first, then recheck for DTC P2279.** **Possible causes:** • Vacuum leaks at the PCV valve, the PCV hose, the purge (PCS) line, the throttle body, the intake manifold, and the brake booster hose • Incorrect camshaft timing
DTC: P2413 **2T PCM, MIL: Yes** **Year:** 2009, 2010 **Model:** MDX, RL, TL, TSX, ZDX **Engine:** 3.5L V6, 3.7L V6 **Transmission:** All	**Exhaust Gas Recirculation (EGR) System Malfunction:** With the engine running at 4,400 rpm, commanded EGR valve lift is 0.040 in. (1 mm). If the actual valve lift is 0.006 in. (0.15 mm) or less for at least 5 seconds, the valve is considered stuck closed. **NOTE: Before you troubleshoot, record all freeze data and any on-board snapshot.** **Possible causes:** • Clogged intake manifold EGR port or EGR valve • Poor connections or loose terminals at the EGR valve and the PCM • "Open" in the wire between the EGR valve and ground • "Open" or "Short" circuit between the PCM and the EGR valve • Faulty EGR valve • PCM needs to be updated with the latest software • Faulty PCM
DTC: P2422 **2T PCM, MIL: Yes** **Year:** 2009, 2010 **Model:** MDX, RDX, TL, TSX, ZDX **Engine:** 2.3L L4, 2.4L L4, 3.5L V6, 3.7L V6 **Transmission:** All	**EVAP Canister Vent Shut Valve Close Malfunction :** The output from the fuel tank pressure sensor is -4 kPa (-1.0 in. Hg, -25 mmHg), -2 kPa (-0.4 in. Hg, -10 mmHg) or less for at least 1.04-8 seconds. Elapsed time after the FTP sensor output exceeds the malfunction threshold 1.04 seconds. Excessive negative pressure is detected 8 seconds. **Possible causes:** • Poor connections or loose terminals at the FTP sensor, the EVAP canister vent shut valve, and the PCM • Blockage in the EVAP canister, canister filter, vent hoses, and drain joint, • Blockage in the FTP sensor air tube or vent • Faulty FTP sensor • Faulty EVAP canister vent shut valve
DTC: P2552 **1T PCM, MIL: Yes** **Year:** 2009, 2010 **Model:** RL **Engine:** 3.7L V6 **Transmission:** All	**Throttle Actuator Control Module Relay Malfunction :** The serial signal is input from the throttle actuator control module for at least 2 seconds after the throttle actuator control module relay is turned OFF. **NOTE: Before you troubleshoot, record all freeze data and any on-board snapshot.** **Possible causes:** • Poor connections or loose terminals at the throttle actuator control module relay, the throttle actuator control module, and the PCM • Faulty throttle actuator control module relay • "Short" circuit between the throttle actuator control module relay and the PCM • PCM may need to be updated with the latest software • Faulty PCM
DTC: P2610 **1T PCM, MIL: Yes** **Year:** 2009, 2010 **Model:** MDX, RDX, RL, TL, TSX, ZDX **Engine:** 2.3L L4, 2.4L L4, 3.5L V6, 3.7L V6 **Transmission:** All	**ECM/PCM Ignition Off Internal Timer Malfunction:** The access process to the ignition off timer fails, or a malfunction is found in the read data for at least 10 seconds. Ignition switch ON (II) when a battery is disconnected and connected again is excluded. **NOTE: Before you troubleshoot, record all freeze data and any on-board snapshot.** **Possible causes:** • ECM/PCM needs to be updated with the latest software • Faulty ECM/PCM

DTC	Trouble Code Title, Conditions & Possible Causes
DTC: P2646 **1T PCM, MIL: Yes** **Year:** 2009, 2010 **Model:** MDX, RDX, RL, TL, TSX, ZDX **Engine:** 2.3L L4, 2.4L L4, 3.5L V6, 3.7L V6 **Transmission:** All	**Rocker Arm Oil Pressure Switch (VTEC Oil Pressure Switch) Circuit Low Voltage :** With the engine running, engine speed a minimum of 4,800 rpm (Variable that is depending on the engine load). The rocker arm oil control solenoid is ON, the rocker arm oil pressure switch remains ON. The execution time is once per driving cycle and DTCs P2648, P2649 are not active. **NOTE: Before you troubleshoot, record all freeze data and any on-board snapshot.** **Possible causes:** • Low engine oil or faulty oil pressure • Poor connections or loose terminals at the rocker arm oil pressure switch, the rocker arm oil control solenoid, and the PCM • "Short" circuit between the PCM and the rocker arm oil pressure switch • Faulty rocker arm oil pressure switch • Faulty rocker arm oil control solenoid • Faulty rocker arm
DTC: P2647 **1T PCM, MIL: Yes** **Year:** 2009, 2010 **Model:** MDX, RDX, RL, TL, TSX, ZDX **Engine:** 2.3L L4, 2.4L L4, 3.5L V6, 3.7L V6 **Transmission:** All	**Rocker Arm Oil Pressure Switch (VTEC Oil Pressure Switch) Circuit High Voltage :** When the rocker arm oil control solenoid is OFF, the rocker arm oil pressure switch remains OFF. (Low lift cam operation) **NOTE: Before you troubleshoot, record all freeze data and any on-board snapshot.** **Possible causes:** • Low oil level • Low oil pressure • "Open" in the wire between the rocker arm oil pressure switch and ground • Poor connections or loose terminals at the rocker arm oil pressure switch, the rocker arm oil control solenoid, and the PCM • "Open" circuit between the PCM and the rocker arm oil pressure switch • Faulty rocker arm oil pressure switch • Faulty rocker arm oil control solenoid assembly • PCM needs to be updated with the latest software • Faulty PCM
DTC: P2648 **1T PCM, MIL: Yes** **Year:** 2009, 2010 **Model:** MDX, RDX, RL, TL, TSX, ZDX **Engine:** 2.3L L4, 2.4L L4, 3.5L V6, 3.7L V6 **Transmission:** All	**Rocker Arm Oil Control Solenoid (VTEC Solenoid Valve) Circuit Low Voltage :** The return signal is OFF (low) for at least 2 seconds when the PCM outputs the ON (high) signal to the rocker arm oil control solenoid. DTC P2649 is not active. **NOTE: Before you troubleshoot, record all freeze data and any on-board snapshot.** **Possible causes:** • Poor connections or loose terminals at the rocker arm oil control solenoid and the PCM • "Short" circuit between the PCM and the rocker arm oil control solenoid • Faulty rocker arm oil control solenoid • PCM needs to be updated with the latest software • Faulty PCM
DTC: P2649 **1T PCM, MIL: Yes** **Year:** 2009, 2010 **Model:** MDX, RDX, RL, TL, TSX, ZDX **Engine:** 2.3L L4, 2.4L L4, 3.5L V6, 3.7L V6 **Transmission:** All	**Rocker Arm Oil Control Solenoid (VTEC Solenoid Valve) Circuit High Voltage :** The return signal is ON (High) for at least 2 seconds when the PCM outputs the OFF (Low) signal to the rocker arm oil control solenoid. With the engine running the execution time is continuous. DTC P2648 is not active. **NOTE: Before you troubleshoot, record all freeze data and any on-board snapshot.** **Possible causes:** • Poor connections or loose terminals at the rocker arm oil control solenoid and the PCM • "Open" circuit between the PCM and the rocker arm oil control solenoid • Faulty rocker arm oil control solenoid • PCM may need to be updated with the latest software • Faulty PCM
DTC: P2714 **2T PCM, MIL: Yes, TCIL: Yes** **Year:** 2010 **Model:** ZDX **Engine:** 3.7L V6 **Transmission:** All	**A/T Clutch Pressure Control Solenoid Valve D Stuck OFF:** Shift lever position D, S, R, with ATF the temperature between -4 °F (-20 °C), 248 °F (120 °C) improper gear ratio output has been detected for the shift schedule. The execution time is continuous and the duration time is between 3-10 seconds or more. **Possible causes:** • Low or dirty transmission fluid • Poor connections or loose terminals • Inspect the strainer for metal debris or excessive clutch material, if present • replace the transmission • Faulty lock-up shift valve, replace the main valve body • Faulty A/T clutch pressure control solenoid valve D

DTC	Trouble Code Title, Conditions & Possible Causes
DTC: P2720 **2T PCM, MIL: Yes, TCIL: Yes** **Year:** 2010 **Model:** ZDX **Engine:** 3.7L V6 **Transmission:** All	**Problem in A/T Clutch Pressure Control Solenoid Valvo D Circuit:** The measured current for the PCM's command value is less than 0.19 with duty cycle at 56.5-89%, with duty cycle at 09% or more it is Less than 0.27. **NOTE: Before you troubleshoot, record all freeze data and any on-board snapshot. This code is caused by an electrical circuit problem and cannot be caused by a mechanical problem in the transmission.** **Possible causes:** • Poor connections and loose terminals between A/T clutch pressure control solenoid valve D and the PCM • "Open" or "Short" between PCM and A/T clutch pressure control solenoid valve D • Faulty A/T clutch pressure control solenoid valve D • Faulty PCM
DTC: P2721 **2T PCM, MIL: Yes, TCIL: Yes** **Year:** 2010 **Model:** ZDX **Engine:** 3.7L V6 **Transmission:** All	**Problem in A/T Clutch Pressure Control Solenoid Valve D:** The measured current for the PCM's command value is more than 0.6 duty cycle less than 13.7%. More than 0.9 with duty cycle at 13.7-27%. (High Input) **NOTE: Before you troubleshoot, record all freeze data and any on-board snapshot. This code is caused by an electrical circuit problem and cannot be caused by a mechanical problem in the transmission.** **Possible causes:** • "Open" in the wires between PCM and body ground, or poor body ground • Faulty secondary valve body (A/T clutch pressure control solenoid valve D) • Faulty PCM
DTC: P2821 **2T PCM, MIL: Yes, TCIL: Yes** **Year:** 2010 **Model:** ZDX **Engine:** 3.7L V6 **Transmission:** All	**Line Pressure Solenoid Valve A Stuck ON:** The input signal from the line pressure switch to the PCM is high during the self shut down mode after the signal is low when the normal line pressure mode was commanded. **NOTE: Before you troubleshoot, record all freeze data and any on-board snapshot. When DTC P2821 is stored in the PCM, engine torque is restricted because of the fail-safe function.** **Possible causes:** • Low or dirty transmission fluid • Inspect the strainer for metal debris or excessive clutch material, if present replace the transmission • Faulty secondary valve body (line pressure solenoid valve A)
DTC: P2826 **2T PCM, TCIL: Yes** **Year:** 2010 **Model:** ZDX **Engine:** 3.7L V6 **Transmission:** All	**Short in Line Pressure Solenoid Valve A Circuit:** The return signal does not match the command to turn on line pressure solenoid valve A for at least 1 second. The circuit detects malfunctions such as a circuit short or open. **NOTE: Before you troubleshoot, record all freeze data and any on-board snapshot. This code is caused by an electrical circuit problem and cannot be caused by a mechanical problem in the transmission.** **Possible causes:** • "Open" or "Short" circuit between PCM and line pressure solenoid valve • Faulty secondary valve body (line pressure solenoid valve A) • Faulty PCM
DTC: P2827 **2T PCM, TCIL: Yes** **Year:** 2010 **Model:** ZDX **Engine:** 3.7L V6 **Transmission:** All	**Open in Line Pressure Solenoid Valve A Circuit:** The return signal does not match the command to turn off line pressure solenoid valve A for at least 1 second. **NOTE: Before you troubleshoot, record all freeze data and any on-board snapshot. This code is caused by an electrical circuit problem and cannot be caused by a mechanical problem in the transmission.** **Possible causes:** • "Open" in the wire between line pressure solenoid valve A and body ground, or poor body ground • "Open" circuit between PCM and line pressure solenoid valve A • Faulty secondary valve body (line pressure solenoid valve A) • Faulty PCM
DTC: P2A00 **2T PCM, MIL: Yes** **Year:** 2009, 2010 **Model:** RDX, TSX **Engine:** 2.3L L4, 2.4L L4 **Transmission:** All	**Air/Fuel Ratio (A/F) Sensor (Sensor 1) Circuit Range/Performance Problem:** The A/F sensor (Sensor 1) output voltage is 3.0 V or less, or 4.8 V or more. The execution time is once per driving cycle and the duration time is 3.5 seconds or more. **NOTE: Before you troubleshoot, record all freeze data and any on-board snapshot.** **Possible causes:** • Poor connections or loose terminals at the A/F sensor (Sensor 1) and the ECM/PCM • Faulty A/F sensor (Sensor 1)
DTC: P2A00 **2T PCM, MIL: Yes** **Year:** 2009, 2010 **Model:** MDX, RL, TL, TSX, ZDX **Engine:** 3.5L V6, 3.7L V6 **Transmission:** All	**Rear Air/Fuel Ratio (A/F) Sensor (Bank 1, Sensor 1) Circuit Range/Performance Problem:** The rear A/F sensor (bank 1, sensor 1) output voltage is 2.55 V or less, or 4.50 V or more. The execution time is once per driving cycle and the duration time is 5.3 seconds or more. **NOTE: Before you troubleshoot, record all freeze data and any on-board snapshot.** **Possible causes:** • Poor connections or loose terminals at the A/F sensor (Sensor 1) and the PCM • Faulty A/F sensor (Sensor 1)

DTC	Trouble Code Title, Conditions & Possible Causes
DTC: P2A03 **2T PCM, MIL: Yes** **Year:** 2009, 2010 **Model:** MDX, RL, TL, TSX, ZDX **Engine:** 3.5L V6, 3.7L V6 **Transmission:** All	**Front Air/Fuel Ratio (A/F) Sensor (Bank 2, Sensor 1) Circuit Range/Performance Problem:** The front A/F sensor (bank 2, sensor 1) output voltage is 2.55 V or less, or 4.50 V or more. The execution time is once per driving cycle and the duration time is 5.3 seconds or more. **Possible causes:** • Poor connections or loose terminals at the A/F sensor (Sensor 1) and the PCM • Faulty A/F sensor (Sensor 1)

Gas Engine OBD II Trouble Code List (U0XXX Codes)

DTC	Trouble Code Title, Conditions & Possible Causes
DTC: U0028 **1T PCM, TCIL: Yes** **Year:** 2009, 2010 **Model:** MDX, RDX **Engine:** 2.3L L4, 3.7L V6 **Transmission:** All	**F-CAN Communication Circuit Error (F-CAN Bus OFF) :** When the powertrain control module (PCM) does not receive the signals via the CAN lines for a set time or more, the PCM detects a malfunction. The PCM does not receive any signals for at least 1 second. **NOTE: Before you troubleshoot, record all freeze data and any on-board snapshot.** **Possible causes:** • PCM needs to be updated with the latest software • Faulty PCM
DTC: U0029 **1T PCM** **Year:** 2009, 2010 **Model:** MDX, TL, TSX, ZDX **Engine:** 3.5L V6, 3.7L V6 **Transmission:** All	**F-CAN A Malfunction (BUS-OFF (PCM)):** The PCM does not receive any signals via the F-CAN A lines for at least 1 second. When a malfunction is detected, the D indicator blinks, and a Pending DTC, a Confirmed DTC, and the freeze data are stored in the PCM memory. The MIL does not come on. **Possible causes:** • Check battery and charging system condition • Loose or poor connections, or worn/shorted wires • "Short" in the F-CAN wires • Faulty gauge control module • PCM may need to be updated with the latest software • Faulty PCM
DTC: U0047 **T PCM** **Year:** 2010 **Model:** ZDX **Engine:** 3.7L V6 **Transmission:** All	**F-CAN B Malfunction (BUS-OFF (PCM):** Adaptive Cruise Control (ACC) malfunction. **NOTE: If DTC U0047 is stored on a vehicle without adaptive cruise control (ACC), no troubleshooting is needed; just clear the DTC with the HDS.** **Possible causes:** • Poor connections or loose terminals at the ACC unit and the PCM • Faulty ACC unit • PCM may need to be updated with the latest software • Faulty PCM
DTC: U0073 **1T PCM, MIL: Yes** **Year:** 2009, 2010 **Model:** RL **Engine:** 3.7L V6 **Transmission:** All	**F-CAN Malfunction (BUS-OFF) :** The PCM does not receive any signals via the F-CAN lines for at least 1 second. **Possible causes:** • PCM may need to be updated with the latest software • Faulty PCM
DTC: U0100 **T PCM** **Year:** 2009, 2010 **Model:** TL, TSX, ZDX **Engine:** 3.5L V6, 3.7L V6 **Transmission:** All	**F-CAN Lost communication with PCM:** Intermittent failure in the F-CAN communication line. **Possible causes:** • Poor connections or loose terminals at the AcuraLink control unit (XM receiver) • "Open" in the wire between the AcuraLink control unit (XM receiver) and the PCM • Faulty gauge control module • Faulty AcuraLink control unit (XM receiver)
DTC: U0100 **1T PCM** **Year:** 2009, 2010 **Model:** TL, TSX **Engine:** 3.5L V6, 3.7L V6 **Transmission:** All	**Gauge Control Module Lost Communication With ECM/PCM:** F-CAN communication line malfunction. **NOTE: If you are troubleshooting multiple DTCs, be sure to follow the instructions in B-CAN System Diagnosis Test Mode A.** **Possible causes:** • Faulty battery or charging system • Perform the gauge control module input test • Loose or poor connections at the gauge control module and the ECM/PCM • Check for faulty inputs • Gauge control module is faulty • Faulty ECM/PCM

DTC	Trouble Code Title, Conditions & Possible Causes
DTC: U0104 **T PCM** **Year:** 2010 **Model:** MDX, ZDX **Engine:** 3.7L V6 **Transmission:** All	**Gauge Control Module Lost Communication With ACC Unit (ACC message):** Poor communication between the Gauge Control Module and ACC Unit. **NOTE: If you are troubleshooting multiple DTCs, be sure to follow the instructions in B-CAN System Diagnosis Test Mode A. If the HDS does not communicate with the ACC unit, the ACC indicator stay on, and no ACC DTCs are stored, go to symptom troubleshooting for ACC indicator does not go off, and no DTCs are stored.** **Possible causes:** • "Open" circuit between gauge control module connector A and ACC unit • Faulty ACC unit • Faulty gauge control module
DTC: U0104 **1T PCM** **Year:** 2009, 2010 **Model:** RL **Engine:** 3.7L V6 **Transmission:** All	**F-CAN Malfunction (Adaptive Cruise Control (ACC) Unit-PCM):** The PCM does not receive any signals via the F-CAN lines for at least 1 second. **NOTE: Before you troubleshoot, record all freeze data and any on-board snapshot.** **Possible causes:** • Poor connections or loose terminals at the ACC unit and the PCM • "Open" or "Short" circuit between the PCM and the ACC unit • Faulty ACC unit
DTC: U0107 **1T PCM, MIL: Yes** **Year:** 2009, 2010 **Model:** RL **Engine:** 3.7L V6 **Transmission:** All	**Lost Communication With Throttle Actuator Control Module :** One of these 2 conditions must be met for at least 250 milliseconds: (1) No serial signals from the throttle actuator control module are detected. (2) The serial signals from the throttle actuator control module are abnormal. **NOTE: Before you troubleshoot, record all freeze data and any on-board snapshot.** **Possible causes:** • Poor connections or loose terminals at the throttle body, the throttle actuator control module relay, the throttle actuator control module, and the PCM • "Open" in the wire between the PCM and ground • "Open" or "Short" circuit between the throttle actuator control module and the PCM • PCM may need to be updated with the latest software • Faulty PCM
DTC: U0114 **1T PCM** **Year:** 2009, 2010 **Model:** MDX, RDX, RL, TL, ZDX **Engine:** 2.3L L4, 3.5L V6, 3.7L V6 **Transmission:** All	**F-CAN Malfunction (PCM-SH-AWD Control Unit) :** No signals via the CAN lines are received for at least 1 second. **NOTE: If DTC U0028 is stored at the same time as DTC U0114, troubleshoot DTC U0028 first, then recheck for DTC U0114.** **Possible causes:** • Blown fuse • Poor connections or loose terminals at the gauge control module, the SH-AWD control unit • "Open" circuit between the SH-AWD control unit and the PCM • "Open" in the wire between the SH-AWD control unit and body ground • Faulty SH-AWD control unit
DTC: U0122 **1T PCM** **Year:** 2009, 2010 **Model:** TL, ZDX **Engine:** 3.5L V6, 3.7L V6 **Transmission:** All	**Lost Communication with VSA Modulator-Control Unit (A/T System):** One of these 2 conditions must be met: (1) The PCM does not receive any signals via the F-CAN A lines for at least 1 second. (2) The information sent from the VSA modulator-control unit is abnormal at least 20 times. DTCs U0029, U0104, U0155 are not active. **NOTE: Before you troubleshoot, record all freeze data and any on-board snapshot. This code is caused by an electrical circuit problem and cannot be caused by a mechanical problem in the transmission.** **Possible causes:** • Poor connections or loose terminals between the VSA modulator-control unit and the PCM • Check for Pending or Confirmed DTCs in the PGM-FI SYSTEM • PCM may need to be updated with the latest software • Faulty PCM
DTC: U0122 **1T PCM** **Year:** 2009, 2010 **Model:** TL, TSX **Engine:** 2.4L L4, 3.5L V6, 3.7L V6 **Transmission:** All	**Gauge Control Module Lost Communication With VSA Modulator-Control Unit (VSA message):** F-CAN communication line Malfunction. **NOTE: If you are troubleshooting multiple DTCs, be sure to follow the instructions in B-CAN System Diagnosis Test Mode A.** **Possible causes:** • Faulty battery and/or charging system • Loose VSA ground or poor connections at the VSA modulator-control unit or gauge control module • Faulty inputs • Faulty gauge control module • Faulty ECM/PCM

DTC	Trouble Code Title, Conditions & Possible Causes
DTC: U0122 **1T PCM, TCIL: Yes** **Year:** 2009, 2010 **Model:** RDX, TL, TSX, ZDX **Engine:** 2.3L L4, 2.4L L4, 3.5L V6, 3.7L V6 **Transmission:** All	**F-CAN A Malfunction (Powertrain Control Module (PCM)-VSA Modulator-Control Unit) (PGM-FI System):** With the ignition switch ON for at least 3 seconds. One of these 2 conditions must be met: (1) The PCM does not receive any signals via the F-CAN A lines for at least 1 second. (2) The information sent from the VSA modulator-control unit is abnormal at least 20 times. **Possible causes:** • Poor connections or loose terminals at the VSA modulator-control unit and the PCM • "Open" circuit between the PCM and the VSA modulator-control unit • VSA modulator-control unit needs to be updated with the latest software • Faulty VSA modulator-control unit
DTC: U0122 **1T PCM, TCIL: Yes** **Year:** 2009, 2010 **Model:** MDX, RL, TSX **Engine:** 2.4L L4, 3.5L V6, 3.7L V6 **Transmission:** All	**Lost Communication with VSA Modulator-Control Unit:** No signals from the VSA modulator-control unit via the CAN lines are received for at least 1 second. The execution time is continuous and the duration time is 1 seconds or more. **NOTE: Before you troubleshoot, record all freeze data and any on-board snapshot.** **Possible causes:** • Poor connections or loose terminals at the gauge control module, the VSA modulator-control unit, and the PCM • "Open circuit between the PCM and the VSA modulator-control • Perform DLC circuit troubleshooting • Faulty VSA modulator-control unit
DTC: U0127 **1T BCM** **Year:** 2009, 2010 **Model:** TL, TSX, ZDX **Engine:** 2.4L L4, 3.5L V6, 3.7L V6 **Transmission:** All	**Gauge Control Module Lost Communication With the TPMS Control Unit (TPMS message):** Poor communication between the Gauge Control Module and TPMS Control Unit. **NOTE: If you are troubleshooting multiple DTCs, be sure to follow the instructions in B-CAN System Diagnosis Test Mode A. If the HDS does not communicate with the TPMS control unit, the TPMS indicator stay on, and no TPMS DTCs are stored, go to symptom troubleshooting for TPMS indicator does not go off, and no DTCs are stored.** **Possible causes:** • Poor connections or loose terminals between gauge control module and TPMS control unit • "Open" circuit between gauge control module and TPMS control unit • Faulty TPMS control unit • Faulty gauge control module
DTC: U0131 **1T PCM** **Year:** 2009, 2010 **Model:** TL **Engine:** 3.5L V6, 3.7L V6 **Transmission:** All	**F-CAN Malfunction (ECM/PCM-EPS Control Unit):** Poor communication between the EPS control unit, and the ECM/PCM. **NOTE: Before you troubleshoot, record all freeze data and any on-board snapshot, and review the general troubleshooting information.** **Possible causes:** • Poor connections or loose terminals at the EPS control unit, and the ECM/PCM • "Open" circuit between the ECM/PCM and the EPS control unit • Faulty EPS control unit
DTC: U0131 **1T PCM** **Year:** 2009, 2010 **Model:** TL **Engine:** 3.5L V6, 3.7L V6 **Transmission:** All	**Gauge Control Module Lost Communication With EPS Control Unit (EPS message):** Poor communication between the Gauge Control Module and the EPS Control Unit. **NOTE: If you are troubleshooting multiple DTCs, be sure to follow the instructions in B-CAN System Diagnosis Test Mode A. If the HDS does not communicate with the EPS control unit, the EPS indicator stay on, and no EPS DTCs are stored, go to symptom troubleshooting for EPS indicator does not go off, and no DTCs are stored.** **Possible causes:** • Loose or poor connections • "Open" circuit between the gauge control module and the EPS control unit • Faulty gauge control module
DTC: U0151 **1T PCM** **Year:** 2009, 2010 **Model:** TL **Engine:** 3.5L V6, 3.7L V6 **Transmission:** All	**Gauge Control Module Lost Communication With SRS Unit:** Poor communication between the gauge control module, and the SRS unit. **NOTE: If you are troubleshooting multiple DTCs, be sure to follow the instructions in B-CAN System Diagnosis Test Mode A. If the HDS does not communicate with the SRS unit, the SRS indicator stay on, go to symptom troubleshooting for SRS indicator stay on, but no DTCs are stored.** **Possible causes:** • Loose or poor connections between the gauge control module, and the SRS unit. • "Open" circuit between the gauge control module, and the SRS unit • Faulty SRS unit • Faulty gauge control module

DTC	Trouble Code Title, Conditions & Possible Causes
DTC: U0155 **1T** **Year:** 2009, 2010 **Model:** TL, ZDX **Engine:** 3.5L V6, 3.7L V6 **Transmission:** All	**Climate Control Unit Lost Communication with Gauge Control Module:** Poor communication between the Climate Control Unit and Gauge Control Module. **NOTE: If you are troubleshooting multiple DTCs, be sure to follow the instructions in B-CAN System Diagnosis Test Mode A.** **Possible causes:** • Loose wires or poor connections on the B-CAN lines between the gauge control module and the climate control unit • Perform the gauge control module input test • "Open" in the wire(s) between the climate control unit and the gauge control module • Faulty climate control unit
DTC: U0155 **1T PCM, TCIL: Yes** **Year:** 2009, 2010 **Model:** MDX, RDX, RL, TL **Engine:** 2.3L L4, 3.5L V6, 3.7L V6 **Transmission:** All	**Lost Communication with Gauge Control Module :** No signals from the gauge control module via the CAN lines are received for at least 1 second. The execution time is continuous and the duration time is 1 second or more. **NOTE: Before you troubleshoot, record all freeze data and any on-board snapshot.** **Possible causes:** • Poor connections and loose terminals at the gauge control module and the PCM • "Open" circuit between the PCM and the gauge control module • Faulty Gauge Control Module • PCM may need to be updated with the latest software • Faulty PCM
DTC: U0155 **1T** **Year:** 2009, 2010 **Model:** TL, TSX, ZDX **Engine:** 2.4L L4, 3.5L V6, 3.7L V6 **Transmission:** All	**Driver's MICU Lost Communication With Gauge Control Module:** Poor communication between the Driver's MICU and Gauge Control Module. **NOTE: If you are troubleshooting multiple DTCs, be sure to follow the instructions in B-CAN System Diagnosis Test Mode A.** **Possible causes:** • Loose or poor connections at the gauge control module and the related units • Perform the gauge control module input test, and do all power, ground, and communication input tests. If the tests prove OK, replace the gauge control module
DTC: U0155 **1T** **Year:** 2009, 2010 **Model:** TL, TSX, ZDX **Engine:** 2.4L L4, 3.5L V6, 3.7L V6 **Transmission:** All	**Lost Communication With the Gauge Control Module:** Communication error between the Gauge Control Module and BSI control unit. **Possible causes:** • "Open" circuit between the gauge control module and the BSI control unit • Faulty BSI control unit
DTC: U0155 **1T** **Year:** 2009, 2010 **Model:** TL, TSX, ZDX **Engine:** 2.4L L4, 3.5L V6, 3.7L V6 **Transmission:** All	**Passenger's MICU Lost Communication With Gauge Control Module:** Poor communication between the Passenger's MICU and Gauge Control Module. **NOTE: If you are troubleshooting multiple DTCs, be sure to follow the instructions in B-CAN System Diagnosis Test Mode A.** **Possible causes:** • Loose or poor connections at the gauge control module and the related units • Perform the gauge control module input test, and do all power, ground, and communication input tests. If the tests prove OK, replace the gauge control module
DTC: U0155 **1T** **Year:** 2009, 2010 **Model:** TL, TSX, ZDX **Engine:** 2.4L L4, 3.5L V6, 3.7L V6 **Transmission:** All	**Immobilizer-Keyless Control Unit Lost Communication With Gauge Control Module:** Poor communication between the Immobilizer-Keyless Control Unit and Gauge Control Module. **NOTE: If you are troubleshooting multiple DTCs, be sure to follow the instructions in B-CAN System Diagnosis Test Mode A.** **Possible causes:** • Perform the gauge control module input test, and do all power, ground, and communication input tests • Loose or poor connections at the gauge control module and the related units • Faulty gauge control module
DTC: U0155 **1T** **Year:** 2009, 2010 **Model:** TL, ZDX **Engine:** 3.5L V6, 3.7L V6 **Transmission:** All	**Lost Communication With Gauge Control Module (VSP/NE Frame):** Poor communication between the Power Seat control Unit and Gauge Control Module. **NOTE: If you are troubleshooting multiple DTCs, be sure to follow the instructions in B-CAN System Diagnosis Test Mode A.** **Possible causes:** • "Open" or poor connection in the wire between the gauge control module and the power seat control unit • Faulty power seat control unit • Faulty gauge control module
DTC: U0155 **1T** **Year:** 2009, 2010 **Model:** TL, ZDX **Engine:** 3.5L V6, 3.7L V6 **Transmission:** All	**Door Multiplex Control Unit Lost Communication With Gauge Control Module:** Poor communication between the Door Multiplex Control Unit and Gauge Control Module. **NOTE: If you are troubleshooting multiple DTCs, be sure to follow the instructions in B-CAN System Diagnosis Test Mode A.** **Possible causes:** • Perform the gauge control module input test, and do all power, ground and communication input tests. If the tests prove OK, replace the gauge control module • Loose or poor connections at the gauge control module and the related units

DTC	Trouble Code Title, Conditions & Possible Causes
DTC: U0155 **1T** **Year:** 2009, 2010 **Model:** TL, ZDX **Engine:** 3.5L V6, 3.7L V6 **Transmission:** All	**Remote Slot Control Unit Lost Communication With Gauge Control Module:** Poor communication between the Remote Slot Control Unit and Gauge Control Module. **NOTE: If you are troubleshooting multiple DTCs, be sure to follow the instructions in B-CAN System Diagnosis Test Mode A.** **Possible causes:** • Loose or poor connections at the gauge control module and the related units • Perform the gauge control module input test, and do all power, ground, and communication input tests. If the tests prove OK, replace the gauge control module
DTC: U0164 **1T** **Year:** 2009, 2010 **Model:** TL, TSX, ZDX **Engine:** 2.4L L4, 3.5L V6, 3.7L V6 **Transmission:** All	**Door Multiplex Control Unit Lost Communication With Climate Control Unit:** Poor communication between the door multiplex control unit and climate control unit. **NOTE: If you are troubleshooting multiple DTCs, be sure to follow the instructions in B-CAN System Diagnosis Test Mode A.** **Possible causes:** • Perform the door multiplex control unit input test and check the power and ground. If OK, replace the driver's power window master switch • Loose or poor connections between the door multiplex control unit and climate control unit.
DTC: U0199 **1T** **Year:** 2010 **Model:** ZDX **Engine:** 3.7L V6 **Transmission:** All	**Lost Communication With P/W (DRLockSW Frame):** Poor communication between the Door Multiplex Control Unit and Power Seat Control Unit. **NOTE: If you are troubleshooting multiple DTCs, be sure to follow the instructions in B-CAN System Diagnosis Test Mode A.** **Possible causes:** • Loose or poor connections between the power seat control unit and the door multiplex control unit • Faulty power seat control unit • "Open" circuit between the door multiplex control unit and the power seat control unit • Faulty door multiplex control unit
DTC: U0199 **1T** **Year:** 2009, 2010 **Model:** TL, ZDX **Engine:** 3.5L V6, 3.7L V6 **Transmission:** All	**Lost Communication With Door Multiplex Control Unit (DRLOCK SW, KLDRLOCK Frame):** Poor communication between the Door Multiplex Control Unit Power Tailgate Control Unit. **NOTE: If you are troubleshooting multiple DTCs, be to follow the instructions in B-CAN System Diagnosis Test Mode A.** **Possible causes:** • "Open" or high resistance between door power window master switch and power tailgate control unit connector • Perform the power window master switch input test, and do all power and ground input tests. If the tests prove OK, replace the power window master switch
DTC: U0199 **1T** **Year:** 2009, 2010 **Model:** TL, TSX, ZDX **Engine:** 2.4L L4, 3.5L V6, 3.7L V6 **Transmission:** All	**Power Control Unit Lost Communication With Door Multiplex Control Unit:** Poor communication between the Power Control Unit and Door Multiplex Control Unit. **NOTE: If you are troubleshooting multiple DTCs, be sure to follow the instructions in B-CAN System Diagnosis Test Mode A.** **Possible causes:** • Perform the door multiplex control unit input test, and do all power, ground, and communication input tests. If the tests prove OK, replace the power window master switch • Loose or poor connections between the power control unit and the door multiplex control unit.
DTC: U0199 **T PCM** **Year:** 2009, 2010 **Model:** TSX **Engine:** 2.4L L4, 3.5L V6 **Transmission:** All	**Power Seat Control Unit Lost Communication With Door Multiplex Control Unit :** Poor communication between the Power Seat Control Unit, and the Door Multiplex Control Unit. **NOTE: If you are troubleshooting multiple DTCs, be sure to follow the instructions in B-CAN System Diagnosis Test Mode A.** **Possible causes:** • Loose or poor connections between the power seat control unit and the door multiplex control unit • Perform the power seat control unit input test • Faulty power seat control unit • Faulty door multiplex control unit
DTC: U0199 **1T** **Year:** 2009, 2010 **Model:** TL, ZDX **Engine:** 3.5L V6, 3.7L V6 **Transmission:** All	**Gauge Control Module Lost Communication With Door Multiplex Control Unit:** Poor communication between the Gauge Control Module and Door Multiplex Control Unit. **NOTE: If you are troubleshooting multiple DTCs, be sure to follow the instructions in B-CAN System Diagnosis Test Mode A.** **Possible causes:** • Perform the door multiplex control unit input test, and do all power, ground, and communication input tests. If the tests prove OK, replace the power window master switch • Loose or poor connections at the door multiplex control unit and the related units
DTC: U0199 **1T** **Year:** 2009, 2010 **Model:** TL, ZDX **Engine:** 3.5L V6, 3.7L V6 **Transmission:** All	**Keyless Access Control Unit Lost Communication With Door Multiplex Control Unit:** Poor communication between the Keyless Access Control Unit and Door Multiplex Control Unit. **NOTE: If you are troubleshooting multiple DTCs, be sure to follow the instructions in B-CAN System Diagnosis Test Mode A.** **Possible causes:** • Perform the door multiplex control unit input test, and do all power, ground, and communication input tests. If the tests prove OK, replace the power window master switch • Loose or poor connections at the door multiplex control unit and the related units

DTC	Trouble Code Title, Conditions & Possible Causes
DTC: U0199 **1T** **Year:** 2009, 2010 **Model:** TL, TSX, ZDX **Engine:** 2.4L L4, 3.5L V6, 3.7L V6 **Transmission:** All	**Driver's MICU Lost Communication With Door Multiplex Control Unit:** Poor communication between the Driver's MICU and Door Multiplex Control Unit. **NOTE: If you are troubleshooting multiple DTCs, be sure to follow the instructions in B-CAN System Diagnosis Test Mode A.** **Possible causes:** • Perform the door multiplex control unit input test, and do all power, ground, and communication input tests. If the tests prove OK, replace the power window master switch • Loose or poor connections at the door multiplex control unit and the related units
DTC: U0199 **T BCM** **Year:** 2009, 2010 **Model:** TSX **Engine:** 2.4L L4, 3.5L V6 **Transmission:** All	**Immobilizer-keyless Control Unit Lost Communication With Door Multiplex Control Unit :** Poor communication between the Immobilizer-keyless Control Unit, and the Door Multiplex Control Unit. **NOTE: If you are troubleshooting multiple DTCs, be sure to follow the instructions in B-CAN System Diagnosis Test Mode A.** **Possible causes:** • Perform the door multiplex control unit input test, and do all power, ground, and communication input tests. If the tests prove OK, replace the power window master switch • Loose or poor connections at the door multiplex control unit and the related units
DTC: U0230 **1T** **Year:** 2010 **Model:** ZDX **Engine:** 3.7L V6 **Transmission:** All	**Lost Communication With Power Tailgate Control Unit:** Poor communication between the Gauge Control Module and Power Tailgate Control Unit. **NOTE: If you are troubleshooting multiple DTCs, be sure to follow the instructions in B-CAN System Diagnosis Test Mode A.** **Possible causes:** • Loose or poor connections between the gauge control module and the power tailgate control unit • "Open" circuit between the gauge control module connector and power tailgate control unit • Faulty power tailgate control unit • Faulty gauge control module
DTC: U0300 **1T PCM, MIL: Yes** **Year:** 2009, 2010 **Model:** MDX, RDX, TSX, ZDX **Engine:** 2.3L L4, 2.4L L4, 3.5L V6, 3.7L V6 **Transmission:** All	**PGM-FI System and A/T System Program Version Mismatch:** The rewriting of the FI CPU or the A/T CPU is done at the PCM, and different data of the serial communication data version is rewritten. The execution time is continuous and the duration time is 500 milliseconds or more. WARNING: Do not turn the ignition switch to LOCK (0) or ACC (I) while updating the PCM. If you turn the ignition switch to LOCK (0) before completion, the PCM will be damaged. **Possible causes:** • PCM may need to be updated with the latest software • Faulty PCM
DTC: U0423 **1T** **Year:** 2010 **Model:** ZDX **Engine:** 3.7L V6 **Transmission:** All	**Invalid Data Received from PCM (Vehicle Speed Data):** Vehicle speed at 6 mph (10 km/h) or more, invalid vehicle speed data received. **Possible causes:** • Troubleshoot the indicated PGM-FI system DTC(s).

Gas Engine OBD II Trouble Code List (U1XXX Codes)

DTC	Trouble Code Title, Conditions & Possible Causes
DTC: U1280 **1T** **Year:** 2009, 2010 **Model:** TL, TSX, ZDX **Engine:** 2.4L L4, 3.5L V6, 3.7L V6 **Transmission:** All	**Communication Bus Line Error (BUS-OFF):** Communication Bus Line Error **NOTE: If you are troubleshooting multiple DTCs, be sure to follow the instructions in B-CAN System Diagnosis Test Mode A.** **Possible causes:** • Check battery and charging system condition • Perform the following input test to help find the faulty unit: • Passenger's MICU input test • Door multiplex control unit (power window master switch) input test • Gauge control unit input test • Power control unit input test • Keyless access control unit input test • Remote slot control unit input test • Immobilizer-keyless control unit input test • Climate control unit power and ground circuit troubleshooting • Power seat control unit input test • HandsFreeLink control unit input test • AcuraLink control unit input test • Audio unit input test • Audio-navigation unit input test • Power tailgate control unit input test • BSI control unit input test • "Open" or "Short" between body ground and driver's under-dash fuse/relay box connector • Faulty driver's MICU
DTC: U1282 **1T PCM** **Year:** 2009, 2010 **Model:** TL **Engine:** 3.5L V6, 3.7L V6 **Transmission:** All	**Door Multiplex Control Unit Lost Communication With Driver's MICU:** Poor communication between the Door Multiplex Control Unit, and the Driver's MICU. **NOTE: If you are troubleshooting multiple DTCs, be sure to follow the instructions in B-CAN System Diagnosis Test Mode A.** **Possible causes:** • Perform the driver's MICU input test, and do all power, ground and communication input tests • Loose or poor connections at driver's under-dash fuse/relay box connector and the related units • Faulty driver's under-dash fuse/relay box
DTC: U1282 **1T** **Year:** 2010 **Model:** ZDX **Engine:** 3.7L V6 **Transmission:** All	**Gauge Control Module Lost Communication With Driver's MICU:** Poor communication between the Gauge Control Module and MICU. **NOTE: If you are troubleshooting multiple DTCs, be sure to follow the instructions in B-CAN System Diagnosis Test Mode A.** **Possible causes:** • Perform the driver's MICU input test, and do all power, ground, and communication input tests. If the tests prove OK, replace the driver's under-dash fuse/relay box • Loose or poor connections at driver's under-dash fuse/relay box and the related units • Faulty gauge control module • Faulty MICU
DTC: U1282 **1T PCM** **Year:** 2009, 2010 **Model:** TL **Engine:** 3.5L V6, 3.7L V6 **Transmission:** All	**Lost Communication With Driver's MICU:** Poor communication between the power seat control unit, and the driver's MICU. **NOTE: If you are troubleshooting multiple DTCs, be sure to follow the instructions in B-CAN System Diagnosis Test Mode A.** **Possible causes:** • Perform the power seat control unit input test • Loose or poor connections between the power seat control unit and the driver's MICU
DTC: U1282 **1T PCM** **Year:** 2009, 2010 **Model:** TL **Engine:** 3.5L V6, 3.7L V6 **Transmission:** All	**Keyless Access Control Unit Lost Communication With Driver's MICU:** Poor communication between the Keyless Access Control Unit, and the Driver's MICU. **NOTE: If you are troubleshooting multiple DTCs, be sure to follow the instructions in B-CAN System Diagnosis Test Mode A.** **Possible causes:** • Perform the driver's MICU input test, and do all power, ground and communication input tests • Loose or poor connections at driver's under-dash fuse/relay box connector and the related units • Faulty driver's under-dash fuse/relay box
DTC: U1282 **1T PCM** **Year:** 2009, 2010 **Model:** TL **Engine:** 3.5L V6, 3.7L V6 **Transmission:** All	**Passenger's MICU Lost Communication With Driver's MICU:** Poor communication between the Passenger's MICU, and the Driver's MICU. **NOTE: If you are troubleshooting multiple DTCs, be sure to follow the instructions in B-CAN System Diagnosis Test Mode A.** **Possible causes:** • Perform the driver's MICU input test, and do all power, ground and communication input tests • Loose or poor connections at driver's under-dash fuse/relay box connector and the related units • Faulty driver's under-dash fuse/relay box

DTC	Trouble Code Title, Conditions & Possible Causes
DTC: U1282 **1T PCM** **Year:** 2009, 2010 **Model:** TL **Engine:** 3.5L V6, 3.7L V6 **Transmission:** All	**Immobilizer-keyless Control Unit Lost Communication With Driver's MICU:** Poor communication between the Immobilizer-keyless Control Unit, and the Driver's MICU. **NOTE: If you are troubleshooting multiple DTCs, be sure to follow the instructions in B-CAN System Diagnosis Test Mode A.** **Possible causes:** • Perform the driver's MICU input test, and do all power, ground and communication input tests • Loose or poor connections at driver's under-dash fuse/relay box connector and the related units • Faulty driver's under-dash fuse/relay box
DTC: U1283 **1T** **Year:** 2009, 2010 **Model:** TSX, ZDX **Engine:** 2.4L L4, 3.5L V6, 3.7L V6 **Transmission:** All	**Lost Communication With Passenger's MICU:** Lost Communication With Passenger's MICU. **NOTE: If you are troubleshooting multiple DTCs, be sure to follow the instructions in B-CAN System Diagnosis Test Mode A.** **Possible causes:** • Check the PCM for DTCs and troubleshoot PCM • Perform the passenger's MICU input test, and do all power, ground and communication input tests. If the tests prove OK, replace the passenger's under-dash fuse/relay box • Loose or poor connections at passenger's under-dash fuse/relay box and the related units
DTC: U1283 **1T** **Year:** 2009, 2010 **Model:** TL, TSX, ZDX **Engine:** 2.4L L4, 3.5L V6, 3.7L V6 **Transmission:** All	**Gauge Control Module Lost Communication With Passenger's MICU:** Poor communication between the Gauge Control Module and MICU. **NOTE: If you are troubleshooting multiple DTCs, be sure to follow the instructions in B-CAN System Diagnosis Test Mode A.** **Possible causes:** • Perform the passenger's MICU input test, and do all power, ground, and communication input tests. If the tests prove OK, replace the passenger's under-dash fuse/relay box • Loose or poor connections at passenger's under-dash fuse/relay box and the related units
DTC: U1283 **1T** **Year:** 2009, 2010 **Model:** TL, TSX, ZDX **Engine:** 2.4L L4, 3.5L V6, 3.7L V6 **Transmission:** All	**Keyless Access Control Unit Lost Communication With Passenger's MICU:** Poor communication between the Keyless Access Control Unit and Passenger's MICU. **NOTE: If you are troubleshooting multiple DTCs, be sure to follow the instructions in B-CAN System Diagnosis Test Mode A.** **Possible causes:** • Perform the passenger's MICU input test, and do all power, ground, and communication input tests. If the tests prove OK, replace the passenger's under-dash fuse/relay box • Loose or poor connections at passenger's under-dash fuse/relay box and the related units
DTC: U1283 **1T** **Year:** 2009, 2010 **Model:** TL, TSX, ZDX **Engine:** 2.4L L4, 3.5L V6, 3.7L V6 **Transmission:** All	**Driver's MICU Lost Communication With Passenger's MICU:** Poor communication between the Driver's MICU and Passenger's MICU. **NOTE: If you are troubleshooting multiple DTCs, be sure to follow the instructions in B-CAN System Diagnosis Test Mode A.** **Possible causes:** • Perform the passenger's MICU input test, and do all power, ground, and communication input tests. If the tests prove OK, replace the driver's under-dash fuse/relay box and the related units • Loose or poor connections at driver's under-dash fuse/relay box and the related units
DTC: U1283 **1T** **Year:** 2009, 2010 **Model:** TSX, ZDX **Engine:** 2.4L L4, 3.5L V6, 3.7L V6 **Transmission:** All	**Door Multiplex Control Unit Lost Communication With Passenger's MICU:** Poor communication between the Door Multiplex Control Unit and Passenger's MICU. **NOTE: If you are troubleshooting multiple DTCs, be sure to follow the instructions in B-CAN System Diagnosis Test Mode A.** **Possible causes:** • Perform the passenger's MICU input test, and do all power, ground and communication input tests. If the tests prove OK, replace the passenger's under-dash fuse/relay box • Loose or poor connections at passenger's under-dash fuse/relay box and the related units
DTC: U1286 **1T** **Year:** 2009, 2010 **Model:** TL, ZDX **Engine:** 3.5L V6, 3.7L V6 **Transmission:** All	**Gauge Control Module Lost Communication With Keyless Access Control Unit:** Poor communication between the Gauge Control Module and Keyless Access Control Unit. **NOTE: If you are troubleshooting multiple DTCs, be sure to follow the instructions in B-CAN System Diagnosis Test Mode A.** **Possible causes:** • Perform the keyless access control unit input test, and do all power, ground, and communication input tests. If the tests prove OK, replace the keyless access control unit • Loose or poor connections at keyless access control unit and the related units
DTC: U1289 **1T** **Year:** 2009, 2010 **Model:** TL, ZDX **Engine:** 3.5L V6, 3.7L V6 **Transmission:** All	**Gauge Control Module Lost Communication With Remote Slot Control Unit:** Poor communication between the Gauge Control Module and Remote Slot Control Unit. **NOTE: If you are troubleshooting multiple DTCs, be sure to follow the instructions in B-CAN System Diagnosis Test Mode A.** **Possible causes:** • Perform the remote slot control unit input test, and do all power, ground, and communication input tests. If the tests prove OK, replace the remote slot control unit • Loose or poor connections at the remote slot control unit and the related units

DTC	Trouble Code Title, Conditions & Possible Causes
DTC: U128B **1T** **Year:** 2009, 2010 **Model:** TL, ZDX **Engine:** 3.5L V6, 3.7L V6 **Transmission:** All	**Gauge Control Module Lost Communication With Power Control Unit:** Poor communication between the Gauge Control Module and Power Control Unit. **NOTE: If you are troubleshooting multiple DTCs, be sure to follow the instructions in B-CAN System Diagnosis Test Mode A.** **Possible causes:** • Perform the power control unit input test, and do all power, ground, and communication input tests. If the tests prove OK, replace the power control unit • Loose or poor connections at power control unit and the related units
DTC: U128D **1T** **Year:** 2009, 2010 **Model:** RDX, TL, TSX, ZDX **Engine:** 2.3L L4, 2.4L L4, 3.5L V6, 3.7L V6 **Transmission:** All	**Lost Communication with Gauge Control Module:** Lost Communication with Gauge Control Module. **NOTE: If you are troubleshooting multiple DTCs, be sure to follow the instructions in B-CAN System Diagnosis Test Mode A.** **Possible causes:** • Poor connections or loose terminals at the AcuraLink control unit (XM receiver) • "Open" circuit between the AcuraLink control unit (XM receiver) and the gauge control module • AcuraLink control unit needs to be updated with the latest software • Faulty AcuraLink control unit
DTC: U128F **1T** **Year:** 2010 **Model:** ZDX **Engine:** 3.7L V6 **Transmission:** All	**Gauge Control Module Lost Communication With BSI Control Unit (BSI Message):** Poor communication between the Gauge Control Module and BSI Control Unit. **NOTE: If you are troubleshooting multiple DTCs, be sure to follow the instructions in B-CAN System Diagnosis Test Mode A.** **Possible causes:** • Loose or poor connections between the BSI control unit and the gauge control module • Perform the gauge control module input test • Faulty BSI control unit

HONDA

Accord

8

BRAKES 8-11

ANTI-LOCK BRAKE SYSTEM (ABS) 8-12
- Wheel Speed Sensors 8-12
  - Removal & Installation..........8-12

BLEEDING THE BRAKE SYSTEM........................8-11
- Bleeding Procedure..................8-11
  - Bleeding Procedure 8-11
  - Bleeding the ABS System 8-11

FRONT DISC BRAKES 8-13
- Brake Caliper..........................8-13
  - Removal & Installation..........8-13
- Disc Brake Pads......................8-13
  - Removal & Installation..........8-13

INFORMATION AND PRECAUTIONS 8-11
  - Anti-lock Systems................8-11
  - Disc and Drum Systems 8-11

PARKING BRAKE 8-15
- Parking Brake Cables...............8-15
  - Adjustment 8-15

REAR DISC BRAKES 8-14
- Brake Caliper..........................8-14
  - Removal & Installation..........8-14
- Disc Brake Pads......................8-14
  - Removal & Installation..........8-14

CHASSIS ELECTRICAL.......8-16

AIR BAG (SUPPLEMENTAL RESTRAINT SYSTEM)........8-16
- General Information..................8-16
  - Arming the System 8-16
  - Disarming the System...........8-16
  - Precautions..........................8-16

DRIVE TRAIN...................8-16
- Clutch.......................................8-16
  - Removal & Installation..........8-16
- Front Halfshaft..........................8-23
  - Removal & Installation..........8-23
- Hydraulic System Bleeding 8-23
  - Bleeding Procedure 8-23
- Intermediate Shaft 8-25
  - Removal & Installation..........8-25

ENGINE COOLING 8-26
- Engine Fan 8-26
  - Removal & Installation..........8-26
- Radiator....................................8-28
  - Removal & Installation..........8-28
- Thermostat 8-30
  - Removal & Installation..........8-30
- Water Pump 8-31
  - Removal & Installation..........8-31

ENGINE ELECTRICAL.........8-32

CHARGING SYSTEM 8-32
- Alternator 8-32
  - Removal & Installation..........8-32

IGNITION SYSTEM 8-34
- Firing Order.............................8-34
- Ignition Coil 8-34
  - Removal & Installation..........8-34
- Ignition Timing.........................8-34
  - Adjustment 8-35
  - Inspection 8-34
- Spark Plugs..............................8-35
  - Removal & Installation..........8-35

STARTING SYSTEM 8-36
- Starter 8-36
  - Removal & Installation..........8-36

ENGINE MECHANICAL 8-37
- Accessory Drive Belts 8-37
  - Accessory Belt Routing............................8-37
  - Adjustment 8-37
  - Inspection 8-37
  - Removal & Installation..........8-37
- Camshaft & Valve Lifters..........8-37
  - Inspection 8-37
  - Removal & Installation..........8-37
- Catalytic Converter...................8-41
  - Removal & Installation..........8-41
- Crankshaft Front Seal...............8-45
  - Removal & Installation..........8-45
- Crankshaft Pulley 8-44
  - Removal & Installation..........8-44
- Cylinder Head 8-45
  - Removal & Installation..........8-45
- Drive Plate...............................8-51
  - Removal & Installation..........8-51
- Intake Manifold 8-51
  - Removal & Installation..........8-51
- Oil Pan 8-56
  - Removal & Installation..........8-56
- Oil Pump..................................8-58
  - Inspection 8-62
  - Removal & Installation..........8-58
- Piston & Ring...........................8-63
  - Positioning 8-63
- Rear Main Seal.........................8-63
  - Removal & Installation..........8-63
- Rocker Arms.............................8-64
  - Removal & Installation..........8-64
- Timing Belt & Sprockets 8-69
  - Removal & Installation..........8-69
- Timing Belt Front Cover............8-68
  - Removal & Installation..........8-68
- Timing Chain & Sprockets 8-72
  - Removal & Installation..........8-72
- Timing Chain Front Cover.........8-71
  - Removal & Installation..........8-71
- Valve Lash................................8-75
  - Adjustment 8-75

ENGINE PERFORMANCE & EMISSION CONTROLS.......8-77
- Accelerator Pedal Position (APP) Sensor 8-77
  - Location.............................8-77
  - Removal & Installation..........8-77
- Camshaft Position (CMP) Sensor 8-77
  - Location...............................8-77
  - Removal & Installation..........8-77
- Crankshaft Position (CKP) Sensor 8-78
  - CKP Pattern Clear/CKP Pattern Learn 8-79
  - Location...............................8-78
  - Removal & Installation..........8-78
- Engine Control Module (ECM)/Powertrain Control Module (PCM).......................8-79
  - CKP Pattern Clear/CKP Pattern Learn 8-81

ECM/PCM Idle Learn Procedure....8-80
ECM/PCM Update....8-80
Removal & Installation....8-79
Engine Coolant Temperature (ECT) Sensor....8-81
Removal & Installation....8-81
Evaporative Emissions (EVAP) Canister....8-82
Location....8-82
Removal & Installation....8-82
Exhaust Gas Recirculation (EGR) Valve....8-83
Location....8-83
Removal & Installation....8-83
Heated Oxygen (HO2S) Sensor....8-84
Location....8-84
Removal & Installation....8-84
Intake Air Temperature (IAT) Sensor....8-87
Location....8-87
Removal & Installation....8-87
Knock Sensor (KS)....8-87
Removal & Installation....8-87
Manifold Absolute Pressure (MAP) Sensor....8-88
Location....8-88
Removal & Installation....8-88
Mass Air Flow (MAF)/Intake Air Temperature (IAT) Sensor (Hot Wire)....8-87
Location....8-87
Removal & Installation....8-87
Positive Crankcase Ventilation (PCV) Valve....8-88
Removal & Installation....8-88
Variable Camshaft Timing Oil Control Solenoid....8-89
Location....8-89
Removal & Installation....8-89

FUEL....8-89

GASOLINE FUEL INJECTION SYSTEM....8-89
Fuel Filter....8-90
Removal & Installation....8-90
Fuel Pump/Fuel Gauge Sending Unit....8-90
Removal & Installation....8-90
Fuel Rail & Injector....8-91
Removal & Installation....8-91
Fuel System Pressure....8-89
Relieving....8-89
Fuel System Service Precautions....8-89
Fuel Tank....8-94
Draining....8-94
Removal & Installation....8-95
Idle Speed....8-97
Adjustment....8-97
Throttle Body....8-97
Removal & Installation....8-97

HEATING & AIR CONDITIONING SYSTEM....8-99

Blower Motor....8-99
Removal & Installation....8-99
Heater Core....8-99
Removal & Installation....8-99

SPECIFICATIONS AND MAINTENANCE CHARTS....8-3

Brake Specifications....8-9
Camshaft and Bearing Specifications....8-5
Capacities....8-4
Crankshaft and Connecting Rod Specifications....8-6
Engine and Vehicle Identification....8-3
Engine Tune-Up Specifications....8-3
Fluid Specifications....8-4
General Engine Specifications....8-3
Maintenance Minder Schedule....8-10
Piston and Ring Specifications....8-6
Tire, Wheel and Ball Joint Specifications....8-9
Torque Specifications....8-7
Valve Specifications....8-5
Wheel Alignment....8-8

STEERING....8-102

Power Rack & Pinion Steering Gear....8-102
Removal & Installation....8-102
Power Steering Pump....8-108
Removal & Installation....8-108

SUSPENSION....8-109

FRONT SUSPENSION....8-109
Control Links....8-109
Removal & Installation....8-109
Lower Ball Joint....8-109
Removal & Installation....8-109
Lower Control Arm....8-110
Removal & Installation....8-110
Stabilizer Bar....8-116
Removal & Installation....8-116
Steering Knuckle....8-111
Removal & Installation....8-111
Strut (Damper/Spring)....8-112
Overhaul....8-114
Removal & Installation....8-112
Upper Ball Joint....8-125
Removal & Installation....8-125
Upper Control Arm....8-125
Removal & Installation....8-125
Wheel Hub & Bearing....8-126
Removal & Installation....8-126

REAR SUSPENSION....8-128
Control Arms/Links....8-128
Removal & Installation....8-128
Knuckle....8-131
Removal & Installation....8-131
Stabilizer Bar....8-133
Removal & Installation....8-133
Strut (Damper/Spring)....8-134
Wheel Hub & Bearing....8-135
Removal & Installation....8-135

SPECIFICATIONS AND MAINTENANCE CHARTS

ENGINE AND VEHICLE IDENTIFICATION

Engine					
Code	Liters	Cu. In. (cc)	Cyl.	Fuel Sys.	Eng. Mfg.
J35Z2	3.5	212.0 (3471)	6	MPFI	Honda
K24Z2	2.4	144.0 (2354)	4	MPFI	Honda
K24Z3	2.4	144.0 (2354)	4	MPFI	Honda

Model Year	
Code	Year
9	2009
A	2010

MPFI: Multiport Fuel Injection

37647_ACCD_C0001

GENERAL ENGINE SPECIFICATIONS

Year	Model	Engine Displacement Liters	Engine ID/VIN	Net Horsepower @ rpm	Net Torque @ rpm (ft. lbs.)	Bore X Stroke (in.)	Com-pression Ratio	Oil Pressure @ rpm
2009	Accord	2.4	K24Z2	177@6500	161@4300	3.43x3.90	10.5:1	44@3000
	Accord	2.4	K24Z3	190@7000	162@4400	3.43x3.90	10.5:1	44@3000
	Accord	3.5	J35Z2	271@6200	254@5000	3.50x3.70	10.5:1	44@3000
2010	Accord	2.4	K24Z2	177@6500	161@4300	3.43x3.90	10.5:1	44@3000
	Accord	2.4	K24Z3	190@7000	162@4400	3.43x3.90	10.5:1	44@3000
	Accord	3.5	J35Z2	271@6200	254@5000	3.50x3.70	10.5:1	44@3000

37647_ACCD_C0002

ENGINE TUNE-UP SPECIFICATIONS

Year	Engine Displacement Liters	Engine ID/VIN	Spark Plugs Gap (in.)	Ignition Timing (deg.)		Fuel Pump (psi)	Idle Speed (rpm)		Valve Clearance	
				MT	AT		MT	AT	In.	Ex.
2009	2.4	K24Z2	0.039-0.043	6-10B	6-10B	48-55	720-830	750-850	0.008-0.010	0.010-0.011
	2.4	K24Z3	0.039-0.043	6-10B	6-10B	48-55	720-830	750-850	0.008-0.010	0.010-0.011
	3.5	J35Z2	0.039-0.043	3-7B	8-12B	57-64	600-700	600-700	0.008-0.009	0.011-0.013
2010	2.4	K24Z2	0.039-0.043	6-10B	6-10B	48-55	720-830	750-850	0.008-0.010	0.010-0.011
	2.4	K24Z3	0.039-0.043	6-10B	6-10B	48-55	720-830	750-850	0.008-0.010	0.010-0.011
	3.5	J35Z2	0.039-0.043	3-7B	8-12B	57-64	600-700	600-700	0.008-0.009	0.011-0.013

NOTE: The Vehicle Emission Control Information label often reflects specification changes made during production.
The label figures must be used if they differ from those in this chart

B: Before Top Dead Center

37647_ACCD_C0003

CAPACITIES

Year	Model	Engine Displacement Liters	Engine ID	Engine Oil with Filter	Transmission (pts.) 5-Spd	Transmission (pts.) Auto.	Drive Axle Front (pts.)	Drive Axle Rear (pts.)	Fuel Tank (gal.)	Cooling System (qts.)
2009	Accord	2.4	K24Z2	4.4	4.0	5.2	—	—	18.5	①
	Accord	2.4	K24Z3	4.4	4.0	5.2	—	—	18.5	①
	Accord	3.5	J35Z2	4.5	4.2	7.0	—	—	18.5	7.0
2010	Accord	2.4	K24Z2	4.4	4.0	5.2	—	—	18.5	①
	Accord	2.4	K24Z3	4.4	4.0	5.2	—	—	18.5	①
	Accord	3.5	J35Z2	4.5	4.2	7.0	—	—	18.5	7.0

NOTE: All capacities are approximate. Add fluid gradually and ensure a proper fluid level is obtained.

NOTE: Capacities given are service, not overhaul capacities

① Automatic Transaxle: 6.2
Manual Transaxle: 6.4

37647_ACCD_C0004

FLUID SPECIFICATIONS

Year	Model	Engine Displ. Liters	Engine Oil	Man. Trans.	Auto. Trans.	Drive Axle Front	Drive Axle Rear	Power Steering Fluid	Brake Master Cylinder	Cooling System
2009	Accord	2.4	5W-20 Honda	Honda MTF	Honda ATF-Z1	—	—	Honda PS Fluid	Honda DOT 3	①
	Accord	3.0	5W-20 Honda	Honda MTF	Honda ATF-Z1	—	—	Honda PS Fluid	Honda DOT 3	①
2010	Accord	2.4	5W-20 Honda	Honda MTF	Honda ATF-Z1	—	—	Honda PS Fluid	Honda DOT 3	①
	Accord	3.5	5W-20 Honda	Honda MTF	Honda ATF-Z1	—	—	Honda PS Fluid	Honda DOT 3	①

DOT: Department Of Transpotation

① Honda Long Life Antifreeze/Coolant-Type2

37647_ACCD_C0005

VALVE SPECIFICATIONS

Year	Engine Displacement Liters	Engine ID/VIN	Seat Angle (deg.)	Face Angle (deg.)	Spring Test Pressure (lbs. @ in.)	Spring Installed Height (in.)	Stem-to-Guide Clearance (in.)		Stem Diameter (in.)	
							Intake	Exhaust	Intake	Exhaust
2009	2.4	K24Z2	45	45	NA	NA	0.0012-0.0022	0.0022-0.0031	0.2156-0.2159	0.2146-0.2150
	2.4	K24Z3	45	45	NA	NA	0.0012-0.0022	0.0022-0.0031	0.2156-0.2159	0.2146-0.2150
	3.5	J35Z2	45	45	NA	NA	0.0008-0.0018	0.0022-0.0031	0.2159-0.2163	0.2146-0.2150
2010	2.4	K24Z2	45	45	NA	NA	0.0012-0.0022	0.0022-0.0031	0.2156-0.2159	0.2146-0.2150
	2.4	K24Z3	45	45	NA	NA	0.0012-0.0022	0.0022-0.0031	0.2156-0.2159	0.2146-0.2150
	3.5	J35Z2	45	45	NA	NA	0.0008-0.0018	0.0022-0.0031	0.2159-0.2163	0.2146-0.2150

NA: Information not available

37647_ACCD_C0007

CAMSHAFT AND BEARING SPECIFICATIONS

All measurements are given in inches.

Year	Engine Displacement Liters	Engine VIN	Journal Diameter	Brg. Oil Clearance	Shaft End-play	Runout	Journal Bore	Lobe Height	
								Intake	Exhaust
2009	2.4	K24Z2	NA	①	0.0020-0.0080	0.0010	NA	②	1.3500
	2.4	K24Z3	NA	①	0.0020-0.0080	0.0010	NA	②	1.3500
	3.5	J35Z2	NA	0.0020-0.0035	0.0020-0.0080	0.0010	NA	③	④
2010	2.4	K24Z2	NA	①	0.0020-0.0080	0.0010	NA	②	1.3500
	2.4	K24Z3	NA	①	0.0020-0.0080	0.0010	NA	②	1.3500
	3.5	J35Z2	NA	0.0020-0.0035	0.0020-0.0080	0.0010	NA	③	④

NA: Information not available

① No. 1 journal: 0.0012-0.0027 inches

No. 2-5 journals: 0.0024-0.0039 inches

② Primary: 1.3285 inches

Mid: 1.3959 inches

Secondary: 1.3285 inches

Mid: 1.4348 in.

Secondary: 1.3891 in.

③ Nos. 1-4 cylinders: 1.3965 in.

Nos 5-6 cylinders: 1.3964 in.

④ Nos. 1-4 cylinders: 1.4481 in

Nos 5-6 cylinders: 1.4472 in.

37047_ACCD_C0006

CRANKSHAFT AND CONNECTING ROD SPECIFICATIONS

All measurements are given in inches.

Year	Engine Displacement Liters	Engine ID/VIN	Crankshaft Main Brg. Journal Dia.	Crankshaft Main Brg. Oil Clearance	Crankshaft Shaft End-play	Crankshaft Thrust on No.	Connecting Rod Journal Diameter	Connecting Rod Oil Clearance	Connecting Rod Side Clearance
2009	2.4	K24Z2	①	②	0.0040-0.0140	4	1.8888-1.8898	0.0013-0.0026	0.0060-0.0140
	2.4	K24Z3	①	②	0.0040-0.0140	4	1.8888-1.8898	0.0013-0.0026	0.0060-0.0140
	3.5	J35Z2	2.8337-2.8346	0.0007-0.0018	0.0040-0.0140	3	2.1644-2.1654	0.0008-0.0017	0.0060-0.0140
2010	2.4	K24Z2	①	②	0.0040-0.0140	4	1.8888-1.8898	0.0013-0.0026	0.0060-0.0140
	2.4	K24Z3	①	②	0.0040-0.0140	4	1.8888-1.8898	0.0013-0.0026	0.0060-0.0140
	3.5	J35Z2	2.8337-2.8346	0.0007-0.0018	0.0040-0.0140	3	2.1644-2.1654	0.0008-0.0017	0.0060-0.0140

① Journals 1, 2 4 and 5: 2.1647-2.1657 inches
Journal 3: 2.1644-2.1654 inches

② Journals 1, 2 4 and 5: 0.0007-0.0016 inches
Journal 3: 0.0010-0.0019 inches

37647_ACCD_C0008

PISTON AND RING SPECIFICATIONS

All measurements are given in inches.

Year	Engine Displacement Liters	Engine ID/VIN	Piston Clearance	Ring Gap Top Compression	Ring Gap Bottom Compression	Ring Gap Oil Control	Ring Side Clearance Top Compression	Ring Side Clearance Bottom Compression	Ring Side Clearance Oil Control
2009	2.4	K24Z2	0.0008-0.0016	0.0080-0.0140	0.0200-0.0260	0.0080-0.0280	0.0024-0.0033	0.0016-0.0026	NA
	2.4	K24Z3	0.0008-0.0016	0.0080-0.0140	0.0200-0.0260	0.0080-0.0280	0.0024-0.0033	0.0016-0.0026	NA
	3.5	J35Z2	0.0006-0.0016	0.0080-0.0140	0.0160-0.0220	0.0080-0.0280	0.0022-0.0031	0.0012-0.0022	NA
2010	2.4	K24Z2	0.0008-0.0016	0.0080-0.0140	0.0200-0.0260	0.0080-0.0280	0.0024-0.0033	0.0016-0.0026	NA
	2.4	K24Z3	0.0008-0.0016	0.0080-0.0140	0.0200-0.0260	0.0080-0.0280	0.0024-0.0033	0.0016-0.0026	NA
	3.5	J35Z2	0.0006-0.0016	0.0080-0.0140	0.0160-0.0220	0.0080-0.0280	0.0022-0.0031	0.0012-0.0022	NA

NA: Information not available

37647_ACCD_C0009

TORQUE SPECIFICATIONS

All readings in ft. lbs.

Year	Engine Displacement Liters	Engine ID/VIN	Cylinder Head Bolts	Main Bearing Bolts	Rod Bearing Bolts	Crankshaft Damper Bolts	Flywheel Bolts	Manifold		Spark Plugs	Oil Pan Drain Plug
								Intake	Exhaust		
2009	2.4	K24Z2	①	②	③	④	⑤	16	25	13	29
	2.4	K24Z3	①	②	③	④	⑤	16	25	13	29
	3.5	J35Z2	⑥	⑦	⑧	⑨	⑩	16	23	13	29
2010	2.4	K24Z2	①	②	③	④	⑤	16	25	13	29
	2.4	K24Z3	①	②	③	④	⑤	16	25	13	29
	3.5	J35Z2	⑥	⑦	⑧	⑨	⑩	16	23	13	29

① Step 1: 29 ft. lbs.
Step 2: Rotate 90 degrees
Step 3: Rotate 90 degrees
Step 4: If new bolt rotate additional 90 degrees

② Step 1: 22 ft. lbs.
Step 2: Rotate 48 degrees

③ Step 1: 30 ft. lbs.
Step 2: Rotate 120 degrees

④ Old bolt:
Step 1: 36 ft. lbs
Step 2: Plus 90 degrees
New bolt:
Step 1: 130 ft. lbs
Step 2: Loosen fully
Step 3: 36 ft. lbs
Step 4: Plus 90 degrees

⑤ Automatic transaxle: 54 ft. lbs.
Manual transaxle: 90 ft. lbs.

⑥ Step 1: 22 ft. lbs.
Step 2: Rotate 90 degrees
Step 3: Rotate 90 degrees
Step 4: If new bolt rotate additional 90 degrees

⑦ Step 1: Cap bolts, 56 ft. lbs.
Step 2: Side bolts, 36 ft. lbs.

⑧ Step 1: 14 ft. lbs.
Step 2: Rotate 90 degrees

⑨ Step 1: 47 ft. lbs.
Step 2: Rotate 60 degrees

⑩ Automatic transaxle: 54 ft. lbs.
Manual transaxle: 76 ft. lbs.

37647_ACCD_C0010

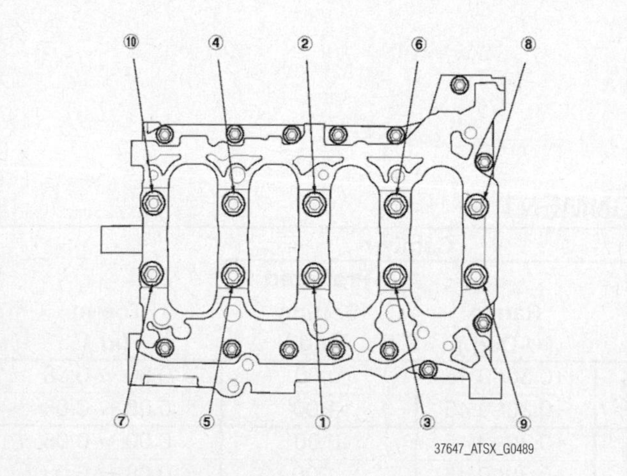
10
4
2
6
8
7
5
1
3
9
37647_ATSX_G0489

Fig. 1 Main bearing torque sequence—4 cylinder engine, 1 of 2

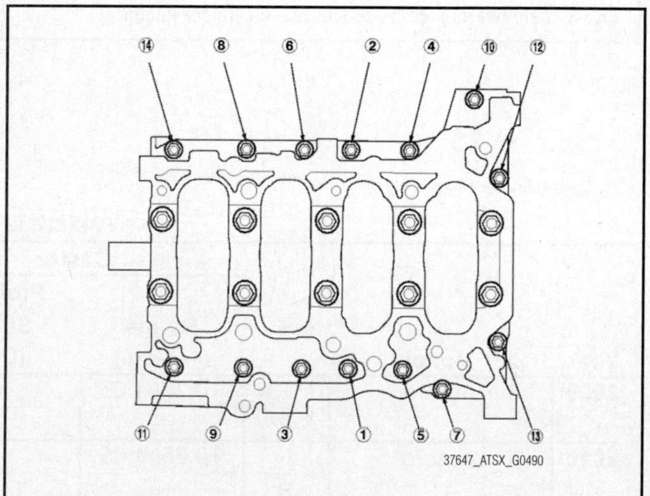
14
8
6
2
4
10
12
11
9
3
1
5
7
13
37647_ATSX_G0490

Fig. 2 Main bearing torque sequence—4 cylinder engine, 2 of 2

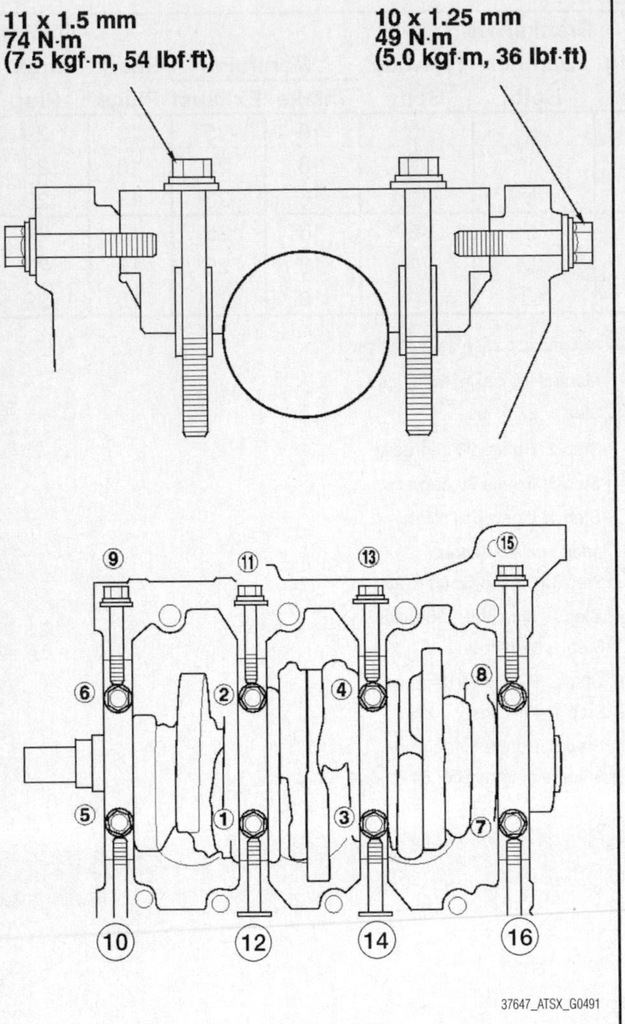
11 x 1.5 mm
74 N·m
(7.5 kgf·m, 54 lbf·ft)
10 x 1.25 mm
49 N·m
(5.0 kgf·m, 36 lbf·ft)
9
11
13
15
6
2
4
8
5
1
3
7
10
12
14
16
37647_ATSX_G0491

Fig. 3 Main bearing torque sequence—6 cylinder engine

WHEEL ALIGNMENT

Year	Model		Caster Range (+/-Deg.)	Caster Preferred Setting (Deg.)	Camber Range (+/-Deg.)	Camber Preferred Setting (Deg.)	Toe-in (in.)
2009	Accord	F	+0.25/-1.05	①	+0.30/-0.45	0.00	0.00 +/-0.08
		R	—	—	+0.30/-0.45	-1 00'	0.08 +/-0.08
2010	Accord	F	+0.25/-1.05	①	+0.30/-0.45	0.00	0.00 +/-0.08
		R	—	—	+0.30/-0.45	-1 00'	0.08 +/-0.08

① 2-door: 3.47; 4-door: 3.48

37647_ACCD_C0011

TIRE, WHEEL AND BALL JOINT SPECIFICATIONS

Year	Model	OEM Tires Standard	OEM Tires Optional	Tire Pressures (psi) Front	Tire Pressures (psi) Rear	Wheel Size	Ball Joint Inspection	Lug Nut (ft. lbs.)
2009	Accord	P215/60R16	P225/50R17	30	30	16x6.5	NA	80
2010	Accord	P215/60R16	P225/50R17	30	30	16x6.5	NA	80

OEM: Original Equipment Manufacturer

PSI: Pounds Per Square Inch

NA: Information not available

37647_ACCD_C0012

BRAKE SPECIFICATIONS

All measurements in inches unless noted

Year	Model		Brake Disc Original Thickness	Brake Disc Minimum Thickness	Brake Disc Maximum Runout	Minimum Lining Thickness Front	Minimum Lining Thickness Rear	Brake Caliper Bracket Bolts (ft. lbs.)	Brake Caliper Mounting Bolts (ft. lbs.)
2009	Accord	F	①	②	0.002	0.060	—	80	37
		R	0.360	0.310	0.002	—	0.060	80	17
2010	Accord	F	①	②	0.002	0.060	—	80	37
		R	0.360	0.310	0.002	—	0.040	80	17

NA: Information not available

F: Front

R: Rear

① Except V6 w/manual transaxle: 0.910
V6 w/automatic transaxle: 1.110

② Except V6 w/manual transaxle: 0.830
V6 w/automatic transaxle: 1.020

37647_ACCD_C0013

MAINTENANCE MINDER SCHEDULE

Honda Accord

All Honda's displays engine oil life and maintenance service items in the information display to indicate when to perform maintenance service. If the engine oil life is 15% or less, based on the onboard computer's caluculations, you will see SERVICE DUE SOON in the information display every time the ignition key is turned to ON. The maintenance minder indicator will also come on and the maintenance code(s) for other scheduled maintenance items needing service will be displayed below the message.

Symbol	Item	Service
A	Engine oil ①	Change
B	Engine oil and filter	Change
	Fluid levels	Inspect
	Brakes	Inspect
	Parking brake adjustment	Check
	Steering gear and linkage	Inspect
	Suspension components	Inspect
	Driveshaft boots	Inspect
	Brake hoses and lines	Inspect
	Exhaust system	Inspect
	Fuel lines and connections	Inspect
1	Tires	Rotate
2	Engine air filter ②	Replace
	Dust and pollen filter ③	Replace
	Accessory drive belt	Inspect
3	Transmission fluid ④	Replace
	Transfer case fluid ④	Replace
4	Spark plugs	Replace
	Timing belt ⑤	Replace
	Water pump	Inspect
	Valve clearance ⑥	Inspect
5	Engine coolant	Replace
6	VTM-4 rear differential fluid	Replace

① If the message SERVICE DUE NOW does not appear more than 12 months after the display is reset, change every year.
② If driven in dusty conditions, replace every 15,000 miles.
③ If driven in urban areas that have a high concentration of soot from industry and diesel, replace every 15,000 miles
④ If regularly driven in mountainous areas at very low speed or trailer towing, change the fluid every 30,000 miles.
⑤ If driven regularly in temperatures over 110 deg.F or below -20 deg.F, or towing a trailer, replace every 60,000 miles.
⑥ Adjust if necessary.

Additionally, replace the brake fluid every 3 years, and inspect the idle speed every 160,000 miles.

To reset the Engine Oil Life Display:

1. Turn the ignition switch to ON.
2. Press the SELECT button repeatedly until the engine oil life display or the service message is displayed.
3. Press the RESET button for about 10 seconds. You will see a MAINT RESET message.
4. Select the appropriate answer, MAINT RESET >N (NO) or MAINT RESET > y (YES) by pressing the SELECT button repeatedly.
   >N or >Y is displayed on the outside temperature >N or >Y is displayed on the outside temperature display.
5. Select the MAINT RESET > Y (YES), and press and hold the RESET button again to reset the engine oil life to 100%.

37647_ACCD_C0014

BRAKES — INFORMATION AND PRECAUTIONS

ANTI-LOCK SYSTEMS

- Certain components within the ABS system are not intended to be serviced or repaired individually.
- Do not use rubber hoses or other parts not specifically specified for and ABS system. When using repair kits, replace all parts included in the kit. Partial or incorrect repair may lead to functional problems and require the replacement of components.
- Lubricate rubber parts with clean, fresh brake fluid to ease assembly. Do not use shop air to clean parts; damage to rubber components may result.
- Use only DOT 3 brake fluid from an unopened container.
- If any hydraulic component or line is removed or replaced, it may be necessary to bleed the entire system.
- A clean repair area is essential. Always clean the reservoir and cap thoroughly before removing the cap. The slightest amount of dirt in the fluid may plug an orifice and impair the system function. Perform repairs after components have been thoroughly cleaned; use only denatured alcohol to clean components. Do not allow ABS components to come into contact with any substance containing mineral oil; this includes used shop rags.
- The Anti-Lock control unit is a microprocessor similar to other computer units in the vehicle. Ensure that the ignition switch is **OFF** before removing or installing controller harnesses. Avoid static electricity discharge at or near the controller.
- If any arc welding is to be done on the vehicle, the control unit should be unplugged before welding operations begin.

DISC AND DRUM SYSTEMS

CAUTION

Dust and dirt accumulating on brake parts during normal use may contain asbestos fibers from production or aftermarket brake linings. Breathing excessive concentrations of asbestos fibers can cause serious bodily harm. Exercise care when servicing brake parts. Do not sand or grind brake lining unless equipment used is designed to contain the dust residue. Do not clean brake parts with compressed air or by dry brushing. Cleaning should be done by dampening the brake components with a fine mist of water, then wiping the brake components clean with a dampened cloth. Dispose of cloth and all residue containing asbestos fibers in an impermeable container with the appropriate label. Follow practices prescribed by the Occupational Safety and Health Administration (OSHA) and the Environmental Protection Agency (EPA) for the handling, processing, and disposing of dust or debris that may contain asbestos fibers.

BRAKES — BLEEDING THE BRAKE SYSTEM

BLEEDING PROCEDURE

BLEEDING PROCEDURE

See Figure 4.

➡Do not reuse the drained fluid. Use only clean Honda DOT 3 Brake Fluid from an unopened container. Using a non-Honda brake fluid can cause corrosion and shorten the life of the system.

➡Do not mix different brands of brake fluid; they may not be compatible.

➡Make sure no dirt or other foreign matter is allowed to contaminate the brake fluid.

WARNING

Do not spill brake fluid on the vehicle, it may damage the paint; if brake fluid does contact the paint, wash it off immediately with water.

➡The reservoir on the master cylinder must be at the MAX (upper) level mark at the start of the bleeding procedure and checked after bleeding each brake caliper. Add fluid as required.

1. Make sure the brake fluid level in the reservoir is at the MAX (upper) level line.
2. Attach a length of clear drain tube to the bleed screw.
3. Have someone slowly pump the brake pedal several times, and then apply steady pressure.
4. Starting at the left-front, loosen the brake bleed screw to allow air to escape from the system. Then tighten the bleed screw securely.
5. Repeat the procedure for each wheel in the sequence shown following until air bubbles no longer appear in the fluid.

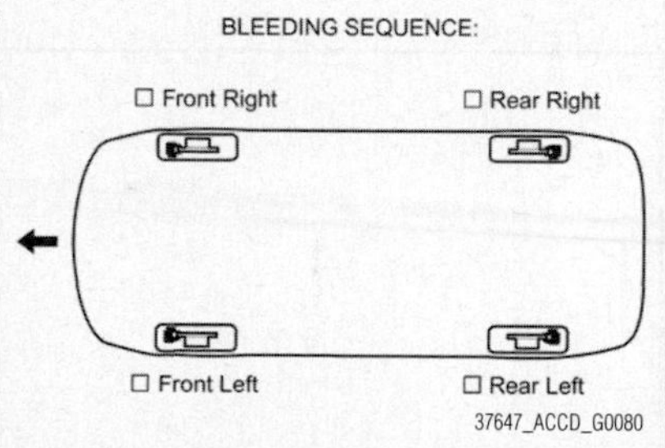
BLEEDING SEQUENCE:
☐ Front Right
☐ Rear Right
☐ Front Left
☐ Rear Left
37647_ACCD_G0080

Fig. 4 Brake bleeding sequence

6. Refill the master cylinder reservoir to the MAX (upper) level line.

BLEEDING THE ABS SYSTEM

The bleeding procedure for the ABS System is the same as the Conventional Bleeding Procedure.

WHEEL SPEED SENSORS

REMOVAL & INSTALLATION

Front

See Figure 5.

1. Turn the ignition switch to LOCK (0).
2. Release the clamp, then disconnect the wheel speed sensor connector.
3. Remove the bolts and the wheel speed sensor.

To install:

4. Install the wheel speed sensor in the reverse order of removal, and note these items:
    - Do not twist the sensor wires.
    - If the wheel speed sensor comes in contact with the wheel bearing, it is faulty.
    - Make sure there is no debris in the sensor mounting hole.
5. Start the engine, and make sure the ABS and the VSA indicators go off.
6. Test-drive the vehicle, and make sure the ABS and the VSA indicators do not come on.

Rear

See Figure 6.

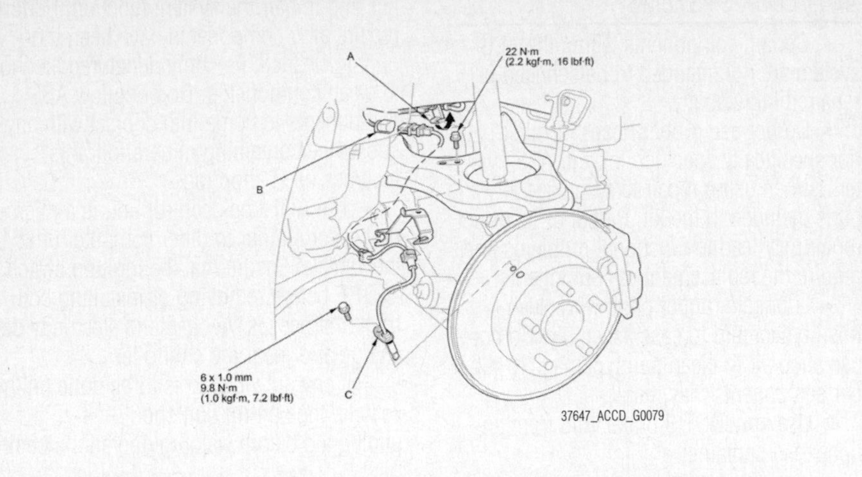
A
22 N·m
(2.2 kgf·m, 16 lbf·ft)
B
6 x 1.0 mm
9.8 N·m
(1.0 kgf·m, 7.2 lbf·ft)
C
37647_ACCD_G0079

Fig. 6 Release the clamp (A), then disconnect the wheel speed sensor connector (B) and remove the wheel speed sensor (C)

1. Turn the ignition switch to LOCK (0).
2. Release the clamp, then disconnect the wheel speed sensor connector.
3. Remove the clamps, the bolt, and the wheel speed sensor.
4. Install the wheel speed sensor in the reverse order of removal, and note these items:
    - Do not twist the sensor wires.
    - If the wheel speed sensor comes in contact with the hub bearing unit, it is faulty.
    - Make sure there is no debris in the sensor mounting hole.
5. Start the engine, and make sure the ABS and the VSA indicators go off.
6. Test-drive the vehicle, and make sure the ABS and the VSA indicators do not come on.

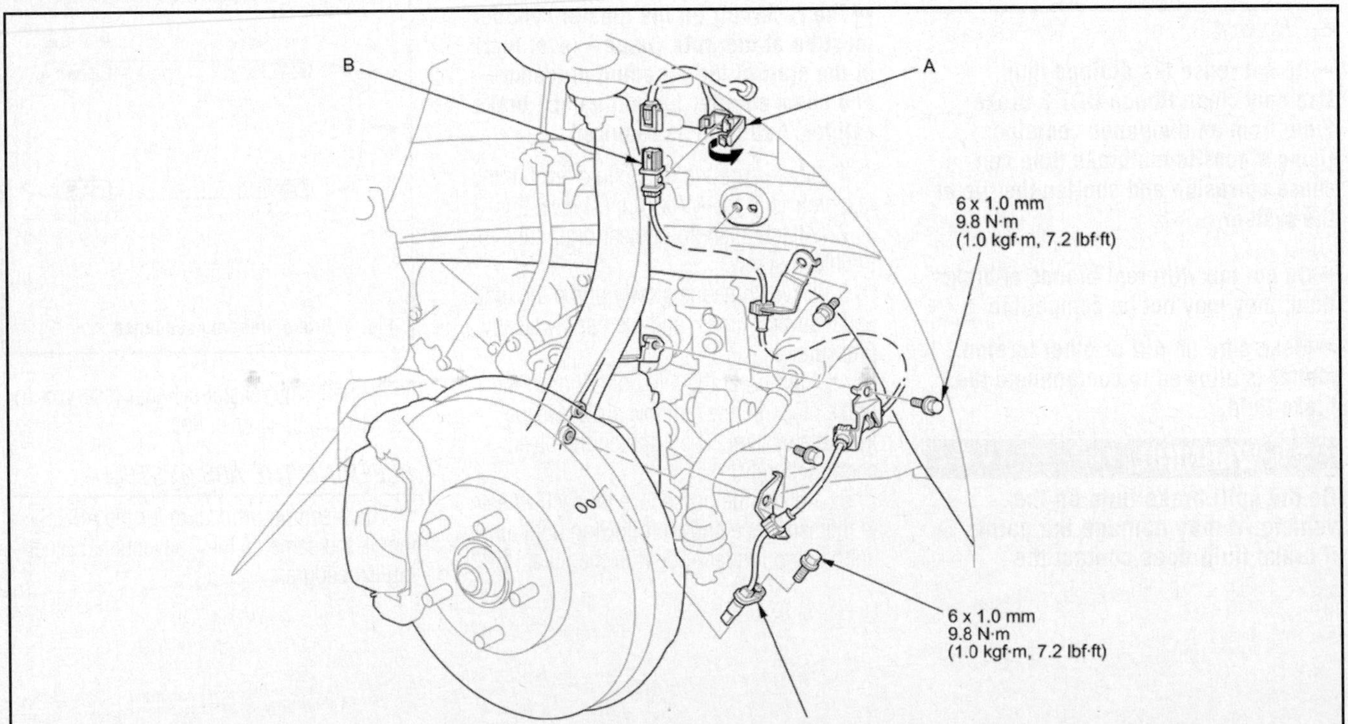
B
A
6 x 1.0 mm
9.8 N·m
(1.0 kgf·m, 7.2 lbf·ft)
6 x 1.0 mm
9.8 N·m
(1.0 kgf·m, 7.2 lbf·ft)
C
37647_ACCD_G0078

Fig. 5 Release the clamp (A), then disconnect the wheel speed sensor connector (B) and remove the wheel speed sensor (C)

BRAKE CALIPER

REMOVAL & INSTALLATION

See Figure 7.

1. Raise the vehicle on a lift.
2. Remove the front wheels.
3. Remove the brake hose mounting bolt.
4. Disconnect the brake hose from the caliper. Place a cap on the end of the brake hose.
5. Remove the brake caliper bracket mounting bolts, then remove the caliper assembly from the knuckle.
6. Installation is the reverse of removal.
7. Tighten the brake caliper bracket mounting bolts to 80 ft. lbs. (108 Nm).

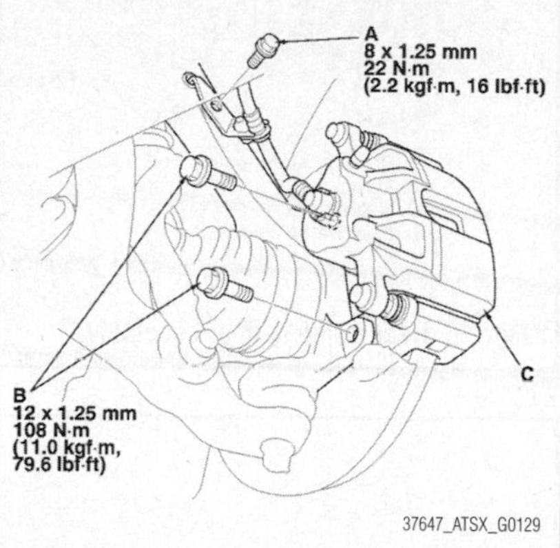
A
8 x 1.25 mm
22 N·m
(2.2 kgf·m, 16 lbf·ft)
B
12 x 1.25 mm
108 N·m
(11.0 kgf·m,
79.6 lbf·ft)
C
37647_ATSX_G0129

Fig. 7 Remove the brake hose mounting bolt

DISC BRAKE PADS

REMOVAL & INSTALLATION

See Figures 8 through 11.

1. Remove some brake fluid from the master cylinder.
2. Raise the vehicle on a lift.
3. Remove the front wheels.
4. Remove the brake hose mounting bolt.
5. Remove the flange bolt while holding the caliper pin with a wrench. Be careful not to damage the pin boot, and pivot the caliper up out of the way. Check the hose and the pin boots for damage and deterioration.
6. Remove the pad shims and the brake pads.
7. Remove the pad retainers.

To install:

8. Clean the caliper bracket thoroughly; remove any rust, and check for grooves and cracks.

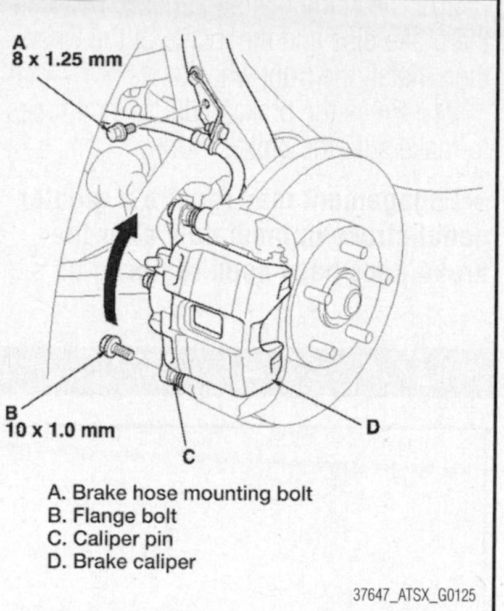
A
8 x 1.25 mm
B
10 x 1.0 mm
C
D
A. Brake hose mounting bolt
B. Flange bolt
C. Caliper pin
D. Brake caliper
37647_ATSX_G0125

Fig. 8 Remove the brake hose mounting bolt

9. Verify that the caliper pins move in and out smoothly. Clean and lube if needed.
10. Inspect the brake disc for runout, thickness, parallelism, and check for damage and cracks.
11. Apply a thin coat of M-77 assembly paste (P/N 08798-9010) to the retainer mating surface of the caliper bracket (indicated by the arrows).
12. Install the pad retainers. Wipe excess assembly paste off the retainers. Keep the assembly paste off the brake disc and the brake pads.
13. Install the brake caliper piston compressor tool on the caliper body.

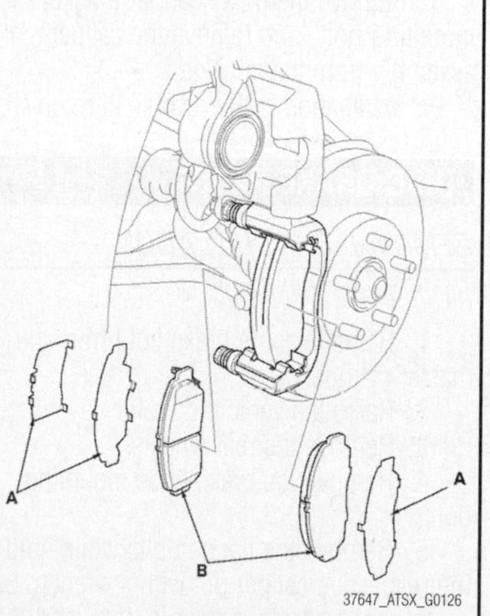
A
A
B
37647_ATSX_G0126

Fig. 9 Remove the pad shims (A) and the brake pads (B)

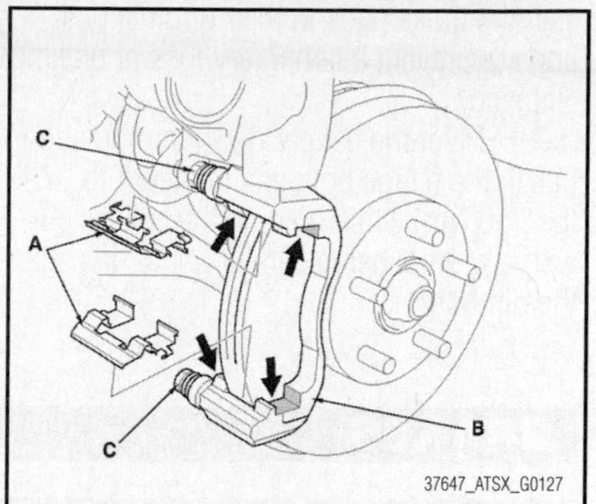
C
A
C
B
37647_ATSX_G0127

Fig. 10 Remove the pad retainers (A)

14. Press in the piston with the brake caliper piston compressor tool so the caliper will fit over the brake pads. Make sure the piston boot is in position to prevent damaging it when pivoting the caliper down.

➡Be careful when pressing in the piston; brake fluid might overflow from the master cylinder's reservoir. If brake fluid gets on any painted surface, wash it off immediately with water.

15. Remove the brake caliper piston compressor tool.
16. Apply a thin coat of M-77 assembly paste (P/N 08798-9010) to the pad side of the shims, the back of the brake pads and the other areas indicated by the arrows. Wipe excess assembly paste off the pad shims and the brake pads friction material. Keep grease and assembly paste off the brake disc and the brake pads. Contaminated brake disc or brake pads reduce stopping ability.
17. Install the brake pads and the pad shims correctly. Install the brake pad with

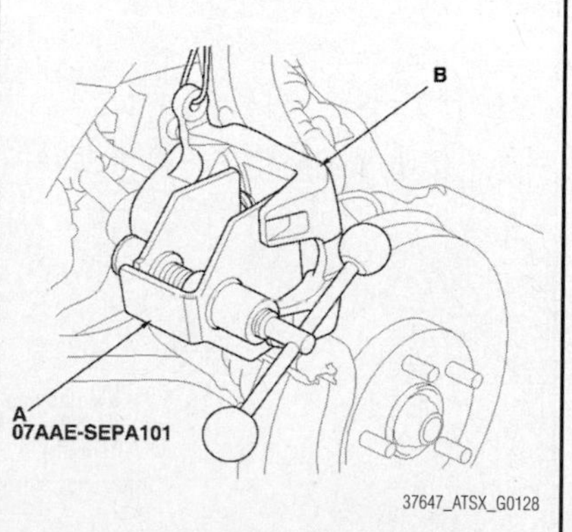
B
A
07AAE-SEPA101
37647_ATSX_G0128

Fig. 11 Install the brake caliper piston compressor tool (A) on the caliper body (B)

the wear indicator on the upper inside. If you are reusing the brake pads, always reinstall the brake pads in their original positions to prevent a temporary loss of braking efficiency.

18. Pivot the caliper down into position. Install the flange bolt, and tighten it to 37 ft. lbs. (50 Nm) while holding the caliper pin with a wrench being careful not to damage the pin boot.

19. Install the brake hose mounting bolt. Tighten the bolt to 16 ft. lbs. (22 Nm).

20. Clean the mating surfaces between the brake disc and the inside of the wheel, then install the front wheels.

21. Press the brake pedal several times to make sure the brakes work.

➡Engagement may require a greater pedal stroke immediately after the brake pads have been replaced as a set. Several applications of the brake pedal will restore the normal pedal stroke.

22. Add brake fluid as needed.

23. After installation, check for leaks at hose and line joints or connections, and retighten if necessary.

24. Test-drive the vehicle, then recheck for leaks.

BRAKES — REAR DISC BRAKES

BRAKE CALIPER

REMOVAL & INSTALLATION

See Figures 12 through 14.

1. Raise and support the vehicle.
2. Remove the rear wheel.
3. Release the parking brake lever fully.
4. Loosen the parking brake cable adjusting nut.

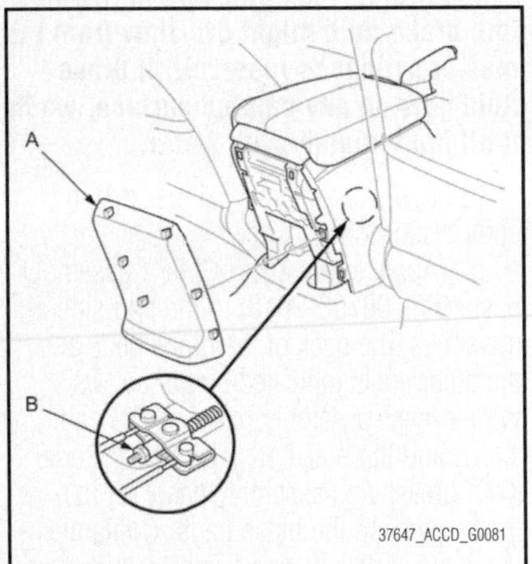
A
B
37647_ACCD_G0081

Fig. 12 Loosen the parking brake cable adjusting nut

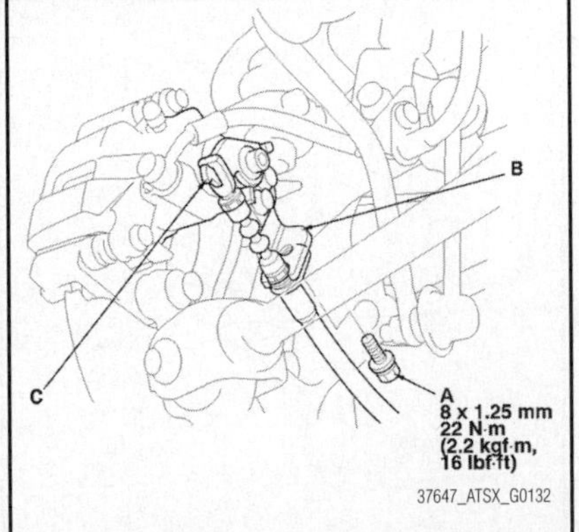
B
C
A
8 x 1.25 mm
22 N·m
(2.2 kgf·m,
16 lbf·ft)
37647_ATSX_G0132

Fig. 13 Remove the flange bolt (A) from the arm (B) and disconnect the parking brake cable from the lever (C)

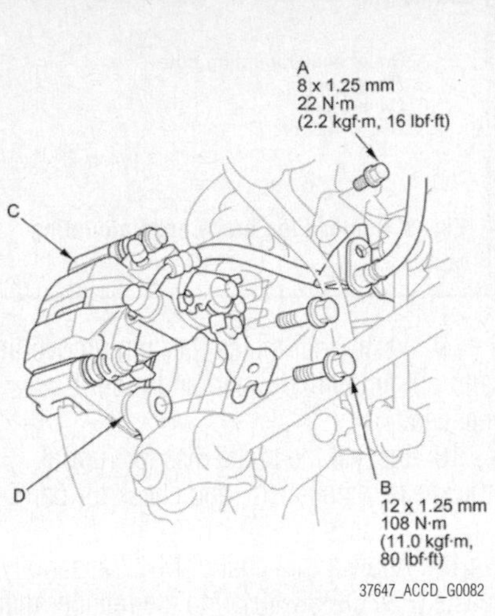
A
8 x 1.25 mm
22 N·m
(2.2 kgf·m, 16 lbf·ft)
C
D
B
12 x 1.25 mm
108 N·m
(11.0 kgf·m,
80 lbf·ft)
37647_ACCD_G0082

Fig. 14 Remove the brake caliper

5. Remove the flange bolt from the arm.
6. Disconnect the parking brake cable from the lever.
7. Remove the brake hose mounting bolt.
8. Remove the brake caliper bracket mounting bolts and remove the caliper assembly from the knuckle.
9. Installation is the reverse of removal.

DISC BRAKE PADS

REMOVAL & INSTALLATION

See Figures 15 through 18.

1. Remove some brake fluid from the master cylinder.
2. Raise the vehicle on a lift.
3. Remove the rear wheels.
4. Remove the brake hose mounting bolt.
5. Remove the flange bolts while holding respective caliper pin with a wrench. Be careful not to damage the pin boot, and remove the caliper. Check the hose, the pin boots, and the parking brake cable boots for damage and deterioration.

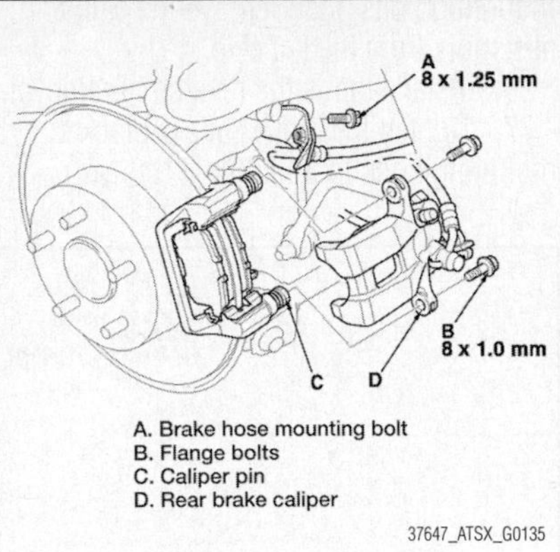
A
8 x 1.25 mm
B
8 x 1.0 mm
C
D
A. Brake hose mounting bolt
B. Flange bolts
C. Caliper pin
D. Rear brake caliper
37647_ATSX_G0135

Fig. 15 Rear brake caliper assembly

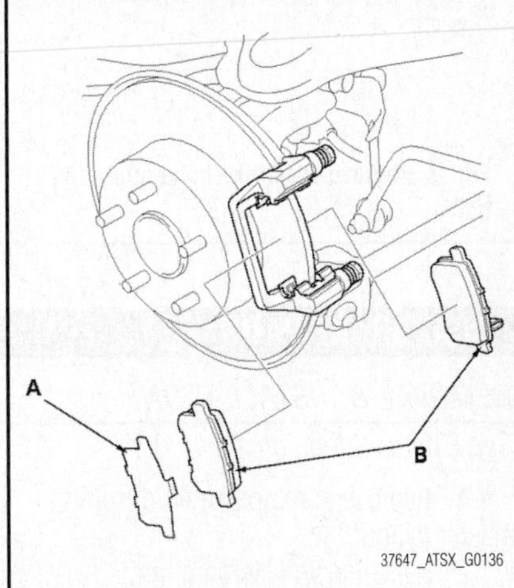
A
B
37647_ATSX_G0136

Fig. 16 Remove the pad shim (A) and the brake pads (B)

➡Do not twist the brake hose and the parking brake cable to prevent damage.

6. Remove the pad shim and the brake pads.
7. Remove the pad retainers (A).

To install:

8. Clean the caliper bracket thoroughly; remove any rust, and check for grooves and cracks.

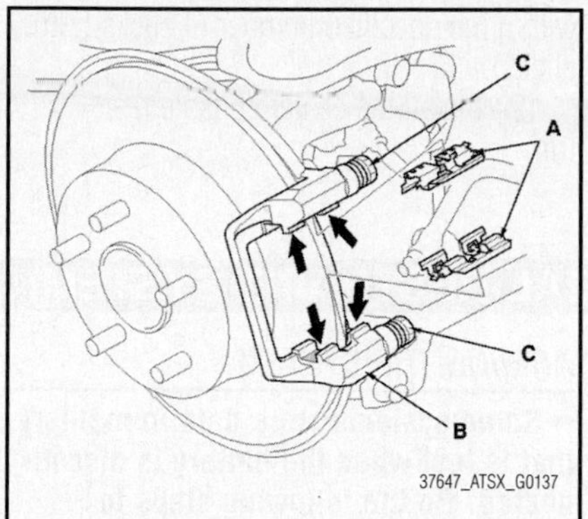
C
A
C
B
37647_ATSX_G0137

Fig. 17 Remove the pad retainers (A)

9. Verify that the caliper pins move in and out smoothly. Clean and lube if needed.

10. Inspect the brake disc for runout, thickness, parallelism, and check for damage and cracks.

11. Apply a thin coat of M-77 assembly paste (P/N 08798-9010) to the retainer mating surface of the caliper bracket (indicated by the arrows).

12. Install the pad retainers. Wipe excess assembly paste off the retainers. Keep the assembly paste off the brake disc and the brake pads.

13. Apply a thin coat of M-77 assembly paste (P/N 08798-9010) to the pad side of the shim, the back of the brake pads, and the other areas indicated by the arrows. Wipe excess assembly paste off the pad shim and the brake pads friction material. Keep grease and assembly paste off the

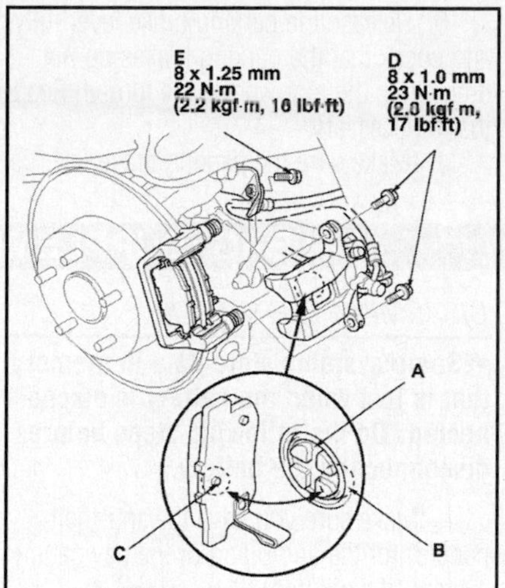
E
8 x 1.25 mm
22 N·m
(2.2 kgf·m, 16 lbf·ft)
D
8 x 1.0 mm
23 N·m
(2.0 kgf·m, 17 lbf·ft)
A
C
B

A. Caliper piston
B. Cutout
C. Tab
D. Flange bolts
E. Brake hose mounting bolt

37647_ATSX_G0138

Fig. 18 Proper installation of pads

brake disc and the brake pads. Contaminated brake disc or brake pads reduce stopping ability.

14. Install the brake pads and pad shim correctly. Install the brake pad with the wear indicator on the bottom inside. If you are reusing the brake pads, always reinstall the brake pads in their original positions to prevent a temporary loss of braking efficiency.

15. Rotate the caliper piston clockwise into the cylinder, then align the cutout in the piston with the tab on the inner pad by turning the piston back. Lubricate the boot with rubber grease to avoid twisting the piston boot. If the piston boot is twisted, back it out so it is positioned properly.

➡**Be careful when moving the piston back in the caliper; brake fluid might overflow from the master cylinder's reservoir. If brake fluid gets on any painted surface, wash it off immediately with water.**

16. Install the caliper. Install the flange bolts, and tighten it to the specified torque while holding the respective caliper pins with a wrench being careful not to damage the pin boots and parking brake cable boots.

17. Install the brake hose mounting bolt.

18. Clean the mating surfaces between the brake disc and the inside of the wheel, then install the rear wheels.

19. Press the brake pedal several times to make sure the brakes work.

➡**Engagement may require a greater pedal stroke immediately after the brake pads have been replaced as a set. Several applications of the brake pedal will restore the normal pedal stroke.**

20. Add brake fluid as needed.

21. After installation, check for leaks at hose and line joints or connections, and retighten if necessary.

22. Test-drive the vehicle, then recheck for leaks.

BRAKES — PARKING BRAKE

PARKING BRAKE CABLES

ADJUSTMENT

See Figures 19 and 20.

1. Release the parking brake lever fully.
2. Pull out the center console rear trim.
3. Loosen the parking brake adjusting nut.
4. Raise and support the vehicle.
5. Remove the rear wheels.
6. Make sure the lever on the rear brake caliper contacts the arm.

➡**The lever will contact the arm when the parking brake adjusting nut is loosened.**

7. Clean the mating surfaces between the brake disc and the inside of the wheel, then install the rear wheels.

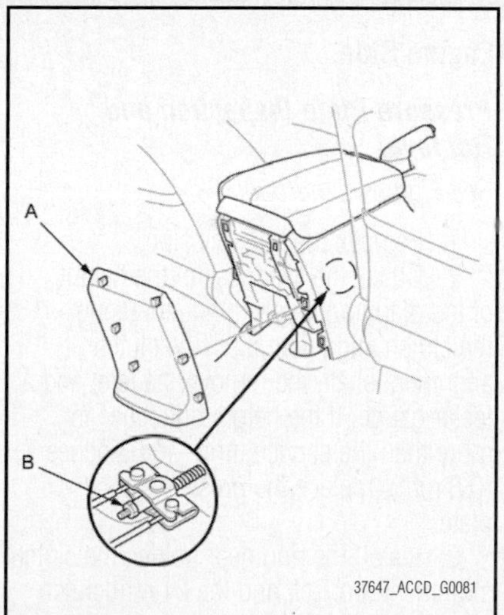
A
B
37647_ACCD_G0081

Fig. 19 Pull out the center console rear trim (A) and loosen the parking brake cable adjusting nut (B)

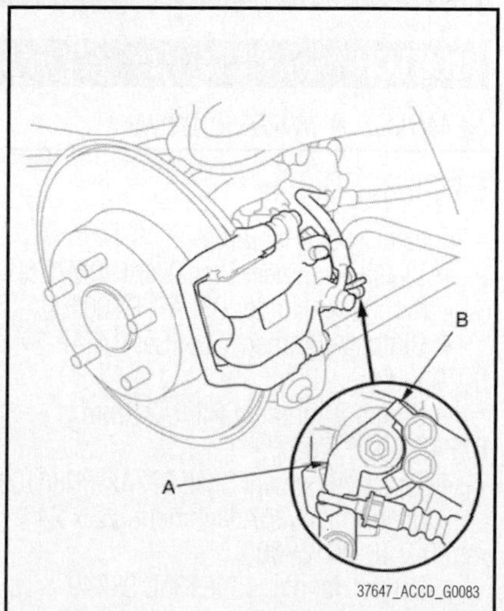
B
A
37647_ACCD_G0083

Fig. 20 Make sure the lever (A) on the rear brake caliper contacts the arm (B)

8. Pull the parking brake lever 1 click.

9. Tighten the parking brake adjusting nut until the parking brakes drag slightly when the rear wheels are turned.

10. Release the parking brake lever fully, and check that the parking brakes do not drag when the rear wheels are turned. Readjust if necessary.

11. Make sure the parking brake lever is within the specified number of clicks (7 to 9 clicks).

12. Install the center console rear trim.

CHASSIS ELECTRICAL

AIR BAG (SUPPLEMENTAL RESTRAINT SYSTEM)

GENERAL INFORMATION

CAUTION

These vehicles are equipped with an air bag system. The system must be disarmed before performing service on, or around, system components, the steering column, instrument panel components, wiring and sensors. Failure to follow the safety precautions and the disarming procedure could result in accidental air bag deployment, possible injury and unnecessary system repairs.

PRECAUTIONS

Disconnect and isolate the battery negative cable before beginning any airbag system component diagnosis, testing, removal, or installation procedures. Allow system capacitor to discharge for two minutes before beginning any component service. This will disable the airbag system. Failure to disable the airbag system may result in accidental airbag deployment, personal injury, or death.

DISARMING THE SYSTEM

➡Some systems store data in memory that is lost when the battery is disconnected. Do the following steps before disconnecting the battery.

1. Make sure you have the anti-theft code(s) for the audio and/or the navigation system (if equipped).
2. For some models or if you're replacing the audio unit, it may be necessary to write down the audio presets (AM and FM), and the XM radio presets (if equipped), because the audio unit does not retain the presets after the battery is disconnected.
3. Make sure the ignition switch is in LOCK (0).
4. Disconnect and isolate the negative cable from the battery.

CAUTION

Always disconnect the negative cable from the battery first.

5. Disconnect the positive cable from the battery.
6. Wait at least 3 minutes for the battery and SRS system to discharge before performing any repairs.

ARMING THE SYSTEM

➡Some systems store data in memory that is lost when the battery is disconnected. Do the following steps to restore the systems back to normal operation.

1. Clean the battery terminals.
2. Reconnect the positive cable to the battery first, then reconnect the negative cable to the battery.

CAUTION

Always connect the positive cable to the battery first.

3. Apply multipurpose grease to the terminals to prevent corrosion.
4. Enter the anti-theft code(s) for the audio system and/or the navigation system (if equipped).
5. Enter the audio presets (if applicable), and enter the XM radio presets (if equipped).
6. Set the clock (for vehicles without navigation).
7. Do the steering column position memorization.

DRIVE TRAIN

CLUTCH

REMOVAL & INSTALLATION

4-Cylinder Engine

Special Tools Required:

- Clutch Alignment Disc 07JAF-PM7011A
- Ring Gear Holder 07LAB-PV00100
- Clutch Alignment Tool Set 07PAF-0020000
- Clutch Alignment Pilot, 21 mm 07PAF-0020370
- Clutch Alignment Shaft 07ZAF-PR8A100
- Bearing Driver Attachment, 22 x 24 mm 07746-0010800
- Driver Handle, 15 x 135L 07749-0010000
- Remover Handle 07936-3710100
- Bearing Remover Shaft Set, 20 mm 07936-3710600
- Slide Hammer 07936-371020A

Engine Side

Pressure Plate Inspection and Removal

See Figures 21 through 23.

1. Remove the transmission.
2. Check the evenness of the height of the diaphragm spring fingers using the clutch alignment disc, the clutch alignment shaft, the remover handle, and a feeler gauge. If the height difference is more than the service limit—0.03inches (0.8 mm), replace the pressure plate.
3. Install the ring gear holder, the clutch alignment tool set, and the 21 mm clutch alignment pilot.
4. To prevent warping, loosen the pressure plate mounting bolts in a crisscross pattern in several steps, then remove the pressure plate.

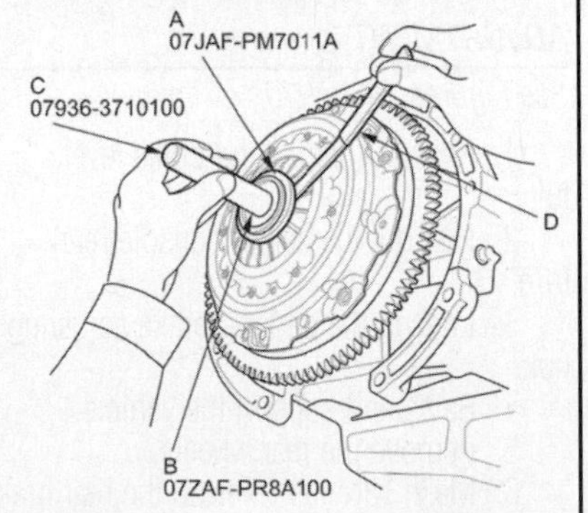
A
07JAF-PM7011A
C
07936-3710100
D
B
07ZAF-PR8A100

A. Clutch alignment disc C. Remover handle
B. Clutch alignment shaft D. Feeler gauge

37647_ACTL_G0595

Fig. 21 Check the evenness of the height of the diaphragm spring fingers using the clutch alignment disc, the clutch alignment shaft, the remover handle, and a feeler gauge

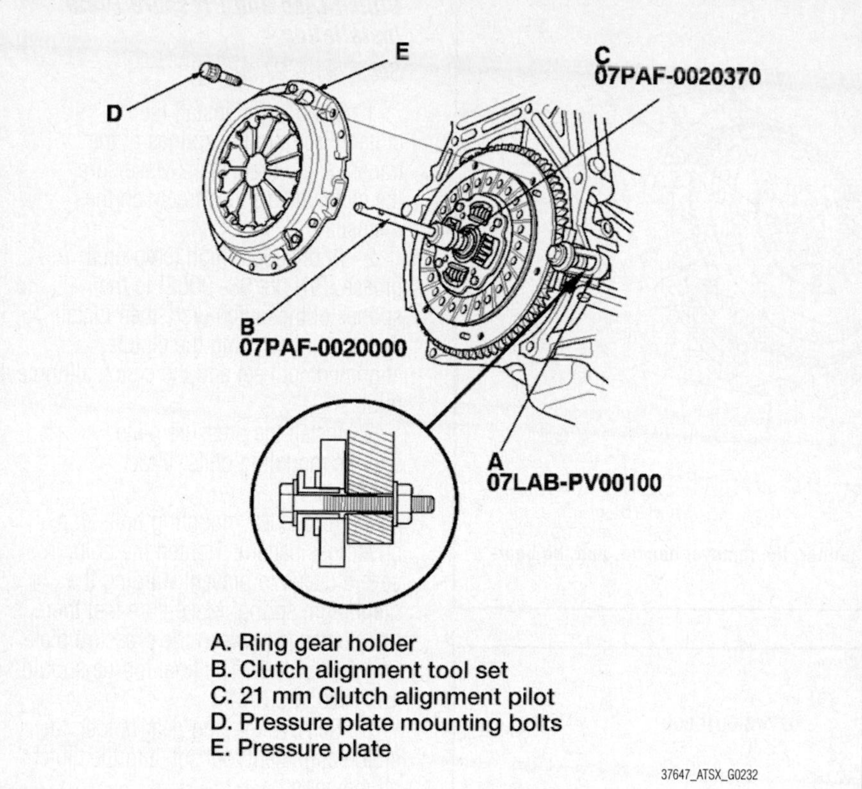
E
C
07PAF-0020370
D
B
07PAF-0020000
A
07LAB-PV00100
A. Ring gear holder
B. Clutch alignment tool set
C. 21 mm Clutch alignment pilot
D. Pressure plate mounting bolts
E. Pressure plate
37647_ATSX_G0232

Fig. 22 Install the ring gear holder, the clutch alignment tool set, and the 21 mm clutch alignment pilot

5. Inspect the fingers of the diaphragm spring for wear at the release bearing contact area.

6. Inspect the pressure plate surface for wear, cracks, and burning.

7. Inspect for warpage using a straight edge and a feeler gauge. Measure across the pressure plate. If the most measurement difference is more than the service limit—0.006 inches (0.15 mm), replace the pressure plate.

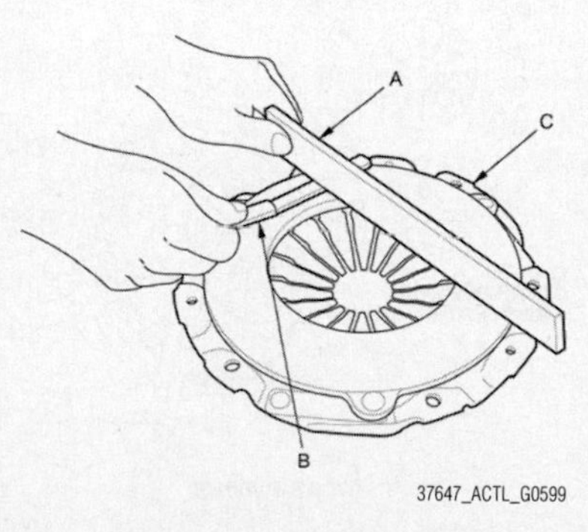
A
C
B
37647_ACTL_G0599

Fig. 23 Inspect for warpage using a straight edge (A) and a feeler gauge (B). Measure across the pressure plate (C)

Clutch Disc Inspection and Removal

1. Remove the clutch disc, the clutch alignment tool set, and the 21 mm clutch alignment pilot.

2. Inspect the lining of the clutch disc for signs of slippage or oil. If the clutch disc looks burnt or is oil soaked, replace the clutch disc. If the clutch disc is oil soaked, find and repair the source of the oil leak.

3. Measure the clutch disc thickness. If the measurement is less than the service limit—0.24 inches (6.0 mm), replace the clutch disc.

4. Measure the depths of the rivets from the clutch disc lining surface to the rivets on both sides. If the measurement is less than the service limit—0.03 inches (0.7 mm), replace the clutch disc.

Flywheel Inspection

1. Remove the ring gear holder.

2. Inspect the ring gear teeth for wear and damage.

3. Inspect the clutch disc mating surface on the flywheel for wear, cracks, and burning.

4. Measure the flywheel runout using a dial indicator. Through at least two full turns with pushing against the flywheel each time you turn it to take up the crankshaft thrust washer clearance. If the measurement is not within the standard (new)—0.002 inches (0.05 mm), replace the flywheel.

Flywheel Replacement

See Figure 24.

1. Install the ring gear holder.

2. Loosen the flywheel mounting bolts in a crisscross pattern in several steps. Remove the bolts, then remove the flywheel and the ring gear holder.

3. Install the flywheel on the crankshaft, and install the mounting bolts finger-tight.

4. Install the ring gear holder, then torque the flywheel mounting bolts in a crisscross pattern in several steps to 90 ft. lbs. (123 Nm).

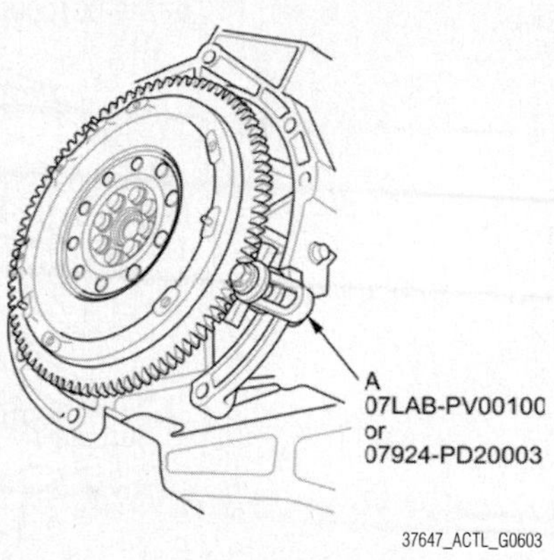
A
07LAB-PV00100
or
07924-PD20003
37647_ACTL_G0603

Fig. 24 Install the ring gear holder (A)

Pilot Bearing Inspection

1. Turn the inner race of the pilot bearing with your finger. The pilot bearing should turn smoothly and quietly. Check that the pilot bearing outer race fits tightly in the flywheel. If the race does not turn smoothly, quietly, or fit tight in the flywheel, replace the pilot bearing.

Pilot Bearing Replacement

See Figures 25 and 26.

1. Remove the pilot bearing using the slide hammer, the remover handle, and the bearing remover shaft set.

2. Install a new pilot bushing into the crankshaft using the 15 x 135L driver handle and the 22 x 24 mm attachment.

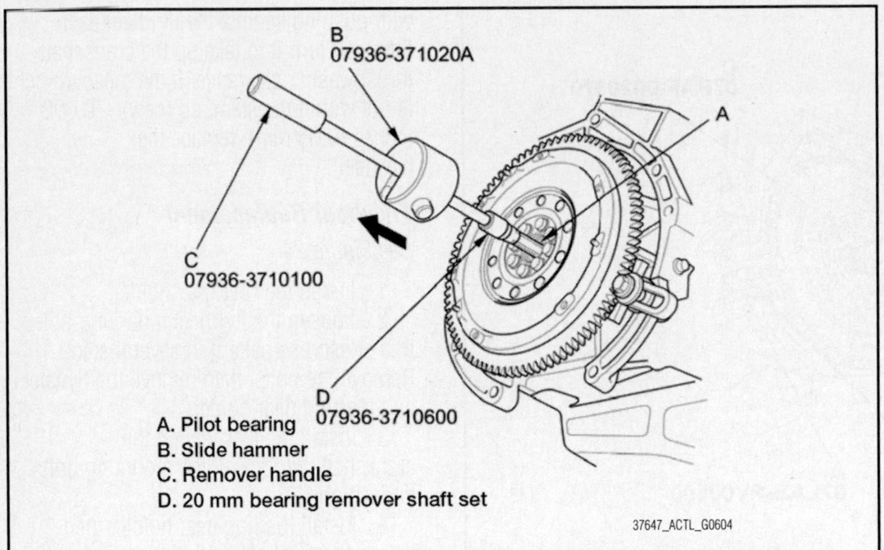
B
07936-371020A
A
C
07936-3710100
D
07936-3710600
A. Pilot bearing
B. Slide hammer
C. Remover handle
D. 20 mm bearing remover shaft set
37647_ACTL_G0604

Fig. 25 Remove the pilot bearing using the slide hammer, the remover handle, and the bearing remover shaft set

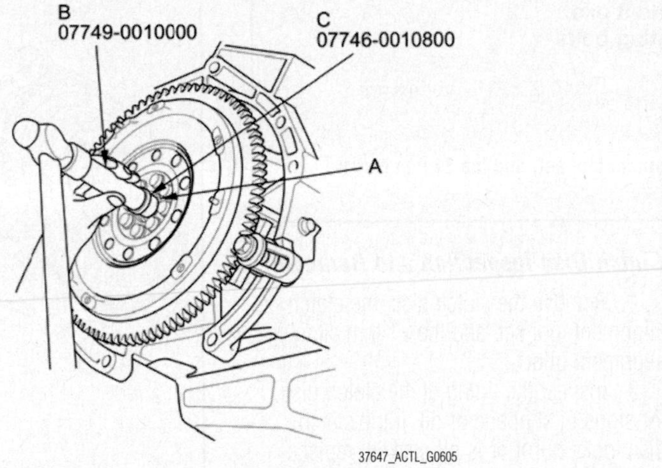
B
07749-0010000
C
07746-0010800
A
37647_ACTL_G0605

Fig. 26 Install a new pilot bushing (A) into the crankshaft using the 15 x 135L driver handle (B) and the 22 x 24 mm attachment (C)

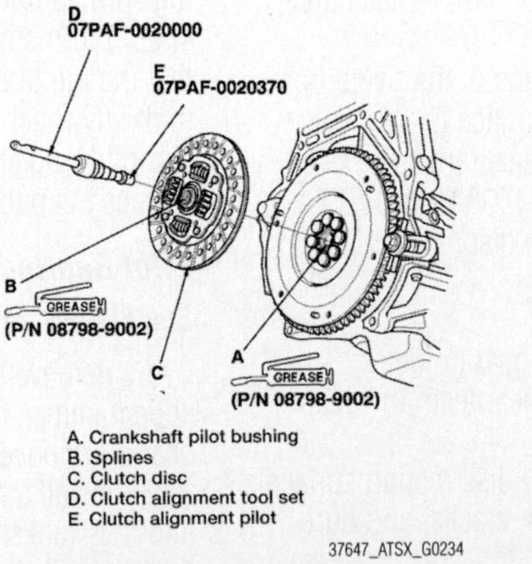
D
07PAF-0020000
E
07PAF-0020370
B
GREASE
(P/N 08798-9002)
C
A
GREASE
(P/N 08798-9002)
A. Crankshaft pilot bushing
B. Splines
C. Clutch disc
D. Clutch alignment tool set
E. Clutch alignment pilot
37647_ATSX_G0234

Fig. 27 Apply super high temp urea grease to the splines of the clutch disc, then install the clutch disc using the clutch alignment tool set and the clutch alignment pilot

Clutch Disc and Pressure Plate Installation

See Figures 27 and 28.

1. Temporarily install the clutch disc onto the splines of the transmission mainshaft. Make sure the clutch disc slides freely on the mainshaft.
2. Apply super high temp urea grease (P/N 08798- 9002) to the splines of the clutch disc, then install the clutch disc using the clutch alignment tool set and the clutch alignment pilot.
3. Install the pressure plate and the mounting bolts, finger-tight.
4. Torque the mounting bolts in a crisscross pattern. Tighten the bolts in several steps to prevent warping the diaphragm spring. Make sure that there is not clearance between the pressure plate and the flywheel. The final torque should be: 19 ft. lbs. (25 Nm)
5. Remove the ring gear holder, the clutch alignment tool set, and the clutch alignment pilot.
6. Make sure the diaphragm spring fingers are all the same height.
7. Do the release bearing inspection, and replace it if necessary.
8. Install the transmission.

Transmission Side

Release Bearing Removal

See Figure 29.

1. Remove the transmission.
2. Remove the release fork boot from the clutch housing.

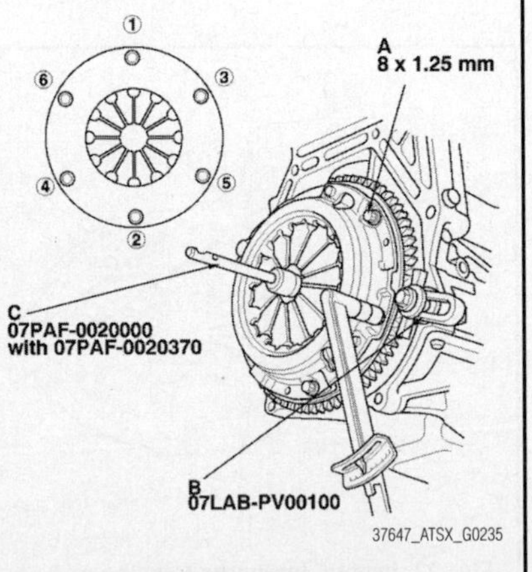
①
⑥
③
④
⑤
②
A
8 x 1.25 mm
C
07PAF-0020000
with 07PAF-0020370
B
07LAB-PV00100
37647_ATSX_G0235

Fig. 28 Torque the mounting bolts in a crisscross pattern

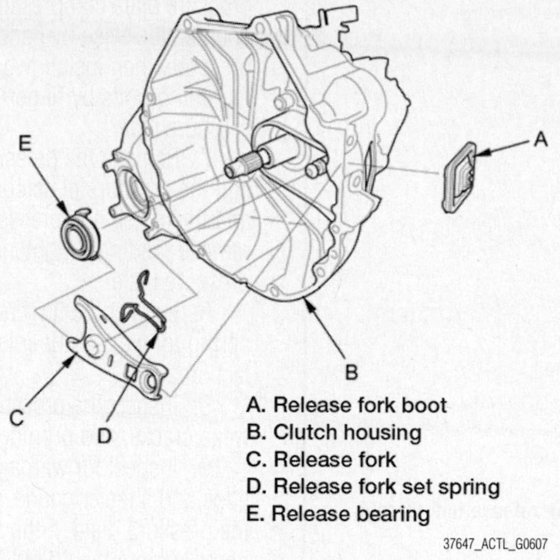
A
B
C
D
E
A. Release fork boot
B. Clutch housing
C. Release fork
D. Release fork set spring
E. Release bearing
37647_ACTL_G0607

Fig. 29 Remove the release fork boot from the clutch housing

3. Remove the release fork from the clutch housing by squeezing the release fork set spring. Remove the release bearing.

Release Bearing Inspection

1. Check the play of the release bearing by spinning it by hand. If there is excessive play or noise, replace the release bearing.

➡**The release bearing is packed with grease. Do not wash it in solvent.**

Release Bearing Installation

See Figures 30 and 31.

1. Apply super high temp urea grease (P/N 08798-9002) to the release fork, the release fork bolt, the release bearing, and the release bearing guide in the shaded areas, then set the release fork set spring.

➡**Replace the release fork bolt if necessary.**

2. With the release fork slid between the release bearing pawls, install the release bearing on the mainshaft while inserting the release fork through the hole in the clutch housing.
3. Align the detent of the release fork with the release fork bolt, then press the detent of the release fork over the release fork bolt squarely.
4. Install the release fork boot. Make sure the boot seals around the release fork and the clutch housing.
5. Move the release fork right and left to make sure that it fits properly against the release bearing and that the release bearing slides smoothly. Wipe off any excess grease.
6. Install the transmission.

6-Cylinder Engine

Special Tools Required:

• Pressure Plate Compressor 07AAE-P8EA000

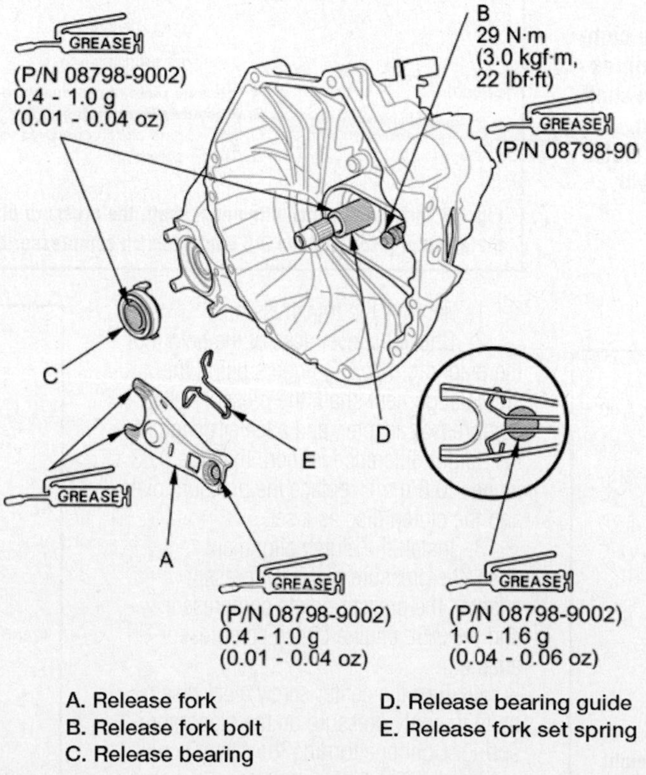
GREASE
(P/N 08798-9002)
0.4 - 1.0 g
(0.01 - 0.04 oz)
B
29 N·m
(3.0 kgf·m,
22 lbf·ft)
GREASE
(P/N 08798-90
C
D
E
GREASE
A
GREASE
(P/N 08798-9002)
0.4 - 1.0 g
(0.01 - 0.04 oz)
GREASE
(P/N 08798-9002)
1.0 - 1.6 g
(0.04 - 0.06 oz)
A. Release fork
B. Release fork bolt
C. Release bearing
D. Release bearing guide
E. Release fork set spring
37647_ACTL_G0608

Fig. 30 Apply super high temp urea grease to the release fork, the release fork bolt, the release bearing, and the release bearing guide in the shaded areas, then set the release fork set spring

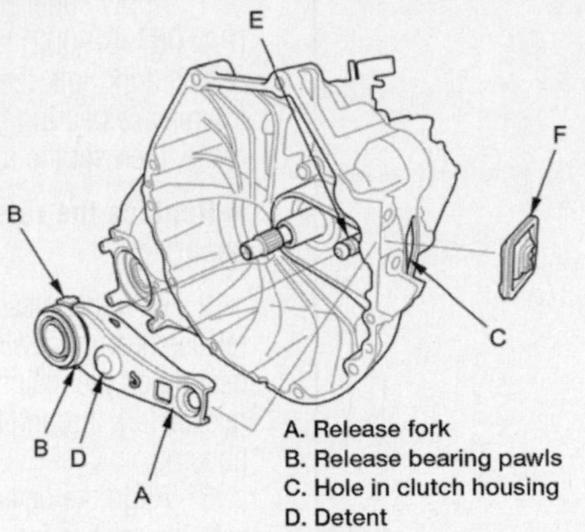
E
F
B
C
B
D
A
A. Release fork
B. Release bearing pawls
C. Hole in clutch housing
D. Detent
E. Release fork bolt
F. Release fork boot
37647_ACTL_G0609

Fig. 31 Install the release bearing on the mainshaft while inserting the release fork through the hole in the clutch housing

- 3.5L Engine Clutch Compressor Adapters 07AAE-P8EA220
- Clutch Alignment Shaft 07AAF-P8EA000
- Pressure Plate Compressor Adapter 07AAK-P8EA000
- Ring Gear Holder 07LAB-PV00100
- Bearing Driver Attachment, 37 x 40 mm 07746-0010200
- Driver Handle, 15 x 135L 07749-0010000

Engine Side

➡You must use the pressure plate compressor, the 3.5 engine clutch compressor adapters, the clutch alignment shaft, and the pressure plate compressor adapter to remove and install the clutch pressure plate or you will damage it.

Pressure Plate Inspection and Removal

See Figures 32 through 37.

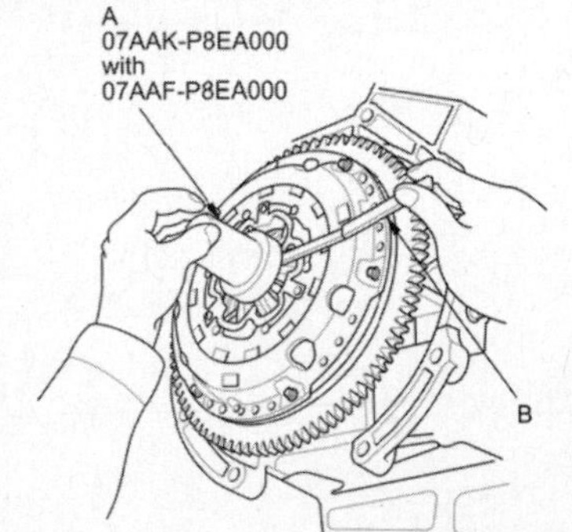
A
07AAK-P8EA000
with
07AAF-P8EA000
B
37647_ACCD_G0178

Fig. 32 Check the evenness of the height of the diaphragm spring fingers using the clutch alignment shaft, the pressure plate compressor adapter (A), and a feeler gauge (B)

1. Remove the transmission.
2. Check the evenness of the height of the diaphragm spring fingers using the clutch alignment shaft, the pressure plate compressor adapter, and a feeler gauge. If the height difference is more than the 0.03 inches (0.8 mm), replace the pressure plate and the clutch disc as a set.
3. Install the clutch alignment shaft, the pressure plate compressor adapter, the pressure plate compressor, and the 3.5L engine clutch compressor adapters.
4. Turn the center screw clockwise by hand to apply pressure on the diaphragm spring. Continue turning the center screw until it stops.
5. Loosen the pressure plate mounting bolts in a crisscross pattern in several steps, then remove the bolts.
6. Turn the center screw on the pressure plate compressor counterclockwise by hand to release the pressure, then install two pressure plate mounting bolts by finger-tight to hold the pressure plate.
7. Remove the pressure plate compressor, the pressure plate compressor adapter, and the clutch compressor adapters, then remove the finger-tightened bolts and the pressure plate.
8. Inspect the fingers of the diaphragm spring for wear at the release bearing contact area.
9. Inspect the pressure plate surface for wear, cracks, and burning.
10. Inspect for warpage using a straight edge and a feeler gauge. Measure across the pressure plate. If the any measured point is more than 0.006 inches (0.15 mm),

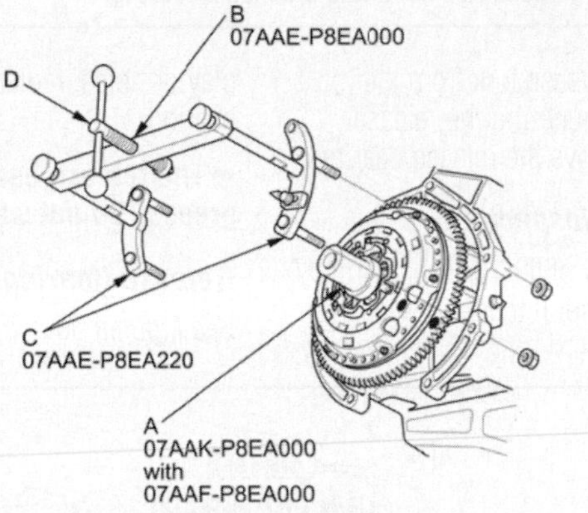
B
07AAE-P8EA000
D
C
07AAE-P8EA220
A
07AAK-P8EA000
with
07AAF-P8EA000
A. Pressure plate compressor adapter
B. Pressure plate compressor
C. 3.5L engine clutch compressor adapters
D. Center screw
37647_ACCD_G0179

Fig. 33 Install the clutch alignment shaft, the pressure plate compressor adapter, the pressure plate compressor, and the 3.5 engine clutch compressor adapters

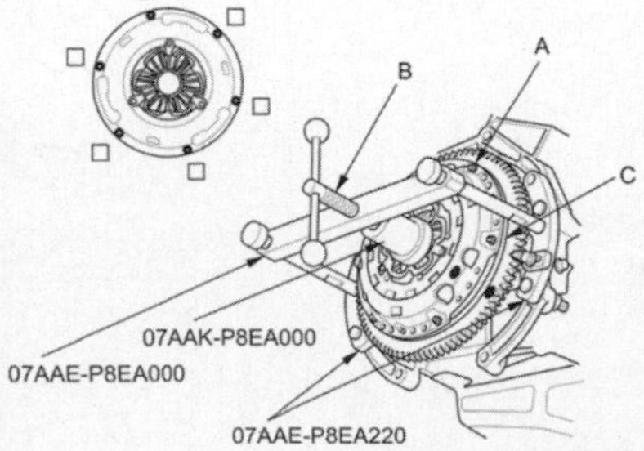
B
A
C
07AAK-P8EA000
07AAE-P8EA000
07AAE-P8EA220
37647_ACCD_G0180

Fig. 34 Loosen the pressure plate mounting bolts (A) in a crisscross pattern in several steps, then remove the bolts

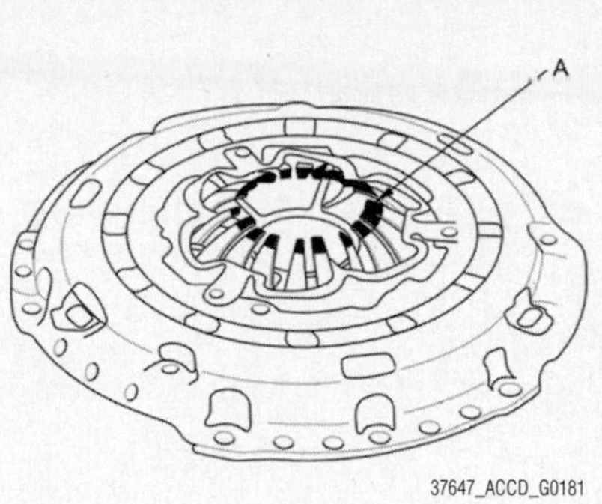
A
37647_ACCD_G0181

Fig. 35 Inspect the fingers of the diaphragm spring (A) for wear at the release bearing contact area

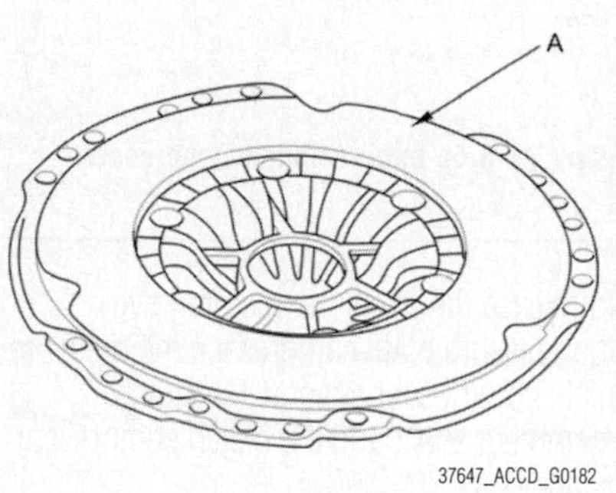
A
37647_ACCD_G0182

Fig. 36 Inspect the pressure plate surface (A) for wear, cracks, and burning

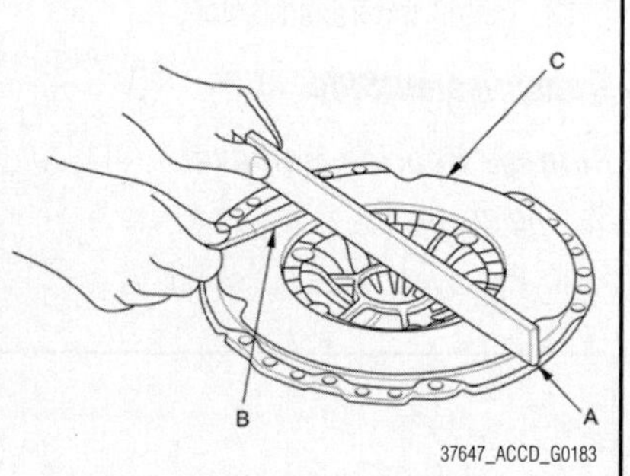
C
B
A
37647_ACCD_G0183

Fig. 37 Inspect for warpage using a straight edge (A) and a feeler gauge (B)

replace the pressure plate and the clutch disc as a set.

Clutch Disc Inspection and Removal

1. Remove the clutch disc and the clutch alignment shaft.
2. Inspect the lining of the clutch disc for signs of slippage or oil. If the clutch disc looks burnt or is oil soaked, replace the clutch disc and the pressure plate as a set. If the clutch disc is oil soaked, find and repair the source of the oil leak.
3. Measure the clutch disc thickness. If the measurement is less than 0.28 inches (7.2 mm), replace the clutch disc and the pressure plate as a set.
4. Measure the depths of the rivets from the clutch disc lining surface to the rivets on both sides. If the measurement is less than 0.01 inches (0.2 mm), replace the clutch disc and the pressure plate as a set.

Flywheel Inspection

See Figure 38.

1. Inspect the ring gear teeth for wear and damage.
2. Inspect the clutch disc mating surface on the flywheel for wear, cracks, and burning.
3. Measure the flywheel runout using a dial indicator. Through at least two full turns with pushing against the flywheel each time you turn it to take up the crankshaft thrust washer clearance. If the measurement is not within the standard, replace the flywheel, and recheck the runout.

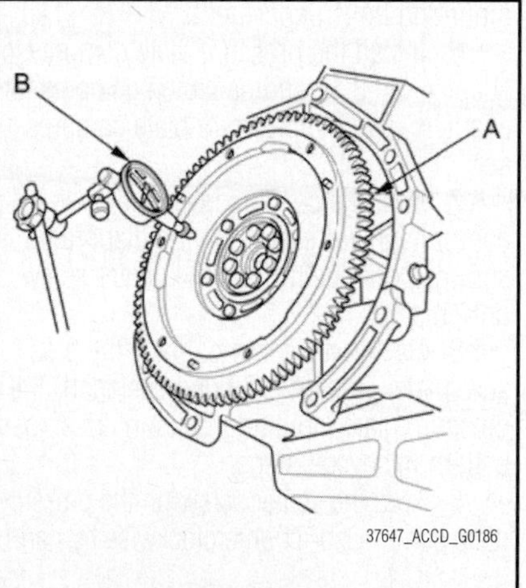
B
A
37647_ACCD_G0186

Fig. 38 Measure the flywheel (A) runout using a dial indicator (B)

Pilot Bearing Inspection

1. Turn the inner race of the pilot bearing with your finger. The pilot bearing should turn smoothly and quietly. Check that the pilot bearing outer race fits tightly in the flywheel. If the race does not turn smoothly, quietly, or fit tight in the flywheel, replace the pilot bearing.

Flywheel and Pilot Bearing Replacement

See Figures 39 through 41.

1. Install the ring gear holder.
2. Loosen the flywheel mounting bolts in a crisscross pattern in several steps. Remove the bolts, then remove the flywheel and the ring gear holder. Do not use impact tools to tighten the flywheel bolts. Using impact tools may permanently damage the flywheel.

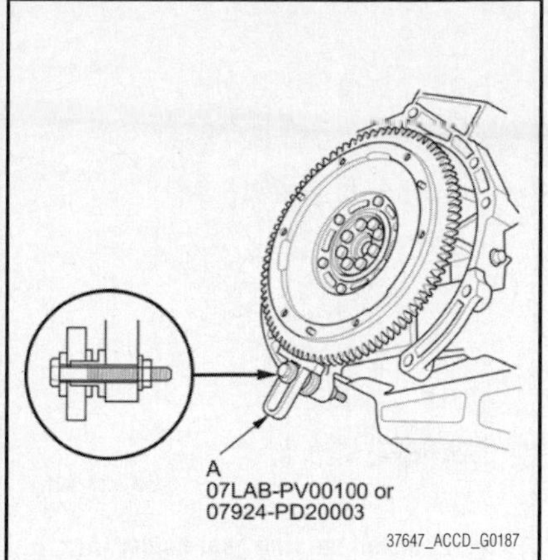
A
07LAB-PV00100 or
07924-PD20003
37647_ACCD_G0187

Fig. 39 Install the ring gear holder (A)

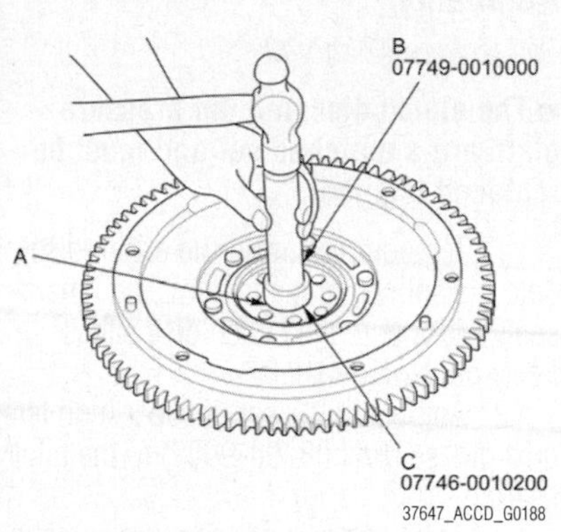
B
07749-0010000
A
C
07746-0010200
37647_ACCD_G0188

Fig. 40 Install a new pilot bearing (A) into the flywheel using the 15 x 135L driver handle (B) and the 37 x 40 mm bearing driver attachment (C)

3. Remove the pilot bearing from the flywheel.
4. Install a new pilot bearing into the flywheel using the 15 x 135L driver handle and the 37 x 40 mm bearing driver attachment. Apply a light coat of oil to the bearing outer surface.
5. Install the flywheel on the crankshaft, and install the mounting bolts finger-tight.
6. Install the ring gear holder, then torque the flywheel mounting bolts in a crisscross pattern in several steps. Do not use impact tools to tighten the flywheel bolts. Using impact tools may permanently damage the flywheel.
7. Remove the ring gear holder.

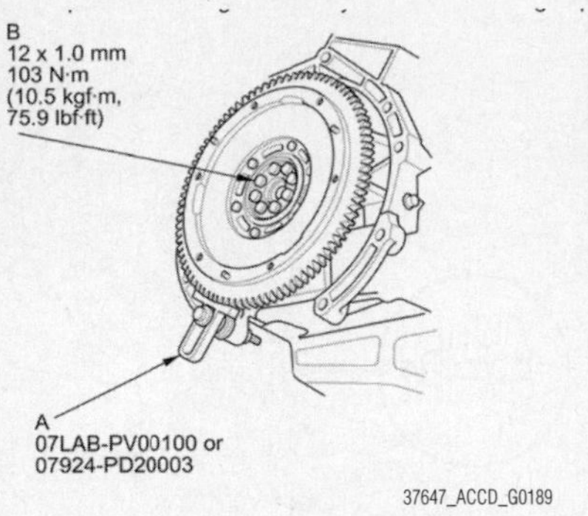
B
12 x 1.0 mm
103 N·m
(10.5 kgf·m,
75.9 lbf·ft)
A
07LAB-PV00100 or
07924-PD20003
37647_ACCD_G0189

Fig. 41 Install the ring gear holder (A), then torque the flywheel mounting bolts (B) in a crisscross pattern in several steps

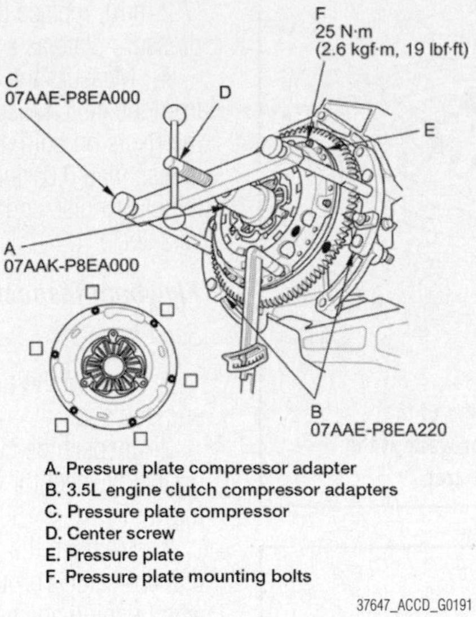
F
25 N·m
(2.6 kgf·m, 19 lbf·ft)
C
07AAE-P8EA000
D
E
A
07AAK-P8EA000
B
07AAE-P8EA220
A. Pressure plate compressor adapter
B. 3.5L engine clutch compressor adapters
C. Pressure plate compressor
D. Center screw
E. Pressure plate
F. Pressure plate mounting bolts
37647_ACCD_G0191

Fig. 43 Install the pressure plate compressor adapter, the 3.5L engine clutch compressor adapters, and the pressure plate compressor

Clutch Disc and Pressure Plate Installation

See Figures 42 and 43.

➡The clutch disc and the pressure plate are a matched set and must be replaced together.

1. Temporarily install the clutch disc onto the splines of the transmission mainshaft. Make sure the clutch disc slides freely on the mainshaft.
2. Apply a light coat of super high temp urea grease (P/N 08798-9002) to the pilot bearing.
3. Apply super high temp urea grease (P/N 08798-9002) to the splines of the clutch disc, then install the clutch disc using the clutch alignment shaft.
4. Align the point marks across the pressure plate and the flywheel.
5. Install the pressure plate and the mounting bolts finger-tight.
6. Install the pressure plate compressor adapter, the 3.5L engine clutch compressor adapters, and the pressure plate compressor.
7. Turn the center screw clockwise by hand to apply pressure on the diaphragm spring. Continue turning the center screw until it stops.
8. Be careful not to damage the pressure plate. Tighten (180 degrees apart) the pressure plate mounting bolts in crisscross pattern in several steps.
9. Turn the center screw on the pressure plate compressor counterclockwise by hand to release the pressure, then remove the pressure plate compressor, the pressure plate compressor adapter, the clutch alignment shaft, and the clutch compressor adapters.
10. Make sure the diaphragm spring fingers are all the same height.
11. Do the release bearing inspection, and replace it if necessary.
12. Install the transmission.

Transmission Side

Release Bearing Removal

See Figure 44.

1. Remove the transmission.

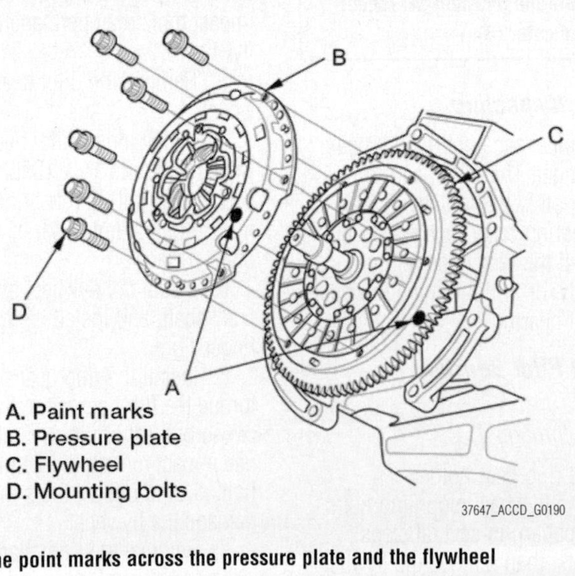
B
C
D
A
A. Paint marks
B. Pressure plate
C. Flywheel
D. Mounting bolts
37647_ACCD_G0190

Fig. 42 Align the point marks across the pressure plate and the flywheel

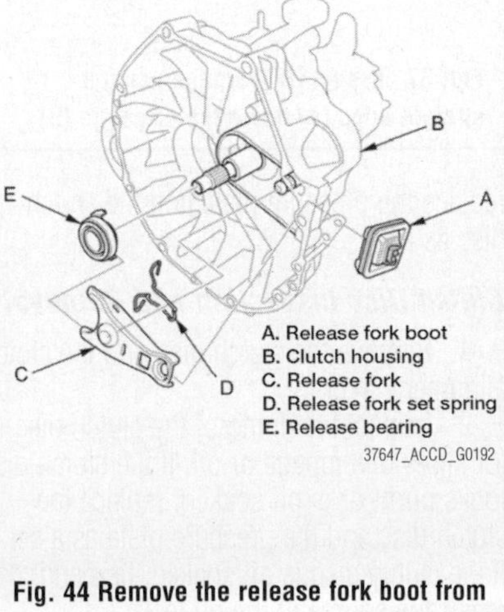
B
E
A
C
D
A. Release fork boot
B. Clutch housing
C. Release fork
D. Release fork set spring
E. Release bearing
37647_ACCD_G0192

Fig. 44 Remove the release fork boot from the clutch housing

2. Remove the release fork boot from the clutch housing.

3. Remove the release fork from the clutch housing by squeezing the release fork set spring. Remove the release bearing.

Release Bearing Inspection

1. Check the play of the release bearing by spinning it by hand. If there is excessive play or noise, replace the release bearing.

➡The release bearing is packed with grease. Do not wash it in solvent.

Release Bearing Installation

See Figure 45.

1. Apply super high temp urea grease (P/N 08798-9002) to the release fork, the release fork bolt, the release bearing, and the release bearing guide, then set the release fork set spring.

➡Replace the release fork bolt if necessary.

2. With the release fork slid between the release bearing pawls, install the release bearing on the mainshaft while inserting the release fork through the hole in the clutch housing.

3. Align the detent of the release fork with the release fork bolt, then press the detent of the release fork over the release fork bolt squarely.

4. Install the release fork boot. Make sure the boot seals around the release fork and the clutch housing.

5. Move the release fork right and left to make sure that it fits properly against the release bearing and that the release bearing slides smoothly. Wipe off any excess grease.

6. Install the transmission.

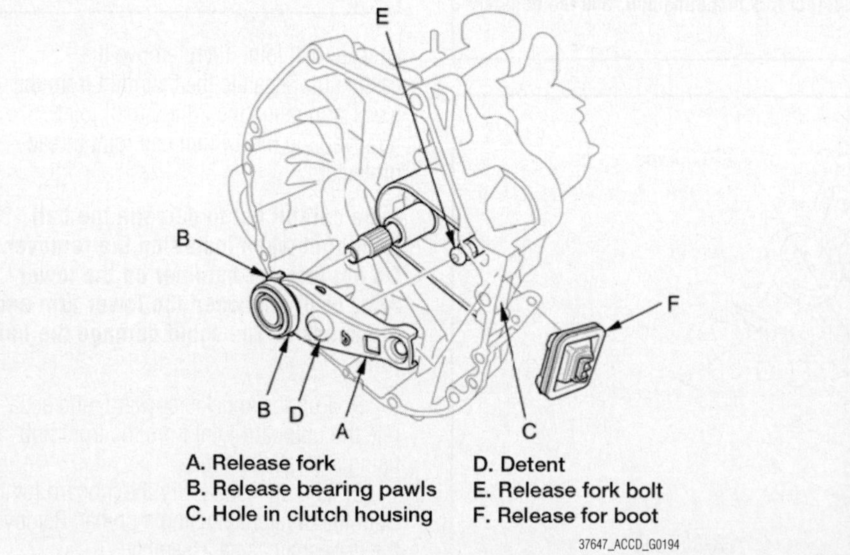
E
B
B
D
A
C
F
A. Release fork
B. Release bearing pawls
C. Hole in clutch housing
D. Detent
E. Release fork bolt
F. Release for boot
37647_ACCD_G0194

Fig. 45 Install the release bearing on the mainshaft while inserting the release fork through the hole in the clutch housing

HYDRAULIC SYSTEM BLEEDING

BLEEDING PROCEDURE

See Figure 46.

➡Precautions to take during bleeding procedure:

- Do not reuse the drained fluid. Always use Honda DOT 3 Brake Fluid from an unopened container. Using a non-Honda brake fluid can cause corrosion and shorten the life of the system.
- Make sure no dirt or other foreign matter is allowed to contaminate the brake fluid.
- Do not spill brake fluid on the vehicle; it may damage the paint or plastic. If brake fluid does contact the paint or plastic, wash it off immediately with water.
- It may be necessary to limit the movement of the release fork with a block of wood to remove all the air from the system.
- Use fender covers to avoid damaging painted surfaces.

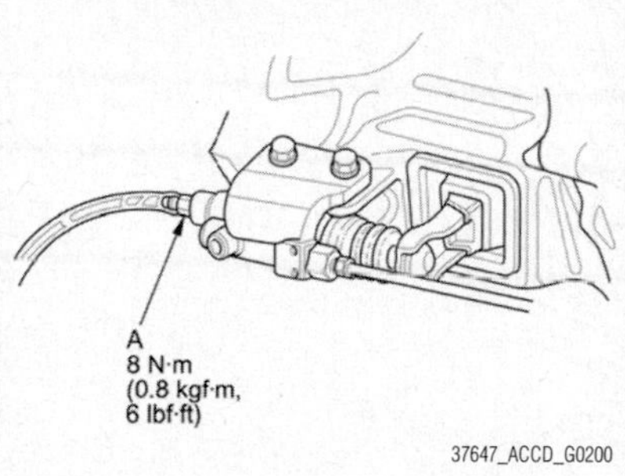
A
8 N·m
(0.8 kgf·m,
6 lbf·ft)
37647_ACCD_G0200

Fig. 46 Attach one end of a clear tube to the bleeder screw (A)

1. Do the battery removal procedure.

2. Make sure the brake fluid level in the clutch reservoir is at the MAX (upper) level line.

3. Attach one end of a clear tube to the bleeder screw, and put the other end into a container. Loosen the bleeder screw to allow air to escape from the system.

4. Make sure there is an adequate supply of fluid in the reservoir, then slowly push the clutch pedal all the way down. Before releasing the pedal, have an assistant temporarily tighten the bleeder screw. Loosen the bleeder screw, and push the clutch pedal down again. Repeat this step until no more bubbles appear at the clear tube.

➡Make sure the fluid level on the reservoir does not go below MIN.

5. Tighten the bleeder screw securely.

6. Refill the brake fluid in the reservoir to the MAX (upper) level line.

7. Do the battery installation procedure.

FRONT HALFSHAFT

REMOVAL & INSTALLATION

See Figures 47 through 53.

Special Tools Required:

- Ball Joint Thread Protector, 14 mm 07AAE-SJAA100
- Ball Joint Remover, 28 mm 07MAC-SL0A202

1. Raise and support the vehicle.

2. Remove the front wheels.

3. Pry up the stake on the spindle nut, then remove the nut.

4. Drain the transmission fluid, then reinstall the drain plug with a new sealing washer:

5. Hold the stabilizer link joint pin using a hex wrench, and remove the flange nut.

6. Separate the front stabilizer link from the lower arm.

7. Remove the damper fork mounting nut, the damper fork mounting bolt, and the damper pinch bolt, then remove the damper fork.

8. Remove the cotter pin from the

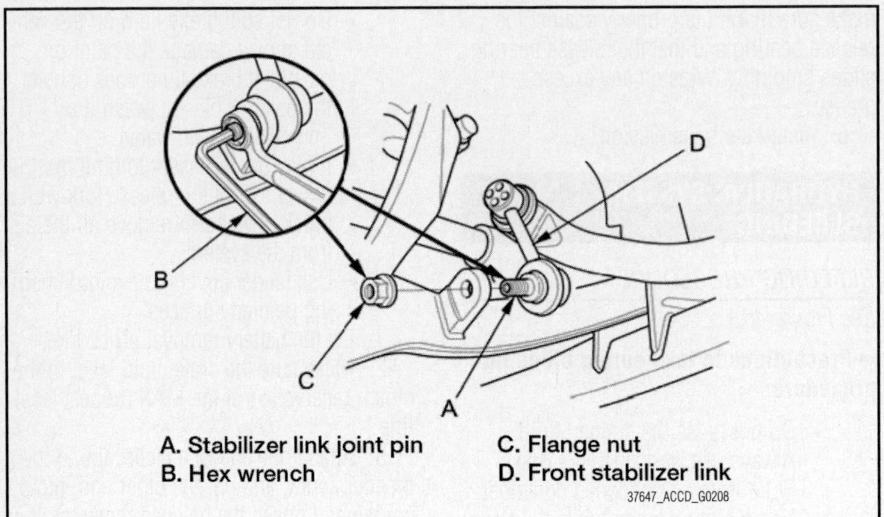

A
B
C
D
A. Stabilizer link joint pin
B. Hex wrench
C. Flange nut
D. Front stabilizer link
37647_ACCD_G0208

Fig. 47 Hold the stabilizer link joint pin using a hex wrench, and remove the flange nut

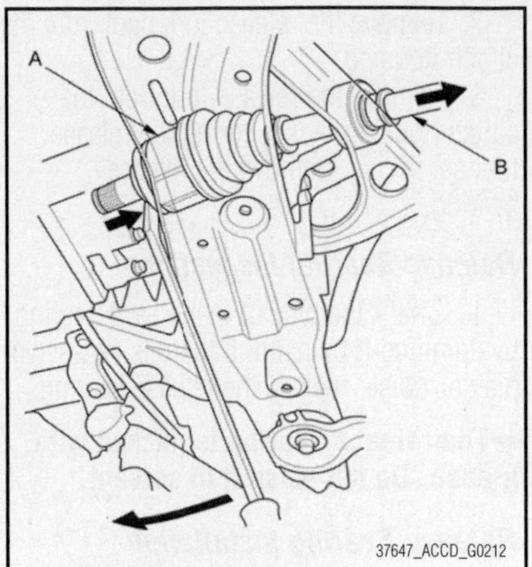

A
B
37647_ACCD_G0212

Fig. 50 Left driveshaft: Pry the inboard joint (A) from the differential using a pry-bar. Remove the driveshaft (B) as an assembly—4 cylinder engine

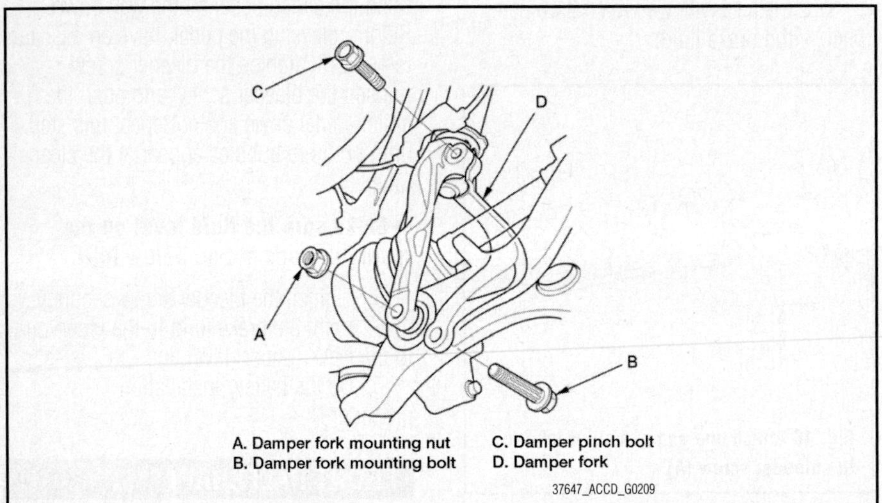

A
B
C
D
A. Damper fork mounting nut
B. Damper fork mounting bolt
C. Damper pinch bolt
D. Damper fork
37647_ACCD_G0209

Fig. 48 Remove the damper fork mounting nut, the damper fork mounting bolt, and the damper pinch bolt, then remove the damper fork

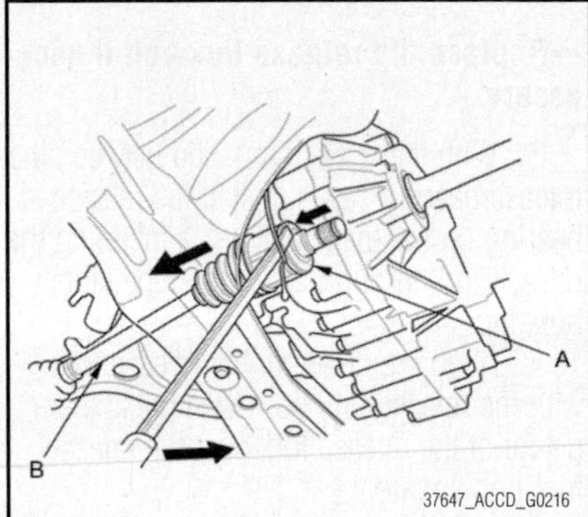

A
B
37647_ACCD_G0216

Fig. 51 Left driveshaft: Pry the inboard joint (A) from the differential using a pry-bar. Remove the driveshaft (B) as an assembly—6 cylinder engine

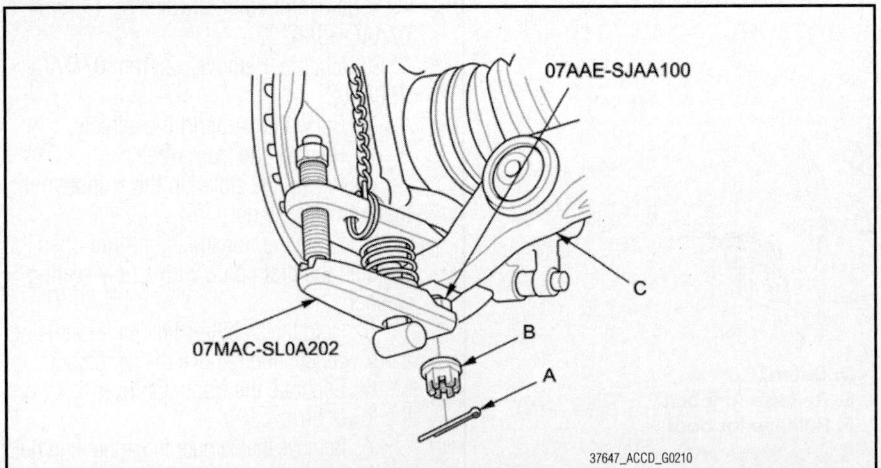

07AAE-SJAA100
07MAC-SL0A202
A
B
C
37647_ACCD_G0210

Fig. 49 Remove the cotter pin (A) from the knuckle ball joint, then remove the castle nut (B). Separate the ball joint from the lower arm (C)

knuckle ball joint, then remove the castle nut. Separate the ball joint from the lower arm using the 28 mm ball joint remover and the 14 mm ball joint thread protector.

➡Be careful not to damage the ball joint boot when installing the remover. Do not force or hammer on the lower arm, or pry between the lower arm and the knuckle. You could damage the ball joint.

9. Pull the knuckle outward, and separate the outboard joint from the front hub using a plastic hammer.
10. Left driveshaft: Pry the inboard joint from the differential using a prybar. Remove the driveshaft as an assembly.

➡Do not pull on the driveshaft, or the inboard joint may come apart. Pull the

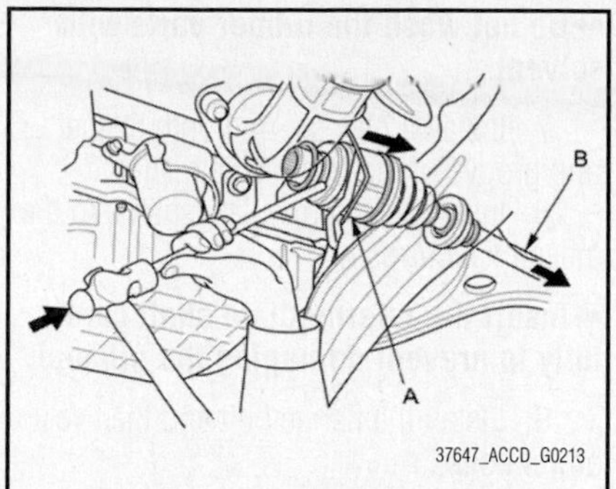
B
A
37647_ACCD_G0213

Fig. 52 Right driveshaft: Drive the inboard joint (A) off of the intermediate shaft using a drift punch and a hammer. Remove the driveshaft (B) as an assembly—4 cylinder engine

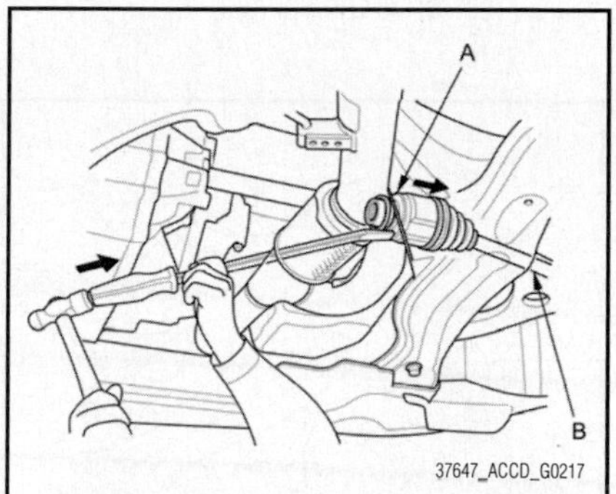
A
B
37647_ACCD_G0217

Fig. 53 Right driveshaft: Drive the inboard joint (A) off of the intermediate shaft using a drift punch and a hammer. Remove the driveshaft (B) as an assembly—6 cylinder engine

inboard joint straight out to avoid damaging the oil seal. Be careful not to damage the oil seal or the end of the inboard joint using the prybar.

11. Right driveshaft: Drive the inboard joint off of the intermediate shaft using a drift punch and a hammer. Remove the driveshaft as an assembly.

➡Do not pull on the driveshaft, or the inboard joint may come apart.

12. Remove the set ring from the left driveshaft inboard joint.
13. Remove the set ring from the intermediate shaft.

To install:

➡Before starting installation, make sure the mating surfaces of the joint and the splined section are clean.

14. Apply Moly 60 paste (P/N 08734-0001) to the contact area of the outboard joint and the front wheel bearing.

➡The paste helps prevent noise and vibration.

15. Install a new set ring into the set ring groove of the left driveshaft inboard joint.
16. Install a new set ring into the set ring groove of the intermediate shaft.
17. Apply super high temp urea grease (P/N 08798-9002) to the whole splined surface of the right driveshaft. After applying grease, remove the grease from the splined grooves at intervals of 2–3 splines and from the set ring groove so that air can bleed from the intermediate shaft.
18. Clean the areas where the driveshaft contacts the differential thoroughly with solvent, and dry then with compressed air.

➡Do not wash the rubber parts with solvent.

19. Insert the inboard end of the driveshaft into the differential or intermediate shaft until the set ring locks in the groove.

➡Insert the driveshaft horizontally to prevent damaging the oil seal.

20. Install the outboard joint into the front hub on the knuckle.
21. Wipe off any grease contamination from the ball joint tapered section and threads, then install the knuckle onto the lower arm. Be careful not to damage the ball joint boot. Wipe off the grease before tightening the nut at the ball joint. Torque the castle nut to the lower torque specification, then tighten it only far enough to align the slot with the ball joint pin hole.

➡Make sure the ball joint boot is not damaged or cracked. Do not align the nut by loosening it.

22. Install a new cotter pin into the ball joint pin hole, and bend the cotter pin.
23. Install the damper fork over the driveshaft and onto the lower arm. Install the damper in the damper fork so the aligning tab is aligned with the slot in the damper fork. Loosely install the damper pinch bolt.
24. Loosely install a new damper fork mounting bolt and a new damper fork mounting nut.
25. Connect the front stabilizer link to the lower arm, and loosely install a new flange nut. Hold the stabilizer link joint pin using a hex wrench, and tighten the flange nut.
26. Place a floor jack under the lower arm, and raise the suspension to load it with the vehicle's weight.

➡Do not put the floor jack under the ball joint.

27. Tighten the damper pinch bolt to 36 ft. lbs. (49 Nm)and tighten the damper fork mounting nut while holding the damper fork mounting bolt to 47 ft. lbs.(64 Nm), then remove the floor jack.
28. Apply a small amount of engine oil to the seating surface of a new spindle nut.
29. Install the spindle nut and tighten it to 242 Ft. Lbs. (329 Nm). After tightening the spindle nut, use a drift to stake the spindle nut shoulder against the driveshaft.
30. Clean the mating surfaces of the brake disc and the wheel, then install the front wheels.
31. Turn the wheel by hand, and make sure there is no interference between the driveshaft and surrounding parts.
32. Refill the transmission with the recommended transmission fluid:
33. Lower the vehicle.
34. Check the wheel alignment, and adjust it if necessary.
35. Test-drive the vehicle.

INTERMEDIATE SHAFT

REMOVAL & INSTALLATION

4-Cylinder Engine

See Figures 54 and 55.

1. Drain the transmission fluid. Reinstall the drain plug using a new sealing washer.
2. Remove the right driveshaft.
3. Remove the flange bolt and the two dowel bolts.
4. Remove the intermediate shaft from the differential. Hold the intermediate shaft horizontally until it is clear of the differential to prevent damaging the oil seal.

To install:

5. Clean the areas where the intermediate shaft contacts the differential thoroughly with solvent, and dry then with compressed air.

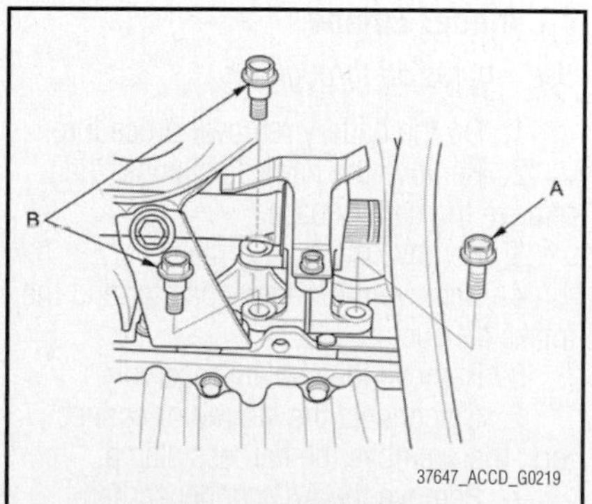
A
B
37647_ACCD_G0219

Fig. 54 Remove the flange bolt (A) and the two dowel bolts (B)

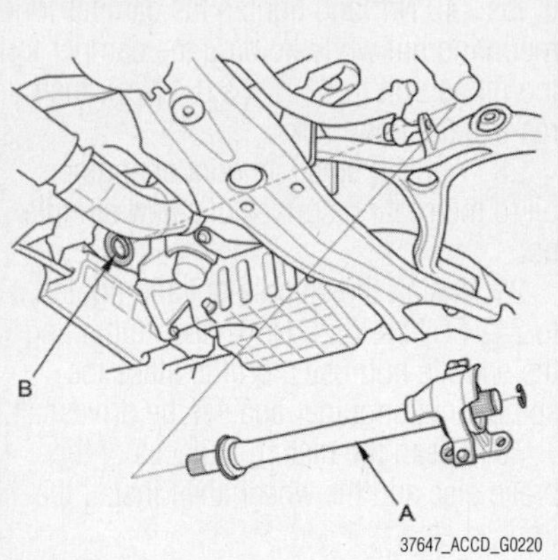
B
A
37647_ACCD_G0220
Fig. 55 Remove the intermediate shaft (A) from the differential

➡Do not wash the rubber parts with solvent.

6. Install a new set ring onto the set ring groove of the intermediate shaft.
7. Insert the intermediate shaft into the differential correctly.

➡Insert the intermediate shaft carefully to prevent damaging the oil seal.

8. Install the flange bolt and the two dowel bolts. Tighten the bolts to 29 ft. lbs. (39 Nm).
9. Install the right driveshaft.
10. Refill the transmission with the recommended transmission fluid.
11. Check the wheel alignment, and adjust it if necessary.
12. Test-drive the vehicle.

6-Cylinder Engine

See Figures 56 and 57.

1. Drain the transmission fluid, then reinstall the drain plug with a new sealing washer.
2. Remove the right driveshaft.
3. Remove exhaust pipe A and the gaskets.
4. Remove the flange bolt and the two dowel bolts.
5. Remove the intermediate shaft from the differential. Hold the intermediate shaft horizontally until it is clear of the differential to prevent damaging the oil seal.

To install:

6. Clean the areas where the intermediate shaft contacts the differential thoroughly with solvent, and dry then with compressed air.

➡Do not wash the rubber parts with solvent.

7. Install a new set ring onto the set ring groove of the intermediate shaft.
8. Insert the intermediate shaft into the differential correctly.

➡Insert the intermediate shaft carefully to prevent damaging the oil seal.

9. Install the flange bolt and the two dowel bolts.
10. Install exhaust pipe A with new gaskets and new nuts.
11. Install the right driveshaft.
12. Refill the transmission with the recommended transmission fluid.
13. Check the wheel alignment, and adjust it if necessary.
14. Test-drive the vehicle.

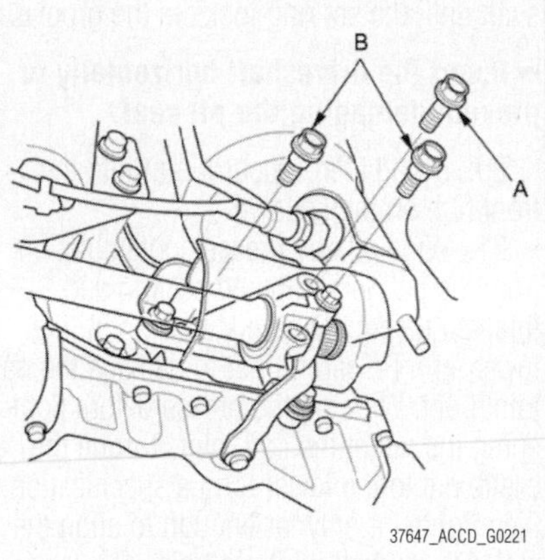
B
A
37647_ACCD_G0221
Fig. 56 Remove the flange bolt (A) and the two dowel bolts (B)

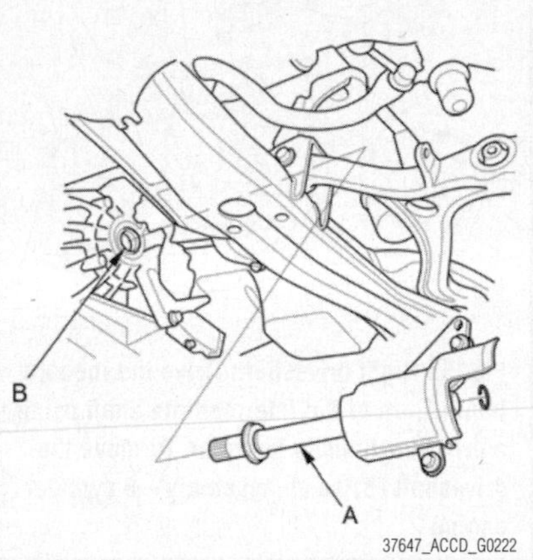
B
A
37647_ACCD_G0222
Fig. 57 Remove the intermediate shaft (A) from the differential

ENGINE COOLING

ENGINE FAN

REMOVAL & INSTALLATION

4-Cylinder Engine

See Figures 58 through 61.

1. Do the battery removal procedure.
2. Remove the harness clamps, then remove the battery base.
3. Remove the front grille cover:
4. Remove the water separator and the intake air duct.
5. Remove the coolant reservoir.
6. Disconnect the fan motor connectors, then remove the harness clamp.
7. Remove the A/C condenser fan shroud assembly, then remove the radiator fan shroud assembly.

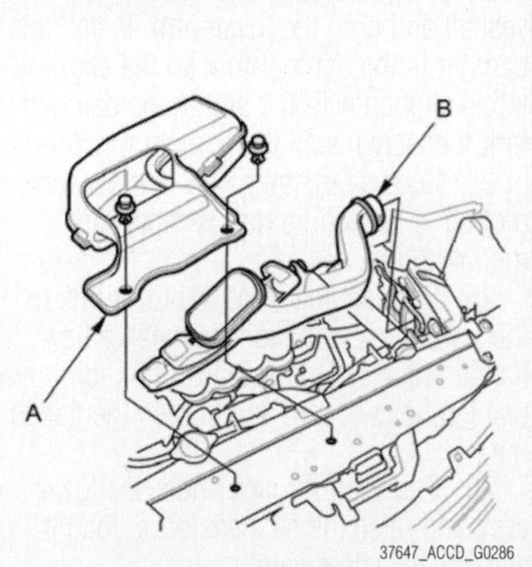
B
A
37647_ACCD_G0286
Fig. 58 Remove the water separator (A) and the intake air duct (B)

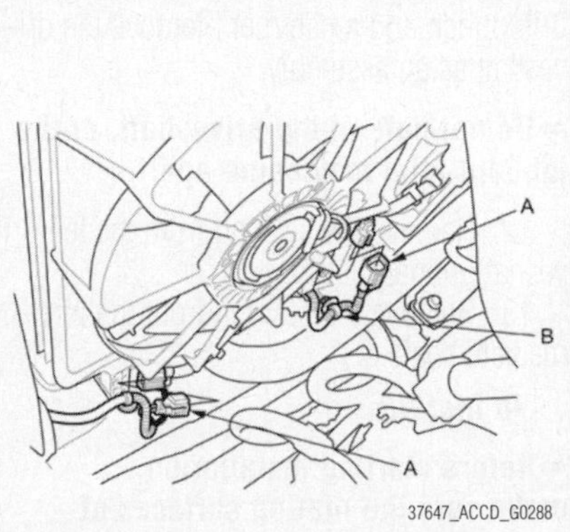
A
B
A
37647_ACCD_G0288
Fig. 59 Disconnect the fan motor connectors (A), then remove the harness clamp (B)

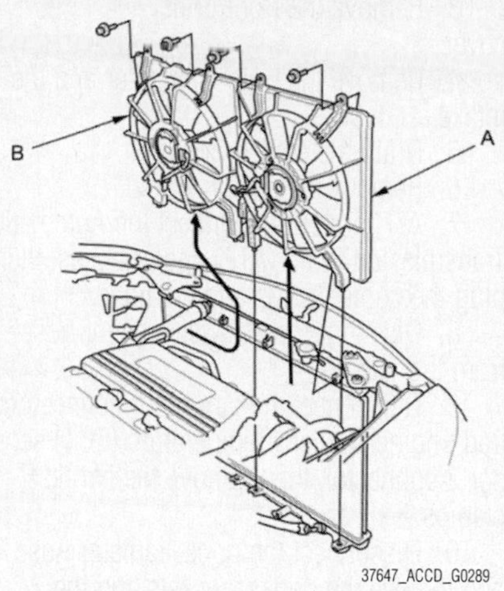
B
A
37647_ACCD_G0289

Fig. 60 Remove the A/C condenser fan shroud assembly (A), then remove the radiator fan shroud assembly (B)

➡Move the radiator fan shroud assembly toward the right side of the vehicle to allow enough space to lift it up and away from the A/C condenser fan shroud assembly.

8. Disassemble the fan shrouds.

To install:

9. Reassemble the fan shrouds.
10. Install the radiator fan shroud assembly, then install the A/C condenser fan shroud assembly.
11. Connect the fan motor connectors, then install the harness clamp.
12. Install the coolant reservoir.
13. Install the water separator and the intake air duct.
14. Install the front grille cover.
15. Install the battery base, then install the harness clamps.
16. Do the battery installation procedure.

37647_ACCD_G0290

Fig. 61 Disassemble the fan shrouds

6-Cylinder Engine

See Figures 62 through 65.

1. Do the battery removal procedure.
2. Remove the battery base.
3. Remove the front grille cover.
4. Remove the splash separator and the intake air ducts.
5. Raise the vehicle on the lift.
6. Remove the splash shield.
7. Disconnect the A/C condenser fan motor connector, and remove the harness clamp.
8. Loosen the A/C condenser fan shroud mounting bolts.
9. Disconnect the radiator fan motor connector.
10. Lower the vehicle on the lift.
11. Remove the coolant reservoir.
12. Remove the upper brackets.

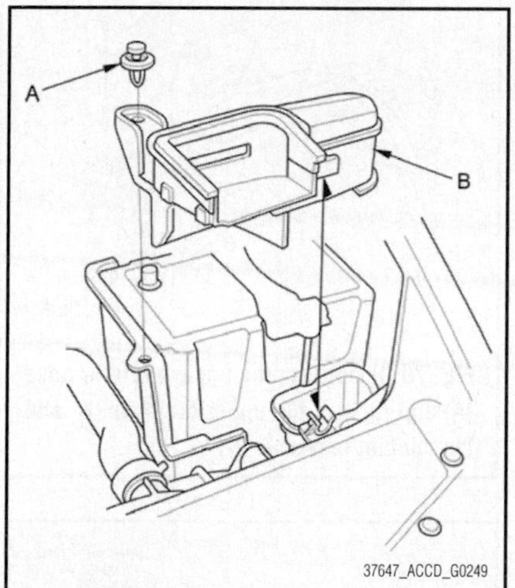
A
B
37647_ACCD_G0249

Fig. 62 Remove the splash separator and the intake air ducts

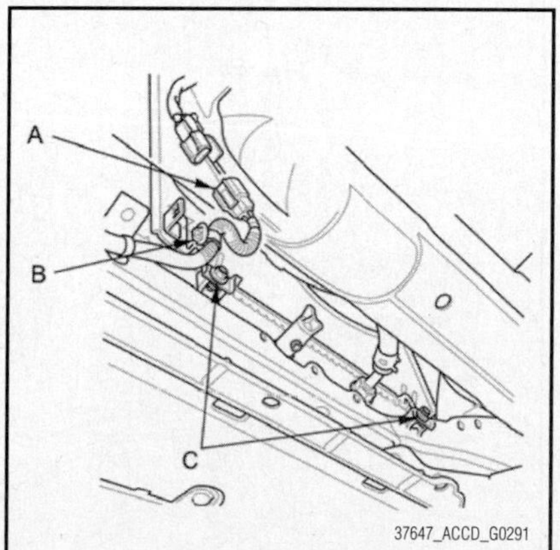
A
B
C
37647_ACCD_G0291

Fig. 63 Disconnect the A/C condenser fan motor connector (A), and remove the harness clamp (B), loosen the A/C condenser fan shroud mounting bolts (C)

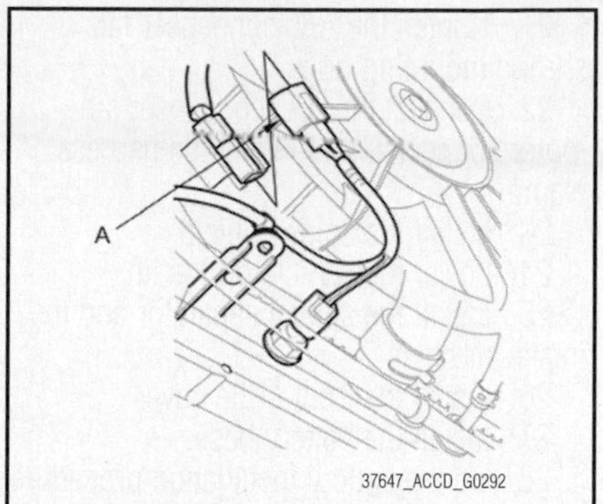
A
37647_ACCD_G0292

Fig. 64 Disconnect the radiator fan motor connector (A)

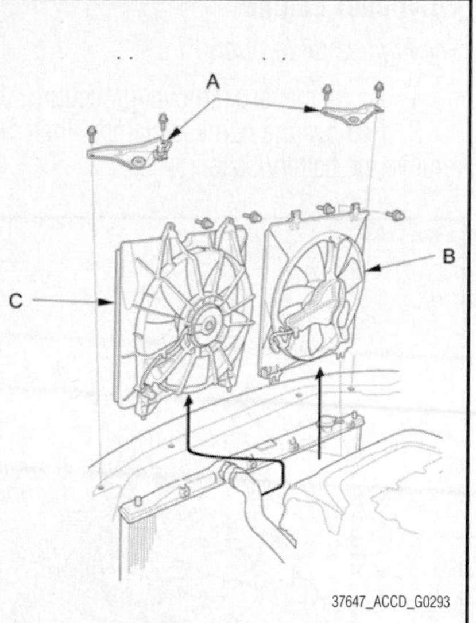
A
B
C
37647_ACCD_G0293

Fig. 65 Remove the upper brackets (A), the A/C condenser fan shroud assembly (B), then remove the radiator fan shroud assembly (C)

13. Remove the A/C condenser fan shroud assembly, then remove the radiator fan shroud assembly.

➡Pull-up the radiator, then move the radiator fan shroud assembly toward the right side of the vehicle to allow for enough space to lift it up and away from the A/C condenser fan shroud assembly.

14. Disassemble the fan shrouds.

To install:

15. Assemble the fan shrouds.
16. Install the radiator fan shroud assembly, then install the A/C condenser fan shroud assembly.
17. Install the upper brackets.
18. Install the coolant reservoir.
19. Raise the vehicle on the lift.
20. Connector the radiator fan motor connector.

21. Tighten the A/C condenser fan shroud mounting bolts.
22. Connect the A/C condenser fan motor connector, and install the harness clamp.
23. Install the splash shield.
24. Lower the vehicle on the lift.
25. Install the splash separator and the intake air ducts.
26. Install the front grille cover.
27. Install the battery base.
28. Do the battery installation procedure.

RADIATOR

REMOVAL & INSTALLATION

4-Cylinder Engine

See Figures 67 through 71.

1. Do the battery removal procedure.
2. Remove the harness clamps, then remove the battery base.

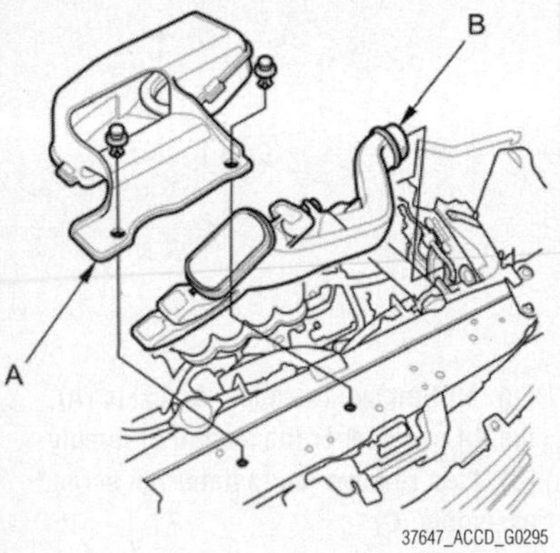
B
A
37647_ACCD_G0295

Fig. 67 Remove the water separator (A) and the intake air duct (B)

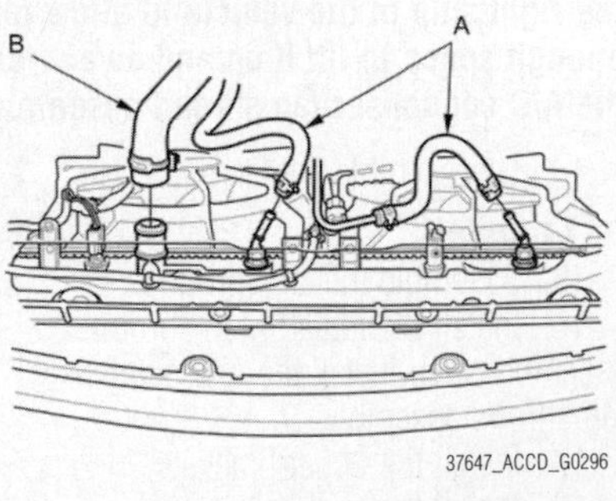
B
A
37647_ACCD_G0296

Fig. 68 A/T model: Disconnect the Automatic Transmission Fluid (ATF) cooler hoses (A), and the lower radiator hose (B)

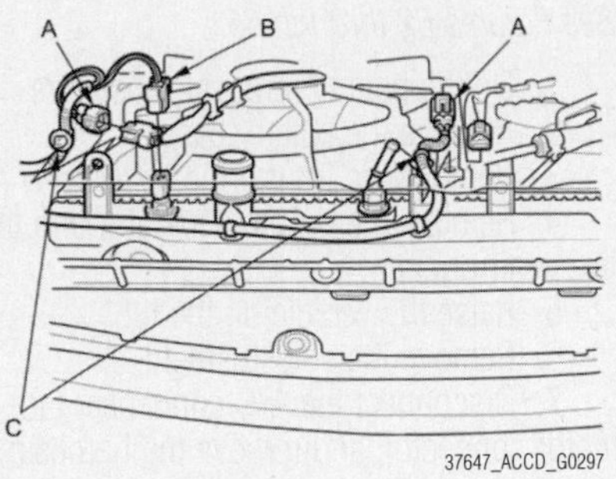
A
B
A
C
37647_ACCD_G0297

Fig. 69 Disconnect the fan motor connectors (A) and ECT sensor 2 connector (B), then remove the harness clamps (C)

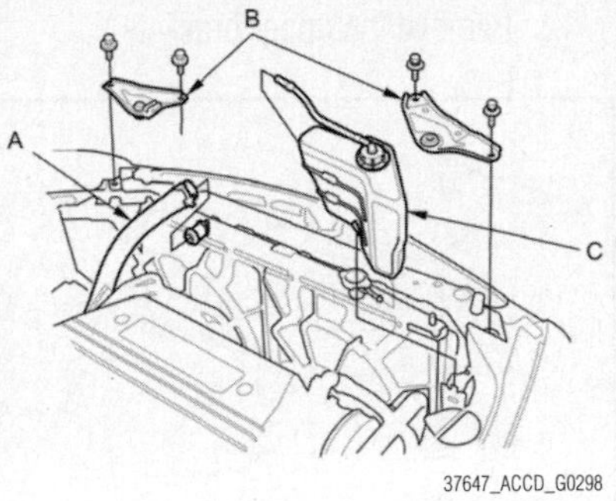
B
A
C
37647_ACCD_G0298

Fig. 70 Disconnect the upper radiator hose (A) and remove the upper brackets (B) and the coolant reservoir (C)

3. Remove the front grille cover.
4. Remove the water separator and the intake air duct.
5. Drain the engine coolant.
6. Remove the splash shield.
7. A/T model: Disconnect the Automatic Transmission Fluid (ATF) cooler hoses, then plug the cooler hoses and the lines.
8. Disconnect the lower radiator hose from the radiator.
9. Disconnect the fan motor connectors and Engine Coolant Temperature (ECT) sensor 2 connector, then remove the harness clamps.
10. Disconnect the upper radiator hose and remove the upper brackets and the coolant reservoir.
11. Remove the radiator.
12. Remove the fan shroud assemblies from the radiator.

To install:

13. Install the radiator in the reverse order of removal. Make sure the upper and lower cushions (4) are set securely.
14. Do the battery installation procedure.
15. A/T model: Refill the transmission with ATF.
16. Fill the radiator with engine coolant, and bleed the air from the cooling system.
17. Clean up any spilled engine coolant.

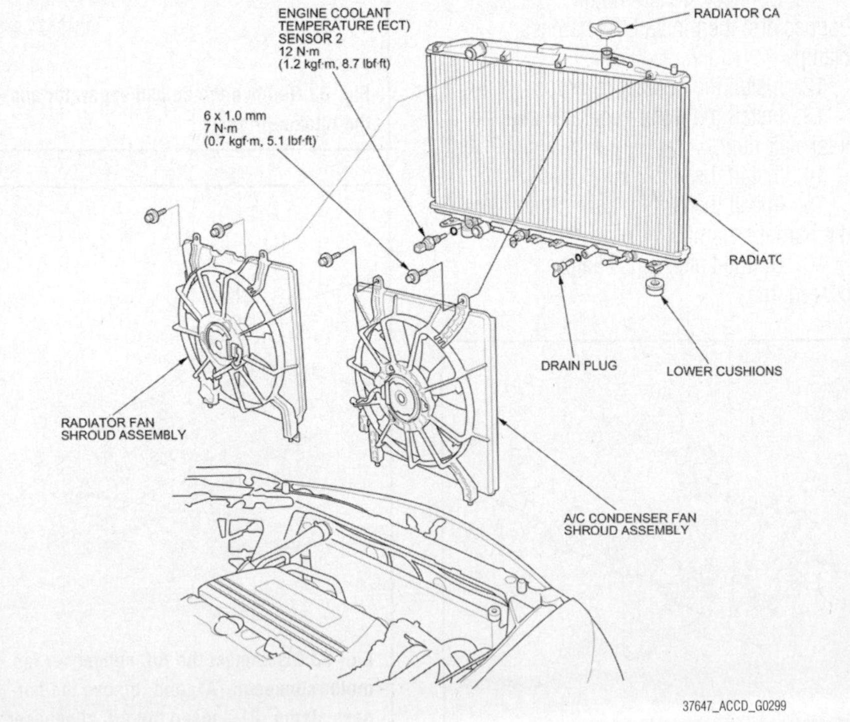
ENGINE COOLANT
TEMPERATURE (ECT)
SENSOR 2
12 N·m
(1.2 kgf·m, 8.7 lbf·ft)
RADIATOR CA
6 x 1.0 mm
7 N·m
(0.7 kgf·m, 5.1 lbf·ft)
RADIATC
DRAIN PLUG
LOWER CUSHIONS
RADIATOR FAN
SHROUD ASSEMBLY
A/C CONDENSER FAN
SHROUD ASSEMBLY
37647_ACCD_G0299

Fig. 71 Exploded view of radiator/shroud/fan assembly

6-Cylinder Engine

See Figures 72 through 77.

1. Remove the front grille cover:
2. Remove the splash separator.
3. Raise the vehicle on the lift.
4. Drain the engine coolant.
5. Remove the splash shield.
6. Disconnect the A/C condenser fan motor connector, and remove the harness clamp.
7. Disconnect the radiator fan motor connector and the Engine Coolant Temperature (ECT) sensor 2 connector, then remove the harness clamp.
8. Disconnect the lower radiator hose from the radiator.
9. A/T model: Disconnect the Automatic Transmission Fluid (ATF) cooler hoses from the radiator, then plug the cooler hoses and the lines.

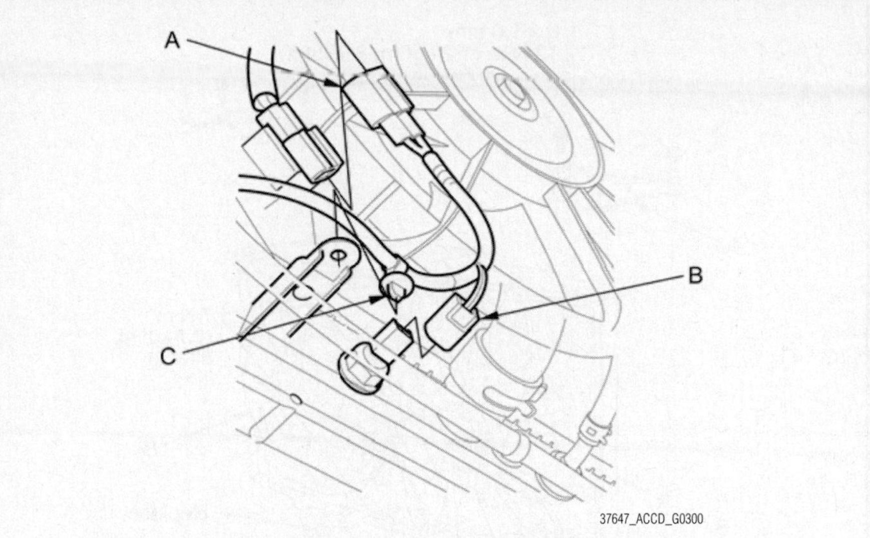

A
B
C
37647_ACCD_G0300

Fig. 74 Disconnect the radiator fan motor connector (A) and the ECT sensor 2 connector (B), then remove the harness clamp (C)

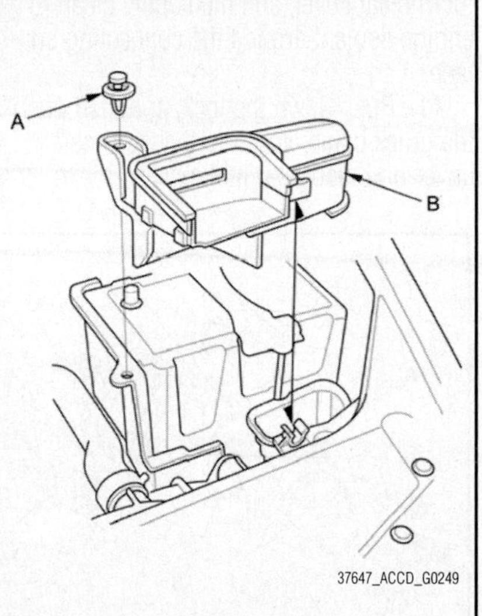

A
B
37647_ACCD_G0249

Fig. 72 Remove the splash separator

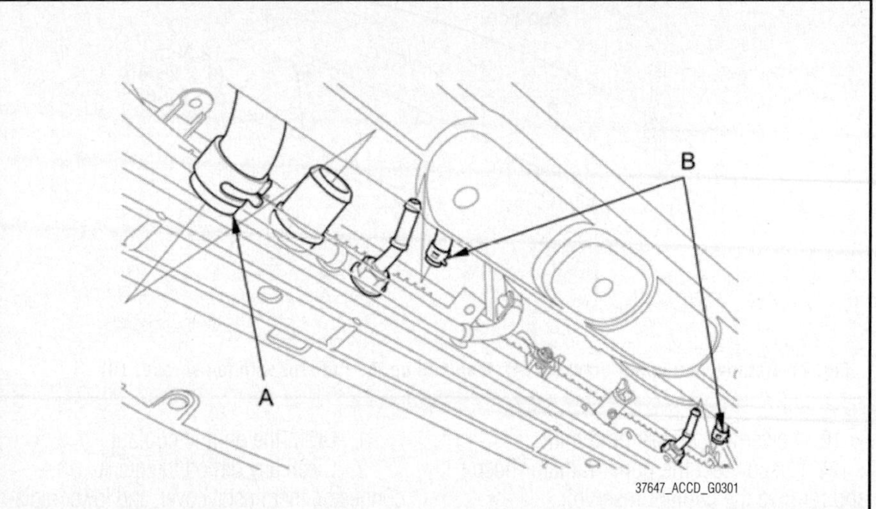

B
A
37647_ACCD_G0301

Fig. 75 Disconnect the lower radiator hose (A) and the ATF cooler hoses (B) from the radiator

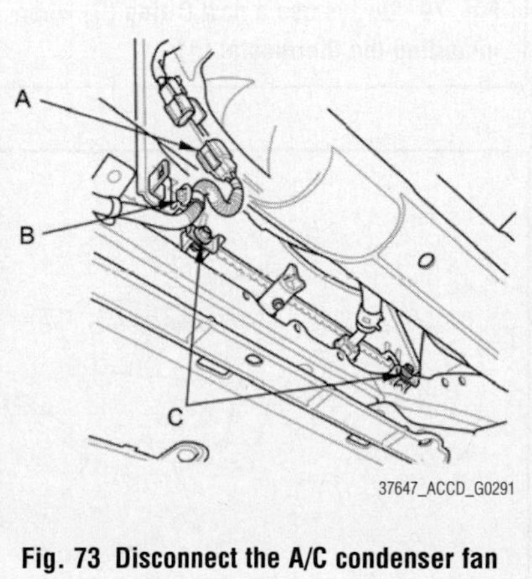

A
B
C
37647_ACCD_G0291

Fig. 73 Disconnect the A/C condenser fan motor connector (A), and remove the harness clamp (B)

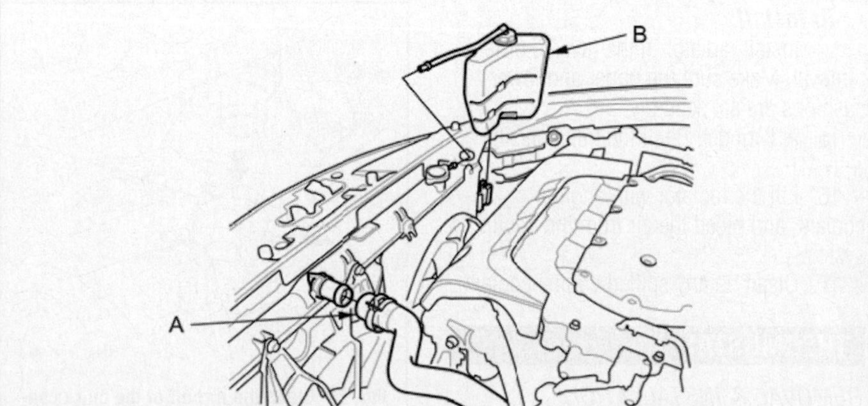

B
A
37647_ACCD_G0302

Fig. 76 Disconnect the upper radiator hose (A) and remove the coolant reservoir (B)

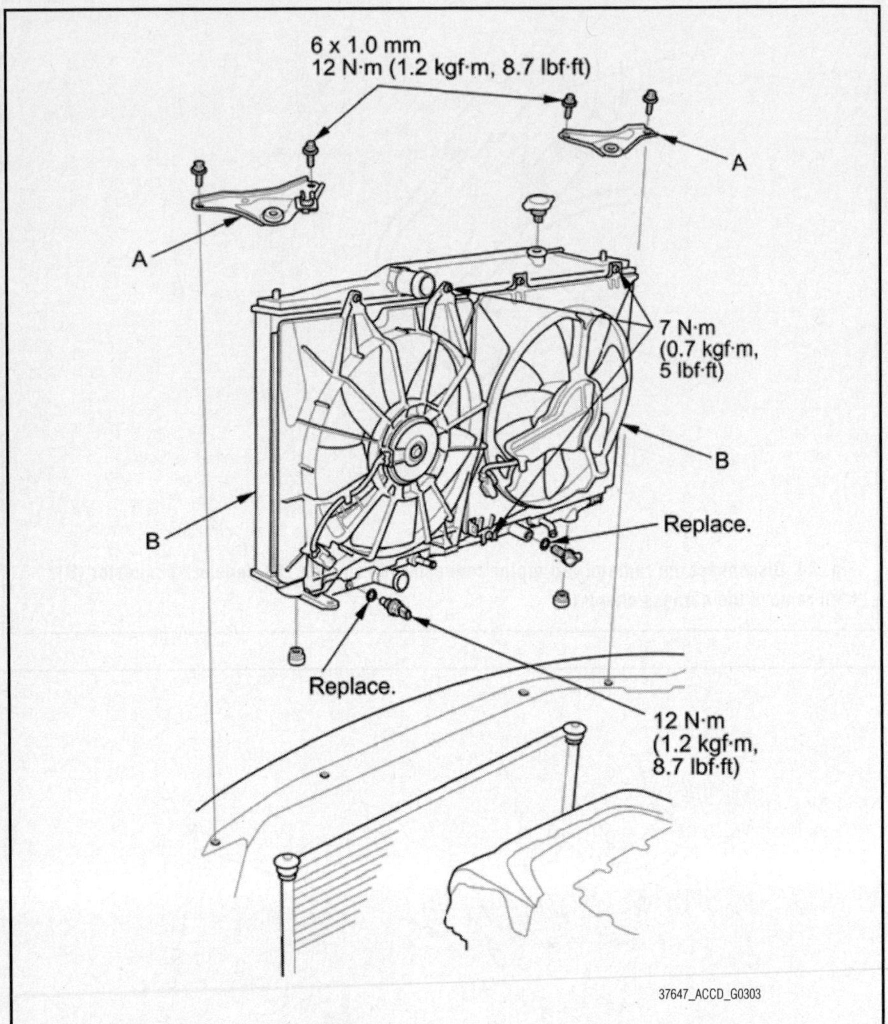
6 x 1.0 mm
12 N·m (1.2 kgf·m, 8.7 lbf·ft)
A
A
7 N·m
(0.7 kgf·m,
5 lbf·ft)
B
B
Replace.
Replace.
12 N·m
(1.2 kgf·m,
8.7 lbf·ft)
37647_ACCD_G0303

Fig. 77 Remove the upper brackets (A), then pull up the radiator with fan shrouds (B)

10. Lower the vehicle on the lift.

11. Disconnect the upper radiator hose and remove the coolant reservoir.

12. Remove the upper brackets, then pull up the radiator with fan shrouds.

13. Remove the fan shrouds and other parts from the radiator.

To install:

14. Install radiator in the reverse order or removal. Make sure the upper and lower cushions are set securely.

15. A/T model: Refill the transmission with ATF.

16. Fill the radiator with engine coolant, and bleed the air from the cooling system.

17. Clean up any spilled engine coolant.

THERMOSTAT

REMOVAL & INSTALLATION

4-Cylinder Engine

See Figures 78 through 80.

1. Drain the engine coolant.

2. Clean any dirt off the quick connector, thermostat cover, and lower radiator hose.

3. Pull out lock by hand, and then wiggle the quick connector loose, and remove it from the thermostat cover. Do not use any tools to remove the quick connector.

4. Remove the thermostat.

To install:

5. Install the thermostat with a new O-ring.

6. Check the quick connector and set ring for cracks or damage. If the connector and/or set ring are cracked or damaged, replace the connector.

7. Make sure the set ring is in place inside the quick connector. If the set ring is off the connector, replace the quick connector.

8. Replace the O-ring in the quick connector.

9. Check the lock. If the lock is damaged or deformed, replace it. When installing the new lock on the connector, press it straight down along the groove.

10. Clean the connecting surface of the thermostat cover, and then apply clean engine coolant around the connecting surface.

11. Push down the lock, and then push the quick connector onto the thermostat cover until you hear it click.

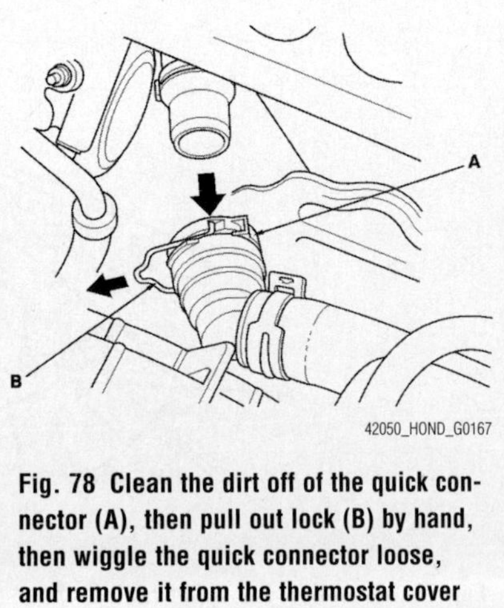
A
B
42050_HOND_G0167

Fig. 78 Clean the dirt off of the quick connector (A), then pull out lock (B) by hand, then wiggle the quick connector loose, and remove it from the thermostat cover

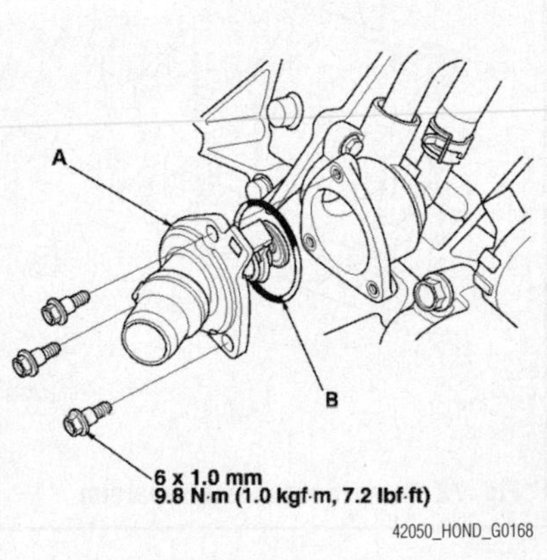
A
B
6 x 1.0 mm
9.8 N·m (1.0 kgf·m, 7.2 lbf·ft)
42050_HOND_G0168

Fig. 79 Always use a new O-ring (B) when installing the thermostat (A)

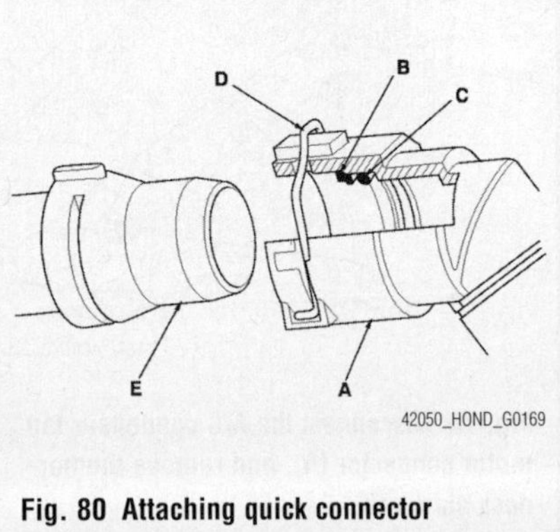
D
B
C
E
A
42050_HOND_G0169

Fig. 80 Attaching quick connector

12. Refill the radiator with engine coolant, and bleed air from the cooling system with the heater valve open.

6-Cylinder Engine

See Figure 81.

1. Drain the engine coolant.
2. Do the battery removal procedure.
3. Remove the thermostat cover, then remove the thermostat.
4. Install the new thermostat with a new rubber seal, then install the thermostat cover.
5. Do the battery installation procedure.
6. Refill the radiator with engine coolant, and bleed the air from the cooling system.
7. Clean up any spilled engine coolant.

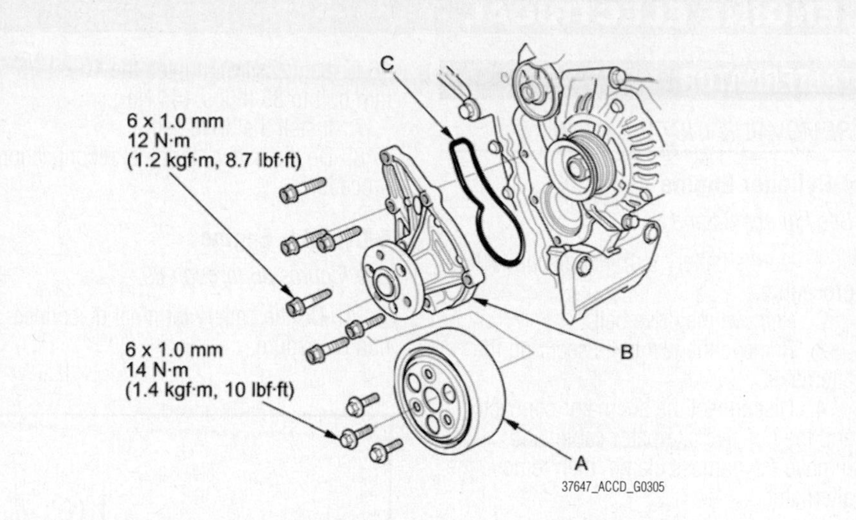
C
6 x 1.0 mm
12 N·m
(1.2 kgf·m, 8.7 lbf·ft)
6 x 1.0 mm
14 N·m
(1.4 kgf·m, 10 lbf·ft)
B
A
37647_ACCD_G0305

Fig. 82 Remove the water pump pulley (A), the water pump)B) and the O-ring (C)

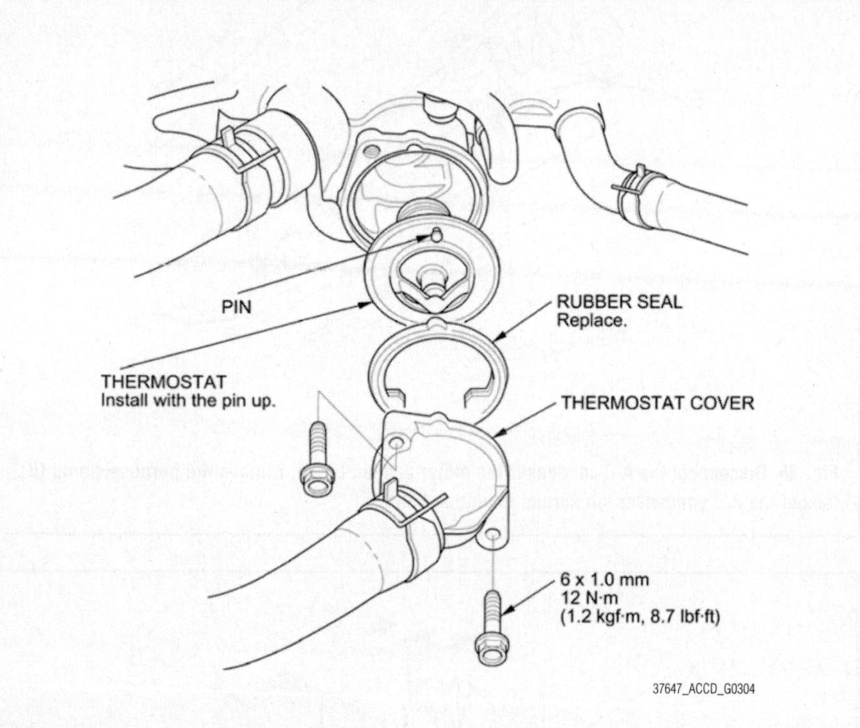
PIN
RUBBER SEAL
Replace.
THERMOSTAT
Install with the pin up.
THERMOSTAT COVER
6 x 1.0 mm
12 N·m
(1.2 kgf·m, 8.7 lbf·ft)
37647_ACCD_G0304

Fig. 81 Exploded view of thermostat housing assembly

WATER PUMP

REMOVAL & INSTALLATION

4-Cylinder Engine

See Figure 82.

1. Remove the drive belt.
2. Drain the engine coolant.
3. Remove the tensioner pulley.
4. Remove the water pump pulley.
5. Remove the six bolts securing the water pump, then remove the water pump.

To install:

6. Inspect and clean the O-ring groove and the mating surface of the water passage.
7. Install the water pump with a new O-ring and the water pump pulley.
8. Clean up any spilled engine coolant.
9. Install the tensioner pulley.
10. Install the drive belt.
11. Refill the radiator with engine coolant, and bleed the air from the cooling system.

6-Cylinder Engine

See Figure 83.

1. Drain the engine coolant.
2. Remove the timing belt:
3. Remove the timing belt adjuster:

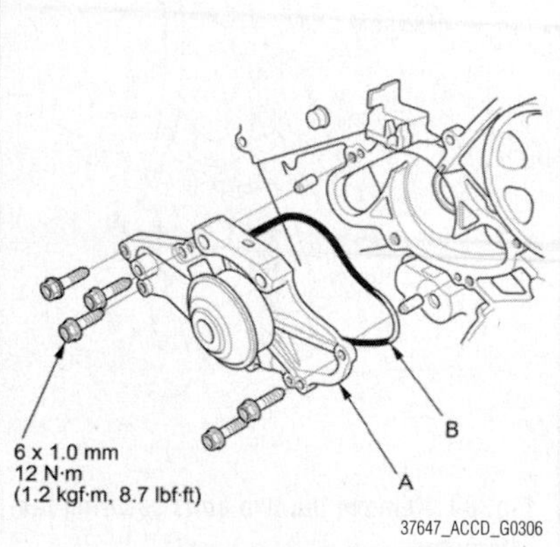
6 x 1.0 mm
12 N·m
(1.2 kgf·m, 8.7 lbf·ft)
B
A
37647_ACCD_G0306

Fig. 83 Remove the five bolts securing the water pump (A), then remove the water pump and O-ring (B)

4. Remove the five bolts securing the water pump, then remove the water pump.
5. Inspect and clean the O-ring groove and the mating surface of the engine block.
6. Install the water pump with a new O-ring.
7. Clean up any spilled engine coolant.
8. Install the timing belt adjuster:
9. Install the timing belt:
10. Refill the radiator with engine coolant, and bleed the air from the cooling system.

ALTERNATOR

REMOVAL & INSTALLATION

4-Cylinder Engine

See Figures 84 and 85.

1. Do the battery terminal disconnection procedure.
2. Remove the drive belt.
3. Remove the two bolts securing the alternator.
4. Disconnect the alternator connector and the positive alternator cable, and remove the harness clamp, then remove the alternator.

To install:

5. Install the alternator, then connect the alternator connector (A) and the positive alternator cable (B), and install the harness clamp (C). Make sure the crimped side of the ring terminal faces away from the alternator when you connect it.
6. Tighten the two bolts securing the alternator. Tighten the 8 x 1.25 mm bolt to 16 ft. lbs. (22 Nm); tighten the 10 x 1.25 mm bolt to 33 ft. lbs. (44 Nm).
7. Install the drive belt.
8. Do the battery terminal reconnection procedure.

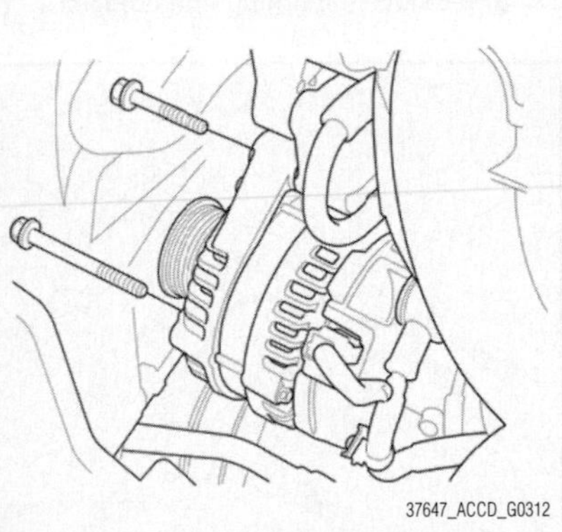
37647_ACCD_G0312

Fig. 84 Remove the two bolts securing the alternator

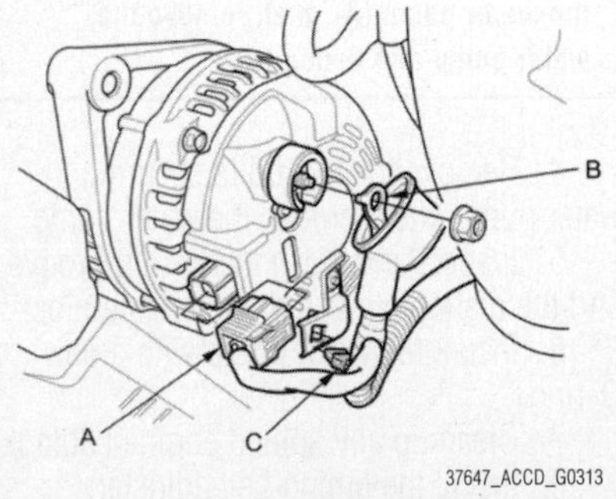
B
A
C
37647_ACCD_G0313

Fig. 85 Disconnect the alternator connector (A) and the positive alternator cable (B), and remove the harness clamp (C)

6-Cylinder Engine

See Figures 86 through 89.

1. Do the battery terminal disconnection procedure.
2. Raise the vehicle on the lift.
3. Remove the engine splash shield.
4. Disconnect the A/C condenser fan motor connector and remove the harness clamp.
5. Loosen the A/C condenser fan shroud mounting bolts.
6. Lower the vehicle on the lift.
7. Remove the A/C condenser fan

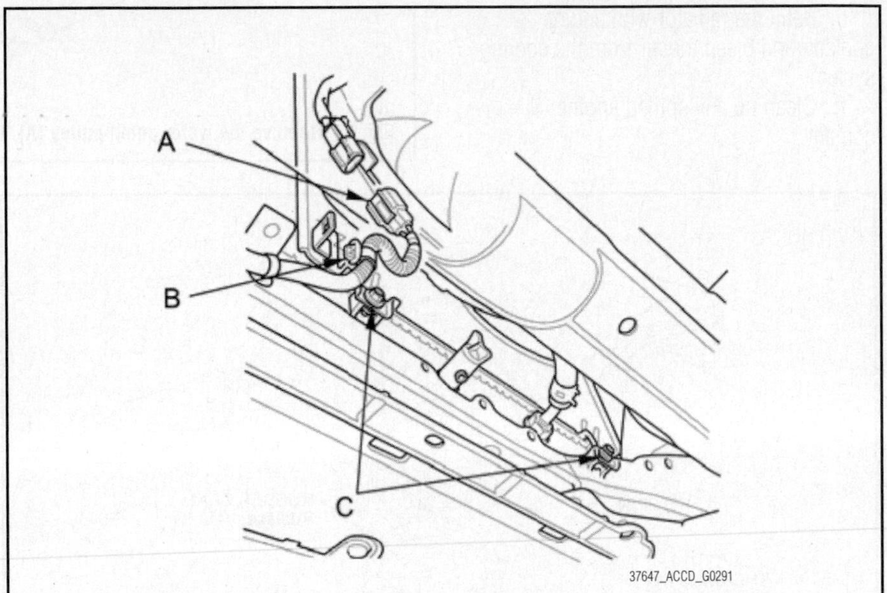
A
B
C
37647_ACCD_G0291

Fig. 86 Disconnect the A/C condenser fan motor connector (A), remove the harness clamp (B), loosen the A/C condenser fan shroud mounting bolts (C)

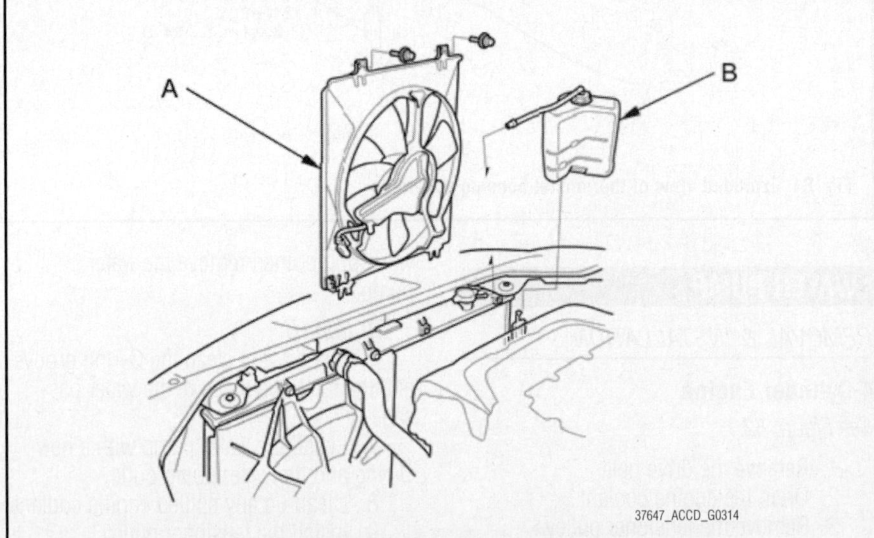
A
B
37647_ACCD_G0314

Fig. 87 Remove the A/C condenser fan shroud assembly (A) and the coolant reservoir (B)

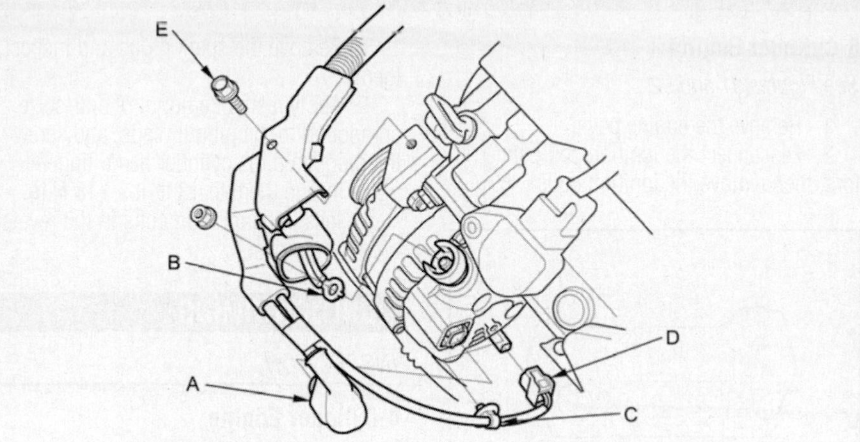
E
B
D
A
C
A. Alternator connector
B. Positive alternator cable
C. Harness clamp
D. A/C compressor clutch connector
E. Bolt
37647_ACCD_G0315

Fig. 88 Disconnect the alternator connector and the positive alternator cable from the alternator

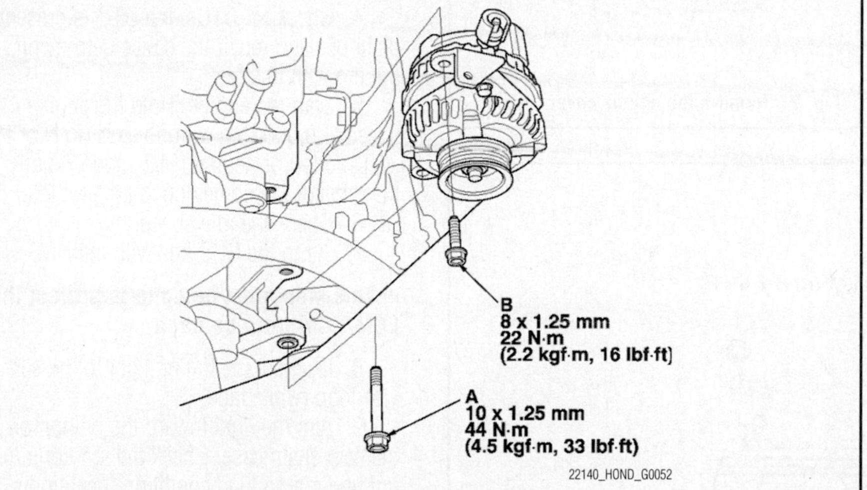
B
8 x 1.25 mm
22 N·m
(2.2 kgf·m, 16 lbf·ft)
A
10 x 1.25 mm
44 N·m
(4.5 kgf·m, 33 lbf·ft)
22140_HOND_G0052

Fig. 89 Remove the mounting bolt (A) and the alternator bracket mounting bolt (B) to remove the alternator—3.5L Engines

shroud assembly and the coolant reservoir.

8. Remove the accessory drive belt.
9. Disconnect the alternator connector and the positive alternator cable from the alternator.
10. Remove the harness clamp from the alternator and disconnect the A/C compressor clutch connector from the A/C compressor.
11. Remove the bolt securing the harness holder.
12. Remove the mounting bolt, the alternator bracket mounting bolt, and the alternator.

To install:

13. Install the alternator, then tighten the mounting bolt and the alternator bracket mounting bolt. Tighten the mounting bolt to 33 ft. lbs. (44 Nm). Tighten the bracket mounting bolt to 16 ft. lbs. (22 Nm).
14. Install the bolt securing the harness holder.
15. Install the harness clamp to the alternator and connect the A/C compressor clutch connector to the A/C compressor.
16. Connect the alternator connecter and the positive alternator cable to the alternator. Make sure the crimped side of the ring terminal faces away from the alternator when you connect it.
17. Install the drive belt.
18. Install the A/C condenser fan shroud assembly and the coolant reservoir.
19. Raise the vehicle on the lift.
20. Tighten the A/C compressor fan shroud mounting bolts.
21. Connect the A/C condenser fan motor connector and install the harness clamp.
22. Install the splash shield.
23. Lower the vehicle on the lift.
24. Do the battery terminal reconnection procedure.

FIRING ORDER

4-Cylinder Engine

The firing order is 1–3–4–2.

6-Cylinder Engine

The firing order is 1–4–2–5–3–6.

IGNITION COIL

REMOVAL & INSTALLATION

4-Cylinder Engine

See Figure 90.

1. Remove the ignition coil cover.
2. Disconnect the ignition coil connectors, then remove the ignition coils.
3. Remove the spark plug and inspect them.
4. Apply a small amount of anti-seize compound to the plugs into the cylinder head, finger tight. Torque them to 13 ft. lbs. (18 Nm).
5. Install the ignition coils, then connect the ignition coil connectors.
6. Install the ignition coil cover.

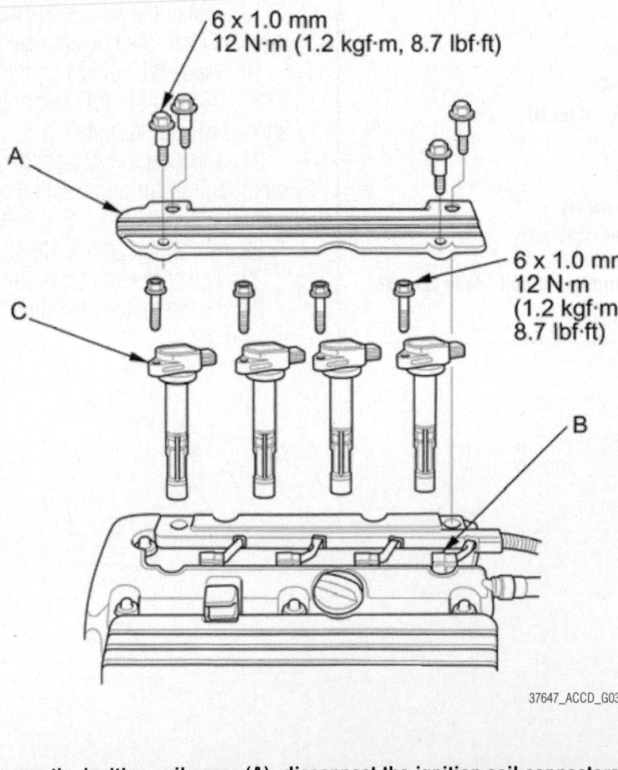
6 x 1.0 mm
12 N·m (1.2 kgf·m, 8.7 lbf·ft)
A
6 x 1.0 mm
12 N·m
(1.2 kgf·m,
8.7 lbf·ft)
C
B
37647_ACCD_G0316

Fig. 90 Remove the ignition coil cover (A), disconnect the ignition coil connectors (B), then remove the ignition coils (C)

6-Cylinder Engine

See Figures 91 and 92.

1. Remove the engine cover.
2. Disconnect the ignition coil connectors, then remove the ignition coils.

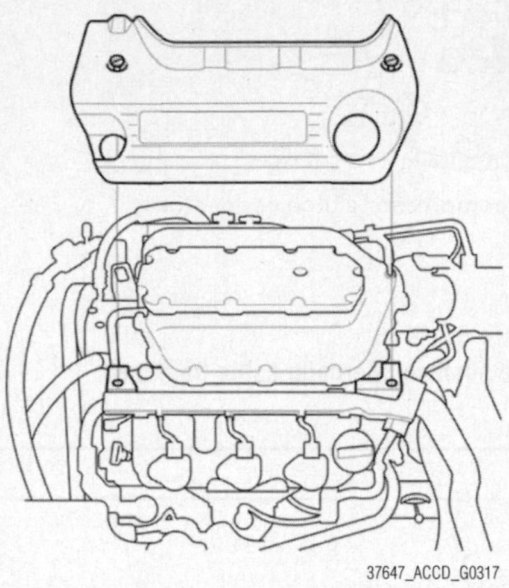
37647_ACCD_G0317

Fig. 91 Remove the engine cover

3. Remove the spark plugs and inspect them.
4. Apply a small amount of anti-seize compound to the plug threads, and screw the plugs into the cylinder head, finger-tight. Torque them to 13 ft. lbs. (18 Nm).
5. Install the ignition coils in the reverse order of removal.

IGNITION TIMING

INSPECTION

4-Cylinder Engine

1. Connect the Honda Diagnostic System (HDS) to the Data Link Connector (DLC).
2. Turn the ignition switch to ON (II).
3. Make sure the HDS communicates with the vehicle and the Engine Control Module (ECM)/Powertrain Control Module (PCM) If it does not communicate, troubleshoot the DLC circuit..
4. Check for DTCs. If a DTC is present, diagnose and repair the cause before continuing with this test.
5. Start the engine. Hold the engine speed at 3,000 rpm with no load (in N or P (A/T model) or neutral (M/T model)) until the radiator fan comes on, then let it idle.
6. Check the idle speed.
7. Jump the SCS line with the HDS.

➡This step must be done to protect the ECM/PCM from damage.

8. Connect the timing light to the service loop (white tape).
9. Aim the light toward the pointer on the cam chain case. Check the ignition timing under a no load condition (headlights, blower fan, rear window defogger, and air conditioner are turned off).

➡The other pointer is not used.

Ignition Timing specifications:

- M/T model: 8°±2° BTDC (RED mark) at idle in neutral
- A/T model: 8°±2° BTDC (RED mark) at idle in N or P

10. If the ignition timing differs from the specification, check the cam timing. If the cam timing is OK, update the ECM/PCM if it does not have the latest software, or substitute a known-good ECM/PCM, then recheck. If the system works properly, and the ECM/PCM was substituted, replace the original ECM/PCM.
11. Disconnect the HDS and the timing light.

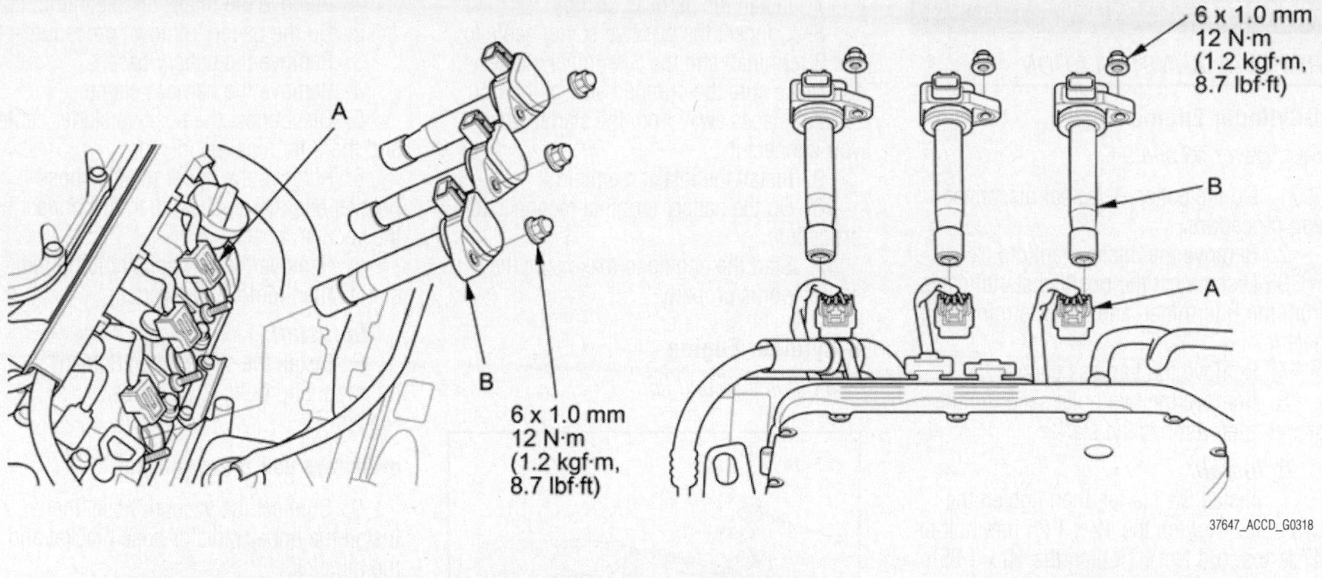
A
B
6 x 1.0 mm
12 N·m
(1.2 kgf·m,
8.7 lbf·ft)
6 x 1.0 mm
12 N·m
(1.2 kgf·m,
8.7 lbf·ft)
B
A
37647_ACCD_G0318

Fig. 92 Disconnect the ignition coil connectors (A), then remove the ignition coils (B)

6-Cylinder Engine

1. Connect the Honda Diagnostic System (HDS) to the Data Link Connector (DLC).
2. Turn the ignition switch to ON (II).
3. Make sure the HDS communicates with the vehicle and the Engine Control Module (ECM)/Powertrain Control Module (PCM). If it does not communicate, troubleshoot the DLC circuit.
4. Check for DTCs. If a DTC is present, diagnose and repair the cause before continuing with this test.
5. Start the engine. Hold the engine speed at 3,000 rpm with no load (in neutral (M/T model) or N or P (A/T model)) until the radiator fan comes on, then let it idle.
6. Check the idle speed.
7. Jump the SCS line with the HDS.
8. Connect the timing light to the No. 1 ignition coil harness.
9. Aim the light toward the pointer on the timing belt cover. Check the ignition timing under a no load condition (headlights, blower fan, rear window defogger, and air conditioner are turned off).

➡The other pointer is not used.

Ignition timing specification: 10 °±2 ° BTDC (RED mark) at idle in N or P

10. If the ignition timing differs from the specification, check the cam timing. If the cam timing is OK, update the ECM/PCM if it does not have the latest software, or substitute a known-good ECM/PCM, then recheck. If the system works properly, and the ECM/PCM was substituted, replace the original ECM/PCM.
11. Disconnect the HDS and the timing light.
12. Install the engine compartment covers.

ADJUSTMENT

Ignition timing is control by the Engine Control Module (ECM)/Power Control Module (PCM). No adjustment is necessary.

SPARK PLUGS

REMOVAL & INSTALLATION

4-Cylinder Engine

Scc Figure 90.

1. Remove the ignition coil cover.
2. Disconnect the ignition coil connectors, then remove the ignition coils.
3. Remove the spark plug and inspect them.
4. Apply a small amount of anti-seize compound to the plugs into the cylinder head, finger tight. Torque them to 13 ft. lbs. (18 Nm).
5. Install the ignition coils, then connect the ignition coil connectors.
6. Install the ignition coil cover.

6-Cylinder Engine

See Figures 91 through 92.

1. Remove the engine cover.
2. Disconnect the ignition coil connectors, then remove the ignition coils.
3. Remove the spark plugs and inspect them.
4. Apply a small amount of anti-seize compound to the plug threads, and screw the plugs into the cylinder head, finger-tight. Torque them to 13 ft. lbs. (18 Nm).
5. Install the ignition coils in the reverse order of removal.

STARTER

REMOVAL & INSTALLATION

4-Cylinder Engine

See Figures 93 and 94.

1. Do the battery terminal disconnection procedure.
2. Remove the intake manifold.
3. Disconnect the positive starter cable from the B terminal, and the S terminal connector.
4. Remove the harness clamp.
5. Remove the two bolts securing the starter, then remove the starter.

To install:

6. Install the starter, then tighten the two bolts. Tighten the 12 x 1.25 mm bolt to 47 ft. lbs. (64 Nm). Tighten the 10 x 1.25 mm bolt to 33 ft. lbs. (44 Nm).

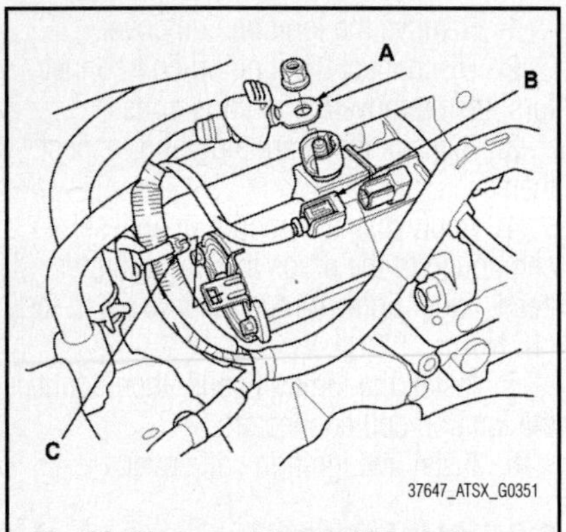
A
B
C
37647_ATSX_G0351
Fig. 93 Disconnect the positive starter cable (A) from the B terminal, and the S terminal connector (B)

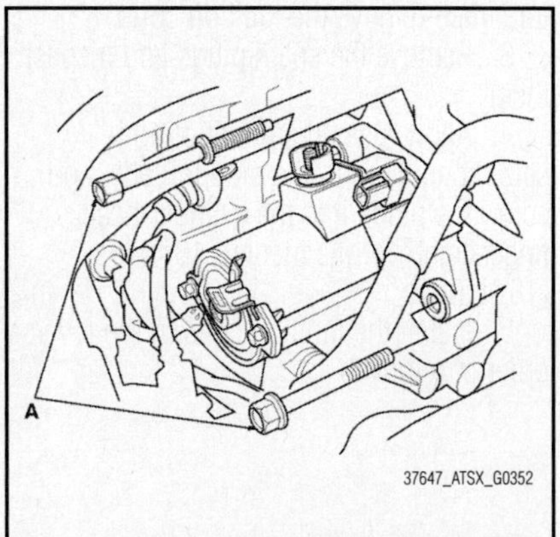
A
37647_ATSX_G0352
Fig. 94 Remove the two bolts (A) securing the starter, then remove the starter

7. Install the harness clamp.
8. Connect the positive starter cable to the B terminal, and the S terminal connector. Make sure the crimped side of the ring terminal faces away from the starter when you connect it.
9. Install the intake manifold.
10. Do the battery terminal reconnection procedure.
11. Start the engine to make sure the starter works properly.

6-Cylinder Engine

See Figures 95 and 96.

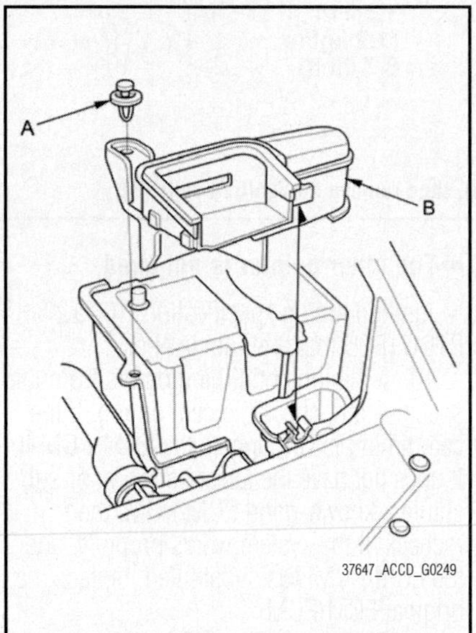
A
B
37647_ACCD_G0249
Fig. 95 Remove the intake air separator

1. Remove the intake air separator.
2. Do the battery removal procedure.
3. Remove the battery base.
4. Remove the harness clamp.
5. Disconnect the positive starter cable and the S terminal connector.
6. Remove the upper radiator hose bracket and the dipstick, then disconnect the vacuum hose.
7. Remove the two bolts holding the starter, then remove the starter.

To install:

8. Install the starter, then tighten the mounting bolts to 33 ft. lbs. (44 Nm).

➡Always use a new gasket.

9. Connect the vacuum hose, then install the upper radiator hose bracket and the dipstick.
10. Connect the positive starter cable and the S terminal connector. Make sure the crimped side of the ring terminal faces away from the starter when you connect it.
11. Install the harness clamp.
12. Install the battery base.
13. Do the battery installation procedure.
14. Start the engine to make sure the starter works properly.
15. Install the intake air separator.
16. Install the engine compartment covers.

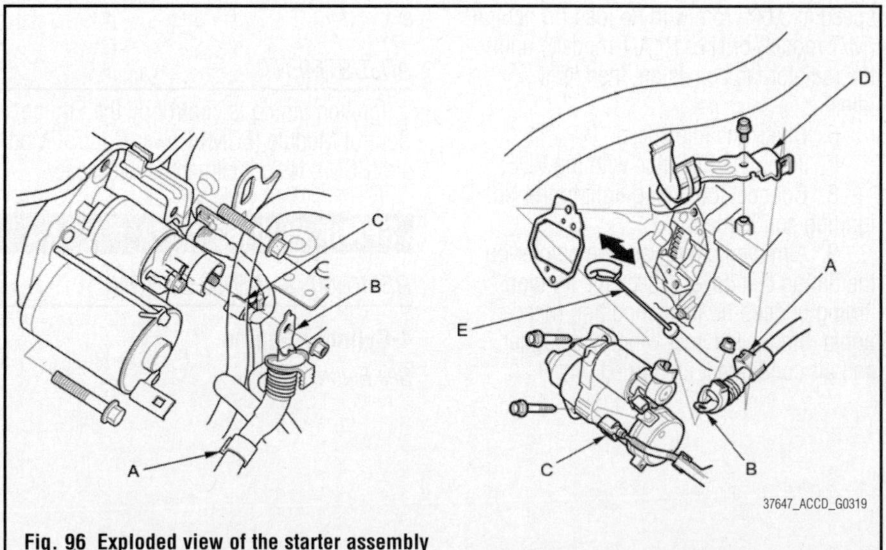
A
B
C
D
E
37647_ACCD_G0319
Fig. 96 Exploded view of the starter assembly

ENGINE MECHANICAL

ACCESSORY DRIVE BELTS

ACCESSORY BELT ROUTING

Refer to the graphic in the Removal and Installation procedure for the proper routing diagram.

INSPECTION

1. Inspect the belt for cracks or damage. If the belt is cracked or damaged, replace it.
2. Check that the position of the auto-tensioner indicator is within the standard range. If it is out of the standard range, replace the drive belt.

ADJUSTMENT

No adjustment is necessary or possible. Proper tension is provided by the auto-tensioner.

REMOVAL & INSTALLATION

4-Cylinder Engine

See Figure 97.

Special Tools Required: Belt tension release tool Snap-on YA9317 or equivalent, commercially available

1. Move the auto-tensioner with the belt tension release tool in the direction shown to relieve tension from the drive belt, then remove the drive belt.
2. Install the new belt in the reverse order of removal.

6-Cylinder Engine

See Figure 98.

1. Set a socket wrench on the drive belt auto-tensioner, and slowly turn the wrench in the direction of the rotation arrow, then remove the drive belt.

➡This is a hydraulic type auto-tensioner, so you must turn the wrench slowly for at least 3 seconds.

2. Install the new belt in the reverse order of removal.

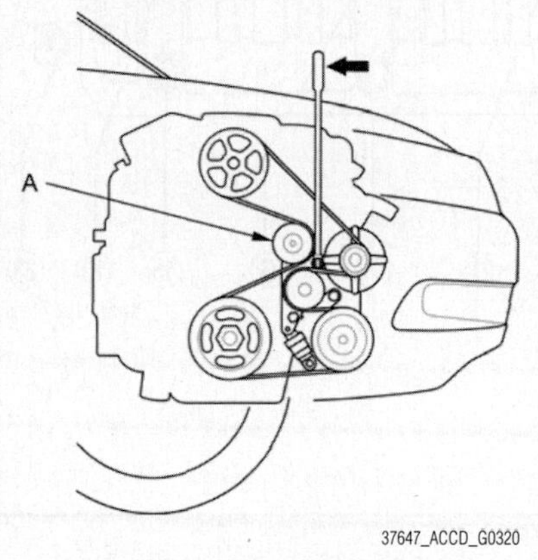
A
37647_ACCD_G0320

Fig. 98 Set a socket wrench on the drive belt auto-tensioner (A), and slowly turn the wrench in the direction of the rotation arrow, then remove the drive belt

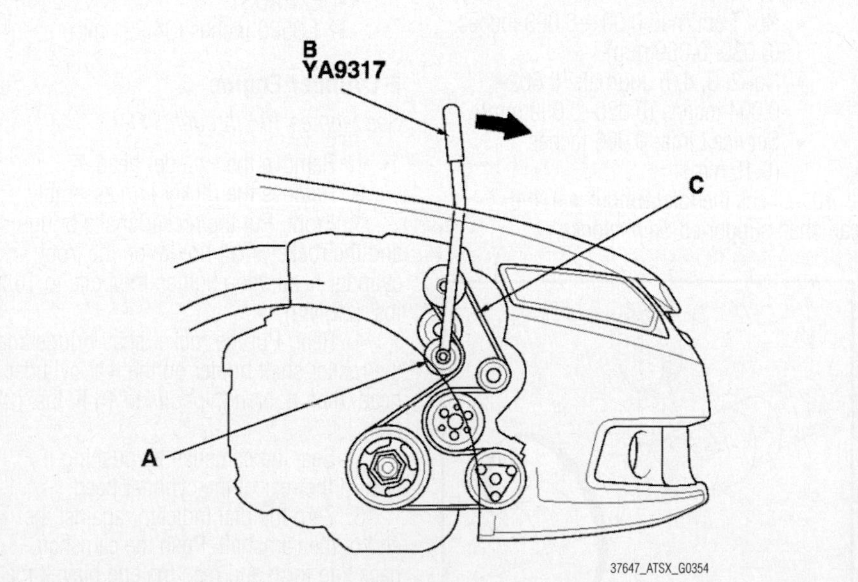
B
YA9317
C
A
37647_ATSX_G0354

Fig 97 Move the auto-tensioner (A) with the belt tension release tool (B) to relieve tension from the drive belt (C), then remove the drive belt

CAMSHAFT & VALVE LIFTERS

INSPECTION

4-Cylinder Engine

See Figures 99 through 103.

➡Do not rotate the camshaft during inspection.

1. Remove the rocker arm assembly.
2. Put the rocker shaft holders, the camshaft, and the camshaft holders on the cylinder head, then tighten the bolts, in sequence, to 16 ft. lbs. (22 Nm) except for bolts 21, 22, and 23. Tighten these three bolts to 9 ft. lbs. (12 Nm).

➡If the engine does not have bolt 21, skip it and continue the torque sequence.

3. Seat the camshaft by pushing it away from the camshaft pulley end of the cylinder head.
4. Zero the dial indicator against the end of the camshaft, then push the camshaft back and forth, and read the end play. If the end play is beyond the service limit, replace the cylinder head and recheck. If it is still beyond the service limit, replace the camshaft.
   a. Camshaft End Play: Standard (New): 0.002–0.008 inches (0.05–0.20 mm)
   b. Service Limit: 0.02 inches (0.4 mm)
5. Loosen the camshaft holder bolts two turns at a time, in a crisscross pattern. Then remove the camshaft holders from the cylinder head.
6. Lift the camshafts out of the cylinder head, wipe them clean, then inspect the lift ramps. Replace the camshaft if any lobes are pitted, scored, or excessively worn.
7. Clean the camshaft journal surfaces in the cylinder head, then set the camshafts back in place. Place a Plastigage strip across each journal.
8. Install the camshaft holders, then tighten the bolts to the specified torque as shown in step 2.
9. Remove the camshaft holders. Measure the widest portion of Plastigage on each journal.
   a. If the camshaft-to-holder clearance is within limits, go to step 11.
   b. If the camshaft-to-holder clearance is beyond the service limit, and the camshaft has been replaced, replace the cylinder head.

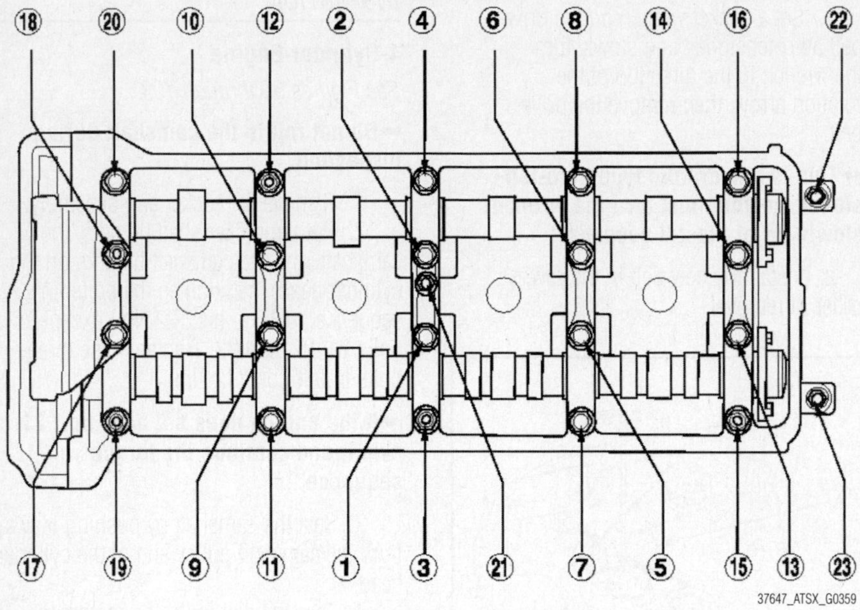
18
20
10
12
2
4
6
8
14
16
22
17
19
9
11
1
3
21
7
5
15
13
23
37647_ATSX_G0359

Fig. 99 Camshaft torque sequence

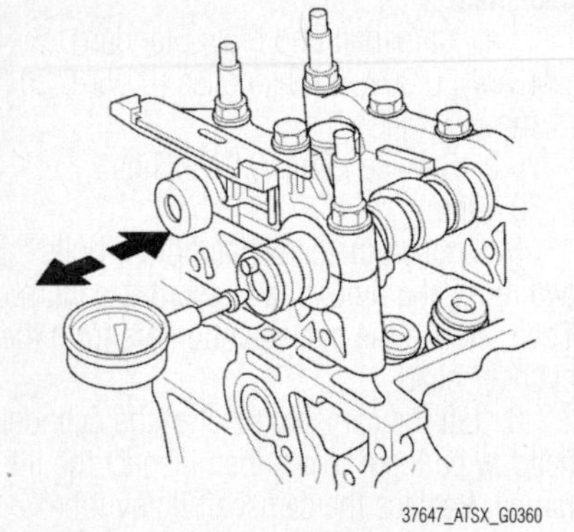
37647_ATSX_G0360

Fig. 100 Checking camshaft end play

c. If the camshaft-to-holder clearance is beyond the service limit, and the camshaft has not been replaced, go to step 10.

➡Camshaft-to-Holder Oil Clearance: Standard (New):

- No. 1 Journal: 0.001–0.003 inches (0.030–0.069 mm)
- No. 2, 3, 4, 5 Journals: 0.002–0.004 inches (0.060–0.099 mm)
- Service Limit: 0.006 inches (0.15 mm)

10. Check the total runout with the camshaft supported on V-blocks.

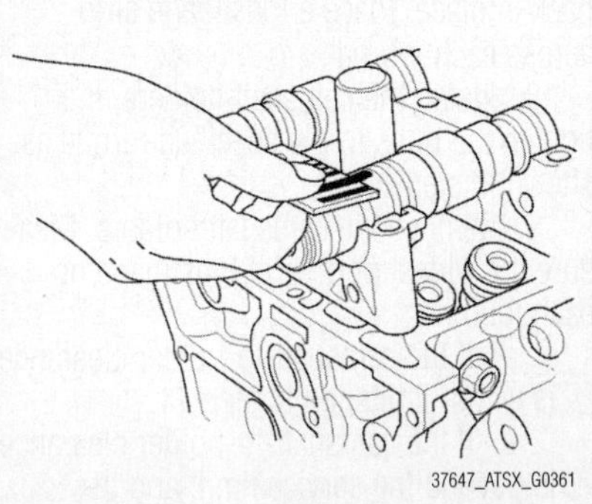
37647_ATSX_G0361

Fig. 101 Measure the widest portion of Plastigage on each journal

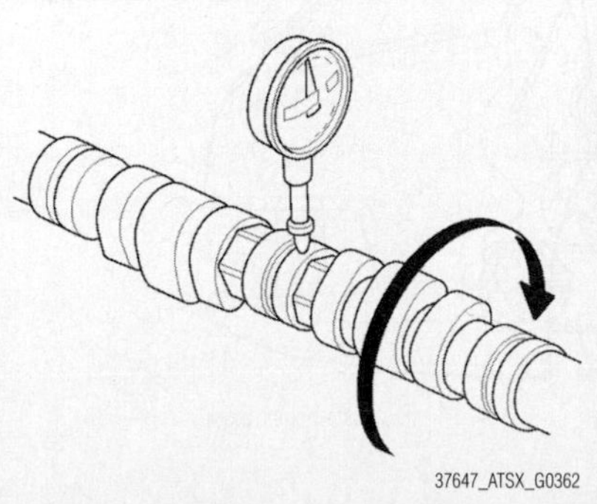
37647_ATSX_G0362

Fig. 102 Check the total runout with the camshaft supported on V-blocks

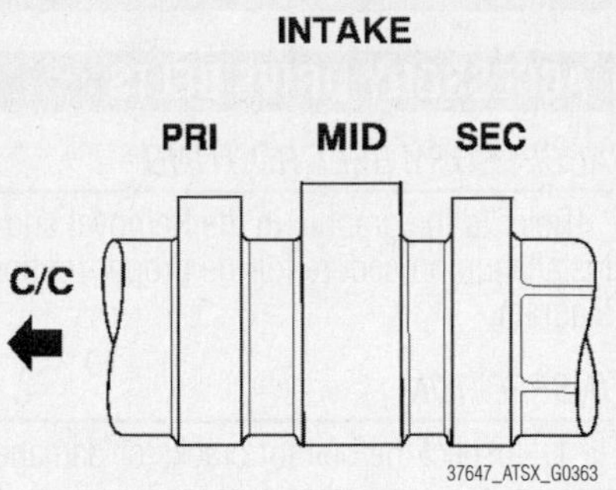
INTAKE
PRI
MID
SEC
C/C
37647_ATSX_G0363

Fig. 103 Measure cam lobe height

a. If the total runout of the camshaft is within the service limit, replace the cylinder head.

b. If the total runout is beyond the service limit, replace the camshaft and recheck the camshaft-to-holder oil clearance. If the oil clearance is still beyond the service limit, replace the cylinder head.

c. Camshaft Total Runout: Standard (New): 0.001 inches (0.03 mm) max.

d. Service Limit: 0.002 inches (0.04 mm)

11. Measure cam lobe height.

➡Cam Lobe Height Standard (New):

- INTAKE
- Primary: 1.3285 inches (33.744 mm)
- Mid: 1.3959 inches (35.456 mm)
- Secondary: 1.3285 inches (33.744 mm)
- EXHAUST
- 1.3500 inches (34.291 mm)

6-Cylinder Engine

See Figures 104 through 111.

1. Remove the cylinder head.
2. Remove the rocker arm assembly.
3. Front: Put the rocker shafts bridge and the rocker shaft holder on the front cylinder head, then tighten the bolts to 16 ft. lbs. (22 Nm).
4. Rear: Put the rocker shaft bridge and the rocker shaft holder on the rear cylinder head, then tighten the bolts to 16 ft. lbs. (22 Nm).
5. Seat the camshaft by pushing it toward the rear of the cylinder head.
6. Zero the dial indicator against the end of the camshaft. Push the camshaft back and forth and read the end play. If the end play is beyond the service limit, replace the thrust cover and recheck. If it is still beyond the service limit, replace the camshaft.

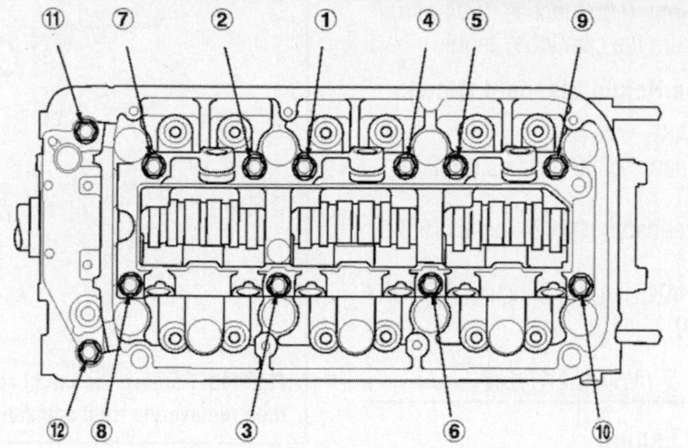

⑪ ⑦ ② ① ④ ⑤ ⑨
⑫ ⑧ ③ ⑥ ⑩
37647_ATSX_G0364

Fig. 104 Front rocker shaft bridge tightening sequence

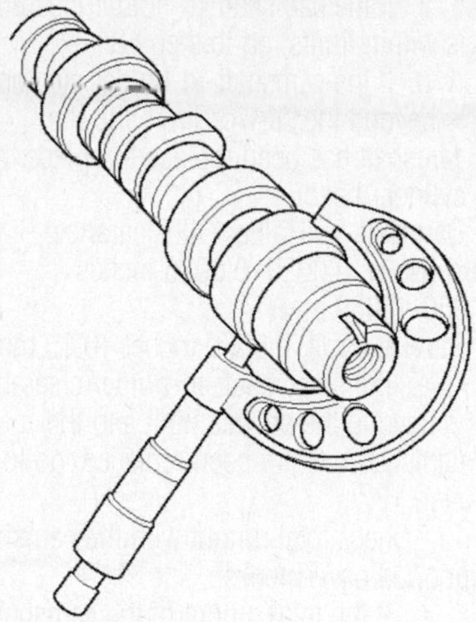

37647_ATSX_G0367

Fig. 107 Measure the diameter of each camshaft journal

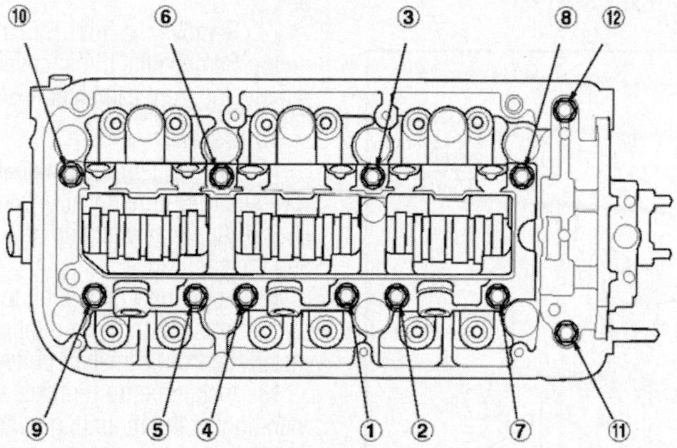

⑩ ⑥ ③ ⑧ ⑫
⑨ ⑤ ④ ① ② ⑦ ⑪
37647_ATSX_G0365

Fig. 105 Rear rocker shaft bridge tightening sequence

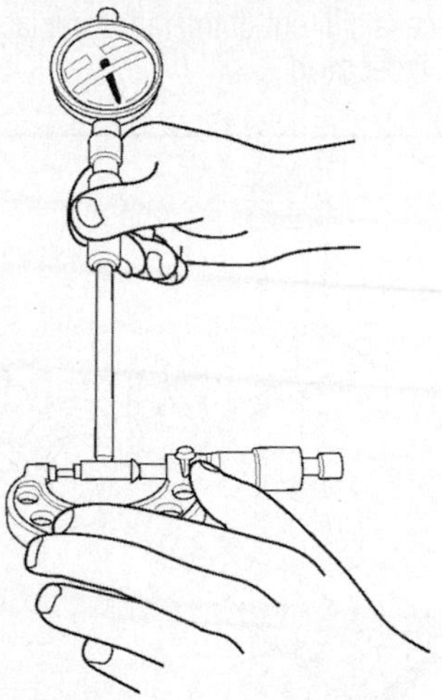

37647_ATSX_G0368

Fig. 108 Zero the gauge to the journal diameter

a. Camshaft End Play: Standard (New): 0.002–0.008 inches (0.05-0.20 mm)

b. Service Limit: 0.008 inches (0.20 mm)

7. Remove the camshaft thrust cover, then pull out the camshaft.

8. Wipe the camshaft clean, then inspect the lift ramps. Replace the camshaft if any lobes are pitted, scored, or excessively worn.

9. Measure the diameter of each camshaft journal.

10. Zero the gauge to the journal diameter.

11. Clean the camshaft bearing surfaces in the cylinder head. Measure the inside diameter of each camshaft bearing surface, and check for an out-of-round condition.

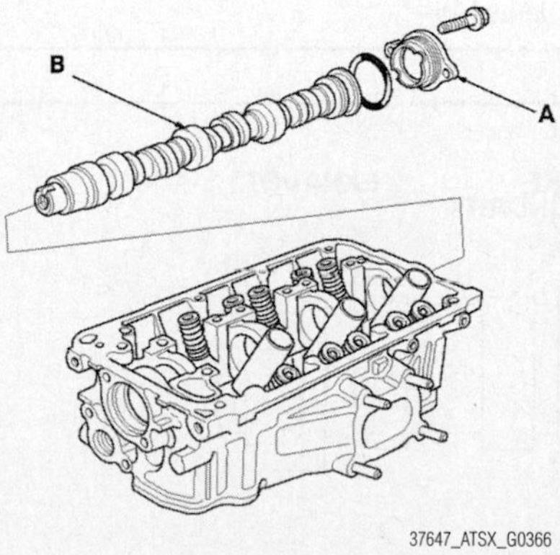

B
A
37647_ATSX_G0366

Fig. 106 Remove the camshaft thrust cover (A), then pull out the camshaft (B)

37647_ATSX_G0369

Fig. 109 Measure the inside diameter of each camshaft bearing surface, and check for an out-of-round condition

a. If the camshaft-to-holder clearance is within limits, go to step 13.

b. If the camshaft-to-holder clearance is beyond the service limit and the camshaft has been replaced, replace the cylinder head.

Camshaft-to-Holder Oil Clearance: Standard (New): 0.0020–0.0035 inches (0.050–0.089 mm)

Service Limit: 0.0060 inches (0.15 mm)

a. If the camshaft-to-holder clearance is beyond the service limit and the camshaft has not been replaced, go to step 12.

12. Check total runout with the camshaft supported on V-blocks.

a. If the total runout of the camshaft is within the service limit, replace the cylinder head.

b. If the total runout is beyond the service limit, replace the camshaft and recheck the oil clearance. If the oil clearance is still out of tolerance, replace the cylinder head.

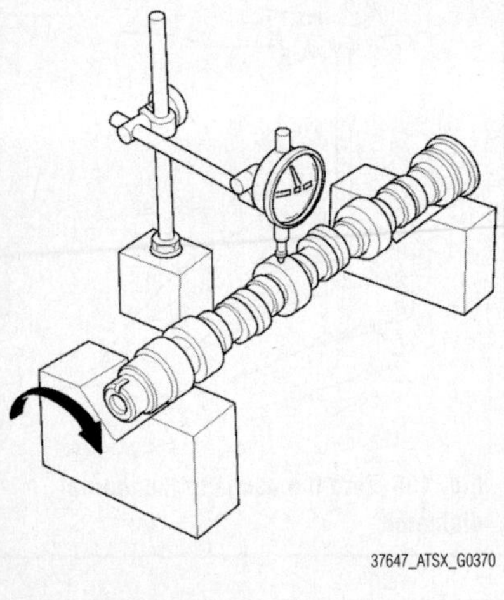
37647_ATSX_G0370

Fig. 110 Check total runout with the camshaft supported on V-blocks

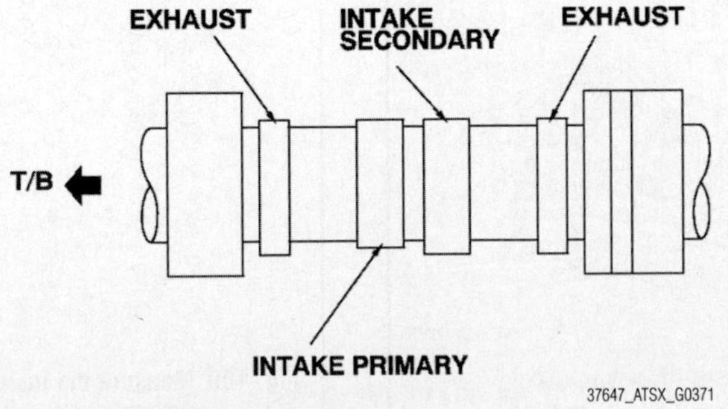
EXHAUST
INTAKE SECONDARY
EXHAUST
T/B
INTAKE PRIMARY
37647_ATSX_G0371

Fig. 111 Measure the cam lobe height

Camshaft Total Runout: Standard (New): 0.0012 inches (0.03 mm) max.

Service Limit: 0.002 inches (0.04 mm)

13. Measure the cam lobe height.

➡Cam Lobe Height Standard (New):

- INTAKE
- Primary: 1.3504 inches (34.299 mm)
- Secondary: 31.4024 inches (5.621 mm)
- EXHAUST: 1.4472 inches (36.760 mm)

REMOVAL & INSTALLATION

6-Cylinder Engine

Front

See Figures 112 and 113.

1. Remove the engine compartment covers.
2. Do the battery removal procedure.
3. Drain the engine coolant.

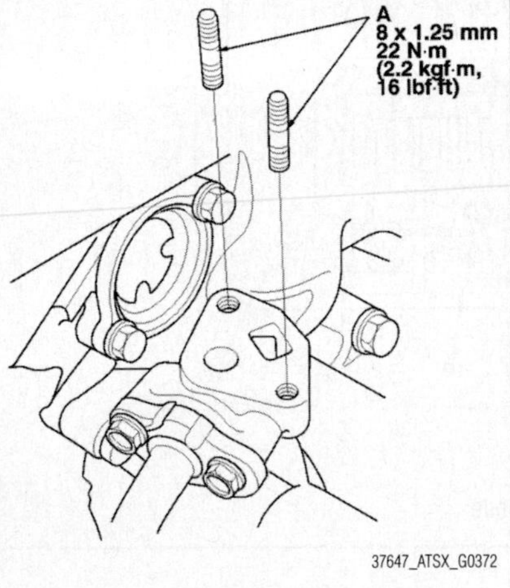
A
8 x 1.25 mm
22 N·m
(2.2 kgf·m,
16 lbf·ft)
37647_ATSX_G0372

Fig. 112 Remove the EGR valve stud bolts (A)

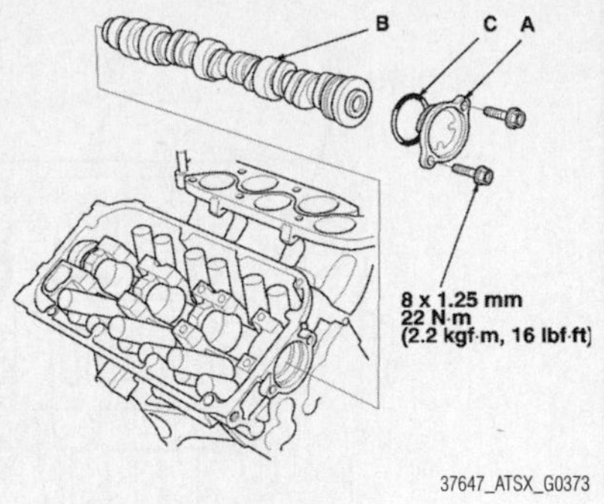
B
C
A
8 x 1.25 mm
22 N·m
(2.2 kgf·m, 16 lbf·ft)
37647_ATSX_G0373

Fig. 113 Remove the thrust cover (A), then remove the front camshaft (B)

4. Disconnect the radiator hoses.
5. Remove the exhaust gas recirculation (EGR) valve.
6. Remove the EGR valve stud bolts.
7. Remove the timing belt.
8. Remove the front rocker arm assembly.
9. Remove the front camshaft pulley.
10. Remove the thrust cover, then remove the front camshaft.

To install:

11. Install the front camshaft in the reverse order of removal. Always use a new O-ring. Apply new engine oil to the journals and the cam lobes.
12. Apply new engine oil to the threads of the camshaft pulley mounting bolt, then install the front camshaft pulley.
13. Install the front rocker arm assembly, then tighten the mounting bolts.
14. Install the timing belt.
15. Adjust the valve clearance.
16. Install the EGR valve stud bolts, then install the EGR valve.
17. Connect the radiator hoses.
18. Do the battery installation procedure.
19. Fill the radiator with engine coolant, and bleed the air from the cooling system with the heater valve open.
20. Install the engine compartment covers.
21. Do the Crankshaft Position (CKP) pattern clear/CKP pattern learn procedure.

Rear

See Figures 114 through 116.

1. Remove the engine compartment covers.
2. Relieve the fuel pressure.
3. Do the battery removal procedure.
4. Drain the engine coolant.
5. Remove the under-hood fuse/relay box from the bracket.
6. Remove the air cleaner assembly.

7. Remove the quick-connect fitting cover, then disconnect the fuel feed hose.
8. Disconnect the heater hoses.
9. Disconnect the Evaporative Emission (EVAP) canister hose, then remove the EVAP canister joint.
10. Remove the upper transmission mount.
11. Remove the timing belt.
12. Remove the rear rocker arm assembly.
13. Remove the rear camshaft pulley.
14. Remove the thrust cover, then remove the rear camshaft.

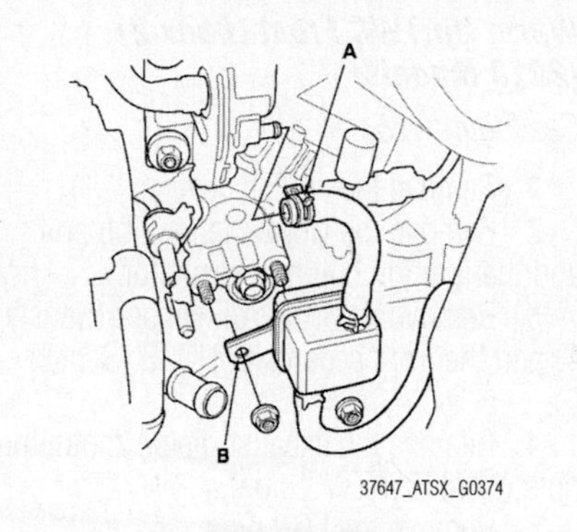
A
B
37647_ATSX_G0374
Fig. 114 Disconnect the EVAP canister hose (A), then remove the EVAP canister joint (B)

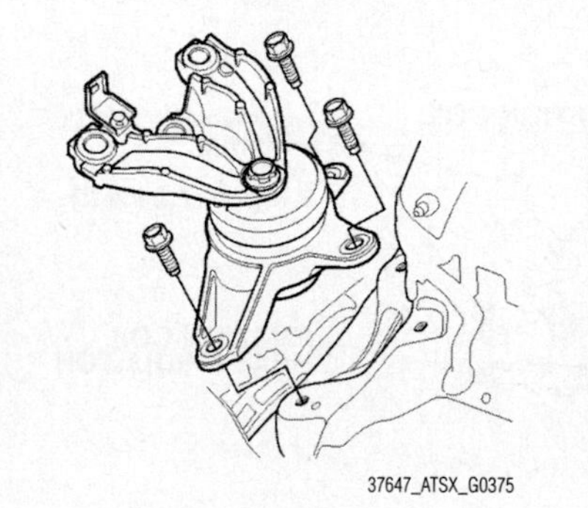
37647_ATSX_G0375
Fig. 115 Remove the upper transmission mount

To install:

15. Install the rear camshaft in the reverse order of removal. Always use a new O-ring. Apply new engine oil to the journals and cam lobes.
16. Apply new engine oil to the threads of the camshaft pulley mounting bolt, then install the rear camshaft pulley.
17. Install the rear rocker arm assembly, then tighten the mounting bolts.
18. Install the timing belt.
19. Install the upper transmission mount.
20. Adjust the valve clearance.
21. Install the EVAP canister joint, then connect the EVAP canister hose.
22. Connect the heater hoses.
23. Connect the fuel feed hose, then install the quick-connect fitting cover.
24. Install the air cleaner assembly.
25. Install the under-hood fuse/relay box to the bracket.
26. Do the battery installation procedure.
27. Inspect for fuel leaks. Turn the ignition switch to ON (II) (do not operate the starter) so the fuel pump runs for about 2 seconds and pressurizes the fuel line. Repeat this operation three times, then check for fuel leakage at any point in the fuel line.
28. Fill the radiator with engine coolant, and bleed the air from the cooling system with the heater valve open.
29. Install the engine compartment covers.
30. Do the Crankshaft Position (CKP) pattern clear/CKP pattern learn procedure.

CATALYTIC CONVERTER

REMOVAL & INSTALLATION

4-Cylinder Engine

Warm Up TWC

See Figures 117 through 122.

1. Raise the vehicle on a lift.
2. Remove the secondary HO2S (Sensor 2).
3. Remove the bolts (A).
4. Remove the WU-TWC bracket.
5. Lower the vehicle.
6. Remove the strut brace.
7. Remove the A/F sensor (Sensor 1).

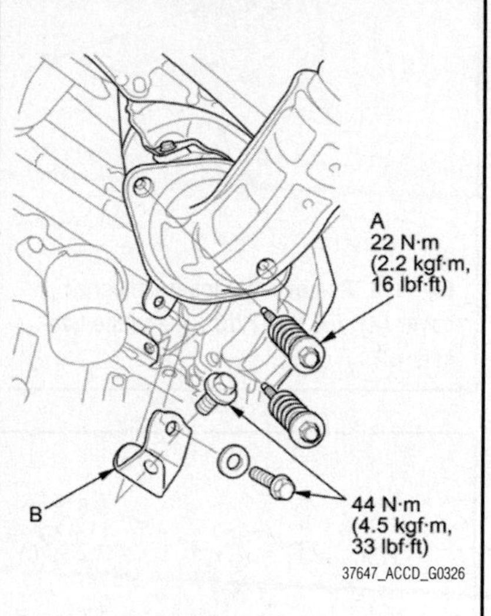
A
22 N·m
(2.2 kgf·m,
16 lbf·ft)
B
44 N·m
(4.5 kgf·m,
33 lbf·ft)
37647_ACCD_G0326
Fig. 118 Remove the bolts (A) and the WU-TWC bracket (B)

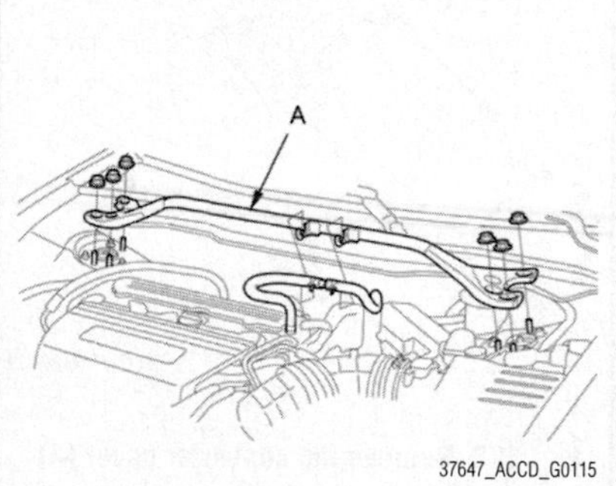
A
37647_ACCD_G0115
Fig. 119 Remove the strut brace

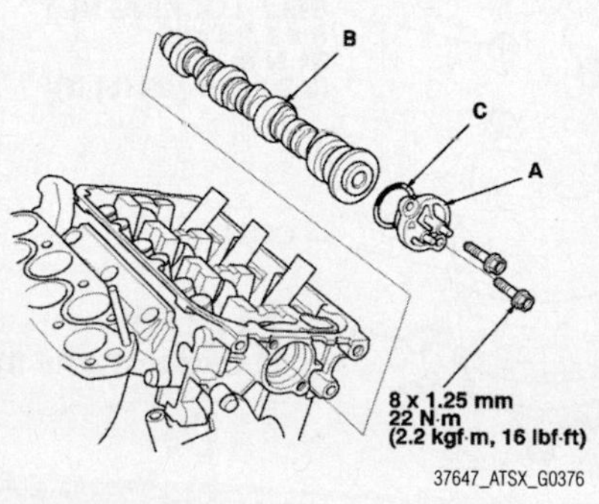
B
C
A
8 x 1.25 mm
22 N·m
(2.2 kgf·m, 16 lbf·ft)
37647_ATSX_G0376
Fig. 116 Remove the thrust cover (A), then remove the rear camshaft (B)

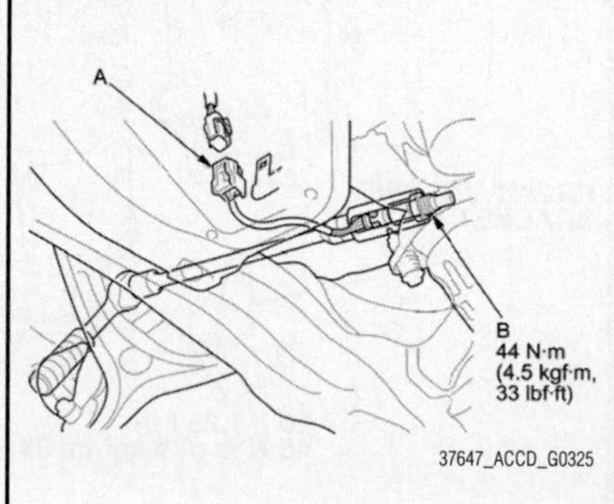
A
B
44 N·m
(4.5 kgf·m,
33 lbf·ft)
37647_ACCD_G0325
Fig. 117 Remove the secondary HO2S (Sensor 2)

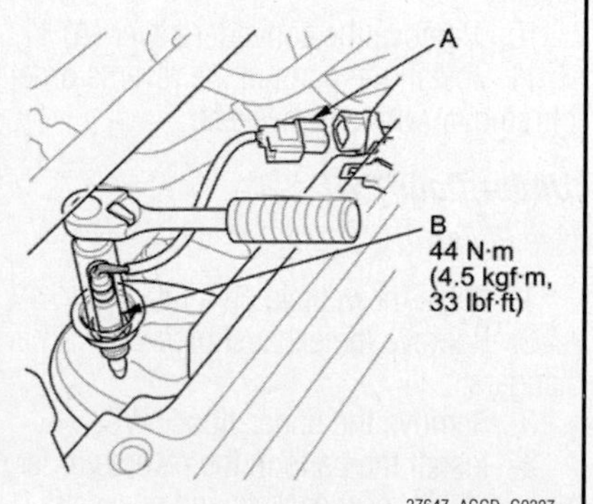
A
B
44 N·m
(4.5 kgf·m,
33 lbf·ft)
37647_ACCD_G0327
Fig. 120 Remove the A/F sensor (Sensor 1)

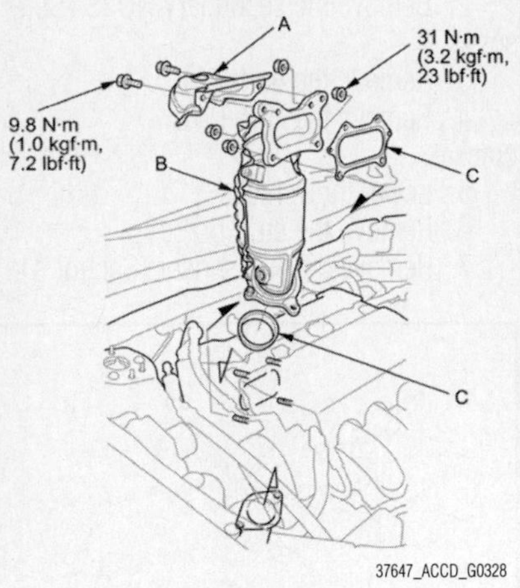
A
31 N·m
(3.2 kgf·m,
23 lbf·ft)
9.8 N·m
(1.0 kgf·m,
7.2 lbf·ft)
B
C
C
37647_ACCD_G0328

Fig. 121 Remove the upper converter cover (A), the WU-TWC (B) and the gaskets (C)

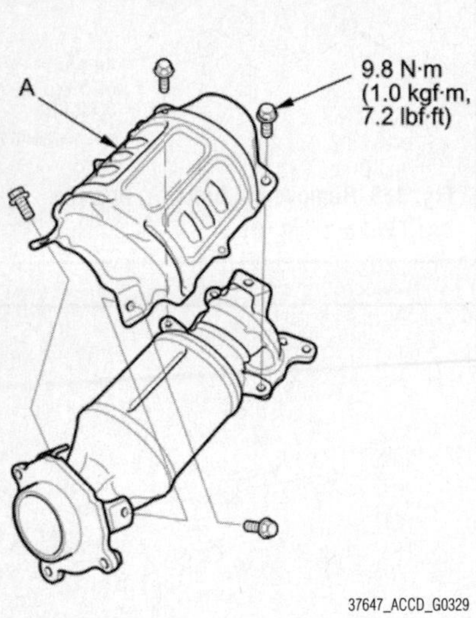
A
9.8 N·m
(1.0 kgf·m,
7.2 lbf·ft)
37647_ACCD_G0329

Fig. 122 Remove the converter cover (A)

8. Remove the upper converter cover.
9. Remove the WU-TWC and the gaskets.
10. Remove the converter cover (A).
11. Install the parts in the reverse order of removal with new gaskets.

Under-floor TWC

See Figure 123.

1. Raise the vehicle on a lift.
2. Remove the exhaust pipe hangers.
3. Remove the under-floor TWC.
4. Install the parts in the reverse order of removal with new gaskets and new self-locking nuts.

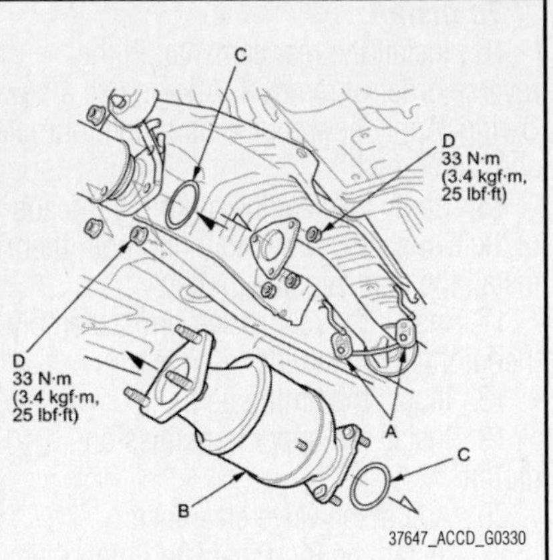
C
D
33 N·m
(3.4 kgf·m,
25 lbf·ft)
D
33 N·m
(3.4 kgf·m,
25 lbf·ft)
A
C
B
37647_ACCD_G0330

Fig. 123 Exploded view of the under-floor TWC assembly

6-Cylinder Engine

Warm Up TWC Front (Bank 2) ('08-09 Models)

See Figure 124.

1. Remove the engine cover.
2. Remove the No. 2 the ignition coil and the ignition coil heat insulator.
3. Remove the front A/F sensor (Sensor 1) and the front secondary HO2S (Sensor 2).
4. Remove the exhaust pipe A mounting nuts (front WU-TWC side).
5. Remove the EGR pipe.
6. Remove the A/C condenser fan assembly and the radiator upper bracket/cushion.
7. Remove the front WU-TWC bracket, then carefully remove the front WU-TWC.
8. Carefully install the front WU-TWC with a new gasket and new self-locking nuts. Tighten the nuts in a crisscross pattern in two or three steps.
9. Install the parts in the reverse order of removal.

Warm Up TWC Front (Bank 2) (2010 Models)

See Figure 125.

1. Remove the engine cover.
2. Remove the No. 2 the ignition coil and the ignition coil heat insulator.
3. Remove the front A/F sensor (Sensor 1) and the front secondary HO2S (Sensor 2).
4. Remove the exhaust pipe A mounting nuts (front WU-TWC side).
5. Remove the EGR pipe.
6. Remove the A/C condenser fan assembly and the radiator upper bracket/cushion.

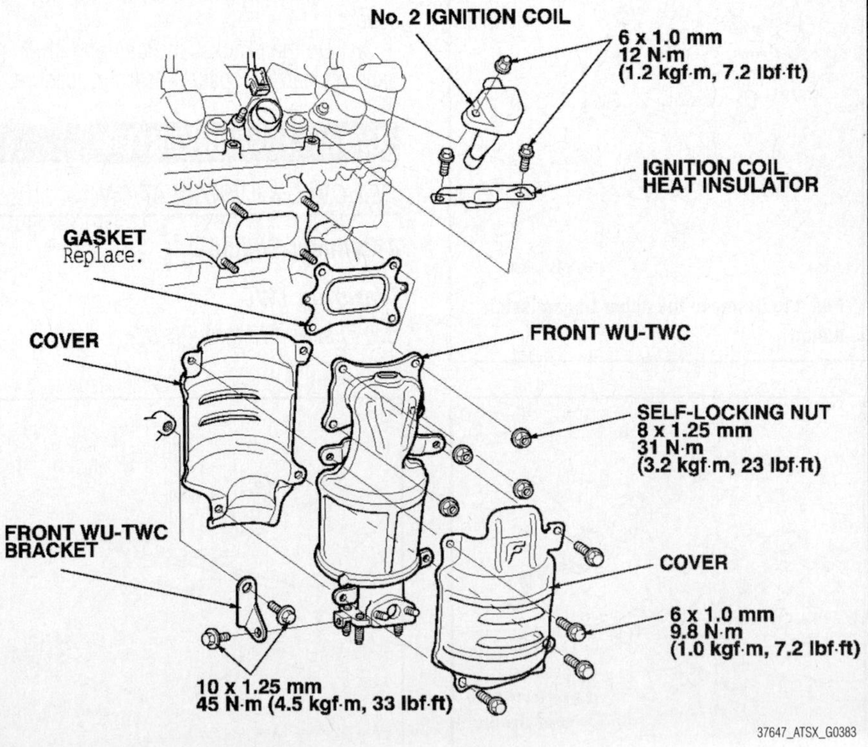
No. 2 IGNITION COIL
6 x 1.0 mm
12 N·m
(1.2 kgf·m, 7.2 lbf·ft)
IGNITION COIL
HEAT INSULATOR
GASKET
Replace.
COVER
FRONT WU-TWC
SELF-LOCKING NUT
8 x 1.25 mm
31 N·m
(3.2 kgf·m, 23 lbf·ft)
FRONT WU-TWC
BRACKET
COVER
6 x 1.0 mm
9.8 N·m
(1.0 kgf·m, 7.2 lbf·ft)
10 x 1.25 mm
45 N·m (4.5 kgf·m, 33 lbf·ft)
37647_ATSX_G0383

Fig. 124 Exploded view of Front WU-TWC assembly

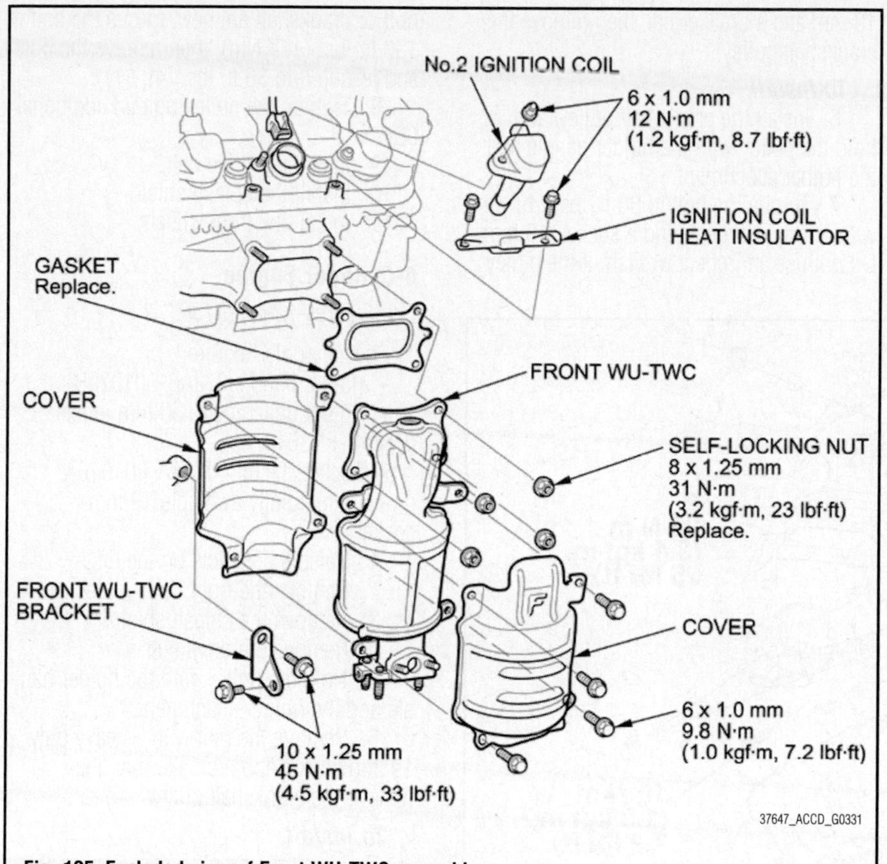
No.2 IGNITION COIL
6 x 1.0 mm
12 N·m
(1.2 kgf·m, 8.7 lbf·ft)
IGNITION COIL
HEAT INSULATOR
GASKET
Replace.
FRONT WU-TWC
COVER
SELF-LOCKING NUT
8 x 1.25 mm
31 N·m
(3.2 kgf·m, 23 lbf·ft)
Replace.
FRONT WU-TWC
BRACKET
COVER
6 x 1.0 mm
9.8 N·m
(1.0 kgf·m, 7.2 lbf·ft)
10 x 1.25 mm
45 N·m
(4.5 kgf·m, 33 lbf·ft)
37647_ACCD_G0331

Fig. 125 Exploded view of Front WU-TWC assembly

7. Carefully remove the front WU-TWC.
8. Carefully install the front WU-TWC with a new gasket and new self-locking nuts. Tighten the nuts in a crisscross pattern in two or three steps.
9. Install the parts in the reverse order of removal.

Warm Up TWC Rear (Bank 1)

See Figure 126.

1. Remove the strut brace.
2. Remove the P/S feed hose clamp.
3. Remove the cowl cover.
4. Remove the No. 4 ignition coil and the ignition coil heat insulator.
5. Remove the rear A/F sensor (Sensor 1) and the rear secondary HO2S (Sensor 2).
6. Remove the exhaust pipe A mounting nuts (rear WU-TWC side).
7. Carefully remove the rear WU-TWC.
8. Carefully install the rear WU-TWC with a new gasket and new self-locking nuts. Tighten the nuts in a crisscross pattern in two or three steps.
9. Install the parts in the reverse order of removal.

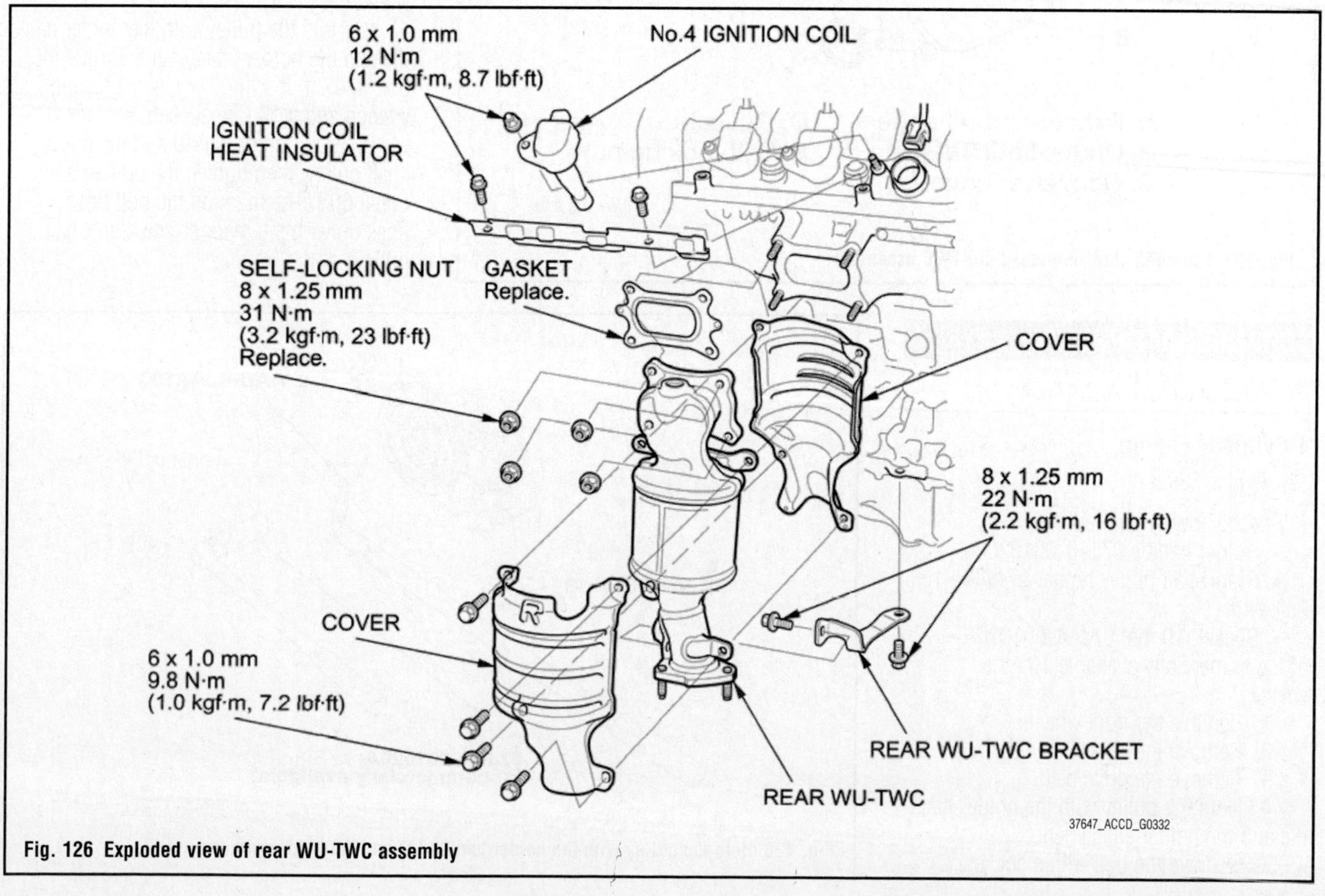
6 x 1.0 mm
12 N·m
(1.2 kgf·m, 8.7 lbf·ft)
No.4 IGNITION COIL
IGNITION COIL
HEAT INSULATOR
SELF-LOCKING NUT
8 x 1.25 mm
31 N·m
(3.2 kgf·m, 23 lbf·ft)
Replace.
GASKET
Replace.
COVER
8 x 1.25 mm
22 N·m
(2.2 kgf·m, 16 lbf·ft)
COVER
6 x 1.0 mm
9.8 N·m
(1.0 kgf·m, 7.2 lbf·ft)
REAR WU-TWC BRACKET
REAR WU-TWC
37647_ACCD_G0332

Fig. 126 Exploded view of rear WU-TWC assembly

Under-floor TWC

See Figure 127.

1. Raise the vehicle on a lift.
2. Remove the exhaust pipe hangers.
3. Remove the under-floor TWC.
4. Remove the converter cover.
5. Install the parts in the reverse order of removal with new gaskets and new self-locking nuts.

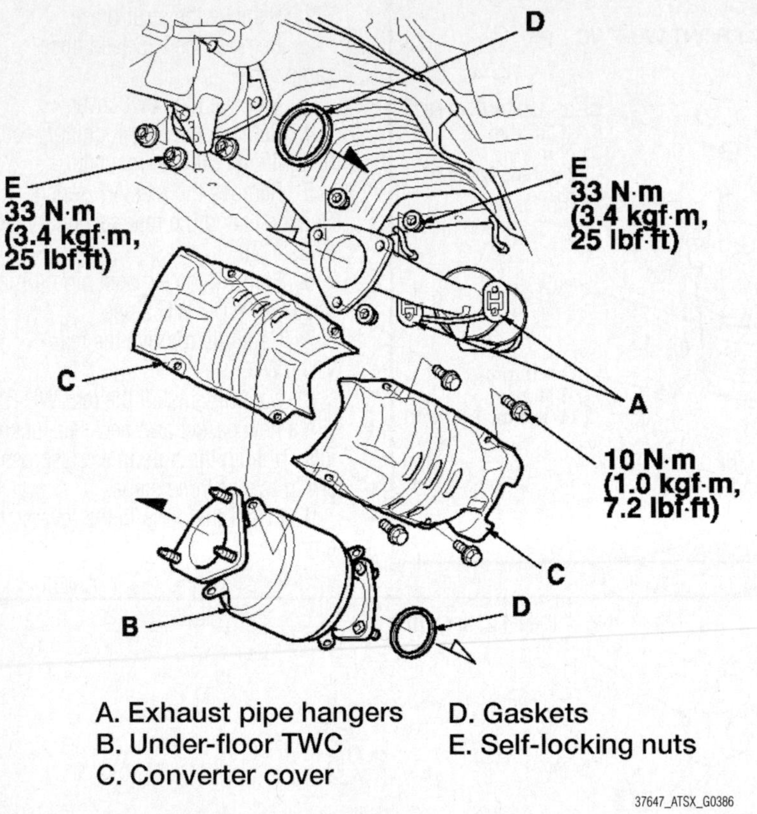
D
E
33 N·m
(3.4 kgf·m,
25 lbf·ft)
E
33 N·m
(3.4 kgf·m,
25 lbf·ft)
C
A
10 N·m
(1.0 kgf·m,
7.2 lbf·ft)
C
D
B
A. Exhaust pipe hangers
B. Under-floor TWC
C. Converter cover
D. Gaskets
E. Self-locking nuts
37647_ATSX_G0386

Fig. 127 Exploded view of under-floor TWC assembly

CRANKSHAFT PULLEY

REMOVAL & INSTALLATION

4-Cylinder Engine

See Figure 128.

Special Tools Required:

- Holder handle 07JAB-001020B
- Crankshaft pulley holder 07AAB-RJAA100
- Socket, 19 mm 07JAA-001020A or a commercially available 19 mm socket

1. Remove the front wheels.
2. Remove the splash shield.
3. Remove the drive belt.
4. Hold the pulley with the holder handle and the holder attachment.
5. Remove the bolt with a socket, 19 mm and a breaker bar, then remove the crankshaft pulley.

To install:

6. Install the crankshaft pulley, and hold the pulley with the holder handle and the holder attachment.
7. Torque the bolt to 36 ft. lbs. (49 Nm) with a torque wrench and a socket, 19 mm. Do not use an impact wrench. If the pulley bolt or crankshaft are new, torque the bolt to 130 ft. lbs. (177 Nm), then remove the bolt and torque it to 36 ft. lbs. (49 Nm).
8. Tighten the pulley bolt an additional 90°.
9. Install the drive belt.
10. Install the splash shield.
11. Install the front wheels.

6-Cylinder Engine

See Figures 129 and 130.

Special Tools Required:

- Holder handle 07JAB-001020B
- Holder attachment, 50 mm, offset 07MAB-PY3010A
- Socket, 19 mm 07JAA-001020A, or a commercially available 19 mm socket

1. Raise the vehicle on the lift.
2. Remove the right front wheel.
3. Remove the splash shield.
4. Remove the drive belt.
5. Hold the pulley with the holder handle and the holder attachment.
6. Remove the bolt with a heavy duty 19 mm socket and a breaker bar, then remove the crankshaft pulley.

To install:

7. Install the crankshaft pulley, and tighten the bolt. Do not use an impact wrench.
8. Hold the pulley with the holder handle and the holder attachment. Torque the bolt to 47 ft. lbs. (64 Nm) with a torque wrench and a 19 mm socket.
9. Mark the bolt head and the crankshaft pulley, then tighten the bolt an additional 60° (The mark on the bolt head lines up with the mark on the crankshaft pulley).

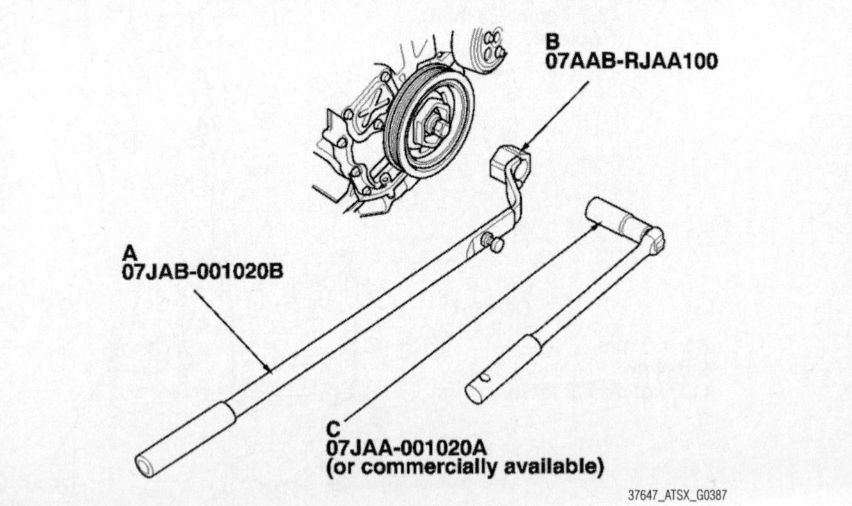
B
07AAB-RJAA100
A
07JAB-001020B
C
07JAA-001020A
(or commercially available)
37647_ATSX_G0387

Fig. 128 Hold the pulley with the holder handle (A) and the holder attachment (B)

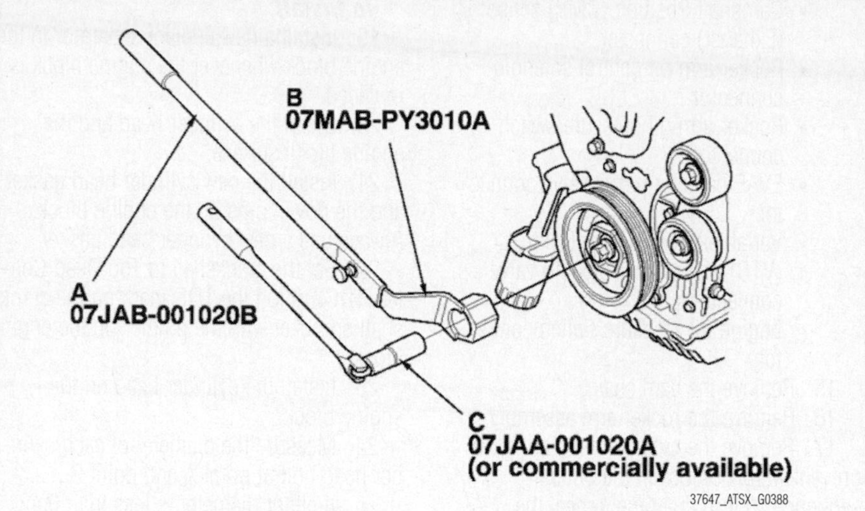
B
07MAB-PY3010A
A
07JAB-001020B
C
07JAA-001020A
(or commercially available)
37647_ATSX_G0388
Fig. 129 Hold the pulley with the holder handle (A) and the holder attachment (B)

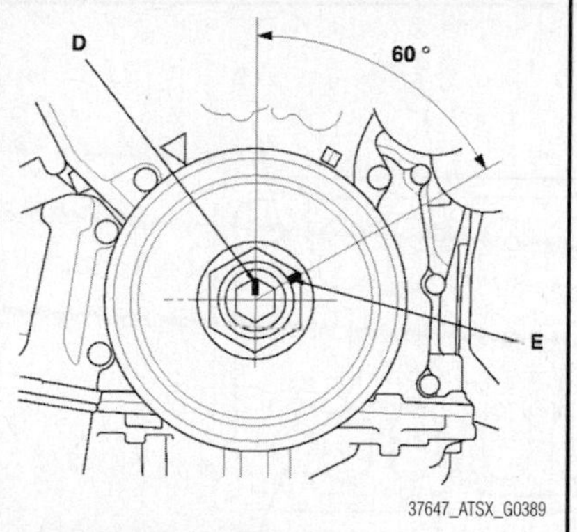
D
60 °
E
37647_ATSX_G0389
Fig. 130 Mark the bolt head (D) and the crankshaft pulley (E)

10. Install the drive belt.
11. Install the splash shield.
12. Install the right front wheel.

CRANKSHAFT FRONT SEAL

REMOVAL & INSTALLATION

6-Cylinder Engine (In Car)

Special Tools Required: Oil seal driver, 64 mm 070AD-RCA0100

1. Remove the timing belt, the timing belt stopper, and the timing belt drive pulley.
2. Remove the pulley end crankshaft oil seal.
3. Clean and dry the crankshaft oil seal housing.
4. Apply a light coat of new engine oil to the lip of the crankshaft oil seal.
5. Using the oil seal driver, 64 mm, drive in the new crankshaft oil seal until the oil seal driver bottoms against the oil pump. When the seal is in place, clean any excess grease off the crankshaft, and check that the oil seal lip is not distorted.
6. Install the timing belt drive pulley, the timing belt stopper, and the timing belt.

CYLINDER HEAD

REMOVAL & INSTALLATION

4-Cylinder Engine

See Figures 131 through 143.

➡Prior to starting the removal of the cylinder head, note the following:

- Use fender covers to avoid damaging painted surfaces.
- To avoid damage, unplug the wiring connectors carefully while holding the connector portion.
- Connect the Honda Diagnostic System (HDS) to the Data Link Connector (DLC), and monitor the Engine Coolant Temperature (ECT) sensor 1. To avoid damaging the cylinder head, wait until the ECT sensor 1 temperature drops below 100°F (38°C) before loosening the cylinder head bolts.
- Mark all wiring and hoses to avoid misconnection. Also, be sure that they do not contact other wiring or hoses, or interfere with other parts.

1. Remove the strut brace.
2. Relieve the fuel pressure.
3. Drain the engine coolant.
4. Remove the drive belt.
5. Remove the intake manifold.
6. Remove the catalytic converter.
7. Remove the Evaporative Emission (EVAP) canister hose.
8. Remove the quick-connect fitting cover, then disconnect the fuel feed hose.
9. Disconnect the four fuel injector connectors, the engine mount control solenoid connector, and remove the ground cables.
10. Remove the four bolts securing the EVAP canister purge valve bracket.
11. Disconnect the upper radiator hose, the heater hoses, and the water bypass hose.

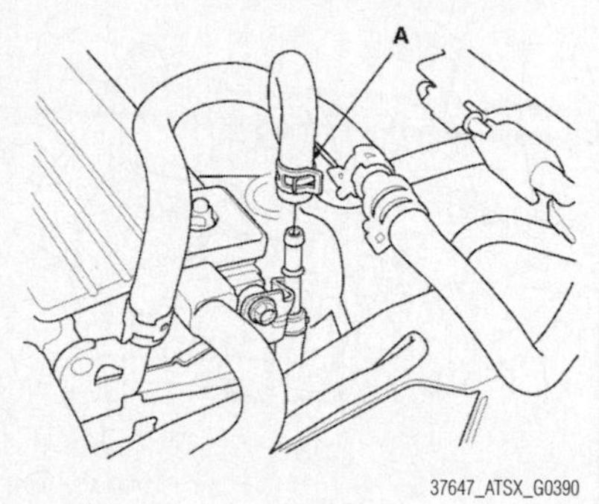
A
37647_ATSX_G0390
Fig. 131 Remove the EVAP canister hose (A)

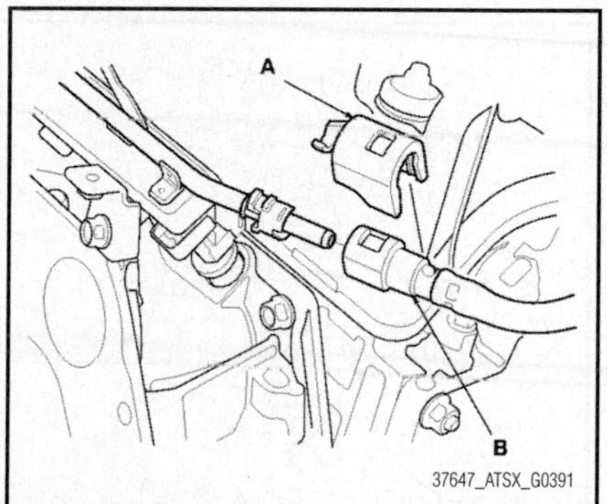
A
B
37647_ATSX_G0391
Fig. 132 Remove the quick-connect fitting cover (A), then disconnect the fuel feed hose (B)

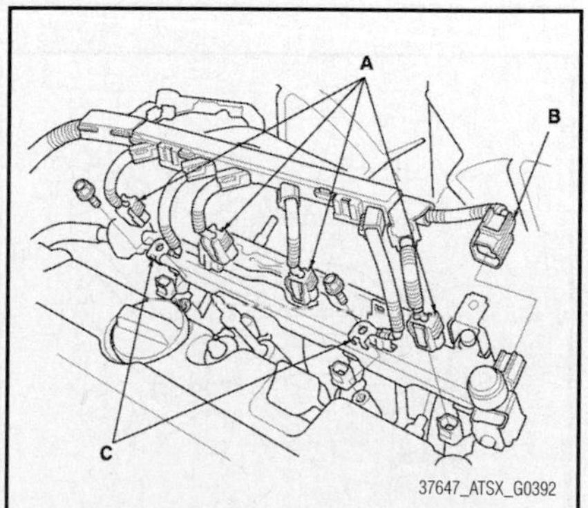
A
B
C
37647_ATSX_G0392
Fig. 133 Disconnect the four fuel injector connectors (A), the engine mount control solenoid connector (B), and remove the ground cables (C)

12. Remove the two bolts securing the connecting pipe.
13. Disconnect the water bypass hose.
14. Disconnect the following engine wire harness connectors, and remove the wire harness clamps from the cylinder head:

- Engine Coolant Temperature (ECT) sensor 1 connector
- Camshaft Position (CMP) sensor A (Intake) connector
- Camshaft Position (CMP) sensor B (Exhaust) connector
- Rocker arm oil control solenoid connector
- Rocker arm oil pressure switch connector
- EVAP canister purge valve connector
- Variable Valve Timing Control (VTC) oil control solenoid valve connector
- Engine oil pressure switch connector

15. Remove the cam chain.
16. Remove the rocker arm assembly.
17. Remove the cylinder head bolts. To prevent warpage, loosen the bolts in sequence ⅓ turn at a time; repeat the sequence until all bolts are loosened.
18. Remove the cylinder head.

To install:

19. Install a new coolant separator in the engine block whenever the engine block is replaced.
20. Clean the cylinder head and the engine block surface.
21. Install the new cylinder head gasket and the dowel pins on the engine block. Always use a new cylinder head gasket.
22. Set the crankshaft to Top Dead Center (TDC). Align the TDC mark on the crankshaft sprocket with the pointer on the engine block.
23. Install the cylinder head on the engine block.
24. Measure the diameter of each cylinder head bolt at point A and point B.
25. If either diameter is less than 0.42 inches (10.6 mm), replace the cylinder head bolt.

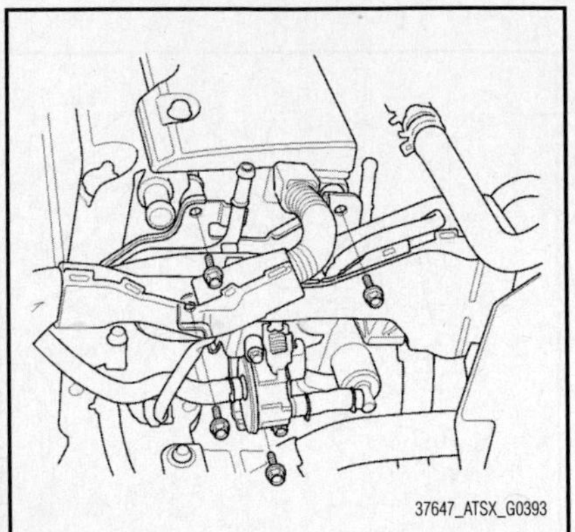
37647_ATSX_G0393

Fig. 134 Remove the four bolts securing the EVAP canister purge valve bracket

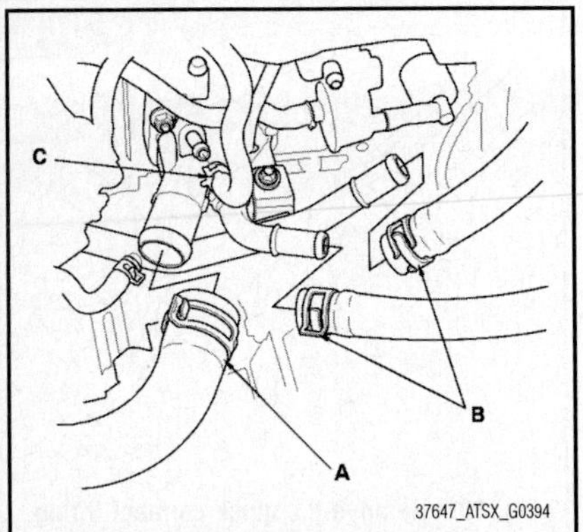
C
B
A
37647_ATSX_G0394

Fig. 135 Disconnect the upper radiator hose (A), the heater hoses (B), and the water bypass hose (C)

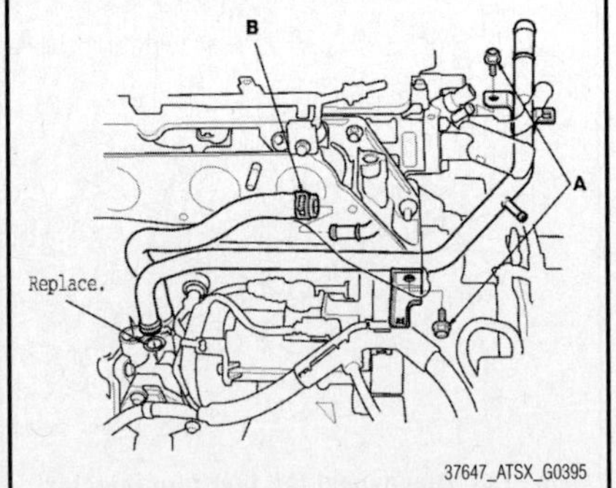
B
A
Replace.
37647_ATSX_G0395

Fig. 136 Remove the two bolts (A) securing the connecting pipe and disconnect the water bypass hose (B)

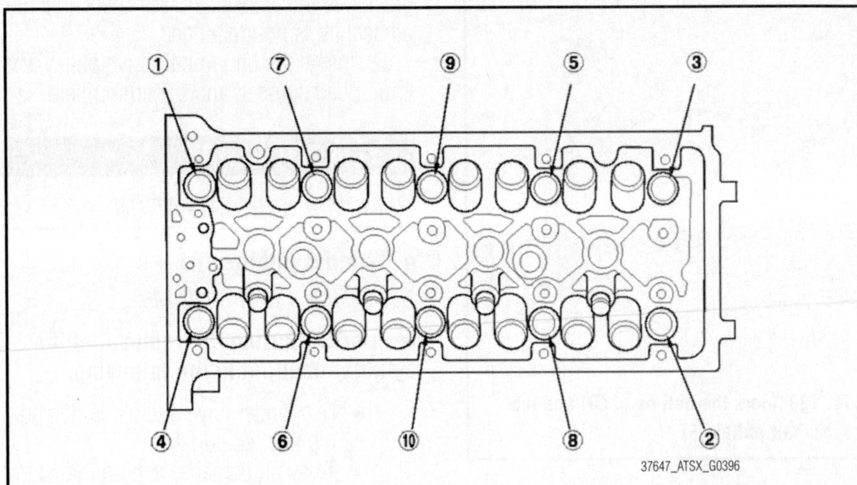
①
⑦
⑨
⑤
③
④
⑥
⑩
⑧
②
37647_ATSX_G0396

Fig. 137 Remove the cylinder head bolts in sequence

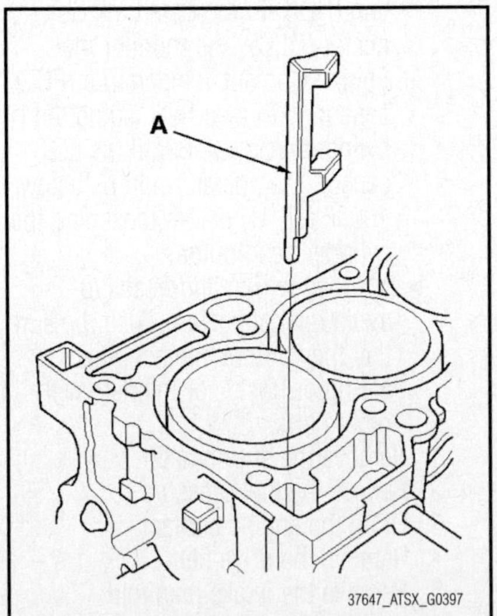
A
37647_ATSX_G0397

Fig. 138 Install a new coolant separator (A) in the engine block

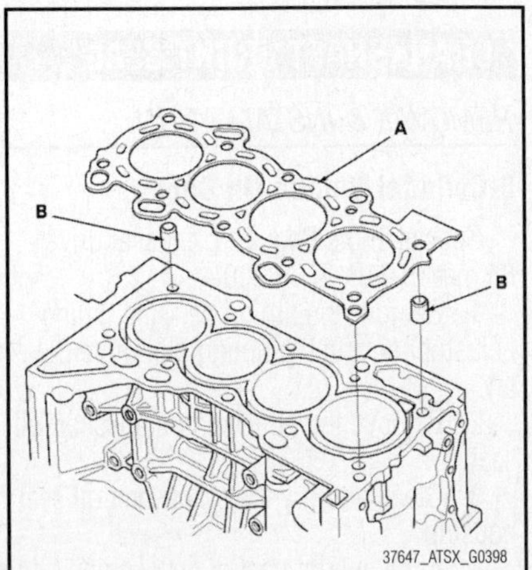
A
B
B
37647_ATSX_G0398

Fig. 139 Install the new cylinder head gasket (A) and the dowel pins (B) on the engine block

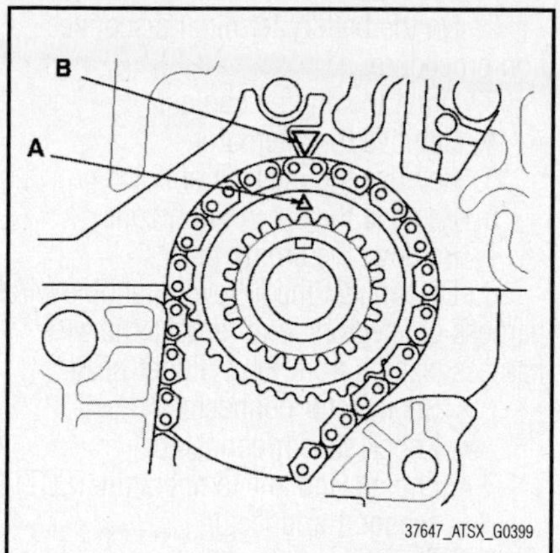
B
A
37647_ATSX_G0399
Fig. 140 Align the TDC mark (A) on the crankshaft sprocket with the pointer (B) on the engine block

26. Apply new engine oil to the threads and under the bolt heads of all cylinder head bolts.

27. Torque the cylinder head bolts in sequence to 29 ft. lbs. (39 Nm). Use a beam-type torque wrench. When using a preset click-type torque wrench, be sure to tighten slowly and do not overtighten. If a bolt makes any noise while you are torquing it, loosen the bolt and retighten it from the first step.

28. After torquing, tighten all cylinder head bolts in two steps (90° per step) using the sequence shown in step 9. If you are using a new cylinder head bolt, tighten the bolt an extra 90°.

➡Remove the cylinder head bolt if you tightened it beyond the specified angle, and go back to step 6 of the procedure. Do not loosen it back to the specified angle.

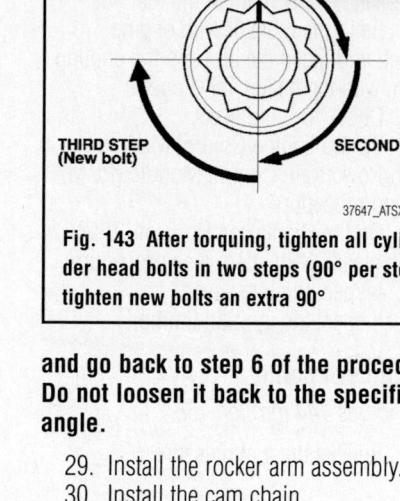

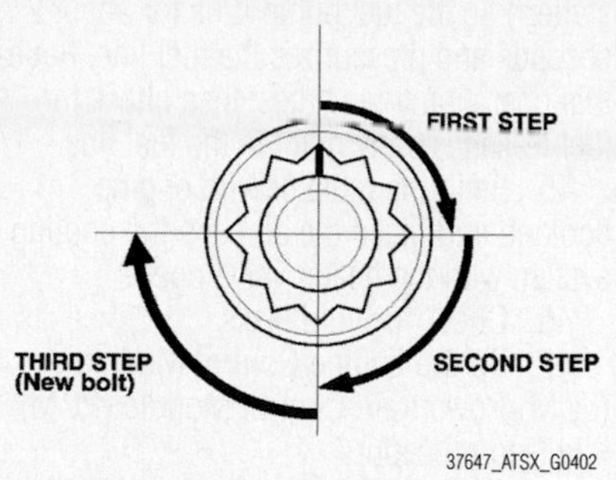
FIRST STEP
THIRD STEP
(New bolt)
SECOND STEP
37647_ATSX_G0402
Fig. 143 After torquing, tighten all cylinder head bolts in two steps (90° per step); tighten new bolts an extra 90°

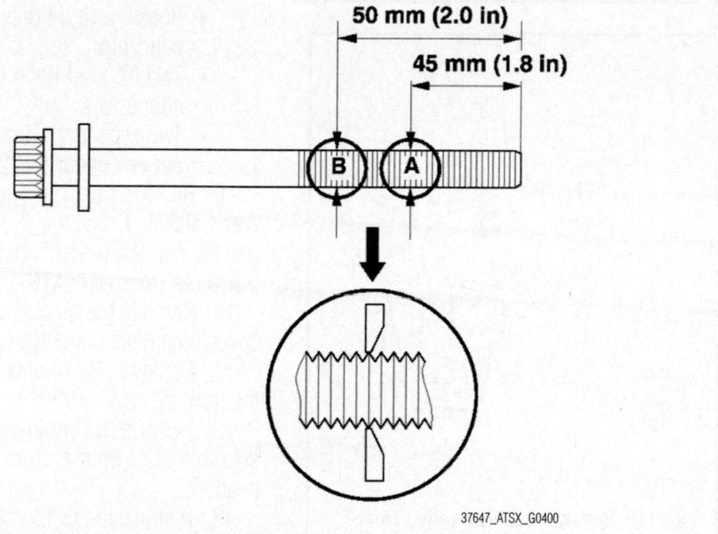
50 mm (2.0 in)
45 mm (1.8 in)
B
A
37647_ATSX_G0400
Fig. 141 Measure the diameter of each cylinder head bolt at point A and point B

29. Install the rocker arm assembly.

30. Install the cam chain.

31. Connect the following engine wire harness connectors, and install the wire harness clamps to the cylinder head:

- Engine Coolant Temperature (ECT) sensor 1 connector
- Camshaft Position (CMP) sensor A (Intake) connector
- Camshaft Position (CMP) sensor B (Exhaust) connector
- Rocker arm oil control solenoid connector
- Rocker arm oil pressure switch connector
- Evaporative Emission (EVAP) canister purge valve connector
- Variable Valve Timing Control (VTC) oil control solenoid valve connector
- Engine oil pressure switch connector

32. Install the two bolts securing the connecting pipe.

33. Install the water bypass hose.

34. Connect the upper radiator hose, the heater hoses, and the water bypass hose.

35. Install the four bolts securing the EVAP canister purge valve bracket.

36. Connect the four fuel injector connectors, the engine mount control solenoid connector, and install the ground cables.

37. Connect the fuel feed hose, then install the quick-connect fitting cover.

38. Connect the EVAP canister hose.

39. Install the catalytic converter.

40. Install the intake manifold.

41. Install the drive belt.

42. Install the strut brace.

43. After installation, check that all tubes, hoses, and connectors are installed correctly.

44. Inspect for fuel leaks. Turn the ignition switch to ON (II) (do not operate the

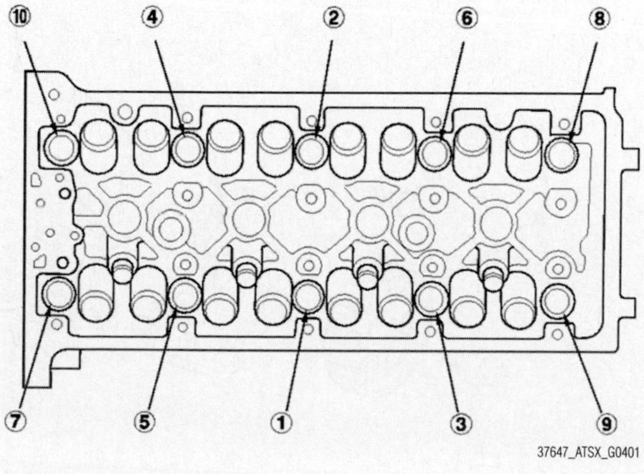
10
4
2
6
8
7
5
1
3
9
37647_ATSX_G0401
Fig. 142 Torque the cylinder head bolts in sequence

starter) so the fuel pump runs for about 2 seconds and pressurizes the fuel line. Repeat this operation three times, then check for fuel leakage at any point in the fuel line.

45. Refill the radiator with engine coolant, and bleed the air from the cooling system with the heater valve open.
46. Check for fluid leaks.
47. Do the Engine Control Module (ECM)/Powertrain Control Module (PCM) idle lean procedure.
48. Do the Crankshaft Position (CKP) pattern clear/CKP pattern lean procedure.
49. Inspect the idle speed.
50. Inspect the ignition timing.

6-Cylinder Engine

See Figures 144 through 156.

1. Relieve the fuel pressure.

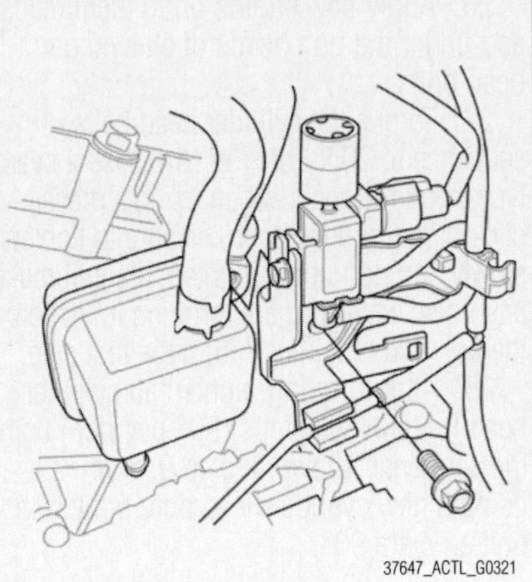
37647_ACTL_G0321

Fig. 145 Remove the engine mount control solenoid valve bracket from the rear cylinder head

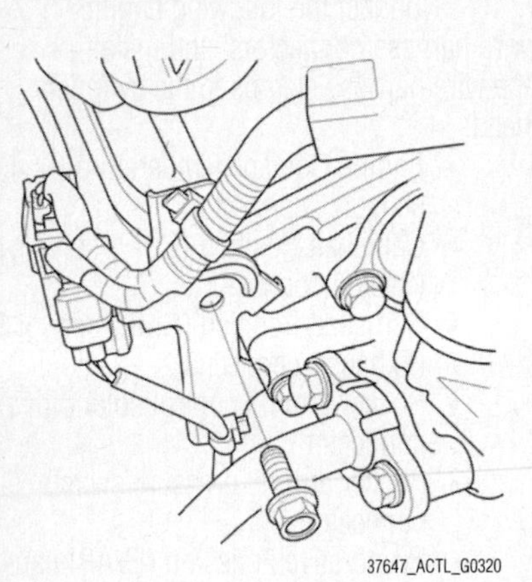
37647_ACTL_G0320

Fig. 144 Remove the connector bracket from the front cylinder head

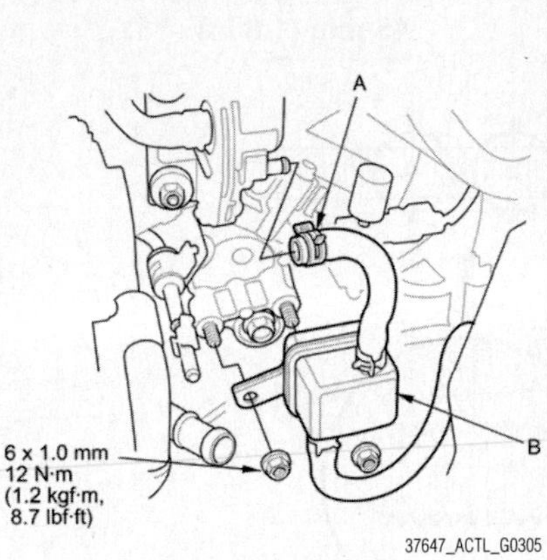
A
B
6 x 1.0 mm
12 N·m
(1.2 kgf·m,
8.7 lbf·ft)
37647_ACTL_G0305

Fig. 146 Remove the Evaporative Emission (EVAP) canister joint with the bracket

2. Do the battery terminal disconnection procedure.
3. Drain the engine coolant.
4. Remove the alternator.
5. Remove the intake manifold.
6. Remove the six ignition coils.
7. Remove the timing belt.
8. Disconnect the following engine wire harness connectors, and remove the wire harness clamps from the cylinder head:
   - Six injector connectors
   - Knock sensor connector
   - Engine Coolant Temperature (ECT) sensor 1 connector
   - Engine mount control solenoid valve connector
   - Camshaft Position (CMP) sensor connector
   - Rocker arm oil control solenoid connector
   - Rocker arm oil pressure switch connector
   - Two Air Fuel Ratio (A/F) sensor connectors
   - Two secondary Heated Oxygen Sensor (secondary HO2S) connectors
9. Remove the front Warm Up Three Way Catalytic Converter (front WU-TWC) and the rear Warm Up Three Way Catalytic Converter (rear WU-TWC).
10. Remove the quick-connect fitting cover, then disconnect the fuel feed hose.
11. Remove the connector bracket from the front cylinder head.
12. Remove the engine mount control solenoid valve bracket from the rear cylinder head.
13. Remove the Evaporative Emission (EVAP) canister joint with the bracket.

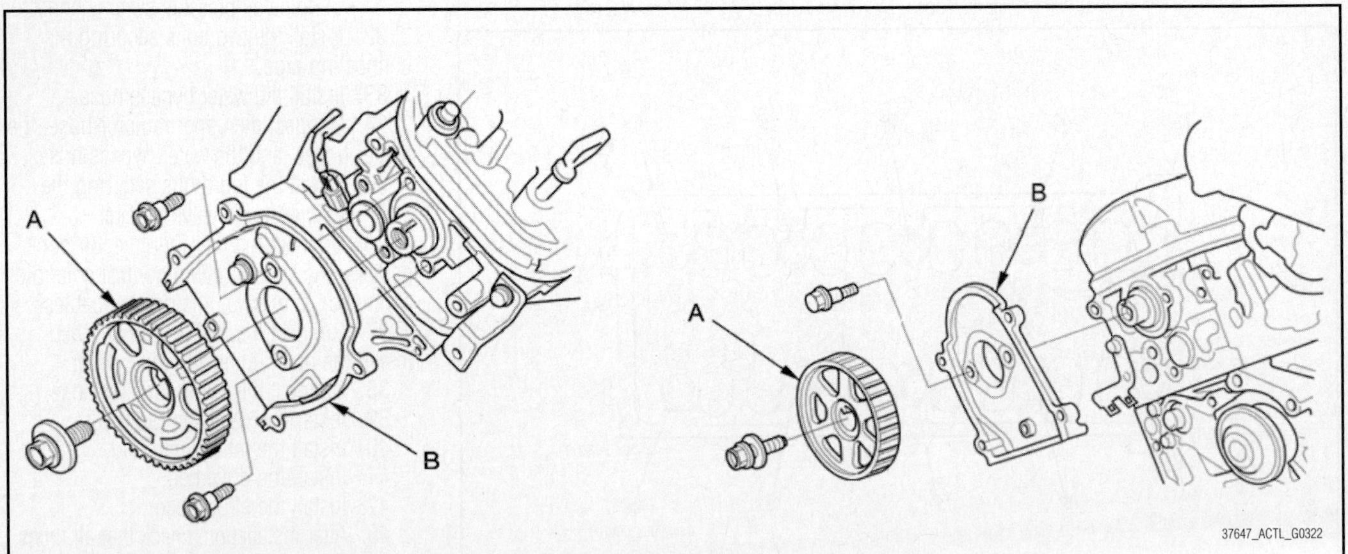
A
B
A
B
37647_ACTL_G0322

Fig. 147 Remove the camshaft pulleys (A) and the back covers (B)

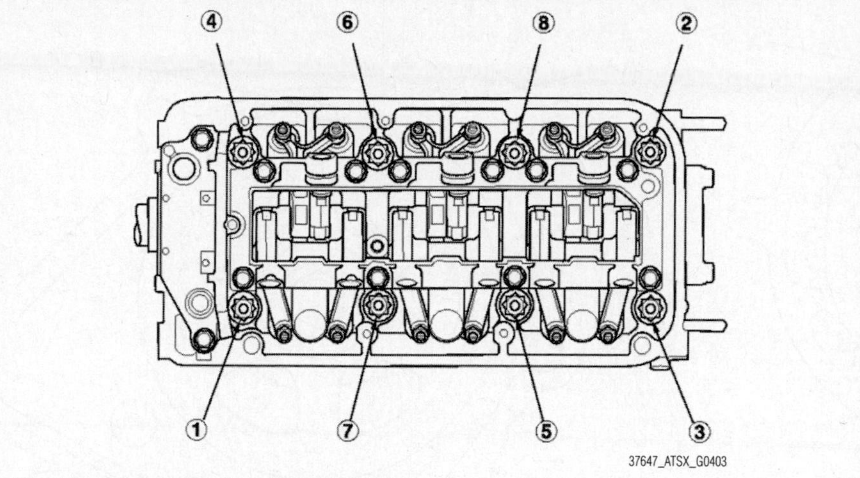
④
⑥
⑧
②
①
⑦
⑤
③
37647_ATSX_G0403

Fig. 148 Remove the front cylinder head bolts

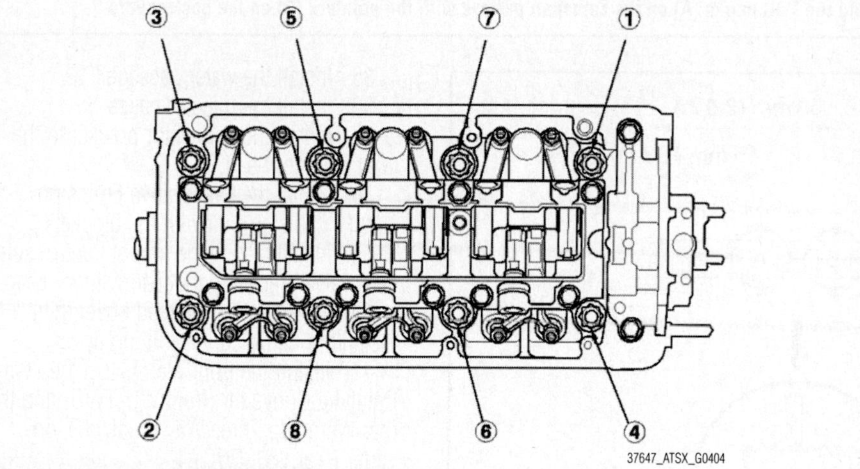
③
⑤
⑦
①
②
⑧
⑥
④
37647_ATSX_G0404

Fig. 149 Remove the rear cylinder head bolts

14. Remove the injector bases.
15. Remove the water passage.
16. Remove the camshaft pulleys and the back covers.
17. Remove the cylinder head covers.
18. Remove the cylinder head bolts. To prevent warpage, loosen the bolts in sequence ⅓ turn at a time; repeat the sequence until all bolts are loosened.
19. Remove the cylinder heads.

To install:

20. Clean the cylinder head and the engine block surface.
21. Clean and install the oil control orifices with new O-rings.
22. Install the dowel pins and the new cylinder head gaskets.
23. Clean the timing belt pulleys, the timing belt guide plate, and the upper and lower covers.
24. Set the timing belt drive pulley to Top Dead Center (TDC) by aligning the TDC mark on the tooth of the timing belt drive pulley with the pointer on the oil pump.
25. Set the camshaft pulleys to TDC by aligning the TDC marks on the camshaft pulleys with the pointers on the back covers.
26. Install the cylinder heads on the engine block.
27. Measure the diameter of each cylinder head bolt at point A and point B.
28. If either diameter is less than 0.44 inches (11.3 mm), replace the cylinder head bolt.
29. Apply new engine oil to the threads and under the bolt heads of all cylinder head bolts.
30. Torque the cylinder head bolts in sequence to 22 ft. lbs. (29 Nm) using a beam-type torque wrench. When using a preset

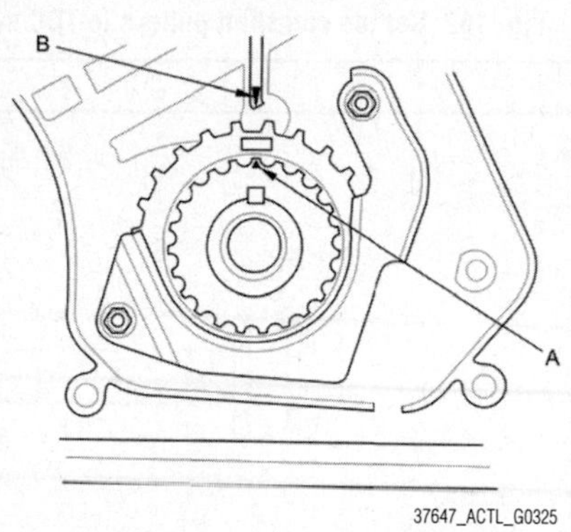
B
A
37647_ACTL_G0325

Fig. 151 Set the timing belt drive pulley to TDC by aligning the TDC mark (A) on the tooth of the timing belt drive pulley with the pointer (B) on the oil pump

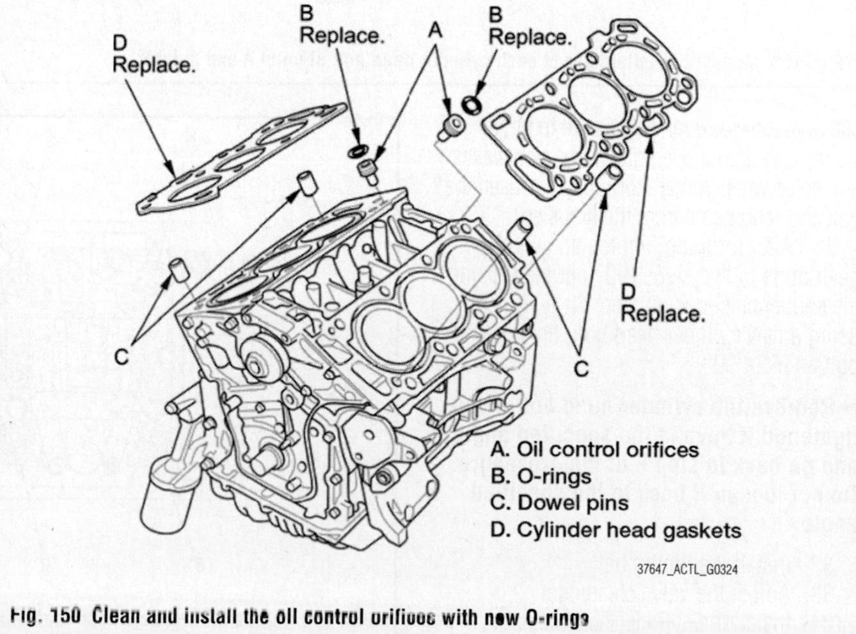
D
Replace.
B
Replace.
A
B
Replace.
D
Replace.
C
C
A. Oil control orifices
B. O-rings
C. Dowel pins
D. Cylinder head gaskets
37647_ACTL_G0324

Fig. 150 Clean and install the oil control orifices with new O-rings

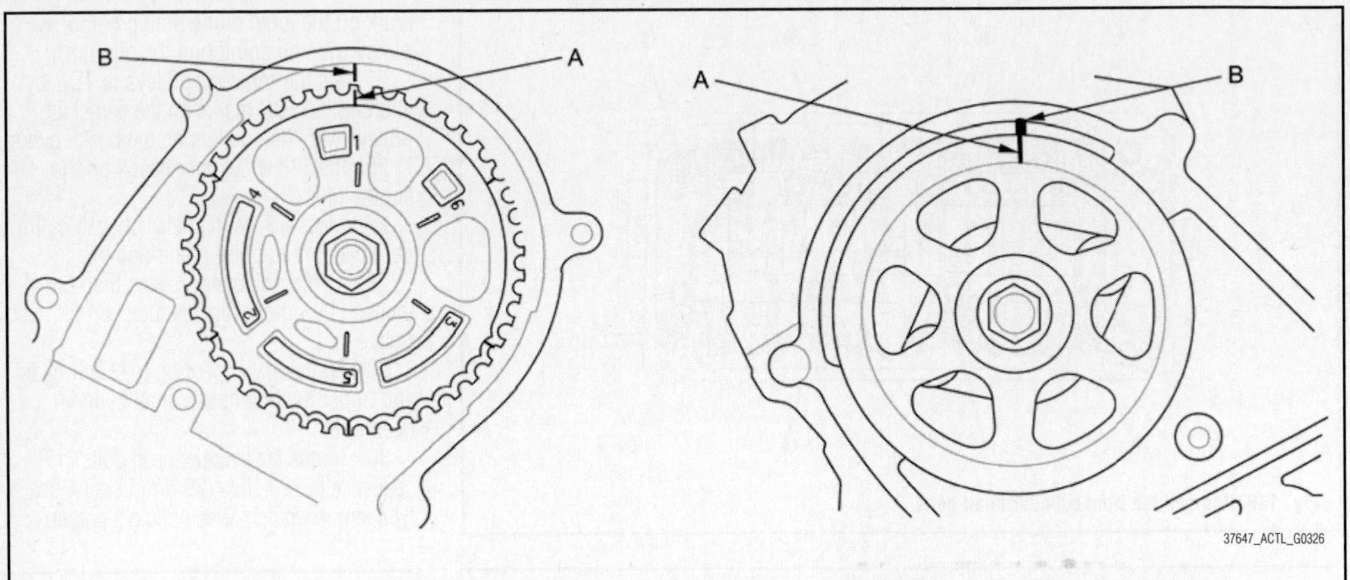
B
A
A
B
37647_ACTL_G0326

Fig. 152 Set the camshaft pulleys to TDC by aligning the TDC marks (A) on the camshaft pulleys with the pointers (B) on the back covers

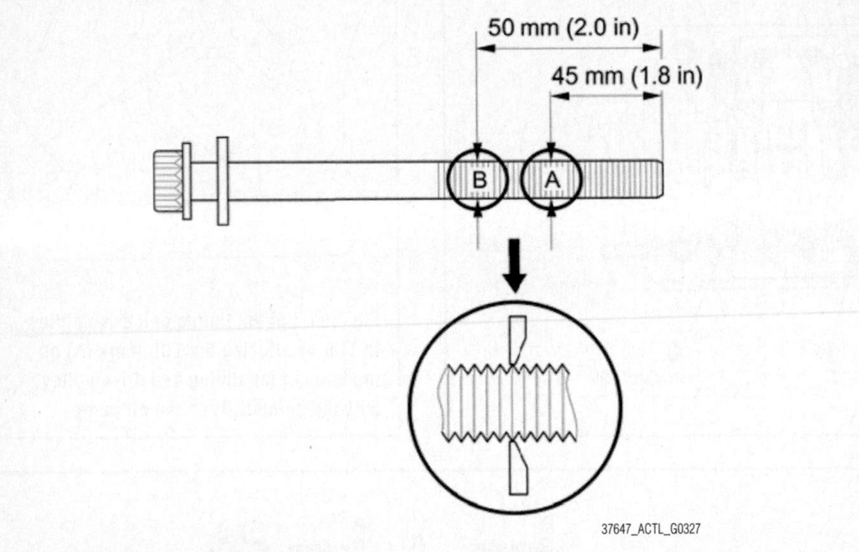
50 mm (2.0 in)
45 mm (1.8 in)
B
A
37647_ACTL_G0327

Fig. 153 Measure the diameter of each cylinder head bolt at point A and point B

click-type torque wrench, be sure to tighten slowly and do not overtighten. If a bolt makes any noise while you are torquing it, loosen the bolt and retighten it from the first step.

31. After torquing, tighten all cylinder head bolts in two steps (90° per step) using the sequence shown in step 11. If you are using a new cylinder head bolt, tighten the bolt an extra 90°.

➡Remove the cylinder head bolt if you tightened it beyond the specified angle, and go back to step 8 of the procedure. Do not loosen it back to the specified angle.

32. Install the timing belt.
33. Adjust the valve clearance.
34. Install the cylinder head covers.
35. Install the water passage.
36. Install the injector bases.
37. Install the connector bracket to the front cylinder head.
38. Install the Evaporative Emission (EVAP) canister joint with the bracket.
39. Install the engine mount control solenoid valve bracket to the rear cylinder head.
40. Connect the fuel feed hose, then install the quick-connect fitting cover.
41. Install the front Warm Up Three Way Catalytic Converter (front WU-TWC) and the rear Warm Up Three Way Catalytic Converter (rear WU-TWC).
42. Connect the following engine wire harness connectors, and install the wire harness clamps to the cylinder head:
   - Six injector connectors
   - Knock sensor connector
   - Engine Coolant Temperature (ECT) sensor 1 connector

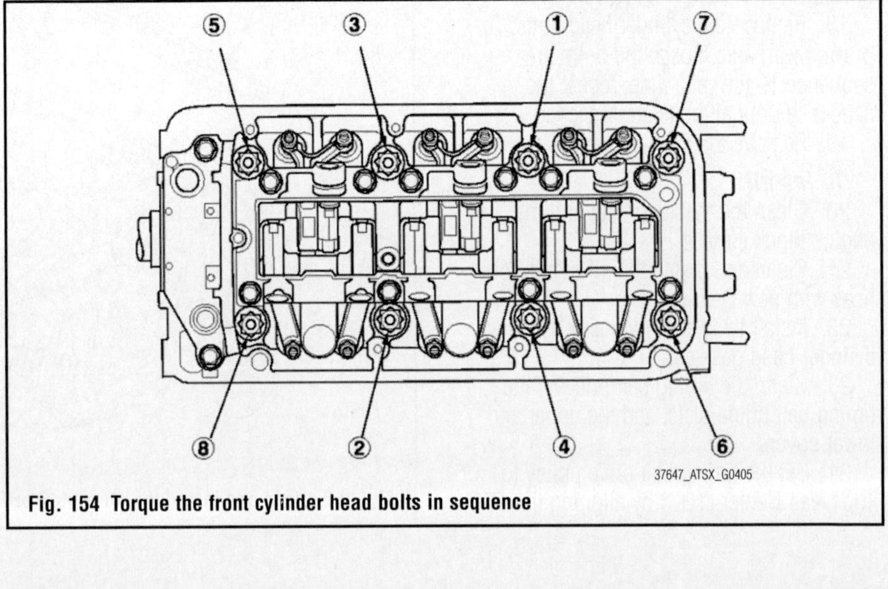
5
3
1
7
8
2
4
6
37647_ATSX_G0405

Fig. 154 Torque the front cylinder head bolts in sequence

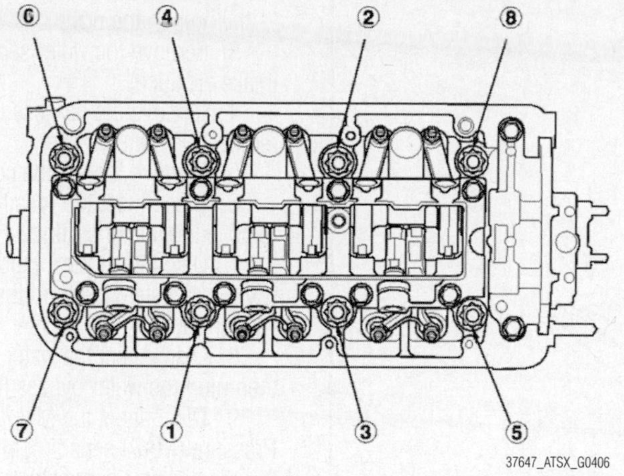
6
4
2
8
7
1
3
5
37647_ATSX_G0406

Fig. 155 Torque the rear cylinder head bolts in sequence

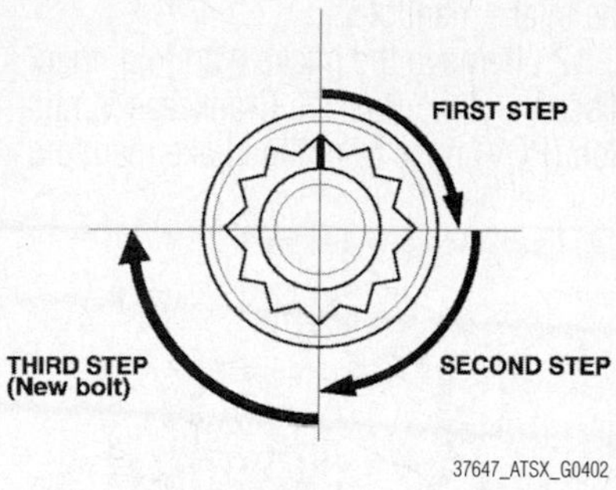
FIRST STEP
THIRD STEP
(New bolt)
SECOND STEP
37647_ATSX_G0402

Fig. 156 After torquing, tighten all cylinder head bolts in two steps (90° per step); tighten new bolts an extra 90°

- Engine mount control solenoid valve connector
- Camshaft Position (CMP) sensor connector
- Rocker arm oil control solenoid connector
- Rocker arm oil pressure switch connector
- Two Air Fuel Ratio (A/F) sensor connectors
- Two secondary Heated Oxygen Sensor (secondary HO2S) connectors

43. Install the six ignition coils.
44. Install the intake manifold.
45. Install the alternator.
46. Do the battery terminal reconnection procedure.
47. After installation, check that all tubes, hoses, and connectors are installed correctly.
48. Inspect for fuel leaks. Turn the ignition switch to ON (II), (do not operate the starter) so the fuel pump runs for about 2 seconds and pressurizes the fuel line. Repeat this operation three times, then check for fuel leakage at any point in the fuel line.
49. Refill the radiator with engine coolant, and bleed the air from the cooling system.
50. Check for fluid leaks.
51. Do the Powertrain Control Module (PCM) idle learn procedure.
52. Do the Crankshaft Position (CKP) pattern clear/CKP pattern learn procedure.
53. Inspect the idle speed.
54. Inspect the ignition timing.

DRIVE PLATE

REMOVAL & INSTALLATION

4-Cylinder Engine

See Figure 157.

1. Remove the transmission assembly.
2. Remove the drive plate and the washer from the engine.

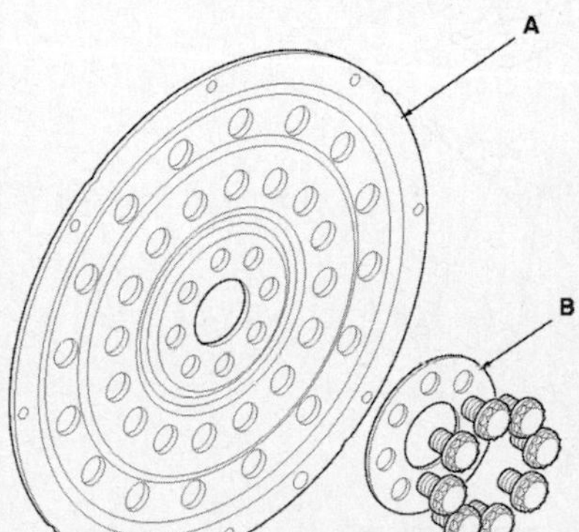
A
B
12 x 1.0 mm
74 N·m (7.5 kgf·m, 54 lbf·ft)
37647_ATSX_G0455

Fig. 157 Remove the drive plate (A) and the washer (B) from the engine

3. Install the drive plate and the washer on the engine, and tighten the eight bolts in a crisscross pattern in at least two steps.
4. Install the transmission assembly.

6-Cylinder Engine

See Figure 158.

1. Remove the transmission assembly.
2. Remove the drive plate and the washer from the engine.
3. Install the drive plate and the washer on the engine, and tighten the eight bolts in a crisscross pattern in at least two steps.

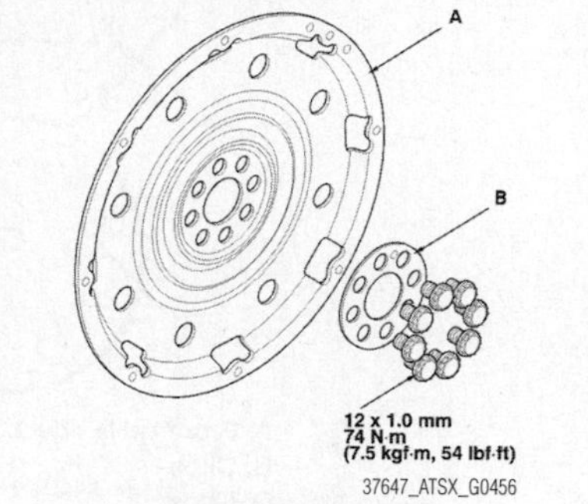
A
B
12 x 1.0 mm
74 N·m
(7.5 kgf·m, 54 lbf·ft)
37647_ATSX_G0456

Fig. 158 Remove the drive plate (A) and the washer (B) from the engine

4. Install the transmission assembly.

INTAKE MANIFOLD

REMOVAL & INSTALLATION

4-Cylinder Engine

See Figures 159 through 169.

1. Do the battery removal procedure.
2. Remove the front grille cover.
   a. Detach the clips.
   b. Pass both hood edge cushions through the holes in the cover by pulling up the rear edge of the cover, and slide the entire cover rearward to release the front edge of it from under the front grille.
   c. Pass the hood latch handle through the hole in the cover.

➡To release the clips, pry up on the center pin at the notch.

   d. Detach the clips.
   e. Release the hooks from both front fender trims.
   f. Pass both hood edge cushions through the holes in the cover by pulling up the rear edge of the cover, and slide the entire cover rearward to release it from the groove of the front grille.

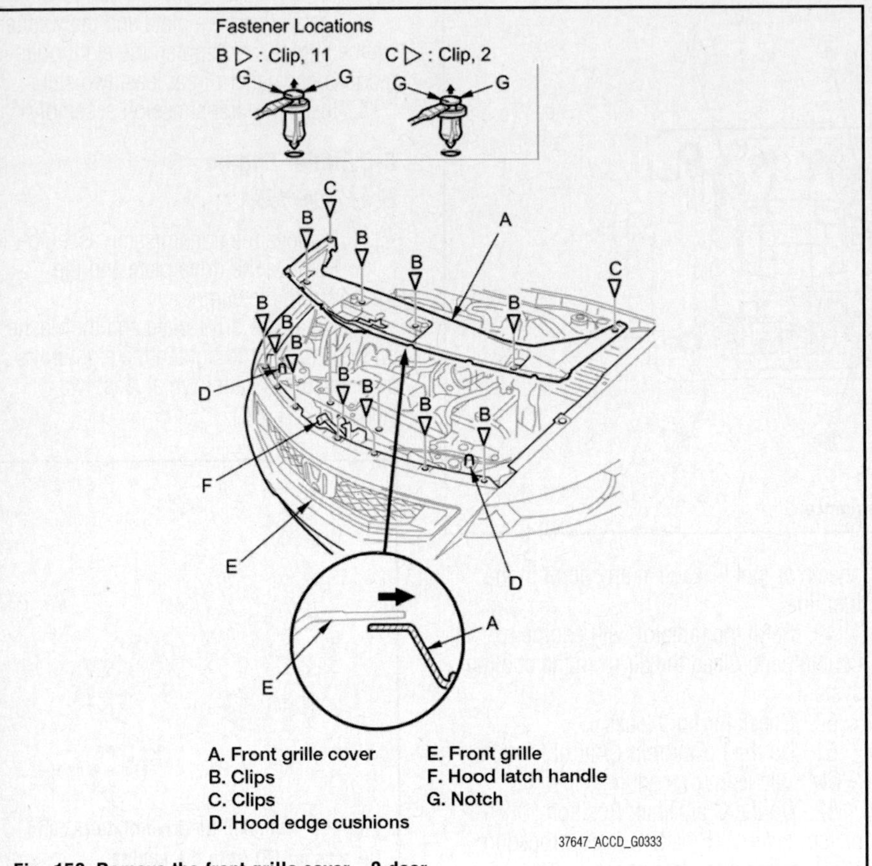
Fastener Locations
B ▷ : Clip, 11
C ▷ : Clip, 2
A. Front grille cover
B. Clips
C. Clips
D. Hood edge cushions
E. Front grille
F. Hood latch handle
G. Notch
37647_ACCD_G0333

Fig. 159 Remove the front grille cover—2-door

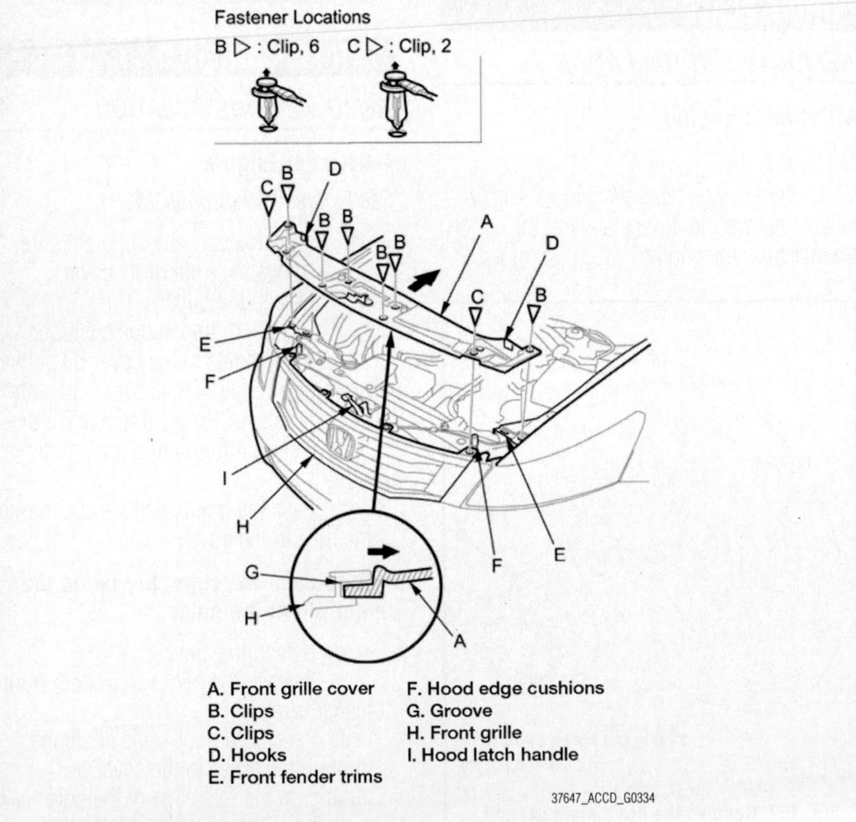
Fastener Locations
B ▷ : Clip, 6
C ▷ : Clip, 2
A. Front grille cover
B. Clips
C. Clips
D. Hooks
E. Front fender trims
F. Hood edge cushions
G. Groove
H. Front grille
I. Hood latch handle
37647_ACCD_G0334

Fig. 160 Remove the front grille cover—4-door

g. Pass the hood latch handle through the hole in the cover.

3. Remove the water separator and the intake air duct.

4. Remove the harness clamps, then remove the battery base.

5. Remove the engine cover.

6. Disconnect the breather pipe, then remove the intake air duct.

7. Disconnect the Evaporative Emission (EVAP) canister hose and the brake booster vacuum hose.

8. Disconnect the water bypass hoses, then plug the water bypass hoses.

9. Disconnect the Manifold Absolute Pressure (MAP) sensor connector and the throttle actuator connector, then remove the wire harness clamps.

10. Remove the intake manifold bracket.

11. Disconnect the vacuum hose from the intake manifold.

12. Remove the intake manifold, then disconnect the Positive Crankcase Ventilation (PCV) hose from the intake manifold.

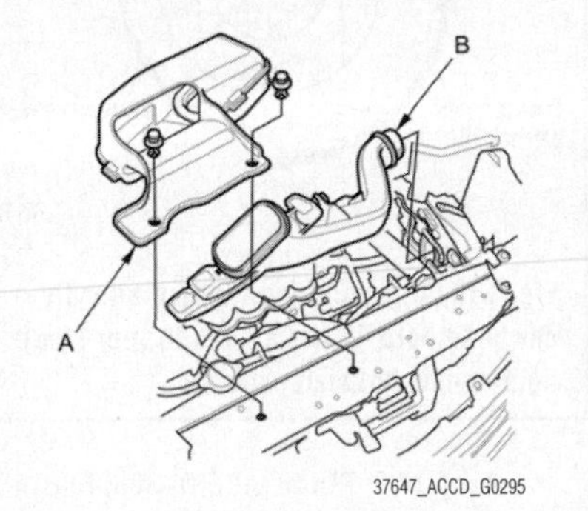
A
B
37647_ACCD_G0295

Fig. 161 Remove the water separator (A) and the intake air duct (B)

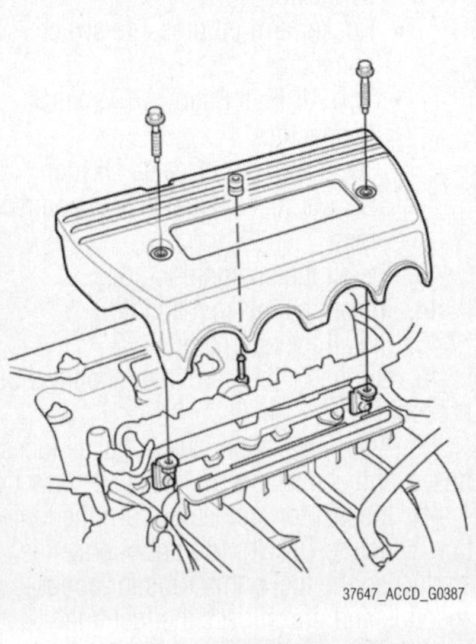
37647_ACCD_G0387

Fig. 162 Remove the engine cover

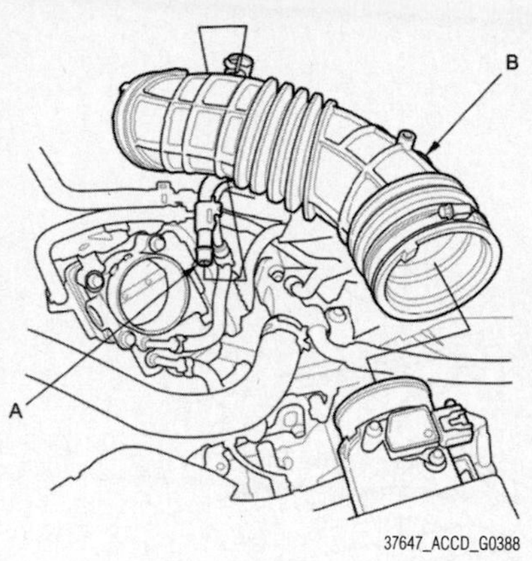
B
A
37647_ACCD_G0388

Fig. 163 Disconnect the breather pipe (A), then remove the intake air duct (B)

To install:

13. Connect the Positive Crankcase Ventilation (PCV) hose to the intake manifold, then install the intake manifold with new gaskets, and tighten the bolts and nuts in a crisscross pattern in three steps, beginning with the inner bolt.
14. Connect the vacuum hose to the intake manifold.
15. Install the intake manifold bracket.
16. Connect the Manifold Absolute Pressure (MAP) sensor connector and the throttle actuator connector, then install the wire harness clamps.
17. Connect the water bypass hoses.
18. Connect the Evaporative Emission (EVAP) canister hose and the brake booster vacuum hose.
19. Install the intake air duct, then connect the breather pipe.

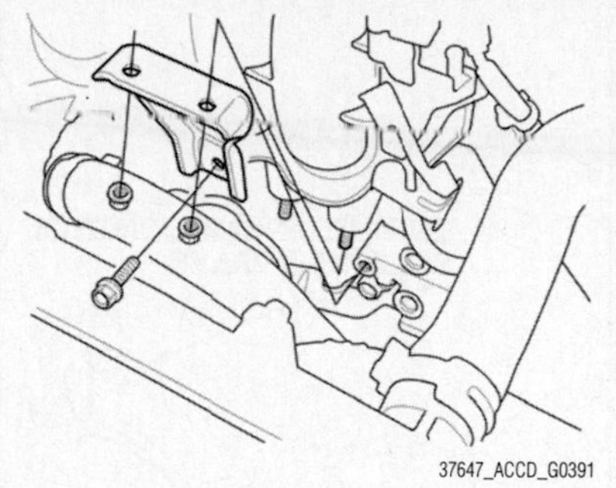
37647_ACCD_G0391

Fig. 166 Remove the intake manifold bracket

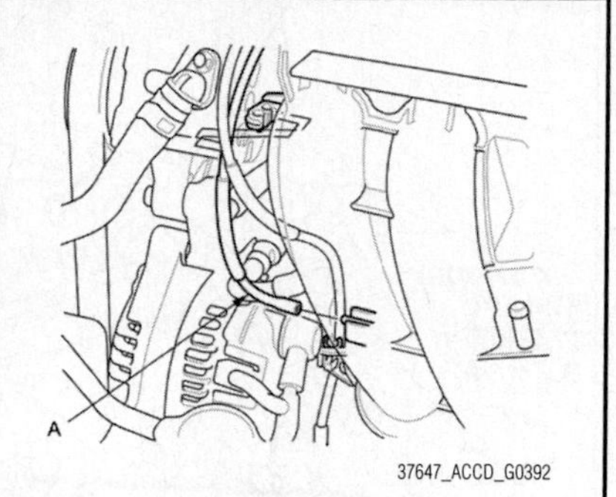
A
37647_ACCD_G0392

Fig. 167 Disconnect the vacuum hose (A) from the intake manifold

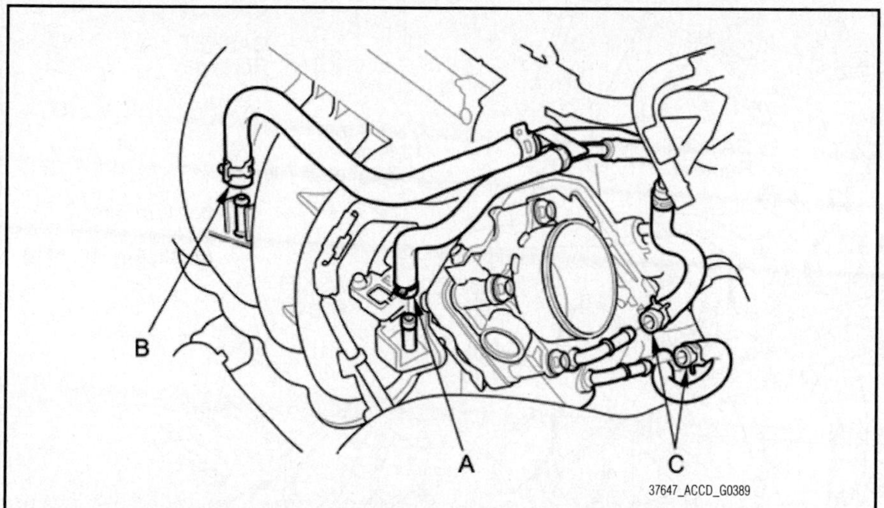
B
A
C
37647_ACCD_G0389

Fig. 164 Disconnect the EVAP canister hose (A), the brake booster vacuum hose (B), and the water bypass hoses (C)

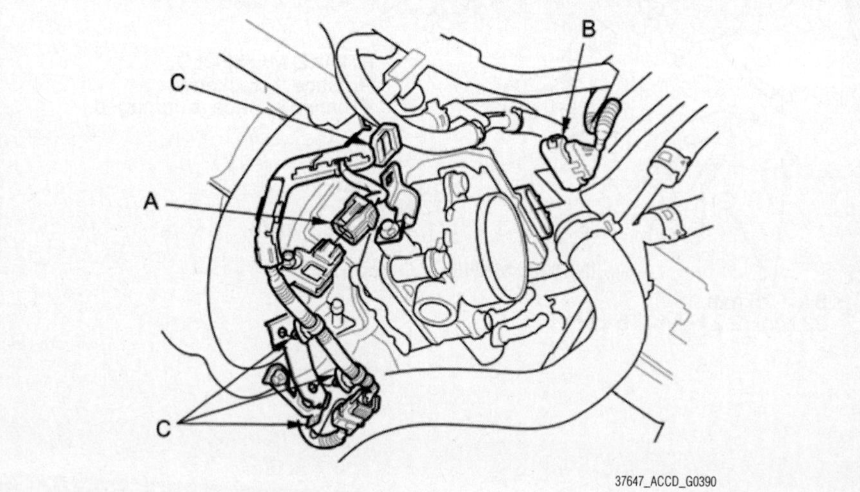
C
B
A
C
37647_ACCD_G0390

Fig. 165 Disconnect the MAP sensor connector (A) and the throttle actuator connector (B), then remove the wire harness clamps (C)

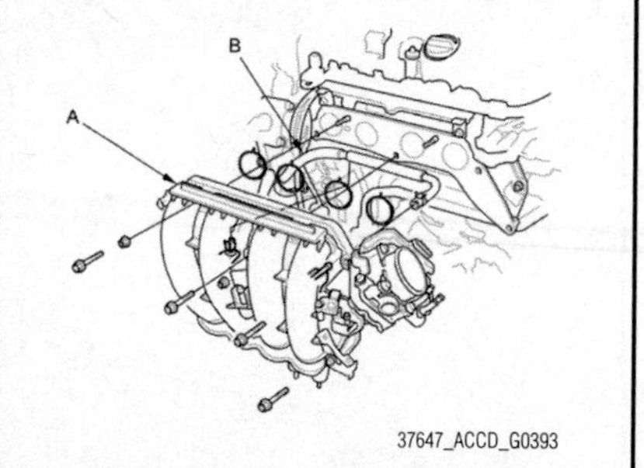
B
A
37647_ACCD_G0393

Fig. 168 Remove the intake manifold (A), then disconnect the PCV hose (B) from the intake manifold

➡When tightening the screw of the hose band, align the edge of the hose band with the mark painted on the hose band. If you tighten the screw over the mark, replace the hose band.

20. Install the engine cover.
21. Install the battery base, then install the harness clamps.
22. Install the water separator and the intake air duct.
23. Install the front grille cover:
24. Do the battery installation procedure.
25. After installation, check that all tubes, hoses, and connectors are installed correctly.

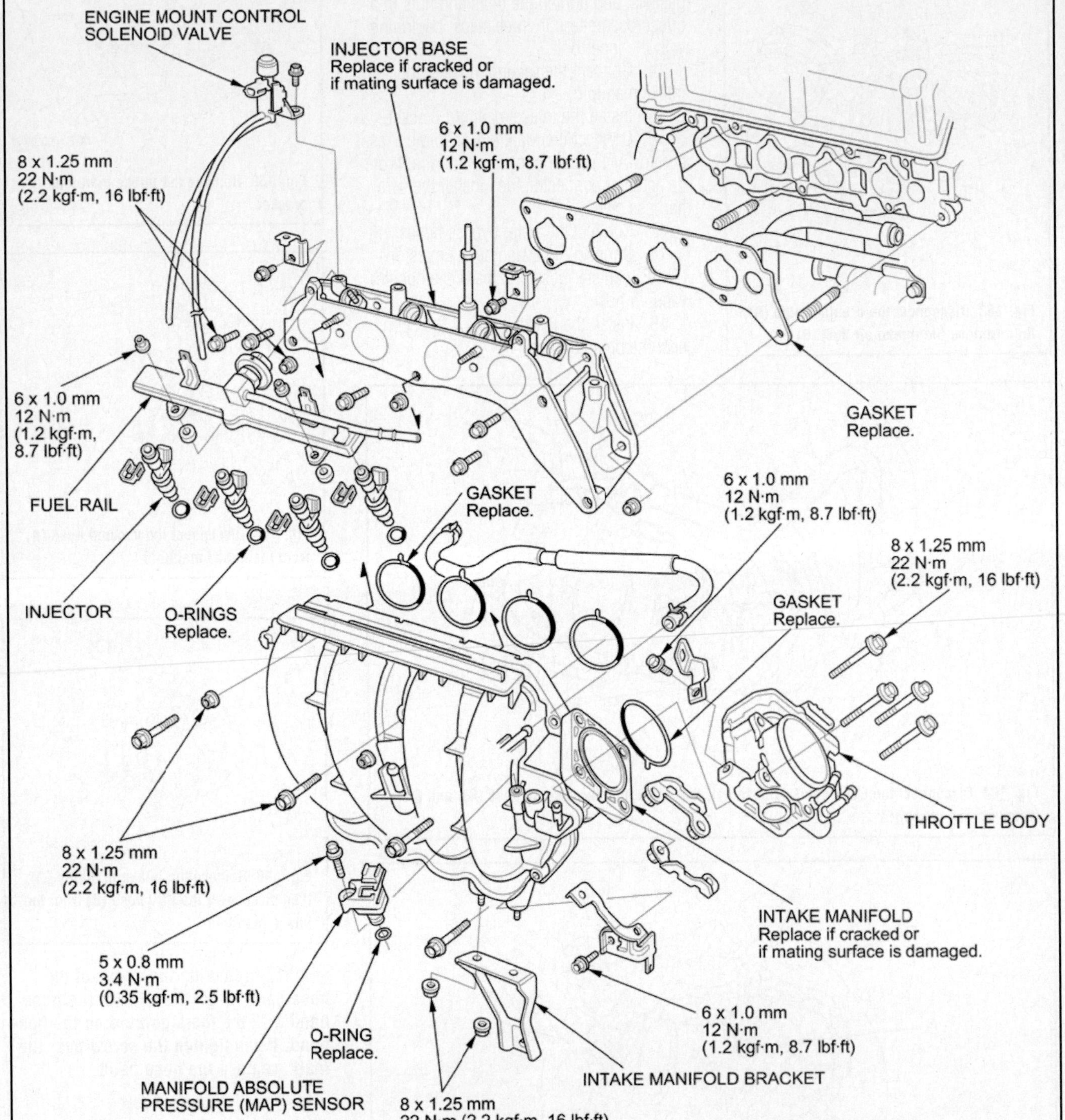
ENGINE MOUNT CONTROL SOLENOID VALVE
INJECTOR BASE
Replace if cracked or if mating surface is damaged.
6 x 1.0 mm
12 N·m
(1.2 kgf·m, 8.7 lbf·ft)
8 x 1.25 mm
22 N·m
(2.2 kgf·m, 16 lbf·ft)
6 x 1.0 mm
12 N·m
(1.2 kgf·m,
8.7 lbf·ft)
GASKET
Replace.
FUEL RAIL
GASKET
Replace.
6 x 1.0 mm
12 N·m
(1.2 kgf·m, 8.7 lbf·ft)
8 x 1.25 mm
22 N·m
(2.2 kgf·m, 16 lbf·ft)
INJECTOR
O-RINGS
Replace.
GASKET
Replace.
THROTTLE BODY
8 x 1.25 mm
22 N·m
(2.2 kgf·m, 16 lbf·ft)
INTAKE MANIFOLD
Replace if cracked or if mating surface is damaged.
5 x 0.8 mm
3.4 N·m
(0.35 kgf·m, 2.5 lbf·ft)
6 x 1.0 mm
12 N·m
(1.2 kgf·m, 8.7 lbf·ft)
O-RING
Replace.
INTAKE MANIFOLD BRACKET
MANIFOLD ABSOLUTE PRESSURE (MAP) SENSOR
8 x 1.25 mm
22 N·m (2.2 kgf·m, 16 lbf·ft)
37647_ACCD_G0394

Fig. 169 Exploded view of intake manifold assembly

26. Clean up any spilled engine coolant.

27. Refill the radiator with engine coolant, and bleed the air from the cooling system.

6-Cylinder Engine

See Figures 170 through 175.

1. Remove the strut brace.
2. Remove the engine cover.
3. Disconnect the Manifold Absolute Pressure (MAP) sensor connector and the breather pipe, and remove the intake air duct.
4. Disconnect the throttle actuator connector, the Evaporative Emission (EVAP) canister purge valve connector, the water bypass hoses, the EVAP hose, and the brake booster vacuum hose.
5. Disconnect the Positive Crankcase Ventilation (PCV) hose, then remove the upper cover mounting bolts and nuts sequentially in three steps, then remove the upper cover.

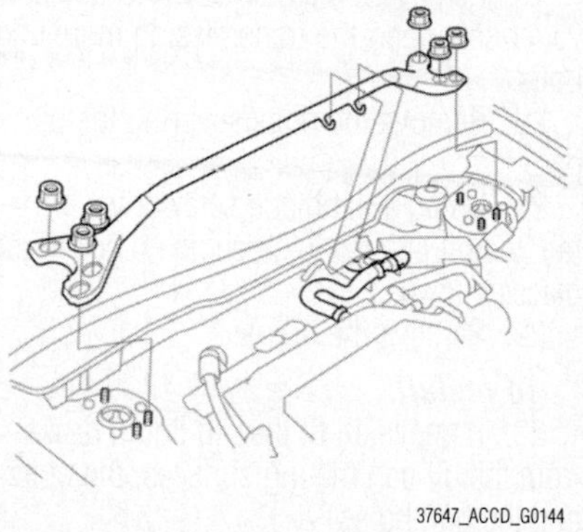
37647_ACCD_G0144

Fig. 170 Remove the strut brace

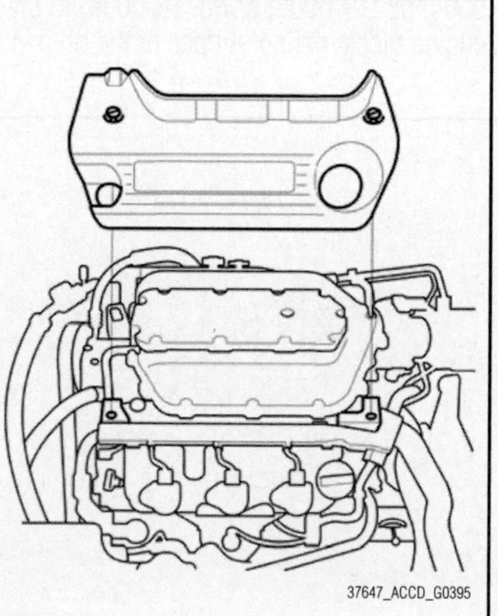
37647_ACCD_G0395

Fig. 171 Remove the engine cover

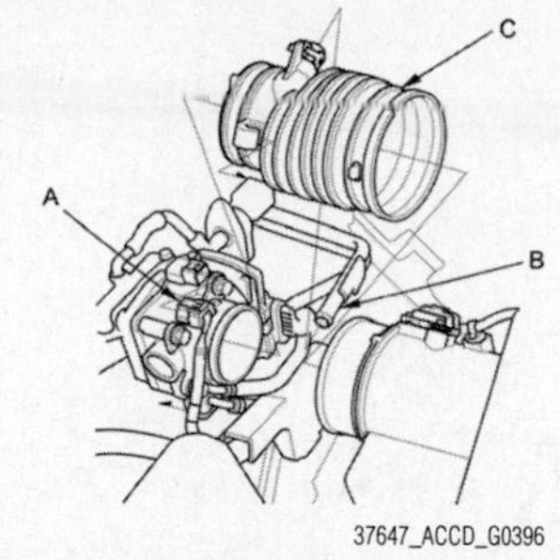
C
A
B
37647_ACCD_G0396

Fig. 172 Disconnect the MAP sensor connector (A) and the breather pipe (B), and remove the intake air duct (C)

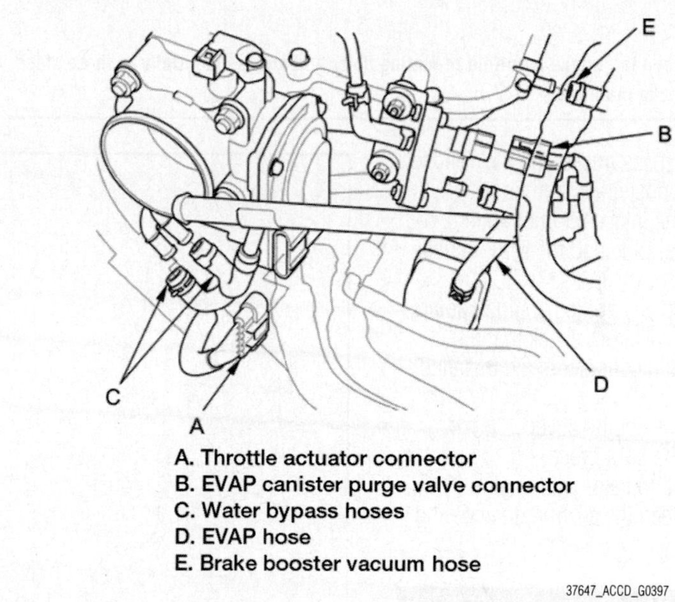
E
B
C
D
A
A. Throttle actuator connector
B. EVAP canister purge valve connector
C. Water bypass hoses
D. EVAP hose
E. Brake booster vacuum hose
37647_ACCD_G0397

Fig. 173 Disconnect the throttle actuator connector, the EVAP canister purge valve connector, the water bypass hoses, the EVAP hose, and the brake booster vacuum hose

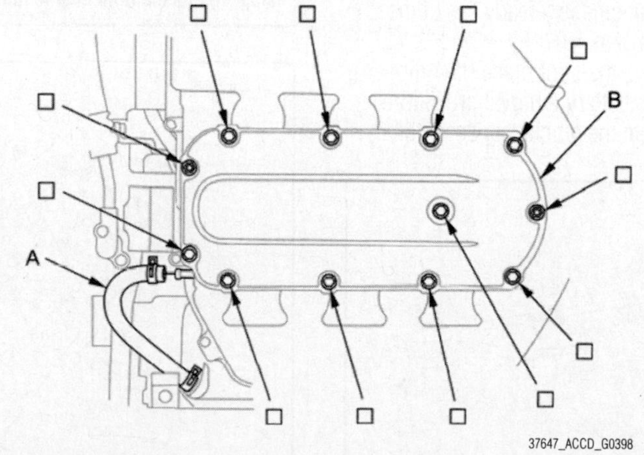
B
A
37647_ACCD_G0398

Fig. 174 Disconnect the PCV hose (A), then remove the upper cover mounting bolts and nuts sequentially in three steps, then remove the upper cover (B)

6. Remove the intake manifold mounting bolts and nuts sequentially in three steps, then remove the intake manifold.

To install:

7. Install the intake manifold. Tighten the bolts and nuts sequentially in three steps. Always use a new intake manifold gasket.
8. Install the upper cover. Tighten the bolts and nuts sequentially in three steps. Always use a new gasket, then connect the PCV hose.
9. Connect the throttle actuator connector, the EVAP canister purge valve connec-

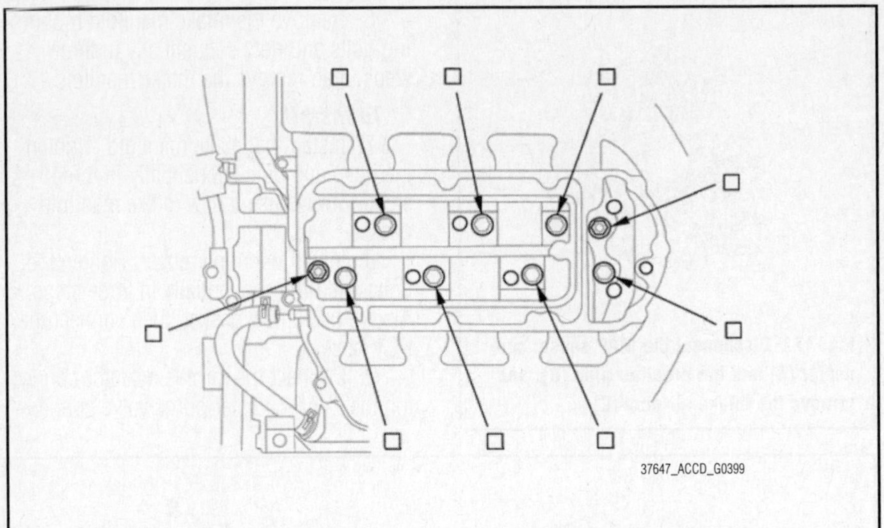
37647_ACCD_G0399

Fig. 175 Remove the intake manifold mounting bolts and nuts sequentially in three steps, then remove the intake manifold

tor, the water bypass hoses, the EVAP hose, and the brake booster vacuum hose.

10. Install the intake air duct, then connect the breather pipe and the MAP sensor connector.
11. Clean up any spilled engine coolant.
12. After installation, check that all tubes, hoses, and connectors are installed correctly.
13. Install the engine cover.
14. Install the strut brace.
15. Refill the radiator with engine coolant, and bleed the air from the cooling system.

OIL PAN

REMOVAL & INSTALLATION

4-Cylinder Engine

See Figures 176 through 182.

1. If the engine is already out of the vehicle, go to step 19.
2. Remove the strut brace (if equipped).
3. Do the battery removal procedure.
4. Remove the air cleaner assembly.

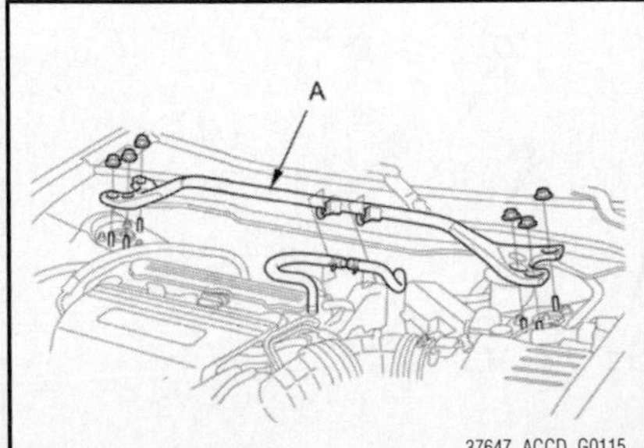
A
37647_ACCD_G0115

Fig. 176 Remove the strut brace

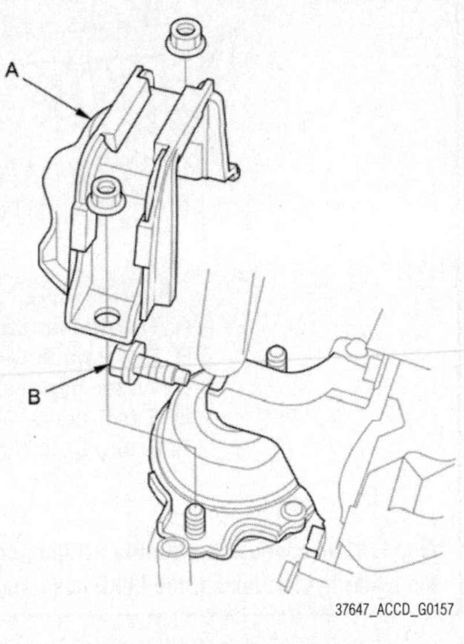
A
B
37647_ACCD_G0157

Fig. 177 Remove the front engine mount stop (A) and the front engine mount bolt (B)

5. Remove the harness clamps, then remove the battery base.
6. Remove the front engine mount stop, then remove the front engine mount bolt.
7. Loosen the rear engine mount mounting bolts.
8. Loosen the upper transmission mount bracket mounting bolts.
9. Raise the vehicle on the lift.
10. Remove the left front wheel.
11. Remove the splash shield.
12. Drain the engine oil.
13. Separate the left side knuckle from the lower arm.
14. Remove the left side damper fork.
15. Remove the left side driveshaft. Coat all precision-finished surface with new engine oil. Tie a plastic bag over the driveshaft end.
16. Remove the nuts securing the lower transmission mount.
17. A/T model: Remove the shift cable bracket.
18. Use a transmission jack to lift the transmission 1.2–1.6 inches (30–40 mm).
19. Remove the clutch cover/torque converter cover and the transmission mounting bolts.
20. Remove the bolts securing the oil pan.
21. Using a flat blade screwdriver, separate the oil pan from the engine block in the places shown.
22. Remove the oil pan.

To install:

23. Remove all of the old liquid gasket from the oil pan mating surfaces, the bolts, and the bolt holes.
24. Clean and dry the oil pan mating surfaces.
25. Apply liquid gasket (P/N 08717-0004, 08718-0003, or 08718-0009) to the engine block mating surface of the oil pan

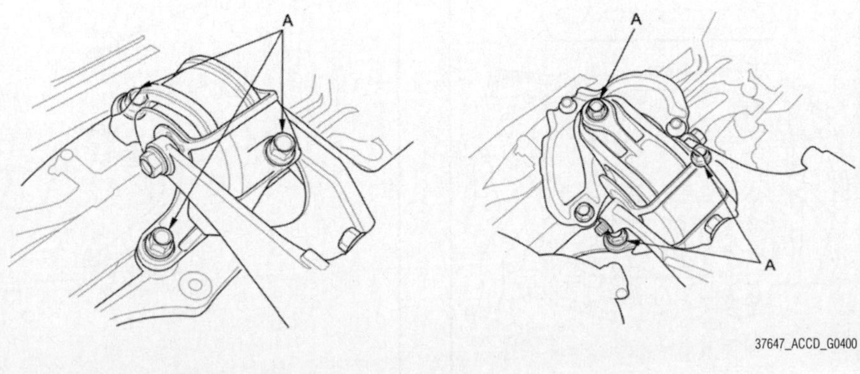
A
A
A
37647_ACCD_G0400

Fig. 178 Loosen the rear engine mount mounting bolts (A)

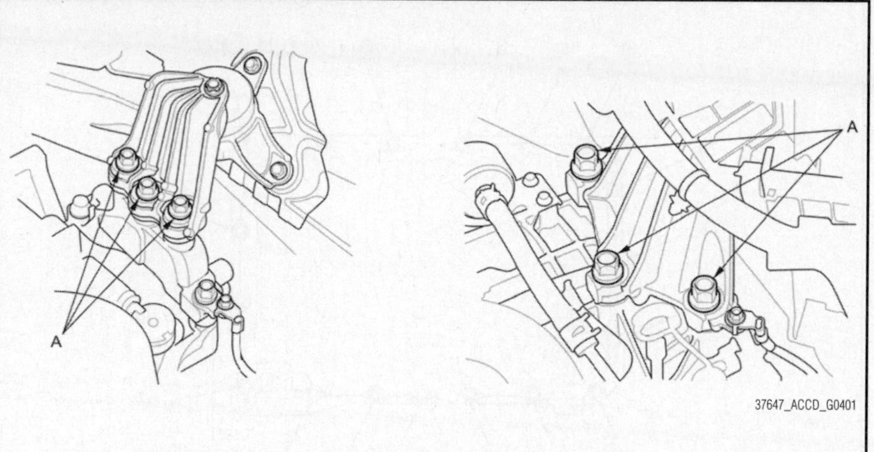
A
A
37647_ACCD_G0401
Fig. 179 Loosen the upper transmission mount bracket mounting bolts (A)

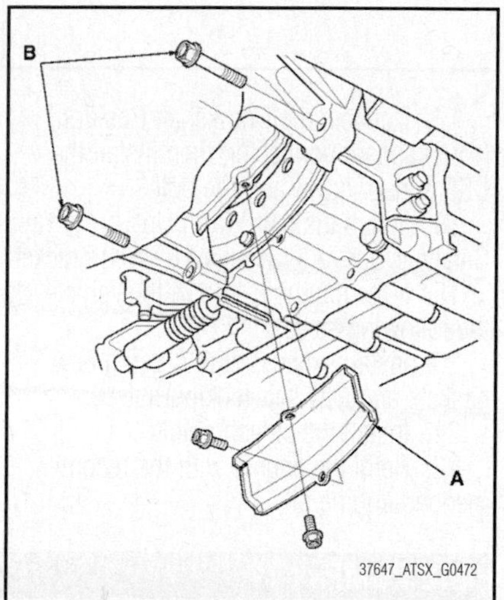
B
A
37647_ATSX_G0472
Fig. 180 Remove the clutch cover/torque converter cover (A) and the transmission mounting bolts (B)

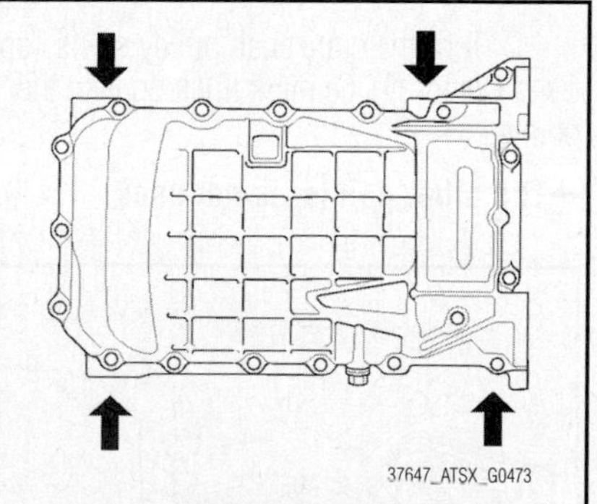
37647_ATSX_G0473
Fig. 181 Separate the oil pan from the engine block in the places shown

and to the inside edge of the threaded bolt holes. Install the component within 5 minutes of applying the liquid gasket.

➡If too much time has passed after applying the liquid gasket, remove the old liquid gasket and residue, then reapply new liquid gasket.

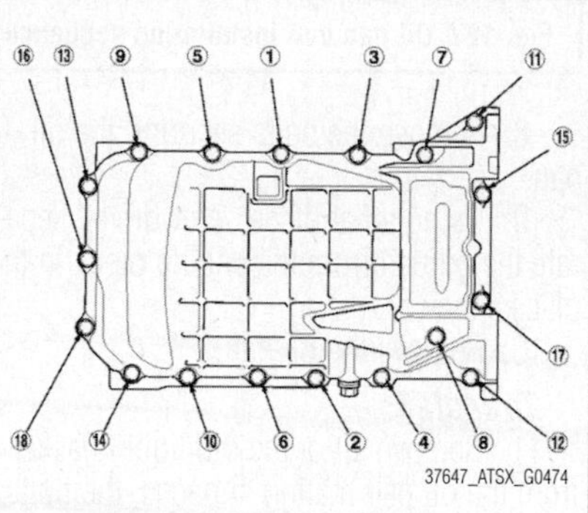
16
13
9
5
1
3
7
11
15
17
12
8
4
2
6
10
14
18
37647_ATSX_G0474
Fig. 182 Oil pan bolt installation sequence

26. Install the oil pan.
27. Tighten the bolts in three steps. In the final step, torque all bolts, in sequence, to 9 ft. lbs. (12 Nm). Wipe off the excess liquid gasket on the each side of crankshaft pulley and the flywheel/drive plate.

➡Wait at least 30 minutes before filling the engine with oil. Do not run the engine for at least 3 hours after installing the oil pan.

28. Install the clutch cover/torque converter cover and the transmission mounting bolts. Tighten the transmission mounting bolts to 47 ft. lbs. (64 Nm).
29. If the engine is still in the vehicle, do steps 8 through 24.
30. Lower the transmission jack from the transmission.
31. A/T model: Install the shift cable bracket.
32. Tighten the nuts securing the lower transmission mount.
33. Install a new set ring on the end of driveshaft, then install the driveshaft. Make sure the ring "clicks" into place in the differential.
34. Connect the lower arm to the left side knuckle.
35. Install the left side damper fork.
36. Install the splash shield.
37. Install the left front wheel.
38. Lower the vehicle on the lift.
39. Tighten the upper transmission mount bracket mounting bolts to 43 ft. lbs. (59 Nm).
40. Tighten the rear engine mount mounting bolts. Tighten the 12 x 1.25 mm bolts to 58 ft. lbs. (78 Nm); tighten the 10 x 1.25 mm bolts to 40 ft. lbs. (54 Nm).
41. Tighten the front engine mount bolt to 40 ft. lbs. (54 Nm), then install the front engine mount stop. Tighten the nuts to 58 ft. lbs. (78 Nm).
42. Install the battery base, then install the harness clamps.
43. Install the air cleaner assembly.
44. Do the battery installation procedure.
45. Install the strut brace (if equipped).
46. Refill the engine with the recommended engine oil.

6-Cylinder Engine

See Figures 183 through 187.

➡If the engine is already out of the vehicle, go to step 6.

1. Raise the vehicle on the lift.
2. Drain the engine oil.
3. Remove the splash shield.
4. Remove exhaust pipe A.
5. Remove the rear Warm Up Three Way Catalytic Converter (rear WU-TWC) bracket.
6. Remove the Crankshaft Position (CKP) sensor cover and the bolt, then disconnect the CKP sensor connector.
7. Remove the clutch/torque converter cover and the four bolts securing the transmission.

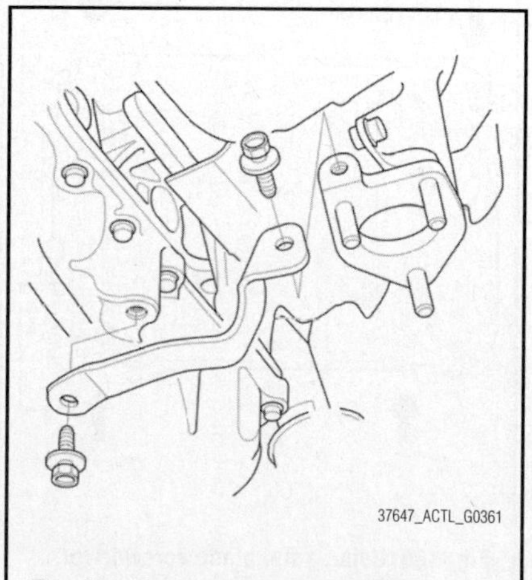
37647_ACTL_G0361
Fig. 183 Remove the rear WU-TWC bracket

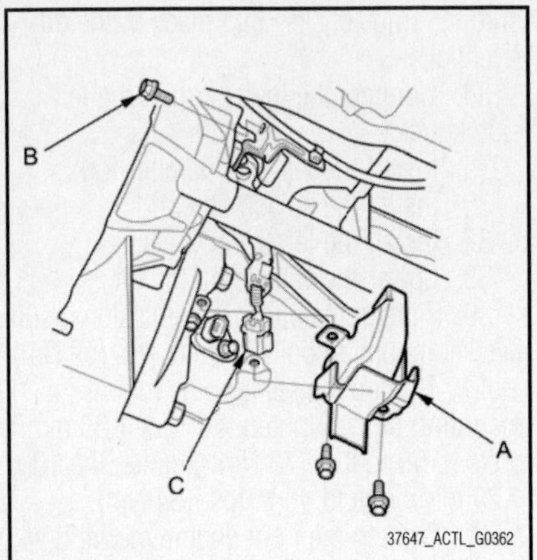
B
C
A
37647_ACTL_G0362

Fig. 184 Remove the CKP sensor cover (A) and the bolt (B), then disconnect the CKP sensor connector (C)

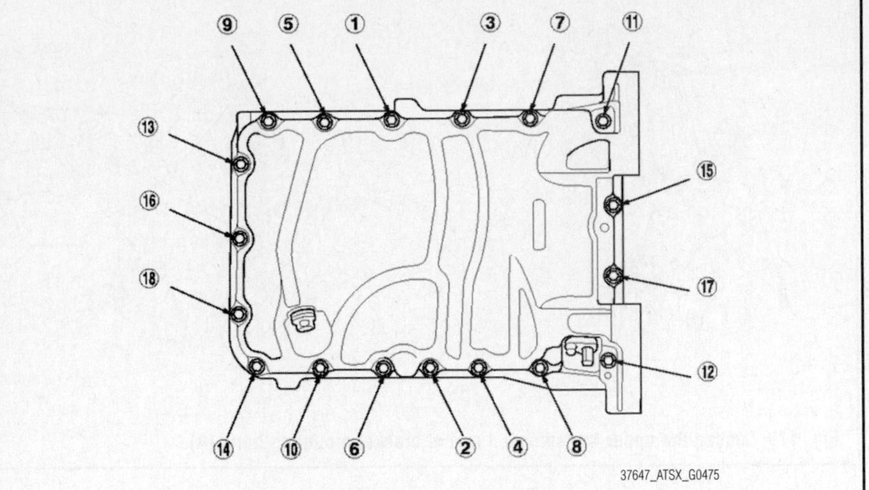
9
5
1
3
7
11
13
16
18
15
17
12
14
10
6
2
4
8
37647_ATSX_G0475

Fig. 187 Oil pan bolt installation sequence

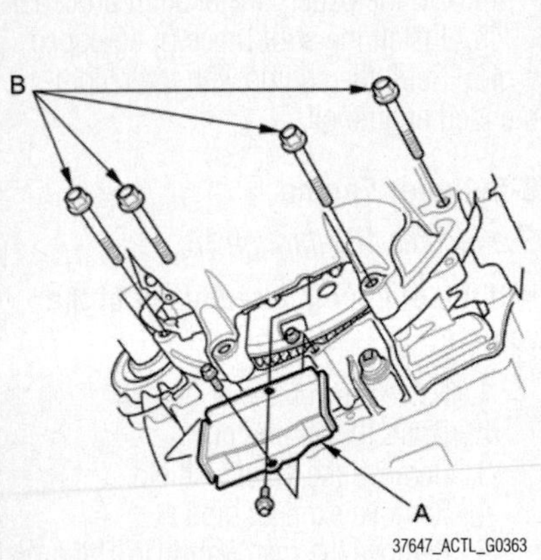
B
A
37647_ACTL_G0363

Fig. 185 Remove the clutch/torque converter cover (A) and the four bolts (B) securing the transmission

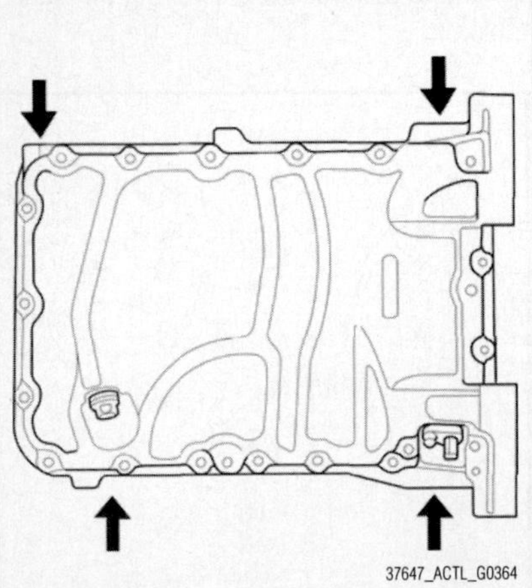
37647_ACTL_G0364

Fig. 186 Using a flat blade screwdriver, separate the oil pan from the engine block in the places shown

8. Remove the bolts securing the oil pan.

9. Using a flat blade screwdriver, separate the oil pan from the engine block in the places shown.

10. Remove the oil pan.

To install:

11. Remove all of the old liquid gasket from the oil pan mating surfaces, the bolts, and the bolt holes.

12. Clean and dry the oil pan mating surfaces.

13. Apply liquid gasket, P/N 08717-0004, 08718-0003, or 08718-0009 to the oil pan mating surface of the engine block and to the inside edge of the threaded bolt holes. Install the component within 5 minutes of applying the liquid gasket.

➡Apply a bead of liquid gasket along the mating edge of the oil pan. If you apply liquid gasket P/N 08718-0012, the component must be installed within 4 minutes. If too much time has passed after applying the liquid gasket, remove the old liquid gasket and residue, then reapply the new liquid gasket.

14. Install the oil pan on the engine block.

15. Tighten the bolts in three steps. In the final step, torque all bolts, in sequence, to 9 ft. lbs. (12 Nm).

➡Wait at least 30 minutes before filling the engine with oil. Do not run the engine for at least 3 hours after installing the oil pan.

16. Tighten the four bolts securing the transmission to 54 ft. lbs. (74 Nm), then install the clutch/torque converter cover.

17. Connect the Crankshaft Position (CKP) sensor connector, then install the CKP sensor cover and the bolt.

18. Install the rear Warm Up Three Way Catalytic Converter (rear WU-TWC) bracket.

19. If the engine is still in the vehicle, do the following steps.

20. Install exhaust pipe A using new gaskets and new self-locking nuts.

21. Install the splash shield.

22. Refill the engine with the recommended engine oil.

OIL PUMP

REMOVAL & INSTALLATION

4-Cylinder Engine

See Figures 188 through 195.

1. Turn the crankshaft pulley so its Top Dead Center (TDC) mark lines up with the pointer.

➡The other pointer is not used.

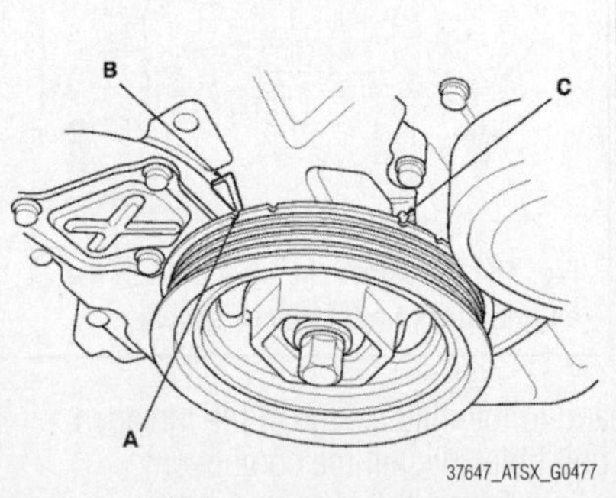
B
C
A
37647_ATSX_G0477

Fig. 188 Turn the crankshaft pulley so its Top Dead Center (TDC) mark (A) lines up with the pointer (B)

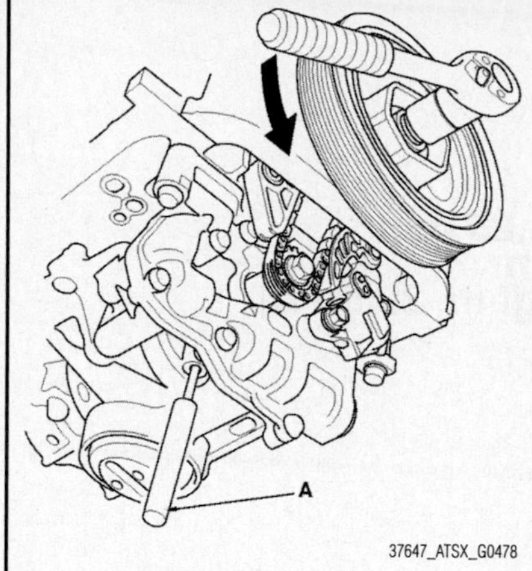
A
37647_ATSX_G0478

Fig. 189 To hold the rear balancer shaft, insert a 6 mm long pin punch (A)

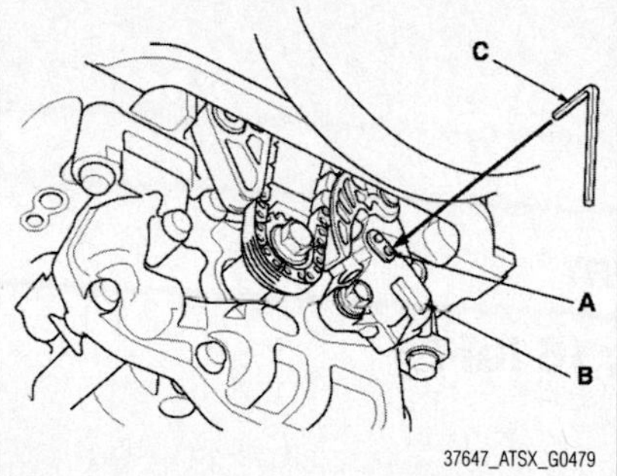
C
A
B
37647_ATSX_G0479

Fig. 190 Align the holes on the lock (A) and the oil pump chain auto-tensioner (B), then insert a 0.12 inches (3.0 mm) diameter pin (C) into the holes

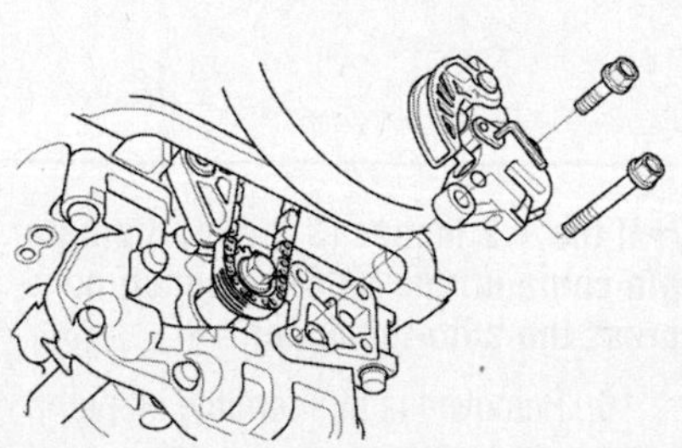
37647_ATSX_G0480

Fig. 191 Remove the oil pump chain auto-tensioner

2. Remove the oil pan.
3. To hold the rear balancer shaft, insert a 6 mm long pin punch (Snap-on PPC108LA or equivalent) into the maintenance hole in the balancer shaft holder and through the rear balancer shaft.

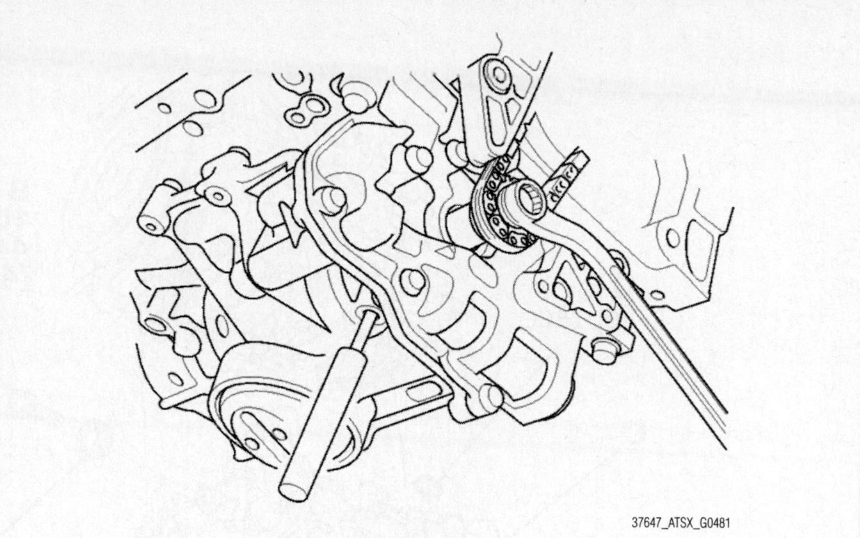
37647_ATSX_G0481

Fig. 192 Loosen the oil pump sprocket mounting bolt.

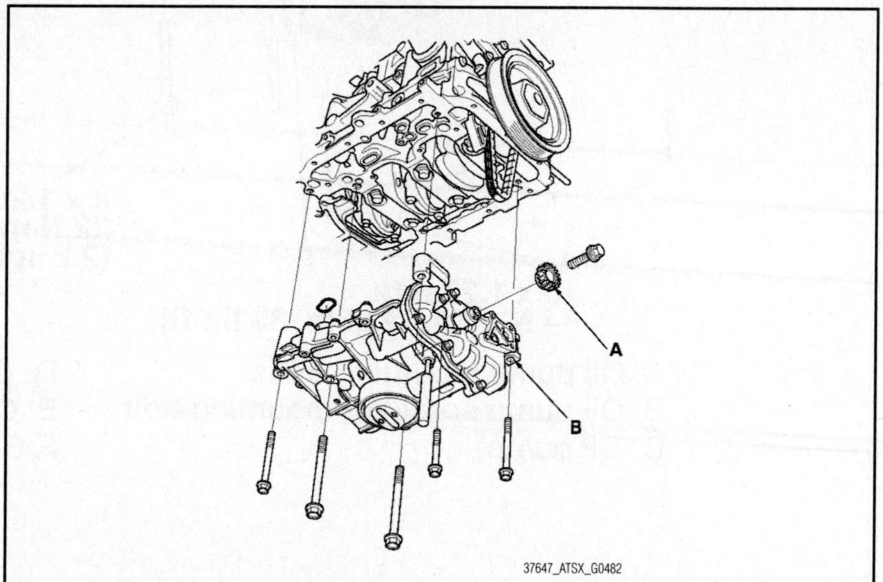
A
B
37647_ATSX_G0482

Fig. 193 Remove the oil pump sprocket (A), then remove the oil pump (B).

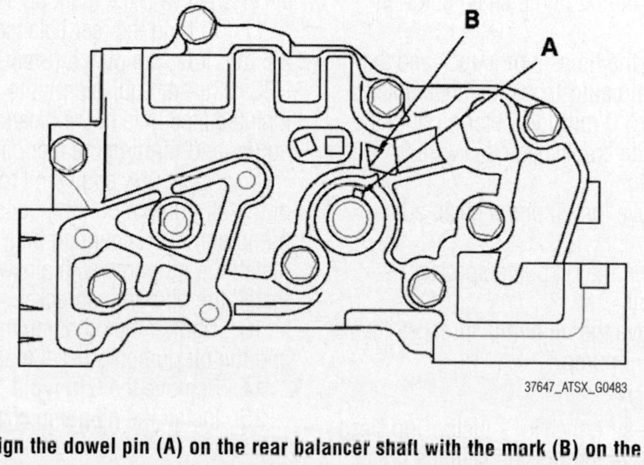
B
A
37647_ATSX_G0483

Fig. 194 Align the dowel pin (A) on the rear balancer shaft with the mark (B) on the oil pump.

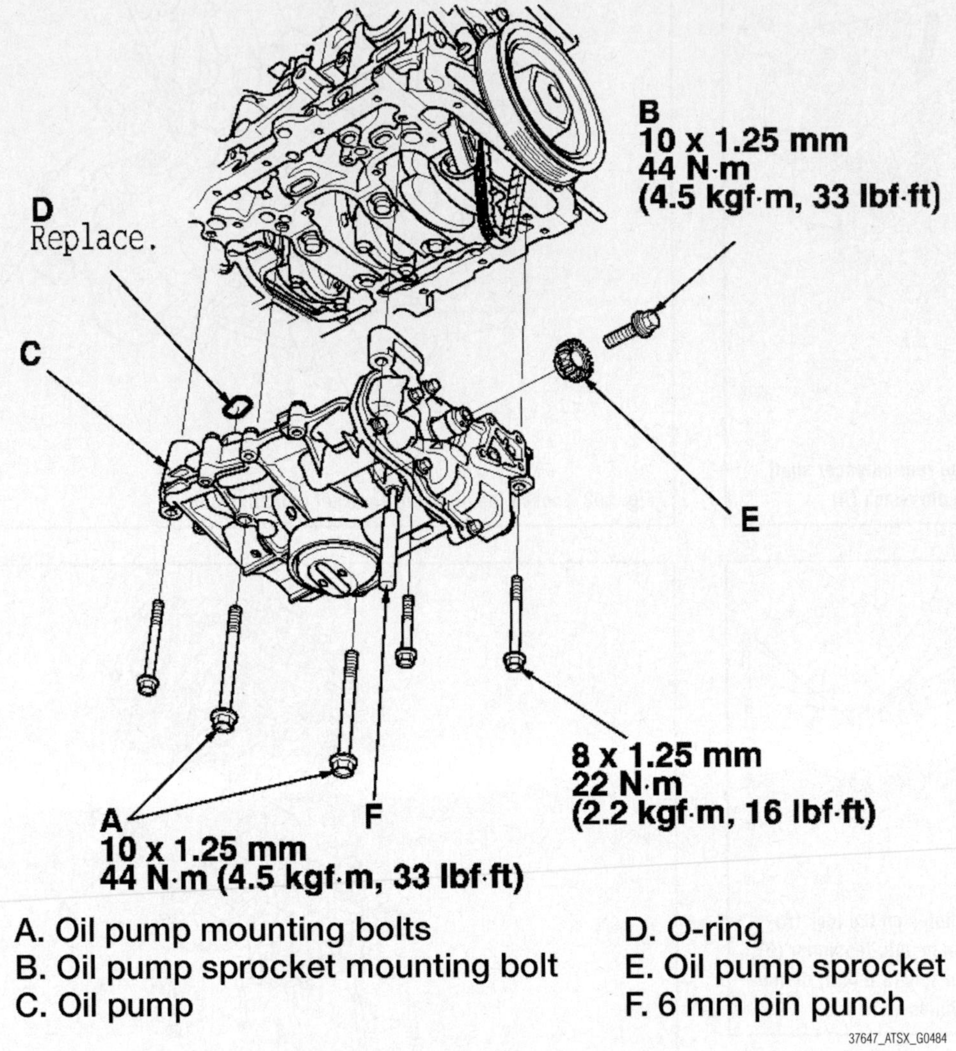
B
10 x 1.25 mm
44 N·m
(4.5 kgf·m, 33 lbf·ft)
D
Replace.
C
E
8 x 1.25 mm
22 N·m
(2.2 kgf·m, 16 lbf·ft)
A
10 x 1.25 mm
44 N·m (4.5 kgf·m, 33 lbf·ft)
F
A. Oil pump mounting bolts
B. Oil pump sprocket mounting bolt
C. Oil pump
D. O-ring
E. Oil pump sprocket
F. 6 mm pin punch
37647_ATSX_G0484

Fig. 195 Install the oil pump with a new O-ring, then install the oil pump sprocket

4. Turn the crankshaft counterclockwise to compress the oil pump chain auto-tensioner.
5. Align the holes on the lock and the oil pump chain auto-tensioner, then insert a 0.12 inches (3.0 mm) diameter pin into the holes. Turn the crankshaft clockwise to secure the pin.
6. Remove the oil pump chain auto-tensioner.
7. Loosen the oil pump sprocket mounting bolt.
8. Remove the oil pump sprocket, then remove the oil pump.

To install:

9. Make sure the No. 1 piston Top Dead Center (TDC) mark lines up with the pointer.
10. Align the dowel pin on the rear balancer shaft with the mark on the oil pump.
11. To hold the rear balancer shaft, insert a 6 mm long pin punch (Snap-on PPC108LA or equivalent) into the maintenance hole in the balancer shaft holder and through the rear balancer shaft.
12. Apply new engine oil to the threads of the oil pump mounting bolts and oil pump sprocket mounting bolt, then loosely install the oil pump with a new O-ring, then install the oil pump sprocket.
13. Tighten the oil pump mounting bolts and the oil pump sprocket mounting bolt.
14. Remove the 6 mm pin punch.
15. Install the oil pump chain auto-tensioner.

➡If the 1.2 inches (3.0 mm) diameter pin come off the auto-tensioner, compress the auto-tensioner.

16. Remove the pin from the oil pump chain auto-tensioner.
17. Install the oil pan.

6-Cylinder Engine

See Figures 196 through 199.

1. Drain the engine oil.
2. Remove the timing belt:
3. Remove the timing belt drive pulley from the crankshaft:
4. Attach the chain hoist to the engine hook on the engine hanger bracket.

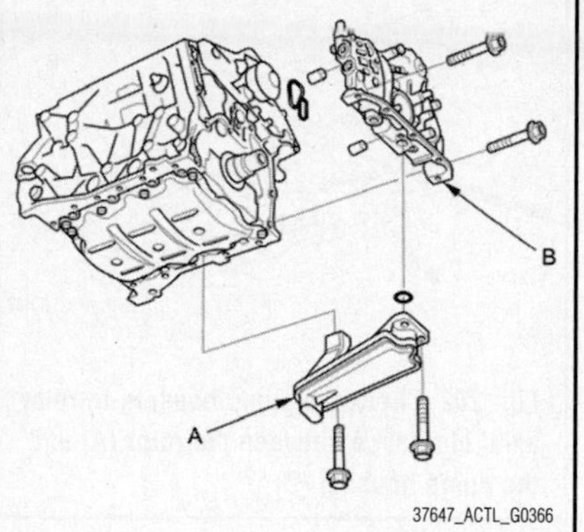
A
B
37647_ACTL_G0366

Fig. 196 Remove the oil strainer (A), then remove the oil pump assembly (B)

5. Remove the rocker arm oil control solenoid/oil filter assembly.
6. Remove the oil pan.
7. Remove the oil strainer.
8. Remove the mounting bolts, then remove the oil pump assembly.

To install:

9. Remove the old oil seal from the oil pump.
10. Clean and dry the crankshaft oil seal housing.
11. Using the oil seal driver, 64 mm, drive in the new crankshaft oil seal until the oil seal driver bottoms on the pump.
12. Remove all of the old liquid gasket from the oil pump mating surfaces, the bolts, and the bolt holes.

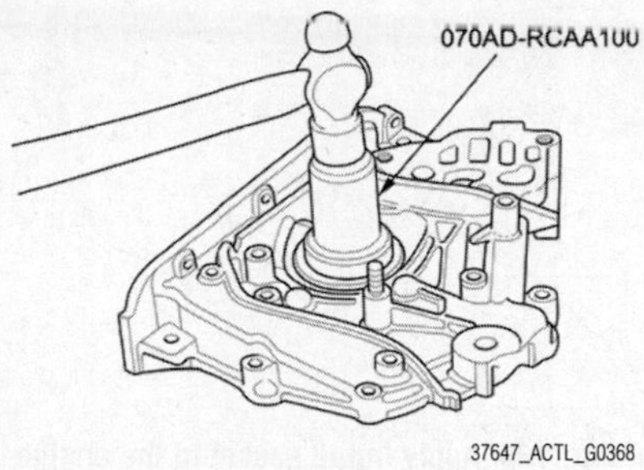
070AD-RCAA100
37647_ACTL_G0368

Fig. 198 Using the oil seal driver, 64 mm, drive in the new crankshaft oil seal until the oil seal driver bottoms on the pump

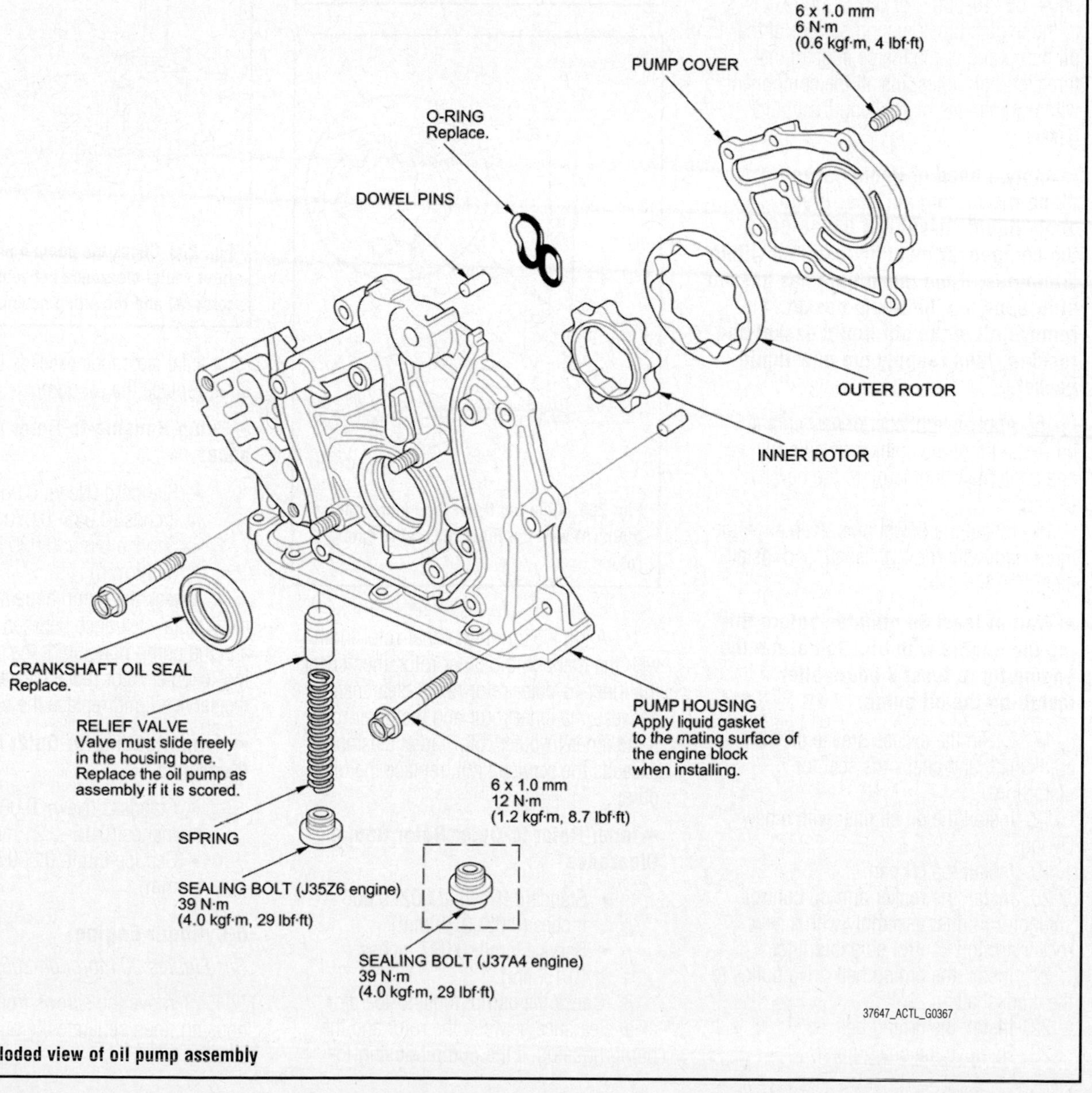
6 x 1.0 mm
6 N·m
(0.6 kgf·m, 4 lbf·ft)
PUMP COVER
O-RING
Replace.
DOWEL PINS
OUTER ROTOR
INNER ROTOR
CRANKSHAFT OIL SEAL
Replace.
RELIEF VALVE
Valve must slide freely in the housing bore. Replace the oil pump as assembly if it is scored.
PUMP HOUSING
Apply liquid gasket to the mating surface of the engine block when installing.
6 x 1.0 mm
12 N·m
(1.2 kgf·m, 8.7 lbf·ft)
SPRING
SEALING BOLT (J35Z6 engine)
39 N·m
(4.0 kgf·m, 29 lbf·ft)
SEALING BOLT (J37A4 engine)
39 N·m
(4.0 kgf·m, 29 lbf·ft)
37647_ACTL_G0367

Fig. 197 Exploded view of oil pump assembly

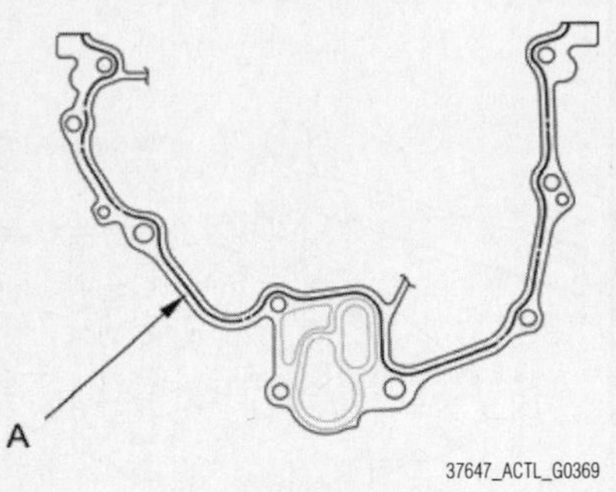
A
37647_ACTL_G0369

Fig. 199 Apply liquid gasket to the engine block mating surface of the oil pump and to the inside edge of the threaded bolt holes

13. Clean and dry the oil pump mating surfaces.

14. Apply liquid gasket, P/N 08717-0004, 08718-0003, or 08718-0009, to the engine block mating surface of the oil pump and to the inside edge of the threaded bolt holes. Install the component within 5 minutes of applying the liquid gasket.

➡Apply a bead of liquid gasket along the mating surface. If you apply liquid gasket P/N 08718-0012, the component must be installed within 5 minutes. If too much time has passed after applying the liquid gasket, remove all of the old liquid gasket and residue, then reapply the new liquid gasket.

15. Apply a light coat of new engine oil to the lip of the crankshaft oil seal, and apply new engine oil to the new O-ring.

16. Install the dowel pins, then align the inner rotor with the crankshaft, and install the oil pump.

➡Wait at least 30 minutes before filling the engine with oil. Do not run the engine for at least 3 hours after installing the oil pump.

17. Clean the excess grease off the crankshaft, and check the seal for distortion.

18. Install the oil strainer with a new O-ring.

19. Install the oil pan.

20. Install the rocker arm oil control solenoid/oil filter assembly with a new rocker arm oil control solenoid filter.

21. Install the timing belt drive pulley to the crankshaft:

22. Install the timing belt:

23. Remove the chain hoist.

INSPECTION

4-Cylinder Engine

See Figures 200 through 203.

1. Remove the pump housing.

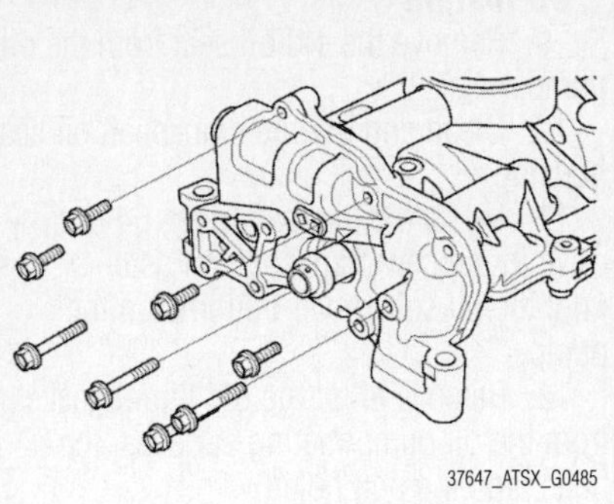
37647_ATSX_G0485

Fig. 200 Remove the pump housing

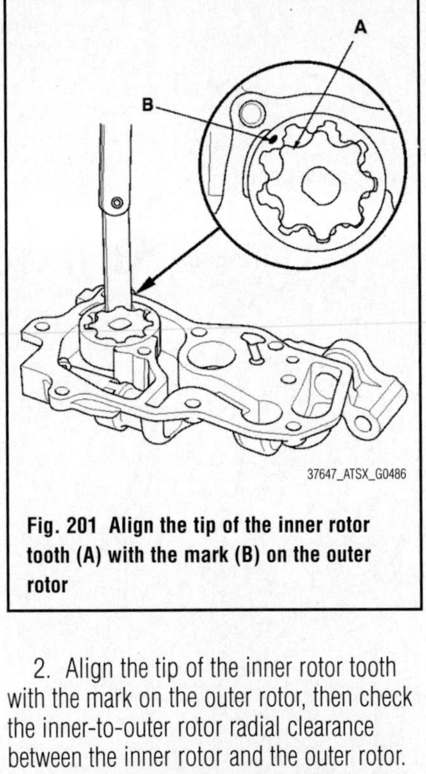
A
B
37647_ATSX_G0486

Fig. 201 Align the tip of the inner rotor tooth (A) with the mark (B) on the outer rotor

2. Align the tip of the inner rotor tooth with the mark on the outer rotor, then check the inner-to-outer rotor radial clearance between the inner rotor and the outer rotor. If the inner-to-outer rotor radial clearance exceeds the service limit, replace the oil pump.

➡Inner Rotor-to-Outer Rotor Radial Clearance

- Standard (New): 0.002–0.006 inches (0.05–0.15 mm)
- Service Limit: 0.007 inches (0.19 mm)

3. Check the pump housing-to-rotor axial clearance between the rotor and the pump housing. If the pump housing-to-rotor axial clearance exceeds the service limit, replace the oil pump.

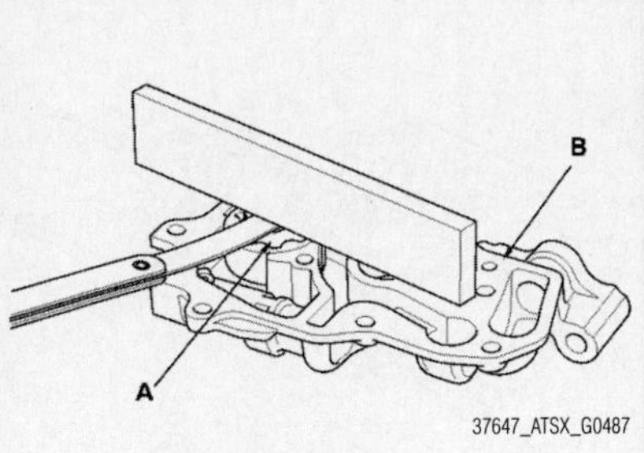
B
A
37647_ATSX_G0487

Fig. 202 Check the pump housing-to-rotor axial clearance between the rotor (A) and the pump housing (B)

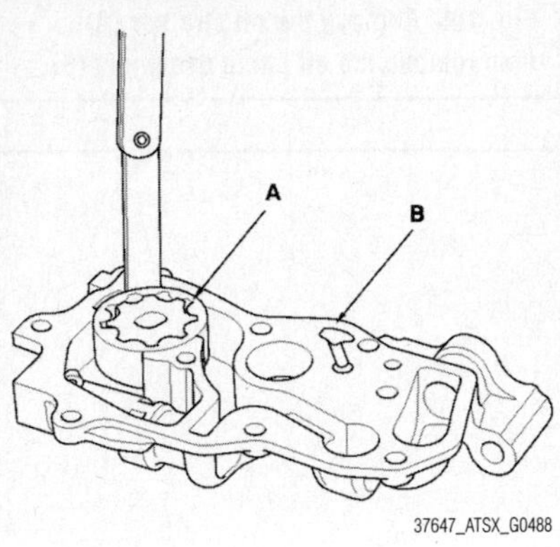
A
B
37647_ATSX_G0488

Fig. 203 Check the pump housing-to-outer rotor radial clearance between the outer rotor (A) and the pump housing (B)

➡Pump Housing-to-Rotor Axial Clearance

- Standard (New): 0.0014–0.0028 inches (0.035–0.070 mm)
- Service Limit: 0.005 inches (0.12 mm)

4. Check the pump housing-to-outer rotor radial clearance between the outer rotor and the pump housing. If the pump housing-to-outer rotor radial clearance exceeds the service limit, replace the oil pump.

➡Pump Housing-to-Outer Rotor Radial Clearance

- Standard (New): 0.0059–0.0083 inches (0.15–0.21 mm)
- Service Limit: 0.009 inches (0.23 mm)

6-Cylinder Engine

See Figures 204 through 206.

1. Remove the screws from the pump housing, then separate the pump housing and the cover.

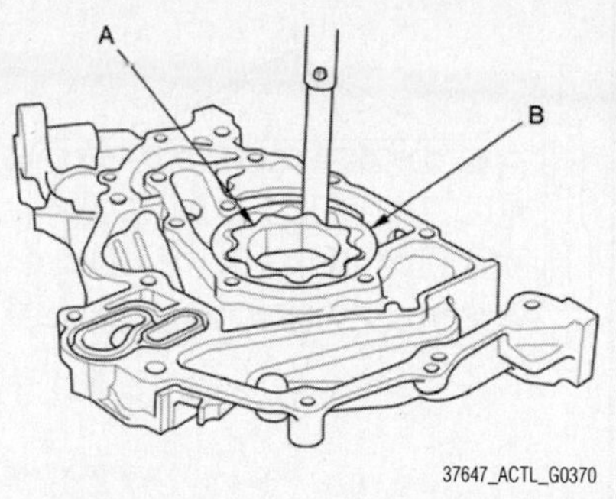
A
B
37647_ACTL_G0370
Fig. 204 Check the inner-to-outer rotor radial clearance between the inner rotor (A) and the outer rotor (B)

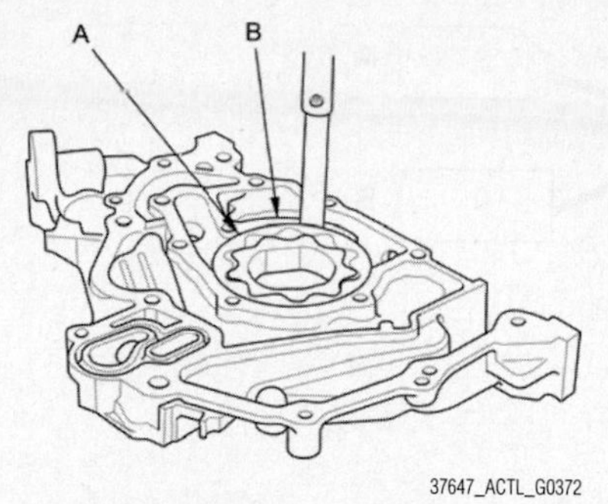
A
B
37647_ACTL_G0372
Fig. 206 Check the pump housing-to-outer rotor radial clearance between the outer rotor (A) and the pump housing (B)

2. Check the inner-to-outer rotor radial clearance between the inner rotor and the outer rotor. If the inner-to-outer rotor radial clearance exceeds the service limit, replace the oil pump assembly.

➡Inner Rotor-to-Outer Rotor Radial Clearance

- Standard (New): 0.002–0.006 inches (0.04–0.16 mm)
- Service Limit: 0.008 inches (0.20 mm)

3. Check the pump housing-to-rotor axial clearance between the rotors and the pump housing. If the pump housing-to-rotor axial clearance exceeds the service limit, replace the oil pump assembly.

➡Pump Housing-to-Rotor Axial Clearance

- Standard (New): 0.001–0.003 inches (0.02–0.07 mm)
- Service Limit: 0.005 inches (0.12 mm)

4. Check the pump housing-to-outer rotor radial clearance between the outer rotor and the pump housing. If the pump housing-to-outer rotor radial clearance exceeds the service limit, replace the oil pump assembly.

➡Pump Housing-to-Outer Rotor Radial Clearance

- Standard (New): 0.004–0.007 inches (0.10–0.19 mm)
- Service Limit: 0.008 inches (0.20 mm)

5. Inspect both rotors and the pump housing for scoring or other damage. Replace the parts, if necessary.

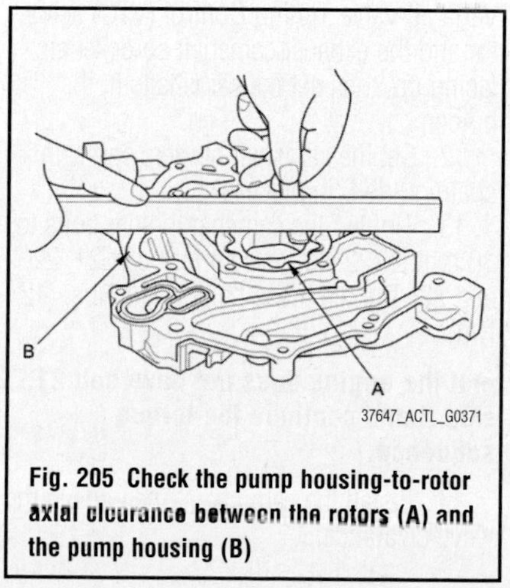
B
A
37647_ACTL_G0371
Fig. 205 Check the pump housing-to-rotor axial clearance between the rotors (A) and the pump housing (B)

6. Apply liquid thread lock to the pump housing screws, then install the oil pump cover.
7. Check that the oil pump turns freely.

PISTON & RING

POSITIONING

4-Cylinder Engine

See Figure 207.

6-Cylinder Engine

See Figure 208.

REAR MAIN SEAL

REMOVAL & INSTALLATION

4-Cylinder Engine

Special Tools Required:

- Driver handle, 15 x 135L 07749-0010000
- Oil seal driver attachment 96 mm 07ZAD-PNAA100

1. Remove the transmission:
2. M/T model: Remove the pressure plate, the clutch disk, and the flywheel.

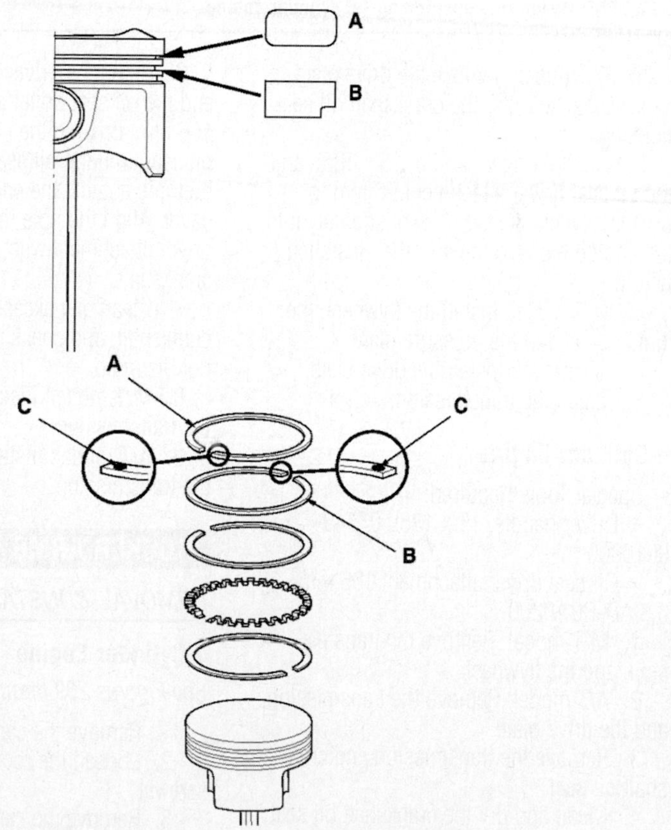
A
B
A
C
C
B
37647_ATSX_G0492
Fig. 207 Piston ring positioning—4 cylinder engine

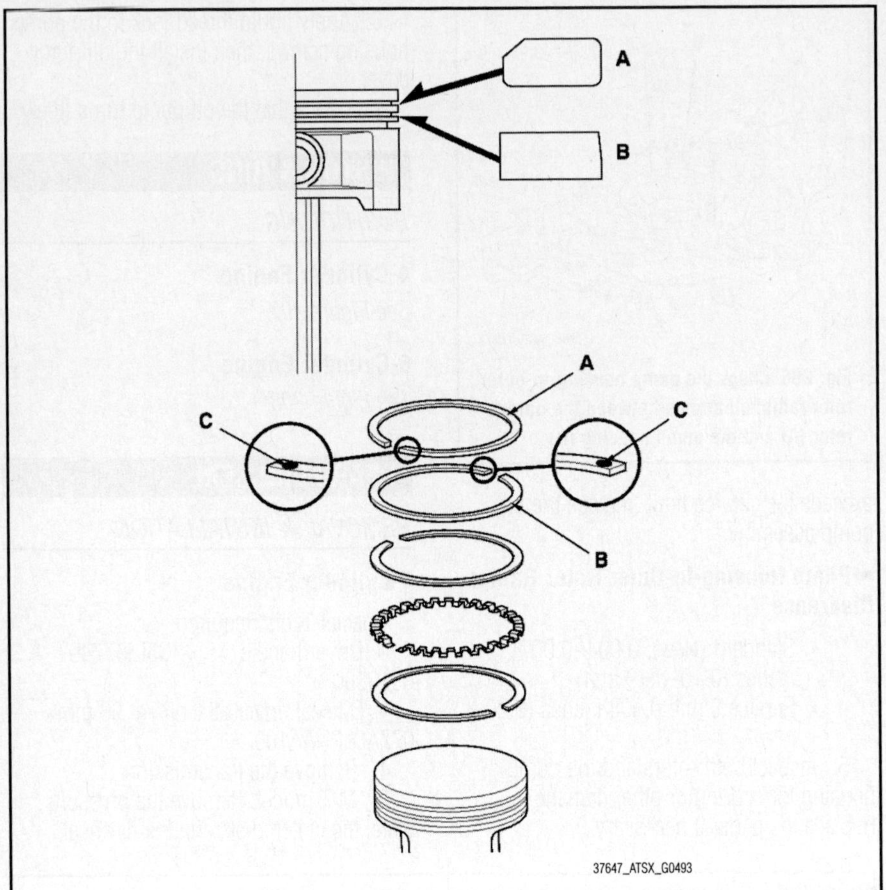
A
B
A
C
C
B
37647_ATSX_G0493

Fig. 208 Piston ring positioning—6 cylinder engine

3. A/T model: Remove the drive plate.
4. Clean and dry the crankshaft oil seal housing.
5. Use the driver handle, 15 x 135L and the oil seal driver attachment 96 mm to drive a new crankshaft oil seal squarely into the engine block to the specified installed height.
6. M/T model: Install the flywheel, the clutch disk, and the pressure plate.
7. A/T model: Install the drive plate.
8. Install the transmission:

6-Cylinder Engine

Special Tools Required:

- Driver handle, 15 x 135L 07749-0010000
- Oil seal driver attachment 106 mm 070AD-RCA0200

1. M/T model: Remove the transmission, and the flywheel.
2. A/T model: Remove the transmission, and the drive plate.
3. Remove the transmission end crankshaft oil seal.
4. Clean and dry the crankshaft oil seal housing.
5. Apply a light coat of new engine oil to the lip of the crankshaft oil seal.
6. Using the driver handle, 15 x 135 L and the oil seal driver attachment, 106 mm, drive in the new crankshaft oil seal until the oil seal driver attachment bottoms against the engine block end cover. Align the hole in the oil seal driver attachment with the pin on the crankshaft.
7. Clean any excess grease off the crankshaft, and check that the oil seal lip is not distorted.
8. M/T model: Install the flywheel, and the transmission.
9. A/T model: Install the drive plate, and the transmission.

ROCKER ARMS

REMOVAL & INSTALLATION

4-Cylinder Engine

See Figures 209 through 213.

1. Remove the cam chain.
2. Loosen the rocker arm adjusting screws.
3. Remove the camshaft holder bolts. To prevent damaging the camshafts, loosen the bolts, in sequence, two turns at a time.

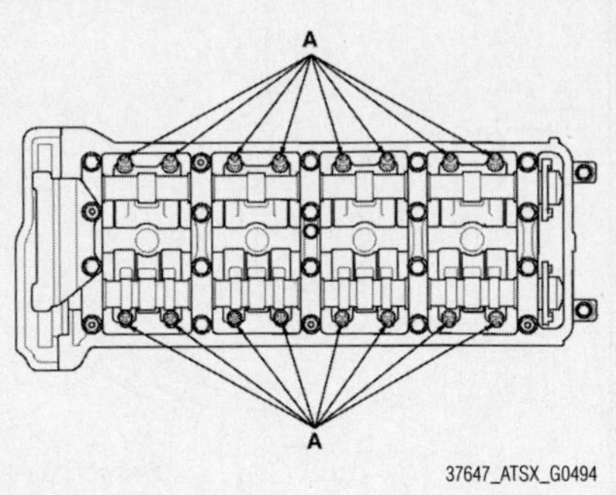
A
A
37647_ATSX_G0494

Fig. 209 Loosen the rocker arm adjusting screws (A)

➡**Bolt #1 is not on all engines.**

4. Remove cam chain guide B, the camshaft holders, and the camshafts.
5. Insert the bolts into the rocker shaft holder, then remove the rocker arm assembly.

To install:

6. Reassemble the rocker arm assembly.
7. Clean and dry the No. 5 rocker shaft holder mating surface.
8. Apply liquid gasket (P/N 08717-0004, 08718-0003, or 08718-0009) to the cylinder head mating surface of the No. 5 rocker shaft holder and to the inside edge of the threaded bolt holes. Install the component within 5 minutes of applying the liquid gasket.

➡ **If too much time has passed after applying the liquid gasket, remove the old liquid gasket and residue, then reapply new liquid gasket.**

9. Insert the bolts into the rocker shaft holder, then install the rocker arm assembly on the cylinder head.
10. Remove the bolts from the rocker shaft holder.
11. Make sure the punch marks on the Variable Valve Timing Control (VTC) actuator and the exhaust camshaft sprocket are facing up, then set the camshafts in the holder.
12. Set the camshaft holders and cam chain guide B in place
13. Tighten the camshaft holder bolts to 16 ft. lbs. (22 Nm) except for bolts 21, 22, and 23. Tighten these bolts to 9 ft. lbs. (12 Nm).

➡**If the engine does not have bolt 21, skip it and continue the torque sequence.**

14. Install the cam chain, then adjust the valve clearance.

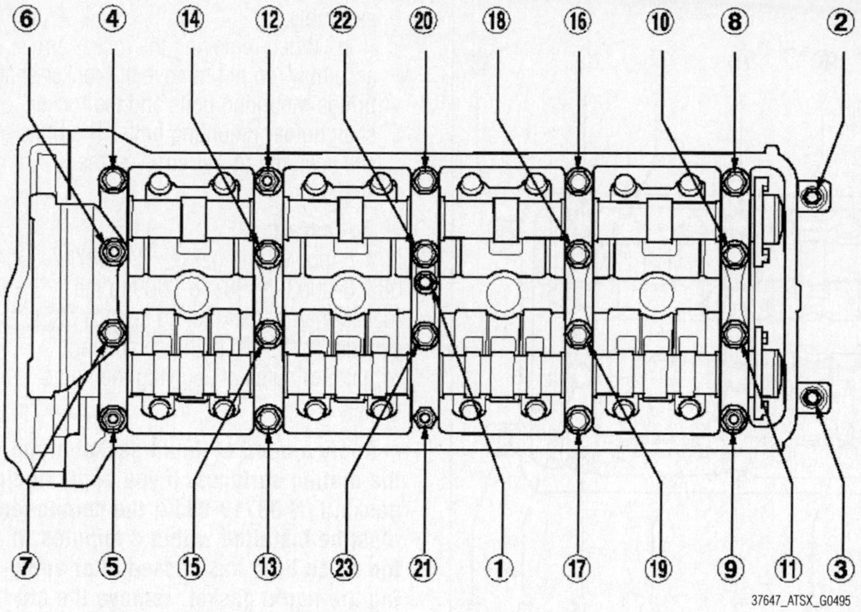
6
4
14
12
22
20
18
16
10
8
2
7
5
15
13
23
21
1
17
19
9
11
3
37647_ATSX_G0495

Fig. 210 Remove the camshaft holder bolts in sequence

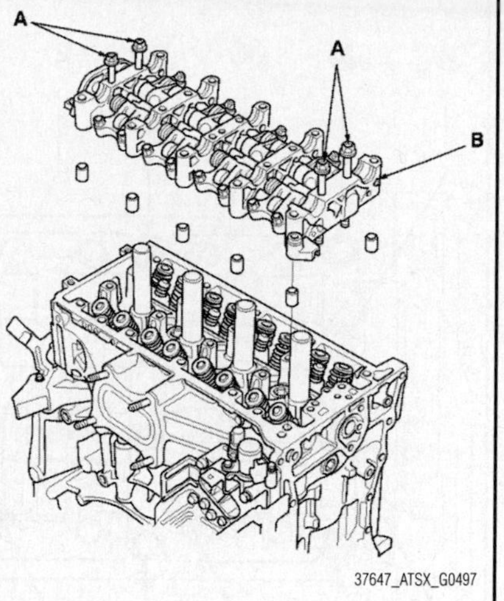
A
A
B
37647_ATSX_G0497

Fig. 212 Insert the bolts (A) into the rocker shaft holder, then remove the rocker arm assembly (B)

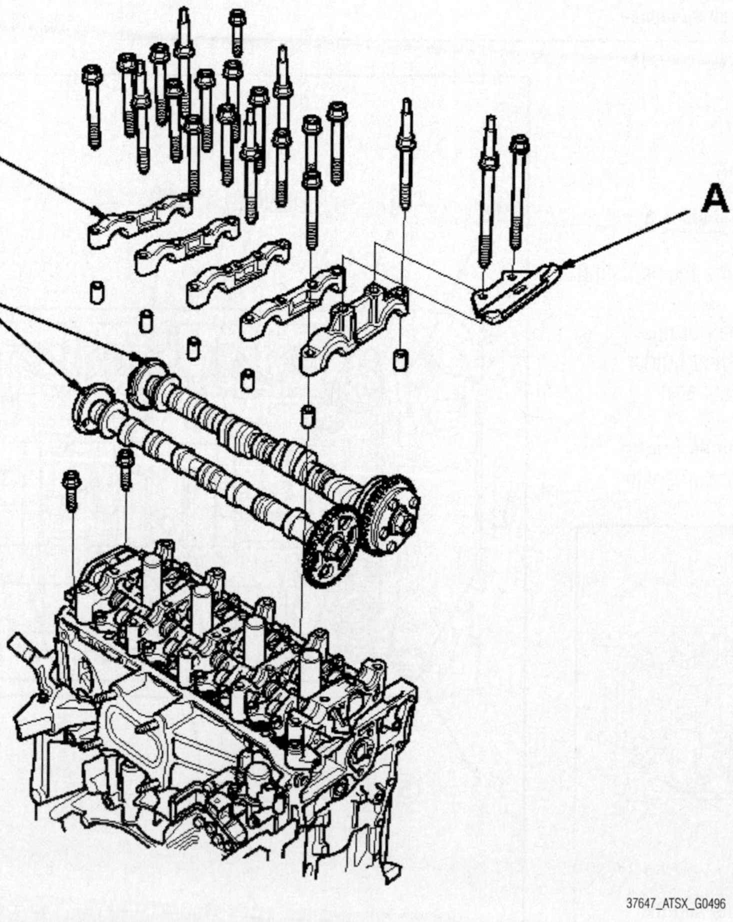
B
A
C
37647_ATSX_G0496

Fig. 211 Remove cam chain guide B (A), the camshaft holders (B), and the camshafts (C)

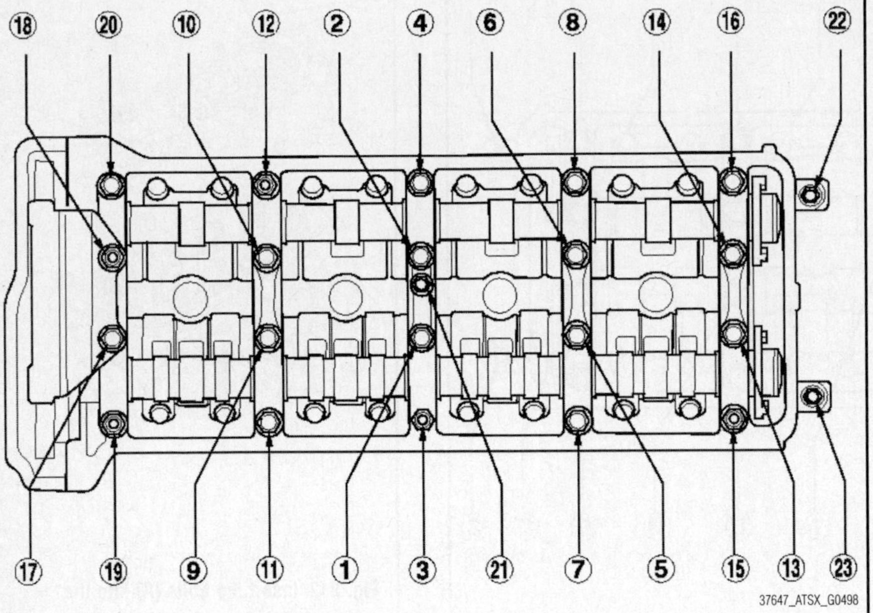
18 20 10 12 2 4 6 8 14 16 22
17 19 9 11 1 3 21 7 5 15 13 23
37647_ATSX_G0498

Fig. 213 Camshaft holder bolt sequence

6-Cylinder Engine

Front Cylinder Head

See Figures 214 through 216.

1. Remove the cylinder head cover.
2. Loosen the locknuts and the adjusting screws
3. Remove the rocker shaft bridge mounting bolts, the rocker shaft holder mounting bolts, and the rocker arm assembly.
   a. Loosen the rocker shaft bridge mounting bolts and the rocker shaft holder mounting bolts in sequence two turns at a time, to prevent damaging the valves or the rocker arm assembly.
   b. When removing the rocker arm assembly, do not remove the rocker shaft bridge mounting bolts and the rocker shaft holder mounting bolts. The bolts will keep the rocker arms on the shafts.

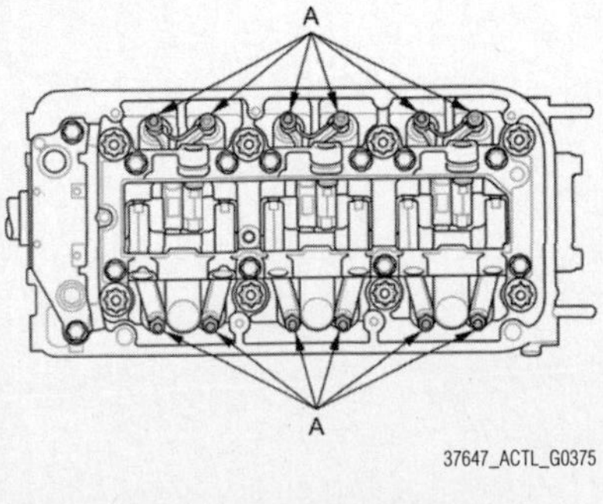
A
A
37647_ACTL_G0375

Fig. 214 Loosen the locknuts and the adjusting screws (A)

To install:

4. Apply liquid gasket, P/N 08717-0004, 08718-0003, or 08718-0009 to the rocker shaft holder mating surface of the cylinder head. Install the component within 5 minutes of applying the liquid gasket.

➡Apply a bead of liquid gasket along the mating surfaces. If you apply liquid gasket P/N 08718-0012, the component must be installed within 4 minutes. If too much time has passed after applying the liquid gasket, remove the old liquid gasket and residue, then reapply the new liquid gasket.

5. Set the rocker arm assembly in place, and loosely install the bolts. Make sure that the rocker arms are properly positioned on the valve stems.

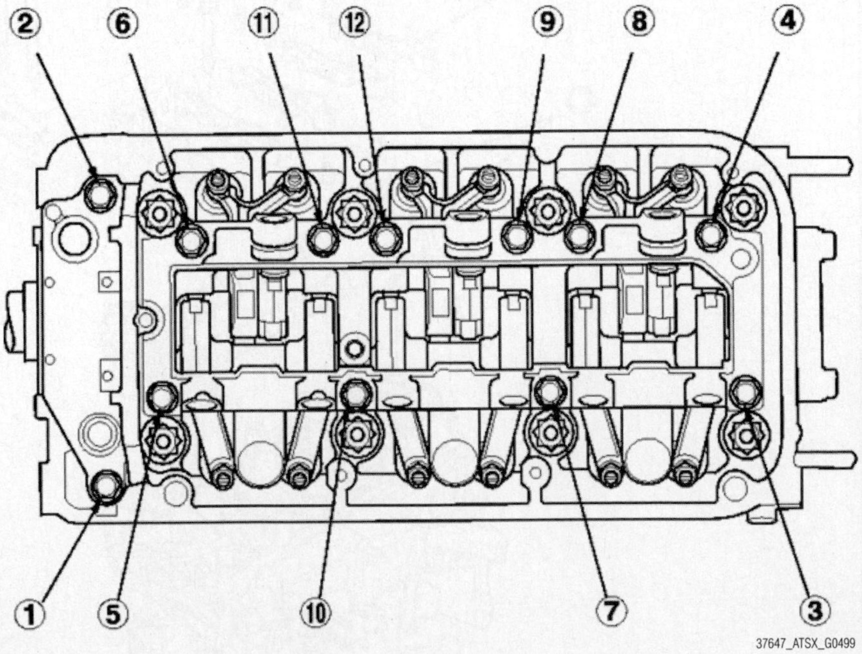
2 6 11 12 9 8 4
1 5 10 7 3
37647_ATSX_G0499

Fig. 215 Front rocker shaft bridge bolt loosening sequence

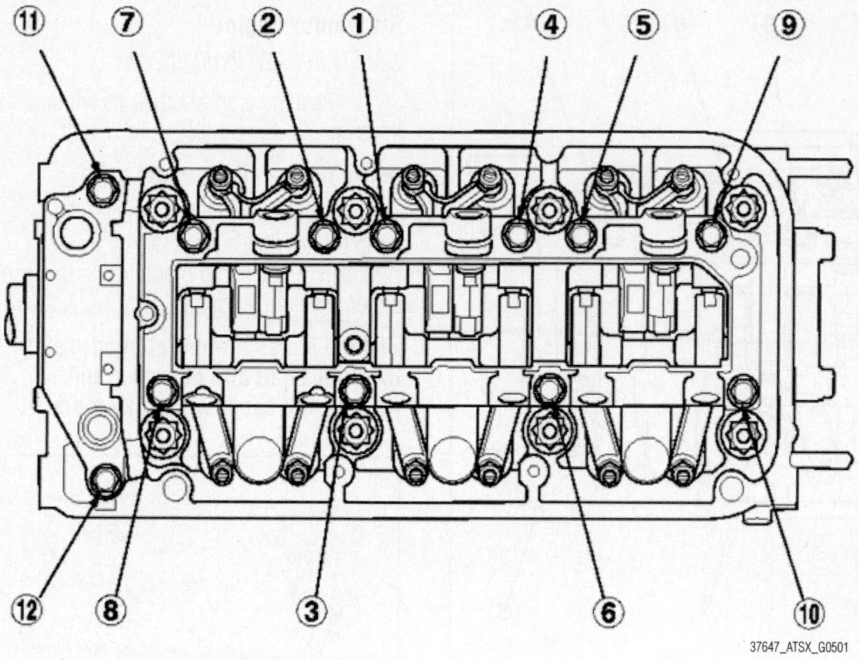
11
7
2
1
4
5
9
12
8
3
6
10
37647_ATSX_G0501

Fig. 216 Tighten each rocker shaft bridge bolt two turns at a time in sequence

To install:

4. Apply liquid gasket, P/N 08717-0004, 08718-0003, or 08718-0009 to the rocker shaft holder mating surface of the cylinder head. Install the component within 5 minutes of applying the liquid gasket.

➡Apply a bead of liquid gasket along the mating surfaces. If you apply liquid gasket P/N 08718-0012, the component must be installed within 4 minutes. If too much time has passed after applying the liquid gasket, remove the old liquid gasket and residue, then reapply the new liquid gasket.

5. Set the rocker arm assembly in place, and loosely install the bolts. Make sure that the rocker arms are properly positioned on the valve stems.

➡Wait at least 30 minutes before filling the engine with oil. Do not run the engine for at least 3 hours after installing the rocker arm assembly.

6. Tighten each bolt two turns at a time in the sequence shown to ensure that the rockers do not bind on the valves.

➡Wait at least 30 minutes before filling the engine with oil. Do not run the engine for at least 3 hours after installing the rocker arm assembly.

6. Tighten each bolt two turns at a time in the sequence shown to ensure that the rockers do not bind on the valves.

Rear Cylinder Head

See Figures 217 through 219.

1. Remove the cylinder head cover.
2. Loosen the locknuts and the adjusting screws.
3. Remove the rocker shaft bridge mounting bolts, the rocker shaft holder mounting bolts, and the rocker arm assembly.
   a. Loosen the rocker shaft bridge mounting bolts and the rocker shaft holder mounting bolts in sequence two turns at a time, to prevent damaging the valves or the rocker arm assembly.
   b. When removing the rocker arm assembly, do not remove the rocker shaft bridge mounting bolts and the rocker shaft holder mounting bolts. The bolts will keep the rocker arms on the shafts.

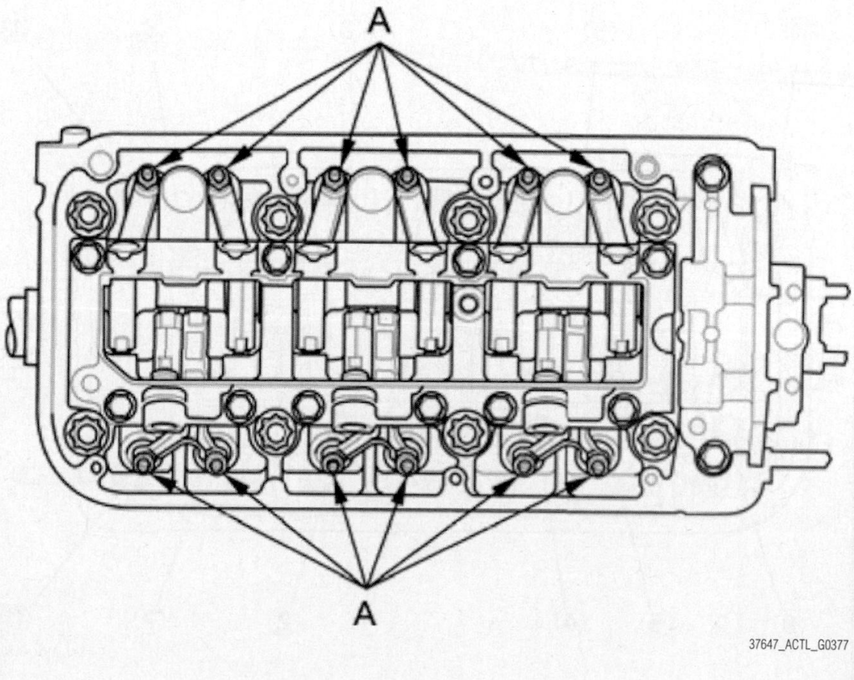
A
A
37647_ACTL_G0377

Fig. 217 Loosen the locknuts and the adjusting screws (A)

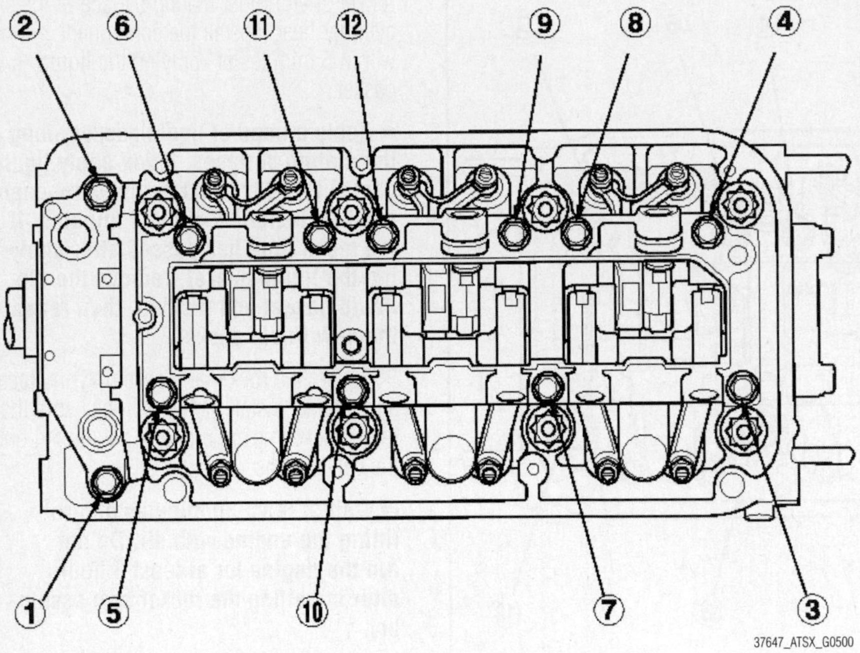
2
6
11
12
9
8
4
1
5
10
7
3
37647_ATSX_G0500

Fig. 218 Rear cylinder head bolt loosening sequence

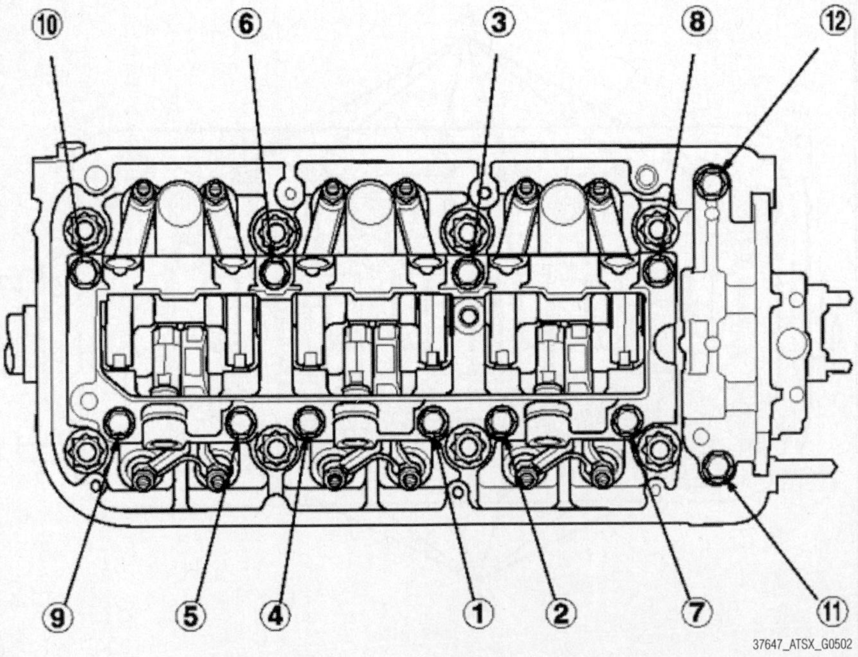
10
6
3
8
12
9
5
4
1
2
7
11
37647_ATSX_G0502

Fig. 219 Tighten each rocker shaft bridge bolt two turns at a time in sequence

TIMING BELT FRONT COVER

REMOVAL & INSTALLATION

6-Cylinder Engine

See Figures 220 through 224.

1. Turn the crankshaft so its white mark lines up with the pointer.

➡The other pointer is not used.

2. Check that the No. 1 piston Top Dead Center (TDC) mark on the front camshaft pulley and the pointer on the front upper cover are aligned.

➡If the marks are not aligned, rotate the crankshaft 360 degrees, and recheck the camshaft pulley mark.

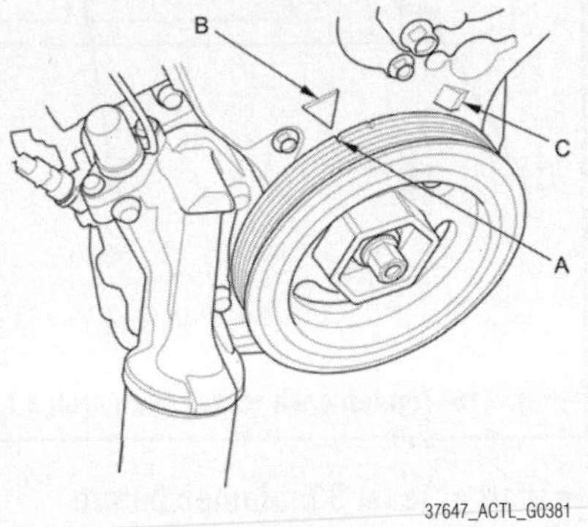
B
C
A
37647_ACTL_G0381

Fig. 220 Turn the crankshaft so its white mark (A) lines up with the pointer (B)

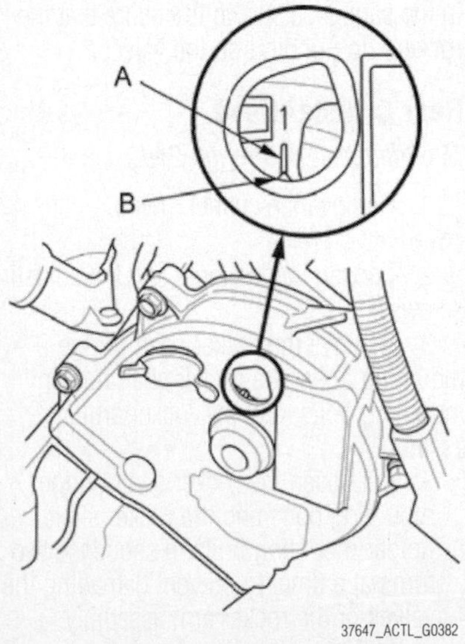
A
1
B
37647_ACTL_G0382

Fig. 221 Check that the No. 1 piston TDC mark (A) on the front camshaft pulley and the pointer (B) on the front upper cover are aligned

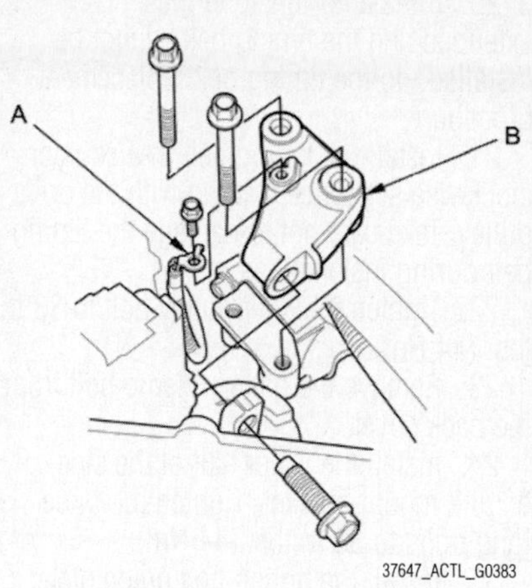
A
B
37647_ACTL_G0383
Fig. 222 Remove the ground cable (A), then remove the upper half of the side engine mount bracket (B)

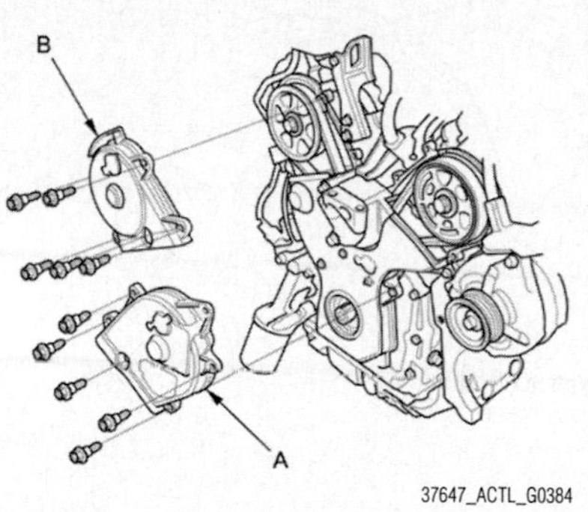
B
A
37647_ACTL_G0384
Fig. 223 Remove the front upper cover (A) and the rear upper cover (B)

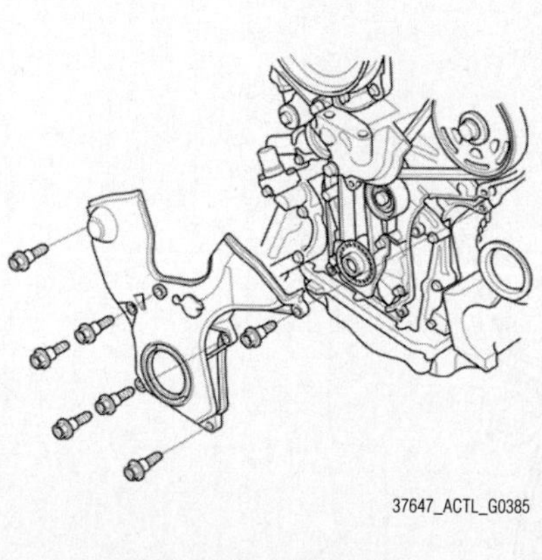
37647_ACTL_G0385
Fig. 224 Remove the lower cover

3. Raise the vehicle on the lift, then remove the right front wheel.
4. Remove the splash shield.
5. Remove the drive belt auto-tensioner.
6. Support the engine with a jack and a wood block under the oil pan.
7. Remove the ground cable, then remove the upper half of the side engine mount bracket.
8. Remove the crankshaft pulley.
9. Remove the front upper cover and the rear upper cover.
10. Remove the lower cover.

To install:

11. Install the lower cover.
12. Install the front upper cover and the rear upper cover.
13. Install the crankshaft pulley.
14. Rotate the crankshaft pulley about six turns clockwise so the timing belt positions itself on the pulleys.
15. Turn the crankshaft pulley so its white mark lines up with the pointer.

➡The other pointer is not used.

16. Check the camshaft pulley marks.

➡If the marks are not aligned, rotate the crankshaft 360 degrees, and recheck the camshaft pulley mark. If the camshaft pulley marks are at TDC, go to step 17. If the camshaft pulley marks are not at TDC, remove the timing belt and repeat steps 2 through 16.

17. Install the upper half of the side engine mount bracket, then tighten the mounting bolts.
18. Install the ground cable.
19. Install the drive belt auto-tensioner.
20. Install the splash shield.
21. Install the right front wheel.

TIMING BELT & SPROCKETS

REMOVAL & INSTALLATION

6-Cylinder Engine

See Figures 220 through 232.

1. Turn the crankshaft so its white mark lines up with the pointer.

➡The other pointer is not used.

2. Check that the No. 1 piston Top Dead Center (TDC) mark on the front camshaft pulley and the pointer on the front upper cover are aligned.

➡If the marks are not aligned, rotate the crankshaft 360 degrees, and recheck the camshaft pulley mark.

3. Raise the vehicle on the lift, then remove the right front wheel.
4. Remove the splash shield.
5. Remove the drive belt auto-tensioner.
6. Support the engine with a jack and a wood block under the oil pan.
7. Remove the ground cable, then remove the upper half of the side engine mount bracket.

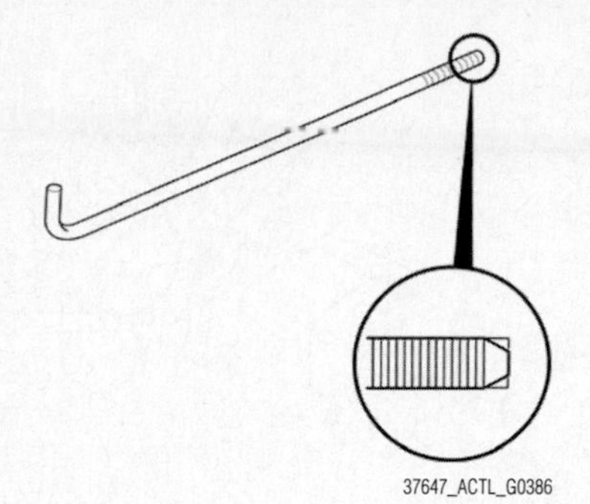
37647_ACTL_G0386
Fig. 225 Remove one of the battery clamp bolts from the battery tray, and grind the end of it

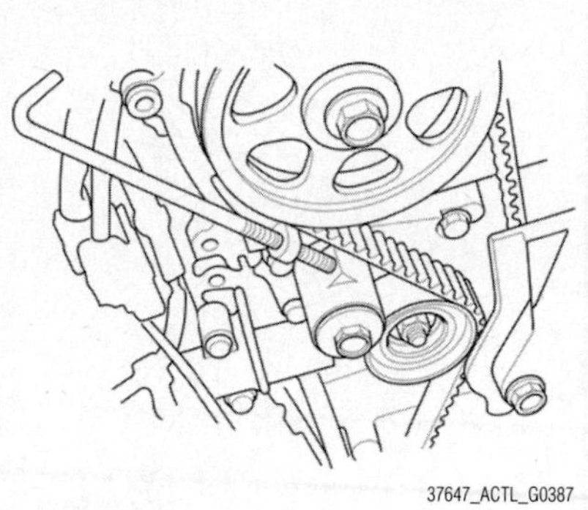
37647_ACTL_G0387
Fig. 226 Thread the battery clamp bolt into hold the timing belt adjuster in its current position

8. Remove the crankshaft pulley.
9. Remove the front upper cover and the rear upper cover.
10. Remove the lower cover.
11. Remove one of the battery clamp bolts from the battery tray, and grind the end of it.
12. Thread the battery clamp bolt into hold the timing belt adjuster in its current

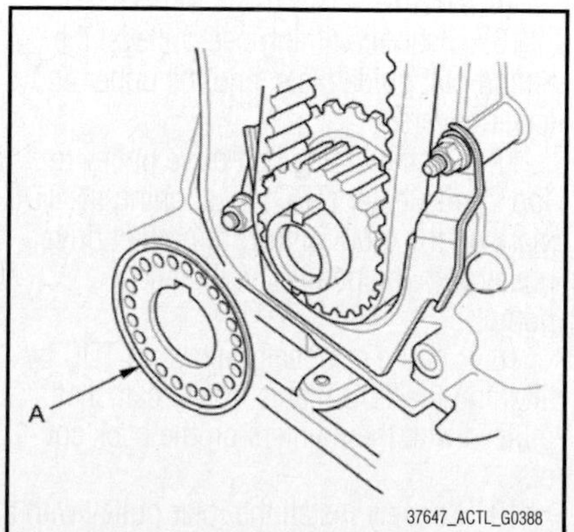
A
37647_ACTL_G0388
Fig. 227 Remove the timing belt guide plate (A)

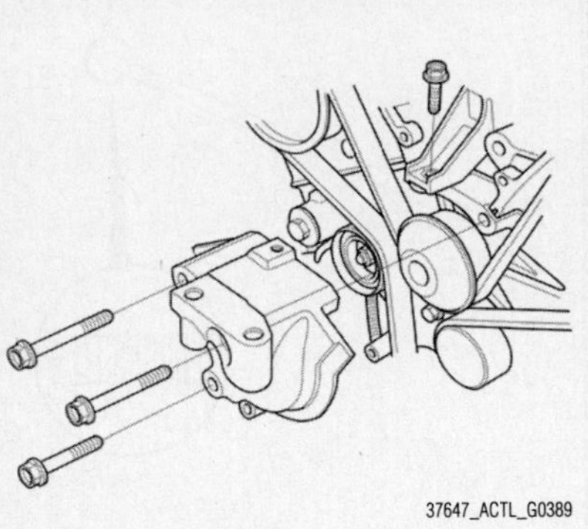
37647_ACTL_G0389

Fig. 228 Remove the lower half of the side engine mount bracket

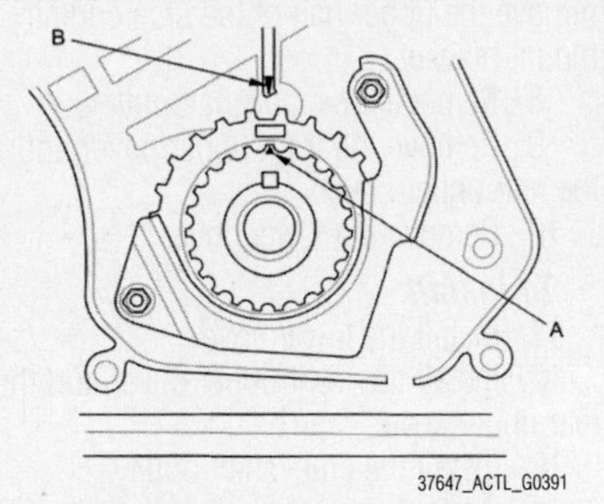
B
A
37647_ACTL_G0391

Fig. 230 Set the timing belt drive pulley to TDC by aligning the TDC mark (A) on the tooth of the timing belt drive pulley with the pointer (B) on the oil pump

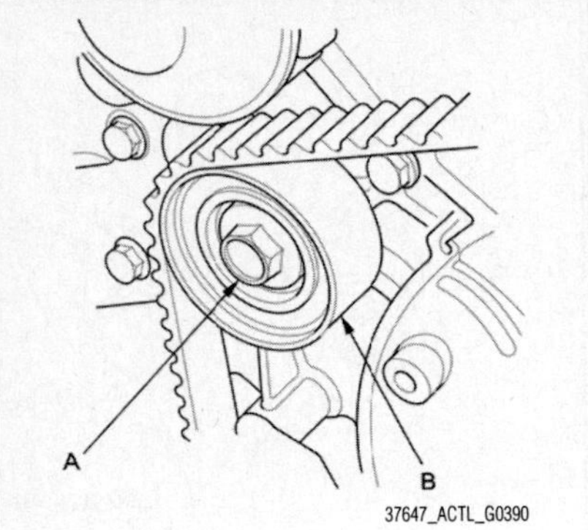
A
B
37647_ACTL_G0390

Fig. 229 Remove the idler pulley bolt (A) and the idler pulley (B), then remove the timing belt

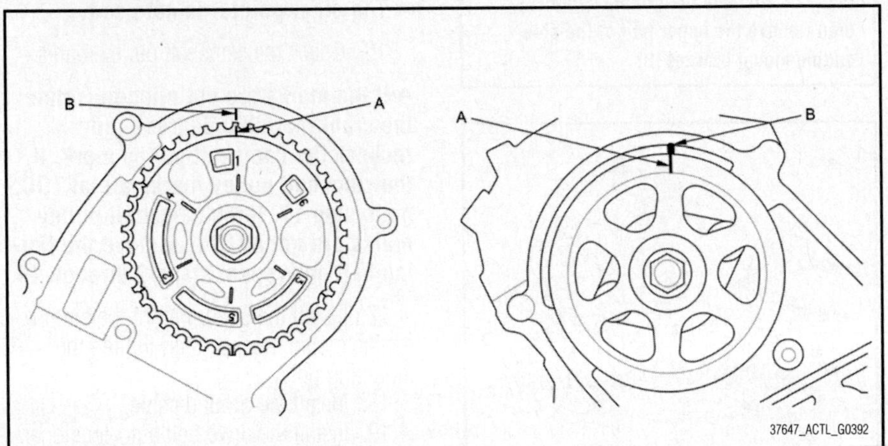
B
A
A
B
37647_ACTL_G0392

Fig. 231 Set the camshaft pulleys to TDC by aligning the TDC marks (A) on the camshaft pulleys with the pointers (B) on the back covers

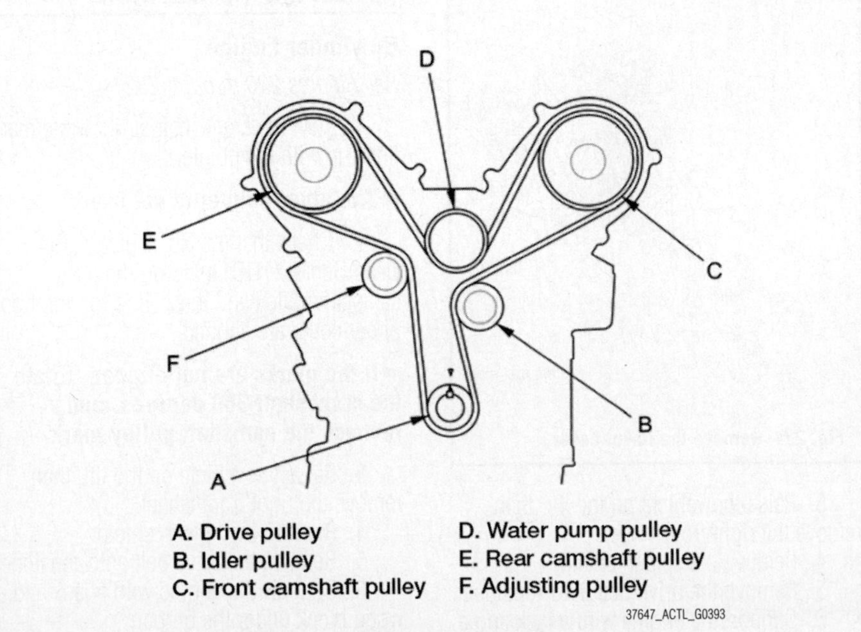
D
E
C
F
B
A
A. Drive pulley
B. Idler pulley
C. Front camshaft pulley
D. Water pump pulley
E. Rear camshaft pulley
F. Adjusting pulley
37647_ACTL_G0393

Fig. 232 Install the timing belt in a counterclockwise sequence starting with the drive pulley

position. Tighten it by hand, do not use a wrench.

13. Remove the timing belt guide plate.

14. Remove the lower half of the side engine mount bracket.

15. Remove the idler pulley bolt and the idler pulley, then remove the timing belt. Discard the idler pulley bolt.

To install:

16. Clean the timing belt pulleys, the timing belt guide plate, and the upper and lower covers.

17. Set the timing belt drive pulley to Top Dead Center (TDC) by aligning the TDC mark on the tooth of the timing belt drive pulley with the pointer on the oil pump.

18. Set the camshaft pulleys to TDC by aligning the TDC marks on the camshaft pulleys with the pointers on the back covers.

19. Loosely install the idler pulley with a new idler pulley bolt so the pulley can move but does not come off.

20. If the auto-tensioner has extended and the timing belt cannot be installed, do the timing belt replacement procedure.

21. Install the timing belt in a counterclockwise sequence starting with the drive pulley. Take care not to damage the timing belt during installation.

22. Tighten the idler pulley bolt to 33 ft. lbs. (44 Nm).

23. Remove the battery clamp bolt from the back cover.

24. Install the lower half of the side engine mount bracket. Tighten the three long bolts to 33 ft. lbs. (44 Nm).

25. Install the timing belt guide plate.

26. Install the lower cover.
27. Install the front upper cover and the rear upper cover.
28. Install the crankshaft pulley.
29. Rotate the crankshaft pulley about six turns clockwise so the timing belt positions itself on the pulleys.
30. Turn the crankshaft pulley so its white mark lines up with the pointer.

➡The other pointer is not used.

31. Check the camshaft pulley marks.

➡If the marks are not aligned, rotate the crankshaft 360 degrees, and recheck the camshaft pulley mark. If the camshaft pulley marks are at TDC, go to step 17. If the camshaft pulley marks are not at TDC, remove the timing belt and repeat steps 2 through 16.

32. Install the upper half of the side engine mount bracket, then tighten the mounting bolts.
33. Install the ground cable.
34. Install the drive belt auto-tensioner.
35. Install the splash shield.
36. Install the right front wheel.

TIMING CHAIN FRONT COVER

REMOVAL & INSTALLATION

4-Cylinder Engine

See Figures 233 through 238.

➡Keep the cam chain away from magnetic fields.

1. Remove the front wheels.
2. Remove the splash shield.
3. Remove the drive belt.
4. Remove the cylinder head cover.
5. Set the No. 1 piston at Top Dead Center (TDC). The punch mark on the Variable Valve Timing Control (VTC) actuator and the punch mark on the exhaust camshaft sprocket should be at the top. Align the TDC marks on the VTC actuator and the exhaust camshaft sprocket.
6. Disconnect the VTC oil control solenoid valve connector and remove the harness clamp.
7. Remove the VTC oil control solenoid valve.
8. Remove the crankshaft pulley.
9. Support the engine with a jack and a wood block under the oil pan.
10. Remove the ground cable, then remove the side engine mount bracket.
11. Remove the side engine mount bracket.
12. Remove the cam chain case and the spacer.

To install:

13. Check the chain case oil seal for damage. If the oil seal is damaged, replace the chain case oil seal.
14. Remove the old liquid gasket from the chain case mating surfaces, the bolts, and the bolt holes.
15. Clean and dry the chain case mating surfaces.
16. Apply liquid gasket (P/N 08717-0004, 08718-0003, or 08718-0009) to the engine block mating surface of the chain case and to the inside edge of the threaded bolt holes. Install the component within 5 minutes of applying the liquid gasket.

➡If too much time has passed after applying the liquid gasket, remove the old liquid gasket and residue, then reapply new liquid gasket.

17. Apply liquid gasket to the engine block upper surface contact areas and the lower block upper surface contact areas on the chain case.

➡Apply about 0.43 inches (11 mm) diameter and about 0.12 inches (3 mm) thickness of liquid gasket to the areas B and C.

18. Apply liquid gasket (P/N 08717-0004, 08718-0003, or 08718-0009) to the oil pan mating surface of the oil pump. Install the component within 5 minutes of applying the liquid gasket.
19. Install the spacer, then install the new O-ring on the chain case. Set the edge of the chain case to the edge of the oil pan, then install the chain case on the engine block. Wipe off the excess liquid gasket on the oil pan and chain case mating area.

➡When installing the chain case, do not slide the bottom surface onto the oil pan mounting surface. Wait at least 30 minutes before filling the engine with oil. Do not run the engine for at least 3 hours after installing the chain case.

20. Install the side engine mount bracket, then tighten the side engine mount bracket mounting bolts to 33 ft. lbs. (44 Nm).

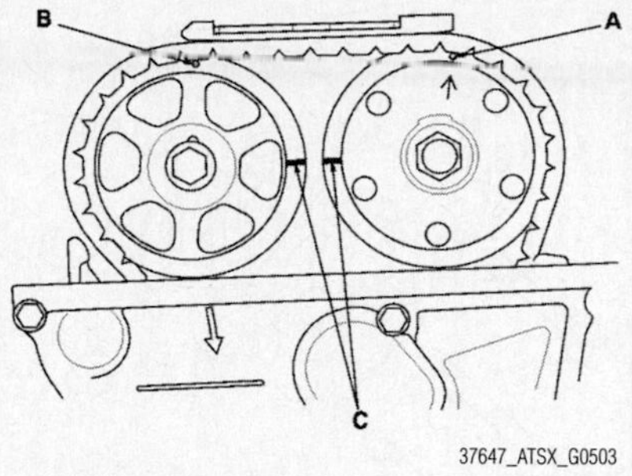
B
A
C
37647_ATSX_G0503

Fig. 233 The punch mark (A) on the VTC actuator and the punch mark (B) on the exhaust camshaft sprocket should be at the top. Align the TDC marks (C) on the VTC actuator and the exhaust camshaft sprocket

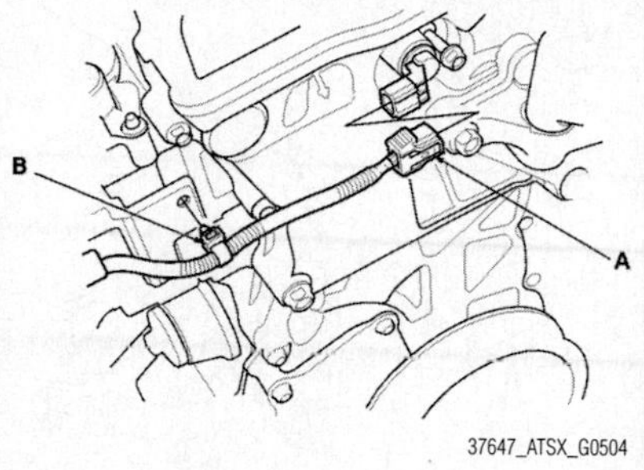
B
A
37647_ATSX_G0504

Fig. 234 Disconnect the VTC oil control solenoid valve connector (A) and remove the harness clamp (B)

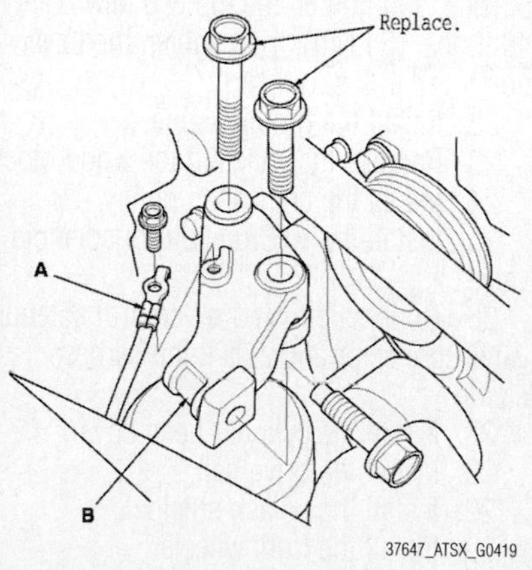
Replace.
A
B
37647_ATSX_G0419

Fig. 235 Remove the ground cable (A), then remove the side engine mount bracket (B)

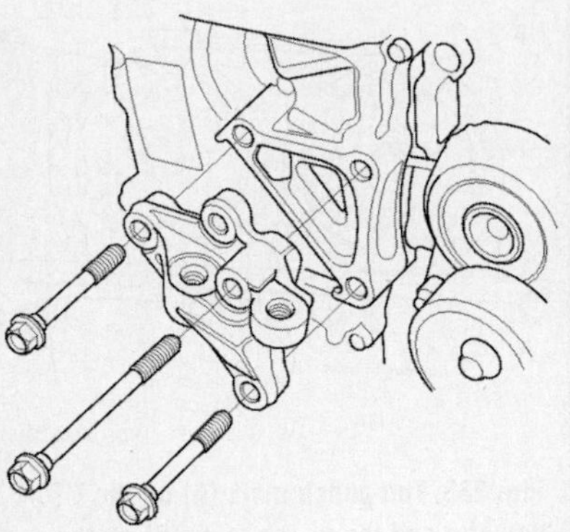

37647_ATSX_G0505

Fig. 236 Remove the side engine mount bracket

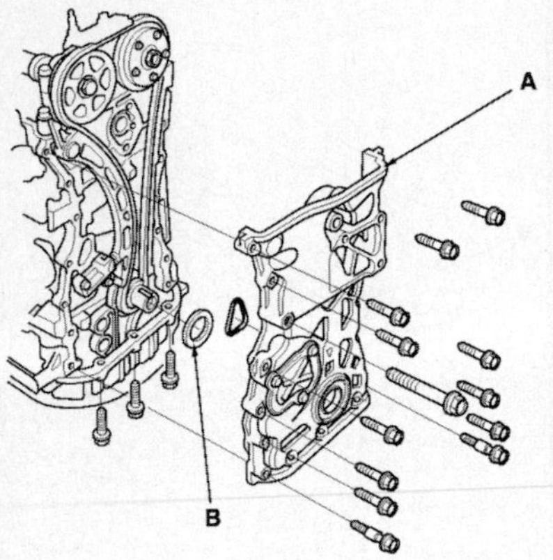
A
B

37647_ATSX_G0506

Fig. 237 Remove the cam chain case (A) and the spacer (B)

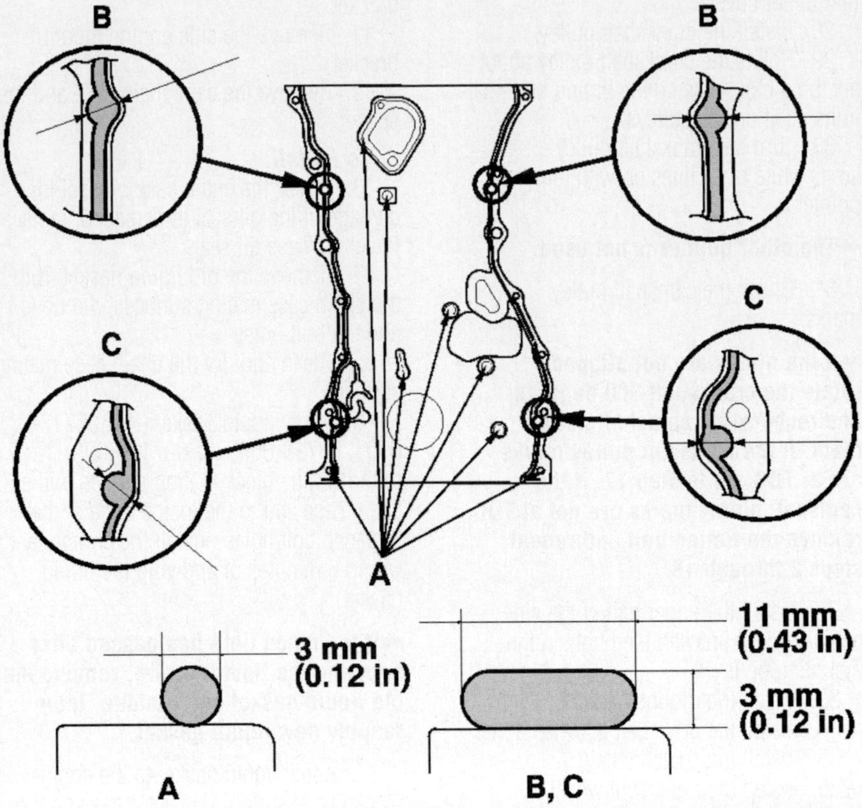
B
B
C
C
A
3 mm
(0.12 in)
11 mm
(0.43 in)
3 mm
(0.12 in)
A
B, C

37647_ATSX_G0516

Fig. 238 Apply about 0.43 inches (11 mm) diameter and about 0.12 inches (3 mm) thickness of liquid gasket to the areas (B) and (C)

21. Install the side engine mount bracket, and tighten the top two new bolts to 40 ft. lbs. (54 Nm), then tighten the lower bolt to 47 ft. lbs. (64 Nm).
22. Install the ground cable.
23. Remove the jack and the wood block.
24. Install the crankshaft pulley.
25. Install the VTC oil control solenoid valve.
26. Connect the VTC oil control solenoid valve connector and install the harness clamp.
27. Install the cylinder head cover.
28. Install the drive belt.
29. Install the splash shield.
30. Install the front wheels.
31. Do the crankshaft position (CKP) pattern clear/CKP pattern learn procedure.

TIMING CHAIN & SPROCKETS

REMOVAL & INSTALLATION

4-Cylinder Engine

See Figures 236, 237 and 239 through 252.

➡Keep the cam chain away from magnetic fields.

1. Remove the front wheels.
2. Remove the splash shield.
3. Remove the drive belt.
4. Remove the cylinder head cover.
5. Set the No. 1 piston at Top Dead Center (TDC). The punch mark on the Variable Valve Timing Control (VTC) actuator and the punch mark on the exhaust camshaft sprocket should be at the top. Align the TDC marks on the VTC actuator and the exhaust camshaft sprocket.
6. Disconnect the VTC oil control solenoid valve connector and remove the harness clamp.
7. Remove the VTC oil control solenoid valve.

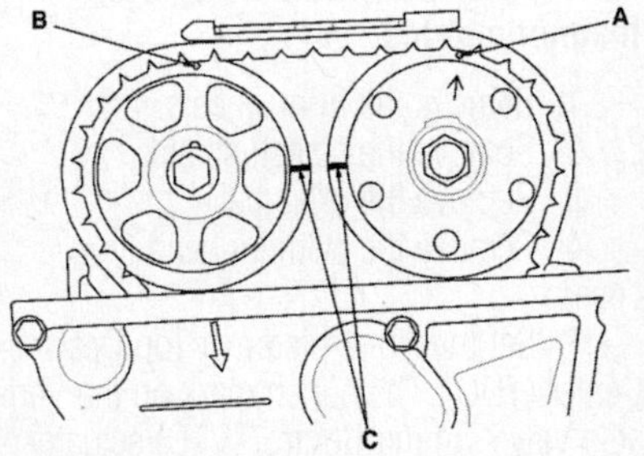
B
A
C

37647_ATSX_G0503

Fig. 239 The punch mark (A) on the VTC actuator and the punch mark (B) on the exhaust camshaft sprocket should be at the top. Align the TDC marks (C) on the VTC actuator and the exhaust camshaft sprocket

8. Remove the crankshaft pulley.
9. Support the engine with a jack and a wood block under the oil pan.
10. Remove the ground cable, then remove the side engine mount bracket.

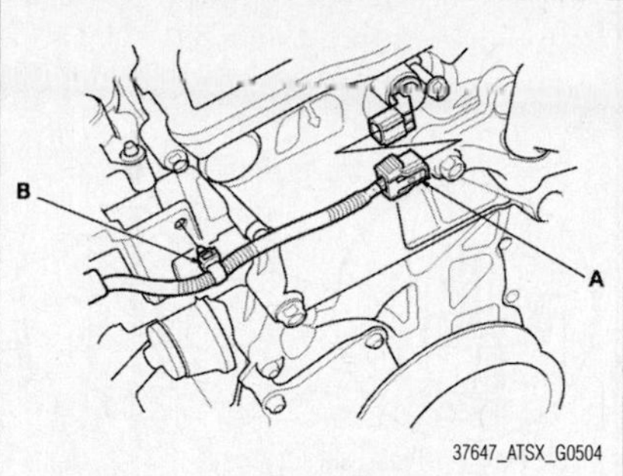
B
A
37647_ATSX_G0504
Fig. 240 Disconnect the VTC oil control solenoid valve connector (A) and remove the harness clamp (B)

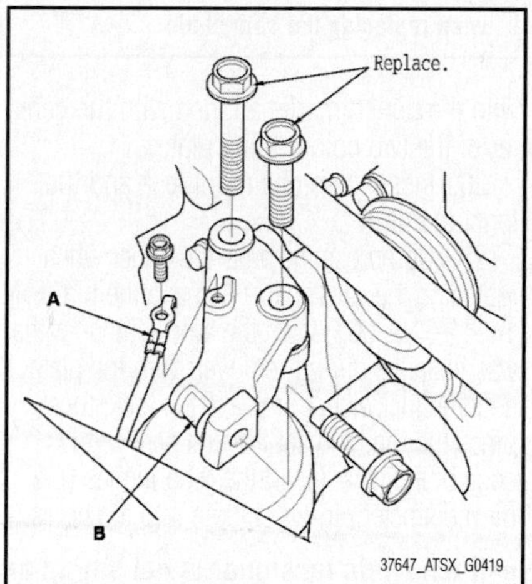
Replace.
A
B
37647_ATSX_G0419
Fig. 241 Remove the ground cable (A), then remove the side engine mount bracket (B)

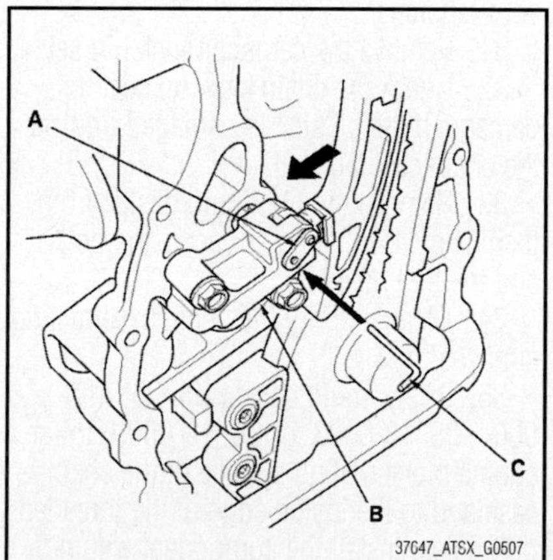
A
C
B
37647_ATSX_G0507
Fig. 242 Align the holes on the lock (A) and the auto-tensioner (B), then insert a 0.05 inches (1.2 mm) diameter pin (C) into the holes

11. Remove the side engine mount bracket.
12. Remove the cam chain case and the spacer.

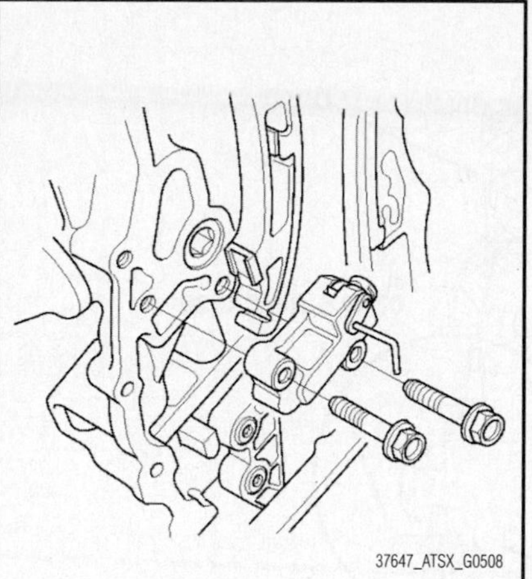
37647_ATSX_G0508
Fig. 243 Remove the auto-tensioner

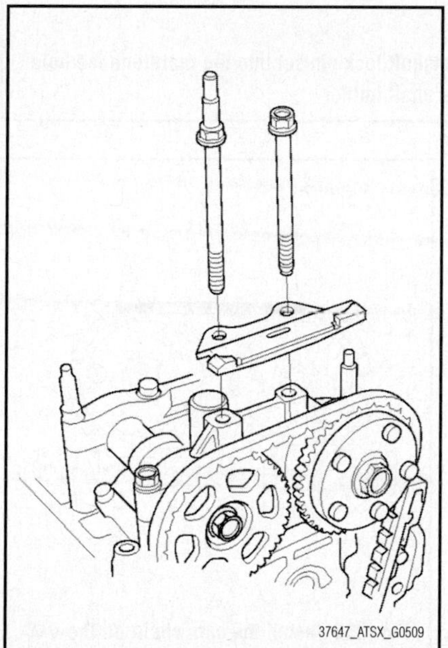
37647_ATSX_G0509
Fig. 244 Remove cam chain guide B

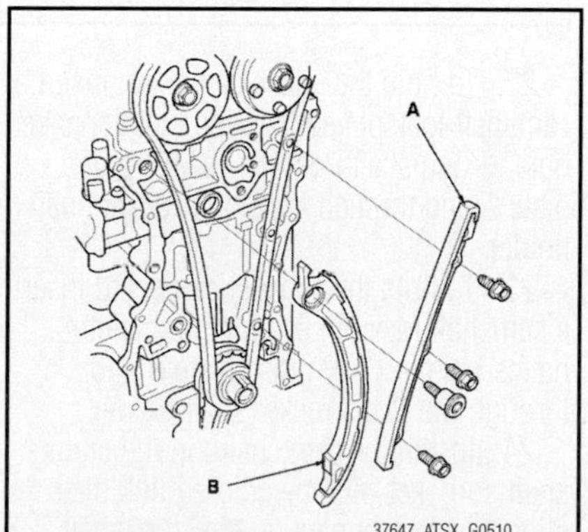
A
B
37647_ATSX_G0510
Fig. 245 Remove cam chain guide A and the tensioner arm (B)

13. Loosely install the crankshaft pulley.
14. Turn the crankshaft counterclockwise to compress the auto-tensioner.
15. Align the holes on the lock and the auto-tensioner, then insert a 0.05 inches (1.2 mm) diameter pin or lock pin (P/N 14511-PNA-003) into the holes. Turn the crankshaft clockwise to secure the pin.
16. Remove the auto-tensioner.
17. Remove cam chain guide B.
18. Remove cam chain guide A and the tensioner arm.
19. Remove the cam chain.

To install:

➡Keep the cam chain away from magnetic fields. Before doing this procedure, check that the Variable Valve Timing Control (VTC) actuator is locked

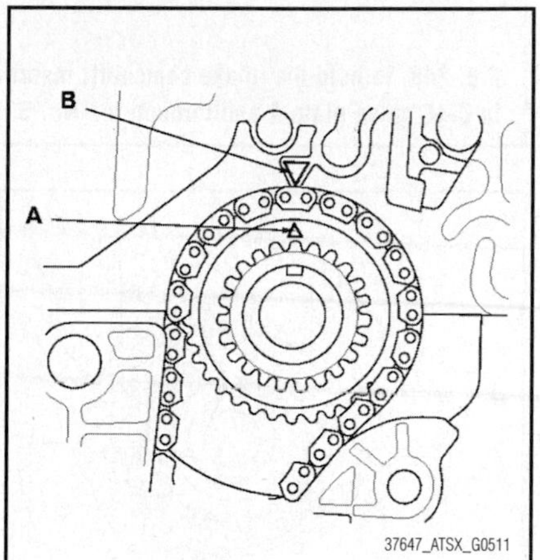
B
A
37647_ATSX_G0511
Fig. 246 Align the TDC mark (A) on the crankshaft sprocket with the pointer (B) on the engine block

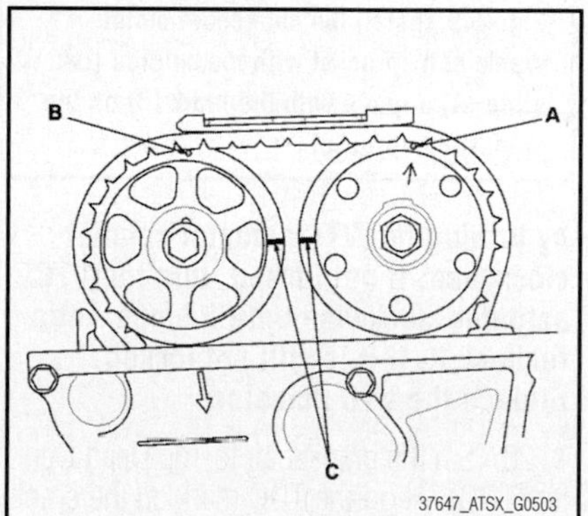
B
A
C
37647_ATSX_G0503
Fig. 247 The punch mark (A) on the VTC actuator and the punch mark (B) on the exhaust camshaft sprocket should be at the top. Align the TDC marks (C) on the VTC actuator and the exhaust camshaft sprocket

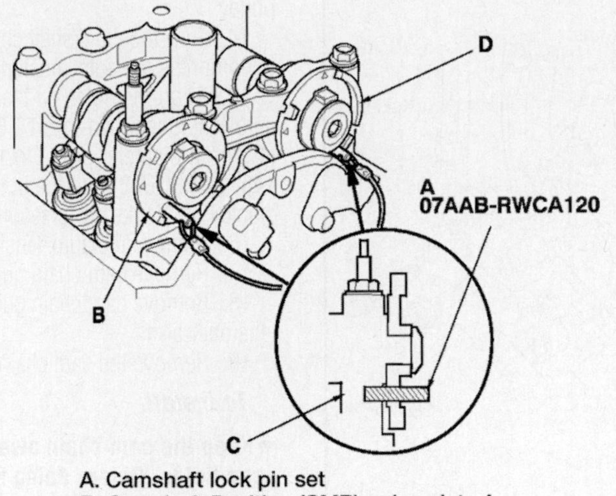
D
A
07AAB-RWCA120
B
C
A. Camshaft lock pin set
B. Camshaft Position (CMP) pulse plate A
C. No. 5 rocker shaft holder
D. Camshaft Position (CMP) pulse plate B
37647_ATSX_G0512

Fig. 248 To hold the intake camshaft, insert a camshaft lock pin set into the maintenance hole in CMP pulse plate A and through the No. 5 rocker shaft holder

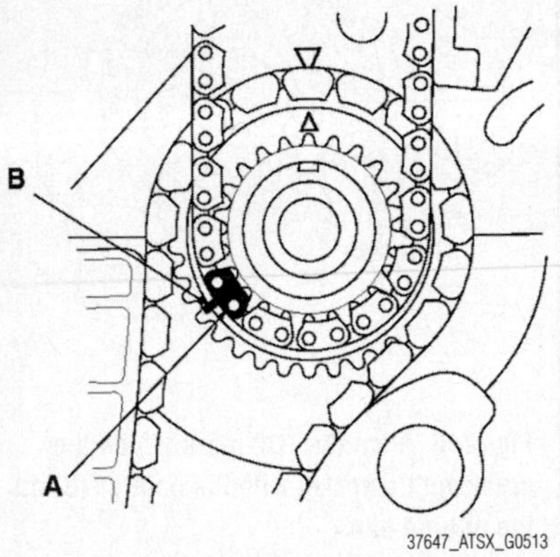
B
A
37647_ATSX_G0513

Fig. 249 Install the cam chain on the crankshaft sprocket with the colored link plate (A) aligned with the mark (B) on the crankshaft sprocket

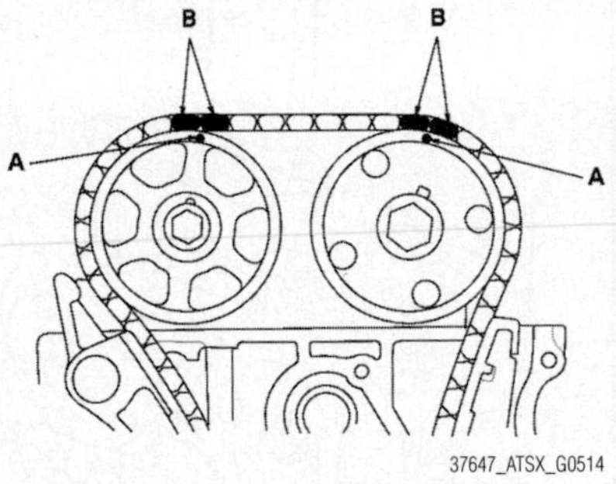
B
B
A
A
37647_ATSX_G0514

Fig. 250 Install the cam chain on the VTC actuator and the exhaust camshaft sprocket with the punch marks (A) aligned with the center of the two colored link plates (B)

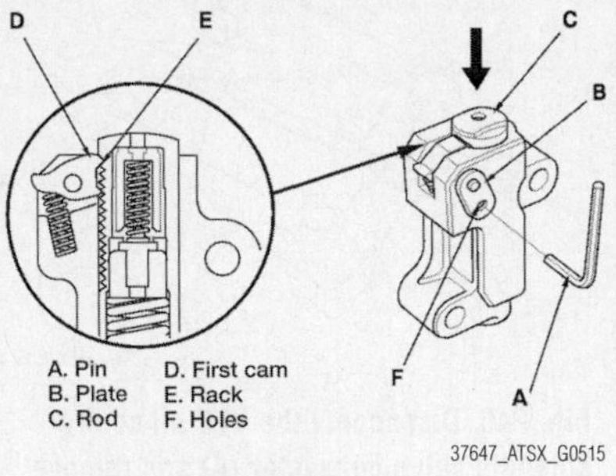
D
E
C
B
F
A
A. Pin
B. Plate
C. Rod
D. First cam
E. Rack
F. Holes
37647_ATSX_G0515

Fig. 251 Compress the auto-tensioner when replacing the cam chain

by turning the VTC actuator counter-clockwise. If not locked, turn the VTC actuator clockwise until it stops, then recheck it. If it is still not locked, replace the VTC actuator.

20. Set the crankshaft to Top Dead Center (TDC). Align the TDC mark on the crankshaft sprocket with the pointer on the engine block.

21. Set the camshafts to TDC. The punch mark on the VTC actuator and the punch mark on the exhaust camshaft sprocket should be at the top. Align the TDC marks on the VTC actuator and the exhaust camshaft sprocket.

22. To hold the intake camshaft, insert a camshaft lock pin set into the maintenance hole in Camshaft Position (CMP) pulse plate A and through the No. 5 rocker shaft holder.

23. To hold the exhaust camshaft, insert a camshaft lock pin set into the maintenance hole in CMP pulse plate B and through the No. 5 rocker shaft holder.

24. Install the cam chain on the crankshaft sprocket with the colored link plate aligned with the mark on the crankshaft sprocket.

25. Install the cam chain on the VTC actuator and the exhaust camshaft sprocket with the punch marks aligned with the center of the two colored link plates.

26. Install cam chain guide A and the tensioner arm.

27. Compress the auto-tensioner when replacing the cam chain. Remove the pin (P/N 14511-PNA-003) from the auto-tensioner that was installed during removal. Turn the plate counterclockwise, to release the lock, then press the rod, and set the first cam to the edge of the rack. Insert the 0.05 inches (1.2 mm) diameter pin or lock pin into the holes.

➡If the chain tensioner is not set up as described, the tensioner will become damaged.

28. Install the auto-tensioner.

29. Install cam chain guide B.

30. Remove the pin or lock pin from the auto-tensioner.

31. Remove the camshaft lock pin set.

32. Check the chain case oil seal for damage. If the oil seal is damaged, replace the chain case oil seal.

33. Remove the old liquid gasket from the chain case mating surfaces, the bolts, and the bolt holes.

34. Clean and dry the chain case mating surfaces.

35. Apply liquid gasket (P/N 08717-0004, 08718-0003, or 08718-0009) to the engine block mating surface of the chain case and to the inside edge of the threaded bolt holes. Install the component within 5 minutes of applying the liquid gasket.

➡If too much time has passed after applying the liquid gasket, remove the old liquid gasket and residue, then reapply new liquid gasket.

36. Apply liquid gasket to the engine block upper surface contact areas and the lower block upper surface contact areas on the chain case.

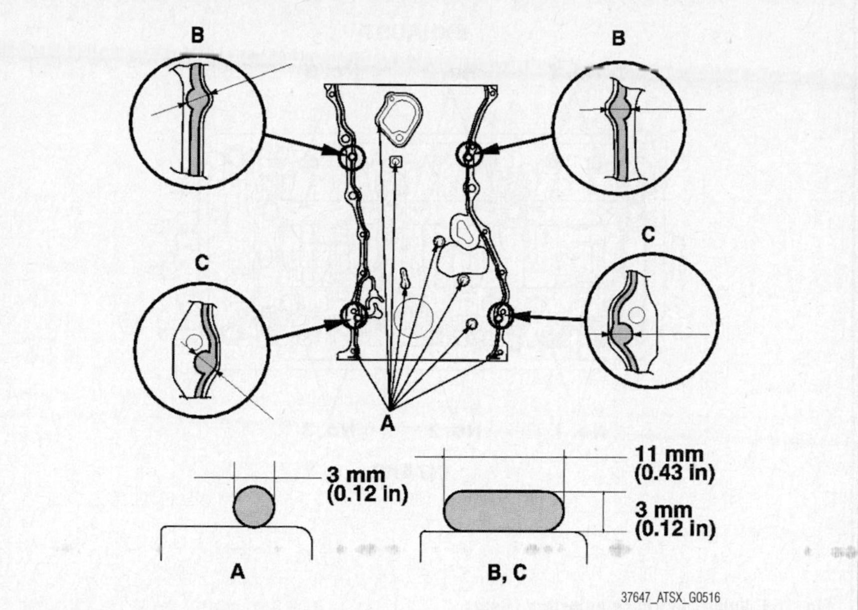
B
B
C
C
A
3 mm
(0.12 in)
11 mm
(0.43 in)
3 mm
(0.12 in)
A
B, C
37647_ATSX_G0516

Fig. 252 Apply about 0.43 inches (11 mm) diameter and about 0.12 inches (3 mm) thickness of liquid gasket to the areas (B) and (C)

➡Apply about 0.43 inches (11 mm) diameter and about 0.12 inches (3 mm) thickness of liquid gasket to the areas B and C.

37. Apply liquid gasket (P/N 08717-0004, 08718-0003, or 08718-0009) to the oil pan mating surface of the oil pump. Install the component within 5 minutes of applying the liquid gasket.

38. Install the spacer, then install the new O-ring on the chain case. Set the edge of the chain case to the edge of the oil pan, then install the chain case on the engine block. Wipe off the excess liquid gasket on the oil pan and chain case mating area.

➡When installing the chain case, do not slide the bottom surface onto the oil pan mounting surface. Wait at least 30 minutes before filling the engine with oil. Do not run the engine for at least 3 hours after installing the chain case.

39. Install the side engine mount bracket, then tighten the side engine mount bracket mounting bolts to 33 ft. lbs. (44 Nm).

40. Install the side engine mount bracket, and tighten the top two new bolts to 40 ft. lbs. (54 Nm), then tighten the lower bolt to 47 ft. lbs. (64 Nm).

41. Install the ground cable.

42. Remove the jack and the wood block.

43. Install the crankshaft pulley.

44. Install the VTC oil control solenoid valve.

45. Connect the VTC oil control solenoid valve connector and install the harness clamp.

46. Install the cylinder head cover.

47. Install the drive belt.

48. Install the splash shield.

49. Install the front wheels.

50. Do the crankshaft position (CKP) pattern clear/CKP pattern learn procedure.

VALVE LASH

ADJUSTMENT

4-Cylinder Engine

See Figures 253 and 254.

Special Tools Required:

- Adjuster 07MAA-PR70110
- Locknut wrench 07MAA-PR70120

➡Connect the Honda Diagnostic System (HDS) to the Data Link Connector (DLC) and monitor the Engine Coolant Temperature (ECT) sensor 1 with the HDS. Adjust the valve clearance only when the ECT sensor 1 temperature is less than 100°F (38°C).

1. Remove the cylinder head cover.
2. Set the No. 1 piston at Top Dead Center (TDC). The punch mark on the Variable Valve Timing Control (VTC) actuator and the punch mark on the exhaust camshaft sprocket should be at the top. Align the TDC marks on the VTC actuator and the exhaust camshaft sprocket.
3. Select the correct feeler gauge for the valve clearance you are going to check.

➡Valve Clearance:

- Intake: 0.008–0.010 inches (0.21–0.25 mm)
- Exhaust: 0.010–0.011 inches (0.25–0.29 mm)

4. Insert the feeler gauge between the adjusting screw and the end of the valve stem, and slide it back and forth; you should feel a slight amount of drag.
5. If you feel too much or too little drag, loosen the locknut with the locknut wrench and the adjuster, and turn the adjusting screw until the drag on the feeler gauge is correct.
6. Tighten the locknut to 10 ft. lbs. (14 Nm), and recheck the clearance. Repeat the adjustment if necessary.
7. Rotate the crankshaft 180° clockwise (camshaft pulley turns 90°).

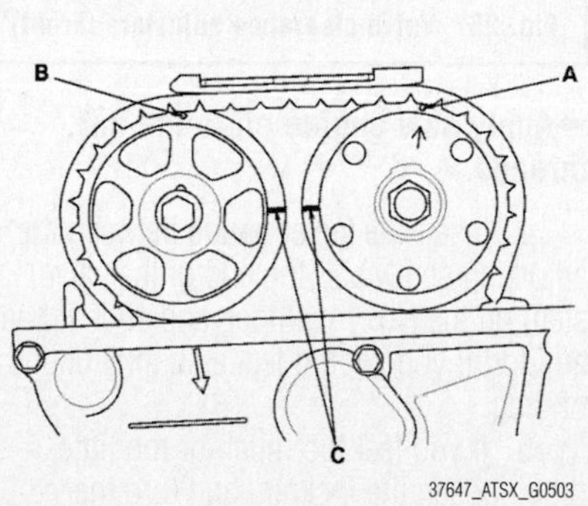
B
A
C
37647_ATSX_G0503

Fig. 253 The punch mark (A) on the VTC actuator and the punch mark (B) on the exhaust camshaft sprocket should be at the top. Align the TDC marks (C) on the VTC actuator and the exhaust camshaft sprocket

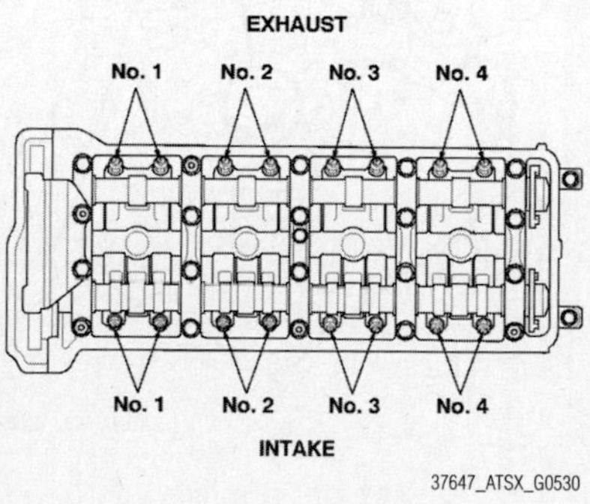
EXHAUST
No. 1
No. 2
No. 3
No. 4
No. 1
No. 2
No. 3
No. 4
INTAKE
37647_ATSX_G0530

Fig. 254 Valve clearance adjusters

8. Check and, if necessary, adjust the valve clearance on the No. 3 cylinder.

9. Rotate the crankshaft 180° clockwise (camshaft pulley turns 90°).

10. Check and, if necessary, adjust the valve clearance on the No. 4 cylinder.

11. Rotate the crankshaft 180° clockwise (camshaft pulley turns 90°).

12. Check and, if necessary, adjust the valve clearance on the No. 2 cylinder.

13. Install the cylinder head cover.

6-Cylinder Engine

See Figures 255 through 258.

➡Connect the Honda Diagnostic System (HDS) to the data link connector (DLC), and monitor the Engine Coolant Temperature (ECT) sensor 1. Adjust the valve clearance only when the ECT sensor 1 is less than 100°F (38°C).

1. Remove the cylinder head covers.

2. Set the No. 1 piston at Top Dead Center (TDC). Align the pointer on the front upper cover with the No. 1 piston TDC mark on the front camshaft pulley.

3. Select the correct feeler gauge for the valve clearance you are going to check.

➡Valve Clearance:

- Intake: 0.008–0.009 inches (0.20–0.24 mm)
- Exhaust: 0.011–0.013 inches (0.28–0.32 mm)

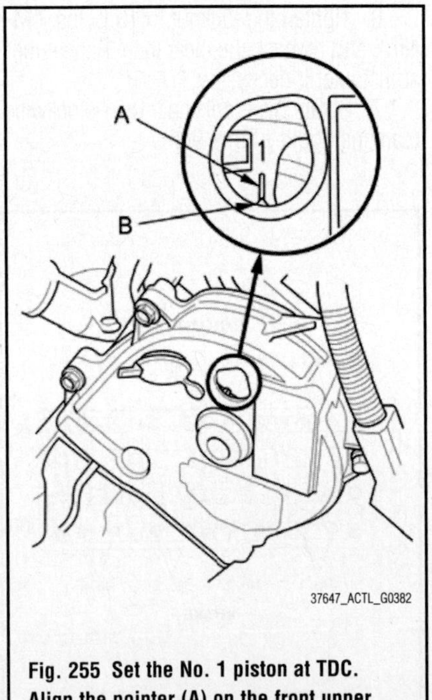
A
1
B
37647_ACTL_G0382

Fig. 255 Set the No. 1 piston at TDC. Align the pointer (A) on the front upper cover with the No. 1 piston TDC mark (B) on the front camshaft pulley

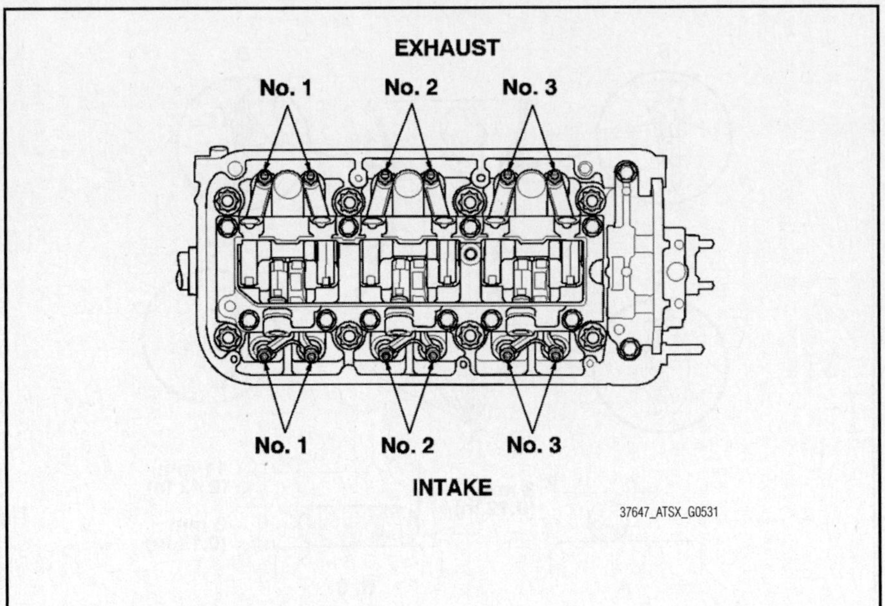
EXHAUST
No. 1
No. 2
No. 3
No. 1
No. 2
No. 3
INTAKE
37647_ATSX_G0531

Fig. 256 Valve clearance adjusters (Rear)

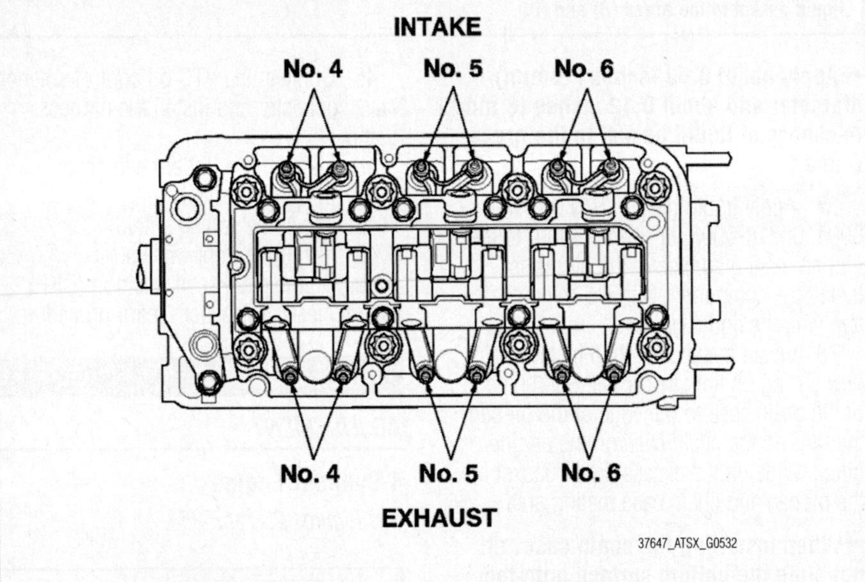
INTAKE
No. 4
No. 5
No. 6
No. 4
No. 5
No. 6
EXHAUST
37647_ATSX_G0532

Fig. 257 Valve clearance adjusters (Front)

➡Apply new engine oil to the nut threads.

4. Insert the feeler gauge between the adjusting screw and the end of the valve stem on the No. 1 cylinder, and slide it back and forth; you should feel a slight amount of drag.

5. If you feel too much or too little drag, loosen the locknut, and turn the adjusting screw until the drag on the feeler gauge is correct.

6. While holding the adjusting screw with the screw driver, tighten the locknut, then recheck the clearance. Repeat the adjustment, if necessary.

➡Specified Torque:

- Intake: 14 ft. lbs. (20 Nm)
- Exhaust: 10 ft. lbs. (14 Nm)

➡Apply new engine oil to the nut threads.

7. Rotate the crankshaft clockwise. Align the pointer on the front upper cover with the No. 4 piston TDC mark on the front camshaft pulley.

8. Check and, if necessary, adjust the valve clearance on the No. 4 cylinder.

9. Rotate the crankshaft clockwise. Align the pointer on the front upper cover with the No. 2 piston TDC mark on the front camshaft pulley.

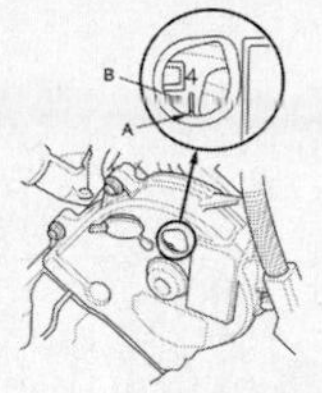
B
A
37647_ACTL_G0394

Fig. 258 Rotate the crankshaft clockwise. Align the pointer (A) on the front upper cover with the No. 4 piston TDC mark (B) on the front camshaft pulley

Check and, if necessary, adjust the valve clearance on the No. 2 cylinder.

10. Rotate the crankshaft clockwise. Align the pointer on the front upper cover with the No. 5 piston TDC mark on the front camshaft pulley.
11. Check and, if necessary, adjust the valve clearance on the No. 5 cylinder.
12. Rotate the crankshaft clockwise. Align the pointer on the front upper cover with the No. 3 piston TDC mark on the front camshaft pulley.
13. Check and, if necessary, adjust the valve clearance on the No. 3 cylinder.
14. Rotate the crankshaft clockwise. Align the pointer on the front upper cover with the No. 6 piston TDC mark on the front camshaft pulley.
15. Check and, if necessary, adjust the valve clearance on the No. 6 cylinder.
16. Install the cylinder head covers.

ENGINE PERFORMANCE & EMISSION CONTROLS

ACCELERATOR PEDAL POSITION (APP) SENSOR

LOCATION

See Figure 259.

Refer to the accompanying illustration.

REMOVAL & INSTALLATION

See Figure 260.

1. Disconnect the APP sensor connector.
2. Remove the accelerator pedal module.

➡The APP sensor is not available separately. Do not disassemble the accelerator pedal module.

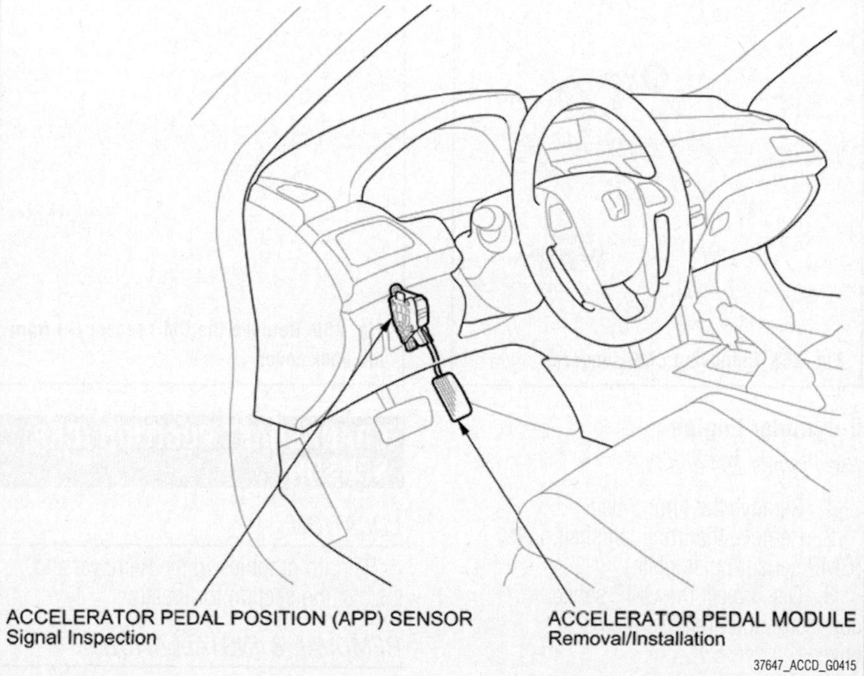
ACCELERATOR PEDAL POSITION (APP) SENSOR
Signal Inspection
ACCELERATOR PEDAL MODULE
Removal/Installation
37647_ACCD_G0415

Fig. 259 Accelerator Pedal Position (APP) sensor location

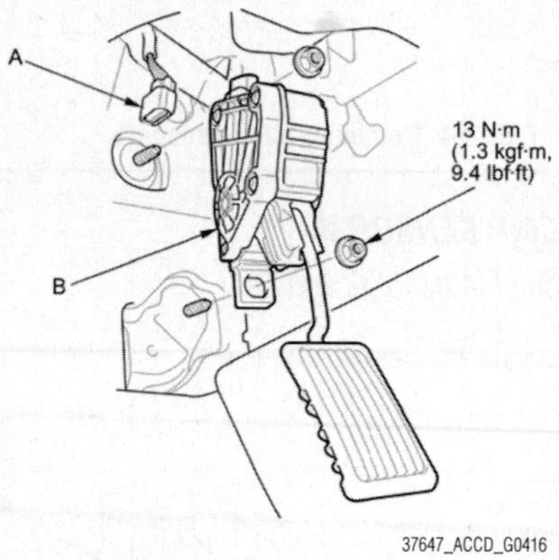
A
B
13 N·m
(1.3 kgf·m,
9.4 lbf·ft)
37647_ACCD_G0416

Fig. 260 Disconnect the APP sensor connector (A)

3. Install the parts in the reverse order of removal.

CAMSHAFT POSITION (CMP) SENSOR

LOCATION

Refer to graphics in the Removal and Installation section for location.

REMOVAL & INSTALLATION

4-Cylinder Engine

CMP Sensor A

See Figure 261.

1. Disconnect the CMP sensor A connector.
2. Remove CMP sensor A from the intake camshaft side of the cylinder head.
3. Install the parts in the reverse order of removal with a new O-ring.

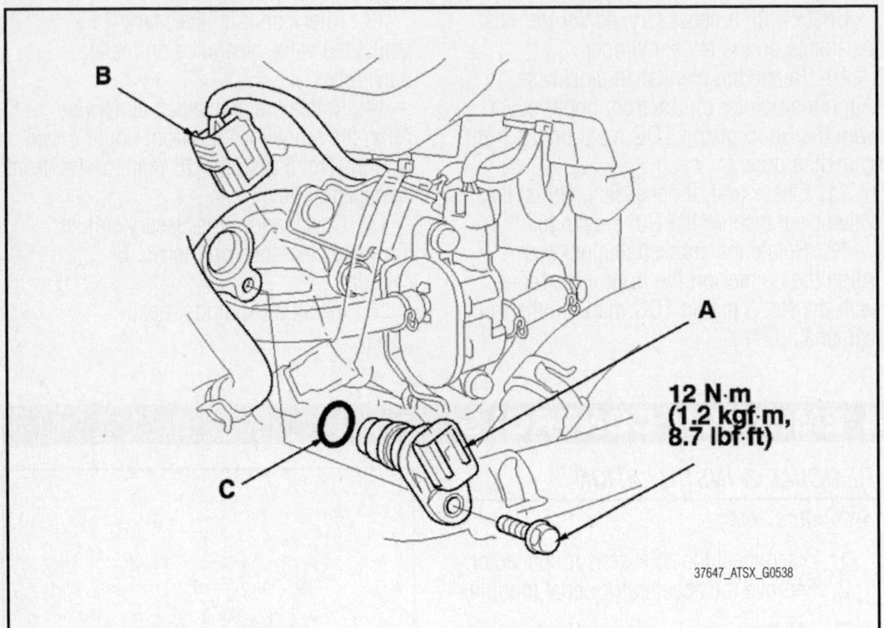
B
A
12 N·m
(1.2 kgf·m,
8.7 lbf·ft)
C
37647_ATSX_G0538

Fig. 261 Removing CMP sensor A

CMP SENSOR B

See Figures 262 and 263.

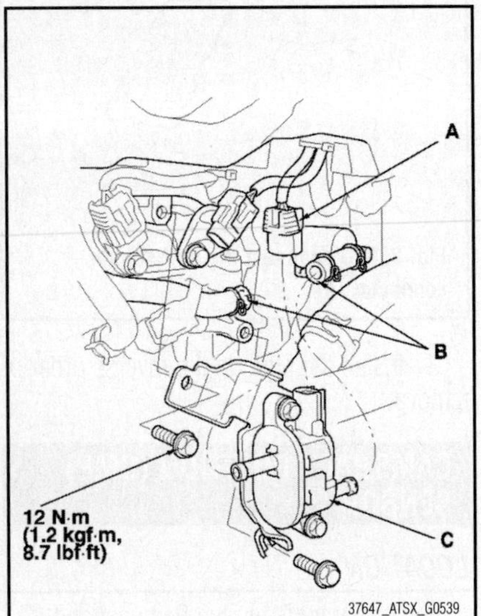
A
B
12 N·m
(1.2 kgf·m,
8.7 lbf·ft)
C
37647_ATSX_G0539

Fig. 262 Disconnect the connector (A) and hoses (B) from the EVAP canister purge valve (C), then remove the EVAP canister purge valve assembly.

1. Disconnect the connector and hoses from the EVAP canister purge valve, then remove the EVAP canister purge valve assembly.
2. Disconnect the CMP sensor B connector.
3. Remove CMP sensor B.
4. Install the parts in the reverse order of removal with a new O-ring.

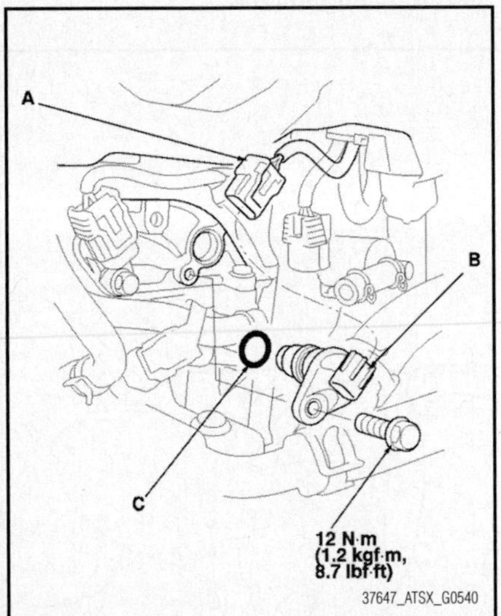
A
B
C
12 N·m
(1.2 kgf·m,
8.7 lbf·ft)
37647_ATSX_G0540

Fig. 263 Removing CMP sensor B

6-Cylinder Engine

See Figures 264 and 265.

1. Remove the timing belt.
2. Remove the front camshaft pulley (CMP sensor pulse plate).
3. Disconnect the CMP sensor connector, then remove the back cover.
4. Remove the CMP sensor from the back cover.
5. Install the parts in the reverse order of removal.
6. Install the timing belt.
7. Do the CKP pattern clear/CKP pattern learn procedure.

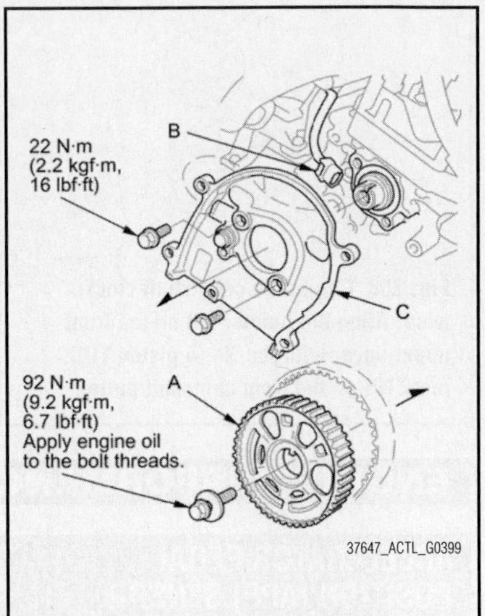
22 N·m
(2.2 kgf·m,
16 lbf·ft)
B
C
92 N·m
(9.2 kgf·m,
6.7 lbf·ft)
Apply engine oil
to the bolt threads.
A
37647_ACTL_G0399

Fig. 264 Remove the front camshaft pulley (CMP sensor pulse plate) (A)

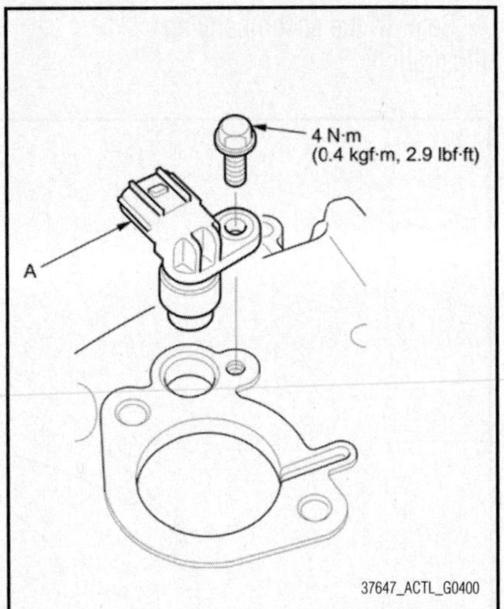
4 N·m
(0.4 kgf·m, 2.9 lbf·ft)
A
37647_ACTL_G0400

Fig. 265 Remove the CMP sensor (A) from the back cover

CRANKSHAFT POSITION (CKP) SENSOR

LOCATION

Refer to graphics in the Removal and Installation section for location.

REMOVAL & INSTALLATION

4-Cylinder Engine

See Figure 266.

1. Raise the vehicle on a lift.

➡Make sure the vehicle is level, because engine oil will drip out when you remove the sensor.

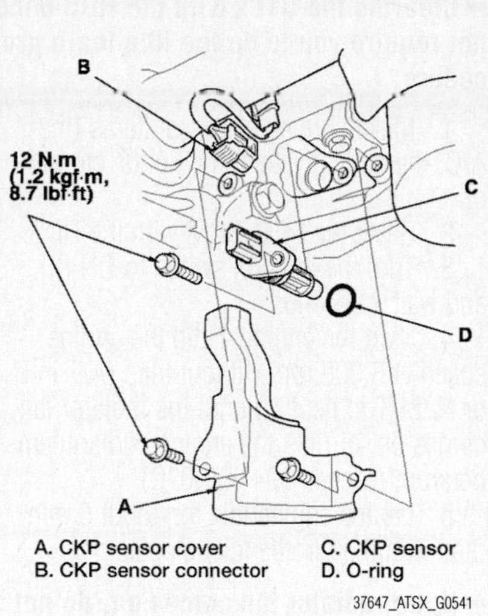
B
12 N·m
(1.2 kgf·m,
8.7 lbf·ft)
C
D
A
A. CKP sensor cover
B. CKP sensor connector
C. CKP sensor
D. O-ring
37647_ATSX_G0541

Fig. 266 Removing CKP sensor

2. Remove the CKP sensor cover.
3. Disconnect the CKP sensor connector.
4. Remove the CKP sensor.
5. Install the parts in the reverse order of removal with a new O-ring.
6. Do the CKP pattern clear/CKP pattern learn procedure.
7. Check the engine oil level, and add more oil if needed.

6-Cylinder Engine

See Figure 267.

1. Raise the vehicle on a lift.

➡**Make sure the vehicle is level, because engine oil will drip out when you remove the sensor.**

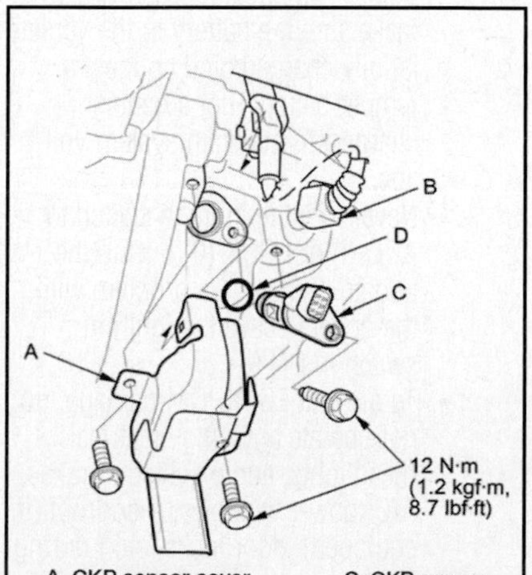
B
D
C
A
12 N·m
(1.2 kgf·m,
8.7 lbf·ft)
A. CKP sensor cover
B. CKP sensor connector
C. CKP sensor
D. O-ring
37647_ACTL_G0401

Fig. 267 Exploded view of CKP sensor assembly

2. Remove the CKP sensor cover.
3. Disconnect the CKP sensor connector.
4. Remove the CKP sensor.
5. Install the parts in the reverse order of removal with a new O-ring.
6. Do the CKP pattern clear/CKP pattern learn procedure.
7. Check for engine oil level, and add more oil if needed.

CKP PATTERN CLEAR/CKP PATTERN LEARN

Clear/Learn Procedure (with the HDS)

1. Connect the HDS to the Data Link Connector (DLC) located under the driver's side of the dashboard.
2. Turn the ignition switch to ON (II).
3. Make sure the HDS communicates with the ECM/PCM and other vehicle systems. If it doesn't, go to the DLC circuit troubleshooting.
4. Select CRANK PATTERN in the ADJUSTMENT MENU with the HDS.
5. Select CRANK PATTERN CLEAR, and clear the CKP pattern.
6. Select CRANK PATTERN LEARNING with the HDS, and follow the screen prompts.

Learn Procedure (without the HDS)

1. Start the engine. Hold the engine speed at 3,000 rpm without load (A/T in P or N, M/T in neutral) until the radiator fan comes on.
2. Test-drive the vehicle on a level road: Decelerate (with the throttle fully closed) from an engine speed of 2,500 rpm down to 1,000 rpm with the Transmission in 2nd.
3. Repeat step 2 several times.
4. Turn the ignition switch to LOCK (0).
5. Turn the ignition switch to ON (II), and wait 30 seconds.

ENGINE CONTROL MODULE (ECM)/POWERTRAIN CONTROL MODULE (PCM)

REMOVAL & INSTALLATION

See Figures 268 and 269.

Special Tools Required:

- Honda Diagnostic System (HDS) tablet tester
- Honda Interface Module (HIM) and an iN workstation with the latest HDS software version
- HDS pocket tester
- GNA600 and an iN workstation with the latest HDS software version

➡**Any one of above updating tools can be used.**

1. Connect the HDS to the Data Link Connector (DLC) located under the driver's side of the dashboard.
2. Turn the ignition switch to ON (II).
3. Make sure the HDS communicates with the ECM/PCM and other vehicle systems. If it doesn't, go to the DLC circuit troubleshooting. If you are returning from the DLC circuit troubleshooting, skip steps 4 through 9, 20 through 25, and 28 through 30, and do this after replacing the ECM/PCM:
   - Replace the engine oil and the engine oil filter.
   - Replace the ATF (A/T model).
   - Clean the throttle body.
4. Select the PGM-FI system with the HDS.
5. Select the INSPECTION MENU with the HDS.
6. Select the ETCS TEST, then select the TP POSITION CHECK, and follow the screen prompts.

➡**If the TP POSITION CHECK indicates FAILED, continue with this procedure.**

7. Select the REPLACE ECM/PCM MENU, then READ DATA, and follow the screen prompts.

➡**Doing this step copies (READS) the engine oil life data from the original ECM/PCM so you can later download (WRITES) it into the new ECM/PCM. If READ DATA indicates FAILED, continue with this procedure.**

8. A/T models: Select the A/T system with the HDS.
9. A/T models: Select the REPLACE TCM/PCM MENU, then select READ DATA, and follow the screen prompts.

➡**Doing this step copies (READS) the ATF life data from the original PCM so you can later download (WRITES) it into the new PCM. If READ DATA indicates FAILED, continue with this procedure.**

10. Turn the ignition switch to LOCK (0).
11. Jump the SCS line with the HDS.
12. Do the battery removal procedure.
13. Remove the bolts.
14. Disconnect ECM/PCM connectors, then remove the ECM/PCM assembly.

➡**ECM/PCM connectors A, B, and C have symbols (A=□, B=△, C=○) embossed on them for identification.**

15. Remove the cover and the bracket from the ECM/PCM.

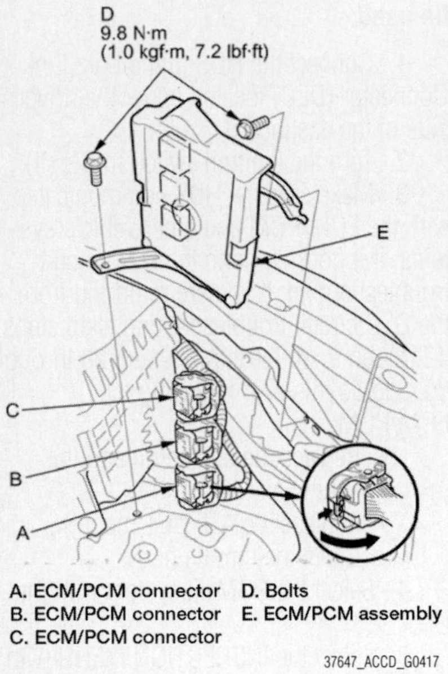
D
9.8 N·m
(1.0 kgf·m, 7.2 lbf·ft)
E
C
B
A
A. ECM/PCM connector
B. ECM/PCM connector
C. ECM/PCM connector
D. Bolts
E. ECM/PCM assembly
37647_ACCD_G0417

Fig. 268 Disconnect ECM/PCM connectors, then remove the ECM/PCM assembly

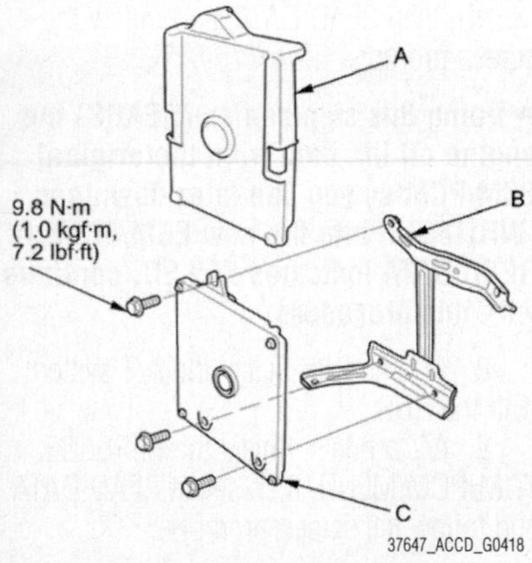
A
9.8 N·m
(1.0 kgf·m,
7.2 lbf·ft)
B
C
37647_ACCD_G0418

Fig. 269 Remove the cover (A) and the bracket (B) from the ECM/PCM (C)

16. Install the ECM/PCM in the reverse order of removal.
17. Do the battery installation procedure.
18. Turn the ignition switch to ON (II).
19. Manually input the VIN to the ECM/PCM with the HDS.

➡DTC P0630 VIN Not Programmed or Mismatch may be stored because the VIN has not been programmed into the ECM/PCM; ignore it, and continue this procedure.

20. If the READ DATA (engine oil life) failed in step 7, go to step 23 (A/T model) or step 26 (M/T model). Otherwise, go to step 21.
21. Select the PGM-FI system with the HDS.
22. Select the REPLACE ECM/PCM MENU, then select WRITE DATA, and follow the screen prompts.

➡If the WRITE DATA indicates FAILED, continue with this procedure.

23. A/T models: If the READ DATA (ATF life) failed in step 9, go to step 26. Otherwise go to step 24.
24. A/T models: Select the A/T SYSTEM with the HDS.
25. A/T models: Select the REPLACE TCM/PCM MENU, then select WRITE DATA, and follow the screen prompts.

➡If the WRITE DATA indicates FAILED, continue with this procedure.

26. Select IMMOBI system with the HDS.
27. Enter the immobilizer ECM/PCM code that you got from the iN, and use the ECM/PCM replacement procedure in the IMMOBI MENU of the HDS; it allows you to start the engine.
28. If the TP POSITION CHECK failed in step 6, clean the throttle body, then go to step 29.
29. If the READ DATA failed in step 7 or the WRITE DATA failed in step 22, replace the engine oil and engine oil filter, then go to step 30 (A/T model) or step 31 (M/T model).
30. If the READ DATA failed in step 9 or the WRITE DATA failed in step 25, replace the ATF, then go to step 31.
31. Select PGM-FI system, and reset the ECM/PCM with the HDS.
32. Update the ECM/PCM if it does not have the latest software.
33. Do the ECM/PCM idle learn procedure.
34. Do the CKP pattern clear/CKP pattern learn procedure.

ECM/PCM IDLE LEARN PROCEDURE

The idle learn procedure must be done so the ECM/PCM can learn the engine idle characteristics.

Do the idle learn procedure whenever you do any of these actions:

- Replace the ECM/PCM.
- Reset the ECM/PCM.
- Update the ECM/PCM.
- Replace or clean the throttle body.
- Disassemble the engine or the transmission.

➡Clearing the DTCs with the HDS does not require you to do the idle learn procedure.

1. Make sure all electrical items (the A/C, the audio system, the lights, etc.) are off.
2. Reset the ECM/PCM with the HDS.
3. Turn the ignition switch to ON (II), and wait 2 seconds.
4. Start the engine. Hold the engine speed at 3,000 rpm without load (A/T in P or N, M/T in neutral) until the radiator fan comes on, or until the engine coolant temperature reaches 194°F (90°C).
5. Let the engine idle for about 5 minutes with the throttle fully closed.

➡If the radiator fan comes on, do not include its running time in the 5 minutes.

ECM/PCM UPDATE

Special Tools Required:

- Honda diagnostic system (HDS) tablet tester
- Honda interface module (HIM) and an iN workstation with the latest HDS software version
- HDS pocket tester
- GNA600 and an iN workstation with the latest HDS software version

➡Any one of the above updating tools can be used.

➡Be aware of the following for update procedures:

- Make sure the HDS/iN workstation has the latest HDS software version.
- Before you update the ECM/PCM, make sure the battery in the vehicle is fully charged, and connect a jumper battery (not a battery charger) to maintain system voltage.
- Never turn the ignition switch to ACC (I) or LOCK (0) during the update. If there is a problem with the update, leave the ignition switch in ON (II).
- To prevent ECM/PCM damage, do not operate anything electrical (headlights, audio system, brakes, A/C, power windows, moonroof (if equipped), door locks, etc.) during the update.
- To ensure the latest program is installed, do an ECM/PCM update whenever the ECM/PCM is substituted or replaced.
- You cannot update an ECM/PCM

with a program it already has. It will only accept a new program.

- High temperature in the engine compartment might cause the ECM/PCM to become too hot to run the update. If the engine has been running before this procedure, open the hood and cool the engine compartment.
- If you need to diagnose the Honda Interface Module (HIM) because the HIM's red (#3) light came on or was flashing during the update, leave the ignition switch in ON (II) when you disconnect the HIM from the Data Link Connector (DLC). This will prevent ECM/PCM damage.

1. Turn the ignition switch to ON (II), but do not start the engine.
2. Connect the HDS to the Data Link Connector (DLC) located under the driver's side of the dashboard.
3. Make sure the HDS communicates with the ECM/PCM and other vehicle systems. If it doesn't, go to the DLC circuit troubleshooting. If you are returning from the DLC circuit troubleshooting, skip steps 4 and 5, and clean the throttle body after updating the ECM/PCM.
4. Select the INSPECTION MENU with the HDS.
5. Select the ETCS TEST, then select the TP POSITION CHECK, and follow the HDS screen prompts.

➡If the TP POSITION CHECK indicates FAILED, continue this procedure.

6. Exit the HDS diagnostic system, then select the update mode, and follow the screen prompts to update the ECM/PCM.
7. If the software in the ECM/PCM is the latest, disconnect the HDS/HIM/GNA600 from the DLC, and go back to the procedure that you were doing. If the software in the ECM/PCM is not the latest, follow the instructions on the screen. If prompted to choose the PGM-FI system or the A/T system (A/T), make sure you update both.

➡If the ECM/PCM update system requires you to cool the ECM/PCM, follow the instructions on the screen. If you run into a problem during the update procedure (programming takes over 15 minutes, status bar goes over 100 %, D (A/T) or immobilizer indicator flashes, HDS tablet freezes, etc.), follow these steps to minimize the chance of damaging the ECM/PCM:

- Leave the ignition switch in ON (II).
- Connect a jumper battery (do not connect a battery charger).
- Shut down the HDS.
- Disconnect the HDS from the DLC.
- Reboot the HDS.
- Reconnect the HDS to the DLC, and do the update again.

8. If the TP POSITION CHECK failed in step 5, clean the throttle body.
9. Do the ECM/PCM idle learn procedure.
10. Do the CKP pattern clear/CKP pattern learn procedure.

CKP PATTERN CLEAR/CKP PATTERN LEARN

Clear/Learn Procedure (with the HDS)

1. Connect the HDS to the Data Link Connector (DLC) located under the driver's side of the dashboard.
2. Turn the ignition switch to ON (II).
3. Make sure the HDS communicates with the ECM/PCM and other vehicle systems. If it doesn't, go to the DLC circuit troubleshooting.
4. Select CRANK PATTERN in the ADJUSTMENT MENU with the HDS.
5. Select CRANK PATTERN LEARNING with the HDS, and follow the screen prompts.

Learn Procedure (without the HDS)

1. Start the engine. Hold the engine speed at 3,000 rpm without load (A/T in P or N, M/T in neutral) until the radiator fan comes on.
2. Test-drive the vehicle on a level road: Decelerate (with the throttle fully closed) from an engine speed of 2,500 rpm down to 1,000 rpm with the A/T in 2, or the M/T in 2nd.
3. Repeat step 2 several times.
4. Turn the ignition switch to LOCK (0).
5. Turn the ignition switch to ON (II), and wait 30 seconds.

How to End a Troubleshooting Session (required after any troubleshooting)

1. Reset the ECM/PCM with the HDS.
2. Do the ECM/PCM idle learn procedure.
3. Turn the ignition switch to LOCK (0).
4. Disconnect the HDS from the DLC.

➡The ECM/PCM is part of the immobilizer system. If you replace the ECM/PCM, for the engine to start, you must use the HDS to instruct the new ECM/PCM and the immobilizer-keyless control unit to recognize each other's unique serial code.

ENGINE COOLANT TEMPERATURE (ECT) SENSOR

REMOVAL & INSTALLATION

4-Cylinder Engine

ECT Sensor 1

See Figure 270.

1. Drain the engine coolant.
2. Disconnect the ECT sensor 1 connector.
3. Remove ECT sensor 1.
4. Install the parts in the reverse order of removal with a new O-ring, then refill the radiator with engine coolant.

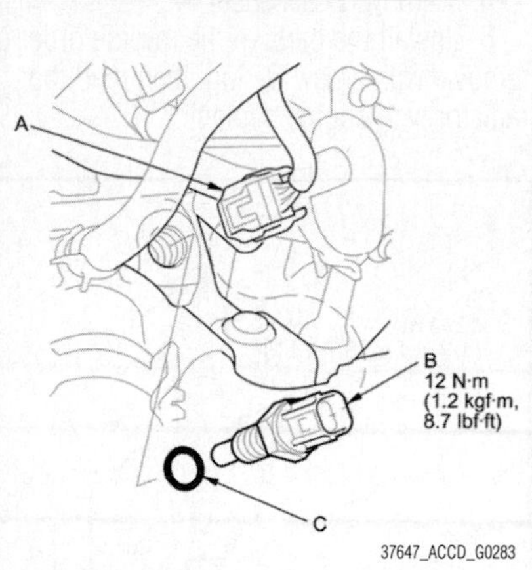
A
B
12 N·m
(1.2 kgf·m,
8.7 lbf·ft)
C
37647_ACCD_G0283

Fig. 270 Disconnect the ECT sensor 1 connector (A), then remove ECT sensor 1 (B) and O-ring (C)

ECT Sensor 2

See Figure 271.

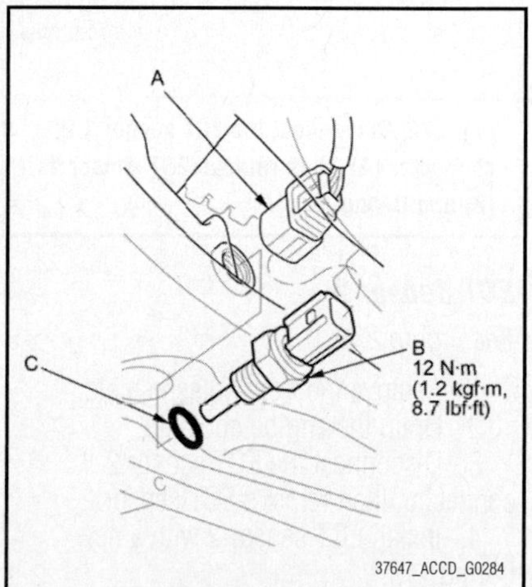
A
B
12 N·m
(1.2 kgf·m,
8.7 lbf·ft)
C
37647_ACCD_G0284

Fig. 271 Disconnect the ECT sensor 2 connector (A), then remove ECT sensor 2 (B) and O-ring (C)

1. Remove the front splash shield.
2. Drain the engine coolant.
3. Disconnect the ECT sensor 2 connector, then remove ECT sensor 2.
4. Install ECT sensor 2 with a new O-ring.
5. Install the front splash shield.
6. Refill the radiator with engine coolant.

6-Cylinder Engine

ECT Sensor 1

See Figure 272.

1. Drain the engine coolant.
2. Remove the engine cover.
3. Disconnect the ECT sensor 1 2P connector.
4. Remove ECT sensor 1.
5. Install the parts in the reverse order of removal with a new O-ring, then refill the radiator with engine coolant.

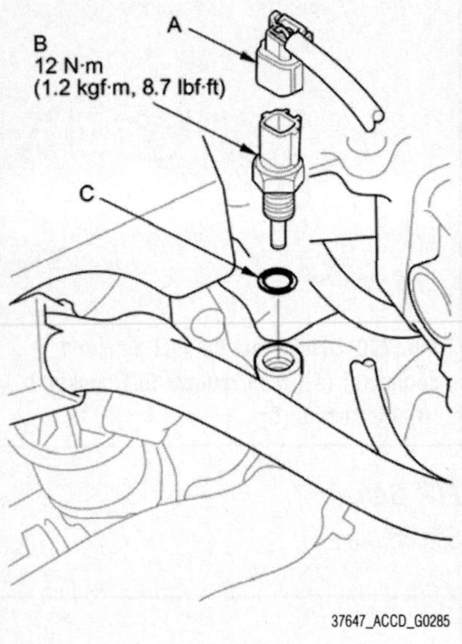
A
B
12 N·m
(1.2 kgf·m, 8.7 lbf·ft)
C
37647_ACCD_G0285

Fig. 272 Disconnect the ECT sensor 1 2P connector (A), then remove ECT sensor 1 (B) and O-ring (C)

ECT Sensor 2

See Figure 273.

1. Remove the front splash shield.
2. Drain the engine coolant.
3. Disconnect the ECT sensor 2 connector, then remove ECT sensor 2.
4. Install ECT sensor 2 with a new O-ring.
5. Install the front splash shield.
6. Refill the radiator with engine coolant.

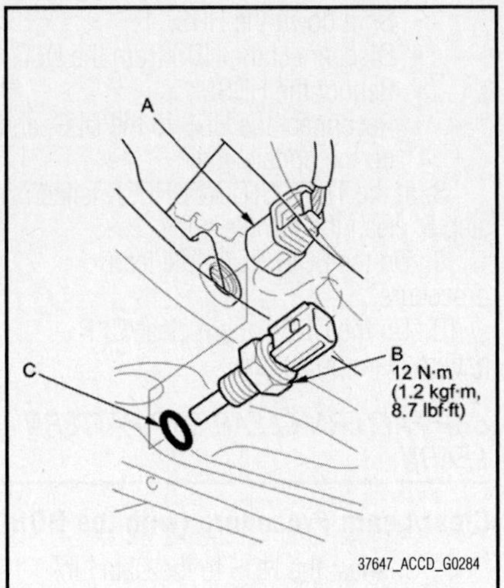
A
B
12 N·m
(1.2 kgf·m,
8.7 lbf·ft)
C
37647_ACCD_G0284

Fig. 273 Disconnect the ECT sensor 2 connector (A), then remove ECT sensor 2 (B) and O-ring (C)

EVAPORATIVE EMISSIONS (EVAP) CANISTER

LOCATION

The Evaporative Emissions (EVAP) canister is located in the rear subframe of the vehicle.

REMOVAL & INSTALLATION

4-Cylinder Engine

See Figures 274 through 276.

1. Raise the vehicle on a lift.
2. Remove the wheel sensor harness clamps.
3. Support the rear subframe with a transmission jack and a wooden block as shown.
4. Remove the rear subframe mounting bolts.
5. Lower the transmission jack and the rear subframe about 50 mm.

➡Be careful not to damage the connecting parts.

6. Remove the bolt, and disconnect the hoses, the EVAP canister vent shut valve 2P connector, and the FTP sensor 3P connector.
7. Remove the bolts, then remove the EVAP canister assembly.
8. Remove the EVAP canister from the EVAP canister bracket.

To install:

9. Install the EVAP canister assembly to the body.

➡Attach the bracket arm to the body.

10. Install the parts in the reverse order of removal. Use new bolts when you install the rear subframe.
11. Check the wheel alignment.

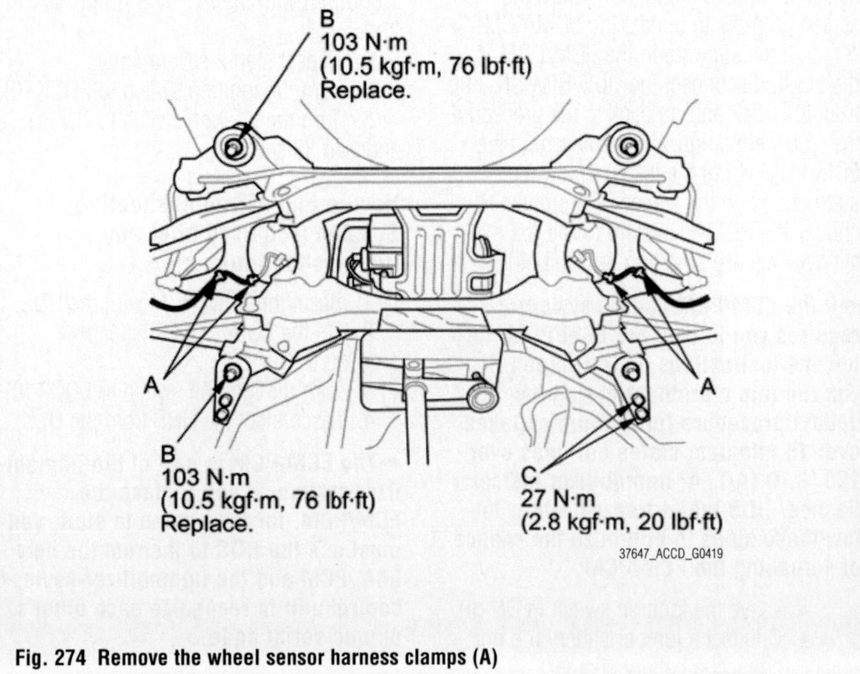
B
103 N·m
(10.5 kgf·m, 76 lbf·ft)
Replace.
A
A
B
103 N·m
(10.5 kgf·m, 76 lbf·ft)
Replace.
C
27 N·m
(2.8 kgf·m, 20 lbf·ft)
37647_ACCD_G0419

Fig. 274 Remove the wheel sensor harness clamps (A)

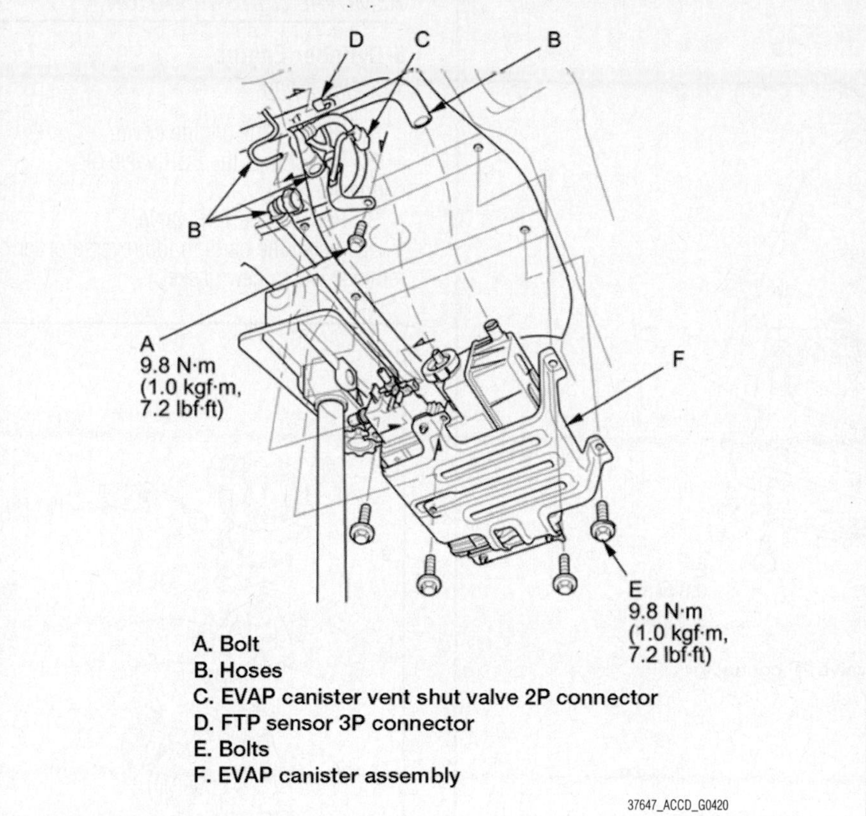
D
C
B
B
A
9.8 N·m
(1.0 kgf·m,
7.2 lbf·ft)
F
E
9.8 N·m
(1.0 kgf·m,
7.2 lbf·ft)
A. Bolt
B. Hoses
C. EVAP canister vent shut valve 2P connector
D. FTP sensor 3P connector
E. Bolts
F. EVAP canister assembly
37647_ACCD_G0420

Fig. 275 Remove the EVAP canister assembly

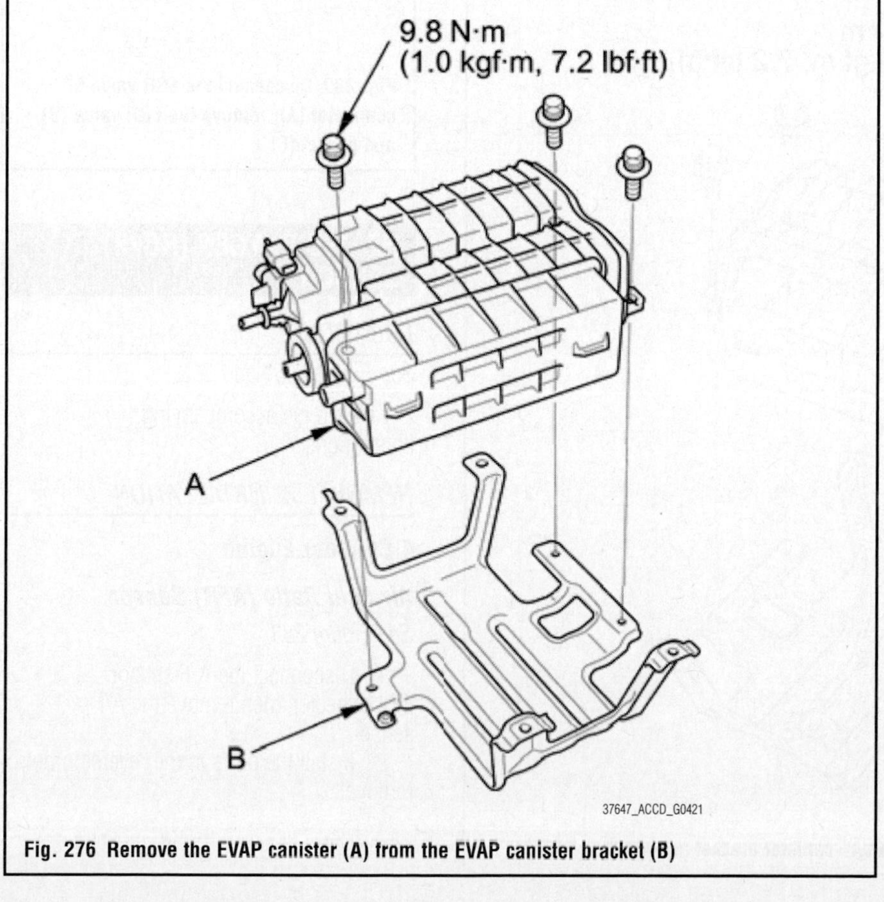
9.8 N·m
(1.0 kgf·m, 7.2 lbf·ft)
A
B
37647_ACCD_G0421

Fig. 276 Remove the EVAP canister (A) from the EVAP canister bracket (B)

6-Cylinder Engine

See Figures 277 through 279.

1. Raise the vehicle on a lift.
2. Remove the wheel sensor harness clamps.
3. Support the rear subframe with a transmission jack and a wooden block.
4. Remove the rear subframe mounting bolts.
5. Lower the transmission jack and the rear subframe about 50 mm.

➡Be careful not to damage the connecting parts.

6. Remove the bolt, and disconnect the hoses, the EVAP canister vent shut valve 2P connector, and the FTP sensor 3P connector.
7. Remove the bolts, then remove the EVAP canister assembly.
8. Remove the EVAP canister from the EVAP canister bracket.

To install:

9. Install the EVAP canister assembly to the body.
10. Install the parts in the reverse order of removal. Use new bolts when you install the rear subframe.
11. Check the wheel alignment.

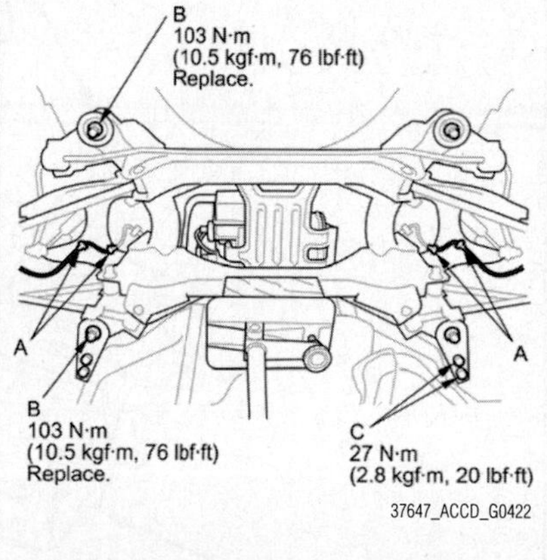
B
103 N·m
(10.5 kgf·m, 76 lbf·ft)
Replace.
A
A
B
103 N·m
(10.5 kgf·m, 76 lbf·ft)
Replace.
C
27 N·m
(2.8 kgf·m, 20 lbf·ft)
37647_ACCD_G0422

Fig. 277 Remove the wheel sensor harness clamps (A) and the rear subframe mounting bolts (B, C)

EXHAUST GAS RECIRCULATION (EGR) VALVE

LOCATION

6-Cylinder Engine

The EGR valve is located at the front left of the engine.

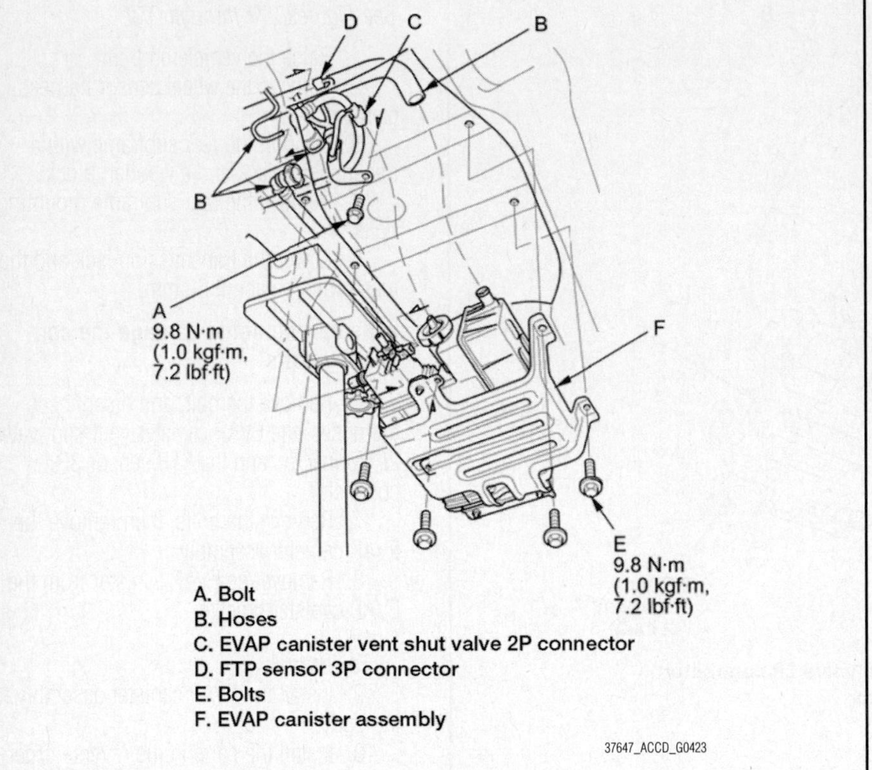
D
C
B
B
A
9.8 N·m
(1.0 kgf·m,
7.2 lbf·ft)
F
E
9.8 N·m
(1.0 kgf·m,
7.2 lbf·ft)
A. Bolt
B. Hoses
C. EVAP canister vent shut valve 2P connector
D. FTP sensor 3P connector
E. Bolts
F. EVAP canister assembly
37647_ACCD_G0423

Fig. 278 Remove the bolt, and disconnect the hoses, the EVAP canister vent shut valve 2P connector, and the FTP sensor 3P connector

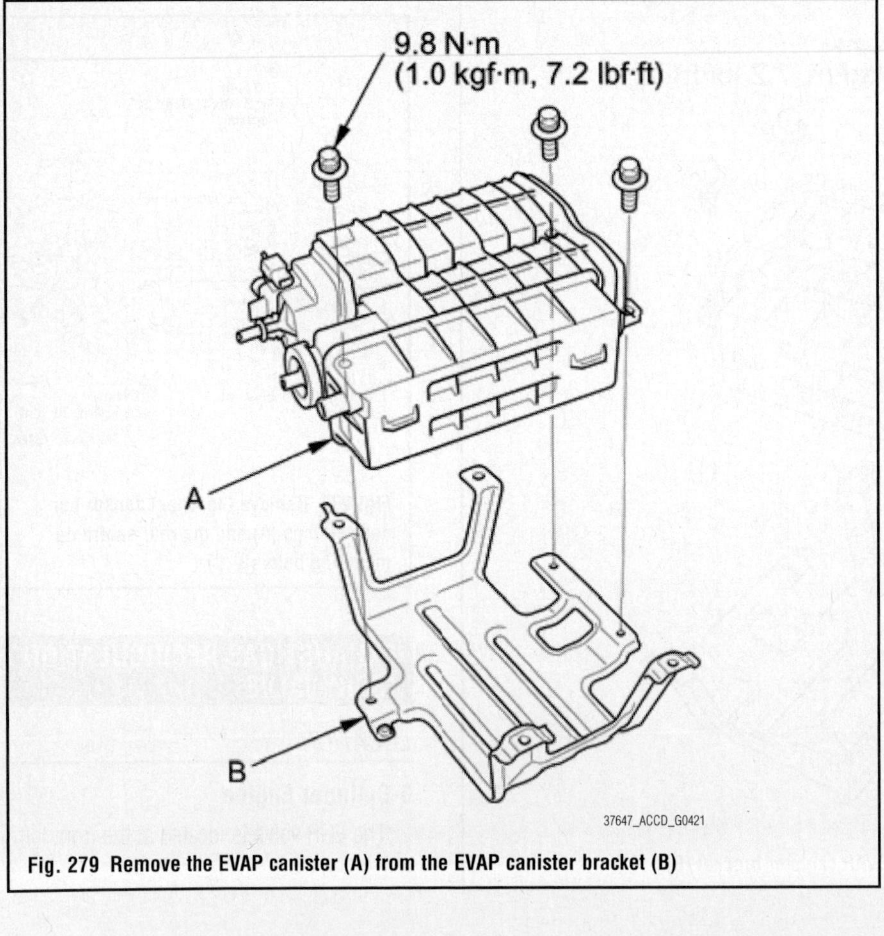
9.8 N·m
(1.0 kgf·m, 7.2 lbf·ft)
A
B
37647_ACCD_G0421

Fig. 279 Remove the EVAP canister (A) from the EVAP canister bracket (B)

REMOVAL & INSTALLATION

6-Cylinder Engine

See Figure 280.

1. Remove the engine cover.
2. Disconnect the EGR valve 5P connector.
3. Remove the EGR valve.
4. Install the parts in the reverse order of removal with a new gasket.

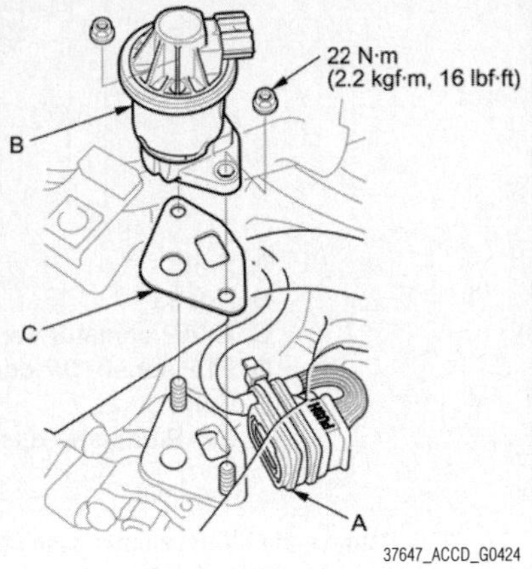
22 N·m
(2.2 kgf·m, 16 lbf·ft)
B
C
A
37647_ACCD_G0424

Fig. 280 Disconnect the EGR valve 5P connector (A), remove the EGR valve (B) and gasket (C)

HEATED OXYGEN (HO2S) SENSOR

LOCATION

See Figures 281 and 282.

Refer to the accompanying illustrations.

REMOVAL & INSTALLATION

4-Cylinder Engine

Air Fuel Ratio (AFR) Sensor

See Figure 283.

1. Disconnect the A/F sensor 4P connector, then remove the A/F sensor.
2. Install the parts in the reverse order of removal.

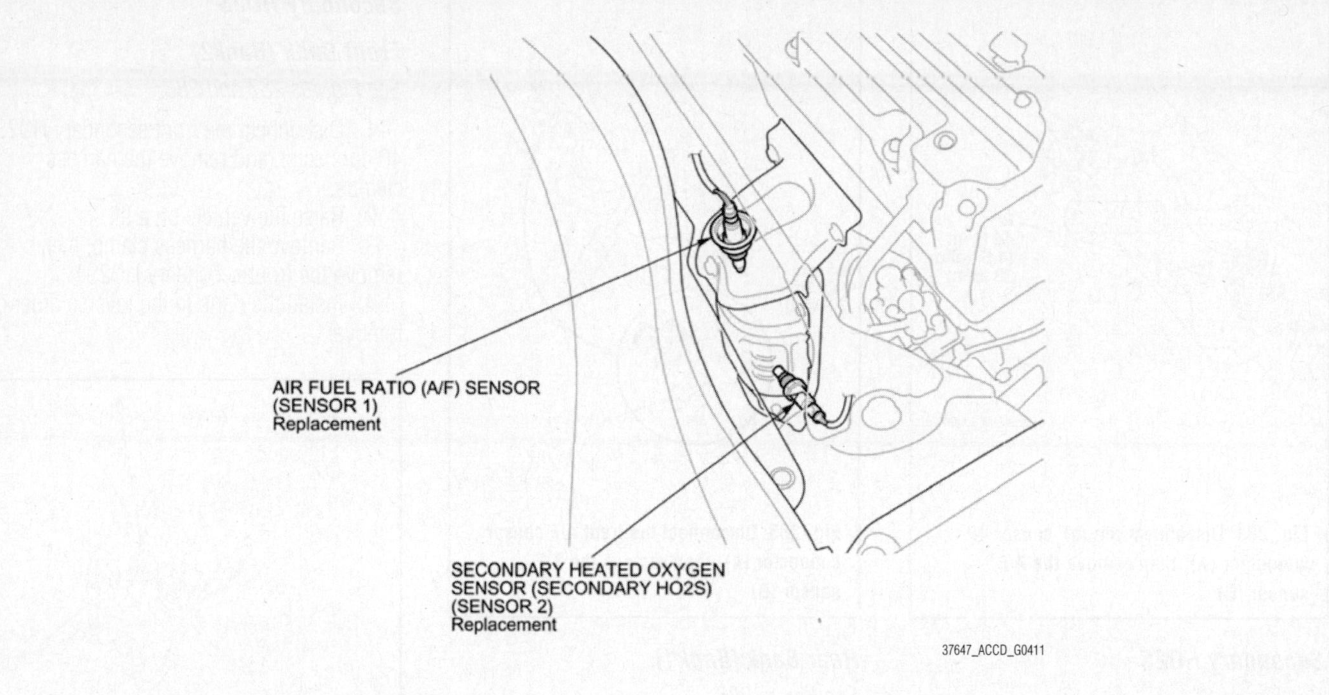
AIR FUEL RATIO (A/F) SENSOR
(SENSOR 1)
Replacement
SECONDARY HEATED OXYGEN
SENSOR (SECONDARY HO2S)
(SENSOR 2)
Replacement
37647_ACCD_G0411

Fig. 281 Heated Oxygen (HO2S) Sensor location—4 cylinder

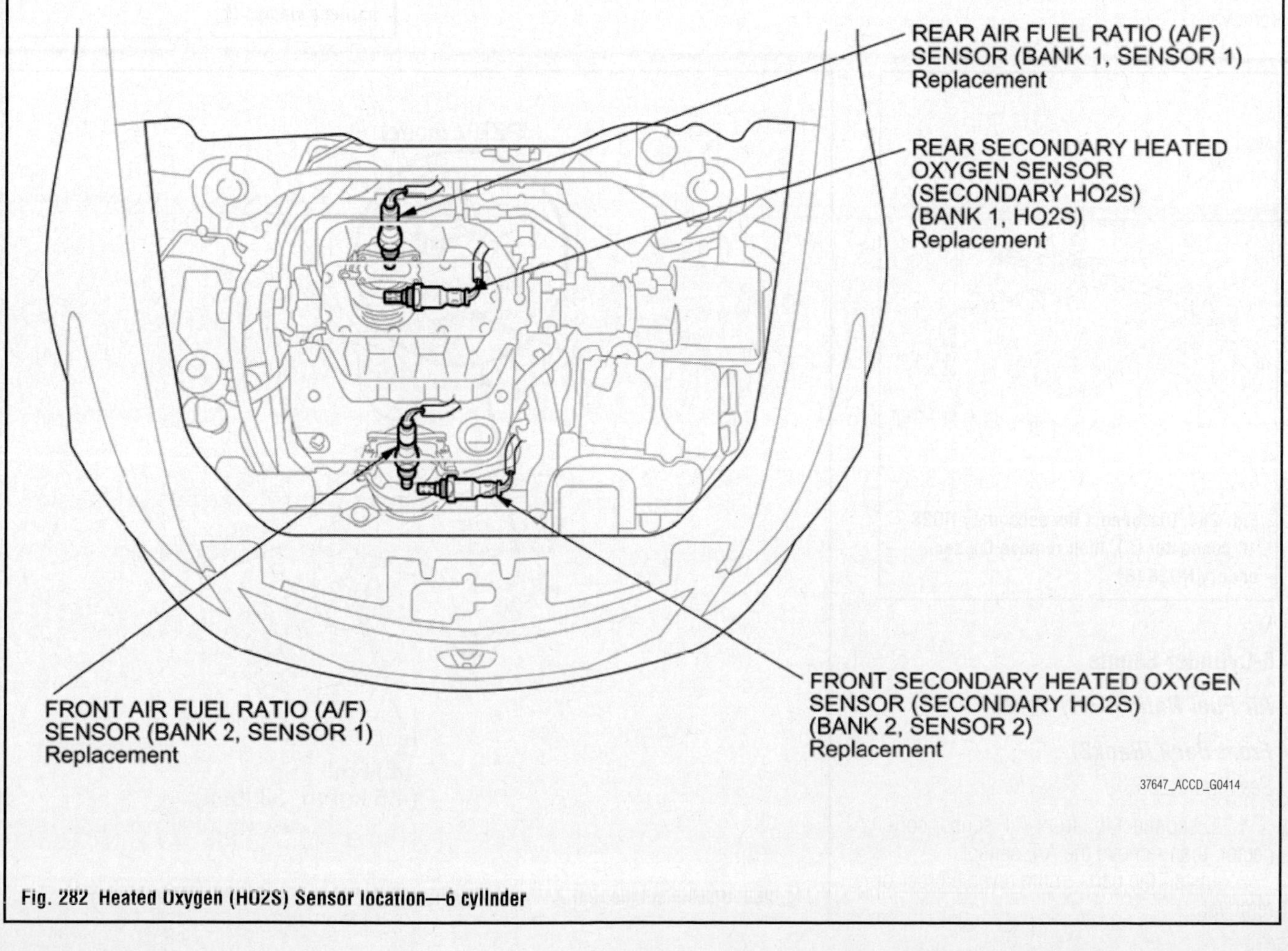
REAR AIR FUEL RATIO (A/F)
SENSOR (BANK 1, SENSOR 1)
Replacement
REAR SECONDARY HEATED
OXYGEN SENSOR
(SECONDARY HO2S)
(BANK 1, HO2S)
Replacement
FRONT AIR FUEL RATIO (A/F)
SENSOR (BANK 2, SENSOR 1)
Replacement
FRONT SECONDARY HEATED OXYGEN
SENSOR (SECONDARY HO2S)
(BANK 2, SENSOR 2)
Replacement
37647_ACCD_G0414

Fig. 282 Heated Oxygen (HO2S) Sensor location—6 cylinder

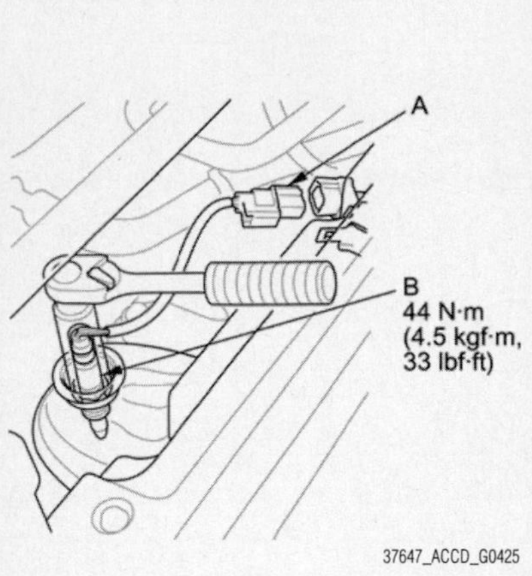
A
B
44 N·m
(4.5 kgf·m,
33 lbf·ft)
37647_ACCD_G0425

Fig. 283 Disconnect the A/F sensor 4P connector (A), then remove the A/F sensor (B)

Secondary HO2S

See Figure 284.

1. Disconnect the secondary HO2S 4P connector, then remove the secondary HO2S.
2. Install the parts in the reverse order of removal.

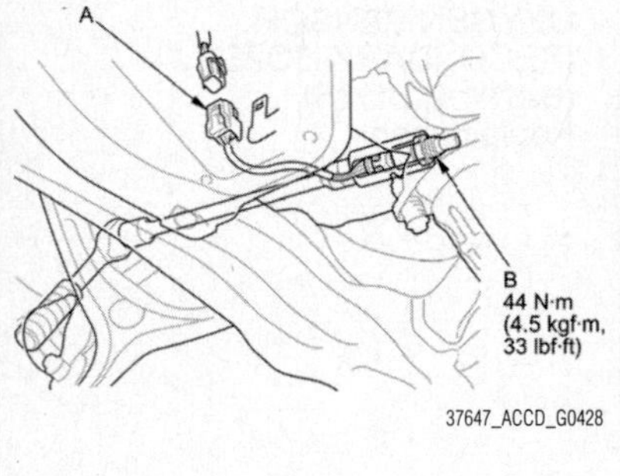
A
B
44 N·m
(4.5 kgf·m,
33 lbf·ft)
37647_ACCD_G0428

Fig. 284 Disconnect the secondary HO2S 4P connector (A), then remove the secondary HO2S (B)

6-Cylinder Engine

Air Fuel Ratio (AFR) Sensor

Front Bank (Bank2)

See Figure 285.

1. Disconnect the front A/F sensor connector, then remove the A/F sensor.
2. Install the parts in the reverse order of removal.

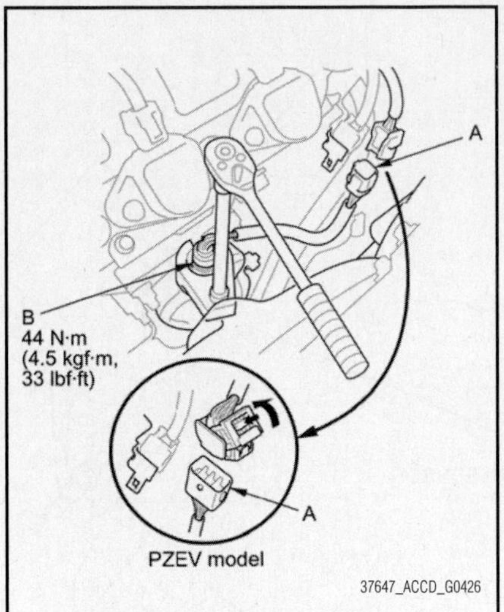
A
B
44 N·m
(4.5 kgf·m,
33 lbf·ft)
A
PZEV model
37647_ACCD_G0426

Fig. 285 Disconnect the front A/F sensor connector (A), then remove the A/F sensor (B)

Rear Bank (Bank1)

See Figure 286.

1. Disconnect the rear A/F sensor connector, then remove the A/F sensor.
2. Install the parts in the reverse order of removal.

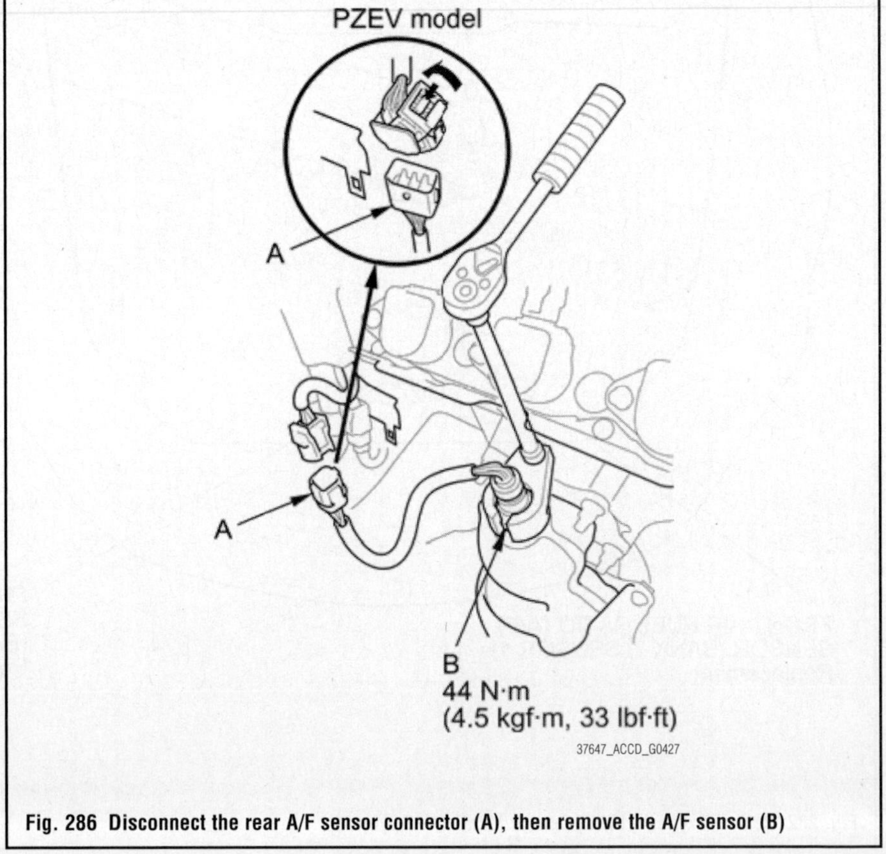
PZEV model
A
A
B
44 N·m
(4.5 kgf·m, 33 lbf·ft)
37647_ACCD_G0427

Fig. 286 Disconnect the rear A/F sensor connector (A), then remove the A/F sensor (B)

Secondary HO2S

Front Bank (Bank2)

See Figures 287 and 288.

1. Disconnect the front secondary HO2S 4P connector, and remove the harness clamps.
2. Raise the vehicle on a lift.
3. Remove the harness clamp, then remove the front secondary HO2S.
4. Install the parts in the reverse order of removal.

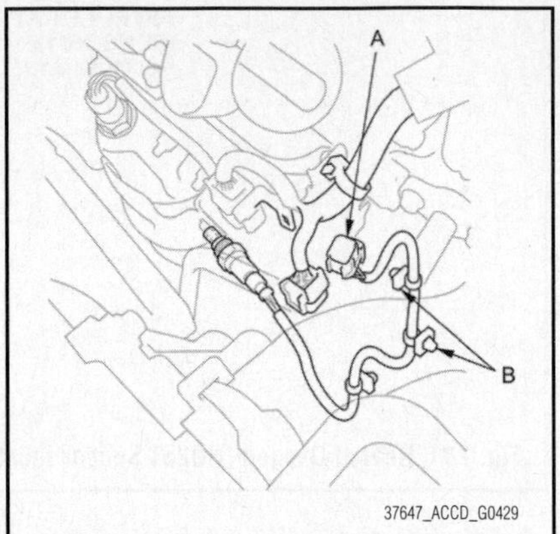
A
B
37647_ACCD_G0429

Fig. 287 Disconnect the front secondary HO2S 4P connector (A), and remove the harness clamps (B)

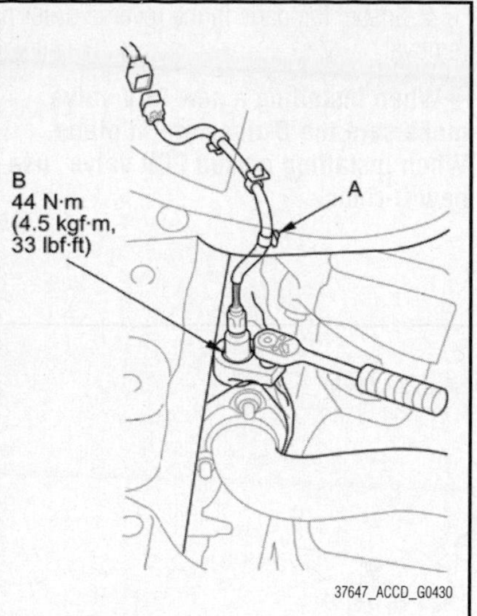
B
44 N·m
(4.5 kgf·m,
33 lbf·ft)
A
37647_ACCD_G0430

Fig. 288 Remove the harness clamp (A), then remove the front secondary HO2S (B)

Rear Bank (Bank 1)

See Figure 289.

1. Raise the vehicle on a lift.
2. Disconnect the rear secondary HO2S 4P connector, and remove the harness clamps.
3. Remove the rear secondary HO2S.
4. Install the parts in the reverse order of removal.

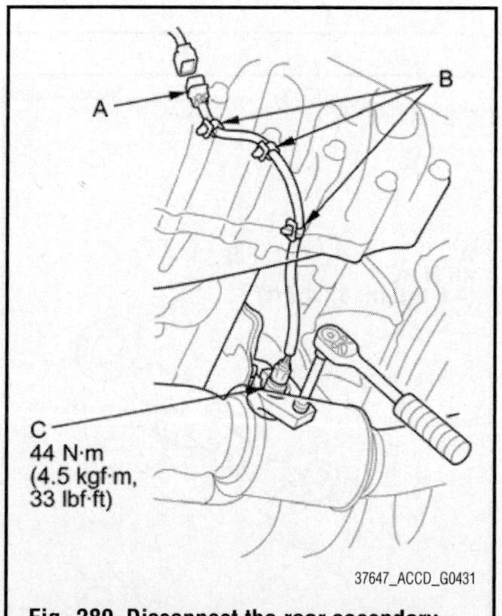
A
B
C
44 N·m
(4.5 kgf·m,
33 lbf·ft)
37647_ACCD_G0431

Fig. 289 Disconnect the rear secondary HO2S 4P connector (A), and remove the harness clamps (B) and the rear secondary HO2S (C)

INTAKE AIR TEMPERATURE (IAT) SENSOR

LOCATION

The Intake Air Temperature (IAT) sensor is an integral part of the Mass Air Flow (MAF) sensor. Refer to the MAF Removal and Installation when needing to service the IAT.

REMOVAL & INSTALLATION

The Intake Air Temperature (IAT) sensor is an integral part of the Mass Air Flow (MAF) sensor. Refer to the MAF Removal and Installation when needing to service the IAT.

KNOCK SENSOR (KS)

REMOVAL & INSTALLATION

4-Cylinder Engine

See Figure 290.

1. Remove the intake manifold.
2. Disconnect the knock sensor connector.
3. Remove the knock sensor.
4. Install the parts in the reverse order of removal.

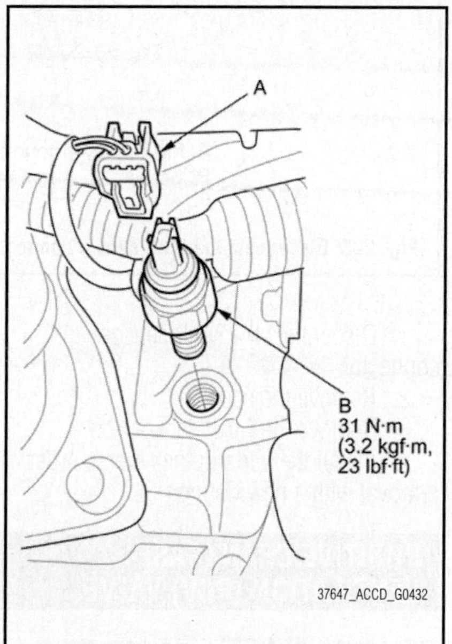
A
B
31 N·m
(3.2 kgf·m,
23 lbf·ft)
37647_ACCD_G0432

Fig. 290 Disconnect the knock sensor connector (A) and remove the knock sensor (B)

6-Cylinder Engine

See Figure 291.

1. Remove the intake manifold.
2. Remove the injector base.
3. Disconnect the knock sensor connector.
4. Remove the knock sensor.
5. Install the parts in the reverse order of removal.

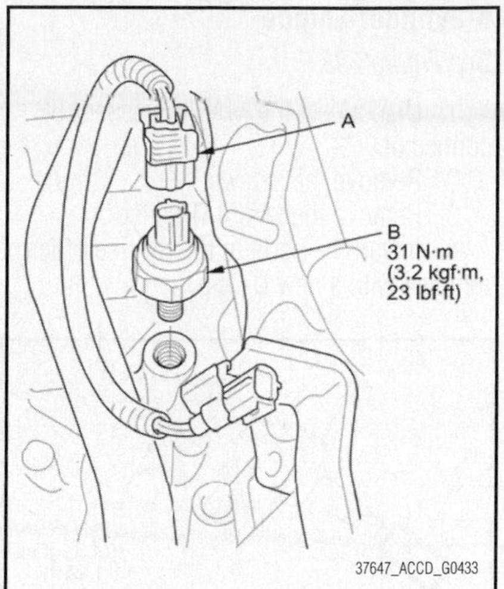
A
B
31 N·m
(3.2 kgf·m,
23 lbf·ft)
37647_ACCD_G0433

Fig. 291 Disconnect the knock sensor connector

MASS AIR FLOW (MAF)/INTAKE AIR TEMPERATURE (IAT) SENSOR (HOT WIRE)

LOCATION

The MAF/IAT sensor is located on the engine air intake.

REMOVAL & INSTALLATION

4-Cylinder Engine

See Figure 292.

1. Disconnect the MAF sensor/IAT sensor connector.
2. Remove the screws.
3. Remove the MAF sensor/IAT sensor.
4. Install the parts in the reverse order of removal with a new O-ring.

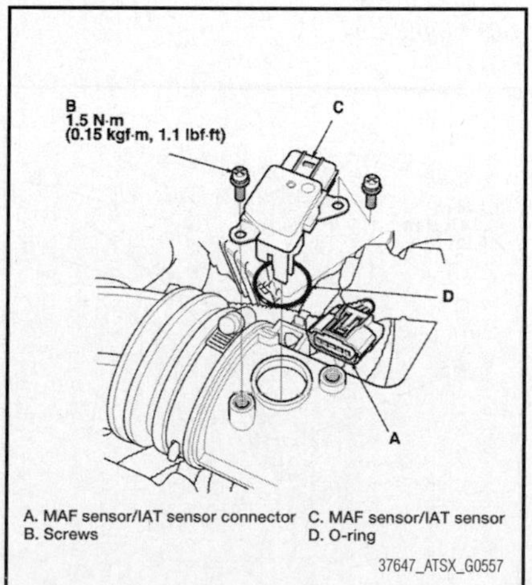
B
1.5 N·m
(0.15 kgf·m, 1.1 lbf·ft)
C
D
A
A. MAF sensor/IAT sensor connector
B. Screws
C. MAF sensor/IAT sensor
D. O-ring
37647_ATSX_G0557

Fig. 292 Disconnect the MAF sensor/IAT sensor connector

6-Cylinder Engine

See Figure 293.

1. Disconnect the MAF/IAT sensor 5P connector.
2. Remove the screws.
3. Remove the MAF/IAT sensor.
4. Install the parts in the reverse order of removal with a new O-ring.

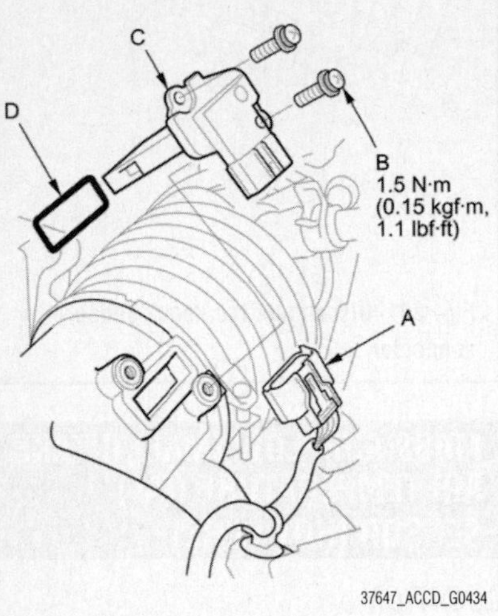
C
D
B
1.5 N·m
(0.15 kgf·m,
1.1 lbf·ft)
A
37647_ACCD_G0434

Fig. 293 Disconnect the MAF/IAT sensor 5P connector

MANIFOLD ABSOLUTE PRESSURE (MAP) SENSOR

LOCATION

Refer to graphic in the Removal and Installation section for location.

REMOVAL & INSTALLATION

4-Cylinder Engine

See Figure 294.

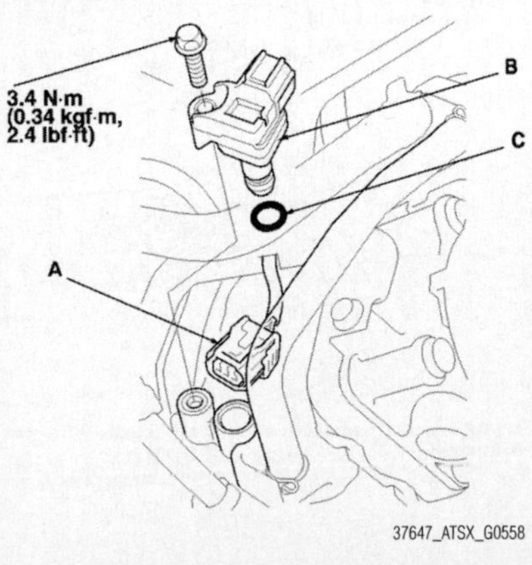
B
3.4 N·m
(0.34 kgf·m,
2.4 lbf·ft)
C
A
37647_ATSX_G0558

Fig. 294 Disconnect the MAP sensor connector (A) and remove the MAP sensor (B)

1. Disconnect the MAP sensor connector.
2. Remove the MAP sensor.
3. Install the parts in the reverse order of removal with a new O-ring.

6-Cylinder Engine

See Figure 295.

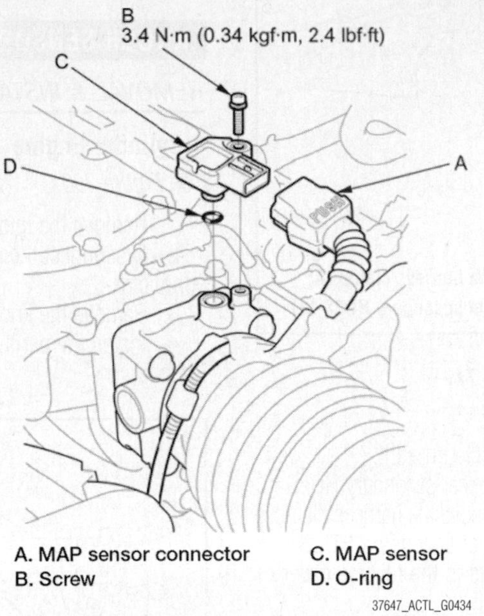
B
3.4 N·m (0.34 kgf·m, 2.4 lbf·ft)
C
D
A
A. MAP sensor connector
B. Screw
C. MAP sensor
D. O-ring
37647_ACTL_G0434

Fig. 295 Disconnect the MAP sensor connector

1. Disconnect the MAP sensor connector.
2. Remove the screw.
3. Remove the MAP sensor.
4. Install the parts in the reverse order of removal with a new O-ring.

POSITIVE CRANKCASE VENTILATION (PCV) VALVE

REMOVAL & INSTALLATION

4-Cylinder Engine

See Figure 296.

1. Disconnect the PCV hose.
2. Remove the PCV valve.
3. Install the parts in the reverse order of removal with a new washer.

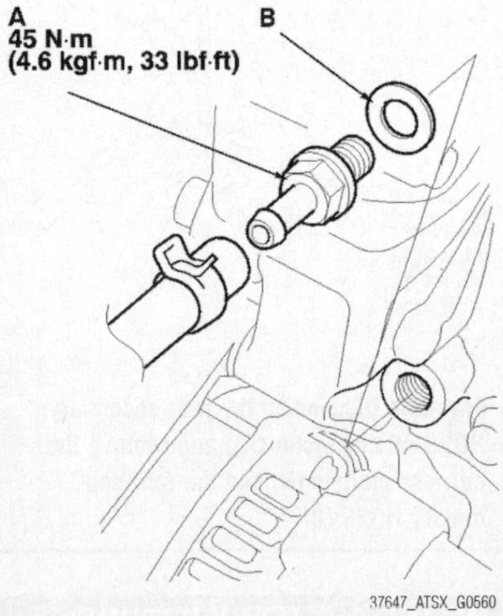
A
45 N·m
(4.6 kgf·m, 33 lbf·ft)
B
37647_ATSX_G0560

Fig. 296 Remove the PCV valve (A)

6-Cylinder Engine

See Figure 297.

1. Remove the engine cover.
2. Remove the bolt.
3. Remove the PCV valve.

➡**Do not to spill oil on the hot exhaust manifold.**

4. Install the parts in the reverse order of removal.

➡**When installing a new PCV valve, make sure the O-rings are in place. When installing a used PCV valve, use new O-rings.**

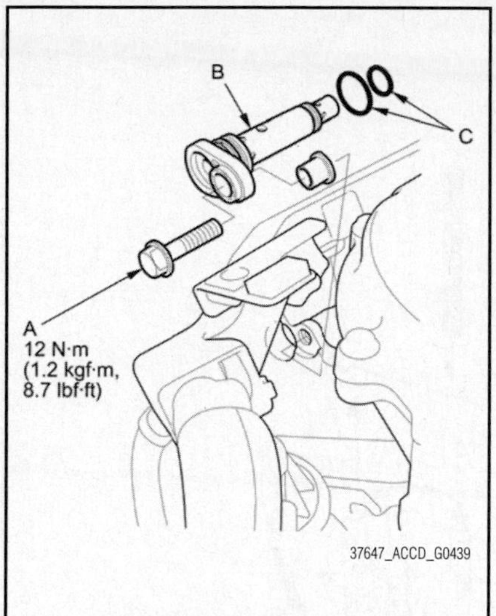
B
C
A
12 N·m
(1.2 kgf·m,
8.7 lbf·ft)
37647_ACCD_G0439

Fig. 297 Remove the bolt (A), the PCV valve (B) and the O-rings (C)

VARIABLE CAMSHAFT TIMING OIL CONTROL SOLENOID

LOCATION

4-Cylinder Engine

The VTC oil control solenoid valve is located on the cylinder head.

REMOVAL & INSTALLATION

4-Cylinder Engine

See Figure 298.

1. Disconnect the VTC oil control solenoid valve 2P connector.
2. Remove the bolt and the VTC oil control solenoid valve.
3. Installation is the reverse of removal.

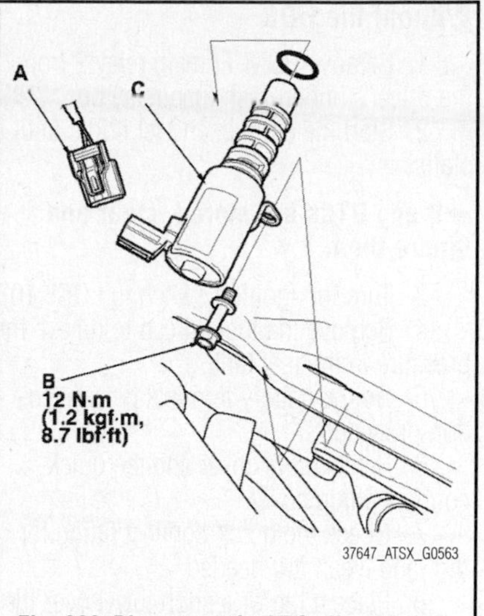
A
C
B
12 N·m
(1.2 kgf·m,
8.7 lbf·ft)
37647_ATSX_G0563

Fig. 298 Disconnect the VTC oil control solenoid valve 2P connector

FUEL — GASOLINE FUEL INJECTION SYSTEM

FUEL SYSTEM SERVICE PRECAUTIONS

Safety is the most important factor when performing not only fuel system maintenance, but any type of maintenance. Failure to conduct maintenance and repairs in a safe manner may result in serious personal injury or death. Work on a vehicle's fuel system components can be accomplished safely and effectively by adhering to the following rules and guidelines.

- To avoid the possibility of fire and personal injury, always disconnect the negative battery cable unless the repair or test procedure requires that battery voltage be applied.
- Always relieve the fuel system pressure prior to disconnecting any fuel system component (injector, fuel rail, pressure regulator, etc.) fitting or fuel line connection. Exercise extreme caution whenever relieving fuel system pressure to avoid exposing skin, face and eyes to fuel spray. Please be advised that fuel under pressure may penetrate the skin or any part of the body that it contacts.
- Always place a shop towel or cloth around the fitting or connection prior to loosening to absorb any excess fuel due to spillage. Ensure that all fuel spillage is quickly removed from engine surfaces. Ensure that all fuel-soaked cloths or towels are deposited into a flame-proof waste container with a lid.
- Always keep a dry chemical (Class B) fire extinguisher near the work area.
- Do not allow fuel spray or fuel vapors to come into contact with a spark or open flame.
- Always use a second wrench when loosening or tightening fuel line connection fittings. This will prevent unnecessary stress and torsion on fuel piping. Always follow the proper torque specifications.
- Always replace worn fuel fitting O-rings with new ones. Do not substitute fuel hose where rigid pipe is installed.

FUEL SYSTEM PRESSURE

RELIEVING

⁂ CAUTION

Before disconnecting fuel lines or hoses, relieve pressure from the system by disabling the fuel pump and disconnecting the fuel line/quick connect fitting in the engine compartment.

With the HDS

1. Connect the HDS to the Data Link Connector (DLC) located under the driver's side of the dashboard.
2. Turn the ignition switch to ON (II).
3. Make sure the HDS communicates with the ECM/PCM. If it doesn't, go to the DLC circuit troubleshooting.
4. Turn the ignition switch to LOCK (0).
5. Remove the fuel fill cap to relieve the pressure in the fuel tank.
6. Turn the ignition switch to ON (II).
7. From the INSPECTION MENU of the HDS, select Fuel Pump OFF, then start the engine, and let it idle until it stalls.

➡Do not allow the engine to idle above 1,000 rpm or the ECM/PCM will continue to operate the fuel pump. Pending or Confirmed DTC may be set during this procedure. Check for DTCs, and clear them as needed.

8. Turn the ignition switch to LOCK (0).
9. Do the battery terminal disconnection procedure.
10. Remove the cover and the quick-connect fitting cover.
11. Check the quick-connect fitting for dirt, and clean it if needed.
12. Place a rag or a shop towel over the quick-connect fitting.
13. Disconnect the quick-connect fitting.
    a. Hold the connector with one hand, and squeeze the retainer tabs with the other hand to release them from the locking tabs.
    b. Pull the connector off.

➡Be careful not to damage the line or other parts. Do not use tools. If the connector does not move, keep the retainer tabs pressed down, and alternately pull and push the connector until it comes off easily. Do not remove the retainer from the line; once removed, the retainer must be replaced with a new one.

14. After disconnecting the quick-connect fitting, check it for dirt or damage.
15. Do the battery terminal reconnection procedure.

Without the HDS

1. Remove PGM-FI main relay 2 from the driver's under-dash fuse/relay box.
2. Start the engine, and let it idle until it stalls.

➡**If any DTCs are stored, clear and ignore them.**

3. Turn the ignition switch to LOCK (0).
4. Remove the fuel fill cap to relieve the pressure in the fuel tank.
5. Do the battery terminal disconnection procedure.
6. Remove the cover and the quick-connect fitting cover.
7. Check the quick-connect fitting for dirt, and clean it if needed.
8. Place a rag or a shop towel over the quick-connect fitting.
9. Disconnect the quick-connect fitting.
   a. Hold the connector with one hand, and squeeze the retainer tabs with the other hand to release them from the locking tabs.
   b. Pull the connector off.

➡**Be careful not to damage the line or other parts. Do not use tools. If the connector does not move, keep the retainer tabs pressed down, and alternately pull and push the connector until it comes off easily. Do not remove the retainer from the line; once removed, the retainer must be replaced with a new one.**

10. After disconnecting the quick-connect fitting, check it for dirt or damage.
11. Do the battery terminal reconnection procedure.

FUEL FILTER

REMOVAL & INSTALLATION

See Figure 299.

➡**The fuel filter should be replaced whenever the fuel pressure drops below the specified value, after making sure that the fuel pump and the fuel pressure regulator are OK.**

1. Remove the fuel tank unit.
2. Remove the fuel filter set.
3. Check these items before installing the fuel tank unit:
   - When connecting the wire harness, make sure the connection is secure and the connectors are firmly locked into place.
   - When installing the fuel gauge sending unit, make sure the connection is secure and the connector is firmly locked into place. Be careful not to bend or twist it excessively.
4. Install the parts in the reverse order of removal with new O-rings and a new bracket. When installing the fuel tank unit, align the marks on the unit and the fuel tank.

➡**Coat the O-rings with clean engine oil; do not use any other oil or fluid. Do not pinch the O-rings during installation. Use all the new parts supplied in the fuel filter replacement kit.**

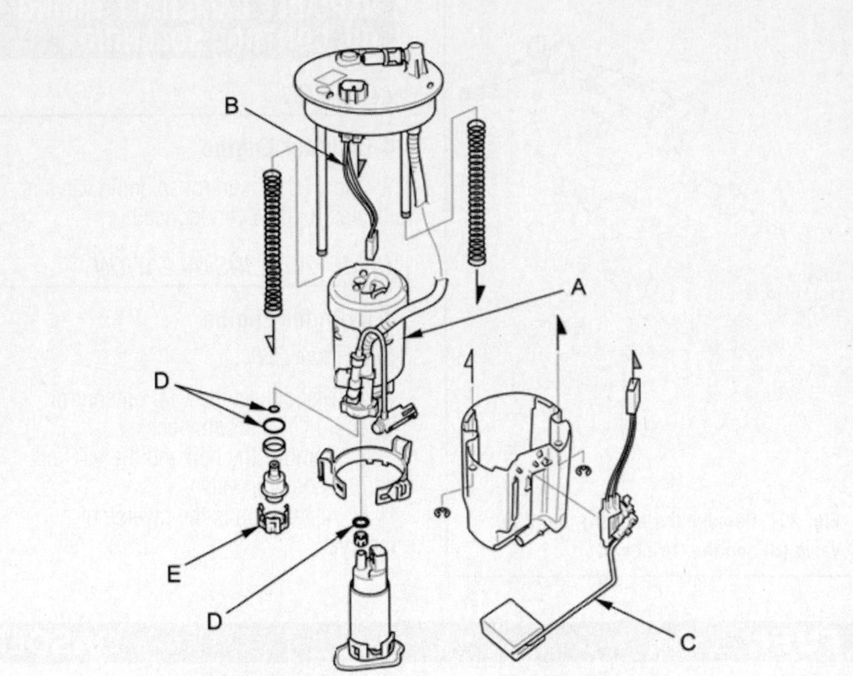
B
A
D
E
D
C
A. Fuel filter set
B. Wire harness
C. Fuel gauge sending unit
D. O-rings
E. Bracket
37647_ACTL_G0444

Fig. 299 Exploded view of fuel pump assembly

FUEL PUMP/FUEL GAUGE SENDING UNIT

REMOVAL & INSTALLATION

See Figures 300 through 303.

1. Remove the fuel tank unit.

   Special Tools Required: Fuel sender wrench 07AAA-S0XA100

   a. Relieve the fuel pressure.
   b. Remove the rear seat cushion cover.
   c. Remove the access panel from the floor.
   d. Disconnect the fuel tank unit 4P connector.

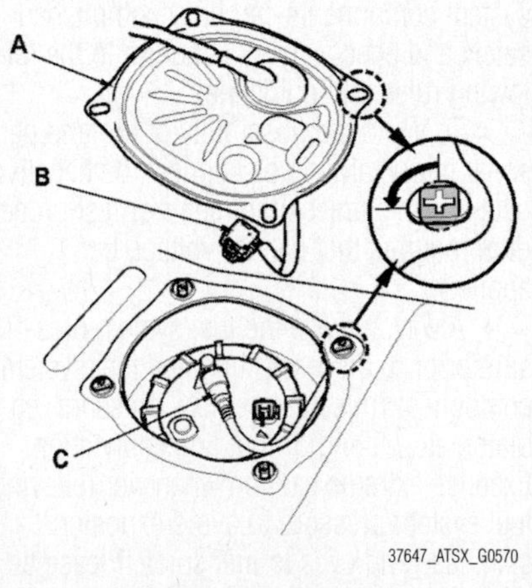
A
B
C
37647_ATSX_G0570

Fig. 300 Remove the access panel (A) and disconnect the fuel tank unit 4P connector (B) and the quick-connect fitting (C)

   e. Disconnect the quick-connect fitting from the fuel tank unit.
   f. Using the special tool, loosen the locknut.
   g. Remove the locknut and the fuel tank unit.
2. Remove the fuel level sensor (fuel gauge sending unit) from the fuel tank unit.

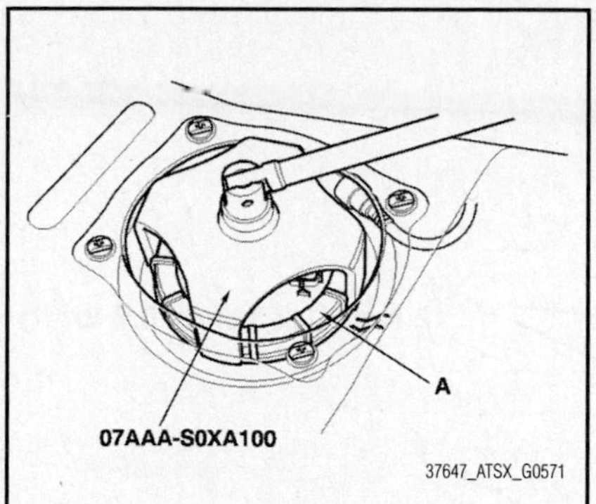
A
07AAA-S0XA100
37647_ATSX_G0571

Fig. 301 Using the special tool, loosen the locknut (A)

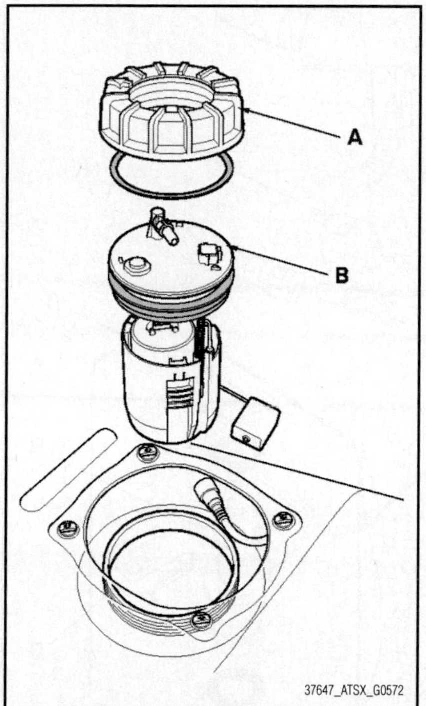
A
B
37647_ATSX_G0572

Fig. 302 Remove the locknut (A) and the fuel tank unit (B)

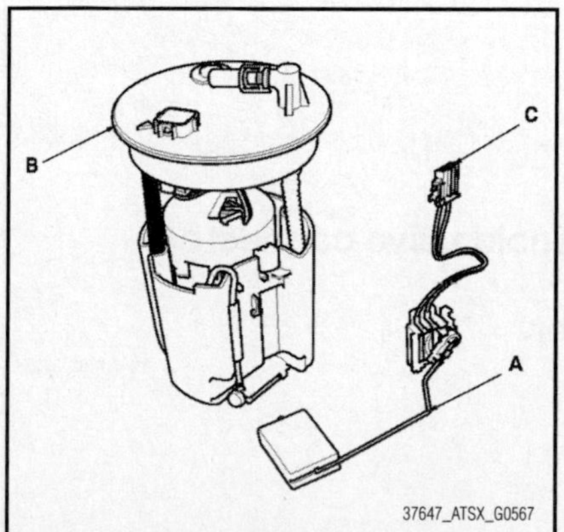
B
C
A
37647_ATSX_G0567

Fig. 303 Remove the fuel level sensor (fuel gauge sending unit) (A) from the fuel tank unit (B)

3. Check these items before installing the fuel tank unit:
   a. When connecting the wire harness, make sure the connection is secure and the connector is firmly locked into place.
   b. When installing the fuel gauge sending unit, make sure the connection is secure. Be careful not to bend or twist it excessively.
4. Install the parts in the reverse order of removal. When installing the fuel tank unit, align the marks on the unit and the fuel tank.

FUEL RAIL & INJECTOR

REMOVAL & INSTALLATION

4-Cylinder Engine

See Figures 304 and 305.

1. Relieve the fuel pressure.
2. Remove the engine cover.
3. Remove the quick-connect fitting cover, then disconnect the quick-connect fitting.
4. Disconnect the injector connectors and the engine mount control solenoid valve connector.
5. Remove the ground cable bolts.
6. Remove the fuel rail mounting nuts from the fuel rail.
7. Remove the fuel rail and the injectors from the injector base.
8. Remove the injector clips from the fuel rail.
9. Remove the injectors from the fuel rail.

To install:

10. Coat the new O-rings (black) with clean engine oil, and insert the injectors into the fuel rail.
11. Install the injectors clips.
12. Coat the new injector O-rings (brown) with clean engine oil.
13. Install the fuel rail and the injectors in the injector base.

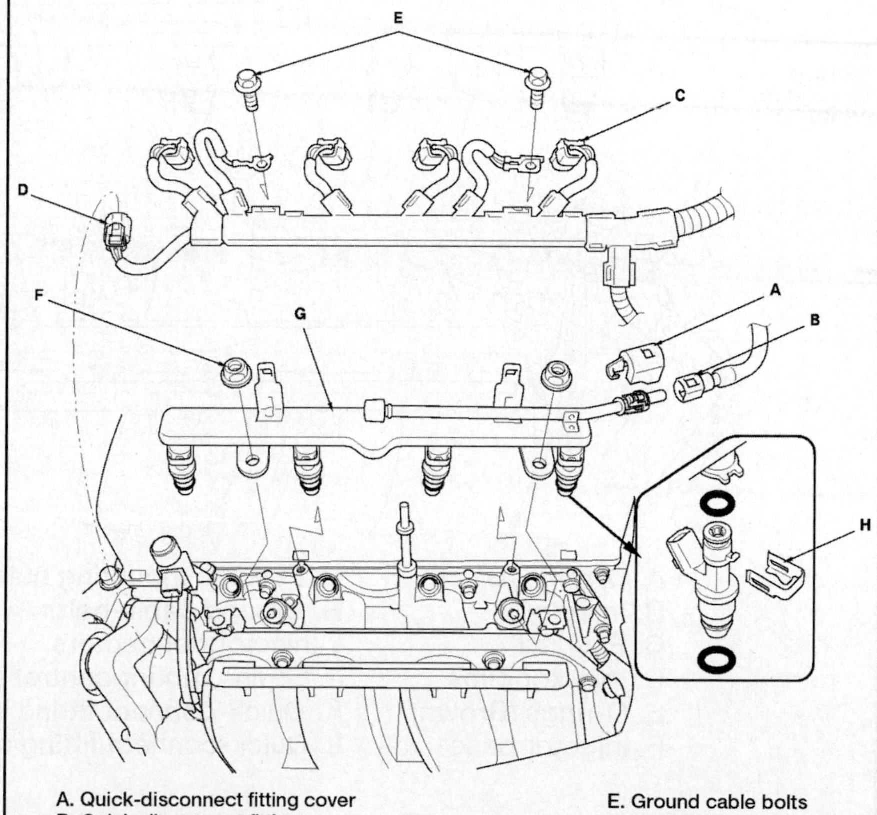
E
C
D
F
G
A
B
H
A. Quick-disconnect fitting cover
B. Quick-disconnect fitting
C. Injector connectors
D. Engine mount control solenoid valve connector
E. Ground cable bolts
F. Fuel rail mounting nuts
G. Fuel rail
H. Injector clips
37647_ATSX_G0568

Fig. 304 Fuel rail assembly

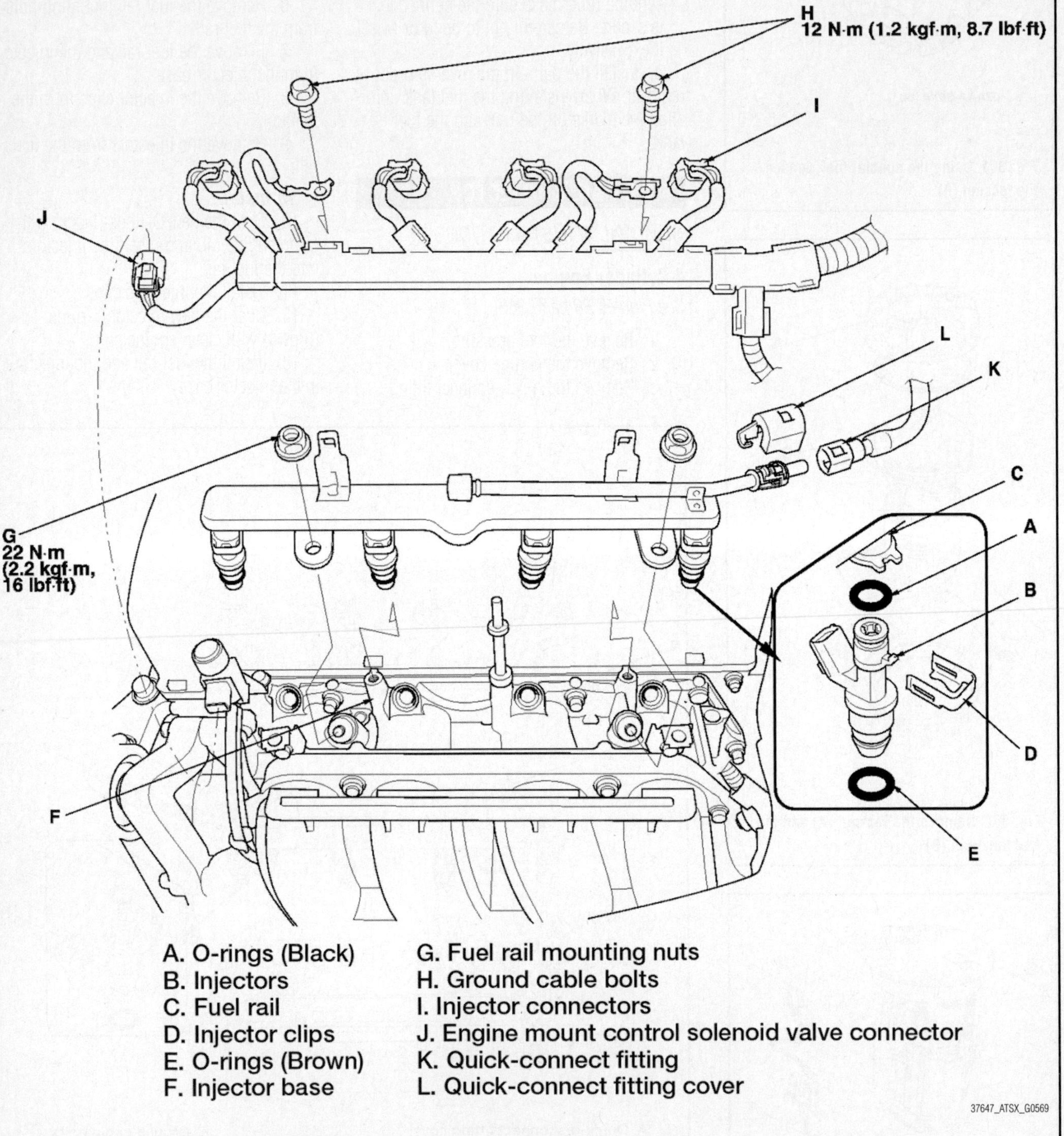
H
12 N·m (1.2 kgf·m, 8.7 lbf·ft)
I
J
L
K
C
A
B
G
22 N·m
(2.2 kgf·m,
16 lbf·ft)
D
F
E
A. O-rings (Black)
B. Injectors
C. Fuel rail
D. Injector clips
E. O-rings (Brown)
F. Injector base
G. Fuel rail mounting nuts
H. Ground cable bolts
I. Injector connectors
J. Engine mount control solenoid valve connector
K. Quick-connect fitting
L. Quick-connect fitting cover
37647_ATSX_G0569

Fig. 305 Installation of injectors and fuel rail

14. Install the fuel rail mounting nuts and the ground cable bolts.

15. Connect the injector connectors and the engine mount control solenoid valve connector.

16. Connect the quick-connect fitting, and quick-connect fitting cover.

17. Turn the ignition switch to ON (II), but do not operate the starter. After the fuel pump runs for about 2 seconds, the fuel rail will be pressurized. Repeat this two or three times, then make sure there are no fuel leaks.

18. Reinstall the engine cover.

6-Cylinder Engine

See Figures 306 and 307.

1. Relieve the fuel pressure.
2. Remove the intake manifold.
3. Disconnect the quick-connect fitting.
4. Remove the fuel joint hose mounting bolt.
5. Disconnect the connectors from the injectors.
6. Remove the fuel rail mounting bolts from the fuel rails.
7. Remove the fuel rails and the injectors from the injector base.
8. Remove the injector clips from the fuel rails.

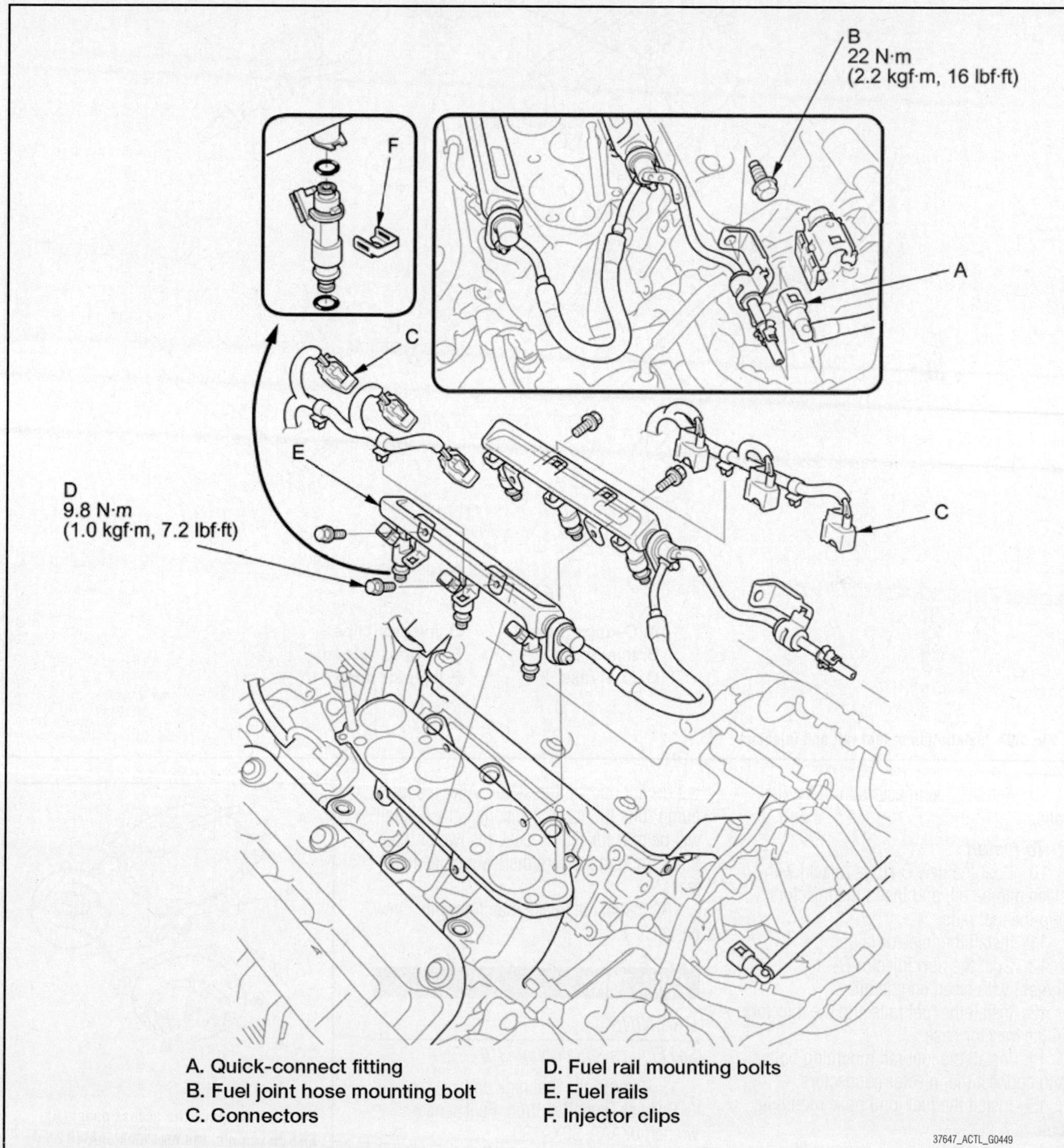
B
22 N·m
(2.2 kgf·m, 16 lbf·ft)
F
A
C
E
D
9.8 N·m
(1.0 kgf·m, 7.2 lbf·ft)
C
A. Quick-connect fitting
B. Fuel joint hose mounting bolt
C. Connectors
D. Fuel rail mounting bolts
E. Fuel rails
F. Injector clips
37647_ACTL_G0449

Fig. 306 Exploded view of fuel rail and injector assembly

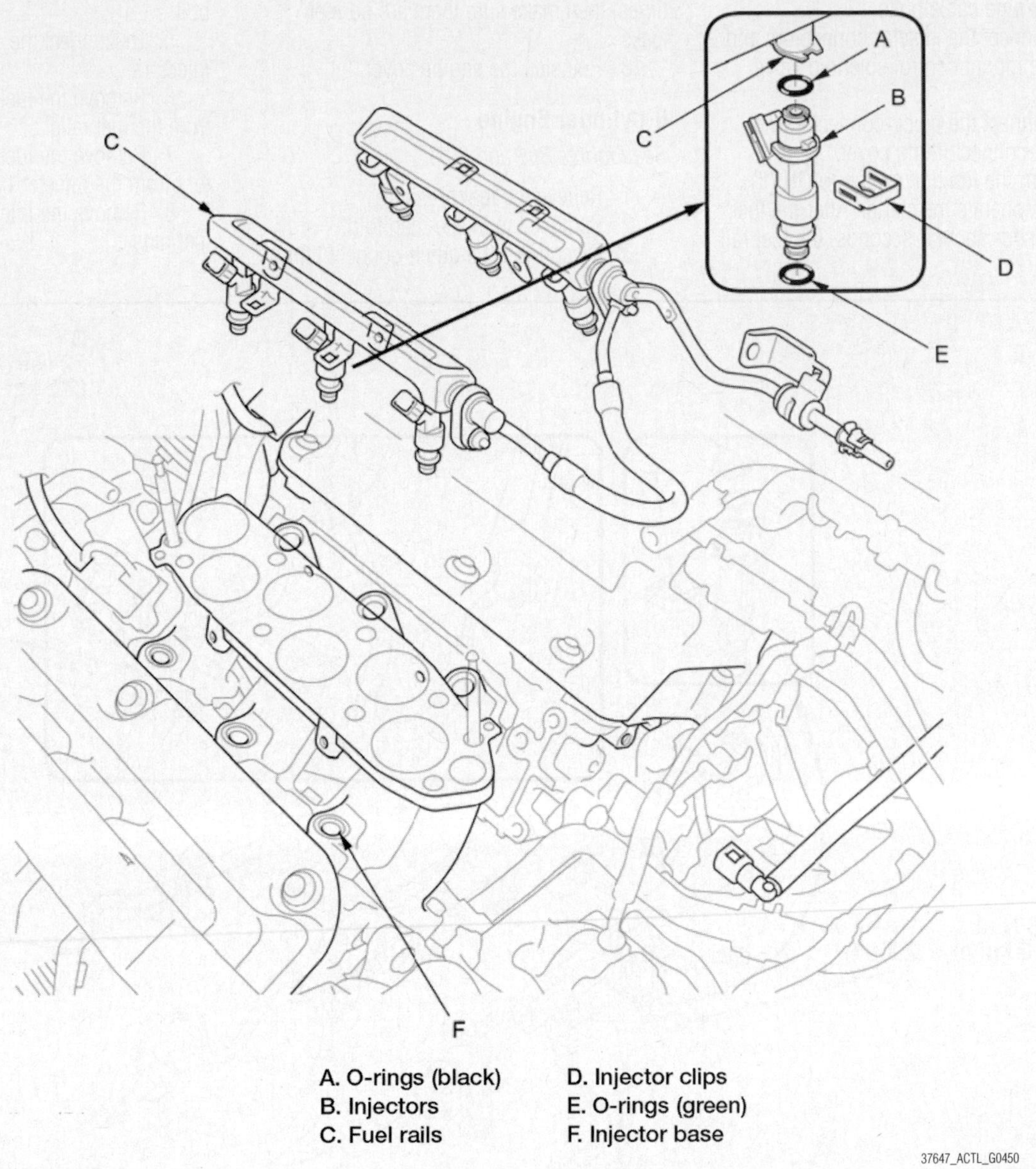
A
B
C
C
D
E
F
A. O-rings (black)
B. Injectors
C. Fuel rails
D. Injector clips
E. O-rings (green)
F. Injector base
37647_ACTL_G0450

Fig. 307 Installation of fuel rail and injectors

9. Remove the injectors from the fuel rails.

To install:

10. Coat the new O-rings (black) with clean engine oil, and insert the injectors into the fuel rails.
11. Install the injector clips.
12. Coat the new injector O-rings (green) with clean engine oil.
13. Install the fuel rails and the injectors in the injector base.
14. Install the fuel rail mounting bolts, and connect the injector connectors.
15. Install the fuel joint hose mounting bolt.
16. Connect the quick-connect fitting.
17. Turn the ignition switch to ON (II), but do not operate the starter. After the fuel pump runs for about 2 seconds, the fuel rail will be pressurized. Repeat this two or three times, then make sure there are no fuel leaks.
18. Install the intake manifold with a new gasket.

FUEL TANK

DRAINING

See Figures 308 through 310.

1. Remove the fuel tank unit.

Special Tools Required: Fuel sender wrench 07AAA-S0XA100

a. Relieve the fuel pressure.

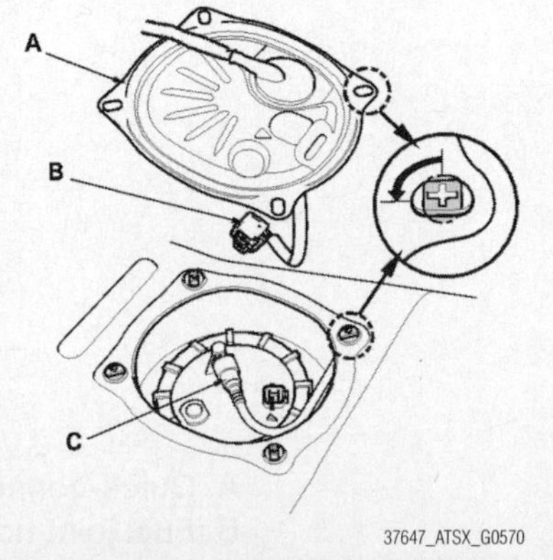
A
B
C
37647_ATSX_G0570

Fig. 308 Remove the access panel (A) and disconnect the fuel tank unit 4P connector (B) and the quick-connect fitting (C)

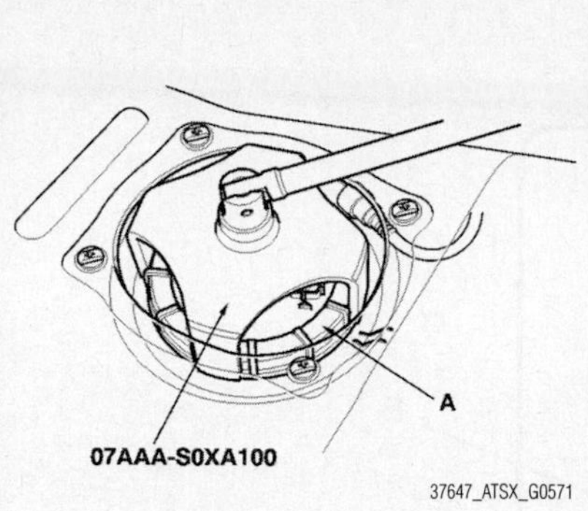
A
07AAA-S0XA100
37647_ATSX_G0571

Fig. 309 Using the special tool, loosen the locknut (A)

b. Remove the rear seat cushion cover.

c. Remove the access panel from the floor.

d. Disconnect the fuel tank unit 4P connector.

e. Disconnect the quick-connect fitting from the fuel tank unit.

f. Using the special tool, loosen the locknut.

g. Remove the locknut and the fuel tank unit.

2. Using a hand pump, a hose, and a container suitable for fuel, draw the fuel from the fuel tank.

3. Reinstall the fuel tank unit.

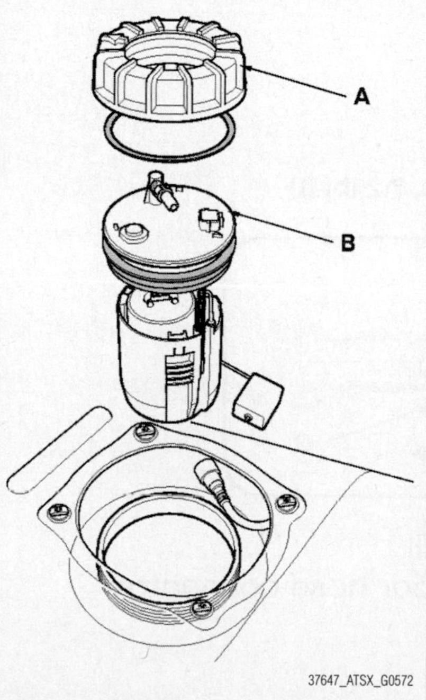
A
B
37647_ATSX_G0572

Fig. 310 Remove the locknut (A) and the fuel tank unit (B)

REMOVAL & INSTALLATION

See Figures 311 through 315.

1. Drain the fuel tank.
2. Reinstall the fuel tank unit without connecting the fuel tank unit 4P connector and the quick-connect fitting.
3. Remove the fuel fill pipe cover.
4. Disconnect the quick-connect fittings and the fuel fill tube from the fuel fill pipe. Slide back the clamps, then twist the hose as you pull to avoid damaging them.
5. Raise the vehicle on a lift.
6. Disconnect the hose from the EVAP canister.
7. Remove the hose from the clamp.

➡**Be careful not to damage the hose.**

8. Remove the exhaust pipe.
9. Remove the middle floor undercover.
10. Remove the fuel tank protector.

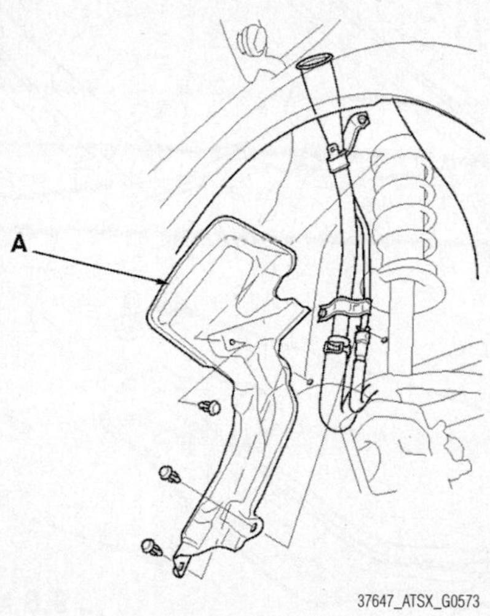
A
37647_ATSX_G0573

Fig. 311 Remove the fuel fill pipe cover (A)

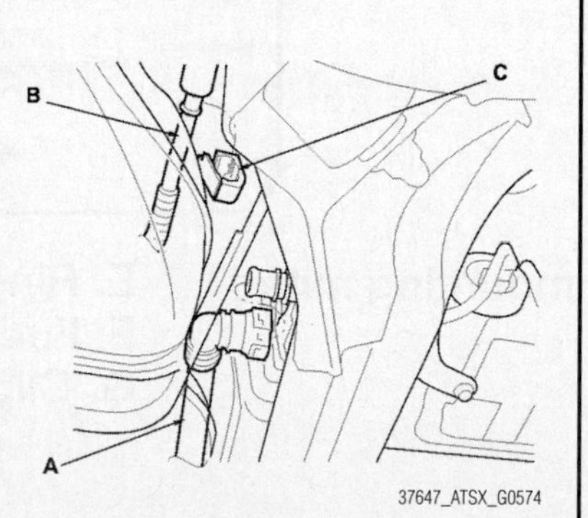
B
C
A
37647_ATSX_G0574

Fig. 312 Disconnect the hose (A) from the EVAP canister

11. Place a jack or other support under the fuel tank.
12. Remove the strap bolts and the straps.
13. Remove the fuel tank.
14. Install the parts in the reverse order of removal.

➡**New fuel tanks have a ring pull at the fuel vapor hose connector. When you connect the hose and confirm that the connection is secure, remove the ring pull by pulling it down.**

Before connecting the fuel fill pipe and the quick-connect fitting, check for dirt, and

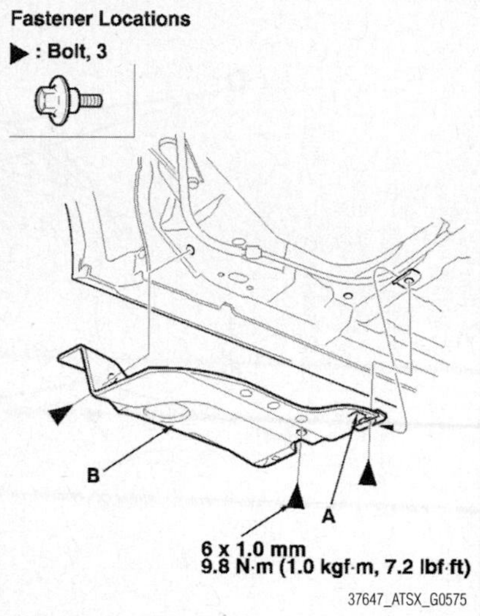
Fastener Locations
▶ : Bolt, 3
B
A
6 x 1.0 mm
9.8 N·m (1.0 kgf·m, 7.2 lbf·ft)
37647_ATSX_G0575

Fig. 313 Remove the middle floor undercover (Left)

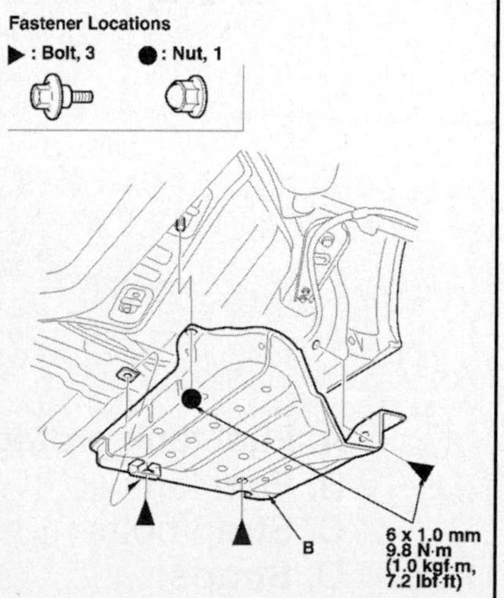
Fastener Locations
▶ : Bolt, 3
● : Nut, 1
B
6 x 1.0 mm
9.8 N·m
(1.0 kgf·m,
7.2 lbf·ft)
37647_ATSX_G0576

Fig. 314 Remove the middle floor undercover (Right)

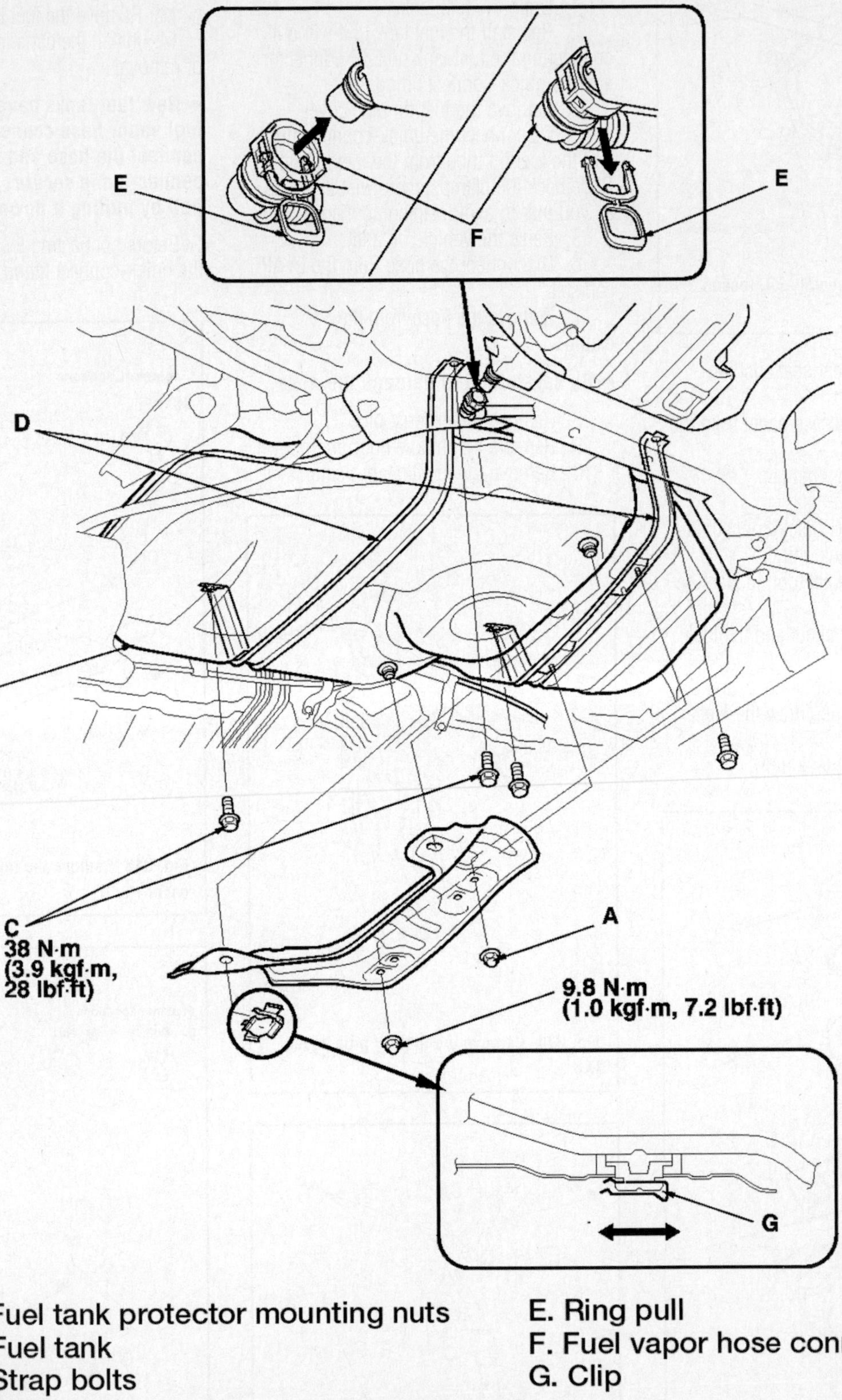
E
F
E
D
B
C
38 N·m
(3.9 kgf·m,
28 lbf·ft)
A
9.8 N·m
(1.0 kgf·m, 7.2 lbf·ft)
G

A. Fuel tank protector mounting nuts
B. Fuel tank
C. Strap bolts
D. Straps
E. Ring pull
F. Fuel vapor hose connector
G. Clip

37647_ATSX_G0577

Fig. 315 Removing the fuel tank

clean them if needed, taking care not to damage the fuel fill pipe or other parts.

When installing the fuel tank protector, make sure to insert it into the clip in the direction shown.

IDLE SPEED

ADJUSTMENT

The idle speed is controlled by the ECM/PCM. No adjustment is possible or necessary.

THROTTLE BODY

REMOVAL & INSTALLATION

4-Cylinder Engine

See Figure 316.

CAUTION

Do not insert your fingers into the installed throttle body when you turn the ignition switch to ON (II) or while the ignition switch is in ON (II). If you do, you will seriously injure your fingers if the throttle valve is activated.

➡If you are replacing or cleaning the throttle body, start at step 1. If you are removing the throttle body, start at step 4.

1. Connect the HDS to the DLC while the engine is stopped.
2. Select the INSPECTION MENU on the HDS.
3. Do the TP POSITION CHECK in the ETCS TEST.
4. Turn the ignition switch to LOCK (0).
5. Remove the intake air duct.
6. Disconnect the throttle body connector.
7. Disconnect and plug the water bypass hoses.
8. Remove the throttle body.
9. Install the parts in the reverse order of removal with a new gasket.

➡When tightening the screw of the hose band, align the edge of the hose band with the mark painted on it.

10. After the installation, refill the radiator with engine coolant.
11. Do the ECM/PCM idle learn procedure.

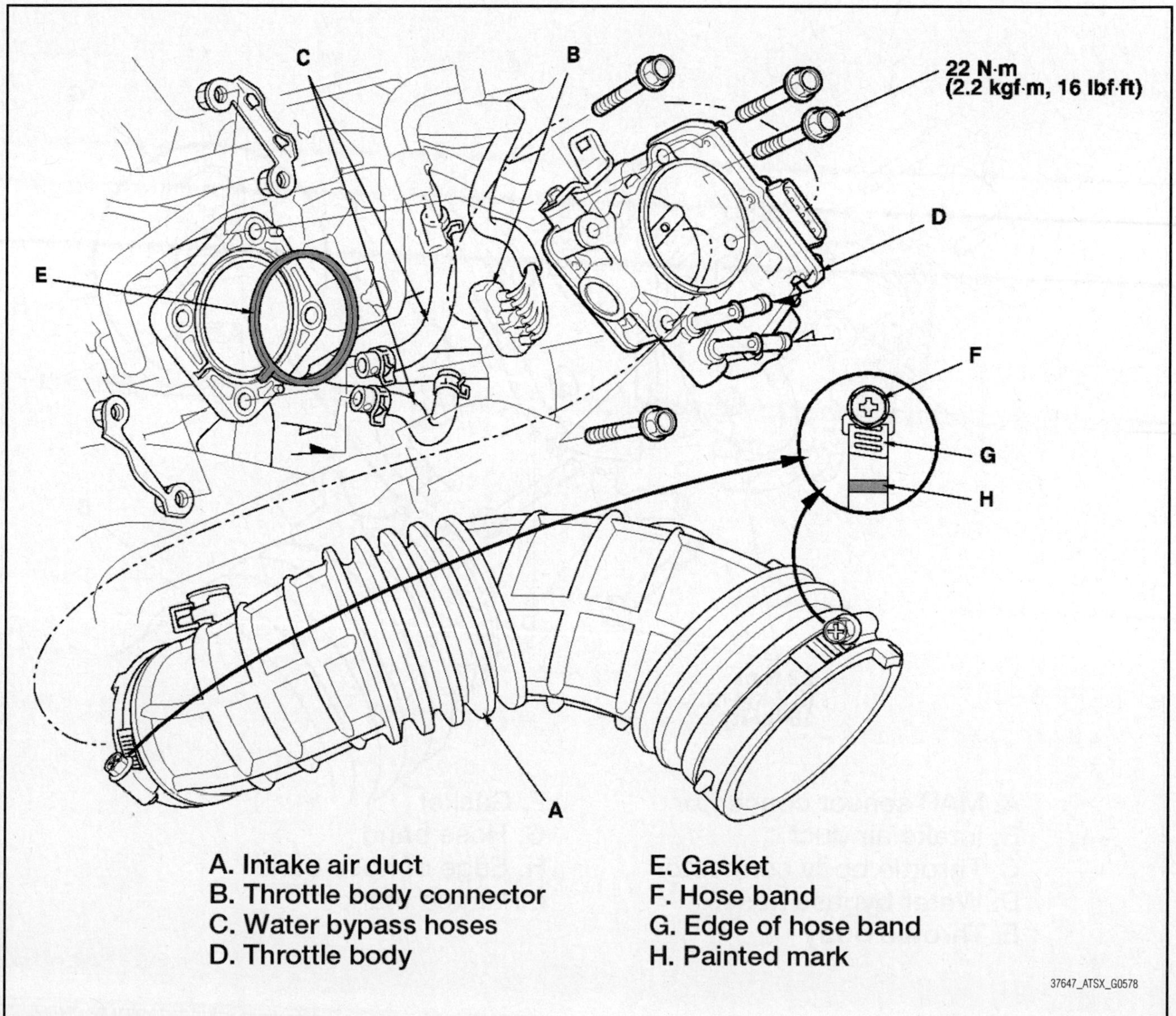
C
B
22 N·m
(2.2 kgf·m, 16 lbf·ft)
D
E
F
G
H
A
A. Intake air duct
B. Throttle body connector
C. Water bypass hoses
D. Throttle body
E. Gasket
F. Hose band
G. Edge of hose band
H. Painted mark
37647_ATSX_G0578

Fig. 316 Exploded view of throttle body removal

6-Cylinder Engine

See Figure 317.

CAUTION

Do not insert your fingers into the installed throttle body when you turn the ignition switch to ON (II) or while the ignition switch is in ON (II). If you do, you will seriously injure your fingers if the throttle valve is activated.

➡If you are replacing or cleaning the throttle body, start at step 1. If you are removing the throttle body, start at step 4.

1. Connect the HDS to the DLC while the engine is stopped.
2. Select the INSPECTION MENU on the HDS.
3. Do the TP POSITION CHECK in the ETCS TEST.
4. Turn the ignition switch to LOCK (0).
5. Disconnect the MAP sensor connector.
6. Remove the intake air duct.
7. Disconnect the throttle body connector.
8. Disconnect and plug the water bypass hoses, then remove the throttle body.
9. Install the parts in the reverse order of removal with a new gasket.

➡When tightening the screw of the hose band, align the edge of the hose band with the mark painted on it.

10. After the installation, refill the radiator with engine coolant.
11. Do the PCM idle learn procedure.

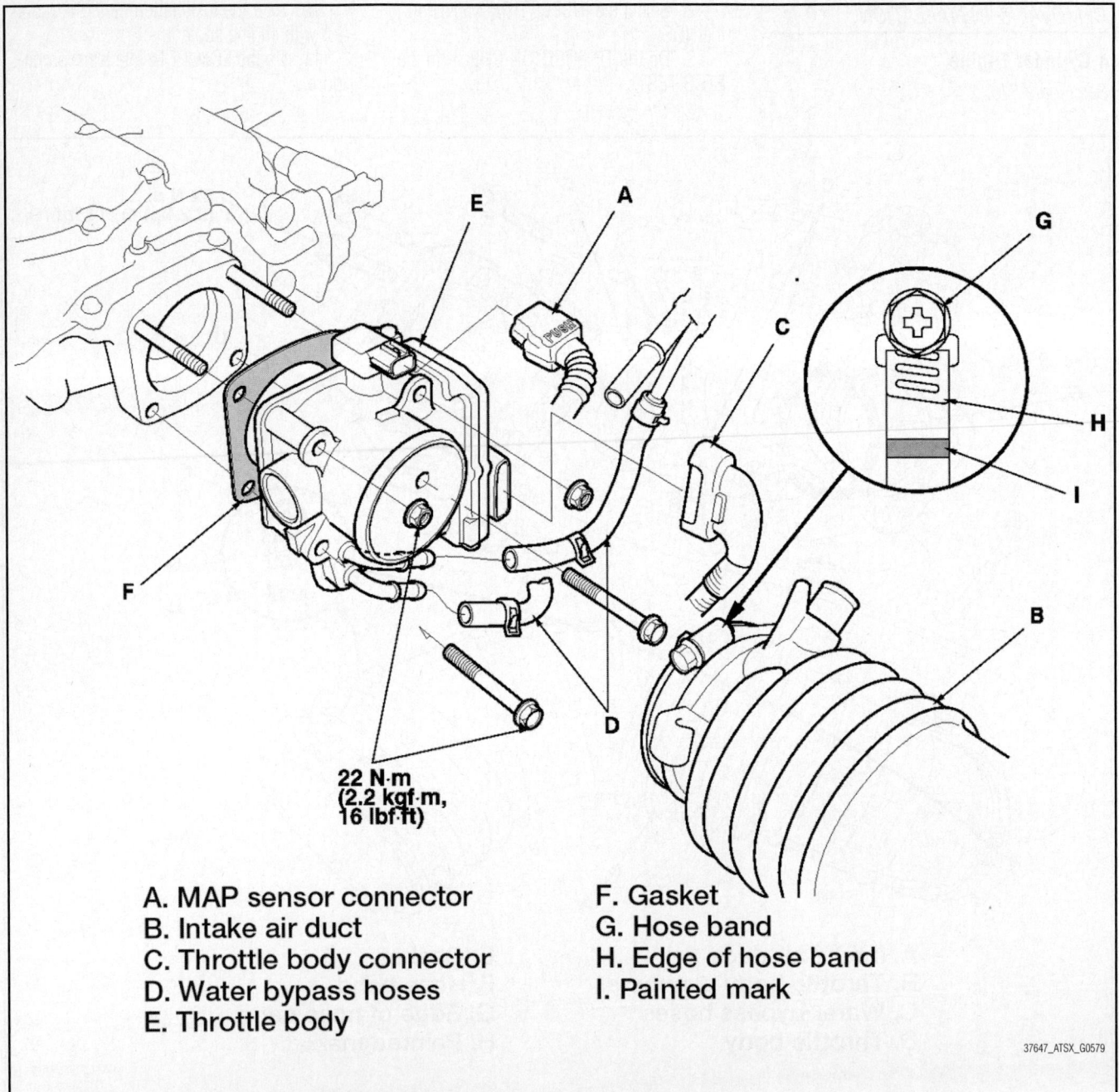

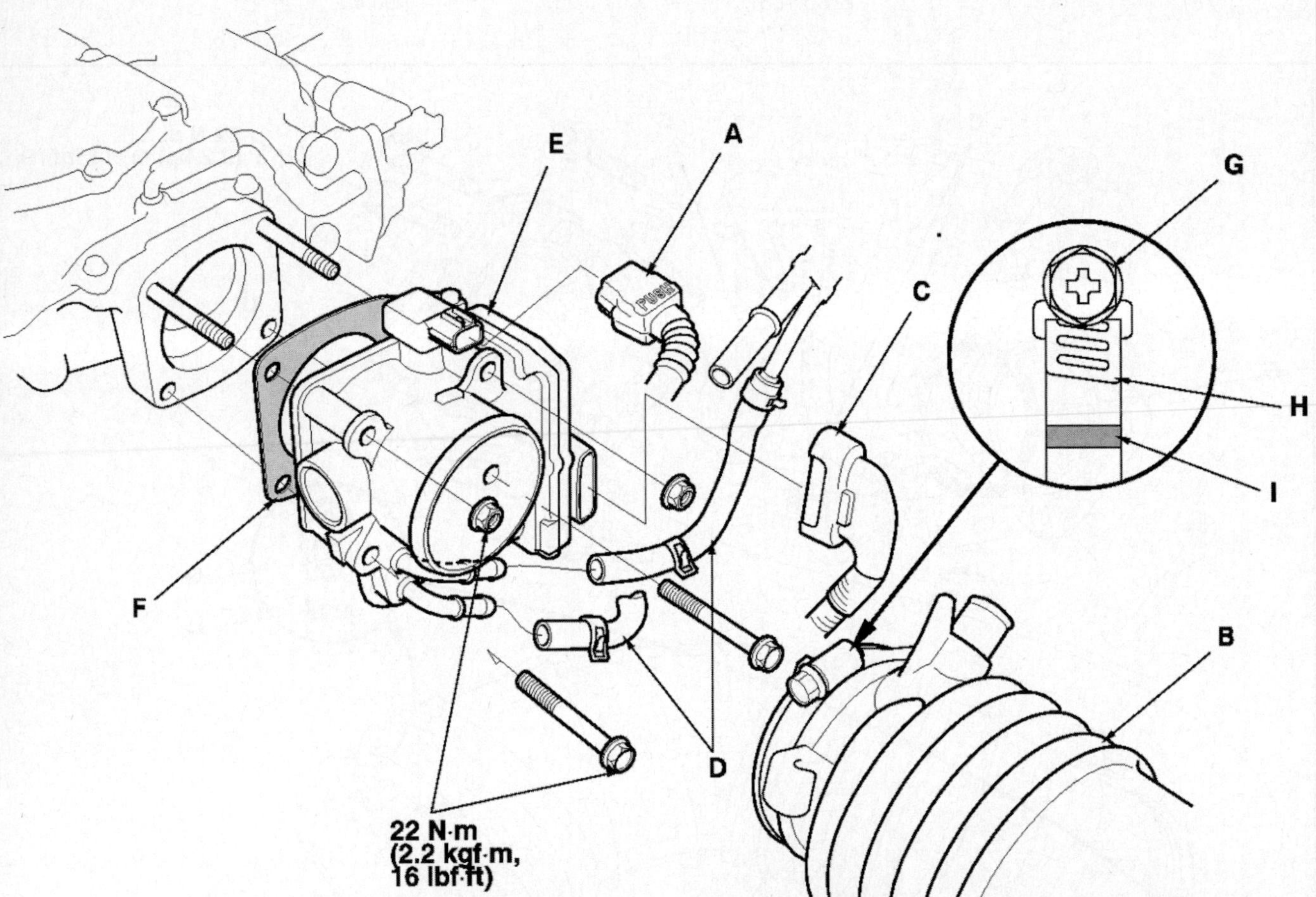
E
A
G
C
H
I
F
B
D
22 N·m
(2.2 kgf·m,
16 lbf·ft)
PUSH

A. MAP sensor connector
B. Intake air duct
C. Throttle body connector
D. Water bypass hoses
E. Throttle body
F. Gasket
G. Hose band
H. Edge of hose band
I. Painted mark

37647_ATSX_G0579

Fig. 317 Exploded view of the throttle body assembly

HEATING & AIR CONDITIONING SYSTEM

BLOWER MOTOR

REMOVAL & INSTALLATION

See Figures 318 through 323.

1. Remove the glove box.
2. Remove the passenger's undercover.
3. Remove the right kick panel.
4. Remove the dust and pollen filter assembly from the blower unit.
5. Remove the bolts and the wire harness clip. Then cut the plastic cross brace in the glove box opening with diagonal cutters in the area shown. Retain the plastic cross brace to be reinstalled later.
6. Disconnect these connectors: Passenger's under-dash fuse/relay box connector D (28P), the stereo amplifier connectors (with premium audio system), the AM/FM antenna lead, and right side wire harness connector C410 (20P). Remove the wire harness clips.
7. Remove the two screws, then remove the cover.

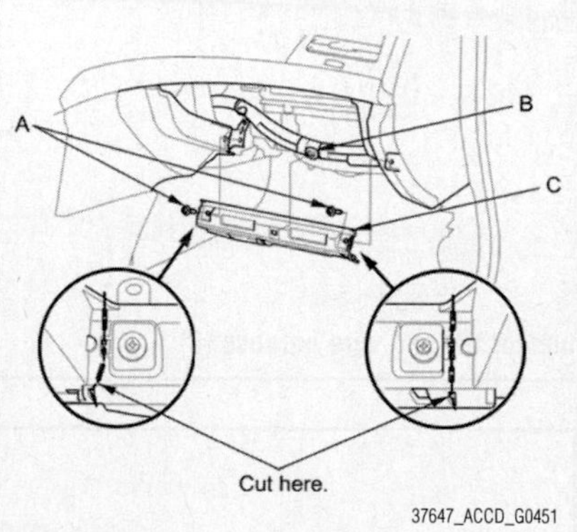
A
B
C
Cut here.
37647_ACCD_G0451

Fig. 318 Remove the bolts (A) and the wire harness clip (B), then cut the plastic cross brace (C)

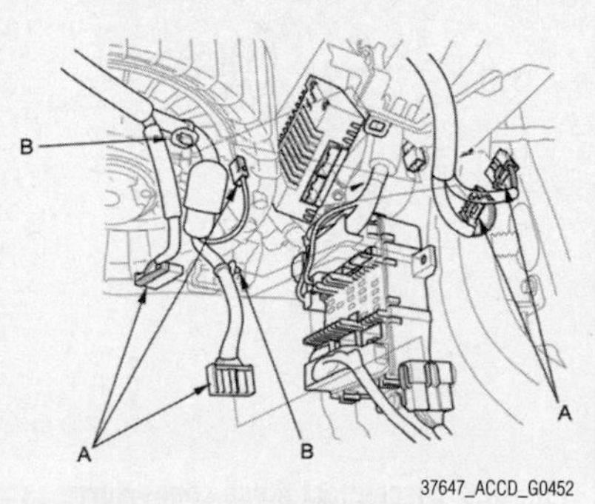
B
A
B
A
37647_ACCD_G0452

Fig. 319 Disconnect the connectors (A), and remove the wire harness clips (B)

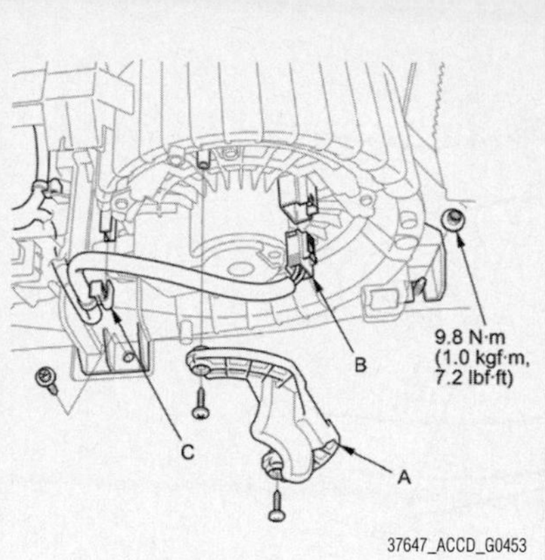
9.8 N·m
(1.0 kgf·m,
7.2 lbf·ft)
B
C
A
37647_ACCD_G0453

Fig. 320 Remove the cover (A), disconnect the connector (B) and wire harness (C)

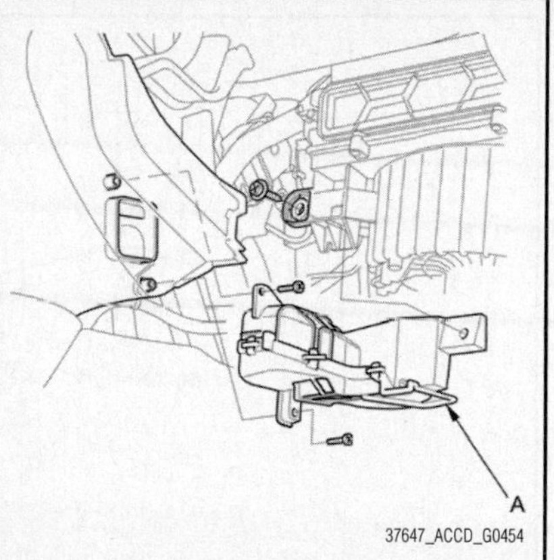
A
37647_ACCD_G0454

Fig. 321 Remove the self-tapping screws, and the passenger's heater duct (A)

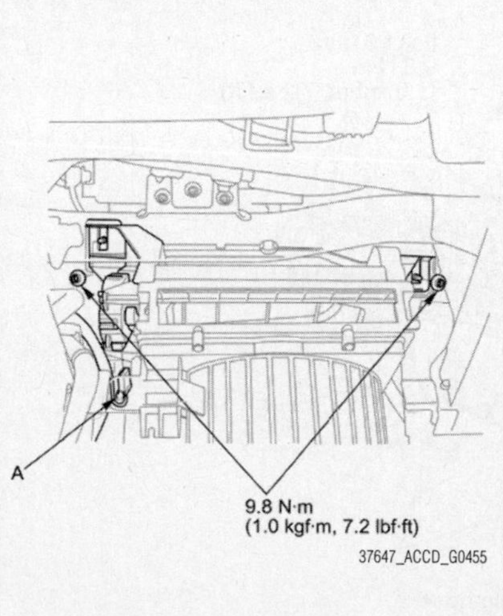
A
9.8 N·m
(1.0 kgf·m, 7.2 lbf·ft)
37647_ACCD_G0455

Fig. 322 Disconnect the connector (A) from the recirculation control motor

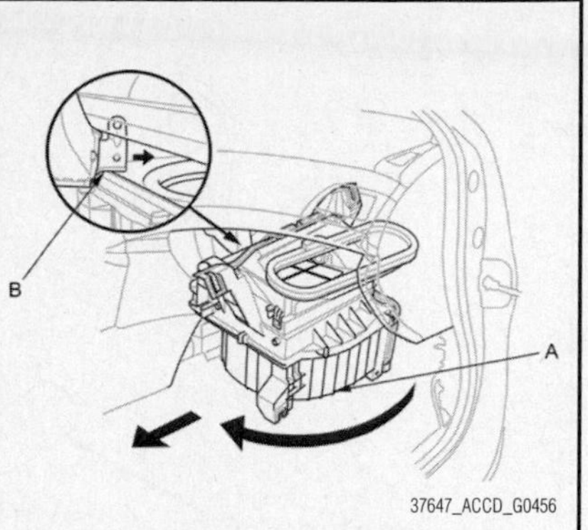
B
A
37647_ACCD_G0456

Fig. 323 Pull the blower unit (A) out while rotating it clockwise, so that the glove box bracket (B) passes through the dust and pollen filter area

8. Disconnect the connector from the blower motor and the wire harness clip. Remove the self-tapping screw and the mounting nut.
9. Remove the self-tapping screws, and the passenger's heater duct.
10. Disconnect the connector from the recirculation control motor. Remove the mounting nuts.
11. Pull the blower unit out while rotating it clockwise as shown, so that the glove box bracket passes through the dust and pollen filter area.
12. Install the unit in the reverse order of removal. Make sure that there is no air leakage.

HEATER CORE

REMOVAL & INSTALLATION

See Figures 324 through 330.

⁂ WARNING

SRS components are located in this area. Review the SRS component locations and the precautions and procedures before doing repairs or service.

1. Do the battery terminal disconnection procedure.
2. Recover the refrigerant with a recovery/recycling/ charging station.
3. Disconnect the A/C line from the evaporator core.
4. When the engine is cool, drain the engine coolant from the radiator.
5. From under the hood, remove the

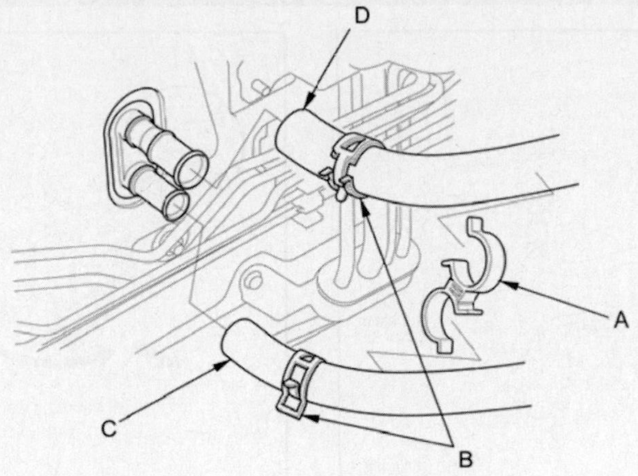
D
A
C
B
A. Clamp
B. Hose clamps
C. Inlet heater hose
D. Outlet heater hose
37647_ACCD_G0470

Fig. 324 Disconnect the inlet heater hose and the outlet heater hose from the heater unit

clamp. Slide the hose clamps back. Disconnect the inlet heater hose and the outlet heater hose from the heater unit. Note the layout of the hoses.

➡Engine coolant will run out when the hoses are disconnected; drain it into a clean drip pan. Be sure not to let coolant spill on the electrical parts or the painted surfaces. If any coolant spills, rinse it off immediately.

6. Remove the mounting nut from the heater unit. Take care not to damage or bend the fuel lines or brake lines, etc..
7. Remove the dashboard.
8. Disconnect the connector. Remove the clips, ducts, and the drain hose. Then remove the mounting bolt, the mounting nuts, and the blower-heater unit.

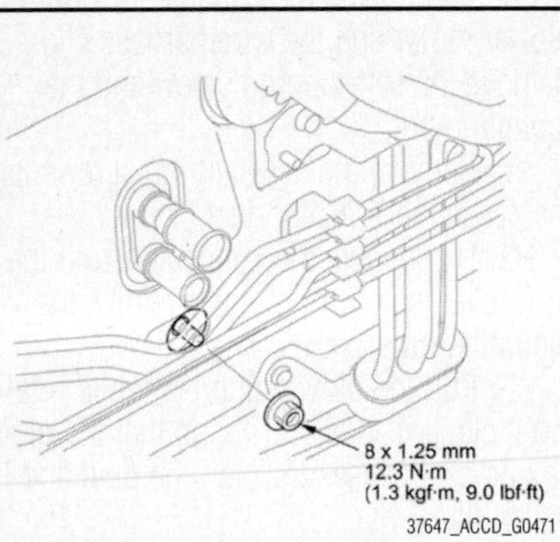
8 x 1.25 mm
12.3 N·m
(1.3 kgf·m, 9.0 lbf·ft)
37647_ACCD_G0471

Fig. 325 Remove the mounting nut from the heater unit

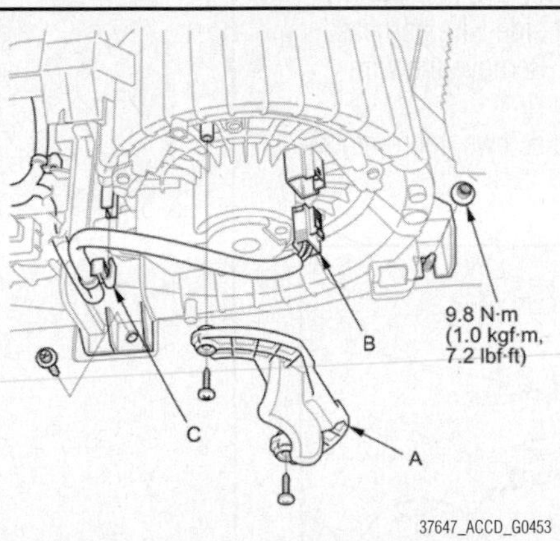
9.8 N·m
(1.0 kgf·m,
7.2 lbf·ft)
B
C
A
37647_ACCD_G0453

Fig. 327 Remove the cover (A), disconnect the connector (B) and wire harness (C)

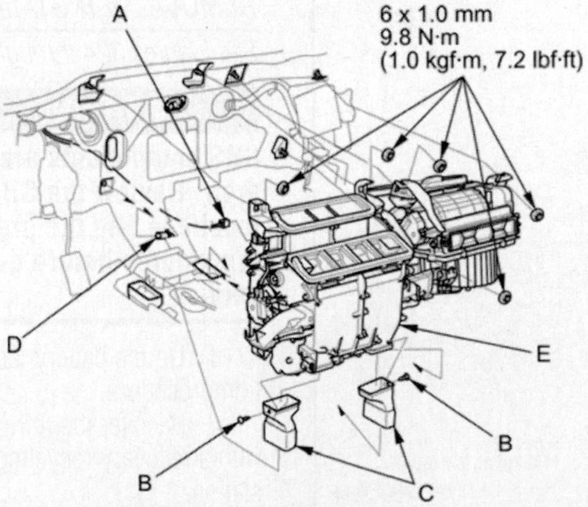
A
6 x 1.0 mm
9.8 N·m
(1.0 kgf·m, 7.2 lbf·ft)
D
E
B
B
C
A. Connector
B. Clips
C. Ducts
D. Drain hose
E. Blower-heater unit
37647_ACCD_G0472

Fig. 326 Remove the blower-heater unit

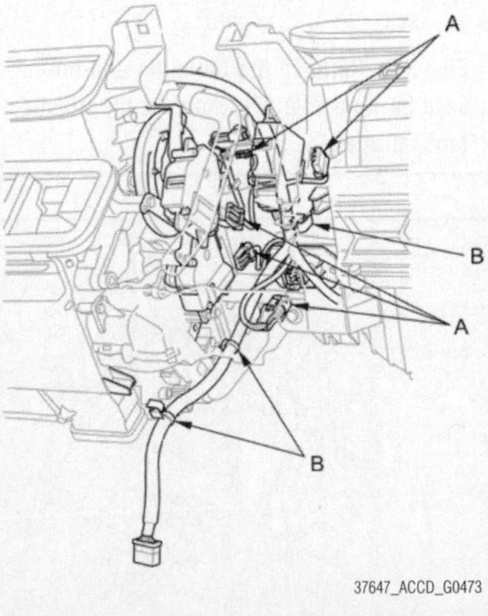
A
B
A
B
37647_ACCD_G0473

Fig. 328 Disconnect these connectors (A) and remove the wire harness clips (B)

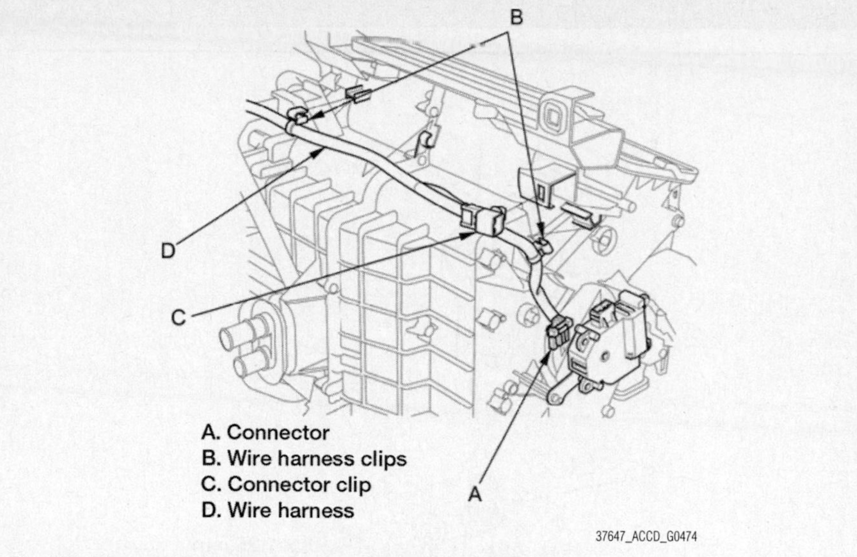
A
B
C
D
A. Connector
B. Wire harness clips
C. Connector clip
D. Wire harness
37647_ACCD_G0474
Fig. 329 Disconnect the connector from the air mix control motor

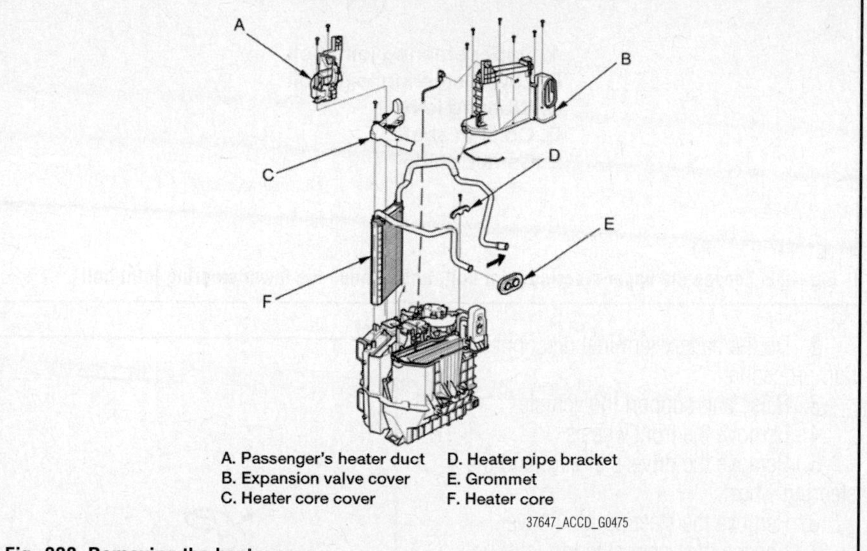
A
B
C
D
E
F
A. Passenger's heater duct
B. Expansion valve cover
C. Heater core cover
D. Heater pipe bracket
E. Grommet
F. Heater core
37647_ACCD_G0475
Fig. 330 Removing the heater core

9. Remove the two screws, then remove the cover.

10. Disconnect the connector from the blower motor. Remove the wire harness clip.

11. Disconnect these connectors: The mode control motor, the power transistor, the evaporator temperature sensor, the passenger's air mix control motor (with climate control), and the recirculation control motor. Remove the wire harness clips.

12. Disconnect the connector from the air mix control motor. Remove the wire harness clips, the connector clip, and the wire harness.

13. Remove the self-tapping screws and the passenger's heater duct. Remove the self-tapping screws and the expansion valve cover. Remove the self-tapping screw and the heater core cover. Remove the self-tapping screws, the heater pipe bracket, and the grommet, and carefully pull out the heater core.

To install:

14. Install the heater core and the evaporator core in the reverse order of removal.

15. Install the heater unit in the reverse order of removal, and note these items:

- Do not interchange the inlet and outlet heater hoses, and install the hose clamps securely.
- Refill the cooling system with engine coolant.
- Make sure that there is no coolant leakage.
- Make sure that there is no air leakage.

16. Do the battery terminal reconnection procedure.

STEERING

POWER RACK & PINION STEERING GEAR

REMOVAL & INSTALLATION

See Figures 331 through 356.

Special Tools Required*:

- Engine Hanger Adapter VSB02C000015
- Engine Support Hanger, A and Reds AAR-T1256
- Ball Joint Remover, 28 mm 07MAC-SL0A202
- Ball Joint Thread Protector, 12 mm 07AAF-SDAA100
- Subframe Adapter VSB02C000016

➡*Available through the Honda Tool and Equipment Program, 888-424-6857.

⁂ WARNING

SRS components are located in this area. Review the SRS component locations: 4-door, 2-door and the precautions and procedures before doing repairs or service.

➡Note these items during removal:

- Using clean solvent and a brush, wash any oil and dirt off the valve body unit, it's lines, and the end of the steering gearbox. Blow dry with compressed air.
- Be sure to remove the steering wheel before disconnecting the steering joint, or damage to the cable reel can occur.
- Lower the front subframe from the body, and remove the steering gearbox through the gap produced by lowering the front subframe.

1. Drain the power steering fluid.

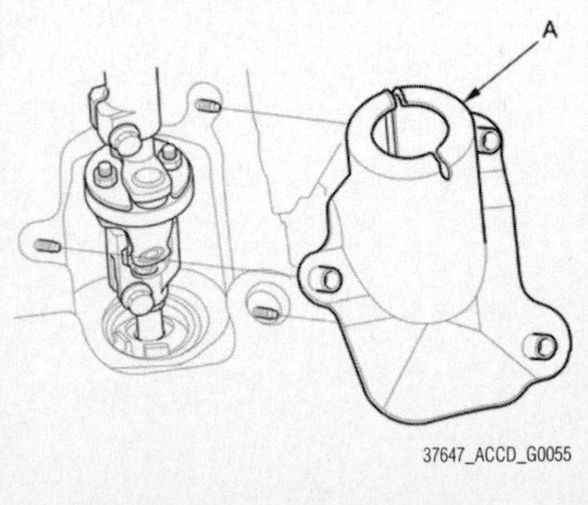
A
37647_ACCD_G0055

Fig. 331 Remove the steering joint cover (A)

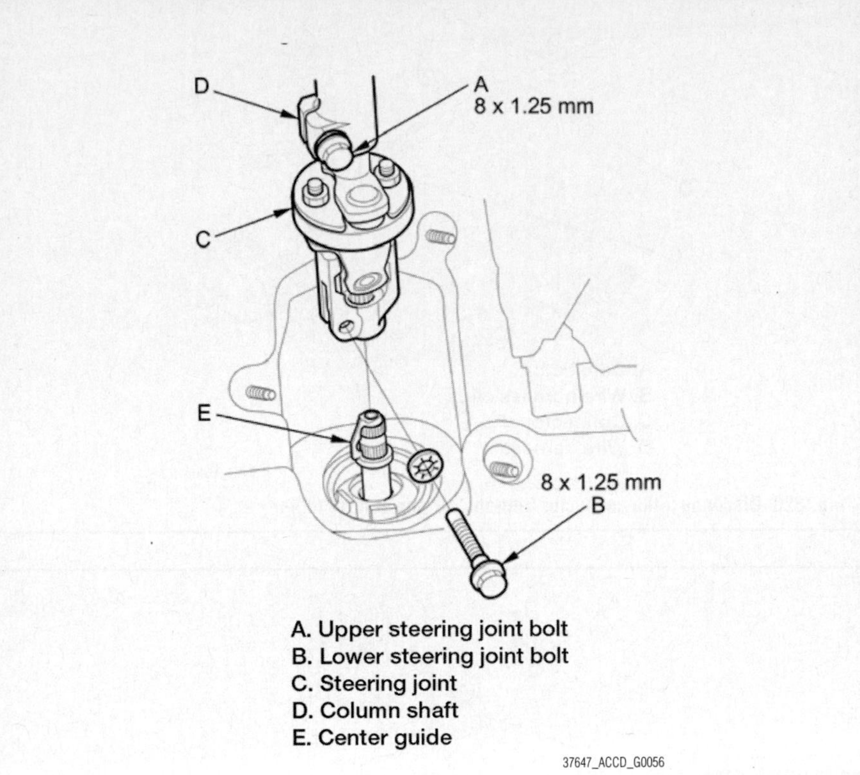
D
A
8 x 1.25 mm
C
E
8 x 1.25 mm
B
A. Upper steering joint bolt
B. Lower steering joint bolt
C. Steering joint
D. Column shaft
E. Center guide
37647_ACCD_G0056

Fig. 332 Loosen the upper steering joint bolt, and remove the lower steering joint bolt

2. Do the battery terminal disconnection procedure.
3. Raise and support the vehicle.
4. Remove the front wheels.
5. Remove the driver's airbag, and the steering wheel.
6. Remove the steering joint cover.
7. Loosen the upper steering joint bolt, and remove the lower steering joint bolt. Disconnect the steering joint by sliding the steering joint into the column shaft. Tighten the upper steering joint bolt to hold the steering joint in place.

➡Do not disconnect the steering joint from the column shaft.

8. Remove the center guide (if equipped) from the top of the pinion shaft, and discard it.
9. Apply vinyl tape to the splines on the pinion shaft.
10. Remove the hood support rod, then use it to prop the hood in the wide-open position.
11. Remove the front grille cover:
12. Remove the strut brace (if equipped).
13. Remove the P/S heat shield.
14. Attach the engine hanger adapter (VSB02C000015) to the threaded hole in the cylinder head.

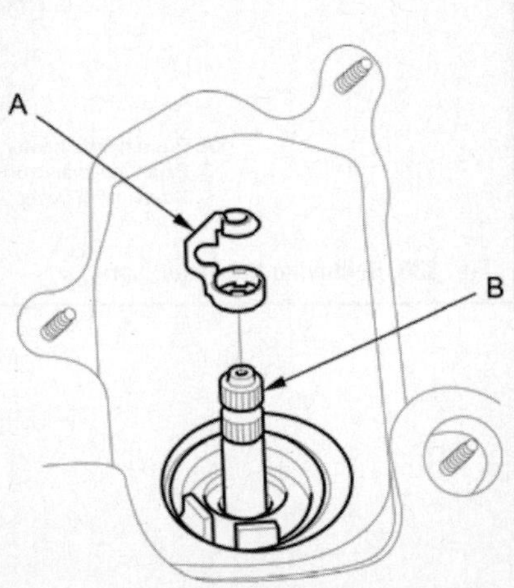
A
B
37647_ACCD_G0479

Fig. 333 Remove the center guide (A) from the top of the pinion shaft (B)

15. Install the engine support hanger (AAR-T1256), then attach the hook to the slotted hole in the hanger adapter. Tighten the wing nut by hand to lift and support the engine/transmission.

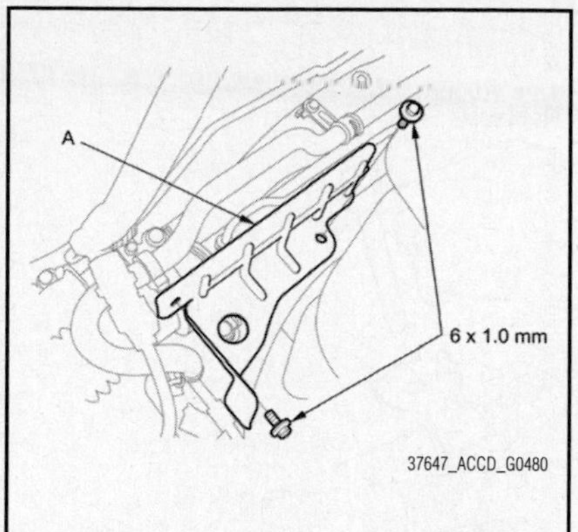
A
6 x 1.0 mm
37647_ACCD_G0480
Fig. 334 Remove the P/S heat shield (A)

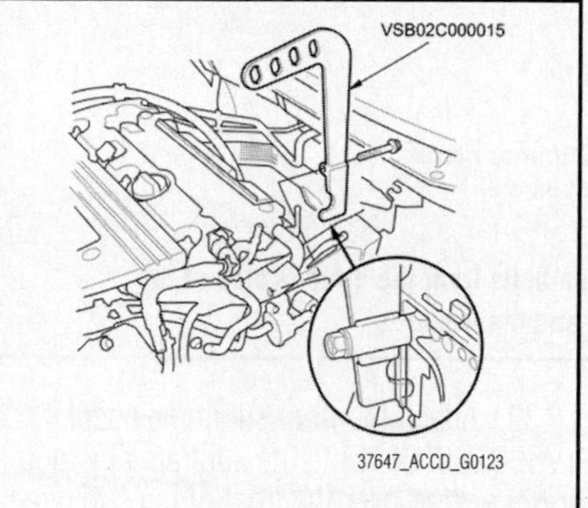
VSB02C000015
37647_ACCD_G0123
Fig. 335 Attach the engine hanger adapter to the threaded hole in the cylinder head

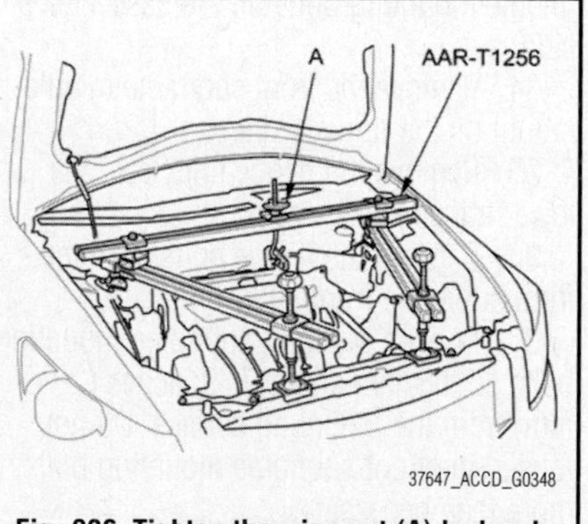
A
AAR-T1256
37647_ACCD_G0348
Fig. 336 Tighten the wing nut (A) by hand to lift and support the engine/transmission

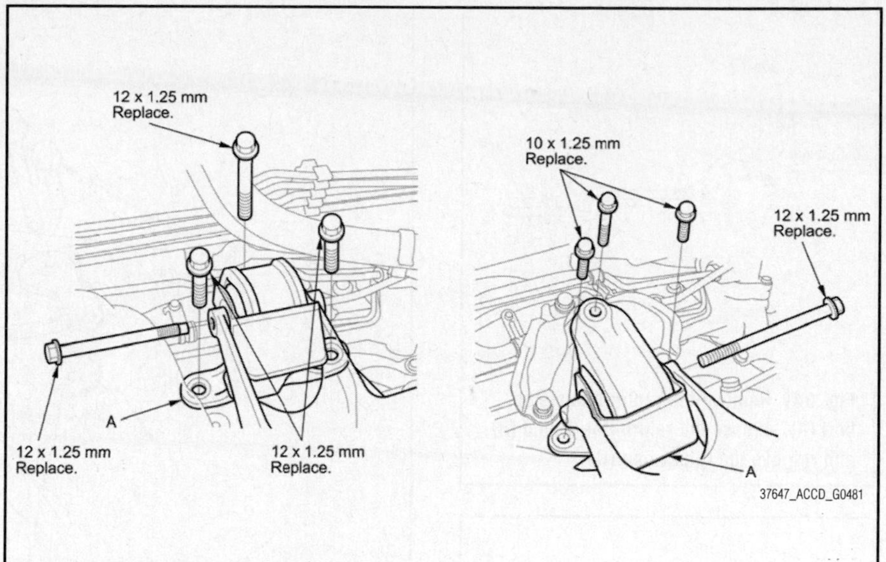
12 x 1.25 mm
Replace.
10 x 1.25 mm
Replace.
12 x 1.25 mm
Replace.
A
12 x 1.25 mm
Replace.
12 x 1.25 mm
Replace.
A
37647_ACCD_G0481
Fig. 337 Remove the rear engine mount (A)

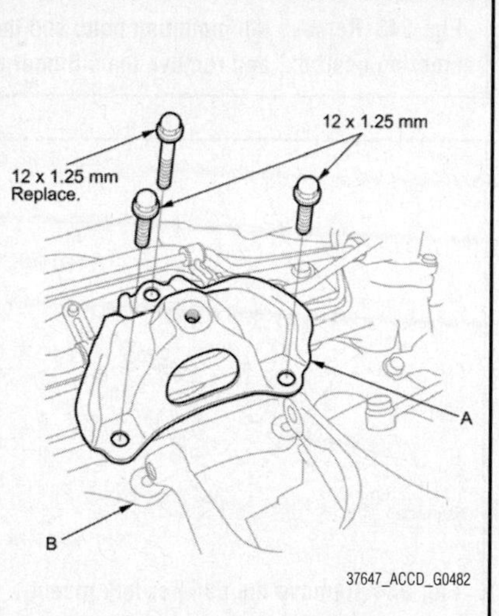
12 x 1.25 mm
12 x 1.25 mm
Replace.
A
B
37647_ACCD_G0482
Fig. 338 A/T: Remove the rear engine mount upper bracket (A) from the base bracket (B)

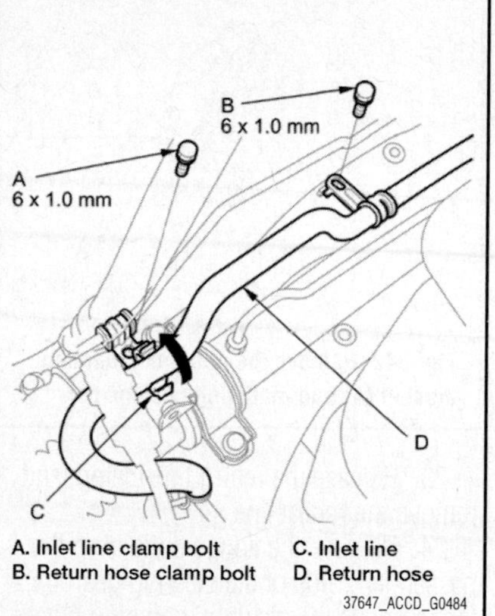
B
6 x 1.0 mm
A
6 x 1.0 mm
D
C
A. Inlet line clamp bolt
B. Return hose clamp bolt
C. Inlet line
D. Return hose
37647_ACCD_G0484
Fig. 340 Remove the inlet line clamp bolt and the return hose clamp bolt

➡**Be careful when working around the windshield.**

16. Remove the rear engine mount.
17. A/T: Remove the rear engine mount upper bracket from the base bracket.
18. Remove the inlet line clamp bolt and the return line clamp bolt.
19. Loosen the flare nuts, and disconnect the inlet line and the return line.
20. Remove the inlet line clamp bolt and the return hose clamp bolt.
21. Release the return hose clamp, and remove the return hose.
22. Remove the return line clamp bolt.

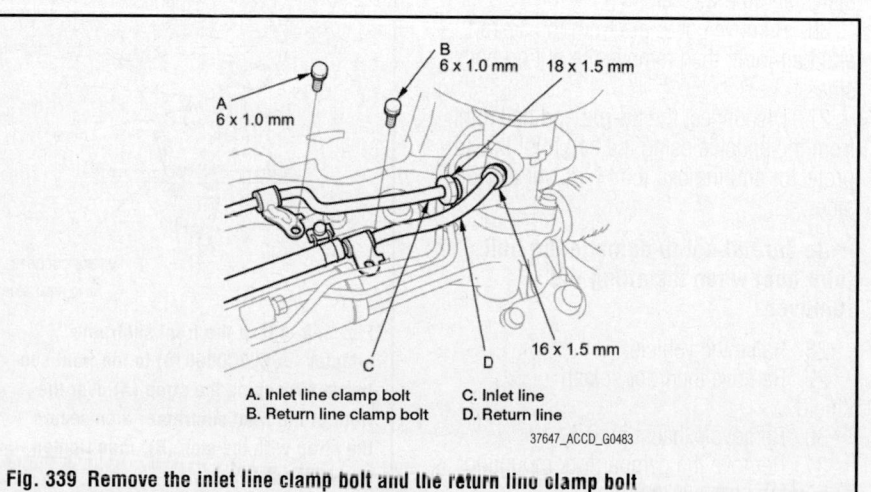

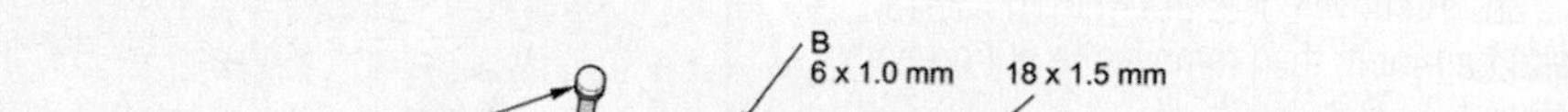
B
6 x 1.0 mm
18 x 1.5 mm
A
6 x 1.0 mm
16 x 1.5 mm
C
D
A. Inlet line clamp bolt
B. Return line clamp bolt
C. Inlet line
D. Return line
37647_ACCD_G0483
Fig. 339 Remove the inlet line clamp bolt and the return line clamp bolt

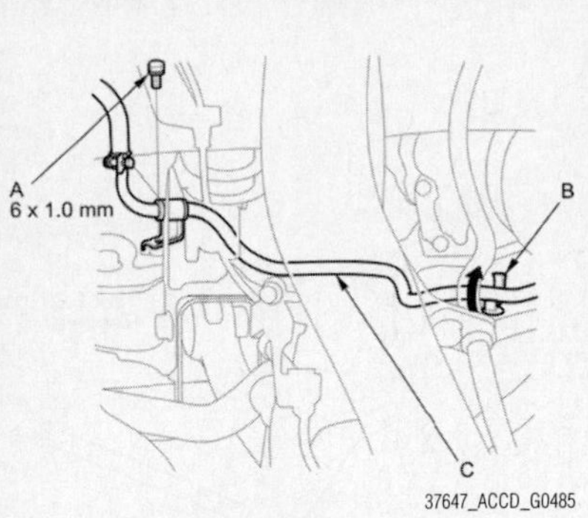
A
6 x 1.0 mm
B
C
37647_ACCD_G0485

Fig. 341 Remove the return line clamp bolt (A), release the return line clamp (B), and remove the return line (C)

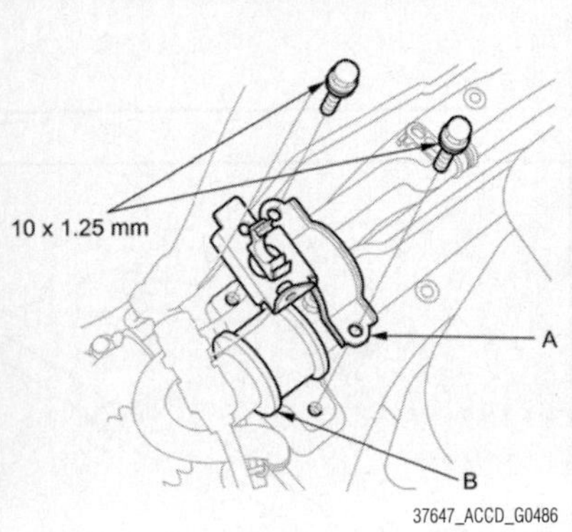
10 x 1.25 mm
A
B
37647_ACCD_G0486

Fig. 342 Remove the gearbox mounting bracket (A) and mounting cushion (B)

23. Release the return line clamp, and remove the return line.
24. Remove the flange bolts from the passenger's side of the steering gearbox, then remove the gearbox mounting bracket and mounting cushion.
25. Remove the mounting bolts and the flange bolts from the driver's side of the steering gearbox, and remove the stiffener plates and the washers.
26. Remove cotter pin from the tie-rod end ball joint, then remove the nut on both sides.
27. Disconnect the tie-rod end ball joint from the knuckle using the ball joint thread protector and the ball joint remover on both side.

➡Be careful not to damage the ball joint boot when installing the remover.

28. Raise the vehicle.
29. Remove the front splash shield.
30. Remove exhaust pipe A.
31. Remove the damper fork mounting nut and the mounting bolt.

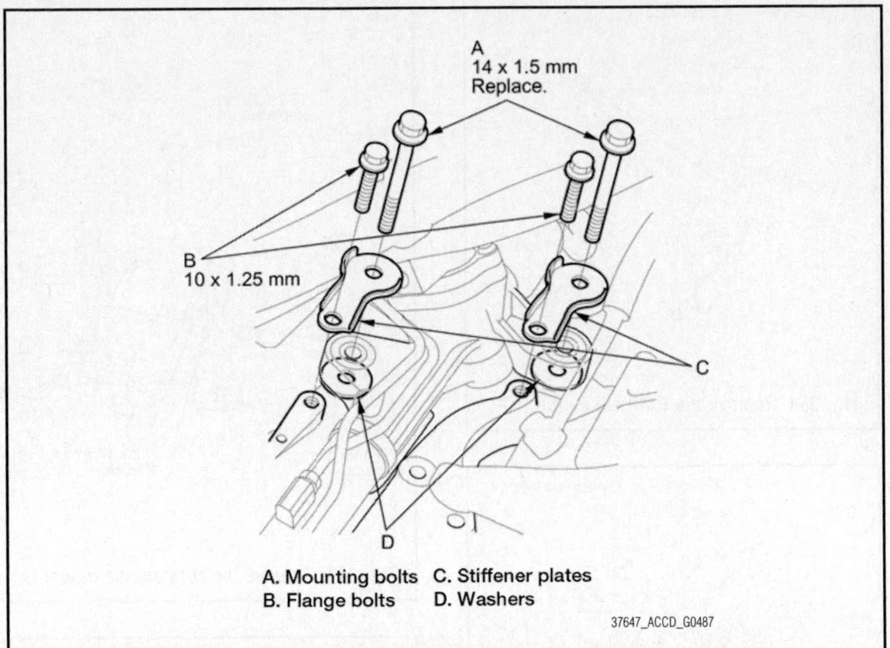
A
14 x 1.5 mm
Replace.
B
10 x 1.25 mm
C
D
A. Mounting bolts
B. Flange bolts
C. Stiffener plates
D. Washers
37647_ACCD_G0487

Fig. 343 Remove the mounting bolts and the flange bolts from the driver's side of the steering gearbox, and remove the stiffener plates and the washers

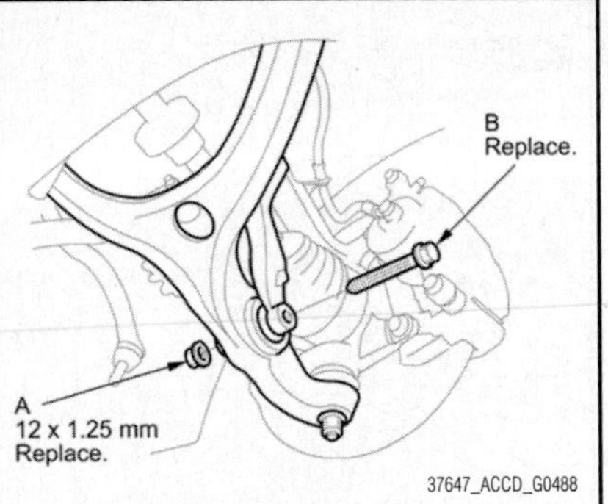
B
Replace.
A
12 x 1.25 mm
Replace.
37647_ACCD_G0488

Fig. 344 Remove the damper fork mounting nut (A) and the mounting bolt (B)

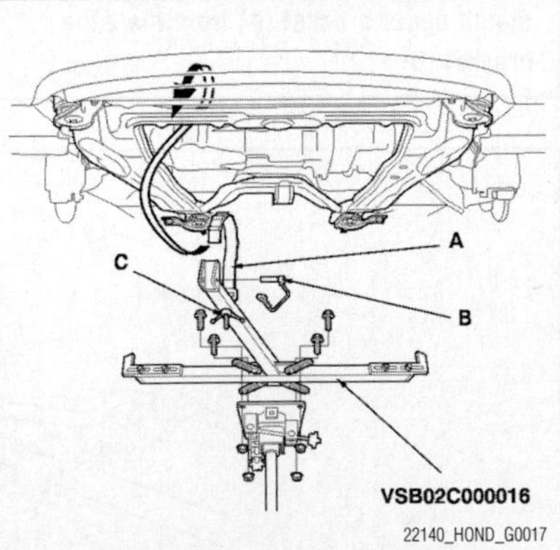
A
C
B
VSB02C000016
22140_HOND_G0017

Fig. 345 Attach the front subframe adapter (VSB02C000016) to the front subframe by looping the strap (A) over the front of the front subframe, then secure the strap with the stop (B), then tighten the wing nut (C)

32. Attach the front subframe adapter (VSB02C000016) to the subframe by looping the strap over the front of the subframe, then secure the strap with the stop, then tighten the wing nut.
33. Remove the front subframe middle mounting bolts on the passenger's side.
34. Remove the front subframe middle mount on the driver's side.
35. Remove the flange nuts from the lower transmission mount.
36. Remove the flange bolts from the front subframe front stiffeners.
37. Loosen the front subframe mounting bolts so they are about 0.79 inches (20 mm) from the mounting surface. Do not loosen the front subframe mounting bolts more than necessary.

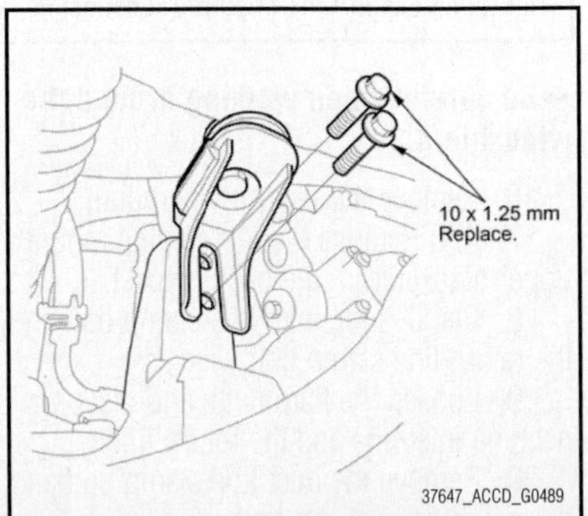
10 x 1.25 mm
Replace.
37647_ACCD_G0489

Fig. 346 Remove the front subframe middle mounting bolts on the passenger's side

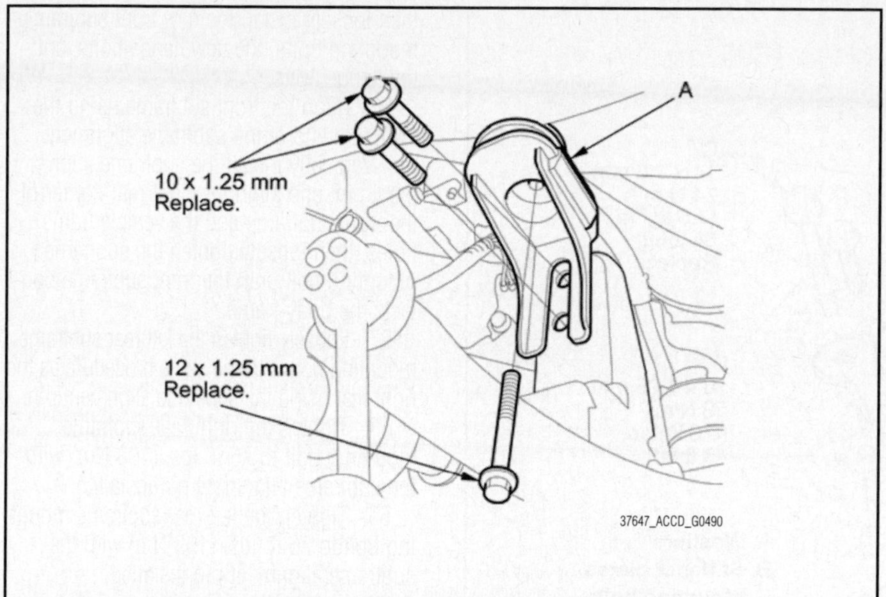
A
10 x 1.25 mm
Replace.
12 x 1.25 mm
Replace.
37647_ACCD_G0490

Fig. 347 Remove the front subframe middle mount (A) on the driver's side

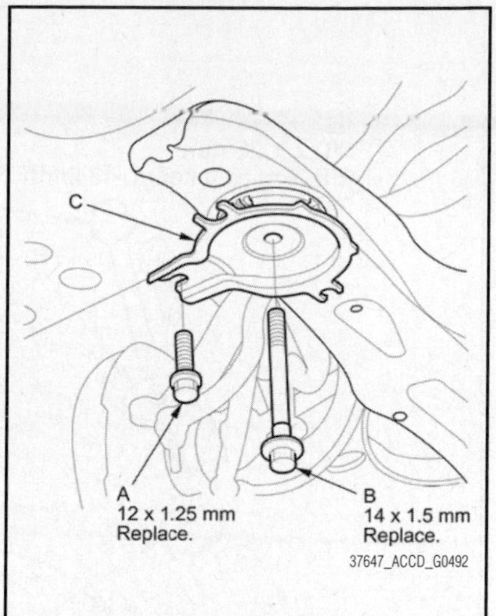
C
A
12 x 1.25 mm
Replace.
B
14 x 1.5 mm
Replace.
37647_ACCD_G0492

Fig. 350 Remove the flange bolts (A) and front subframe mounting bolts (B) from the front subframe rear stiffeners (C)

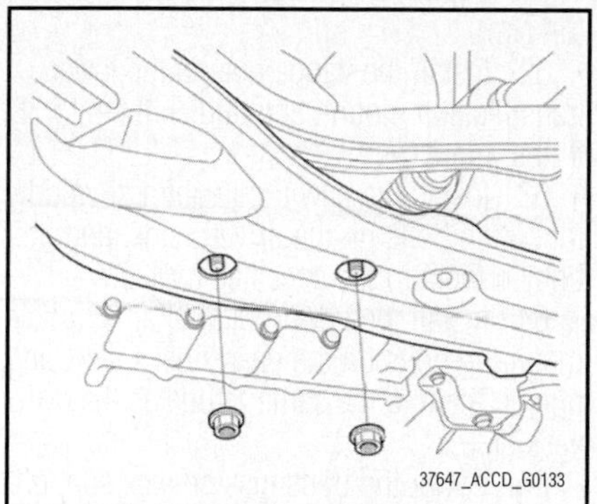
37647_ACCD_G0133

Fig. 348 Remove the flange nuts (A) from the lower transmission mount

38. Remove the flange bolts and front subframe mounting bolts from the front subframe rear stiffeners.

39. Lower the jack slowly until the front subframe has dropped about 2.71 inches (69 mm).

40. Carefully move the steering gearbox toward the driver's side until the pinion shaft clears the fenderwell opening on the body.

41. Remove the steering gearbox through the fenderwell opening on the driver's side.

42. Remove the pinion shaft grommet from the top of the valve housing.

43. After removing the steering gearbox, make sure that no power steering fluid gets on the gearbox mount cushions, the gearbox housing, the surface of the front subframe, and stiffener. Wipe off any spilled fluid at once.

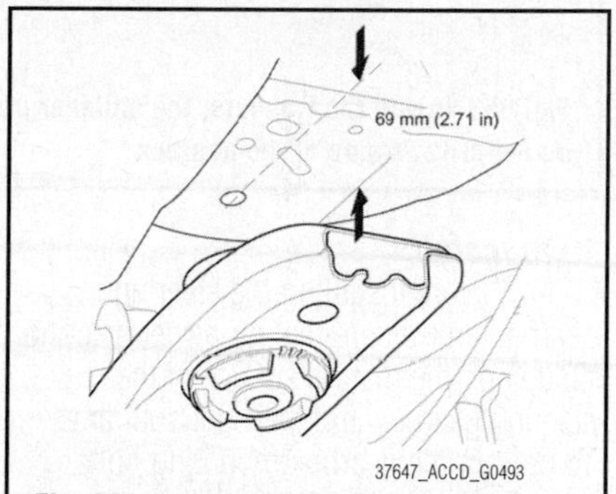
69 mm (2.71 in)
37647_ACCD_G0493

Fig. 351 Lower the jack slowly until the front subframe has dropped about 2.71 inches (69 mm)

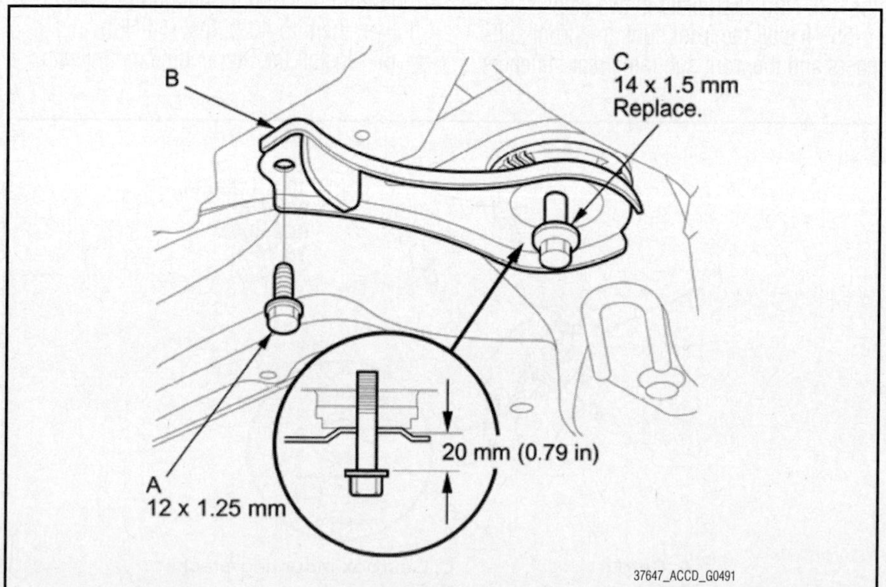
B
C
14 x 1.5 mm
Replace.
20 mm (0.79 in)
A
12 x 1.25 mm
37647_ACCD_G0491

Fig. 349 Remove the flange bolts (A) from the front subframe front stiffeners (B), and loosen the front subframe mounting bolts (C)

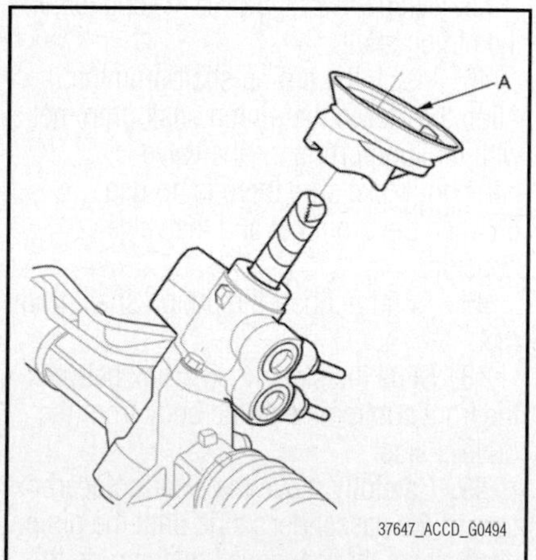
A
37647_ACCD_G0494

Fig. 352 Remove the pinion shaft grommet (A) from the top of the valve housing

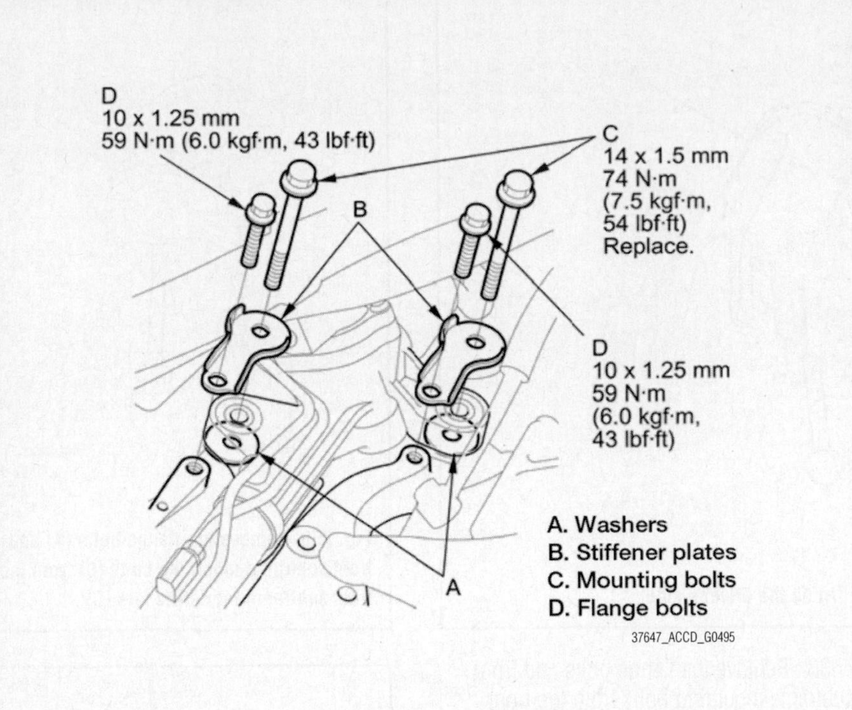
D
10 x 1.25 mm
59 N·m (6.0 kgf·m, 43 lbf·ft)
C
14 x 1.5 mm
74 N·m
(7.5 kgf·m,
54 lbf·ft)
Replace.
B
D
10 x 1.25 mm
59 N·m
(6.0 kgf·m,
43 lbf·ft)
A
A. Washers
B. Stiffener plates
C. Mounting bolts
D. Flange bolts
37647_ACCD_G0495

Fig. 353 Install the washers, the stiffener plates, the new mounting bolts, and the flange bolts on the driver's side of the gearbox

To install:

44. Before installing the steering gearbox, make sure that no power steering fluid is on the mating surface of the steering gearbox and the front subframe. To prevent the gearbox mounting bolts from loosening after the installation, remove any power steering fluid from the mount cushions and the bolt holes.

45. Wrap vinyl tape over the splines on the pinion shaft.

46. Install the pinion shaft grommet. Align the slot in the pinion shaft grommet with the lug portion on the valve housing. Make sure there is no gap between the grommet and the valve housing.

47. Turn the lip of the pinion shaft grommet.

48. Slide the steering gearbox between the front subframe and the body from the driver's side.

49. Carefully move the steering gearbox toward the passenger's side until the pinion shaft clears the fenderwell opening on the body.

50. Continue moving the gearbox toward the passenger's side until the steering gearbox is in position.

51. Install the washers, the stiffener plates, the new mounting bolts, and the flange bolts on the driver's side of the gearbox. Then loosely install the mounting bolts and the flange bolts.

52. Position the cutout on the mounting cushion as shown, and install it on the passenger's side of the steering gearbox.

53. Install the gearbox mounting bracket over the mounting cushion, and tighten the flange bolts to 28 ft. lbs. (38 Nm).

54. Tighten the flange bolts on the driver's side of the steering gearbox to 28 ft. lbs. (38 Nm) alternately in two steps.

55. Install the front subframe front stiffeners and the front subframe rear stiffeners, then loosely install the new front subframe mounting bolts, the new flange bolts, and the flange bolts.

56. Align the front subframe using the 15.7 mm end of the subframe alignment pin. Vertically install the subframe alignment pin, and align the right rear corner of the front subframe and the vehicle frame holes, then loosely tighten the subframe mounting bolt until the front subframe contacts the body frame.

57. Loosely tighten the left rear subframe mounting bolt with the same procedure as the right rear using the subframe alignment pin.

58. Tighten the right rear subframe mounting bolt to 76 ft. lbs. (103 Nm) with the subframe alignment pin installed.

59. Tighten the left rear subframe mounting bolt to 76 ft. lbs. (103 Nm) with the subframe alignment pin installed.

60. Tighten the front subframe mounting bolts to 76 ft. lbs. (103 Nm).

61. Tighten the flange bolts to 40 ft. lbs. (54 Nm).

62. Install the flange nuts to the lower transmission mount, and tighten them to 33 ft. lbs. (44 Nm).

63. Install the new front subframe middle mounting bolts on the driver's side, and tighten them to the specified torque.

64. Install the new front subframe middle mounting bolts on the passenger's side, and tighten them to the same torque as the driver's side.

65. Lower the transmission jack supporting the front subframe.

66. A/T: Install the rear engine mount upper bracket to the base bracket with a new mounting bolt and mounting bolts, and tighten them to 40 ft. lbs. (54 Nm).

67. Install the rear engine mount with

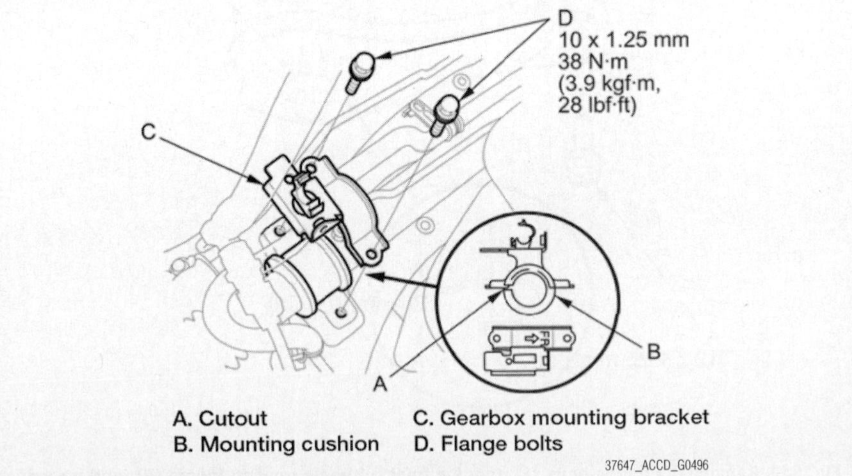
D
10 x 1.25 mm
38 N·m
(3.9 kgf·m,
28 lbf·ft)
C
B
A
A. Cutout
B. Mounting cushion
C. Gearbox mounting bracket
D. Flange bolts
37647_ACCD_G0496

Fig. 354 Position the cutout on the mounting cushion

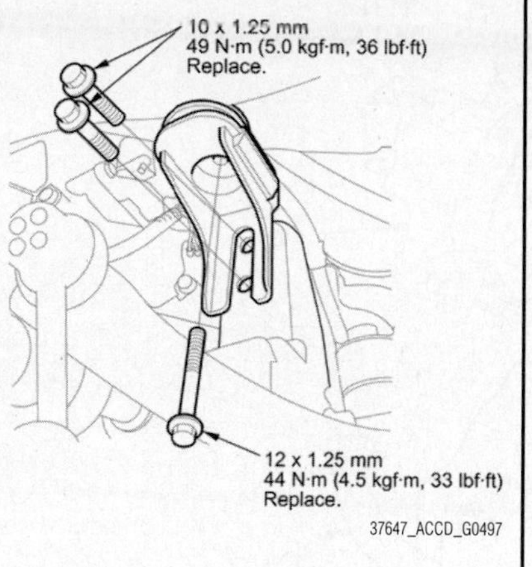
10 x 1.25 mm
49 N·m (5.0 kgf·m, 36 lbf·ft)
Replace.
12 x 1.25 mm
44 N·m (4.5 kgf·m, 33 lbf·ft)
Replace.
37647_ACCD_G0497

Fig. 355 Install the new front subframe middle mounting bolts on the driver's side

new mounting bolts, and lightly tighten them.

68. Remove the engine support hanger, the hanger balance bar, and the hanger adapter set.
69. Tighten the rear engine mount mounting bolts to the specified torque.
70. Install the new damper fork mounting bolt and the new mounting nut, and loosely tighten the nut.
71. Install exhaust pipe A.
72. Install the front splash shield.
73. Lower the vehicle.
74. On both sides, wipe off any grease contamination from the ball joint tapered section and threads. Reconnect the tie-rod end ball joint to the knuckle. Install the nut, and tighten it to 40 ft. lbs. (54 Nm).
75. Install a new cotter pin.
76. Loosely connect the return line and inlet line to the valve housing by hand.
77. Install the return line clamp bolt.
78. Install the return line to the return hose clamp, and clamp it.
79. Install the inlet line clamp bracket bolt and the return hose clamp bracket bolt.
80. Install the return hose to the return hose clamp, and clamp it.
81. Install the inlet line clamp bracket bolt and the return line clamp bolt.
82. Tighten the flare nuts.
83. Install the P/S heat shield with the flange bolts.
84. Install the strut brace.
85. Install the front grille cover.
86. Place a floor jack under the lower arm, and raise the suspension to load it with the vehicle's weight. Do not place the jack against the ball joint pin of the knuckle.
87. Tighten the damper fork mounting nut while holding the mounting bold to 47 ft. lbs. (64 Nm).
88. Install the front wheels, then set the wheels in the straight ahead position.

➡Before installing the wheel, clean the mating surfaces of the brake disc and the inside of the wheel.

89. Center the steering rack within its stroke.
90. Loosen the upper steering joint bolt, and slip the lower end of the steering joint onto the pinion shaft taking care to align the gap within the angle of 89 +/- 20 degrees.

➡Pick up the tabs of the pinion shaft grommet, and turn up the lip of the pinion shaft grommet securely in place. Make sure that light does not enter from the space between the pinion shaft grommet and the body.

91. Align the bolt hole on the steering joint with the groove around the pinion shaft, then loosely install the lower steering joint bolt. Be sure that the joint bolt is securely in the groove in the pinion shaft.
92. Pull on the steering joint to make sure that the steering joint is fully seated, then tighten the lower joint bolt to 21 ft. lbs. (28 Nm).
93. Tighten the upper steering joint bolt to 21 ft. lbs. (28 Nm).
94. Install the steering joint cover.
95. Install the steering wheel, and the driver's airbag.
96. Do the battery terminal reconnection procedure, and check these items:

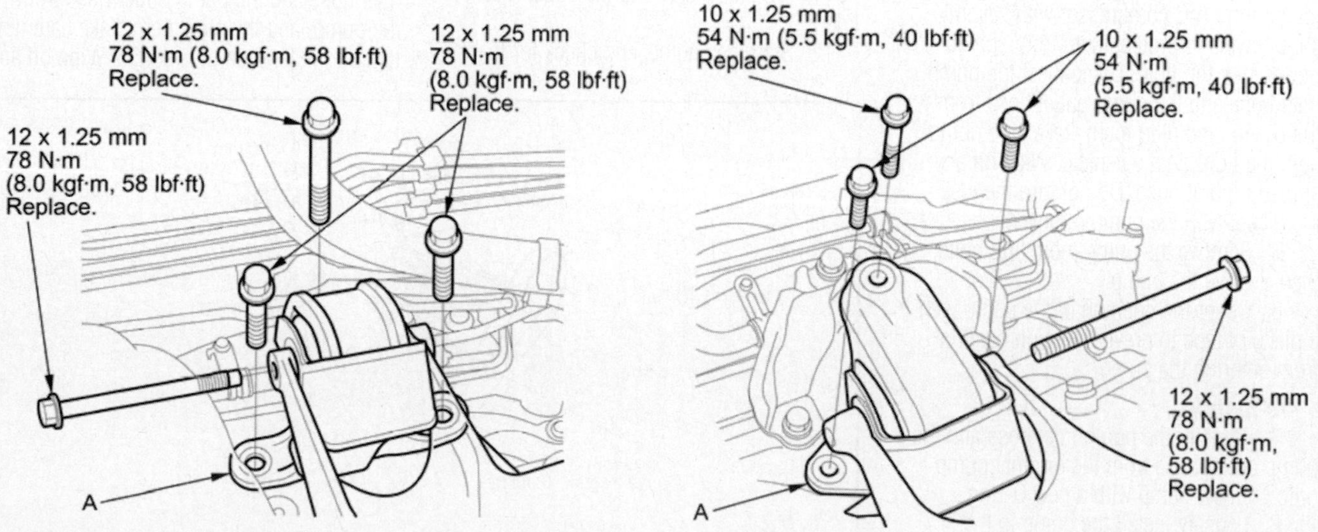
12 x 1.25 mm
78 N·m (8.0 kgf·m, 58 lbf·ft)
Replace.
12 x 1.25 mm
78 N·m
(8.0 kgf·m, 58 lbf·ft)
Replace.
12 x 1.25 mm
78 N·m
(8.0 kgf·m, 58 lbf·ft)
Replace.
A
10 x 1.25 mm
54 N·m (5.5 kgf·m, 40 lbf·ft)
Replace.
10 x 1.25 mm
54 N·m
(5.5 kgf·m, 40 lbf·ft)
Replace.
12 x 1.25 mm
78 N·m
(8.0 kgf·m,
58 lbf·ft)
Replace.
A
37647_ACCD_G0498

Fig. 356 Install the rear engine mount (A) with new mounting bolts

a. Turn the ignition switch to ON (II) and check that the SRS indicator comes on for about 6 seconds, and then goes off.

b. Make sure the horn and turn signal switches work properly.

c. Make sure the steering wheel switches work properly.

97. Fill the system with power steering fluid, and bleed air from the system.

98. After installation, check these items.

a. Start the engine, allow it to idle, and turn the steering wheel from lock to lock several times to warm up the fluid. Check the gearbox for leaks.

b. Check the steering wheel spoke angle.

c. If steering spoke angles to the right and left are not equal (steering wheel and rack are not centered), correct the engagement of the joint/pinion shaft serrations.

d. Set the steering column to the center tilt position, and to the center telescopic position, then do the front toe inspection

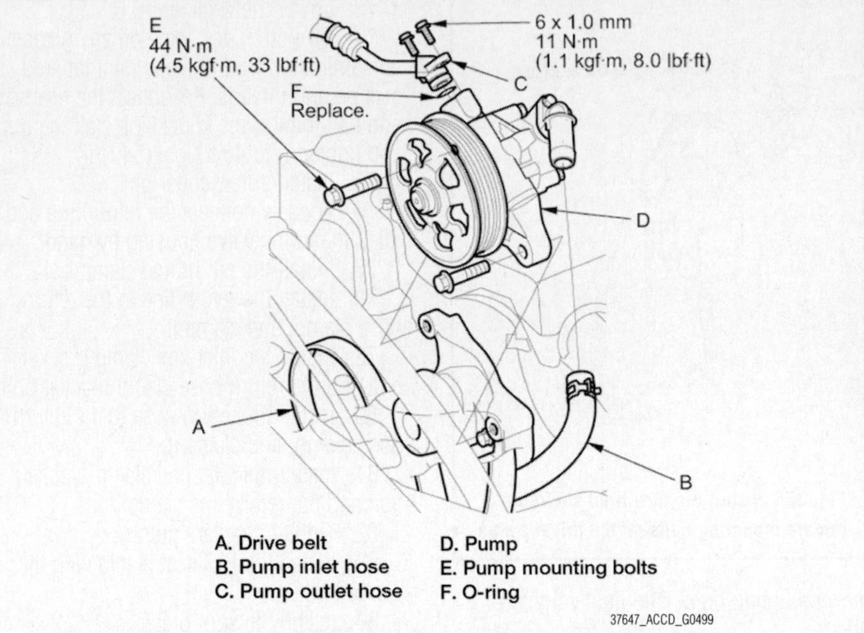
E
44 N·m
(4.5 kgf·m, 33 lbf·ft)
6 x 1.0 mm
11 N·m
(1.1 kgf·m, 8.0 lbf·ft)
C
F
Replace.
D
A
B
A. Drive belt
B. Pump inlet hose
C. Pump outlet hose
D. Pump
E. Pump mounting bolts
F. O-ring
37647_ACCD_G0499

Fig. 357 Disconnect the pump inlet hose and the pump outlet hose from the pump

POWER STEERING PUMP

REMOVAL & INSTALLATION

4-Cylinder Engine

See Figure 357.

1. Place a suitable container under the vehicle to catch any spilled fluid.
2. Drain the power steering fluid from the reservoir.
3. Remove the drive belt from the pump pulley.
4. Cover the auto-tensioner, the alternator, and the A/C compressor with several shop towels to protect them from spilled power steering fluid. Disconnect the pump inlet hose and the pump outlet hose from the pump, and plug them. Take care not to spill the fluid on the vehicle. Wipe off any spilled fluid at once. Do not turn the steering wheel with the pump removed.
5. Remove the pump mounting bolts, then remove the pump.
6. Cover the opening of the pump with a piece of tape to prevent foreign material from entering the pump.

To install:

7. Transfer the pump inlet hose and the pump outlet hose from the original pump onto the new pump with a new O-ring.
8. Loosely install the pump in the pump bracket with the mounting bolts, then tighten the pump fittings.
9. Tighten the pump mounting bolts to 33 ft. lbs. (44 Nm).
10. Install the drive belt.

➡Note these items during drive belt installation:

- Inspect the belt for wear and cracks. Replace the belt if necessary.
- Make sure that the belt is properly positioned on the pulleys.
- Do not get power steering fluid or grease on the auto-tensioner, the alternator, the A/C compressor, and the drive belt, or the pulley faces. Clean off any fluid or grease before installation.

11. Fill the reservoir to the upper level line.
12. Start the engine, and check for leaks.

6-Cylinder Engine

See Figure 358.

1. Place a suitable container under the vehicle to catch any spilled fluid.
2. Drain the power steering fluid from the reservoir.
3. Remove the engine cover.
4. Remove the drive belt from the pump pulley.
5. Cover the auto-tensioner, the alternator, and the A/C compressor with several shop towels to protect them from spilled power steering fluid. Disconnect the pump inlet hose and the pump outlet hose from the pump, and the plug them. Take care not to spill the fluid on the vehicle. Wipe off any

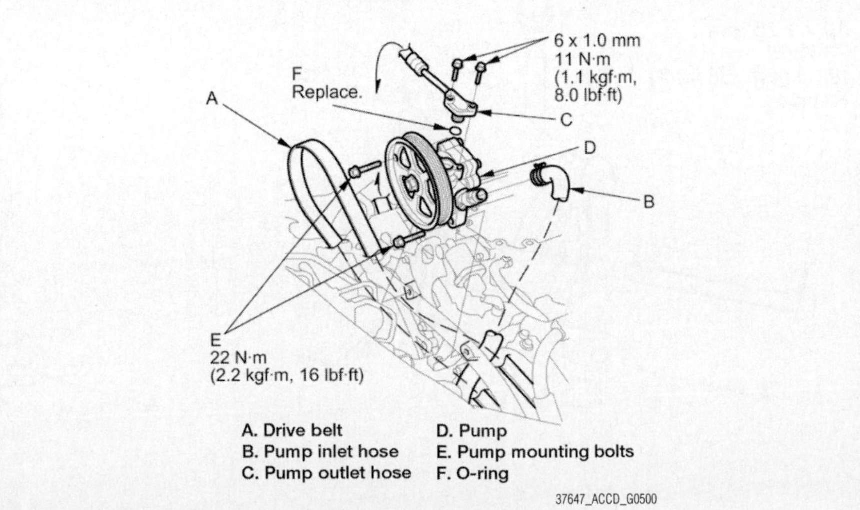
6 x 1.0 mm
11 N·m
(1.1 kgf·m,
8.0 lbf·ft)
F
Replace.
A
C
D
B
E
22 N·m
(2.2 kgf·m, 16 lbf·ft)
A. Drive belt
B. Pump inlet hose
C. Pump outlet hose
D. Pump
E. Pump mounting bolts
F. O-ring
37647_ACCD_G0500

Fig. 358 Disconnect the pump inlet hose and the pump outlet hose from the pump

spilled fluid at once. Do not turn the steering wheel with the pump removed.

6. Remove the pump mounting bolts, then remove the pump.

7. Cover the opening of the pump with a piece of tape to prevent foreign material from entering the pump.

To install:

8. Transfer the pump inlet hose and the pump outlet hose from the original pump onto the new pump with a new O-ring.

9. Loosely install the pump in the pump bracket with the mounting bolts, then tighten the pump fittings.

10. Tighten the pump mounting bolts to 16 ft. lbs. (22 Nm).

11. Install the drive belt.

➡Note these items during drive belt installation:

- Inspect the belt for wear and cracks. Replace the belt if necessary.
- Make sure that the belt is properly positioned on the pulleys.
- Do not get power steering fluid or grease on the auto-tensioner, the alternator, the A/C compressor, and the drive belt or pulley faces. Clean off any fluid or grease before installation.

12. Install the engine cover.

13. Fill the reservoir to the upper level line.

14. Start the engine, and check for leaks.

SUSPENSION

FRONT SUSPENSION

CONTROL LINKS

REMOVAL & INSTALLATION

Front Stabilizer Link

See Figure 359.

1. Raise the vehicle on a lift.
2. Remove the front wheel.
3. Remove the self-locking nut and the flange nut while holding the respective joint pin with a hex wrench, then remove the stabilizer link.

To install:

4. Install the stabilizer link on the stabilizer bar and the lower arm with the joint pins set at the center of their range of movement.

➡The stabilizer link has a paint mark. The left stabilizer link is marked with yellow paint, and the right stabilizer link is marked with white paint.

5. Install the new self-locking nut and the new flange nut, and tighten them to the specified torque while holding the respective joint pin with a hex wrench.
6. Clean the mating surfaces of the brake disc and the inside of the wheel, then install the front wheel.
7. Test-drive the vehicle.
8. After 5 minutes of driving, tighten the self-locking nut again to the specified torque.

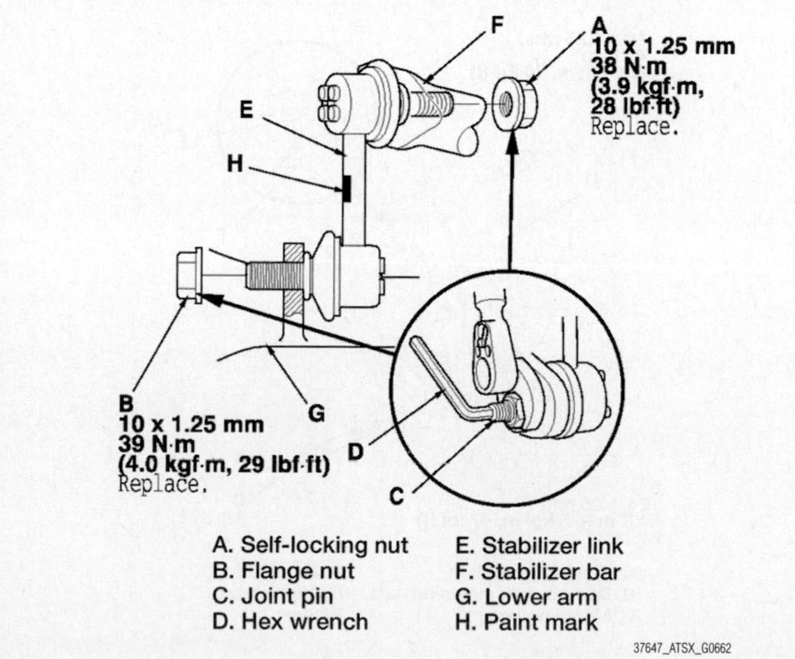
F
A
10 x 1.25 mm
38 N·m
(3.9 kgf·m,
28 lbf·ft)
Replace.
E
H
B
10 x 1.25 mm
39 N·m
(4.0 kgf·m, 29 lbf·ft)
Replace.
G
D
C
A. Self-locking nut
B. Flange nut
C. Joint pin
D. Hex wrench
E. Stabilizer link
F. Stabilizer bar
G. Lower arm
H. Paint mark
37647_ATSX_G0662

Fig. 359 Front stabilizer link

LOWER BALL JOINT

REMOVAL & INSTALLATION

See Figures 360 through 362.

Special Tools Required:

- Ball joint thread protector, 14 mm 07AAE-SJAA100
- Ball joint thread protector, 12 mm 07AAF-SDAA100
- Ball joint thread protector, 10 mm 07AAF-SECA120
- Ball joint remover, 32 mm 07MAC-SL0A102
- Ball joint remover, 28 mm 07MAC-SL0A202

➡Always use a ball joint remover to disconnect a ball joint. Do not strike the housing or any other part of the ball joint connection to disconnect it.

1. Install a hex nut or the ball joint thread protector onto the threads of the ball joint.

➡Using a hex nut, make sure the nut is flush with the ball joint pin end to prevent damage to the threaded end of the ball joint pin.

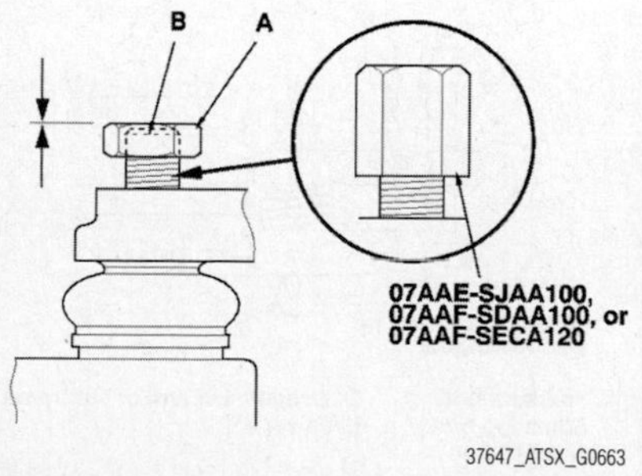
B
A
07AAE-SJAA100,
07AAF-SDAA100, or
07AAF-SECA120
37647_ATSX_G0663

Fig. 360 Install a hex nut (A) or the ball joint thread protector onto the threads of the ball joint (B)

2. Apply grease to the ball joint remover on the areas shown. This will ease the installation of the tool, and prevent damage to the pressure bolt threads.

3. Loosen the pressure bolt, and install the ball joint remover as shown. Insert the jaws carefully, making sure not to damage the ball joint boot. Adjust the jaw spacing by turning the adjusting bolt.

➡Fasten the safety chain securely to a suspension arm or the subframe. Do not fasten it to a brake line or wire harness.

4. After adjusting the adjusting bolt, make sure the head of the adjusting bolt is in the position shown to allow the jaw to pivot.

5. With a wrench, tighten the pressure bolt until the ball joint pin pops loose from the ball joint connecting hole. If necessary, apply penetrating type lubricant to loosen the ball joint pin.

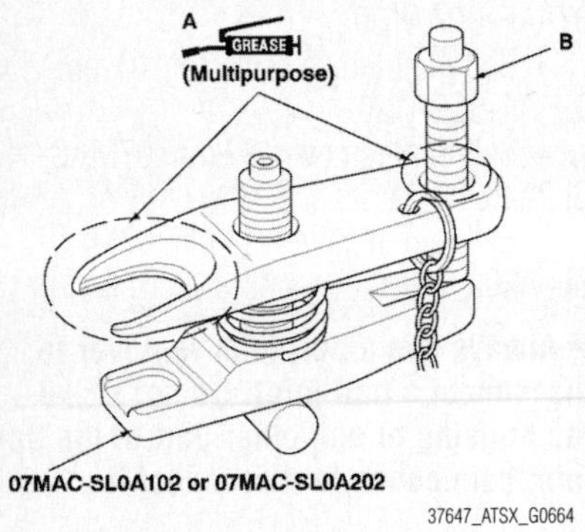
A
GREASE
(Multipurpose)
B
07MAC-SL0A102 or 07MAC-SL0A202
37647_ATSX_G0664

Fig. 361 Apply grease to the ball joint remover on the areas shown (A)

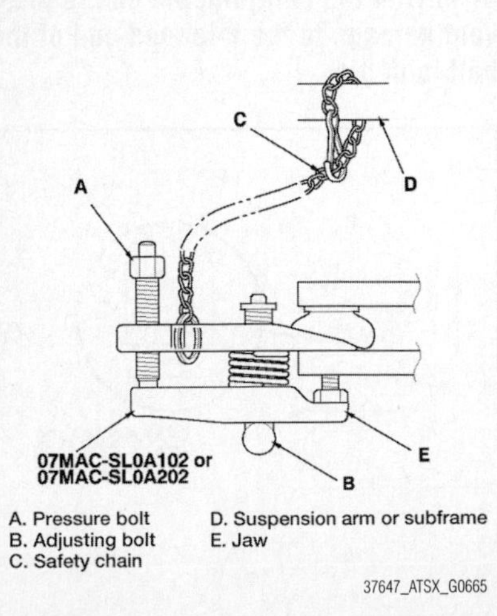
C
A
D
07MAC-SL0A102 or
07MAC-SL0A202
E
B
A. Pressure bolt
B. Adjusting bolt
C. Safety chain
D. Suspension arm or subframe
E. Jaw
37647_ATSX_G0665

Fig. 362 Loosen the pressure bolt, and install the ball joint remover as shown

➡Do not use pneumatic or electric tools on the pressure bolt.

6. Remove the ball joint remover, then remove the nut or the ball joint thread protector from the end of the ball joint pin, and pull the ball joint out of the ball joint connecting hole. Inspect the ball joint boot, and replace it if damaged.

LOWER CONTROL ARM

REMOVAL & INSTALLATION

See Figures 363 and 364.

Special Tools Required:

- Ball joint thread protector, 14 mm 07AAE-SJAA100
- Ball joint remover, 28 mm 07MAC-SL0A202
- Bushing driver 070AF-TA0A100
- Bushing receiver set 070AF-TA0A220

1. Raise the vehicle on a lift.
2. Remove the front wheel.
3. Remove the damper pinch bolt and the damper fork mounting nut while holding the mounting bolt, then remove the damper fork from the damper and the lower arm.

➡During installation, insert the aligning tab on the damper unit into the slot of the damper fork. Use the new damper fork mounting bolt and the new mounting nut, and torque the nut while holding the bolt during reassembly.

4. Disconnect the stabilizer link from the lower arm.
5. Remove the cotter pin from the knuckle ball joint, then remove the castle nut.

➡During installation, insert the new cotter pin into the ball joint pin hole from the front to the rear of the vehicle, and bend it. Check the ball joint pin hole direction before connecting the ball joint.

6. Disconnect the knuckle ball joint from the lower arm using the ball joint thread protector and the ball joint remover.

➡Be careful not to damage the ball joint boot when installing the remover. Do not force or hammer on the lower arm, or pry between the lower arm and the knuckle. You could damage the ball joint.

7. Remove the lower arm mounting bolts, and remove the lower arm.

➡Use new lower arm mounting bolts during reassembly.

To install:

8. Install the lower arm in the reverse order of removal, and note these items:
   - First install all of the components, and lightly tighten the bolts and the nuts, then raise the suspension to load it with the vehicle's weight before fully tightening it to the

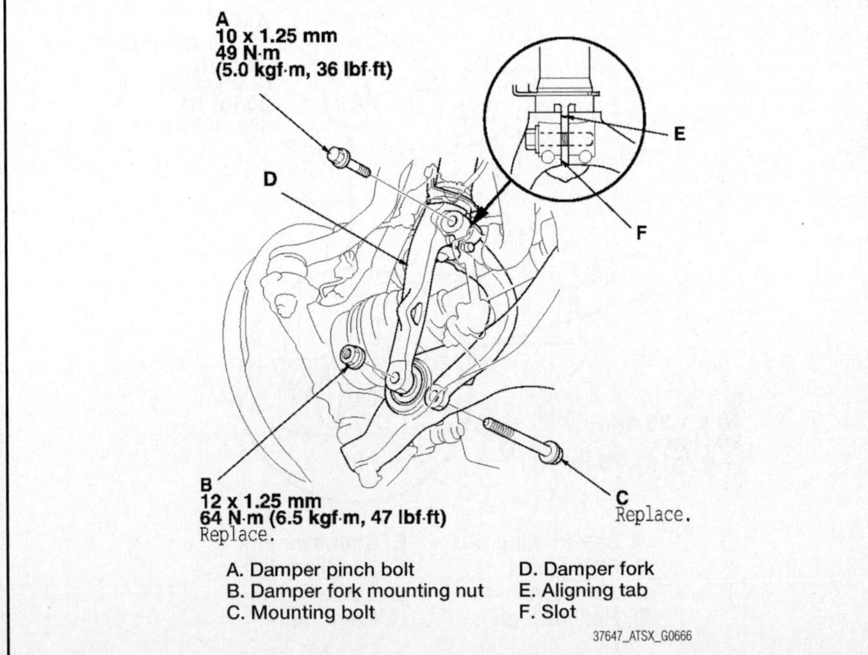
A
10 x 1.25 mm
49 N·m
(5.0 kgf·m, 36 lbf·ft)
D
E
F
B
12 x 1.25 mm
64 N·m (6.5 kgf·m, 47 lbf·ft)
Replace.
C
Replace.
A. Damper pinch bolt
B. Damper fork mounting nut
C. Mounting bolt
D. Damper fork
E. Aligning tab
F. Slot
37647_ATSX_G0666

Fig. 363 Remove the damper pinch bolt and the damper fork mounting nut while holding the mounting bolt

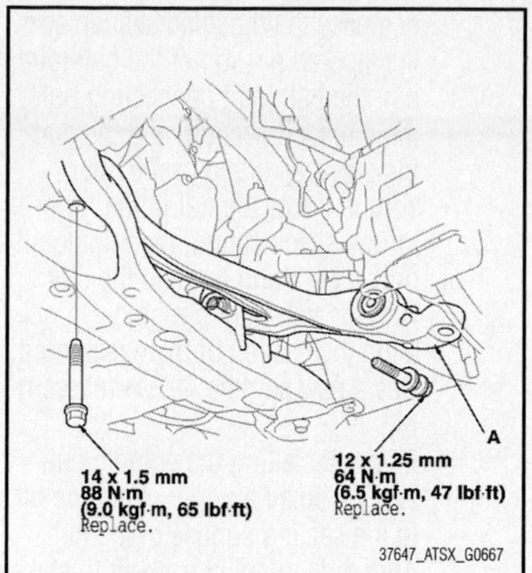
A
14 x 1.5 mm
88 N·m
(9.0 kgf·m, 65 lbf·ft)
Replace.
12 x 1.25 mm
64 N·m
(6.5 kgf·m, 47 lbf·ft)
Replace.
37647_ATSX_G0667

Fig. 364 Remove the lower arm mounting bolts, and remove the lower arm (A)

specified torque. Do not place the jack against the ball joint pin of the knuckle.

- Be careful not to damage the ball joint boot when connecting the knuckle.
- Before connecting the ball joint, degrease the threaded section and the tapered portion of the ball joint pin, the ball joint connecting hole, the threaded section, and the mating surfaces of the castle nut.
- Torque the castle nut to the lower torque specification, 58–65 ft. lbs. (78–88 Nm), then tighten it only far enough to align the slot with the ball joint pin hole. Do not align the castle nut by loosening it.
- Before installing the wheel, clean the mating surfaces of the brake disc and the inside of the wheel.

9. Check the wheel alignment, and adjust it if necessary.

STEERING KNUCKLE

REMOVAL & INSTALLATION

See Figures 365 through 369.

Special Tools Required:

- Ball joint thread protector, 14 mm 07AAE-SJAA100
- Ball joint thread protector, 12 mm 07AAF-SDAA100
- Ball joint thread protector, 10 mm 07AAF-SECA120
- Ball joint remover, 28 mm 07MAC-SL0A202
- Hub dis/assembly tool 07GAF-SD4A100
- Attachment, 72 x 75 mm 07746-0010600

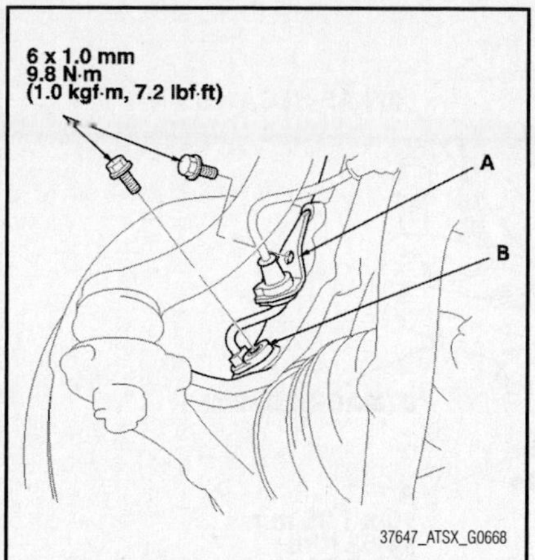
6 x 1.0 mm
9.8 N·m
(1.0 kgf·m, 7.2 lbf·ft)
A
B
37647_ATSX_G0668

Fig. 365 Remove the wheel speed sensor harness bracket (A) and the wheel speed sensor (B)

- Driver handle 07749-0010000
- Attachment, 80 x 96 mm 07ZAD-PNA0100
- Support base 07965-SD90100

1. Raise the vehicle on a lift.
2. Remove the wheel nuts and the front wheel.
3. Remove the brake hose bracket mounting bolt.
4. Remove the brake caliper bracket mounting bolts, then remove the caliper assembly from the knuckle. To prevent damage to the caliper assembly or the brake hose, use a short piece of wire to hang the caliper assembly from the undercarriage. Do not twist the brake hose excessively.
5. Remove the wheel speed sensor harness bracket and the wheel speed sensor from the knuckle. Do not disconnect the wheel speed sensor connector.
6. Pry up the stake on the spindle nut, then remove the nut.
7. Remove the front brake disc.
8. Check the front hub for damage and cracks.
9. Remove the cotter pin from the tie-rod end ball joint, then remove the nut.

➡**During installation, install the new cotter pin after tightening the nut, and bend its end.**

10. Disconnect the tie-rod end ball joint from the knuckle using the ball joint thread protector and the ball joint remover.
11. Remove the cotter pin from the knuckle ball joint, then remove the castle nut.

➡**During installation, insert the new cotter pin into the ball joint pin hole from the front to the rear of the vehicle, and bend its end as shown. Check the**

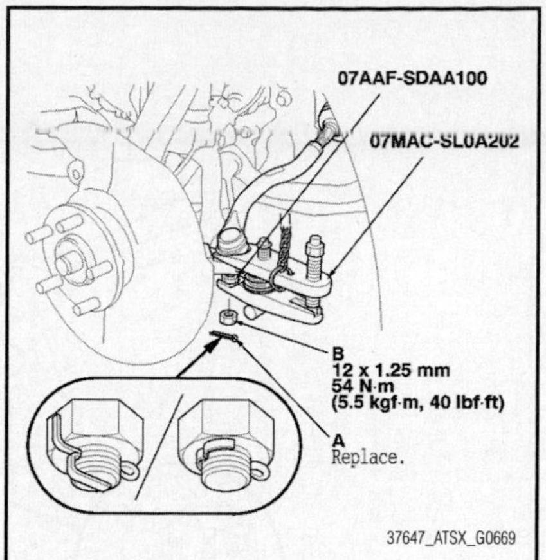
07AAF-SDAA100
07MAC-SL0A202
B
12 x 1.25 mm
54 N·m
(5.5 kgf·m, 40 lbf·ft)
A
Replace.
37647_ATSX_G0669

Fig. 366 Remove the cotter pin (A) from the tie-rod end ball joint, then remove the nut (B)

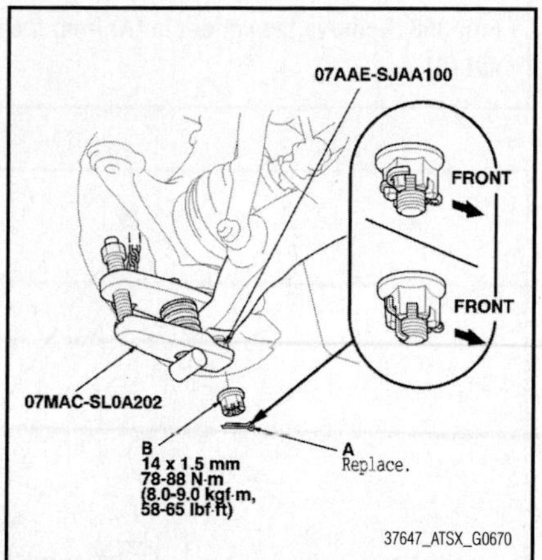
07AAE-SJAA100
FRONT
FRONT
07MAC-SL0A202
B
14 x 1.5 mm
78-88 N·m
(8.0-9.0 kgf·m,
58-65 lbf·ft)
A
Replace.
37647_ATSX_G0670

Fig. 367 Remove the cotter pin (A) from the knuckle ball joint, then remove the castle nut (B)

ball joint pin hole direction before connecting the ball joint.

12. Disconnect the knuckle ball joint from the lower arm using the ball joint thread protector and the ball joint remover.

➡**Be careful not to damage the ball joint boot when installing the remover. Do not force or hammer on the lower arm, or pry between the lower arm and the knuckle. You could damage the ball joint.**

13. Remove the cotter pin from the upper arm ball joint, then remove the castle nut.

➡**During installation, insert the new cotter pin into the ball joint pin hole from the front to the rear of the vehicle, and bend its end as shown. Check the ball joint pin hole direction before connecting the ball joint.**

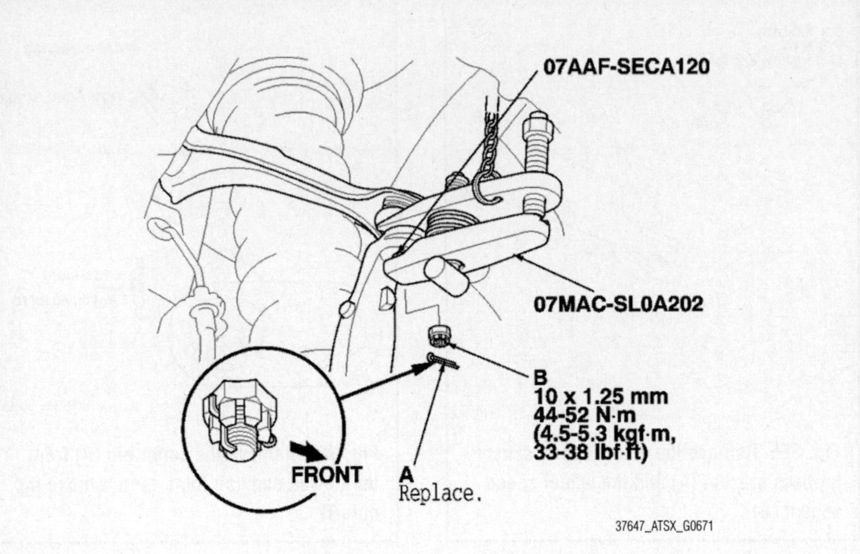
07AAF-SECA120
07MAC-SL0A202
B
10 x 1.25 mm
44-52 N·m
(4.5-5.3 kgf·m,
33-38 lbf·ft)
FRONT
A
Replace.
37647_ATSX_G0671

Fig. 368 Remove the cotter pin (A) from the upper arm ball joint, then remove the castle nut (B)

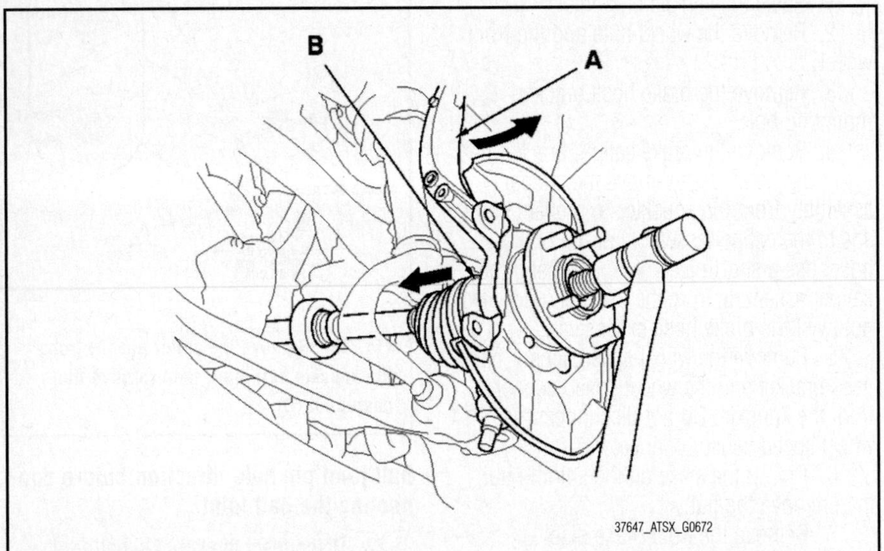
B
A
37647_ATSX_G0672

Fig. 369 Pull the knuckle (A) outward, and separate the outboard joint (B) from the front hub using a plastic hammer

14. Disconnect the upper arm ball joint from the knuckle using the ball joint thread protector and the ball joint remover.
15. Pull the knuckle outward, and separate the outboard joint from the front hub using a plastic hammer.

➡Do not pull the driveshaft end outward. The driveshaft inboard joint may come apart. During installation, apply grease to the mating surfaces of the wheel bearing and the driveshaft outboard joint.

To install:

16. Install the knuckle/hub in the reverse order of removal, and note these items:
  - First install all of the components, and lightly tighten the bolts and the nuts, then raise the suspension to load it with the vehicle's weight before fully tightening to the specified torque. Do not place the jack against the ball joint pin of the knuckle.
  - Be careful not to damage the ball joint boot when connecting the knuckle.
  - Before connecting the ball joint, degrease the threaded section and the tapered portion of the ball joint pin, the ball joint connecting hole, the threaded section, and the mating surfaces of the castle nut.
  - Torque the castle nut to the lower torque specification, then tighten it only far enough to align the slot with the ball joint pin hole. Do not align the castle nut by loosening it.
  - Use a new spindle nut on reassembly.
  - Before installing the spindle nut, apply a small amount of engine oil to the seating surface of the nut. After tightening, use a drift to stake the spindle nut shoulder against the driveshaft.
  - Before installing the brake disc, clean the mating surfaces of the front hub and the inside of the brake disc.
  - Before installing the wheel, clean the mating surfaces of the brake disc and the inside of the wheel.
17. Check the wheel alignment, and adjust it if necessary.

STRUT (DAMPER/SPRING)

REMOVAL & INSTALLATION

See Figures 370 through 375.

1. Raise the vehicle on a lift.
2. Remove the front wheel.
3. Remove the wheel speed sensor harness bracket mounting bolt.
4. Remove the damper pinch bolt and the damper fork mounting nut while holding the mounting bolt, then remove the damper fork from the damper and the lower arm.
5. Remove the front strut brace mounting nuts.

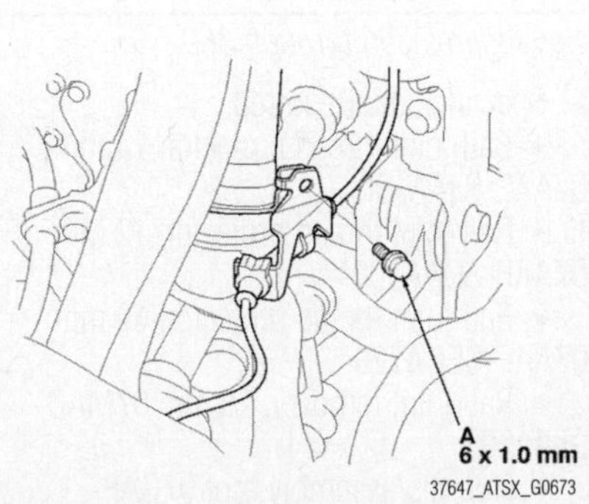
A
6 x 1.0 mm
37647_ATSX_G0673

Fig. 370 Remove the wheel speed sensor harness bracket mounting bolt (A)

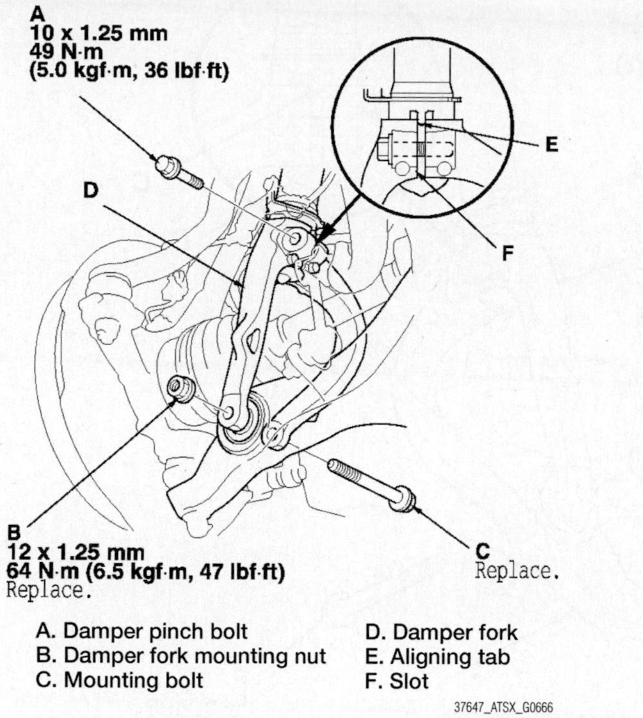
A
10 x 1.25 mm
49 N·m
(5.0 kgf·m, 36 lbf·ft)
D
E
F
B
12 x 1.25 mm
64 N·m (6.5 kgf·m, 47 lbf·ft)
Replace.
C
Replace.
A. Damper pinch bolt
B. Damper fork mounting nut
C. Mounting bolt
D. Damper fork
E. Aligning tab
F. Slot
37647_ATSX_G0666

Fig. 371 Remove the damper pinch bolt and the damper fork mounting nut while holding the mounting bolt

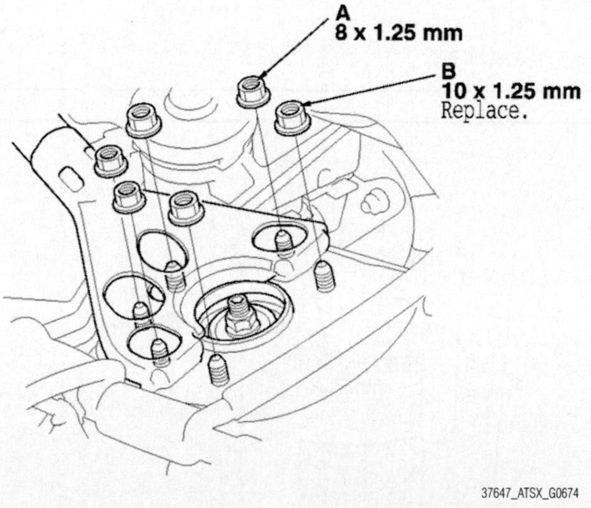
A
8 x 1.25 mm
B
10 x 1.25 mm
Replace.
37647_ATSX_G0674

Fig. 372 Remove the front strut brace mounting nuts (A) and the damper mounting nuts (B)

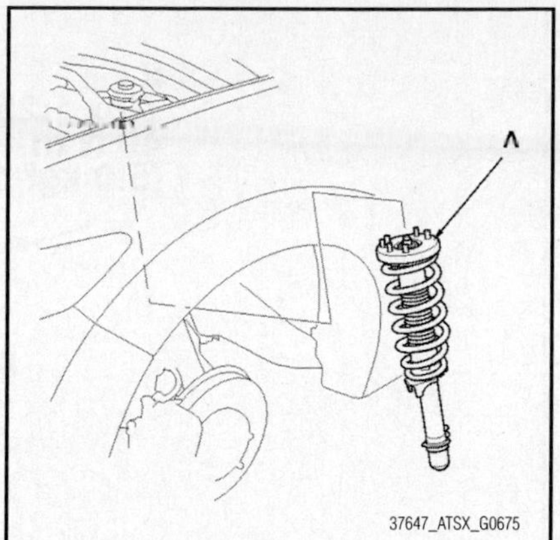
A
37647_ATSX_G0675

Fig. 373 Remove the damper/spring (A)

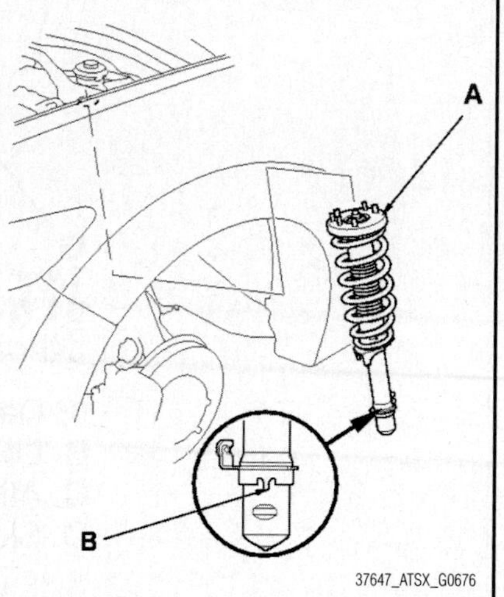
A
B
37647_ATSX_G0676

Fig. 374 Position the damper/spring (A) in the body with the aligning tab (B) facing inside

6. Remove the damper mounting nuts from the top of the damper. Do not let the damper/spring drop down under its own weight.
7. Remove the damper/spring.

➡**Be careful not to damage the body.**

To install:

8. Position the damper/spring in the body with the aligning tab facing inside.
9. Loosely install the new damper mounting nuts to the top of the damper.
10. Loosely install the front strut brace mounting nuts.
11. Install the damper fork over the driveshaft and onto the lower arm. Install the aligning tab on the damper unit into the slot of the damper fork.
12. Loosely install the damper pinch bolt into the damper fork.
13. Connect the damper fork and the lower arm with the new damper fork mounting bolt, then lightly tighten the new mounting nut.
14. Place a floor jack under the lower arm, and raise the suspension to load it with the vehicle's weight.
15. Tighten the damper pinch bolt and the damper fork mounting nut while holding the mounting bolt to the specified torque. Damper pinch bolt: tighten to 36 ft. lbs. (49 Nm); damper fork mounting nut: tighten to 47 ft. lbs. (64 Nm).
16. Tighten the damper mounting nuts and front strut brace mounting nuts on top of the damper to the specified torque.

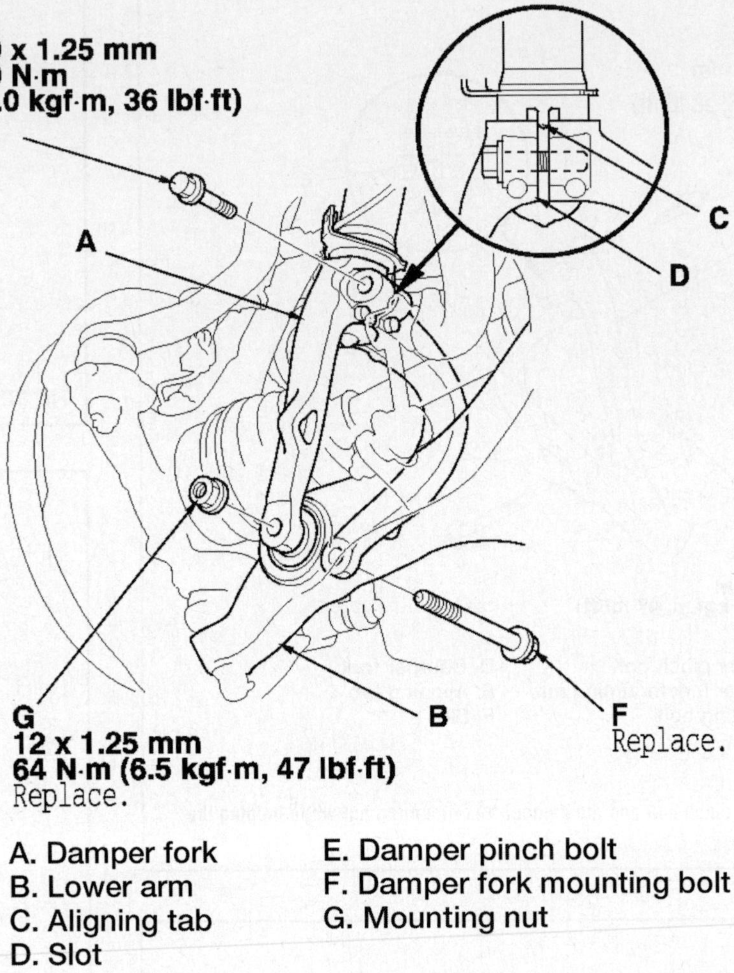
E
10 x 1.25 mm
49 N·m
(5.0 kgf·m, 36 lbf·ft)
A
C
D
G
12 x 1.25 mm
64 N·m (6.5 kgf·m, 47 lbf·ft)
Replace.
B
F
Replace.
A. Damper fork
B. Lower arm
C. Aligning tab
D. Slot
E. Damper pinch bolt
F. Damper fork mounting bolt
G. Mounting nut
37647_ATSX_G0677

Fig. 375 Installing damper fork

Damper mounting nuts: tighten to 41 ft. lbs. (55 Nm); strut brace mounting nuts: tighten to 16 ft. lbs. (22 Nm).

17. Install the wheel speed sensor harness bracket.
18. Clean the mating surfaces of the brake disc and the inside of the wheel, then install the front wheel.
19. Check the wheel alignment, and adjust it if necessary.

OVERHAUL

Disassembly

See Figures 376 and 377.

1. Compress the damper spring, then remove the self-locking nut while holding the damper shaft with a hex wrench. Do not compress the damper spring more than is necessary to remove the self-locking nut.
2. Release the pressure from the strut spring compressor, then disassemble the damper as shown in the Exploded View.

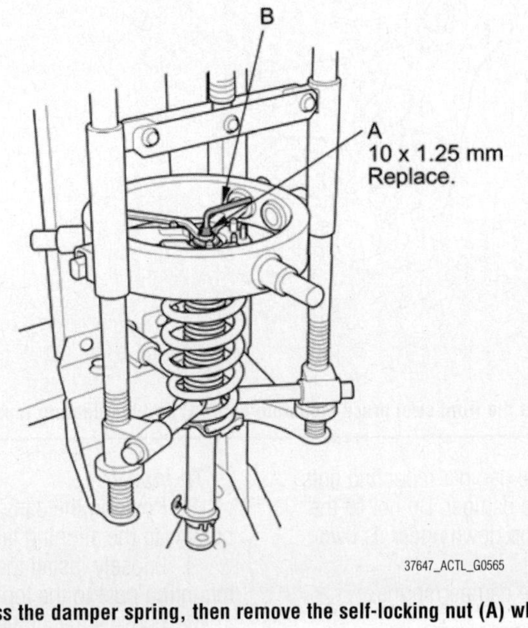
B
A
10 x 1.25 mm
Replace.
37647_ACTL_G0565

Fig. 376 Compress the damper spring, then remove the self-locking nut (A) while holding the damper shaft with a hex wrench (B)

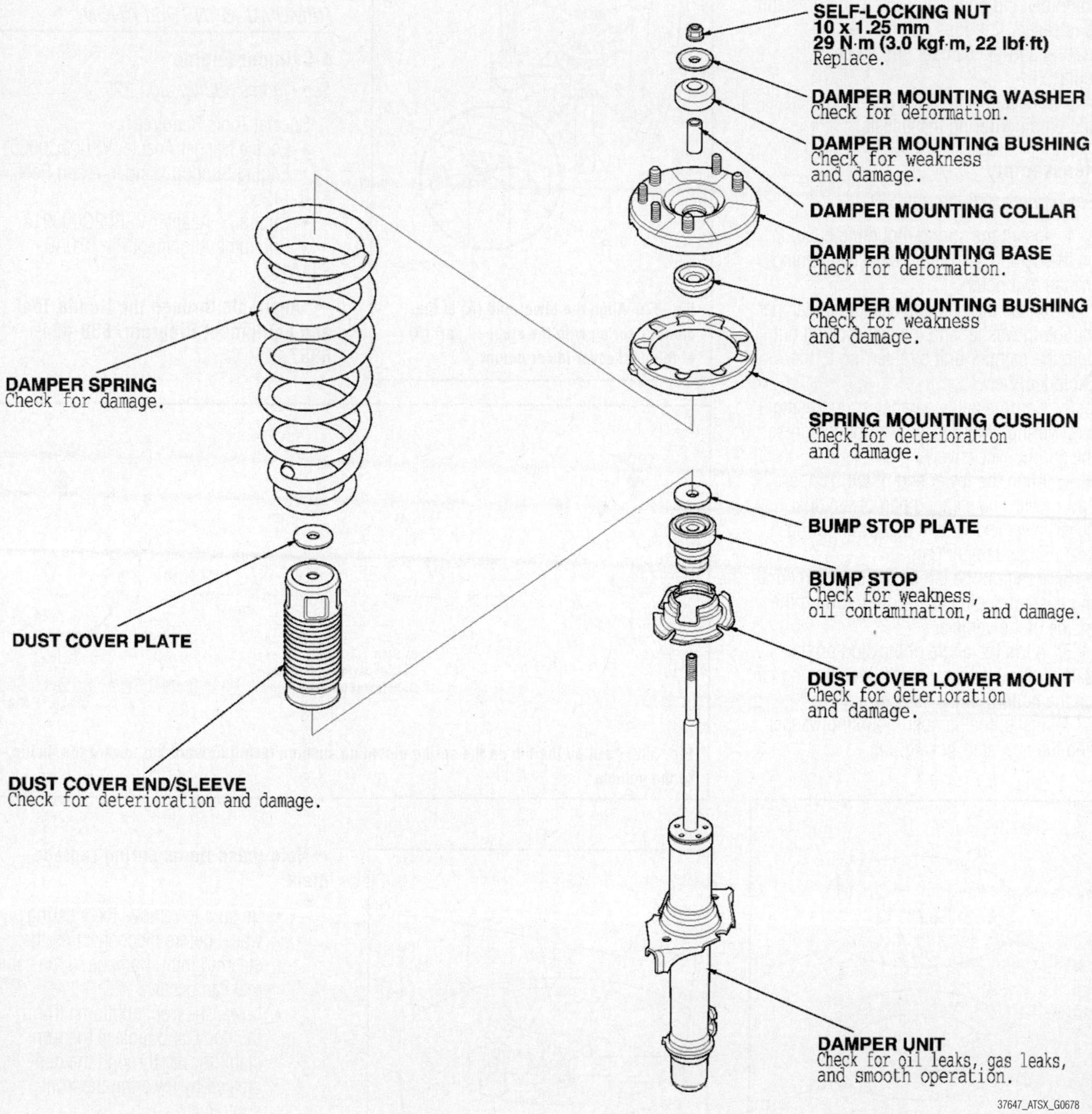
SELF-LOCKING NUT
10 x 1.25 mm
29 N·m (3.0 kgf·m, 22 lbf·ft)
Replace.
DAMPER MOUNTING WASHER
Check for deformation.
DAMPER MOUNTING BUSHING
Check for weakness and damage.
DAMPER MOUNTING COLLAR
DAMPER MOUNTING BASE
Check for deformation.
DAMPER MOUNTING BUSHING
Check for weakness and damage.
DAMPER SPRING
Check for damage.
SPRING MOUNTING CUSHION
Check for deterioration and damage.
BUMP STOP PLATE
BUMP STOP
Check for weakness, oil contamination, and damage.
DUST COVER PLATE
DUST COVER LOWER MOUNT
Check for deterioration and damage.
DUST COVER END/SLEEVE
Check for deterioration and damage.
DAMPER UNIT
Check for oil leaks, gas leaks, and smooth operation.
37647_ATSX_G0678

Fig. 377 Exploded view of damper/spring assembly

Inspection

1. Reassemble all the parts, except for the damper spring.
2. Compress the damper assembly by hand, and check for smooth operation through a full stroke, both compression and extension. The damper should extend smoothly and constantly when compression is released. If it does not, the gas is leaking and the damper should be replaced.
3. Check for oil leaks, abnormal noises, and binding during these tests.

Reassembly

See Figures 378 through 381.

1. Install the spring mounting cushion on the damper mounting base by aligning the tab and notch.
2. Install all the parts except the damper mounting washer and the self-locking nut onto the damper unit by referring to the Exploded View.
3. Compress the damper spring using a strut spring compressor. Do not compress the spring excessively.
4. Align the lower end of the damper spring with the stepped part of the dust cover lower mount and the lower spring seat on the damper unit.
5. Position the tab on the spring mounting cushion facing forward but toward the inside of the vehicle.
6. Align the angle of the stud on the damper mounting base with the aligning tab on the bottom of the damper unit.
7. Install the damper mounting washer and the new self-locking nut.

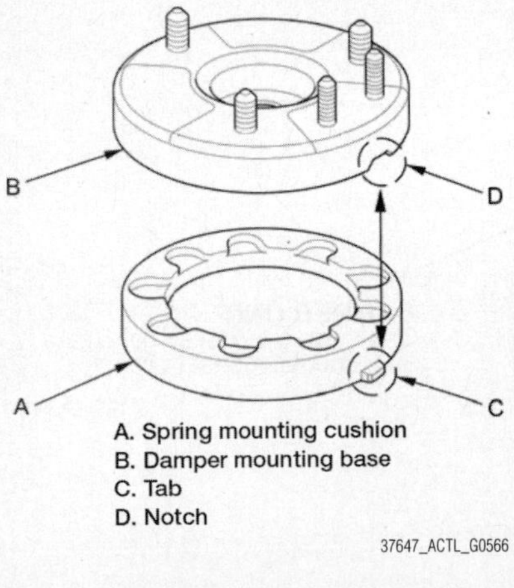
A
B
C
D
A. Spring mounting cushion
B. Damper mounting base
C. Tab
D. Notch
37647_ACTL_G0566

Fig. 378 Install the spring mounting cushion on the damper mounting base by aligning the tab and notch

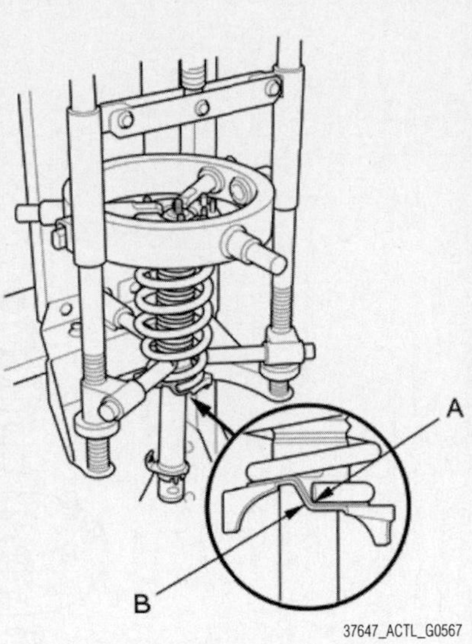
A
B
37647_ACTL_G0567

Fig. 379 Align the lower end (A) of the damper spring with the stepped part (B) of the dust cover lower mount

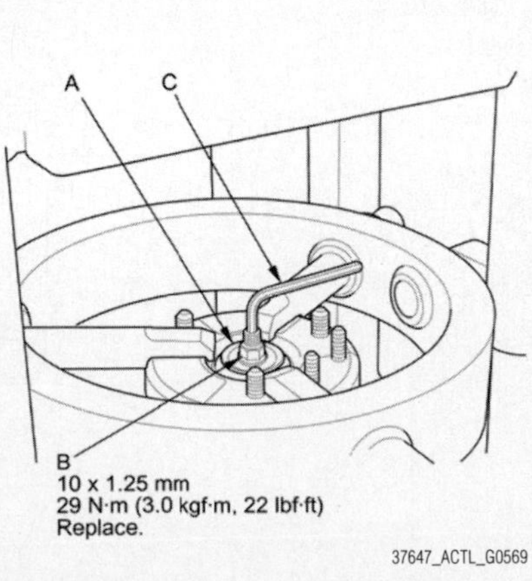
A
C
B
10 x 1.25 mm
29 N·m (3.0 kgf·m, 22 lbf·ft)
Replace.
37647_ACTL_G0569

Fig. 381 Install the damper mounting washer (A) and the new self-locking nut (B)

8. Hold the damper shaft with a hex wrench, and tighten the self-locking nut to 22 ft. lbs. (29 Nm).
9. Remove the damper/spring from the strut spring compressor.

STABILIZER BAR

REMOVAL & INSTALLATION

4-Cylinder Engine

See Figures 382 through 395.

Special Tools Required*:

- Engine Hanger Adapter VSB02C000015
- Engine Support Hanger, A and Reds AAR-T1256
- Subframe Adapter VSB02C000016
- Subframe Alignment Pin 070AG-SJAA10S

➡*Available through the Honda Tool and Equipment Program, 888-424-6857.

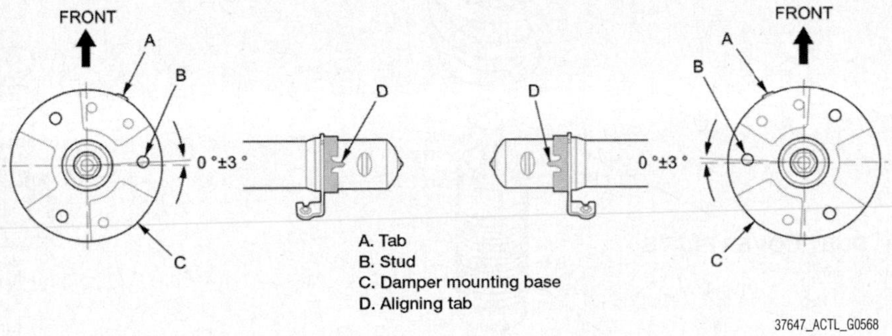
FRONT
A
B
C
D
0 °±3 °
FRONT
A
B
C
D
0 °±3 °
A. Tab
B. Stud
C. Damper mounting base
D. Aligning tab
37647_ACTL_G0568

Fig. 380 Position the tab on the spring mounting cushion facing forward but toward the inside of the vehicle

➡Note these items during replacement:

- Be sure to remove the steering wheel before disconnecting the steering joint. Damage to the cable reel can occur.
- Lower the front subframe from the body, and replace the front stabilizer bar through the gap created by lowering the front subframe.

1. Remove the hood support rod, then use it to prop the hood in the wide-open position.
2. Remove the front grille cover:
3. Do the battery terminal disconnection procedure.
4. Raise and support the vehicle.
5. Remove the front wheels.

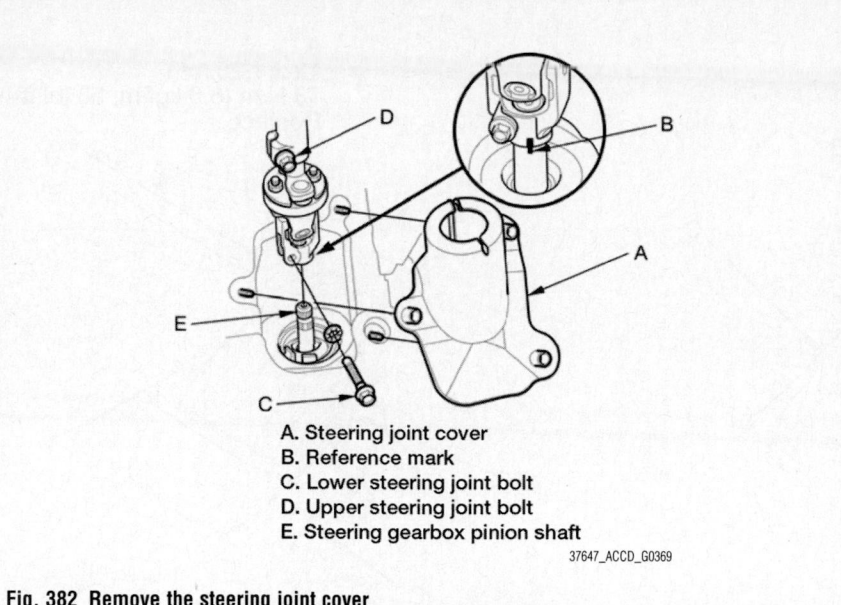

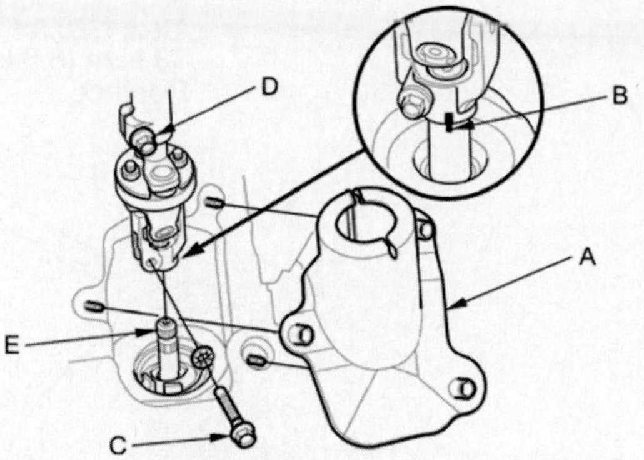
D
B
A
E
C
A. Steering joint cover
B. Reference mark
C. Lower steering joint bolt
D. Upper steering joint bolt
E. Steering gearbox pinion shaft
37647_ACCD_G0369

Fig. 382 Remove the steering joint cover

6. Remove the driver's airbag and the steering wheel.

7. Remove the steering joint cover.

8. Loosen the steering joint upper bolt, and remove the steering joint lower bolt. Disconnect the steering joint by sliding the steering joint into the column shaft. Tighten the steering joint upper bolt to hold the column shaft.

➡Do not disconnect the steering joint from the column shaft. If the center guide is in place and has not moved, leave it in place. If the center guide has moved or been removed, discard it.

9. Disconnect the Power Steering Pressure (PSP) switch connector.

10. Remove the power steering pump outlet hose mounting bolt.

11. Remove the front strut brace (if equipped).

12. Attach the engine hanger adapter

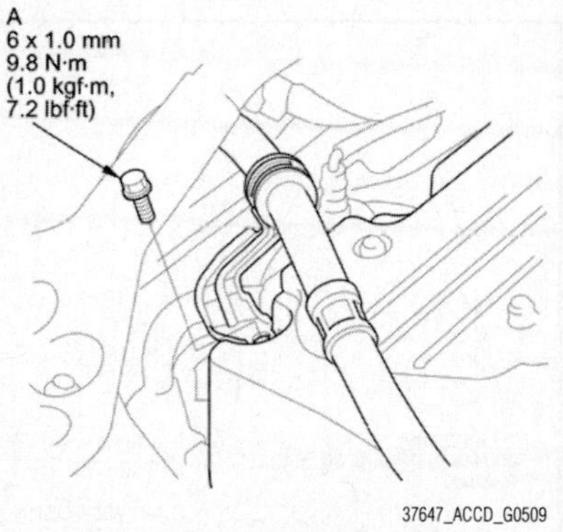
A
6 x 1.0 mm
9.8 N·m
(1.0 kgf·m,
7.2 lbf·ft)
37647_ACCD_G0509

Fig. 384 Remove the power steering pump outlet hose mounting bolt (A)

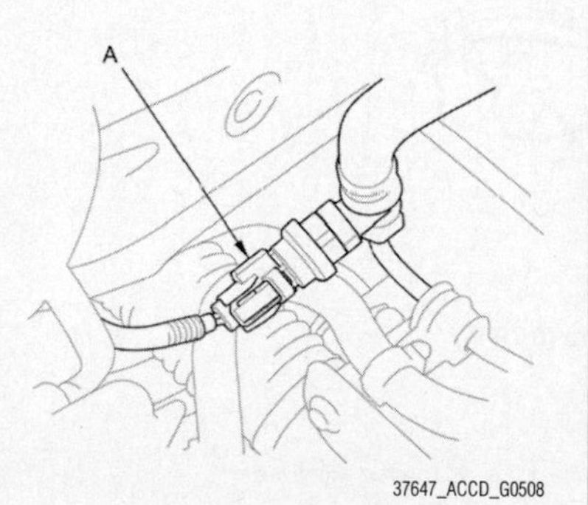
A
37647_ACCD_G0508

Fig. 383 Disconnect the Power Steering Pressure (PSP) switch connector (A)

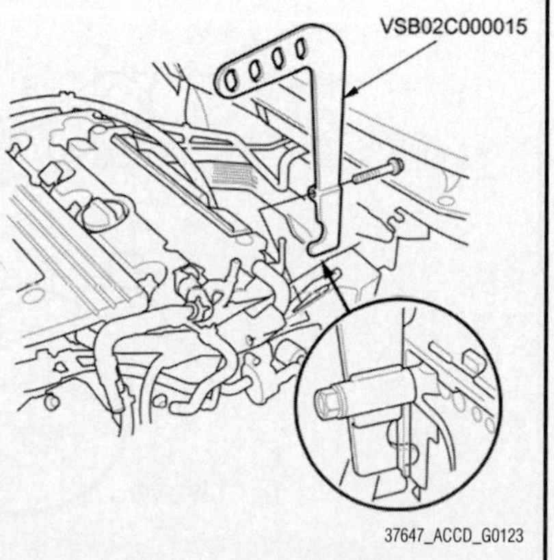
VSB02C000015
37647_ACCD_G0123

Fig. 385 Attach the engine hanger adapter to the threaded hole in the cylinder head

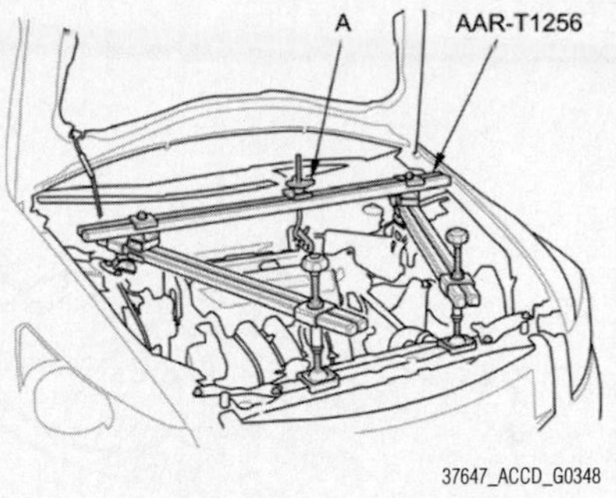
A
AAR-T1256
37647_ACCD_G0348

Fig. 386 Tighten the wing nut (A) by hand to lift and support the engine/transmission

(VSB02C000015) to the threaded hole in the cylinder head.

13. Install the engine support hanger (AAR-T1256), then attach the hook to the slotted hole in the engine hanger adapter. Tighten the wing nut by hand to lift and support the engine/transmission.

➡Be careful when working around the windshield.

14. Remove the engine mount bolt from the rear engine mount and the rear engine mount bracket.

➡Use a new engine mount bolt during reassembly.

15. Raise the vehicle on the lift to full height.

16. Remove the front splash shield.

17. Remove the nuts securing of the lower transmission mount.

18. Remove the exhaust pipe A hanger from the front subframe.

19. Attach the subframe adapter (VSB02C000016) to the subframe, hang the belt of the subframe adapter over the front of the subframe, then secure the belt with its stop.

20. Raise the jack, line up the slots in the front subframe adapter arms with the bolt holes on the jack base, then securely attach them with four bolts.

21. Remove the front subframe mounting bolts on both sides of the middle mount.

➡Use new mounting bolts during reassembly.

22. Disconnect both sides of the stabilizer link from the stabilizer bar.

23. Remove the flange bolts on both sides of the front subframe front stiffener.

24. Loosen the front side of the subframe mounting bolts to obtain a 0.79

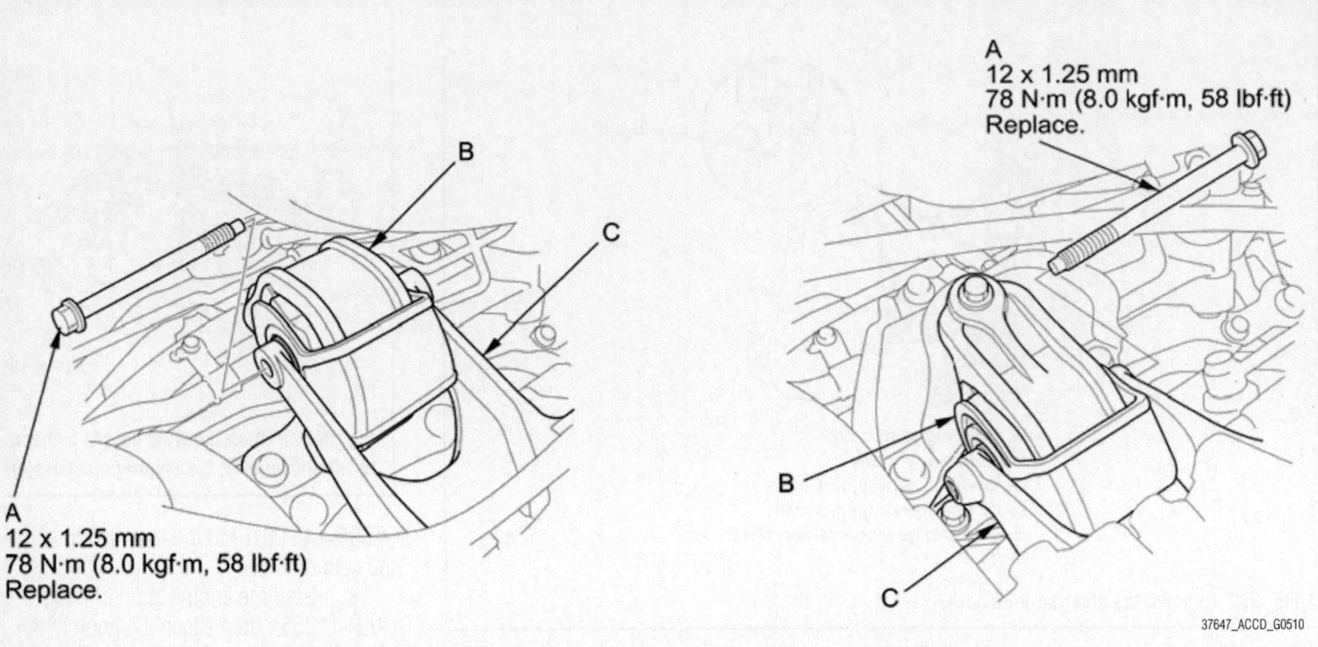
A
12 x 1.25 mm
78 N·m (8.0 kgf·m, 58 lbf·ft)
Replace.
B
C
A
12 x 1.25 mm
78 N·m (8.0 kgf·m, 58 lbf·ft)
Replace.
B
C
37647_ACCD_G0510

Fig. 387 Remove the engine mount bolt (A) from the rear engine mount (B) and the rear engine mount bracket (C)

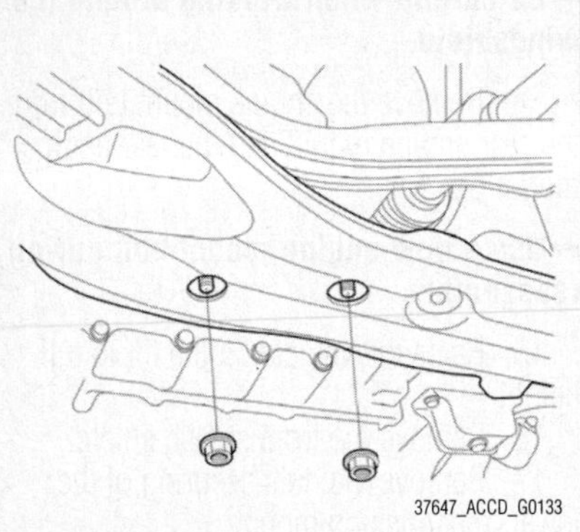
37647_ACCD_G0133

Fig. 388 Remove the nuts (A) securing of the lower transmission mount

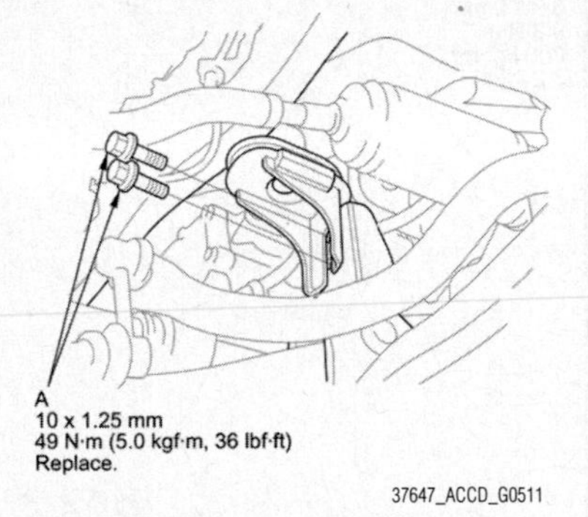
A
10 x 1.25 mm
49 N·m (5.0 kgf·m, 36 lbf·ft)
Replace.
37647_ACCD_G0511

Fig. 390 Remove the front subframe mounting bolts (A) on both sides of the middle mount

inches (20 mm) distance between the bolt seat and the mounting surface. Do not loosen the mounting bolts more than necessary.

25. Remove the flange bolts on both sides of the front subframe rear stiffener.

26. Loosen the rear side of the subframe mounting bolts to obtain a 1.18 inches (30 mm) distance between the bolt seat and the mounting surface. Do not loosen the mounting bolts more than necessary.

27. Lower the transmission jack with the front subframe adapter slowly until the front subframe has dropped about 1.18 inches (30 mm).

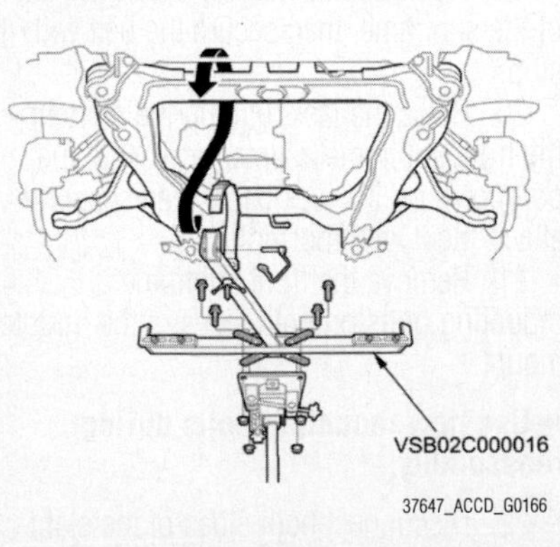
VSB02C000016
37647_ACCD_G0166

Fig. 389 Attach the subframe adapter to the subframe by looping the belt over the front of the subframe, then secure the belt with its stop

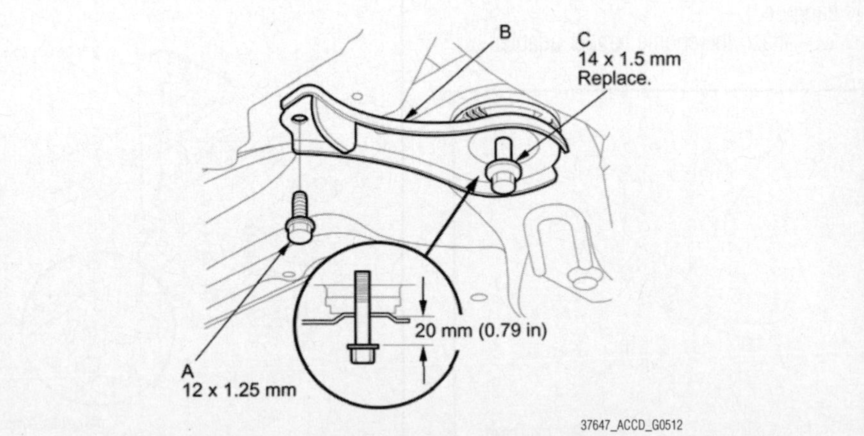
B
C
14 x 1.5 mm
Replace.
20 mm (0.79 in)
A
12 x 1.25 mm
37647_ACCD_G0512

Fig. 391 Remove the flange bolts (A) of the front subframe front stiffener (B), loosen subframe mounting bolts (C)

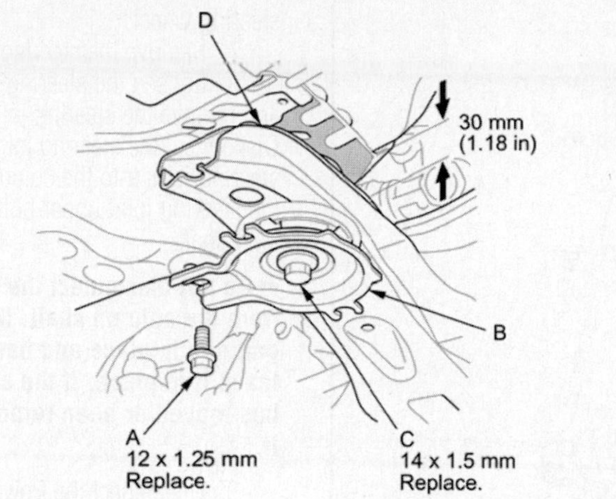
D
30 mm
(1.18 in)
B
A
12 x 1.25 mm
Replace.
C
14 x 1.5 mm
Replace.
A. Flange bolts
B. Front subframe rear stiffener
C. Subframe mounting bolts
D. Front subframe
37647_ACCD_G0513

Fig. 392 Remove the flange bolts of the front subframe rear stiffener, loosen the subframe mounting bolts

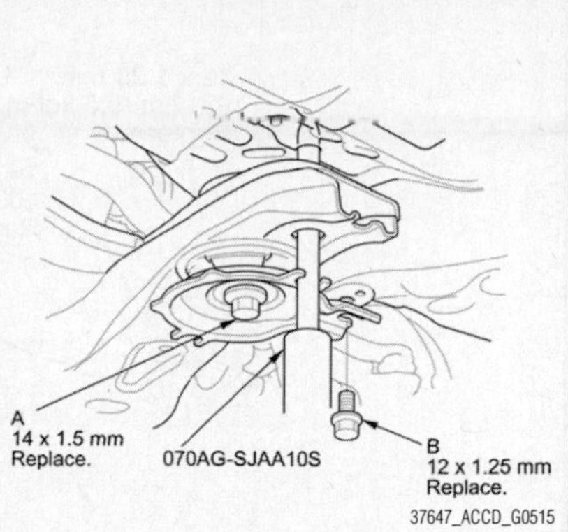
A
14 x 1.5 mm
Replace.
070AG-SJAA10S
B
12 x 1.25 mm
Replace.
37647_ACCD_G0515

Fig. 394 Align the front subframe using the subframe alignment pin

➡Do not lower the front subframe beyond the loosened subframe mounting bolts clearance.

28. Remove the flange bolts and the bushing holders, then remove the bushings.

➡During installation, align the paint marks on the stabilizer bar with the side of the bushings.

29. Move the stabilizer bar toward the passenger's side, and remove the stabilizer bar.

To install:

30. Install the stabilizer bar.

➡Note the right and left direction of the stabilizer bar. Note the direction of installation for the bushings.

31. Align the front subframe using the subframe alignment pin. Vertically install the subframe alignment pin, and align the right-rear corner of the front subframe and vehicle frame holes, then loosely tighten the new subframe mounting bolt until the front subframe contacts the body frame.

32. Loosely tighten the left-rear subframe mounting bolt using the same procedure as the right-rear with the subframe alignment pin.

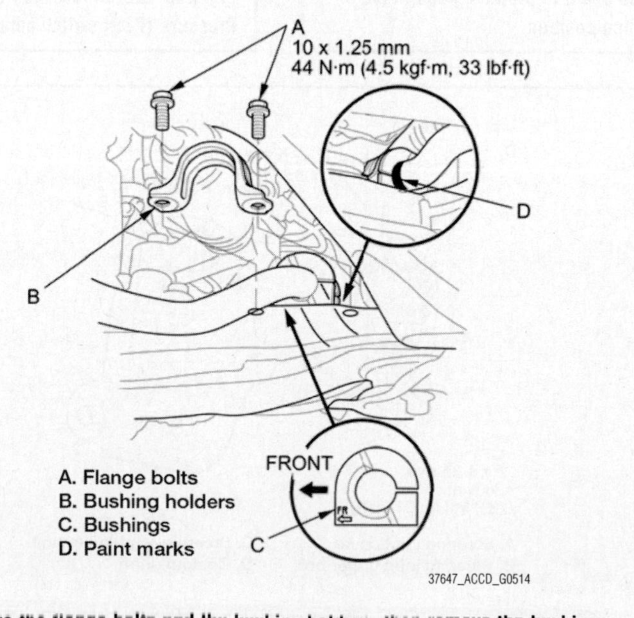
A
10 x 1.25 mm
44 N·m (4.5 kgf·m, 33 lbf·ft)
D
B
FRONT
C
A. Flange bolts
B. Bushing holders
C. Bushings
D. Paint marks
37647_ACCD_G0514

Fig. 393 Remove the flange bolts and the bushing holders, then remove the bushings

33. Loosely install the new 12 mm flange bolts to the subframe rear stiffener.

34. Torque the subframe mounting bolts to 76 ft. lbs. (103 Nm) starting with the right-rear bolt. Use the subframe alignment pin when tightening the rear side bolts.

➡Before tightening the new front side subframe mounting bolts, raise the jack and loosely install the 12 mm flange bolts to align the subframe front stiffener.

35. Check all of the front subframe mounting bolts, and retighten if necessary.

36. Install all of the removed parts in the reverse order of removal, and note these items:

- Refer to stabilizer link removal/installation to connect the stabilizer bar to the links.
- If the center guide is in place, use it to determine the steering joint installation angle.
- If the center guide is gone, check the steering joint installation angle.
- Check the steering wheel installation.
- When connecting the rear engine mount to the rear engine mount bracket, first lightly tighten the mounting bolt, then remove the engine support hanger, and tighten it to the specified torque.
- Before installing the wheel, clean the mating surfaces of the brake disc and inside of the wheel.

37. Do the battery terminal reconnection procedure, then turn the ignition switch to ON (II) and check that the SRS indicator should come on for about 6 seconds and then go off.

38. Check the wheel alignment, and adjust it if necessary.

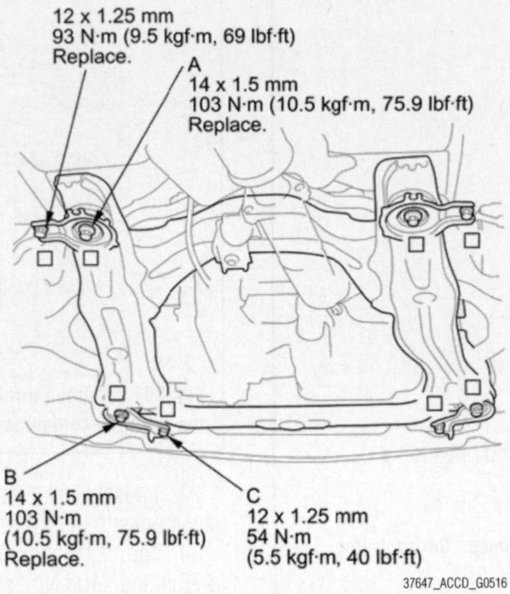
12 x 1.25 mm
93 N·m (9.5 kgf·m, 69 lbf·ft)
Replace.
A
14 x 1.5 mm
103 N·m (10.5 kgf·m, 75.9 lbf·ft)
Replace.
B
14 x 1.5 mm
103 N·m
(10.5 kgf·m, 75.9 lbf·ft)
Replace.
C
12 x 1.25 mm
54 N·m
(5.5 kgf·m, 40 lbf·ft)
37647_ACCD_G0516

Fig. 395 Install the rear side bolts (A), the front side subframe mounting bolts (B) and the flange bolts (C)

6-Cylinder Engine

See Figures 396 through 412.

Special Tools Required*:

- 2008 V6 Attachment Arm SIL02C000033
- Engine Hanger Balance Bar VSB02C000019
- Engine Support Hanger, A and Reds AAR-T1256
- Subframe Adapter VSB02C000016
- Subframe Alignment Pin 070AG-SJAA10S

➡*Available through the Honda Tool and Equipment Program, 888-424-6857.

➡Note these items during replacement:

- Be sure to remove the steering wheel before disconnecting the steering joint. Damage to the cable reel can occur.
- Lower the front subframe from the body, and replace the front stabilizer bar through the gap created by lowering the front subframe.

1. Remove the hood support rod, then use it as shown to prop the hood in the wide-open position.
2. Remove the front grille cover:
3. Do the battery terminal disconnection procedure.
4. Raise and support the vehicle.
5. Remove the front wheels.
6. Remove the driver's airbag and the steering wheel.
7. Remove the steering joint cover.
8. Loosen the steering joint upper bolt, and remove the steering joint lower bolt. Disconnect the steering joint by sliding the steering joint into the column shaft. Tighten the steering joint upper bolt to hold the column shaft.

➡Do not disconnect the steering joint from the column shaft. If the center guide is in place and has not moved, leave it in place. If the center guide has moved or been removed, discard it.

9. Disconnect the Power Steering Pressure (PSP) switch connector.
10. Remove the power steering pump outlet hose mounting bolt.
11. Remove the front strut brace.
12. Remove the connector bracket from the front cylinder head; use the bracket bolt hole to attach the engine hanger balance bar front arm.

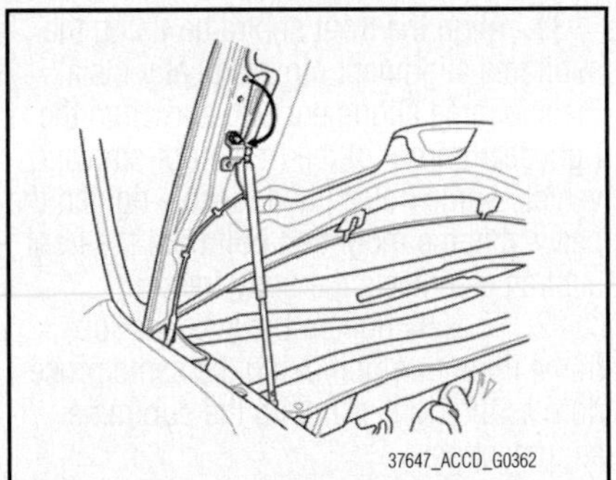
37647_ACCD_G0362

Fig. 396 Remove the hood support rod, and then use it to prop the hood in the wide-open position

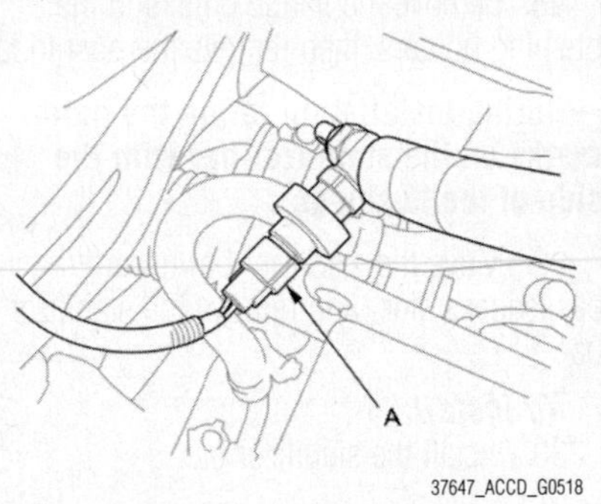
A
37647_ACCD_G0518

Fig. 398 Disconnect the Power Steering Pressure (PSP) switch connector (A)

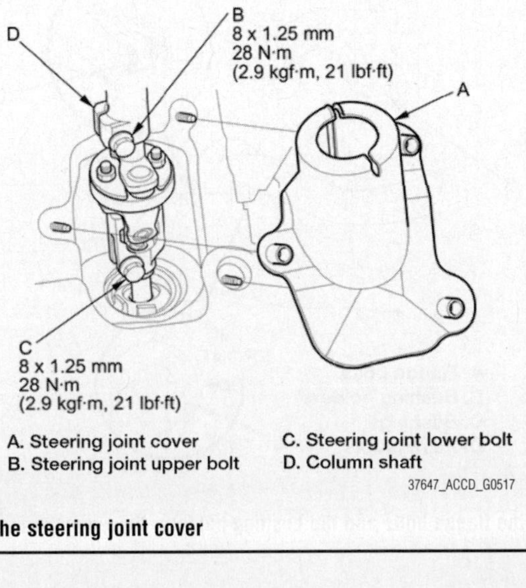
B
8 x 1.25 mm
28 N·m
(2.9 kgf·m, 21 lbf·ft)
D
A
C
8 x 1.25 mm
28 N·m
(2.9 kgf·m, 21 lbf·ft)
A. Steering joint cover
B. Steering joint upper bolt
C. Steering joint lower bolt
D. Column shaft
37647_ACCD_G0517

Fig. 397 Remove the steering joint cover

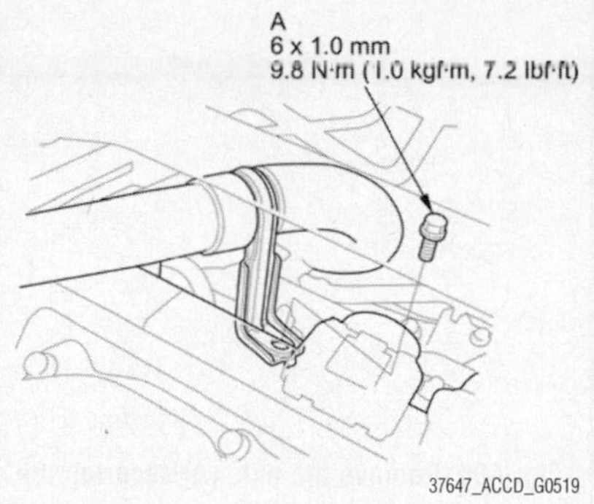
A
6 x 1.0 mm
9.8 N·m (1.0 kgf·m, 7.2 lbf·ft)
37647_ACCD_G0519

Fig. 399 Remove the power steering pump outlet hose mounting bolt (A)

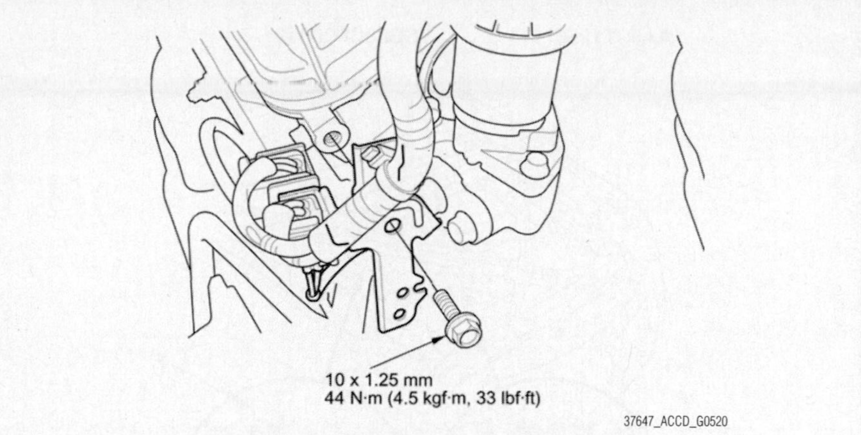
10 x 1.25 mm
44 N·m (4.5 kgf·m, 33 lbf·ft)
37647_ACCD_G0520

Fig. 401 Remove the connector bracket from the front cylinder head

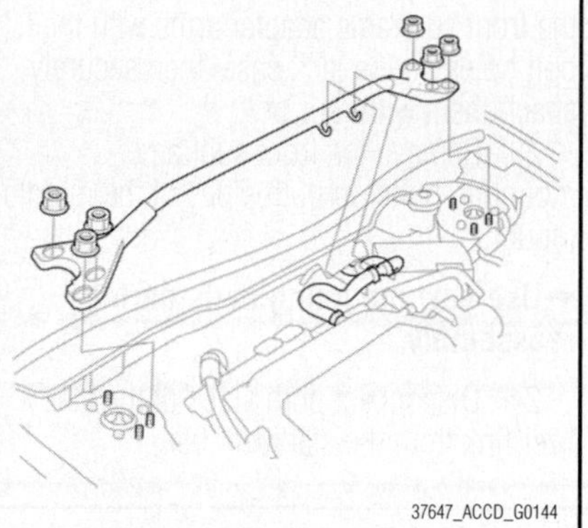
37647_ACCD_G0144

Fig. 400 Remove the front strut brace

13. Remove the harness clamp bracket from the rear cylinder head; use the bracket bolt hole to attach the 2008 V6 attachment arm (SIL02C000033).

14. Lift and support the engine with engine support hanger (AAR-T1256) and the engine hanger balance bar (VSB02C000019). Attach the front arm to the front cylinder head with several spacers and the bolt (10 x 1.25 mm). Attach the 2008 V6 attachment arm to the rear cylinder head and the bolt (8 x 1.25 mm), then attach the rear arm to the 2008 V6 attachment arm.

15. A/T: Remove the rear engine mount stop.

➡Use new nuts during reassembly.

16. Remove the engine mount bolt from the rear engine mount and the rear engine mount bracket.

➡Use a new engine mount bolt during reassembly.

17. Raise the vehicle on the lift to full height.

18. Remove the front splash shield.

19. Remove the nuts securing the lower transmission mount.

20. Remove the exhaust pipe A hanger from the front subframe.

21. Attach the subframe adapter (VSB02C000016) to the subframe, hang the

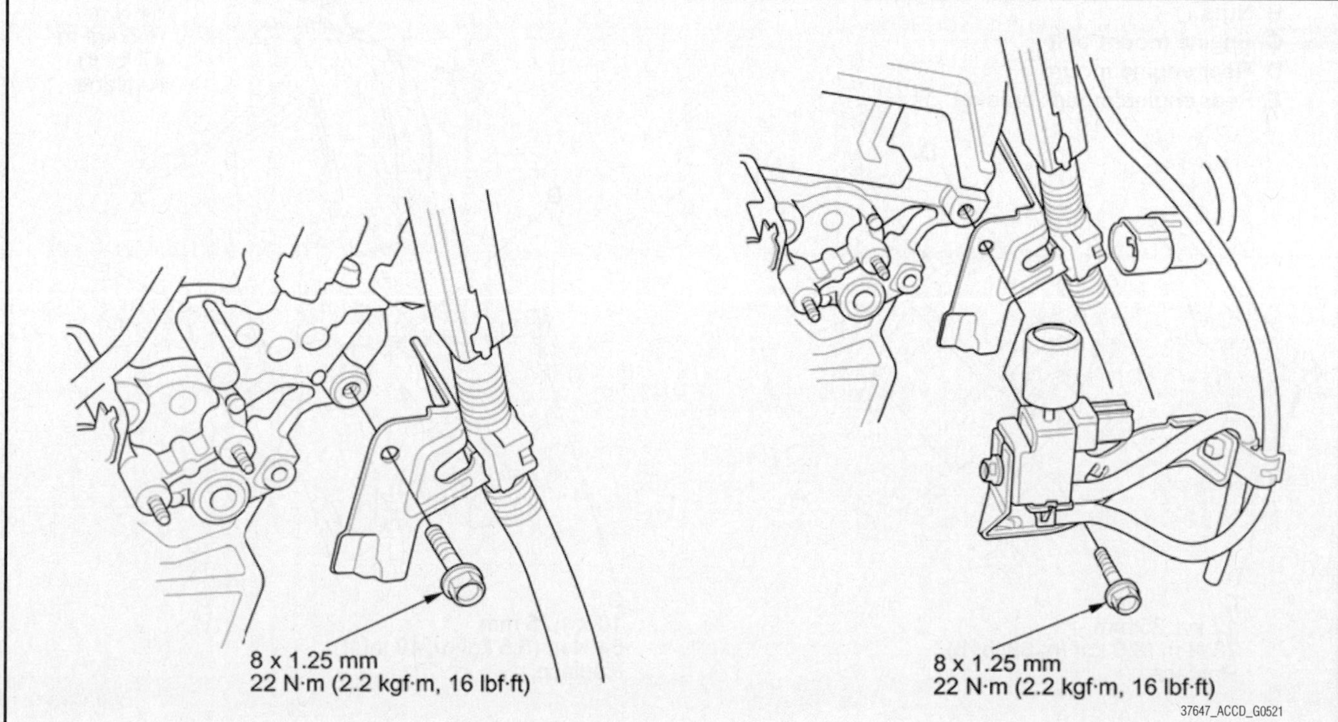
8 x 1.25 mm
22 N·m (2.2 kgf·m, 16 lbf·ft)
8 x 1.25 mm
22 N·m (2.2 kgf·m, 16 lbf·ft)
37647_ACCD_G0521

Fig. 402 Remove the harness clamp bracket from the rear cylinder head

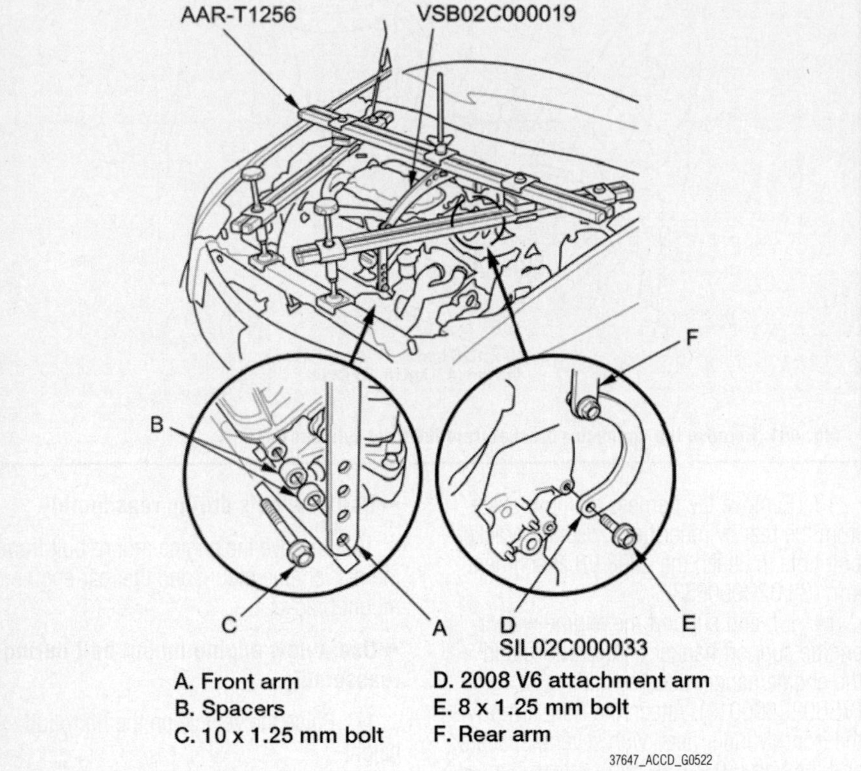
AAR-T1256
VSB02C000019
F
B
C
A
D
E
SIL02C000033
A. Front arm
B. Spacers
C. 10 x 1.25 mm bolt
D. 2008 V6 attachment arm
E. 8 x 1.25 mm bolt
F. Rear arm
37647_ACCD_G0522

Fig. 403 Lift and support the engine with engine support hanger and the engine hanger balance bar

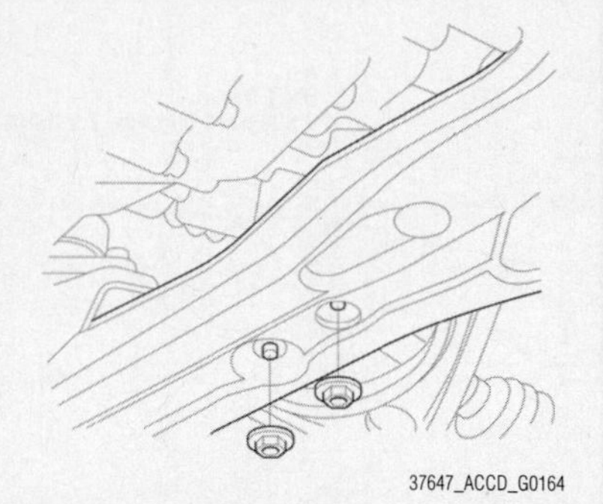
37647_ACCD_G0164

Fig. 405 Remove the nuts (A) securing the lower transmission mount

belt of the subframe adapter over the front of the subframe, then secure the belt with its stop.

22. Raise the jack, line up the slots in the front subframe adapter arms with the bolt holes on the jack base, then securely attach them with four bolts.

23. Remove the front subframe mounting bolts on both sides of the middle mount.

➡Use new mounting bolts during reassembly.

24. Disconnect both sides of the stabilizer link from the stabilizer bar.

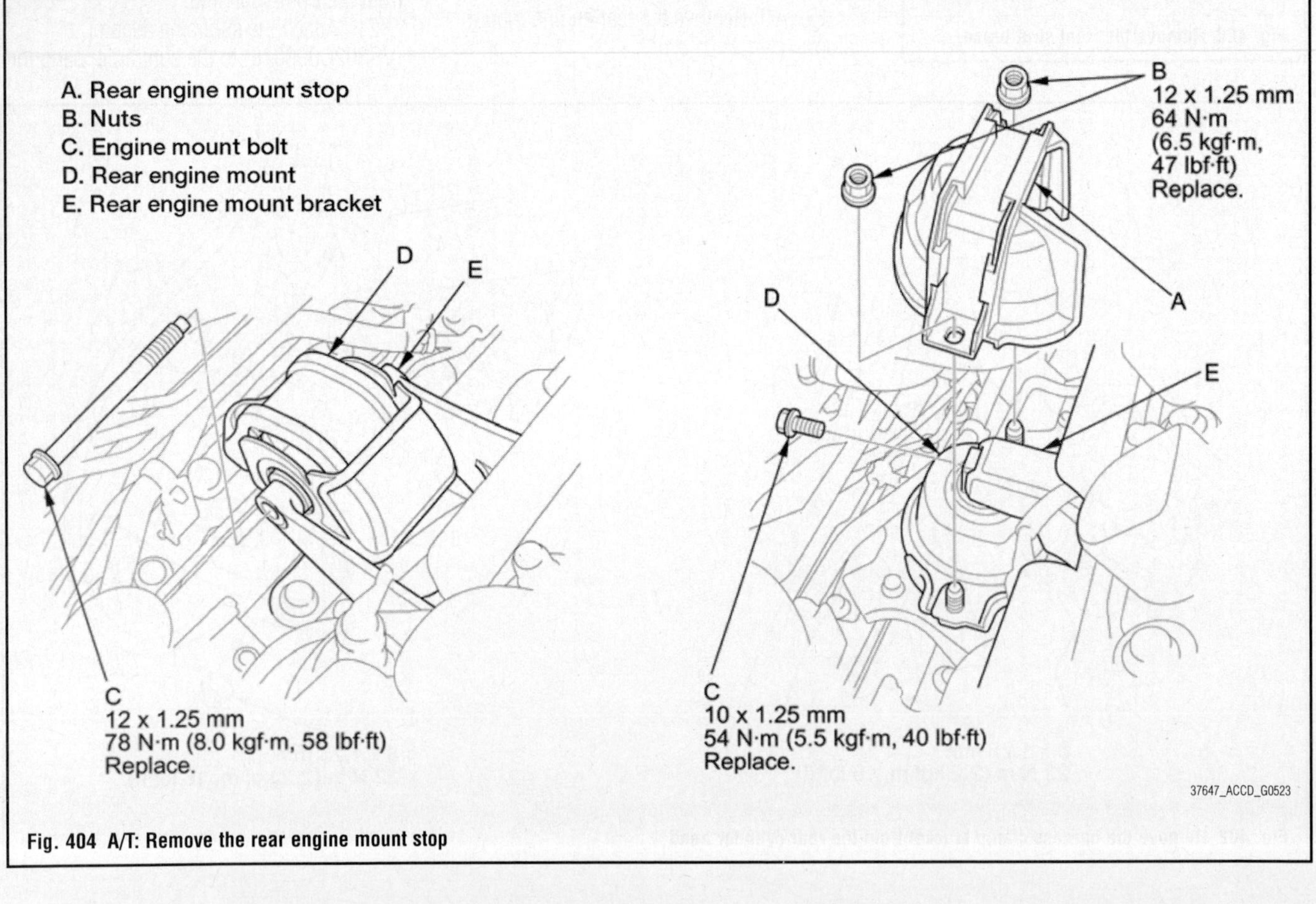
A. Rear engine mount stop
B. Nuts
C. Engine mount bolt
D. Rear engine mount
E. Rear engine mount bracket
D
E
C
12 x 1.25 mm
78 N·m (8.0 kgf·m, 58 lbf·ft)
Replace.
B
12 x 1.25 mm
64 N·m
(6.5 kgf·m,
47 lbf·ft)
Replace.
A
D
E
C
10 x 1.25 mm
54 N·m (5.5 kgf·m, 40 lbf·ft)
Replace.
37647_ACCD_G0523

Fig. 404 A/T: Remove the rear engine mount stop

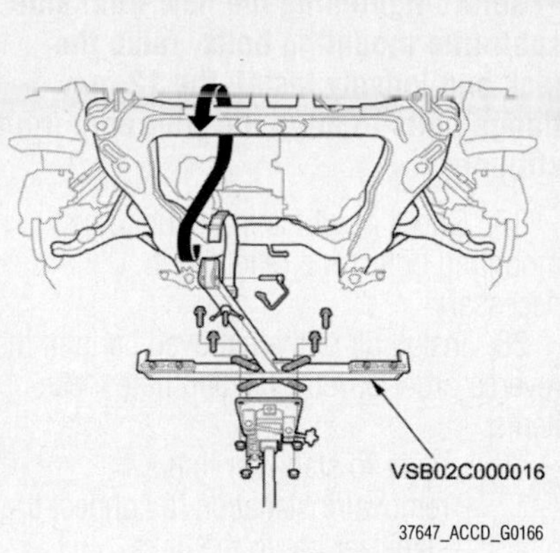
VSB02C000016
37647_ACCD_G0166

Fig. 406 Attach the subframe adapter to the subframe by looping the belt over the front of the subframe, then secure the belt with its stop

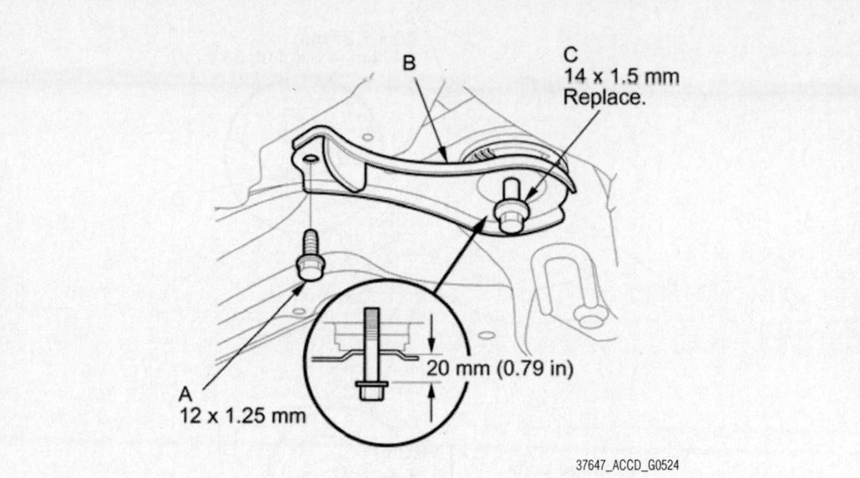
B
C
14 x 1.5 mm
Replace.
20 mm (0.79 in)
A
12 x 1.25 mm
37647_ACCD_G0524

Fig. 408 Remove the flange bolts (A) on both sides of the front subframe front stiffener (B); loosen the front side of the front subframe mounting bolts (C)

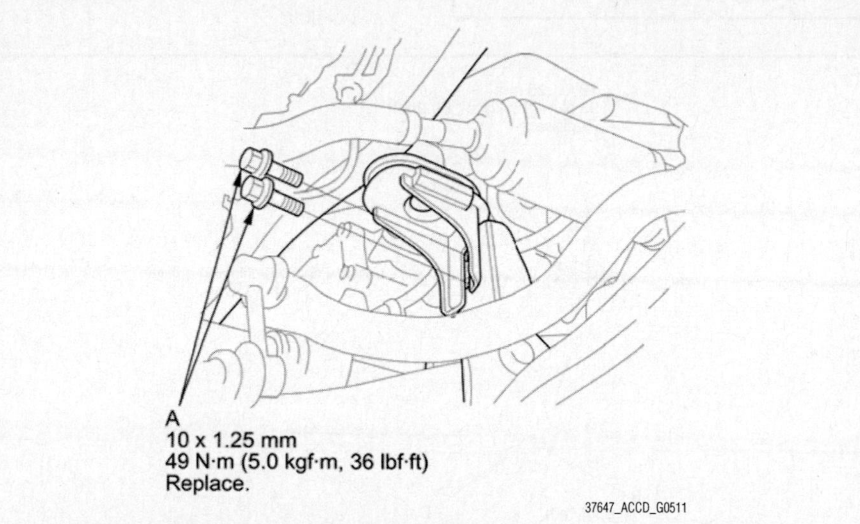
A
10 x 1.25 mm
49 N·m (5.0 kgf·m, 36 lbf·ft)
Replace.
37647_ACCD_G0511

Fig. 407 Remove the front subframe mounting bolts (A) on both sides of the middle mount

25. Remove the flange bolts on both sides of the front subframe front stiffener.
26. Loosen the front side of the front subframe mounting bolts to obtain a 0.79 inches (20 mm) distance between the bolt seat and the mounting surface. Do not loosen the mounting bolts more than necessary.
27. Remove the flange bolts on both sides of the front subframe rear stiffener.
28. Loose the rear side of the subframe mounting bolts to obtain a 1.18 inches (30 mm) distance between the bolt seat and the mounting surface. Do not loosen the mounting bolts more than necessary.
29. Lower the transmission jack with the front subframe adapter slowly until the front subframe has dropped about 1.18 inches (30 mm).

➡Do not lower the front subframe beyond the loosened subframe mounting bolts clearance.

30. Remove the flange bolts and the bushing holders, then remove the bushings.

➡During installation, align the paint marks on the stabilizer bar with the side of the bushings.

31. Move the stabilizer bar toward the passenger's side, and remove the stabilizer bar.

To install:

32. Install the stabilizer bar.

➡Note the right and left direction of the stabilizer bar. Note the direction of installation for the bushings.

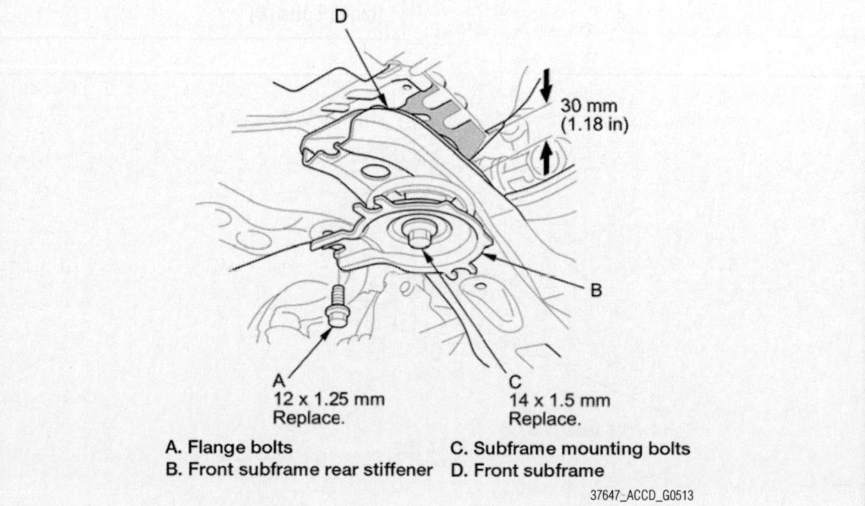
D
30 mm
(1.18 in)
B
A
12 x 1.25 mm
Replace.
C
14 x 1.5 mm
Replace.
A. Flange bolts
B. Front subframe rear stiffener
C. Subframe mounting bolts
D. Front subframe
37647_ACCD_G0513

Fig. 409 Remove the flange bolts on both sides of the front subframe rear stiffener

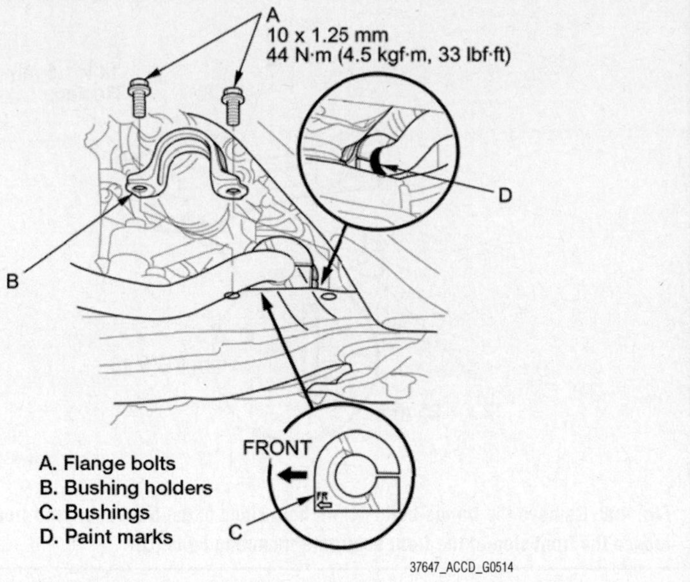
A
10 x 1.25 mm
44 N·m (4.5 kgf·m, 33 lbf·ft)
D
B
FRONT
C
A. Flange bolts
B. Bushing holders
C. Bushings
D. Paint marks
37647_ACCD_G0514

Fig. 410 Remove the flange bolts and the bushing holders, then remove the bushings

33. Align the front subframe using the subframe alignment pin. Vertically install the subframe alignment pin, and align the right-rear corner of the front subframe and vehicle frame holes, then loosely tighten the new subframe mounting bolt until the front subframe contacts the body frame.

34. Loosely tighten the left-rear subframe mounting bolt using the same procedure as the right-rear with the subframe alignment pin.

35. Loosely install the new 12 mm flange bolts to the subframe rear stiffener.

36. Torque the subframe mounting bolts to the specified torque starting with the right-rear bolt. Use the subframe alignment pin when tightening the rear side bolts.

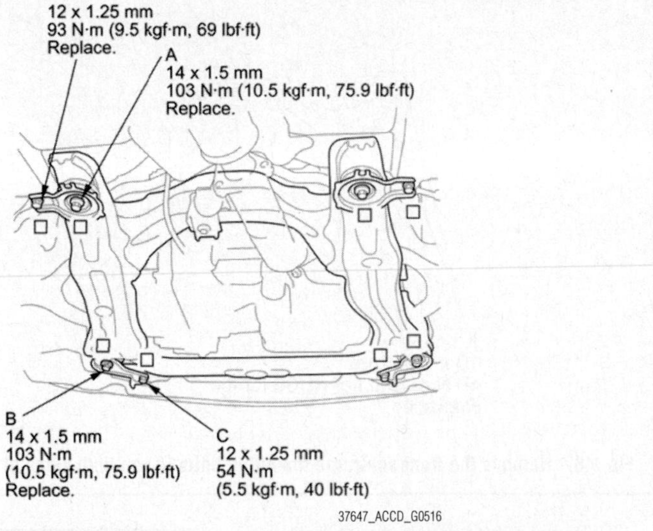
12 x 1.25 mm
93 N·m (9.5 kgf·m, 69 lbf·ft)
Replace.
A
14 x 1.5 mm
103 N·m (10.5 kgf·m, 75.9 lbf·ft)
Replace.
B
14 x 1.5 mm
103 N·m
(10.5 kgf·m, 75.9 lbf·ft)
Replace.
C
12 x 1.25 mm
54 N·m
(5.5 kgf·m, 40 lbf·ft)
37647_ACCD_G0516

Fig. 412 Install the rear side bolts (A), the front side subframe mounting bolts (B) and the flange bolts (C)

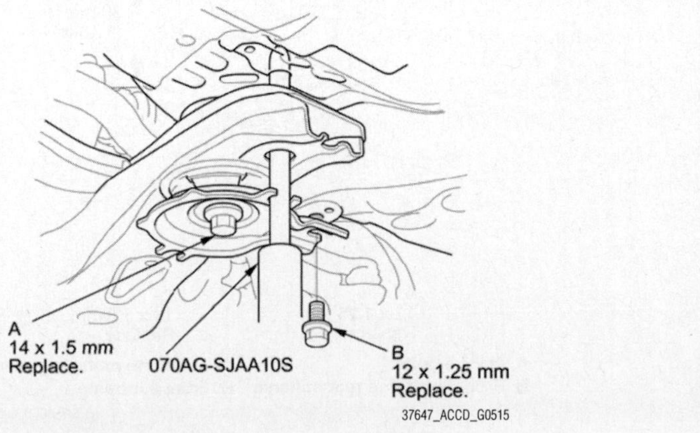
A
14 x 1.5 mm
Replace.
070AG-SJAA10S
B
12 x 1.25 mm
Replace.
37647_ACCD_G0515

Fig. 411 Align the front subframe using the subframe alignment pin

➡Before tightening the new front side subframe mounting bolts, raise the jack and loosely install the 12 mm flange bolts to align the subframe front stiffener.

37. Check all of the front subframe mounting bolts, and retighten if necessary.

38. Install all of the removed parts in the reverse order of removal, and note these items:

- Refer to stabilizer link removal/installation to connect the stabilizer bar to the links.
- If the center guide is in place, use it to determine the steering joint installation angle.
- If the center guide is gone, check the steering joint installation angle.
- Check the steering wheel installation.
- When connecting the rear engine mount to the rear engine mount bracket, first lightly tighten the mounting bolt, then remove the engine support hanger, and tighten it to the specified torque.
- Before installing the wheel, clean the mating surfaces of the brake disc and inside of the wheel.

39. Do the battery terminal reconnection procedure, then turn the ignition switch to ON (II) and check that the SRS indicator should comes on for about 6 seconds and then goes off.

40. Check the wheel alignment, and adjust it if necessary.

UPPER BALL JOINT

REMOVAL & INSTALLATION

See Figures 413 through 415.

Special Tools Required:

- Ball joint thread protector, 14 mm 07AAE-SJAA100
- Ball joint thread protector, 12 mm 07AAF-SDAA100
- Ball joint thread protector, 10 mm 07AAF-SECA120
- Ball joint remover, 32 mm 07MAC-SL0A102
- Ball joint remover, 28 mm 07MAC-SL0A202

➡Always use a ball joint remover to disconnect a ball joint. Do not strike the housing or any other part of the ball joint connection to disconnect it.

1. Install a hex nut or the ball joint thread protector onto the threads of the ball joint.

➡Using a hex nut, make sure the nut is flush with the ball joint pin end to prevent damage to the threaded end of the ball joint pin.

2. Apply grease to the ball joint remover on the areas shown. This will ease the installation of the tool, and prevent damage to the pressure bolt threads.
3. Loosen the pressure bolt, and install the ball joint remover as shown. Insert the jaws carefully, making sure not to damage the ball joint boot. Adjust the jaw spacing by turning the adjusting bolt.

➡Fasten the safety chain securely to a suspension arm or the subframe. Do not fasten it to a brake line or wire harness.

4. After adjusting the adjusting bolt, make sure the head of the adjusting bolt is in the position shown to allow the jaw to pivot.
5. With a wrench, tighten the pressure bolt until the ball joint pin pops loose from the ball joint connecting hole. If necessary, apply penetrating type lubricant to loosen the ball joint pin.

➡Do not use pneumatic or electric tools on the pressure bolt.

6. Remove the ball joint remover, then remove the nut or the ball joint thread protector from the end of the ball joint pin, and pull the ball joint out of the ball joint connecting hole. Inspect the ball joint boot, and replace it if damaged.

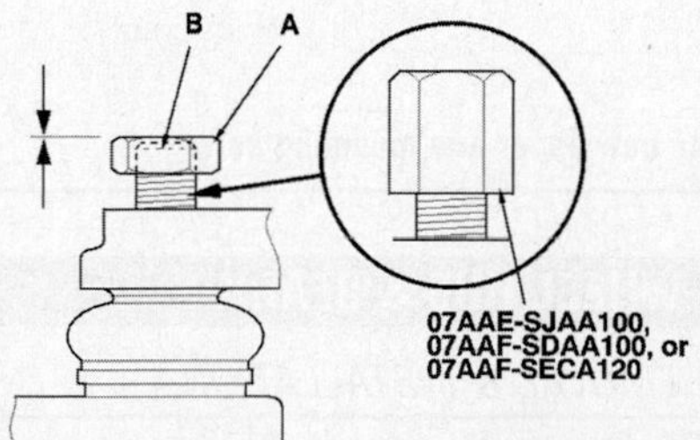
B
A
07AAE-SJAA100,
07AAF-SDAA100, or
07AAF-SECA120

37647_ATSX_G0663

Fig. 413 Install a hex nut (A) or the ball joint thread protector onto the threads of the ball joint (B)

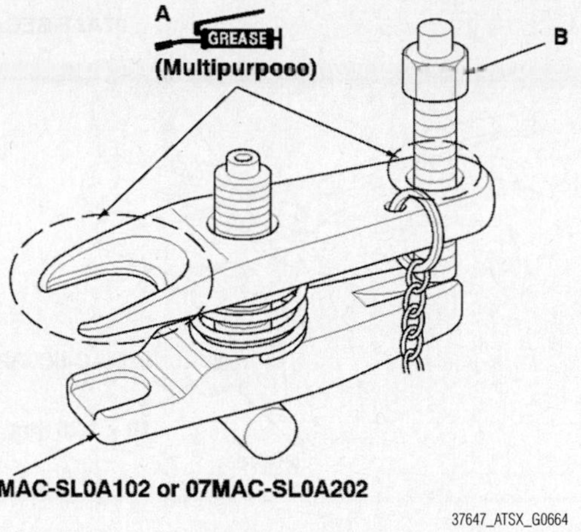
A
GREASE
(Multipurpose)
B
07MAC-SL0A102 or 07MAC-SL0A202

37647_ATSX_G0664

Fig. 414 Apply grease to the ball joint remover on the areas shown (A)

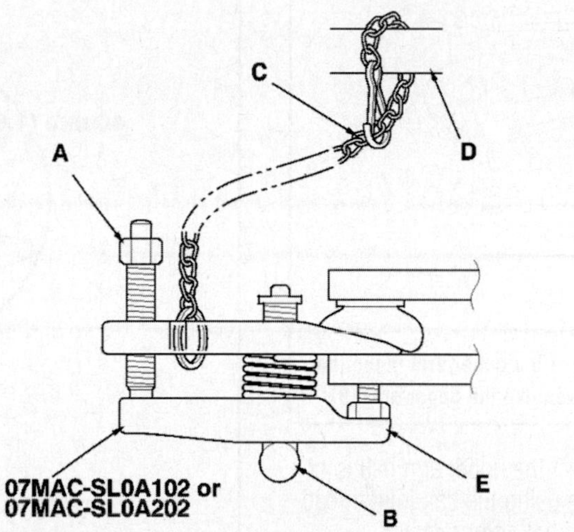
C
A
D
07MAC-SL0A102 or
07MAC-SL0A202
E
B

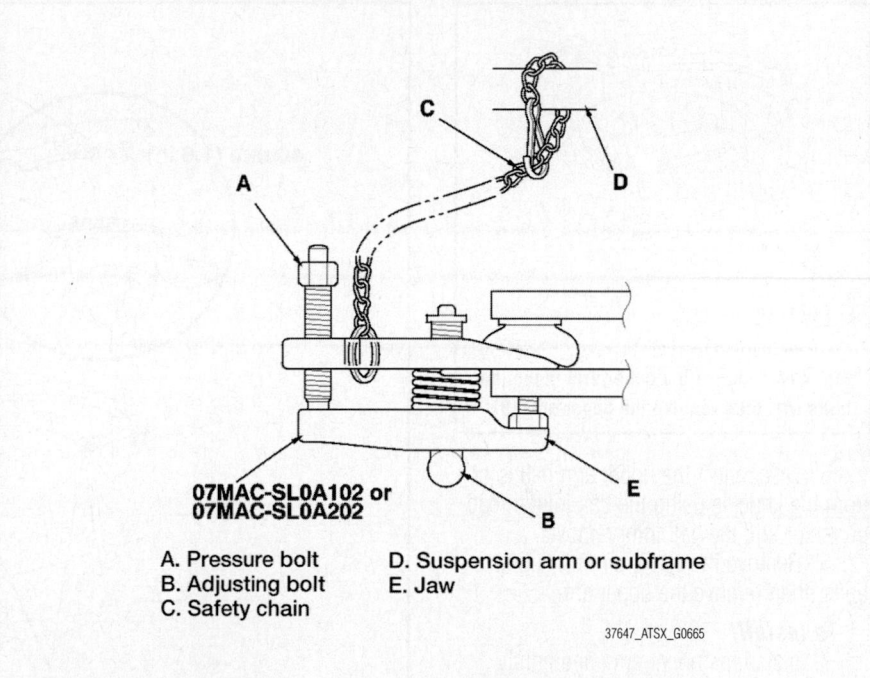

A. Pressure bolt
B. Adjusting bolt
C. Safety chain
D. Suspension arm or subframe
E. Jaw

37647_ATSX_G0665

Fig. 415 Loosen the pressure bolt, and install the ball joint remover as shown

UPPER CONTROL ARM

REMOVAL & INSTALLATION

See Figures 416 through 418.

Special Tools Required:

- Ball joint thread protector, 10 mm 07AAF-SECA120
- Ball joint remover, 28 mm 07MAC-SL0A202

1. Raise the vehicle on a lift.
2. Remove the front wheel.
3. Remove the front damper/spring.
4. Remove the cotter pin from the upper arm ball joint, then remove the castle nut.

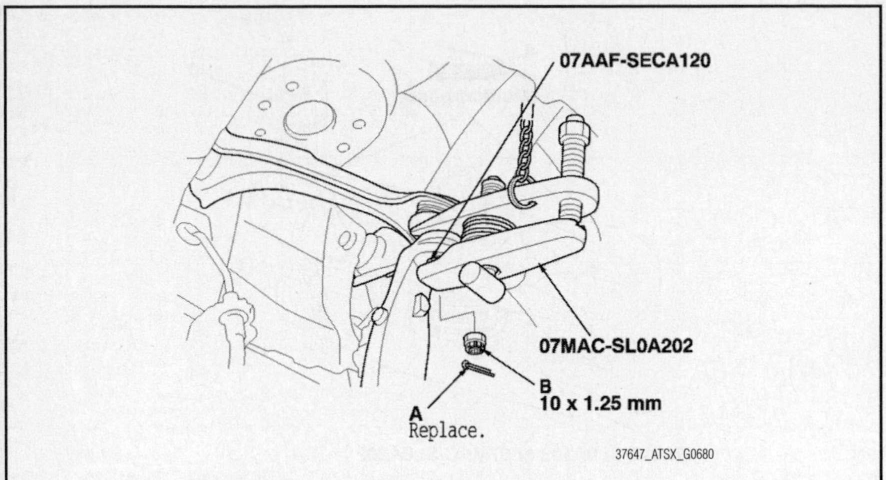
07AAF-SECA120
07MAC-SL0A202
B
10 x 1.25 mm
A
Replace.
37647_ATSX_G0680

Fig. 416 Remove the cotter pin (A) from the upper arm ball joint, then remove the castle nut (B)

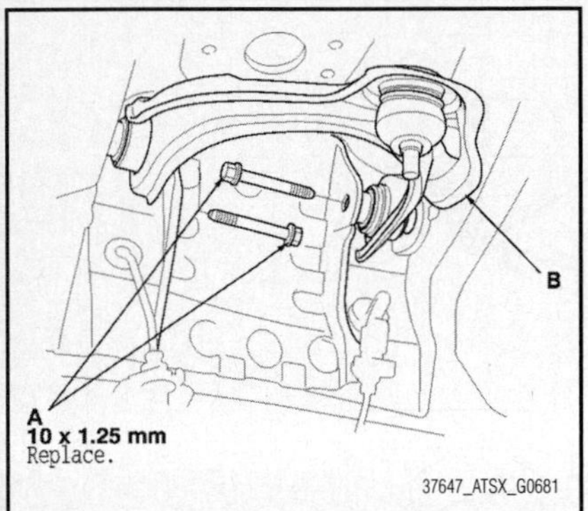
B
A
10 x 1.25 mm
Replace.
37647_ATSX_G0681

Fig. 417 Remove the upper arm mounting bolts (A), then remove the upper arm (B)

5. Disconnect the upper arm ball joint from the knuckle using the ball joint thread protector and the ball joint remover.
6. Remove the upper arm mounting bolts, then remove the upper arm.

To install:

7. Install the upper arm, and lightly tighten the new upper arm mounting bolts, then connect the knuckle, and lightly tighten the castle nut.

➡Be careful not to damage the ball joint boot when connecting the knuckle. Before connecting the ball joint, degrease the threaded section and the tapered portion of the ball joint pin, the ball joint connecting hole, the threaded section and the mating surfaces of the castle nut.

8. Place a floor jack under the lower arm, and raise the suspension until the clearance between the top of the upper arm ball joint and the backside of the fender cut out point is 1.6 inches (40 mm), then tighten the upper arm mounting bolts to 23 ft. lbs. (31 Nm).

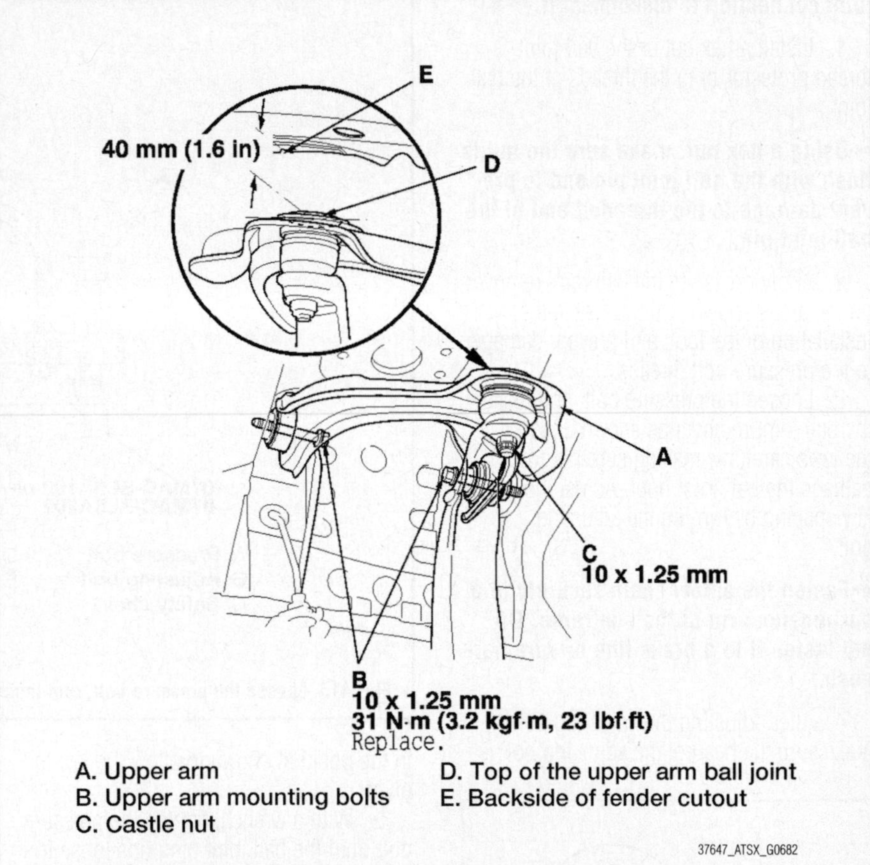
E
40 mm (1.6 in)
D
A
C
10 x 1.25 mm
B
10 x 1.25 mm
31 N·m (3.2 kgf·m, 23 lbf·ft)
Replace.
A. Upper arm
B. Upper arm mounting bolts
C. Castle nut
D. Top of the upper arm ball joint
E. Backside of fender cutout
37647_ATSX_G0682

Fig. 418 Install the upper arm, and lightly tighten the new upper arm mounting bolts

➡To measure the specified clearance, temporarily remove the front inner fender.

9. Lower the floor jack
10. Install the front damper/spring.
11. Place the floor jack under the lower arm, and raise the suspension to load it with the vehicle's weight.
12. Tighten the castle nut on the upper arm ball joint to 33–38 ft. lbs. (44–52 Nm).

➡Torque the castle nut to the lower torque specification, then tighten it only far enough to align the slot with the ball joint pin hole. Do not align the castle nut by loosening it. Insert the new cotter pin into the ball joint pin hole from the front to the rear of the vehicle, and bend its end. Check the ball joint pin hole direction before connecting the ball joint.

13. Clean the mating surfaces of the brake disc and the inside of the wheel, then install the front wheel.
14. Check the wheel alignment, and adjust it if necessary.

WHEEL HUB & BEARING

REMOVAL & INSTALLATION

See Figures 419 through 425.

1. Remove the knuckle/hub.
2. Separate the hub from the knuckle using the hub dis/assembly pin and a hydraulic press. Hold the knuckle with the

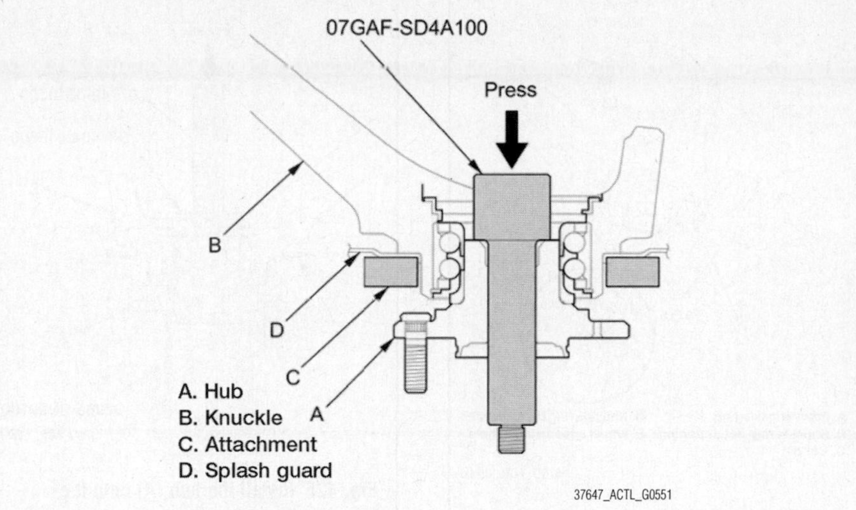
07GAF-SD4A100
Press
B
D
C
A
A. Hub
B. Knuckle
C. Attachment
D. Splash guard
37647_ACTL_G0551

Fig. 419 Separate the hub from the knuckle using the hub dis/assembly pin and a hydraulic press

attachment of the hydraulic press or equivalent tool. Be careful not to damage or deform the splash guard. Hold onto the hub to keep it from falling when pressed clear.

3. Press the wheel bearing inner race off of the hub using the hub dis/assembly pin, a commercially available bearing separator, and a press.

4. Remove the snap ring and the splash guard from the knuckle.

5. Press the wheel bearing out of the knuckle using the attachment, the driver handle, and a press.

6. Wash the knuckle and the hub thoroughly in high flash point solvent before reassembly.

7. Press a new wheel bearing into the knuckle using the old bearing, a steel plate, the attachment, the support base, and a press.

➡Install the wheel bearing with the wheel speed sensor magnetic encoder (brown color), toward the inside of the knuckle. Remove any oil, grease, dust, metal debris, and other foreign material from the magnetic encoder surface.

Keep any magnetic tools away from the magnetic encoder surface. Be careful not to damage the magnetic encoder surface when you insert the wheel bearing.

8. Check the front knuckle ring for damage or deformation, and replace it if necessary.

➡When installing the new front knuckle ring, position the knuckle ring notch portion toward the cut out near

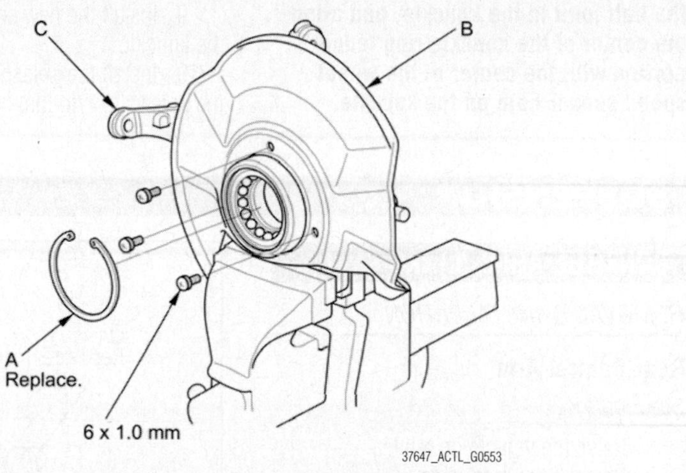
C
B
A
Replace.
6 x 1.0 mm
37647_ACTL_G0553

Fig. 421 Remove the snap ring (A) and the splash guard (B) from the knuckle (C)

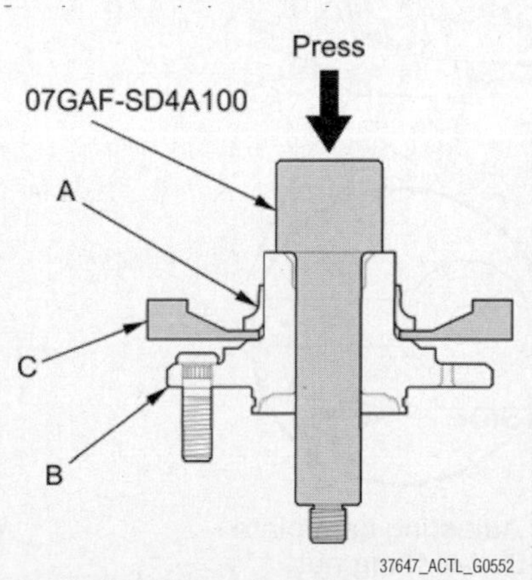
Press
07GAF-SD4A100
A
C
B
37647_ACTL_G0552

Fig. 420 Press the wheel bearing inner race (A) off of the hub (B) using the hub dis/assembly pin, a commercially available bearing separator (C), and a press

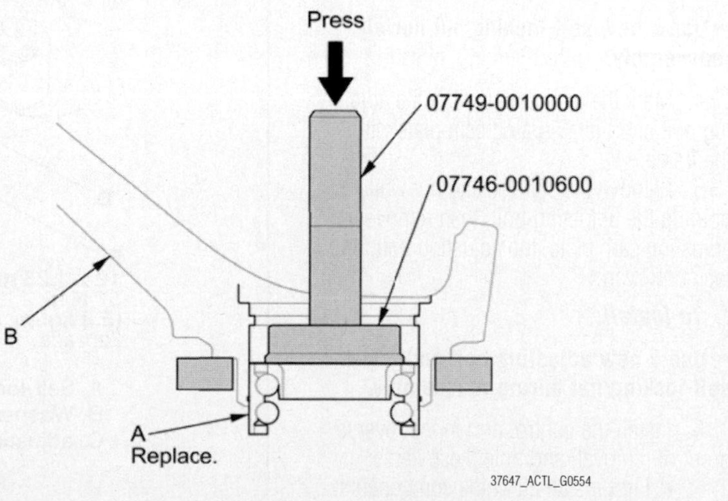
Press
07749-0010000
07746-0010600
B
A
Replace.
37647_ACTL_G0554

Fig. 422 Press the wheel bearing (A) out of the knuckle (B) using the attachment, the driver handle, and a press

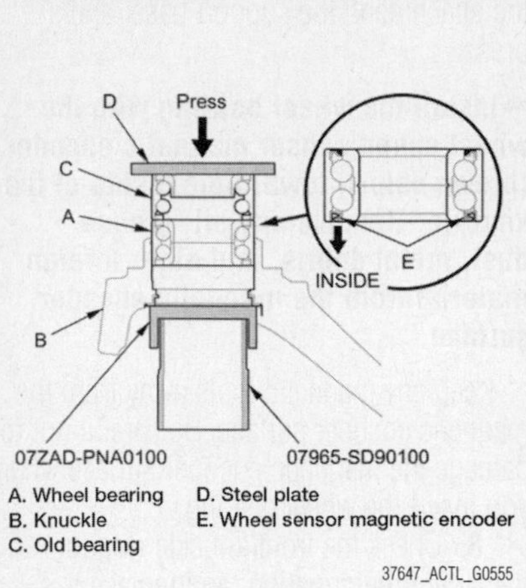

Press
D
C
A
B
E
INSIDE
07ZAD-PNA0100
07965-SD90100
A. Wheel bearing
B. Knuckle
C. Old bearing
D. Steel plate
E. Wheel sensor magnetic encoder
37647_ACTL_G0555

Fig. 423 Press a new wheel bearing into the knuckle using the old bearing, a steel plate, the attachment, the support base, and a press

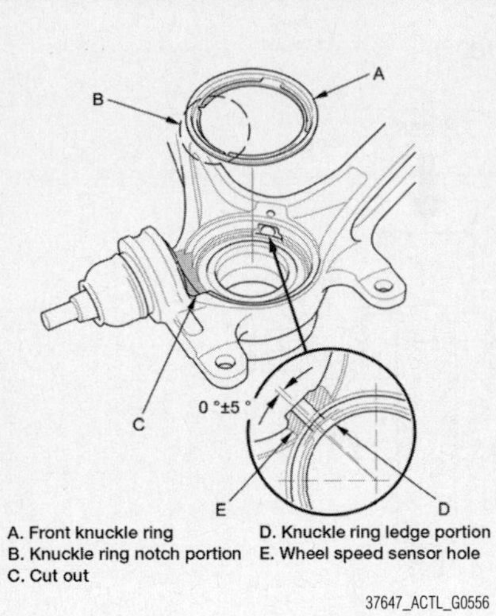

A
B
C
0 °±5 °
E
D
A. Front knuckle ring
B. Knuckle ring notch portion
C. Cut out
D. Knuckle ring ledge portion
E. Wheel speed sensor hole
37647_ACTL_G0556

Fig. 424 Check the front knuckle ring for damage or deformation, and replace it if necessary

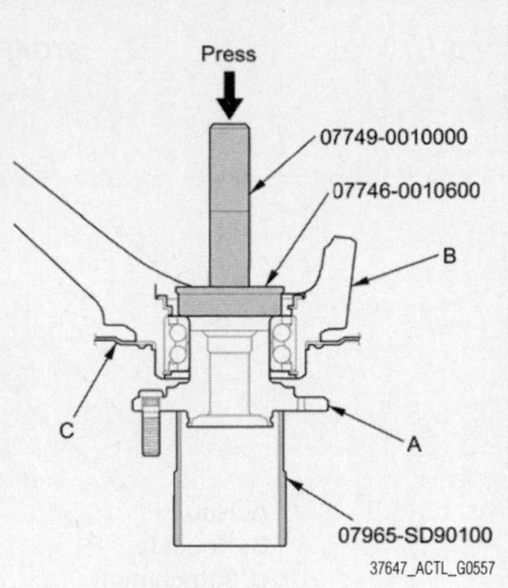

Press
07749-0010000
07746-0010600
B
C
A
07965-SD90100
37647_ACTL_G0557

Fig. 425 Install the hub (A) onto the knuckle (B) using the attachment, the driver handle, the support base, and a hydraulic press

the ball joint in the knuckle, and align the center of the knuckle ring ledge portion with the center of the wheel speed sensor hole on the knuckle.

9. Install the new snap ring securely in the knuckle.
10. Install the splash guard, and tighten the screws to 7 ft. lbs. (10 Nm).
11. Install the hub onto the knuckle using the attachment, the driver handle, the support base, and a hydraulic press. Be careful not to damage the splash guard.
12. Install the knuckle/hub.

SUSPENSION — REAR SUSPENSION

CONTROL ARMS/LINKS

REMOVAL & INSTALLATION

Rear Control Arm

See Figure 426.

1. Raise the vehicle on a lift.
2. Remove the rear wheel.
3. Remove the control arm mounting self-locking nut and the washer from the knuckle side.

➡Use a new self-locking nut during reassembly.

4. Mark the cam positions of the adjusting bolt and the adjusting cam plate with the frame.
5. Remove the self-locking nut while holding the adjusting bolt, then remove the adjusting cam plate, the adjusting bolt, and the control arm.

To install:

➡Use a new adjusting bolt and a new self-locking nut during reassembly.

6. Install the control arm in the reverse order of removal, and note these items:
   - First install all of the components, and lightly tighten the bolts and the nuts, then raise the suspension to

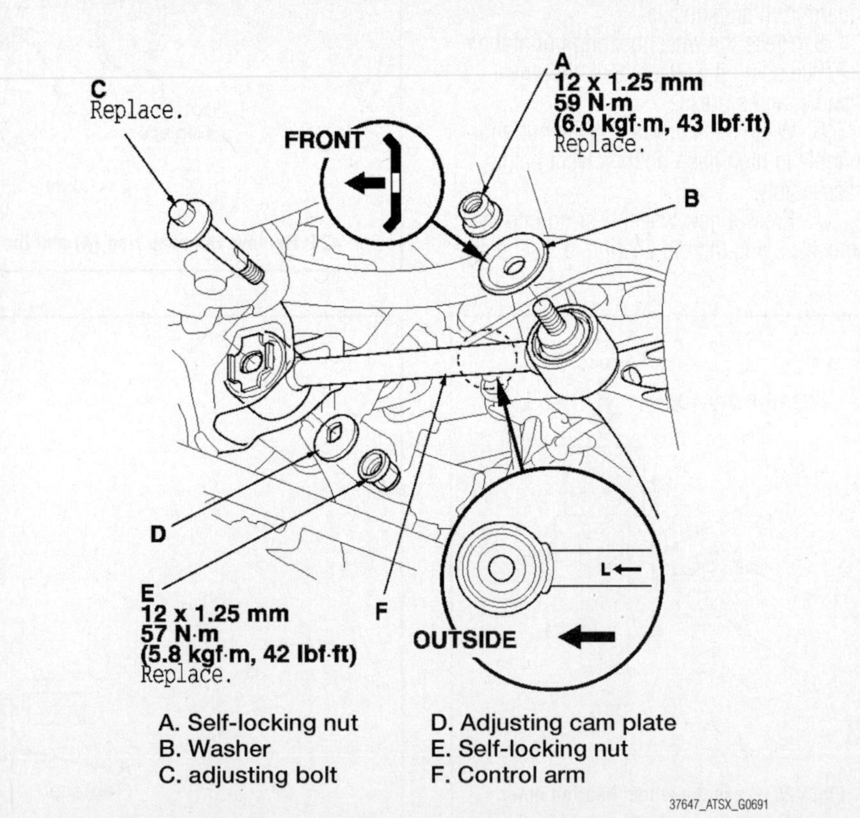

C
Replace.
FRONT
A
12 x 1.25 mm
59 N·m
(6.0 kgf·m, 43 lbf·ft)
Replace.
B
D
E
12 x 1.25 mm
57 N·m
(5.8 kgf·m, 42 lbf·ft)
Replace.
F
OUTSIDE
L←
A. Self-locking nut
B. Washer
C. adjusting bolt
D. Adjusting cam plate
E. Self-locking nut
F. Control arm
37647_ATSX_G0691

Fig. 426 Remove the self-locking nut while holding the adjusting bolt, then remove the adjusting cam plate, the adjusting bolt, and the control arm

load it with the vehicle's weight before fully tightening to the specified torque.

- Position the extended surfaces of the cam on the adjusting bolt and the adjusting cam plate facing down.
- Align the cam positions of the adjusting bolt and the adjusting cam plate with the marked positions on the frame when tightening the self-locking nut.
- Before installing the wheel, clean the mating surfaces on the brake disc and the inside of the wheel.

7. Check the wheel alignment, and adjust it if necessary.

Rear Lower Arm A

See Figure 427.

1. Raise the vehicle on a lift.
2. Remove the rear wheel.
3. Remove the parking brake cable mounting bolt.
4. Remove the lower arm A mounting bolts, then remove lower arm A.

To install:

➡Use new mounting bolts during reassembly.

5. Install lower arm A in the reverse order of removal, and note these items:
   - First install all of the components, and lightly tighten the bolts, then raise the suspension to load it with the vehicle's weight before fully tightening to the specified torque.
   - Before installing the wheel, clean the mating surfaces on the brake disc and the inside of the wheel.
6. Check the wheel alignment, and adjust it if necessary.

Rear Lower Arm B

See Figure 428.

1. Raise the vehicle on a lift.
2. Remove the rear wheel.
3. Remove the lower arm B mounting bolts, then remove lower arm B.

To install:

➡Use new mounting bolts during reassembly.

4. Install lower arm B in the reverse order of removal, and note these items:
   - First install all of the components, and lightly tighten the bolts, then raise the suspension to load it with the vehicle's weight before fully tightening to the specified torque.
   - Make sure the clearance between lower arm B and the parking brake cable is more than 0.20 inches (5 mm).
   - Before installing the wheel, clean the mating surfaces on the brake disc and the inside of the wheel.
5. Check the wheel alignment, and adjust it if necessary.

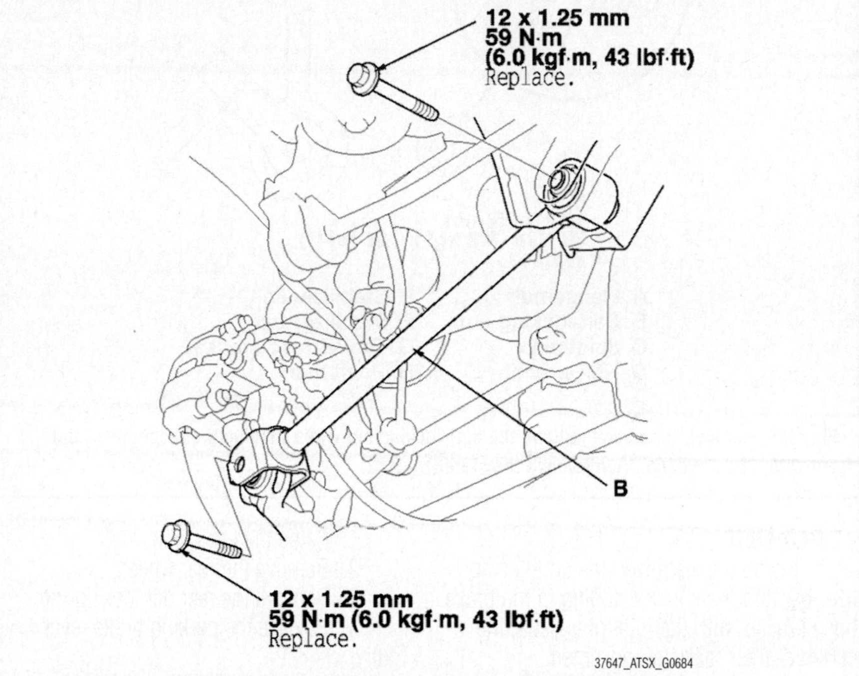
12 x 1.25 mm
59 N·m
(6.0 kgf·m, 43 lbf·ft)
Replace.
B
12 x 1.25 mm
59 N·m (6.0 kgf·m, 43 lbf·ft)
Replace.
37647_ATSX_G0684

Fig. 428 Remove the lower arm B mounting bolts, then remove lower arm B

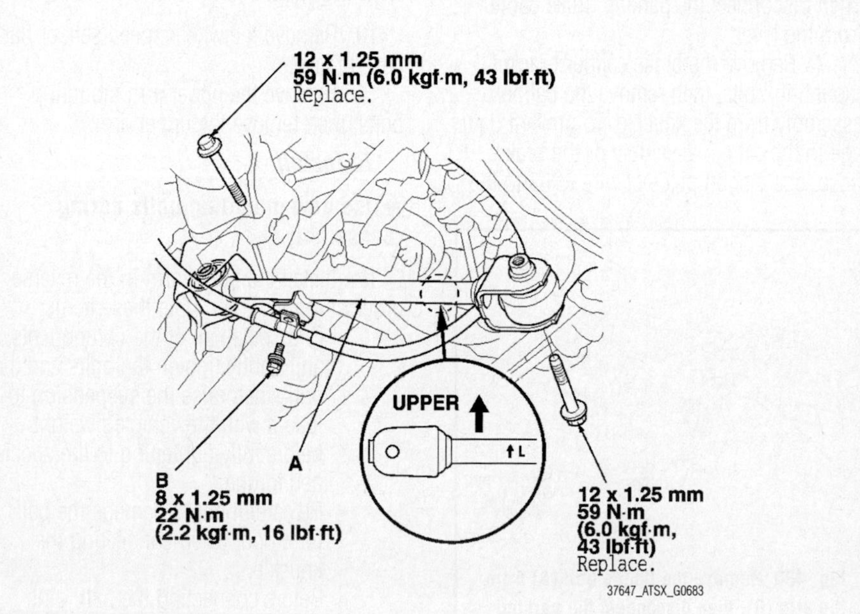
12 x 1.25 mm
59 N·m (6.0 kgf·m, 43 lbf·ft)
Replace.
UPPER
L
A
B
8 x 1.25 mm
22 N·m
(2.2 kgf·m, 16 lbf·ft)
12 x 1.25 mm
59 N·m
(6.0 kgf·m,
43 lbf·ft)
Replace.
37647_ATSX_G0683

Fig. 427 Remove the lower arm A mounting bolts, then remove lower arm A

Rear Stabilizer Link

See Figure 436.

1. Raise the vehicle on a lift.
2. Remove the rear wheel.
3. Remove the flange nut and the self-locking nut while holding the respective joint pin with a hex wrench, then remove the stabilizer link.

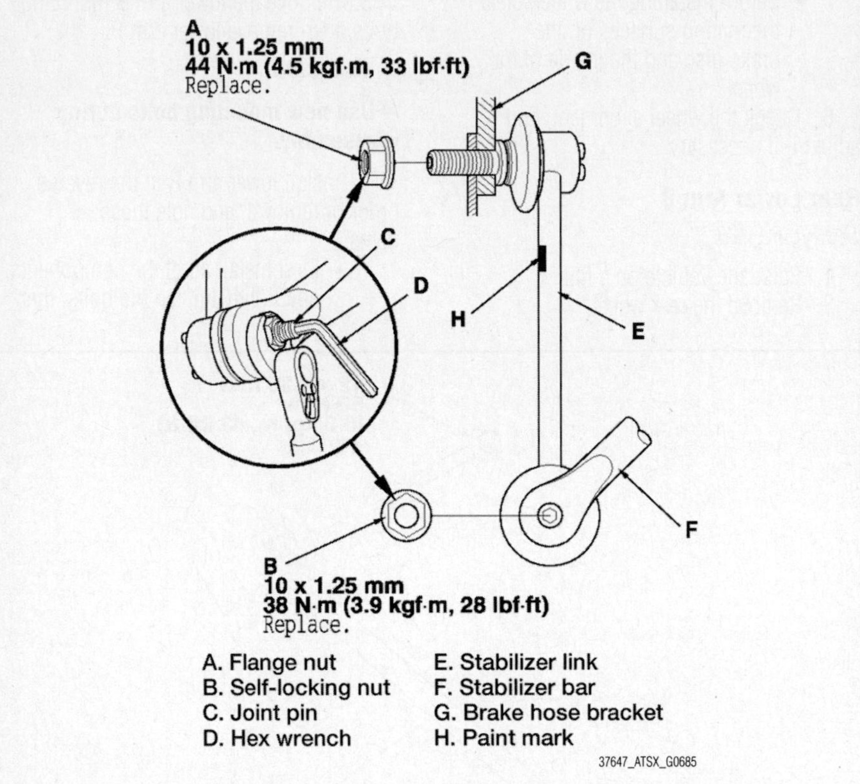
A
10 x 1.25 mm
44 N·m (4.5 kgf·m, 33 lbf·ft)
Replace.
G
C
D
H
E
F
B
10 x 1.25 mm
38 N·m (3.9 kgf·m, 28 lbf·ft)
Replace.
A. Flange nut
B. Self-locking nut
C. Joint pin
D. Hex wrench
E. Stabilizer link
F. Stabilizer bar
G. Brake hose bracket
H. Paint mark
37647_ATSX_G0685

Fig. 429 Remove the flange nut and the self-locking nut while holding the respective joint pin with a hex wrench, then remove the stabilizer link

To install:

4. Install the stabilizer link on the stabilizer bar and the knuckle adding in the brake hose bracket with the joint pins set at the center of their range of movement.

➡The stabilizer link has a paint mark. The paint mark indicates the difference between the left and right stabilizer links. Install the end of the stabilizer link with the paint mark in the upper position.

5. Install the new flange nut and the new self-locking nut, and tighten them to the specified torque while holding the respective joint pin with a hex wrench.

6. Clean the mating surfaces of the brake disc and the inside of the wheel, then install the rear wheel.

7. Test-drive the vehicle.

8. After 5 minutes of driving, tighten the self-locking nuts again to the specified torque.

Rear Upper Arm

See Figures 430 through 434.

Special Tools Required:

- Ball joint thread protector, 14 mm 07AAE-SJAA100
- Ball joint remover, 32 mm 07MAC-SL0A102

1. Raise the vehicle on a lift.
2. Remove the rear wheel.
3. Remove the rear damper/spring.
4. Release the parking brake lever fully.
5. Loosen the parking brake cable adjusting nut.
6. Remove the flange bolt from the arm. Then disconnect the parking brake cable from the lever.
7. Remove the brake caliper bracket mounting bolts, then remove the caliper assembly from the knuckle. To prevent damage to the caliper assembly or the brake hose, use a short piece of wire to hang the caliper assembly from the undercarriage. Do not twist the brake hose excessively.

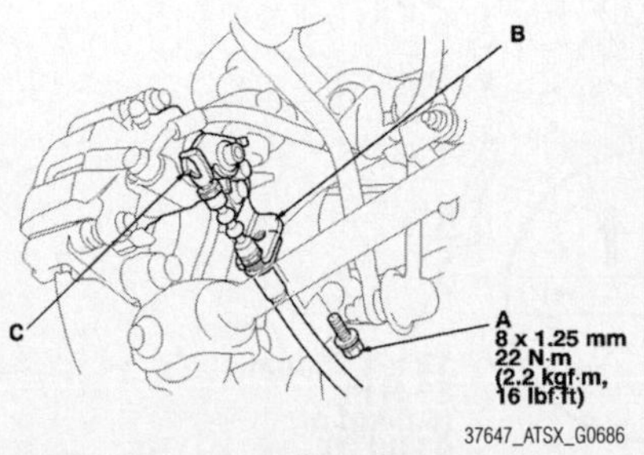
B
C
A
8 x 1.25 mm
22 N·m
(2.2 kgf·m,
16 lbf·ft)
37647_ATSX_G0686

Fig. 430 Remove the flange bolt (A) from the arm (B), then disconnect the parking brake cable from the lever (C)

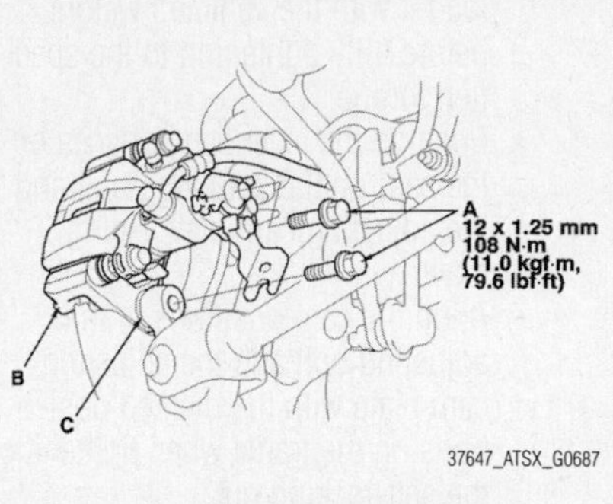
A
12 x 1.25 mm
108 N·m
(11.0 kgf·m,
79.6 lbf·ft)
B
C
37647_ATSX_G0687

Fig. 431 Remove the brake caliper bracket mounting bolts (A), then remove the caliper assembly (B) from the knuckle

➡Make sure the washers are reinstalled between the brake caliper bracket and the knuckle, if they are removed.

8. Remove the lock pin from the upper arm ball joint, then remove the castle nut.

➡During installation, install the lock pin as shown after tightening the new castle nut.

9. Disconnect the upper arm ball joint from the knuckle using the ball joint thread protector and the ball joint remover.

➡Be careful not to damage the ball joint boot when installing the remover. During installation, to connect the ball joint, position a floor jack under the connecting point of the knuckle and lower arm A, and raise the suspension with the jack.

10. Remove the wheel speed sensor harness bracket.

11. Remove the upper arm mounting bolts, then remove the upper arm.

To install:

➡Use new mounting bolts during reassembly.

12. Install the upper arm in the reverse order of removal, and note these items:

- First install all of the components, and lightly tighten the bolts and the nuts, then raise the suspension to load it with the vehicle's weight before fully tightening to the specified torque.
- Be careful not to damage the ball joint boot when connecting the knuckle.
- Before connecting the ball joint, degrease the threaded section and the tapered portion of the ball joint

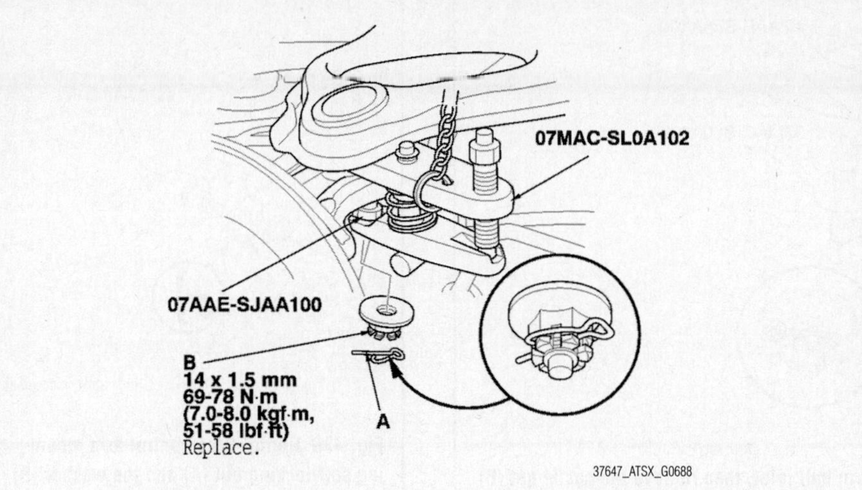
07MAC-SL0A102
07AAE-SJAA100
B
14 x 1.5 mm
69-78 N·m
(7.0-8.0 kgf·m,
51-58 lbf·ft)
Replace.
A
37647_ATSX_G0688

Fig. 432 Remove the lock pin (A) from the upper arm ball joint, then remove the castle nut (B)

pin, the ball joint connecting hole, the threaded section, and the mating surfaces of the castle nut.

- Torque the castle nut to the lower torque specification, then tighten it only far enough to align the slot with the ball joint pin hole. Do not align the castle nut by loosening it.
- After installing the brake caliper, make sure the clearance between lower arm B and the parking brake cable is more than 0.20 inches (5 mm).
- Do the parking brake adjustment.
- Before installing the wheel, clean the mating surfaces on the brake disc and the inside of the wheel.

13. Check the wheel alignment, and adjust it if necessary.

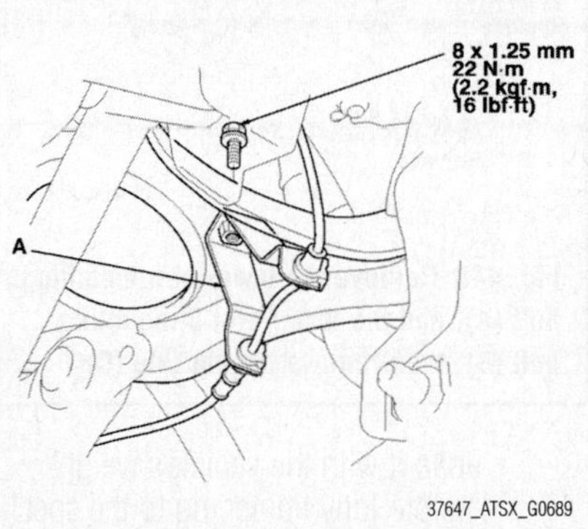
8 x 1.25 mm
22 N·m
(2.2 kgf·m,
16 lbf·ft)
A
37647_ATSX_G0689

Fig. 433 Remove the wheel speed sensor harness bracket (A)

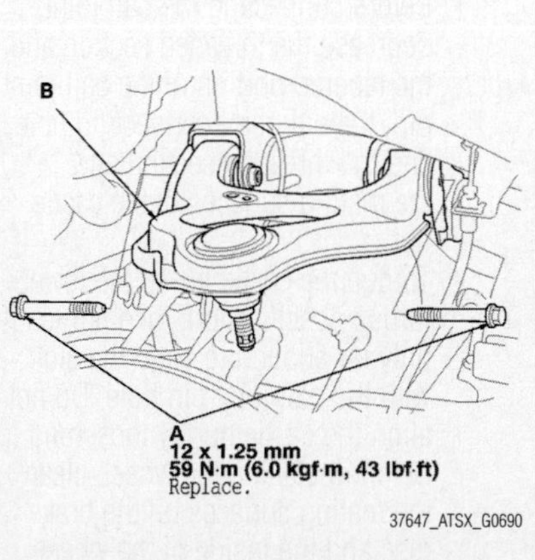
B
A
12 x 1.25 mm
59 N·m (6.0 kgf·m, 43 lbf·ft)
Replace.
37647_ATSX_G0690

Fig. 434 Remove the upper arm mounting bolts (A), then remove the upper arm (B)

KNUCKLE

REMOVAL & INSTALLATION

See Figures 435 through 441.

1. Remove the hub bearing unit.
2. Remove the splash guard.
3. Remove the lock pin from the upper arm ball joint, then remove the castle nut.

➡During installation, install the lock pin after tightening the new castle nut.

4. Disconnect the upper arm ball joint from the knuckle using the ball joint thread protector and the ball joint remover.

➡Be careful not to damage the ball joint boot when installing the remover. During installation, to connect the ball joint, raise the suspension with a jack.

5. Remove the wheel speed sensor from the knuckle. Do not disconnect the wheel speed sensor connector.
6. Remove the flange nut while holding the joint pin with a hex wrench, then disconnect the stabilizer link from the knuckle, and remove the brake hose bracket.

➡Use the new flange nut during reassembly.

7. Remove the damper lower mounting bolt.

➡Use the new mounting bolt during reassembly.

8. Remove the control arm mounting self-locking nut and the washer.

➡Use a new self-locking nut during reassembly.

9. Remove the lower arm mounting bolt, and the lower arm B mounting bolt, then remove the knuckle.

➡Use new mounting bolts during reassembly.

To install:

10. Install the knuckle in the reverse order of removal, and note these items:
    - First install all of the components, and lightly tighten the bolts and the nuts, then raise the suspension to

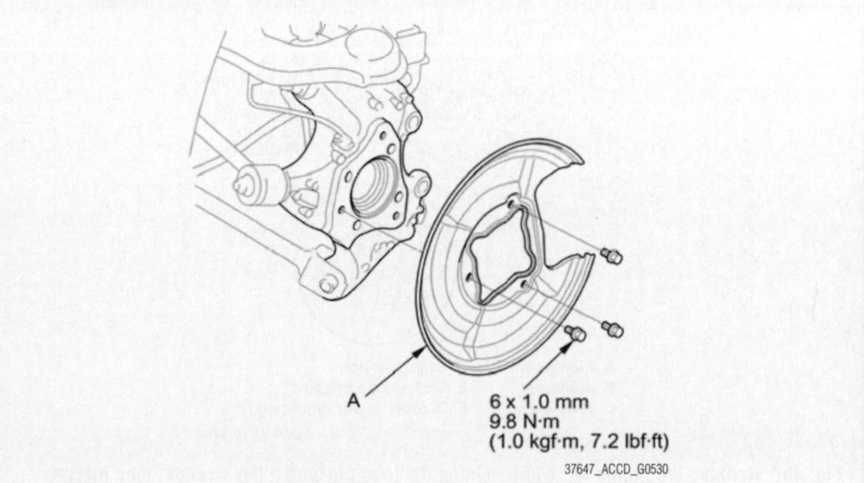
A
6 x 1.0 mm
9.8 N·m
(1.0 kgf·m, 7.2 lbf·ft)
37647_ACCD_G0530

Fig. 435 Remove the splash guard (A)

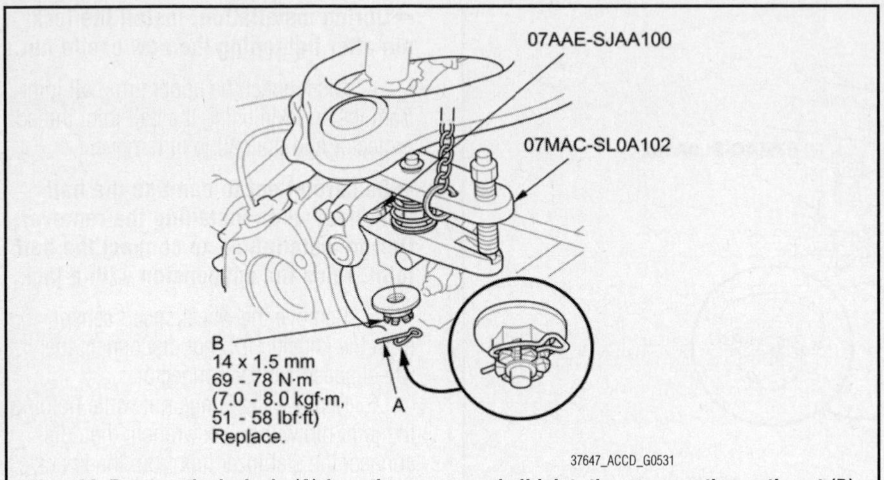
07AAE-SJAA100
07MAC-SL0A102
B
14 x 1.5 mm
69 - 78 N·m
(7.0 - 8.0 kgf·m,
51 - 58 lbf·ft)
Replace.
A
37647_ACCD_G0531

Fig. 436 Remove the lock pin (A) from the upper arm ball joint, then remove the castle nut (B)

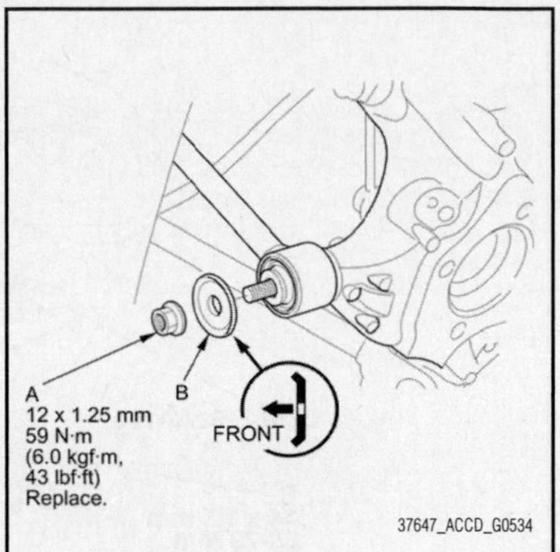
A
12 x 1.25 mm
59 N·m
(6.0 kgf·m,
43 lbf·ft)
Replace.
B
FRONT
37647_ACCD_G0534

Fig. 439 Remove the control arm mounting self-locking nut (A) and the washer (B)

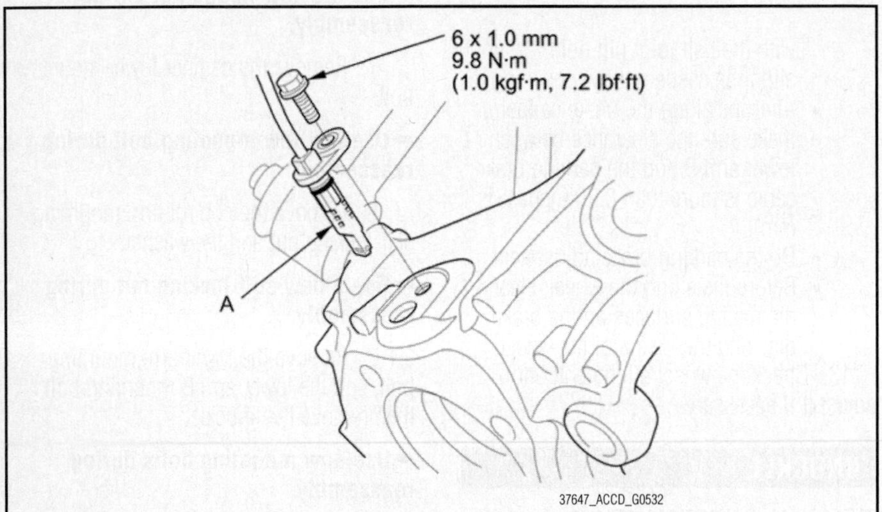
6 x 1.0 mm
9.8 N·m
(1.0 kgf·m, 7.2 lbf·ft)
A
37647_ACCD_G0532

Fig. 437 Remove the wheel speed sensor (A) from the knuckle

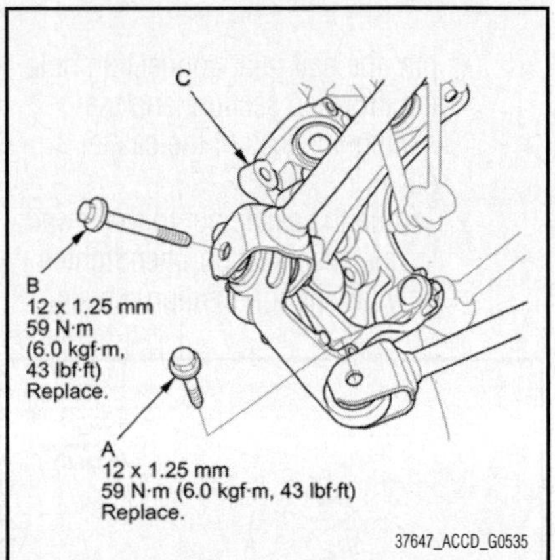
C
B
12 x 1.25 mm
59 N·m
(6.0 kgf·m,
43 lbf·ft)
Replace.
A
12 x 1.25 mm
59 N·m (6.0 kgf·m, 43 lbf·ft)
Replace.
37647_ACCD_G0535

Fig. 440 Remove the lower arm mounting bolt (A), and the lower arm B mounting bolt (B), then remove the knuckle (C)

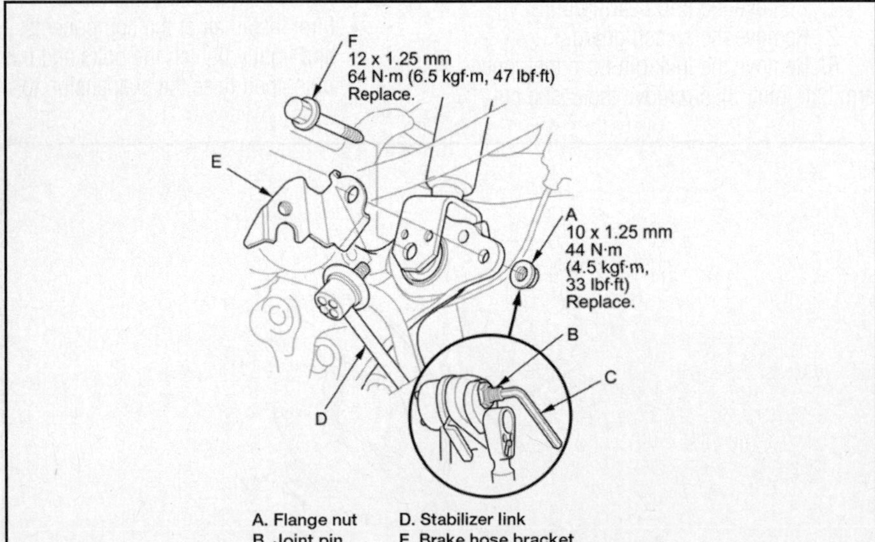
F
12 x 1.25 mm
64 N·m (6.5 kgf·m, 47 lbf·ft)
Replace.
E
A
10 x 1.25 mm
44 N·m
(4.5 kgf·m,
33 lbf·ft)
Replace.
B
C
D
A. Flange nut
B. Joint pin
C. Hex wrench
D. Stabilizer link
E. Brake hose bracket
F. Damper lower mounting bolt
37647_ACCD_G0533

Fig. 438 Remove the flange nut while holding the joint pin with a hex wrench, then disconnect the stabilizer link from the knuckle, and remove the brake hose bracket

load it with the vehicle's weight before fully tightening to the specified torque.
- Be careful not to damage the ball joint boot when connecting the knuckle.
- Before connecting the ball joint, degrease the threaded section and the tapered portion of the ball joint pin, the ball joint connecting hole, and the threaded section and the mating surfaces of the castle nut.
- Torque the castle nut to the lower torque specification, then tighten it only far enough to align the slot with the ball joint pin hole. Do not align the castle nut by loosening it.
- Before installing the wheel, clean the mating surfaces on the brake disc and the inside of the wheel.

11. Check the wheel alignment, and adjust it if necessary.

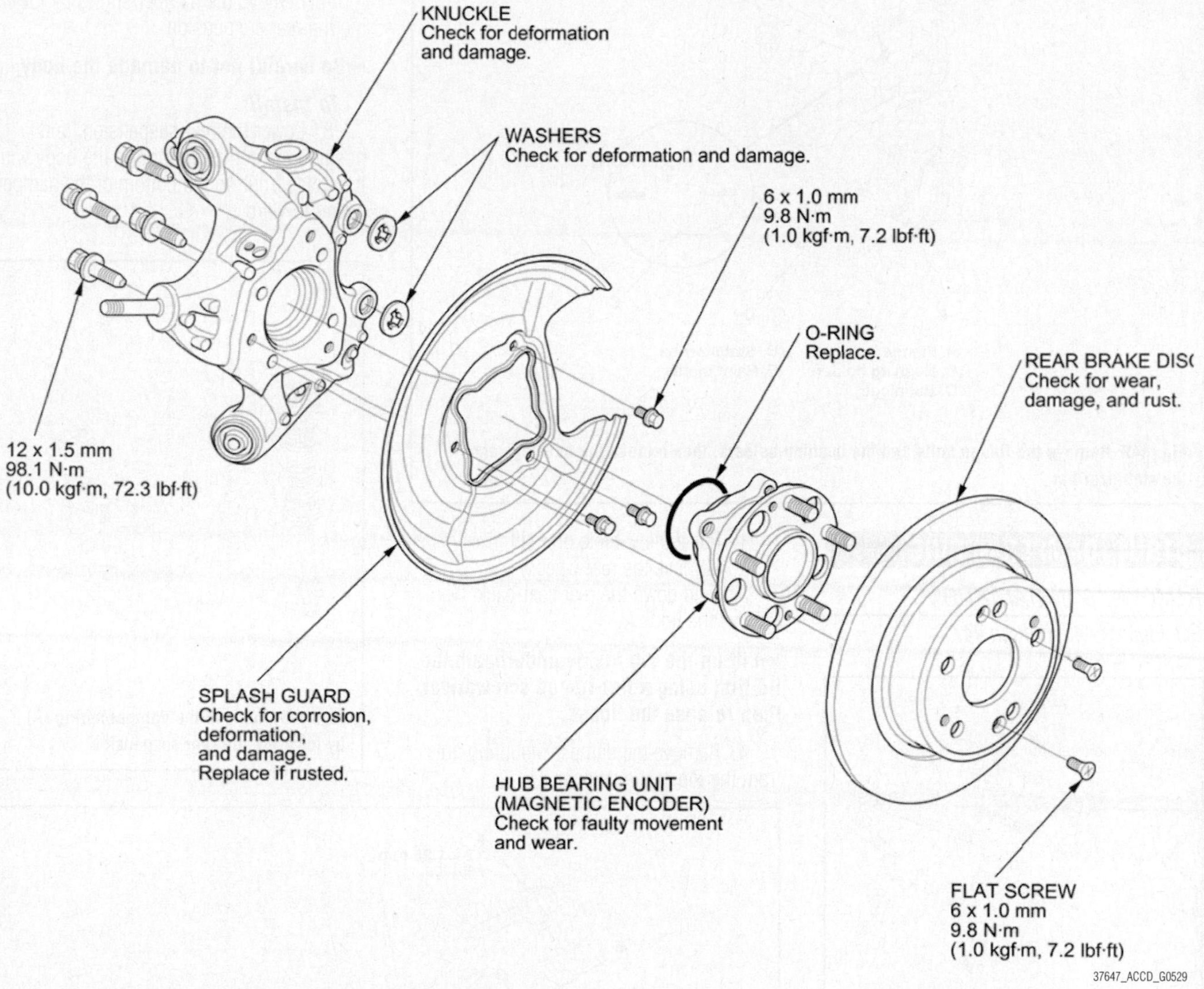
KNUCKLE
Check for deformation and damage.
WASHERS
Check for deformation and damage.
6 x 1.0 mm
9.8 N·m
(1.0 kgf·m, 7.2 lbf·ft)
O-RING
Replace.
REAR BRAKE DISC
Check for wear, damage, and rust.
12 x 1.5 mm
98.1 N·m
(10.0 kgf·m, 72.3 lbf·ft)
SPLASH GUARD
Check for corrosion, deformation, and damage.
Replace if rusted.
HUB BEARING UNIT
(MAGNETIC ENCODER)
Check for faulty movement and wear.
FLAT SCREW
6 x 1.0 mm
9.8 N·m
(1.0 kgf·m, 7.2 lbf·ft)
37647_ACCD_G0529

Fig. 441 Exploded view of rear knuckle and hub assembly

STABILIZER BAR

REMOVAL & INSTALLATION

See Figure 442.

1. Raise the vehicle on a lift.
2. Remove the rear wheels.
3. Disconnect both stabilizer links from the stabilizer bar.
4. Remove the flange bolts and the bushing holders, then remove the bushings and the stabilizer bar.

To install:

➡During installation, align the paint marks on the stabilizer bar with the side of the bushings.

5. Install the stabilizer bar in the reverse order of removal, and note these items:

- Note the right and left direction of the stabilizer bar.
- Note the direction of installation for the bushing.
- Refer to the stabilizer link removal/installation to connect the stabilizer bar to the links.
- Before installing the wheel, clean the mating surfaces of the brake disc and the inside of the wheel.

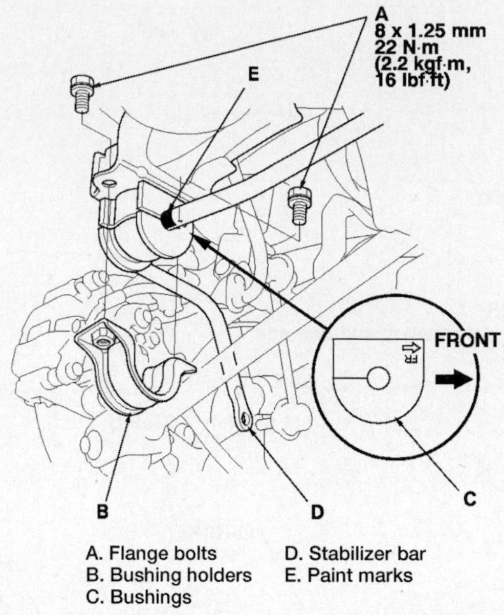
A
8 x 1.25 mm
22 N·m
(2.2 kgf·m,
16 lbf·ft)
E
FRONT
B
D
C
A. Flange bolts
B. Bushing holders
C. Bushings
D. Stabilizer bar
E. Paint marks
37647_ATSX_G0692

Fig. 442 Remove the flange bolts and the bushing holders, then remove the bushings and the stabilizer bar

STRUT (DAMPER/SPRING)

REMOVAL & INSTALLATION

See Figures 443 through 447.

1. Raise the vehicle on a lift.
2. Remove the rear wheel.
3. Fold down the rear seat-back, then remove the lid.

➡Lift up the tab inside underneath the lid first using a flat-tipped screwdriver, then release the hooks.

4. Remove the damper mounting nuts from the top of the damper.
5. Remove the flange nut while holding the joint pin with a hex wrench, then disconnect the stabilizer link from the knuckle, and remove the brake hose bracket.
6. Remove the damper lower mounting bolt.
7. Remove the damper/spring by lowering the rear suspension.

➡Be careful not to damage the body.

To install:

8. Lower the rear suspension, and position the damper/spring in the body with the welded nut on the bottom of the damper facing forward.

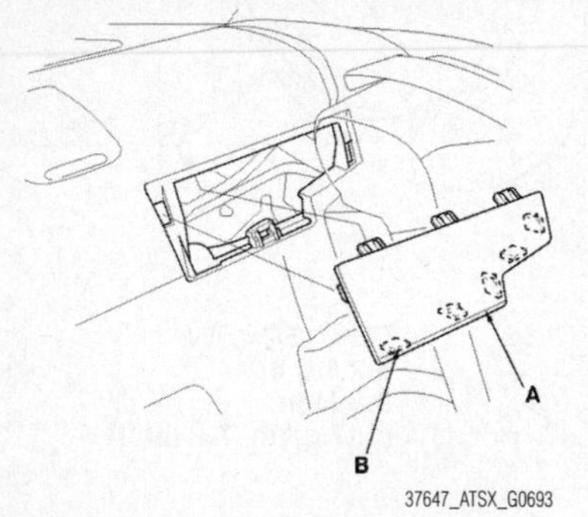
A
B
37647_ATSX_G0693

Fig. 443 Remove the lid (A) and lift up the tab (B)

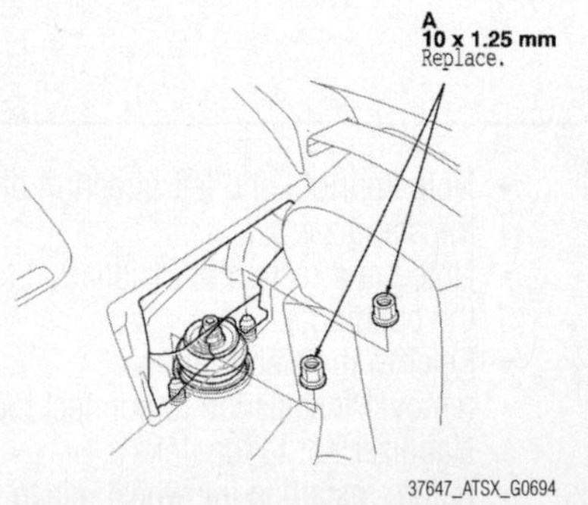
A
10 x 1.25 mm
Replace.
37647_ATSX_G0694

Fig. 444 Remove the damper mounting nuts (A) from the top of the damper

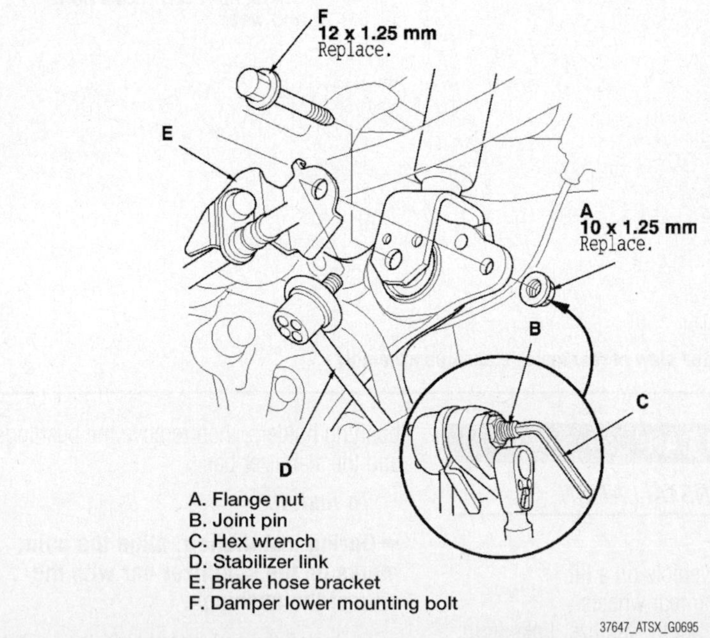
F
12 x 1.25 mm
Replace.
E
A
10 x 1.25 mm
Replace.
B
C
D
A. Flange nut
B. Joint pin
C. Hex wrench
D. Stabilizer link
E. Brake hose bracket
F. Damper lower mounting bolt
37647_ATSX_G0695

Fig. 445 Remove the flange nut while holding the joint pin with a hex wrench, then disconnect the stabilizer link from the knuckle, and remove the brake hose bracket

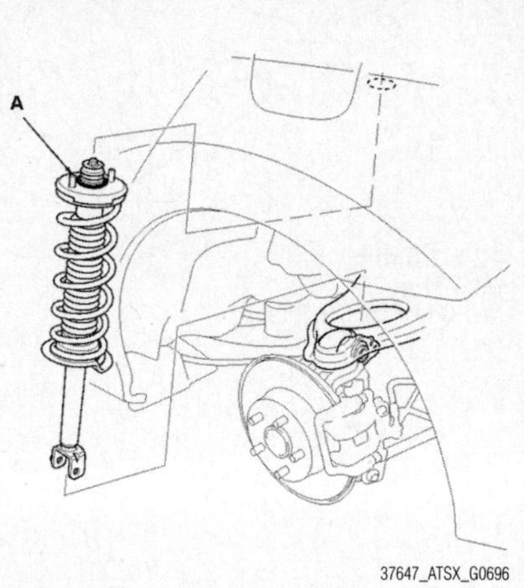
A
37647_ATSX_G0696

Fig. 446 Remove the damper/spring (A) by lowering the rear suspension

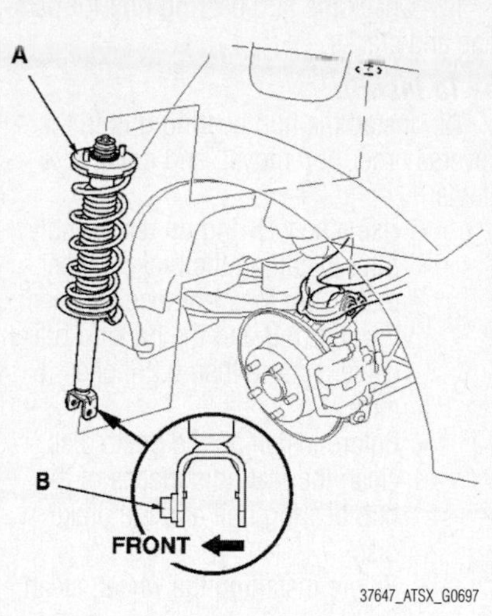
A
B
FRONT
37647_ATSX_G0697

Fig. 447 Lower the rear suspension, and position the damper/spring (A) in the body with the welded nut (B) on the bottom of the damper facing forward

➡Be careful not to damage the body. Make sure the damper is installed in the correct direction.

9. Loosely install the new damper mounting nuts to the top of the damper.
10. Loosely install the new damper lower mounting bolt on the bottom of the damper. Connect the stabilizer link to the brake hose bracket to the knuckle, and loosely install the new flange nut.
11. Place a floor jack under the connecting point of the knuckle and lower arm A, and raise the suspension to load with the vehicle's weight.
12. Tighten the damper lower mounting bolt and the flange nut while holding the joint pin with the hex wrench to 33 ft. lbs. (44 Nm).
13. Tighten the damper mounting nuts on top of the damper to 41 ft. lbs. (55 Nm).
14. Install the lid, and set the rear seatback to the original position.
15. Clean the mating surfaces of the brake disc and the inside of the wheel, then install the rear wheel.
16. Check the wheel alignment, and adjust it if necessary.

WHEEL HUB & BEARING

REMOVAL & INSTALLATION

See Figures 448 through 452.

1. Raise and support the vehicle.
2. Remove the wheel nuts, and the rear wheel.
3. Release the parking brake lever fully.
4. Loosen the parking brake cable adjusting nut.
5. Remove the flange bolt from the arm. Then disconnect the parking brake cable from the lever.
6. Remove the brake hose mounting bolt.
7. Remove the brake caliper bracket mounting bolts, then remove the caliper assembly from the knuckle. To prevent damage to the caliper assembly or the brake hose, use a short piece of wire to hang the caliper assembly from the undercarriage. Do not twist the brake hose excessively.
8. Remove the two washers.

➡During installation, make sure the washers are installed between the brake caliper bracket and the knuckle.

9. Remove the rear brake disc.
10. Remove the hub bearing unit and the O-ring.

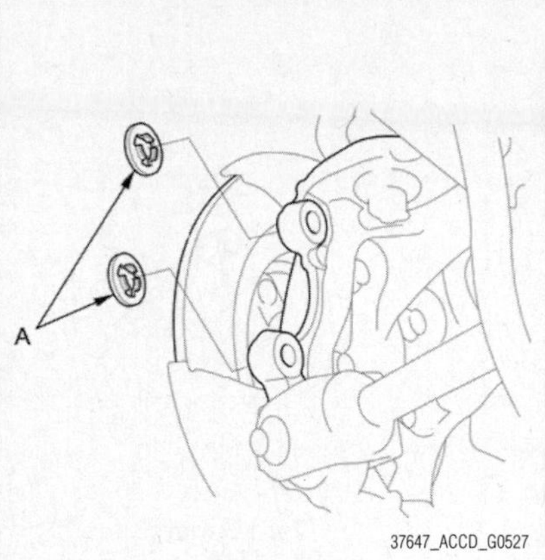
A
37647_ACCD_G0527

Fig. 450 Remove the two washers (A)

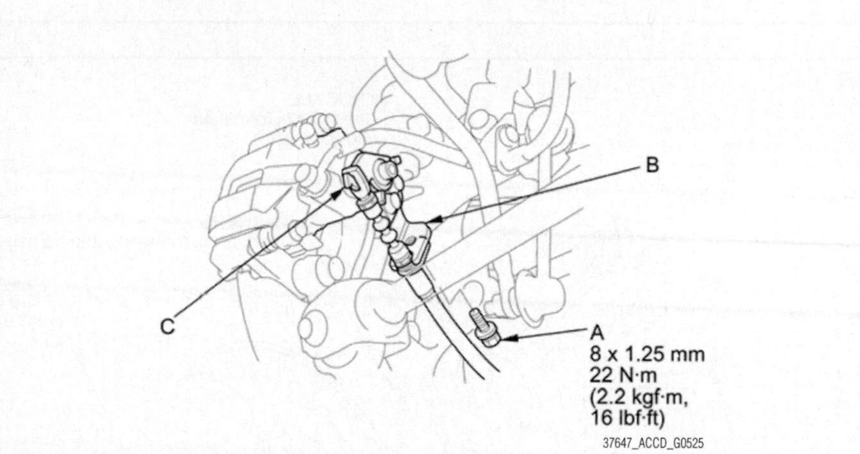
B
C
A
8 x 1.25 mm
22 N·m
(2.2 kgf·m,
16 lbf·ft)
37647_ACCD_G0525

Fig. 448 Remove the flange bolt (A) from the arm (B), then disconnect the parking brake cable from the lever (C)

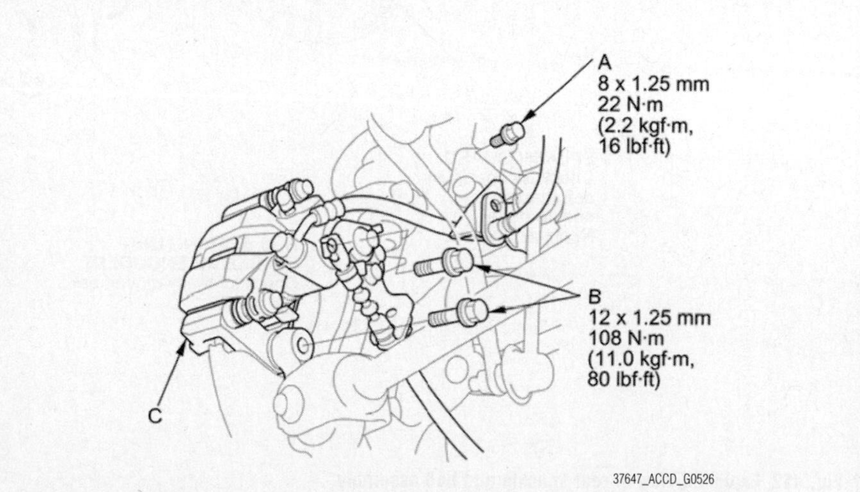
A
8 x 1.25 mm
22 N·m
(2.2 kgf·m,
16 lbf·ft)
B
12 x 1.25 mm
108 N·m
(11.0 kgf·m,
80 lbf·ft)
C
37647_ACCD_G0526

Fig. 449 Remove the brake hose mounting bolt (A), the brake caliper mounting bolts (B) and the caliper assembly (C)

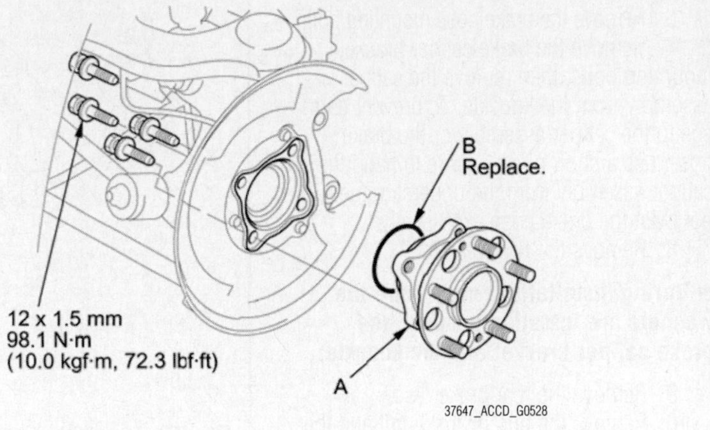

B
Replace.
12 x 1.5 mm
98.1 N·m
(10.0 kgf·m, 72.3 lbf·ft)
A
37647_ACCD_G0528

Fig. 451 Remove the hub bearing unit (A) and the O-ring (B)

11. Check the hub bearing unit for damage and cracks.

To install:

12. Install the hub bearing unit in the reverse order of removal, and note these items:

- Use a new O-ring on reassembly.
- After installing the brake caliper, make sure the clearance between lower arm B and the parking brake cable is more than 0.2 inches (5 mm).
- Before installing the brake disc, clean the mating surfaces of the hub bearing unit and the brake disc.
- Before installing the wheel, clean the mating surfaces of the brake disc and the inside of the wheel.

13. Check the wheel alignment, and adjust it if necessary.

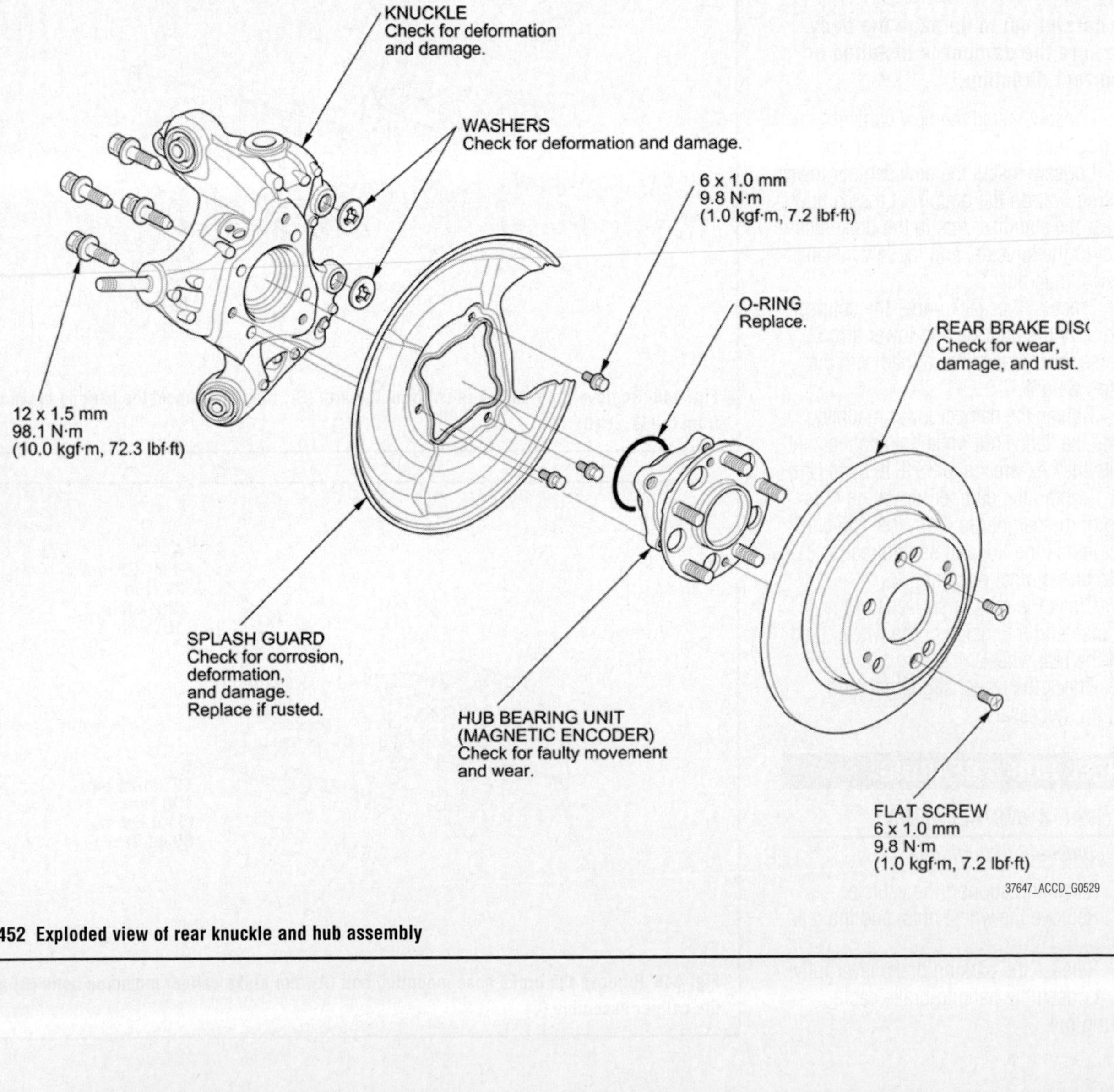

KNUCKLE
Check for deformation
and damage.
WASHERS
Check for deformation and damage.
6 x 1.0 mm
9.8 N·m
(1.0 kgf·m, 7.2 lbf·ft)
O-RING
Replace.
REAR BRAKE DISC
Check for wear,
damage, and rust.
12 x 1.5 mm
98.1 N·m
(10.0 kgf·m, 72.3 lbf·ft)
SPLASH GUARD
Check for corrosion,
deformation,
and damage.
Replace if rusted.
HUB BEARING UNIT
(MAGNETIC ENCODER)
Check for faulty movement
and wear.
FLAT SCREW
6 x 1.0 mm
9.8 N·m
(1.0 kgf·m, 7.2 lbf·ft)
37647_ACCD_G0529

Fig. 452 Exploded view of rear knuckle and hub assembly

HONDA

Civic

9

BRAKES 9-12

ANTI-LOCK BRAKE SYSTEM (ABS) 9-13
- Wheel Speed Sensors 9-13
  - Removal & Installation 9-13

BLEEDING THE BRAKE SYSTEM 9-12
- Bleeding Procedure 9-12
  - Bleeding Procedure 9-12
  - Bleeding the ABS System 9-12

FRONT DISC BRAKES 9-14
- Brake Caliper 9-14
  - Removal & Installation 9-14
- Disc Brake Pads 9-16
  - Removal & Installation 9-16

INFORMATION AND PRECAUTIONS 9-12
- Anti-lock Systems 9-12
- Disc and Drum Systems 9-12

PARKING BRAKE 9-21
- Parking Brake Cables 9-21
  - Adjustment 9-21
- Parking Brake Shoes 9-21
  - Adjustment 9-21
  - Removal & Installation 9-21

REAR DISC BRAKES 9-17
- Brake Caliper 9-17
  - Removal & Installation 9-17
- Disc Brake Pads 9-18
  - Removal & Installation 9-18

Rear Drum Brakes 9-18
- Brake Drum 9-18
  - Removal & Installation 9-18
- Brake Shoes 9-19
  - Removal & Installation 9-19

CHASSIS ELECTRICAL 9-22

AIR BAG (SUPPLEMENTAL RESTRAINT SYSTEM) 9-22
- General Information 9-22
  - Arming the System 9-22
  - Disarming the System 9-22
  - Precautions 9-22

DRIVE TRAIN 9-22
- Clutch 9-22
  - Removal & Installation 9-22
- Front Driveshaft 9-23
  - Removal & Installation 9-23
- Hydraulic System
  - Bleeding 9-22
    - Fluid Fill Procedure 9-22

ENGINE COOLING 9-24
- Engine Fan 9-24
  - Removal & Installation 9-24
- Radiator & Fan 9-24
  - Removal & Installation 9-24
- Thermostat 9-25
  - Removal & Installation 9-25
- Water Pump 9-25
  - Removal & Installation 9-25

ENGINE ELECTRICAL 9-25

CHARGING SYSTEM 9-25
- Alternator 9-25
  - Removal & Installation 9-25

IGNITION SYSTEM 9-26
- Firing Order 9-26
- Idle Speed 9-26
  - Inspection 9-26
- Ignition Coil 9-26
  - Removal & Installation 9-26
- Ignition Timing 9-27
  - Adjustment 9-27
  - Inspection 9-27
- Spark Plugs 9-27
  - Removal & Installation 9-27

STARTING SYSTEM 9-27
- Starter 9-27
  - Removal & Installation 9-27

ENGINE MECHANICAL 9-28
- Accessory Drive Belts 9-28
  - Accessory Belt Routing 9-28
  - Adjustment 9-28
  - Inspection 9-28
  - Removal & Installation 9-28
- Camshaft 9-29
  - Inspection 9-29
  - Removal & Installation 9-30
- Catalytic Converter 9-32
  - Removal & Installation 9-32
- Crankshaft Damper (Pulley) 9-33
  - Removal & Installation 9-33
- Crankshaft Front Seal 9-34
  - Removal & Installation 9-34
- Cylinder Head 9-34
  - Removal & Installation 9-34
- Exhaust Manifold 9-39
  - Removal & Installation 9-39
- Flywheel 9-41
  - Removal & Installation 9-41
- Intake Manifold 9-41
  - Removal & Installation 9-41
- Oil Pan 9-43
  - Removal & Installation 9-43
- Oil Pump 9-44
  - Inspection 9-46
  - Removal & Installation 9-44
- Piston and Ring 9-49
  - Positioning 9-49
- Rear Main Seal 9-49
  - Removal & Installation 9-49
- Rocker Arms 9-50
  - Removal & Installation 9-50
- Timing Chain & Sprockets 9-52
  - Removal & Installation 9-52
- Valve Lash 9-57
  - Adjustment 9-57

ENGINE PERFORMANCE & EMISSION CONTROLS 9-58
- Accelerator Pedal Position (APP) Sensor 9-58
  - Location 9-58
  - Removal & Installation 9-58
- Camshaft Position (CMP) Sensor 9-58
  - Location 9-58
  - Removal & Installation 9-59

Crankshaft Position (CKP) Sensor 9-59
CKP Pattern Clear/CKP Pattern Lean Procedure 9-59
Location 9-59
Removal & Installation 9-59
Engine Control Module/ Powertrain Control Module (ECM/PCM) 9-60
ECM/PCM Idle Learn Procedure 9-61
ECM/PCM Update 9-61
Location 9-60
Removal & Installation 9-60
Engine Coolant Temperature (ECT) Sensor 9-62
Location 9-62
Removal & Installation 9-62
Evaporative Emissions (EVAP) Canister 9-63
Location 9-63
Removal & Installation 9-63
Exhaust Gas Recirculation (EGR) Valve 9-63
Location 9-63
Removal & Installation 9-63
Heated Oxygen Sensor (HO2S) 9-64
Location 9-64
Removal & Installation 9-64
Intake Air Temperature/Mass Air Flow (IAT/MAF) Sensor 9-64
Location 9-64
Removal & Installation 9-64
Knock Sensor (KS) 9-65
Location 9-65
Removal & Installation 9-65
Malfunction Indicator Light (MIL) 9-65
Reset Procedure 9-65
Manifold Absolute Pressure (MAP) Sensor 9-65
Location 9-65
Removal & Installation 9-65
Positive Crankcase Ventilation (PCV) Valve 9-66
Location 9-66
Removal & Installation 9-66
Throttle Actuator Control (TAC) 9-66
Location 9-66
Removal & Installation 9-66
Throttle Position (TP) Sensor 9-66
Location 9-66
Removal & Installation 9-66

FUEL 9-68

GASOLINE FUEL INJECTION SYSTEM 9-68
Fuel Filter 9-69
Removal & Installation 9-69
Fuel Pump/ Fuel Gauge Sending Unit 9-70
Removal & Installation 9-70
Fuel Rail and Injector 9-70
Removal & Installation 9-70
Fuel System Pressure 9-68
Relieving 9-68
Fuel System Service Precautions 9-68
Fuel Tank 9-71
Draining 9-71
Removal & Installation 9-71
Idle Speed 9-74
Adjustment 9-74
Throttle Body 9-74
Removal & Installation 9-74

HEATING & AIR CONDITIONING SYSTEM 9-76
Blower Motor 9-76
Removal & Installation 9-76
Heater Unit & Core 9-76
Removal & Installation 9-76

SPECIFICATIONS AND MAINTENANCE CHARTS 9-3
Brake Specifications 9-9
Camshaft and Bearing Specifications 9-5
Capacities 9-4
Crankshaft and Connecting Rod Specifications 9-6
Engine and Vehicle Identification 9-3
Engine Tune-Up Specifications 9-3
Fluid Specifications 9-4
General Engine Specifications 9-3
Piston and Ring Specifications 9-6
Maintenance Minder Schedule 9-10,11
Tire, Wheel and Ball Joint Specifications 9-8
Torque Specifications 9-7
Valve Specifications 9-5
Wheel Alignment 9-8

STEERING 9-77
Electronic Power Steering (EPS) Control Unit 9-77
Removal & Installation 9-77
Electronic Power Steering (EPS) Motor 9-77
Removal & Installation 9-77
Power Rack & Pinion Steering Gear 9-78
Removal & Installation 9-78
Torque Sensor Neutral Position Memorization 9-82
Power Steering Pump 9-82
Bleeding 9-83
Removal & Installation 9-83

SUSPENSION 9-84

FRONT SUSPENSION 9-84
Lower Ball Joint 9-84
Removal & Installation 9-84
Lower Control Arm 9-84
Removal & Installation 9-84
MacPherson Strut 9-84
Overhaul 9-84
Removal & Installation 9-84
Stabilizer Bar 9-84
Removal & Installation 9-84
Steering Knuckle 9-84
Removal & Installation 9-84
Wheel Bearings 9-85
Adjustment 9-86
Removal & Installation 9-85

REAR SUSPENSION 9-86
Knuckle/Hub Bearing 9-86
Removal & Installation 9-86
MacPherson Struts 9-86
Overhaul 9-86
Removal & Installation 9-86
Stabilizer Bar 9-87
Removal & Installation 9-87
Upper Control Arm 9-88
Removal & Installation 9-88

SPECIFICATIONS AND MAINTENANCE CHARTS

ENGINE AND VEHICLE IDENTIFICATION

Engine					
Code	Liters	Cu. In. (cc)	Cyl.	Fuel Sys.	Eng. Mfg.
R18A1	1.8	110.0 (1798)	4	PGM-FI	Honda
R18A4	1.8	110.0 (1798)	4	CNG	Honda
K20Z3	2.0	121.9 (1997)	4	PGM-FI	Honda

Model Year	
Code	Year
9	2009
A	2010

PGM-FI: Programmed Fuel Injection

CNG: Compressed Natural Gas

37647_CIVC_C0001

GENERAL ENGINE SPECIFICATIONS

Year	Model	Engine Displacement Liters	Engine ID/VIN	Net Horsepower @ rpm	Net Torque @ rpm (ft. lbs.)	Bore X Stroke (in.)	Com-pression Ratio	Oil Pressure @ rpm
2009	Civic	1.8	R18A1	140@4800	128@4800	3.19x3.44	10.5:1	50@3000
	Civic GX	1.8	R18A4	113@6300	109@4300	3.19x3.44	10.5:1	50@3000
	Civic	2.0	K20Z3	197@7800	139@6100	3.39x3.39	11.0:1	44@3000
2010	Civic	1.8	R18A1	140@4800	128@4800	3.19x3.44	10.5:1	50@3000
	Civic GX	1.8	R18A4	113@6300	109@4300	3.19x3.44	10.5:1	50@3000
	Civic	2.0	K20Z3	197@7800	139@6100	3.39x3.39	11.0:1	44@3000

37647_CIVC_C0002

ENGINE TUNE-UP SPECIFICATIONS

Year	Engine Displacement Liters	Engine ID/VIN	Spark Plugs Gap (in.)	Ignition Timing (deg.)		Fuel Pump (psi)	Idle Speed (rpm)		Valve Clearance	
				MT	AT		MT	AT	In.	Ex.
2009	1.8	R18A1	0.039-0.043	6-10B	6-10B	48-55	620-720	620-720	0.007-0.009	0.009-0.011
	1.8	R18A4	0.039-0.043	6-10B	6-10B	—	①	①	0.007-0.009	0.009-0.011
	2.0	K20Z3	0.039-0.043	6-10B	—	48-55	700-800	—	0.008-0.010	0.010-0.011
2010	1.8	R18A1	0.039-0.043	6-10B	6-10B	48-55	620-720	620-720	0.007-0.009	0.009-0.011
	1.8	R18A4	0.039-0.043	6-10B	6-10B	—	①	①	0.007-0.009	0.009-0.011
	2.0	K20Z3	0.039-0.043	6-10B	—	48-55	700-800	—	0.008-0.010	0.010-0.011

NOTE: The Vehicle Emission Control Information label often reflects specification changes made during production. The label figures must be used if they differ from those in this chart

B: Before Top Dead Center

① Shown with no load; with full electrical load: 660-760 rpm

37647_CIVC_C0003

CAPACITIES

Year	Model	Engine Displacement Liters	Engine ID	Engine Oil with Filter	Transmission (pts.) 5-Spd	6-Spd	Auto.	Fuel Tank (gal.)	Cooling System (qts.)
2009	Civic	1.8	R18A1	3.9	3.0	3.2	5.0	13.2	①
	Civic GX	1.8	R18A4	3.9	3.0	3.2	5.0	②	①
	Civic	2.0	K20Z3	4.6	3.2	3.2	—	13.2	③
2010	Civic	1.8	R18A1	3.9	3.0	3.2	5.0	13.2	①
	Civic GX	1.8	R18A4	3.9	3.0	3.2	5.0	②	①
	Civic	2.0	K20Z3	4.6	3.2	3.2	—	13.2	③

NOTE: All capacities are approximate. Add fluid gradually and ensure a proper fluid level is obtained.

NOTE: Capacities given are service, not overhaul capacities

① M/T 4 door with coolant change: 5.48
M/T 4 door with engine overhaul: 6.88
A/T 4 door with coolant change: 5.60
A/T 4 door with engine overhaul: 7.10
M/T 2 door with coolant change: 5.48
M/T 2 door with engine overhaul: 6.88
A/T 2 door with coolant change: 6.80
A/T 2 door with engine overhaul: 7.52

② Fuel system is CNG

③ With coolant change: 4.76
With engine overhaul: 7.20

37647_CIVC_C0004

FLUID SPECIFICATIONS

Year	Model	Engine Displacement Liters	Engine Oil	Man. Trans.	Auto. Trans.	Power Steering Fluid	Brake Master Cylinder	Cooling System
2009	Civic	1.8	5W-20 Honda	Honda MTF	Honda ATF-Z1	Honda PS Fluid	Honda DOT 3	①
	Civic	2.0	5W-20 Honda	Honda MTF	—	Honda PS Fluid	Honda DOT 3	①
2010	Civic	1.8	5W-20 Honda	Honda MTF	Honda ATF-Z1	Honda PS Fluid	Honda DOT 3	①
	Civic	2.0	5W-20 Honda	Honda MTF	—	Honda PS Fluid	Honda DOT 3	①

DOT: Department Of Transpotation

① Honda Long Life Antifreeze/Coolant-Type2

② Hypoid gear oil SAE 90, API classified GL4 or GL5 only

37647_CIVC_C0005

VALVE SPECIFICATIONS

Year	Engine Displacement Liters	Engine ID/VIN	Seat Angle (deg.)	Face Angle (deg.)	Spring Test Pressure (lbs. @ in.)	Spring Installed Height (in.)	Stem-to-Guide Clearance (in.)		Stem Diameter (in.)	
							Intake	Exhaust	Intake	Exhaust
2009	1.8	R18A1	45	45	NA	NA	0.0008-0.0020	0.0020-0.0031	0.2157-0.2161	0.2146-0.2150
	1.8	R18A4	45	45	NA	NA	0.0008-0.0020	0.0020-0.0031	0.2157-0.2161	0.2146-0.2150
	2.0	K20Z3	45	45	NA	NA	0.0012-0.0022	0.0022-0.0031	0.2156-0.2159	0.2146-0.2150
2010	1.8	R18A1	45	45	NA	NA	0.0008-0.0020	0.0020-0.0031	0.2157-0.2161	0.2146-0.2150
	1.8	R18A4	45	45	NA	NA	0.0008-0.0020	0.0020-0.0031	0.2157-0.2161	0.2146-0.2150
	2.0	K20Z3	45	45	NA	NA	0.0012-0.0022	0.0022-0.0031	0.2156-0.2159	0.2146-0.2150

NA: Information not available

37647_CIVC_C0007

CAMSHAFT AND BEARING SPECIFICATIONS

All measurements are given in inches.

Year	Engine Displacement Liters	Engine VIN	Journal Diameter	Brg. Oil Clearance	Shaft End-play	Runout	Journal Bore	Lobe Height	
								Intake	Exhaust
2009	1.8	R18A1	NA	0.0018-0.0033	0.0020-0.0100	0.0010	NA	①	1.4100
	1.8	R18A4	NA	0.0018-0.0033	0.0020-0.0100	0.0010	NA	②	1.3538
	2.0	K20Z3	NA	③	0.0020-0.0080	0.0010	NA	④	⑤
2010	1.8	R18A1	NA	0.0018-0.0033	0.0020-0.0100	0.0010	NA	①	1.4100
	1.8	R18A4	NA	0.0018-0.0033	0.0020-0.0100	0.0010	NA	②	1.3538
	2.0	K20Z3	NA	③	0.0020-0.0080	0.0010	NA	④	⑤

NA: Information not available

① Primary: 1.4076 in.
Secondary A: 1.3920 in.
Secondary B: 1.4184 in.

② Primary: 1.3473 in.
Secondary A: 1.3338 in.
Secondary B: 1.3537 in.

③ No. 1 journal: 0.001-0.003 in.
No. 2-5 journals: 0.002-0.004 in

④ Primary: 1.2910 in.
Mid: 1.3990 in.
Secondary: 1.2865 in.

⑤ Primary: 1.2902 in.
Mid: 1.3688 in.
Secondary: 1.2859 in.

37647_CIVC_C0006

CRANKSHAFT AND CONNECTING ROD SPECIFICATIONS

All measurements are given in inches.

Year	Engine Displacement Liters	Engine ID/VIN	Crankshaft Main Brg. Journal Dia.	Crankshaft Main Brg. Oil Clearance	Crankshaft Shaft End-play	Crankshaft Thrust on No.	Connecting Rod Journal Diameter	Connecting Rod Oil Clearance	Connecting Rod Side Clearance
2009	1.8	R18A1	2.1644-2.1654	0.0007-0.0013	0.0040-0.0140	4	1.7707-1.7716	0.0009-0.0017	0.0060-0.0140
	1.8	R18A4	2.1644-2.1654	0.0007-0.0013	0.0040-0.0140	4	1.7707-1.7716	0.0009-0.0017	0.0060-0.0140
	2.0	K20Z3	①	②	0.0040-0.0140	4	1.7707-1.7717	0.0013-0.0026	0.0060-0.0120
2010	1.8	R18A1	2.1644-2.1654	0.0007-0.0013	0.0040-0.0140	4	1.7707-1.7716	0.0009-0.0017	0.0060-0.0140
	1.8	R18A4	2.1644-2.1654	0.0007-0.0013	0.0040-0.0140	4	1.7707-1.7716	0.0009-0.0017	0.0060-0.0140
	2.0	K20Z3	①	②	0.0040-0.0140	4	1.7707-1.7717	0.0013-0.0026	0.0060-0.0120

① Journals 1, 2 4 and 5: 2.1648-2.1657
Journal 3: 2.1644-2.1654

② Journals 1, 2 4 and 5: 0.0007-0.0016
Journal 3: 0.0010-0.0019

37647_CIVC_C0008

PISTON AND RING SPECIFICATIONS

All measurements are given in inches.

Year	Engine Displacement Liters	Engine ID/VIN	Piston Clearance	Ring Gap Top Compression	Ring Gap Bottom Compression	Ring Gap Oil Control	Ring Side Clearance Top Compression	Ring Side Clearance Bottom Compression	Ring Side Clearance Oil Control
2009	1.8	R18A1	0.0004-0.0014	0.0080-0.0140	0.0160-0.0020	①	0.0018-0.0028	0.0014-0.0024	NA
	1.8	R18A4	0.0004-0.0014	0.0080-0.0140	0.0160-0.0220	0.0080-0.0200	0.0018-0.0028	0.0012-0.0022	NA
	2.0	K20Z3	0.0008-0.0016	0.0080-0.0140	0.0200-0.0260	0.0080-0.0280	0.0018-0.0028	0.0016-0.0026	NA
2010	1.8	R18A1	0.0004-0.0014	0.0080-0.0140	0.0160-0.0020	①	0.0018-0.0028	0.0014-0.0024	NA
	1.8	R18A4	0.0004-0.0014	0.0080-0.0140	0.0160-0.0220	0.0080-0.0200	0.0018-0.0028	0.0014-0.0024	NA
	2.0	K20Z3	0.0008-0.0016	0.0080-0.0140	0.0200-0.0260	0.0080-0.0280	0.0018-0.0028	0.0016-0.0026	NA

NA: Information not available

① Riken: 0.008-0.020 in.
Except RIKEN: 0.008-0.028 in.

37647_CIVC_C0009

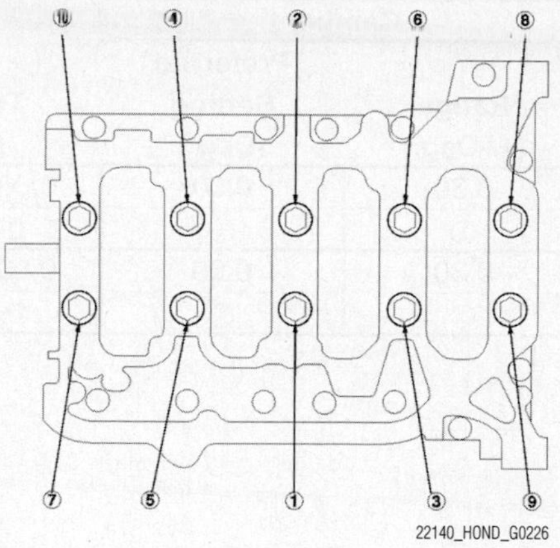
10
4
2
6
8
7
5
1
3
9
22140_HOND_G0226

Fig. 1 Main bearing torque sequence—1.8L, 2.0L Engines

TORQUE SPECIFICATIONS

All readings in ft. lbs.

Year	Engine Disp. Liters	Engine ID/VIN	Cylinder Head Bolts	Main Bearing Bolts	Rod Bearing Bolts	Crankshaft Damper Bolts	Flywheel Bolts	Manifold Intake	Manifold Exhaust	Spark Plugs	Oil Pan Drain Plug
2009	1.8	R18A1	①	②	③	④	⑤	17	23	18	29
	1.8	R18A4	①	②	③	④	⑤	17	23	18	29
	2.0	K20Z3	①	②	⑥	④	⑤	16	23	18	29
2010	1.8	R18A1	①	②	③	④	⑤	17	23	18	29
	1.8	R18A4	①	②	③	④	⑤	17	23	18	29
	2.0	K20Z3	①	②	⑥	④	⑤	16	23	18	29

① Step 1: 29 ft. lbs.
Step 2: Rotate 90 degrees
Step 3: Rotate 90 degrees
Step 4: If new bolt rotate additional 60 degrees

② Step 1: 18 ft. lbs.
Step 2: Rotate 57 degrees

③ Step 1: 14 ft. lbs.
Step 2: Rotate 90 degrees

④ Old bolt:
Step 1: 27 ft. lbs
Step 2: Plus 90 degrees
New bolt:
Step 1: 130 ft. lbs
Step 2: Loosen fully
Step 3: 27 ft. lbs
Step 4: Plus 90 degrees

⑤ Automatic transaxle: 54 ft. lbs.
Manual transaxle: 76 ft. lbs.

⑥ Step 1: 22 ft. lbs.
Step 2: Rotate 90 degrees

37647_CIVC_C0010

WHEEL ALIGNMENT

Year	Model		Caster Range (+/-Deg.)	Caster Preferred Setting (Deg.)	Camber Range (+/-Deg.)	Camber Preferred Setting (Deg.)	Toe-in (in.)
2009	Civic	F	1.00	+7.00	0.30	0.00	0.00+/-0.08
		R	—	—	①	①	0.08 ②
2010	Civic	F	1.00	+7.00	0.30	0.00	0.00+/-0.08
		R	—	—	①	①	0.08 ②

① With "C" marks on upper arms: -0 degrees 45 minutes (preferred)
Range:
+1 degree 5 minutes
-0 degrees 45 minutes
Without "C" marks on upper arms: -1 degree 30 minutes (preferred)
Range:
+1 degree 5 minutes
-0 degrees 45 minutes

② Range: +0.08 to -0.45 in.

37647_CIVC_C0011

TIRE, WHEEL AND BALL JOINT SPECIFICATIONS

Year	Model	OEM Tires Standard	OEM Tires Optional	Tire Pressures (psi) Front	Tire Pressures (psi) Rear	Wheel Size	Ball Joint Inspection	Lug Nut (ft. lbs.)
2009	Civic	①	②	③	③	NA	NA	80
2010	Civic	①	②	③	③	NA	NA	80

OEM: Original Equipment Manufacturer
PSI: Pounds Per Square Inch
NA: Information not available

① DX, DX-VP, DX-G: P195/65R15
LX, LX-S, EX, EX-L: P205/55R16
Si: P215/45R17

② Multiple tire and wheel options available through aftermarket dealers.

③ 4-door:
DX, DX-VP, DX-G: 30 psi
LX, LX-S, EX, EX-L: 32 psi
Si with ABS: Front 32, Rear 29 psi
Si with VSA: Front 32, Rear 33 psi
2-door:
Except Si Model: 32 psi
Si with ABS: Front 32, Rear 29 psi
Si with VSA: 32 psi

37647_CIVC_C0012

BRAKE SPECIFICATIONS

All measurements in inches unless noted

			Brake Disc			Brake Drum Diameter		Minimum Lining Thickness		Brake Caliper	
Year	Model		Original Thickness	Minimum Thickness	Maximum Runout	Original Inside Diameter	Max. Wear Limit	Front	Rear	Bracket Bolts (ft. lbs.)	Mounting Bolts (ft. lbs.)
2009	Civic	F	0.82-0.83	0.750	0.0016	—	—	0.38-0.40	—	80	25
		R	0.35-0.36	0.310	0.0016	7.870-7.874	7.910	—	①	54	17
	Civic GX	F	0.90-0.91	0.830	0.0016	—	—	0.38-0.40	—	80	25
		R	—	—	—	8.657-8.661	8.700	—	0.180	54	17
	Civic Si	F	0.98-0.99	0.910	0.0016			0.35-0.38		80	25
		R	0.35-0.36	0.310	0.0016	—	—	—	0.33-0.37	54	17
2010	Civic	F	0.82-0.83	0.750	0.0016	—	—	0.38-0.40	—	80	25
		R	0.35-0.36	0.310	0.0016	7.870-7.874	7.910	—	①	54	17
	Civic GX	F	0.90-0.91	0.830	0.0016	—	—	0.38-0.40	—	80	25
		R	—	—	—	8.657-8.661	8.700	—	0.180	54	17
	Civic Si	F	0.98-0.99	0.910	0.0016			0.35-0.38		80	25
		R	0.35-0.36	0.310	0.0016	—	—	—	0.33-0.37	54	17

F: Front

R: Rear

① With rear disc: 0.33-0.37 in.
With rear drum: 0.160

37647_CIVC_C0013

MAINTENANCE MINDER SCHEDULE

Honda—Civic (Except GX)

All Honda's displays engine oil life and maintenance service items in the information display to indicate when to perform maintenance service. If the engine oil life is 15% or less, based on the onboard computer's caluculations, you will see SERVICE DUE SOON in the information display every time the ignition key is turned to ON. The maintenance minder indicator will also come on and the maintenance code(s) for other scheduled maintenance items needing service will be displayed below the message.

Symbol	Item	Service
A	Engine oil ①	Change
B	Engine oil and filter	Change
	Fluid levels	Inspect
	Brakes	Inspect
	Parking brake adjustment	Check
	Steering gear and linkage	Inspect
	Suspension components	Inspect
	Driveshaft boots	Inspect
	Brake hoses and lines	Inspect
	Exhaust system	Inspect
	Fuel lines and connections	Inspect
1	Tires	Rotate
2	Engine air filter ②	Replace
	Dust and pollen filter ③	Replace
	Accessory drive belt	Inspect
3	Transmission fluid ④	Replace
4	Spark plugs	Replace
	Valve clearance ⑥	Inspect
5	Engine coolant	Replace

① If the message SERVICE DUE NOW does not appear more than 12 months after the display is reset, change every year.

② If driven in dusty conditions, replace every 15,000 miles.

③ If driven in urban areas that have a high concentration of soot from industry and diesel, replace every 15,000 miles

④ If regularly driven in mountainous areas at very low speed or trailer towing, change the fluid every 30,000 miles.

⑤ If driven regularly in temperatures over 110 deg.F or below -20 deg.F, or towing a trailer, replace every 60,000 miles.

⑥ Adjust if necessary.

Additionally, replace the brake fluid every 3 years, and inspect the idle speed every 160,000 miles.

To reset the Engine Oil Life Display:

1. Turn the ignition switch to ON.
2. Press the SELECT button repeatedly until the engine oil life display or the service message is displayed.
3. Press the RESET button for about 10 seconds. You will see a MAINT RESET message.
4. Select the appropriate answer, MAINT RESET >N (NO) or MAINT RESET > y (YES) by pressing the SELECT button repeatedly.

   >N or >Y is displayed on the outside temperature >N or >Y is displayed on the outside temperature display.
5. Select the MAINT RESET > Y (YES), and press and hold the RESET button again to reset the engine oil life to 100%.

37647_CIVC_C0014

MAINTENANCE MINDER SCHEDULE

Honda—Civic GX

All Honda's displays engine oil life and maintenance service items In the information display to indicate when to perform maintenance service. If the engine oil life is 15% or less, based on the onboard computer's caluculations, you will see SERVICE DUE SOON in the information display every time the ignition key is turned to ON. The maintenance minder indicator will also come on and the maintenance code(s) for other scheduled maintenance items needing service will be displayed below the message.

Symbol	Item	Service
A	Engine oil ①	Change
B	Engine oil and filter	Change
	Fuel filter element B (low pressure)	Change
	Fuel filter element A (high pressure)	Drain
	Fluid levels	Inspect
	Brakes	Inspect
	Parking brake adjustment	Inspect
	Steering gear and linkage	Inspect
	Suspension components	Inspect
	Driveshaft boots	Inspect
	Brake hoses and lines	Inspect
	Exhaust system	Inspect
	Fuel lines and connections	Inspect
1	Tires	Rotate
2	Engine air filter ②	Replace
	Dust and pollen filter ③	Replace
	Fuel filter element A (high pressure)	Replace
	Fuel tank	Inspect
	Valve clearance ④	Inspect
	Accessory drive belt	Inspect
3	Transmission fluid	Replace
4	Spark plugs	Replace
5	Engine coolant	Replace

① If the message SERVICE DUE NOW does not appear more than 12 months after the display is reset, change every year.

② If driven in dusty conditions, replace every 15,000 miles.

③ If driven in urban areas that have a high concentration of soot from industry and diesel, replace every 15,000 miles

④ Adjust if necessary.

Additionally, replace the brake fluid every 3 years, and inspect the idle speed every 160,000 miles.

To reset the Engine Oil Life Display:

1. Turn the ignition switch to ON.
2. Press the SELECT button repeatedly until the engine oil life display or the service message is displayed.
3. Press the RESET button for about 10 seconds. You will see a MAINT RESET message.
4. Select the appropriate answer, MAINT RESET >N (NO) or MAINT RESET > y (YES) by pressing the SELECT button repeatedly.
   >N or >Y is displayed on the outside temperature >N or >Y is displayed on the outside temperature display.
5. Select the MAINT RESET > Y (YES), and press and hold the RESET button again to reset the engine oil life to 100%.

37647_CIVC_C0015

BRAKES — INFORMATION AND PRECAUTIONS

ANTI-LOCK SYSTEMS

- Certain components within the ABS system are not intended to be serviced or repaired individually.
- Do not use rubber hoses or other parts not specifically specified for and ABS system. When using repair kits, replace all parts included in the kit. Partial or incorrect repair may lead to functional problems and require the replacement of components.
- Lubricate rubber parts with clean, fresh brake fluid to ease assembly. Do not use shop air to clean parts; damage to rubber components may result.
- Use only DOT 3 brake fluid from an unopened container.
- If any hydraulic component or line is removed or replaced, it may be necessary to bleed the entire system.
- A clean repair area is essential. Always clean the reservoir and cap thoroughly before removing the cap. The slightest amount of dirt in the fluid may plug an orifice and impair the system function. Perform repairs after components have been thoroughly cleaned; use only denatured alcohol to clean components. Do not allow ABS components to come into contact with any substance containing mineral oil; this includes used shop rags.
- The Anti-Lock control unit is a microprocessor similar to other computer units in the vehicle. Ensure that the ignition switch is **OFF** before removing or installing controller harnesses. Avoid static electricity discharge at or near the controller.
- If any arc welding is to be done on the vehicle, the control unit should be unplugged before welding operations begin.

DISC AND DRUM SYSTEMS

CAUTION

Dust and dirt accumulating on brake parts during normal use may contain asbestos fibers from production or aftermarket brake linings. Breathing excessive concentrations of asbestos fibers can cause serious bodily harm. Exercise care when servicing brake parts. Do not sand or grind brake lining unless equipment used is designed to contain the dust residue. Do not clean brake parts with compressed air or by dry brushing. Cleaning should be done by dampening the brake components with a fine mist of water, then wiping the brake components clean with a dampened cloth. Dispose of cloth and all residue containing asbestos fibers in an impermeable container with the appropriate label. Follow practices prescribed by the Occupational Safety and Health Administration (OSHA) and the Environmental Protection Agency (EPA) for the handling, processing, and disposing of dust or debris that may contain asbestos fibers.

BRAKES — BLEEDING THE BRAKE SYSTEM

BLEEDING PROCEDURE

BLEEDING PROCEDURE

➡**See "Bleeding the ABS System" in this section.**

BLEEDING THE ABS SYSTEM

See Figure 2.

➡**Do not reuse the drained fluid. Use only new Honda DOT 3 Brake Fluid from an unopened container. Using a non-Honda brake fluid can cause corrosion and shorten the life of the system.**

➡**Make sure no dirt or other foreign matter gets in the brake fluid.**

1. The reservoir connected to the master cylinder must be at the MAX (upper) level mark at the start of the bleeding procedure, and checked after bleeding each wheel location. Add fluid as required.
2. Make sure the brake fluid level in the reservoir is at the MAX (upper) level line.
3. Have someone slowly pump the brake pedal several times, then apply steady pressure.
4. Start the bleeding at the driver's side of the front brake system. Bleed the calipers or the wheel cylinders in the sequence shown.
5. Attach a length of clear drain tube to the bleed screw, then loosen the bleed screw to allow air to escape from the system. Then tighten the bleed screw securely.
6. Refill the master cylinder reservoir to the MAX (upper) level line.
7. Repeat the procedure for each brake circuit until there are no air bubbles in the fluid.

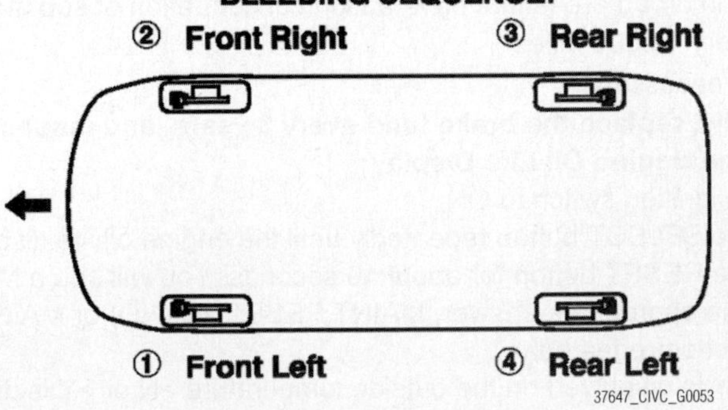
BLEEDING SEQUENCE:
② Front Right
③ Rear Right
① Front Left
④ Rear Left
37647_CIVC_G0053

Fig. 2 Start the bleeding at the driver's side of the front brake system. Bleed the calipers or the wheel cylinders in the sequence shown.

WHEEL SPEED SENSORS

REMOVAL & INSTALLATION

See Figures 3 and 4.

➡Required: guide pin tool (07AAG-SVBA100, or equivalent).

1. Turn the ignition switch to LOCK (0).
2. Release the clamp, then disconnect the wheel speed sensor connector.
3. Remove the clips, the bolt, and the wheel speed sensor.

To install:

4. Install the wheel speed sensor in the reverse order of removal, and note these items:
5. Do not twist the sensor wires.
6. If the wheel speed sensor comes in contact with the wheel bearing, it is faulty.
7. Make sure there is no debris in the sensor mounting hole.
8. Start the engine, and make sure the ABS or VSA indicator goes off.
9. Test-drive the vehicle, and make sure the ABS or VSA indicator does not come on.

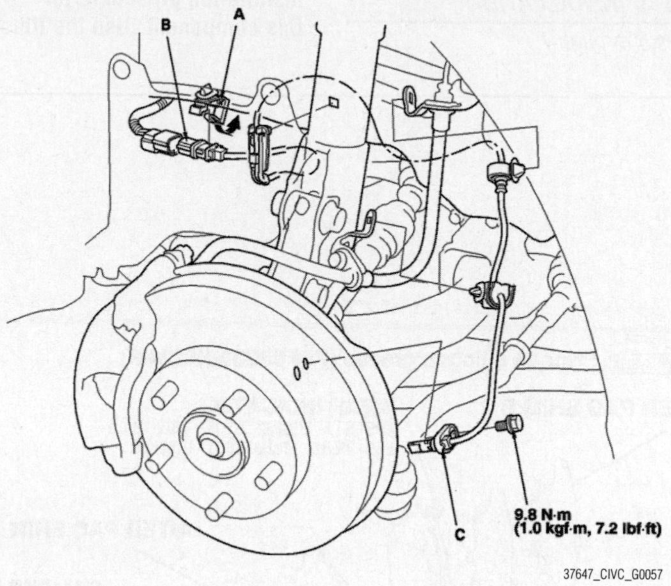
B
A
C
9.8 N·m
(1.0 kgf·m, 7.2 lbf·ft)
37647_CIVC_G0057

Fig. 3 Removing the front wheel speed sensor: Release the clamp (A), then disconnect the wheel speed sensor connector (B), then remove the clips, the bolt, and the wheel speed sensor (C)—front wheel

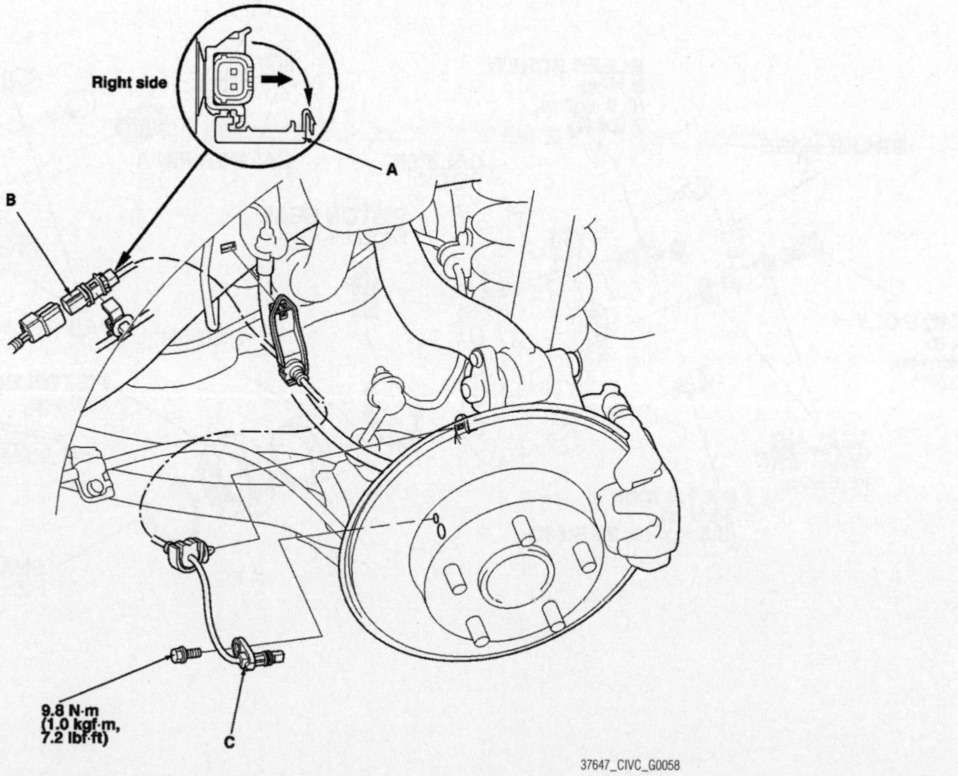
Right side
A
B
9.8 N·m
(1.0 kgf·m,
7.2 lbf·ft)
C
37647_CIVC_G0058

Fig. 4 Removing the front wheel speed sensor: Release the clamp (A), then disconnect the wheel speed sensor connector (B), then remove the clips, the bolt, and the wheel speed sensor (C)—rear wheel

BRAKE CALIPER

REMOVAL & INSTALLATION

See Figures 5 through 7.

➡Manufacturer does not provide a specific removal and installation procedure for this component. Use the illustration as a guide when servicing this component.

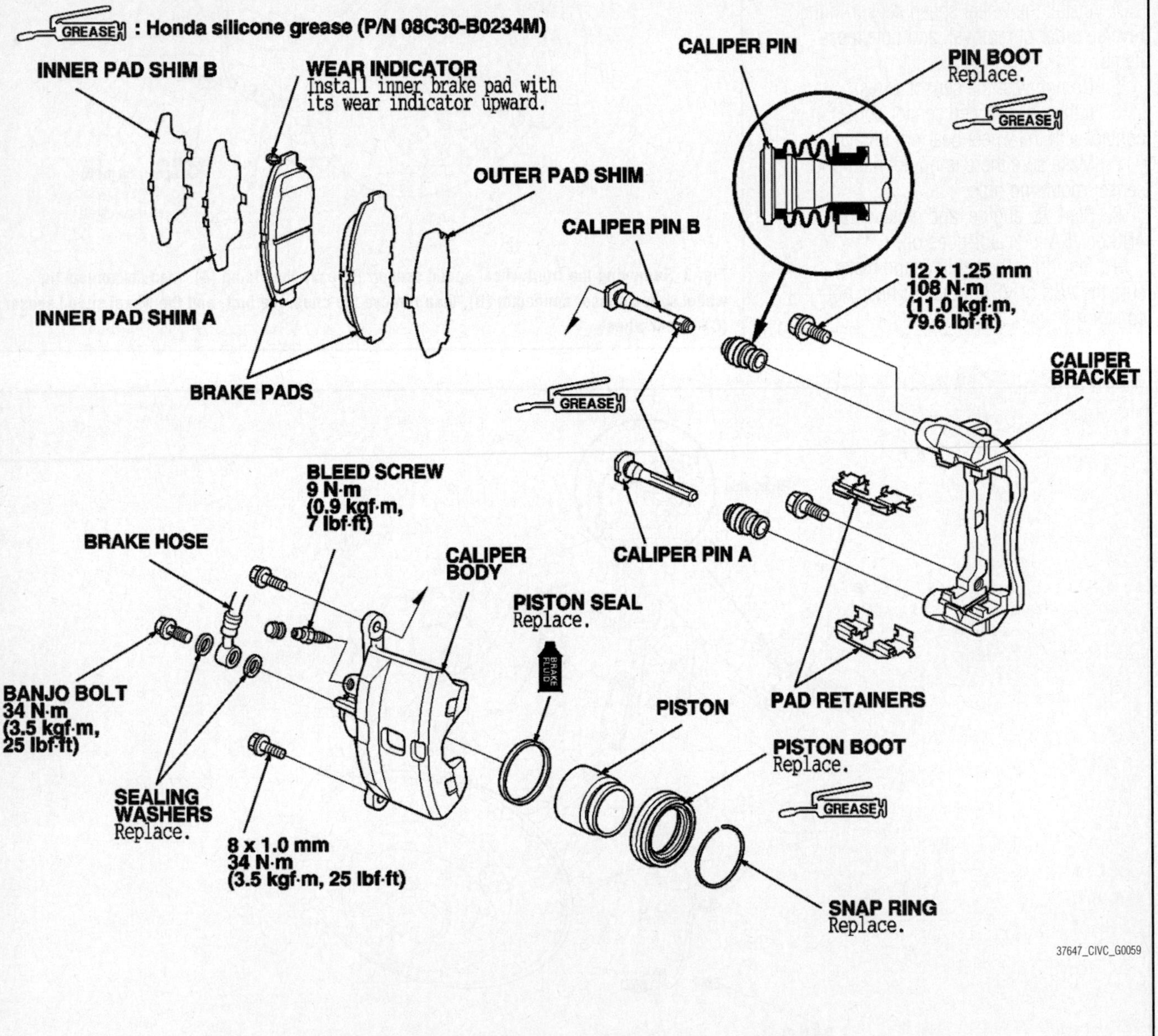
GREASE : Honda silicone grease (P/N 08C30-B0234M)
INNER PAD SHIM B
WEAR INDICATOR
Install inner brake pad with its wear indicator upward.
CALIPER PIN
PIN BOOT
Replace.
GREASE
OUTER PAD SHIM
CALIPER PIN B
12 x 1.25 mm
108 N·m
(11.0 kgf·m,
79.6 lbf·ft)
INNER PAD SHIM A
CALIPER
BRACKET
BRAKE PADS
GREASE
BLEED SCREW
9 N·m
(0.9 kgf·m,
7 lbf·ft)
BRAKE HOSE
CALIPER
BODY
CALIPER PIN A
PISTON SEAL
Replace.
BRAKE FLUID
BANJO BOLT
34 N·m
(3.5 kgf·m,
25 lbf·ft)
PISTON
PAD RETAINERS
PISTON BOOT
Replace.
GREASE
SEALING
WASHERS
Replace.
8 x 1.0 mm
34 N·m
(3.5 kgf·m, 25 lbf·ft)
SNAP RING
Replace.
37647_CIVC_G0059

Fig. 5 Exploded view of the front brake caliper—with 1.8L except DX model

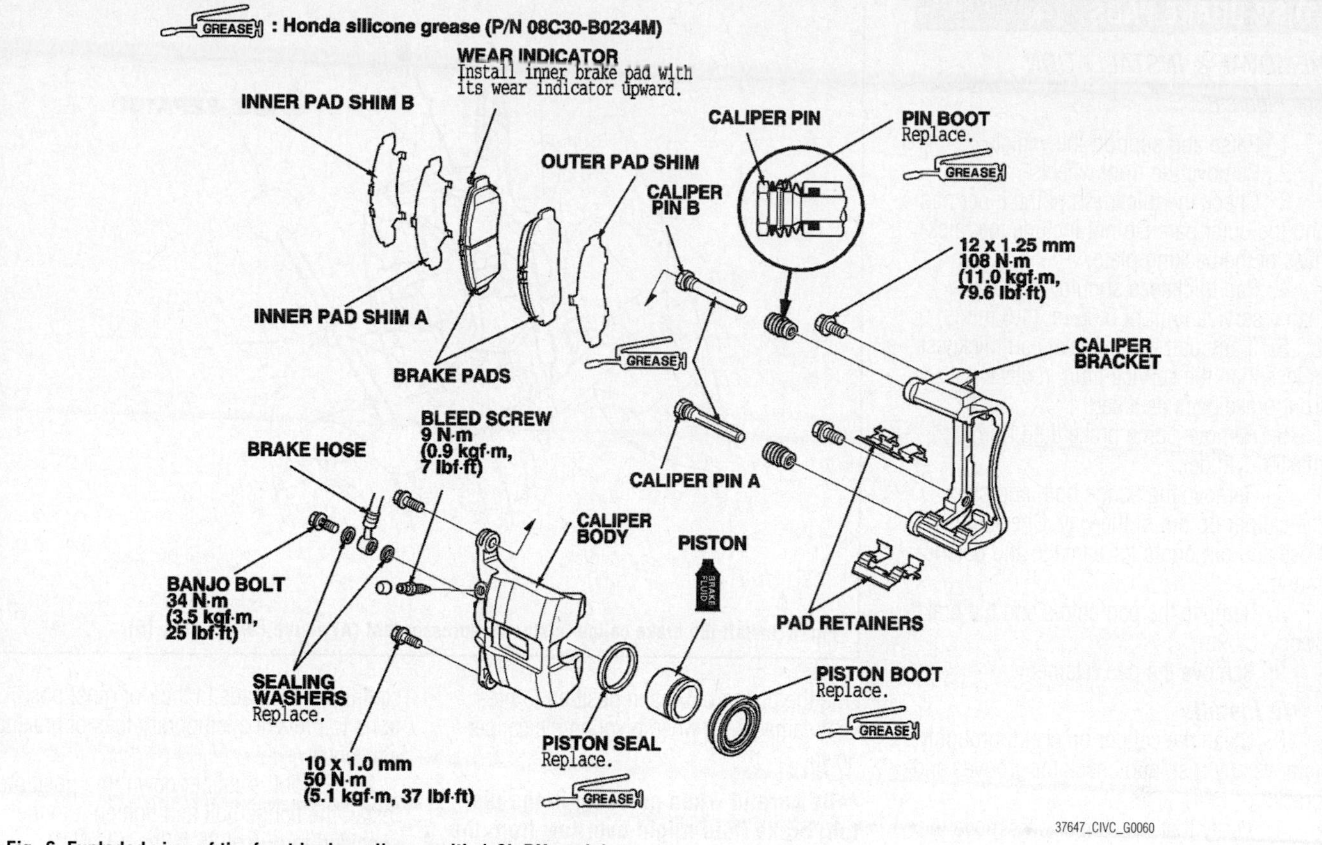
GREASE : Honda silicone grease (P/N 08C30-B0234M)
WEAR INDICATOR
Install inner brake pad with its wear indicator upward.
INNER PAD SHIM B
CALIPER PIN
PIN BOOT
Replace.
OUTER PAD SHIM
CALIPER PIN B
12 x 1.25 mm
108 N·m
(11.0 kgf·m,
79.6 lbf·ft)
INNER PAD SHIM A
CALIPER BRACKET
BRAKE PADS
BLEED SCREW
9 N·m
(0.9 kgf·m,
7 lbf·ft)
BRAKE HOSE
CALIPER PIN A
CALIPER BODY
PISTON
BANJO BOLT
34 N·m
(3.5 kgf·m,
25 lbf·ft)
PAD RETAINERS
SEALING WASHERS
Replace.
PISTON BOOT
Replace.
PISTON SEAL
Replace.
10 x 1.0 mm
50 N·m
(5.1 kgf·m, 37 lbf·ft)
37647_CIVC_G0060

Fig. 6 Exploded view of the front brake caliper—with 1.8L DX model

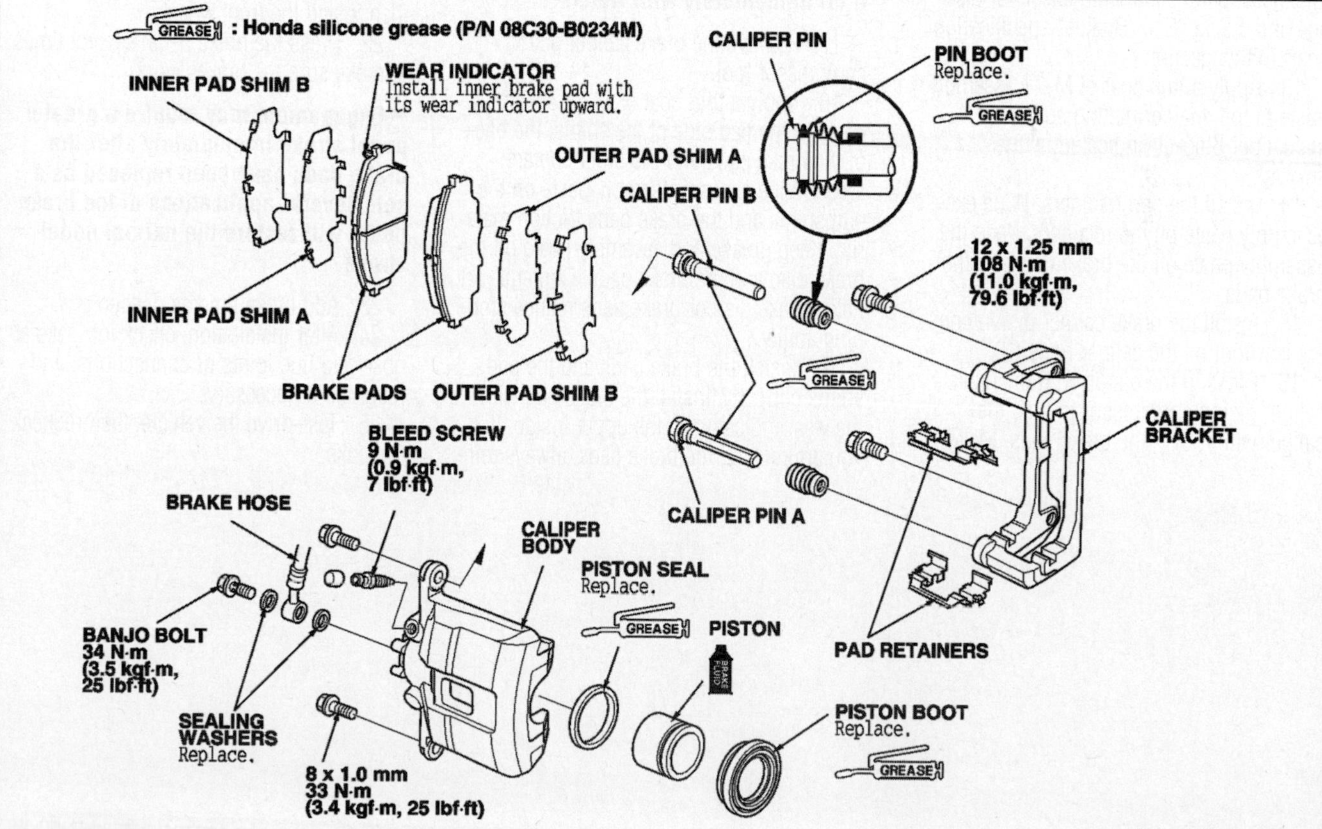
GREASE : Honda silicone grease (P/N 08C30-B0234M)
CALIPER PIN
INNER PAD SHIM B
WEAR INDICATOR
Install inner brake pad with its wear indicator upward.
PIN BOOT
Replace.
OUTER PAD SHIM A
CALIPER PIN B
12 x 1.25 mm
108 N·m
(11.0 kgf·m,
79.6 lbf·ft)
INNER PAD SHIM A
BRAKE PADS
OUTER PAD SHIM B
CALIPER BRACKET
BLEED SCREW
9 N·m
(0.9 kgf·m,
7 lbf·ft)
BRAKE HOSE
CALIPER PIN A
CALIPER BODY
PISTON SEAL
Replace.
BANJO BOLT
34 N·m
(3.5 kgf·m,
25 lbf·ft)
PISTON
PAD RETAINERS
SEALING WASHERS
Replace.
PISTON BOOT
Replace.
8 x 1.0 mm
33 N·m
(3.4 kgf·m, 25 lbf·ft)
37647_CIVC_G0061

Fig. 7 Exploded view of the front brake caliper—with 2.0L Si model

DISC BRAKE PADS

REMOVAL & INSTALLATION

See Figure 8.

1. Raise and support the vehicle.
2. Remove the front wheels.
3. Check the thickness of the inner pad and the outer pad. Do not include the thickness of the backing plate.
4. Pad thickness should not be less than a service limit of 0.06 in. (1.6 mm).
5. If any part of the brake pad thickness is less than the service limit, replace the front brake pads as a set.
6. Remove some brake fluid from the master cylinder.
7. Remove the flange bolt, and pivot the caliper up out of the way. Check the hose and pin boots for damage and deterioration.
8. Remove the pad shims and the brake pads.
9. Remove the pad retainers.

To install:

10. Clean the caliper bracket thoroughly; remove any rust, and check for grooves and cracks.
11. Verify that the caliper pins move in and out smoothly. Clean and lube if needed.
12. Inspect the brake disc for runout, thickness, parallelism, and check for damage and cracks. See "Brakes" specification chart in this section.
13. Apply a thin coat of M-77 assembly paste to the retainer mating surface of the caliper bracket (indicated by the arrows).
14. Install the pad retainers. Wipe excess assembly paste off the retainers. Keep the assembly paste off the brake disc and the brake pads.
15. Install the brake caliper piston compressor tool on the caliper body (B).
16. Press in the piston with the brake caliper piston compressor tool so the caliper will fit over the brake pads. Make sure the piston boot is in position to prevent damaging it when pivoting the caliper down.

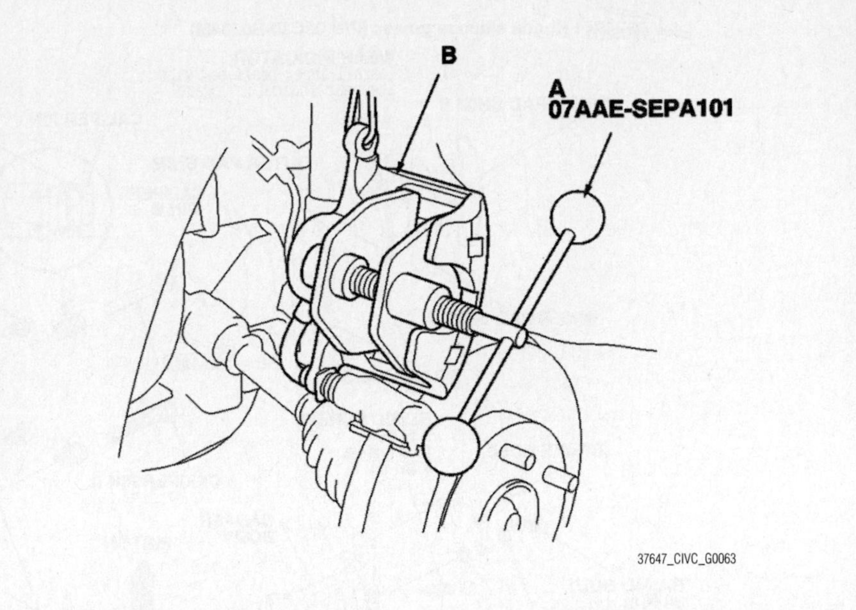
B
A
07AAE-SEPA101
37647_CIVC_G0063

Fig. 8 Install the brake caliper piston compressor tool (A) on the caliper body (B).

➡Be careful when pressing in the piston; brake fluid might overflow from the master cylinder's reservoir. If brake fluid gets on any painted surface, wash it off immediately with water.

17. Remove the brake caliper piston compressor tool.
18. Apply a thin coat of M-77 assembly paste to the pad side of the shims, the back of the brake pads and the other contact areas. Wipe excess assembly paste off the pad shims and the brake pads friction material. Keep grease and assembly paste off the brake disc and the brake pads. Contaminated brake disc or brake pads reduce stopping ability.
19. Install the brake pads and the pad shims correctly. Install the brake pad with the wear indicator on the upper inside. If you are reusing the brake pads, always reinstall the brake pads in their original positions to prevent a temporary loss of braking efficiency.
20. Pivot the caliper down into position. Install the flange bolt and tighten it to the specified torque of 25 ft. lbs. (34 Nm).
21. Clean the mating surfaces between the brake disc and the inside of the wheel, then install the front wheels.
22. Press the brake pedal several times to make sure the brakes work.

➡Engagement may require a greater pedal stroke immediately after the brake pads have been replaced as a set. Several applications of the brake pedal will restore the normal pedal stroke.

23. Add brake fluid as needed.
24. After installation, check for leaks at hose and line joints or connections, and retighten if necessary.
25. Test-drive the vehicle, then recheck for leaks.

BRAKE CALIPER

REMOVAL & INSTALLATION

See Figures 9 and 10.

➡Keep any grease off the brake disc and the brake pads.

1. Raise and support the vehicle.
2. Remove the rear wheel.
3. Release thc parking brake lever fully.
4. Remove the brake hose mounting bolt from the bracket.
5. Remove the brake caliper bracket

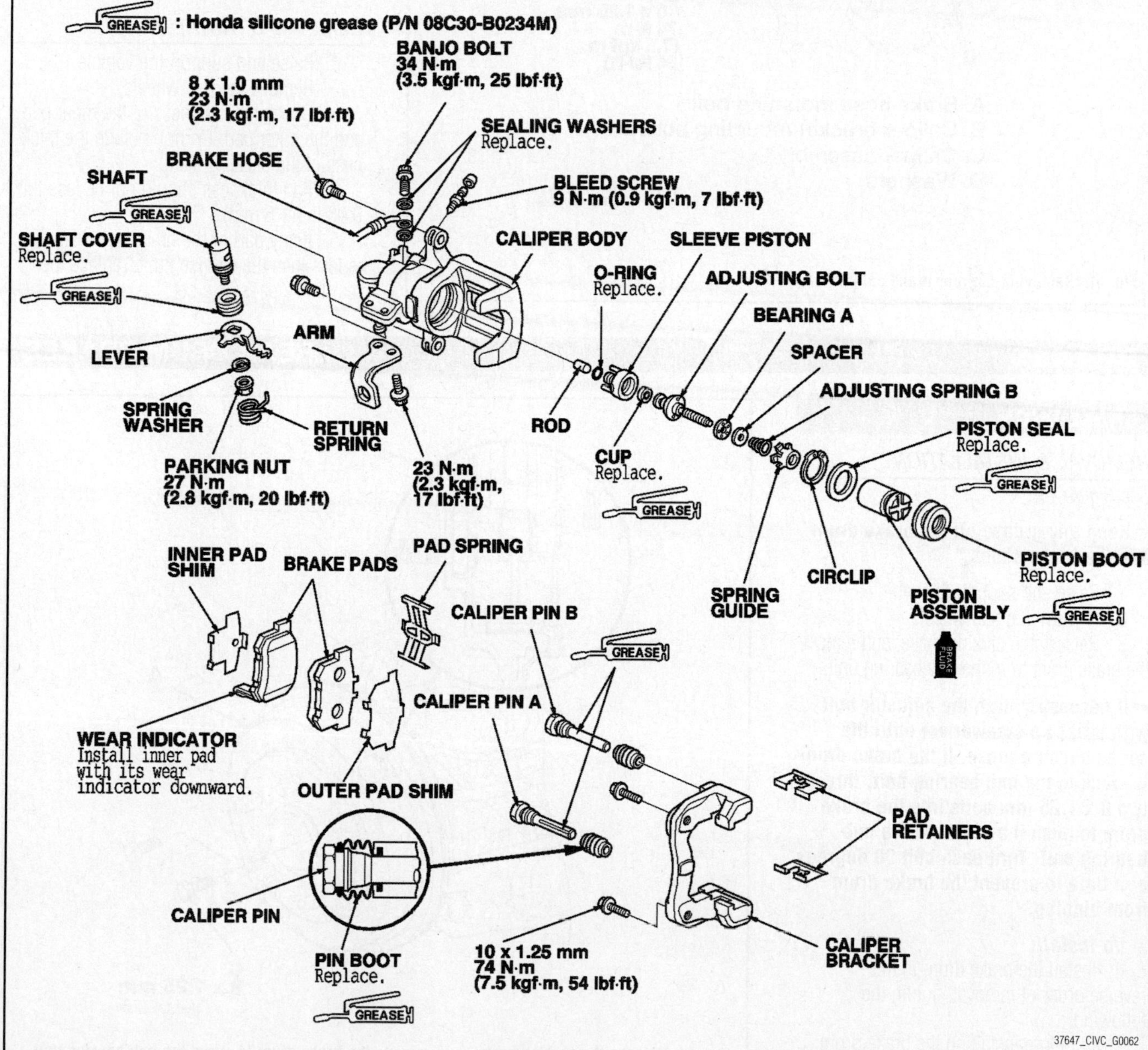
GREASE : Honda silicone grease (P/N 08C30-B0234M)
BANJO BOLT 34 N·m (3.5 kgf·m, 25 lbf·ft)
8 x 1.0 mm 23 N·m (2.3 kgf·m, 17 lbf·ft)
SEALING WASHERS Replace.
BRAKE HOSE
SHAFT
BLEED SCREW 9 N·m (0.9 kgf·m, 7 lbf·ft)
SHAFT COVER Replace.
CALIPER BODY
SLEEVE PISTON
O-RING Replace.
ADJUSTING BOLT
BEARING A
ARM
LEVER
SPACER
ADJUSTING SPRING B
SPRING WASHER
RETURN SPRING
ROD
PISTON SEAL Replace.
PARKING NUT 27 N·m (2.8 kgf·m, 20 lbf·ft)
23 N·m (2.3 kgf·m, 17 lbf·ft)
CUP Replace.
INNER PAD SHIM
BRAKE PADS
PAD SPRING
PISTON BOOT Replace.
CIRCLIP
SPRING GUIDE
PISTON ASSEMBLY
CALIPER PIN B
BRAKE FLUID
CALIPER PIN A
WEAR INDICATOR Install inner pad with its wear indicator downward.
OUTER PAD SHIM
PAD RETAINERS
CALIPER PIN
PIN BOOT Replace.
10 x 1.25 mm 74 N·m (7.5 kgf·m, 54 lbf·ft)
CALIPER BRACKET
37647_CIVC_G0062

Fig. 9 Exploded view of the rear brake caliper

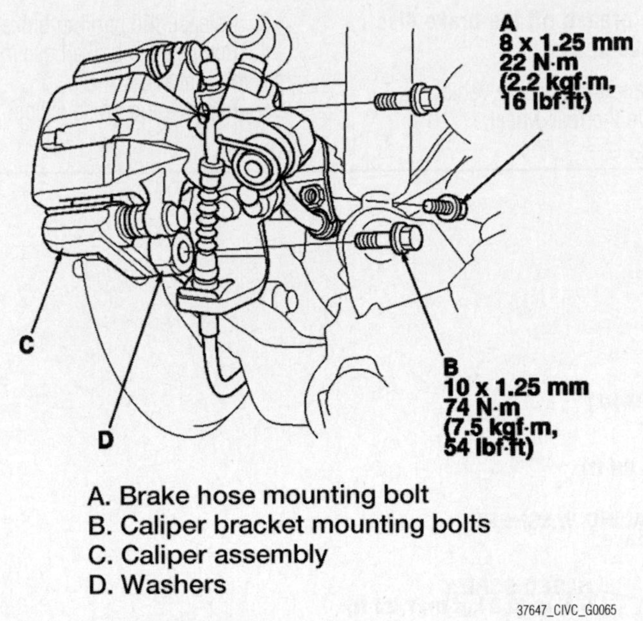
A
8 x 1.25 mm
22 N·m
(2.2 kgf·m,
16 lbf·ft)
B
10 x 1.25 mm
74 N·m
(7.5 kgf·m,
54 lbf·ft)
C
D
A. Brake hose mounting bolt
B. Caliper bracket mounting bolts
C. Caliper assembly
D. Washers
37647_CIVC_G0065

Fig. 10 Removing the rear brake caliper

mounting bolts, then remove the caliper assembly from the knuckle. To prevent damage to the caliper assembly or brake hose, use a short piece of wire to hang the caliper assembly from the undercarriage. Do not twist the brake hose and the parking brake cable excessively.

➡Make sure the washers are in position on reassembly, if they are removed.

6. Installation is the reverse of the removal procedure.

DISC BRAKE PADS

REMOVAL & INSTALLATION

1. Raise and support the vehicle.
2. Remove the rear wheels.
3. Check the thickness of the inner pad and the outer pad. Do not include the thickness of the backing plate.
4. Pad thickness should not be less than 0.06 in. (1.6 mm).
5. If any part of the brake pad thickness is less than the service limit, replace the rear brake pads as a set.

BRAKES — REAR DRUM BRAKES

BRAKE DRUM

REMOVAL & INSTALLATION

See Figure 11.

➡Keep any grease off the brake drum and the brake shoes.

1. Raise and support the vehicle.
2. Remove the rear wheel.
3. Release the parking brake, and remove the brake drum from the hub bearing unit.

➡If necessary, turn the adjuster bolt with a flat-tip screwdriver until the shoes become loose. If the brake drum is stuck to the hub bearing unit, thread two 8 x 1.25 mm bolts into the brake drum to push it away from the hub bearing unit. Turn each bolt 90 degrees at a time to prevent the brake drum from binding.

To install:

4. Install the brake drum in the reverse order of removal, noting the following:
   a. Before installing the brake drum, clean the mating surfaces between the hub bearing unit and the inside of the brake drum.
   b. After installation, press the brake pedal several times to make sure the

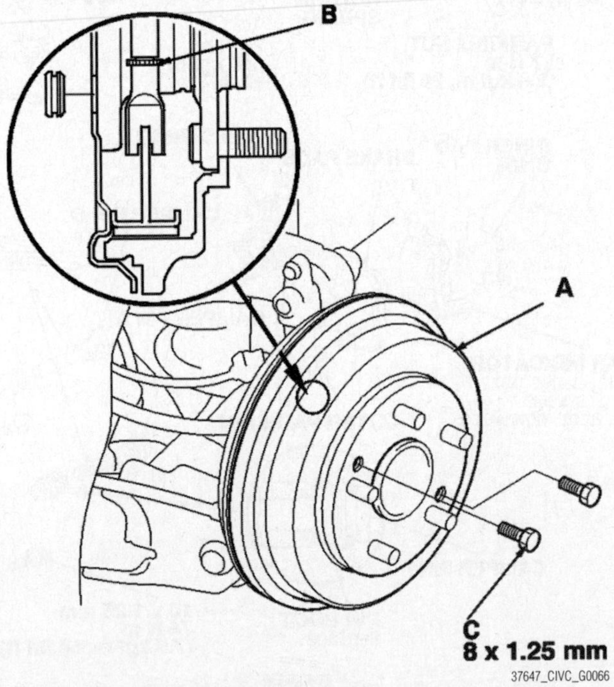
B
A
C
8 x 1.25 mm
37647_CIVC_G0066

Fig. 11 Release the parking brake, and remove the brake drum (A) from the hub bearing unit. If necessary, turn the adjuster bolt (B) with a flat-tip screwdriver until the shoes become loose. If the brake drum is stuck to the hub bearing unit, thread two 8 x 1.25 mm bolts (C) into the brake drum to push it away from the hub bearing unit. Turn each bolt 90 degrees at a time to prevent the brake drum from binding.

brakes work and self adjust the brake shoes. Do not drive before doing this procedure.

5. Clean the mating surfaces between the brake drum and the inside of the wheel, then install the rear wheel.

BRAKE SHOES

REMOVAL & INSTALLATION

See Figures 12 through 15.

1. Raise and support the vehicle.
2. Remove the rear wheels.
3. Release the parking brake, and remove the brake drum.
4. Remove the tension pins by pushing the respective retainer springs, and turning the pin.
5. Remove the lower return spring, and remove the brake shoe assembly over the hub.
6. Remove the forward brake shoe by removing the upper return spring, and disassemble the brake shoe assembly.
7. Remove the rearward brake shoe by disconnecting the parking brake cable from the parking brake lever.
8. Remove the U-clip, the wave washer, and the pivot pin, and separate the parking brake lever from the brake shoe.

To install:

9. Apply Molykote®-44MA grease to the sliding surface of the pivot pin for the rearward brake shoe.
10. Install the parking brake lever and the wave washer on the pivot pin, and secure the pin with a new U-clip.

➡Pinch the U-clip securely to prevent the parking brake lever from coming out of the brake shoe.

11. Connect the parking brake cable to the parking brake lever.
12. Apply a thin coat of Molykote-44MA grease to the connecting rod ends, and the sliding surfaces. Wipe off any excess. Keep grease off the brake linings.
13. Apply a thin coat of Molykote-44MA grease to the shoe ends and to the edge of the shoe surfaces that make contact with the backing plate. Wipe off any excess. Keep grease off the brake linings.
14. Install the two connecting rods on the adjuster bolt.
15. Clean the threaded portions of longer connecting rod and the sliding surface of short connecting rod, then coat them with Molykote 44MA grease.
16. Shorten connecting rod by fully turning in the adjuster bolt.
17. Assemble the brake shoes with the upper return spring, with the connecting rods and the adjuster bolt onto the backing plate. Reconnect the parking brake cable to the parking brake lever, then install the self-adjuster lever and the self-adjuster spring on the forward brake shoe.
18. Install the tension pins and the retainer springs by pushing in the respective spring and turning each pin.

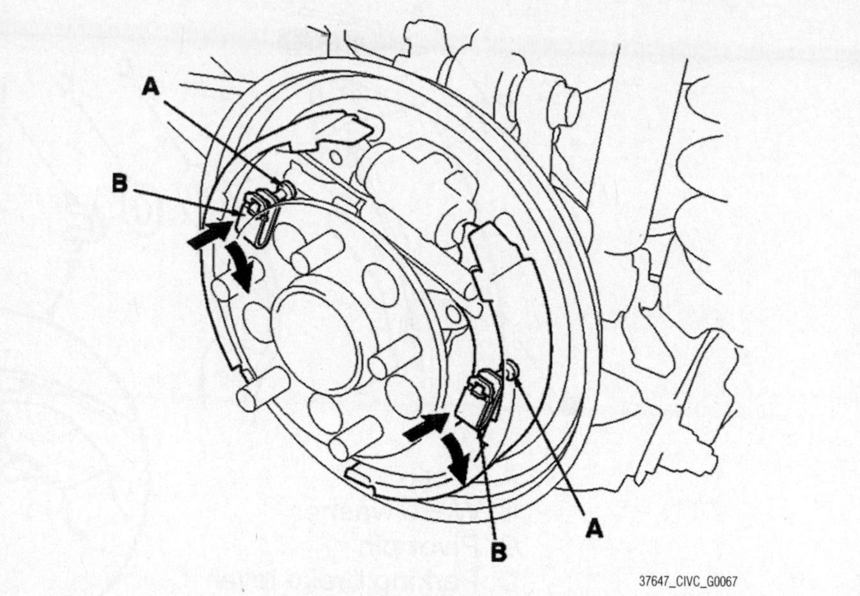
A
B
A
B
37647_CIVC_G0067

Fig. 12 Remove the tension pins (A) by pushing the respective retainer springs (B), and turning the pin.

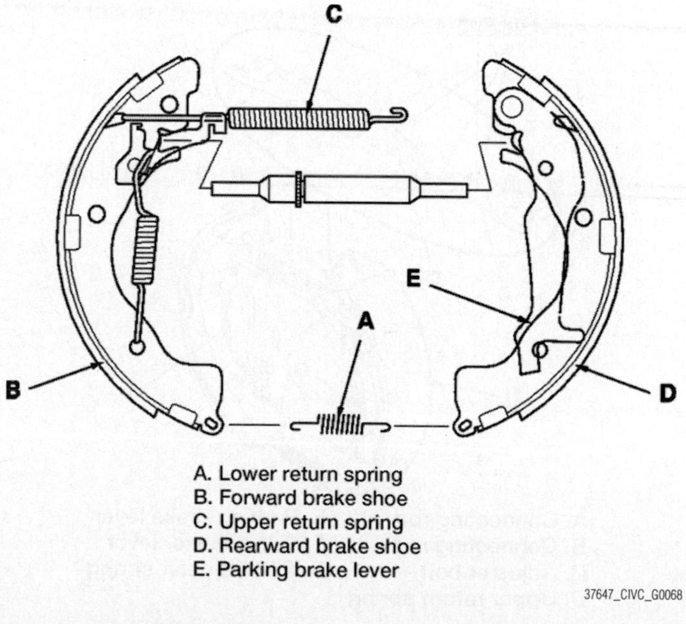
C
E
A
B
D
A. Lower return spring
B. Forward brake shoe
C. Upper return spring
D. Rearward brake shoe
E. Parking brake lever
37647_CIVC_G0068

Fig. 13 Showing the brake shoe components

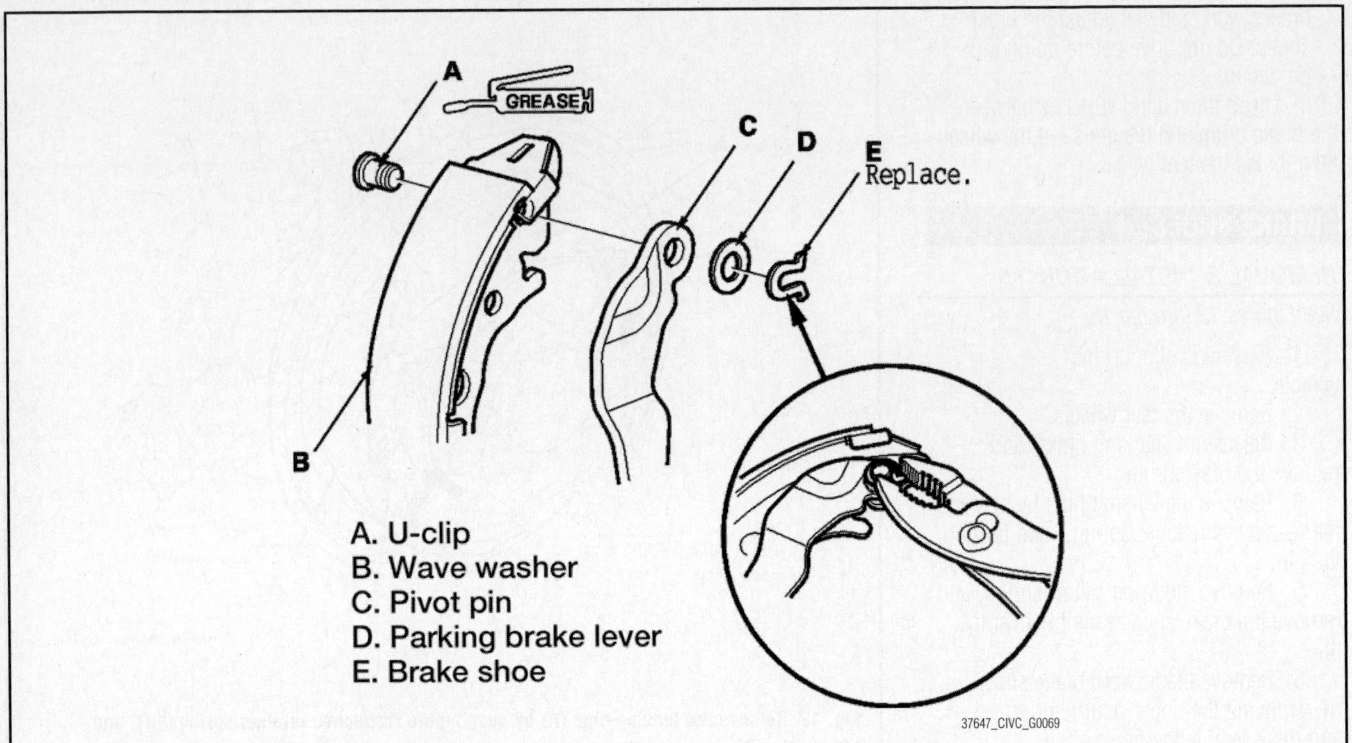
A
GREASE
C
D
E
Replace.
B
A. U-clip
B. Wave washer
C. Pivot pin
D. Parking brake lever
E. Brake shoe
37647_CIVC_G0069

Fig. 14 Removing hardware from the brake shoe

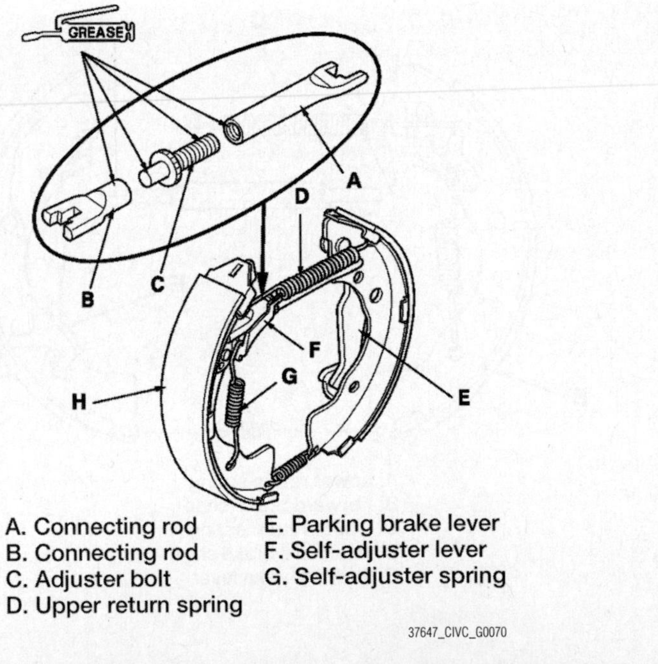
GREASE
A
D
B
C
F
G
H
E
A. Connecting rod
B. Connecting rod
C. Adjuster bolt
D. Upper return spring
E. Parking brake lever
F. Self-adjuster lever
G. Self-adjuster spring
37647_CIVC_G0070

Fig. 15 Assemble the brake shoes with the upper return spring, with the connecting rods and the adjuster bolt onto the backing plate. Reconnect the parking brake cable to the parking brake lever, then install the self-adjuster lever and the self-adjuster spring on the forward brake shoe.

19. Install the lower return spring.

➡Make sure the brake shoes are positioned on the brake shoe bosses on the backing plate, and the fitting on the top of the brake shoes are fitted into the wheel cylinder pistons.

20. Install the brake drum.

➡Before installing the brake drum, clean the mating surfaces between the hub bearing unit and the inside of the brake drum.

21. Clean the mating surfaces between the brake drum and the inside of the wheel, then install the rear wheels.

22. Press the brake pedal several times to make sure the brakes work and to set the self-adjusting brake.

➡Engagement of the brakes may require a greater pedal stroke immediately after the brake shoes have been replaced as a set. Several applications of the brake pedal will restore the normal pedal stroke.

23. Adjust the parking brake.

PARKING BRAKE CABLES

ADJUSTMENT

Rear Disc Brake Type

See Figure 16.

1. Remove the center console panel.
2. Release the parking brake lever fully.
3. Loosen the parking brake adjusting nut at the lower end of the parking brake handle.
4. Raise and support the vehicle.
5. Remove the rear wheels.
6. Make sure the lever on the rear brake caliper contacts the stop pin.

➡The lever will only contact the stop pin when the parking brake adjusting nut is loosened.

7. Clean the mating surfaces between the brake disc and the inside of the wheel, then install the rear wheels.
8. Pull the parking brake lever 1 click.
9. Tighten the parking brake adjusting nut until the parking brakes drag slightly when the rear wheels are turned.
10. Release the parking brake lever fully, and check that the parking brakes do not drag when the rear wheels are turned. Readjust if necessary.
11. Make sure the parking brake lever is within the specified number of clicks (8 to 10 clicks).
12. Install the center console panel.

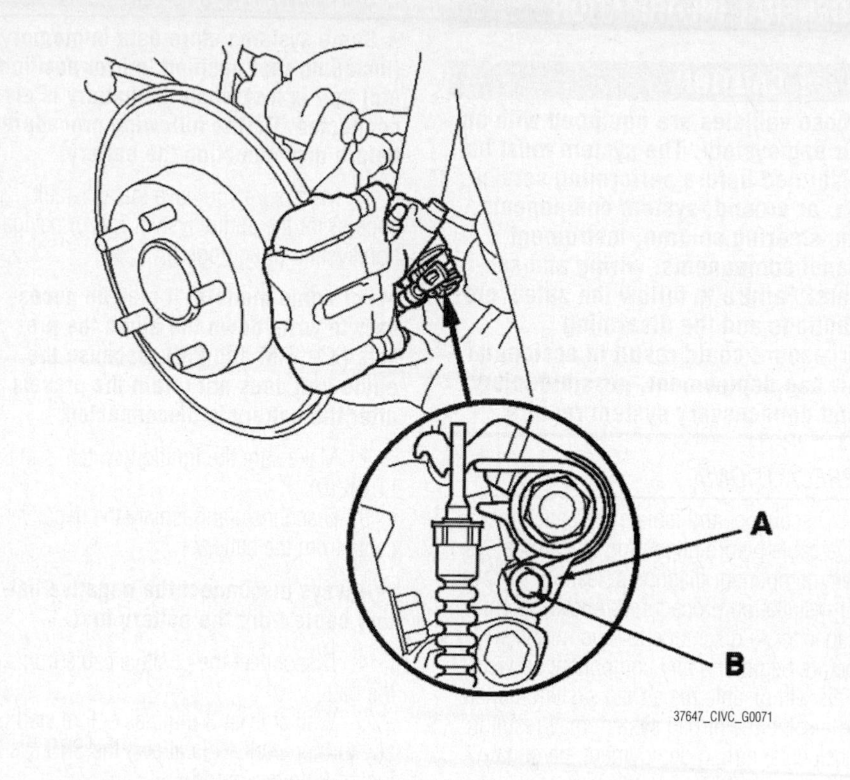
A
B
37647_CIVC_G0071

Fig. 16 Make sure the lever (A) on the rear brake caliper contacts the stop pin (B).

Adjustment - Rear Drum Brake Type

1. Remove the center console panel.
2. Release the parking brake lever fully.
3. Loosen the parking brake adjusting nut at the lower end of the parking brake handle.
4. Press the brake pedal several times to set the self-adjusting brake before adjusting the parking brake.
5. Pull the parking brake lever 1 click. Raise and support the vehicle.
6. Tighten the parking brake adjusting nut until the parking brakes drag slightly when the rear wheels are turned.
7. Release the parking brake lever fully, and check that the parking brakes do not drag when the rear wheels are turned. Readjust if necessary.
8. Make sure the parking brakes are fully applied when the parking brake lever is pulled all the way (8 to 10 clicks).
9. Install the center console panel.

PARKING BRAKE SHOES

REMOVAL & INSTALLATION

➡See "Brake Shoes" in this section.

ADJUSTMENT

➡See "Brake Shoes" in this section.

CHASSIS ELECTRICAL

AIR BAG (SUPPLEMENTAL RESTRAINT SYSTEM)

GENERAL INFORMATION

CAUTION

These vehicles are equipped with an air bag system. The system must be disarmed before performing service on, or around, system components, the steering column, instrument panel components, wiring and sensors. Failure to follow the safety precautions and the disarming procedure could result in accidental air bag deployment, possible injury and unnecessary system repairs.

PRECAUTIONS

Disconnect and isolate the battery negative cable before beginning any airbag system component diagnosis, testing, removal, or installation procedures. Allow system capacitor to discharge for two minutes before beginning any component service. This will disable the airbag system. Failure to disable the airbag system may result in accidental airbag deployment, personal injury, or death.

DISARMING THE SYSTEM

➡Some systems store data in memory (including seat position, mirror position, etc) that is lost when the battery is disconnected. Do the following procedures before disconnecting the battery.

1. Make sure you have the anti-theft code(s) for the audio system and/or navigation system (if equipped).

➡For some models, it may be necessary to write down the audio the presets (XM, AM and FM), because the audio unit does not retain the presets after the battery is disconnected.

2. Make sure the ignition switch is in LOCK (0).
3. Disconnect and isolate the negative cable from the battery.

➡Always disconnect the negative battery cable from the battery first.

4. Disconnect the positive cable from the battery.
5. Wait at least 3 minutes before starting any further work. This allows the SRS system to fully discharge.

ARMING THE SYSTEM

➡Some systems store data in memory (including seat position, mirror position, etc) that is lost when the battery is disconnected. Do the following procedures to restore the system back to normal operation.

1. Clean the battery terminals.
2. Test the battery, if necessary.
3. Reconnect the positive cable to the battery first, then reconnect the negative cable to the battery.

CAUTION

Always connect the positive cable to the battery first.

4. Apply multipurpose grease to the terminals to prevent corrosion.
5. Enter the anti-theft code(s) for the audio system and/or navigation system (if equipped).
6. Enter the audio presets.
7. Set the clock (for vehicles without navigation).

DRIVE TRAIN

CLUTCH

REMOVAL & INSTALLATION

See Figures 17 and 18.

1. Remove the transmission.
2. Check the evenness of the height of the diaphragm spring fingers using the clutch alignment disc, the clutch alignment shaft, the remover handle, and a feeler gauge. If the height difference is more than the service limit of 0.03 in. (0.8 mm), replace the pressure plate.
3. Install the ring gear holder, the clutch alignment shaft, and the remover handle.
4. To prevent warping, loosen the pressure plate mounting bolts in a crisscross pattern in several steps, then remove the pressure plate.

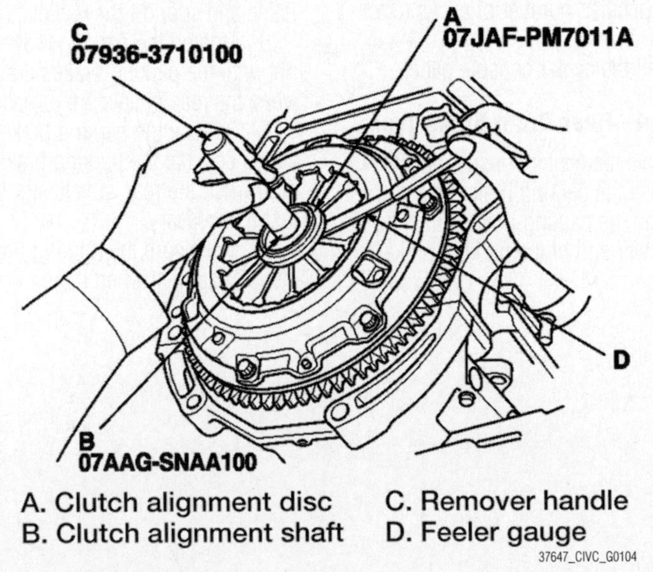
C
07936-3710100
A
07JAF-PM7011A
D
B
07AAG-SNAA100
A. Clutch alignment disc
B. Clutch alignment shaft
C. Remover handle
D. Feeler gauge
37647_CIVC_G0104

Fig. 17 Check the evenness of the height of the diaphragm spring fingers using the clutch alignment disc (A), the clutch alignment shaft (B), the remover handle (C), and a feeler gauge (D). If the height difference is more than the service limit of 0.03 in. (0.8 mm), replace the pressure plate.

HYDRAULIC SYSTEM BLEEDING

FLUID FILL PROCEDURE

1. Make sure the brake fluid level in the clutch reservoir is at the MAX (upper) level line.

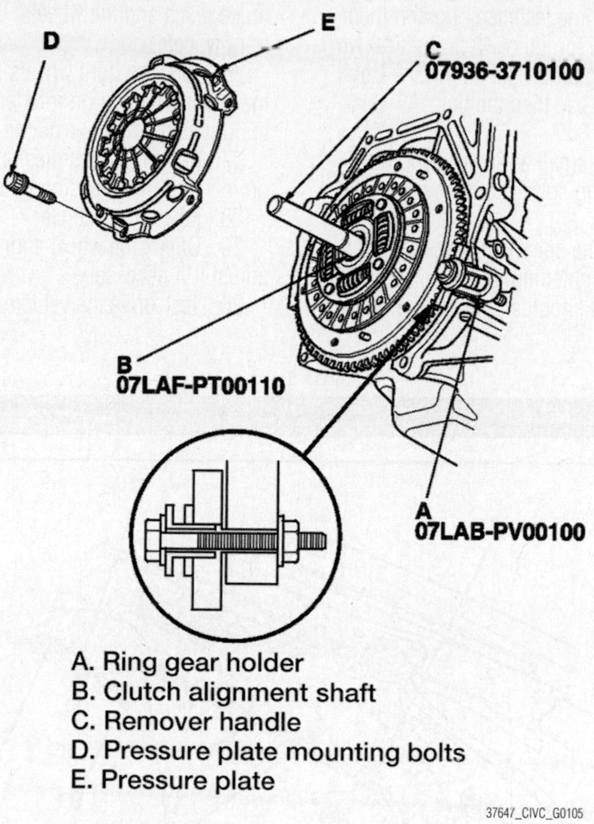
D
E
C
07936-3710100
B
07LAF-PT00110
A
07LAB-PV00100
A. Ring gear holder
B. Clutch alignment shaft
C. Remover handle
D. Pressure plate mounting bolts
E. Pressure plate
37647_CIVC_G0105

Fig. 18 Install the ring gear holder, the clutch alignment shaft, and the remover handle. To prevent warping, loosen the pressure plate mounting bolts in a crisscross pattern in several steps, then remove the pressure plate.

2. Attach one end of a clear tube to the bleeder screw and put the other end into a container. Loosen the bleeder screw to allow air to escape from the system.
3. Make sure there is an adequate supply of fluid in the reservoir, then slowly push the clutch pedal all the way down. Before releasing the pedal, have an assistant temporarily tighten the bleeder screw. Loosen the bleeder screw, and push the pedal down again. Repeat this step until no more bubbles appear in the clear tube.

➡**Make sure the fluid level on the reservoir does not go below MIN (lower).**

4. Tighten the bleeder screw securely.
5. Refill the brake fluid in the reservoir to the MAX (upper) level line.

FRONT DRIVESHAFT

REMOVAL & INSTALLATION

See Figure 19.

1. Raise and support the vehicle.
2. Remove the front wheels.
3. Pry up the stake on the spindle nut, then remove the nut.
4. Drain the transmission fluid, then reinstall the drain plug with a new washer.
5. Remove the nuts and the bolt, then separate the lower arm using a prybar.
6. Separate the driveshaft outboard joint from the front hub using a soft face hammer.
7. Pull the knuckle outward, and remove the driveshaft outboard joint from the front hub.
8. On the left driveshaft, pry the inboard joint from the differential using the prybar. Remove the driveshaft as an assembly.

➡**Do not pull on the driveshaft or the inboard joint may come apart. Pull the inboard joint straight out to avoid damaging the oil seal. Be careful not to damage the oil seal or the end of the inboard joint using the prybar.**

9. On the right driveshaft, drive the inboard joint off of the intermediate shaft using a drift punch and a hammer (M/T model). Pry the inboard joint from the differential using the prybar (A/T model)

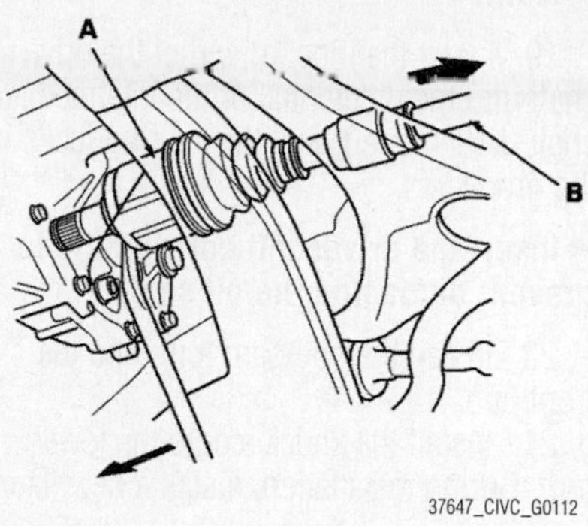
A
B
37647_CIVC_G0112

Fig. 19 On the left driveshaft, pry the inboard joint (A) from the differential using the prybar. Remove the driveshaft (B) as an assembly.

10. Remove the driveshaft as an assembly.
11. Do not pull on the driveshaft or the inboard joint may come apart. Pull the inboard joint straight out to avoid damaging the oil seal.

➡**Be careful not to damage the oil seal or the end of the inboard using the pry-bar.**

12. Remove the set ring from the inboard joint (except right driveshaft with M/T model).
13. Remove the set ring from the intermediate shaft (M/T model).

To install:

➡**Before starting installation, make sure the mating surfaces of the joint and the splined section are clean.**

14. Apply about 5 g (0.18 oz) of moly 60 paste to the contact area of the outboard joint and the front wheel bearing.

➡**The paste helps prevent noise and vibration.**

15. Install a new set ring into the set ring groove of the driveshaft.
16. Install a new set ring into the set ring groove of the intermediate shaft.
17. On M/T model, apply specified super high temp urea grease to the whole splined surface of the right driveshaft. After applying grease, remove the grease from the splined grooves at intervals of 2-3 splines and from the set ring groove so that air can bleed from the intermediate shaft.
18. Clean the areas where the driveshaft contacts the differential thoroughly with solvent, and dry them with compressed air.

➡**Do not wash the rubber parts with solvent.**

19. Insert the inboard end of the driveshaft into the differential or the intermediate shaft (M/T model) until the set ring locks in the groove.

➡**Insert the driveshaft horizontally to prevent damaging the oil seal.**

20. Install the outboard joint into the front hub.
21. Install the knuckle onto the lower arm. During installation, install a new flange bolt and new self-locking nuts. After lightly tightening all three fasteners, tighten them to the specified torque of 43 ft. lbs. (59 Nm) in the following order: the nut on the front, the nut on the rear, then the bolt.
22.
23. Apply a small amount of engine oil to the seating surface of a new spindle nut.
24. Install the spindle nut, then tighten the nut. After tightening, use a drift to stake the spindle nut shoulder against the driveshaft.
25. Clean the mating surfaces of the brake discs and the wheels, then install the front wheels.
26. Turn the front wheels by hand, and make sure there is no interference between the driveshaft and surrounding parts.
27. Refill the transmission with the recommended transmission fluid:
28. Lower the vehicle.
29. Check the wheel alignment, and adjust it if necessary.
30. Test-drive the vehicle.

ENGINE COOLING

ENGINE FAN

REMOVAL & INSTALLATION

➡**See "Radiator & Fan" in this section.**

RADIATOR & FAN

REMOVAL & INSTALLATION

1.8L Engine

See Figures 20 and 21.

1. Do the battery removal procedure.
2. Drain the engine coolant.
3. Remove the front grille cover.
4. Disconnect the fan motor connectors and the hood switch connector, then remove the harness clamps.
5. Disconnect the reservoir hose and remove the radiator cap base mounting bolts, then the radiator upper bracket bolts and brackets.
6. Disconnect the upper radiator hose.
7. Raise the vehicle on the lift.
8. Remove the splash shield.
9. Disconnect the ECT sensor 2 connector, and remove the harness clamp, then disconnect the lower radiator hose from the radiator.
10. Lower the vehicle on the lift.
11. Remove the condenser bracket mounting bolts, then remove the upper radiator support.
12. Pull up the radiator, then remove the fan shroud assemblies and other parts from the radiator.

To install:

13. Install the radiator in the reverse order of removal. Make sure the upper and lower cushions are set securely.
14. Install the bulkhead in the reverse order of removal. Apply body paint to the bulkhead mounting bolts.
15. Do the battery installation procedure.

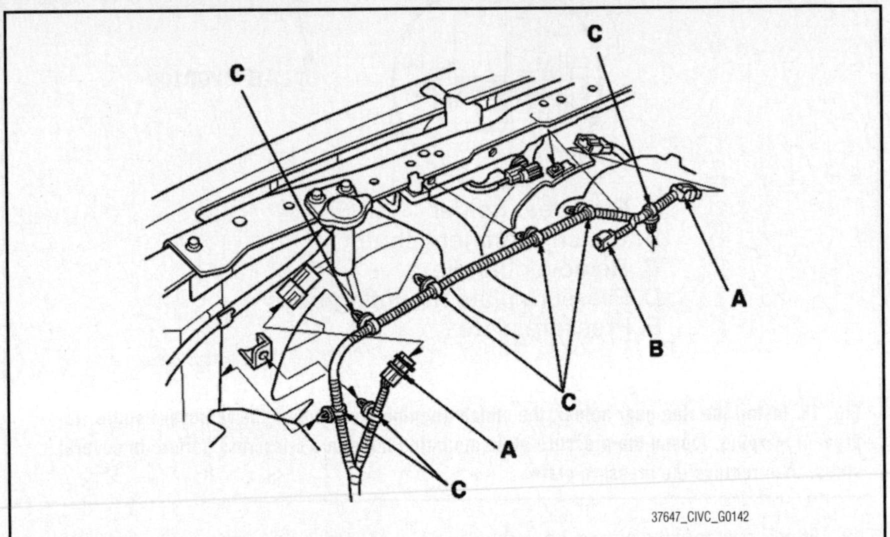
C
C
A
B
C
A
C
37647_CIVC_G0142

Fig. 20 Disconnect the fan motor connectors (A) and the hood switch connector (B), then remove the harness clamps (C).

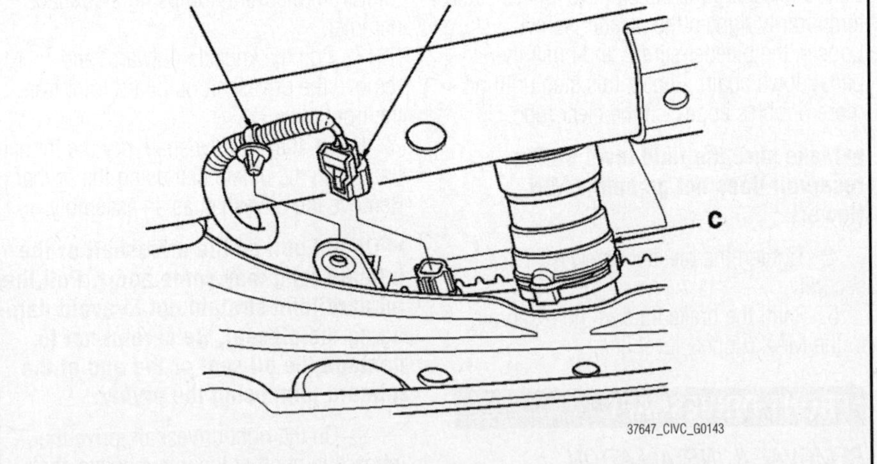
B
A
C
37647_CIVC_G0143

Fig. 21 Disconnect the ECT sensor 2 connector (A), and remove the harness clamp (B), then disconnect the lower radiator hose (C) from the radiator.

16. Fill the radiator with engine coolant, and bleed the air from the cooling system.

THERMOSTAT

REMOVAL & INSTALLATION

1.8L Engine

1. Drain the engine coolant.
2. Remove the harness clamp bracket and the thermostat cover, then remove the thermostat.
3. Install the thermostat with a new rubber seal, then install the harness clamp bracket.
4. Refill the radiator with engine coolant, and bleed the air from the cooling system.

2.0L Engine

1. Drain the engine coolant.
2. Remove the splash shield.
3. Disconnect the lower radiator hose, then remove the thermostat.
4. Install the new thermostat assembly with a new O-ring, then install the lower radiator hose.
5. Install the splash shield.
6. Refill the radiator with engine coolant, and bleed the air from the cooling system.

WATER PUMP

REMOVAL & INSTALLATION

1.8L Engine

1. Drain the engine coolant.
2. Remove the drive belt auto-tensioner.
3. Remove the water pump by removing the five bolts.
4. Clean and inspect the O-ring groove and the mating surface of the engine block.
5. Install the water pump with a new O-ring.
6. Clean up any spilled engine coolant.
7. Install the drive belt auto-tensioner.
8. Refill the radiator with engine coolant, and bleed the air from the cooling system.

2.0L Engine

1. Remove the drive belt.
2. Drain the engine coolant.
3. Remove the drive belt auto-tensioner pulley.
4. Remove the crankshaft pulley.
5. Remove the oil cooler joint pipe, then remove the seven bolts securing the water pump. Remove the water pump.

To install:

6. Inspect, and clean the O-ring groove and mating surface of the water passage.
7. Install the water pump with a new O-rings and the oil cooler joint pipe.
8. Clean up any spilled engine coolant.
9. Install the crankshaft pulley.
10. Install the drive belt auto-tensioner pulley.
11. Install the drive belt.
12. Refill the radiator with engine coolant, and bleed the air from the cooling system.

ENGINE ELECTRICAL

CHARGING SYSTEM

ALTERNATOR

REMOVAL & INSTALLATION

1.8L Engine

1. Do the battery terminal disconnect procedure, then wait at least 3 minutes before beginning work.
2. Remove the drive belt.
3. Disconnect the alternator connector and the positive alternator cable from the alternator.
4. Remove the harness connector and the harness clamps.
5. Remove the retaining bolts and remove the alternator.
6. Installation is the reverse of the removal procedure.
7. Tighten the alternator mounting bolts to 17 ft. lbs. (24 Nm).

2.0L Engine

See Figure 20.

1. Do the battery terminal disconnect procedure, then wait at least 3 minutes before beginning work.
2. Remove the drive belt.
3. Remove the front grille cover.
4. Disconnect the fan motor connectors and the hood switch connector, then remove the harness clamps.
5. Remove the coolant reservoir hose, the radiator cap base mounting bolts, the clips and the radiator upper brackets.
6. Remove the condenser bracket mounting bolts, then remove the upper radiator support.
7. Remove the three bolts securing the alternator.
8. Disconnect the alternator connector and the positive alternator cable, and remove the harness clamp, then remove the alternator.
9. Installation is the reverse of the removal procedure.
10. Tighten the alternator mounting bolts to 16 ft. lbs. (22 Nm).

FIRING ORDER

Firing order for all engines is 1-3-4-2.

IGNITION COIL

REMOVAL & INSTALLATION

1.8L Engine

See Figure 22.

1. Disconnect the ignition coil connectors, then remove the ignition coils.
2. Remove the spark plugs and inspect them.
3. Apply a small amount of anti-seize compound to the plug threads, and screw the plugs into the cylinder head, finger tight. Torque them to 18 ft. lbs. (25 Nm).
4. Install the ignition coils in the reverse order of removal.

2.0L Engine

See Figure 23.

1. Remove the ignition coil cover.
2. Disconnect the ignition coil connectors, then remove the ignition coils.
3. Remove the spark plugs and inspect them.
4. Apply a small amount of anti-seize compound to the plug threads, and screw the plugs into the cylinder head, finger tight. Torque them to 18 ft. lbs. (25 Nm).
5. Install the ignition coils in the reverse order of removal.

IDLE SPEED

INSPECTION

1. Before checking the idle speed, check these items:
   - The malfunction indicator lamp (MIL) has not been reported on, and there are no DTCs.
   - Ignition timing
   - Spark plugs
   - Air cleaner
   - PCV system
2. Apply the parking brake, and make sure the headlights are off.
3. Disconnect the evaporative emission (EVAP) canister purge valve connector.
4. Connect the scan tool to the data link

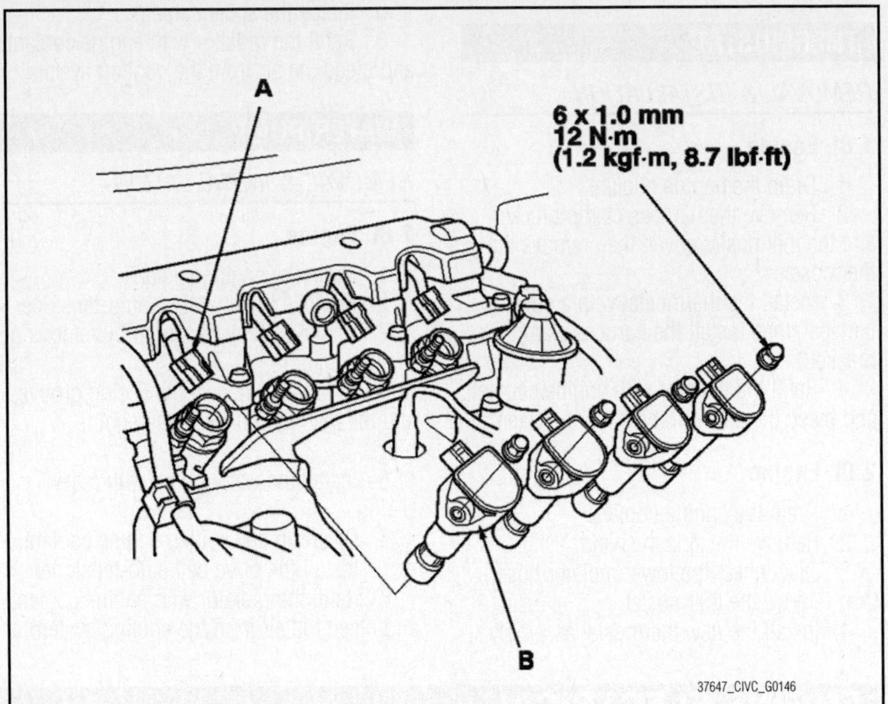
A
6 x 1.0 mm
12 N·m
(1.2 kgf·m, 8.7 lbf·ft)
B
37647_CIVC_G0146

Fig. 22 Disconnect the ignition coil connectors (A), then remove the ignition coils (B).

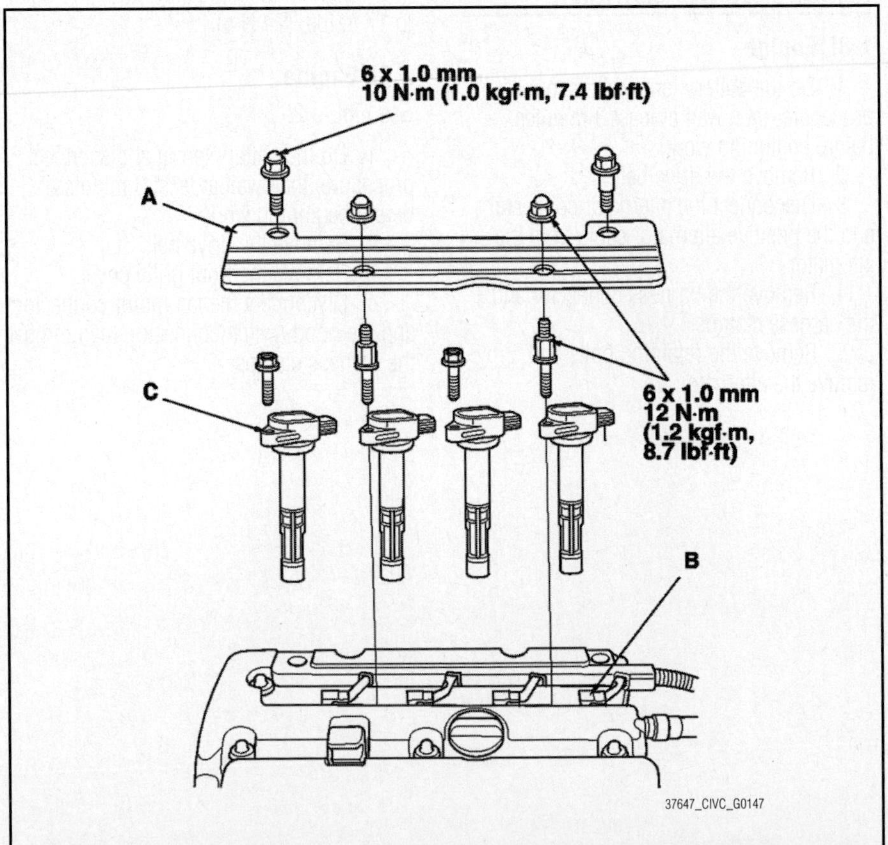
6 x 1.0 mm
10 N·m (1.0 kgf·m, 7.4 lbf·ft)
A
C
6 x 1.0 mm
12 N·m
(1.2 kgf·m,
8.7 lbf·ft)
B
37647_CIVC_G0147

Fig. 23 Disconnect the ignition coil cover (A) and the connectors (B), then remove the ignition coils (C).

connector (DLC) located under the driver's side of the dashboard.

5. Make sure the scan tool communicates with the ECM/PCM. If it doesn't, perform further DLC circuit troubleshooting.

6. Start the engine. Hold the engine speed at 3,000 RPM without load (in neutral) until the radiator fan comes on, then let it idle.

7. Check the idle speed without load conditions: headlights, blower fan, radiator fan, and air conditioner off. Idle speed should be 750±50 RPM.

8. Let the engine idle for 1 minute with high electric load (A/C on, temperature set to max cool, blower fan on high, headlights on high beam). Idle speed should be 750±50 RPM.

a. If the idle speed is not within specification, do the "ECM/PCM Idle Learn Procedure". See "Engine Control Module (ECM/PCM)" in "ENGINE PERFORMANCE & EMISSION CONTROLS" section.

b. If the idle speed is still not within specification, go to the symptom troubleshooting.

9. Reconnect the EVAP canister purge valve connector.

IGNITION TIMING

INSPECTION

1. Connect the Honda Diagnostic System (HDS), or equivalent scan tool, to the data link connector (DLC).

2. Turn the ignition switch to ON (II).

3. Make sure the scan tool communicates with the vehicle and the engine control module (ECM/PCM)/powertrain control module (PCM). If it does not communicate, troubleshoot the DLC circuit.

4. Check for DTCs. If a DTC is present, diagnose and repair the cause before continuing with this test.

5. Start the engine. Hold the engine speed at 3,000 RPM with no load (neutral (M/T model or in N or P A/T model) until the radiator fan comes on, then let it idle.

6. Check the idle speed. See "Idle Speed" in this section.

7. Jump the SCS line with the scan tool.

8. Connect the timing light to the No. 1 ignition coil harness.

9. Aim the light toward the pointer on the cam chain case. Check the ignition timing under a no load condition (headlights, blower fan, rear window defogger, and air conditioner are turned off).

10. Ignition timing should be:

- M/I model: 8°±2° BTDC (RED mark) at idle in Neutral
- A/T model: 8°±2° BTDC (RED mark) at idle in N or P

➡On 2.0L engine, the other pointer on the right side is not used.

11. If the ignition timing differs from the specification, check the cam timing. If the cam timing is OK, update the ECM/PCM if it does not have the latest software, or substitute a known-good ECM/PCM, then recheck. If the system works properly, and the ECM/PCM was substituted, replace the original ECM/PCM.

12. Disconnect the scan tool and the timing light.

ADJUSTMENT

➡Ignition timing is electronically controlled by the ECM/PCM. No adjustment is possible.

SPARK PLUGS

REMOVAL & INSTALLATION

➡See "Ignition Coil" in this section.

ENGINE ELECTRICAL

STARTING SYSTEM

STARTER

REMOVAL & INSTALLATION

1.8L Engine

1. Disconnect the battery cables.

2. Remove the exhaust pipe.

3. Remove the intake manifold bracket.

4. Remove the wiring harness clamps and the harness connector from each clamp.

5. Remove the two bolts securing the starter, and then remove the starter from the engine.

6. Disconnect the starter wiring harnesses.

To install:

7. Connect the positive starter cable and the connector.

➡Make sure the starter cable crimped side of the ring terminal faces away from the starter when you connect it.

8. Install the starter, and then loosely install the upper mounting bolt and the lower mounting bolt.

9. Tighten the upper mounting bolt to 33 ft. lbs. (44 Nm), and then tighten the lower mounting bolt to 33 ft. lbs. (44 Nm).

10. The remainder of the installation is the reverse order of removal.

2.0L Engine

See Figure 24.

1. Disconnect the negative battery cable.

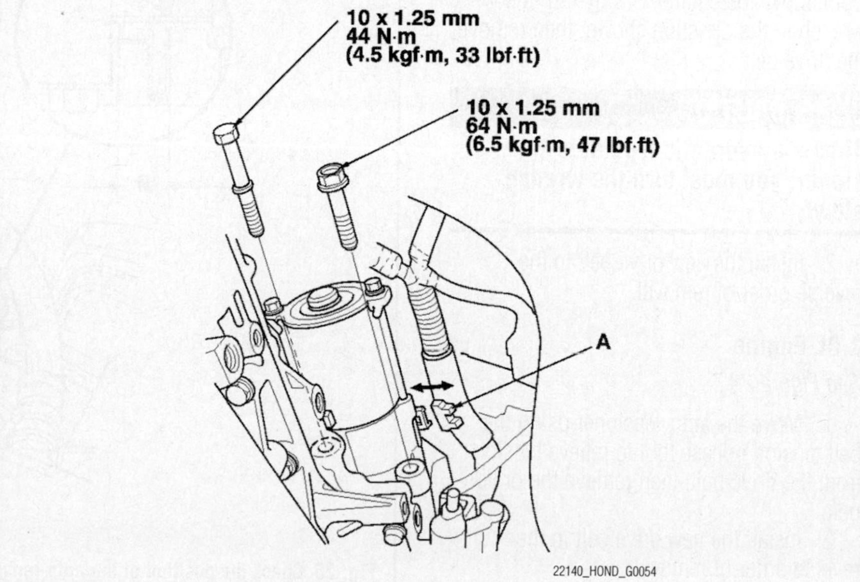
10 x 1.25 mm
44 N·m
(4.5 kgf·m, 33 lbf·ft)
10 x 1.25 mm
64 N·m
(6.5 kgf·m, 47 lbf·ft)
A
22140_HOND_G0054

Fig. 24 Starter mounting bolt torque specifications—2.0L Civic

2. Remove the engine splash shield.
3. Remove the intake manifold bracket.
4. Remove the harness clamp, and the two bolts securing the starter, and then remove the starter from the engine.
5. Disconnect the positive starter cable and the S terminal connector.
6. Remove the harness clamp and remove the starter.

To install:

7. Install the wiring harness clamp.
8. Connect the positive starter cable and S terminal connector.

➡Make sure the starter cable crimped side of the ring terminal faces away from the starter when you connect it.

9. Install the starter and tighten the mounting bolts.
10. The remainder of the installation is the reverse order of removal.

ENGINE MECHANICAL

ACCESSORY DRIVE BELTS

ACCESSORY BELT ROUTING

➡See "Removal & Installation" for belt routing figures.

INSPECTION

See Figures 25 and 26.

1. Inspect the belt for cracks or damage If the belt is cracked or damaged, replace it..
2. Check the position of the auto-tensioner indicator pointer is within the standard range. If it is out of the standard range, replace the drive belt.

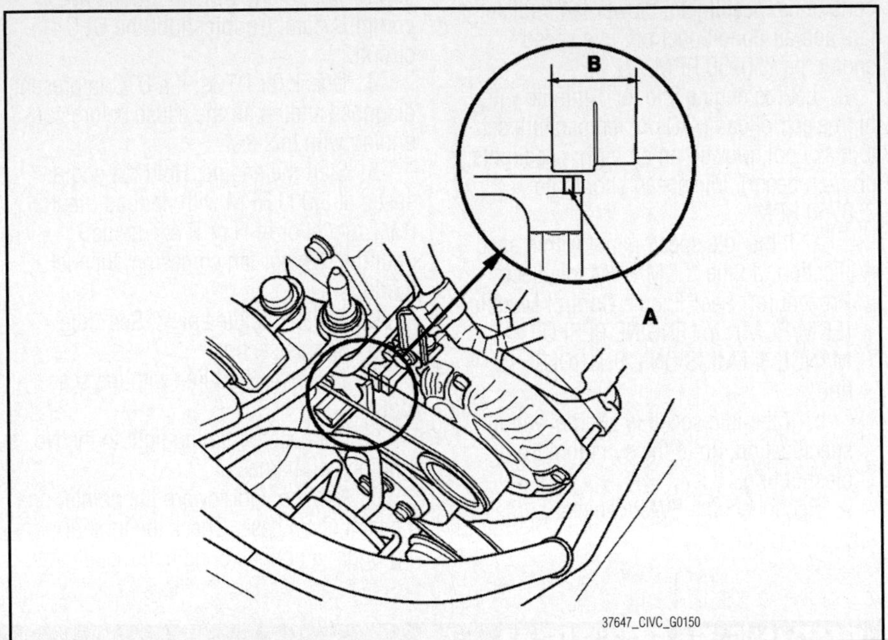
B
A
37647_CIVC_G0150
Fig. 25 Check the position of the auto-tensioner indicator pointer (A) is within the standard range (B) as shown.—1.8L engine

ADJUSTMENT

➡These belt systems are equipped with automatic tensioners and no adjustment is required.

REMOVAL & INSTALLATION

1.8L Engine

See Figure 27.

1. Set a long-handled, boxed-end wrench on the drive belt auto-tensioner from above the engine. Slowly turn the wrench in the direction shown, then remove the drive belt.

⁂ CAUTION

This is a hydraulic type auto-tensioner; you must turn the wrench slowly.

2. Install the new drive belt in the reverse order of removal.

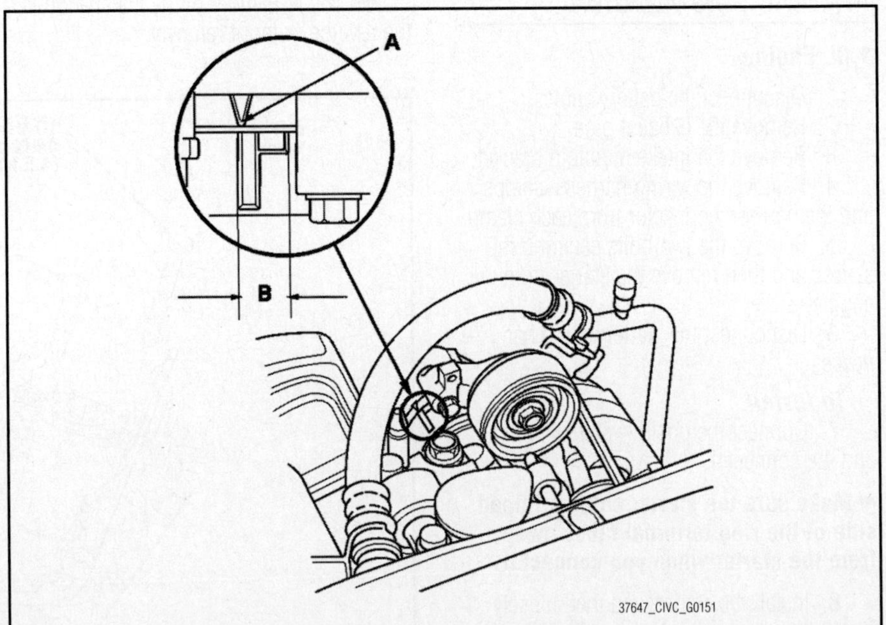
A
B
37647_CIVC_G0151
Fig. 26 Check the position of the auto-tensioner indicator pointer (A) is within the standard range (B) as shown.—2.0L engine

2.0L Engine

See Figure 28.

1. Move the auto-tensioner using the belt tension release tool to relieve tension from the drive belt, then remove the drive belt.
2. Install the new drive belt in the reverse order of removal.

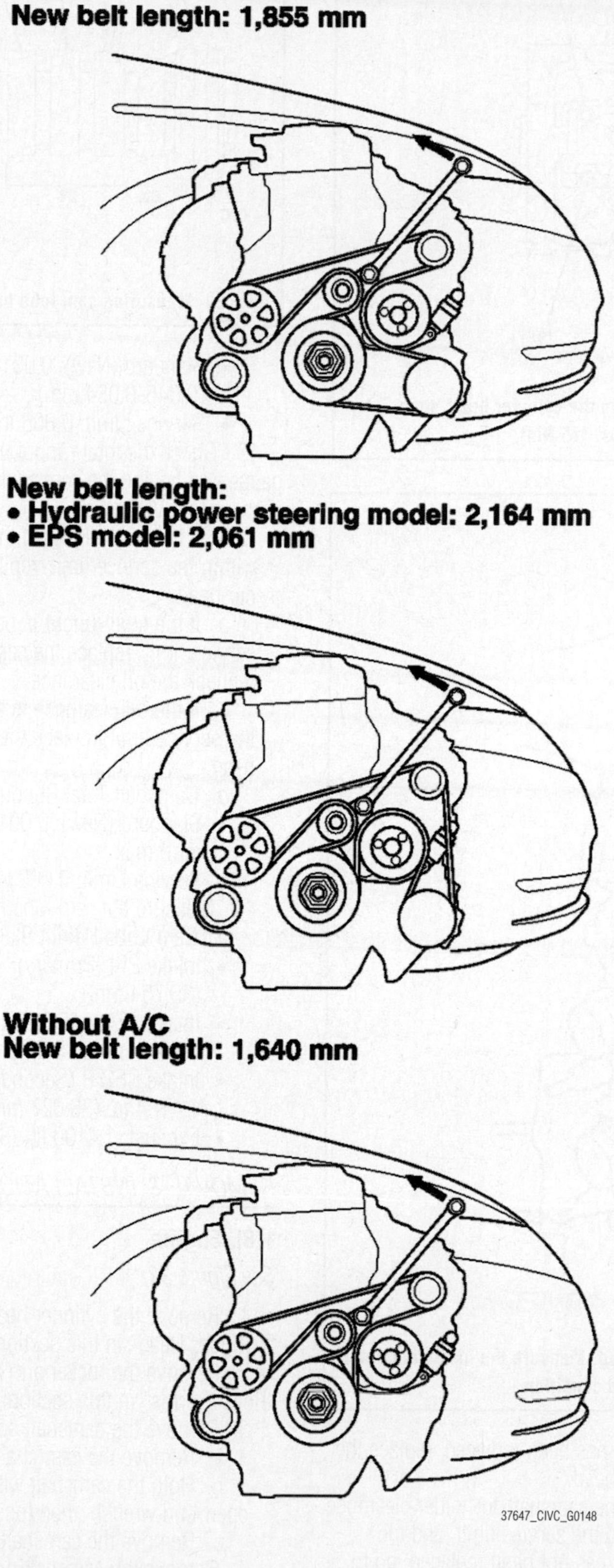
New belt length: 1,855 mm
New belt length:
• Hydraulic power steering model: 2,164 mm
• EPS model: 2,061 mm
Without A/C
New belt length: 1,640 mm
37647_CIVC_G0148

Fig. 27 Removing accessory drive belt—1.8L engine

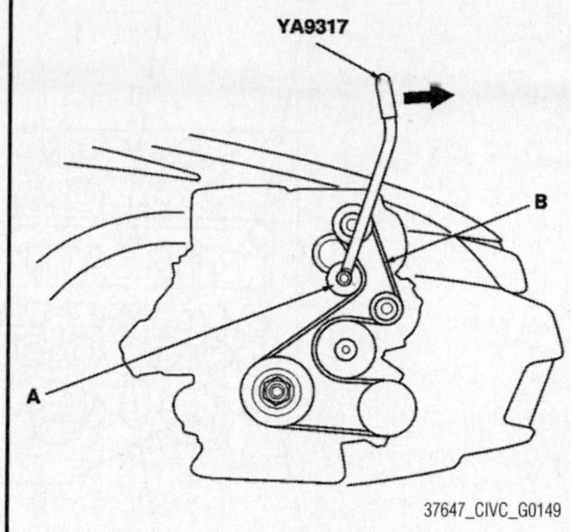
YA9317
B
A
37647_CIVC_G0149

Fig. 28 Removing accessory drive belt—2.0L engine

CAMSHAFT

INSPECTION

1.8L Engine

See Figures 29 through 31.

1. Remove the camshaft sprocket.
2. Remove the rocker arm assembly. See "Rocker Arms" in this section.
3. Disassemble the rocker arm assembly.
4. Put the rocker shaft holders and the lost motion holder on the cylinder head, then tighten the bolts, in sequence, to the specified torque of 11 ft. lbs. (15 Nm).
5. Seat the camshaft by pushing it away from the camshaft pulley end of the cylinder head.
6. Zero a dial indicator against the end of the camshaft, then push the camshaft back and forth, and read the end play. If the end play is beyond the service limit, replace the thrust cover and recheck. If it is still beyond the service limit, replace the camshaft".
7. Camshaft End Play should be:
   - Standard (New): 0.002-0.010 in. (0.050-0.250 mm)
   - Service Limit: 0.02 in. (0.4 mm)
8. Remove the camshaft. See "Removal & Installation".
9. Wipe the camshaft clean, then inspect the lift ramps. Replace the camshaft if any lobes are pitted, scored, or excessively worn.
10. Measure the diameter of each camshaft journal.
11. Zero the gauge to the journal diameter.
12. Clean the camshaft bearing surfaces

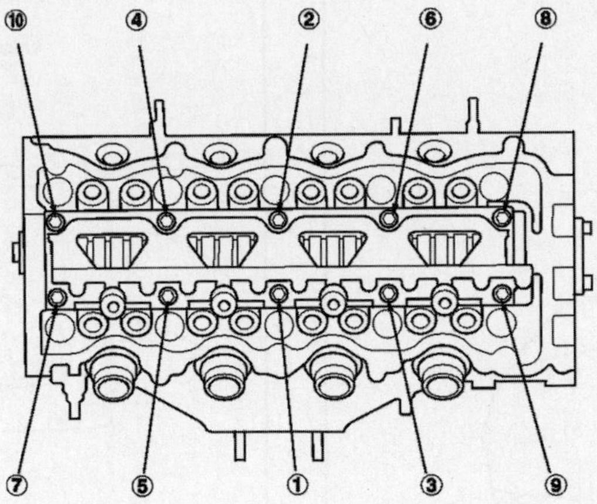
37647_CIVC_G0154

Fig. 29 Put the rocker shaft holders and the lost motion holder on the cylinder head, then tighten the bolts, in sequence, to the specified torque of 11 ft. lbs. (15 Nm).

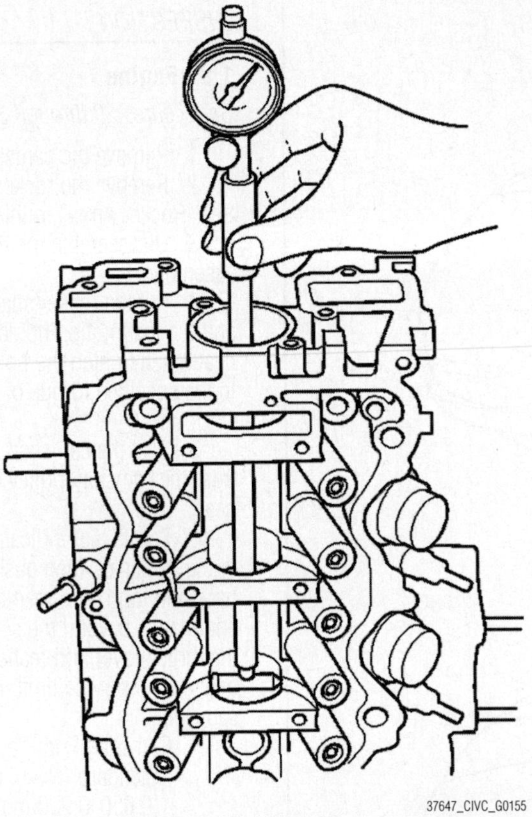
37647_CIVC_G0155

Fig. 30 Clean the camshaft bearing surfaces in the cylinder head. Measure the inside diameter of each camshaft bearing surface, and check for an out-of-round condition.

in the cylinder head. Measure the inside diameter of each camshaft bearing surface, and check for an out-of-round condition.

a. If the camshaft-to-holder clearance is within limits, go to step 13.

b. If the camshaft-to-holder clearance is beyond the service limit, and the camshaft has been replaced, replace the cylinder head.

c. If the camshaft-to-holder clearance is beyond the service limit, and the camshaft has not been replaced, go to step 12.

Camshaft-to-Holder Oil Clearance should be:

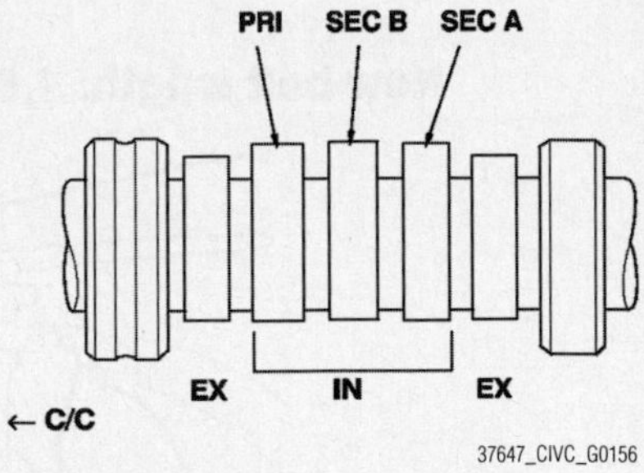
PRI
SEC B
SEC A
EX
IN
EX
← C/C
37647_CIVC_G0156

Fig. 31 Measuring cam lobe height

- Standard (New): 0.0018-0.0033 in. (0.045-0.084 mm)
- Service Limit: 0.006 in. (0.15 mm)

13. Check the total runout using the dial gauge and having the camshaft supported at each end on V-blocks.

a. If the total runout of the camshaft is within the service limit, replace the cylinder head.

b. If the total runout is beyond the service limit, replace the camshaft, and recheck the oil clearance.

c. If the oil clearance is still beyond the service limit, replace the cylinder head.

d. Camshaft Total Runout should be:

- Standard (New): 0.001 in. (0.03 mm) max.
- Service Limit: 0.002 in. (0.04 mm)

14. Measure the cam lobe height.

a. Cam Lobe Height Standard (New):

- Intake PRI (Primary): 1.4076 in. (35.754 mm)
- Intake SEC A (Secondary A): 1.3920 in. (35.358 mm)
- Intake SEC B (Secondary B): 1.4184 in. (36.027 mm)
- Exhaust: 1.4100 in. (35.813 mm)

REMOVAL & INSTALLATION

1.8L Engine

See Figure 32.

1. Remove the cylinder head. See "Cylinder Head" in this section.
2. Remove the rocker arm assembly. See "Rocker Arms" in this section.
3. Remove the camshaft sprocket:

a. Remove the cam chain.

b. Hold the camshaft with a 27 mm open-end wrench, then loosen the bolt.

c. Remove the camshaft sprocket.

4. Remove the camshaft position (CMP) sensor.
5. Remove the camshaft thrust cover, then pull out the camshaft.

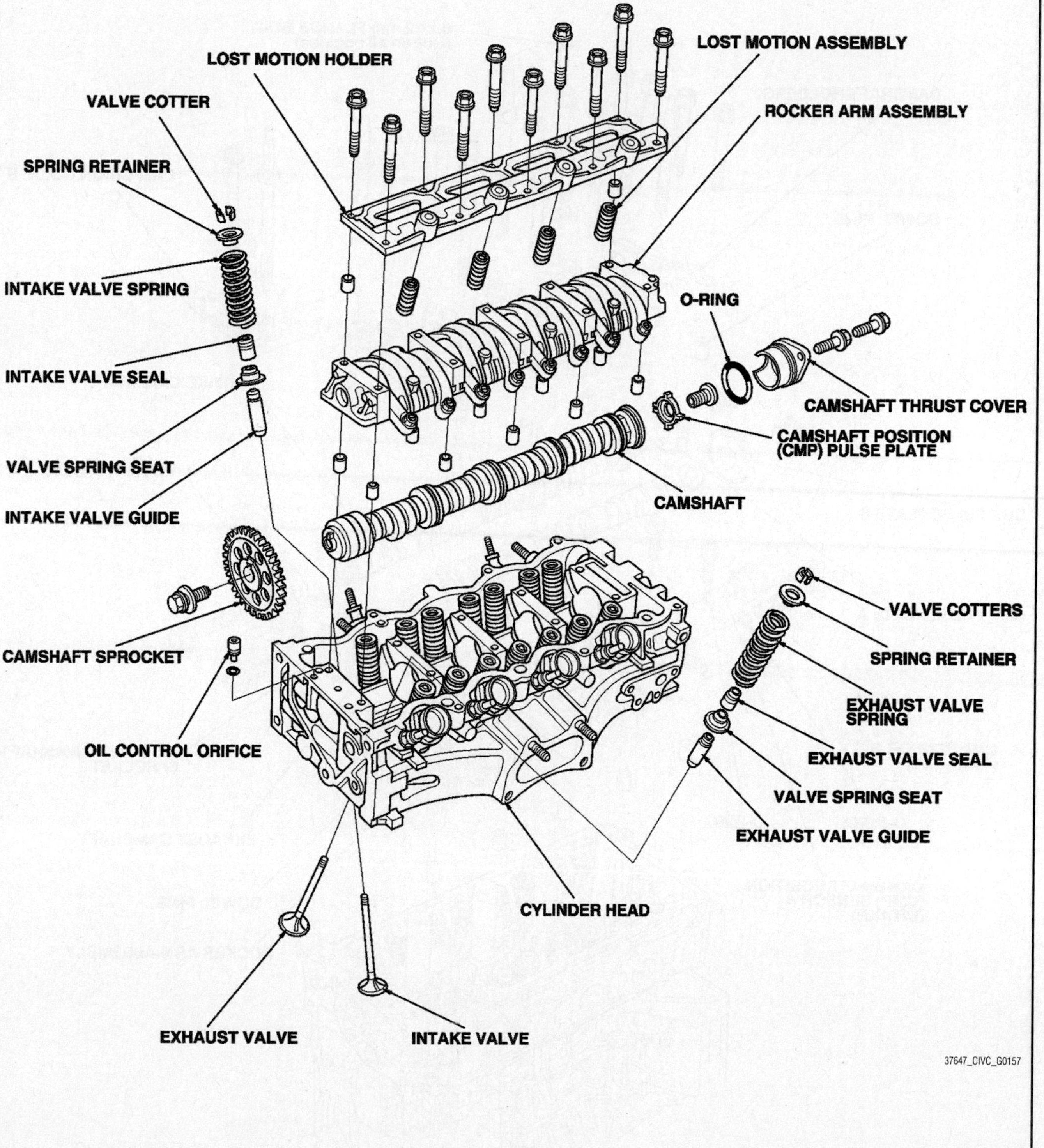
LOST MOTION HOLDER
LOST MOTION ASSEMBLY
VALVE COTTER
ROCKER ARM ASSEMBLY
SPRING RETAINER
INTAKE VALVE SPRING
O-RING
INTAKE VALVE SEAL
CAMSHAFT THRUST COVER
CAMSHAFT POSITION (CMP) PULSE PLATE
VALVE SPRING SEAT
INTAKE VALVE GUIDE
CAMSHAFT
VALVE COTTERS
CAMSHAFT SPROCKET
SPRING RETAINER
EXHAUST VALVE SPRING
OIL CONTROL ORIFICE
EXHAUST VALVE SEAL
VALVE SPRING SEAT
EXHAUST VALVE GUIDE
CYLINDER HEAD
EXHAUST VALVE
INTAKE VALVE
37647_CIVC_G0157

Fig. 32 Exploded view of the camshaft and related components—1.8L engine

6. Installation is the reverse of the removal procedure.

2.0L Engine

See Figure 33.

➡**Manufacturer does not provide a specific removal and installation procedure for this component. Use the illustration as a guide when servicing this component.**

CATALYTIC CONVERTER

REMOVAL & INSTALLATION

See Figures 34 and 35.

1. Remove the A/F sensor (sensor 1).

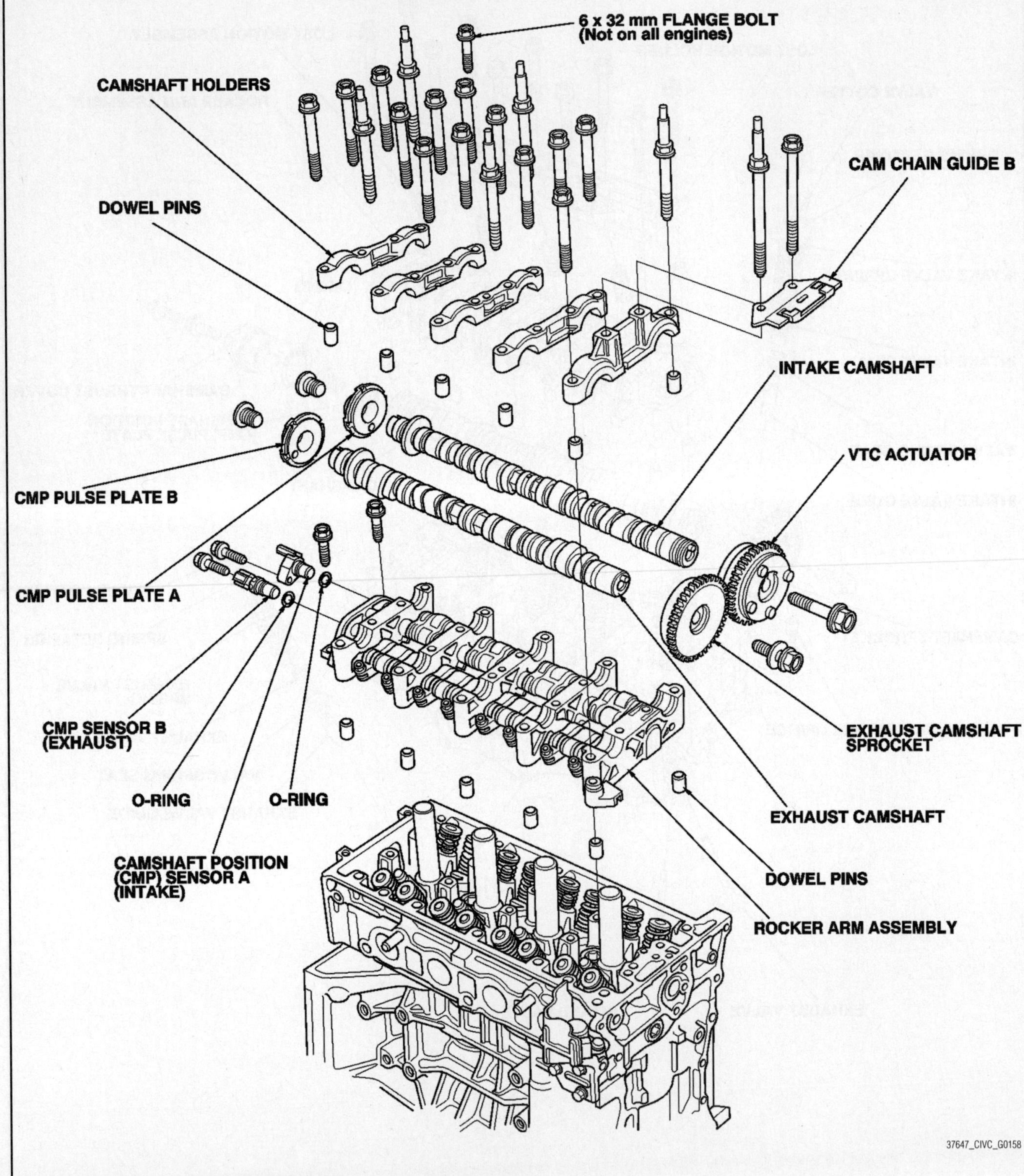
6 x 32 mm FLANGE BOLT
(Not on all engines)
CAMSHAFT HOLDERS
CAM CHAIN GUIDE B
DOWEL PINS
INTAKE CAMSHAFT
VTC ACTUATOR
CMP PULSE PLATE B
CMP PULSE PLATE A
CMP SENSOR B
(EXHAUST)
EXHAUST CAMSHAFT
SPROCKET
O-RING
O-RING
EXHAUST CAMSHAFT
CAMSHAFT POSITION
(CMP) SENSOR A
(INTAKE)
DOWEL PINS
ROCKER ARM ASSEMBLY
37647_CIVC_G0158

Fig. 33 Exploded view of the camshaft and related components—2.0L engine

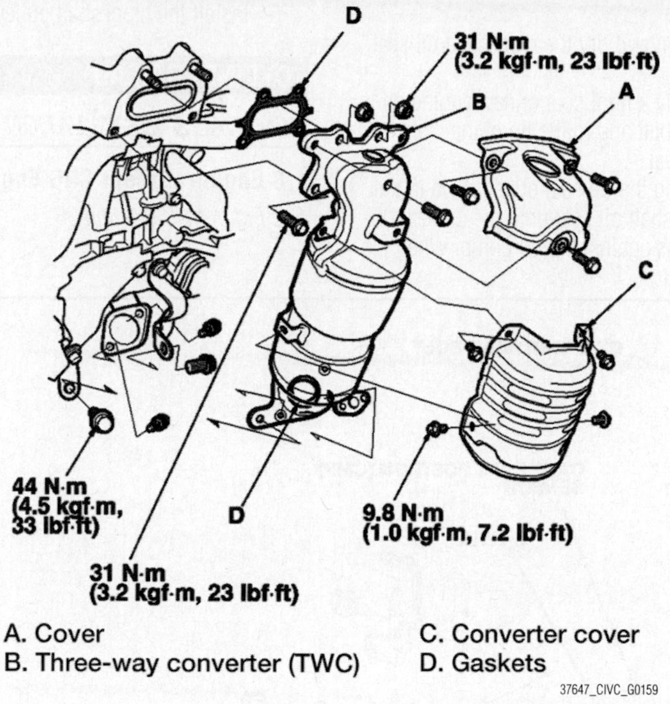
D
31 N·m
(3.2 kgf·m, 23 lbf·ft)
B
A
C
44 N·m
(4.5 kgf·m,
33 lbf·ft)
D
9.8 N·m
(1.0 kgf·m, 7.2 lbf·ft)
31 N·m
(3.2 kgf·m, 23 lbf·ft)
A. Cover
B. Three-way converter (TWC)
C. Converter cover
D. Gaskets
37647_CIVC_G0159

Fig. 34 Removing the catalytic converter—1.8L engine

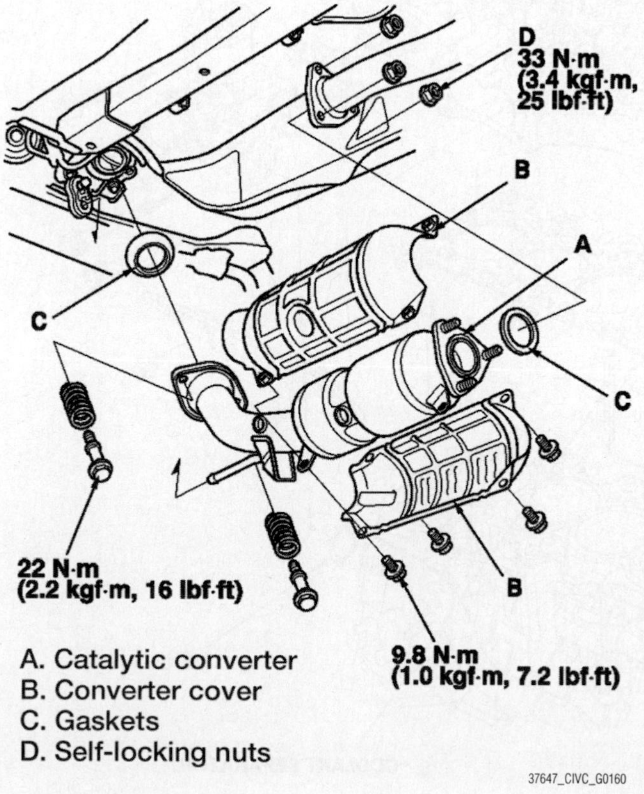
D
33 N·m
(3.4 kgf·m,
25 lbf·ft)
B
A
C
C
22 N·m
(2.2 kgf·m, 16 lbf·ft)
B
9.8 N·m
(1.0 kgf·m, 7.2 lbf·ft)
A. Catalytic converter
B. Converter cover
C. Gaskets
D. Self-locking nuts
37647_CIVC_G0160

Fig. 35 Removing the catalytic converter—2.0L engine

2. Remove secondary HO2S (sensor 2).
3. Remove the EGR pipe (1.8L engine).
4. Remove the cover (1.8L engine).
5. Remove the TWC.
6. Remove the converter cover and gaskets.
7. Install the parts in the reverse order of removal with new gaskets.

CRANKSHAFT DAMPER (PULLEY)

REMOVAL & INSTALLATION

1. Remove the right front wheel.
2. Remove the drive belt.
3. Counterhold the pulley and remove the bolt, then remove the crankshaft pulley.
4. Clean the crankshaft pulley, the crankshaft mating surface, the bolt, and the washer.
5. Lubricate the bolt with engine oil prior to installation.
6. Install the crankshaft pulley onto the crankshaft by aligning the flat sides of the pulley with the flat sides of the inner oil pump gear.
7. On 1.8L engine, do the following:
   a. Counterhold the crankshaft pulley and install and torque the bolt to 51 ft. lbs. (69 Nm).

**** CAUTION**

Do not use an impact wrench.

   b. If the pulley bolt or crankshaft are new, torque the bolt to 130 ft. lbs. (177 Nm), then remove the bolt and re-install it and torque it to 51 ft. lbs. (69 Nm).
   c. Mark the bolt head at 12 o'clock and then mark the crankshaft pulley with a reference line 90° clockwise from the bolt mark, then tighten the bolt an additional 90°. The mark on the bolt head lines up with the mark on the crankshaft pulley.
8. On 2.0L engine, do the following:
   a. Counterhold the crankshaft pulley and install and torque the bolt to 36 ft. lbs. (49 Nm).

**** CAUTION**

Do not use an impact wrench.

   b. If the pulley bolt or crankshaft are new, torque the bolt to 130 ft. lbs. (177 Nm), then remove the bolt and re-install it and torque it to 36 ft. lbs. (49 Nm).
   c. Mark the bolt head at 12 o'clock and then mark the crankshaft pulley with a reference line 90° clockwise from the bolt mark, then tighten the bolt an

additional 90°. The mark on the bolt head lines up with the mark on the crankshaft pulley.

9. Install the drive belt.
10. Install the right front wheel.

CRANKSHAFT FRONT SEAL

REMOVAL & INSTALLATION

➡This procedure is performed with the engine in the vehicle.

1. Remove the crankshaft pulley as described in this section.
2. Remove the pulley end crankshaft oil seal.
3. Clean and dry the crankshaft oil seal housing.
4. Apply a light coat of new engine oil to the crankshaft and to the lip of the crankshaft oil seal.
5. Using the oil seal driver, drive in the new crankshaft oil seal until the oil seal driver bottoms against the oil pump. When the seal is in place, clean any excess grease off the crankshaft, and check that the oil seal lip is not distorted.
6. Install the crankshaft pulley.

CYLINDER HEAD

REMOVAL & INSTALLATION

1.8 Engine (Except CNG Engine)

See Figures 36 through 38.

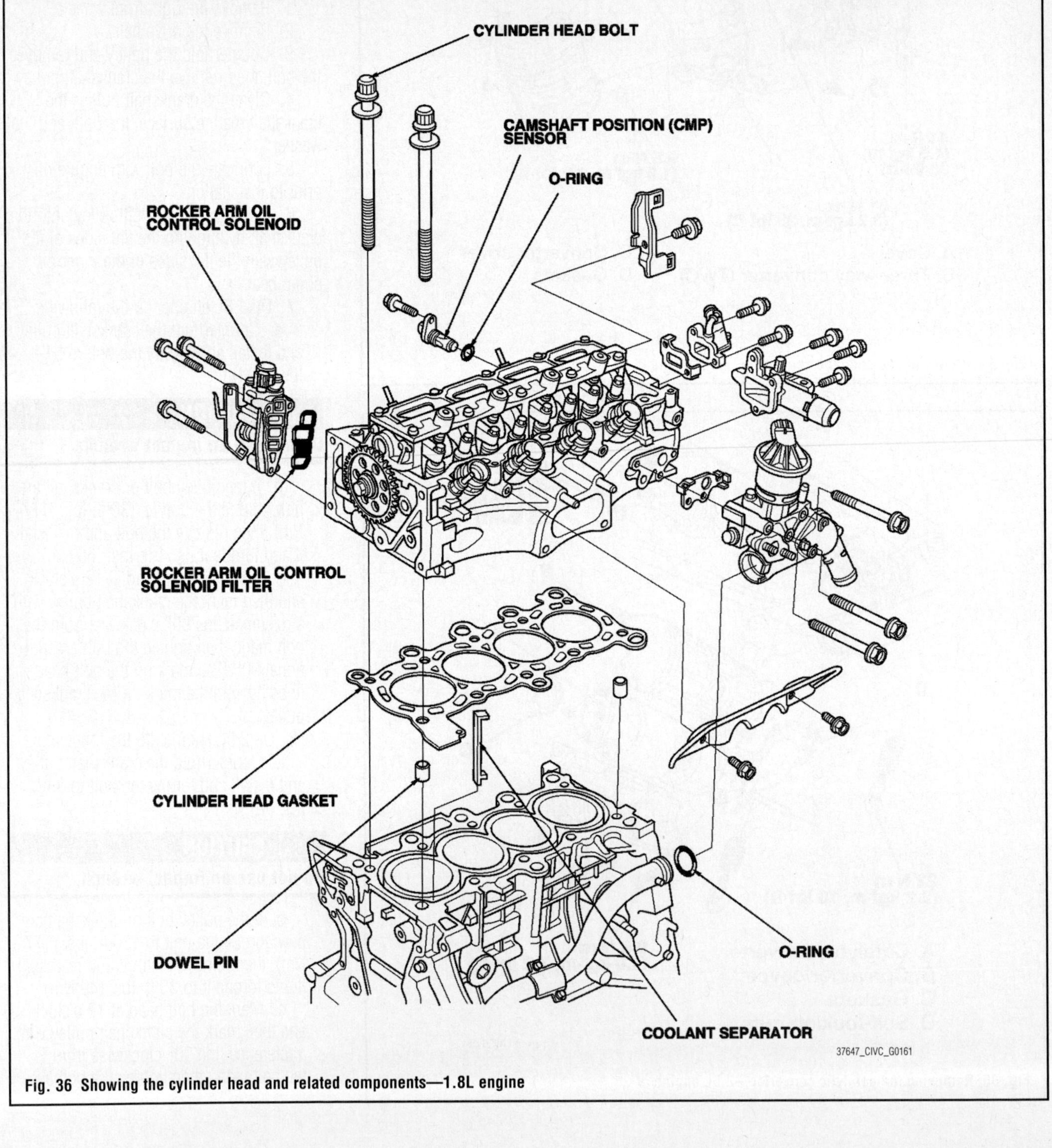
CYLINDER HEAD BOLT
CAMSHAFT POSITION (CMP) SENSOR
O-RING
ROCKER ARM OIL CONTROL SOLENOID
ROCKER ARM OIL CONTROL SOLENOID FILTER
CYLINDER HEAD GASKET
DOWEL PIN
O-RING
COOLANT SEPARATOR
37647_CIVC_G0161

Fig. 36 Showing the cylinder head and related components—1.8L engine

1. Before starting work, refer to the "Precautions" in this section.
2. Connect the Honda Diagnostic System (HDS), or equivalent scan tool, to the data link connecter (DLC), and monitor the engine coolant temperature (ECT) sensor 1. To avoid damaging the cylinder head, wait until the engine coolant temperature drops below 100°F (38°C) before loosening the cylinder head bolts.
3. Mark all wiring and hoses to avoid misconnection. Also, be sure that they do not contact other wiring or hoses, or interfere with other parts.
4. Relieve the fuel pressure. See "Relieving Fuel System Pressure" in the "FUEL SYSTEMS" section.
5. Do the battery terminal disconnect procedure, then wait at least 3 minutes before beginning work.
6. Drain the engine coolant.
7. Remove the air cleaner assembly.
8. Remove the drive belt.
9. Remove the intake manifold. See "Intake Manifold" in this section.
10. Remove the harness clamps and disconnect the positive crankcase ventilation (PCV) hose from the clamp.
11. Remove the air cleaner bracket, then remove the harness holder from the cylinder head.
12. On M/T model, disconnect the upper radiator hose and the heater hose from the cylinder head.
13. On A/T model, disconnect the upper radiator hose, the heater hose, and the water bypass hose.
14. Disconnect the following engine wire harness connectors, and remove the wire harness clamps from the cylinder head:

- Four fuel injector connectors
- Engine coolant temperature (ECT) sensor 1 connector
- Air fuel ratio (A/F) sensor connector
- Secondary heated oxygen sensor (secondary HO2S) connector
- Exhaust gas recirculation (EGR) valve connector
- Rocker arm oil control valve connector
- Rocker arm oil pressure sensor (EOP sensor) connector
- EVAP canister purge valve connector
- Camshaft position (CMP) sensor connector

15. Remove the four ignition coils.
16. Remove the three way catalytic converter (TWC).
17. Remove the thermostat housing.
18. Remove the cam chain. See "Timing Chain & Sprockets" in this section.
19. Remove the cylinder head bolts. To prevent warpage, loosen the bolts, in sequence, 1/3 turn at a time.
20. Repeat the sequence until all bolts are loosened.
21. Remove the cylinder head.

To install:

22. Clean the cylinder head and the engine block surface.
23. Install a new coolant separator in the engine block whenever the engine block is replaced.
24. Install the new cylinder head gasket and the dowel pins on the engine block. Always use a new cylinder head gasket.
25. Set the crankshaft to top dead center (TDC). Align the TDC mark on the crankshaft sprocket with the pointer on the engine block.
26. Set the camshaft to TDC. The "UP" mark on the camshaft sprocket should be at the top, and the TDC grooves on the camshaft sprocket should line up with the top edge of the head.
27. Install the cylinder head on the engine block.
28. Measure the diameter of each cylinder head bolt at multiple points.
29. If either diameter is less than 0.42 in. (10.6 mm), replace the cylinder head bolt.
30. Apply new engine oil to the threads and under the bolt heads of all cylinder head bolts.
31. Torque the cylinder head bolts in sequence to 29 ft. lbs. (39 Nm). Use a beam-type torque wrench. When using a preset click-type torque wrench, be sure to tighten slowly and do not overtighten. If a bolt makes any noise while you are torquing it, loosen the bolt and retighten it from the first step.
32. After torquing, tighten all cylinder head bolts in two steps, an additional 90° per step, using the sequence shown.

➡If you are using a new cylinder head bolt, tighten the bolt an extra 60°.

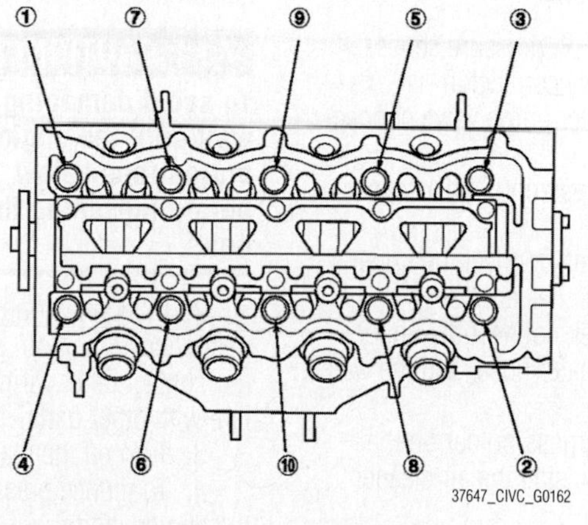
1
7
9
5
3
4
6
10
8
2
37647_CIVC_G0162

Fig. 37 Cylinder head bolt loosening sequence—1.8L engine

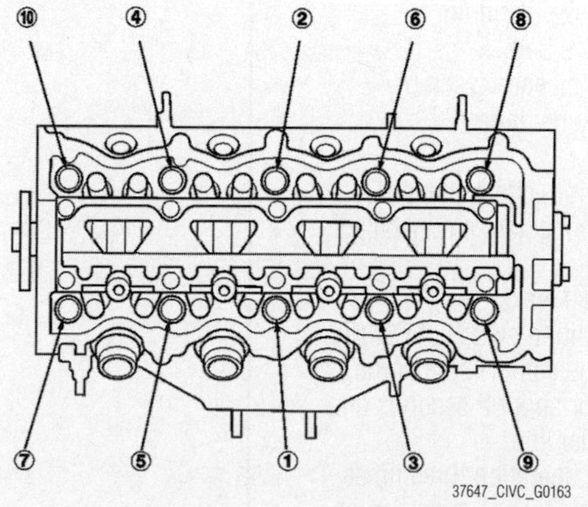
10
4
2
6
8
7
5
1
3
9
37647_CIVC_G0163

Fig. 38 Cylinder head bolt tightening sequence—1.8L engine

➡Remove the cylinder head bolt if you tightened it beyond the specified angle, and go back to the procedure. Do not loosen it back to the specified angle.

33. Install the cam chain. See "Timing Chain & Sprockets" in this section.
34. Adjust the valve clearance.
35. Install the cylinder head cover.
36. Install the thermostat housing.
37. Install the three way catalytic converter (TWC).
38. Install the four ignition coils.
39. Connect the following engine wire harness connectors, and install the wire harness clamps to the cylinder head:
    - Four fuel injector connectors
    - Engine coolant temperature (ECT) sensor 1 connector
    - Air/Fuel ratio (A/F) sensor connector
    - Secondary heated oxygen sensor (secondary HO2S) connector
    - Exhaust gas recirculation (EGR) valve connector
    - Rocker arm oil control valve connector
    - Rocker arm oil pressure sensor (EOP sensor) connector
    - EVAP canister purge valve connector
    - Camshaft position (CMP) sensor connector
40. On M/T model, connect the upper radiator hose and the heater hose.
41. On A/T model, connect the upper radiator hose, the heater hose, and the water bypass hose.
42. Install the harness holder on the cylinder head, then install the air cleaner bracket.
43. Install the harness clamps, and connect the positive crankcase ventilation (PCV) hose to the clamp.
44. Install the intake manifold.
45. Install the drive belt.
46. Install the air cleaner assembly.
47. Do the battery reconnect procedure.
48. After installation, check that all tubes, hoses, and connectors are installed correctly.
49. Inspect for fuel leaks:
    a. Turn the ignition switch to ON (II) (do not operate the starter) so that the fuel pump runs for about 2 seconds and pressurizes the fuel line.
    b. Repeat this operation three times, then check for fuel leakage at any point in the fuel line.
50. Refill the radiator with engine coolant, and bleed the air from the cooling system.
51. Check for fluid leaks.
52. Do the Engine Control Module (ECM/PCM) Idle Learn Procedure.
53. Do the Crankshaft Position (CKP) Pattern Clear/CKP Pattern Learn Procedure. See "Crankshaft Position (CKP) Sensor" in "ENGINE PERFORMANCE & EMISSION CONTROLS" section.
54. Inspect the idle speed.
55. Inspect the ignition timing.

1.8L CNG Engine

See Figures 39 through 42.

⁂ CAUTION

To avoid damaging the wiring and terminals, unplug the wiring connectors carefully while holding the connector portion.

1. Connect the Honda Diagnostic System (HDS), or equivalent scan tool, to the data link connector (DLC), and monitor the engine coolant temperature (ECT) sensor 1.

⁂ CAUTION

To avoid damaging the cylinder head, wait until the engine coolant temperature drops below 100°F (38°C) before loosening the cylinder head bolts.

2. Mark all wiring and hoses to avoid misconnection. Also, be sure that they do not contact other wiring or hoses, or interfere with other parts.
3. Turn off the manual shut-off valve.
4. To reduce pressure in the lines, start the engine, and run it until it stalls.
5. Do the battery terminal disconnect procedure, then wait at least 3 minutes before beginning work.
6. Drain the engine coolant.
7. Remove the air cleaner assembly.
8. Remove the drive belt.
9. Remove the intake manifold. See "Intake Manifold" in this section.
10. Remove the harness clamps and remove the positive crankcase ventilation (PCV) hose from the clamp.
11. Remove the air cleaner bracket, then remove the harness holder from the cylinder head.
12. Disconnect the upper radiator hose, the heater hose, and the water bypass hose.
13. Disconnect the following engine wire harness connectors, and remove the wire harness clamps from the cylinder head:
    - Four fuel injector connectors
    - Engine coolant temperature (ECT) sensor 1 connector
    - Air fuel ratio (A/F) sensor connector
    - Secondary heated oxygen sensor (secondary HO2S) connector
    - Rocker arm oil control valve connector
    - Rocker arm oil pressure sensor (EOP sensor) connector
    - Camshaft position (CMP) sensor connector
    - Remove the four ignition coils.
    - Remove the three way catalytic converter (TWC).
    - Remove the thermostat housing.
    - Remove the cam chain. See "Timing Chain & Sprockets" in this section.
14. Remove the cylinder head bolts. To prevent warpage, unscrew the bolts, in sequence, 1/3 turn at a time. Repeat the sequence until all bolts are loosened.
15. Remove the cylinder head.

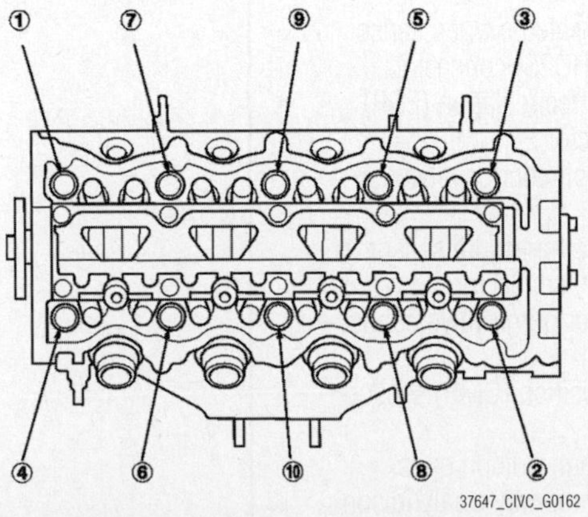

①
⑦
⑨
⑤
③
④
⑥
⑩
⑧
②
37647_CIVC_G0162

Fig. 39 Cylinder head bolt loosening sequence

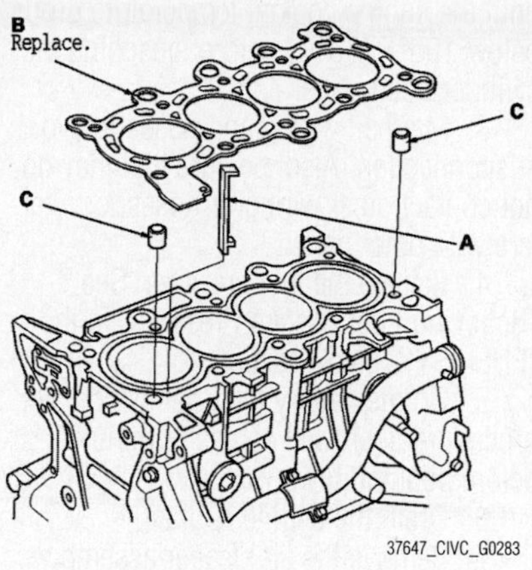
B
Replace.
C
C
A
37647_CIVC_G0283

Fig. 40 Install the new coolant separator (A) in the engine block whenever the engine block is replaced. Install the new cylinder head gasket (B) and the dowel pins (C) on the engine block.

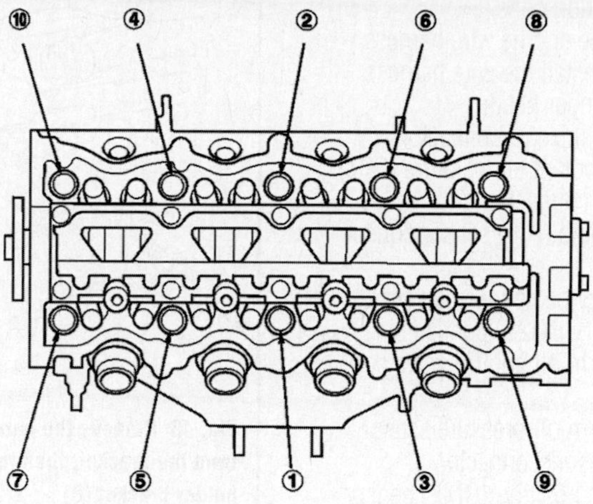
10
4
2
6
8
7
5
1
3
9
37647_CIVC_G0163

Fig. 42 Cylinder head bolt tightening sequence

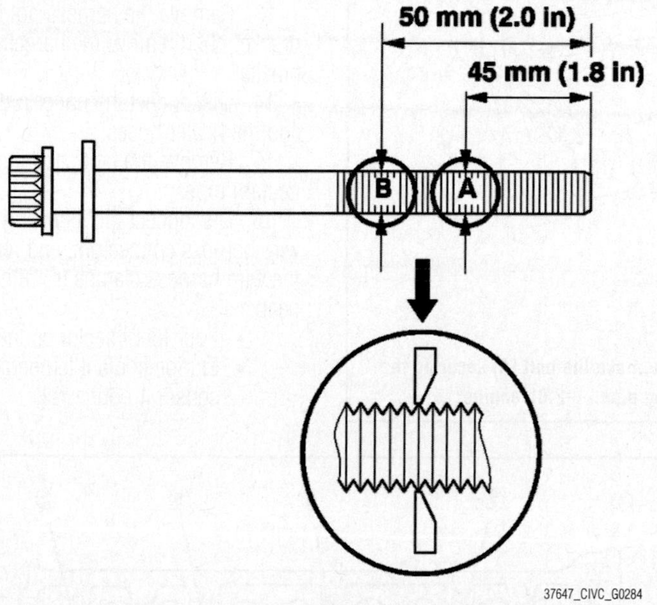
50 mm (2.0 in)
45 mm (1.8 in)
B
A
37647_CIVC_G0284

Fig. 41 Measure the diameter of each cylinder head bolt at point A and point B. If either diameter is less than 10.6 mm (0.42 in), replace the cylinder head bolt.

To install:

16. Clean the cylinder head and the block surface.
17. Install the new coolant separator in the engine block whenever the engine block is replaced.
18. Install the new cylinder head gasket and the dowel pins on the engine block. Always use a new cylinder head gasket.
19. Set the crankshaft to top dead center (TDC): Align the TDC mark on the crankshaft sprocket with the pointer on the engine block.
20. Set the camshaft TDC. The "UP" mark on the camshaft sprocket should be at the top, and the TDC grooves on the camshaft sprocket should line up with the top edge of the head.
21. Install the cylinder head on the block.
22. Measure the diameter of each cylinder head bolt at point A and point B, as shown. If either diameter is less than 10.6 mm (0.42 in), replace the cylinder head bolt.
23. Apply new engine oil to the threads and under the bolt heads of all cylinder head bolts.
24. Torque the cylinder head bolts in sequence to 29 ft. lbs. (39 Nm). Use a beam-type torque wrench. When using a preset click-type torque wrench, be sure to tighten slowly and do not overtighten. If a bolt makes any noise while you are torquing it, loosen the bolt and retighten it from the first step.
25. After torquing, tighten all cylinder head bolts in two steps (90° per step) using the sequence shown. If you are using a new cylinder head bolt, tighten the bolt an extra 60°.

➡Remove the cylinder head bolt if you tightened it beyond the specified angle, and go back and repeat the entire tightening procedure. Do not loosen it back to the specified angle.

26. Install, adjust or connect the following:
    - Cam chain
    - Valve clearance
    - Cylinder head cover

- Thermostat housing
- Three way catalytic converter (TWC)
- Four ignition coils

27. Connect the engine wire harness connectors, and install the wire harness clamps to the cylinder head:
    - Four fuel injector connectors
    - Engine coolant temperature (ECT) sensor 1 connector
    - Air fuel ratio (A/F) sensor connector
    - Secondary heated oxygen sensor (secondary HO2S) connector
    - Rocker arm oil control valve connector
    - Rocker arm oil pressure sensor (EOP sensor) connector
    - Camshaft position (CMP) sensor connector
28. Connect the upper radiator hose, the heater hose, and the water bypass hose.
29. Install the harness holder on the cylinder head, then Install the air cleaner bracket.
30. Install the harness clamps and connect the positive crankcase ventilation (PCV) hose to the clamp.
31. Install the intake manifold.
32. Install the drive belt.
33. Install the air cleaner assembly.
34. Do the battery terminal reconnection procedure.
35. After installation, check that all tubes, hoses and connectors are installed correctly.
36. Do the leak inspection.
37. Refill the radiator with engine coolant, and bleed the air from the cooling system.
38. Do the Powertrain Control Module (PCM) Idle Learn Procedure. See "Engine Control Module/Powertrain Control Module (ECM/PCM)" in "ENGINE PERFORMANCE & EMISSION CONTROLS" section.
39. Do the Crankshaft Position (CKP) Pattern Clear/CKP Pattern Learn Procedure. See "Crankshaft Position (CKP) Sensor" in "ENGINE PERFORMANCE & EMISSION CONTROLS" section.
40. Inspect the idle speed.
41. Inspect the ignition timing.

2.0L Engine

See Figures 43 through 46.

1. Before starting work, refer to the "Precautions" in this section.
2. Connect the Honda Diagnostic System (HDS), or equivalent scan tool, to the data link connecter (DLC), and monitor the engine coolant temperature (ECT) sensor 1. To avoid damaging the cylinder head, wait until the engine coolant temperature drops below 100°F (38°C) before loosening the cylinder head bolts.
3. Mark all wiring and hoses to avoid misconnection. Also, be sure that they do not contact other wiring or hoses, or interfere with other parts.
4. Relieve the fuel pressure. See "Relieving Fuel System Pressure" in the "FUEL SYSTEMS" section.
5. Do the battery terminal disconnect procedure, then wait at least 3 minutes before beginning work.
6. Drain the engine coolant.
7. Remove the air cleaner assembly.
8. Remove the drive belt.
9. Remove the intake manifold. See "Intake Manifold" in this section.
10. Remove the exhaust manifold. See "Exhaust Manifold" in this section.
11. Disconnect the evaporative emission (EVAP) canister hose and the brake booster vacuum hose.
12. Remove the quick-connect fitting cover, then disconnect the fuel feed hose.
13. Remove the harness holder from the bracket, then remove the harness holder bracket.
14. Disconnect the upper radiator hose and the heater hoses.
15. Remove the bolt securing the connecting pipe.
16. Disconnect the following engine wire harness connectors, and remove the wire harness clamps from the cylinder head:
    - Four fuel injector connectors
    - Engine coolant temperature (ECT) sensor 1 connector

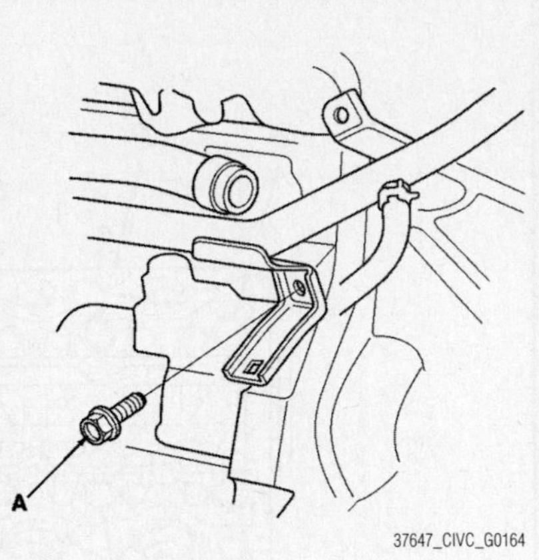
A
37647_CIVC_G0164

Fig. 43 Remove the harness holder (A) from the bracket, then remove the harness holder bracket (B). —2.0L engine

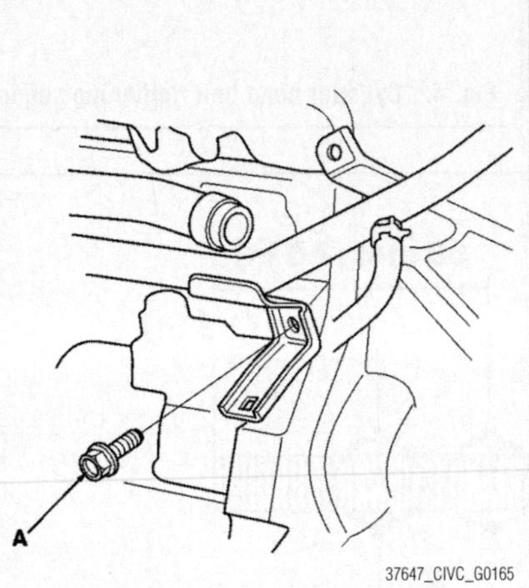
A
37647_CIVC_G0165

Fig. 44 Remove the bolt (A) securing the connecting pipe. —2.0L engine

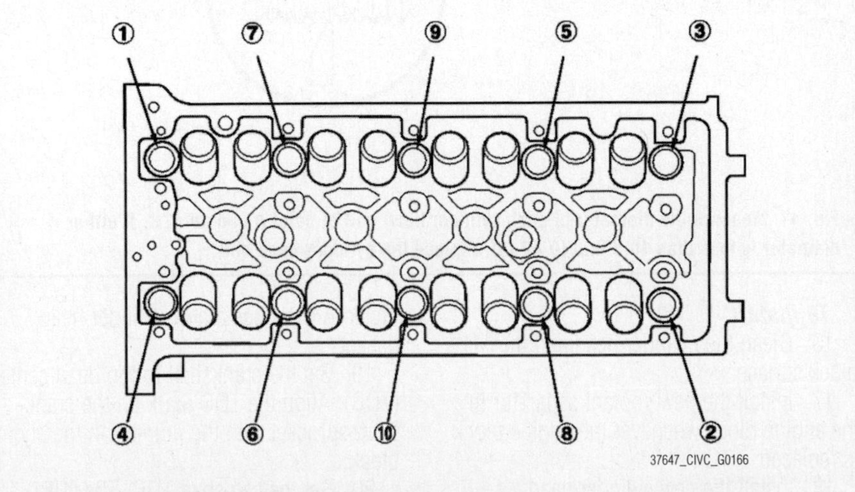
①
⑦
⑨
⑤
③
④
⑥
⑩
⑧
②
37647_CIVC_G0166

Fig. 45 Cylinder head bolt loosening sequence—2.0L engine

- Camshaft position (CMP) sensor A (Intake) connector
- Camshaft position (CMP) sensor B (Exhaust) connector
- Rocker arm oil control valve connector
- Rocker arm oil pressure switch connector
- EVAP canister purge valve connector
- Engine oil pressure switch connector

17. Remove the cam chain. See "Timing Chain & Sprockets" in this section.
18. Remove the rocker arm assembly. See "Rocker Arms" in this section.
19. Remove the cylinder head bolts. To prevent warpage, loosen the bolts in sequence 1/3 turn at a time.
20. Repeat the sequence until all bolts are loosened.
21. Remove the cylinder head.

To install:

22. Install a new coolant separator in the engine block whenever the engine block is replaced.
23. Clean the cylinder head and the engine block surface.
24. Install the new cylinder head gasket and the dowel pins on the engine block. Always use a new cylinder head gasket.
25. Set the crankshaft to top dead center (TDC). Align the TDC mark on the crankshaft sprocket with the pointer on the engine block.
26. Set the camshaft pulley to TDC.
27. Install the cylinder head on the engine block.
28. Measure the diameter of each cylinder head bolt at multiple points.
29. If either diameter is less than 0.42 in. (10.6 mm), replace the cylinder head bolt.
30. Apply new engine oil to the threads and under the bolt heads of all cylinder head bolts.
31. Torque the cylinder head bolts in sequence to 29 ft. lbs. (39 Nm). Use a beam-type torque wrench. When using a preset click-type torque wrench, be sure to tighten slowly and do not overtighten. If a bolt makes any noise while you are torquing it, loosen the bolt and retighten it from the first step.
32. After torquing, tighten all cylinder head bolts in two steps, an additional 90° per step, using the sequence shown.

➡If you are using a new cylinder head bolt, tighten the bolt an extra 60°.

➡Remove the cylinder head bolt if you tightened it beyond the specified angle, and go back to the procedure. Do not loosen it back to the specified angle.

33. Install the rocker arm assembly.
34. Install the cam chain.
35. Connect the following engine wire harness connectors, and install the wire harness clamps to the cylinder head:
    - Four fuel injector connectors
    - Engine coolant temperature (ECT) sensor 1 connector
    - Camshaft position (CMP) sensor A (Intake) connector
    - Camshaft position (CMP) sensor B (Exhaust) connector
    - Rocker arm oil control valve connector
    - Rocker arm oil pressure switch connector
    - EVAP canister purge valve connector
    - Engine oil pressure switch connector
36. Install the bolt securing the connecting pipe.
37. Connect the upper radiator hose and the heater hoses.
38. Install the harness holder bracket, then install the harness holder.
39. Connect the fuel feed hose, then install the quick-connect fitting cover.
40. Connect the evaporative emission (EVAP) canister hose and the brake booster vacuum hose.
41. Install the exhaust manifold.
42. Install the intake manifold.
43. Install the drive belt.
44. Install the air cleaner assembly
45. Do the battery reconnect procedure.
46. After installation, check that all tubes, hoses, and connectors are installed correctly.
47. Inspect for fuel leaks:
    a. Turn the ignition switch to ON (II) (do not operate the starter) so that the fuel pump runs for about 2 seconds and pressurizes the fuel line.
    b. Repeat this operation three times, then check for fuel leakage at any point in the fuel line.
48. Refill the radiator with engine coolant, and bleed the air from the cooling system.
49. Check for fluid leaks.
50. Do the Engine Control Module (ECM/PCM) Idle Learn Procedure.
51. Do the Crankshaft Position (CKP) Pattern Clear/CKP Pattern Learn Procedure. See "Crankshaft Position (CKP) Sensor" in "ENGINE PERFORMANCE & EMISSION CONTROLS" section.
52. Inspect the idle speed.
53. Inspect the ignition timing.

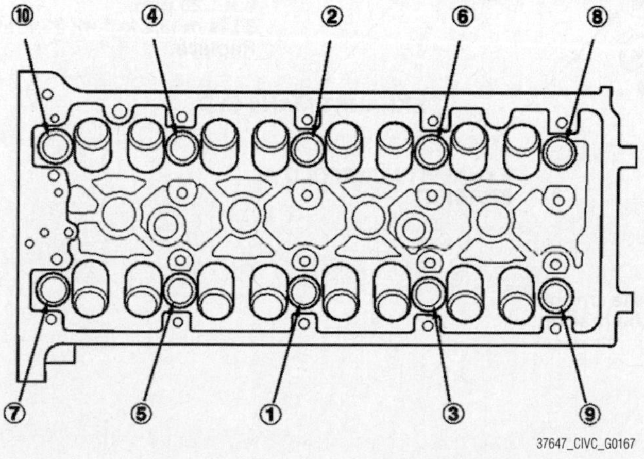
10
4
2
6
8
7
5
1
3
9
37647_CIVC_G0167
Fig. 46 Cylinder head bolt tightening sequence—2.0L engine

EXHAUST MANIFOLD

REMOVAL & INSTALLATION

2.0L Engine

See Figure 47.

1. Before servicing the vehicle, refer to the precautions in the beginning of this section.
2. Drain the cooling system and relieve the fuel system pressure.
3. Remove or disconnect the following:
    - Negative battery cable
    - Catalytic converter

NOTE: Use new gaskets and self-locking nuts when reassembling.

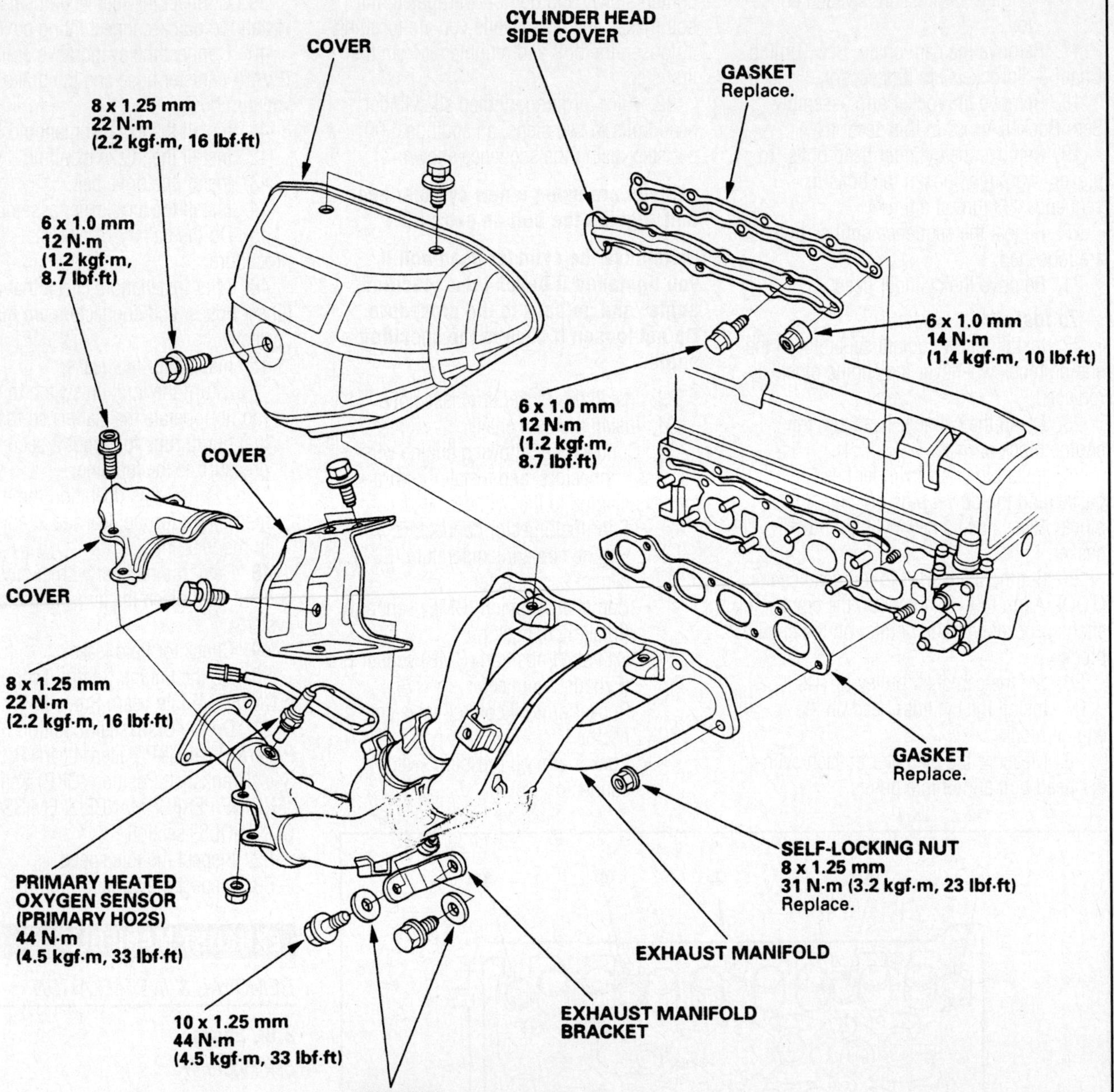
COVER
CYLINDER HEAD SIDE COVER
GASKET
Replace.
8 x 1.25 mm
22 N·m
(2.2 kgf·m, 16 lbf·ft)
6 x 1.0 mm
12 N·m
(1.2 kgf·m,
8.7 lbf·ft)
6 x 1.0 mm
14 N·m
(1.4 kgf·m, 10 lbf·ft)
6 x 1.0 mm
12 N·m
(1.2 kgf·m,
8.7 lbf·ft)
COVER
COVER
8 x 1.25 mm
22 N·m
(2.2 kgf·m, 16 lbf·ft)
GASKET
Replace.
SELF-LOCKING NUT
8 x 1.25 mm
31 N·m (3.2 kgf·m, 23 lbf·ft)
Replace.
PRIMARY HEATED
OXYGEN SENSOR
(PRIMARY HO2S)
44 N·m
(4.5 kgf·m, 33 lbf·ft)
EXHAUST MANIFOLD
10 x 1.25 mm
44 N·m
(4.5 kgf·m, 33 lbf·ft)
EXHAUST MANIFOLD
BRACKET
WASHERS
Make sure the smooth
side faces the bracket.
9347FG10

Fig. 47 Exploded view of the exhaust manifold—2.0L engines

- Exhaust manifold heat shields
- Heated Oxygen (HO_2S) sensor connector
- Exhaust manifold bracket
- Exhaust manifold

4. Installation is the reverse of the removal procedure.

FLYWHEEL

REMOVAL & INSTALLATION

With M/T

➡See "Clutch" in "DRIVETRAIN" section for removal procedure.

With A/T

1. Remove the transmission assembly.
2. Remove the drive plate and the washer from the engine crankshaft.
3. Install the drive plate and the washer on the engine crankshaft, and tighten the eight bolts in a crisscross pattern in at least two steps to 54 ft. lbs. (74 Nm).
4. Install the transmission assembly.

INTAKE MANIFOLD

REMOVAL & INSTALLATION

1.8L Engine

See Figures 48 and 49.

1. Remove the cowl cover and the under-cowl panel.
2. Relieve the fuel pressure. See "Fuel Pressure Relieving" in "FUEL SYSTEMS" section.
3. Remove the air cleaner assembly.
4. Remove the intake air duct.
5. Remove the injector cover.
6. Disconnect the evaporative emission (EVAP) canister hose, the brake booster vacuum hose, and the positive crankcase ventilation (PCV) hose, and remove the power steering (P/S) hose clamp.
7. Remove the quick-connect fitting cover, then disconnect the fuel feed hose.
8. Remove the following engine wire harness connectors, and remove the wire harness clamps from the intake manifold:

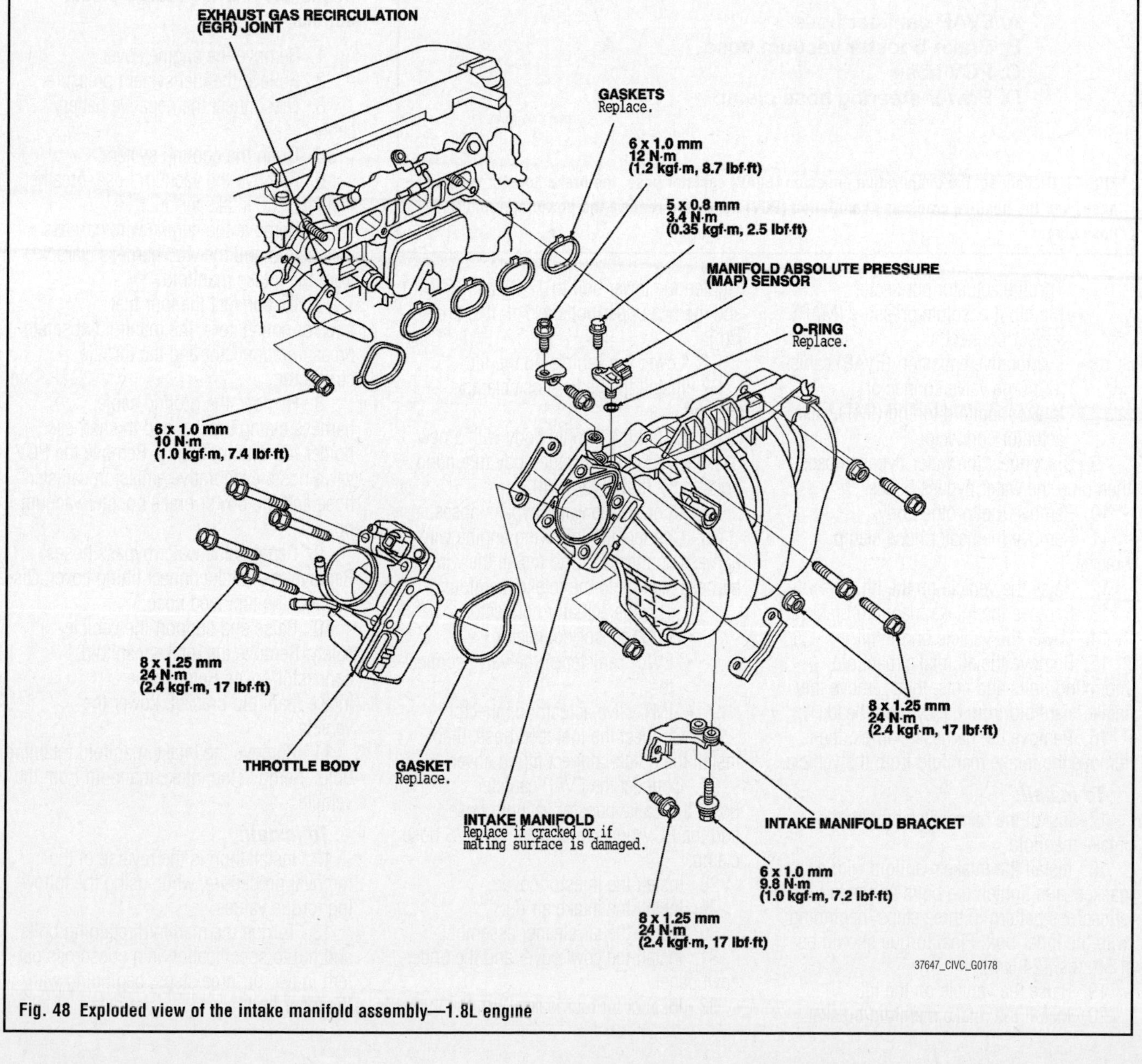

EXHAUST GAS RECIRCULATION (EGR) JOINT
GASKETS Replace.
6 x 1.0 mm 12 N·m (1.2 kgf·m, 8.7 lbf·ft)
5 x 0.8 mm 3.4 N·m (0.35 kgf·m, 2.5 lbf·ft)
MANIFOLD ABSOLUTE PRESSURE (MAP) SENSOR
O-RING Replace.
6 x 1.0 mm 10 N·m (1.0 kgf·m, 7.4 lbf·ft)
8 x 1.25 mm 24 N·m (2.4 kgf·m, 17 lbf·ft)
8 x 1.25 mm 24 N·m (2.4 kgf·m, 17 lbf·ft)
THROTTLE BODY
GASKET Replace.
INTAKE MANIFOLD Replace if cracked or if mating surface is damaged.
INTAKE MANIFOLD BRACKET
6 x 1.0 mm 9.8 N·m (1.0 kgf·m, 7.2 lbf·ft)
8 x 1.25 mm 24 N·m (2.4 kgf·m, 17 lbf·ft)
37647_CIVC_G0178

Fig. 48 Exploded view of the intake manifold assembly—1.8L engine

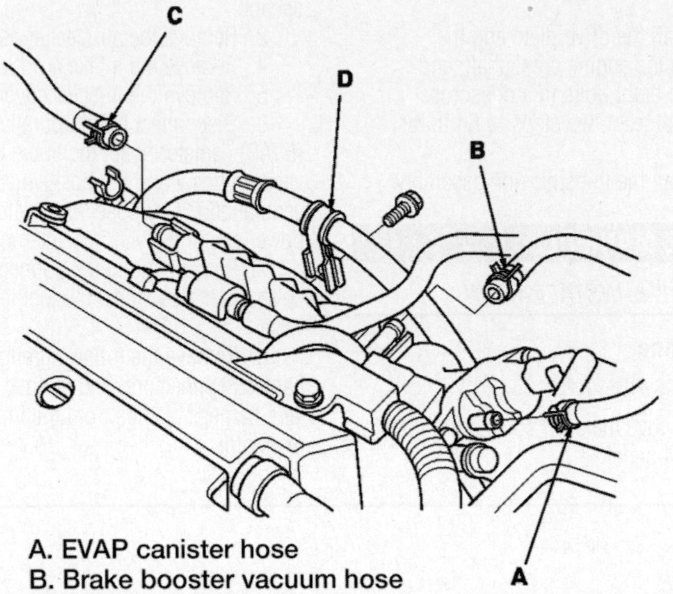
C
D
B
A
A. EVAP canister hose
B. Brake booster vacuum hose
C. PCV hose
D. Power steering hose clamp
37647_CIVC_G0179

Fig. 49 Disconnect the evaporative emission (EVAP) canister hose, the brake booster vacuum hose, and the positive crankcase ventilation (PCV) hose, and remove the power steering (P/S) hose clamp

- Throttle actuator connector
- Manifold absolute pressure (MAP) sensor connector
- Evaporative emission (EVAP) canister purge valve connector
- Intake manifold tuning (IMT) valve actuator connector

9. Disconnect the water bypass hoses, then plug the water bypass hoses.
10. Remove the throttle body.
11. Remove the heater hose clamp bracket.
12. Raise the vehicle on the lift.
13. Remove the intake manifold bracket.
14. Lower the vehicle on the lift.
15. Remove the all intake manifold mounting bolts and nuts, then remove the intake manifold from the cylinder head.
16. Remove the harness clamps, then remove the intake manifold from the vehicle.

To install:

17. Install the harness clamps to the intake manifold.
18. Install the intake manifold with new gaskets and tighten the bolts and nuts in a crisscross pattern in three steps, beginning with the inner bolt. Final torque should be 17 ft. lbs. (24 Nm).
19. Raise the vehicle on the lift.
20. Install the intake manifold bracket. Tighten the lower bolt to 17 ft. lbs. (24 Nm) and the two upper bolts to 7 ft. lbs. (10 Nm).
21. Lower the vehicle on the lift.
22. Install the heater hose clamp bracket.
23. Install the throttle body with a new gasket. Torque the throttle body mounting bolts to 17 ft. lbs. (24 Nm).
24. Connect the water bypass hoses.
25. Connect the following engine wire harness connectors, and install the wire harness clamps to the intake manifold:
    - Throttle actuator connector
    - MAP sensor connector
    - EVAP canister purge valve connector
    - IMT valve actuator connector
26. Connect the fuel feed hose, then install the quick-connect fitting cover.
27. Connect the EVAP canister hose, the brake booster vacuum hose, and the PCV hose, and install the P/S hose clamp.
28. Install the injector cover.
29. Install the intake air duct.
30. Install the air cleaner assembly.
31. Install the cowl cover and the under-cowl panel.
32. Inspect for fuel leaks. Turn the ignition switch to ON (II) (do not operate the starter) so the fuel pump runs for about 2 seconds and pressurizes the fuel line. Repeat this operation three times, then check for fuel leakage at any point in the fuel line.
33. After installation, check that all tubes, hoses, and connectors are installed correctly.
34. Clean up any spilled engine coolant.
35. Refill the radiator with engine coolant, and bleed the air from the cooling system.

2.0L Engine

See Figure 50.

➡Make sure to acquire the anti-theft code from the radio and write down the frequencies for the radio's preset buttons.

1. Remove the engine cover.
2. Relieve the fuel system pressure.
3. Disconnect the negative battery cable.
4. Drain the cooling system.
5. Remove the vacuum hose, breather pipe and the intake air duct.
6. Remove the engine wire harness connectors and the wire harness clamps from the intake manifold.
7. Disconnect the four fuel injector connectors, the manifold absolute pressure connector and the throttle connector.
8. Remove the ground cable, harness clamp bracket and the harness holder from its mounting. Remove the PCV valve hose, evaporative emission canister hose and the power brake booster vacuum hose.
9. Remove the water bypass hoses. Remove the quick connect fitting cover. Disconnect the fuel feed hose.
10. Raise and support the vehicle safely. Remove the intake manifold connector cover. Remove the intake manifold bracket. Lower the vehicle.
11. Remove the intake manifold retaining bolts. Remove the intake manifold from the vehicle.

To install:

12. Installation is the reverse of the removal procedure, while using the following torque values:
13. Torque the manifold retaining bolts and nut to specification in a crisscross pattern in two or three steps, beginning with the inner bolt.

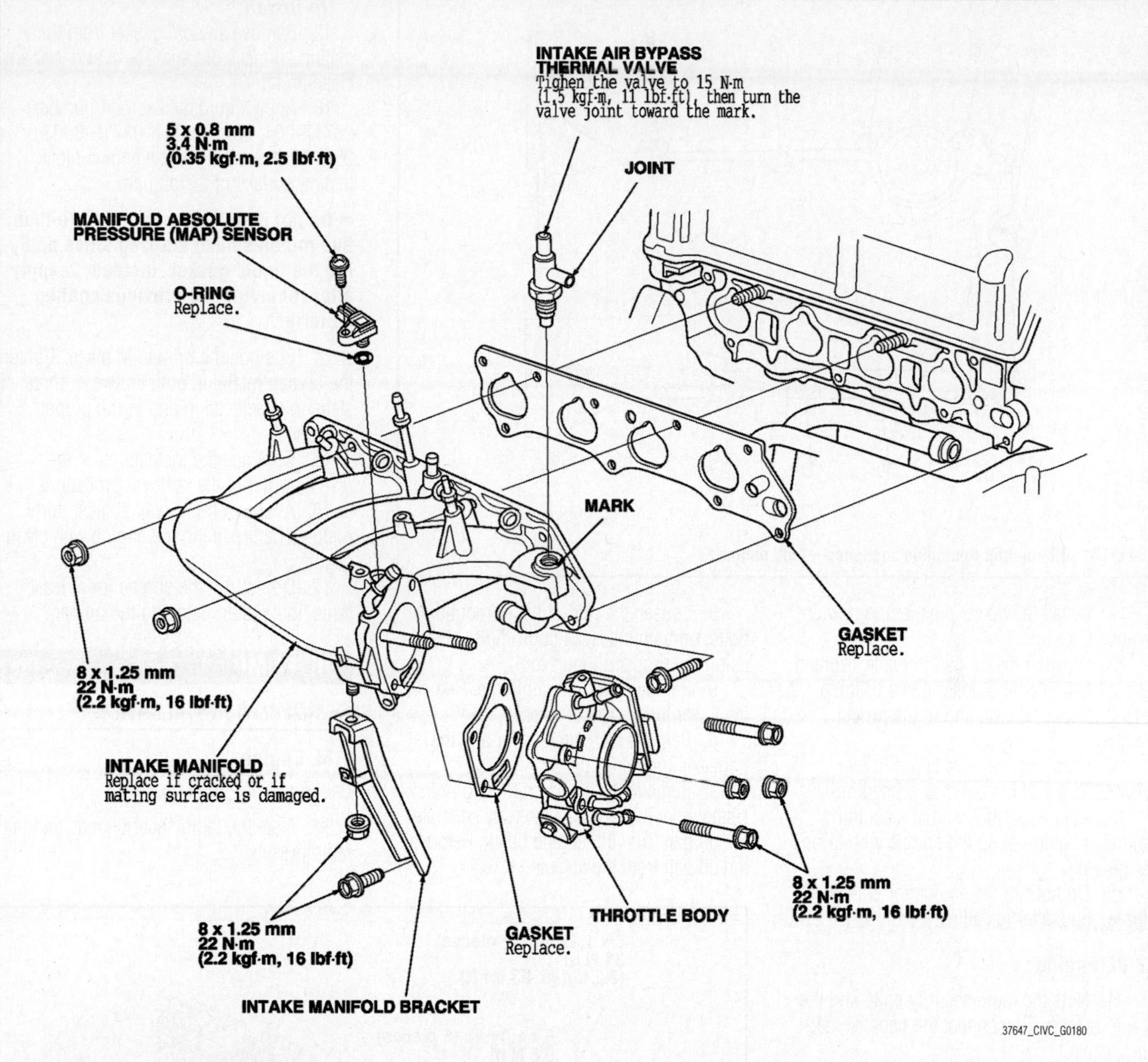
INTAKE AIR BYPASS THERMAL VALVE
Tighen the valve to 15 N·m (1.5 kgf·m, 11 lbf·ft), then turn the valve joint toward the mark.
5 x 0.8 mm
3.4 N·m
(0.35 kgf·m, 2.5 lbf·ft)
JOINT
MANIFOLD ABSOLUTE PRESSURE (MAP) SENSOR
O-RING
Replace.
MARK
GASKET
Replace.
8 x 1.25 mm
22 N·m
(2.2 kgf·m, 16 lbf·ft)
INTAKE MANIFOLD
Replace if cracked or if mating surface is damaged.
8 x 1.25 mm
22 N·m
(2.2 kgf·m, 16 lbf·ft)
THROTTLE BODY
GASKET
Replace.
8 x 1.25 mm
22 N·m
(2.2 kgf·m, 16 lbf·ft)
INTAKE MANIFOLD BRACKET
37647_CIVC_G0180

Fig. 50 Exploded view of the intake manifold assembly—2.0L engine

OIL PAN

REMOVAL & INSTALLATION

1.8L Engine

See Figure 51.

1. Note the radio security code and the radio presets. Disconnect the negative battery cable.
2. Remove the drive belt. Remove the air conditioning condenser fan shroud.
3. Disconnect the compressor electrical connectors. Remove the compressor retaining bolts, and position it to the side without discharging the system.
4. Raise and support the vehicle safely. Remove the splash shield. Drain the engine oil.
5. Remove the exhaust pipe. Properly support the oil pan. Remove the lower torque rod. Remove the oil pan support tool.
6. Remove the lower torque rod bracket. Remove the air conditioning compressor bracket.
7. If equipped with automatic transaxle, remove the shift cable cover. Remove the torque converter cover.
8. Remove the clutch cover if equipped with manual transaxle.
9. Remove the oil pan retaining bolts. Using a flat bladed tool, carefully separate the oil pan from the engine block. Remove the oil pan from the vehicle.

To install:

10. Remove any old gasket from the mating surfaces. Be sure these surfaces are clean and dry.
11. Apply liquid gasket, part number 08717-004, 08718-0001, 08718-0003 or 08718-0009 evenly to the engine block mating surface of the oil pan.

➡Do not install the parts if more than five minutes have elapsed since applying the liquid gasket. Instead, reapply after removing the previous coating material.

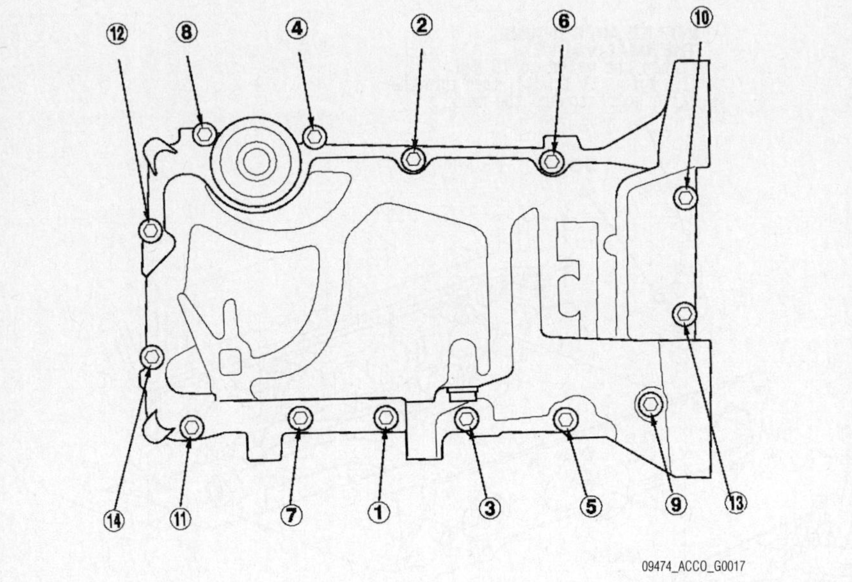
12
8
4
2
6
10
14
11
7
1
3
5
9
13
09474_ACCO_G0017

Fig. 51 Oil pan bolt tightening sequence—1.8L engine

12. Install the dowel pins, using new O-rings.
13. Position the oil pan in place. Tighten the oil pan retaining bolts in two or three steps to specification and in the proper sequence.
14. Continue the installation in the reverse order of the removal procedure.
15. After assembly, wait at least thirty minutes before filling the engine with clean engine oil.
16. Do not run the engine for at least three hours after installing the oil pan.

2.0L Engine

1. Note the radio security code and the radio presets. Disconnect the negative battery cable.
2. Raise and support the vehicle safely. Drain the engine oil. Remove the front tires.
3. Remove the splash shield. Separate the stabilizer links. Separate the knuckles from the lower arms.
4. Remove the steering gearbox bracket. Remove the steering gearbox mounting bolt, stiffener mounting bolt and stiffener.
5. Remove the gearbox mounting bolt, stiffener mounting bolt and stiffener. Remove the harness clamp from the subframe
6. Remove the lower torque rod. Remove the front mount mounting bolt.
7. Use a marker and make alignment marks on the reference lines that align with the centers of the rear subframe mounting bolts.
8. Loosen the mid-stiffener mounting bolts, on both sides. Support the subframe using the proper support tool.
9. Remove the front subframe. Remove the lower torque rod bracket.
10. Remove the clutch cover and the transaxle mounting bolts.
11. Remove the oil pan retaining bolts. Using a flat bladed tool, carefully separate the oil pan from the engine block. Remove the oil pan from the vehicle.

To install:

12. Remove any old gasket from the mating surfaces. Be sure these surfaces are clean and dry.
13. Apply liquid gasket, part number 08717-004, 08718-0001, 08718-0003 or 08718-0009 evenly to the engine block mating surface of the oil pan.

➡Do not install the parts if more than five minutes have elapsed since applying the liquid gasket. Instead, reapply after removing the previous coating material.

14. Position the oil pan in place. Tighten the oil pan retaining bolts in two or three steps to specification and in the proper sequence.
15. Continue the installation in the reverse order of the removal procedure.
16. After assembly, wait at least thirty minutes before filling the engine with clean engine oil.
17. Do not run the engine for at least three hours after installing the oil pan.

OIL PUMP

REMOVAL & INSTALLATION

1.8L Engine

See Figures 52 through 54.

1. Note the radio security code and the radio presets.

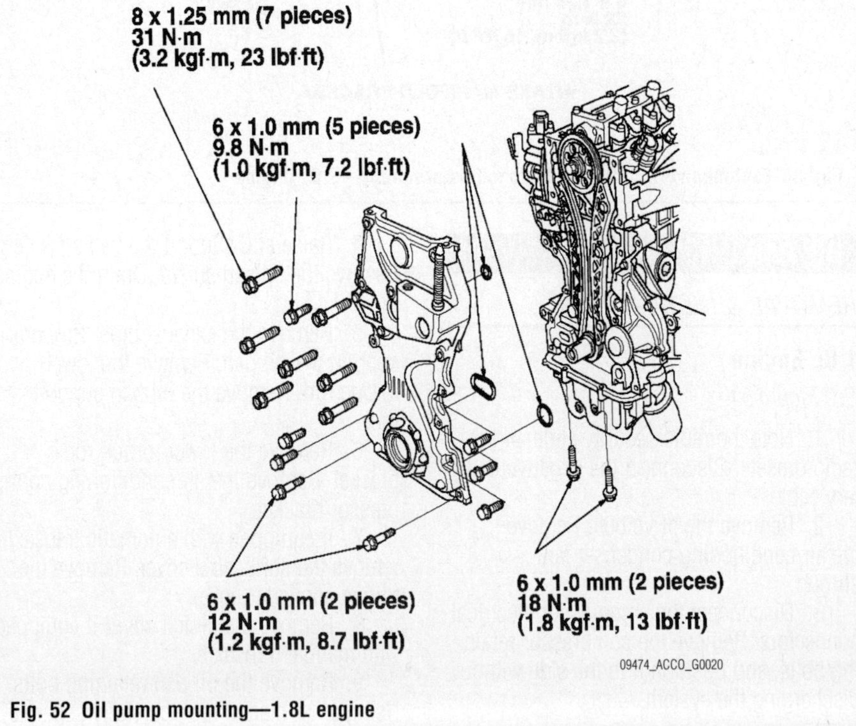
8 x 1.25 mm (7 pieces)
31 N·m
(3.2 kgf·m, 23 lbf·ft)
6 x 1.0 mm (5 pieces)
9.8 N·m
(1.0 kgf·m, 7.2 lbf·ft)
6 x 1.0 mm (2 pieces)
12 N·m
(1.2 kgf·m, 8.7 lbf·ft)
6 x 1.0 mm (2 pieces)
18 N·m
(1.8 kgf·m, 13 lbf·ft)
09474_ACCO_G0020

Fig. 52 Oil pump mounting—1.8L engine

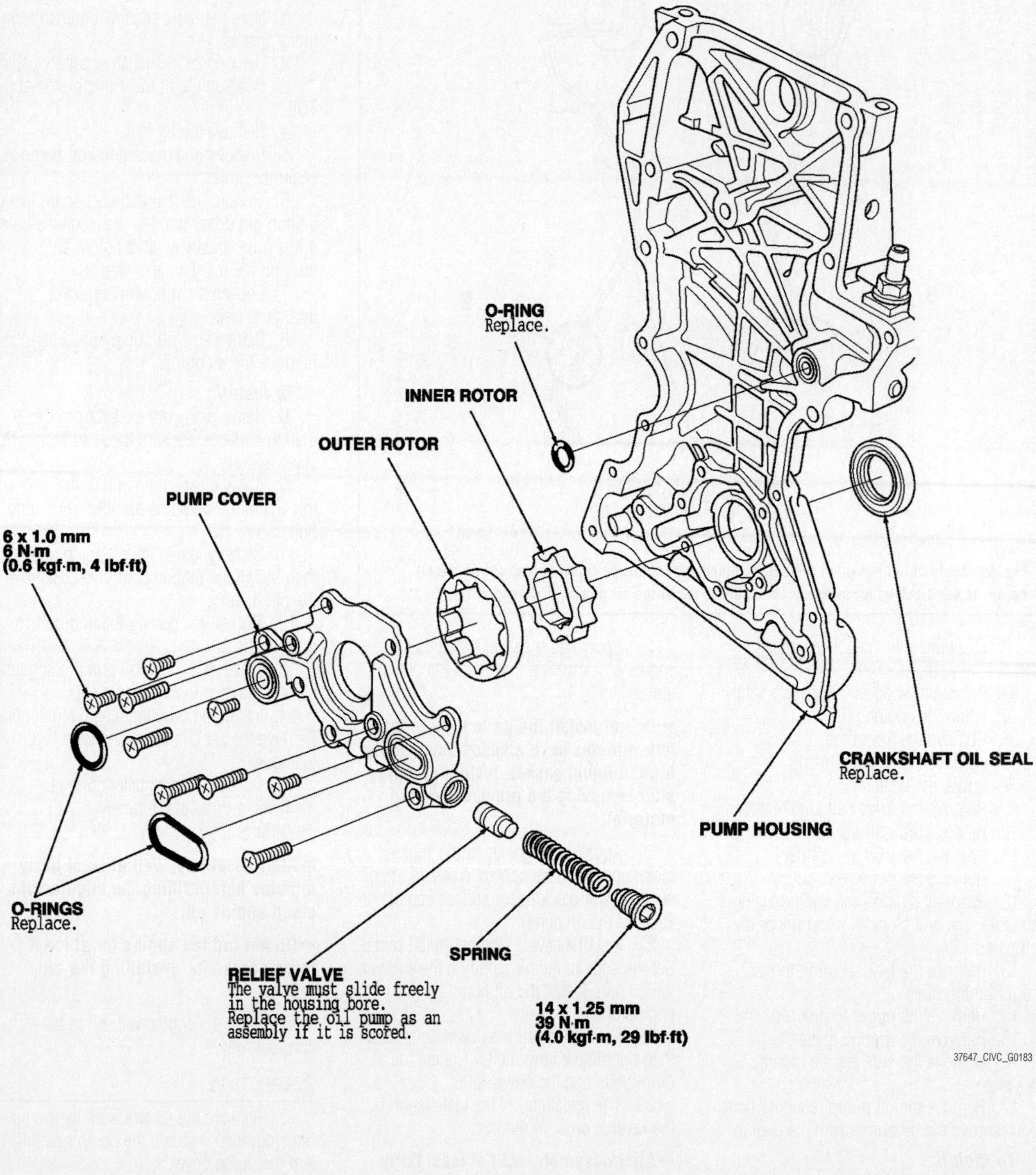
O-RING
Replace.
INNER ROTOR
OUTER ROTOR
PUMP COVER
6 x 1.0 mm
6 N·m
(0.6 kgf·m, 4 lbf·ft)
CRANKSHAFT OIL SEAL
Replace.
PUMP HOUSING
O-RINGS
Replace.
SPRING
RELIEF VALVE
The valve must slide freely
in the housing bore.
Replace the oil pump as an
assembly if it is scored.
14 x 1.25 mm
39 N·m
(4.0 kgf·m, 29 lbf·ft)
37647_CIVC_G0183

Fig. 53 Exploded view of the oil pump assembly—1.8L engine

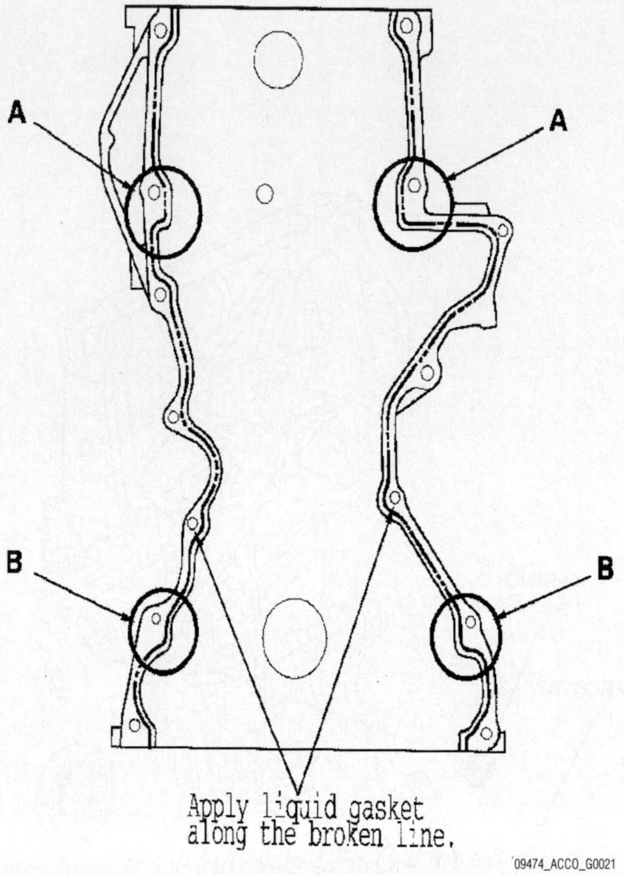
A
A
B
B
Apply liquid gasket
along the broken line.
09474_ACCO_G0021

Fig. 54 Apply liquid gasket to the engine block upper surface contact areas (A) on the oil pump, lower block upper surface contact areas (B) of the oil pump—1.8L engine

2. Disconnect the negative battery cable.
3. Raise and support the vehicle safely.
4. Drain the engine oil.
5. Remove the front tires.
6. Remove the splash shield.
7. Lower the vehicle.
8. Remove the drive belt auto tensioner.
9. Remove the cylinder head cover.
10. Remove the PCV valve hose.
11. Remove the crankshaft pulley.
12. Properly support the engine, using a suitable jack and block of wood under the oil pan.
13. Remove the bolt securing the air conditioning line.
14. Remove the upper torque rod.
15. Remove the ground cable.
16. Remove the side engine mount bracket.
17. Remove the oil pump retaining bolts and remove the oil pump from the engine.

To install:

18. Remove any old gasket from the mating surfaces. Be sure these surfaces are clean and dry.
19. Apply liquid gasket evenly to the engine block mating surface of the oil pump.

➡Do not install the parts if more than four minutes have elapsed since applying the liquid gasket. Instead, reapply after removing the previous coating material.

20. Apply liquid gasket to the engine block upper surface contact areas on the oil pump, lower block upper surface contact areas of the oil pump.
21. Install a new O-ring on the oil pump. Set the edge of the oil pump on the edge of the oil pan. Install the oil pump on the engine block.
22. Loosely install the dowel bolts, and then tighten the 8mm bolts. Tighten the 6mm bolts and the dowel bolts.
23. The remainder of the installation is the reverse order of removal.

➡After assembly, wait at least thirty minutes before filling the engine with clean engine oil.

➡Do not run the engine for at least three hours after installing the oil pan.

24. Refill the engine with oil to the correct level.

2.0L Engine

See Figures 55 through 59.

1. Note the radio security code and the radio presets.
2. Disconnect the negative battery cable.
3. Position the number one piston at TDC.
4. Remove the oil pan.
5. Remove and discard the oil pump chain tensioner.
6. To hold the rear balancer shaft, insert a 6mm pin driver into the maintenance hole in the lower balancer shaft holder and through the rear balancer shaft.
7. Loosen the oil pump sprocket mounting bolt.
8. Remove the oil pump sprocket. Remove the oil pump.

To install:

9. Remove any old gasket from the mating surfaces. Be sure these surfaces are clean and dry.
10. Apply clean engine oil to the threads of the oil pump sprocket mounting bolt.
11. Loosely install the oil pump, and then install the oil pump sprocket. Remove the pin driver.
12. Tighten the pump retaining bolts to specification.
13. Squeeze the new oil pump chain tensioner and then install the set clip.
14. Install the oil pump chain tensioner. Remove the set clip from the pump chain tensioner.
15. Continue the installation in the reverse order of the removal procedure.

➡After assembly, wait at least thirty minutes before filling the engine with clean engine oil.

➡Do not run the engine for at least three hours after installing the oil pan.

16. Refill the engine with oil to the correct level.

INSPECTION

1. Remove the screws from the pump housing, then separate the pump housing and the pump cover.
2. Check the inner-to-outer rotor radial clearance between the inner rotor and the outer rotor. If the inner-to-outer rotor clear-

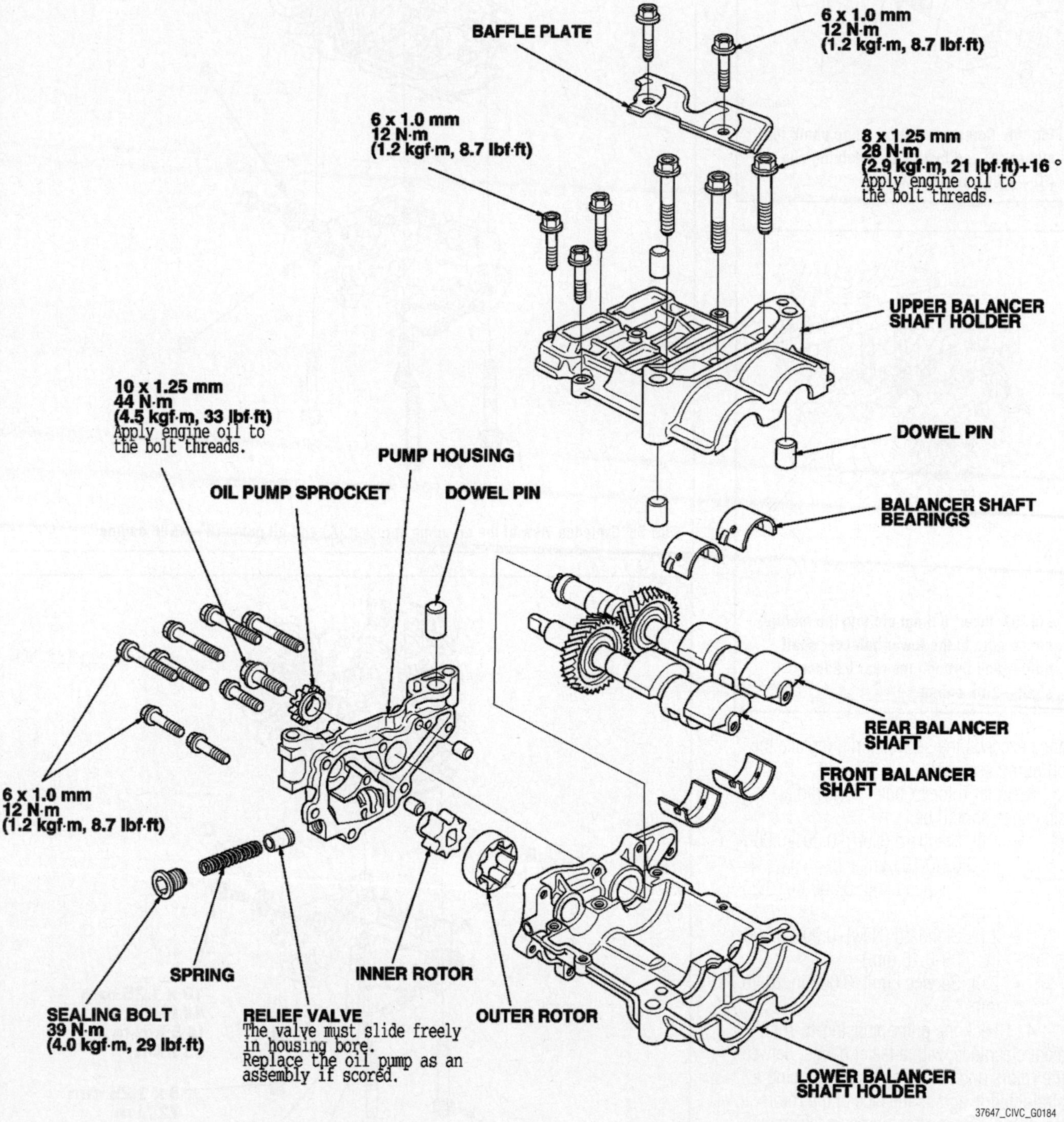
BAFFLE PLATE
6 x 1.0 mm
12 N·m
(1.2 kgf·m, 8.7 lbf·ft)
6 x 1.0 mm
12 N·m
(1.2 kgf·m, 8.7 lbf·ft)
8 x 1.25 mm
28 N·m
(2.9 kgf·m, 21 lbf·ft)+16 °
Apply engine oil to
the bolt threads.
UPPER BALANCER
SHAFT HOLDER
10 x 1.25 mm
44 N·m
(4.5 kgf·m, 33 lbf·ft)
Apply engine oil to
the bolt threads.
DOWEL PIN
PUMP HOUSING
OIL PUMP SPROCKET
DOWEL PIN
BALANCER SHAFT
BEARINGS
REAR BALANCER
SHAFT
FRONT BALANCER
SHAFT
6 x 1.0 mm
12 N·m
(1.2 kgf·m, 8.7 lbf·ft)
SPRING
INNER ROTOR
SEALING BOLT
39 N·m
(4.0 kgf·m, 29 lbf·ft)
RELIEF VALVE
The valve must slide freely
in housing bore.
Replace the oil pump as an
assembly if scored.
OUTER ROTOR
LOWER BALANCER
SHAFT HOLDER
37647_CIVC_G0184

Fig. 55 Exploded view of the oil pump assembly—2.0L engine

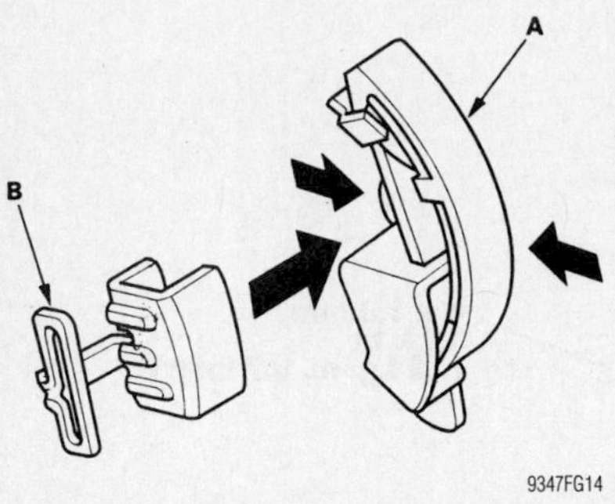
A
B
9347FG14

Fig. 56 Compress the oil pump chain tensioner (A) and install the retaining clip (B)—2.0L engine

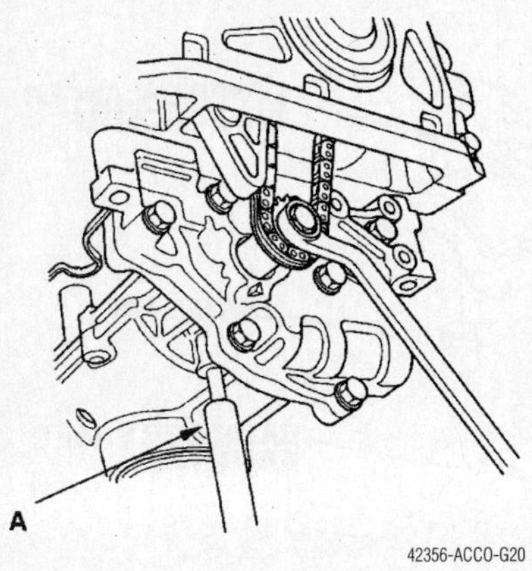
A
42356-ACCO-G20

Fig. 57 Insert a 6mm pin into the maintenance hole in the lower balancer shaft holder and through the rear balancer shaft—2.0L engine

ance exceeds the service limit, replace the oil pump assembly.

3. Inner rotor-to-outer rotor radial clearance should be:

- 1.8L Standard (New): 0.001-0.006 in. 0.02-0.16 mm)
- 1.8L Service Limit: 0.008 in. (0.20 mm)
- 2.0L Standard (New): 0.002-0.006 in. 0.03-0.16 mm)
- 2.0L Service Limit: 0.008 in. (0.20 mm)

4. Check the pump housing-to-rotor axial clearance, with a feeler gauge, between the rotors and the pump housing, using a straightedge across the face of the rotors. If the pump housing-to-rotor axial clearance exceeds the service limit, replace the oil pump assembly.

5. Pump housing-to-rotor axial clearance should be:

- 1.8L Standard (New): 0.001-0.003 in. (0.02-0.07 mm)

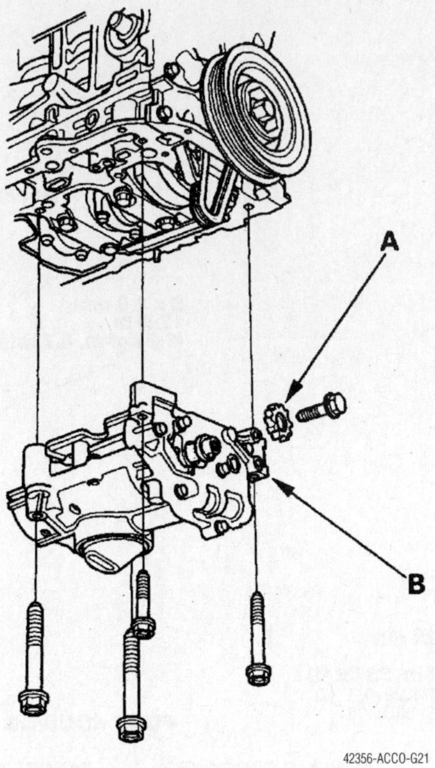
A
B
42356-ACCO-G21

Fig. 58 Exploded view of the oil pump sprocket (A) and oil pump (B)—2.0L engine

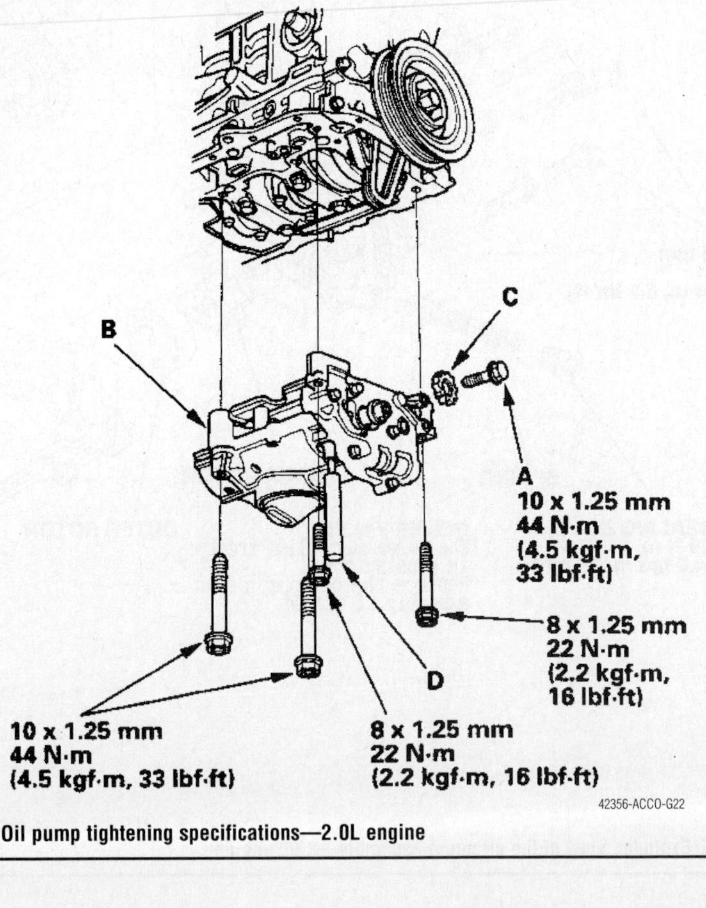
C
B
A
10 x 1.25 mm
44 N·m
(4.5 kgf·m,
33 lbf·ft)
8 x 1.25 mm
22 N·m
(2.2 kgf·m,
16 lbf·ft)
D
10 x 1.25 mm
44 N·m
(4.5 kgf·m, 33 lbf·ft)
8 x 1.25 mm
22 N·m
(2.2 kgf·m, 16 lbf·ft)
42356-ACCO-G22

Fig. 59 Oil pump tightening specifications—2.0L engine

- 1.8L Service Limit: 0.005 in. (0.12 mm)
- 2.0L Standard (New): 0.0014-0.0028 in. (0.035-0.070 mm)
- 2.0L Service Limit: 0.005 in. (0.12 mm)

6. Check the pump housing-to-outer rotor radial clearance between the outer rotor and the pump housing. If the pump housing-to-outer rotor radial clearance exceeds the service limit, replace the oil pump assembly.
7. Pump housing-to-outer rotor radial clearance should be:
   - 1.8L Standard (New): 0.0039-0.0069 in. (0.100-0.175 mm)
   - 1.8L Service Limit: 0.008 in. (0.20 mm)
   - 2.0L Standard (New): 0.006-0.008 in. (0.15-0.21 mm)
   - 2.0L Service Limit: 0.009 in. (0.23 mm)
8. Inspect both rotors and the pump housing for scoring or other damage. Replace the parts, if necessary.
9. Apply liquid thread lock to the pump housing screws, then install the oil pump cover.
10. Check that the oil pump turns freely.

PISTON AND RING

POSITIONING

See Figures 60 and 61.

REAR MAIN SEAL

REMOVAL & INSTALLATION

1. Remove the transmission and, on M/T, the pressure plate, the clutch disc, and the flywheel; on A/T model, remove the drive plate.
2. Remove the transmission end crankshaft oil seal.
3. Clean and dry the crankshaft oil seal housing.
4. Apply a light coat of new engine oil to the crankshaft and to the lip of the crankshaft oil seal.
5. Use the driver handle, 15 x 135L and oil seal driver attachment, 96 mm to drive a new crankshaft oil seal squarely into the engine block to the specified installed height of 0.001–0.047 in. (0.2–1.2 mm)
6. Install the driveplate or pressure plate, clutch disc and flywheel, and the transmission.

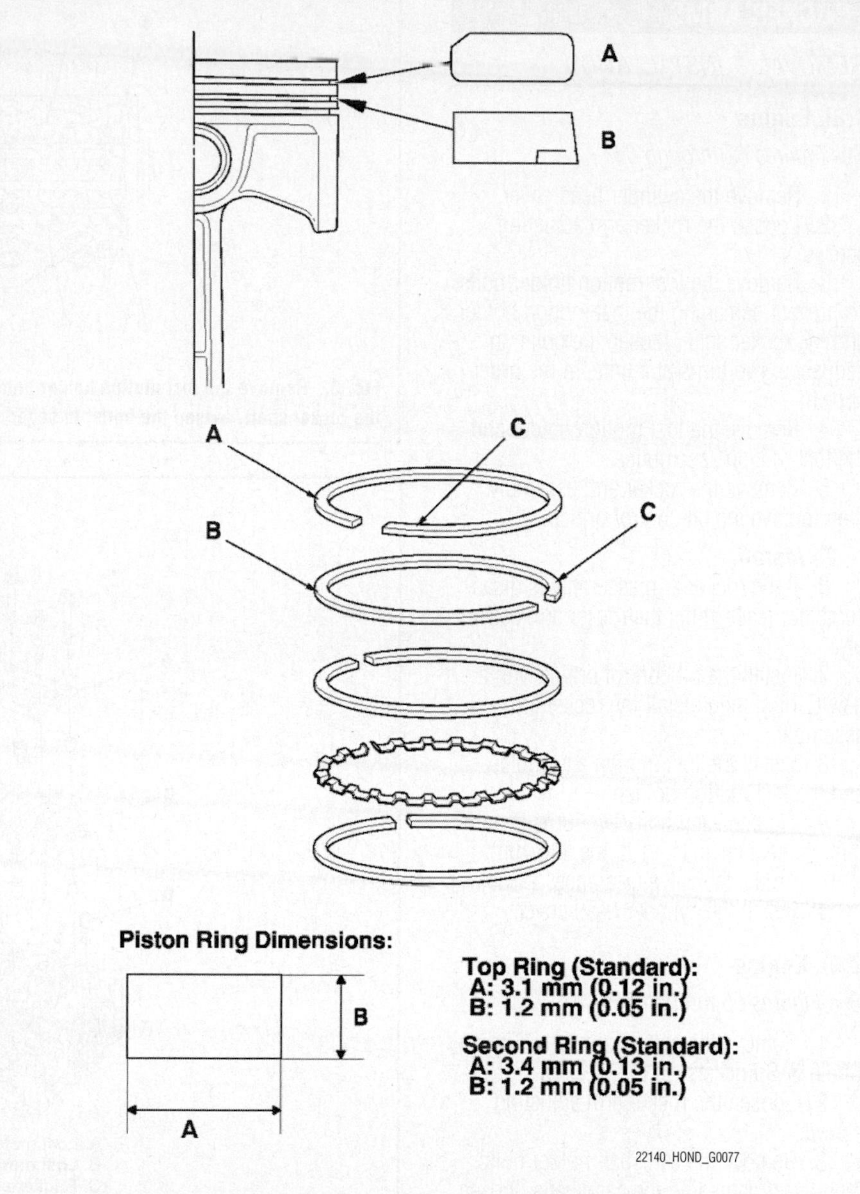
A
B
A
C
B
C
Piston Ring Dimensions:
B
A
Top Ring (Standard):
A: 3.1 mm (0.12 in.)
B: 1.2 mm (0.05 in.)
Second Ring (Standard):
A: 3.4 mm (0.13 in.)
B: 1.2 mm (0.05 in.)
22140_HOND_G0077

Fig. 60 Top ring (A), second ring (B) and the manufacturing marks (C) must face upward

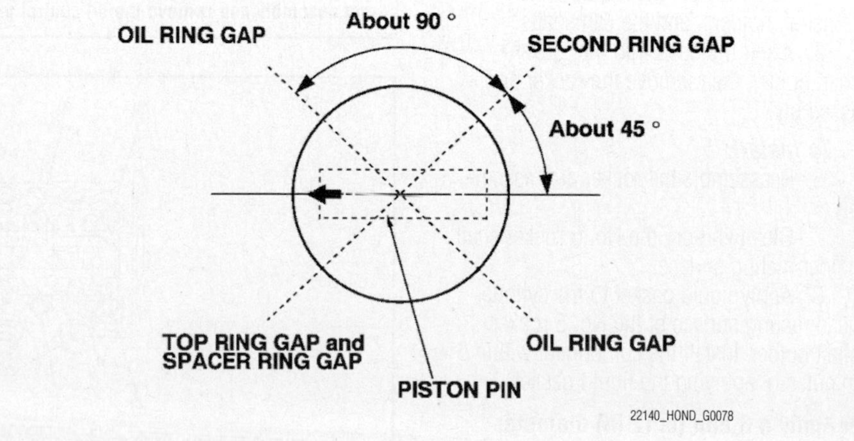
OIL RING GAP
About 90 °
SECOND RING GAP
About 45 °
TOP RING GAP and SPACER RING GAP
OIL RING GAP
PISTON PIN
22140_HOND_G0078

Fig. 61 Piston ring positioning

ROCKER ARMS

REMOVAL & INSTALLATION

1.8L Engine

See Figures 62 through 64.

1. Remove the cylinder head cover.
2. Loosen the rocker arm adjusting screws.
3. Remove the lost motion holder bolts. To prevent damaging the lost motion holder and the rocker shaft, loosen the bolts, in sequence, two turns at a time, in the order shown.
4. Remove the lost motion holder and the lost motion assemblies.
5. Remove the rocker arm assembly, then remove the oil control orifice.

To install:

6. If the rocker arm assembly is disassembled, reassemble the rocker arm assembly.
7. Install the oil control orifice with a new O-ring, then install the rocker arm assembly.
8. Install the lost motion assembles and the lost motion holder.
9. Tighten each bolts two turns at a time, in sequence, to 11 ft. lbs. (15 Nm).
10. Adjust the valve clearance.
11. Install the cylinder head cover.

2.0L Engine

See Figures 65 through 68.

1. Remove the cam chain. See "Timing Chain & Sprockets" in this section.
2. Loosen the rocker arm adjusting screws.
3. Remove the camshaft holder bolts. To prevent damaging the camshafts, loosen the bolts, in sequence, two turns at a time.

➡Bolt "1" is not on all engines.

4. Remove cam chain guide, the camshaft holders, and the camshafts.
5. Insert the bolts into the rocker shaft holder, then remove the rocker arm assembly.

To install:

6. Reassemble the rocker arm assembly.
7. Clean and dry the No. 5 rocker shaft holder mating surface.
8. Apply liquid gasket to the cylinder head mating surface of the No. 5 rocker shaft holder. Install the component within 5 minutes of applying the liquid gasket.

➡Apply a 3 mm (0.12 in) diameter bead of liquid gasket along the cylinder

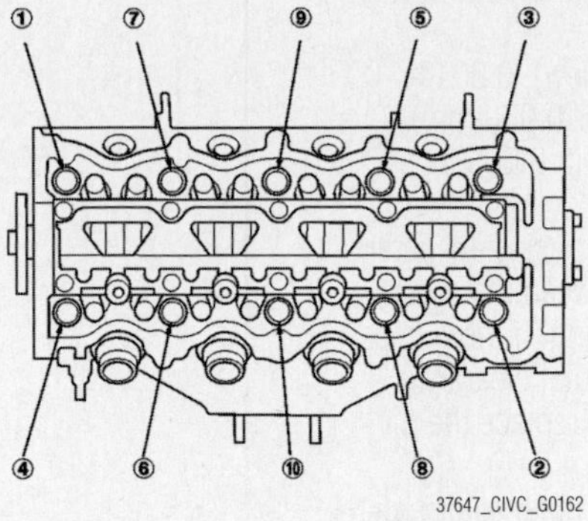
1
7
9
5
3
4
6
10
8
2
37647_CIVC_G0162

Fig. 62 Remove the lost motion holder bolts. To prevent damaging the lost motion holder and the rocker shaft, loosen the bolts, in sequence, two turns at a time, in the order shown.

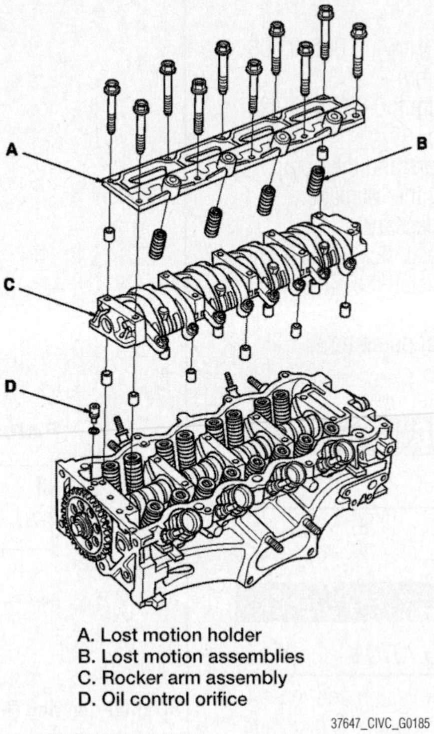
A
B
C
D
A. Lost motion holder
B. Lost motion assemblies
C. Rocker arm assembly
D. Oil control orifice
37647_CIVC_G0185

Fig. 63 Remove the lost motion holder and the lost motion assemblies, then remove the rocker arm assembly and remove the oil control orifice.

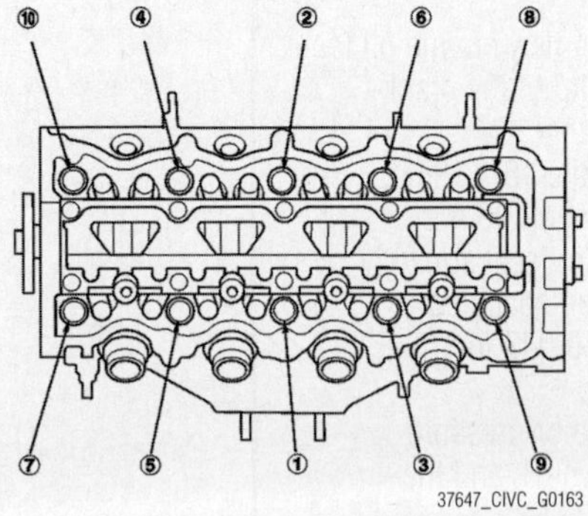
10
4
2
6
8
7
5
1
3
9
37647_CIVC_G0163

Fig. 64 Tighten each bolts two turns at a time, in sequence, to 11 ft. lbs. (15 Nm).

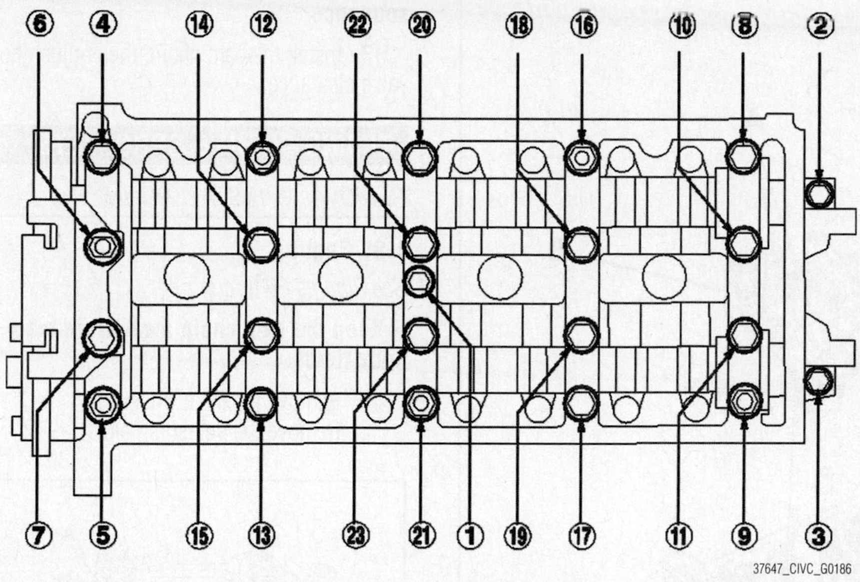
6
4
14
12
22
20
18
16
10
8
2
7
5
15
13
23
21
1
19
17
11
9
3
37647_CIVC_G0186

Fig. 65 Remove the camshaft holder bolts. To prevent damaging the camshafts, loosen the bolts, in sequence, two turns at a time.

head-to-rocker shaft holder mating surface. If too much time has passed after applying the liquid gasket, remove the old liquid gasket and residue, then reapply new liquid gasket.

9. Insert the bolts used for removal into the rocker shaft holder, then install the rocker arm assembly on the cylinder head. Remove these bolts from the rocker shaft holder.

10. Make sure the punch marks on the VTC actuator and the exhaust camshaft sprocket are facing up, then set the camshafts in the holder. Apply new engine oil to the camshaft journals and lobes.

11. Set the camshaft holders and cam chain guide in place.

12. Tighten the bolts to the specified torque, in sequence, to the following:

- 8x1.25 mm bolts: 16 ft. lbs. (22 Nm)
- 6x1.0 mm bolts: 8.7 ft. lbs. (12 Nm)
- 6 x 1.0 mm bolts: 21, 22, 23 in sequence

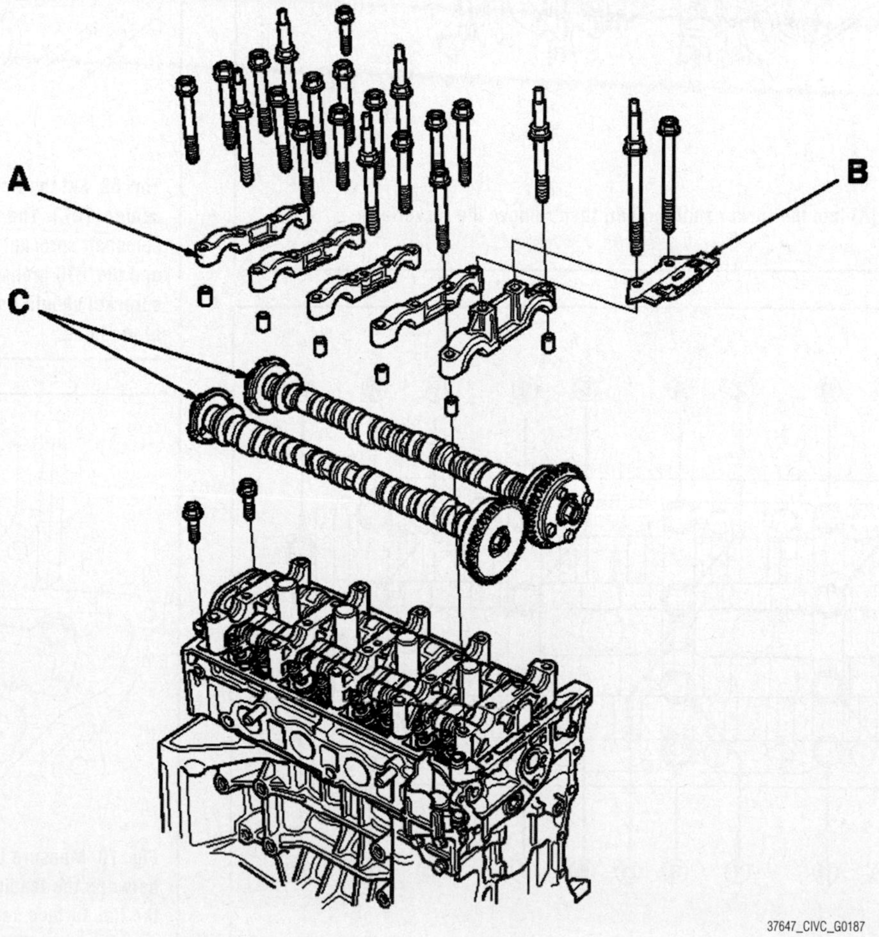
A
B
C
37647_CIVC_G0187

Fig. 66 Remove cam chain guide (B), the camshaft holders (A), and the camshafts (C).

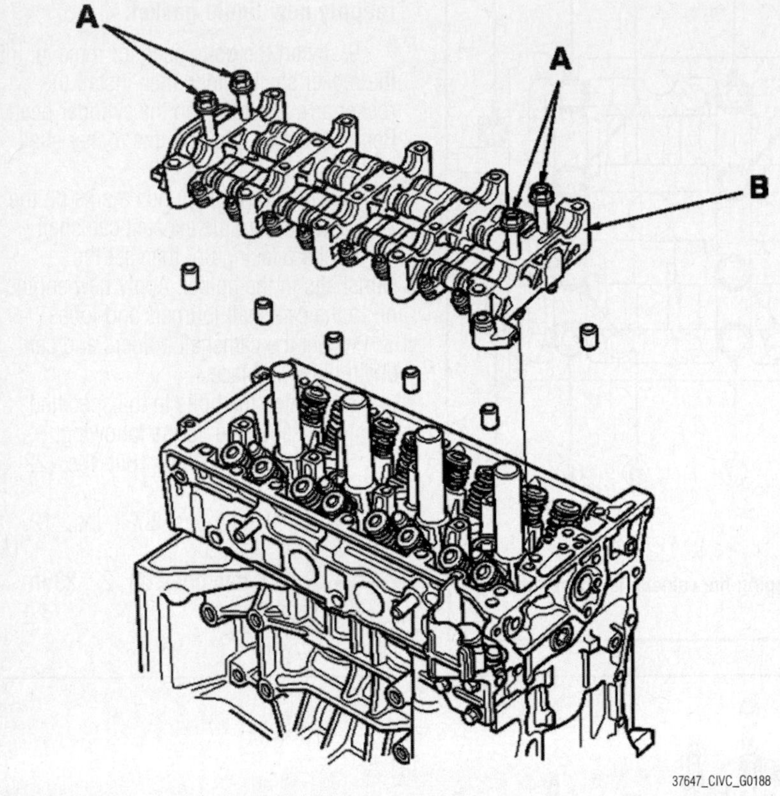
A
A
B
37647_CIVC_G0188

Fig. 67 Insert the bolts (A) into the rocker shaft holder, then remove the rocker arm assembly (B).

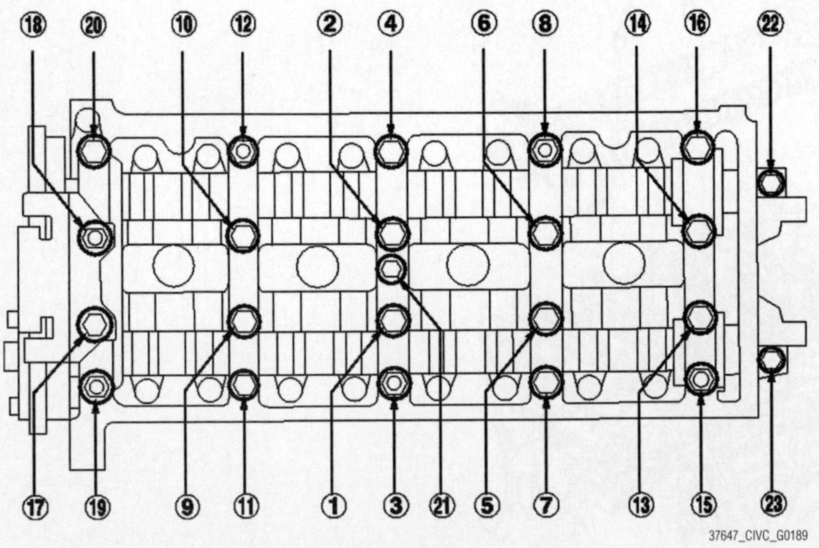
18 20 10 12 2 4 6 8 14 16 22
17 19 9 11 1 3 21 5 7 13 15 23
37647_CIVC_G0189

Fig. 68 Rocker arm bolt tightening sequence

➡If the engine does not have bolt 21, skip it and continue the torque sequence.

13. Install the cam chain, then adjust the valve clearance.

TIMING CHAIN & SPROCKETS

REMOVAL & INSTALLATION

1.8L Engine

See Figures 69 through 74.

➡Keep the cam chain away from magnetic fields.

1. Remove the front wheels.
2. Remove the splash shield.

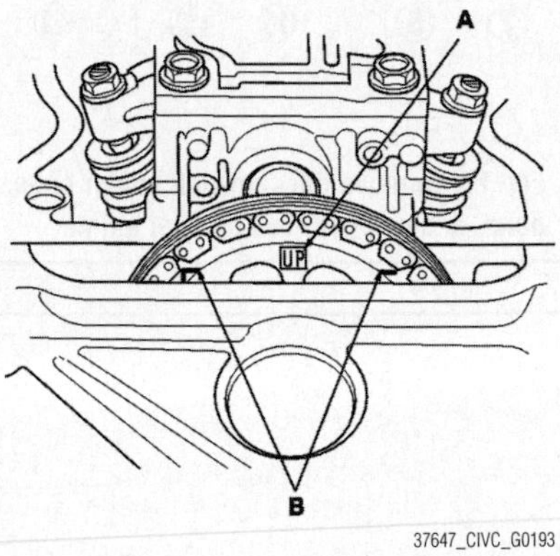
A
UP
B
37647_CIVC_G0193

Fig. 69 Set the No. 1 piston at top dead center (TDC). The "UP" mark (A) on the camshaft sprocket should be at the top, and the TDC grooves (B) on the camshaft sprocket should line up with the top edge of the head.

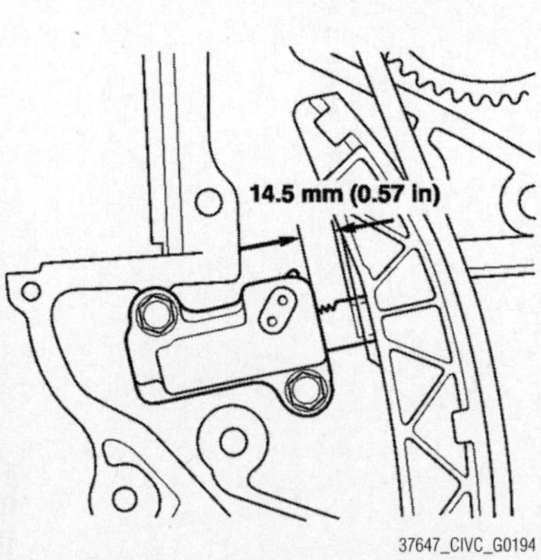
14.5 mm (0.57 in)
37647_CIVC_G0194

Fig. 70 Measure the tensioner rod length between the tensioner body and bottom of the flat surface section on the tensioner rod. If the length is more than the service limit, replace the cam chain.

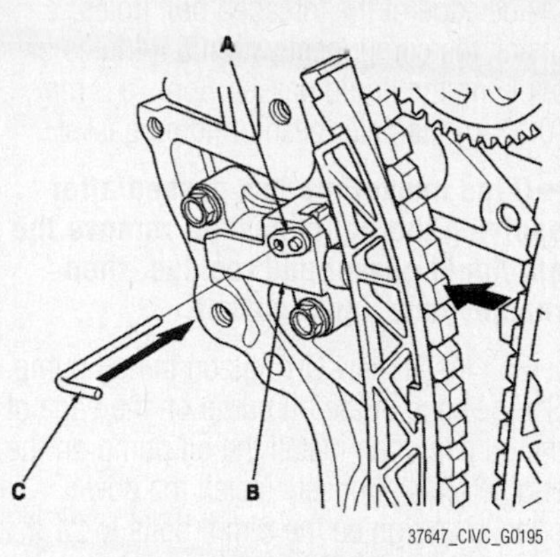
A
C
B
37647_CIVC_G0195

Fig. 71 Align the holes on the lock (A) and the auto-tensioner (B), then insert a 1.0 mm (0.04 in) diameter pin (C) into the holes. Turn the crankshaft clockwise to secure the pin.

3. Remove the drive belt auto-tensioner.
4. Remove the cylinder head cover.
5. Set the No. 1 piston at top dead center (TDC). The "UP" mark on the camshaft sprocket should be at the top, and the TDC grooves on the camshaft sprocket should line up with the top edge of the head.
6. Disconnect the positive crankcase ventilation (PCV) hose.
7. Remove the crankshaft pulley.
8. Support the engine with a jack and a wood block under the oil pan.
9. Remove the bolt securing the A/C line, then remove the upper torque rod.
10. Remove the ground cable, then remove the side engine mount bracket.
11. Remove the oil pump.
12. Measure the tensioner rod length between the tensioner body and bottom of the flat surface section on the tensioner rod. If the length is more than the service limit, replace the cam chain.
13. Tensioner rod length service limit is 0.57 in. (14.5 mm).
14. Loosely install the crankshaft pulley.
15. Turn the crankshaft counterclockwise to compress the auto-tensioner.
16. Align the holes on the lock and the auto-tensioner, then insert a 0.04 in. (1.0 mm) diameter pin into the holes. Turn the crankshaft clockwise to secure the pin.

➡Check the auto-tensioner cam position If the position is not aligned, set the first cam to the first edge of the rack.

17. Remove the auto-tensioner.

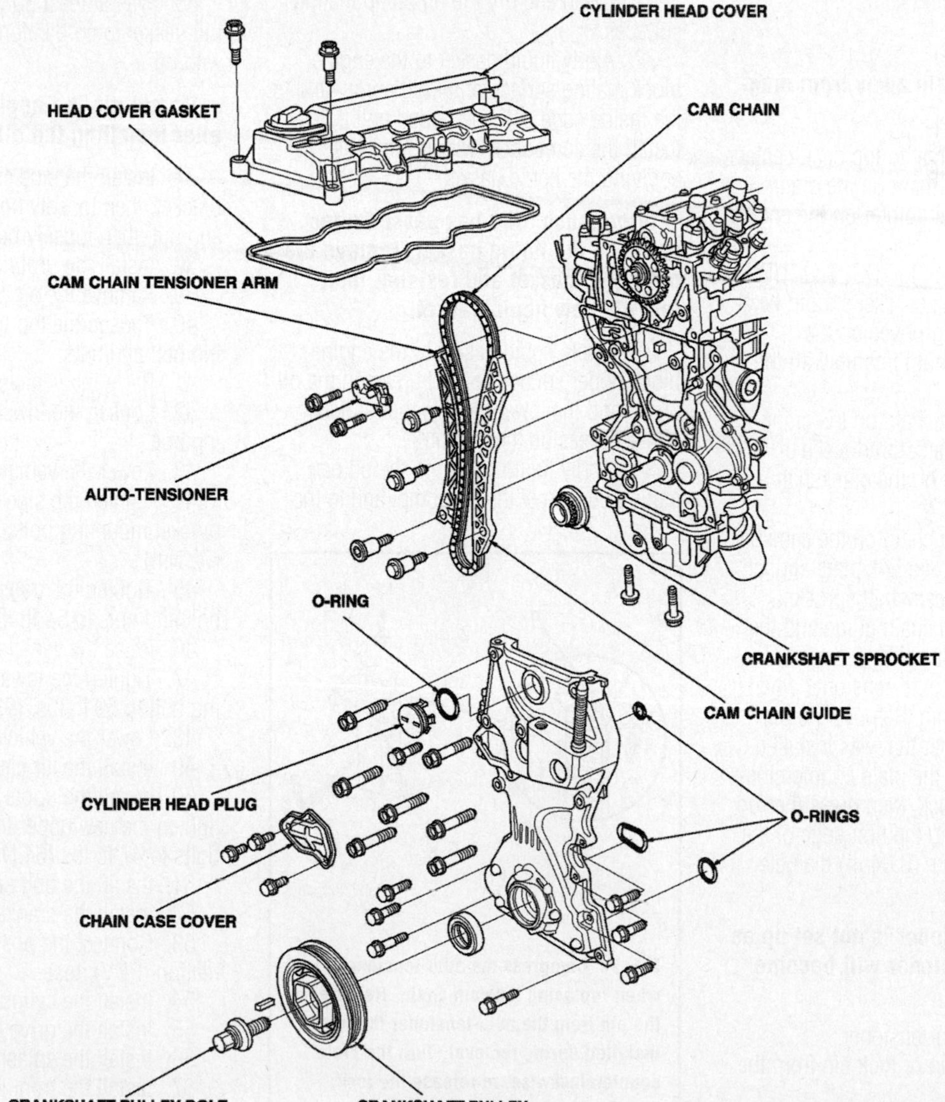
CYLINDER HEAD COVER
HEAD COVER GASKET
CAM CHAIN
CAM CHAIN TENSIONER ARM
AUTO-TENSIONER
O-RING
CRANKSHAFT SPROCKET
CAM CHAIN GUIDE
CYLINDER HEAD PLUG
O-RINGS
CHAIN CASE COVER
CRANKSHAFT PULLEY BOLT
CRANKSHAFT PULLEY
37647_CIVC_G0196

Fig. 72 Exploded view of the timing chain (cam chain) and related components—1.8L engine

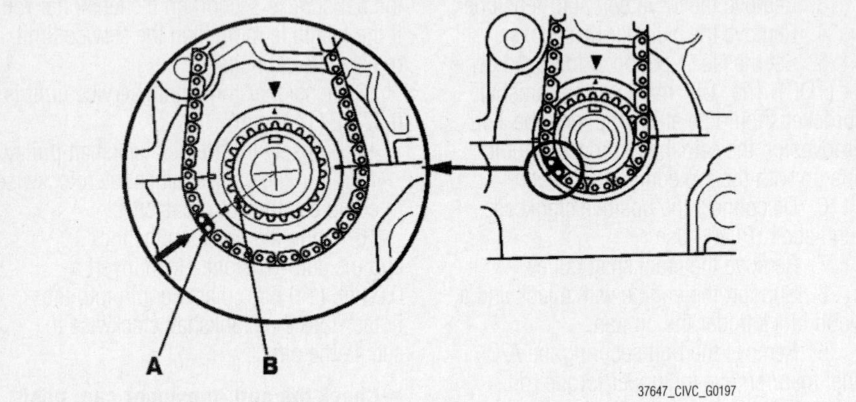
A
B
37647_CIVC_G0197

Fig. 73 Install the cam chain on the crankshaft sprocket with the colored piece (A) aligned with the mark (B) on the crankshaft sprocket.

18. Remove the crankshaft pulley.
19. Remove the cam chain guide and the cam chain tensioner arm.
20. Remove the cam chain.

To install:

➡Keep the cam chain away from magnetic fields.

21. Set the crankshaft to top dead center (TDC). Align the TDC mark on the crankshaft sprocket with the pointer on the engine block.
22. Set the camshaft to TDC. The "UP" mark on the camshaft sprocket should be at the top, and the TDC grooves on the camshaft sprocket should line up with the top edge of the head.
23. Install the cam chain on the crankshaft sprocket with the colored piece aligned with the mark on the crankshaft sprocket.
24. Install the cam chain on the camshaft sprocket with the colored link plate aligned with the mark on the camshaft sprocket.
25. Install the cam chain guide and the cam chain tensioner arm.
26. Compress the auto-tensioner when replacing the cam chain. Remove the pin from the auto-tensioner that was installed during removal. Turn the plate counterclockwise, to release the lock, then press the rod, and set the first cam to the first edge of the rack. Insert the 1.0 mm (0.04 in) diameter pin into the holes.

➡If the chain tensioner is not set up as described, the tensioner will become damaged.

27. Install the auto-tensioner.
28. Remove the pin or lock pin from the auto-tensioner.
29. Check the oil pump oil seal for damage If the oil seal is damaged, replace the oil seal..
30. Remove all of the old liquid gasket from the oil pump mating surfaces, the bolts, and the bolt holes.
31. Clean and dry the oil pump mating surfaces.
32. Apply liquid gasket to the engine block mating surface of the oil pump, and to the inside edge of the threaded bolt holes. Install the component within 5 minutes of applying the liquid gasket.

➡If too much time has passed after applying the liquid gasket, remove the old liquid gasket and residue, then reapply new liquid gasket.

33. Apply liquid gasket to the engine block upper surface contact areas on the oil pump and the lower block upper surface contact areas on the oil pump.
34. Apply liquid gasket to the oil pan mating surface of the oil pump, and to the inside edge of the threaded bolt holes. Install the component within 5 minutes of applying the liquid gasket. Apply a 3 mm (0.12 in) diameter bead of liquid gasket.

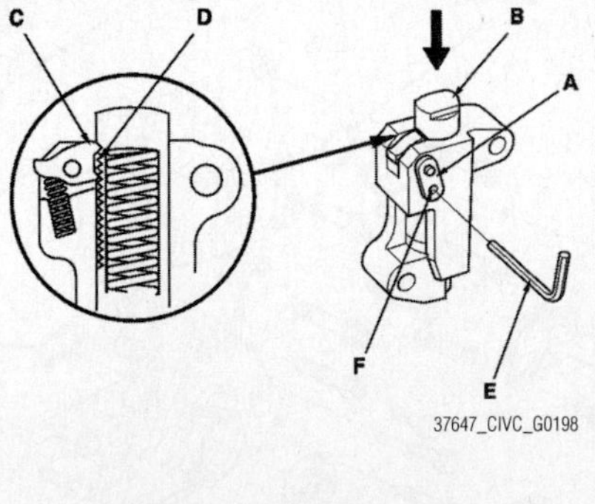
C
D
B
A
F
E
37647_CIVC_G0198

Fig. 74 Compress the auto-tensioner when replacing the cam chain. Remove the pin from the auto-tensioner that was installed during removal. Turn the plate counterclockwise, to release the lock, then press the rod, and set the first cam to the first edge of the rack. Insert the 1.0 mm (0.04 in) diameter pin into the holes.

➡If too much time has passed after applying the liquid gasket, remove the old liquid gasket and residue, then reapply new liquid gasket.

35. Install new O-rings on the oil pump. Set the edge of the oil pump on the edge of the oil pan, then install the oil pump on the engine block. Loosely install the dowel bolts, then tighten the 8 mm bolts to 21 ft. lbs. (33 Nm), the 6 mm bolts and the dowel bolts. Wipe off the excess liquid gasket on the oil pan and oil pump mating surface.

➡When installing the oil pump, do not slide the bottom surface onto the oil pan mounting surface.

36. Wait at least 30 minutes to allow liquid gasket to cure before filling the engine with oil.

➡Do not run the engine within 3 hours after installing the oil pump.

37. Install the side engine mount bracket, then loosely tighten the new bolt and nut, then loosely tighten the bolt.
38. Install the ground cable.
39. Remove the air cleaner assembly.
40. Loosen the top transmission mounting bolt and nuts.
41. Raise the vehicle on the lift.
42. Loosen the lower torque rod mounting bolt.
43. Lower the vehicle on the lift.
44. Tighten the side engine mount bracket mounting bolts and nut to 52 ft. lbs. (72 Nm).
45. Tighten the transmission mounting bolt and nuts to 54 ft. lbs. (74 Nm).
46. Raise the vehicle on the lift.
47. Tighten the lower torque rod mounting bolt to 69 ft. lbs. (93 Nm).
48. Lower the vehicle on the lift.
49. Install the air cleaner assembly.
50. Install the upper torque rod, then tighten the new upper torque rod mounting bolts to 47 ft. lbs (64 Nm).
51. Install the bolt securing the A/C line.
52. Install the crankshaft pulley.
53. Connect the positive crankcase ventilation (PCV) hose.
54. Install the cylinder head cover.
55. Install the drive belt auto-tensioner.
56. Install the splash shield.
57. Install the front wheels.
58. Do the Crankshaft Position (CKP) Pattern Clear/CKP Pattern Learn Procedure. See "Crankshaft Position (CKP) Sensor" in

"ENGINE PERFORMANCE & EMISSION CONTROLS" section.

2.0L Engine

See Figures 75 through 78.

➡Keep the cam chain away from magnetic fields.

1. Remove the front wheels.
2. Remove the splash shield.
3. Drain the engine coolant.
4. Remove the drive belt.
5. Remove the cylinder head cover.
6. Set the No. 1 piston at top dead center (TDC). The punch mark on the variable valve timing control (VTC) actuator and the punch mark on the exhaust camshaft sprocket should be at the top. Align the TDC marks on the VTC actuator and the exhaust camshaft sprocket.
7. Remove the oil cooler hose joint pipe from the water pump.
8. Disconnect the crankshaft position (CKP) sensor connector and the VTC oil control solenoid valve connector and

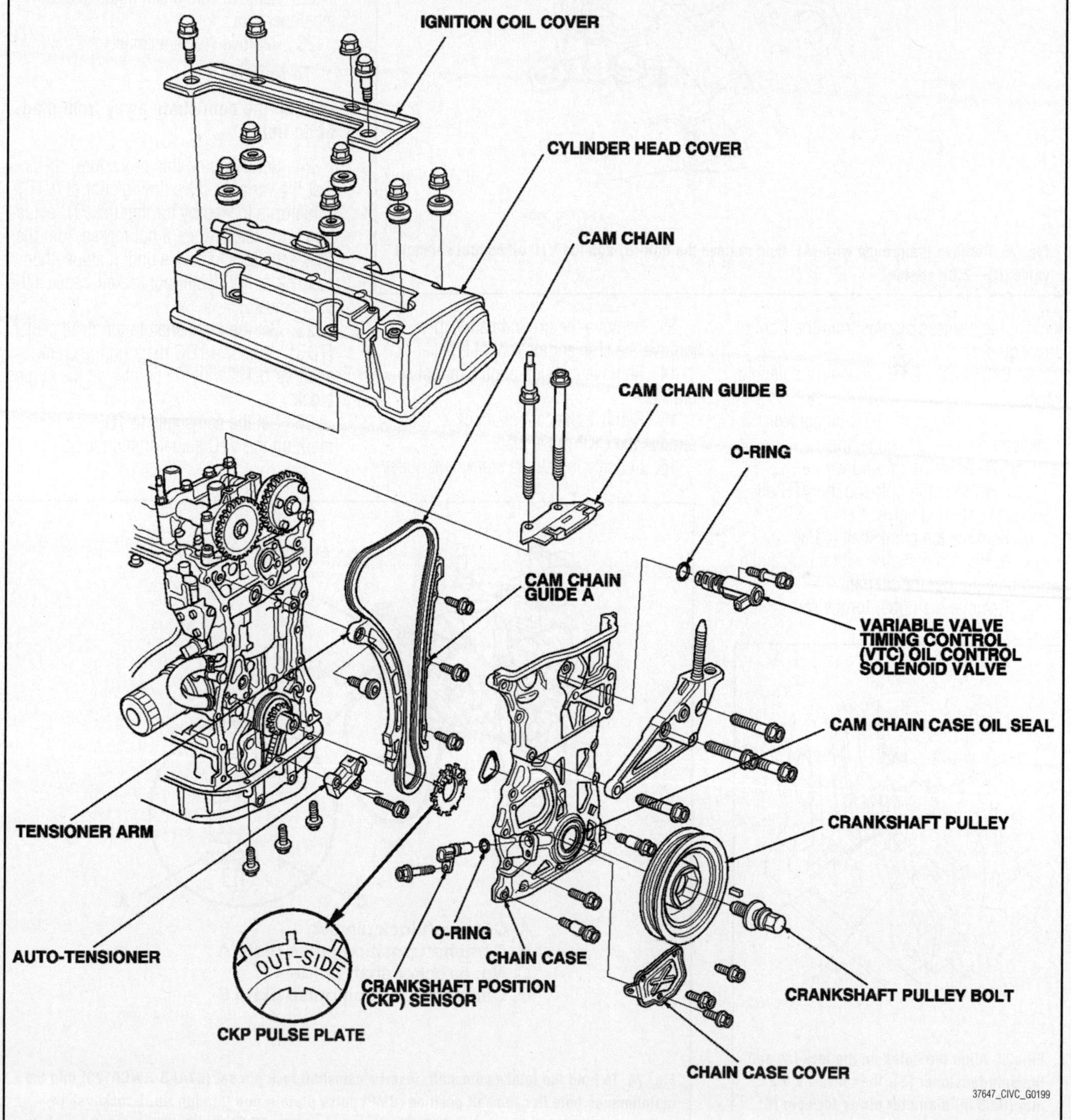
IGNITION COIL COVER
CYLINDER HEAD COVER
CAM CHAIN
CAM CHAIN GUIDE B
O-RING
CAM CHAIN GUIDE A
VARIABLE VALVE TIMING CONTROL (VTC) OIL CONTROL SOLENOID VALVE
CAM CHAIN CASE OIL SEAL
CRANKSHAFT PULLEY
TENSIONER ARM
AUTO-TENSIONER
OUT-SIDE
O-RING
CHAIN CASE
CRANKSHAFT POSITION (CKP) SENSOR
CRANKSHAFT PULLEY BOLT
CKP PULSE PLATE
CHAIN CASE COVER
37647_CIVC_G0199

Fig. 75 Exploded view of the timing chain (cam chain) and related engine components—2.0L engine

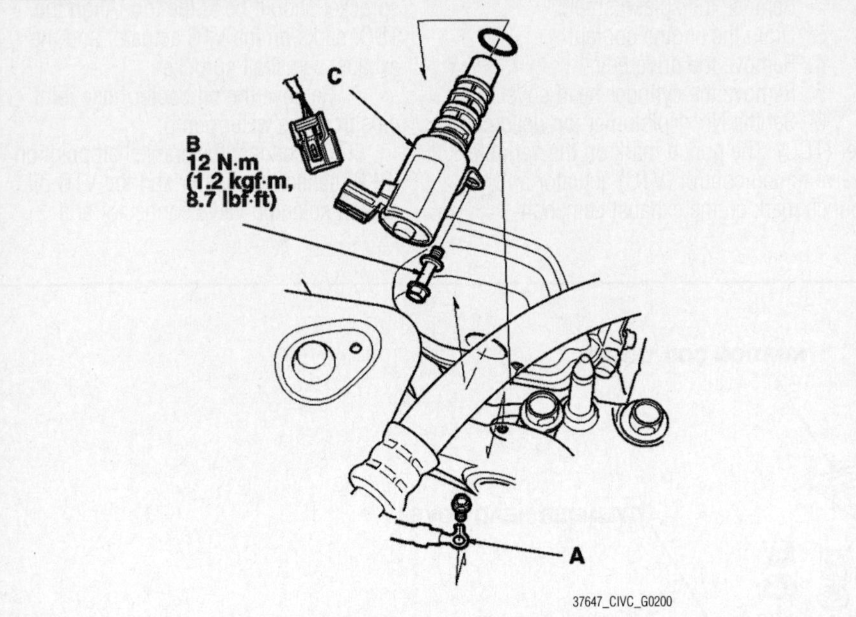
C
B
12 N·m
(1.2 kgf·m,
8.7 lbf·ft)
A
37647_CIVC_G0200

Fig. 76 Remove the ground wire (A), then remove the bolt (B) and the VTC oil control solenoid valve (C)—2.0L engine

remove the harness clamps from the front of the engine.

9. Remove the VTC oil control solenoid valve:

a. Disconnect the VTC oil control solenoid valve 2P connector.

b. Remove the ground wire.

c. Remove the bolt and the VTC oil control solenoid valve.

10. Remove the crankshaft pulley.

11. Support the engine with a jack and a wood block under the oil pan.

12. Remove the upper torque rod.

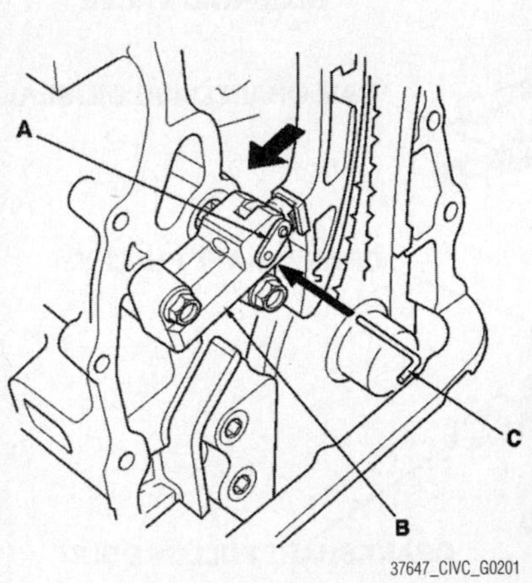
A
C
B
37647_CIVC_G0201

Fig. 77 Align the holes on the lock (A) and the auto-tensioner (B), then insert a 1.2 mm (0.05 in) diameter pin or lock pin (C) into the holes. Turn the crankshaft clockwise to secure the pin—2.0L engine

13. Remove the ground cable, then remove the side engine mount bracket.

14. Remove the side engine mount bracket.

15. Remove the cam chain case, then remove the CKP pulse plate.

16. Loosely install the crankshaft pulley.

17. Turn the crankshaft counterclockwise to compress the auto-tensioner.

18. Align the holes on the lock and the auto-tensioner, then insert a 1.2 mm (0.05 in) diameter pin or lock pin into the holes. Turn the crankshaft clockwise to secure the pin.

19. Remove the auto-tensioner.

20. Remove the crankshaft pulley.

21. Remove cam chain guide from on top of the camshaft pulleys.

22. Remove cam chain guide and the tensioner arm.

23. Remove the cam chain.

To install:

➡Keep the cam chain away from magnetic fields.

24. Before doing this procedure, check that the variable valve timing control (VTC) actuator is locked by turning the VTC actuator counterclockwise. If not locked, turn the VTC actuator clockwise until it stops, then recheck it. If it is still not locked, replace the VTC actuator.

25. Set the crankshaft to top dead center (TDC). Align the TDC mark on the crankshaft sprocket with the pointer on the engine block.

26. Set the camshafts to TDC. The punch mark on the VTC actuator and the punch

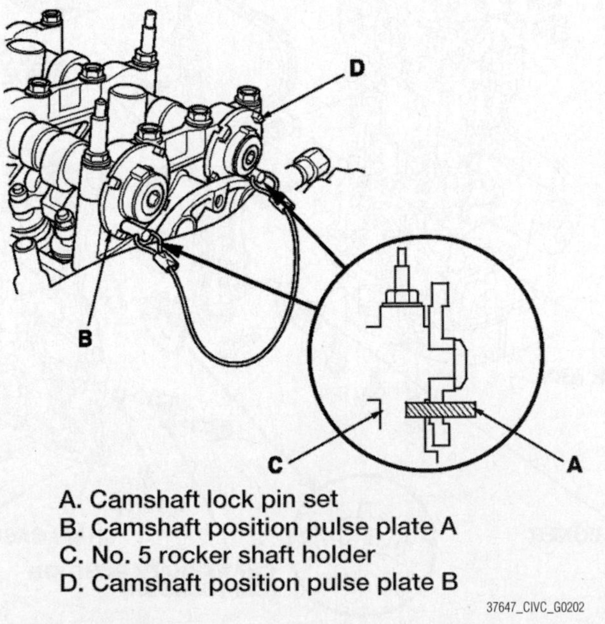
D
B
C
A
A. Camshaft lock pin set
B. Camshaft position pulse plate A
C. No. 5 rocker shaft holder
D. Camshaft position pulse plate B
37647_CIVC_G0202

Fig. 78 To hold the intake camshaft, insert a camshaft lock pin set (07AAB-RWCA120) into the maintenance hole in camshaft position (CMP) pulse plate A and through No. 5 rocker shaft holder. To hold the exhaust camshaft, insert a camshaft lock pin set into the maintenance hole in CMP pulse plate B and through No. 5 rocker shaft holder.

mark on the exhaust camshaft sprocket should be at the top. Align the TDC marks on the VTC actuator and the exhaust camshaft sprocket.

27. To hold the intake camshaft, insert a camshaft lock pin set (07AAB-RWCA120) into the maintenance hole in camshaft position (CMP) pulse plate A and through No. 5 rocker shaft holder.

28. To hold the exhaust camshaft, insert a camshaft lock pin set into the maintenance hole in CMP pulse plate B and through No. 5 rocker shaft holder.

29. Install the cam chain on the crankshaft sprocket with the colored link plate aligned with the mark on the crankshaft sprocket.

30. Install the cam chain on the VTC actuator and the exhaust camshaft sprocket with the punch marks aligned with the two colored link plates.

31. Install cam chain guide A and the tensioner arm.

32. Compress the auto-tensioner when replacing the cam chain. Remove the pin from the auto-tensioner that was installed during removal. Turn the plate counterclockwise, to release the lock, then press the rod, and set the first cam to the first edge of the rack. Insert the 1.2 mm (0.05 in) diameter pin into the holes.

➡If the chain tensioner is not set up as described, the tensioner will be damaged.

33. Install cam chain guide B on top of the cam pulleys.

34. Install the auto-tensioner.

35. Remove the pin or lock pin from the auto-tensioner.

36. Remove the camshaft lock pin set.

37. Install the crankshaft position (CKP) pulse plate.

38. Check the chain case oil seal for damage. If the oil seal is damaged, replace the chain case oil seal.

39. Remove all of the old liquid gasket from the chain case mating surfaces, the bolts, and the bolt holes.

40. Clean and dry the chain case mating surfaces.

41. Apply liquid gasket to the engine block mating surface of the chain case, and to the inside edge of the threaded bolt holes. Install the component within 5 minutes of applying the liquid gasket.

42. Apply a 3 mm (0.12 in) diameter bead of liquid gasket along these areas.

➡If too much time has passed after applying the liquid gasket, remove the old liquid gasket and residue, then reapply new liquid gasket.

43. Apply liquid gasket to the engine block upper surface contact areas and the lower block upper surface contact areas on the chain case.

44. Apply liquid gasket to the oil pan mating surface of the chain case, and to the inside edge of the threaded bolt holes. Install the component within 5 minutes of applying the liquid gasket.

45. Apply a 3 mm (0.12 in) diameter bead of liquid gasket along this area.

➡If too much time has passed after applying the liquid gasket, remove the old liquid gasket and residue, then reapply new liquid gasket.

46. Install a new O-ring on the chain case. Set the edge of the chain case on the edge of the oil pan, then install the chain case on the engine block. Wipe off the excess liquid gasket on the oil pan and chain case mating surface.

➡When installing the chain case, do not slide the bottom surface onto the oil pan mounting surface. Wait at least 30 minutes before filling the engine with oil. Do not run the engine within 3 hours after installing the chain case.

47. Install the side engine mount bracket. Tighten the bolts to 33 ft. lbs. (44 Nm).

48. Install the side engine mount bracket, then loosely tighten the new bolt and nut, and loosely tighten the bolt.

49. Install the ground cable.

50. Remove the air cleaner assembly.

51. Loosen the transmission mounting bolt and nuts.

52. Raise the vehicle on the lift.

53. Loosen the lower torque rod mounting bolt.

54. Loosen the front mount mounting bolt.

55. Lower the vehicle on the lift.

56. Tighten the side engine mount mounting bolts and nut to 52 ft. lbs. (72 Nm).

57. Tighten the transmission mounting bolt and nuts to 54 ft. lbs. (74 Nm).

58. Raise the vehicle on the lift.

59. Tighten the lower torque rod mounting bolt to 69 ft. lbs. (93 Nm).

60. Lower the vehicle on the lift.

61. Install the air cleaner assembly.

62. Install the upper torque rod, then tighten the new upper torque rod mounting bolts to 47 ft. lbs. (64 Nm).

63. Raise the vehicle on the lift.

64. Tighten the front mount mounting bolt to 47 ft. lbs. (64 Nm).

65. Install the crankshaft pulley.

66. Install the splash shield.

67. Lower the vehicle on the lift.

68. Install the variable valve timing control (VTC) oil control solenoid valve.

69. Install the harness clamps and connect the CKP sensor connector and the VTC oil control solenoid valve connector.

70. Install the oil cooler hose joint pipe with a new O-ring.

71. Install the cylinder head cover.

72. Install the drive belt.

73. Refill the radiator with engine coolant, and bleed the air from the cooling system.

74. Do the Crankshaft Position (CKP) Pattern Clear/CKP Pattern Learn Procedure. See "Crankshaft Position (CKP) Sensor" in "ENGINE PERFORMANCE & EMISSION CONTROLS" section.

VALVE LASH

ADJUSTMENT

1.8L Engine

➡Adjust valves only when the cylinder head temperature is less than 100 degrees F.

1. Before servicing the vehicle, refer to the precautions in the beginning of this section. Disconnect the negative battery cable.

2. Note the radio security code and the radio presets.

3. Remove the cylinder head cover retaining bolts. Remove the cylinder head cover from the engine.

4. Set the number one piston at TDC. The UP mark on the camshaft sprocket should be at the top, and the TDC grooves on the camshaft sprocket should line up with the top edge of the cylinder head.

5. Using the proper gauge feeler gauge, adjust the valves on cylinder number one.

6. Rotate the crankshaft clockwise. Align the number three piston TDC groove on the camshaft sprocket with the top edge of the cylinder head.

7. Using the proper gauge feeler gauge, adjust the valves on cylinder number three.

8. Rotate the crankshaft clockwise. Align the number four piston TDC groove on the camshaft sprocket with the top edge of the cylinder head.

9. Using the proper gauge feeler gauge, adjust the valves on cylinder number four.

10. Rotate the crankshaft clockwise. Align the number two piston TDC groove on the camshaft sprocket with the top edge of the cylinder head.

11. Using the proper gauge feeler gauge, adjust the valves on cylinder number two.

12. Install the cylinder head cover.

2.0L Engine

➡Adjust valves only when the cylinder head temperature is less than 100 degrees F.

1. Before servicing the vehicle, refer to the precautions in the beginning of this section. Disconnect the negative battery cable.

2. Note the radio security code and the radio presets.

3. Remove the cylinder head cover retaining bolts. Remove the cylinder head cover from the engine.

4. Set the number one piston at TDC. The punch mark on the Variable Timing Control (VTC) actuator and the punch mark on the exhaust camshaft sprocket should be at the top. Align the TDC marks on the VTC actuator and exhaust camshaft sprocket.

5. Using the proper gauge feeler gauge, adjust the valves on cylinder number one.

6. Rotate the crankshaft 180 degrees. Using the proper gauge feeler gauge, adjust the valves on cylinder number three.

7. Rotate the crankshaft 180 degrees. Using the proper gauge feeler gauge, adjust the valves on cylinder number four.

8. Rotate the crankshaft 180 degrees. Using the proper gauge feeler gauge, adjust the valves on cylinder number two.

9. Install the cylinder head cover.

ENGINE PERFORMANCE & EMISSION CONTROLS

ACCELERATOR PEDAL POSITION (APP) SENSOR

LOCATION

The Accelerator Pedal Position (APP) sensor is located on the top of the accelerator pedal assembly.

REMOVAL & INSTALLATION

See Figures 79 and 80.

1. Disconnect the APP sensor 6P connector.

2. Remove the clip.

➡Do not reuse the clip once it is removed.

3. Push the tab, and remove the accelerator pedal pad from the pedal stop.

4. Remove the accelerator pedal module.

➡The APP sensor is not available separately. Do not disassemble the accelerator pedal module.

To install:

5. Set the accelerator pedal pad to the pedal stop.

6. Install the accelerator pedal module with a new clip.

7. Reconnect the APP sensor 6P connector.

CAMSHAFT POSITION (CMP) SENSOR

LOCATION

➡See "Removal & Installation" for location reference figures.

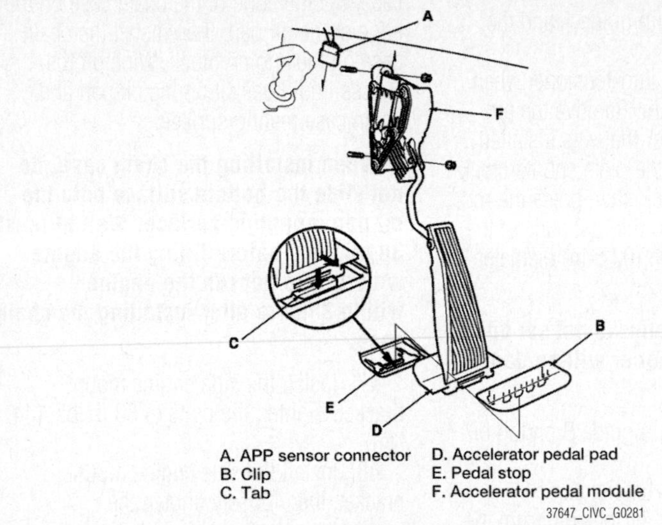
A
F
B
C
E
D
A. APP sensor connector
B. Clip
C. Tab
D. Accelerator pedal pad
E. Pedal stop
F. Accelerator pedal module
37647_CIVC_G0281

Fig. 79 Removing the accelerator pedal assembly—1.8L engine

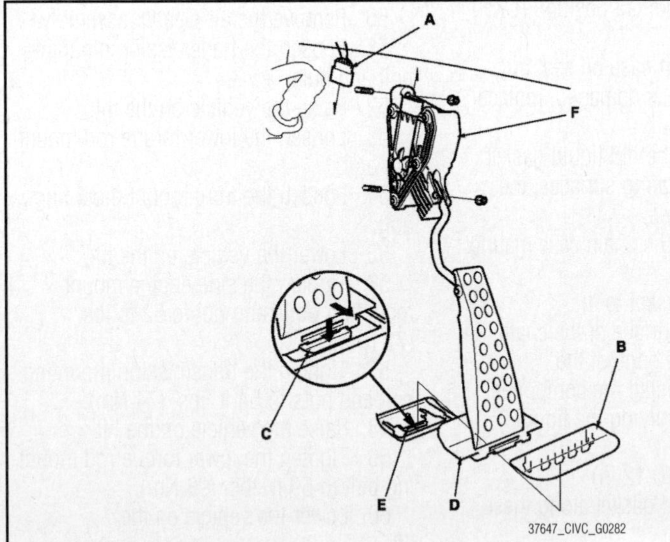
A
F
B
C
E
D
37647_CIVC_G0282

Fig. 80 Removing the accelerator pedal assembly—2.0L engine

REMOVAL & INSTALLATION

1.8L Engine

See Figure 81.

1. Remove the cowl cover and the under-cowl panel.
2. Disconnect the CMP sensor 3P connector.
3. Remove the CMP sensor.
4. Install the parts in the reverse order of removal with a new O-ring.

2.0L Engine

See Figures 82 and 83.

1. Remove the air cleaner.
2. Disconnect the CMP sensor A 3P connector.
3. Remove CMP sensor A from the intake camshaft side of the cylinder head.
4. Install the parts in the reverse order of removal with a new O-ring.

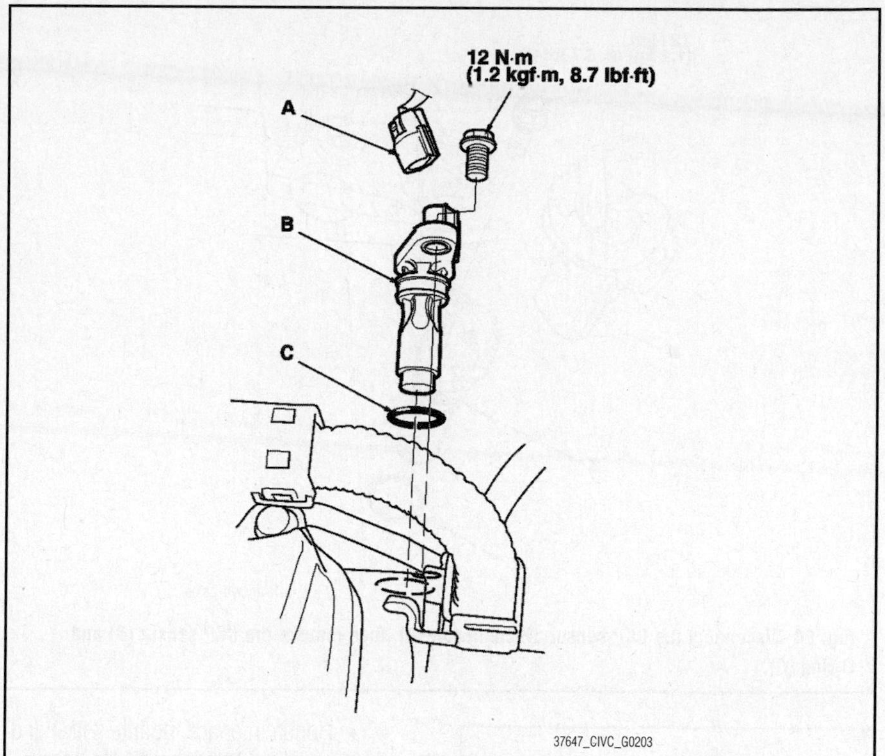

12 N·m
(1.2 kgf·m, 8.7 lbf·ft)
A
B
C
37647_CIVC_G0203

Fig. 81 Showing the CMP sensor connector (A), CMP sensor (B), and the O-ring (C)

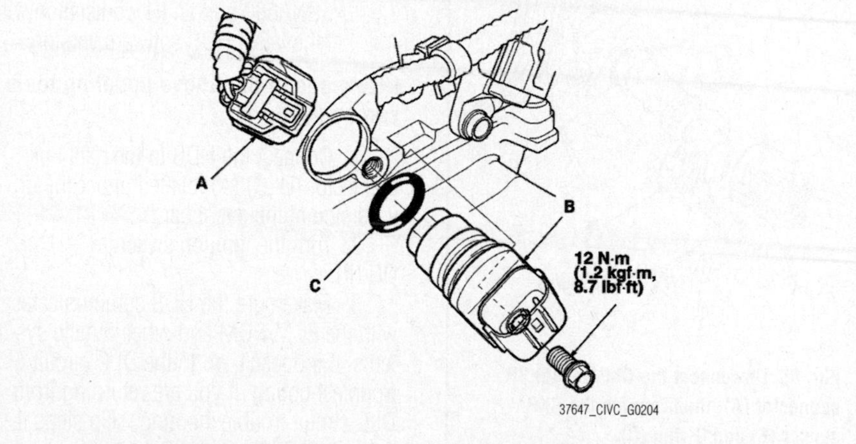

A
B
C
12 N·m
(1.2 kgf·m,
8.7 lbf·ft)
37647_CIVC_G0204

Fig. 82 Showing the CMP sensor A connector (A), CMP sensor A (B), and the O-ring (C)

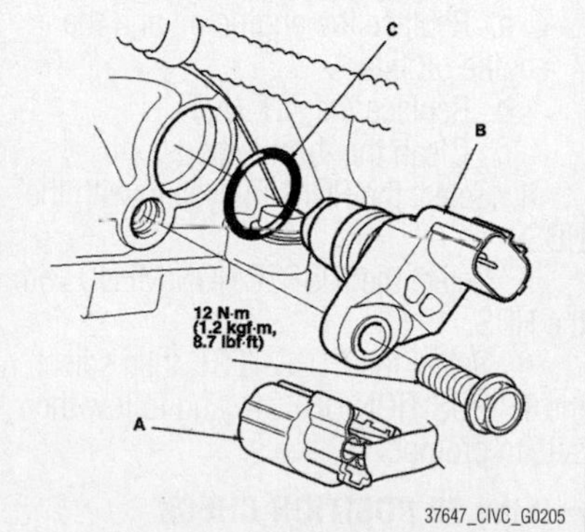

C
B
12 N·m
(1.2 kgf·m,
8.7 lbf·ft)
A
37647_CIVC_G0205

Fig. 83 Showing the CMP sensor B connector (A), CMP sensor B (B), and the O-ring (C)

CRANKSHAFT POSITION (CKP) SENSOR

LOCATION

➡The Crankshaft Position (CKP) sensor is located on the engine block near the crankshaft pulley. See "Removal & Installation" for figure references of component locations.

REMOVAL & INSTALLATION

1.8L Engine

See Figure 84.

1. Remove the splash shield.
2. Disconnect the CKP sensor 3P connector.
3. Remove the CKP sensor and O-ring.
4. Install the parts in the reverse order of removal with a new O-ring.
5. Do the "CKP Pattern Clear/CKP Pattern Learn Procedure".

2.0L Engine

See Figure 85.

1. Disconnect the CKP sensor 3P connector.
2. Remove the CKP sensor and O-ring.
3. Install the parts in the reverse order of removal with a new O-ring.
4. Do the "CKP Pattern Clear/CKP Pattern Learn Procedure".

CKP PATTERN CLEAR/CKP PATTERN LEAN PROCEDURE

1. Start the engine. Hold the engine speed at 3,000 RPM without load (in neutral) until the radiator fan comes on.
2. Test-drive the vehicle on a level road: Decelerate (with the throttle fully closed) from an engine speed of 2,500 RPM down to 1,000 RPM with the transmission in 1st.
3. Test-drive the vehicle on a level road: Decelerate (with the throttle fully closed) from an engine speed of 5,000 RPM down to 3,000 RPM with the transmission in 1st.
4. Repeat previous two steps several times.
5. Turn the ignition switch to LOCK (0).
6. Turn the ignition switch to ON (II), and wait 30 seconds.

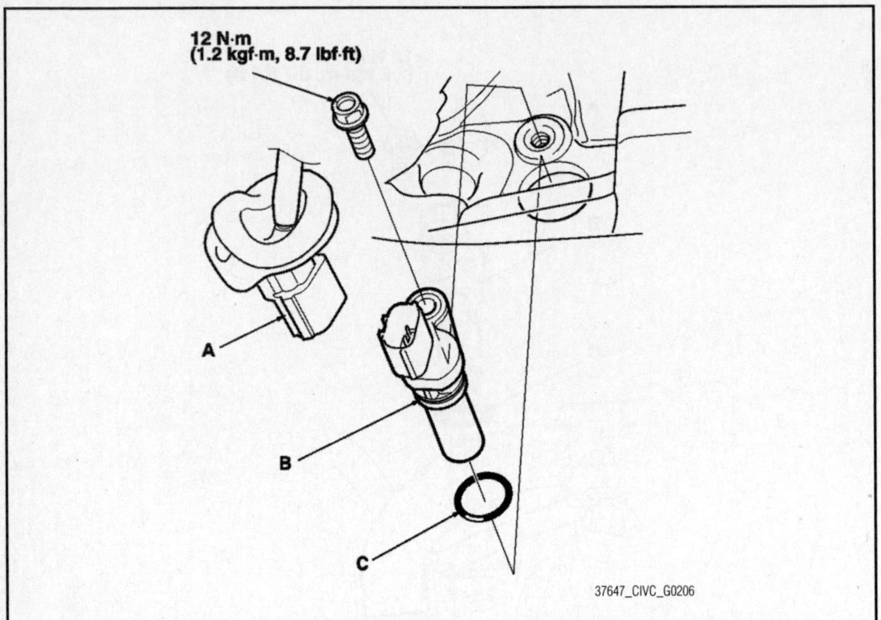
12 N·m
(1.2 kgf·m, 8.7 lbf·ft)
A
B
C
37647_CIVC_G0206

Fig. 84 Disconnect the CKP sensor 3P connector (A), then remove the CKP sensor (B) and O-ring (C).

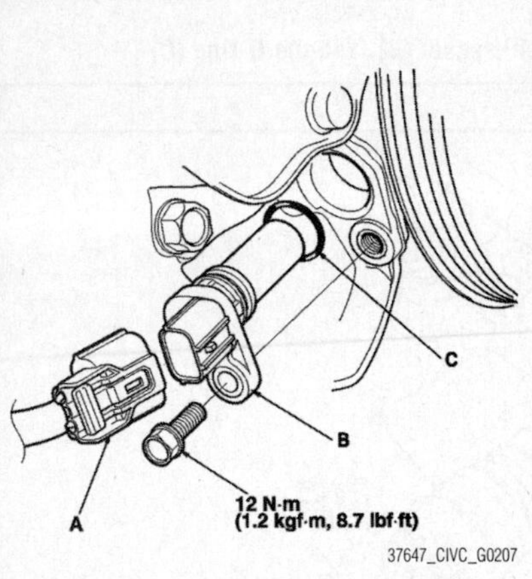
C
B
A
12 N·m
(1.2 kgf·m, 8.7 lbf·ft)
37647_CIVC_G0207

Fig. 85 Disconnect the CKP sensor 3P connector (A), then remove the CKP sensor (B) and O-ring (C).

ENGINE CONTROL MODULE/POWERTRAIN CONTROL MODULE (ECM/PCM)

LOCATION

➡The ECM/PCM is located in the engine compartment. See "Removal & Installation" for figure references of component locations.

REMOVAL & INSTALLATION

➡The following special tools are needed to perform this operation:

- Honda diagnostic system (HDS) tablet tester
- Honda interface module (HIM) and an iN workstation with the latest HDS software version
- HDS pocket tester
- GNA600 and an iN workstation with the latest HDS software version

➡Any one of the above updating tools can be used.

1. Connect the HDS to the data link connector (DLC) (A) located under the driver's side of the dashboard.
2. Turn the ignition switch to ON (II).
3. Make sure the HDS communicates with the ECM/PCM and other vehicle systems. If it doesn't, go to the DLC circuit troubleshooting. If you are returning from DLC circuit troubleshooting, skip steps 4 through 9, 20 through 25, and 28 through 30, and do this after replacing the ECM/PCM:
   a. Replace the engine oil and the engine oil filter.
   b. Replace the ATF (A/T).
   c. Clean the throttle body.
4. Select the PGM-FI system with the HDS.
5. Select the INSPECTION MENU with the HDS.
6. Select the ETCS TEST, then select the TP POSITION CHECK, and follow the screen prompts.

➡If the TP POSITION CHECK indicates FAILED, continue with this procedure.

7. Select the REPLACE ECM/PCM MENU, then select READ DATA, and follow the screen prompts.

➡Doing this step copies (READS) the engine oil life data from the original ECM/PCM so you can later download (WRITES) it into the new ECM/PCM. If READ DATA indicates FAILED, continue with this procedure.

8. A/T: Select the A/T system with the HDS.
9. A/T: Select the REPLACE TCM/PCM MENU, then select READ DATA, and follow the screen prompts.

➡Doing this step copies (READS) the ATF life data from the original PCM so you can later download (WRITES) it into the new PCM. If READ DATA indicates FAILED, continue with this procedure.

10. Turn the ignition switch to LOCK (0).
11. Jump the SCS line with the HDS.
12. Do the battery removal procedure.
13. Remove the ECM/PCM cover.
14. Remove the bolts (D), then remove the ECM/PCM (E).
15. Disconnect ECM/PCM connectors A, B, and C.

➡ECM/PCM connectors A, B, and C have symbols (A=□, B=△, C=○) embossed on them for identification.

16. Install the ECM/PCM in the reverse order of removal.
17. Do the battery installation procedure.
18. Turn the ignition switch to ON (II).
19. Manually input the VIN to the ECM/PCM with the HDS.

➡DTC P0630 VIN Not Programmed or Mismatch may be stored because the VIN has not been programmed into the ECM/PCM; ignore it, and continue this procedure.

20. If the READ DATA (engine oil life) failed in step 7, go to step 23 (A/T) or step 26 (M/T). Otherwise, go to step 21.
21. Select the PGM-FI system with the HDS.
22. Select the REPLACE ECM/PCM MENU, then select WRITE DATA, and follow the screen prompts.

➡If the WRITE DATA indicates FAILED, continue with this procedure.

23. A/T: If the READ DATA (ATF life) failed in step 9, go to step 26. Otherwise go to step 24.
24. A/T: Select the A/T SYSTEM with the HDS.

25. A/T: Select the REPLACE TCM/PCM MENU, then select WRITE DATA, and follow the screen prompts.

➡If the WRITE DATA indicates FAILED, continue with this procedure.

26. Select IMMOBI system with the HDS.

27. Enter the immobilizer ECM/PCM code that you got from the iN, and use the ECM/PCM replacement procedure in the IMMOBI MENU of the HDS; it allows you to start the engine.

28. If the TP POSITION CHECK failed in step 6, clean the throttle body, then go to step 29.

29. If the READ DATA failed in step 7 or the WRITE DATA failed in step 22, replace the engine oil and engine oil filter, then go to step 30 (A/T) or step 31 (M/T).

30. A/T: If the READ DATA failed in step 9 or the WRITE DATA failed in step 25, replace the ATF, then go to step 31.

31. Select PGM-FI system, and reset the ECM/PCM with the HDS.

32. Update the ECM/PCM if it does not have the latest software.

33. Do the "CKP Pattern Clear/CKP Pattern Learn Procedure".

ECM/PCM UPDATE

➡The following special tools are required to properly update the ECM/PCM:

- Honda diagnostic system (HDS) tablet tester
- Honda interface module (HIM) and an iN workstation with the latest HDS software version HDS pocket tester
- GNA600 and an iN workstation with the latest HDS software version

➡Any one of the above updating tools can be used.

➡Pay attention to the following before starting the update:

- Make sure the HDS/iN workstation has the latest HDS software version.
- Before you update the ECM/PCM, make sure the battery in the vehicle is fully charged, and connect a jumper battery (not a battery charger) to maintain system voltage.
- Never turn the ignition switch to ACC (I) or LOCK (0) during the update. If there is a problem with the update, leave the ignition switch in ON (II).
- To prevent ECM/PCM damage, do not operate anything electrical (headlights, audio system, brakes, A/C, power windows, moonroof (if equipped), door locks, etc.) during the update.
- To ensure the latest program is installed, do an ECM/PCM update whenever the ECM/PCM is substituted or replaced.
- You cannot update an ECM/PCM with a program it already has. It will only accept a new program.
- High temperature in the engine compartment might cause the ECM/PCM to become too hot to run the update. If the engine was running before this procedure, open the hood, and cool the engine compartment.
- If you need to diagnose the Honda interface module (HIM) because the HIM's red (#3) light came on or was flashing during the update, leave the ignition switch in ON (II) when you disconnect the HIM from the data link connector (DLC). This will prevent ECM/PCM damage.

1. Turn the ignition switch to ON (II), but do not start the engine.

2. Connect the HDS to the data link connector (DLC) (A) located under the driver's side of the dashboard.

3. Make sure the HDS communicates with the ECM/PCM and other vehicle systems. If it doesn't, go to the DLC circuit troubleshooting. If you are returning from the DLC circuit troubleshooting, skip steps 4 and 5 and clean the throttle body after updating the ECM/PCM.

4. Select the INSPECTION MENU with the HDS.

5. Select the ETCS TEST, then select the TP POSITION CHECK, and follow the HDS screen prompts.

➡If the TP POSITION CHECK indicates FAILED, continue this procedure.

6. Exit the HDS diagnostic system, then select the update mode, and follow the screen prompts to update the ECM/PCM.

7. If the software in the ECM/PCM is the latest, disconnect the HDS/HIM/GNA600 from the DLC, and go back to the procedure that you were doing. If the software in the ECM/PCM is not the latest, follow the instructions on the screen. If prompted to choose the PGM-FI system or the A/T system (A/T), make sure you update both.

➡NOTE: If the ECM/PCM update system requires you to cool the ECM/PCM, follow the instructions on the screen. If you run into a problem during the update procedure (programming takes over 15 minutes, status bar goes over 100 %, D (A/T) or immobilizer indicator flashes, HDS tablet freezes, etc.), follow these steps to minimize the chance of damaging the ECM/PCM:

- Leave the ignition switch in ON (II).
- Connect a jumper battery (do not connect a battery charger).
- Shut down the HDS.
- Disconnect the HDS from the DLC.
- Reboot the HDS.
- Reconnect the HDS to the DLC, and do the update again.

8. If the TP POSITION CHECK failed in step 5, clean the throttle body.

9. Do the "ECM/PCM Idle Learn Procedure".

10. Do the "CKP Pattern Clear/CKP Pattern Learn Procedure".

ECM/PCM IDLE LEARN PROCEDURE

➡The idle learn procedure must be done so the ECM/PCM can learn the engine idle characteristics.

1. Do the idle learn procedure whenever you do any of these actions:
Replace the ECM/PCM.
- Reset the ECM/PCM.
- Update the ECM/PCM.
- Replace or clean the throttle body.
- Disassemble the engine or the transmission.

➡Erasing DTCs with the HDS does not require you to do the idle learn procedure.

2. Make sure all electrical items (the A/C, the audio system, the lights, etc.) are off.

3. Reset the ECM/PCM with the HDS.

4. Turn the ignition switch to ON (II), and wait 2 seconds.

5. Start the engine. Hold the engine speed at 3,000 RPM without load (A/T in P or N, M/T in neutral) until the radiator fan comes on, or until the engine coolant temperature reaches 194°F (90°C).

6. Let the engine idle for about 5 minutes with the throttle fully closed.

➡If the radiator fan comes on, do not include its running time in the 5 minutes.

ENGINE COOLANT TEMPERATURE (ECT) SENSOR

LOCATION

These engine use two Engine Coolant Temperature (ECT) sensors. See figures in "Removal & Installation" for identification of locations.

REMOVAL & INSTALLATION

1.8L Engine

ECT sensor 1

See Figure 86.

1. Drain the engine coolant.
2. Disconnect the ECT sensor 1 2P connector.
3. Remove ECT sensor 1 and O-ring.
4. Install the parts in the reverse order of removal with a new O-ring, then refill the radiator with engine coolant.

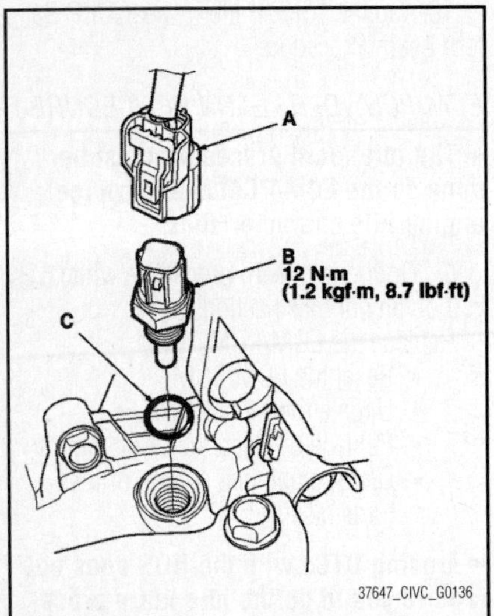
A
B
12 N·m
(1.2 kgf·m, 8.7 lbf·ft)
C
37647_CIVC_G0136

Fig. 86 Disconnect the ECT sensor 1 2P connector (A), then remove the ECT sensor 1 (B) and O-ring (C)

ECT sensor 2

See Figure 87.

1. Drain the engine coolant.
2. Remove the splash shield.
3. Disconnect the ECT sensor 2 2P connector.
4. Remove ECT sensor 2 and O-ring.
5. Install the parts in the reverse order of removal with a new O-ring, then refill the radiator with engine coolant.
6. Install the splash shield.

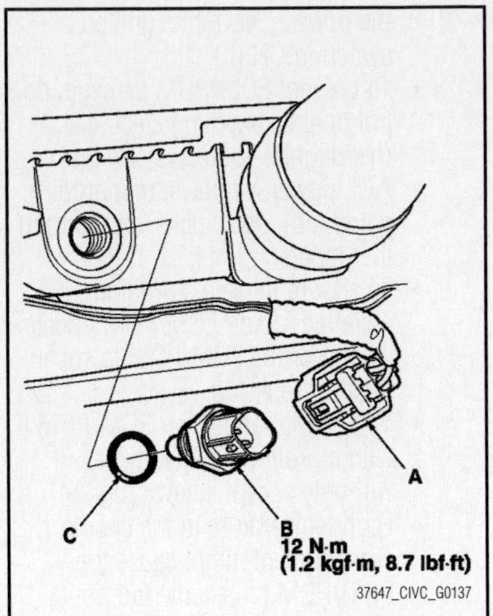
A
B
12 N·m
(1.2 kgf·m, 8.7 lbf·ft)
C
37647_CIVC_G0137

Fig. 87 Disconnect the ECT sensor 2 2P connector (A), then remove the ECT sensor 2 (B) and O-ring (C).

2.0L Engine

ECT sensor 1

See Figures 88 and 89.

1. Drain the engine coolant.
2. Remove the air cleaner.
3. Disconnect the CMP sensor A 3P connector, and the CMP sensor B 3P connector, then remove the harness cover.
4. Disconnect the ECT sensor 1 2P connector, then remove the ECT sensor 1 and O-ring.
5. Install the parts in the reverse order of removal with a new O-ring, then refill the radiator with engine coolant.

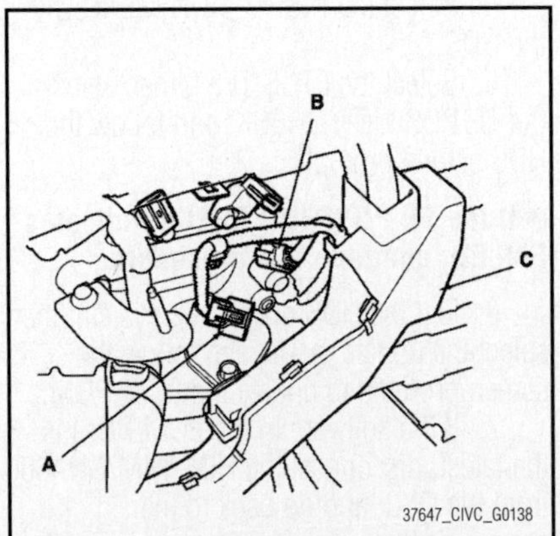
B
C
A
37647_CIVC_G0138

Fig. 88 Disconnect the CMP sensor A 3P connector (A), and the CMP sensor B 3P connector (B), then remove the harness cover (C).

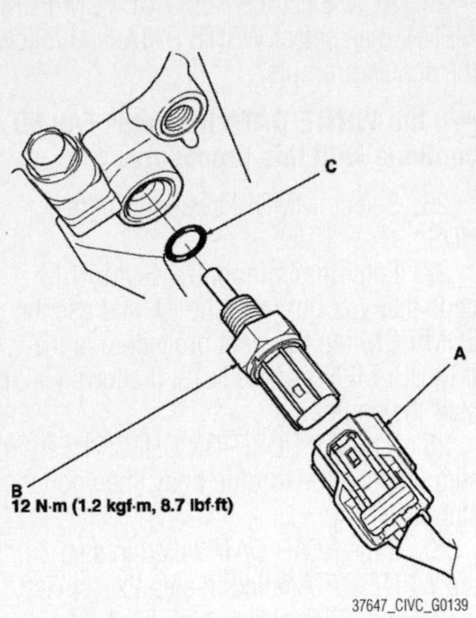
C
A
B
12 N·m (1.2 kgf·m, 8.7 lbf·ft)
37647_CIVC_G0139

Fig. 89 Disconnect the ECT sensor 1 2P connector (A), then remove the ECT sensor 1 (B) and O-ring (C).

ECT sensor 2

See Figure 90.

1. Drain the engine coolant.
2. Remove the splash shield.
3. Disconnect the ECT sensor 2 2P connector, then remove ECT sensor 2 and O-ring.
4. Install parts in the reverse order of removal with a new O-ring.
5. Install the splash shield.
6. Refill the radiator with engine coolant.

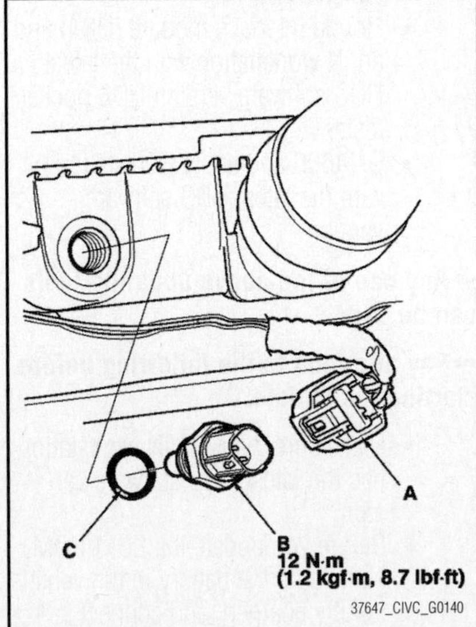
A
C
B
12 N·m
(1.2 kgf·m, 8.7 lbf·ft)
37647_CIVC_G0140

Fig. 90 Disconnect the ECT sensor 2 2P connector (A), then remove the ECT sensor 2 (B) and O-ring (C).

EVAPORATIVE EMISSIONS (EVAP) CANISTER

LOCATION

The Evaporative Emissions (EVAP) canister is located under the vehicle, just forward of the fuel tank. See "Removal & Installation" for figure references of component locations.

REMOVAL & INSTALLATION

See Figures 91 and 92.

1. Raise the vehicle on a lift.
2. Remove the undercover at the fuel tank.
3. Remove the EVAP canister guard pipe.
4. Remove the hoses, the FTP sensor connector, the EVAP canister vent shut valve connector, and the bolts.
5. Remove the EVAP canister.
6. Remove the cap.
7. Remove the EVAP canister vent shut valve and O-rings.
8. Install the EVAP canister vent shut valve in the new EVAP canister with new O-rings.

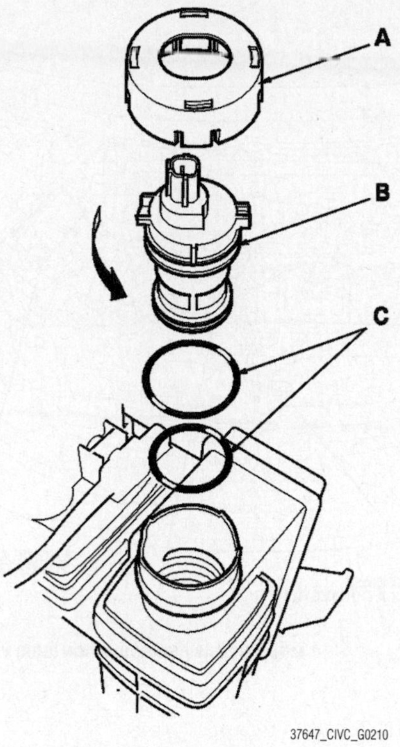
A
B
C
37647_CIVC_G0210

Fig. 92 Remove the cap (A), then remove the EVAP canister vent shut valve (B) and O-rings (C).

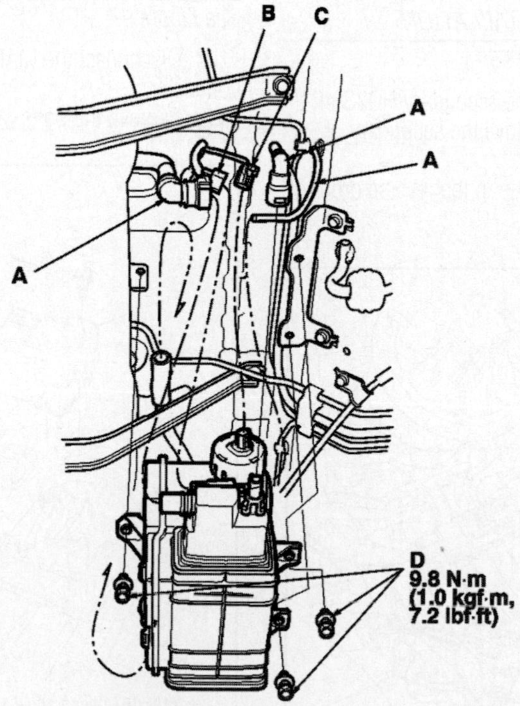
B
C
A
A
A
D
9.8 N·m
(1.0 kgf·m,
7.2 lbf·ft)
A. Hoses
B. FTP sensor connector
C. Vent shut valve connector
D. Bolts
37647_CIVC_G0209

Fig. 91 Removing the EVAP canister

➡**Do not coat the O-rings with engine oil.**

9. Install the parts in the reverse order of removal.

EXHAUST GAS RECIRCULATION (EGR) VALVE

LOCATION

See Figure 93.

The Exhaust Gas Recirculation (EGR) valve is used on 1.8L engine only and is located in the engine compartment, near the top front of the engine.

REMOVAL & INSTALLATION

1.8L Engine

See Figure 94.

1. Remove the injector cover.
2. Disconnect the EGR valve 6P connector.
3. Remove the EGR valve and gasket.
4. Install the parts in the reverse order of removal with a new gasket.
5. Install the injector cover.

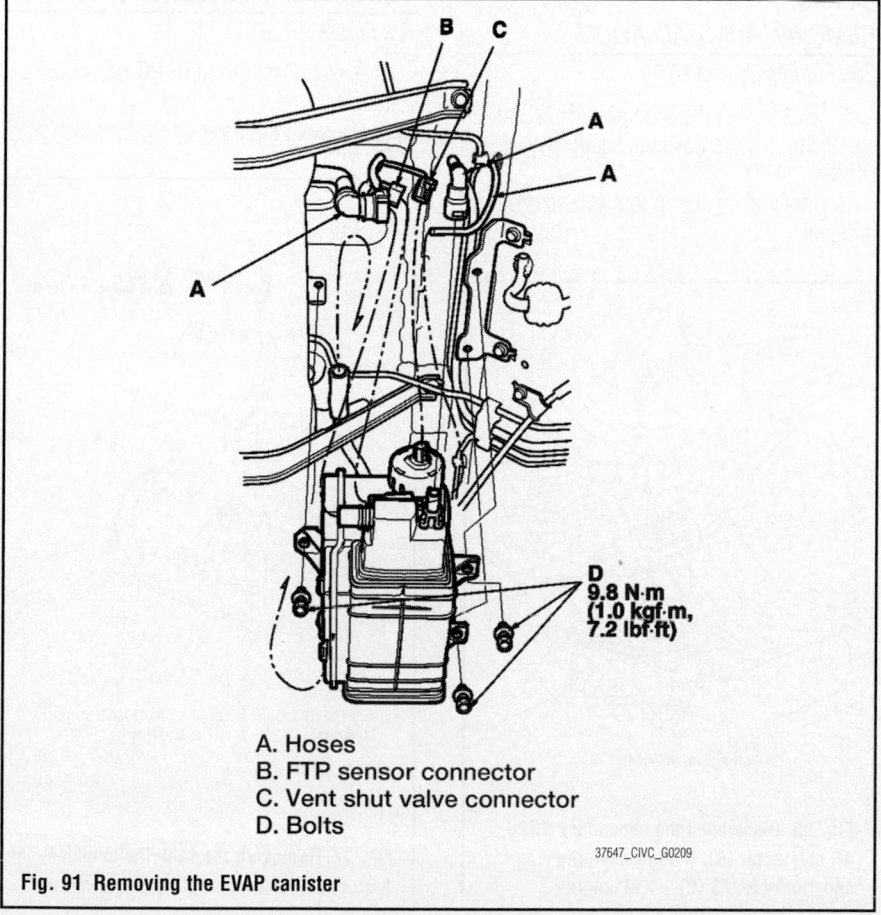

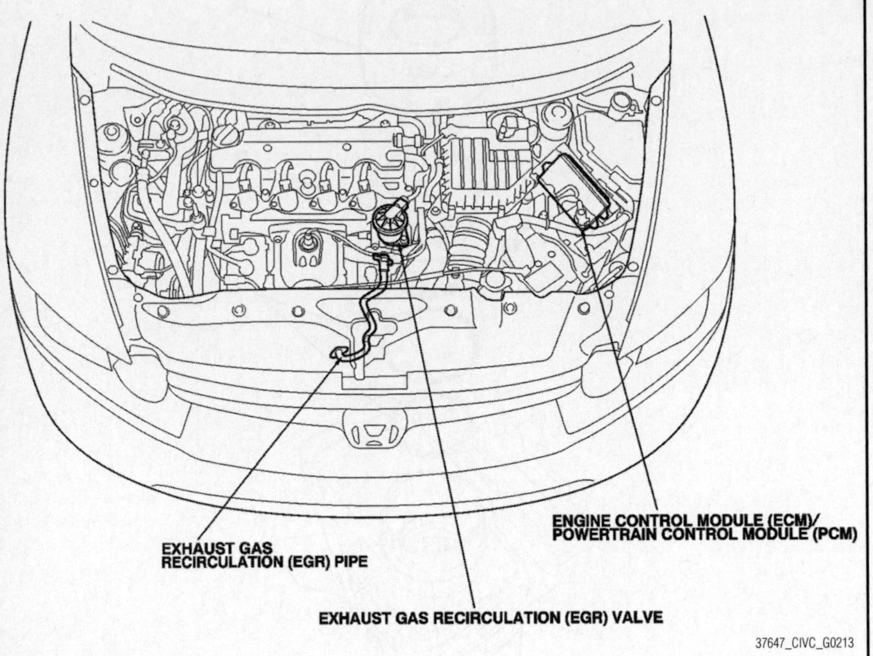

ENGINE CONTROL MODULE (ECM)/ POWERTRAIN CONTROL MODULE (PCM)
EXHAUST GAS RECIRCULATION (EGR) PIPE
EXHAUST GAS RECIRCULATION (EGR) VALVE
37647_CIVC_G0213

Fig. 93 Showing the location of the EGR valve and pipe

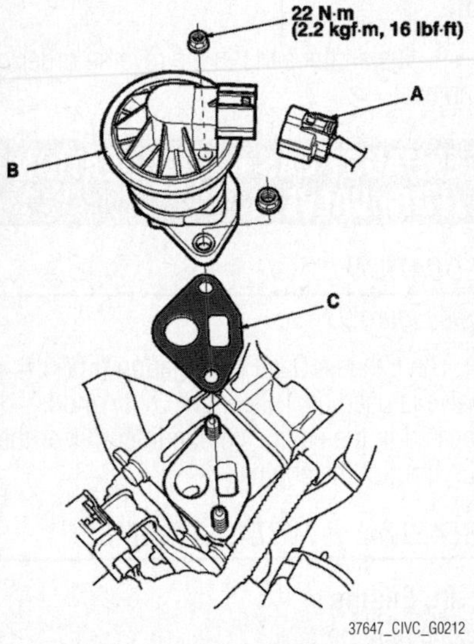

22 N·m (2.2 kgf·m, 16 lbf·ft)
A
B
C
37647_CIVC_G0212

Fig. 94 Disconnect the EGR valve 6P connector (A), then remove the EGR valve (B) and gasket (C).

HEATED OXYGEN SENSOR (HO2S)

LOCATION

The secondary Heated Oxygen Sensor (HO2S) is located on the side of the catalytic converter.

REMOVAL & INSTALLATION

See Figures 95 and 96.

1. Disconnect the secondary HO2S 4P connector, then remove the secondary HO2S.
2. Install the parts in the reverse order of removal.

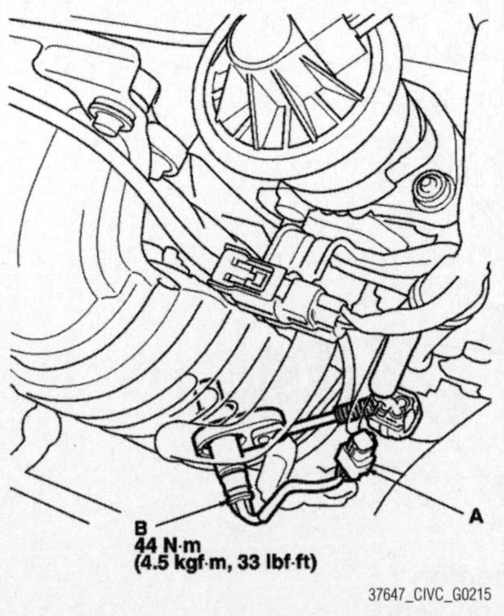

B
44 N·m (4.5 kgf·m, 33 lbf·ft)
A
37647_CIVC_G0215

Fig. 95 Disconnect the secondary HO2S 4P connector (A), then remove the secondary HO2S (B)—1.8L engine

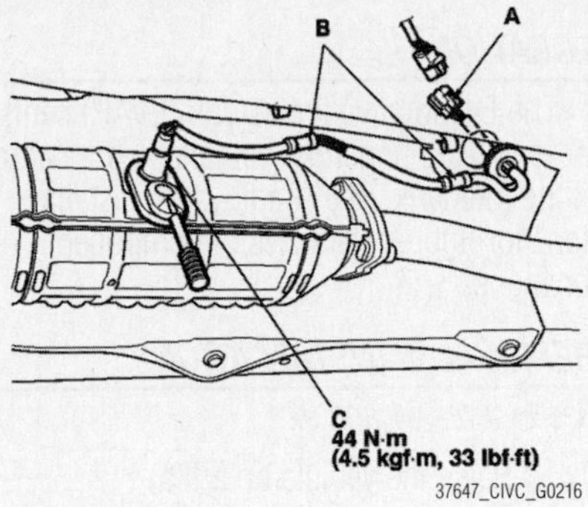

B
A
C
44 N·m (4.5 kgf·m, 33 lbf·ft)
37647_CIVC_G0216

Fig. 96 Disconnect the secondary HO2S 4P connector (A), then remove the secondary HO2S (B)—2.0L engine

INTAKE AIR TEMPERATURE/MASS AIR FLOW (IAT/MAF) SENSOR

LOCATION

The IAT/MAF sensor is located on the end of the intake duct, near the throttle body.

REMOVAL & INSTALLATION

See Figure 97.

1. Disconnect the MAF/IAT sensor 5P connector.
2. Remove the bolts.

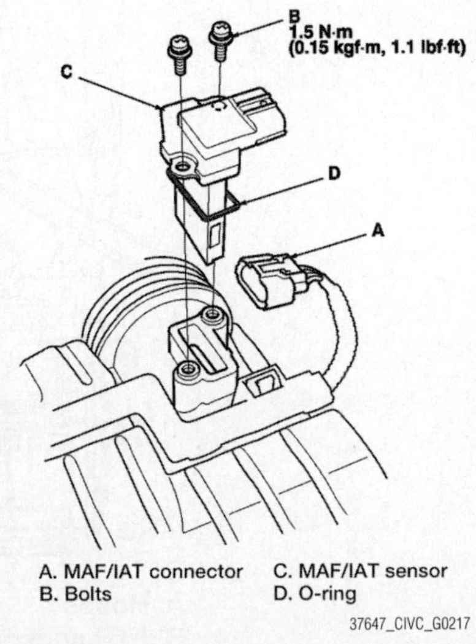

B
1.5 N·m (0.15 kgf·m, 1.1 lbf·ft)
C
D
A
A. MAF/IAT connector
B. Bolts
C. MAF/IAT sensor
D. O-ring
37647_CIVC_G0217

Fig. 97 Removing the MAF/IAT sensor—1.8L engine shown; 2.0L engine similar

3. Remove the MAF/IAT sensor and O ring.
4. Install the parts in the reverse order of removal with a new O-ring.

KNOCK SENSOR (KS)

LOCATION

The Knock Sensor is threaded into the side of the engine block. For 1.8L engine, this is accessible when the intake manifold is removed.

REMOVAL & INSTALLATION

1.8L Engine

See Figure 98.

1. Remove the intake manifold.
2. Disconnect the knock sensor 1P connector.
3. Remove the knock sensor.
4. Install the parts in the reverse order of removal.

2.0L Engine

See Figure 99.

1. Disconnect the knock sensor 1P connector.
2. Remove the knock sensor.
3. Install the parts in the reverse order of removal.

MALFUNCTION INDICATOR LIGHT (MIL)

RESET PROCEDURE

➡**No specific reset procedure is provided.**

MANIFOLD ABSOLUTE PRESSURE (MAP) SENSOR

LOCATION

The Manifold Absolute Pressure (MAP) sensor is located on the firewall for the 1.8L engine. It is located near the front center of the engine for the 2.0L application.

REMOVAL & INSTALLATION

See Figure 100.

1. Remove the cowl cover and the under-cowl panel.
2. Disconnect the MAP sensor 3P connector.
3. Remove the MAP sensor and O-ring.
4. Install the parts in the reverse order of removal with a new O-ring

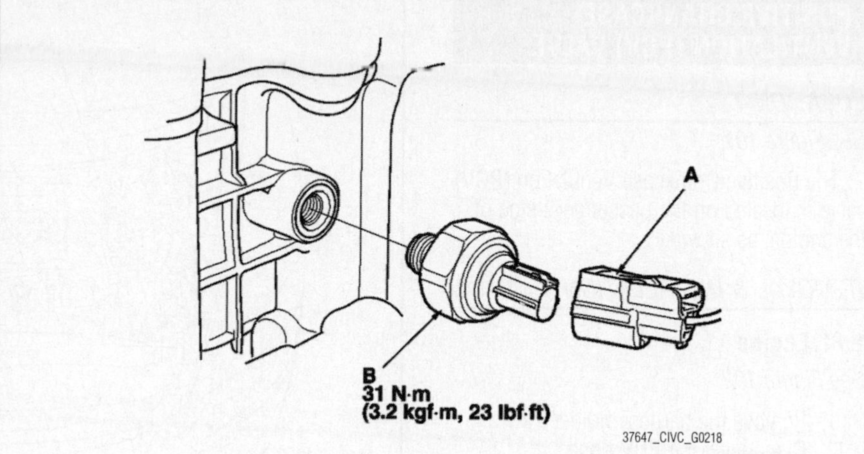
A
B
31 N·m
(3.2 kgf·m, 23 lbf·ft)
37647_CIVC_G0218

Fig. 98 Removing the connector (A) and the knock sensor (B)—1.8L engine

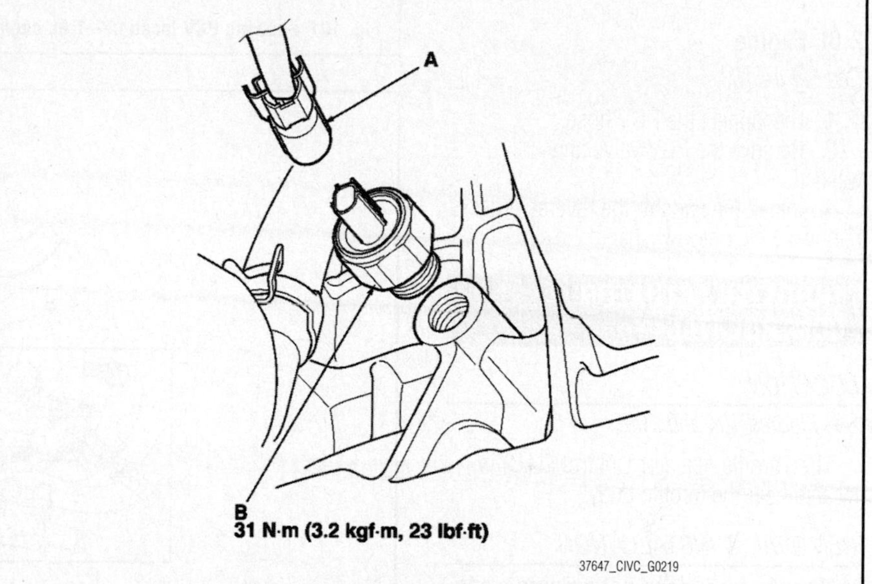
A
B
31 N·m (3.2 kgf·m, 23 lbf·ft)
37647_CIVC_G0219

Fig. 99 Removing the connector (A) and the knock sensor (B)—2.0L engine

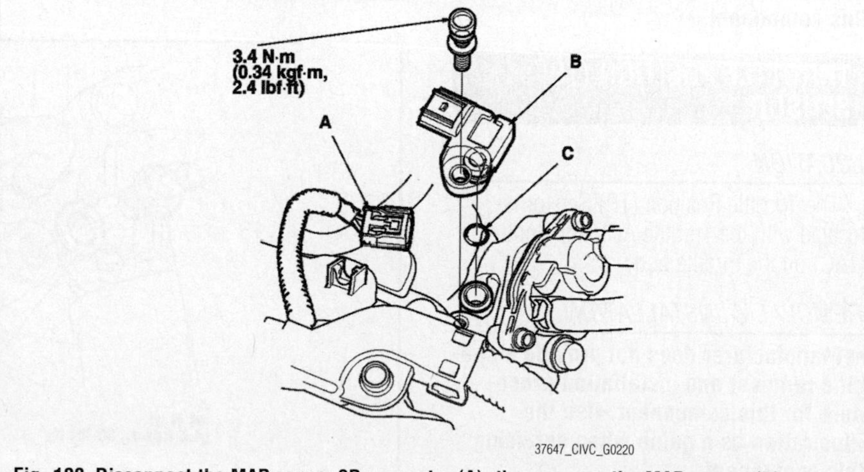
3.4 N·m
(0.34 kgf·m,
2.4 lbf·ft)
B
A
C
37647_CIVC_G0220

Fig. 100 Disconnect the MAP sensor 3P connector (A), then remove the MAP sensor (B) and O-ring (C)—1.8L shown; 2.0L similar

POSITIVE CRANKCASE VENTILATION (PCV) VALVE

LOCATION

See Figure 101.

The Positive Crankcase Ventilation (PCV) valve is located on the passenger's side of the engine, as shown.

REMOVAL & INSTALLATION

1.8L Engine

See Figure 102.

1. Remove the harness holder.
2. Disconnect the PCV hose.
3. Remove the PCV valve and washer.
4. Install the parts in the reverse order of removal with a new washer.

2.0L Engine

See Figure 103.

1. Disconnect the PCV hose.
2. Remove the PCV valve and washer.
3. Install the parts in the reverse order of removal with a new washer.

THROTTLE ACTUATOR CONTROL (TAC)

LOCATION

See Figures 104 and 105.

The Throttle Actuator Control (TAC) is located near the throttle body.

REMOVAL & INSTALLATION

➡Manufacturer does not provide a specific removal and installation procedure for this component. Use the illustration as a guide when servicing this component.

THROTTLE POSITION (TP) SENSOR

LOCATION

The Throttle Position (TP) Sensor is located with the Throttle Actuator Control (TAC) on the throttle body.

REMOVAL & INSTALLATION

➡Manufacturer does not provide a specific removal and installation procedure for this component. Use the illustration as a guide when servicing this component.

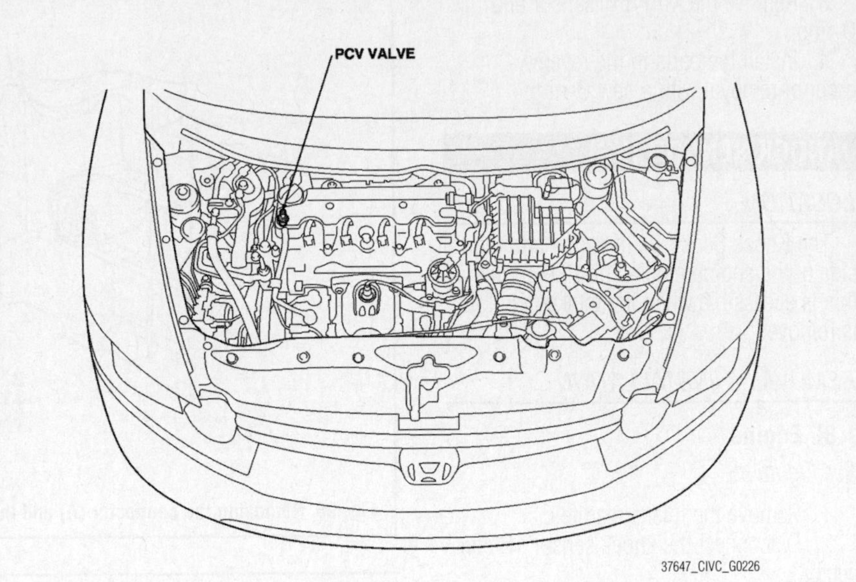
PCV VALVE
37647_CIVC_G0226

Fig. 101 Showing PCV location—1.8L engine shown; 2.0L engine location similar

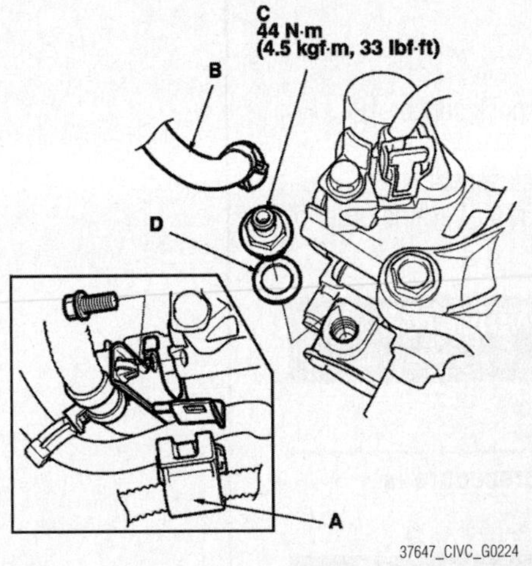
C
44 N·m
(4.5 kgf·m, 33 lbf·ft)
B
D
A
37647_CIVC_G0224

Fig. 102 Remove the harness holder (A), disconnect the PCV hose (B), and remove the PCV valve (C) and washer (D).

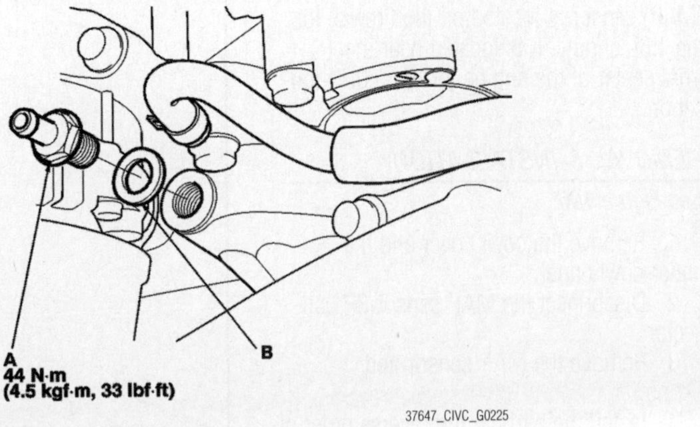
A
44 N·m
(4.5 kgf·m, 33 lbf·ft)
B
37647_CIVC_G0225

Fig. 103 Remove the PCV valve (A) and washer (B).

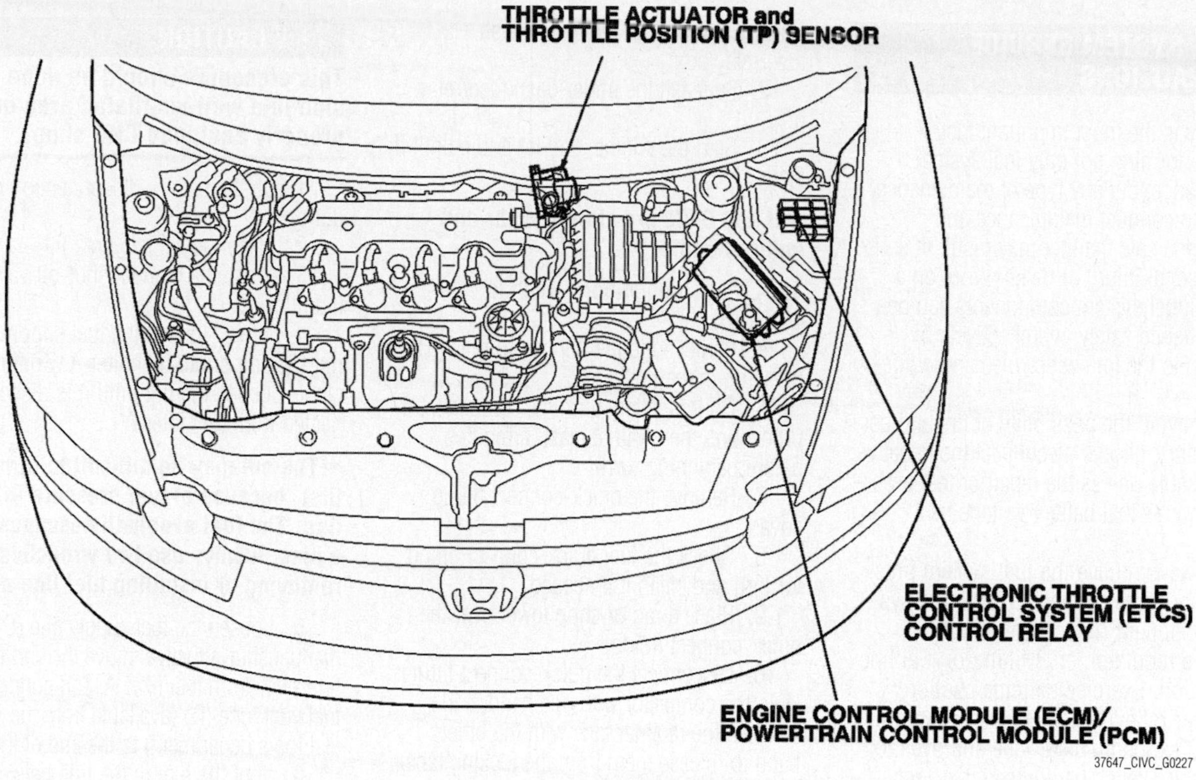
THROTTLE ACTUATOR and THROTTLE POSITION (TP) SENSOR
ELECTRONIC THROTTLE CONTROL SYSTEM (ETCS) CONTROL RELAY
ENGINE CONTROL MODULE (ECM)/ POWERTRAIN CONTROL MODULE (PCM)
37647_CIVC_G0227

Fig. 104 Showing the location of the Throttle Actuator Control (TAC)—1.8L engine

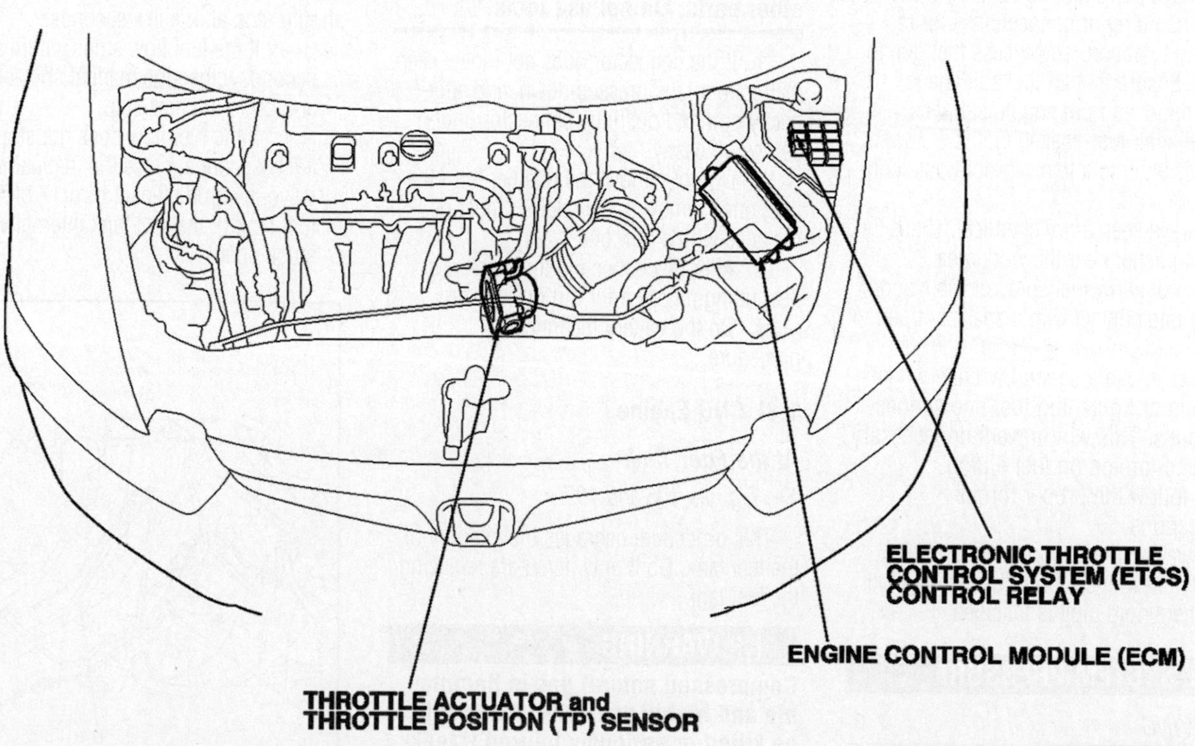
ELECTRONIC THROTTLE CONTROL SYSTEM (ETCS) CONTROL RELAY
ENGINE CONTROL MODULE (ECM)
THROTTLE ACTUATOR and THROTTLE POSITION (TP) SENSOR
37647_CIVC_G0228

Fig. 105 Showing the location of the Throttle Actuator Control (TAC)—2.0L engine

FUEL SYSTEM SERVICE PRECAUTIONS

Safety is the most important factor when performing not only fuel system maintenance, but any type of maintenance. Failure to conduct maintenance and repairs in a safe manner may result in serious personal injury or death. Work on a vehicle's fuel system components can be accomplished safely and effectively by adhering to the following rules and guidelines.

- To avoid the possibility of fire and personal injury, always disconnect the negative battery cable unless the repair or test procedure requires that battery voltage be applied.
- Always relieve the fuel system pressure prior to disconnecting any fuel system component (injector, fuel rail, pressure regulator, etc.) fitting or fuel line connection. Exercise extreme caution whenever relieving fuel system pressure to avoid exposing skin, face and eyes to fuel spray. Please be advised that fuel under pressure may penetrate the skin or any part of the body that it contacts.
- Always place a shop towel or cloth around the fitting or connection prior to loosening to absorb any excess fuel due to spillage. Ensure that all fuel spillage is quickly removed from engine surfaces. Ensure that all fuel-soaked cloths or towels are deposited into a flame-proof waste container with a lid.
- Always keep a dry chemical (Class B) fire extinguisher near the work area.
- Do not allow fuel spray or fuel vapors to come into contact with a spark or open flame.
- Always use a second wrench when loosening or tightening fuel line connection fittings. This will prevent unnecessary stress and torsion on fuel piping. Always follow the proper torque specifications.
- Always replace worn fuel fitting O-rings with new ones. Do not substitute fuel hose where rigid pipe is installed.

FUEL SYSTEM PRESSURE

RELIEVING

Except 1.8L CNG Engine

1. Remove the under-dash fuse/relay box, then remove PGM-FI main relay 2 (FUEL PUMP) from the under-dash fuse/relay box.
2. Reinstall the under-dash fuse/relay box.
3. Start the engine, and let it idle until it stalls.

➡If any DTCs are stored, clear and ignore them.

4. Turn the ignition switch to LOCK (0).
5. Remove the fuel fill cap to relieve the pressure in the fuel tank.
6. Do the battery terminal disconnect procedure, then wait at least 3 minutes before beginning work.
7. Remove the quick-connect fitting cover.
8. Check the fuel quick-connect fitting for dirt, and clean it if needed.
9. Place a rag or shop towel over the quick-connect fitting.
10. Disconnect the quick-connect fitting: Hold the connector with one hand, and squeeze the retainer tabs with the other hand to release them from the locking tabs. Pull the connector off.

✻✻ CAUTION

Be careful not to damage the line or other parts. Do not use tools.

11. If the connector does not move, keep the retainer tabs pressed down, and alternately pull and push the connector until it comes off easily.
12. Do not remove the retainer from the line; once removed, the retainer must be replaced with a new one.
13. After disconnecting the quick-connect fitting, check it for dirt or damage.
14. Do the battery reconnect procedure.

1.8L CNG Engine

At the Fuel Tank

See Figures 106 and 107.

This procedure degrades the integrity of the fuel tank. Do it only if you are replacing the fuel tank.

✻✻ WARNING

Compressed natural gas is flammable and highly explosive. You could be killed or seriously injured if leaking natural gas is ignited. Stop the engine, and keep heat, sparks, and flames away.

✻✻ CAUTION

This procedure should be done outside in a well-ventilated area or in a properly equipped CNG shop.

1. Do the battery disconnection procedure.
2. Raise the vehicle on a lift.
3. Close the manual shut-off valve at the filler door.
4. Slowly loosen the fuel supply line nut (B) going into fuel filter A. Continue to slowly loosen the nut until it is disconnected from fuel filter A.

➡The nut may be difficult to turn at first, because of fuel pressure in the line. The fuel eventually escapes with a hiss. Always use two wrenches when removing or installing fuel line nuts.

5. Loosen the fuel supply line (C) at the manual shut-off valve, move the end of the line away from fuel filter A, and connect a fuel vent tube (D) (available from the AH special tools department) to the end of the line.
6. Vent the fuel in the line between the fuel tank and the manual shut-off valve by fully opening the manual shut-off valve. The small amount of fuel in the line should vent quickly with a hissing sound. The fuel flow should stop after a few seconds.
   a. If the fuel flow stops within a few seconds, close the manual shut-off valve, and go to the next step.
   b. If the fuel flow does not stop within a few seconds, close the manual shut-off valve, then do steps 7 thru 17 of At the fuel tank (If the fuel tank internal valve is stuck open).

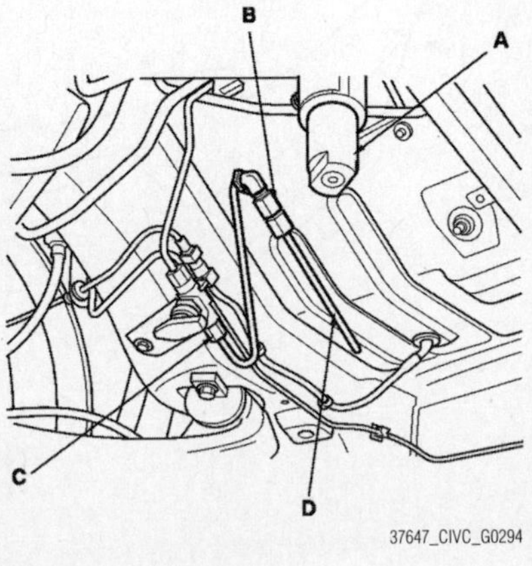

B
A
C
D
37647_CIVC_G0294

Fig. 106 Loosen the fuel supply line

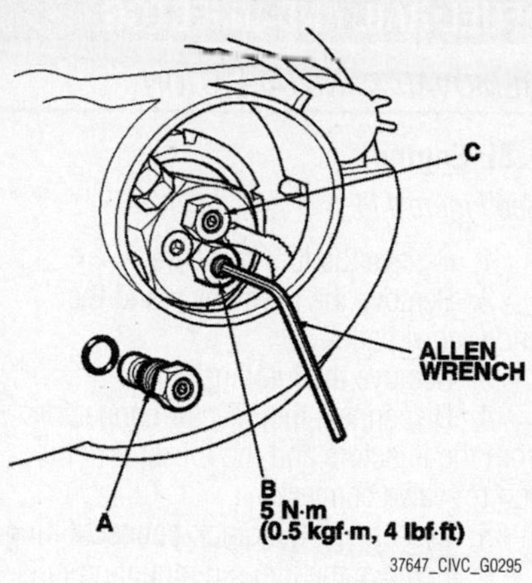
C
ALLEN WRENCH
A
B
5 N·m
(0.5 kgf·m, 4 lbf·ft)
37647_CIVC_G0295

Fig. 107 Slowly remove the manual lock-down valve (A) from the fuel line side of the fuel tank, and install a manual override vent tool (B) (available from the AH special tools department) in its place. Do not remove the valve (C) from the receptacle side of the fuel tank.

7. Attach one end of the ground wire (included with the new fuel tank) to the fuel supply line. Route the other end of the ground wire into the vehicle, where it will not interfere with vehicle movement.
8. Lower the vehicle to the ground.
9. Slowly remove the manual lock-down valve (A) from the fuel line side of the fuel tank, and install a manual override vent tool (B) (available from the AH special tools department) in its place. Do not remove the valve (C) from the receptacle side of the fuel tank.
10. With the help of an assistant, push the vehicle outside, near a metal water pipe or a chain-link fence, where the fuel can be safely vented.
11. Attach the other end of the ground wire to the water pipe or to a chain-link fence post buried in the ground. This disperses any static electricity into the ground.
12. Using an Allen wrench, slowly turn the bolt in the manual override vent tool clockwise until the bolt stops.
13. Vent the fuel from the tank by opening the manual shut-off valve. The fuel tank is empty when the hissing is gone, or when fuel cannot be felt coming from the vent pipe. Once the tank is empty, close the manual shut-off valve.

➡**Depending on the amount of fuel in the tank, fuel venting can take several minutes, up to a few hours.**

14. Remove the manual override vent tool, and reinstall the manual lock-down valve.
15. Reinstall the fuel line to fuel filter A with a new O-ring, then retighten the fuel line at the manual shut-off valve.
16. Do a leak inspection after replacing the fuel tank.
17. Do the battery reconnection procedure.

Between Fuel Tank & Engine

See Figure 108.

This procedure will allow you to safely work on any part of the fuel system downstream of the fuel tank such as the fuel joint block or the manual shut-off valve.

⁂ CAUTION

This procedure should be done outside in a well-ventilated area or in a properly equipped CNG shop.

➡**Make sure the manual shut-off valve is open.**

1. Remove the rear seat.
2. Disconnect the fuel subharness 6P connector (this prevents the tank from supplying fuel to the system).
3. Start the engine, and let it idle. After a few minutes, the engine will stall.
4. Turn the ignition switch to LOCK (0).

Between the Engine & Manual Shut-off Valve

This procedure will allow you to safely work on any part of the fuel system downstream of the manual shut-off valve, such as fuel pressure regulator P1 or the fuel injectors.

1. Raise the vehicle on a lift.
2. Close the manual shut-off valve.
3. Start the engine, and let it idle. After a few minutes, the engine will stall.
4. Turn the ignition switch to LOCK (0).

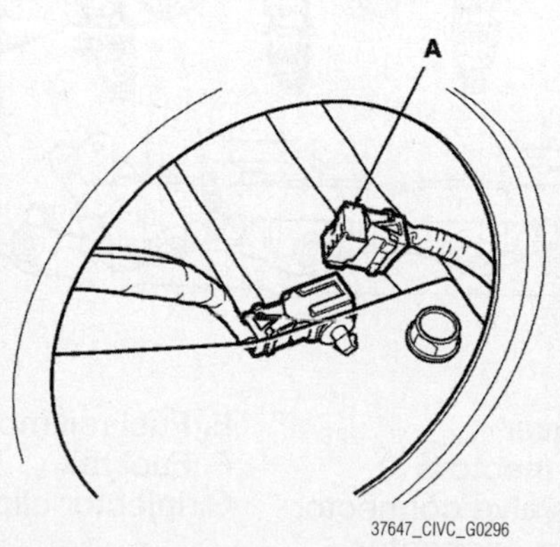
A
37647_CIVC_G0296

Fig. 108 Disconnect the fuel subharness 6P connector (A) (this prevents the tank from supplying fuel to the system).

FUEL FILTER

REMOVAL & INSTALLATION

See Figure 109.

The fuel filter should be replaced whenever the fuel pressure drops below the specified value, after making sure that the fuel pump and the fuel pressure regulator are OK.

1. Remove the fuel tank unit.
2. Remove the fuel filter.
3. Check these items before installing the fuel tank unit:
   a. When connecting the wire harness, make sure the connection is secure and the connectors are firmly locked into place.
   b. When installing the fuel gauge sending unit, make sure the connection is secure and the connector is firmly locked into place. Be careful not to bend or twist it excessively.

To install:

4. Install the parts in the reverse order of removal with new O-rings.
5. When installing the fuel tank unit, align the marks on the unit and the fuel tank.

➡**Coat the O-rings with clean engine oil; do not use any other type oils or fluids. Do not pinch the O-rings during installation.**

6. Use all new parts supplied in the fuel filter replacement kit.

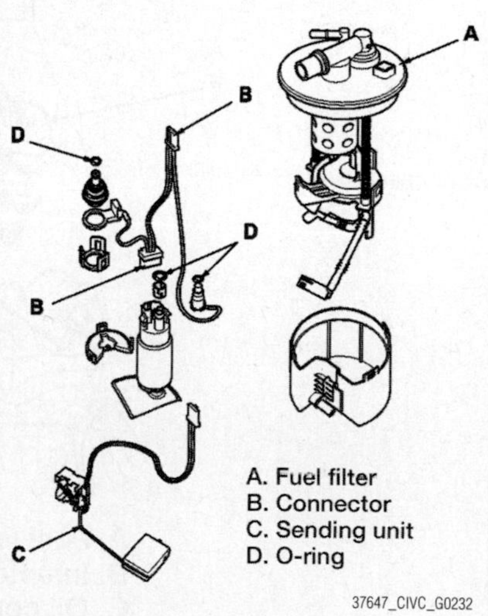
A
B
D
D
B
C
A. Fuel filter
B. Connector
C. Sending unit
D. O-ring
37647_CIVC_G0232

Fig. 109 Exploded view of the fuel pump assembly, showing the fuel filter

FUEL PUMP/ FUEL GAUGE SENDING UNIT

REMOVAL & INSTALLATION

See Figure 110.

1. Remove the fuel tank unit.
2. Remove the fuel level sensor (fuel sending unit) from the fuel tank unit.
3. Check these items before installing the fuel tank unit:
   a. When connecting the wire harness, make sure the connection is secure and the connector is firmly locked into place.
   b. When installing the fuel gauge sending unit, make sure the connection is secure. Be careful not to bend or twist it excessively.
4. Install the parts in the reverse order of removal. When installing the fuel tank unit, align the marks on the unit and the fuel tank.

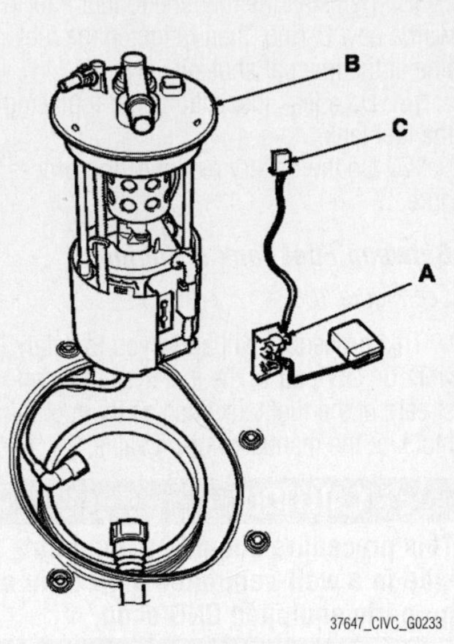
B
C
A
37647_CIVC_G0233
Fig. 110 Remove the connector (C), fuel level sensor (fuel sending unit) (A) from the fuel tank unit (B).

FUEL RAIL AND INJECTOR

REMOVAL & INSTALLATION

1.8L Engine

See Figure 111.

1. Relieve the fuel pressure.
2. Remove the cowl cover and the under-cowl panel.
3. Remove the fuel line cover.
4. Disconnect the injector connectors from the injectors and the rocker arm oil control valve connector.
5. Disconnect the quick-connect fitting.
6. Remove the fuel rail mounting nuts from the fuel rail.
7. Remove the injector clips from the injectors.
8. Remove the injectors from the fuel rail.

To install:

9. Coat the new O-rings with clean engine oil, and insert the injectors into the fuel rail.

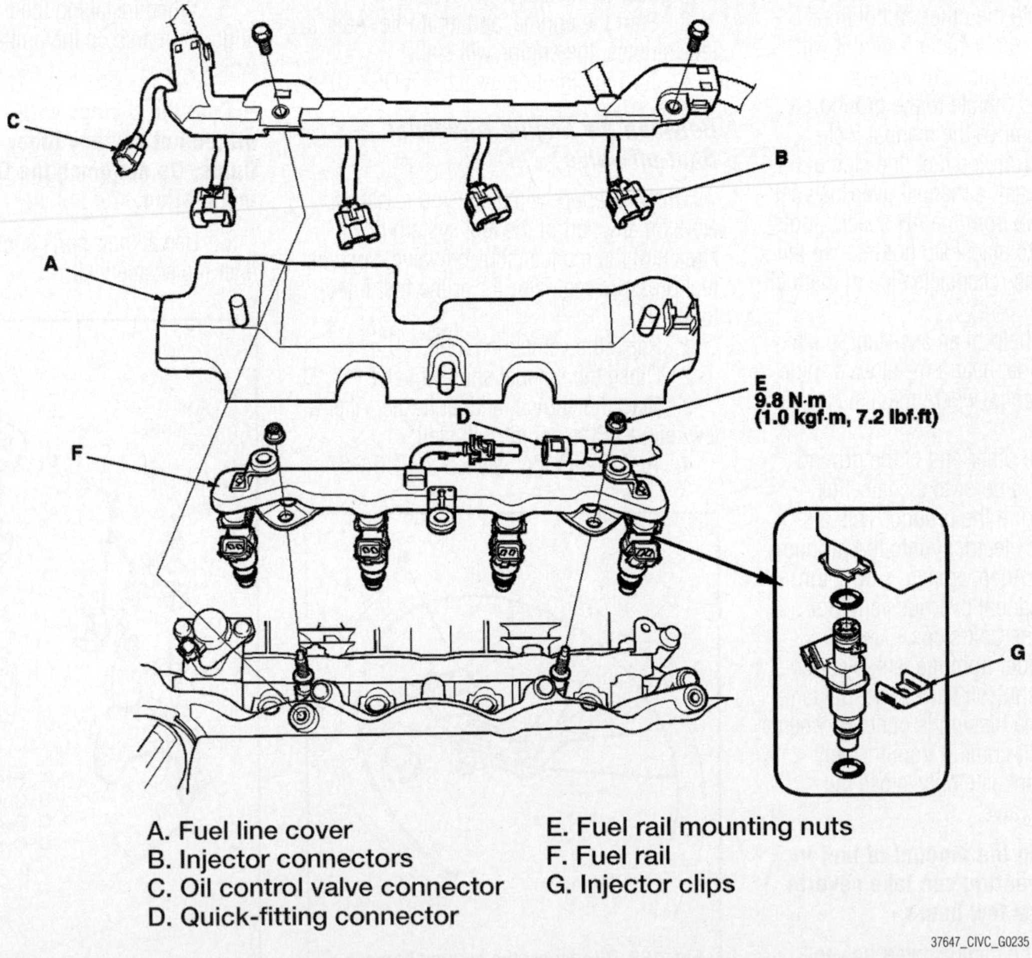
C
B
A
D
E
9.8 N·m
(1.0 kgf·m, 7.2 lbf·ft)
F
G
A. Fuel line cover
B. Injector connectors
C. Oil control valve connector
D. Quick-fitting connector
E. Fuel rail mounting nuts
F. Fuel rail
G. Injector clips
37647_CIVC_G0235
Fig. 111 Showing the fuel rail and injector components—1.8L engine

10. Install the injector clips.
11. Coat the new injector O-rings with clean engine oil.
12. Install the fuel rail and the injectors into the cylinder head.
13. Install the fuel rail mounting nuts.
14. Connect the connectors on the injectors.
15. Connect the quick-connect fitting.
16. Turn the ignition switch to ON (II), but do not operate the starter. After the fuel pump runs for about 2 seconds, the fuel rail is pressurized. Repeat this two or three times, then make sure there are no fuel leaks.
17. Install the fuel line cover.
18. Install the cowl cover and the under-cowl panel.

2.0L Engine

See Figure 112.

1. Relieve fuel pressure.
2. Remove the engine cover.
3. Disconnect the injector connectors from the injectors.
4. Remove the ground cable bolt.
5. Disconnect the quick-connect fitting.
6. Remove the fuel rail mounting nuts from the fuel rail.
7. Remove the fuel rail and the injectors from the injector base.
8. Remove the injector clips from the injectors.
9. Remove the injectors from the fuel rail.
10. Coat the new O-rings with clean engine oil, and insert the injectors into the fuel rail.
11. Install the injector clips.
12. Coat the injector O-rings with clean engine oil.
13. Install the fuel rail and the injectors in the injector base.
14. Install the fuel rail mounting nuts and tighten to 16 ft. lbs. (22 Nm).
15. Connect the connectors on the injectors, and reinstall the ground cable bolt.
16. Connect the quick-connect fitting.
17. Turn the ignition switch to ON (II), but do not operate the starter. After the fuel pump runs for about 2 seconds, the fuel rail is pressurized. Repeat this two or three times, then make sure there are no fuel leaks.
18. Install the engine cover.

FUEL TANK

DRAINING

1. Remove the fuel tank unit. See "Fuel Pump/Fuel Gauge Sending Unit" in this section.
2. Using a hand pump, a hose, and a container suitable for fuel, draw the fuel from the fuel tank.
3. Reinstall the fuel tank unit.

REMOVAL & INSTALLATION

Except 1.8L CNG Engine

See Figure 113.

1. Drain the fuel tank, then disconnect the fuel tank unit 4P connector and the quick-connect fitting.
2. Raise the vehicle on a lift.
3. Remove the cover and the EVAP canister guard pipe.
4. Disconnect the fuel fill tube. Slide back the clamps, then twist the hoses as you pull to avoid damaging them.
5. Disconnect the quick-connect fitting and the fuel tank vapor recirculation tube.
6. Place a jack or other support under the fuel tank.
7. Remove the strap bolts and the strap.
8. Remove the fuel tank.
9. Install the parts in the reverse order of removal.

1.8L CNG Engine

See Figures 114 through 117.

⁂ WARNING

Compressed natural gas is flammable and highly explosive. You could be killed or seriously injured if leaking natural gas is ignited. Stop the engine, and keep heat, sparks, and flames away.

➡You must be an SCI-certified technician to do fuel tank inspection or replacement work.

⁂ CAUTION

This procedure degrades the integrity of the fuel tank. If you do it, do not reinstall the original fuel tank. Install a new one.

1. Relieve fuel pressure between the fuel tank and the engine.
2. Raise the vehicle on a lift.
3. Remove the under-floor cover.
4. Remove the three fuel pipes at the fuel joint block and the two fuel joint block mounting bolts.
5. Lower the vehicle.
6. Remove the left-rear door, then disconnect the connector.

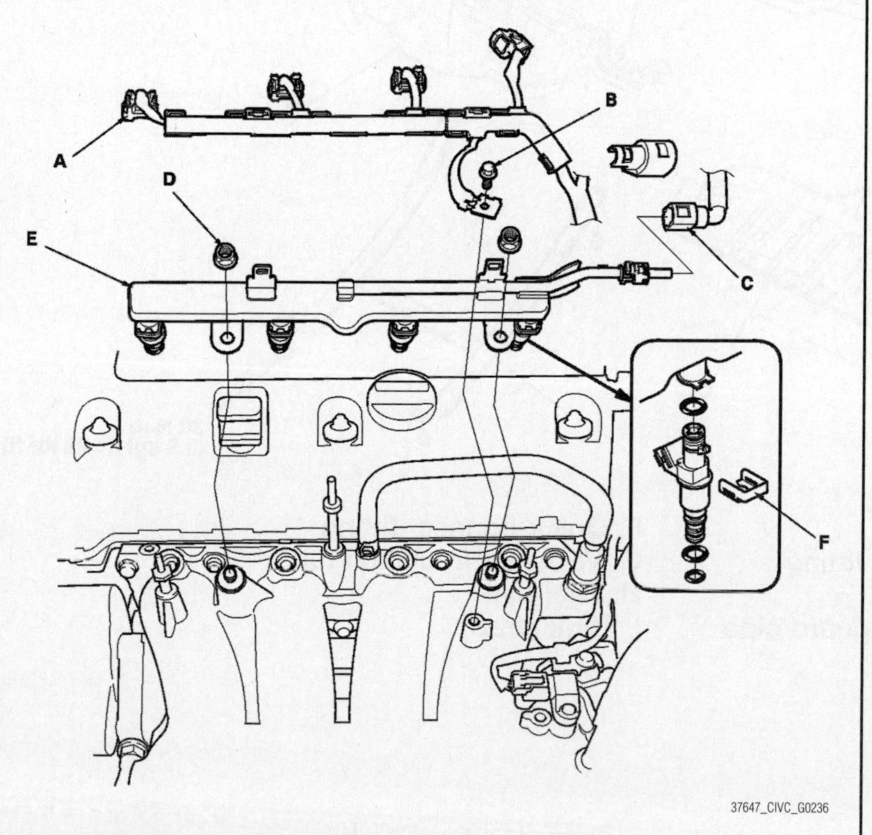
A
B
C
D
E
F
37647_CIVC_G0236

Fig. 112 Showing the fuel rail and injector components—2.0L engine

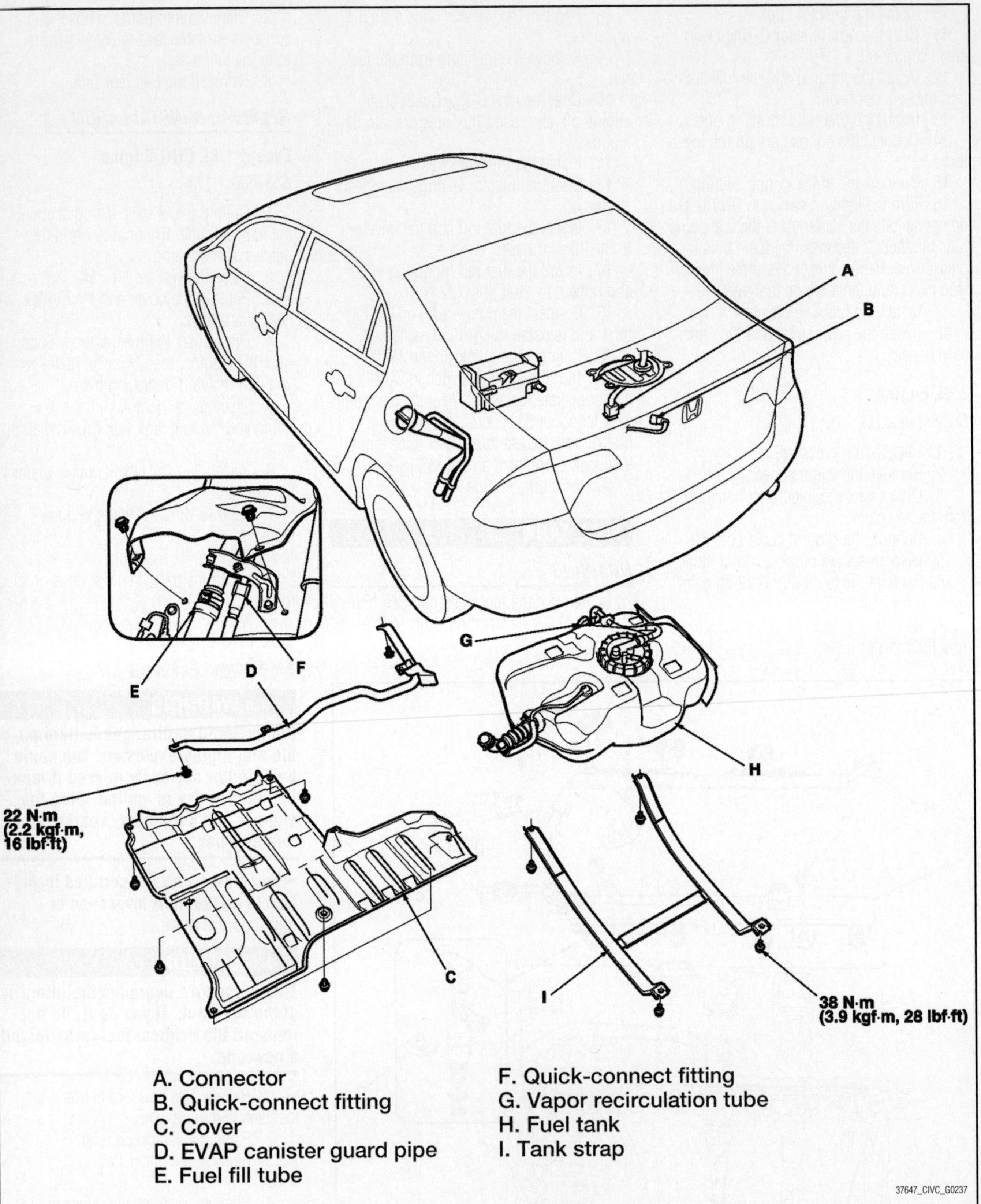
A
B
G
D
E
F
H
22 N·m
(2.2 kgf·m,
16 lbf·ft)
C
I
38 N·m
(3.9 kgf·m, 28 lbf·ft)
A. Connector
B. Quick-connect fitting
C. Cover
D. EVAP canister guard pipe
E. Fuel fill tube
F. Quick-connect fitting
G. Vapor recirculation tube
H. Fuel tank
I. Tank strap
37647_CIVC_G0237

Fig. 113 Showing the fuel tank and related components

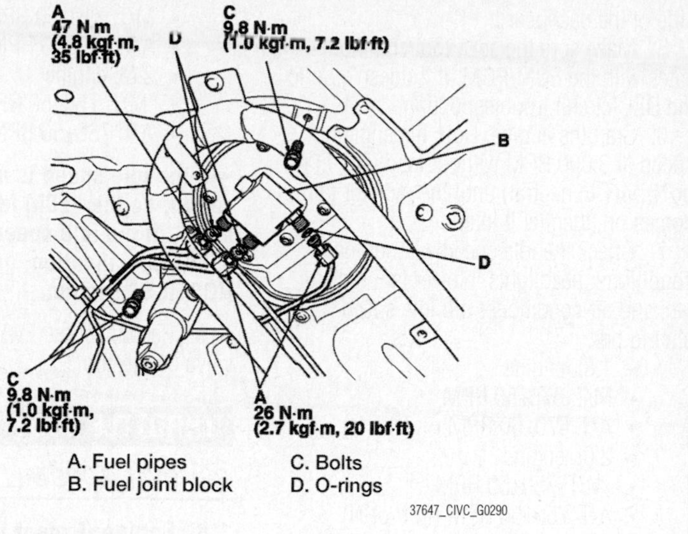
A
47 N·m
(4.8 kgf·m,
35 lbf·ft)
D
C
9.8 N·m
(1.0 kgf·m, 7.2 lbf·ft)
B
D
C
9.8 N·m
(1.0 kgf·m,
7.2 lbf·ft)
A
26 N·m
(2.7 kgf·m, 20 lbf·ft)
A. Fuel pipes
B. Fuel joint block
C. Bolts
D. O-rings
37647_CIVC_G0290

Fig. 114 Removing the fuel pipes

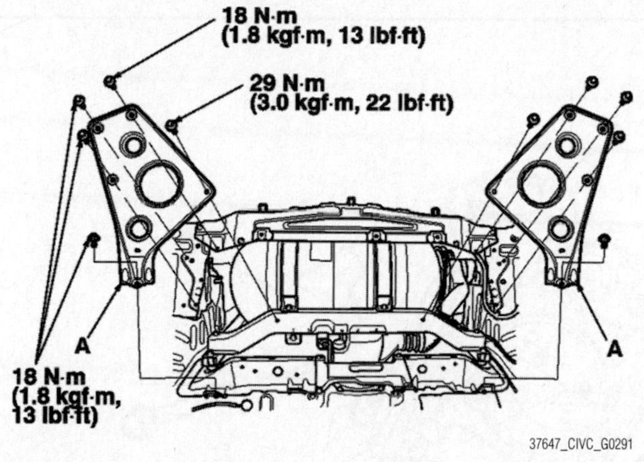
18 N·m
(1.8 kgf·m, 13 lbf·ft)
29 N·m
(3.0 kgf·m, 22 lbf·ft)
A
A
18 N·m
(1.8 kgf·m,
13 lbf·ft)
37647_CIVC_G0291

Fig. 115 Remove the gussets (A).

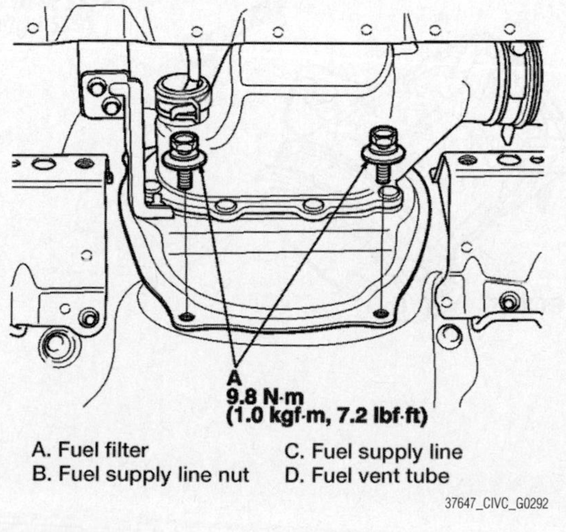
A
9.8 N·m
(1.0 kgf·m, 7.2 lbf·ft)
A. Fuel filter
B. Fuel supply line nut
C. Fuel supply line
D. Fuel vent tube
37647_CIVC_G0292

Fig. 116 Remove the bolts indicated (A).

7. Remove the rear seat.
8. Remove the rear seat belt.
9. Remove the C-pillar trim.
10. Remove the rear shelf and the trunk floor lid.
11. Remove the trunk partition.
12. Remove the gussets.
13. Remove the harness clip and the rear parcel cover.
14. Remove the harness clip and the rear shelf gussets.
15. Remove the rear floor upper cross-member gusset.
16. Remove the bolts indicated (A).
17. Disconnect the vent hose from the fuel tank.
18. Remove the tank frame mounting bolts on either side of the tank.
19. Push the front seats forward, then cover the seats and the floor with a tarp or similar material.
20. With the help of two other technicians, lift the frame, and slide the entire fuel tank assembly (A) into the passenger compartment. Be careful not to damage the fuel joint block.
21. Rotate the fuel tank so the fuel joint block and the frame can clear the door opening, then carefully remove the fuel tank assembly from the vehicle.

To install:

22. Install the new fuel tank assembly in the reverse order of removal.
23. Do this when you install the left rear door:
    a. Tighten the door checker mounting bolt.
    b. Adjust the door position.
24. Open the manual lock-down valves on the new tank (new tank come with these valves closed).
25. Do the leak inspection procedure.

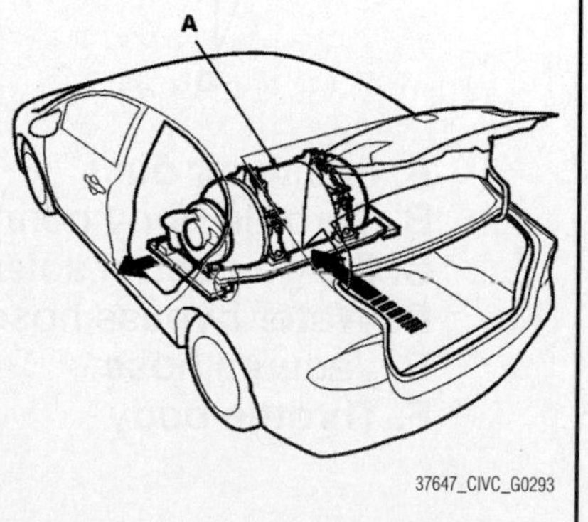
A
37647_CIVC_G0293

Fig. 117 Showing the CNG fuel tank location

IDLE SPEED

ADJUSTMENT

➡The idle speed is not adjustable except by performing the "ECM/PCM Idle Learn Procedure". The following information will help determine if the idle speed is not to specification.

1. Before checking the idle speed, check these items:
   - The malfunction indicator lamp (MIL) has not been reported on, and there are no DTCs.
   - Ignition timing
   - Spark plugs
   - Air cleaner
   - PCV system
2. Apply the parking brake, and make sure the headlights are off.
3. Disconnect the evaporative emission (EVAP) canister purge valve connector.
4. Connect the scan tool to the data link connector (DLC) located under the driver's side of the dashboard.
5. Make sure the scan tool communicates with the ECM/PCM. If it doesn't, go to the DLC circuit troubleshooting.
6. Start the engine. Hold the engine speed at 3,000 RPM without load (A/T in P or N, M/T in neutral) until the radiator fan comes on, then let it idle.
7. Check the idle speed without load conditions: headlights, blower fan, radiator fan, and air conditioner off. Idle speed should be:
   - 1.8L engine
   - M/T: 670±50 RPM
   - A/T: 670±50 RPM (in P or N)
   - 2.0L engine
   - M/T: 750±50 RPM
   - A/T: 750±50 RPM (in P or N)
8. Let the engine idle for 1 minute with high electric load (A/C on, temperature set to max cool, blower fan on high, headlights on high beam). Idle speed should be:
   - 1.8L engine
   - M/T: 710±50 RPM
   - A/T: 710±50 RPM (in P or N)
   - 2.0L engine
   - M/T: 750±50 RPM
   - A/T: 750±50 RPM (in P or N)

➡If the idle speed is not within specification, do the "ECM Idle Learn Procedure". If the idle speed is still not within specification, go to the "DIAGNOSTICS" section.

9. Reconnect the EVAP canister purge valve connector.

THROTTLE BODY

REMOVAL & INSTALLATION

1.8L Engine (Except CNG) & 2.0L

See Figures 118 and 119.

1. Turn the ignition switch to LOCK (0).

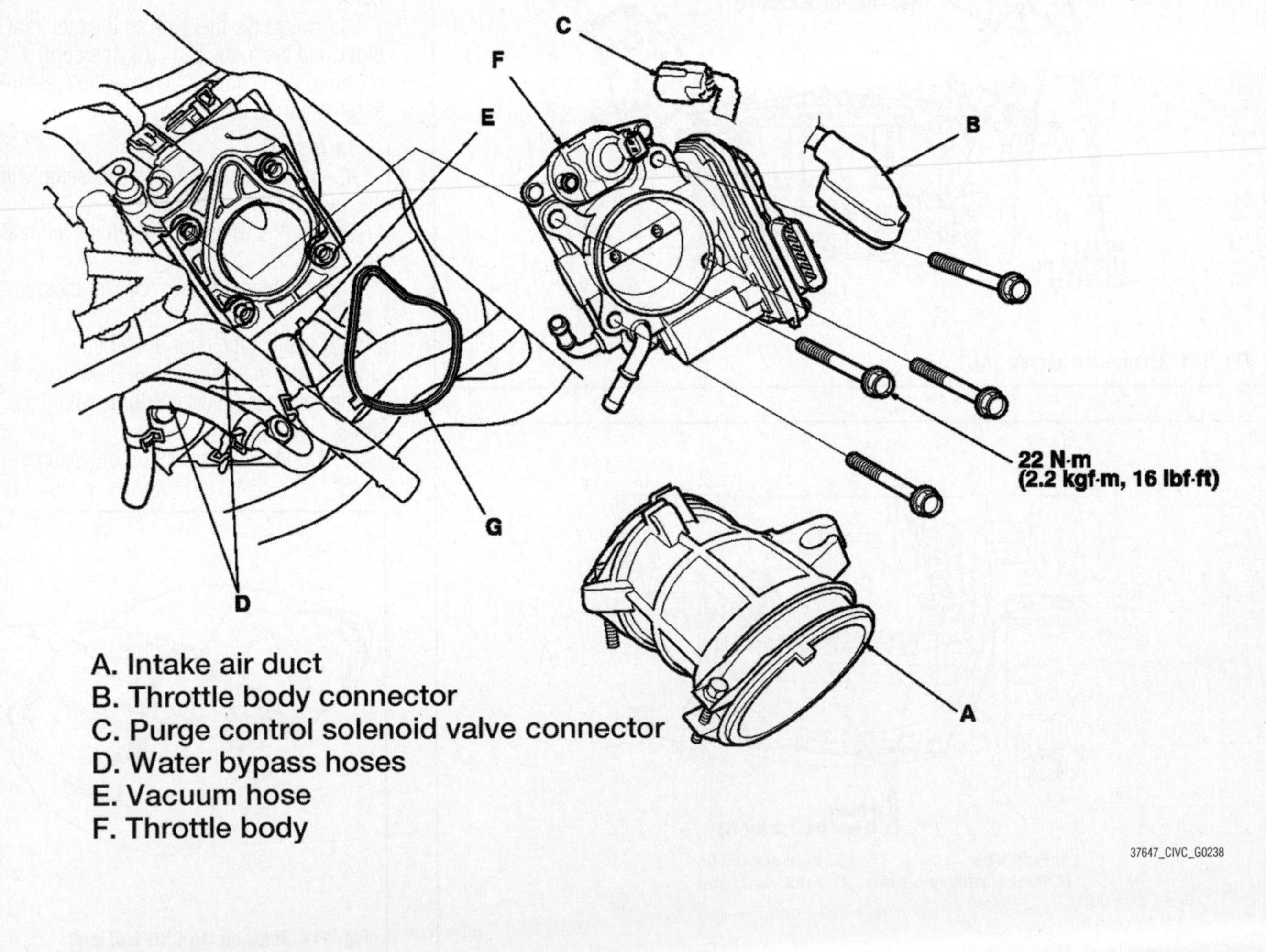
C
F
E
B
22 N·m
(2.2 kgf·m, 16 lbf·ft)
G
D
A
A. Intake air duct
B. Throttle body connector
C. Purge control solenoid valve connector
D. Water bypass hoses
E. Vacuum hose
F. Throttle body
37647_CIVC_G0238

Fig. 118 Showing the throttle body assembly components—1.8L engine (except CNG)

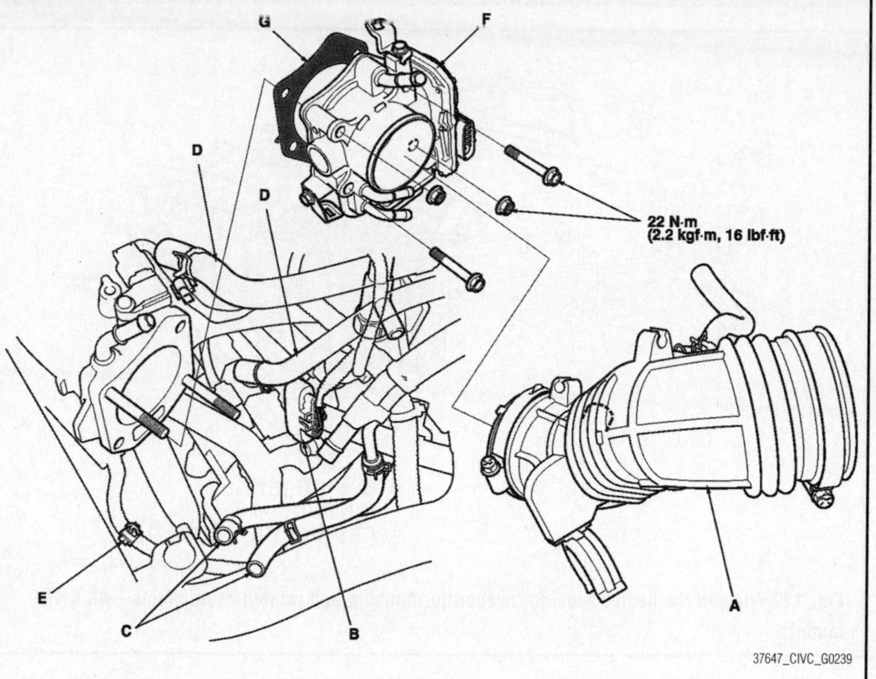
G
F
D
D
22 N·m
(2.2 kgf·m, 16 lbf·ft)
E
C
B
A
37647_CIVC_G0239

Fig. 119 Showing the throttle body assembly components—2.0L engine

2. Remove the cowl cover and the under-cowl panel.

3. Remove the air cleaner and the intake air duct.

4. Disconnect the throttle body connector and the EVAP purge control solenoid valve connector.

5. Disconnect and plug the water bypass hoses.

6. Disconnect the vacuum hose.

7. Remove the throttle body.

8. Install the parts in the reverse order of removal with a new gasket, then do this.

9. Refill the radiator with engine coolant.

10. Do the “ECM Idle Learn Procedure”. See “Engine Control Module/Powertrain Control Module (ECM/PCM)” in this section.

1.8L CNG Engine

See Figure 120.

⁂ WARNING

Do not insert your fingers into the installed throttle body when you turn the ignition switch to ON (II) or while the ignition switch is in ON (II). If you do, you will seriously injure your fingers if the throttle valve is activated.

➡If replacing or cleaning the throttle body, start at step 1. If removing the throttle body, start at step 4.

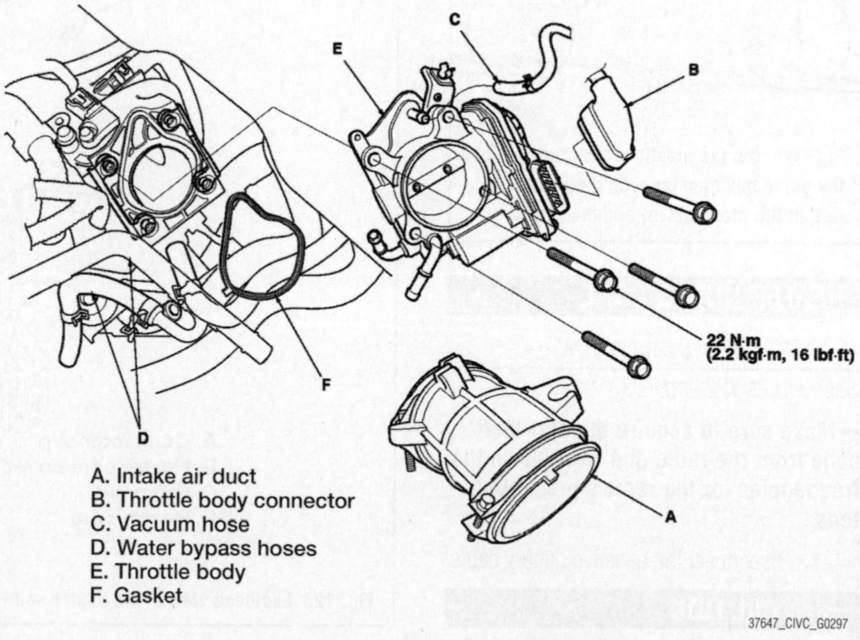
C
E
B
22 N·m
(2.2 kgf·m, 16 lbf·ft)
F
D
A
A. Intake air duct
B. Throttle body connector
C. Vacuum hose
D. Water bypass hoses
E. Throttle body
F. Gasket
37647_CIVC_G0297

Fig. 120 Removing/installing the throttle body assembly—1.8L CNG engine

1. Connect the HDS, or equivalent scan tool, to the DLC while the engine is stopped.

2. Select the INSPECTION MENU on the HDS.

3. Do the TP POSITION CHECK in the ETCS TEST.

4. Turn the ignition switch to LOCK (0).

5. Remove the cowl cover and the under-cowl panel.

6. Remove the air cleaner, then remove the intake air duct (A).

7. Disconnect the throttle body connector (B) and the vacuum hose (C).

8. Disconnect and plug the water bypass hoses (D).

9. Remove the throttle body (E) and gasket (F).

10. Install the parts in the reverse order of removal with a new gasket.

11. Refill the radiator with engine coolant.

12. Do the PCM idle learn procedure after replacing throttle body.

HEATING & AIR CONDITIONING SYSTEM

BLOWER MOTOR

REMOVAL & INSTALLATION

See Figure 121.

1. Remove the glove box.
2. Cut the plastic cross brace in the glove box opening with diagonal cutters in the area shown, and discard it.
3. Remove the bolts and the glove box frame.
4. Remove the wire harness clip, the self-tapping screws, and the passenger's heater duct.
5. Disconnect the connector from the blower motor. Remove the wire harness clip.
6. Disconnect the connector from the recirculation control motor. Remove the self-tapping screws, the bolt, the mounting nuts, and the blower unit.
7. Installation is the reverse order of removal. Make sure that there is no air leakage.

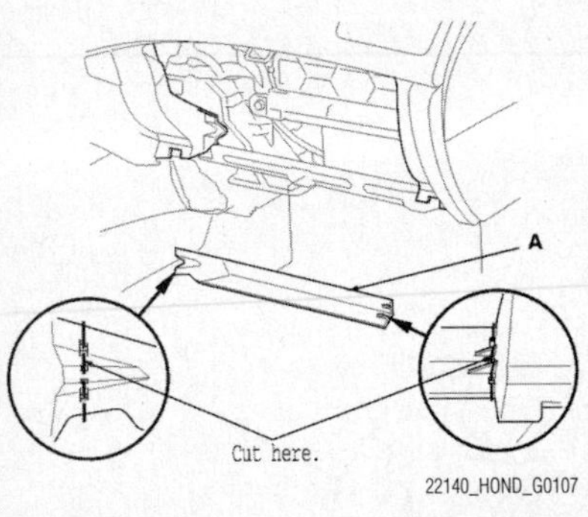
A
Cut here.
22140_HOND_G0107

Fig. 121 Cut the plastic cross brace (A) in the glove box opening with diagonal cutters in the area shown, and discard it

HEATER UNIT & CORE

REMOVAL & INSTALLATION

See Figures 122 and 123.

➡Make sure to acquire the anti-theft code from the radio and write down the frequencies for the radio's preset buttons.

1. Disconnect the negative battery cable.

✱✱ CAUTION

After disconnecting the negative battery cable, wait for at least 3 minutes for the air bag module to deplete its energy before working the on the instrument panel or steering wheel.

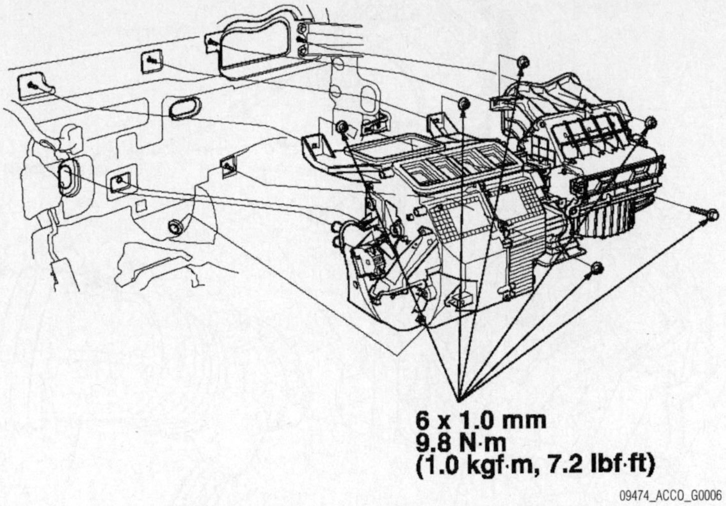
6 x 1.0 mm
9.8 N·m
(1.0 kgf·m, 7.2 lbf·ft)
09474_ACCO_G0006

Fig. 122 View of the heater housing, evaporator housing and related components—All Civic Models

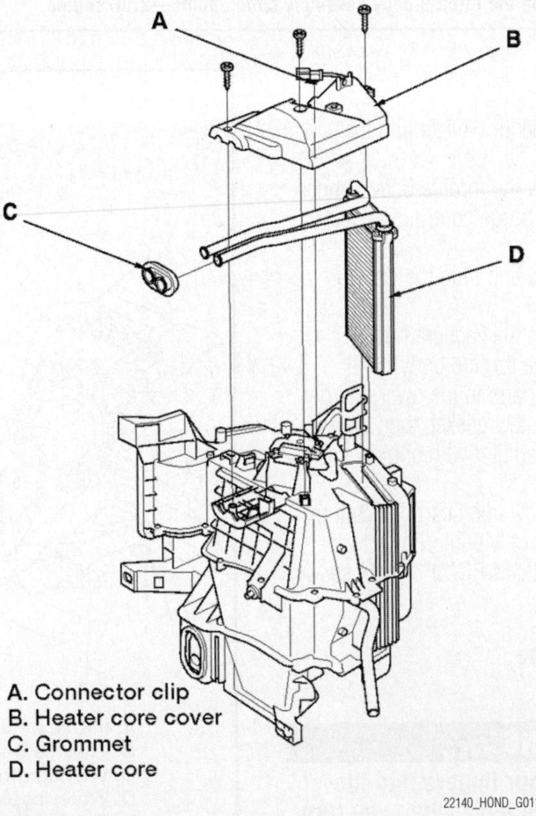
A
B
C
D
A. Connector clip
B. Heater core cover
C. Grommet
D. Heater core
22140_HOND_G0111

Fig. 123 Exploded view of the heater unit—All Civic models

2. Discharge the air conditioning system.
3. Remove the air cleaner assembly. Disconnect the heater hoses. Remove the mounting nut from the heater unit.
4. Remove the sub-display visor. Remove the navigation system, if equipped. On vehicles without navigation system remove the radio.

5. Remove the dashboard retaining screws. Detach the retaining clips along the lower edge of the instrument panel.

6. Detach the retaining clips along the upper edge of the instrument panel. Gently pull forward to release the hooks from the holder of the gauge module.

7. Remove the instrument panel.

8. Disconnect the electrical connector from the blower motor. Remove the wire harness clip. Disconnect the connector from the recirculation control motor.

9. Disconnect the connectors from the mode control motor, the evaporator temperature sensor and the power transistor. Remove the wire clip harness.

10. Disconnect the connectors from the air mix control motor and the air conditioning wire harness. Remove the connector clip and the wire harness clips.

11. Remove the mounting bolt, mounting nuts and the heater/blower unit from the vehicle.

12. Remove the self taping screws, remove the grommet and carefully pry out the heater core.

To install:

13. Installation is the reverse of the removal procedure.

14. Evacuate, charge and leak test the air conditioning system refrigerant, as required.

15. Operate the engine to normal operating temperatures; then, check the climate control operation and check for leaks.

16. Enter the antitheft codes for the radio and the navigation system.

STEERING

ELECTRONIC POWER STEERING (EPS) CONTROL UNIT

REMOVAL & INSTALLATION

See Figure 124.

1. Do the battery terminal disconnection procedure and wait 3 minutes before starting work.

2. Remove the passenger's dashboard under cover.

3. Remove the passenger's side kick panel.

4. Disconnect EPS control unit connector A (2P), connector B (2P), connector C (2P), and connector D (28P). Remove the nuts from the EPS control unit. Remove the EPS control unit.

To install:

6. Install the EPS control unit in the reverse order of removal.

7. Do the battery terminal reconnection procedure.

8. Do the "Torque Sensor Neutral Position Memorization". See procedure under "Power Rack & Pinion Steering Gear" for EPS application.

9. After installation, start the engine, allow it to idle. Turn the steering wheel from lock to lock several times. Check that the EPS indicator does not come on.

ELECTRONIC POWER STEERING (EPS) MOTOR

REMOVAL & INSTALLATION

See Figures 125 and 126.

✲✲ CAUTION

Do not allow dust, dirt, or other foreign materials to enter the steering gearbox. Make sure not to get any silicone grease on the terminal part of the connectors and switches, especially if you have silicone grease on your hands or gloves.

1. Remove the steering gearbox. See "Power Rack & Pinion Steering Gear" in this section.

2. Disconnect the torque sensor 3P connector from the steering gearbox, then remove the wire harness clamp bolts and the ground terminal.

3. Remove the EPS motor from the steering gearbox, then remove the O-ring.

To install:

4. Clean the mating surface of the EPS motor and the steering gearbox.

5. Apply a thin coat of silicone grease to the O-ring and carefully fit it on the EPS motor.

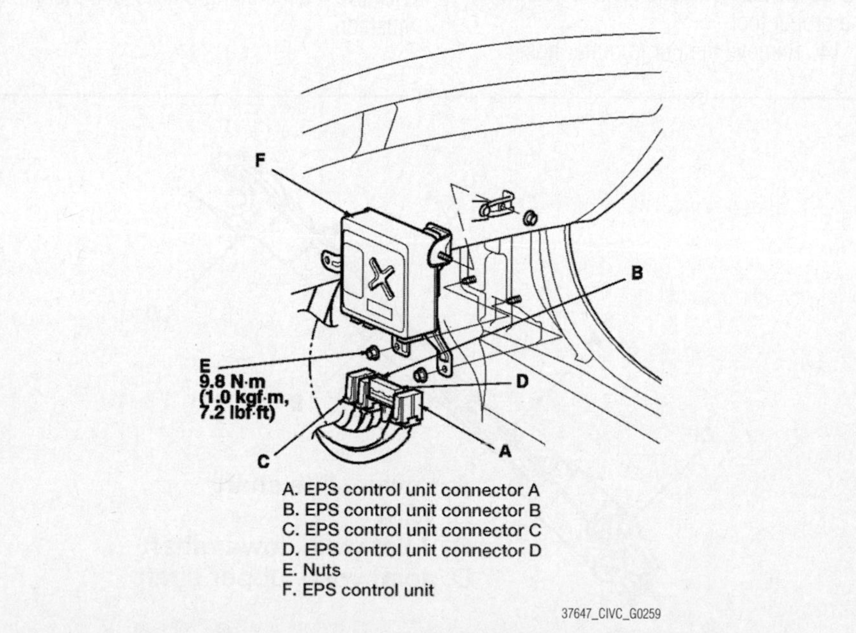
F
B
E
9.8 N·m
(1.0 kgf·m,
7.2 lbf·ft)
D
C
A
A. EPS control unit connector A
B. EPS control unit connector B
C. EPS control unit connector C
D. EPS control unit connector D
E. Nuts
F. EPS control unit
37647_CIVC_G0259

Fig. 124 Disconnect EPS control unit connector A (2P), connector B (2P), connector C (2P), and connector D (28P). Remove the nuts from the EPS control unit. Remove the EPS control unit.

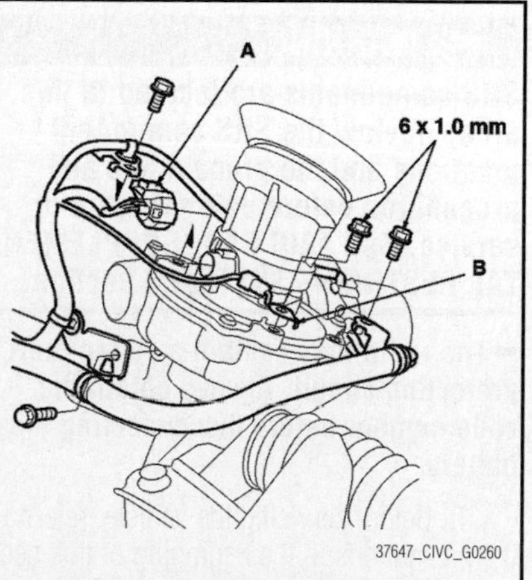
A
6 x 1.0 mm
B
37647_CIVC_G0260

Fig. 125 Disconnect the torque sensor 3P connector (A) from the steering gearbox, then remove the wire harness clamp bolts and the ground terminal (B).

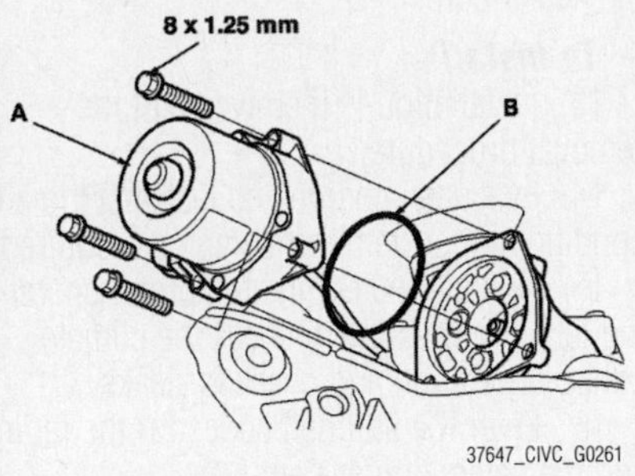
8 x 1.25 mm
A
B
37647_CIVC_G0261

Fig. 126 Remove the EPS motor (A) from the steering gearbox, then remove the O-ring (B).

6. Apply steering grease into the EPS motor shaft.
7. Set the motor on the steering gearbox by engaging the EPS motor shaft and the worm shaft.
8. Turn the EPS motor two or three times to the right and left about 45 degrees. Make sure the EPS motor is evenly seated on the steering gearbox, and that the O-ring is not pinched between the mating surfaces.
9. Loosely install the EPS motor mount bolts, then turn the steering wheel two or three times to the right and left about 45 degrees.
10. Tighten the EPS motor mount bolts to 14 ft. lbs. (20 Nm).

POWER RACK & PINION STEERING GEAR

REMOVAL & INSTALLATION

Without EPS

See Figures. 127 and 128.

WARNING

SRS components are located in this area. Review the SRS component locations and the precautions and procedures before doing repairs or service. See "AIR BAG (SUPPLEMENTAL RESTRAINT SYSTEM)" section.

➡The radio may contain a coded theft protection circuit. Always obtain the code number before disconnecting the battery.

1. Before servicing the vehicle, refer to the precautions in the beginning of this section.
2. Disconnect the negative and positive battery cables. Wait 3 minutes before working around the air bags.
3. Raise and support the vehicle safely. Remove the front tires. Drain the power steering fluid.
4. Remove the driver's side airbag.
5. Remove the steering wheel.

➡Be sure to remove the steering wheel before disconnecting the steering joint as damage to the cable reel can occur.

6. Remove the driver's dashboard lower cover. Remove the driver's dashboard under cover.
7. Remove the steering joint bolts. Disconnect the steering joint by moving the joint toward the column.
8. Hold the lower slide shaft on the column with a piece of wire between the joint yoke on the lower slide shaft to the joint yoke on the upper shaft.
9. Remove the center guide, if equipped and discard it.

➡The center guide is for factory assembly only.

10. Remove the harness holder and the air cleaner housing bracket.
11. Remove the breather pipe and the harness holder bracket.
12. Attach special tool 07AAK-SNAA120, or equivalent to the cylinder head. Install the engine support hanger, AAR-12566 or equivalent. Attach the hook to the tool and tighten the wing nut. Lift and support the engine.
13. Remove the cotter pin from the tie rod ball joint and loosen the nut. Separate the tie rod ball joint and the knuckle, using the proper tool.
14. Remove the pump outlet hose clamp from the intake manifold. Remove the inlet line clamp bolt. Open the return line holder and remove the return line clamp bolt.
15. Loosen the adjustable hose clamp and disconnect the return hose. Loosen the flare nut and disconnect the inlet line.
16. Remove the splash shield. Remove the lower ball joint mounting bolt and flange nuts from the lower arm. Disconnect the lower arm from the lower ball housing.
17. Note the reference marks on both sides of the subframe that line up with the body (these are for installation reference).
18. Attach tool VSB02C000016 to the front subframe and a transaxle jack or powertrain lift tool. Make sure that the subframe is securely supported.
19. Remove the front subframe middle mount bolt from the left side. Remove the front subframe middle mount bolt from the right side. Remove the two 12mm flange bolts from the lower torque rod and bracket.
20. Remove the front subframe front mounting bolts from the right and left sides of the vehicle. Discard them.
21. Remove the front subframe rear mounting bolts from the right and left sides of the vehicle. Discard them.
22. Lower the front subframe and steering gearbox as an assembly. Remove the pinion shaft grommet from the top of the valve body unit.
23. Remove the two 10mm bolts from the right side of the steering gearbox. Remove the mounting bracket and mounting cushion.

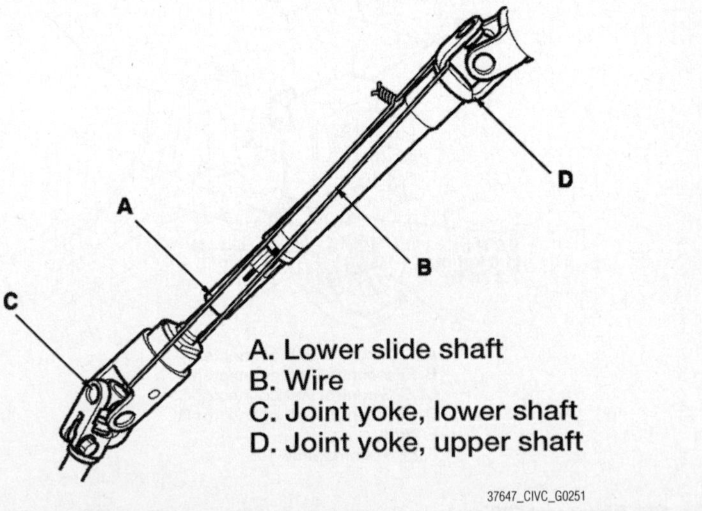
A
B
C
D
A. Lower slide shaft
B. Wire
C. Joint yoke, lower shaft
D. Joint yoke, upper shaft
37647_CIVC_G0251

Fig. 127 Hold the lower slide shaft on the column with a piece of wire between the joint yoke on the lower slide shaft to the joint yoke on the upper shaft.

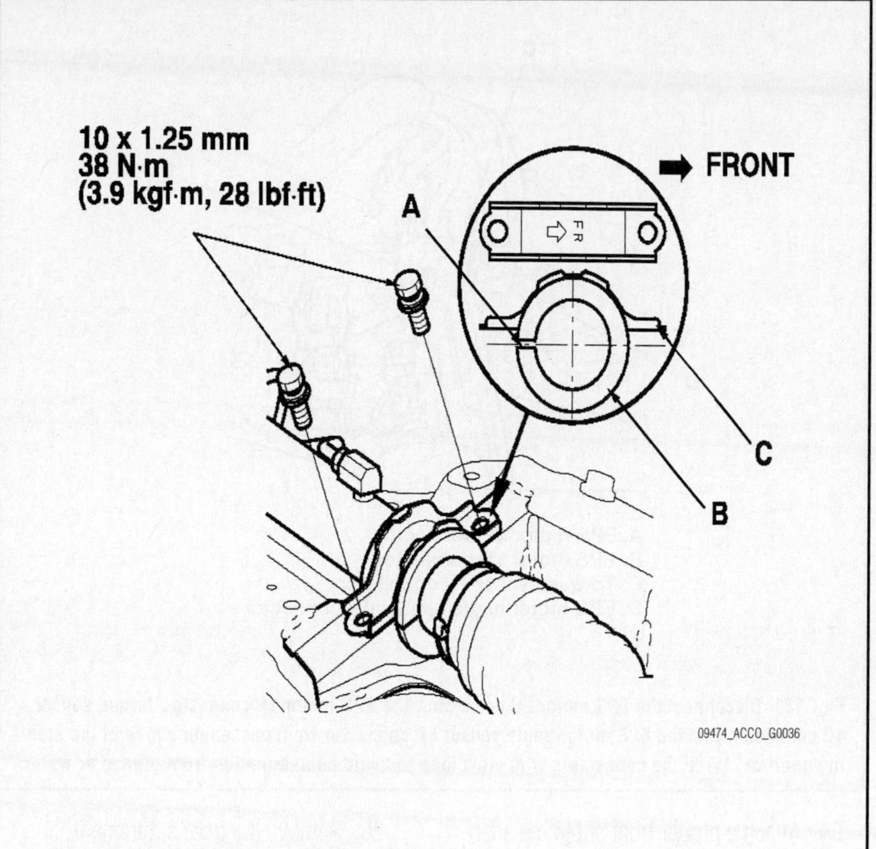
10 x 1.25 mm
38 N·m
(3.9 kgf·m, 28 lbf·ft)
FRONT
A
B
C
09474_ACCO_G0036

Fig. 128 Cutout positioning on mounting cushion

24. Remove the four 10mm flange bolts from the left side of the steering gearbox. Remove the stiffener plates.
25. Remove the steering gearbox from the front subframe

To install:

26. Position the assembly to its mounting on the subframe.
27. Loosely install the stiffener plates and gearbox mounting bolts on the left side of the steering gearbox.
28. Position the cutout on the mounting cushion and install it on the right side of the steering gearbox.
29. Install the gearbox mounting bracket over the mounting cushion and loosely install the two 10mm bolts.
30. Tighten the 10mm bolts on both sides of the steering gearbox to 40 ft. lbs. right side and 28 ft. lbs. left side, alternately in two or more steps.
31. Install the pinion shaft grommet. Align the slot in the pinion shaft grommet with the lug portion on the valve housing. The grommet must not have a gap at the mating surface of the grommet and valve housing.
32. Carefully raise the front subframe in position.
33. Continue the installation in the reverse order of the removal procedure.
34. Be sure to center the cable reel by first rotating it clockwise until it stops. Then rotate it counterclockwise (approximately two and a half turns) until the arrow mark on the label points straight up. Install the steering wheel.
35. Fill the system with the proper grade and type power steering fluid.
36. Start the engine, allow it to idle, and turn the steering wheel from lock-to-lock several times to warm up the fluid. Check and adjust the fluid level.
37. Check the gearbox for fluid leaks, correct as necessary.
38. Check front end alignment.
39. Check the steering wheel spoke angle. Adjust by turning the right and left tie rods equally, if necessary.

With EPS

See Figures 127, 129 through 136.

➡Note these items during removal:

- Use solvent and a brush, wash any oil and dirt off the end of the steering gearbox. Avoid any electrical parts. Blow dry with compressed air. Wear eye protection.
- Make sure to remove the steering wheel before disconnecting the steering joint, or damage to the cable reel occur.

1. Do the battery terminal disconnect procedure, then wait at least 3 minutes before beginning work.
2. Raise and support the vehicle.
3. Remove the front wheels.
4. Remove the driver's airbag, and the steering wheel.
5. Remove the driver's dashboard undercover.
6. Remove the steering joint cover at the floorboard.
7. Release the lock lever, and adjust the steering column to the full tilt up position, and to the full telescopic in position. Tighten the lock lever.
8. Hold the lower slide shaft on the column with a piece of wire between the joint yoke on the lower slide shaft to the joint yoke on the upper shaft.
9. Remove the steering joint bolt and disconnect the steering joint by moving the steering joint toward the column.
10. Remove the center guide (if equipped) from the top of the pinion shaft and discard it.

➡The center guide is for factory assembly use only.

11. Remove the cowl cover and the under-cowl panel.
12. Remove the air cleaner housing.
13. On non-Si models, remove the air cleaner mounting bracket and install the 1.8 support eyelet (07AAK-SNAA400) behind the breather pipe and down to the threaded hole on the cylinder head. Install a retaining bolt.

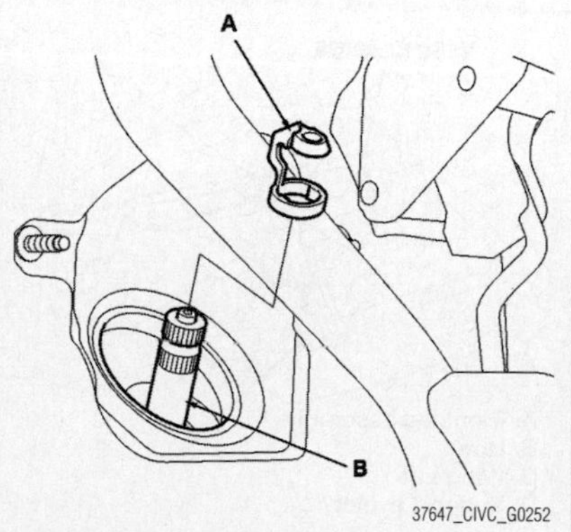
A
B
37647_CIVC_G0252

Fig. 129 Remove the center guide (A) (if equipped) from the top of the pinion shaft (B), and discard it.

14. On Si models, attach the engine hanger adapter (VSB02C000015) to the threaded hole in the cylinder head.

15. Install the front leg assembly, hook, and the wing nut to an "A and Reds" engine support hanger (AAR-T1256) onto the engine hanger. Carefully position the engine hanger on the vehicle, and attach the hook to the forward hole in the engine hanger adapter (Si model) or the slotted hole in the 1.8 support eyelet (Except Si model). Tighten the wing nut by hand to lift and support the engine/transmission.

16. Remove the cotter pin from the tie-rod end ball joint, then remove the nut on both sides.

17. Disconnect the tie-rod end ball joint from the knuckle using the ball joint thread protector and the ball joint remover.

➡Be careful not to damage the ball joint boot when installing the remover.

18. Remove the front splash shield.

19. Disconnect the EPS motor 2P connector, the EPS motor 1P connector, torque sensor 4P connector, the EPS motor angle sensor 6P connector from passenger's side of the steering gearbox. Wrap the connectors with vinyl tape to avoid contamination from grease or water.

20. Remove the lower ball joint mounting bolt and the self-locking nuts from the lower arm on both sides.

21. Disconnect the lower arm from the lower ball joint housing.

22. On Si models, remove the exhaust hanger from the three way catalytic converter (TWC).

23. Note the reference marks on both sides of the subframe that line up with the body (these are for installation reference).

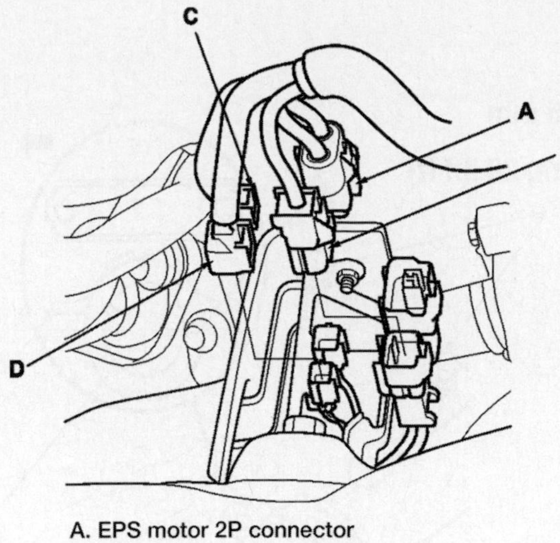
C
A
B
D
A. EPS motor 2P connector
B. EPS motor 1P connector
C. Torque sensor 4P connector
D. EPS motor angle sensor 6P connector
37647_CIVC_G0253

Fig. 131 Disconnect the EPS motor 2P connector, the EPS motor 1P connector, torque sensor 4P connector, and the EPS motor angle sensor 6P connector from passenger's side of the steering gearbox. Wrap the connectors with vinyl tape to avoid contamination from grease or water.

24. Attach a proper front subframe adaptor to the front subframe and the transmission jack or powertrain lift, then tighten the front subframe adaptor screw. Make sure the front subframe is securely supported by the jack with the front subframe adaptor.

25. Remove the front subframe middle mount bolt from the left and right sides.

26. Remove the lower torque rod mounting bolts.

27. On Si models, remove the lower radiator hose from the radiator hose stay, then remove the front mounting bolt and nut.

28. Remove the front subframe front mounting bolts on both sides.

29. Remove the front subframe rear mounting bolts on both sides.

30. Lower the front subframe and steering gearbox as an assembly by lowering the jack slowly.

31. On non-Si models, remove the harness clips from the harness clamp brackets on the steering gear, then remove the harness clamp brackets.

32. On Si models, remove the harness clamp bracket from the front subframe, then remove the harness clips.

33. Remove the EPS motor 2P connector, the EPS motor 1P connector, torque sensor 4P connector, the EPS motor angle sensor 6P connector from the passenger's side of the gearbox mounting bracket.

34. Remove the pinion shaft grommet from the top of torque sensor.

35. Remove the two 10 mm bolts from the right side of the steering gearbox, then remove the gearbox mounting bracket and the mounting cushion.

36. Remove the four 10 mm flange bolts from the left side of the steering gearbox, then remove the stiffener plates.

37. Remove the steering gearbox from the front subframe.

To install:

38. Place the steering gearbox in position on the front subframe.

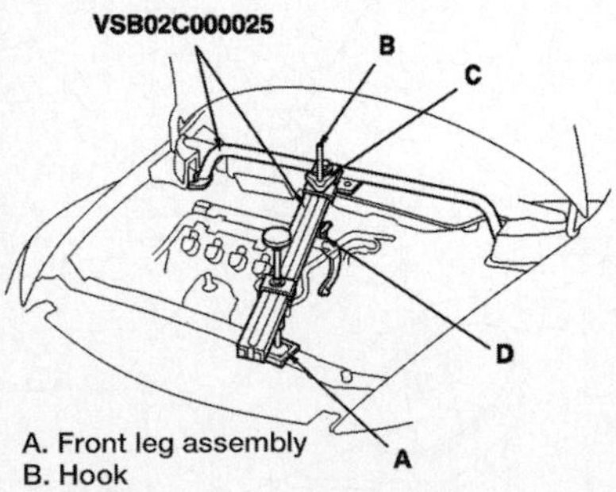
VSB02C000025
B
C
D
A
A. Front leg assembly
B. Hook
C. Wing nut
D. Support eyelet
37647_CIVC_G0097

Fig. 130 Installed position of the engine support hanger

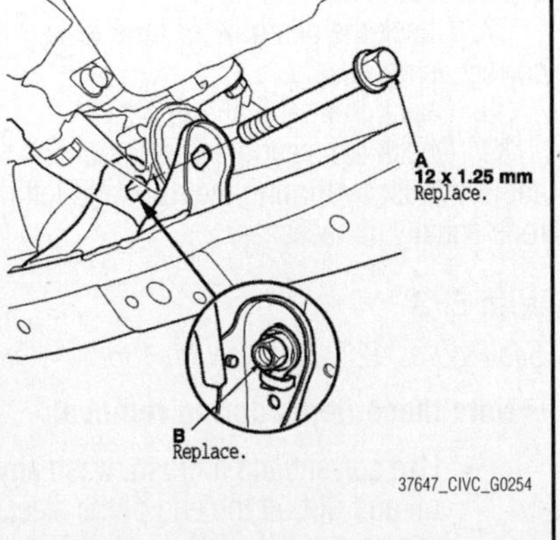
A
12 x 1.25 mm
Replace.
B
Replace.
37647_CIVC_G0254

Fig. 132 Remove the front mounting bolt (A) and nut (B)—Si models

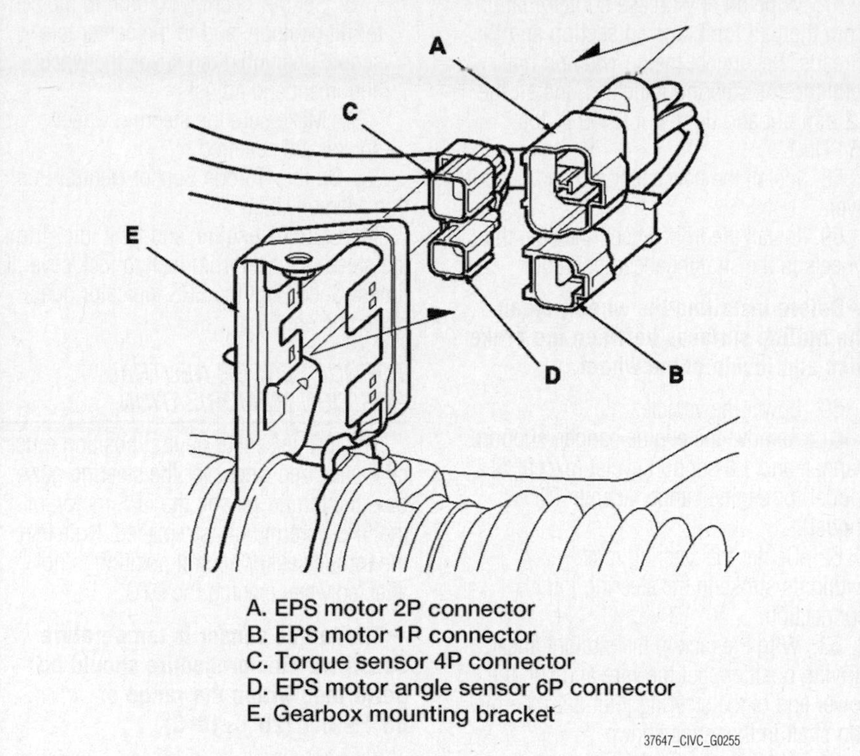
A
C
E
D
B

A. EPS motor 2P connector
B. EPS motor 1P connector
C. Torque sensor 4P connector
D. EPS motor angle sensor 6P connector
E. Gearbox mounting bracket

37647_CIVC_G0255

Fig. 133 Remove the EPS motor 2P connector, the EPS motor 1P connector, torque sensor 4P connector, the EPS motor angle sensor 6P connector from the passenger's side of the gearbox mounting bracket.

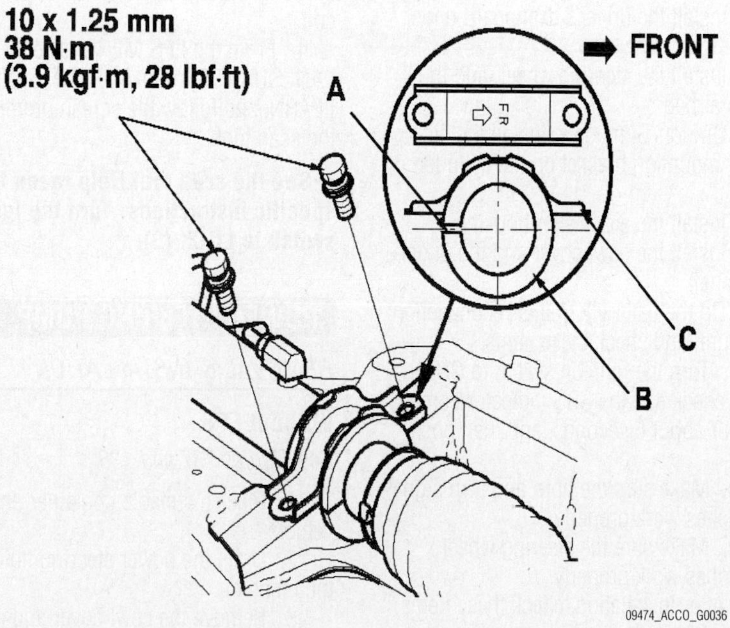
10 x 1.25 mm
38 N·m
(3.9 kgf·m, 28 lbf·ft)
FRONT
A
FR
C
B
09474_ACCO_G0036

Fig. 134 Position the cutout (A) on the mounting cushion (B), as shown, and install it on the right side of the steering gearbox securely. Install the gearbox mounting bracket (C) over the mounting cushion, and loosely install the two 10 mm bolts.

39. Loosely install the stiffener plates and the gearbox mounting bolts on the left side of the steering gearbox.

40. Position the cutout on the mounting cushion, as shown, and install it on the right side of the steering gearbox securely. Install the gearbox mounting bracket over the mounting cushion, and loosely install the two 10 mm bolts.

41. Tighten the 10 mm bolts on both sides of the steering gearbox to 28 ft. lbs. (38 Nm), alternately in two or more steps.

42. Install the pinion shaft grommet. Align the slot in the pinion shaft grommet with the lug portion on the torque sensor. The grommet must not have a gap at the mating surface of the grommet and the torque sensor.

43. Install the EPS motor 2P connector, the EPS motor 1P connector, torque sensor 4P connector, the EPS motor angle sensor 6P connector on the right side of the gearbox mounting bracket.

44. On non-Si models, install the harness clamp bracket to the front subframe, then install the harness clips to the harness clamp bracket.

45. On Si models, install the harness clips to the harness clamp bracket, then install the harness clamp bracket to the front subframe.

46. Carefully raise the front subframe with the front subframe adapter and the transmission jack or the powertrain lift until the front subframe is in position, then loosely install new front subframe mounting bolts.

➡Be sure that the pinion shaft grommet is in place securely. Make sure the pinion shaft grommet is not turned up. Incorrect installation can cause leakage of water, mud, and noise.

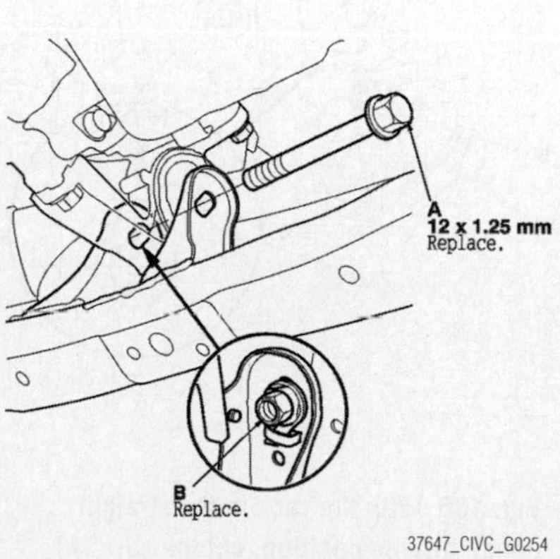
A
12 x 1.25 mm
Replace.
B
Replace.
37647_CIVC_G0254

Fig. 135 Install and tighten new front mounting bolt (A) and nut (B)—Si models

47. Position the subframe, then align the front subframe reference marks to the body, as noted during removal.
48. Tighten the new front and rear subframe mounting bolts to the 76 ft. lbs. (103 Nm) on both sides.
49. On Si models, install the new nut and the new front mounting bolt and tighten the mounting bolt to 47 ft. lbs. (64 Nm). Install the lower radiator hose to the radiator hose stay.
50. Install the new lower torque rod mounting bolts and tighten them to 47 ft. lbs. (64 Nm).
51. Install the new front subframe middle mount bolt on the left and right sides and tighten it to 47 ft. lbs. (64 Nm).
52. Lower the transmission jack supporting the front subframe.
53. On Si models, install the exhaust hanger to the three way catalytic converter.
54. Connect the lower arm to the lower ball joint. Install a new flange bolt and the new self-locking nuts. After lightly tightening all three fasteners, tighten them to 47 ft. lbs. (64 Nm) in the following order; the flange nut on the front, the flange nut on the rear, then the flange bolt.
55. Remove the vinyl tape, then connect the EPS motor 2P connector, the EPS motor 1P connector, torque sensor 4P connector, the EPS motor angle sensor 6P connector to the steering gearbox. Make sure to push these connectors until you hear a click so that the connectors are secured.
56. Install the front splash shield.

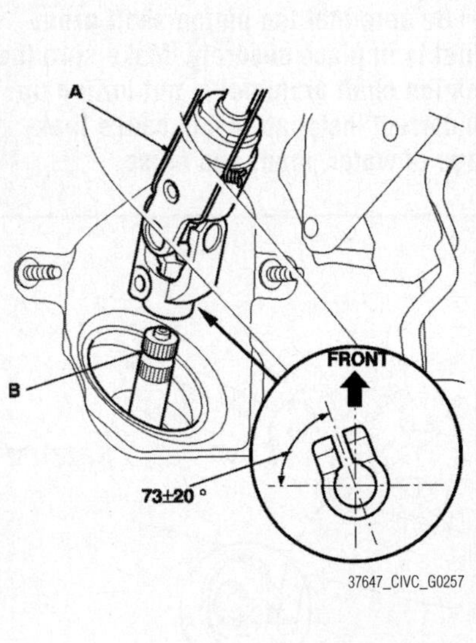
A
B
FRONT
73±20 °
37647_CIVC_G0257

Fig. 136 With the rack in the straight ahead driving position, cut the wire (A) and slip the lower end of the steering joint onto the pinion shaft (B) in the range shown.

57. Wipe off any grease contamination from the ball joint tapered section and the threads. Reconnect the tie-rod end ball joints to the steering knuckles. Install the 12 mm nut and tighten it to 40 ft. lbs. (54 Nm).
58. Install the new cotter pin and bend it over.
59. Install the front wheel, then set the wheels in the straight ahead position.

➡Before installing the wheel, clean the mating surfaces between the brake disc and inside of the wheel.

60. Lower the vehicle.
61. Remove the engine hanger, support hanger, and 1.8 support eyelet (except Si model) or engine hanger adapter (Si model).
62. Center the steering rack within its stroke in the steering joint connection.
63. With the rack in the straight ahead driving position, cut the wire and slip the lower end of the steering joint onto the pinion shaft in the range shown.
64. Align the bolt hole on the steering joint with the groove around the pinion shaft. Then loosely install the joint bolt. Be sure that the joint bolt is securely in the groove in the pinion shaft. Pull on the steering joint to make sure that the steering joint is fully seated. Tighten the steering joint bolt to 21 ft. lbs. (28 Nm).
65. Install the steering joint cover.
66. Install the driver's dashboard under cover.
67. Install the steering wheel, and the driver's airbag.
68. On non-Si models, install the air cleaner mounting bracket on the cylinder head.
69. Install the air cleaner housing.
70. Install the cowl cover and the under-cowl panel.
71. Do the battery terminal reconnection procedure, and check these items:
    a. Turn the ignition switch to ON (II) and check that the SRS indicator comes on for about 6 seconds, and then goes off.
    b. Make sure the horn and turn signal switches work properly.
    c. Make sure the steering wheel switches work properly.
72. After installation, check these items:
    a. Check the steering wheel spoke angle. If steering spoke angles to the right and left are not equal (steering wheel and rack are not centered), correct the engagement of the joint/pinion shaft splines.
    b. Set the steering column to the center tilt position, and to the center telescopic position, then check the wheel alignment and adjust.
    c. Make sure the steering wheel spokes are centered.
73. Do the "Torque Sensor Neutral Position Memorization".
74. Start the engine, and let it idle. Turn the steering wheel from lock to lock several times. Check that the EPS indicator does not come on.

TORQUE SENSOR NEUTRAL POSITION MEMORIZATION

The torque sensor neutral position must be memorized whenever the steering gearbox, the torque sensor, the EPS motor, or the EPS control unit is replaced. Note that the torque sensor neutral position is not affected when erasing the DTC.

➡The torque sensor is temperature sensitive. This procedure should be performed within the range of 68°F±18°F (20°C±10°C).

1. With the ignition switch in LOCK (0), connect the Honda Diagnostic System (HDS), or equivalent scan tool and software, to the data link connector (DLC), located under the driver's side of the dashboard.
2. Turn the ignition switch to ON (II).
3. Make sure the scan tool communicates with the vehicle and the EPS control unit. If it doesn't, troubleshoot the DLC circuit.
4. From the EPS MENU, select MISCELLANEOUS TEST, then TORQUE SENSOR LEARN, and follow the screen prompts on the scan tool.

➡See the scan tool Help menu for specific instructions. Turn the ignition switch to LOCK (0).

POWER STEERING PUMP

REMOVAL & INSTALLATION

Without EPS

See Figures 137 and 138.

1. Place a suitable container under the vehicle.
2. Drain the power steering fluid from the reservoir.
3. Remove the cowl cover and under-cowl panel.
4. Remove the air cleaner.
5. Remove the front splash shield.
6. If equipped with a manual transaxle, remove the shift cable bracket.

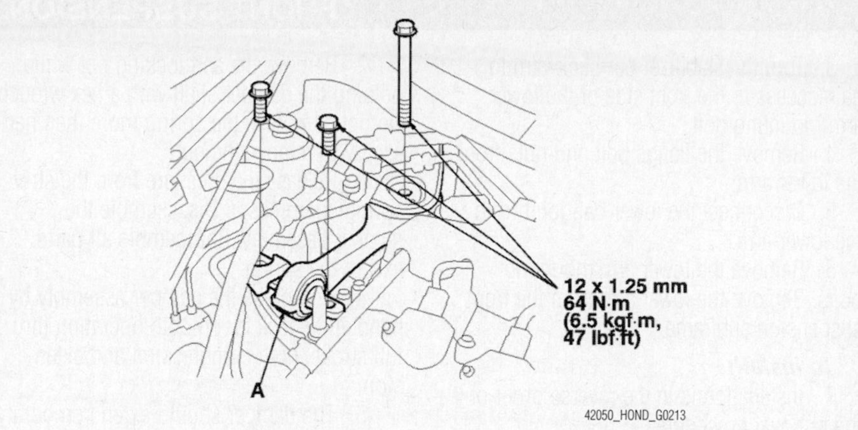
12 x 1.25 mm
64 N·m
(6.5 kgf·m,
47 lbf·ft)
A
42050_HOND_G0213

Fig. 137 Remove the upper torque rod (A)

7. If equipped with an automatic transaxle, disconnect the shift cable from the control lever.
8. Remove the upper torque rod from the body.
9. Remove the drive belt from the pump pulley.
10. Cover the parts around the power steering pump with several shop towels to protect them from spilled power steering fluid.
11. Disconnect the pump inlet hose and pump outlet hose from the pump, and plug them.

⁂ WARNING

Take care not to spill the fluid on the body or any parts. Wipe off any spilled fluid at once. Do not turn the steering wheel with the pump removed.

12. Remove the pump outlet hose O-ring, and discard it.
13. Remove the pump mounting bolts.
14. Cover the opening of the pump with a piece of tape to prevent foreign material from entering the pump.

To install:

15. Move the power steering pump toward the driver's side, and then raise it.
16. Connect the pump inlet hose and pump outlet hose onto the new pump with new O-ring.
17. Loosely install the pump in the pump bracket with the mounting bolts, and then tighten the pump fittings securely.
18. Tighten the pump mounting bolts to the specified torque shown in the accompanying illustration.

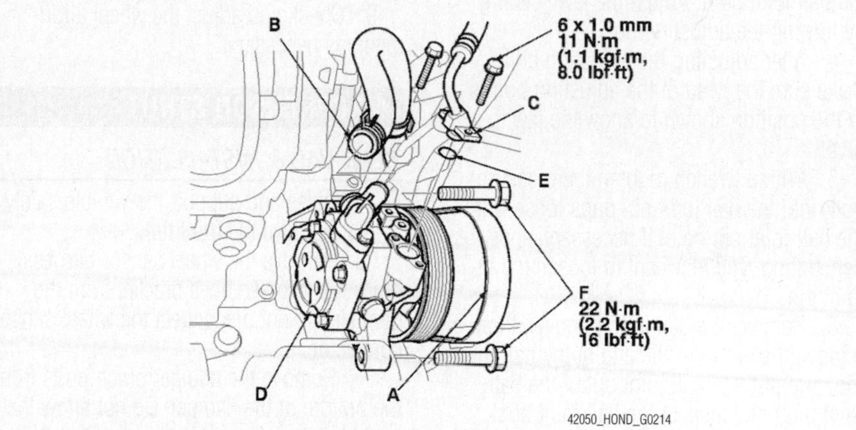
B
6 x 1.0 mm
11 N·m
(1.1 kgf·m,
8.0 lbf·ft)
C
E
F
22 N·m
(2.2 kgf·m,
16 lbf·ft)
D
A
42050_HOND_G0214

Fig. 138 Power steering pump mounting

19. Install the drive belt. Make sure that the belt is properly positioned on the pulleys.

⁂ WARNING

Do not get power steering fluid or grease on any parts around the power steering pump, drive belt, or pulley faces. Clean off any fluid or grease before installation.

20. Fill the reservoir to the upper level line and bleed the system.

BLEEDING

Check the reservoir at regular intervals, and add the recommended fluid as necessary. Always use Honda Power Steering Fluid. Using any other type of power steering fluid or automatic transmission fluid can cause increased wear and poor steering in cold weather.

➡If the fluid is contaminated, the screen in the reservoir may be partially blocked. Replace the reservoir if necessary.

1. Remove the reservoir from its holder. Raise the reservoir, and then disconnect the return hose to drain the reservoir. Take care not to spill the fluid on the body and parts. Wipe off any spilled fluid at once.

➡Inspect the reservoir screen for any debris. If the reservoir screen is clogged, replace the reservoir.

2. Connect a hose of suitable diameter to the disconnected return hose, and put the hose end in a suitable container.
3. Start the engine, let it run at idle, and turn the steering wheel from lock-to-lock several times. When fluid stops running out of the hose, shut off the engine. Discard the fluid.
4. Reinstall the return hose on the reservoir.
5. Fill the reservoir to the upper level line.
6. Start the engine and run it at fast idle, and then turn the steering from lock-to-lock several times to bleed air from the system.
7. Recheck the fluid level and add some if necessary. Do not fill the reservoir beyond the upper level line.
8. If the fluid is contaminated, dark, or discolored, repeat the procedure as necessary.

LOWER BALL JOINT

REMOVAL & INSTALLATION

See Figure 139.

1. Install a hex nut onto the threads of the ball joint. Make sure the nut is flush with the ball joint pin end to prevent damage to the threaded end of the ball joint pin.
2. Apply grease to the ball joint remover on the areas shown. This will ease installation of the tool and prevent damage to the pressure bolt threads.
3. Loosen the pressure bolt, and install the ball joint remover as shown. Insert the jaws carefully, making sure not to damage the ball joint boot. Adjust the jaw spacing by turning the adjusting bolt.
4. After adjusting the adjusting bolt, make sure the head of the adjusting bolt is in the position shown to allow the jaw to pivot.
5. With a wrench, tighten the pressure bolt until the ball joint pin pops loose from the ball joint pin hole. If necessary, apply penetrating type lubricant to loosen the ball joint pin.
6. Remove the ball joint remover, then remove the nut from the end of the ball joint pin, and pull the ball joint out of the ball joint pin hole. Inspect the ball joint boot, and replace it if damaged.

LOWER CONTROL ARM

REMOVAL & INSTALLATION

1. Raise and support the vehicle safely. Remove the front tires.
2. Remove the flange nut while holding the ball joint pin. Disconnect the stabilizer links from the lower arm.

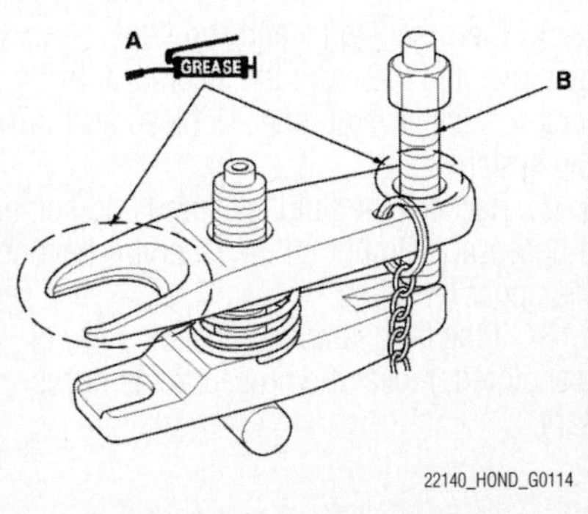
A
GREASE
B
22140_HOND_G0114

Fig. 139 Apply grease to the ball joint remover on the areas shown (A). This will ease installation of the tool and prevent damage to the pressure bolt (B) threads.

3. Turn the stabilizer bar backward to gain access to the front side of the lower arm mounting bolt.
4. Remove the flange bolt and nuts from the lower arm.
5. Disconnect the lower ball joint from the lower arm.
6. Remove the lower arm mounting bolts. Remove the lower arm from the front suspension subframe.

To install:

7. Installation is in the reverse order of the removal procedure.
8. Be sure to use new flange bolts, castle nut and lock pin. Tighten the flange nut to 25 ft. lbs. (34 Nm).
9. Check and adjust the wheel alignment, as necessary.

MACPHERSON STRUT

REMOVAL & INSTALLATION

1. Raise and support the vehicle safely.
2. Remove the front tires.
3. Remove the wheel sensor harness bracket and brake hose bracket from the damper. Do not disconnect the wheel sensor connector.
4. Remove the damper pinch bolts from the bottom of the damper. Do not allow the knuckle to rotate too far outward as this may cause the inner CV joint bearing to unseat.
5. Turn the ignition switch to the ON position. Turn the windshield wipers on; turn the ignition switch off leaving the wipers near the A-pillars.
6. Remove the service cap and lid. Remove the three flange nuts at the top of the damper.

➡Damper springs are different, left and right. Mark the springs before continuing.

7. Remove the strut assembly from the vehicle.

To install:

8. Position the assembly to its mounting on the vehicle.
9. Install the mounting bolts.
10. Continue the installation in the reverse order of the removal procedure.
11. Check front end alignment.

OVERHAUL

1. Compress the damper spring with a commercially available strut spring compressor according to the manufacturer's instructions.
2. Remove the self locking nut while holding the damper shaft with a hex wrench. Do not compress the spring more than necessary to remove the nut.
3. Release the pressure from the strut spring compressor. Disassemble the damper assembly. Reassemble all parts, except the spring.
4. Compress the damper assembly by hand and check for smooth operation thru full stroke, both compression and extension.
5. The damper should extend smoothly and constantly when compression is released. If not, gas is leaking and the damper should be replaced.
6. Install all parts except the self locking nut onto the damper unit.
7. Align the bottom of the spring and the stepped part of the lower spring seat. The hole in the upper spring seat and the arrow on the damper mounting base must point toward the knuckle mounting area.
8. Install the damper assembly on a commercially available strut spring compressor.
9. Compress the damper spring with the strut spring compressor. Install a new self locking nut on the damper shaft.
10. Hold the damper shaft with a hex wrench and tighten the self locking nut.

STEERING KNUCKLE

REMOVAL & INSTALLATION

➡The steering knuckle is removed with the wheel bearings as an assembly. Refer to "Wheel Bearings" in this section.

STABILIZER BAR

REMOVAL & INSTALLATION

1. Raise and support the vehicle safely. Remove the front tires.
2. Disconnect both stabilizer links from the stabilizer bar.
3. Remove the flange bolts and the bushing holders. Remove the bushings and the stabilizer bar from the front suspension subframe.

To install:

4. Position the assembly to its mounting on the vehicle.
5. Install the mounting bolts.
6. Continue the installation in the reverse order of the removal procedure.

WHEEL BEARINGS

REMOVAL & INSTALLATION

See Figure 140.

➡A hydraulic press and several bearing drivers and attachments are needed to remove and install the hub and bearing.

1. Before servicing the vehicle, refer to the precautions in the beginning of this section.
2. Pry the spindle nut stake away from the spindle, and then loosen the nut.
3. Raise and safely support the vehicle.
4. Remove or disconnect the following:
   - Front wheel and the spindle nut
   - Wheel sensor wire bracket from the knuckle, but don't disconnect it.
   - Caliper mounting bolts and the caliper. Support the caliper out of the way with a length of wire. Do not let the caliper hang from the brake hose.
   - 6mm brake disc retaining screws. Screw two 12mm bolts into the disc to push it away from the hub.
   - Tie rod castle nut
   - Tie rod ball joint using a suitable ball joint remover.
   - Cotter pin and loosen the lower arm ball joint nut half the length of the joint threads.
   - Ball joint and lower arm using a suitable puller with the pawls applied to the lower arm.

➡Avoid damaging the ball joint boot. If necessary, apply penetrating type lubricant to loosen the ball joint.

   - Ball joint nut cover.
   - Cotter pin and the upper ball joint nut.
   - Upper ball joint and knuckle using a ball joint remover.
5. Use a plastic mallet to free the halfshaft from the knuckle. Pull the knuckle out to remove it.

➡A new wheel bearing must be used when the hub is removed.

6. Place the knuckle in a press and use a base and pilot to press the hub assembly out of the wheel bearing.
7. Remove the knuckle ring seal and circlip. Remove the splash guard from the knuckle.
8. Press the wheel bearing out of the knuckle using a driving attachment.

To install:

9. Clean the knuckle and hub assembly and inspect them for damage.
10. Install or connect the following:

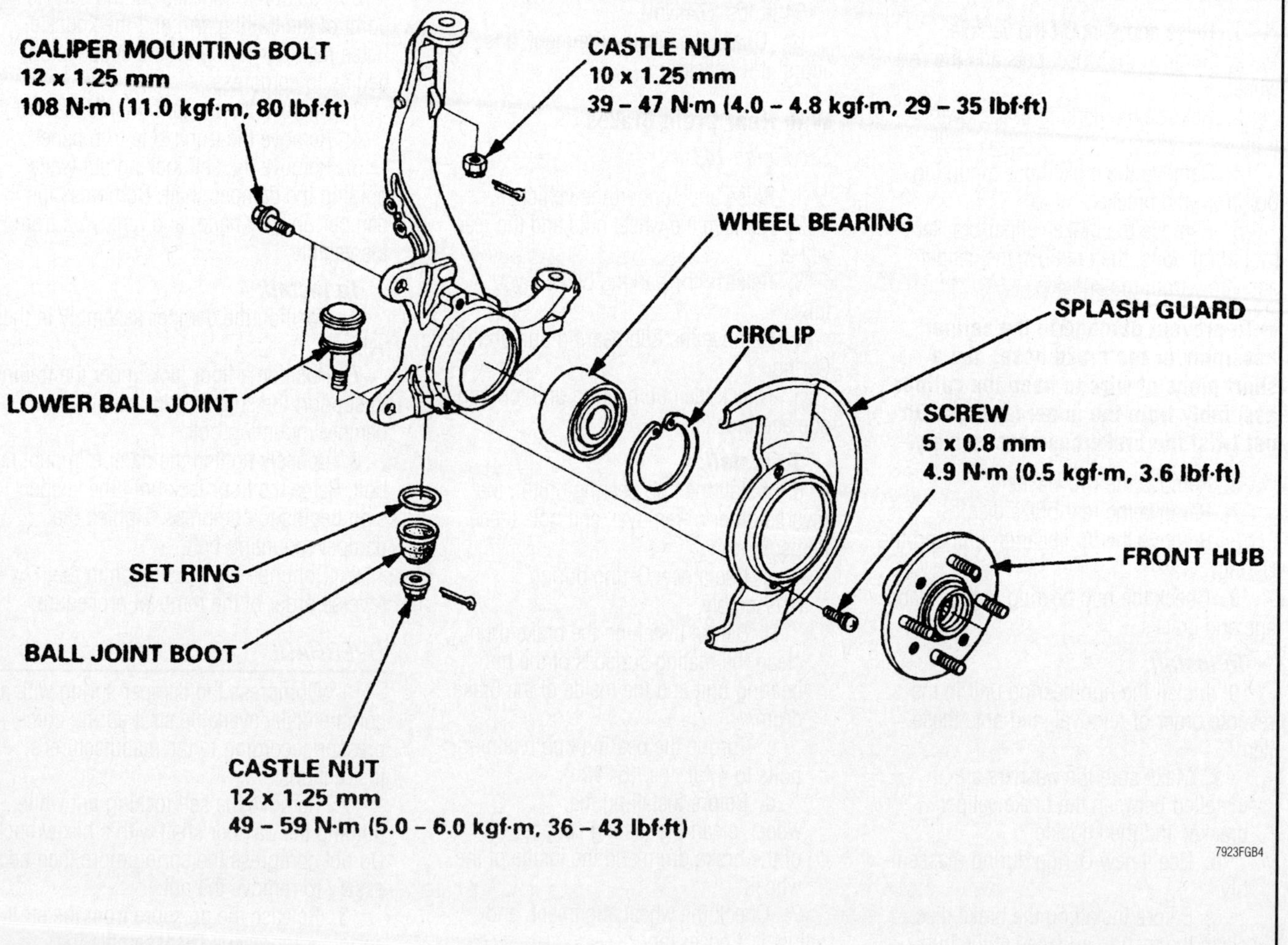

CALIPER MOUNTING BOLT
12 x 1.25 mm
108 N·m (11.0 kgf·m, 80 lbf·ft)
CASTLE NUT
10 x 1.25 mm
39 – 47 N·m (4.0 – 4.8 kgf·m, 29 – 35 lbf·ft)
WHEEL BEARING
SPLASH GUARD
CIRCLIP
LOWER BALL JOINT
SCREW
5 x 0.8 mm
4.9 N·m (0.5 kgf·m, 3.6 lbf·ft)
FRONT HUB
SET RING
BALL JOINT BOOT
CASTLE NUT
12 x 1.25 mm
49 – 59 N·m (5.0 – 6.0 kgf·m, 36 – 43 lbf·ft)
7923FGB4

Fig. 140 Knuckle components

- New wheel bearing into the hub using a driving tool
- Circlip in the outer groove of the knuckle
- Splash guard
- Hub assembly into the steering knuckle using a base and a driving and guide tool
- Knuckle ring seal
- Knuckle onto the spindle
- Knuckle onto the upper and lower ball joints and tighten the castle nuts
- Tie rod ball joint onto the steering knuckle

11. Tighten the upper ball joint nut and tie rod nut to 29–35 ft. lbs. (40–48 Nm) and the lower ball joint castle nut to 36–43 ft. lbs. (50–60 Nm).
12. Install or connect the following:
    - Anti-lock Brake System (ABS) wheel sensor wire brackets onto the knuckle. Tighten the mounting bolts to 84 inch lbs. (10 Nm).
    - Brake disc; use 2 lug nuts to evenly draw the disc onto the hub
    - Retainer screws: 84 inch lbs. (10 Nm)
    - Spindle washer and nut. Don't tighten the nut until the vehicle is on the ground.
    - Brake caliper and tighten the bolts to 80 ft. lbs. (110 Nm)
    - Front wheels and lower the vehicle
13. Tighten the spindle nut to 134 ft. lbs. (185 Nm), stake the nut, and install the grease cap.
14. Check and adjust the vehicle's front wheel alignment.

➡**Avoid damaging the ball joint boot. If necessary, apply penetrating-type lubricant to loosen the ball joint.**

ADJUSTMENT

➡**The wheel bearings are not adjustable. If they are not within specification, the wheel bearings must be replaced.**

SUSPENSION — REAR SUSPENSION

KNUCKLE/HUB BEARING

REMOVAL & INSTALLATION

With Rear Disc Brakes

See Figures 141 and 142.

1. Raise and support the vehicle.
2. Remove the wheel nuts and the rear wheel.
3. Release the parking brake lever fully.
4. Remove the brake hose mounting bolt from the bracket.
5. Remove the brake caliper bracket mounting bolts, then remove the caliper assembly from the knuckle.

➡**To prevent damage to the caliper assembly or the brake hose, use a short piece of wire to hang the caliper assembly from the undercarriage. Do not twist the brake hose excessively.**

6. Remove the two washers.
7. Remove the rear brake disc.
8. Remove the hub bearing unit and the O-ring.
9. Check the hub bearing unit for damage and cracks.

To install:

10. Install the hub bearing unit in the reverse order of removal, and note these items:
    a. Make sure the washers are installed between the brake caliper bracket and the knuckle.
    b. Use a new O-ring during reassembly.
    c. Before installing the brake disc, clean the mating surfaces of the hub bearing unit and the inside of the brake disc.
    d. Torque the bearing hub retaining bolts to 47 ft. lbs. (64 Nm).
    e. Before installing the wheel, clean the mating surfaces of the brake disc and the inside of the wheel.
    f. Torque the caliper retaining bolts to 54 ft. lbs. (74 Nm).
11. Check the wheel alignment, and adjust it if necessary.

With Rear Drum Brakes

See Figure 143.

1. Raise and support the vehicle.
2. Remove the wheel nuts and the rear wheel.
3. Release the parking brake lever fully.
4. Remove the hub bearing unit and the O-ring.
5. Check the hub bearing unit for damage and cracks.

To install:

6. Install the hub bearing unit in the reverse order of removal, and note these items:
    a. Use a new O-ring during reassembly.
    b. Before installing the brake drum, clean the mating surfaces of the hub bearing unit and the inside of the brake drum.
    c. Torque the bearing hub retaining bolts to 47 ft. lbs. (64 Nm).
    d. Before installing the wheel, clean the mating surfaces of the brake drum and the inside of the wheel.
7. Check the wheel alignment, and adjust it if necessary.

MACPHERSON STRUTS

REMOVAL & INSTALLATION

1. Raise and support the vehicle safely. Remove the rear tires.
2. Position a floor jack at the connecting point of the trailing arm and the knuckle. Raise the floor jack until the suspension begins to compress.
3. Remove the flange bolt and discard it.
4. Remove the trunk side trim panel.
5. Remove the self locking nut while holding the damper shaft. Compress the damper unit, by hand, and remove it from the vehicle.

To install:

6. Position the damper assembly in the vehicle.
7. Position a floor jack under the trailing to support the suspension. Install a new damper mounting bolt.
8. Loosely tighten the damper mounting bolt. Raise the floor jack until the suspension begins to compress. Tighten the damper mounting bolt.
9. Continue the installation in the reverse order of the removal procedure.

OVERHAUL

1. Compress the damper spring with a commercially available strut spring compressor according to the manufacturer's instructions.
2. Remove the self locking nut while holding the damper shaft with a hex wrench. Do not compress the spring more than necessary to remove the nut.
3. Release the pressure from the strut spring compressor. Disassemble the damper assembly. Reassemble all parts, except the spring.

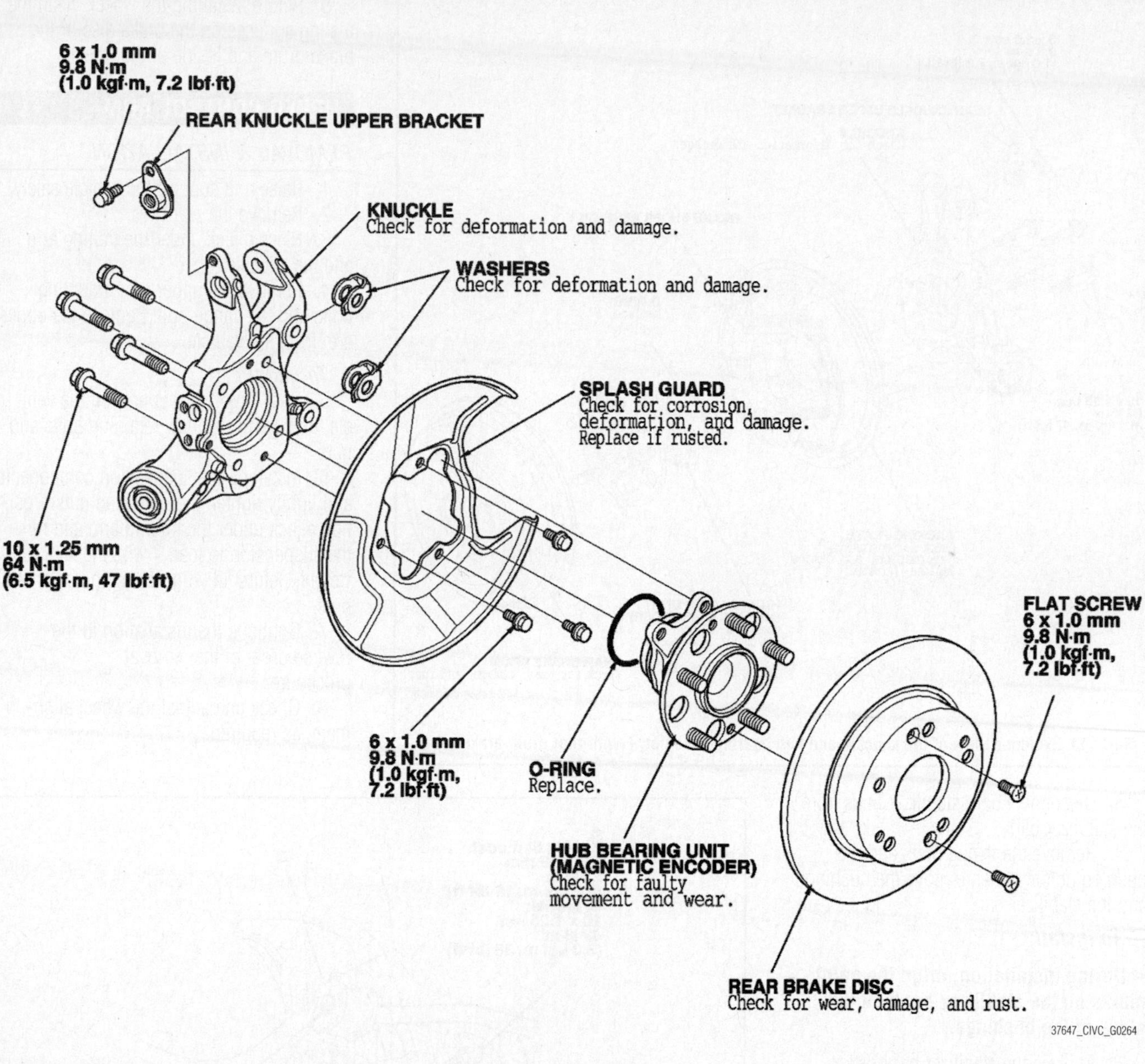
6 x 1.0 mm
9.8 N·m
(1.0 kgf·m, 7.2 lbf·ft)
REAR KNUCKLE UPPER BRACKET
KNUCKLE
Check for deformation and damage.
WASHERS
Check for deformation and damage.
SPLASH GUARD
Check for corrosion, deformation, and damage. Replace if rusted.
10 x 1.25 mm
64 N·m
(6.5 kgf·m, 47 lbf·ft)
FLAT SCREW
6 x 1.0 mm
9.8 N·m
(1.0 kgf·m,
7.2 lbf·ft)
6 x 1.0 mm
9.8 N·m
(1.0 kgf·m,
7.2 lbf·ft)
O-RING
Replace.
HUB BEARING UNIT
(MAGNETIC ENCODER)
Check for faulty movement and wear.
REAR BRAKE DISC
Check for wear, damage, and rust.
37647_CIVC_G0264

Fig. 141 Exploded view of the knuckle and hub bearing assembly—with rear disc brakes

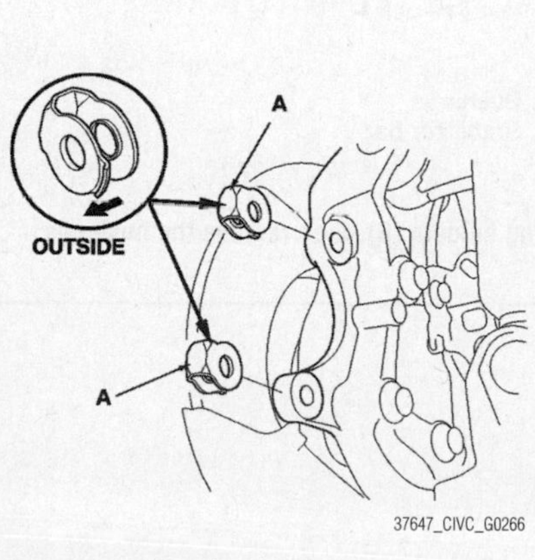
A
OUTSIDE
A
37647_CIVC_G0266

Fig. 142 Remove the two washers (A).

4. Compress the damper assembly by hand and check for smooth operation thru full stroke, both compression and extension.
5. The damper should extend smoothly and constantly when compression is released. If not, gas is leaking and the damper should be replaced.
6. Install all parts except the self locking nut onto the damper unit.
7. Align the bottom of the spring and the stepped part of the lower spring seat. The hole in the upper spring seat and the arrow on the damper mounting base must point toward the knuckle mounting area.
8. Install the damper assembly on a commercially available strut spring compressor.
9. Compress the damper spring with the strut spring compressor. Install a new self locking nut on the damper shaft.
10. Hold the damper shaft with a hex wrench and tighten the self locking nut.

STABILIZER BAR

REMOVAL & INSTALLATION

See Figure 144.

1. Raise and support the vehicle.
2. Remove the rear wheels.

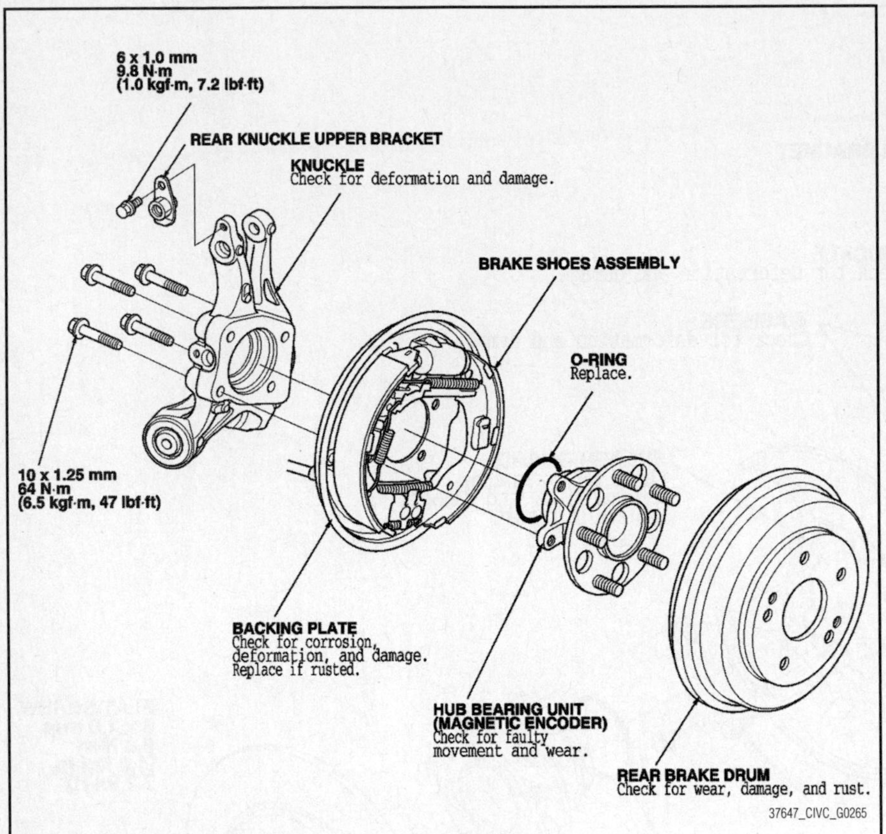
6 x 1.0 mm
9.8 N·m
(1.0 kgf·m, 7.2 lbf·ft)
REAR KNUCKLE UPPER BRACKET
KNUCKLE
Check for deformation and damage.
BRAKE SHOES ASSEMBLY
O-RING
Replace.
10 x 1.25 mm
64 N·m
(6.5 kgf·m, 47 lbf·ft)
BACKING PLATE
Check for corrosion, deformation, and damage.
Replace if rusted.
HUB BEARING UNIT
(MAGNETIC ENCODER)
Check for faulty movement and wear.
REAR BRAKE DRUM
Check for wear, damage, and rust.
37647_CIVC_G0265

Fig. 143 Exploded view of the knuckle and hub bearing assembly—with rear drum brakes

3. Disconnect both stabilizer links from the stabilizer bar.

4. Remove the flange bolts and the bushing holders, then remove the bushings and the stabilizer bar.

To install:

➡During installation, align the paint marks on the stabilizer bar with the sides of the bushings.

5. Replace the stabilizer bar bracket, if necessary. Tighten the bracket bolts to 29 ft. lbs. (39 Nm).

6. Install the stabilizer bar in the reverse order of removal.

7. Tighten the stabilizer bar bolts as follows:

- Except Si: 16 ft. lbs. (22 Nm)
- Si: 36 ft. lbs. (49 Nm)

➡Note the right and left direction of the stabilizer bar.

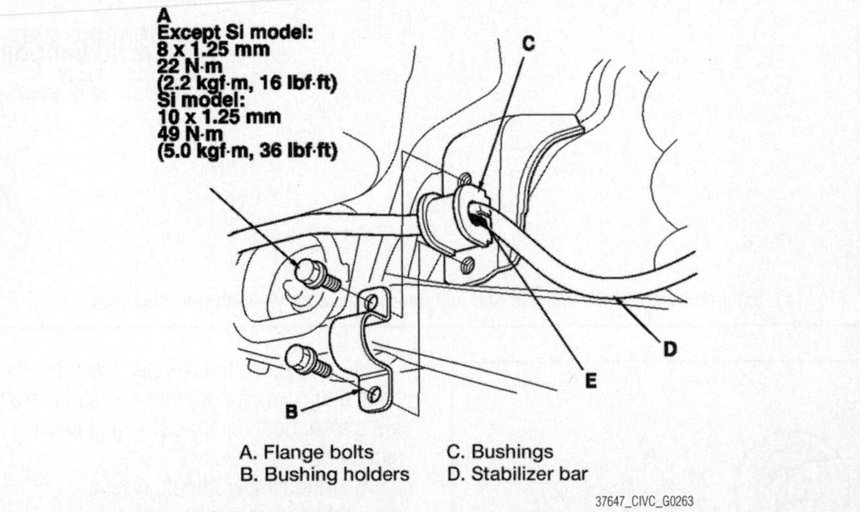
A
Except Si model:
8 x 1.25 mm
22 N·m
(2.2 kgf·m, 16 lbf·ft)
Si model:
10 x 1.25 mm
49 N·m
(5.0 kgf·m, 36 lbf·ft)
C
D
E
B
A. Flange bolts
B. Bushing holders
C. Bushings
D. Stabilizer bar
37647_CIVC_G0263

Fig. 144 Remove the flange bolts (A) and the bushing holders (B), then remove the bushings (C) and the stabilizer bar (D).

8. Before installing the wheel, clean the mating surfaces on the brake disc or the brake drum and inside of the wheel.

UPPER CONTROL ARM

REMOVAL & INSTALLATION

1. Raise and support the vehicle safely.
2. Remove the rear tires.
3. Place a jack under the trailing arm and the knuckle.
4. Remove the upper arm mounting bolts and the flange bolt. Remove the upper arm from the vehicle.

To install:

5. Position the upper arm on the vehicle. Be sure to use new retaining bolts and nuts.
6. Install all the suspension components and lightly tighten the bolts and nuts. Position a jack under the trailing arm and raise the suspension to load it with the vehicles weight, before fully tightening the bolts and nuts.
7. Continue the installation in the reverse order of the removal procedure.
8. Check and adjust the wheel alignment, as required.

HONDA

Civic Hybrid

10

BRAKES 10-8

ANTI-LOCK BRAKE SYSTEM (ABS) 10-9
Wheel Speed Sensors 10-9
Fluid Fill Procedure 10-9
Removal & Installation 10-9
BLEEDING THE BRAKE SYSTEM 10-8
Bleeding Procedure 10-8
Bleeding Procedure 10-8
Bleeding the ABS System 10-8
FRONT DISC BRAKES 10-10
Brake Caliper 10-10
Removal & Installation 10-10
Disc Brake Pads 10-10
Removal & Installation 10-10
INFORMATION AND PRECAUTIONS 10-8
Anti-lock Systems 10-8
Disc and Drum Systems 10-8
PARKING BRAKE 10-13
Parking Brake Cables 10-13
Adjustment 10-13
Parking Brake Shoes 10-14
Adjustment 10-14
Removal & Installation 10-14
REAR DISC BRAKES 10-11
Brake Caliper 10-11
Removal & Installation 10-11
Disc Brake Pads 10-12
Removal & Installation 10-12
REAR DRUM BRAKES 10-12
Brake Drum 10-12
Removal & Installation 10-12
Brake Shoes 10-12
Removal & Installation 10-12

CHASSIS ELECTRICAL 10-14

AIR BAG (SUPPLEMENTAL RESTRAINT SYSTEM) 10-14
General Information 10-14
Arming the System 10-14
Disarming the System 10-14
Precautions 10-14

DRIVE TRAIN 10-15

Driveshaft 10-15
CV-boots Inspection 10-15
Removal & Installation 10-15

ENGINE COOLING 10-16

Engine Fan 10-16
Removal & Installation 10-16
Hoses 10-16
Removal & Installation 10-16
Radiator & Fan 10-16
Removal & Installation 10-16
Thermostat 10-16
Removal & Installation 10-16
Water Pump 10-16
Removal & Installation 10-16

ENGINE ELECTRICAL 10-17

IGNITION SYSTEM 10-17
Firing Order 10-17
Idle Speed 10-17
Inspection 10-17
Ignition Coil 10-17
Removal & Installation 10-17
Ignition Timing 10-17
Adjustment 10-17
Inspection 10-17
STARTING SYSTEM 10-18
Spark Plug Wires 10-17
Removal & Installation 10-17
Spark Plugs 10-17
Removal & Installation 10-17
Starter 10-18
Removal & Installation 10-17

ENGINE MECHANICAL 10-18

Accessory Drive Belts 10-18
Accessory Belt Routing 10-18
Adjustment 10-18
Inspection 10-18
Removal & Installation 10-18
Camshaft 10-18
Inspection 10-18
Removal & Installation 10-19
Catalytic Converter 10-19
Removal & Installation 10-19
Crankshaft Damper (Pulley) 10-19
Removal & Installation 10-19
Cylinder Head 10-21
Removal & Installation 10-21
Exhaust Manifold 10-22
Removal & Installation 10-22
Flywheel 10-23
Removal & Installation 10-23
Intake Manifold 10-23
Removal & Installation 10-23
Oil Pan 10-23
Removal & Installation 10-23
Oil Pump 10-24
Inspection 10-24
Removal & Installation 10-23
Piston and Ring Positioning 10-24
Rear Main Seal 10-24
Removal & Installation 10-24
Rocker Arms 10-25
Removal & Installation 10-25
Timing Chain & Sprockets 10-25
Removal & Installation 10-25
Valve Cover (Cylinder Head Cover) 10-25
Removal & Installation 10-25
Valve Lash 10-26
Adjustment 10-26

ENGINE PERFORMANCE & EMISSION CONTROLS 10-26

Accelerator Pedal Position (APP) Sensor 10-26
Location 10-26
Removal & Installation 10-26
Camshaft Position (CMP) Sensor 10-26
Location 10-26
Removal & Installation 10-26
Crankshaft Position (CKP) Sensor 10-26
CKP Pattern Clear/CKP Pattern Lean Procedure 10-26
Location 10-26
Removal & Installation 10-26

Engine Coolant Temperature (ECT) Sensor....10-27
Location....10-27
Removal & Installation....10-27
Evaporative Emissions (EVAP) Canister....10-27
Location....10-27
Removal & Installation....10-27
Exhaust Gas Recirculation (EGR) Valve....10-28
Location....10-28
Removal & Installation....10-28
Intake Air Temperature (IAT) Sensor....10-28
Location....10-28
Knock Sensor (KS)....10-28
Location....10-28
Removal & Installation....10-28
Malfunction Indicator Light (MIL)....10-28
Reset Procedure....10-28
Manifold Absolute Pressure (MAP) Sensor....10-29
Location....10-29
Removal & Installation....10-29
Mass Air Flow/Intake Air Temperature (MAF/IAT) Sensor....10-28
Location....10-28
Removal & Installation....10-28
Positive Crankcase Ventilation (PCV) Valve....10-29
Location....10-29
Removal & Installation....10-29
Powertrain Control Module (PCM)....10-29
Location....10-29
PCM Idle Learn Procedure....10-30
Removal & Installation....10-29
Throttle Position (TP) Sensor....10-30
Location....10-30
Removal & Installation....10-30
Vehicle Speed Sensor (VSS)....10-30
Location....10-30
Removal & Installation....10-30

FUEL....10-31

GASOLINE FUEL INJECTION SYSTEM....10-31
Fuel Filter....10-31
Removal & Installation....10-31
Fuel Pump/ Fuel Gauge Sending Unit....10-31
Removal & Installation....10-31
Fuel Rail and Injector....10-31
Removal & Installation....10-31
Fuel System Pressure....10-31
Relieving....10-31
Fuel System Service Precautions....10-31
Fuel Tank....10-32
Draining....10-32
Removal & Installation....10-32
Idle Speed....10-32
Inspection & Adjustment....10-32
Throttle Body....10-33
Removal & Installation....10-33

HEATING & AIR CONDITIONING SYSTEM....10-34
Blower Motor....10-34
Removal & Installation....10-34
Heater Unit & Core....10-34
Removal & Installation....10-34

SPECIFICATIONS AND MAINTENANCE CHARTS....10-3
Brake Specifications....10-7
Camshaft and Bearing Specifications....10-5
Capacities....10-4
Crankshaft and Connecting Rod Specifications....10-5
Engine and Vehicle Identification....10-3
Engine Tune-Up Specifications....10-3
Fluid Specifications....10-4
General Engine Specifications....10-3
Piston and Ring Specifications....10-5
Maintenance Minder Schedule....10-7
Tire, Wheel and Ball Joint Specifications....10-6
Torque Specifications....10-6
Valve Specifications....10-4
Wheel Alignment....10-6

STEERING....10-35
Electronic Power Steering (EPS) Control Unit....10-35
Removal & Installation....10-35
Electronic Power Steering (EPS) Motor....10-35
Removal & Installation....10-35
Power Rack & Pinion Steering Gear....10-36
Removal & Installation....10-36
Torque Sensor Neutral Position Memorization....10-38

SUSPENSION....10-39

FRONT SUSPENSION....10-39
Lower Ball Joint....10-39
Removal & Installation....10-39
Lower Control Arm....10-39
Removal & Installation....10-39
MacPherson Strut....10-39
Overhaul....10-39
Removal & Installation....10-39
Stabilizer Bar....10-40
Removal & Installation....10-40
Steering Knuckle....10-40
Removal & Installation....10-40
Wheel Bearings....10-40
Adjustment....10-40
Removal & Installation....10-40

REAR SUSPENSION....10-41
Knuckle/Hub Bearing....10-41
Removal & Installation....10-41
MacPherson Struts....10-43
Overhaul....10-43
Removal & Installation....10-43
Stabilizer Bar....10-43
Removal & Installation....10-43
Upper Control Arm....10-43
Removal & Installation....10-43

SPECIFICATIONS AND MAINTENANCE CHARTS

ENGINE AND VEHICLE IDENTIFICATION

Engine						Model Year	
Code	Liters	Cu. In. (cc)	Cyl.	Fuel Sys.	Eng. Mfg.	Code	Year
LDA2	1.3	82.0 (1339)	4	PGM-FI	Honda	9	2009
						A	2010

PGM-FI: Programmed Fuel Injection

37647_CHYB_C0001

GENERAL ENGINE SPECIFICATIONS

Year	Model	Engine Displacement Liters	Engine ID/VIN	Net Horsepower @ rpm	Net Torque @ rpm (ft. lbs.)	Bore X Stroke (in.)	Com-pression Ratio	Oil Pressure @ rpm
2009	Civic Hybrid	1.3	LDA2	110@6000	123@2500	2.87x3.15	10.8:1	50@3000
2010	Civic Hybrid	1.3	LDA2	110@6000	123@2500	2.87x3.15	10.8:1	50@3000

37647_CHYB_C0002

ENGINE TUNE-UP SPECIFICATIONS

Year	Engine Displacement Liters	Engine ID/VIN	Spark Plugs Gap (in.)	Ignition Timing (deg.)		Fuel Pump (psi)	Idle Speed (rpm)		Valve Clearance	
				MT	AT		MT	AT	In.	Ex.
2009	1.3	LDA2	0.039-0.043	—	8-12B	38-46	—	770-870	0.006-0.007	0.009-0.011
2010	1.3	LDA2	0.039-0.043	—	8-12B	38-46	—	770-870	0.006-0.007	0.009-0.011

NOTE: The Vehicle Emission Control Information label often reflects specification changes made during production.

The label figures must be used if they differ from those in this chart

B: Before Top Dead Center

37647_CHYB_C0003

CAPACITIES

Year	Model	Engine Displacement Liters	Engine ID	Engine Oil with Filter	Transmission (pts.) 5-Spd	Transmission (pts.) Auto.	Drive Axle Front (pts.)	Drive Axle Rear (pts.)	Fuel Tank (gal.)	Cooling System (qts.)
2009	Civic Hybrid	1.3	LDA2	3.4	3.2	6.0	—	—	12.3	5.0
2010	Civic Hybrid	1.3	LDA2	3.4	3.2	6.0	—	—	12.3	5.0

NOTE: All capacities are approximate. Add fluid gradually and ensure a proper fluid level is obtained.

NOTE: Capacities given are service, not overhaul capacities

37647_CHYB_C0004

FLUID SPECIFICATIONS

Year	Model	Engine Displ. Liters	Engine Oil	Auto. Trans.	Power Steering Fluid	Brake Master Cylinder	Cooling System
2009	Civic Hybrid	1.3	0W-20 Honda	Honda CVTF	NA	Honda DOT 3	①
2010	Civic Hybrid	1.3	0W-20 Honda	Honda CVTF	NA	Honda DOT 3	①

DOT: Department Of Transpotation

① Honda Long Life Antifreeze/Coolant-Type2

37647_CHYB_C0005

VALVE SPECIFICATIONS

Year	Engine Displacement Liters	Engine ID/VIN	Seat Angle (deg.)	Face Angle (deg.)	Spring Test Pressure (lbs. @ in.)	Spring Installed Height (in.)	Stem-to-Guide Clearance (in.) Intake	Stem-to-Guide Clearance (in.) Exhaust	Stem Diameter (in.) Intake	Stem Diameter (in.) Exhaust
2009	1.3	LDA2	45	45	NA	NA	0.0008-0.0020	0.0020-0.0031	0.2157-0.2161	0.2146-0.2150
2010	1.3	LDA2	45	45	NA	NA	0.0008-0.0020	0.0020-0.0031	0.2157-0.2161	0.2146-0.2150

NA: Information not available

37647_CHYB_C0007

CAMSHAFT AND BEARING SPECIFICATIONS

All measurements are given in inches.

Year	Engine Displacement Liters	Engine VIN	Journal Diameter	Brg. Oil Clearance	Shaft End-play	Runout	Journal Bore	Lobe Height	
								Intake	Exhaust
2009	1.3	LDA2	NA	0.0020-0.0035	0.0020-0.0060	0.0010	NA	①	②
2010	1.3	LDA2	NA	0.0020-0.0035	0.0020-0.0060	0.0010	NA	①	②

NA: Information not available

① 1st: 1.1692 in.
2nd: 1.4003 in.
3rd: 1.4196 in.

② 1st: 1.1771 in.
2nd: 1.4054 in.

37647_CHYB_C0006

CRANKSHAFT AND CONNECTING ROD SPECIFICATIONS

All measurements are given in inches.

Year	Engine Displacement Liters	Engine ID/VIN	Crankshaft				Connecting Rod		
			Main Brg. Journal Dia.	Main Brg. Oil Clearance	Shaft End-play	Thrust on No.	Journal Diameter	Oil Clearance	Side Clearance
2009	1.3	LDA2	1.9676-1.9685	0.0007-0.0014	0.0040-0.0140	4	1.5739-1.5748	0.0008-0.0015	0.0060-0.0120
2010	1.3	LDA2	1.9676-1.9685	0.0007-0.0014	0.0040-0.0140	4	1.5739-1.5748	0.0008-0.0015	0.0060-0.0120

37647_CHYB_C0008

PISTON AND RING SPECIFICATIONS

All measurements are given in inches.

Year	Engine Disp. Liters	Engine ID/VIN	Piston Clearance	Ring Gap			Ring Side Clearance		
				Top Compression	Bottom Compression	Oil Control	Top Compression	Bottom Compression	Oil Control
2009	1.3	LDA2	0.0004-0.0016	0.0060-0.0120	0.0140-0.0200	0.0080-0.0280	0.0025-0.0035	0.0012-0.0022	NA
2010	1.3	LDA2	0.0004-0.0016	0.0060-0.0120	0.0140-0.0200	0.0080-0.0280	0.0025-0.0035	0.0012-0.0022	NA

NA: Information not available

37647_CHYB_C0009

TORQUE SPECIFICATIONS

All readings in ft. lbs.

Year	Engine Displacement Liters	Engine ID/VIN	Cylinder Head Bolts	Main Bearing Bolts	Rod Bearing Bolts	Crankshaft Damper Bolts	Flywheel Bolts	Manifold Intake	Manifold Exhaust	Spark Plugs	Oil Pan Drain Plug
2009	1.3	LDA2	①	②	③	④	33	17	16	13	29
2010	1.3	LDA2	①	②	③	④	33	17	16	13	29

① Step 1: 22 ft. lbs.
Step 2: Rotate 130 degrees
② Step 1: 18 ft. lbs.
Step 2: Rotate 40 degrees
③ Step 1: 7.2 ft. lbs.
Step 2: Rotate 90 degrees
④ Old bolt:
Step 1: 27 ft. lbs
Step 2: Plus 90 degrees
New bolt:
Step 1: 130 ft. lbs
Step 2: Loosen fully
Step 3: 27 ft. lbs
Step 4: Plus 90 degrees

37647_CHYB_C0010

WHEEL ALIGNMENT

Year	Model		Caster Range (+/-Deg.)	Caster Preferred Setting (Deg.)	Camber Range (+/-Deg.)	Camber Preferred Setting (Deg.)	Toe-in (in.)
2009	Civic	F	1.00	+7.10	0.50	0.05	0.00+/-0.08
	Hybrid	R	—	—	0.183	-0.90	0.00+/-0.08
2010	Civic	F	1.00	+7.10	0.50	0.05	0.00+/-0.08
	Hybrid	R	—	—	0.183	-0.90	0.00+/-0.08

37647_CHYB_C0011

TIRE, WHEEL AND BALL JOINT SPECIFICATIONS

Year	Model	OEM Tires Standard	OEM Tires Optional	Tire Pressures (psi) Front	Tire Pressures (psi) Rear	Wheel Size	Ball Joint Inspection	Lug Nut (ft. lbs.)
2009	Civic Hybrid	195/65R15	None	32	32	NA	NA	80
2010	Civic Hybrid	195/65R15	None	32	32	NA	NA	80

OEM: Original Equipment Manufacturer
PSI: Pounds Per Square Inch
NA: Information not available

37647_CHYB_C0012

BRAKE SPECIFICATIONS

All measurements in inches unless noted

			Brake Disc			Brake Drum Diameter			Minimum Pad Thickness		Brake Caliper	
Year	Model		Original Thickness	Minimum Thickness	Maximum Runout	Original Inside Diameter	Max. Wear Limit	Max. Machine Diameter	Front	Rear	Bracket Bolts (ft. lbs.)	Mounting Bolts (ft. lbs.)
2009	Civic	F	0.82-0.83	0.750	0.002	NA	NA	NA	0.37-0.41	—	80	17
	Hybrid	R	0.35-0.36	0.310	0.002	NA	NA	NA	—	0.33-0.37	54	17
2010	Civic	F	0.82-0.83	0.750	0.002	NA	NA	NA	0.37-0.41	—	80	17
	Hybrid	R	0.35-0.36	0.310	0.002	NA	NA	NA	—	0.33-0.37	54	17

NA: Information not available

F: Front

R: Rear

37647_CHYB_C0013

MAINTENANCE MINDER SCHEDULE

Honda—Civic Hybrid

All Honda's displays engine oil life and maintenance service items in the information display to indicate when to perform maintenance service. If the engine oil life is 15% or less, based on the onboard computer's caluculations, you will see SERVICE DUE SOON in the information display every time the ignition key is turned to ON. The maintenance minder indicator will also come on and the maintenance code(s) for other scheduled maintenance items needing service will be displayed below the message.

The maintenance minder is an important feature of the information display. Based on engine and transmission operating conditions, and accumulated engine revolutions, the Civic Hybrid's onboard computer (PCM) calculates the remaining engine oil and the CVT fluid life. The system also displays the remaining engine oil life along with the code(s) for other scheduled maintenance items needing service.

Symbol	Item	Service
A	Engine oil ①	Change
B	Engine oil and filter	Change
	Fluid levels	Inspect
	Brakes	Inspect
	Parking brake adjustment	Check
	Steering gear and linkage	Inspect
	Suspension components	Inspect
	Driveshaft boots	Inspect
	Brake hoses and lines	Inspect
	Exhaust system	Inspect
	Fuel lines and connections	Inspect
1	Tires	Rotate
2	Engine air filter ②	Replace
	Dust and pollen filter ③	Replace
	Accessory drive belt	Inspect
3	CVT fluid	Replace
4	Spark plugs	Replace
	Valve clearance ④	Inspect
5	Engine coolant	Replace

① If the message SERVICE DUE NOW does not appear more than 12 months after the display is reset, change every year.

② If driven in dusty conditions, replace every 15,000 miles.

③ If driven in urban areas that have a high concentration of dust, pollen or soot, replace every 15,000 miles.

④ Adjust if necessary, otherwise, adjust valves during service only if they are noisy.

Additionally, replace the brake fluid every 3 years, and inspect the idle speed every 160,000 miles.

37647_CHYB_C0014

BRAKES — INFORMATION AND PRECAUTIONS

ANTI-LOCK SYSTEMS

- Certain components within the ABS system are not intended to be serviced or repaired individually.
- Do not use rubber hoses or other parts not specifically specified for and ABS system. When using repair kits, replace all parts included in the kit. Partial or incorrect repair may lead to functional problems and require the replacement of components.
- Lubricate rubber parts with clean, fresh brake fluid to ease assembly. Do not use shop air to clean parts; damage to rubber components may result.
- Use only DOT 3 brake fluid from an unopened container.
- If any hydraulic component or line is removed or replaced, it may be necessary to bleed the entire system.
- A clean repair area is essential. Always clean the reservoir and cap thoroughly before removing the cap. The slightest amount of dirt in the fluid may plug an orifice and impair the system function. Perform repairs after components have been thoroughly cleaned; use only denatured alcohol to clean components. Do not allow ABS components to come into contact with any substance containing mineral oil; this includes used shop rags.
- The Anti-Lock control unit is a microprocessor similar to other computer units in the vehicle. Ensure that the ignition switch is **OFF** before removing or installing controller harnesses. Avoid static electricity discharge at or near the controller.
- If any arc welding is to be done on the vehicle, the control unit should be unplugged before welding operations begin.

DISC AND DRUM SYSTEMS

CAUTION

Dust and dirt accumulating on brake parts during normal use may contain asbestos fibers from production or aftermarket brake linings. Breathing excessive concentrations of asbestos fibers can cause serious bodily harm. Exercise care when servicing brake parts. Do not sand or grind brake lining unless equipment used is designed to contain the dust residue. Do not clean brake parts with compressed air or by dry brushing. Cleaning should be done by dampening the brake components with a fine mist of water, then wiping the brake components clean with a dampened cloth. Dispose of cloth and all residue containing asbestos fibers in an impermeable container with the appropriate label. Follow practices prescribed by the Occupational Safety and Health Administration (OSHA) and the Environmental Protection Agency (EPA) for the handling, processing, and disposing of dust or debris that may contain asbestos fibers.

BRAKES — BLEEDING THE BRAKE SYSTEM

BLEEDING PROCEDURE

BLEEDING PROCEDURE

➡See "Bleeding the ABS System" in this section.

BLEEDING THE ABS SYSTEM

See Figure 1.

➡Do not reuse the drained fluid. Use only new Honda DOT 3 Brake Fluid from an unopened container. Using a non-Honda brake fluid can cause corrosion and shorten the life of the system.

➡Make sure no dirt or other foreign matter gets in the brake fluid.

1. The reservoir connected to the master cylinder must be at the MAX (upper) level mark at the start of the bleeding procedure, and checked after bleeding each wheel location. Add fluid as required.
2. Make sure the brake fluid level in the reservoir is at the MAX (upper) level line.
3. Have someone slowly pump the brake pedal several times, then apply steady pressure.
4. Start the bleeding at the driver's side of the front brake system. Bleed the calipers or the wheel cylinders in the sequence shown.
5. Attach a length of clear drain tube to the bleed screw, then loosen the bleed screw to allow air to escape from the system. Then tighten the bleed screw securely.
6. Refill the master cylinder reservoir to the MAX (upper) level line.
7. Repeat the procedure for each brake circuit until there are no air bubbles in the fluid.

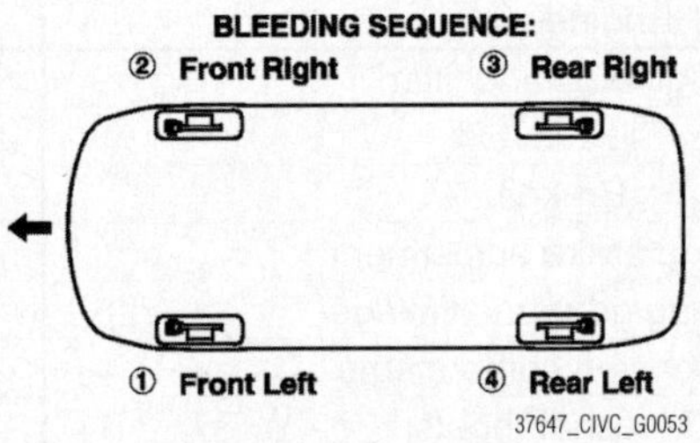
BLEEDING SEQUENCE:
② Front Right
③ Rear Right
① Front Left
④ Rear Left
37647_CIVC_G0053

Fig. 1 Start the bleeding at the driver's side of the front brake system. Bleed the calipers or the wheel cylinders in the sequence shown.

WHEEL SPEED SENSORS

REMOVAL & INSTALLATION

See Figures 2 and 3.

➡Required: guide pin tool (07AAG-SVBA100, or equivalent).

1. Turn the ignition switch to LOCK (0).
2. Release the clamp, then disconnect the wheel speed sensor connector.
3. Remove the clips, the bolt, and the wheel speed sensor.

To install:

4. Install the wheel speed sensor in the reverse order of removal, and note these items:
5. Do not twist the sensor wires.
6. If the wheel speed sensor comes in contact with the wheel bearing, it is faulty.
7. Make sure there is no debris in the sensor mounting hole.

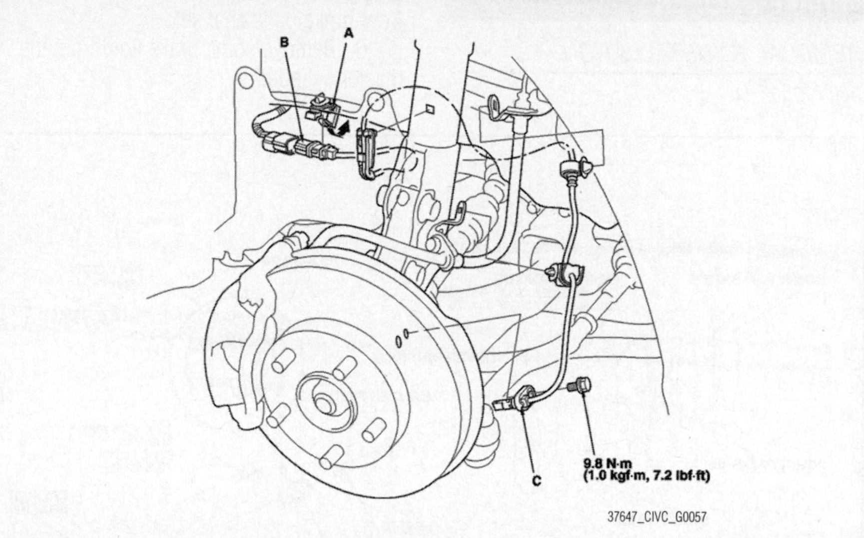
B
A
9.8 N·m
(1.0 kgf·m, 7.2 lbf·ft)
C
37647_CIVC_G0057

Fig. 2 Removing the front wheel speed sensor: Release the clamp (A), then disconnect the wheel speed sensor connector (B), then remove the clips, the bolt, and the wheel speed sensor (C).—front wheel

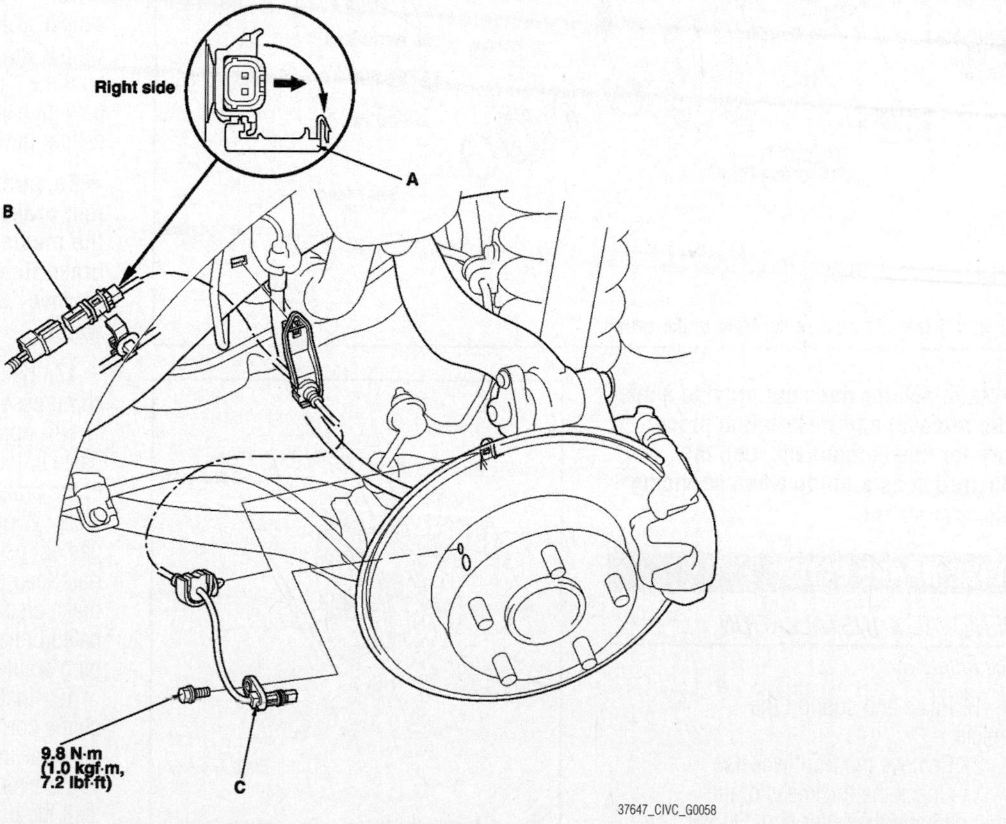
Right side
A
B
9.8 N·m
(1.0 kgf·m,
7.2 lbf·ft)
C
37647_CIVC_G0058

Fig. 3 Removing the front wheel speed sensor: Release the clamp (A), then disconnect the wheel speed sensor connector (B), then remove the clips, the bolt, and the wheel speed sensor (C).—rear wheel

8. Start the engine, and make sure the ABS or VSA indicator goes off.
9. Test-drive the vehicle, and make sure the ABS or VSA indicator does not come on.

FLUID FILL PROCEDURE

➡See "Bleeding the ABS System" in this section.

BRAKE CALIPER

REMOVAL & INSTALLATION

See Figure 4.

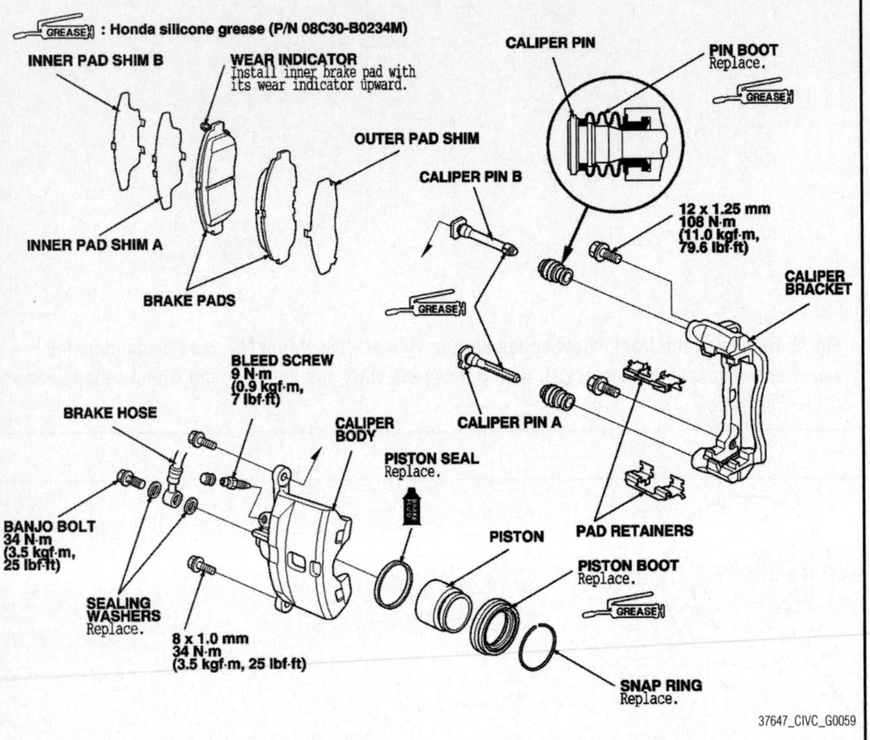
GREASE : Honda silicone grease (P/N 08C30-B0234M)
INNER PAD SHIM B
WEAR INDICATOR
Install inner brake pad with its wear indicator upward.
CALIPER PIN
PIN BOOT
Replace.
GREASE
OUTER PAD SHIM
CALIPER PIN B
12 x 1.25 mm
108 N·m
(11.0 kgf·m,
79.6 lbf·ft)
INNER PAD SHIM A
CALIPER BRACKET
BRAKE PADS
GREASE
BLEED SCREW
9 N·m
(0.9 kgf·m,
7 lbf·ft)
BRAKE HOSE
CALIPER BODY
CALIPER PIN A
PISTON SEAL
Replace.
BANJO BOLT
34 N·m
(3.5 kgf·m,
25 lbf·ft)
PISTON
PAD RETAINERS
PISTON BOOT
Replace.
GREASE
SEALING WASHERS
Replace.
8 x 1.0 mm
34 N·m
(3.5 kgf·m, 25 lbf·ft)
SNAP RING
Replace.
37647_CIVC_G0059

Fig. 4 Exploded view of the front brake caliper

➡Manufacturer does not provide a specific removal and installation procedure for this component. Use the illustration as a guide when servicing this component.

DISC BRAKE PADS

REMOVAL & INSTALLATION

See Figure 5.

1. Raise and support the vehicle.
2. Remove the front wheels.
3. Check the thickness of the inner pad and the outer pad. Do not include the thickness of the backing plate.
4. Pad thickness should not be less than a service limit of 0.06 in. (1.6 mm).
5. If any part of the brake pad thickness is less than the service limit, replace the front brake pads as a set.
6. Remove some brake fluid from the master cylinder.

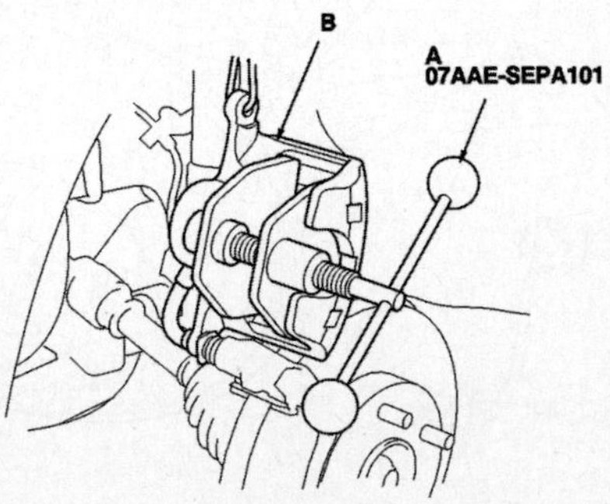
B
A
07AAE-SEPA101
37647_CIVC_G0063

Fig. 5 Install the brake caliper piston compressor tool (A) on the caliper body (B).

7. Remove the flange bolt, and pivot the caliper up out of the way. Check the hose and pin boots for damage and deterioration.
8. Remove the pad shims and the brake pads.
9. Remove the pad retainers.

To install:

10. Clean the caliper bracket thoroughly; remove any rust, and check for grooves and cracks.
11. Verify that the caliper pins move in and out smoothly. Clean and lube if needed.
12. Inspect the brake disc for runout, thickness, parallelism, and check for damage and cracks. See "Brakes" specification chart in this section.
13. Apply a thin coat of M-77 assembly paste to the retainer mating surface of the caliper bracket (indicated by the arrows).
14. Install the pad retainers. Wipe excess assembly paste off the retainers. Keep the assembly paste off the brake disc and the brake pads.
15. Install the brake caliper piston compressor tool on the caliper body.
16. Press in the piston with the brake caliper piston compressor tool so the caliper will fit over the brake pads. Make sure the piston boot is in position to prevent damaging it when pivoting the caliper down.

➡Be careful when pressing in the piston; brake fluid might overflow from the master cylinder's reservoir. If brake fluid gets on any painted surface, wash it off immediately with water.

17. Remove the brake caliper piston compressor tool.
18. Apply a thin coat of M-77 assembly paste to the pad side of the shims, the back of the brake pads and the other contact areas. Wipe excess assembly paste off the pad shims and the brake pads friction material. Keep grease and assembly paste off the brake disc and the brake pads. Contaminated brake disc or brake pads reduce stopping ability.
19. Install the brake pads and the pad shims correctly. Install the brake pad with the wear indicator on the upper inside. If you are reusing the brake pads, always reinstall the brake pads in their original positions to prevent a temporary loss of braking efficiency.
20. Pivot the caliper down into position. Install the flange bolt and tighten it to the specified torque of 25 ft. lbs. (34 Nm).

21. Clean the mating surfaces between the brake disc and the inside of the wheel, then install the front wheels.

22. Press the brake pedal several times to make sure the brakes work.

➡**Engagement may require a greater pedal stroke immediately after the brake pads have been replaced as a set. Several applications of the brake pedal will restore the normal pedal stroke.**

23. Add brake fluid as needed.

24. After installation, check for leaks at hose and line joints or connections, and retighten if necessary.

25. Test-drive the vehicle, then recheck for leaks.

BRAKES

REAR DISC BRAKES

BRAKE CALIPER

REMOVAL & INSTALLATION

See Figure 6.

➡**Keep any grease off the brake disc and the brake pads.**

1. Raise and support the vehicle.
2. Remove the rear wheel.
3. Release the parking brake lever fully.
4. Remove the brake hose mounting bolt from the bracket.
5. Remove the brake caliper bracket mounting bolts, then remove the caliper assembly from the knuckle. To prevent damage to the caliper assembly or brake hose, use a short piece of wire to hang the caliper assembly from the undercarriage. Do

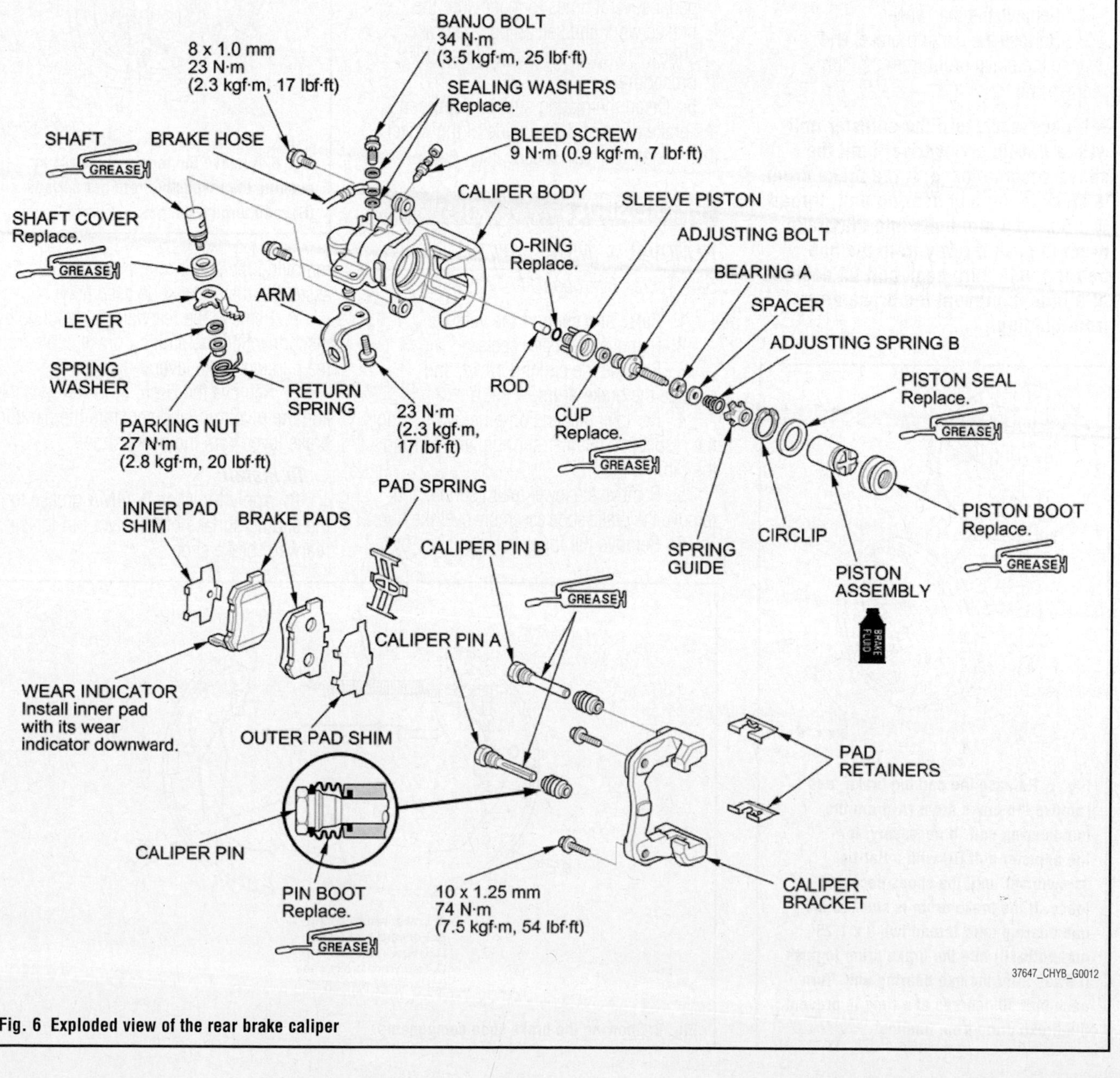
BANJO BOLT
34 N·m
(3.5 kgf·m, 25 lbf·ft)
8 x 1.0 mm
23 N·m
(2.3 kgf·m, 17 lbf·ft)
SEALING WASHERS
Replace.
SHAFT
BRAKE HOSE
BLEED SCREW
9 N·m (0.9 kgf·m, 7 lbf·ft)
GREASE
CALIPER BODY
SLEEVE PISTON
SHAFT COVER
Replace.
ADJUSTING BOLT
O-RING
Replace.
BEARING A
GREASE
ARM
LEVER
SPACER
ADJUSTING SPRING B
SPRING
WASHER
RETURN
SPRING
ROD
PISTON SEAL
Replace.
23 N·m
(2.3 kgf·m,
17 lbf·ft)
CUP
Replace.
GREASE
PARKING NUT
27 N·m
(2.8 kgf·m, 20 lbf·ft)
GREASE
PAD SPRING
INNER PAD
SHIM
BRAKE PADS
PISTON BOOT
Replace.
CIRCLIP
CALIPER PIN B
SPRING
GUIDE
PISTON
ASSEMBLY
GREASE
GREASE
BRAKE FLUID
CALIPER PIN A
WEAR INDICATOR
Install inner pad
with its wear
indicator downward.
OUTER PAD SHIM
PAD
RETAINERS
CALIPER PIN
CALIPER
BRACKET
PIN BOOT
Replace.
10 x 1.25 mm
74 N·m
(7.5 kgf·m, 54 lbf·ft)
GREASE
37647_CHYB_G0012

Fig. 6 Exploded view of the rear brake caliper

not twist the brake hose and the parking brake cable excessively.

➡Make sure the washers are in position on reassembly, if they are removed.

6. Installation is the reverse of the removal procedure.

DISC BRAKE PADS

REMOVAL & INSTALLATION

1. Raise and support the vehicle.
2. Remove the rear wheels.
3. Check the thickness of the inner pad and the outer pad. Do not include the thickness of the backing plate.
4. Pad thickness should not be less than 0.06 in. (1.6 mm).
5. If any part of the brake pad thickness is less than the service limit, replace the rear brake pads as a set.

BRAKES

REAR DRUM BRAKES

BRAKE DRUM

REMOVAL & INSTALLATION

See Figure 7.

➡Keep any grease off the brake drum and the brake shoes.

1. Raise and support the vehicle.
2. Remove the rear wheel.
3. Release the parking brake, and remove the brake drum from the hub bearing unit.

➡If necessary, turn the adjuster bolt with a flat-tip screwdriver until the shoes become loose. If the brake drum is stuck to the hub bearing unit, thread two 8 x 1.25 mm bolts into the brake drum to push it away from the hub bearing unit. Turn each bolt 90 degrees at a time to prevent the brake drum from binding.

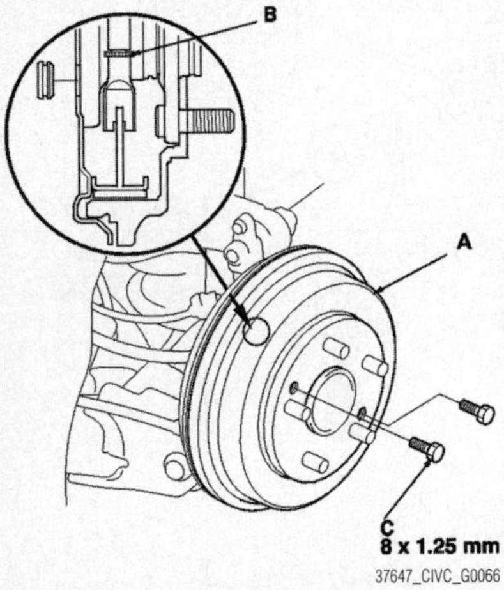
B
A
C
8 x 1.25 mm
37647_CIVC_G0066

Fig. 7 Release the parking brake, and remove the brake drum (A) from the hub bearing unit. If necessary, turn the adjuster bolt (B) with a flat-tip screwdriver until the shoes become loose. If the brake drum is stuck to the hub bearing unit, thread two 8 x 1.25 mm bolts (C) into the brake drum to push it away from the hub bearing unit. Turn each bolt 90 degrees at a time to prevent the brake drum from binding.

To install:

4. Install the brake drum in the reverse order of removal, noting the following:
   a. Before installing the brake drum, clean the mating surfaces between the hub bearing unit and the inside of the brake drum.
   b. After installation, press the brake pedal several times to make sure the brakes work and self adjust the brake shoes. Do not drive before doing this procedure.
5. Clean the mating surfaces between the brake drum and the inside of the wheel, then install the rear wheel.

BRAKE SHOES

REMOVAL & INSTALLATION

See Figures 8 through 11.

1. Raise and support the vehicle.
2. Remove the rear wheels.
3. Release the parking brake, and remove the brake drum.
4. Remove the tension pins by pushing the respective retainer springs, and turning the pin.
5. Remove the lower return spring, and remove the brake shoe assembly over the hub.
6. Remove the forward brake shoe by removing the upper return spring, and disassemble the brake shoe assembly.
7. Remove the rearward brake shoe by disconnecting the parking brake cable from the parking brake lever.
8. Remove the U-clip, the wave washer, and the pivot pin, and separate the parking brake lever from the brake shoe.

To install:

9. Apply Molykote®-44MA grease to the sliding surface of the pivot pin for the rearward brake shoe.

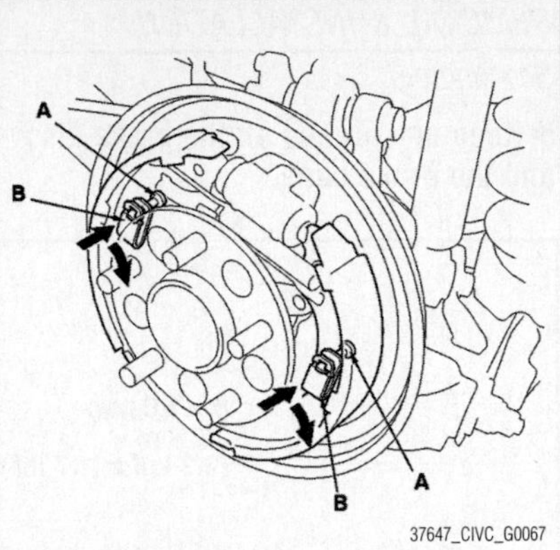
A
B
B
A
37647_CIVC_G0067

Fig. 8 Remove the tension pins (A) by pushing the respective retainer springs (B), and turning the pin.

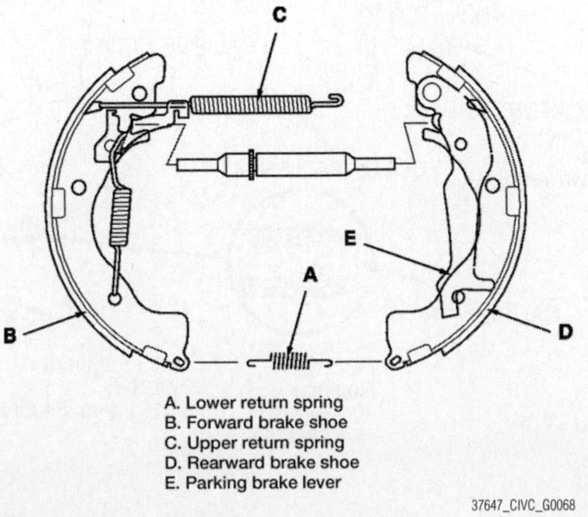
C
E
A
B
D
A. Lower return spring
B. Forward brake shoe
C. Upper return spring
D. Rearward brake shoe
E. Parking brake lever
37647_CIVC_G0068

Fig. 9 Showing the brake shoe components

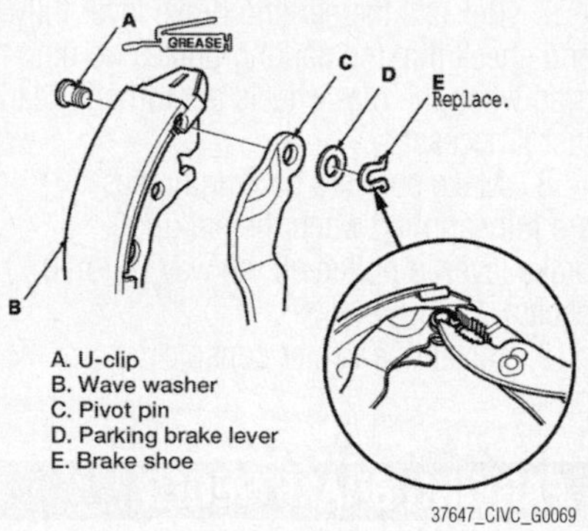
A
GREASE
C
D
E
Replace.
B
A. U-clip
B. Wave washer
C. Pivot pin
D. Parking brake lever
E. Brake shoe
37647_CIVC_G0069

Fig. 10 Removing hardware from the brake shoe

10. Install the parking brake lever and the wave washer on the pivot pin, and secure the pin with a new U-clip.

➡Pinch the U-clip securely to prevent the parking brake lever from coming out of the brake shoe.

11. Connect the parking brake cable to the parking brake lever.

12. Apply a thin coat of Molykote-44MA grease to the connecting rod ends, and the sliding surfaces. Wipe off any excess. Keep grease off the brake linings.

13. Apply a thin coat of Molykote-44MA grease to the shoe ends and to the edge of the shoe surfaces that make contact with the backing plate. Wipe off any excess. Keep grease off the brake linings.

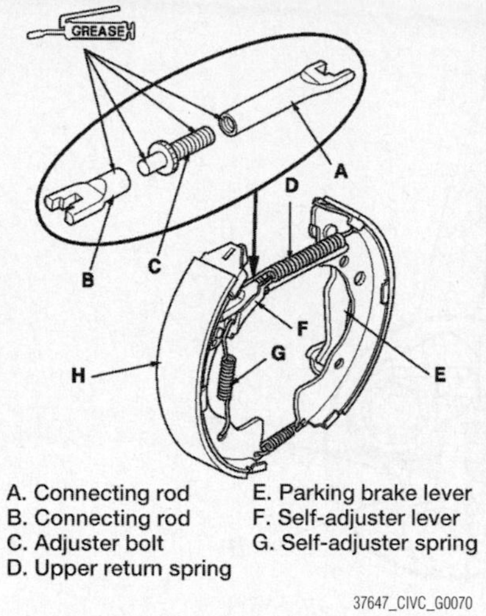
GREASE
A
D
B
C
F
G
H
E
A. Connecting rod
B. Connecting rod
C. Adjuster bolt
D. Upper return spring
E. Parking brake lever
F. Self-adjuster lever
G. Self-adjuster spring
37647_CIVC_G0070

Fig. 11 Assemble the brake shoes with the upper return spring, with the connecting rods and the adjuster bolt onto the backing plate. Reconnect the parking brake cable to the parking brake lever, then install the self-adjuster lever and the self-adjuster spring on the forward brake shoe.

14. Install the two connecting rods on the adjuster bolt.

15. Clean the threaded portions of longer connecting rod and the sliding surface of short connecting rod, then coat them with Molykote 44MA grease.

16. Shorten connecting rod by fully turning in the adjuster bolt.

17. Assemble the brake shoes with the upper return spring, with the connecting rods and the adjuster bolt onto the backing plate. Reconnect the parking brake cable to the parking brake lever, then install the self-adjuster lever and the self-adjuster spring on the forward brake shoe.

18. Install the tension pins and the retainer springs by pushing in the respective spring and turning each pin.

19. Install the lower return spring.

➡Make sure the brake shoes are positioned on the brake shoe bosses on the backing plate, and the fitting on the top of the brake shoes are fitted into the wheel cylinder pistons.

20. Install the brake drum.

➡Before installing the brake drum, clean the mating surfaces between the hub bearing unit and the inside of the brake drum.

21. Clean the mating surfaces between the brake drum and the inside of the wheel, then install the rear wheels.

22. Press the brake pedal several times to make sure the brakes work and to set the self-adjusting brake.

➡Engagement of the brakes may require a greater pedal stroke immediately after the brake shoes have been replaced as a set. Several applications of the brake pedal will restore the normal pedal stroke.

23. Adjust the parking brake.

BRAKES — PARKING BRAKE

PARKING BRAKE CABLES

ADJUSTMENT

Rear Disc Brake Type

See Figure 12.

1. Remove the center console panel.
2. Release the parking brake lever fully.
3. Loosen the parking brake adjusting nut at the lower end of the parking brake handle.
4. Raise and support the vehicle.
5. Remove the rear wheels.
6. Make sure the lever on the rear brake caliper contacts the stop pin.

➡The lever will only contact the stop pin when the parking brake adjusting nut is loosened.

7. Clean the mating surfaces between the brake disc and the inside of the wheel, then install the rear wheels.
8. Pull the parking brake lever 1 click.
9. Tighten the parking brake adjusting nut until the parking brakes drag slightly when the rear wheels are turned.
10. Release the parking brake lever fully, and check that the parking brakes do not drag when the rear wheels are turned. Readjust if necessary.
11. Make sure the parking brake lever is within the specified number of clicks (8 to 10 clicks).
12. Install the center console panel.

Adjustment - Rear Drum Brake Type

1. Remove the center console panel.
2. Release the parking brake lever fully.
3. Loosen the parking brake adjusting nut at the lower end of the parking brake handle.
4. Press the brake pedal several times to set the self-adjusting brake before adjusting the parking brake.
5. Pull the parking brake lever 1 click. Raise and support the vehicle.

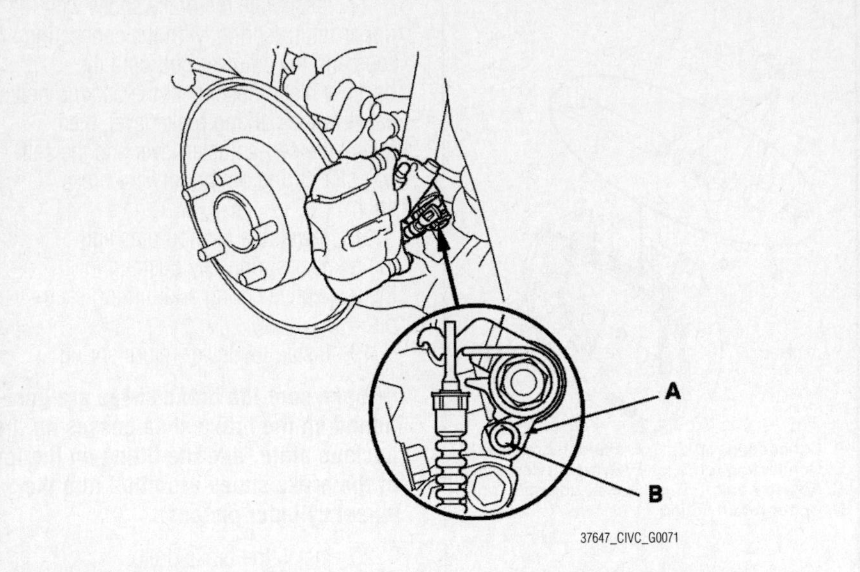
A
B
37647_CIVC_G0071

Fig. 12 Make sure the lever (A) on the rear brake caliper contacts the stop pin (B).

6. Tighten the parking brake adjusting nut until the parking brakes drag slightly when the rear wheels are turned.
7. Release the parking brake lever fully, and check that the parking brakes do not drag when the rear wheels are turned. Readjust if necessary.
8. Make sure the parking brakes are fully applied when the parking brake lever is pulled all the way (8 to 10 clicks).
9. Install the center console panel.

PARKING BRAKE SHOES

REMOVAL & INSTALLATION

➡See "Brake Shoes" in this section.

ADJUSTMENT

➡See "Brake Shoes" in this section.

CHASSIS ELECTRICAL

AIR BAG (SUPPLEMENTAL RESTRAINT SYSTEM)

GENERAL INFORMATION

⁂ CAUTION

These vehicles are equipped with an air bag system. The system must be disarmed before performing service on, or around, system components, the steering column, instrument panel components, wiring and sensors. Failure to follow the safety precautions and the disarming procedure could result in accidental air bag deployment, possible injury and unnecessary system repairs.

PRECAUTIONS

Disconnect and isolate the battery negative cable before beginning any airbag system component diagnosis, testing, removal, or installation procedures. Allow system capacitor to discharge for two minutes before beginning any component service. This will disable the airbag system. Failure to disable the airbag system may result in accidental airbag deployment, personal injury, or death.

DISARMING THE SYSTEM

➡Some systems store data in memory (including seat position, mirror position, etc) that is lost when the battery is disconnected. Do the following procedures before disconnecting the battery.

1. Make sure you have the anti-theft code(s) for the audio system and/or navigation system (if equipped).

➡For some models, it may be necessary to write down the audio the presets (XM, AM and FM), because the audio unit does not retain the presets after the battery is disconnected.

2. Make sure the ignition switch is in LOCK (0).
3. Disconnect and isolate the negative cable from the battery.

➡Always disconnect the negative battery cable from the battery first.

4. Disconnect the positive cable from the battery.
5. Wait at least 3 minutes before starting any further work. This allows the SRS system to fully discharge.

ARMING THE SYSTEM

➡Some systems store data in memory (including seat position, mirror position, etc) that is lost when the battery is disconnected. Do the following procedures to restore the system back to normal operation.

1. Clean the battery terminals.
2. Test the battery, if necessary.
3. Reconnect the positive cable to the battery first, then reconnect the negative cable to the battery.

⁂ CAUTION

Always connect the positive cable to the battery first.

4. Apply multipurpose grease to the terminals to prevent corrosion.
5. Enter the anti-theft code(s) for the audio system and/or navigation system (if equipped).
6. Enter the audio presets.
7. Set the clock (for vehicles without navigation).

DRIVE TRAIN

DRIVESHAFT

REMOVAL & INSTALLATION

See Figure 13.

1. Raise and support the vehicle.
2. Remove the front wheels.
3. Pry up the stake on the spindle nut, then remove the nut.
4. Drain the transmission fluid, then reinstall the drain plug with a new washer.
5. Remove the nuts and the bolt, then separate the lower arm using a prybar.
6. Separate the driveshaft outboard joint from the front hub using a soft face hammer.
7. Pull the knuckle outward, and remove the driveshaft outboard joint from the front hub.
8. On the left driveshaft, pry the inboard joint from the differential using the prybar. Remove the driveshaft as an assembly.

➡Do not pull on the driveshaft or the inboard joint may come apart. Pull the inboard joint straight out to avoid damaging the oil seal. Be careful not to damage the oil seal or the end of the inboard joint using the prybar.

9. On the right driveshaft, drive the inboard joint off of the intermediate shaft using a drift punch and a hammer (M/T model). Pry the inboard joint from the differential using the prybar (A/T model).
10. Remove the driveshaft as an assembly.
11. Do not pull on the driveshaft or the inboard joint may come apart. Pull the inboard joint straight out to avoid damaging the oil seal.

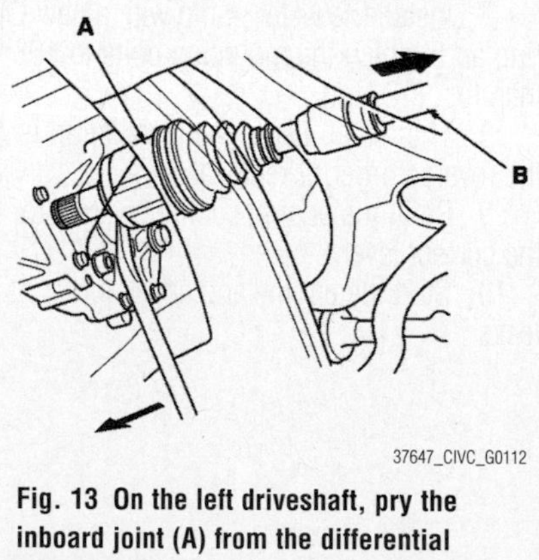
A
B
37647_CIVC_G0112

Fig. 13 On the left driveshaft, pry the inboard joint (A) from the differential using the prybar. Remove the driveshaft (B) as an assembly.

➡Be careful not to damage the oil seal or the end of the inboard using the prybar.

12. Remove the set ring from the inboard joint (except right driveshaft with M/T model).
13. Remove the set ring from the intermediate shaft (M/T model).

To install:

➡Before starting installation, make sure the mating surfaces of the joint and the splined section are clean.

14. Apply about 5 g (0.18 oz) of moly 60 paste to the contact area of the outboard joint and the front wheel bearing.

➡The paste helps prevent noise and vibration.

15. Install a new set ring into the set ring groove of the driveshaft.
16. Install a new set ring into the set ring groove of the intermediate shaft.
17. On M/T model, apply specified super high temp urea grease to the whole splined surface of the right driveshaft. After applying grease, remove the grease from the splined grooves at intervals of 2-3 splines and from the set ring groove so that air can bleed from the intermediate shaft.
18. Clean the areas where the driveshaft contacts the differential thoroughly with solvent, and dry them with compressed air.

➡Do not wash the rubber parts with solvent.

19. Insert the inboard end of the driveshaft into the differential or the intermediate shaft (M/T model) until the set ring locks in the groove.

➡Insert the driveshaft horizontally to prevent damaging the oil seal.

20. Install the outboard joint into the front hub.
21. Install the knuckle onto the lower arm. During installation, install a new flange bolt and new self-locking nuts. After lightly tightening all three fasteners, tighten them to the specified torque of 43 ft. lbs. (59 Nm) in the following order: the nut on the front, the nut on the rear, then the bolt.
22. Apply a small amount of engine oil to the seating surface of a new spindle nut.
23. Install the spindle nut, then tighten the nut. After tightening, use a drift to stake the spindle nut shoulder against the driveshaft.
24. Clean the mating surfaces of the brake discs and the wheels, then install the front wheels.
25. Turn the front wheels by hand, and make sure there is no interference between the driveshaft and surrounding parts.
26. Refill the transmission with the recommended transmission fluid:
27. Lower the vehicle.
28. Check the wheel alignment, and adjust it if necessary.
29. Test-drive the vehicle.

CV-BOOTS INSPECTION

See Figure 14.

1. Check the inboard boot and the outboard boot on the driveshaft for cracks, damage, leaking grease, and the loose boot bands. If any damage is found, replace the boot and the boot bands.
2. Check the driveshaft for cracks and damage. If any damage is found, replace the driveshaft.
3. Check the inboard joint and the outboard joint for cracks and damage. If any damage is found, replace the inboard joint or the outboard joint as an assembly.
4. Hold the inboard joint and turn the front wheel by hand, then make sure the joint is not excessively loose. If necessary, replace the inboard joint or the outboard joint as an assembly.

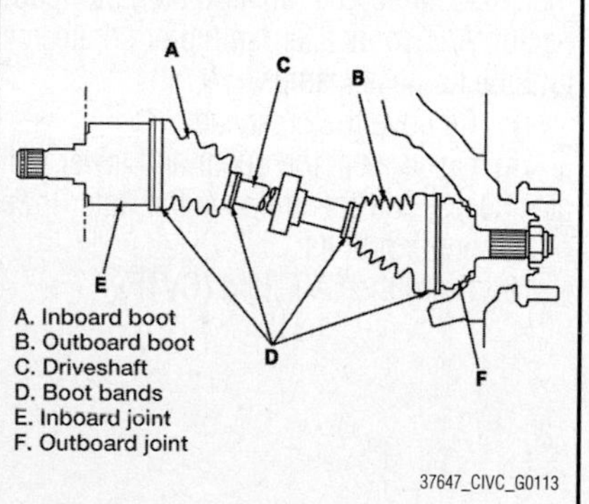
A
C
B
E
D
F
A. Inboard boot
B. Outboard boot
C. Driveshaft
D. Boot bands
E. Inboard joint
F. Outboard joint
37647_CIVC_G0113

Fig. 14 Inspecting the driveshaft and boots

ENGINE COOLING

ENGINE FAN

REMOVAL & INSTALLATION

➡See "Radiator & Fan" in this section.

HOSES

REMOVAL & INSTALLATION

➡Manufacturer does not provide a specific removal and installation procedure for the cooling system hoses. Use the illustrations in this section as a guide when servicing these components.

RADIATOR & FAN

REMOVAL & INSTALLATION

1. Drain the engine cooling system.
2. Raise and safely support the vehicle.
3. Remove the front wheels.
4. Remove the engine undercover and splash shield.
5. Remove the engine appearance cover.
6. Remove the resonator.
7. Remove the air intake assembly.
8. Remove the front grille cover.
9. Remove the clips and the radiator mount upper bracket/cushion.
10. Disconnect the hood latch switch connector, and then remove the mounting bolts and the clamps.
11. Disconnect the reservoir hose and remove the mounting bolts, and then remove the water filler.
12. Remove the harness clamps, and then disconnect the radiator fan motor connector, A/C condenser fan motor connector, and the harness clamps.
13. Remove the bulkhead.
14. Disconnect the upper and lower radiator hoses, and then remove the water lower pipe mounting bolts.
15. Drain the CVT fluid (CVTF).
16. Disconnect the transmission fluid cooler hoses and the lower radiator hose.
17. Disconnect the Engine Coolant Temperature (ECT) sensor 2 connector, and then remove the clamp.
18. Disconnect the upper radiator hose and water filler hose.
19. Remove the mounting bolts, and then pull up the radiator fan shroud assembly and the A/C condenser fan shroud assembly.

To install:

20. Installation is the reverse order of removal.
21. Adjust the hood latch alignment if necessary.
22. Refill the engine cooling system to the correct level.

THERMOSTAT

REMOVAL & INSTALLATION

1. Drain the engine cooling system.
2. Remove the air intake assembly.
3. Disconnect the upper radiator hose, lower radiator hose, heater hoses and water bypass hose.
4. Remove the water passage and remove the thermostat.

To install:

5. Remove all of the old liquid gasket from the water passage, thermostat housing mating surfaces, the bolts, and the bolt holes.
6. Apply a 1.5 mm wide bead of the liquid gasket on the water passage, P/N 08717-0004, 08718-0001, 08718-0003, or 08718-0009, along the broken lines. Install the component within 5 minutes of applying the liquid gasket.
7. Install the rubber seal on the thermostat, and then install the thermostat with the pin up.
8. Install the water passage and tighten the mounting bolts to 108 inch lbs. (12 Nm).
9. The remainder of the installation is the reverse order of removal.
10. Refill the engine cooling system to the correct level.

WATER PUMP

REMOVAL & INSTALLATION

See Figure 15.

1. Drain the engine cooling system.
2. Remove the engine undercover and splash shield.
3. Remove the accessory drive belt.
4. Remove the water pump pulley mounting bolts.
5. Remove the auto tensioner.
6. Remove the five bolts securing the water pump, and then remove the water pump with water pump pulley.

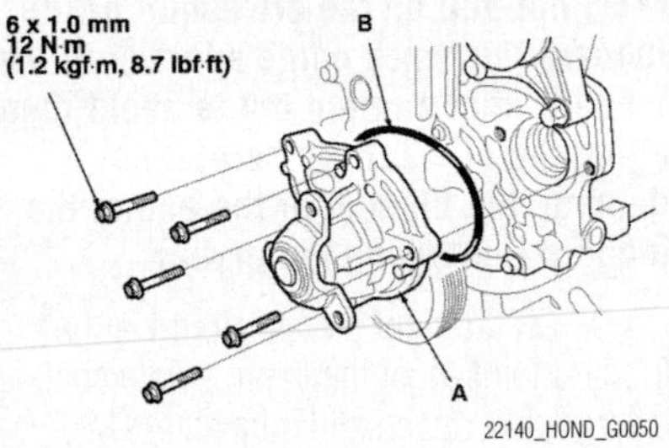
6 x 1.0 mm
12 N·m
(1.2 kgf·m, 8.7 lbf·ft)
B
A
22140_HOND_G0050

Fig. 15 Water pump (A) and O-ring (B) mounting

To install:

7. Install the water pump with a new O-ring and tighten the mounting bolts to 108 inch lbs. (12 Nm).
8. The remainder of the installation is the reverse order of removal.
9. Refill the engine cooling system to the correct level.
10. Start the engine and check for leaks.

FIRING ORDER

Firing order for all engines is 1-3-4-2.

IGNITION COIL

REMOVAL & INSTALLATION

See Figure 16.

1. Remove the engine appearance cover.
2. Disconnect the ignition coil connectors, and then remove the intake side ignition coils and the exhaust side ignition coils.
3. Install the ignition coils in the reverse order of removal.

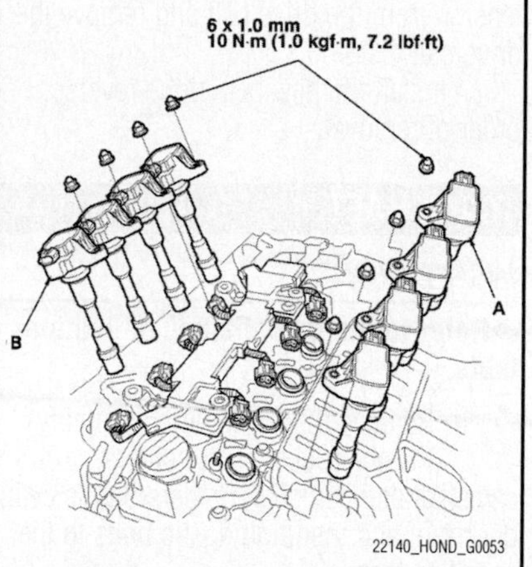
6 x 1.0 mm
10 N·m (1.0 kgf·m, 7.2 lbf·ft)
A
B
22140_HOND_G0053

Fig. 16 Exploded view of the intake side (A) and exhaust side (B) ignition coils

IDLE SPEED

INSPECTION

1. Before checking the idle speed, check these items:
   - The malfunction indicator lamp (MIL) has not been reported on, and there are no DTCs.
   - Ignition timing
   - Spark plugs
   - Air cleaner
   - PCV system
2. Apply the parking brake, and make sure the headlights are off.
3. Disconnect the evaporative emission (EVAP) canister purge valve connector.
4. Connect the scan tool to the data link connector (DLC) located under the driver's side of the dashboard.
5. Make sure the scan tool communicates with the PCM. If it doesn't, perform further DLC circuit troubleshooting.
6. Start the engine. Hold the engine speed at 3,000 RPM without load (in neutral) until the radiator fan comes on, then let it idle.
7. Check the idle speed without load conditions: headlights, blower fan, radiator fan, and air conditioner off. Idle speed should be 820±50 RPM.
8. Let the engine idle for 1 minute with high electric load (A/C on, temperature set to max cool, blower fan on high, headlights on high beam). Idle speed should be 820±50 RPM.
   a. If the idle speed is not within specification, do the "PCM Idle Learn Procedure". See "Engine Control Module (PCM)" in "ENGINE PERFORMANCE & EMISSION CONTROLS" section.
   b. If the idle speed is still not within specification, go to the symptom troubleshooting.
9. Reconnect the EVAP canister purge valve connector.

IGNITION TIMING

INSPECTION

1. Connect the Honda Diagnostic System (HDS), or equivalent scan tool, to the data link connector (DLC).
2. Turn the ignition switch to ON (II).
3. Make sure the HDS communicates with the vehicle and the Powertrain Control Module (PCM) If it does not communicate, troubleshoot the DLC circuit.
4. Check for DTCs. If a DTC is present, diagnose and repair the cause before continuing with this test.
5. Start the engine. Hold the engine speed at 3,000 RPM with no load (in N or P (A/T model) or neutral (M/T model)) until the radiator fan comes on, and then let it idle.
6. Check the idle speed.
7. Jump the SCS line with the HDS.

➡This step must be done to protect the PCM from damage.

8. Connect the timing light to the service loop (white tape).
9. Aim the light toward the pointer on the cam chain case. Check the ignition timing under a no load condition (headlights, blower fan, rear window defogger, and air conditioner are turned off).
10. If the ignition timing differs from the specification, check the cam timing. If the cam timing is OK, update the PCM if it does not have the latest software, or substitute a known-good PCM, and then recheck. If the system works properly, and the PCM was substituted, replace the original PCM.
11. Disconnect the HDS and the timing light.

ADJUSTMENT

➡Ignition timing is electronically controlled by the PCM. No adjustment is possible.

SPARK PLUGS

REMOVAL & INSTALLATION

1. Remove the ignition coil. For additional information, refer to the following section, "Ignition Coil, Removal and Installation."
2. Using a spark plug socket equipped with a rubber insert to properly hold the plug, turn the spark plug counterclockwise to loosen and remove the spark plug from the threaded hole in the cylinder head.

To install:

3. Apply a light coating of an anti-seize compound to the spark plug threads.
4. Carefully thread the plug into the threaded spark plug hole by hand. If resistance is felt before the plug is almost completely threaded, back the plug out and begin threading again. In tight, hard to reach areas, a small piece of rubber hose pressed onto the spark plug can be used as a threading tool. The rubber hose will hold the plug and while twisting the end of the hose, and the hose will be flexible enough to twist before allowing the plug to cross thread.
5. Carefully tighten the spark plug to 13 ft. lbs. (18 Nm).
6. The remainder of the installation is the reverse order of removal.

SPARK PLUG WIRES

REMOVAL & INSTALLATION

➡See "Ignition Coil" in this section.

STARTER

REMOVAL & INSTALLATION

1. Disconnect the negative battery cable.
2. Remove the air intake assembly.
3. Disconnect the positive starter cable and the S terminal connector from the starter, and remove the upper radiator hose bracket.
4. Remove the two bolts holding the starter and remove the starter.

To install:

5. Install the starter, and tighten the mounting bolts to 33 ft. lbs. (44 Nm).
6. Connect the positive starter cable and the S terminal connector to the starter, and install the upper radiator hose bracket. Make sure the crimped side of the ring terminal faces away from the starter when you connect it.
7. Install the air intake assembly.
8. Connect the negative battery cable.
9. Turn the IMA battery module switch OFF.
10. Start the engine to make sure the starter works properly.
11. Turn the IMA battery module switch ON.
12. If the IMA battery level gauge (BAT) displays no segments, start the engine, and hold it between 3,500–4,000 RPM without load (in N or P) until the BAT displays at least three segments.

ENGINE MECHANICAL

ACCESSORY DRIVE BELTS

ACCESSORY BELT ROUTING

See Figure 17.

Refer to the accompanying illustration for belt routing.

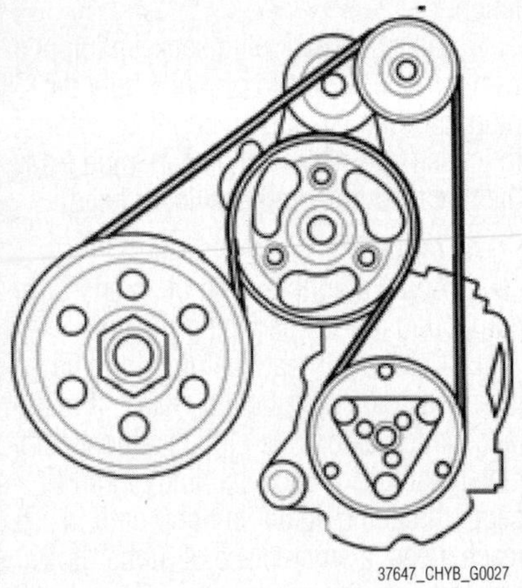
37647_CHYB_G0027

Fig. 17 Showing the accessory drive belt routing

INSPECTION

See Figure 18.

1. Inspect the belt for cracks or damage If the belt is cracked or damaged, replace it..
2. Check the position of the auto-tensioner indicator pointer is within the standard range. If it is out of the standard range, replace the drive belt.

ADJUSTMENT

➡These belt systems are equipped with automatic tensioners and no adjustment is required.

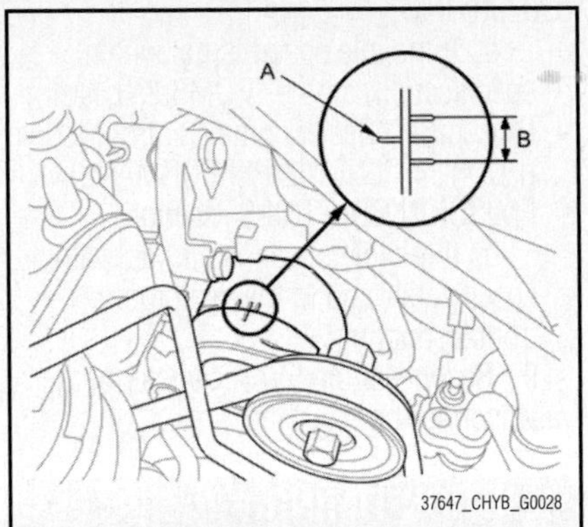
A
B
37647_CHYB_G0028

Fig. 18 Check the position of the auto-tensioner indicator pointer (A) is within the standard range (B) as shown

REMOVAL & INSTALLATION

See Figure 19.

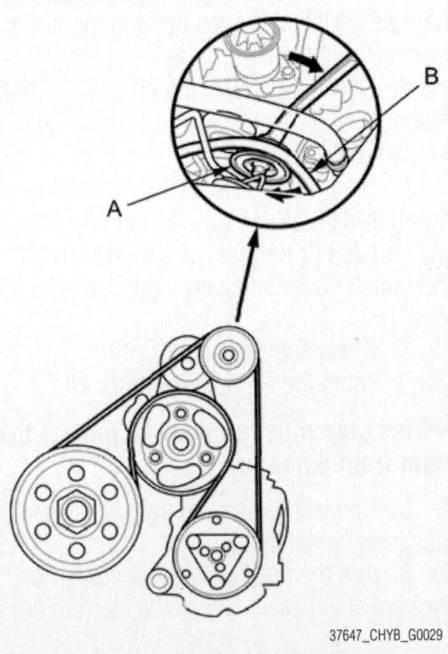
B
A
37647_CHYB_G0029

Fig. 19 Move the auto-tensioner (A) to relieve tension from the drive belt (B), and remove the drive belt.

1. Move the auto-tensioner to relieve tension from the drive belt and remove the drive belt.
2. Install the new belt in the reverse order of removal.

CAMSHAFT

INSPECTION

➡Refer to the camshaft specifications table.

1. Remove the rocker arm assembly.
2. Put the rocker shaft holders, camshaft and camshaft holders on the cylinder head, and then tighten the bolts to the specified torque.
3. Seat the camshaft by pushing it away from the camshaft pulley end of the cylinder head.
4. Zero the dial indicator against the end of the camshaft, and then push the camshaft back and forth and read the end play. If the end play is beyond the service limit, replace the cylinder head and recheck. If it is still beyond the service limit, replace the camshaft.
5. Loosen the camshaft holder bolts two turns at a time, in a crisscross pattern. Then remove the camshaft holders from the cylinder head.
6. Lift the camshafts out of the cylinder head, wipe them clean, and then inspect the lift ramps. Replace the camshaft if any lobes are pitted, scored, or excessively worn.
7. Clean the camshaft journal surfaces in the cylinder head, and then set the camshafts back in place. Place a Plastigage® strip across each journal.
8. Install the camshaft holders, and then tighten the bolts to the specified torque.
9. Remove the camshaft holders. Measure the widest portion of Plastigage® on each journal. If the camshaft-to-holder

clearance is beyond the service limit and the camshaft has been replaced, replace the cylinder head.

10. Check the total runout with the camshaft supported on V-blocks. If the total runout is beyond the service limit, replace the camshaft and recheck the camshaft-to-holder oil clearance. If the oil clearance is still beyond the service limit, replace the cylinder head.

11. Measure the cam lobe height.

REMOVAL & INSTALLATION

See Figures 20 and 21.

1. Disconnect the negative battery cable.
2. Remove the engine appearance cover.
3. Remove the ignition coils.
4. Remove the ignition coil wiring harness holder, and then disconnect the breather hose.
5. Remove the cylinder head cover.
6. Make a reference mark across the camshaft sprocket and the timing chain.
7. Apply new engine oil to the sliding surface of the cam chain tensioner through the oil return hole in the cylinder head.
8. Remove the cylinder head plug.
9. Hold the crankshaft pulley and set the socket wrench on the camshaft sprocket bolt.
10. Remove the maintenance bolt, and turn the camshaft clockwise to compress the cam chain tensioner, and then install the 6 x 1.0 mm bolt in the bolt hole in the engine block through the maintenance hole and the cam chain tensioner.
11. Hold the camshaft with a 27 mm open-end wrench, and then loosen the camshaft sprocket bolt.
12. Remove the camshaft sprocket bolt, and then remove the camshaft sprocket.
13. Loosen the locknuts and the adjusting screws.
14. Remove the lost motion holder bolts and the camshaft holder bolts. To prevent damaging the camshaft, loosen the bolts in sequence two turns at a time, in a criss-cross pattern.
15. Remove the lost motion holder, the lost motion assembly, and the rocker arm assembly, and then remove the camshaft.

To install:

16. Put the camshaft, the camshaft holders, and the lost motion holder on the cylinder head, and then tighten the bolts as follows:
    - 8mm bolts: 16 ft. lbs. (22 Nm)
    - 8mm bolts 11 and 13: 14 ft. lbs. (20 Nm)
    - 6mm bolt: 9 ft. lbs. (12 Nm)
17. Seat the camshaft by pushing it away from the camshaft pulley end of the cylinder head.
18. Install the cam chain around the camshaft sprocket aligned with the reference mark, and then install the camshaft sprocket on the camshaft.
19. Hold the camshaft with a 27 mm open-end wrench, and then tighten the bolt to 41 ft. lbs. (56 Nm).
20. Apply new engine oil to the sliding surface of the cam chain tensioner through the oil return hole in the cylinder head.
21. Hold the crankshaft pulley and set the socket wrench on the camshaft sprocket bolt.
22. Turn the camshaft clockwise to compress the cam chain tensioner, and then remove the 6 x 1.0 mm bolt.
23. Install the maintenance bolt and new washer and tighten to 15 ft. lbs. (20 Nm).
24. Install the new cylinder head plug.
25. The remainder of the installation is the reverse order of removal.

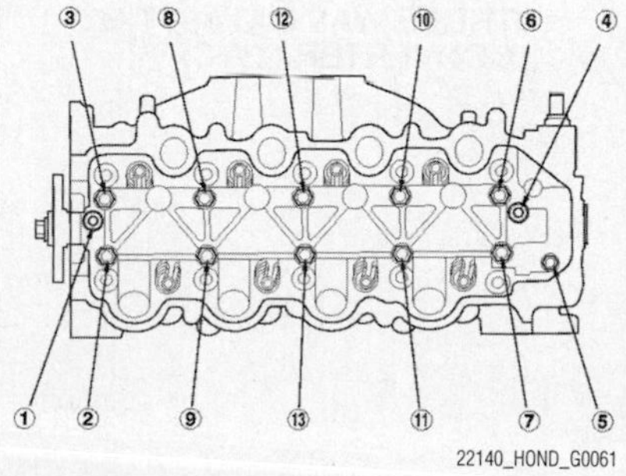
③
⑧
⑫
⑩
⑥
④
①
②
⑨
⑬
⑪
⑦
⑤
22140_HOND_G0061

Fig. 20 Lost motion holder bolt removal sequence

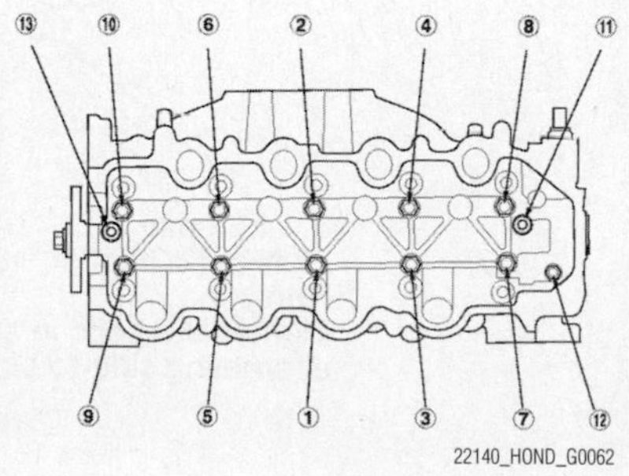
⑬
⑩
⑥
②
④
⑧
⑪
⑨
⑤
①
③
⑦
⑫
22140_HOND_G0062

Fig. 21 Camshaft holder torque sequence—1.3L Engine

CATALYTIC CONVERTER

REMOVAL & INSTALLATION

See Figure 22.

➡Manufacturer does not provide a specific removal and installation procedure for this component. Use the illustration as a guide when servicing this component.

CRANKSHAFT DAMPER (PULLEY)

REMOVAL & INSTALLATION

1. Remove the front wheels.
2. Remove the front undercover and the splash shield.
3. Remove the drive belt.
4. Hold the crankshaft pulley with the holder handle and the holder attachment (50mm).
5. Remove the crankshaft pulley bolt with a socket (19 mm) and a breaker bar.
6. Remove the under-floor three way catalytic converter (TWC).
7. Support the engine with a jack and a wood block under the oil pan.
8. Remove the lower torque rod mounting bolts.
9. Remove the crankshaft pulley while moving the engine forward.

To install:

10. Clean the crankshaft pulley (A), the crankshaft (B), the bolt (C), and the washer (D). Lubricate with new engine oil as shown.
11. Install the crankshaft pulley while moving the engine forward and loosely install the crankshaft pulley.
12. Install the lower torque rod mounting bolts.
13. Remove the jack and the wood block.
14. Install the under-floor TWC.
15. Tighten the crankshaft pulley bolt. Do not use an impact wrench.
16. Hold the pulley with the holder handle and the holder attachment, then torque the bolt to 27 ft. lbs. (37 Nm) with a torque wrench and a heavy duty 19 mm socket.
    a. If the pulley bolt or crankshaft are new, torque the bolt to 130 ft. lbs. (177 Nm), then remove the bolt and torque it to 27 ft. lbs. (37 Nm).
17. Mark the bolt head and the crankshaft pulley with a reference line, then tighten the bolt an additional 90°.
18. Install the drive belt.
19. Install the splash shield and the front undercover.
20. Install the front wheels.

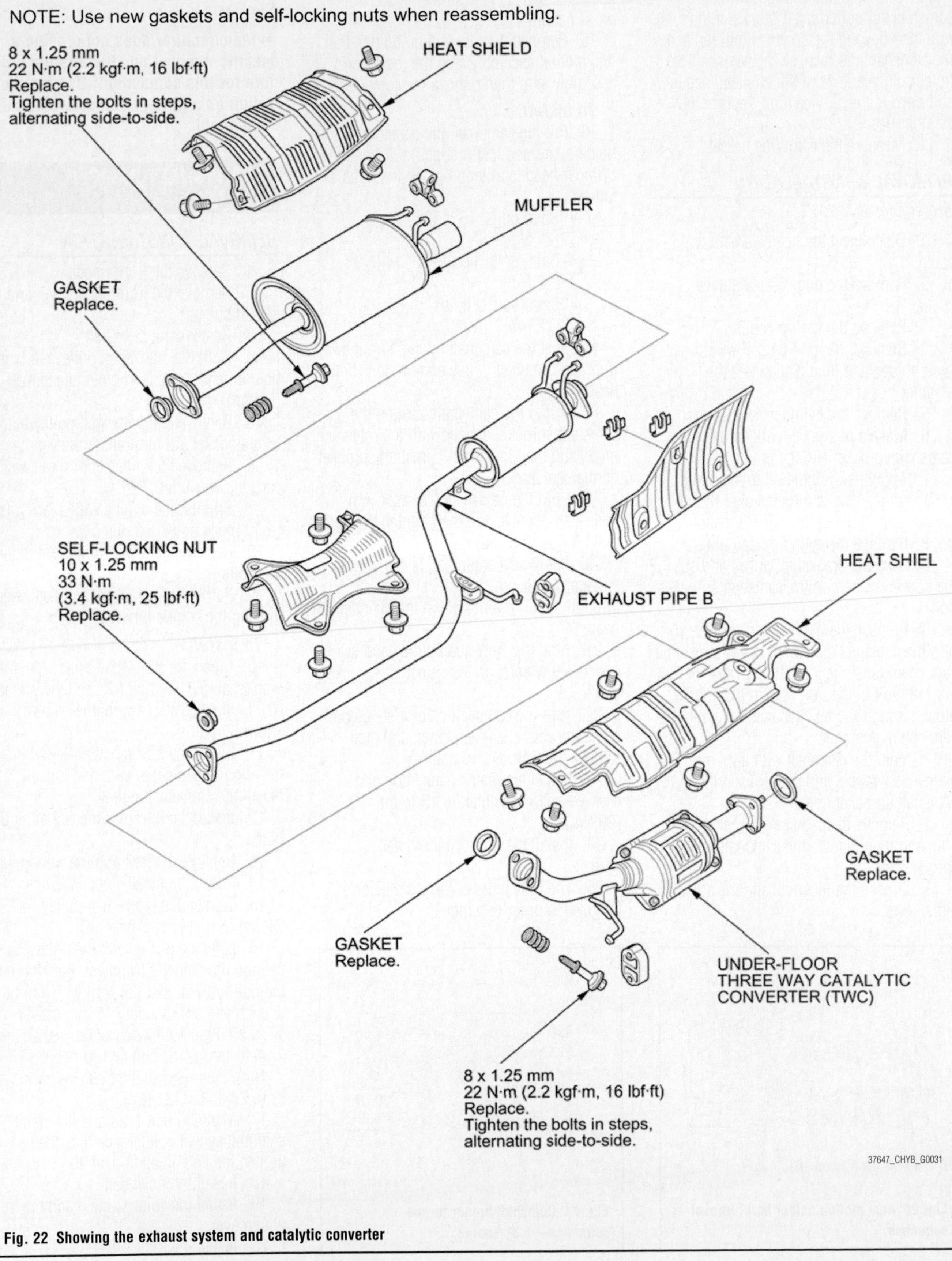

NOTE: Use new gaskets and self-locking nuts when reassembling.
8 x 1.25 mm
22 N·m (2.2 kgf·m, 16 lbf·ft)
Replace.
Tighten the bolts in steps, alternating side-to-side.
HEAT SHIELD
MUFFLER
GASKET
Replace.
SELF-LOCKING NUT
10 x 1.25 mm
33 N·m
(3.4 kgf·m, 25 lbf·ft)
Replace.
EXHAUST PIPE B
HEAT SHIEL
GASKET
Replace.
GASKET
Replace.
UNDER-FLOOR THREE WAY CATALYTIC CONVERTER (TWC)
8 x 1.25 mm
22 N·m (2.2 kgf·m, 16 lbf·ft)
Replace.
Tighten the bolts in steps, alternating side-to-side.
37647_CHYB_G0031

Fig. 22 Showing the exhaust system and catalytic converter

CYLINDER HEAD

REMOVAL & INSTALLATION

See Figures 23 and 24.

➡Connect the Honda Diagnostic System (HDS) to the data link connector (DLC), and monitor the engine coolant temperature (ECT) sensor 1. To avoid damaging the cylinder head, wait until the ECT 1 temperature drops below 100°F (38°C) before loosening the cylinder head bolts.

1. Properly relieve the fuel system pressure.
2. Remove the engine appearance cover.
3. Drain the engine cooling system.
4. Remove the air intake assembly.
5. Remove the intake manifold. For additional information, refer to the following section, "Intake Manifold, Removal and Installation."
6. Disconnect the knock sensor connector, and then remove the harness clamp and the connecting pipe.
7. Disconnect the Camshaft Position (CMP) sensor and the Engine Coolant Temperature (ECT) sensor 1 connector, and then remove the harness clamp.
8. Remove the eight ignition coils.
9. Remove the harness holder and disconnect the breather hose.
10. Disconnect the Evaporative Emission (EVAP) canister purge valve connector, and then remove the harness clamp.
11. Disconnect the purge joint from the bracket, and then remove the fuel pipe nut and remove the EVAP canister purge valve bracket bolts.
12. Disconnect the upper radiator hose, the lower radiator hose, the water bypass hose, and the heater hoses.
13. Remove the accessory drive belt.
14. Turn the crankshaft pulley so its Top Dead Center (TDC) mark lines up with the pointer.
15. Remove the water pump.
16. Remove the cylinder head cover.
17. Remove the warm-up three way catalytic converter (WU-TWC).
18. Remove the crankshaft pulley.
19. Support the engine with a suitable jack and a wood block under the oil pan.
20. Remove the ground cable, and then remove the side engine mount bracket.
21. Disconnect the Crankshaft Position (CKP) sensor and the harness clamp, and then remove the dipstick tube.
22. Remove the timing chain case.
23. Make a reference mark across the camshaft sprocket and the timing chain.
24. Loosely install the crankshaft pulley.
25. Apply new engine oil to the sliding surface of the cam chain tensioner through the oil return hole in the cylinder head.
26. Hold the crankshaft pulley and set the socket wrench on the camshaft sprocket bolt.
27. Turn the camshaft clockwise to compress the cam chain tensioner, and then install the 6 x 1.0 mm bolt in the bolt hole in the engine block through the timing chain tensioner.
28. Hold the camshaft with a 27 mm open-end wrench, and then remove the camshaft sprocket.
29. Remove the top bolt that secures the timing chain guide.
30. Remove the cylinder head bolts. To prevent warpage, loosen the bolts in sequence ⅓ turn at a time; repeat the sequence until all bolts are loosened.
31. Remove the cylinder head.

To install:

32. Clean the cylinder head and the engine block surface.
33. Install the new cylinder head g asket and the dowel pins on the engine block. Always use a new cylinder head gasket.
34. Set the crankshaft to TDC. Align the TDC mark on the crankshaft sprocket with the pointer on the oil pump.
35. Set the No. 1 piston at TDC. The "UP" mark on the camshaft sprocket should be at the top, and the TDC grooves on the camshaft sprocket should line up with the top edge of the cylinder head.
36. Install the cylinder head on the engine block.
37. Replace any stretched cylinder head bolts.
38. Apply new engine oil to the threads and flange of all cylinder head bolts. Be sure to install the 165 mm long head bolt (A) in the location shown.
39. Tighten the cylinder head bolts in sequence to 22 ft. lbs. (29 Nm), use a beam-type torque wrench. When using a preset-click-type torque wrench, be sure to tighten slowly and do not overtighten. If a bolt makes any noise while you are torquing it, loosen the bolt and retighten it from the first step. Then tighten the bolts an additional 130°.
40. Install the timing chain guide mounting top bolt.
41. Install the cam chain around the camshaft sprocket aligned with the reference mark, and then install the camshaft sprocket on the camshaft.
42. Apply new engine oil to the bolt threads and flange. Hold the camshaft with a 27 mm open-end wrench, and then tighten the bolt to 41 ft. lbs. (56 Nm).
43. Loosely install the crankshaft pulley.
44. Apply new engine oil to the sliding surface of the cam chain tensioner through the oil return hole in the cylinder head.
45. Hold the crankshaft pulley and set the socket wrench on the camshaft sprocket bolt.
46. Turn the camshaft clockwise to compress the cam chain tensioner, and then remove the 6 x 1.0 mm bolt.
47. Remove all of the old liquid gasket from the chain case mating surfaces, the bolts, and the bolt holes.
48. Clean and dry the chain case mating surfaces.
49. Apply liquid gasket, P/N 08717-0004, 08718-0001, 08718-0003, or 08718-0009, evenly to the cylinder head and engine block mating surface of the timing chain case. Install the component within 5 minutes of applying the liquid gasket.
50. Apply liquid gasket, P/N 08718-0001, or 08718-0009 evenly to the oil pan mating surface of the timing chain case.

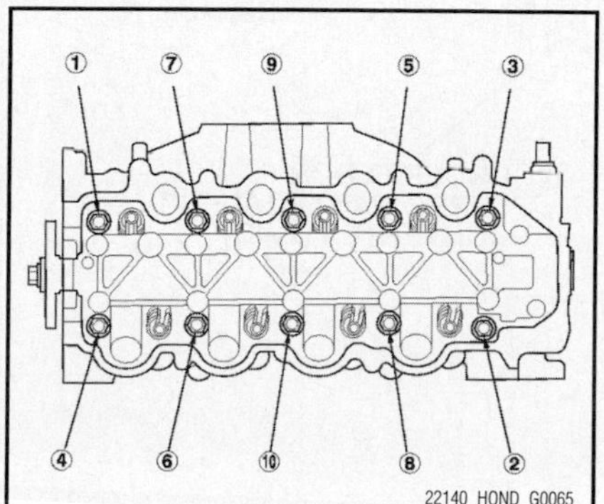
1
7
9
5
3
4
6
10
8
2
22140_HOND_G0065

Fig. 23 Cylinder head bolt removal sequence

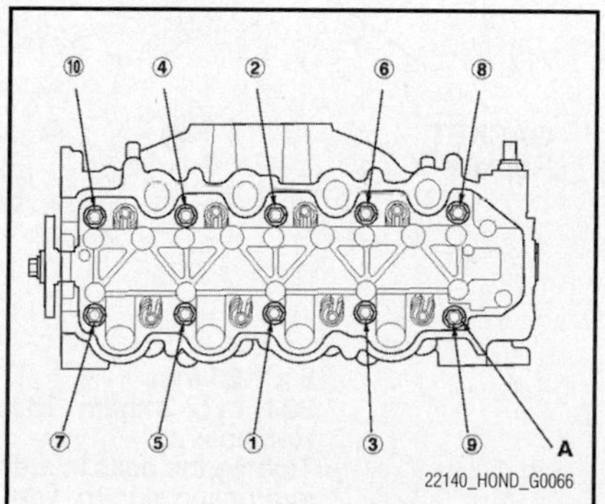
10
4
2
6
8
7
5
1
3
9
A
22140_HOND_G0066

Fig. 24 Cylinder head bolt torque sequence

Install the component within 5 minutes of applying the liquid gasket.

51. Set the edge of the chain case to the edge of the oil pan, and then install the chain case on the engine block. Wipe off the excess liquid gasket on the oil pan and chain case mating area.

52. The remainder of the installation is the reverse order of removal.

53. Inspect for fuel leaks.

54. Refill the engine cooling system to the correct level.

55. Start the engine and check for leaks.

EXHAUST MANIFOLD

REMOVAL & INSTALLATION

See Figure 25.

➡**Manufacturer does not provide a specific removal and installation**

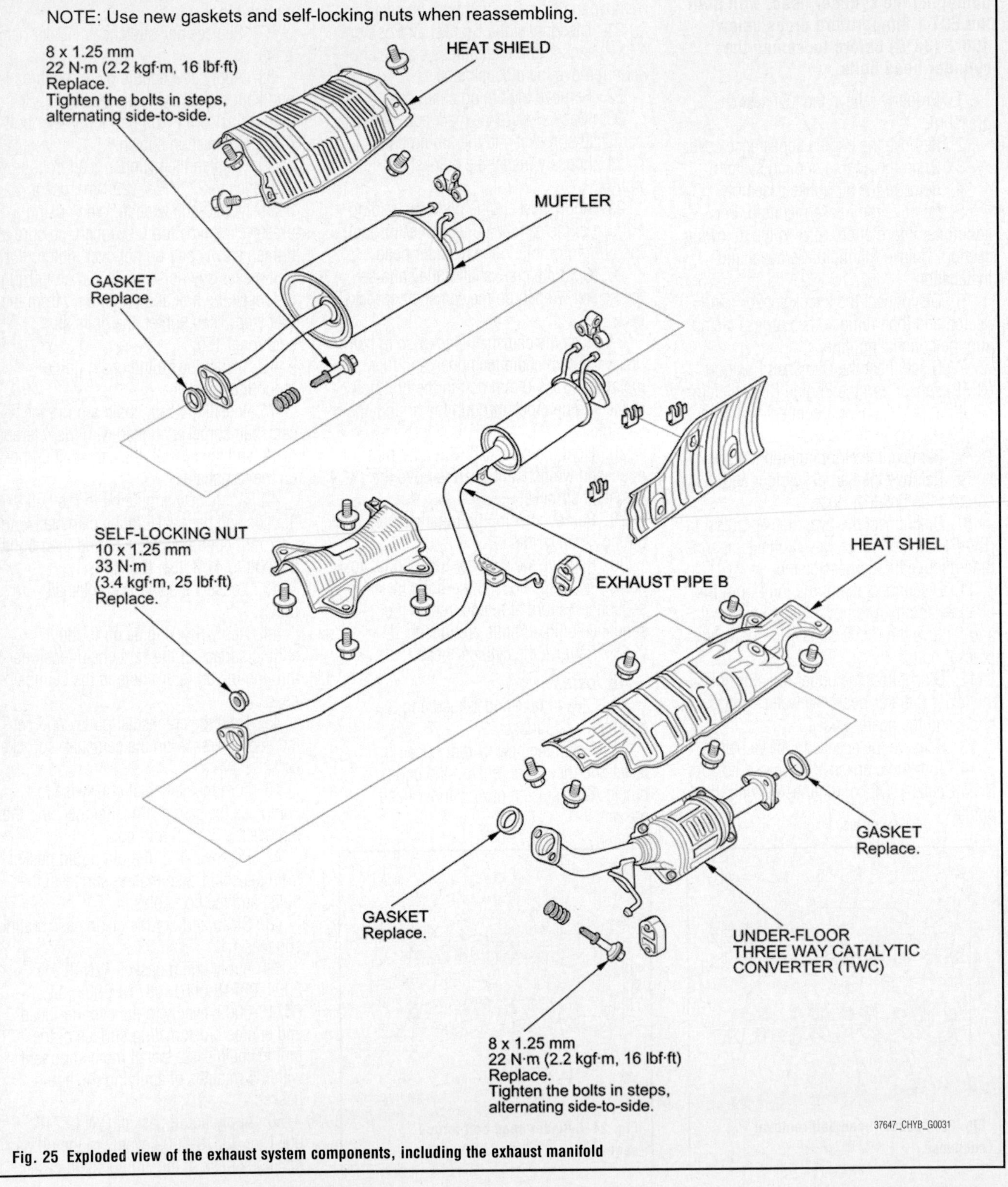
NOTE: Use new gaskets and self-locking nuts when reassembling.
8 x 1.25 mm
22 N·m (2.2 kgf·m, 16 lbf·ft)
Replace.
Tighten the bolts in steps, alternating side-to-side.
HEAT SHIELD
MUFFLER
GASKET
Replace.
SELF-LOCKING NUT
10 x 1.25 mm
33 N·m
(3.4 kgf·m, 25 lbf·ft)
Replace.
EXHAUST PIPE B
HEAT SHIEL
GASKET
Replace.
GASKET
Replace.
UNDER-FLOOR THREE WAY CATALYTIC CONVERTER (TWC)
8 x 1.25 mm
22 N·m (2.2 kgf·m, 16 lbf·ft)
Replace.
Tighten the bolts in steps, alternating side-to-side.
37647_CHYB_G0031

Fig. 25 Exploded view of the exhaust system components, including the exhaust manifold

procedure for this component. Use the illustration as a guide when servicing this component.

FLYWHEEL

REMOVAL & INSTALLATION

1. Remove the transmission assembly.
2. Remove the drive plate and the washer from the engine crankshaft.
3. Install the drive plate and the washer on the engine crankshaft, and tighten the eight bolts in a crisscross pattern in at least two steps to 54 ft. lbs. (74 Nm).
4. Install the transmission assembly.

INTAKE MANIFOLD

REMOVAL & INSTALLATION

See Figure 26.

1. Disconnect the negative battery cable.
2. Drain the engine cooling system.
3. Remove the engine appearance cover.
4. Remove the resonator.
5. Remove the intake air duct.
6. Remove the air cleaner assembly.
7. Remove the front grille cover.
8. Raise the vehicle on the lift.
9. Remove the front wheels.
10. Remove the front undercover and the splash shield.
11. Remove the intake manifold bracket bolts and the A/C compressor harness clamp.
12. Lower the vehicle on the lift.
13. Disconnect the water bypass hose and the vacuum hose, and the throttle actuator connector.
14. Disconnect the water bypass hose.
15. Disconnect the EGR valve connector, the intake manifold sub-harness connector, the clamp, and then remove the ground cables.
16. Disconnect the fuel injector connectors, rocker arm oil control valve connectors, rocker arm oil pressure sensor connector, and then remove the engine wire harness from the brackets.
17. Disconnect the positive crankcase ventilation (PCV) hose.
18. Remove the intake manifold assembly.

To install:

19. Install the intake manifold assembly and tighten the bolts/nuts in a crisscross pattern in three steps, beginning with the inner bolt. Use a new gasket.

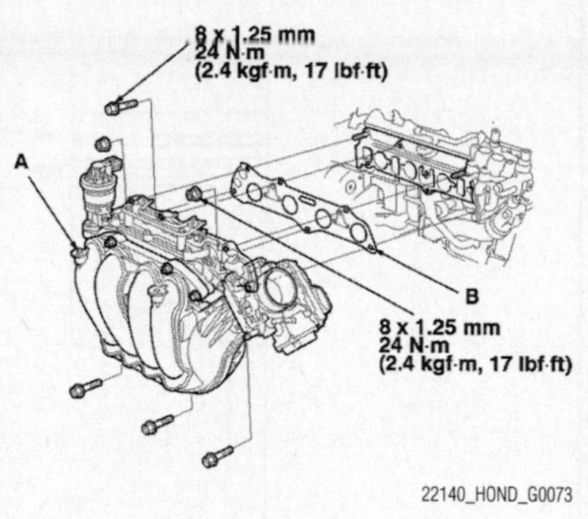
8 x 1.25 mm
24 N·m
(2.4 kgf·m, 17 lbf·ft)
A
B
8 x 1.25 mm
24 N·m
(2.4 kgf·m, 17 lbf·ft)
22140_HOND_G0073

Fig. 26 Install the intake manifold (A) using a new gasket (B)

20. The remainder of the installation is the reverse order of removal.
21. Refill the engine cooling system to the correct level.

OIL PAN

REMOVAL & INSTALLATION

See Figure 27.

1. Drain the engine oil.
2. Remove the steering wheel and the steering joint bolt.
3. Remove the front wheels.
4. Remove the front undercover and the splash shield.
5. Remove the A/C condenser fan shroud assembly.
6. Disconnect the A/C compressor clutch connector, and then remove the A/C compressor.
7. Remove the intake manifold bracket and the harness clamp.
8. Remove the under-floor Three Way Catalytic Converter (TWC).
9. Disconnect the front stabilizer links.
10. Disconnect the suspension lower arm ball joints.
11. Disconnect the steering gearbox harness connectors.
12. Attach the universal lifting eyelet, and the engine support hanger.
13. Remove the lower torque rod mounting bolts.
14. Make the appropriate reference line at the both side front subframe that line up with the edge on the body.
15. Loosen the mid stiffener mounting bolts.
16. Using the front subframe adapter, remove the front subframe.
17. Remove the harness clamps, the mounting bolt, and the dipstick tube.
18. Remove the lower torque rod bracket mounting bolts.
19. Remove the two bolts securing the transmission.
20. Remove the bolts securing the oil pan.
21. Insert a flat blade screwdriver where shown, and separate the oil pan from the engine block.
22. Remove the oil pan.

To install:

23. Remove all of the old liquid gasket from the oil pan mating surfaces, the bolts, and the bolt holes.
24. Clean and dry the oil pan mating surfaces and the O-ring groove.
25. Install the dowel pins, and install the new O-ring and the oil pan gasket on the oil pan.
26. Apply liquid gasket, P/N 08718-0001, or 08718-0009, evenly to the engine block mating surface of the oil pan and to the inside edge of the threaded bolt holes. Install the component within 5 minutes of applying the liquid gasket.
27. Install the oil pan.
28. Tighten all the bolts in three steps to 8.7 ft. lbs. (12 Nm), except for bolt No. 1 which is tighten to 17 ft. lbs. (24 Nm). Wipe off the excess liquid gasket on the each side of the crankshaft pulley and the flywheel.
29. Tighten the two bolts securing the transmission to 47 ft. lbs. (64 Nm).
30. Install the lower torque rod bracket mounting bolts and tighten to 54 ft. lbs. (74 Nm).
31. The remainder of the installation is the reverse order of removal.
32. Wait at least 30 minutes before filling the engine oil.
33. Check and adjust the wheel alignment as necessary.

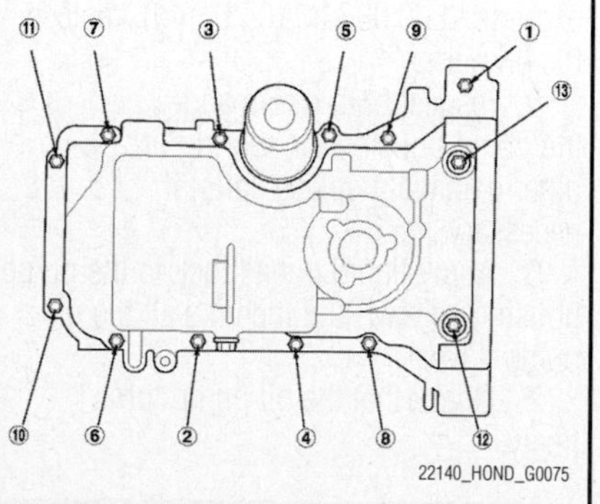
⑪ ⑦ ③ ⑤ ⑨ ①
⑬
⑩ ⑥ ② ④ ⑧ ⑫
22140_HOND_G0075

Fig. 27 Oil pan mounting bolt torque sequence

OIL PUMP

REMOVAL & INSTALLATION

1. Disconnect the negative battery cable.
2. Drain the engine oil.
3. Remove the timing chain. For additional information, refer to "Timing Chain & Sprockets" in this section.
4. Remove the oil screen.
5. Remove the oil pump.

To install:

6. Install the dowel pins and a new O-ring on the oil pump, and then align the inner rotor with the crankshaft, and install the oil pump. Tighten the mounting bolts to 86 inch lbs. (10 Nm).
7. Install the oil screen with a new gasket.
8. Install the timing chain.
9. Refill the engine with oil to the correct level.
10. Connect the negative battery cable.

INSPECTION

1. Remove the screws from the pump housing, and then separate the housing and the cover.
2. Check the inner-to-outer rotor radial clearance between the inner rotor and the outer rotor. If the inner-to-outer rotor radial clearance exceeds 0.008 in. (0.20 mm), replace the oil pump.
3. Check the pump housing-to-rotor axial clearance between the rotor and the pump housing. If the pump housing-to-rotor axial clearance exceeds the 0.006 in. (0.15 mm), replace the oil pump.
4. Check the pump housing-to-outer rotor radial clearance between the outer rotor and the pump housing. If the pump housing-to-outer rotor radial clearance exceeds the 0.008 in. (0.20 mm), replace the oil pump.
5. Inspect both rotors and the pump housing for scoring or other damage. Replace parts, if necessary.
6. Apply liquid thread lock to the pump housing screws, and then install the oil pump cover.
7. Check that the oil pump turns freely.

PISTON AND RING

POSITIONING

See Figures 28 and 29.

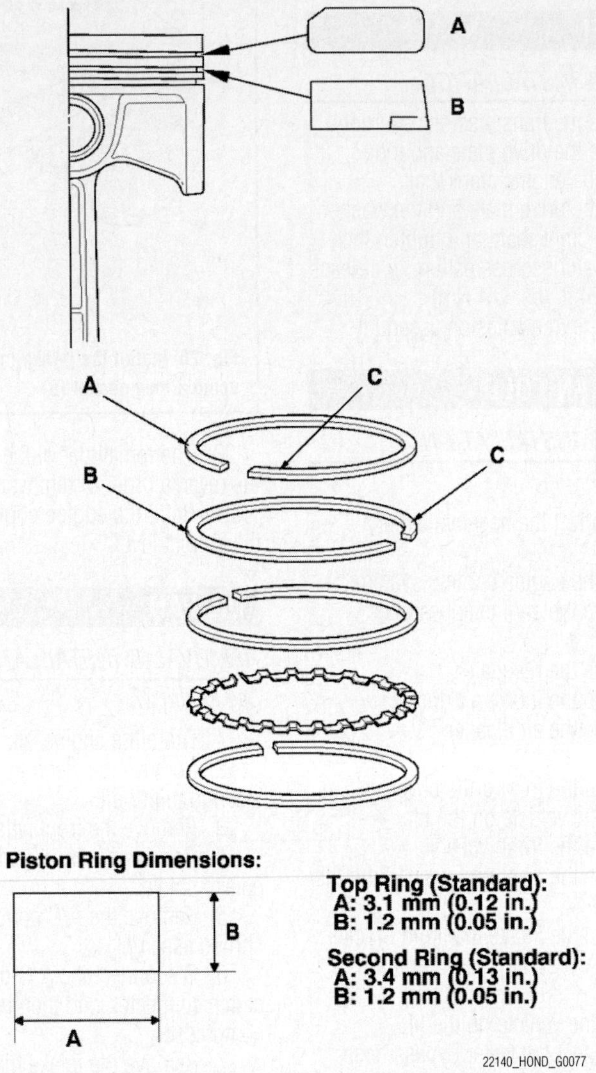
A
B
A
C
B
C
Piston Ring Dimensions:
B
A
Top Ring (Standard):
A: 3.1 mm (0.12 in.)
B: 1.2 mm (0.05 in.)
Second Ring (Standard):
A: 3.4 mm (0.13 in.)
B: 1.2 mm (0.05 in.)
22140_HOND_G0077

Fig. 28 Top ring (A), second ring (B) and the manufacturing marks (C) must face upward

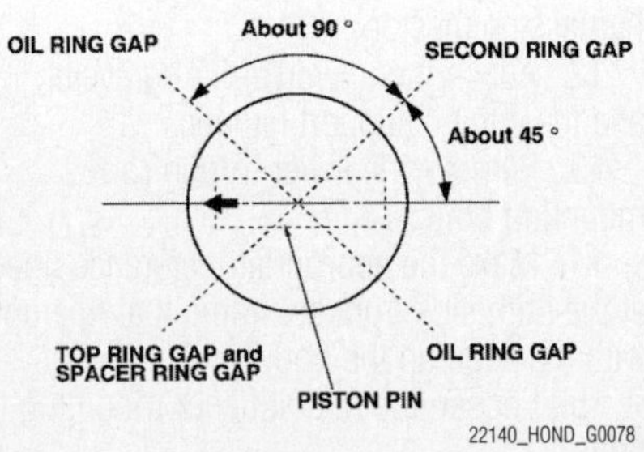
OIL RING GAP
About 90 °
SECOND RING GAP
About 45 °
TOP RING GAP and SPACER RING GAP
OIL RING GAP
PISTON PIN
22140_HOND_G0078

Fig. 29 Piston ring positioning

REAR MAIN SEAL

REMOVAL & INSTALLATION

1. Remove the transaxle from the vehicle.
2. Remove the IMA motor, if equipped.
3. Remove the driveplate from the crankshaft.
4. Carefully pry the crankshaft seal out of the retainer.

To install:

5. Apply clean engine oil to the lip of the new seal.
6. Install the seal onto the crankshaft and into the retainer using the appropriate seal driver.

7. Install the IMA motor, if equipped.
8. Install the driveplate and the transmission.

ROCKER ARMS

REMOVAL & INSTALLATION

➡See "Camshaft" in this section.

TIMING CHAIN & SPROCKETS

REMOVAL & INSTALLATION

See Figures 30 through 32.

1. Disconnect the negative battery cable.
2. Raise and safely support the vehicle.
3. Remove the front wheels.
4. Remove the front undercover and the splash shield.
5. Remove the accessory drive belt.
6. Turn the crankshaft pulley so its Top Dead Center (TDC) mark lines up with the pointer.
7. Remove the water pump pulley.
8. Remove the cylinder head cover.
9. Remove the crankshaft pulley.
10. Remove the oil pan.
11. Support the engine with a suitable jack and a wood block under the engine block.
12. Remove the ground cable and the side engine mount bracket.
13. Disconnect the Crankshaft Position (CKP) sensor connector, and then remove the dipstick tube mounting bolt and the harness clamps.
14. Remove the timing chain case, and then remove the CKP pulse plate.
15. Apply new engine oil to the sliding surface of the cam chain tensioner slider.
16. Hold the cam chain tensioner slider with a screwdriver, then remove the bolt, and loosen the bolt.
17. Remove the timing chain tensioner slider.
18. Remove the timing chain tensioner and the cam chain guide.
19. Remove the timing chain.

To install:

20. Set the crankshaft to TDC. Align the TDC mark on the crankshaft sprocket with the pointer on the oil pump.
21. Set the No. 1 piston at TDC. The "UP" mark on the camshaft sprocket should be at the top, and the TDC grooves on the camshaft sprocket should line up with the top edge of the cylinder head.
22. Install the cam chain on the crankshaft sprocket with the colored piece aligned with the TDC mark on the crankshaft sprocket.

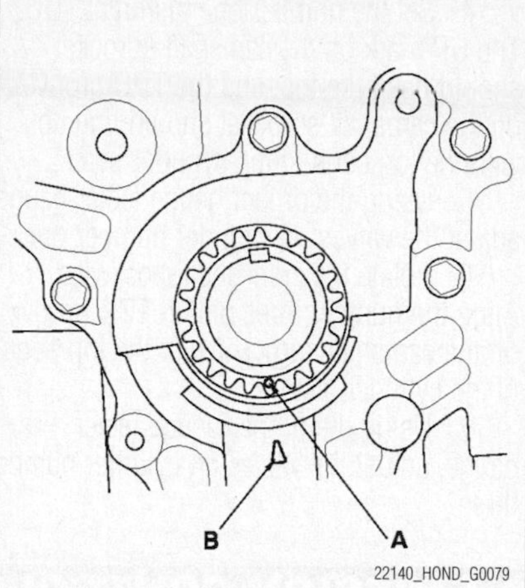
B
A
22140_HOND_G0079

Fig. 30 Set the crankshaft to TDC. Align the TDC mark (A) on the crankshaft sprocket with the pointer (B) on the oil pump

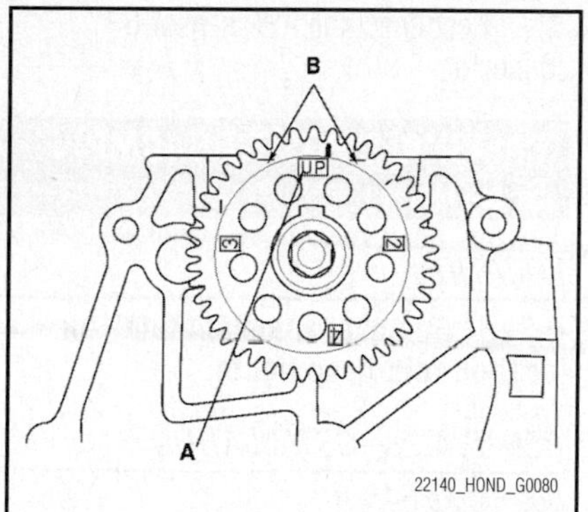
B
UP
A
22140_HOND_G0080

Fig. 31 Set the No. 1 piston at TDC. The "UP" mark (A) on the camshaft sprocket should be at the top, and the TDC grooves (B) on the camshaft sprocket should line up with the top edge of the cylinder head

23. Install the cam chain on the camshaft sprocket with the pointer aligned with the center of the two colored pieces.
24. Apply new engine oil to the threads of the cam chain tensioner mounting bolt.
25. Install the cam chain tensioner and the cam chain guide.
26. Install the cam chain tensioner slider, and tighten the lower side bolt loosely.
27. Apply new engine oil to the sliding surface of the cam chain tensioner slider.
28. Turn the cam chain tensioner clockwise to compress the cam chain tensioner slider. Install the remaining bolt, and then tighten the two bolts.
29. Check the chain case oil seal for damage If the oil seal is damaged, replace the chain case oil seal.
30. Remove all of the old liquid gasket

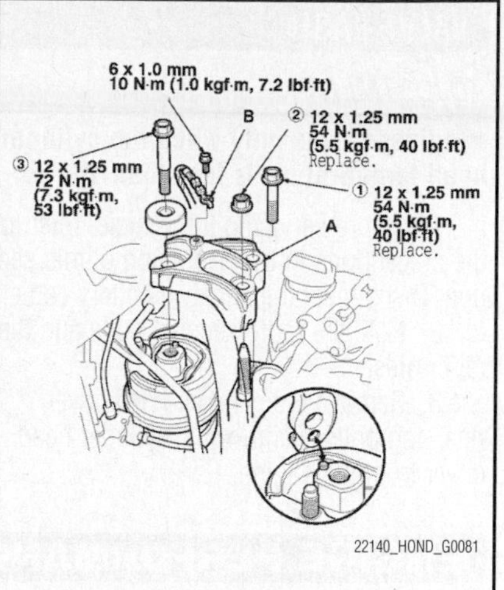
6 x 1.0 mm
10 N·m (1.0 kgf·m, 7.2 lbf·ft)
B
② 12 x 1.25 mm
54 N·m
(5.5 kgf·m, 40 lbf·ft)
Replace.
③ 12 x 1.25 mm
72 N·m
(7.3 kgf·m,
53 lbf·ft)
① 12 x 1.25 mm
54 N·m
(5.5 kgf·m,
40 lbf·ft)
Replace.
A
22140_HOND_G0081

Fig. 32 Install the side engine mount bracket (A), then tighten the mounting bolts and nut in the numbered sequence shown

from the chain case mating surfaces, the bolts, and the bolt holes.
31. Clean and dry the chain case mating surfaces.
32. Apply liquid gasket, P/N 08717-0004, 08718-0001, 08718-0003, or 08718-0009, evenly to the cylinder head and the engine block mating surface of the chain case. Install the component within 5 minutes of applying the liquid gasket.
33. Install the CKP pulse plate and the timing chain case. Tighten the mounting bolts to 23 ft. lbs. (31 Nm).
34. Install the harness clamps and the dipstick tube mounting bolt, and then connect the CKP sensor connector.
35. Install the side engine mount bracket, and then tighten the mounting bolts and nut in the numbered sequence shown.
36. Install the ground cable.
37. Remove the jack and the wood block.
38. Install the oil pan.
39. Install the crankshaft pulley.
40. Install the cylinder head cover.
41. Install the water pump pulley.
42. Install the accessory drive belt.
43. Install the splash shield and the front undercover.
44. Install the front wheels.
45. Connect the negative battery cable.

VALVE COVER (CYLINDER HEAD COVER)

REMOVAL & INSTALLATION

➡See "Cylinder Head" in this section.

VALVE LASH

ADJUSTMENT

➡Adjust valves only when the cylinder head temperature is less than 100°F.

1. Before servicing the vehicle, refer to the precautions in the beginning of this section. Disconnect the negative battery cable.
2. Note the radio security code and the radio presets.
3. Remove the cylinder head cover retaining bolts. Remove the cylinder head cover from the engine.
4. Set the number one piston at TDC. The UP mark on the camshaft sprocket should be at the top, and the TDC grooves on the camshaft sprocket should line up with the top edge of the cylinder head.
5. Using the proper gauge feeler gauge, adjust the valves on cylinder number one.
6. Rotate the crankshaft clockwise. Align the number three piston TDC groove on the camshaft sprocket with the top edge of the cylinder head.
7. Using the proper gauge feeler gauge, adjust the valves on cylinder number three.
8. Rotate the crankshaft clockwise. Align the number four piston TDC groove on the camshaft sprocket with the top edge of the cylinder head.
9. Using the proper gauge feeler gauge, adjust the valves on cylinder number four.
10. Rotate the crankshaft clockwise. Align the number two piston TDC groove on the camshaft sprocket with the top edge of the cylinder head.
11. Using the proper gauge feeler gauge, adjust the valves on cylinder number two.
12. Install the cylinder head cover.

ENGINE PERFORMANCE & EMISSION CONTROLS

ACCELERATOR PEDAL POSITION (APP) SENSOR

LOCATION

The Accelerator Pedal Position (APP) sensor is located on the top of the accelerator pedal assembly.

REMOVAL & INSTALLATION

See Figure 33.

1. Disconnect the APP sensor 6P connector.
2. Remove the clip.

➡Do not reuse the clip once it is removed.

3. Push the tab, and remove the accelerator pedal pad from the pedal stop.
4. Remove the accelerator pedal module.

➡The APP sensor is not available separately. Do not disassemble the accelerator pedal module.

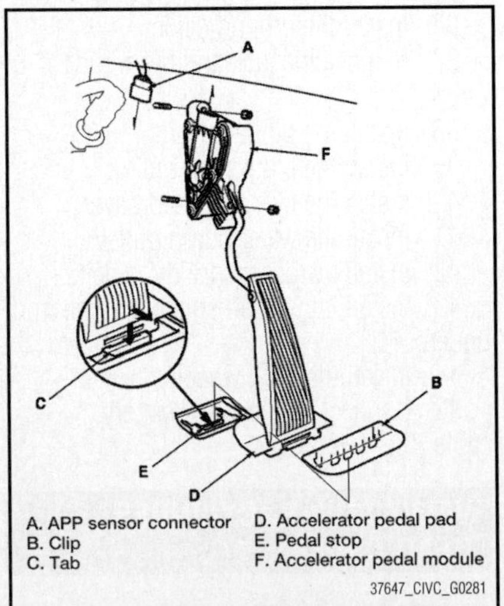
A
F
C
B
E
D
A. APP sensor connector
B. Clip
C. Tab
D. Accelerator pedal pad
E. Pedal stop
F. Accelerator pedal module
37647_CIVC_G0281

Fig. 33 Removing the accelerator pedal assembly

To install:

5. Set the accelerator pedal pad to the pedal stop.
6. Install the accelerator pedal module with a new clip.
7. Reconnect the APP sensor 6P connector.

CAMSHAFT POSITION (CMP) SENSOR

LOCATION

➡See "Removal & Installation" for location reference figure.

REMOVAL & INSTALLATION

See Figure 34.

1. Remove the air cleaner.
2. Disconnect the CMP sensor A 3P connector.
3. Remove CMP sensor A from the intake camshaft side of the cylinder head.

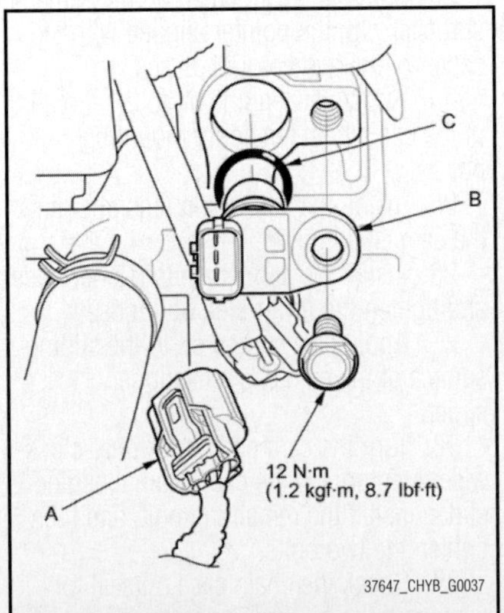
C
B
A
12 N·m
(1.2 kgf·m, 8.7 lbf·ft)
37647_CHYB_G0037

Fig. 34 Showing the CMP sensor connector (A), CMP sensor (B), and the O-ring (C)

4. Install the parts in the reverse order of removal with a new O-ring.

CRANKSHAFT POSITION (CKP) SENSOR

LOCATION

➡The Crankshaft Position (CKP) sensor is located on the engine block near the crankshaft pulley. See "Removal & Installation" for figure references of component locations.

REMOVAL & INSTALLATION

See Figure 35.

1. Disconnect the CKP sensor 3P connector.
2. Remove the CKP sensor and O-ring.
3. Install the parts in the reverse order of removal with a new O-ring.
4. Do the "CKP Pattern Clear/CKP Pattern Learn Procedure".

CKP PATTERN CLEAR/CKP PATTERN LEAN PROCEDURE

1. Start the engine. Hold the engine speed at 3,000 RPM without load (in neutral) until the radiator fan comes on.

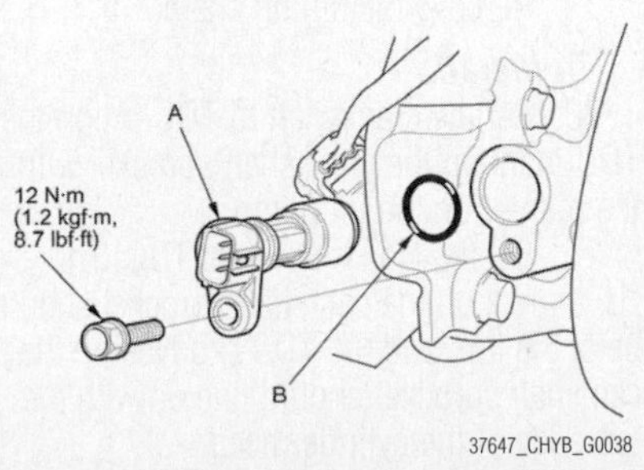
A
12 N·m
(1.2 kgf·m,
8.7 lbf·ft)
B
37647_CHYB_G0038

Fig. 35 Disconnect the CKP sensor 3P connector, then remove the CKP sensor (A) and O-ring (B).

2. Test-drive the vehicle on a level road: Decelerate (with the throttle fully closed) from an engine speed of 2,500 RPM down to 1,000 RPM with the transmission in 1st.

3. Test-drive the vehicle on a level road: Decelerate (with the throttle fully closed) from an engine speed of 5,000 RPM down to 3,000 RPM with the transmission in 1st.

4. Repeat previous two steps several times.

5. Turn the ignition switch to LOCK (0).

6. Turn the ignition switch to ON (II), and wait 30 seconds.

ENGINE COOLANT TEMPERATURE (ECT) SENSOR

LOCATION

These engine use two Engine Coolant Temperature (ECT) sensors. See figures in "Removal & Installation" for identification of locations.

REMOVAL & INSTALLATION

See Figures 36 and 37.

1. Drain the engine coolant.
2. Remove the air cleaner, if removing ECT sensor 1, or the splash shield if removing ECT sensor 2.
3. Disconnect the ECT sensor 2P connector.
4. Remove ECT sensor and O-ring.
5. Install the parts in the reverse order of removal with a new O-ring, then refill the radiator with engine coolant.

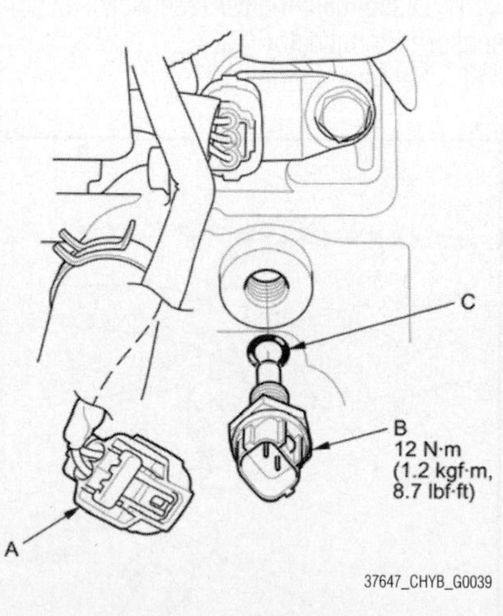
C
B
12 N·m
(1.2 kgf·m,
8.7 lbf·ft)
A
37647_CHYB_G0039

Fig. 36 Disconnect the ECT sensor 1 2P connector (A), then remove the ECT sensor 1 (B) and O-ring (C)

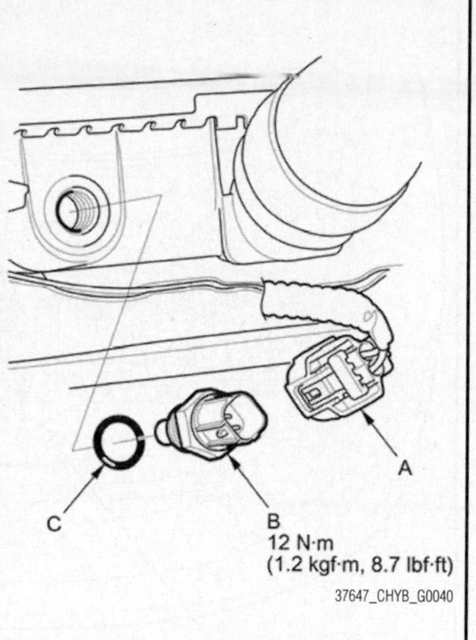
A
C
B
12 N·m
(1.2 kgf·m, 8.7 lbf·ft)
37647_CHYB_G0040

Fig. 37 Disconnect the ECT sensor 2 2P connector (A), then remove the ECT sensor 2 (B) and O-ring (C).

EVAPORATIVE EMISSIONS (EVAP) CANISTER

LOCATION

The Evaporative Emissions (EVAP) canister is located under the vehicle, just forward of the fuel tank. See "Removal & Installation" for figure references of component locations.

REMOVAL & INSTALLATION

See Figure 38.

1. Raise the vehicle on a lift.
2. Remove the undercover at the fuel tank.
3. Remove the EVAP canister guard pipe.
4. Remove the hoses, the FTP sensor connector, the EVAP canister vent shut valve connector, and the bolts.
5. Remove the EVAP canister.

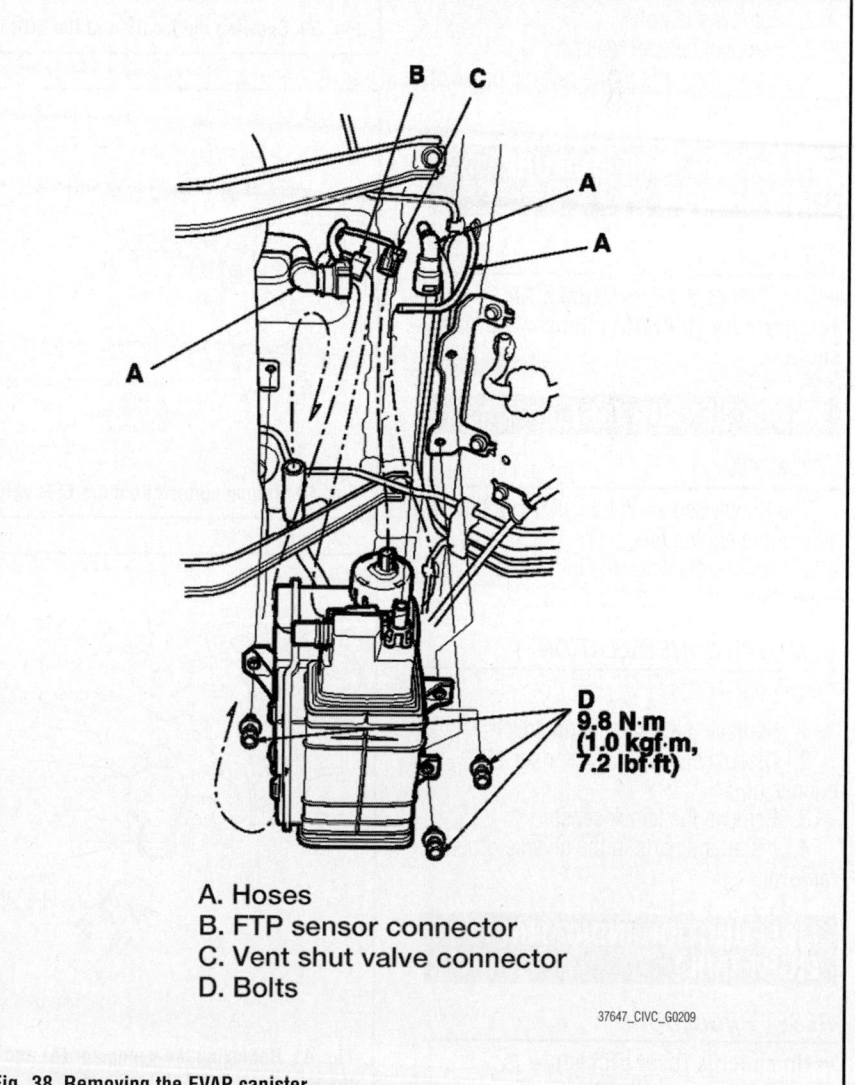

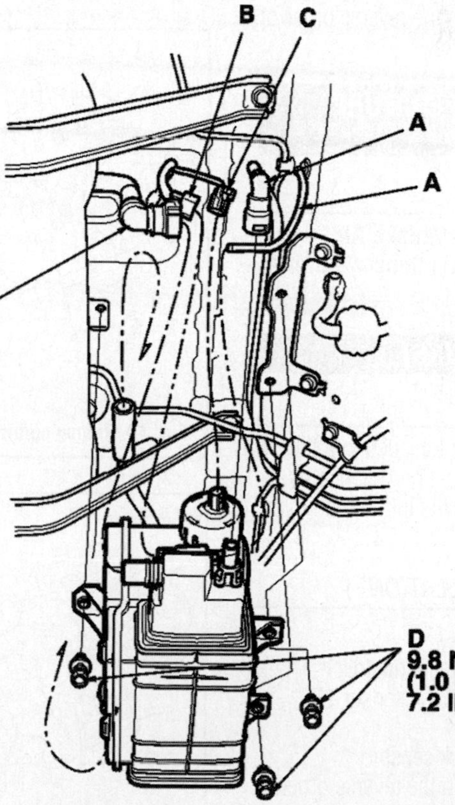
B
C
A
A
A
D
9.8 N·m
(1.0 kgf·m,
7.2 lbf·ft)
A. Hoses
B. FTP sensor connector
C. Vent shut valve connector
D. Bolts
37647_CIVC_G0209

Fig. 38 Removing the EVAP canister

6. Install the EVAP canister vent shut valve in the new EVAP canister with new O-rings.

➡**Do not coat the O-rings with engine oil.**

7. Install the parts in the reverse order of removal.

EXHAUST GAS RECIRCULATION (EGR) VALVE

LOCATION

See Figure 39.

The Exhaust Gas Recirculation (EGR) valve is used on 1.8L engine only and is located in the engine compartment, near the top front of the engine.

REMOVAL & INSTALLATION

See Figure 40.

1. Remove the EGR valve 6P connector.
2. Remove the bolts.
3. Remove the EGR valve.
4. Install the parts in the reverse order of removal with a new gasket.

INTAKE AIR TEMPERATURE (IAT) SENSOR

LOCATION

➡**See "Mass Air Flow/Intake Air Temperature (MAF/IAT) Sensor" in this section.**

KNOCK SENSOR (KS)

LOCATION

The Knock Sensor is threaded into the side of the engine block. For 1.8L engine, this is accessible when the intake manifold is removed.

REMOVAL & INSTALLATION

See Figure 41.

1. Remove the intake manifold.
2. Disconnect the knock sensor 1P connector.
3. Remove the knock sensor.
4. Install the parts in the reverse order of removal.

MALFUNCTION INDICATOR LIGHT (MIL)

RESET PROCEDURE

➡**No specific reset procedure is provided.**

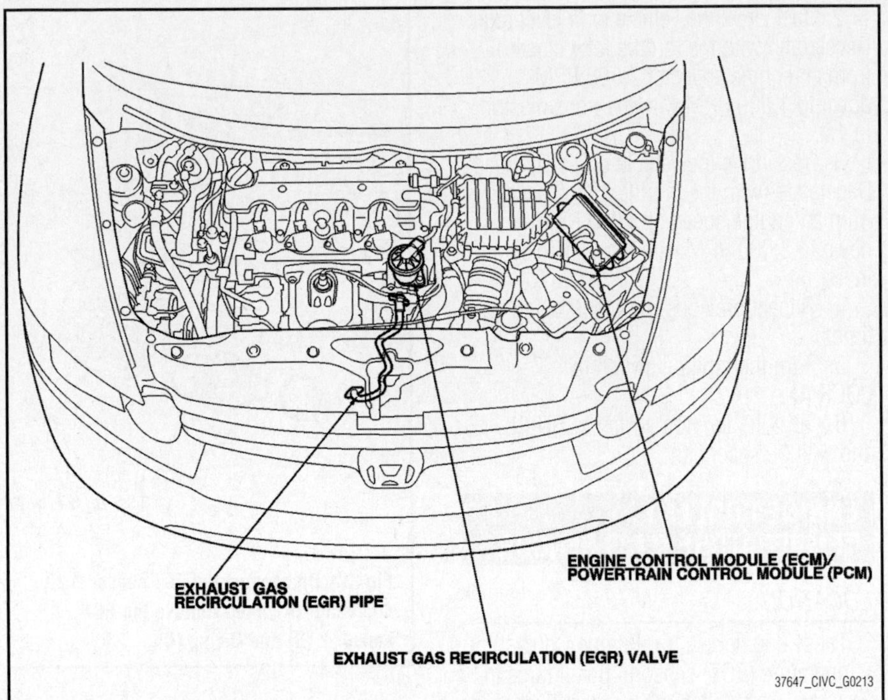
ENGINE CONTROL MODULE (ECM)/
POWERTRAIN CONTROL MODULE (PCM)
EXHAUST GAS
RECIRCULATION (EGR) PIPE
EXHAUST GAS RECIRCULATION (EGR) VALVE
37647_CIVC_G0213

Fig. 39 Showing the location of the EGR valve and pipe

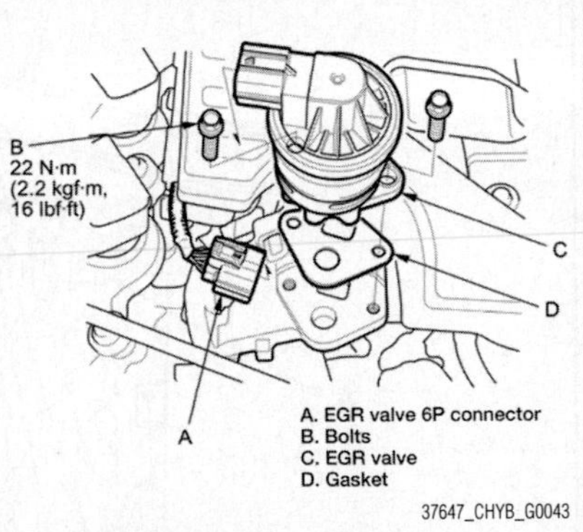
B
22 N·m
(2.2 kgf·m,
16 lbf·ft)
C
D
A
A. EGR valve 6P connector
B. Bolts
C. EGR valve
D. Gasket
37647_CHYB_G0043

Fig. 40 Removing/installing the EGR valve

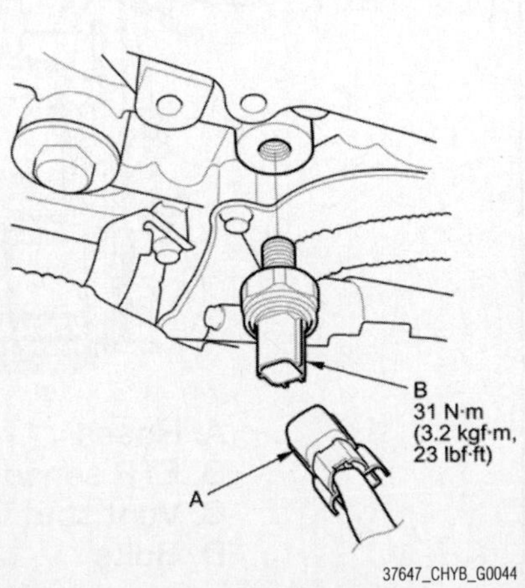
B
31 N·m
(3.2 kgf·m,
23 lbf·ft)
A
37647_CHYB_G0044

Fig. 41 Removing the connector (A) and the knock sensor (B)

MASS AIR FLOW/INTAKE AIR TEMPERATURE (MAF/IAT) SENSOR

LOCATION

The Mass Air Flow/Intake Air Temperature (MAF/IAT) sensor is located on the intake air duct. See "Removal & Installation" for location reference figure.

REMOVAL & INSTALLATION

See Figure 42.

1. Disconnect the MAF sensor/IAT sensor 5P connector (A).
2. Remove the bolts (B).

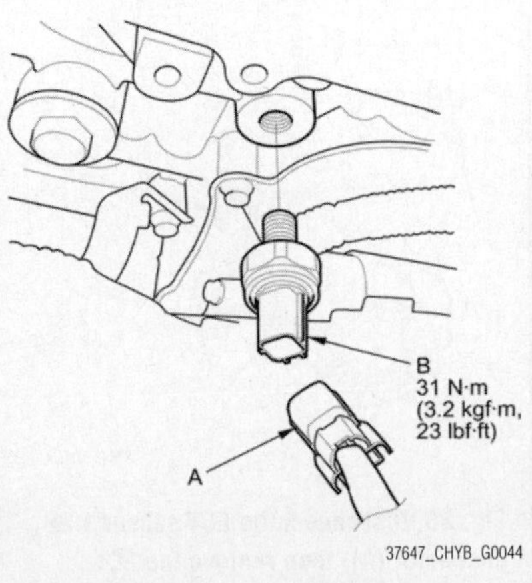
B
31 N·m
(3.2 kgf·m,
23 lbf·ft)
A
37647_CHYB_G0044

Fig. 42 Removing the MAF/IAT sensor

3. Remove the MAF sensor/IAT sensor (C).

4. Install the parts in the reverse order of removal with a new O-ring (D).

MANIFOLD ABSOLUTE PRESSURE (MAP) SENSOR

LOCATION

The Manifold Absolute Pressure (MAP) sensor is located on the firewall for the 1.8L engine. It is located near the front center of the engine for the 2.0L application.

REMOVAL & INSTALLATION

See Figure 43.

1. Disconnect the MAP sensor 3P connector.
2. Remove the MAP sensor and O-ring.
3. Install the parts in the reverse order of removal with a new O-ring

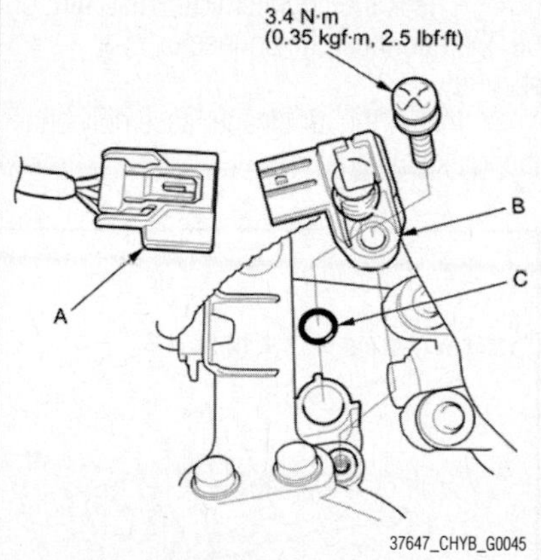
3.4 N·m
(0.35 kgf·m, 2.5 lbf·ft)
A
B
C
37647_CHYB_G0045

Fig. 43 Disconnect the MAP sensor 3P connector (A), then remove the MAP sensor (B) and O-ring (C)

POSITIVE CRANKCASE VENTILATION (PCV) VALVE

LOCATION

See Figure 44.

The Positive Crankcase Ventilation (PCV) valve is located on the passenger's side of the engine, as shown.

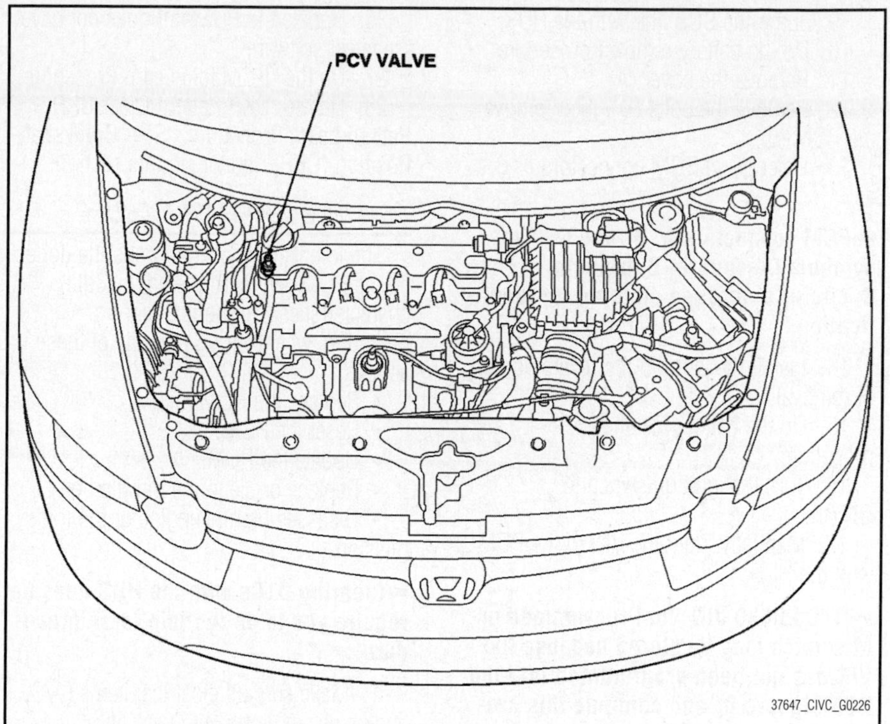
PCV VALVE
37647_CIVC_G0226

Fig. 44 Showing PCV location—1.8L engine shown; 2.0L engine location similar

REMOVAL & INSTALLATION

See Figure 45.

1. Remove the intake manifold. See "Intake Manifold" in "ENGINE MECHANICAL" section.
2. Disconnect the PCV hose (A), then unscrew the PCV valve (B), and remove it.

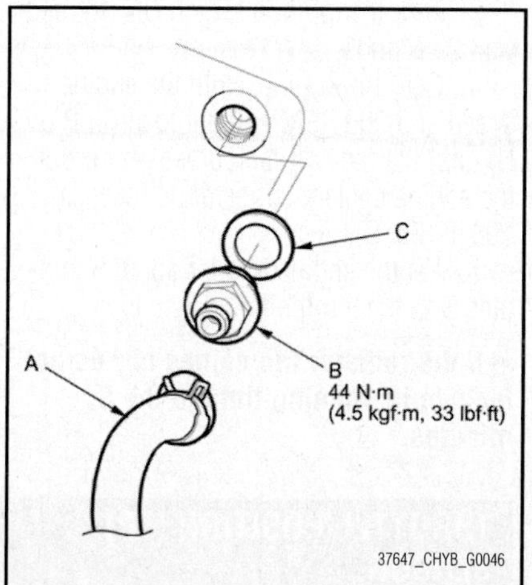
C
A
B
44 N·m
(4.5 kgf·m, 33 lbf·ft)
37647_CHYB_G0046

Fig. 45 Removing the PCV valve hose (A), PCV valve (B) and washer (C)

3. Install the parts in the reverse order of removal with a new washer (C).

POWERTRAIN CONTROL MODULE (PCM)

LOCATION

The Powertrain Control Module (PCM) is located on the driver's side of the engine compartment, mounted to the inner fender area.

REMOVAL & INSTALLATION

1. Connect the HDS, or equivalent scan tool, to the data link connector (DLC) located under the driver's side of the dashboard.
2. Turn the ignition switch to ON (II).
3. Make sure the HDS communicates with the PCM and other vehicle systems. If it doesn't, perform further DLC circuit troubleshooting.
   a. If you are returning from DLC circuit troubleshooting, skip steps 4 through 7, 18 through 20, 23 through 24, and do this after replacing the PCM:
      - Replace the engine oil and the engine oil filter.
      - Clean the throttle body.
4. Select the PGM-FI system with the HDS.
5. Select the INSPECTION MENU with the HDS.
6. Select the ETCS TEST, then select the TP POSITION CHECK, and follow the screen prompts.

➡If the TP POSITION CHECK indicates FAILED, continue with this procedure.

7. Select the REPLACE PCM MENU, then select READ DATA and follow the screen prompts.

➡Doing this step copies (READS) the engine oil life data from the original PCM so you can later download (WRITES) it into the new PCM. If READ DATA indicates FAILED, continue with this procedure.

8. Turn the ignition switch to LOCK (0).

9. Jump the SCS line with the HDS.
10. Do the battery removal procedure.
11. Remove the cover (A).
12. Remove the bolts (D), then remove the PCM (E).
13. Disconnect PCM connectors A, B, and C.

➡PCM connectors A, B, and C have symbols (A=square, B=triangle, C=circle) embossed on them for identification.

14. Install the parts in the reverse order of removal.
15. Do the battery installation procedure.
16. Turn the ignition switch to ON (II).
17. Manually input the VIN to the PCM with the HDS.

➡DTC P0630 VIN Not Programmed or Mismatch may be stored because the VIN has not been programmed into the PCM; ignore it, and continue this procedure.

18. If the READ DATA (engine oil life) failed in step 7, go to step 23; otherwise, go to step 19.
19. Select the PGM-FI system with the HDS.
20. Select the REPLACE PCM MENU, then select WRITE DATA and follow the screen prompts.

➡If the WRITE DATA indicates FAILED, continue with this procedure.

21. Select IMMOBI system with the HDS.
22. Enter the immobilizer PCM code that you got from the iN, and use the PCM replacement procedure in the IMMOBI MENU of the HDS; it allows you to start the engine.
23. If the TP POSITION CHECK failed in step 6, clean the throttle body, then go to step 24.
24. If the READ DATA failed in step 8, or the WRITE DATA failed in step 20, replace the engine oil and engine oil filter, then go to step 25.
25. Select PGM-FI system and reset the PCM with the HDS.
26. Update the PCM if it does not have the latest software.
27. Do the "PCM Idle Learn Procedure."
28. Do the "CKP Pattern Clear/CKP Pattern Learn Procedure." See "Crankshaft Position (CKP) Sensor" in this section.

PCM IDLE LEARN PROCEDURE

The idle learn procedure must be done so the PCM can learn the engine idle characteristics. Do the idle learn procedure whenever you do any of these actions:

- Replace the PCM.
- Reset the PCM.
- Update the PCM.
- Replace or clean the throttle body.
- Disassemble the engine or the transmission.

➡Clearing DTCs with the HDS does not require you to do the idle learn procedure.

1. Make sure all electrical items (A/C, audio system, lights, etc.) are off.
2. Reset the PCM with the HDS, or equivalent scan tool.
3. Turn the ignition switch ON (II), and wait 2 seconds.
4. Start the engine. Hold the engine speed at 3,000 RPM without load (in P or N) until the radiator fan comes on, or until the engine coolant temperature reaches 194°F (90°C).
5. Let the engine idle for about 5 minutes with the throttle fully closed.

➡If the radiator fan comes on, do not include its running time in the 5 minutes.

THROTTLE POSITION (TP) SENSOR

LOCATION

The Throttle Position (TP) Sensor is located with the Throttle Actuator Control (TAC) on the throttle body.

REMOVAL & INSTALLATION

➡Manufacturer does not provide a specific removal and installation procedure for this component. Use the illustration as a guide when servicing this component.

VEHICLE SPEED SENSOR (VSS)

LOCATION

The Vehicle Speed Sensor (VSS) is located on the CVT assembly. See "Removal & Installation" for additional location reference.

REMOVAL & INSTALLATION

See Figure 46.

1. Remove the air cleaner assembly and the intake air duct.
2. Disconnect the vehicle speed sensor connector, then remove the vehicle speed sensor.
3. Install a new O-ring on a new vehicle speed sensor, then install the vehicle speed sensor in the transmission housing.
4. Check the connector for rust, dirt, or oil, then connect the connector securely.
5. Install the air cleaner assembly and the intake air duct.

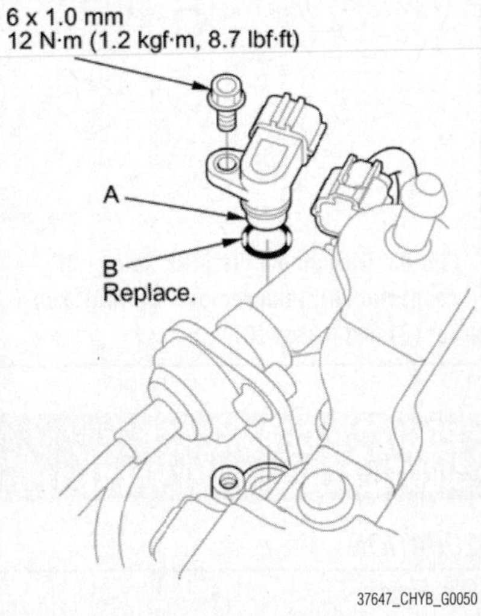
6 x 1.0 mm
12 N·m (1.2 kgf·m, 8.7 lbf·ft)
A
B
Replace.
37647_CHYB_G0050

Fig. 46 Removing/installing the vehicle speed sensor (A) and O-ring (B)

FUEL SYSTEM SERVICE PRECAUTIONS

Safety is the most important factor when performing not only fuel system maintenance, but any type of maintenance. Failure to conduct maintenance and repairs in a safe manner may result in serious personal injury or death. Work on a vehicle's fuel system components can be accomplished safely and effectively by adhering to the following rules and guidelines.

- To avoid the possibility of fire and personal injury, always disconnect the negative battery cable unless the repair or test procedure requires that battery voltage be applied.
- Always relieve the fuel system pressure prior to disconnecting any fuel system component (injector, fuel rail, pressure regulator, etc.) fitting or fuel line connection. Exercise extreme caution whenever relieving fuel system pressure to avoid exposing skin, face and eyes to fuel spray. Please be advised that fuel under pressure may penetrate the skin or any part of the body that it contacts.
- Always place a shop towel or cloth around the fitting or connection prior to loosening to absorb any excess fuel due to spillage. Ensure that all fuel spillage is quickly removed from engine surfaces. Ensure that all fuel-soaked cloths or towels are deposited into a flame-proof waste container with a lid.
- Always keep a dry chemical (Class B) fire extinguisher near the work area.
- Do not allow fuel spray or fuel vapors to come into contact with a spark or open flame.
- Always use a second wrench when loosening or tightening fuel line connection fittings. This will prevent unnecessary stress and torsion on fuel piping. Always follow the proper torque specifications.
- Always replace worn fuel fitting O-rings with new ones. Do not substitute fuel hose where rigid pipe is installed.

FUEL SYSTEM PRESSURE

RELIEVING

1. Remove PGM-FI main relay 2 from the auxiliary under-hood fuse/relay box.
2. Start the engine, and let it idle until it stalls.

➡If any DTCs are stored, clear and ignore them.

3. Turn the ignition switch to LOCK (0)
4. Remove the fuel fill cap.
5. Do the battery terminal reconnection procedure.
6. Remove the air intake assembly.
7. Remove the cover and quick-connect fitting cover.
8. Check the fuel quick-connect fitting for dirt, and clean it if needed.
9. Place a rag or shop towel over the quick-connect fitting.
10. Disconnect the quick-connect fitting: Hold the connector with one hand, and squeeze the retainer tabs with the other hand to release them from the locking tabs. Pull the connector off.

FUEL FILTER

REMOVAL & INSTALLATION

➡The fuel filter is an integrated part of the fuel pump assembly. For additional information, refer to "Fuel Pump" in this section.

FUEL PUMP/ FUEL GAUGE SENDING UNIT

REMOVAL & INSTALLATION

See Figure 47.

⁑ CAUTION

The fuel injection system remains under pressure, even after the engine has been turned OFF. The fuel system pressure must be relieved before disconnecting any fuel lines. Failure to follow this procedure may result in fire, explosion, or personal injury.

1. Before servicing the vehicle, refer to the precautions in the beginning of this section.
2. Disconnect the negative battery cable.
3. Relieve the fuel pressure.
4. Remove or disconnect the following:
   - Rear seat cushions
   - Remove the fuel filler cap
   - Rear floor upper crossmember
   - Fuel pump access panel
   - Fuel pump electrical harness

➡Clean the fuel line fittings before disconnecting them.

5. Disconnect the quick connect fitting from the pump assembly.
6. Using Special Tool 07AAA-SOXA100, or equivalent, loosen the fuel tank unit locknut. Remove the locknut and the fuel tank unit. Allow the fuel in the pump drain into the tank before removing the pump from the vehicle.

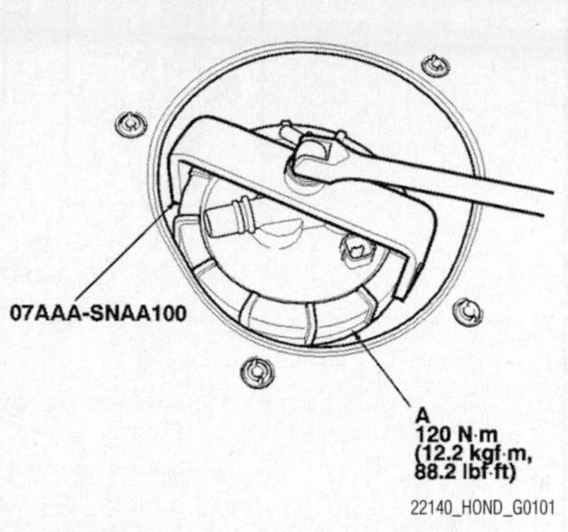
07AAA-SNAA100
A
120 N·m
(12.2 kgf·m,
88.2 lbf·ft)
22140_HOND_G0101

Fig. 47 Using a suitable fuel sender wrench, tighten a new locknut to 88 ft. lbs. (120 Nm)

To install:

7. Be sure to use a new base gasket and new O-rings.
8. When installing the fuel tank unit, align the marks on the fuel tank and the fuel tank unit.
9. Make sure the electrical connections are secure and the connector firmly locked in place.
10. Make sure the fuel line connections are secure and the connector firmly locked in place.
11. Using Special Tool 07AAA-SOXA100, or equivalent, tighten the fuel tank unit locknut to 88 ft. lbs. (120 Nm).
12. The remainder of the installation is the reverse order of removal.

FUEL RAIL AND INJECTOR

REMOVAL & INSTALLATION

See Figure 48.

1. Relieve the fuel pressure, as shown in this section.
2. Remove the intake air resonator.
3. Remove the air cleaner.
4. Remove the nut and bolt.
5. Disconnect the connectors from the injectors, the intake side ignition coils, the EGR valve, the rocker arm oil control valve, the rocker arm oil pressure sensor, the MAP sensor, and the rocker arm oil pressure switch.
6. Remove the ground terminals.
7. Disconnect the hoses from the fuel rail.
8. Remove the fuel rail mounting nuts

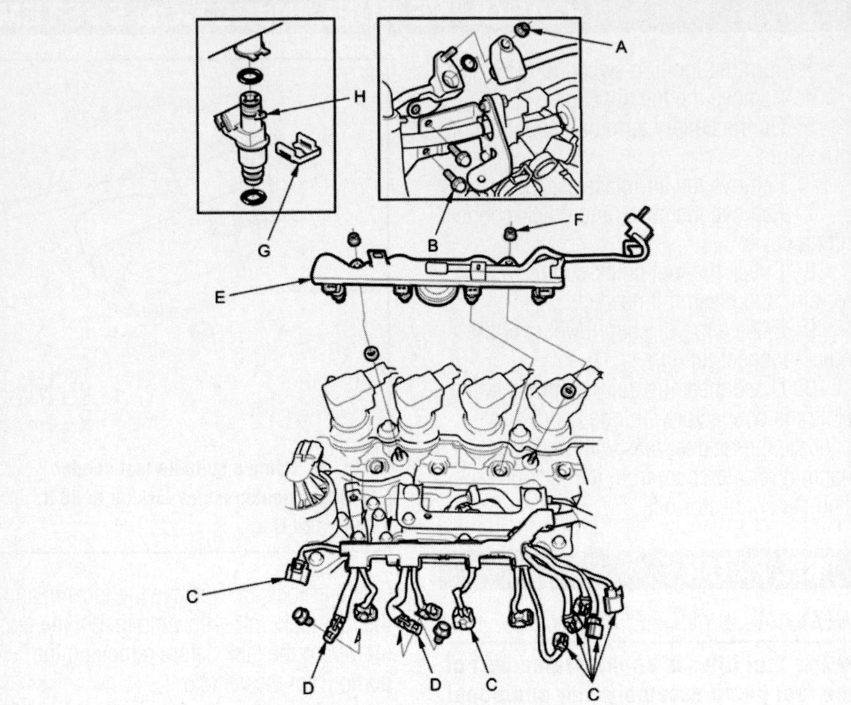

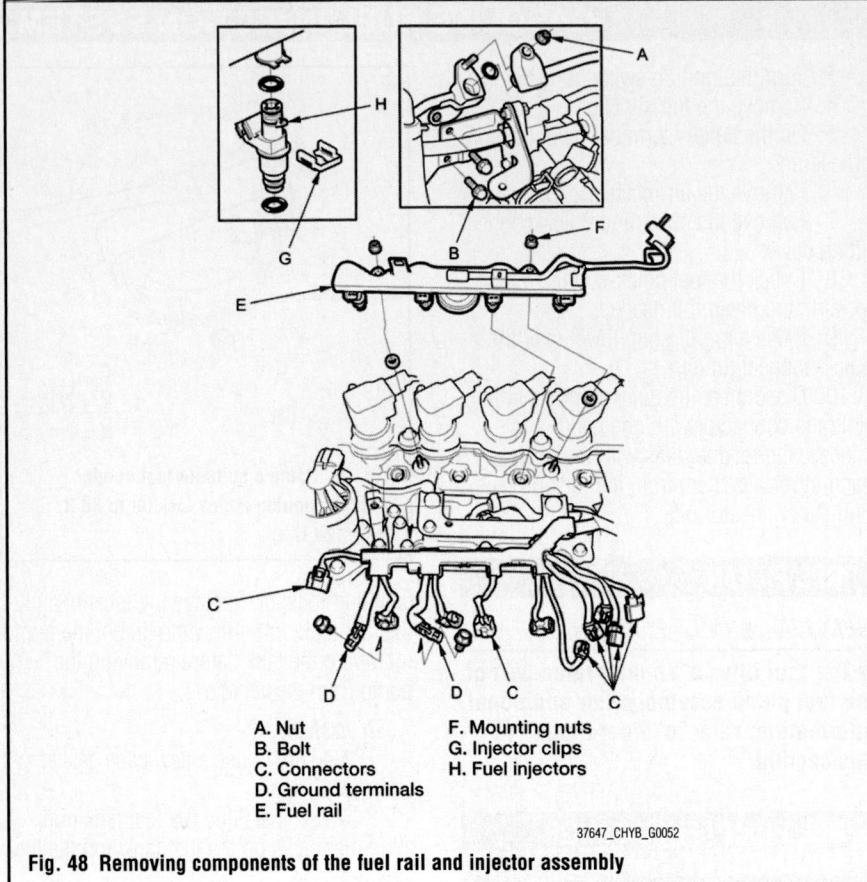

A
H
F
G
B
E
C
D
D
C
C
A. Nut
B. Bolt
C. Connectors
D. Ground terminals
E. Fuel rail
F. Mounting nuts
G. Injector clips
H. Fuel injectors
37647_CHYB_G0052

Fig. 48 Removing components of the fuel rail and injector assembly

from the fuel rail, then remove the injectors and fuel rail together.

9. Remove the injector clips from the injectors.
10. Remove the injectors from the fuel rail.

To install:

11. Coat the new O-rings with clean engine oil, and insert the injectors into the fuel rail.
12. Install the injector clips.
13. Coat the injector O-rings with clean engine oil.
14. Install the injectors into the injector base.
15. Install the fuel rail mounting nuts and the bolt.
16. Install the nut with a new O-ring.
17. Reconnect the fuel hoses.
18. Connect the ground terminal.
19. Connect the connectors on the injectors, the intake side ignition coils, the EGR valve, the rocker arm oil control valve, the rocker arm oil pressure sensor, the MAP sensor, and the rocker arm oil pressure switch.
20. Install the air cleaner.
21. Install the intake air resonator.
22. Turn the ignition switch to ON (II), but do not operate the starter. After the fuel pump runs for about 2 seconds, the fuel pressure in the fuel line rises.
23. Repeat this two or three times, then make sure there are no fuel leaks.

FUEL TANK

DRAINING

1. Remove the fuel tank unit. See "Fuel Pump/Fuel Gauge Sending Unit" in this section.
2. Using a hand pump, a hose, and a container suitable for fuel, draw the fuel from the fuel tank.
3. Reinstall the fuel tank unit.

REMOVAL & INSTALLATION

See Figure 49.

1. Drain the fuel tank, then disconnect the fuel tank unit 4P connector and the quick-connect fitting.
2. Raise the vehicle on a lift.
3. Remove the cover and the EVAP canister guard pipe.
4. Disconnect the fuel fill tube. Slide back the clamps, then twist the hoses as you pull to avoid damaging them.
5. Disconnect the quick-connect fitting and the fuel tank vapor recirculation tube.
6. Place a jack or other support under the fuel tank.
7. Remove the strap bolts and the strap.
8. Remove the fuel tank.
9. Install the parts in the reverse order of removal.

IDLE SPEED

INSPECTION & ADJUSTMENT

➡The idle speed is not adjustable except by performing the "PCM Idle Learn Procedure". The following information will help determine if the idle speed is not to specification.

1. Before checking the idle speed, check these items:
   - The malfunction indicator lamp (MIL) has not been reported on, and there are no DTCs.
   - Ignition timing
   - Spark plugs
   - Air cleaner
   - PCV system
2. Apply the parking brake, and make sure the headlights are off.
3. Disconnect the evaporative emission (EVAP) canister purge valve connector.
4. Connect the scan tool to the data link connector (DLC) located under the driver's side of the dashboard.
5. Make sure the scan tool communicates with the PCM. If it doesn't, go to the DLC circuit troubleshooting.
6. Start the engine. Hold the engine speed at 3,000 RPM without load (A/T in P or N, M/T in neutral) until the radiator fan comes on, then let it idle.
7. Check the idle speed without load conditions: headlights, blower fan, radiator fan, and air conditioner off. Idle speed should be 820±50 RPM (in P or N)
8. Let the engine idle for 1 minute with high electric load (A/C on, temperature set to max cool, blower fan on high, headlights on high beam). Idle speed should be 820±50 RPM (in P or N)

➡If the idle speed is not within specification, do the "ECM Idle Learn Procedure". If the idle speed is still not within specification, go to the "DIAGNOSTICS" section.

9. Reconnect the EVAP canister purge valve connector.

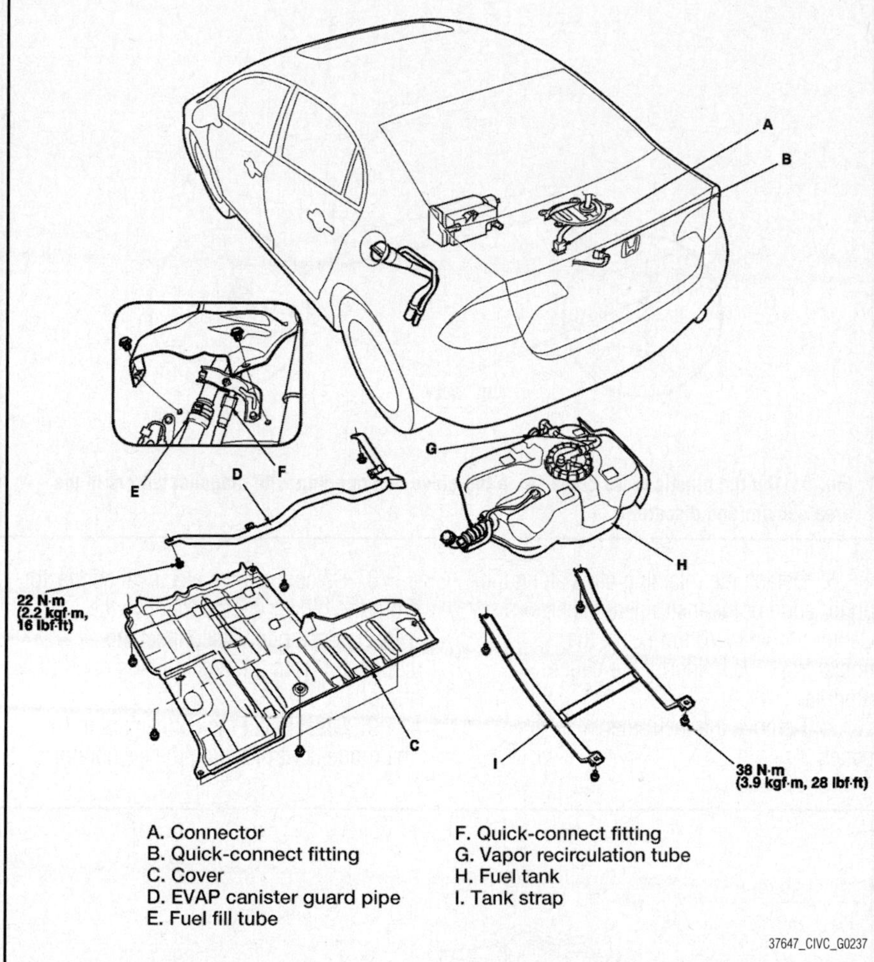
A
B
G
D
F
E
H
22 N·m
(2.2 kgf·m,
16 lbf·ft)
C
I
38 N·m
(3.9 kgf·m, 28 lbf·ft)
A. Connector
B. Quick-connect fitting
C. Cover
D. EVAP canister guard pipe
E. Fuel fill tube
F. Quick-connect fitting
G. Vapor recirculation tube
H. Fuel tank
I. Tank strap
37647_CIVC_G0237

Fig. 49 Showing the fuel tank and related components

THROTTLE BODY

REMOVAL & INSTALLATION

See Figure 50

1. Connect the HDS, or equivalent scan tool, to the DLC while the engine is stopped.
2. Select the INSPECTION MENU on the HDS.
3. Do the TP POSITION CHECK in the ETCS TEST.
4. Turn the ignition switch to LOCK (0).
5. Remove the cowl cover and under-cowl panel.
6. Remove the air cleaner, and the intake air duct.
7. Disconnect the throttle body connector.
8. Disconnect and plug the water bypass hoses and vacuum hose.
9. Remove the throttle body.
10. Install the parts in the reverse order of removal with a new gasket.

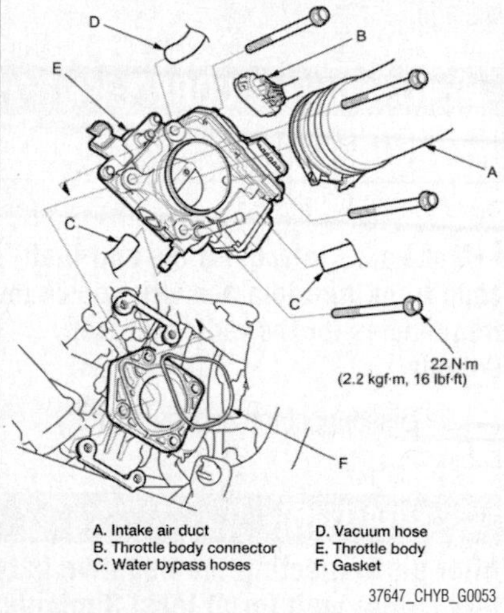
D
B
E
A
C
C
22 N·m
(2.2 kgf·m, 16 lbf·ft)
F
A. Intake air duct
B. Throttle body connector
C. Water bypass hoses
D. Vacuum hose
E. Throttle body
F. Gasket
37647_CHYB_G0053

Fig. 50 Exploded view of the throttle body assembly

HEATING & AIR CONDITIONING SYSTEM

BLOWER MOTOR

REMOVAL & INSTALLATION

See Figure 51.

1. Remove the glove box.
2. Cut the plastic cross brace in the glove box opening with diagonal cutters in the area shown, and discard it.
3. Remove the bolts and the glove box frame.
4. Remove the wire harness clip, the self-tapping screws, and the passenger's heater duct.
5. Disconnect the connector from the blower motor. Remove the wire harness clip.
6. Disconnect the connector from the recirculation control motor. Remove the self-tapping screws, the bolt, the mounting nuts, and the blower unit.
7. Installation is the reverse order of removal. Make sure that there is no air leakage.

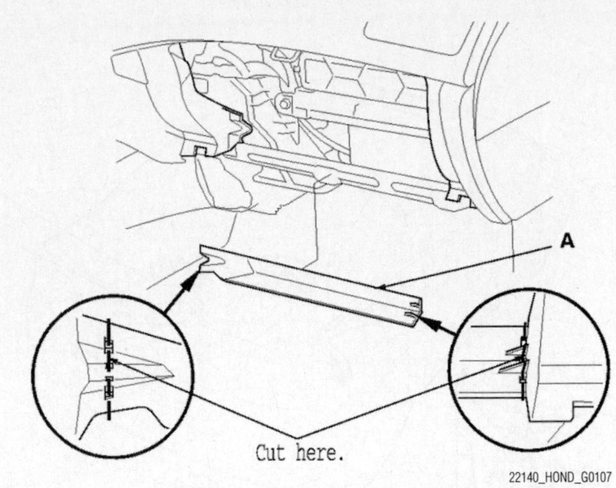
A
Cut here.
22140_HOND_G0107

Fig. 51 Cut the plastic cross brace (A) in the glove box opening with diagonal cutters in the area shown, and discard it

HEATER UNIT & CORE

REMOVAL & INSTALLATION

See Figures 52 and 53.

➡Make sure to acquire the anti-theft code from the radio and write down the frequencies for the radio's preset buttons.

1. Disconnect the negative battery cable.

⁂ CAUTION

After disconnecting the negative battery cable, wait for at least 3 minutes for the air bag module to deplete its energy before working the on the instrument panel or steering wheel.

2. Discharge the air conditioning system.
3. Remove the air cleaner assembly. Disconnect the heater hoses. Remove the mounting nut from the heater unit.
4. Remove the sub-display visor. Remove the navigation system, if equipped. On vehicles without navigation system remove the radio.
5. Remove the dashboard retaining screws. Detach the retaining clips along the lower edge of the instrument panel.
6. Detach the retaining clips along the upper edge of the instrument panel. Gently pull forward to release the hooks from the holder of the gauge module.
7. Remove the instrument panel.
8. Disconnect the electrical connector from the blower motor. Remove the wire harness clip. Disconnect the connector from the recirculation control motor.
9. Disconnect the connectors from the mode control motor, the evaporator

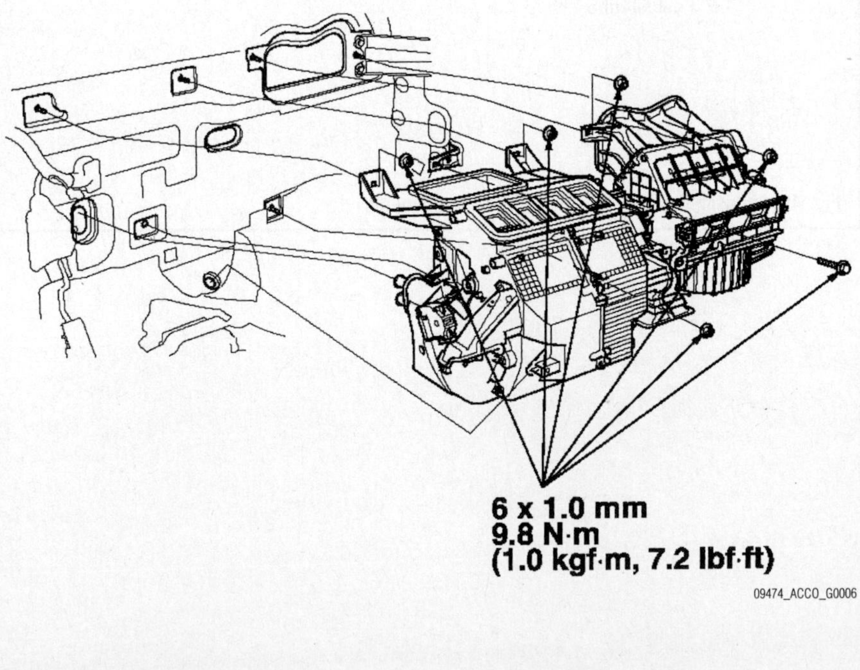
6 x 1.0 mm
9.8 N·m
(1.0 kgf·m, 7.2 lbf·ft)
09474_ACCO_G0006

Fig. 52 View of the heater housing, evaporator housing and related components

temperature sensor and the power transistor. Remove the wire clip harness.

10. Disconnect the connectors from the air mix control motor and the air conditioning wire harness. Remove the connector clip and the wire harness clips.

11. Remove the mounting bolt, mounting nuts and the heater/blower unit from the vehicle.

12. Remove the self taping screws, remove the grommet and carefully pry out the heater core.

To install:

13. Installation is the reverse of the removal procedure.

14. Evacuate, charge and leak test the air conditioning system refrigerant, as required.

15. Operate the engine to normal operating temperatures; then, check the climate control operation and check for leaks.

16. Enter the antitheft codes for the radio and the navigation system.

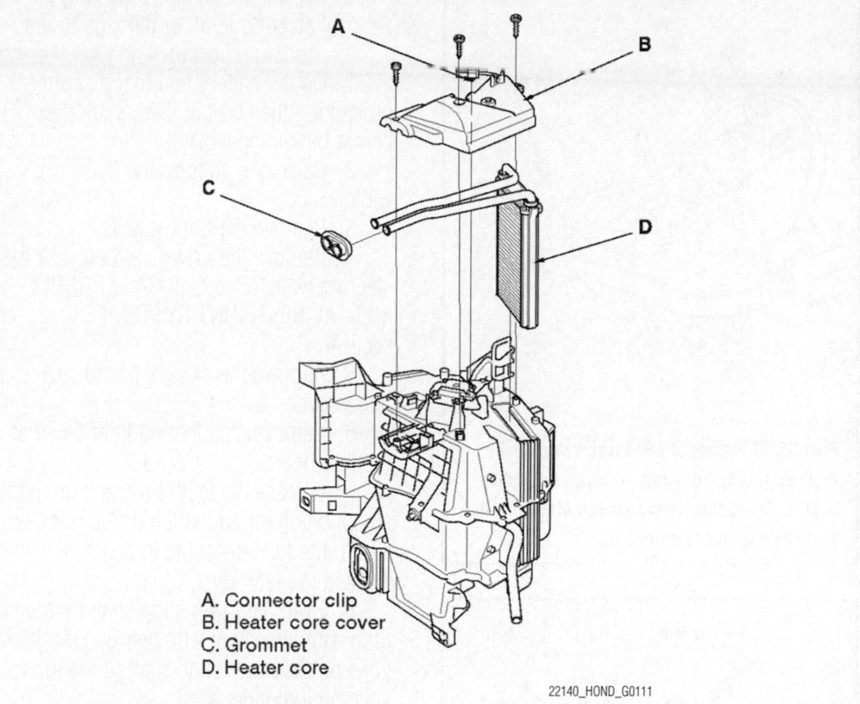
A
B
C
D
A. Connector clip
B. Heater core cover
C. Grommet
D. Heater core
22140_HOND_G0111

Fig. 53 Exploded view of the heater unit

STEERING

ELECTRONIC POWER STEERING (EPS) CONTROL UNIT

REMOVAL & INSTALLATION

See Figure 54.

1. Do the battery terminal disconnection procedure and wait 3 minutes before starting work.

2. Remove the passenger's dashboard under cover.

3. Remove the passenger's side kick panel.

4. Disconnect EPS control unit connector A (2P), connector B (2P), connector C (2P), and connector D (28P). Remove the nuts from the EPS control unit. Remove the EPS control unit.

To install:

5. Install the EPS control unit in the reverse order of removal.

6. Do the battery terminal reconnection procedure.

7. Do the "Torque Sensor Neutral Position Memorization". See procedure under "Power Rack & Pinion Steering Gear" for EPS application.

8. After installation, start the engine, allow it to idle. Turn the steering wheel from lock to lock several times. Check that the EPS indicator does not come on.

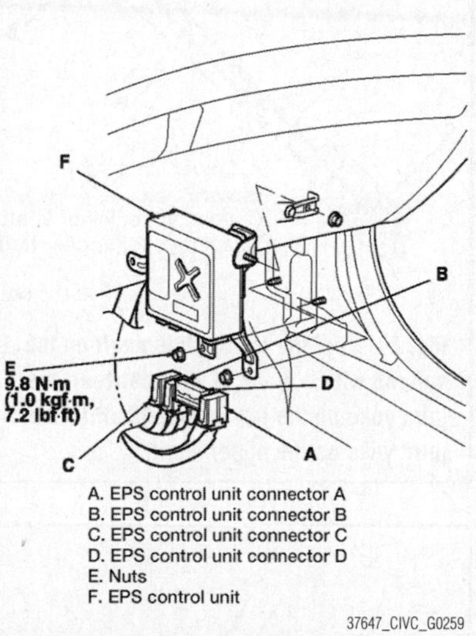
F
B
E
9.8 N·m
(1.0 kgf·m,
7.2 lbf·ft)
D
C
A
A. EPS control unit connector A
B. EPS control unit connector B
C. EPS control unit connector C
D. EPS control unit connector D
E. Nuts
F. EPS control unit
37647_CIVC_G0259

Fig. 54 Disconnect EPS control unit connector A (2P), connector B (2P), connector C (2P), and connector D (28P). Remove the nuts from the EPS control unit. Remove the EPS control unit.

ELECTRONIC POWER STEERING (EPS) MOTOR

REMOVAL & INSTALLATION

See Figures 55 and 56.

⁂ CAUTION

Do not allow dust, dirt, or other foreign materials to enter the steering gearbox. Make sure not to get any silicone grease on the terminal part of the connectors and switches, especially if you have silicone grease on your hands or gloves.

1. Remove the steering gearbox. See "Power Rack & Pinion Steering Gear" in this section.

2. Disconnect the torque sensor 3P connector from the steering gearbox, then remove the wire harness clamp bolts and the ground terminal.

3. Remove the EPS motor from the steering gearbox, then remove the O-ring.

To install:

4. Clean the mating surface of the EPS motor and the steering gearbox.

5. Apply a thin coat of silicone grease to the O-ring and carefully fit it on the EPS motor.

6. Apply steering grease into the EPS motor shaft.

7. Set the motor on the steering gearbox by engaging the EPS motor shaft and the worm shaft.

8. Turn the EPS motor two or three times to the right and left about 45 degrees.

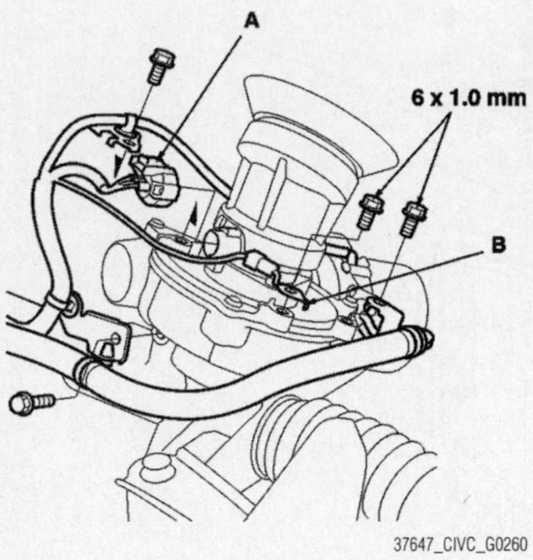
A
6 x 1.0 mm
B
37647_CIVC_G0260

Fig. 55 Disconnect the torque sensor 3P connector (A) from the steering gearbox, then remove the wire harness clamp bolts and the ground terminal (B).

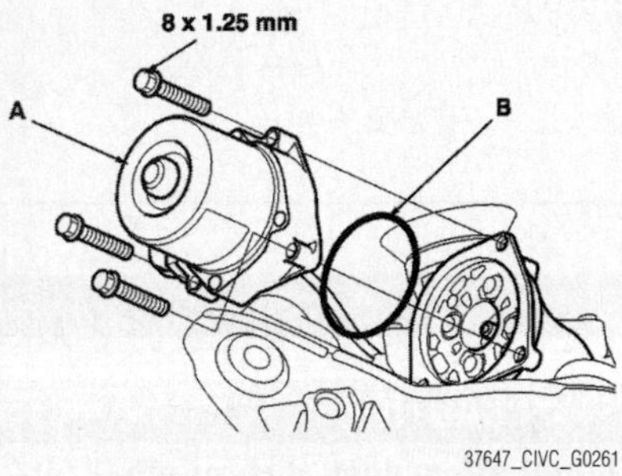
8 x 1.25 mm
A
B
37647_CIVC_G0261

Fig. 56 Remove the EPS motor (A) from the steering gearbox, then remove the O-ring (B).

Make sure the EPS motor is evenly seated on the steering gearbox, and that the O-ring is not pinched between the mating surfaces.

9. Loosely install the EPS motor mount bolts, then turn the steering wheel two or three times to the right and left about 45 degrees.

10. Tighten the EPS motor mount bolts to 14 ft. lbs. (20 Nm).

POWER RACK & PINION STEERING GEAR

REMOVAL & INSTALLATION

See Figures 57 through 63.

➡Note these items during removal:

- Use solvent and a brush, wash any oil and dirt off the end of the steering gearbox. Avoid any electrical parts. Blow dry with compressed air. Wear eye protection.
- Make sure to remove the steering wheel before disconnecting the steering joint, or damage to the cable reel occur.

1. Do the battery terminal disconnect procedure, then wait at least 3 minutes before beginning work.
2. Raise and support the vehicle.
3. Remove the front wheels.
4. Remove the driver's airbag, and the steering wheel. See "AIR BAG (SUPPLEMENTAL RESTRAINT SYSTEM)" section.
5. Remove the driver's dashboard undercover.
6. Remove the steering joint cover at the floorboard.
7. Release the lock lever, and adjust the steering column to the full tilt up position, and to the full telescopic in position. Tighten the lock lever.
8. Hold the lower slide shaft on the column with a piece of wire between the joint yoke on the lower slide shaft to the joint yoke on the upper shaft.

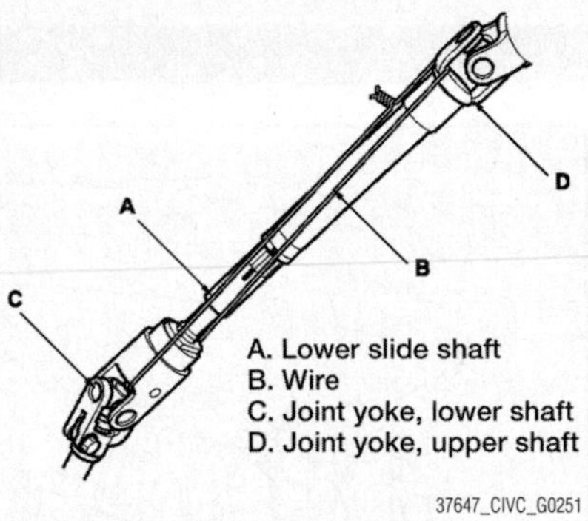
A
B
C
D
A. Lower slide shaft
B. Wire
C. Joint yoke, lower shaft
D. Joint yoke, upper shaft
37647_CIVC_G0251

Fig. 57 Hold the lower slide shaft on the column with a piece of wire between the joint yoke on the lower slide shaft to the joint yoke on the upper shaft.

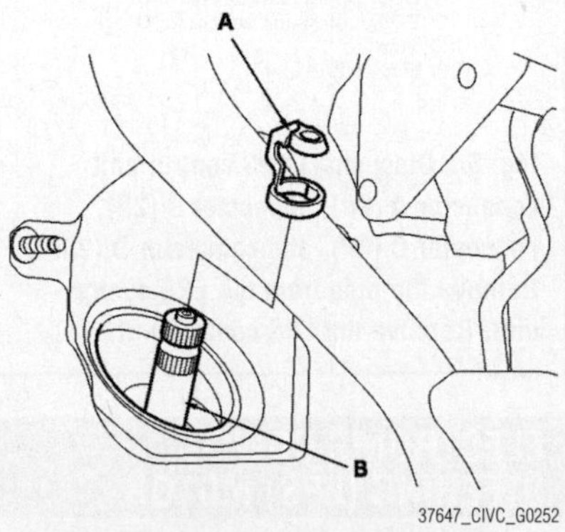
A
B
37647_CIVC_G0252

Fig. 58 Remove the center guide (A) (if equipped) from the top of the pinion shaft (B), and discard it.

9. Remove the steering joint bolt and disconnect the steering joint by moving the steering joint toward the column.
10. Remove the center guide (if equipped) from the top of the pinion shaft and discard it.

➡The center guide is for factory assembly use only.

11. Remove the cowl cover and the under-cowl panel.
12. Remove the air cleaner housing.
13. Remove the air cleaner mounting bracket and install the 1.8 support eyelet (07AAK-SNAA400) behind the breather pipe and down to the threaded hole on the cylinder head. Install a retaining bolt.
14. Attach the engine hanger adapter (VSB02C000015) to the threaded hole in the cylinder head.
15. Install the front leg assembly, hook, and the wing nut to an "A and Reds" engine support hanger (AAR-T1256) onto the engine hanger. Carefully position the engine hanger on the vehicle, and attach the hook to the forward hole in the engine hanger adapter in the 1.8 support eyelet. Tighten the wing nut by hand to lift and support the engine/transmission.
16. Remove the cotter pin from the tie-rod end ball joint, then remove the nut on both sides.
17. Disconnect the tie-rod end ball joint from the knuckle using the ball joint thread protector and the ball joint remover.

➡Be careful not to damage the ball joint boot when installing the remover.

18. Remove the front splash shield.
19. Disconnect the EPS motor 2P connector, the EPS motor 1P connector, torque sensor 4P connector, the EPS motor angle

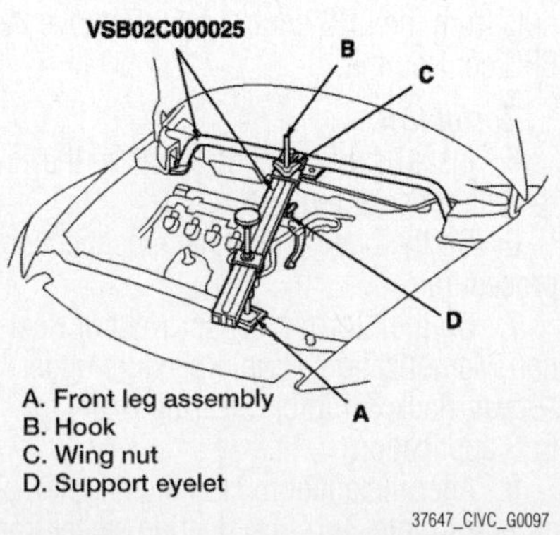
VSB02C000025
B
C
D
A
A. Front leg assembly
B. Hook
C. Wing nut
D. Support eyelet
37647_CIVC_G0097

Fig. 59 Installed position of the engine support hanger

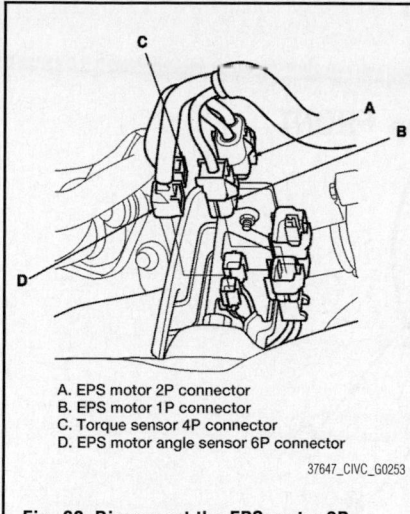

C
A
B
D
A. EPS motor 2P connector
B. EPS motor 1P connector
C. Torque sensor 4P connector
D. EPS motor angle sensor 6P connector
37647_CIVC_G0253

Fig. 60 Disconnect the EPS motor 2P connector, the EPS motor 1P connector, torque sensor 4P connector, and the EPS motor angle sensor 6P connector from passenger's side of the steering gearbox. Wrap the connectors with vinyl tape to avoid contamination from grease or water.

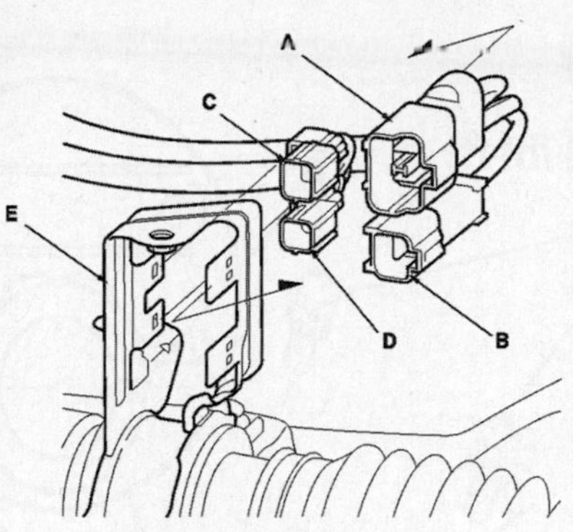
A
C
E
D
B
A. EPS motor 2P connector
B. EPS motor 1P connector
C. Torque sensor 4P connector
D. EPS motor angle sensor 6P connector
E. Gearbox mounting bracket
37647_CIVC_G0255

Fig. 61 Remove the EPS motor 2P connector, the EPS motor 1P connector, torque sensor 4P connector, the EPS motor angle sensor 6P connector from the passenger's side of the gearbox mounting bracket.

sensor 6P connector from passenger's side of the steering gearbox. Wrap the connectors with vinyl tape to avoid contamination from grease or water.

20. Remove the lower ball joint mounting bolt and the self-locking nuts from the lower arm on both sides.

21. Disconnect the lower arm from the lower ball joint housing.

22. Note the reference marks on both sides of the subframe that line up with the body (these are for installation reference).

23. Attach a proper front subframe adaptor to the front subframe and the transmission jack or powertrain lift, then tighten the front subframe adaptor screw. Make sure the front subframe is securely supported by the jack with the front subframe adaptor.

24. Remove the front subframe middle mount bolt from the left and right sides.

25. Remove the lower torque rod mounting bolts.

26. Remove the front subframe front mounting bolts on both sides.

27. Remove the front subframe rear mounting bolts on both sides.

28. Lower the front subframe and steering gearbox as an assembly by lowering the jack slowly.

29. Remove the harness clips from the harness clamp brackets on the steering gear, then remove the harness clamp brackets.

30. Remove the EPS motor 2P connector, the EPS motor 1P connector, torque sensor 4P connector, the EPS motor angle sensor 6P connector from the passenger's side of the gearbox mounting bracket.

31. Remove the pinion shaft grommet from the top of torque sensor.

32. Remove the two 10 mm bolts from the right side of the steering gearbox, then remove the gearbox mounting bracket and the mounting cushion.

33. Remove the four 10 mm flange bolts from the left side of the steering gearbox, then remove the stiffener plates.

34. Remove the steering gearbox from the front subframe.

To install:

35. Place the steering gearbox in position on the front subframe.

36. Loosely install the stiffener plates and the gearbox mounting bolts on the left side of the steering gearbox.

37. Position the cutout on the mounting cushion, as shown, and install it on the right side of the steering gearbox securely. Install the gearbox mounting bracket over the mounting cushion, and loosely install the two 10 mm bolts.

38. Tighten the 10 mm bolts on both sides of the steering gearbox to 28 ft. lbs. (38 Nm), alternately in two or more steps.

39. Install the pinion shaft grommet. Align the slot in the pinion shaft grommet with the lug portion on the torque sensor. The grommet must not have a gap at the mating surface of the grommet and the torque sensor.

40. Install the EPS motor 2P connector, the EPS motor 1P connector, torque sensor 4P connector, the EPS motor angle sensor 6P connector on the right side of the gearbox mounting bracket.

41. Install the harness clamp bracket to the front subframe, then install the harness clips to the harness clamp bracket.

42. Carefully raise the front subframe with the front subframe adapter and the transmission jack or the powertrain lift until the front subframe is in position, then loosely install new front subframe mounting bolts.

➡Be sure that the pinion shaft grommet is in place securely. Make sure the pinion shaft grommet is not turned up. Incorrect installation can cause leakage of water, mud, and noise.

43. Position the subframe, then align the front subframe reference marks to the body, as noted during removal.

44. Tighten the new front and rear subframe mounting bolts to the 76 ft. lbs. (103 Nm) on both sides.

45. Install the new lower torque rod mounting bolts and tighten them to 47 ft. lbs. (64 Nm).

46. Install the new front subframe middle mount bolt on the left and right sides and tighten it to 47 ft. lbs. (64 Nm).

47. Lower the transmission jack supporting the front subframe.

48. Connect the lower arm to the lower ball joint. Install a new flange bolt and the new self-locking nuts. After lightly tightening all three fasteners, tighten them to 47 ft. lbs. (64 Nm) in the following order; the flange nut on the front, the flange nut on the rear, then the flange bolt.

49. Remove the vinyl tape, then connect the EPS motor 2P connector, the EPS motor 1P connector, torque sensor 4P connector, the EPS motor angle sensor 6P connector to the steering gearbox. Make sure to push these connectors until you hear a click so that the connectors are secured.

50. Install the front splash shield.

51. Wipe off any grease contamination from the ball joint tapered section and the threads. Reconnect the tie-rod end ball joints to the steering knuckles. Install the 12 mm nut and tighten it to 40 ft. lbs. (54 Nm).

52. Install the new cotter pin and bend it over.

53. Install the front wheel, then set the wheels in the straight ahead position.

➡Before installing the wheel, clean the mating surfaces between the brake disc and inside of the wheel.

54. Lower the vehicle.

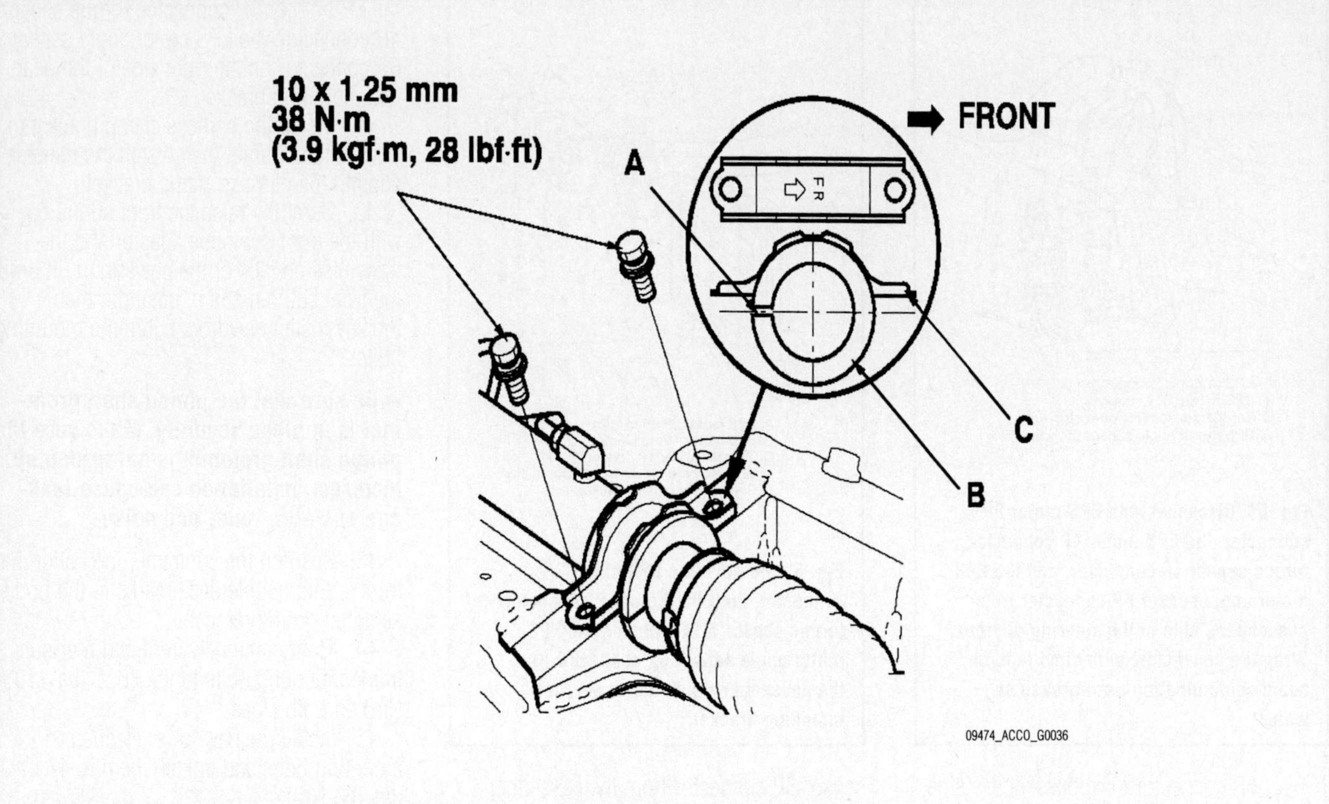
10 x 1.25 mm
38 N·m
(3.9 kgf·m, 28 lbf·ft)
FRONT
A
B
C
FR
09474_ACCO_G0036

Fig. 62 Position the cutout (A) on the mounting cushion (B), as shown, and install it on the right side of the steering gearbox securely. Install the gearbox mounting bracket (C) over the mounting cushion, and loosely install the two 10 mm bolts.

55. Remove the engine hanger, support hanger, and 1.8 support eyelet.

56. Center the steering rack within its stroke in the steering joint connection.

57. With the rack in the straight ahead driving position, cut the wire and slip the lower end of the steering joint onto the pinion shaft in the range shown.

58. Align the bolt hole on the steering joint with the groove around the pinion shaft. Then loosely install the joint bolt. Be sure that the joint bolt is securely in the groove in the pinion shaft. Pull on the steering joint to make sure that the steering joint is fully seated. Tighten the steering joint bolt to 21 ft. lbs. (28 Nm).

59. Install the steering joint cover.

60. Install the driver's dashboard under cover.

61. Install the steering wheel, and the driver's airbag. See "AIR BAG (SUPPLEMENTAL RESTRAINT SYSTEM)" section.

62. Install the air cleaner mounting bracket on the cylinder head.

63. Install the air cleaner housing.

64. Install the cowl cover and the undercowl panel.

65. Do the battery terminal reconnection procedure, and check these items:

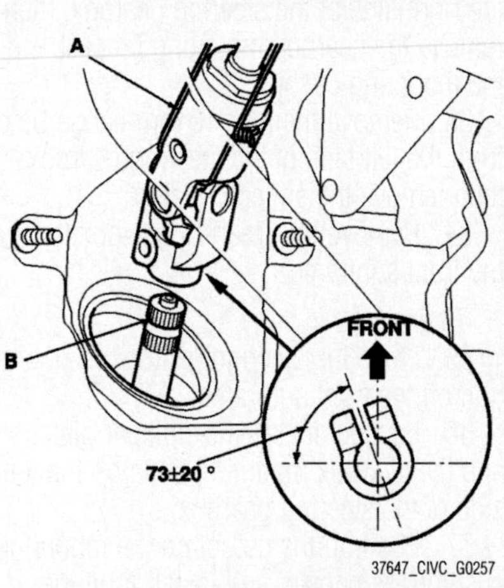
A
B
FRONT
73±20°
37647_CIVC_G0257

Fig. 63 With the rack in the straight ahead driving position, cut the wire (A) and slip the lower end of the steering joint onto the pinion shaft (B) in the range shown.

a. Turn the ignition switch to ON (II) and check that the SRS indicator comes on for about 6 seconds, and then goes off.

b. Make sure the horn and turn signal switches work properly.

c. Make sure the steering wheel switches work properly.

66. After installation, check these items:

a. Check the steering wheel spoke angle. If steering spoke angles to the right and left are not equal (steering wheel and rack are not centered), correct the engagement of the joint/pinion shaft splines.

b. Set the steering column to the center tilt position, and to the center telescopic position, then check the wheel alignment and adjust.

c. Make sure the steering wheel spokes are centered.

67. Do the "Torque Sensor Neutral Position Memorization".

68. Start the engine, and let it idle. Turn the steering wheel from lock to lock several times. Check that the EPS indicator does not come on.

TORQUE SENSOR NEUTRAL POSITION MEMORIZATION

The torque sensor neutral position must be memorized whenever the steering gearbox, the torque sensor, the EPS motor, or the EPS control unit is replaced. Note that the torque sensor neutral position is not affected when erasing the DTC.

➡The torque sensor is temperature sensitive. This procedure should be

performed within the range of 68°F±18°F (20°C±10°C).

1. With the ignition switch in LOCK (0), connect the Honda Diagnostic System (HDS), or equivalent scan tool and software, to the data link connector (DLC), located under the driver's side of the dashboard.

2. Turn the ignition switch to ON (II).

3. Make sure the scan tool communicates with the vehicle and the EPS control unit. If it doesn't, troubleshoot the DLC circuit.

4. From the EPS MENU, select MISCELLANEOUS TEST, then TORQUE SENSOR LEARN, and follow the screen prompts on the scan tool.

➡See the scan tool Help menu for specific instructions. Turn the ignition switch to LOCK (0).

SUSPENSION

FRONT SUSPENSION

LOWER BALL JOINT

REMOVAL & INSTALLATION

See Figure 64.

1. Install a hex nut onto the threads of the ball joint. Make sure the nut is flush with the ball joint pin end to prevent damage to the threaded end of the ball joint pin.

2. Apply grease to the ball joint remover on the areas shown. This will ease installation of the tool and prevent damage to the pressure bolt threads.

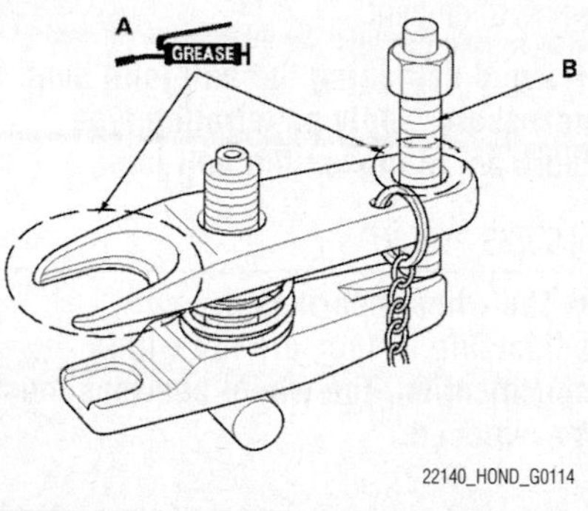
A
GREASE
B
22140_HOND_G0114

Fig. 64 Apply grease to the ball joint remover on the areas shown (A). This will ease installation of the tool and prevent damage to the pressure bolt (B) threads.

3. Loosen the pressure bolt, and install the ball joint remover as shown. Insert the jaws carefully, making sure not to damage the ball joint boot. Adjust the jaw spacing by turning the adjusting bolt.

4. After adjusting the adjusting bolt, make sure the head of the adjusting bolt is in the position shown to allow the jaw to pivot.

5. With a wrench, tighten the pressure bolt until the ball joint pin pops loose from the ball joint pin hole. If necessary, apply penetrating type lubricant to loosen the ball joint pin.

6. Remove the ball joint remover, then remove the nut from the end of the ball joint pin, and pull the ball joint out of the ball joint pin hole. Inspect the ball joint boot, and replace it if damaged.

LOWER CONTROL ARM

REMOVAL & INSTALLATION

1. Raise and support the vehicle safely. Remove the front tires.

2. Remove the flange nut while holding the ball joint pin. Disconnect the stabilizer links from the lower arm.

3. Turn the stabilizer bar backward to gain access to the front side of the lower arm mounting bolt.

4. Remove the flange bolt and nuts from the lower arm.

5. Disconnect the lower ball joint from the lower arm.

6. Remove the lower arm mounting bolts. Remove the lower arm from the front suspension subframe.

To install:

7. Installation is in the reverse order of the removal procedure.

8. Be sure to use new flange bolts, castle nut and lock pin. Tighten the flange nut to 25 ft. lbs. (34 Nm).

9. Check and adjust the wheel alignment, as necessary.

MACPHERSON STRUT

REMOVAL & INSTALLATION

1. Raise and support the vehicle safely.

2. Remove the front tires.

3. Remove the wheel sensor harness bracket and brake hose bracket from the damper. Do not disconnect the wheel sensor connector.

4. Remove the damper pinch bolts from the bottom of the damper. Do not allow the knuckle to rotate too far outward as this may cause the inner CV joint bearing to unseat.

5. Turn the ignition switch to the ON position. Turn the windshield wipers on; turn the ignition switch off leaving the wipers near the A-pillars.

6. Remove the service cap and lid. Remove the three flange nuts at the top of the damper.

➡Damper springs are different, left and right. Mark the springs before continuing.

7. Remove the strut assembly from the vehicle.

To install:

8. Position the assembly to its mounting on the vehicle.

9. Install the mounting bolts.

10. Continue the installation in the reverse order of the removal procedure.

11. Check front end alignment.

OVERHAUL

1. Compress the damper spring with a commercially available strut spring compressor according to the manufacturer's instructions.

2. Remove the self locking nut while holding the damper shaft with a hex wrench. Do not compress the spring more than necessary to remove the nut.

3. Release the pressure from the strut spring compressor. Disassemble the damper assembly. Reassemble all parts, except the spring.

4. Compress the damper assembly by hand and check for smooth operation thru full stroke, both compression and extension.

5. The damper should extend smoothly and constantly when compression is released. If not, gas is leaking and the damper should be replaced.

6. Install all parts except the self locking nut onto the damper unit.

7. Align the bottom of the spring and the stepped part of the lower spring seat. The hole in the upper spring seat and the arrow on the damper mounting base must point toward the knuckle mounting area.

8. Install the damper assembly on a commercially available strut spring compressor.

9. Compress the damper spring with the strut spring compressor. Install a new self locking nut on the damper shaft.

10. Hold the damper shaft with a hex wrench and tighten the self locking nut.

STEERING KNUCKLE

REMOVAL & INSTALLATION

➡The steering knuckle is removed with the wheel bearings as an assembly. Refer to "Wheel Bearings" in this section.

STABILIZER BAR

REMOVAL & INSTALLATION

1. Raise and support the vehicle safely. Remove the front tires.
2. Disconnect both stabilizer links from the stabilizer bar.
3. Remove the flange bolts and the bushing holders. Remove the bushings and the stabilizer bar from the front suspension subframe.

To install:

4. Position the assembly to its mounting on the vehicle.
5. Install the mounting bolts.
6. Continue the installation in the reverse order of the removal procedure.

WHEEL BEARINGS

REMOVAL & INSTALLATION

See Figure 65.

➡A hydraulic press and several bearing drivers and attachments are needed to remove and install the hub and bearing.

1. Before servicing the vehicle, refer to the precautions in the beginning of this section.
2. Pry the spindle nut stake away from the spindle, and then loosen the nut.
3. Raise and safely support the vehicle.
4. Remove or disconnect the following:
   - Front wheel and the spindle nut
   - Wheel sensor wire bracket from the knuckle, but don't disconnect it.
   - Caliper mounting bolts and the caliper. Support the caliper out of the way with a length of wire. Do not let the caliper hang from the brake hose.
   - 6mm brake disc retaining screws. Screw two 12mm bolts into the disc to push it away from the hub.
   - Tie rod castle nut
   - Tie rod ball joint using a suitable ball joint remover.
   - Cotter pin and loosen the lower arm ball joint nut half the length of the joint threads.
   - Ball joint and lower arm using a suitable puller with the pawls applied to the lower arm.

➡Avoid damaging the ball joint boot. If necessary, apply penetrating type lubricant to loosen the ball joint.

   - Ball joint nut cover.
   - Cotter pin and the upper ball joint nut.
   - Upper ball joint and knuckle using a ball joint remover.
5. Use a plastic mallet to free the half-shaft from the knuckle. Pull the knuckle out to remove it.

➡A new wheel bearing must be used when the hub is removed.

6. Place the knuckle in a press and use a base and pilot to press the hub assembly out of the wheel bearing.
7. Remove the knuckle ring seal and circlip. Remove the splash guard from the knuckle.
8. Press the wheel bearing out of the knuckle using a driving attachment.

To install:

9. Clean the knuckle and hub assembly and inspect them for damage.
10. Install or connect the following:
    - New wheel bearing into the hub using a driving tool
    - Circlip in the outer groove of the knuckle
    - Splash guard
    - Hub assembly into the steering knuckle using a base and a driving and guide tool
    - Knuckle ring seal
    - Knuckle onto the spindle
    - Knuckle onto the upper and lower ball joints and tighten the castle nuts
    - Tie rod ball joint onto the steering knuckle
11. Tighten the upper ball joint nut and tie rod nut to 29–35 ft. lbs. (40–48 Nm) and the lower ball joint castle nut to 36–43 ft. lbs. (50–60 Nm).
12. Install or connect the following:
    - Anti-lock Brake System (ABS) wheel sensor wire brackets onto the knuckle. Tighten the mounting bolts to 84 inch lbs. (10 Nm).
    - Brake disc; use 2 lug nuts to evenly draw the disc onto the hub
    - Retainer screws: 84 inch lbs. (10 Nm)
    - Spindle washer and nut. Don't tighten the nut until the vehicle is on the ground.
    - Brake caliper and tighten the bolts to 80 ft. lbs. (110 Nm)
    - Front wheels and lower the vehicle
13. Tighten the spindle nut to 134 ft. lbs. (185 Nm), stake the nut, and install the grease cap.
14. Check and adjust the vehicle's front wheel alignment.

➡Avoid damaging the ball joint boot. If necessary, apply penetrating-type lubricant to loosen the ball joint.

ADJUSTMENT

➡The wheel bearings are not adjustable. If they are not within specification, the wheel bearings must be replaced.

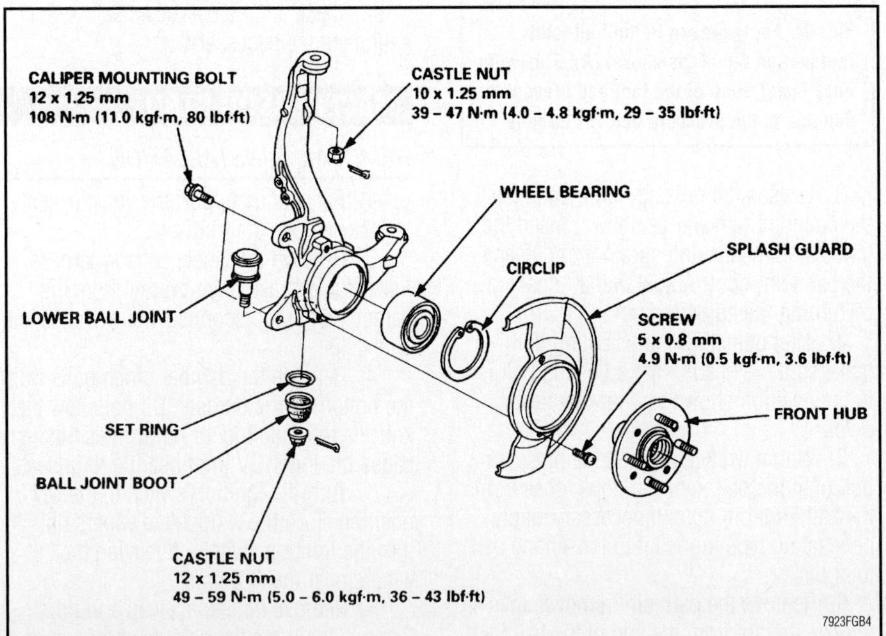
CALIPER MOUNTING BOLT
12 x 1.25 mm
108 N·m (11.0 kgf·m, 80 lbf·ft)
CASTLE NUT
10 x 1.25 mm
39 – 47 N·m (4.0 – 4.8 kgf·m, 29 – 35 lbf·ft)
WHEEL BEARING
SPLASH GUARD
CIRCLIP
LOWER BALL JOINT
SCREW
5 x 0.8 mm
4.9 N·m (0.5 kgf·m, 3.6 lbf·ft)
FRONT HUB
SET RING
BALL JOINT BOOT
CASTLE NUT
12 x 1.25 mm
49 – 59 N·m (5.0 – 6.0 kgf·m, 36 – 43 lbf·ft)
7923FGB4

Fig. 65 Knuckle components

KNUCKLE/HUB BEARING

REMOVAL & INSTALLATION

With Rear Disc Brakes

See Figures 66 and 67.

1. Raise and support the vehicle.
2. Remove the wheel nuts and the rear wheel.
3. Release the parking brake lever fully.
4. Remove the brake hose mounting bolt from the bracket.
5. Remove the brake caliper bracket mounting bolts, then remove the caliper assembly from the knuckle.

➡To prevent damage to the caliper assembly or the brake hose, use a short piece of wire to hang the caliper assembly from the undercarriage. Do not twist the brake hose excessively.

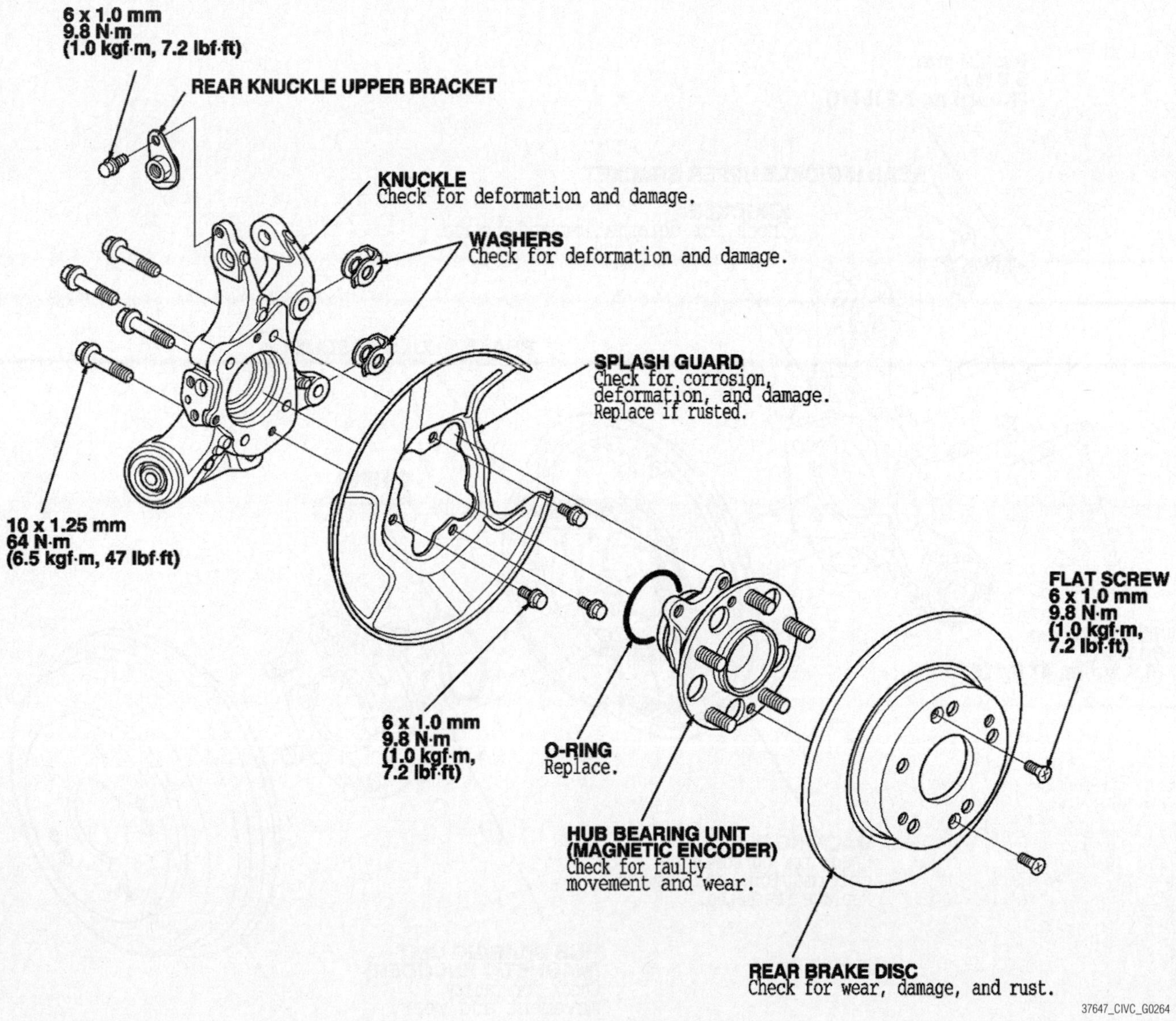
6 x 1.0 mm
9.8 N·m
(1.0 kgf·m, 7.2 lbf·ft)
REAR KNUCKLE UPPER BRACKET
KNUCKLE
Check for deformation and damage.
WASHERS
Check for deformation and damage.
SPLASH GUARD
Check for corrosion, deformation, and damage.
Replace if rusted.
10 x 1.25 mm
64 N·m
(6.5 kgf·m, 47 lbf·ft)
FLAT SCREW
6 x 1.0 mm
9.8 N·m
(1.0 kgf·m, 7.2 lbf·ft)
6 x 1.0 mm
9.8 N·m
(1.0 kgf·m, 7.2 lbf·ft)
O-RING
Replace.
HUB BEARING UNIT
(MAGNETIC ENCODER)
Check for faulty movement and wear.
REAR BRAKE DISC
Check for wear, damage, and rust.
37647_CIVC_G0264

Fig. 66 Exploded view of the knuckle and hub bearing assembly—with rear disc brakes

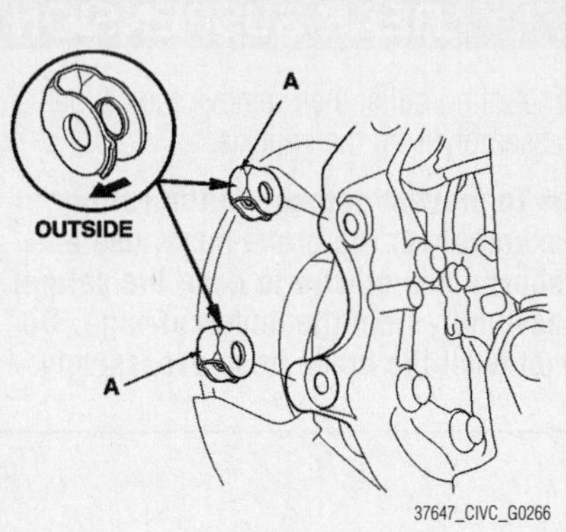
A
OUTSIDE
A
37647_CIVC_G0266

Fig. 67 Remove the two washers (A).

6. Remove the two washers.
7. Remove the rear brake disc.
8. Remove the hub bearing unit and the O-ring.
9. Check the hub bearing unit for damage and cracks.

To install:

10. Install the hub bearing unit in the reverse order of removal, and note these items:
   a. Make sure the washers are installed between the brake caliper bracket and the knuckle.
   b. Use a new O-ring during reassembly.
   c. Before installing the brake disc, clean the mating surfaces of the hub bearing unit and the inside of the brake disc.
   d. Torque the bearing hub retaining bolts to 47 ft. lbs. (64 Nm).
   e. Before installing the wheel, clean the mating surfaces of the brake disc and the inside of the wheel.
   f. Torque the caliper retaining bolts to 54 ft. lbs. (74 Nm).
11. Check the wheel alignment, and adjust it if necessary.

With Rear Drum Brakes

See Figure 68.

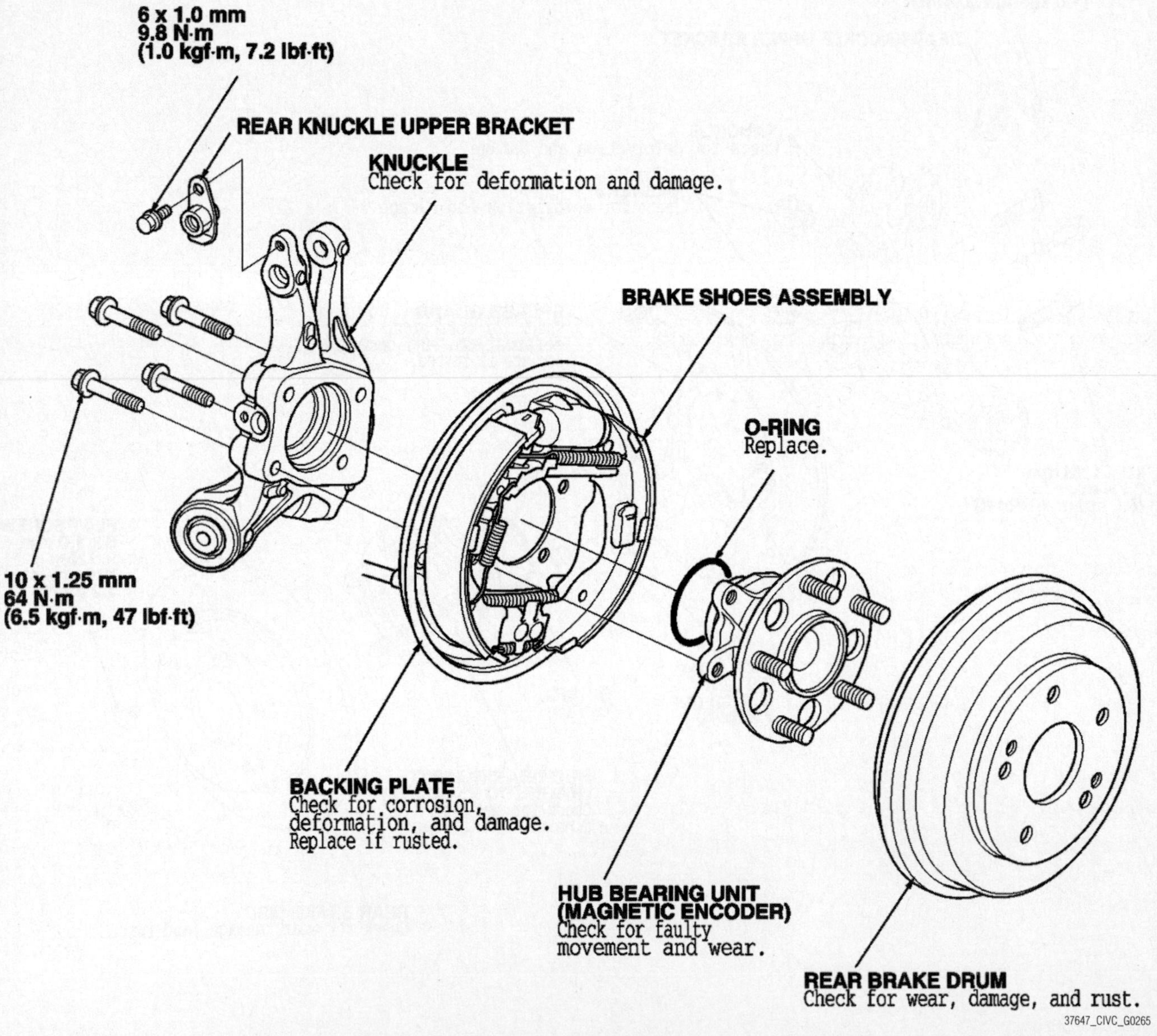
6 x 1.0 mm
9.8 N·m
(1.0 kgf·m, 7.2 lbf·ft)
REAR KNUCKLE UPPER BRACKET
KNUCKLE
Check for deformation and damage.
BRAKE SHOES ASSEMBLY
O-RING
Replace.
10 x 1.25 mm
64 N·m
(6.5 kgf·m, 47 lbf·ft)
BACKING PLATE
Check for corrosion, deformation, and damage.
Replace if rusted.
HUB BEARING UNIT (MAGNETIC ENCODER)
Check for faulty movement and wear.
REAR BRAKE DRUM
Check for wear, damage, and rust.
37647_CIVC_G0265

Fig. 68 Exploded view of the knuckle and hub bearing assembly—with rear drum brakes

1. Raise and support the vehicle.
2. Remove the wheel nuts and the rear wheel.
3. Release the parking brake lever fully.
4. Remove the hub bearing unit and the O-ring.
5. Check the hub bearing unit for damage and cracks.

To install:

6. Install the hub bearing unit in the reverse order of removal, and note these items:
   a. Use a new O-ring during reassembly.
   b. Before installing the brake drum, clean the mating surfaces of the hub bearing unit and the inside of the brake drum.
   c. Torque the bearing hub retaining bolts to 47 ft. lbs. (64 Nm).
   d. Before installing the wheel, clean the mating surfaces of the brake drum and the inside of the wheel.
7. Check the wheel alignment, and adjust it if necessary.

MACPHERSON STRUTS

REMOVAL & INSTALLATION

1. Raise and support the vehicle safely. Remove the rear tires.
2. Position a floor jack at the connecting point of the trailing arm and the knuckle. Raise the floor jack until the suspension begins to compress.
3. Remove the flange bolt and discard it.
4. Remove the trunk side trim panel.
5. Remove the self locking nut while holding the damper shaft. Compress the damper unit, by hand, and remove it from the vehicle.

To install:

6. Position the damper assembly in the vehicle.
7. Position a floor jack under the trailing to support the suspension. Install a new damper mounting bolt.
8. Loosely tighten the damper mounting bolt. Raise the floor jack until the suspension begins to compress. Tighten the damper mounting bolt.
9. Continue the installation in the reverse order of the removal procedure.

OVERHAUL

1. Compress the damper spring with a commercially available strut spring compressor according to the manufacturer's instructions.
2. Remove the self locking nut while holding the damper shaft with a hex wrench. Do not compress the spring more than necessary to remove the nut.
3. Release the pressure from the strut spring compressor. Disassemble the damper assembly. Reassemble all parts, except the spring.
4. Compress the damper assembly by hand and check for smooth operation thru full stroke, both compression and extension.
5. The damper should extend smoothly and constantly when compression is released. If not, gas is leaking and the damper should be replaced.
6. Install all parts except the self locking nut onto the damper unit.
7. Align the bottom of the spring and the stepped part of the lower spring seat. The hole in the upper spring seat and the arrow on the damper mounting base must point toward the knuckle mounting area.
8. Install the damper assembly on a commercially available strut spring compressor.
9. Compress the damper spring with the strut spring compressor. Install a new self locking nut on the damper shaft.
10. Hold the damper shaft with a hex wrench and tighten the self locking nut.

STABILIZER BAR

REMOVAL & INSTALLATION

See Figure 69.

1. Raise and support the vehicle.
2. Remove the rear wheels.
3. Disconnect both stabilizer links from the stabilizer bar.
4. Remove the flange bolts and the bushing holders, then remove the bushings and the stabilizer bar.

To install:

➡**During installation, align the paint marks on the stabilizer bar with the sides of the bushings.**

6. Replace the stabilizer bar bracket, if necessary. Tighten the bracket bolts to 29 ft. lbs. (39 Nm).
7. Install the stabilizer bar in the reverse order of removal.
8. Tighten the stabilizer bar bolts to 16 ft. lbs. (22 Nm).

➡**Note the right and left direction of the stabilizer bar.**

9. Before installing the wheel, clean the mating surfaces on the brake disc or the brake drum and inside of the wheel.

UPPER CONTROL ARM

REMOVAL & INSTALLATION

1. Raise and support the vehicle safely.
2. Remove the rear tires.

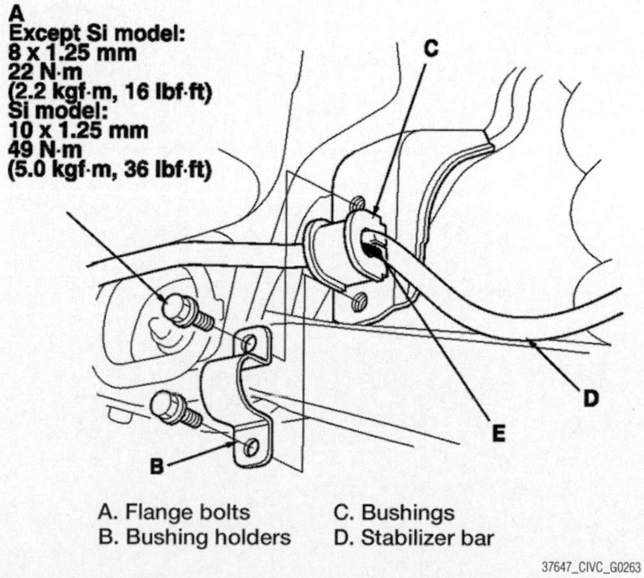
A
Except Si model:
8 x 1.25 mm
22 N·m
(2.2 kgf·m, 16 lbf·ft)
Si model:
10 x 1.25 mm
49 N·m
(5.0 kgf·m, 36 lbf·ft)
C
D
E
B
A. Flange bolts
B. Bushing holders
C. Bushings
D. Stabilizer bar
37647_CIVC_G0263

Fig. 69 Remove the flange bolts (A) and the bushing holders (B), then remove the bushings (C) and the stabilizer bar (D).

3. Place a jack under the trailing arm and the knuckle.

4. Remove the upper arm mounting bolts and the flange bolt. Remove the upper arm from the vehicle.

To install:

5. Position the upper arm on the vehicle. Be sure to use new retaining bolts and nuts.

6. Install all the suspension components and lightly tighten the bolts and nuts. Position a jack under the trailing arm and raise the suspension to load it with the vehicles weight, before fully tightening the bolts and nuts.

7. Continue the installation in the reverse order of the removal procedure.

8. Check and adjust the wheel alignment, as required.

HONDA

CR-V

11

BRAKES........................11-9

ANTI-LOCK BRAKE SYSTEM (ABS) 11-10
Wheel Speed Sensors................11-10
Removal & Installation 11-10
BLEEDING THE BRAKE SYSTEM........................11-9
Bleeding Procedure..................11-9
Bleeding Procedure 11-9
Bleeding the ABS System 11-9
FRONT DISC BRAKES........11-11
Brake Caliper..........................11-11
Removal & Installation........11-11
Disc Brake Pads.....................11-11
Removal & Installation........11-11
INFORMATION AND PRECAUTIONS 11-9
Anti-lock Systems................11-9
Disc and Drum Systems.............................11-9
PARKING BRAKE.............11-14
Parking Brake Cables.............11-14
Adjustment 11-14
Parking Brake Shoes..............11-14
Adjustment 11-15
Removal & Installation........11-14
REAR DISC BRAKES 11-12
Brake Caliper..........................11-12
Removal & Installation........11-12
Disc Brake Pads.....................11-12
Removal & Installation........11-12

CHASSIS ELECTRICAL 11-16

AIR BAG (SUPPLEMENTAL RESTRAINT SYSTEM) 11-16
General Information................11-16
Arming the System 11-17
Clockspring Centering........11-17
Disarming the System.........11-16
Precautions.......................11-16

DRIVE TRAIN 11-18

Front Halfshaft........................11-18
CV-boots Inspection 11-19
Removal & Installation........11-18
Intermediate Shaft 11-19
Removal & Installation........11-19
Rear Differential......................11-22
Removal & Installation........11-22
Rear Driveshaft (Propeller Shaft) 11-19
Removal & Installation........11-19
Rear Halfshaft.........................11-20
CV-boots Inspection 11-20
Removal & Installation........11-20
Rear Pinion Seal.....................11-20
Removal & Installation........11-20
Transfer Case Assembly 11-23
Removal & Installation........11-23

ENGINE COOLING 11-24

Engine Fan 11-24
Removal & Installation........11-24
Radiator..................................11-25
Removal & Installation........11-25
Thermostat 11-26
Removal & Installation........11-26
Water Pump 11-26
Removal & Installation........11-26

ENGINE ELECTRICAL 11-28

CHARGING SYSTEM 11-28
Alternator 11-28
Removal & Installation........11-28
IGNITION SYSTEM 11-28
Firing Order............................11-28
Ignition Coil 11-28
Removal & Installation........11-28
Ignition Timing........................11-28
Adjustment 11-28
Spark Plugs............................11-28
Removal & Installation........11-28
STARTING SYSTEM 11-29
Starter 11-29
Removal & Installation........11-29

ENGINE MECHANICAL......11-30

Accessory Drive Belts 11-30
Accessory Belt Routing.......11-30
Adjustment 11-30
Inspection 11-30
Removal & Installation........11-30
Camshaft and Valve Lifters......11-30
Removal & Installation........11-30
Catalytic Converter 11-33
Removal & Installation........11-33
Crankshaft Front Seal 11-35
Removal & Installation........11-35
Crankshaft Pulley 11-33
Removal & Installation........11-33
Cylinder Head 11-35
Removal & Installation........11-35
Drive Plate..............................11-39
Removal & Installation........11-39
Exhaust Manifold 11-39
Removal & Installation........11-39
Intake Manifold 11-40
Removal & Installation........11-40
Oil Pan 11-43
Removal & Installation........11-43
Oil Pump................................11-44
Inspection 11-47
Removal & Installation........11-44
Piston and Ring.......................11-48
Positioning 11-48
Rear Main Seal.......................11-49
Removal & Installation........11-49
Rocker Arms...........................11-49
Removal & Installation........11-49
Timing (Cam) Chain & Sprockets 11-49
Removal & Installation........11-49
Timing (Cam) Chain Front Cover 11-49
Removal & Installation........11-49
Valve Lash..............................11-50
Adjustment 11-50

ENGINE PERFORMANCE & EMISSION CONTROLS 11-54

Accelerator Pedal Position (APP) Sensor 11-54
Location.............................11-54
Removal & Installation........11-54
Camshaft Position (CMP) Sensor 11-54

Location....11-54
Removal & Installation....11-54
Crankshaft Position (CKP) Sensor....11-55
CKP Pattern Clear/CKP Pattern Learn....11-55
Location....11-55
Removal & Installation....11-55
Engine Coolant Temperature (ECT) Sensor....11-55
Location....11-55
Removal & Installation....11-56
Evaporative Emissions (EVAP) Canister....11-56
Location....11-56
Removal & Installation....11-56
Exhaust Gas Recirculation (EGR) Valve....11-57
Location....11-57
Removal & Installation....11-57
Heated Oxygen (HO2S) Sensor....11-58
Location....11-58
Removal & Installation....11-58
Intake Air Temperature (IAT) Sensor....11-59
Location....11-59
Removal & Installation....11-59
Knock Sensor (KS)....11-59
Location....11-59
Removal & Installation....11-59
Malfunction Indicator Light (MIL)....11-59
Reset Procedure....11-59
Manifold Absolute Pressure (MAP) Sensor....11-60
Location....11-60
Removal & Installation....11-60
Mass Air Flow (MAF) Sensor....11-60
Location....11-60
Removal & Installation....11-60
Positive Crankcase Ventilation (PCV) Valve....11-60
Location....11-60
Removal & Installation....11-60
Powertrain Control Module (PCM)....11-61
Location....11-61
Removal & Installation....11-61
Throttle Position Sensor (TPS)....11-62
Removal & Installation....11-62
Variable Camshaft Timing Oil Control Solenoid....11-62
Location....11-62
Removal & Installation....11-63

FUEL....11-63

GASOLINE FUEL INJECTION SYSTEM....11-63
Fuel Filter....11-64
Removal & Installation....11-64
Fuel Level Sending Unit....11-65
Location....11-65
Removal & Installation....11-65
Fuel Rail and Injector....11-66
Removal & Installation....11-66
Fuel System Pressure....11-63
Relieving....11-63
Fuel System Service Precautions....11-63
Fuel Tank....11-67
Draining....11-67
Removal & Installation....11-67
Fuel Tank Unit....11-65
Removal & Installation....11-65
Idle Speed....11-67
Adjustment....11-67
Throttle Body....11-67
Removal & Installation....11-67

HEATING & AIR CONDITIONING SYSTEM....11-69
Blower Motor....11-69
Removal & Installation....11-69
Heater Core....11-69
Removal and Installation....11-69
Heater Unit....11-71
Removal & Installation....11-71

SPECIFICATIONS AND MAINTENANCE CHARTS....11-3
Brake Specifications....11-7
Camshaft and Bearing Specifications Chart....11-5
Capacities....11-4
Crankshaft and Connecting Rod Specifications....11-5
Engine and Vehicle Identification....11-3
Fluid Specifications....11-3
Gasoline Engine Tune-Up Specifications....11-4
General Engine Specifications....11-3
Piston and Ring Specifications....11-5
Maintenance Minder Schedule....11-8
Tire, Wheel and Ball Joint Specifications....11-7
Torque Specifications....11-6
Valve Specifications....11-4
Wheel Alignment....11-6

STEERING....11-72
Power Rack & Pinion Steering Gear....11-72
Removal & Installation....11-72
Power Steering Pump....11-74
Bleeding....11-75
Removal & Installation....11-74

SUSPENSION....11-75

FRONT SUSPENSION....11-75
Control Links....11-75
Removal & Installation....11-75
Lower Ball Joint....11-75
Removal & Installation....11-75
Lower Control Arm....11-76
Removal & Installation....11-76
Stabilizer Bar....11-79
Removal & Installation....11-79
Steering Knuckle....11-76
Removal & Installation....11-76
Strut (Damper/Spring)....11-77
Overhaul....11-78
Removal & Installation....11-77
Wheel Hub & Bearing....11-81
Adjustment....11-81
Removal & Installation....11-81

REAR SUSPENSION....11-82
Control Arms/Links....11-82
Removal & Installation....11-82
Stabilizer Bar....11-85
Removal & Installation....11-85
Strut (Damper/Spring)....11-83
Overhaul....11-83
Removal & Installation....11-83
Wheel Hub & Bearing....11-85
Adjustment....11-86
Removal & Installation....11-85

SPECIFICATIONS AND MAINTENANCE CHARTS

ENGINE AND VEHICLE IDENTIFICATION

Engine							Model Year	
Code ①	Liters (cc)	Cu. In.	Cyl.	Fuel Sys.	Engine Type	Eng. Mfg.	Code ②	Year
K24Z1	2.4 (2,354)	144	4	MPFI	DOHC	Honda	9	2009
K24Z6	2.4 (2,354)	144	4	MPFI	DOHC	Honda	A	2010

MPFI: Multi-Port Fuel Injection

DOHC: Double Overhead Camshaft

① Stamped into the front of the engine block and can be seen through the window next to the "H" logo on the front grill

② 10th digit of the Vehicle Identification Number (VIN)

37647_HCRV_C0001

GENERAL ENGINE SPECIFICATIONS

All measurements are given in inches.

Year	Model	Engine Displacement Liters	Engine Series ID/VIN	Net Horsepower @ rpm	Net Torque @ rpm (ft. lbs.)	Bore x Stroke (in.)	Com-pression Ratio	Oil Pressure @ rpm
2009	CR-V	2.4	K24Z1	166@5800	161@4200	3.43 x 3.90	9.7:1	44@3000
2010	CR-V	2.4	K24Z6	180@6800	N/A	3.43 x 3.90	10.5:1	44@3000

37647_HCRV_C0002

GASOLINE ENGINE TUNE-UP SPECIFICATIONS

Year	Engine Displacement Liters	Engine Series ID/VIN	Spark Plug Gap (in.)	Ignition Timing (deg.)		Fuel Pump (psi)	Idle Speed (rpm)		Valve Clearance (in.)	
				MT	AT		MT	AT	Intake	Exhaust
2009	2.4	K24Z1	0.039-0.043	—	6-10B	47-54	—	600-700	0.008-0.010	0.011-0.013
2010	2.4	K24Z6	0.039-0.043	—	6-10B	47-54	—	600-700	0.008-0.010	0.010-0.011

NOTE: The Vehicle Emission Control Information label reflects specification changes made during production.

Follow the figures on the label if they differ from those in this chart.

B: Before top dead center

37647_HCRV_C0003

CAPACITIES

Year	Model	Engine Displacement Liters	Engine Series ID/VIN	Engine Oil with Filter (qts.)	Transmission (pts.) Manual	Transmission (pts.) Auto. ①	Transfer Case (pts.)	Drive Axle Front (pts.)	Drive Axle Rear (pts.) ①	Fuel Tank (gal.)	Cooling System (qts.)
2009	CR-V	2.4	K24Z1	4.4	—	②	③	③	2.6	15.3	5.3
2010	CR-V	2.4	K24Z6	4.4	—	5.2	③	③	2.6	15.3	6.2

NOTE: All capacities are approximate. Add fluid gradually and check to be sure a proper fluid level is obtained.

① Drain and refill

② 2WD: 5.4 pts.
4WD: 5.2 pts.

③ Included in transaxle refill figure

37647_HCRV_C0004

FLUID SPECIFICATIONS

Year	Model	Engine Displacement Liters	Engine Series ID/VIN	Engine Oil	Manual Trans.	Auto. Trans.	Drive Axle	Transfer Case	Power Steering Fluid	Brake Master Cylinder	Cooling System
2009	CR-V	2.4	K24Z1	①	—	②	③	②	④	⑤	⑥
2010	CR-V	2.4	K24Z6	①	—	②	③	②	④	⑤	⑥

DOT: Department Of Transportation

① Honda Motor Oil: American Honda P/N 08798-9023 (5W-20), Honda Canada P/N CA66806 (5W-20)

② Honda Automatic Transmission Fluid (ATF-Z1): American Honda P/N 08200-9001, Honda Canada P/N CA66689

③ Rear differential (4WD) Honda Dual Pump Fluid II: P/N 08200-9007

④ Honda Power Steering Fluid: P/N 08206-9002

⑤ Honda DOT 3 Brake Fluid: P/N 08798-9008

⑥ Honda Long Life Antifreeze/Coolant Type 2: P/N OL 999-9001; Honda Coolant Concentrate: P/N OL 999-9020

37647_HCRV_C0005

VALVE SPECIFICATIONS

Year	Engine Displacement Liters	Engine Series ID/VIN	Seat Angle (deg.)	Face Angle (deg.)	Spring Test Pressure (lbs. @ in.)	Spring Installed Height (in.)	Stem-to-Guide Clearance (in.) Intake	Stem-to-Guide Clearance (in.) Exhaust	Stem Diameter (in.) Intake	Stem Diameter (in.) Exhaust
2009	2.4	K24Z1	NA	NA	NA	①	0.0012-0.0022	0.0022-0.0031	0.2156-0.2159	0.2146-0.2150
2010	2.4	K24Z6	NA	NA	NA	①	0.0012-0.0022	0.0022-0.0031	0.2156-0.2159	0.2146-0.2150

NA: Information not available

① Valve spring free length:
Intake: 1.873 in.
Exhaust: 1.954 in.

37647_HCRV_C0006

CAMSHAFT AND BEARING SPECIFICATIONS CHART

All measurements are given in inches.

Year	Engine Displ. Liters	Engine Series ID/VIN	Journal Dia.	Brg. Oil Clearance	Shaft End-play	Runout	Journal Bore	Lobe Height Intake	Lobe Height Exhaust
2009	2.4	K24Z1	NA	①	0.002-0.008	0.001 max.	NA	②	1.3422
2010	2.4	K24Z6	NA	③	0.002-0.008	0.001 max.	NA	④	1.3500

NA: Information not available

① Journal No. 1: 0.002-0.003 in.
Journals No. 2, 3, 4, 5: 0.002-0.004 in.

② Intake, primary: 1.3489 in.
Intake, secondary: 1.1668 in.

③ Journal No. 1: 0.0012-0.0027 inches
Journals No. 2, 3, 4, 5: 0.0024-0.0039 inches

④ Intake, primary: 1.3285 inches
Intake, Mid: 1.3959 inches
Intake, secondary: 1.3285 inches

37647_HCRV_C0007

CRANKSHAFT AND CONNECTING ROD SPECIFICATIONS

All measurements are given in inches.

Year	Engine Displacement Liters	Engine Series ID/VIN	Crankshaft Main Brg. Journal Dia.	Crankshaft Main Brg. Oil Clearance	Crankshaft Shaft End-play	Crankshaft Thrust on No.	Connecting Rod Journal Diameter	Connecting Rod Oil Clearance	Connecting Rod Side Clearance
2009	2.4	K24Z1	①	②	0.0040-0.0140	3	2.0100	0.0002-0.0006	0.0060-0.0120
2010	2.4	K24Z6	①	②	0.0040-0.0140	3	2.0100	0.0002-0.0006	0.0060-0.0140

① Except No. 3: 2.1648-2.1657
No. 3: 2.1644-2.1654

② Except No. 3: 0.0007-0.0016
No. 3: 0.0010-0.0019

37647_HCRV_C0008

PISTON AND RING SPECIFICATIONS

All measurements are given in inches.

Year	Engine Displ. Liters	Engine Series ID/VIN	Piston Clearance	Ring Gap Top Compression	Ring Gap Bottom Compression	Ring Gap Oil Control	Ring Side Clearance Top Compression	Ring Side Clearance Bottom Compression	Ring Side Clearance Oil Control
2009	2.4	K24Z1	0.0008-0.0016	0.0080-0.0140	0.0160-0.0220	0.0100-0.0260	0.0014-0.0024	0.0012-0.0022	NA
2010	2.4	K24Z6	0.0008-0.0016	0.0080-0.0140	0.0200-0.0260	0.0080-0.0280	0.0024-0.0033	0.0016-0.0026	NA

37647_HCRV_C0009

TORQUE SPECIFICATIONS

All readings in ft. lbs.

Year	Engine Displacement Liters	Engine Series ID/VIN	Cylinder Head Bolts	Main Bearing Bolts	Rod Bearing Bolts	Crankshaft Damper Bolts	Flywheel Bolts	Manifold Intake	Manifold Exhaust	Spark Plugs	Oil Pan Drain Plug
2009	2.4	K24Z1	①	②	③	④	⑤	⑥	⑦	13	29
2010	2.4	K24Z6	①	⑧	⑨	④	⑤	⑥	⑦	13	29

NOTE: Dip cylinder head bolts, main bearing bolts, and crankshaft damper bolt in clean engine oil prior to tightening.

① Step 1: 29 ft. lbs.
Step 2: plus 90 degrees
Step 3: plus 90 degrees
Step 4: NEW BOLT ONLY plus 90 degrees

② Tighten bearing cap bolts in sequence:
Step 1: 22 ft. lbs.
Step 2: plus 56 degrees
Tighten the 8mm bolts to 16 ft. lbs., in sequence

③ Step 1: Tighten to 14 ft. lbs.
Step 2: plus 90 degrees

④ Step 1: Tighten to 36 ft. lbs.
Step 2: plus 90 degrees

⑤ Tighten in sequence to 54 ft. lbs. in at least 2 steps

⑥ Tighten in sequence to 16 ft. lbs. in 3 steps

⑦ Tighten in sequence to 33 ft. lbs. in 3 steps

⑧ Tighten bearing cap bolts in sequence:
Step 1: 22 ft. lbs.
Step 2: plus 48 degrees
Tighten the 8mm bolts to 16 ft. lbs., in sequence

⑨ Step 1: Tighten to 30 ft. lbs.
Step 2: plus 120 degrees

37647_HCRV_C0010

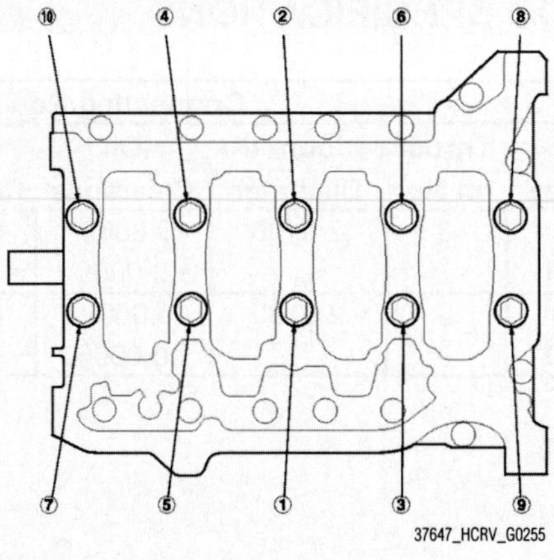
⑩
④
②
⑥
⑧
⑦
⑤
①
③
⑨
37647_HCRV_G0255

Fig. 1 Main bearing torque sequence

WHEEL ALIGNMENT

Year	Model		Caster Range (+/-Deg.)	Caster Preferred Setting (Deg.)	Camber Range (+/-Deg.)	Camber Preferred Setting (Deg.)	Toe-in (Deg.)
2009	CR-V	Front	1.00	+3.02	0.30	0	0+/-0.08
		Rear	—	—	0.45	-1.00	0.08+/-0.08
2010	CR-V	Front	1.00	+2.44	0.45	0	0+/-0.08
		Rear	—	—	0.45	-1.00	0.08+0.08/-0.04

NOTE: Measurements are given for unladen vehicle: fuel, engine coolant, and fluid levels are full. Spare tire, jack, hand tools, and mats are in designated posit

37647_HCRV_C0011

TIRE, WHEEL AND BALL JOINT SPECIFICATIONS

Year	Model	OEM Tires Standard	OEM Tires Optional	Tire Pressures (psi) Front	Tire Pressures (psi) Rear	Wheel Size	Ball Joint Inspection	Lug Nut Torque (ft. lbs.)
2009	CR-V	P225/65R17	—	30	30	NA	NA	80
2010	CR-V	P225/65R17	—	30	30	NA	NA	80

OEM: Original Equipment Manufacturer

PSI: Pounds Per Square Inch

NA: Information not available

37647_HCRV_C0012

BRAKE SPECIFICATIONS

All measurements in inches unless noted

Year	Model		Brake Disc Original Thickness	Brake Disc Minimum Thickness	Brake Disc Maximum Runout	Brake Drum Diameter Original Inside Diameter	Brake Drum Diameter Max. Wear Limit	Brake Drum Diameter Maximum Machine Diameter	Minimum Lining Thickness	Brake Caliper Bracket Bolts (ft. lbs.)	Brake Caliper Mounting Bolts (ft. lbs.)
2009	CR-V	F	1.10-1.11	1.020	0.0016	—	—	—	0.060	80	25
		R	0.35-0.36	0.300	0.0016	—	—	—	0.060	41	16
2010	CR-V	F	1.09-1.11	1.020	0.0016	—	—	—	0.060	101	37
		R	0.35-0.36	0.300	0.0016	—	—	—	0.060	80	17

F: Front

R: Rear

37647_HCRV_C0013

MAINTENANCE MINDER SCHEDULE
Honda CR-V

All Honda's displays engine oil life and maintenance service items in the information display to indicate when to perform maintenance service. If the engine oil life is 15% or less, based on the onboard computer's calculations, you will see SERVICE DUE SOON in the information display every time the ignition key is turned to ON. The maintenance minder indicator will also come on and the maintenance code(s) for other scheduled maintenance items needing service will be displayed below the message.

Symbol	Item	Service
A	Engine oil ①	Change
B	Engine oil and filter	Change
	Fluid levels	Inspect
	Brakes	Inspect
	Parking brake adjustment	Check
	Steering gear and linkage	Inspect
	Suspension components	Inspect
	Driveshaft boots	Inspect
	Brake hoses and lines	Inspect
	Exhaust system	Inspect
	Fuel lines and connections	Inspect
1	Tires	Rotate
2	Engine air filter ②	Replace
	Dust and pollen filter ③	Replace
	Accessory drive belt	Inspect
3	Transmission fluid ④	Replace
	Transfer case fluid ④	Replace
4	Spark plugs	Replace
	Timing belt ⑤	Replace
	Water pump	Inspect
	Valve clearance ⑥	Inspect
5	Engine coolant	Replace
6	VTM-4 rear differential fluid	Replace

① If the message SERVICE DUE NOW does not appear more than 12 months after the display is reset, change every year.

② If driven in dusty conditions, replace every 15,000 miles.

③ If driven in urban areas that have a high concentration of soot from industry and diesel, replace every 15,000 miles

④ If regularly driven in mountainous areas at very low speed or trailer towing, change the fluid every 30,000 miles.

⑤ If driven regularly in temperatures over 110 deg. F or below -20 deg. F, or towing a trailer, replace every 60,000 miles.

⑥ Adjust if necessary.

Additionally, replace the brake fluid every 3 years, and inspect the idle speed every 160,000 miles.

To reset the Engine Oil Life Display:

1. Turn the ignition switch to ON.
2. Press the SELECT button repeatedly until the engine oil life display or the service message is displayed.
3. Press the RESET button for about 10 seconds. You will see a MAINT RESET message.
4. Select the appropriate answer, MAINT RESET >N (NO) or MAINT RESET > y (YES) by pressing the SELECT button repeatedly.
   >N or >Y is displayed on the outside temperature >N or >Y is displayed on the outside temperature display.
5. Select the MAINT RESET > Y (YES), and press and hold the RESET button again to reset the engine oil life to 100%.

37647_HCRV_C0014

BRAKES — INFORMATION AND PRECAUTIONS

ANTI-LOCK SYSTEMS

- Certain components within the ABS system are not intended to be serviced or repaired individually.
- Do not use rubber hoses or other parts not specifically specified for and ABS system. When using repair kits, replace all parts included in the kit. Partial or incorrect repair may lead to functional problems and require the replacement of components.
- Lubricate rubber parts with clean, fresh brake fluid to ease assembly. Do not use shop air to clean parts; damage to rubber components may result.
- Use only DOT 3 brake fluid from an unopened container.
- If any hydraulic component or line is removed or replaced, it may be necessary to bleed the entire system.
- A clean repair area is essential. Always clean the reservoir and cap thoroughly before removing the cap. The slightest amount of dirt in the fluid may plug an orifice and impair the system function. Perform repairs after components have been thoroughly cleaned; use only denatured alcohol to clean components. Do not allow ABS components to come into contact with any substance containing mineral oil; this includes used shop rags.
- The Anti-Lock control unit is a microprocessor similar to other computer units in the vehicle. Ensure that the ignition switch is **OFF** before removing or installing controller harnesses. Avoid static electricity discharge at or near the controller.
- If any arc welding is to be done on the vehicle, the control unit should be unplugged before welding operations begin.

DISC AND DRUM SYSTEMS

CAUTION

Dust and dirt accumulating on brake parts during normal use may contain asbestos fibers from production or aftermarket brake linings. Breathing excessive concentrations of asbestos fibers can cause serious bodily harm. Exercise care when servicing brake parts. Do not sand or grind brake lining unless equipment used is designed to contain the dust residue. Do not clean brake parts with compressed air or by dry brushing. Cleaning should be done by dampening the brake components with a fine mist of water, then wiping the brake components clean with a dampened cloth. Dispose of cloth and all residue containing asbestos fibers in an impermeable container with the appropriate label. Follow practices prescribed by the Occupational Safety and Health Administration (OSHA) and the Environmental Protection Agency (EPA) for the handling, processing, and disposing of dust or debris that may contain asbestos fibers.

BRAKES — BLEEDING THE BRAKE SYSTEM

BLEEDING PROCEDURE

BLEEDING PROCEDURE

➡Note the following:

- Do not reuse the drained fluid. Use only clean Honda DOT 3 Brake Fluid from an unopened container. Using a non-Honda brake fluid can cause corrosion and shorten the life of the system.
- Make sure no dirt or other foreign matter is allowed to contaminate the brake fluid.
- Do not spill brake fluid on the vehicle, it may damage the paint. If brake fluid does contact the paint, wash it off immediately with water.
- The reservoir connected to the master cylinder must be at the MAX (upper) level mark at the start of the bleeding procedure and checked after bleeding each brake system. Add fluid as required.

1. Make sure the brake fluid level in the reservoir is at the MAX (upper) level line.
2. Have someone slowly pump the brake pedal several times, then apply steady pressure.
3. Start the bleeding at the driver's side of the front brake system.
4. Bleed the calipers or the wheel cylinders in the following sequence:
   - Front left
   - Front right
   - Rear right
   - Rear left
5. Attach a length of clear drain tube to the bleed screw, then loosen the bleed screw to allow air to escape from the system. Then tighten the bleed screw securely.

➡Do not loosen the special bolt on the rear caliper.

6. Refill the master cylinder reservoir to the MAX (upper) level line.
7. Repeat the procedure for each brake circuit until no air bubbles are in the fluid.

BLEEDING THE ABS SYSTEM

The bleeding procedure for the ABS System is the same as the conventional bleeding procedure. Refer to Brakes, Bleeding the Brake System, Bleeding Procedure.

WHEEL SPEED SENSORS

REMOVAL & INSTALLATION

Front Wheels

See Figure 2.

1. Before servicing the vehicle, refer to the Precautions Section.
2. Turn the ignition switch to LOCK (0).
3. Release the connector holding clamps, then disconnect the wheel speed sensor connector.
4. Remove the clips, the bolt, and the wheel speed sensor.

To install:

5. Install the wheel speed sensor in the reverse order of removal.
6. Install the sensor carefully to avoid twisting the wires.
7. If the wheel speed sensor comes in contact with the wheel bearing, it is faulty. Investigate the cause before replacing the sensor.
8. Start the engine and check that the ABS and the VSA indicators go off.
9. Test-drive the vehicle, and check that the ABS and the VSA indicators do not come on.

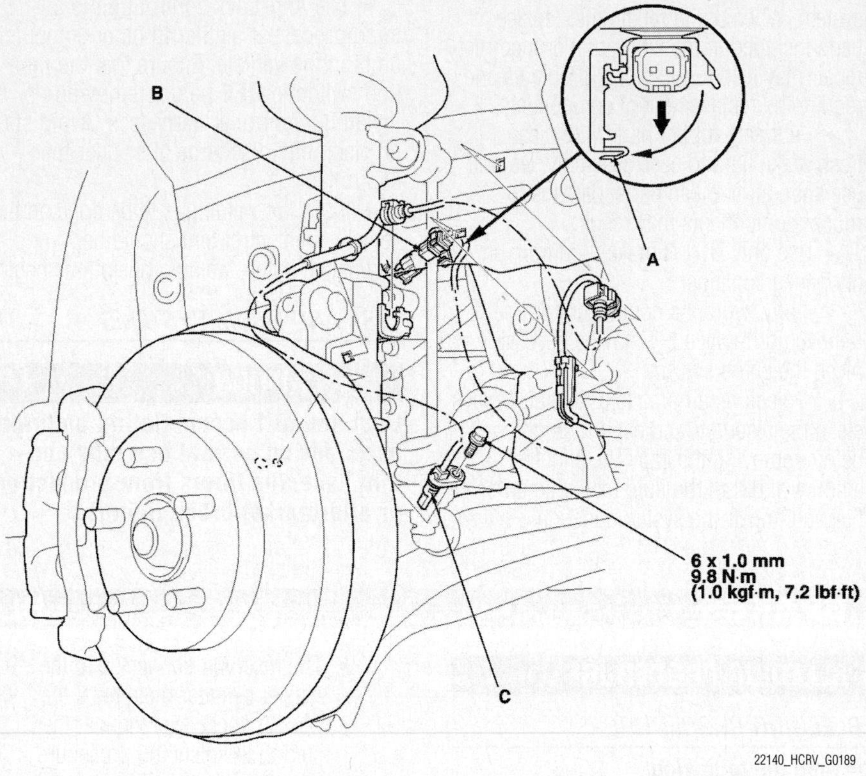
A
B
C
6 x 1.0 mm
9.8 N·m
(1.0 kgf·m, 7.2 lbf·ft)
22140_HCRV_G0189

Fig. 2 View of front wheel speed sensor

Rear Wheels

See Figure 3.

1. Before servicing the vehicle, refer to the Precautions Section.
2. Turn the ignition switch to LOCK (0).
3. Release the connector holding clamps, then disconnect the wheel speed sensor connector.

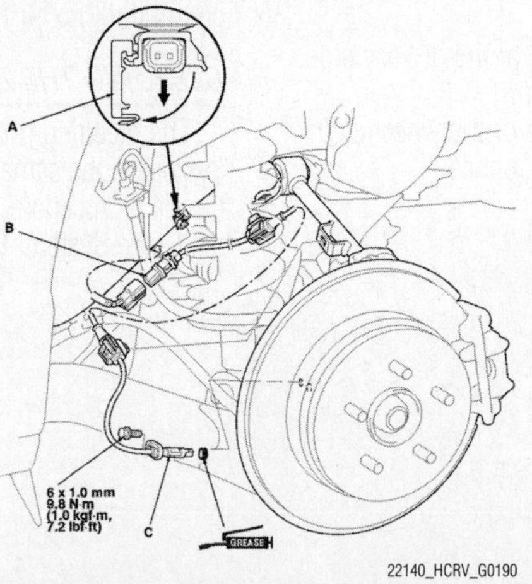
A
B
C
6 x 1.0 mm
9.8 N·m
(1.0 kgf·m,
7.2 lbf·ft)
GREASE
22140_HCRV_G0190

Fig. 3 View of rear wheel speed sensor

4. Remove the clips, the bolt, and the wheel speed sensor.

To install:

5. Install the wheel speed sensor in the reverse order of removal.
6. Apply multipurpose grease to the O-ring.
7. Install the sensor carefully to avoid twisting the wires.
8. If the wheel speed sensor comes in contact with the hub bearing unit, it is faulty. Investigate the cause before replacing the sensor.
9. Start the engine and check that the ABS and the VSA indicators go off.
10. Test-drive the vehicle and check that the ABS and the VSA indicators do not come on.

BRAKES

BRAKE CALIPER

REMOVAL & INSTALLATION

See Figure 4.

➡Keep any grease off the brake disc and brake pads.

1. Raise and support the vehicle.
2. Remove the front wheels.
3. Remove the brake hose mounting bolt.
4. Remove the brake caliper bracket mounting bolts, then remove the caliper assembly from the knuckle.

To install:

5. Install the brake caliper in the reverse order of removal.
6. Clean the mating surfaces of the brake disc and the inside of the wheel, then install the front wheel.

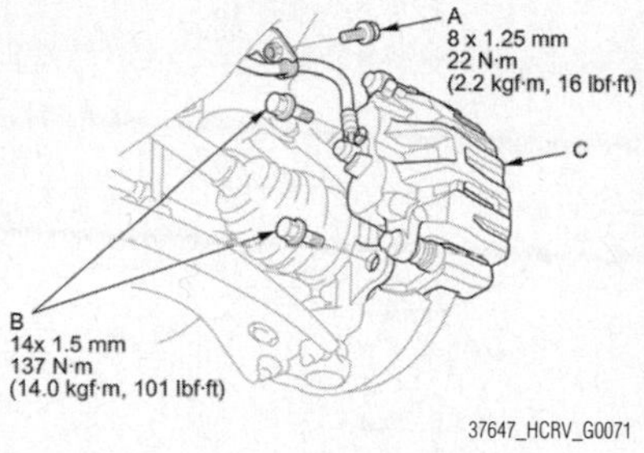
A
8 x 1.25 mm
22 N·m
(2.2 kgf·m, 16 lbf·ft)
C
B
14x 1.5 mm
137 N·m
(14.0 kgf·m, 101 lbf·ft)
37647_HCRV_G0071

Fig. 4 Remove the brake hose mounting bolt (A), the brake caliper bracket mounting bolts (B), then remove the caliper assembly (C)

DISC BRAKE PADS

REMOVAL & INSTALLATION

See Figures 5 through 9.

1. Remove some of the brake fluid from the master cylinder.
2. Raise and support the vehicle.
3. Remove the front wheels.
4. Remove the flange bolt while holding the caliper pin with a wrench. Be careful not to damage the pin boot, and pivot the caliper up out of the way. Check the hose and the pin boots for damage and deterioration.
5. Remove the brake pads and the pad shims.
6. Remove the pad retainers.

To install:

7. Clean the caliper bracket thoroughly; remove any rust, and check for grooves and cracks.

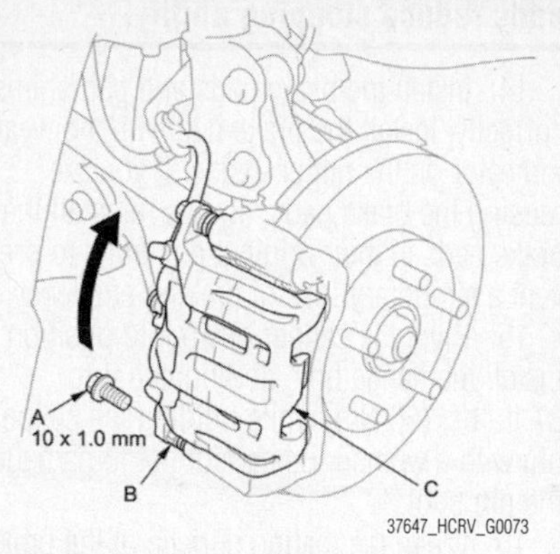
A
10 x 1.0 mm
B
C
37647_HCRV_G0073

Fig. 5 Remove the flange bolt (A) while holding the caliper pin (B) with a wrench, and pivot the caliper (C)

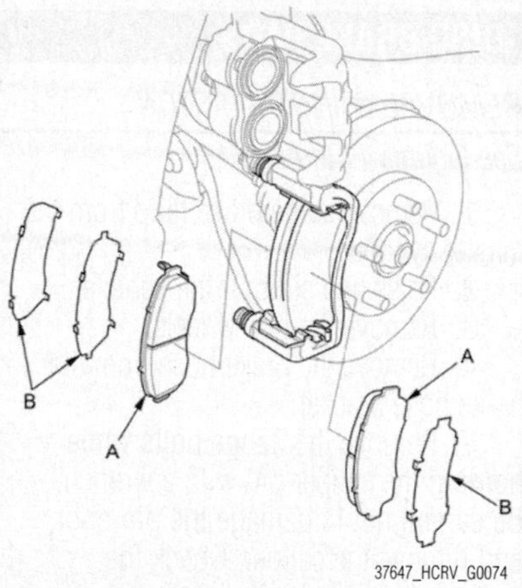
B
A
A
B
37647_HCRV_G0074

Fig. 6 Remove the brake pads (A) and the pad shims (B)

➡Verify that the caliper pins move in and out smoothly. Clean and lube as needed.

8. Apply a thin coat of M-77 assembly paste (P/N 08798-9010) to the retainer mating surface of the caliper bracket (indicated by the arrows and shaded area).
9. Install the pad retainers. Wipe excess assembly paste off the retainers. Keep any assembly paste off the brake discs and the brake pads.
10. Mount the brake caliper piston compressor on the caliper body.
11. Press in the piston with the brake caliper piston compressor tool so the caliper will fit over the brake pads. Make sure the piston boot is in position to prevent damaging it when pivoting the caliper down.

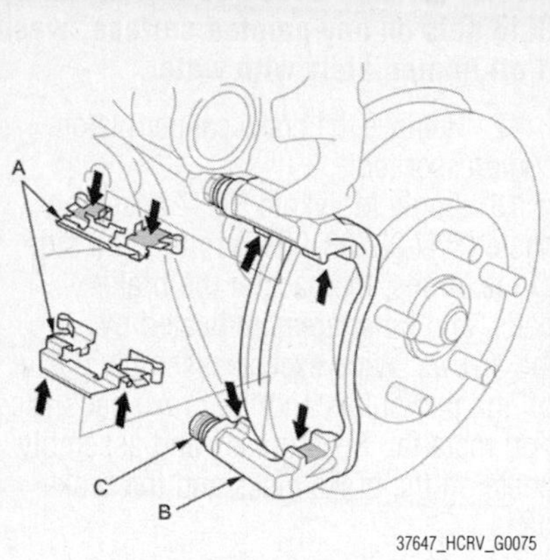
A
C
B
37647_HCRV_G0075

Fig. 7 Remove the pad retainers (A)

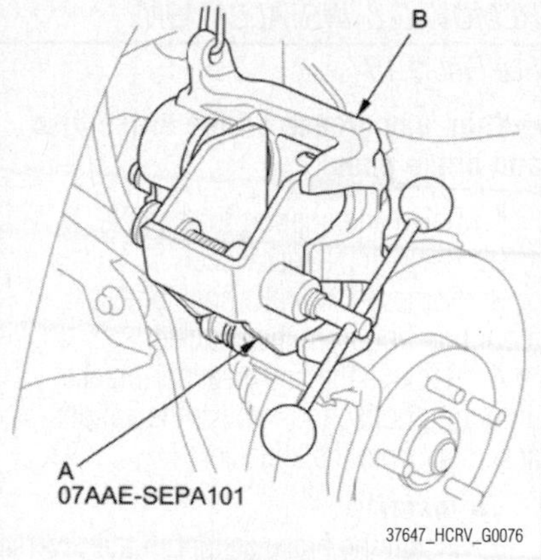
B
A
07AAE-SEPA101
37647_HCRV_G0076

Fig. 8 Mount the brake caliper piston compressor (A) on the caliper body (B)

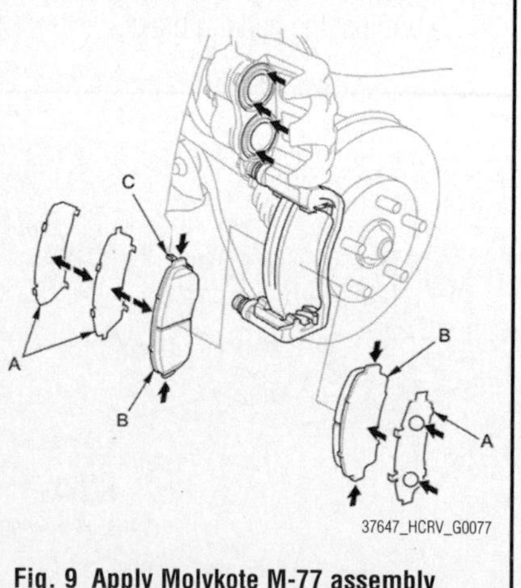
C
A
B
B
A
37647_HCRV_G0077

Fig. 9 Apply Molykote M-77 assembly paste (P/N 08798-9010) to the pad side of the shims (A), the back of the brake pads (B), and other areas indicated by the arrows

➡Be careful when pressing in the piston; brake fluid might overflow from the master cylinder's reservoir. If brake fluid gets on any painted surface, wash it off immediately with water.

12. Remove the brake caliper piston compressor tool.

13. Apply Molykote M-77 assembly paste (P/N 08798-9010) to the pad side of the shims, the back of the brake pads, and other areas indicated by the arrows. Wipe excess assembly paste off the pad shims and the brake pads friction material. Keep grease and assembly paste off the brake discs and the brake pads.

⁂ CAUTION

Contaminated brake discs or brake pads reduce stopping ability.

14. Install the brake pads and pad shims correctly. Install the brake pad with the wear indicator on the upper inside. If you are reusing the brake pads, always reinstall the brake pads in their original positions to prevent a temporary loss of braking efficiency.

15. Pivot the caliper down into position. Install the flange bolt, and tighten it to 37 ft. lbs. (50 Nm) while holding the caliper pin with a wrench. Be careful not to damage the pin boot.

16. Clean the mating surface of the brake disc and the inside of the wheel, then install the front wheels.

17. Press the brake pedal several times to make sure the brakes work.

➡Engagement may require a greater pedal stroke immediately after the brake pads have been replaced as a set. Several applications of the brake pedal will restore the normal pedal stroke.

18. Add brake fluid as needed.

19. After installation, check for leaks at hose and line joints or connections, and retighten if necessary.

20. Test-drive the vehicle, then check for leaks.

BRAKES

REAR DISC BRAKES

BRAKE CALIPER

REMOVAL & INSTALLATION

See Figure 10.

➡Keep any grease off the brake disc and brake pads.

1. Raise and support the vehicle.
2. Remove the rear wheel.
3. Remove the brake hose bracket mounting bolt from the knuckle.
4. Remove the brake caliper bracket mounting bolts, and remove the caliper assembly from the knuckle.

To install:

5. Install the brake caliper in the reverse order of removal, and note these items:
   - Before installing the brake disc/drum, clean the mating surface of the hub bearing unit and the inside of the brake disc/drum.
   - Adjust the parking brake.

6. Clean the mating surfaces of the brake disc/drum and the inside of the wheel, then install the rear wheel.

DISC BRAKE PADS

REMOVAL & INSTALLATION

See Figures 11 through 18.

1. Remove some brake fluid from the master cylinder.
2. Raise and support the vehicle.
3. Remove the rear wheels.
4. Remove the brake hose from the brake hose bracket.
5. Remove the flange bolts while holding the caliper pin with a wrench. Be careful not to damage the pin boot, and remove the caliper. Check the hose and pin boots for damage and deterioration.
6. Remove the outer pad and the pad shim.

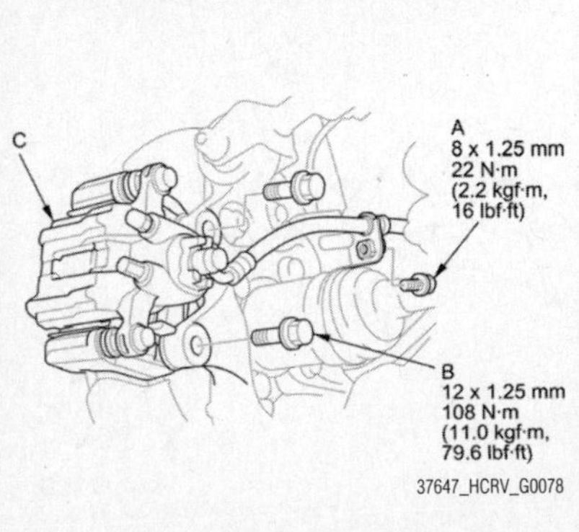
C
A
8 x 1.25 mm
22 N·m
(2.2 kgf·m,
16 lbf·ft)
B
12 x 1.25 mm
108 N·m
(11.0 kgf·m,
79.6 lbf·ft)
37647_HCRV_G0078

Fig. 10 Remove the brake hose bracket mounting bolt (A), the brake caliper bracket mounting bolts (B), and remove the caliper assembly (C) from the knuckle

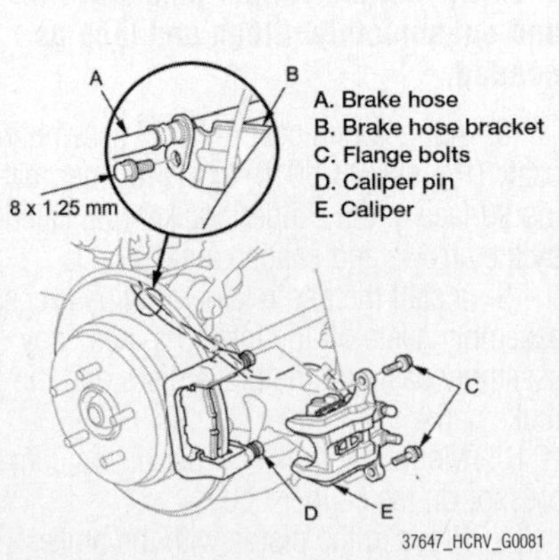
A
B
A. Brake hose
B. Brake hose bracket
C. Flange bolts
D. Caliper pin
E. Caliper
8 x 1.25 mm
C
D
E
37647_HCRV_G0081

Fig. 11 Remove the brake hose from the brake hose bracket

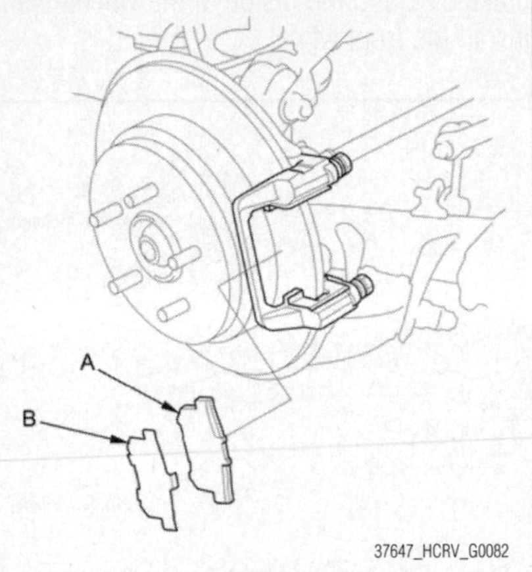
A
B
37647_HCRV_G0082

Fig. 12 Remove the outer pad (A) and the pad shim (B)

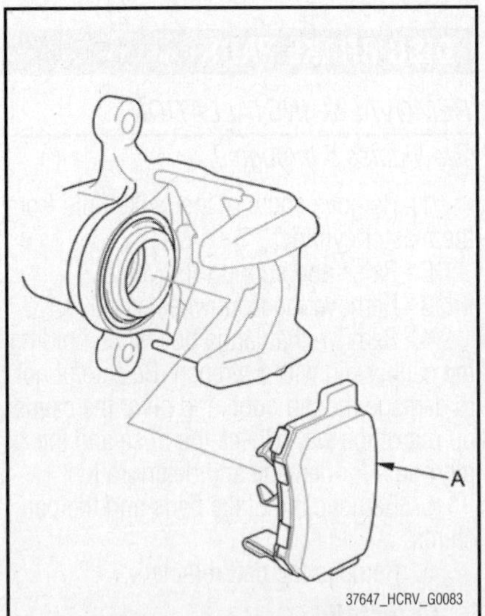
A
37647_HCRV_G0083

Fig. 13 Remove the inner pad (A) from the caliper

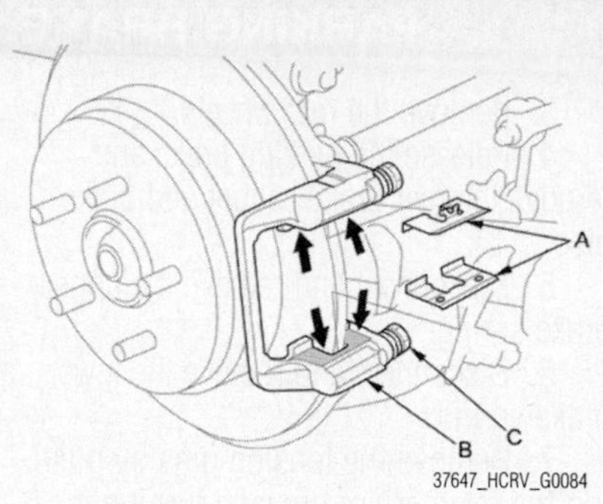
A
B
C
37647_HCRV_G0084

Fig. 14 Remove the pad retainers (A)

7. Remove the inner pad from the caliper.

8. Remove the pad retainers.

To install:

9. Clean the caliper bracket thoroughly; remove any rust, and check for grooves and cracks. Verify that the caliper pins move in and out smoothly. Clean and lube as needed.

10. Inspect the brake disc/drum for damage and cracks. Inspect for runout, thickness, and parallelism.

11. Apply Molykote M-77 assembly paste (P/N 08798-9010) to the retainers on their mating surfaces (indicated by the arrows) against the caliper bracket.

12. Install the pad retainers. Wipe excess assembly paste off the retainers. Keep any assembly paste off the brake discs and the brake pads.

13. Mount the brake caliper piston compressor on the caliper body.

14. Press in the piston with the brake caliper piston compressor tool so the caliper will fit over the brake pads. Make sure the piston boot is in position to prevent damaging it when pivoting the caliper down.

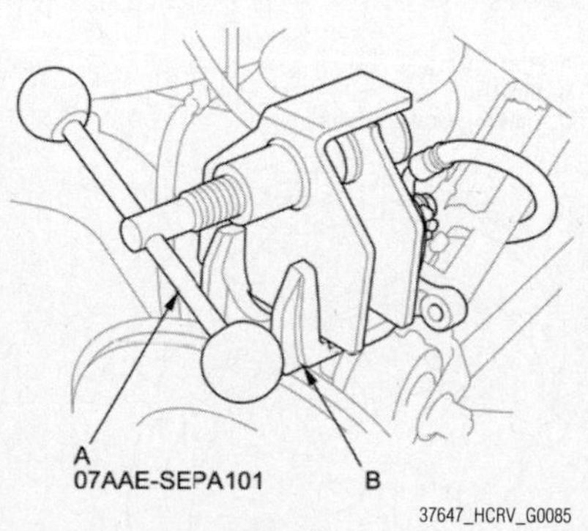
A
07AAE-SEPA101
B
37647_HCRV_G0085

Fig. 15 Mount the brake caliper piston compressor (A) on the caliper body (B)

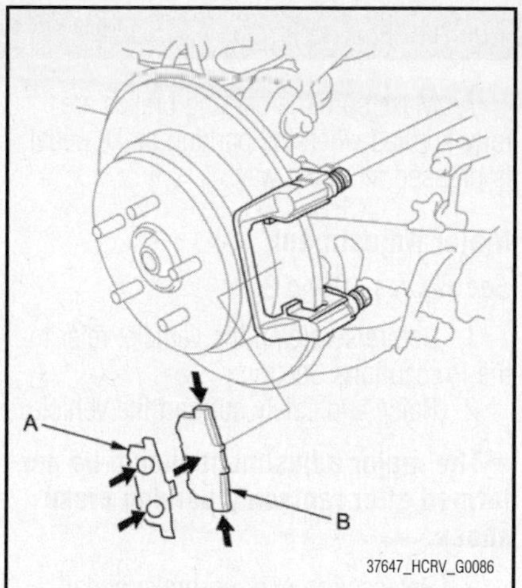
A
B
37647_HCRV_G0086

Fig. 16 Apply Molykote M-77 assembly paste to the pad side of the shims (A), the back of the brake pads (B), and other areas indicated by the arrows

➡Be careful when pressing in the piston; brake fluid might overflow from the master cylinder's reservoir. If brake fluid gets on any painted surface, wash it off immediately with water.

15. Remove the brake caliper piston compressor tool.

16. Apply Molykote M-77 assembly paste (P/N 08798-9010) to the pad side of the shims, the back of the brake pads, and other areas indicated by the arrows. Wipe excess assembly paste off the pad shims and the brake pads friction material. Keep grease and assembly paste off the brake discs and the brake pads.

⁂ CAUTION

Contaminated brake discs or brake pads reduce stopping ability.

17. Install the brake pads and pad shims correctly. Install the brake pad with the wear indicator on the bottom inside. If you are reusing the brake pads, always reinstall the brake pads in their original positions to prevent a temporary loss of braking efficiency.

18. Install the caliper into position. Install the flange bolts, and tighten it to 17 ft. lbs. (23 Nm) while holding the caliper pin with a wrench. Be careful not to damage the pin boots.

19. Install the brake hose to the brake hose bracket.

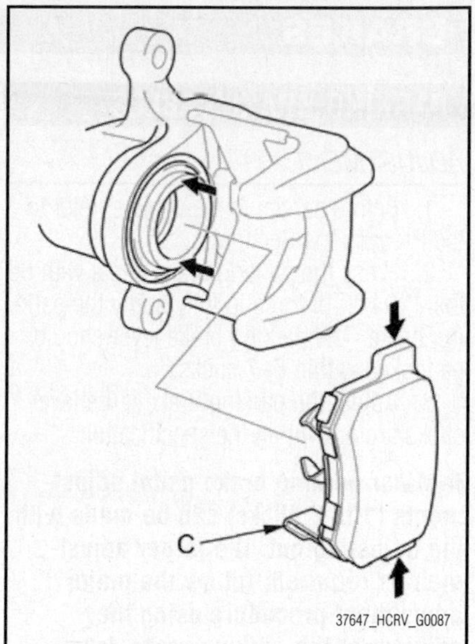
C
37647_HCRV_G0087

Fig. 17 Install the brake pad with the wear indicator (C) on the bottom inside

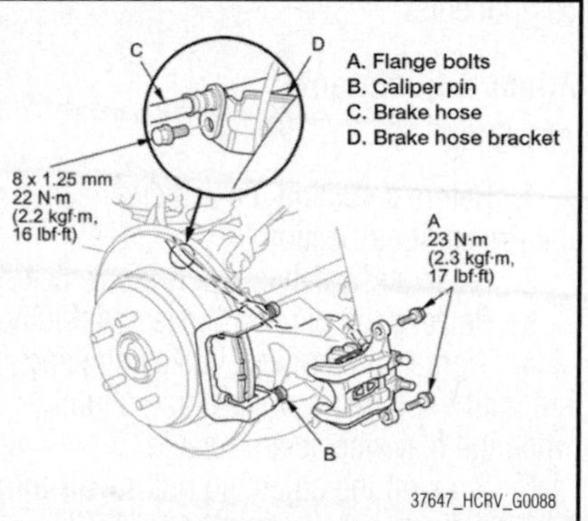
C
D
A. Flange bolts
B. Caliper pin
C. Brake hose
D. Brake hose bracket
8 x 1.25 mm
22 N·m
(2.2 kgf·m,
16 lbf·ft)
A
23 N·m
(2.3 kgf·m,
17 lbf·ft)
B
37647_HCRV_G0088

Fig. 18 Install the caliper into position

20. Clean the mating surface of the brake disc/drum and the inside of the wheel, then install the rear wheels.

21. Press the brake pedal several times to make sure the brakes work.

➡Engagement may require a greater pedal stroke immediately after the brake pads have been replaced as a set. Several applications of the brake pedal will restore the normal pedal stroke.

22. Add brake fluid as needed.

23. After installation, check for leaks at hose and line joints or connections, and retighten if necessary.

24. Test-drive the vehicle, then check for leaks.

PARKING BRAKE CABLES

ADJUSTMENT

1. Before servicing the vehicle, refer to the Precautions Section.
2. Press the parking brake pedal with 66 lbs. (294 N) of force to fully apply the parking brake. The parking brake lever should be locked within 6–7 clicks.
3. Adjust the parking brake if the lever clicks are not within the specification.

➡Minor parking brake pedal adjustments (1 to 2 clicks) can be made with the adjusting nut. If a larger adjustment is required, follow the major adjustment procedure using the adjuster at the parking brake drum.

4. After installing new parking brake shoes or a new rear brake disc/drum, make sure to drive the vehicle to break-in the components.

Minor Adjustment

See Figure 19.

1. Before servicing the vehicle, refer to the Precautions Section.
2. Raise and safely support the vehicle.
3. Release the parking brake pedal fully.
4. Tighten the parking brake adjusting nut until the parking brakes drag slightly when the rear wheels are rotated.
5. Back off the adjusting nut in half-turn increments.
6. Release the parking brake pedal fully, and check that the parking brakes do not drag when the rear wheels are rotated. Readjust if necessary.

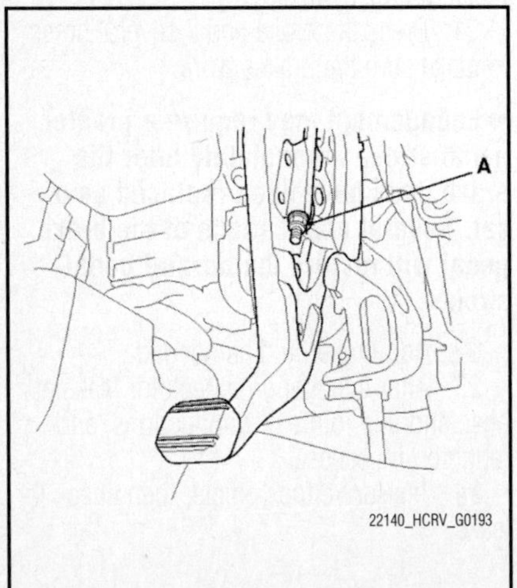
A
22140_HCRV_G0193

Fig. 19 View of parking brake adjusting nut (A)

7. Make sure the parking brakes are fully applied when the parking brake pedal is pressed in all the way.

Major Adjustment

See Figures 19 and 20.

1. Before servicing the vehicle, refer to the Precautions Section.
2. Raise and safely support the vehicle.

➡The major adjustment should be performed after replacing parking brake shoes.

3. Release the parking brake pedal fully.
4. Back off the parking brake adjusting nut on the parking brake pedal.
5. Remove the rear wheels.
6. Remove the access plug.
7. Turn the adjuster with a flat-tip screwdriver until the shoes lock against the parking brake drum. Then back off the adjuster 8 clicks and install the access plug.
8. Clean the mating surface of the brake disc/drum and the inside of the wheel, then install the rear wheels.
9. Do the minor adjustment procedure, as necessary.

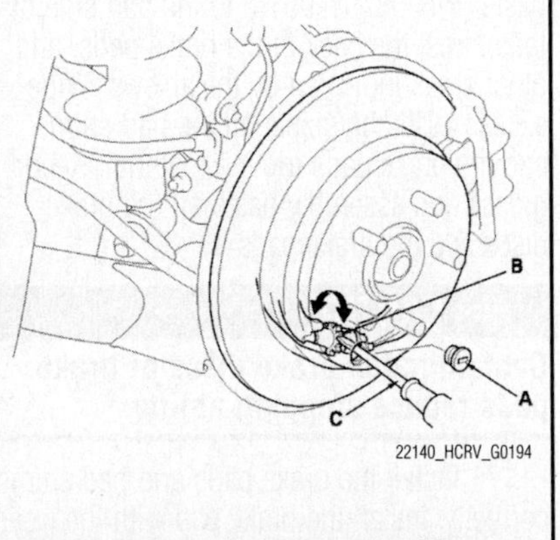
B
A
C
22140_HCRV_G0194

Fig. 20 Remove the access plug (A), turn the adjuster (B) with a flat-tip screwdriver (C) until the shoes lock against the parking brake drum

PARKING BRAKE SHOES

REMOVAL & INSTALLATION

See Figures 21 through 27.

1. Before servicing the vehicle, refer to the Precautions Section.
2. Raise and safely support the vehicle.
3. Remove the rear wheels.
4. Release the parking brake and remove the rear brake caliper and brake disc.
5. Disconnect and remove the upper brake spring.
6. Disconnect and remove the lower brake spring.
7. Remove the tension pins by pushing the retainer springs and turning the pins.
8. Remove the adjuster assembly by moving the forward brake shoe.
9. Remove the rearward brake shoe by disconnecting the parking brake cable from the parking brake lever.

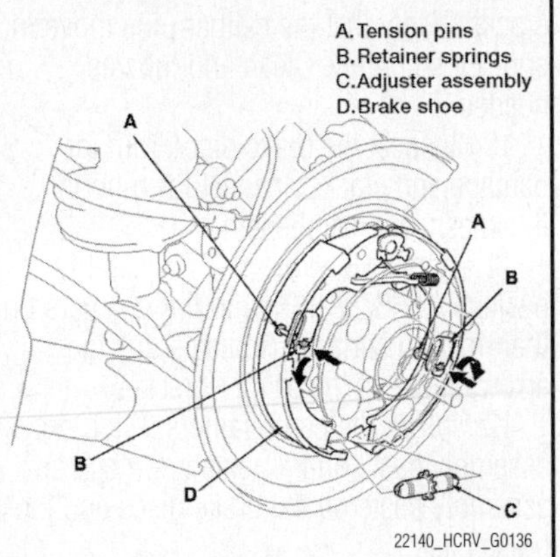
A. Tension pins
B. Retainer springs
C. Adjuster assembly
D. Brake shoe
A
A
B
B
D
C
22140_HCRV_G0136

Fig. 21 Remove the tension pins by pushing the retainer springs and turning the pins. Remove the adjuster assembly by moving the forward brake shoe

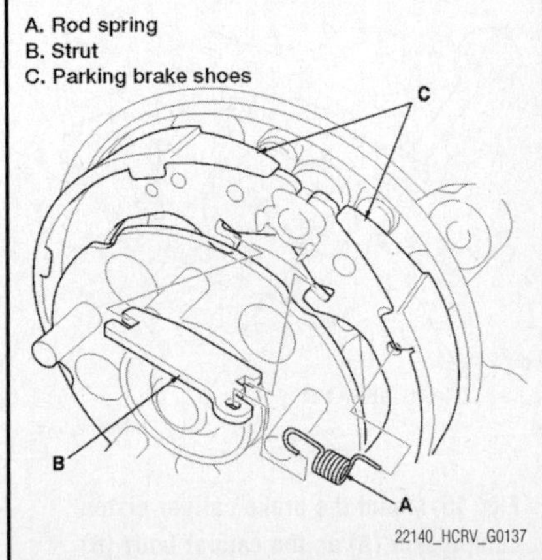
A. Rod spring
B. Strut
C. Parking brake shoes
C
B
A
22140_HCRV_G0137

Fig. 22 Disconnect the rod spring (A) and remove the strut (B). Remove the parking brake shoes (C)

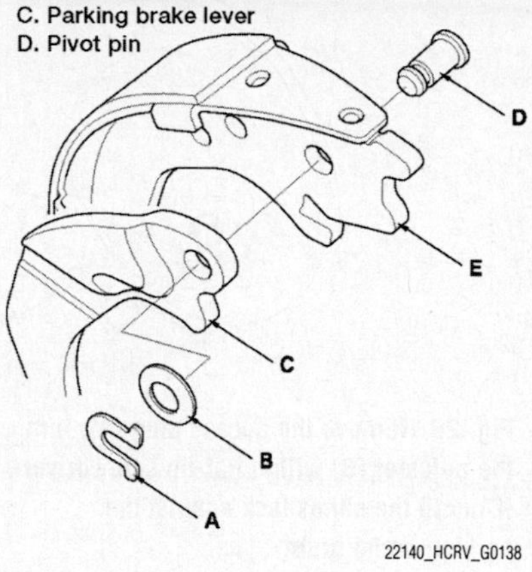
A. U-clip
B. Wave washer
C. Parking brake lever
D. Pivot pin
A
B
C
D
E
22140_HCRV_G0138

Fig. 23 Remove the U-clip, wave washer, parking brake lever, and pivot pin from the brake shoe

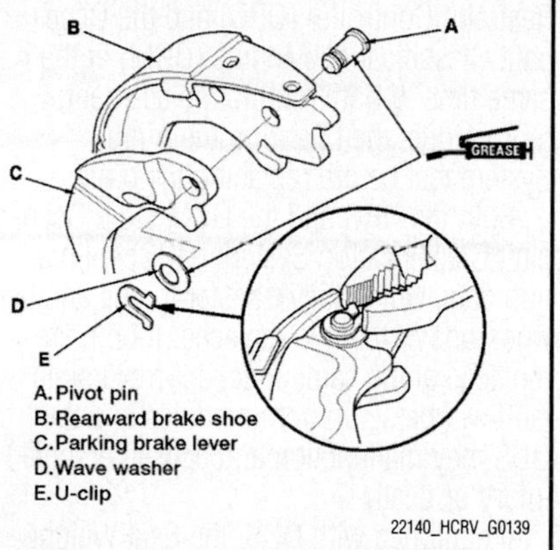
A
B
C
D
GREASE
A. Pivot pin
B. Rearward brake shoe
C. Parking brake lever
D. Wave washer
E. U-clip
22140_HCRV_G0139

Fig. 24 Apply Molykote 44 MA grease to the sliding surface of the pivot pin and insert the pin into the rearward brake shoe. Install the parking brake lever and wave washer on the pivot pin and secure with a new U-clip

10. Disconnect the rod spring, and remove the strut.

11. Remove the parking brake shoes.

12. Remove the U-clip, wave washer, parking brake lever, and pivot pin from the brake shoe.

To install:

13. Apply Molykote 44 MA grease to the sliding surface of the pivot pin and insert the pin into the rearward brake shoe.

14. Install the parking brake lever and wave washer on the pivot pin and secure with a new U-clip.

15. Install the wave washer with its convex side facing out.

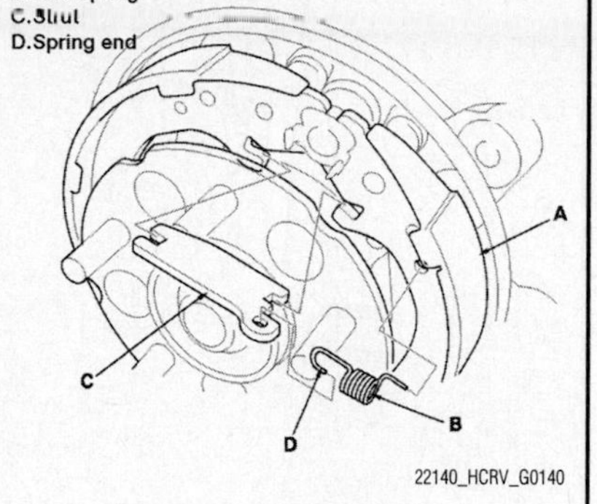
A. Position the parking brake shoes
B. Rod spring
C. Strut
D. Spring end
A
B
C
D
22140_HCRV_G0140

Fig. 25 Position the parking brake shoes, then hook the rod spring on the strut first with the spring end pointing downward. Then, hook the rod spring to the parking brake shoe and install the strut on the parking brake shoes

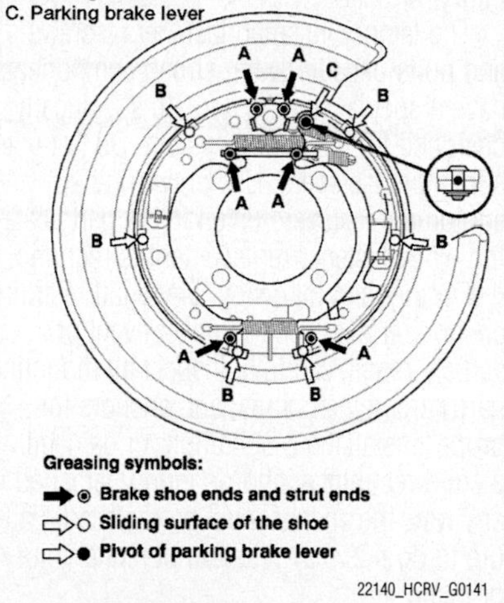
A. Sliding surfaces
B. Parking brake shoe
C. Parking brake lever
Greasing symbols:
Brake shoe ends and strut ends
Sliding surface of the shoe
Pivot of parking brake lever
22140_HCRV_G0141

Fig. 26 Apply Molykote 44 MA grease to the sliding surfaces, the opposite edges of the parking brake shoe, and the pivot of the parking brake lever

16. Pinch the U-clip securely to prevent the pivot pin from coming out of the brake shoe.

17. Position the parking brake shoes, then hook the rod spring on the strut first with the spring end pointing downward. Then, hook the rod spring to the parking brake shoe and install the strut on the parking brake shoes.

18. Apply Molykote 44 MA grease to the sliding surfaces, the opposite edges of the parking brake shoe, and the pivot of the parking brake lever as shown. Wipe off any excess. Keep grease off the brake linings.

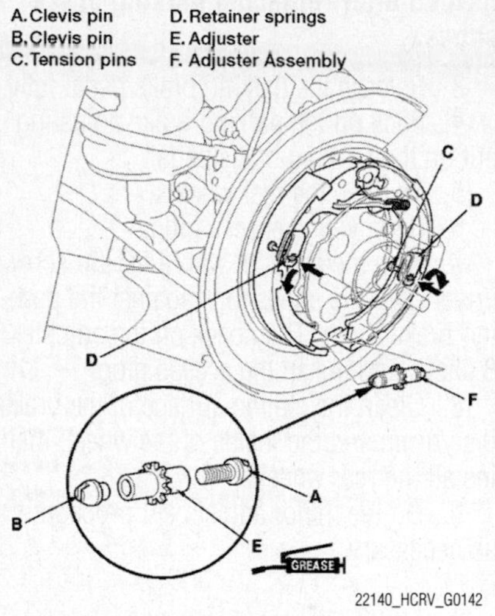
A. Clevis pin
B. Clevis pin
C. Tension pins
D. Retainer springs
E. Adjuster
F. Adjuster Assembly
A
B
C
D
E
F
GREASE
22140_HCRV_G0142

Fig. 27 Reinstall the tension pins and retainer springs. Make sure the tension pin does not contact the parking brake lever. Clean the threaded portions of the clevis A, and coat the threads of the clevis with grease. Clean the sliding surface of the clevis B, and coat the sliding surface of the clevis B with grease. Install the clevis A and B on the adjuster and shorten the clevis A by turning the adjuster

19. Connect the parking brake cable to the parking brake lever. Apply silicone grease to the cable contact surface on the backing plate.

20. Reinstall the tension pins and retainer springs. Make sure the tension pin does not contact the parking brake lever.

21. Clean the threaded portions of the clevis A, and coat the threads of the clevis with grease. Clean the sliding surface of the clevis B, and coat the sliding surface of the clevis B with grease. Install the clevis A and B on the adjuster and shorten the clevis A by turning the adjuster.

22. Position the brake shoe adjuster assembly on the parking brake shoes.

23. Reinstall the lower brake spring.

24. Reinstall the upper brake spring.

25. Install the rear brake disc/drum and rear brake caliper.

26. Adjust the parking brake.

ADJUSTMENT

See Figures 28 and 29.

1. Before servicing the vehicle, refer to the Precautions Section.

2. Raise and safely support the vehicle.

➡The major adjustment should be performed after replacing parking brake shoes.

3. Release the parking brake pedal fully.
4. Back off the parking brake adjusting nut on the parking brake pedal.
5. Remove the rear wheels.
6. Remove the access plug.
7. Turn the adjuster with a flat-tip screwdriver until the shoes lock against the parking brake drum. Then back off the adjuster 8 clicks and install the access plug.
8. Clean the mating surface of the brake disc/drum and the inside of the wheel, then install the rear wheels.
9. Do the minor adjustment procedure, as necessary.

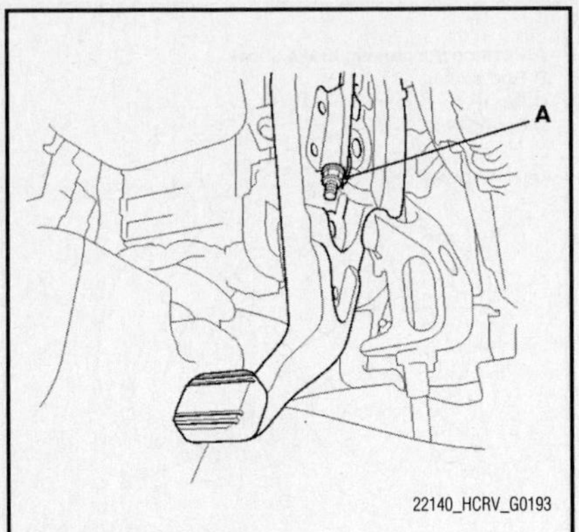
A
22140_HCRV_G0193

Fig. 28 View of parking brake adjusting nut (A)

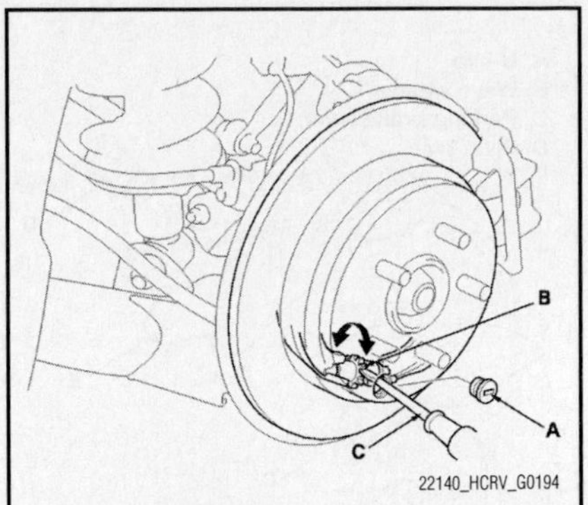
B
A
C
22140_HCRV_G0194

Fig. 29 Remove the access plug (A), turn the adjuster (B) with a flat-tip screwdriver (C) until the shoes lock against the parking brake drum

CHASSIS ELECTRICAL

AIR BAG (SUPPLEMENTAL RESTRAINT SYSTEM)

GENERAL INFORMATION

PRECAUTIONS

Disconnect and isolate the battery negative cable before beginning any airbag system component diagnosis, testing, removal, or installation procedures. Allow system capacitor to discharge for two minutes before beginning any component service. This will disable the airbag system. Failure to disable the airbag system may result in accidental airbag deployment, personal injury, or death.

Do not place an intact undeployed airbag face down on a solid surface. The airbag will propel into the air if accidentally deployed and may result in personal injury or death.

When carrying or handling an undeployed airbag, the trim side (face) of the airbag should be pointing towards the body to minimize possibility of injury if accidental deployment occurs. Failure to do this may result in personal injury or death.

Replace airbag system components with OEM replacement parts. Substitute parts may appear interchangeable, but internal differences may result in inferior occupant protection. Failure to do so may result in occupant personal injury or death.

Wear safety glasses, rubber gloves, and long sleeved clothing when cleaning powder residue from vehicle after an airbag deployment. Powder residue emitted from a deployed airbag can cause skin irritation. Flush affected area with cool water if irritation is experienced. If nasal or throat irritation is experienced, exit the vehicle for fresh air until the irritation ceases. If irritation continues, see a physician.

Do not use a replacement airbag that is not in the original packaging. This may result in improper deployment, personal injury, or death.

The factory installed fasteners, screws and bolts used to fasten airbag components have a special coating and are specifically designed for the airbag system. Do not use substitute fasteners. Use only original equipment fasteners listed in the parts catalog when fastener replacement is required.

During, and following, any child restraint anchor service, due to impact event or vehicle repair, carefully inspect all mounting hardware, tether straps, and anchors for proper installation, operation, or damage. If a child restraint anchor is found damaged in any way, the anchor must be replaced. Failure to do this may result in personal injury or death.

Deployed and non-deployed airbags may or may not have live pyrotechnic material within the airbag inflator.

Do not dispose of driver/passenger/curtain airbags or seat belt tensioners unless you are sure of complete deployment. Refer to the Hazardous Substance Control System for proper disposal.

Dispose of deployed airbags and tensioners consistent with state, provincial, local, and federal regulations.

After any airbag component testing or service, do not connect the battery negative cable. Personal injury or death may result if the system test is not performed first.

If the vehicle is equipped with the Occupant Classification System (OCS), do not connect the battery negative cable before performing the OCS Verification Test using the scan tool and the appropriate diagnostic information. Personal injury or death may result if the system test is not performed properly.

Never replace both the Occupant Restraint Controller (ORC) and the Occupant Classification Module (OCM) at the same time. If both require replacement, replace one, then perform the Airbag System test before replacing the other.

Both the ORC and the OCM store Occupant Classification System (OCS) calibration data, which they transfer to one another when one of them is replaced. If both are replaced at the same time, an irreversible fault will be set in both modules and the OCS may malfunction and cause personal injury or death.

If equipped with OCS, the Seat Weight Sensor is a sensitive, calibrated unit and must be handled carefully. Do not drop or handle roughly. If dropped or damaged, replace with another sensor. Failure to do so may result in occupant injury or death.

If equipped with OCS, the front passenger seat must be handled carefully as well. When removing the seat, be careful when setting on floor not to drop. If dropped, the sensor may be inoperative, could result in occupant injury, or possibly death.

If equipped with OCS, when the passenger front seat is on the floor, no one should sit in the front passenger seat. This uneven force may damage the sensing ability of the seat weight sensors. If sat on and damaged, the sensor may be inoperative, could result in occupant injury, or possibly death.

DISARMING THE SYSTEM

1. Before servicing the vehicle, refer to the Precautions Section.
2. Turn the ignition switch to **OFF**.
3. Disconnect the negative battery cable

and isolate it from accidental reconnection. Insulate the cable end with high-quality electrical tape or a similar non-conductive wrapping.

4. Wait at least 3 minutes for the system capacitor to discharge before performing any service. The air bag system is designed to retain enough voltage to deploy the air bag for a short period of time after the battery has been disconnected.

ARMING THE SYSTEM

1. Before servicing the vehicle, refer to the Precautions Section.
2. Reconnect the negative battery cable.
3. To confirm proper system operation, turn the ignition switch to the **ON** position. The SRS indicator light should light for at least 7 seconds and then go off.

CLOCKSPRING CENTERING

See Figures 30 through 34.

CAUTION

Before servicing, or working around, the SRS system, turn the ignition switch OFF, disconnect both battery cables and wait at least 3 minutes. When servicing, or working around, the SRS system, do not work directly in front of the air bag module.

1. Before servicing the vehicle, refer to the Precautions Section.
2. Position the front wheels in the straight ahead position.
3. Remove the driver's airbag.
4. Disconnect the connector from the clockspring (also called a cable reel), then remove the steering wheel bolt.

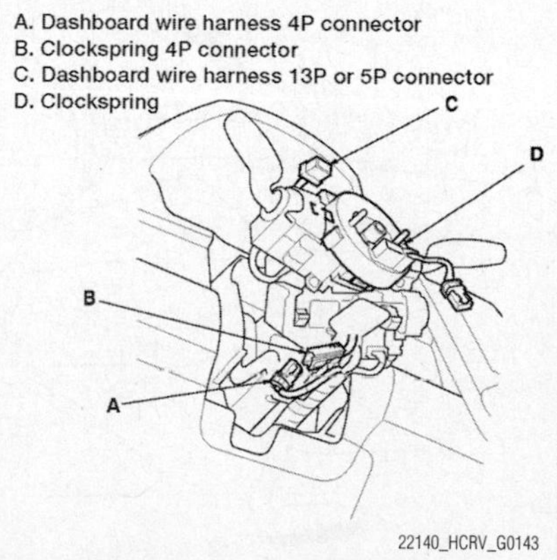
A. Dashboard wire harness 4P connector
B. Clockspring 4P connector
C. Dashboard wire harness 13P or 5P connector
D. Clockspring
C
D
B
A
22140_HCRV_G0143

Fig. 30 Disconnect the dashboard wire harness 4P connector from the clockspring 4P connector, then disconnect the dashboard wire harness 13P or 5P connector from the clockspring

5. Confirm that the front wheels point straight ahead, then remove the steering wheel with a steering wheel puller .Do not tap on the steering wheel or steering column shaft when removing the steering wheel.
6. Remove the driver's dashboard undercover.
7. Remove the column cover screws, then remove the column covers.
8. Disconnect the dashboard wire harness 4P connector from the clockspring 4P connector, then disconnect the dashboard wire harness 13P or 5P connector from the clockspring.
9. Release the lock tab (A) under the clockspring connector with a 90° hook shaped tool (B). Slide the tool below the clockspring connector just above the lock tab. Release the lower lock tab (C) and slide the clockspring off the column.

To install:

10. Before installing the steering wheel, make sure the front wheels are aligned straight ahead.
11. If not already done, perform the battery terminal disconnection procedure.

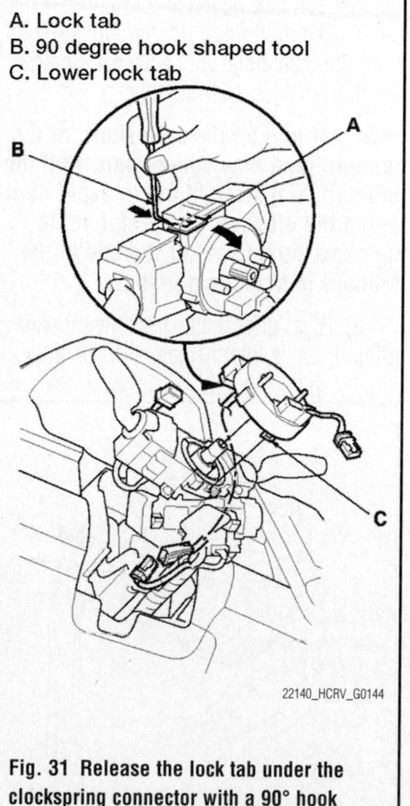
A. Lock tab
B. 90 degree hook shaped tool
C. Lower lock tab
A
B
C
22140_HCRV_G0144

Fig. 31 Release the lock tab under the clockspring connector with a 90° hook shaped tool. Slide the tool below the clockspring connector just above the lock tab. Release the lower lock tab and slide the clockspring off the column

Some systems store data in memory that is lost when the battery is disconnected. Do thc following steps before disconnecting the battery:

a. Make sure to record the anti-theft code(s) for the audio and/or the navigation system (if equipped).
b. Write down the audio presets (AM and FM) and the XM audio presets (if equipped), because the audio unit does not retain the presets after the battery is disconnected.
c. Make sure the ignition switch is in LOCK position.
d. Disconnect and isolate the negative cable from the battery.

➡**Always disconnect the negative cable from the battery first.**

e. Disconnect the positive cable from the battery.

12. Wait for 3 minutes for the SRS system to discharge before continuing the procedure.
13. Set the turn signal canceling sleeve so that the projections are aligned vertically.
14. Carefully install the clockspring on the steering column shaft. Then connect 13P or 5P connector to the clockspring, and connect the 4P connector to the dashboard wire harness 4P connector.
15. Install the steering column covers.
16. If necessary, center the clockspring. Do this by first rotating the clockspring clockwise until it stops. Then, rotate it counterclockwise (about 3 turns) until the arrow mark on the clockspring label points straight up.

➡**New replacement clocksprings come centered.**

17. Position the 2 tabs of the turn signal canceling sleeve as shown, and install the steering wheel on to the steering column

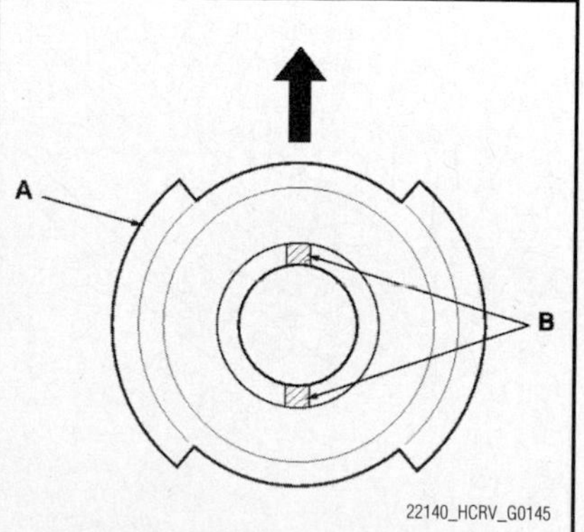
A
B
22140_HCRV_G0145

Fig. 32 Set the turn signal canceling sleeve (A) so that the projections (B) are aligned vertically

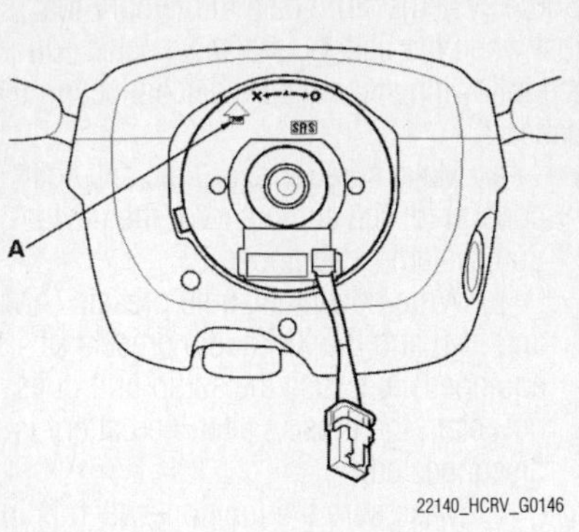
A
22140_HCRV_G0146

Fig. 33 Center the clockspring by rotating the clockspring clockwise until it stops. Then, rotate it counterclockwise (about 3 turns) until the arrow mark (A) on the clockspring label points straight up

shaft, making sure the steering wheel hub engages the pins of the clockspring and tabs of the turn signal canceling sleeve. Do not tap on the steering wheel or steering column shaft when installing the steering wheel.

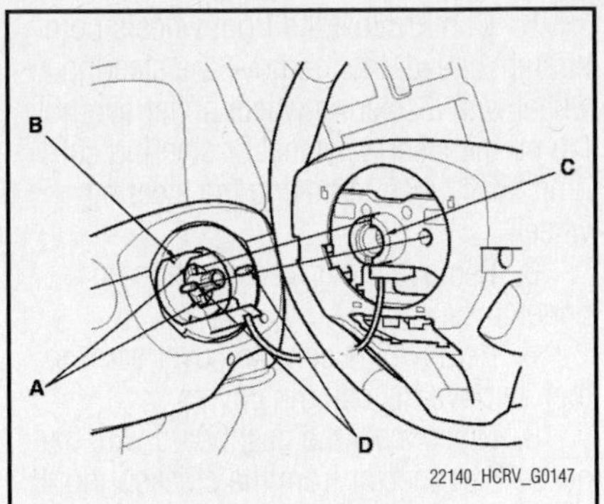
B
C
A
D
22140_HCRV_G0147

Fig. 34 Position the 2 tabs of the turn signal canceling sleeve as shown, and install the steering wheel on to the steering column shaft, making sure the steering wheel hub engages the pins of the clockspring and tabs of the turn signal canceling sleeve

18. Install a new steering wheel bolt and tighten to 29 ft. lbs. (39 Nm).
19. Reconnect the connectors.
20. Install the driver's airbag.
21. Connect the battery terminals.
22. Clear any DTC's.
23. After installing the clockspring, confirm proper system operation:
    - Turn the ignition switch to ON (II); the SRS indicator should come on for about 6 seconds and then go off
    - After the SRS indicator has turned off, turn the steering wheel fully left and right to confirm the SRS indicator does not come on
    - Make sure the horn works
    - Make sure the cruise control buttons work
    - Make sure the switches near the steering wheel all work properly

DRIVE TRAIN

FRONT HALFSHAFT

REMOVAL & INSTALLATION

See Figures 35 through 38.

1. Raise and support the vehicle.
2. Remove the front wheels.
3. Pry up the stake on the spindle nut, then remove the nut.
4. Drain the transmission fluid, then reinstall the drain plug with a new sealing washer.
5. Remove the nuts and bolt, then separate the lower arm using a prybar.
6. Pull the knuckle outward, and separate the outboard joint from the front hub using a soft face hammer.
7. Left driveshaft: Pry the inboard joint from the differential with a prybar. Remove the driveshaft as an assembly.

➡Do not pull on the driveshaft, or the inboard joint may come apart. Pull the inboard joint straight out to avoid damaging the oil seal. Be careful not to damage the oil seal or the end of the inboard joint using a prybar.

8. Right driveshaft: Drive the inboard joint off of the intermediate shaft using a drift punch and a hammer. Remove the driveshaft as an assembly.

➡Do not pull on the driveshaft or the inboard joint may come apart. Be careful not to damage the end of the inboard joint with the drift.

9. Remove the set ring from the left driveshaft inboard joint.
10. Remove the set ring from the intermediate shaft.

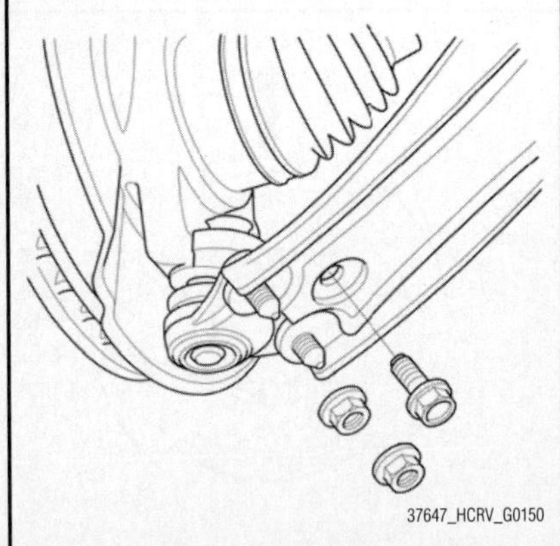
37647_HCRV_G0150

Fig. 35 Remove the nuts and bolt, then separate the lower arm using a prybar

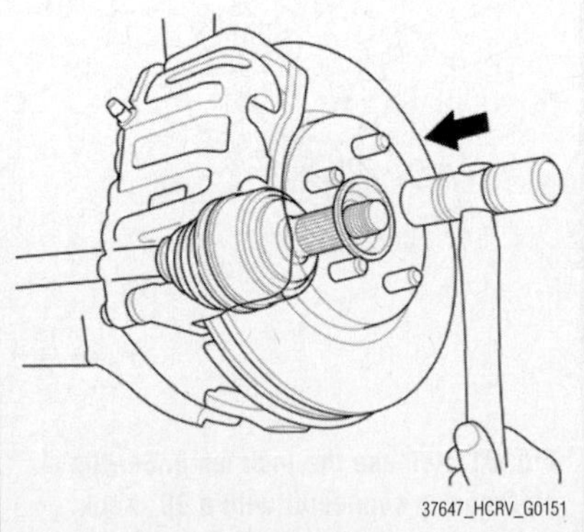
37647_HCRV_G0151

Fig. 36 Pull the knuckle outward, and separate the outboard joint from the front hub using a soft face hammer

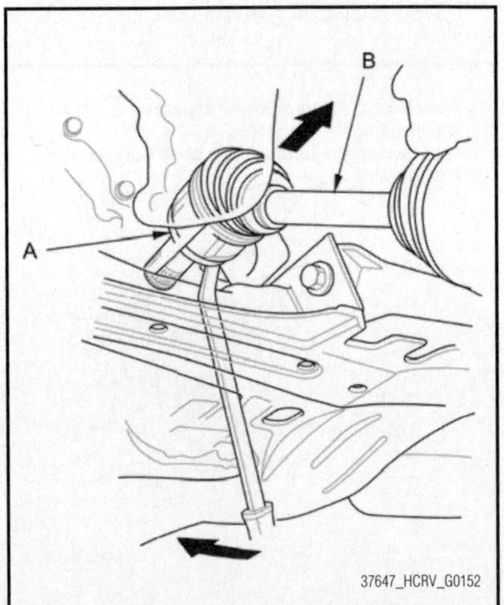
B
A
37647_HCRV_G0152

Fig. 37 Left driveshaft: Pry the inboard joint (A) from the differential with a prybar; remove the driveshaft (B) as an assembly

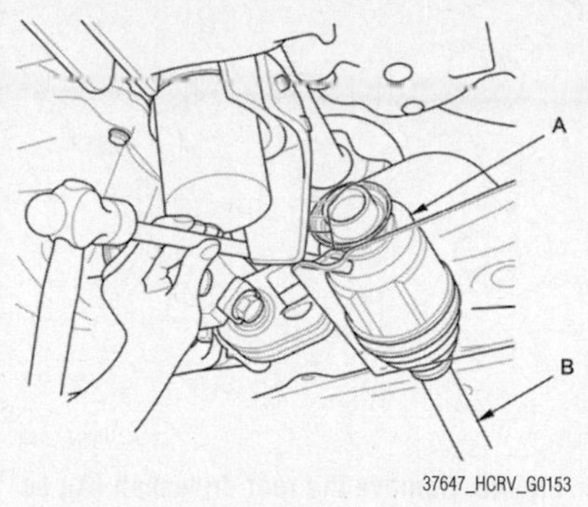
A
B
37647_HCRV_G0153

Fig. 38 Right driveshaft: Drive the inboard joint (A) off of the intermediate shaft using a drift punch and a hammer. Remove the driveshaft (B) as an assembly

To install:

➡Before starting installation, make sure the mating surfaces of the joint and the splined section are clean.

11. Apply grease 0.18 oz (5 g) of Moly 60 paste (P/N 08734-0001) to the contact area of the outboard joint and the front wheel bearing.

➡The paste helps to prevent noise and vibration.

12. Install a new set ring into the set ring groove of the left driveshaft inboard joint.
13. Install a new set ring into the set ring groove of the intermediate shaft.
14. Apply 0.02–0.04 oz (0.5–1.0 g) of super high temp urea grease (P/N 08798-9002) to the whole splined surface of the right driveshaft. After applying grease, remove the grease from the splined grooves at intervals of 2–3 splines and from the set ring groove so that air can bleed from the intermediate shaft.
15. Clean the areas where the driveshaft contacts the differential thoroughly with solvent, and dry with compressed air.

➡Do not wash the rubber parts with solvent.

16. Insert the inboard end of the driveshaft into the differential or the intermediate shaft until the set ring locks in the groove.

➡Insert the driveshaft horizontally to prevent damaging the oil seal.

17. Install the outboard joint into the front hub on the knuckle.
18. Install the knuckle onto the lower arm. Then tighten the nuts and bolt to 38 ft. lbs. (52 Nm).

➡During installation, loosely install a new flange bolt and new self-locking nuts. After lightly tightening all three fasteners, tighten them to the specified torque in the following order; the nut on the front, the nut on the rear, then the bolt.

19. Apply a small amount of engine oil to the seating surface of a new spindle nut.
20. Install the spindle nut, then tighten it to 242 ft. lbs. (328 Nm). After tightening, use a drift to stake the spindle nut shoulder against the driveshaft.
21. Clean the mating surfaces of the brake discs and the front wheels, then install the front wheels.
22. Turn the front wheels by hand, and make sure there is no interference between the driveshaft and the surrounding parts.
23. Lower the vehicle.
24. Refill the transmission with the recommended automatic transmission fluid.
25. Check the wheel alignment, and adjust it if necessary.
26. Test-drive the vehicle.

CV-BOOTS INSPECTION

Check the inboard boot and the outboard boot on the driveshaft for cracks, damage, leaking grease, and loose boot bands. If any damage is found, replace the boot and the boot bands.

INTERMEDIATE SHAFT

REMOVAL & INSTALLATION

See Figure 39.

1. Drain the automatic transmission fluid. Reinstall the drain plug with a new sealing washer.
2. Remove the front right driveshaft.
3. Remove the flange bolt and the two dowel bolts.

➡Remove the set ring from the intermediate shaft if it is installed.

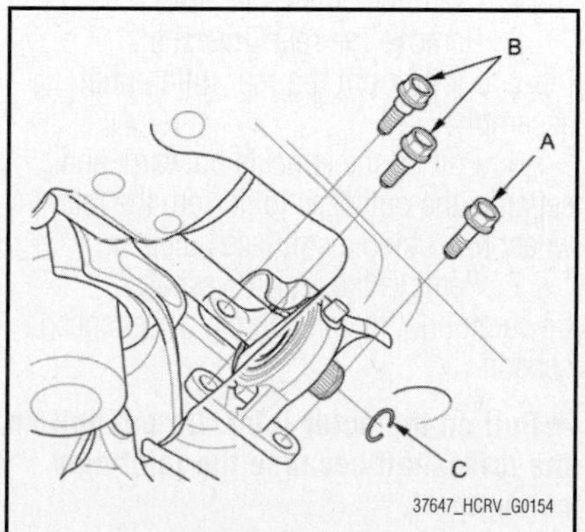
B
A
C
37647_HCRV_G0154

Fig. 39 Remove the flange bolt (A) and the two dowel bolts (B)

4. Remove the intermediate shaft from the differential. Hold the intermediate shaft horizontal until it is clear of the differential to prevent damaging the oil seal.

To install:

5. Clean the areas where the intermediate shaft contact the differential thoroughly with solvent, and dry with compressed air.

➡Do not wash the rubber parts with solvent.

6. Insert the intermediate shaft into differential correctly.

➡Hold the intermediate shaft horizontally to prevent damaging the oil seal.

7. Install the flange bolt and two dowel bolts, then install a new set ring. Tighten the bolts to 29 ft. lbs. (39 Nm).
8. Install the front right driveshaft.
9. Refill the transmission with the recommended automatic transmission fluid.
10. Check the wheel alignment, and adjust it if necessary.
11. Test-drive the vehicle.

REAR DRIVESHAFT (PROPELLER SHAFT)

REMOVAL & INSTALLATION

See Figures 40 through 44.

1. Raise and support the vehicle.
2. Remove the No. 1 propeller shaft protector.
3. Remove the No. 2 propeller shaft protector.
4. Make reference marks across the No. 1 propeller shaft and the transfer companion flange.
5. Remove the flange bolts, then separate the No. 1 propeller shaft from the transfer assembly.
6. Remove the center support bearing mounting bolts.

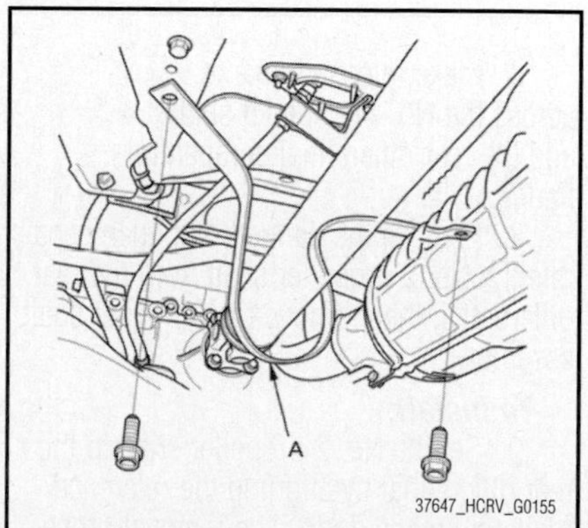
A
37647_HCRV_G0155

Fig. 40 Remove the No. 1 propeller shaft protector (A)

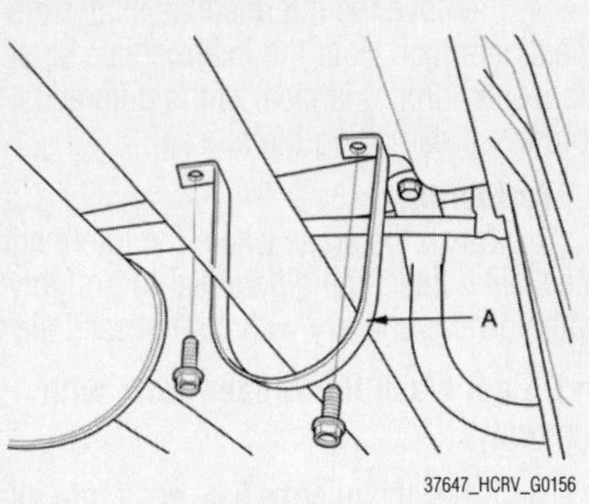
A
37647_HCRV_G0156

Fig. 41 Remove the No. 2 propeller shaft protector (A)

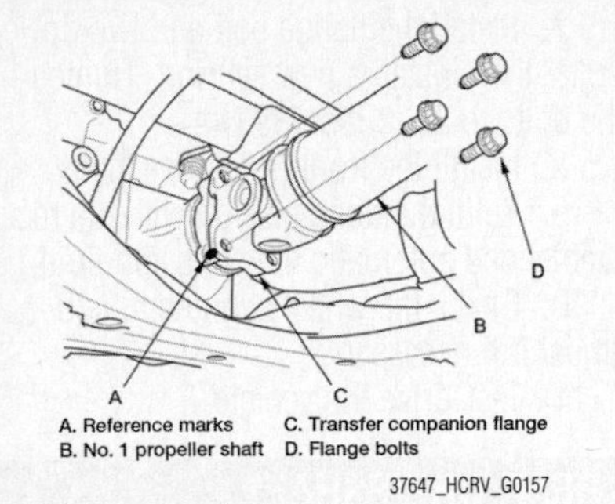
A
B
C
D
A. Reference marks
B. No. 1 propeller shaft
C. Transfer companion flange
D. Flange bolts
37647_HCRV_G0157

Fig. 42 Make reference marks across the No. 1 propeller shaft and the transfer companion flange

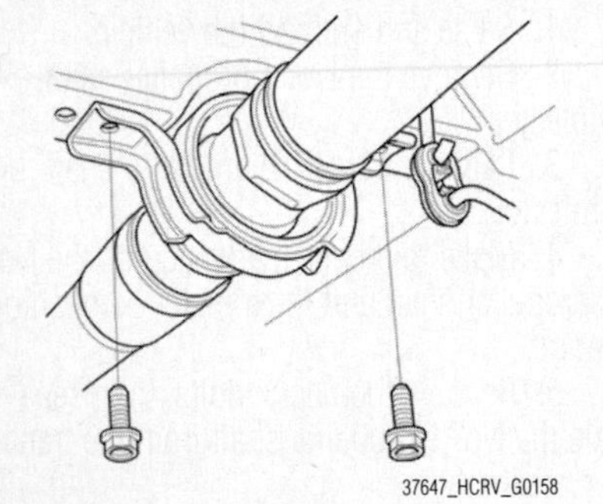
37647_HCRV_G0158

Fig. 43 Remove the center support bearing mounting bolts

7. Make reference marks across the No. 2 propeller shaft and the rear differential companion flange.
8. Remove the flange bolts, then separate the No. 2 propeller shaft from the rear differential, then remove the propeller shaft assembly.

To install:

9. Set the No. 2 propeller shaft to the rear differential by aligning the reference mark you made during the removal procedure. Then install new flange bolts and tighten to 24 ft. lbs. (32 Nm).

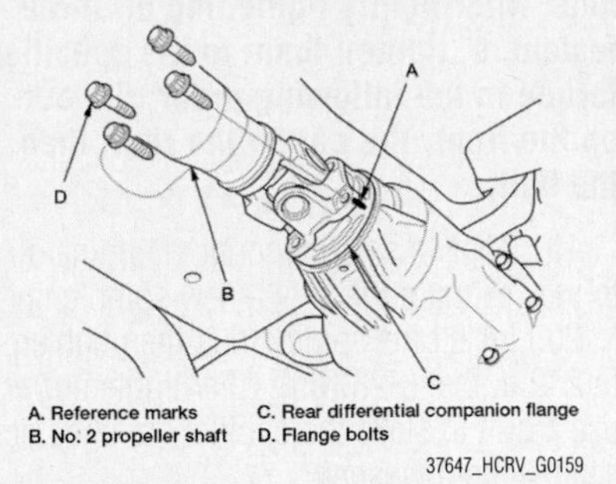
A
B
C
D
A. Reference marks
B. No. 2 propeller shaft
C. Rear differential companion flange
D. Flange bolts
37647_HCRV_G0159

Fig. 44 Make reference marks across the No. 2 propeller shaft and the rear differential companion flange

10. Install the center support bearing with new bolts. Tighten the bolts to 29 ft. lbs. (39 Nm).
11. Set the No. 1 propeller shaft to the transfer companion flange by aligning the reference mark you made during the removal procedure. Then install new flange bolts and tighten to 24 ft. lbs. (32 Nm).
12. Install the No. 2 propeller shaft protector.
13. Install the No. 1 propeller shaft protector.
14. If you installed a new propeller shaft assembly, test-drive the vehicle at 55 mph (88 km/h), and check for noise or vibration. If there is a noise or vibration, rotate the propeller shaft 180 degrees from its current alignment with the rear differential companion flange, then recheck.

REAR HALFSHAFT

REMOVAL & INSTALLATION

See Figure 45.

1. Raise and support the vehicle.
2. Remove the rear wheels.
3. Pry up the stake on the spindle nut, then remove the nut.
4. Drain the differential fluid.
5. Remove the rear driveshaft inboard joint from the rear differential assembly.
6. Pull up the knuckle outward, and separate the outboard joint from the rear wheel hub using a soft face hammer.
7. Remove the rear driveshaft. Be careful not to damage the wheel speed sensor.

➡Pull on the outer joint. Do not pull on the driveshaft because the joint may come apart.

8. Remove the set ring from inboard joint.

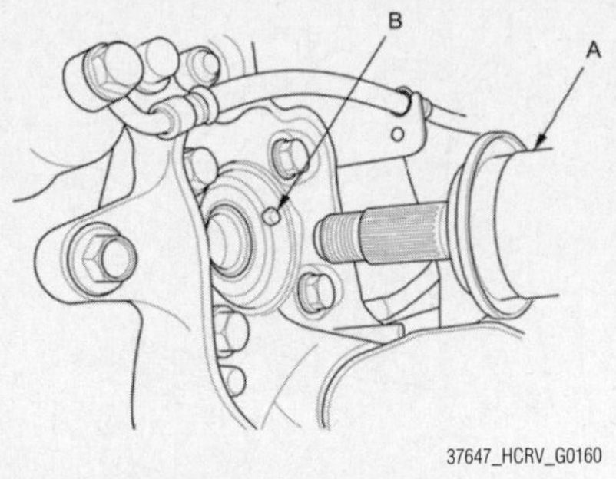
B
A
37647_HCRV_G0160

Fig. 45 Remove the rear driveshaft (A); be careful not to damage the wheel speed sensor (B)

To install:

➡Before starting installation, make sure the mating surfaces of the joint and the splined section are clean.

9. Install a new set ring onto the set ring groove of the rear driveshaft inboard joint.
10. Install the outboard joint into the rear hub.

➡Be careful not to damage the wheel speed sensor.

11. Clean the areas where the driveshaft contacts the differential thoroughly with solvent, and dry with compressed air.

➡Do not wash the rubber parts with solvent.

12. Insert the inboard end of the driveshaft into the differential until the set ring locks in the groove.

➡Insert the driveshaft horizontally to prevent damaging the oil seal.

13. Apply a small amount of engine oil to the seating surface of a new spindle nut.
14. Install the new spindle nut, then tighten it to 181 ft. lbs. (245 Nm). After tightening, use a drift to stake the spindle nut shoulder against the driveshaft.
15. Clean the mating surfaces of the brake discs and the rear wheels, then install the rear wheels.
16. Turn the rear wheel by hand, and make sure there is no interference between the driveshaft and the surrounding parts.
17. Refill the differential fluid.
18. Lower the vehicle.
19. Test-drive the vehicle.

CV-BOOTS INSPECTION

Check the inboard boot and the outboard boot on the driveshaft for cracks, damage, leaking grease, and loose boot bands. If any damage is found, replace the boot and the boot bands.

REAR PINION SEAL

REMOVAL & INSTALLATION

Special Tools Required:
- Oil Seal Driver Attachment 07NAD-P200100
- Oil Seal Driver, 65 mm 07JAD-PL90100
- Attachment, 78 x 80 mm 07NAD-PX40100
- Driver Handle, 15 x 135L 07749-0010000

Rear Driveshaft Side

See Figures 46 and 47.

1. Remove the rear differential.
2. Remove the oil seals from the differential housing.

➡Be careful not to damage the differential carrier while prying out the seals.

3. Install a new right side oil seal squarely using the 15 x 135L driver handle and 78 x 80 mm attachment. Installation depth of the oil seal is 0.35 inches (9 mm) below the machined edge of the rear differential carrier assembly. Be careful not to damage the lip of the oil seal.
4. Install a new left side oil seal even and flush with the machined edge of the rear differential carrier assembly using the 15 x 135L driver handle and the oil seal driver attachment. Be careful not to damage the lip of the oil seal.
5. Install the rear differential.

Propeller Shaft Side

See Figures 48 through 54.

1. Drain the differential fluid.

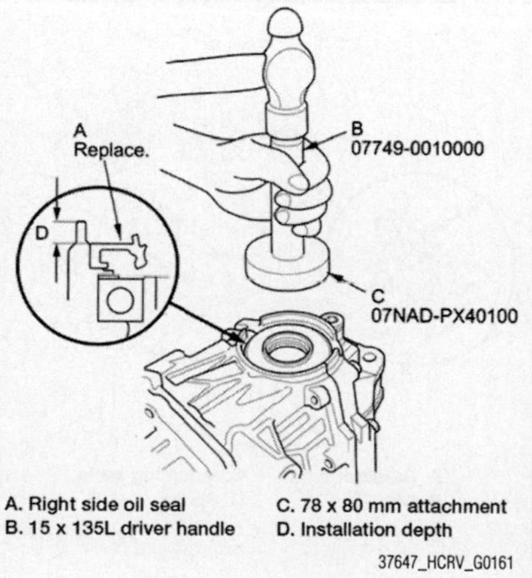
A Replace.
B 07749-0010000
C 07NAD-PX40100
D
A. Right side oil seal
B. 15 x 135L driver handle
C. 78 x 80 mm attachment
D. Installation depth
37647_HCRV_G0161

Fig. 46 Install a new right side oil seal

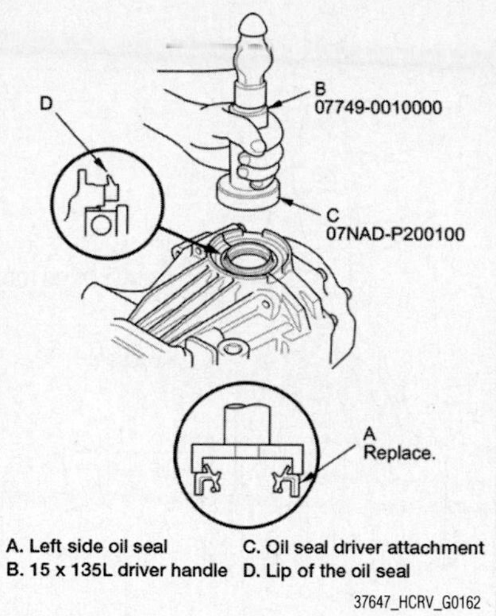
B 07749-0010000
D
C 07NAD-P200100
A Replace.
A. Left side oil seal
B. 15 x 135L driver handle
C. Oil seal driver attachment
D. Lip of the oil seal
37647_HCRV_G0162

Fig. 47 Install a new left side oil seal

2. Separate the propeller shaft from the rear differential assembly.
3. Remove the companion flange.
   a. Pry up the stake on the lock nut from the groove of the clutch guide, making sure that the tab completely clears the groove to prevent damaging the clutch guide.
   b. Install the companion flange holder and holder handle on the companion flange.
   c. Loosen the locknut counterclockwise so that its tab comes out from the groove in the clutch guide.
   d. Tighten the locknut until its tab aligns with the groove.

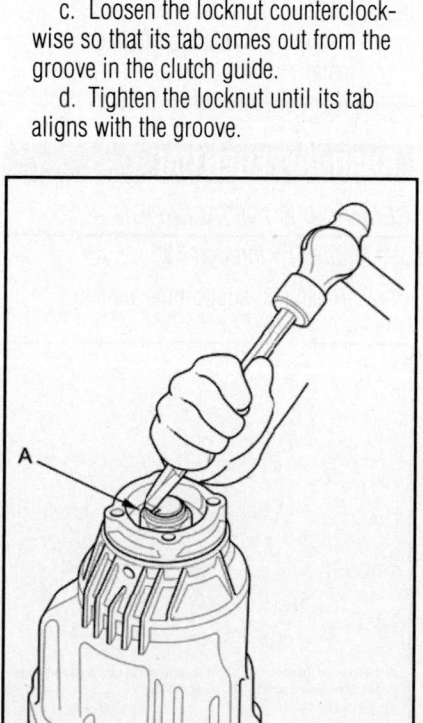

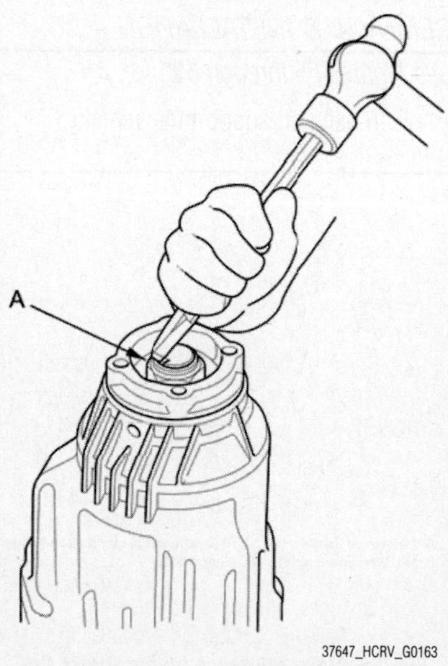
A
37647_HCRV_G0163

Fig. 48 Pry up the stake on the lock nut (A) from the groove of the clutch guide

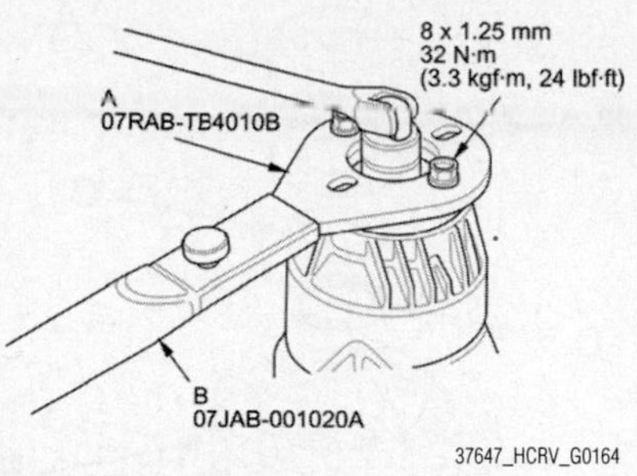
8 x 1.25 mm
32 N·m
(3.3 kgf·m, 24 lbf·ft)
A 07RAB-TB4010B
B 07JAB-001020A
37647_HCRV_G0164

Fig. 49 Install the companion flange holder (A) and holder handle (B) on the companion flange

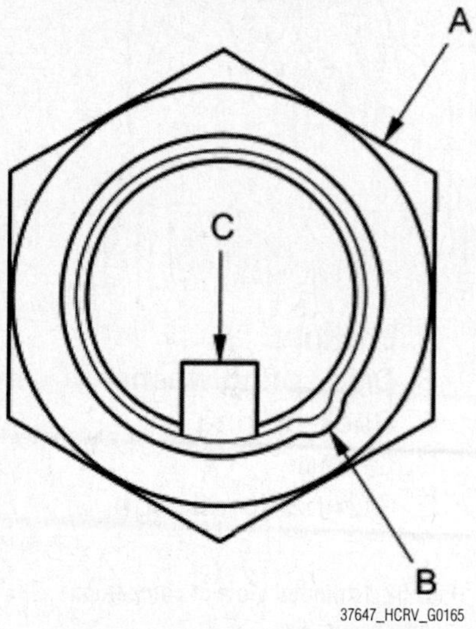
A
C
B
37647_HCRV_G0165

Fig. 50 Loosen the locknut (A) counterclockwise so that its tab (B) comes out from the groove (C) in the clutch guide

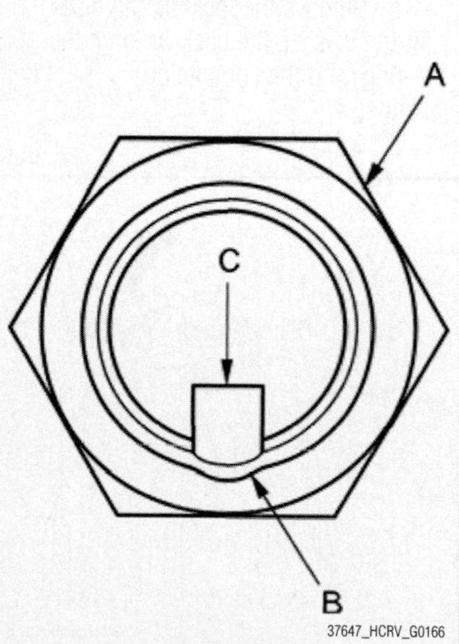
A
C
B
37647_HCRV_G0166

Fig. 51 Tighten the locknut (A) until its tab (B) aligns with the groove (C)

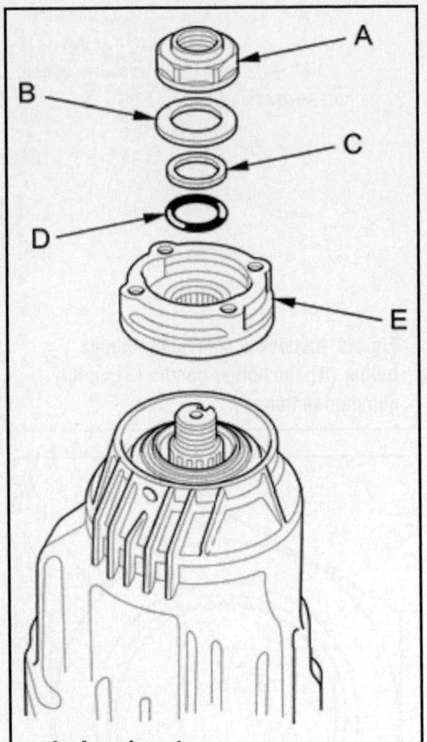
A
B
C
D
E
A. Locknut
B. Disc spring washer
C. Back-up ring
D. O-ring
E. Companion flange
37647_HCRV_G0167

Fig. 52 Exploded view of companion flange assembly

e. Clean any dirt from inside of the groove in the clutch guide, then loosen the locknut.

f. Remove the locknut, the disc spring washer, the back-up ring, the O-ring, and the companion flange.

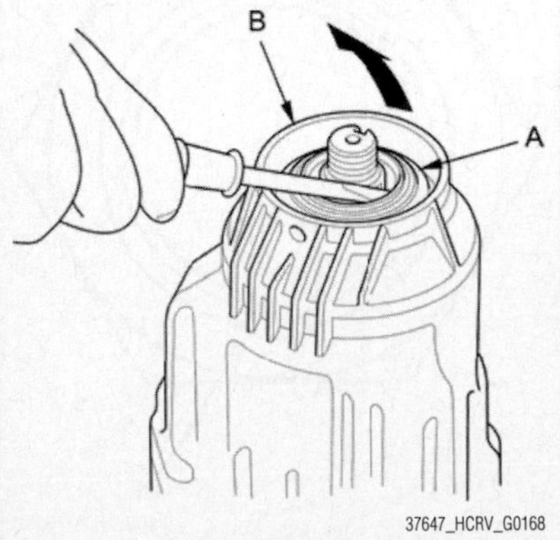
B
A
37647_HCRV_G0168

Fig. 53 Remove the oil seal (A) from the Torque Control Differential (TCD) case (B)

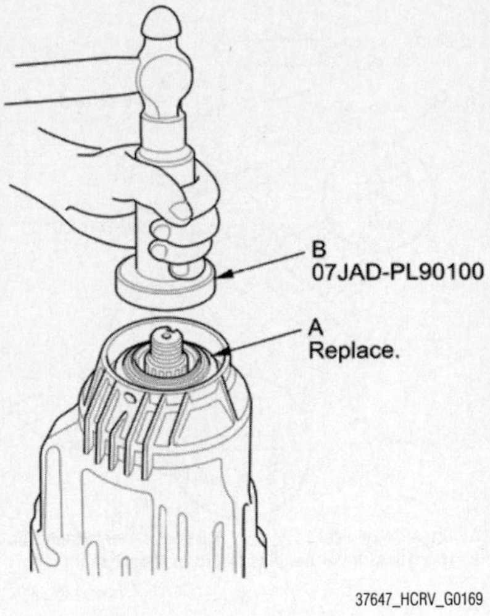
B
07JAD-PL90100
A
Replace.
37647_HCRV_G0169

Fig. 54 Install a new oil seal (A) even and flush with the case using the 65 mm oil seal driver (B)

4. Remove the oil seal from the Torque Control Differential (TCD) case.

➡**Be careful not to damage the shaft or case while prying out the seal.**

To install:

5. Install a new oil seal even and flush with the case using the 65 mm oil seal driver. Be careful not to damage the lip of the oil seal.

6. Install the companion flange.

7. Install the No. 2 propeller shaft to the rear differential assembly .

REAR DIFFERENTIAL

REMOVAL & INSTALLATION

See Figures 55 through 60.

1. Raise and support the vehicle.

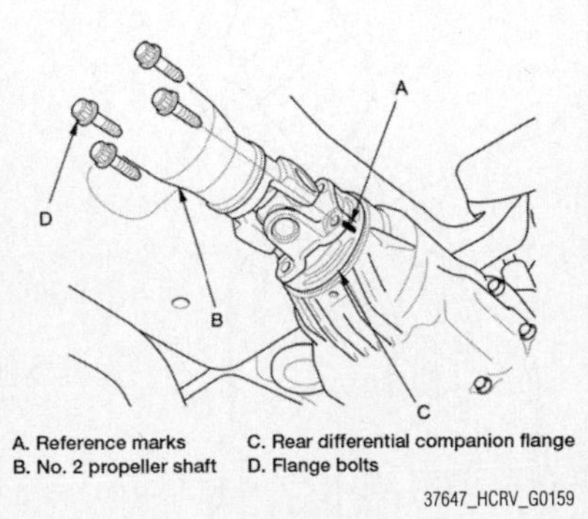
A
D
B
C
A. Reference marks
B. No. 2 propeller shaft
C. Rear differential companion flange
D. Flange bolts
37647_HCRV_G0159

Fig. 55 Make reference marks across the No. 2 propeller shaft and the rear differential companion flange

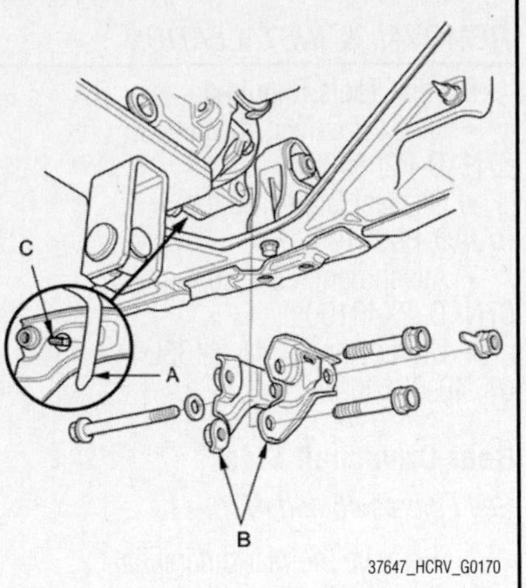
C
A
B
37647_HCRV_G0170

Fig. 56 Remove the breather tube (A) from the clip (C); remove the right and left rear differential mount brackets (B)

2. Drain the differential fluid. Reinstall the drain plug with a new sealing washer.

3. Mark the propeller shaft and the companion flange of the rear differential assembly so they can be reinstalled in their original positions.

4. Separate the propeller shaft from the rear differential assembly.

➡**Suspend the propeller shaft with an appropriate size wire or nylon strap.**

5. Remove the breather tube from the clip.

6. Place a transmission jack under the rear differential assembly, then remove the right and left rear differential mount brackets.

7. Remove the breather tubes from the clip.

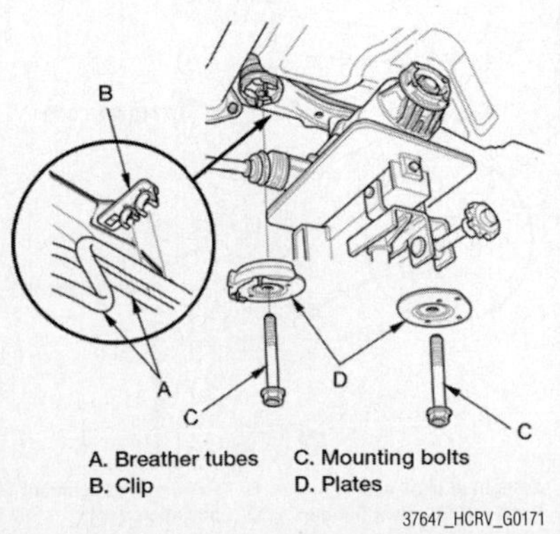
B
A
D
C
C
A. Breather tubes
B. Clip
C. Mounting bolts
D. Plates
37647_HCRV_G0171

Fig. 57 Remove the breather tubes from the clip

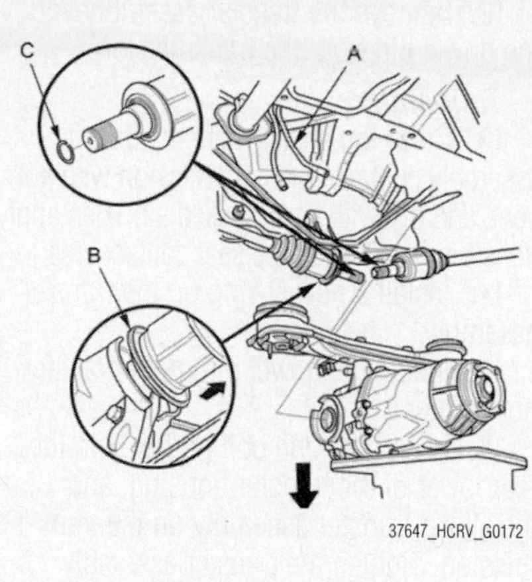
C
A
B
37647_HCRV_G0172

Fig. 58 Lowering the differential assembly showing the breather tube (A), driveshaft ring (B), and set rings (C)

8. Remove the mounting bolts and the plates.
9. Disconnect the breather tube from the breather tube fitting.
10. Lower the rear differential assembly while pulling both driveshaft inboard joints out of the rear differential assembly.

➡Be careful not to damage the driveshaft ring when prying out the driveshaft inboard joints.

11. Remove the set rings.
12. Remove the rear differential mount assembly from the rear differential assembly.

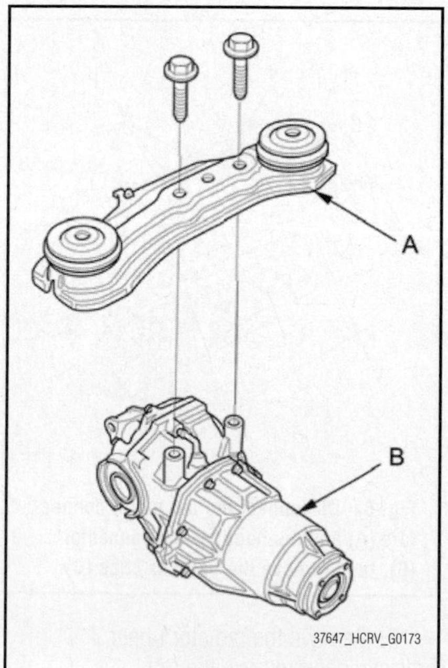
A
B
37647_HCRV_G0173

Fig. 59 Remove the rear differential mount assembly (A) from the rear differential assembly (B)

To install:

13. Install the rear differential mount assembly to the rear differential assembly with new bolts. Tighten the bolts to 51 ft. lbs. (69 Nm).
14. Raise the rear differential with the transmission jack.
15. Install new set rings into the groove of the rear driveshaft inboard joints, then insert the driveshafts into the rear differential.
16. Lift the rear differential up into position, then push on both driveshafts to lock the set rings into place. Connect the breather tube.
17. Align the tab of the rubber mount with the hole of the plates, then install the plates and torque the rear differential mount assembly mounting bolts.

➡The rubber heat insulator is installed only in the right side.

18. Install the breather tubes to the clips.
19. Install the right and left rear differential mount brackets, then torque the new bolts. Tighten the two 12 x 1.25 mm bolts to 44 ft. lbs. (59 Nm) and the 10 x 1.25 mm bolt to 47 ft. lbs. (64 Nm).
20. Install the breather tube to the clip.
21. Install the No. 2 propeller shaft onto the rear differential by aligning the reference marks made during removal. Make sure you use new mounting bolts. Tighten the bolts to 24 ft. lbs. (32 Nm).
22. Refill the rear differential with the recommended fluid.
23. Test-drive the vehicle.

TRANSFER CASE ASSEMBLY

REMOVAL & INSTALLATION

See Figures 61 and 62.

1. Shift the transmission into N.
2. Raise the vehicle on a lift, and make sure it is supported securely.
3. Remove the drain plug, and drain the Automatic Transmission Fluid (ATF).
4. Reinstall the drain plug with a new sealing washer.
5. Disconnect the A/F sensor connector (K24Z1 engine) and the secondary HO2S connector.
6. K24Z1 engine: Remove the A/F sensor harness and the secondary HO2S harness from the harness clamps.
7. K24Z6 engine: Remove the secondary HO2S harness from the harness clamp.
8. Remove the three-way catalytic converter (K24Z1 engine) or the under-floor three-way catalytic converter (K24Z6 engine).

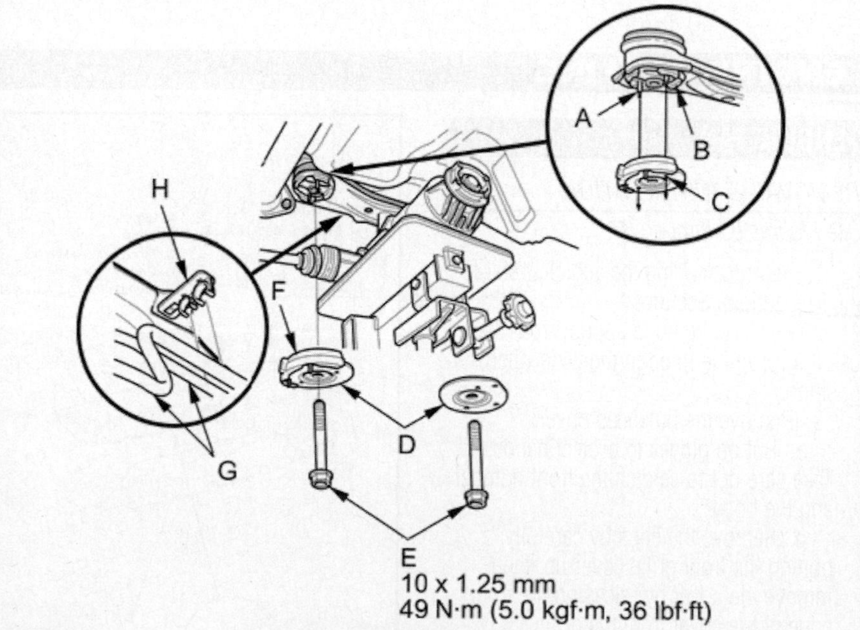
A
B
C
H
F
G
D
E
10 x 1.25 mm
49 N·m (5.0 kgf·m, 36 lbf·ft)
A. Tab
B. Rubber mount
C. Hole
D. Plates
E. Rear differential mount assembly mounting bolts
F. Rubber heat insulator
G. Breather tubes
H. Clips
37647_HCRV_G0174

Fig. 60 Aligning the rear differential mount assembly

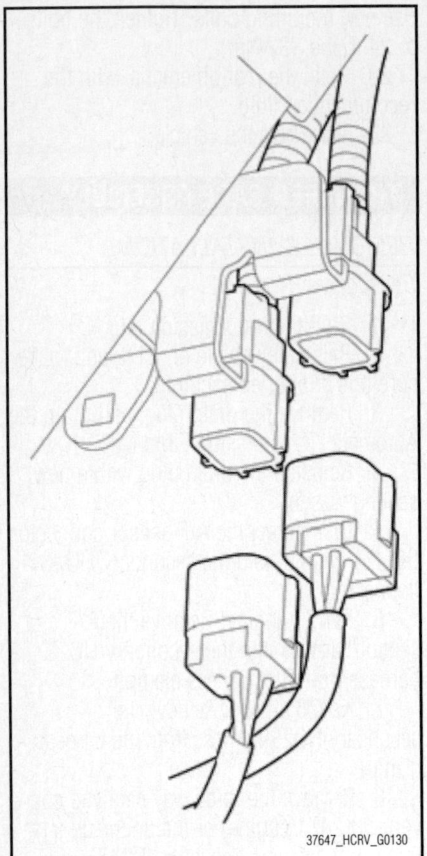
37647_HCRV_G0130

Fig. 61 Disconnect the A/F sensor connector (K24Z1 engine) and the secondary HO2S connector

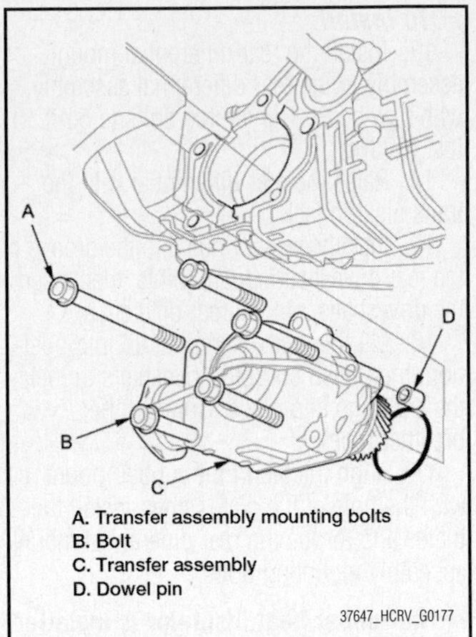
A
B
C
D
A. Transfer assembly mounting bolts
B. Bolt
C. Transfer assembly
D. Dowel pin
37647_HCRV_G0177

Fig. 62 Remove the transfer assembly mounting bolts (4), and pull out the bolt to the limit of travel

9. Make a reference mark across the propeller shaft and the transfer companion flange.

10. Separate the propeller shaft from the transfer companion flange.

11. Remove the transfer assembly mounting bolts (4), and pull out the bolt to the limit of travel.

12. Remove the transfer assembly and the dowel pin from the transmission.

To install:

13. Clean the areas where the transfer assembly contacts the transmission with solvent, and dry with compressed air. Then apply transmission fluid to the seal contact area.

14. Install a new O-ring on the transfer assembly.

15. Install the dowel pin in the transfer housing.

16. Install the one bolt part-way in the rear lower of the transfer housing, and install the transfer assembly on the transmission. Tighten the transfer assembly mounting bolts to 33 ft. lbs. (44 Nm).

17. Install the propeller shaft to the transfer companion flange by aligning the reference mark.

18. Install the three-way catalytic converter (K24Z1 engine) or the under-floor three-way catalytic converter (K24Z6 engine) with the bolts, new self-locking nuts, and new gaskets (A).

19. Connect the A/F sensor connector (K24Z1 engine) and the secondary HO2S connector.

20. K24Z1 engine: Install the A/F sensor harness and the secondary HO2S harness in the harness clamps.

21. K24Z6 engine: Install the secondary HO2S harness in the harness clamp.

22. Refill the transmission with ATF.

ENGINE COOLING

ENGINE FAN

REMOVAL & INSTALLATION

See Figures 63 through 66.

1. Before servicing the vehicle, refer to the Precautions Section.

2. Remove the hood support rod, then use it to prop the hood in the wide-open position.

3. Remove the bulkhead cover:

 a. Put on gloves to protect hands. Take care not to scratch the front bumper and the body.

 b. Remove the clips by carefully pulling the front grille cover up, then remove the cover by releasing the front edge of the cover from the grille.

➡To remove the clips, pry the inner clip up at the edge near the line on its head.

4. Disconnect the fan motor connectors and the hood switch connector, then remove the harness clips.

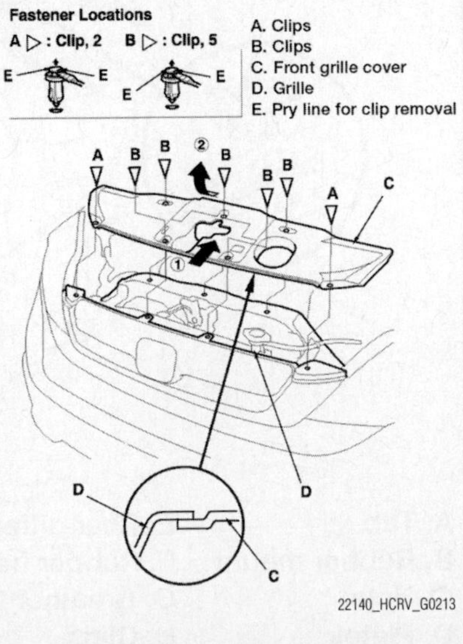
Fastener Locations
A ▷ : Clip, 2 B ▷ : Clip, 5
A. Clips
B. Clips
C. Front grille cover
D. Grille
E. Pry line for clip removal
E
E
E
E
A
B
B
B
B
B
A
C
D
D
C
22140_HCRV_G0213

Fig. 63 Remove the bulkhead cover

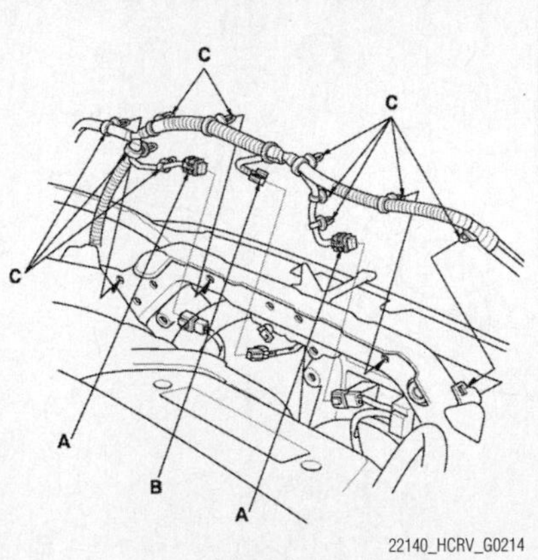
C
C
C
A
B
A
22140_HCRV_G0214

Fig. 64 Disconnect the fan motor connectors (A) and the hood switch connector (B), then remove the harness clips (C)

5. Remove the radiator upper brackets, then remove the front bulkhead.

6. Remove the condenser fan shroud assembly, then remove the radiator fan

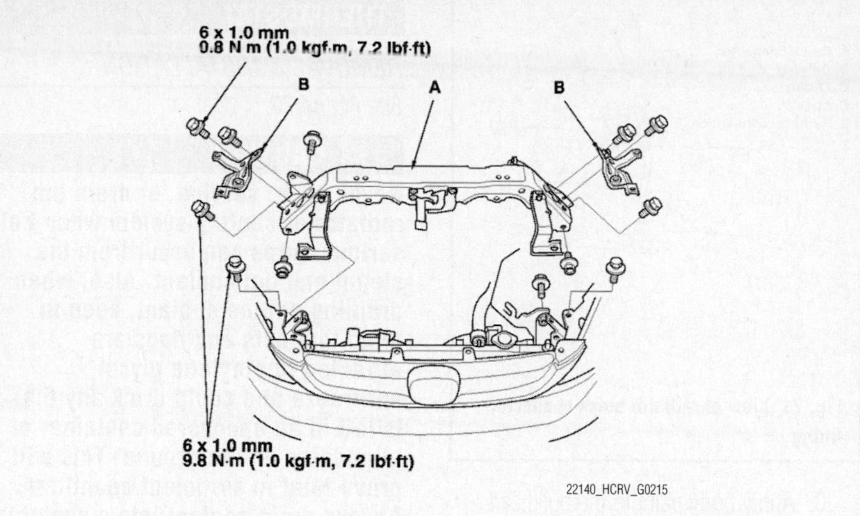
6 x 1.0 mm
0.8 N m (1.0 kgf·m, 7.2 lbf·ft)
B
A
B
6 x 1.0 mm
9.8 N·m (1.0 kgf·m, 7.2 lbf·ft)
22140_HCRV_G0215

Fig. 65 Remove the radiator upper brackets (A), then remove the front bulkhead (B)

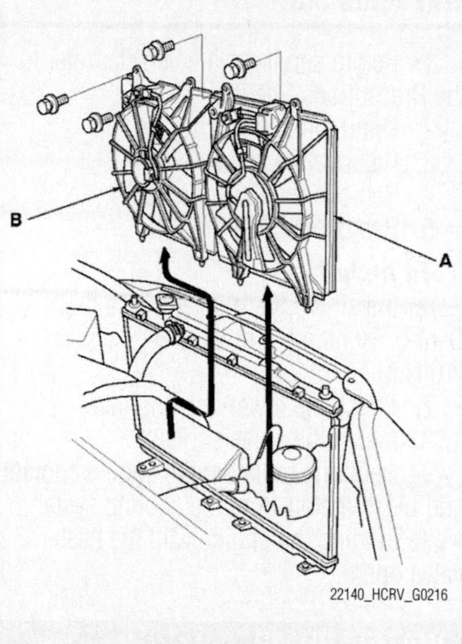
B
A
22140_HCRV_G0216

Fig. 66 Remove the condenser fan shroud assembly (A), then remove the radiator fan shroud assembly (B) from the condenser fan shroud side

shroud assembly from the condenser fan shroud side.

7. Disassemble the fan shrouds, as necessary.

To install:

8. Assemble the fan shrouds.

9. Install the radiator fan shroud assembly, then install the condenser fan shroud assembly.

10. Install the front bulkhead, then install the radiator upper brackets.

11. Apply body paint to the bulkhead mounting bolts.

12. Connect the fan motor connectors and the hood switch connector, then install the harness clips.

13. Install the bulkhead cover in the reverse order of removal, and note these items:

- If the clips are damaged or stress-whitened, replace them with new ones
- Push the clips and the hooks into place securely

RADIATOR

REMOVAL & INSTALLATION

See Figures 63, 67 through 71.

✱✱ CAUTION

Never open a radiator cap or cooling system when the coolant temperature is above 100°F (37°C). Always drain coolant into a sealable container. If spillage occurs, take care to clean the spill as quickly as possible.

1. Before servicing the vehicle, refer to the Precautions Section.

2. Disconnect the negative battery cable.

3. Drain the engine coolant.

4. Remove the hood support rod, then use it to prop the hood in the wide-open position.

5. Remove the bulkhead cover:

a. Put on gloves to protect hands. Take care not to scratch the front bumper and the body.

b. Remove the clips by carefully pulling the front grille cover up, then remove the cover by releasing the front edge of the cover from the grille.

➡To remove the clips, pry the inner clip up at the edge near the line on its head.

6. Disconnect the coolant reservoir hose and the upper radiator hose.

7. Raise and safely support the vehicle.

8. Remove the splash shield.

9. Disconnect the Engine Coolant Temperature (ECT) sensor 2 connector, and remove the harness clip.

10. Clean any dirt off the quick connector, the radiator, and the lower radiator hose.

11. Pull out the lock by hand, then wiggle the quick connector loose, and remove it from the radiator. Do not use any tools to remove the quick connector.

12. Disconnect the Automatic Transmission Fluid (ATF) cooler hoses, then plug the hoses and the lines.

13. Lower the vehicle.

14. Disconnect the fan motor connectors and the hood switch connector, then remove the harness clips.

15. Remove the radiator upper brackets, then remove the front bulkhead.

16. Pull up the radiator, then remove the radiator fan shroud assembly, the A/C condenser fan shroud assembly, the radiator cap, the ETC sensor 2, and the drain plug.

To install:

17. Reassemble the radiator with new O-rings.

18. Install the radiator. Make sure the lower cushions are set securely.

19. Install the front bulkhead, then install the radiator upper brackets.

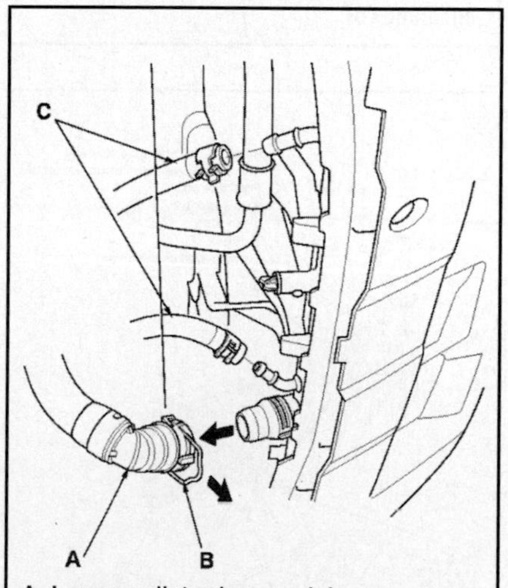
C
A
B
A. Lower radiator hose quick connector
B. Quick connector lock
C. Automatic Transmission Fluid (ATF) cooler hoses
22140_HCRV_G0217

Fig. 67 View of the lower radiator hose quick connector

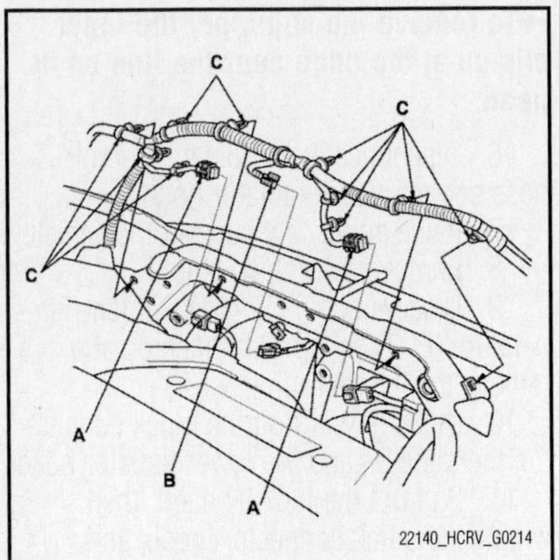
C
C
C
A
B
A
22140_HCRV_G0214

Fig. 68 Disconnect the fan motor connectors (A) and the hood switch connector (B), then remove the harness clips (C)

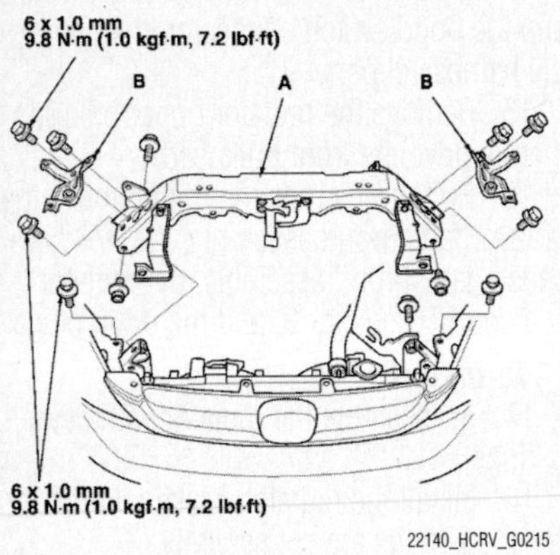
6 x 1.0 mm
9.8 N·m (1.0 kgf·m, 7.2 lbf·ft)
B
A
B
6 x 1.0 mm
9.8 N·m (1.0 kgf·m, 7.2 lbf·ft)
22140_HCRV_G0215

Fig. 69 Remove the radiator upper brackets (A), then remove the front bulkhead (B)

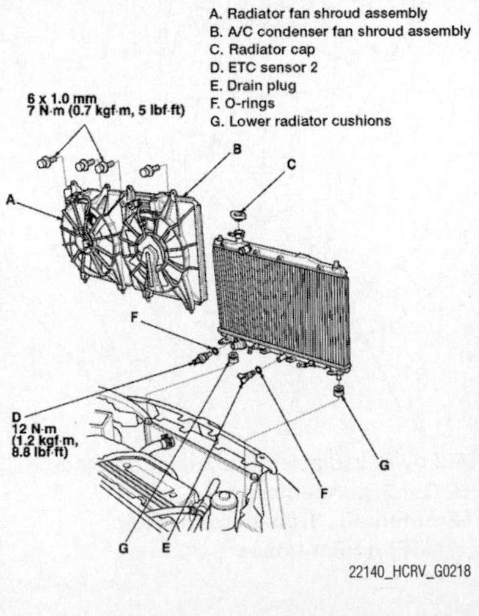
A. Radiator fan shroud assembly
B. A/C condenser fan shroud assembly
C. Radiator cap
D. ETC sensor 2
E. Drain plug
F. O-rings
G. Lower radiator cushions
6 x 1.0 mm
7 N·m (0.7 kgf·m, 5 lbf·ft)
B
C
A
F
D
12 N·m
(1.2 kgf·m,
8.8 lbf·ft)
G
F
G
E
22140_HCRV_G0218

Fig. 70 Expanded view of radiator removal

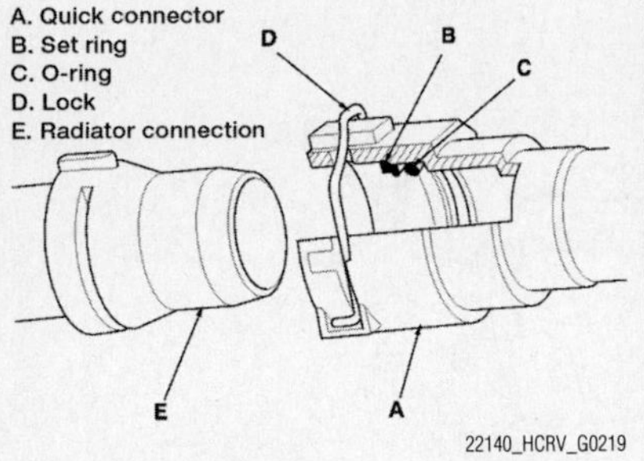
A. Quick connector
B. Set ring
C. O-ring
D. Lock
E. Radiator connection
D
B
C
E
A
22140_HCRV_G0219

Fig. 71 View of radiator quick connector fitting

20. Apply body paint to the bulkhead mounting bolts.
21. Connect the fan motor connectors and the hood switch connector, then install the harness clips.
22. Raise and safely support the vehicle.
23. Check the quick connector and the set ring for cracks or damage. If the connector and/or set ring are cracked or damaged, replace the connector.
24. Make sure the set ring is in place inside the quick connector. If the set ring is displaced or not properly seated in the connector, replace the quick connector.
25. Replace the O-ring in the quick connector.
26. Check the lock. If the lock is damaged or deformed, replace it. When installing the new lock to the connector, slide it straight down along the groove.
27. Install a new lower radiator hose on the quick connector and install the clamp.
28. Clean the connecting surface of the radiator, then apply clean engine coolant around the connecting surface.
29. Push down the lock, then push the quick connector onto the radiator until you hear it click.
30. Install the ATF cooler hoses.
31. Connect the ECT sensor 2 connector and install the harness clip.
32. Install the splash shield.
33. Lower the vehicle.
34. Install the coolant reservoir hose and the upper radiator hose.
35. Install the bulkhead cover.
36. Refill the radiator with engine coolant and bleed the air from the cooling system with the heater valve open.

THERMOSTAT

REMOVAL & INSTALLATION

See Figure 72.

CAUTION

Never open, service, or drain the radiator or cooling system when hot; serious burns can occur from the steam and hot coolant. Also, when draining engine coolant, keep in mind that cats and dogs are attracted to ethylene glycol antifreeze and could drink any that is left in an uncovered container or in puddles on the ground. This will prove fatal in sufficient quantities. Always drain coolant into a sealable container. Coolant should be reused unless it is contaminated or is several years old.

1. Before servicing the vehicle, refer to the Precautions Section.
2. Drain the engine coolant.
3. Remove the splash shield.
4. Remove the lower coolant hose.
5. Remove the thermostat.

To install:

6. Install the thermostat with a new O-ring. Tighten the bolts to 84 inch lbs. (10 Nm).
7. Install the lower coolant hose.
8. Install the splash shield.
9. Refill the radiator with engine coolant and bleed the air from the cooling system while running the engine with the heater valve open.

WATER PUMP

REMOVAL & INSTALLATION

See Figure 73.

CAUTION

Never open, service, or drain the radiator or cooling system when hot; serious burns can occur from the steam and hot coolant. Also, when draining engine coolant, keep in mind that cats and dogs are attracted to ethylene glycol antifreeze and could drink any that is left in an uncovered container or in puddles on the ground. This will prove fatal in sufficient quantities. Always drain coolant into a sealable container. Coolant should be reused unless it is contaminated or is several years old.

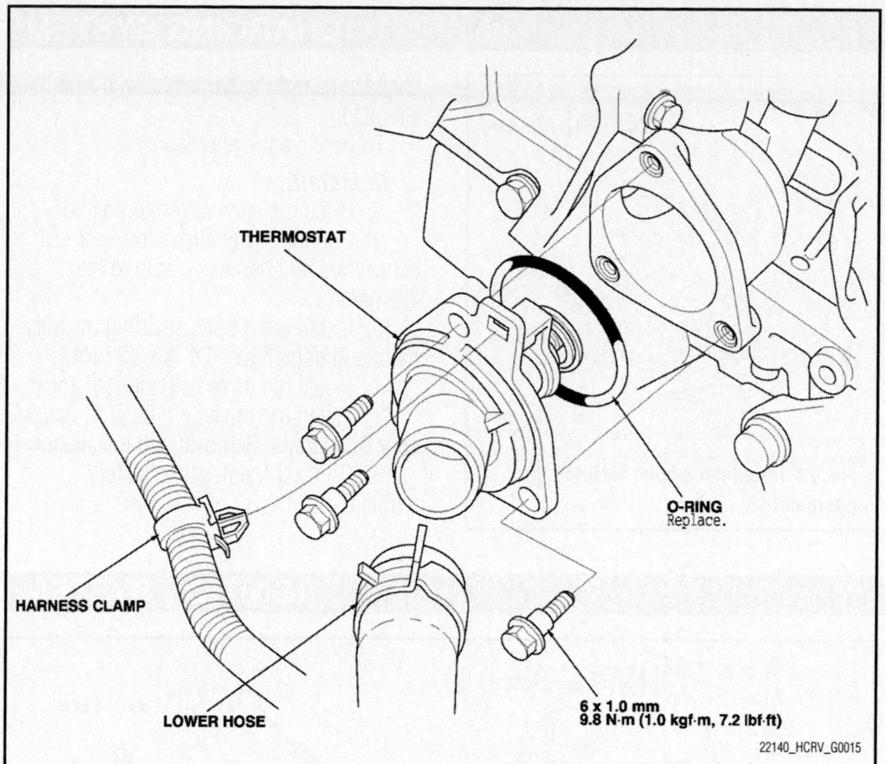
THERMOSTAT
HARNESS CLAMP
LOWER HOSE
O-RING
Replace.
6 x 1.0 mm
9.8 N·m (1.0 kgf·m, 7.2 lbf·ft)
22140_HCRV_G0015

Fig. 72 Exploded view of the thermostat and related components

1. Before servicing the vehicle, refer to the Precautions Section.
2. Remove the drive belt.
3. Drain the engine coolant.
4. Remove the drive belt auto-tensioner pulley.
5. K24Z6 engine: Remove the water pump pulley.
6. Remove the 6 bolts securing the water pump, then remove the water pump.

To install:

7. Inspect and clean the O-ring groove and mating surface of the water passage.
8. Install the water pump with new O-ring and tighten the 6 bolts to 106 inch lbs. (12 Nm).
9. Clean up any spilled engine coolant.
10. Install the drive belt auto-tensioner pulley.
11. Refill the radiator with engine coolant and bleed the air from the cooling system while running the engine with the heater valve open.

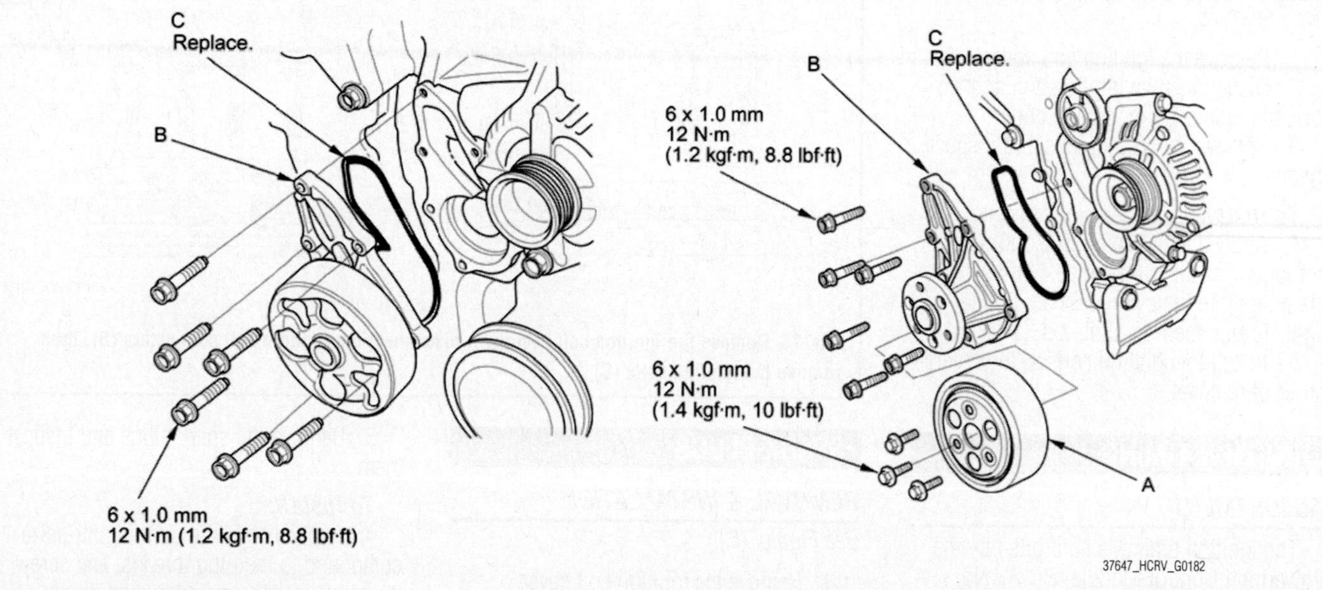
C
Replace.
B
6 x 1.0 mm
12 N·m (1.2 kgf·m, 8.8 lbf·ft)
B
C
Replace.
6 x 1.0 mm
12 N·m
(1.2 kgf·m, 8.8 lbf·ft)
6 x 1.0 mm
12 N·m
(1.4 kgf·m, 10 lbf·ft)
A
37647_HCRV_G0182

Fig. 73 Removing the water pump

ENGINE ELECTRICAL — CHARGING SYSTEM

ALTERNATOR

REMOVAL & INSTALLATION

See Figure 74.

1. Before servicing the vehicle, refer to the Precautions Section.
2. Disconnect the negative battery terminal.
3. Remove the accessory drive belt. Refer to Accessory Drive Belts, Removal and Installation.
4. Remove the drive belt auto-tensioner.
5. Remove the 3 bolts securing the alternator.
6. Disconnect the alternator electrical connectors and the harness clamp from the alternator.
7. Remove the alternator.

To install:

8. Place the alternator into position.
9. Connector the alternator electrical connectors and harness clamp to the alternator.
10. Install the 3 bolts securing the alternator and tighten to 16 ft. lbs. (22 Nm).
11. Install the drive belt auto-tensioner.
12. Install the drive belt. Refer to Accessory Drive Belts, Removal and Installation.
13. Connect the negative battery terminal.

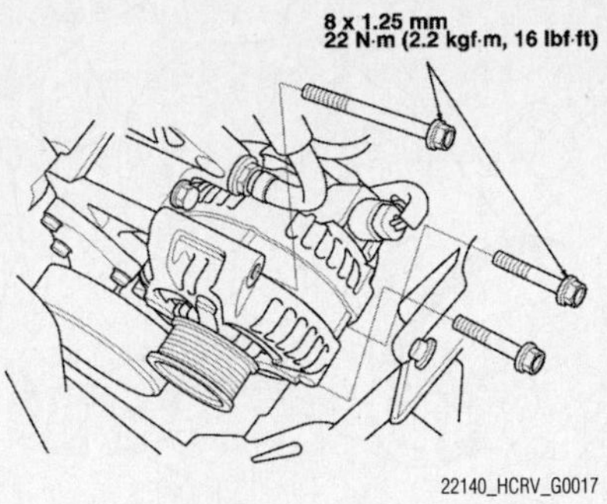
8 x 1.25 mm
22 N·m (2.2 kgf·m, 16 lbf·ft)
22140_HCRV_G0017

Fig. 74 Install the 3 bolts securing the alternator

ENGINE ELECTRICAL — IGNITION SYSTEM

FIRING ORDER

2.4L Engine, Firing order: 1–3–4–2

IGNITION COIL

REMOVAL & INSTALLATION

See Figure 75.

1. Remove the ignition coil cover.
2. Disconnect the ignition coil connectors, then remove the ignition coils.
3. Remove the spark plugs and inspect them.

To install:

4. Apply a small amount of anti-seize compound to the plug threads, and screw the plugs into the cylinder head, finger-tight. Torque them to 13 ft. lbs. (18 Nm).
5. Install the ignition coils in the reverse order of removal.

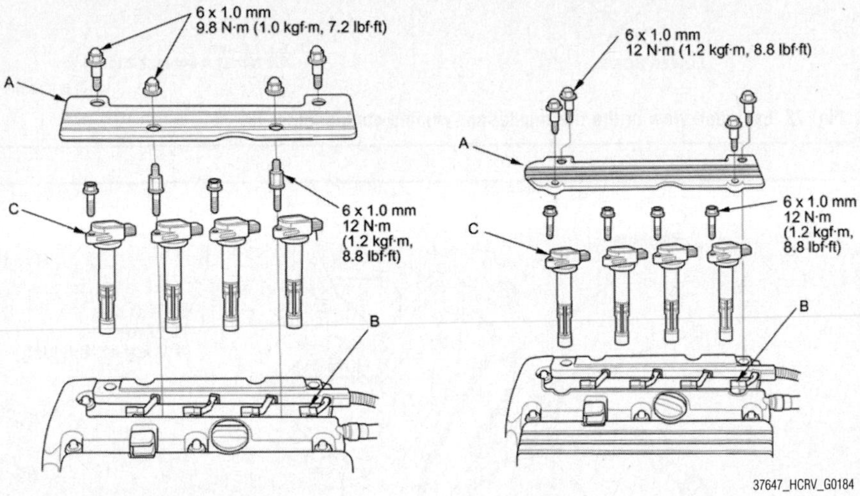
6 x 1.0 mm
9.8 N·m (1.0 kgf·m, 7.2 lbf·ft)
A
C
6 x 1.0 mm
12 N·m
(1.2 kgf·m,
8.8 lbf·ft)
B
6 x 1.0 mm
12 N·m (1.2 kgf·m, 8.8 lbf·ft)
A
C
6 x 1.0 mm
12 N·m
(1.2 kgf·m,
8.8 lbf·ft)
B
37647_HCRV_G0184

Fig. 75 Remove the ignition coil cover (A), disconnect the ignition coil connectors (B), then remove the ignition coils (C)

IGNITION TIMING

ADJUSTMENT

The ignition timing is controlled by the Powertrain Control Module (PCM). No adjustment is necessary or possible.

SPARK PLUGS

REMOVAL & INSTALLATION

See Figure 75.

1. Remove the ignition coil cover.
2. Disconnect the ignition coil connectors, then remove the ignition coils.
3. Remove the spark plugs and inspect them.

To install:

4. Apply a small amount of anti-seize compound to the plug threads, and screw the plugs into the cylinder head, finger-tight. Torque them to 13 ft. lbs. (18 Nm).

STARTER

REMOVAL & INSTALLATION

See Figures 76 through 80.

1. Do the battery terminal disconnection procedure.
2. Remove the splash shield.
3. Remove the intake manifold bracket.
4. Disconnect the knock sensor connector.
5. Remove the harness clamp and remove the two bolts securing the starter, then remove the starter from the engine.
6. Disconnect the starter cable from the B terminal, then disconnect the BLK/WHT wire from the S terminal.
7. Remove the harness clamp, then remove the starter.

To install:

8. Install the starter cable and BLK/WHT wire. Make sure the starter cable crimped side of the ring terminal faces away from the starter when you connect it.
9. Install the harness clamp.
10. Install the starter, tighten the bolts, then install the harness clamp.
11. Reconnect the knock sensor connector.
12. Install the intake manifold bracket.
13. Install the splash shield.
14. Do the battery terminal reconnection procedure.
15. Start the engine to make sure the starter works properly.

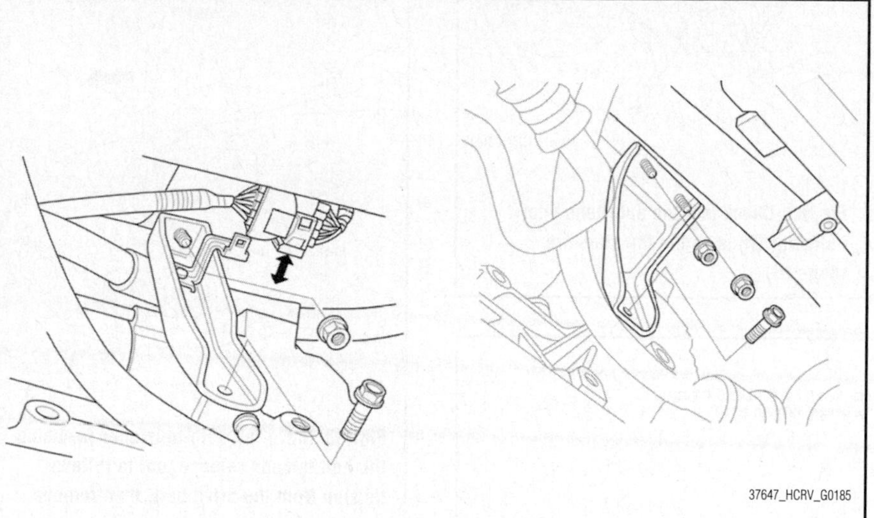
37647_HCRV_G0185

Fig. 76 Remove the intake manifold bracket

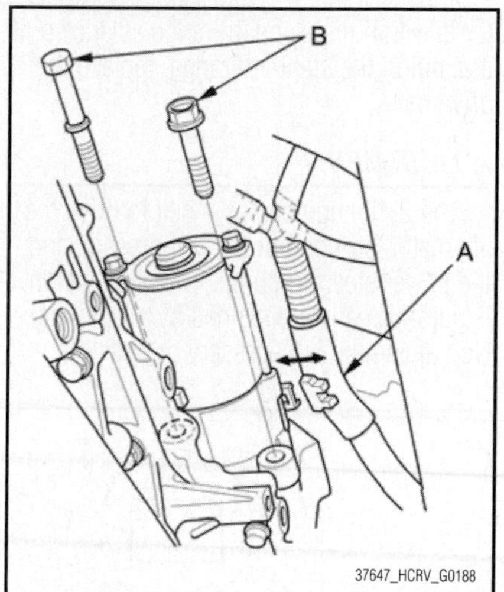
B
A
37647_HCRV_G0188

Fig. 79 Remove the harness clamp (A), and remove the two bolts (B) securing the starter

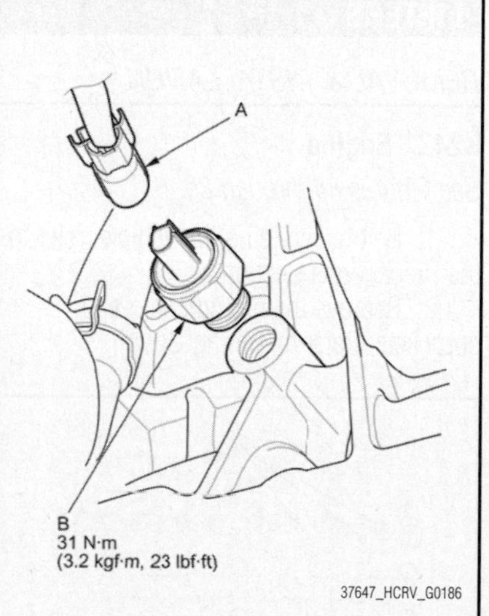
A
B
31 N·m
(3.2 kgf·m, 23 lbf·ft)
37647_HCRV_G0186

Fig. 77 K24Z1 engine knock sensor

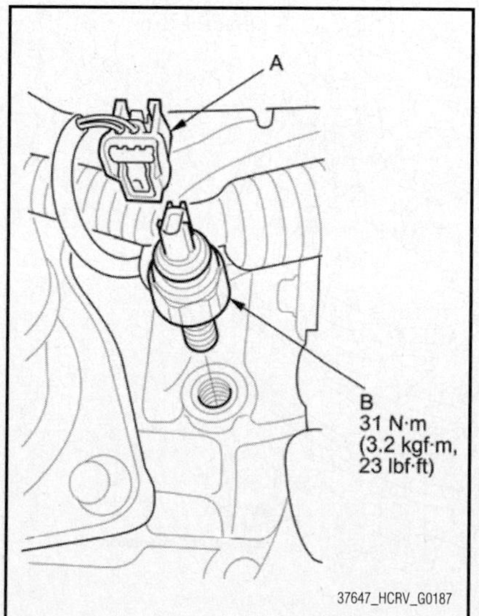
A
B
31 N·m
(3.2 kgf·m,
23 lbf·ft)
37647_HCRV_G0187

Fig. 78 K24Z6 engine knock sensor

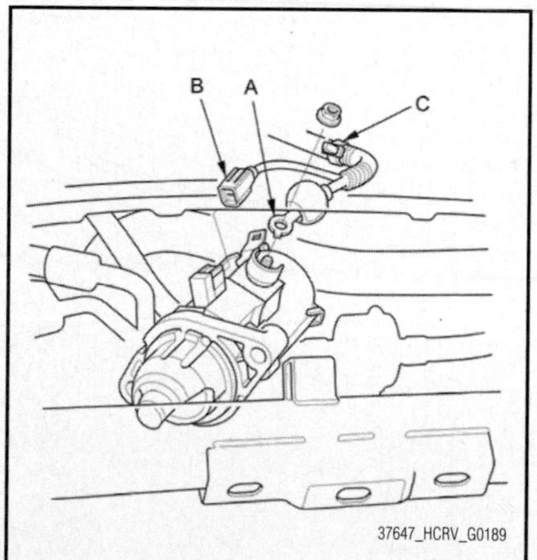
B
A
C
37647_HCRV_G0189

Fig. 80 Disconnect the starter cable (A) from the B terminal, then disconnect the BLK/WHT wire (B) from the S terminal; remove the harness clamp (C)

ENGINE MECHANICAL

ACCESSORY DRIVE BELTS

ACCESSORY BELT ROUTING

See Figure 81.

Refer to the accompanying illustration for belt routing.

INSPECTION

See Figure 85.

1. Inspect the belt for cracks or damage. If the belt is cracked or damaged, replace it.
2. Check that the auto-tensioner indicator is within the standard range as shown. If it is out of the standard range, replace the drive belt.

ADJUSTMENT

The 2.4L engine uses 1 belt to drive the alternator, air conditioner compressor, and the power steering pump. The belt tension is automatically maintained by a tensioner. No adjustment is necessary or possible.

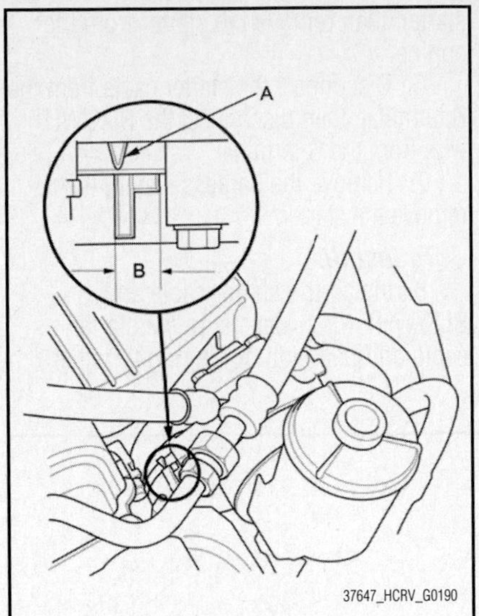
A
B
37647_HCRV_G0190

Fig. 82 Check that the auto-tensioner indicator (A) is within the standard range (B)

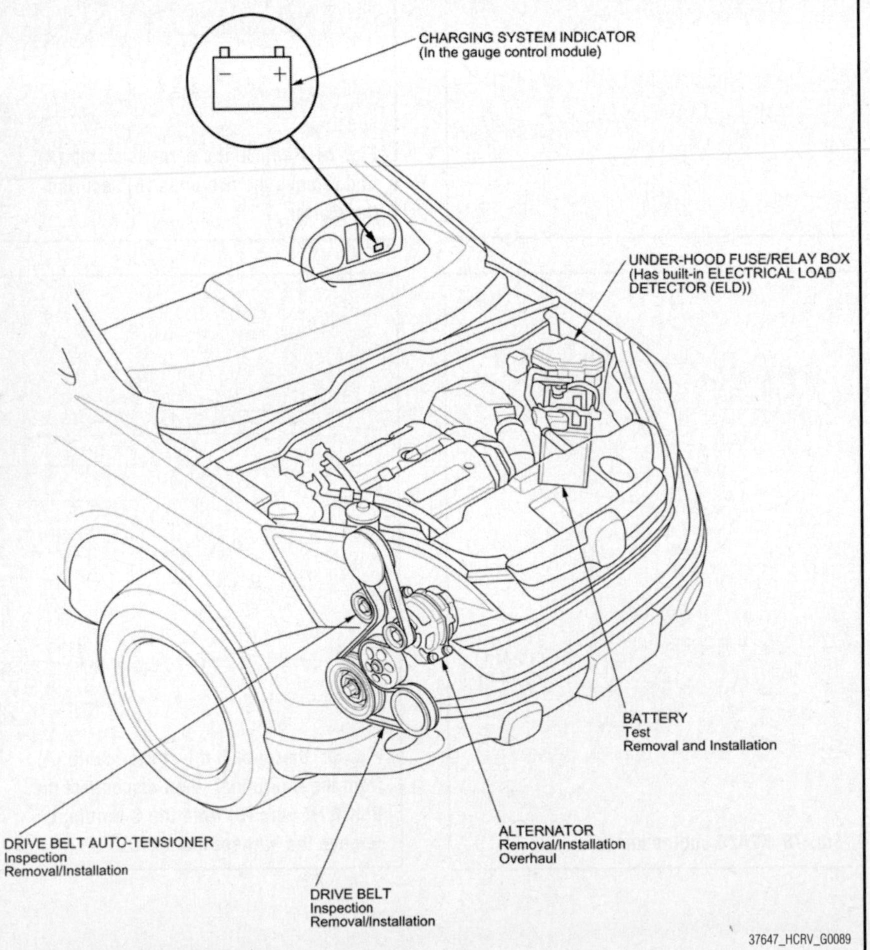
CHARGING SYSTEM INDICATOR
(In the gauge control module)
UNDER-HOOD FUSE/RELAY BOX
(Has built-in ELECTRICAL LOAD DETECTOR (ELD))
BATTERY
Test
Removal and Installation
ALTERNATOR
Removal/Installation
Overhaul
DRIVE BELT AUTO-TENSIONER
Inspection
Removal/Installation
DRIVE BELT
Inspection
Removal/Installation
37647_HCRV_G0089

Fig. 81 Accessory belt routing

REMOVAL & INSTALLATION

See Figure 83.

Special Tools Required: Belt Tension Release Tool Snap-on YA9317 or equivalent, commercially available

1. Move the auto-tensioner using the belt tension release tool to relieve tension from the drive belt, then remove the drive belt.
2. Install the new belt in the reverse order of removal.

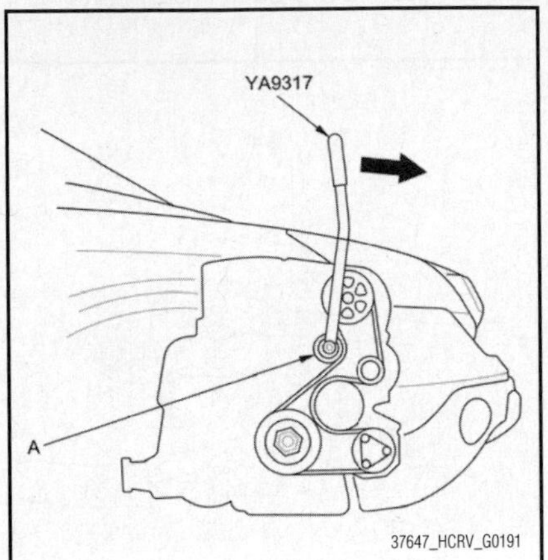
YA9317
A
37647_HCRV_G0191

Fig. 83 Move the auto-tensioner (A) using the belt tension release tool to relieve tension from the drive belt, then remove the drive belt

CAMSHAFT AND VALVE LIFTERS

REMOVAL & INSTALLATION

K24Z1 Engine

See Figures 84 through 89.

1. Before servicing the vehicle, refer to the Precautions Section.
2. Remove the timing chain and sprockets. Refer to Timing Chain,

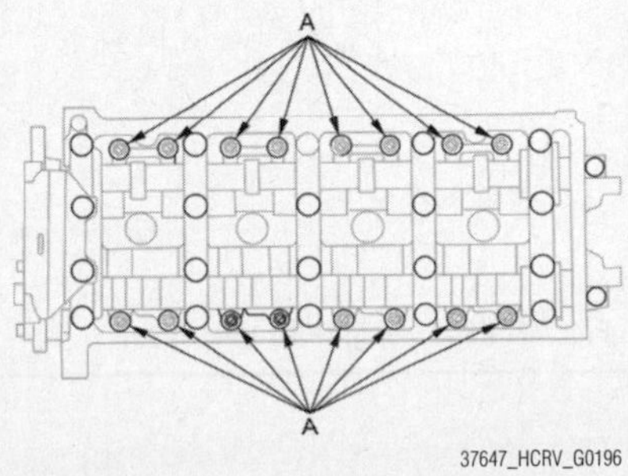
A
A
37647_HCRV_G0196

Fig. 84 Loosen the rocker arm adjusting screws (A)

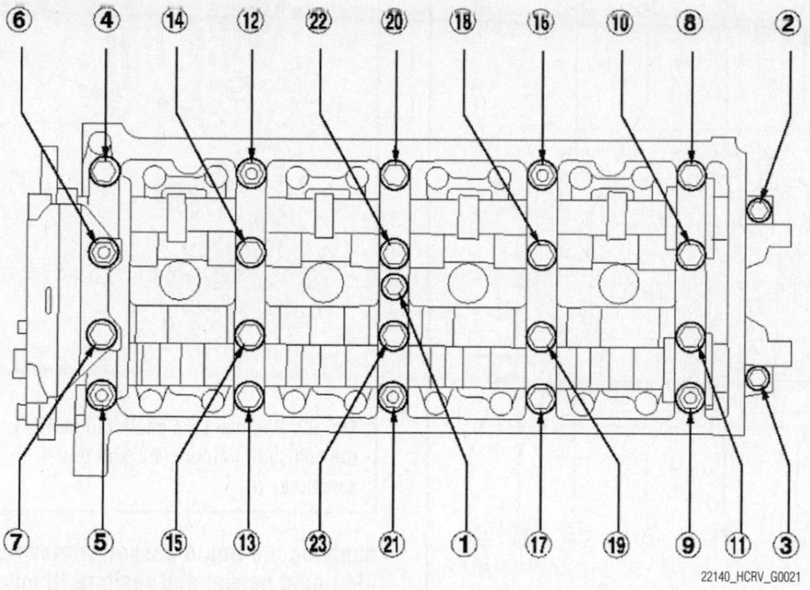
6
4
14
12
22
20
18
16
10
8
2
7
5
15
13
23
21
1
17
19
9
11
3
22140_HCRV_G0021

Fig. 85 Remove the camshaft holder bolts. To prevent damage to the camshafts, loosen the bolts 2 turns at a time in the sequence illustrated

Sprockets, Front Cover and Seal, Removal and Installation.

3. Loosen the rocker arm adjusting screws.

4. Remove the camshaft holder bolts. To prevent damaging the camshafts, loosen the bolts, in sequence, 2 turns at a time.

➡Mark the camshafts, camshaft brackets, and bolts so they are placed in the same position and direction at installation.

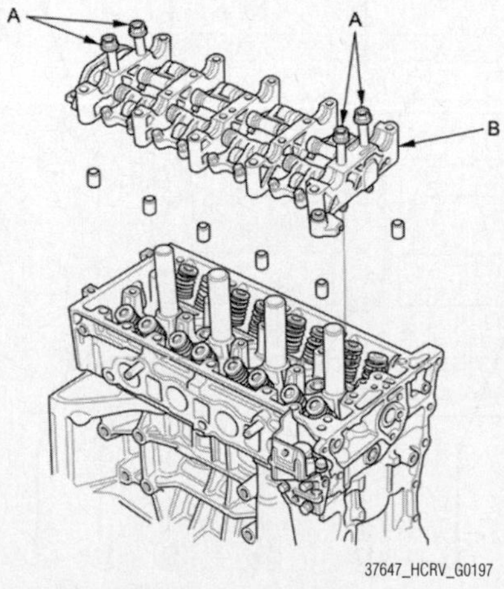
A
A
B
37647_HCRV_G0197

Fig. 86 Insert the bolts (A) into the rocker shaft holder, then remove the rocker arm assembly (B)

5. Remove the camshaft chain guide, the camshaft holders, and the camshafts.

6. Insert the bolts into the rocker shaft holder, then remove the rocker arm assembly.

To install:

7. Reassemble the rocker arm assembly.

8. Clean and dry the No. 5 rocker shaft holder mating surface.

9. Apply liquid gasket (P/N 08717-0004, 08718-0001, 08718-0003, or 08718-0009) to the cylinder head mating surface of the No. 5 rocker shaft holder and to the inside edge of the bolt holes. Install the component within 5 minutes of applying the liquid gasket.

a. Apply a bead of liquid gasket about 0.12 inches (3mm) in diameter along the broken line.

b. If applying liquid gasket P/N 08718-0012, the component must be installed within 4 minutes.

c. If too much time has passed after applying the liquid gasket, remove the old liquid gasket and residue, then reapply new liquid gasket.

10. Insert the bolts into the rocker shaft holder, then install the rocker arm assembly on the cylinder head.

11. Remove the bolts from the rocker shaft holder.

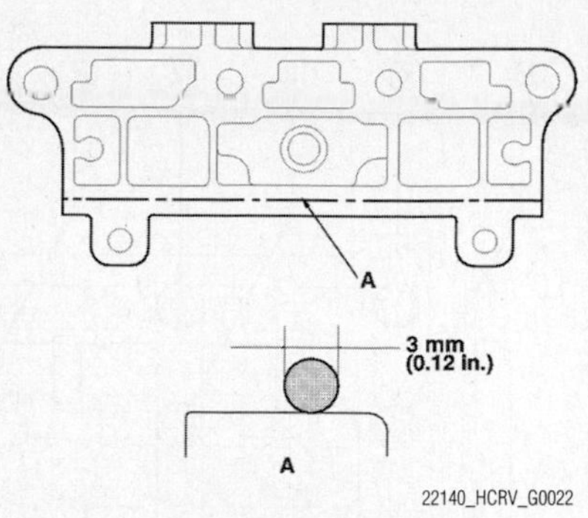
A
3 mm
(0.12 in.)
A
22140_HCRV_G0022

Fig. 87 Apply a bead of liquid gasket about 0.12 inches (3mm) in diameter along the broken line (A)

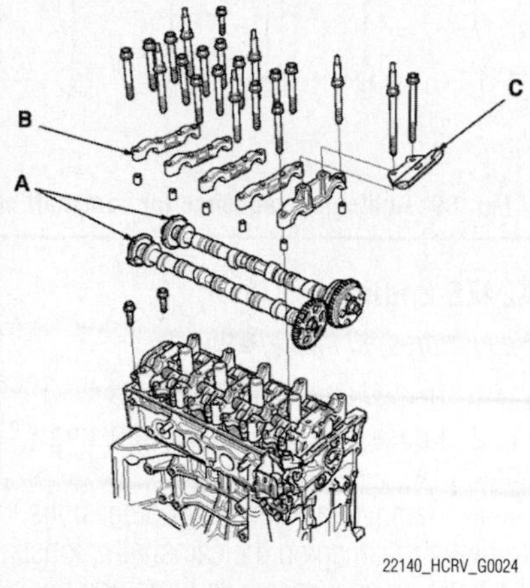
C
B
A
22140_HCRV_G0024

Fig. 88 Set the camshafts (A) in the holder. Set the camshaft holders (B) and camshaft chain guide (C) in place

12. Make sure the punch marks on the variable valve timing control actuator and exhaust camshaft sprocket are facing up, then set the camshafts in the holder.

13. Set the camshaft holders and camshaft chain guide B in place.

14. Tighten the camshaft holder bolts to the specified torque in the sequence shown.

a. Bolts 8 x 1.25mm: 16 ft. lbs. (22 Nm).

b. Bolts 6 x 1.0mm: 106 inch lbs. (12 Nm)—bolts 21, 22, and 23.

➡If the engine does not have bolt 21, skip it and continue the torque sequence.

15. Install the timing chain and sprockets. Refer to Timing Chain, Sprockets, Front Cover and Seal, Removal and Installation.

16. Adjust the valve clearance as needed.

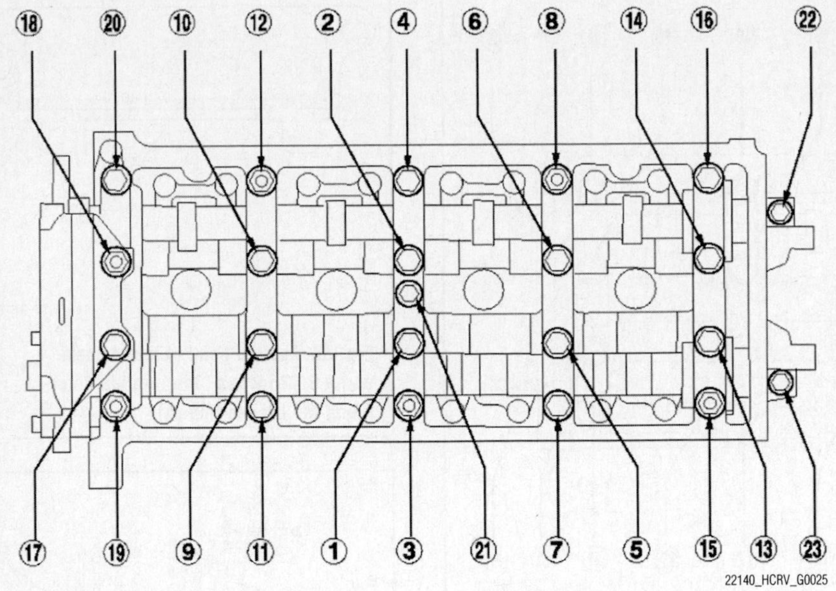
18
20
10
12
2
4
6
8
14
16
22
17
19
9
11
1
3
21
7
5
15
13
23
22140_HCRV_G0025

Fig. 89 Tightening sequence for camshaft holder bolts (caps)

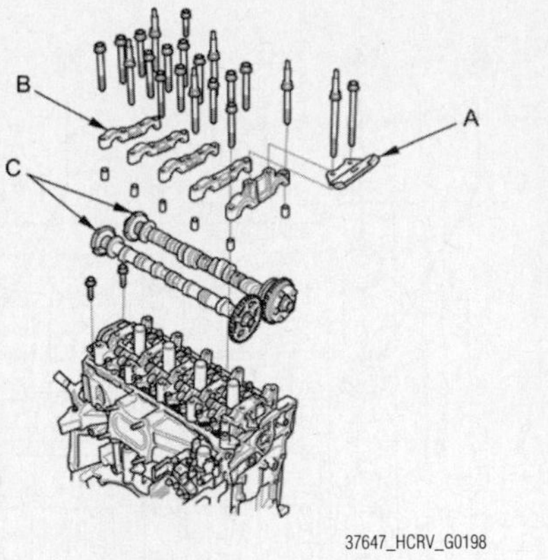
B
A
C
37647_HCRV_G0198

Fig. 92 Remove cam chain guide B (A), the camshaft holders (B), and the camshafts (C)

K24Z6 Engine

See Figures 90 through 95.

1. Remove the cam chain.
2. Loosen the rocker arm adjusting screws (A).
3. Remove the camshaft holder bolts. To prevent damaging the camshafts, loosen the bolts, in sequence, two turns at a time.

➡**Bolt 21 is not on all engines.**

4. Remove cam chain guide B, the camshaft holders, and the camshafts.
5. Insert the bolts into the rocker shaft holder, then remove the rocker arm assembly.

To install:

6. Reassemble the rocker arm assembly. Clean and dry the No. 5 rocker shaft holder mating surface.
7. Apply liquid gasket, P/N 08717-0004, 08718-0001, 08718-0003, or 08718-0009, evenly to the cylinder head mating surface of the No. 5 rocker shaft holder, and to the inside edge of the threaded bolt holes. Install the component within 5 minutes of applying the liquid gasket.

➡**Apply a 0.12 inches (3 mm) diameter bead of liquid gasket along the broken line. If too much time has passed after applying the liquid gasket, remove the old liquid gasket and residue, then reapply new liquid gasket.**

8. Insert the bolts into the rocker shaft holder, then install the rocker arm assembly on the cylinder head.
9. Remove the bolts from the rocker shaft holder.
10. Make sure the punch marks on the Variable Valve Timing Control (VTC) actuator and the exhaust camshaft sprocket are facing up, then set the camshafts in the holder. Apply new engine oil to the camshaft journals and lobes.

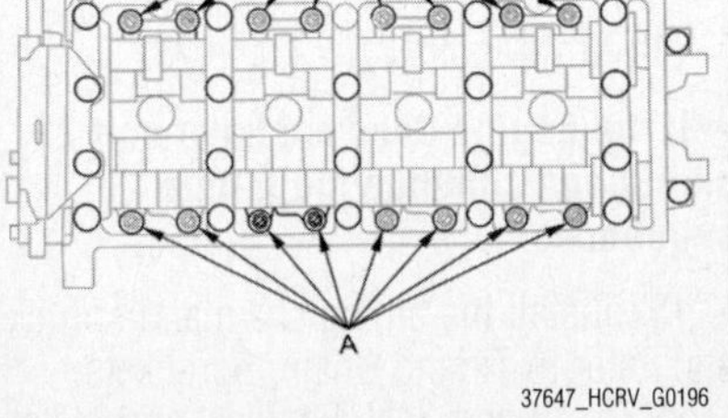
A
A
37647_HCRV_G0196

Fig. 90 Loosen the rocker arm adjusting screws (A)

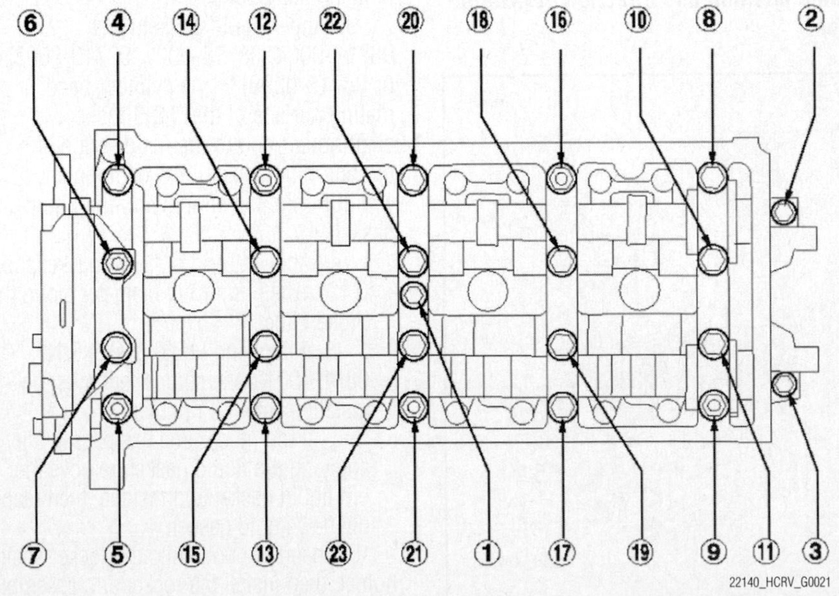
6
4
14
12
22
20
18
16
10
8
2
7
5
15
13
23
21
1
17
19
9
11
3
22140_HCRV_G0021

Fig. 91 Remove the camshaft holder bolts; to prevent damage to the camshafts, loosen the bolts , in sequence, 2 turns at a time

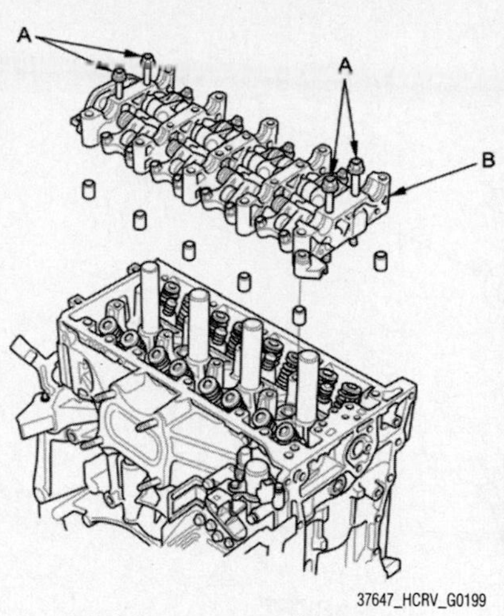
A
A
B
37647_HCRV_G0199
Fig. 93 Insert the bolts (A) into the rocker shaft holder, then remove the rocker arm assembly (B)

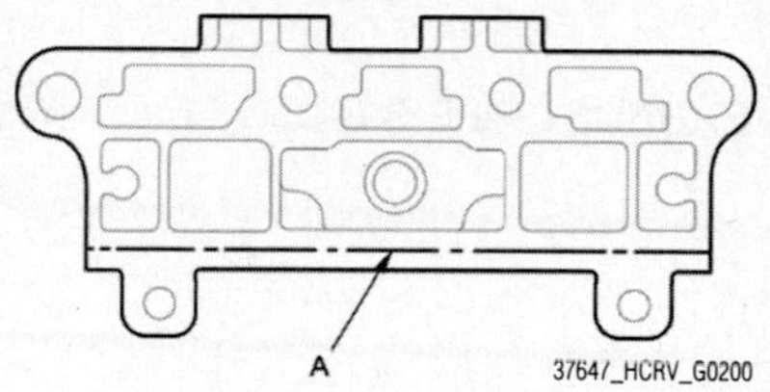
A
37647_HCRV_G0200
Fig. 94 Apply a 0.12 inches (3 mm) diameter bead of liquid gasket along the broken line (A)

11. Set the camshaft holders and the cam chain guide B in place.
12. Tighten the bolts to the specified torque.
    a. Bolts 8 x 1.25mm: 16 ft. lbs. (22 Nm).
    b. Bolts 6 x 1.0mm: 9 ft. lbs. (12 Nm)—bolts 21, 22, and 23.

➡If the engine does not have bolt 21, skip it and continue the torque sequence.

13. Install the cam chain, then adjust the valve clearance.

CATALYTIC CONVERTER

REMOVAL & INSTALLATION

K24Z1 Engine

See Figure 96.

➡The manufacturer does not provide a specific Removal and Installation procedure for this component. Refer to the graphic(s) when servicing this component.

K24Z6 Engine

Warm Up Three Way Catalytic Converter (WU-TWC)

See Figures 97 and 98.

1. Remove the A/F sensor (Sensor 1).
2. Remove the under-cowl panel.
3. Raise the vehicle on a lift.
4. Remove the bolts and the WU-TWC bracket.
5. Remove the bolts.
6. Lower the vehicle.
7. Remove the upper converter cover.
8. Remove the WU-TWC and the gaskets.
9. Install the parts in the reverse order of removal with new gaskets.

Under-Floor TWC

See Figure 99.

1. Remove the secondary HO2S (Sensor 2).
2. Remove the bolts.
3. Remove the nuts, then remove the under-floor TWC.
4. Install the parts in the reverse order of removal with new gaskets and new self-locking nuts.

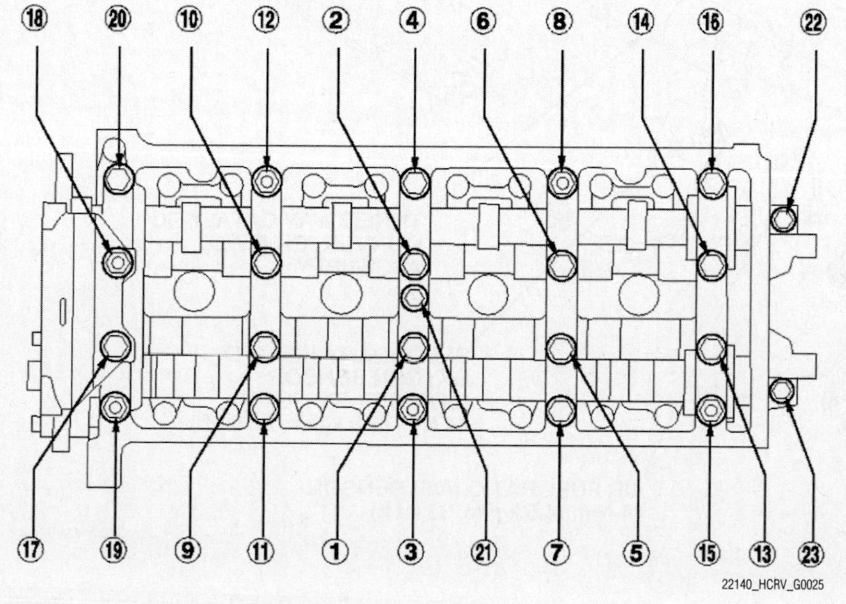
18
20
10
12
2
4
6
8
14
16
22
17
19
9
11
1
3
21
7
5
15
13
23
22140_HCRV_G0025
Fig. 95 Tightening sequence for camshaft holder bolts

CRANKSHAFT PULLEY

REMOVAL & INSTALLATION

See Figure 100.

Special Tools Required:

- Holder handle 07JAB-001020B
- Crankshaft pulley holder 07AAB-RJAA100
- Socket, 19mm 07JAA-001020A or a commercially available 19mm socket

1. Before servicing the vehicle, refer to the Precautions Section.
2. Remove the front wheels.
3. Remove the splash shield.
4. Remove the accessory drive belt. Refer to Accessory Drive Belts, Removal and Installation.
5. Hold the pulley with the holder handle and the holder attachment.
6. Remove the bolt with a 19mm socket and breaker bar, then remove the crankshaft pulley.

To install:

7. Clean the crankshaft damper pulley, crankshaft, bolt, and washer. Lubricate crankshaft damper bolt with new engine oil.
8. Install the crankshaft damper pulley and hold the pulley with holder handle and holder attachment.
9. Tighten the crankshaft damper bolt to 36 ft. lbs. (49 Nm).
10. Tighten the crankshaft damper bolt an additional 90°.
11. Install the accessory drive belt. Refer to Accessory Drive Belts, Removal and Installation.
12. Install the splash shield.
13. Install the front wheels.

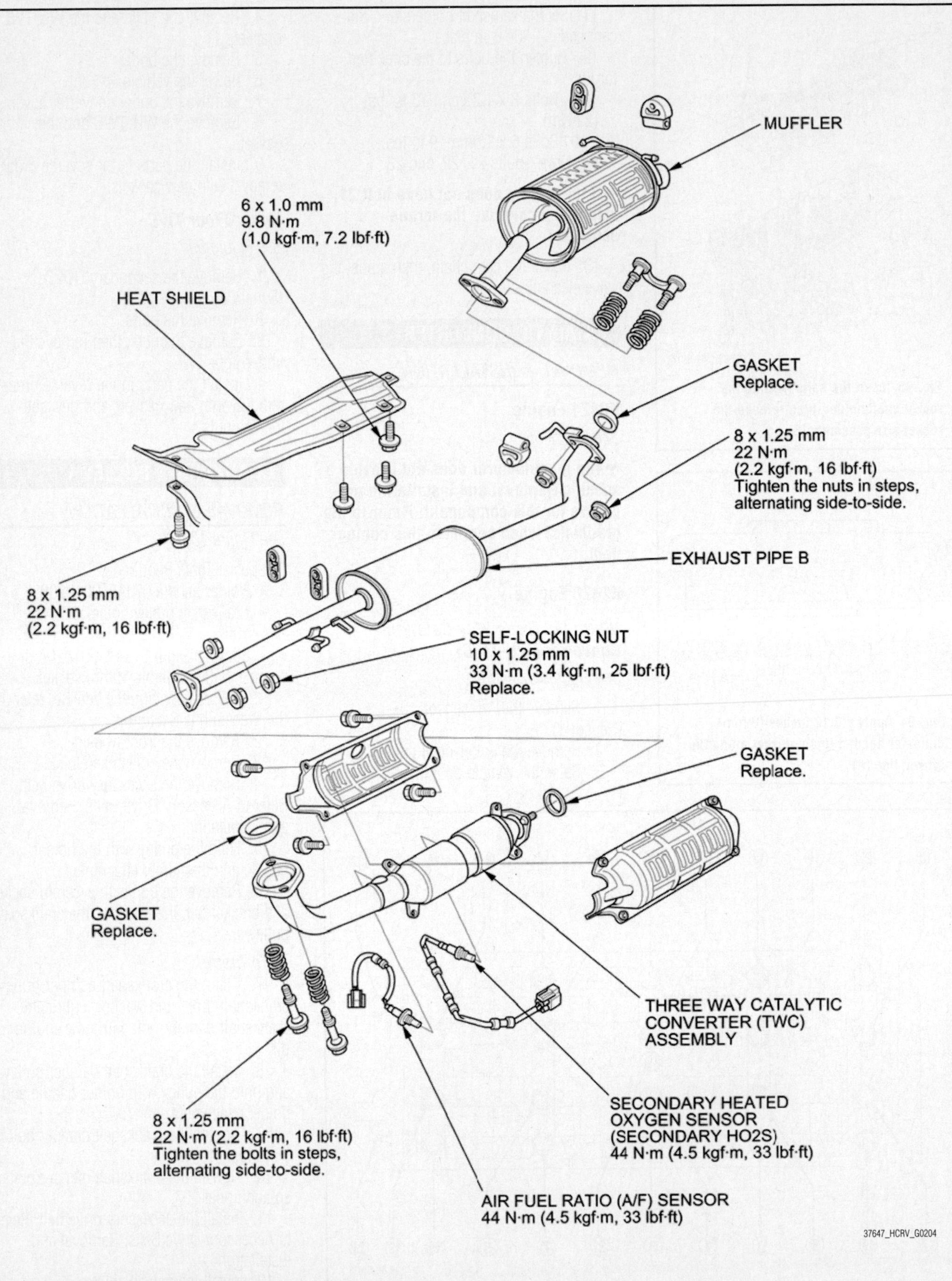
MUFFLER
6 x 1.0 mm
9.8 N·m
(1.0 kgf·m, 7.2 lbf·ft)
HEAT SHIELD
GASKET
Replace.
8 x 1.25 mm
22 N·m
(2.2 kgf·m, 16 lbf·ft)
Tighten the nuts in steps, alternating side-to-side.
EXHAUST PIPE B
8 x 1.25 mm
22 N·m
(2.2 kgf·m, 16 lbf·ft)
SELF-LOCKING NUT
10 x 1.25 mm
33 N·m (3.4 kgf·m, 25 lbf·ft)
Replace.
GASKET
Replace.
GASKET
Replace.
THREE WAY CATALYTIC CONVERTER (TWC) ASSEMBLY
8 x 1.25 mm
22 N·m (2.2 kgf·m, 16 lbf·ft)
Tighten the bolts in steps, alternating side-to-side.
SECONDARY HEATED OXYGEN SENSOR (SECONDARY HO2S)
44 N·m (4.5 kgf·m, 33 lbf·ft)
AIR FUEL RATIO (A/F) SENSOR
44 N·m (4.5 kgf·m, 33 lbf·ft)
37647_HCRV_G0204

Fig. 96 Exploded view of exhaust system

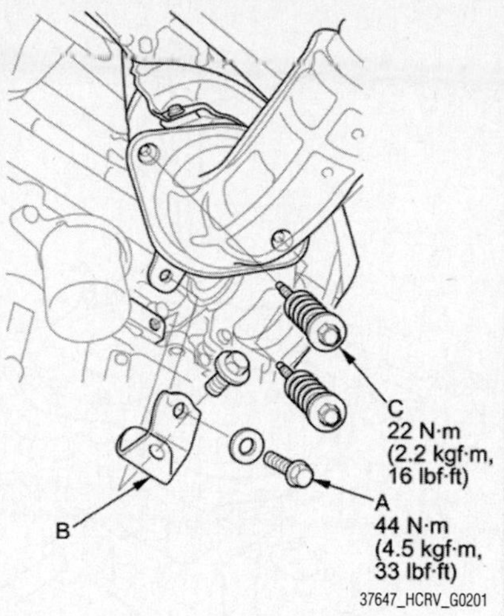
C
22 N·m
(2.2 kgf·m,
16 lbf·ft)
B
A
44 N·m
(4.5 kgf·m,
33 lbf·ft)
37647_HCRV_G0201

Fig. 97 Remove the bolts (A), the WU-TWC bracket (B) and the bolts (C)

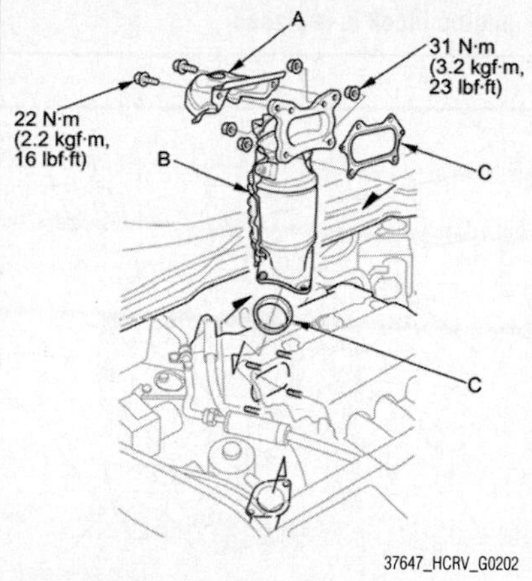
A
31 N·m
(3.2 kgf·m,
23 lbf·ft)
22 N·m
(2.2 kgf·m,
16 lbf·ft)
B
C
C
37647_HCRV_G0202

Fig. 98 Remove the upper converter cover (A), the WU-TWC (B) and the gaskets (C)

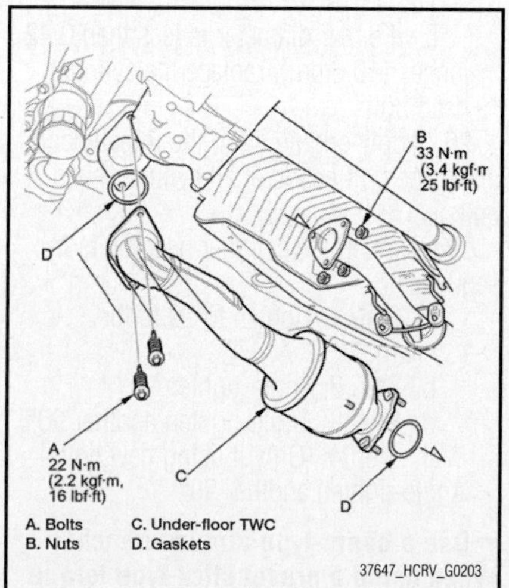
B
33 N·m
25 lbf·ft)
D
A
22 N·m
(2.2 kgf·m,
16 lbf·ft)
C
D
A. Bolts
B. Nuts
C. Under-floor TWC
D. Gaskets
37647_HCRV_G0203

Fig. 99 Exploded view of the under-floor TWC assembly

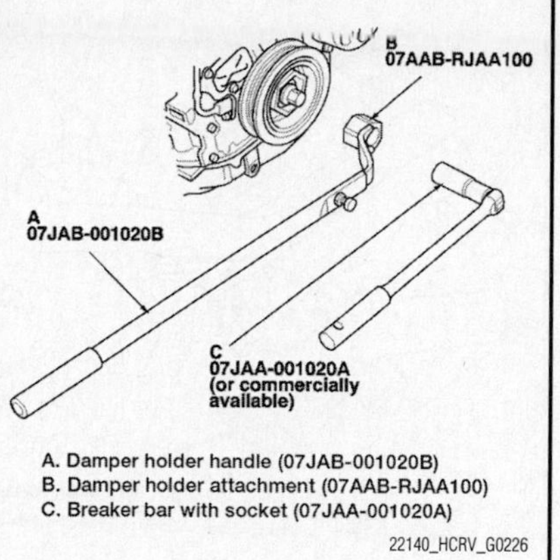
B
07AAB-RJAA100
A
07JAB-001020B
C
07JAA-001020A
(or commercially
available)
A. Damper holder handle (07JAB-001020B)
B. Damper holder attachment (07AAB-RJAA100)
C. Breaker bar with socket (07JAA-001020A)
22140_HCRV_G0226

Fig. 100 View of special tools used to remove the crankshaft damper

CRANKSHAFT FRONT SEAL

REMOVAL & INSTALLATION

The 2.4L engine utilizes a camshaft timing chain case seal. Refer to Timing Chain, Sprockets, Front Cover and Seal, Removal and Installation.

CYLINDER HEAD

REMOVAL & INSTALLATION

K24Z1 Engine

See Figures 101 through 111.

1. Before servicing the vehicle, refer to the Precautions Section.

⁂ WARNING

Use fender covers to avoid damaging painted surfaces. To avoid damage, unplug the wiring connectors carefully while holding the connector portion. To avoid damaging the cylinder head, wait until the engine coolant temperature drops below 100°F (38°C) before loosening the cylinder head bolts.

➡Mark all wiring and hoses to avoid misconnection.

2. Relieve fuel pressure.
3. Drain the engine coolant.
4. Remove the air cleaner housing.
5. Remove the accessory drive belt.
6. Remove the intake manifold.
7. Remove the exhaust manifold.
8. Disconnect the Evaporative Emission (EVAP) canister hose and the brake booster vacuum hose.

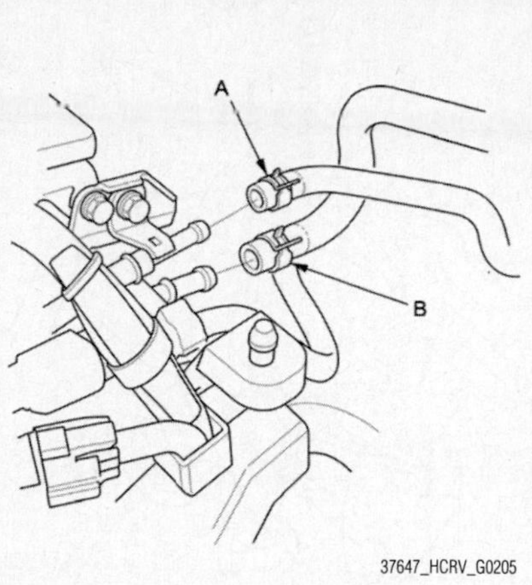
A
B
37647_HCRV_G0205

Fig. 101 Disconnect the Evaporative Emission (EVAP) canister hose (A) and the brake booster vacuum hose (B)

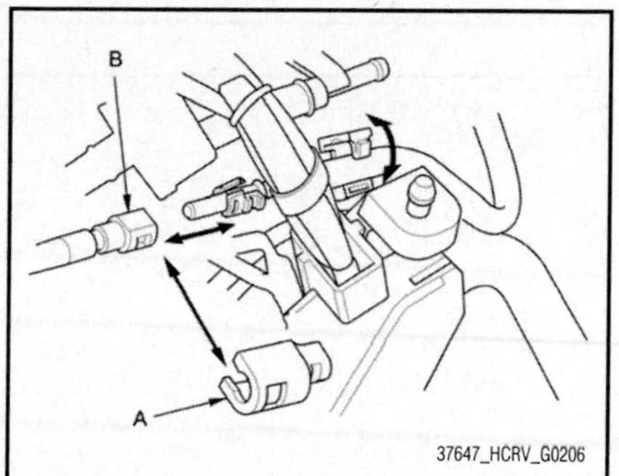
B
A
37647_HCRV_G0206

Fig. 102 Remove the quick-connect fitting cover (A), then disconnect the fuel feed hose (B)

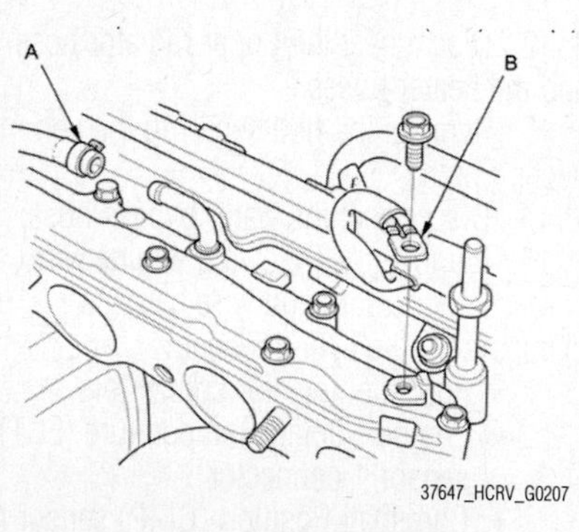
A
B
37647_HCRV_G0207

Fig. 103 Disconnect the PCV hose (A) and remove the ground cable (B)

9. Remove the quick-connect fitting cover, then disconnect the fuel feed hose.
10. Disconnect the Positive Crankcase Ventilation (PCV) hose and remove the ground cable.
11. Remove the harness holder from the bracket, then remove the harness holder bracket.

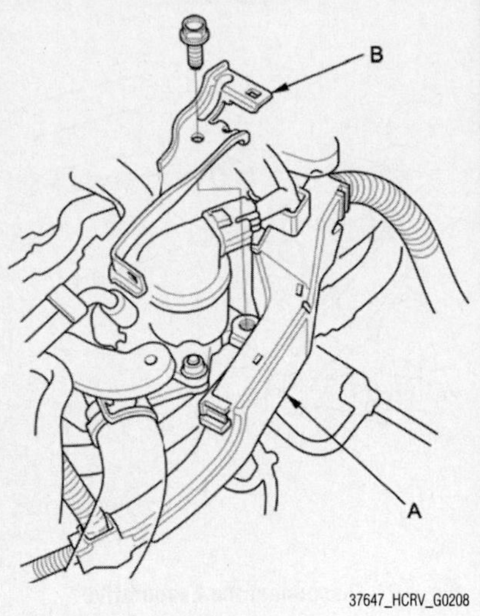
B
A
37647_HCRV_G0208

Fig. 104 Remove the harness holder (A) from the bracket, then remove the harness holder bracket (B)

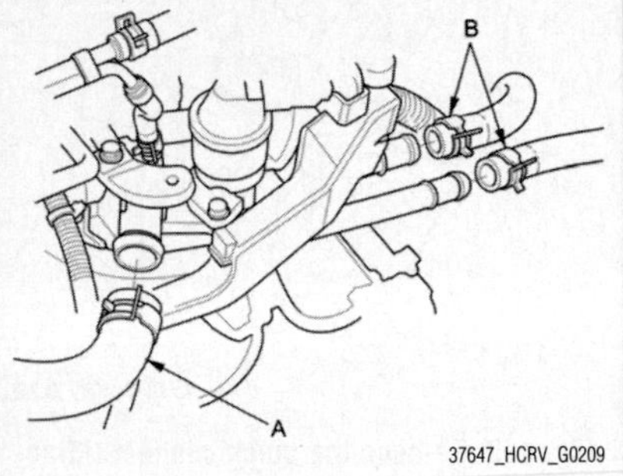
B
A
37647_HCRV_G0209

Fig. 105 Disconnect the upper radiator hose (A) and the heater hoses (B)

12. Disconnect the upper radiator hose and the heater hoses.
13. Remove the bolt securing the connecting pipe.
14. Disconnect the water bypass hose.
15. Remove the following engine wire harness connectors and wire harness clamps from the cylinder head:
- Four fuel injector connectors
- Engine Coolant Temperature (ECT) sensor 1 connector
- Camshaft Position (CMP) sensor A (Intake) connector
- Camshaft Position (CMP) sensor B (Exhaust) connector
- Rocker arm oil control solenoid connector
- Rocker arm oil pressure switch connector
- EVAP canister purge valve connector
- Exhaust Gas Recirculation (EGR) valve connector

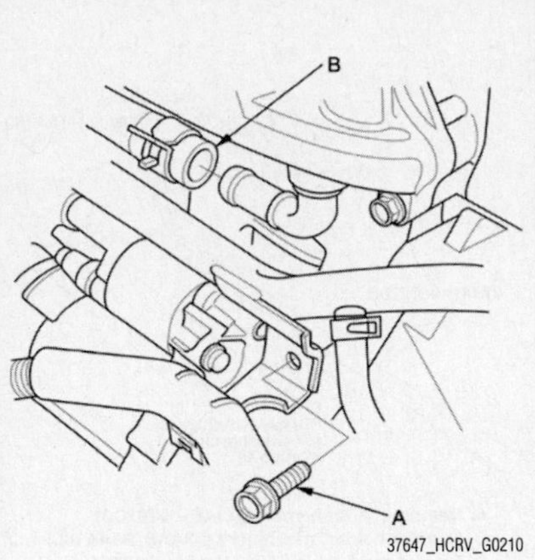
B
A
37647_HCRV_G0210

Fig. 106 Remove the bolt (A) securing the connecting pipe, then disconnect the water bypass hose (B)

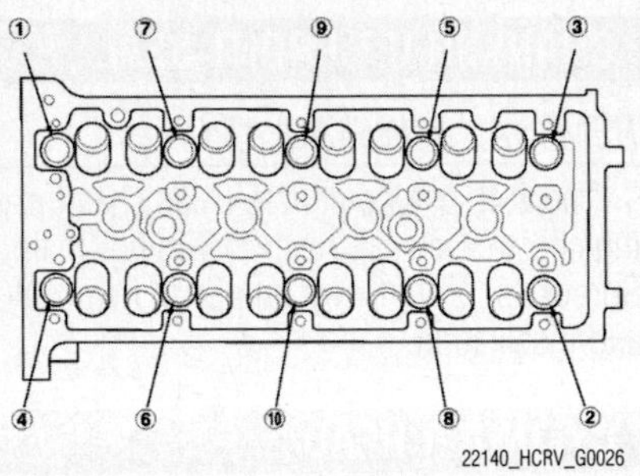
1
7
9
5
3
4
6
10
8
2
22140_HCRV_G0026

Fig. 107 Loosen the cylinder head bolts in sequence

16. Remove the camshaft chain.
17. Remove the rocker arm assembly.
18. Remove the cylinder head bolts. To prevent warpage, loosen the bolts in sequence 1/3 turn at a time; repeat the sequence until all bolts are loosened.
19. Remove the cylinder head.

To install:

20. Install a new coolant separator in the engine block whenever the engine block is replaced.
21. Clean the cylinder head and block surface.
22. Install a new cylinder head gasket and dowel pins on the engine block.

➡Always use a new cylinder head gasket.

23. Set the crankshaft to Top Dead Center (TDC). Align the TDC mark on the crankshaft sprocket with the pointer on the engine block.
24. Install the cylinder head on the block.
25. If reusing the cylinder head bolts:
    a. Measure the diameter of each cylinder head bolt at point A and point B.

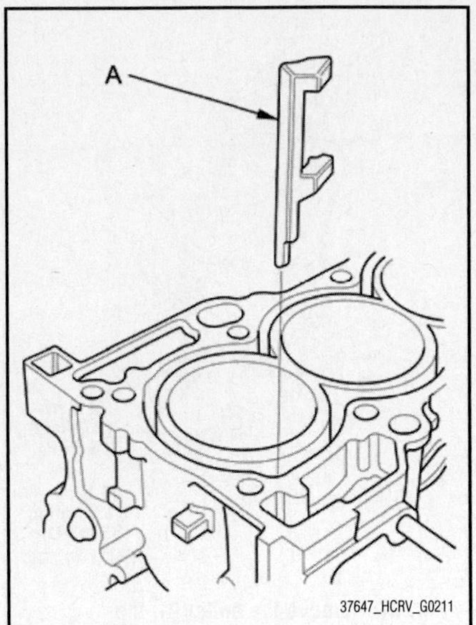
A
37647_HCRV_G0211

Fig. 108 Install a new coolant separator (A) in the engine block whenever the engine block is replaced

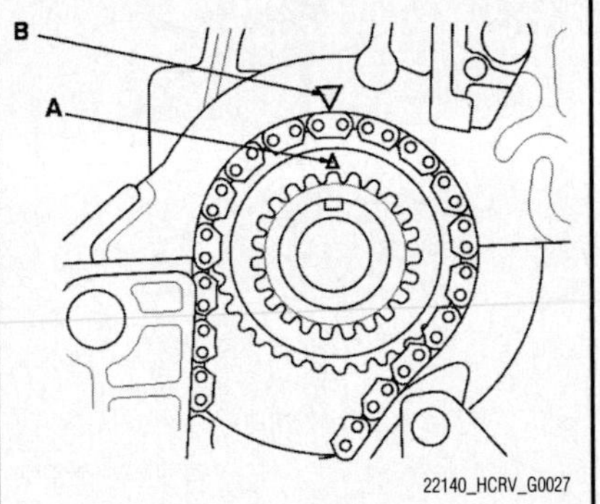
B
A
22140_HCRV_G0027

Fig. 109 Align the TDC mark (A) on the crankshaft sprocket with the pointer (B) on the engine block

    b. If either diameter is less than 0.42 inches (10.6mm), replace the cylinder head bolt.
26. Apply engine oil to the threads and under the bolt heads of all cylinder head bolts.
27. Tighten the cylinder head bolts in sequence:
    a. Step 1: Tighten to 29 ft. lbs. (39 Nm).
    b. Step 2: Angle-tighten 90°.
    c. Step 3: Angle-tighten another 90°.
    d. Step 4: (Only if using new bolts) Angle-tighten another 90°.

➡Use a beam-type torque wrench. When using a preset click-type torque wrench, be sure to tighten slowly and do not over-tighten. If a bolt makes any noise while you are tightening it,

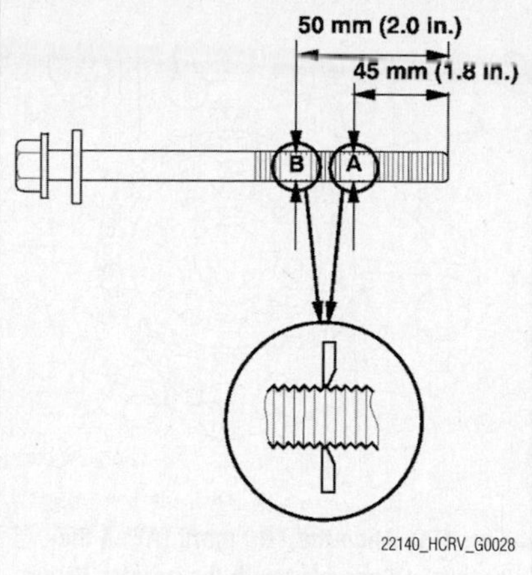
50 mm (2.0 in.)
45 mm (1.8 in.)
B
A
22140_HCRV_G0028

Fig. 110 Measure the diameter of each cylinder head bolt at point A and point B. Replace with new bolts if required

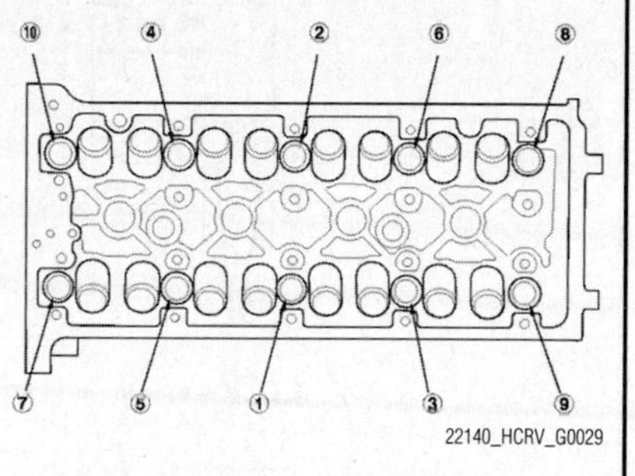
22140_HCRV_G0029

Fig. 111 Cylinder head bolt tightening sequence

loosen the bolt and retighten it from the first step.

28. Install the rocker arm assembly.
29. Install the camshaft chain.
30. Connect the following engine wire harness connectors and install the wire harness clamps to the cylinder head:
    - Four fuel injector connectors
    - ECT sensor 1 connector
    - CMP sensor A (Intake) connector
    - CMP sensor B (Exhaust) connector
    - Rocker arm oil control solenoid connector
    - Rocker arm oil pressure switch connector
    - EVAP canister purge valve connector
    - EGR valve connector
31. Install the bolt securing the connecting pipe.
32. Install the water bypass hose.
33. Install the upper radiator hose and the heater hoses.
34. Install the harness holder bracket, then install the harness holder.
35. Install the PCV hose and the ground cable.
36. Connect the fuel feed hose, then install the quick-connect fitting cover.
37. Install the EVAP canister hose and the brake booster vacuum hose.
38. Install the exhaust manifold.
39. Install the intake manifold.
40. Install the accessory drive belt.
41. Install the air cleaner housing.
42. After installation, check that all tubes, hoses, and connectors are installed correctly.
43. Inspect for fuel leaks. Turn the ignition switch **ON** (do not operate the starter) so the fuel pump runs for about 2 seconds and pressurizes the fuel line. Repeat this operation 3 times, then check for fuel leakage at all points in the fuel line.
44. Refill the radiator with engine coolant and bleed the air from the cooling system with the heater valve open and the engine running.
45. Inspect the idle speed.
46. Inspect the ignition timing.

K24Z6 Engine

See Figures 112 through 122.

1. Relieve the fuel pressure.
2. Drain the engine coolant.
3. Remove the drive belt.
4. Remove the intake manifold.
5. Remove the warm up TWC.
6. Disconnect the Evaporative Emission (EVAP) canister hose.
7. Remove the quick-connect fitting cover, then disconnect the fuel feed hose.
8. Disconnect the four fuel injector connectors and remove the ground cables.
9. Remove the four bolts securing the EVAP canister purge valve bracket.
10. Disconnect the upper radiator hose, the heater hoses, and the water bypass hose.

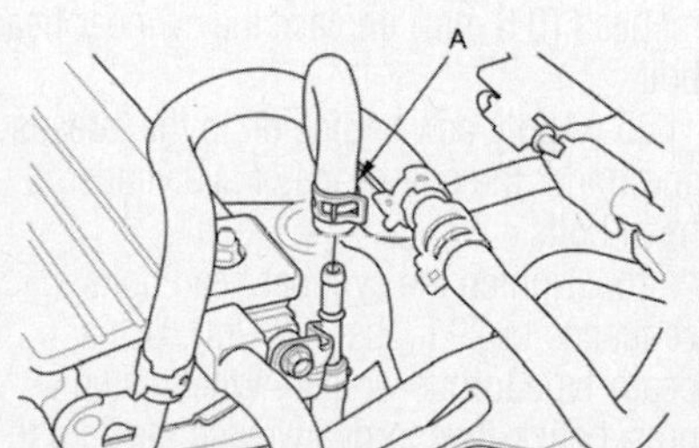
A
37647_HCRV_G0212

Fig. 112 Disconnect the Evaporative Emission (EVAP) canister hose (A)

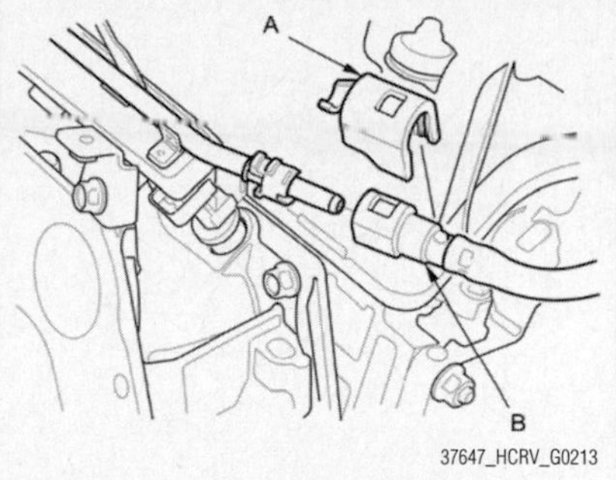
A
B
37647_HCRV_G0213

Fig. 113 Remove the quick-connect fitting cover (A), then disconnect the fuel feed hose (B)

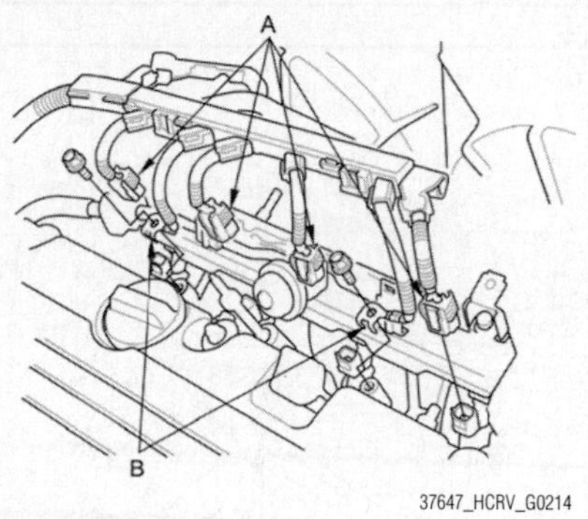
A
B
37647_HCRV_G0214

Fig. 114 Disconnect the four fuel injector connectors (A), remove the ground cables (B)

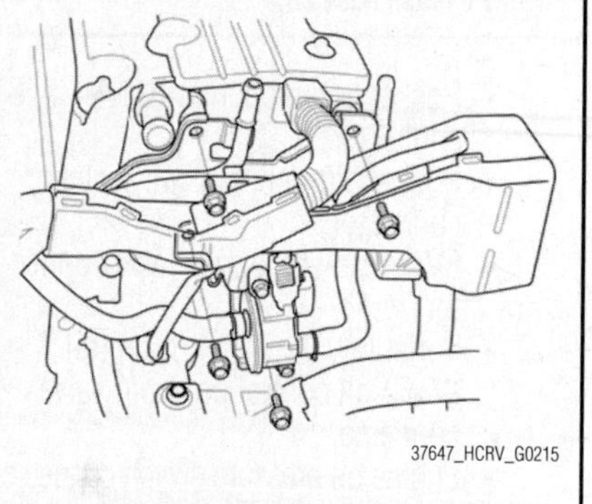
37647_HCRV_G0215

Fig. 115 Remove the four bolts securing the EVAP canister purge valve bracket

11. Remove the two bolts securing the connecting pipe.
12. Disconnect the water bypass hose.
13. Disconnect the following engine wire harness connectors and remove the wire harness clamps from the cylinder head:
    - Engine Coolant Temperature (ECT) sensor 1 connector
    - Camshaft Position (CMP) sensor A (Intake) connector
    - Camshaft Position (CMP) sensor B (Exhaust) connector

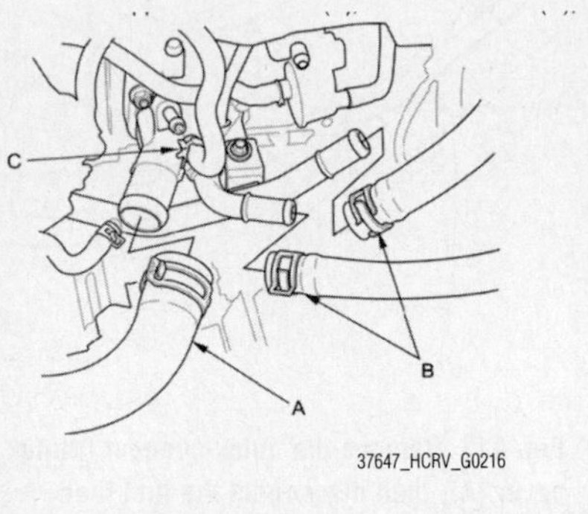
C
B
A
37647_HCRV_G0216

Fig. 116 Disconnect the upper radiator hose (A), the heater hoses (B), and the water bypass hose (C)

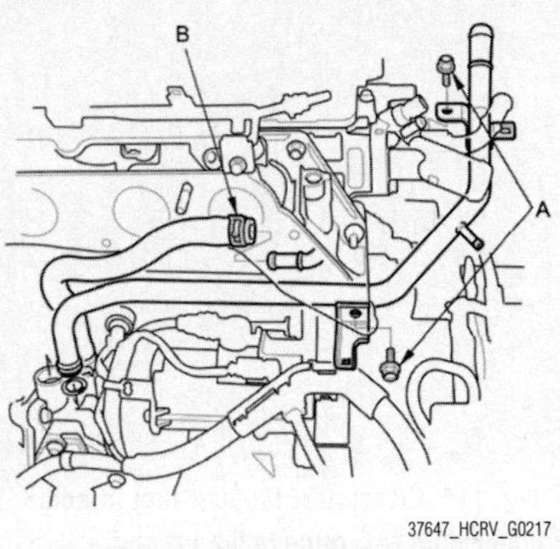
B
A
37647_HCRV_G0217

Fig. 117 Remove the two bolts (A) securing the connecting pipe, disconnect the water bypass hose (B)

- Rocker arm oil control valve connector
- Rocker arm oil pressure switch connector
- EVAP canister purge valve connector
- Variable Valve Timing Control (VTC) oil control solenoid valve connector
- Engine oil pressure switch connector

14. Remove the cam chain.
15. Remove the rocker arm assembly.
16. Remove the cylinder head bolts. To prevent warpage, loosen the bolts in sequence ⅓ turn at a time; repeat the sequence until all bolts are loosened.
17. Remove the cylinder head.

To install:

18. Clean the cylinder head and the engine block surface.
19. Install a new coolant separator in the engine block whenever the engine block is replaced.
20. Install the new cylinder head gasket and the dowel pins on the engine block. Always use a new cylinder head gasket.

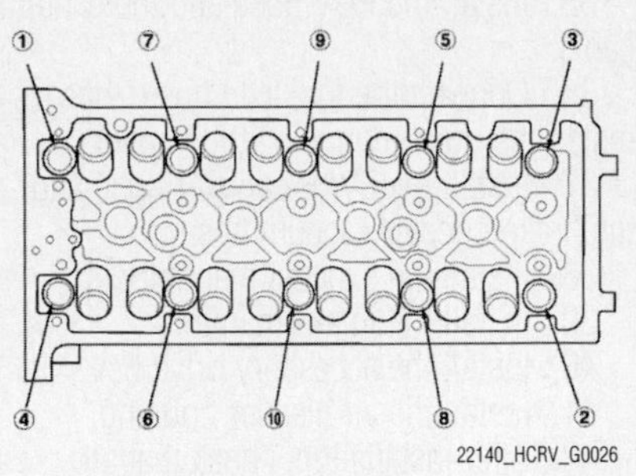
①
⑦
⑨
⑤
③
④
⑥
⑩
⑧
②
22140_HCRV_G0026

Fig. 118 Remove the cylinder head bolts, loosen the bolts in sequence

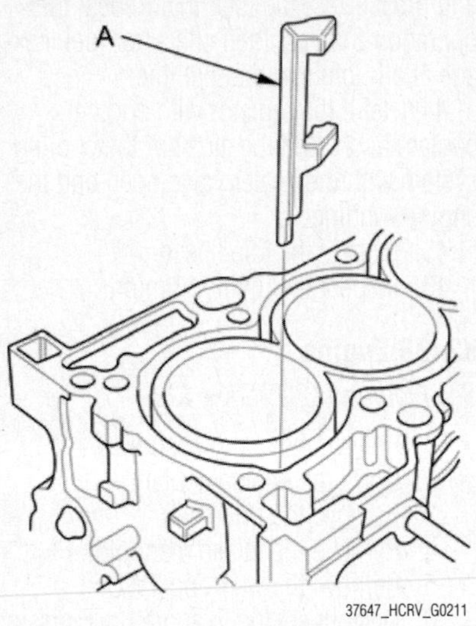
A
37647_HCRV_G0211

Fig. 119 Install a new coolant separator (A) in the engine block whenever the engine block is replaced

21. Set the crankshaft to Top Dead Center (TDC). Align the TDC mark on the crankshaft sprocket with the pointer on the engine block.
22. Install the cylinder head on the engine block.
23. Measure the diameter of each cylinder head bolt at point A and point B.
24. If either diameter is less than 0.42 inches (10.6 mm), replace the cylinder head bolt.
25. Apply new engine oil to the threads and under the bolt heads of all cylinder head bolts.
26. Tighten the cylinder head bolts in sequence to 29 ft. lbs. (39 Nm). Use a beam-type torque wrench. When using a preset click-type torque wrench, be sure to tighten slowly and do not overtighten. If a bolt makes any noise while you are torquing it, loosen the bolt and retighten it from the first step.
27. After torquing, tighten all cylinder

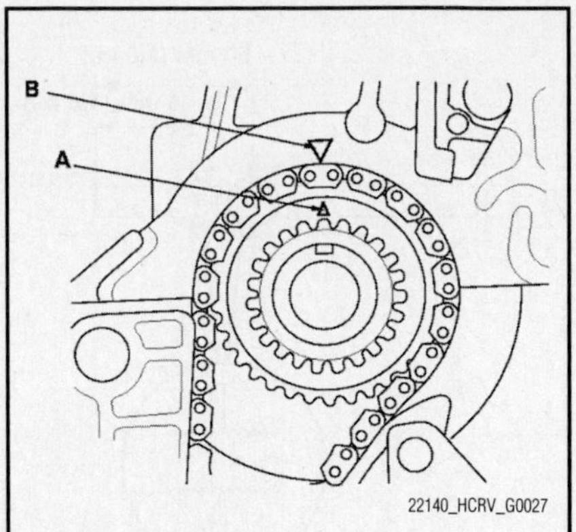
B
A
22140_HCRV_G0027

Fig. 120 Align the TDC mark (A) on the crankshaft sprocket with the pointer (B) on the engine block

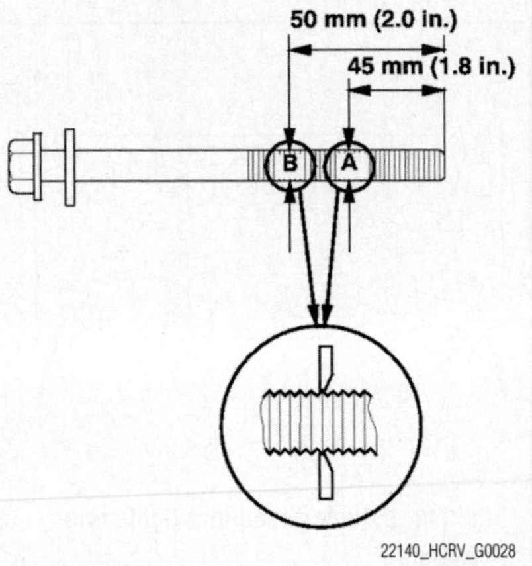
50 mm (2.0 in.)
45 mm (1.8 in.)
B
A
22140_HCRV_G0028

Fig. 121 Measure the diameter of each cylinder head bolt at point A and point B. Replace with new bolts if required

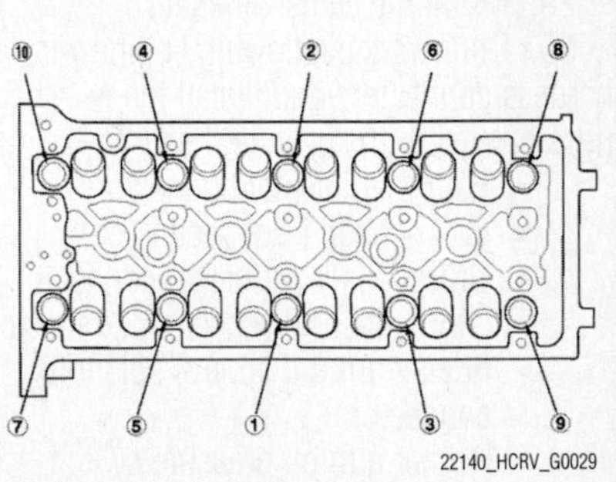
⑩
④
②
⑥
⑧
⑦
⑤
①
③
⑨
22140_HCRV_G0029

Fig. 122 Cylinder head bolt tightening sequence

head bolts in two steps (90° per step) using the sequence shown in step 9. If you are using a new cylinder head bolt, tighten the bolt an extra 90°.

➡Remove the cylinder head bolt if you tightened it beyond the specified angle,

and go back to step 6 of the procedure. Do not loosen it back to the specified angle.

28. Install the rocker arm assembly.
29. Install the cam chain.
30. Connect the following engine wire harness connectors, and install the wire harness clamps to the cylinder head:
    - Engine Coolant Temperature (ECT) sensor 1 connector
    - Camshaft Position (CMP) sensor A (Intake) connector
    - Camshaft Position (CMP) sensor B (Exhaust) connector
    - Rocker arm oil control solenoid connector
    - Rocker arm oil pressure switch connector
    - Evaporative Emission (EVAP) canister purge valve connector
    - Variable Valve Timing Control (VTC) oil control solenoid valve connector
    - Engine oil pressure switch connector
31. Install the two bolts (A) securing the connecting pipe.
32. Connect the water bypass hose (B).
33. Connect the upper radiator hose (A), the heater hoses (B), and the water bypass hose (C).
34. Install the four bolts securing the EVAP canister purge valve bracket.
35. Connect the four fuel injector connectors (A), install the ground cables (B).
36. Connect the fuel feed hose (A), then install the quick-connect fitting cover (B).
37. Connect the EVAP canister hose (A).
38. Install the warm up TWC.
39. Install the intake manifold.
40. Install the drive belt.
41. After installation, check that all tubes, hoses, and connectors are installed correctly.
42. Inspect for fuel leaks. Turn the ignition switch to ON (II) (do not operate the starter) so the fuel pump runs for about 2 seconds and pressurizes the fuel line. Repeat this operation three times, then check for fuel leakage at any point in the fuel line.
43. Refill the radiator with engine coolant, and bleed the air from the cooling system.
44. Check for fluid leaks.
45. Do the Powertrain Control Module (PCM) idle learn procedure.
46. Do the Crankshaft Position (CKP) pattern clear/CKP pattern learn procedure.
47. Inspect the idle speed.
48. Inspect the ignition timing.

EXHAUST MANIFOLD

REMOVAL & INSTALLATION

K24Z1 Engine

See Figure 123.

1. Before servicing the vehicle, refer to the Precautions Section.
2. Disconnect the air fuel ratio sensor connector and secondary heated oxygen sensor connector
3. Remove the 3-way catalytic converter.
4. Remove the under-cowl cover, then remove the strut brace.
5. Remove the rocker arm oil control solenoid.
6. Remove the cover and exhaust manifold bracket.
7. Remove the exhaust manifold.

To install:

8. Install the exhaust manifold with a new gasket.
9. Tighten the bolts and nuts in a crisscross pattern in 3 steps, beginning with the inner nut. Torque specification: 33 ft. lbs. (44 Nm).
10. Install the other parts in the reverse order of removal and torque according to the illustration.

DRIVE PLATE

REMOVAL & INSTALLATION

1. Remove the transmission assembly.
2. Remove the drive plate and the washer from the engine crankshaft.
3. Install the drive plate and the washer on the engine crankshaft, and tighten the eight bolts in a crisscross pattern in at least two steps.
4. Install the transmission assembly.

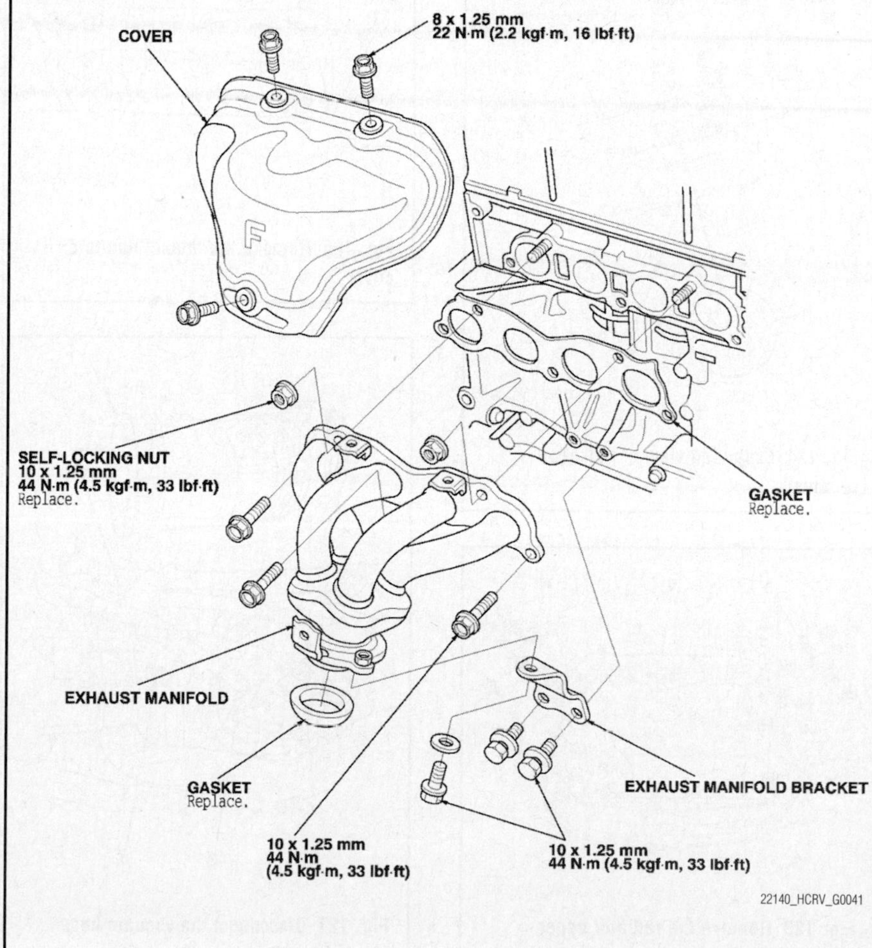

COVER
8 x 1.25 mm
22 N·m (2.2 kgf·m, 16 lbf·ft)
SELF-LOCKING NUT
10 x 1.25 mm
44 N·m (4.5 kgf·m, 33 lbf·ft)
Replace.
GASKET
Replace.
EXHAUST MANIFOLD
GASKET
Replace.
EXHAUST MANIFOLD BRACKET
10 x 1.25 mm
44 N·m
(4.5 kgf·m, 33 lbf·ft)
10 x 1.25 mm
44 N·m (4.5 kgf·m, 33 lbf·ft)
22140_HCRV_G0041

Fig. 123 Exploded view of the exhaust manifold and related components

INTAKE MANIFOLD

REMOVAL & INSTALLATION

K24Z1 Engine

See Figures 124 through 131.

1. Before servicing the vehicle, refer to the Precautions Section.
2. Remove the hood support rod, then use it to prop the hood in the wide-open position.
3. Remove the bulkhead cover:
   a. Put on gloves to protect your hands. Take care not to scratch the front bumper and the body.
   b. Remove the clips by carefully pulling the front grille cover up, then remove the cover by releasing the front edge of the cover from the grille.
   c. To remove the clips, pry the inner clip up at the edge near the line on its head.

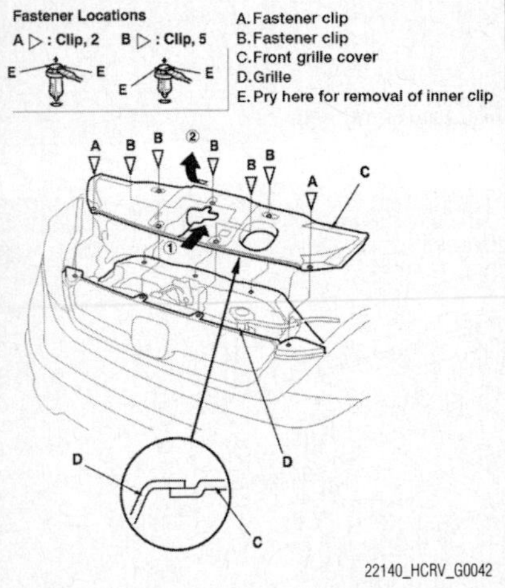

Fastener Locations
A ▷: Clip, 2 B ▷: Clip, 5
A. Fastener clip
B. Fastener clip
C. Front grille cover
D. Grille
E. Pry here for removal of inner clip
22140_HCRV_G0042

Fig. 124 Expanded view of bulkhead removal

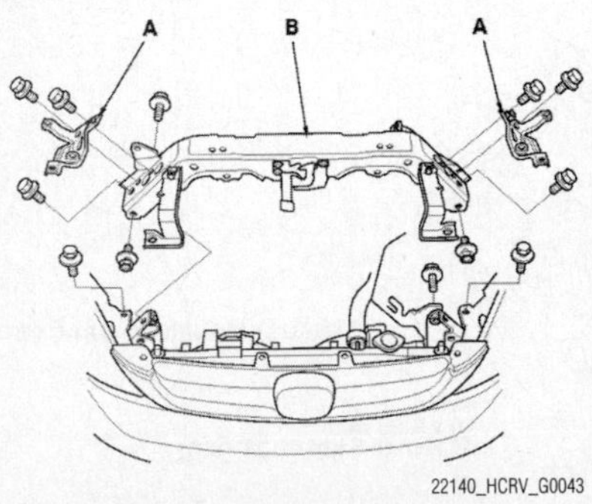

A
B
A
22140_HCRV_G0043

Fig. 125 Remove the radiator upper brackets (A), then remove the front bulkhead (B)

4. Disconnect the fan motor connectors and the hood switch connector, then remove the harness clips.
5. Remove the radiator upper brackets, then remove the front bulkhead.
6. Remove the intake manifold cover.
7. Disconnect the vacuum hose and breather pipe, then remove the intake air duct.
8. Remove the following engine wire harness connectors and wire harness clamps from the intake manifold:
   - Manifold Absolute Pressure (MAP) sensor connector
   - Throttle actuator connector
9. Disconnect the Evaporative Emission (EVAP) canister hose and brake booster vacuum hose.

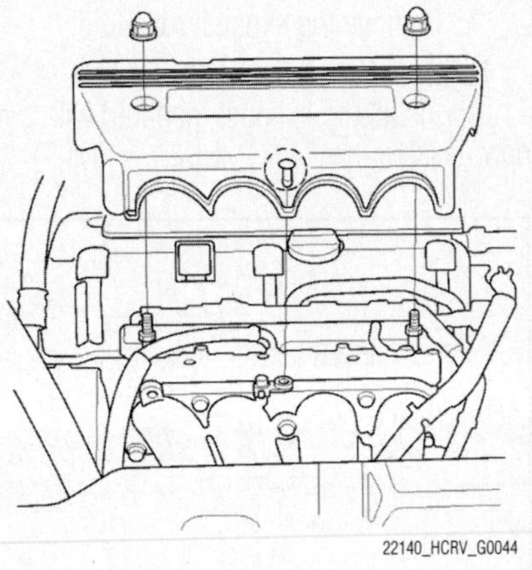

22140_HCRV_G0044

Fig. 126 Remove the intake manifold cover

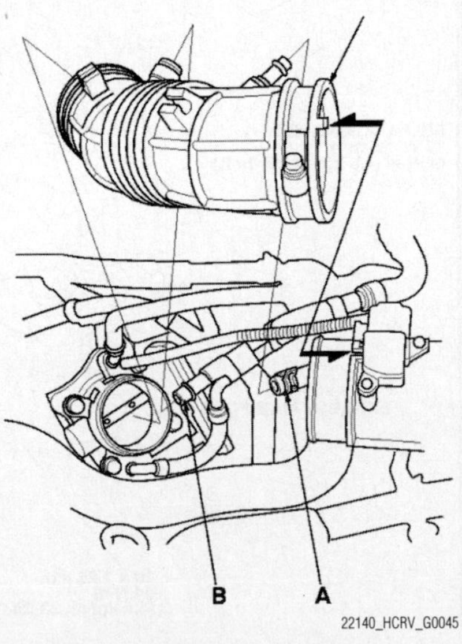

B
A
22140_HCRV_G0045

Fig. 127 Disconnect the vacuum hose (A) and breather pipe (B), then remove the intake air duct (C)

10. Remove the harness bracket mounting bolts.
11. Disconnect the water bypass hoses, then plug the water bypass hoses.
12. Raise and safely support the vehicle.
13. Remove the splash shield.
14. Remove the connector, then remove the intake manifold bracket.
15. Lower the vehicle.
16. Remove the intake manifold.

To install:

17. Install the intake manifold with a new gasket, and tighten the bolts and nuts in a crisscross pattern in 3 steps, beginning with the inner bolt. Tighten the bolts and nuts to 16 ft. lbs. (22 Nm).
18. Raise and safely support the vehicle.
19. Install the intake manifold bracket, then install the connector. Tighten the bracket bolt and nut to 16 ft. lbs. (22 Nm).
20. Install the splash shield.
21. Lower the vehicle.
22. Install the water bypass hoses.
23. Install the EVAP canister hose and the brake booster vacuum hose.
24. Install the harness bracket mounting bolts.

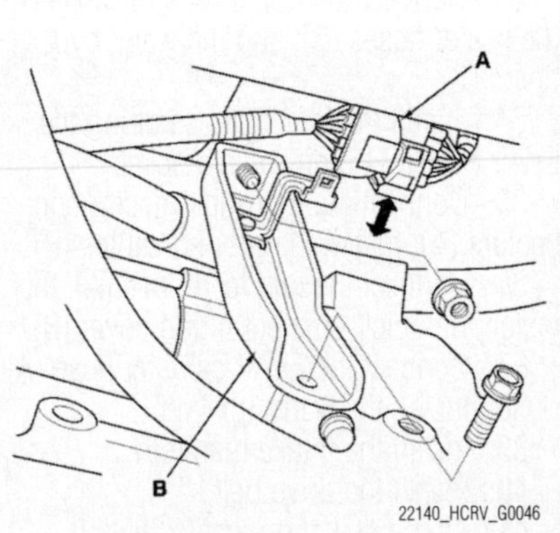

A
B
22140_HCRV_G0046

Fig. 128 Remove the connector (A), then remove the intake manifold bracket (B)

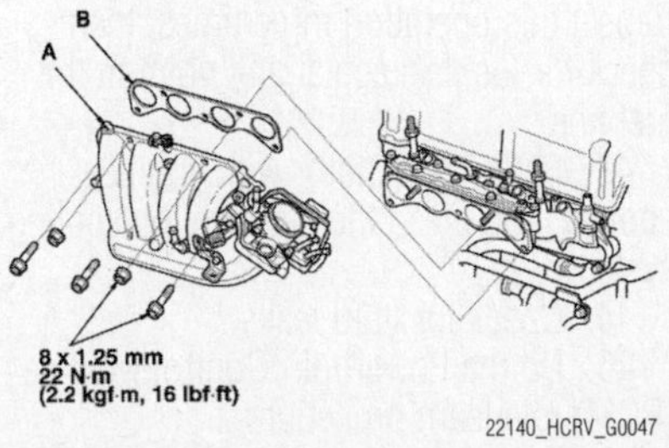

B
A
8 x 1.25 mm
22 N·m
(2.2 kgf·m, 16 lbf·ft)
22140_HCRV_G0047

Fig. 129 Install the intake manifold (A) with a new gasket (B)

25. Connect the engine wire harness connectors and install the wire harness clamps to the intake manifold including the:

- MAP sensor connector
- Throttle actuator connector

26. Install the intake air duct, then install the vacuum hose and the breather pipe.
27. Install the intake manifold cover and tighten the nuts to 106 inch lbs. (12 Nm).
28. Install the front bulkhead, then install the radiator upper brackets. Tighten the bolts to 86 inch lbs. (10 Nm).
29. Apply body paint to the bulkhead mounting bolts.
30. Connect the fan motor connectors and the hood switch connector, then install the harness clips.
31. Install the bulkhead cover.
32. Clean up any spilled engine coolant.
33. After installation, check that all tubes, hoses, and connectors are installed correctly.
34. Refill the radiator with the proper type and amount of engine coolant and bleed the air from the cooling system with the heater valve open.

K24Z6 Engine

See Figures 132 through 140.

1. Remove the hood support rod, then use it to prop the hood in the wide-open position.
2. Remove the bulkhead cover.
3. Disconnect the fan motor connectors and the hood switch connector, then remove the harness clips.
4. Remove the radiator upper brackets, then remove the front bulkhead.
5. Remove the engine cover.
6. Disconnect the breather pipe, then remove the air flow tube.
7. Disconnect the Evaporative Emission (EVAP) canister hose and the brake booster vacuum hose.
8. Disconnect the water bypass hoses, then plug the water bypass hoses.
9. Disconnect the Manifold Absolute Pressure (MAP) sensor connector and the throttle actuator connector, then remove the wire harness clamps.
10. Remove the intake manifold bracket.
11. Remove the intake manifold, then disconnect the Positive Crankcase Ventilation (PCV) hose from the intake manifold.

To install:

12. Connect the Positive Crankcase Ventilation (PCV) hose to the intake

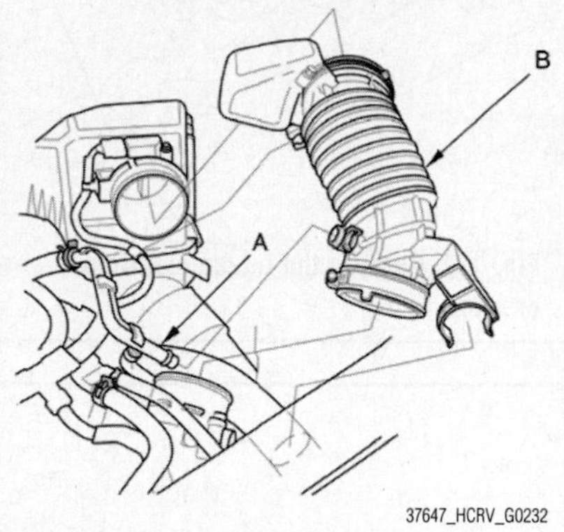
B
A
37647_HCRV_G0232

Fig. 134 Disconnect the breather pipe (A), then remove the air flow tube (B)

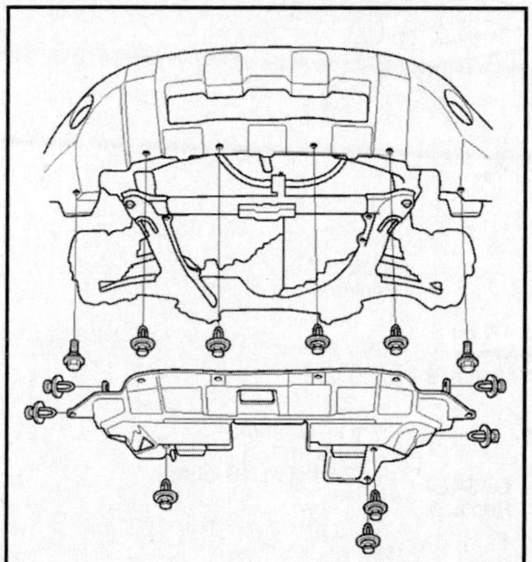
22140_HCRV_G0048

Fig. 130 Install the splash shield

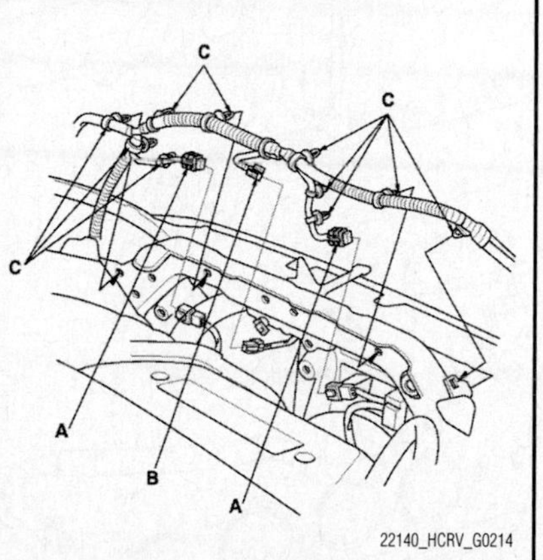
C
C
C
A
B
A
22140_HCRV_G0214

Fig. 132 Disconnect the fan motor connectors (A) and the hood switch connector (B), then remove the harness clips (C)

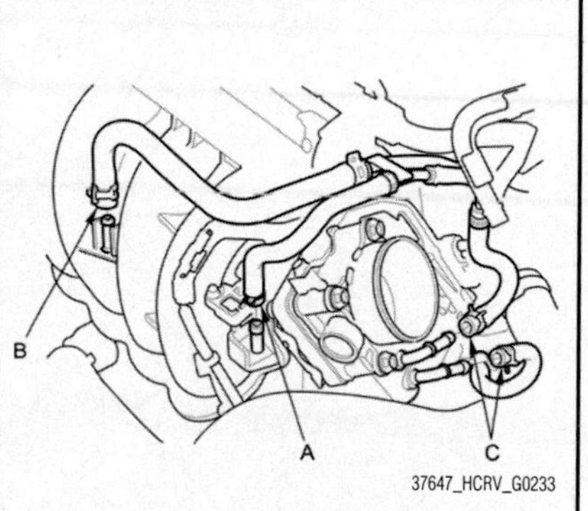
B
A
C
37647_HCRV_G0233

Fig. 135 Disconnect the EVAP canister hose (A) and the brake booster vacuum hose (B); disconnect the water bypass hoses (C)

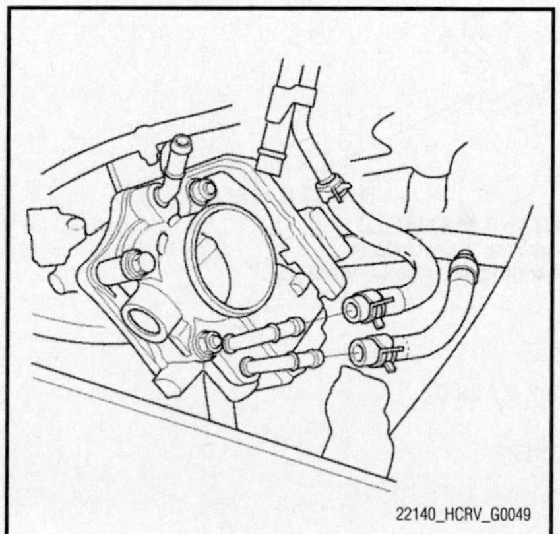
22140_HCRV_G0049

Fig. 131 Install the water bypass hoses

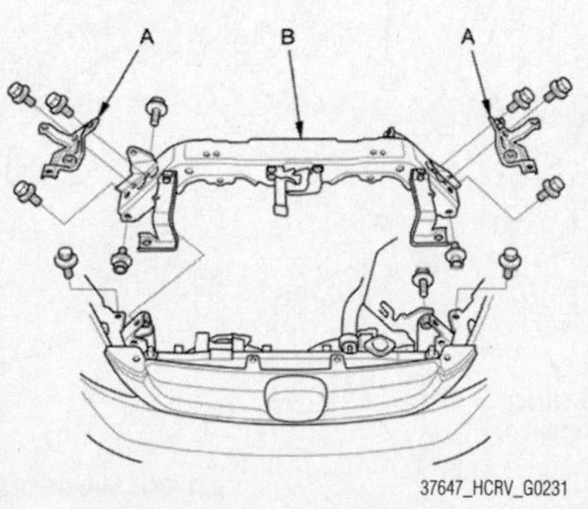
A
B
A
37647_HCRV_G0231

Fig. 133 Remove the radiator upper brackets (A), then remove the front bulkhead (B)

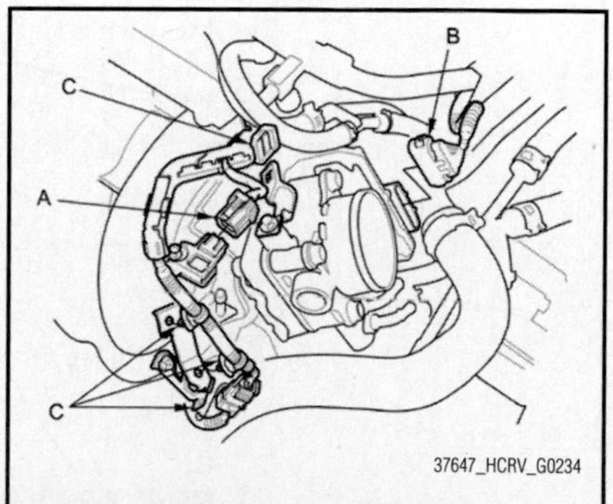
B
C
A
C
37647_HCRV_G0234

Fig. 136 Disconnect the Manifold Absolute Pressure (MAP) sensor connector (A) and the throttle actuator connector (B), then remove the wire harness clamps (C)

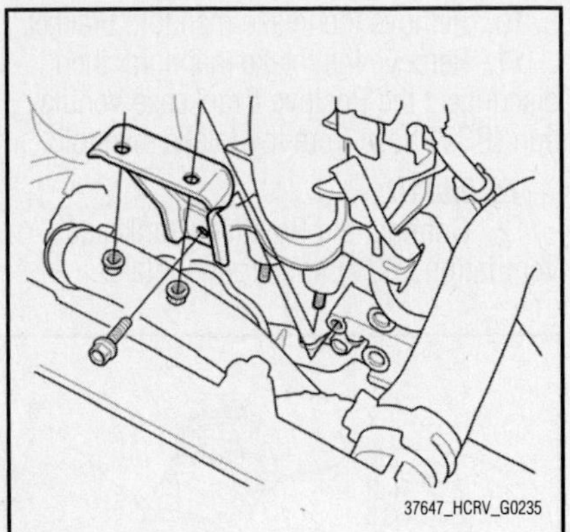

Fig. 137 Remove the intake manifold bracket

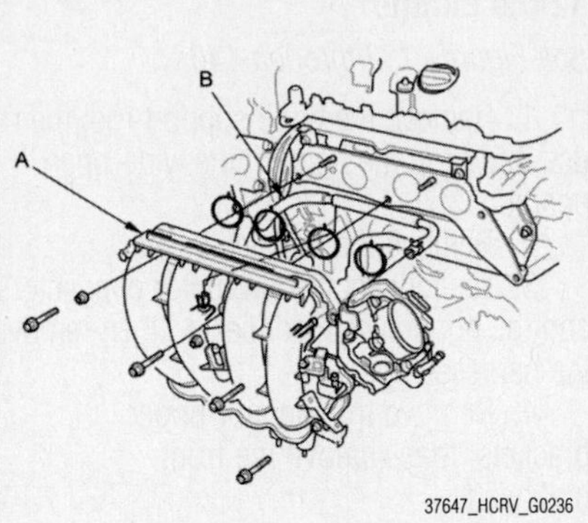
A
B

Fig. 138 Remove the intake manifold (A), then disconnect the PCV hose (B) from the intake manifold

manifold, then install the intake manifold with new gaskets, and tighten the bolts and nuts in a crisscross pattern in three steps, beginning with the inner bolt.

13. Install the intake manifold bracket.
14. Connect the Manifold Absolute Pressure (MAP) sensor connector and the throttle actuator connector, then install the wire harness clamps.
15. Connect the water bypass hoses.
16. Connect the Evaporative Emission (EVAP) canister hose and the brake booster vacuum hose.
17. Install the air flow tube, then connect the breather pipe.

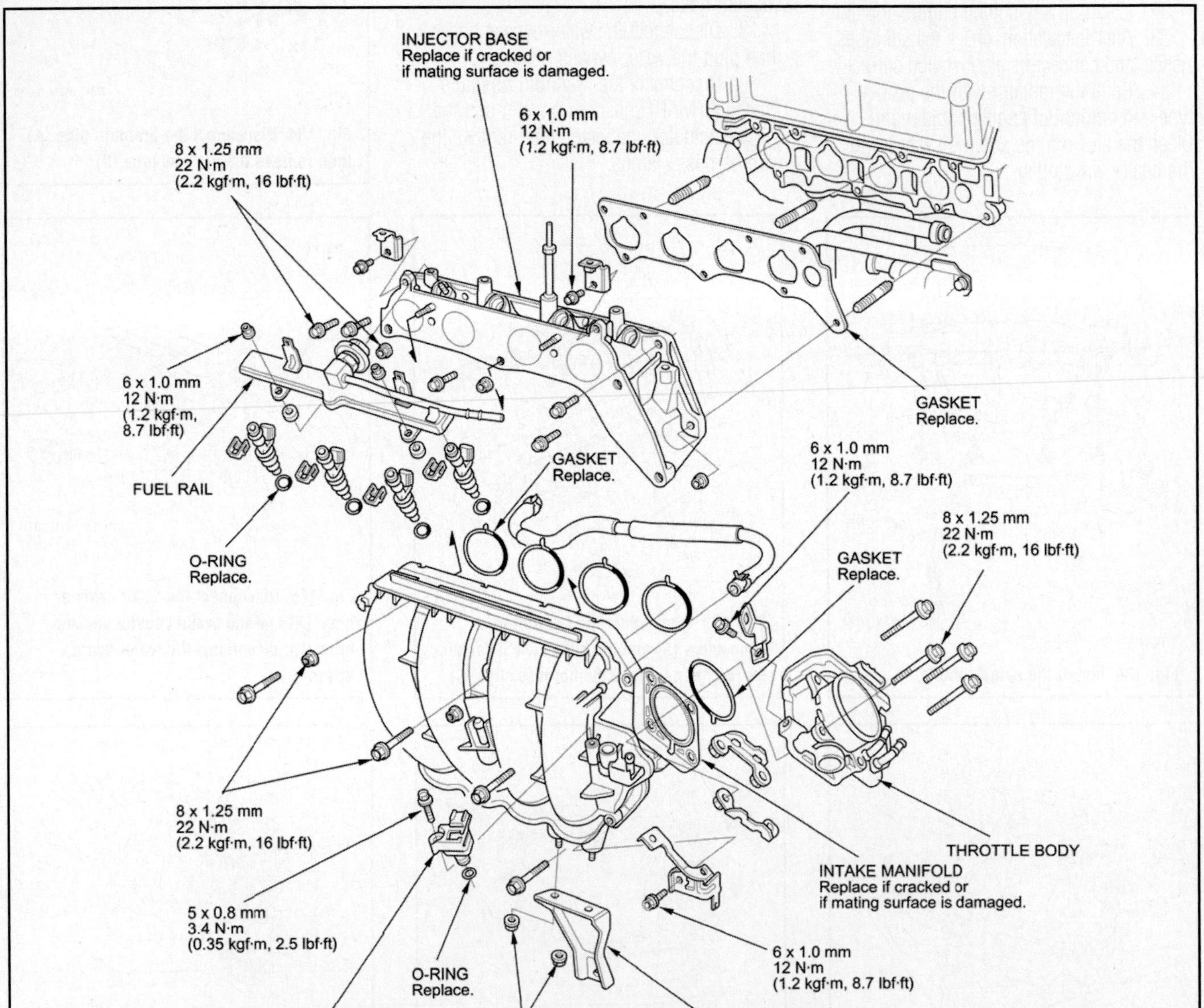
INJECTOR BASE
Replace if cracked or if mating surface is damaged.
6 x 1.0 mm 12 N·m (1.2 kgf·m, 8.7 lbf·ft)
8 x 1.25 mm 22 N·m (2.2 kgf·m, 16 lbf·ft)
6 x 1.0 mm 12 N·m (1.2 kgf·m, 8.7 lbf·ft)
FUEL RAIL
GASKET Replace.
GASKET Replace.
O-RING Replace.
6 x 1.0 mm 12 N·m (1.2 kgf·m, 8.7 lbf·ft)
8 x 1.25 mm 22 N·m (2.2 kgf·m, 16 lbf·ft)
GASKET Replace.
8 x 1.25 mm 22 N·m (2.2 kgf·m, 16 lbf·ft)
THROTTLE BODY
INTAKE MANIFOLD
Replace if cracked or if mating surface is damaged.
5 x 0.8 mm 3.4 N·m (0.35 kgf·m, 2.5 lbf·ft)
6 x 1.0 mm 12 N·m (1.2 kgf·m, 8.7 lbf·ft)
O-RING Replace.
MANIFOLD ABSOLUTE PRESSURE (MAP) SENSOR
8 x 1.25 mm 22 N·m (2.2 kgf·m, 16 lbf·ft)
INTAKE MANIFOLD BRACKET

Fig. 139 Exploded view of the intake manifold assembly

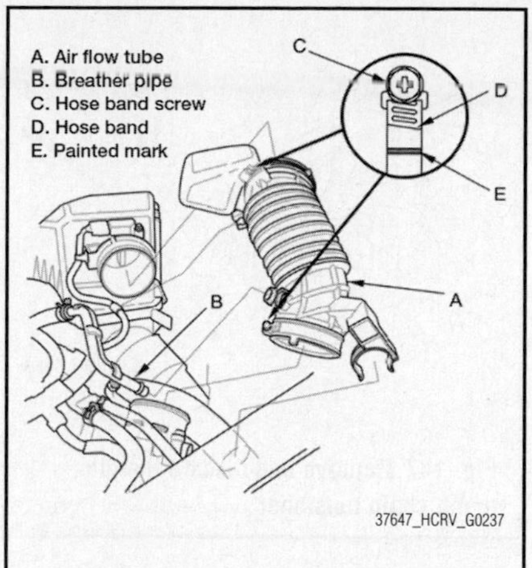
A. Air flow tube
B. Breather pipe
C. Hose band screw
D. Hose band
E. Painted mark
C
D
E
B
A
37647_HCRV_G0237

Fig. 140 Install the air flow tube, then connect the breather pipe

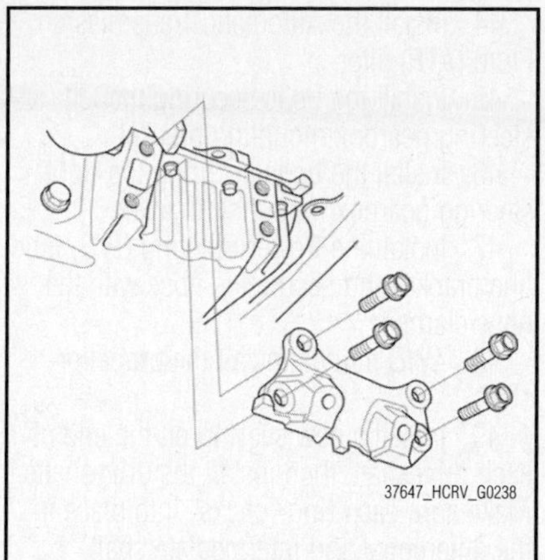
37647_HCRV_G0238

Fig. 141 Remove the lower torque rod bracket

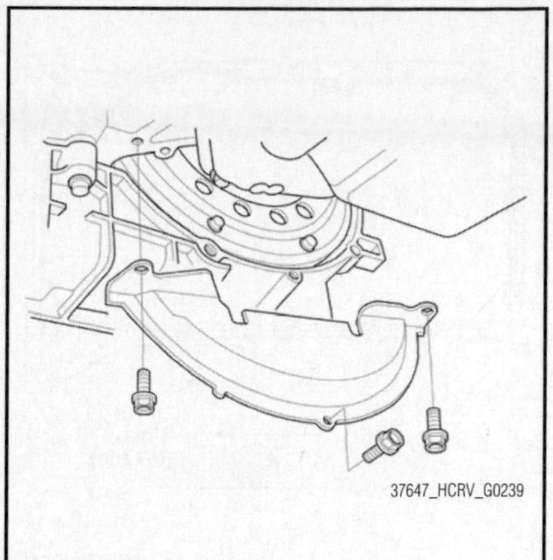
37647_HCRV_G0239

Fig. 142 Remove the torque converter cover

➡When tightening the screw of the hose band, align the edge of the hose band with the mark painted on the hose band. If you tighten the screw over the mark, replace the hose band.

18. Install the engine cover.
19. Install the front bulkhead, then install the radiator upper brackets.
20. Apply body paint to the bulkhead mounting bolts.
21. Connect the fan motor connectors and the hood switch connector, then install the harness clips.
22. Install the bulkhead cover.
23. Clean up any spilled engine coolant.
24. After installation, check that all tubes, hoses, and connectors are installed correctly.
25. Refill the radiator with engine coolant, and bleed the air from the cooling system.

OIL PAN

REMOVAL & INSTALLATION

See Figures 141 through 145.

1. If the engine is already out of the vehicle, go to step 21. If the engine is still mounted in the vehicle, follow steps 1–20.
2. Raise the vehicle on the lift.
3. Drain the engine oil.
4. Remove the front wheels.
5. Remove the splash shield.
6. Disconnect the Air Fuel Ratio (A/F) sensor connector and secondary Heated Oxygen Sensor (secondary HO2S) connector, then remove the Three Way Catalytic Converter (TWC).
7. Remove the shift cable.
8. Separate the stabilizer links from the stabilizer bar.
9. Separate the knuckles from the lower arms.
10. 4WD model: Remove the propeller shaft.
11. Remove a bolt securing the P/S fluid line bracket, and unclamp the P/S fluid line clamps on the front subframe.
12. Remove the bolts securing the left steering gearbox mounting bracket.
13. Remove the bolts securing the right steering gearbox mounting brackets.
14. Remove the bolt securing the Automatic Transmission Fluid (ATF) filter.
15. Remove the lower torque rod.
16. Make the appropriate reference lines at both ends of the subframe that line up with the body.
17. Remove the subframe mounting bolts on both side.
18. Attach the subframe adapter to the subframe and hang the belt of the subframe adapter over the front of the subframe, then secure the belt with its stop.
19. Raise the jack and line up the slots in the arms with the bolt holes on the corner of the jack base, then attach them with bolts securely.
20. Remove the subframe.
21. Remove the lower torque rod bracket.
22. Remove the torque converter cover.
23. Remove the bolts securing the oil pan.
24. Using a flat blade screwdriver, separate the oil pan from the block in the places shown.
25. Remove the oil pan.

To install:

26. Remove all of the old liquid gasket from the oil pan mating surfaces, bolts, and bolt holes.
27. Clean and dry the oil pan mating surfaces.

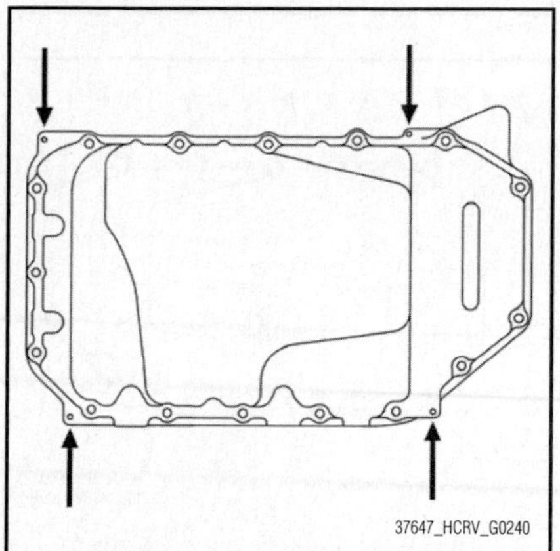
37647_HCRV_G0240

Fig. 143 Using a flat blade screwdriver, separate the oil pan from the block in the places shown

28. Apply liquid gasket (P/N 08717-0004, 08718-0003, or 08718-0009) to the engine block mating surface of the oil pan and to the inside edge of the bolt holes. Install the component within 5 minutes of applying the liquid gasket.

➡Apply a bead of liquid gasket along the broken line. If too much time has passed after applying the liquid gasket, remove the old liquid gasket and residue, then reapply new liquid gasket.

29. Install the oil pan.
30. Tighten the bolts in three steps. In the final step, tighten all bolts, in sequence, to 9 ft. lbs. (12 Nm). Wipe off the excess liquid gasket on the each side of crankshaft pulley and drive plate.

➡Wait at least 30 minutes to allow the liquid gasket to cure before filling the engine with oil. Do not run the engine for at least 3 hours after installing the oil pan.

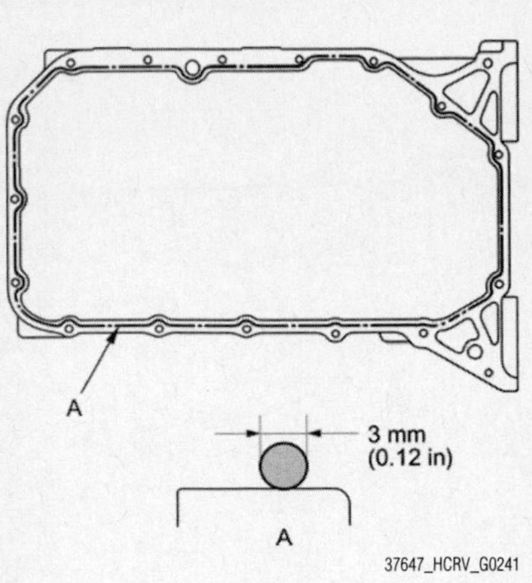
A
3 mm
(0.12 in)
A
37647_HCRV_G0241

Fig. 144 Apply a bead of liquid gasket along the broken line (A)

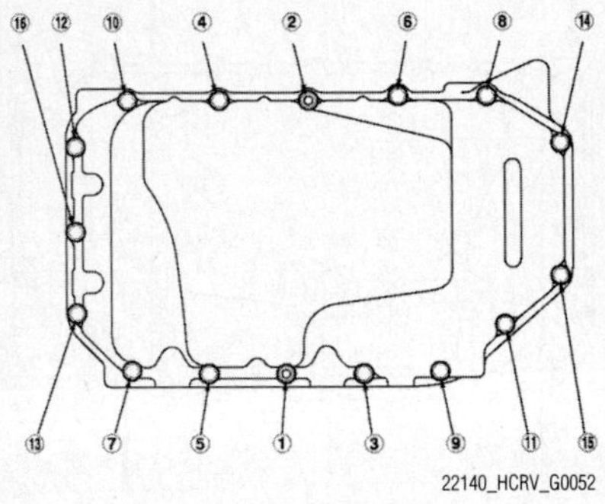
16 12 10 4 2 6 8 14
13 7 5 1 3 9 11 15
22140_HCRV_G0052

Fig. 145 Oil pan bolt tightening sequence

31. Install the torque converter cover.
32. Install the lower torque rod bracket. Tighten the bolts to 60 ft. lbs. (81 Nm).
33. If the engine is still in the vehicle, do steps 9 through 31.
34. Using the subframe adapter and a jack, raise the subframe up to body.
35. Loosely install the new 14 x 1.5 mm bolts.
36. Align all reference marks on the front subframe with the body, then tighten the bolts on the front subframe to 76 ft. lbs. (103 Nm).
37. Tighten the new subframe mounting bolts on both sides to 76 ft. lbs. (103 Nm).
38. Lower the vehicle on the lift.
39. Loosen the upper torque rod mounting bolt.
40. Raise the vehicle on the lift.
41. Install the lower torque rod, then tighten the new lower torque rod mounting bolts in the numbered sequence shown.
42. Lower the vehicle on the lift.
43. Tighten the upper torque rod mounting bolt.
44. Install the Automatic Transmission Fluid (ATF) filter.
45. Install the bolts securing the left steering gearbox mounting bracket.
46. Install the bolts securing the right steering gearbox mounting bracket.
47. Install the Power Steering (P/S) fluid line bracket and secure the hose with the hose clamps.
48. 4WD model: Install the propeller shaft.
49. Install a new set ring on the end of each driveshaft, then install the driveshafts. Make sure each ring "clicks" into place in the differential and intermediate shaft.
50. Connect the lower arms to the knuckles.
51. Connect the stabilizer links to the stabilizer bar.
52. Install the shift cable.
53. Install the Three Way Catalytic Converter (TWC). Use new gaskets and new self-locking nuts. Connect the Air Fuel Ratio (A/F) sensor connector and secondary Heated Oxygen Sensor (secondary HO2S) connector.
54. Install the splash shield.
55. Install the front wheels.
56. Check the wheel alignment.

OIL PUMP

REMOVAL & INSTALLATION

K24Z1 Engine

See Figures 146 through 151.

1. Before servicing the vehicle, refer to the Precautions Section.
2. Turn the crankshaft pulley so the Top Dead Center (TDC) mark lines up with the pointer.

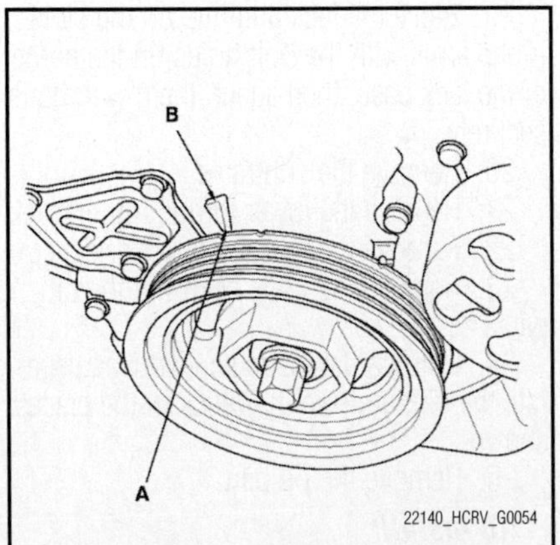
B
A
22140_HCRV_G0054

Fig. 146 Turn the crankshaft pulley so the Top Dead Center (TDC) mark (A) lines up with the pointer (B)

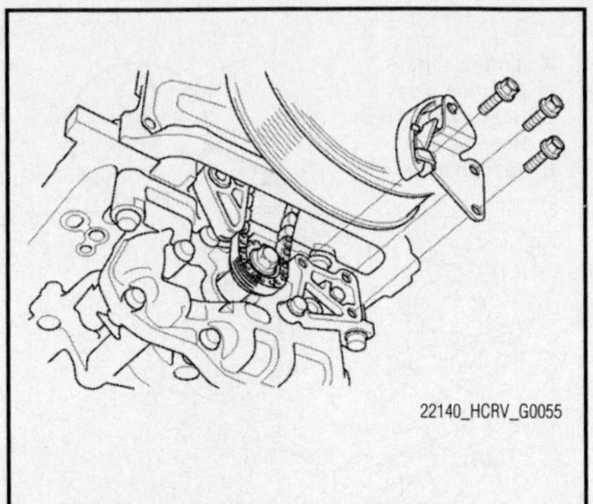
22140_HCRV_G0055

Fig. 147 Remove and discard the oil pump chain tensioner

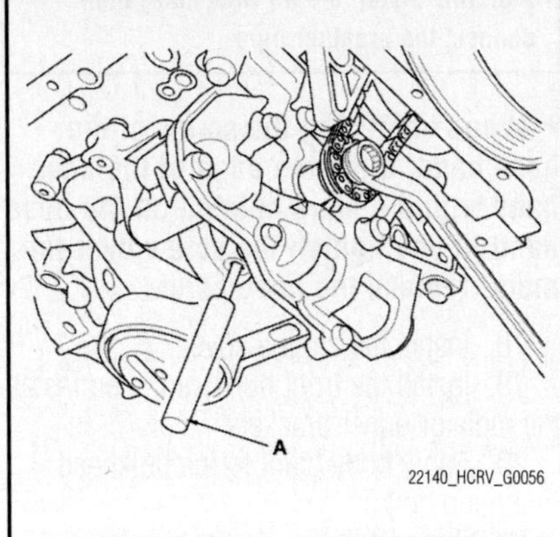
A
22140_HCRV_G0056

Fig. 148 Hold the rear balancer shaft with a 6mm pin driver (A) inserted into the maintenance hole in the lower balancer shaft holder and through the rear balancer shaft

3. Remove the oil pan. Refer to Oil Pan, removal & installation.
4. Remove and discard the oil pump chain tensioner.
5. To hold the rear balancer shaft, insert a 6mm pin driver into the maintenance hole in the lower balancer shaft holder and through the rear balancer shaft.
6. Loosen the oil pump sprocket mounting bolt.
7. Remove the oil pump sprocket, then remove the oil pump.

To install:

8. Make sure the No. 1 piston TDC mark lines up with the pointer.
9. Align the dowel pin on the rear balancer shaft with the mark on the oil pump.
10. To hold the rear balancer shaft, insert a 6mm pin driver into the maintenance hole in the lower balancer shaft holder and through the rear balancer shaft.
11. Apply engine oil to the threads of the oil pump mounting bolts.
12. Loosely install the oil pump, then install the oil pump sprocket.

13. Remove the 6mm pin driver.
14. Tighten the oil pump mounting bolts:
a. Bolts 10 x 1.25mm: 33 ft. lbs. (44 Nm).
b. Bolts 8 x 1.25mm: 16 ft. lbs. (22 Nm).
15. Squeeze the new oil pump chain tensioner, then install the set clip.

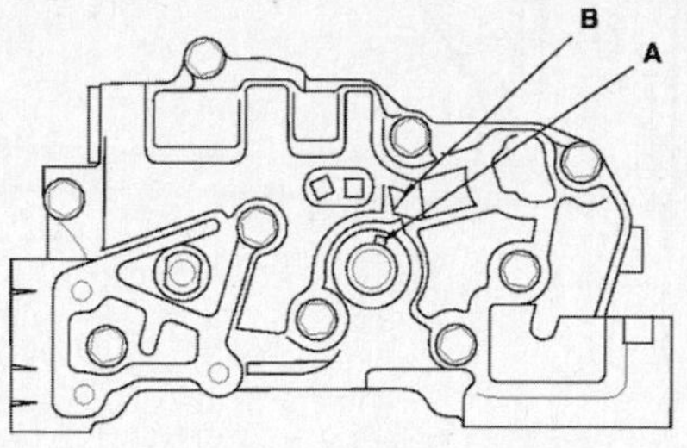
B
A
22140_HCRV_G0057

Fig. 149 Align the dowel pin (A) on the rear balancer shaft with the mark (B) on the oil pump

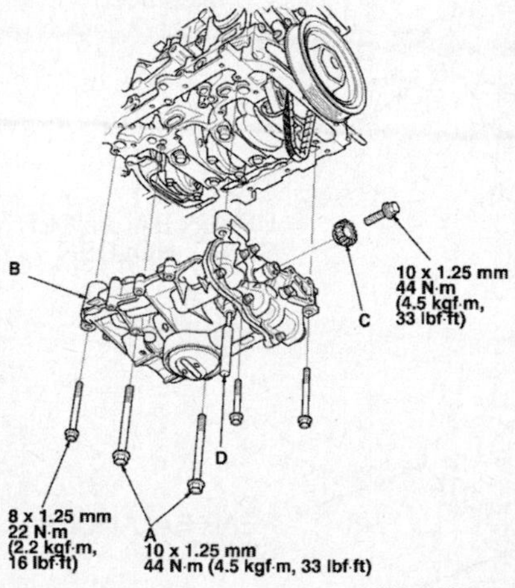
B
C
10 x 1.25 mm
44 N·m
(4.5 kgf·m,
33 lbf·ft)
D
8 x 1.25 mm
22 N·m
(2.2 kgf·m,
16 lbf·ft)
A
10 x 1.25 mm
44 N·m (4.5 kgf·m, 33 lbf·ft)
22140_HCRV_G0058

Fig. 150 Exploded view of the oil pump and related components

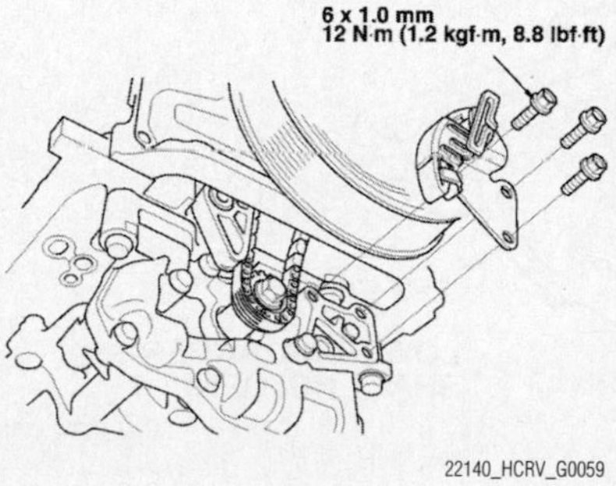
6 x 1.0 mm
12 N·m (1.2 kgf·m, 8.8 lbf·ft)
22140_HCRV_G0059

Fig. 151 Install the new oil pump chain tensioner and remove the set clip

➡The set clip is supplied with the oil pump chain tensioner.

16. Install the new oil pump chain tensioner. Tighten to 106 inch lbs. (12 Nm).
17. Remove the set clip from the oil pump chain tensioner.
18. Install the oil pan. Refer to Oil Pan, removal & installation.
19. Run the engine and check for oil leakage.

K24Z6 Engine

See Figures 152 through 160.

1. Turn the crankshaft pulley so its Top Dead Center (TDC) mark lines up with the pointer.
2. Remove the oil pan.
3. To hold the rear balancer shaft, insert a 6 mm long pin punch (Snap-On® PPC108LA or equivalent) into the maintenance hole in the balancer shaft holder and through the rear balancer shaft.
4. Turn the crankshaft counterclockwise to compress the oil pump chain auto-tensioner.
5. Align the holes on the lock and the oil pump chain auto-tensioner, then insert a 0.12 inches (3.0 mm) diameter pin into the holes. Turn the crankshaft clockwise to secure the pin.

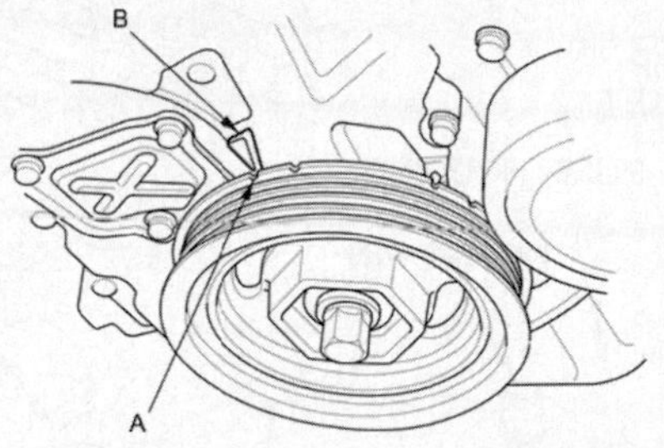
B
A
37647_HCRV_G0243

Fig. 152 Turn the crankshaft pulley so its Top Dead Center (TDC) mark (A) lines up with the pointer (B)

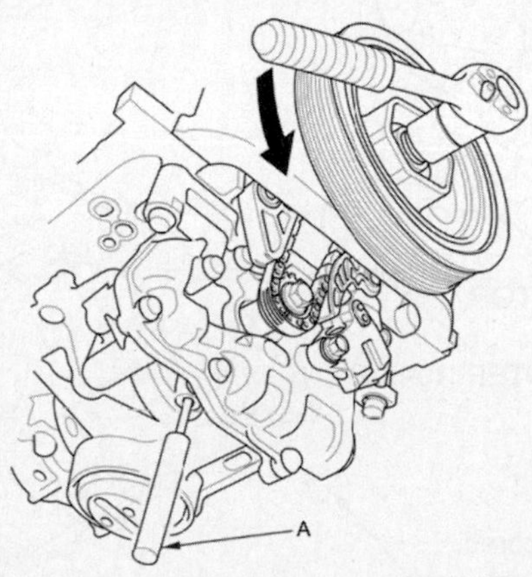
A
37647_HCRV_G0244

Fig. 153 To hold the rear balancer shaft, insert a 6 mm long pin punch (A) into the maintenance hole in the balancer shaft holder and through the rear balancer shaft

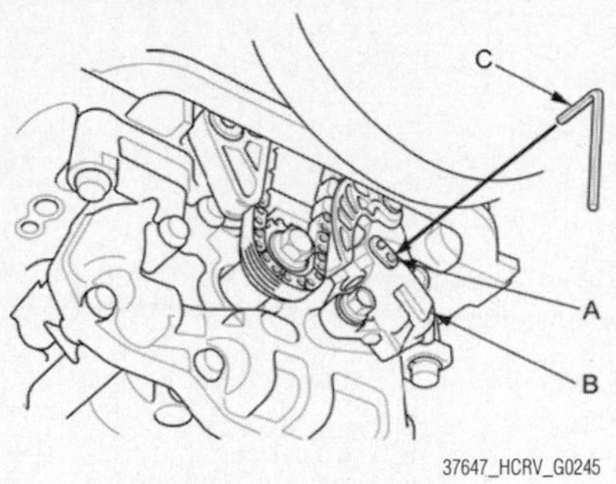
C
A
B
37647_HCRV_G0245

Fig. 154 Align the holes on the lock (A) and the oil pump chain auto-tensioner (B), then insert a 0.12 inches (3.0 mm) diameter pin (C) into the holes

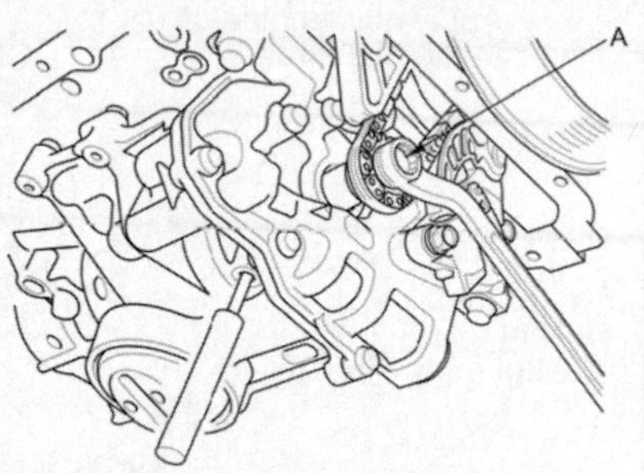
A
37647_HCRV_G0246

Fig. 155 Loosen the oil pump sprocket mounting bolt (A)

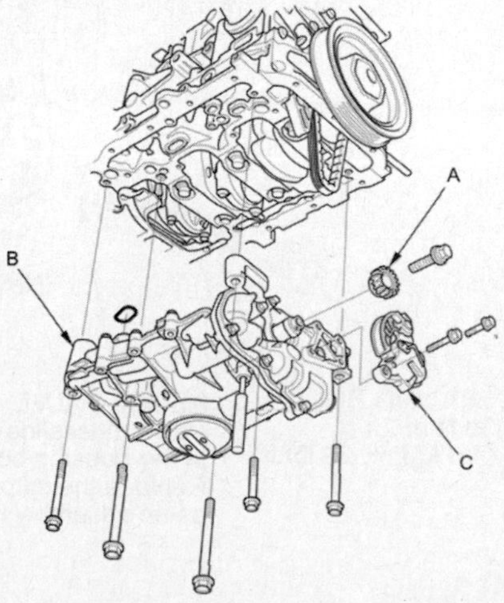
A
B
C
37647_HCRV_G0247

Fig. 156 Remove the oil pump sprocket (A) and the oil pump (B), then remove the oil pump chain auto-tensioner (C)

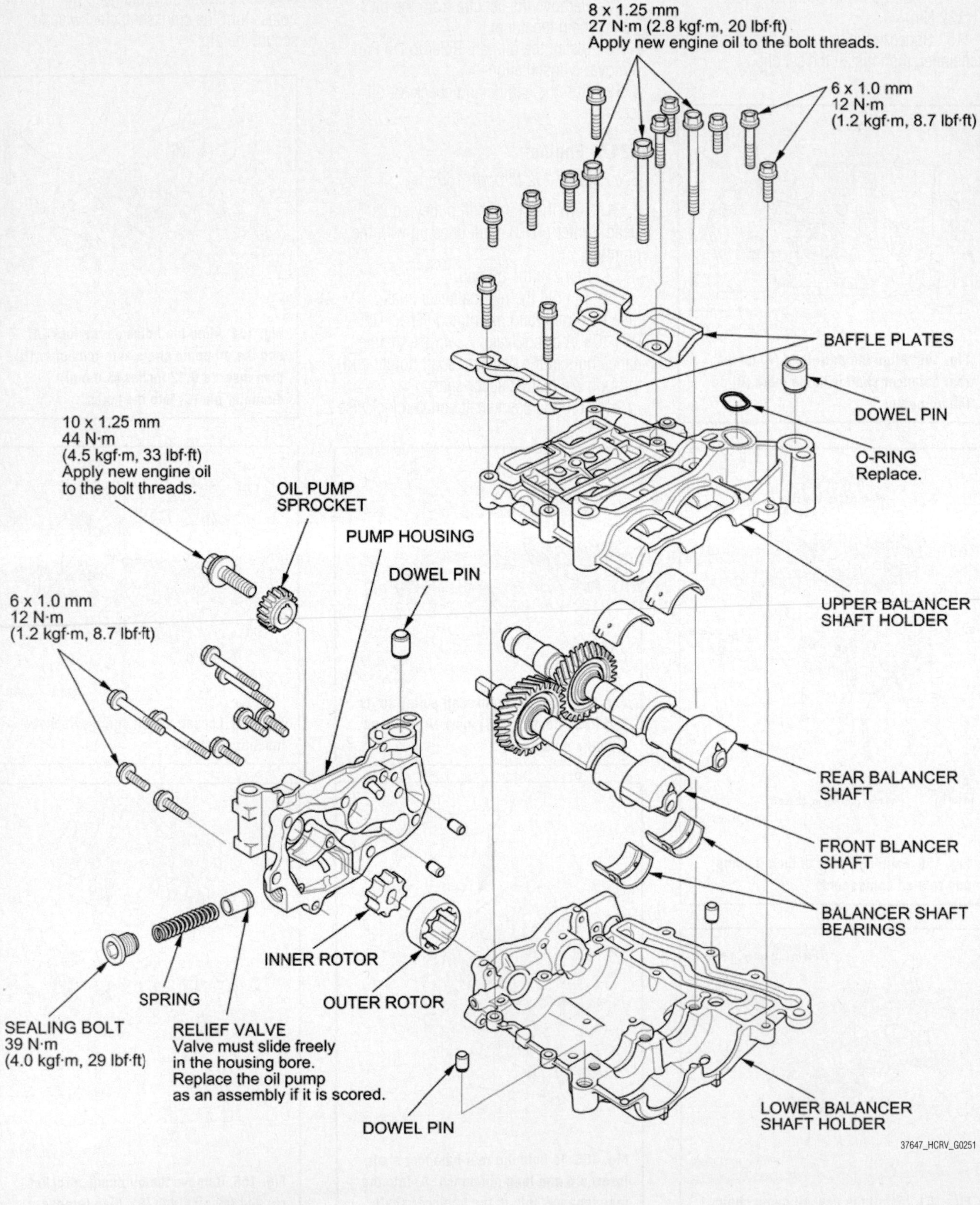
8 x 1.25 mm
27 N·m (2.8 kgf·m, 20 lbf·ft)
Apply new engine oil to the bolt threads.
6 x 1.0 mm
12 N·m
(1.2 kgf·m, 8.7 lbf·ft)
BAFFLE PLATES
DOWEL PIN
O-RING
Replace.
10 x 1.25 mm
44 N·m
(4.5 kgf·m, 33 lbf·ft)
Apply new engine oil
to the bolt threads.
OIL PUMP
SPROCKET
PUMP HOUSING
DOWEL PIN
UPPER BALANCER
SHAFT HOLDER
6 x 1.0 mm
12 N·m
(1.2 kgf·m, 8.7 lbf·ft)
REAR BALANCER
SHAFT
FRONT BLANCER
SHAFT
BALANCER SHAFT
BEARINGS
INNER ROTOR
SPRING
OUTER ROTOR
SEALING BOLT
39 N·m
(4.0 kgf·m, 29 lbf·ft)
RELIEF VALVE
Valve must slide freely
in the housing bore.
Replace the oil pump
as an assembly if it is scored.
DOWEL PIN
LOWER BALANCER
SHAFT HOLDER
37647_HCRV_G0251

Fig. 157 Exploded view of oil pump assembly

➡Check the oil pump chain auto-tensioner cam position. If the position is not aligned, set the first cam to the first edge of the rack.

6. Loosen the oil pump sprocket mounting bolt.
7. Remove the oil pump sprocket and the oil pump, then remove the oil pump chain auto-tensioner.

To install:

8. Make sure the No. 1 piston Top Dead Center (TDC) mark lines up with the pointer.
9. Align the dowel pin on the rear balancer shaft with the mark on the oil pump.
10. To hold the rear balancer shaft, insert a 6 mm long pin punch (Snap-On® PPC108LA or equivalent) into the maintenance hole in the balancer shaft holder and through the rear balancer shaft.
11. Turn the plate counterclockwise, to release the lock, then push the oil pump chain auto-tensioner arm, and set the first cam to the first edge of the rack. Insert a 0.12 inches (3.0 mm) diameter pin into the hole.

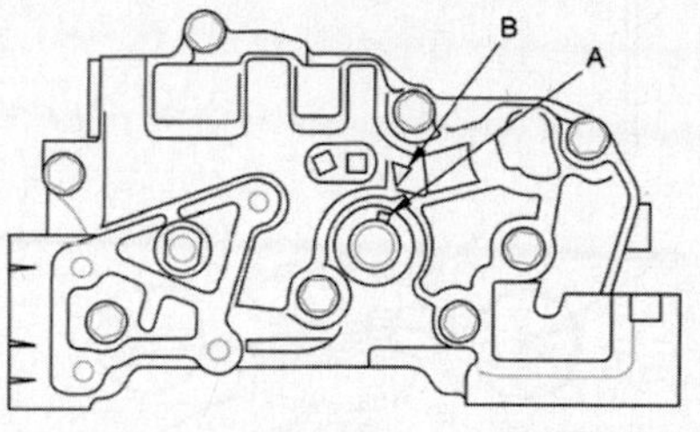

B
A
37647_HCRV_G0248

Fig. 158 Align the dowel pin (A) on the rear balancer shaft with the mark (B) on the oil pump

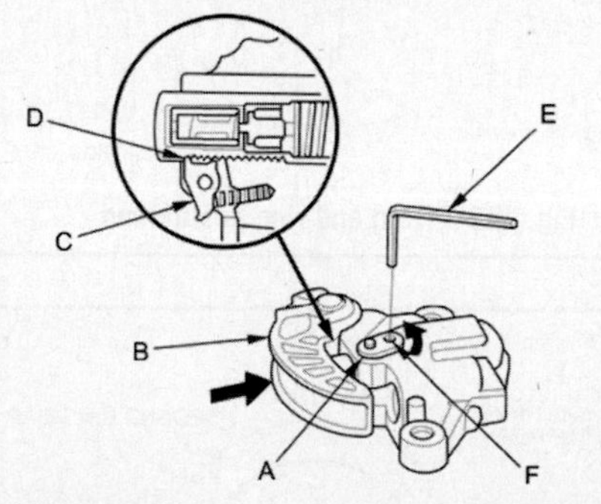

D
E
C
B
A
F
A. Plate
B. Oil pump chain auto-tensioner arm
C. First cam
D. Rack
E. Pin
F. Hole
37647_HCRV_G0249

Fig. 159 Turn the plate counterclockwise, to release the lock, then push the oil pump chain auto-tensioner arm, and set the first cam to the first edge of the rack; insert a 0.12 inches (3.0 mm) diameter pin into the hole

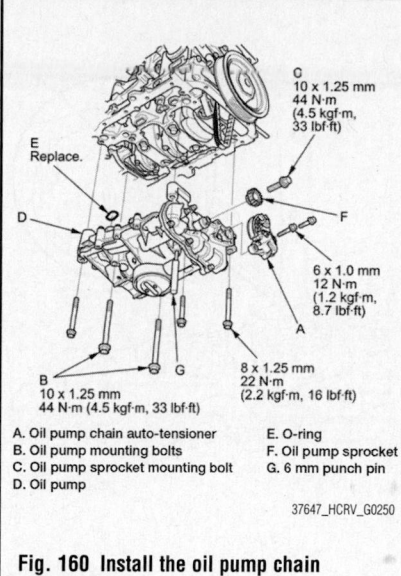

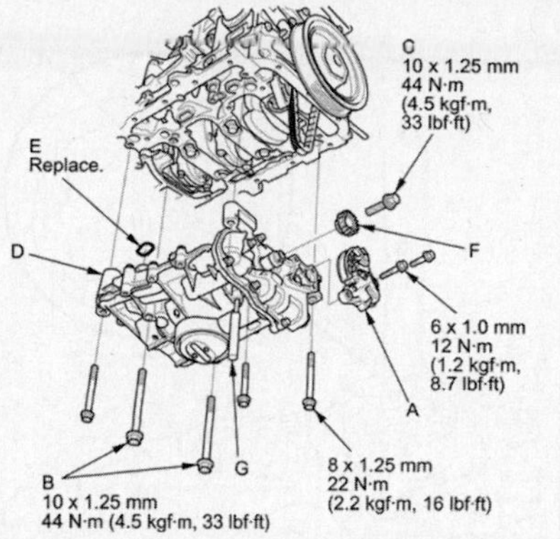

C
10 x 1.25 mm
44 N·m
(4.5 kgf·m,
33 lbf·ft)
E
Replace.
D
F
6 x 1.0 mm
12 N·m
(1.2 kgf·m,
8.7 lbf·ft)
A
B
10 x 1.25 mm
44 N·m (4.5 kgf·m, 33 lbf·ft)
G
8 x 1.25 mm
22 N·m
(2.2 kgf·m, 16 lbf·ft)
A. Oil pump chain auto-tensioner
B. Oil pump mounting bolts
C. Oil pump sprocket mounting bolt
D. Oil pump
E. O-ring
F. Oil pump sprocket
G. 6 mm punch pin
37647_HCRV_G0250

Fig. 160 Install the oil pump chain auto-tensioner

➡If the chain tensioner is not set up as described, the tensioner will become damaged.

12. Install the oil pump chain auto-tensioner.
13. Apply new engine oil to the threads of the oil pump mounting bolts and the oil pump sprocket mounting bolt, then Loosely install the oil pump with a new O-ring, then install the oil pump sprocket.
14. Tighten the oil pump mounting bolts and the oil pump sprocket mounting bolt.
15. Remove the 6 mm pin punch.
16. Remove the 0.12 inches (3.0 mm) diameter pin from the oil pump chain auto-tensioner.
17. Install the oil pan.

INSPECTION

K24Z1 Engine

See Figures 161 through 164.

1. Before servicing the vehicle, refer to the Precautions Section.
2. Remove the oil pump housing.
3. Check the inner-to-outer rotor radial clearance between the inner rotor and the outer rotor.
    a. If the inner-to-outer rotor radial clearance exceeds the service limit, replace the oil pump.
    b. Inner rotor-to-outer rotor radial clearance standard (new): 0.002–0.006 inches (0.06–0.16mm).
    c. Service Limit: 0.008 inch (0.20mm).

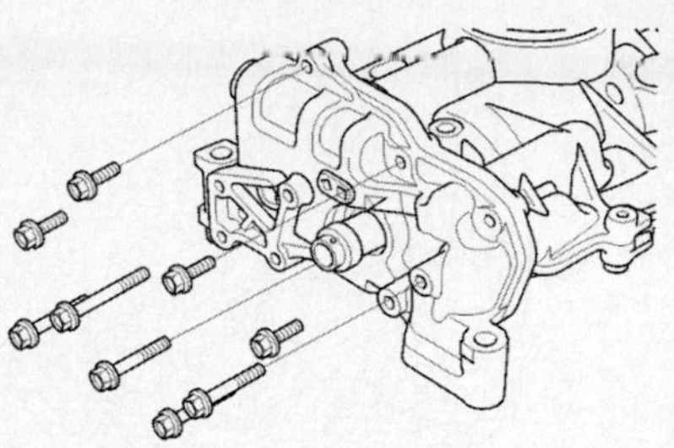

22140_HCRV_G0228

Fig. 161 Remove the oil pump housing

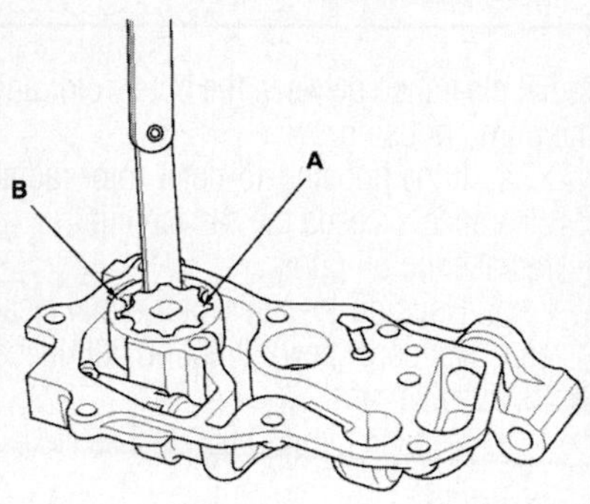

B
A
22140_HCRV_G0229

Fig. 162 Check the inner-to-outer rotor radial clearance between the inner rotor (A) and the outer rotor (B)

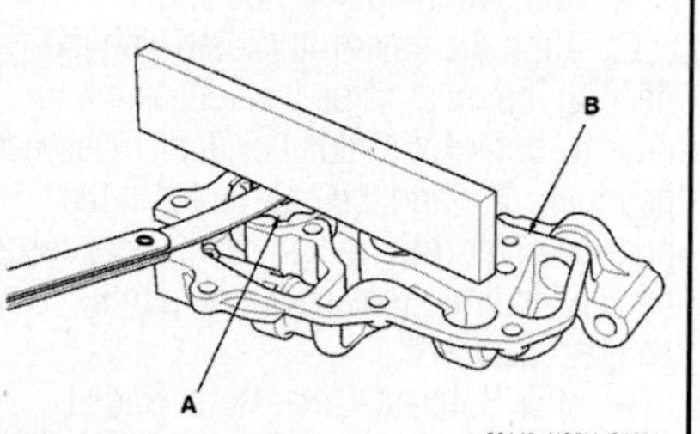

B
A
22140_HCRV_G0230

Fig. 163 Check the housing-to-rotor axial clearance between the rotor (A) and the pump housing (B)

4. Check the housing-to-rotor axial clearance between the rotor and the pump housing.
    a. If the housing-to-rotor axial clearance exceeds the service limit, replace the oil pump.
    b. Housing-to-rotor axial clearance Standard (New): 0.0014–0.0028 inches (0.035–0.070mm).
    c. Service Limit: 0.005 inches (0.12mm).
5. Check the housing-to-outer rotor

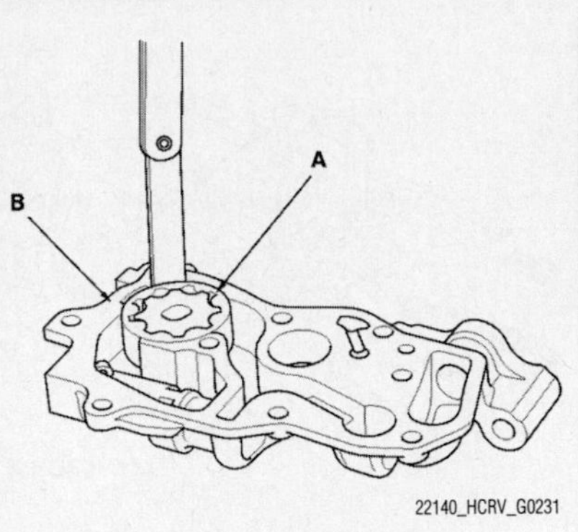
A
B
22140_HCRV_G0231
Fig. 164 Check the housing-to-outer rotor radial clearance between the outer rotor (A) and the pump housing (B)

radial clearance between the outer rotor and the pump housing.

a. If the housing-to-outer rotor radial clearance exceeds the service limit, replace the oil pump.

b. Housing-to-outer rotor radial clearance standard (new): 0.006–0.008 inches (0.15–0.21mm).

c. Service Limit: 0.009 inches (0.23mm).

6. Inspect both rotors and the pump housing for scoring or other damage. Replace parts if necessary.

K24Z6 Engine

See Figures 165 through 167.

1. Remove the pump housing.
2. Align the inner rotor tooth with the mark on the outer rotor, then check the inner-to-outer rotor radial clearance between the inner rotor and the outer rotor. If the inner-to-outer rotor radial clearance exceeds the service limit, replace the oil pump.

Specifications:

- Inner Rotor-to-Outer Rotor Radial Clearance
- Standard (New): 0.002–0.006 inches (0.05–0.15 mm)
- Service Limit: 0.007 inches (0.19 mm)

3. Check the pump housing-to-rotor axial clearance between the rotor and the pump housing. If the pump housing-to-rotor axial clearance exceeds the service limit, replace the oil pump.

Specifications:

- Pump Housing-to-Rotor Axial Clearance
- Standard (New): 0.0014–0.0028 inches (0.035–0.070 mm)
- Service Limit: 0.005 inches (0.12 mm)

4. Check the pump housing-to-outer rotor radial clearance between the outer rotor and the pump housing. If the pump housing-to-outer rotor radial clearance exceeds the service limit, replace the oil pump.

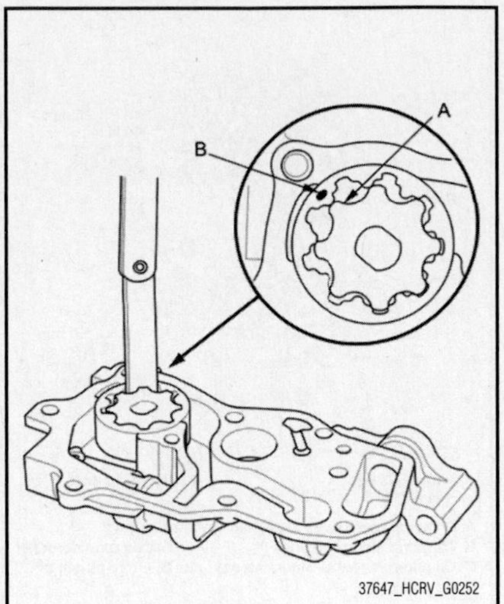
A
B
37647_HCRV_G0252
Fig. 165 Align the inner rotor tooth (A) with the mark (B) on the outer rotor, then check the inner-to-outer rotor radial clearance between the inner rotor and the outer rotor

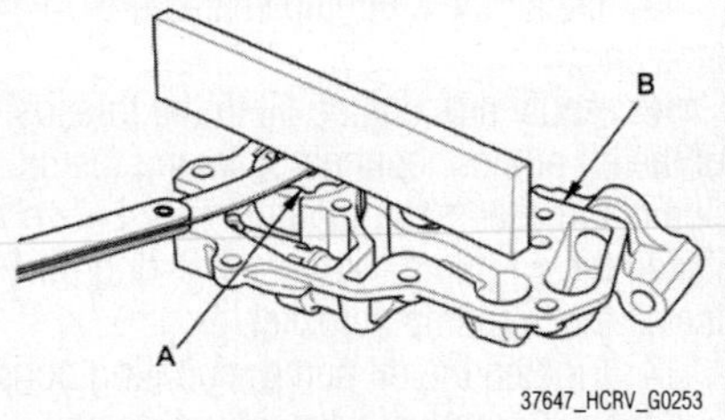
B
A
37647_HCRV_G0253
Fig. 166 Check the pump housing-to-rotor axial clearance between the rotor (A) and the pump housing (B)

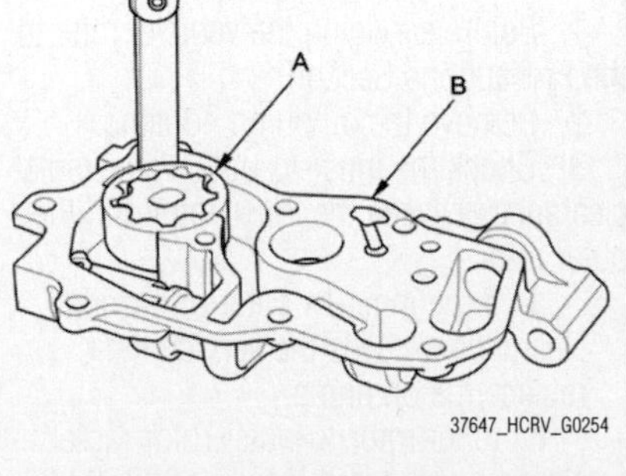
A
B
37647_HCRV_G0254
Fig. 167 Check the pump housing-to-outer rotor radial clearance between the outer rotor (A) and the pump housing (B)

Specifications:

- Pump Housing-to-Outer Rotor Radial Clearance
- Standard (New): 0.006–0.008 inches (0.15–0.21 mm)
- Service Limit: 0.009 inches (0.23 mm)

5. Inspect both rotors and the pump housing for scoring or other damage. Replace parts if necessary.

PISTON AND RING

POSITIONING

See Figures 168 and 169.

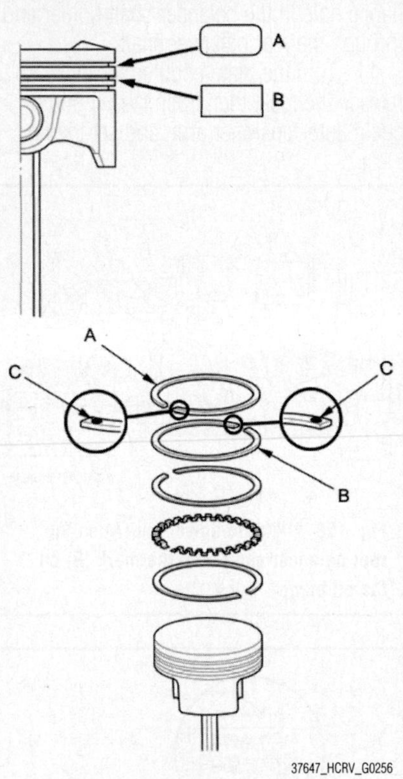
A
B
A
C
C
B
37647_HCRV_G0256
Fig. 168 Piston and ring positioning

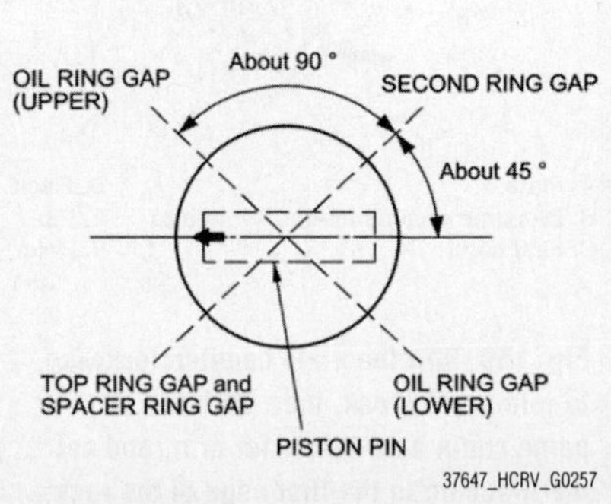
About 90 °
OIL RING GAP (UPPER)
SECOND RING GAP
About 45 °
TOP RING GAP and SPACER RING GAP
OIL RING GAP (LOWER)
PISTON PIN
37647_HCRV_G0257
Fig. 169 Piston ring end gap positioning

REAR MAIN SEAL

REMOVAL & INSTALLATION

See Figures 170 and 171.

Special Tools Required:

- Driver handle, 15 x 135L 07749-0010000
- Oil seal driver attachment 96mm 07ZAD-PNAA100

1. Before servicing the vehicle, refer to the Precautions Section.
2. Remove the transaxle.
3. Remove the drive plate.
4. Use a suitable tool to remove the old crankshaft oil seal.

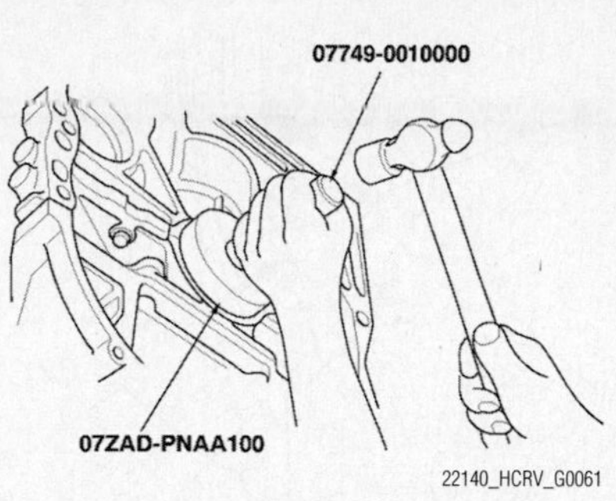
07749-0010000
07ZAD-PNAA100
22140_HCRV_G0061

Fig. 170 Using special tools 07749-0010000 and 07ZAD-PNAA100 to drive in the new crankshaft rear main seal

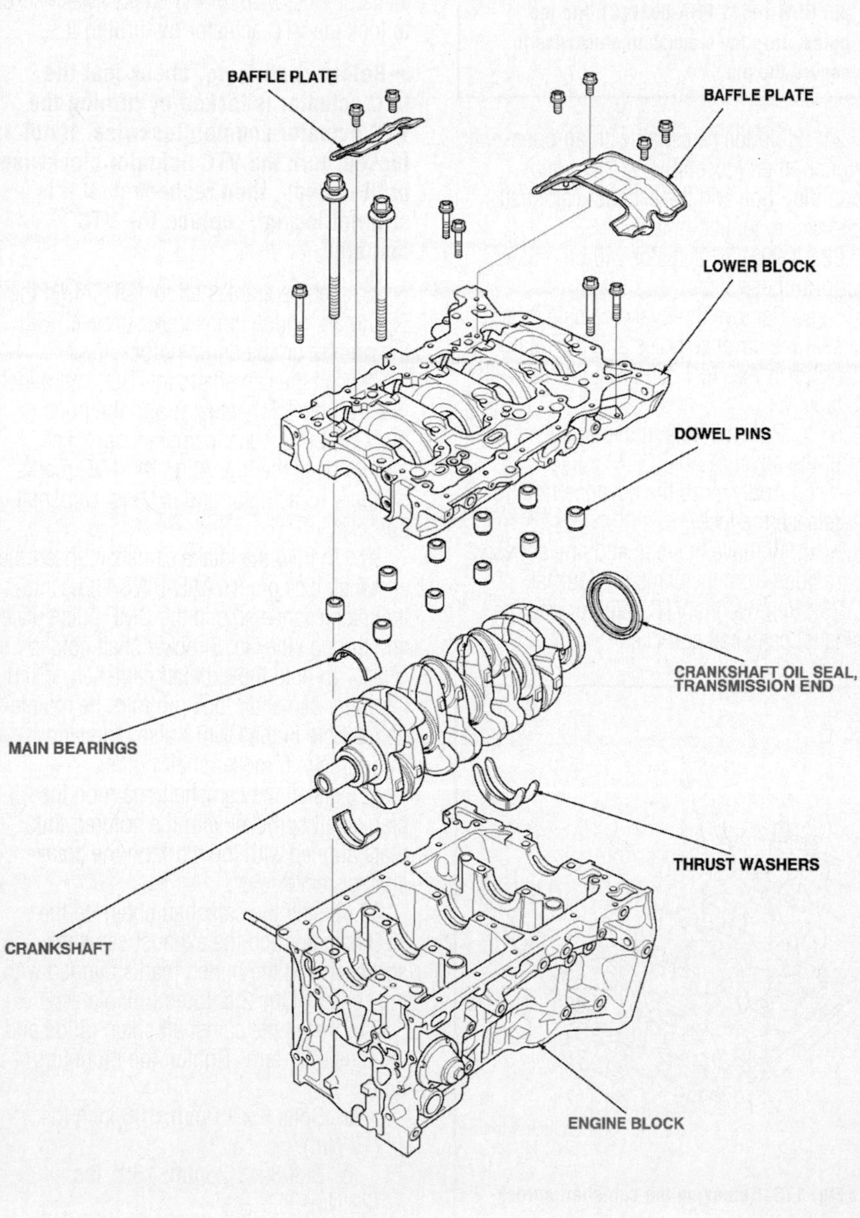
BAFFLE PLATE
BAFFLE PLATE
LOWER BLOCK
DOWEL PINS
CRANKSHAFT OIL SEAL, TRANSMISSION END
MAIN BEARINGS
THRUST WASHERS
CRANKSHAFT
ENGINE BLOCK
22140_HCRV_G0062

Fig. 171 Exploded view of engine block assembly

To install:

5. Clean and dry the crankshaft oil seal housing.
6. Use the driver and attachment to drive a new oil seal squarely into the block to the specified installed height of 0.001–0.047 inches (0.2–1.2mm) from flush.
7. Install the drive plate.
8. Install the transaxle.

ROCKER ARMS

REMOVAL & INSTALLATION

For service information, refer to Camshaft and Valve Lifters, Removal and Installation.

TIMING (CAM) CHAIN FRONT COVER

REMOVAL & INSTALLATION

For service information, refer to Cam Chain and Sprockets, Removal and Installation.

TIMING (CAM) CHAIN & SPROCKETS

REMOVAL & INSTALLATION

See Figures 172 through 184.

1. Before servicing the vehicle, refer to the Precautions Section.

➡Keep the camshaft chain away from magnetic fields.

2. Remove the front wheels.
3. Remove the splash shield.
4. Remove the accessory drive belt.
5. Remove the cylinder head cover.
6. Set the No. 1 piston at Top Dead Center (TDC). The punch mark on the Variable Valve Timing Control (VTC) actuator and the punch mark on the exhaust camshaft sprocket should be at the top. Align the TDC marks on the VTC actuator and exhaust camshaft sprocket.
7. Disconnect the Crankshaft Position (CKP) sensor connector and the VTC oil control solenoid valve connector.
8. Remove the VTC oil control solenoid valve.
9. Remove the crankshaft damper pulley.
10. Support the engine with a jack and a wood block under the oil pan.
11. Remove the upper torque rod.
12. Remove the ground cable, then remove the side engine mount bracket.

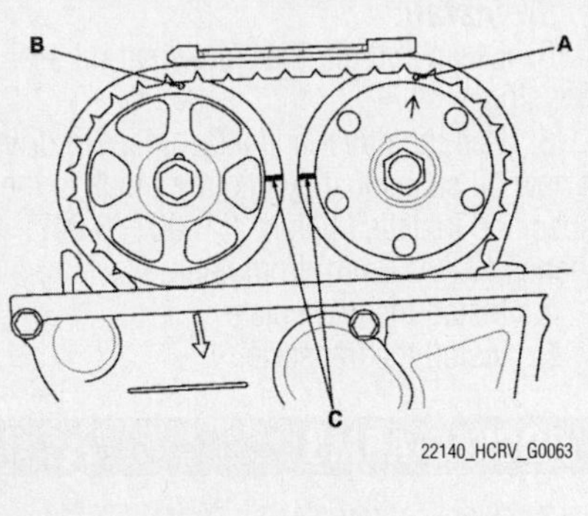
B
A
C
22140_HCRV_G0063

Fig. 172 Set the No. 1 piston at TDC. The punch mark (A) on the VTC actuator and the punch mark (B) on the exhaust camshaft sprocket should be at the top. Align the TDC marks (C) on the VTC actuator and exhaust camshaft sprocket.

13. Remove the side engine mount bracket mounting bolts.
14. Remove the camshaft chain case and the side engine mount bracket.
15. Loosely install the crankshaft damper pulley.
16. Turn the crankshaft counterclockwise to compress the auto-tensioner.
17. Align the holes on the lock and the auto-tensioner, then insert a 0.05 inches (1.2mm) diameter pin or lock pin (P/N 14511-PNA-003) into the holes. Turn the crankshaft clockwise to secure the pin.
18. Remove the auto-tensioner.
19. Remove camshaft chain guides and the tensioner arm.
20. Remove the camshaft chain.

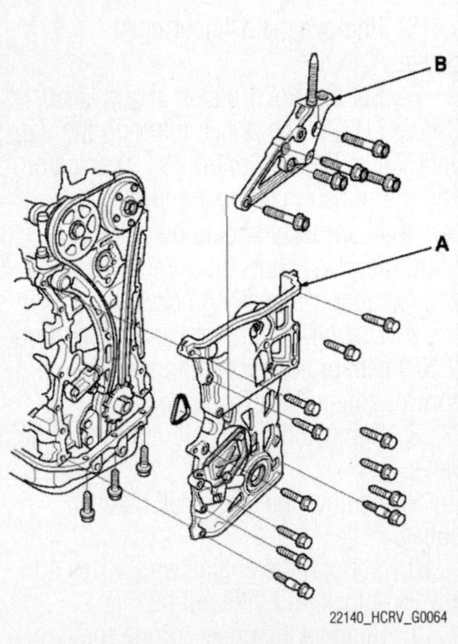
B
A
22140_HCRV_G0064

Fig. 173 Remove the camshaft chain case (A) and the side engine mount bracket (B)

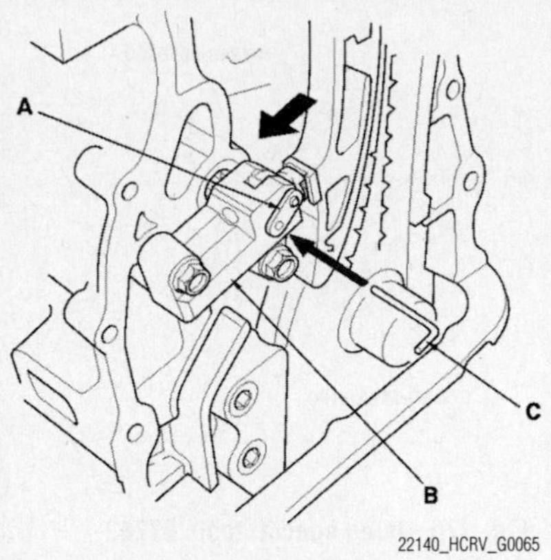
A
C
B
22140_HCRV_G0065

Fig. 174 Align the holes on the lock (A) and the auto-tensioner (B), then insert a 0.05 inches (1.2mm) diameter pin or lock pin (P/N 14511-PNA-003) (C) into the holes. Turn the crankshaft clockwise to secure the pin

21. Hold the camshaft with an open-end wrench, then loosen the VTC actuator mounting bolt and the exhaust camshaft sprocket mounting bolt.
22. If the VTC actuator will be reused, perform these steps:
  a. Remove the intake camshaft, and seal the advance holes and the retard holes in the No 1 camshaft journal with tape.
  b. Punch a hole in the tape over one of the advance holes.
  c. Apply air to the advance hole to release the lock.
  d. Remove the tape and any adhesive residue from the camshaft journal.
23. Remove the VTC actuator and the exhaust camshaft sprocket.

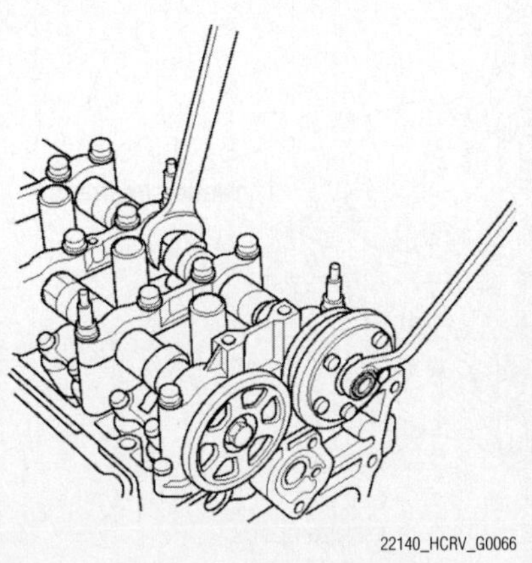
22140_HCRV_G0066

Fig. 175 Removing the camshaft sprockets holding the camshaft with an open-end wrench

To install:

24. Install the VTC actuator and the exhaust camshaft sprocket.

➡Install the VTC actuator to unlock position.

25. Apply engine oil to the threads of the VTC actuator mounting bolt and exhaust camshaft mounting bolt, then install the bolts.
26. Hold the camshaft with an open-end wrench, then tighten the bolts:
  a. VTC actuator mounting bolt 12 x 1.25mm: 83 ft. lbs. (113 Nm).
  b. Exhaust camshaft sprocket mounting bolt 10 x 1.25mm: 53 ft. lbs. (72 Nm).
27. Hold the camshaft and turn the VTC actuator clockwise until it clicks. Make sure to lock the VTC actuator by turning it.

➡Before continuing, check that the VTC actuator is locked by turning the VTC actuator counterclockwise. If not locked, turn the VTC actuator clockwise until it stops, then recheck it. If it is still not locked, replace the VTC actuator.

28. Set the crankshaft to TDC. Align the TDC mark on the crankshaft sprocket with the pointer on the engine block.
29. Set the camshafts to TDC. The punch mark on the VTC actuator and the punch mark on the exhaust camshaft sprocket should be at the top. Align the TDC marks on the VTC actuator and exhaust camshaft sprocket.
30. To hold the intake camshaft, insert the camshaft lock pin (07AAB-RWCA120) into the maintenance hole in the CMP pulse plate and through the No. 5 rocker shaft holder.
31. To hold the exhaust camshaft, insert the other camshaft lock pin into the maintenance hole in the CMP pulse plate and through No. 5 rocker shaft holder.
32. Install the camshaft chain on the crankshaft sprocket with the colored link plate aligned with the mark on the crankshaft sprocket.
33. Install the camshaft chain on the VTC actuator and the exhaust camshaft sprocket with the punch marks aligned with the center of the 2 colored link plates.
34. Install the camshaft chain guide and the tensioner arm. Tighten the mounting bolts:
  a. Bolts 6 x 1.0mm: 106 inch lbs. (12 Nm).
  b. Bolt 8 x 1.25mm: 16 ft. lbs. (22 Nm).
35. Compress the auto-tensioner when replacing the camshaft chain. Remove the

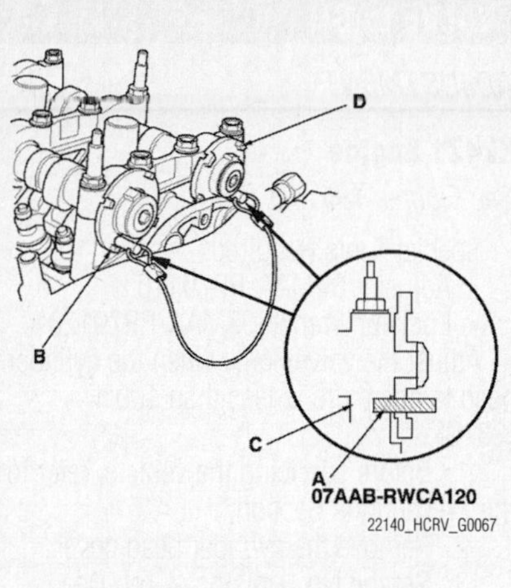
D
B
C
A
07AAB-RWCA120
22140_HCRV_G0067

Fig. 176 Insert a camshaft lock pin (07AAB-RWCA120) (A) into the maintenance hole in the CMP pulse plate (B) and through the No. 5 rocker shaft holder (C)

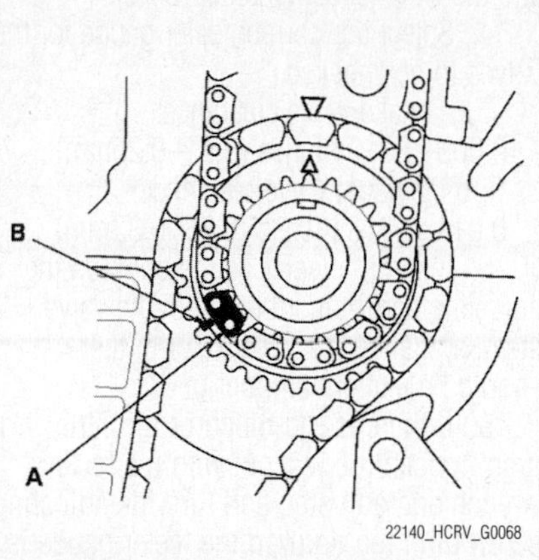
B
A
22140_HCRV_G0068

Fig. 177 Install the camshaft chain on the crankshaft sprocket with the colored link plate (A) aligned with the mark (B) on the crankshaft sprocket

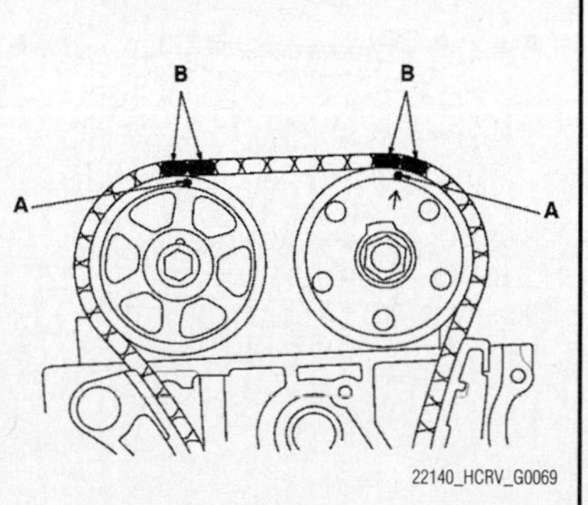
B
B
A
A
22140_HCRV_G0069

Fig. 178 Install the camshaft chain on the VTC actuator and the exhaust camshaft sprocket with the punch marks (A) aligned with the center of the 2 colored link plates (B)

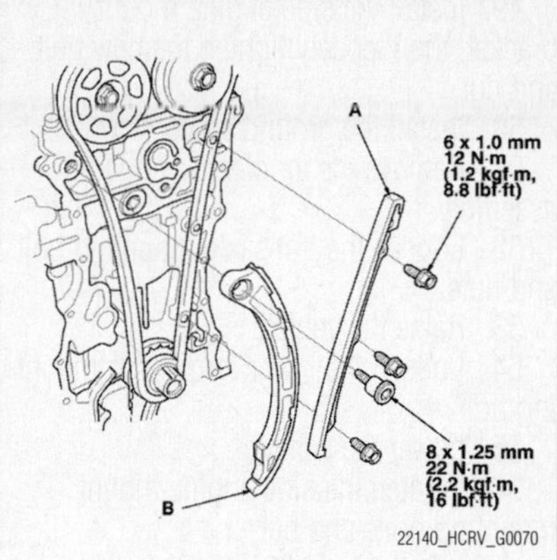
A
6 x 1.0 mm
12 N·m
(1.2 kgf·m,
8.8 lbf·ft)
8 x 1.25 mm
22 N·m
(2.2 kgf·m,
16 lbf·ft)
B
22140_HCRV_G0070

Fig. 179 Install the camshaft chain guide (A) and the tensioner arm (B)

pin (P/N 14511-PNA-003) from the auto-tensioner that was installed during removal. Turn the plate counterclockwise, to release the lock, then press the rod, and set the first cam to the edge of the rack. Insert the 0.05 inches (1.2mm) diameter pin or lock pin into the holes.

⁂ WARNING

If the chain tensioner is not set up as described, the tensioner will become damaged.

36. Install the auto-tensioner and tighten the bolts to 106 inch lbs. (12 Nm).
37. Install the camshaft chain guide and tighten the mounting bolts to 16 ft. lbs. (22 Nm).
38. Remove the pin or lock pin (P/N 14511-PNA-003) from the auto-tensioner.
39. Remove the camshaft lock pin set.
40. Check the chain case oil seal for damage. If the oil seal is damaged, replace the chain case oil seal.
41. Remove old liquid gasket from the chain case mating surfaces, bolts, and bolt holes.

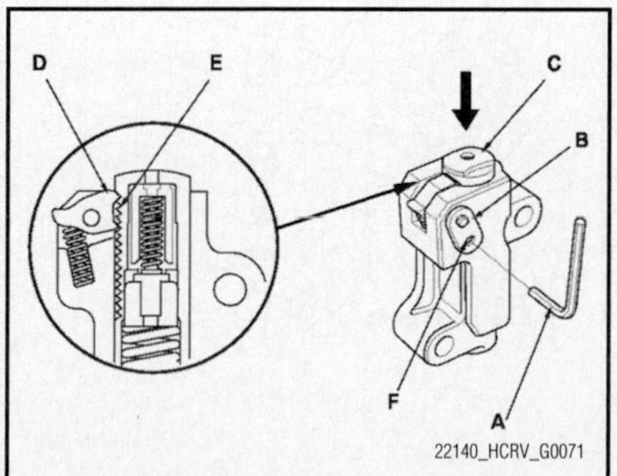
D
E
C
B
F
A
22140_HCRV_G0071

Fig. 180 Compress the auto-tensioner as illustrated using a lock pin

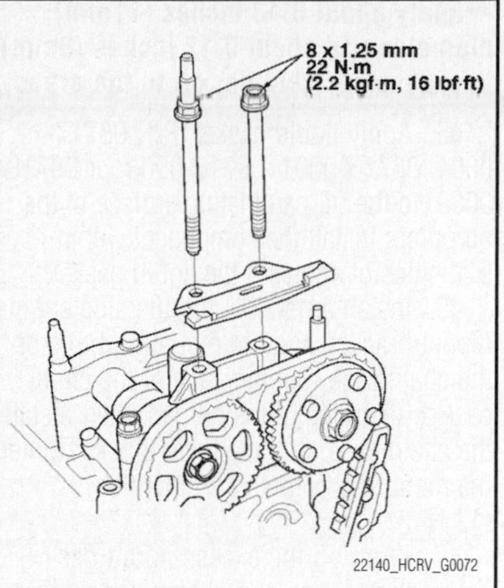
8 x 1.25 mm
22 N·m
(2.2 kgf·m, 16 lbf·ft)
22140_HCRV_G0072

Fig. 181 Install the camshaft chain guide

42. Clean and dry the chain case mating surfaces.
43. Apply liquid gasket (P/N 08717-0004, 08718-0001, 08718-0003, or 08718-0009) to the engine block mating surface of the chain case and to the inside edge of the bolt holes. Install the component within 5 minutes of applying the liquid gasket.
   - Apply a bead of liquid gasket about 0.12 inches (3mm) in diameter along the broken line.
   - If you apply liquid gasket P/N 08718-0012, the component must be installed within 4 minutes.
   - If too much time has passed after applying the liquid gasket, remove the old liquid gasket and residue, then reapply new liquid gasket.
44. Apply liquid gasket to the engine block upper surface contact areas and the lower block upper surface contact areas on the chain case.

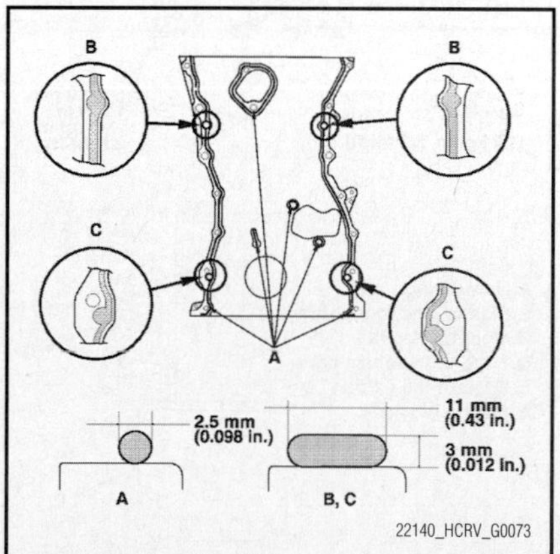
B
B
C
C
A
2.5 mm
(0.098 in.)
11 mm
(0.43 in.)
3 mm
(0.012 in.)
A
B, C
22140_HCRV_G0073

Fig. 182 Apply liquid gasket to the engine block mating surface of the chain case and to the inside edge of the bolt holes as illustrated

➡Apply about 0.43 inches (11mm) diameter and about 0.12 inches (3mm) thickness of liquid gasket to the areas.

45. Apply liquid gasket (P/N 08717-0004, 08718-0001, 08718-0003, or 08718-0009) to the oil pan mating surface of the oil pump. Install the component within 5 minutes of applying the liquid gasket.

46. Install a new O-ring, the side engine mount bracket, and the mounting bolts on the chain case. Set the edge of the chain case to the edge of the oil pan, then install the chain case on the engine block. Tighten the mounting bolts to 106 inch lbs. (12 Nm).

47. Wipe off the excess liquid gasket on the oil pan and chain case mating area.

- When installing the chain case, do not slide the bottom surface onto the oil pan mounting surface.
- Wait at least 30 minutes to allow the liquid gasket to cure before filling the engine with oil.
- Do not run the engine for at least 3 hours after installing the chain case.

48. Tighten the side engine mount bracket mounting bolts to 33 ft. lbs. (44 Nm).

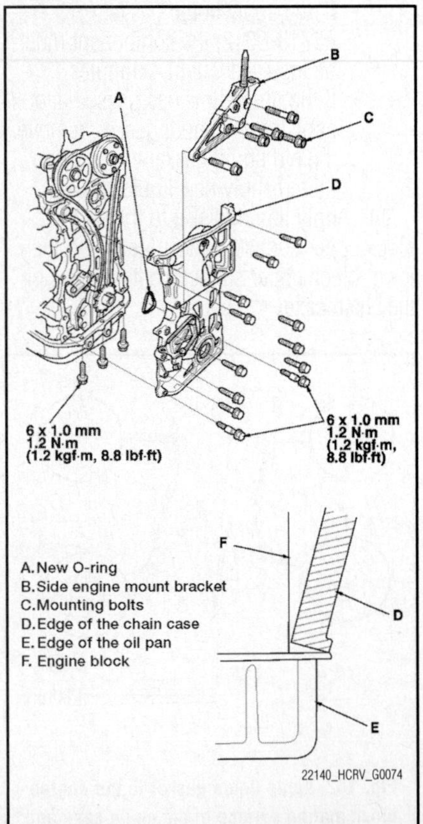
B
A
C
D
6 x 1.0 mm
1.2 N·m
(1.2 kgf·m, 8.8 lbf·ft)
6 x 1.0 mm
1.2 N·m
(1.2 kgf·m,
8.8 lbf·ft)
F
D
E
A. New O-ring
B. Side engine mount bracket
C. Mounting bolts
D. Edge of the chain case
E. Edge of the oil pan
F. Engine block
22140_HCRV_G0074

Fig. 183 Expanded view of the timing chain case and related components

49. Install the side engine mount bracket, then loosely tighten the new bolt and nut.

50. Install the ground cable.

51. Remove the air cleaner housing assembly.

52. Loosen the transaxle mounting bolt and nuts.

53. Raise the vehicle.

54. Loosen the lower torque rod mounting bolt.

55. Lower the vehicle.

56. Tighten the side engine mount mounting bolts and nut.

a. Bolt 12 x 1.25mm: 52 ft. lbs. (72 Nm).

b. Bolt 14 x 1.5mm and nut: 54 ft. lbs. (74 Nm).

57. Tighten the transaxle mounting bolt and nuts to 54 ft. lbs. (74 Nm).

58. Raise the vehicle.

59. Tighten the lower torque rod mounting bolt to 69 ft. lbs. (93 Nm).

60. Lower the vehicle.

61. Install the air cleaner housing assembly.

62. Install the upper torque rod, then tighten the new upper torque rod mounting bolts in the numbered sequence shown to the torque illustrated.

63. Install the crankshaft damper pulley.

64. Install the VTC oil control solenoid valve.

65. Connect the CKP sensor connector and VTC oil control solenoid valve connector.

66. Install the cylinder head cover.

67. Install the accessory drive belt.

68. Perform the CKP pattern clear/CKP learn procedure.

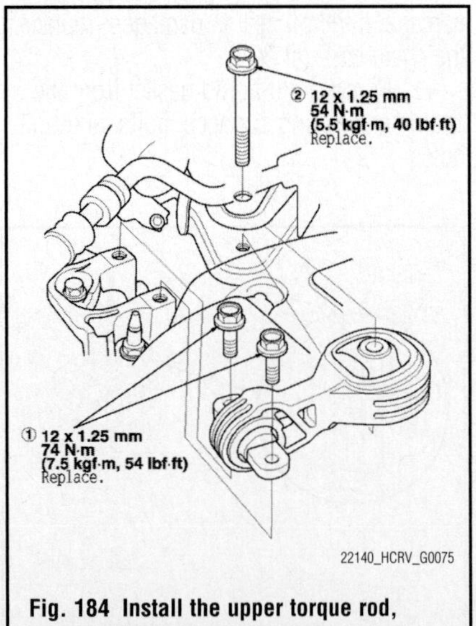
② 12 x 1.25 mm
54 N·m
(5.5 kgf·m, 40 lbf·ft)
Replace.
① 12 x 1.25 mm
74 N·m
(7.5 kgf·m, 54 lbf·ft)
Replace.
22140_HCRV_G0075

Fig. 184 Install the upper torque rod, then tighten the mounting bolts in the numbered sequence shown

VALVE LASH

ADJUSTMENT

K24Z1 Engine

See Figures 185 and 186.

Special Tools Required:

- Adjuster 07MAA-PR70110
- Locknut wrench 07MAA-PR70120

Adjust the valves only when the cylinder head temperature is less than 100°F (38°C).

1. Before servicing the vehicle, refer to the Precautions Section.

2. Remove the cylinder head cover.

3. Set the No. 1 piston at Top Dead Center (TDC). The punch mark on the Variable Valve Timing Control (VTC) actuator and the punch mark on the exhaust camshaft sprocket should be at the top. Align the TDC marks on the VTC actuator and the exhaust camshaft sprocket.

4. Select the correct feeler gauge for the valves to be checked.

a. Intake valve clearance: 0.008–0.010 inches (0.21–0.25mm).

b. Exhaust valve clearance: 0.011–0.013 inches (0.28–0.32mm).

5. Insert the feeler gauge between the adjusting screw and the end of the valve stem, and slide it back and forth; there should be a slight amount of drag.

6. If there is too much or too little drag, loosen the locknut with the locknut wrench and adjuster, and turn the adjusting screw until the drag on the feeler gauge is correct.

7. Tighten the locknut to the specified torque and recheck the clearance. Repeat

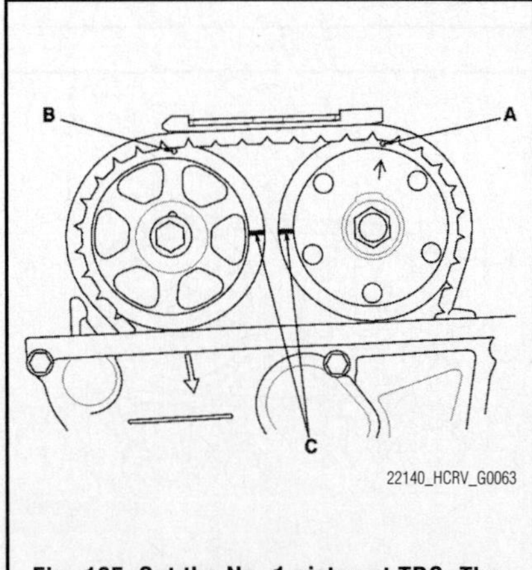
B
A
C
22140_HCRV_G0063

Fig. 185 Set the No. 1 piston at TDC. The punch mark (A) on the VTC actuator and the punch mark (B) on the exhaust camshaft sprocket should be at the top. Align the TDC marks (C) on the VTC actuator and exhaust camshaft sprocket.

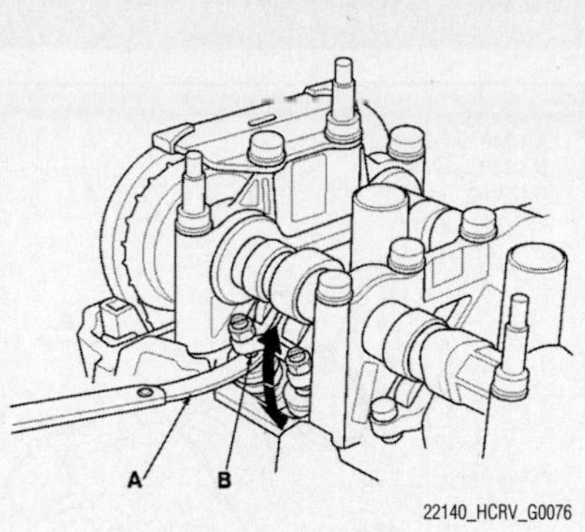
A
B
22140_HCRV_G0076

Fig. 186 Insert the feeler gauge (A) between the adjusting screw (B) and the end of the valve stem, and slide it back and forth; there should be a slight amount of drag

the adjustment if necessary. Apply engine oil to the nut threads.

a. Intake locknut torque: 14 ft. lbs. (20 Nm).

b. Exhaust locknut torque: 10 ft. lbs. (14 Nm).

8. Rotate the crankshaft 180° clockwise (the camshaft pulley turns 90°).
9. Check and, if necessary, adjust the valve clearance on the No. 3 cylinder.
10. Rotate the crankshaft 180° clockwise (the camshaft pulley turns 90°).
11. Check and, if necessary, adjust the valve clearance on the No. 4 cylinder.
12. Rotate the crankshaft 180° clockwise (the camshaft pulley turns 90°).
13. Check and, if necessary, adjust the valve clearance on the No. 2 cylinder.
14. Install the cylinder head cover.

K24Z6 Engine

See Figures 185, 187 and 188.

Special Tools Required:

- Locknut Wrench 07MAA-PR70120
- Adjuster 07MAA-PR70110

➡Connect the Honda Diagnostic System (HDS) to the Data Link Connector (DLC) and monitor the Engine Coolant Temperature (ECT) sensor 1 with the HDS. Adjust the valve clearance only when the ECT sensor 1 temperature is less than 100°F(38°C).

1. Remove the cylinder head cover.
2. Set the No. 1 piston at Top Dead Center (TDC). The punch mark on the Variable Valve Timing Control (VTC) actuator and the punch mark on the exhaust camshaft sprocket should be at the top. Align the TDC marks on the VTC actuator and the exhaust camshaft sprocket.
3. Select the correct feeler gauge for the valve clearance you are going to check.

Valve Clearance:

- Intake: 0.008–0.010 inches (0.21–0.25 mm)
- Exhaust: 0.010–0.011 inches (0.25–0.29 mm)

4. Insert the feeler gauge between the adjusting screw and the end of the valve stem, and slide it back and forth; you should feel a slight amount of drag.
5. If you feel too much or too little drag, loosen the locknut with the locknut wrench and adjuster, and turn the adjusting screw until the drag on the feeler gauge is correct.
6. Tighten the locknut to the specified torque, and recheck the clearance. Repeat the adjustment if necessary.

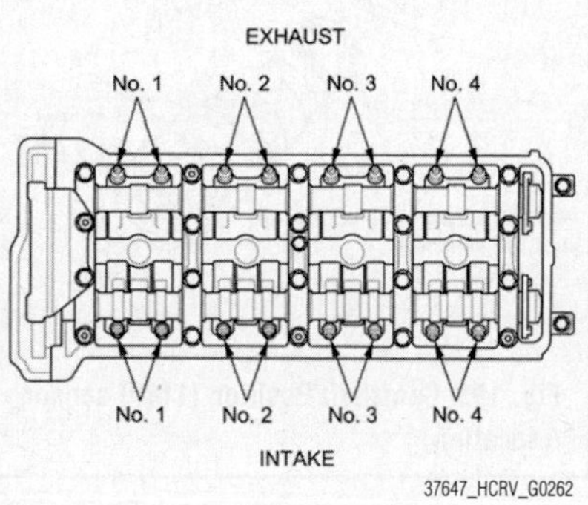
EXHAUST
No. 1
No. 2
No. 3
No. 4
No. 1
No. 2
No. 3
No. 4
INTAKE
37647_HCRV_G0262

Fig. 187 Valve clearance adjusting screws

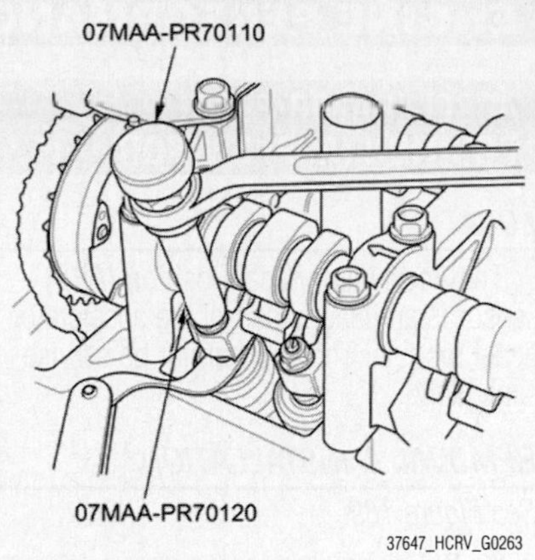
07MAA-PR70110
07MAA-PR70120
37647_HCRV_G0263

Fig. 188 Loosen the locknut with the locknut wrench and adjuster, and turn the adjusting screw

Specified Torque:

- Intake: 7 x 0.75 mm: 10 ft. lbs. (14 Nm)

➡Apply new engine oil to the nut threads.

- Exhaust: 7 x 0.75 mm: 10 ft. lbs. (14 Nm)

➡Apply new engine oil to the nut threads.

7. Rotate the crankshaft 180 degrees clockwise (camshaft pulley turns 90 degrees).
8. Check and, if necessary, adjust the valve clearance on the No. 3 cylinder.
9. Rotate the crankshaft 180 degrees clockwise (camshaft pulley turns 90 degrees).
10. Check and, if necessary, adjust the valve clearance on the No. 4 cylinder.
11. Rotate the crankshaft 180 degrees clockwise (camshaft pulley turns 90 degrees).
12. Check and, if necessary, adjust the valve clearance on the No. 2 cylinder.
13. Install the cylinder head cover.

ACCELERATOR PEDAL POSITION (APP) SENSOR

LOCATION

The Accelerator Pedal Position (APP) sensor is an integral part of the accelerator pedal module. The APP cannot be service separately.

REMOVAL & INSTALLATION

See Figure 189.

1. Disconnect the accelerator pedal module connector.
2. Remove the accelerator pedal module.

➡The APP sensor is not available separately. Do not disassemble the accelerator pedal module.

3. Install the parts in the reverse order of removal.

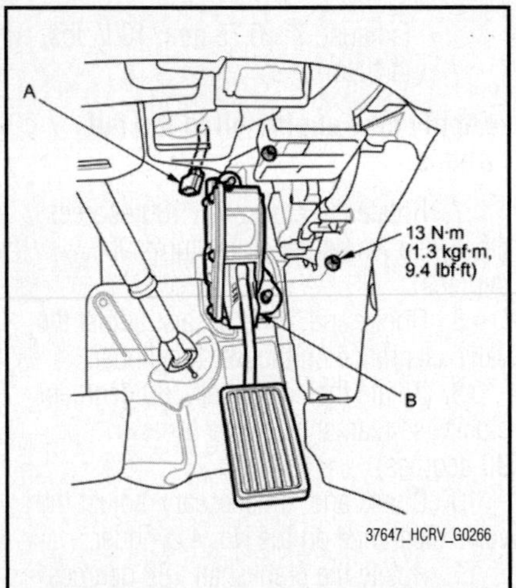
A
13 N·m
(1.3 kgf·m,
9.4 lbf·ft)
B
37647_HCRV_G0266

Fig. 189 Disconnect the accelerator pedal module connector (A); remove the accelerator pedal module (B)

CAMSHAFT POSITION (CMP) SENSOR

LOCATION

Refer to the graphics in the Removal and Installation section for the location(s).

REMOVAL & INSTALLATION

K24Z1 Engine

See Figures 190 through 192.

1. Before servicing the vehicle, refer to the Precautions Section.
2. Remove the air cleaner.

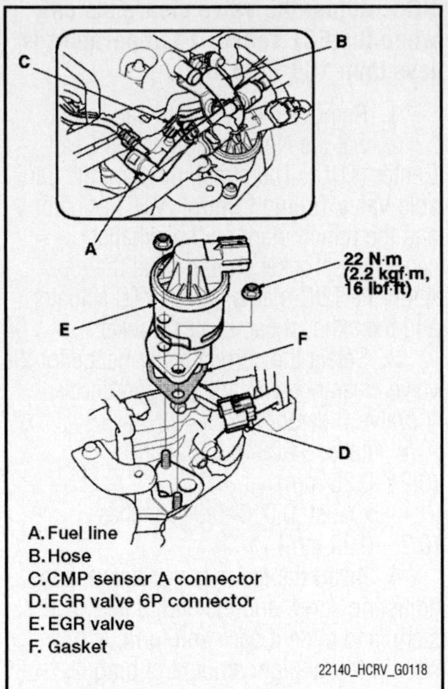
B
C
A
22 N·m
(2.2 kgf·m,
16 lbf·ft)
E
F
D
A. Fuel line
B. Hose
C. CMP sensor A connector
D. EGR valve 6P connector
E. EGR valve
F. Gasket
22140_HCRV_G0118

Fig. 190 Exhaust Gas Recirculation (EGR) valve location

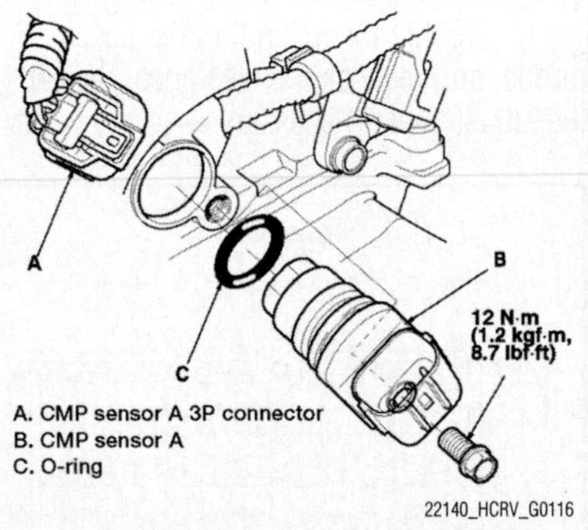
A
B
12 N·m
(1.2 kgf·m,
8.7 lbf·ft)
C
A. CMP sensor A 3P connector
B. CMP sensor A
C. O-ring
22140_HCRV_G0116

Fig. 191 Camshaft Position (CMP) sensor A location

3. Relieve the fuel pressure.
4. Remove the fuel line and disconnect the hose.
5. Disconnect the CMP sensor A connector.
6. Remove the EGR valve 6P connector.
7. Remove the EGR valve.
8. Disconnect the CMP sensor A 3P connector.
9. Remove CMP sensor A from the intake camshaft side of the cylinder head.
10. Disconnect the CMP sensor B connector.
11. Remove CMP sensor B.

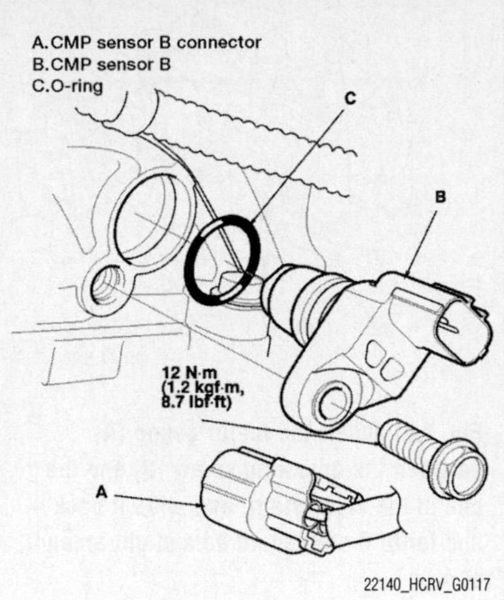
A. CMP sensor B connector
B. CMP sensor B
C. O-ring
C
B
12 N·m
(1.2 kgf·m,
8.7 lbf·ft)
A
22140_HCRV_G0117

Fig. 192 Camshaft Position (CMP) sensor B location

To install:

12. Install the CMP sensor B with a new O-ring and tighten the mounting bolt to 104 inch lbs. (12 Nm).
13. Connect the CMP sensor B connector.
14. Install CMP sensor A to the intake camshaft side of the cylinder head with a new O-ring and tighten the mounting bolt to 104 inch lbs. (12 Nm).
15. Connect the CMP sensor A 3P connector.
16. Install the EGR valve with a new gasket and tighten the mounting nuts to 16 ft. lbs. (22 Nm).
17. Connect the EGR valve 6P connector.
18. Connect the fuel line and the hose.
19. Install the air cleaner.

K24Z6 Engine

CMP Sensor A

See Figure 193.

1. Disconnect the CMP sensor A 3P connector.
2. Remove CMP sensor A from the intake camshaft side of the cylinder head.
3. Install the parts in the reverse order of removal with a new O-ring.

CMP Sensor B

See Figures 194 and 195.

1. Disconnect the connector and the hoses from the EVAP canister purge valve, then remove the EVAP canister purge valve assembly.
2. Disconnect the CMP sensor B connector.

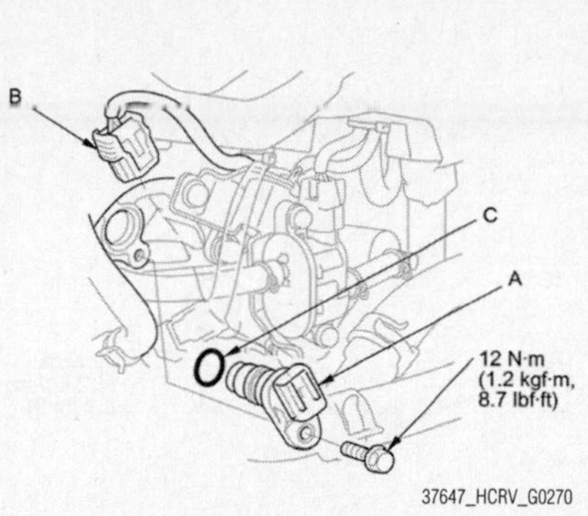
B
C
A
12 N·m
(1.2 kgf·m,
8.7 lbf·ft)
37647_HCRV_G0270

Fig. 193 Disconnect the CMP sensor A 3P connector (B); remove the CMP sensor A (A) and the O-ring (C)

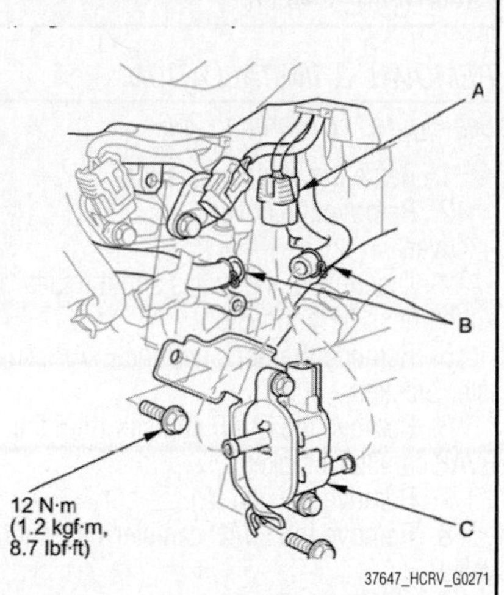
A
B
12 N·m
(1.2 kgf·m,
8.7 lbf·ft)
C
37647_HCRV_G0271

Fig. 194 Disconnect the connector (A) and the hoses (B) from the EVAP canister purge valve (C)

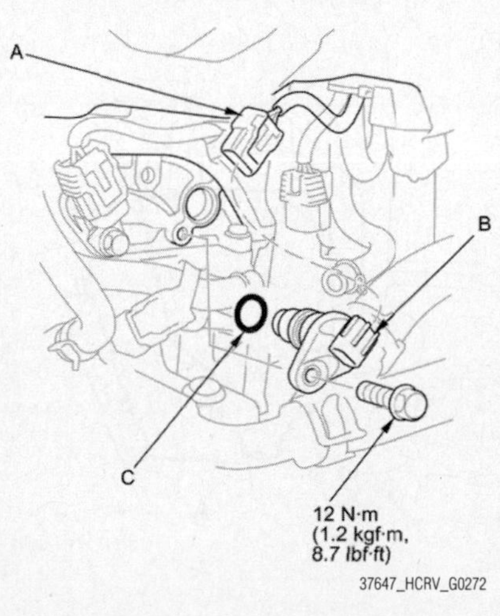
A
B
C
12 N·m
(1.2 kgf·m,
8.7 lbf·ft)
37647_HCRV_G0272

Fig. 195 Disconnect the CMP sensor B connector (A); remove the CMP sensor B (B) and the O-ring (C)

3. Remove CMP sensor B.
4. Install the parts in the reverse order of removal with a new O-ring.

CRANKSHAFT POSITION (CKP) SENSOR

LOCATION

Refer to the graphics in the Removal and Installation section for the location(s).

REMOVAL & INSTALLATION

K24Z1 Engine

See Figure 196.

1. Disconnect the CKP sensor connector.
3. Install the parts in the reverse order of removal with a new O-ring.
4. Do the CKP pattern clear/CKP pattern learn procedure.

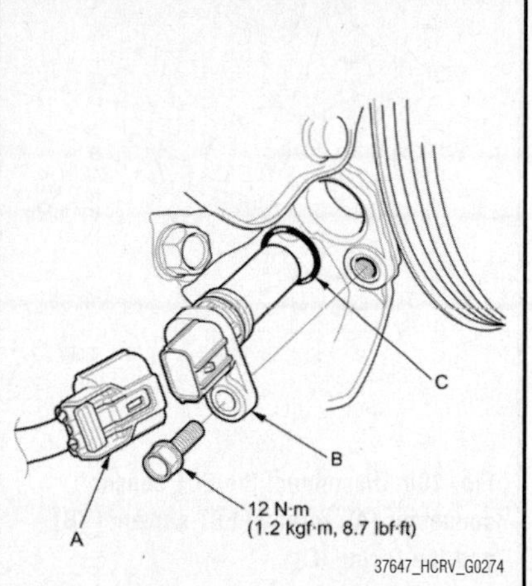
C
B
A
12 N·m
(1.2 kgf·m, 8.7 lbf·ft)
37647_HCRV_G0274

Fig. 196 Disconnect the CKP sensor connector (A); remove the CKP sensor (B) and the O-ring (C)

K24Z6 Engine

See Figure 197.

1. Raise the vehicle on a lift.

➡Make sure the vehicle is level, because engine oil will drip out when you remove the sensor.

2. Remove the CKP sensor cover.
3. Disconnect the CKP sensor connector.
4. Remove the CKP sensor.
5. Install the parts in the reverse order of removal with a new O-ring.
6. Do the CKP pattern clear/CKP pattern learn procedure.
7. Check the engine oil level, and add more oil if needed.

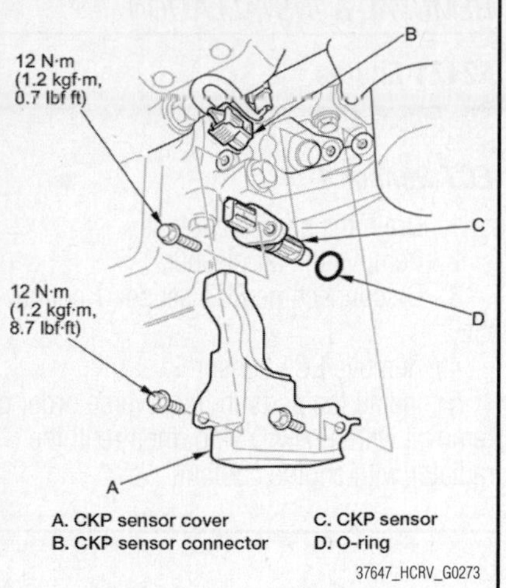
12 N·m
(1.2 kgf·m,
0.7 lbf·ft)
B
C
12 N·m
(1.2 kgf·m,
8.7 lbf·ft)
D
A
A. CKP sensor cover
B. CKP sensor connector
C. CKP sensor
D. O-ring
37647_HCRV_G0273

Fig. 197 Removing the CKP sensor

CKP PATTERN CLEAR/CKP PATTERN LEARN

Clear/Learn Procedure (with the HDS)

1. Connect the HDS to the Data Link Connector (DLC) located under the driver's side of the dashboard.
2. Turn the ignition switch to ON (II).
3. Make sure the HDS communicates with the PCM and other vehicle systems. If it doesn't, go to the DLC circuit troubleshooting.
4. Select the CRANK PATTERN in the ADJUSTMENT MENU with the HDS.
5. Select the CRANK PATTERN CLEAR, and clear the CKP pattern.
6. Select the CRANK PATTERN LEARNING with the HDS, and follow the screen prompts.

Learn Procedure (without the HDS)

1. Start the engine. Hold the engine speed at 3,000 rpm without load (in P or N) until the radiator fan comes on.
2. Test-drive the vehicle on a level road: Decelerate (with the throttle fully closed) from an engine speed of 2,500 rpm down to 1,000 rpm with the transmission in 2 position.
3. Repeat the previous step several times.
4. Turn the ignition switch to LOCK (0).
5. Turn the ignition switch to ON (II), and wait 30 seconds.

ENGINE COOLANT TEMPERATURE (ECT) SENSOR

LOCATION

Refer to the graphics in the Removal and Installation section for the location(s).

REMOVAL & INSTALLATION

K24Z1 Engine

See Figure 198.

ECT Sensor 1

1. Drain the engine coolant.
2. Remove the air cleaner.
3. Disconnect the ECT sensor 1 connector.
4. Remove ECT sensor 1.
5. Install the parts in the reverse order of removal with a new O-ring, then refill the radiator with engine coolant.

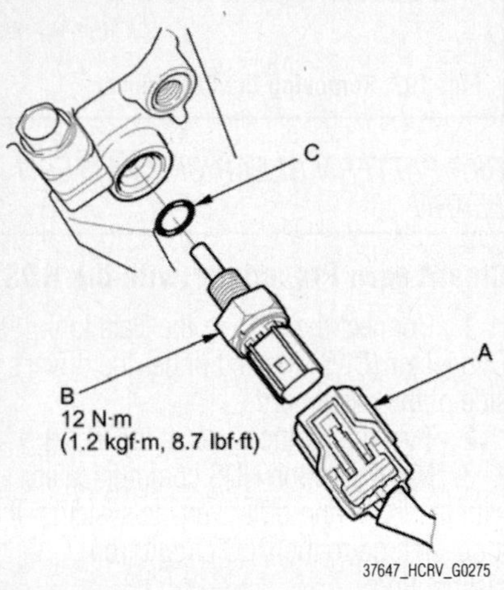
C
A
B
12 N·m
(1.2 kgf·m, 8.7 lbf·ft)
37647_HCRV_G0275

Fig. 198 Disconnect the ECT sensor 1 connector (A); remove the ECT sensor 1 (B) and the O-ring (C)

ECT Sensor 2

See Figure 199.

1. Remove the splash shield.
2. Drain the engine coolant.
3. Disconnect the ECT sensor 2 connector, then remove ECT sensor 2.
4. Install ECT sensor 2 with a new O-ring.
5. Install the splash shield.
6. Refill the radiator with engine coolant.

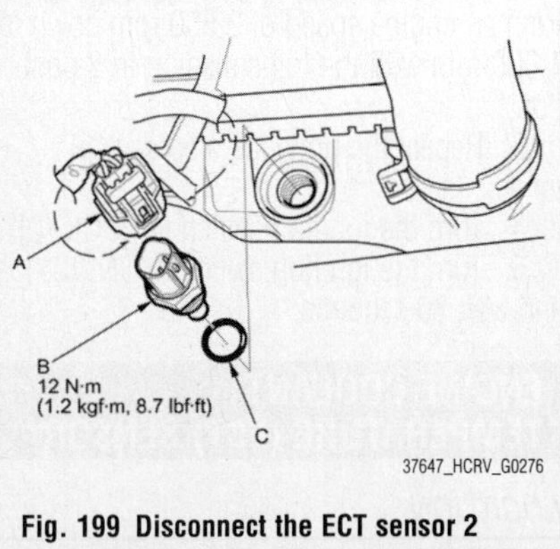
A
B
12 N·m
(1.2 kgf·m, 8.7 lbf·ft)
C
37647_HCRV_G0276

Fig. 199 Disconnect the ECT sensor 2 connector (A), then remove ECT sensor 2 (B) and the O-ring (C)

K24Z6 Engine

ECT Sensor 1

See Figure 200.

1. Drain the engine coolant.
2. Disconnect the ECT sensor 1 connector (A).
3. Remove ECT sensor 1.
4. Install the parts in the reverse order of removal with a new O-ring, then refill the radiator with engine coolant.

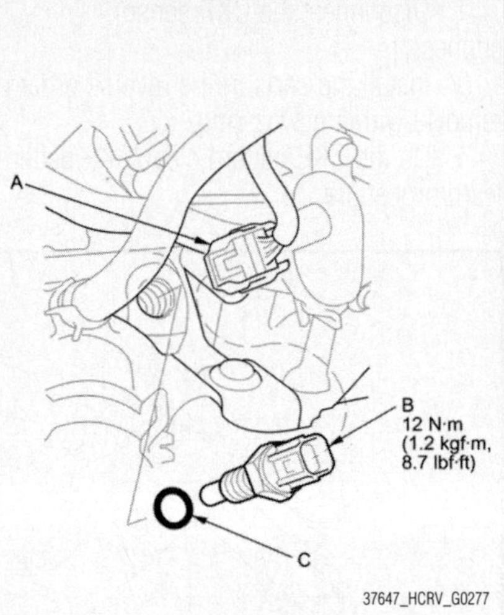
A
B
12 N·m
(1.2 kgf·m,
8.7 lbf·ft)
C
37647_HCRV_G0277

Fig. 200 Disconnect the ECT sensor 1 connector (A); remove ECT sensor 1 (B) and the O-ring (C)

ECT Sensor 2

See Figure 201.

1. Remove the front splash shield.
2. Drain the engine coolant.
3. Disconnect the ECT sensor 2 connector, then remove ECT sensor 2.
4. Install ECT sensor 2 with a new O-ring.
5. Install the front splash shield.
6. Refill the radiator with engine coolant.

EVAPORATIVE EMISSIONS (EVAP) CANISTER

LOCATION

The EVAP canister is mounted under the vehicle between the muffler and the rear wheel.

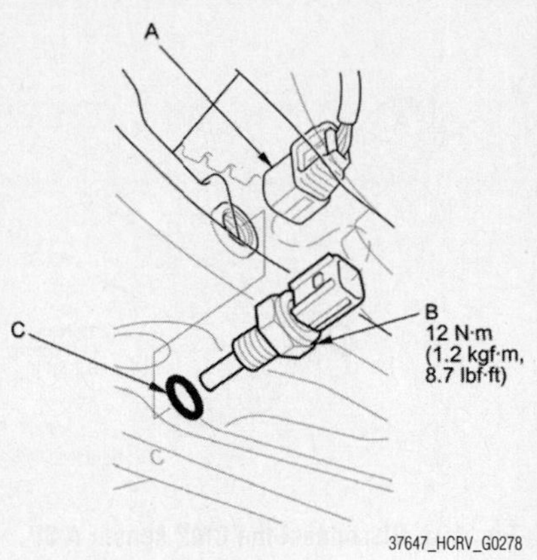
A
B
12 N·m
(1.2 kgf·m,
8.7 lbf·ft)
C
37647_HCRV_G0278

Fig. 201 Disconnect the ECT sensor 2 connector (A), then remove ECT sensor 2 (B)and the O-ring (C)

REMOVAL & INSTALLATION

See Figures 201 through 206.

1. Raise the vehicle on a lift.
2. Remove the cover.

cover.

4. Disconnect the hoses and the fuel subharness 6P connector.
5. Remove the bolts and the EVAP canister bracket.
6. Remove the EVAP canister from the EVAP canister bracket.
7. Remove the cap (A).
8. Remove the EVAP canister vent shut valve.
9. Install the EVAP canister vent shut valve into the new EVAP canister with new O-rings.

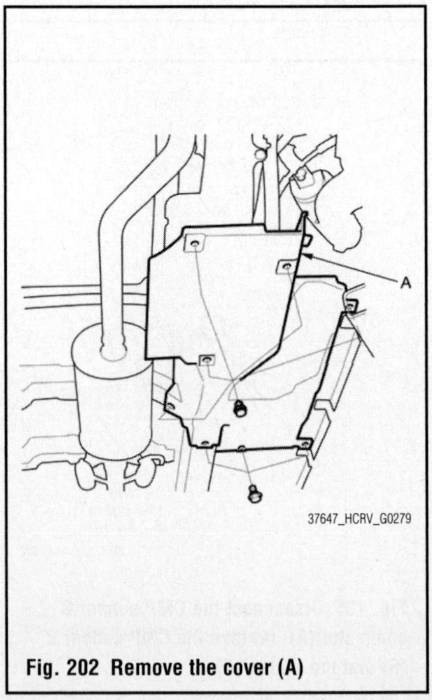
A
37647_HCRV_G0279

Fig. 202 Remove the cover (A)

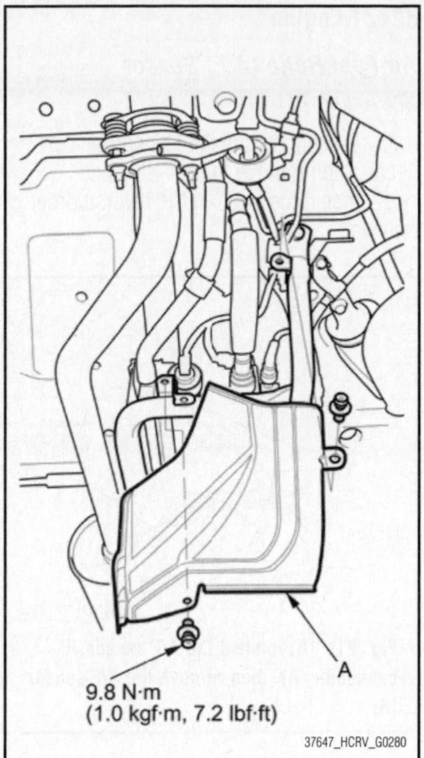
A
9.8 N·m
(1.0 kgf·m, 7.2 lbf·ft)
37647_HCRV_G0280

Fig. 203 Remove the EVAP canister baffle cover (A)

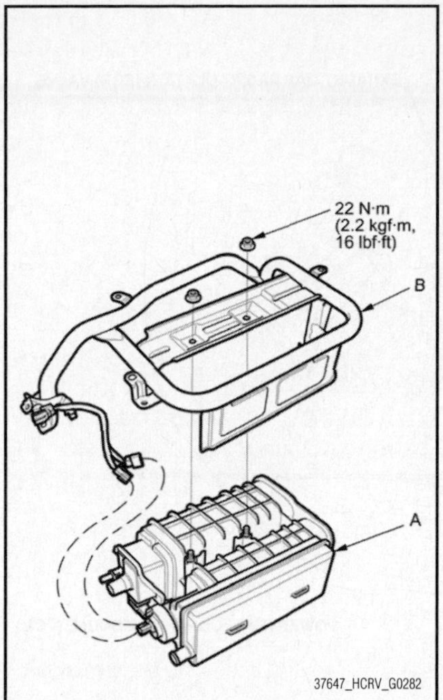
22 N·m
(2.2 kgf·m,
16 lbf·ft)
B
A
37647_HCRV_G0282

Fig. 205 Remove the EVAP canister (A) from the EVAP canister bracket (B)

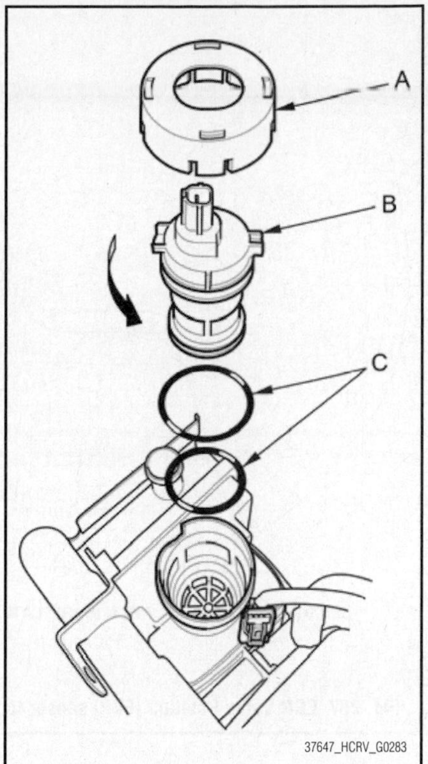
A
B
C
37647_HCRV_G0283

Fig. 206 Remove the cap (A), the EVAP canister vent shut valve (B), and the O-rings (C)

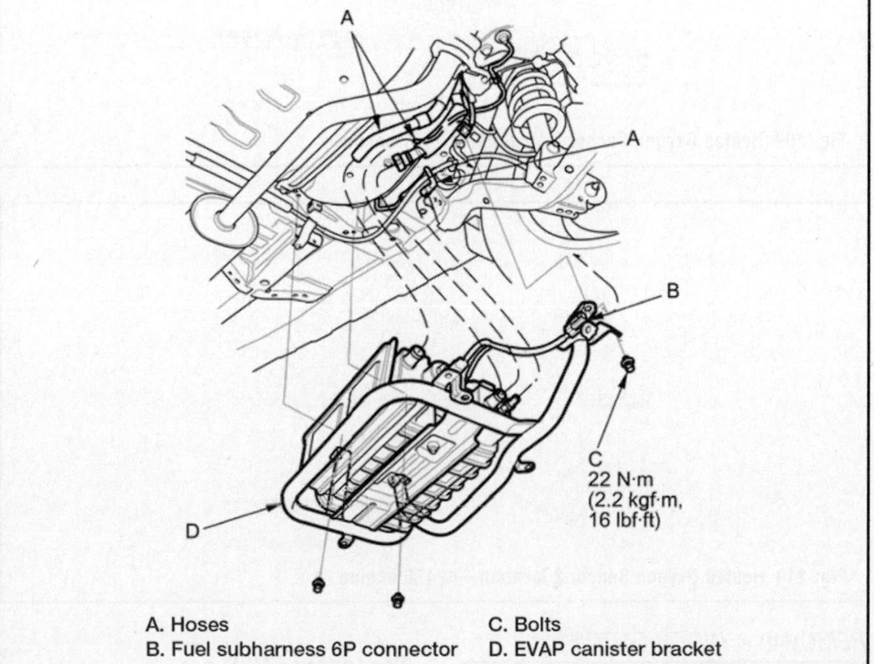
A
A
B
C
22 N·m
(2.2 kgf·m,
16 lbf·ft)
D
A. Hoses
B. Fuel subharness 6P connector
C. Bolts
D. EVAP canister bracket
37647_HCRV_G0281

Fig. 204 Disconnect the hoses and the fuel subharness 6P connector

➡**Do not coat the O-rings with oil.**

10. Install the parts in the reverse order of removal.

EXHAUST GAS RECIRCULATION (EGR) VALVE

LOCATION

See Figure 207.

Refer to the accompanying illustration.

REMOVAL & INSTALLATION

See Figure 208.

1. Before servicing the vehicle, refer to the Precautions Section.
2. Remove the air cleaner.
3. Relieve the fuel pressure.
4. Remove the fuel line. Disconnect the hose.
5. Disconnect the CMP sensor A connector.
6. Remove the EGR valve 6P connector.
7. Remove the EGR valve.

To install:

8. Installation is the reverse of the removal procedure.
9. Use a new gasket during installation.

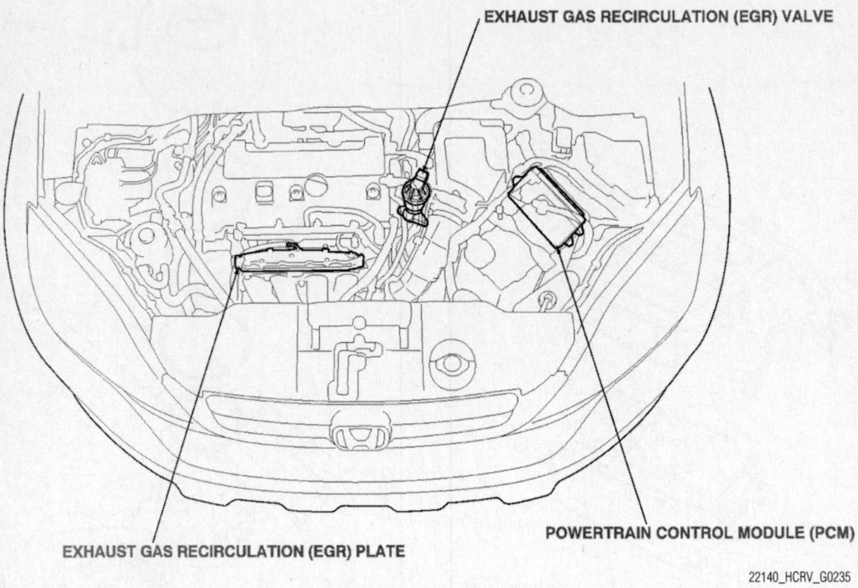
EXHAUST GAS RECIRCULATION (EGR) VALVE
EXHAUST GAS RECIRCULATION (EGR) PLATE
POWERTRAIN CONTROL MODULE (PCM)
22140_HCRV_G0235

Fig. 207 EGR Valve Position (EVP) sensor location

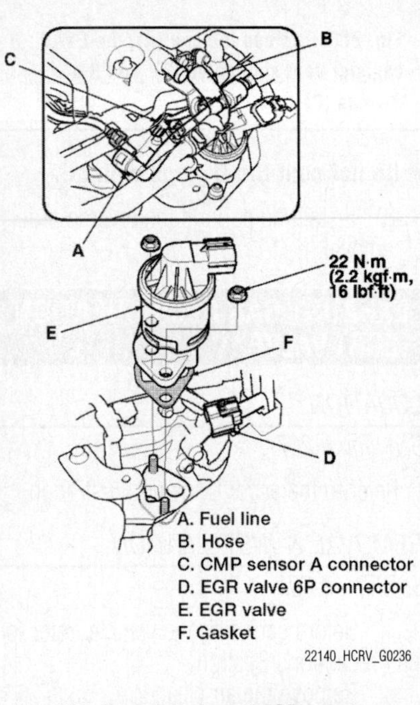
B
C
A
22 N·m
(2.2 kgf·m,
16 lbf·ft)
E
F
D
A. Fuel line
B. Hose
C. CMP sensor A connector
D. EGR valve 6P connector
E. EGR valve
F. Gasket
22140_HCRV_G0236

Fig. 208 Exploded view of EGR valve

10. Tighten the mounting nuts to 16 ft. lbs. (22 Nm).

HEATED OXYGEN (HO2S) SENSOR

LOCATION

See Figures 209 and 210.

Refer to the accompanying illustrations.

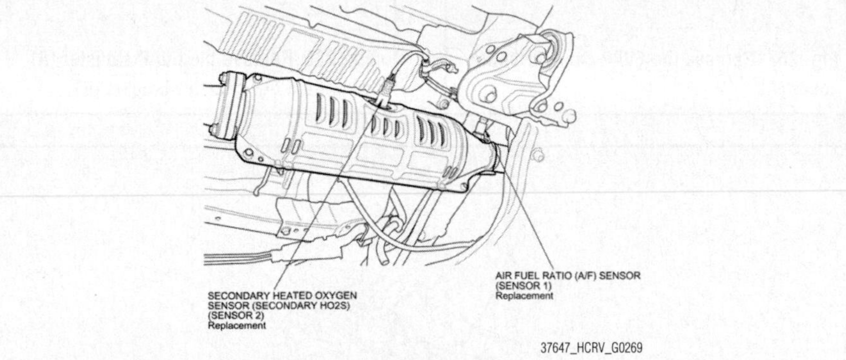
SECONDARY HEATED OXYGEN SENSOR (SECONDARY HO2S) (SENSOR 2) Replacement
AIR FUEL RATIO (A/F) SENSOR (SENSOR 1) Replacement
37647_HCRV_G0269

Fig. 209 Heated Oxygen Sensor 2 location—K24Z1 engine

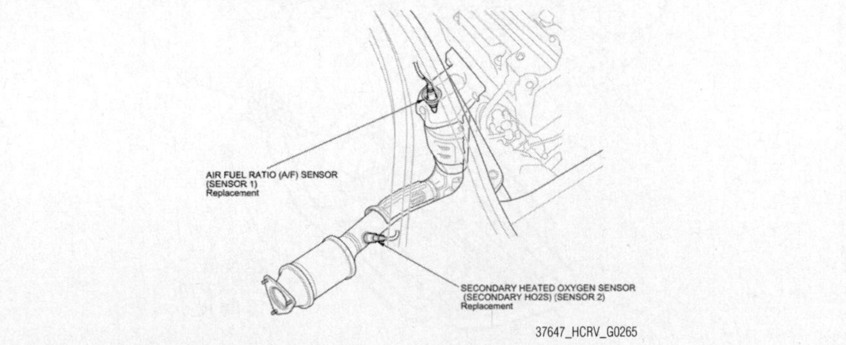
AIR FUEL RATIO (A/F) SENSOR (SENSOR 1) Replacement
SECONDARY HEATED OXYGEN SENSOR (SECONDARY HO2S) (SENSOR 2) Replacement
37647_HCRV_G0265

Fig. 210 Heated Oxygen Sensor 2 location—K24Z6 engine

REMOVAL & INSTALLATION

Special Tools Required: O2 Sensor Wrench Snap-On® S3176 or equivalent, commercially available

K24Z1 Engine

Air Fuel Ratio (A/F) Sensor

See Figure 211.

1. Disconnect the A/F sensor 4P connector, then remove the A/F sensor.
2. Install the parts in the reverse order of removal.

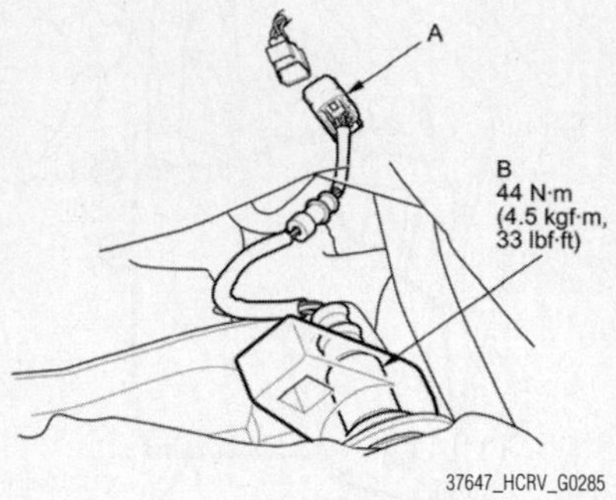
A
B
44 N·m
(4.5 kgf·m,
33 lbf·ft)
37647_HCRV_G0285

Fig. 211 Disconnect the A/F sensor 4P connector (A), then remove the A/F sensor (B)

Secondary HO2S

See Figure 212.

1. Disconnect the secondary HO2S 4P connector, then remove the secondary HO2S.

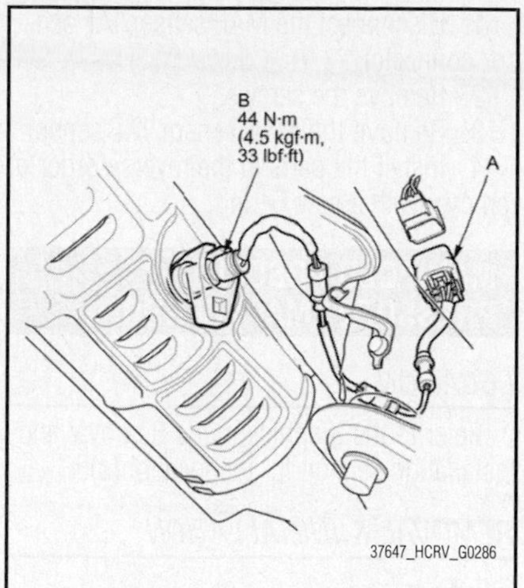
B
44 N·m
(4.5 kgf·m,
33 lbf·ft)
A
37647_HCRV_G0286

Fig. 212 Disconnect the secondary HO2S 4P connector (A), then remove the secondary HO2S (B)

2. Install the parts in the reverse order of removal.

K24Z6 Engine

Air Fuel Ratio (A/F) Sensor

See Figure 213.

1. Disconnect the A/F sensor 4P connector, then remove the A/F sensor.
2. Install the parts in the reverse order of removal.

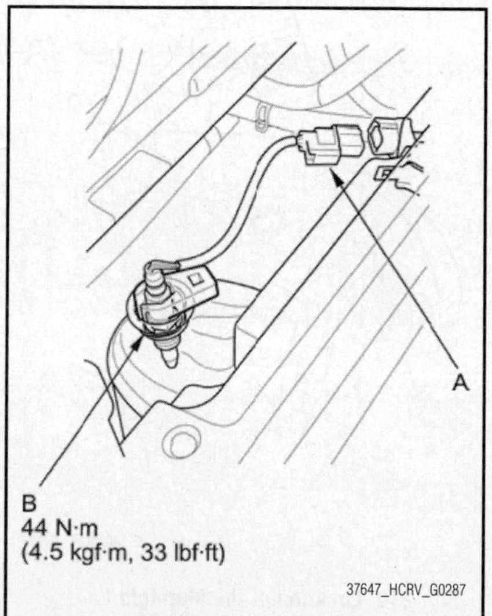
A
B
44 N·m
(4.5 kgf·m, 33 lbf·ft)
37647_HCRV_G0287

Fig. 213 Disconnect the A/F sensor 4P connector (A), then remove the A/F sensor (B)

Secondary HO2S

See Figure 214.

1. Raise the vehicle on a lift.
2. Disconnect the secondary HO2S 4P connector.

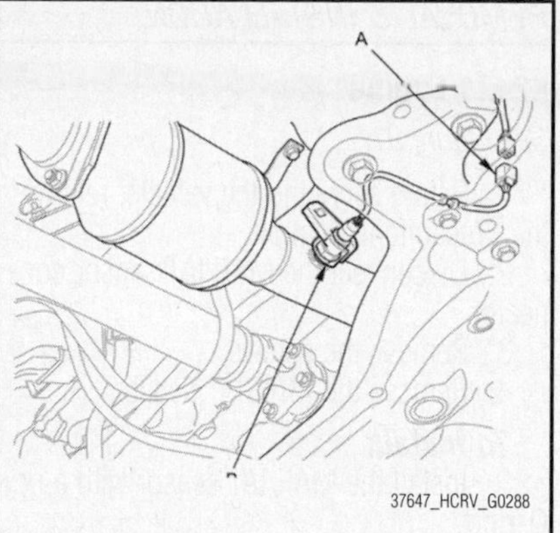
A
B
37647_HCRV_G0288

Fig. 214 Disconnect the secondary HO2S 4P connector (A) and remove the secondary HO2S (Sensor 2) (B)

3. Remove the secondary HO2S (Sensor 2).
4. Install the parts in the reverse order of removal.

INTAKE AIR TEMPERATURE (IAT) SENSOR

LOCATION

The Intake Air Temperature (IAT) sensor is an integral part of the Mass Air Flow (MAF) sensor/Intake Air Temperature (IAT) sensor assembly which is mounted on the air intake duct. Refer to the Mass Air Flow section for information regarding servicing this component.

REMOVAL & INSTALLATION

The Intake Air Temperature (IAT) sensor is an integral part of the Mass Air Flow (MAF) sensor/Intake Air Temperature (IAT) sensor assembly which is mounted on the air intake duct. Refer to the Mass Air Flow section for information regarding servicing this component.

KNOCK SENSOR (KS)

LOCATION

Refer to the graphics in the Removal and Installation section for the location(s).

REMOVAL & INSTALLATION

K24Z1 Engine

See Figure 215.

1. Before servicing the vehicle, refer to the Precautions Section.
2. Raise and safely support the vehicle.
3. Disconnect the knock sensor connector.

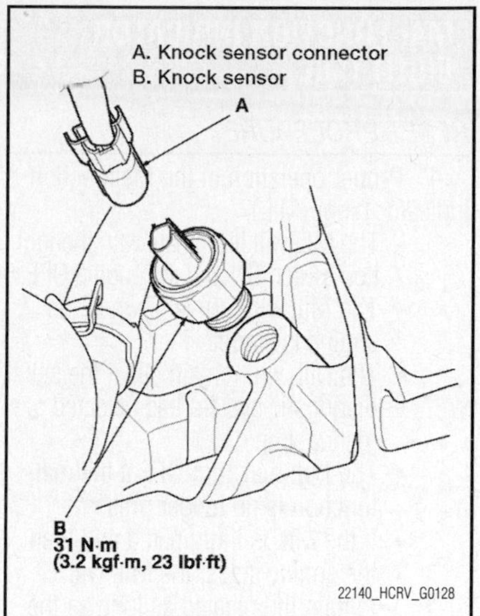
A. Knock sensor connector
B. Knock sensor
A
B
31 N·m
(3.2 kgf·m, 23 lbf·ft)
22140_HCRV_G0128

Fig. 215 Location of the Knock Sensor (KS)

4. Remove the knock sensor.

To install:

5. Install the knock sensor and tighten to 23 ft. lbs. (31 Nm).
6. Connect the knock sensor connector.

K24Z6 Engine

See Figure 216.

1. Remove the intake manifold.
2. Disconnect the knock sensor connector.
3. Remove the knock sensor.
4. Install the parts in the reverse order of removal.

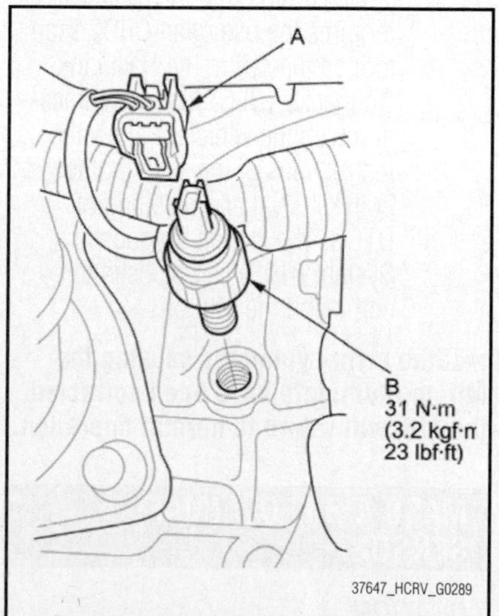
A
B
31 N·m
23 lbf·ft)
37647_HCRV_G0289

Fig. 216 Disconnect the knock sensor connector (A); remove the knock sensor (B)

MALFUNCTION INDICATOR LIGHT (MIL)

RESET PROCEDURE

1. Proper operation of the Malfunction Indicator Light (MIL):
   - The MIL will illuminate with the ignition switch ON and the engine OFF
   - The MIL will turn OFF when the engine is started
   - The MIL will remain ON if the self-diagnostic system has detected a malfunction
   - The MIL may turn OFF if the malfunction is no longer present
   - If the MIL is illuminated and then the engine stalls, the MIL will remain illuminated as long as the ignition switch is ON
   - If the MIL is not illuminated and the engine stalls, the MIL will not illuminate until the ignition switch is cycled OFF, then ON
2. Resetting the MIL:
   - The control module turns OFF the MIL after 3 consecutive ignition cycles that the diagnostic system runs and does not fail
   - The control module turns OFF the MIL after a current Diagnostic Trouble Code (DTC) clears when the diagnostic cycle runs and passes
   - There may still be a history of DTC's stored in the system. These will clear after 40 consecutive warm-up cycles, if no failures are reported by any other related diagnostic system
   - Manual resetting of the MIL and any DTC stored in the system, requires the use of an OBD2 scan tool connected to the Data Link Connector (DLC) for communication with the vehicle. Follow the instructions of the scan tool for both retrieval and resetting of DTC's. The Honda Diagnostic System (HDS) can be used to command the MIL off.

➡**If the error symptoms causing the MIL to illuminate have been corrected, the MIL will return to normal operation.**

MASS AIR FLOW (MAF) SENSOR

LOCATION

The Mass Air Flow (MAF) sensor/Intake Air Temperature (IAT) sensor assembly is mounted on the air intake duct.

REMOVAL & INSTALLATION

K24Z1 Engine

See Figure 217.

1. Before servicing the vehicle, refer to the Precautions Section.
2. Disconnect the MAF/IAT sensor connector.
3. Remove the screws.
4. Remove the MAF/IAT sensor.

To install:

5. Install the MAF/IAT sensor with a new O-ring.
6. Tighten the screws (B) to 13 inch lbs. (1.5 Nm).

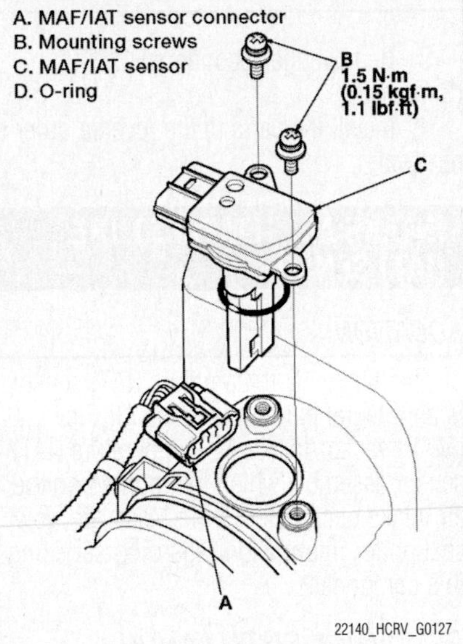
A. MAF/IAT sensor connector
B. Mounting screws
C. MAF/IAT sensor
D. O-ring
B
1.5 N·m
(0.15 kgf·m,
1.1 lbf·ft)
C
A
22140_HCRV_G0127

Fig. 217 Location of Mass Air Flow (MAF) and Intake Air Temperature (IAT) sensor

K24Z6 Engine

See Figure 218.

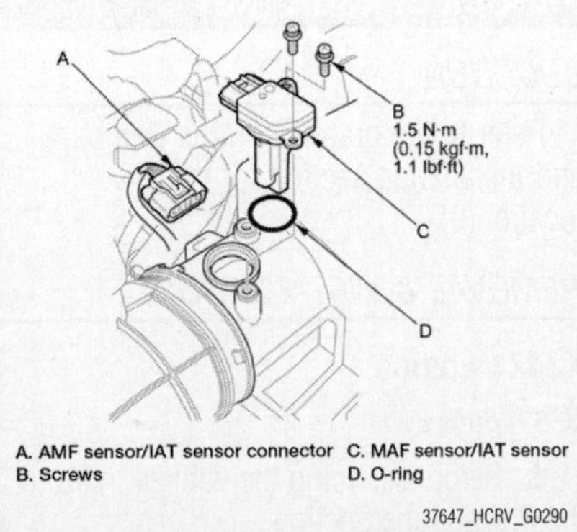
A
B
1.5 N·m
(0.15 kgf·m,
1.1 lbf·ft)
C
D
A. AMF sensor/IAT sensor connector
B. Screws
C. MAF sensor/IAT sensor
D. O-ring
37647_HCRV_G0290

Fig. 218 Disconnect the MAF sensor/IAT sensor connector

1. Disconnect the MAF sensor/IAT sensor connector.
2. Remove the screws.
3. Remove the MAF sensor/IAT sensor.
4. Install the parts in the reverse order of removal with a new O-ring.

MANIFOLD ABSOLUTE PRESSURE (MAP) SENSOR

LOCATION

Refer to the graphics in the Removal and Installation section for the location(s).

REMOVAL & INSTALLATION

K24Z1 Engine

See Figure 219.

1. Before servicing the vehicle, refer to the Precautions Section.
2. Disconnect the MAP sensor connector.
3. Remove the MAP sensor.

To install:

4. Install the MAP sensor with a new O-ring.
5. Connect the MAP sensor connector.

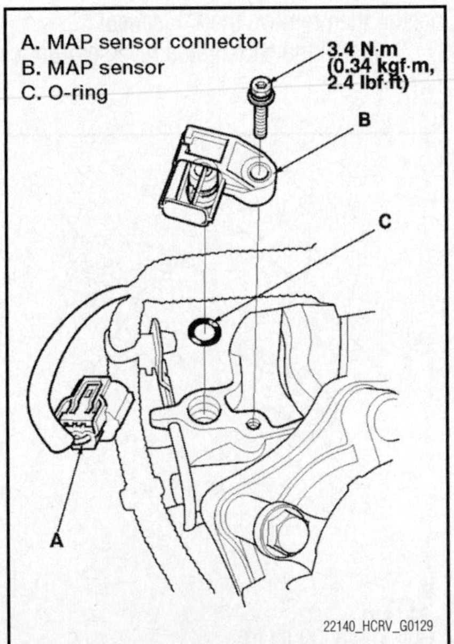
A. MAP sensor connector
B. MAP sensor
C. O-ring
3.4 N·m
(0.34 kgf·m,
2.4 lbf·ft)
B
C
A
22140_HCRV_G0129

Fig. 219 Location of the Manifold Absolute Pressure (MAP) Sensor

K24Z6 Engine

See Figure 220.

1. Disconnect the MAP sensor connector.
2. Remove the MAP sensor.
3. Install the parts in the reverse order of removal with a new O-ring.

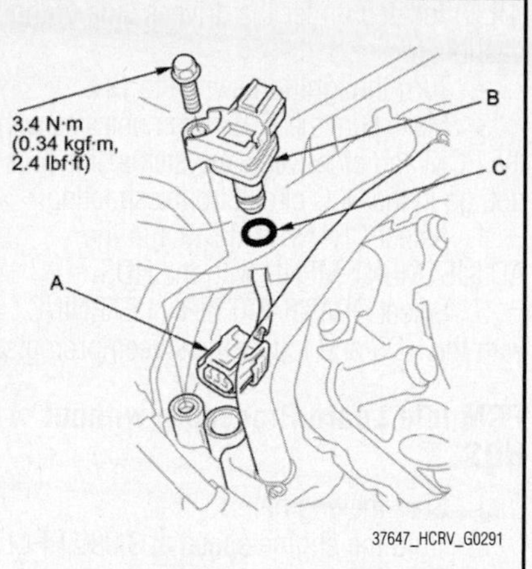
3.4 N·m
(0.34 kgf·m,
2.4 lbf·ft)
B
C
A
37647_HCRV_G0291

Fig. 220 Disconnect the MAP sensor connector (A); remove the MAP sensor (B) and the O-ring (C)

POSITIVE CRANKCASE VENTILATION (PCV) VALVE

LOCATION

Refer to the graphics in the Removal and Installation section for the location(s).

REMOVAL & INSTALLATION

See Figures 221 and 222.

1. Disconnect the PCV hose.
2. Remove the PCV valve.
3. Install the parts in the reverse order of removal with a new washer.

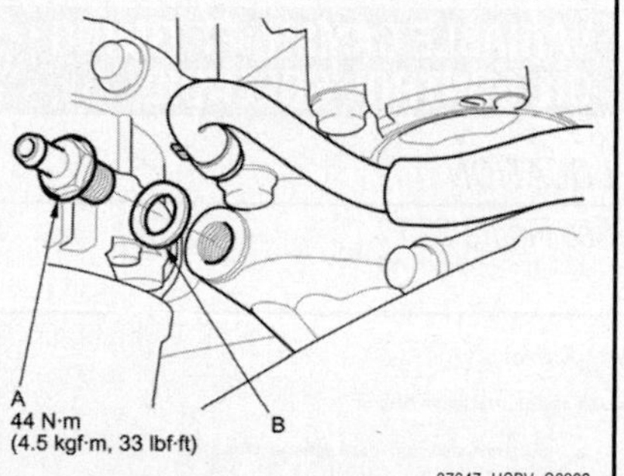
A
44 N·m
(4.5 kgf·m, 33 lbf·ft)
B
37647_HCRV_G0293

Fig. 221 Remove the PCV valve (A) and the washer (B)—K24Z1 engine

POWERTRAIN CONTROL MODULE (PCM)

LOCATION

See Figure 223.

The PCM is located on the driver's side in the engine compartment on the inner fender.

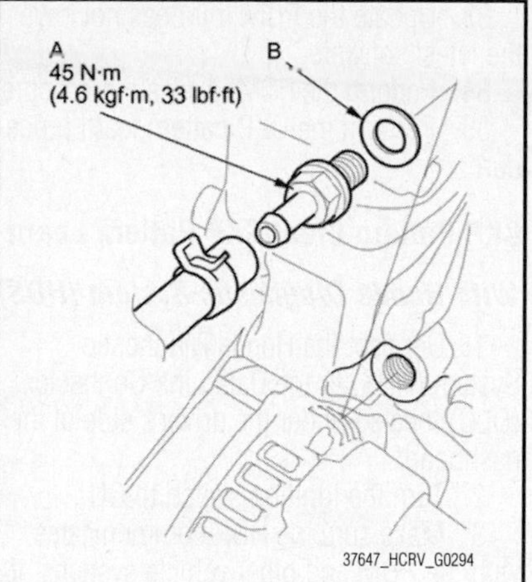
A
45 N·m
(4.6 kgf·m, 33 lbf·ft)
B
37647_HCRV_G0294

Fig. 222 Remove the PCV valve (A) and the washer (B)—K24Z6 engine

REMOVAL & INSTALLATION

See Figures 223 and 224.

Special tools required (one of the following):

- Honda Diagnostic System (HDS) tablet tester
- Honda Interface Module (HIM) and an iN workstation with the latest software version
- HDS pocket tester
- GNA600 and an iN workstation with the latest software version

➡Make sure the HDS is loaded with the latest software version. If you are replacing the PCM after substituting a known-good PCM, reinstall the original PCM, then do this procedure. During the procedure, if any READ DATA, WRITE DATA, or other data checks fail, note the failure, then continue.

1. Before servicing the vehicle, refer to the Precautions Section.
2. Connect the HDS to the Data Link Connector (DLC) located under the driver's side of the dashboard.
3. Turn the ignition switch to ON (II).
4. Make sure the HDS communicates with the PCM and other vehicles systems.
5. Select the PGM-FI system with the HDS.
6. Select the INSPECTION MENU with the HDS.
7. Select the ETCS TEST, then select the TP POSITION CHECK, and follow the screen prompts.

➡If the TP POSITION CHECK indicates FAILED, continue with this procedure.

8. Select the REPLACE PCM MENU, then select READ DATA and follow the screen prompts.

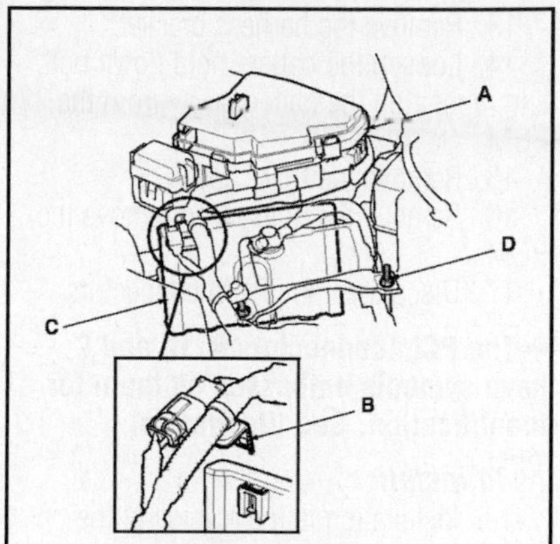
A
D
C
B

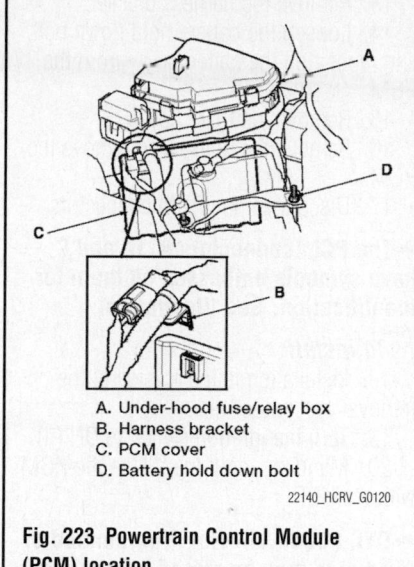

A. Under-hood fuse/relay box
B. Harness bracket
C. PCM cover
D. Battery hold down bolt

22140_HCRV_G0120

Fig. 223 Powertrain Control Module (PCM) location

➡Doing this step copies (READS) the engine oil life data from the original PCM so you can later download (WRITES) it into the new PCM. If READ DATA indicates FAILED, continue with this procedure.

9. Select the A/T system with the HDS.
10. Select the REPLACE TCM/PCM MENU, then select READ DATA and follow the screen prompts.

➡Doing this step copies (READS) the ATF life data from the original PCM so you can later download (WRITES) it into the new PCM. If READ DATA indicates FAILED, continue with this procedure.

11. Turn the ignition switch to LOCK (0).
12. Remove the under-hood fuse/relay box.

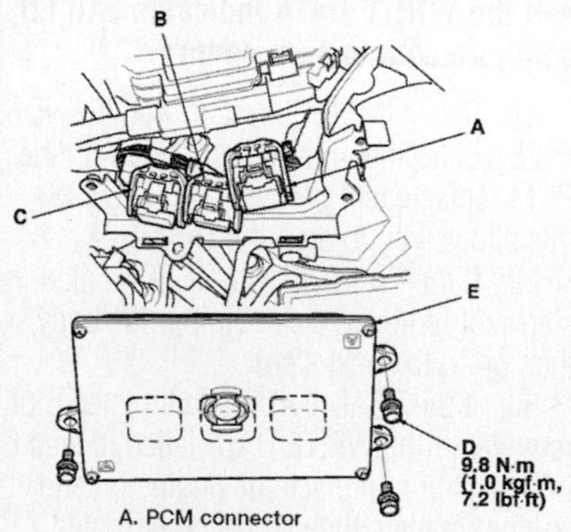
B
A
C
E
D
9.8 N·m
(1.0 kgf·m,
7.2 lbf·ft)

A. PCM connector
B. PCM connector
C. PCM connector
D. Mounting bolts
E. PCM

22140_HCRV_G0121

Fig. 224 Disconnect the PCM connectors and remove the PCM

13. Remove the harness bracket.
14. Loosen the battery hold down bolt, and reposition the battery away from the PCM.
15. Remove the PCM cover.
16. Remove the bolts, then remove the PCM.
17. Disconnect the PCM connectors.

➡The PCM connectors A, B, and C have symbols embossed on them for identification. See illustration.

To install:

18. Installation is the reverse of the removal procedure.
19. Turn the ignition switch to ON (II).
20. Manually input the VIN to the PCM with the HDS.

➡DTC P0630 VIN Not Programmed or Mismatch may be stored because the VIN has not been programmed into the PCM; ignore it, and continue this procedure.

21. If the READ DATA (engine oil life) failed in step 9 of removal, go to step 7 below. Otherwise, go to the next step.
22. Select the PGM-FI system with the HDS.
23. Select the REPLACE PCM MENU, then select WRITE DATA and follow the screen prompts.

➡If the WRITE DATA indicates FAILED, continue with this procedure.

24. If the READ DATA (ATF life) failed in step 11 of removal, go to step 9 below. Otherwise go to the next step.
25. Select the A/T SYSTEM with the HDS.
26. Select the REPLACE TCM/PCM MENU, then select WRITE DATA and follow the screen prompts.

➡If the WRITE DATA indicates FAILED, continue with this procedure.

27. Select IMMOBI system with the HDS.
28. Enter the immobilizer code with the PCM replacement procedure in the HDS; this allows you to start the engine.
29. If the TP POSITION CHECK failed in step 7 of removal, clean the throttle body, then go to the next step.
30. If the READ DATA failed in step 8 of removal or the WRITE DATA failed in step 6 in installation, replace the engine oil and engine oil filter, then go to the next step.
31. If the READ DATA failed in step 11 of removal or the WRITE DATA failed in step 9 of installation, replace the ATF, then go to the next step.
32. Select PGM-FI system and reset the PCM with the HDS.
33. Update the PCM if it does not have the latest software.
34. Perform the PCM idle learn procedure.
35. Perform the CKP pattern learn procedure.

CKP Pattern Clear/CKP Pattern Learn

With Honda Diagnostic System (HDS)

1. Connect the Honda Diagnostic System (HDS) to the Data Link Connector (DLC) located under the driver's side of the dashboard.
2. Turn the ignition switch to ON.
3. Make sure the HDS communicates with the PCM and other vehicle systems. If it does not, go to the DLC circuit troubleshooting.
4. Select CRANK PATTERN in the ADJUSTMENT MENU with the HDS.
5. Select CRANK PATTERN LEARNING with the HDS and follow the screen prompts.

Without Honda Diagnostic System (HDS)

1. Start the engine.
2. Hold the engine speed at 3,000 RPM without load (in P or N) until the radiator fan comes on.
3. Test-drive the vehicle on a level road: decelerate (with the throttle fully closed) from an engine speed of 2,500 RPM down to 1,000 RPM with the transaxle in 2 position.
4. Repeat step 3 several times.
5. Turn the ignition switch to LOCK.
6. Turn the ignition switch to ON and wait 30 seconds.

PCM Idle Learn Procedure with HDS

1. Connect the Honda Diagnostic System (HDS) to the Data Link Connector (DLC) located under the driver's side of the dashboard.
2. Turn the ignition switch to ON.
3. Make sure the HDS communicates with the PCM and other vehicle systems. If it does not, go to the DLC circuit troubleshooting.
4. Select CRANK PATTERN in the ADJUSTMENT MENU with the HDS.
5. Select CRANK PATTERN LEARNING with the HDS and follow the screen prompts.

PCM Idle Learn Procedure without HDS

1. Start the engine.
2. Hold the engine speed at 3,000 RPM without load (in P or N) until the radiator fan comes on.
3. Test-drive the vehicle on a level road: decelerate (with the throttle fully closed) from an engine speed of 2,500 RPM down to 1,000 RPM with the transaxle in 2 position.
4. Repeat step 3 several times.
5. Turn the ignition switch to LOCK.
6. Turn the ignition switch to ON and wait 30 seconds.

THROTTLE POSITION SENSOR (TPS)

REMOVAL & INSTALLATION

The TPS is a component of the electronic throttle control system. This sensor is located with the throttle actuator and is integral to the throttle body. Refer to Throttle Body, removal & installation.

VARIABLE CAMSHAFT TIMING OIL CONTROL SOLENOID

LOCATION

See Figure 225.

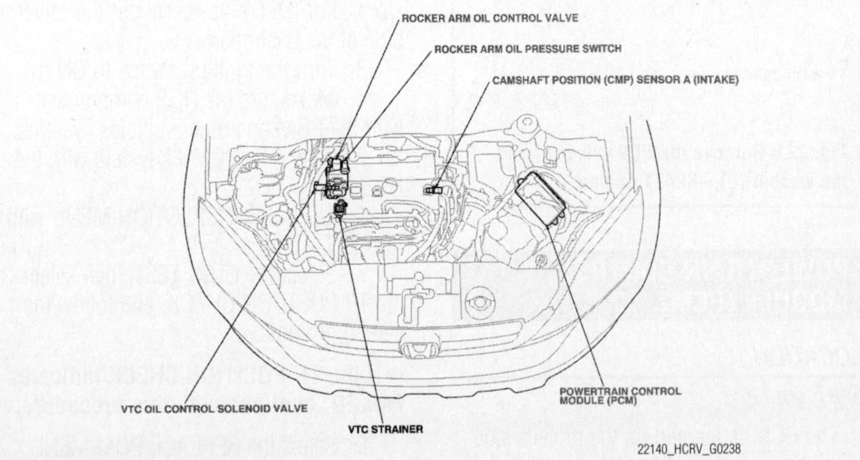
ROCKER ARM OIL CONTROL VALVE
ROCKER ARM OIL PRESSURE SWITCH
CAMSHAFT POSITION (CMP) SENSOR A (INTAKE)
VTC OIL CONTROL SOLENOID VALVE
VTC STRAINER
POWERTRAIN CONTROL MODULE (PCM)
22140_HCRV_G0238

Fig. 225 Location of the variable camshaft timing oil control solenoid valve and related components—K24Z1 engine

REMOVAL & INSTALLATION

K24Z1 Engine

See Figure 226.

1. Before servicing the vehicle, refer to the Precautions Section.
2. Disconnect the Variable Timing Control (VTC) oil control solenoid valve 2P connector.
3. Remove the mounting bolt and the VTC oil control solenoid valve.

To install:

4. Replace the VTC valve O-ring. Coat a new O-ring with clean engine oil, then install it.
5. Clean and dry the mating surface of the valve.
6. Install the VTC solenoid valve. Tighten the mounting bolt to 9 ft. lbs. (12 Nm).
7. Connect the 2P connector.

➡Do not install the valve while wearing cloth fibrous gloves. Be careful not to contaminate the cylinder head opening.

K24Z6 Engine

See Figure 227.

1. Disconnect the VTC oil control solenoid valve 2P connector.
2. Remove the bolt and the VTC oil control solenoid valve.

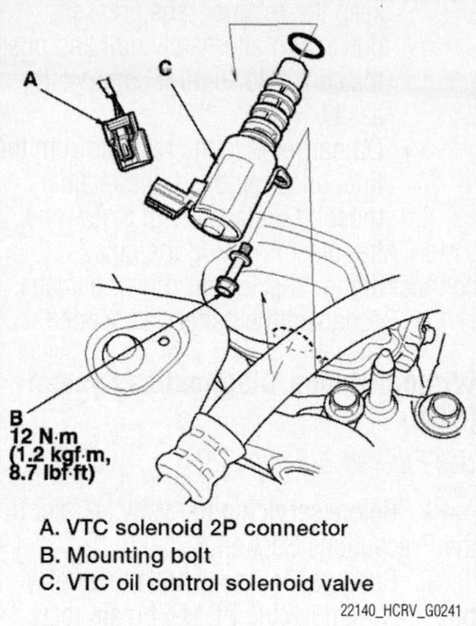
A
C
B
12 N·m
(1.2 kgf·m,
8.7 lbf·ft)
A. VTC solenoid 2P connector
B. Mounting bolt
C. VTC oil control solenoid valve
22140_HCRV_G0241

Fig. 226 View of the VTC oil control solenoid during removal

3. Check the VTC oil control solenoid valve strainer for clogging. If the strainer is clogged, replace the VTC oil control solenoid valve.
4. Note the amount of valve opening by observing the position of the piston shoulder through the valve retard drain port. If you see the shoulder of the piston, the valve is open and must be replaced.

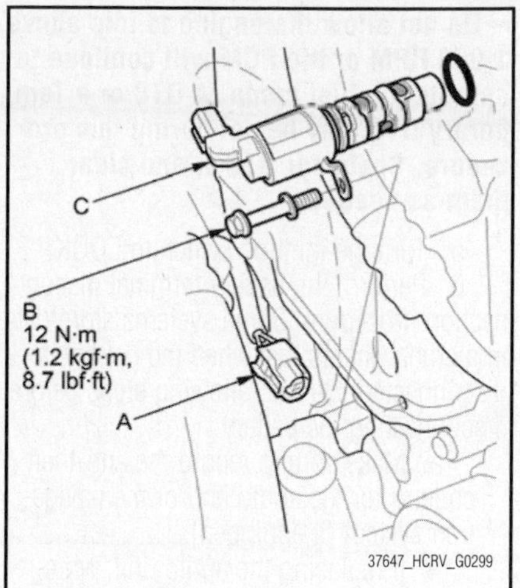
C
B
12 N·m
(1.2 kgf·m,
8.7 lbf·ft)
A
37647_HCRV_G0299

Fig. 227 Disconnect the VTC oil control solenoid valve 2P connector (A); remove the bolt (B) and the VTC oil control solenoid valve (C)

To install:

5. Replace the VTC oil control valve O-ring.
6. Coat a new O-ring with clean engine oil, then install it on the valve.
7. Clean and dry the mating surface of the valve.
8. Install the VTC oil control valve.

➡Do not install the valve while wearing cloth fibrous gloves. Be careful not to contaminate the cylinder head opening.

FUEL

GASOLINE FUEL INJECTION SYSTEM

FUEL SYSTEM SERVICE PRECAUTIONS

Safety is the most important factor when performing not only fuel system maintenance, but any type of maintenance. Failure to conduct maintenance and repairs in a safe manner may result in serious personal injury or death. Work on a vehicle's fuel system components can be accomplished safely and effectively by adhering to the following rules and guidelines.

- To avoid the possibility of fire and personal injury, always disconnect the negative battery cable unless the repair or test procedure requires that battery voltage be applied.
- Always relieve the fuel system pressure prior to disconnecting any fuel system component (injector, fuel rail, pressure regulator, etc.) fitting or fuel line connection. Exercise extreme caution whenever relieving fuel system pressure to avoid exposing skin, face and eyes to fuel spray. Please be advised that fuel under pressure may penetrate the skin or any part of the body that it contacts.
- Always place a shop towel or cloth around the fitting or connection prior to loosening to absorb any excess fuel due to spillage. Ensure that all fuel spillage is quickly removed from engine surfaces. Ensure that all fuel-soaked cloths or towels are deposited into a flame-proof waste container with a lid.
- Always keep a dry chemical (Class B) fire extinguisher near the work area.
- Do not allow fuel spray or fuel vapors to come into contact with a spark or open flame.
- Always use a second wrench when loosening or tightening fuel line connection fittings. This will prevent unnecessary stress and torsion on fuel piping. Always follow the proper torque specifications.
- Always replace worn fuel fitting O-rings with new ones. Do not substitute fuel hose where rigid pipe is installed.

FUEL SYSTEM PRESSURE

RELIEVING

Before disconnecting fuel lines or hoses, relieve pressure from the system by disabling the fuel pump and then disconnecting the fuel tube/quick connect fitting in the engine compartment.

With Honda Diagnostics System (HDS)

See Figure 228.

1. Before servicing the vehicle, refer to the Precautions Section.
2. Remove the fuel fill cap, to relieve the pressure in the fuel tank.
3. Turn the ignition switch to ON.
4. From the INSPECTION MENU of the Honda Diagnostics System (HDS), select Fuel Pump OFF, then start the engine, and let it idle until it stalls.

➡Do not allow the engine to idle above 1,000 RPM or the PCM will continue to operate the fuel pump. A DTC or a Temporary DTC may be set during this procedure. Check for DTC's, and clear them as needed.

5. Turn the ignition switch to LOCK.
6. Perform the battery terminal disconnection procedure. Some systems store data in memory that is lost when the battery is disconnected. Do the following steps before disconnecting the battery:
   a. Make sure to record the anti-theft code(s) for the audio and/or the navigation system (if equipped).
   b. If replacing the audio unit, write down the audio presets (AM and FM), and the XM audio presets (if equipped), because the audio unit does not retain the presets after the battery is disconnected.
   c. Make sure the ignition switch is in LOCK position.
   d. Disconnect and isolate the negative cable from the battery.

➡Always disconnect the negative cable from the battery first.

   e. Disconnect the positive cable from the battery.
7. Remove the quick-connect fitting cover.
8. Check the fuel quick-connect fitting for dirt, and clean it if needed.
9. Place a rag or shop towel over the quick-connect fitting.
10. Disconnect the quick-connect fitting: hold the connector with one hand and squeeze the retainer tabs with the other hand to release them from the locking tabs. Pull the connector off.
   - Be careful not to damage the line or other parts.
   - Do not use tools.
   - If the connector does not move, keep the retainer tabs pressed down, and alternately pull and push the connector until it comes off easily.
   - Do not remove the retainer from the line; once removed, the retainer must be replaced with a new one.
11. After disconnecting the quick-connect fitting, check it for dirt or damage.
12. Reconnect the battery as needed.

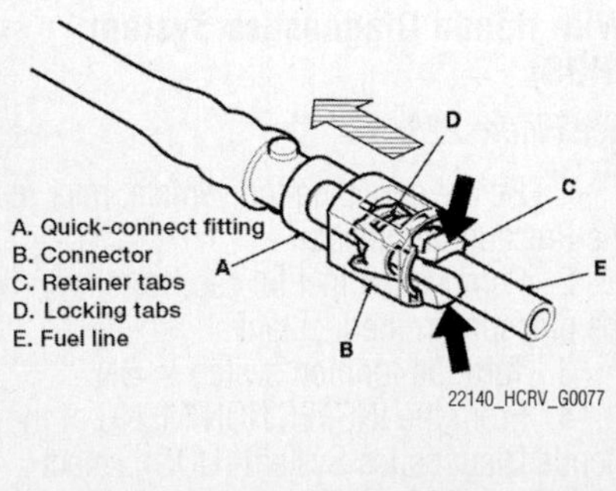
A. Quick-connect fitting
B. Connector
C. Retainer tabs
D. Locking tabs
E. Fuel line
A
B
C
D
E
22140_HCRV_G0077

Fig. 228 Quick-connect fuel connection illustrated

Without Honda Diagnostics System (HDS)

See Figures 228 and 229.

1. Before servicing the vehicle, refer to the Precautions Section.
2. Remove the under-dash fuse/relay box, then remove the PGM-FI main relay 2 (FUEL PUMP) (A) from the under-dash fuse/relay box.
3. Reinstall the under-dash fuse/relay box.
4. Start the engine, and let it idle until it stalls.

➡If any DTC's are stored, clear and ignore them.

5. Turn the ignition switch to LOCK.
6. Remove the fuel fill cap.
7. Perform the battery terminal disconnection procedure. Some systems store data in memory that is lost when the battery is disconnected. Do the following steps before disconnecting the battery:
   a. Make sure to record the anti-theft code(s) for the audio and/or the navigation system (if equipped).
   b. If replacing the audio unit, write down the audio presets (AM and FM), and the XM audio presets (if equipped), because the audio unit does not retain the presets after the battery is disconnected.
   c. Make sure the ignition switch is in LOCK position.
   d. Disconnect and isolate the negative cable from the battery.

➡Always disconnect the negative cable from the battery first.

   e. Disconnect the positive cable from the battery.
   f. Disconnect the positive cable from the battery.
8. Remove the quick-connect fitting cover.
9. Check the fuel quick-connect fitting for dirt, and clean it if needed.
10. Place a rag or shop towel over the quick-connect fitting.
11. Disconnect the quick-connect fitting: hold the connector with one hand and squeeze the retainer tabs with the other hand to release them from the locking tabs. Pull the connector off.
   - Be careful not to damage the line or other parts
   - Do not use tools
   - If the connector does not move, keep the retainer tabs pressed down, and alternately pull and push the connector until it comes off easily
   - Do not remove the retainer from the line; once removed, the retainer must be replaced with a new one
12. After disconnecting the quick-connect fitting, check it for dirt or damage.
13. Reconnect the battery as needed.

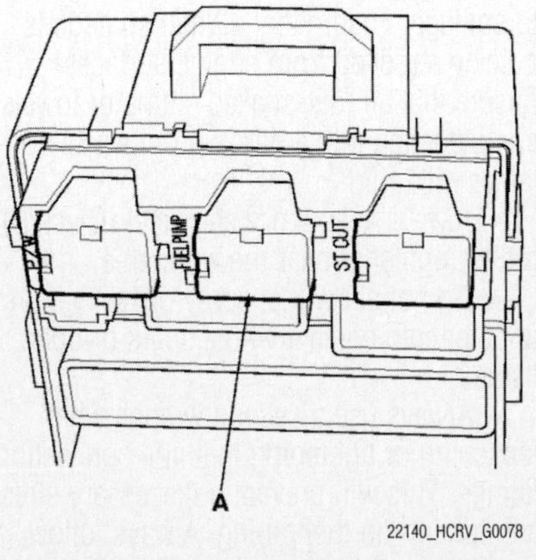
FUEL PUMP
ST CUT
A
22140_HCRV_G0078

Fig. 229 Remove the PGM-FI main relay 2 (FUEL PUMP) (A) from the under-dash fuse/relay box

FUEL FILTER

REMOVAL & INSTALLATION

See Figure 230.

The fuel filter should be replaced whenever the fuel pressure drops below the specified value (47–54 psi), after making sure that the fuel pump and the fuel pressure regulator are functioning properly.

1. Before servicing the vehicle, refer to the Precautions Section.
2. Relieve the fuel pressure. Refer to Relieving Fuel System Pressure.
3. Remove the fuel pump unit. Refer to Fuel Pump, removal & installation.
4. Remove the fuel filter set.

To install:

5. Check these items before installing the fuel pump tank unit:
   a. When connecting the wire harness, make sure the connection is secure and

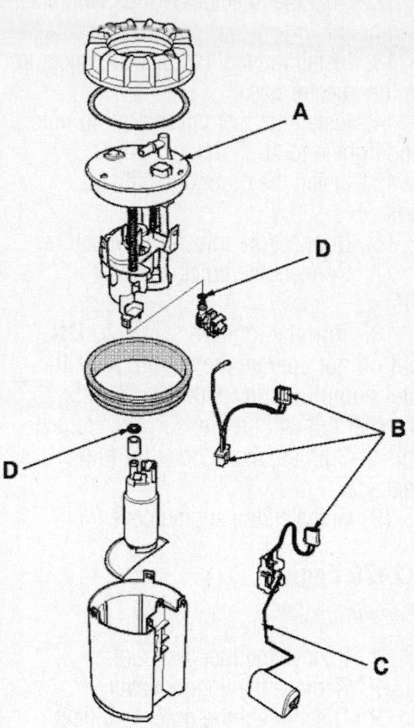
A
D
B
D
C

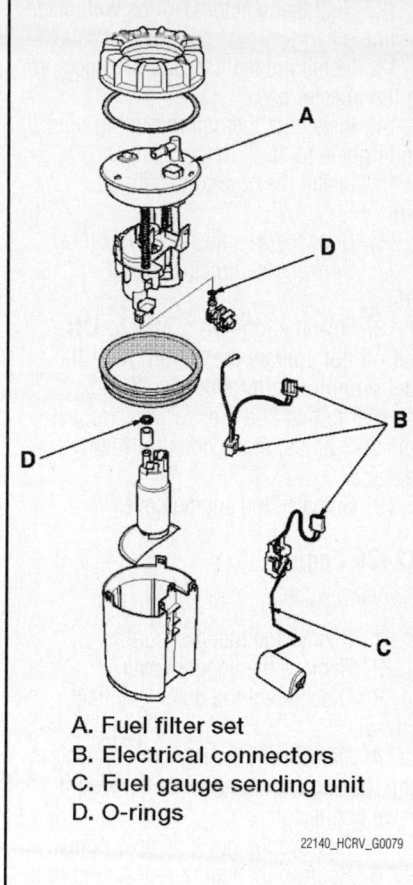

A. Fuel filter set
B. Electrical connectors
C. Fuel gauge sending unit
D. O-rings

22140_HCRV_G0079

Fig. 230 Exploded view of the fuel pump tank unit

the connectors are firmly locked into place.

b. When installing the fuel gauge sending unit, make sure the connection is secure and the connector is firmly locked into place. Be careful not to bend or twist it excessively.

6. Install the parts in the reverse order of removal with new O-rings.

➡**Coat the O-rings with clean engine oil.**

7. Install the fuel pump tank unit.

FUEL LEVEL SENDING UNIT

LOCATION

The fuel level sensor is part of the fuel tank unit.

REMOVAL & INSTALLATION

See Figure 231.

1. Before servicing the vehicle, refer to the Precautions Section.

2. Remove the fuel tank unit. Refer to Fuel Pump, removal & installation.

3. Remove the fuel level sensor (fuel sending unit) from the fuel tank unit.

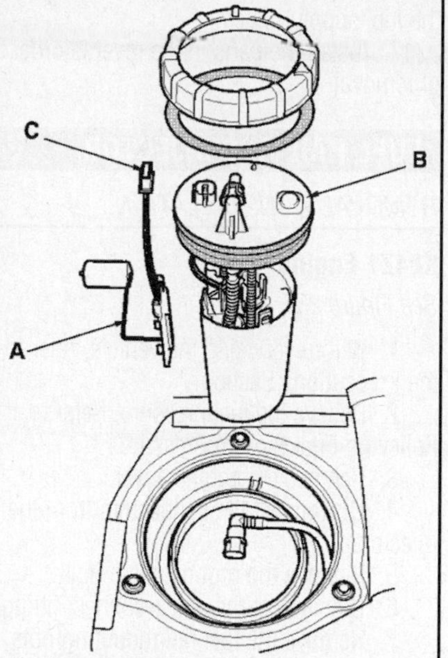
C
B
A

A. Fuel level sending unit
B. Fuel tank unit
C. 4P connector

22140_HCRV_G0245

Fig. 231 Exploded view of fuel pump and fuel level sending unit

To install:

4. When connecting the wire harness, make sure the connection is secure and the connector is firmly locked into place.

5. When installing the fuel gauge sending unit, make sure the connection is secure and the connector is firmly locked into place. Be careful not to bend or twist it excessively.

6. Install the parts in the reverse order of removal.

7. Install the fuel tank unit. Refer to Fuel Pump, removal & installation.

FUEL TANK UNIT

REMOVAL & INSTALLATION

See Figures 232 and 233.

Special tools required: Fuel sender wrench 07AAA-S0XA100

1. Before servicing the vehicle, refer to the Precautions Section.

2. Relieve the fuel pressure. Refer to Relieving Fuel System Pressure.

3. Disconnect the negative battery cable.

4. Remove the fuel fill cap.

5. Fold the left side rear seat forward and pull back the carpet to expose the access panel.

6. Remove the access panel from the floor.

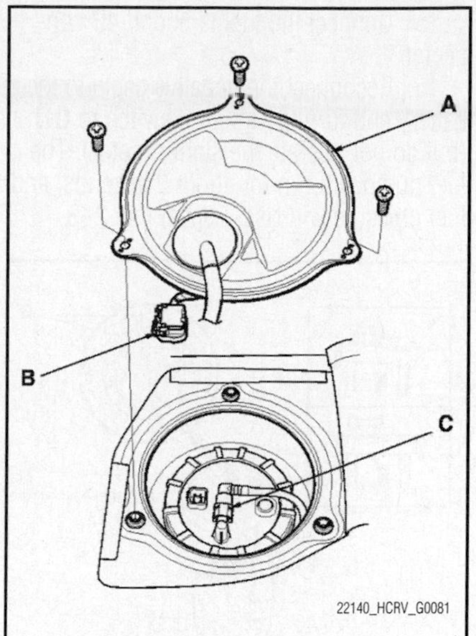
A
B
C

22140_HCRV_G0081

Fig. 232 Remove the access panel (A) from the floor, disconnect the 4P connector (B) and the quick-connect fitting (C) from the fuel pump tank unit

7. Disconnect the fuel tank unit 4P connector.

8. Disconnect the quick-connect fitting from the fuel tank unit.

9. Using the special tool, loosen the locknut.

10. Remove the locknut, the base gasket, and the fuel pump tank unit.

To install:

11. Temporarily attach a new base gasket to the fuel pump tank unit, then insert the fuel tank unit partially into the fuel tank.

- Be careful not to damage the new base gasket
- Be careful not to bend the fuel gauge sending unit
- Do not coat the base gasket with oil

12. Transfer the base gasket from the fuel tank unit to the fuel tank.

13. Align the marks on the fuel tank and the fuel tank unit, then insert the fuel tank unit until it sits on the base gasket.

⁂ CAUTION

To prevent a fuel leak, check the base gasket, visually or by hand, to make sure it is not pinched.

14. Using the special tool, tighten the new locknut to 52 ft. lbs. (70 Nm).

- After tightening, make sure the marks are still aligned
- After installation, check the base gasket visually or by hand to be sure it is not pinched

15. Connect the fuel tank unit 4P connector.

16. Reconnect the negative cable to the battery and turn the ignition switch to **ON** (but do not operate the starter motor). The fuel pump will run for about 2 seconds, and fuel pressure will rise. Repeat this 2–3 times, and check that there is no leakage in the fuel supply system.

17. Install the parts in the reverse order of removal.

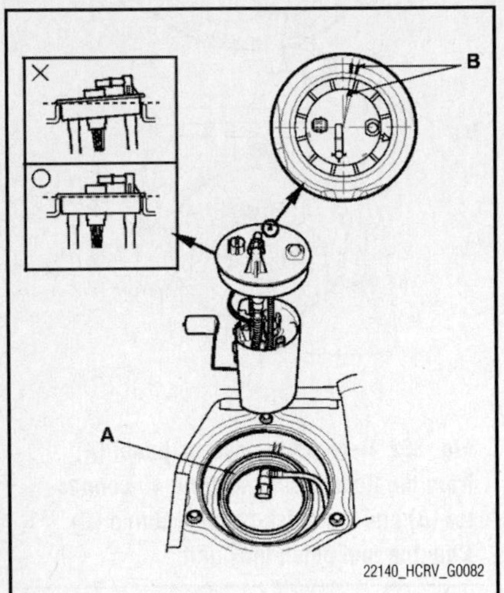
B
A
22140_HCRV_G0082

Fig. 233 Install the base gasket (A). Align the marks (B) on the fuel tank and the fuel tank unit, then insert the fuel tank unit until it sits on the base gasket

FUEL RAIL AND INJECTOR

REMOVAL & INSTALLATION

K24Z1 Engine

See Figure 234.

1. Before servicing the vehicle, refer to the Precautions Section.
2. Relieve the fuel pressure. Refer to Relieving Fuel System Pressure.
3. Remove the engine cover.
4. Disconnect the connectors from the injectors.
5. Remove the ground cable bolt.
6. Disconnect the quick-connect fitting.
7. Remove the fuel rail mounting nuts from the fuel rail.
8. Remove the injector clips from the injectors.
9. Remove the injectors from the fuel rail.

To install:

10. Coat the new O-rings with clean engine oil and insert the injectors into the fuel rail.
11. Install the injector clips.
12. Coat the injector O-rings with clean engine oil.
13. Install the fuel rail and the injectors in the injector base.
14. Install the fuel rail mounting nuts and tighten to 16 ft. lbs. (22 Nm).
15. Install the ground cable and bolt.
16. Connect the injector connectors.
17. Connect the quick-connect fitting.
18. Turn the ignition switch to **ON**, but do not operate the starter. After the fuel pump runs for about 2 seconds, the fuel rail will be pressurized. Repeat this 2–3 times, then check for fuel leakage.
19. Reinstall the engine cover.

K24Z6 Engine

See Figure 235.

1. Relieve the fuel pressure.
2. Remove the engine cover.
3. Disconnect the quick-connect fitting.
4. Disconnect the injector connectors and the engine mount control solenoid valve connector.
5. Remove the ground cable bolts.
6. Remove the fuel rail mounting nuts from the fuel rail.
7. Remove the fuel rail and the injectors from the injector base.
8. Remove the injector clips from the fuel rail.
9. Remove the injectors from the fuel rail.

To install:

10. Coat the new O-rings (black) with clean engine oil, and insert the injectors into the fuel rail.
11. Install the injector clips.
12. Coat the new injector O-rings (brown) with clean engine oil.
13. Install the fuel rail and the injectors in the injector base.
14. Install the fuel rail mounting nuts and the ground cable bolts.
15. Connect the injector connectors and the engine mount control solenoid valve connector.
16. Connect the quick-connect fitting.
17. Turn the ignition switch to ON (II), but do not operate the starter. After the fuel pump runs for about 2 seconds, the fuel rail is pressurized. Repeat this two or three times, then make sure there are no fuel leaks.
18. Reinstall the engine cover.

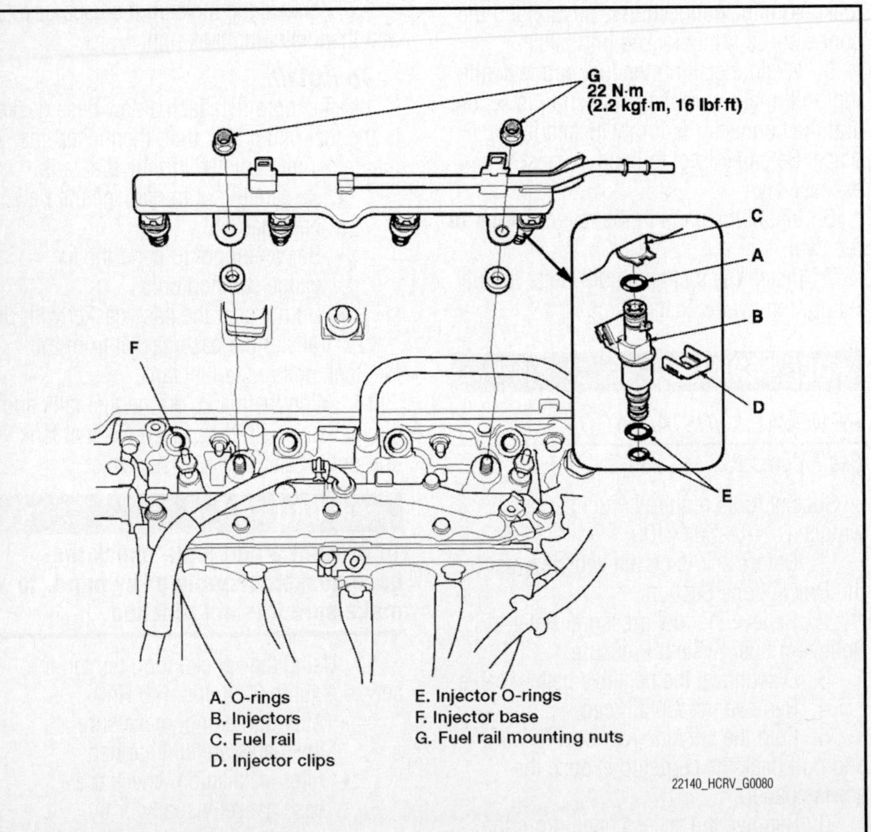
G
22 N·m
(2.2 kgf·m, 16 lbf·ft)
C
A
B
D
E
F
A. O-rings
B. Injectors
C. Fuel rail
D. Injector clips
E. Injector O-rings
F. Injector base
G. Fuel rail mounting nuts
22140_HCRV_G0080

Fig. 234 Exploded view of the fuel injector/fuel rail

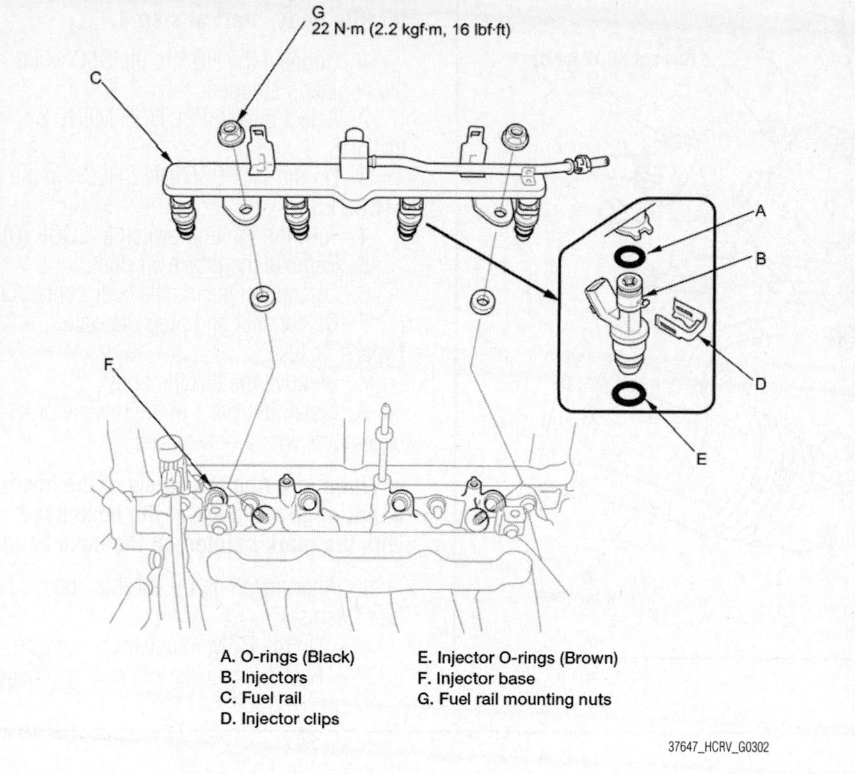
G
22 N·m (2.2 kgf·m, 16 lbf·ft)
C
A
B
D
E
F
A. O-rings (Black)
B. Injectors
C. Fuel rail
D. Injector clips
E. Injector O-rings (Brown)
F. Injector base
G. Fuel rail mounting nuts
37647_HCRV_G0302

Fig. 235 Exploded view of fuel rail and injector assembly

IDLE SPEED

ADJUSTMENT

Idle speed is maintained by the Powertrain Control Module (PCM). No adjustment is necessary or possible.

THROTTLE BODY

REMOVAL & INSTALLATION

K24Z1 Engine

See Figure 237.

CAUTION

Do not insert your fingers into the installed throttle body when you turn the ignition switch to ON or while the ignition switch is ON. Serious injury may result if the throttle valve is activated.

➡If replacing or cleaning the throttle body, start at step 2. If removing the throttle body, start at step 5.

1. Before servicing the vehicle, refer to the Precautions Section.

FUEL TANK

DRAINING

1. Remove the fuel tank unit.
2. Using a hand pump, a hose, and a container suitable for fuel, draw the fuel from the fuel tank.
3. Reinstall the fuel tank unit.

REMOVAL & INSTALLATION

See Figure 236.

1. Remove the fuel tank unit, then drain the fuel tank.
2. Raise the vehicle on a lift.
3. Remove the EVAP canister.
4. Remove the cover and the fuel tank guard.
5. 4WD model: Remove the propeller shaft.
6. Remove the exhaust pipe.
7. Disconnect the hoses. Slide back the clamps, then twist the hoses as you pull to avoid damaging them.
8. Place a jack or other support under the fuel tank.
9. Remove the strap bolts and the straps.
10. Remove the fuel tank.
11. Install the parts in the reverse order of removal.

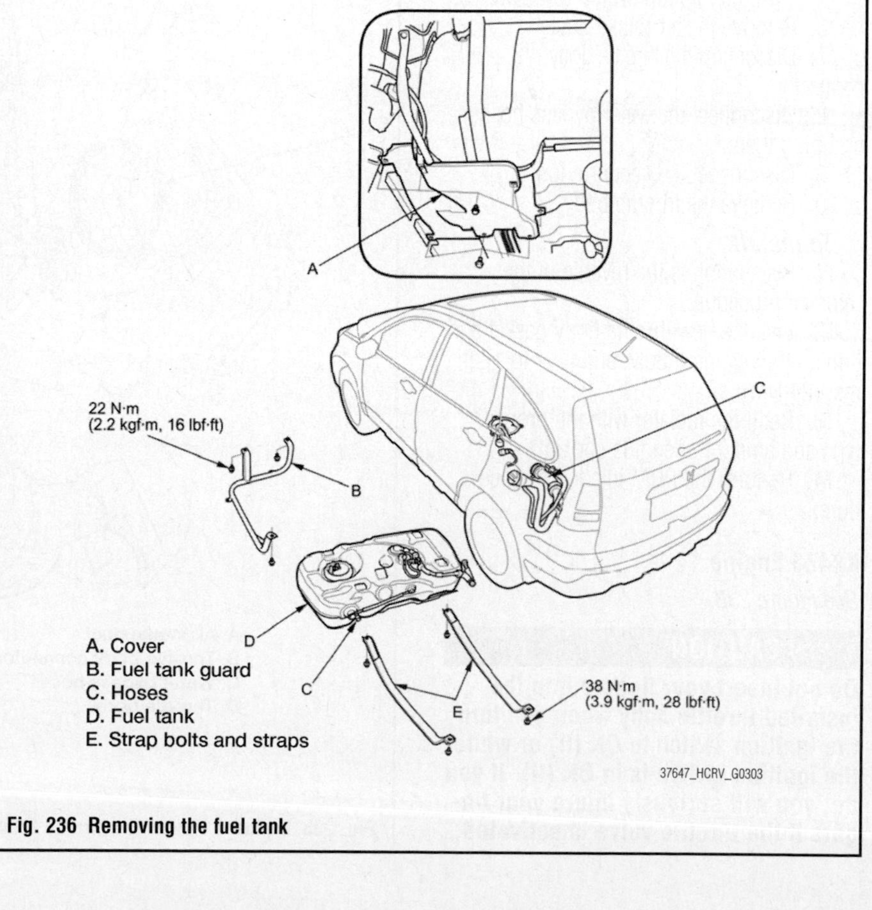
A
C
22 N·m
(2.2 kgf·m, 16 lbf·ft)
B
D
C
E
38 N·m
(3.9 kgf·m, 28 lbf·ft)
A. Cover
B. Fuel tank guard
C. Hoses
D. Fuel tank
E. Strap bolts and straps
37647_HCRV_G0303

Fig. 236 Removing the fuel tank

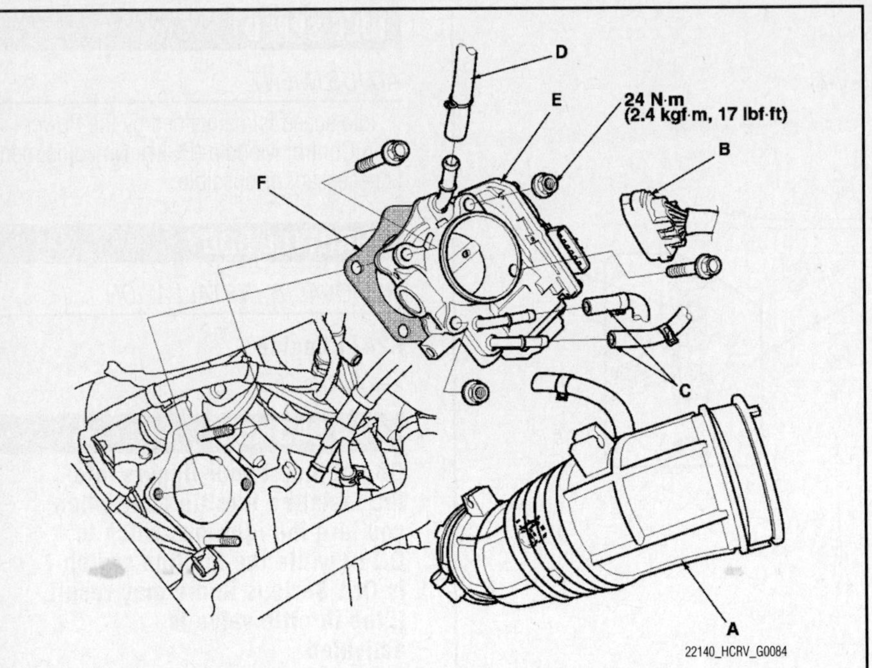
D
E
24 N·m
(2.4 kgf·m, 17 lbf·ft)
B
F
C
A
22140_HCRV_G0084

Fig. 237 View of throttle body components

2. Connect the Honda Diagnostics System (HDS) to the Data Link Connector (DLC) while the engine is stopped.
3. Select the INSPECTION MENU on the HDS.
4. Perform the TP POSITION CHECK in the ETCS TEST.
5. Turn the ignition switch to LOCK (0).
6. Remove the air intake duct.
7. Disconnect the throttle body connector.
8. Disconnect the water bypass hoses and plug them.
9. Disconnect the vacuum hose.
10. Remove the throttle body.

To install:

11. Installation is the reverse of the removal procedure.
12. Install a new throttle body gasket and tighten the retaining bolts and nuts to 17 ft. lbs. (24 Nm).
13. Refill the radiator with the proper type and amount of engine coolant.
14. Perform the PCM idle learn procedure.

K24Z6 Engine

See Figure 238.

CAUTION

Do not insert your fingers into the installed throttle body when you turn the ignition switch to ON (II) or while the ignition switch is in ON (II). If you do, you will seriously injure your fingers if the throttle valve is activated.

➡If you are replacing the throttle body, start at step 1. If you are removing the throttle body, start at step 4.

1. Connect the HDS to the DLC while the engine is stopped.
2. Select the INSPECTION MENU on the HDS.
3. Do the TP POSITION CHECK in the ETCS TEST.
4. Turn the ignition switch to LOCK (0).
5. Remove the intake air duct.
6. Disconnect the throttle body connector.
7. Disconnect and plug the water bypass hoses.
8. Remove the throttle body.
9. Install the parts in the reverse order of removal with a new gasket.

➡When torquing the screw of the hose band, align the edge of the hose band with the mark painted on the hose band.

10. After installing the throttle body, do these items:
    - Do the PCM idle learn procedure.
    - Refill the radiator with engine coolant.

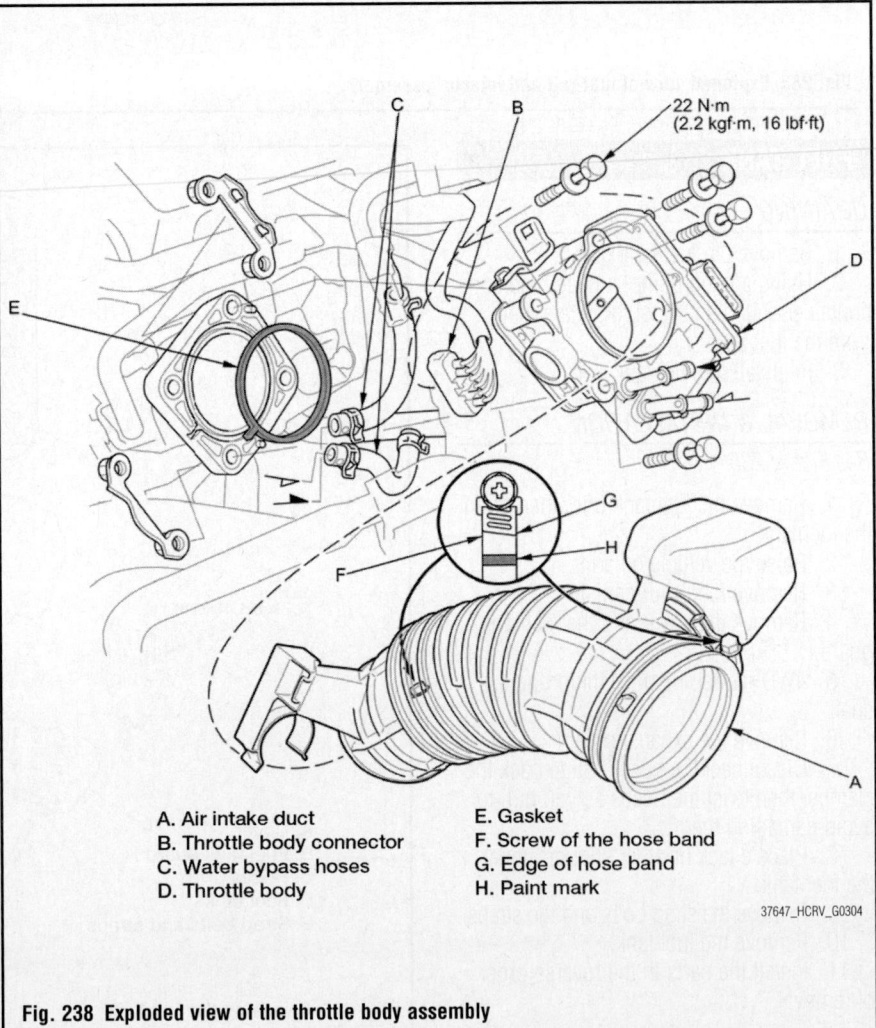

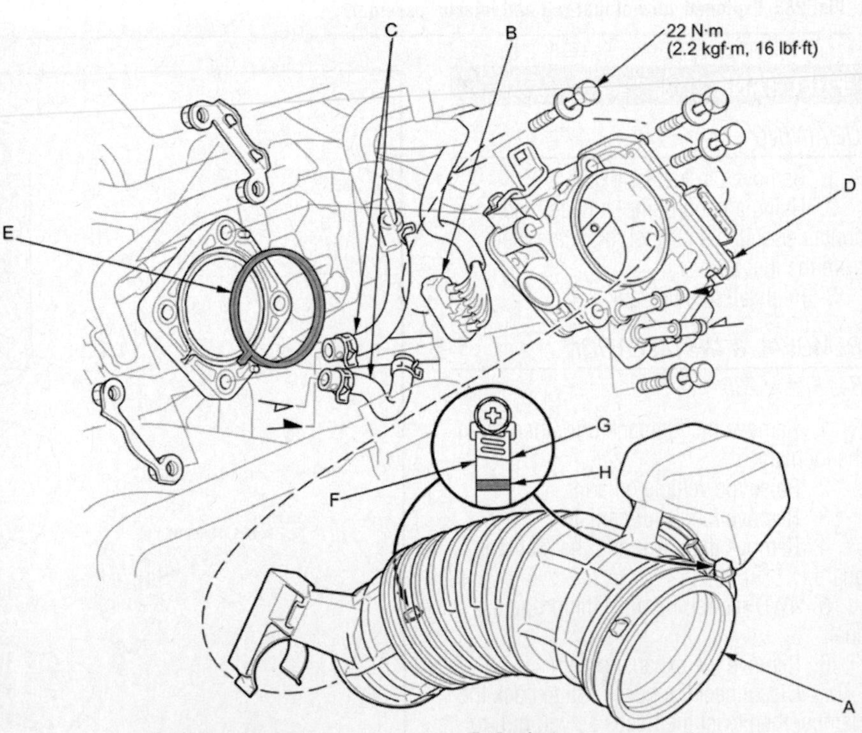
C
B
22 N·m
(2.2 kgf·m, 16 lbf·ft)
D
E
G
H
F
A

A. Air intake duct
B. Throttle body connector
C. Water bypass hoses
D. Throttle body
E. Gasket
F. Screw of the hose band
G. Edge of hose band
H. Paint mark

37647_HCRV_G0304

Fig. 238 Exploded view of the throttle body assembly

HEATING & AIR CONDITIONING SYSTEM

BLOWER MOTOR

REMOVAL & INSTALLATION

See Figures 239 through 243.

1. Remove the glove box.
2. Disconnect the connector, then remove the connector clip, the wire harness clips, the bolts, and the glove box frame.
3. Cut the plastic cross brace in the glove box opening with diagonal cutters in the area shown. Retain the plastic cross brace it will be reinstalled.
4. Remove the wire harness clips, the self-tapping screws, and the passenger's heater duct.
5. Disconnect the connector from the blower motor. Remove the wire harness clip.

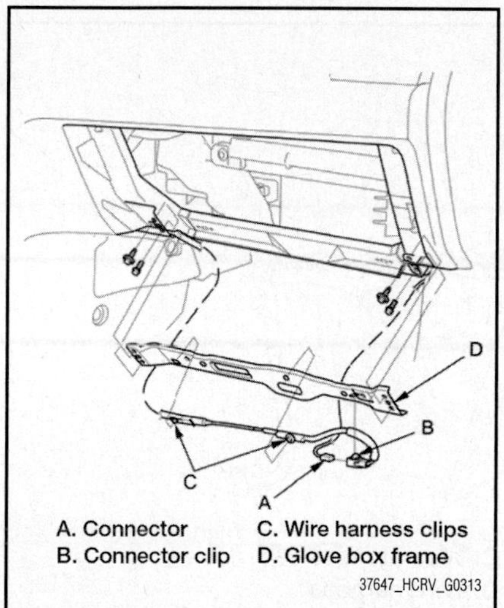

D
B
C
A
A. Connector
B. Connector clip
C. Wire harness clips
D. Glove box frame
37647_HCRV_G0313

Fig. 239 Disconnect the connector, then remove the connector clip, the wire harness clips, the bolts, and the glove box frame

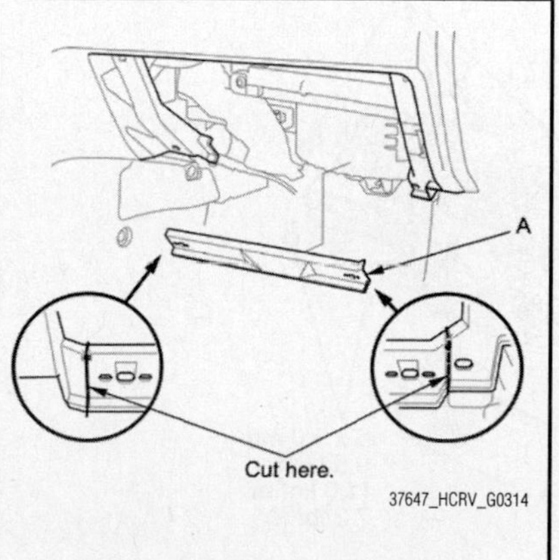

A
Cut here.
37647_HCRV_G0314

Fig. 240 Cut the plastic cross brace (A) in the glove box opening

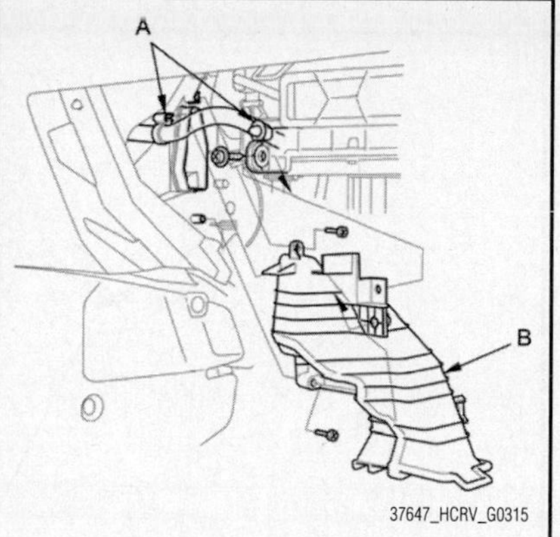

A
B
37647_HCRV_G0315

Fig. 241 Remove the wire harness clips (A), the self-tapping screws, and the passenger's heater duct (B)

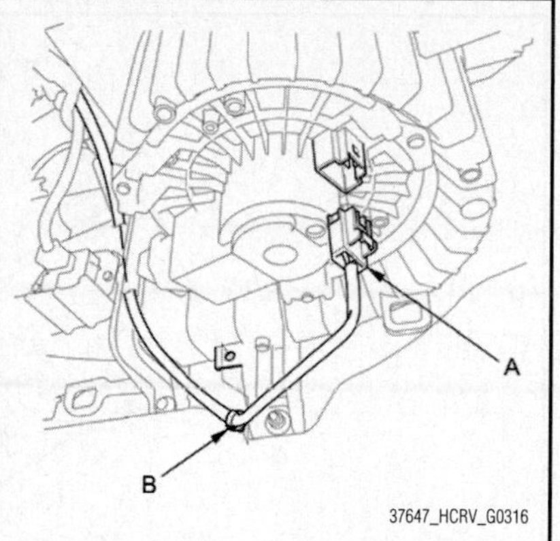

A
B
37647_HCRV_G0316

Fig. 242 Disconnect the connector (A) from the blower motor; remove the wire harness clip (B)

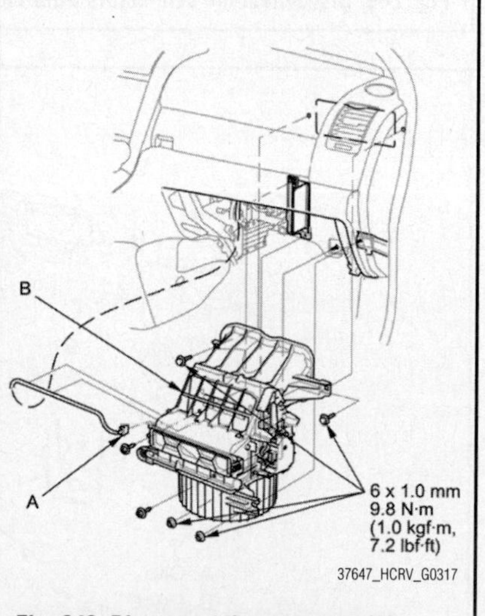

B
A
6 x 1.0 mm
9.8 N·m
(1.0 kgf·m,
7.2 lbf·ft)
37647_HCRV_G0317

Fig. 243 Disconnect the connector (A) from the recirculation control motor; remove the self-tapping screws, the bolt, the mounting nuts, and the blower unit (B)

6. Disconnect the connector from the recirculation control motor. Remove the self-tapping screws, the bolt, the mounting nuts, and the blower unit.
7. Install the unit in the reverse order of removal. Make sure that there are no air leaks.

HEATER CORE

REMOVAL & INSTALLATION

See Figures 244 through 251.

**** WARNING**

SRS components are located in this area. Review the SRS component locations and the precautions and procedures before doing repairs or service.

1. Do the battery terminal disconnection procedure.
2. Recover the refrigerant with a recovery/recycling/ charging station
3. Disconnect the A/C line from the evaporator core.
4. When the engine is cool, drain the engine coolant from the radiator.
5. From under the hood, slide the hose clamps back. Disconnect the inlet heater hose and the outlet heater hose from the heater unit. Note the layout of the hoses.

➡Engine coolant will run out when the hoses are disconnected; drain it into a clean drip pan. Be sure not to let coolant spill on the electrical parts or the painted surfaces. If any coolant spills, rinse it off immediately.

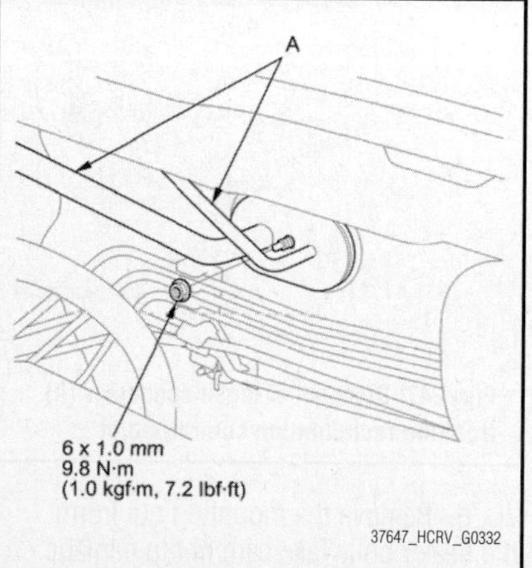

A
6 x 1.0 mm
9.8 N·m
(1.0 kgf·m, 7.2 lbf·ft)
37647_HCRV_G0332

Fig. 244 Remove the nut, then disconnect the A/C lines (A) from the evaporator core

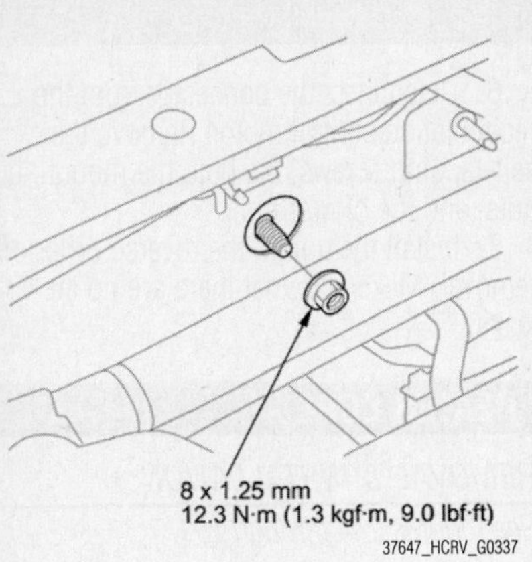
8 x 1.25 mm
12.3 N·m (1.3 kgf·m, 9.0 lbf·ft)
37647_HCRV_G0337

Fig. 245 Remove the mounting nut from the heater unit

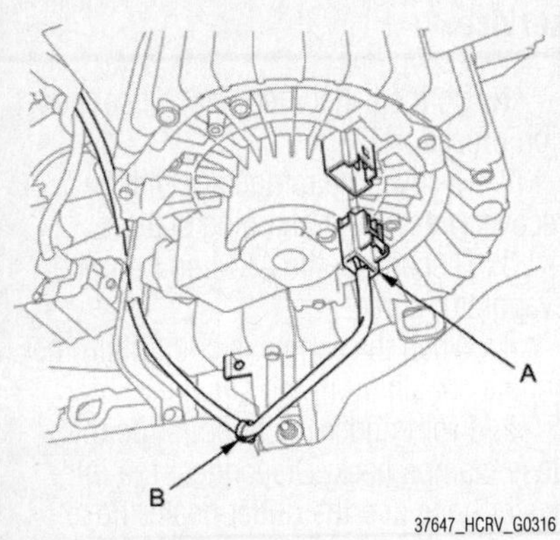
A
B
37647_HCRV_G0316

Fig. 246 Disconnect the connector (A) from the blower motor; remove the wire harness clip (B)

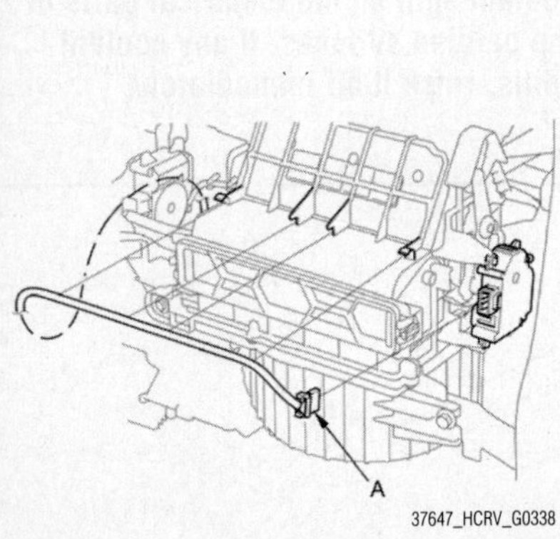
A
37647_HCRV_G0338

Fig. 247 Disconnect these connector (A) from the recirculation control motor

6. Remove the mounting nut from the heater unit. Take care not to damage or bend the fuel lines or the brake lines.
7. Remove the dashboard.

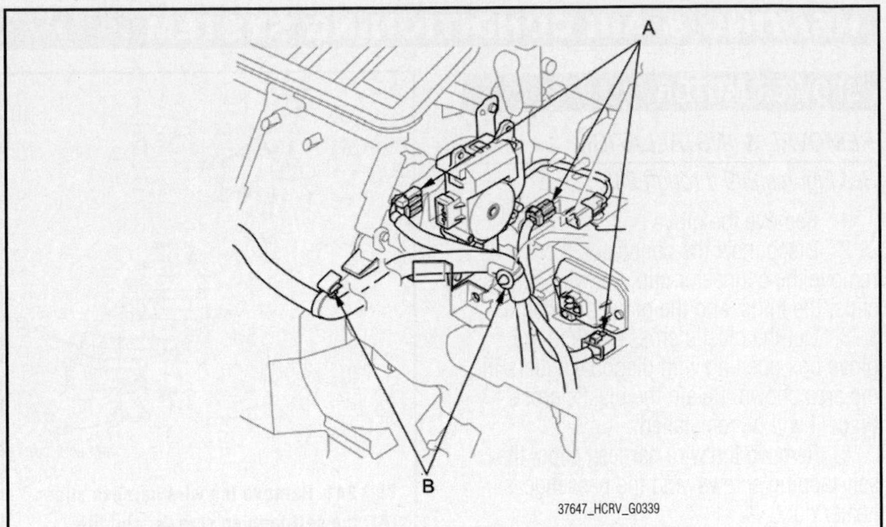
A
B
37647_HCRV_G0339

Fig. 248 Disconnect these connectors (A) from the mode control motor, the evaporator temperature sensor, and the power transistor; remove the wire harness clips (B)

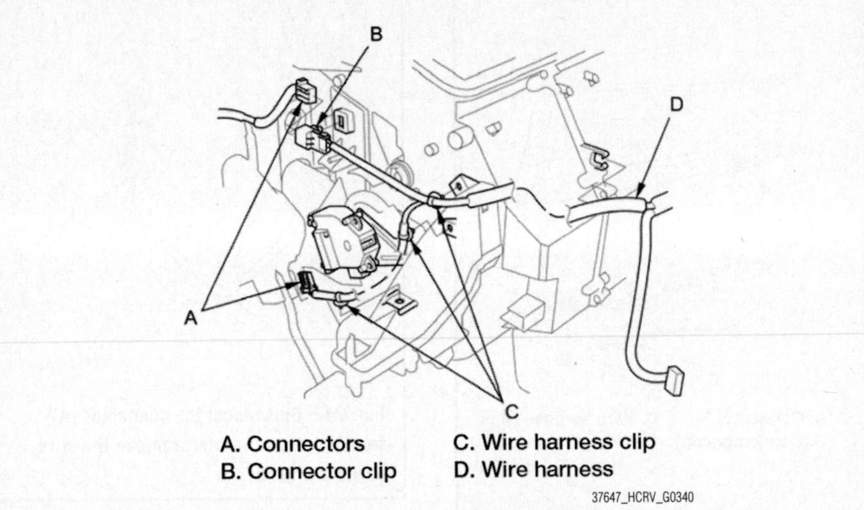
B
D
A
C
A. Connectors
B. Connector clip
C. Wire harness clip
D. Wire harness
37647_HCRV_G0340

Fig. 249 Disconnect the connectors from the air mix control motor and A/C wire harness

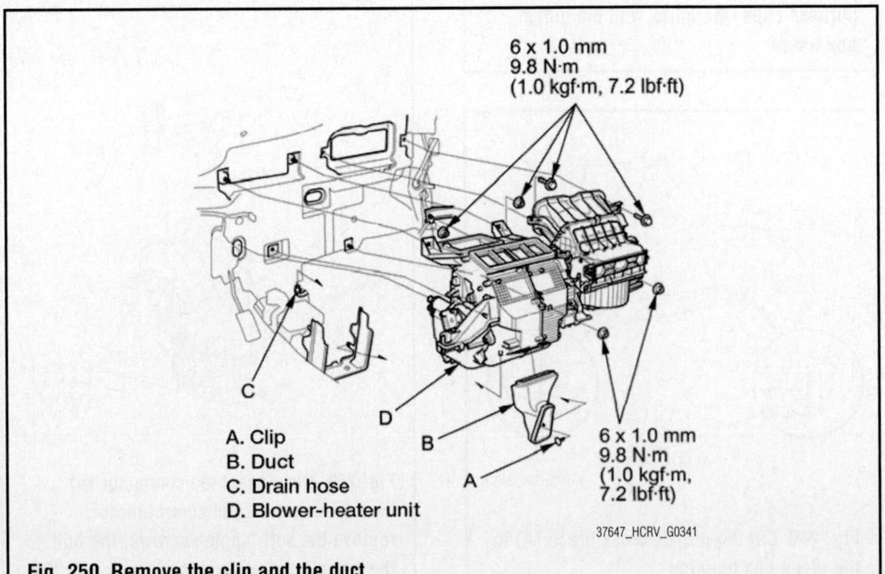
6 x 1.0 mm
9.8 N·m
(1.0 kgf·m, 7.2 lbf·ft)
C
D
B
A
A. Clip
B. Duct
C. Drain hose
D. Blower-heater unit
6 x 1.0 mm
9.8 N·m
(1.0 kgf·m,
7.2 lbf·ft)
37647_HCRV_G0341

Fig. 250 Remove the clip and the duct

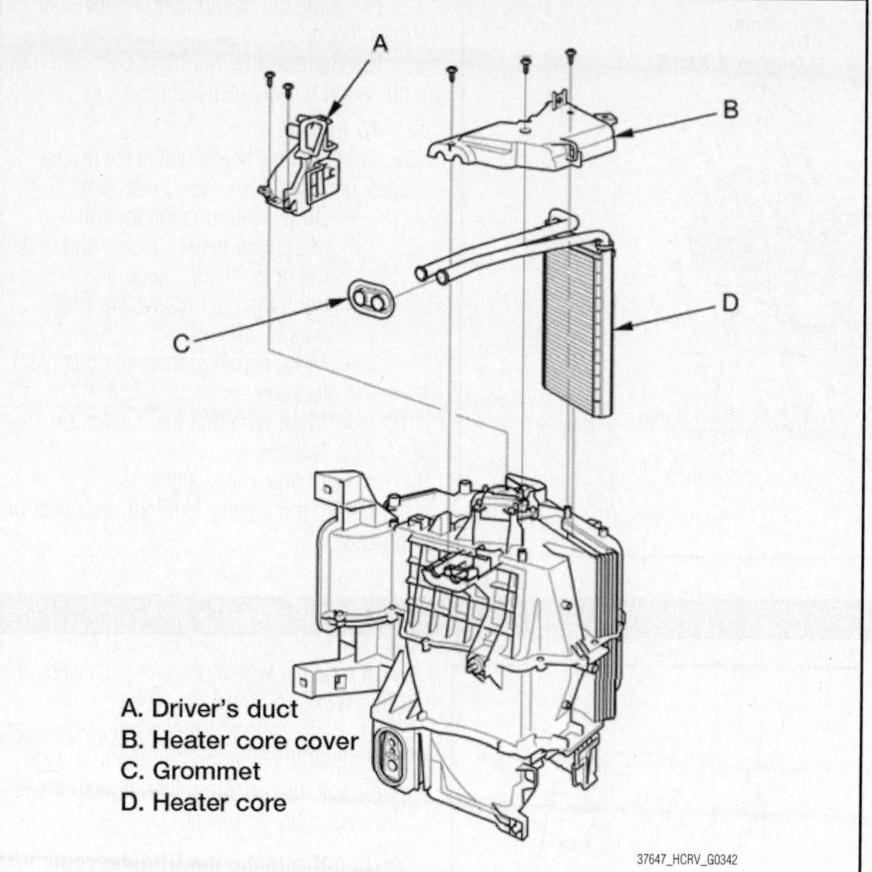
A
B
C
D
A. Driver's duct
B. Heater core cover
C. Grommet
D. Heater core
37647_HCRV_G0342

Fig. 251 Removing the heater core

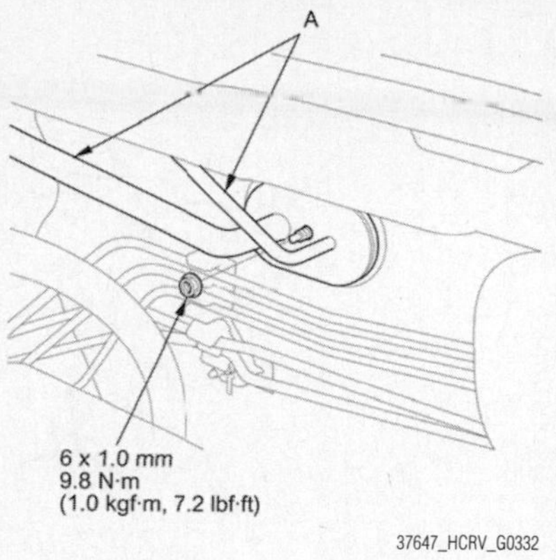
A
6 x 1.0 mm
9.8 N·m
(1.0 kgf·m, 7.2 lbf·ft)
37647_HCRV_G0332

Fig. 252 Remove the nut, then disconnect the A/C lines (A) from the evaporator core

8. Disconnect the connector from the blower motor. Remove the wire harness clip.
9. Disconnect these connector from the recirculation control motor.
10. Disconnect these connectors from the mode control motor, the evaporator temperature sensor, and the power transistor. Remove the wire harness clips.
11. Disconnect the connectors from the air mix control motor and A/C wire harness.
12. Remove the connector clip, the wire harness clips, and the wire harness.
13. Remove the clip and the duct.
14. Slide the clamp, then remove the drain hose.
15. Remove the mounting bolt, mounting nuts, and blower-heater unit.
16. Remove the self-tapping screws and the driver's duct.
17. Remove the self-tapping screws, the heater core cover, the grommet, and carefully pull out the heater core.

To install:

18. Install the heater core and the evaporator core in the reverse order of removal.
19. Install the heater unit in the reverse order of removal, and note these items:
    - Do not interchange the inlet and outlet heater hoses, and install the hose clamps securely.
    - Refill the cooling system with engine coolant.
    - Make sure that there is no coolant leakage.
    - Make sure that there is no air leakage.
    - Charge the system.
20. Do the battery terminal reconnection procedure.

HEATER UNIT

REMOVAL & INSTALLATION

See Figures 245 through 250, 252 and 253.

⁂ WARNING

SRS components are located in this area. Review the SRS component locations and the precautions and procedures before doing repairs or service.

1. Do the battery terminal disconnection procedure.
2. Recover the refrigerant with a recovery/recycling/ charging station
3. Disconnect the A/C line from the evaporator core.
4. When the engine is cool, drain the engine coolant from the radiator.
5. From under the hood, slide the hose clamps back. Disconnect the inlet heater hose and the outlet heater hose from the heater unit. Note the layout of the hoses.

➡Engine coolant will run out when the hoses are disconnected; drain it into a clean drip pan. Be sure not to let coolant spill on the electrical parts or the painted surfaces. If any coolant spills, rinse it off immediately.

6. Remove the mounting nut from the heater unit. Take care not to damage or bend the fuel lines or the brake lines.
7. Remove the dashboard.
8. Disconnect the connector from the blower motor. Remove the wire harness clip.
9. Disconnect these connector from the recirculation control motor.
10. Disconnect these connectors from the mode control motor, the evaporator temperature sensor, and the power transistor. Remove the wire harness clips.
11. Disconnect the connectors from the air mix control motor and A/C wire harness.
12. Remove the connector clip, the wire harness clips, and the wire harness.
13. Remove the clip and the duct.

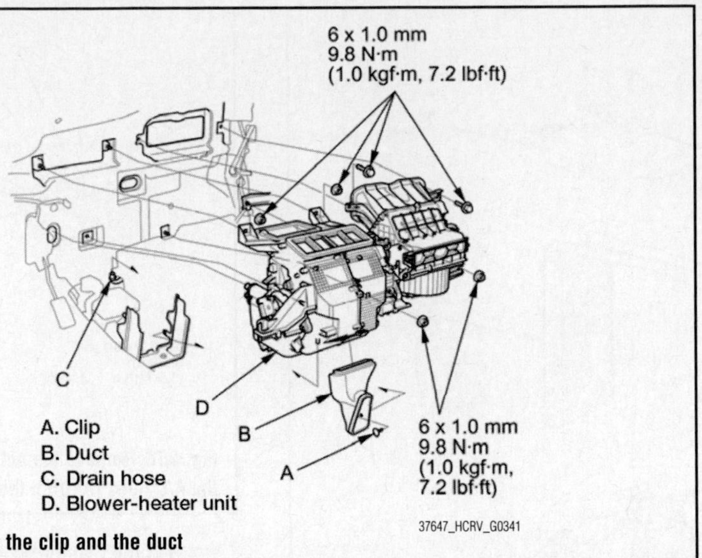

6 x 1.0 mm
9.8 N·m
(1.0 kgf·m, 7.2 lbf·ft)
C
D
B
A
A. Clip
B. Duct
C. Drain hose
D. Blower-heater unit
6 x 1.0 mm
9.8 N·m
(1.0 kgf·m,
7.2 lbf·ft)
37647_HCRV_G0341

Fig. 253 Remove the clip and the duct

14. Slide the clamp, then remove the drain hose.

15. Remove the mounting bolt, mounting nuts, and blower-heater unit.

To install:

16. Install the heater unit in the reverse order of removal, and note these items:

- Do not interchange the inlet and outlet heater hoses, and install the hose clamps securely.
- Refill the cooling system with engine coolant.
- Make sure that there is no coolant leakage.
- Make sure that there is no air leakage.
- Charge the system.

17. Do the battery terminal reconnection procedure.

STEERING

POWER RACK & PINION STEERING GEAR

REMOVAL & INSTALLATION

See Figures 254 through 265.

Special Tools Required:

- Ball Joint Remover, 28 mm 07MAC-SL0A202
- Ball Joint Thread Protector, 12 mm 07AAF-SDAA100

➡Note these items during removal:

- Using solvent and a brush, wash any oil and dirt off the valve body unit, its lines, and the end of the steering gearbox. Blow dry with compressed air.
- Be sure to remove the steering wheel before disconnecting the steering joint. Damage to the cable reel can occur.
- The illustrations show a 2WD model.

1. Do the battery terminal disconnection procedure.
2. Drain the power steering fluid.
3. Raise and support the vehicle.
4. Remove the front wheels.
5. Remove the steering wheel.
6. Remove the driver's dashboard undercover.
7. Remove the steering joint cover.
8. Release the tilt/telescopic lock lever, and adjust the steering column to the full tilt up position, and to the full telescopic in position.
9. Tighten the tilt/telescopic lock lever.

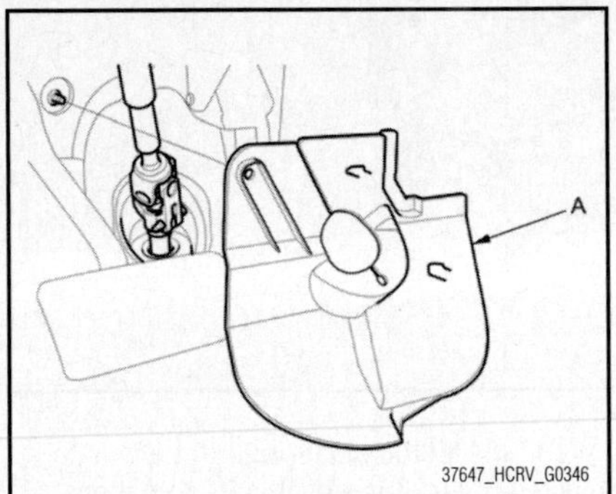

A
37647_HCRV_G0346

Fig. 254 Remove the steering joint cover (A)

10. Hold the steering column lower slide shaft together with a piece wire between the upper yoke and the lower yoke.

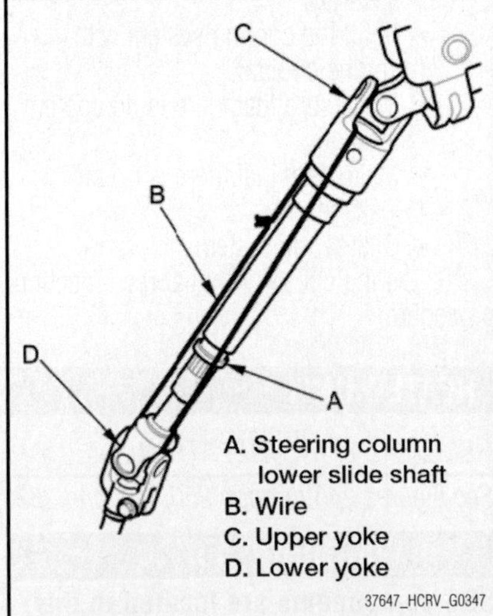

C
B
D
A
A. Steering column lower slide shaft
B. Wire
C. Upper yoke
D. Lower yoke
37647_HCRV_G0347

Fig. 255 Hold the steering column lower slide shaft together with a piece wire between the upper yoke and the lower yoke

11. Release the tilt/telescopic lock lever, and adjust the steering column to the full telescopic out position, then tighten the tilt/telescopic lock lever.

➡Do not release the tilt/telescopic lock lever when removing the steering column from the frame.

12. Remove the steering joint bolt. Disconnect the steering joint by sliding the steering joint into the column. Do not disconnect the steering joint from the column shaft.

➡If the center guide is in place and has not moved, leave it in place. If the center guide has come off, discard it.

13. Apply vinyl tape to the splines on the pinion shaft.
14. Remove the cotter pin from the 12 mm nut, and loosen the nut.
15. Disconnect the tie-rod end ball joint from the knuckle using the ball joint thread protector and the ball joint remover.
16. Remove the air cleaner housing.
17. For 2010 model: Remove the underfloor TWC.
18. Remove the P/S heat shield.
19. Loosen the adjustable hose clamp and disconnect the return hose.
20. Remove the inlet line clamp bolt and return hose clamp bolt.
21. Loosen the 18 mm flare nut and disconnect the inlet line.
22. Remove the return hose clamp bolt and open the return hose clamp.
23. Remove the pump outlet hose clamp bolt, then open the return line clamps.

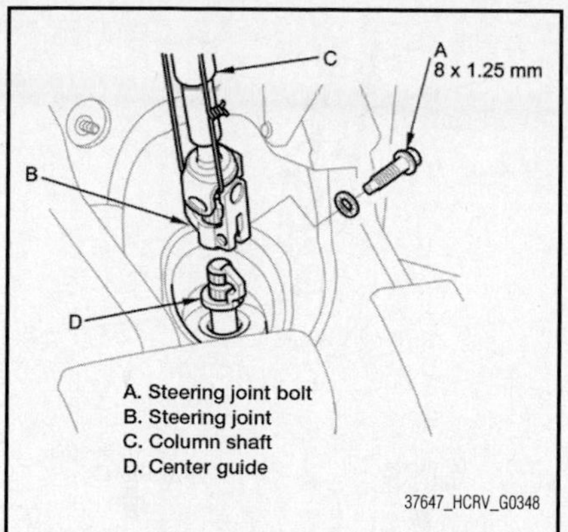
A
8 x 1.25 mm
B
C
D
A. Steering joint bolt
B. Steering joint
C. Column shaft
D. Center guide
37647_HCRV_G0348

Fig. 256 Remove the steering joint bolt; disconnect the steering joint by sliding the steering joint into the column

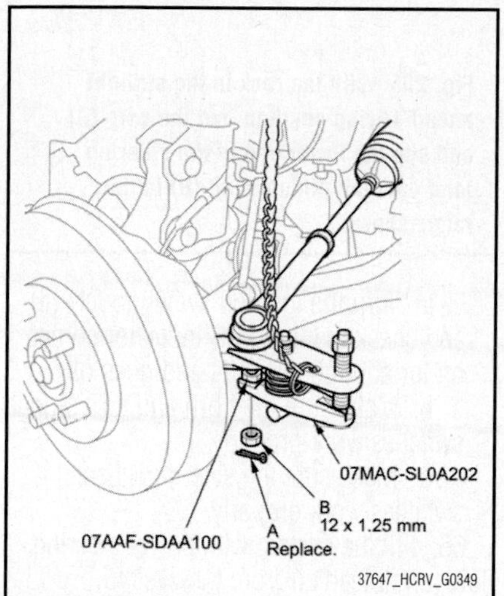
07MAC-SL0A202
07AAF-SDAA100
A
Replace.
B
12 x 1.25 mm
37647_HCRV_G0349

Fig. 257 Remove the cotter pin (A) from the 12 mm nut (B), and loosen the nut

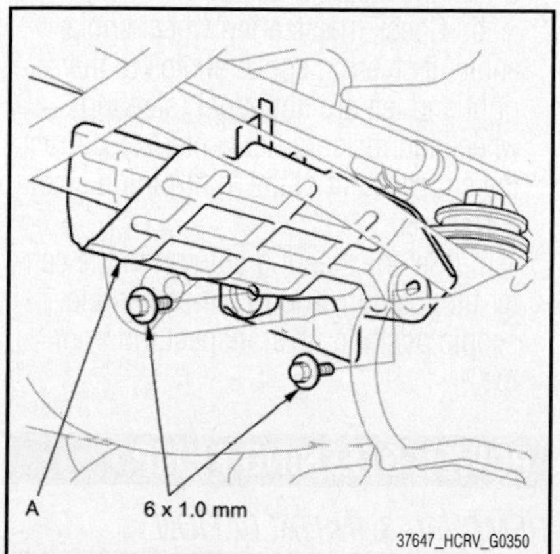
A
6 x 1.0 mm
37647_HCRV_G0350

Fig. 258 Remove the P/S heat shield (A)

24. Remove the 10 mm flange bolts and washers from the driver's side of the steering gearbox.

25. Remove the 8 mm flange bolts, then remove the stabilizer bar bushing holder.

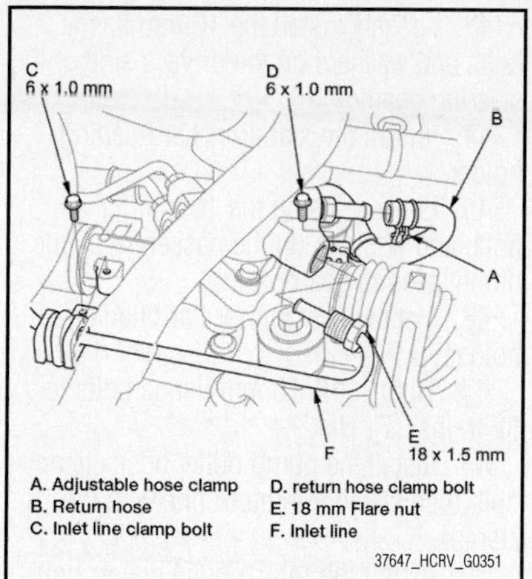
C
6 x 1.0 mm
D
6 x 1.0 mm
B
A
F
E
18 x 1.5 mm
A. Adjustable hose clamp
B. Return hose
C. Inlet line clamp bolt
D. return hose clamp bolt
E. 18 mm Flare nut
F. Inlet line
37647_HCRV_G0351

Fig. 259 Loosen the adjustable hose clamp and disconnect the return hose

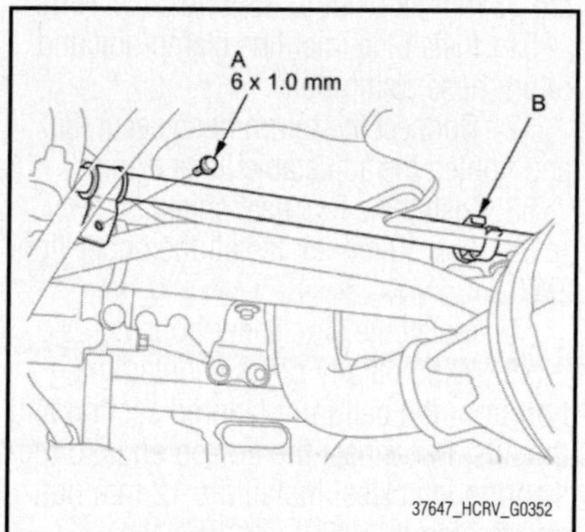
A
6 x 1.0 mm
B
37647_HCRV_G0352

Fig. 260 Remove the return hose clamp bolt (A) and open the return hose clamp (B)

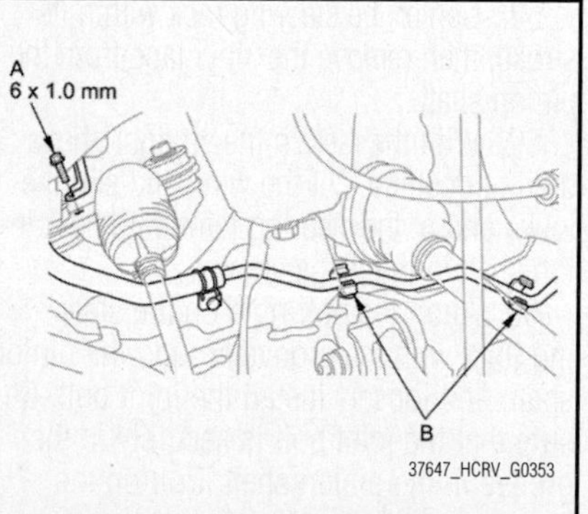
A
6 x 1.0 mm
B
37647_HCRV_G0353

Fig. 261 Remove the pump outlet hose clamp bolt (A), then open the return line clamps (B)

26. Remove the 10 mm flange bolts and washers from the passenger's side of the steering gearbox.

27. Remove the 8 mm flange bolts, then remove the stabilizer bar bushing holder.

28. Remove the 10 mm flange bolts from both of the gearbox brackets, then remove the brackets.

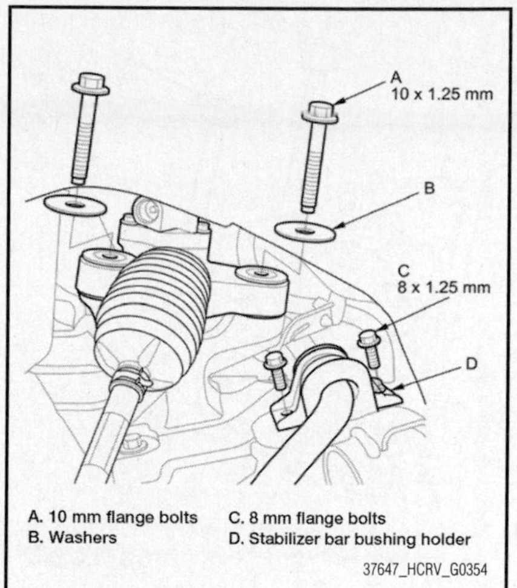
A
10 x 1.25 mm
B
C
8 x 1.25 mm
D
A. 10 mm flange bolts
B. Washers
C. 8 mm flange bolts
D. Stabilizer bar bushing holder
37647_HCRV_G0354

Fig. 262 Remove the 10 mm flange bolts and washers from the driver's side of the steering gearbox

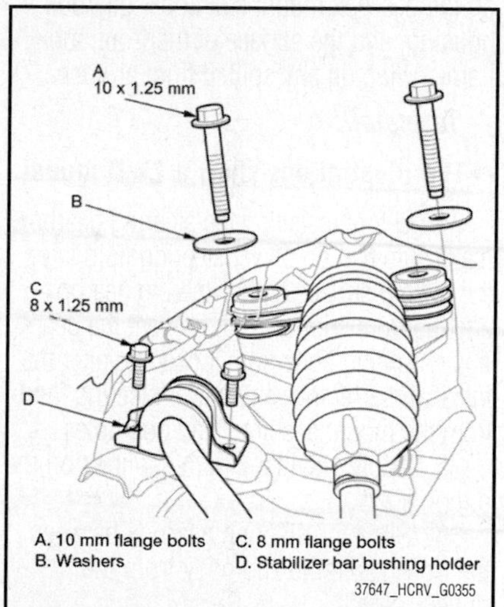
A
10 x 1.25 mm
B
C
8 x 1.25 mm
D
A. 10 mm flange bolts
B. Washers
C. 8 mm flange bolts
D. Stabilizer bar bushing holder
37647_HCRV_G0355

Fig. 263 Remove the 10 mm flange bolts and washers from the passenger's side of the steering gearbox

29. Move the steering gearbox toward the front, and remove the pinion shaft grommet from the top of the valve housing.

30. Move the steering gearbox to the driver's side, and rotate it so the pinion shaft points toward the rear of the vehicle.

31. Carefully move the steering gearbox as an assembly toward the driver's side of the vehicle until the pinion shaft clears the wheelwell opening.

32. Remove the steering gearbox through the wheelwell opening on the driver's side.

33. After removing the steering gearbox, make sure that no power steering fluid gets

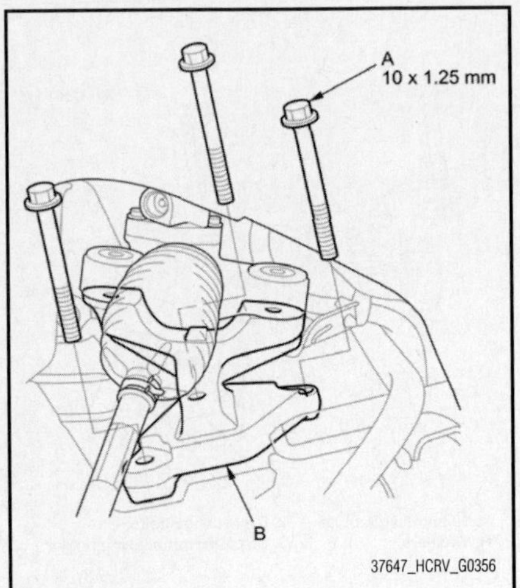
A
10 x 1.25 mm
B
37647_HCRV_G0356

Fig. 264 Remove the 10 mm flange bolts (A) from both of the gearbox brackets (B), then remove the brackets

on the gearbox mount cushions, gearbox housing, and the surface of the front subframe. Wipe off any spilled fluid at once.

To install:

➡The illustrations show a 2WD model.

34. Before installing the steering gearbox, make sure that no power steering fluid is on the mating surface of the steering gearbox and the front subframe. To prevent the gearbox mounting bolts from loosening after the installation, remove any power steering fluid from the mount cushions and bolt holes.
35. Apply vinyl tape to the splines on the pinion shaft.
36. Slide the steering gearbox between the front subframe and body from the driver's side.
37. Carefully move the steering gearbox toward the passenger's side until the pinion shaft clears the wheelwell opening on the body.
38. Rotate the steering gearbox so the pinion shaft points upward.
39. Continue moving the steering gearbox toward the passenger's side until the steering gearbox is in position.
40. Remove the vinyl tape from pinion shaft, and install the pinion shaft grommet. Align the slot in the pinion shaft grommet with the lug portion on the valve housing. The grommet must not have a gap at the mating surface of the grommet and valve housing.
41. Install the gearbox brackets, and loosely install the steering gearbox bracket mounting bolts.
42. Install the gearbox brackets with the 10 mm flange bolts. Tighten the bolts to 48 ft. lbs. (66 Nm).
43. Loosely install the 10 mm flange bolts and washers on the driver's side of the steering gearbox.
44. Install the stabilizer bar bushing holder.
45. Loosely install the 10 mm flange bolts and washers on the passenger's side of the steering gearbox.
46. Install the stabilizer bar bushing holder.
47. Tighten the 10 mm flange bolts to 52 ft. lbs. (71 Nm).
48. Install the pump outlet hose clamp bolt, then clamp the return line with the clamps.
49. Install the return hose clamp bolt, then clamp the return hose with the clamp.
50. Connect the inlet line, and tighten the 18 mm flare nut to 35 ft. lbs. (47 Nm).
51. Install the inlet line clamp bolt and return hose clamp bolt.
52. Connect the return hose securely, and tighten the adjustable hose clamp.
53. Install the P/S heat shield.
54. For '10 model: Install the under-floor TWC.
55. Install the air cleaner housing.
56. Wipe off any grease contamination from the ball joint tapered section and threads. Reconnect the tie-rod ends to the steering knuckles. Install the 12 mm nut, and tighten it to 40 ft. lbs. (54 Nm).
57. Install a new cotter pin, and bend it.

➡Check the boot for damage and deterioration. If there is damage or deterioration, replace the boot.

58. Center the steering rack within its stroke, then remove the vinyl tape from the pinion shaft.
59. With the rack in the straight ahead driving position, cut the wire, and slip the lower end of the steering joint onto the pinion shaft in the range shown.
60. Align the bolt hole on the steering joint with the groove around the pinion shaft, and loosely install the joint bolt. Be sure that the joint bolt is securely in the groove in the pinion shaft. Pull on the steering joint to make sure that the steering joint is fully seated. Tighten the steering joint bolt to 21 ft. lbs. (28 Nm).
61. Install the steering joint cover.
62. Install the driver's dashboard undercover.
63. Install the front wheels, then set the wheels in the straight ahead position.
64. Install the steering wheel and the driver's airbag.
65. Do the battery terminal reconnection procedure, and do these tasks:

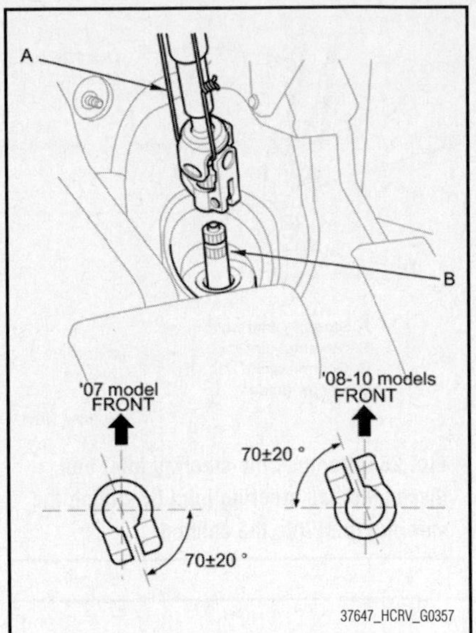
A
B
'07 model
FRONT
'08-10 models
FRONT
70±20 °
70±20 °
37647_HCRV_G0357

Fig. 265 With the rack in the straight ahead driving position, cut the wire (A), and slip the lower end of the steering joint onto the pinion shaft (B) in the range shown

a. Turn the ignition switch to ON (II) and check that the SRS indicator comes on for about 6 seconds and goes off.
b. Make sure the horn and turn signal switches work properly.
c. Make sure the steering wheel switches work properly.

66. Fill the system with power steering fluid, and bleed air from the system.
67. After installation, and do these tasks:
a. Start the engine, allow it to idle, and turn the steering wheel from lock to lock several times to warm up the fluid. Check the steering gearbox for leaks.
b. Check the steering wheel spoke angle. If steering spoke angles to the right and left are not equal (steering wheel and rack are not centered), correct the engagement of the joint/pinion shaft serrations.
c. Set the steering column to the center tilt position, and to the center telescopic position, then inspect the front toe.

POWER STEERING PUMP

REMOVAL & INSTALLATION

See Figure 266.

1. Place a suitable container under the vehicle to catch any spilled fluid.
2. Drain the power steering fluid from the reservoir.
3. Remove the drive belt from the pump pulley.

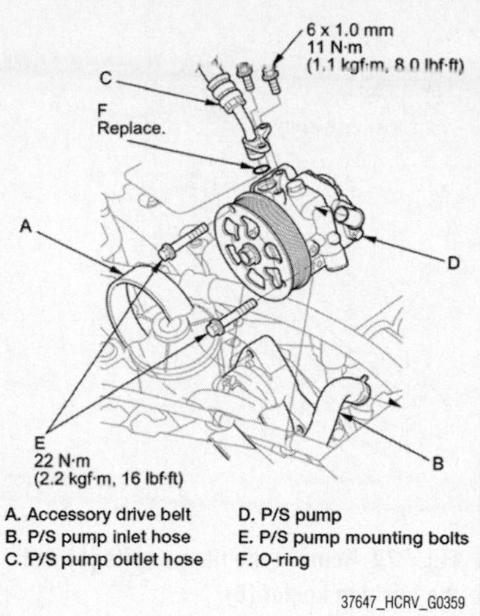
6 x 1.0 mm
11 N·m
(1.1 kgf·m, 8.0 lbf·ft)
C
F
Replace.
A
D
E
22 N·m
(2.2 kgf·m, 16 lbf·ft)
B
A. Accessory drive belt
B. P/S pump inlet hose
C. P/S pump outlet hose
D. P/S pump
E. P/S pump mounting bolts
F. O-ring
37647_HCRV_G0359

Fig. 266 Removing the hydraulic power steering pump

4. Cover the auto-tensioner, the alternator, and the A/C compressor with several shop towels to protect them from spilled power steering fluid. Disconnect the pump inlet hose and pump outlet hose from the pump, and plug them. Take care not to spill the fluid on the vehicle. Wipe off any spilled fluid at once. Do not turn the steering wheel with the pump removed.

5. Remove the pump mounting bolts.

6. Cover the opening of the pump to prevent spillage.

7. Transfer the pump inlet hose and the pump outlet hose from the original pump onto the new pump with a new O-ring.

8. Loosely install the pump in the pump bracket with the mounting bolts, then tighten the pump fittings to the specified torque.

9. Tighten the pump mounting bolts to 16 ft. lbs. (22 Nm).

10. Install the drive belt.

➡Note these items during drive belt installation:

- Inspect the drive belt for wear and cracks. Replace the belt if necessary.
- Make sure that the drive belt is properly positioned on the pulleys.
- Do not get power steering fluid or grease on the auto-tensioner, the alternator, the A/C compressor, and drive belt, or pulley faces. Clean off any fluid or grease before installation.

11. Fill the reservoir to the upper level line.

12. Start the engine, and check for fluid leaks.

BLEEDING

1. Before servicing the vehicle, refer to the Precautions Section.

2. Stop the engine.

3. Turn the steering wheel fully to the right and left several times.

➡Do not allow the fluid level in the reservoir tank to go below the MIN level line. Check and add fluid as needed.

4. Run the engine at idle speed. Turn the steering wheel fully to the right and then fully to the left. Hold for about 3 seconds. Check for fluid leakage.

5. Repeat the above step several times at 3 second intervals.

⁂ WARNING

Do not hold the steering wheel in the locked position for more than 10 seconds. Damage to the pump may occur.

6. Check for air bubbles or cloudy fluid. If found, repeat the bleeding procedure.

7. Stop the engine and check the fluid level. Correct as required.

SUSPENSION

FRONT SUSPENSION

CONTROL LINKS

REMOVAL & INSTALLATION

Front Stabilizer Link

See Figure 267.

1. Raise and support the vehicle.
2. Remove the front wheel.
3. Remove the flange nuts while holding the respective joint pin with a hex wrench, then remove the stabilizer link.

To install:

4. Install the stabilizer link on the stabilizer bar and the damper with the joint pins set at the center of their range of movement.

➡The stabilizer link has a paint mark. Align the paint mark on the stabilizer link facing inward.

5. Install the flange nuts, and lightly tighten them.

6. Place a jack under the lower arm, and raise the suspension to load it with the vehicle's weight.

7. Tighten the flange nuts to 58 ft. lbs. (78 Nm) while holding the respective joint pin with a hex wrench.

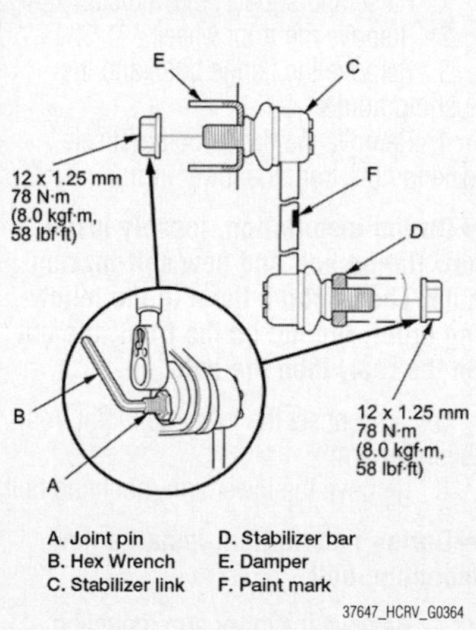
E
C
12 x 1.25 mm
78 N·m
(8.0 kgf·m,
58 lbf·ft)
F
D
B
A
12 x 1.25 mm
78 N·m
(8.0 kgf·m,
58 lbf·ft)
A. Joint pin
B. Hex Wrench
C. Stabilizer link
D. Stabilizer bar
E. Damper
F. Paint mark
37647_HCRV_G0364

Fig. 267 Front stabilizer link components

8. Clean the mating surface of the brake disc and the inside of the wheel, then install the front wheel.

LOWER BALL JOINT

REMOVAL & INSTALLATION

See Figures 268 and 269.

Special Tools Required:

- Ball Joint Remover, 32 mm 07MAC-SL0A102
- Ball Joint Thread Protector, 14 mm 07AAE-SJAA100

1. Raise and support the vehicle.
2. Remove the front wheel.
3. Remove the flange bolt and flange nuts from the lower arm.

➡During installation, install a new flange bolt and new flange nuts. After lightly tightening all three fasteners, tighten them to 38 ft. lbs. (52 Nm) in the following order; the nut on the front, the nut on the rear, then the bolt.

4. Disconnect the lower ball joint from the lower arm.

5. Remove the lock pin from the lower ball joint pin, then remove the castle nut.

➡During installation, install the lock pin after tightening a new castle nut.

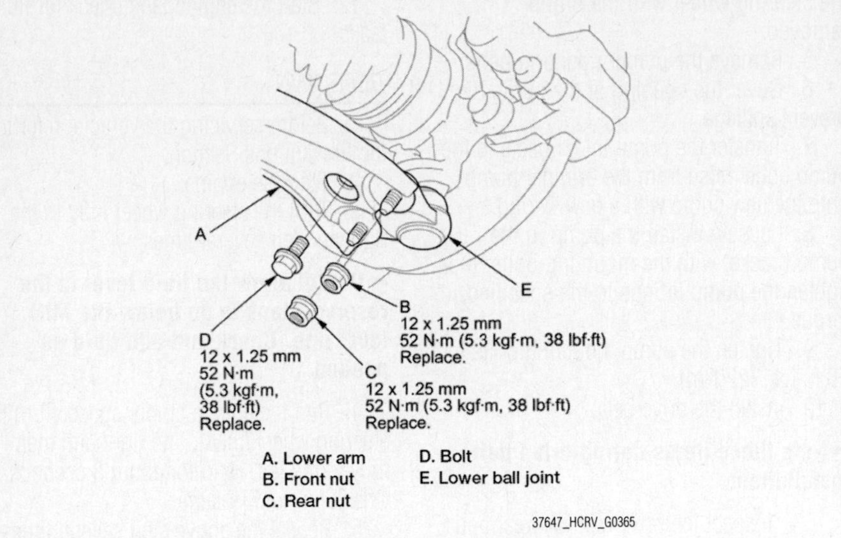
A
B
12 x 1.25 mm
52 N·m (5.3 kgf·m, 38 lbf·ft)
Replace.
C
12 x 1.25 mm
52 N·m (5.3 kgf·m, 38 lbf·ft)
Replace.
D
12 x 1.25 mm
52 N·m
(5.3 kgf·m,
38 lbf·ft)
Replace.
E
A. Lower arm
B. Front nut
C. Rear nut
D. Bolt
E. Lower ball joint
37647_HCRV_G0365

Fig. 268 Remove the flange bolt and flange nuts from the lower arm

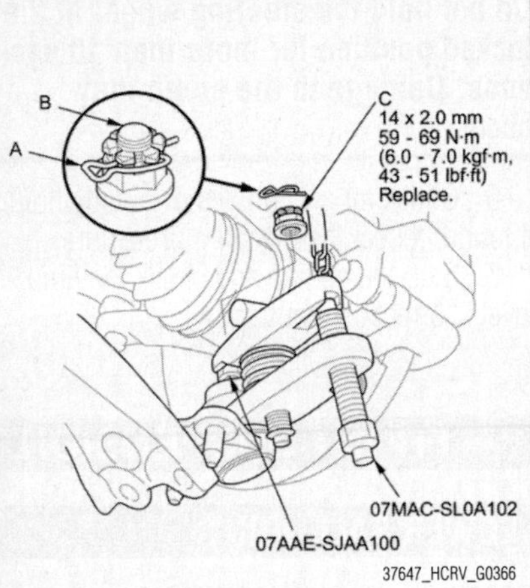
B
A
C
14 x 2.0 mm
59 - 69 N·m
(6.0 - 7.0 kgf·m,
43 - 51 lbf·ft)
Replace.
07MAC-SL0A102
07AAE-SJAA100
37647_HCRV_G0366

Fig. 269 Remove the lock pin (A) from the lower ball joint pin (B), then remove the castle nut (C)

6. Disconnect the lower ball joint from the knuckle using the ball joint thread protector and the ball joint remover.

To install:

7. Install the lower ball joint in the reverse order of removal, and note these items:
   - First install all the components, and lightly tighten the bolts and nuts, then tighten the lower ball joint to the lower arm to the specified torque. Raise the suspension to load it with the vehicle's weight before fully tightening the lower ball joint to the knuckle to the specified torque.
   - Torque the castle nut to the lower torque specification, then tighten it only far enough to align the slot with the ball joint pin hole. Do not align the castle nut by loosening it.
   - Before installing the wheel, clean the mating surface of the brake disc and the inside of the wheel.
   - Check the wheel alignment, and adjust it if necessary.

LOWER CONTROL ARM

REMOVAL AND & INSTALLATION

See Figures 268, 270 and 271.

1. Raise and support the vehicle.
2. Remove the front wheel.
3. Remove the flange bolts and the bushing holder.
4. Remove the flange bolt and self-locking nuts from the lower arm.

➡During installation, loosely install a new flange bolt and new self-locking nuts. Then tighten them in the following order; the nut on the front, the nut on the rear, then the bolt.

5. Disconnect the lower ball joint from the lower arm.
6. Remove the lower arm mounting bolt.

➡During installation, install a new mounting bolt.

7. Remove the lower arm mounting bolts, then remove the lower arm from the front suspension subframe.

To install:

➡During installation, install new mounting bolts.

8. Install the lower arm in the reverse order of removal, and note these items:

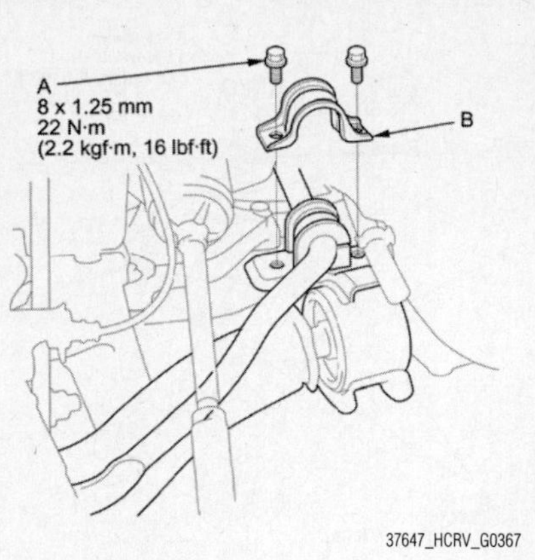
A
8 x 1.25 mm
22 N·m
(2.2 kgf·m, 16 lbf·ft)
B
37647_HCRV_G0367

Fig. 270 Remove the flange bolts (A) and the bushing holder (B)

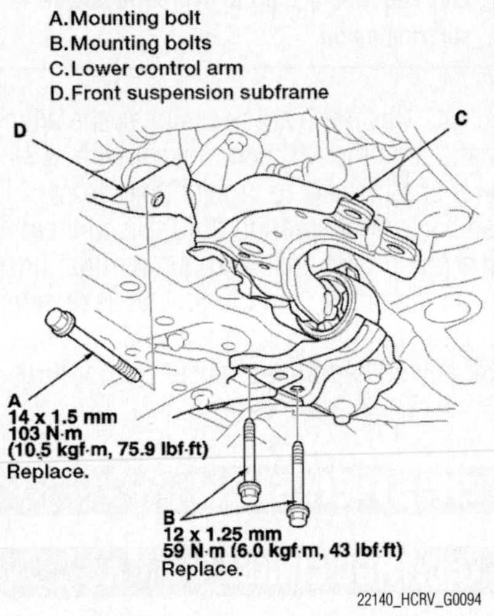
A.Mounting bolt
B.Mounting bolts
C.Lower control arm
D.Front suspension subframe
D
C
A
14 x 1.5 mm
103 N·m
(10.5 kgf·m, 75.9 lbf·ft)
Replace.
B
12 x 1.25 mm
59 N·m (6.0 kgf·m, 43 lbf·ft)
Replace.
22140_HCRV_G0094

Fig. 271 Remove the lower arm mounting bolts, then remove the lower arm from the front suspension subframe

   - First install all the components, and lightly tighten the bolts and nuts, then raise the suspension to load it with the vehicle's weight before fully tightening to the specified torque.
   - Before installing the wheel, clean the mating surface of the brake disc and the inside of the wheel.
   - Check the wheel alignment, and adjust it if necessary.

STEERING KNUCKLE

REMOVAL & INSTALLATION

See Figure 272.

Special Tools Required:

- Ball joint remover, 32mm 07MAC-SL0A102

• Ball joint remover, 28mm 07MAC-SL0A202

• Ball joint thread protector, 14mm 071AF-S3VA000

1. Before servicing the vehicle, refer to the Precautions Section.
2. Raise and safely support the vehicle.
3. Remove the front wheels.
4. Remove the brake hose mounting bolt.
5. Remove the brake caliper bracket mounting bolts and the caliper assembly from the knuckle.

➡To prevent damage to the caliper assembly or the brake hose, use a short piece of wire to hang the caliper assembly from the undercarriage. Do not twist the brake hose excessively.

6. Remove the wheel speed sensor from the knuckle. Do not disconnect the wheel speed sensor connector.
7. Raise the stake, then remove the spindle nut.
8. Remove the front brake disc rotor.
9. Check the front hub for damage and cracks.
10. Remove the cotter pin from the tie-rod end ball joint, then remove the nut.
11. Disconnect the tie-rod end ball joint from the knuckle using a ball joint remover.
12. Remove the lock pin from the lower ball joint pin, then remove the castle nut.
13. Disconnect the lower ball joint from the lower arm.
14. Remove the strut mounting bolts and the self-locking nuts from the strut.
15. Remove the driveshaft outboard joint from the knuckle by tapping the driveshaft end with a plastic hammer while drawing the hub outward.

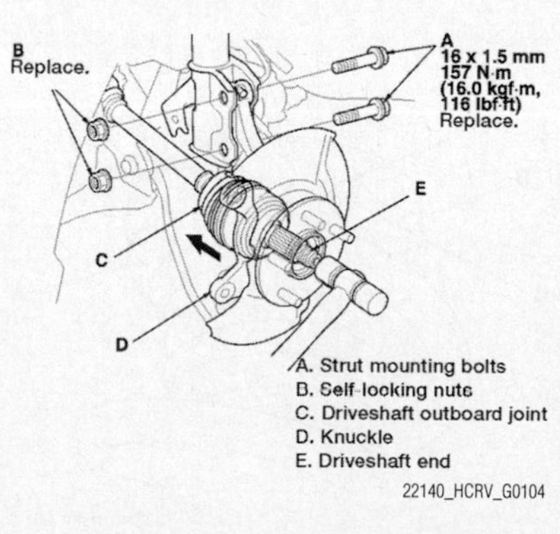
B
Replace.
A
16 x 1.5 mm
157 N·m
(16.0 kgf·m,
116 lbf·ft)
Replace.
E
C
D
A. Strut mounting bolts
B. Self-locking nuts
C. Driveshaft outboard joint
D. Knuckle
E. Driveshaft end
22140_HCRV_G0104

Fig. 272 Remove the strut mounting bolts and the self-locking nuts from the strut. Remove the driveshaft outboard joint from the knuckle by tapping the driveshaft end with a plastic hammer while drawing the hub outward

16. Remove the knuckle/hub.

➡Do not pull the driveshaft end outward as the driveshaft inboard joint may come apart.

To install:

Install all the components and lightly tighten the bolts and nuts, then raise the suspension to load it with the vehicle weight before fully tightening to the specified torque values.

• Be careful not to damage the ball joint boot when installing the knuckle

• Before connecting the lower ball joint to the knuckle, degrease the threaded section and tapered portion of the ball joint pin, the knuckle connecting hole, the threaded section, and the mating surface of the castle nut

• Torque the castle nut to the lower torque specification, then tighten it only far enough to align the slot with the ball joint pin hole. Do not align the castle nut by loosening it

• Use a new spindle nut during reassembly

• Before installing the spindle nut, apply a small amount of engine oil to the seating surface of the nut. After tightening, use a drift to stake the spindle nut shoulder against the driveshaft.

• Before installing the brake disc, clean the mating surface of the front hub and the inside of the brake disc rotor.

17. Apply grease to the mating surface of the wheel bearing and the driveshaft outboard joint.
18. Install the knuckle/hub into position.
19. Install the driveshaft outboard joint to the knuckle.
20. Loosely install the strut mounting bolts and the self-locking nuts to the strut.
21. With the weight of the vehicle on the suspension, tighten the strut mounting bolts and nuts to 116 ft. lbs. (157 Nm).
22. Connect the lower ball joint to the lower arm. Tighten the castle nut to 43–51 ft. lbs. (59–69 Nm).
23. Install the lock pin into the lower ball joint.
24. Connect the tie-rod end ball joint to the knuckle and tighten the nut to 40 ft. lbs. (54 Nm). Install the cotter pin.
25. Install the front brake disc rotor.
26. Install the spindle nut and tighten to 242 ft. lbs. (328 Nm).
27. Stake the spindle nut with a drift tool.
28. Install the wheel speed sensor to the knuckle and tighten to 86 inch lbs. (10 Nm).
29. Install the caliper assembly to the knuckle and tighten the brake caliper bracket mounting bolts to 101 ft. lbs. (137 Nm).
30. Install the brake hose mounting bolt and tighten to 16 ft. lbs. (22 Nm).
31. Install the wheel and tighten to 80 ft. lbs. (108 Nm).
32. Check the wheel alignment and adjust as necessary.

STRUT (DAMPER/SPRING)

REMOVAL & INSTALLATION

See Figures 73 through 75.

1. Before servicing the vehicle, refer to the Precautions Section.
2. Turn the ignition switch to ON, then turn on the windshield wipers. Turn the ignition switch to LOCK when the wipers are near the A-pillars.
3. Raise and safely support the vehicle.
4. Remove the front wheel.
5. Remove the wheel speed sensor harness guide, the harness clip, and the brake hose from the strut. Do not disconnect the wheel speed sensor connector.
6. Make an alignment marking on the camber adjusting bolt and strut for approximate installation alignment later.
7. Remove the flange nut, while holding the joint pin with a hex wrench, and disconnect the stabilizer link from the strut.
8. Remove the strut mounting bolts and self-locking nuts from the strut.
9. Remove the lid by releasing the hooks.
10. Remove the flange nuts from the top of the strut.
11. Remove the strut assembly.

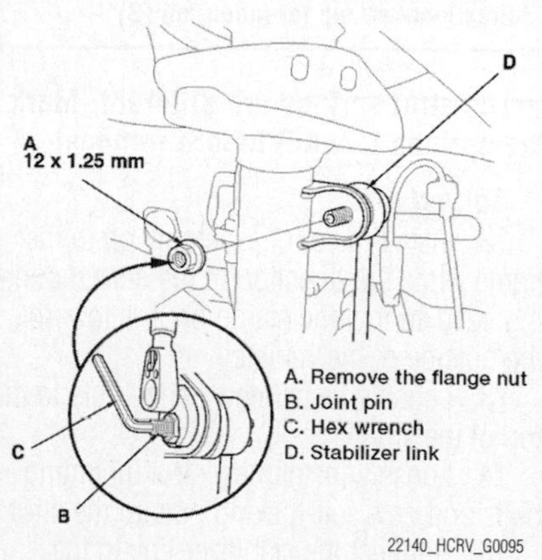
D
A
12 x 1.25 mm
C
B
A. Remove the flange nut
B. Joint pin
C. Hex wrench
D. Stabilizer link
22140_HCRV_G0095

Fig. 273 Remove the flange nut, while holding the joint pin with a hex wrench, and disconnect the stabilizer link from the strut

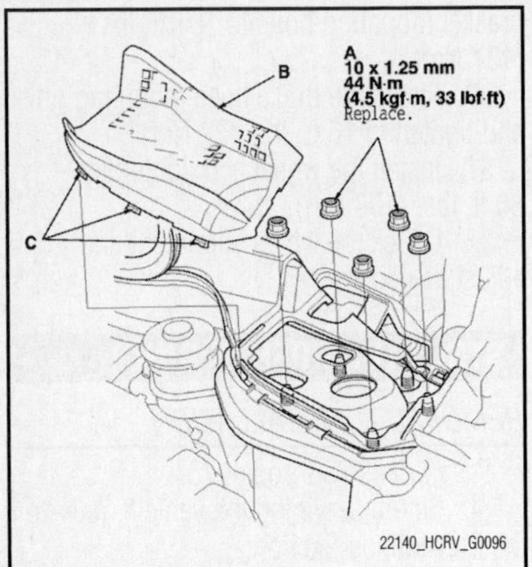
A
10 x 1.25 mm
44 N·m
(4.5 kgf·m, 33 lbf·ft)
Replace.
B
C
22140_HCRV_G0096

Fig. 274 Remove the lid (B) by releasing the hooks (C) and remove the flange nuts (A) from the top of the strut

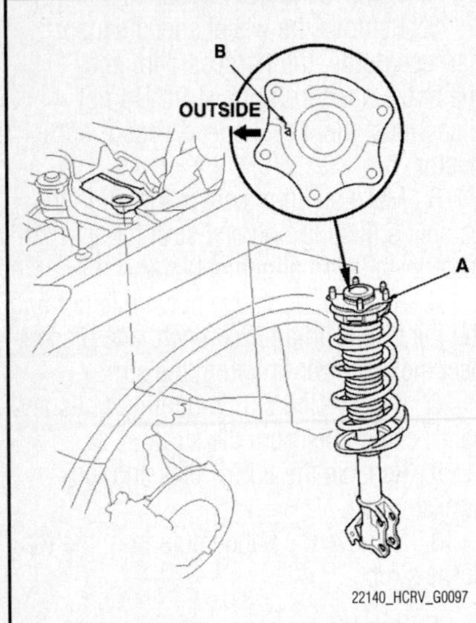
B
OUTSIDE
A
22140_HCRV_G0097

Fig. 275 View of strut assembly (A) and directional stamp for mounting (B)

➡The strut springs are different. Mark the springs L and R before removal.

To install:

12. Install the strut assembly on to the frame. Note the direction of the strut mounting base so that the stamp on it is toward the outside of the vehicle.

13. Loosely install new flange nuts to the top of the strut.

14. Loosely install new strut mounting bolts and new self-locking nuts to the strut.

15. Connect the stabilizer link to the strut, and loosely install the flange nut, while holding the joint pin with a hex wrench.

16. Install the wheel speed sensor harness guide, the harness clip, and the brake hose to the strut.

17. Raise the front suspension with a floor jack to load the suspension with the vehicle's weight.

18. Tighten the strut mounting bolts and the self-locking nuts to 116 ft. lbs. (157 Nm).

19. Tighten the flange nuts on top of the strut to 33 ft. lbs. (44 Nm).

20. Install the lid by pushing the hooks into place securely.

21. Clean the mating surface of the brake disc and the inside of the wheel, then install the front wheel.

22. Check the wheel alignment and adjust if necessary.

OVERHAUL

See Figures 276 through 280.

Special Tools Required:

- Strut nut adapter 07AAA-SVAA100
- Strut spring compressor: Branick MST-580A or Model 7200, or equivalent

1. Before servicing the vehicle, refer to the Precautions Section.

2. Remove the strut assembly from the vehicle.

3. Remove the cap from the top of the strut damper.

4. Compress the strut damper spring, then remove the self-locking nut using the strut nut adapter and a ratchet or breaker bar while holding the strut damper shaft with a hex wrench.

➡Do not compress the spring more than necessary to remove the nut.

5. Release the pressure from the strut spring compressor, then disassemble the strut damper.

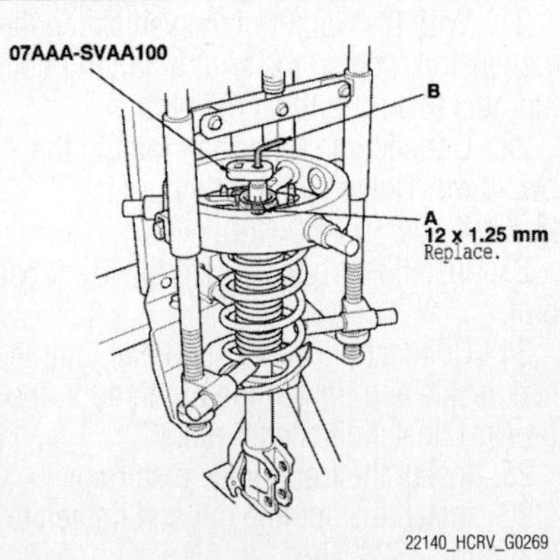
07AAA-SVAA100
B
A
12 x 1.25 mm
Replace.
22140_HCRV_G0269

Fig. 276 Compress the strut damper spring, then remove the self-locking nut (A) using the strut nut adapter and a ratchet or breaker bar while holding the strut damper shaft with a hex wrench (B)

To assemble:

6. Check for smooth operation through a full stroke, both compression and extension of the strut. The strut damper should extend smoothly and constantly when compression is released. If it does not, the gas may be leaking and the strut damper should be replaced.

7. Check for oil leaks, abnormal noises, or binding.

8. Install the upper spring mounting cushion on the upper spring seat by aligning the ledge portion.

9. Install the strut damper spring on the upper spring mounting cushion, then align the upper end of the strut damper spring with the cushion stop.

10. Compress the strut damper spring.

11. Install all the parts except the strut damper mounting washer and self-locking nut onto the strut damper unit.

12. Align the bottom of the strut damper spring with the stepped part of the lower spring seat.

13. Align the center of raised portion on the upper spring seat and the strut damper unit.

14. Align the strut damper bracket and the strut damper mounting base so that stamp on it is toward the outside of the vehicle.

15. Install a new self-locking nut.

16. Hold the strut damper shaft using a hex wrench, and tighten the new self-locking nut using the strut nut adapter and a torque wrench.

17. Tighten the self-locking strut nut to 33 ft. lbs. (44 Nm).

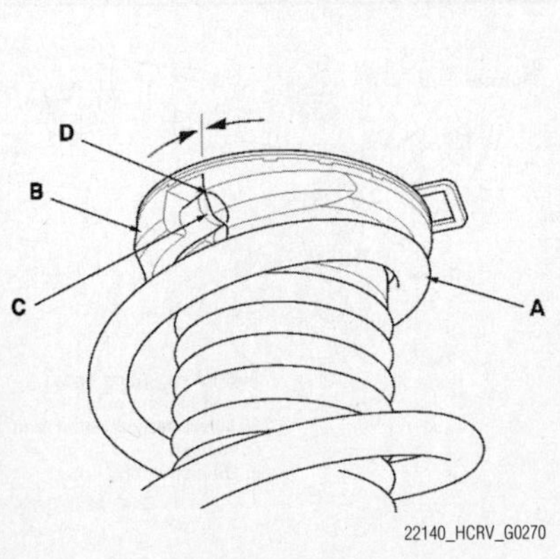
D
B
C
A
22140_HCRV_G0270

Fig. 277 Install the strut damper spring on the upper spring mounting cushion, then align the upper end of the strut damper spring with the cushion stop

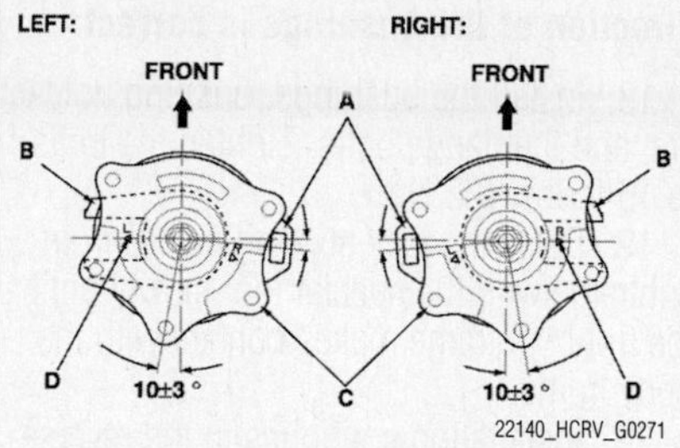
LEFT:
RIGHT:
FRONT
FRONT
A
B
B
D
10±3 °
C
10±3 °
D
22140_HCRV_G0271

Fig. 278 Align the center of raised portion on the upper spring seat and the strut damper unit. Align the strut damper bracket and the strut damper mounting base so that stamp is toward the outside of the vehicle

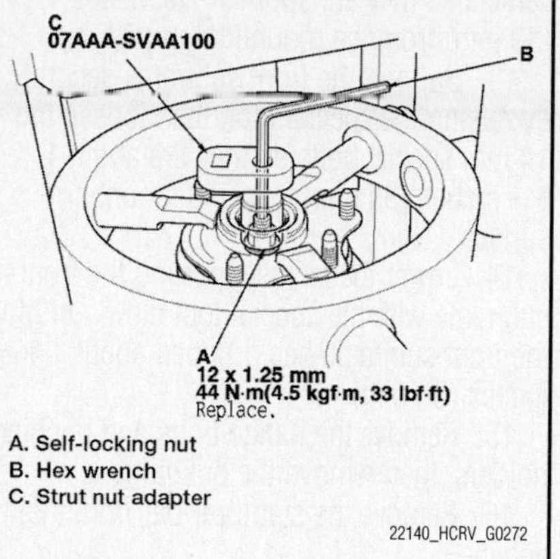
C
07AAA-SVAA100
B
A
12 x 1.25 mm
44 N·m(4.5 kgf·m, 33 lbf·ft)
Replace.
A. Self-locking nut
B. Hex wrench
C. Strut nut adapter
22140_HCRV_G0272

Fig. 279 Install a new self-locking nut on the strut damper as illustrated

STABILIZER BAR

REMOVAL & INSTALLATION

See Figures 281 through 286.

Special Tools Required:

- Universal eyelet 07AAK-SNAA120
- Front subframe adapter VSB02C000016
- Engine support hanger, A and Reds AAR-T-12566

1. Before servicing the vehicle, refer to the Precautions Section.
2. Matchmark the stabilizer bar for proper reinstallation.
3. Raise and safely support the vehicle.
4. Remove the front wheels.

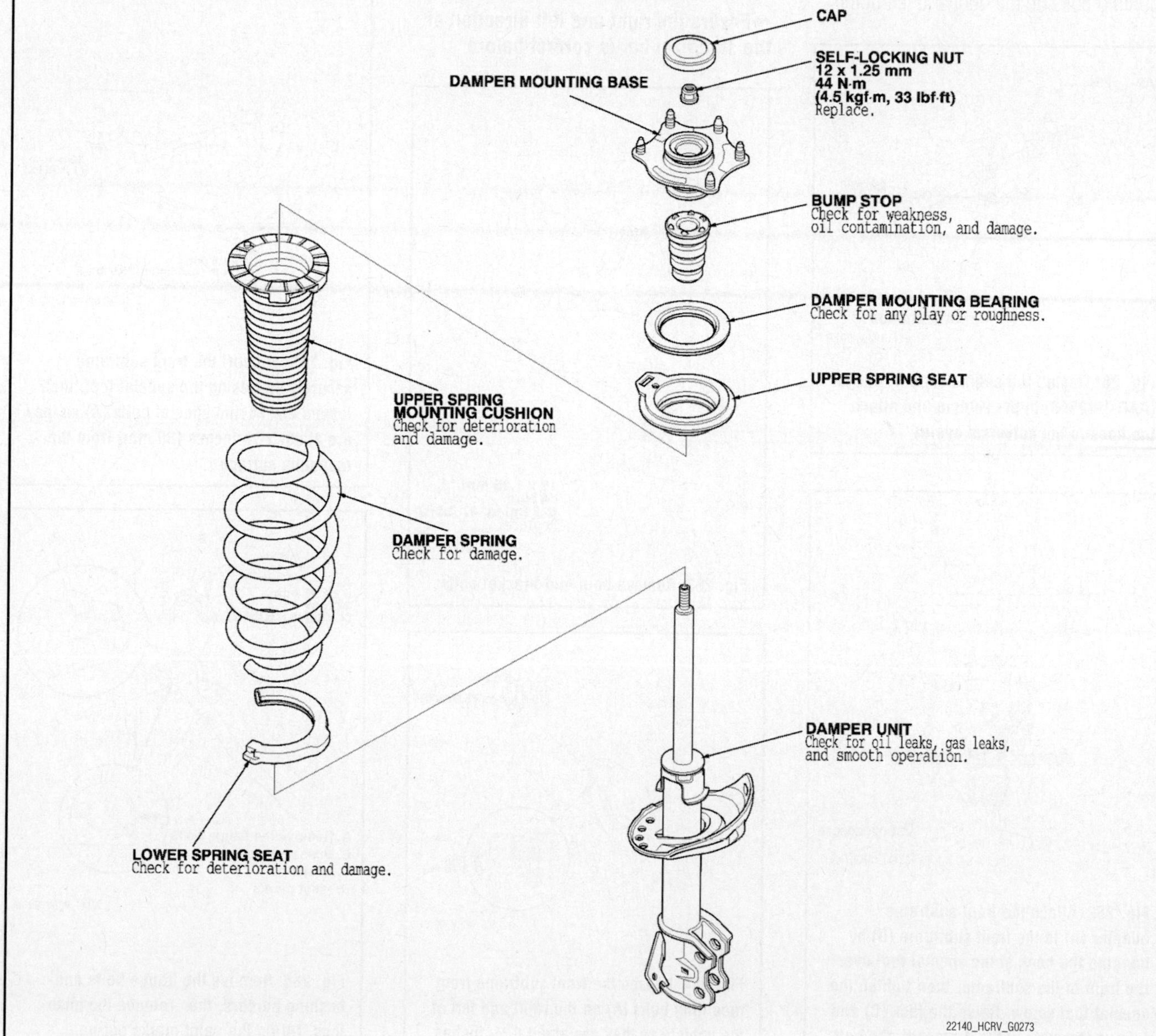
CAP
SELF-LOCKING NUT
12 x 1.25 mm
44 N·m
(4.5 kgf·m, 33 lbf·ft)
Replace.
DAMPER MOUNTING BASE
BUMP STOP
Check for weakness, oil contamination, and damage.
DAMPER MOUNTING BEARING
Check for any play or roughness.
UPPER SPRING SEAT
UPPER SPRING MOUNTING CUSHION
Check for deterioration and damage.
DAMPER SPRING
Check for damage.
DAMPER UNIT
Check for oil leaks, gas leaks, and smooth operation.
LOWER SPRING SEAT
Check for deterioration and damage.
22140_HCRV_G0273

Fig. 280 Exploded view of front strut components

5. Disconnect both stabilizer control links from the stabilizer bar.

6. Remove the cowl cover.

7. Attach the universal eyelet to the cylinder head.

8. Install the engine support hanger (AAR-T-12566) to the vehicle and attach the hook to the universal eyelet.

9. Attach the front subframe adapter to the front subframe by hanging the hook of the special tool over the front of the subframe, then tighten the special tool screw.

10. Raise the jack and line up the slots in the arms with the bolt holes on the corner of the jack base, then securely attach them with bolts.

11. Remove both mid-bracket bolts.

12. Loosen the front subframe front mounting bolts on the right and left of the vehicle so they are about 1 3⁄16 inches (30 mm) from the mounting surface.

13. Support the front subframe securely by raising the special tool, then loosen the 14 mm special bolts so they are about 1 3⁄16 inches (30 mm) from the mounting surface.

14. Lower the jack supporting the front subframe with the special tool slowly until the front subframe has dropped about 1 3⁄16 inches (30 mm).

15. Remove the flange bolts and bushing holders, then remove the bushings.

16. Remove the stabilizer bar from the vehicle.

To install:

17. Install the stabilizer bar into position aligning the paint marks on the stabilizer bar with the sides of the bushings.

➡Ensure the right and left direction of the stabilizer bar is correct before installation. Make sure the fore and aft direction of the bushings is correct.

18. Install the bushings, bushing holders, and the flange bolts. Tighten the bolts to 16 ft. lbs. (22 Nm).

19. Raise the jack supporting the front subframe with the special tool slowly until the front subframe makes contact with the body frame.

20. Installation continues in the reverse of the removal procedure.

21. Tighten the bolts to the specified torque:

a. Bolts 14 x 1.5mm: 76 ft. lbs. (103 Nm).

b. Bolts 12 x 1.25mm: 47 ft. lbs. (64 Nm).

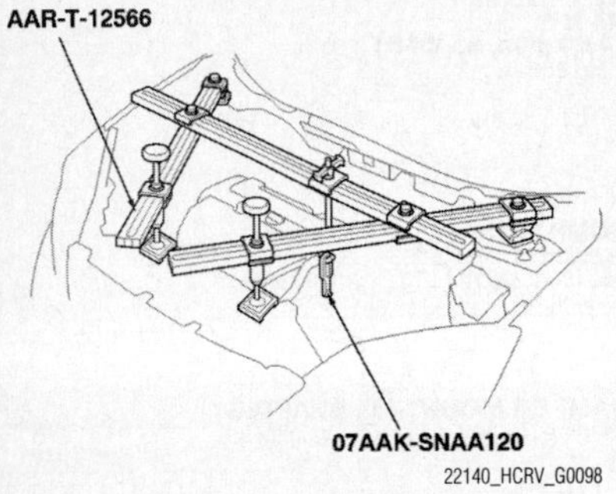
AAR-T-12566
07AAK-SNAA120
22140_HCRV_G0098

Fig. 281 Install the engine support hanger (AAR-T-12566) to the vehicle and attach the hook to the universal eyelet

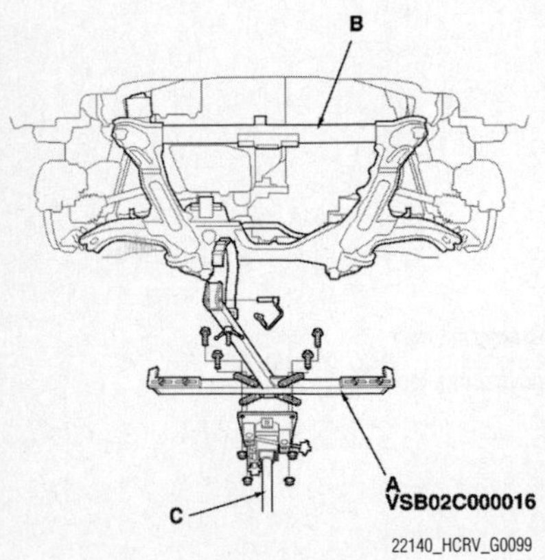
B
A
VSB02C000016
C
22140_HCRV_G0099

Fig. 282 Attach the front subframe adapter (A) to the front subframe (B) by hanging the hook of the special tool over the front of the subframe, then tighten the special tool screw. Raise the jack (C) and line up the slots in the arms with the bolt holes on the corner of the jack base

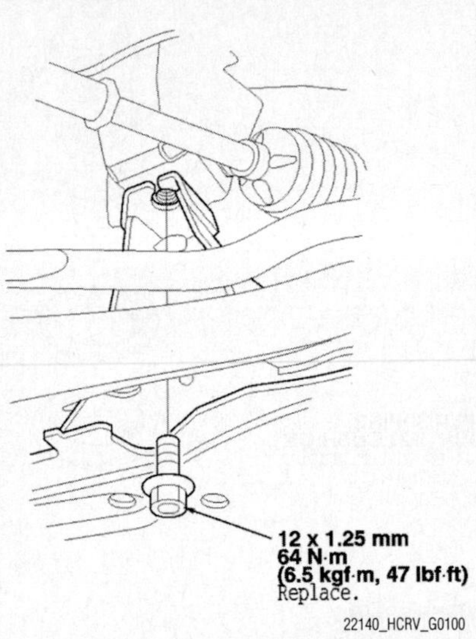
12 x 1.25 mm
64 N·m
(6.5 kgf·m, 47 lbf·ft)
Replace.
22140_HCRV_G0100

Fig. 283 Remove both mid-bracket bolts

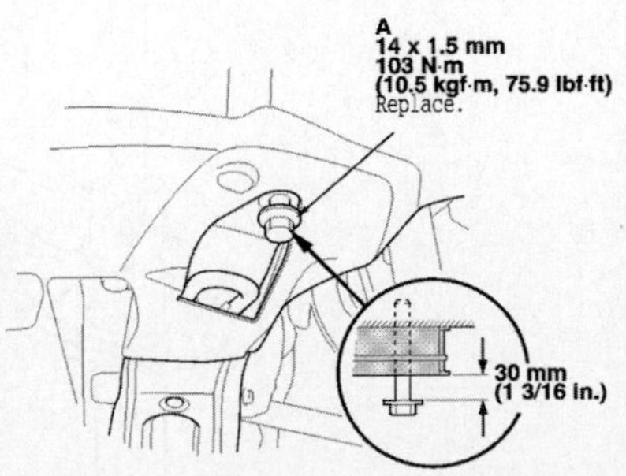
A
14 x 1.5 mm
103 N·m
(10.5 kgf·m, 75.9 lbf·ft)
Replace.
30 mm
(1 3/16 in.)
22140_HCRV_G0101

Fig. 284 Loosen the front subframe front mounting bolts (A) on the right and left of the vehicle so they are about 1 3⁄16 inches (30 mm) from the mounting surface

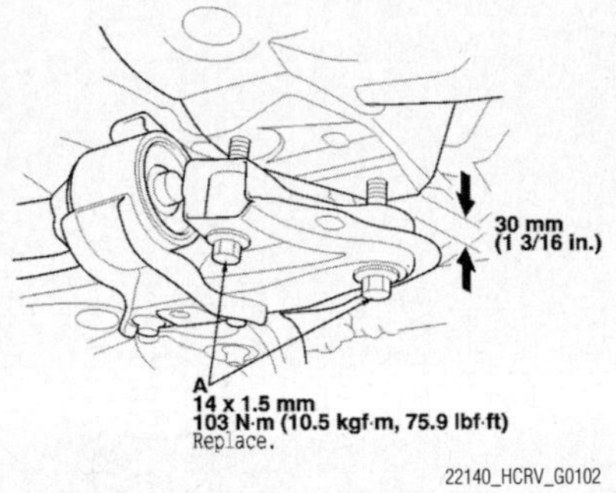
30 mm
(1 3/16 in.)
A
14 x 1.5 mm
103 N·m (10.5 kgf·m, 75.9 lbf·ft)
Replace.
22140_HCRV_G0102

Fig. 285 Support the front subframe securely by raising the special tool, then loosen the 14 mm special bolts (A) so they are about 1 3⁄16 inches (30 mm) from the mounting surface

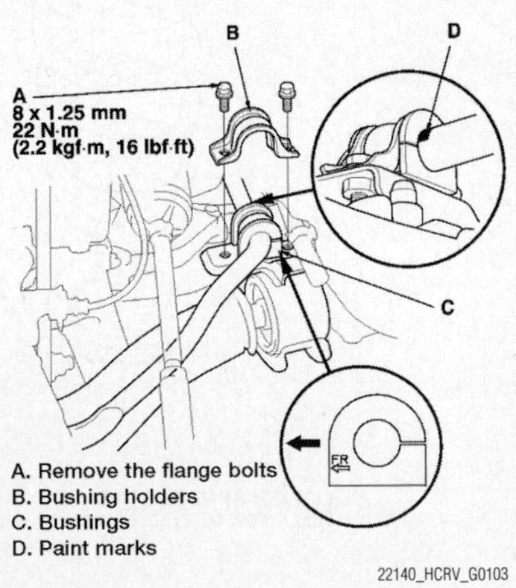
B
D
A
8 x 1.25 mm
22 N·m
(2.2 kgf·m, 16 lbf·ft)
C
FR
A. Remove the flange bolts
B. Bushing holders
C. Bushings
D. Paint marks
22140_HCRV_G0103

Fig. 286 Remove the flange bolts and bushing holders, then remove the bushings. (Align the paint marks during installation)

22. Check the front wheel alignment and make adjustments as necessary.

WHEEL HUB & BEARING

REMOVAL & INSTALLATION

See Figures 287 through 290.

Special Tools Required:

- Hub disassembly/assembly tool 07GAF-SD40100
- Attachment, 72 x 75mm 07746-0010600
- Driver 07749-0010000
- Attachment, 96mm 07948-SB00101
- Support base 07965-SD90100

1. Before servicing the vehicle, refer to the Precautions Section.
2. Remove the knuckle/hub assembly. Refer to Steering Knuckle, removal & installation.
3. Separate the hub from the knuckle using the hub disassembly/assembly tool and a hydraulic press. Hold the knuckle with the attachment of the hydraulic press or equivalent tool. Be careful not to deform the splash guard. Hold onto the hub to keep it from falling when pressed clear.
4. Press the wheel bearing inner race off of the hub using the hub disassembly/assembly tool, a commercially available bearing separator, and a press.

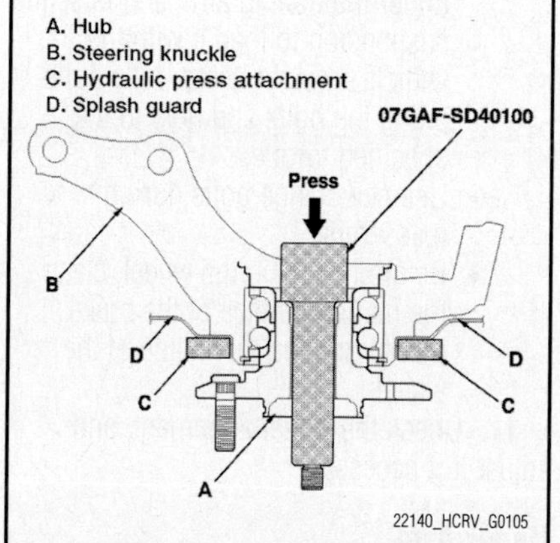
A. Hub
B. Steering knuckle
C. Hydraulic press attachment
D. Splash guard
07GAF-SD40100
Press
B
D
C
D
C
A
22140_HCRV_G0105

Fig. 287 Separate the hub from the knuckle using the hub disassembly/assembly tool and a hydraulic press. Hold the knuckle with the attachment of the hydraulic press or equivalent tool. Be careful not to deform the splash guard

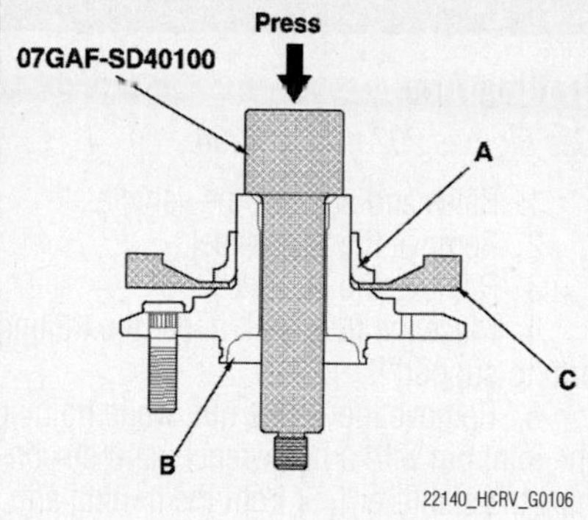
Press
07GAF-SD40100
A
C
B
22140_HCRV_G0106

Fig. 288 Press the wheel bearing inner race (A) off of the hub (B) using the hub disassembly/assembly tool, a commercially available bearing separator (C), and a press

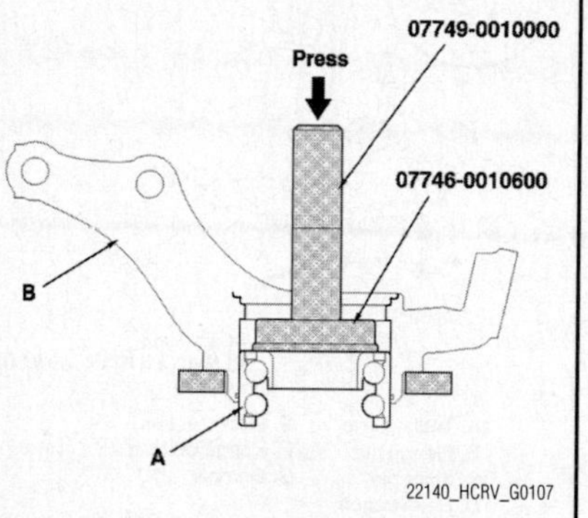
07749-0010000
Press
07746-0010600
B
A
22140_HCRV_G0107

Fig. 289 Press the wheel bearing (A) out of the knuckle (B) using the attachment, the driver, and a press

5. Remove the splash guard and the snap ring from the knuckle.
6. Press the wheel bearing out of the knuckle using the attachment, the driver, and a press.

To install:

7. Wash the knuckle and hub thoroughly in high flash-point solvent before reassembly.

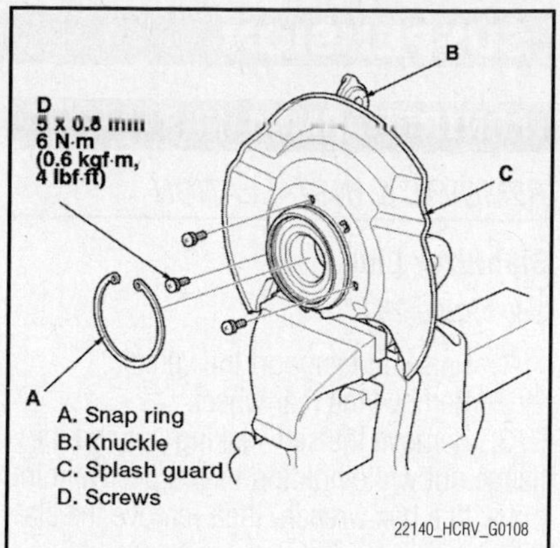
B
D
5 x 0.8 mm
6 N·m
(0.6 kgf·m,
4 lbf·ft)
C
A
A. Snap ring
B. Knuckle
C. Splash guard
D. Screws
22140_HCRV_G0108

Fig. 290 Install the snap ring securely in the knuckle. Install the splash guard and tighten the screws

8. Press a new wheel bearing into the knuckle using the old bearing, a steel plate, the attachment, the support base, and a press.
   - Install the wheel bearing with the wheel speed sensor magnetic encoder (brown color) toward the inside of the knuckle
   - Remove any oil, grease, dust, metal debris, and other foreign material from the encoder surface
   - Keep all magnetic tools away from the encoder surface
   - Be careful not to damage the encoder surface when inserting the wheel bearing
9. Install the snap ring securely in the knuckle.
10. Install the splash guard and tighten the screws to 48 inch lbs. (6 Nm).
11. Install the hub onto the knuckle using the attachment, the driver, the support base, and a hydraulic press. Be careful not to distort the splash guard.
12. Install the knuckle/hub assembly. Refer to Steering Knuckle, removal & installation.

ADJUSTMENT

The front wheel bearings are not adjustable. If the bearings are noisy or become loose, they must be replaced.

CONTROL ARMS/LINKS

REMOVAL & INSTALLATION

Stabilizer Link

See Figure 291.

1. Raise and support the vehicle.
2. Remove the rear wheel.
3. Remove the self-locking nut and the flange nut while holding the respective joint pin with a hex wrench, then remove the stabilizer link.

To install:

4. Install the stabilizer link on the stabilizer bar and trailing arm with the joint pins set at the center of their range of movement.

➡The left stabilizer link has a yellow paint mark, while the right stabilizer link has a white paint mark.

5. Install a new self-locking nut and a new flange nut, and lightly tighten them.
6. Place the floor jack under the trailing arm, and raise the suspension to load it with the vehicle's weight.
7. Tighten the self-locking nut and the flange nut to the specified torque while holding the respective joint pin with a hex wrench.
8. Clean the mating surface of the brake disc/drum and the inside of the wheel, then install the rear wheel.
9. Test-drive the vehicle.

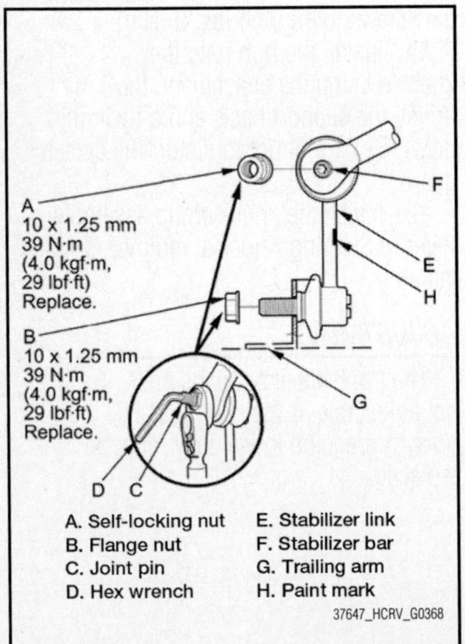

A
10 x 1.25 mm
39 N·m
(4.0 kgf·m,
29 lbf·ft)
Replace.
B
10 x 1.25 mm
39 N·m
(4.0 kgf·m,
29 lbf·ft)
Replace.
C
D
E
F
G
H
A. Self-locking nut
B. Flange nut
C. Joint pin
D. Hex wrench
E. Stabilizer link
F. Stabilizer bar
G. Trailing arm
H. Paint mark
37647_HCRV_G0368

Fig. 291 Removing the stabilizer link

Trailing Arm

See Figures 292 through 294.

1. Raise and support the vehicle.
2. Remove the rear wheel.
3. Remove the knuckle.
4. Place the floor jack under the trailing arm to support it.
5. Remove the flange nut, while holding the joint pin with a hex wrench, and disconnect the stabilizer link from the trailing arm.

➡During installation, install a new flange nut.

6. Remove the flange bolt, and disconnect the damper from the trailing arm.
7. Remove the trailing arm front mounting bolts.

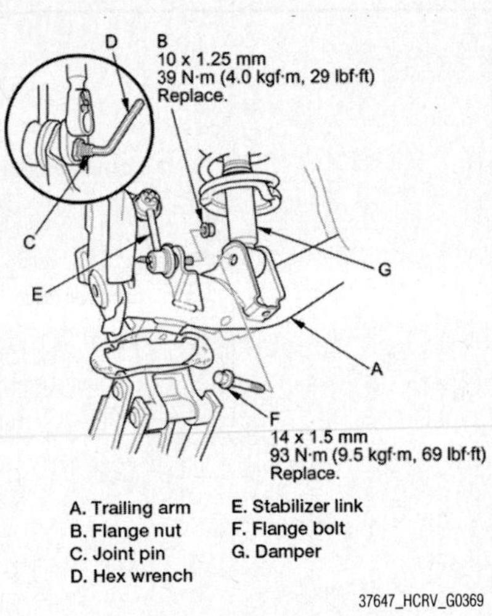

D
B
10 x 1.25 mm
39 N·m (4.0 kgf·m, 29 lbf·ft)
Replace.
C
E
G
A
F
14 x 1.5 mm
93 N·m (9.5 kgf·m, 69 lbf·ft)
Replace.
A. Trailing arm
B. Flange nut
C. Joint pin
D. Hex wrench
E. Stabilizer link
F. Flange bolt
G. Damper
37647_HCRV_G0369

Fig. 292 Removing the rear trailing arm

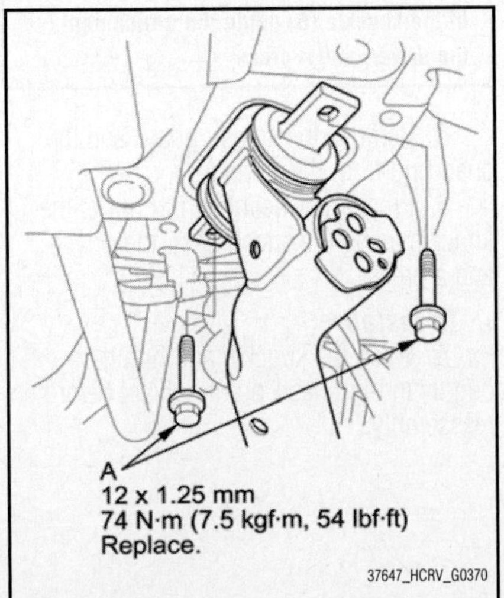

A
12 x 1.25 mm
74 N·m (7.5 kgf·m, 54 lbf·ft)
Replace.
37647_HCRV_G0370

Fig. 293 Remove the trailing arm front mounting bolts (A)

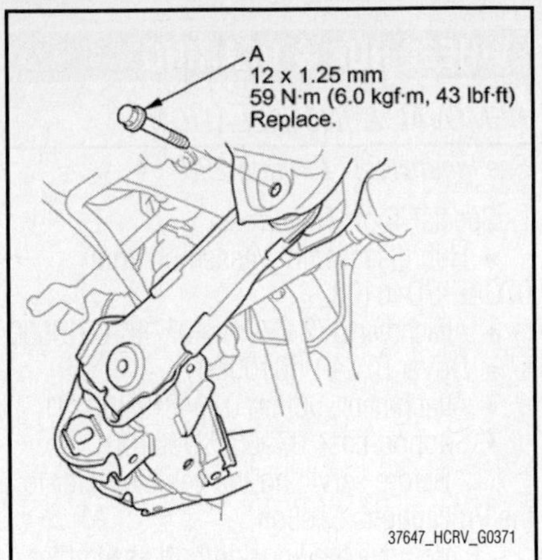

A
12 x 1.25 mm
59 N·m (6.0 kgf·m, 43 lbf·ft)
Replace.
37647_HCRV_G0371

Fig. 294 Remove the trailing arm rear mounting bolt (A)

8. Remove the trailing arm rear mounting bolt.
9. Lower the jack, and remove the trailing arm.

To install:

10. Install the trailing arm in the reverse order of removal, and note these items:
    - First install all the suspension components, and lightly tighten the bolts and nuts, then place a jack under the trailing arm, and raise the suspension to load it with the vehicle's weight before fully tightening the bolts and nuts to the specified torque.
    - Use new flange bolts during reassembly.
    - Before installing the wheel, clean the mating surface of the brake disc/drum and the inside of the wheel.
11. Check the wheel alignment, and adjust it if necessary.

Upper Arm

See Figure 295.

1. Before servicing the vehicle, refer to the Precautions Section.
2. Raise and safely support the vehicle.
3. Remove the rear wheel.
4. Place a floor jack under the trailing arm, and support the suspension.
5. Remove the wheel speed sensor harness bracket from the upper arm.
6. Remove the flange bolts and remove the upper arm.

To install:

7. Install the upper arm and loosely install the flange bolts.

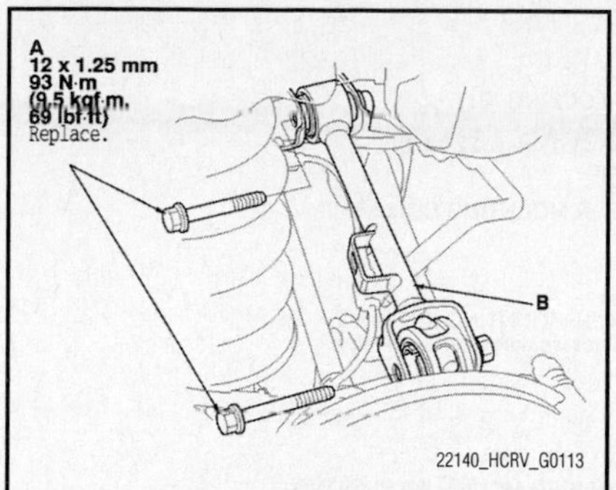
A
12 x 1.25 mm
93 N·m
(9.5 kgf·m,
69 lbf·ft)
Replace.
B
22140_HCRV_G0113

Fig. 295 Remove the flange bolts (A) and remove the upper arm (B)

8. Install the wheel speed sensor harness bracket to the upper arm.

9. Place a floor jack under the trailing arm and raise the suspension to load it with the vehicle weight before tightening the flange bolts.

10. Tighten the new flange bolts to 69 ft. lbs. (93 Nm).

11. Install the rear wheel.

12. Check the wheel alignment and adjust it if necessary.

STRUT (DAMPER/SPRING)

REMOVAL & INSTALLATION

See Figures 296 through 298.

1. Before servicing the vehicle, refer to the Precautions Section.

2. Raise and safely support the vehicle.

3. Remove the rear wheel.

4. Remove the lid on the cargo area side trim panel by releasing the hooks.

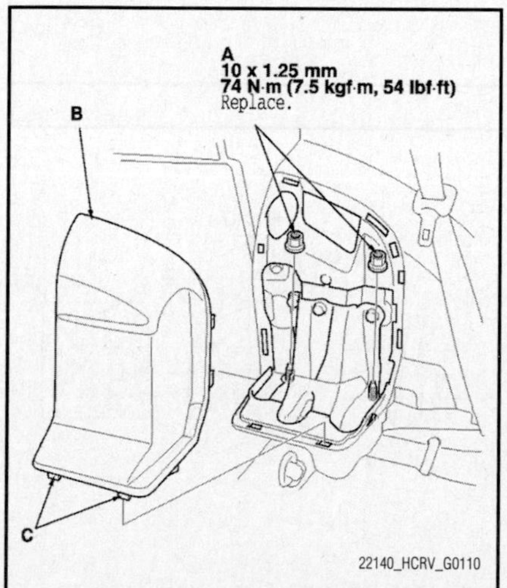
A
10 x 1.25 mm
74 N·m (7.5 kgf·m, 54 lbf·ft)
Replace.
B
C
22140_HCRV_G0110

Fig. 296 Remove the lid (B) on the cargo area side trim panel by releasing the hooks (C) and remove the flange nuts (A) from the top of the strut

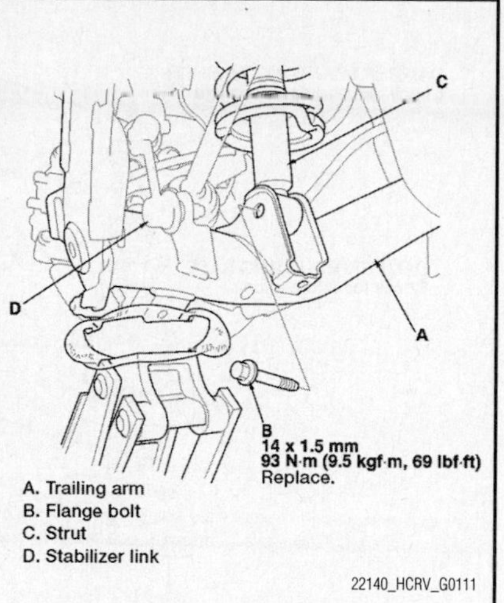
C
D
A
B
14 x 1.5 mm
93 N·m (9.5 kgf·m, 69 lbf·ft)
Replace.
A. Trailing arm
B. Flange bolt
C. Strut
D. Stabilizer link
22140_HCRV_G0111

Fig. 297 Position a floor jack under the trailing arm and raise it until the suspension begins to compress. Disconnect the stabilizer link from the trailing arm. Remove the flange bolt from the bottom of the strut

5. Remove the flange nuts from the top of the strut.

6. Remove the flange bolts, then remove the brake hose bracket.

7. Position a floor jack under the trailing arm. Raise the floor jack until the suspension begins to compress.

8. Disconnect the stabilizer link from the trailing arm.

9. Remove the flange bolt from the bottom of the strut.

10. Remove the flange bolt, then remove the parking brake cable.

11. Remove the trailing arm mounting bolts.

12. Lower the rear suspension, then remove the strut assembly from the vehicle.

To install:

13. Position the strut assembly between the body and the trailing arm. Note the direction of the strut mounting base so that the hook or the paint mark on it is toward the outside of the vehicle.

14. Loosely install new flange nuts on the top of the strut.

15. Loosely install new trailing arm mounting bolts.

16. Install the parking brake cable, then install the flange bolt.

17. Position a floor jack under the trailing arm. Raise the floor jack until the hole in the trailing arm aligns with the hole in the strut.

18. Loosely install a new flange bolt on the bottom of the strut.

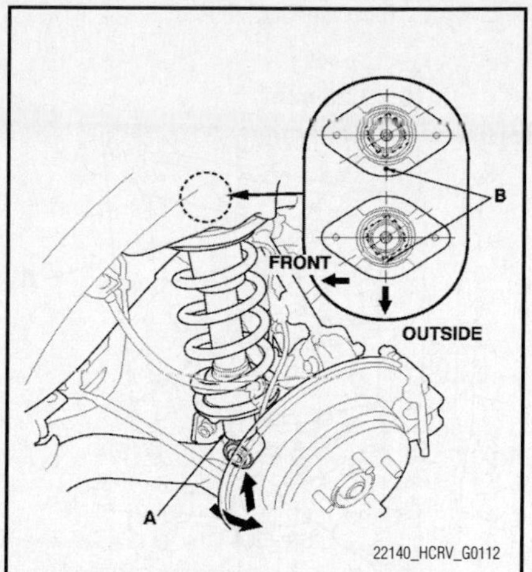
B
FRONT
OUTSIDE
A
22140_HCRV_G0112

Fig. 298 Position the strut assembly (A) between the body and the trailing arm. Note the direction of the strut mounting base and the hook or the paint mark (B)

19. Connect the stabilizer link to the trailing arm.

20. Raise the rear suspension with a floor jack to load the suspension with the vehicle's weight.

21. Tighten new flange nuts, new flange bolts, and other fasteners to the specified torque values.

a. Flange nuts and bolts: 54 ft. lbs. (74 Nm).

b. Parking brake cable bolt: 16 ft. lbs. (22 Nm).

c. New flange bolt at bottom of strut: 69 ft. lbs. (93 Nm).

d. Brake hose bracket bolts: 16 ft. lbs. (22 Nm).

22. Install the strut cover lid.

23. Clean the mating surface of the brake disc and the inside of the wheel, then install the rear wheel.

24. Check the wheel alignment and adjust it if necessary.

OVERHAUL

See Figures 299 through 303.

Special tool required: Strut spring compressor: Branick MST-580A or Model 7200, or equivalent

1. Before servicing the vehicle, refer to the Precautions Section.

2. Remove the strut assembly from the vehicle.

3. Compress the damper spring, then remove the 10 mm nut while holding the damper shaft with a hex wrench.

➡Do not compress the damper spring more than necessary to remove the nut.

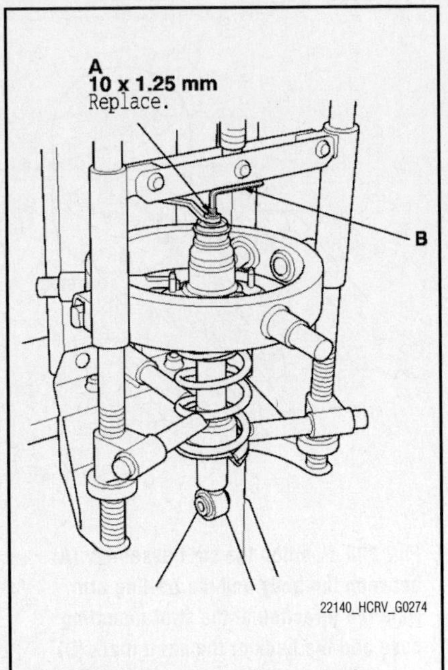
A
10 x 1.25 mm
Replace.
B
22140_HCRV_G0274

Fig. 299 Compress the damper spring, then remove the 10 mm nut (A) while holding the damper shaft with a hex wrench (B)

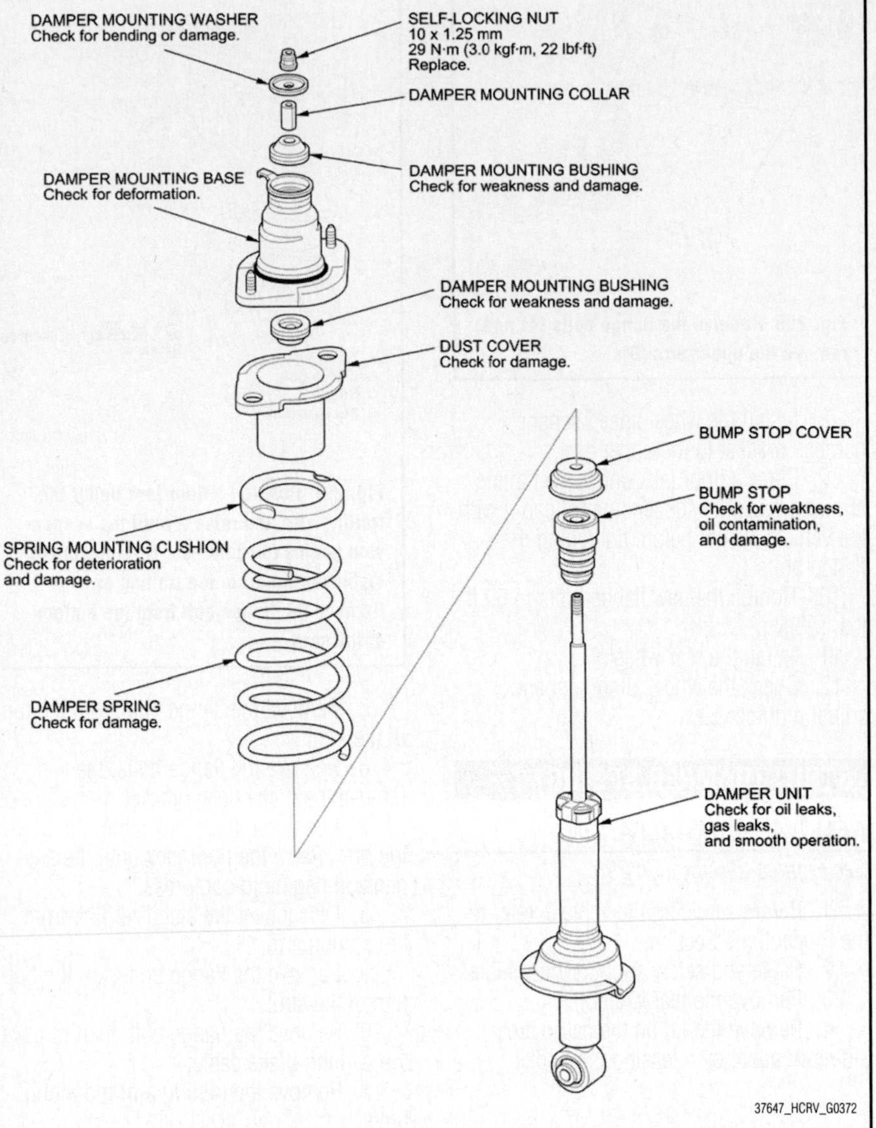
DAMPER MOUNTING WASHER
Check for bending or damage.
SELF-LOCKING NUT
10 x 1.25 mm
29 N·m (3.0 kgf·m, 22 lbf·ft)
Replace.
DAMPER MOUNTING COLLAR
DAMPER MOUNTING BASE
Check for deformation.
DAMPER MOUNTING BUSHING
Check for weakness and damage.
DAMPER MOUNTING BUSHING
Check for weakness and damage.
DUST COVER
Check for damage.
BUMP STOP COVER
BUMP STOP
Check for weakness, oil contamination, and damage.
SPRING MOUNTING CUSHION
Check for deterioration and damage.
DAMPER SPRING
Check for damage.
DAMPER UNIT
Check for oil leaks, gas leaks, and smooth operation.
37647_HCRV_G0372

Fig. 300 Exploded view of rear strut assembly

To install:

5. Reassemble all the parts, except the upper spring mounting cushion, the bump stop, and the damper spring.

6. Compress the damper assembly by hand, and check for smooth operation through a full stroke, both compression and extension.

➡The damper should extend smoothly and constantly when compression is released. If it does not, the gas is leaking, and the damper should be replaced.

7. Check for oil leaks, abnormal noises, and binding.

8. Compress the damper spring with the spring compressor.

9. Install all the parts, except the damper mounting washer and self-locking nut, onto the damper unit.

10. Align the bottom of the damper spring with the stepped part of the lower spring seat.

11. Align the bottom of the damper and the damper mounting base as shown so that the hook or the paint mark is toward the outside of the vehicle.

12. Install the damper mounting washer and a new self-locking nut.

13. Hold the damper shaft using a hex wrench and tighten the new self-locking nut to 22 ft. lbs. (29 Nm).

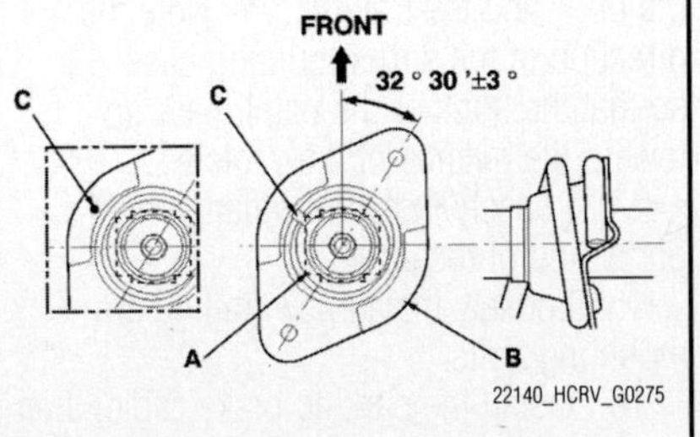
FRONT
32 ° 30 '±3 °
C
C
A
B
22140_HCRV_G0275

Fig. 301 Align the bottom (A) of the damper and the damper mounting base (B) so that the hook or the paint mark (C) is toward the outside of the vehicle—left side illustrated

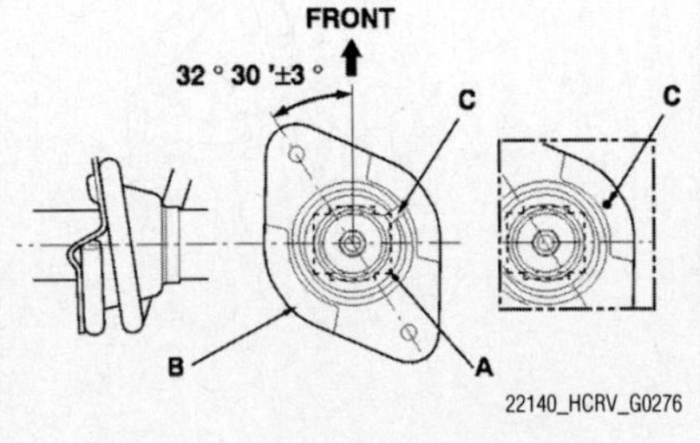
FRONT
32 ° 30 '±3 °
C
C
B
A
22140_HCRV_G0276

Fig. 302 Align the bottom (A) of the damper and the damper mounting base (B) so that the hook or the paint mark (C) is toward the outside of the vehicle—right side illustrated

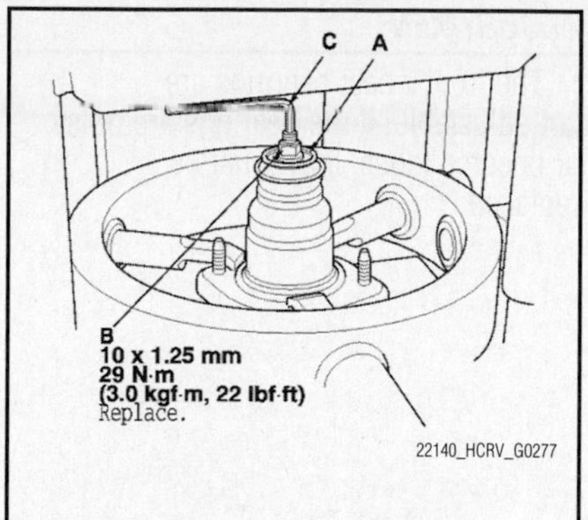
C A
B
10 x 1.25 mm
29 N·m
(3.0 kgf·m, 22 lbf·ft)
Replace.
22140_HCRV_G0277

Fig. 303 Install the damper mounting washer (A) and a new self-locking nut (B). Hold the damper shaft using a hex wrench (C)

STABILIZER BAR

REMOVAL & INSTALLATION

See Figure 316.

1. Raise and support the vehicle.
2. Remove the rear wheels.
3. Disconnect both stabilizer links from the stabilizer bar.
4. Remove the flange bolts and the bushing holders, then remove the bushings and the stabilizer bar.

To install:

5. Install the stabilizer bar in the reverse order of removal, and note these items:
   - Note the right and left direction of the stabilizer bar.
   - Align the paint marks (E) on the stabilizer bar with the sides of the bushings.
   - Note the fore/aft direction of the bushings.
   - Refer to the stabilizer link removal/installation to connect the stabilizer bar to the links.

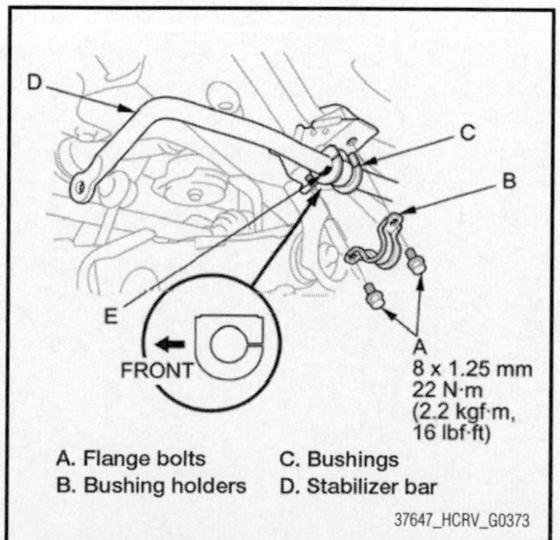
D
C
B
E
FRONT
A
8 x 1.25 mm
22 N·m
(2.2 kgf·m,
16 lbf·ft)
A. Flange bolts
B. Bushing holders
C. Bushings
D. Stabilizer bar
37647_HCRV_G0373

Fig. 304 Remove the flange bolts and the bushing holders, then remove the bushings and the stabilizer bar

   - Clean the mating surface of the brake disc/drum and the inside of the wheel, then install the rear wheel.
6. Check the wheel alignment, and adjust it if necessary.

WHEEL HUB & BEARING

REMOVAL & INSTALLATION

See Figures 305 through 308.

1. Raise and support the vehicle.
2. Remove the wheel nuts and the rear wheel.
3. Remove the brake hose bracket mounting bolt from the knuckle.
4. Remove the brake caliper bracket mounting bolts, and remove the caliper assembly from the knuckle. To prevent damage to the caliper assembly or brake hose, use a short piece of wire to hang the caliper assembly from the undercarriage. Do not twist the brake hose excessively.

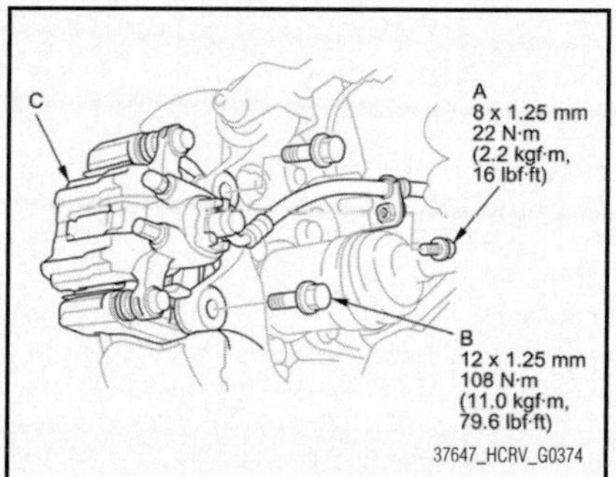
C
A
8 x 1.25 mm
22 N·m
(2.2 kgf·m,
16 lbf·ft)
B
12 x 1.25 mm
108 N·m
(11.0 kgf·m,
79.6 lbf·ft)
37647_HCRV_G0374

Fig. 305 Remove the brake hose bracket mounting bolt (A), the brake caliper bracket mounting bolts (B), and the caliper assembly (C)

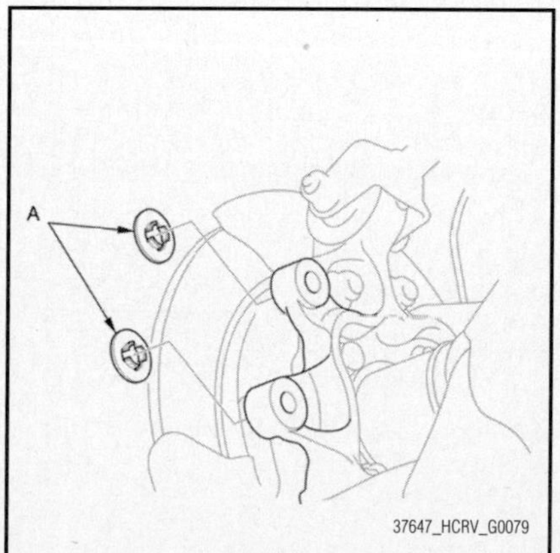
A
37647_HCRV_G0079

Fig. 306 Remove the two washers (A)

5. Remove the two washers.
6. 4WD model: Pry up the stake on the spindle nut, then remove the nut.
7. Release the parking brake, and remove the brake disc/drum.
8. 4WD model: Remove the flange bolts, and remove the hub bearing unit by tapping the driveshaft end with a soft face hammer while drawing the hub bearing unit outward.
9. 2WD model: Remove the hub bearing unit and the O-ring.
10. Check the hub bearing unit for damage and cracks.
11. Install the hub bearing unit in the reverse order of removal, and note these items:
    - 2WD model: Use a new O-ring on reassembly.
    - 4WD model: Use a new spindle nut on reassembly.
    - 4WD model: Before installing the spindle nut, apply a small amount of engine oil to the seating surface of the nut. After tightening, use a drift to stake the spindle nut shoulder against the driveshaft.

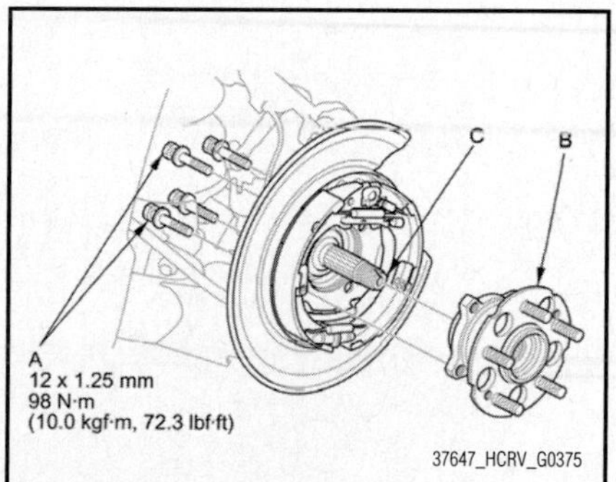
C
B
A
12 x 1.25 mm
98 N·m
(10.0 kgf·m, 72.3 lbf·ft)
37647_HCRV_G0375

Fig. 307 4WD model: Remove the flange bolts (A), and remove the hub bearing unit (B) by tapping the driveshaft end (C) with a soft face hammer while drawing the hub bearing unit outward

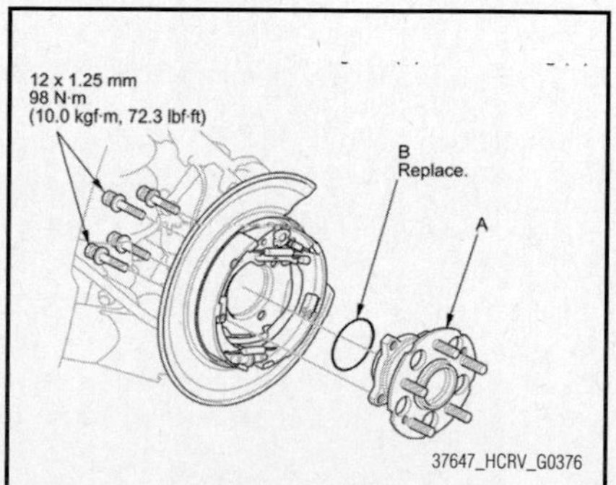
12 x 1.25 mm
98 N·m
(10.0 kgf·m, 72.3 lbf·ft)
B
Replace.
A
37647_HCRV_G0376

Fig. 308 2WD model: Remove the hub bearing unit (A) and the O-ring (B)

- Before installing the brake disc/drum, clean the mating surface of the hub bearing unit and the inside of the brake disc/drum.
- Before installing the wheel, clean the mating surface of the brake disc/drum and the inside of the wheel.
- After installation, press the brake pedal several times to make sure the brakes work.

12. Check the wheel alignment, and adjust it if necessary.

ADJUSTMENT

The front wheel bearings are not adjustable. If the bearings are noisy or become loose, they must be replaced.

HONDA

Element

12

BRAKES........................12-9

BLEEDING THE BRAKE SYSTEM........................12-9
- Bleeding Procedure..................12-9
  - Bleeding Procedure 12-9

FRONT DISC BRAKES........12-10
- Brake Caliper..........................12-10
  - Removal & Installation........12-10
- Disc Brake Pads......................12-10
  - Removal & Installation........12-10

INFORMATION AND PRECAUTIONS 12-9
  - Anti-lock Systems................12-9
  - Disc and Drum Systems.......12-9

PARKING BRAKE.............12-10
- Parking Brake Cables.............12-10
  - Adjustment 12-10
- Parking Brake Shoes...............12-11
  - Adjustment 12-11

REAR DISC BRAKES 12-10
- Brake Caliper..........................12-10
  - Removal & Installation........12-10
- Disc Brake Pads......................12-10
  - Removal & Installation........12-10

CHASSIS ELECTRICAL 12-11

AIR BAG (SUPPLEMENTAL RESTRAINT SYSTEM) 12-11
- General Information.................12-11
  - Arming the System 12-12
  - Disarming the System.........12-11
  - Service Precautions 12-11

DRIVETRAIN 12-12

- Clutch Driven Disc and Pressure Plate......................12-12
  - Adjustments........................12-12
  - Removal & Installation........12-12
- Clutch Master Cylinder 12-13
  - Removal & Installation........12-13
- Clutch Slave Cylinder.............12-14
  - Removal & Installation........12-14
- Front Halfshaft........................12-15
  - CV-joint Overhaul 12-15
  - Removal & Installation........12-15
- Hydraulic Clutch System.........12-14
  - Bleeding 12-14
- Rear Halfshaft.........................12-15
  - CV-joint Overhaul 12-16
  - Removal & Installation........12-15
- Rear Pinion Seal.....................12-17
  - Removal & Installation........12-17
- Transfer Assembly..................12-15
  - Removal & Installation........12-15

ENGINE COOLING 12-17

- Thermostat 12-17
  - Removal & Installation........12-17
- Water Pump 12-18
  - Removal & Installation........12-18

ENGINE ELECTRICAL 12-18

CHARGING SYSTEM 12-18
- Alternator 12-18
  - Removal & Installation........12-18

IGNITION SYSTEM 12-18
- Ignition Coil 12-18
  - Removal & Installation........12-18
- Ignition Timing........................12-18
  - Adjustment 12-19
  - Inspection 12-18
- Spark Plugs............................12-19
  - Removal & Installation........12-19

STARTING SYSTEM 12-20
- Starter 12-20
  - Removal & Installation........12-20

ENGINE MECHANICAL......12-20

- Accessory Drive Belts 12-20
  - Accessory Belt Routing........12-20
  - Adjustment 12-21
  - Inspection 12-20
  - Removal & Installation........12-21
- Camshaft and Valve Lifters......12-21
  - Removal & Installation........12-21
- Crankshaft Damper.................12-21
  - Removal & Installation........12-21
- Crankshaft Front Seal.............12-21
  - Removal & Installation........12-21
- Cylinder Head 12-21
  - Removal & Installation........12-21
- Driveplate 12-23
  - Removal & Installation........12-23
- Exhaust Manifold 12-23
  - Removal & Installation........12-23
- Flywheel 12-25
  - Removal & Installation........12-25
- Intake Manifold 12-25
  - Removal & Installation........12-25
- Oil Pan 12-25
  - Removal & Installation........12-25
- Oil Pump 12-25
  - Removal & Installation........12-25
- Piston and Ring.......................12-27
  - Positioning 12-27
- Rear Main Seal.......................12-27
  - Removal & Installation........12-27
- Rocker Arms/Shafts................12-28
  - Removal & Installation........12-28
- Timing Chain & Front Seal 12-30
  - Removal & Installation........12-30
- Valve (Rocker Arm) Covers 12-32
  - Removal & Installation........12-32
- Valve Lash..............................12-33
  - Adjustment 12-33

ENGINE PERFORMANCE & EMISSION CONTROL 12-33

- Air Fuel (A/F) Ratio Sensor.....12-33
  - Location.............................12-33
  - Removal & Installation........12-33
- Camshaft Position (CMP) Sensor 12-34
  - Location.............................12-34
  - Removal & Installation........12-34
- Crankshaft Position (CKP) Sensor 12-34
  - Location.............................12-34
  - Removal & Installation........12-34
- Electronic Control Module (ECM) Powertrain Control Module.....................12-34
  - Location.............................12-34
  - Removal & Installation........12-35
- Engine Coolant Temperature (ECT) Sensor 12-37
  - Location.............................12-37
  - Removal & Installation........12-37

Heated Oxygen (HO2S) Sensor 12-38
Location 12-38
Removal & Installation 12-38
Intake Air Temperature (IAT) Sensor 12-38
Location 12-38
Removal & Installation 12-38
Knock Sensor (KS) 12-38
Location 12-38
Removal & Installation 12-38
Manifold Absolute Pressure (MAP) Sensor 12-38
Location 12-38
Removal & Installation 12-39
Mass Air Flow (MAF) Sensor ..12-39
Location 12-39
Removal & Installation 12-39
Output Shaft Speed (OSS) Sensor 12-39
Location 12-39
Removal & Installation 12-39

FUEL 12-40

GASOLINE FUEL INJECTION SYSTEM 12-40
Fuel Filter 12-40
Removal & Installation 12-40
Fuel Pump 12-41
Removal & Installation 12-41
Fuel Rail & Injectors 12-41
Removal & Installation 12-41
Fuel System Service Precautions 12-40
Idle Speed 12-42
Inspection 12-42
Relieving Fuel System Pressure 12-40
Throttle Body 12-41
Removal & Installation 12-41

HEATING & AIR CONDITIONING SYSTEM 12-43

Blower Motor 12-43
Removal & Installation 12-43
Heater Core 12-44
Removal & Installation 12-44

SPECIFICATIONS AND MAINTENANCE CHARTS 12-3

Brake Specifications 12-6
Camshaft and Bearing Specifications 12-5
Capacities 12-4
Crankshaft and Connecting Rod Specifications 12-5
Engine and Vehicle Identification 12-3
Engine Tune-Up Specifications 12-3
Fluid Specifications 12-4
General Engine Specifications ...12-3
Piston and Ring Specifications ...12-5
Scheduled Maintenance Intervals 12-8
Tire, Wheel and Ball Joint Specifications 12-7
Torque Specifications 12-6
Valve Specifications 12-4
Wheel Alignment 12-6

STEERING 12-48

Power Rack & Pinion Steering Gear 12-48
Removal & Installation 12-48
Power Steering Pump 12-49
Removal & Installation 12-49

SUSPENSION 12-50

FRONT SUSPENSION 12-50
Coil Spring 12-50
Removal & Installation 12-50
Lower Ball Joint 12-51
Removal & Installation 12-51
Lower Control Arm 12-51
Control Arm Bushing Replacement 12-51
Removal & Installation 12-51
Stabilizer Bar 12-51
Removal & Installation 12-51
Stabilizer Link 12-51
Removal & Installation 12-51
Steering Knuckle 12-52
Removal & Installation 12-52
Strut/Damper 12-52
Removal & Installation 12-52
Wheel Bearings 12-52
Adjustment 12-52
Removal & Installation 12-52

REAR SUSPENSION 12-55
Coil Spring 12-55
Removal & Installation 12-55
Strut/Damper 12-55
Removal & Installation 12-55
Upper Ball Joint 12-56
Removal & Installation 12-56
Upper Control Arm 12-56
Control Arm Bushing Replacement 12-56
Removal & Installation 12-56
Wheel Bearings 12-56
Adjustment 12-56
Removal & Installation 12-56

SPECIFICATIONS AND MAINTENANCE CHARTS

ENGINE AND VEHICLE IDENTIFICATION CHART

Engine Code						
Code	Liters (cc)	Cu. In.	Cyl.	Fuel Sys.	Engine Type	Eng. Mfg.
K24A8	2.4 (2354)	144	4	SMFI	DOHC	Honda

Model Year	
Code ①	Year
9	2009
A	2010

DOHC: Double Overhead Cam

SMFI: Sequential Multi-port Fuel Injection

① 10th position of VIN

37647_ELEM_C0001

GENERAL ENGINE SPECIFICATIONS

Year	Model	Engine Displacement Liters (VIN)	Net Horsepower @ rpm	Net Torque @ rpm (ft. lbs.)	Bore x Stroke (in.)	Com-pression Ratio	Oil Pressure @ rpm
2009	Element	2.4 (K24A8)	166@5800	161@4500	3.42x3.90	9.7:1	44@3000
2010	Element	2.4 (K24A8)	166@5800	161@4500	3.42x3.90	9.7:1	44@3000

SMFI: Sequential Multi-port Fuel Injection

37647_ELEM_C0002

ENGINE TUNE-UP SPECIFICATIONS

Year	Engine Displacement Liters (VIN)	Spark Plug Gap (in.)	Ignition Timing (deg.)		Fuel Pump (psi)	Idle Speed (rpm)		Valve Clearance (in.)	
			MT	AT		MT	AT	In.	Ex.
2009	2.4 (K24A8)	0.039-0.043	6-10B	6-10B	48-55	650-750	650-750	0.008-0.010	0.011-0.013
2010	2.4 (K24A8)	0.039-0.043	6-10B	6-10B	48-55	650-750	650-750	0.008-0.010	0.011-0.013

NOTE: The Vehicle Emission Control Information label often reflects changes made during production and must be used if they differ from this chart.

NOTE: The fuel pressure readings are given with the vacuum hose connected to the regulator and the engine running

B: Before top dead center

HYD: Hydraulic

37647_ELEM_C0003

CAPACITIES

Year	Model	Engine Displacement Liters (VIN)	Engine Oil with Filter (qts.)	Transmission (pts.) 5-Spd	Transmission (pts.) Auto.	Transfer Case (pts.)	Drive Axle Front (pts.)	Drive Axle Rear (pts.)	Fuel Tank (gal.)	Cooling System (qts.)
2009	Element	2.4 (K24A8)	4.4	4.0	①	②	②	2.2	15.9	③
2010	Element	2.4 (K24A8)	4.4	4.0	①	②	②	2.2	15.9	③

NOTE: All capacities are approximate. Add fluid gradually and check to be sure a proper fluid level is obtained.

① 2WD: 6.6 pts. for fluid change, 15.2 for overhaul
4WD: 6.3 pts. for fluid change, 13.8 for overhaul

② Included in transaxle refill figure

③ Manual trans: 7.6 qts
Auto trans: 7.5 qts.

37647_ELEM_C0004

FLUID SPECIFICATIONS

Year	Model	Engine Displ. Liters	Engine Oil	Man. Trans.	Auto. Trans.	Drive Axle Front	Drive Axle Rear	Transfer Case	Power Steering Fluid	Brake Master Cylinder	Cooling System
2009	Element	2.4	5W-20 Honda	Honda MTF	Honda ATF-Z1	—	—	—	Honda PS Fluid	Honda DOT 3	①
2010	Element	2.4	5W-20 Honda	Honda MTF	Honda ATF-Z1	—	—	—	Honda PS Fluid	Honda DOT 3	①

DOT: Department Of Transpotation

① Honda Long Life Antifreeze/Coolant-Type2

37647_ELEM_C0014

VALVE SPECIFICATIONS

Year	Engine Displacement Liters (VIN)	Seat Angle (deg.)	Face Angle (deg.)	Spring Test Pressure (lbs. @ in.)	Spring Installed Height (in.)	Stem-to-Guide Clearance (in.) Intake	Stem-to-Guide Clearance (in.) Exhaust	Stem Diameter (in.) Intake	Stem Diameter (in.) Exhaust
2009	2.4 (K24A8)	NA	NA	NA	①	0.0012-0.0022	0.0022-0.0031	0.2156-0.2159	0.2146-0.2150
2010	2.4 (K24A8)	NA	NA	NA	①	0.0012-0.0022	0.0022-0.0031	0.2156-0.2159	0.2146-0.2150

NA: Not Available

① Valve spring free length:
Intake: 1.668 in.
Exhaust: 1.745 in.

37647_ELEM_C0005

CAMSHAFT AND BEARING SPECIFICATIONS

All measurements are given in inches.

Year	Engine Displacement Liters	Engine VIN	Journal Diameter	Brg. Oil Clearance	Shaft End-play	Runout	Journal Bore	Lobe Height Intake	Lobe Height Exhaust
2009	2.4	K24A8	NA	①	0.0020-0.0080	0.0010	NA	②	1.3422
2010	2.4	K24A8	NA	①	0.0020-0.0080	0.0010	NA	②	1.3422

NA: Information not available

① No. 1 journal: 0.001-0.003 in.
No. 2-5 journals: 0.002-0.004 in

② Primary: 1.3356 in.
Secondary: 1.1668 in.

37647_ELEM_C0013

CRANKSHAFT AND CONNECTING ROD SPECIFICATIONS

All measurements are given in inches

Year	Engine Displacement Liters (VIN)	Crankshaft Main Brg. Journal Dia.	Crankshaft Main Brg. Oil Clearance	Crankshaft Shaft End-play	Crankshaft Thrust on No.	Connecting Rod Journal Diameter	Connecting Rod Oil Clearance	Connecting Rod Side Clearance
2009	2.4 (K24A8)	①	②	0.0040-0.0140	3	1.8888-1.8898	0.0008-0.0019	0.016
2010	2.4 (K24A8)	①	②	0.0040-0.0140	3	1.8888-1.8898	0.0008-0.0019	0.016

① Except No. 3: 2.1648-2.1657
No. 3: 2.1644-2.1654

② Except No. 3: 0.0007-0.0016
No. 3: 0.0010-0.0019

37647_ELEM_C0006

PISTON AND RING SPECIFICATIONS

All measurements are given in inches

Year	Engine Displacement Liters (VIN)	Piston Clearance	Ring Gap Top Compression	Ring Gap Bottom Compression	Ring Gap Oil Control	Ring Side Clearance Top Compression	Ring Side Clearance Bottom Compression	Ring Side Clearance Oil Control
2009	2.4 (K24A8)	0.0008-0.0016	0.0080-0.0140	0.0160-0.0220	0.0080-0.0280	0.0018-0.0028	0.0020-0.0030	NA
2010	2.4 (K24A8)	0.0008-0.0016	0.0080-0.0140	0.0160-0.0220	0.0080-0.0280	0.0018-0.0028	0.0020-0.0030	NA

NA: Not Applicable

37647_ELEM_C0007

TORQUE SPECIFICATIONS

All readings in ft. lbs.

Year	Engine Displacement Liters (VIN)	Cylinder Head Bolts	Main Bearing Bolts	Rod Bearing Bolts	Crankshaft Damper Bolts	Flywheel Bolts	Manifold Intake	Manifold Exhaust	Spark Plugs	Oil Pan Drain Plug
2009	2.4 (K24A8)	①	②	③	④	76	16	33	13	33
2010	2.4 (K24A8)	①	②	③	④	76	16	33	13	33

NOTE: Dip main bearing bolts and crankshaft damper bolt in clean engine oil prior to tightening.

① Step 1: 29 ft. lbs.
Step 2: +90 degrees
Step 3: +90 degrees
Step 4: NEW BOLT ONLY +90 degrees

② 22 ft. lbs. +56 degrees

③ 14 ft. lbs. +90 degrees

④ 36 ft. lbs. +90 degrees

37647_ELEM_C0008

WHEEL ALIGNMENT

Year	Model		Caster Range (+/-Deg.)	Caster Preferred Setting (Deg.)	Camber Range (+/-Deg.)	Camber Preferred Setting (Deg.)	Toe-in (in.)
2009	Element	F	1.00	+1.75	0.75	0	0+/-0.08
		R	—	—	0.75	-1.00	0.08+/-0.08
2010	Element	F	1.00	+1.75	0.75	0	0+/-0.08
		R	—	—	0.75	-1.00	0.08+/-0.08

37647_ELEM_C0009

TIRE, WHEEL AND BALL JOINT SPECIFICATIONS

Year	Model	OEM Tires Standard	OEM Tires Optional	Tire Pressures (psi) Front	Tire Pressures (psi) Rear	Wheel Size	Ball Joint Inspection	Lug Nut (ft. lbs.)
2009	Element	P215/70R16	None	26	26	6JJ	NS	80
2010	Element	P215/70R16	None	26	26	6JJ	NS	80

OEM: Original Equipment Manufacturer

PSI: Pounds Per Square Inch

NS: Not specified by manufacturer

37647_ELEM_C0010

BRAKE SPECIFICATIONS

All measurements in inches unless noted

Year	Model		Brake Disc Original Thickness	Brake Disc Minimum Thickness	Brake Disc Maximum Runout	Brake Drum Diameter Original Inside Diameter	Brake Drum Diameter Max. Wear Limit	Brake Drum Diameter Maximum Machine Diameter	Minimum Lining Thickness Front	Minimum Lining Thickness Rear	Brake Caliper Bracket Bolts (ft. lbs.)	Brake Caliper Mounting Bolts (ft. lbs.)
2009	Element	F	0.910	0.830	0.004	—	—	—	0.060	—	80	25
		R	0.350	0.300	0.004	—	—	—	0.060	—	41	16
2010	Element	F	0.910	0.830	0.004	—	—	—	0.060	—	80	25
		R	0.350	0.300	0.004	—	—	—	0.060	—	41	16

F: Front

R: Rear

37647_ELEM_C0011

SCHEDULED MAINTENANCE INTERVALS

HONDA—ELEMENT

TO BE SERVICED	TYPE OF SERVICE	VEHICLE MILEAGE INTERVAL (x1000)											
		10	20	30	40	50	60	70	80	90	100	110	120
Accessory drive belts	I & A			✓			✓			✓			✓
Air cleaner element	R			✓			✓			✓			✓
Air conditioning filter	R			✓			✓			✓			✓
Brake fluid	R											✓	
Brake hoses & lines (including ABS)	I		✓		✓		✓		✓		✓		
Cooling system hoses & connections	I		✓		✓		✓		✓		✓		
Engine coolant	R												✓
Engine oil	R	✓	✓	✓	✓	✓	✓	✓	✓	✓	✓	✓	✓
Engine oil and coolant levels	I	Inspect at each fuel stop											
Engine oil filter	R		✓		✓		✓		✓		✓		
Exhaust system	I		✓		✓		✓		✓		✓		
Fluid levels and condition	I		✓		✓		✓		✓		✓		
Front and rear brakes	I		✓		✓		✓		✓		✓		
Fuel lines & connection	I		✓		✓		✓		✓		✓		
Halfshaft boots	I		✓		✓		✓		✓		✓		
Idle speed	I & A											✓	
Parking brake system	I & A		✓		✓		✓		✓		✓		
Rear differential fluid	R										✓		
Rotate and inspect tires	I	✓	✓	✓	✓	✓	✓	✓	✓	✓	✓	✓	✓
Spark plugs	R											✓	
Suspension components	I		✓		✓		✓		✓		✓		
Tie rod ends, steering gear box & boots	I		✓		✓		✓		✓		✓		
Transmission fluid	R												✓
Valve clearance	I											✓	

R: Replace I: Inspect A: Adjust

FREQUENT OPERATION MAINTENANCE (SEVERE SERVICE)

If a vehicle is operated under any of the following conditions it is considered severe service:

- Towing a trailer or using a camper or car-top carrier.

- Repeated short trips of less than 5 miles in temperatures below freezing, or trips of less than 10 miles in any temperature.

- Extensive idling or low-speed driving for long distances as in heavy commercial use, such as delivery, taxi or police cars.

- Operating on rough, muddy or salt-covered roads.

- Operating on unpaved or dusty roads.

- Driving in extremely hot (over 90°) conditions.

Air cleaner element: replace every 15,000 miles

Engine oil and filter: replace every 3750 miles or 6 months, whichever occurs first.

Timing belt: replace every 60,000 miles if the vehicle is regularly driven in temperatures above 110°F or below -20°F.

Transmission fluid: replace every 30,000 miles.

Rear differential fluid: replace every 60,000 miles.

Front and rear brakes: inspect every 7500 miles or 6 months, whichever occurs first.

Locks and hinges: lubricate every 15,000 miles.

Tie rods, steering gear box, boots: inspect every 7500 miles or 6 months, whichever occurs first.

Suspension components: inspect every 7500 miles or 6 months, whichever occurs first.

Halfshaft boots: inspect every 7500 miles or 6 months, whichever occurs first.

37647_ELEM_C0012

BRAKES — INFORMATION AND PRECAUTIONS

ANTI-LOCK SYSTEMS

- Certain components within the ABS system are not intended to be serviced or repaired individually.
- Do not use rubber hoses or other parts not specifically specified for and ABS system. When using repair kits, replace all parts included in the kit. Partial or incorrect repair may lead to functional problems and require the replacement of components.
- Lubricate rubber parts with clean, fresh brake fluid to ease assembly. Do not use shop air to clean parts; damage to rubber components may result.
- Use only DOT 3 brake fluid from an unopened container.
- If any hydraulic component or line is removed or replaced, it may be necessary to bleed the entire system.
- A clean repair area is essential. Always clean the reservoir and cap thoroughly before removing the cap. The slightest amount of dirt in the fluid may plug an orifice and impair the system function. Perform repairs after components have been thoroughly cleaned; use only denatured alcohol to clean components. Do not allow ABS components to come into contact with any substance containing mineral oil; this includes used shop rags.
- The Anti-Lock control unit is a microprocessor similar to other computer units in the vehicle. Ensure that the ignition switch is **OFF** before removing or installing controller harnesses. Avoid static electricity discharge at or near the controller.
- If any arc welding is to be done on the vehicle, the control unit should be unplugged before welding operations begin.

DISC AND DRUM SYSTEMS

CAUTION

Dust and dirt accumulating on brake parts during normal use may contain asbestos fibers from production or aftermarket brake linings. Breathing excessive concentrations of asbestos fibers can cause serious bodily harm. Exercise care when servicing brake parts. Do not sand or grind brake lining unless equipment used is designed to contain the dust residue. Do not clean brake parts with compressed air or by dry brushing. Cleaning should be done by dampening the brake components with a fine mist of water, then wiping the brake components clean with a dampened cloth. Dispose of cloth and all residue containing asbestos fibers in an impermeable container with the appropriate label. Follow practices prescribed by the Occupational Safety and Health Administration (OSHA) and the Environmental Protection Agency (EPA) for the handling, processing, and disposing of dust or debris that may contain asbestos fibers.

BRAKES — BLEEDING THE BRAKE SYSTEM

BLEEDING PROCEDURE

BLEEDING PROCEDURE

See Figure 1.

When bleeding the brake system, observe the following:

- Do not reuse the drained fluid. Use only clean Honda DOT 3 Brake Fluid from an unopened container. Using a non-Honda brake fluid can cause corrosion and shorten the life of the system. Do not mix different brands of brake fluid; they may not be compatible.
- Make sure no dirt or other foreign matter is allowed to contaminate the brake fluid.
- Do not spill brake fluid on the vehicle, it may damage the paint; if brake fluid does contact the paint, wash it off immediately with water.

1. The reservoir on the master cylinder must be at the MAX (upper) level mark at the start of the bleeding procedure and checked after bleeding each brake caliper. Add fluid as required.
2. Make sure the brake fluid level in the reservoir is at the MAX (upper) level line.
3. Slide a piece of clear plastic hose over the bleed screw, and submerge the other end in a container of new brake fluid.
4. Have someone slowly pump the brake pedal several times, then apply steady pressure.
5. Loosen the left-front brake bleed screw to allow air to escape from the system. Then tighten the bleed screw securely.
6. Repeat the procedure for each caliper until no air bubbles are in the fluid. Bleed the calipers in the sequence shown.
7. Refill the master cylinder reservoir to the MAX (upper) level line.

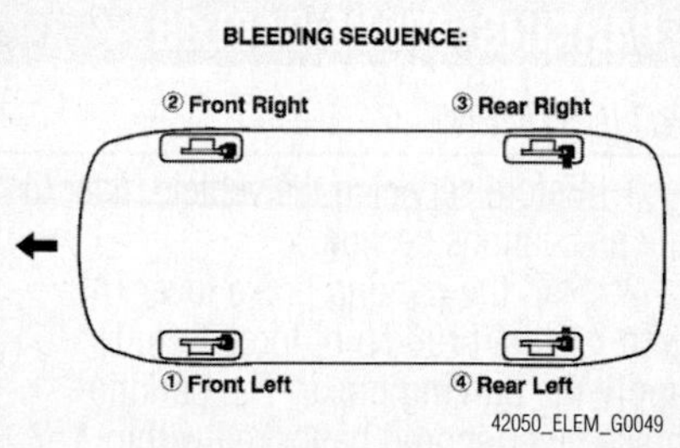
BLEEDING SEQUENCE:
② Front Right
③ Rear Right
① Front Left
④ Rear Left
42050_ELEM_G0049

Fig. 1 Proper brake bleeding sequence

BRAKES — FRONT DISC BRAKES

BRAKE CALIPER

REMOVAL & INSTALLATION

1. Remove the wheel
2. Remove the brake hose banjo bolt and washers. Discard the washers.
3. Remove the upper and lower bolts.
4. Lift off the caliper.
5. Remove the pad springs.
6. Remove the pads and shims.
7. Remove the pad retainers.
8. Installation is the reverse of removal. Coat both sides of the shims and the backs of the pads with brake grease. Torque the bolts to 25 ft. lbs. (34 Nm). Install new washers and torque the banjo bolt to 25 ft. lbs. (34 Nm).

DISC BRAKE PADS

REMOVAL & INSTALLATION

1. Remove the lower bolt.
2. Pivot the caliper up and hold the pads.
3. Remove the pad springs.
4. Remove the pads and shims.
5. Remove the pad retainers.
6. Installation is the reverse of removal. Coat both sides of the shims and the backs of the pads with brake grease. Torque the lower bolt to 25 ft. lbs. (34 Nm).

BRAKES — REAR DISC BRAKES

BRAKE CALIPER

REMOVAL & INSTALLATION

1. Remove the wheel
2. Remove the brake hose banjo bolt and washers. Discard the washers.
3. Remove the caliper pin bolts.
4. Lift off the caliper and suspend it safely.
5. Remove the pads and shims.
6. Remove the pad retainers.
7. Installation is the reverse of removal. Coat both sides of the shims and the backs of the pads with brake grease. Torque the bolts to 16 ft. lbs. (22 Nm). Install new washers and torque the banjo bolt to 25 ft. lbs. (34 Nm).

DISC BRAKE PADS

REMOVAL & INSTALLATION

1. Remove the caliper pin bolts.
2. Lift off the caliper and suspend it safely.
3. Remove the pads and shims.
4. Remove the pad retainers.
5. Installation is the reverse of removal. Coat both sides of the shims and the backs of the pads with brake grease. Torque the bolts to 16 ft. lbs. (22 Nm).

BRAKES — PARKING BRAKE

PARKING BRAKE CABLES

ADJUSTMENT

1. Before servicing the vehicle, refer to the precautions section.
2. Pull the parking brake lever (A) with 44 lbs. (196 N) of force to fully apply the parking brake. The parking brake lever should be locked within 4–7 clicks.
3. Adjust the parking brake if the lever clicks are not within the specification.

Minor Adjustment

See Figure 2.

1. Raise the rear of the vehicle, and support it with safety stands in the proper locations.
2. Release the parking brake lever fully.
3. Remove the center console, by pulling the front edge of the center console to release the clips, and remove the console by releasing the hooks.
4. Pull the parking brake lever 1 click.
5. Tighten the adjusting nut (A) until the parking brakes drag slightly when the rear wheels are turned.
6. Release the parking brake lever fully, and check that the parking brakes do not drag when the rear wheels are turned. Readjust if necessary.
7. Make sure the parking brakes are fully applied when the parking brake lever is pulled all the way.
8. Install the console in the reverse order of removal. During installation, check for damaged or stress-whitened clips, and replace them with new ones. Push the clip portions into place securely.

Major Adjustment

See Figures 3 and 4.

➡This adjustment should be done when replacing parking brake shoes and after lining surface break-in.

1. Raise the rear of the vehicle, and support it with safety stands in the proper locations.
2. Release the parking brake lever fully.

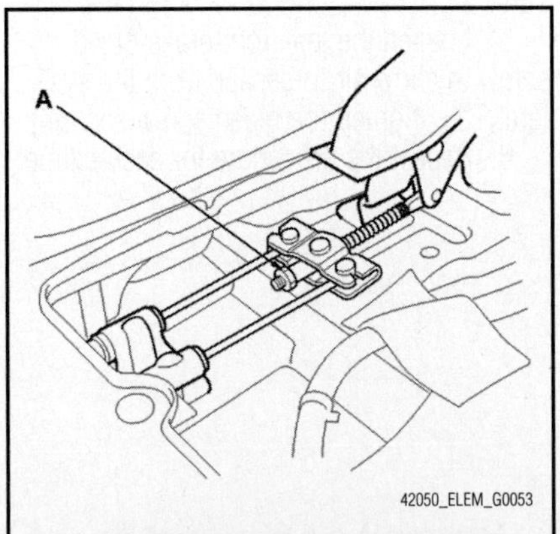
A
42050_ELEM_G0053

Fig. 2 Tighten the adjusting nut (A) until the parking brakes drag slightly when the rear wheels are turned

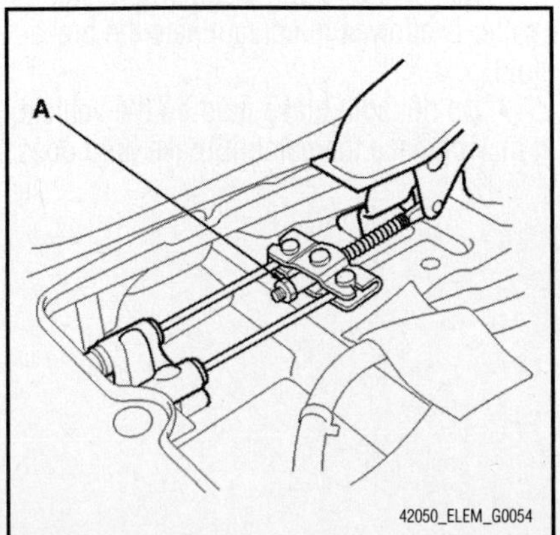
A
42050_ELEM_G0054

Fig. 3 Back off the adjusting nut (A) in the equalizer

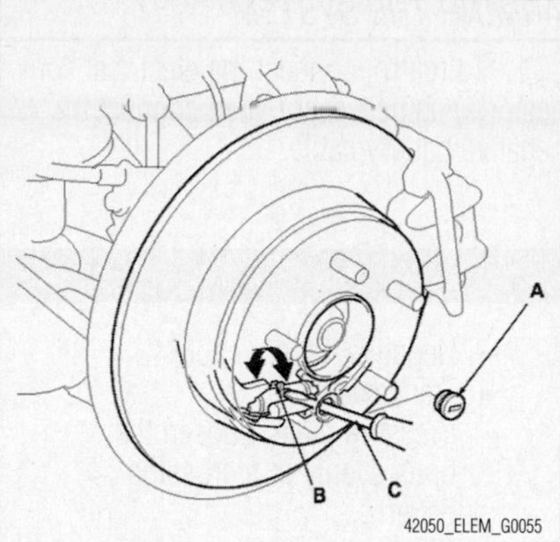
A
B
C
42050_ELEM_G0055

Fig. 4 Remove the access plug (A). Turn the adjuster (B) with a flat-tip screwdriver (C) until the shoes lock against the drum. Then back off 8 clicks, and install the access plug.

3. Remove the center console, by pulling the front edge of the center console to release the clips, and remove the console by releasing the hooks.
4. Back off the adjusting nut (A) in the equalizer.
5. Remove the rear wheels.
6. Remove the access plug (A).
7. Turn the adjuster (B) with a flat-tip screwdriver (C) until the shoes lock against the drum. Then back off 8 clicks, and install the access plug.
8. Perform the minor adjustment procedure.
9. Install the rear wheels.
10. Install the console in the reverse order of removal. During installation, check for damaged or stress-whitened clips, and replace them with new ones. Push the clip portions into place securely.

PARKING BRAKE SHOES

ADJUSTMENT

1. Before servicing the vehicle, refer to the precautions section.
2. Pull the parking brake lever (A) with 44 lbs. (196 N) of force to fully apply the parking brake. The parking brake lever should be locked within 4–7 clicks.
3. Adjust the parking brake if the lever clicks are not within the specification.

Minor Adjustment

See Figure 2.

1. Raise the rear of the vehicle, and support it with safety stands in the proper locations.
2. Release the parking brake lever fully.
3. Remove the center console, by pulling the front edge of the center console to release the clips, and remove the console by releasing the hooks.
4. Pull the parking brake lever 1 click.
5. Tighten the adjusting nut (A) until the parking brakes drag slightly when the rear wheels are turned.
6. Release the parking brake lever fully, and check that the parking brakes do not drag when the rear wheels are turned. Readjust if necessary.
7. Make sure the parking brakes are fully applied when the parking brake lever is pulled all the way.
8. Install the console in the reverse order of removal. During installation, check for damaged or stress-whitened clips, and replace them with new ones. Push the clip portions into place securely.

Major Adjustment

See Figures 3 and 4.

➡This adjustment should be done when replacing parking brake shoes and after lining surface break-in.

1. Raise the rear of the vehicle, and support it with safety stands in the proper locations.
2. Release the parking brake lever fully.
3. Remove the center console, by pulling the front edge of the center console to release the clips, and remove the console by releasing the hooks.
4. Back off the adjusting nut (A) in the equalizer.
5. Remove the rear wheels.
6. Remove the access plug (A).
7. Turn the adjuster (B) with a flat-tip screwdriver (C) until the shoes lock against the drum. Then back off 8 clicks, and install the access plug.
8. Perform the minor adjustment procedure.
9. Install the rear wheels.
10. Install the console in the reverse order of removal. During installation, check for damaged or stress-whitened clips, and replace them with new ones. Push the clip portions into place securely.

CHASSIS ELECTRICAL

GENERAL INFORMATION

⁂ CAUTION

These vehicles are equipped with an air bag system. The system must be disarmed before performing service on, or around, system components, the steering column, instrument panel components, wiring and sensors. Failure to follow the safety precautions and the disarming procedure could result in accidental air bag deployment, possible injury and unnecessary system repairs.

AIR BAG (SUPPLEMENTAL RESTRAINT SYSTEM)

SERVICE PRECAUTIONS

⁂ CAUTION

Disconnect and isolate the battery negative cable before beginning any airbag system component diagnosis, testing, removal, or installation procedures. Allow system capacitor to discharge for two minutes before beginning any component service. This will disable the airbag system. Failure to disable the airbag system may result in accidental airbag deployment, personal injury, or death.

DISARMING THE SYSTEM

1. Disconnect and isolate the negative battery cable. Wait 3 minutes for the system capacitor to discharge before performing any service.
2. To disarm the driver's airbag, remove the access panel from the steering wheel, then disconnect the driver's airbag 4P connector from the cable reel.
3. To disarm the front passenger's airbag, remove the glove box, then disconnect the passenger's airbag 4P connector from dashboard wire harness B.
4. To disarm the side airbag, disconnect the side airbag 2P connector from the floor wire harness.

5. To disarm the seat belt tensioner, disconnect the seat belt tensioner 2P connector from the rear door wire harness.

6. To disarm the seat belt buckle tensioner, disconnect the seat belt buckle tensioner 4P connector.

7. To disarm the SRS unit, disconnect the SRS unit connector A, B or C, as applicable.

ARMING THE SYSTEM

1. To rearm, connect the electrical connector(s) as necessary, then connect the negative battery cable.

DRIVETRAIN

CLUTCH DRIVEN DISC AND PRESSURE PLATE

ADJUSTMENTS

The Element is equipped with a hydraulic clutch system. No adjustment is necessary.

REMOVAL & INSTALLATION

See Figure 5.

1. Before servicing the vehicle, refer to the precautions section.
2. Remove or disconnect the following:

- Negative battery cable
- Transaxle
- Pressure plate. Loosen the bolts evenly in a crossing pattern.
- Clutch disc

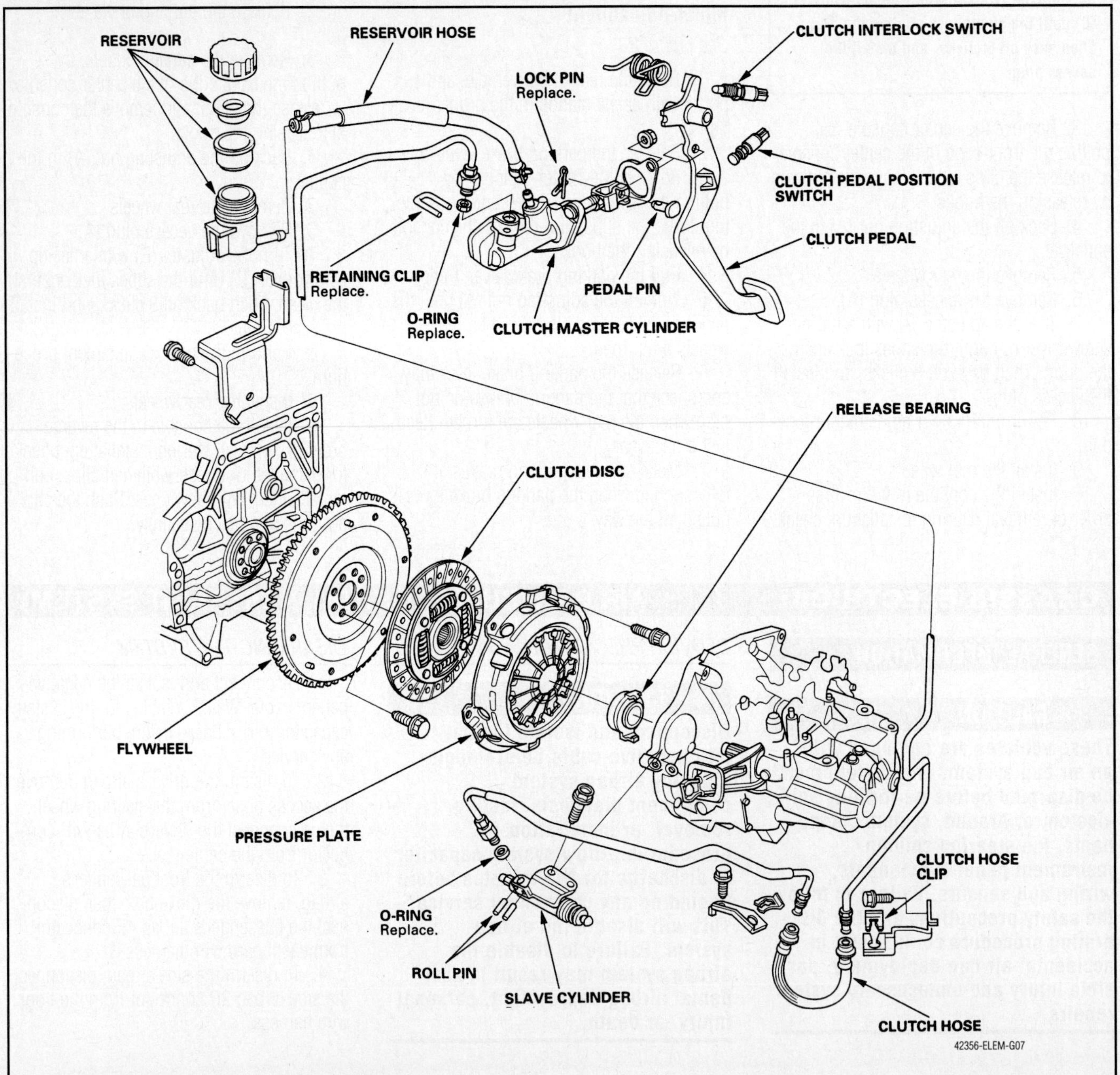

RESERVOIR
RESERVOIR HOSE
CLUTCH INTERLOCK SWITCH
LOCK PIN
Replace.
CLUTCH PEDAL POSITION SWITCH
CLUTCH PEDAL
RETAINING CLIP
Replace.
PEDAL PIN
O-RING
Replace.
CLUTCH MASTER CYLINDER
RELEASE BEARING
CLUTCH DISC
FLYWHEEL
PRESSURE PLATE
CLUTCH HOSE CLIP
O-RING
Replace.
ROLL PIN
SLAVE CYLINDER
CLUTCH HOSE
42356-ELEM-G07

Fig. 5 Exploded view of the clutch system components

To install:

3. Install the clutch disc and pressure plate. Tighten the pressure plate bolts in a crisscross pattern, in several steps to 19 ft. lbs. (26 Nm).
4. Install or connect the following:
   - Transaxle
   - Negative battery cable

CLUTCH MASTER CYLINDER

REMOVAL & INSTALLATION

See Figures 6 through 10.

⁂ WARNING

Always use fender covers to avoid damaging painted surfaces. Do not spill brake fluid on the vehicle; it may damage the paint; if brake fluid does contact the paint, wash it off immediately with water.

1. Remove the brake fluid from the clutch master cylinder reservoir with a syringe.
2. Make sure you have the anti-theft code for the radio, then write down the audio presets. Disconnect the negative cable from the battery first, then disconnect the positive cable. Remove the battery.
3. Remove the air cleaner housing.
4. Remove the battery base.
5. Pry out the lock pin, and pull the pedal pin out of the yoke. Remove the master cylinder mounting nuts.
6. Remove the reservoir mounting bolt.
7. Remove the clutch line bracket.
8. Remove the clutch master cylinder from the vehicle.
9. Disconnect the reservoir hose, then remove the retaining clip and clutch line from the clutch master cylinder. Plug the end of the reservoir hose and clutch line with a shop towel to prevent brake fluid from coming out.
10. Remove the O-ring and clutch master cylinder seal from the clutch master cylinder.

To install:

11. Install the clutch master cylinder in the reverse order of removal, and note these items:
    a. Apply brake assembly lube to the clutch line, and install a new O-ring.
    b. Tighten the master cylinder mounting nuts to 9 ft. lbs. (13 Nm).
12. Install the battery base.
13. Install the air cleaner housing.
14. Install the battery. Clean the battery posts and cable terminals with sandpaper. Connect the positive cable to the battery first, then connect the negative cable, and apply grease to prevent corrosion.
15. Enter the anti-theft code for the radio, then enter the audio presets, and set the clock.
16. Perform the power window control unit reset procedure, as follows.

➡Resetting the power window control unit is required after performing the following procedures: Loss of battery power, Loss of power from the No. 23 (20 A) fuse in the under-dash fuse/relay box, Open circuit caused by disconnecting the 14P connector from the power window master switch

a. Make sure the driver's window does not work in AUTO with the ignition switch ON (II).
b. Start the engine.
c. Lower the driver's window all the way down by pushing the driver's power window switch to the second detent (AUTO DOWN); when the window reaches the bottom, hold the switch in the AUTO DOWN position for 2 seconds.
d. Raise the driver's window all the

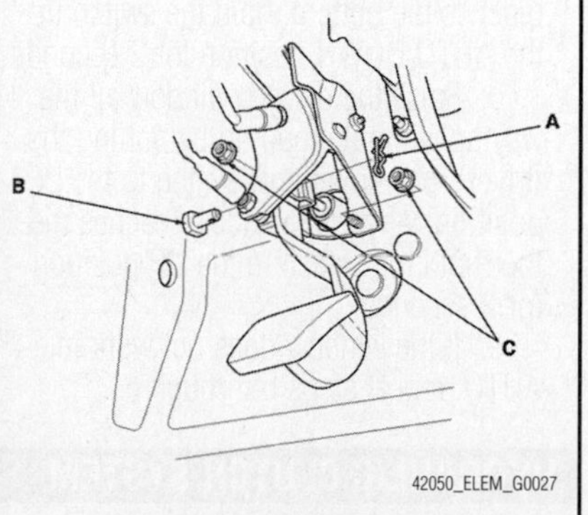
A
B
C
42050_ELEM_G0027
Fig. 6 Pry out the lock pin (A), and pull the pedal pin (B) out of the yoke. Remove the master cylinder mounting nuts (C)

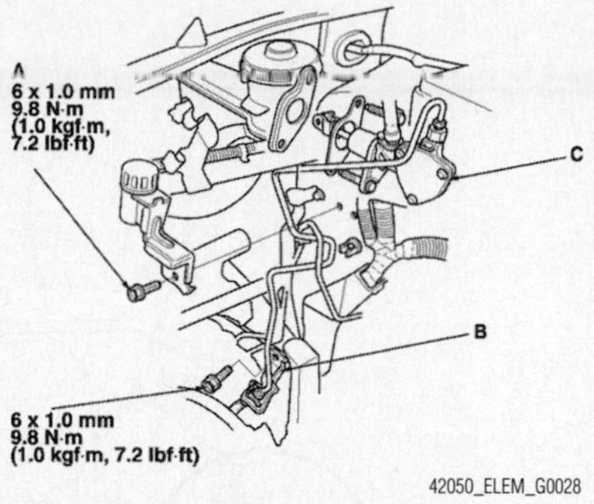
A
6 x 1.0 mm
9.8 N·m
(1.0 kgf·m,
7.2 lbf·ft)
C
B
6 x 1.0 mm
9.8 N·m
(1.0 kgf·m, 7.2 lbf·ft)
42050_ELEM_G0028
Fig. 7 Remove the reservoir mounting bolt (A), the clutch line bracket (B), then remove the clutch master cylinder (C)

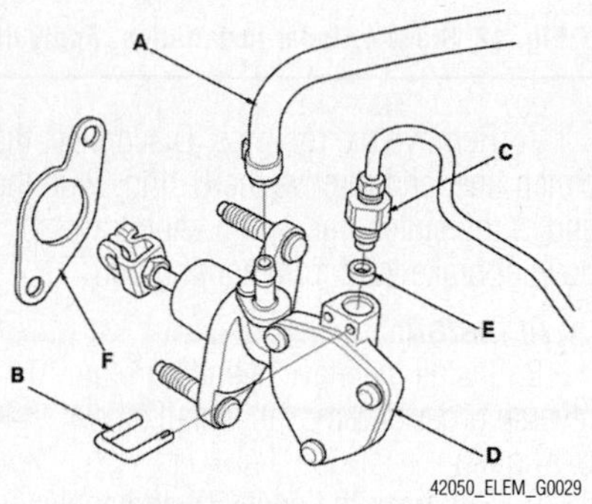
A
C
E
F
B
D
42050_ELEM_G0029
Fig. 8 View of the reservoir hose (A), retaining clips (B), clutch line (C), clutch master cylinder (D), O-ring (E) and clutch master cylinder seal (F)

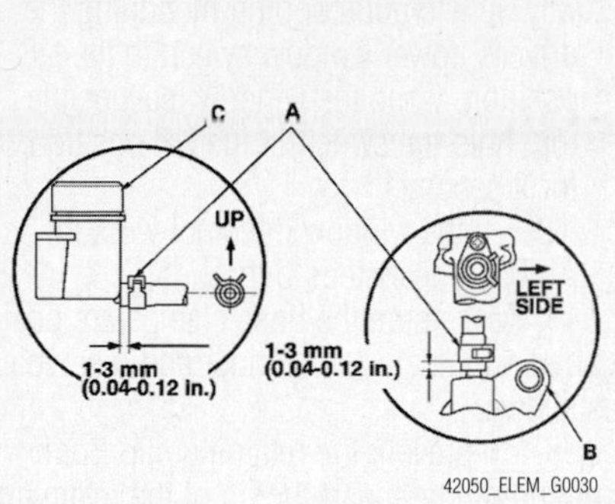
C
A
UP
LEFT SIDE
1-3 mm
(0.04-0.12 in.)
1-3 mm
(0.04-0.12 in.)
B
42050_ELEM_G0030
Fig. 9 The hose clamps (A) must be properly positioned on the master cylinder (B) and reservoir (C)

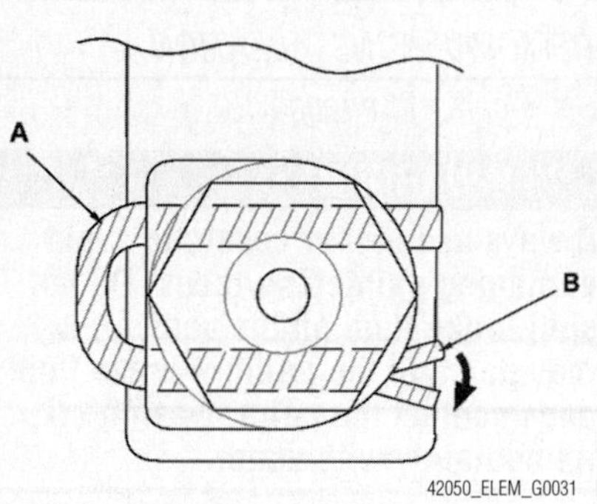
A
B
42050_ELEM_G0031
Fig. 10 Pry the tip of the retaining clips (B) apart with a screwdriver to prevent the clip (A) from coming off

way up without stopping by pulling the driver's power window switch to the UP position; when the window reaches the top, hold the switch in the UP position for 2 seconds.

e. If the window does not work in AUTO, repeat steps b through e.

17. Make sure the hose clamps are positioned on the master cylinder and reservoir as shown.

18. To prevent the retaining clip from coming off, pry apart the tip of the retaining clip with a screwdriver.

➡ **Reservoir filling is covered in the bleeding procedure.**

19. Bleed the clutch hydraulic system.

CLUTCH SLAVE CYLINDER

REMOVAL & INSTALLATION

See Figures 11 through 13.

WARNING

Always use fender covers to avoid damaging painted surfaces. Do not spill brake fluid on the vehicle; it may damage the paint; if brake fluid does contact the paint, wash it off immediately with water.

1. Remove the brake fluid from the clutch master cylinder reservoir with a syringe.
2. Make sure you have the anti-theft code for the radio, then write down the audio presets. Disconnect the negative cable from the battery first, then disconnect the positive cable. Remove the battery.
3. Remove the air cleaner housing.
4. Remove the battery base.
5. Remove the clutch line bracket.
6. Remove the mounting bolts and the slave cylinder.

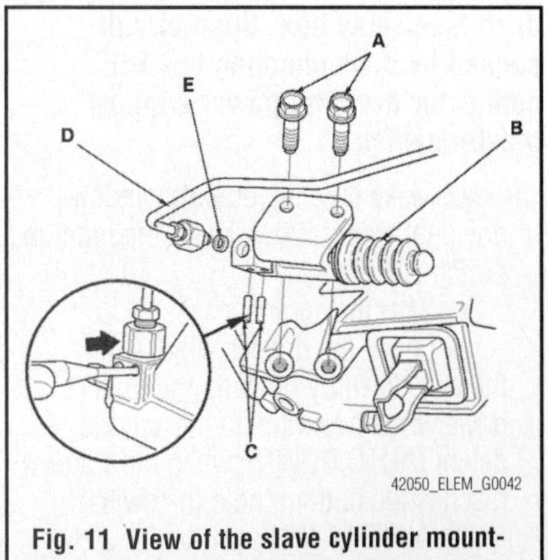
A
E
D
B
C
42050_ELEM_G0042

Fig. 11 View of the slave cylinder mounting bolts (A), slave cylinder (B), roll pins (C), clutch line (D) and O-ring (E)

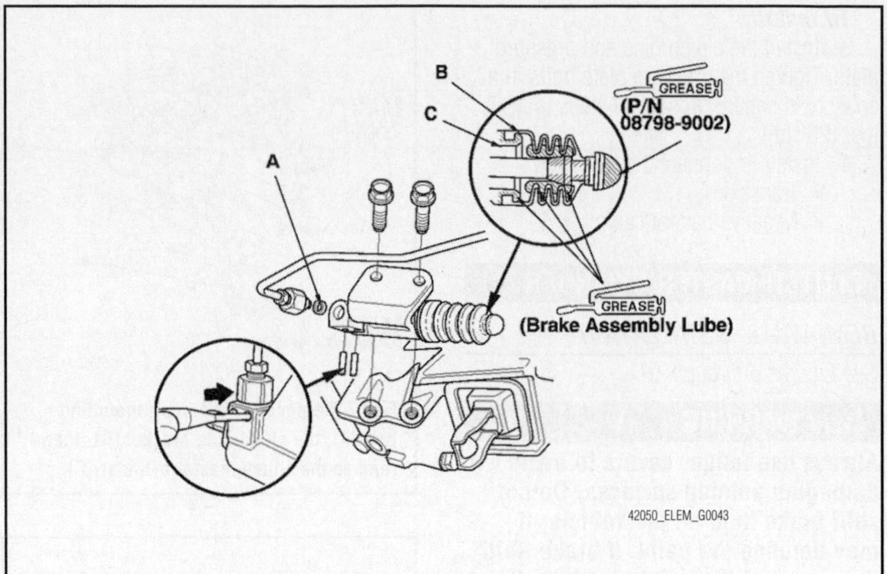
B
C
A
GREASE
(P/N 08798-9002)
GREASE
(Brake Assembly Lube)
42050_ELEM_G0043

Fig. 12 Slave cylinder installation. Apply the proper type of grease where shown

7. Remove the roll pins. Disconnect the clutch line, and remove the O-ring. Plug the end of the clutch line with a shop towel to prevent brake fluid from coming out.

To install:

8. Install the slave cylinder in the reverse order of removal. Install a new O-ring (A).
9. Pull back the boot (B), and apply brake assembly lube to the boot and slave cylinder rod (C). Reinstall the boot.
10. Apply super high temp urea grease (P/N 08798-9002) to the slave cylinder push rod. Tighten the slave cylinder mounting bolts to 16 ft. lbs. (22 Nm).
11. Attach a hose (A) to the bleeder screw, and suspend the hose in a container of brake fluid.
12. Make sure there is enough fluid in the clutch master cylinder, then slowly pump the clutch pedal until no more bubbles appear at the bleeder hose.
13. It may be necessary to limit the movement of the release fork (B) with a block of wood to remove all the air from the system.
14. Tighten the bleeder screw to 70 inch lbs. (8 Nm); do not overtighten the screw.
15. Refill the clutch master cylinder with fluid when done. Use only Honda DOT 3 Brake Fluid from an unopened container.
16. Make sure the fluid level in the reservoir is at the MAX (upper) level line.
17. Install the air cleaner housing.
18. Install the battery base.
19. Install the battery. Clean the battery posts and cable terminals with sandpaper. Connect the positive cable to the battery first, then connect the negative cable, and apply grease to prevent corrosion.
20. Enter the anti-theft code for the radio, then enter the audio presets, and set the clock.
21. Perform the power window control unit reset procedure, as follows.

➡**Resetting the power window control unit is required after performing the following procedures: Loss of battery power, Loss of power from the No. 23 (20 A) fuse in the under-dash fuse/relay box, Open circuit caused by disconnecting the 14P connector from the power window master switch**

a. Make sure the driver's window does not work in AUTO with the ignition switch ON (II).

b. Start the engine.

c. Lower the driver's window all the way down by pushing the driver's power window switch to the second detent (AUTO DOWN); when the window reaches the bottom, hold the switch in the AUTO DOWN position for 2 seconds.

d. Raise the driver's window all the way up without stopping by pulling the driver's power window switch to the UP position; when the window reaches the top, hold the switch in the UP position for 2 seconds.

e. If the window does not work in AUTO, repeat steps b through e.

HYDRAULIC CLUTCH SYSTEM

BLEEDING

See Figure 13.

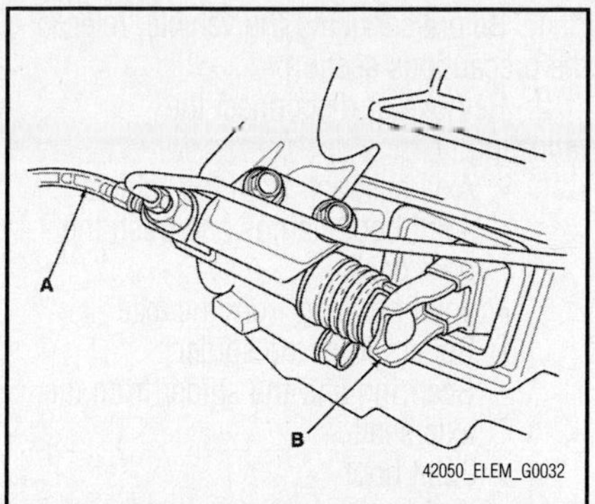
A
B
42050_ELEM_G0032

Fig. 13 View of the hose (A) attached to the bleeder screw and the release fork (B)

1. Before servicing the vehicle, refer to the precautions section.
2. Attach a hose (A) to the bleeder screw, and suspend the hose in a container of brake fluid.
3. Make sure there is enough fluid in the clutch master cylinder, then slowly pump the clutch pedal until no more bubbles appear at the bleeder hose.
4. It may be necessary to limit the movement of the release fork (B) with a block of wood to remove all the air from the system.
5. Tighten the bleeder screw to 70 inch lbs. (8 Nm); do not overtighten the screw.
6. Refill the clutch master cylinder with fluid when done. Use only Honda DOT 3 Brake Fluid from an unopened container.
7. Make sure the fluid level in the reservoir is at the MAX (upper) level line.

TRANSFER ASSEMBLY

REMOVAL & INSTALLATION

See Figure 14.

1. Before servicing the vehicle, refer to the precautions section.

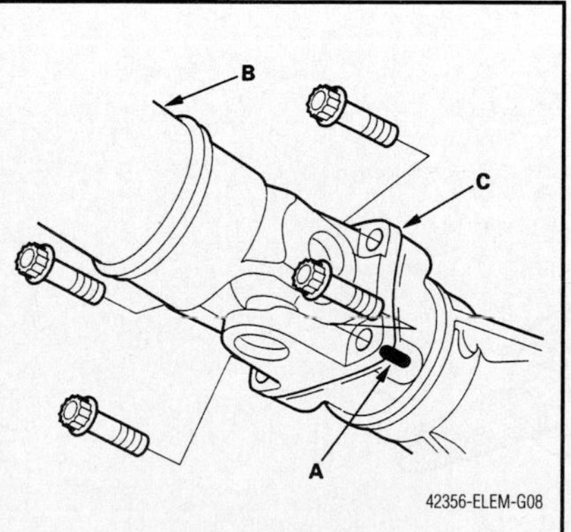
B
C
A
42356-ELEM-G08

Fig. 14 Matchmark (A) the installed position of the propeller shaft (B) and transfer companion flange (C)

2. Drain the transaxle fluid. Install the drain plug with a new gasket and tighten to 36 ft. lbs. (49 Nm).
3. Disconnect the negative battery cable.
4. Matchmark the installed position of the propeller shaft and transfer companion flange.
5. Remove or disconnect the following:
   - Propeller shaft from the transfer assembly
   - Mounting bolts and transfer assembly

To install:

6. Clean the transfer assembly mating surfaces, then apply clean transmission fluid to the mating surfaces.
7. Install or connect the following:
   - New O-ring seal on the transfer assembly
   - 4 bolts in the transfer housing, then the transfer assembly with the dowel pin. Tighten the 10mm bolts to 33 ft. lbs. (44 Nm).
   - Propeller shaft to the transfer companion flange, aligning the mark made during removal. Tighten the 8mm bolts to 24 ft. lbs. (33 Nm).
   - Negative battery cable
8. Fill the transaxle to the correct level and check for leaks.

FRONT HALFSHAFT

REMOVAL & INSTALLATION

1. Before servicing the vehicle, refer to the precautions section.
2. Drain the transaxle.
3. Remove or disconnect the following:
   - Negative battery cable
   - Front wheels
   - Spindle nut
   - Stabilizer bar
   - Lower ball joint from the control arm
4. On the left side, pry the inboard joint from the case with a prybar.
5. On the right side, drive the inboard shaft off the intermediate shaft with a drift and hammer.
6. Installation is the reverse of removal. Observe the following torques:
   - Ball stud nuts: 40 ft. lbs. (54 Nm)
   - Stabilizer link nuts: 29 ft. lbs. (39 Nm)
   - Spindle nut: 181 ft. lbs. (245 Nm)

CV-JOINT OVERHAUL

Outboard Joint

1. Before servicing the vehicle, refer to the precautions section.
2. Remove or disconnect the following:
   - Axle halfshaft from the vehicle and place it in a vise
   - Outboard joint boot clamps and push the boot back
   - Outboard joint by driving it off the axle shaft with a brass drift and hammer
   - Outboard joint boot

To install:

➡Use new circlips and boot clamps for assembly.

3. Install the outboard joint boot and clamps to the axle shaft.
4. Fill the outboard joint with grease. Install the outboard joint to the axle shaft. Tap the stub shaft with a brass hammer to seat the circlip.
5. Fill the outboard joint boot with grease and install the boot clamps.
6. Install the axle halfshaft to the vehicle.

Inboard Joint

1. Before servicing the vehicle, refer to the precautions section.
2. Remove or disconnect the following:
   - Axle halfshaft from the vehicle.
   - Inboard joint boot clamps and push the boot back
   - Inboard joint housing from the axle
   - Rollers from the spider
   - Snapring and the spider from the axle shaft
   - Inboard joint boot

To install:

➡Use new circlips and boot clamps for assembly.

3. Install or connect the following:
   - Inboard joint boot and clamps to the axle shaft
   - Spider with a new snapring
   - Rollers to the spider
4. Fill the joint housing with grease and install it.
5. Fill the inboard joint boot with grease and install the boot clamps.
6. Install the axle halfshaft to the vehicle.

REAR HALFSHAFT

REMOVAL & INSTALLATION

1. Before servicing the vehicle, refer to the precautions section.
2. Drain the differential.
3. Remove or disconnect the following:
   - Negative battery cable
   - Rear wheels
   - Spindle nut
4. Pry the inboard joint from the differential.

5. Remove the outer CV-joint stub shaft from the hub by tapping the stub shaft with a plastic hammer.

To install:

➡**Use new circlips and self-locking nuts for assembly.**

6. Install the outer CV-joint stub shaft into the hub.

7. Install the inner CV-joint to the differential until the circlip locks in the retaining groove.

8. Install or connect the following:
- Spindle nut. Tighten the nut to 134 ft. lbs. (181 Nm).
- Rear wheels
- Negative battery cable

9. Fill the differential to the correct level and check for leaks.

CV-JOINT OVERHAUL

See Figure 15.

1. Before servicing the vehicle, refer to the precautions section.

2. Remove or disconnect the following:
- Axle halfshaft from the vehicle
- Joint boot clamps and push the boot back
- Joint housing from the axle
- Rollers from the spider
- Snapring and the spider from the axle shaft
- Joint boot

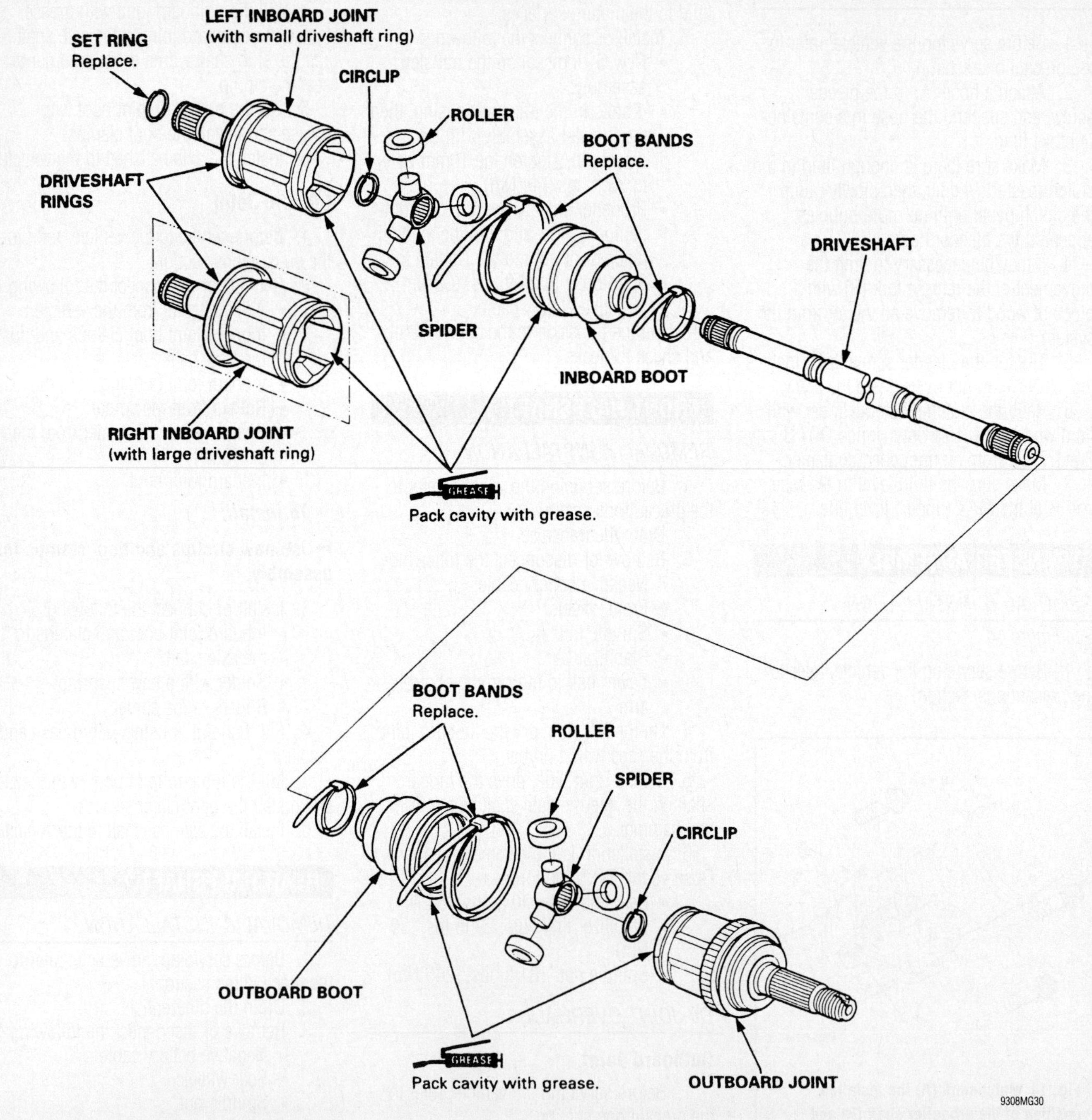
LEFT INBOARD JOINT
(with small driveshaft ring)
SET RING
Replace.
CIRCLIP
ROLLER
BOOT BANDS
Replace.
DRIVESHAFT RINGS
DRIVESHAFT
SPIDER
INBOARD BOOT
RIGHT INBOARD JOINT
(with large driveshaft ring)
GREASE
Pack cavity with grease.
BOOT BANDS
Replace.
ROLLER
SPIDER
CIRCLIP
OUTBOARD BOOT
GREASE
Pack cavity with grease.
OUTBOARD JOINT
9308MG30

Fig. 15 Exploded view of the rear axle

To install:

➡**Use new circlips and boot clamps for assembly.**

3. Install or connect the following:
    - Joint boot and clamps to the axle shaft
    - Spider with a new snapring
    - Rollers to the spider
4. Fill the joint housing with grease and install it.
5. Fill the joint boot with grease and install the boot clamps.
6. Install the axle halfshaft to the vehicle.

REAR PINION SEAL

REMOVAL & INSTALLATION

See Figure 16.

1. Before servicing the vehicle, refer to the precautions section.
2. Remove or disconnect the following:
    - Driveshaft
    - Companion flange
    - Pinion seal

To install:

➡**Use a new locknut and O-ring for assembly.**

3. Install or connect the following:
    - Pinion seal. Drive the seal square into the bore.
    - Companion flange. Tighten the locknut to 108 ft. lbs. (147 Nm).
    - Driveshaft. Tighten the flange bolts to 24 ft. lbs. (32 Nm).

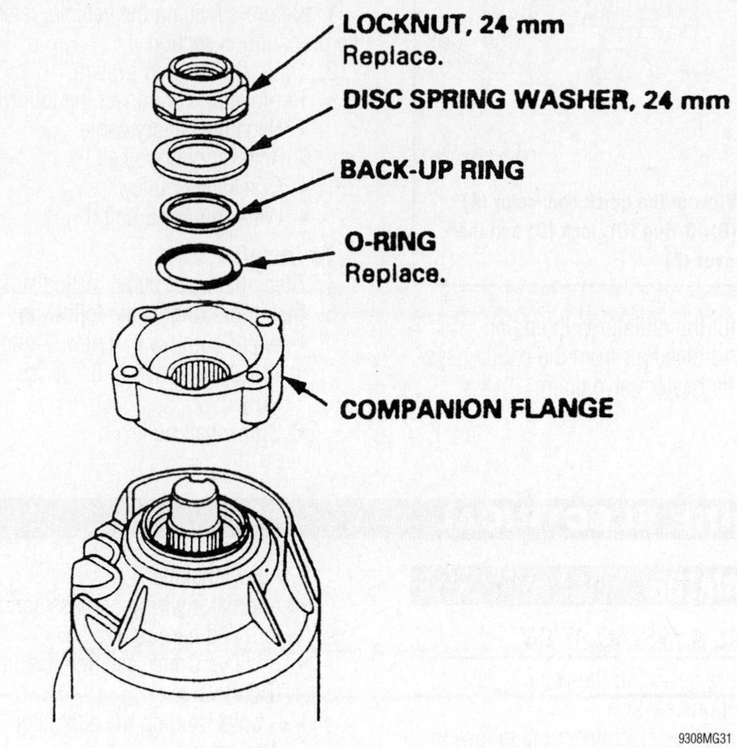
LOCKNUT, 24 mm
Replace.
DISC SPRING WASHER, 24 mm
BACK-UP RING
O-RING
Replace.
COMPANION FLANGE
9308MG31

Fig. 16 Exploded view of the rear differential pinion components

ENGINE COOLING

THERMOSTAT

REMOVAL & INSTALLATION

See Figures 17 and 18.

1. Before servicing the vehicle, refer to the precautions section.
2. Drain the engine coolant.
3. Disconnect the negative battery cable.
4. Clean any dirt from quick connector, thermostat cover, and lower radiator hose.
5. Pull the lock out by hand, then wiggle the quick connector loose, and remove it from the thermostat cover. Do not use any tools to remove the quick connector.
6. Remove the thermostat.

To install:

7. Install the thermostat with a new O-ring.
8. Check the quick connector (A) and set ring (B) for cracks or damage. If the connector and/or set ring are cracked or damaged, replace the connector.

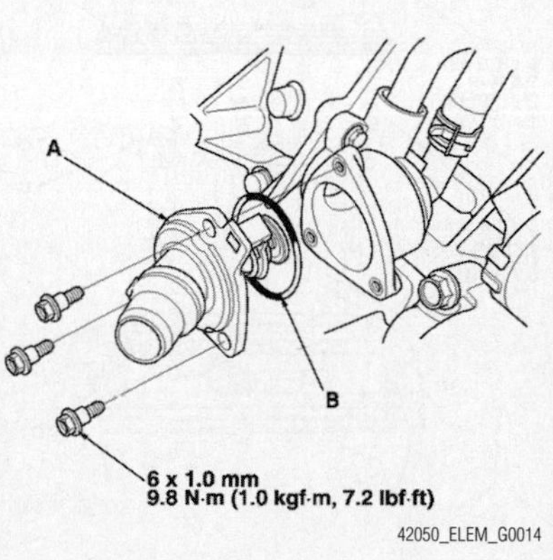
A
B
6 x 1.0 mm
9.8 N·m (1.0 kgf·m, 7.2 lbf·ft)
42050_ELEM_G0014

Fig. 17 View of the thermostat (A) and O-ring (B). Always use a new O-ring during thermostat installation

➡**Make sure the set ring is in place inside the quick connector. If the set ring is off the connector, replace the quick connector.**

9. Replace the O-ring (C) in the quick connector.
10. Check the lock (D). If the lock is damaged or deformed, replace it. When installing the new lock to the connector, push it straight down along the groove.
11. Clean the connecting surface of the thermostat cover (E), then apply clean engine coolant around the connecting surface.
12. Push the lock down, then push the quick connector onto the thermostat cover until you hear an audible click.

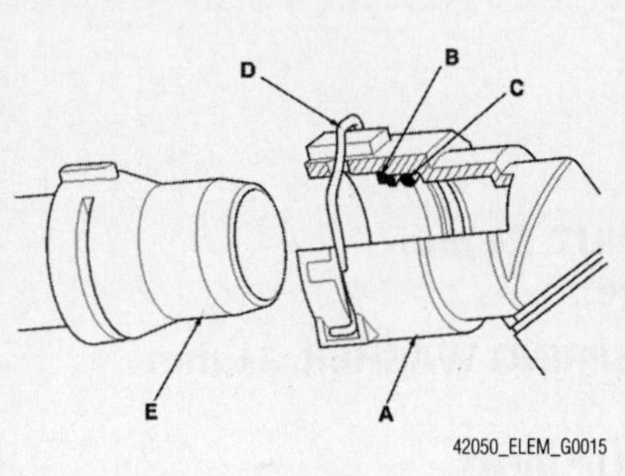
D
B
C
E
A
42050_ELEM_G0015

Fig. 18 View of the quick connector (A), set ring (B), O-ring (C), lock (D) and thermostat cover (E)

13. Refill the radiator with engine coolant, and bleed air from the cooling system with the heater valve open.

WATER PUMP

REMOVAL & INSTALLATION

See Figure 19.

1. Before servicing the vehicle, refer to the precautions section.
2. Drain the cooling system.
3. Remove or disconnect the following:
   - Negative battery cable
   - Accessory drive belt
   - Crankshaft pulley
   - Water pump (6 bolts)

To install:

4. Clean the water pump mating surfaces.
5. Install or connect the following:
   - Water pump with a new O-ring. Torque the bolts to 8.7 ft. lbs. (12 Nm).
   - Crankshaft pulley

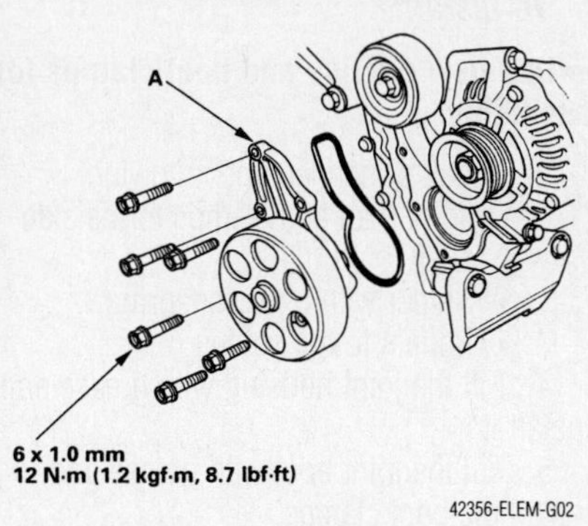
A
6 x 1.0 mm
12 N·m (1.2 kgf·m, 8.7 lbf·ft)
42356-ELEM-G02

Fig. 19 Exploded view of the water pump mounting

   - Accessory drive belt
   - Negative battery cable
6. Refill the engine cooling system.

ENGINE ELECTRICAL — CHARGING SYSTEM

ALTERNATOR

REMOVAL & INSTALLATION

1. Before servicing the vehicle, refer to the precautions section.
2. Remove or disconnect the following:
   - Negative, then the positive battery cables
   - Accessory drive belt
   - Auto-tensioner
   - Alternator wiring harness connectors and harness clamp
   - Positive Crankcase Ventilation (PCV) valve
   - 3 bolts holding the alternator
   - Alternator

To install:

3. Install or connect the following:
   - Alternator. Tighten the bolts to 16 ft. lbs. (22 Nm).
   - PCV valve
   - Alternator wiring harness connectors and harness clamp
   - Auto tensioner
   - Accessory drive belt
   - Negative battery cable

ENGINE ELECTRICAL — IGNITION SYSTEM

IGNITION COIL

REMOVAL & INSTALLATION

See Figure 20.

1. Before servicing the vehicle, refer to the precautions section.
2. Disconnect the negative battery cable.
3. Remove the ignition coil cover, disconnect the ignition coil connectors, then remove the ignition coils.

To install:

4. Install the ignition coils and tighten the retainers to 8.7 ft. lbs. (12 Nm).
5. Attach the ignition coil electrical connectors.
6. Install the ignition coil cover and tighten the retainers to 7.2 ft. lbs. (9.8 Nm).
7. Connect the negative battery cable.

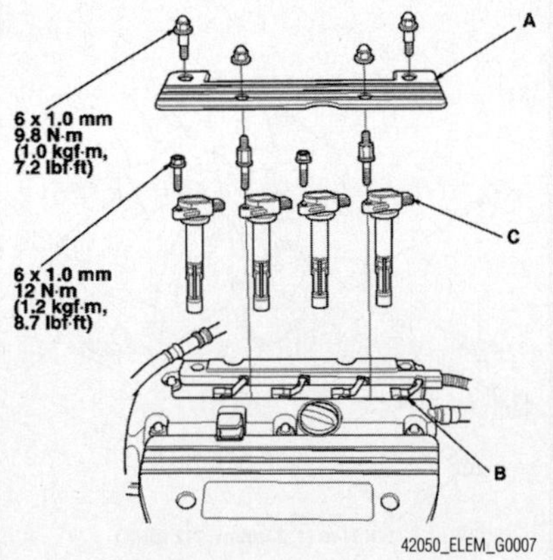
A
6 x 1.0 mm
9.8 N·m
(1.0 kgf·m,
7.2 lbf·ft)
C
6 x 1.0 mm
12 N·m
(1.2 kgf·m,
8.7 lbf·ft)
B
42050_ELEM_G0007

Fig. 20 Exploded view of the ignition coil cover (A), coil connectors (B) and ignition coils (C)

IGNITION TIMING

INSPECTION

See Figure 21.

1. Connect the Honda Diagnostic System (HDS) to the data link connector (DLC), and check for DTCs. If a DTC is present, diagnose and repair the cause before inspecting the ignition timing.
2. Start the engine. Hold the engine speed at 3,000 rpm without load (in Park or Neutral) until the radiator fan comes on, then let it idle.
3. Check the idle speed, as outlined in the Fuel System Section.
4. Jump the SCS line with the HDS.
5. Free the service loop from the wire harness, then connect the timing light to the service loop.
6. Aim the light toward the pointer (A) on the cam chain case. Check the ignition timing under a no load condition (headlights, blower fan, rear window defogger,

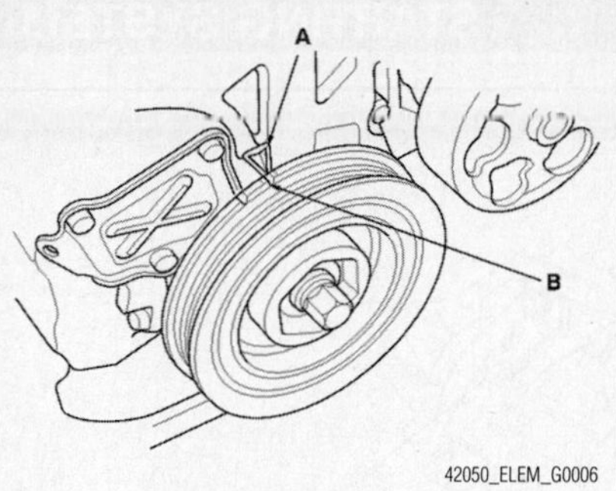
A
B
42050_ELEM_G0006

Fig. 21 Aim the light toward the pointer (A) on the cam chain case. Check the ignition timing under a no load condition (headlights, blower fan, rear window defogger, and air conditioner are turned off)

and air conditioner are turned off). The ignition timing should be:

a. M/T: 6–10° BTDC (RED mark B) at idle in Neutral

b. A/T: 6–10° BTDC (RED mark B) at idle in Park or Neutral

7. If the ignition timing differs from the specification, update the engine control module (ECM)/powertrain control module (PCM) if it does not have the latest software, or substitute a known-good ECM/PCM, then recheck. If the system works properly, and the ECM/PCM was substituted, replace the original ECM/PCM.
8. Disconnect the HDS and the timing light.
9. Secure the service loop to the wire harness with wire ties.

ADJUSTMENT

The ignition timing is controlled by the Powertrain Control Module (PCM). No adjustment is necessary or possible.

SPARK PLUGS

REMOVAL & INSTALLATION

See Figure 22.

1. Disconnect the negative battery cable.
2. Remove the ignition coil, as outlined later in this section.
3. Remove the spark plug.
4. Inspect the spark plug.

To install:

5. Install the spark plug and tighten to 13 ft. lbs. (18 Nm).
6. Install the ignition coil.
7. Connect the negative battery cable.

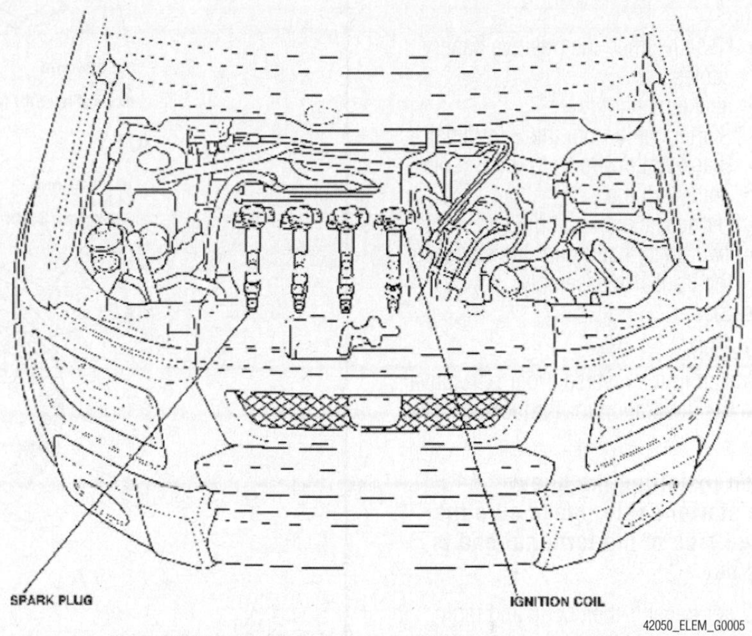
SPARK PLUG
IGNITION COIL
42050_ELEM_G0005

Fig. 22 Spark plug and ignition coil locations

ENGINE ELECTRICAL

STARTER

REMOVAL & INSTALLATION

See Figure 23.

➡The factory sound system has a coded theft protection system. It is recommended that you know your reset code before you begin.

1. Before servicing the vehicle, refer to the precautions section.
2. Remove or disconnect the following:
   - Negative then the positive battery cables
   - Intake manifold
   - Starter cable from the B terminal
   - Black/white wire from the S (solenoid) terminal
   - Harness clamp and holder
   - Two bolts that mount the starter to the transaxle assembly
   - Starter

To install:

3. Install in the reverse order of removal. Refer to the illustration for torque specifications.

➡When installing the heavy gauge starter cable, make sure the crimped side of the terminal end is facing out.

4. Enter the anti-theft code and radio presets.

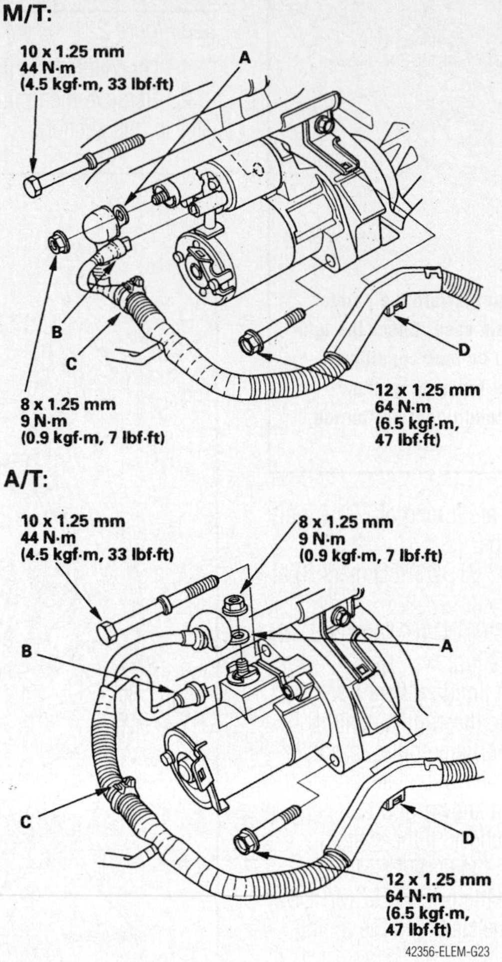
M/T:
10 x 1.25 mm
44 N·m
(4.5 kgf·m, 33 lbf·ft)
A
B
C
D
8 x 1.25 mm
9 N·m
(0.9 kgf·m, 7 lbf·ft)
12 x 1.25 mm
64 N·m
(6.5 kgf·m,
47 lbf·ft)
A/T:
10 x 1.25 mm
44 N·m
(4.5 kgf·m, 33 lbf·ft)
8 x 1.25 mm
9 N·m
(0.9 kgf·m, 7 lbf·ft)
A
B
C
D
12 x 1.25 mm
64 N·m
(6.5 kgf·m,
47 lbf·ft)
42356-ELEM-G23

Fig. 23 Starter mounting

ENGINE MECHANICAL

➡Disconnecting the negative battery cable may interfere with the functions of the on board computer systems and may require the computer to undergo a relearning process, once the negative battery cable is reconnected.

ACCESSORY DRIVE BELTS

ACCESSORY BELT ROUTING

See Figure 24.

Refer to the accompanying figure for drive belt routing.

INSPECTION

See Figure 25.

1. Inspect the drive belt for signs of glazing or cracking. A glazed belt will be perfectly smooth from slippage, while a good belt will have a slight texture of fabric visible. Cracks

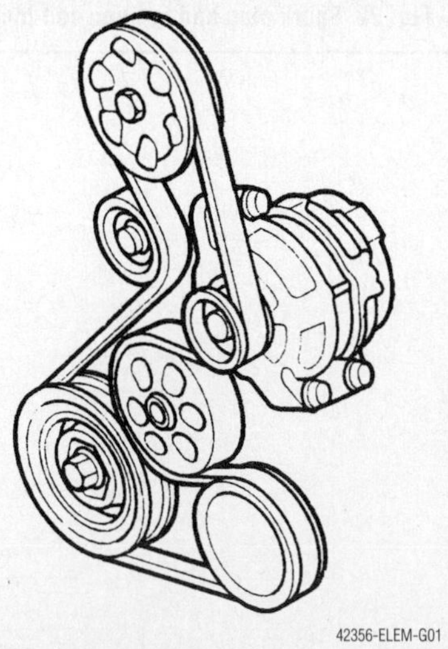
42356-ELEM-G01

Fig. 24 Accessory drive belt routing

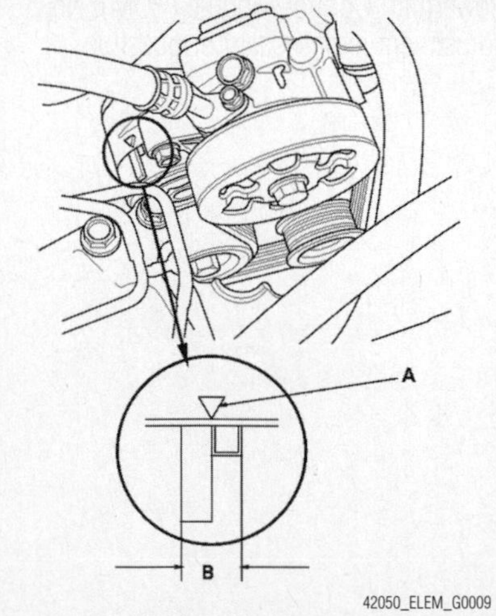
A
B
42050_ELEM_G0009

Fig. 25 When inspecting the belt, check that the auto-tensioner indicator (A) is within the standard range (B)

will usually start at the inner edge of the belt and run outward. All worn or damaged drive belts should be replaced immediately.

2. Check that the auto-tensioner indicator (A) is within the standard range (B) as shown. If it is out of the standard range, replace the drive belt.

ADJUSTMENT

The belt tension maintained by an automatic tensioner. No adjustment is necessary or possible.

REMOVAL & INSTALLATION

See Figures 24 and 26.

➡This procedure requires the use of a special Belt tension release tool, Snap-on YA9317 or equivalent tool.

1. Move the auto-tensioner (A) with the belt tension release tool (B) in the direction shown to relieve tension from the drive belt, and remove the drive belt.
2. Install the new belt in the reverse order of removal.

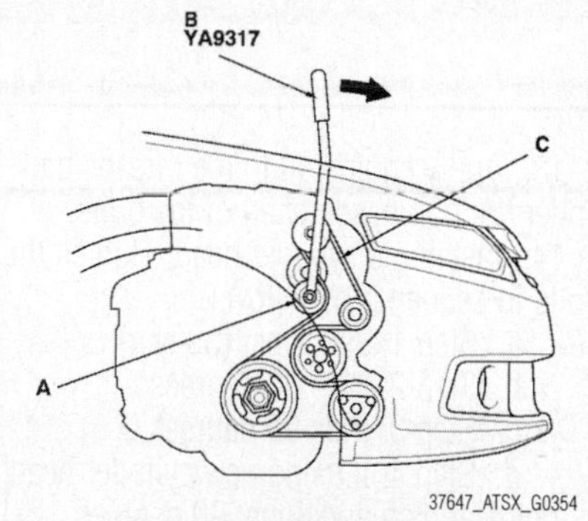
B
YA9317
C
A
37647_ATSX_G0354

Fig. 26 To remove the belt, move the auto-tensioner (A) with the belt tension release tool (B) in the direction shown to relieve tension from the drive belt, and remove the drive belt

CAMSHAFT AND VALVE LIFTERS

REMOVAL & INSTALLATION

See the Rocker Arm Shaft Removal & Installation procedure.

CRANKSHAFT DAMPER

REMOVAL & INSTALLATION

See Figures 27 through 29.

➡This procedure requires the use of the following special tools, or their equivalents: Holder handle 07JAB-00102JB, 50mm Holder attachment

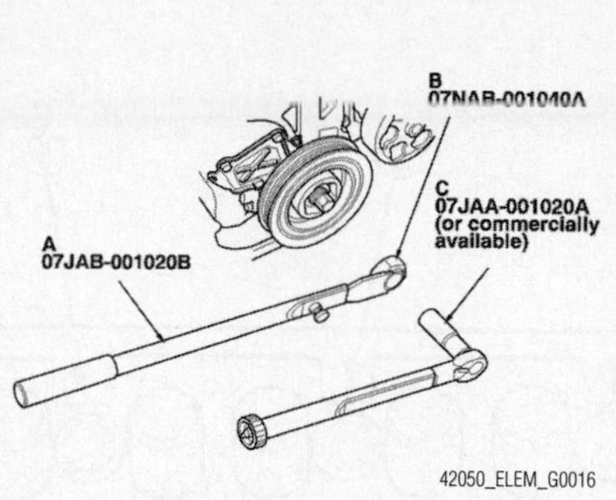
B
07NAB-001040A
C
07JAA-001020A
(or commercially available)
A
07JAB-001020B
42050_ELEM_G0016

Fig. 27 To remove the crankshaft pulley, hold the pulley with holder handle (A) and holder attachment (B), then remove the bolt with a 19mm socket (C) and breaker bar

07NAB-001040A, and 19mm Socket 07JAA-001020A.

1. Before servicing the vehicle, refer to the precautions section.
2. Remove the right front wheel.
3. Remove the splash shield.
4. Remove the drive belt.
5. Hold the pulley with holder handle (A) and holder attachment (B).
6. Remove the bolt with a 19mm socket (C) and breaker bar, then remove the crankshaft pulley.

To install:

7. Clean the crankshaft pulley (A), crankshaft (B), bolt (C), and washer (D). Lubricate with the new engine oil as shown.
8. Install the crankshaft pulley, and hold the pulley with holder handle (A) and holder attachment (B).
9. Tighten the bolt to 36 ft. lbs. (49 Nm) with a torque wrench and 19mm socket (C). Never use an impact wrench to tighten the crankshaft pulley bolt.

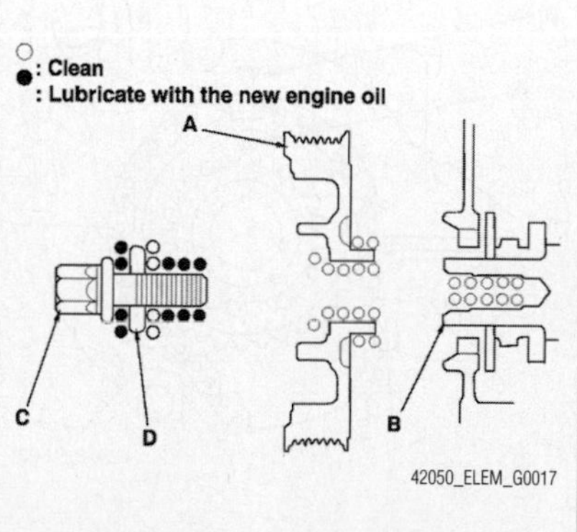
○: Clean
●: Lubricate with the new engine oil
A
C
D
B
42050_ELEM_G0017

Fig. 28 Clean the crankshaft pulley (A), crankshaft (B), bolt (C), and washer (D). Lubricate with the new engine oil as shown

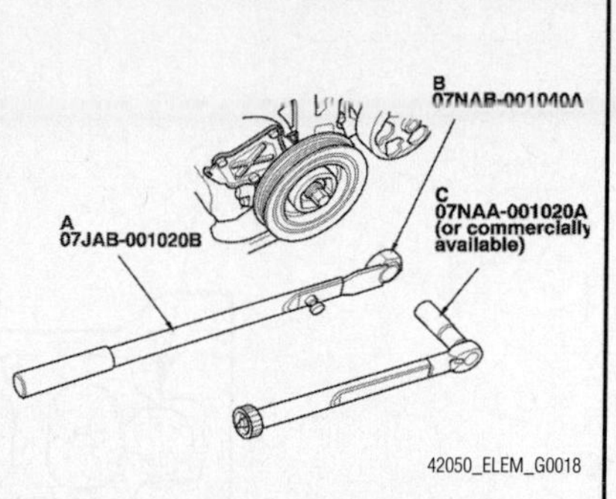
B
07NAB-001040A
A
07JAB-001020B
C
07NAA-001020A
(or commercially available)
42050_ELEM_G0018

Fig. 29 Install the crankshaft pulley, and hold the pulley with holder handle (A) and holder attachment (B). Tighten the bolt to specifications with a torque wrench and 19mm socket

10. Tighten the pulley bolt an additional 90°.
11. Install the drive belt.
12. Install the splash shield.
13. Install the right front wheel.

CRANKSHAFT FRONT SEAL

REMOVAL & INSTALLATION

For the 2.4L engine, see the Timing Chain Removal & Installation procedure.

CYLINDER HEAD

REMOVAL & INSTALLATION

See Figures 30 through 33.

1. Before servicing the vehicle, refer to the precautions section.
2. Drain the cooling system.
3. Relieve the fuel system pressure.
4. Remove or disconnect the following:
   - Negative battery cable
   - Accessory drive belt
   - Intake Air Temperature (IAT) sensor connector
   - Vacuum hoses and breather pipe and air intake duct
   - Fuel feed hose
   - Bolt securing the connecting pipe support bracket to the engine block
   - Evaporative emission (EVAP) canister hose and brake booster vacuum hose
   - Intake manifold
   - Exhaust manifold
   - Cam chain
   - Positive Crankcase Ventilation (PCV) hose and ground cable
   - Upper radiator hose, heater hoses and water bypass hose

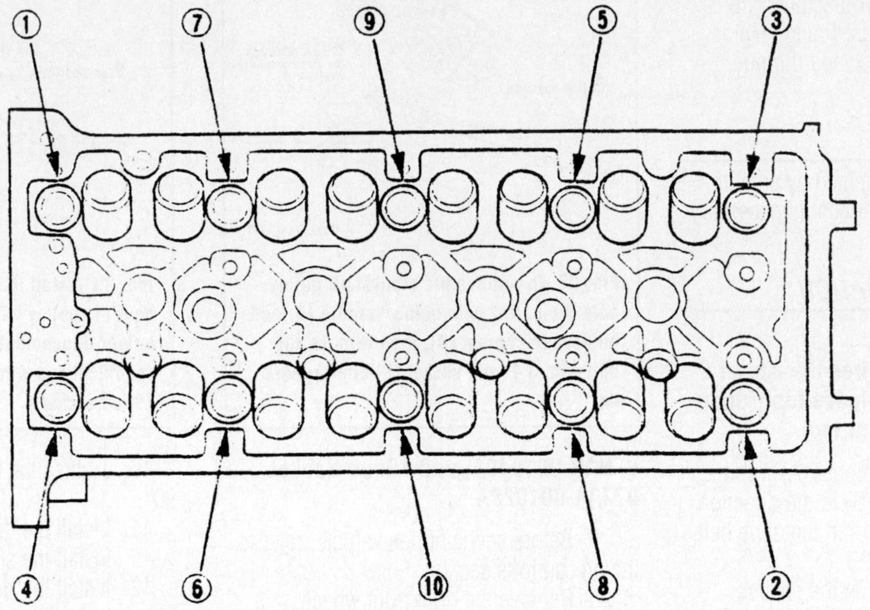
1
7
9
5
3
4
6
10
8
2
9359MG01

Fig. 30 Cylinder head bolt loosening sequence

5. Remove the following engine wire harness connectors and wire harness clamps from the cylinder head:
   - Four injector connector
   - Engine Coolant Temperature (ECT) sensor connector
   - Camshaft Position (CMP) sensor A & B (intake & exhaust) connectors
   - VTEC solenoid valve connector
   - Engine Oil Pressure (EOP) sensor connector
6. Remove or disconnect the following:
   - 3 bolts holding the EVAP canister purge valve bracket and remove the two bolts (B) securing the harness bracket
   - Timing (cam) chain
   - Rocker arm assembly
7. Loosen the cylinder head bolts in sequence and ⅓ turns until all bolts are loose.
8. Remove the cylinder head.

To install:

9. Be sure all cylinder head and block gasket surfaces are clean. Check the cylinder head for warpage. If warpage is less than 0.002 in. (0.05mm), cylinder head resurfacing is not required. Maximum resurface limit is 0.008 in. (0.2mm) based on a cylinder head height of 3.94 in. (100mm).
10. Install or connect the following:
    - New gasket and dowel pins on the cylinder block
11. Set the crankshaft to Top Dead Center (TDC). Align the TDC mark (A) on the crankshaft sprocket with the pointer (B) on the cylinder block.
12. Measure the diameter of each cylinder head bolt at points A & B, as shown in the illustration. If either diameter is less than 0.42 in. (10.6mm), replace the head bolt

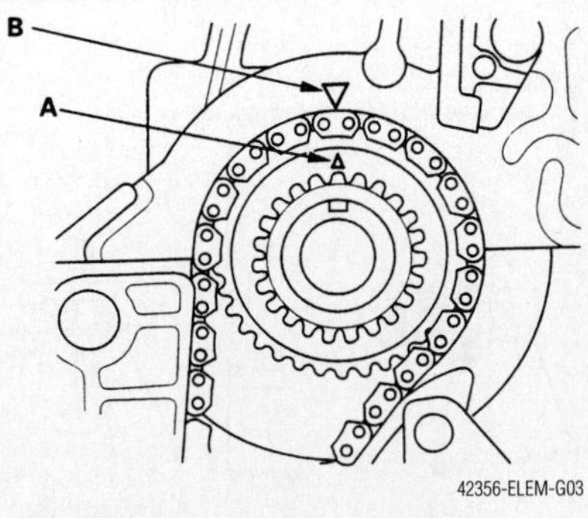
B
A
42356-ELEM-G03

Fig. 31 Set the crankshaft to TDC by aligning the mark (A) on the crankshaft sprocket with the pointer (B) on the cylinder block

13. Apply engine oil to the threads and under the bolt heads of all of the bolts.
14. Install the cylinder head. Tighten the bolts in sequence as follows:
    a. Step 1: 29 ft. lbs. (39 Nm).
    b. Step 2: Plus 90 degrees.
    c. Step 3: Plus 90 degrees.
    d. Step 4: If using new cylinder head bolts, add an additional 90 degrees.
15. The remainder of installation is the reverse of removal.

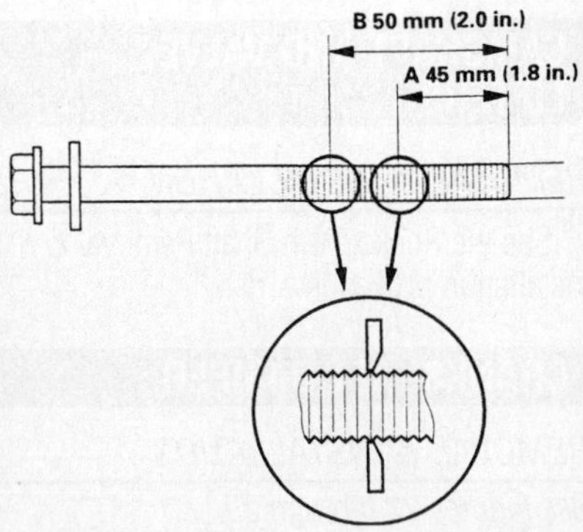
B 50 mm (2.0 in.)
A 45 mm (1.8 in.)
9359MG02

Fig. 32 Cylinder head bolt inspection

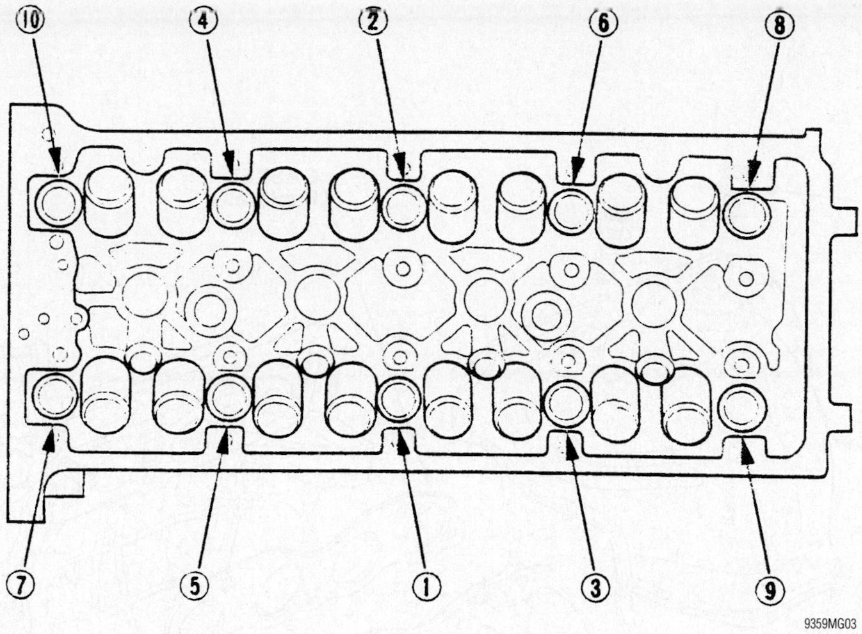
10
4
2
6
8
7
5
1
3
9
9359MG03

Fig. 33 Cylinder head bolt torque sequence

16. Fill the cooling system.
17. Connect the negative battery cable and enter the radio security code.
18. Start the engine and check carefully for any leaks.

DRIVEPLATE

REMOVAL & INSTALLATION

With Automatic Transmission

See Figure 34.

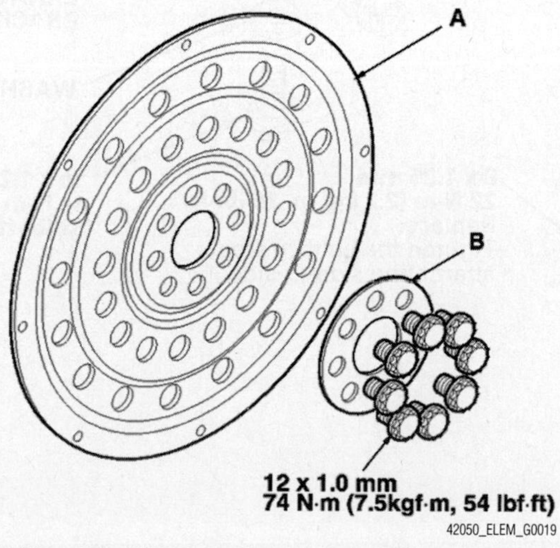
A
B
12 x 1.0 mm
74 N·m (7.5kgf·m, 54 lbf·ft)
42050_ELEM_G0019

Fig. 34 View of the drive plate (A) and washer

1. Before servicing the vehicle, refer to the precautions section.
2. Remove the transmission assembly, as outlined in the Drive Train Section.
3. Remove the drive plate (A) and washer (B) from the engine crankshaft.

To install:

4. Install the drive plate and washer on the engine crankshaft, and tighten the eight bolts in a crisscross pattern in two or more steps to a final torque of 54 ft. lbs. (74 Nm).
5. Install the transmission assembly, as outlined in the Drive Train Section.

EXHAUST MANIFOLD

REMOVAL & INSTALLATION

See Figure 35.

1. Before servicing the vehicle, refer to the precautions section.
2. Raise and safely support the vehicle.
3. Remove or disconnect the following:
   - VTEC solenoid valve
   - Intermediate shaft heat cover
   - Cover and exhaust manifold bracket
   - Exhaust manifold

To install:

4. Clean the mounting surfaces.
5. Install or connect the following:
   - New gasket on the cylinder head
   - Exhaust manifold. Tighten the nuts, in a criss-cross pattern starting with the inner nut, to 33 ft. lbs. (45 Nm).
   - Exhaust manifold bracket and cover
   - Intermediate shaft heat cover
   - VTEC solenoid valve

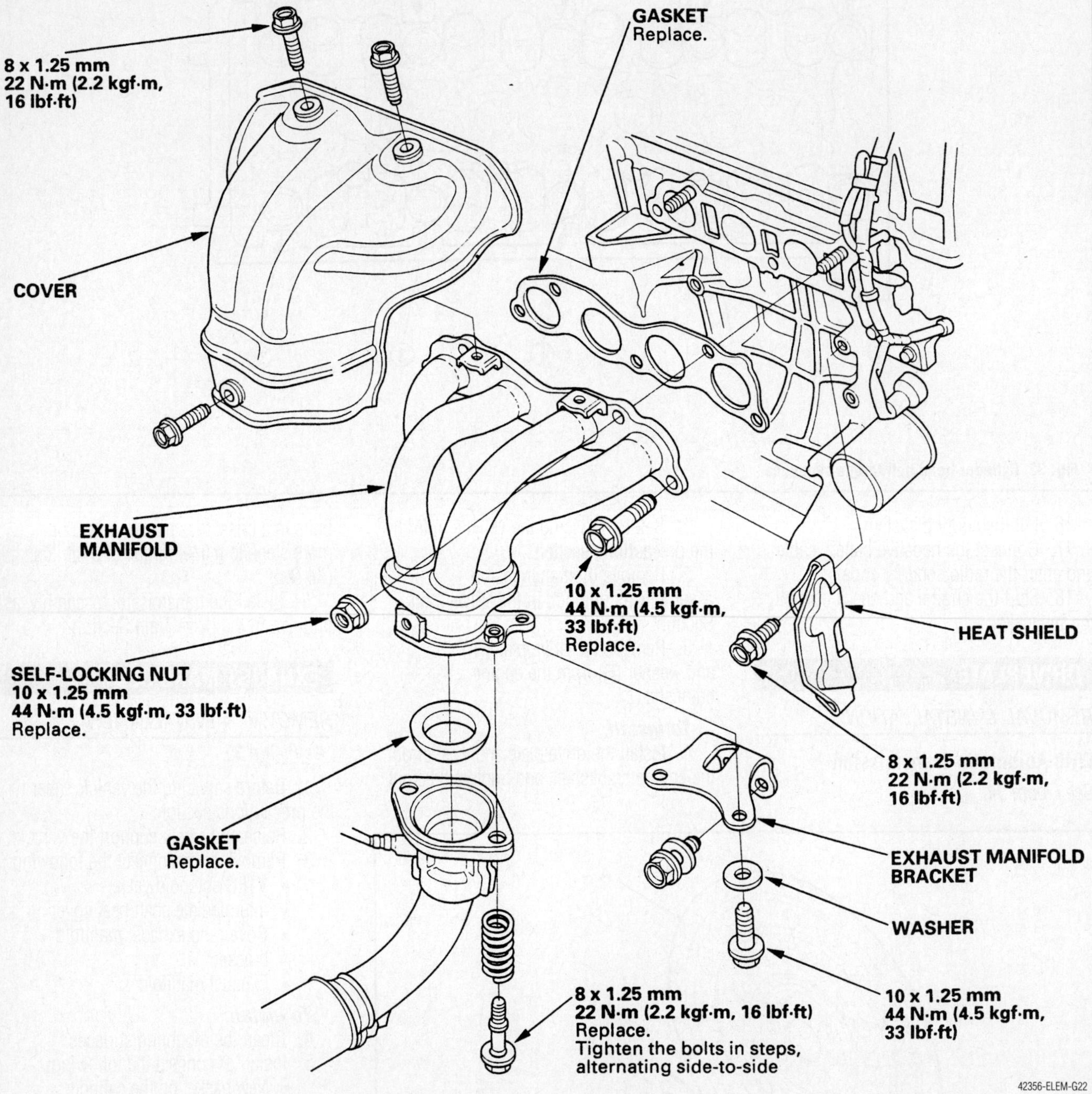
8 x 1.25 mm
22 N·m (2.2 kgf·m,
16 lbf·ft)
GASKET
Replace.
COVER
EXHAUST
MANIFOLD
10 x 1.25 mm
44 N·m (4.5 kgf·m,
33 lbf·ft)
Replace.
HEAT SHIELD
SELF-LOCKING NUT
10 x 1.25 mm
44 N·m (4.5 kgf·m, 33 lbf·ft)
Replace.
8 x 1.25 mm
22 N·m (2.2 kgf·m,
16 lbf·ft)
GASKET
Replace.
EXHAUST MANIFOLD
BRACKET
WASHER
8 x 1.25 mm
22 N·m (2.2 kgf·m, 16 lbf·ft)
Replace.
Tighten the bolts in steps,
alternating side-to-side
10 x 1.25 mm
44 N·m (4.5 kgf·m,
33 lbf·ft)
42356-ELEM-G22

Fig. 35 Exploded view of the exhaust manifold and related components

FLYWHEEL

REMOVAL & INSTALLATION

With Manual Transmission

See Figures 36 and 37.

1. Before servicing the vehicle, refer to the precautions section.
2. Remove the pressure plate and clutch disc, as outlined in the Drive Train Section.
3. Install the special tool on the flywheel, as shown in the accompanying illustration.
4. Remove the flywheel mounting bolts in a crisscross pattern in several steps, then remove the flywheel.

To install:

5. Install the flywheel on the crankshaft, and install the mounting bolts, finger-tight.

Install the special tool, then torque the flywheel mounting bolts in a crisscross pattern in several steps to 76 ft. lbs. (103 Nm).

6. Install the clutch disc and pressure plate, as outlined in the Drive Train Section.

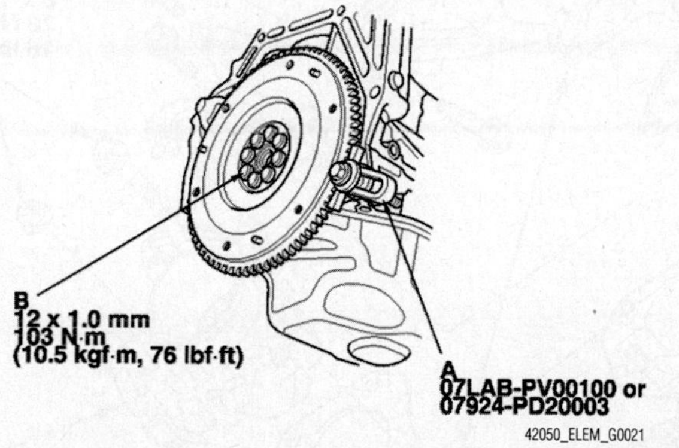
B
12 x 1.0 mm
103 N·m
(10.5 kgf·m, 76 lbf·ft)
A
07LAB-PV00100 or
07924-PD20003
42050_ELEM_G0021

Fig. 36 View of the special tool (A) installed on the flywheel

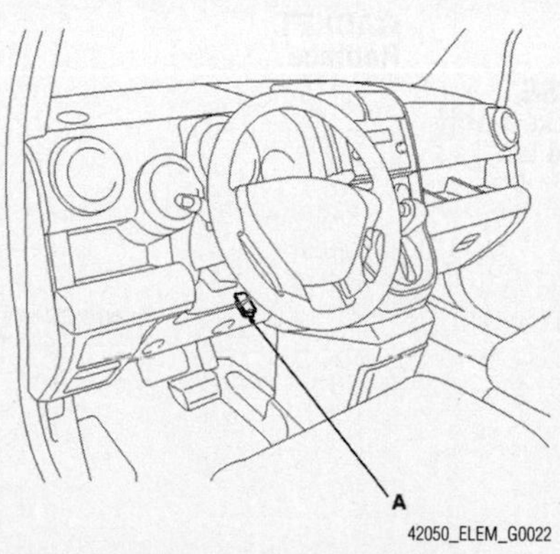
A
42050_ELEM_G0022

Fig. 37 After installing the special tool (A), tighten the flywheel mounting bolts (B) to specifications in several steps, using a criss-cross pattern

INTAKE MANIFOLD

REMOVAL & INSTALLATION

See Figure 38.

1. Before servicing the vehicle, refer to the precautions section.
2. Disconnect the negative battery cable.
3. Drain the engine coolant into a sealable container.
4. Remove or disconnect the following:
   - Intake Air Temperature (IAT) sensor electrical connector
   - Vacuum hose and breather pipe and the air intake duct
   - Intake manifold cover
   - Throttle and cruise control cables by loosening the locknuts, then slipping the cable ends out of the accelerator linkage.

➡Do not bend the cables during removal. Always replace any throttle or cruise control cables that get kinked during removal.

   - Evaporative emission (EVAP) canister hose and brake booster vacuum hose
   - Idle Air Control (IAC) valve connectors
   - Throttle Position (TP) sensor connector
   - Manifold Absolute Pressure (MAP) sensor connector
   - Necessary engine wire harness connectors and wire harness clamps from the intake manifold
   - Bolt securing the harness holder and remove the harness clamps
   - Water bypass hoses, then plug them
   - Harness clamp and harness connector from the intake manifold bracket
   - Intake manifold bracket
   - A/T vacuum hose
   - Retainer and intake manifold

To install:

5. Clean the mounting surfaces.
6. Install or connect the following:
   - New gasket
   - Intake manifold. Tighten the bolts, in a criss-cross pattern beginning with the inner bolt, to 16 ft. lbs. (22 Nm).
   - A/T vacuum hose
   - Intake manifold bracket
   - Harness clamp and connector to the intake manifold bracket
   - Water bypass hoses
   - Bolt securing the harness holder and tighten to 8.7 ft. lbs. (12 Nm)
   - Harness clamps
   - EVAP canister hose and brake booster vacuum hose
   - Throttle and cruise control cables
   - Intake manifold cover
   - Intake air duct
   - IAT sensor connector, vacuum hose and breather pipe
7. Refill the cooling system.
8. Connect the negative battery cable, start the engine, and check for leaks.

OIL PAN

REMOVAL & INSTALLATION

See Figure 39.

1. Before servicing the vehicle, refer to the precautions section.
2. Drain the engine oil.
3. Remove or disconnect the following:
   - Subframe. See Engine Removal and Installation.
   - With a manual transmission, remove the stiffener
   - Oil pan bolts
   - Oil pan. A gasket cutter will be needed.

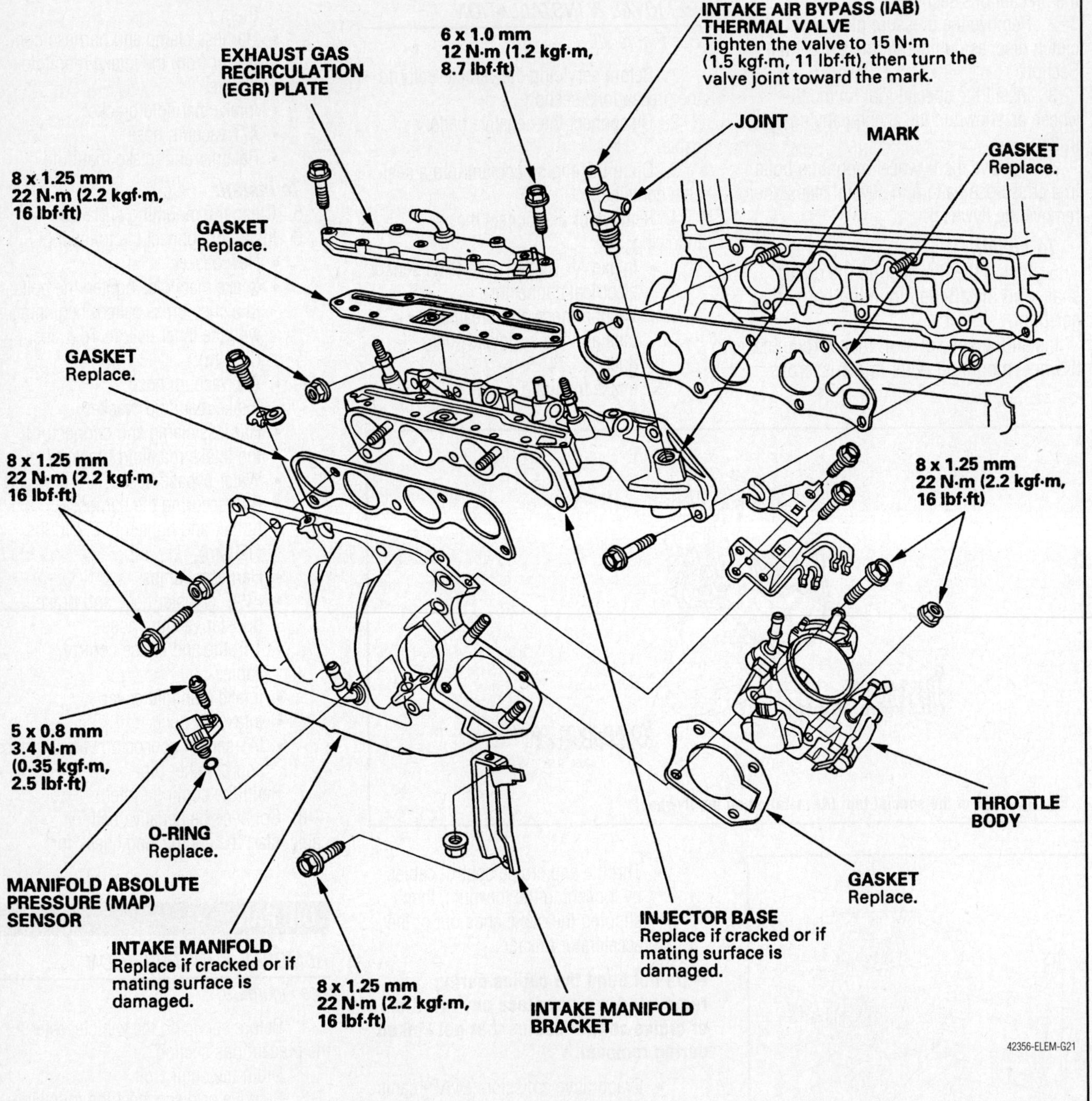

EXHAUST GAS RECIRCULATION (EGR) PLATE
6 x 1.0 mm
12 N·m (1.2 kgf·m, 8.7 lbf·ft)
INTAKE AIR BYPASS (IAB) THERMAL VALVE
Tighten the valve to 15 N·m (1.5 kgf·m, 11 lbf·ft), then turn the valve joint toward the mark.
JOINT
MARK
GASKET
Replace.
8 x 1.25 mm
22 N·m (2.2 kgf·m, 16 lbf·ft)
GASKET
Replace.
GASKET
Replace.
8 x 1.25 mm
22 N·m (2.2 kgf·m, 16 lbf·ft)
8 x 1.25 mm
22 N·m (2.2 kgf·m, 16 lbf·ft)
5 x 0.8 mm
3.4 N·m
(0.35 kgf·m, 2.5 lbf·ft)
O-RING
Replace.
MANIFOLD ABSOLUTE PRESSURE (MAP) SENSOR
INTAKE MANIFOLD
Replace if cracked or if mating surface is damaged.
8 x 1.25 mm
22 N·m (2.2 kgf·m, 16 lbf·ft)
INTAKE MANIFOLD BRACKET
INJECTOR BASE
Replace if cracked or if mating surface is damaged.
GASKET
Replace.
THROTTLE BODY
42356-ELEM-G21

Fig. 38 Exploded view of the intake manifold and related components

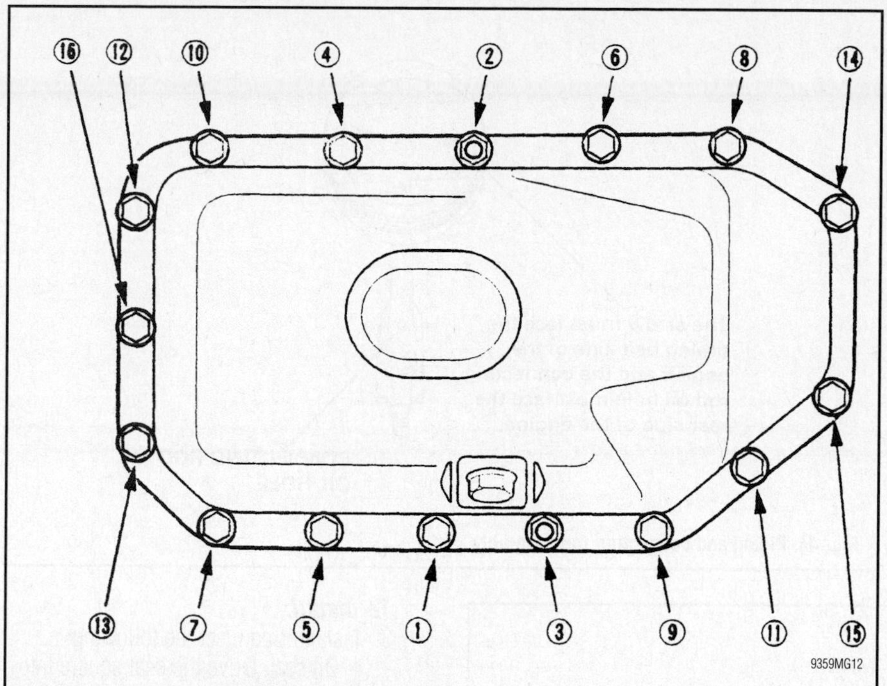
16
12
10
4
2
6
8
14
13
7
5
1
3
9
11
15
9359MG12

Fig. 39 Oil pan fastener tightening sequence

To install:

4. Apply a bead of liquid gasket to the oil pan mating surface. make sure to install the pan within 4 minutes of applying the gasket maker.

5. Installation is the reverse of removal. Torque the bolts, in sequence, in 2 or 3 steps, to 9 ft. lbs. (12 Nm).

OIL PUMP

REMOVAL & INSTALLATION

See Figures 40 and 41.

1. Before servicing the vehicle, refer to the precautions section.

2. Drain the engine oil.

3. Set the No. 1 piston to Top Dead Center (TDC).

4. Remove or disconnect the following:
   - Negative battery cable
   - Oil pan
   - Oil pump chain tensioner and discard

5. Insert a 6mm pin driver into the maintenance hole in the lower balance shaft holder and through the rear balancer shaft to hold the rear balancer shaft.

6. Loosen the oil pump sprocket mounting bolt.
   - Oil pump sprocket
   - Oil pump

To install:

7. Make sure that No.1 piston is at TDC.

8. Align the dowel pin on the rear balance shaft with the mark on the pump.

9. Insert a 6mm pin into the maintenance hole in the lower balance shaft holder, through the rear balancer shaft to hold the shaft.

10. Install or connect the following:

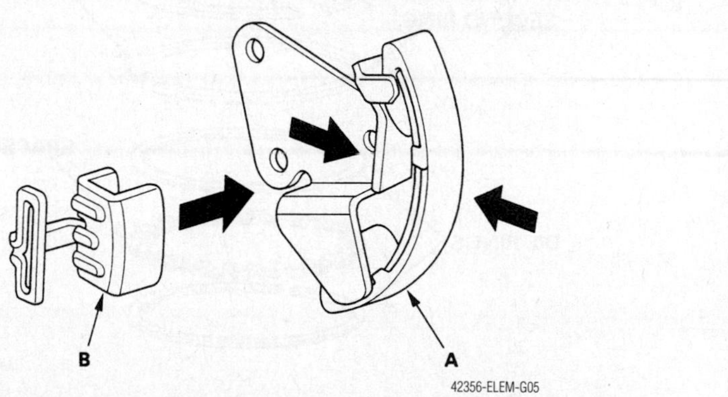
B
A
42356-ELEM-G05

Fig. 41 Squeeze the new oil pump chain tensioner (A) then install the set clip (A) on it as shown. The clip is supplied with the new tensioner

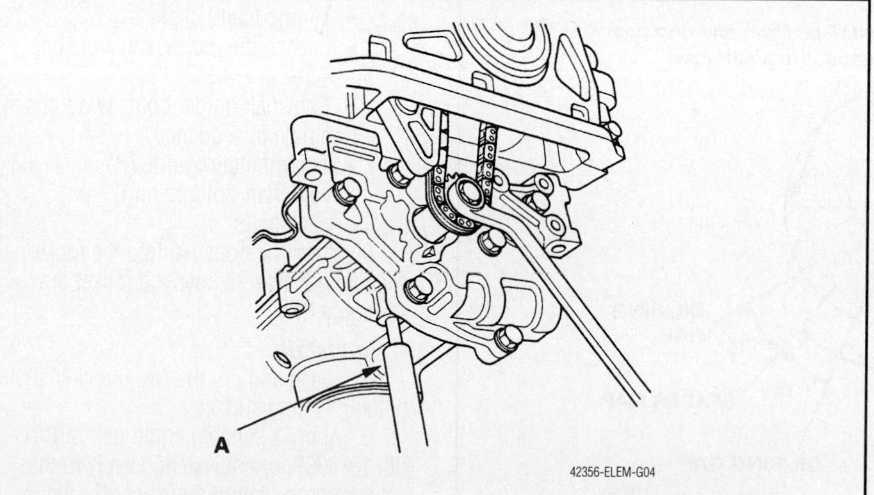
A
42356-ELEM-G04

Fig. 40 Insert a 6mm pin into the maintenance hole in the lower balance shaft holder, through the rear balancer shaft to hold the shaft, then loosen the sprocket mounting bolt

   - Engine oil to the threads of the oil pump sprocket mounting bolt
   - Oil pump and sprocket loosely

11. Remove the balance shaft holding pin.

12. Torque the 10mm mounting bolts to 33 ft. lbs. (44 Nm); the 8mm bolts to 16 ft. lbs. (22 Nm).

13. Torque the pulley bolt to 33 ft. lbs. (44 Nm).

14. Squeeze the new oil pump chain tensioner then install the set clip on it as shown in the illustration.

15. Install or connect the following:
   - New oil pump chain tensioner and torque the bolts to 9 ft. lbs. (12 Nm). Remove the set clip from the tensioner.
   - Oil pan

16. Fill the engine with oil.

PISTON AND RING

POSITIONING

See Figures 42 through 44.

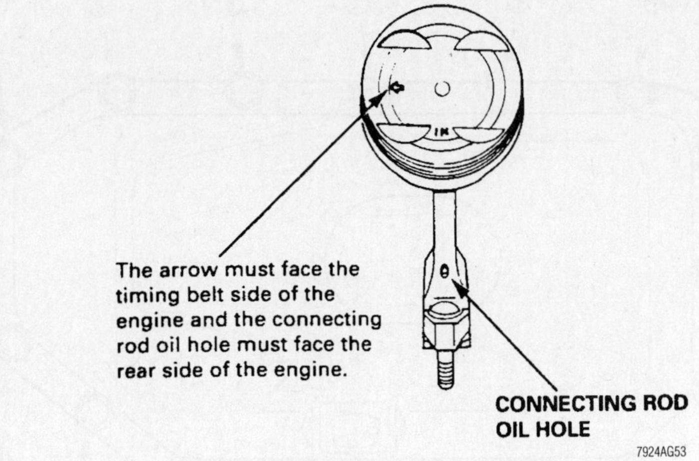
The arrow must face the timing belt side of the engine and the connecting rod oil hole must face the rear side of the engine.
CONNECTING ROD OIL HOLE
7924AG53

Fig. 44 Piston and connecting rod assembly

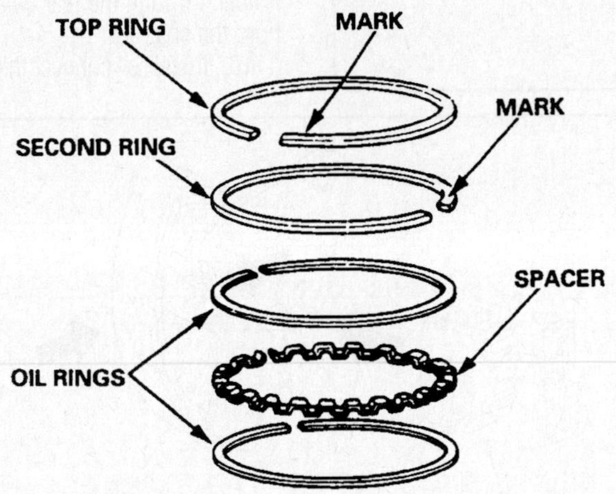
TOP RING
MARK
MARK
SECOND RING
SPACER
OIL RINGS
7924AG55

Fig. 42 Piston ring positioning and top mark location

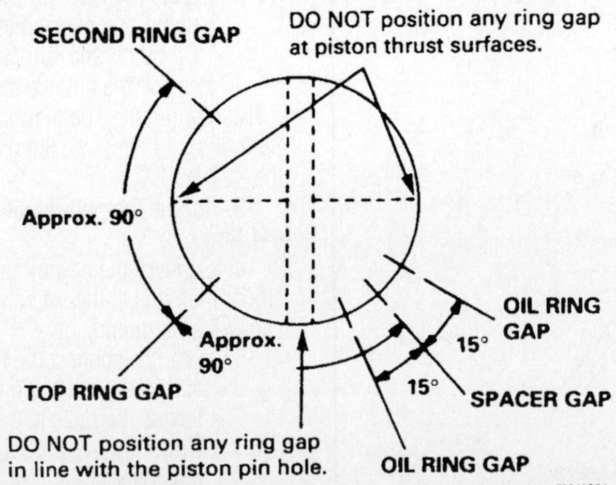
SECOND RING GAP
DO NOT position any ring gap at piston thrust surfaces.
Approx. 90°
OIL RING GAP
Approx. 90°
15°
15°
SPACER GAP
TOP RING GAP
DO NOT position any ring gap in line with the piston pin hole.
OIL RING GAP
7924AG54

Fig. 43 Piston ring end-gap spacing

REAR MAIN SEAL

REMOVAL & INSTALLATION

1. Before servicing the vehicle, refer to the precautions section.
2. Remove or disconnect the following:
   - Transaxle
   - Clutch pressure plate and disc, if equipped
   - Flywheel
   - Oil seal

To install:

3. Install or connect the following:
   - Oil seal. Drive the seal square into the seal case.
   - Flywheel. Tighten the bolts in a crossing pattern to 76 ft. lbs. (103 Nm).
   - Clutch pressure plate and disc, if equipped
   - Transaxle
4. Check the fluid levels.
5. Start the engine and check for leaks.

ROCKER ARMS/SHAFTS

REMOVAL & INSTALLATION

See Figures 45 through 49.

1. Before servicing the vehicle, refer to the precautions section.
2. Remove or disconnect the following:
   - Timing (cam) chain
   - Loosen the rocker arm adjusting screws
   - Camshaft holder bolts, two turns at a time in sequence
   - Timing chain guide (B), camshaft holders and camshafts
3. Insert the bolts (A) into the rocker shaft holder, then remove the rocker arm assembly (B)

To install:

4. Clean and dry the No. 5 rocker shaft holding mating surface.
5. Apply a suitable liquid gasket P/N 08718-0009, or equivalent, evenly to the cylinder head mating surface of the No. 5 rocker shaft holder.

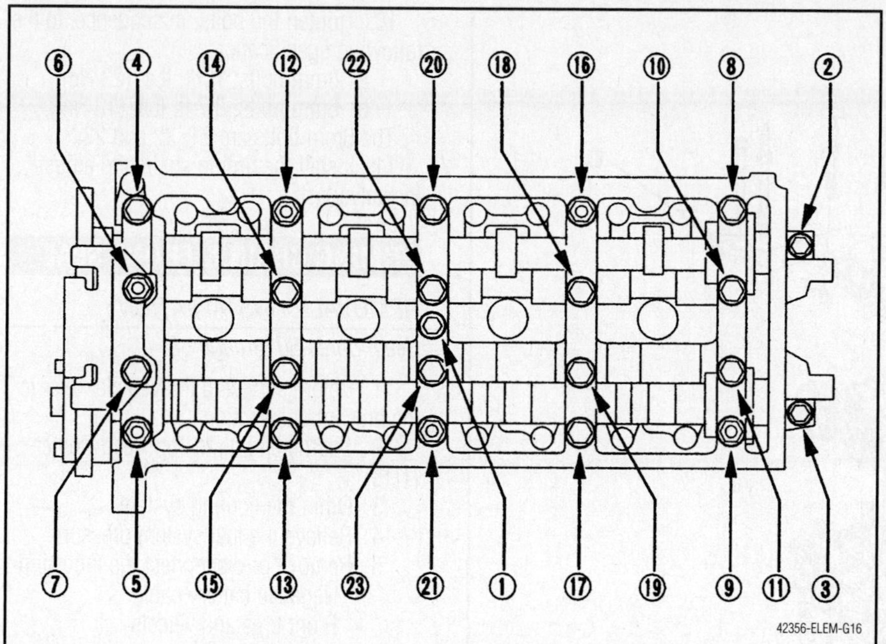

6 4 14 12 22 20 18 16 10 8 2
7 5 15 13 23 21 1 17 19 9 11 3
42356-ELEM-G16

Fig. 45 Camshaft holder bolt loosening sequence. Note that bolt 1 in the illustration is not on all engines.

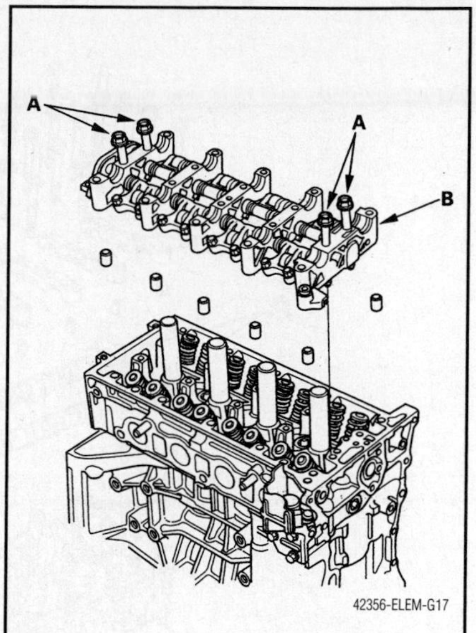

A
A
B
42356-ELEM-G17

Fig. 46 Insert the bolts (A) into the rocker shaft holder, then remove the rocker arm assembly (B)

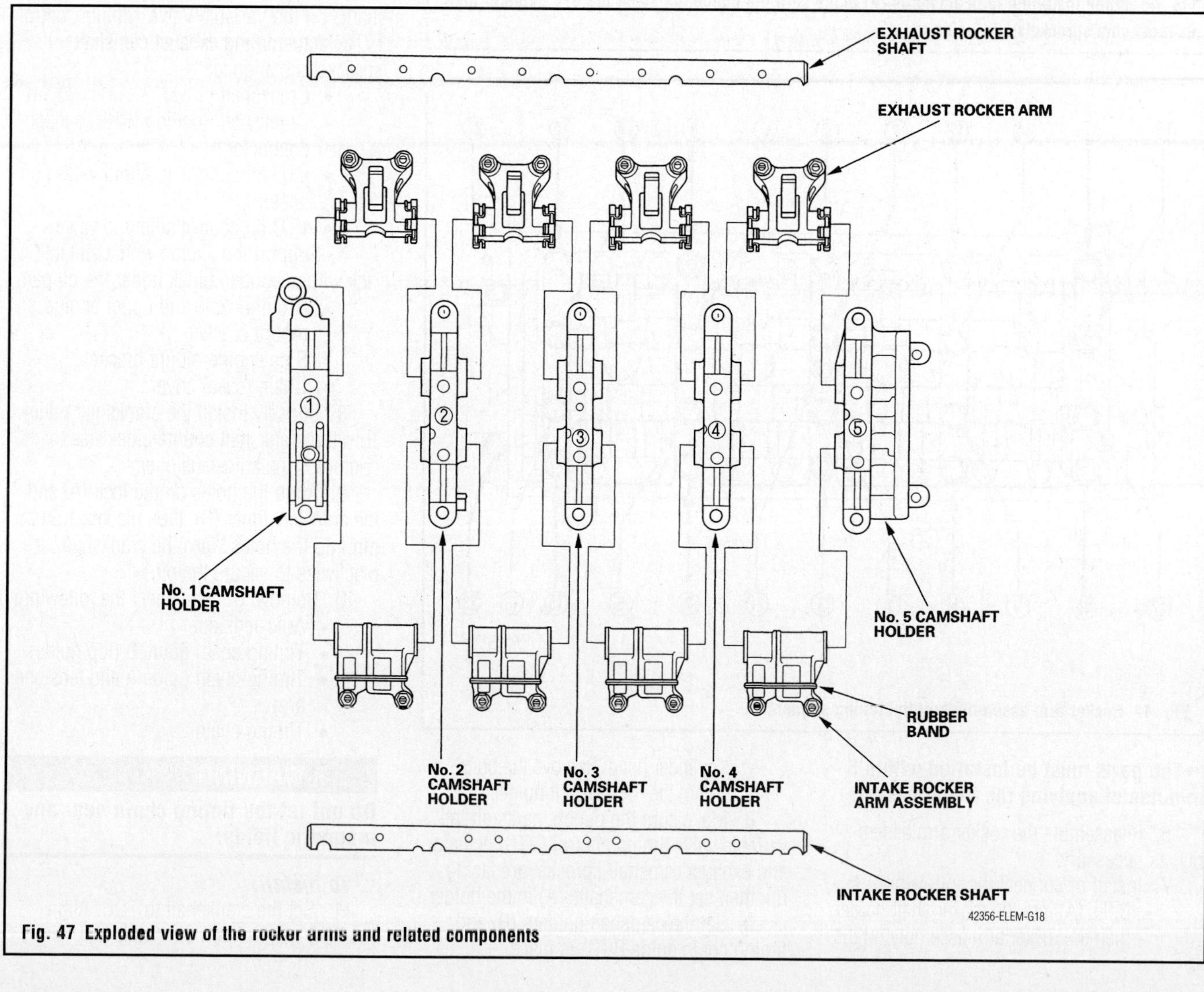

EXHAUST ROCKER SHAFT
EXHAUST ROCKER ARM
No. 1 CAMSHAFT HOLDER
No. 5 CAMSHAFT HOLDER
RUBBER BAND
No. 2 CAMSHAFT HOLDER
No. 3 CAMSHAFT HOLDER
No. 4 CAMSHAFT HOLDER
INTAKE ROCKER ARM ASSEMBLY
INTAKE ROCKER SHAFT
42356-ELEM-G18

Fig. 47 Exploded view of the rocker arms and related components

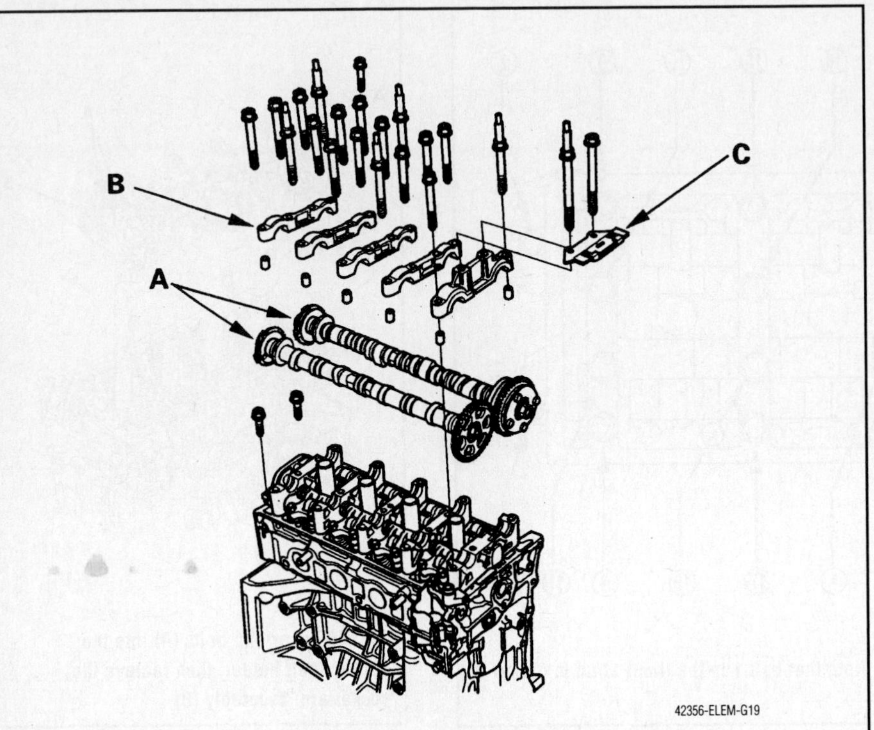
B
C
A
42356-ELEM-G19

Fig. 48 When installing the camshafts (A) make sure the punch marks on the VTC actuator and exhaust cam sprockets are facing up

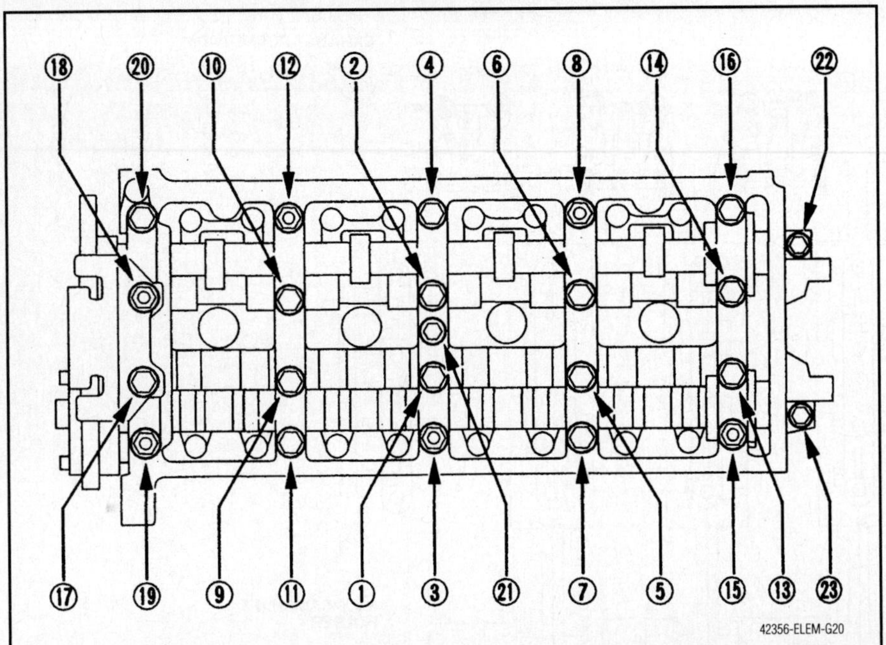
18 20 10 12 2 4 6 8 14 16 22
17 19 9 11 1 3 21 7 5 15 13 23
42356-ELEM-G20

Fig. 49 Rocker arm assembly bolt tightening sequence

➡The parts must be installed within 5 minutes of applying the liquid gasket.

6. Reassemble the rocker arm assembly, as necessary.
7. Install or connect the following
   - Bolts (A) into the rocker shaft holder, then the rocker arm assembly on the cylinder head. Remove the bolts from the rocker shaft holder.
8. Make sure the punch marks on the variable valve timing control (VTC) actuator and exhaust camshaft sprocket are facing up, then set the camshafts (A) in the holder
9. Set the camshaft holders (B) and timing chain guide B (C) in place.
10. Tighten the bolts, in sequence, to the following specification:
    a. 8mm bolts: 16 ft. lbs. (22 Nm).
    b. 6mm bolts: 8.7 ft. lbs. (12 Nm).

    The 6mm bolts are 21, 22 and 23.
11. Install the timing chain and adjust the valve lash.

TIMING CHAIN & FRONT SEAL

REMOVAL & INSTALLATION

See Figures 50 through 58.

1. Before servicing the vehicle, refer to the precautions section.
2. Set the engine to Top Dead Center (TDC).
3. Drain the cooling system.
4. Relieve the fuel system pressure.
5. Remove or disconnect the following:
   - Negative battery cable
   - Front tires and wheels
   - Splash shield
   - Drive belt
   - Cylinder head cover.
6. Check that the No. 1 piston TDC marks on the Variable Valve Timing Control (VTC) actuator and exhaust camshaft sprocket are aligned.
   - Crankshaft pulley
   - Crankshaft Position (CKP) sensor connector
   - VTC oil control solenoid valve connector
   - VTC oil control solenoid valve
7. Support the engine with a suitable jack with a wooden block under the oil pan.
   - Ground cable and upper engine mount bracket
   - Side engine mount bracket
   - Chain (case) cover
8. Loosely install the crankshaft pulley. Turn the crankshaft counterclockwise to compress the auto-tensioner.
9. Align the holes on the lock (A) and the auto-tensioner (B), then place a 1.5mm pin into the holes. Turn the crankshaft clockwise to secure the pin.
10. Remove or disconnect the following:
    - Auto-tensioner
    - Timing chain guide B (top guide)
    - Timing chain guide A and tensioner arm
    - Timing chain

⁂ WARNING

Do not let the timing chain near any magnetic fields.

To install:

11. Set the crankshaft to TDC. Align the TDC mark (A) on the crankshaft

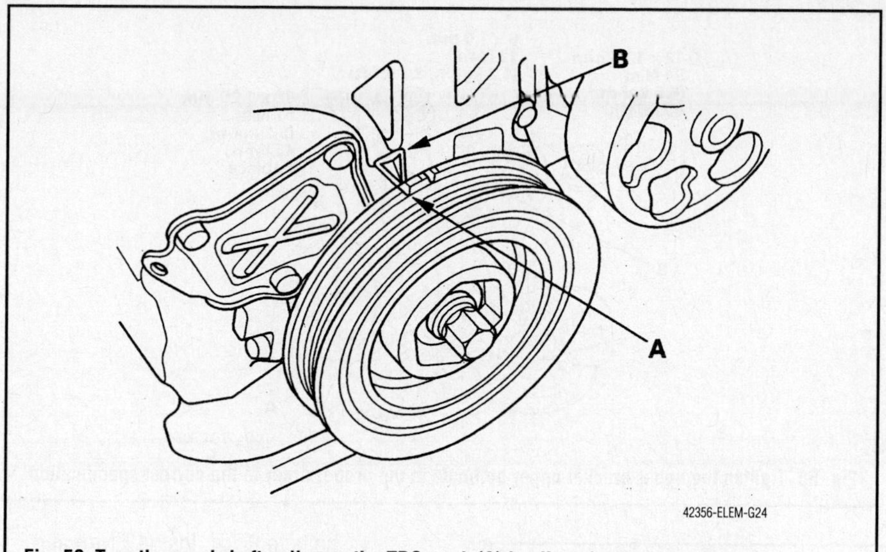
B
A
42356-ELEM-G24

Fig. 50 Turn the crankshaft pulley so the TDC mark (A) is aligned with the pointer (B)

sprocket with the pointer (B) on the cylinder block.

12. Set the camshafts to TDC. The punch mark (A) on the VTC actuator and the punch mark (B) on the exhaust camshaft (C) should be at the top. Align the TDC marks (C) on the VTC actuator and exhaust camshaft sprockets.

13. Install or connect the following:

- Timing chain on the crankshaft sprocket with the colored link of the chain aligned with the mark on the crank sprocket
- Timing chain on the VTC actuator and exhaust camshaft sprocket with the punch marks aligned with the center of the 2 colored links
- Timing chain guide A and tensioner arm. Tighten the guide

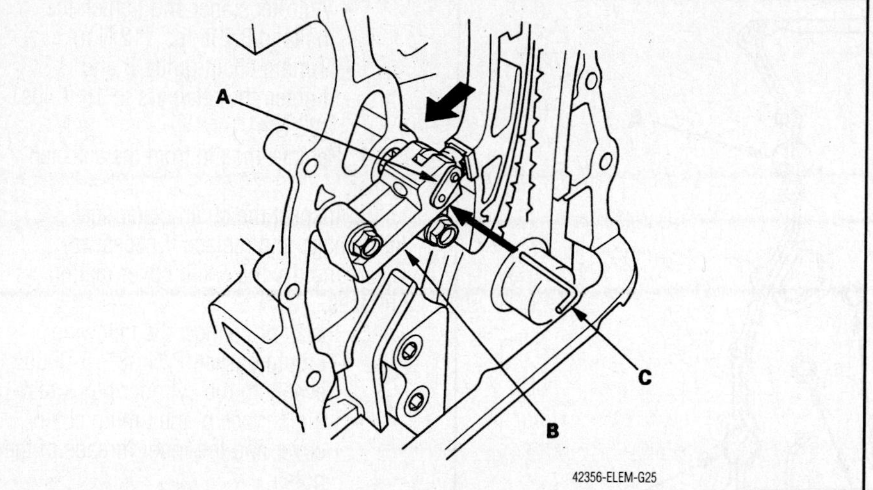
A
C
B
42356-ELEM-G25

Fig. 51 Align the holes on the lock (A) and the auto-tensioner (B), then place a 1.5mm pin into the holes. Turn the crankshaft clockwise to secure the pin

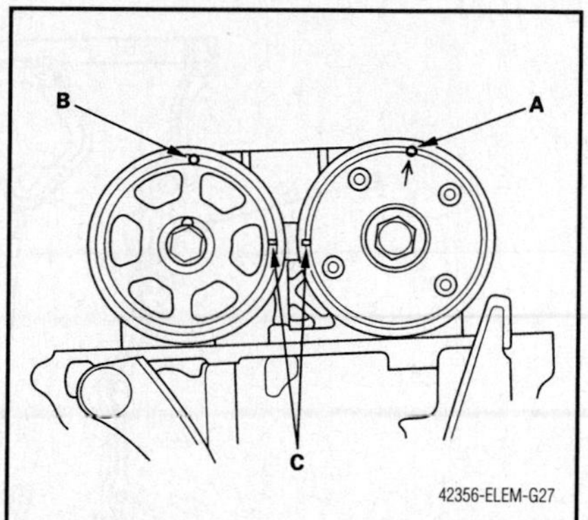
B
A
C
42356-ELEM-G27

Fig. 53 The mark (A) on the VTC actuator and the mark (B) on the exhaust cam (C) should be at the top. Align the TDC marks (C) on the VTC actuator and exhaust cam sprockets.

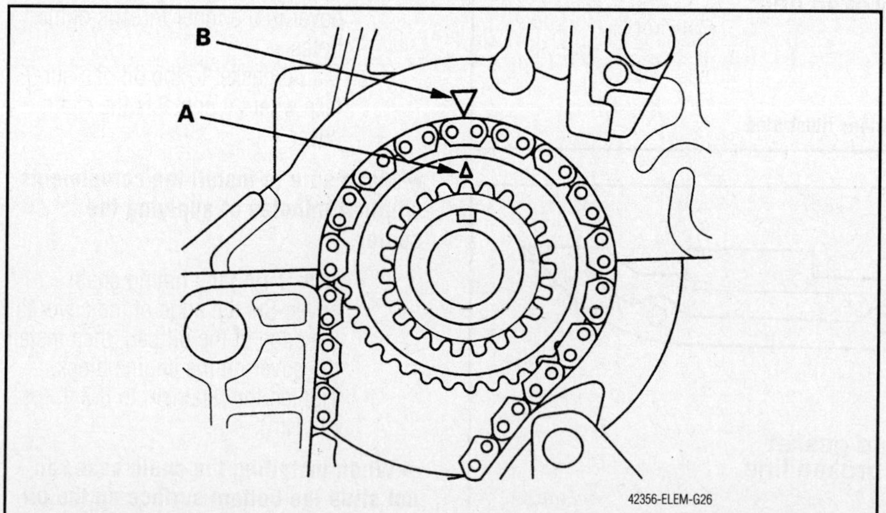
B
A
42356-ELEM-G26

Fig. 52 Set the crankshaft to TDC. Align the TDC mark (A) on the crankshaft sprocket with the pointer (D) on the cylinder block.

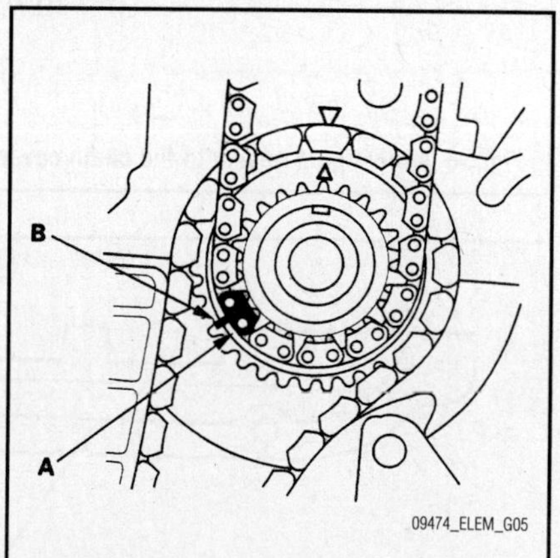
B
A
09474_ELEM_G05

Fig. 54 Install the timing chain on the crankshaft sprocket with the colored link of the chain aligned with the mark on the crank sprocket

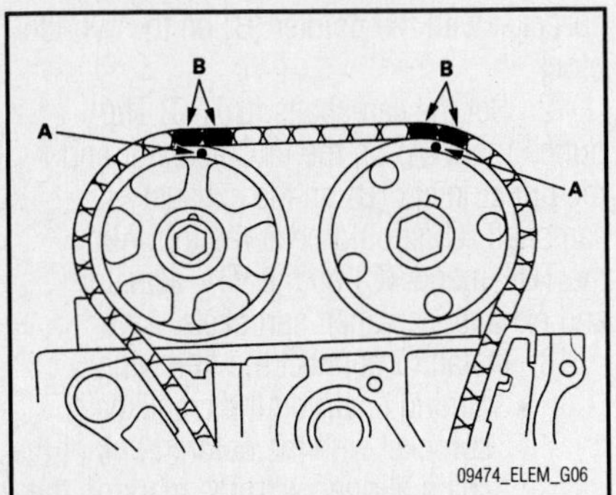
B
B
A
A
09474_ELEM_G06
Fig. 55 Install the timing chain on the VTC actuator and exhaust camshaft sprocket with the punch marks aligned with the center of the 2 colored links

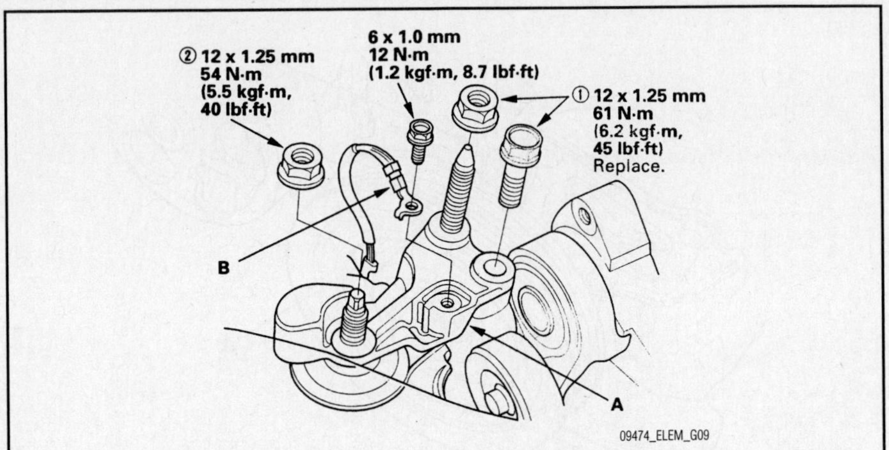
② 12 x 1.25 mm
54 N·m
(5.5 kgf·m,
40 lbf·ft)
6 x 1.0 mm
12 N·m
(1.2 kgf·m, 8.7 lbf·ft)
① 12 x 1.25 mm
61 N·m
(6.2 kgf·m,
45 lbf·ft)
Replace.
B
A
09474_ELEM_G09
Fig. 58 Tighten the upper bracket upper bolt/nuts in the proper order to the correct specification

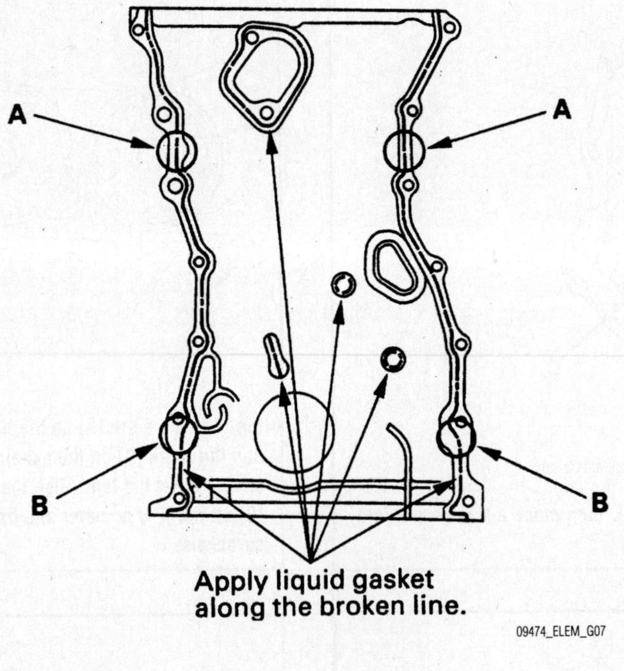
A
A
B
B
Apply liquid gasket
along the broken line.
09474_ELEM_G07
Fig. 56 Apply liquid gasket to the chain cover locations illustrated

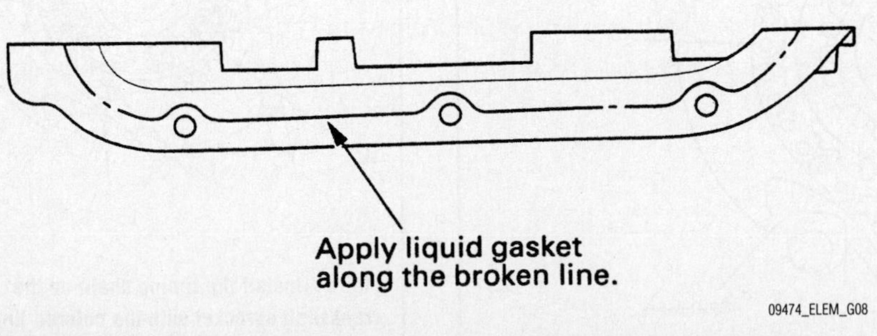
Apply liquid gasket
along the broken line.
09474_ELEM_G08
Fig. 57 Apply liquid gasket to the oil pan surface where it contacts the chain cover

bolts to 8.7 ft. lbs. (12 Nm) and the tensioner arm retainer to 16 ft. lbs. (22 Nm).

- Auto-tensioner and tighten the bolts to 8.7 ft. lbs. (12 Nm)
- Timing chain guide B and tighten the retainers to 16 ft. lbs. (22 Nm)

14. Remove the pin from the auto-tensioner.

15. Inspect the chain cover seal for damage and replace if necessary. Clean and dry the chain cover mating surfaces.

16. Install or connect the following:

- Liquid gasket, P/N 08718-0009 evenly to the cylinder block mating surface of the timing chain cover and the inner threads of the holes
- Liquid gasket to the cylinder block upper surface contact areas on the chain cover and the oil pan mating surface of the chain cover in the inner threads of the holes
- Liquid gasket to the oil pan surface where it contacts the chain cover

➡Make sure to install the components within 4 minutes of applying the sealer.

- New O-ring the timing chain cover. Set the edge of the cover to the edge of the oil pan, then install the cover on the engine block. Tighten the retainers to 8.7 ft. lbs. (12 Nm).

➡When installing the chain case, do not slide the bottom surface on the oil pan mounting surface.

- Side engine mounting bracket and

tighten the retainers to 33 ft. lbs. (44 Nm)
- Upper mount, then tighten the bolts/nuts as shown in the illustration
- Ground cable
- VTC oil control solenoid valve
- CKP sensor and VTC oil control solenoid valve connectors
- Crankshaft pulley. Tighten the bolt to 36 ft. lbs. (49 Nm), then tighten an additional 90 degrees.
- Cylinder head cover
- Drive belt
- Splash shield

17. Fill the engine cooling system and connect the negative battery cable.

VALVE (ROCKER ARM) COVERS

REMOVAL & INSTALLATION

1. Before servicing the vehicle, refer to the precautions section.
2. Disconnect the negative battery cable.
3. Remove the intake manifold cover.
4. Remove the four ignition coils.
5. Remove the two bolts securing the vacuum line.
6. Remove the bolt securing the power steering hose bracket.
7. Remove the dipstick and breather hose.
8. Remove the retainers and the cylinder head cover.
9. Installation is the reverse of the removal procedure.

VALVE LASH

ADJUSTMENT

See Figure 59.

Adjust the valves only when the cylinder head temperature is less than 100°F (38°C).

1. Before servicing the vehicle, refer to the precautions section.
2. Remove or disconnect the following:
   - Negative battery cable
   - Valve (rocker arm) cover
3. Set the timing marks as shown in the illustration with N0.1 at TDC. Check all clearances. Intake should be 0.008–0.010 in.; exhaust should be 0.011–0.013 in. Intake locknut torque is 14 ft. lbs. (19 Nm); exhaust is 10 ft. lbs. (14 Nm).
4. Rotate the crankshaft 180 degrees clockwise and recheck No.3.
5. Rotate the crankshaft 180 degrees clockwise and recheck No.4.
6. Rotate the crankshaft 180 degrees clockwise and recheck No.2.

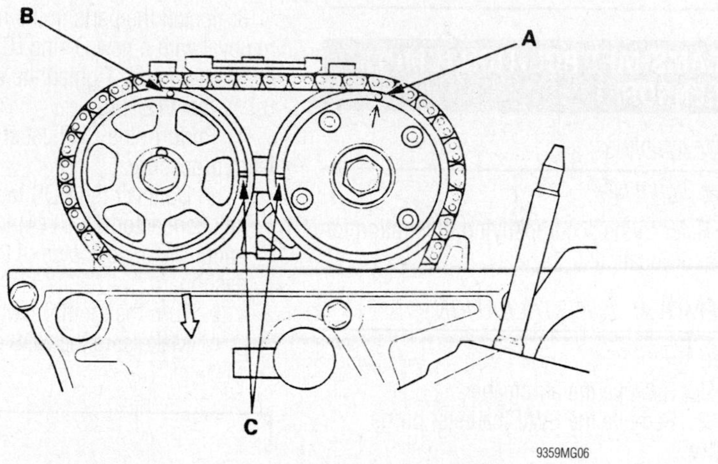
B
A
C
9359MG06
Fig. 59 Align the timing marks

ENGINE PERFORMANCE & EMISSION CONTROL

AIR FUEL (A/F) RATIO SENSOR

LOCATION

See Figure 60.

Refer to the accompanying illustration for sensor location.

REMOVAL & INSTALLATION

See Figure 61.

➡This procedure requires the use of O2 sensor socket wrench, Snap-on YA8875, SP Tools 93750, or equivalent, commercially available O2 sensor removal tool.

1. Disconnect the A/F sensor 4P connector (A), then remove the A/F sensor (B), using the proper tool.
2. Install the parts in the reverse order of removal. Tighten the sensor to 33 ft. lbs. (44 Nm).

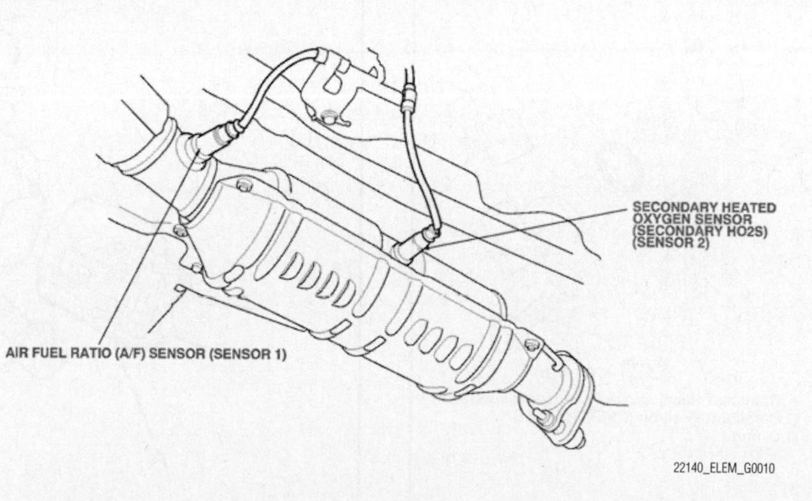
SECONDARY HEATED OXYGEN SENSOR (SECONDARY HO2S) (SENSOR 2)
AIR FUEL RATIO (A/F) SENSOR (SENSOR 1)
22140_ELEM_G0010
Fig. 60 A/F ratio sensor and Secondary HO2S locations

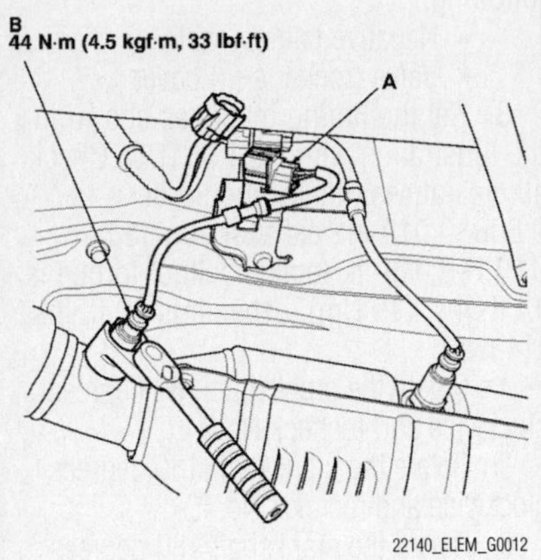
B
44 N·m (4.5 kgf·m, 33 lbf·ft)
A
22140_ELEM_G0012

Fig. 61 View of the A/F sensor connector (A) and sensor (B)

CAMSHAFT POSITION (CMP) SENSOR

LOCATION

See Figure 62.

Refer to the accompanying illustration for sensor location.

REMOVAL & INSTALLATION

See Figure 62.

1. Remove the air cleaner.
2. Remove the EVAP canister purge valve.
3. Disconnect the CMP sensor B connector (A).
4. Remove CMP sensor B (B).
5. Install the parts in the reverse order of removal with a new O-ring (C) coated with clean engine oil. Tighten the sensor bolt to 8.7 ft. lbs. (12 Nm).

CRANKSHAFT POSITION (CKP) SENSOR

LOCATION

See Figures 63 and 64.

Refer to the accompanying illustration for sensor location.

REMOVAL & INSTALLATION

See Figures 63 and 64.

1. Disconnect the CKP sensor connector (A).
2. Remove the CKP sensor (B).

To install:

3. Install the parts in the reverse order of removal with a new O-ring (C) coated with clean engine oil. Tighten the sensor bolt to 8.7 ft. lbs. (12 Nm).
4. Perform the CKP pattern clear/pattern learn procedure:
   a. Connect the HDS to the data link connector (DLC) (A) located under the driver's side of the dashboard.
   b. Turn the ignition switch ON (II).
   c. Make sure the HDS communicates with the ECM/PCM and other vehicle system.
   d. Select CRANK PATTERN in the ADJUSTMENT MENU with the HDS.
   e. Select CRANK PATTERN LEARNING with the HDS, and follow the screen prompts.

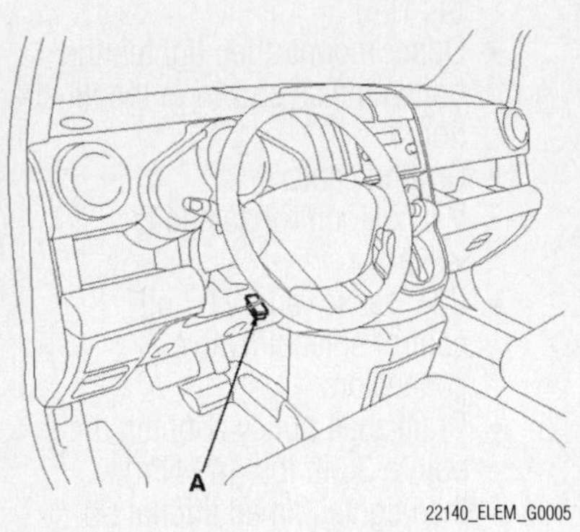
A
22140_ELEM_G0005

Fig. 64 Location of the HDS DLC (A)

ELECTRONIC CONTROL MODULE (ECM) POWERTRAIN CONTROL MODULE

LOCATION

See Figure 65.

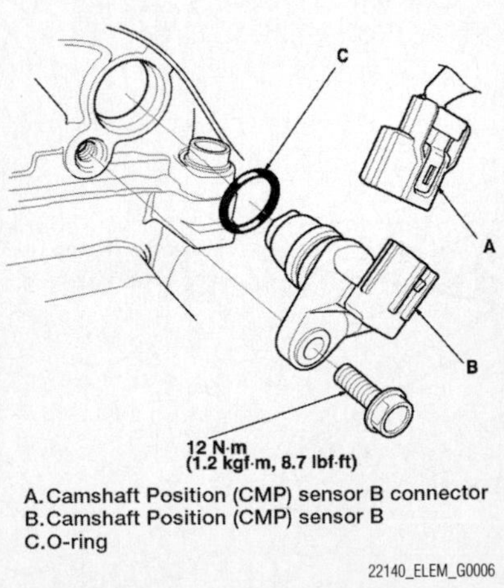
C
A
B
12 N·m
(1.2 kgf·m, 8.7 lbf·ft)
A. Camshaft Position (CMP) sensor B connector
B. Camshaft Position (CMP) sensor B
C. O-ring
22140_ELEM_G0006

Fig. 62 Exploded view of the CMP sensor B connector (A) and sensor (B)

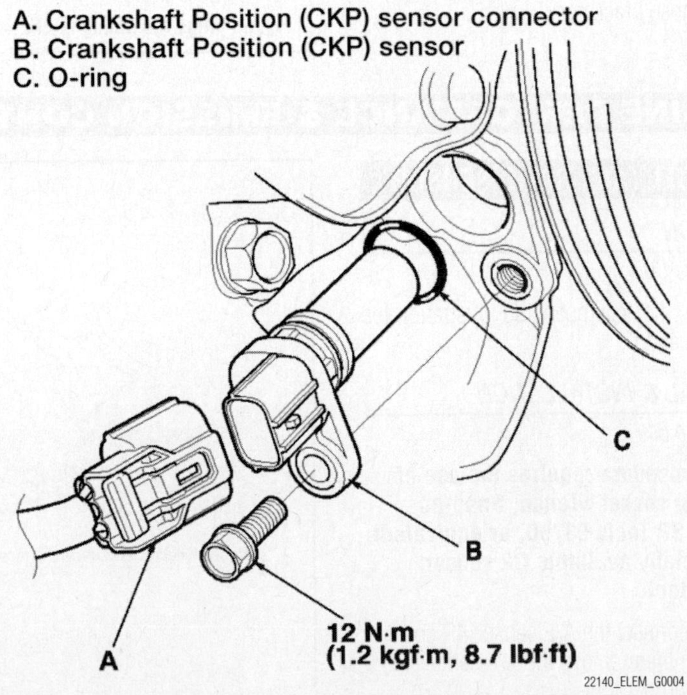
A. Crankshaft Position (CKP) sensor connector
B. Crankshaft Position (CKP) sensor
C. O-ring
C
B
A
12 N·m
(1.2 kgf·m, 8.7 lbf·ft)
22140_ELEM_G0004

Fig. 63 View of the CKP sensor connector (A) and sensor (B)

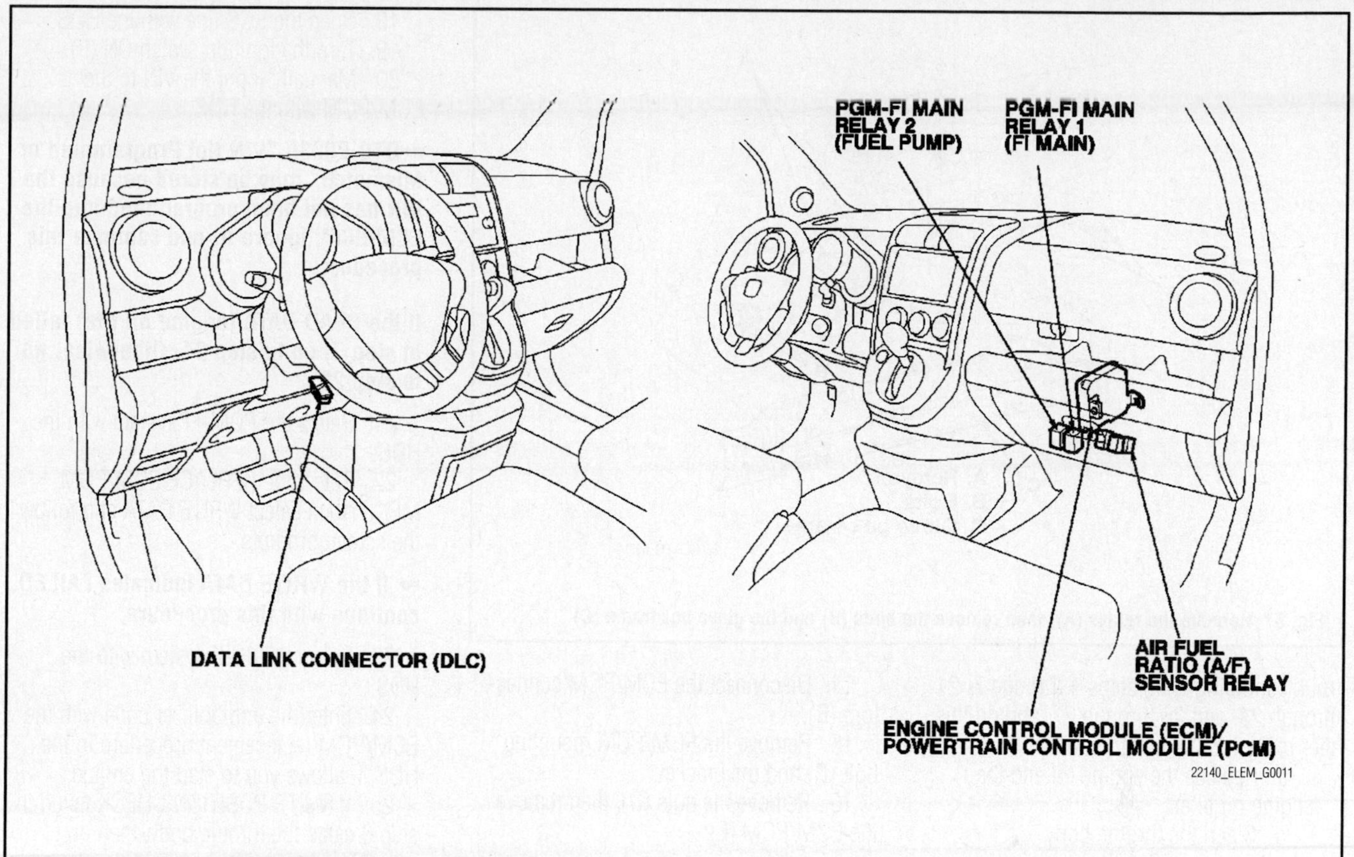
PGM-FI MAIN RELAY 2 (FUEL PUMP)
PGM-FI MAIN RELAY 1 (FI MAIN)
DATA LINK CONNECTOR (DLC)
AIR FUEL RATIO (A/F) SENSOR RELAY
ENGINE CONTROL MODULE (ECM)/ POWERTRAIN CONTROL MODULE (PCM)
22140_ELEM_G0011

Fig. 65 Location of the ECM/PCM and related components

➡On manual transaxle equipped vehicles, it is referred to as the Engine Control Module (ECM). On automatic transaxle equipped vehicles, it is referred to as the Powertrain Control Module (PCM).

Refer to the accompanying illustration for ECM/PCM location.

REMOVAL & INSTALLATION

See Figures 64, 66 through 68.

➡On manual transaxle equipped vehicles, it is referred to as the Engine Control Module (ECM). On automatic transaxle equipped vehicles, it is referred to as the Powertrain Control Module (PCM).

This procedure requires the following special tools (or their equivalents):

- Honda diagnostics system (HDS) tablet tester
- Honda interface module (HIM) and an iN workstation with HDS and CM update software
- HDS pocket tester
- GNA 600 and an iN workstation with HDS and CM update software

➡Make sure the HDS is loaded with the latest software version.

- If you are replacing the ECM/PCM after substituting a known-good ECM/PCM, reinstall the original ECM/PCM, then do this procedure.
- During the procedure, if any READ DATA, WRITE DATA, or other data checks fail, note the failure, then continue.

1. Connect the HDS to the data link connector (DLC) (A) located under the driver's side of the dashboard.
2. Turn the ignition switch ON (II).
3. Make sure the HDS communicates with the ECM/PCM and other vehicle system. If you are returning from DLC circuit

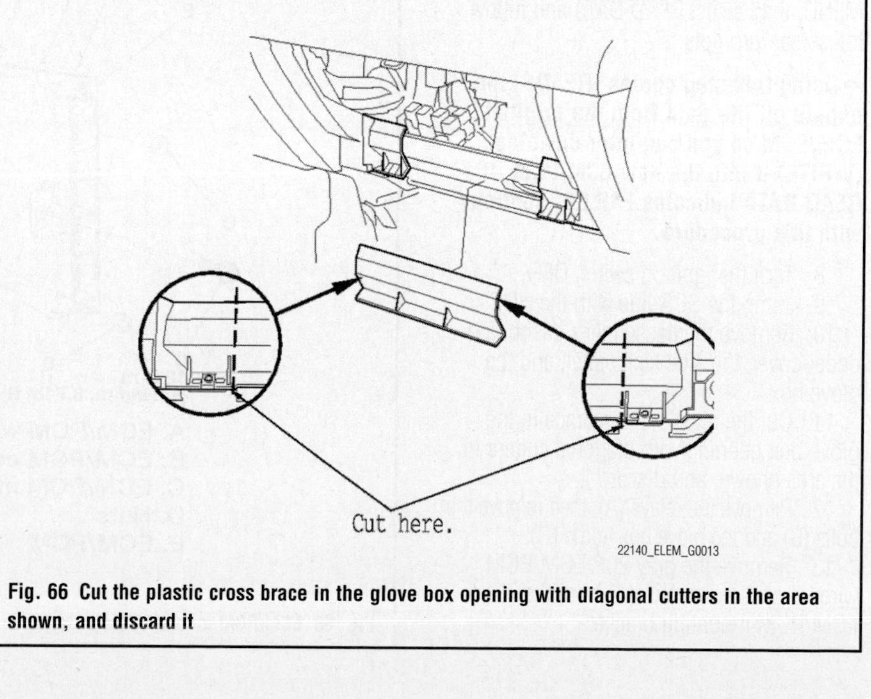
Cut here.
22140_ELEM_G0013

Fig. 66 Cut the plastic cross brace in the glove box opening with diagonal cutters in the area shown, and discard it

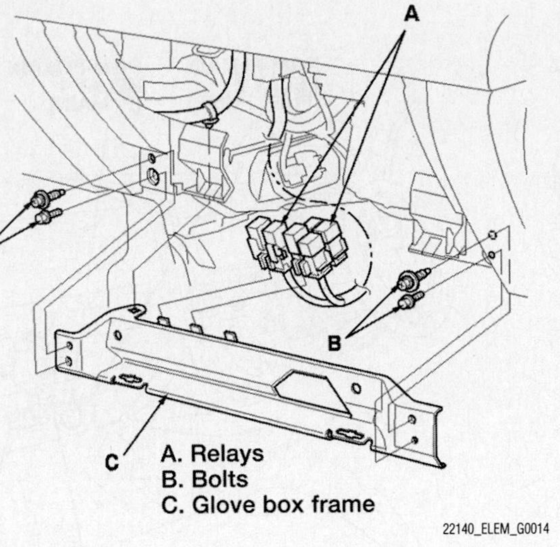
A
B
B
C
A. Relays
B. Bolts
C. Glove box frame
22140_ELEM_G0014

Fig. 67 Remove the relays (A), then remove the bolts (B) and the glove box frame (C)

troubleshooting, skip steps 4 through 7, 21 through 23, and 26 through 27, and do this after replacing the ECM/PCM:

a. Replace the engine oil and the engine oil filter.

b. Clean the throttle body.

4. Select the PGM-FI system with the HDS.

5. Select the INSPECTION MENU with the HDS.

6. Select the ETCS TEST, then select the TP POSITION CHECK, and follow the screen prompts.

➡If the TP POSITION CHECK indicates FAILED, continue with this procedure.

7. Select the REPLACE ECM/PCM MENU, then select READ DATA and follow the screen prompts.

➡Doing this step copies (READS) the engine oil life data from the original ECM/PCM so you can later download (WRITE) it into the new ECM/PCM. If READ DATA indicates FAILED, continue with this procedure.

8. Turn the ignition switch OFF.

9. Jump the SCS line with the HDS.

10. Remove the passenger's dashboard under cover, the side kick panel, and the glove box.

11. Cut the plastic cross brace in the glove box opening with diagonal cutters in the area shown, and discard it.

12. Remove the relays (A), then remove the bolts (B) and the glove box frame (C).

13. Remove the gray 20P ECM/PCM wire harness connector (A) from the ECM/PCM mounting bracket.

14. Disconnect the ECM/PCM connectors (B).

15. Remove the ECM/PCM mounting bolt (C) and the bracket.

16. Remove the nuts (D), then remove the ECM/PCM (E).

To install:

17. Install the parts in the reverse order of removal.

18. Open the SCS line with the HDS.

19. Turn the ignition switch ON (II).

20. Manually input the VIN to the ECM/PCM with the HDS.

➡DTC P0630 "VIN Not Programmed or Mismatch" may be stored because the VIN has not been programmed into the ECM/PCM; ignore it, and continue this procedure.

If the READ DATA (engine oil life) failed in step 7, go to step 24. Otherwise, go to step 22.

21. Select the PGM-FI system with the HDS.

22. Select the REPLACE ECM/PCM MENU, then select WRITE DATA and follow the screen prompts.

➡ If the WRITE DATA indicates FAILED, continue with this procedure.

23. Select IMMOBI system with the HDS.

24. Enter the immobilizer code with the ECM/PCM replacement procedure in the HDS; it allows you to start the engine.

25. If the TP POSITION CHECK failed in step 6 clean the throttle body, then go to step 27.

26. If the READ DATA failed in step 7 or the WRITE DATA failed in step 23, replace

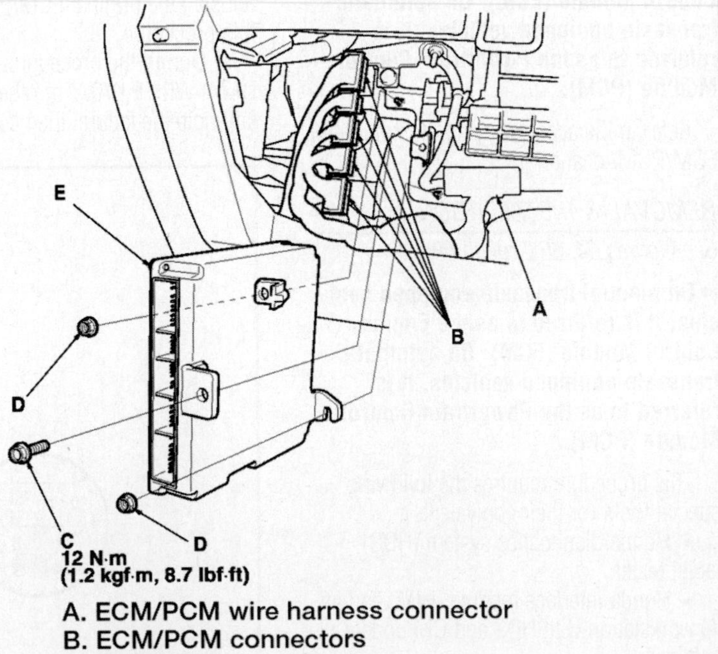
E
A
B
D
C
12 N·m
(1.2 kgf·m, 8.7 lbf·ft)
D
A. ECM/PCM wire harness connector
B. ECM/PCM connectors
C. ECM/PCM mounting bolt
D. Nuts
E. ECM/PCM
22140_ELEM_G0015

Fig. 68 ECM/PCM mounting (refer to procedure for component identification)

the engine oil and engine oil filter, then go to step 28.

27. Select PGM-FI system, and reset the ECM/PCM with thc HDS.

28. Update the ECM/PCM if it does not have the latest software.

29. Perform the ECM/PCM idle learn procedure, as follows:

a. Make sure all electrical items (A/C, audio, lights, etc.) are off.

b. Reset the ECM/PCM with the HDS.

c. Turn the ignition switch ON (II), and wait 2 seconds.

d. Start the engine. Hold the engine speed at 3,000 rpm without load (in Park or neutral) until the radiator fan comes on, or until the engine coolant temperature reaches 194ºF (90ºC).

e. Let the engine idle for about 5 minutes with the throttle fully closed.

➡If the radiator fan comes on, do not include its running time in the 5 minutes

30. Perform the CKP pattern learn procedure, as follows:

a. Connect the HDS to the data link connector (DLC) (A) located under the driver's side of the dashboard.

b. Turn the ignition switch ON (II).

c. Make sure the HDS communicates with the ECM/PCM and other vehicle system.

d. Select CRANK PATTERN in the ADJUSTMENT MENU with the HDS.

e. Select CRANK PATTERN LEARNING with the HDS, and follow the screen prompts.

ENGINE COOLANT TEMPERATURE (ECT) SENSOR

LOCATION

See Figures 69 and 70.

Refer to the accompanying illustration for sensor locations.

REMOVAL & INSTALLATION

Sensor 1

See Figure 69.

1. Remove the air cleaner.
2. Remove the EVAP canister purge valve.
3. Unbolt the under-hood fuse/relay box bolt, and move the assembly aside.
4. Drain the engine coolant.
5. Disconnect the ECT sensor 1 connector (A).
6. Remove the ECT sensor 1 (B).
7. Install the parts in the reverse order of removal with a new O-ring (C) coated with clean engine oil, then refill the radiator with engine coolant.

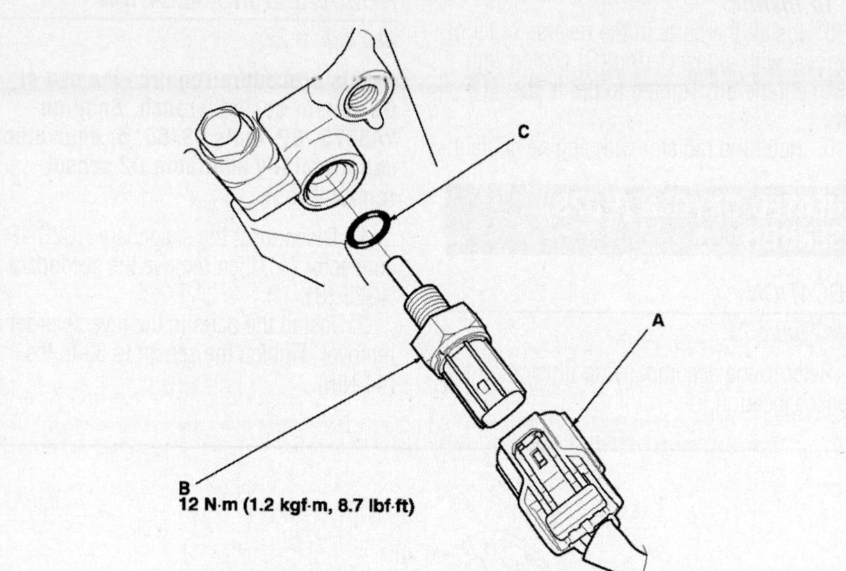
C
A
B
12 N·m (1.2 kgf·m, 8.7 lbf·ft)
A. Engine Coolant Temperature (ECT) sensor 1 connector
B. Engine Coolant Temperature (ECT) 1 sensor
C. O-ring
22140_ELEM_G0007

Fig. 69 View of the ECT sensor 1 connector (A) and sensor (B). Use a new O-ring (C) coated with clean engine oil during installation

To install:

8. Install the parts in the reverse order of removal with a new O-ring (C) coated with clean engine oil. Tighten to 8.7 ft. lbs. (12 Nm).
9. Refill the radiator with engine coolant.

Sensor 2

See Figure 70.

1. Drain the engine coolant.
2. Remove the splash shield.
3. Disconnect the ECT sensor 2 connector (A).
4. Remove ECT sensor 2 (B).

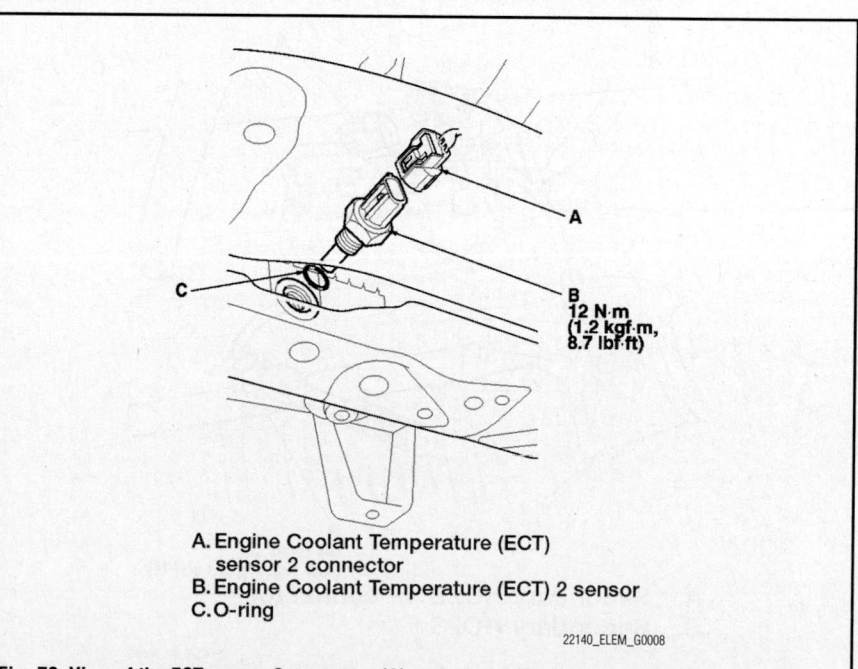

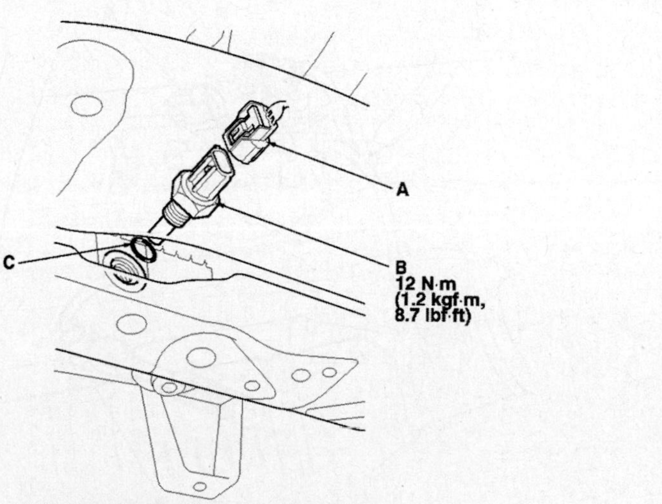
A
C
B
12 N·m
(1.2 kgf·m,
8.7 lbf·ft)
A. Engine Coolant Temperature (ECT) sensor 2 connector
B. Engine Coolant Temperature (ECT) 2 sensor
C. O-ring
22140_ELEM_G0008

Fig. 70 View of the ECT sensor 2 connector (A) and sensor (B). Use a new O-ring (C) coated with clean engine oil during installation

To install:

5. Install the parts in the reverse order of removal with a new O-ring (C) coated with clean engine oil. Tighten to 8.7 ft. lbs. (12 Nm).
6. Refill the radiator with engine coolant.

HEATED OXYGEN (HO2S) SENSOR

LOCATION

See Figure 71.

Refer to the accompanying illustration for sensor location.

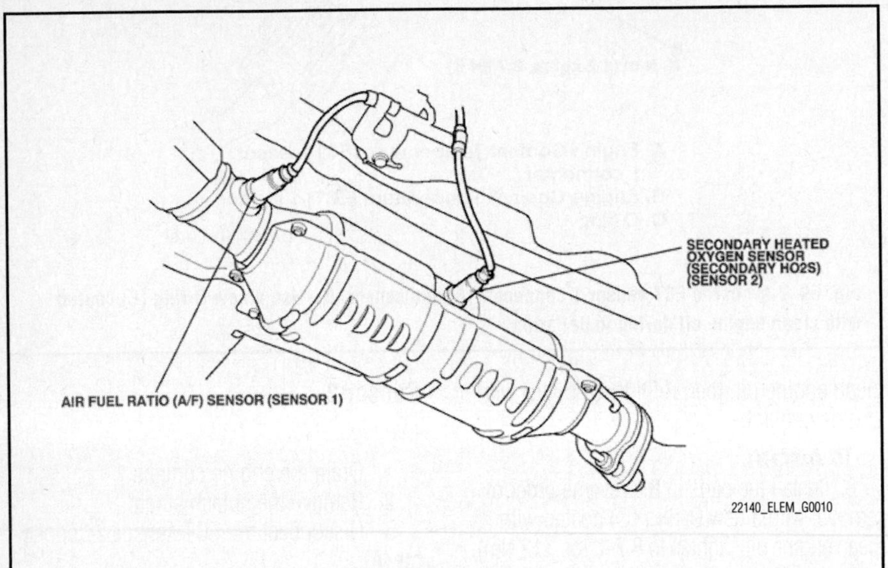
SECONDARY HEATED OXYGEN SENSOR (SECONDARY HO2S) (SENSOR 2)
AIR FUEL RATIO (A/F) SENSOR (SENSOR 1)
22140_ELEM_G0010

Fig. 71 A/F ratio sensor and Secondary HO2S locations

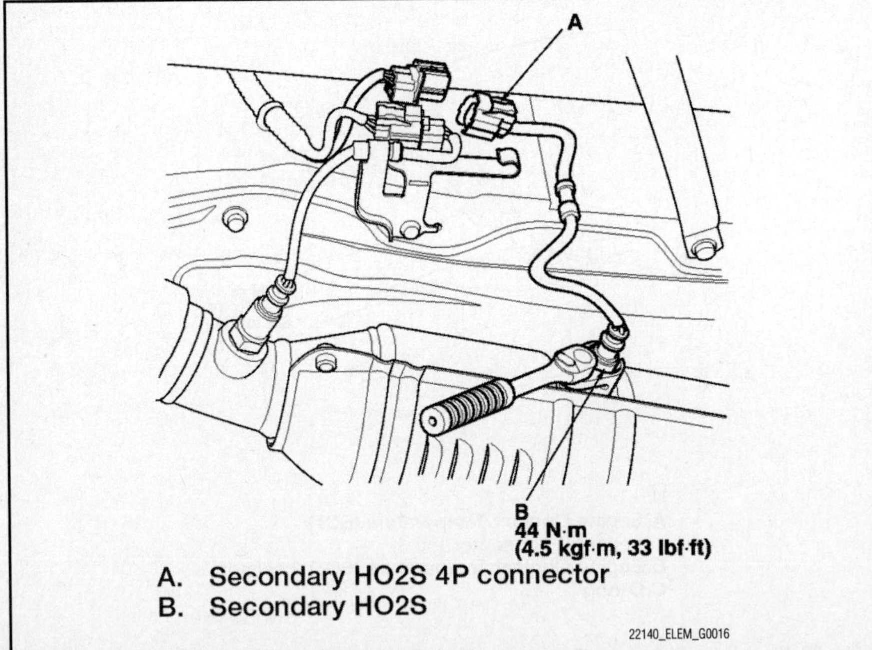
A
B
44 N·m
(4.5 kgf·m, 33 lbf·ft)
A. Secondary HO2S 4P connector
B. Secondary HO2S
22140_ELEM_G0016

Fig. 72 Disconnect the secondary HO2S 4P connector (A), then remove the secondary HO2S (B)

REMOVAL & INSTALLATION

See Figure 72.

➡This procedure requires the use of O2 sensor socket wrench, Snap-on YA8875, SP Tools 93750, or equivalent, commercially available O2 sensor removal tool.

1. Disconnect the secondary HO2S 4P connector (A), then remove the secondary HO2S (B).
2. Install the parts in the reverse order of removal. Tighten the sensor to 33 ft. lbs. (44 Nm).

INTAKE AIR TEMPERATURE (IAT) SENSOR

LOCATION

Refer to Mass Air Flow (MAF) sensor.

REMOVAL & INSTALLATION

Refer to Mass Air Flow (MAF) sensor.

KNOCK SENSOR (KS)

LOCATION

See Figure 73.

Refer to the accompanying illustration for sensor location.

REMOVAL & INSTALLATION

See Figures 73.

1. Disconnect the Knock Sensor (KS) connector (A).
2. Remove the KS (B).
3. Install the parts in the reverse order of removal. Tighten the KS to 23 ft. lbs. (32 Nm).

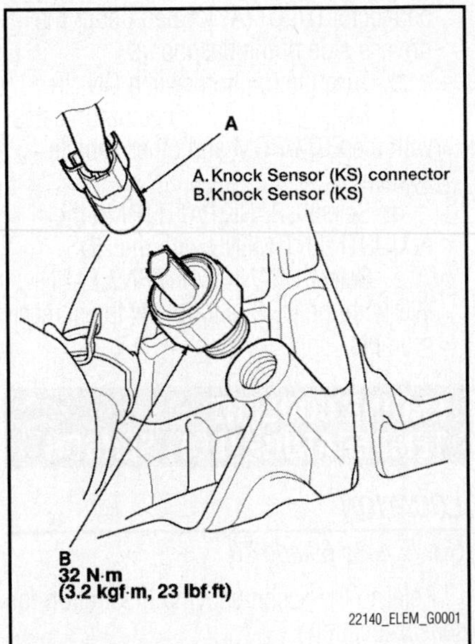
A
A. Knock Sensor (KS) connector
B. Knock Sensor (KS)
B
32 N·m
(3.2 kgf·m, 23 lbf·ft)
22140_ELEM_G0001

Fig. 73 View of the Knock Sensor (KS) connector (A) and KS (B)

MANIFOLD ABSOLUTE PRESSURE (MAP) SENSOR

LOCATION

See Figure 74.

Refer to the accompanying illustration for sensor location.

REMOVAL & INSTALLATION

See Figure 74.

1. Disconnect the MAP sensor connector (A).
2. Remove the MAP sensor (B).
3. Install the parts in the reverse order of removal with a new O-ring (C) coated with clean engine oil. Tighten the sensor retainer to 2.4 ft. lbs. (3.4 Nm).

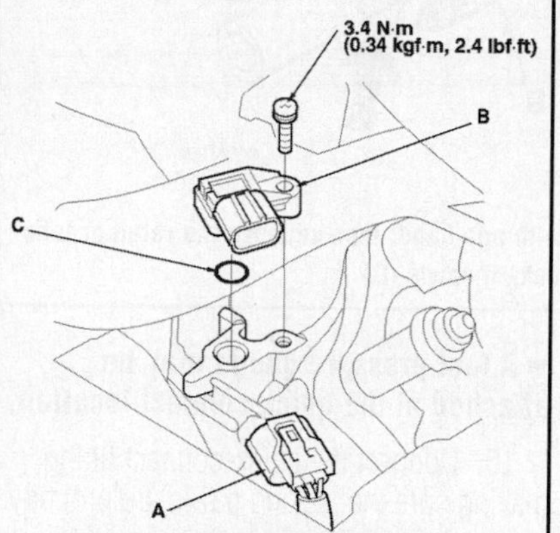
3.4 N·m
(0.34 kgf·m, 2.4 lbf·ft)
B
C
A

A. Manifold Absolute Pressure (MAP) sensor connector
B. Manifold Absolute Pressure (MAP) sensor
C. O-ring

22140_ELEM_G0003

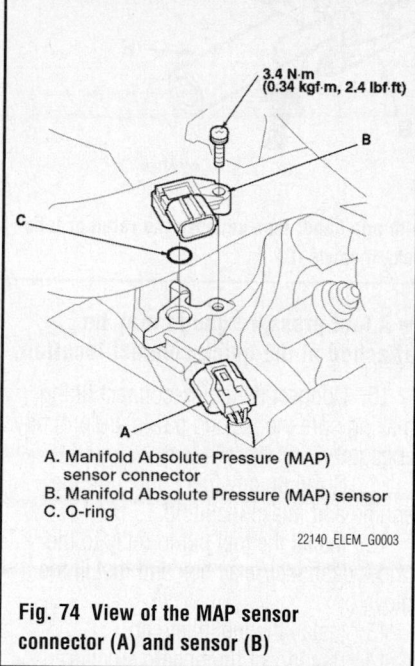

Fig. 74 View of the MAP sensor connector (A) and sensor (B)

MASS AIR FLOW (MAF) SENSOR

LOCATION

See Figure 75.

Refer to the accompanying illustration for sensor location.

REMOVAL & INSTALLATION

See Figure 75.

➡The Intake Air Temperature Sensor is integrated with the Mass Air Flow (MAF) sensor.

1. Disconnect the MAF sensor/IAT sensor connector (A).
2. Remove the bolts (B).
3. Remove the MAF sensor/IAT sensor (C).
4. Install the parts in the reverse order of removal with a new O-ring (D) coated with clean engine oil.

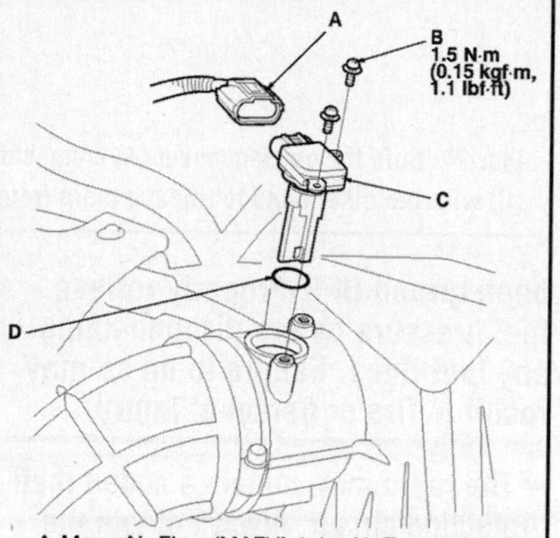
A
B
1.5 N·m
(0.15 kgf·m,
1.1 lbf·ft)
C
D

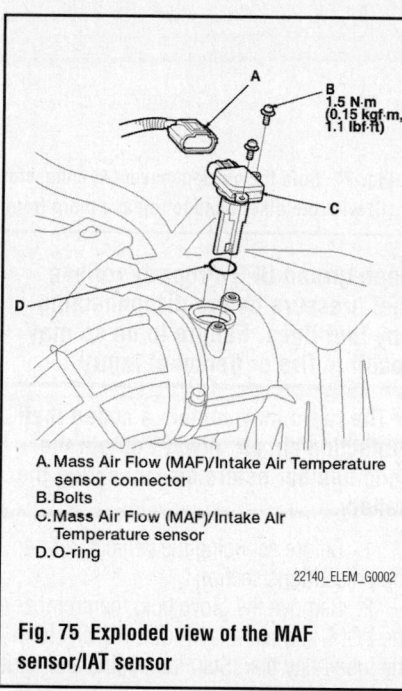

A. Mass Air Flow (MAF)/Intake Air Temperature sensor connector
B. Bolts
C. Mass Air Flow (MAF)/Intake Air Temperature sensor
D. O-ring

22140_ELEM_G0002

Fig. 75 Exploded view of the MAF sensor/IAT sensor

OUTPUT SHAFT SPEED (OSS) SENSOR

LOCATION

With Manual Transaxle

See Figure 76.

Refer to the accompanying illustration sensor location.

REMOVAL & INSTALLATION

With Manual Transaxle

See Figure 76.

1. Remove the air cleaner.
2. Disconnect the output shaft (countershaft) speed sensor connector (A).
3. Remove the output shaft (countershaft) speed sensor (B).
4. Install the parts in the reverse order of removal with a new O-ring (C) coated with clean engine oil. Tighten the sensor retainer to 8.7 ft. lbs. (12 Nm).

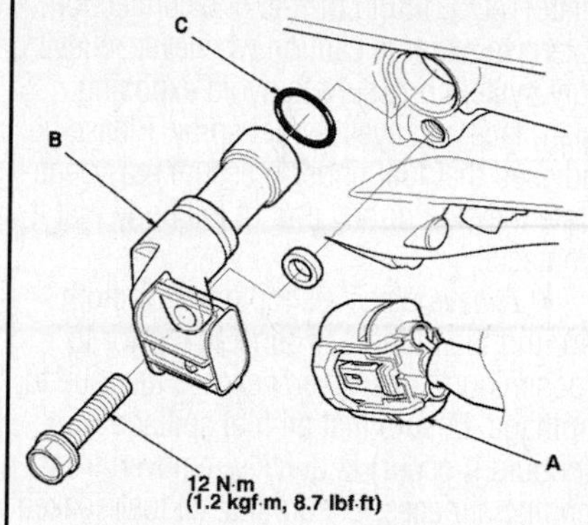
C
B
A
12 N·m
(1.2 kgf·m, 8.7 lbf·ft)

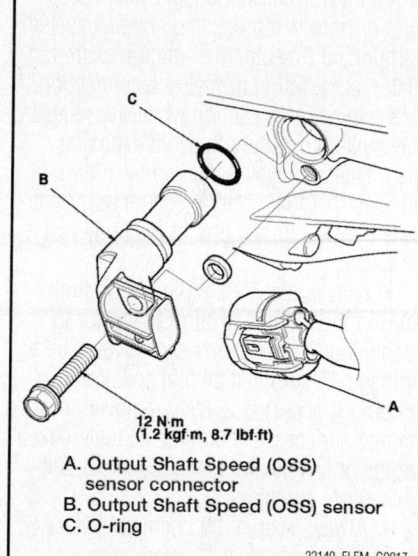

A. Output Shaft Speed (OSS) sensor connector
B. Output Shaft Speed (OSS) sensor
C. O-ring

22140_ELEM_G0017

Fig. 76 View of the Output Shaft Speed (OSS) connector (A), sensor (B) and O-ring (C)

FUEL SYSTEM SERVICE PRECAUTIONS

Safety is the most important factor when performing not only fuel system maintenance but any type of maintenance. Failure to conduct maintenance and repairs in a safe manner may result in serious personal injury or death. Maintenance and testing of the vehicle's fuel system components can be accomplished safely and effectively by adhering to the following rules and guidelines.

- To avoid the possibility of fire and personal injury, always disconnect the negative battery cable unless the repair or test procedure requires that battery voltage be applied.
- Always relieve the fuel system pressure prior to disconnecting any fuel system component (injector, fuel rail, pressure regulator, etc.), fitting or fuel line connection. Exercise extreme caution whenever relieving fuel system pressure to avoid exposing skin, face and eyes to fuel spray. Please be advised that fuel under pressure may penetrate the skin or any part of the body that it contacts.
- Always place a shop towel or cloth around the fitting or connection prior to loosening to absorb any excess fuel due to spillage. Ensure that all fuel spillage (should it occur) is quickly removed from engine surfaces. Ensure that all fuel soaked cloths or towels are deposited into a suitable waste container.
- Always keep a dry chemical (Class B) fire extinguisher near the work area.
- Do not allow fuel spray or fuel vapors to come into contact with a spark or open flame.
- Always use a back-up wrench when loosening and tightening fuel line connection fittings. This will prevent unnecessary stress and torsion to fuel line piping.
- Always replace worn fuel fitting O-rings with new Do not substitute fuel hose or equivalent where fuel pipe is installed.

Before servicing the vehicle, make sure to also refer to the precautions in the beginning of this section as well.

RELIEVING FUEL SYSTEM PRESSURE

See Figure 77.

⁂ CAUTION

The fuel injection system remains under pressure after the engine has been turned OFF. Properly relieve fuel pressure before disconnecting any fuel lines. Failure to do so may result in fire or personal injury.

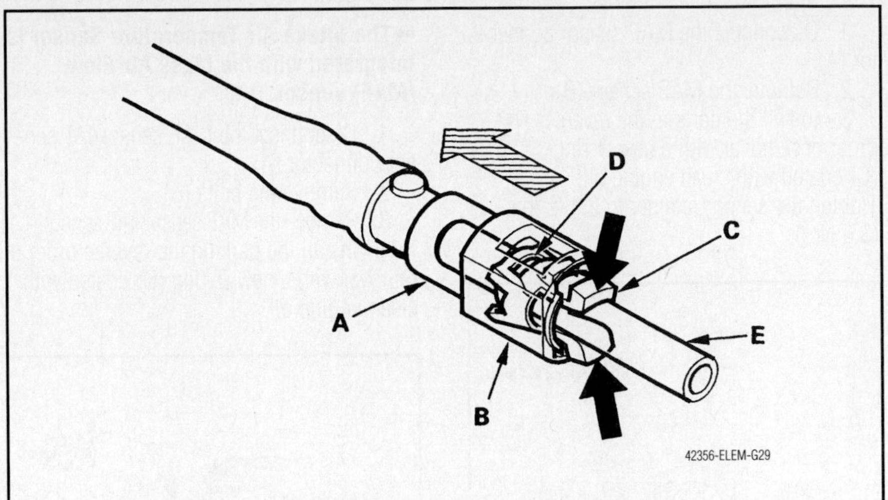

D
C
A
E
B
42356-ELEM-G29

Fig. 77 Hold the quick-connect (A) connector (B) with one hand, then squeeze the retainer tabs (C) with the other hand to release them from the locking pawls (D)

➡The radio may contain a coded theft protection circuit. Always obtain the code number before disconnecting the battery.

1. Before servicing the vehicle, refer to the precautions section.
2. Remove the glove box, then remove the PGM-FI main relay (FUEL PUMP) from the fuse/relay box. Start the engine and let it run until it stalls.
3. Turn the engine OFF.
4. Disconnect the negative battery cable.
5. Remove the fuel filler cap.
6. Remove the quick-connect fitting cover.
7. Clean any dirt from the quick-connect fitting.
8. Place a rag or shop towel over quick-connect fitting.
9. Detach the quick-connect fitting by holding the connector with one hand, then squeeze the retainer tabs with the other hand to release them from the locking pawls. Pull the connector off.

⁂ CAUTION

Do not allow fuel spray or fuel vapors to come in contact with a spark or open flame. Keep a dry chemical fire extinguisher nearby. Never store fuel in an open container due to risk of fire or explosion.

➡A fuel pressure gauge may be attached at the quick-connect location.

10. Connect the quick-connect fitting, making sure the locking pawls are properly engaged,
11. Clean up any fuel spilled on the engine and intake manifold.
12. Install the fuel pump relay to the under dash fuel/relay box and install the glove box.
13. Install the fuel filler cap.
14. Reconnect the negative battery cable.
15. Turn the ignition **ON**, but don't start the engine. Repeat this 2 or 3 times to pressurize the fuel system. Check for fuel leaks.
16. Enter the radio security code.

FUEL FILTER

REMOVAL & INSTALLATION

See Figure 78.

➡The fuel filter should be replaced whenever the fuel pressure drops below 48 psi, after making sure that the fuel pump and fuel pressure regulator are okay.

1. Before servicing the vehicle, refer to the precautions section.
2. Relieve the fuel system pressure.
3. Remove or disconnect the following:
   - Negative battery cable
   - Fuel pump
   - Fuel filter carrier (A)
   - Fuel filter

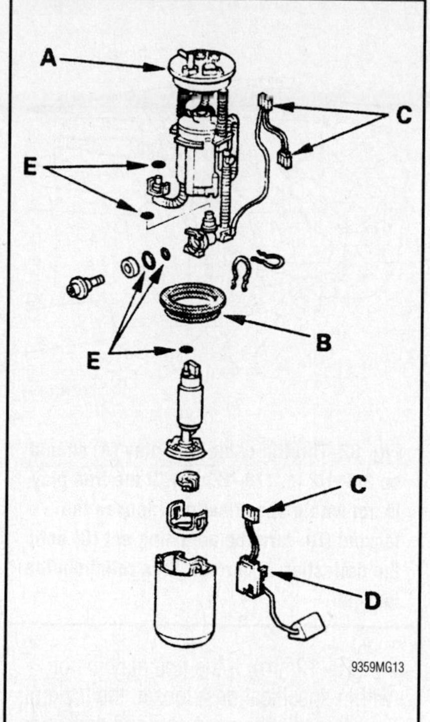
A
C
E
E
B
C
D
9359MG13

Fig. 78 Exploded view of the fuel filter mounting

To install:

4. Install or connect the following:
   - Fuel filter
   - Fuel lines
   - New gasket (B)
   - New o-rings (E)
   - Connectors (C)
   - Sending unit (D)
5. Start the engine and check for leaks.

FUEL PUMP

REMOVAL & INSTALLATION

1. Before servicing the vehicle, refer to the precautions section.
2. Relieve the fuel system pressure.
3. Remove or disconnect the following:
   - Negative battery cable
   - Fuel filler cap
   - Center console, then both track floor covers and sill trims.
4. Fold back the floor covering until you can get to the access panel
   - Access panel from the floor
   - Fuel pump connector
   - Fuel supply and return line quick-connect fittings
   - Fuel pump locknut, using special tool No. 07XAA-001010A
   - Fuel pump/sending unit assembly
5. Installation is the reverse of removal.

FUEL RAIL & INJECTORS

REMOVAL & INSTALLATION

See Figure 79.

1. Before servicing the vehicle, refer to the precautions section.
2. Relieve the fuel system pressure.
3. Remove or disconnect the following:
   - Negative battery cable
   - Engine cover
   - Injector connectors, ground cable and harness holder
   - Fuel line quick-connect fittings
   - Fuel rail mounting nuts
   - Injector clip(s) from the injector(s)
   - Fuel injectors from the fuel rail

To install:

4. Install or connect the following:
   - Injectors to the fuel rail with new O-rings coated with clean engine oil.
   - Injector clips
   - Injectors in the injector base
   - Fuel rail and injector assembly. Tighten the nuts to 16 ft. lbs. (22 Nm).
   - Ground cable bolt
   - Injector connectors
   - Fuel lines
   - Negative battery cable
5. Start the engine and check for leaks.

THROTTLE BODY

REMOVAL & INSTALLATION

See Figures 80 through 82.

1. Before servicing the vehicle, refer to the precautions section.
2. Relieve the fuel system pressure.
3. Disconnect the negative battery cable.
4. Remove the engine cover.

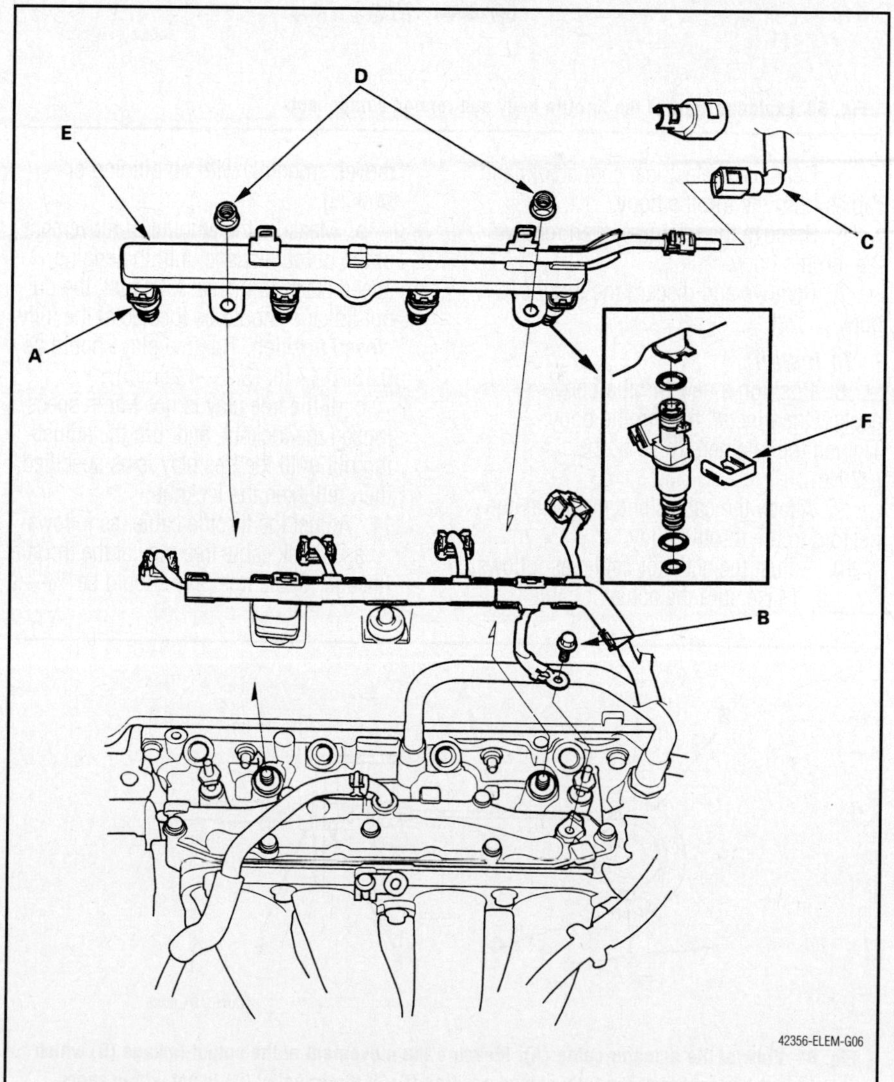
D
E
C
A
F
B
42356-ELEM-G06

Fig. 79 Exploded view of the fuel rail (F), injectors (A) and related components

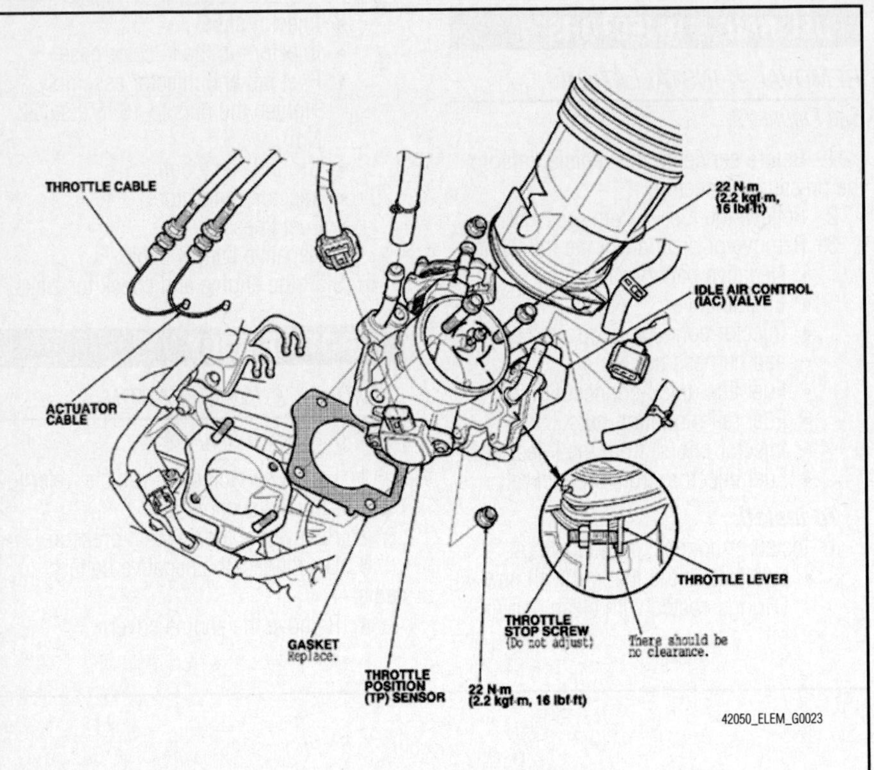
THROTTLE CABLE
22 N·m
(2.2 kgf·m,
16 lbf·ft)
IDLE AIR CONTROL
(IAC) VALVE
ACTUATOR
CABLE
THROTTLE LEVER
THROTTLE
STOP SCREW
(Do not adjust)
There should be
no clearance.
GASKET
Replace.
THROTTLE
POSITION
(TP) SENSOR
22 N·m
(2.2 kgf·m, 16 lbf·ft)
42050_ELEM_G0023

Fig. 80 Exploded view of the throttle body and related components

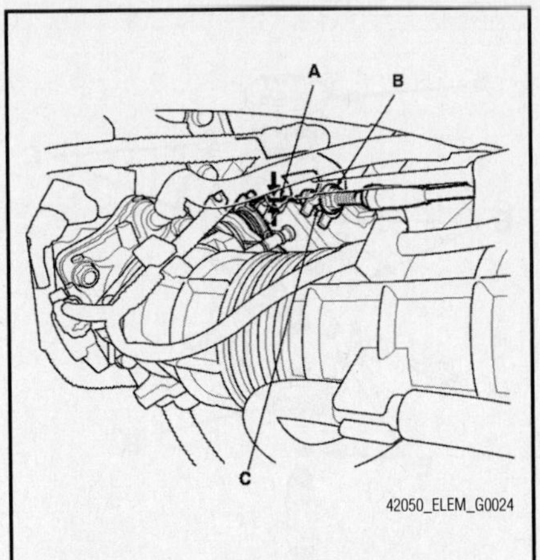
A
B
C
42050_ELEM_G0024

Fig. 82 Throttle cable free play (A) should be 3/8–1/2 in. (10–12mm). If the free play is not within specifications, loosen the locknut (B), turn the adjusting nut (C) until the deflection is correct, then retighten the locknut

5. Detach the electrical connectors and cables from the throttle body.

6. Remove the retainers, then remove the throttle body.

7. Remove and discard the throttle body gasket.

To install:

8. Position a new throttle body gasket, then install the throttle body. Tighten the retainers to 16 ft. lbs. (22 Nm).

9. Attach the cables and electrical connectors to the throttle body.

10. Adjust the actuator cable, as follows:

a. Make sure the actuator cable moves smoothly with no binding or sticking.

b. Measure the amount of movement of the output linkage until the engine speed starts to increase. At first, the output linkage should be located at the fully closed position. The free play should be 0.13–0.17 in. (3.25–4.25mm).

c. If the free play is not within specs, loosen the locknut, and turn the adjusting nut until the free play is as specified, then retighten the locknut.

11. Adjust the throttle cable, as follows:

a. Check cable free play at the throttle linkage. Cable free play should be 3/8–1/2 in. (10–12mm). If the free play is not within specifications, loosen the locknut, turn the adjusting nut until the deflection is as specified, then retighten the locknut.

b. With the cable properly adjusted, check the throttle valve to be sure it opens fully when you push the accelerator pedal to the floor. Also check the throttle valve to be sure it returns to the idle position whenever you release the accelerator pedal.

12. Connect the negative battery cable.

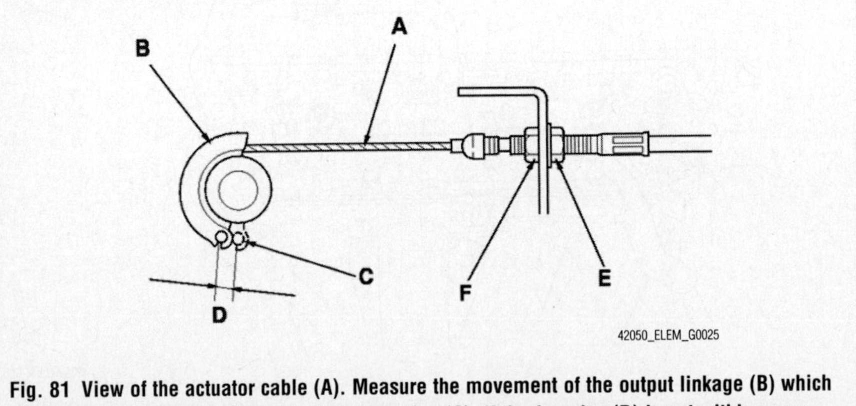
A
B
C
D
F
E
42050_ELEM_G0025

Fig. 81 View of the actuator cable (A). Measure the movement of the output linkage (B) which should first be located at the fully closed position (C). If the freeplay (D) is not within specs, loosen the locknut (E) and turn the adjusting nut (F) until the free play is correct

IDLE SPEED

INSPECTION

See Figure 83.

➡Leave the idle air control (IAC) valve connected.

Before checking the idle speed, check these items:

- The malfunction indicator lamp (MIL) has not been reported on.
- Ignition timing
- Spark plugs
- Air cleaner
- PCV system

1. Pull the parking brake lever up.

2. Disconnect the evaporative emission (EVAP) canister purge valve connector.

3. Connect a suitable OBD II compliant scan tool to the Data Link Connector (DLC) located under the driver's side of the dashboard.

4. Start the engine. Hold the engine speed at 3,000 rpm without load (in Park or neutral) until the radiator fan comes on, then let it idle.

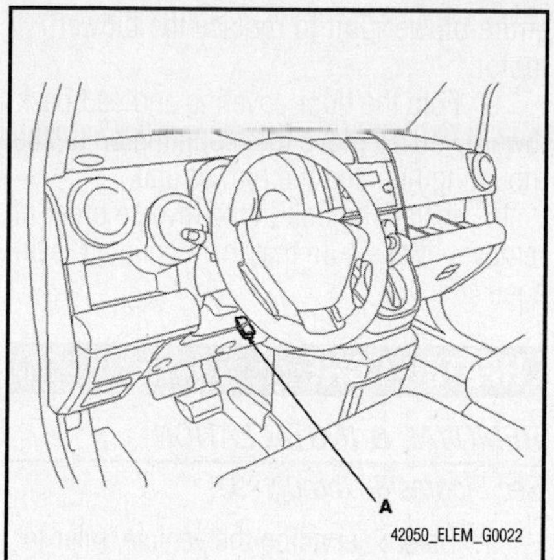
A
42050_ELEM_G0022

Fig. 83 The Data Link Connector (DLC) is located under the driver's side of the dashboard

5. Check the idle speed without load conditions: headlights, blower fan, radiator fan, and air conditioner off.

6. Idle speed should be:
   a. M/T: 650–750 rpm
   b. A/T: 650–750 rpm (in Park or neutral)
7. Let the engine idle for 1 minute with a high electrical load (A/C switch on, temperature set to MAX cool, blower fan on high, rear window defogger on, and headlights on high beam).
8. Idle speed should be:
   a. M/T: 670–770 rpm
   b. A/T: 670–770 rpm (in Park or neutral)
9. If the idle speed is not within specification, do the ECM/PCM idle learn procedure. The idle learn procedure must be done so the ECM/PCM can learn the engine idle characteristics.

The idle learn procedure must be performed whenever you do any of the following:

- Replace ECM/PCM.
- Reset ECM/PCM.
- Update ECM/PCM.

➡Erasing DTCs with the HDS does not require you to do the idle learn procedure.

- Clean or replace the throttle body.

a. Make sure all electrical items (A/C, audio, rear window defogger, lights, etc.) are off.
b. Reset the ECM/PCM with the HDS.
c. Turn the ignition switch ON (II), and wait 2 seconds.
d. Start the engine. Hold the engine speed at 3,000 rpm without load (in Park or neutral) until the radiator fan comes on, or until the engine coolant temperature reaches 194 °F (90 °C).
e. Let the engine idle for about 5 minutes with the throttle fully closed.

➡If the radiator fan comes on, do not include its running time in the 5 minutes.

10. Reconnect the EVAP canister purge valve connector.

HEATING & AIR CONDITIONING SYSTEM

BLOWER MOTOR

REMOVAL & INSTALLATION

See Figures 84 through 88.

1. Remove the passenger's dashboard under cover, as follows:
   a. Gently pull down the rear edge to release the clips.
   b. Pull the cover away to release the pins (B) from the holders (C).
2. Use a suitable trim panel removal tool to remove the passenger's kick panel.

⁂ WARNING

Be careful not to scratch or damage the dash when removing the glove box.

3. Remove the glove box, as follows:
   a. While holding the glove box, remove the glove box stop on each side.
   b. Remove the bolts, then remove the glove box.
4. Cut the plastic cross brace in the glove box opening with diagonal cutters in the area shown. Remove and discard the plastic cross brace.
5. Remove the relays (A) and the wire harness clip (B), then remove the bolts and the glove box frame (C).
6. Remove the ECM/PCM.

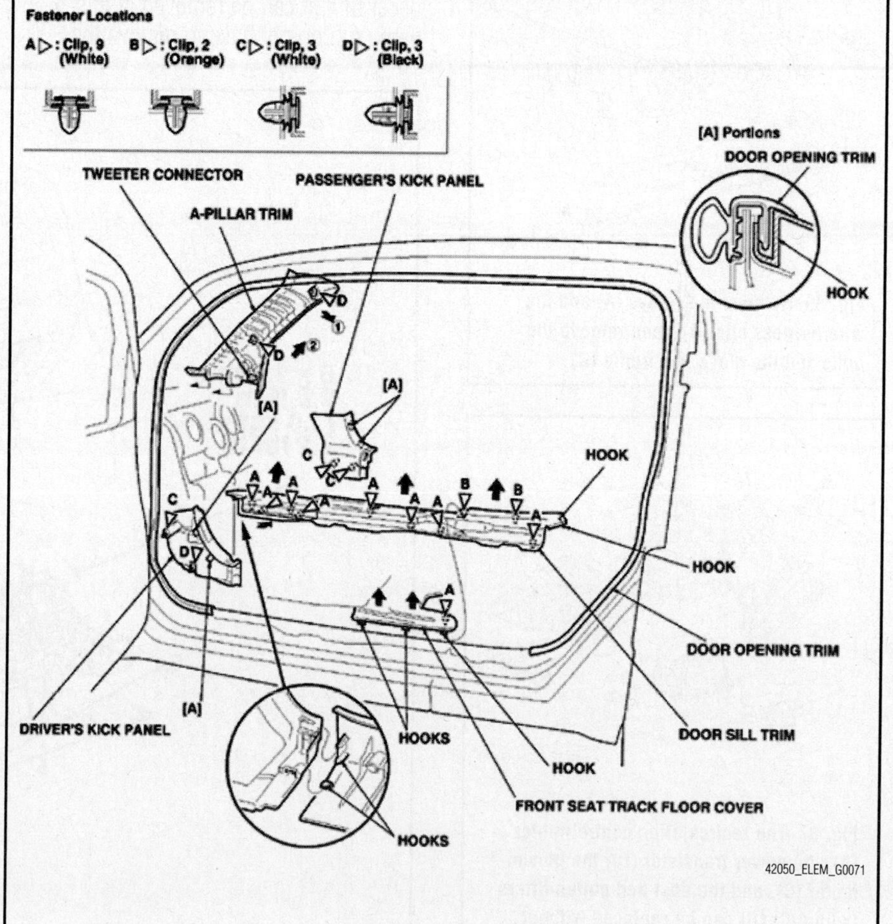
Fastener Locations
A: Clip, 9 (White)
B: Clip, 2 (Orange)
C: Clip, 3 (White)
D: Clip, 3 (Black)
[A] Portions
DOOR OPENING TRIM
HOOK
TWEETER CONNECTOR
PASSENGER'S KICK PANEL
A-PILLAR TRIM
HOOK
HOOK
DOOR OPENING TRIM
DRIVER'S KICK PANEL
HOOKS
DOOR SILL TRIM
HOOK
FRONT SEAT TRACK FLOOR COVER
HOOKS
42050_ELEM_G0071

Fig. 84 Passenger side kick panel and related interior trim panels

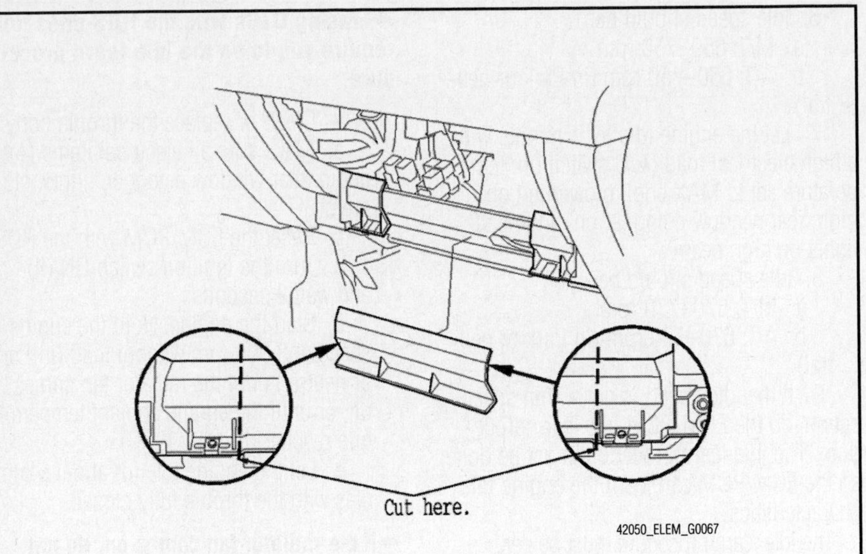
Cut here.
42050_ELEM_G0067

Fig. 85 Cut the plastic cross brace in the glove box opening with diagonal cutters in the area shown. Remove and discard the plastic cross brace

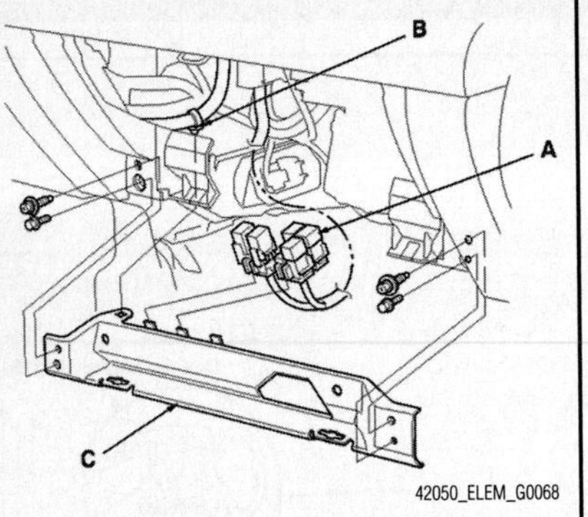
B
A
C
42050_ELEM_G0068

Fig. 86 Remove the relays (A) and the wire harness clip (B), then remove the bolts and the glove box frame (C)

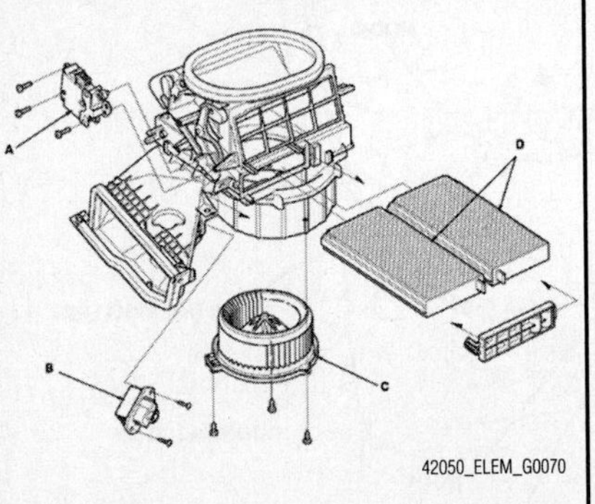
A
D
B
C
42050_ELEM_G0070

Fig. 87 The recirculation control motor (A), the power transistor (B), the blower motor (C), and the dust and pollen filters (with A/C) (D) can be replaced without removing the blower unit

7. Disconnect the connectors from the blower motor, the power transistor, and the recirculation control motor, then remove the wire harness clips.

8. If blower motor replacement is necessary, it can be removed at this time. You do not have to remove the entire blower unit to replace the blower motor.

9. Fold the floor covering and pad back toward you. Remove the mounting bolts, the mounting nut, and the blower unit.

10. Install the unit in the reverse order of removal. Make sure that there is no air leakage.

HEATER CORE

REMOVAL & INSTALLATION

See Figures 88 through 92.

1. Before servicing the vehicle, refer to the precautions section.
2. Disconnect the negative battery cable.
3. Drain the cooling system into a clean container for reuse.
4. Disconnect the A/C lines from the evaporator core it equipped with A/C.
5. Open the cable clamp (A) and disconnect the heater valve cable (B) from the heater valve arm (C). Turn the heater valve arm to the full open position as illustrated.
6. Remove the heater hoses from the heater core
7. Remove the mounting bolt and heater valve. Remove the nut from the

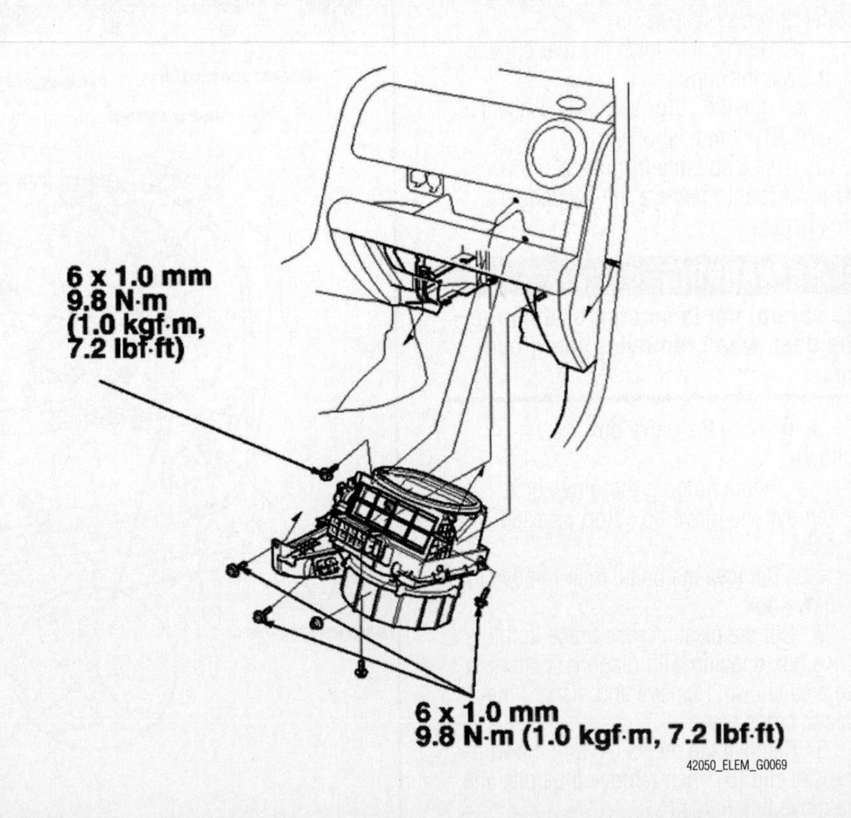
6 x 1.0 mm
9.8 N·m
(1.0 kgf·m,
7.2 lbf·ft)
6 x 1.0 mm
9.8 N·m (1.0 kgf·m, 7.2 lbf·ft)
42050_ELEM_G0069

Fig. 88 Blower unit mounting and fastener tightening specifications.

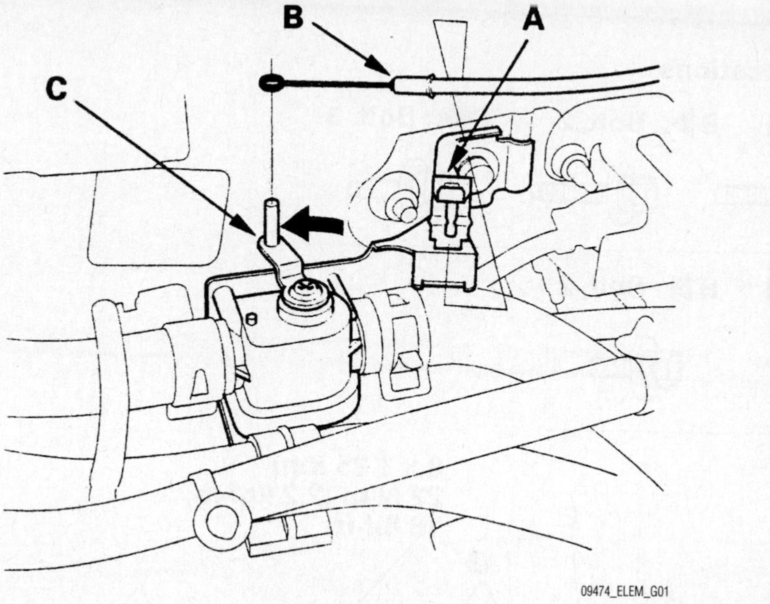
B
A
C
09474_ELEM_G01

Fig. 89 Open the cable clamp (A) and disconnect the heater valve cable (B) from the heater valve arm (C). Turn the heater valve arm to the full open position

heater unit being careful of any lines, hoses and wiring in the vicinity.

8. Remove the dashboard as follows:
   a. Remove the driver's side lower instrument panel cover clips and remove the lower cover.
   b. Remove the glove box stops from each side of the glove box.
   c. Remove the glove box-to-instrument panel bolts and the glove box.
   d. Remove the passenger's side lower instrument panel cover clips and remove the lower cover.
   e. Remove the center lower cover clips and remove the lower cover.
   f. Remove the passenger vent.
   g. Remove the A–trim on both sides.
   h. Remove the passenger side kick panel.
   i. Remove the steering wheel.
   j. Remove the steering column covers.
   k. Disconnect the wiring from the combination switch and remove the assembly by removing the screw on top of the switch.
   l. Disconnect the ignition switch con-

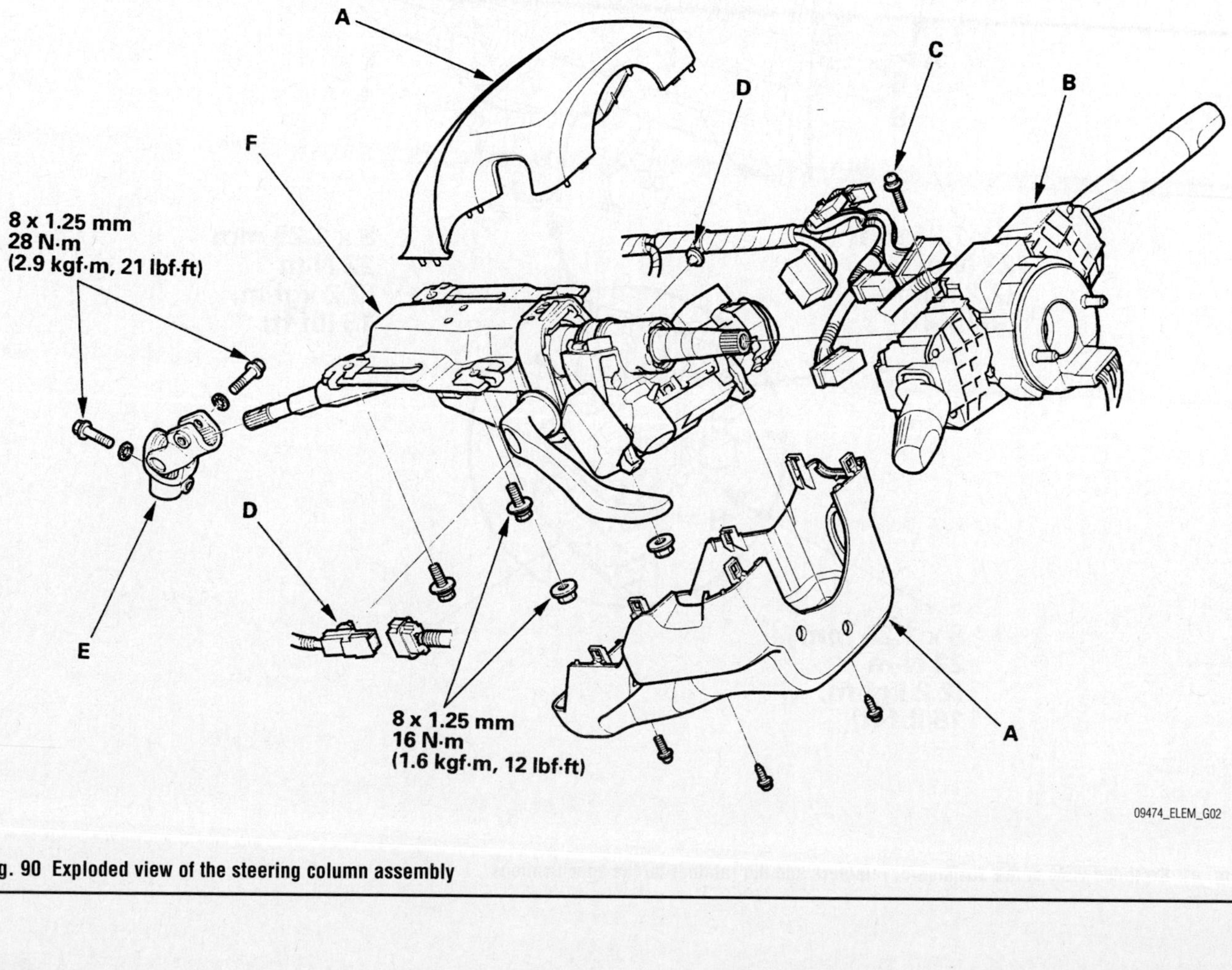
A
D
C
B
F
8 x 1.25 mm
28 N·m
(2.9 kgf·m, 21 lbf·ft)
D
E
8 x 1.25 mm
16 N·m
(1.6 kgf·m, 12 lbf·ft)
A
09474_ELEM_G02

Fig. 90 Exploded view of the steering column assembly

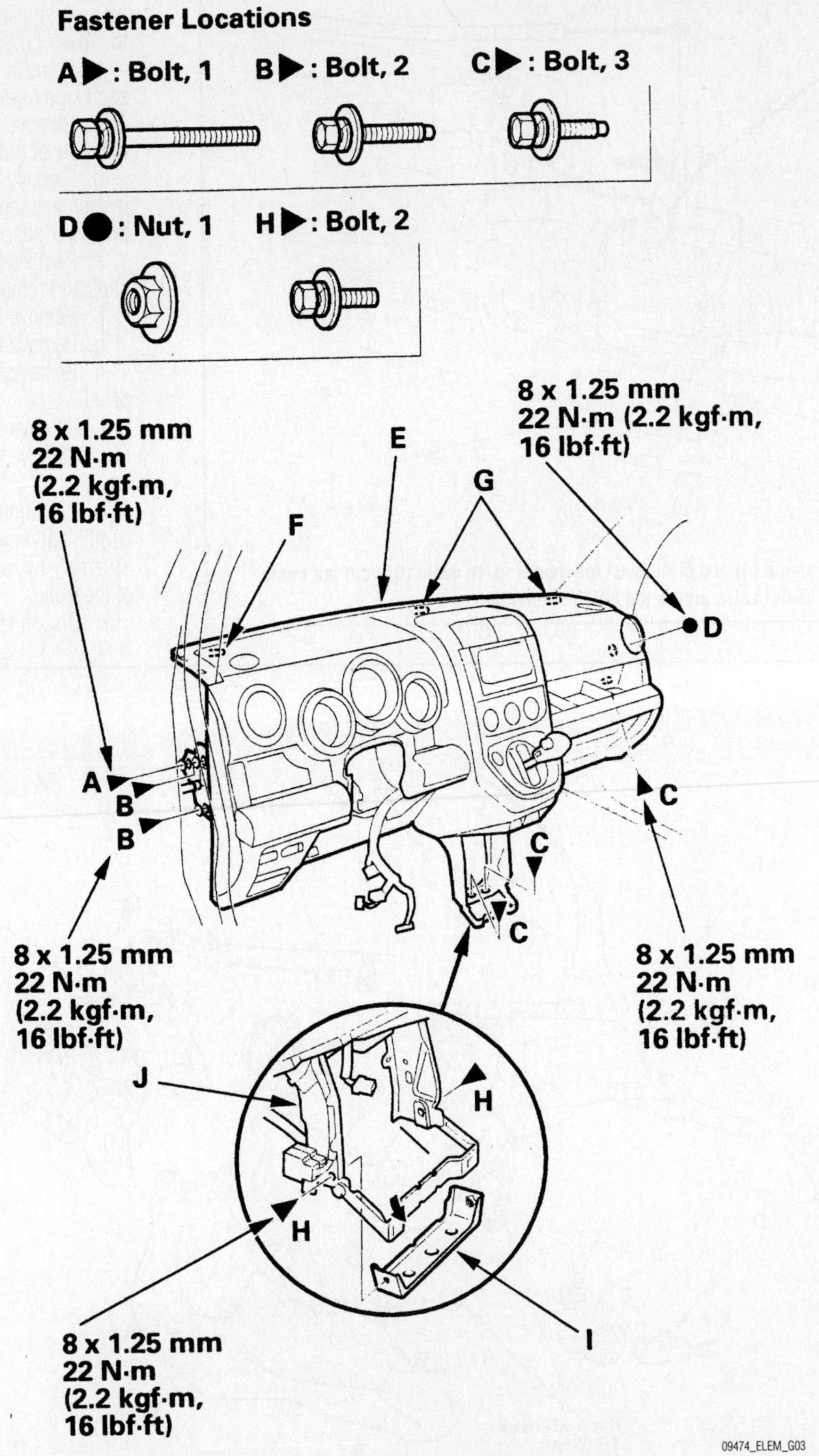
Fastener Locations
A: Bolt, 1
B: Bolt, 2
C: Bolt, 3
D: Nut, 1
H: Bolt, 2
8 x 1.25 mm
22 N·m (2.2 kgf·m,
16 lbf·ft)
E
G
F
D
8 x 1.25 mm
22 N·m
(2.2 kgf·m,
16 lbf·ft)
A
B
B
C
C
C
8 x 1.25 mm
22 N·m
(2.2 kgf·m,
16 lbf·ft)
8 x 1.25 mm
22 N·m
(2.2 kgf·m,
16 lbf·ft)
J
H
H
I
8 x 1.25 mm
22 N·m
(2.2 kgf·m,
16 lbf·ft)
09474_ELEM_G03

Fig. 91 Exploded view of the dashboard, retainers and the retainer torque specifications

nectors and release the wire harness clips from the column.

m. Disconnect the steering joint bolt and disconnect it from the column shaft.

n. Remove the steering column retainers and the column.

o. Control cable if equipped with an automatic transmission or shift cable if equipped with a manual transaxle.

p. Woofer, if equipped.

q. On the driver's side disconnect the following:

- Tweeter connector
- Drivers door wiring connectors
- Brake switch connector
- Clutch switch connector, if equipped
- Engine compartment harness connectors from the fuse/relay box

r. In the middle of the dashboard, disconnect the SRS control unit connector, floor harness connector and engine compartment harness connectors.

s. On the passenger side, disconnect the following:

- Passenger door wiring connectors
- Antenna lead
- PCM connectors
- Engine wire harness connectors
- Heater sub-harness connectors
- Passenger airbag connectors.
- Amplifier connectors
- Wire harness protector from the amplifier, if equipped

t. Remove any remaining harness and connector clips.

u. Dashboard bolts. Refer to the exploded view for bolt location and torque values.

v. Remove the dashboard.

9. Remove the PCM.
10. Disconnect the following connectors:
    - Dashboard wiring harness
    - Air mixture control motor
    - Evaporator temperature sensor
    - Power transistor
    - Mode control motor
    - Blower motor

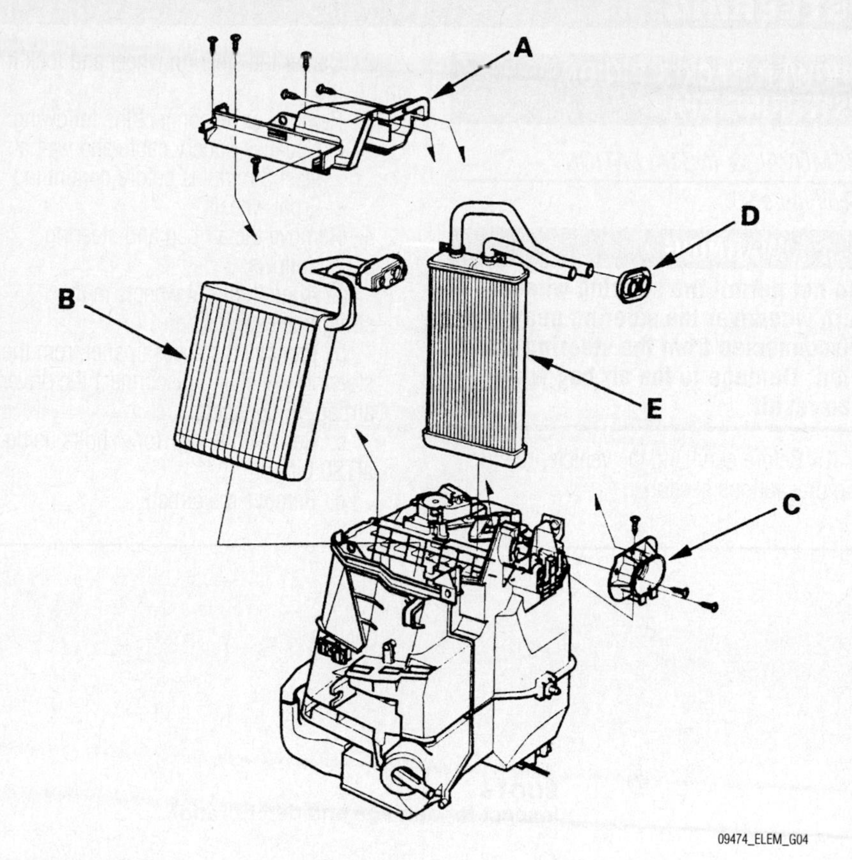
A
B
C
D
E
09474_ELEM_G04

Fig. 92 Exploded view of the heater unit assembly

11. Disconnect the following clips:
    - Wire harness clips
    - Connector clips
12. Remove the wire harness, heater duct and clip.
13. Remove the drain hose and the mounting nuts and the heater unit.
14. Remove the screws and the expansion valve cover (A).
15. If equipped with A/C, remove the evaporator core (B).
16. Remove the screws and the flange cover (C).
17. Remove the grommet (D) and the heater core being careful not to damage any lines.

To install:

18. Installation is the reverse of removal. Refer to the exploded views of the heater unit assembly, dashboard and steering column assembly for component location, fastener location and torque specifications.
19. Refill the cooling system.
20. Connect the negative battery cable.
21. Evacuate and charge and leak test the air conditioning system refrigerant.
22. Run the engine to normal operating temperatures; then, check the climate control operation and check for leaks.

STEERING

POWER RACK & PINION STEERING GEAR

REMOVAL & INSTALLATION

See Figure 93.

⁂ WARNING

Do not permit the steering wheel to turn whenever the steering gear is disconnected from the steering column. Damage to the air bag wiring can result.

1. Before servicing the vehicle, refer to the precautions section.
2. Center the steering wheel and lock it in position.
3. Remove or disconnect the following:
   - Negative battery cable and wait at least 3 minutes before continuing
   - Front wheels
4. Remove the air bag and steering wheel as follows:
   a. Align the front wheels in the straight ahead position
   b. Remove the access panel from the steering wheel and disconnect the drivers airbag 4P connector.
   c. Remove the two Torx® bolts using a T30 bit.
   d. Remove the airbag.
   e. Disconnect the cruise control connector and horn switch connector.
   f. Loosen the steering wheel bolt and using a suitable puller, free the steering wheel.

➡Do not tap on the steering wheel or column shaft during removal. If you thread the puller bolts more than 5 threads into the wheel hub you will hit the cable reel and damage it. To prevent damage, insert a pair of jam nuts 5 threads up on each puller bolt.

   g. Remove the puller, steering wheel bolt and wheel.

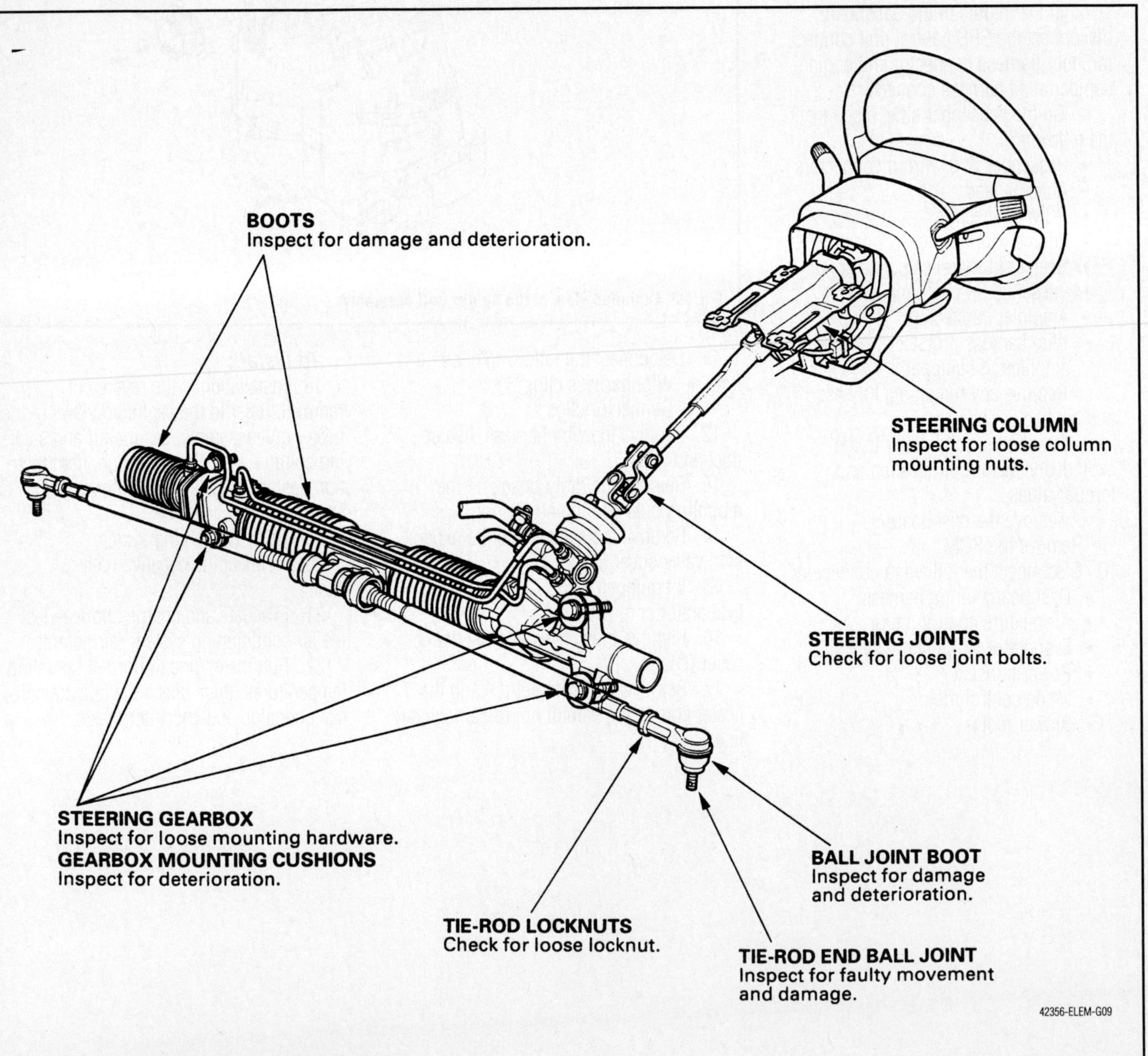
BOOTS
Inspect for damage and deterioration.
STEERING COLUMN
Inspect for loose column mounting nuts.
STEERING JOINTS
Check for loose joint bolts.
STEERING GEARBOX
Inspect for loose mounting hardware.
GEARBOX MOUNTING CUSHIONS
Inspect for deterioration.
BALL JOINT BOOT
Inspect for damage and deterioration.
TIE-ROD LOCKNUTS
Check for loose locknut.
TIE-ROD END BALL JOINT
Inspect for faulty movement and damage.
42356-ELEM-G09

Fig. 93 Power steering gear and related components

5. Remove or disconnect the following:
   - Driver's side dashboard lower cover and undercover
   - Steering joint bolts and disconnect the joint by moving the joint towards the column
   - Center pin from the top of the pinion shaft, if equipped and discard the pin
   - Tie rod ends
   - Power steering heat baffle plate
   - Engine wiring harness clamp and clip from their brackets
   - Loosen the adjustable hose clamp and disconnect the return hose
   - Loosen the 14mm flare nut and disconnect the feed line
   - Open the hose holders on the return hose and remove the clamp
   - Power steering pressure switch connector
   - Feed line on the power steering line mounting bracket and set it aside
   - Body stiffener
   - Left, then right side flange bolts and washers
   - Mounting brackets

6. Lower the unit so the pinion shaft points upward. Remove the pinion shaft grommet. The steering gear is removed through the driver's side.

To install:

7. Installation of the steering gear is the reverse of removal. Observe the following torques:
   - Mounting bracket and side flange bolts: 46 ft. lbs. (62 Nm)
   - Supply line flare nut: 27 ft. lbs. (37 Nm)
   - Tie rod ball stud nuts: 40 ft. lbs. (54 Nm)
   - Steering joint bolts: 21 ft. lbs. (28 Nm)

8. Install the steering wheel and air bag as follows:

a. First make sure the front wheels are aligned straight ahead. Center the cable reel by rotating the cable reel clockwise until it stops, then rotate it counterclockwise about 2 ½ turns. The arrow mark on the cable reel should point straight up.

b. Position the tabs on the turn signal canceling sleeve, install the steering wheel and make sure the wheel hub engages the pins of the cable reel and tabs of the canceling sleeve. Do not tap on the wheel or column.

c. Install the steering wheel bolt and tighten to 29 ft. lbs. (39 Nm). Connect the horn switch, cruise control switch and ensure the wiring is routed correctly and properly secured.

d. Install the driver's side air bag and tighten the Torx ® bolts to 7 ft. lbs. (9 Nm).

e. Connect the cable reel to the airbag 4P connector and install the access panel.

f. Connect the negative battery cable.

g. Turn the ignition switch on and ensure the airbag light illuminates for about 6 seconds and then goes out.

h. Ensure proper operation of the horn and cruise control.

POWER STEERING PUMP

REMOVAL & INSTALLATION

See Figures 95 through 96.

1. Place a suitable container under the vehicle.

2. Drain the power steering fluid from the reservoir.

3. Remove the drive belt (A) from the pump pulley, as follows.

➡This procedure requires the use of a special Belt tension release tool, Snap-on YA9317 or equivalent tool.

a. Move the auto-tensioner (A) with the belt tension release tool (B) in the direction shown to relieve tension from the drive belt, and remove the drive belt.

b. Install the new belt in the reverse order of removal.

4. Cover the auto-tensioner, alternator, and A/C compressor with several shop towels to protect them from spilled power steering fluid. Disconnect the pump inlet

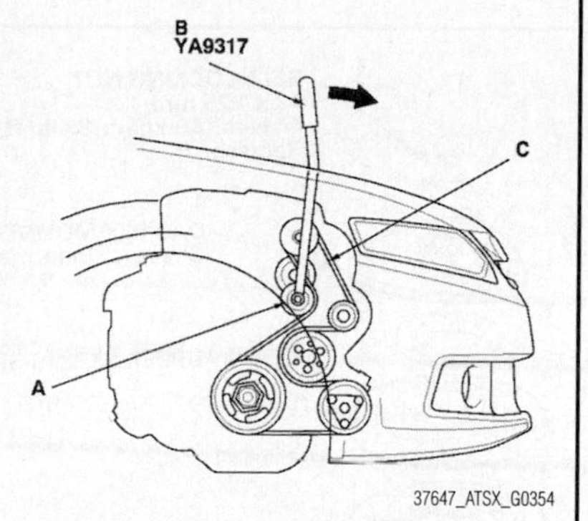
B
YA9317
C
A
37647_ATSX_G0354

Fig. 94 To remove the belt, move the auto-tensioner (A) with the belt tension release tool (B) in the direction shown to relieve tension from the drive belt, and remove the drive belt

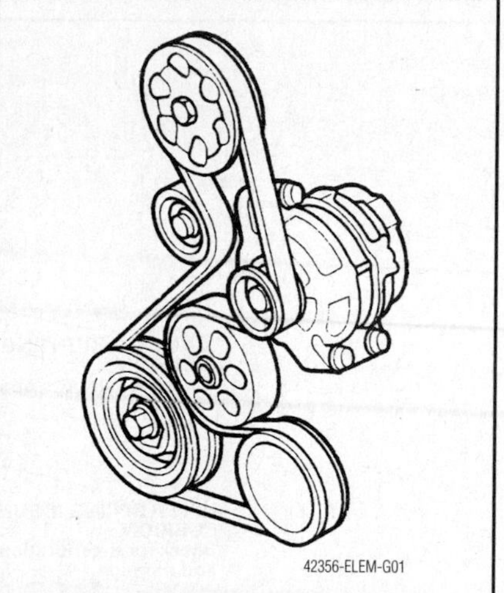
42356-ELEM-G01

Fig. 95 Accessory drive belt routing

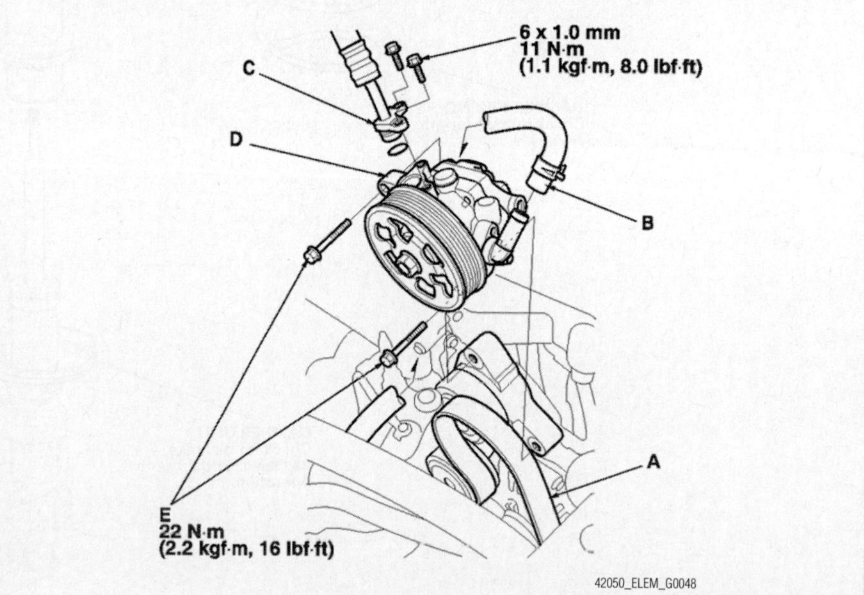
6 x 1.0 mm
11 N·m
(1.1 kgf·m, 8.0 lbf·ft)
C
D
B
A
E
22 N·m
(2.2 kgf·m, 16 lbf·ft)
42050_ELEM_G0048

Fig. 96 View of the driver belt (A), power steering pump inlet hose (B), outlet hose (C), power steering pump (D) and mounting bolts (E)

hose (B) and the pump outlet hose (C) from the pump (D), and plug them. Take care not to spill the fluid on the body or parts. Wipe off any spilled fluid at once. Do not turn the steering wheel with the pump removed.

5. Remove the pump mounting bolts (E).

6. Cover the opening of the pump with a piece of tape to prevent foreign material from entering the pump.

To install:

7. Connect the pump inlet hose and the pump outlet hose with a new O-ring.

8. Loosely install the pump in the pump bracket with the mounting bolts, then tighten the pump fittings securely.

✲✲ WARNING

Do not get power steering fluid or grease on the auto-tensioner, alternator, A/C compressor, and drive belt or pulley faces. Clean off any fluid or grease before installation.

9. Install the drive belt. Make sure that the belt is properly positioned on the pulleys.

10. Tighten the pump mounting bolts to 16 ft. lbs. (22 Nm).

11. Fill the power steering fluid reservoir to the upper level line.

SUSPENSION

FRONT SUSPENSION

COIL SPRING

REMOVAL & INSTALLATION

See Figure 97.

1. Before servicing the vehicle, refer to the precautions section.

2. Remove the strut from the vehicle and install in a strut spring compressor. Compress the spring until the end of the spring comes away from the spring seat.

3. Remove the upper strut mount, spring seat and related components.

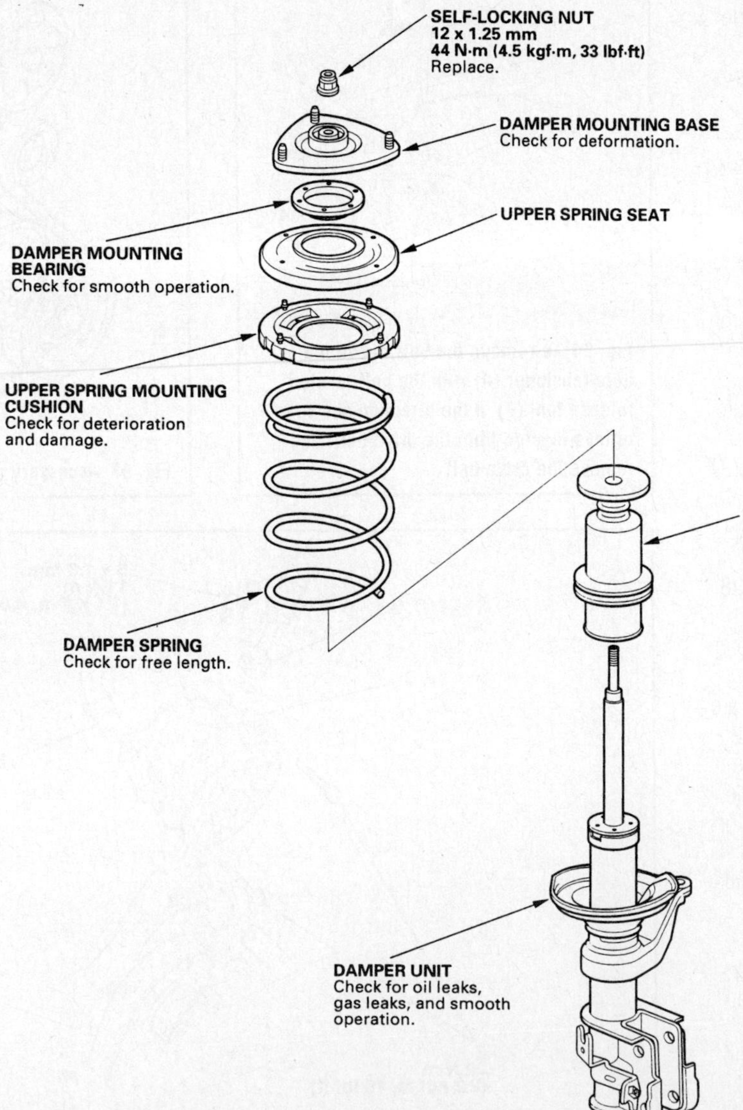
SELF-LOCKING NUT
12 x 1.25 mm
44 N·m (4.5 kgf·m, 33 lbf·ft)
Replace.
DAMPER MOUNTING BASE
Check for deformation.
UPPER SPRING SEAT
DAMPER MOUNTING BEARING
Check for smooth operation.
UPPER SPRING MOUNTING CUSHION
Check for deterioration and damage.
BUMP STOP
Check for weakness and damage.
DAMPER SPRING
Check for free length.
DAMPER UNIT
Check for oil leaks, gas leaks, and smooth operation.
42356-ELEM-G11

Fig. 97 Exploded view of the front strut (damper and spring) assembly

4. Remove the coil spring from the strut spring compressor.

To install:

➡**Use a new self-locking nut.**

5. Compress the spring and position the strut so that the end of the spring aligns with the notch in the spring seat.
6. Install the upper strut mounting components and tighten the nut to 33 ft. lbs. (44 Nm).
7. Install the strut to the vehicle.
8. Check the wheel alignment and adjust as necessary.

LOWER BALL JOINT

REMOVAL & INSTALLATION

The ball joint is not replaceable.

LOWER CONTROL ARM

REMOVAL & INSTALLATION

1. Before servicing the vehicle, refer to the precautions section.
2. Remove or disconnect the following:
   - Front wheel
   - Stabilizer link
   - Lower arm from the knuckle
   - Lower arm

To install:

3. Install all suspension components and fasteners and hand tighten them.. Place a jack under the suspension, raise the suspension with the jack and load the jack with the vehicle weight.
4. Installation is the reverse of removal. Observe the following torques:
   - Lower arm bolts: 61 ft. lbs. (83 Nm)
   - Ball stud nut: 51 ft. lbs. (69 Nm). Install the cotter pin into the ball joint from the inside to the outside of the vehicle.
   - Stabilizer link: 29 ft. lbs. (39 Nm)

CONTROL ARM BUSHING REPLACEMENT

The lower control arm front inner bushing and the damper fork bushing are serviced with the control arm as an assembly.

STABILIZER BAR

REMOVAL & INSTALLATION

See Figures 98 and 99.

1. Raise the front of the vehicle, and support it with safety stands in the proper locations.
2. Remove the front wheels.

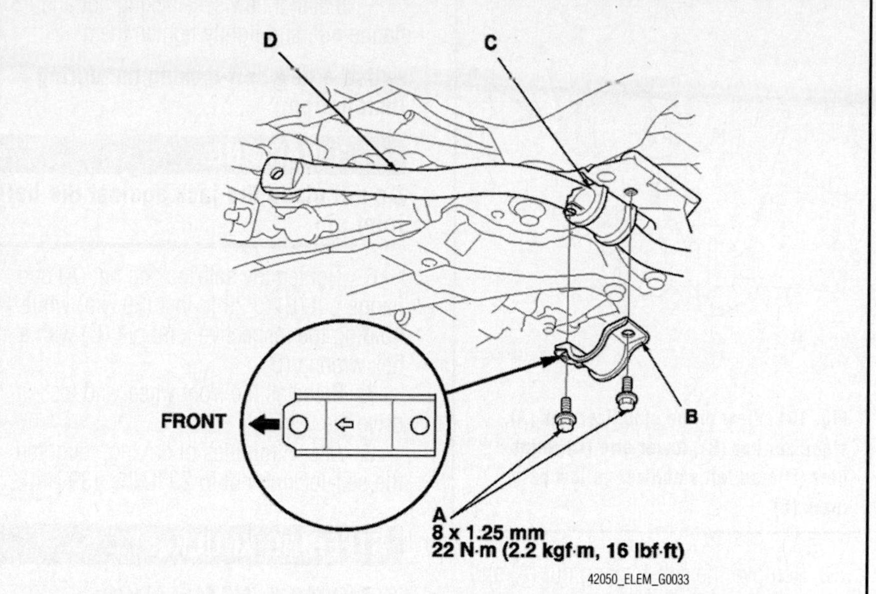
D
C
B
FRONT
A
8 x 1.25 mm
22 N·m (2.2 kgf·m, 16 lbf·ft)
42050_ELEM_G0033

Fig. 98 Remove the flange bolts (A) and bushing holders (B), then remove the bushings (C) and the stabilizer bar (D)

3. Disconnect the stabilizer links from the stabilizer bar on the right and left sides. Refer to the Stabilizer Link procedure in this section.
4. Remove the flange bolts (A) and bushing holders (B), then remove the bushings (C) and the stabilizer bar (D).

To install:

5. Install the stabilizer bar in the reverse order of removal, and note these items:
   a. Note the right and left direction of the stabilizer bar. The paint mark (A) on the stabilizer bar shows the right side.
   b. Do not set the bushings on the bent or curved part of the stabilizer bar.
   c. Note the fore/aft direction of the bushing holders.

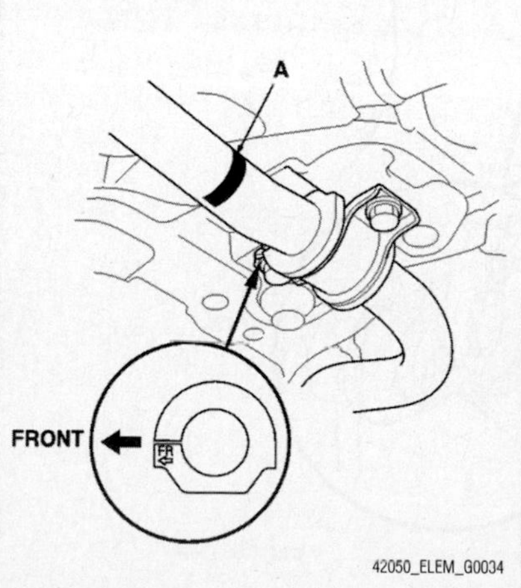
A
FRONT
42050_ELEM_G0034

Fig. 99 The paint mark (A) indicates the right side of the stabilizer bar

   d. Refer to stabilizer link removal/installation to connect the stabilizer bar to the links.

STABILIZER LINK

REMOVAL & INSTALLATION

See Figures 100 through 102.

1. Raise the front of the vehicle, and support it with safety stands in the proper locations.
2. Remove the front wheel.

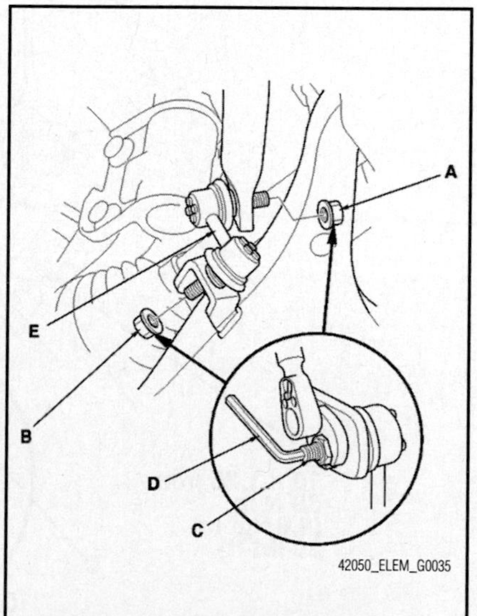
A
E
B
D
C
42050_ELEM_G0035

Fig. 100 Remove the self-locking nut (A) and flange nut (B) while holding the respective joint pin (C) with a hex wrench (D), and remove the stabilizer link (E)

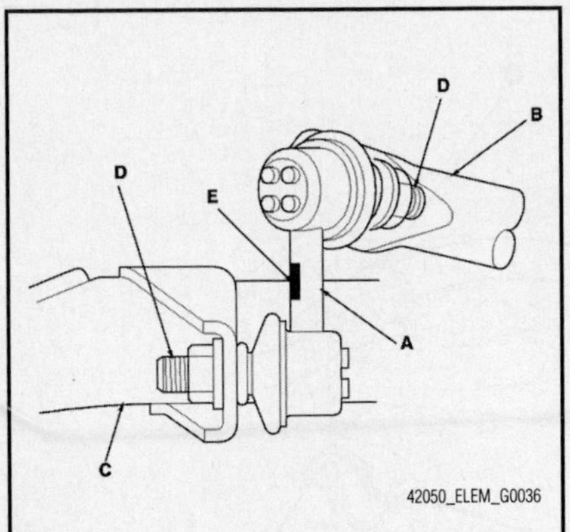
D
B
D
E
A
C
42050_ELEM_G0036

Fig. 101 View of the stabilizer link (A), stabilizer bar (B), lower arm (C), joint pins (D) and left stabilizer yellow paint mark (E)

3. Remove the self-locking nut (A) and flange nut (B) while holding the respective joint pin (C) with a hex wrench (D), and remove the stabilizer link (E).

To install:

4. Install the stabilizer link (A) on the stabilizer bar (B) and lower arm (C) with the joint pins (D) set at the center of their range of movement.

➡The left stabilizer has a yellow paint mark (E), while the right stabilizer link has a white paint mark.

5. Install a new self-locking nut and flange nut, and lightly tighten them.

➡Use a new self-locking nut during installation.

⁂ WARNING

Do not place the jack against the ball joint pin.

6. Tighten the self-locking nut (A) and flange nut (B) to 29 ft. lbs. (39 Nm) while holding the respective joint pin (C) with a hex wrench (D).
7. Reinstall the front wheel and test-drive the vehicle.
8. After 5 minutes of driving, retighten the self-locking nut to 29 ft. lbs. (39 Nm).

STEERING KNUCKLE

REMOVAL & INSTALLATION

Refer to the Wheel Bearing Removal & Installation procedure.

STRUT/DAMPER

REMOVAL & INSTALLATION

See Figure 103.

1. Before servicing the vehicle, refer to the precautions section.
2. Remove or disconnect the following:
   - Front wheel
   - Tie rod end
   - Brake hose retainer
   - ABS sensor harness bracket and brake hose bracket. Do not disconnect the wheel sensor connector.
   - Pinch bolts from the damper, while holding the nuts
   - Flange nuts from the top of the damper
   - Strut (damper), after lowering the lower control arm

To install:

➡Use new self-locking fasteners for assembly.

3. Install or connect the following:
   - Strut (damper). Tighten the upper mounting nuts to 33 ft. lbs. (44 Nm).
   - Tighten the pinch bolts to 116 ft. lbs. (157 Nm)
   - ABS sensor
   - Tie rod end
   - Brake hose retainer
   - Front wheel

WHEEL BEARINGS

ADJUSTMENT

The wheel bearings are sealed units and are not adjustable.

REMOVAL & INSTALLATION

See Figure 104.

1. Before servicing the vehicle, refer to the precautions section.
2. Remove or disconnect the following:
   - Front wheel
   - Brake hose bracket
   - Brake caliper and rotor. Forcing screws are needed to remove the rotor.
   - Spindle nut
   - ABS sensor
   - Stabilizer link
   - Lower arm from the knuckle
   - Strut-to-knuckle bolts
   - Steering hub/knuckle assembly
3. Press the hub from the knuckle. The bearings and races can now be pressed out and replaced.

➡With ABS, install the bearing with the magnetic encoder (brown color) toward the inside of the knuckle.

4. Observe the following torques:
   - Strut bolts: 116 ft. lbs. (157 Nm)
   - Ball stud nuts: 51 ft. lbs. (69 Nm)
   - Stabilizer bar link: 29 ft. lbs. (39 Nm)
   - Spindle nut: 181 ft. lbs. (245 Nm)

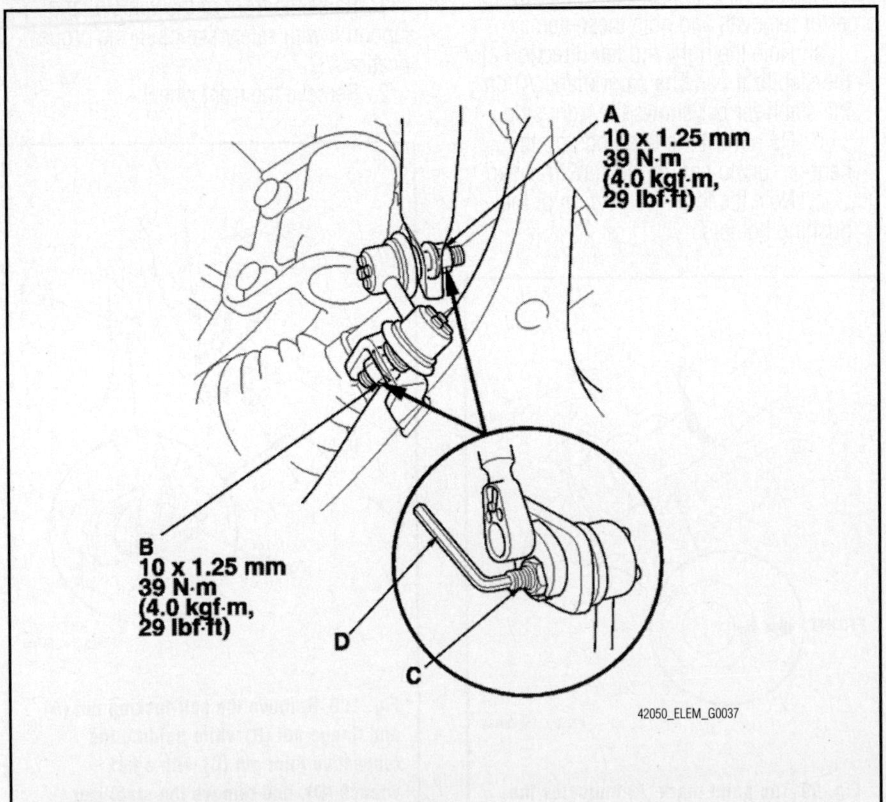
A
10 x 1.25 mm
39 N·m
(4.0 kgf·m,
29 lbf·ft)
B
10 x 1.25 mm
39 N·m
(4.0 kgf·m,
29 lbf·ft)
D
C
42050_ELEM_G0037

Fig. 102 View of the self-locking nut (A) and flange nut (B), joint pin (C) and hex wrench (D)

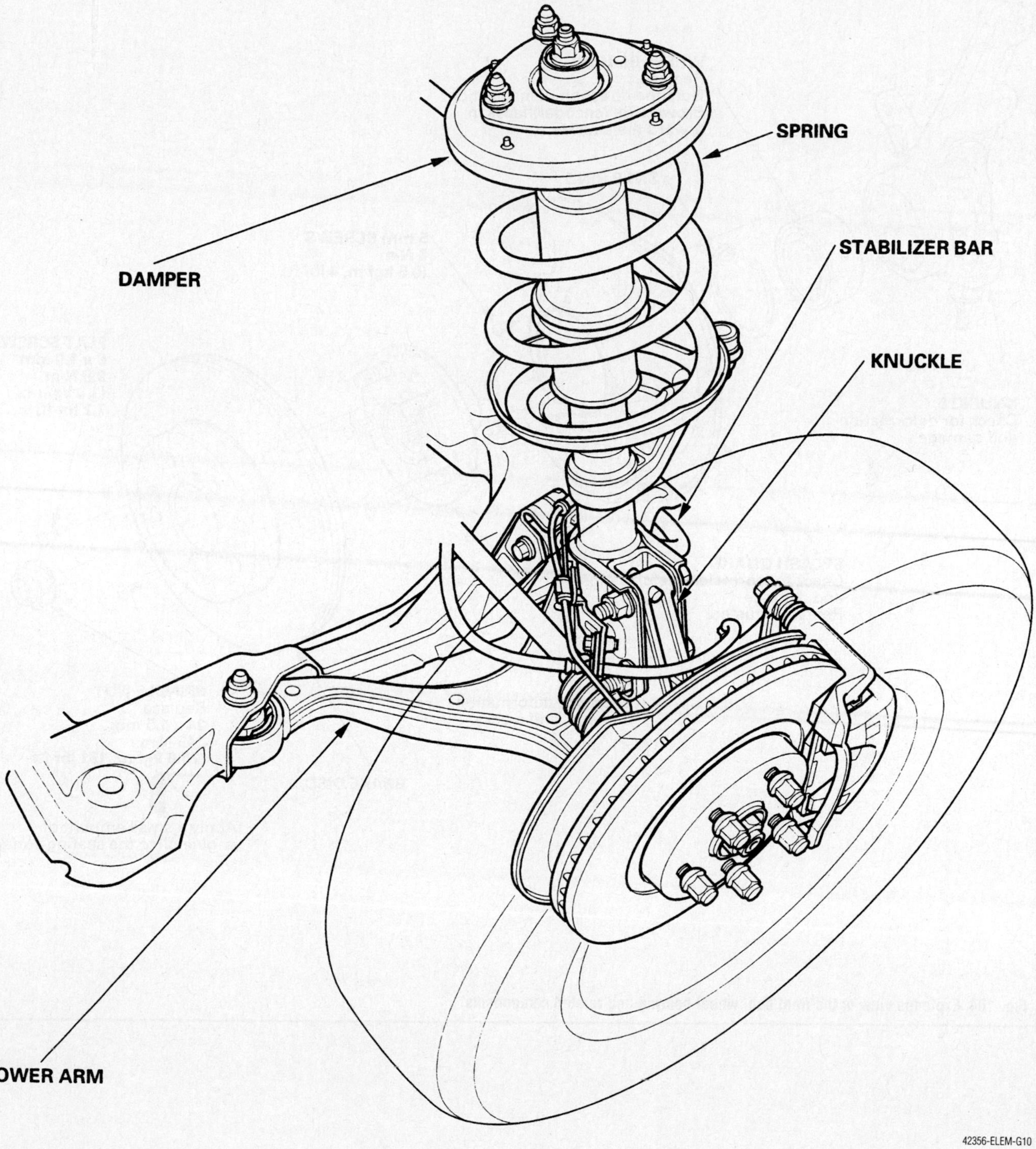
SPRING
DAMPER
STABILIZER BAR
KNUCKLE
LOWER ARM
42356-ELEM-G10

Fig. 103 Front suspension components

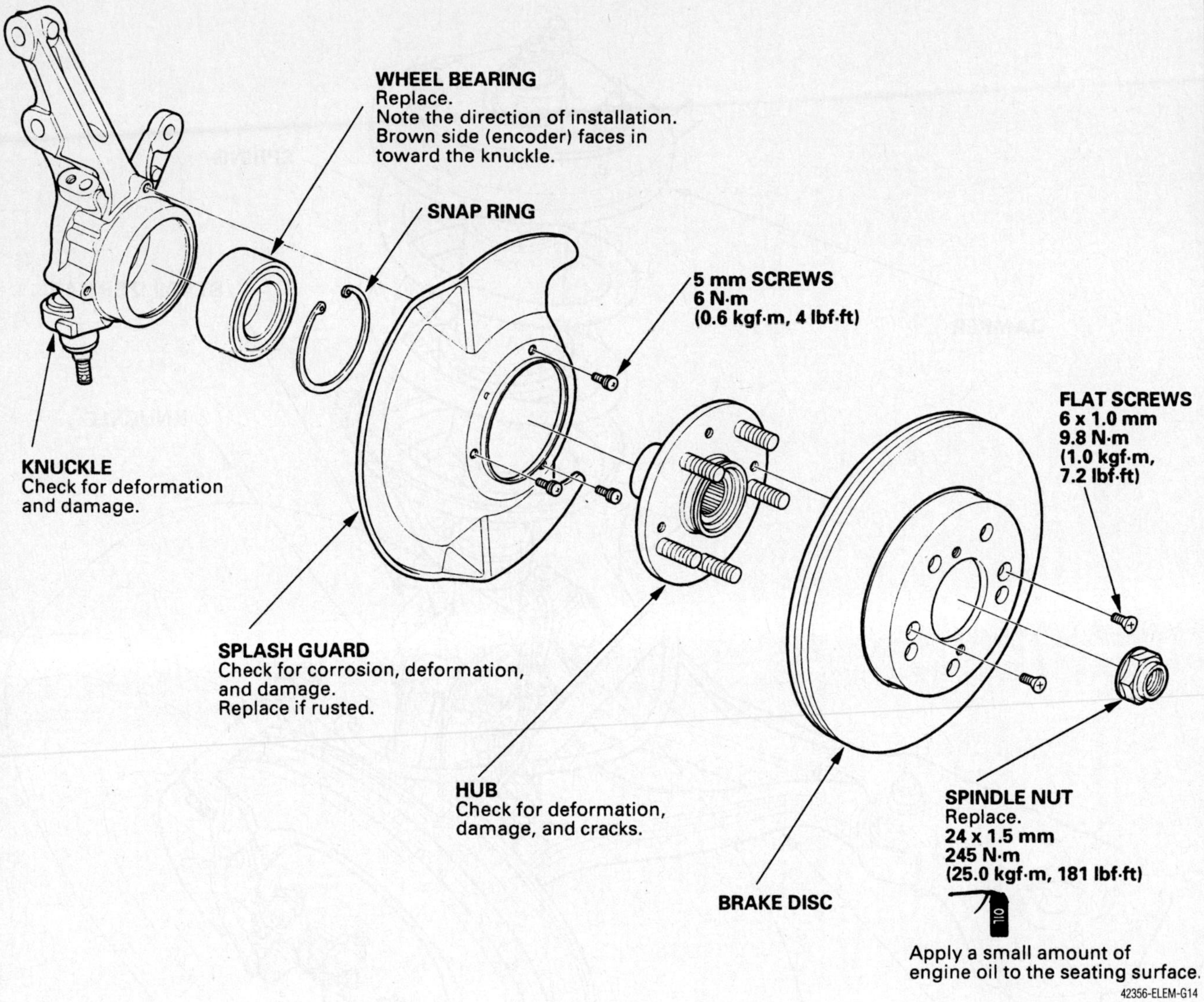
WHEEL BEARING
Replace.
Note the direction of installation.
Brown side (encoder) faces in toward the knuckle.
SNAP RING
5 mm SCREWS
6 N·m
(0.6 kgf·m, 4 lbf·ft)
FLAT SCREWS
6 x 1.0 mm
9.8 N·m
(1.0 kgf·m,
7.2 lbf·ft)
KNUCKLE
Check for deformation and damage.
SPLASH GUARD
Check for corrosion, deformation, and damage.
Replace if rusted.
HUB
Check for deformation, damage, and cracks.
SPINDLE NUT
Replace.
24 x 1.5 mm
245 N·m
(25.0 kgf·m, 181 lbf·ft)
BRAKE DISC
OIL
Apply a small amount of engine oil to the seating surface.
42356-ELEM-G14

Fig. 104 Exploded view of the front hub, wheel bearing and related components

COIL SPRING

REMOVAL & INSTALLATION

See Figure 105.

1. Before servicing the vehicle, refer to the precautions section.
2. Remove the strut from the vehicle and install in a strut spring compressor. Compress the spring until the end of the spring comes away from the spring seat.
3. Remove or disconnect the following:
   - Upper strut mount, spring seat and related components
   - Coil spring from the strut spring compressor

To install:

➡Use a new self-locking nut.

4. Compress the spring and position the strut so that the end of the spring aligns with the notch in the spring seat.
5. Install or connect the following:
   - Upper strut mounting components and tighten the nut to 22 ft. lbs. (29 Nm).
   - Strut to the vehicle
6. Check the wheel alignment and adjust as necessary.

STRUT/DAMPER

REMOVAL & INSTALLATION

See Figure 106.

1. Before servicing the vehicle, refer to the precautions section.
2. Support the vehicle under the lower control arm.
3. Romove or disconnect the following:
   - Rear wheel
   - Flange bolt from the bottom of the damper (strut)
   - Evaporative emission (EVAP) canister bolts, and loosen the EVAP canister mounting (left side only)
   - Interior access panel, if necessary
   - Flange nuts from the top of the damper in the cargo area
   - Strut

To install:

4. Install or connect the following:
   - Strut. Position the damper mounting base so the indent mark is toward the inside of the vehicle,
   - Upper flange nuts, hand-tight only
   - Bottom flange bolt, hand-tight only
5. With the suspension raised with a jack to load it with the vehicles weight,

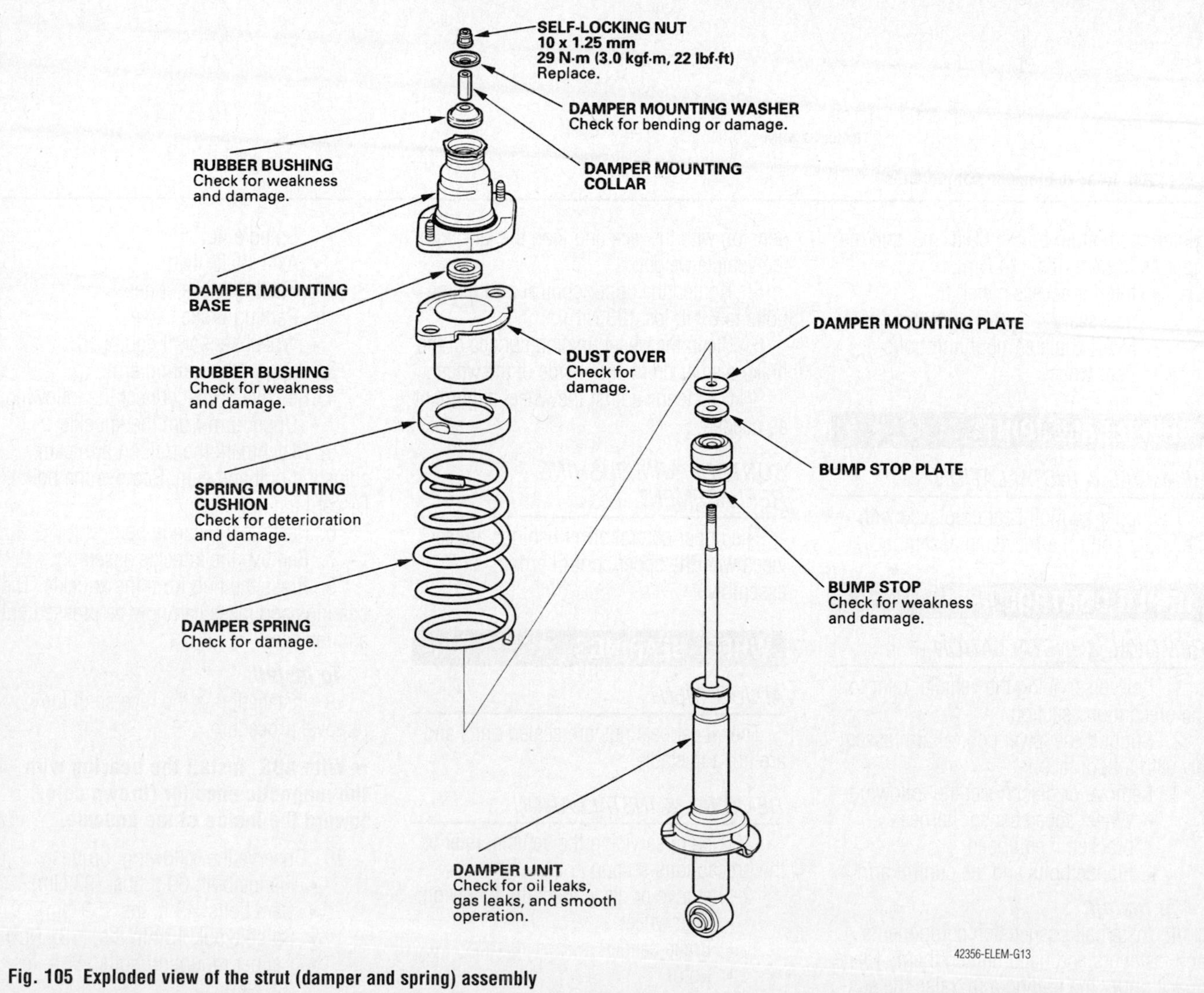
SELF-LOCKING NUT
10 x 1.25 mm
29 N·m (3.0 kgf·m, 22 lbf·ft)
Replace.
DAMPER MOUNTING WASHER
Check for bending or damage.
RUBBER BUSHING
Check for weakness and damage.
DAMPER MOUNTING COLLAR
DAMPER MOUNTING BASE
RUBBER BUSHING
Check for weakness and damage.
DUST COVER
Check for damage.
DAMPER MOUNTING PLATE
BUMP STOP PLATE
SPRING MOUNTING CUSHION
Check for deterioration and damage.
BUMP STOP
Check for weakness and damage.
DAMPER SPRING
Check for damage.
DAMPER UNIT
Check for oil leaks, gas leaks, and smooth operation.
42356-ELEM-G13

Fig. 105 Exploded view of the strut (damper and spring) assembly

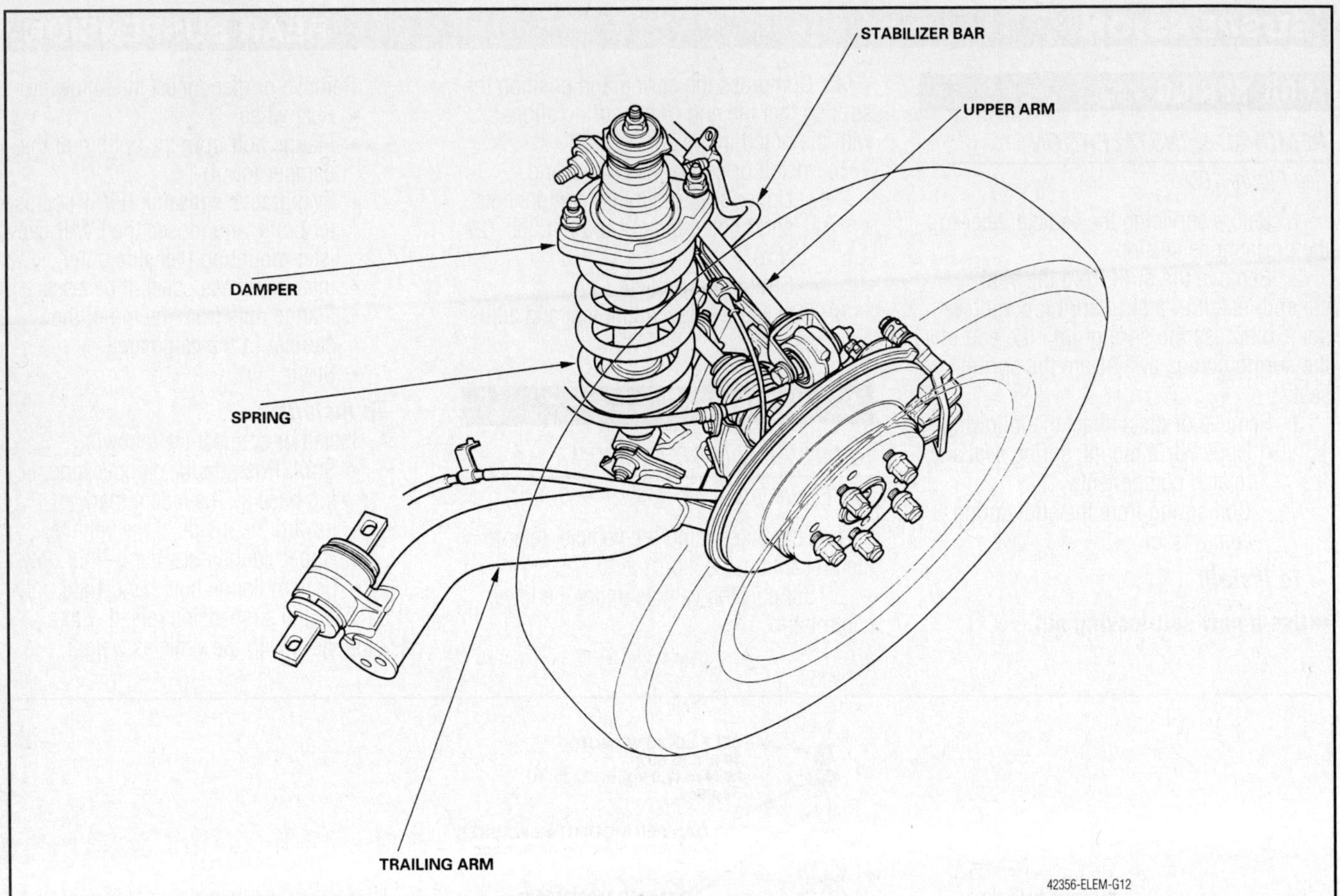

STABILIZER BAR
UPPER ARM
DAMPER
SPRING
TRAILING ARM
42356-ELEM-G12

Fig. 106 Rear suspension components

tighten the bottom bolt to 69 ft. lbs. and the top nuts to 54 ft. lbs. (74 Nm).

- Interior access panel, if necessary
- EVAP canister mounting bolts
- Rear wheel

UPPER BALL JOINT

REMOVAL & INSTALLATION

The upper ball joints are replaced with the upper control arms as an assembly.

UPPER CONTROL ARM

REMOVAL & INSTALLATION

1. Before servicing the vehicle, refer to the precautions section.
2. Support the lower control arm assembly with a floor jack.
3. Remove or disconnect the following:
   - Wheel speed sensor harness bracket, if equipped
   - Flange bolts and the control arm

To install:

4. Install all suspension components and fasteners and hand tighten them.. Place a jack under the trailing arm, raise the suspension with the jack and load the jack with the vehicle weight.
5. Tighten the upper control arm flange bolts to 69 ft. lbs. (93 Nm).
6. Clean the wheel, mating surface of the brake disc or drum and inside of the wheel.
7. Check and adjust the wheel alignment as needed.

CONTROL ARM BUSHING REPLACEMENT

The upper control arm bushings are serviced with the upper control arm as an assembly.

WHEEL BEARINGS

ADJUSTMENT

The wheel bearings are sealed units and are not adjustable.

REMOVAL & INSTALLATION

1. Before servicing the vehicle, refer to the precautions section.
2. Remove or disconnect the following:
   - Rear wheel
   - Brake caliper
   - Rotor
   - Spindle nut
   - Axle shaft (4wd)
   - Parking brake shoes
   - Parking brake cable
   - Wheel sensor, if equipped
3. Support the trailing arm.
4. Remove or disconnect the following:
   - Upper arm from the knuckle
5. Matchmark the trailing arm cam adjusting bolt and cam. Remove the bolt. Discard the nut.
6. Remove the flange bolt.
7. Remove the knuckle assembly.
8. Press the hub from the knuckle. The bearings and races can now be pressed out and replaced.

To install:

9. Installation is the reverse of the removal procedure.

➡With ABS, install the bearing with the magnetic encoder (brown color) toward the inside of the knuckle.

10. Observe the following torques:
    - Flange bolt: 69 ft. lbs. (93 Nm)
    - Cam bolts: 43 ft. lbs. (59 Nm)
    - Spindle nut: 134 ft. lbs. (181 Nm)
    - Caliper mounting bolts: 41 ft. lbs. (55 Nm)

HONDA

Fit

13

BRAKES 13-8

ANTI-LOCK BRAKE SYSTEM (ABS) 13-9
Wheel Speed Sensors 13-9
Removal & Installation 13-9
BLEEDING THE BRAKE SYSTEM 13-8
Bleeding Procedure 13-8
Bleeding Procedure 13-8
Bleeding the ABS System 13-8
FRONT DISC BRAKES 13-11
Brake Caliper 13-11
Removal & Installation 13-11
Disc Brake Pads 13-11
Removal & Installation 13-11
INFORMATION AND PRECAUTIONS 13-8
Anti-lock Systems 13-8
Disc and Drum Systems 13-8
PARKING BRAKE 13-13
Parking Brake Cables 13-13
Adjustment 13-13
REAR DRUM BRAKES 13-11
Brake Drum 13-11
Removal & Installation 13-11
Brake Shoes 13-11
Removal & Installation 13-11

CHASSIS ELECTRICAL 13-13

AIR BAG (SUPPLEMENTAL RESTRAINT SYSTEM) 13-13
General Information 13-13
Arming the System 13-13
Cable Reel Centering 13-13
Disarming the System 13-13
Precautions 13-13

DRIVE TRAIN 13-14

Clutch Driven Disc & Pressure Plate 13-14
Removal & Installation 13-14
Front Halfshaft 13-15
CV-boots Inspection 13-16
Removal & Installation 13-15
Hydraulic System Bleeding 13-14
Bleeding Procedure 13-14

ENGINE COOLING 13-16

Engine Fan 13-16
Removal & Installation 13-16
Radiator 13-17
Removal & Installation 13-17
Thermostat 13-18
Removal & Installation 13-18
Water Pump 13-18
Removal & Installation 13-18

ENGINE ELECTRICAL 13-20

CHARGING SYSTEM 13-20
Alternator 13-20
Removal & Installation 13-20
IGNITION SYSTEM 13-20
Firing Order 13-20
Ignition Coil 13-20
Removal & Installation 13-20
Ignition Timing 13-20
Adjustment 13-20
Inspection 13-20
Spark Plugs 13-20
Removal & Installation 13-20
STARTING SYSTEM 13-21
Starter 13-21
Removal & Installation 13-21

ENGINE MECHANICAL 13-21

Accessory Drive Belts 13-21
Accessory Belt Routing 13-21
Adjustment 13-22
Inspection 13-21
Removal & Installation 13-22
Camshaft and Valve Lifters 13-22
Inspection 13-22
Removal & Installation 13-23
Catalytic Converter 13-23
Removal & Installation 13-23
Crankshaft Pulley 13-24
Removal & Installation 13-24
Cylinder Head 13-25
Removal & Installation 13-25
Flywheel/Drive Plate 13-27
Removal & Installation 13-27
Intake Manifold 13-27
Removal & Installation 13-27
Oil Pan 13-28
Removal & Installation 13-28
Oil Pump 13-30
Inspection 13-30
Removal & Installation 13-30
Piston and Ring 13-31
Positioning 13-31
Rear Main Seal 13-31
Removal & Installation 13-31
Rocker Arms 13-31
Removal & Installation 13-31
Timing (Cam) Chain Front Case, Chain & Sprockets 13-33
Removal & Installation 13-33
Valve Lash 13-36
Adjustment 13-36

ENGINE PERFORMANCE & EMISSION CONTROLS 13-37

Accelerator Pedal Position (APP) Sensor 13-37
Location 13-37
Removal & Installation 13-37
Camshaft Position (CMP) Sensor 13-37
Location 13-37
Removal & Installation 13-37
Crankshaft Position (CKP) Sensor 13-38
CKP Pattern Clear/CKP Pattern Learn 13-38
Location 13-38
Removal & Installation 13-38
Engine Control Module (ECM)/Powertrain Control Module (PCM) 13-38
ECM/PCM Idle Learn Procedure 13-40
ECM/PCM Update 13-40
Removal & Installation 13-38
Engine Coolant Temperature (ECT) Sensor 13-40
Location 13-40
Removal & Installation 13-40
Evaporative Emissions (EVAP) Canister 13-41
Location 13-41
Removal & Installation 13-41

Exhaust Gas Recirculation (EGR) Valve....13-42
Location....13-42
Removal & Installation....13-42
Heated Oxygen (HO2S) Sensor....13-42
Location....13-42
Removal & Installation....13-42
Intake Air Temperature (IAT) Sensor....13-42
Location....13-42
Removal & Installation....13-42
Knock Sensor (KS)....13-42
Location....13-42
Removal & Installation....13-42
Manifold Absolute Pressure (MAP) Sensor....13-43
Location....13-43
Removal & Installation....13-43
Mass Air Flow (MAF) Sensor/ Intake Air Temperature (IAT) Sensor (Hot Wire)....13-43
Location....13-43
Removal & Installation....13-43
Positive Crankcase Ventilation (PCV) Valve....13-43
Location....13-43
Removal & Installation....13-43

FUEL....13-44

GASOLINE FUEL INJECTION SYSTEM....13-44
Fuel Filter....13-45
Removal & Installation....13-45
Fuel Level Sending Unit....13-45
Location....13-45
Removal & Installation....13-45
Fuel Rail and Injector....13-45
Removal & Installation....13-45
Fuel System Pressure....13-44
Relieving....13-44
Fuel System Service Precautions....13-44
Fuel Tank....13-47
Draining....13-47
Removal & Installation....13-47
Fuel Tank Unit....13-48
Removal & Installation....13-48
Idle Speed....13-49
Adjustment....13-49
Throttle Body....13-49
Removal & Installation....13-49

HEATING & AIR CONDITIONING SYSTEM....13-50
Blower Motor....13-50
Removal & Installation....13-50
Heater Core....13-51
Removal & Installation....13-51
Heater Unit....13-52
Removal & Installation....13-52

SPECIFICATIONS AND MAINTENANCE CHARTS....13-3
Brake Specifications....13-7
Camshaft Gasoline Specifications and Bearing....13-5
Capacities....13-4
Crankshaft and Connecting Rod Specifications....13-5
Engine and Vehicle Identification....13-3
Engine Tune-Up Specifications....13-3
Fluid Specifications....13-4
General Engine Specifications....13-3
Piston and Ring Specifications....13-5
Scheduled Maintenance Intervals....13-7
Tire, Wheel and Ball Joint Specifications....13-7
Torque Specifications....13-6
Valve Specifications....13-4
Wheel Alignment....13-6

STEERING....13-53
Electrical Power Steering (EPS) Control Unit....13-53
Removal & Installation....13-53
Electrical Power Steering (EPS) Motor....13-53
Removal & Installation....13-53
Power Rack & Pinion Steering Gear....13-54
Removal & Installation....13-54

SUSPENSION....13-58

FRONT SUSPENSION....13-58
Control Links....13-58
Removal & Installation....13-58
Lower Ball Joint....13-58
Removal & Installation....13-58
Lower Control Arm....13-59
Removal & Installation....13-59
Stabilizer Bar....13-66
Removal & Installation....13-66
Steering Knuckle....13-60
Removal & Installation....13-60
Strut (Damper/Spring)....13-62
Overhaul....13-62
Removal & Installation....13-62
Wheel Hub & Bearing....13-67
Removal & Installation....13-67

REAR SUSPENSION....13-69
Coil Spring....13-69
Removal & Installation....13-69
Rear Axle Beam....13-70
Removal & Installation....13-70
Shock Absorber (Damper)....13-71
Removal & Installation....13-71
Testing....13-71
Wheel Hub & Bearing....13-72
Removal & Installation....13-72

SPECIFICATIONS AND MAINTENANCE CHARTS

ENGINE AND VEHICLE IDENTIFICATION

Engine							Model Year	
Code ①	Liters (cc)	Cu. In.	Cyl.	Fuel Sys.	Engine Type	Eng. Mfg.	Code ②	Year
CD3	1.5 (1496)	91.00	I4	MPFI	SOHC	Honda	9	2009
							A	2010

MPFI: Multi-Point Fuel Injection

DOHC: Double Overhead Camshafts

① 4th-6th digits of VIN

② 10th digit of VIN

37647_HFIT_C0001

GENERAL ENGINE SPECIFICATIONS

All measurements are given in inches.

Year	Model	Engine Displacement Liters	Engine Series VIN	Net Horsepower @ rpm	Net Torque @ rpm (ft. lbs.)	Bore x Stroke (in.)	Com-pression Ratio	Oil Pressure @ rpm
2009	Fit	1.5	CD3	109@5800	105@4800	3.11 x 3.52	10.4:1	15.6@Idle
2010	Fit	1.5	CD3	109@5800	105@4800	3.11 x 3.52	10.4:1	15.6@Idle

NA: Not Available

37647_HFIT_C0002

GASOLINE ENGINE TUNE-UP SPECIFICATIONS

Year	Engine Displacement Liters	Engine VIN	Spark Plug Gap (in.)	Ignition Timing (deg.)		Fuel Pump (psi)	Idle Speed (rpm)		Valve Clearance	
				MT	AT		MT	AT	In.	Ex.
2009	1.5	CD3	0.042-0.051	①	①	47-54	②	②	HYD	HYD
2010	1.5	CD3	0.042-0.051	①	①	47-54	②	②	HYD	HYD

NOTE: The Vehicle Emission Control Information label reflects specification changes made during production.
Follow the figures on the label if they differ from those in this chart.

HYD: Hydraulic

① Ignition timing is preset and cannot be adjusted

② Idle speed is maintained by the Electronic Control Module (ECM)

37647_HFIT_C0003

CAPACITIES

Year	Model	Engine Displacement Liters	Engine VIN	Engine Oil with Filter (qts.)	Transmission (pts.) Manual	Transmission (pts.) Auto. ①	Fuel Tank (gal.)	Cooling System (qts.)
2009	Fit	1.5	CD3	3.8	3.4	12.6	10.8	4.0
2010	Fit	1.5	CD3	3.8	3.4	12.6	10.8	4.0

NOTE: All capacities are approximate. Add fluid gradually and check to be sure a proper fluid level is obtained.

① Drain and refill

37647_HFIT_C0004

FLUID SPECIFICATIONS

Year	Model	Engine Displacement Liters	Engine ID/VIN	Engine Oil	Auto. Trans.	Manual Trans.	Power Steering Fluid	Brake Master Cylinder
2009	Fit	1.5	CD3	5W-20	ATF-Z1	①	—	②
2010	Fit	1.5	CD3	5W-20	ATF-Z1	①	—	②

DOT: Department Of Transportation

① Honda Manual Transmission Fluid (MTF): P/N 08798-9031

② DOT 3

37647_HFIT_C0005

VALVE SPECIFICATIONS

Year	Engine Displacement Liters	Engine VIN	Seat Angle (deg.)	Face Angle (deg.)	Spring Test Pressure (lbs. @ in.)	Spring Installed Height (in.)	Stem-to-Guide Clearance (in.) Intake	Stem-to-Guide Clearance (in.) Exhaust	Stem Diameter (in.) Intake	Stem Diameter (in.) Exhaust
2009	1.5	CD3	45	45	—	①	0.0008-0.0020	0.0020-0.0031	0.2157-0.2161	0.2146-0.2150
2010	1.5	CD3	45	45	—	①	0.0008-0.0020	0.0020-0.0031	0.2157-0.2161	0.2146-0.2150

① Free length - intake: 1.189 in.; exhaust: 2.259in.

37647_HFIT_C0006

CAMSHAFT AND BEARING SPECIFICATIONS CHART

All measurements are given in inches.

Year	Engine Displ. Liters	Engine ID/VIN	Journal Dia.	Brg. Oil Clearance	Shaft End-play	Runout	Journal Bore	Lobe Height Intake	Lobe Height Exhaust
2009	1.5	CD3	NA	0.0018-0.0033	0.002-0.0100	0.0010	NA	①	①
2010	1.5	CD3	NA	0.0018-0.0033	0.002-0.0100	0.0010	NA	①	①

NA: Not Available

① Intake Primary: 1.39291 inch

Intake Secondary: 1.20193 inch

Exhaust: 1.39321 inch

37647_HFIT_C0007

CRANKSHAFT AND CONNECTING ROD SPECIFICATIONS

All measurements are given in inches.

Year	Engine Displacement Liters	Engine VIN	Crankshaft Main Brg. Journal Dia.	Crankshaft Main Brg. Oil Clearance	Crankshaft Shaft End-play	Crankshaft Thrust on No.	Connecting Rod Journal Diameter	Connecting Rod Oil Clearance	Connecting Rod Side Clearance
2009	1.5	CD3	1.9676-1.9685	0.0020	0.018	—	1.5739-1.5748	0.0008-0.0015	0.006-0.0120
2010	1.5	CD3	1.9676-1.9685	0.0020	0.018	—	1.5739-1.5748	0.0008-0.0015	0.006-0.0120

37647_HFIT_C0008

PISTON AND RING SPECIFICATIONS

All measurements are given in inches.

Year	Engine Displ. Liters	Engine VIN	Piston Clearance	Ring Gap Top Compression	Ring Gap Bottom Compression	Ring Gap Oil Control	Ring Side Clearance Top Compression	Ring Side Clearance Bottom Compression	Ring Side Clearance Oil Control
2009	1.5	CD3	0.0004-0.0016	0.006-0.0120	0.014-0.0200	0.008-0.0280	0.0026-0.0035	0.0012-0.0022	NA
2010	1.5	CD3	0.0004-0.0016	0.006-0.0120	0.014-0.0200	0.008-0.0280	0.0026-0.0035	0.0012-0.0022	NA

NA: Not Available

37647_HFIT_C0009

TORQUE SPECIFICATIONS

All readings in ft. lbs.

Year	Engine Displacement Liters	Engine VIN	Cylinder Head Bolts	Main Bearing Bolts	Rod Bearing Bolts	Crankshaft Damper Bolts	Flywheel Bolts	Manifold Intake	Manifold Exhaust	Spark Plugs	Oil Pan Drain Plug
2009	1.5	CD3	①	②	③	④	NA	17	33	13	NA
2010	1.5	CD3	①	②	③	④	NA	17	33	13	NA

NA: Not Available

① Step 1: 22 ft. lbs.
Step 2: Plus 130 degrees

② Step 1: 18 ft. lbs.
Step 2: Plus 40 degrees

③ Step 1: 7.2 ft. lbs.
Step 2: Plus 90 degrees

④ Step 1: New bolt: 130 ft. lbs.
Step 2: 27 ft. lbs.

37647_HFIT_C0010

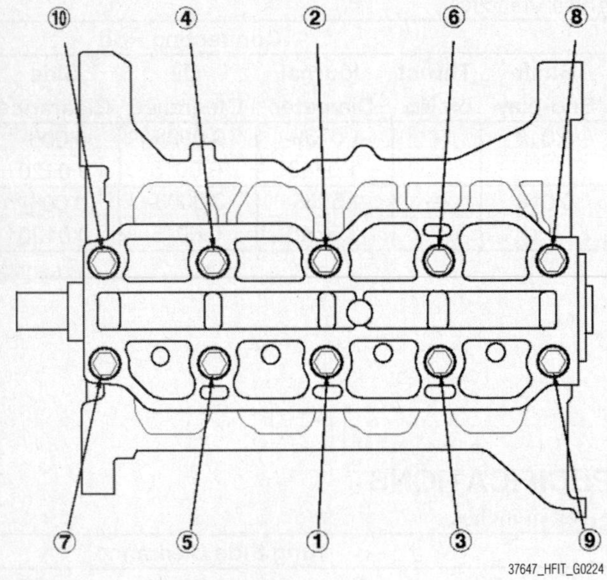
10
4
2
6
8
7
5
1
3
9
37647_HFIT_G0224

Fig. 1 Main bearing torque sequence

WHEEL ALIGNMENT

Year	Model		Caster Range (+/-Deg.)	Caster Preferred Setting (Deg.)	Camber Range (+/-Deg.)	Camber Preferred Setting (Deg.)	Toe-in (mm)
2009	Fit	Front	3.75 +/- 1.0	0.0	0.0 +/- 1.0	0.0	0.0 +/- 3.0
		Rear	—	—	-1.5 +/- 1.0	—	2.5 +/- 2.5
2010	Fit	Front	3.75 +/- 1.0	0.0	0.0 +/- 1.0	0.0	0.0 +/- 3.0
		Rear	—	—	-1.5 +/- 1.0	—	2.5 +/- 2.5

37647_HFIT_C0011

TIRE, WHEEL AND BALL JOINT SPECIFICATIONS

Year	Model	OEM Tires Standard	OEM Tires Optional	Tire Pressures (psi) Front	Tire Pressures (psi) Rear	Wheel Size	Ball Joint Inspection	Lug Nut Torque (ft. lbs.)
2009	Fit	P175/65R14	P195/55R15	①	①	N/A	②	80
2010	Fit	P175/65R14	P195/55R15	①	①	N/A	②	80

① Refer to placard on vehicle for proper inflation pressure.

② Replace if any measurable movement is found.

37647_HFIT_C0012

BRAKE SPECIFICATIONS

All measurements in inches unless noted

Year	Model		Brake Disc Original Thickness	Brake Disc Minimum Thickness	Brake Disc Maximum Runout	Brake Drum Diameter Original Inside Diameter	Brake Drum Diameter Max. Wear Limit	Brake Drum Diameter Maximum Machine Diameter	Minimum Lining Thickness	Brake Caliper Bracket Bolts (ft. lbs.)	Brake Caliper Mounting Bolts (ft. lbs.)
2009	Fit	F	0.830	0.750	0.004	—	—	—	0.060	62-69	17
		R	—	—	—	7.870-7.874	7.91	7.91	0.080	—	—
2010	Fit	F	0.830	0.750	0.004	—	—	—	0.060	62-69	17
		R	—	—	—	7.870-7.874	7.91	7.91	0.080	—	—

37647_HFIT_C0013

SCHEDULED MAINTENANCE INTERVALS

HONDA - FIT

NOTE: HONDA FIT uses a Maintenance Service light system. There are few items associated with specific mileage intervals; most are based on SYMBOLS that appear with the Maintenance Service light; such as: If Service light shows symbol "A" at 15%, service is due soon; at 5%, service is due immediately; at 0%, service is past due.

37647_HFIT_C0014

BRAKES — INFORMATION AND PRECAUTIONS

ANTI-LOCK SYSTEMS

- Certain components within the ABS system are not intended to be serviced or repaired individually.
- Do not use rubber hoses or other parts not specifically specified for and ABS system. When using repair kits, replace all parts included in the kit. Partial or incorrect repair may lead to functional problems and require the replacement of components.
- Lubricate rubber parts with clean, fresh brake fluid to ease assembly. Do not use shop air to clean parts; damage to rubber components may result.
- Use only DOT 3 brake fluid from an unopened container.
- If any hydraulic component or line is removed or replaced, it may be necessary to bleed the entire system.
- A clean repair area is essential. Always clean the reservoir and cap thoroughly before removing the cap. The slightest amount of dirt in the fluid may plug an orifice and impair the system function. Perform repairs after components have been thoroughly cleaned; use only denatured alcohol to clean components. Do not allow ABS components to come into contact with any substance containing mineral oil; this includes used shop rags.
- The Anti-Lock control unit is a microprocessor similar to other computer units in the vehicle. Ensure that the ignition switch is **OFF** before removing or installing controller harnesses. Avoid static electricity discharge at or near the controller.
- If any arc welding is to be done on the vehicle, the control unit should be unplugged before welding operations begin.

DISC AND DRUM SYSTEMS

CAUTION

Dust and dirt accumulating on brake parts during normal use may contain asbestos fibers from production or aftermarket brake linings. Breathing excessive concentrations of asbestos fibers can cause serious bodily harm. Exercise care when servicing brake parts. Do not sand or grind brake lining unless equipment used is designed to contain the dust residue. Do not clean brake parts with compressed air or by dry brushing. Cleaning should be done by dampening the brake components with a fine mist of water, then wiping the brake components clean with a dampened cloth. Dispose of cloth and all residue containing asbestos fibers in an impermeable container with the appropriate label. Follow practices prescribed by the Occupational Safety and Health Administration (OSHA) and the Environmental Protection Agency (EPA) for the handling, processing, and disposing of dust or debris that may contain asbestos fibers.

BRAKES — BLEEDING THE BRAKE SYSTEM

BLEEDING PROCEDURE

BLEEDING PROCEDURE

WARNING

Do not reuse the drained fluid. Use only clean Honda DOT 3 Brake Fluid from an unopened container. Using a non-Honda brake fluid can cause corrosion and shorten the life of the system. Do not mix different brands of brake fluid; they may not be compatible. Make sure no dirt or other foreign matter is allowed to contaminate the brake fluid. Do not spill brake fluid on the vehicle, it may damage the paint; if brake fluid does contact the paint, wash it off immediately with water.

1. Ensure the reservoir connected to the master cylinder is at the MAX (upper) level mark at the start of the bleeding procedure and checked after bleeding each brake system. Add fluid as required.
2. Have someone slowly pump the brake pedal several times, then apply steady pressure.
3. Start the bleeding at the driver's side of the front brake system.
4. Bleed the calipers or the wheel cylinders in the following sequence:
   - Left front
   - Right front
   - Right rear
   - Left rear
5. Attach a length of clear drain tube to the bleed screw, then, loosen the bleed screw to allow air to escape from the system. Then tighten the bleed screw securely.
6. Refill the master cylinder reservoir to the MAX (upper) level line.
7. Repeat the procedure for each brake circuit until no air bubbles are in the fluid.

BLEEDING THE ABS SYSTEM

➡Brake fluid replacement and air bleeding procedures are identical to the procedures used on vehicles without ABS. Refer to the Bleeding Procedure above.

WHEEL SPEED SENSORS

REMOVAL & INSTALLATION

Front

See Figure 2.

1. Turn the ignition switch to LOCK (0).
2. Remove the grommet, then disconnect the wheel speed sensor connector.
3. Remove the bracket, the clip, and the wire guide grommet.
4. Remove the bolts and the wheel speed sensor.
5. Install the wheel speed sensor in the reverse order of removal, and note these items:
   - Do not twist the sensor wires.
   - If the wheel speed sensor comes in contact with the hub bearing unit, it is faulty.
   - Make sure the grommet is installed properly.
   - Make sure there is no debris in the sensor mounting hole.
6. Start the engine, and make sure the ABS and the VSA indicators go off.
7. Test-drive the vehicle, and make sure the ABS and the VSA indicators do not come on.

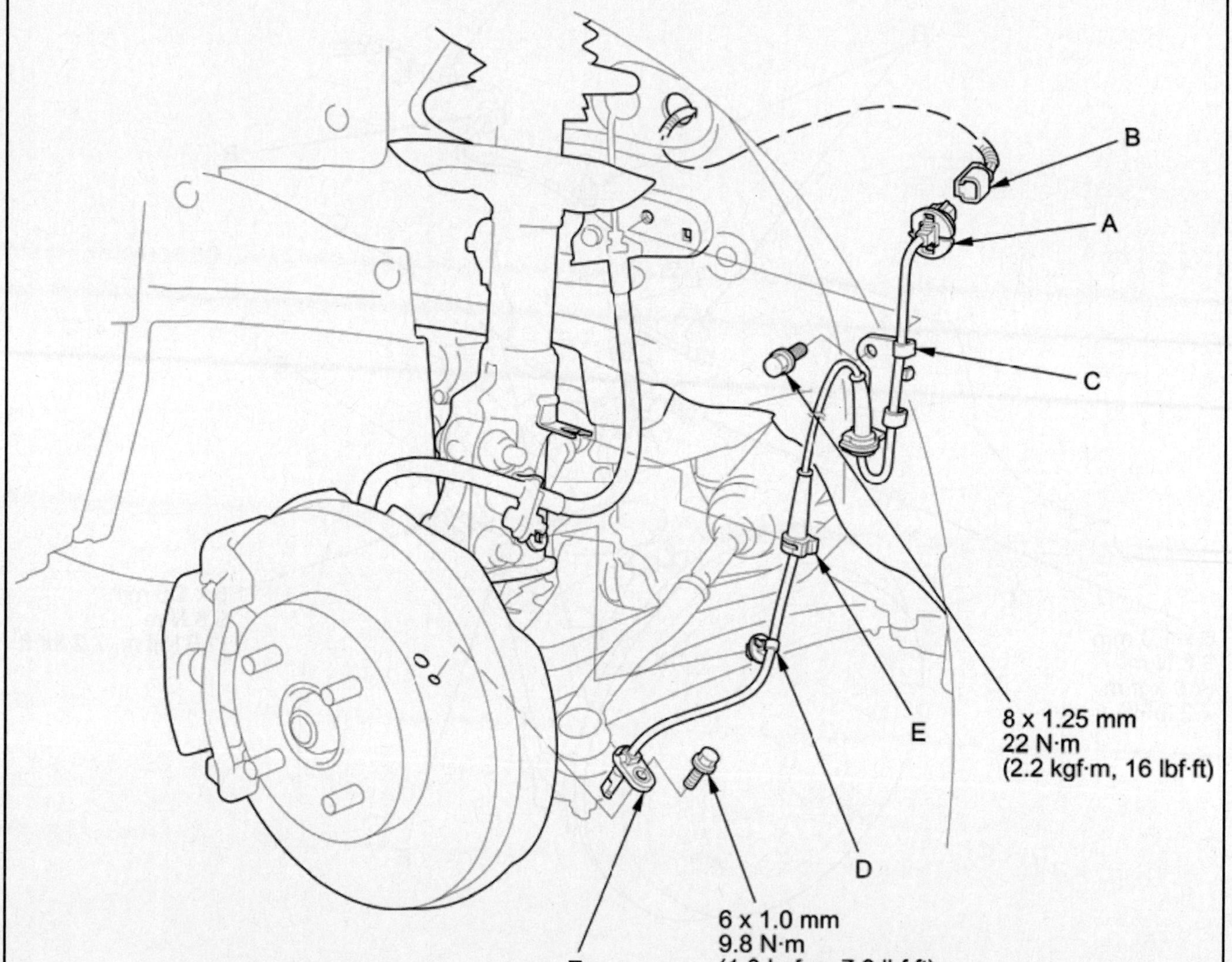
B
A
C
8 x 1.25 mm
22 N·m
(2.2 kgf·m, 16 lbf·ft)
E
D
6 x 1.0 mm
9.8 N·m
(1.0 kgf·m, 7.2 lbf·ft)
F
A. Grommet
B. Wheel speed sensor connector
C. Bracket
D. Clip
E. Wire guide grommet
F. Wheel speed sensor
37647_HFIT_G0067

Fig. 2 Front wheel speed sensor assembly

Rear

See Figure 3.

1. Turn the ignition switch to LOCK (0).
2. Pull back the carpet under the rear seat, then disconnect the wheel speed sensor connector.
3. Remove the grommet, the bracket, the clips, and the wire guide grommet.
4. Remove the bolt and the wheel speed sensor.
5. Install the wheel speed sensor in the reverse order of removal, and note these items:
   - Do not twist the sensor wires.
   - If the wheel speed sensor comes in contact with the hub bearing unit, it is faulty.
   - Make sure the grommet is installed properly.
   - Make sure there is no debris in the sensor mounting hole.
6. Start the engine, and make sure the ABS and the VSA indicators go off.
7. Test-drive the vehicle, and make sure the ABS and the VSA indicators do not come on.

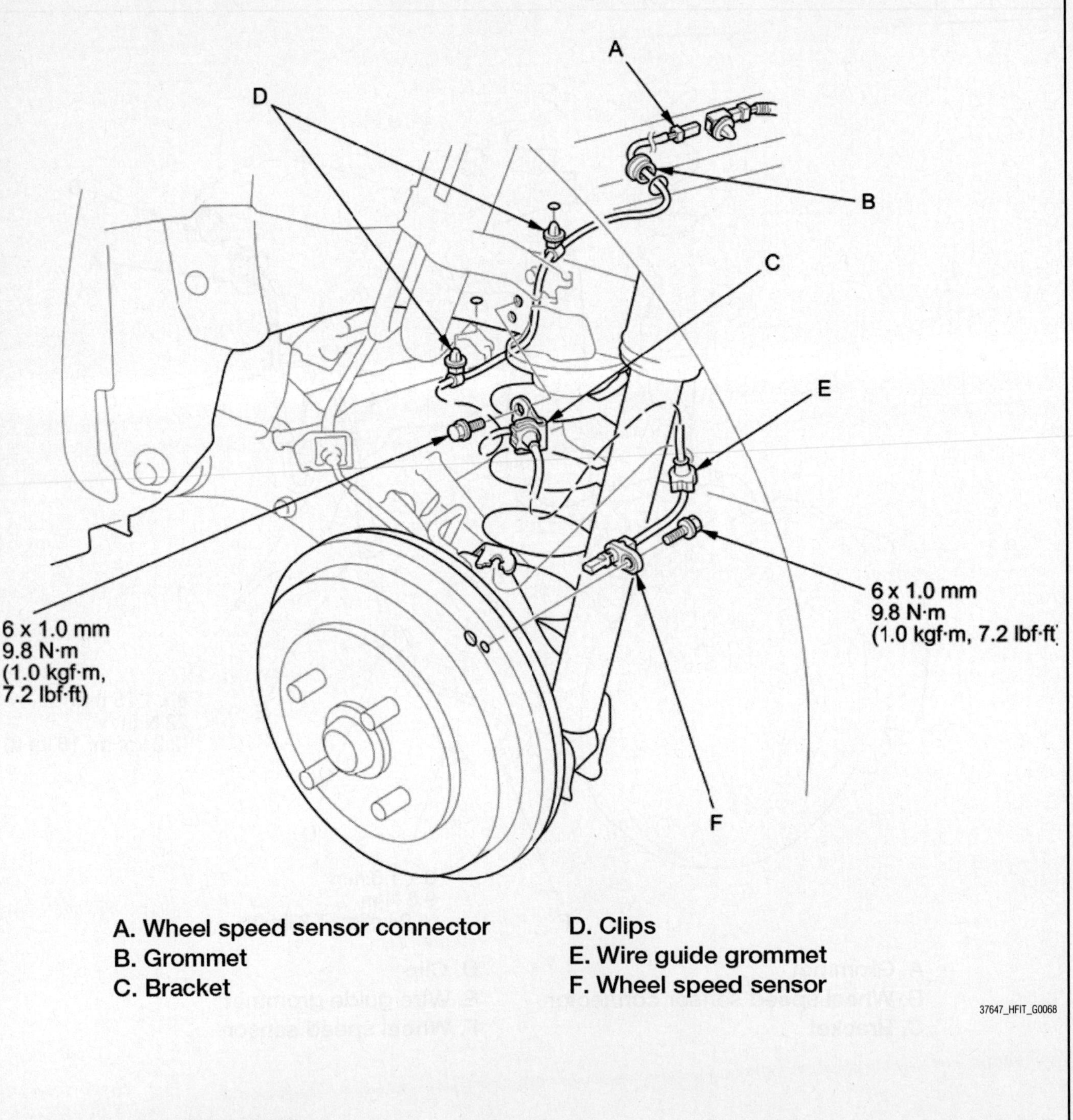
A
B
C
D
E
F
6 x 1.0 mm
9.8 N·m
(1.0 kgf·m, 7.2 lbf·ft)
6 x 1.0 mm
9.8 N·m
(1.0 kgf·m,
7.2 lbf·ft)
A. Wheel speed sensor connector
B. Grommet
C. Bracket
D. Clips
E. Wire guide grommet
F. Wheel speed sensor
37647_HFIT_G0068

Fig. 3 Rear wheel speed sensor assembly

BRAKES — FRONT DISC BRAKES

BRAKE CALIPER

REMOVAL & INSTALLATION

See Figure 4.

1. Remove the wheel nuts and front wheel.
2. Remove the brake hose mounting bolt.
3. Remove the brake caliper bracket mounting bolts, and remove the caliper assembly from the knuckle.
4. Detach the brake caliper from the brake hose, if the caliper is being replaced.
5. If the caliper is not being replaced, it can be moved out of the way and suspended with wire, so the hose is not under strain.

WARNING

To prevent damage to the caliper assembly or brake hose, use a short piece of wire to hang the caliper assembly from the undercarriage. Do not twist the brake hose excessively.

To install:

6. Position the caliper in place.

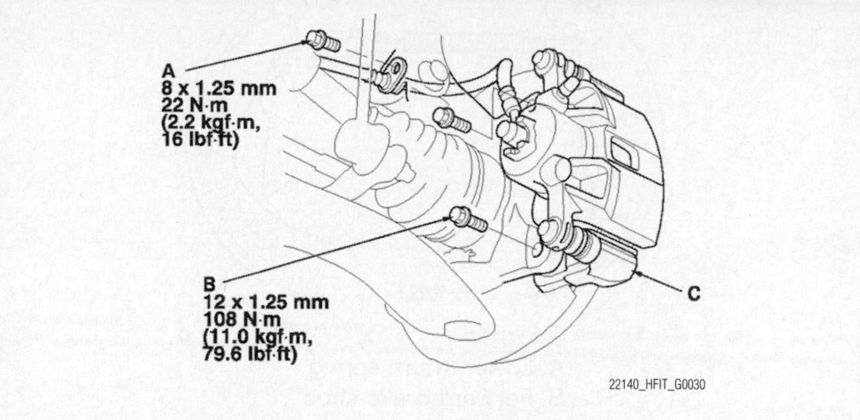
A
8 x 1.25 mm
22 N·m
(2.2 kgf·m,
16 lbf·ft)
B
12 x 1.25 mm
108 N·m
(11.0 kgf·m,
79.6 lbf·ft)
C
22140_HFIT_G0030

Fig. 4 Showing the front brake caliper brake hose mounting bolt (A), caliper mounting bolts (B), and caliper assembly (C)

7. If removed, reattach the brake line to the caliper.
8. Install and tighten the caliper mounting bolts to 80 ft. lbs. (108 Nm).
9. Install the brake hose clip retaining bolt.
10. If the brake hose was removed from the caliper, the system, perform "Bleeding Procedure" in this section.

DISC BRAKE PADS

REMOVAL & INSTALLATION

➡Manufacturer does not provide a separate procedure for disc brake pad removal and installation.

BRAKES — REAR DRUM BRAKES

BRAKE DRUM

REMOVAL & INSTALLATION

See Figure 5.

CAUTION

Keep any grease off the brake drum and brake shoes.

1. Raise the rear of the vehicle, and support it with safety stands in the proper locations.
2. Remove the rear wheel.
3. Remove the parking brake, and remove the brake drum from the hub bearing unit. If necessary, turn the adjuster bolt with a flat-tip screwdriver until the shoes become loose.
4. If the brake drum has clung to the hub bearing unit, thread two 8 x 1.25 mm bolts into the brake drum to push it away from the hub bearing unit. Turn each bolt 90 degrees at a time to prevent cocking the brake drum.

To install:

5. Install the brake drum in the reverse order of removal.
   a. After installation, press the brake pedal several times to make sure the brakes work and self adjust the brake shoes.
   b. Before installing the brake drum, clean the mating surfaces of the rear hub and the inside of the brake drum.
   c. Clean the mating surfaces of the brake drum and the inside of the wheel, then install the rear wheel.

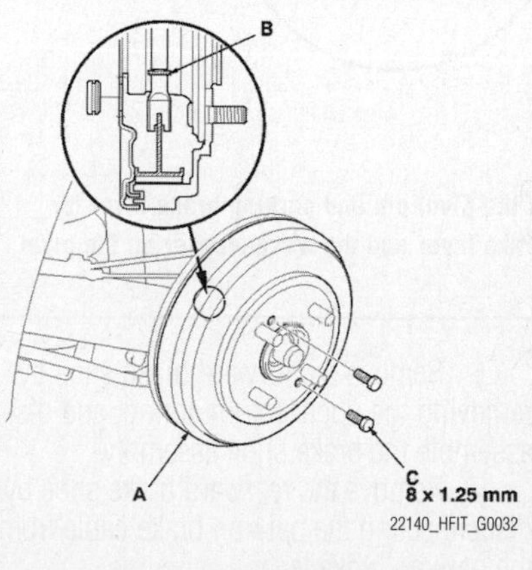
B
A
C
8 x 1.25 mm
22140_HFIT_G0032

Fig. 5 Remove the parking brake, then the brake drum (A), turn the adjuster bolt (B) until the shoes become loose

BRAKE SHOES

REMOVAL & INSTALLATION

See Figures 6 through 9.

1. Raise the rear of the vehicle, and support it with safety stands in the proper locations.

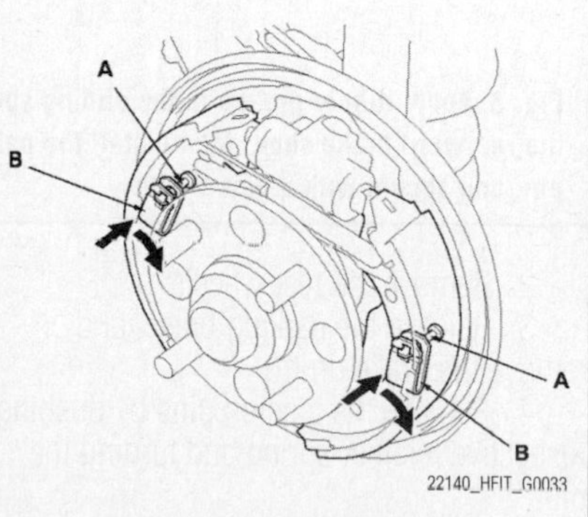
A
B
A
B
22140_HFIT_G0033

Fig. 6 Removing the tension pins (A) by pushing the retainer spring (B)

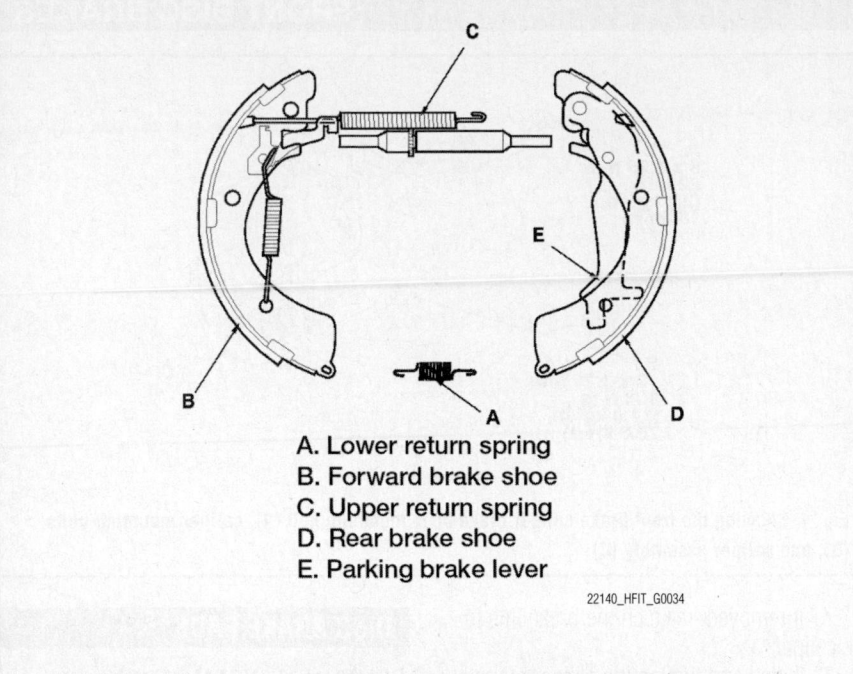

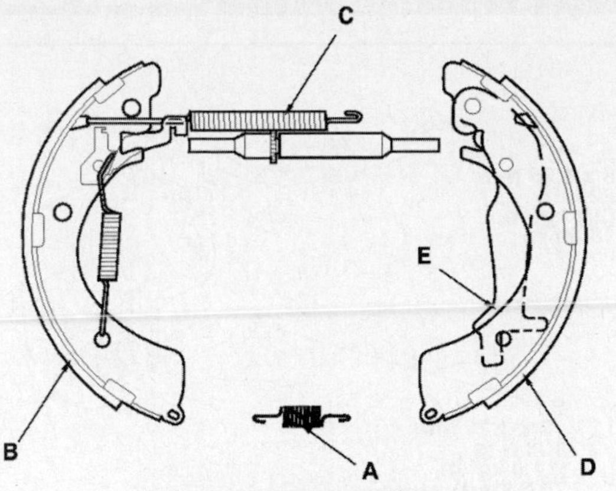
C
E
B
A
D
A. Lower return spring
B. Forward brake shoe
C. Upper return spring
D. Rear brake shoe
E. Parking brake lever
22140_HFIT_G0034

Fig. 7 Showing the lower return spring, forward brake shoe, upper return spring, and rear brake shoe, and parking brake lever locations

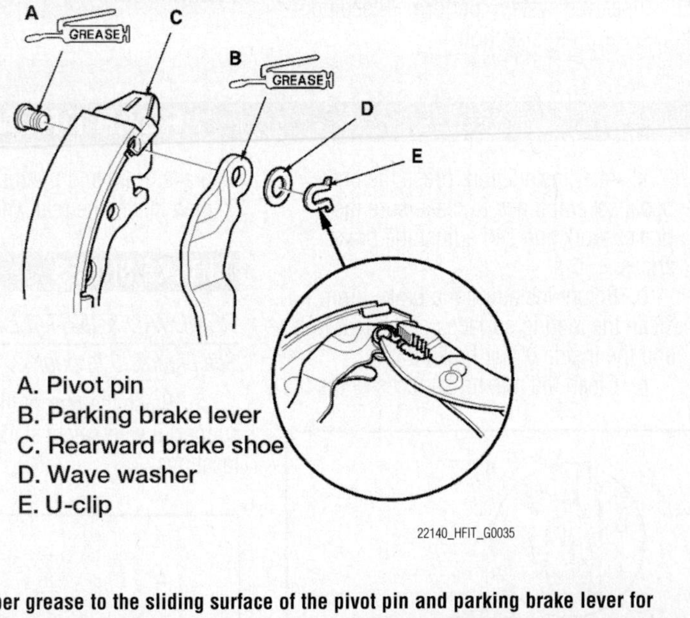
A
GREASE
C
B
GREASE
D
E
A. Pivot pin
B. Parking brake lever
C. Rearward brake shoe
D. Wave washer
E. U-clip
22140_HFIT_G0035

Fig. 8 Apply rubber grease to the sliding surface of the pivot pin and parking brake lever for the rearward brake shoe, then install the parking brake lever and the wave washer on the pivot pin, and secure with a new U-clip

2. Remove the rear wheels.
3. Release the parking brake, and remove the brake drum.
4. Remove the tension pins by pushing respective retainer spring and turning the pin.
5. Remove the lower return spring, and remove the brake shoe assembly over the hub.
6. Remove the forward brake shoe by removing the upper return spring, and disassemble the brake shoe assembly.
7. Remove the rearward brake shoe by disconnecting the parking brake cable from the parking brake lever.
8. Remove the U-clip, wave washer, and pivot pin, and separate the parking brake lever from the brake shoe.

To install:

9. Apply rubber grease to the sliding surface of the pivot pin and parking brake lever for the rearward brake shoe.
10. Install the parking brake lever and the wave washer on the pivot pin, and secure with a new U-clip.

➡Pinch the U-clip securely to prevent the parking brake lever from coming out of the brake shoe.

11. Connect the parking brake cable to the parking brake lever.
12. Apply a thin coat of rubber grease to the connecting rod ends and the sliding surfaces. Wipe off any excess. Keep grease off the brake linings.
13. Apply a thin coat of Molykote 44 MA grease to the shoe ends and the edge of the shoe surfaces that contact the backing plate. Wipe off any excess. Keep grease off the brake linings.
14. Install connecting rods A and B on the adjuster bolt.

➡Clean the threaded portions of connecting rod A and the sliding surface of connecting rod B, then coat them with rubber grease.

15. Shorten connecting rod A by fully turning the adjuster bolt.
16. Assemble the brake shoes, the upper return spring, and with the connecting rods the adjuster bolt on the backing plate, then install the self-adjuster lever and the self-adjuster spring on the forward brake shoe.
17. Install the tension pins and the retainer springs by pushing in respective spring and turning each pin. Install the lower return spring.

➡Make sure the brake shoe positioning on the brake shoe bosses of the backing

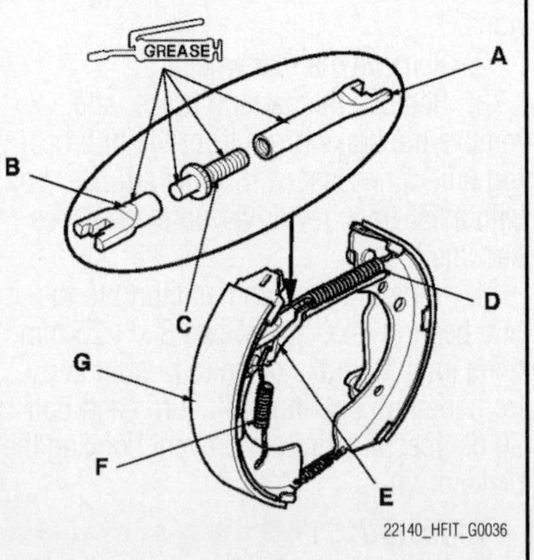
GREASE
A
B
C
D
G
F
E
22140_HFIT_G0036

Fig. 9 Reassembling the rear brake shoes

plate, and fitting the top of the brake shoes onto the wheel cylinder pistons.

18. Before installing the brake drum, clean the mating surface of the rear hub and the inside of the brake drum.
19. Install the brake drum.
20. Install the rear wheels.
21. Press the brake pedal several times to make sure the brakes work and to set the self-adjusting brake.

➡Engagement of the brakes may require a greater pedal stroke immediately after the brake shoes have been replaced as a set. Several applications of the brake pedal will restore the normal pedal stroke.

22. Adjust the parking brake.

BRAKES — PARKING BRAKE

PARKING BRAKE CABLES

ADJUSTMENT

See Figure 10.

1. Pull the parking brake lever with about 44 lbs. of force to fully apply the parking brake. The parking brake lever should be locked within 6–8 clicks. If the number of lever clicks is excessive, adjust the parking brake.
2. Remove the center console.
3. Release the parking brake lever fully.
4. Loosen the parking brake cable adjusting nut.

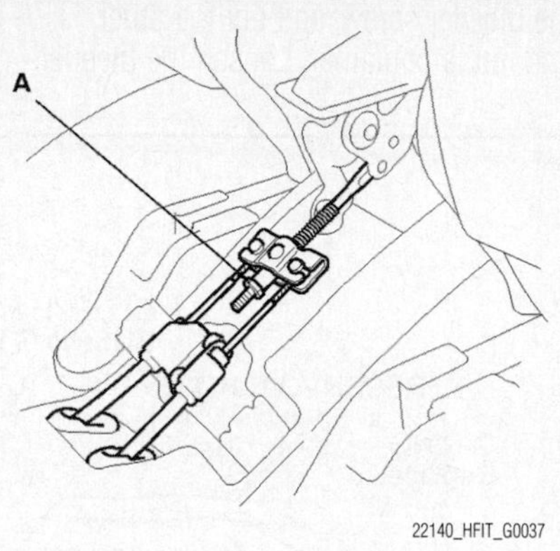
A
22140_HFIT_G0037

Fig. 10 Showing the parking brake cable adjusting nut (A)

5. Press the brake pedal several times to set the self-adjusting brake before adjusting the parking brake.
6. Pull the parking brake lever 1 click.
7. Tighten the parking brake cable adjusting nut until the parking brakes drag slightly when the rear wheels are turned.
8. Release the parking brake lever fully, and check that the parking brakes do not drag when the rear wheels are turned. Readjust if necessary.
9. Make sure the parking brakes are fully applied when the parking brake lever is pulled all the way.
10. Install the center console.

CHASSIS ELECTRICAL — AIR BAG (SUPPLEMENTAL RESTRAINT SYSTEM)

GENERAL INFORMATION

CAUTION

These vehicles are equipped with an air bag system. The system must be disarmed before performing service on, or around, system components, the steering column, instrument panel components, wiring and sensors. Failure to follow the safety precautions and the disarming procedure could result in accidental air bag deployment, possible injury and unnecessary system repairs.

PRECAUTIONS

CAUTION

Disconnect and isolate the battery negative cable before beginning any airbag system component diagnosis, testing, removal, or installation procedures. Wait at least 90 seconds after the ignition switch is turned off and the negative (-) terminal cable is disconnected from the battery before starting the operation. The SRS is equipped with a backup power source, so if work is started within 90 seconds after disconnecting the negative (-) terminal cable from the battery, the SRS may be deployed. Failure to disable the airbag system may result in accidental airbag deployment, personal injury, or death.

DISARMING THE SYSTEM

1. Before servicing the vehicle, refer to the Precautions Section.
2. Turn the ignition switch to the **LOCK** position.
3. Disconnect the negative battery cable.
4. Wait three minutes for the battery power to fully discharge from the system.

ARMING THE SYSTEM

1. Before servicing the vehicle, refer to the Precautions Section.
2. Connect the negative battery cable.
3. Turn the ignition switch **ON**.
4. Verify that the air bag indicator illuminates for 4–8 seconds, then goes off.

CABLE REEL CENTERING

See Figure 11.

Center the cable reel by first rotating the cable reel clockwise until it stops. Then rotate it counterclockwise (about three turns) until the arrow mark on the cable reel label points straight up.

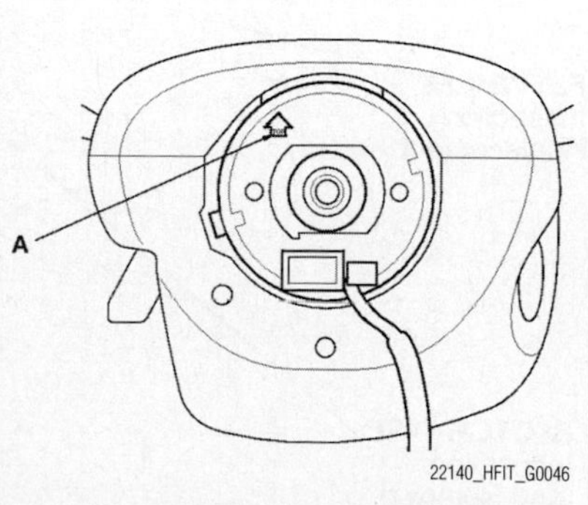
A
22140_HFIT_G0046

Fig. 11 Showing the cable reel and arrow mark

DRIVE TRAIN

CLUTCH DRIVEN DISC & PRESSURE PLATE

REMOVAL & INSTALLATION

See Figure 12.

➡The manufacturer does not provide a specific Removal and Installation procedure for this component. Refer to the graphic(s) when servicing this component.

HYDRAULIC SYSTEM BLEEDING

BLEEDING PROCEDURE

1. Make sure the brake fluid level in the clutch reservoir is at the MAX (upper) level line.
2. Attach one end of a clear tube to the bleeder screw, and put the other end into a container. Loosen the bleeder screw to allow air to escape from the system.
3. Make sure there is an adequate supply of fluid in the reservoir, then slowly push the clutch pedal all the way down. Before releasing the pedal, have an assistant temporarily tighten the bleeder screw. Loosen the bleeder screw and push the clutch pedal down again. Repeat this step until no more bubbles appear at the clear tube.

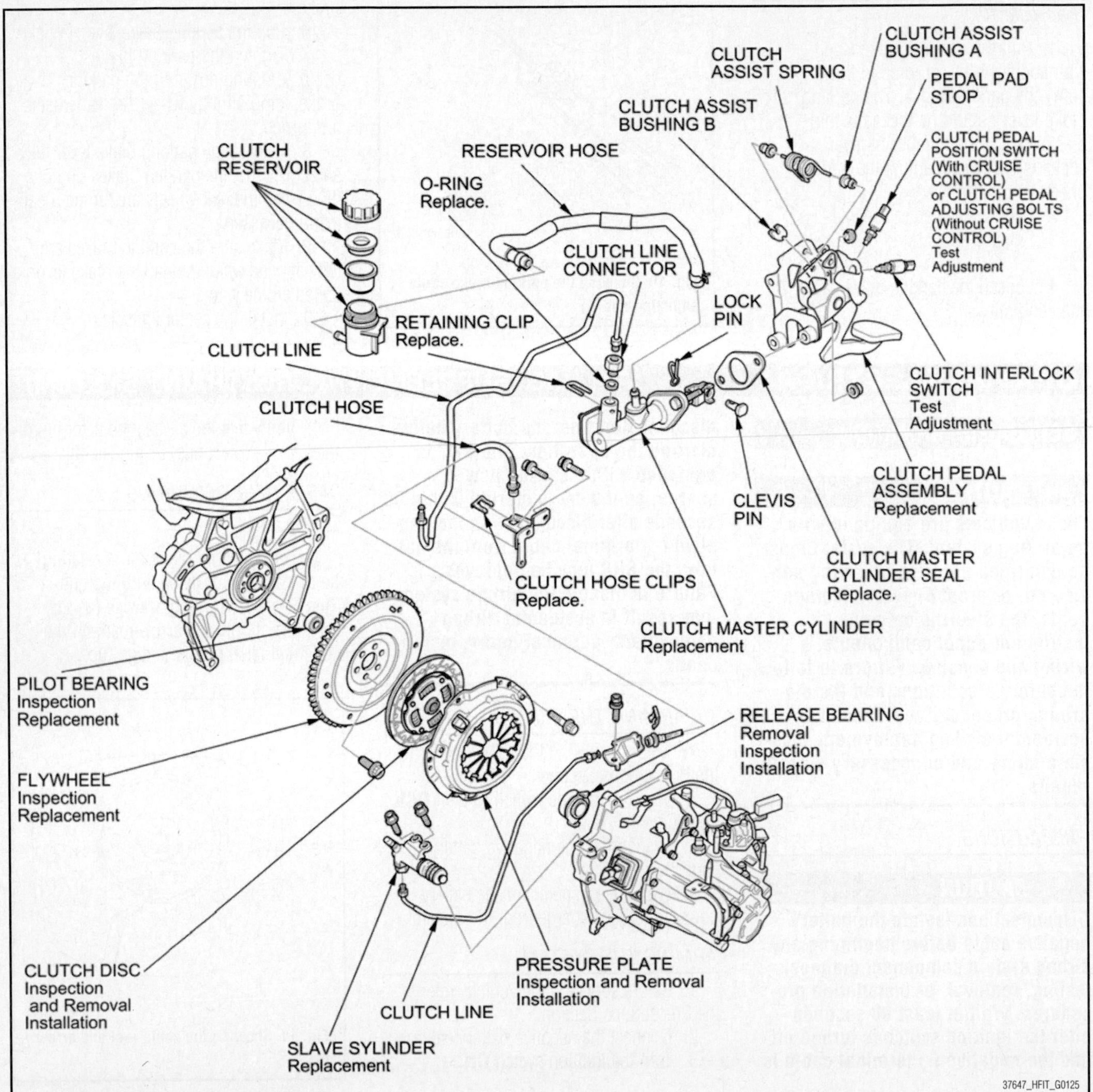
CLUTCH ASSIST BUSHING A
CLUTCH ASSIST SPRING
PEDAL PAD STOP
CLUTCH ASSIST BUSHING B
CLUTCH PEDAL POSITION SWITCH (With CRUISE CONTROL) or CLUTCH PEDAL ADJUSTING BOLTS (Without CRUISE CONTROL) Test Adjustment
CLUTCH RESERVOIR
RESERVOIR HOSE
O-RING Replace.
CLUTCH LINE CONNECTOR
LOCK PIN
RETAINING CLIP Replace.
CLUTCH LINE
CLUTCH INTERLOCK SWITCH Test Adjustment
CLUTCH HOSE
CLUTCH PEDAL ASSEMBLY Replacement
CLEVIS PIN
CLUTCH MASTER CYLINDER SEAL Replace.
CLUTCH HOSE CLIPS Replace.
CLUTCH MASTER CYLINDER Replacement
PILOT BEARING Inspection Replacement
RELEASE BEARING Removal Inspection Installation
FLYWHEEL Inspection Replacement
CLUTCH DISC Inspection and Removal Installation
PRESSURE PLATE Inspection and Removal Installation
CLUTCH LINE
SLAVE SYLINDER Replacement
37647_HFIT_G0125

Fig. 12 Exploded view of clutch assembly

➡Make sure the fluid level on the reservoir does not go below MIN.

4. Tighten the bleeder screw securely.
5. Refill the brake fluid in the reservoir to the MAX (upper) level line.

FRONT HALFSHAFT

REMOVAL & INSTALLATION

See Figures 13 and 14.

1. Raise and support the vehicle.
2. Remove the front wheels.
3. Pry up the stake on the spindle nut, then remove the nut.
4. Drain the transmission fluid. Reinstall the drain plug using a new sealing washer:
5. Remove the lock pin from the lower arm ball joint, then remove the castle nut. Separate the ball joint from the knuckle using the 14 mm ball joint thread protector and the 28 mm ball joint remover .

➡Be careful not to damage the ball joint boot when installing the remover. Do not hammer or pry on the lower arm to disconnect the ball joint or you will damage the ball joint.

6. Pull the knuckle outward, and separate the outboard joint from the front hub using a soft face hammer.
7. Left and right driveshaft (M/T model)/left driveshaft (A/T model): Pry the inboard joint from the differential with a prybar. Remove the driveshaft as an assembly.

➡Do not pull the assembly by the driveshaft, or the inboard joint may come apart. Pull the inboard joint straight out to avoid damaging the oil seal. Be careful not to damage the oil seal or the end of the inboard joint with the prybar.

8. Right driveshaft (A/T model): Drive the inboard joint off of the intermediate shaft using a drift punch and a hammer. Remove the driveshaft as an assembly.

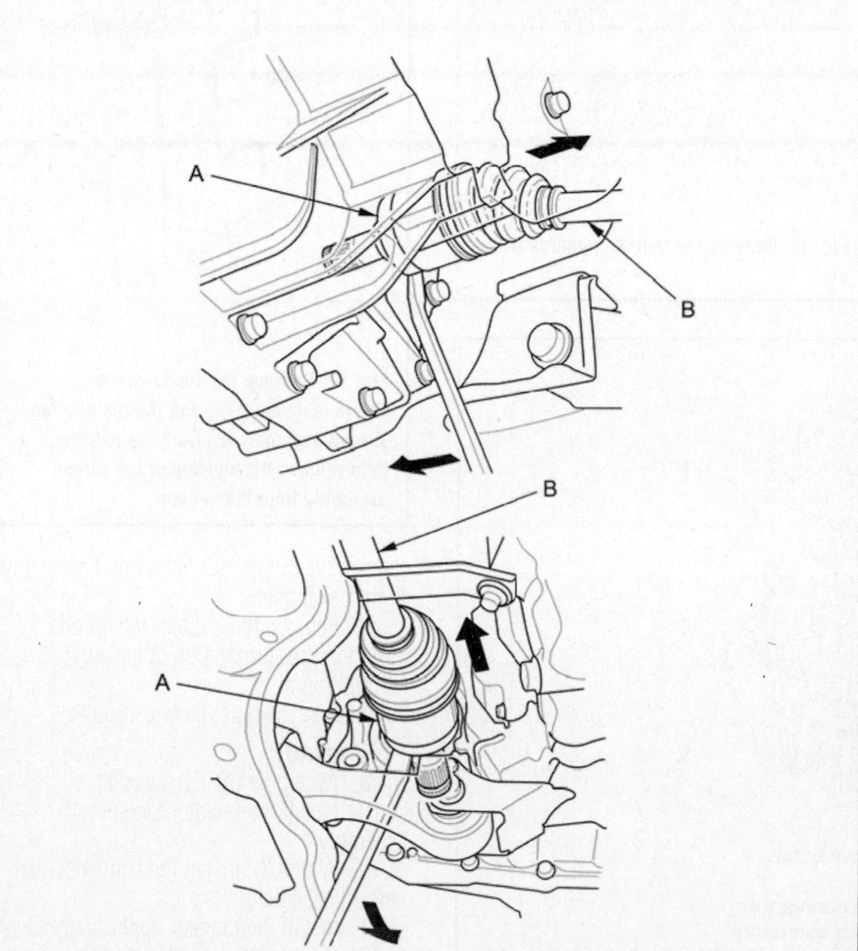
A
B
A
B
37647_HFIT_G0133

Fig. 13 Left and right driveshaft (M/T model)/left driveshaft (A/T model)

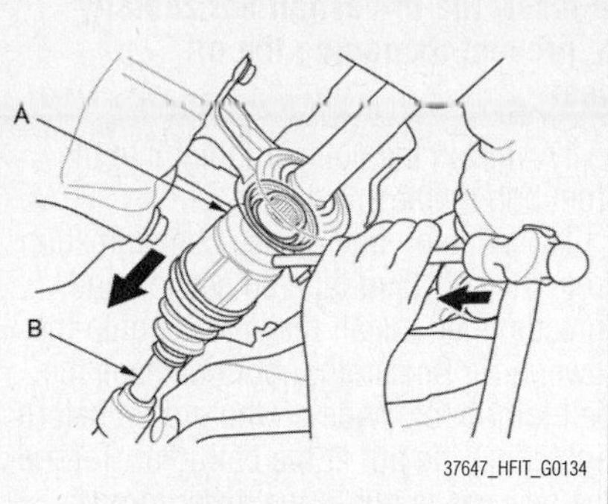
A
B
37647_HFIT_G0134

Fig. 14 Right driveshaft (A/T model)

➡Do not pull the assembly by the driveshaft, or the inboard joint may come apart.

9. Remove the set ring from the inboard joint (Except A/T model right driveshaft).
10. Remove the set ring from the intermediate shaft (A/T model right driveshaft).

To install:

NOTE: Before starting installation, make sure the mating surfaces of the joint and the splined section are clean.

11. Apply Moly 60 paste (P/N 08734-0001) to the contact area of the outboard joint and the front wheel bearing.

➡The paste helps to prevent noise and vibration.

12. Install a new set ring onto the set ring groove of the driveshaft inboard joint (except A/T model right driveshaft).
13. Install a new set ring onto the set ring groove of the intermediate shaft (A/T model).
14. Apply super high temp urea grease (P/N 08798-9002) to the whole splined surface of the right driveshaft. After applying grease, remove the grease from the splined grooves at intervals of 2–3 splines and from the set ring groove so that air can bleed from the intermediate shaft.
15. Clean the areas where the driveshaft contacts the differential thoroughly with solvent or brake cleaner, and dry them with compressed air.

➡Do not wash the rubber parts with solvent.

16. Insert the inboard end of the driveshaft into the differential or intermediate shaft until the set ring locks in the groove.

➡Insert the driveshaft horizontally to prevent damaging the oil seal.

17. Install the outboard joint into the front hub on the knuckle.
18. Wipe off any grease contamination from the ball joint tapered section and threads, then install the knuckle onto the lower arm. Be careful not to damage the ball joint boot. Wipe off the grease before tightening the nut at the ball joint. Torque the new castle nut to the lower torque specification, then tighten it only far enough to align the slot with the ball joint pin hole.

➡Make sure the ball joint boot is not damaged or cracked. Do not align the nut by loosening it.

19. Install the lock pin into the ball joint pin hole.
20. Apply a small amount of engine oil to the seating surface of a new spindle nut.
21. Install the spindle nut, then tighten it. After tightening, use a drift to stake the spindle nut shoulder against the driveshaft.
22. Clean the mating surfaces of the brake disc and the wheel, then install the front wheels.
23. Turn the front wheel by hand, and make sure there is no interference between the driveshaft and surrounding parts.
24. Refill the transmission with the recommended transmission fluid:
25. Lower the vehicle.
26. Check the wheel alignment, and adjust it if necessary.
27. Test-drive the vehicle.

CV-BOOTS INSPECTION

Inspect the CV boots for cracks, wear, leakage and for damaged mounting bands.

ENGINE COOLING

ENGINE FAN

REMOVAL & INSTALLATION

See Figures 15 through 18.

1. Remove the coolant reservoir.
2. Disconnect the radiator fan motor connector, then remove the harness clamp.
3. With A/C: Disconnect the condenser fan motor connector and the A/C compressor clutch connector, then remove the harness clamps.
4. Remove the radiator upper brackets.
5. With A/C: Remove the condenser fan shroud assembly and the radiator fan shroud assembly from the radiator, then remove the condenser fan shroud assembly from the vehicle.
6. Remove the radiator fan shroud assembly from right side of the engine compartment.
7. Disassemble the fan shrouds.

To install:

8. Assemble the fan shrouds.
9. Install the radiator fan shroud assembly.
10. With A/C: Install the condenser fan shroud assembly.
11. Install the radiator upper brackets.
12. Connect the radiator fan motor connector, then install the harness clamp.
13. With A/C: Connect the condenser fan motor connector and the A/C compressor

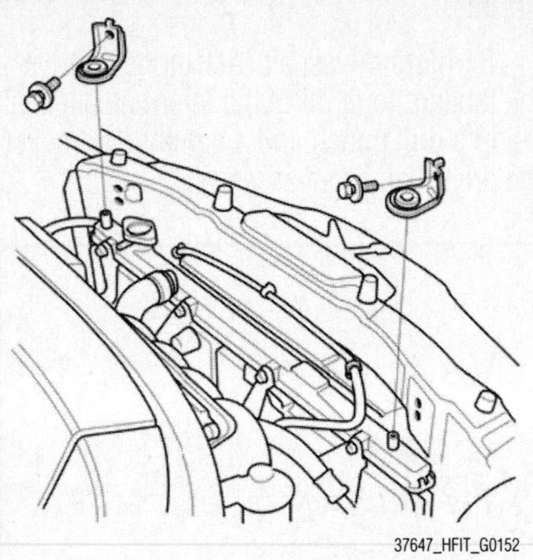
37647_HFIT_G0152

Fig. 16 Remove the radiator upper brackets

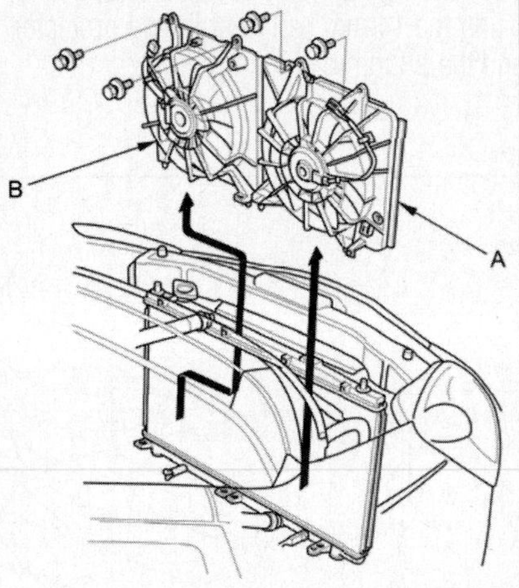
B
A
37647_HFIT_G0153

Fig. 17 Remove the condenser fan shroud assembly (A) and the radiator fan shroud assembly (B) from the radiator, then remove the condenser fan shroud assembly from the vehicle

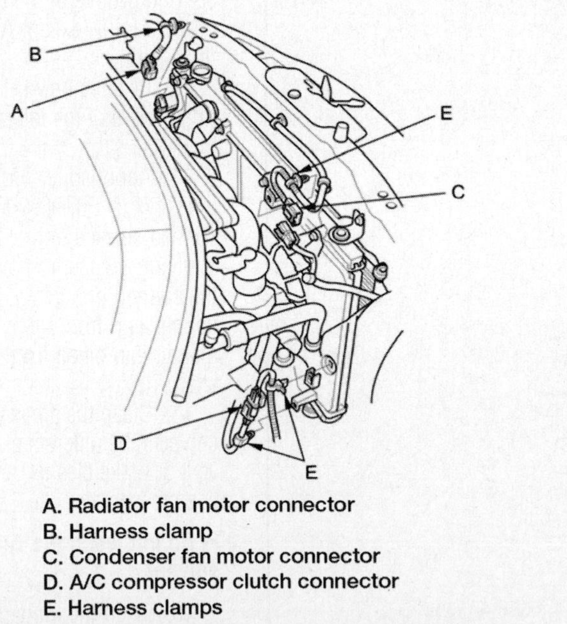
B
A
E
C
D
E
A. Radiator fan motor connector
B. Harness clamp
C. Condenser fan motor connector
D. A/C compressor clutch connector
E. Harness clamps
37647_HFIT_G0151

Fig. 15 Disconnect the radiator fan motor connector, then remove the harness clamp

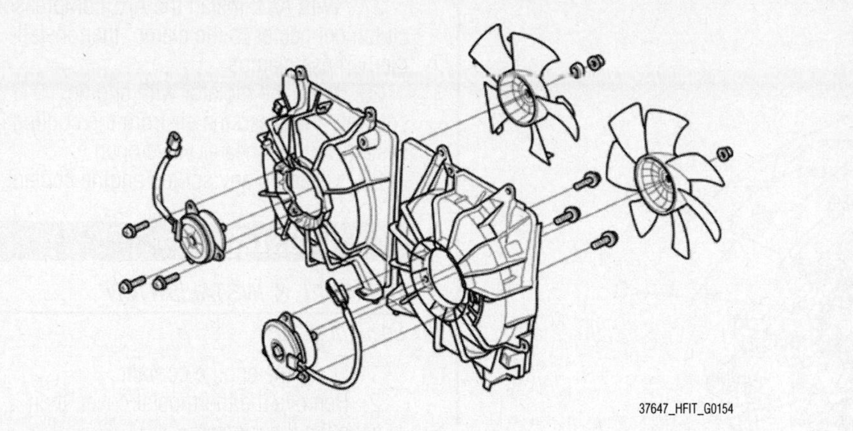
37647_HFIT_G0154
Fig. 18 Disassemble the fan shrouds

clutch connector, then install the harness clamps.

14. Install the coolant reservoir.

RADIATOR

REMOVAL & INSTALLATION

See Figures 19 through 24.

1. Drain the engine coolant.
2. Raise the vehicle on the lift to full height.
3. Disconnect the Engine Coolant Temperature (ECT) sensor 2 connector, then remove the harness clamp.
4. With A/C: Remove the A/C compressor clutch connector from the clamp, then remove the harness clamps.
5. Disconnect the lower radiator hose.

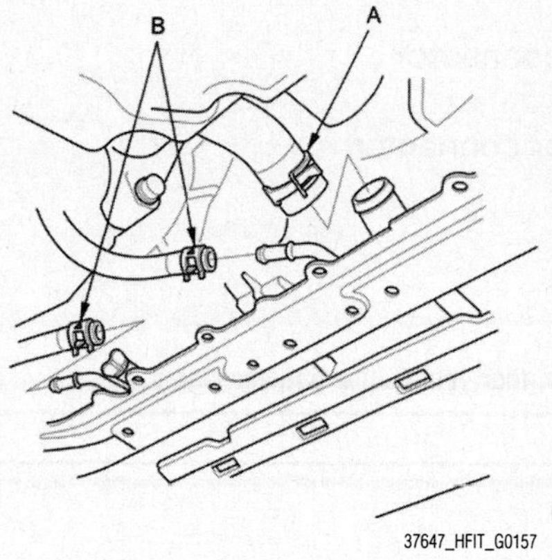
B
A
37647_HFIT_G0157
Fig. 20 Disconnect the lower radiator hose (A) and the ATF cooler hoses (B)

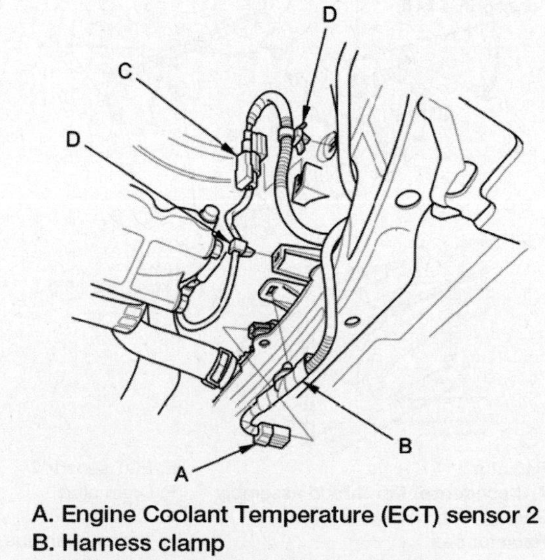
D
C
D
B
A
A. Engine Coolant Temperature (ECT) sensor 2
B. Harness clamp
C. A/C compressor clutch connector
D. Harness clamps
37647_HFIT_G0156
Fig. 19 Disconnect the ECT sensor 2 connector, then remove the harness clamp

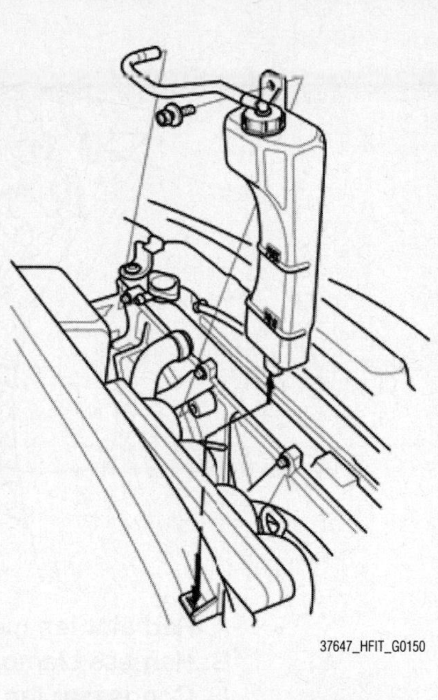
37647_HFIT_G0150
Fig. 21 Remove the coolant reservoir

6. A/T model: Remove the Automatic Transmission Fluid (ATF) cooler hoses, then plug the hose and line.
7. Lower the vehicle on the lift.
8. Remove the coolant reservoir.
9. Disconnect the radiator fan motor connector, then remove the harness clamp.
10. With A/C: Disconnect the condenser fan motor connector, then remove the harness clamp.
11. Remove the upper radiator hose.
12. Remove the radiator upper brackets.
13. Pull up the radiator.
14. With A/C: Remove the A/C condenser fan shroud assembly.
15. Remove the radiator fan shroud assembly, radiator cap, the ECT sensor 2, and the drain plug.

To install:

16. Reassemble the radiator with new O-rings.
17. Install the radiator. Make sure the lower cushions are set securely.
18. Install the radiator upper brackets.
19. Install the upper radiator hose.
20. Connect the radiator fan motor connector, then install the harness clamp.
21. With A/C: Connect the condenser fan motor connector, then install the harness clamp.
22. Install the coolant reservoir.

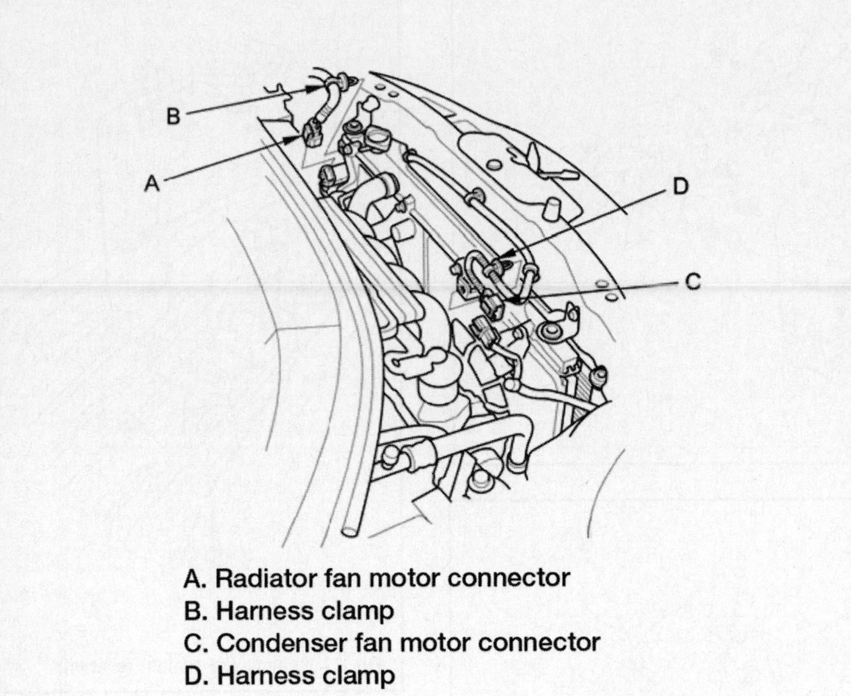
B
A
D
C
A. Radiator fan motor connector
B. Harness clamp
C. Condenser fan motor connector
D. Harness clamp
37647_HFIT_G0158

Fig. 22 Disconnect the radiator fan motor connector, then remove the harness clamp

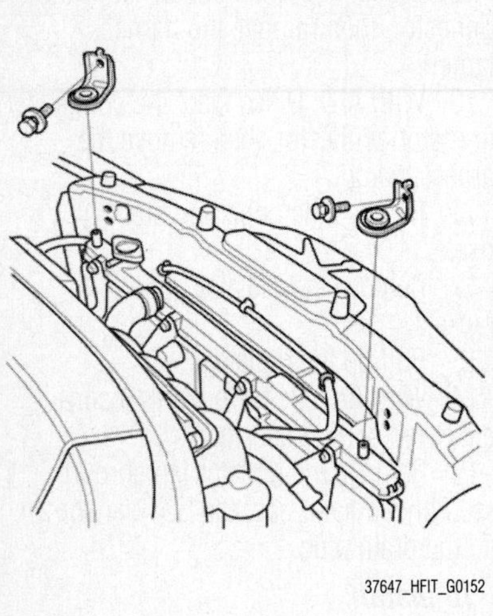
37647_HFIT_G0152

Fig. 23 Remove the radiator upper brackets

23. Raise the vehicle on the lift to full height.
24. Install the lower radiator hose.
25. A/T model: Remove the plug from the hose and the line, then install the Automatic Transmission Fluid (ATF) cooler hoses.
26. Connect the Engine Coolant Temperature (ECT) sensor 2 connector, then install the harness clamp.

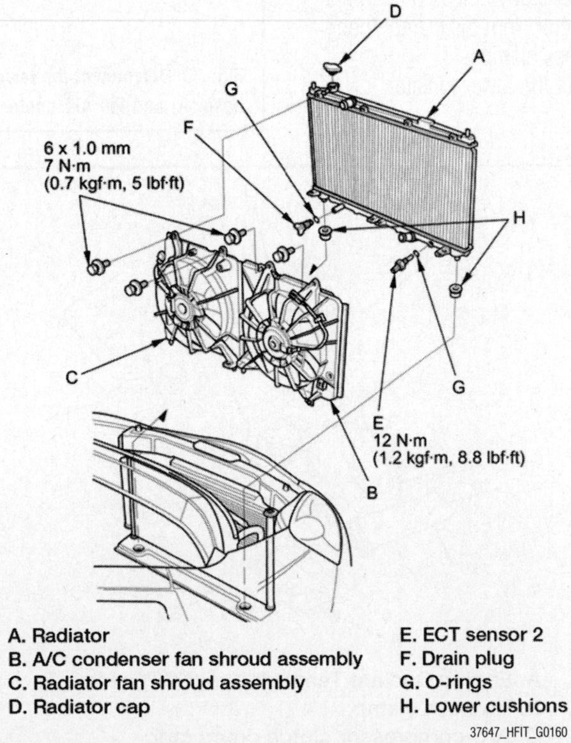
D
A
G
F
6 x 1.0 mm
7 N·m
(0.7 kgf·m, 5 lbf·ft)
H
C
G
E
12 N·m
(1.2 kgf·m, 8.8 lbf·ft)
B
A. Radiator
B. A/C condenser fan shroud assembly
C. Radiator fan shroud assembly
D. Radiator cap
E. ECT sensor 2
F. Drain plug
G. O-rings
H. Lower cushions
37647_HFIT_G0160

Fig. 24 Pull up the radiator

27. With A/C: Install the A/C compressor clutch connector to the clamp, then install the harness clamps.
28. Refill the radiator with engine coolant, and bleed the air from the cooling system with the heater valve open.
29. Clean up any spilled engine coolant.

THERMOSTAT

REMOVAL & INSTALLATION

See Figure 25.

1. Drain the engine coolant.
2. Remove the thermostat cover, then remove the thermostat.
3. Install the new rubber seal onto the thermostat, then install the thermostat with the pin up, and install the thermostat cover.
4. Refill the radiator with engine coolant, then bleed the air from the cooling system.
5. Clean up any spilled engine coolant.

WATER PUMP

REMOVAL & INSTALLATION

See Figures 26 and 27.

1. Drain the engine coolant.
2. Remove the right front wheel.

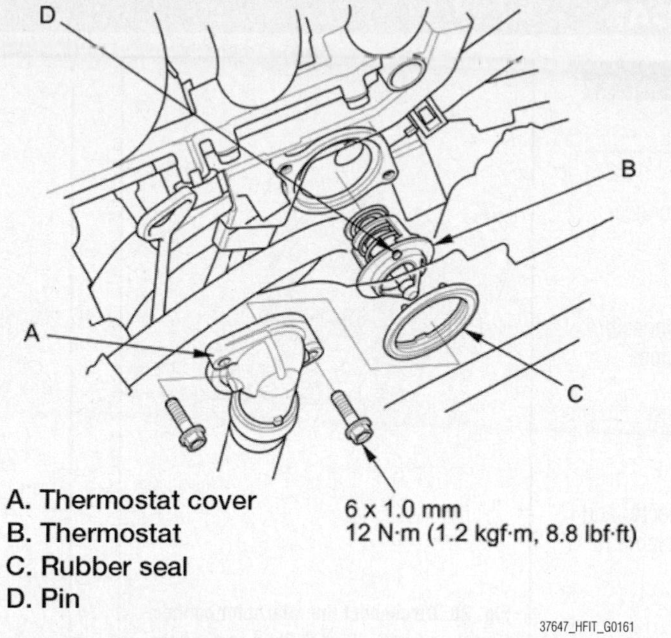
D
B
A
C
A. Thermostat cover
B. Thermostat
C. Rubber seal
D. Pin
6 x 1.0 mm
12 N·m (1.2 kgf·m, 8.8 lbf·ft)
37647_HFIT_G0161

Fig. 25 Remove the thermostat cover, then remove the thermostat

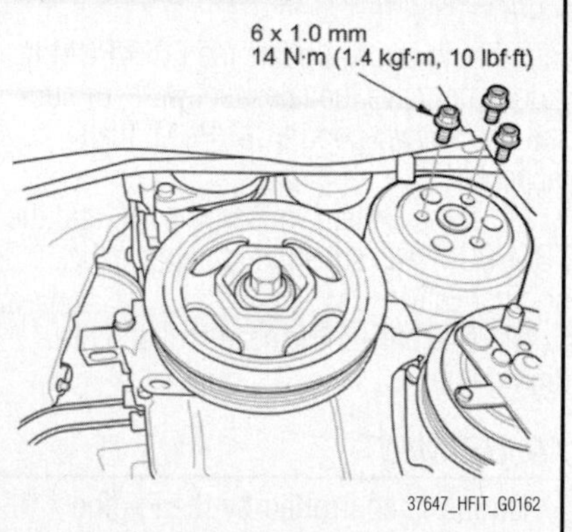
6 x 1.0 mm
14 N·m (1.4 kgf·m, 10 lbf·ft)
37647_HFIT_G0162

Fig. 26 Remove the water pump pulley

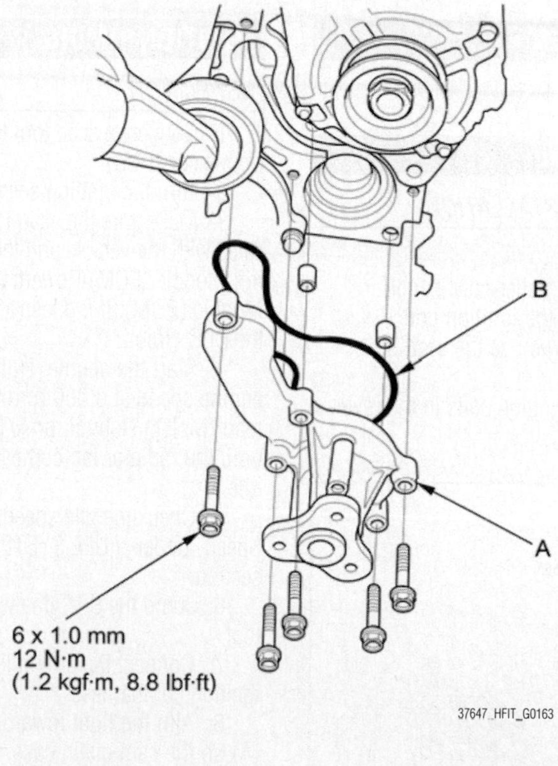
B
A
6 x 1.0 mm
12 N·m
(1.2 kgf·m, 8.8 lbf·ft)
37647_HFIT_G0163

Fig. 27 Remove the water pump (A) and O-ring (B) by removing the five bolts

3. Remove the splash shield.
4. Loosen the water pump pulley mounting bolts.
5. Remove the drive belt.
6. Remove the water pump pulley.
7. Remove the water pump and O-ring by removing the five bolts.
8. Inspect and clean the O-ring groove and the mating surface of the engine block.
9. Install the water pump with a new O-ring in the reverse order of removal.
10. Clean up any spilled engine coolant.
11. Install the water pump pulley.
12. Install the drive belt.
13. Tighten the water pump pulley mounting bolts.
14. Install the splash shield.
15. Install the right front wheel.
16. Refill the radiator with engine coolant, and bleed the air from the cooling system with the heater valve open.

ENGINE ELECTRICAL — CHARGING SYSTEM

ALTERNATOR

REMOVAL & INSTALLATION

See Figures 28 and 29.

1. Do the battery terminal disconnection procedure.
2. Remove the drive belt.
3. Remove the intake manifold.
4. Disconnect the alternator connector and BLK wire, then remove the harness clamp from the alternator.
5. Remove the alternator.

To install:

6. Install the alternator.
7. Connect the alternator connector and BLK wire, then install the harness clamp to the alternator.
8. Install the intake manifold.
9. Install the drive belt.
10. Do the battery terminal reconnection procedure.

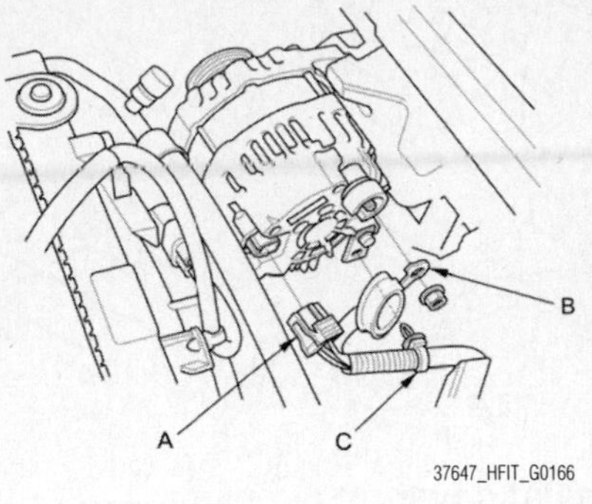
B
A
C
37647_HFIT_G0166

Fig. 28 Disconnect the alternator connector (A) and BLK wire (B), then remove the harness clamp (C) from the alternator

37647_HFIT_G0167

Fig. 29 Remove the alternator

ENGINE ELECTRICAL — IGNITION SYSTEM

FIRING ORDER

Firing order is: 1–3–4–2

IGNITION COIL

REMOVAL & INSTALLATION

See Figure 30.

1. Remove the under-cowl panel.
2. Disconnect the ignition coil connectors, then remove the ignition coils.
3. Install the ignition coils in the reverse order of removal.

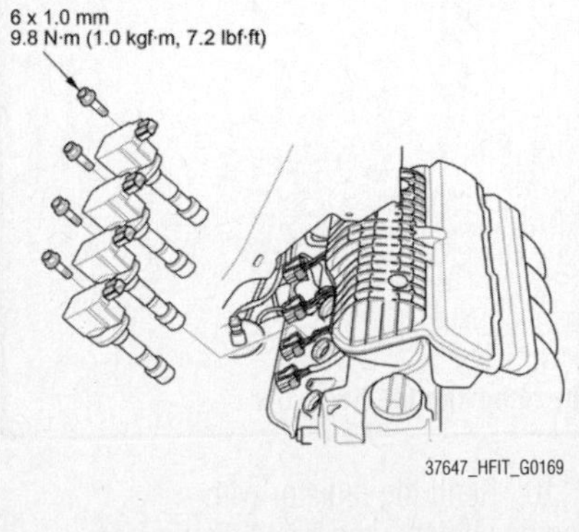
6 x 1.0 mm
9.8 N·m (1.0 kgf·m, 7.2 lbf·ft)
37647_HFIT_G0169

Fig. 30 Disconnect the ignition coil connectors, then remove the ignition coils

IGNITION TIMING

INSPECTION

1. Connect a scan tool to the Data Link Connector (DLC).
2. Turn the ignition switch ON (II).
3. Make sure the scan tool communicates with the vehicle and the Engine Control Module (ECM)/Powertrain Control Module (PCM). If it doesn't, troubleshoot the DLC circuit.
4. Start the engine. Hold the engine speed at 3,000 rpm with no load (M/T in Neutral, or A/T in P or N) until the radiator fan comes on, then let it idle.
5. Check the idle speed. See "Idle Speed" under "FUEL SYSTEMS" section.
6. Jump the SCS line with the scan tool.
7. Connect the timing light to the No. 1 ignition coil harness.
8. Aim the light toward the pointer (A) on the cam chain case. Check the ignition timing under a no load condition (headlights, blower fan, rear window defogger, and air conditioner are turned off).
9. Ignition timing should be 8°, plus/minus 2° BTDC.
10. If the ignition timing differs from the specification, check the cam timing. If the cam timing is OK, update the ECM/PCM if it does not have the latest software, or substitute a known-good ECM/PCM, then recheck.
11. If the system works properly, and the ECM/PCM was substituted, replace the original ECM/PCM.
12. Disconnect the scan tool and the timing light.

ADJUSTMENT

Timing is controlled by the Engine Control Module (ECM)/Powertrain Control Module (PCM). No timing adjustment is possible or necessary.

SPARK PLUGS

REMOVAL & INSTALLATION

See Figure 30.

1. Remove the under-cowl panel.
2. Disconnect the ignition coil connectors, then remove the ignition coils.
3. Remove the spark plugs and inspect them.
4. Apply a small amount of anti-seize compound to the plug threads, and screw the plugs into the cylinder head, finger tight. Torque them to 13 ft. lbs. (18 Nm).
5. Install the ignition coils in the reverse order of removal.

ENGINE ELECTRICAL

STARTER

REMOVAL & INSTALLATION

See Figures 31 through 33.

1. Do the battery terminal disconnection procedure.
2. Remove the coolant reservoir.
3. Remove the intake manifold.

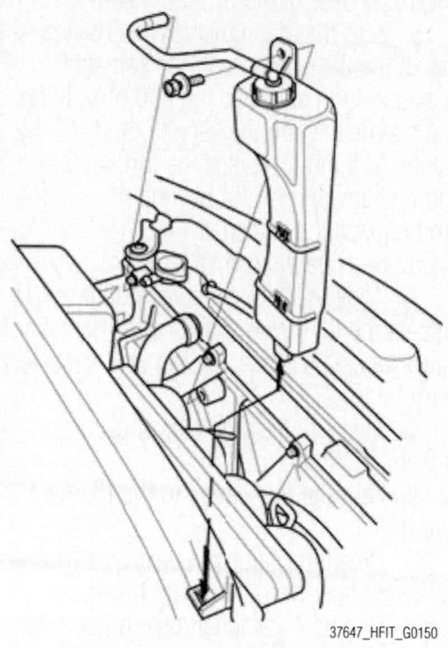
37647_HFIT_G0150

Fig. 31 Remove the coolant reservoir

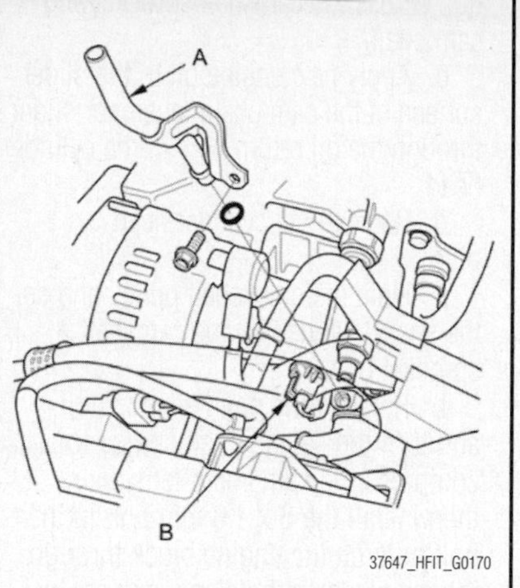
A
B
37647_HFIT_G0170

Fig. 32 Remove the dipstick, then remove the dipstick tube (A); disconnect the oil pressure switch connector (B)

4. Remove the dipstick, then remove the dipstick tube.
5. Disconnect the oil pressure switch connector.
6. Disconnect the starter cable from the B terminal and disconnect the connector from the S terminal.
7. Remove the two bolts securing the starter, then remove the starter from under the vehicle.

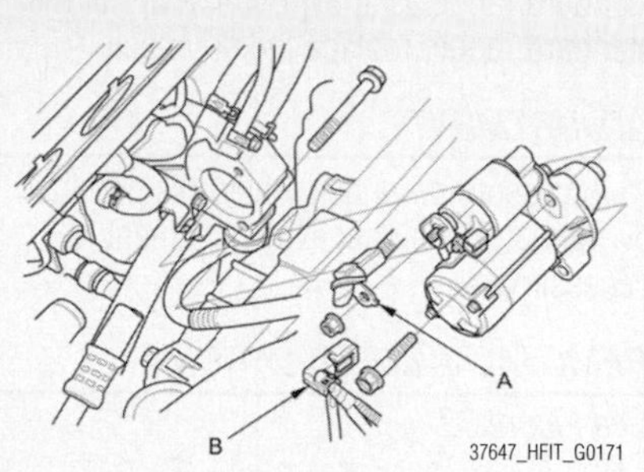
A
B
37647_HFIT_G0171

Fig. 33 Disconnect the starter cable (A) from the B terminal and disconnect the connector (B) from the S terminal

To install:

8. Install the starter to the engine. Connect the starter cable and connector. Make sure the crimped side of the starter cable terminal faces away from the starter when you connect it.
9. Install the dipstick tube with a new O-ring.
10. Connect the oil pressure switch connector.
11. Install the coolant reservoir.
12. Install the intake manifold.
13. Do the battery terminal reconnection procedure.
14. Start the engine to make sure the starter works properly.

ENGINE MECHANICAL

ACCESSORY DRIVE BELTS

ACCESSORY BELT ROUTING

See Figures 34 and 35.

Refer to the accompanying illustrations.

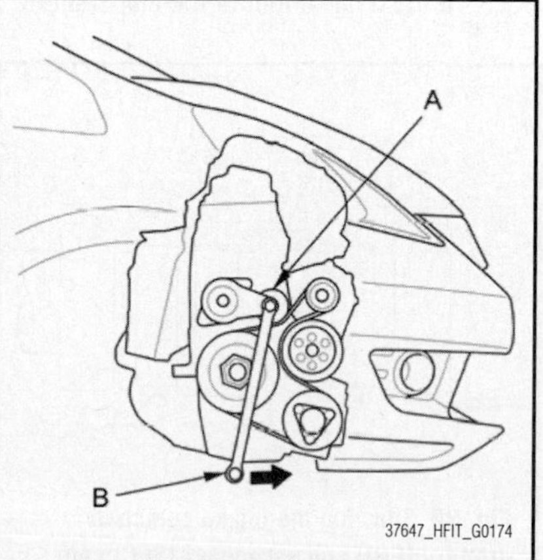
A
B
37647_HFIT_G0174

Fig. 34 Accessory drive belt routing with A/C

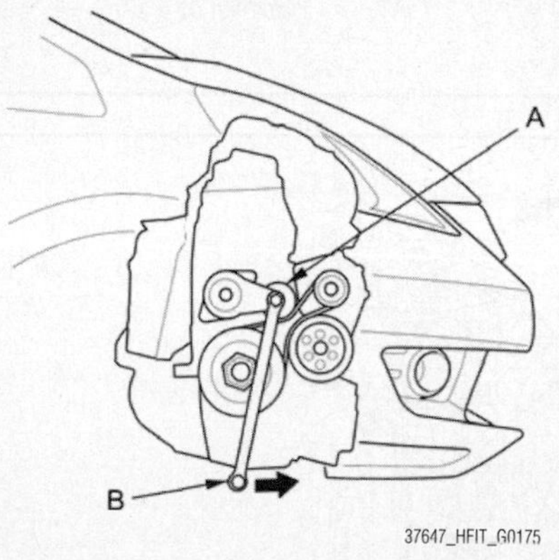
A
B
37647_HFIT_G0175

Fig. 35 Accessory drive belt routing without A/C

INSPECTION

See Figure 36.

1. Inspect the belt for cracks or damage. If the belt is cracked or damaged, replace it.

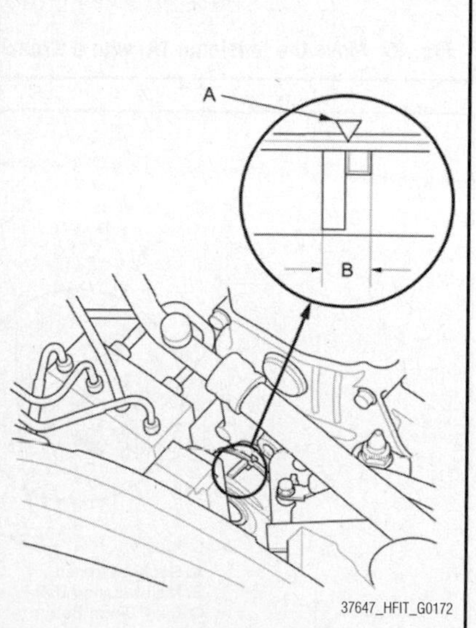
A
B
37647_HFIT_G0172

Fig. 36 Auto-tensioner indicator's pointer (A) is within the standard range (B)

2. Check that the position of the auto-tensioner indicator's pointer is within the standard range as shown. If it is out of the standard range, replace the drive belt.

ADJUSTMENT

Accessory drive belt tension is provided by the auto-tensioner. No adjustment is necessary.

REMOVAL & INSTALLATION

See Figure 37.

1. Remove the splash shield.
2. Move the tensioner with a wrench in the direction shown to relieve tension from the drive belt, then remove the drive belt.
3. Install the new belt in the reverse order of removal.

CAMSHAFT AND VALVE LIFTERS

INSPECTION

See Figures 38 and 39.

1. Remove the camshaft sprocket as follows:

a. Remove the cylinder head cover.

b. Make a reference mark in one position across the camshaft sprocket and cam chain.

c. Apply new engine oil to the slider surface of the cam chain tensioner slider through the oil return hole in the cylinder head.

d. Remove the cylinder head plug.

e. Hold the crankshaft pulley and set the socket wrench on the camshaft sprocket bolt.

f. Remove the maintenance bolt, and turn the camshaft clockwise to compress the cam chain tensioner, then install the 6 x 1.0 mm bolt in the bolt hole on the engine block through the maintenance hole and cam chain tensioner.

WARNING

Turning torque should not exceed 41 ft. lbs. (56 Nm), when turning the camshaft. Do not turn the camshaft counterclockwise.

g. Hold the camshaft with a 27 mm open-end wrench, then remove the camshaft sprocket.

➡Hang the cam chain with a wire.

2. Remove the rocker arm assembly. See "Rocker Arm" in this section.
3. Put the rocker shaft on the cylinder head, then tighten the bolts, in an alternating pattern from the center outward, to 22 ft. lbs. (29 Nm).
4. Seat the camshaft by pushing it toward the rear of the cylinder head.
5. Zero the dial indicator against the end of the camshaft. Push the camshaft back and forth and read the end play. If the end play is beyond the service limit of 0.02 inches (0.5 mm), replace the thrust cover and recheck. If it is still beyond the service limit replace the camshaft.
6. Remove the camshaft.
7. Wipe the camshaft clean, then inspect the lift ramps. Replace the camshaft it any lobes are pitted, scored or excessively worn.
8. Measure the diameter of each camshaft journal.
9. Zero the dial gauge to the journal diameter.
10. Clean the camshaft bearing surfaces in the cylinder head. Measure the inside diameter of each camshaft bearing surface, and check for an out-of-round condition.
11. If the camshaft-to-holder clearance is within limits, go to step 12.
12. If the camshaft-to-holder clearance is beyond the service limit of 0.004 inches (0.100 mm) and the camshaft has been replaced, replace the cylinder head.
13. If the camshaft-to-holder clearance is beyond the service limit 0.004 inches (0.100 mm) and the camshaft has not been replaced, go to the next step.
14. Check the total runout with the camshaft supported on V-blocks.
15. If the total runout of the camshaft is

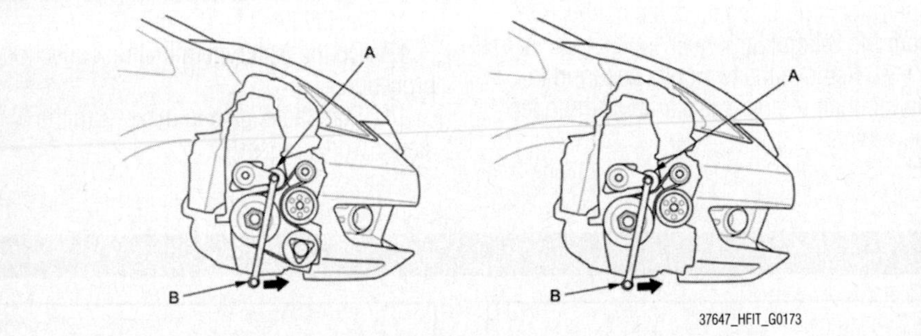
A
B
A
B
37647_HFIT_G0173

Fig. 37 Move the tensioner (A) with a wrench (B)

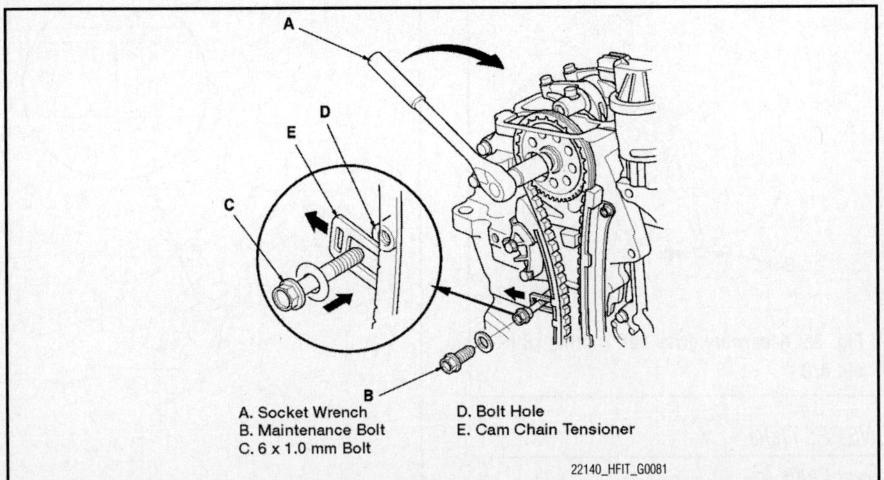
A
D
E
C
B
A. Socket Wrench
B. Maintenance Bolt
C. 6 x 1.0 mm Bolt
D. Bolt Hole
E. Cam Chain Tensioner
22140_HFIT_G0081

Fig. 38 Remove the maintenance bolt, turn the camshaft clockwise to compress the cam chain tensioner, then install a 6 x 1.0 mm bolt in the bolt hole on the engine block through the maintenance hole and cam chain tensioner

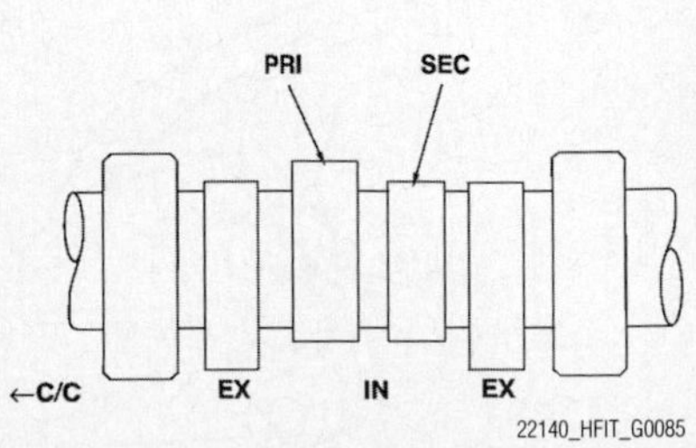
PRI
SEC
←C/C
EX
IN
EX
22140_HFIT_G0085

Fig. 39 Showing the intake camshaft primary (PRI) and secondary (SEC) cam lobes (C/C=cam chain)

within the service limit of 0.002 inches (0.04 mm), replace the cylinder head.

16. If the total runout is beyond the service limit of 0.002 inches (0.04 mm), replace the camshaft and recheck the camshaft-to-holder oil clearance. If the oil clearance is still beyond the service limit, replace the cylinder head.

17. Measure the cam lobe height. It should be:

- Intake Primary: 1.39291 inches (35.3799 mm)
- Intake Secondary: 1.20193 inches (30.5291 mm)
- Exhaust: 1.39321 inches (35.3877 mm)

REMOVAL & INSTALLATION

See Figures 38 and 40.

1. Remove the air cleaner assembly.
2. Remove the camshaft sprocket as follows:
   a. Remove the cylinder head cover.
   b. Make a reference mark in one position across the camshaft sprocket and cam chain.
   c. Apply new engine oil to the slider surface of the cam chain tensioner slider through the oil return hole in the cylinder head.
   d. Remove the cylinder head plug.
   e. Hold the crankshaft pulley and set the socket wrench on the camshaft sprocket bolt.
   f. Remove the maintenance bolt, and turn the camshaft clockwise to compress the cam chain tensioner, then install the 6 x 1.0 mm bolt in the bolt hole on the engine block through the maintenance hole and cam chain tensioner.

WARNING

Turning torque should not exceed 41 ft. lbs. (56 Nm), when turning the camshaft. Do not turn the camshaft counterclockwise.

   g. Hold the camshaft with a 27 mm open-end wrench, then remove the camshaft sprocket.
3. Remove the rocker arm assembly. See "Rocker Arm" in this section.
4. Remove the air cleaner housing bracket, ground cable and harness clamps, then remove the harness holder from the bracket.
5. Disconnect the Camshaft Position (CMP) sensor connector, then remove the CMP sensor.

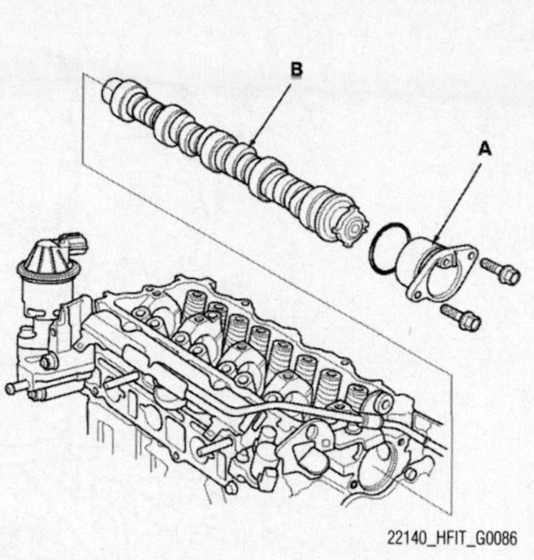

B
A
22140_HFIT_G0086

Fig. 40 Remove the camshaft thrust cover (A), then pull out the camshaft (B)

6. Remove the camshaft thrust cover, then pull out the camshaft.

To install:

7. Install the camshaft into the cylinder head, then install the camshaft thrust cover with new O-ring. Tighten the bolts to 7.2 ft. lbs. (9.8 Nm).
8. Install the CMP sensor with new O-ring, then connect the CMP sensor connector.
9. Install the harness holder, then install harness clamps, ground cable and air cleaner housing bracket.
10. Install the camshaft sprocket as follows:

WARNING

Keep the cam chain away from magnetic fields.

   a. Install the cam chain to the camshaft sprocket by alignment the reference mark made during removal, then install the camshaft sprocket on the camshaft.
   b. Hold the camshaft with a 27 mm open-end wrench, then tighten the bolt to 41 ft. lbs. (56 Nm).
   c. Apply new engine oil to the slider surface of the cam chain tensioner slider through the oil return hole in the cylinder head.
   d. Hold the crankshaft pulley and set the socket wrench on the camshaft sprocket bolt.
   e. Turn the camshaft clockwise to compress the cam chain tensioner, then remove the 6 x 1.0 mm bolt.

WARNING

Turning torque should not exceed 41 ft. lbs. (56 Nm), when turning the camshaft. Do not turn the camshaft counterclockwise.

   f. Install the maintenance bolt with a new washer to 14 ft. lbs. (20 Nm).
   g. Install the new cylinder head plug.
   h. Install the cylinder head cover.
11. Install the rocker arm assembly. See "Rocker Arm" in this section.
12. Install the air cleaner assembly.

CATALYTIC CONVERTER

REMOVAL & INSTALLATION

Warm Up Three Way Catalytic Converter (WU-TWC)

See Figures 41 and 42.

1. Raise the vehicle on a lift.
2. Remove the bolts.
3. Lower the vehicle.

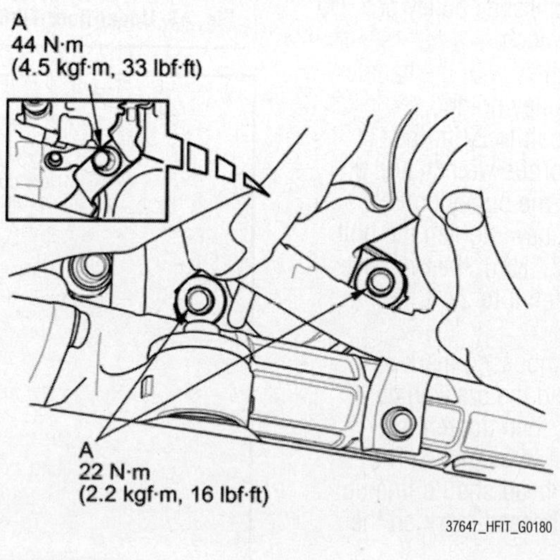

A
44 N·m
(4.5 kgf·m, 33 lbf·ft)
A
22 N·m
(2.2 kgf·m, 16 lbf·ft)
37647_HFIT_G0180

Fig. 41 Remove the bolts (A)

4. Remove the A/F sensor (Sensor 1).
5. Remove the EGR pipe.
6. Remove the cover.
7. Remove the bolts and nuts.
8. Remove the WU-TWC.
9. Install the parts in the reverse order of removal with new gaskets.

Under-Floor TWC

See Figure 43.

1. Raise the vehicle on a lift.
2. Remove the secondary HO2S (Sensor 2).
3. Remove the nuts and the bolts.
4. Remove the under-floor TWC.
5. Remove the cover.
6. Install the parts in the reverse order of removal with new gaskets.

CRANKSHAFT PULLEY

REMOVAL & INSTALLATION

See Figures 44 and 45.

Special Tools Required:

- Crankshaft Pulley Holder 07AAB-RJAA100
- Socket, 19 mm 07JAA-001020A or equivalent
- Holder Handle 07JAB-001020B

1. Remove the drive belt.
2. Hold the pulley with the handle and the crankshaft pulley holder.
3. Remove the bolt with a 19 mm socket and a breaker bar, then remove the crankshaft pulley.

Installation

4. Clean the crankshaft pulley, the crankshaft, the bolt, and the washer. Lubricate with new engine oil.
5. Install the crankshaft pulley.
6. Tighten the crankshaft pulley bolt. Do not use an impact wrench.
   a. Hold the pulley with the handle and crankshaft pulley holder, then tighten the bolt to 27 ft. lbs. (37 Nm) with a torque wrench and a 19 mm socket. If the pulley bolt or crankshaft are new, tighten the bolt to 130 ft. lbs. (177 Nm), then remove the bolt and tighten it to 27 ft. lbs. (37 Nm).
   b. Mark the embossed mark on the bolt flange and the crankshaft pulley as shown, then tighten the bolt an additional 90 degrees (The mark on the bolt head should line up with the next embossed mark on the bolt flange).
7. Install the drive belt.

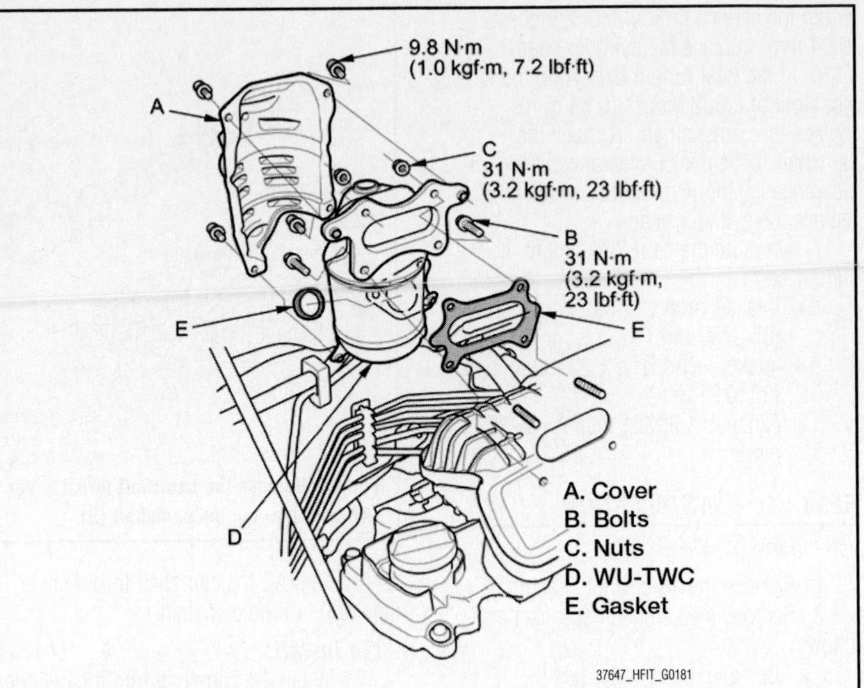
9.8 N·m
(1.0 kgf·m, 7.2 lbf·ft)
A
C
31 N·m
(3.2 kgf·m, 23 lbf·ft)
B
31 N·m
(3.2 kgf·m,
23 lbf·ft)
E
E
D
A. Cover
B. Bolts
C. Nuts
D. WU-TWC
E. Gasket
37647_HFIT_G0181

Fig. 42 Exploded view of the WU-TWC

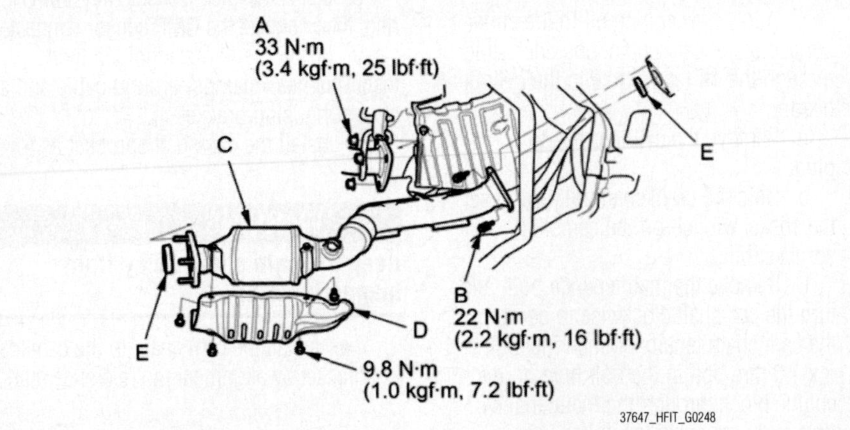
A
33 N·m
(3.4 kgf·m, 25 lbf·ft)
C
E
B
22 N·m
(2.2 kgf·m, 16 lbf·ft)
D
E
9.8 N·m
(1.0 kgf·m, 7.2 lbf·ft)
37647_HFIT_G0248

Fig. 43 Under-floor TWC assembly

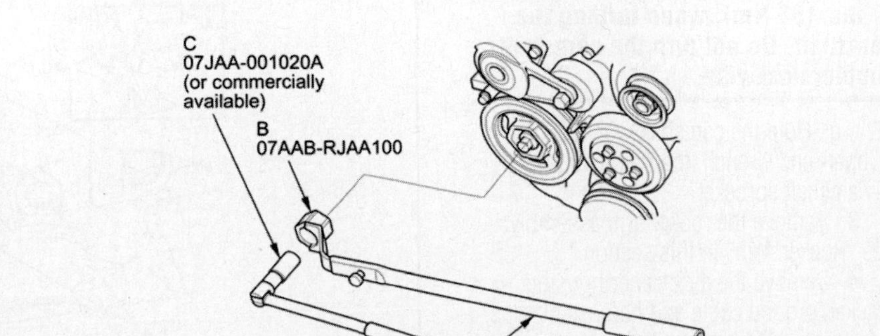
C
07JAA-001020A
(or commercially
available)
B
07AAB-RJAA100
A
07JAB-001020B
37647_HFIT_G0182

Fig. 44 Hold the pulley with the handle (A) and the crankshaft pulley holder (B)

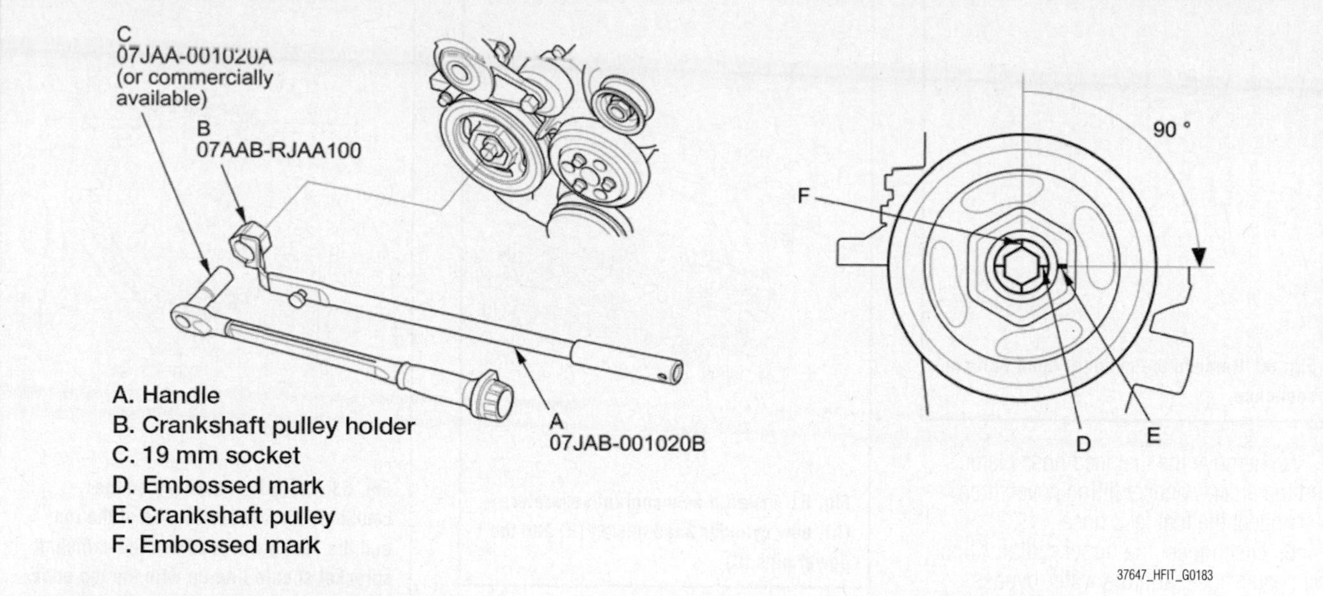
C
07JAA-001020A
(or commercially
available)
B
07AAB-RJAA100
A
07JAB-001020B
90 °
F
D
E
A. Handle
B. Crankshaft pulley holder
C. 19 mm socket
D. Embossed mark
E. Crankshaft pulley
F. Embossed mark
37647_HFIT_G0183

Fig. 45 Tighten the crankshaft pulley bolt

CYLINDER HEAD

REMOVAL & INSTALLATION

See Figures 46 through 55.

➡To avoid damaging the cylinder head, wait until the engine coolant temperature drops below 100°F (38°C) before loosening the cylinder head bolts.

➡Mark all wiring and hoses to avoid misconnection. Also, be sure that they do not contact other wiring or hoses, or interfere with other parts.

1. Relieve fuel pressure.
2. Drain the engine coolant.
3. Do the battery removal procedure.
4. Remove the air cleaner housing assembly.

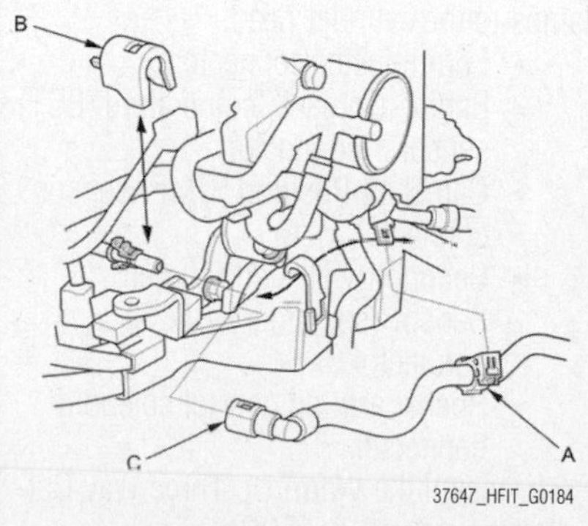
B
C
A
37647_HFIT_G0184

Fig. 46 Remove the fuel feed hose clamp (A) and the quick-connect fitting cover (B), then disconnect the fuel feed hose (C)

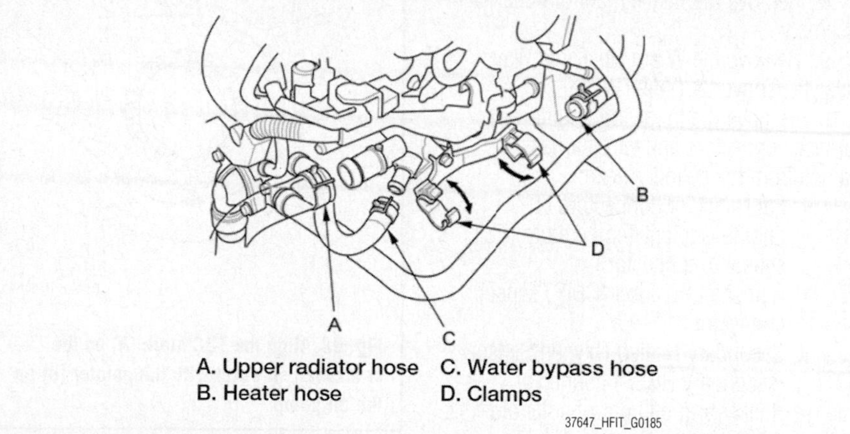
B
D
A
C
A. Upper radiator hose
B. Heater hose
C. Water bypass hose
D. Clamps
37647_HFIT_G0185

Fig. 47 Disconnect the upper radiator hose, the heater hose, and the water bypass hose

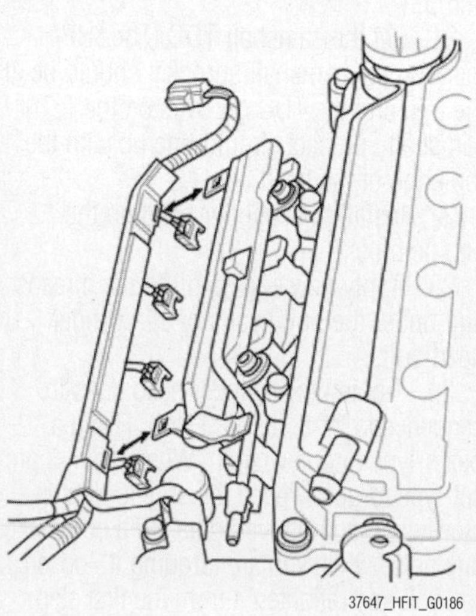
37647_HFIT_G0186

Fig. 48 Remove the harness holder

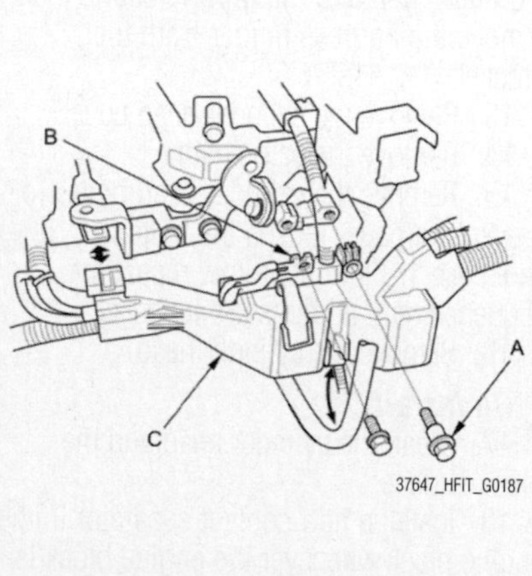
B
A
C
37647_HFIT_G0187

Fig. 49 Remove the harness holder mounting bolt (A) and the ground cable (B), then remove the harness holder (C) from the bracket

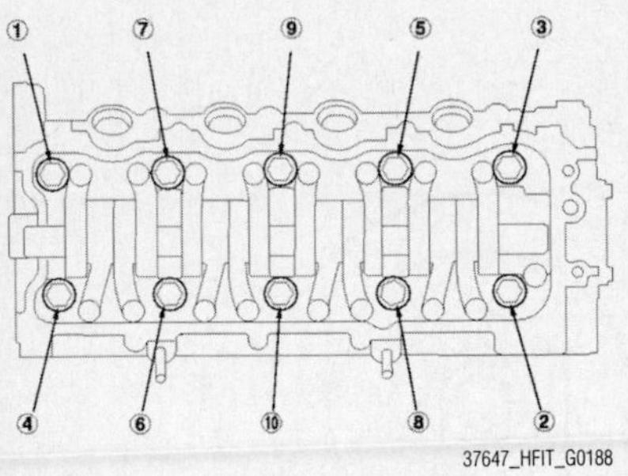
①
⑦
⑨
⑤
③
④
⑥
⑩
⑧
②
37647_HFIT_G0188

Fig. 50 Remove the cylinder head bolts in sequence

5. Remove the fuel feed hose clamp and the quick-connect fitting cover, then disconnect the fuel feed hose.

6. Disconnect the upper radiator hose, the heater hose, and the water bypass hose.

7. Remove the heater hose from the clamps.

8. Remove the intake manifold/chamber assembly.

9. Remove the Warm Up Three Way Catalytic Converter (WU-TWC).

10. Remove the following engine wire harness connectors and wire harness clamps from the cylinder head:

- Four injector connectors
- Engine Coolant Temperature (ECT) sensor 1 connector
- Camshaft Position (CMP) sensor connector
- Secondary Heated Oxygen Sensor (secondary HO2S) connector
- Rocker arm oil control solenoid connector

11. Remove the harness holder.

12. Remove the harness holder mounting bolt and the ground cable, then remove the harness holder from the bracket.

13. Remove the cylinder head cover.

14. Remove the cam chain.

15. Remove the cylinder head bolts. To prevent warpage, loosen the bolts in sequence 1/3 turn at a time; repeat the sequence until all bolts are loosened.

16. Remove the cylinder head.

To install:

17. Clean the cylinder head and the block surface.

18. Install a new coolant separator in the engine block whenever the engine block is replaced.

19. Install the new cylinder head gasket and the dowel pins on the engine block. Always use a new cylinder head gasket.

20. Set the crankshaft to Top Dead Center (TDC). Align the TDC mark on the crankshaft sprocket with the pointer on the oil pump.

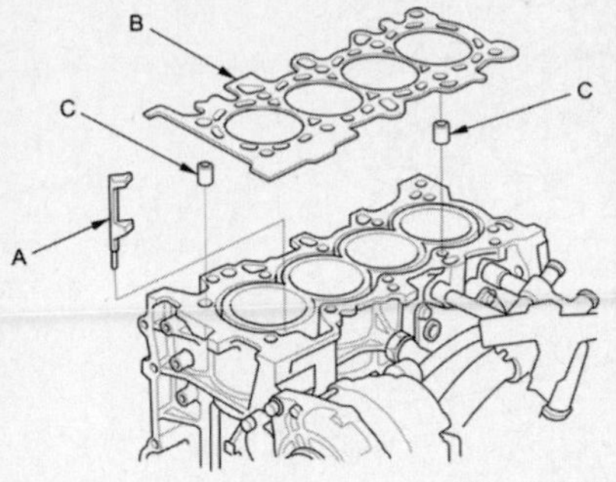
B
C
C
A
37647_HFIT_G0190

Fig. 51 Install a new coolant separator (A), new cylinder head gasket (B) and the dowel pins (C)

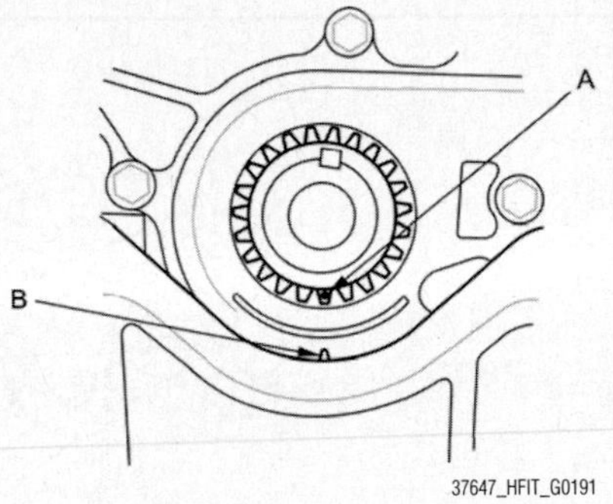
A
B
37647_HFIT_G0191

Fig. 52 Align the TDC mark (A) on the crankshaft sprocket with the pointer (B) on the oil pump

21. Set the camshaft TDC. The "UP" mark on the camshaft sprocket should be at the top, and the TDC grooves on the camshaft sprocket should line up with the top edge of the head.

22. Install the cylinder head on the engine block.

23. Apply new engine oil to the threads and under the bolt heads of all cylinder head bolts.

24. Tighten the cylinder head bolts in sequence to 22 ft. lbs. (29 Nm). Using a beam-type torque wrench. When using a preset-type torque wrench, be sure to tighten slowly and do not overtighten. If a bolt makes any noise while you are torquing it, loosen the bolt and retighten it from the first step.

25. Tighten all cylinder head bolts an additional 130°.

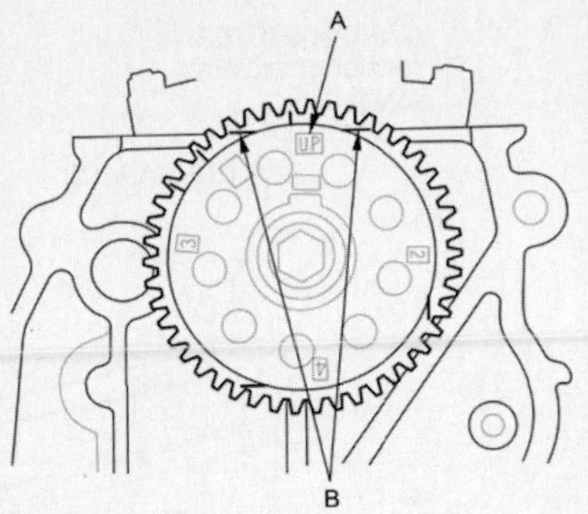
A
UP
B
37647_HFIT_G0192

Fig. 53 The "UP" mark (A) on the camshaft sprocket should be at the top, and the TDC grooves (B) on the camshaft sprocket should line up with the top edge of the head

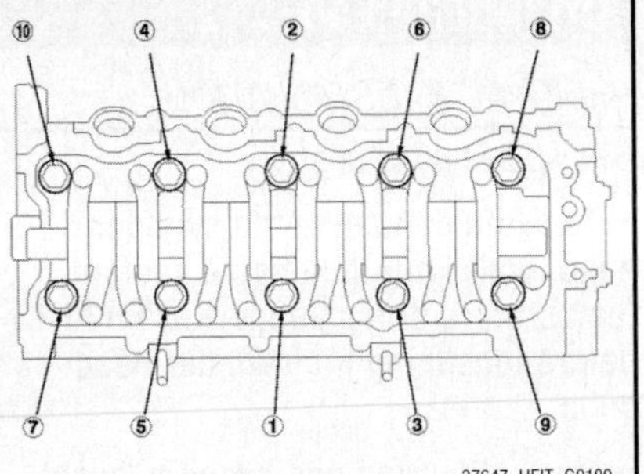
⑩
④
②
⑥
⑧
⑦
⑤
①
③
⑨
37647_HFIT_G0189

Fig. 54 Tighten the cylinder head bolts in sequence

26. Install the cam chain.

27. Install the cylinder head cover.

28. Install the harness holder, then install the ground cable.

29. Install the harness holder.

30. Connect the engine wire harness connectors, and install the wire harness clamps to the cylinder head:

- Four injector connectors
- Engine Coolant Temperature (ECT) sensor 1 connector
- Camshaft Position (CMP) sensor connector
- Secondary Heated Oxygen Sensor (secondary HO2S) connector
- Rocker arm oil control solenoid connector

31. Install the Warm Up Three Way Catalytic Converter (WU-TWC).

32. Install the intake manifold/chamber assembly.

33. Install the upper radiator hose,

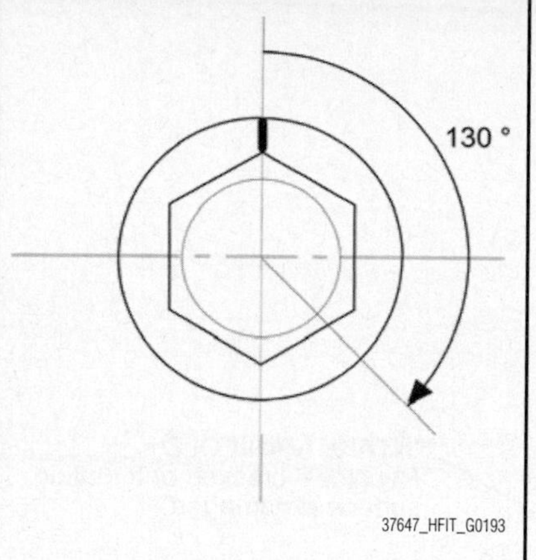
130 °
37647_HFIT_G0193

Fig. 55 Tighten all cylinder head bolts an additional 130°

the heater hose, and the water bypass hose.

34. Install the heater hose to the clamps.
35. Connect the fuel feed hose, then install the quick-connect fitting cover and the fuel feed hose clamp.
36. Install the air cleaner housing assembly.
37. Do the battery installation procedure.
38. Inspect for fuel leaks. Turn the ignition switch to ON (II) (do not operate the starter) so the fuel pump runs for about 2 seconds and pressurizes the fuel line. Repeat this operation three times, then check for fuel leakage at any point in the fuel line.
39. Refill the radiator with engine coolant, and bleed air from the cooling system with the heater valve open.
40. Do the Crankshaft Position (CKP) pattern clear/CKP pattern learn procedure.
41. Inspect the idle speed.
42. Inspect the ignition timing.

FLYWHEEL/DRIVE PLATE

REMOVAL & INSTALLATION

Flywheel

1. Remove the transaxle.
2. Remove the clutch.
3. Install the ring gear holder.
4. Loosen the flywheel mounting bolts in a crisscross pattern in several steps. Remove the bolts, then remove the flywheel and the ring gear holder.
5. Install the flywheel on the crankshaft, and install the mounting bolts finger-tight.
6. Install the ring gear holder, then torque the flywheel mounting bolts in a crisscross pattern in several steps. Tighten to 87 ft. lbs. (118 N m).
7. Install the clutch.
8. Install the transaxle.

Drive Plate

1. Remove the transaxle assembly.
2. Remove the drive plate and the washer from the engine crankshaft.
3. Install the drive plate and the washer on the engine crankshaft, and tighten the six bolts in a crisscross pattern in at least two steps. Tighten the bolts to 54 ft. lbs. (74 N m).
4. Install the transaxle assembly.

INTAKE MANIFOLD

REMOVAL & INSTALLATION

See Figures 56 through 58.

1. Remove the harness holder mounting bolt, then remove the harness holder from the bracket.
2. Remove the Positive Crankcase Ventilation (PCV) hose.
3. Disconnect the Exhaust Gas Recirculation (EGR) valve connector.
4. Remove the dipstick.
5. Remove the intake manifold.

To install:

6. Install the intake manifold with new gaskets, and tighten the bolts and nuts in a crisscross pattern in three steps, beginning with the inner bolt. Tighten to 17 ft. lbs. (24 N m).

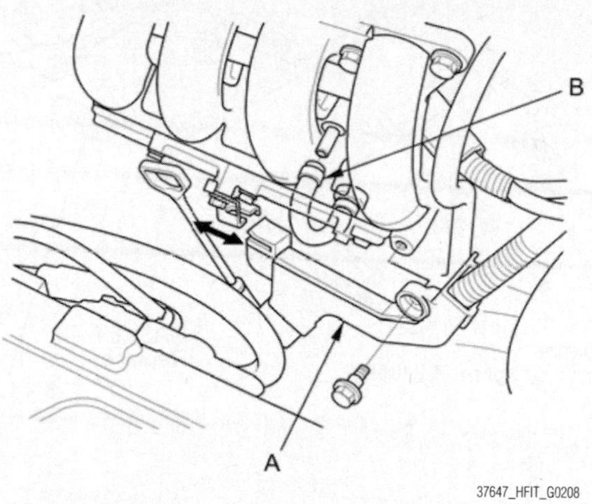
B
A
37647_HFIT_G0208

Fig. 56 Remove the harness holder mounting bolt, then remove the harness holder (A) and the PCV hose (B)

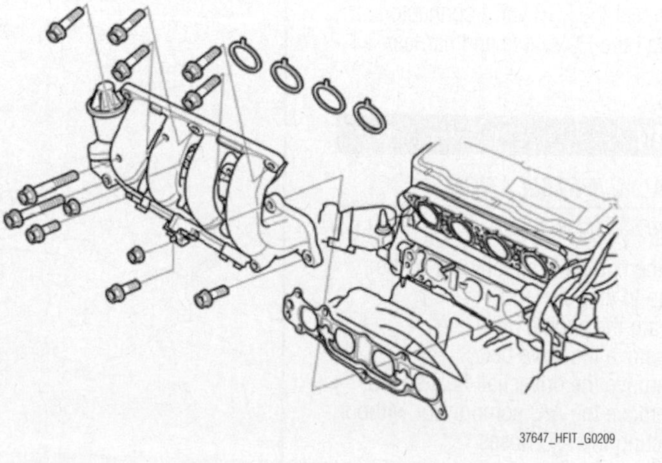
37647_HFIT_G0209

Fig. 57 Remove the intake manifold

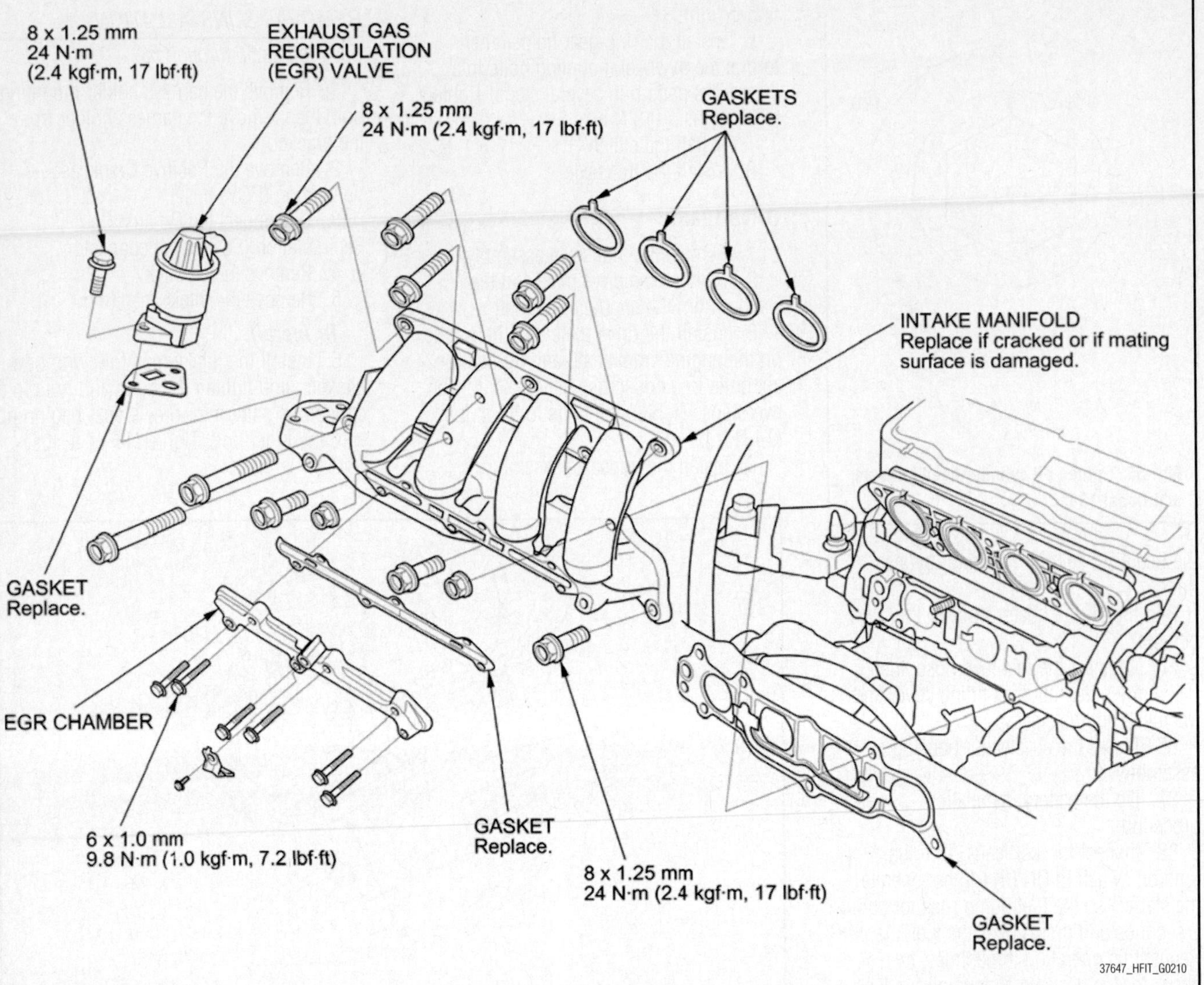
8 x 1.25 mm
24 N·m
(2.4 kgf·m, 17 lbf·ft)
EXHAUST GAS
RECIRCULATION
(EGR) VALVE
8 x 1.25 mm
24 N·m (2.4 kgf·m, 17 lbf·ft)
GASKETS
Replace.
INTAKE MANIFOLD
Replace if cracked or if mating surface is damaged.
GASKET
Replace.
EGR CHAMBER
6 x 1.0 mm
9.8 N·m (1.0 kgf·m, 7.2 lbf·ft)
GASKET
Replace.
8 x 1.25 mm
24 N·m (2.4 kgf·m, 17 lbf·ft)
GASKET
Replace.
37647_HFIT_G0210

Fig. 58 Exploded view of the intake manifold assembly

7. Install the dipstick.
8. Connect the EGR valve connector.
9. Install the PCV hose and harness holder.

OIL PAN

REMOVAL & INSTALLATION

See Figures 59 through 65.

1. If the engine is already out of the vehicle, go to step 7.
2. Drain the engine oil.
3. Remove the drive belt.
4. Remove the driveshaft heat shield.
5. Remove the A/C compressor without disconnecting the A/C hoses.
6. A/T model: Remove the shift cable cover.

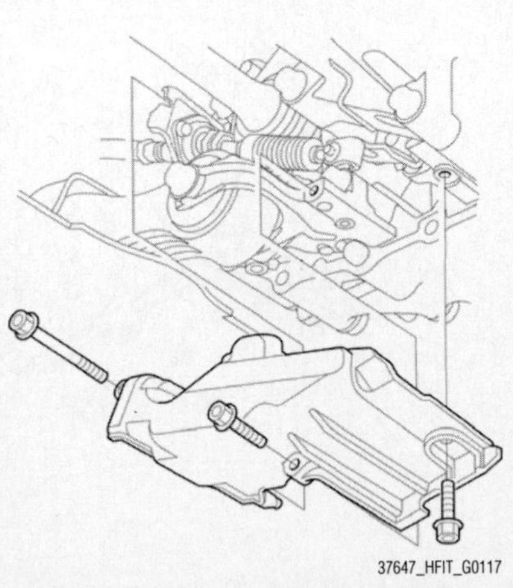
37647_HFIT_G0117

Fig. 59 Remove the shift cable cover

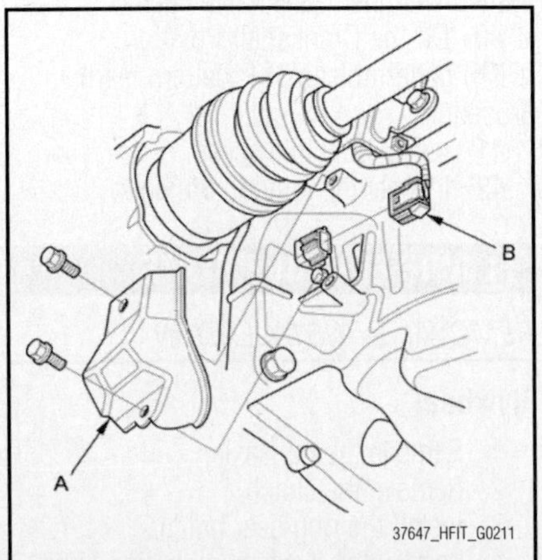
B
A
37647_HFIT_G0211

Fig. 60 Remove the CKP sensor cover (A), then disconnect the CKP sensor connector (B)

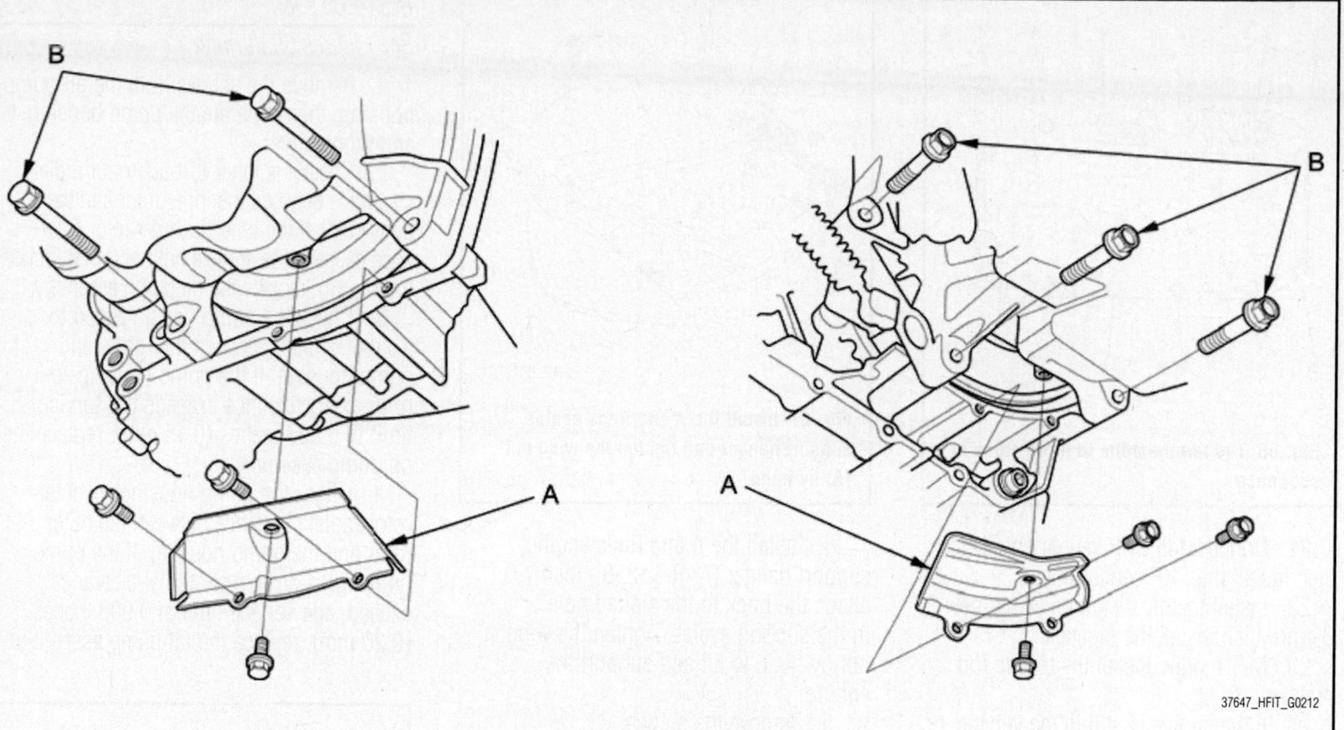
B
A
A
B
37647_HFIT_G0212

Fig. 61 Remove the clutch cover/torque converter cover (A), and transaxle mounting bolts (B)

7. M/T model: Remove the torque rod bracket.

8. Remove the dipstick, then remove the dipstick tube.

9. Remove the crankshaft position (CKP) sensor cover, then disconnect the CKP sensor connector.

10. Remove the clutch cover/torque converter cover, and transaxle mounting bolts.

11. Remove the oil pan bolts. Note the bolt locations by their size.

12. Using a flat blade screwdriver, separate the oil pan from the block in the places shown.

13. Remove the oil pan.

To install:

14. Remove any old liquid gasket from the oil pan mating surfaces bolts, and bolt holes.

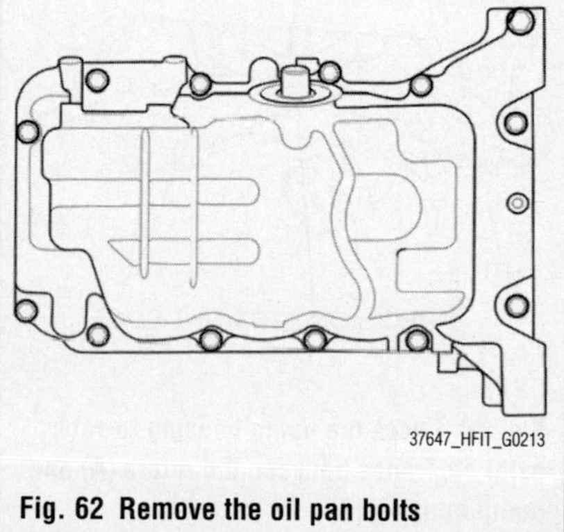
37647_HFIT_G0213

Fig. 62 Remove the oil pan bolts

15. Clean and dry the oil pan mating surfaces.

16. Install the new oil pan gasket, new O-ring, and dowel pins on the oil pan.

17. Apply liquid gasket (P/N 08717-0004, 08718-0003, or 08718-0009) to the engine block mating surface of the oil pan and to the inside edge of the bolt holes. Install the component within 5 minutes of applying the liquid gasket.

18. Install the oil pan.

➡Note the following:

- Wait at least 30 minutes before filling the engine with oil.
- Do not run the engine for at least 3 hours after installing the oil pan.
- Make sure to install the bolts in the correct locations according to size.

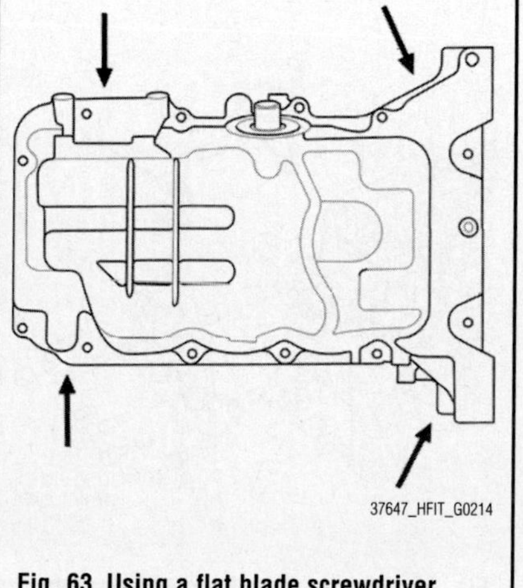
37647_HFIT_G0214

Fig. 63 Using a flat blade screwdriver, separate the oil pan from the block

19. Tighten the bolts in three steps. Wipe off the excess liquid gasket on the each side of crankshaft pulley and the flywheel/drive plate.

20. Install the clutch cover/torque converter cover, and install the transaxle mounting bolts.

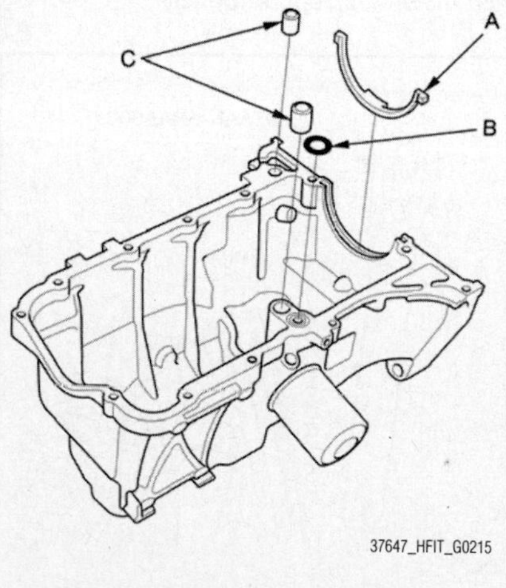
A
C
B
37647_HFIT_G0215

Fig. 64 Install the new oil pan gasket (A), new O-ring (B), and dowel pins (C) on the oil pan

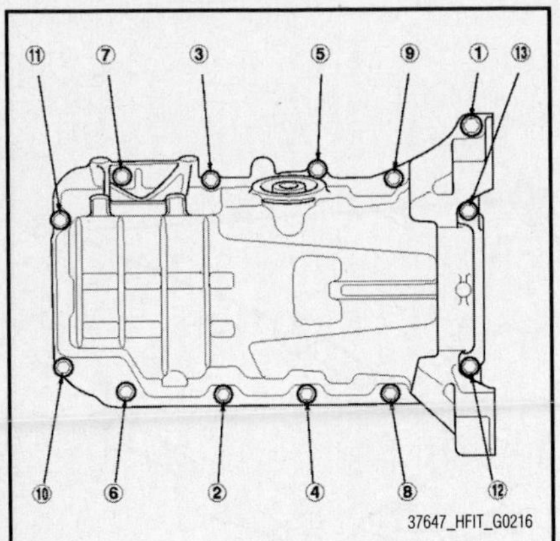
11
7
3
5
9
1
13
10
6
2
4
8
12
37647_HFIT_G0216

Fig. 65 Tighten the bolts in three steps in sequence

21. Connect the CKP sensor connector, then install the CKP sensor cover.
22. Install the dipstick tube with a new O-ring, then install the dipstick.
23. M/T model: Install the torque rod bracket.
24. If the engine is still in the vehicle, do the remaining steps.
25. A/T model: Install the shift cable cover.
26. Install the A/C compressor.
27. Install the driveshaft heat shield.
28. Install the drive belt.
29. Refill the engine with engine oil.

OIL PUMP

REMOVAL & INSTALLATION

See Figures 66 through 68.

1. Remove the under-cowl panel.
2. Remove the cam chain.
3. Remove the alternator mounting bolt, then install the support eyelet.

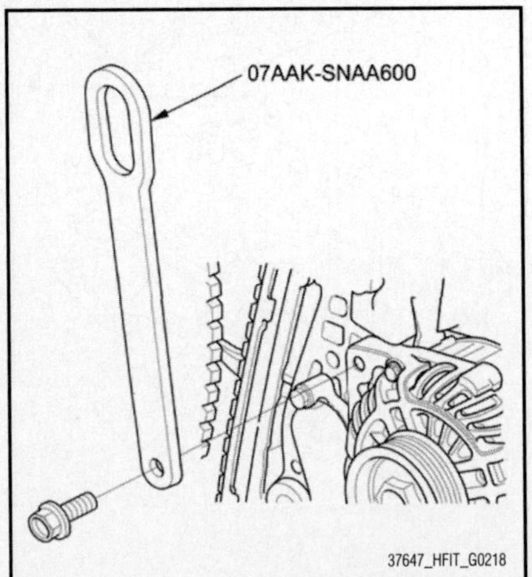
07AAK-SNAA600
37647_HFIT_G0218

Fig. 66 Remove the alternator mounting bolt, then install the support eyelet

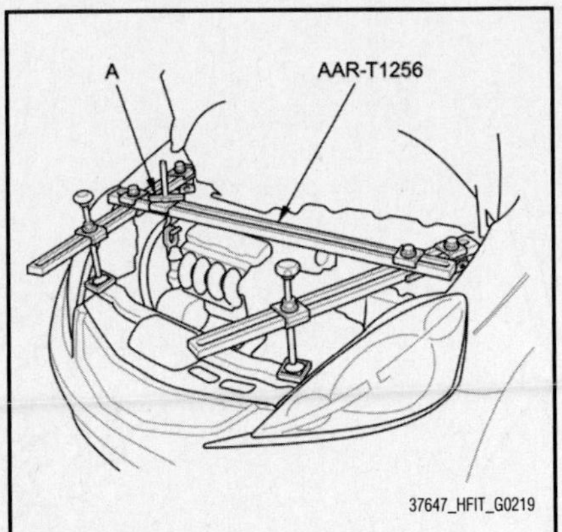
A
AAR-T1256
37647_HFIT_G0219

Fig. 67 Install the A and Reds engine support hanger and tighten the wing nut (A) by hand

4. Install the A and Reds engine support hanger (AAR-T1256), then attach the hook to the slotted hole in the support eyelet. Tighten the wing nut by hand to lift and support the engine.
5. Remove the oil pan.
6. Remove the oil screen, then remove the oil pump.

To install:

7. Clean the O-ring groove and the mating surface of the engine block.
8. Install the oil pump with a new O-ring.
9. Install the oil screen with a new gasket.
10. Install the oil pan.
11. Support the engine with a jack and a wood block under the oil pan.
12. Remove the engine hanger and support eyelet, then tighten the alternator mounting bolt.
13. Install the cam chain.

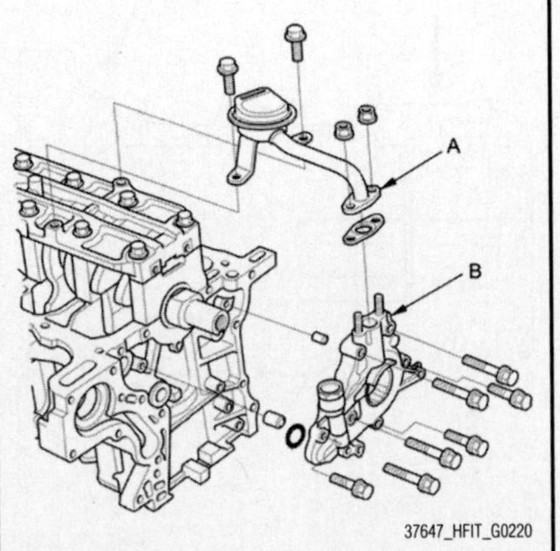
A
B
37647_HFIT_G0220

Fig. 68 Remove the oil screen (A), then remove the oil pump (B)

INSPECTION

See Figures 69 through 71.

1. Remove the screws from the oil pump housing, then separate the pump housing and pump cover.
2. Check the inner-to-outer rotor radial clearance between the inner rotor and the outer rotor. If the inner-to-outer rotor clearance exceeds the service limit of 0.008 inches (0.20 mm), replace the oil pump assembly.
3. Check the pump housing-to-rotor axial clearance between the rotors and pump housing. If the pump housing-to-rotor axial clearance exceeds the service limit of 0.006 inches (0.15 mm), replace the oil pump assembly.
4. Check the pump housing-to-outer rotor radial clearance between the outer rotor and the pump housing. If the pump housing-to-outer rotor radial clearance exceeds the service limit of 0.008 inches (0.20 mm), replace the oil pump assembly.

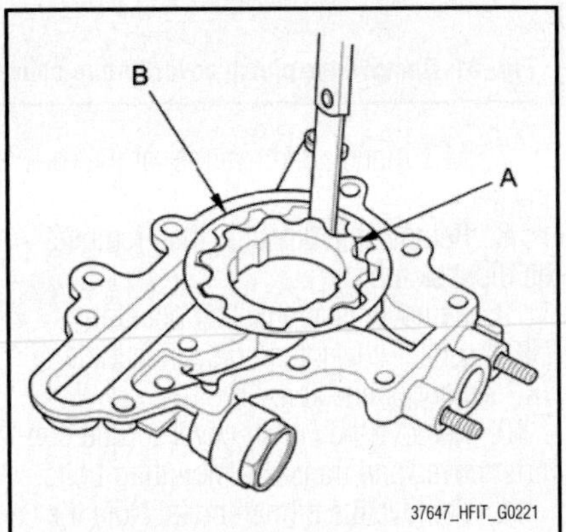
B
A
37647_HFIT_G0221

Fig. 69 Check the inner-to-outer rotor radial clearance between the inner rotor (A) and the outer rotor (B)

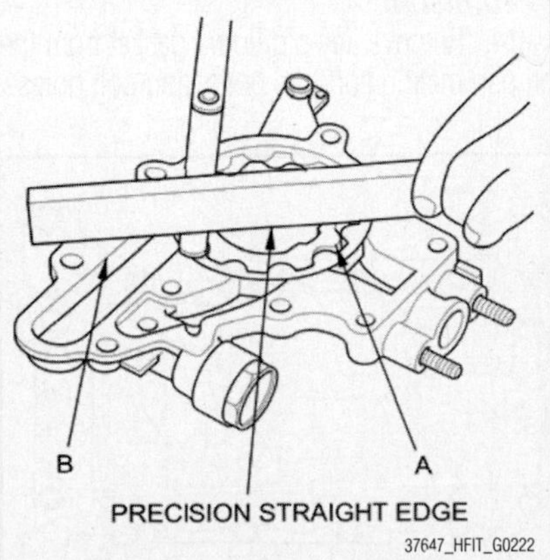
B
A
PRECISION STRAIGHT EDGE
37647_HFIT_G0222

Fig. 70 Check the pump housing-to-rotor axial clearance between the rotors (A) and pump housing (B)

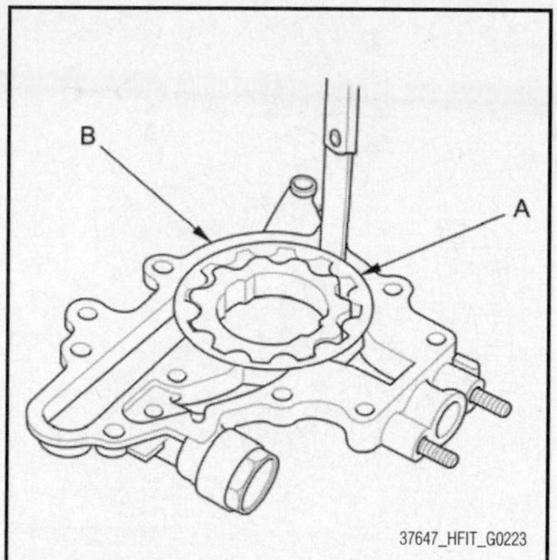
B
A
37647_HFIT_G0223

Fig. 71 Check the pump housing-to-outer rotor radial clearance between the outer rotor (A) and the pump housing (B)

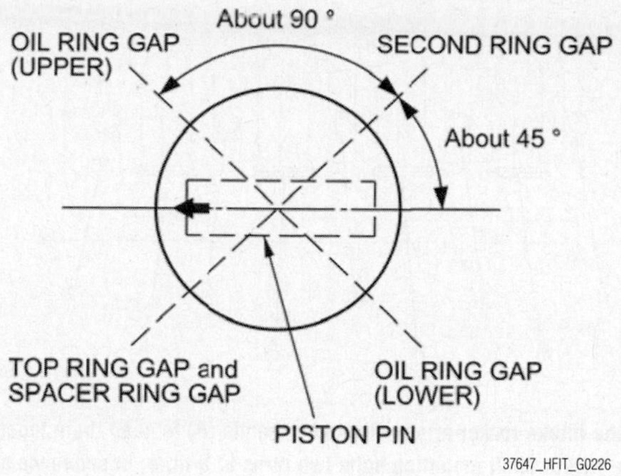
About 90 °
OIL RING GAP (UPPER)
SECOND RING GAP
About 45 °
TOP RING GAP and SPACER RING GAP
OIL RING GAP (LOWER)
PISTON PIN
37647_HFIT_G0226

Fig. 73 Piston ring end gap positioning

5. Inspect both rotors and the pump housing for scoring or other damage. Replace the parts, if necessary.
6. Check that the oil pump turns freely.

PISTON AND RING

POSITIONING

See Figures 72 and 73.

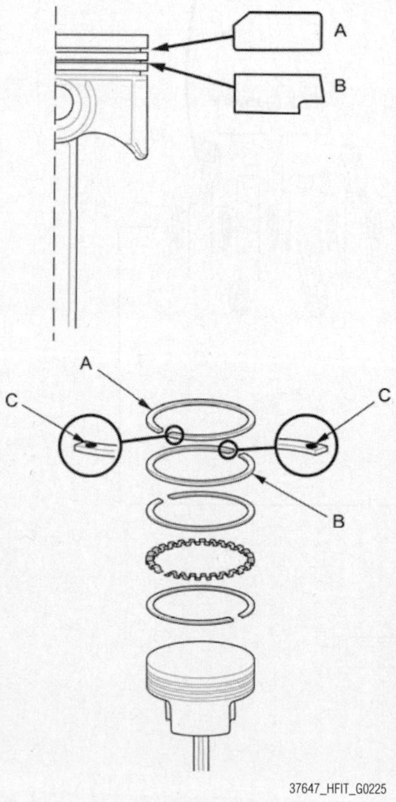
A
B
A
C
C
B
37647_HFIT_G0225

Fig. 72 Piston ring positioning

REAR MAIN SEAL

REMOVAL & INSTALLATION

With Manual Transaxle

1. Remove the ball bearing from the clutch housing, using an adjustable bearing puller and a slide hammer.
2. Remove the oil seal from the transaxle side.

To install:

3. Drive in the new oil seal from the transaxle side using the driver and 37 x 40 mm attachment.
4. Drive in the new ball bearing from the transaxle side using the driver and 52 x 55 mm attachment.

With Automatic Transaxle

1. Remove the mainshaft bearing and the oil seal using the adjustable bearing puller and a slide hammer.

To install:

2. Install the new mainshaft bearing until it bottoms in the torque converter housing using the driver and the attachment (62 x 68 mm).
3. Install the new oil seal flush with the housing using the driver and the attachment (72 x 75 mm).

ROCKER ARMS

REMOVAL & INSTALLATION

See Figures 74 through 78.

1. Remove the cylinder head cover.
2. Loosen the rocker arm adjusting screws.
3. Bundle the intake rocker arms with

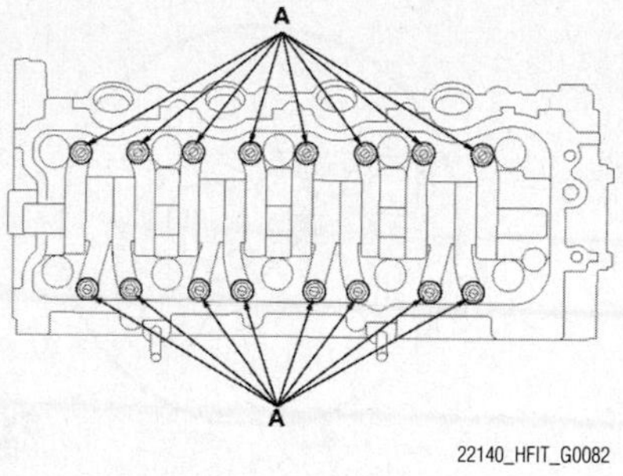
A
A
22140_HFIT_G0082

Fig. 74 Showing the rocker arm adjusting screws (A)

rubber bands to keep them together as a set. Unscrew the rocker shaft mounting bolts two turns at a time, in sequence shown.

4. Remove the rocker shaft mounting bolts, then remove the rocker arm assembly.
5. When disassembling the rocker arms, remove the dowel pin, then remove the rocker arm from cam chain side on the rocker shaft.

➡Identify parts as they are removed to ensure reinstallation in original location.

To install:

➡When disassembling or reassembling the rocker arms, remove or install the rocker arm from cam chain side on the rocker shaft.

6. Inspect the rocker arm shaft and rocker arms.
7. If reused, the rocker arms must be installed in the same positions.
8. Prior to reassembling, clean all the parts in solvent, dry them, and apply lubricant to any contact points.

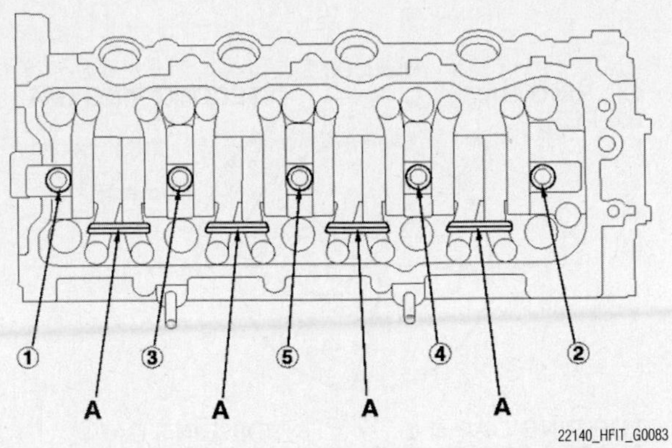

1
3
5
4
2
A
A
A
A
22140_HFIT_G0083

Fig. 75 Bundle the intake rocker arms with rubber bands (A) to keep them together as a set, then unscrew the rocker shaft mounting bolts two turns at a time, in sequence shown

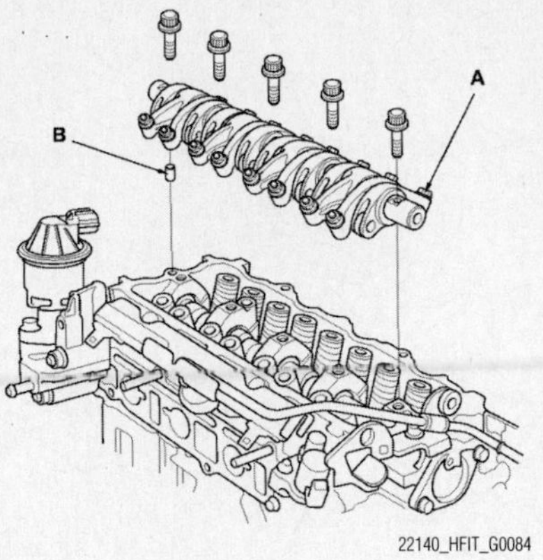

A
B
22140_HFIT_G0084

Fig. 76 Removing the dowel pin (B), then the rocker arm assembly (A)

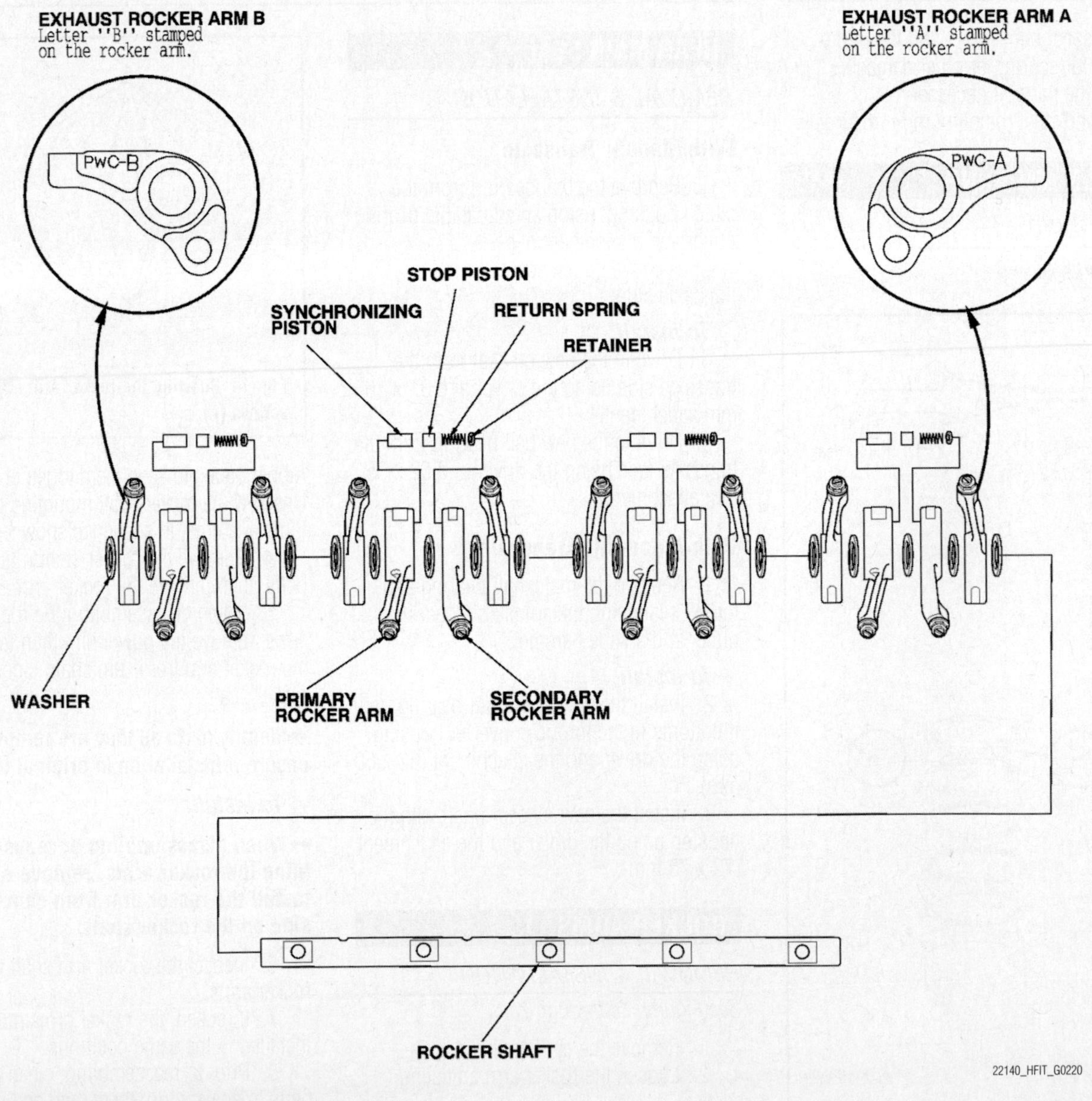

EXHAUST ROCKER ARM B
Letter ''B'' stamped on the rocker arm.
PWC-B
EXHAUST ROCKER ARM A
Letter ''A'' stamped on the rocker arm.
PWC-A
STOP PISTON
SYNCHRONIZING PISTON
RETURN SPRING
RETAINER
WASHER
PRIMARY ROCKER ARM
SECONDARY ROCKER ARM
ROCKER SHAFT
22140_HFIT_G0220

Fig. 77 Exploded view of rocker arm and shaft assemblies

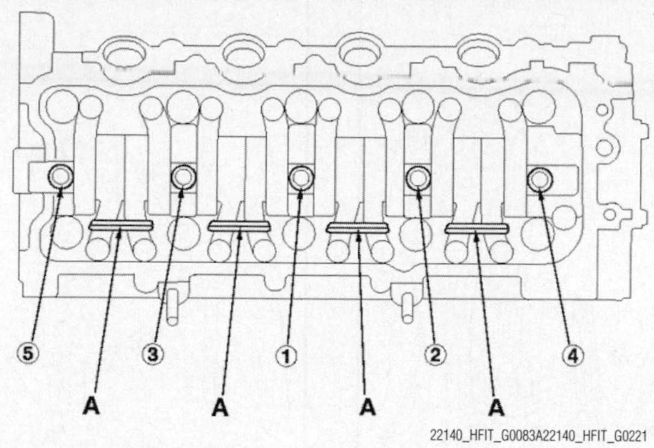
5
3
1
2
4
A
A
A
A
22140_HFIT_G0083A22140_HFIT_G0221

Fig. 78 Showing rocker arm bolt tightening sequence

9. Install the rocker arm assembly with dowel pin into position on the cylinder head.

➡When reassembling the rocker arms, Install the rocker arm from cam chain side on the rocker shaft

10. Tighten each bolt, two turns at a time, in the sequence shown. Torque bolts to 22 ft. lbs. (29 Nm).

➡Apply new engine oil to the bolt threads and flange.

11. Remove the rubber bands from the intake rocker arms.
12. Adjust the valve clearance.
13. Install the cylinder head cover.

TIMING (CAM) CHAIN FRONT CASE, CHAIN & SPROCKETS

REMOVAL & INSTALLATION

See Figures 79 through 95.

➡Keep the cam chain away from magnetic fields.

1. Remove the cylinder head cover.
2. Set the No. 1 piston at Top Dead Center (TDC). The "UP" mark on the camshaft sprocket should be at the top, and the TDC grooves on the camshaft sprocket should line up with the top edge of the head.
3. Remove the right front wheel.
4. Remove the splash shield.
5. Loosen the water pump pulley mounting bolts.
6. Remove the drive belt.
7. Remove the water pump pulley.
8. Remove the crankshaft pulley.
9. Remove the drive belt auto-tensioner.

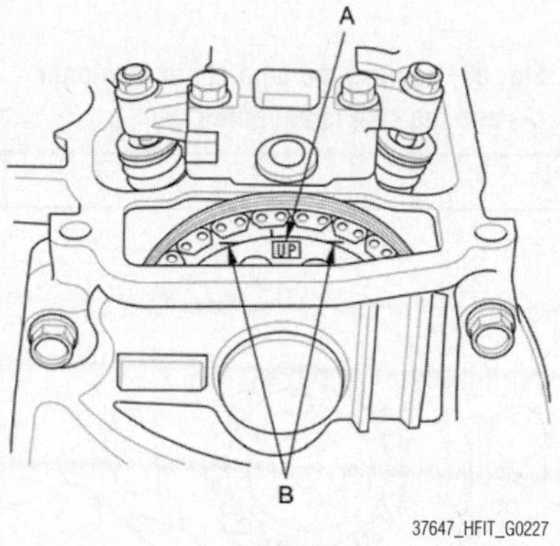
A
UP
B
37647_HFIT_G0227

Fig. 79 Set the No. 1 piston at Top Dead Center (TDC). The "UP" mark (A) on the camshaft sprocket should be at the top, and the TDC grooves (B) on the camshaft sprocket should line up with the top edge of the head

10. Support the engine with a jack and a wood block under the oil pan.
11. Remove the ground cable, then remove the side engine mount/bracket assembly.
12. Remove the cam chain case.
13. Measure the cam chain separation. If the distance is less than the service limit of 0.59 inches (15 mm), replace the cam chain and cam chain tensioner.
14. Apply new engine oil to the sliding surface of the cam chain tensioner slider.
15. Hold the cam chain tensioner slider with the screwdriver, then remove the bolt, and loosen the bolt.
16. Remove the cam chain tensioner slider.
17. Remove the cam chain tensioner and the cam chain guide.
18. Remove the cam chain.

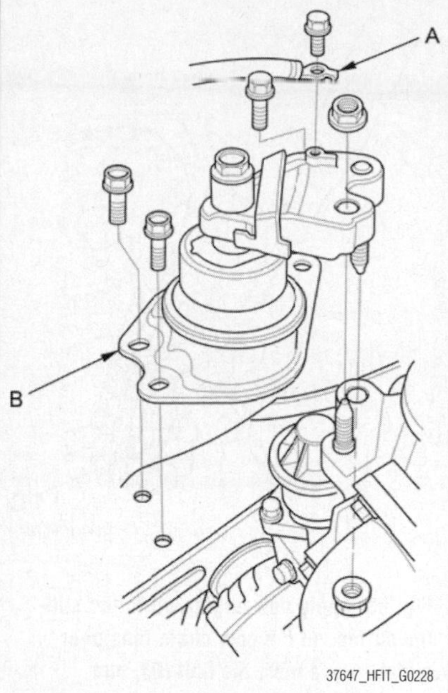
A
B
37647_HFIT_G0228

Fig. 80 Remove the ground cable (A), then remove the side engine mount/bracket assembly (B)

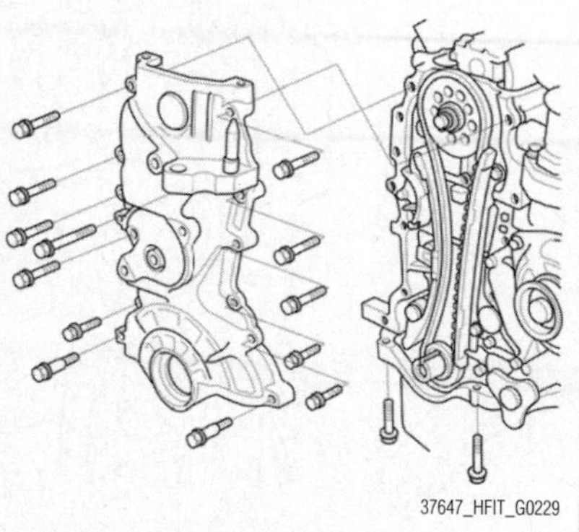
37647_HFIT_G0229

Fig. 81 Remove the cam chain case

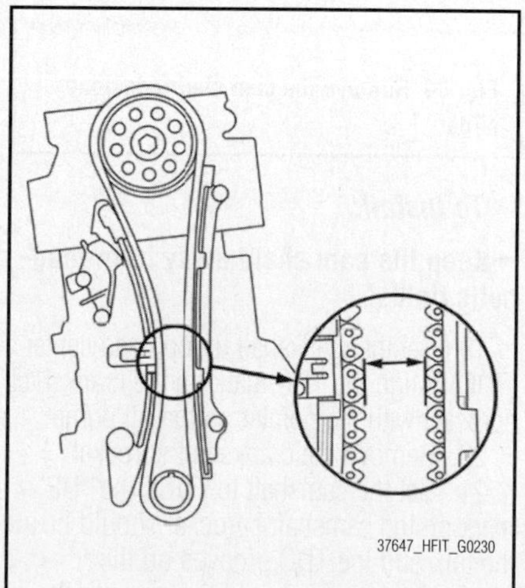
37647_HFIT_G0230

Fig. 82 Measure the cam chain separation

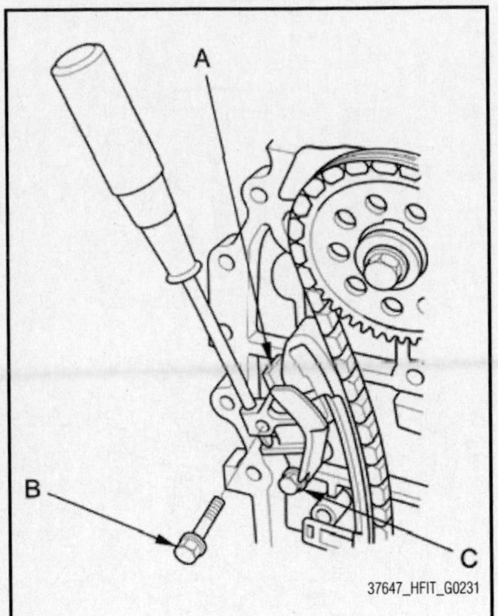

A
B
C
37647_HFIT_G0231

Fig. 83 Apply new engine oil to the sliding surface of the cam chain tensioner slider (A); remove the bolt (B), and loosen the bolt (C)

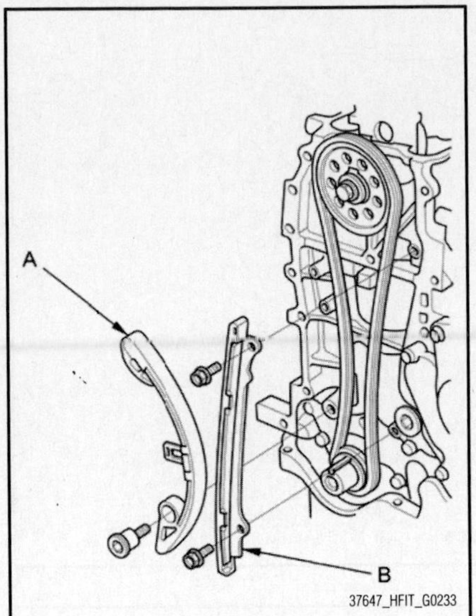

A
B
37647_HFIT_G0233

Fig. 85 Remove the cam chain tensioner (A) and the cam chain guide (B)

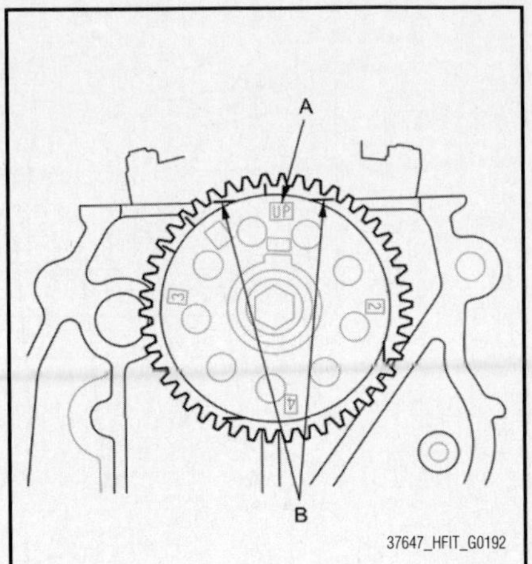

A
UP
B
37647_HFIT_G0192

Fig. 87 The UP mark (A) on the camshaft sprocket should be at the top, and the TDC grooves (B) on the camshaft sprocket should line up with the top edge of the head

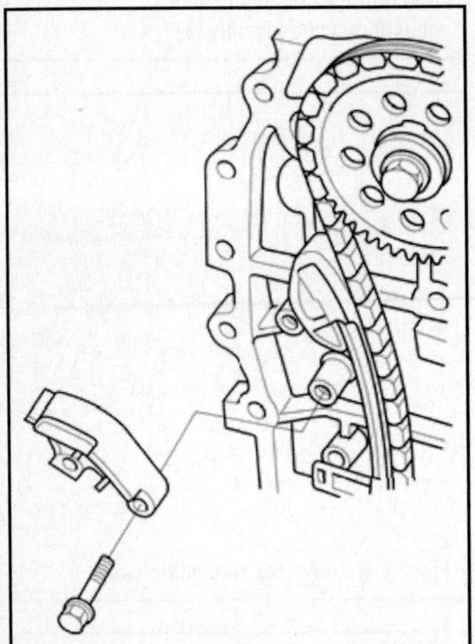

37647_HFIT_G0232

Fig. 84 Remove the cam chain tensioner slider

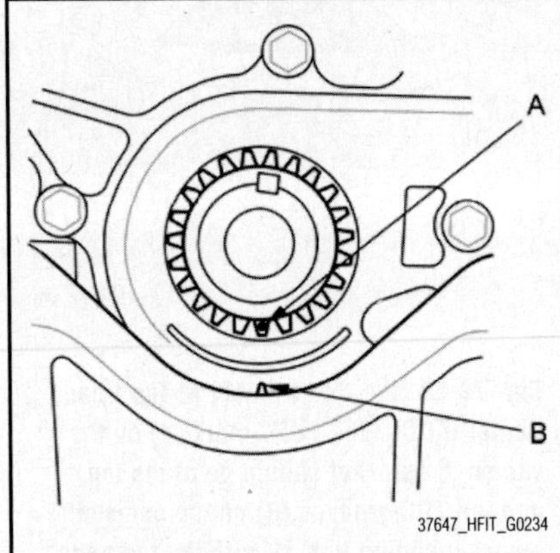

A
B
37647_HFIT_G0234

Fig. 86 Align the TDC mark (A) on the crankshaft sprocket with the pointer (B) on the oil pump

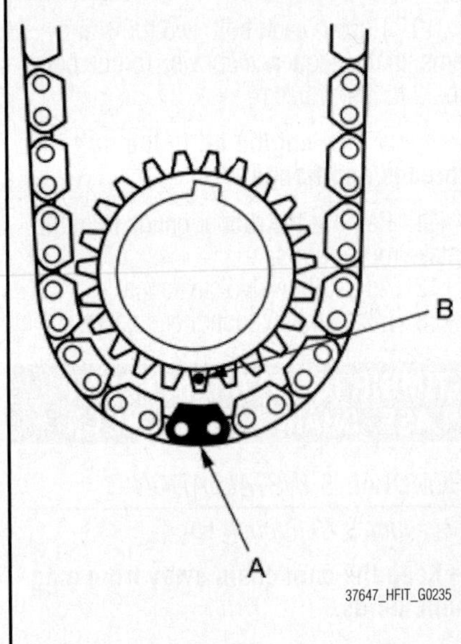

B
A
37647_HFIT_G0235

Fig. 88 Install the cam chain on the crankshaft sprocket with the colored piece (A) aligned with the TDC mark (B) on the crankshaft sprocket, then install the crankshaft sprocket to the crankshaft

To install:

➡Keep the cam chain away from magnetic fields.

19. Set the crankshaft to top dead center (TDC). Align the TDC mark on the crankshaft sprocket with the pointer on the oil pump.
20. Remove the crankshaft sprocket.
21. Set the camshaft to TDC. The "UP" mark on the camshaft sprocket should be at the top, and the TDC grooves on the camshaft sprocket should line up with the top edge of the head.
22. Install the cam chain on the crankshaft sprocket with the colored piece aligned with the TDC mark on the crankshaft sprocket, then install the crankshaft sprocket to the crankshaft.
23. Install the cam chain on the camshaft sprocket with the pointers aligned with the three colored pieces as shown.
24. Install the cam chain tensioner and the cam chain guide.
25. Install the cam chain tensioner slider, and loosely tighten the bolt.
26. Apply new engine oil to the sliding surface of the cam chain tensioner slider.
27. Rotate the cam chain tensioner slider clockwise to compress the cam chain tensioner, and install the remaining bolt, then tighten the bolts.
28. Check the chain case oil seal for damage. If the oil seal is damaged, replace the chain case oil seal.
29. Remove the all of the old liquid gasket from the chain case mating surfaces, the bolts, and the bolt holes.
30. Clean and dry the chain case mating surfaces.
31. Apply liquid gasket (P/N 08717-0004, 08718-0003, or 08718-0009) to the cylinder head and the engine block mating

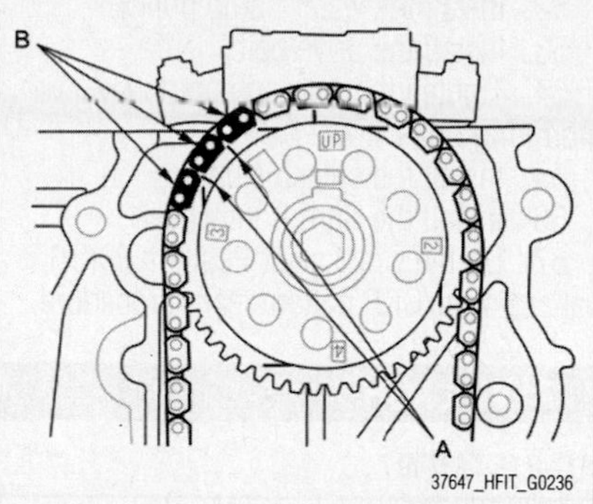
B
A
37647_HFIT_G0236

Fig. 89 Install the cam chain on the camshaft sprocket with the pointers (A) aligned with the three colored pieces (B)

surfaces of the cam chain case and to the inside edge of the bolt holes. Install the component within 5 minutes of applying the liquid gasket.

➡If you apply liquid gasket P/N 08718-0012, the component must be installed within 4 minutes. If too much time has passed after applying the liquid gasket, remove the old liquid gasket and residue, then reapply new liquid gasket.

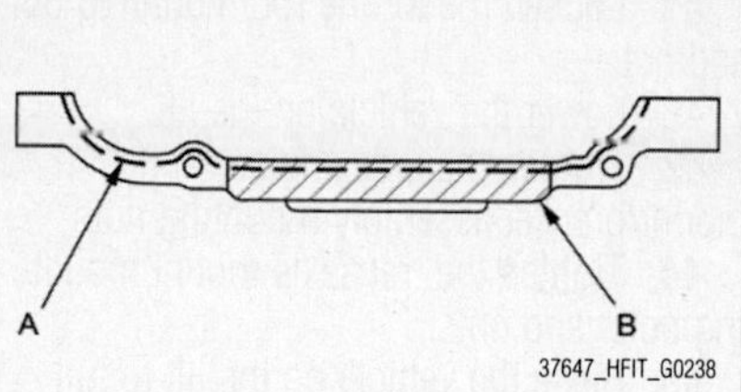
A
B
37647_HFIT_G0238

Fig. 91 Apply liquid to the oil pan mating surface of the cam chain case and to the inside edge of the bolt holes

32. Apply liquid gasket (P/N 08717-0004, 08718-0003, or 08718-0009) to the oil pan mating surface of the cam chain case and to the inside edge of the bolt holes. Install the component within 5 minutes of applying the liquid gasket.

➡If you apply liquid gasket P/N 08718-0012, the component must be installed within 4 minutes. If too much time has passed after applying the liquid gasket, remove the old liquid gasket and residue, then reapply new liquid gasket.

33. Set the edge of the chain case to the edge of the oil pan, then install the chain case on the engine block.

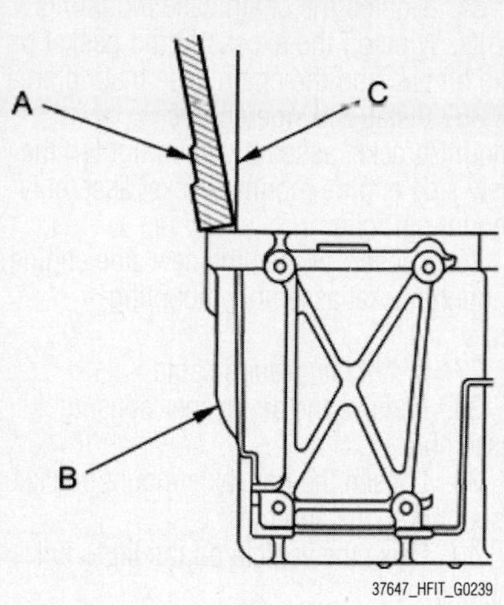
A
C
B
37647_HFIT_G0239

Fig. 93 Set the edge of the chain case (A) to the edge of the oil pan (B), then install the chain case on the engine block (C)

➡When installing the chain case, do not slide the bottom surface onto the oil pan mounting surface. Wait at least 30 minutes before filling the engine with oil. Do not run the engine for at least 3 hours after installing the chain case.

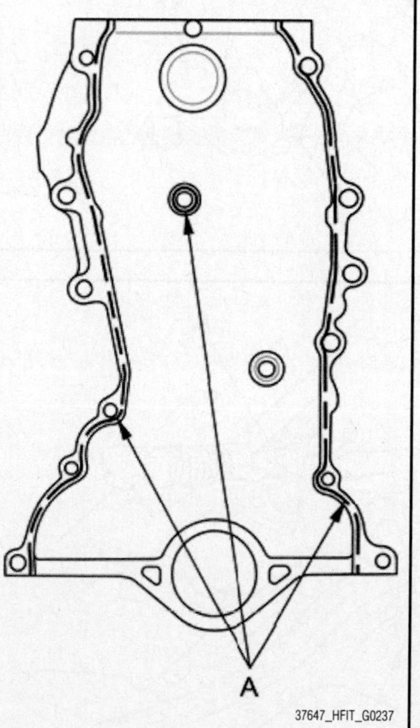
A
37647_HFIT_G0237

Fig. 90 Apply liquid gasket to the cylinder head and the engine block mating surfaces of the cam chain case and to the inside edge of the bolt holes

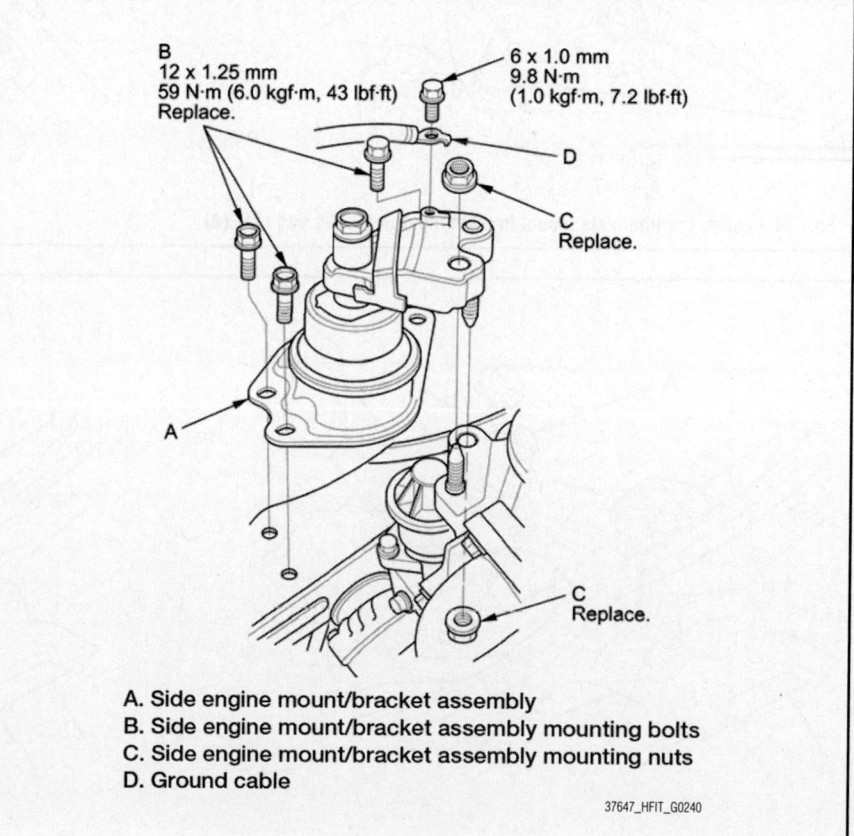

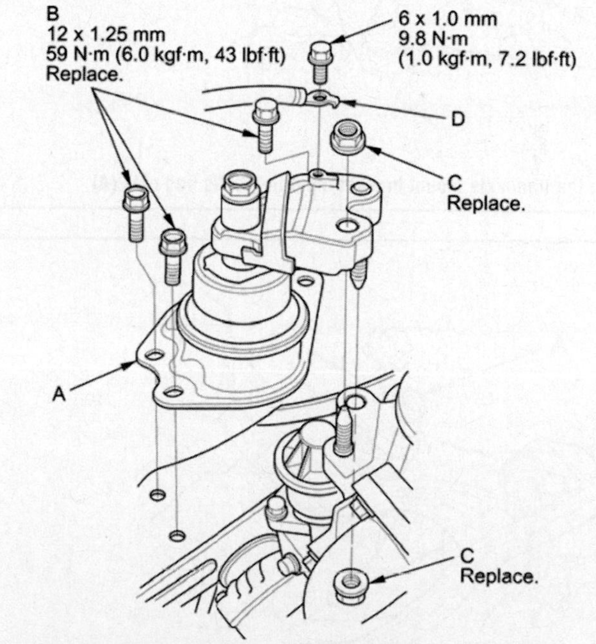
B
12 x 1.25 mm
59 N·m (6.0 kgf·m, 43 lbf·ft)
Replace.
6 x 1.0 mm
9.8 N·m
(1.0 kgf·m, 7.2 lbf·ft)
D
C
Replace.
A
C
Replace.
A. Side engine mount/bracket assembly
B. Side engine mount/bracket assembly mounting bolts
C. Side engine mount/bracket assembly mounting nuts
D. Ground cable
37647_HFIT_G0240

Fig. 92 Install the side engine mount/bracket assembly

34. Tighten the chain case mounting bolts. Wipe off the excess liquid gasket on the oil pan and the chain case mating area.

35. Install the side engine mount/bracket assembly, then tighten the new side engine mount/bracket assembly mounting bolts.

36. Loosely tighten the new side engine mount/bracket assembly mounting nuts.

37. Install the ground cable.

38. Remove the air cleaner housing assembly.

39. Loosen the transaxle mount bracket mounting bolts and nuts.

40. Raise the vehicle on the lift to full height.

41. Loosen the torque rod mounting bolt and nut.

42. Lower the vehicle on the lift.

43. Tighten the side engine mount/bracket assembly mounting nuts.

44. Tighten the transaxle mount mounting bolts and nuts.

45. Raise the vehicle on the lift to full height.

46. Tighten the torque rod mounting bolt and nut.

47. Lower the vehicle on the lift.

48. Install the air cleaner housing assembly.

49. Install the cylinder head cover.

50. Install the drive belt auto-tensioner.

51. Install the crankshaft pulley.

52. Install the water pump pulley.

53. Install the drive belt.

54. Tighten the water pump pulley mounting bolts.

55. Install the splash shield.

56. Install the right front wheel.

57. Do the Crankshaft Position (CKP) pattern clear/CKP pattern learn procedure.

VALVE LASH

ADJUSTMENT

See Figures 96 and 97.

➡**Valves should be adjusted only when the cylinder head temperature is less than 100°F (38°C).**

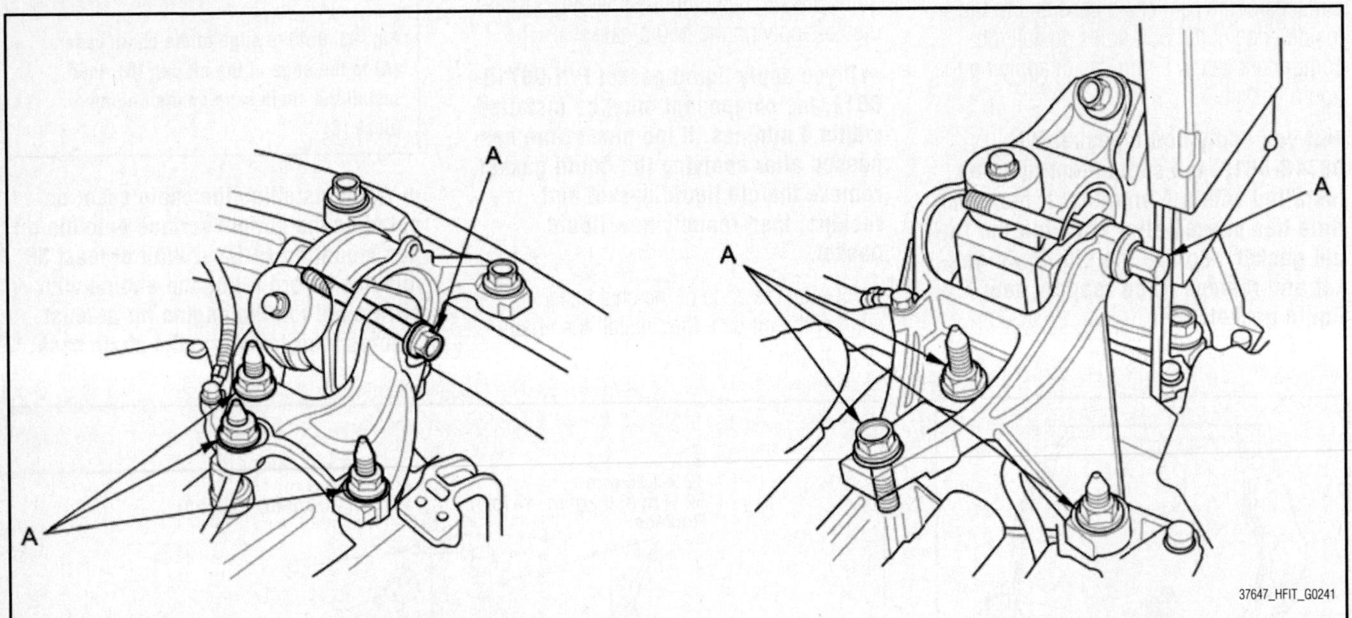
A
A
A
A
37647_HFIT_G0241

Fig. 94 Loosen the transaxle mount bracket mounting bolts and nuts (A)

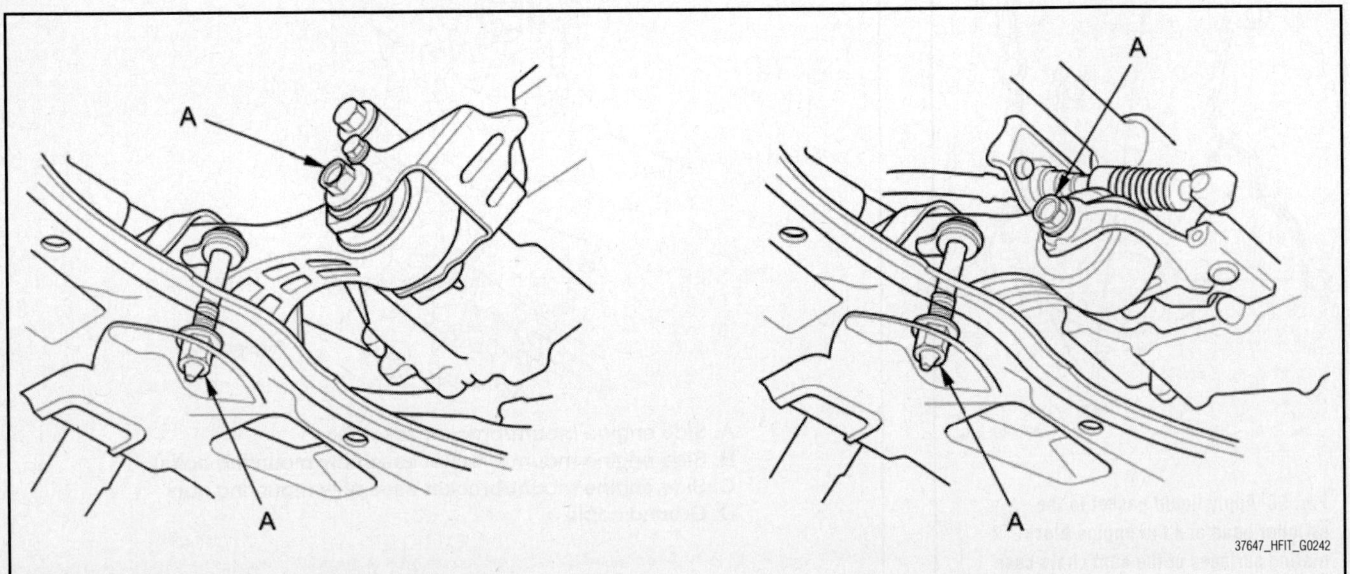
A
A
A
A
37647_HFIT_G0242

Fig. 95 Loosen the torque rod mounting bolt and nut (A)

1. Remove the cylinder head cover.

2. Set the No. 1 piston at top dead center (TDC). The "UP" mark on the camshaft sprocket should be at the top, and the TDC grooves on the camshaft sprocket should line up with the top edge of the head.

3. Select the correct thickness feeler gauge for the valves to check.

4. Proper valve clearance should be:
   - Intake: 0.006–0.007 inches (0.15–0.19 mm)
   - Exhaust: 0.010–0.012 inches (0.26–0.30 mm)

5. Insert the feeler gauge between the adjusting screw and the end of the valve stem and slide it back and forth; a slight amount of drag should be felt.

6. If too much or too little drag is present, loosen the locknut, and turn the adjusting screw until the drag on the feeler gauge is correct.

7. Tighten the locknut to 10 ft. lbs. (14 Nm) and recheck the clearance. Repeat the adjustment if necessary.

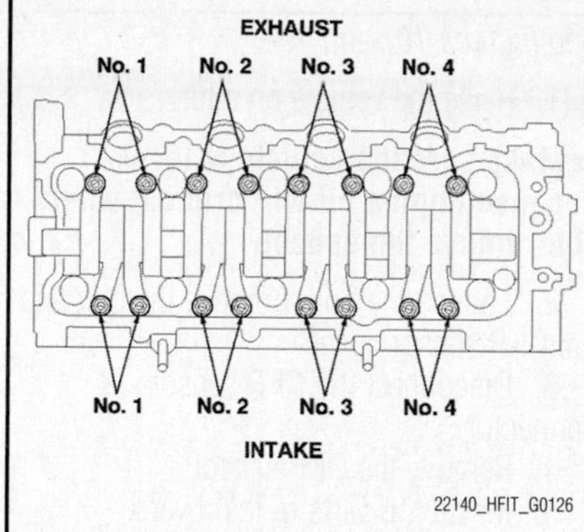
EXHAUST
No. 1
No. 2
No. 3
No. 4
No. 1
No. 2
No. 3
No. 4
INTAKE
22140_HFIT_G0126

Fig. 96 Showing valve positions for reference during adjustment

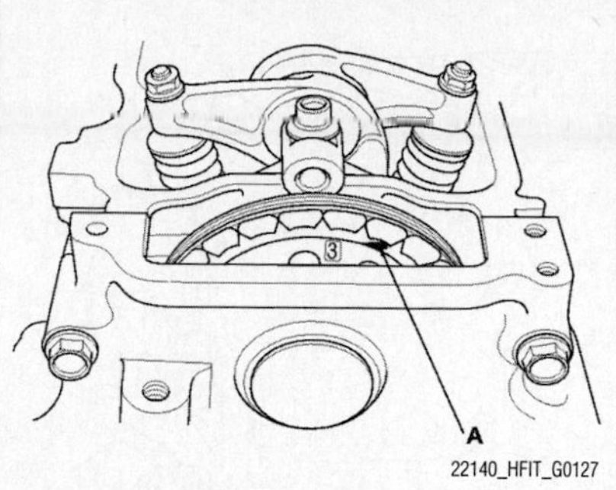
A
22140_HFIT_G0127

Fig. 97 Rotate the crankshaft clockwise. Align the No. 3 piston TDC groove (A) on the camshaft sprocket with the top edge of the head

8. Rotate the crankshaft clockwise. Align the No. 3 piston TDC groove on the camshaft sprocket with the top edge of the head.

9. Check and if necessary, adjust the valve clearance on No. 3 cylinder.

10. Repeat this procedure for cylinder No. 4 and then for cylinder No. 2.

ENGINE PERFORMANCE & EMISSION CONTROLS

ACCELERATOR PEDAL POSITION (APP) SENSOR

LOCATION

The accelerator Pedal Position (APP) sensor is an integral part of the accelerator pedal module.

REMOVAL & INSTALLATION

See Figures 98 and 99.

1. Remove the driver's dashboard lower cover.

2. Disconnect the APP sensor connector (A).

3. Remove the accelerator pedal module (B).

➡The APP sensor is not available separately. Do not disassemble the accelerator pedal module.

4. Install the parts in the reverse order of removal.

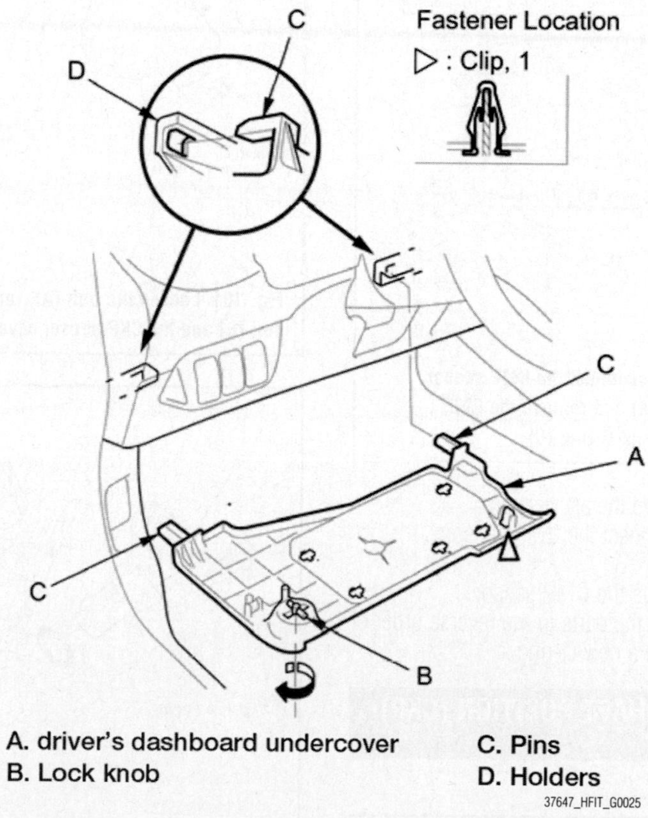
Fastener Location
▷ : Clip, 1
C
D
C
A
C
B
A. driver's dashboard undercover
B. Lock knob
C. Pins
D. Holders
37647_HFIT_G0025

Fig. 98 Remove the driver's dashboard lower cover

CAMSHAFT POSITION (CMP) SENSOR

LOCATION

Refer to graphics in the Removal and Installation section for location.

REMOVAL & INSTALLATION

See Figure 100.

1. Remove the cowl cover and the under-cowl panel.

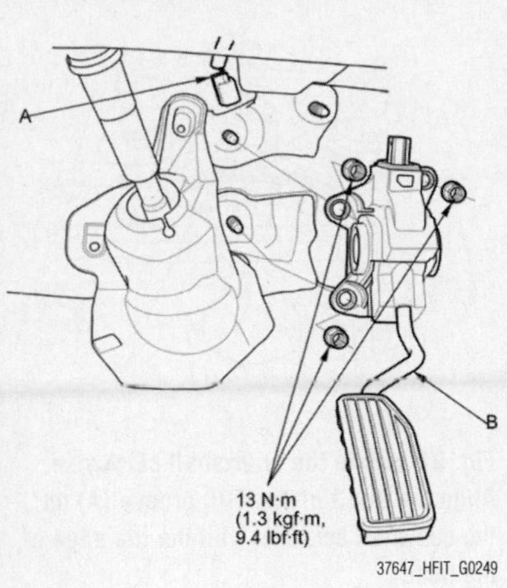
A
B
13 N·m
(1.3 kgf·m,
9.4 lbf·ft)
37647_HFIT_G0249

Fig. 99 APP sensor connector (A) and accelerator pedal module (B)

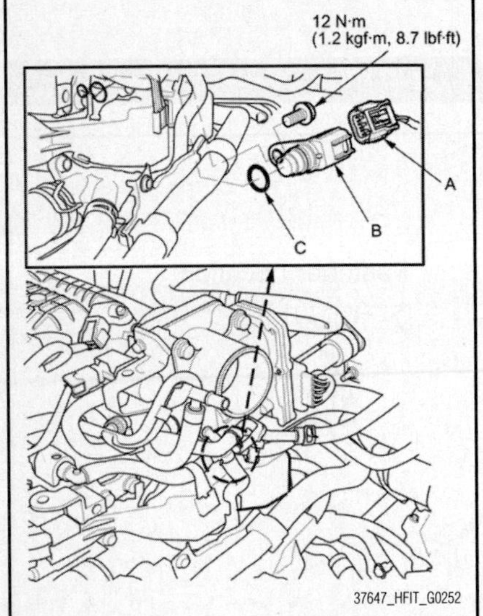
12 N·m
(1.2 kgf·m, 8.7 lbf·ft)
A
B
C
37647_HFIT_G0252

Fig. 100 Disconnect the CMP sensor connector (A) and remove the CMP sensor (B) and O-ring (C)

2. Remove the air cleaner.
3. Disconnect the CMP sensor connector.
4. Remove the CMP sensor.
5. Install the parts in the reverse order of removal with a new O-ring.

CRANKSHAFT POSITION (CKP) SENSOR

LOCATION

Refer to graphics in the Removal and Installation section for location.

REMOVAL & INSTALLATION

See Figures 101 and 102.

1. Raise the vehicle on a lift.

➡Make sure the vehicle is level, because engine oil will drip out when you remove the sensor.

2. Loosen the bolt. Remove the bolt and the CKP sensor cover.
3. Disconnect the CKP sensor connector.
4. Remove the CKP sensor.
5. Install the parts in the reverse order of removal with a new O-ring.
6. Do the CKP pattern clear/CKP pattern learn procedure.
7. Check the engine oil level, and add more oil if needed.

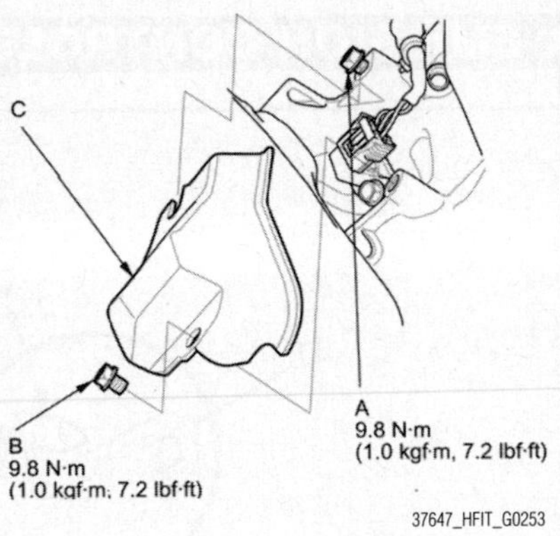
C
A
9.8 N·m
(1.0 kgf·m, 7.2 lbf·ft)
B
9.8 N·m
(1.0 kgf·m, 7.2 lbf·ft)
37647_HFIT_G0253

Fig. 101 Loosen the bolt (A), remove the bolt (B) and the CKP sensor cover (C)

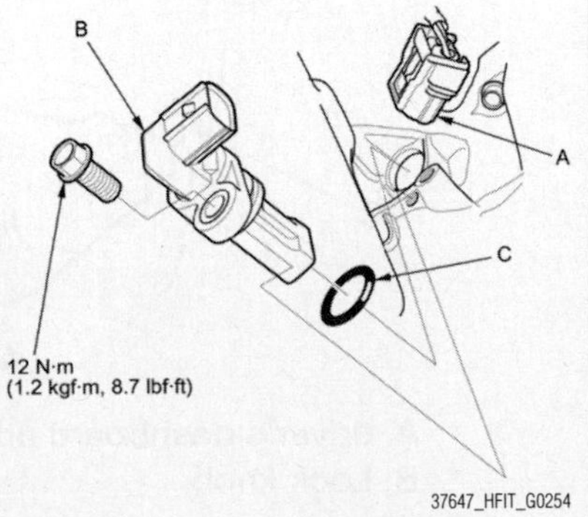
B
A
C
12 N·m
(1.2 kgf·m, 8.7 lbf·ft)
37647_HFIT_G0254

Fig. 102 Disconnect the CKP sensor connector (A), remove the CKP sensor (B) and O-ring (C)

CKP PATTERN CLEAR/CKP PATTERN LEARN

Clear/Learn Procedure (with the HDS)

1. Connect the HDS to the Data Link Connector (DLC) located under the driver's side of the dashboard.
2. Turn the ignition switch to ON (II).
3. Make sure the HDS communicates with the ECM/PCM and other vehicle systems. If it doesn't, go to the DLC circuit troubleshooting.
4. Select CRANK PATTERN in the ADJUSTMENT MENU with the HDS.
5. Select CRANK PATTERN LEARNING with the HDS, and follow the screen prompts.

Learn Procedure (without the HDS)

1. Start the engine. Hold the engine speed at 3,000 rpm without load (A/T in P or N, M/T in neutral) until the radiator fan comes on.
2. Test-drive the vehicle on a level road: Decelerate (with the throttle fully closed) from an engine speed of 2,500 rpm down to 1,000 rpm with the A/T in 2, or the M/T in 2nd.
3. Repeat step 2 several times.
4. Turn the ignition switch to LOCK (0).
5. Turn the ignition switch to ON (II), and wait 30 seconds.

ENGINE CONTROL MODULE (ECM)/POWERTRAIN CONTROL MODULE (PCM)

REMOVAL & INSTALLATION

See Figures 103 and 104.

Special Tools Required:

- Honda Diagnostic System (HDS) tablet tester
- Honda Interface Module (HIM) and an iN workstation with the latest HDS software version
- HDS pocket tester
- GNA600 and an iN workstation with the latest HDS software version
- Any one of the above updating tools can be used.

➡Make sure the HDS/iN workstation has the latest HDS software version.

1. Connect the HDS to the Data Link Connector (DLC) located under the driver's side of the dashboard.
2. Turn the ignition switch to ON (II).
3. Make sure the HDS communicates with the ECM/PCM and other vehicle sys-

tems. If it doesn't, go to the DLC circuit troubleshooting. If you are returning from the DLC circuit troubleshooting, skip steps 4 through 9, 19 through 24, and 27 through 29, and do these procedures after replacing the ECM/PCM:

- Replace the engine oil and the engine oil filter.
- Replace the ATF (A/T).
- Clean the throttle body.

4. Select the PGM-FI system with the HDS.
5. Select the INSPECTION MENU with the HDS.
6. Select the ETCS TEST, then select the TP POSITION CHECK, and follow the screen prompts.

➡If the TP POSITION CHECK indicates FAILED, continue with this procedure.

7. Select the REPLACE ECM/PCM MENU, then select READ DATA, and follow the screen prompts.

➡Doing this step copies (READS) the engine oil life data from the original ECM/PCM so you can later download (WRITES) it into the new ECM/PCM. If READ DATA indicates FAILED, continue with this procedure.

8. A/T: Select the A/T system with the HDS.
9. A/T: Select the REPLACE TCM/PCM MENU, then READ DATA, and follow the screen prompts.

➡A/T: Doing this step copies (READS) the ATF life data from the original PCM so you can later download (WRITES) it into the new PCM. A/T: If READ DATA indicates FAILED, continue with this procedure.

10. Jump the SCS line with the HDS.
11. Turn the ignition switch to LOCK (0).
12. Remove the ECM/PCM cover.
13. Remove the battery holder, and reposition the battery away from the ECM/PCM.

➡Do not disconnect the battery terminals.

14. Remove the bolts, and loosen the bolt.
15. Disconnect ECM/PCM connectors A, B, and C, then remove the ECM/PCM.

➡ECM/PCM connectors A, B, and C have symbols (A=_□, B=_△, C=_○) embossed on them for identification.

16. Install a known-good ECM/PCM in the reverse order of removal.
17. Turn the ignition switch to ON (II).
18. Manually input the VIN to the ECM/PCM with the HDS.

➡DTC P0630 (VIN Not Programmed or Mismatch) may be stored because the VIN has not been programmed into the ECM/PCM; ignore it, and continue this procedure.

19. If the READ DATA (engine oil life) failed in step 7, go to step 22 (A/T) or step 25 (M/T). Otherwise, go to step 20.
20. Select the PGM-FI system with the HDS.
21. Select the REPLACE ECM/PCM MENU, then WRITE DATA, and follow the screen prompts.

➡If the WRITE DATA indicates FAILED, continue with this procedure.

22. A/T: If the READ DATA (ATF life) failed in step 8, go to step 25. Otherwise go to step 23.
23. A/T: Select the A/T SYSTEM with the HDS.
24. A/T: Select the REPLACE ECM/PCM MENU, then WRITE DATA, and follow the screen prompts.

➡A/T: If the WRITE DATA indicates FAILED, continue with this procedure.

25. Select IMMOBI system with the HDS.
26. Enter the immobilizer ECM/PCM code that you got from the iN, and use the

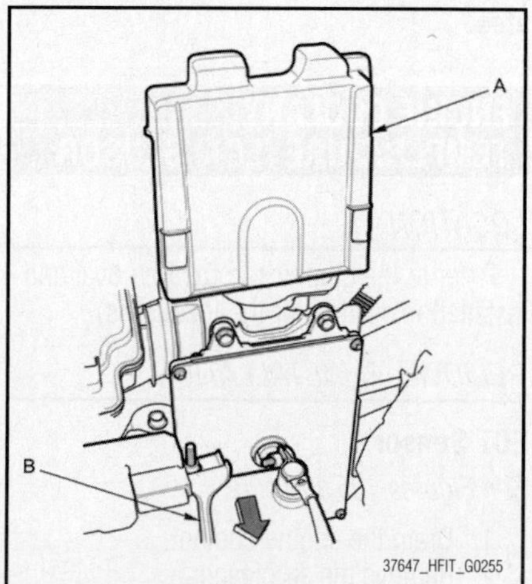
A
B
37647_HFIT_G0255

Fig. 103 Remove the ECM/PCM cover (A); remove the battery holder (B)

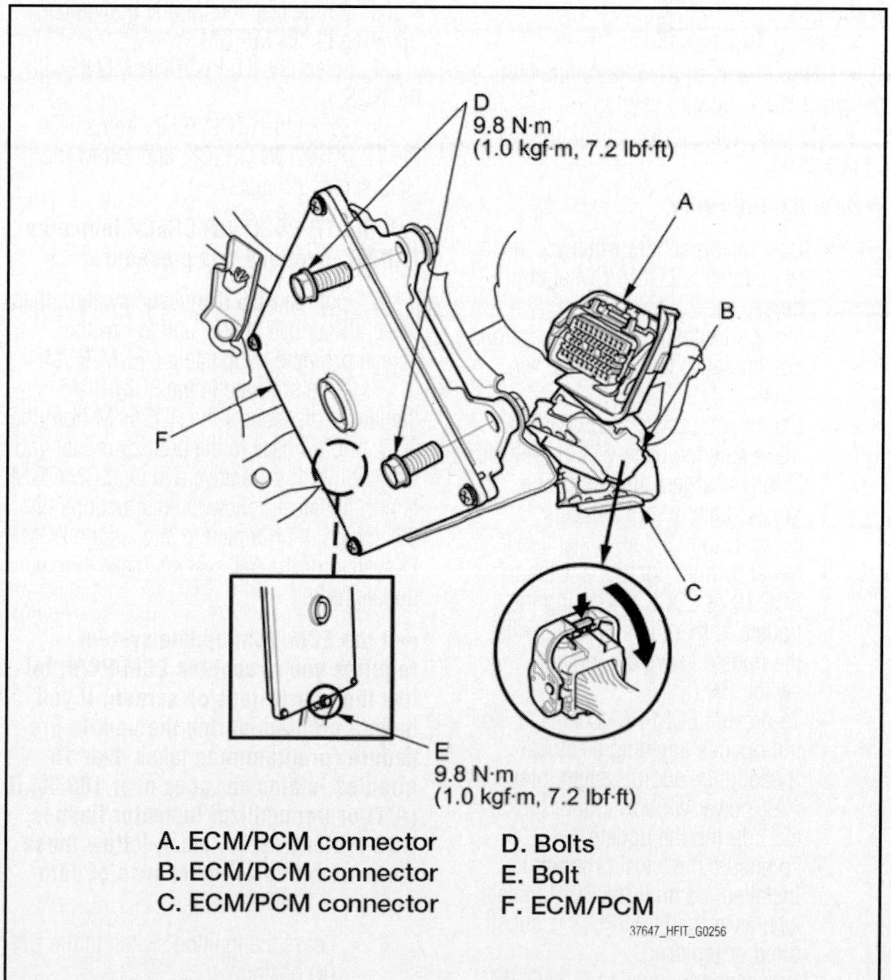
D
9.8 N·m
(1.0 kgf·m, 7.2 lbf·ft)
A
B
F
C
E
9.8 N·m
(1.0 kgf·m, 7.2 lbf·ft)
A. ECM/PCM connector
B. ECM/PCM connector
C. ECM/PCM connector
D. Bolts
E. Bolt
F. ECM/PCM
37647_HFIT_G0256

Fig. 104 ECM/PCM removal

ECM/PCM replacement procedure in the HDS; it allows you to start the engine.

27. If the TP POSITION CHECK failed in step 6 clean the throttle body, then go to step 28.

28. If the READ DATA failed in step 7 or the WRITE DATA failed in step 21, replace the engine oil and the engine oil filter, then go to step 29 (A/T) or step 30 (M/T).

29. If the READ DATA failed in step 9 or the WRITE DATA failed in step 24, replace the ATF, then go to step 30.

30. Select the PGM-FI system, and reset the ECM/PCM with the HDS.

31. Update the ECM/PCM if it does not have the latest software.

32. Do the ECM/PCM idle learn procedure.

33. Do the CKP pattern learn procedure.

ECM/PCM UPDATE

Special Tools Required:

- Honda Diagnostic System (HDS) tablet tester
- Honda Interface Module (HIM) and an iN workstation with the latest HDS software version
- HDS pocket tester
- GNA600 and an iN workstation with the latest HDS software version
- Any one of the above updating tools can be used.

➡**Note the following:**

- Use this procedure when you need to update the ECM/PCM at any time.
- Make sure the HDS/iN workstation has the latest HDS software version.
- Before you update the ECM/PCM, make sure the battery in the vehicle is fully charged, and connect a jumper battery (not a battery charger) to maintain system voltage.
- Never turn the ignition switch to ACC (I) or LOCK (0) during the update. If there is a problem with the update, leave the ignition switch ON (II).
- To prevent ECM/PCM damage, do not operate anything electrical (headlights, audio system, brakes, A/C, power windows, door locks, etc.) during the update.
- To ensure the latest program is installed, do an ECM/PCM update whenever the ECM/PCM is substituted or replaced.
- You cannot update an ECM/PCM with a program it already has. It will only accept a new program.
- High temperature in the engine compartment might cause the ECM/PCM to become too hot to run the update. If the engine has been running before this procedure, open the hood and cool the engine compartment.
- If you need to diagnose the Honda interface module (HIM) because the HIM's red (#3) light came on or was flashing during the update, leave the ignition switch in ON (II) when you disconnect the HIM from the Data Link Connector (DLC). This prevents damage to the ECM/PCM.

1. Turn the ignition switch to ON (II), but do not start the engine.

2. Connect the HDS to the Data Link Connector (DLC) located under the driver's side of the dashboard.

3. Make sure the HDS communicates with the ECM/PCM and other vehicle systems. If it doesn't, go to the DLC circuit troubleshooting. If you are returning from the DLC circuit troubleshooting, skip steps 4 and 5, and clean the throttle body after updating the ECM/PCM.

4. Select the INSPECTION MENU with the HDS.

5. Select the ETCS TEST, then select the TP POSITION CHECK, and follow the HDS screen prompts.

➡**If the TP POSITION CHECK indicates FAILED, continue this procedure.**

6. Exit the HDS diagnostic system, then select the update mode, and follow the screen prompts to update the ECM/PCM.

7. If the software in the ECM/PCM is the latest, disconnect the HDS/HIM from the DLC, and go back to the procedure that you were doing. If the software in the ECM/PCM is not the latest, follow the instructions on the screen. If prompted to choose the PGM-FI system or the A/T system, make sure you update both.

➡**If the ECM/PCM update system requires you to cool the ECM/PCM, follow the instructions on screen. If you have a problem during the update procedure (programming takes over 15 minutes, status bar goes over 100 %, D (A/T) or immobilizer indicator flashes, HDS tablet freezes, etc.), follow these steps to minimize the chance of damaging the ECM/PCM:**

- Leave the ignition switch in the ON (II) position.
- Connect a jumper battery (do not connect a battery charger).
- Shut down the HDS.
- Disconnect the HDS from the DLC.
- Reboot the HDS.
- Reconnect the HDS to the DLC, and try the update procedure again.

8. If the TP POSITION CHECK failed in step 5, clean the throttle body.

9. Do the ECM/PCM idle learn procedure.

10. Do the CKP pattern learn procedure.

ECM/PCM IDLE LEARN PROCEDURE

The idle learn procedure must be done so the ECM/PCM can learn the engine idle characteristics.

Do the idle learn procedure whenever you do any of these actions:

- Replace ECM/PCM.
- Reset ECM/PCM.
- Update ECM/PCM.
- Replace or clean the throttle body.
- Disassemble the engine or the transaxle.

➡**Clearing DTCs with the HDS does not require that you to do the idle learn procedure.**

1. Make sure all electrical items (A/C, audio, lights, etc.) are off.

2. Reset the ECM/PCM with the HDS.

3. Turn the ignition switch to ON (II), and wait 2 seconds.

4. Start the engine. Hold the engine speed at 3,000 rpm without load (A/T in P or N, M/T in neutral) until the radiator fan comes on, or until the engine coolant temperature reaches 194°F (90°C).

5. Let the engine idle for about 5 minutes with the throttle fully closed.

➡**If the radiator fan comes on, do not include its running time in the 5 minutes.**

ENGINE COOLANT TEMPERATURE (ECT) SENSOR

LOCATION

Refer to the graphics in the Removal and Installation section for the location(s).

REMOVAL & INSTALLATION

ECT Sensor 1

See Figures 105 and 106.

1. Drain the engine coolant.
2. Remove the air cleaner.
3. Remove the hose clamp stay bolt.
4. Disconnect the ECT sensor 1 connector.
5. Remove ECT sensor 1.

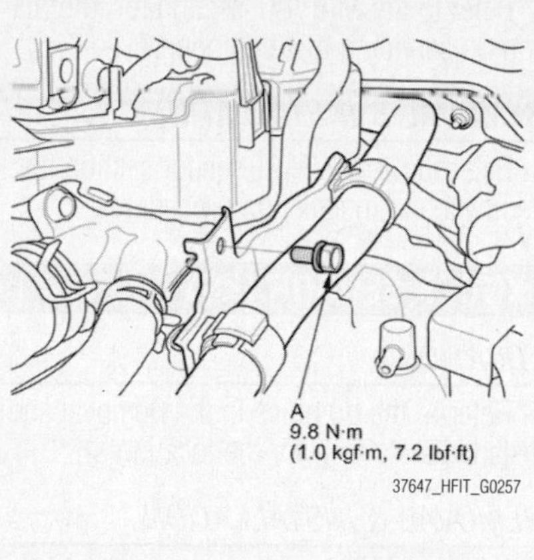
A
9.8 N·m
(1.0 kgf·m, 7.2 lbf·ft)
37647_HFIT_G0257

Fig. 105 Remove the hose clamp stay bolt (A)

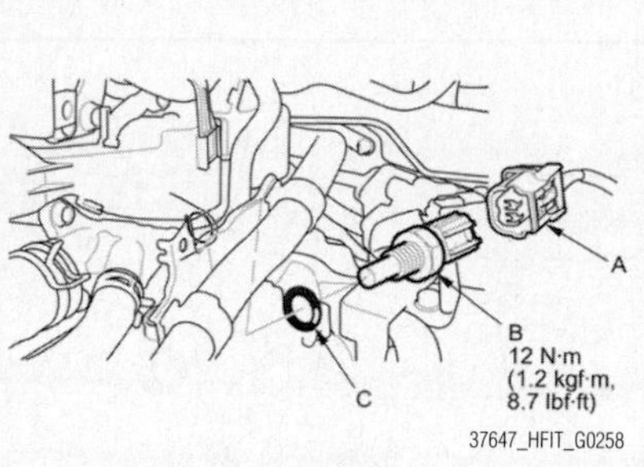
A
B
12 N·m
(1.2 kgf·m,
8.7 lbf·ft)
C
37647_HFIT_G0258

Fig. 106 Disconnect the ECT sensor 1 connector (A); remove the ECT sensor 1 (B) and the O-ring (C)

6. Install the parts in the reverse order of removal with a new O-ring, then refill the radiator with engine coolant.

ECT Sensor 2

See Figure 107.

1. Drain the engine coolant.
2. Raise the vehicle on a lift.

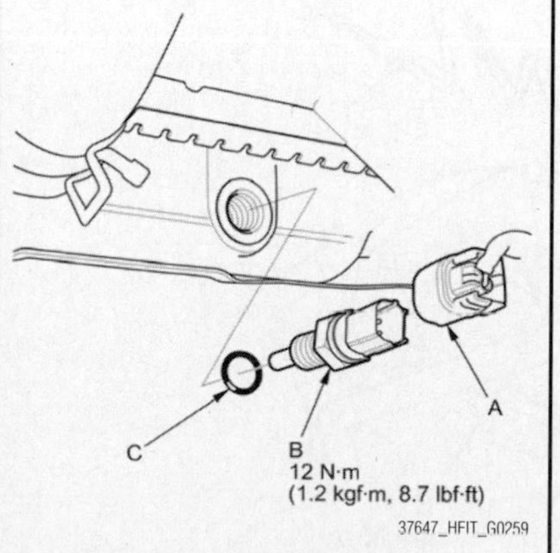
A
C
B
12 N·m
(1.2 kgf·m, 8.7 lbf·ft)
37647_HFIT_G0259

Fig. 107 Disconnect the ECT sensor 2 connector (A), then remove ECT sensor 2 (B) and the O-ring (C)

3. Disconnect the ECT sensor 2 connector, then remove ECT sensor 2.
4. Install the parts in the reverse order of removal with a new O-ring, then refill the radiator with engine coolant.

EVAPORATIVE EMISSIONS (EVAP) CANISTER

LOCATION

The EVAP canister is located forward of the rear suspension on the passenger side.

REMOVAL & INSTALLATION

See Figures 108 through 112.

1. Raise the vehicle on a lift.
2. Except LX (A/T model): Remove the fuel tank cover
3. LX (A/T model): Remove the floor under cover assembly.

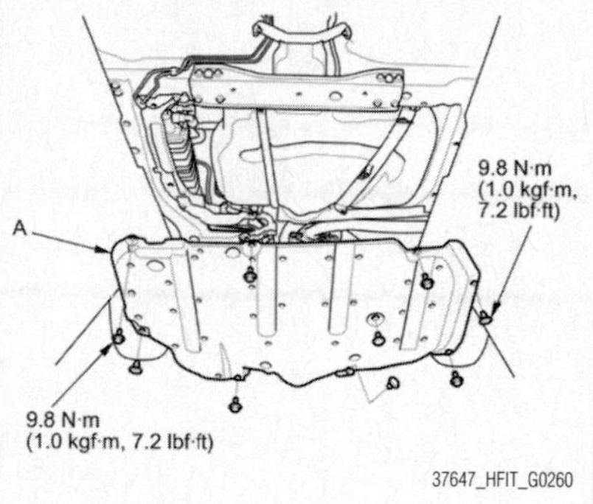
9.8 N·m
(1.0 kgf·m,
7.2 lbf·ft)
A
9.8 N·m
(1.0 kgf·m, 7.2 lbf·ft)
37647_HFIT_G0260

Fig. 108 Except LX (A/T model): Remove the fuel tank cover (A)

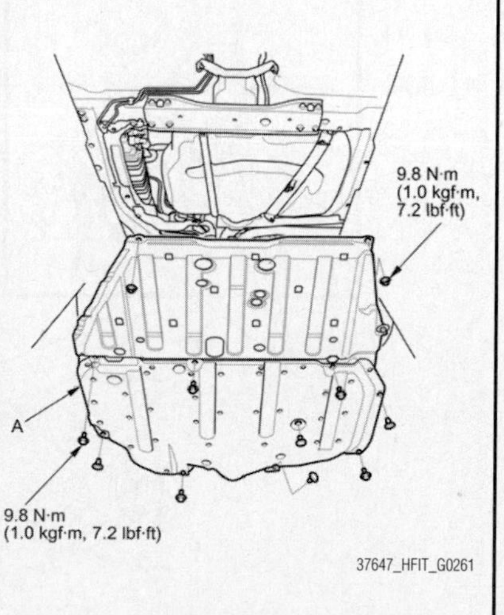
9.8 N·m
(1.0 kgf·m,
7.2 lbf·ft)
A
9.8 N·m
(1.0 kgf·m, 7.2 lbf·ft)
37647_HFIT_G0261

Fig. 109 LX (A/T model): Remove the floor under cover assembly (A)

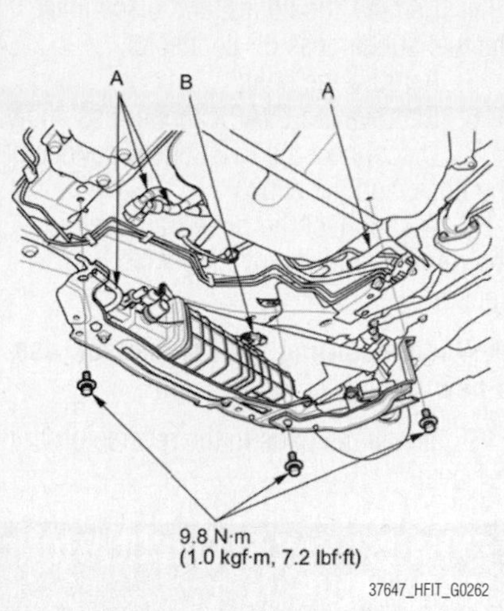
A
B
A
9.8 N·m
(1.0 kgf·m, 7.2 lbf·ft)
37647_HFIT_G0262

Fig. 110 Remove the hoses (A), and disconnect the fuel subharness 6P connector (B)

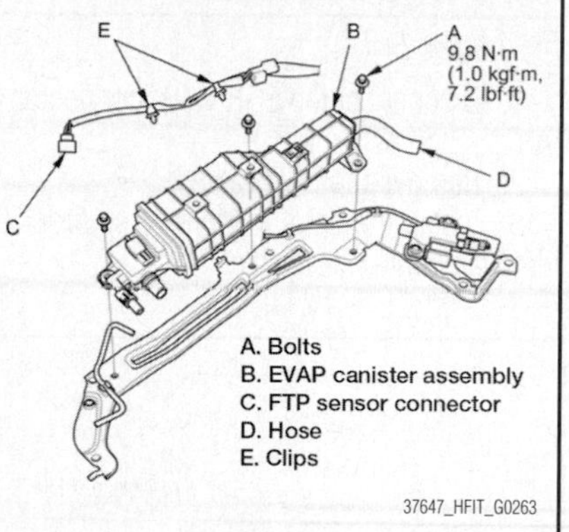
E
B
A
9.8 N·m
(1.0 kgf·m,
7.2 lbf·ft)
D
C
A. Bolts
B. EVAP canister assembly
C. FTP sensor connector
D. Hose
E. Clips
37647_HFIT_G0263

Fig. 111 EVAP canister assembly

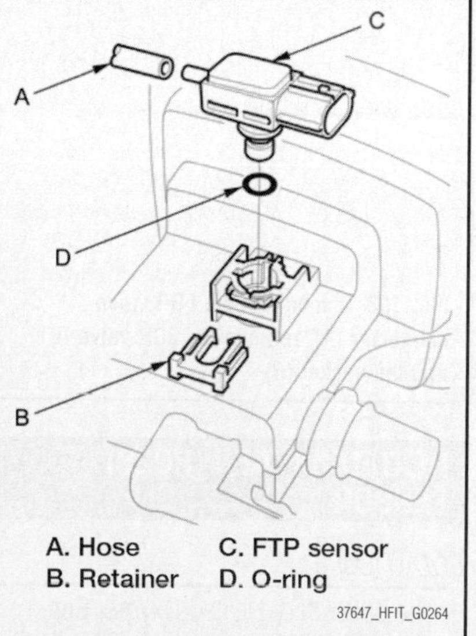
C
A
D
B
A. Hose
B. Retainer
C. FTP sensor
D. O-ring
37647_HFIT_G0264

Fig. 112 Disconnect the hose, remove the retainer and remove the FTP sensor

4. Remove the hoses, and disconnect the fuel subharness 6P connector.
5. Remove the bolts.
6. Remove the EVAP canister assembly.
7. Disconnect the FTP sensor connector, the hose, and the clips.
8. Disconnect the hose, remove the retainer and remove the FTP sensor.

➡When installing the FTP sensor, use a new O-ring.

9. Install the parts in the reverse order of removal.

EXHAUST GAS RECIRCULATION (EGR) VALVE

LOCATION

The Exhaust Gas Recirculation (EGR) valve is located on the intake manifold just above the alternator.

REMOVAL & INSTALLATION

See Figure 113.

1. Disconnect the EGR valve connector (A).
2. Remove the EGR valve.
3. Install the parts in the reverse order of removal with a new gasket.

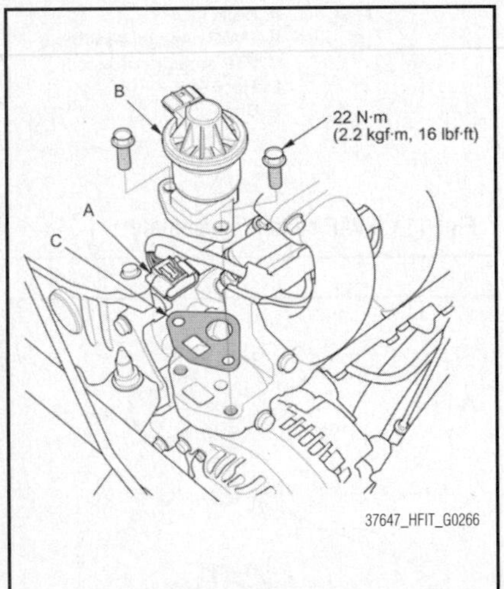
B
22 N·m
(2.2 kgf·m, 16 lbf·ft)
A
C
37647_HFIT_G0266

Fig. 113 Disconnect the EGR valve connector (A); remove the EGR valve (B) and the gasket (C)

HEATED OXYGEN (HO2S) SENSOR

LOCATION

The secondary Heated Oxygen Sensor (HO2S) is located on the under-floor TWC.

REMOVAL & INSTALLATION

See Figures 114 and 115.

Special Tools Required: O2 Sensor Wrench, Snap-on S6176 or equivalent, commercially available

1. Disconnect the secondary HO2S connector.
2. Raise the vehicle on a lift.
3. Remove the supporters, then remove the secondary HO2S.
4. Install the parts in the reverse order of removal.

INTAKE AIR TEMPERATURE (IAT) SENSOR

LOCATION

The Intake Air Temperature (IAT) sensor is an integral part of the Mass Air Flow (MAF/Intake Air Temperature (IAT) located on the engine air intake.

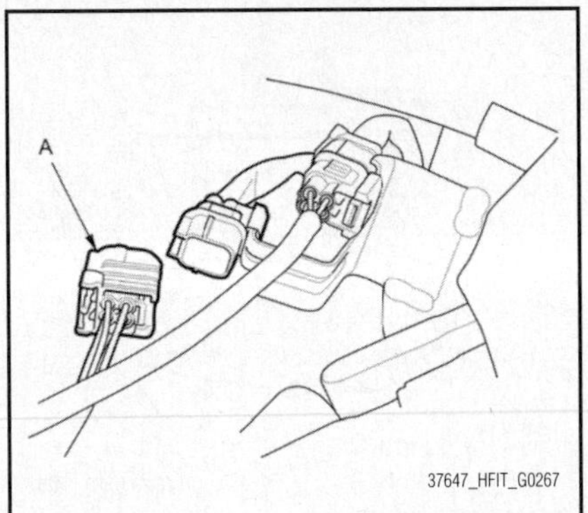
A
37647_HFIT_G0267

Fig. 114 Disconnect the secondary HO2S connector (A)

Refer to the MAF/IAT section for information on servicing this component.

REMOVAL & INSTALLATION

Refer to the MAF/IAT sensor section for Removal and Installation procedures.

KNOCK SENSOR (KS)

LOCATION

Refer to the graphics in the Removal and Installation section for the location(s).

REMOVAL & INSTALLATION

See Figure 116.

1. Disconnect the knock sensor connector.
2. Remove the knock sensor.

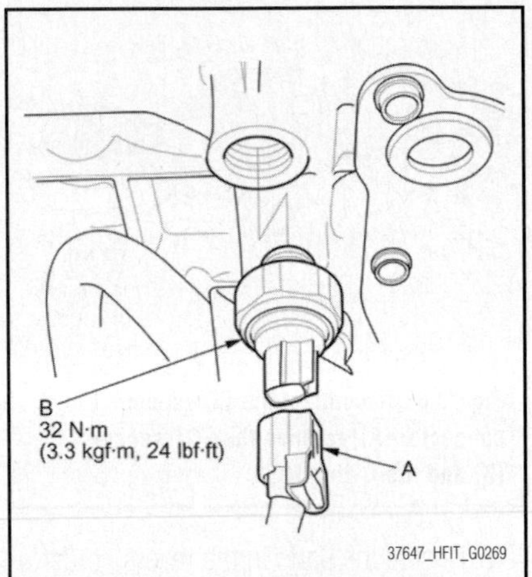
B
32 N·m
(3.3 kgf·m, 24 lbf·ft)
A
37647_HFIT_G0269

Fig. 116 Disconnect the knock sensor connector (A) and remove the knock sensor (B)

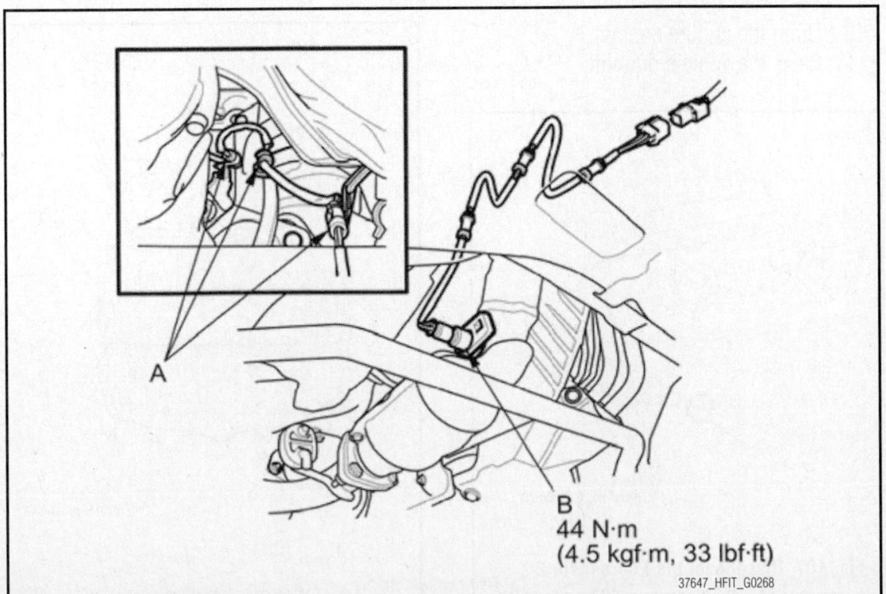
A
B
44 N·m
(4.5 kgf·m, 33 lbf·ft)
37647_HFIT_G0268

Fig. 115 Remove the supporters (A), then remove the secondary HO2S (B)

3. Install the parts in the reverse order of removal.

MASS AIR FLOW (MAF) SENSOR/INTAKE AIR TEMPERATURE (IAT) SENSOR (HOT WIRE)

LOCATION

The Intake Air Temperature (IAT) sensor is an integral part of the Mass Air Flow (MAF/Intake Air Temperature (IAT) located on the engine air intake.

REMOVAL & INSTALLATION

See Figure 117.

1. Disconnect the MAF sensor/IAT sensor connector.
2. Remove the screws.
3. Remove the MAF sensor/IAT sensor.
4. Install the parts in the reverse order of removal with a new gasket.

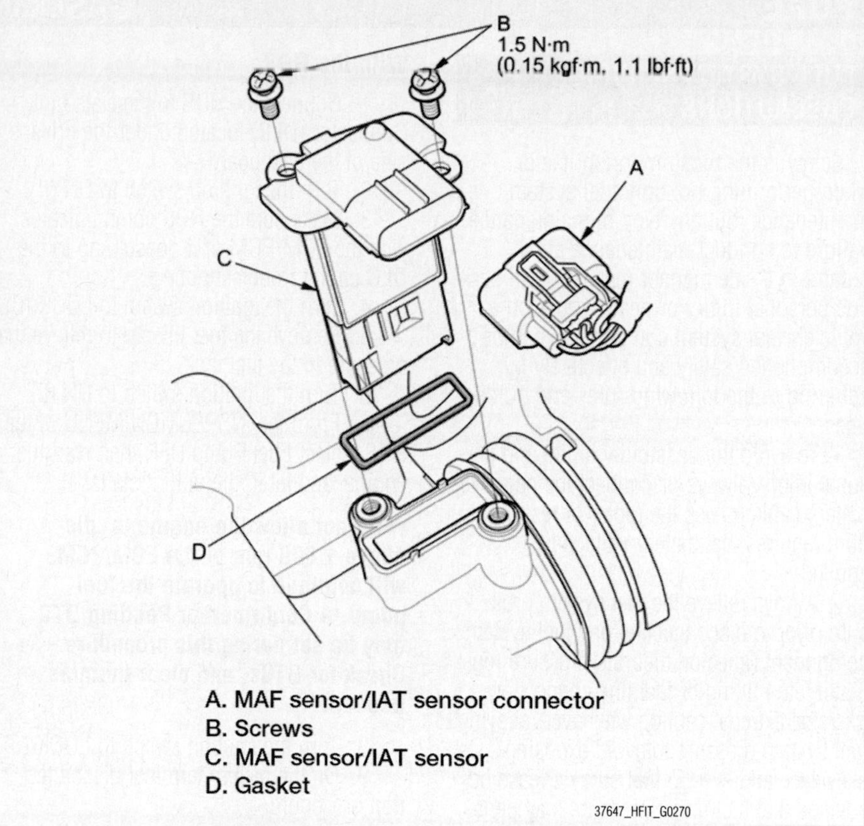
B
1.5 N·m
(0.15 kgf·m, 1.1 lbf·ft)
A
C
D
A. MAF sensor/IAT sensor connector
B. Screws
C. MAF sensor/IAT sensor
D. Gasket
37647_HFIT_G0270

Fig. 117 Disconnect the MAF sensor/IAT sensor connector

MANIFOLD ABSOLUTE PRESSURE (MAP) SENSOR

LOCATION

Refer to the graphics in the Removal and Installation section for the location(s).

REMOVAL & INSTALLATION

See Figure 118.

1. Disconnect the MAP sensor connector.
2. Remove the MAP sensor.
3. Install the parts in the reverse order of removal with a new O-ring.

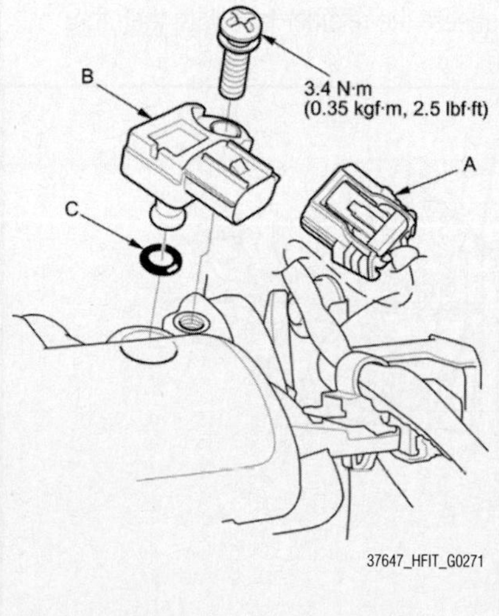
B
3.4 N·m
(0.35 kgf·m, 2.5 lbf·ft)
A
C
37647_HFIT_G0271

Fig. 118 Disconnect the MAP sensor connector (A), remove the MAP sensor (B) and the O-ring (C)

POSITIVE CRANKCASE VENTILATION (PCV) VALVE

LOCATION

Refer to the graphics in the Removal and Installation section for the location(s).

REMOVAL & INSTALLATION

See Figures 119 and 120.

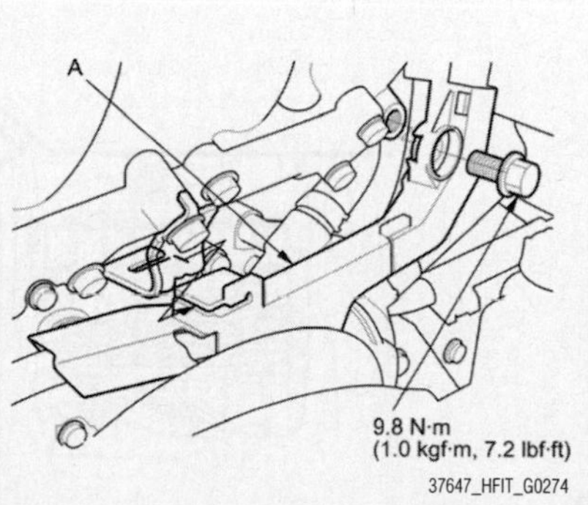
A
9.8 N·m
(1.0 kgf·m, 7.2 lbf·ft)
37647_HFIT_G0274

Fig. 119 Remove the harness holder (A)

1. Remove the harness holder.
2. Disconnect the hose.
3. Remove the PCV valve.
4. Install the parts in the reverse order of removal with a new 14 mm washer.

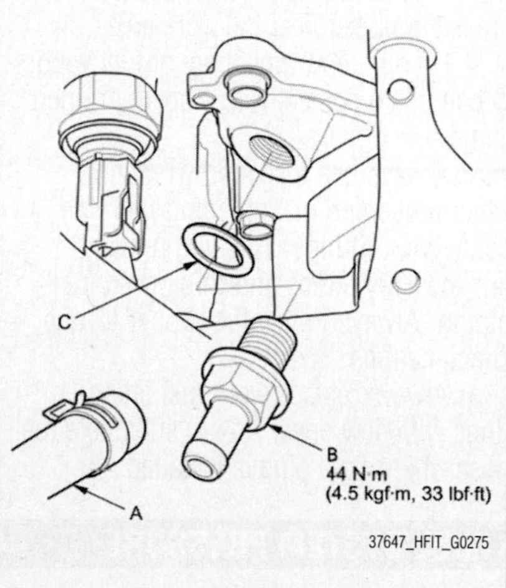
C
B
44 N·m
(4.5 kgf·m, 33 lbf·ft)
A
37647_HFIT_G0275

Fig. 120 Disconnect the hose (A); remove the PCV valve (B) and washer (C)

FUEL SYSTEM SERVICE PRECAUTIONS

Safety is the most important factor when performing not only fuel system maintenance, but any type of maintenance. Failure to conduct maintenance and repairs in a safe manner may result in serious personal injury or death. Work on a vehicle's fuel system components can be accomplished safely and effectively by adhering to the following rules and guidelines.

- To avoid the possibility of fire and personal injury, always disconnect the negative battery cable unless the repair or test procedure requires that battery voltage be applied.
- Always relieve the fuel system pressure prior to disconnecting any fuel system component (injector, fuel rail, pressure regulator, etc.) fitting or fuel line connection. Exercise extreme caution whenever relieving fuel system pressure to avoid exposing skin, face and eyes to fuel spray. Please be advised that fuel under pressure may penetrate the skin or any part of the body that it contacts.
- Always place a shop towel or cloth around the fitting or connection prior to loosening to absorb any excess fuel due to spillage. Ensure that all fuel spillage is quickly removed from engine surfaces. Ensure that all fuel-soaked cloths or towels are deposited into a flame-proof waste container with a lid.
- Always keep a dry chemical (Class B) fire extinguisher near the work area.
- Do not allow fuel spray or fuel vapors to come into contact with a spark or open flame.
- Always use a second wrench when loosening or tightening fuel line connection fittings. This will prevent unnecessary stress and torsion on fuel piping. Always follow the proper torque specifications.
- Always replace worn fuel fitting O-rings with new ones. Do not substitute fuel hose where rigid pipe is installed.

FUEL SYSTEM PRESSURE

RELIEVING

Before disconnecting fuel lines or hoses, relieve pressure from the system by disabling the fuel pump and then disconnecting the fuel tube/quick connect fitting in the engine compartment.

With the HDS

1. Connect the HDS to the Data Link Connector (DLC) located under the driver's side of the dashboard.
2. Turn the ignition switch to ON (II).
3. Make sure the HDS communicates with the ECM/PCM. If it doesn't, go to the DLC circuit troubleshooting.
4. Turn the ignition switch to LOCK (0).
5. Remove the fuel fill cap to relieve the pressure in the fuel tank.
6. Turn the ignition switch to ON (II).
7. From the INSPECTION MENU of the HDS, select Fuel Pump OFF, then start the engine, and let it idle until it stalls.

➡Do not allow the engine to idle above 1,000 rpm or the ECM/PCM will continue to operate the fuel pump. A Confirmed or Pending DTC may be set during this procedure. Check for DTCs, and clear them as needed.

8. Turn the ignition switch to LOCK (0).
9. Do the battery terminal disconnection procedure.
10. Remove the quick-connect fitting cover.
11. Check the fuel quick-connect fitting for dirt, and clean it if needed.
12. Place a rag or shop towel over the quick-connect fitting.
13. Disconnect the quick-connect fitting: Hold the connector with one hand, and squeeze the retainer tabs with the other hand to release them from the locking tabs. Pull the connector off.

➡Be careful not to damage the line or other parts. Do not use tools. If the connector does not move, keep the retainer tabs pressed down, and alternately pull and push the connector until it comes off easily. Do not remove the retainer from the line; once removed, the retainer must be replaced with a new one.

14. After disconnecting the quick-connect fitting, check it for dirt or damage.
15. Do the battery terminal reconnection procedure.

Without the HDS

See Figure 121.

1. Open the fuse access panel.
2. Remove PGM-FI main relay 2 from the under-dash fuse/relay box.
3. Start the engine, and let it idle until it stalls.

➡If any DTCs are stored, clear and ignore them.

4. Turn the ignition switch to LOCK (0).
5. Remove the fuel fill cap to relieve the pressure in the fuel tank.
6. Do the battery terminal disconnection procedure.
7. Remove the quick-connect fitting cover.
8. Check the fuel quick-connect fitting for dirt, and clean it if needed.
9. Place a rag or shop towel over the quick-connect fitting.
10. Disconnect the quick-connect fitting: Hold the connector with one hand, and squeeze the retainer tabs with the other

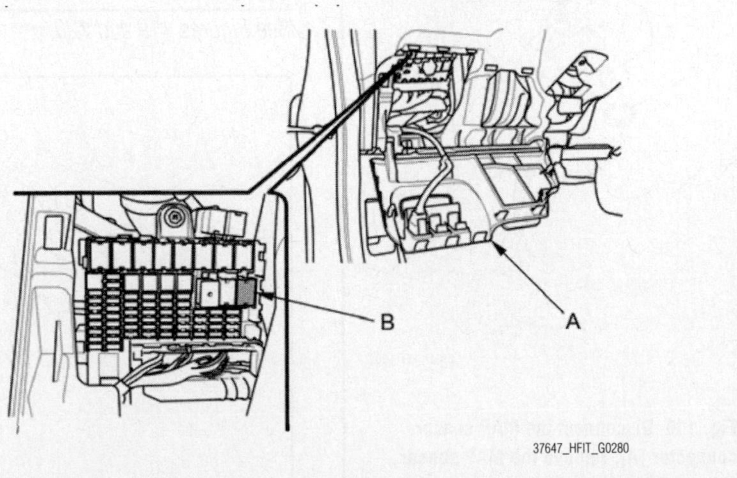
B
A
37647_HFIT_G0280

Fig. 121 Open the fuse access panel (A) and remove PGM-FI main relay 2 (B) from the under-dash fuse/relay box

hand to release them from the locking tabs. Pull the connector off.

➡Be careful not to damage the line or other parts. Do not use tools. If the connector does not move, keep the retainer tabs pressed down, and alternately pull and push the connector until it comes off easily. Do not remove the retainer from the line; once removed, the retainer must be replaced with a new one.

11. After disconnecting the quick-connect fitting, check it for dirt or damage.
12. Do the battery terminal reconnection procedure.

FUEL FILTER

REMOVAL & INSTALLATION

See Figure 122.

The fuel filter should be replaced whenever the fuel pressure drops below 47–54 psi (320–370 kPa), after making sure that the fuel pump and the fuel pressure regulator are OK.

1. Remove the fuel tank unit.
2. Remove the fuel gauge sending unit, and the reservoir.
3. Remove the fuel pressure regulator, the fuel pump, and the wire harness from the fuel filter set.
4. Check these items before installing the fuel tank unit:
   a. When connecting the wire harness, make sure the connection is secure and the connectors are firmly locked into place.
   b. When installing the fuel gauge sending unit, make sure the connection is secure and the connector is firmly locked into place. Be careful not to bend or twist it excessively.
5. Install the parts in the reverse order of removal with new O-rings. When installing the fuel tank unit, align the marks on the unit and the fuel tank.

➡Coat the O-rings with clean engine oil; do not use any other oil or fluid. Do not pinch the O-rings during installation. Use all the new parts supplied in the fuel filter replacement kit.

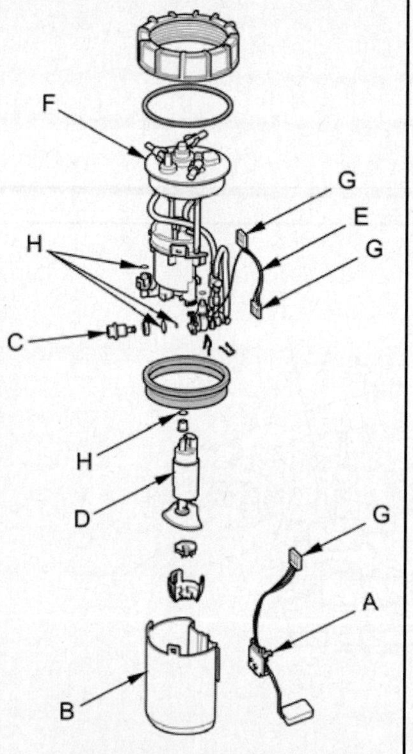
F
G
E
G
H
C
H
D
G
A
B
A. Fuel gauge sending unit
B. Reservoir
C. Fuel pressure regulator
D. Fuel pump
E. Wire harness
F. Fuel filter set
G. Connectors
H. O-rings
37647_HFIT_G0281

Fig. 122 Exploded view of the fuel tank unit

FUEL LEVEL SENDING UNIT

LOCATION

The fuel gauge sending unit is an integral part of the fuel tank unit.

REMOVAL & INSTALLATION

See Figure 123.

1. Remove the fuel tank unit.
2. Remove the fuel level sensor (fuel sending unit) from the fuel tank unit.
3. Check these items before installing the fuel tank unit:
   a. When connecting the wire harness, make sure the connection is secure and the connector is firmly locked into place.
   b. When installing the fuel gauge sending unit, make sure the connection is secure. Be careful not to bend or twist it excessively.
4. Install the parts in the reverse order of removal. When installing the fuel tank unit, align the marks on the unit and the fuel tank.

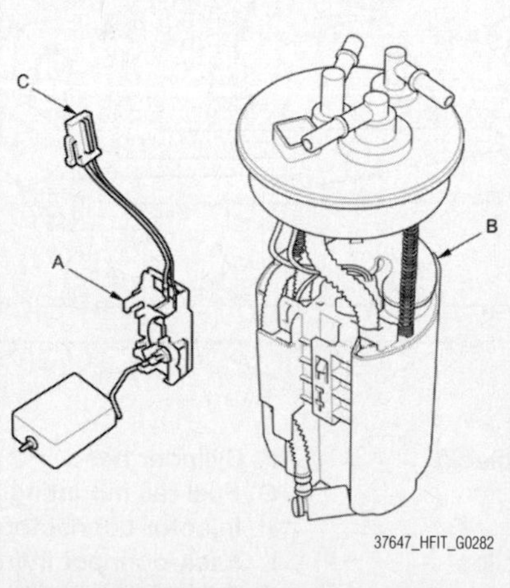
C
A
B
37647_HFIT_G0282

Fig. 123 Remove the fuel level sensor (fuel sending unit) (A) from the fuel tank unit (B)

FUEL RAIL AND INJECTOR

REMOVAL & INSTALLATION

See Figures 124 and 125.

1. Relieve the fuel pressure.
2. Remove the intake manifold and the intake manifold chamber.
3. Remove the quick-connect fitting cover, then disconnect the quick-connect fitting.
4. Disconnect the injector connectors.
5. Remove the fuel rail mounting nuts from the fuel rail.
6. Remove the fuel rail and the injectors from the cylinder head.
7. Remove the injector clips from the fuel rail.
8. Remove the injectors from the fuel rail.

To install:

9. Coat the new O-rings (black) with clean engine oil, and insert the injectors into the fuel rail.
10. Install the injector clips.
11. Coat the new injector O-rings (brown) with clean engine oil.
12. Install the fuel rail and the injectors in the cylinder head.
13. Install the fuel rail mounting nuts.
14. Connect the injector connectors.
15. Connect the quick-connect fitting and the quick-connect fitting cover.
16. Turn the ignition switch to ON (II), but do not operate the starter. After the fuel pump runs for about 2 seconds, the fuel rail is pressurized. Repeat this two or three times, then check for fuel leaks.
17. Reinstall the intake manifold and the intake manifold chamber.

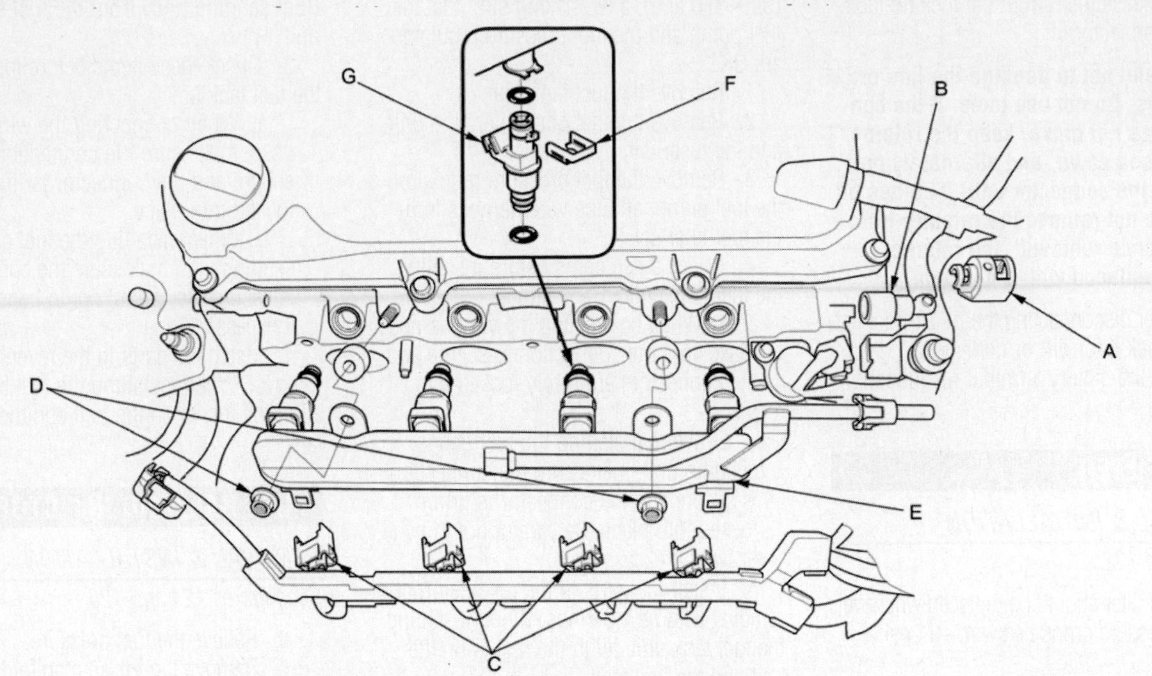
G
F
B
A
D
E
C
37647_HFIT_G0292

A. Quick-connect fitting cover
B. Quick-connect fitting
C. Injector connectors
D. Fuel rail mounting nuts
E. Fuel rail
F. Injector clips
G. Injectors

Fig. 124 Exploded view of the fuel rail and injector assembly

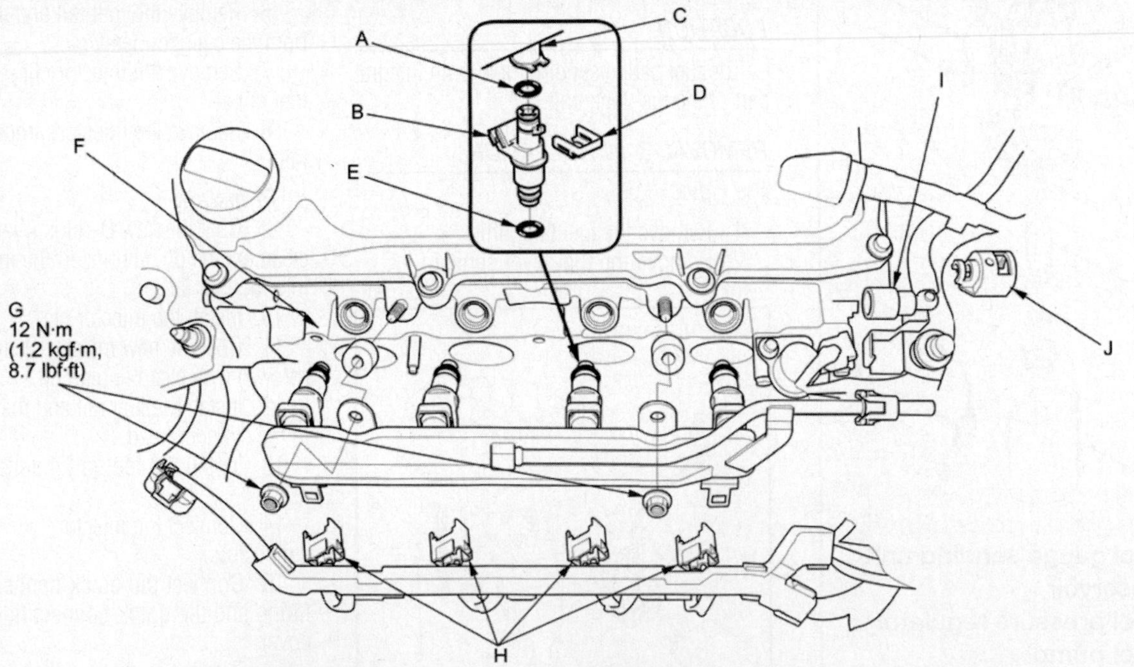
C
A
D
B
F
E
I
G
12 N·m
(1.2 kgf·m,
8.7 lbf·ft)
J
H
37647_HFIT_G0293

A. O-rings (black)
B. Injectors
C. Fuel rail
D. Injector clips
E. O-rings (brown)
F. Cylinder head
G. Fuel rail mounting nuts
H. Injector connectors
I. Quick-connect fitting
J. Quick-connect fitting cover

Fig. 125 Installation of injectors and fuel rail

FUEL TANK

DRAINING

1. Remove the fuel tank unit.
2. Using a hand pump, a hose, and a container suitable for fuel, draw the fuel from the fuel tank.
3. Reinstall the fuel tank unit.

REMOVAL & INSTALLATION

See Figures 126 through 132.

1. Relieve the fuel pressure.
2. Drain the fuel tank.
3. Reinstall the fuel tank unit without connecting the fuel tank unit 4P connector and the quick-connect fitting (feed line).
4. Raise the vehicle on a lift.
5. Remove the fuel tank guard, and the fuel tank protector.
6. Except LX (A/T model): Remove the fuel tank cover.

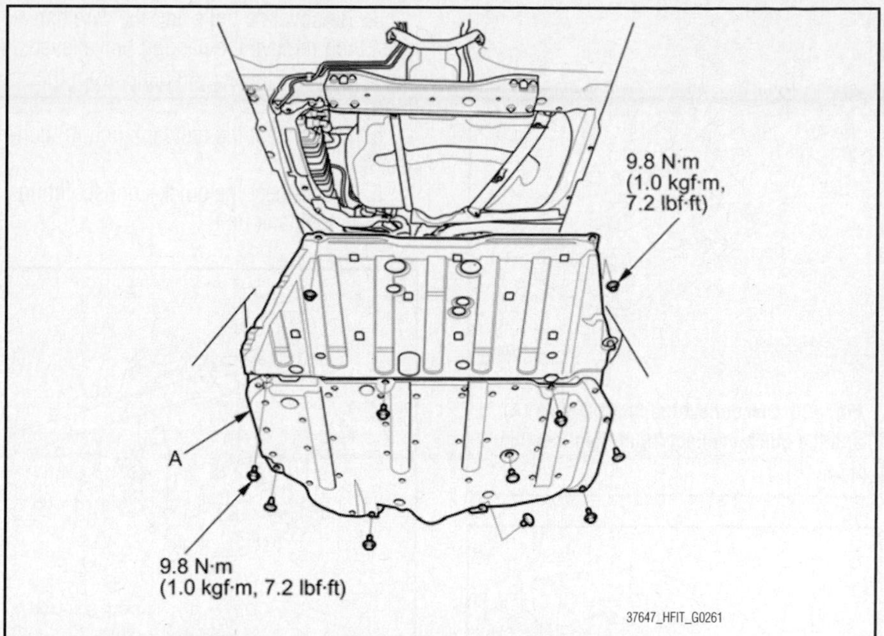
9.8 N·m
(1.0 kgf·m,
7.2 lbf·ft)
A
9.8 N·m
(1.0 kgf·m, 7.2 lbf·ft)
37647_HFIT_G0261

Fig. 128 LX (A/T model): Remove the floor under cover assembly (A)

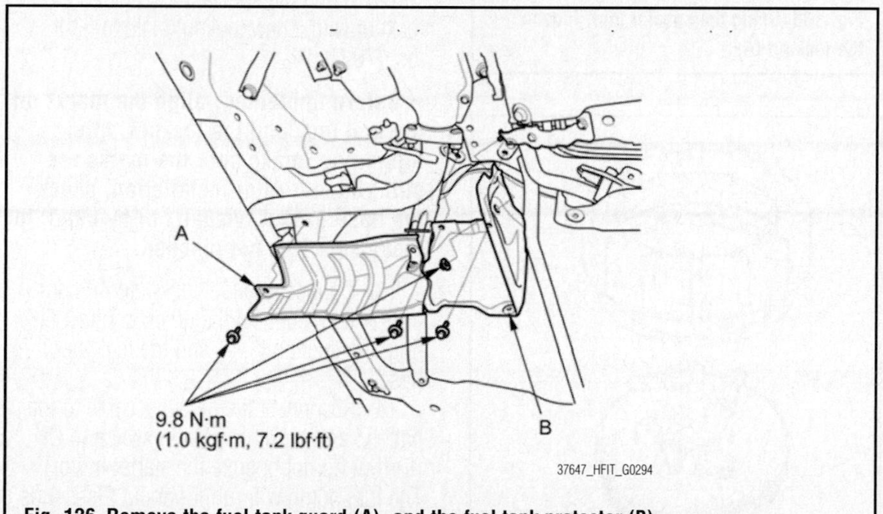
A
9.8 N·m
(1.0 kgf·m, 7.2 lbf·ft)
B
37647_HFIT_G0294

Fig. 126 Remove the fuel tank guard (A), and the fuel tank protector (B)

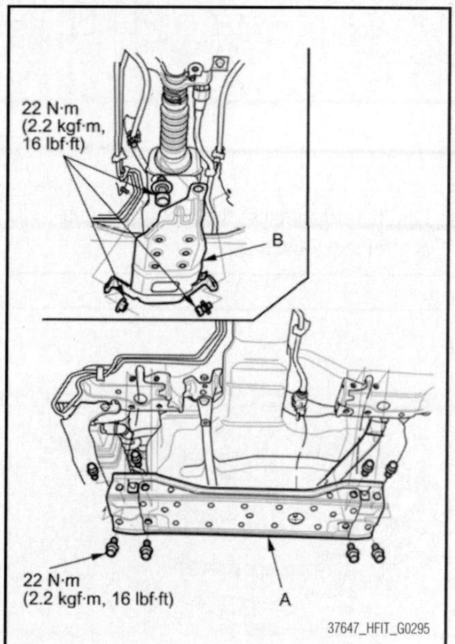
22 N·m
(2.2 kgf·m,
16 lbf·ft)
B
22 N·m
(2.2 kgf·m, 16 lbf·ft)
A
37647_HFIT_G0295

Fig. 129 Remove the front floor cross beam (A), and the tank mount bracket (B)

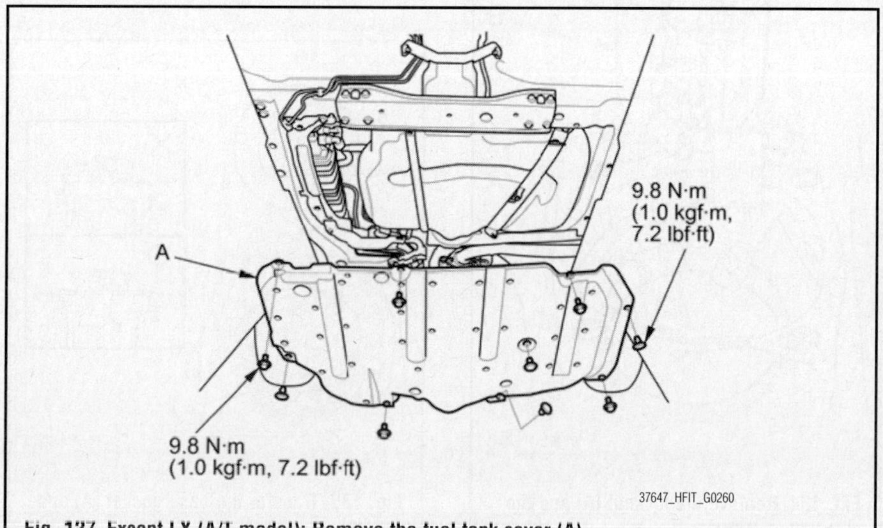
9.8 N·m
(1.0 kgf·m,
7.2 lbf·ft)
A
9.8 N·m
(1.0 kgf·m, 7.2 lbf·ft)
37647_HFIT_G0260

Fig. 127 Except LX (A/T model): Remove the fuel tank cover (A)

7. LX (A/T model): Remove the floor under cover assembly.
8. Remove the front floor cross beam and the tank mount bracket.
9. Disconnect the fuel fill hose and the quick-connect fitting (fuel suction tube).
10. Disconnect the fuel tank vapor control valve hose.
11. Place a jack or other support under the fuel tank, then remove the strap bolts.
12. Remove the fuel tank.
13. Install the parts in the reverse order of removal.

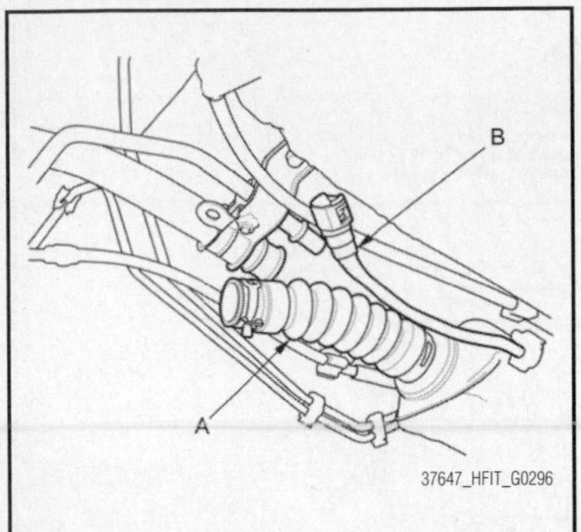
B
A
37647_HFIT_G0296

Fig. 130 Disconnect the fuel fill hose (A), and the quick-connect fitting (fuel suction tube) (B)

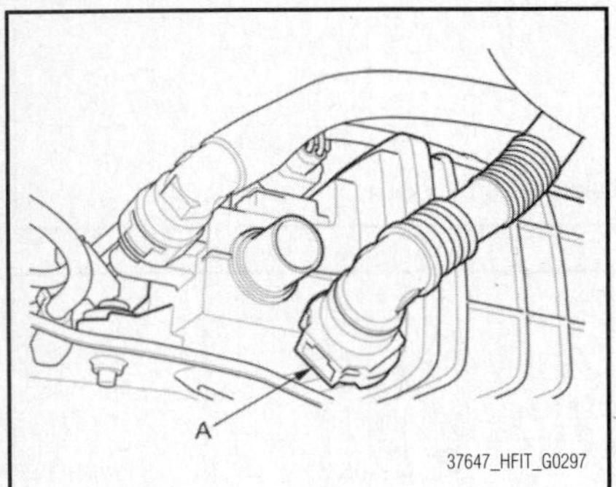
A
37647_HFIT_G0297

Fig. 131 Disconnect the fuel tank vapor control valve hose (A)

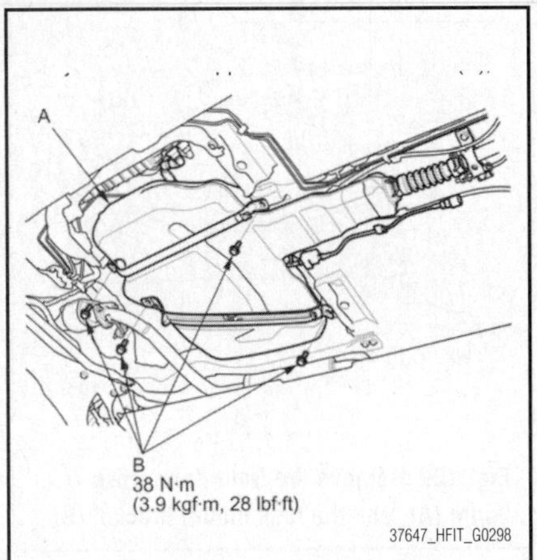
A
B
38 N·m
(3.9 kgf·m, 28 lbf·ft)
37647_HFIT_G0298

Fig. 132 Place a jack or other support under the fuel tank (A), then remove the strap bolts (B)

FUEL TANK UNIT

REMOVAL & INSTALLATION

See Figures 133 through 135.

Special Tools Required: Fuel Sender Wrench 07AAA-S0XA100

1. Relieve the fuel pressure.
2. Remove the center console.
3. Remove the bolts and the wire harness, then remove the parking brake lever.
4. Remove the access panel from the floor.
5. Disconnect the fuel tank unit 4P connector.
6. Disconnect the quick-connect fittings from the fuel tank unit.

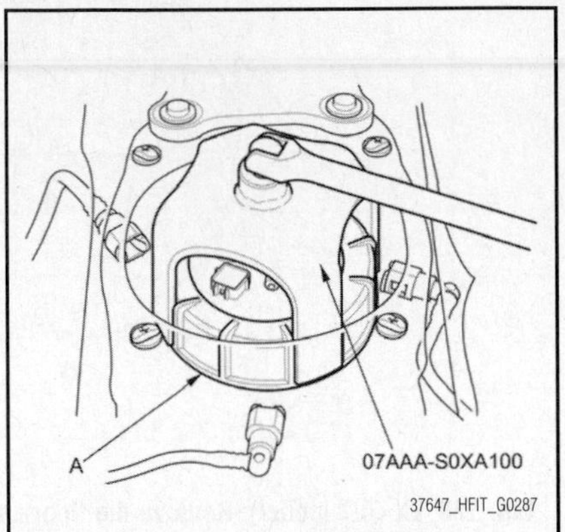
A
07AAA-S0XA100
37647_HFIT_G0287

Fig. 133 Using the special tool, loosen the locknut (A)

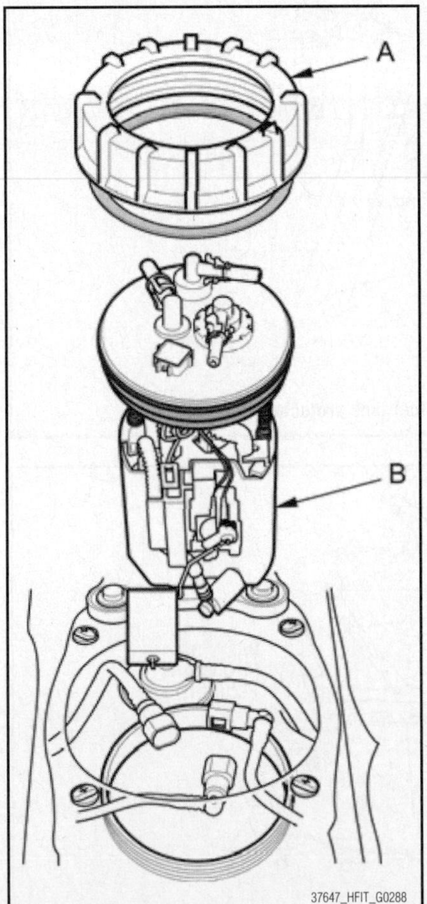
A
B
37647_HFIT_G0288

Fig. 134 Remove the locknut (A) and the fuel tank unit (B)

7. Using the special tool, loosen the locknut.
8. Remove the locknut and the fuel tank unit.

To install:

9. Temporarily attach a new base gasket to the fuel tank unit, then insert the fuel tank unit partially into the fuel tank.

➡Be careful not to damage the new base gasket. Be careful not to bend the fuel gauge sending unit. Do not coat the base gasket with oil.

10. Transfer the base gasket from the fuel tank unit to the fuel tank.
11. Align the marks on the fuel tank and fuel tank unit, then insert the fuel tank unit into the fuel tank until the fuel tank unit rests on top of the base gasket.

➡To avoid a fuel leak, check the base gasket, visually or by hand, to make sure it is not pinched.

12. Using the special tool, tighten a new locknut with a new locknut plate to 51 ft. lbs. (70 N m).

➡Before tightening, align the marks on the fuel tank and the locknut. After tightening, make sure the marks are still aligned. After installation, check the base gasket, visually or by hand, to make sure it is not pinched.

13. Connect the fuel tank unit 4P connector, then connect the quick-connect fitting, the suction hose, and the fuel vapor hose.
14. Reconnect the negative cable to the battery, and turn the ignition switch to ON (II) (but do not operate the starter motor). The fuel pump will run for about 2 seconds, and fuel pressure rises. Repeat this two or three times, then check for fuel leaks.

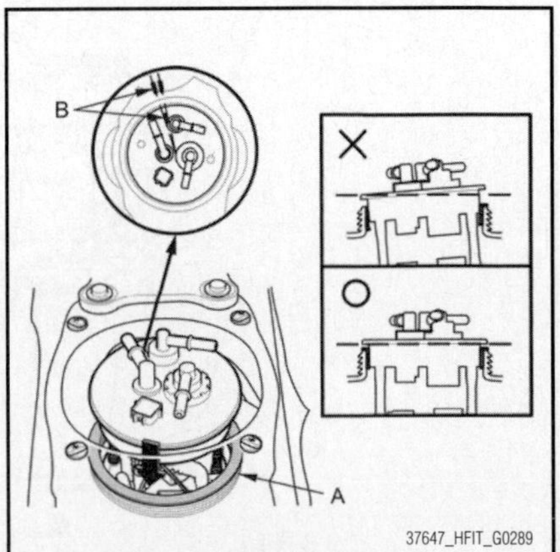
B
A
37647_HFIT_G0289

Fig. 135 Transfer the base gasket (A) and align the marks (B)

15. Reinstall the access panel, the parking brake lever, and the center console.

IDLE SPEED

ADJUSTMENT

Idle speed is controlled by the ECM/PCM. No adjustment is necessary or possible.

THROTTLE BODY

REMOVAL & INSTALLATION

See Figure 136.

CAUTION

Do not insert your fingers into the installed throttle body when you turn the ignition switch to ON (II) or while the ignition switch is in ON (II). If you do, you will seriously injure your fingers if the throttle valve is activated.

➡If you are replacing or cleaning the throttle body, start at step 1. If you are removing the throttle body, start at step 4.

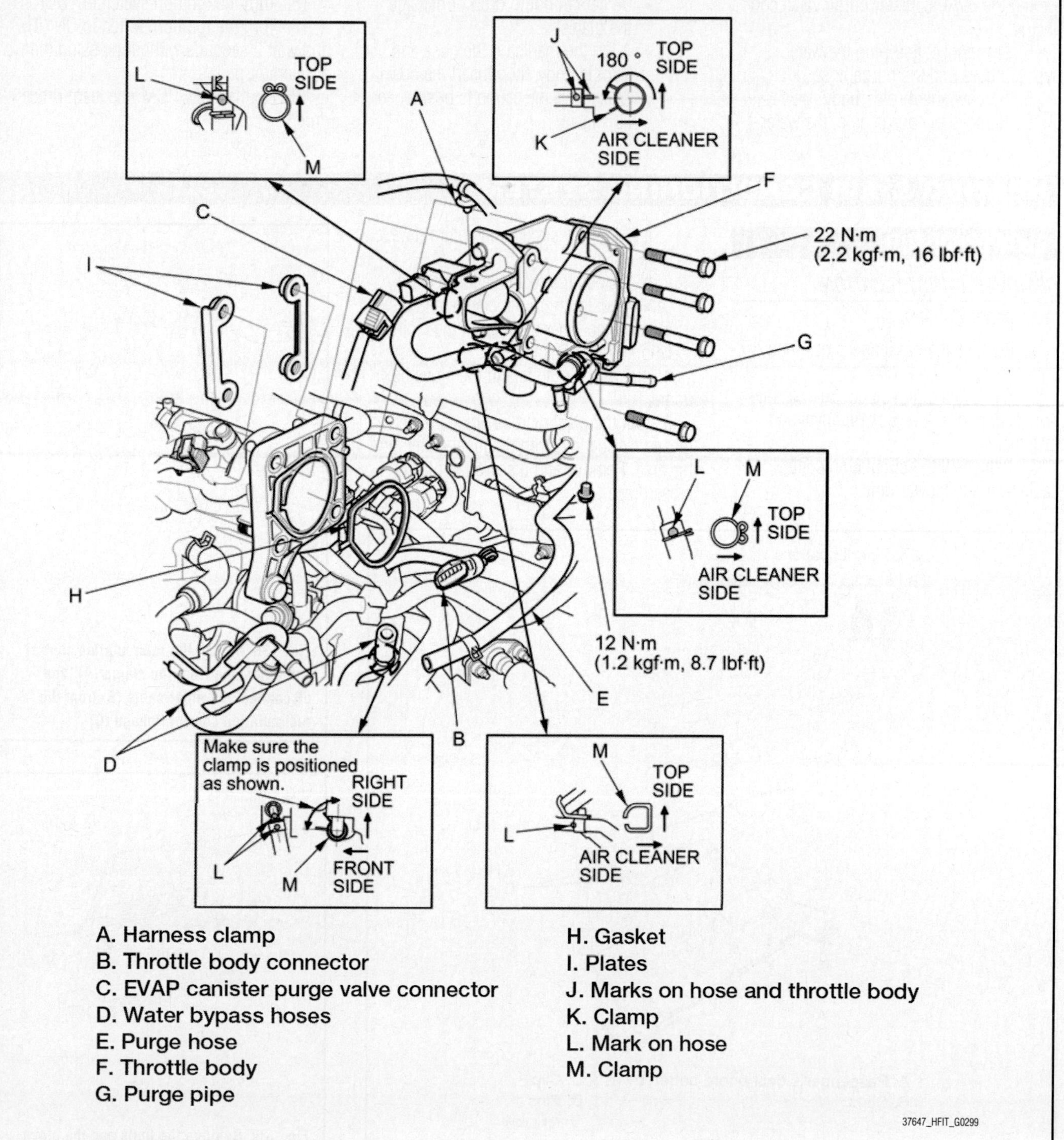

TOP SIDE
L
M
A
J
180 °
TOP SIDE
K
AIR CLEANER SIDE
F
C
I
22 N·m
(2.2 kgf·m, 16 lbf·ft)
G
L
M
TOP SIDE
AIR CLEANER SIDE
H
12 N·m
(1.2 kgf·m, 8.7 lbf·ft)
E
B
D
Make sure the clamp is positioned as shown.
RIGHT SIDE
FRONT SIDE
L
M
M
TOP SIDE
L
AIR CLEANER SIDE
A. Harness clamp
B. Throttle body connector
C. EVAP canister purge valve connector
D. Water bypass hoses
E. Purge hose
F. Throttle body
G. Purge pipe
H. Gasket
I. Plates
J. Marks on hose and throttle body
K. Clamp
L. Mark on hose
M. Clamp
37647_HFIT_G0299

Fig. 136 Exploded view of the throttle body assembly

1. Connect the HDS to the DLC while the engine is stopped.
2. Select the INSPECTION MENU on the HDS.
3. Do the TP POSITION CHECK in the ETCS TEST.
4. Turn the ignition switch to LOCK (0).
5. Remove the air cleaner.
6. Remove the harness clamp.
7. Disconnect the throttle body connector and the EVAP canister purge valve connector.
8. Disconnect and plug the water bypass hoses and the purge hose.
9. Remove the throttle body.
10. Remove the purge pipe, the water bypass hoses, and the purge hose from the throttle body.
11. Install the parts in the reverse order of removal with a new gasket, then refill the radiator with engine coolant. Note the following during installation:
    - If you replace or clean the throttle body, go to the next step.
    - If you did not replace or clean the throttle body, this procedure is complete.
    - Be careful not to drop or damage the plates.
    - Align the marks on the hose and throttle body, then insert the hose. Make sure the clamp is positioned as shown.
    - Align the mark on the hose as shown, then insert the hose. Make sure the clamp is positioned as shown.
12. Turn the ignition switch to ON (II).
13. Reset the ECM/PCM with the HDS.
14. Select the ETCS TEST in the INSPECTION MENU with the HDS.
15. Select the TP POSITION CHECK, then clear the Throttle Position (TP) learned value.
16. Turn the ignition switch to LOCK (0).
17. Turn the ignition switch to ON (II), and wait 2 seconds without pressing the accelerator pedal.
18. Do the ECM/PCM idle learn procedure.

HEATING & AIR CONDITIONING SYSTEM

BLOWER MOTOR

REMOVAL & INSTALLATION

See Figures 137 through 143.

1. Remove the passenger's dashboard undercover.
2. Remove the glove box. Refer to Interior section for Glove Box Removal and Installation.
3. Remove the recirculation control cable from the blower unit.
   a. Set the recirculation control lever to FRESH.
   b. Detach the recirculation control cable housing from the clamp, and disconnect the inner cable from the recirculation control linkage.
4. Remove the bolts and the glove box frame.
5. Cut the plastic cross brace in the glove box opening with diagonal cutters in the area. Remove and discard the plastic cross brace.

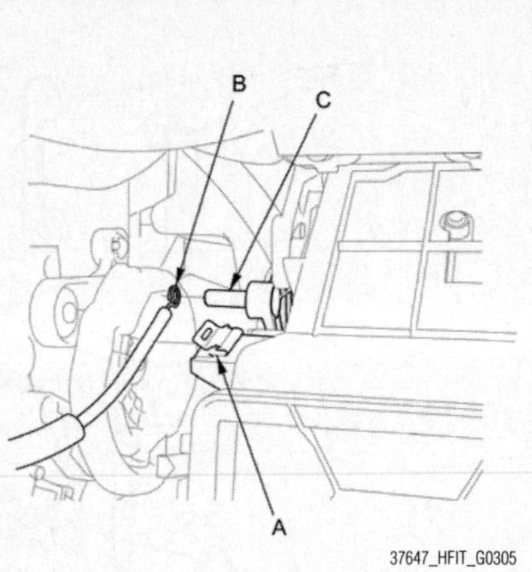
B
C
A
37647_HFIT_G0305

Fig. 138 Detach the recirculation control cable housing from the clamp (A), and disconnect the inner cable (B) from the recirculation control linkage (C)

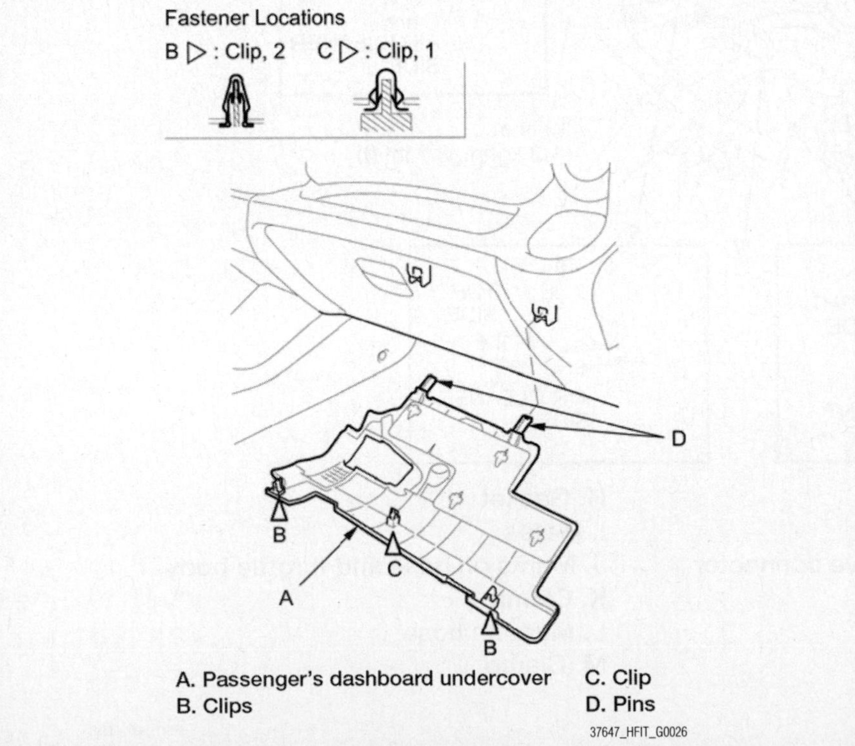
Fastener Locations
B ▷ : Clip, 2
C ▷ : Clip, 1
D
B
C
A
B
A. Passenger's dashboard undercover
B. Clips
C. Clip
D. Pins
37647_HFIT_G0026

Fig. 137 Remove the passenger's dashboard undercover

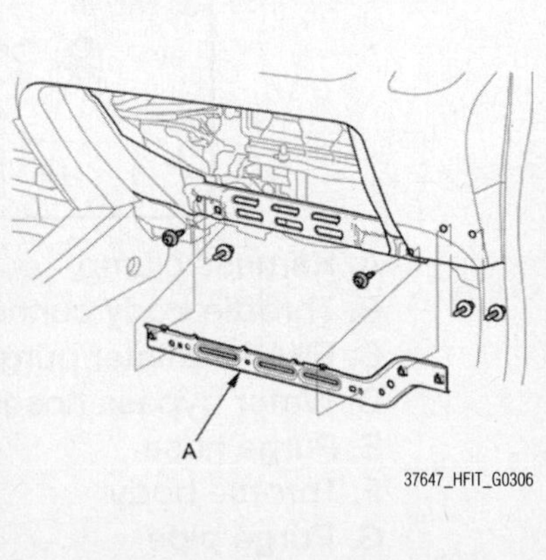
A
37647_HFIT_G0306

Fig. 139 Remove the bolts and the glove box frame (A)

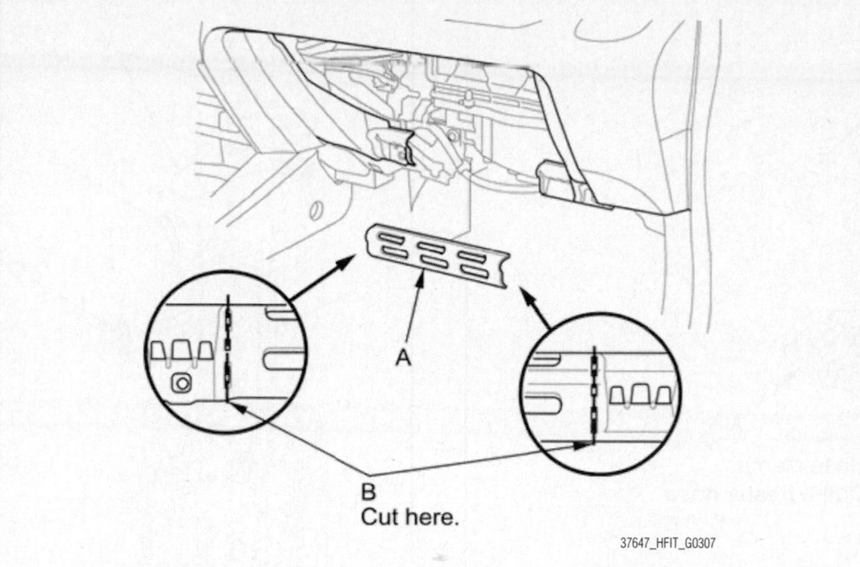
A
B
Cut here.
37647_HFIT_G0307

Fig. 140 Cut the plastic cross brace (A) in the glove box opening with diagonal cutters in the area (B)

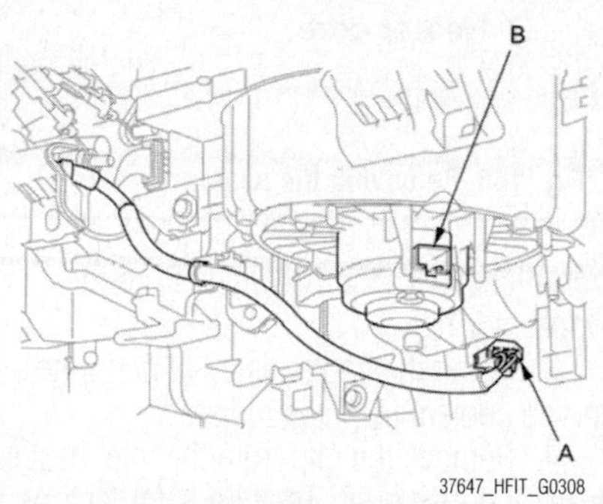
B
A
37647_HFIT_G0308

Fig. 141 Disconnect the connector (A) from the blower motor (B)

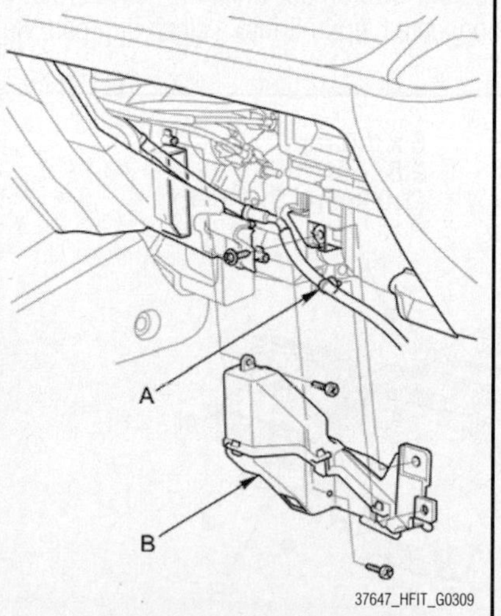
A
B
37647_HFIT_G0309

Fig. 142 Remove the self-tapping screws and the harness clip (A), then remove the passenger's heater duct (B)

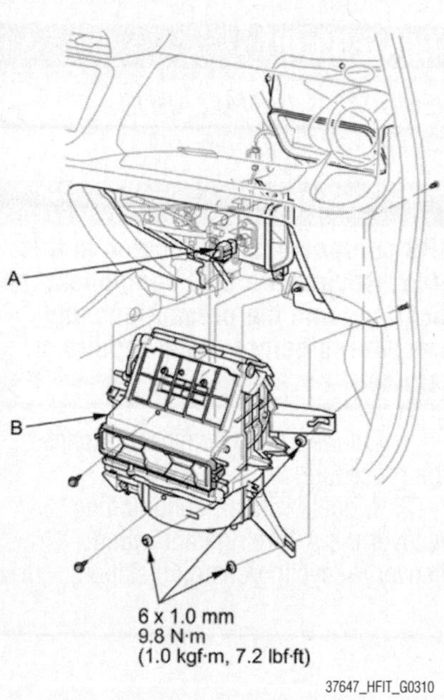
A
B
6 x 1.0 mm
9.8 N·m
(1.0 kgf·m, 7.2 lbf·ft)
37647_HFIT_G0310

Fig. 143 Remove the connector clip (A), the self-tapping screws and the mounting nuts, then pull the blower unit (B) out

6. Disconnect the connector from the blower motor.

7. Remove the self-tapping screws and the harness clip, then remove the passenger's heater duct.

8. Remove the connector clip, the self-tapping screws and the mounting nuts, then pull the blower unit out.

9. Install the unit in the reverse order of removal. Make sure that there is no air leakage.

HEATER CORE

REMOVAL & INSTALLATION

See Figures 144 through 148.

⁂ WARNING

SRS components are located in this area. Review the SRS component locations and the precautions and procedures before doing repairs or service.

1. Do the battery terminal disconnection procedure.
2. Models with air conditioning: Recover the A/C refrigerant with a recovery/recycling/charging station, then disconnect the A/C lines from the evaporator core.
3. When the engine is cool, drain the engine coolant from the radiator.
4. Remove the clip from the inlet heater hose, and discard it. The clip is for factory assembly use only. Slide the hose clamps back, then disconnect the inlet heater hose and the outlet heater hose from the heater unit. Note the layout of the hoses. Engine coolant will run out when the hoses are disconnected; drain it into a clean drip pan. Be sure not to let coolant spill on electrical parts or painted surfaces. If any coolant spills, rinse it off immediately.
5. Remove the mounting nut from the heater unit. Take care not to damage or bend the fuel lines or the brake lines.
6. Remove the dashboard.
7. Disconnect the connector from the evaporator sensor, and remove the blower resistor wire harness.
8. Remove the drain hose, then remove the mounting nuts and blower-heater unit.
9. Remove the self-tapping screws and the heater core cover.
10. Remove the holder and the grommet, then carefully pull out the heater core.

To install:

11. Install the heater core in the reverse order of removal.
12. Install the heater unit in the reverse order of removal, and note these items:
    - Do not interchange the inlet and outlet heater hoses, and install the hose clamps securely.
    - Refill the cooling system with engine coolant.
    - If necessary, recharge the A/C system.
    - Make sure that there is no coolant leakage.
    - Make sure that there is no air leakage.

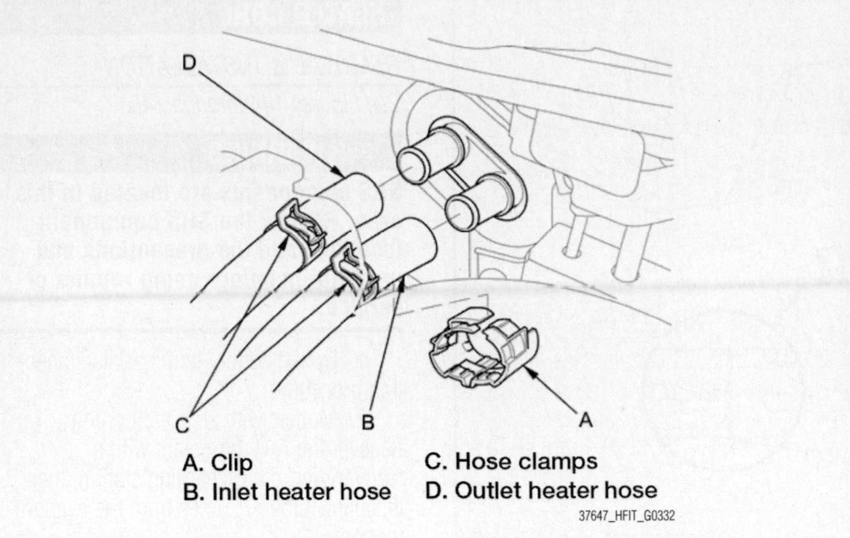
D
C
B
A
A. Clip
B. Inlet heater hose
C. Hose clamps
D. Outlet heater hose
37647_HFIT_G0332

Fig. 144 Disconnect the heater hoses

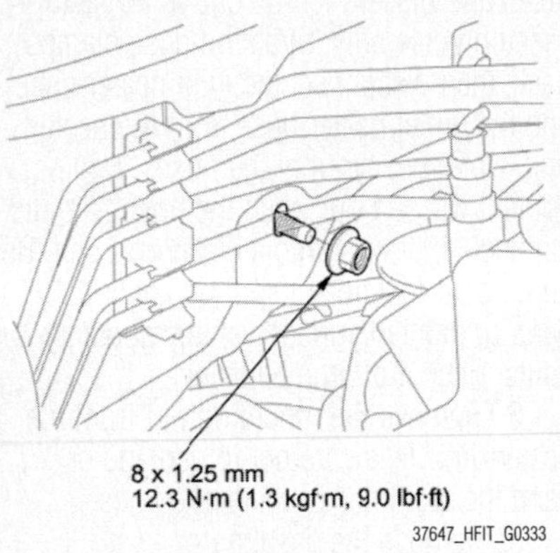
8 x 1.25 mm
12.3 N·m (1.3 kgf·m, 9.0 lbf·ft)
37647_HFIT_G0333

Fig. 145 Remove the mounting nut from the heater unit

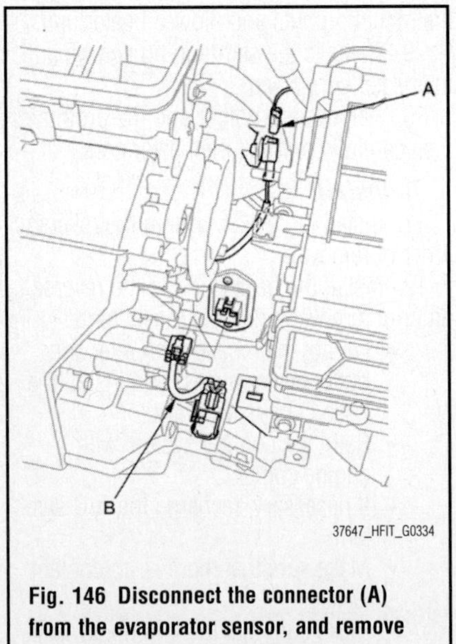
A
B
37647_HFIT_G0334

Fig. 146 Disconnect the connector (A) from the evaporator sensor, and remove the blower resistor wire harness (B)

13. Do the battery terminal reconnection procedure.

HEATER UNIT

REMOVAL & INSTALLATION

See Figures 144 through 148.

⁂ WARNING

SRS components are located in this area. Review the SRS component locations and the precautions and procedures before doing repairs or service.

1. Do the battery terminal disconnection procedure.
2. Models with air conditioning: Recover the A/C refrigerant with a recovery/recycling/charging station, then disconnect the A/C lines from the evaporator core.
3. When the engine is cool, drain the engine coolant from the radiator.
4. Remove the clip from the inlet heater hose, and discard it. The clip is for factory assembly use only. Slide the hose clamps back, then disconnect the inlet heater hose and the outlet heater hose from the heater unit. Note the layout of the hoses. Engine coolant will run out when the hoses are disconnected; drain it into a clean drip pan. Be

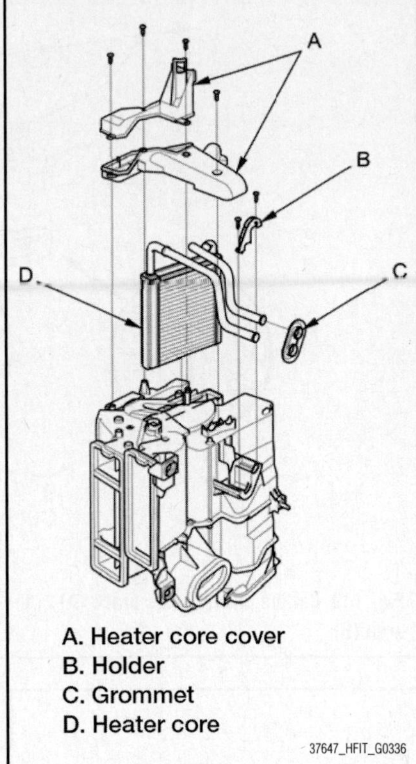
A
B
C
D
A. Heater core cover
B. Holder
C. Grommet
D. Heater core
37647_HFIT_G0336

Fig. 148 Removing the heater core

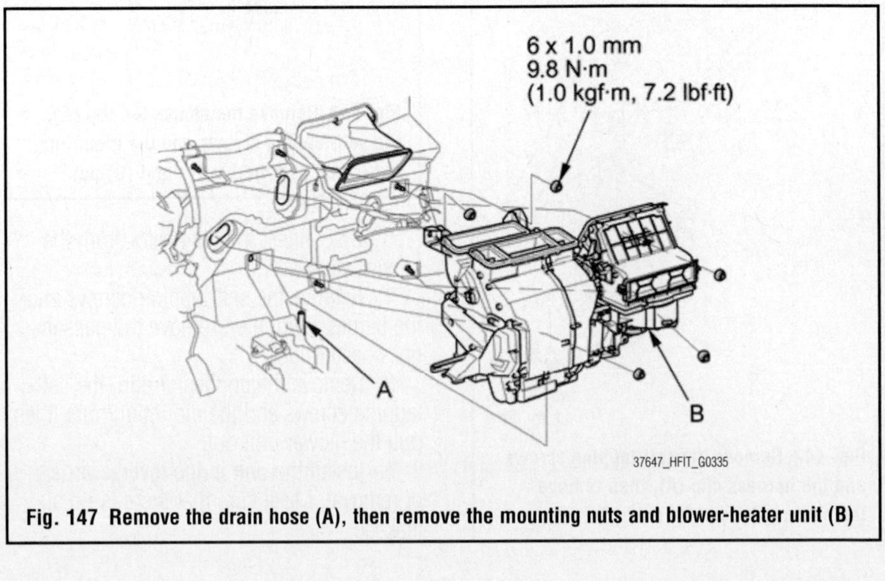
6 x 1.0 mm
9.8 N·m
(1.0 kgf·m, 7.2 lbf·ft)
A
B
37647_HFIT_G0335

Fig. 147 Remove the drain hose (A), then remove the mounting nuts and blower-heater unit (B)

sure not to let coolant spill on electrical parts or painted surfaces. If any coolant spills, rinse it off immediately.

5. Remove the mounting nut from the heater unit. Take care not to damage or bend the fuel lines or the brake lines.

6. Remove the dashboard.

7. Disconnect the connector from the evaporator sensor, and remove the blower resistor wire harness.

8. Remove the drain hose, then remove the mounting nuts and blower-heater unit.

To install:

9. Install the heater unit in the reverse order of removal, and note these items:

- Do not interchange the inlet and outlet heater hoses, and install the hose clamps securely.
- Refill the cooling system with engine coolant.
- If necessary, recharge the A/C system.
- Make sure that there is no coolant leakage.
- Make sure that there is no air leakage.

10. Do the battery terminal reconnection procedure.

STEERING

ELECTRICAL POWER STEERING (EPS) CONTROL UNIT

REMOVAL & INSTALLATION

See Figure 149.

1. Do the battery terminal disconnection procedure.
2. Remove the under-dash fuse/relay box.
3. Disconnect EPS control unit connector A (11P) and connector B (16P).
4. Loosen the bolt.
5. Remove the bolt and nut from the EPS control unit bracket.
6. Remove the EPS control unit with the bracket from the body.

To install:

7. Install the EPS control unit in the reverse order of removal.

➡Install the bracket with the EPS control unit as shown in position. Connect EPS control unit connector A (11P) and connector B (16P), then confirm the connectors are fully seated.

8. Do the battery terminal reconnection procedure.
9. After installation, start the engine, allow it to idle, and turn the steering wheel from lock to lock several times. Make sure that the EPS indicator does not come on.

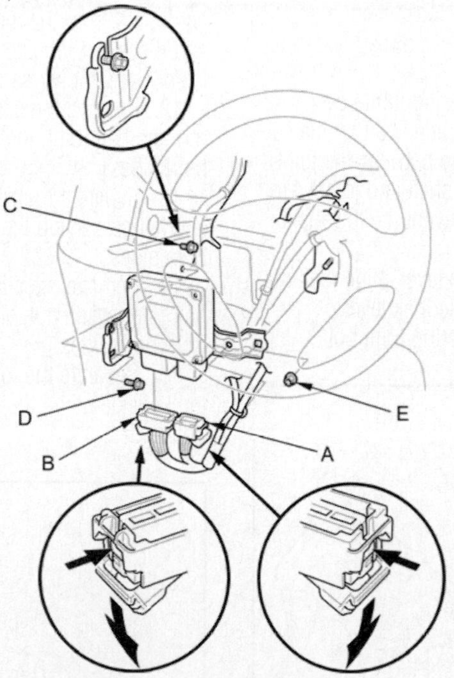
C
D
E
B
A
A. EPS control unit connector A
B. EPS control unit connector B
C. Bolt
D. Bolt
E. Nut
37647_HFIT_G0338

Fig. 149 Disconnect EPS control unit connector A (11P) and connector B (16P)

ELECTRICAL POWER STEERING (EPS) MOTOR

REMOVAL & INSTALLATION

See Figures 150 and 151.

➡Do not allow dust, dirt, or other foreign materials to enter the steering gearbox.

1. Remove the steering gearbox.
2. Remove the EPS motor 3P connector and EPS motor angle sensor 8P connector from the connector bracket, then remove clips.
3. Remove the EPS motor from the steering gearbox, then remove the O-rings.

➡Do not discard the O-ring C.

To install:

4. Clean the mating surface of the EPS motor and the steering gearbox.

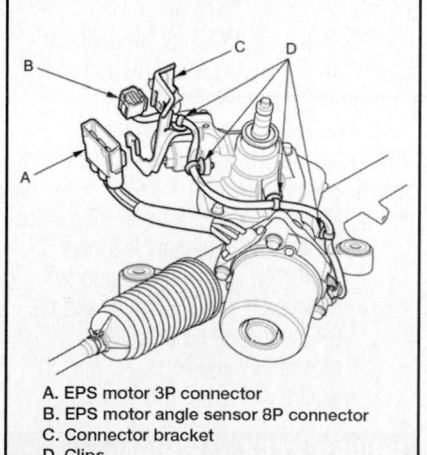

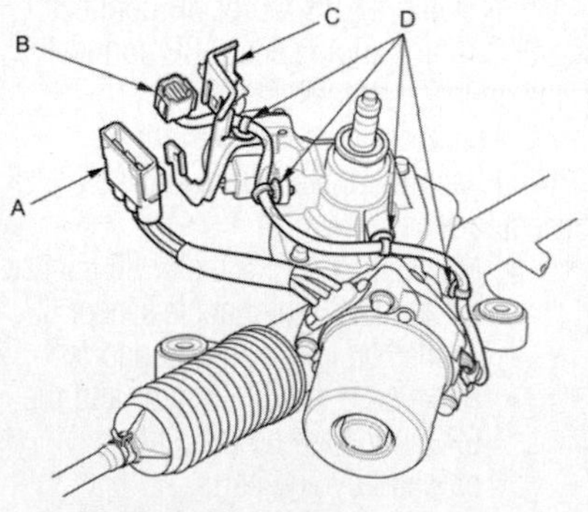
B
C
D
A
A. EPS motor 3P connector
B. EPS motor angle sensor 8P connector
C. Connector bracket
D. Clips
37647_HFIT_G0339

Fig. 150 Remove the EPS motor 3P connector and EPS motor angle sensor 8P connector from the connector bracket

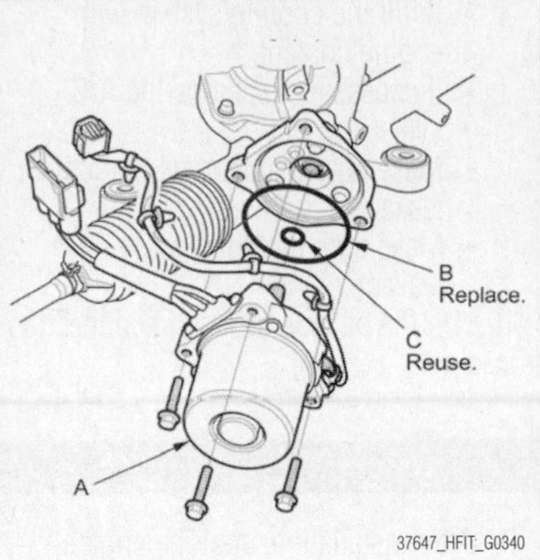
B
Replace.
C
Reuse.
A
37647_HFIT_G0340

Fig. 151 Remove the EPS motor (A) from the steering gearbox, then remove the O-rings (B, C)

5. Apply grease included in the motor set to the new O-ring, and carefully fit it on the EPS motor.
6. Install the O-ring to the EPS motor shaft.
7. Apply steering grease to the EPS motor shaft.
8. Install the EPS motor on the steering gearbox by engaging the EPS motor shaft and the worm shaft.
9. Before tightening the bolts, turn the motor two or three times to the right and left about 45 degrees. Make sure the EPS motor is evenly seated on the steering gearbox, and that the O-ring is not pinched between the mating surfaces.
10. Loosely install the EPS motor mounting bolts, then turn the pinion shaft two or three times to the right and left about 45 degrees.
11. Tighten the EPS motor mounting bolts to 14 ft. lbs. (20 N m).
12. Install the EPS motor 3P connector and EPS motor angle sensor 8P connector to the connector bracket.
13. Install the wire harness clips.
14. Finish the installation, and note these items:
   - Make sure the EPS motor 3P connector and EPS motor angle sensor 8P connector are properly connected.
   - Make sure the EPS motor and the EPS wires are not caught or pinched by any parts.
15. Install the steering gearbox.

POWER RACK & PINION STEERING GEAR

REMOVAL & INSTALLATION

See Figures 152 through 169.

Special Tools Required*:

- Ball Joint Remover, 28 mm 07MAC-SL0A202
- Universal Lifting Eyelet 07AAK-SNAA120
- 1.8 Support Bolt 07AAK-SNAA500
- Engine Support Hanger, A and Reds AAR-T1256

➡* Available through the Honda Tool and Equipment Program 888-424-6857.

➡Note these items during removal:

- Use solvent and a brush, wash any oil and dirt off the end of the steering gearbox, but avoid any electrical parts. Blow dry with compressed air.
- Be sure to remove the steering wheel before disconnecting the steering joint or damage to the cable reel can occur.
- Lower the front subframe from the body then remove the steering gearbox.

1. Do the battery terminal disconnection procedure.
2. Raise the vehicle on a lift.
3. Remove the front wheels.
4. Release the lock lever, tilt the steering column all the way up slide it all the way in, then tighten the lock lever.
5. Remove the driver's airbag and the steering wheel.
6. Remove the steering joint cover.
7. Hold the lower slide shaft on the column with a piece of wire between the joint yoke of the lower slide shaft and joint yoke of the upper shaft to prevent the lower slide from pulling out.
8. Release the lock lever, slide it all the way out, then tighten the lock lever.
9. Remove the steering joint bolt.

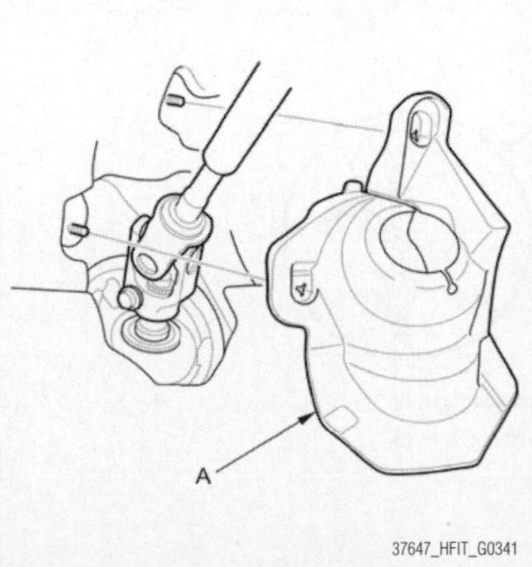
A
37647_HFIT_G0341

Fig. 152 Remove the steering joint cover (A)

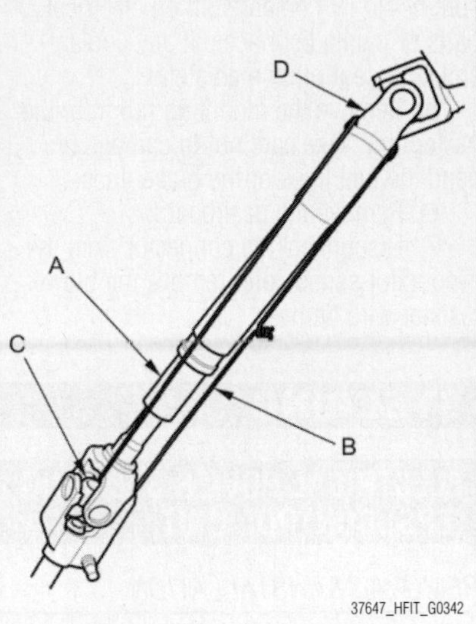
D
A
C
B
37647_HFIT_G0342

Fig. 153 Hold the lower slide shaft on the column with a piece of wire between the joint yoke of the lower slide shaft and joint yoke of the upper shaft to prevent the lower slide from pulling out

10. Disconnect the steering joint by sliding the steering joint into the column.
11. Remove the center guide (if equipped) from the top of the pinion shaft, and discard it. The center guide is for factory assembly use only.
12. Remove the cotter pin from the tie-rod end ball joint, then remove the nut on both sides.
13. Separate the tie-rod end ball joint and knuckle using the ball joint remover on both sides.
14. Remove the clip from the lower arm ball joint castle nut, and remove the nut from both sides.
15. Separate the lower arm ball joint and the knuckle using the ball joint remover on both sides.

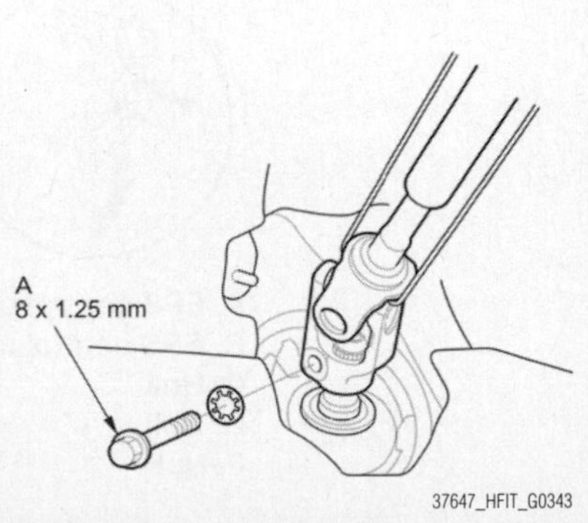
A
8 x 1.25 mm
37647_HFIT_G0343

Fig. 154 Remove the steering joint bolt (A)

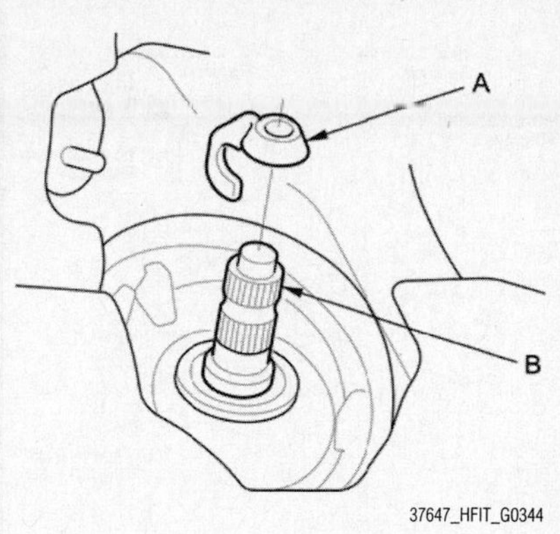
A
B
37647_HFIT_G0344

Fig. 155 Remove the center guide (A) (if equipped) from the top of the pinion shaft (B), and discard it

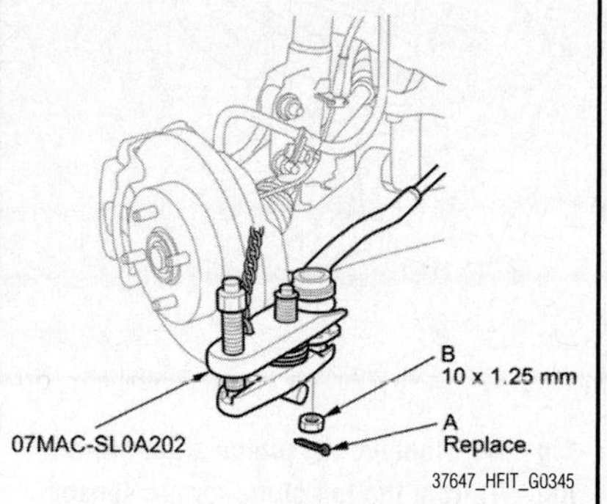
B
10 x 1.25 mm
07MAC-SL0A202
A
Replace.
37647_HFIT_G0345

Fig. 156 Remove the cotter pin (A) from the tie-rod end ball joint, then remove the nut (B) on both sides

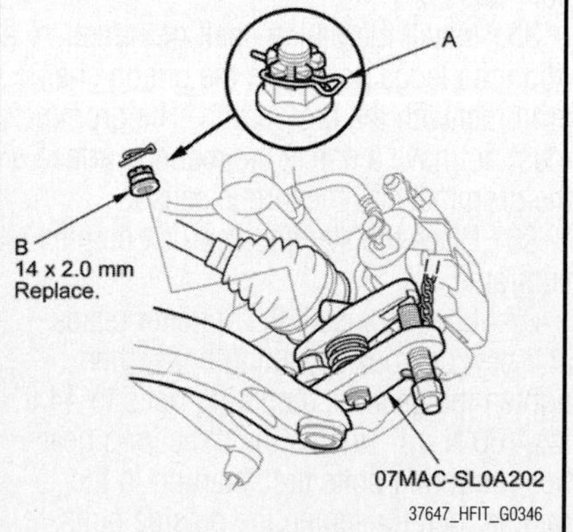
A
B
14 x 2.0 mm
Replace.
07MAC-SL0A202
37647_HFIT_G0346

Fig. 157 Remove the clip (A) from the lower arm ball joint castle nut (B), and remove the nut from both sides

16. Remove the stabilizer link from the stabilizer bar on both sides.
17. Remove the air cleaner housing.
18. Remove the cowl cover and the under-cowl panel.
19. Install the universal lifting eyelet (07AAK-SNAA120) to the bolt hole (A) at the air cleaner housing mounting bracket with the 1.8 support bolt (07AAK-SNAA500).
20. Remove the harness clamp from its clamp bracket located in front of the left damper top.
21. Set up the engine support hanger (AAR-T1256). Carefully position the engine support hanger to the vehicle; position both cross-arm foot bases over the harness clamp brackets on both sides, and position both front stands on the front bulkhead.

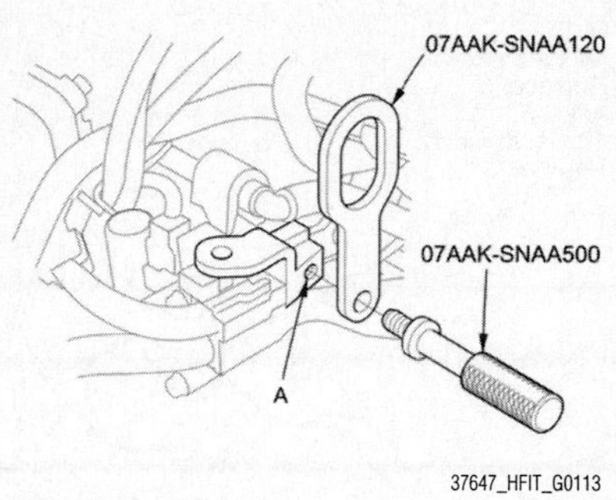
07AAK-SNAA120
07AAK-SNAA500
A
37647_HFIT_G0113

Fig. 158 Install the universal lifting eyelet to the bolt hole (A)

Attach the hook to the universal eyelet, tighten the wing nut by hand, and lift and support the engine.

22. Disconnect the EPS motor angle sensor 8P connector and torque sensor 6P connector from the steering gearbox.

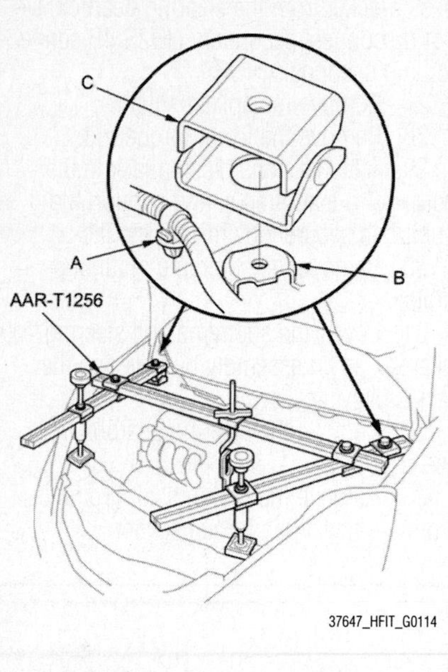
C
A
B
AAR-T1256
37647_HFIT_G0114

Fig. 160 Remove the harness clamp (A) from its clamp bracket (B) located in front of the left damper top; position both cross-arm foot bases (C) over the harness clamp brackets on both sides

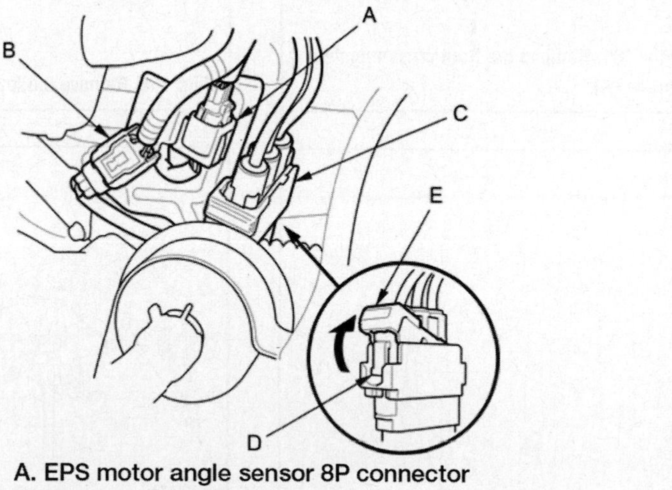
A
B
C
E
D

A. EPS motor angle sensor 8P connector
B. Torque sensor 6P connector
C. EPS motor 3P connector
D. Lock
E. Lever

37647_HFIT_G0347

Fig. 159 Disconnect the EPS motor angle sensor 8P connector and torque sensor 6P connector from the steering gearbox

23. Disconnect the EPS motor 3P connector by pushing the lock and pulling up the lever.

24. Wrap the connectors with the vinyl tape to avoid contamination from grease or water.

25. Remove the front cross-member brace.

26. Remove the secondary HO2S harness bracket from the steering gearbox. Do not disconnect secondary HO2S 4P connector and secondary HO2S.

27. Remove the splash shield.

28. Remove the lower torque rod.

29. Attach a transmission jack to the center of the subframe, and support the subframe securely by raising the jack.

30. Remove the subframe mounting bolts.

31. Lower the subframe and steering gearbox as an assembly by lowering the jack slowly.

32. Remove the steering gearbox from the subframe.

33. Remove the pinion shaft grommet from the top of the torque sensor.

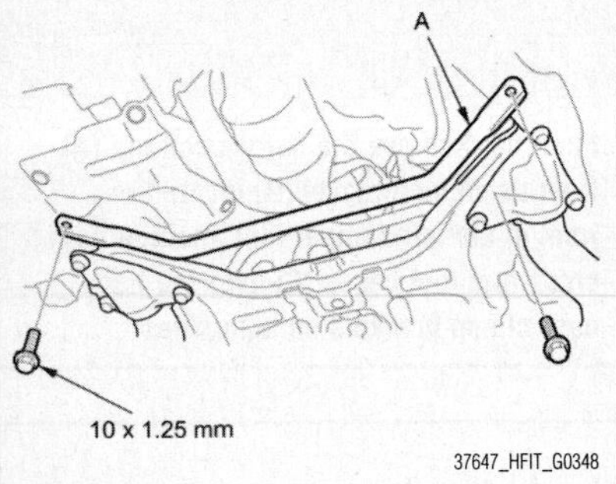
A
10 x 1.25 mm
37647_HFIT_G0348

Fig. 161 Remove the front cross-member brace (A)

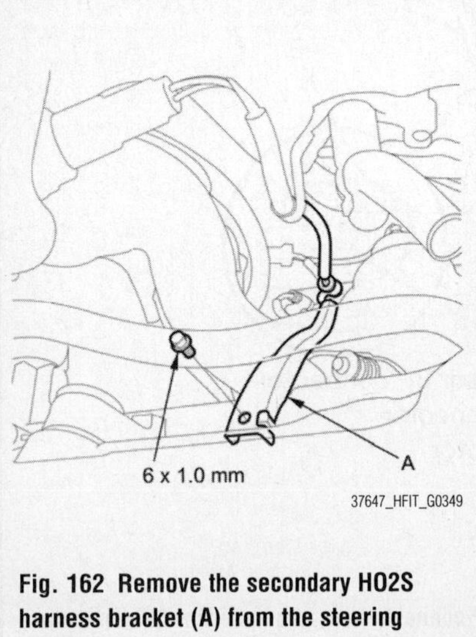
6 x 1.0 mm
A
37647_HFIT_G0349

Fig. 162 Remove the secondary HO2S harness bracket (A) from the steering gearbox

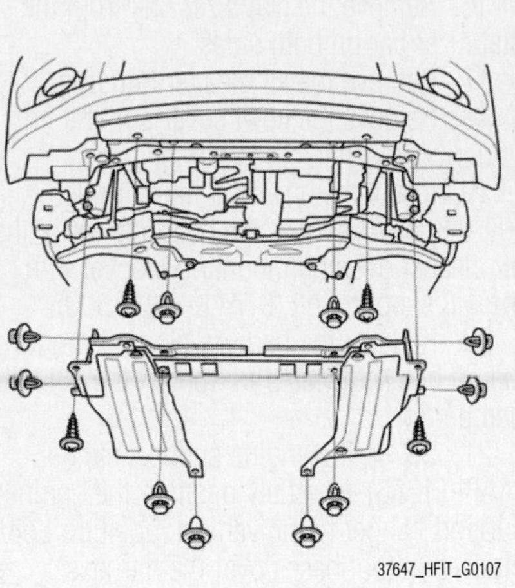
37647_HFIT_G0107

Fig. 163 Remove the splash shield

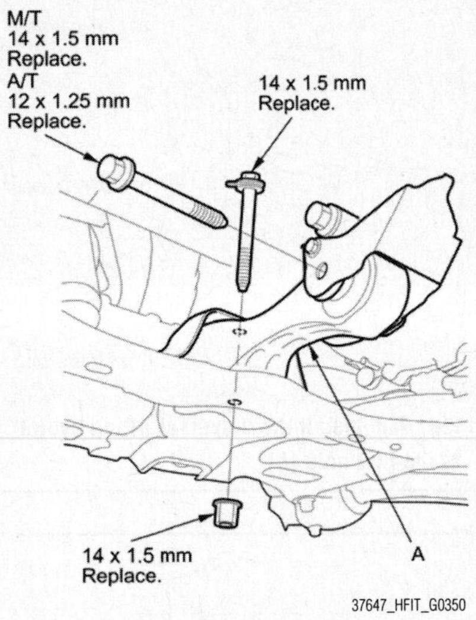
M/T
14 x 1.5 mm
Replace.
A/T
12 x 1.25 mm
Replace.
14 x 1.5 mm
Replace.
14 x 1.5 mm
Replace.
A
37647_HFIT_G0350

Fig. 164 Remove the lower torque rod (A)

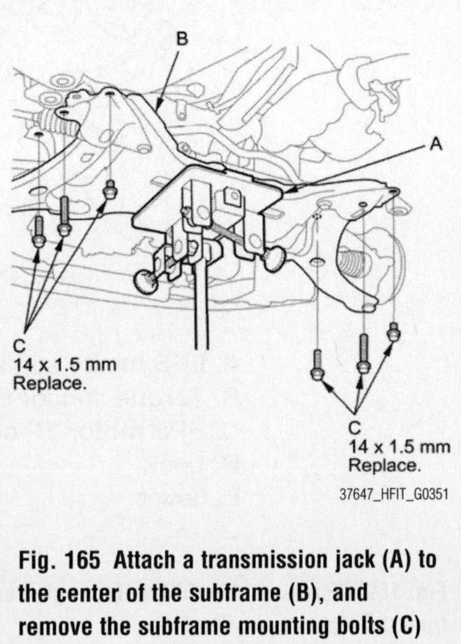
B
A
C
14 x 1.5 mm
Replace.
C
14 x 1.5 mm
Replace.
37647_HFIT_G0351

Fig. 165 Attach a transmission jack (A) to the center of the subframe (B), and remove the subframe mounting bolts (C)

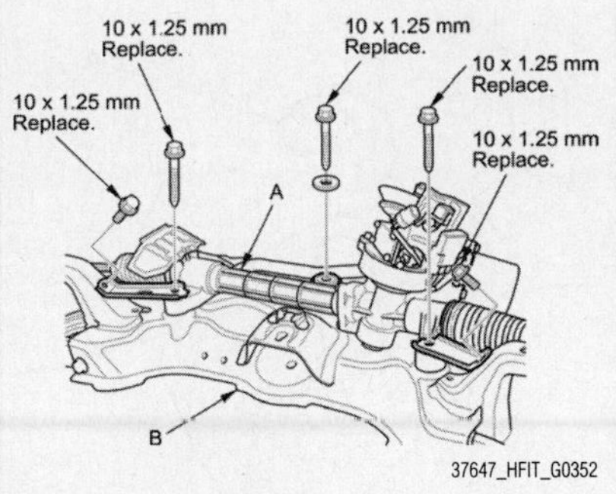
10 x 1.25 mm
Replace.
10 x 1.25 mm
Replace.
10 x 1.25 mm
Replace.
10 x 1.25 mm
Replace.
10 x 1.25 mm
Replace.
A
B
37647_HFIT_G0352

Fig. 166 Remove the steering gearbox (A) from the subframe (B)

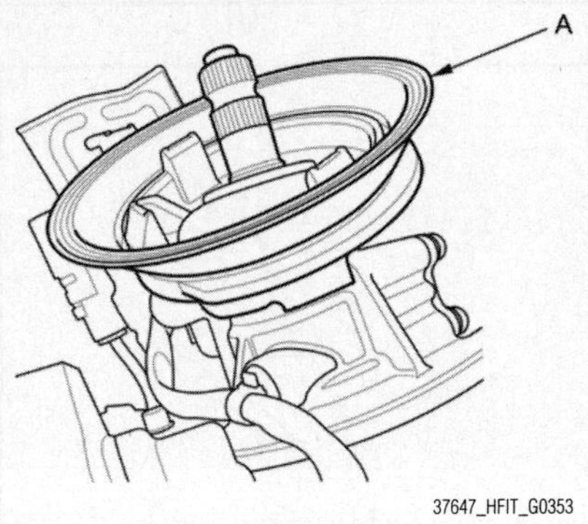
A
37647_HFIT_G0353

Fig. 167 Remove the pinion shaft grommet (A) from the top of the torque sensor

To install:

34. Before installing the steering gearbox, make sure that no grease is on the mating surface of the steering gearbox and the front subframe. To prevent the gearbox mounting bolts from loosening after the installation, remove any grease from the bolt holes.

35. Install the pinion shaft grommet. Align the ledge portion in the pinion shaft grommet with the lug portion. The grommet must not have a gap at the mating surface of the grommet and the torque sensor.

36. Place the steering gearbox on the subframe.

37. Loosely install the stiffener plates, and new gearbox mounting bolts, then tighten the gearbox mounting bolts to 44 ft. lbs. (60 N m). Tighten the three long gearbox mounting bolts first, starting in the middle then tightening the outside bolts. After the gearbox mounting bolts are tightened, then tighten the stiffener mounting bolts to 43 ft. lbs. (59 N m).

38. Turn the lip of the pinion shaft grommet to ease installation.

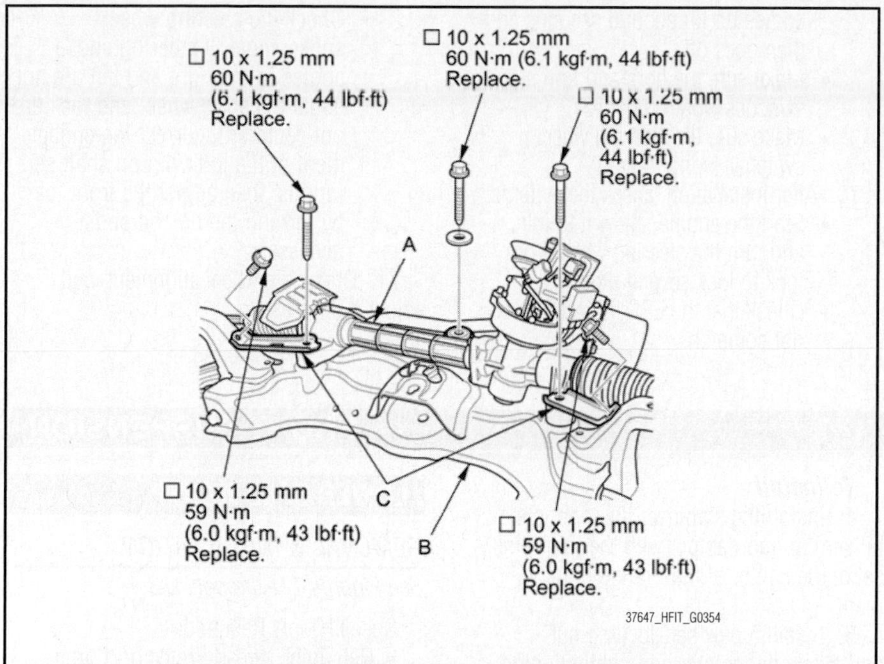
10 x 1.25 mm
60 N·m
(6.1 kgf·m, 44 lbf·ft)
Replace.
10 x 1.25 mm
60 N·m (6.1 kgf·m, 44 lbf·ft)
Replace.
10 x 1.25 mm
60 N·m
(6.1 kgf·m,
44 lbf·ft)
Replace.
A
10 x 1.25 mm
59 N·m
(6.0 kgf·m, 43 lbf·ft)
Replace.
C
B
10 x 1.25 mm
59 N·m
(6.0 kgf·m, 43 lbf·ft)
Replace.
37647_HFIT_G0354

Fig. 168 Place the steering gearbox (A) on the subframe (B) and install the stiffener plates (C)

39. Set the subframe with the steering gearbox on the transmission jack and support it.

40. Carefully raise the subframe with the transmission jack, and pass the pinion shaft into the passenger's compartment. Return the lip of the pinion shaft grommet.

➡Be sure that the pinion shaft grommet is securely in place. Make sure the lip of the pinion shaft grommet is not turned up. Incorrect installation can cause leakage of water or mud, and noise. Take care not to damage the lower arm ball joint boot with the edge of the knuckle, etc.

41. Loosely install new subframe mounting bolts, then tighten the bolts to 69 ft. lbs. (93 N m).

42. Install the lower torque rod with new mounting bolts and a new mounting nut, and tighten the bolt to 61 ft. lbs. (83 N m) and tighten the nut to 69 ft. lbs. (93 N m).

➡Make sure the stopper of the flange bolt is securely installed.

43. Install the splash shield.

44. Lower the vehicle.

45. Remove the engine support hanger (AAR-T1256) from the vehicle.

46. Remove the cross arm foot, then install the harness clip to the harness clamp bracket on the left side.

47. Remove the universal lifting eyelet (07AAK-SNAA120) and 1.8 support bolt (07AAK-SNAA500).

48. Install the secondary HO2S harness bracket to the steering gearbox.

49. Install the front cross-member brace. Tighten the mounting bolts to 40 ft. lbs. (54 N m).

50. Remove the vinyl tape from the connectors.

51. Connect the EPS motor angle sensor 8P connector, torque sensor 6P connector to the steering gearbox.

52. Pull down the lever of the EPS motor 3P connector, then confirm the connector is fully seated.

53. Install the air cleaner housing.

54. Install the under cowl panel and cowl cover.

55. Connect the stabilizer links.

56. Wipe off any grease contamination from the ball joint tapered section and threads. Reconnect the tie-rod end ball joints to the knuckles. Install the nuts, and tighten to 32 ft. lbs. (43 N m) on both sides.

57. Install a new cotter pin and bend it.

58. Wipe off any grease contamination from the lower arm ball joint tapered section and thread. Then reconnect the lower arm to the knuckle. Install the new castle nut and tighten it to 47–54 ft. lbs. (64–74 N m) on both sides.

➡Be careful not to damage the lower ball joint boot. Check the ball joint boot for deformation before connecting the knuckle. Torque the castle nut to the lower torque specification, then tighten it only far enough to align the slot with the joint pin clip hole. Do not align the castle nut by loosening it.

59. Install the clip.

60. Install the front wheels, then set the wheels in the straight ahead position.

➡Before installing the wheel, clean the mating surfaces of the brake disc and the inside of the wheel.

61. Cut the wire, while holding the lower slide shaft on the steering column.

62. Slip the lower end of the steering joint onto the pinion shaft taking care to align the gap within the angle shown.

63. Release the lock lever, tilt the steering column all the way down, side it all the way out, then tighten the lock lever.

64. Align the bolt hole on the steering joint with the groove around the pinion shaft, then loosely install the lower steering joint bolt. Be sure that the joint bolt is securely in the groove in the pinion shaft.

65. Pull on the steering joint to make sure that the steering joint is fully seated, then tighten the lower joint bolt to the specified torque.

66. Install the steering joint cover.

67. Install the steering wheel and the driver's airbag .

68. With the tires raised off the ground, check for the following symptoms and probable causes by turning the steering wheel fully to the right and left several times.

- Rubbing sound coming from the lower steering column area. Steering column joint is contacting the cover.
- Grating sound from the lower steering column area, or a rough feeling

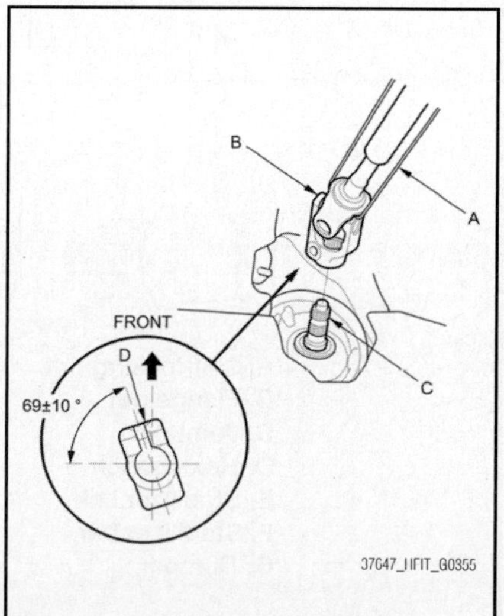
B
A
FRONT
D
C
69±10 °
37647_HFIT_G0355

Fig. 169 Cut the wire, while holding the lower slide shaft on the steering column

during steering. Poor engagement of the pinion shaft serrations.

- Noise from around the steering wheel during steering. Poor engagement of the SRS cable reel with the steering wheel, or a damaged cable reel.

69. Do the battery terminal reconnection procedure, and check these items:

- Turn the ignition switch to ON (II), and check that the SRS indicator comes on for about 6 seconds and then goes off.
- Make sure the horn and turn signal switches work properly.
- Make sure the steering wheel switches work properly.

70. After installation, check these items:

- Start the engine, allow it to idle, and turn the steering wheel from lock to lock several times.
- Check that the EPS indicator does not come on.
- Check the steering wheel spoke angle. If steering spoke angles to the right and left are not equal (steering wheel and rack are not centered), correct the engagement of the joint/pinion shaft serrations, then adjust the front toe by turning the tie-rod ends, if necessary.

71. Check the wheel alignment, and adjust it if necessary.

SUSPENSION

FRONT SUSPENSION

CONTROL LINKS

REMOVAL & INSTALLATION

Front Stabilizer Link

See Figure 170.

1. Raise the front of the vehicle, and support it with safety stands in the proper locations.
2. Remove the front wheel.
3. Remove the self-locking nut and the flange nut while holding the respective joint pin with a hex wrench, then remove the stabilizer link.

To install:

4. Install the stabilizer link on the stabilizer bar and damper with the joint pins set at the center of their range of movement.
5. Install a new self-locking nut and a new flange nut, and lightly tighten them.
6. Tighten the lower self-locking nut to 28 ft. lbs. (38 Nm) and the upper flange nut to 22 ft. lbs. (29 Nm), while holding the respective joint pin with a hex wrench.
7. Reinstall all removed parts, and test-drive the vehicle.

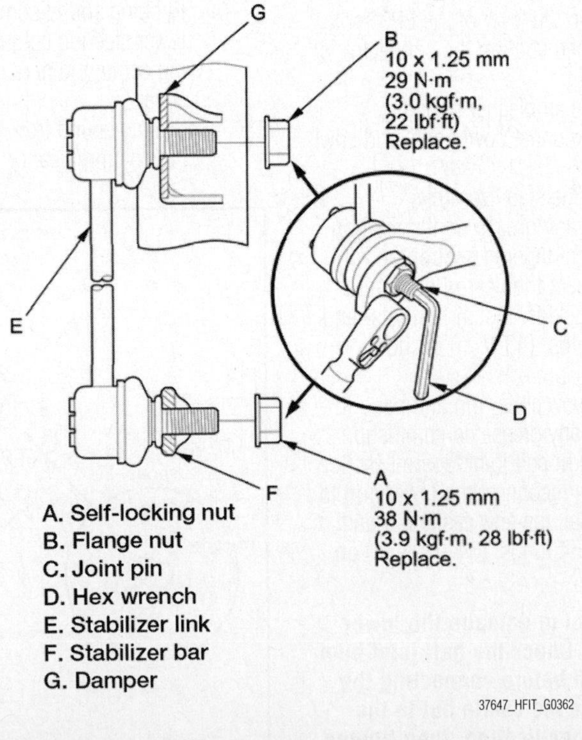
G
B
10 x 1.25 mm
29 N·m
(3.0 kgf·m,
22 lbf·ft)
Replace.
C
E
D
F
A
10 x 1.25 mm
38 N·m
(3.9 kgf·m, 28 lbf·ft)
Replace.
A. Self-locking nut
B. Flange nut
C. Joint pin
D. Hex wrench
E. Stabilizer link
F. Stabilizer bar
G. Damper
37647_HFIT_G0362

Fig. 170 Removing/installing the stabilizer link

LOWER BALL JOINT

REMOVAL & INSTALLATION

See Figures 171 through 173.

Special Tools Required:

- Ball Joint Thread Protector, 10 mm 07AAF-SECA120
- Ball Joint Thread Protector, 14 mm 071AF-S3VA000
- Ball Joint Remover, 28 mm 07MAC-SL0A202

➡**Always use a ball joint remover to disconnect a ball joint. Do not strike the housing or any other part of the ball joint connection to disconnect it.**

1. Install a hex nut or the ball joint thread protector onto the threads of the ball joint.

➡**Using a hex nut, make sure the nut is flush with the ball joint pin end to prevent damage to the threaded end of the ball joint pin.**

2. Apply grease to the ball joint remover on the areas shown. This will ease installation of the tool and prevent damage to the pressure bolt threads.

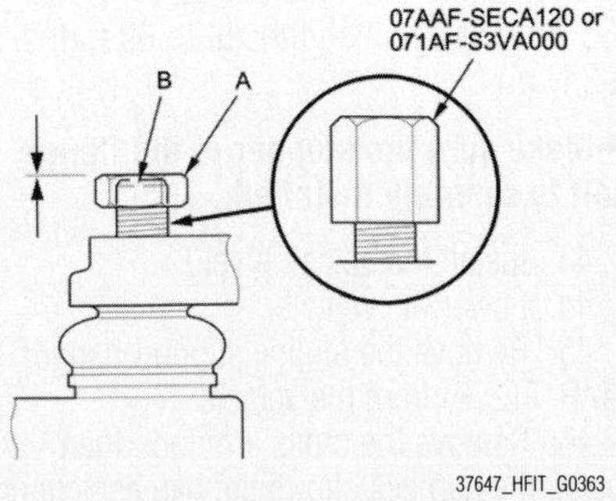
07AAF-SECA120 or
071AF-S3VA000
B
A
37647_HFIT_G0363

Fig. 171 Install a hex nut (A) or the ball joint thread protector onto the threads of the ball joint (B)

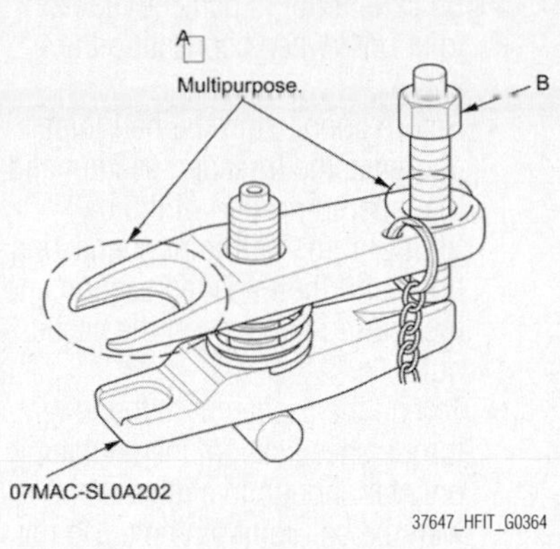
A
Multipurpose.
B
07MAC-SL0A202
37647_HFIT_G0364

Fig. 172 Apply grease to the ball joint remover on the areas shown (A) to ease installation of the tool and prevent damage to the pressure bolt (B) threads

3. Loosen the pressure bolt, and install the ball joint remover as shown. Insert the jaws carefully, making sure not to damage the ball joint boot. Adjust the jaw spacing by turning the adjusting bolt.

➡Fasten the safety chain securely to a suspension arm or the subframe. Do not fasten it to a brake line or wire harness.

4. After adjusting the adjusting bolt, make sure the head of the adjusting bolt is in the position shown to allow the jaw to pivot.

5. With a wrench, tighten the pressure bolt until the ball joint pin pops loose from the ball joint connecting hole. If necessary, apply penetrating type lubricant to loosen the ball joint pin.

➡Do not use pneumatic or electric tools on the pressure bolt.

6. Remove the ball joint remover, then remove the nut from the end of the ball joint pin, and pull the ball joint out of the ball joint connecting hole. Inspect the ball joint boot, and replace it if damaged.

LOWER CONTROL ARM

REMOVAL & INSTALLATION

See Figures 174 and 175.

➡Do not remove the lower arm from both sides at the same time. The lower arm mounting bolts also secure the subframe to the vehicle.

1. Raise and support the vehicle.
2. Remove the front wheel.
3. Remove the lock pin from the lower arm ball joint, then remove the castle nut.

➡During installation, install the lock pin as shown after tightening the new castle nut.

4. Disconnect the lower ball joint from the knuckle using the ball joint remover.

➡Be careful not to damage the ball joint boot when installing the remover.

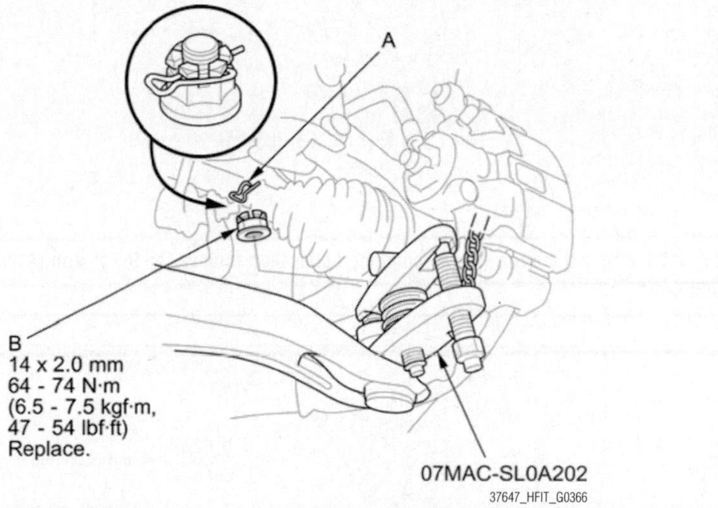
A
B
14 x 2.0 mm
64 - 74 N·m
(6.5 - 7.5 kgf·m,
47 - 54 lbf·ft)
Replace.
07MAC-SL0A202
37647_HFIT_G0366

Fig. 174 Remove the lock pin (A) from the lower arm ball joint, then remove the castle nut (B)

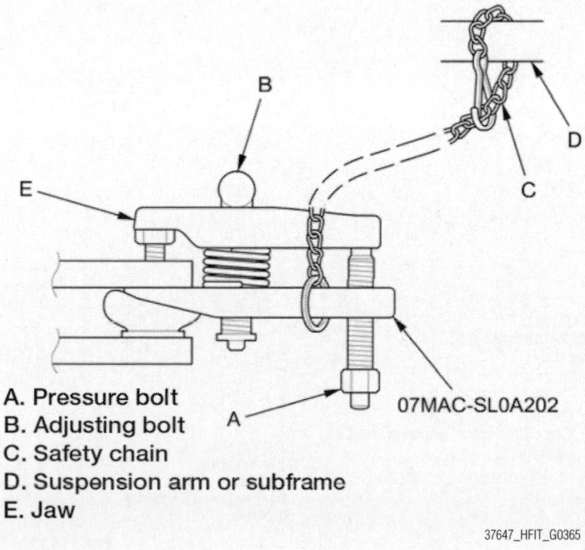
B
D
E
C
A
07MAC-SL0A202
A. Pressure bolt
B. Adjusting bolt
C. Safety chain
D. Suspension arm or subframe
E. Jaw
37647_HFIT_G0365

Fig. 173 Loosen the pressure bolt, and install the ball joint remover

Do not force or hammer on the lower arm, or pry between the lower arm and the knuckle. You could damage the ball joint.

5. Remove the lower arm mounting bolts, then remove the lower arm from the front subframe.

To install:

➡Use new lower arm mounting bolts during reassembly.

6. Install the lower arm in the reverse order of removal, and note these items:
   - First install all the components, and lightly tighten the bolts and the nuts, then raise the suspension to load it with the vehicle's weight before fully tightening it to the specified torque values. Do not place the jack against the ball joint of the lower arm.

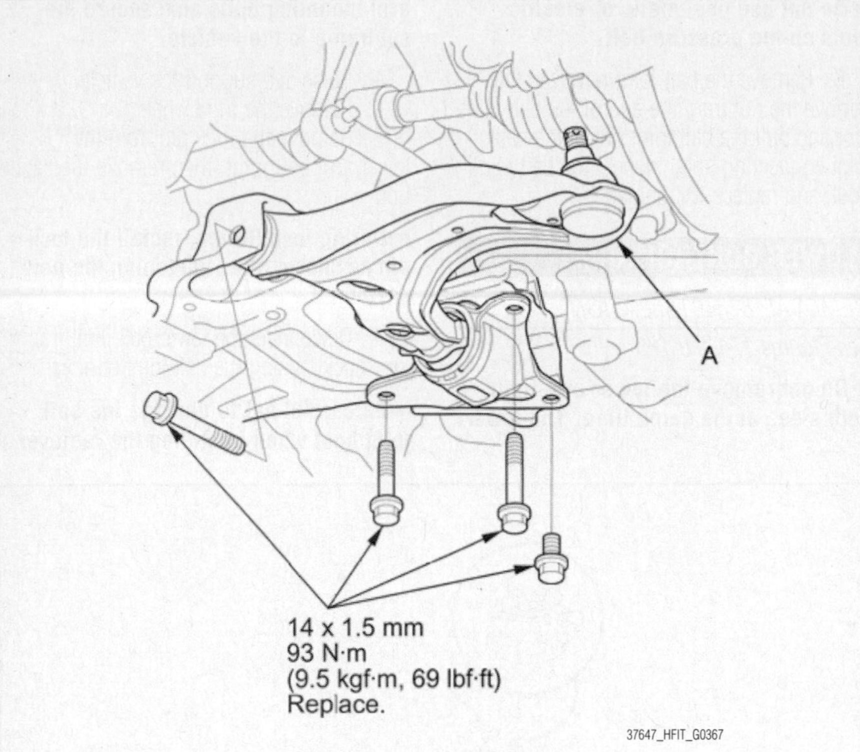
A
14 x 1.5 mm
93 N·m
(9.5 kgf·m, 69 lbf·ft)
Replace.
37647_HFIT_G0367

Fig. 175 Remove the lower arm mounting bolts, then remove the lower arm (A) from the front subframe

- Be careful not to damage the ball joint boot when connecting the knuckle.
- Before connecting the ball joint, degrease the threaded section and the tapered portion of the ball joint pin, the ball joint connecting hole, and the threaded section and the mating surfaces of the castle nut.
- Torque the castle nut to the lower torque specification, then tighten it only far enough to align the slot with the ball joint pin hole. Do not align the castle nut by loosening it.
- Before installing the wheel, clean the mating surfaces of the brake disc and the inside of the wheel.

7. Check the wheel alignment, and adjust it if necessary.

STEERING KNUCKLE

REMOVAL & INSTALLATION

See Figures 176 through 181.

Special Tools Required:

- Ball Joint Remover, 28 mm 07MAC-SL0A202

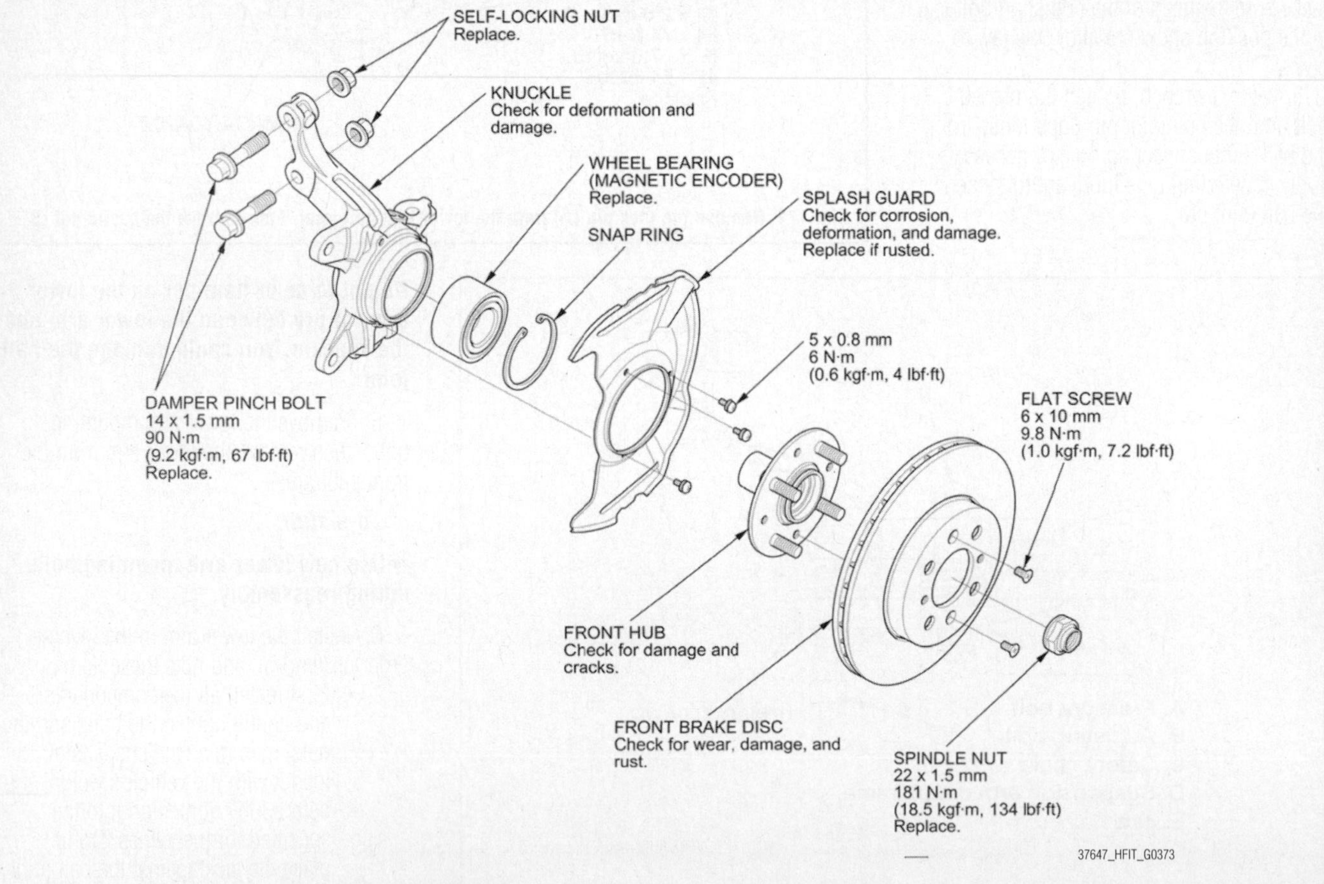
SELF-LOCKING NUT
Replace.
KNUCKLE
Check for deformation and damage.
WHEEL BEARING
(MAGNETIC ENCODER)
Replace.
SNAP RING
SPLASH GUARD
Check for corrosion, deformation, and damage.
Replace if rusted.
5 x 0.8 mm
6 N·m
(0.6 kgf·m, 4 lbf·ft)
FLAT SCREW
6 x 10 mm
9.8 N·m
(1.0 kgf·m, 7.2 lbf·ft)
DAMPER PINCH BOLT
14 x 1.5 mm
90 N·m
(9.2 kgf·m, 67 lbf·ft)
Replace.
FRONT HUB
Check for damage and cracks.
FRONT BRAKE DISC
Check for wear, damage, and rust.
SPINDLE NUT
22 x 1.5 mm
181 N·m
(18.5 kgf·m, 134 lbf·ft)
Replace.
37647_HFIT_G0373

Fig. 176 Exploded view of the steering knuckle assembly

- Hub Dis/Assembly Pin, 40 mm 07GAF-SE00100
- Driver Handle, 15 x 135L 07749-0010000
- Bearing Driver Attachment, 52 x 55 mm 07746-0010400
- Support Base 07965-SD90100
- Ball Joint Thread Protector, 10 mm 07AAF-SECA120
- Ball Joint Thread Protector, 14 mm 071AF-S3VA000

1. Raise and support the vehicle.
2. Remove the wheel nuts and the front wheel.
3. Remove the brake hose mounting bolt from the damper bracket.
4. Remove the brake caliper bracket mounting bolts, then remove the caliper assembly from the knuckle.

➡To prevent damage to the caliper assembly or the brake hose, use a short piece of wire to hang the caliper assembly from the undercarriage. Do not twist the brake hose excessively.

5. Pry up the stake on the spindle nut, then remove the nut.
6. Remove the front brake disc.
7. Check the front hub for damage and cracks.
8. Remove the wheel speed sensor from the knuckle. Do not disconnect the wheel speed sensor connector.
9. Remove the cotter pin from the tie-rod end ball joint, then remove the nut.

➡During installation, install the new cotter pin after tightening the nut, and bend its end as shown.

10. Disconnect the tie-rod end ball joint from the knuckle using the ball joint remover.
11. Remove the lock pin from the lower arm ball joint, then remove the castle nut.

➡During installation, install the lock pin as shown after tightening the new castle nut.

12. Disconnect the lower ball joint from the knuckle using the ball joint remover.

➡Be careful not to damage the ball joint boot when installing the remover. Do not force or hammer on the lower arm, or pry between the lower arm and the knuckle. You could damage the ball joint.

13. Remove the damper pinch bolts and the self-locking nuts from the damper.

➡Use new damper pinch bolts and new self-locking nuts during reassembly.

14. Pull the knuckle outward, and separate the outboard joint from the front hub using a soft face hammer.

➡Do not pull the driveshaft end outward. The driveshaft inboard joint may come apart. During installation, apply grease to the mating surfaces of the wheel bearing and the driveshaft outboard joint.

To install:

15. Install the knuckle/hub in the reverse order of removal, and note these items:
   - First install all the components, and lightly tighten the bolts and the nuts, then raise the suspension to load it with the vehicle's weight before fully tightening to the specified torque values. Do not place the jack against the ball joint of the lower arm.
   - Be careful not to damage the ball joint boot when connecting the knuckle.
   - Before connecting the ball joint, degrease the threaded section and the tapered portion of the ball joint pin, the ball joint connecting hole, and the threaded section and the mating surfaces of the castle nut.
   - Torque the castle nut to the lower torque specification, then tighten it only far enough to align the slot with the ball joint pin hole. Do not align the castle nut by loosening it.
   - Use a new spindle nut on reassembly.

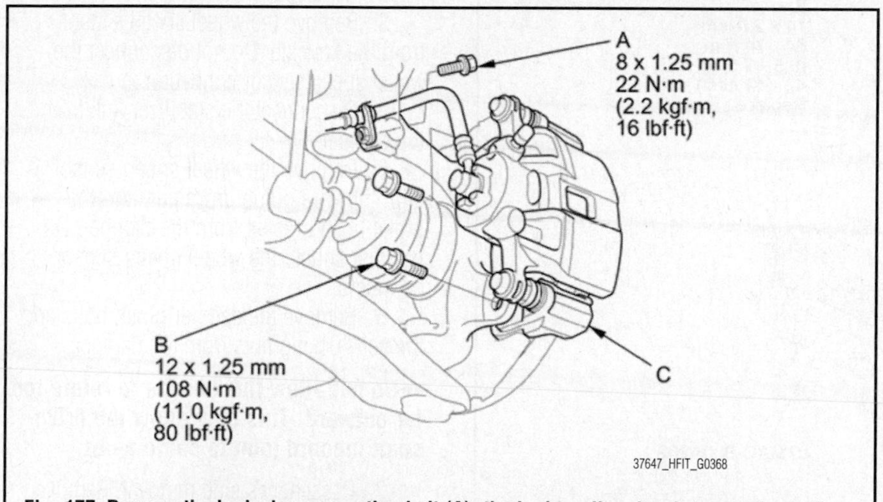
A
8 x 1.25 mm
22 N·m
(2.2 kgf·m,
16 lbf·ft)
B
12 x 1.25 mm
108 N·m
(11.0 kgf·m,
80 lbf·ft)
C
37647_HFIT_G0368

Fig. 177 Remove the brake hose mounting bolt (A), the brake caliper bracket mounting bolts (B), then remove the caliper assembly (C)

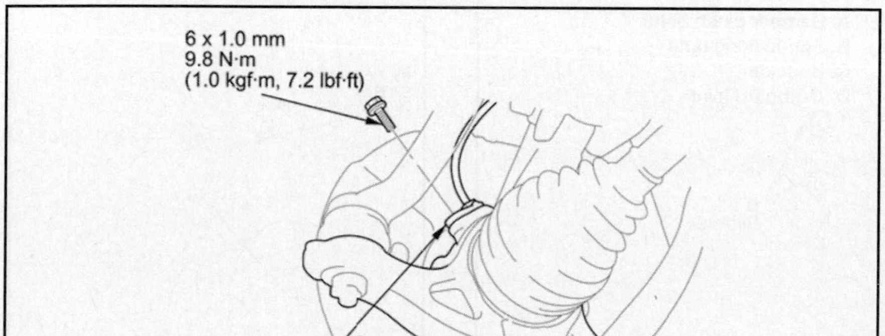
6 x 1.0 mm
9.8 N·m
(1.0 kgf·m, 7.2 lbf·ft)
A
B
37647_HFIT_G0369

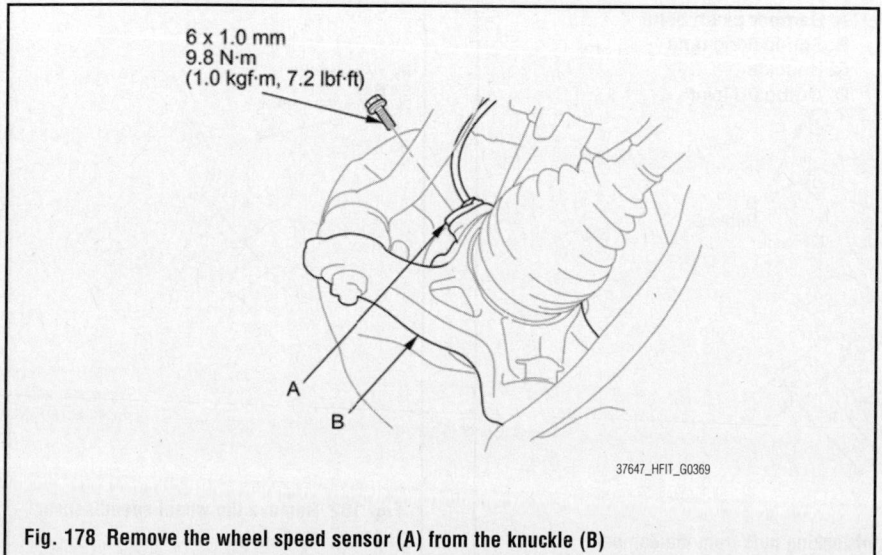

Fig. 178 Remove the wheel speed sensor (A) from the knuckle (B)

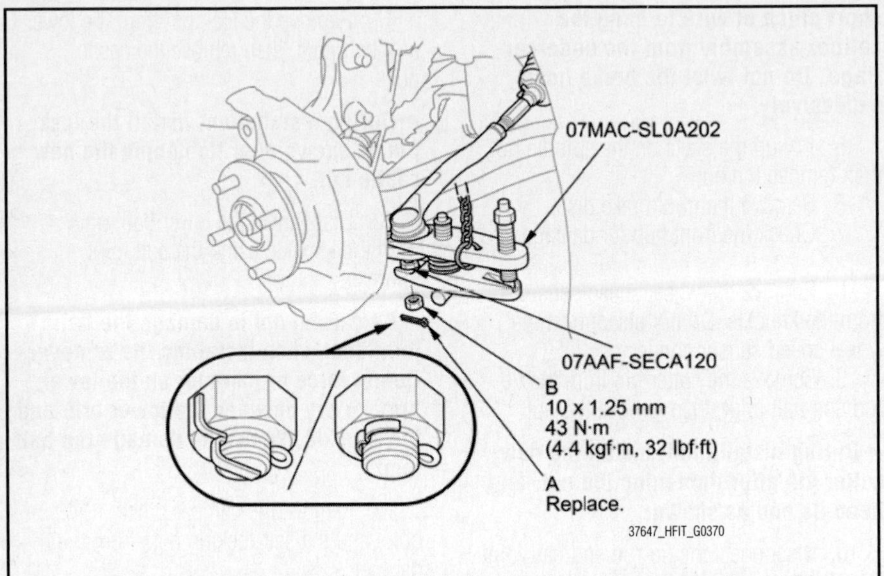
07MAC-SL0A202
07AAF-SECA120
B
10 x 1.25 mm
43 N·m
(4.4 kgf·m, 32 lbf·ft)
A
Replace.
37647_HFIT_G0370

Fig. 179 Remove the cotter pin (A) from the tie-rod end ball joint, then remove the nut (B)

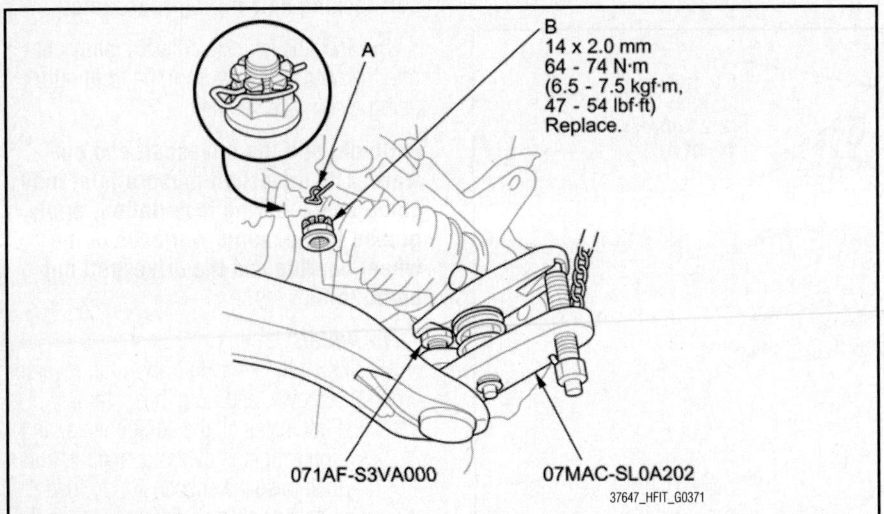
A
B
14 x 2.0 mm
64 - 74 N·m
(6.5 - 7.5 kgf·m,
47 - 54 lbf·ft)
Replace.
071AF-S3VA000
07MAC-SL0A202
37647_HFIT_G0371

Fig. 180 Remove the lock pin (A) from the lower arm ball joint, then remove the castle nut (B)

- Before installing the spindle nut, apply a small amount of engine oil to the seating surface of the nut. After tightening, use a drift to stake the spindle nut shoulder against the driveshaft.
- Before installing the brake disc, clean the mating surfaces of the front hub and the inside of the brake disc.
- Before installing the wheel, clean the mating surfaces of the brake disc and the inside of the wheel.

16. Check the wheel alignment, and adjust it if necessary.

STRUT (DAMPER/SPRING)

REMOVAL & INSTALLATION

See Figures 182 through 187.

1. Raise and support the vehicle.
2. Remove the front wheel.
3. Remove the wheel speed sensor from the knuckle. Do not disconnect the wheel speed sensor connector.
4. Disconnect the stabilizer link from the damper.
5. Remove the wheel speed sensor clip, the wire guide grommet, and the brake hose bracket from the damper. Do not disconnect the wheel speed sensor connector.
6. Remove the damper pinch bolts and the self-locking nuts from the damper.

➡Do not allow the knuckle to rotate too far outward. This may allow the driveshaft inboard joint to come apart.

7. Passenger's side removal: Remove the lid.

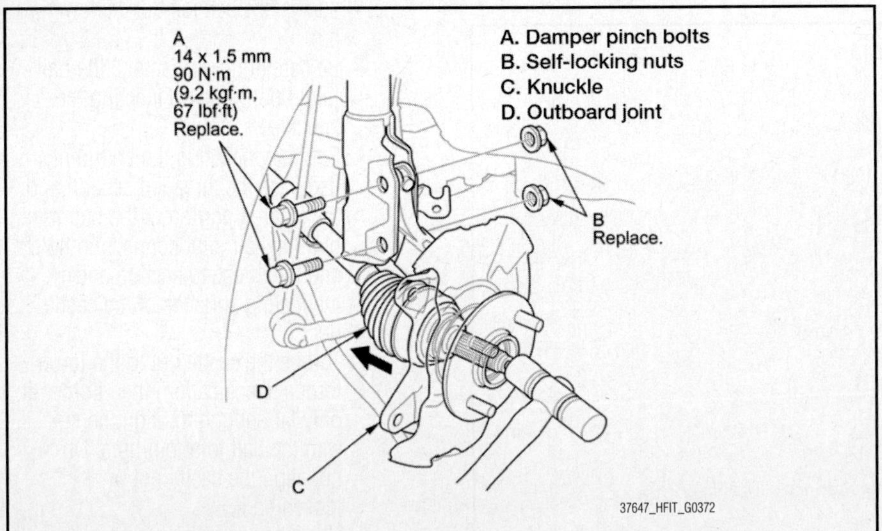
A
14 x 1.5 mm
90 N·m
(9.2 kgf·m,
67 lbf·ft)
Replace.
A. Damper pinch bolts
B. Self-locking nuts
C. Knuckle
D. Outboard joint
B
Replace.
D
C
37647_HFIT_G0372

Fig. 181 Remove the damper pinch bolts and the self-locking nuts from the damper

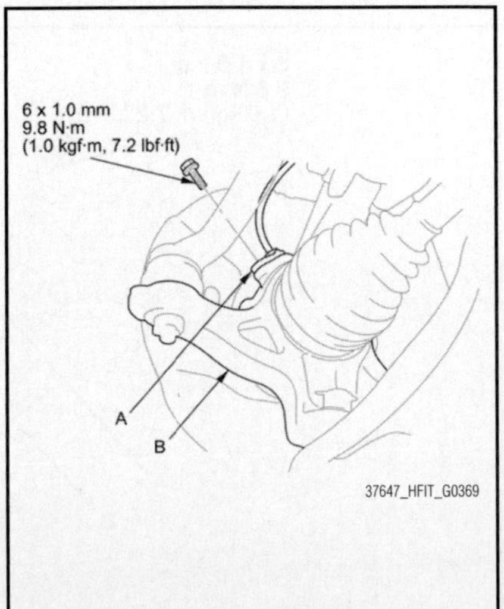
6 x 1.0 mm
9.8 N·m
(1.0 kgf·m, 7.2 lbf·ft)
A
B
37647_HFIT_G0369

Fig. 182 Remove the wheel speed sensor (A) from the knuckle

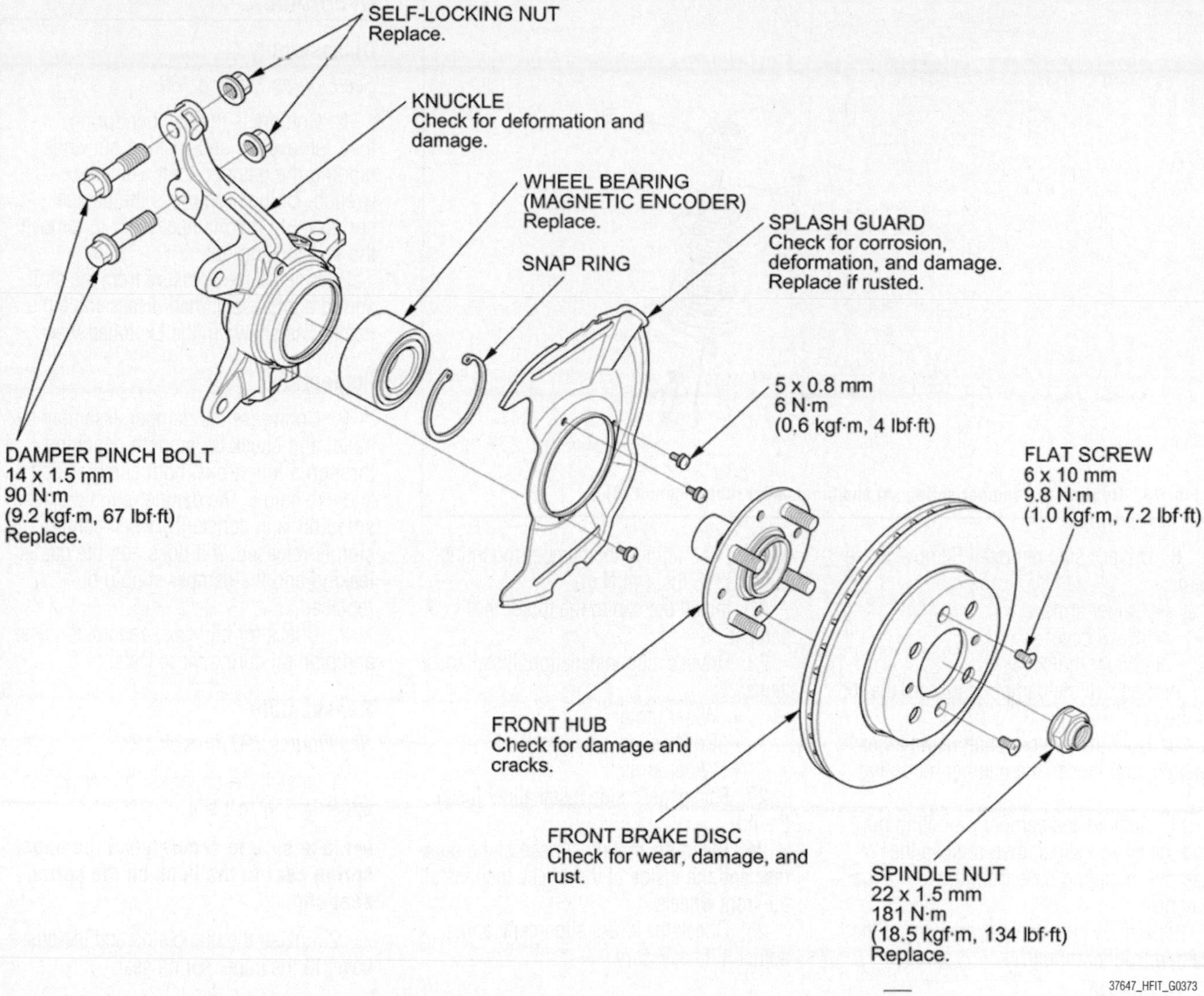
SELF-LOCKING NUT
Replace.
KNUCKLE
Check for deformation and damage.
WHEEL BEARING (MAGNETIC ENCODER)
Replace.
SNAP RING
SPLASH GUARD
Check for corrosion, deformation, and damage. Replace if rusted.
5 x 0.8 mm
6 N·m
(0.6 kgf·m, 4 lbf·ft)
DAMPER PINCH BOLT
14 x 1.5 mm
90 N·m
(9.2 kgf·m, 67 lbf·ft)
Replace.
FLAT SCREW
6 x 10 mm
9.8 N·m
(1.0 kgf·m, 7.2 lbf·ft)
FRONT HUB
Check for damage and cracks.
FRONT BRAKE DISC
Check for wear, damage, and rust.
SPINDLE NUT
22 x 1.5 mm
181 N·m
(18.5 kgf·m, 134 lbf·ft)
Replace.
37647_HFIT_G0373

Fig. 183 Remove the wheel speed sensor clip, the wire guide grommet, and the brake hose bracket from the damper

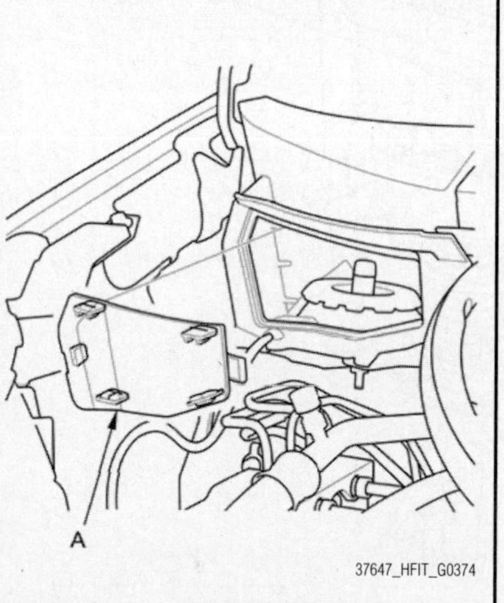
A
37647_HFIT_G0374

Fig. 184 Passenger's side removal: Remove the lid (A)

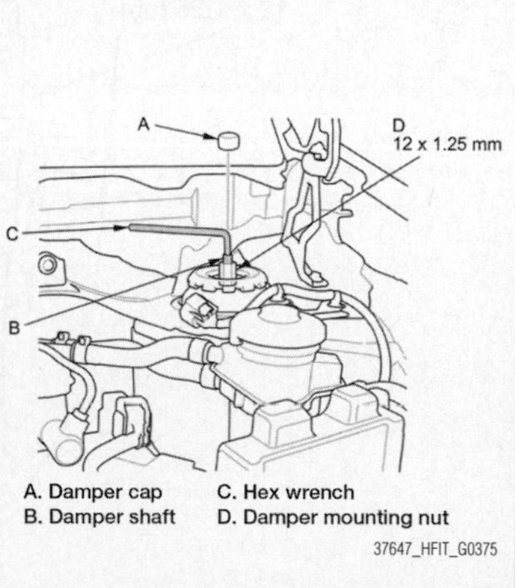
A
D
12 x 1.25 mm
C
B
A. Damper cap
B. Damper shaft
C. Hex wrench
D. Damper mounting nut
37647_HFIT_G0375

Fig. 185 Remove the damper cap from the top of the damper

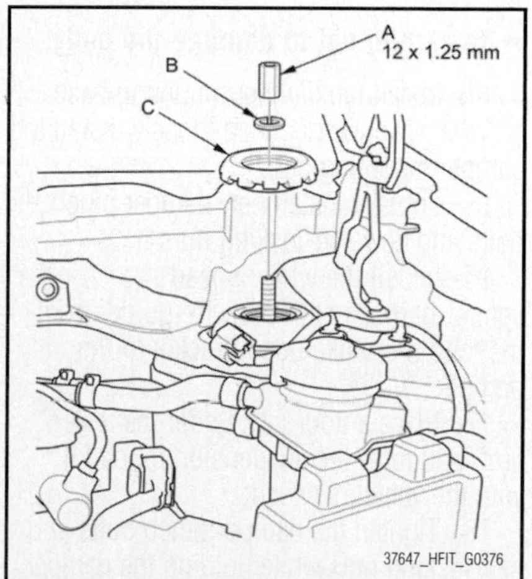
A
12 x 1.25 mm
B
C
37647_HFIT_G0376

Fig. 186 Remove the damper mounting nut (A) and the wave washer (B), then remove the damper mounting base (C) from the top of the damper

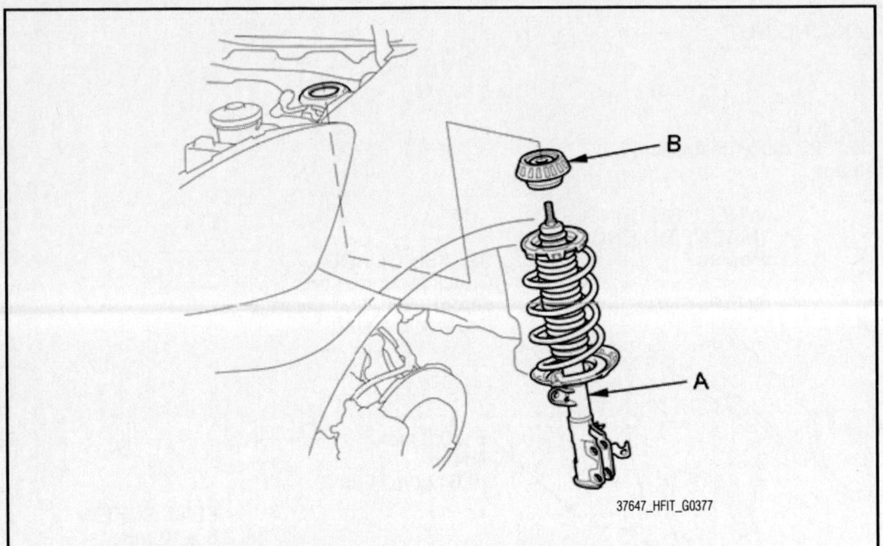
B
A
37647_HFIT_G0377

Fig. 187 Remove the damper/spring (A) and the damper rubber mount (B)

8. Driver's side removal: Remove these items:
- Wiper arms
- Cowl cover
- Wiper motor

9. Remove the damper cap from the top of the damper.
10. Hold the damper shaft using a hex wrench, and loosen the damper mounting nut.
11. Remove the damper mounting nut and the wave washer, then remove the damper mounting base from the top of the damper.
12. Remove the damper/spring and the damper rubber mount.

To install:

13. Install the damper rubber mount and the damper/spring onto the upper mount.

➡Be careful not to damage the body.

14. Install the damper mounting base and the wave washer, then loosely install the damper mounting nut.
15. Loosely install new damper pinch bolts and new self-locking nuts.
16. Install the wheel speed sensor harness clip, the wire guide grommet and the brake hose bracket to the damper.
17. Place a floor jack under the lower arm, and raise the suspension to load it with the vehicle's weight.
18. Tighten the damper pinch bolts and self-locking nuts while holding the damper pinch bolt to 67 ft. lbs. (90 N m).
19. Install the stabilizer link to the damper and tighten to the specified torque.
20. Hold the damper shaft with a hex wrench, and tighten the damper mounting nut to 33 ft. lbs. (44 N m).
21. Install the cap to the top of the damper.
22. Driver's side installation: Install these items:
- Wiper motor
- Cowl cover
- Wiper arms

23. Passenger's side installation: Install the lid.
24. Clean the mating surface of the brake disc and the inside of the wheel, then install the front wheels.
25. Check the wheel alignment, and adjust it if necessary.

OVERHAUL

Disassembly

See Figures 188 and 189.

1. Compress the damper spring, then remove the self-locking nut while holding the damper shaft with a hex wrench. Do not compress the damper spring more than is necessary to remove the self-locking nut.
2. Release the pressure from the strut spring compressor, then disassemble the damper as shown in the Exploded View.

Inspection

1. Compress the damper assembly by hand, and check for smooth operation through a full stroke, both compression and extension. The damper should extend smoothly and constantly when compression is released. If it does not, the gas is leaking and the damper should be replaced.
2. Check for oil leaks, abnormal noises, and binding during these tests.

Reassembly

See Figures 190 through 193.

1. Install the upper spring seat to the upper spring seat clip.

➡Make sure to securely set the upper spring seat to the hook on the spring seat clip.

2. Install the bump stop and the dust cover to the upper spring seat.

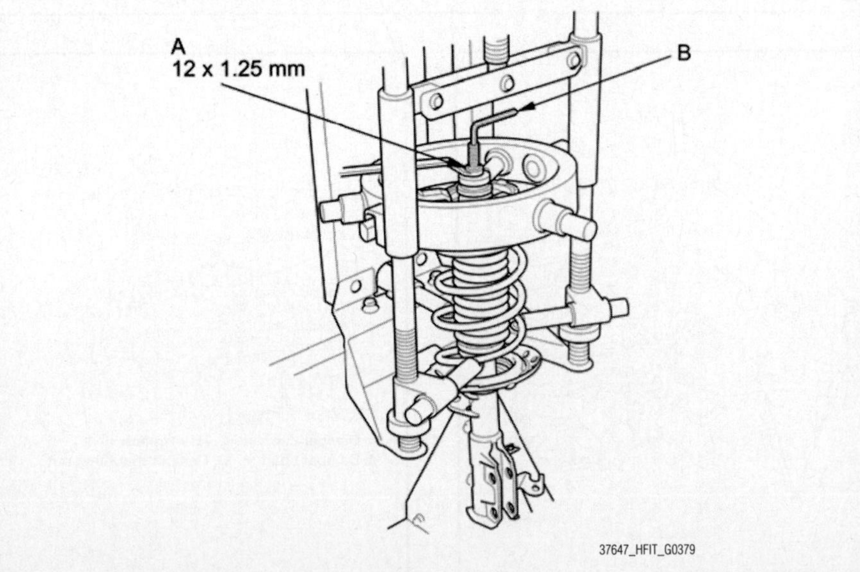
A
12 x 1.25 mm
B
37647_HFIT_G0379

Fig. 188 Compress the damper spring, then remove the self-locking nut (A) while holding the damper shaft with a hex wrench (B)

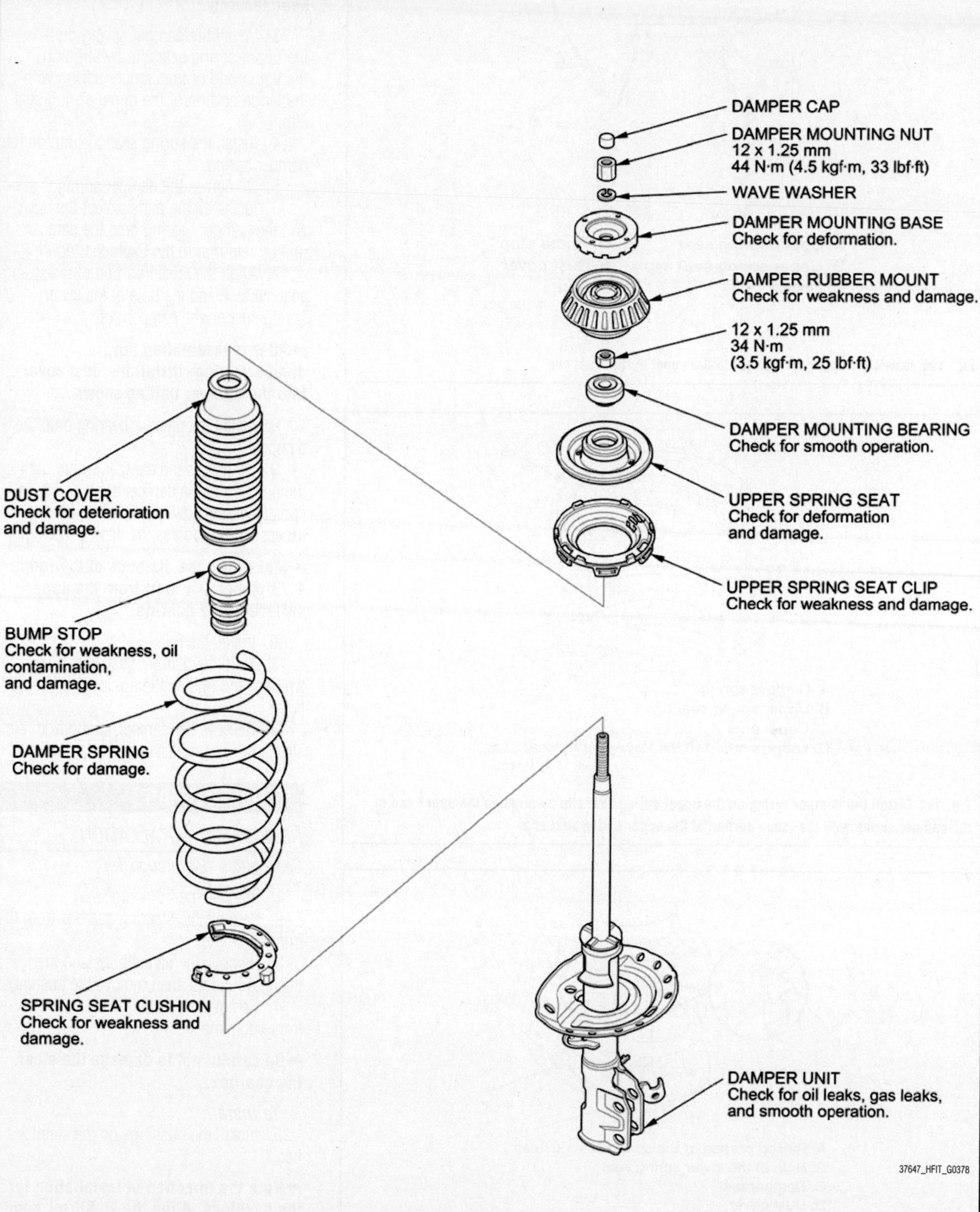
DAMPER CAP
DAMPER MOUNTING NUT
12 x 1.25 mm
44 N·m (4.5 kgf·m, 33 lbf·ft)
WAVE WASHER
DAMPER MOUNTING BASE
Check for deformation.
DAMPER RUBBER MOUNT
Check for weakness and damage.
12 x 1.25 mm
34 N·m
(3.5 kgf·m, 25 lbf·ft)
DAMPER MOUNTING BEARING
Check for smooth operation.
UPPER SPRING SEAT
Check for deformation and damage.
UPPER SPRING SEAT CLIP
Check for weakness and damage.
DUST COVER
Check for deterioration and damage.
BUMP STOP
Check for weakness, oil contamination, and damage.
DAMPER SPRING
Check for damage.
SPRING SEAT CUSHION
Check for weakness and damage.
DAMPER UNIT
Check for oil leaks, gas leaks, and smooth operation.
37647_HFIT_G0378

Fig. 189 Exploded view of damper/spring assembly

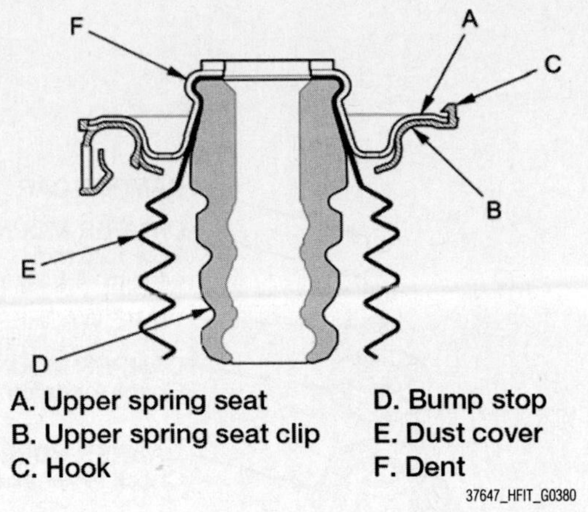
F
A
C
B
E
D
A. Upper spring seat
B. Upper spring seat clip
C. Hook
D. Bump stop
E. Dust cover
F. Dent
37647_HFIT_G0380

Fig. 190 Install the upper spring seat to the upper spring seat clip

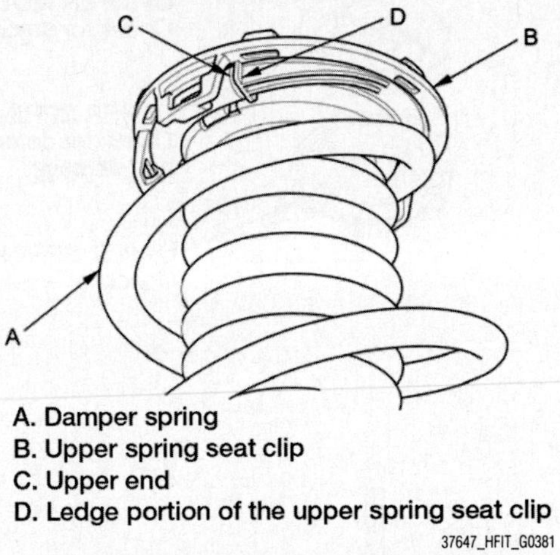
C
D
B
A
A. Damper spring
B. Upper spring seat clip
C. Upper end
D. Ledge portion of the upper spring seat clip
37647_HFIT_G0381

Fig. 191 Install the damper spring on the upper spring seat clip by aligning the upper end of the damper spring with the ledge portion of the upper spring seat clip

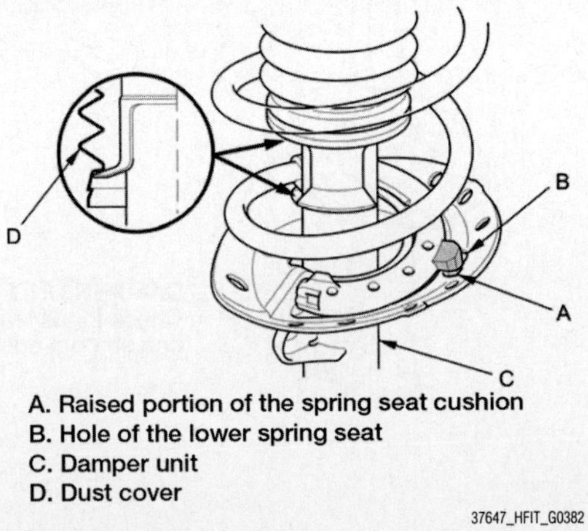
B
D
A
C
A. Raised portion of the spring seat cushion
B. Hole of the lower spring seat
C. Damper unit
D. Dust cover
37647_HFIT_G0382

Fig. 192 Align the raised portion of the spring seat cushion and the hole of the lower spring seat on the damper unit

➡Push the bump stop and the dust cover into the dent of the upper spring seat securely.

3. Install the damper spring on the upper spring seat clip by aligning the upper end of the damper spring with the ledge portion of the upper spring seat clip.
4. Install the spring seat cushion on the damper spring.
5. Compress the damper spring.
6. Install all the parts except the nut and the damper bearing onto the damper unit by referring to the Exploded View.
7. Align the raised portion of the spring seat cushion and the hole of the lower spring seat on the damper unit.

➡After reassembling the damper/spring, install the dust cover into the damper unit as shown.

8. Install the damper bearing onto the damper shaft.
9. Compress the damper spring until the position of the damper shaft comes in contact with the damper bearing. Do not excessively compress the damper spring.

➡Make sure the distance of the rod is 1.73 inches (44 mm) from the upper surface of the bearing.

10. Install the nut.
11. Hold the damper shaft with a hex wrench, and tighten the nut to 25 ft. lbs. (34 N m).
12. Remove the damper/spring from the strut spring compressor.

STABILIZER BAR

REMOVAL & INSTALLATION

See Figures 194 through 196.

1. Remove the front subframe.
2. Remove the steering gearbox from the front subframe.
3. Remove the flange bolts and the bushing holders, then remove the bushings.
4. Remove the stabilizer bar from the front subframe.

➡Be careful not to damage the steering gearbox.

To install:

5. Install the bushings on the stabilizer bar.

➡Note the direction of installation for the bushings. Align the stabilizer band clamp with the side of the bushings.

6. Install the stabilizer bar with the bushings on the front subframe.

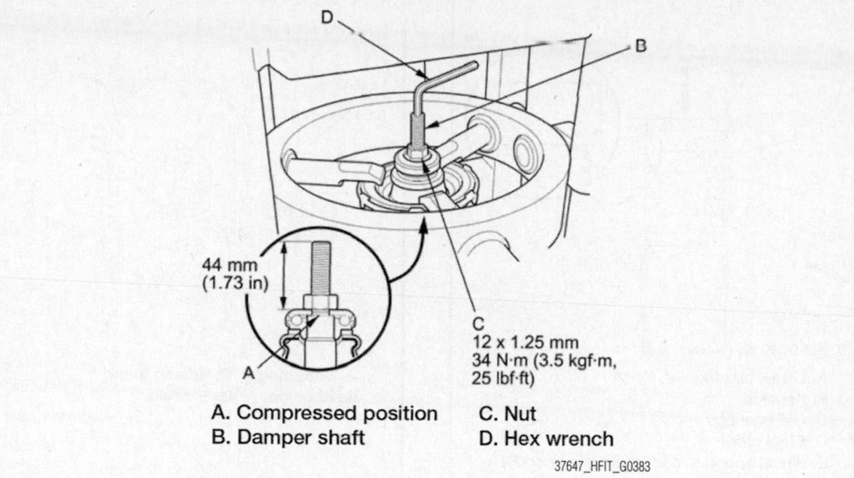
D
B
44 mm
(1.73 in)
A
C
12 x 1.25 mm
34 N·m (3.5 kgf·m,
25 lbf·ft)
A. Compressed position
B. Damper shaft
C. Nut
D. Hex wrench
37647_HFIT_G0383

Fig. 193 Compress the damper spring until the position of the damper shaft comes in contact with the damper bearing

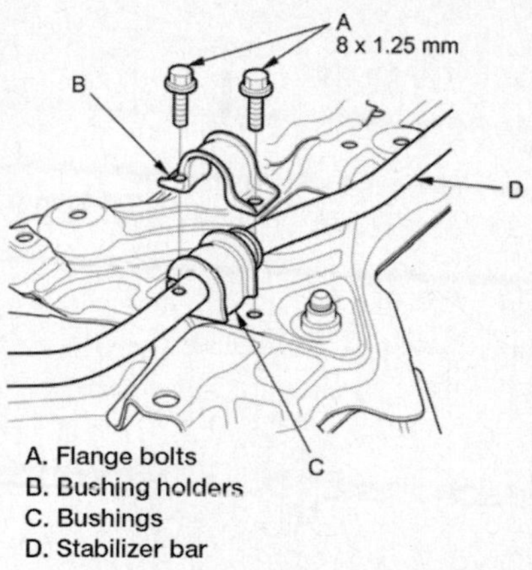
A
8 x 1.25 mm
B
D
C
A. Flange bolts
B. Bushing holders
C. Bushings
D. Stabilizer bar
37647_HFIT_G0384

Fig. 194 Remove the flange bolts and the bushing holders, then remove the bushings

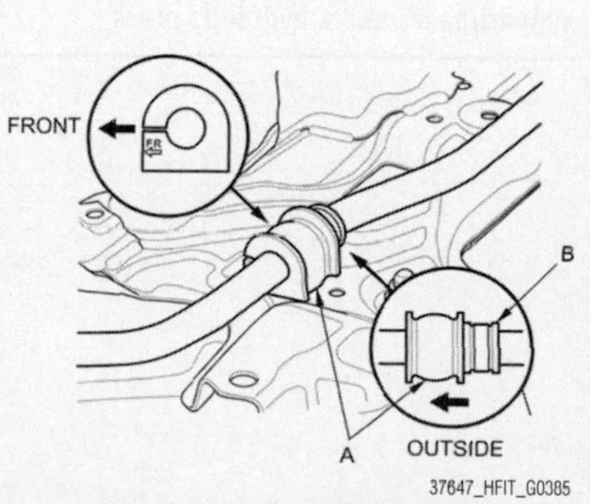
FRONT
B
A
OUTSIDE
37647_HFIT_G0385

Fig 195 Install the bushings (A) on the stabilizer bar and align the stabilizer band clamp (B) with the side of the bushings

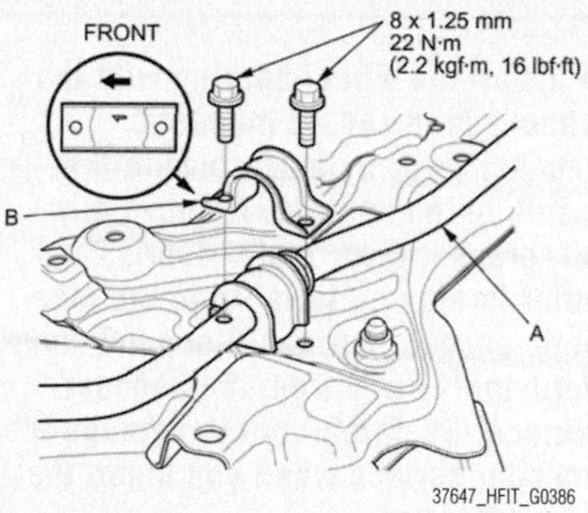
FRONT
8 x 1.25 mm
22 N·m
(2.2 kgf·m, 16 lbf·ft)
B
A
37647_HFIT_G0386

Fig. 196 Install the stabilizer bar (A) with the bushings on the front subframe

➡Note the right and left direction of the stabilizer bar. Note the direction of installation for the bushing holders.

7. Install the steering gearbox on the front subframe.
8. Install the front subframe.

WHEEL HUB & BEARING

REMOVAL & INSTALLATION

See Figures 197 through 202.

Special Tools Required:

- Ball Joint Remover, 28 mm 07MAC-SL0A202
- Hub Dis/Assembly Pin, 40 mm 07GAF-SE00100
- Driver Handle, 15 x 135L 07749-0010000
- Bearing Driver Attachment, 52 x 55 mm 07746-0010400
- Support Base 07965-SD90100

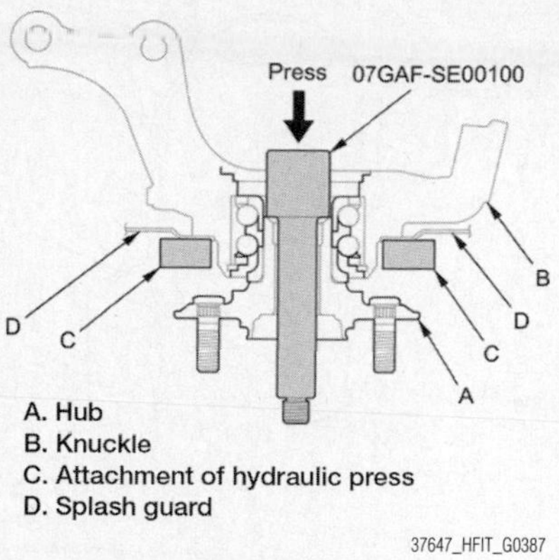
Press
07GAF-SE00100
B
D
C
A
D
C
A. Hub
B. Knuckle
C. Attachment of hydraulic press
D. Splash guard
37647_HFIT_G0387

Fig. 197 Separate the hub from the knuckle using the hub dis/assembly tool and a hydraulic press

- Ball Joint Thread Protector, 10 mm 07AAF-SECA120
- Ball Joint Thread Protector, 14 mm 071AF-S3VA000

1. Remove the steering knuckle.
2. Separate the hub from the knuckle using the hub dis/assembly tool and a hydraulic press. Hold the knuckle with the attachment of the hydraulic press or equivalent tool. Be careful not to damage or deform the splash guard. Hold onto the hub to keep it from falling when pressed clear.
3. Press the wheel bearing inner race off of the hub using the hub dis/assembly tool, a commercially available bearing separator, and a press.
4. Remove the splash guard and the snap ring from the knuckle.

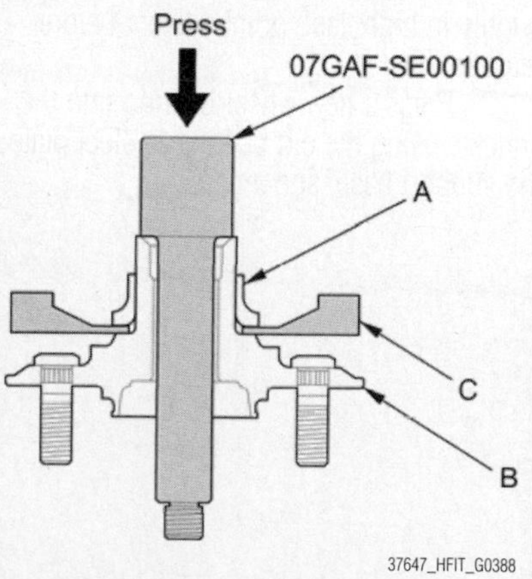
Press
07GAF-SE00100
A
C
B
37647_HFIT_G0388

Fig. 198 Press the wheel bearing inner race (A) off of the hub (B) using the hub dis/assembly tool, a commercially available bearing separator (C), and a press

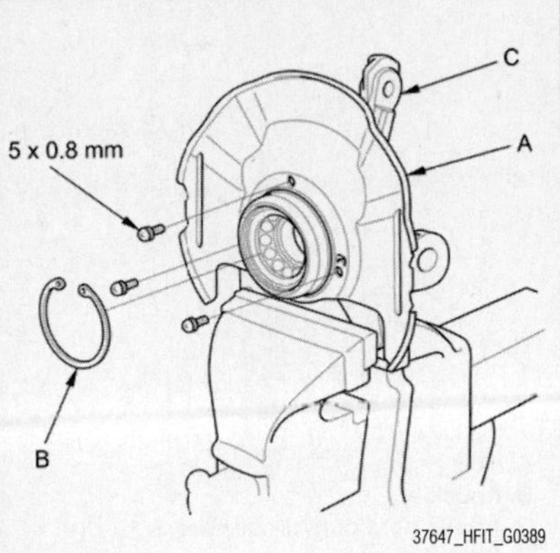
C
A
5 x 0.8 mm
B
37647_HFIT_G0389

Fig. 199 Remove the splash guard (A) and the snap ring (B) from the knuckle (C)

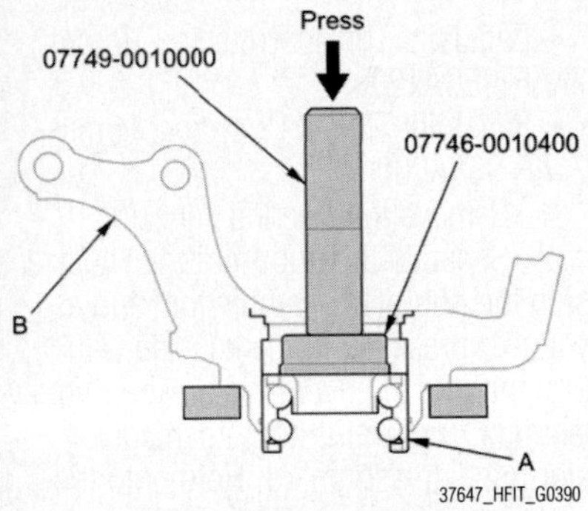
Press
07749-0010000
07746-0010400
B
A
37647_HFIT_G0390

Fig. 200 Press the wheel bearing (A) out of the knuckle (B) using the bearing driver attachment, the driver handle, and a press

5. Press the wheel bearing out of the knuckle using the bearing driver attachment, the driver handle, and a press.

6. Wash the knuckle and the hub thoroughly in high flash point solvent before reassembly.

7. Press a new wheel bearing into the knuckle using the old bearing, a steel plate, the support base, and a press.

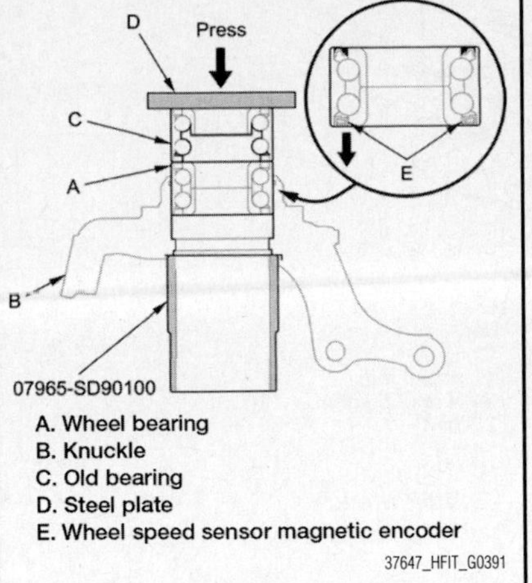
D
Press
C
A
E
B
07965-SD90100
A. Wheel bearing
B. Knuckle
C. Old bearing
D. Steel plate
E. Wheel speed sensor magnetic encoder
37647_HFIT_G0391

Fig. 201 Press a new wheel bearing into the knuckle using the old bearing, a steel plate, the support base, and a press

➡Install the wheel bearing with the wheel speed sensor magnetic encoder (brown color), toward the inside of the knuckle. Remove any oil, grease, dust, metal debris, and other foreign material from the magnetic encoder surface. Keep any magnetic tools away from the encoder surface. Be careful not to damage the encoder surface when you insert the wheel bearing.

8. Install the snap ring securely in the knuckle.

9. Install the splash guard, and tighten the screws to the specified torque value.

10. Install the hub onto the knuckle using the bearing driver attachment, the driver handle, the support base, and a hydraulic press. Be careful not to damage the splash guard.

11. Install the steering knuckle.

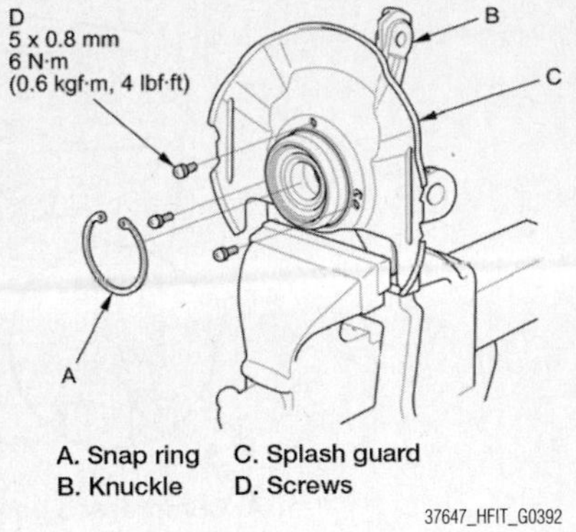
D
5 x 0.8 mm
6 N·m
(0.6 kgf·m, 4 lbf·ft)
B
C
A
A. Snap ring
B. Knuckle
C. Splash guard
D. Screws
37647_HFIT_G0392

Fig. 202 Install the snap ring securely in the knuckle

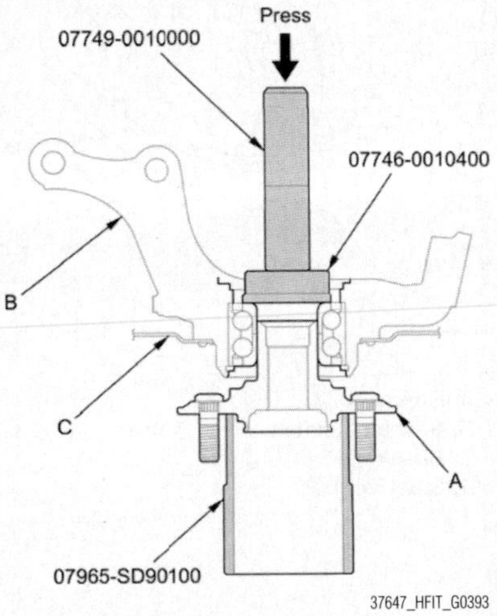
Press
07749-0010000
07746-0010400
B
C
A
07965-SD90100
37647_HFIT_G0393

Fig. 203 Install the hub (A) onto the knuckle (B) using the bearing driver attachment, the driver handle, the support base, and a hydraulic press

COIL SPRING

REMOVAL & INSTALLATION

See Figures 203 through 209.

1. Raise and support the vehicle.
2. Remove the rear wheel.
3. Remove the wheel speed sensor and the wire guide grommet from both sides of the axle beam. Do not disconnect the wheel speed sensor connector.
4. Position the floor jack under spring seat on both sides of the axle beam. Raise the floor jack until the suspension begins to compress.
5. Remove the damper mounting bolt from both sides.
6. Lower the floor jack gradually.
7. Remove the spring, the upper rubber mount, and the lower rubber mount. Do not lower the jack more than necessary.

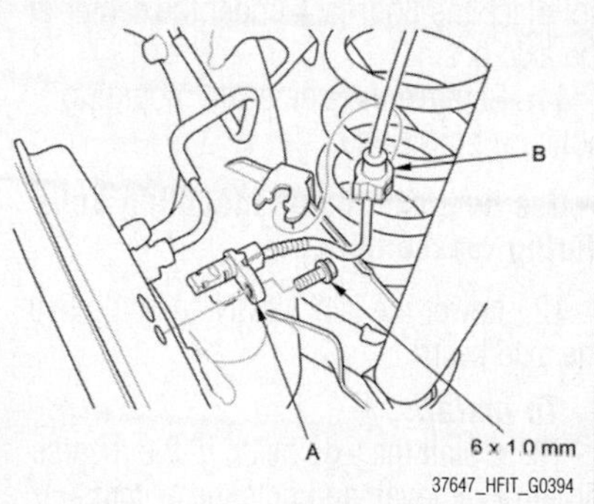
B
A
6 x 1.0 mm
37647_HFIT_G0394

Fig. 204 Remove the wheel speed sensor (A) and the wire guide grommet (B) from both sides of the axle beam

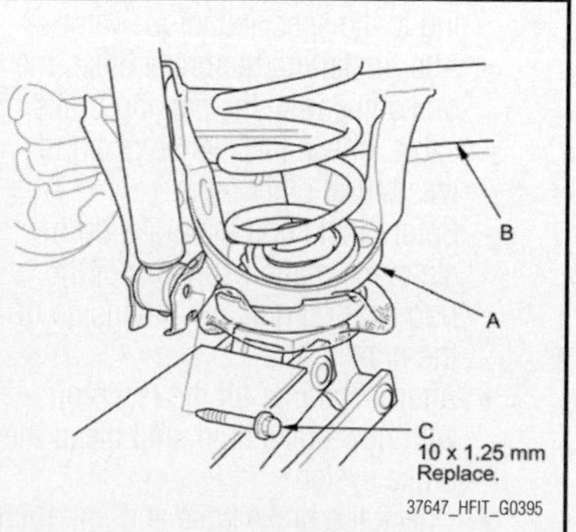
B
A
C
10 x 1.25 mm
Replace.
37647_HFIT_G0395

Fig. 205 Position the floor jack under spring seat (A) on both sides of the axle beam (B); remove the damper mounting bolt (C) from both sides

To install:

8. Install the upper rubber mount on the spring by aligning the upper end of the spring with the stop on the upper rubber mount.
9. Install the lower rubber mount on the spring by aligning the lower end of the spring with the stop on the lower rubber mount.
10. Install the tab of the lower rubber mount into the groove of the lower spring seat.

➡Make sure that the tab of the lower rubber mount is properly installed into the axle beam. Make sure that the spring is installed correctly.

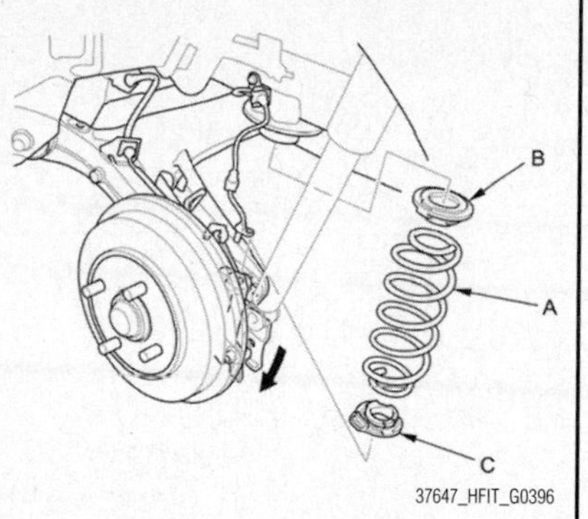
B
A
C
37647_HFIT_G0396

Fig. 206 Remove the spring (A), the upper rubber mount (B), and the lower rubber mount (C)

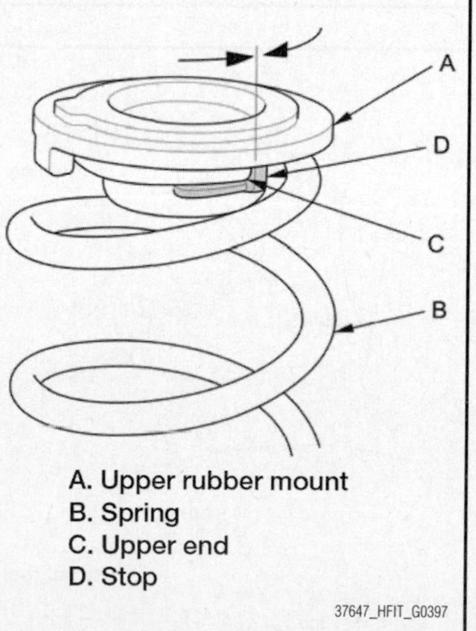
A
D
C
B
A. Upper rubber mount
B. Spring
C. Upper end
D. Stop
37647_HFIT_G0397

Fig. 207 Install the upper rubber mount on the spring by aligning the upper end of the spring with the stop on the upper rubber mount

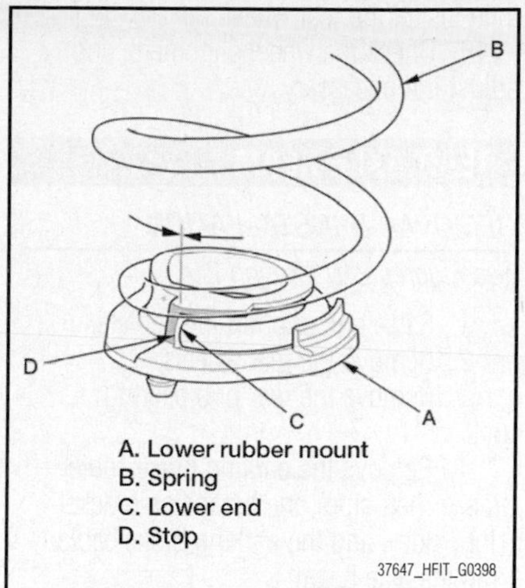
B
D
C
A
A. Lower rubber mount
B. Spring
C. Lower end
D. Stop
37647_HFIT_G0398

Fig. 208 Install the lower rubber mount on the spring by aligning the lower end of the spring with the stop on the lower rubber mount

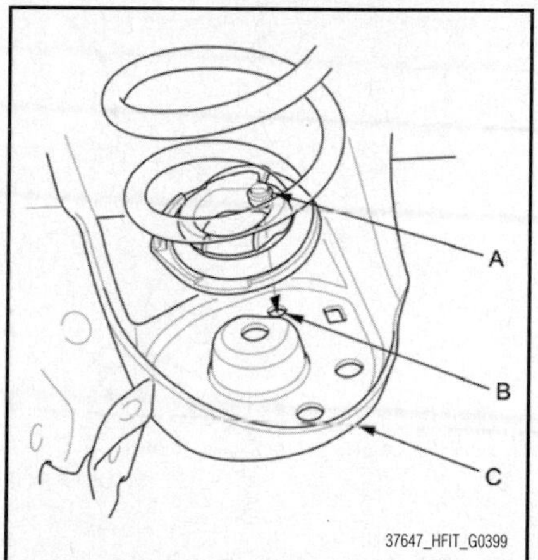
A
B
C
37647_HFIT_G0399

Fig. 209 Install the tab (A) of the lower rubber mount into the groove (B) of the lower spring seat

11. Position a floor jack under the lower spring seat on both sides of the axle beam.
12. Slowly raise the floor jacks until you can align the bolt hole with the holes in the axle beam and the damper, then loosely tighten the new damper mounting bolt on both sides.
13. Raise the rear suspension with the floor jacks until the vehicle just lifts off of the lift, then tighten the damper mounting bolts to the specified torque.
14. Install the wheel speed sensor and the wire guide grommet on both sides of the axle beam.

15. Clean the mating surfaces of the brake drum and the inside of the wheel, then install the rear wheel.

16. Check the wheel alignment, and adjust it if necessary.

REAR AXLE BEAM

REMOVAL & INSTALLATION

See Figures 210 through 214.

1. Raise and support the vehicle.
2. Remove the rear wheels.
3. Remove the rear hub bearing unit.
4. Remove the parking brake cable bracket (left side), the brake line bracket (right side), and the parking brake cable from the axle beam.

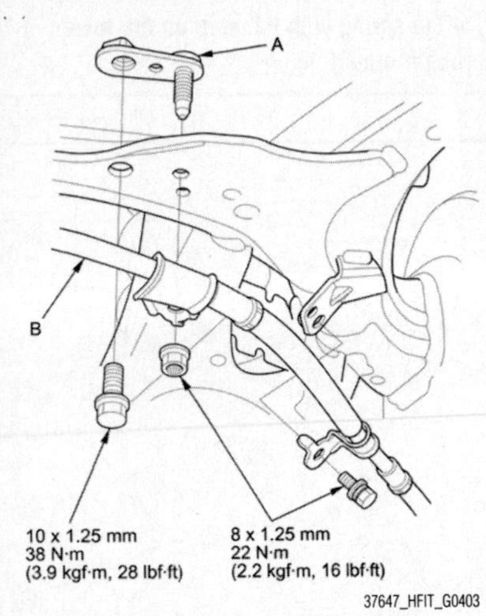
A
B
10 x 1.25 mm
38 N·m
(3.9 kgf·m, 28 lbf·ft)
8 x 1.25 mm
22 N·m
(2.2 kgf·m, 16 lbf·ft)
37647_HFIT_G0403

Fig. 210 Remove the parking brake cable bracket (left side) (A), the brake line bracket (right side), and the parking brake cable (B) from the axle beam

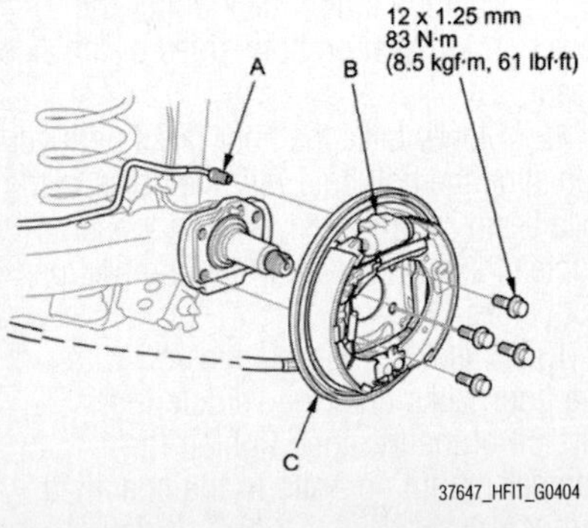
12 x 1.25 mm
83 N·m
(8.5 kgf·m, 61 lbf·ft)
A
B
C
37647_HFIT_G0404

Fig. 211 Disconnect the brake line (A) from the wheel cylinder (B); remove the backing plate (C)

5. Disconnect the brake line from the wheel cylinder, and plug the line with a shop towel.
6. Remove the backing plate with the brake shoes assembly from the spindle.
7. Remove the wheel speed sensor and the wire guide grommet from the axle beam. Do not disconnect the wheel speed sensor connector.
8. Remove the rear spring.
9. Disconnect the brake hose from both sides of the brake line, then remove the brake line by removing the brake hose clip.

➡Do not spill brake fluid on the vehicle; it may damage the paint; if brake fluid gets on the paint, wash it off immediately with water. Plug the end of a hose and joints to prevent spilling brake fluid. During installation, install new brake hose clips.

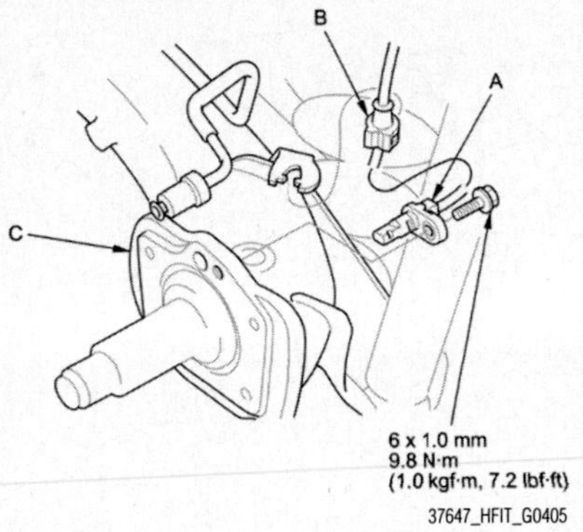
B
A
C
6 x 1.0 mm
9.8 N·m
(1.0 kgf·m, 7.2 lbf·ft)
37647_HFIT_G0405

Fig. 212 Remove the wheel speed sensor (A) and the wire guide grommet (B) from the axle beam (C)

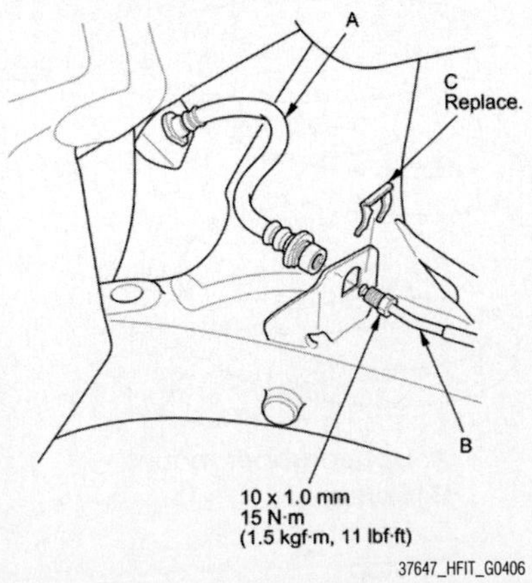
A
C
Replace.
B
10 x 1.0 mm
15 N·m
(1.5 kgf·m, 11 lbf·ft)
37647_HFIT_G0406

Fig. 213 Disconnect the brake hose (A) from both sides of the brake line (B), then remove the brake line by removing the brake hose clip (C)

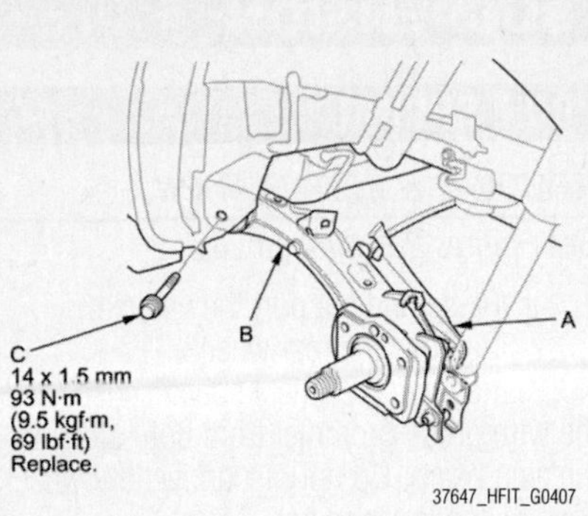
A
B
C
14 x 1.5 mm
93 N·m
(9.5 kgf·m,
69 lbf·ft)
Replace.
37647_HFIT_G0407

Fig. 214 Place a floor jack under the lower spring seat (A) on both sides of the axle beam (B); remove the axle beam mounting bolts (C) on both sides

10. Place a floor jack under the lower spring seat on both sides of the axle beam, and support it by raising the floor jack. Do not place the floor jack under the center of the axle beam.
11. Remove the axle beam mounting bolts on both sides.

➡Use new axle beam mounting bolts during reassembly.

12. Lower the jack slowly, then remove the axle beam.

To install:

13. Install the axle beam in the reverse order of removal, and note these items:
    - First install all the components, and lightly tighten the bolts, and place a jack under the lower spring seat of the axle beam on both sides, then raise the suspension to load with the vehicle's weight before fully tightening to the specified torque values.
    - After installing the brake hose, the brake line, and the parking brake cable, check for interference and twisting of other parts.
    - Before installing the brake drum, clean the mating surfaces of the hub bearing unit and the inside of the brake drum.
    - After installing, fill the reservoir with new brake fluid, and bleed the brake system.
    - Check the brake hose and line joint for leaks, and tighten if necessary.
    - Before installing the wheel, clean the mating surfaces of the brake drum and the inside of the wheel.
14. Check the wheel alignment, and adjust it if necessary.

SHOCK ABSORBER (DAMPER)

REMOVAL & INSTALLATION

See Figures 215 through 218.

1. Raise and support the vehicle.
2. Remove the rear wheel.
3. Remove the wire guide grommet (A) from both sides of the axle beam.
4. Position the floor jack under lower spring seat on both sides of the axle beam. Raise the floor jack until the suspension begins to compress.
5. Remove the damper mounting bolt from both sides.
6. Remove the access panel from the cargo area side trim.
7. Remove the self-locking nut while holding the damper shaft with a hex wrench.
8. Remove the damper mount washer and the damper rubber mount from the top of the damper.

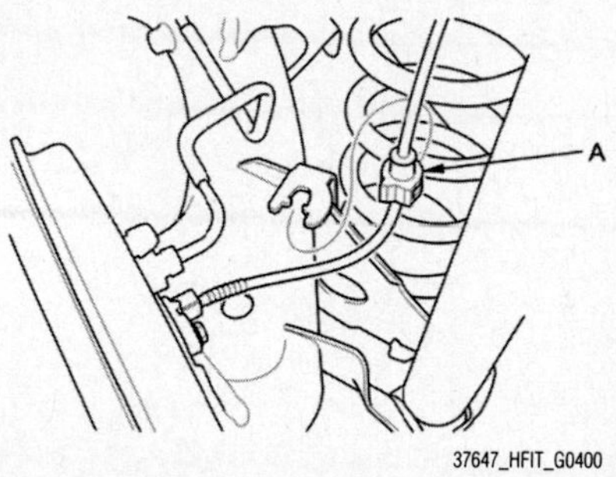
A
37647_HFIT_G0400

Fig. 215 Remove the wire guide grommet (A) from both sides of the axle beam

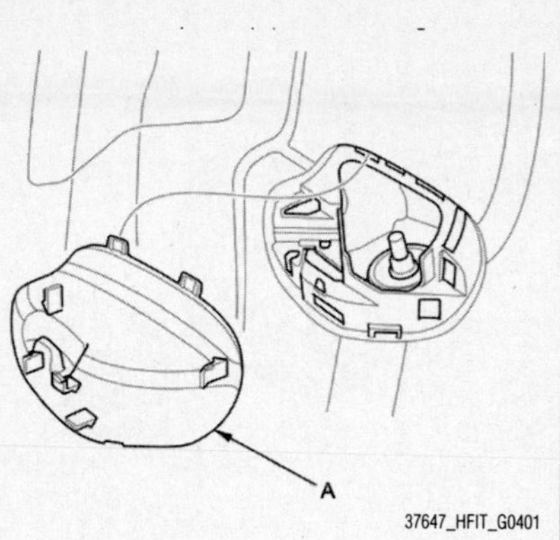
A
37647_HFIT_G0401

Fig. 217 Remove the access panel (A) from the cargo area side trim

9. Compress the damper unit (A) by hand, and remove it from the vehicle.

➡Be careful not to damage the body0>Remove the damper rubber mount (B).

To install:

10. Install the damper rubber mount onto the damper unit. Position the damper assembly between the body and the axle beam.

➡Be careful not to damage the body.

11. Position the floor jack under lower spring seat on both sides of the axle beam.
12. Slowly raise the jack until you can align the bolt hole with the holes in the axle beam and the damper, then loosely tighten the new damper mounting bolt on both sides.
13. Raise the rear suspension with the jack until the vehicle just lifts off of the lift, then tighten the damper mounting bolts to the specified torque.
14. Install the damper rubber mount, the damper mounting washer, and the new self-locking nut on the damper shaft.

➡During installation, note the direction of the damper rubber mount and the damper mounting washer.

15. Tighten the self-locking nut to 22 ft. lbs. (29 N m) while holding the damper shaft with a hex wrench.
16. Install the access panel on the cargo area side trim.
17. Install the wire guide grommet on both sides of the axle beam.
18. Clean the mating surfaces of the brake drum and the inside of the wheel, then install the rear wheel.

TESTING

1. Install the flange nut on the damper shaft end, and set the socket wrench and T-handle on the nut.
2. Compress the damper assembly by hand, and check for smooth operation through a full stroke, both compression and extension. The damper should extend smoothly and constantly when compression is released. If it does not, the gas is leaking and the damper should be replaced.
3. Check for oil leaks, abnormal noises, and binding during these tests.

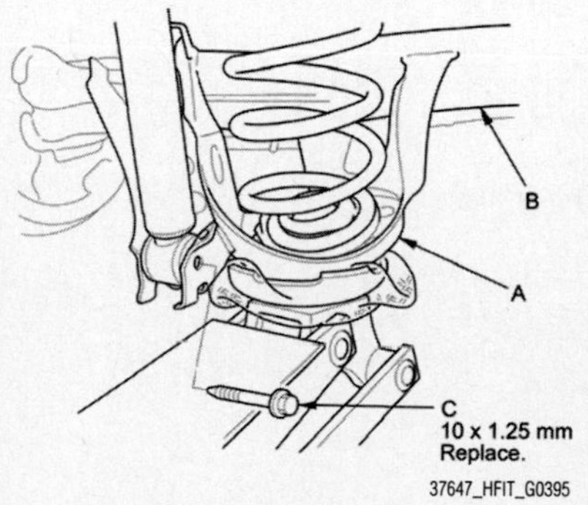
B
A
C
10 x 1.25 mm
Replace.
37647_HFIT_G0395

Fig. 216 Position the floor jack under spring seat (A) on both sides of the axle beam (B); remove the damper mounting bolt (C) from both sides

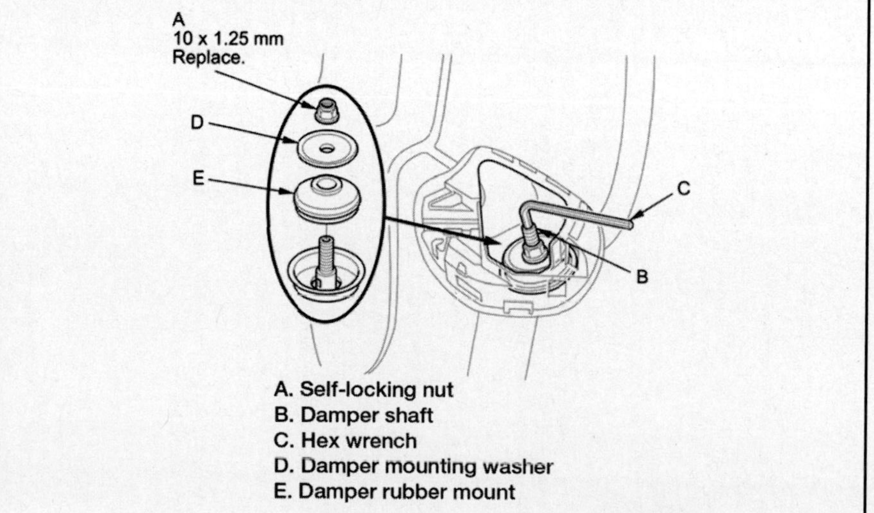
A
10 x 1.25 mm
Replace.
D
E
C
B
A. Self-locking nut
B. Damper shaft
C. Hex wrench
D. Damper mounting washer
E. Damper rubber mount
37647_HFIT_G0402

Fig. 218 Remove the self-locking nut while holding the damper shaft with a hex wrench

WHEEL HUB & BEARING

REMOVAL & INSTALLATION

See Figure 219.

1. Raise and support the vehicle.
2. Remove the wheel nuts and the rear wheel.
3. Remove the brake drum.
4. Remove the hub cap. Raise the stake, then remove the spindle nut.
5. Remove the hub bearing unit from the spindle.
6. Check the hub bearing unit for damage and cracks.
7. Install the hub bearing unit in the reverse order of removal, and note these items:

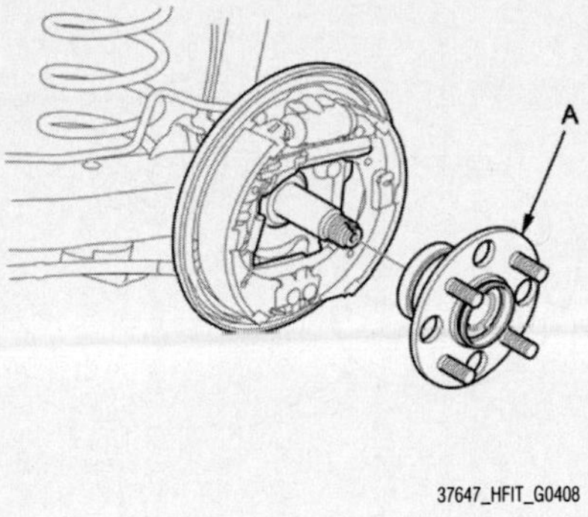
A
37647_HFIT_G0408
Fig. 219 Remove the hub bearing unit (A) from the spindle

- Tighten all mounting hardware to the specified torque values.
- Use a new spindle nut and hub cap on reassembly.
- Before installing the spindle nut, apply a small amount of engine oil to the seating surface of the nut. After tightening, use a drift to stake the spindle nut shoulder against the spindle.
- Before installing the brake drum, clean the mating surface of the hub bearing unit and the inside of the brake drum.
- Before installing the wheel, clean the mating surface of the brake drum and the inside of the wheel.

HONDA

Insight Hybrid

14

BRAKES........................14-9

ANTI-LOCK BRAKE SYSTEM (ABS) 14-10
Wheel Speed Sensor Rings.....14-11
Removal & Installation........14-11
Wheel Speed Sensors 14-10
Removal & Installation........14-10
BLEEDING THE BRAKE SYSTEM........................14-9
Bleeding Procedure...................14-9
Bleeding Procedure 14-9
Bleeding the ABS System 14-9
FRONT DISC BRAKES........14-12
Brake Caliper..........................14-12
Removal & Installation........14-12
Disc Brake Pads.......................14-12
Removal & Installation........14-12
INFORMATION AND PRECAUTIONS 14-9
Anti-lock Systems................14-9
Disc and Drum Systems 14-9
PARKING BRAKE.............14-16
Parking Brake Cables 14 16
Adjustment 14-16
Parking Brake Shoes 14-16
Removal & Installation........14-16
REAR DRUM BRAKES........14-13
Brake Drum 14-13
Removal & Installation........14-13
Brake Shoes 14-13
Removal & Installation........14-13

CHASSIS ELECTRICAL 14-16

AIR BAG (SUPPLEMENTAL RESTRAINT SYSTEM) 14-16
General Information.................14-16
Precautions.........................14-16

DRIVE TRAIN 14-17

Front Halfshaft........................14-17
CV-boots Inspection..........14-18
Removal & Installation........14-17

ENGINE COOLING..........14-19

Engine Fan 14-19
Removal & Installation........14-19
Radiator.................................14-20
Removal & Installation........14-20
Thermostat 14-21
Removal & Installation........14-21
Water Pump 14-21
Removal & Installation........14-21

ENGINE ELECTRICAL 14-22

CHARGING SYSTEM 14-22
Alternator 14-22
HYBRID INTEGRATED MOTOR ASSISTED (IMA) SYTEM.......................14-22
DC-DC Converter..................14-24
Removal & Installation........14-24
IMA Motor Drain Cover...........14-24
Removal & Installation........14-24
IMA Motor Housing 14-24
Removal & Installation........14-24
IMA Motor Power Cable..........14-25
Removal & Installation........14-25
IMA Motor Rotor....................14-27
IMA Rotor Position Calibration.......................14-29
Removal & Installation........14-27
IMA Motor Rotor Position Sensor 14-27
IMA Rotor Position Calibration.......................14-27
Removal & Installation........14-27
IMA Service Precautions.........14-22
Disconnecting the Motor Power Cable Connector From the Motor Stator......14-23
Turning Off and on Power to the High Voltage Circuit ...14-23
IMA System Description..........14-22
Operating Conditions..........14-22
IPU Case 14-29
Removal & Installation........14-29
IPU Cover...............................14-30
Removal & Installation........14-30
IPU Module Air Duct..............14-30
Removal & Installation........14-30
IPU Module Fan 14-31
Removal & Installation........14-31
Motor Control Module (MCM) 14-31
IMA Rotor Position Calibration.......................14-31
Motor Control Module (MCM) Update 14-31
Removal & Installation........14-31
Motor Power Inverter (MPI) Module................................14-32
IMA Rotor Position Calibration.......................14-33
Removal & Installation........14-32
IGNITION SYSTEM 14-33
Firing Order............................14-33
Ignition Coil 14-33
Removal & Installation........14-33
Ignition Timing.......................14-33
Adjustment 14-33
Inspection 14-33
Spark Plugs............................14-33
Removal & Installation........14-33
STARTING SYSTEM 14-34
Starter 14-34
Removal & Installation........14-34

ENGINE MECHANICAL......14-34

Accessory Drive Belts 14-34
Accessory Belt Routing.......14-34
Adjustment 14-34
Inspection 14-34
Removal & Installation........14-34
Cam Chain Case Oil Seal........14-38
Removal & Installation........14-38
Camshaft and Valve Lifters......14-34
Inspection 14-34
Removal & Installation........14-35
Catalytic Converter.................14-37
Removal & Installation........14-37
Crankshaft Pulley 14-37
Removal & Installation........14-37
Cylinder Head 14-38
CKP Pattern Clear/CKP Pattern Learn Procedure...14-40
Removal & Installation........14-38
Drive Plate.............................14-40
Removal & Installation........14-40
Intake Manifold 14-40
Removal & Installation........14-40
Oil Pan 14-42
Removal & Installation........14-42

Oil Pump 14-43
Inspection 14-44
Removal & Installation 14-43
Piston and Ring 14-45
Positioning 14-45
Rocker Arms 14-45
Removal & Installation 14-45
Timing (Cam) Chain & Sprockets 14-46
Removal & Installation 14-46
Valve Lash 14-49
Adjustment 14-49

ENGINE PERFORMANCE & EMISSION CONTROLS...14-51

Accelerator Pedal Position (APP) Sensor 14-51
Location 14-51
Removal & Installation 14-51
Camshaft Position (CMP) Sensor 14-51
Location 14-51
Removal & Installation 14-51
Crankshaft Position (CKP) Sensor 14-51
CKP Pattern Clear/CKP Pattern Learn Procedure ... 14-51
Location 14-51
Removal & Installation 14-51
Engine Coolant Temperature (ECT) Sensor 14-52
Location 14-52
Removal & Installation 14-52
Evaporative Emissions (EVAP) Canister 14-52
Location 14-52
Removal & Installation 14-52
Exhaust Gas Recirculation (EGR) Valve 14-53
Location 14-53
Removal & Installation 14-53
Heated Oxygen (HO2S) Sensor 14-53
Location 14-53
Removal & Installation 14-53
Intake Air Temperature (IAT) Sensor 14-53
Location 14-53
Removal & Installation 14-53
Knock Sensor (KS) 14-53
Location 14-53
Removal & Installation 14-53
Manifold Absolute Pressure (MAP) Sensor 14-54
Location 14-54
Removal & Installation 14-54
Mass Air Flow (MAF) Sensor 14-53
Location 14-53
Removal & Installation 14-53
Positive Crankcase Ventilation (PCV) Valve 14-54
Location 14-54
Removal & Installation 14-54
Powertrain Control Module (PCM) 14-55
Location 14-55
PCM Update 14-56
Removal & Installation 14-55
Throttle Position (TP) Sensor 14-56
Location 14-56
Removal & Installation 14-56
Vehicle Speed Sensor (VSS) 14-57
Removal & Installation 14-57

FUEL 14-58

GASOLINE FUEL INJECTION SYSTEM 14-58
Fuel Filter 14-59
Removal & Installation 14-59
Fuel Level Sensor/Fuel Gauge Sending Unit 14-60
Removal & Installation 14-60
Fuel Rail and Injector 14-60
Removal & Installation 14-60
Fuel System Pressure 14-58
Relieving 14-58
Fuel System Service Precautions 14-58
Fuel Tank 14-62
Draining 14-62
Removal & Installation 14-62
Fuel Tank Unit 14-61
Removal & Installation 14-61
Idle Speed 14-63
Adjustment 14-63
Throttle Body 14-63
Removal & Installation 14-63

HEATING & AIR CONDITIONING SYSTEM...14-65

Blower Motor 14-65
Removal & Installation 14-65
Heater Core 14-65
Removal and Installation 14-65
Heater Unit 14-66
Removal & Installation 14-66

SPECIFICATIONS AND MAINTENANCE CHARTS 14-3

Brake Specifications 14-7
Camshaft and Bearing Specifications Chart 14-5
Capacities 14-4
Crankshaft and Connecting Rod Specifications 14-5
Engine and Vehicle Identification 14-3
Gasoline Engine Tune-Up Specifications 14-3
Fluid Specifications 14-4
General Engine Specifications 14-3
Piston and Ring Specifications 14-5
Maintenance Minder Schedule 14-8
Tire, Wheel and Ball Joint Specifications 14-7
Torque Specifications 14-6
Valve Specifications 14-4
Wheel Alignment 14-6

STEERING 14-67

Power Rack & Pinion Steering Gear 14-67
Removal & Installation 14-67
Torque Sensor Neutral Position Memorization 14-70

SUSPENSION 14-70

FRONT SUSPENSION 14-70
Control Links 14-70
Removal & Installation 14-70
Lower Ball Joint 14-70
Removal & Installation 14-70
Lower Control Arm 14-71
Removal & Installation 14-71
Stabilizer Bar 14-75
Removal & Installation 14-75
Steering Knuckle 14-71
Removal & Installation 14-71
Strut (Damper/Spring) 14-73
Overhaul 14-74
Removal & Installation 14-73
Wheel Hub & Bearing 14-76
Removal & Installation 14-76
REAR SUSPENSION 14-78
Axle Beam 14-78
Removal & Installation 14-78
Coil Spring 14-78
Removal & Installation 14-78
Shock Absorber (Damper) 14-79
Removal & Installation 14-79
Testing 14-79
Wheel Hub & Bearing 14-80
Adjustment 14-80
Removal & Installation 14-80

SPECIFICATIONS AND MAINTENANCE CHARTS

ENGINE AND VEHICLE IDENTIFICATION

Engine							Model Year	
Code ①	Liters (cc)	Cu. In.	Cyl.	Fuel Sys.	Engine Type	Eng. Mfg.	Code ②	Year
LDA3	1.3 (1.339)	82	4	MPFI	SOHC	Honda	A	2010

MPFI: Multi-Port Fuel Injection

DOHC: Double Overhead Camshaft

① Stamped into the front of the engine block and can be seen through the window next to the "H" logo on the front grill

② 10th digit of the Vehicle Identification Number (VIN)

37647_INHY_C0001

GENERAL ENGINE SPECIFICATIONS

All measurements are given in inches.

Year	Model	Engine Displacement Liters	Engine Series ID/VIN	Net Horsepower @ rpm	Net Torque @ rpm (ft. lbs.)	Bore x Stroke (in.)	Com-pression Ratio	Oil Pressure @ rpm
2010	Insight Hybrid	1.3	LDA3	①	②	NA	NA	NA

NA: Not Available

① Engine: 98@5800

Motor: 13@1500

② Engine: 123@1000

Motor: 58@1000

37647_INHY_C0002

GASOLINE ENGINE TUNE-UP SPECIFICATIONS

Year	Engine Displacement Liters	Engine Series ID/VIN	Spark Plug Gap (in.)	Ignition Timing (deg.)		Fuel Pump (psi)	Idle Speed (rpm)		Valve Clearance (in.)	
				MT	AT		MT	AT	Intake	Exhaust
2010	1.3	LDA3	0.039-0.043	—	8-12B	38-46	—	700-800	0.006-0.007	0.009-0.011

NOTE: The Vehicle Emission Control Information label reflects specification changes made during production.

Follow the figures on the label if they differ from those in this chart.

B: Before top dead center

37647_INHY_C0003

CAPACITIES

Year	Model	Engine Displacement Liters	Engine Series ID/VIN	Engine Oil with Filter (qts.)	Transmission (pts.)		Drive Axle		Fuel Tank (gal.)	Cooling System (qts.)
					Manual	Auto. ①	Front (pts.)	Rear (pts.) ①		
2010	Insight Hybrid	1.3	LDA3	3.2	—	3.0	②	NA	10.6	4.8

NOTE: All capacities are approximate. Add fluid gradually and check to be sure a proper fluid level is obtained.

NA: Not Available

① Drain and refill

② Included in transaxle refill figure

37647_INHY_C0004

FLUID SPECIFICATIONS

Year	Model	Engine Displacement Liters	Engine Series ID/VIN	Engine Oil	Auto. Trans.	Drive Axle	Transfer Case	Brake Master Cylinder	Cooling System
2010	Insight	1.3	LDA3	①	②	NA	NA	③	④

DOT: Department Of Transportation

NA: Not Available

① Honda Motor Oil: American Honda P/N 08798-9022 (0W-20), Honda Canada P/N 08798-8023C (0W-20)

② Honda CVTF: Honda P/N 08200-9006

③ Honda DOT 3 Brake Fluid: P/N 08798-9008

④ Honda Long Life Antifreeze/Coolant Type 2: P/N OL 999-9001

37647_INHY_C0005

VALVE SPECIFICATIONS

Year	Engine Displacement Liters	Engine Series ID/VIN	Seat Angle (deg.)	Face Angle (deg.)	Spring Test Pressure (lbs. @ in.)	Spring Installed Height (in.)	Stem-to-Guide Clearance (in.)		Stem Diameter (in.)	
							Intake	Exhaust	Intake	Exhaust
2010	1.3	LDA3	NA	NA	NA	①	0.0008-0.0020	0.0020-0.0031	0.2157-0.2161	0.2146-0.2150

NA: Information not available

① Valve spring free length:
Intake: 2.096-2.097 inches
Exhaust: 2.256-2.257 inches

37647_INHY_C0006

CAMSHAFT AND BEARING SPECIFICATIONS CHART

All measurements are given in inches.

Year	Engine Displ. Liters	Engine Series ID/VIN	Journal Dia.	Brg. Oil Clearance	Shaft End-play	Runout	Journal Bore	Lobe Height Intake	Lobe Height Exhaust
2010	1.3	LDA3	NA	0.0020-0.0035	0.002-0.006	0.001 max.	NA	①	②

NA: Not Available

① Intake PRI: 1.1693 inches

Intake SEC: 1.4116 inches

② Exhaust PRI: 1.1772 inches

Exhaust SEC: 1.3965 inches

37647_INHY_C0007

CRANKSHAFT AND CONNECTING ROD SPECIFICATIONS

All measurements are given in inches.

	Engine	Engine	Crankshaft				Connecting Rod		
Year	Disp. Liters	Series ID/VIN	Main Brg. Journal Dia.	Main Brg. Oil Clearance	Shaft End-play	Thrust on No.	Journal Diameter	Oil Clearance	Side Clearance
2010	1.3	LDA3	1.9676-1.9685	0.0007-0.0014	0.0040-0.0140	4	1.5739-1.5748	0.0002-0.0006	0.0060-0.0140

37647_INHY_C0008

PISTON AND RING SPECIFICATIONS

All measurements are given in inches.

	Engine	Engine		Ring Gap			Ring Side Clearance		
Year	Displ. Liters	Series ID/VIN	Piston Clearance	Top Compression	Bottom Compression	Oil Control	Top Compression	Bottom Compression	Oil Control
2010	1.3	LDA3	0.0008-0.0018	0.0060-0.0120	0.0120-0.0170	0.0080-0.0280	0.0026-0.0035	0.0012-0.0022	NA

37647_INHY_C0009

TORQUE SPECIFICATIONS

All readings in ft. lbs.

Year	Engine Disp. Liters	Engine Series ID/VIN	Cylinder Head Bolts	Main Bearing Bolts	Rod Bearing Bolts	Crankshaft Damper Bolts	Flywheel Bolts	Manifold Intake	Manifold Exhaust	Spark Plugs	Oil Pan Drain Plug
2010	1.3	LDA3	①	②	③	④	33	17	NA	13	⑤

NOTE: Dip cylinder head bolts, main bearing bolts, and crankshaft damper bolt in clean engine oil prior to tightening.

NA: Not Available

① Step 1: 22 ft. lbs.
Step 2: plus 130 degrees

② Tighten bearing cap bolts in sequence:
Step 1: 18 ft. lbs.
Step 2: plus 40 degrees

③ Step 1: Tighten to 7.2 ft. lbs.
Step 2: plus 90 degrees

④ Step 1: Tighten to 27 ft. lbs.
Step 2: plus 90 degrees

⑤ Tighten in sequence to 17 ft. lbs. in 3 steps

37647_INHY_C0010

WHEEL ALIGNMENT

Year	Model		Caster Range (+/-Deg.)	Caster Preferred Setting (Deg.)	Camber Range (+/-Deg.)	Camber Preferred Setting (Deg.)	Toe-in (Deg.)
2010	Insight Hybrid	F	1.00	+3.20	1.0	0	0+/-0.12
		R	—	—	1.0	-1.00	0.10+/-0.10

NOTE: Measurements are given for unladen vehicle: fuel, engine coolant, and fluid levels are full. Spare tire, jack, hand tools, and mats are in designated positior

37647_INHY_C0011

TIRE, WHEEL AND BALL JOINT SPECIFICATIONS

Year	Model	OEM Tires Standard	OEM Tires Optional	Tire Pressures (psi) Front	Tire Pressures (psi) Rear	Wheel Size	Lug Nut Torque (ft. lbs.)
2010	Insight Hybrid	SBRP175/65R15	—	33	33	15	80

OEM: Original Equipment Manufacturer

PSI: Pounds Per Square Inch

NA: Information not available

37647_INHY_C0012

BRAKE SPECIFICATIONS

All measurements in inches unless noted

Year	Model		Brake Disc Original Thickness	Brake Disc Minimum Thickness	Brake Disc Maximum Runout	Brake Drum Diameter Original Inside Diameter	Brake Drum Diameter Max. Wear Limit	Brake Drum Diameter Maximum Machine Diameter	Minimum Lining Thickness	Brake Caliper Bracket Bolts (ft. lbs.)	Brake Caliper Mounting Bolts (ft. lbs.)
2010	Insight	F	0.830	0.750	0.0016	—	—	—	0.060	80	25
		R	—	—	—	7.874	7.91	NA	0.040	—	—

F: Front

R: Rear

NA: Not Available

37647_INHY_C0013

MAINTENANCE MINDER SCHEDULE

Honda Insight Hybrid

All Honda's displays engine oil life and maintenance service items in the information display to indicate when to perform maintenance service. If the engine oil life is 15% or less, based on the onboard computer's caluculations, you will see SERVICE DUE SOON in the information display every time the ignition key is turned to ON. The maintenance minder indicator will also come on and the maintenance code(s) for other scheduled maintenance items needing service will be displayed below the message.

Symbol	Item	Service
A	Engine oil ①	Change
B	Engine oil and filter	Change
	Fluid levels	Inspect
	Brakes	Inspect
	Parking brake adjustment	Check
	Steering gear and linkage	Inspect
	Suspension components	Inspect
	Driveshaft boots	Inspect
	Brake hoses and lines	Inspect
	Exhaust system	Inspect
	Fuel lines and connections	Inspect
1	Tires	Rotate
2	Engine air filter ②	Replace
	Dust and pollen filter ③	Replace
	Accessory drive belt	Inspect
3	CVT fluid	Replace
4	Spark plugs	Replace
	Valve clearance ④	Inspect
5	Engine coolant	Replace

① If the message SERVICE DUE NOW does not appear more than 12 months after the display is reset, change every year.

② If driven in dusty conditions, replace every 15,000 miles.

③ If driven in urban areas that have a high concentration of soot from industry and diesel, replace every 15,000 miles

⑥ Adjust if necessary.

Additionally, replace the brake fluid every 3 years, and inspect the idle speed every 160,000 miles.

To reset the Engine Oil Life Display:

1. Turn the ignition switch to ON.
2. Press the SELECT button repeatedly until the engine oil life display or the service message is displayed.
3. Press the RESET button for about 10 seconds. You will see a MAINT RESET message.
4. Select the appropriate answer, MAINT RESET >N (NO) or MAINT RESET > y (YES) by pressing the SELECT button repeatedly.

   >N or >Y is displayed on the outside temperature >N or >Y is displayed on the outside temperature display.
5. Select the MAINT RESET > Y (YES), and press and hold the RESET button again to reset the engine oil life to 100%.

37647_INHY_C0014

BRAKES — INFORMATION AND PRECAUTIONS

ANTI-LOCK SYSTEMS

- Certain components within the ABS system are not intended to be serviced or repaired individually.
- Do not use rubber hoses or other parts not specifically specified for and ABS system. When using repair kits, replace all parts included in the kit. Partial or incorrect repair may lead to functional problems and require the replacement of components.
- Lubricate rubber parts with clean, fresh brake fluid to ease assembly. Do not use shop air to clean parts; damage to rubber components may result.
- Use only DOT 3 brake fluid from an unopened container.
- If any hydraulic component or line is removed or replaced, it may be necessary to bleed the entire system.
- A clean repair area is essential. Always clean the reservoir and cap thoroughly before removing the cap. The slightest amount of dirt in the fluid may plug an orifice and impair the system function. Perform repairs after components have been thoroughly cleaned; use only denatured alcohol to clean components. Do not allow ABS components to come into contact with any substance containing mineral oil; this includes used shop rags.
- The Anti-Lock control unit is a microprocessor similar to other computer units in the vehicle. Ensure that the ignition switch is **OFF** before removing or installing controller harnesses. Avoid static electricity discharge at or near the controller.
- If any arc welding is to be done on the vehicle, the control unit should be unplugged before welding operations begin.

DISC AND DRUM SYSTEMS

CAUTION

Dust and dirt accumulating on brake parts during normal use may contain asbestos fibers from production or aftermarket brake linings. Breathing excessive concentrations of asbestos fibers can cause serious bodily harm. Exercise care when servicing brake parts. Do not sand or grind brake lining unless equipment used is designed to contain the dust residue. Do not clean brake parts with compressed air or by dry brushing. Cleaning should be done by dampening the brake components with a fine mist of water, then wiping the brake components clean with a dampened cloth. Dispose of cloth and all residue containing asbestos fibers in an impermeable container with the appropriate label. Follow practices prescribed by the Occupational Safety and Health Administration (OSHA) and the Environmental Protection Agency (EPA) for the handling, processing, and disposing of dust or debris that may contain asbestos fibers.

BRAKES — BLEEDING THE BRAKE SYSTEM

BLEEDING PROCEDURE

BLEEDING PROCEDURE

➡**Note the following while bleeding the brake system:**

- Do not reuse the drained fluid. Use only new Honda DOT 3 Brake Fluid from an unopened container. Using a non-Honda brake fluid can cause corrosion and shorten the life of the system.
- Make sure no dirt or other foreign matter is allowed to contaminate the brake fluid.
- Do not spill brake fluid on the vehicle; it may damage the paint. If brake fluid does contact the paint, wash it off immediately with water.
- The reservoir connected to the master cylinder must be at the MAX (upper) level mark at the start of the bleeding procedure and checked after bleeding each wheel location. Add fluid as required.

1. Make sure the brake fluid level in the reservoir is at the MAX (upper) level line.
2. Have someone slowly pump the brake pedal several times, then apply steady pressure.
3. Start the bleeding at the driver's side of the front brake system.

➡**Bleed the calipers or the wheel cylinders in the following sequence:**

- Front left
- Front right
- Rear right
- Rear left

4. Attach a length of clear drain tube to the bleed screw, then loosen the bleed screw to allow air to escape from the system. Then tighten the bleed screw securely.
5. Refill the master cylinder reservoir to the MAX (upper) level line.
6. Repeat the procedure for each brake circuit until there are no air bubbles are in the fluid.

BLEEDING THE ABS SYSTEM

Bleeding the ABS system is done the same as a conventional brake system. Refer to Bleeding the Brake System in this section.

WHEEL SPEED SENSORS

REMOVAL & INSTALLATION

Front

See Figure 1.

1. Turn the ignition switch to LOCK (0).
2. Remove the front wheels.
3. Remove the grommet, then disconnect the wheel speed sensor connector.
4. Remove the bolt and the bracket, the clip, and the grommet.
5. Remove the bolt and the wheel speed sensor.

To install:

6. Install the wheel speed sensor in the reverse order of removal, and note these items:
   - Do not twist the sensor wires.
   - If the wheel speed sensor comes in contact with the wheel bearing unit, it is faulty.
   - Make sure the grommet is installed properly.
   - Make sure there is no debris in the sensor mounting hole.
7. Start the engine, and make sure the ABS, indicator go off.
8. Test-drive the vehicle, and make sure the ABS, indicator do not come on.

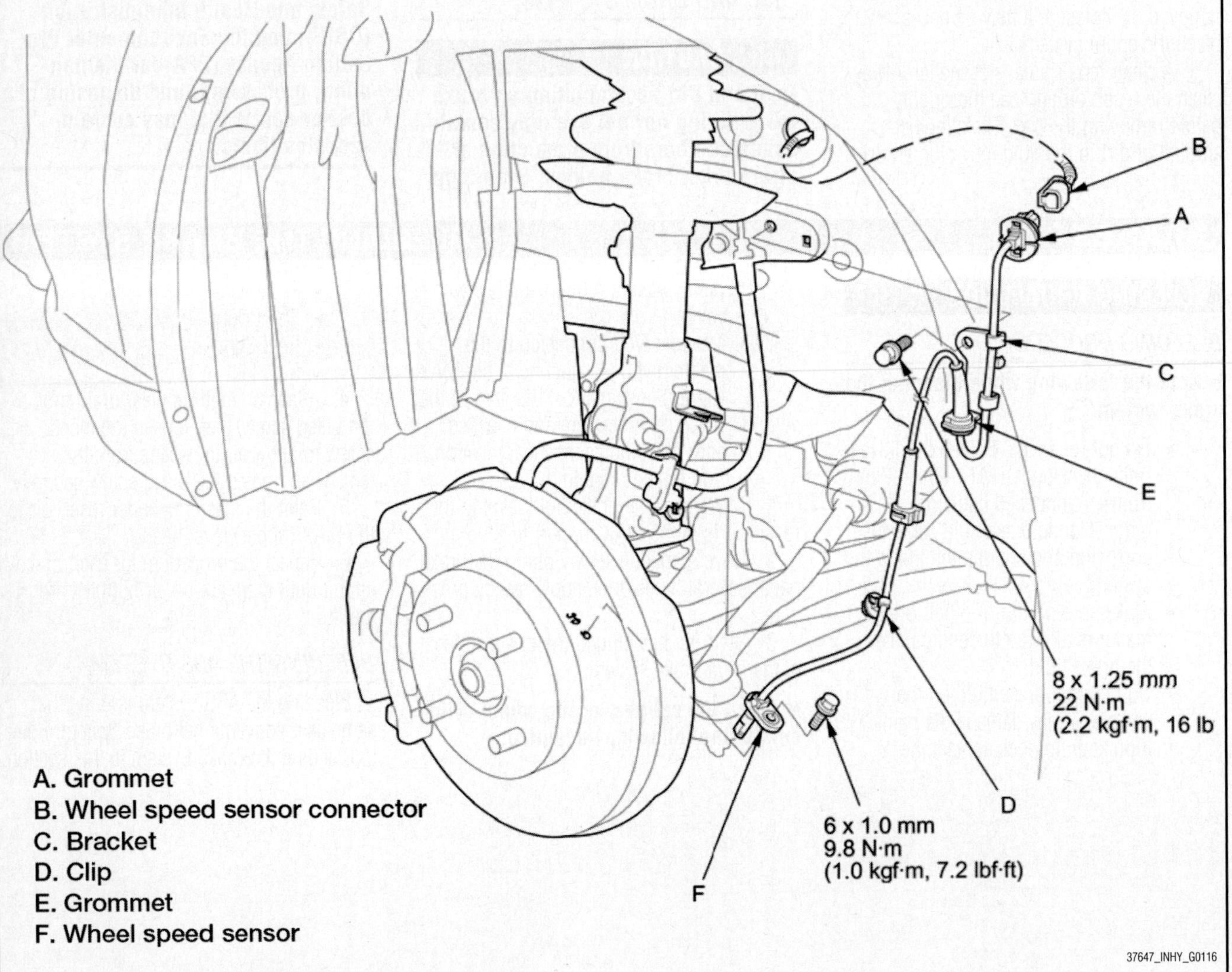
B
A
C
E
8 x 1.25 mm
22 N·m
(2.2 kgf·m, 16 lb
D
6 x 1.0 mm
9.8 N·m
(1.0 kgf·m, 7.2 lbf·ft)
F
A. Grommet
B. Wheel speed sensor connector
C. Bracket
D. Clip
E. Grommet
F. Wheel speed sensor
37647_INHY_G0116

Fig. 1 Exploded view of front wheel speed sensor assembly

Rear

See Figure 2.

1. Turn the ignition switch to LOCK (0).
2. Remove the rear wheels.
3. Release the connector holding clamps, then disconnect the wheel sensor connector.
4. Remove the clip and the grommets.
5. Remove the bolt and the wheel speed sensor.

To install:

6. Install the wheel speed sensor in the reverse order of removal, and note these items:
   - Do not twist the sensor wires.
   - If the wheel speed sensor comes in contact with the hub bearing unit, it is faulty.
   - Make sure the grommet is installed properly.
   - Make sure there is no debris in the sensor mounting hole.
7. Start the engine, and make sure the ABS indicator go off.
8. Test-drive the vehicle, and make sure the ABS indicator do not come on.

WHEEL SPEED SENSOR RINGS

REMOVAL & INSTALLATION

Refer to wheel bearing Removal and Installation when servicing this component.

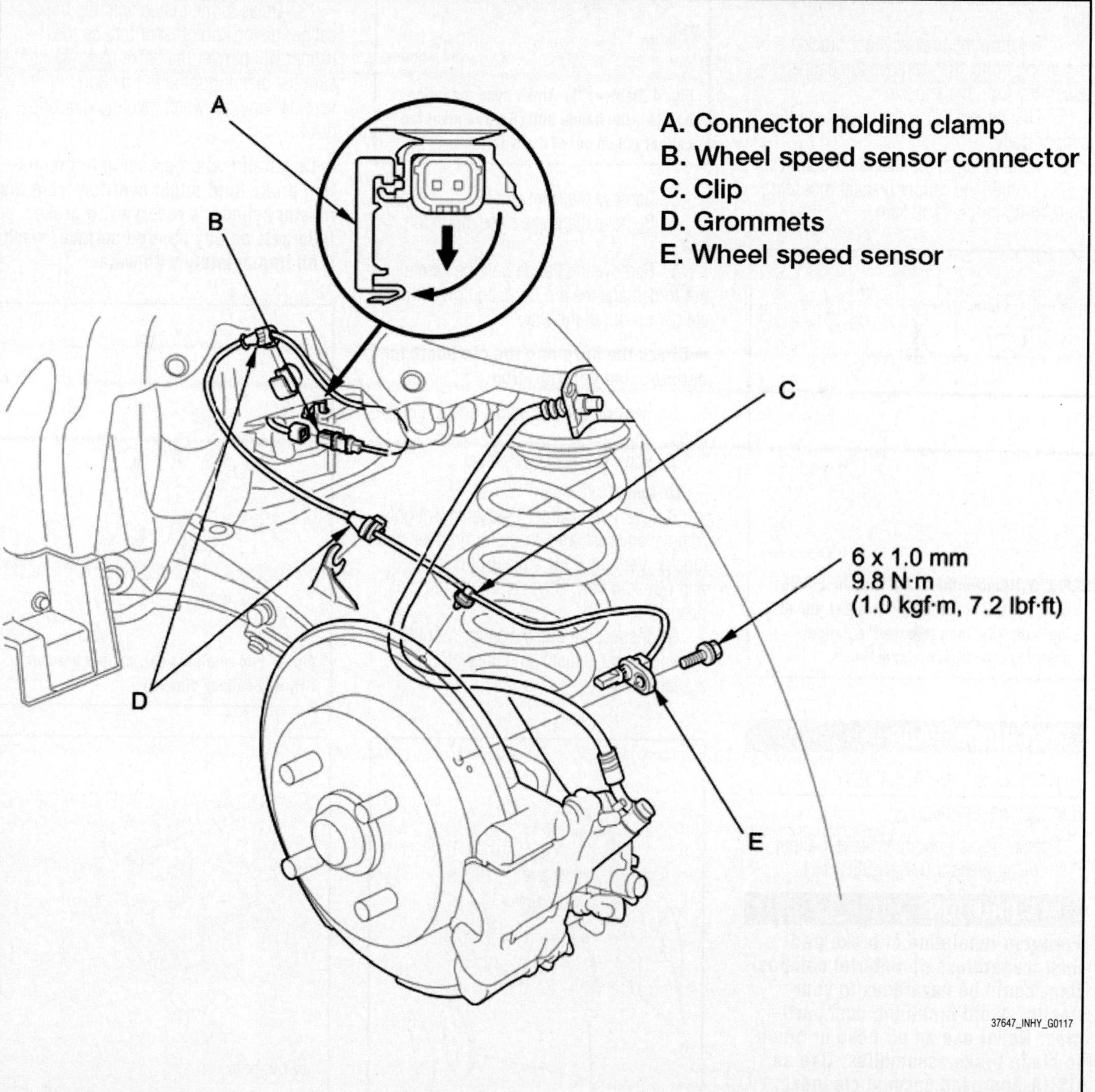
A
B
C
D
E
A. Connector holding clamp
B. Wheel speed sensor connector
C. Clip
D. Grommets
E. Wheel speed sensor
6 x 1.0 mm
9.8 N·m
(1.0 kgf·m, 7.2 lbf·ft)
37647_INHY_G0117

Fig. 2 Exploded view of rear wheel speed sensor assembly—vehicle with rear disc brakes shown, Insight similar

BRAKE CALIPER

REMOVAL & INSTALLATION

See Figure 3.

➡Keep any grease off the brake disc and brake pads.

1. Raise the vehicle on a lift.
2. Remove the front wheel.
3. Remove the brake hose mounting bolt.
4. Remove the brake caliper bracket mounting bolts, then remove the caliper assembly from the knuckle.
5. Disconnect the brake hose from the caliper body.
6. Installation is the reverse of removal.
7. Tighten the caliper bracket mounting bolts to 80 ft. lbs. (108 Nm).

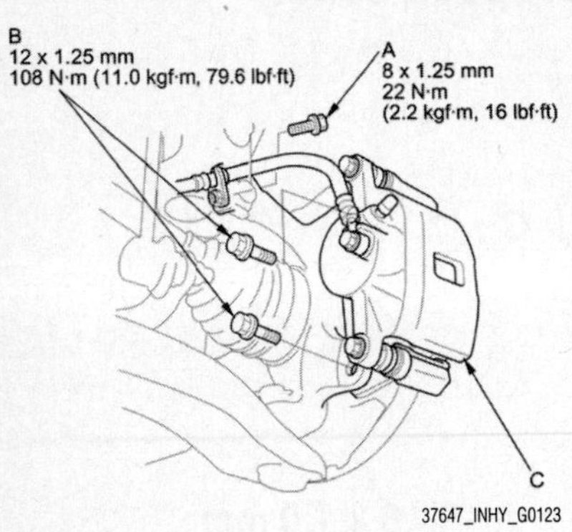
B
12 x 1.25 mm
108 N·m (11.0 kgf·m, 79.6 lbf·ft)
A
8 x 1.25 mm
22 N·m
(2.2 kgf·m, 16 lbf·ft)
C
37647_INHY_G0123

Fig. 3 Remove the brake hose mounting bolt (A), the brake caliper bracket mounting bolts (B), then remove the caliper assembly (C) from the knuckle

DISC BRAKE PADS

REMOVAL & INSTALLATION

See Figures 4 through 8.

Special Tools Required: Brake Caliper Piston Compressor 07AAE-SEPA101

CAUTION

Frequent inhalation of brake pad dust, regardless of material composition, could be hazardous to your health. Avoid breathing dust particles. Never use an air hose or brush to clean brake assemblies. Use an OSHA-approved vacuum cleaner.

1. Remove some of the brake fluid from the master cylinder.
2. Raise the vehicle on a lift.

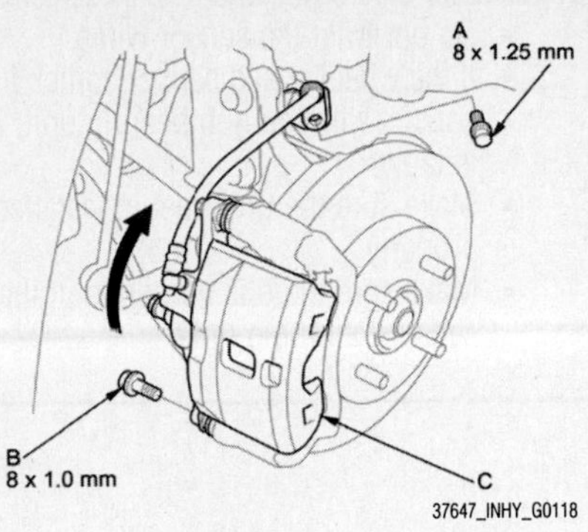
A
8 x 1.25 mm
B
8 x 1.0 mm
C
37647_INHY_G0118

Fig. 4 Remove the brake hose mounting bolt (A), the flange bolt (B), and pivot the caliper (C) up out of the way

3. Remove the front wheels.
4. Remove the brake hose mounting bolt.
5. Remove the flange bolt, be careful not to damage the pin boot, and pivot the caliper up out of the way.

➡Check the hose and the pin boots for damage and deterioration.

6. Remove the pad shims and the brake pads.
7. Remove the pad retainers.

To install:

8. Clean the caliper bracket thoroughly; remove any rust, and check for grooves and cracks. Verify that the caliper pins move in and out smoothly. Clean and lube the pins if needed.
9. Inspect the brake disc for runout, thickness, parallelism, and check for damage and cracks.

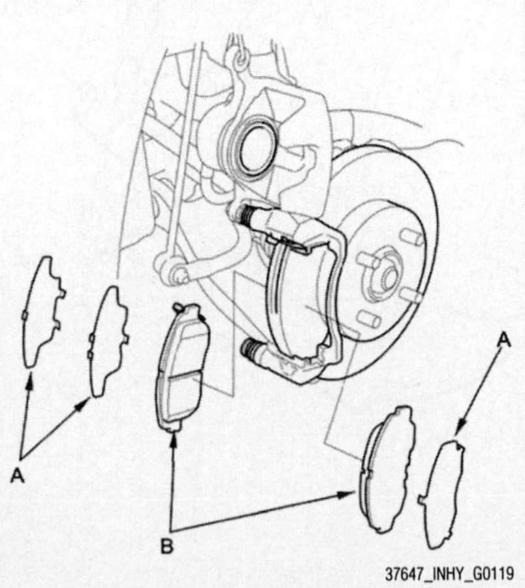
A
A
B
37647_INHY_G0119

Fig. 5 Remove the pad shims (A) and the brake pads (B)

10. Apply a thin coat of M-77 assembly paste (P/N 08798-9010) to the retainer mating surface of the caliper bracket (indicated by the arrows and shaded area).
11. Install the pad retainers. Wipe excess assembly paste off the retainers. Keep the assembly paste off the brake disc and the brake pads.
12. Mount a brake caliper piston compressor on the caliper body.
13. Press in the piston with the brake caliper piston compressor tool so the caliper will fit over the brake pads. Make sure the piston boot is in position to prevent damaging it when pivoting the caliper down.

➡Be careful when pressing in the piston; brake fluid might overflow from the master cylinder's reservoir. If brake fluid gets on any painted surface, wash it off immediately with water.

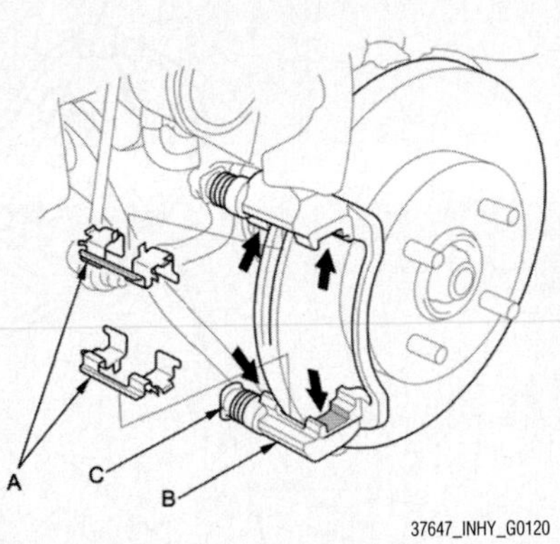
A
C
B
37647_INHY_G0120

Fig. 6 Pad retainers (A), caliper bracket (B), and caliper pins (C)

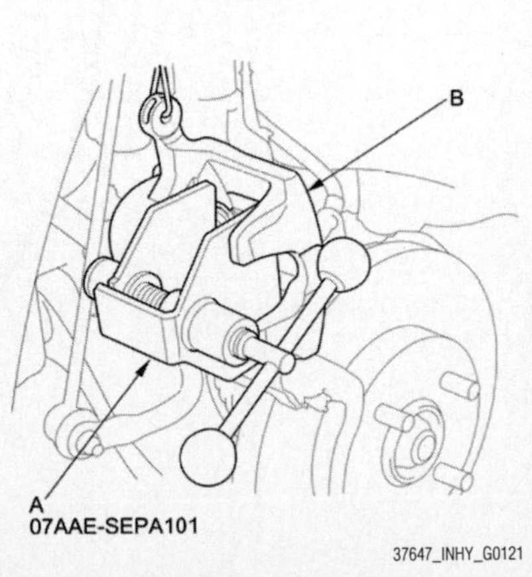
B
A
07AAE-SEPA101
37647_INHY_G0121

Fig. 7 Mount a brake caliper piston compressor (A) on the caliper body (B)

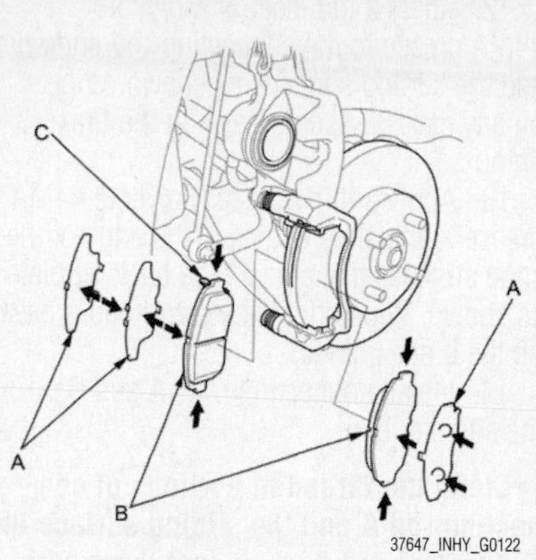
C
A
A
B
37647_INHY_G0122

Fig. 8 Apply Molykote M-77 assembly paste to the pad side of the shims (A), the back of the brake pads (B), and other areas indicated by the arrows

14. Remove the brake caliper piston compressor tool.

15. Apply Molykote M-77 assembly paste (P/N 08798-9010) to the pad side of the shims, the back of the brake pads, and other areas indicated by the arrows. Wipe excess assembly paste off the pad shims and the brake pads friction material. Keep grease and assembly paste off the brake discs and the brake pads. Contaminated brake discs or brake pads reduce stopping ability.

16. Install the brake pads and the pad shims correctly. Install the brake pad with the wear indicator on the upper inside. If you are reusing the brake pads, always reinstall the brake pads in their original positions to prevent a temporary loss of braking efficiency.

17. Pivot the caliper down into position. Install the flange bolt and tighten it to 25 ft. lbs. (34 Nm).

18. Install the brake hose mounting bolt.

19. Clean the mating surfaces of the brake disc and the inside of the wheel, then install the front wheels.

20. Press the brake pedal several times to make sure the brakes work.

➡**Engagement may require a greater pedal stroke immediately after the brake pads have been replaced as a set. Several applications of the brake pedal will restore the normal pedal stroke.**

21. Add brake fluid as needed.

22. After installation, check for leaks at hose and line joints or connections, and retighten if necessary.

23. Test-drive the vehicle, then check for leaks.

BRAKES

REAR DRUM BRAKES

BRAKE DRUM

REMOVAL & INSTALLATION

➡**Keep any grease off the brake drum and brake shoes.**

1. Raise the vehicle on a lift .
2. Remove the rear wheel.
3. Remove the parking brake, and remove the brake drum from the hub bearing unit.

➡**If necessary, turn the adjuster bolt with a flat-tip screwdriver until the shoes become loose. If the brake drum has clung to the hub bearing unit. Thread two 8 x 1.25 mm bolts into the brake drum to push it away from the hub bearing unit. Turn each bolt 90 degrees at a time to prevent cocking the brake drum.**

To install:

4. Install the brake drum in the reverse order of removal.

➡**Before installing the brake drum, clean the mating surfaces between the rear hub and the inside of the brake drum. After installation, press the brake pedal several times to make sure the brakes work and self-adjust the brake shoes.**

5. Clean the mating surfaces between the brake drum and the inside of the wheel, then install the rear wheel.

BRAKE SHOES

REMOVAL & INSTALLATION

See Figures 9 through 16.

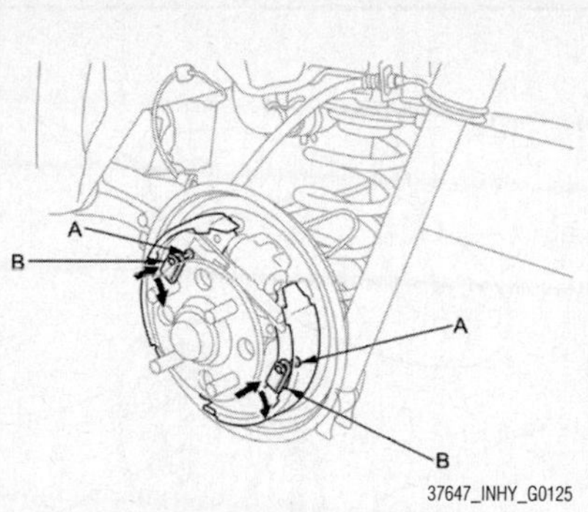
A
B
A
B
37647_INHY_G0125

Fig. 9 Remove the tension pins (A) by pushing the respective retainer spring (B) and turning the pin

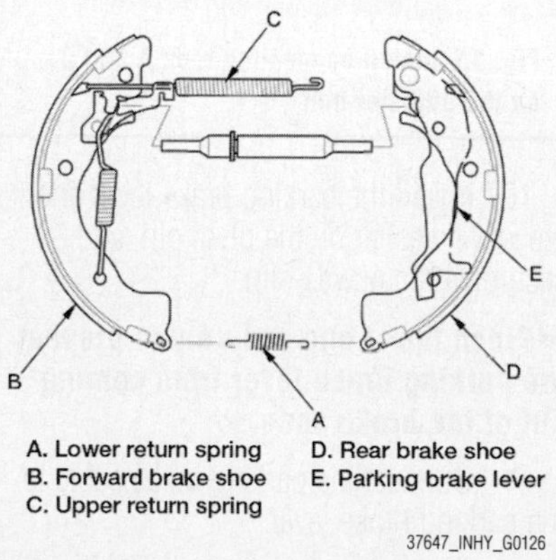
C
E
B
D
A
A. Lower return spring
B. Forward brake shoe
C. Upper return spring
D. Rear brake shoe
E. Parking brake lever
37647_INHY_G0126

Fig. 10 Remove the lower return spring, and remove the brake shoe assembly over the hub

1. Raise the vehicle on a lift.
2. Remove the rear wheels.
3. Release the parking brake, and remove the brake drum.
4. Remove the tension pins by pushing the respective retainer spring and turning the pin.
5. Remove the lower return spring, and remove the brake shoe assembly over the hub.
6. Remove the forward brake shoe by removing the upper return spring, and disassemble the brake shoe assembly.
7. Remove the rearward brake shoe by disconnecting the parking brake cable from the parking brake lever.

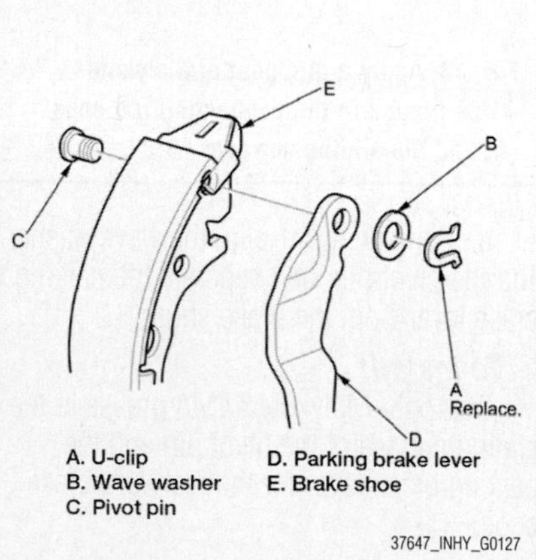
E
B
C
A
Replace.
D
A. U-clip
B. Wave washer
C. Pivot pin
D. Parking brake lever
E. Brake shoe
37647_INHY_G0127

Fig. 11 Remove the U-clip, the wave washer, and the pivot pin, and separate the parking brake lever from the brake shoe

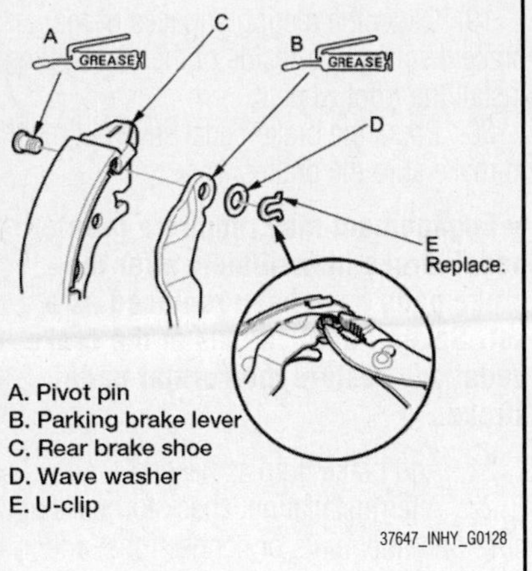
A
C
B
GREASE
GREASE
D
E
Replace.
A. Pivot pin
B. Parking brake lever
C. Rear brake shoe
D. Wave washer
E. U-clip
37647_INHY_G0128

Fig. 12 Apply Molykote 44MA grease to the sliding surface of the pivot pin and the parking brake lever for the rearward brake shoe

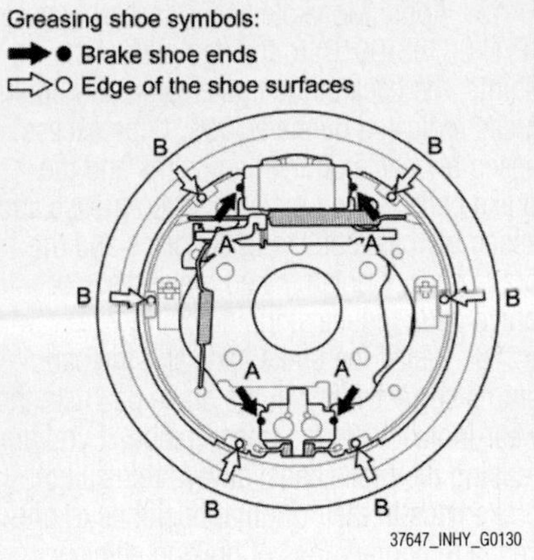
Greasing shoe symbols:
Brake shoe ends
Edge of the shoe surfaces
B
A
37647_INHY_G0130

Fig. 14 Apply a thin coat of Molykote 44MA grease to the shoe ends (A) and the edge of the shoe surfaces (B) that contact the backing plate

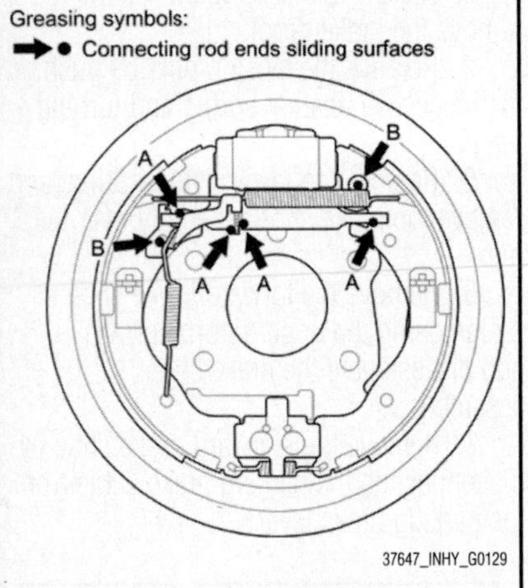
Greasing symbols:
Connecting rod ends sliding surfaces
A
B
37647_INHY_G0129

Fig. 13 Apply a thin coat of Molykote 44MA grease to the connecting rod ends (A) and the sliding surfaces (B)

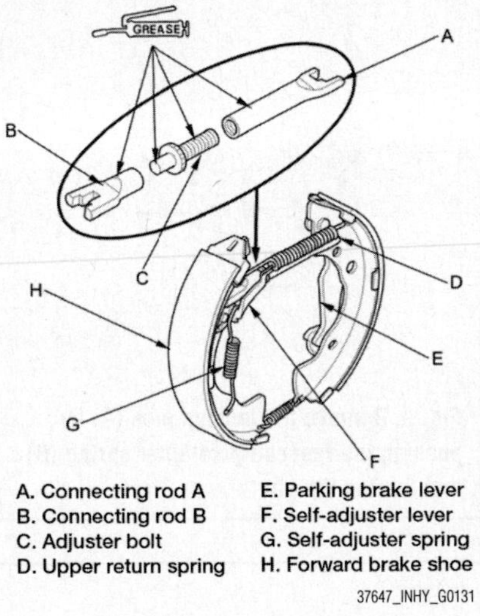
GREASE
A
B
C
D
E
F
G
H
A. Connecting rod A
B. Connecting rod B
C. Adjuster bolt
D. Upper return spring
E. Parking brake lever
F. Self-adjuster lever
G. Self-adjuster spring
H. Forward brake shoe
37647_INHY_G0131

Fig. 15 Install connecting rods A and B on the adjuster bolt

8. Remove the U-clip, the wave washer, and the pivot pin, and separate the parking brake lever from the brake shoe.

To install:

9. Apply Molykote 44MA grease to the sliding surface of the pivot pin and the parking brake lever for the rearward brake shoe.

10. Install the parking brake lever and the wave washer on the pivot pin, and secure with a new U-clip.

➡Pinch the U-clip securely to prevent the parking brake lever from coming out of the brake shoe.

11. Connect the parking brake cable to the parking brake lever.

12. Apply a thin coat of Molykote 44MA grease to the connecting rod ends and the sliding surfaces as shown. Wipe off any excess. Keep grease off the brake linings.

13. Apply a thin coat of Molykote 44MA grease to the shoe ends and the edge of the shoe surfaces that contact the backing plate as shown. Wipe off any excess. Keep grease off the brake linings.

14. Install connecting rods A and B on the adjuster bolt.

➡Clean the threaded portions of connecting rod A and the sliding surface of connecting rod B, then coat them with Molykote 44MA grease. Shorten connecting rod A by fully turning in the adjuster bolt.

15. Assemble the brake shoes with the upper return spring, and with the connecting rods and the adjuster bolt onto the backing plate. Reconnect the parking brake cable to the parking brake lever, then install the self-adjuster lever and the self-adjuster spring on the forward brake shoe.

16. Install the tension pins and the retainer springs by pushing in the respective spring and turning each pin.

17. Install the lower return spring.

➡Make sure the brake shoes are positioned on the brake shoe bosses on the backing plate, and the fittings on the top of the brake shoes are fitted into the wheel cylinder pistons.

18. Install the brake drum.

➡Before installing the brake drum, clean the mating surface between the rear hub and the inside of the brake drum.

19. Clean the mating surfaces between the brake drum and the inside of the wheel, then install the rear wheels.

20. Press the brake pedal several times to make sure the brakes work and to set the self-adjusting brake.

➡Engagement of the brakes may require a greater pedal stroke immediately after the brake shoes have been replaced as a set. Several applications of the brake pedal will restore the normal pedal stroke.

21. Do the parking brake adjustment.

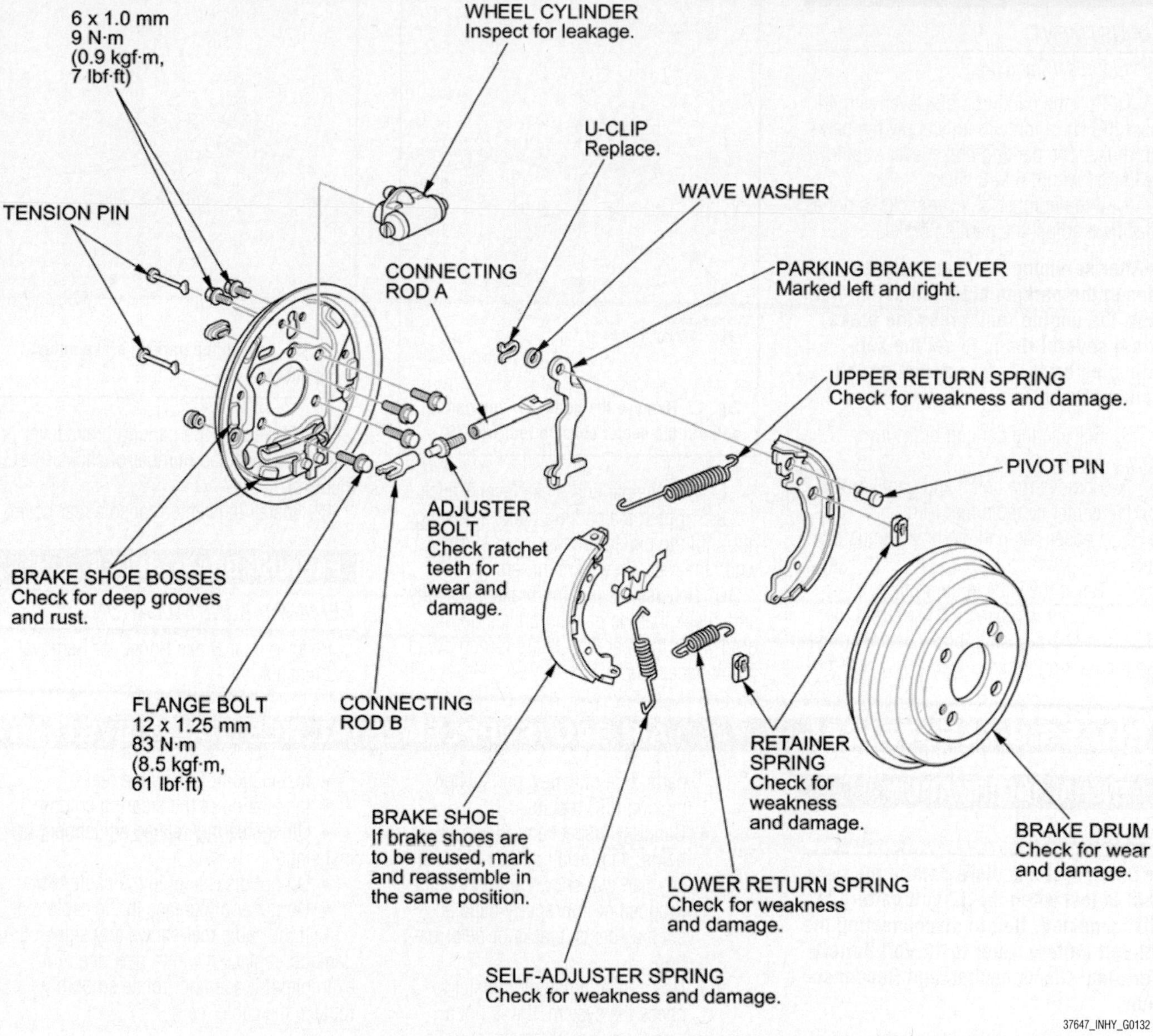
6 x 1.0 mm
9 N·m
(0.9 kgf·m,
7 lbf·ft)
WHEEL CYLINDER
Inspect for leakage.
U-CLIP
Replace.
WAVE WASHER
TENSION PIN
CONNECTING
ROD A
PARKING BRAKE LEVER
Marked left and right.
UPPER RETURN SPRING
Check for weakness and damage.
PIVOT PIN
ADJUSTER
BOLT
Check ratchet
teeth for
wear and
damage.
BRAKE SHOE BOSSES
Check for deep grooves
and rust.
FLANGE BOLT
12 x 1.25 mm
83 N·m
(8.5 kgf·m,
61 lbf·ft)
CONNECTING
ROD B
RETAINER
SPRING
Check for
weakness
and damage.
BRAKE DRUM
Check for wear
and damage.
BRAKE SHOE
If brake shoes are
to be reused, mark
and reassemble in
the same position.
LOWER RETURN SPRING
Check for weakness
and damage.
SELF-ADJUSTER SPRING
Check for weakness and damage.
37647_INHY_G0132

Fig. 16 Exploded view of rear drum brake assembly

BRAKES — PARKING BRAKE

PARKING BRAKE CABLES

ADJUSTMENT

See Figures 17 and 18.

1. Pull the parking brake lever with 44 lbs. (196 N) of force to fully apply the parking brake. The parking brake lever should be locked within 6 to 8 clicks.
2. If the number of lever clicks is not as specified, adjust the parking brake.

➡After servicing the rear brake shoes, loosen the parking brake adjusting nut, start the engine, and press the brake pedal several times to set the self-adjusting brake before adjusting the parking brake.

3. Release the parking brake lever fully.
4. Remove the bolts, and gently pull out the center console rear trim.
5. Loosen the parking brake adjusting nut.
6. Raise the vehicle on a lift.
7. Press the brake pedal several times to set the self-adjusting brake before adjusting the parking brake.

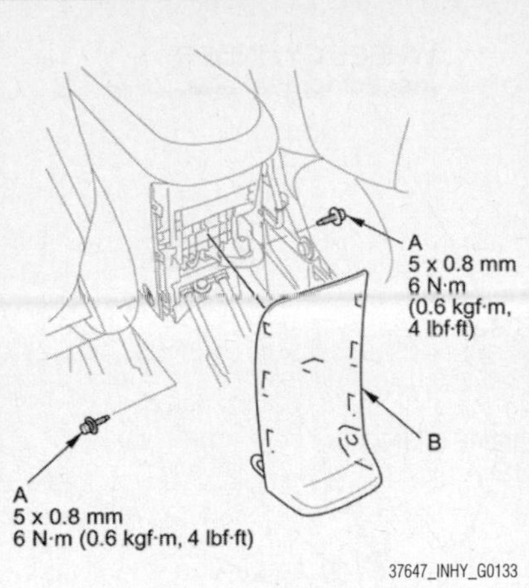
A
5 x 0.8 mm
6 N·m
(0.6 kgf·m,
4 lbf·ft)
B
A
5 x 0.8 mm
6 N·m (0.6 kgf·m, 4 lbf·ft)
37647_INHY_G0133

Fig. 17 Remove the bolts (A), and gently pull out the center console rear trim (B)

8. Pull the parking brake lever 1 click.
9. Tighten the parking brake adjusting nut until the parking brakes drag slightly when the rear wheels are turned.
10. Release the parking brake lever fully, and check that the parking brakes do not drag when the rear wheels are turned. Readjust if necessary.

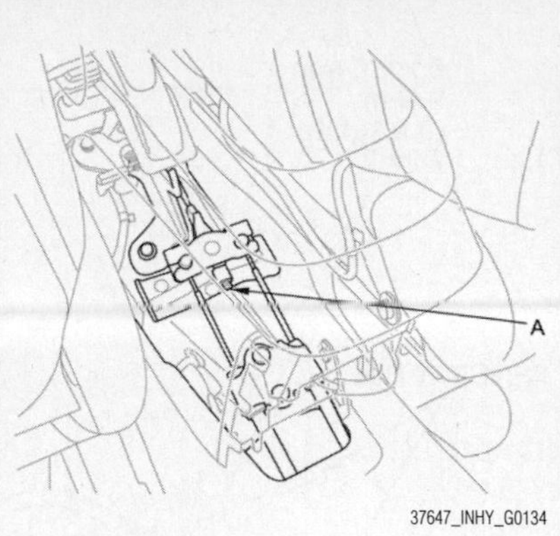
A
37647_INHY_G0134

Fig. 18 Loosen the parking brake adjusting nut (A)

11. Make sure the parking brake lever is within the specified number of clicks (6 to 8 clicks).
12. Install the center console rear cover.

PARKING BRAKE SHOES

REMOVAL & INSTALLATION

Refer to Rear Brake Shoes for Removal and Installation.

CHASSIS ELECTRICAL — AIR BAG (SUPPLEMENTAL RESTRAINT SYSTEM)

GENERAL INFORMATION

PRECAUTIONS

➡Some systems store data in memory that is lost when the 12 volt battery is disconnected. Before disconnecting the 12 volt battery, refer to 12 Volt Battery Terminal Disconnection and Reconnection.

Please read the following precautions carefully before servicing the airbag system. Observe the instructions described in this manual, or the airbags could accidentally deploy and cause damage or injuries.

- Except when doing electrical inspections, always turn the ignition switch to LOCK (0), disconnect the negative cable from the 12 volt battery, then wait at least 3 minutes before starting work.

NOTE: The SRS memory is not erased even if the ignition switch is turned to LOCK (0) or the battery cables are disconnected from the 12 volt battery.

- Use replacement parts which are manufactured to the same standards and quality as the original parts. Do not install used SRS parts. Use only new parts when making SRS repairs.
- Carefully inspect any SRS part before you install it. Do not install any part that shows signs of being dropped or improperly handled, such as dents, cracks or deformation.
- Use only a digital multimeter to check the system. If it is not a Honda multimeter, make sure its output is 10 mA (0.01 A) or less when switched to the lowest value in the ohmmeter range. A tester with a higher output could cause accidental deployment and possible injury.
- Do not put objects on the front passenger's airbag.

Steering-Related Precautions

Cable Reel Alignment:

- Misalignment of the cable reel could cause an open in the wiring, making the SRS system, remote steering wheel controls, and the horn inoperative. Center the cable reel whenever you do the following.
- Installation of the steering wheel
- Installation of the cable reel
- Installation of the steering column
- Other steering-related adjustment or installation
- Do not disassemble the cable reel.
- Do not apply grease to the cable reel.
- If the cable reel shows any signs of damage, replace it with a new one. For example, if it does not rotate smoothly, replace the cable reel.

Airbag Handling and Storage

Do not disassemble an airbag. It has no serviceable parts. Once an airbag has been deployed, it cannot be repaired or reused.

For temporary storage of an airbag during service, observe the following precautions.

- Store the removed airbag with the pad surface up. Never put anything on the airbag.
- To prevent damage to the airbag, keep it away from any oil, grease, detergent, or water.
- Store the removed airbag on a secure, flat surface away from any high heat source (exceeding 200 °F/93 °C).
- Never do electrical tests on the airbags, such as measuring resistance.

- Do not position yourself in front of the airbag during removal, inspection, or replacement.
- For proper disposal of a damaged airbag, refer to airbag disposal.

The side curtain airbag module assembly is a long, jointed part containing an inflator, a flexible bag, and brackets. When removing or installing the side curtain airbag inflator assembly, never do these things:

- Drop the curtain airbag.
- Cut, tear, or unwrap the tape strips.
- Handle the flexible bag.

SRS Unit, Front and Side Impact Sensors, Rear Safing Sensor, Driver's Seat Position Sensor, and Front Passenger's Weight Sensors

Some systems store data in memory that is lost when the 12 volt battery is disconnected. Before disconnecting the 12 volt battery, refer to 12 Volt Battery Terminal Disconnection and Reconnection.

- Turn the ignition switch to LOCK (0), disconnect the negative cable from the 12 volt battery, then wait at least 3 minutes before beginning installation or replacement of the SRS unit or disconnecting the connectors from the SRS unit.
- Be careful not to bump or impact the SRS unit, front impact sensors or side impact sensors when the ignition switch is at ON (II), or for at least 3 minutes after the ignition switch is turned to LOCK (0).
- During installation or replacement, be careful not to bump (by impact wrench, hammer, etc.) the area around the SRS unit, front impact sensors or side impact sensors. The airbags could accidentally deploy and cause damage or injury.
- After a collision where a front airbag, side airbags, side curtain airbags, or a seat belt tensioner deployed, go to Component Replacement/Inspection after Deployment. After a collision where the airbags or the side airbags did not deploy, inspect for any damage or any deformation on the SRS unit, front impact sensors and side impact sensors. Replace all damaged parts.
- Do not disassemble the SRS unit, front impact sensors and side impact sensors.
- Be sure the SRS unit, front impact sensors and side impact sensors are installed securely with the mounting bolts torqued to 7 ft. lbs. (9.8 Nm) whenever you remove or replace the SRS unit, all impact sensors, always install the components with new TORX;rM bolts.
- Do not spill water or oil on the SRS unit or the side impact sensors.

Wiring Precautions

Some of the SRS wiring can be identified by special yellow outer covering, and the SRS connectors can be identified by their yellow color. Observe the instructions.

- Never attempt to modify, splice, or repair SRS wiring. If there is an open or damage in SRS wiring, replace the harness.
- Be sure to install the harness wires so they do not get pinched or interfere with other parts.
- Make sure all SRS ground locations are clean, and grounds are securely fastened for optimum metal-to-metal contact. Poor grounds can cause intermittent problems that are difficult to diagnose.
- Do not use any silicone based cleaners or lubricants on any SRS connectors or terminals.

Precautions for Electrical Inspections

Special Tools Required: Back Probe Adapter, 17 mm 07TAZ-001020A

Make sure the 12 volt battery is fully charged when doing electrical tests. If the 12 volt battery is not fully charged, the results of the tests may not be accurate.

When using electrical test equipment, insert the probe of the tester into the wire side of the connector (except waterproof connector). Do not insert the probe of the tester into the terminal side of the connector, and do not tamper with the connector.

Use back probe adapter 07TAZ-001020A. Do not insert the probe forcibly.

Use specified service connectors in troubleshooting.

Using improper tools could cause an error in inspection due to poor metal-to-metal contact.

DRIVE TRAIN

FRONT HALFSHAFT

REMOVAL & INSTALLATION

See Figures 19 through 22.

Special Tools Required:

- Ball Joint Remover, 28 mm 07MAC-SL0A202
- Ball Joint Thread Protector, 14 mm 071AF-S3VA000

1. Raise the vehicle on a lift.
2. Remove the front wheels.
3. Pry up the stake on the spindle nut, then remove the nut.
4. Drain the transmission fluid, then reinstall the drain plug with a new sealing washer.
5. Remove the lock pin from the lower arm ball joint, then remove the castle nut. Separate the ball joint from the knuckle

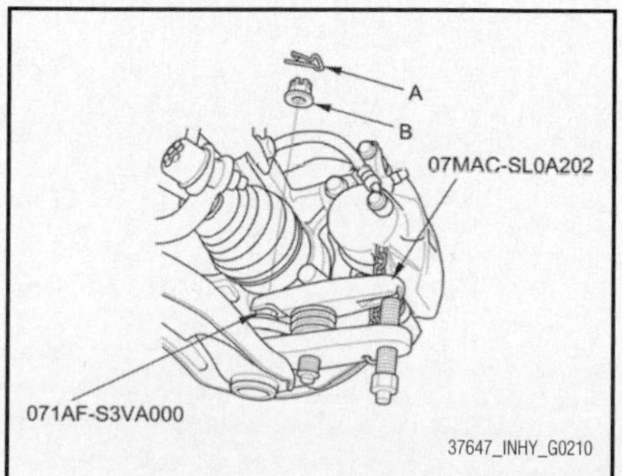
A
B
07MAC-SL0A202
071AF-S3VA000
37647_INHY_G0210

Fig. 19 Remove the lock pin (A) from the lower arm ball joint, then remove the castle nut (B)

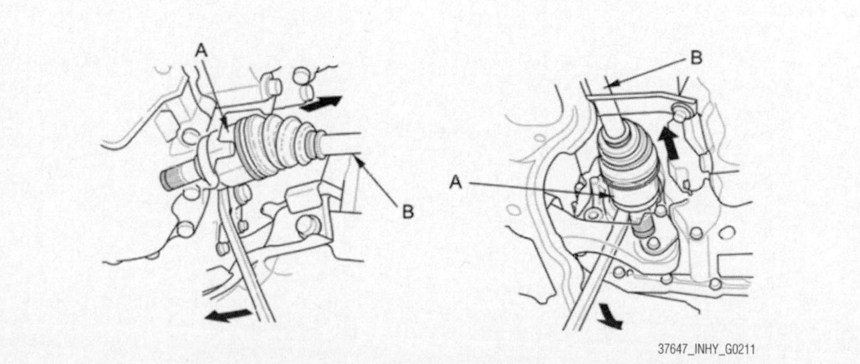
A
B
B
A
37647_INHY_G0211

Fig. 20 Pry the inboard joint (A) from the differential using a prybar: do not pull on the driveshaft (B)

using the 14 mm ball joint thread protector and the 28 mm ball joint remover.

➡Be careful not to damage the ball joint boot when installing the remover. Do not force or hammer on the lower arm, or pry between the lower arm and the knuckle. You could damage the ball joint.

6. Pull the knuckle outward, and separate the outboard joint from the front hub using a plastic hammer.

7. Pry the inboard joint from the differential using a prybar. Remove the driveshaft as an assembly.

➡Do not pull on the driveshaft, or the inboard joint may come apart. Pull the inboard joint straight out to avoid damaging the oil seal. Be careful not to damage the oil seal or the end of the inboard joint with the prybar.

8. Remove the set ring from the inboard joint.

To install:

➡Before starting installation, make sure the mating surfaces of the joint and the splined section are clean.

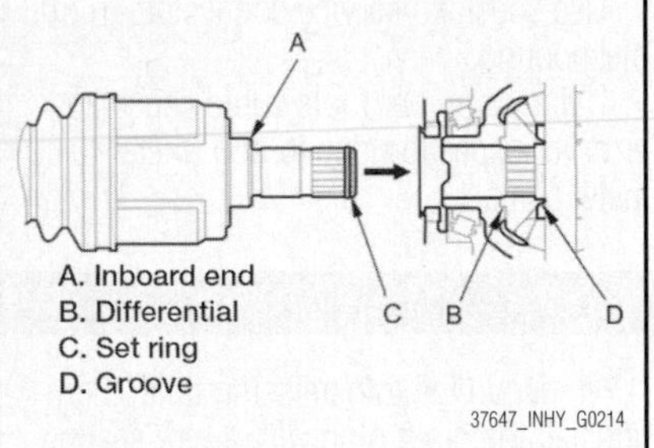
A
A. Inboard end
B. Differential
C. Set ring
D. Groove
C
B
D
37647_INHY_G0214

Fig. 21 Insert the inboard end of the driveshaft into the differential until the set ring locks in the groove

9. Apply Moly 60 paste (P/N 08734-0001) to the contact area of the outboard joint and the front wheel bearing.

➡The paste helps to prevent noise and vibration.

10. Install a new set ring into the set ring groove of the driveshaft inboard joint.

11. Clean the areas where the driveshaft contacts the differential thoroughly with solvent or brake cleaner, and dry with compressed air.

➡Do not wash the rubber parts with solvent.

12. Insert the inboard end of the driveshaft into the differential until the set ring locks in the groove.

➡Insert the driveshaft horizontally to prevent damaging the oil seal.

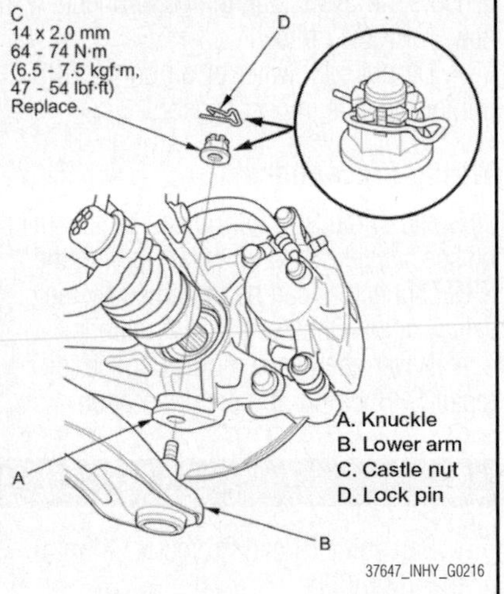
C
14 x 2.0 mm
64 - 74 N·m
(6.5 - 7.5 kgf·m,
47 - 54 lbf·ft)
Replace.
D
A. Knuckle
B. Lower arm
C. Castle nut
D. Lock pin
A
B
37647_INHY_G0216

Fig. 22 Install the knuckle onto the lower arm

13. Install the outboard joint into the front hub on the knuckle.

14. Wipe off any grease contamination from the ball joint tapered section and threads, then install the knuckle onto the lower arm. Be careful not to damage the ball joint boot. Wipe off the grease before tightening the nut at the ball joint. Torque the new castle nut to the lower torque specification, then tighten it only far enough to align the slot with the ball joint pin hole.

➡Make sure the ball joint boot is not damaged or cracked. Do not align the nut by loosening it.

15. Install the lock pin into the ball joint pin hole.

16. Apply a small amount of engine oil to the seating surface of a new spindle nut.

17. Install the spindle nut, then tighten it. After tightening, use a drift to stake the spindle nut shoulder against the driveshaft.

18. Clean the mating surfaces of the brake discs and the wheels, then install the front wheels.

19. Turn the front wheel by hand, and make sure there is no interference between the driveshaft and the surrounding parts.

20. Lower the vehicle on the lift.

21. Refill the transmission with the recommended transmission fluid.

22. Check the wheel alignment, and adjust it if necessary.

23. Test-drive the vehicle.

CV-BOOTS INSPECTION

Check the inboard boot and the outboard boot on the driveshaft for cracks, damage, leaking grease, and loose boot bands. If any damage is found, replace the boot and the boot bands.

ENGINE COOLING

ENGINE FAN

REMOVAL & INSTALLATION

See Figures 23 through 29.

1. Remove the splash shield.
2. Raise the vehicle on the lift to full height.
3. Remove the A/C compressor clutch connector from the clamp, then remove the harness clamps.
4. Lower the vehicle on the lift.
5. Remove the air cleaner.
6. Remove the air duct bracket.
7. Disconnect the fan motor connectors and the harness clamp.
8. Remove the coolant reservoir.
9. Remove the A/C condenser fan shroud assembly and the radiator fan shroud assembly from the radiator, then remove the A/C condenser fan shroud assembly from the vehicle.

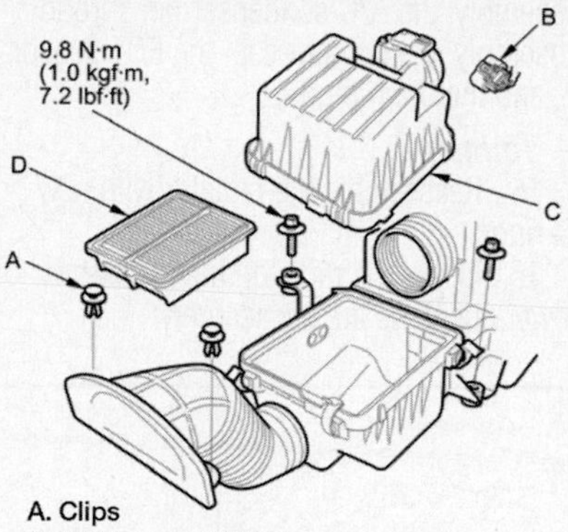
9.8 N·m
(1.0 kgf·m,
7.2 lbf·ft)
A
B
C
D

A. Clips
B. MAF sensor/IAT sensor connector
C. Air cleaner housing cover
D. Air cleaner element

37647_INHY_G0232

Fig. 25 Exploded view of air cleaner assembly

10. Move the radiator fan shroud assembly to the right side of the engine compartment to clear the upper radiator hose, then lift it out.
11. Disassemble the fan shrouds.

To install:

12. Assemble the fan shrouds.
13. Install the radiator fan shroud assembly, then install the A/C condenser fan shroud assembly.
14. Install the coolant reservoir.
15. Connect the fan motor connectors and the harness clamp.
16. Install the air duct bracket.
17. Install the air cleaner.
18. Raise the vehicle on the lift to full height.
19. Install the A/C compressor clutch connector to the clamp, then install the harness clamps.
20. Lower the vehicle on the lift.
21. Install the splash shields.

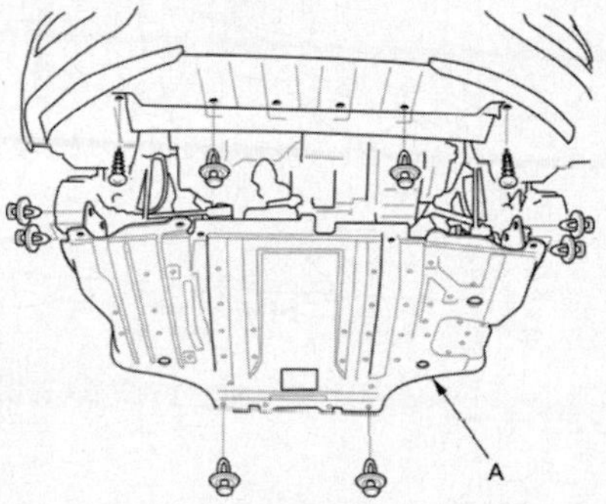
A

37647_INHY_G0183

Fig. 23 Remove the splash shield

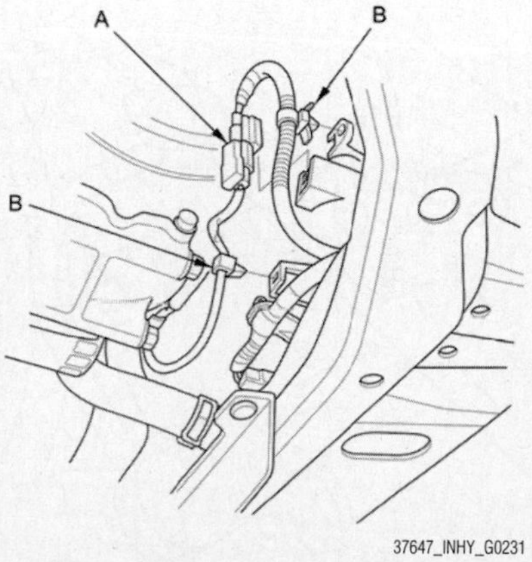
A
B
B

37647_INHY_G0231

Fig. 24 Remove the A/C compressor clutch connector (A) from the clamp, then remove the harness clamps (B)

37647_INHY_G0233

Fig. 26 Remove the air duct bracket

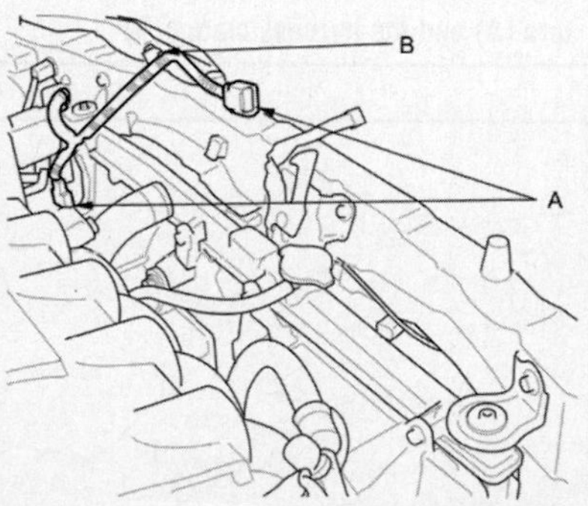
B
A

37647_INHY_G0234

Fig. 27 Disconnect the fan motor connectors (A) and the harness clamp (B)

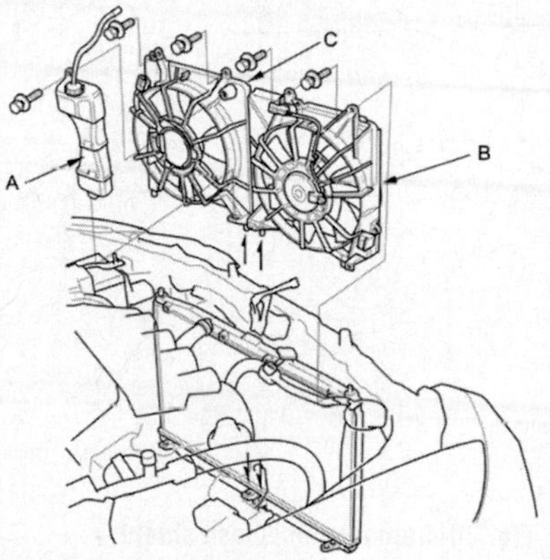
C
A
B

37647_INHY_G0235

Fig. 28 Remove the coolant reservoir (A), the A/C condenser fan shroud assembly (B) and the radiator fan shroud assembly (C)

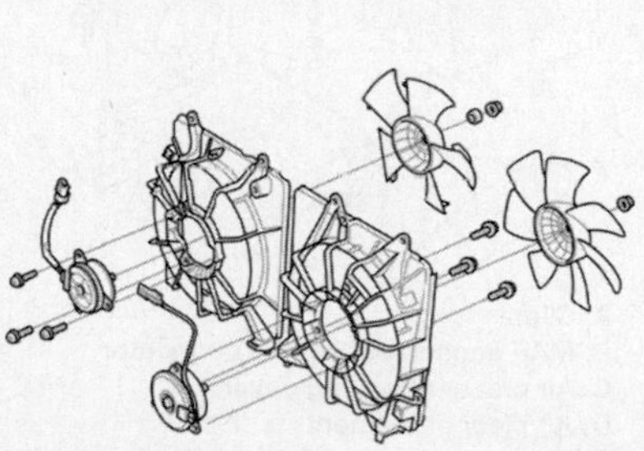

37647_INHY_G0236

Fig. 29 Disassemble the fan shrouds

RADIATOR

REMOVAL & INSTALLATION

See Figures 30 through 38.

1. Drain the engine coolant.
2. Remove the splash shield.
3. Remove the air cleaner.
4. Remove the air duct bracket.
5. Disconnect the fan motor connectors and the harness clamp.
6. Remove the radiator brackets.
7. Raise the vehicle on the lift to full height.
8. Disconnect the engine coolant temperature (ECT) sensor 2 connector, and remove the A/C compressor clutch connector from the clamp.
9. Remove the harness clamps.
10. Disconnect the lower radiator hose.
11. Remove the CVT fluid (CVTF) cooler hoses, then plug the hose and line.
12. Lower the vehicle on the lift.
13. Remove the upper radiator hose.
14. Pull up the radiator, then remove the coolant reservoir, the radiator fan shroud assembly, the A/C condenser fan shroud assembly, the radiator cap, the ECT sensor 2, and the drain plug.

To install:

15. Reassemble the radiator with new O-rings.
16. Install the radiator. Make sure the lower cushions are set securely.

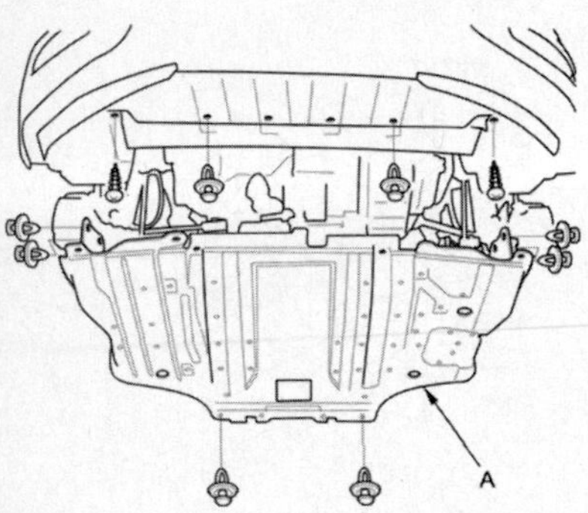
A

37647_INHY_G0183

Fig. 30 Remove the splash shield

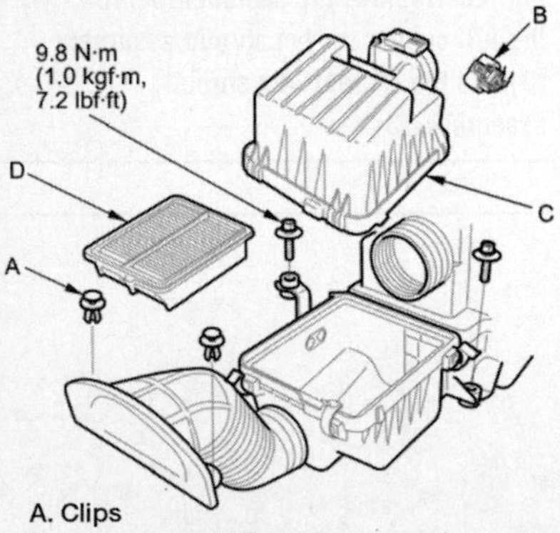
9.8 N·m
(1.0 kgf·m,
7.2 lbf·ft)
A
B
C
D

A. Clips
B. MAF sensor/IAT sensor connector
C. Air cleaner housing cover
D. Air cleaner element

37647_INHY_G0232

Fig. 31 Exploded view of air cleaner assembly

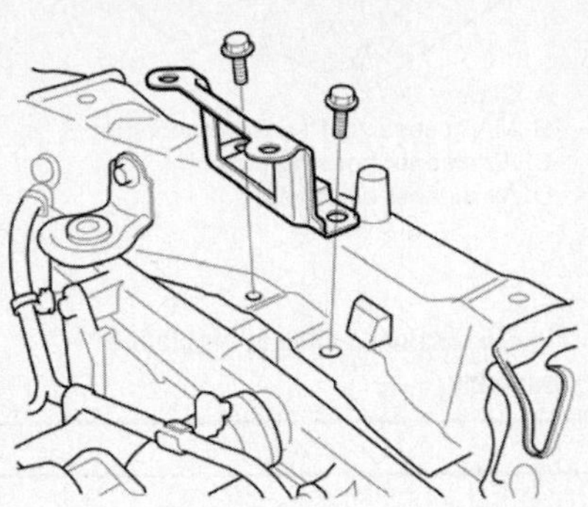

37647_INHY_G0233

Fig. 32 Remove the air duct bracket

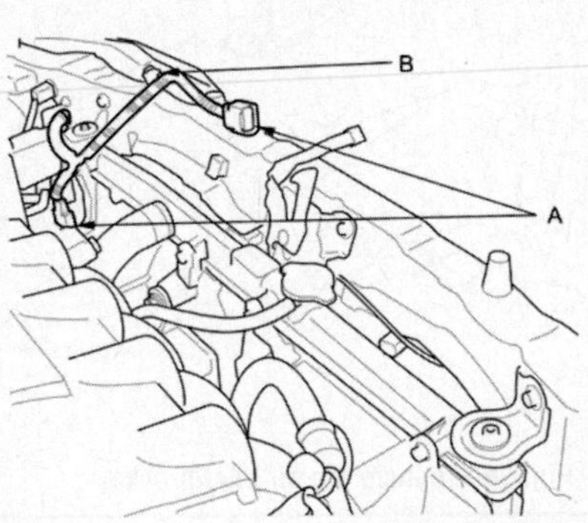
B
A

37647_INHY_G0234

Fig. 33 Disconnect the fan motor connectors (A) and the harness clamp (B)

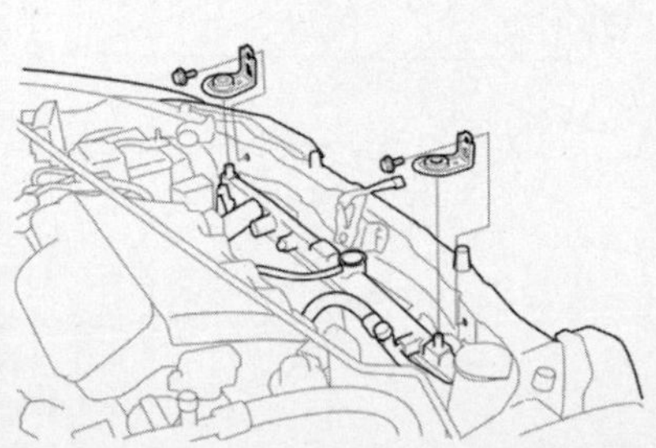

37647_INHY_G0237

Fig. 34 Remove the radiator brackets

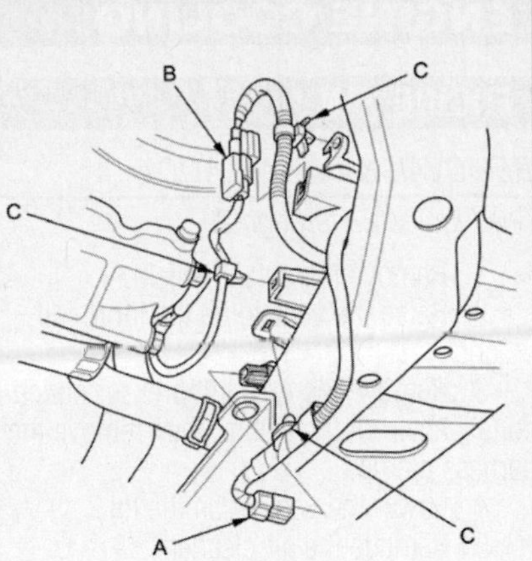
B
C
C
A
C

37647_INHY_G0238

Fig. 35 Disconnect the ECT sensor 2 connector (A), and remove the A/C compressor clutch connector (B) from the clamp; remove the harness clamps (C)

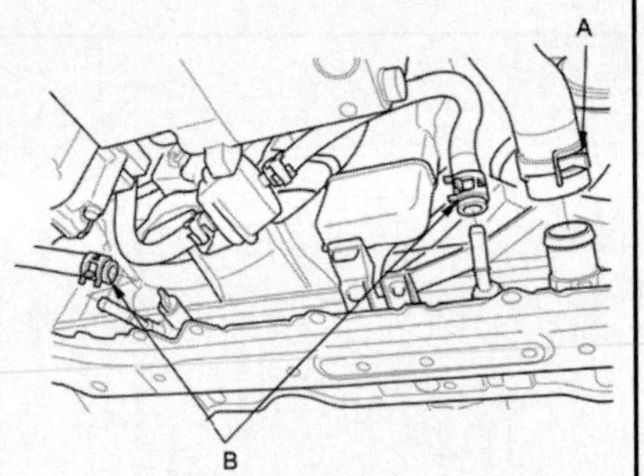
A
B

37647_INHY_G0239

Fig. 36 Disconnect the lower radiator hose (A); remove the CVT fluid (CVTF) cooler hoses (B)

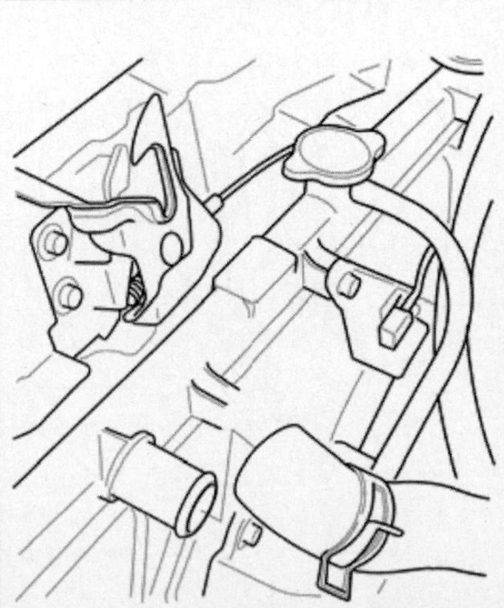

37647_INHY_G0240

Fig. 37 Remove the upper radiator hose

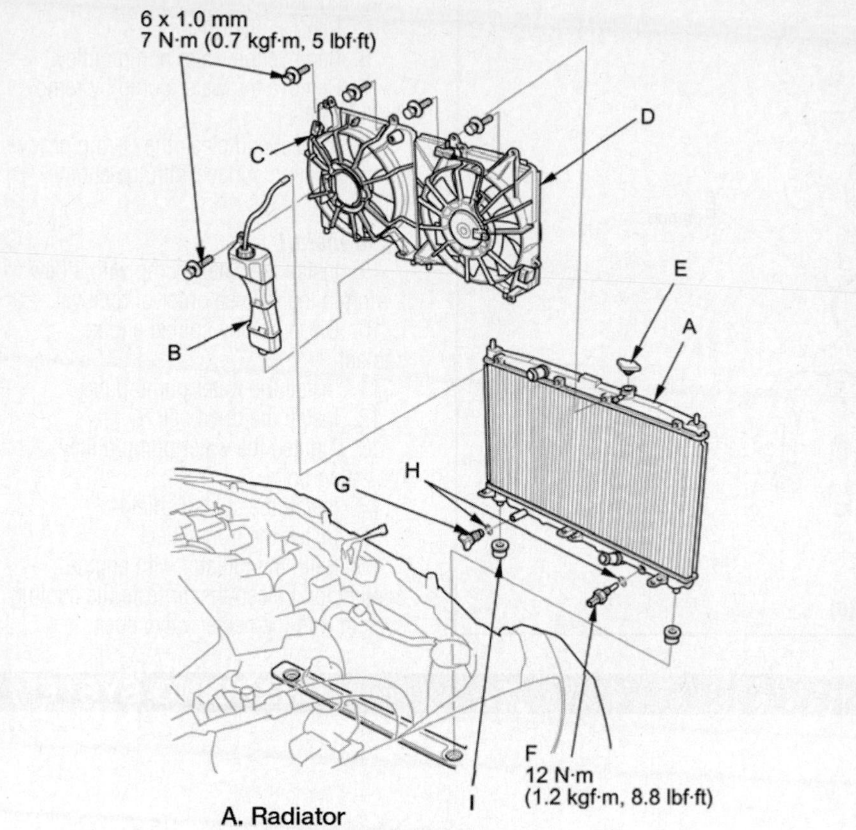
6 x 1.0 mm
7 N·m (0.7 kgf·m, 5 lbf·ft)
C
D
E
A
B
H
G
F
12 N·m
(1.2 kgf·m, 8.8 lbf·ft)
I

A. Radiator
B. Coolant reservoir
C. Radiator fan shroud assembly
D. A/C condenser fan shroud assembly
E. Radiator cap
F. ECT sensor 2
G. Drain plug
H. O-rings
I. Lower cushions

37647_INHY_G0241

Fig. 38 Pull up the radiator, then remove the coolant reservoir, the radiator fan shroud assembly, the A/C condenser fan shroud assembly, the radiator cap, the ECT sensor 2, and the drain plug

17. Install the upper radiator hose.
18. Raise the vehicle on the lift to full height.
19. Install the lower radiator hose.
20. Install the CVTF cooler hoses.
21. Connect the ECT sensor 2 connector, and install the A/C compressor clutch connector to the clamp.
22. Install the harness clamps.
23. Lower the vehicle on the lift.
24. Install the radiator brackets.
25. Connect the fan motor connectors and the harness clamp.
26. Install the air duct bracket.
27. Install the air cleaner.
28. Install the splash shields.
29. Refill the radiator with engine coolant, and bleed the air from the cooling system.
30. Clean up any spilled engine coolant.

THERMOSTAT

REMOVAL & INSTALLATION

See Figure 39.

1. Drain the engine coolant.
2. Remove the thermostat cover, then remove the thermostat.
3. Install the new rubber seal on the thermostat, then install the thermostat with the pin up, and install the thermostat cover.
4. Refill the radiator with engine coolant, then bleed the air from the cooling system.
5. Clean up any spilled engine coolant.

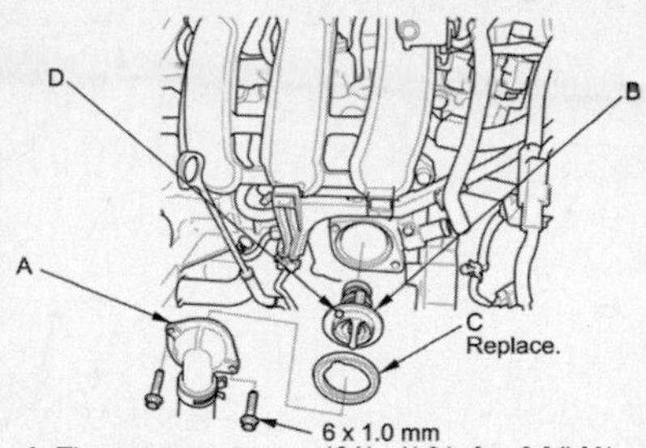
D
B
A
C
Replace.
6 x 1.0 mm
12 N·m(1.2 kgf·m, 8.8 lbf·ft)

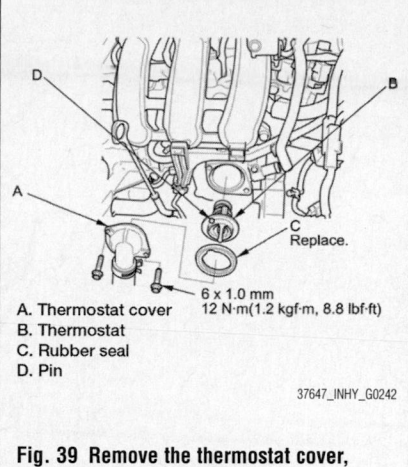

A. Thermostat cover
B. Thermostat
C. Rubber seal
D. Pin

37647_INHY_G0242

Fig. 39 Remove the thermostat cover, then remove the thermostat

WATER PUMP

REMOVAL & INSTALLATION

See Figures 40 through 42.

1. Remove the front wheel.
2. Remove the splash shields.
3. Drain the engine coolant.

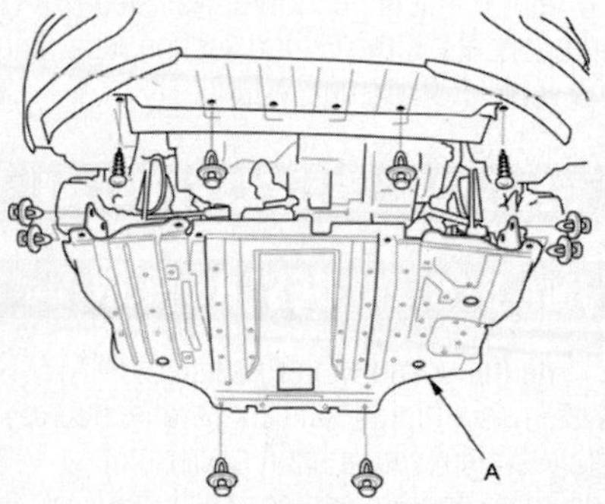
A

37647_INHY_G0183

Fig. 40 Remove the splash shields

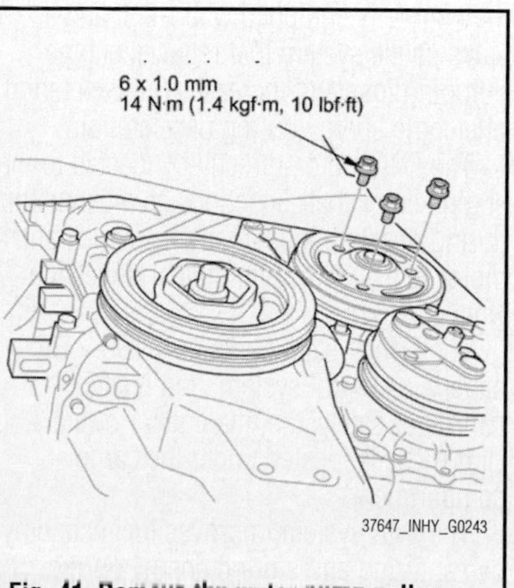
6 x 1.0 mm
14 N·m (1.4 kgf·m, 10 lbf·ft)

37647_INHY_G0243

Fig. 41 Remove the water pump pulley

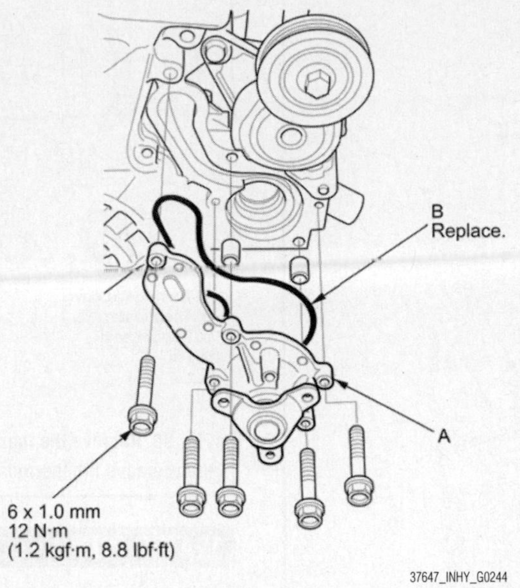

B
Replace.
A
6 x 1.0 mm
12 N·m
(1.2 kgf·m, 8.8 lbf·ft)
37647_INHY_G0244

Fig. 42 Remove the water pump (A) and the O-ring (B)

4. Loosen the water pump pulley mounting bolts.
5. Remove the drive belt.
6. Remove the water pump pulley.
7. Remove the water pump by removing the five bolts.
8. Inspect and clean the O-ring groove and the mating surface with the engine block.

To install:

9. Install the water pump with a new O-ring in the reverse order of removal.
10. Clean up any spilled engine coolant.
11. Install the water pump pulley.
12. Install the drive belt.
13. Tighten the water pump pulley mounting bolts.
14. Install the splash shields.
15. Install the front wheel.
16. Refill the radiator with engine coolant, and bleed the air from the cooling system with the heater valve open.

ENGINE ELECTRICAL — CHARGING SYSTEM

ALTERNATOR

Refer to Integrated Motor Assisted (IMA) components in the Hybrid section regarding the vehicle Charging System.

ENGINE ELECTRICAL — HYBRID INTEGRATED MOTOR ASSISTED (IMA) SYTEM

IMA SYSTEM DESCRIPTION

The Integrated Motor Assisted (IMA) system is a highly-efficient parallel hybrid drive system with a main power unit (gasoline engine) and an assist unit (electric motor).

The engine is an in-line, 4-cylinder, 8-valve power plant with a displacement of 82 cu. in (1.339 liters). To reduce fuel use, the engine is equipped with i-DSI and a valve pause system that reduces engine pumping loss and increases the generation of electric energy during deceleration.

The IMA motor, directly connected to the engine crankshaft, functions as a generator during deceleration, an engine starter, and a motor to assist the engine that drives the wheels.

The IMA system contains a 100 V DC battery, a control system, and related parts. For safety, the Intelligent Power Unit (IPU) is located under the cargo compartment.

The IMA system improves fuel economy by capturing and storing energy during deceleration.

OPERATING CONDITIONS

1. Engine start: The IMA system drives the IMA motor and starts the engine during normal starts and when re-starting from an idle stop. The IMA motor is linked directly to the engine crankshaft.
2. Start running: The IMA motor assists the engine.
3. Slow acceleration: Runs only with the engine.
4. Low speed cruise: At a constant speed of about 25 mph (40 km/h) with light load, all engine cylinders are deactivated, and the vehicle runs only with the IMA motor. When the State-Of-Charge (SOC) drops, the engine starts and the IMA motor begins to charge the battery module.
5. Acceleration: The IMA motor assists the engine.
6. High acceleration: The IMA motor assists the engine.
7. High speed cruise: Runs only with the engine. When the state-of-charge drops, the IMA motor begins to charge the battery module.
8. Deceleration: All engine cylinders are deactivated, and the IMA motor captures the deceleration energy and charges the battery module.
9. Stop: When conditions are satisfied, the PCM stops the engine automatically (auto stop).

IMA SERVICE PRECAUTIONS

The IMA (integrated motor assisted) system uses high voltage (100 V) circuits. Be sure to shut off the electrical circuits and isolate the IMA system and related parts before servicing the IMA system.

The high voltage cables and their covers are identified by orange coloring. The caution labels are attached to high voltage and other related parts. When the system is energized, be careful not to touch these cables and parts without adequate protective gear. The front floor under-cover protecting the high voltage cables is marked .

If the 12 V battery is discharged, its cables have been disconnected, or the MCM (Motor Control Module) has been reset, the IMA battery level indicator does not display the State-Of-Charge (SOC) when the engine is first started. To display the level in the indicator, start the engine, and hold it between 3,500 and 4,000 rpm without load (in P or N) until the level in the indicator is half full.

Observe the following instructions when inspecting or servicing the IMA system:

When the system is energized, servicing, disassembling, or replacing items marked in each procedure requires insulated tools.

When the IMA system indicator is on, do the IMA system troubleshooting first.

Wear insulated gloves whenever you inspect or service the IMA system. Be sure to check the gloves for pin holes, tears, or other damage.

To make sure the system is not energized, turn the battery module switch OFF, and secure the switch in the OFF position with the locking cover before servicing the IMA system (see Turning off and on power to the high voltage circuit).

Wait at least 5 minutes after turning off the battery module switch, then disconnect the negative cable from the 12 V battery (it takes about 5 minutes for the PDU capacitor to discharge).

Before disconnecting the high voltage cable terminals, use a voltmeter to make sure that the voltage between the terminals is below 30 V.

When the system is energized and you are servicing parts without an insulated sheath, be sure to use insulated tools to prevent short circuiting.

The rotor assembly contains very strong magnets and should be handled with special care. People with pacemakers or other magnetically sensitive medical devices should not handle the rotor assembly.

Use the special tool (rotor puller) to remove or install the rotor assembly.

⁂ WARNING

If the rotor is installed by hand, it may suddenly be pulled toward the stator with great force, causing serious hand or finger injury. Always use the special tool (rotor puller) to remove or install the rotor assembly.

Keep the rotor assembly away from magnetically sensitive devices.

After disconnecting the high voltage terminals, busbars, etc., insulate the parts with insulated tape.

As a safety warning, attach a sign saying, WORKING ON HIGH VOLTAGE PARTS. DO NOT TOUCH! to the steering wheel.

DISCONNECTING THE MOTOR POWER CABLE CONNECTOR FROM THE MOTOR STATOR

See Figure 43.

⁂ CAUTION

IMA components are located in this area. The IMA is a high-voltage system. You must be familiar with the IMA system before working on or around it. Make sure you have read the IMA service precautions before doing repairs or service.

1. Turn the ignition switch to LOCK (0). Turn the battery module switch OFF.
2. Slide the protector in the direction of the arrow. Push the tab, then raise the lever. Remove the IMA motor power cable from the motor stator.

➡Note the following:

- If the outside of the IMA motor power cable connector is dirty, clean it before you disconnect it.
- Cover the disconnected connector with a plastic bag, and wrap the IMA motor power cable terminals with insulating tape.
- If the IMA motor power cable is wet, dry them with a clean shop towel. Do not use compressed air.

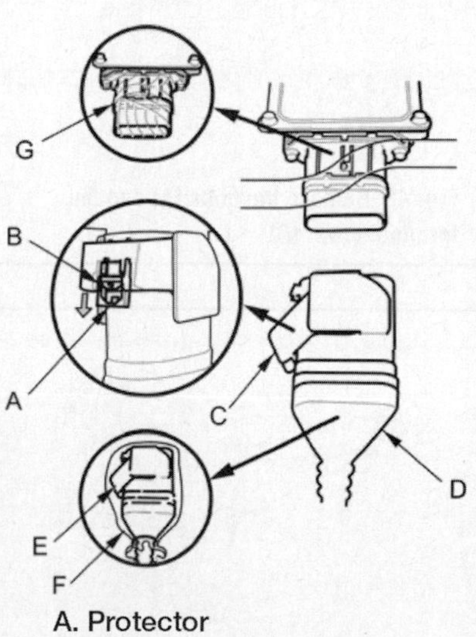
G
B
A
C
E
F
D
A. Protector
B. Tab
C. Lever
D. IMA motor power cable
E. Connector
F. Plastic bag
G. Insulating tape
37647_INHY_G0255

Fig. 43 Remove the IMA motor power cable from the motor stator

TURNING OFF AND ON POWER TO THE HIGH VOLTAGE CIRCUIT

See Figures 44 through 47.

⁂ CAUTION

IMA components are located in this area. The IMA is a high-voltage system. You must be familiar with the IMA system before working on or around it. Make sure you have read the IMA service precautions before doing repairs or service.

The following procedure should be done before you work on or near any energized high voltage components. Follow the procedure exactly. Otherwise, you may be injured or may damage equipment.

1. Turn the ignition switch to LOCK (0), then remove the key from the ignition switch.

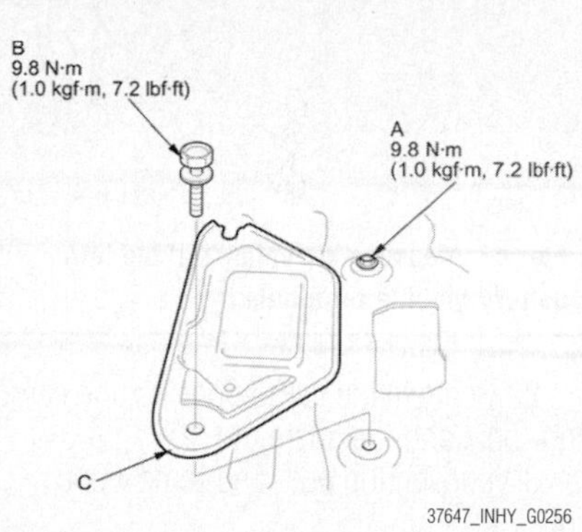
B
9.8 N·m
(1.0 kgf·m, 7.2 lbf·ft)
A
9.8 N·m
(1.0 kgf·m, 7.2 lbf·ft)
C
37647_INHY_G0256

Fig. 44 Loosen the bolt (A), and remove the bolt (B) and the battery module switch lid (C)

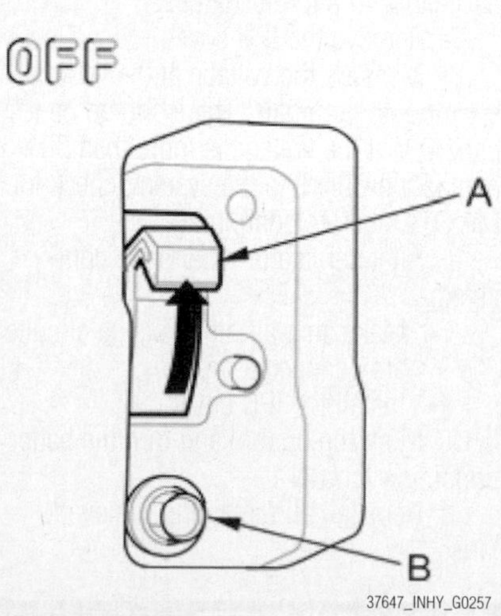
OFF
A
B
37647_INHY_G0257

Fig. 45 Turn the battery module switch (A) OFF, then check that the bolt (B) is showing

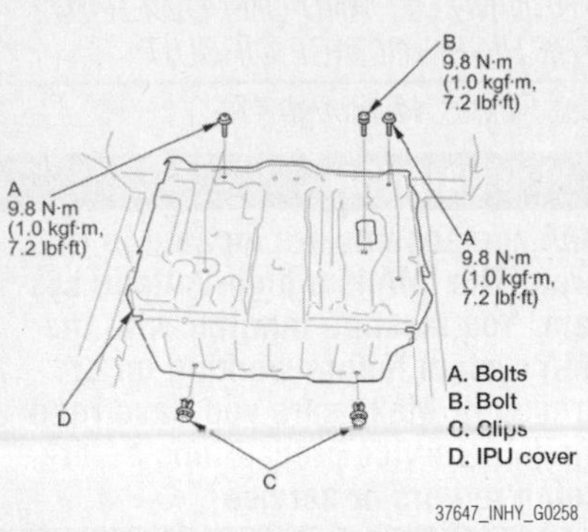
B
9.8 N·m
(1.0 kgf·m,
7.2 lbf·ft)
A
9.8 N·m
(1.0 kgf·m,
7.2 lbf·ft)
A
9.8 N·m
(1.0 kgf·m,
7.2 lbf·ft)
D
C
A. Bolts
B. Bolt
C. Clips
D. IPU cover
37647_INHY_G0258

Fig. 46 Remove the IPU cover

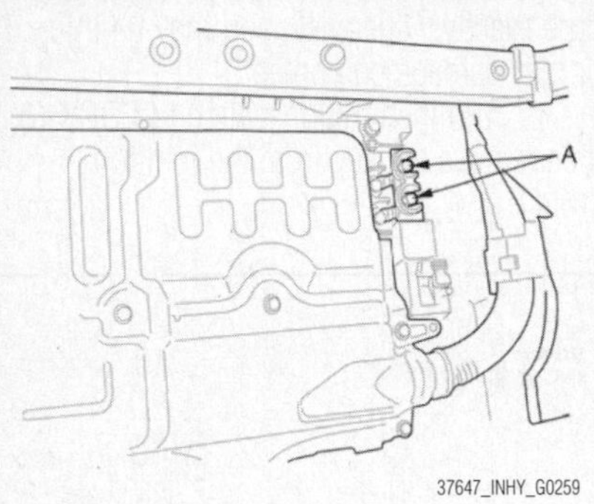
A
37647_INHY_G0259

Fig. 47 Measure the voltage at the IMA battery module terminals (A)

2. Remove the cargo floor lid, the cargo floor box, and the spare tire.
3. Loosen the bolt, and remove the bolt.
4. Remove the battery module switch lid from the IPU cover.
5. Turn the battery module switch OFF, then check that the bolt is showing.
6. Wait at least 5 minutes to allow the PDU capacitors to discharge.
7. Remove the IPU cover.
8. Measure the voltage at the IMA battery module terminals. There should be less than 30 V. If the voltage is more than 30 V, there is a problem in the system; check for IMA DTCs before continuing.
9. After service or repairs are completed:
   - Make sure all high voltage circuits are connected properly.
   - Install the IPU cover.
10. Push the button, and turn the battery module switch ON.
11. Reinstall all remaining removed parts.

DC-DC CONVERTER

REMOVAL & INSTALLATION

See Figures 48 and 49.

CAUTION

IMA components are located in this area. The IMA is a high-voltage system. You must be familiar with the IMA system before working on or around it. Make sure you have read the IMA service precautions before doing repairs or service.

1. Make sure the ignition switch is in LOCK (0).
2. Do the 12 volt battery terminal disconnection procedure.
3. Remove the MCM.
4. Remove the bolts and the terminal cover.
5. Disconnect the DC-DC converter connector.
6. Remove the bolts and the busbar.
7. Remove the bolts and the DC-DC converter.
8. Install the parts in the reverse order of removal.

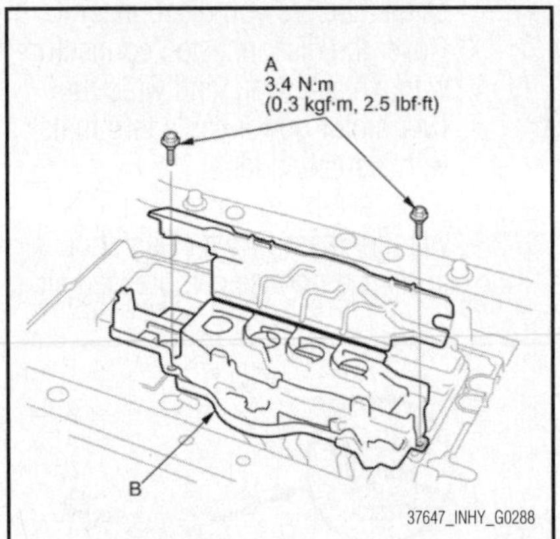
A
3.4 N·m
(0.3 kgf·m, 2.5 lbf·ft)
B
37647_INHY_G0288

Fig. 48 Remove the bolts (A) and the terminal cover (B)

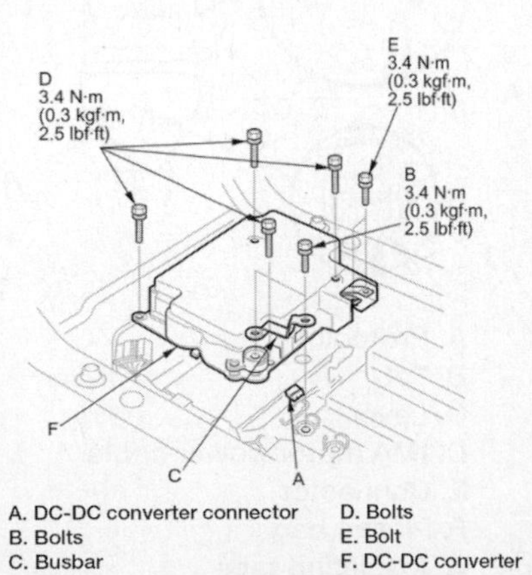
E
3.4 N·m
(0.3 kgf·m,
2.5 lbf·ft)
D
3.4 N·m
(0.3 kgf·m,
2.5 lbf·ft)
B
3.4 N·m
(0.3 kgf·m,
2.5 lbf·ft)
F
C
A
A. DC-DC converter connector
B. Bolts
C. Busbar
D. Bolts
E. Bolt
F. DC-DC converter
37647_INHY_G0289

Fig. 49 Disconnect the DC-DC converter connector

9. Do the 12 volt battery terminal reconnection procedure.

IMA MOTOR DRAIN COVER

REMOVAL & INSTALLATION

See Figures 50 and 51.

CAUTION

IMA components are located in this area. The IMA is a high-voltage system. You must be familiar with the IMA system before working on or around it. Make sure you have read the IMA service precautions before doing repairs or service.

1. Remove the splash shields.
2. Remove the drain cover.
3. Install the parts in the reverse order of removal.

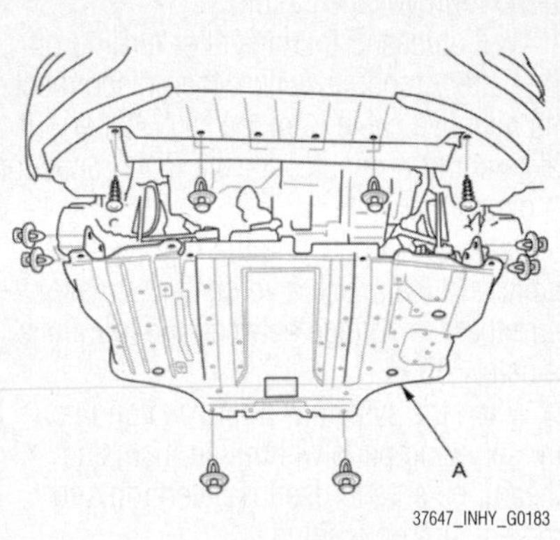
A
37647_INHY_G0183

Fig. 50 Remove the splash shields

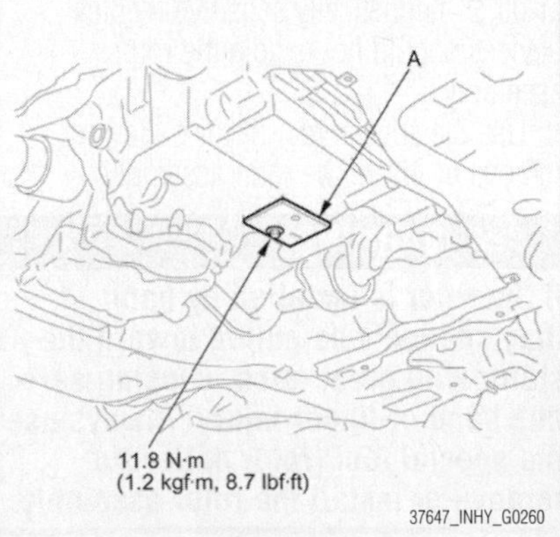
A
11.8 N·m
(1.2 kgf·m, 8.7 lbf·ft)
37647_INHY_G0260

Fig. 51 Remove the drain cover (A)

IMA MOTOR HOUSING

REMOVAL & INSTALLATION

See Figure 52.

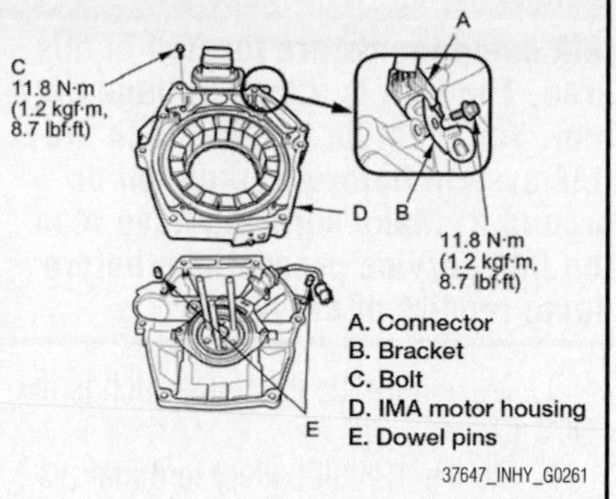
A
C
11.8 N·m
(1.2 kgf·m,
8.7 lbf·ft)
D
B
11.8 N·m
(1.2 kgf·m,
8.7 lbf·ft)
E
A. Connector
B. Bracket
C. Bolt
D. IMA motor housing
E. Dowel pins
37647_INHY_G0261

Fig. 52 Remove the connector and the bracket

⁂ CAUTION

IMA components are located in this area. The IMA is a high-voltage system. You must be familiar with the IMA system before working on or around it. Make sure you have read the IMA service precautions before doing repairs or service.

1. Remove the IMA motor rotor.
2. Remove the connector and the bracket.
3. Remove the bolt and the IMA motor housing.

➡Set the dowel pins in the IMA motor housing before installing the motor stator on the engine.

4. Install the parts in the reverse order of removal.
5. Install the IMA motor rotor.
6. Do the motor rotor position calibration procedure.

IMA MOTOR POWER CABLE

REMOVAL & INSTALLATION

See Figures 53 through 64.

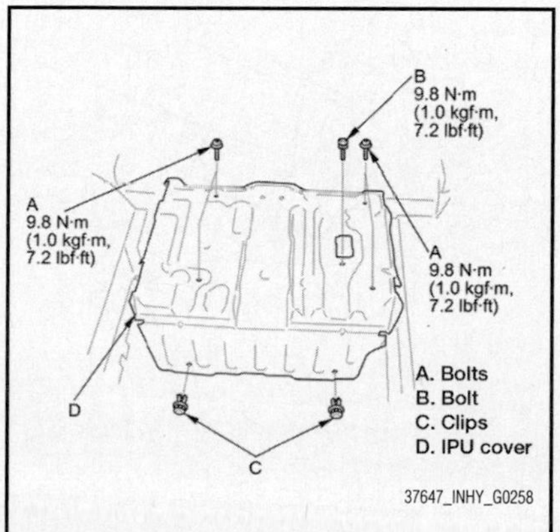
B
9.8 N·m
(1.0 kgf·m,
7.2 lbf·ft)
A
9.8 N·m
(1.0 kgf·m,
7.2 lbf·ft)
A
9.8 N·m
(1.0 kgf·m,
7.2 lbf·ft)
D
C
A. Bolts
B. Bolt
C. Clips
D. IPU cover
37647_INHY_G0258

Fig. 53 Remove the IPU cover

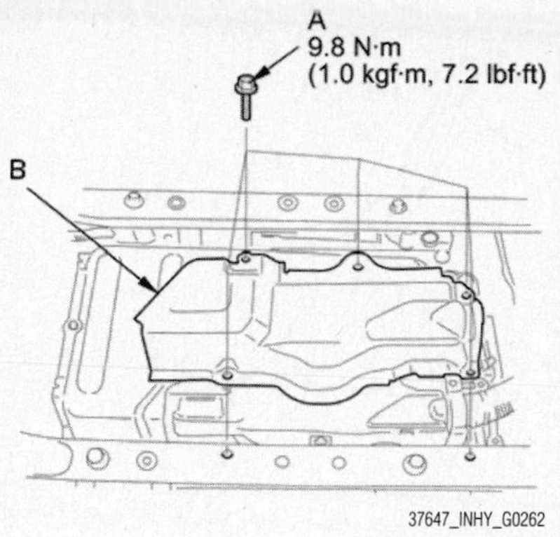
A
9.8 N·m
(1.0 kgf·m, 7.2 lbf·ft)
B
37647_INHY_G0262

Fig. 54 Remove the bolts (A) and the PCU lid (B)

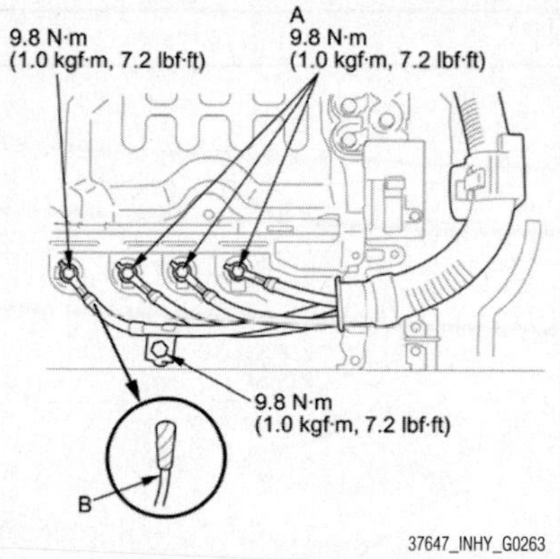
9.8 N·m
(1.0 kgf·m, 7.2 lbf·ft)
A
9.8 N·m
(1.0 kgf·m, 7.2 lbf·ft)
9.8 N·m
(1.0 kgf·m, 7.2 lbf·ft)
B
37647_INHY_G0263

Fig. 55 Remove the IMA motor power cables (A)

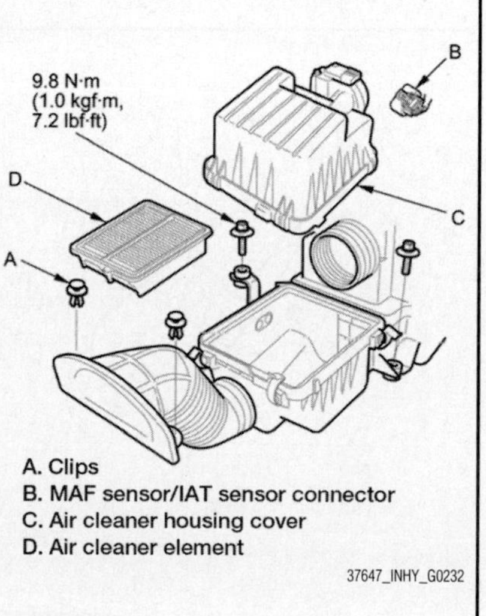
9.8 N·m
(1.0 kgf·m,
7.2 lbf·ft)
B
D
C
A
A. Clips
B. MAF sensor/IAT sensor connector
C. Air cleaner housing cover
D. Air cleaner element
37647_INHY_G0232

Fig. 56 Remove the air cleaner

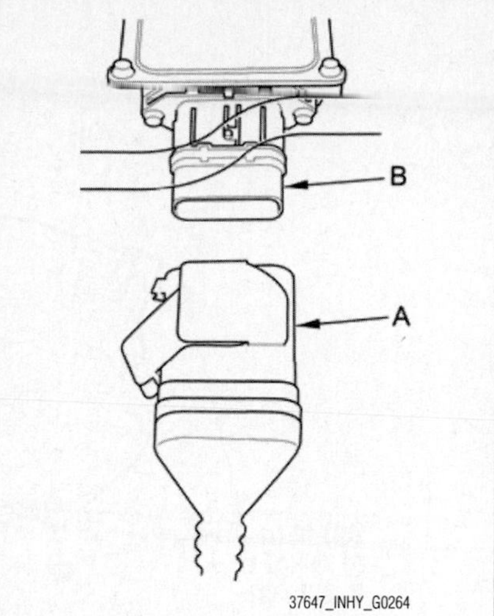
B
A
37647_INHY_G0264

Fig. 57 Disconnect the IMA motor power cable connector (A) from the motor stator (B)

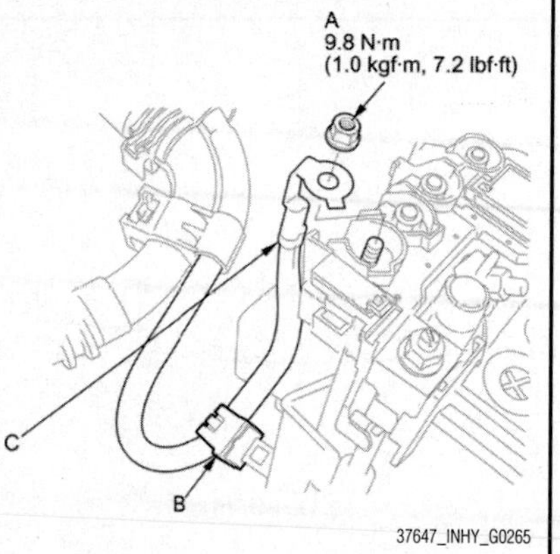
A
9.8 N·m
(1.0 kgf·m, 7.2 lbf·ft)
C
B
37647_INHY_G0265

Fig. 58 Remove the nut (A), the clamp (B), and the 12 V power cable (C) from the battery terminal fuse box

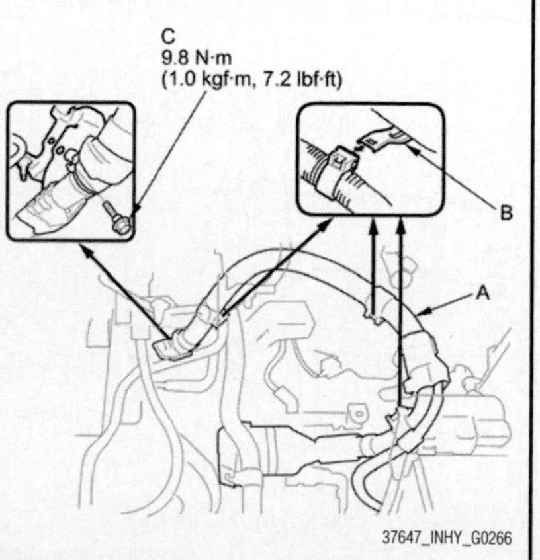
C
9.8 N·m
(1.0 kgf·m, 7.2 lbf·ft)
B
A
37647_INHY_G0266

Fig. 59 Remove the IMA motor power cable (A) from the brackets (B); remove the bolt (C)

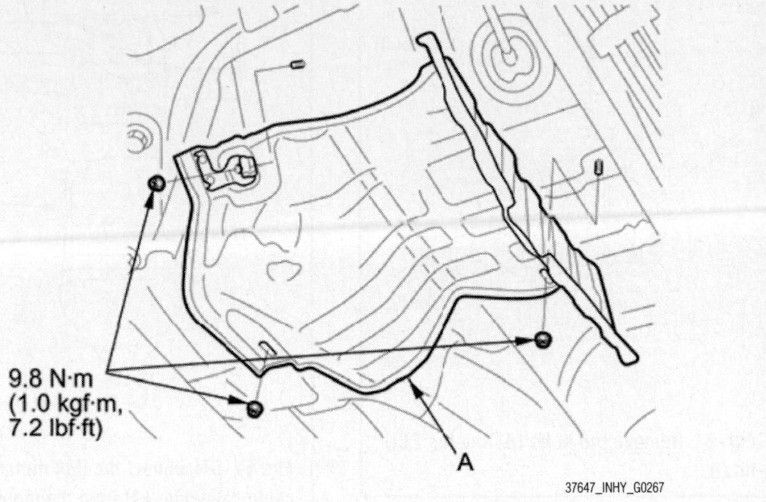
9.8 N·m
(1.0 kgf·m,
7.2 lbf·ft)
A
37647_INHY_G0267

Fig. 60 Remove the heat shield (A)

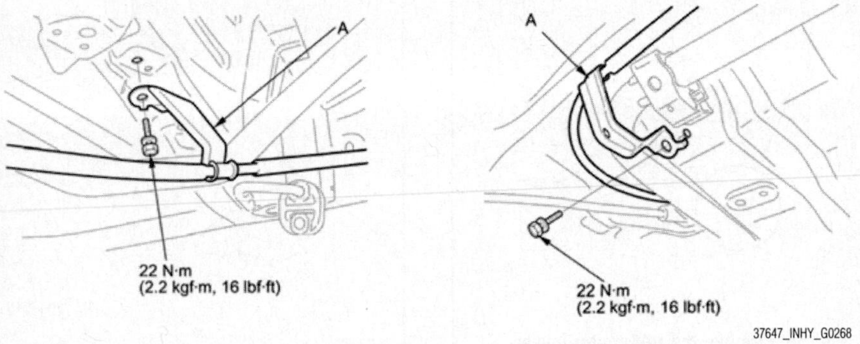
A
22 N·m
(2.2 kgf·m, 16 lbf·ft)
A
22 N·m
(2.2 kgf·m, 16 lbf·ft)
37647_INHY_G0268

Fig. 61 Remove the parking brake cable brackets (A)

CAUTION

IMA components are located in this area. The IMA is a high-voltage system. You must be familiar with the IMA system before working on or around it. Make sure you have read the IMA service precautions before doing repairs or service.

1. Make sure the ignition switch is in LOCK (0).
2. Do the 12 volt battery terminal disconnection procedure.
3. Remove the IPU cover.
4. Remove the bolts and the PCU lid.
5. Remove the IMA motor power cables.

➡Check the position of the U phase, V phase, and W phase cables before disconnecting the IMA motor power cables.

6. Remove the 12 V power cable, and wrap it with insulating tape.
7. Remove the air cleaner.
8. Disconnect the IMA motor power cable connector from the motor stator.

➡Refer to disconnecting the IMA motor power cable connector from the motor stator. If the IMA motor power cable terminals are wet, dry them with a clean shop towel. Do not use compressed air.

9. Remove the nut, the clamp, and the 12 V power cable from the battery terminal fuse box.
10. Remove the cowl cover and the under-cowl panel.
11. Remove the IMA motor power cable from the brackets.
12. Remove the bolt.

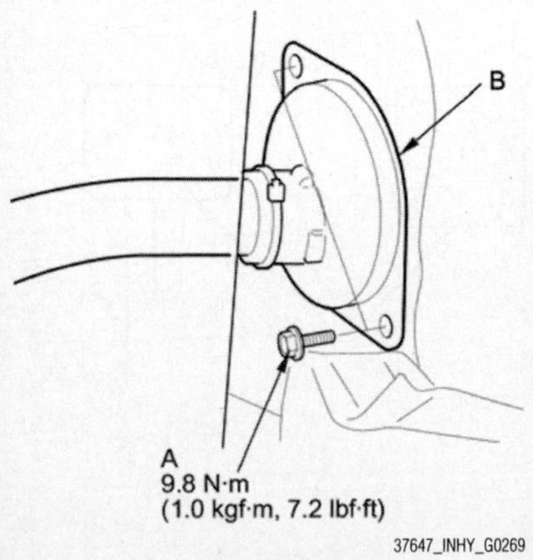
B
A
9.8 N·m
(1.0 kgf·m, 7.2 lbf·ft)
37647_INHY_G0269

Fig. 62 Remove the bolts (A) and the cover (B)

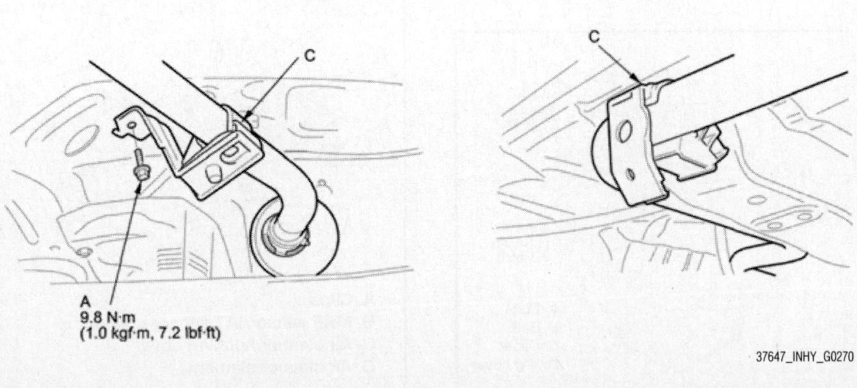
C
A
9.8 N·m
(1.0 kgf·m, 7.2 lbf·ft)
C
37647_INHY_G0270

Fig. 63 Remove the bolts (A) and the nut (B) then remove the clamps (C)—1 of 2

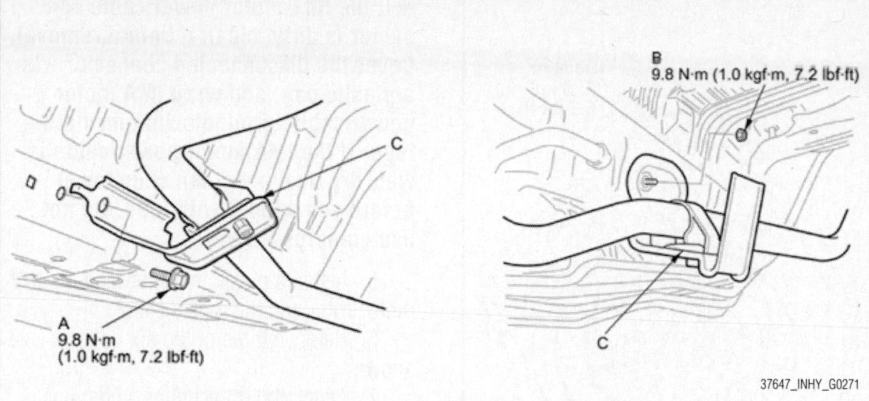
B
9.8 N·m (1.0 kgf·m, 7.2 lbf·ft)
C
A
9.8 N·m
(1.0 kgf·m, 7.2 lbf·ft)
C
37647_INHY_G0271

Fig. 64 Remove the bolts (A) and the nut (B) then remove the clamps (C)—2 of 2

13. Remove the fuel tank.
14. Remove the exhaust pipe and the muffler.
15. Remove the under-floor TWC.
16. Remove the heat shield.
17. Remove the parking brake cable brackets.
18. Remove the bolts and the cover.
19. Remove the bolts and the nut then remove the clamps.

➡Do not reuse the clamps. Replace them with new ones.

To install:

20. Install the parts in the reverse order of removal.
21. Do the 12 volt battery terminal reconnection procedure.

IMA MOTOR ROTOR POSITION SENSOR

REMOVAL & INSTALLATION

See Figure 65.

1. Remove the IMA motor housing.
2. Remove the bolts and the IMA motor rotor position sensor.

To install:

3. Install the parts in the reverse order of removal.

➡Tighten the A bolts first, then tighten the B bolts.

4. Do the motor rotor position calibration procedure.

IMA ROTOR POSITION CALIBRATION

Do the IMA motor rotor position calibration whenever any of these actions are done:

- The MCM is replaced.
- The IMA motor rotor position sensor is replaced or removed during service.
- The IMA motor is replaced.

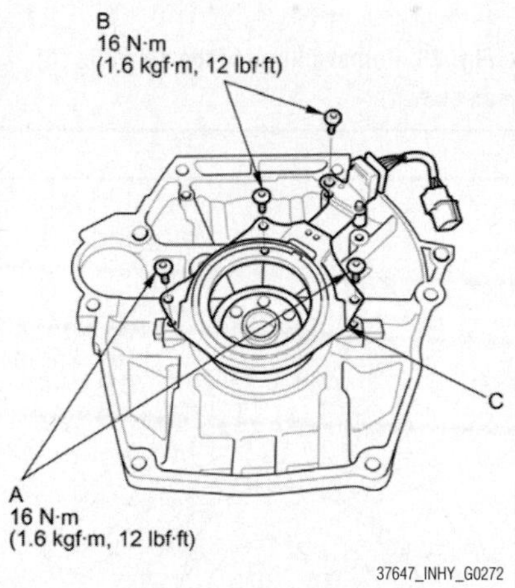
B
16 N·m
(1.6 kgf·m, 12 lbf·ft)
C
A
16 N·m
(1.6 kgf·m, 12 lbf·ft)
37647_INHY_G0272

Fig. 65 Remove the bolts (A, B) and the IMA motor rotor position sensor (C)

- The engine assembly is replaced.
- The transmission is replaced.

1. Connect the HDS to the Data Link Connector (DLC) located under the driver's side of the dashboard.
2. Turn the ignition switch to ON (II).
3. Make sure the HDS communicates with the vehicle and the MCM (IMA system). If it doesn't, troubleshoot the DLC circuit.
4. Select IMA SYSTEM on the HDS.
5. Select the MOTOR ROTOR POSITION CALIBRATION in the ADJUSTMENT MENU of the HDS.
6. Turn the ignition switch to LOCK (0), and disconnect the HDS from the DLC.

IMA MOTOR ROTOR

REMOVAL & INSTALLATION

See Figures 66 through 73.

Special Tools Required: Rotor Puller 07YAC-PHM010C

⁂ CAUTION

IMA components are located in this area. The IMA is a high-voltage system. You must be familiar with the IMA system before working on or around it. Make sure you have read the IMA service precautions before doing repairs or service.

⁂ CAUTION

The motor rotor contains very strong magnets and should be handled with special care. People with pacemakers or other sensitive medical devices should not handle the IMA motor rotor.

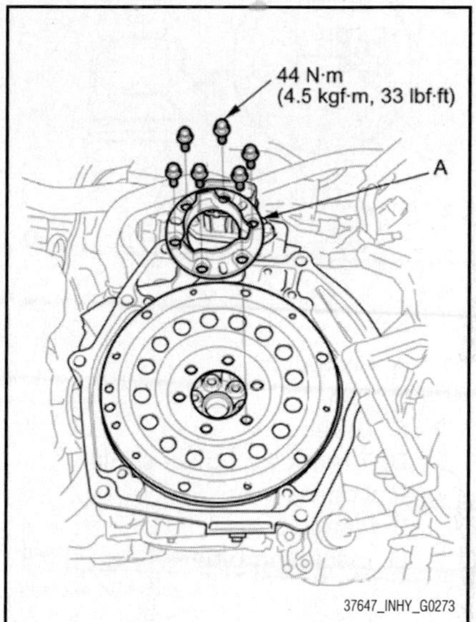
44 N·m
(4.5 kgf·m, 33 lbf·ft)
A
37647_INHY_G0273

Fig. 66 Remove the support (A)

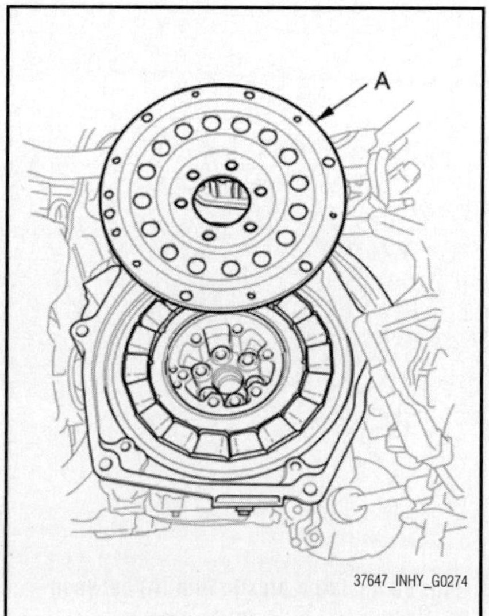
A
37647_INHY_G0274

Fig. 67 Remove the drive plate (A)

⁂ WARNING

If the rotor is installed by hand, it may suddenly be pulled toward the stator with great force, causing serious hand or finger injury. Always use the special tool (rotor puller) to remove or install the rotor assembly.

- Keep the motor rotor away from magnetically sensitive devices.

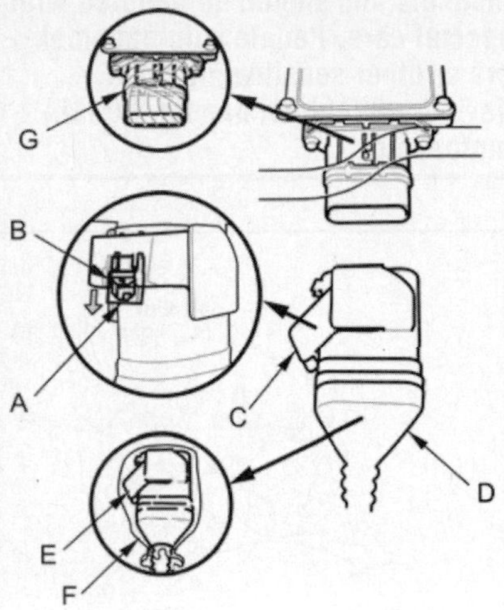
G
B
A
C
D
E
F

A. Protector
B. Tab
C. Lever
D. IMA motor power cable
E. Connector
F. Plastic bag
G. Insulating tape

37647_INHY_G0255

Fig. 68 Remove the IMA motor power cable from the motor stator

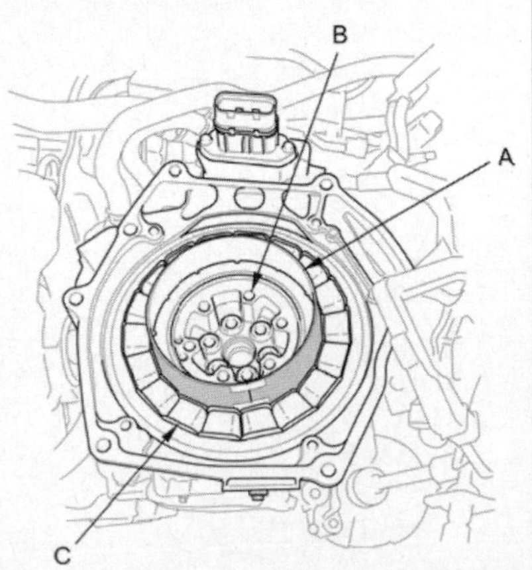
B
A
C

37647_INHY_G0275

Fig. 69 Install a plastic film (A) between the IMA motor rotor (B) and the motor stator (C)

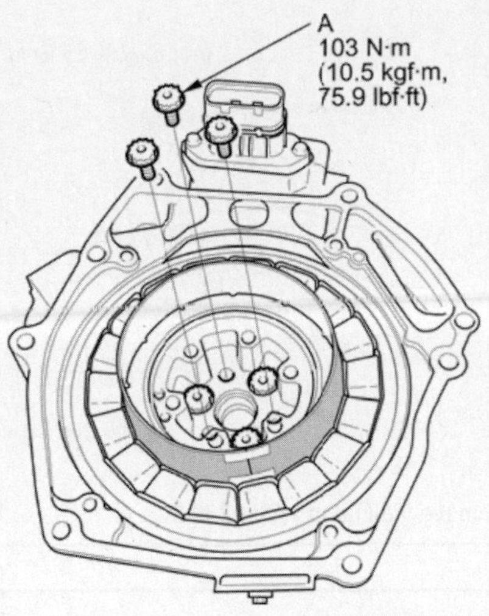
A
103 N·m
(10.5 kgf·m,
75.9 lbf·ft)

37647_INHY_G0276

Fig. 70 Remove three of the six bolts (A) as shown

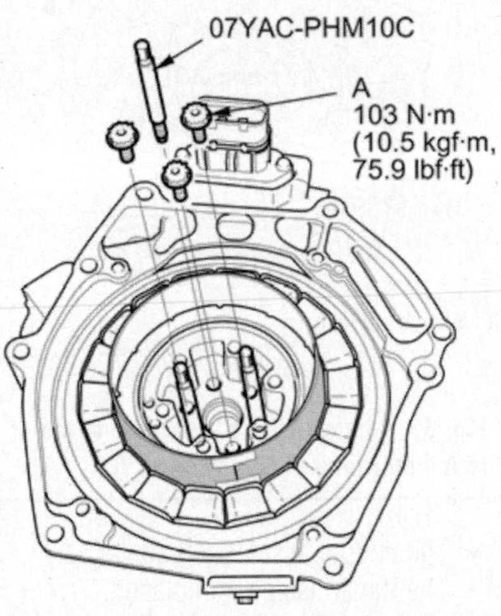
07YAC-PHM10C
A
103 N·m
(10.5 kgf·m,
75.9 lbf·ft)

37647_INHY_G0277

Fig. 71 Install the rotor puller guide pins, then remove the remaining three bolts (A)

- Do not blow air near the rotor, as metal particles may get on the magnets.
- Store the rotor in the designated storage box, and keep it away from sensitive devices during storage.

1. Remove the transmission.
2. Remove the support.
3. Remove the drive plate.
4. Turn the ignition switch to LOCK (0). Turn the battery module switch OFF. Slide the protector in the direction of the arrow. Push the tab, then raise the lever. Remove the IMA motor power cable from the motor stator.

➡If the IMA motor power cable connector is dirty, clean it before removal. Cover the disconnected connector with a plastic bag, and wrap IMA motor power cable terminals with insulating tape. If the IMA motor power cable is wet, dry it with a clean shop towel before you wrap it with tape. Do not use compressed air.

5. Install a plastic film between the IMA motor rotor and the motor stator.
6. Remove three of the six bolts as shown.
7. Install the rotor puller guide pins, then remove the remaining three bolts.
8. Attach the rotor puller with its supplied bolts.

➡When installing the rotor puller, position the puller to fit over the guide pins.

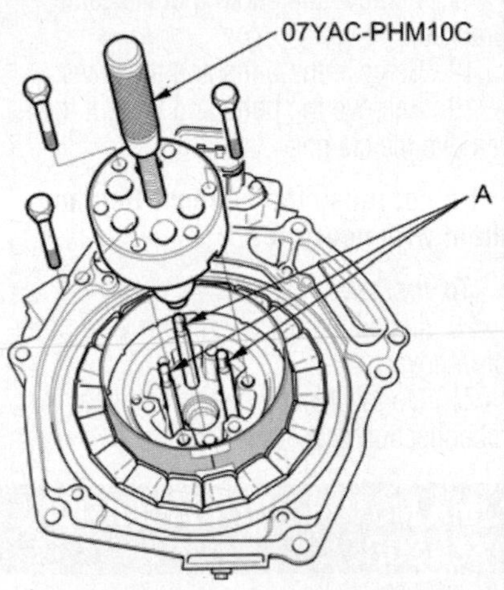
07YAC-PHM10C
A

37647_INHY_G0278

Fig. 72 Position the puller to fit over the guide pins (A)

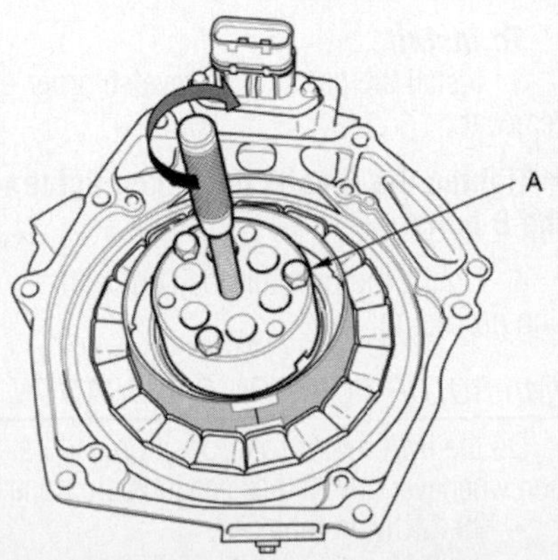
A

37647_INHY_G0279

Fig. 73 Remove the IMA motor rotor (A)

9. Remove the IMA motor rotor.

10. To prevent damage to the rotor magnets while working on the stator, place the rotor, with puller attached, into the puller case.

To install:

11. Install the parts in the reverse order of removal.

➡When installing the rotor, install the special tool to the rotor and set the rotor with the tool tip stretched out. Turn the handle of the special tool slowly when inserting the rotor into the stator. The rotor is drawn into the stator by magnetic force.

12. Remove the plastic film.

13. Reconnect the IMA motor power cable to the motor stator.

14. Reinstall the drive plate.

15. Reinstall the support.

16. Install the transmission.

17. Do the motor rotor position calibration procedure.

IMA ROTOR POSITION CALIBRATION

Do the IMA motor rotor position calibration whenever any of these actions are done:

- The MCM is replaced.
- The IMA motor rotor position sensor is replaced or removed during service.
- The IMA motor is replaced.
- The engine assembly is replaced.
- The transmission is replaced.

1. Connect the HDS to the Data Link Connector (DLC) located under the driver's side of the dashboard.

2. Turn the ignition switch to ON (II).

3. Make sure the HDS communicates with the vehicle and the MCM (IMA system). If it doesn't, troubleshoot the DLC circuit.

4. Select IMA SYSTEM on the HDS.

5. Select the MOTOR ROTOR POSITION CALIBRATION in the ADJUSTMENT MENU of the HDS.

6. Turn the ignition switch to LOCK (0), and disconnect the HDS from the DLC.

IPU CASE

REMOVAL & INSTALLATION

See Figures 74 through 79.

⁂ CAUTION

IMA components are located in this area. The IMA is a high-voltage system. You must be familiar with the IMA system before working on or around it. Make sure you have read the IMA service precautions before doing repairs or service.

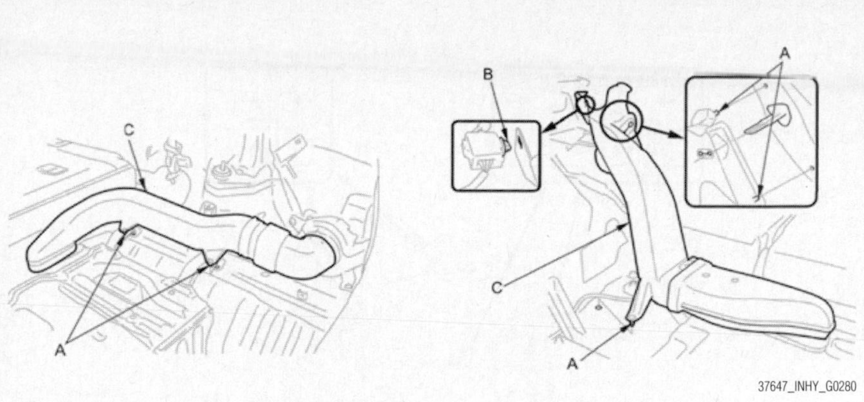
A
B
C
C
A
A
37647_INHY_G0280

Fig. 74 Remove the clips (A, B) and the IPU module air duct (C)

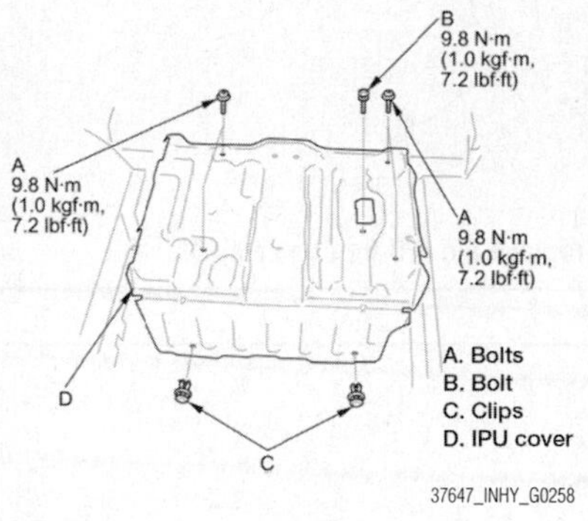
B
9.8 N·m
(1.0 kgf·m,
7.2 lbf·ft)
A
9.8 N·m
(1.0 kgf·m,
7.2 lbf·ft)
A
9.8 N·m
(1.0 kgf·m,
7.2 lbf·ft)
D
C
A. Bolts
B. Bolt
C. Clips
D. IPU cover
37647_INHY_G0258

Fig. 75 Remove the IPU cover

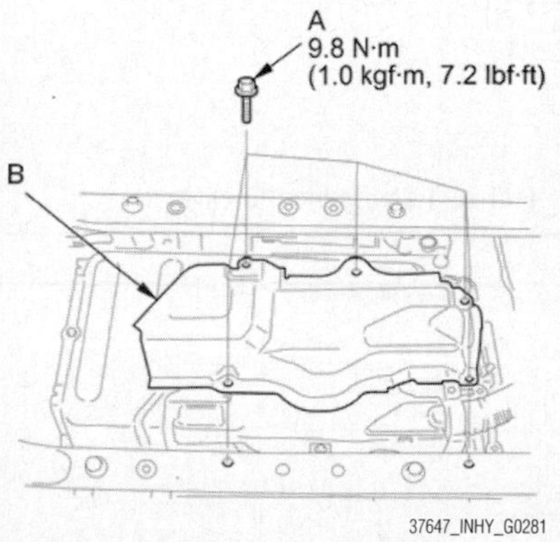
A
9.8 N·m
(1.0 kgf·m, 7.2 lbf·ft)
B
37647_INHY_G0281

Fig. 76 Remove the bolts (A) and the PCU lid (B)

1. Make sure the ignition switch to LOCK (0).

2. Do the 12 volt battery terminal disconnection procedure.

3. Remove the IPU module air duct.

a. Remove the cargo floor lid, the IPU duct cover, and the cargo area side trim panel.

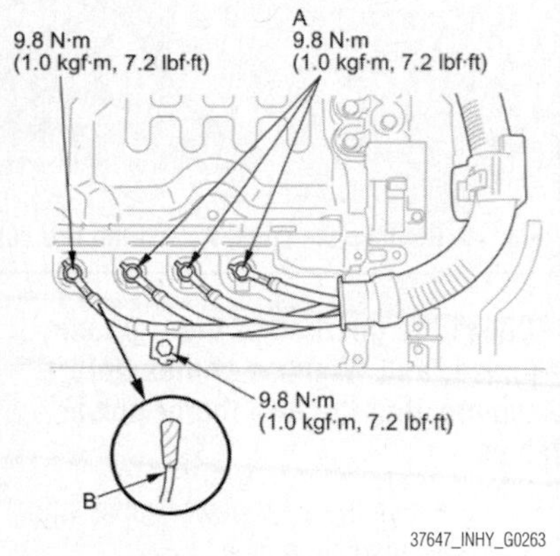
9.8 N·m
(1.0 kgf·m, 7.2 lbf·ft)
A
9.8 N·m
(1.0 kgf·m, 7.2 lbf·ft)
9.8 N·m
(1.0 kgf·m, 7.2 lbf·ft)
B
37647_INHY_G0263

Fig. 77 Remove the IMA motor power cables

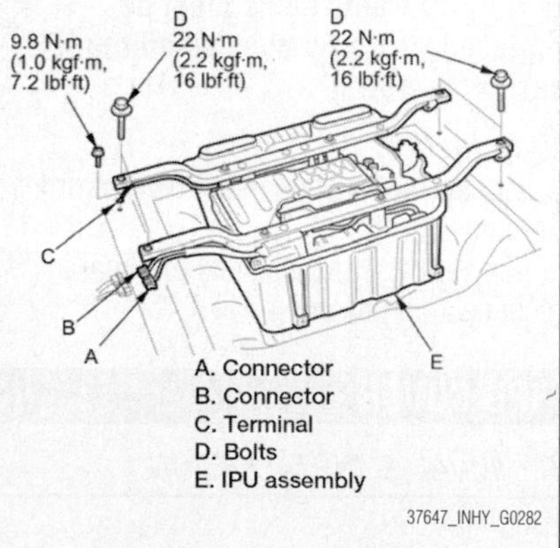
9.8 N·m
(1.0 kgf·m,
7.2 lbf·ft)
D
22 N·m
(2.2 kgf·m,
16 lbf·ft)
D
22 N·m
(2.2 kgf·m,
16 lbf·ft)
C
B
A
E
A. Connector
B. Connector
C. Terminal
D. Bolts
E. IPU assembly
37647_INHY_G0282

Fig. 78 Disconnect the connectors

b. Remove the clips and the IPU module air duct.

4. Remove the IPU cover.

5. Remove the PCU lid.

6. Remove the IMA motor power cables (A).

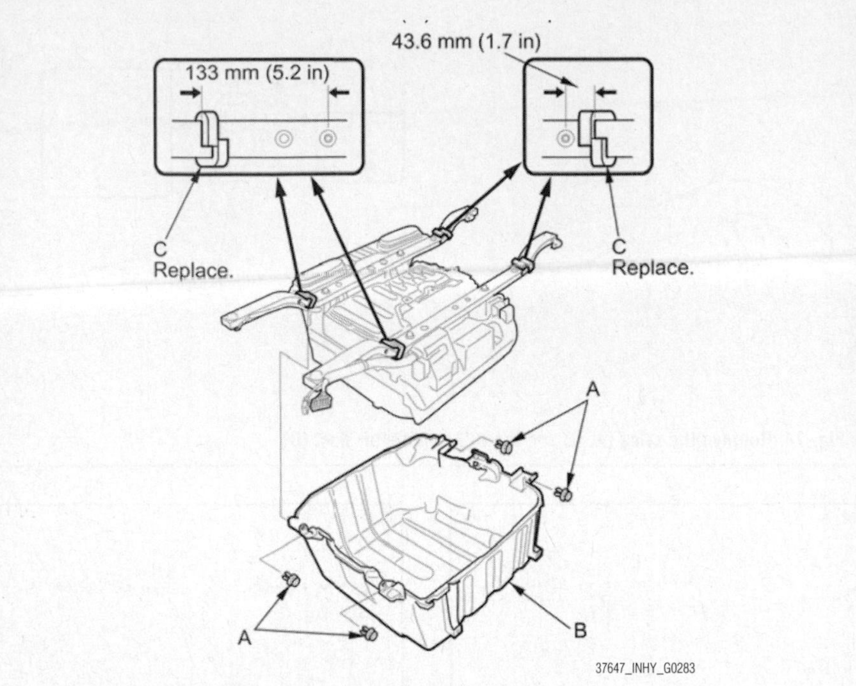
43.6 mm (1.7 in)
133 mm (5.2 in)
C
Replace.
C
Replace.
A
A
B
37647_INHY_G0283

Fig. 79 Remove the clips (A) and the IPU case (B); replace the IPU frame seals (C)

➡Check the position of the U phase, V phase, and W phase cables before disconnecting the IMA motor power cables.

7. Remove the 12 V power cable, and wrap it with insulating tape.
8. Disconnect the connectors.
9. Remove the terminal.
10. Remove the bolts and the IPU assembly.
11. Remove the clips and the IPU case.

➡The IPU frame seals must be replaced with new ones when the IPU case is removed.

To install:

12. Install the parts in the reverse order of removal.
13. Do the 12 volt battery terminal reconnection procedure.

IPU COVER

REMOVAL & INSTALLATION

See Figure 80.

CAUTION

IMA components are located in this area. The IMA is a high-voltage system. You must be familiar with the IMA system before working on or around it. Make sure you have read the IMA service precautions before doing repairs or service.

1. Remove the cargo floor lid, the IPU duct cover, and the spare tire beam.
2. Turn the battery module switch OFF.
3. Remove the bolts and the clips.
4. Remove the IPU cover.
5. Install the parts in the reverse order of removal.

➡Before the battery module switch is turned ON, make sure all the high voltage circuits are connected properly. Then push the button, and turn the battery module switch ON.

IPU MODULE AIR DUCT

REMOVAL & INSTALLATION

See Figure 81.

1. Remove the cargo floor lid, the IPU duct cover, and the cargo area side trim panel.
2. Remove the clips and the IPU module air duct.
3. Installation is the reverse of removal.

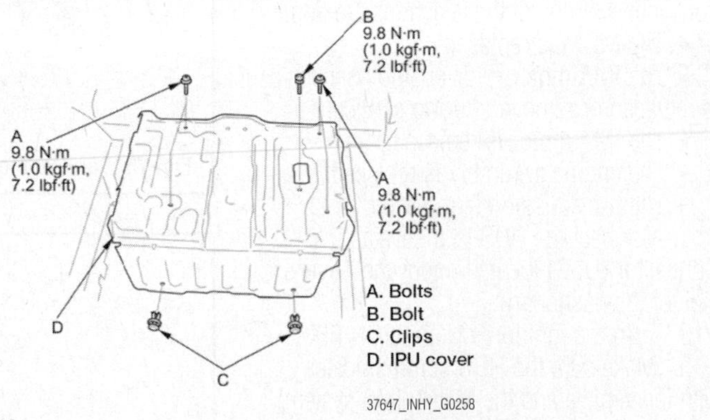
B
9.8 N·m
(1.0 kgf·m,
7.2 lbf·ft)
A
9.8 N·m
(1.0 kgf·m,
7.2 lbf·ft)
A
9.8 N·m
(1.0 kgf·m,
7.2 lbf·ft)
D
C
A. Bolts
B. Bolt
C. Clips
D. IPU cover
37647_INHY_G0258

Fig. 80 Remove the IPU cover

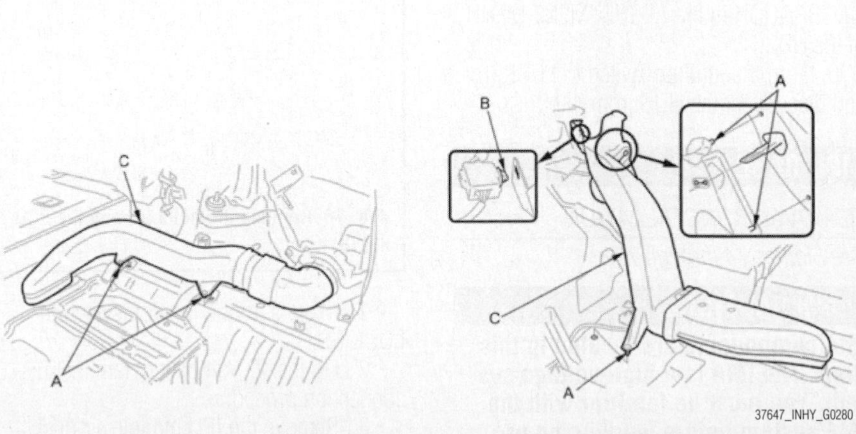
B
A
C
C
A
A
37647_INHY_G0280

Fig. 81 Remove the clips (A, B) and the IPU module air duct (C)

IPU MODULE FAN

REMOVAL & INSTALLATION

See Figure 82.

1. Remove the right cargo area side trim panel.
2. Disconnect the IPU module fan connector.
3. Remove the bolts, the nut, and the IPU module fan.
4. Install the parts in the reverse order of removal.

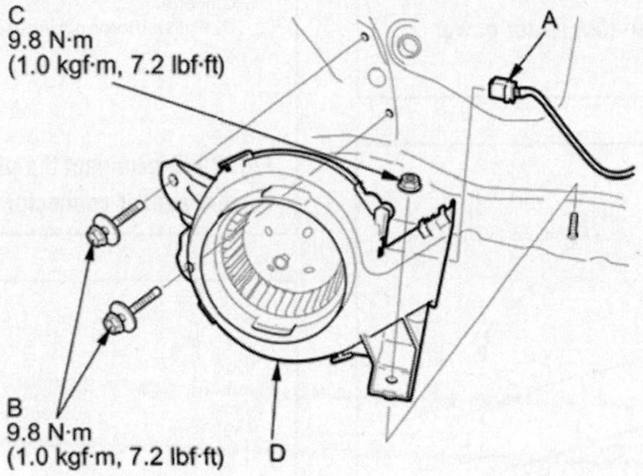
C
9.8 N·m
(1.0 kgf·m, 7.2 lbf·ft)
A
B
9.8 N·m
(1.0 kgf·m, 7.2 lbf·ft)
D

A. IPU module fan connector
B. Bolts
C. Nut
D. IPU module fan

37647_INHY_G0284

Fig. 82 Disconnect the IPU module fan connector

MOTOR CONTROL MODULE (MCM)

REMOVAL & INSTALLATION

See Figures 80, 83 through 86.

⁂ CAUTION

IMA components are located in this area. The IMA is a high-voltage system. You must be familiar with the IMA system before working on or around it. Make sure you have read the IMA service precautions before doing repairs or service.

1. Remove the IPU cover.
2. Remove the bolts and the PCU lid.
3. Remove the bolts and the PCU busplate.
4. Remove the bolts and the PCU cover.
5. Disconnect the MCM connectors.
6. Remove the bolts and the MCM.
7. Install the parts in the reverse order of removal.
8. Do the motor rotor position calibration procedure.

IMA ROTOR POSITION CALIBRATION

Do the IMA motor rotor position calibration whenever any of these actions are done:

- The MCM is replaced.
- The IMA motor rotor position sensor is replaced or removed during service.
- The IMA motor is replaced.
- The engine assembly is replaced.
- The transmission is replaced.

1. Connect the HDS to the Data Link Connector (DLC) located under the driver's side of the dashboard.
2. Turn the ignition switch to ON (II).
3. Make sure the HDS communicates with the vehicle and the MCM (IMA system). If it doesn't, troubleshoot the DLC circuit.
4. Select IMA SYSTEM on the HDS.
5. Select the MOTOR ROTOR POSITION CALIBRATION in the ADJUSTMENT MENU of the HDS.

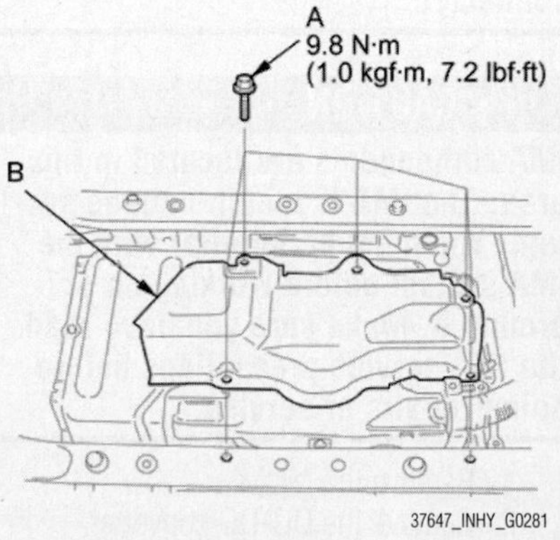
A
9.8 N·m
(1.0 kgf·m, 7.2 lbf·ft)
B
37647_INHY_G0281

Fig. 83 Remove the bolts (A) and the PCU lid (B)

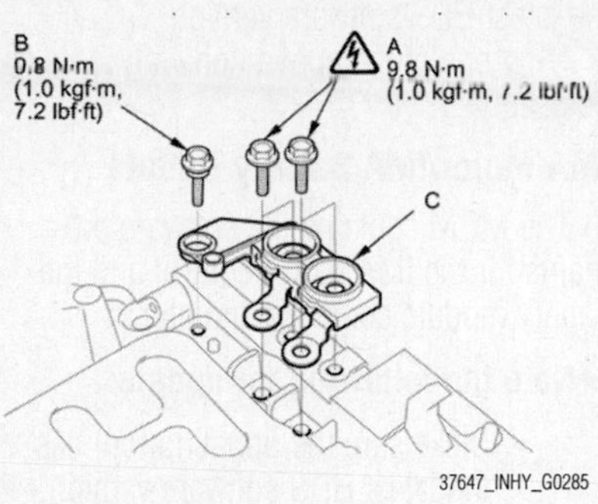
B
0.8 N·m
(1.0 kgf·m,
7.2 lbf·ft)
A
9.8 N·m
(1.0 kgf·m, 7.2 lbf·ft)
C
37647_INHY_G0285

Fig. 84 Remove the bolts (A, B) and the PCU busplate (C)

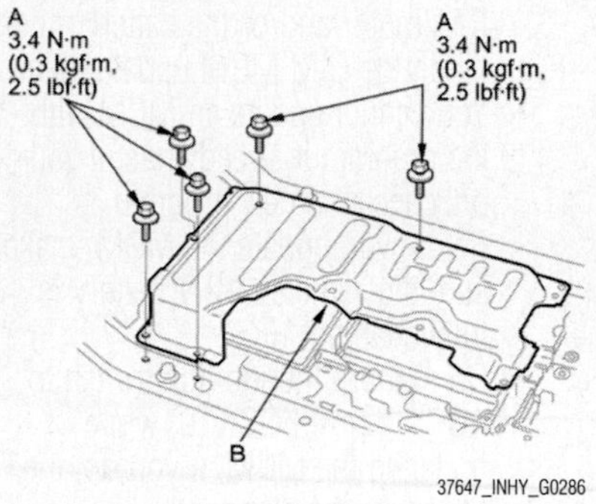
A
3.4 N·m
(0.3 kgf·m,
2.5 lbf·ft)
A
3.4 N·m
(0.3 kgf·m,
2.5 lbf·ft)
B
37647_INHY_G0286

Fig. 85 Remove the bolts (A) and the PCU cover (B)

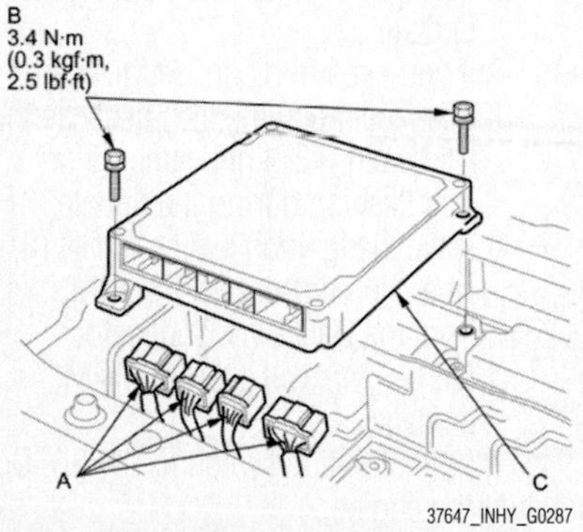
B
3.4 N·m
(0.3 kgf·m,
2.5 lbf·ft)
A
C
37647_INHY_G0287

Fig. 86 Disconnect the MCM connectors (A); remove the bolts (B) and the MCM (C)

6. Turn the ignition switch to LOCK (0), and disconnect the HDS from the DLC.

MOTOR CONTROL MODULE (MCM) UPDATE

Special Tools Required*:

- Honda Diagnostic System (HDS) tablet tester
- Honda Interface Module (HIM) and an iN workstation with the latest HDS software version
- HDS pocket tester

- GNA600 and an iN workstation with the latest HDS software version

*Any one of the above updating tools can be used.

IMA Motor/IMA Battery Update

The MCM contains the software programs for the IMA motor control and the battery module condition monitor.

➡Note the following for update:

- Make sure the updating tool has the latest HDS software version.
- To ensure the latest programs are installed, do an MCM update whenever the MCM is substituted or replaced.
- If you are using the HIM, select the IMA motor and/or the battery module in the HIM MCM update menu.
- You cannot update an MCM with the program it already has. It will only accept a new program.
- Before you update the MCM, make sure the vehicle's 12 V battery is fully charged.
- Do not turn the ignition switch to ACC (I) or to LOCK (0) while updating the MCM. If you do, the MCM can be damaged.
- To prevent MCM damage, do not operate anything electrical (audio system, brakes, A/C, power windows, door locks, etc.) during the update.
- If you need to diagnose the Honda interface module (HIM) because the HIM's red (#3) light came on or was flashing during the update, leave the ignition switch in ON (II) when you disconnect the HIM from the Data Link Connector (DLC). This will prevent MCM damage.

1. Turn the ignition switch to ON (II). Do not start the engine.
2. Connect the updating tool to the Data Link Connector (DLC) located under the driver's side of dashboard.
3. Make sure the updating tool communicates with the vehicle and the MCM. If it doesn't, troubleshoot the DLC circuit.
4. Do the MCM update procedure as described on the HIM label and in the MCM update system.

MOTOR POWER INVERTER (MPI) MODULE

REMOVAL & INSTALLATION

See Figures 87 through 91.

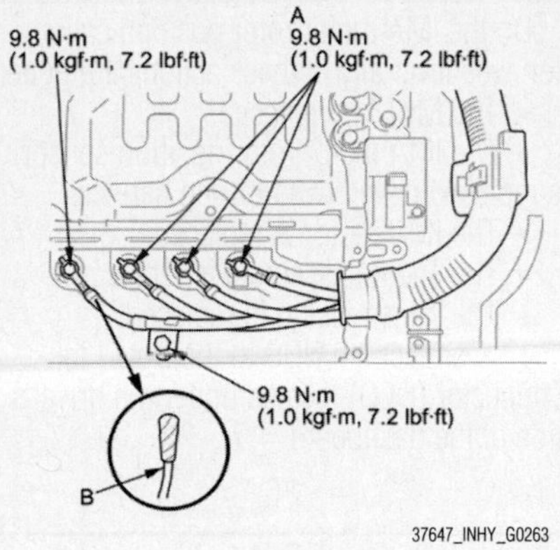
9.8 N·m
(1.0 kgf·m, 7.2 lbf·ft)
A
9.8 N·m
(1.0 kgf·m, 7.2 lbf·ft)
9.8 N·m
(1.0 kgf·m, 7.2 lbf·ft)
B
37647_INHY_G0263

Fig. 87 Remove the IMA motor power cables

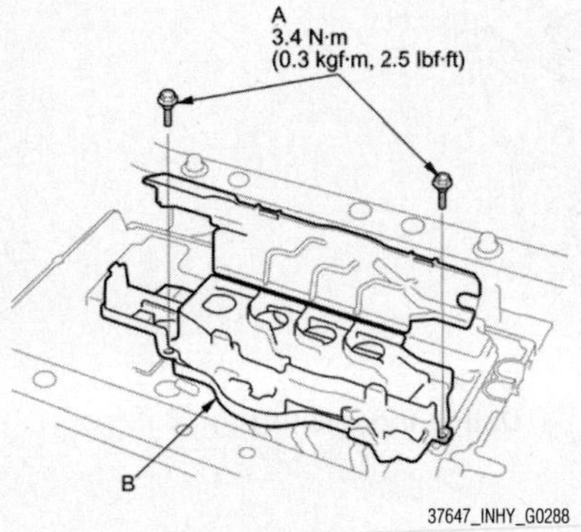
A
3.4 N·m
(0.3 kgf·m, 2.5 lbf·ft)
B
37647_INHY_G0288

Fig. 88 Remove the bolts (A) and the terminal cover (B)

⁂ WARNING

The IMA motor power cables carry high voltage when the engine is running or the IMA system is energized. To avoid serious injury from electrical shock, do not start the engine with the IMA motor power cables disconnected.

⁂ CAUTION

IMA components are located in this area. The IMA is a high-voltage system. You must be familiar with the IMA system before working on or around it. Make sure you have read the IMA service precautions before doing repairs or service.

1. Remove the MCM.
2. Remove the DC-DC converter.
3. Remove the IMA motor power cables.

➡Check the position of the U phase, V phase, and W phase cables before

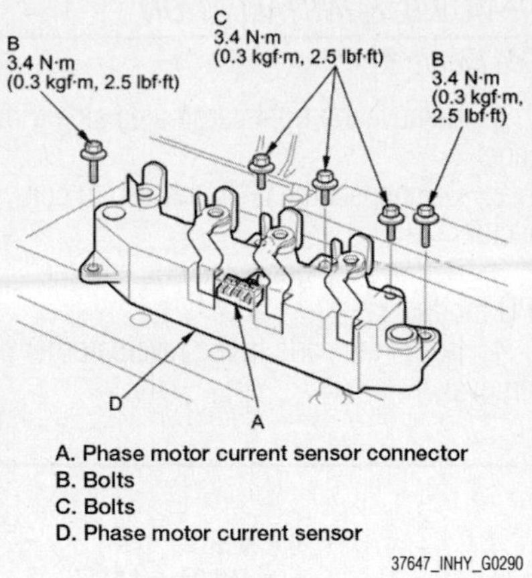
B
3.4 N·m
(0.3 kgf·m, 2.5 lbf·ft)
C
3.4 N·m
(0.3 kgf·m, 2.5 lbf·ft)
B
3.4 N·m
(0.3 kgf·m,
2.5 lbf·ft)
D
A
A. Phase motor current sensor connector
B. Bolts
C. Bolts
D. Phase motor current sensor
37647_INHY_G0290

Fig. 89 Disconnect the phase motor current sensor connector

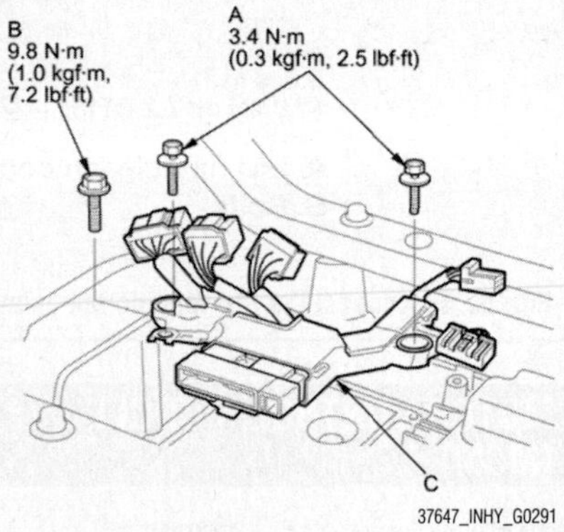
B
9.8 N·m
(1.0 kgf·m,
7.2 lbf·ft)
A
3.4 N·m
(0.3 kgf·m, 2.5 lbf·ft)
C
37647_INHY_G0291

Fig. 90 Remove the bolts (A, B) and the PCU wire harness (C)

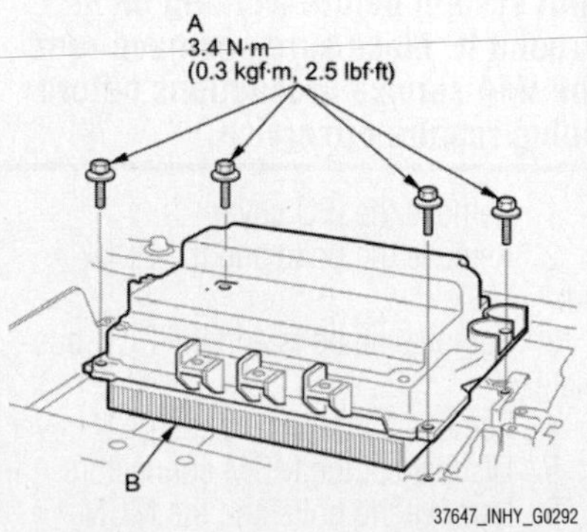
A
3.4 N·m
(0.3 kgf·m, 2.5 lbf·ft)
B
37647_INHY_G0292

Fig. 91 Remove the bolts (A) and the MPI module (B)

disconnecting the IMA motor power cables.

4. Remove the 12 V power cable, and wrap it with insulating tape.
5. Remove the bolts and the terminal cover.
6. Disconnect the phase motor current sensor connector.
7. Remove the bolts and the phase motor current sensor.
8. Remove the bolts and the PCU wire harness.
9. Remove the bolts and the MPI module.
10. Install the parts in the reverse order of removal.
11. Do the motor rotor position calibration procedure.

IMA ROTOR POSITION CALIBRATION

Do the IMA motor rotor position calibration whenever any of these actions are done:

- The MCM is replaced.
- The IMA motor rotor position sensor is replaced or removed during service.
- The IMA motor is replaced.
- The engine assembly is replaced.
- The transmission is replaced.

1. Connect the HDS to the Data Link Connector (DLC) located under the driver's side of the dashboard
2. Turn the ignition switch to ON (II).
3. Make sure the HDS communicates with the vehicle and the MCM (IMA system). If it doesn't, troubleshoot the DLC circuit.
4. Select IMA SYSTEM on the HDS.
5. Select the MOTOR ROTOR POSITION CALIBRATION in the ADJUSTMENT MENU of the HDS.
6. Turn the ignition switch to LOCK (0), and disconnect the HDS from the DLC.

ENGINE ELECTRICAL

IGNITION SYSTEM

IGNITION COIL

REMOVAL & INSTALLATION

See Figures 92 and 93.

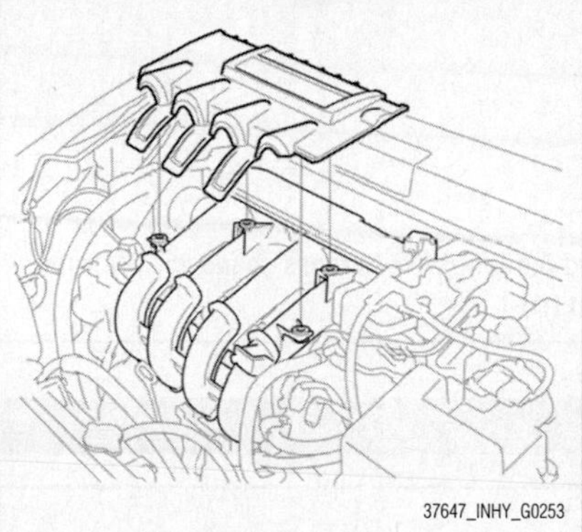
37647_INHY_G0253

Fig. 92 Remove the engine cover

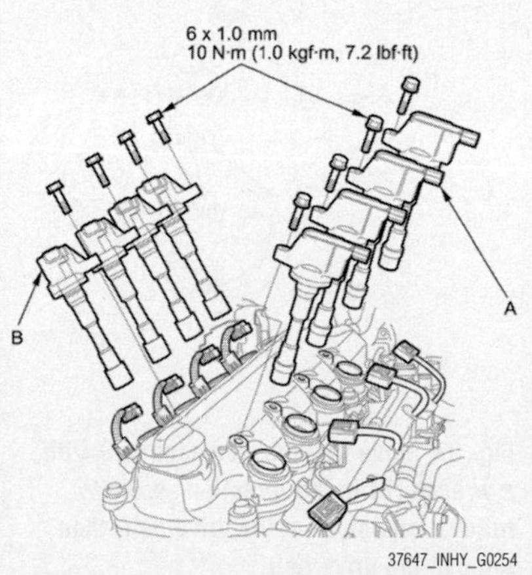
6 x 1.0 mm
10 N·m (1.0 kgf·m, 7.2 lbf·ft)
A
B
37647_INHY_G0254

Fig. 93 Disconnect the ignition coil connectors, then remove the intake side ignition coils (A) and the exhaust side ignition coils (B)

1. Remove the under-cowl panel.
2. Remove the engine cover.
3. Disconnect the ignition coil connectors, then remove the intake side ignition coils and the exhaust side ignition coils.
4. Install the ignition coils in the reverse order of removal.

FIRING ORDER

Firing order for the 1.3L LDA3 engine is 1–3–4–2.

IGNITION TIMING

INSPECTION

See Figure 94.

1. Connect the Honda Diagnostic System (HDS) to the Data Link Connector (DLC).
2. Turn the ignition switch to ON (II).
3. Make sure the HDS communicates with the vehicle and the Powertrain Control Module (PCM). If it does not communicate, troubleshoot the DLC circuit.

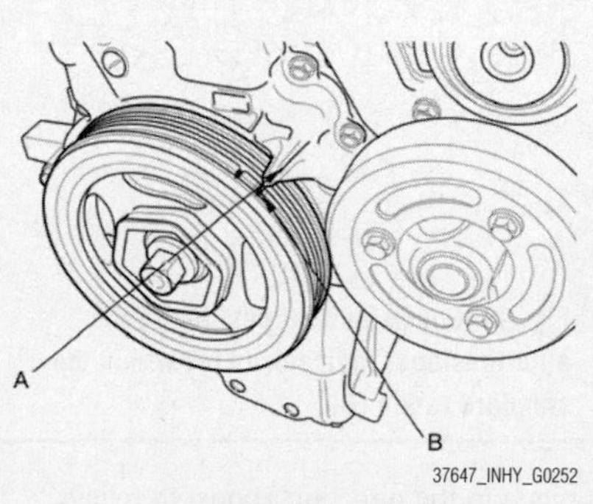
A
B
37647_INHY_G0252

Fig. 94 Aim the light toward the pointer (A) on the cam chain case; RED mark (B)

4. Check for DTCs. If a DTC is present, diagnose and repair the cause before inspecting the ignition timing.
5. Start the engine. Hold the engine speed at 3,000 rpm with no load (in P or N) until the radiator fan comes on, then let it idle.
6. Check the idle speed.
7. Jump the SCS line with the HDS.
8. Connect the timing light to the exhaust side No. 1 ignition coil harness.
9. Aim the light toward the pointer on the cam chain case. Check the ignition timing under no load condition (headlights, blower fan, rear window defogger, and air conditioner are turned off).
10. If the ignition timing differs from 8–12° BTDC, check the cam timing. If the cam timing is OK, update the PCM if it does not have the latest software, or substitute a known-good PCM, then recheck. If the system works properly, and the PCM was substituted, replace the original PCM.
11. Disconnect the HDS and the timing light.

ADJUSTMENT

Ignition timing is controlled by the Powertrain Control Module (PCM). No adjustment is necessary or possible.

SPARK PLUGS

REMOVAL & INSTALLATION

1. Remove the ignition coils.
2. Remove the spark plugs.
3. Installation is the reverse of removal.

ENGINE ELECTRICAL

STARTER

REMOVAL & INSTALLATION

See Figure 95.

1. Do the 12 volt battery terminal disconnection procedure.
2. Remove the air cleaner.
3. Disconnect the positive starter cable and the BLK/WHT harness connector from the S terminal, then remove the heater hose bracket.
4. Remove the two bolts securing the starter, then remove the starter.

To install:

5. Install the starter, then tighten the starter mounting bolts to 33 ft. lbs. (44 Nm).
6. Connect the positive starter cable and the BLK/WHT harness connector to the S terminal, then install the heater hose bracket. Make sure the crimped side of the ring terminal is facing out.
7. Install the air cleaner.
8. Do the 12 volt battery terminal reconnection procedure.
9. If the IMA battery level indicator displays no level, start the engine, and hold it between 3,500 and 4,000 rpm without load (in P or N) until the level in the indicator is half full.

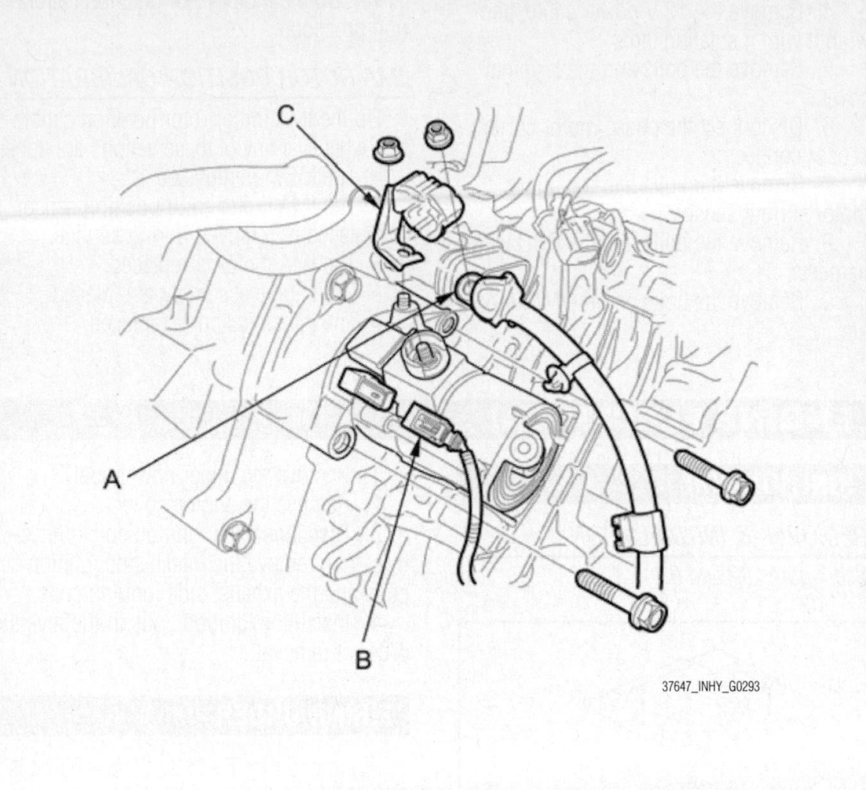
C
A
B
37647_INHY_G0293

Fig. 95 Disconnect the positive starter cable (A) and the BLK/WHT harness connector (B) from the S terminal, then remove the heater hose bracket (C)

ENGINE MECHANICAL

ACCESSORY DRIVE BELTS

ACCESSORY BELT ROUTING

Refer to the graphic in the Removal and Installation section for proper routing of the accessory drive belt.

INSPECTION

See Figure 96.

1. Inspect the belt for cracks or damage. If the belt is cracked or damaged, replace it.
2. Check the that position of the auto-tensioner indicator is within the standard range as shown. If it is out of the standard range, replace the drive belt.

ADJUSTMENT

The accessory drive belt tension is set by the drive belt auto-tensioner. No adjustment is needed.

REMOVAL & INSTALLATION

See Figure 97.

1. Move the auto-tensioner with a wrench in the direction shown to relieve tension from the drive belt, then remove the drive belt.
2. Install the new belt in the reverse order of removal.

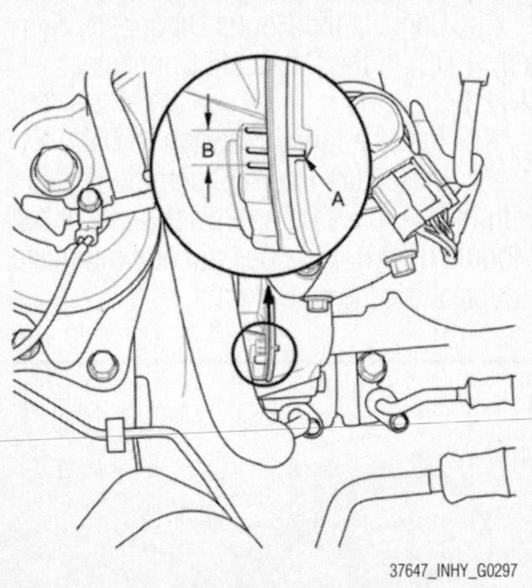
B
A
37647_INHY_G0297

Fig. 96 Check the that position of the auto-tensioner indicator (A) is within the standard range (B)

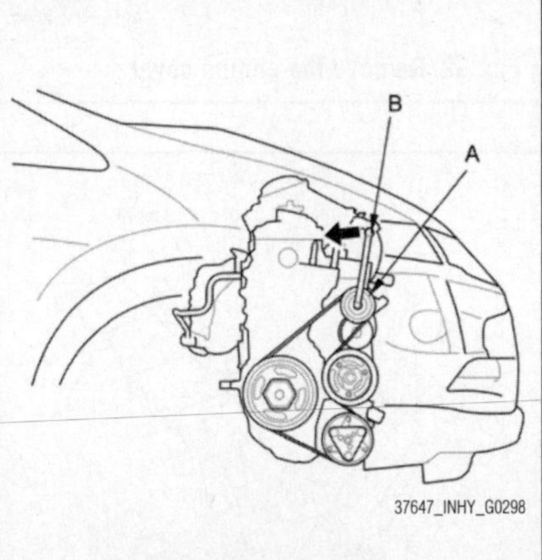
B
A
37647_INHY_G0298

Fig. 97 Move the auto-tensioner (A) with a wrench (B) in the direction shown to relieve tension from the drive belt, then remove the drive belt

CAMSHAFT AND VALVE LIFTERS

INSPECTION

See Figures 98 through 101.

➡Do not rotate the camshaft during inspection.

1. Remove the rocker arm assembly.

Specified Torque:

1. Put the camshaft and camshaft holders on the cylinder head, then tighten the bolts to the specified torque.
   - 8 mm Bolts: 16 ft. lbs. (22 Nm)
   - 8 mm Bolts: 14 ft. lbs. (20 Nm)
   - 6 mm Bolt: 9 ft. lbs. (12 Nm)

➡Apply new engine oil to the bolt threads and flange.

2. Torque the bolts in a crisscross pattern starting in the middle working to the ends.
3. Seat the camshaft by pushing it toward the rear of the cylinder head.

Camshaft End Play

1. Zero the dial indicator against the end of the camshaft, then push the camshaft back and forth, and read the end play. If the end play is beyond the service limit, replace the cylinder head and recheck. If it is still beyond the service limit, replace the camshaft.
   - Standard (New): 0.002–0.006 inches (0.05–0.15 mm)
   - Service Limit: 0.01 inches (0.3 mm)
2. Loosen the camshaft holder bolts two turns at a time, in a crisscross pattern. Then remove the camshaft holders from the cylinder head.
3. Lift the camshaft out of the cylinder head, wipe it clean, then inspect the lift ramps. Replace the camshaft if any lobes are pitted, scored, or excessively worn.

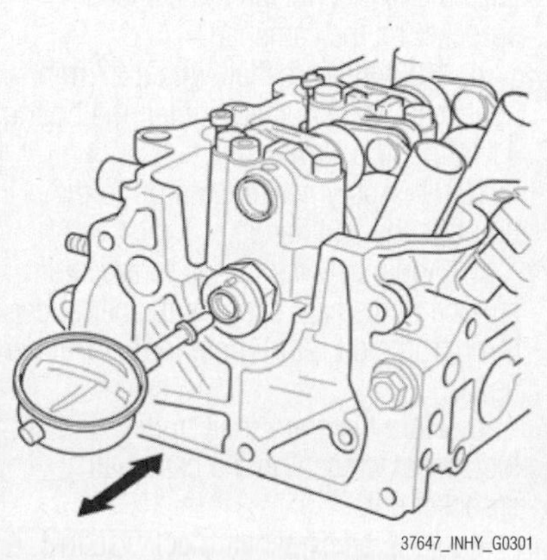
37647_INHY_G0301
Fig. 98 Checking camshaft end play

Camshaft-to-Holder Oil Clearance

1. Clean the camshaft journal surfaces in the cylinder head, then set the camshaft back in place. Place a Plastigage strip across each journal.
2. Install the camshaft holders, then tighten the bolts to the specified torque as shown in step 2.
3. Remove the camshaft holders. Measure the widest part of Plastigage on each journal.
   a. If the camshaft-to-holder clearance is within limits, go to step 11.
   b. If the camshaft-to-holder clearance is beyond the service limit and the camshaft has been replaced, replace the cylinder head.
   c. If the camshaft-to-holder clearance is beyond the service limit and the camshaft has not been replaced, go to step 10.
   - Standard (New): 0.0020–0.0035 inches (0.050–0.089 mm)
   - Service Limit: 0.004 inches (0.10 mm)

Camshaft Total Runout

1. Check the total runout with the camshaft supported on V-blocks.
   a. If the total runout of the camshaft is within the service limit, replace the cylinder head.
   b. If the total runout is beyond the service limit, replace the camshaft and recheck the camshaft-to-holder oil clearance. If the oil clearance is still out of tolerance, replace the cylinder head.
   - Standard (New): 0.001 inches (0.03 mm) max.
   - Service Limit: 0.002 inches (0.04 mm)

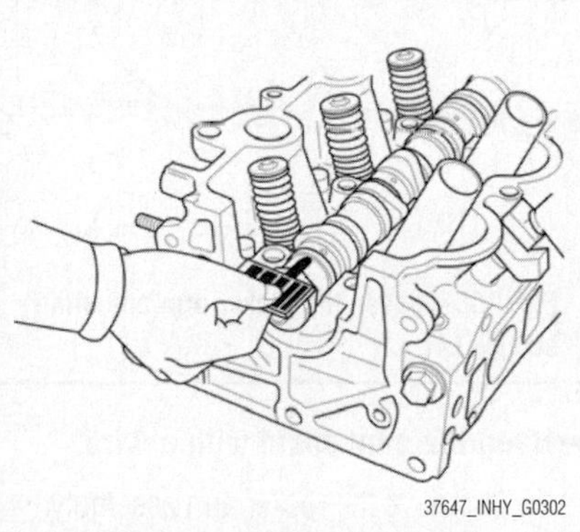
37647_INHY_G0302
Fig. 99 Measure the widest part of Plastigage on each journal

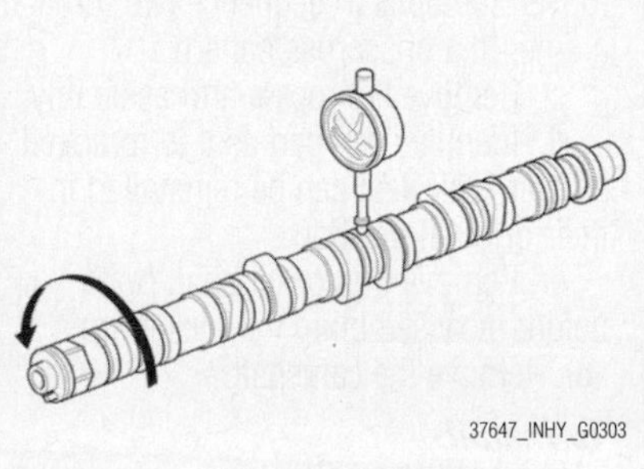
37647_INHY_G0303
Fig. 100 Checking total runout

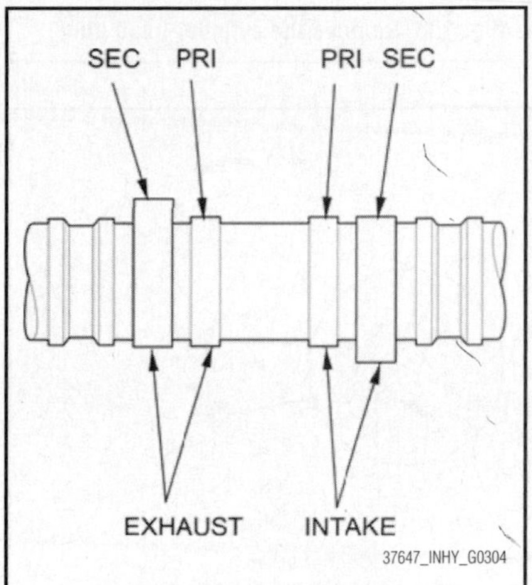
SEC
PRI
PRI
SEC
EXHAUST
INTAKE
37647_INHY_G0304
Fig. 101 Measure the cam lobe height

Cam Lobe Height Standard (New):

1. Measure the cam lobe height.
   - PRI: Primary; SEC: Secondary
   - Intake: SEC: 1.4116 inches (35.854 mm); PRI: 1.1693 inches (29.700 mm)
   - Exhaust: SEC: 1.3965 inches (35.470 mm); PRI: 1.1772 inches (29.900 mm)

REMOVAL & INSTALLATION

See Figures 102 through 107.

1. Remove the air cleaner.
2. Remove the intake manifold.

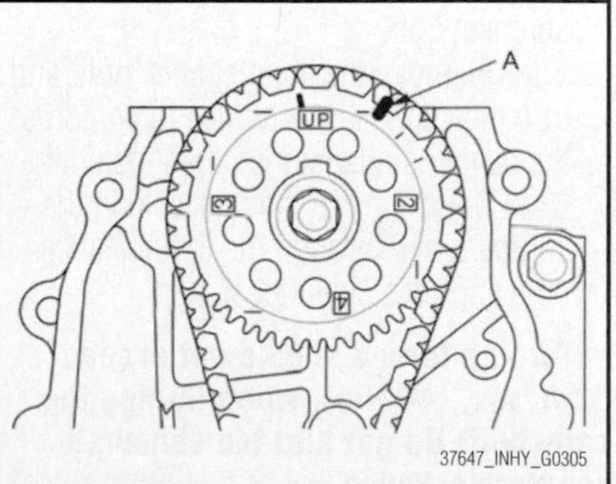
A
UP
37647_INHY_G0305
Fig. 102 Make a reference mark (A) across the camshaft sprocket and cam chain

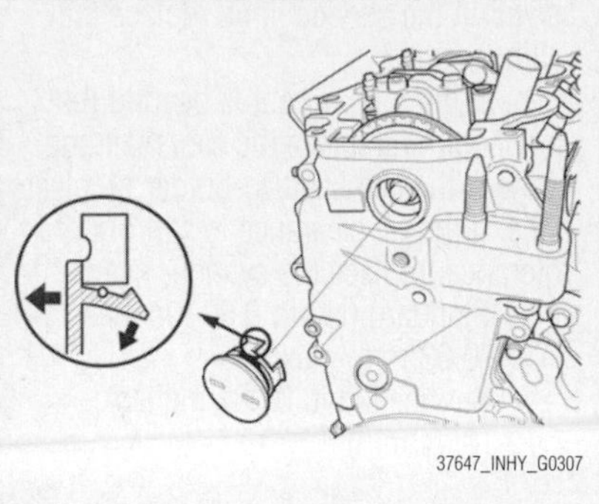
37647_INHY_G0307

Fig. 103 Remove the cylinder head plug

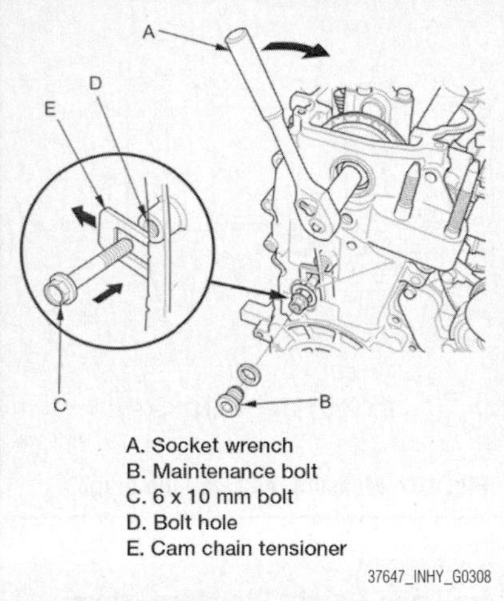
A
D
E
C
B
A. Socket wrench
B. Maintenance bolt
C. 6 x 10 mm bolt
D. Bolt hole
E. Cam chain tensioner
37647_INHY_G0308

Fig. 104 Hold the crankshaft pulley, and set the socket wrench on the camshaft sprocket bolt

3. Remove the cylinder head cover.
4. Remove the camshaft sprocket.
   a. Make a reference mark across the camshaft sprocket and cam chain.
   b. Apply new engine oil to the slider surface of the cam chain tensioner slider through the oil return hole in the cylinder head.
   c. Remove the cylinder head plug.
   d. Hold the crankshaft pulley, and set the socket wrench on the camshaft sprocket bolt.
   e. Remove the maintenance bolt, and turn the camshaft clockwise to compress the cam chain tensioner, then install the 6 x 1.0 mm bolt in the bolt hole in the engine block through the maintenance hole and cam chain tensioner.

➡**Turning torque should not exceed 33 ft. lbs. (44 Nm), when turning the camshaft. Do not turn the camshaft counterclockwise.**

   f. Hold the camshaft with a 27 mm open-end wrench, then remove the camshaft sprocket.

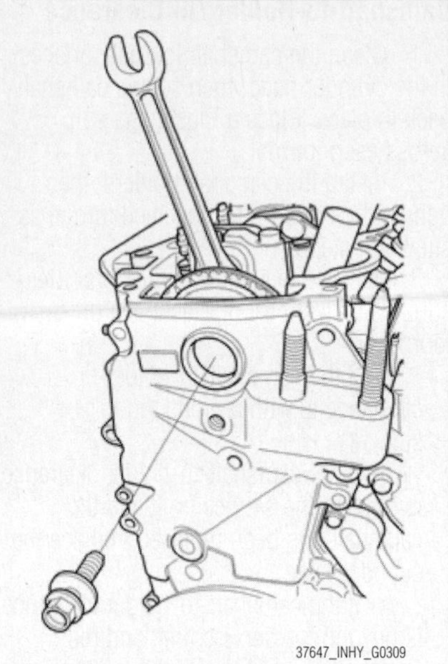
37647_INHY_G0309

Fig. 105 Hold the camshaft with a 27 mm open-end wrench, then remove the camshaft sprocket

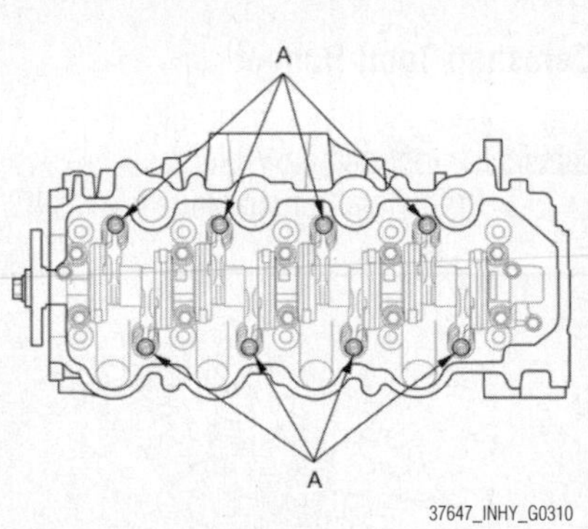
A
A
37647_INHY_G0310

Fig. 106 Loosen the rocker arm adjusting screws (A)

➡**Hang the cam chain with a wire.**

5. Remove the rocker arm assembly.
   a. Loosen the rocker arm adjusting screws.
   b. Remove the camshaft holder bolts. To prevent damaging the camshaft, loosen the bolts in sequence two turns at a time, in a crisscross pattern.
   c. Remove the rocker arm assembly.
   d. Identify each part as it is removed so that each item can be reinstalled in their original location.
   e. Remove the rocker shaft bolts before disassembling the rocker arms.
6. Remove the camshaft.

To install:

7. Install the camshaft.
8. Install the rocker arm assembly.

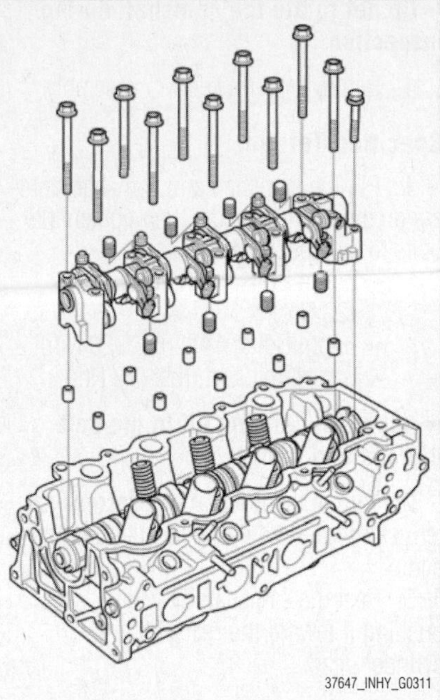
37647_INHY_G0311

Fig. 107 Remove the rocker arm assembly

   a. Reassemble the rocker arm assembly.
   b. If reused, the rocker arms must be installed in their original location.
   c. Prior to reassembling, clean all the parts in solvent, dry them, and apply new engine oil to any contact points.
   d. Apply new engine oil to the threads of the rocker shaft bolts when installing them.
   e. When replacing the rocker arm assembly, remove the fastening hardware from the new rocker arm assembly.
9. Install the camshaft sprocket.

NOTE: Keep the cam chain away from magnetic fields.

   a. Install the cam chain around the camshaft sprocket by alignment the reference mark, then install the camshaft sprocket on the camshaft.
   b. Hold the camshaft with a 27 mm open-end wrench, then tighten the bolt to 41 ft. lbs. (56 Nm).

   NOTE: Apply new engine oil to the bolt threads and flange.

   c. Apply new engine oil to the slider surface of the cam chain tensioner slider through the oil return hole in the cylinder head.
   d. Hold the crankshaft pulley, and set the socket wrench on the camshaft sprocket bolt.
   e. Turn the camshaft clockwise to compress the cam chain tensioner, then remove the 6 x 1.0 mm bolt.

➡Turning torque should not exceed 33 ft. lbs. (44 Nm), when turning the camshaft. Do not turn the camshaft counterclockwise.

f. Install the maintenance bolt with a new washer.

g. Install the new cylinder head plug.

10. Install the cylinder head cover.
11. Install the intake manifold.
12. Install the air cleaner.

CATALYTIC CONVERTER

REMOVAL & INSTALLATION

Warm Up Three Way Catalytic Converter (WU-TWC)

See Figures 108 and 109.

1. Raise the vehicle on a lift.
2. Remove the bolts.
3. Remove the cowl cover and the under-cowl panel.
4. Remove the EGR pipe.
5. Remove the A/F sensor (Sensor 1).
6. Remove the cover.
7. Remove the bolts and the nuts.
8. Remove the WU-TWC.

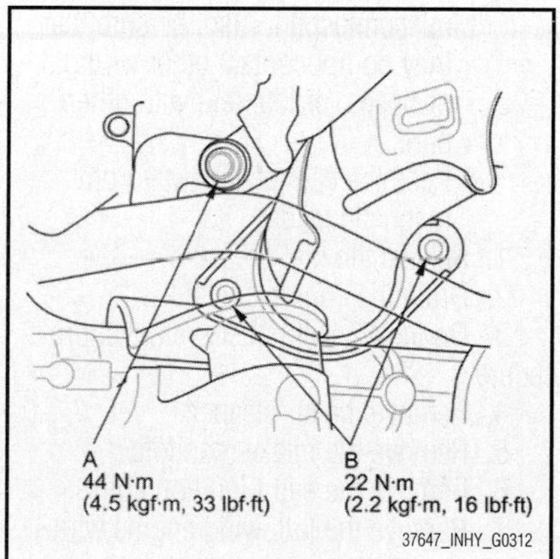
A
44 N·m
(4.5 kgf·m, 33 lbf·ft)
B
22 N·m
(2.2 kgf·m, 16 lbf·ft)
37647_INHY_G0312

Fig. 108 Remove the bolts (A, B)

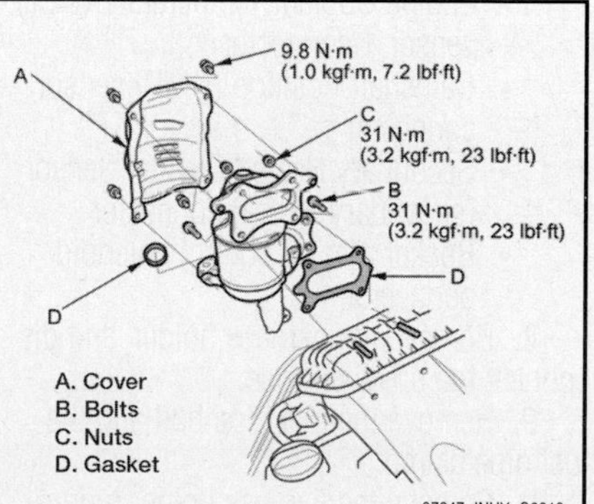
9.8 N·m
(1.0 kgf·m, 7.2 lbf·ft)
A
C
31 N·m
(3.2 kgf·m, 23 lbf·ft)
B
31 N·m
(3.2 kgf·m, 23 lbf·ft)
D
D
A. Cover
B. Bolts
C. Nuts
D. Gasket
37647_INHY_G0313

Fig. 109 Remove the WU-TWC

9. Remove the gaskets from the WU-TWC.
10. Install the parts in the reverse order of removal with new gaskets.

Under-Floor TWC

See Figure 110.

1. Raise the vehicle on a lift.
2. Remove secondary HO2S (Sensor 2).
3. Remove the nuts and the bolts.
4. Remove the under-floor TWC.
5. Remove the cover.
6. Install the parts in the reverse order of removal with new gaskets.

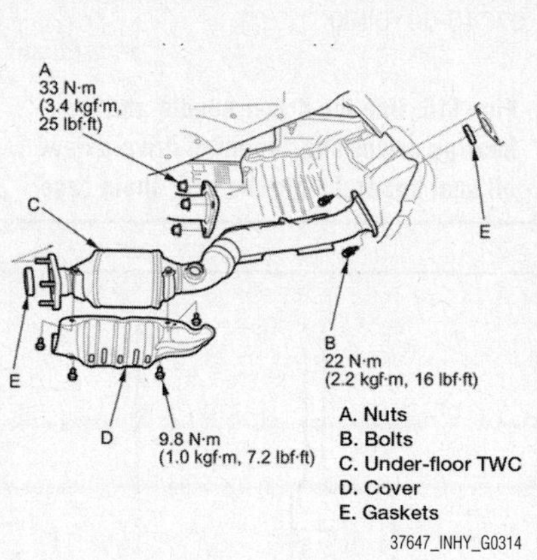
A
33 N·m
(3.4 kgf·m,
25 lbf·ft)
C
E
E
B
22 N·m
(2.2 kgf·m, 16 lbf·ft)
D
9.8 N·m
(1.0 kgf·m, 7.2 lbf·ft)
A. Nuts
B. Bolts
C. Under-floor TWC
D. Cover
E. Gaskets
37647_INHY_G0314

Fig. 110 Remove the under-floor TWC

CRANKSHAFT PULLEY

REMOVAL & INSTALLATION

See Figures 112 through 115.

Special Tools Required:

- Holder Handle 07JAB-001020A
- Holder Attachment, 50 mm 07NAB-001040A
- Socket, 19 mm 07JAA-001020A or equivalent

1. Remove the front wheels.

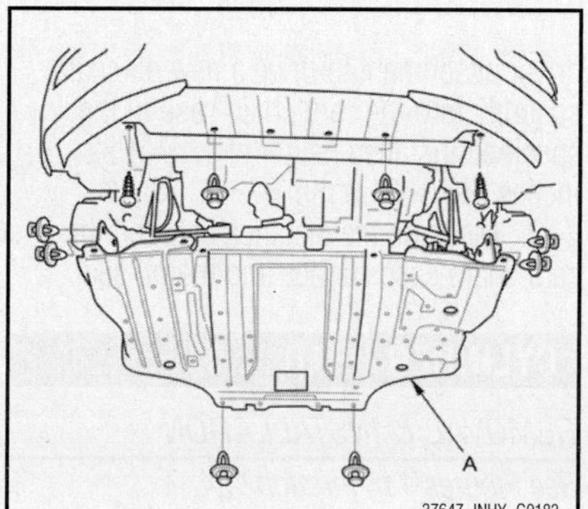
A
37647_INHY_G0183

Fig. 111 Remove the splash shields

2. Remove the splash shields.
3. Remove the accessory drive belt.
4. Hold the pulley with the holder handle and the pulley holder attachment.
5. Remove the bolt with a 19 mm socket wrench and a breaker bar, then remove the crankshaft pulley.

To install:

6. Remove any oil and clean the crankshaft pulley, the crankshaft, the bolt, and the washer. Lubricate with new engine oil as shown.
7. Install the crankshaft pulley.
8. Tighten the crankshaft pulley bolt. Do not use an impact wrench.

a. Hold the pulley with the holder handle and the pulley holder attachment, then tighten the bolt to 27 ft. lbs. (37 Nm) with a torque wrench and heavy duty 19 mm socket wrench. If the pulley bolt or crankshaft is new, tighten the bolt to

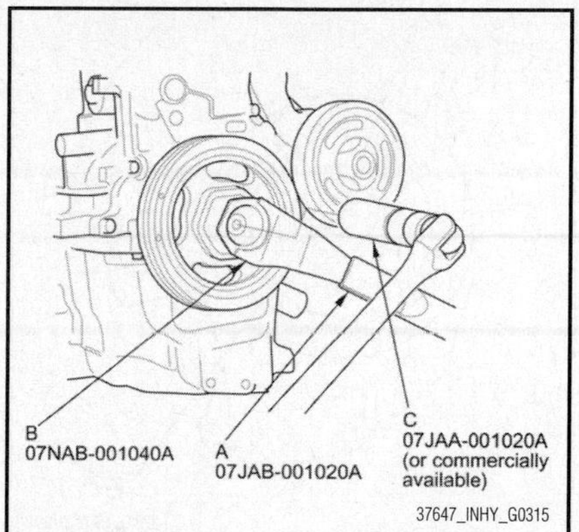
B
07NAB-001040A
A
07JAB-001020A
C
07JAA-001020A
(or commercially
available)
37647_INHY_G0315

Fig. 112 Hold the pulley with the holder handle (A) and the pulley holder attachment (B); remove the bolt with a 19 mm socket wrench (C) and a breaker bar

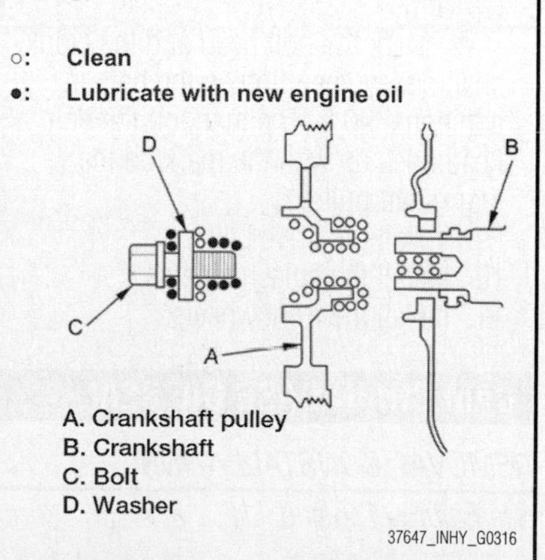
○: Clean
●: Lubricate with new engine oil
D
B
C
A
A. Crankshaft pulley
B. Crankshaft
C. Bolt
D. Washer
37647_INHY_G0316

Fig. 113 Remove any oil and clean the crankshaft pulley, the crankshaft, the bolt, and the washer; lubricate with new engine oil as shown

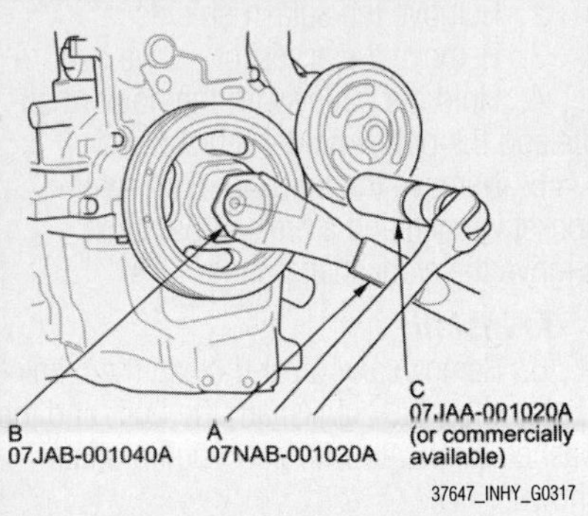
B
07JAB-001040A
A
07NAB-001020A
C
07JAA-001020A
(or commercially available)
37647_INHY_G0317

Fig. 114 Hold the pulley with the holder handle (A) and the pulley holder attachment (B), then tighten the bolt with a torque wrench and heavy duty 19 mm socket wrench (C)

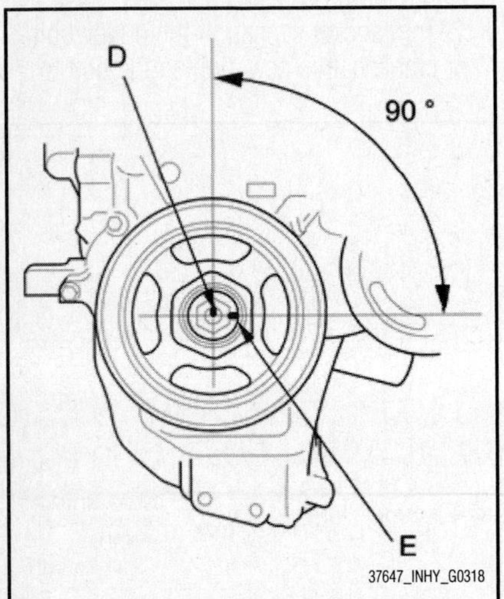
D
90 °
E
37647_INHY_G0318

Fig. 115 Mark the bolt head (D) and the crankshaft pulley (E), then tighten the bolt an additional 90 degrees

130 ft. lbs. (177 Nm), then remove the bolt and tighten it to 27 ft. lbs. (37 Nm).

b. Mark the bolt head and the crankshaft pulley, then tighten the bolt an additional 90 ° (The mark on the bolt head lines up with the mark on the crankshaft pulley).

9. Install the drive belt.
10. Install the splash shields.
11. Install the front wheels.

CAM CHAIN CASE OIL SEAL

REMOVAL & INSTALLATION

See Figures 116 and 117.

Special Tools Required:

- Driver Handle, 15 x 135L 07749-0010000 Bearing Driver Attachment, 52 x 55 mm 07746-0010400

1. Use the driver handle and the bearing driver attachment to drive a new oil seal squarely into the cam chain case to the specified installed height of 1.19–1.22 inches (30.3–31.0 mm).

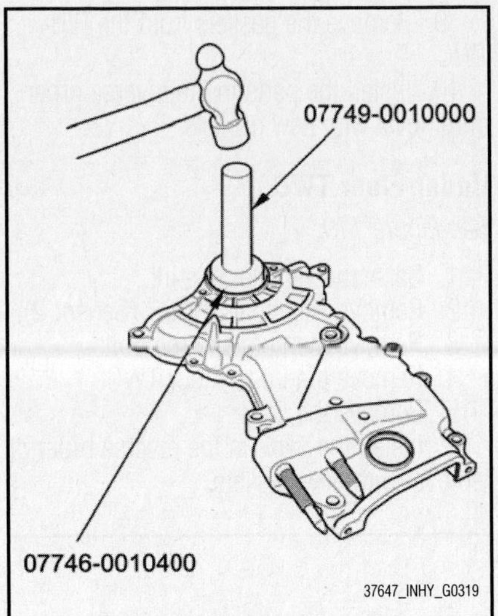
07749-0010000
07746-0010400
37647_INHY_G0319

Fig. 116 Use the driver handle and the bearing driver attachment to drive a new oil seal squarely into the cam chain case

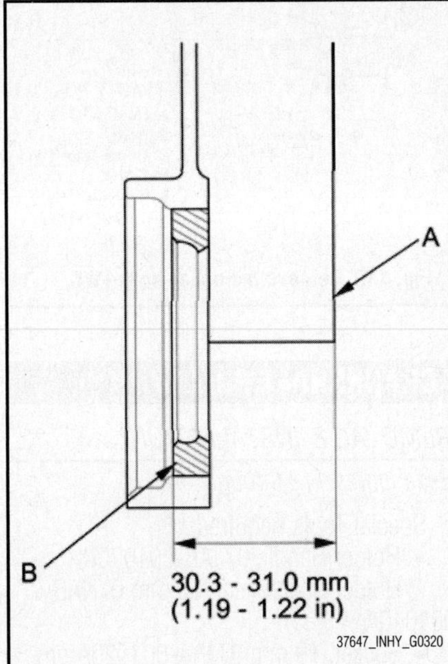
A
B
30.3 - 31.0 mm
(1.19 - 1.22 in)
37647_INHY_G0320

Fig. 117 Measure the distance between the cam chain case surface (A) and the oil seal (B)

2. Measure the distance between the cam chain case surface and the oil seal.

CYLINDER HEAD

REMOVAL & INSTALLATION

See Figures 118 through 124.

➡Note the following when servicing this component:

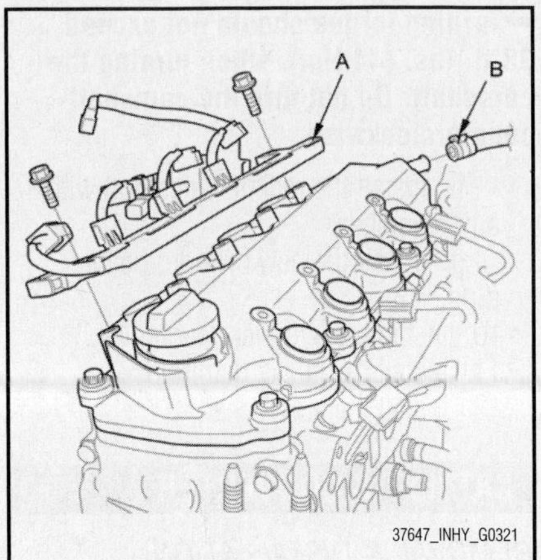
A
B
37647_INHY_G0321

Fig. 118 Remove the harness holder (A), and disconnect the breather hose (B)

- Use fender covers to avoid damaging painted surfaces.
- To avoid damage, unplug the wiring connectors carefully while holding the connector portion.
- To avoid damaging the cylinder head, wait until the engine coolant temperature drops below 100 °F (38 °C) before loosening the cylinder head bolts.
- Mark all wiring and hoses to avoid misconnection. Also, be sure that they do not contact other wiring or hoses, or interfere with other parts.
- Keep the cam chain away from magnetic fields.

1. Relieve the fuel pressure.
2. Drain the engine coolant.
3. Do the 12 volt battery removal procedure.
4. Remove the air cleaner.
5. Remove the intake manifold.
6. Remove the eight ignition coils.
7. Remove the following engine wire harness connectors and wire harness clamps from the cylinder head:
   - Four injector connectors
   - Engine Coolant Temperature (ECT) sensor 1 connector
   - Camshaft Position (CMP) sensor connector
   - Secondary Heated Oxygen Sensor (secondary HOS2) connector
   - Rocker arm oil control solenoid connector
8. Remove the harness holder, and disconnect the breather hose.
9. Remove the fuel pipe bolt and the fuel pipe clamp.
10. Remove the harness holder mounting bolt and the ground cable, then remove the harness holder.

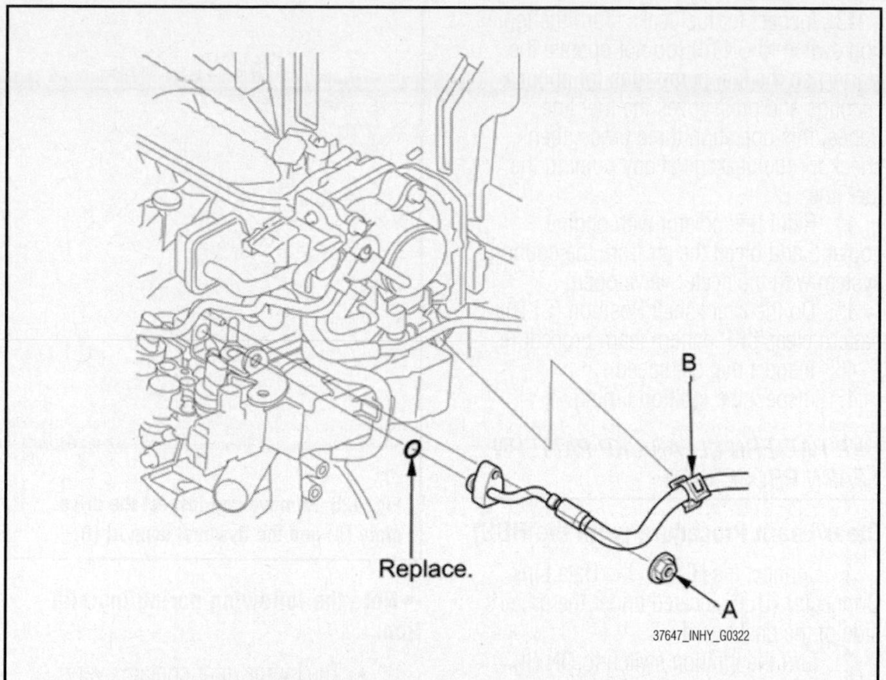
B
Replace.
A
37647_INHY_G0322

Fig. 119 Remove the fuel pipe bolt (A) and the fuel pipe clamp (B)

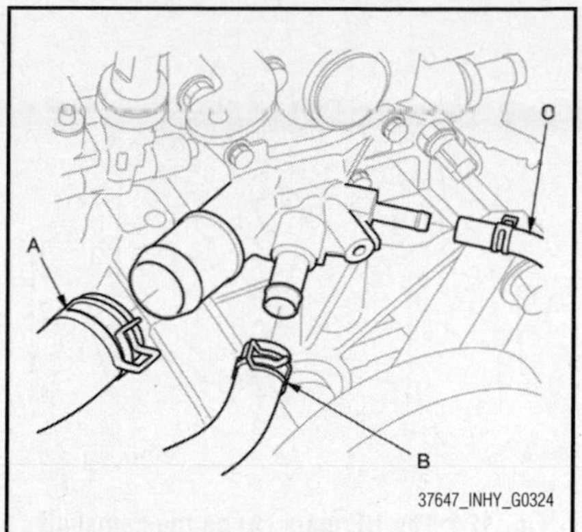
C
A
B
37647_INHY_G0324

Fig. 121 Remove the upper radiator hose (A), the water bypass hose (B), and the heater hose (C)

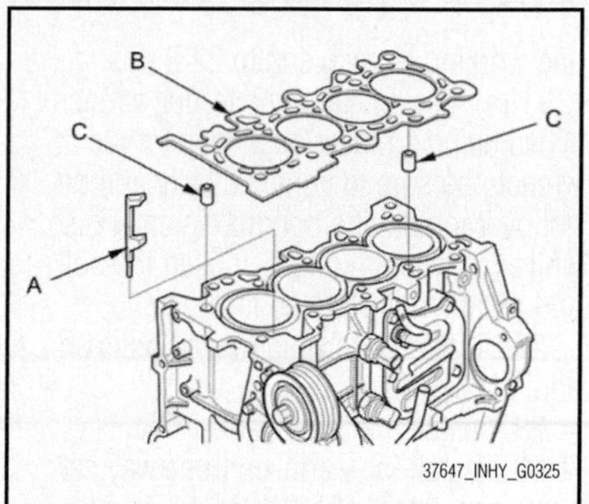
B
C
C
A
37647_INHY_G0325

Fig. 122 Install a new coolant separator (A), new cylinder head gasket (B) and the dowel pins (C)

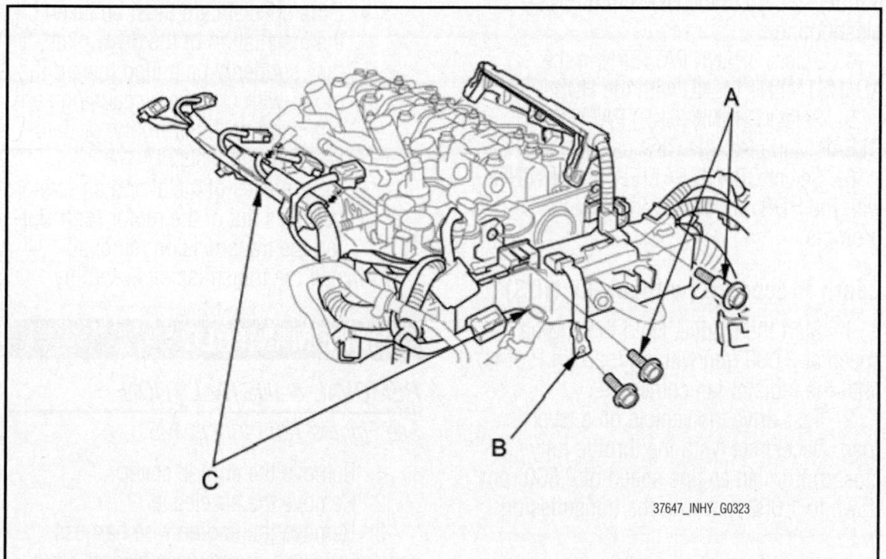
A
B
C
37647_INHY_G0323

Fig. 120 Remove the harness holder mounting bolt (A) and the ground cable (B), then remove the harness holder (C)

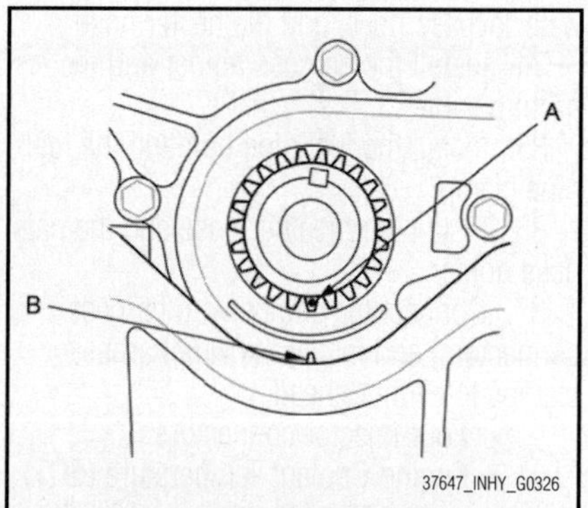
A
B
37647_INHY_G0326

Fig. 123 Align the TDC mark (A) on the crankshaft sprocket with the pointer (B) on the oil pump

11. Remove the upper radiator hose, the water bypass hose, and the heater hose.
12. Remove the drive belt.
13. Remove the water pump.
14. Remove the cylinder head cover.
15. Remove the warm-up three way catalytic converter (WU-TWC).
16. Remove the cam chain.
17. Remove the cylinder head bolts. To prevent warpage, loosen the bolts in a crisscross sequence ⅓ turn at a time starting from each end; repeat the sequence until all bolts are loosened.
18. Remove the cylinder head.

To install:

19. Clean the cylinder head and the engine block surface.
20. Install a new coolant separator in the engine block whenever the engine block is replaced.
21. Install the new cylinder head gasket and the dowel pins on the engine block. Always use a new cylinder head gasket.
22. Set the crankshaft to Top Dead Center (TDC). Align the TDC mark on the crankshaft sprocket with the pointer on the oil pump.
23. Set the camshaft TDC. The UP mark on the camshaft sprocket should be at the top, and the TDC grooves on the camshaft sprocket should line up with the top edge of the head.
24. Install the cylinder head on the engine block.
25. Apply new engine oil to the threads and flange of all cylinder head bolts.
26. Tighten the cylinder head bolts in a criss cross sequence staring in the middle

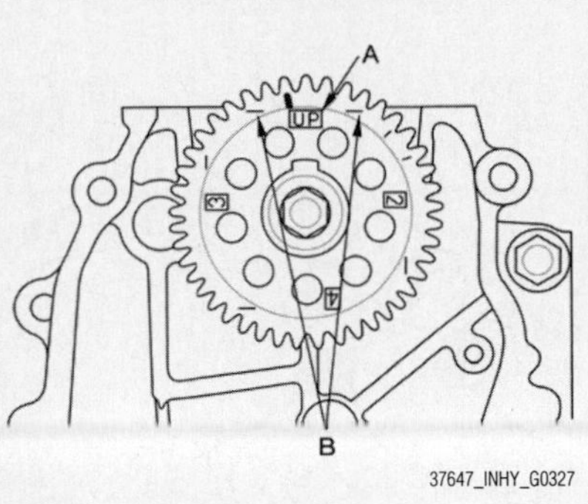
A
UP
B
37647_INHY_G0327

Fig. 124 The UP mark (A) on the camshaft sprocket should be at the top, and the TDC grooves (B) on the camshaft sprocket should line up with the top edge of the head

and working to each end to 22 ft. lbs. (29 Nm), use a beam-type torque wrench. When using a preset-click-type torque wrench, be sure to tighten slowly and do not overtighten. If a bolt makes any noise while you are torquing it, loosen the bolt and retighten it from the first step.

27. Tighten all cylinder head bolts an additional 130°.
28. Install the cam chain.
29. Install the warm-up three way catalytic converter (WU-TWC).
30. Install the cylinder head cover.
31. Install the water pump.
32. Install the drive belt.
33. Install the upper radiator hose, the water bypass hose, and the heater hose.
34. Install the harness holder and the ground cable.
35. Install the fuel pipe bolt and the fuel pipe clamp.
36. Install the breather hose and the harness holder.
37. Connect the engine wire harness connectors, and install the wire harness clamps to cylinder head.
    - Four injector connectors
    - Engine Coolant Temperature (ECT) sensor 1 connector
    - Camshaft Position (CMP) sensor connector
    - Secondary Heated Oxygen Sensor (secondary HO2S) connector
    - Rocker arm oil control solenoid connector
38. Install the eight ignition coils.
39. Install the intake manifold.
40. Install the air cleaner.
41. After installation, check that all tubes, hoses, and connectors are installed correctly.
42. Do the 12 volt battery installation procedure.
43. Inspect for fuel leaks. Turn the ignition switch to ON (II) (do not operate the starter) so the fuel pump runs for about 2 seconds and pressurizes the fuel line. Repeat this operation three times, then check for fuel leakage at any point in the fuel line.
44. Refill the radiator with engine coolant, and bleed the air from the cooling system with the heater valve open.
45. Do the Crankshaft Position (CKP) pattern clear/CKP pattern learn procedure.
46. Inspect the idle speed.
47. Inspect the ignition timing.

CKP PATTERN CLEAR/CKP PATTERN LEARN PROCEDURE

Clear/Learn Procedure (with the HDS)

1. Connect the HDS to the Data Link Connector (DLC) located under the driver's side of the dashboard.
2. Turn the ignition switch to ON (II).
3. Make sure the HDS communicates with the PCM and all other vehicle systems. If it doesn't, go to the DLC circuit troubleshooting.
4. Select CRANK PATTERN in the ADJUSTMENT MENU with the HDS.
5. Select CRANK (CKP) PATTERN CLEAR, and clear the CKP pattern.
6. Select CRANK PATTERN LEARNING with the HDS, and follow the screen prompts.

Learn Procedure (without the HDS)

1. Start the engine. Hold the engine speed at 3,000 rpm without load (in P or N) until the radiator fan comes on.
2. Test-drive the vehicle on a level road: Decelerate (with the throttle fully closed) from an engine speed of 2,500 rpm down to 1,000 rpm with the transmission in L.
3. Repeat step 2 several times.
4. Turn the ignition switch to LOCK (0).
5. Turn the ignition switch to ON (II), and wait 30 seconds.

DRIVE PLATE

REMOVAL & INSTALLATION

See Figure 125.

1. Remove the transmission assembly.
2. Remove and inspect the drive plate (A) and the flywheel support (B). Replace it if it is damaged.

To install:

3. Install the drive plate and the flywheel support on the motor rotor (C), and secure them with the bolts, if removed.

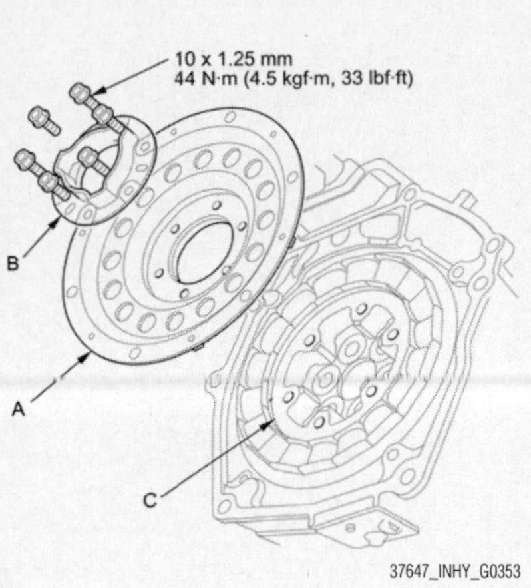
10 x 1.25 mm
44 N·m (4.5 kgf·m, 33 lbf·ft)
B
A
C
37647_INHY_G0353

Fig. 125 Remove and inspect the drive plate (A) and the flywheel support (B)

➡**Note the following during installation:**

- The motor rotor contains very strong magnets and should be handled with special care.
- Extra special care must be taken on the installation of the drive plate, it may suddenly be pulled toward the motor with great force causing serious hand or finger injury, or damaged components.
- Keep pieces of metal and all foreign particles out of the motor rotor during the transmission removal.

4. Install the transmission assembly.

INTAKE MANIFOLD

REMOVAL & INSTALLATION

See Figures 126 through 135.

1. Remove the engine cover.
2. Remove the air cleaner.
3. Remove the engine wire harness connectors and wire harness clamps from the intake manifold:

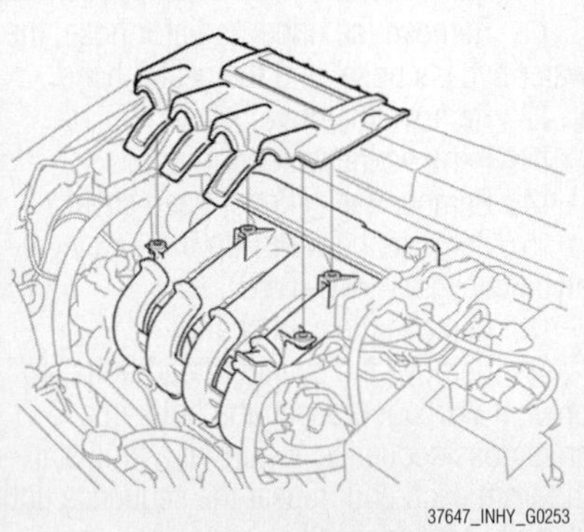
37647_INHY_G0253

Fig. 126 Remove the engine cover

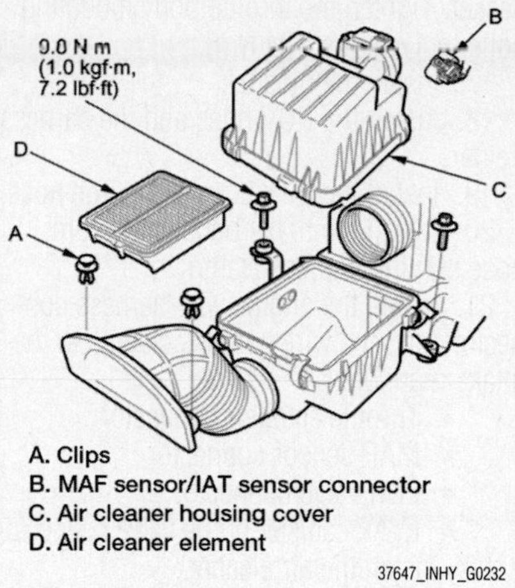
9.0 N·m
(1.0 kgf·m,
7.2 lbf·ft)
A
B
C
D
A. Clips
B. MAF sensor/IAT sensor connector
C. Air cleaner housing cover
D. Air cleaner element
37647_INHY_G0232

Fig. 127 Exploded view of air cleaner assembly

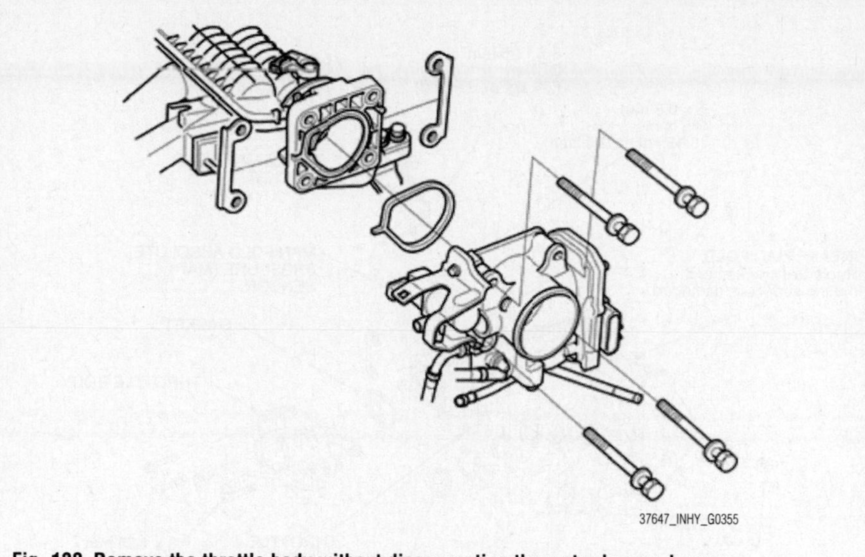
37647_INHY_G0355

Fig. 130 Remove the throttle body without disconnecting the water bypass hoses

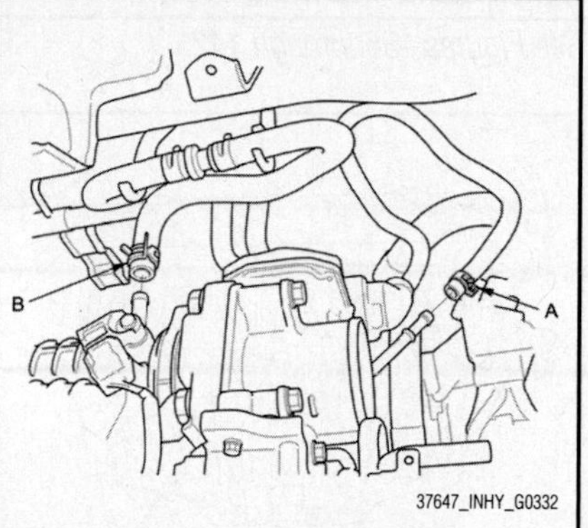
A
B
37647_INHY_G0332

Fig. 128 Disconnect the Evaporative Emission (EVAP) canister hose (A) and the brake booster vacuum hose (B)

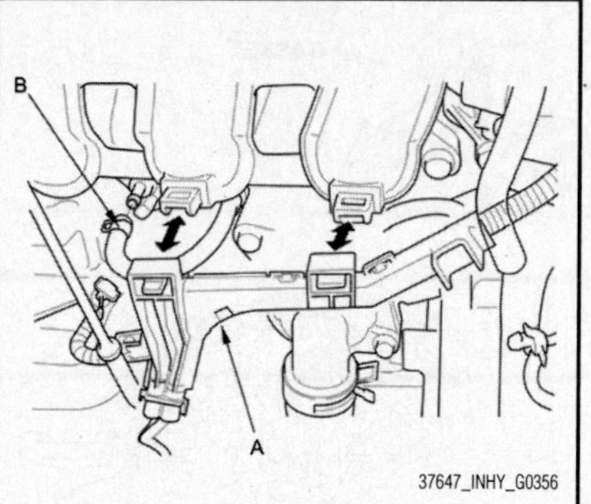
A
B
37647_INHY_G0356

Fig. 131 Remove the harness holder (A) and the Positive Crankcase Ventilation (PCV) hose (B)

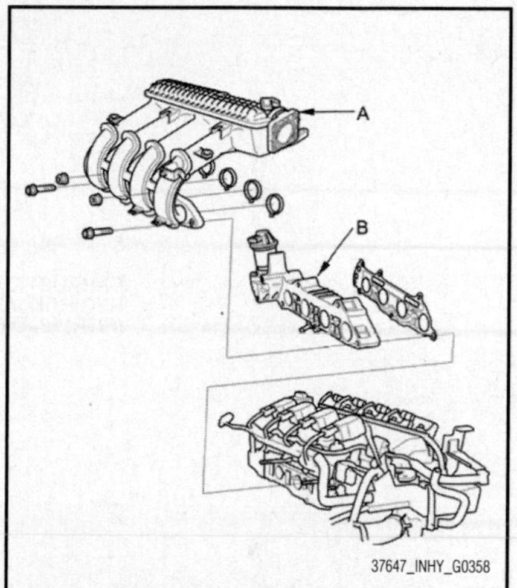
A
B
37647_INHY_G0358

Fig. 133 Remove the intake manifold (A) and the EGR plate (B)

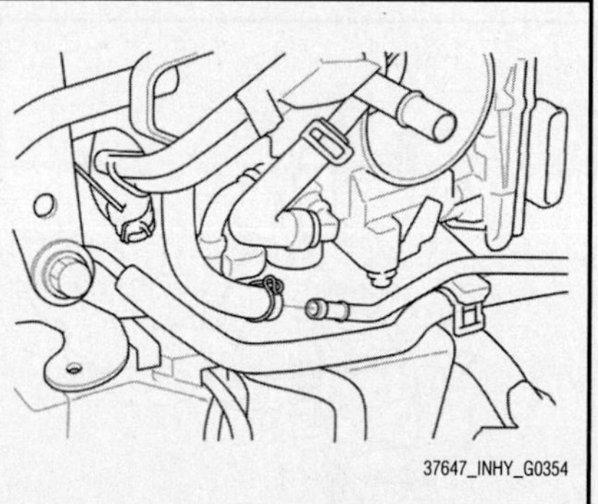
37647_INHY_G0354

Fig. 129 Remove the EVAP canister purge hose

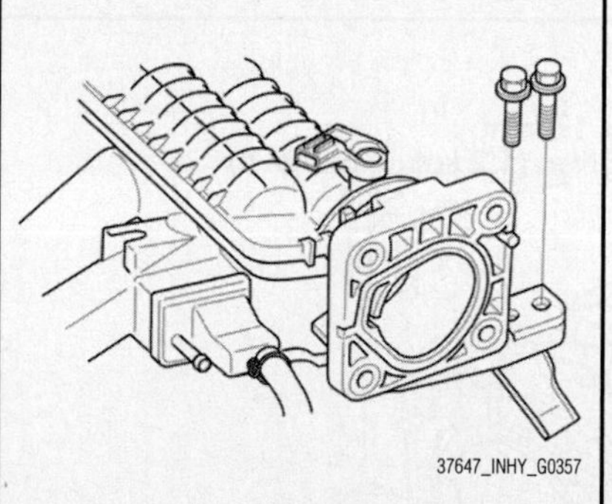
37647_INHY_G0357

Fig. 132 Remove the intake manifold bracket mounting bolts

- Throttle actuator connector
- Manifold Absolute Pressure (MAP) sensor connector
- Exhaust Gas Recirculation (EGR) valve connector
- Evaporative Emission (EVAP) canister purge valve connector

4. Disconnect the Evaporative Emission (EVAP) canister hose and the brake booster vacuum hose.
5. Remove the EVAP canister purge hose.
6. Remove the throttle body without disconnecting the water bypass hoses.
7. Remove the harness holder from the intake manifold.
8. Remove the Positive Crankcase Ventilation (PCV) hose.
9. Remove the dipstick.
10. Remove the intake manifold bracket mounting bolts.
11. Remove the intake manifold.
12. Remove the EGR plate.

To install:

13. Install the EGR plate with a new gasket.
14. Install the intake manifold with new gaskets, and tighten the bolts/nuts in a crisscross pattern in three steps, beginning with the inner bolt.
15. Loosen the intake manifold bracket mounting bolt. Tighten the mounting bolts, then tighten the mounting bolt.

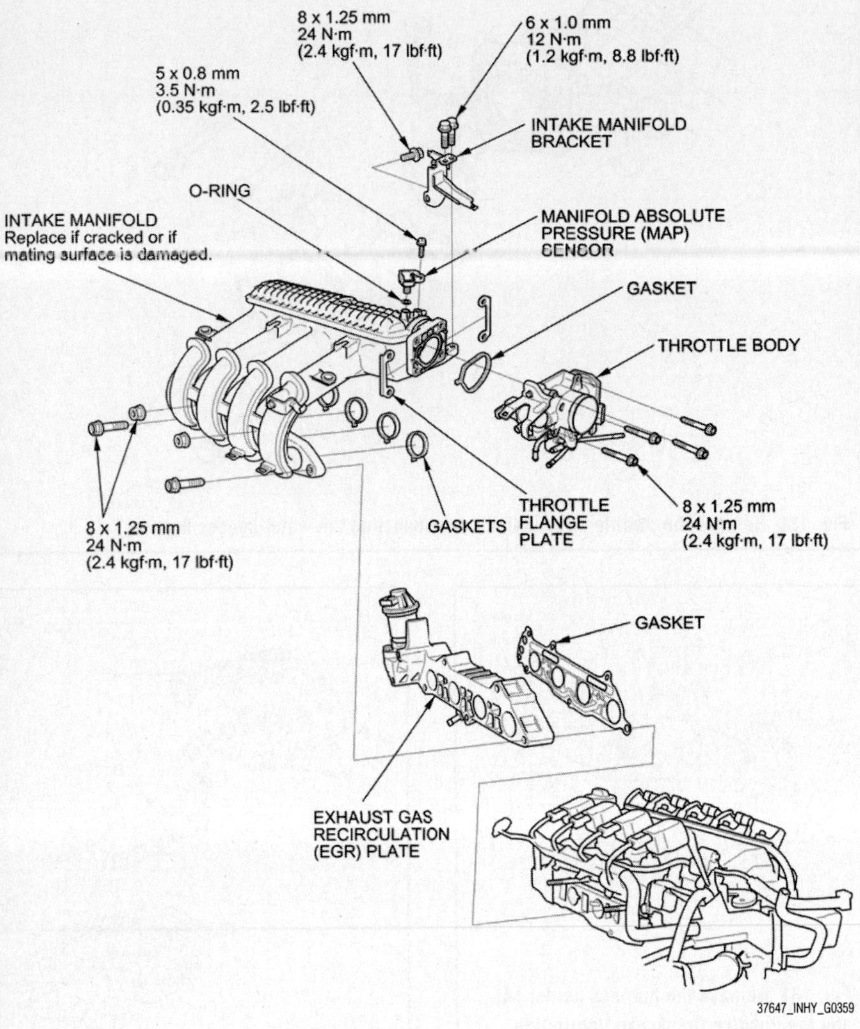
8 x 1.25 mm
24 N·m
(2.4 kgf·m, 17 lbf·ft)
6 x 1.0 mm
12 N·m
(1.2 kgf·m, 8.8 lbf·ft)
5 x 0.8 mm
3.5 N·m
(0.35 kgf·m, 2.5 lbf·ft)
INTAKE MANIFOLD
BRACKET
O-RING
INTAKE MANIFOLD
Replace if cracked or if
mating surface is damaged.
MANIFOLD ABSOLUTE
PRESSURE (MAP)
SENSOR
GASKET
THROTTLE BODY
8 x 1.25 mm
24 N·m
(2.4 kgf·m, 17 lbf·ft)
GASKETS
THROTTLE
FLANGE
PLATE
8 x 1.25 mm
24 N·m
(2.4 kgf·m, 17 lbf·ft)
GASKET
EXHAUST GAS
RECIRCULATION
(EGR) PLATE
37647_INHY_G0359

Fig. 134 Exploded view of the intake manifold assembly

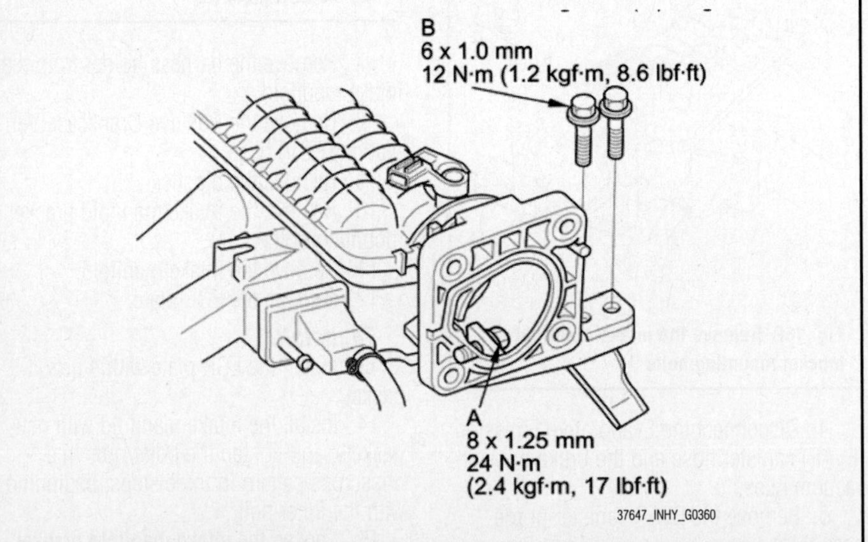
B
6 x 1.0 mm
12 N·m (1.2 kgf·m, 8.6 lbf·ft)
A
8 x 1.25 mm
24 N·m
(2.4 kgf·m, 17 lbf·ft)
37647_INHY_G0360

Fig. 135 Loosen the intake manifold bracket mounting bolt (A); tighten the mounting bolts (B), then tighten the mounting bolt (A)

16. Install the throttle body with a new gasket. Tighten the throttle body mounting bolts to 17 ft. lbs. (24 Nm).
17. Install the dipstick.
18. Install the PCV hose and the harness holder.
19. Install the EVAP canister purge hose.
20. Install the brake booster vacuum hose and the harness clamp.
21. Install the engine wire harness connectors and the wire harness clamps to the intake manifold:
    - Throttle actuator connector
    - MAP sensor connector
    - EGR valve connector
    - EVAP canister purge valve connector
22. Install the air cleaner.
23. Install the engine cover.

OIL PAN

REMOVAL & INSTALLATION

See Figures 136 through 142.

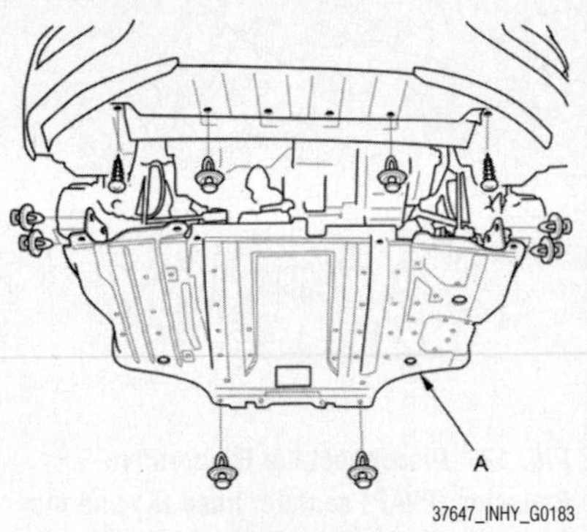
A
37647_INHY_G0183

Fig. 136 Remove the splash shields

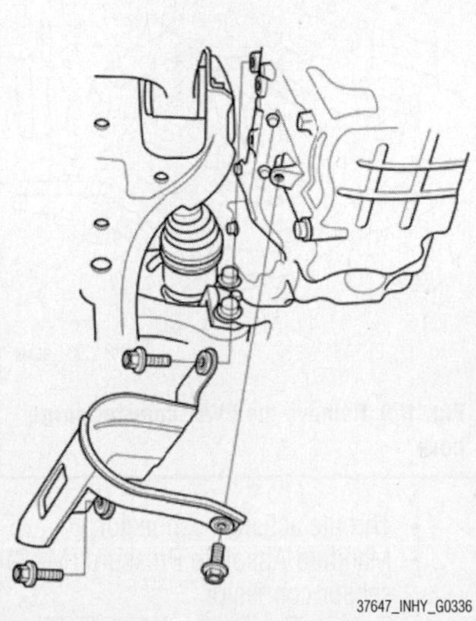
37647_INHY_G0336

Fig. 137 Remove the driveshaft heat shield

1. If the engine is already out of the vehicle, go to step 7.
2. Remove the splash shields.
3. Drain the engine oil.
4. Remove the drive belt.

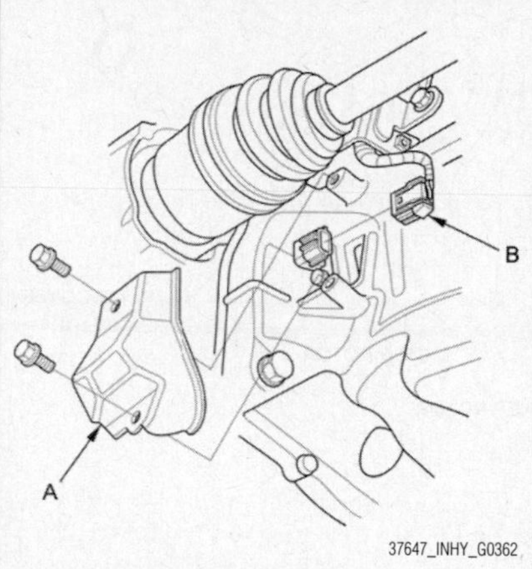
B
A
37647_INHY_G0362

Fig. 138 Remove the CKP sensor cover (A), then disconnect the CKP sensor connector (B)

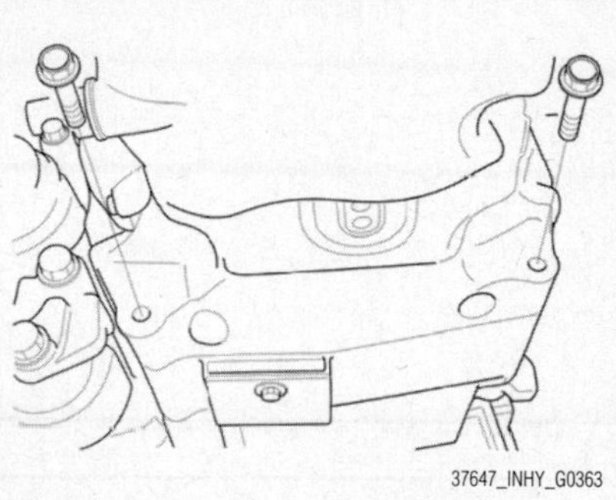
37647_INHY_G0363

Fig. 139 Remove the transmission mounting bolts

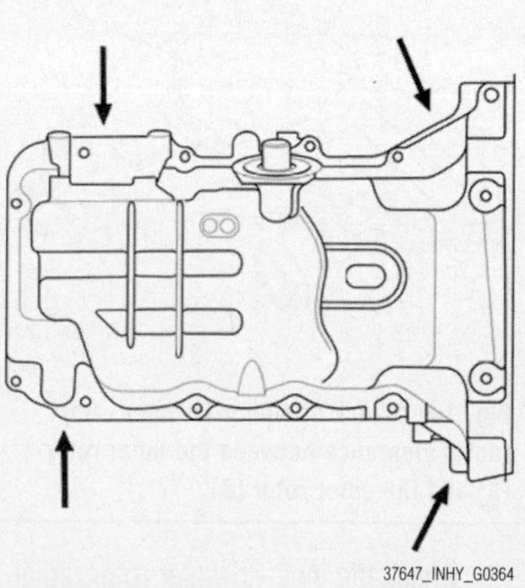
37647_INHY_G0364

Fig. 140 Insert a flat blade screwdriver where shown, and separate the oil pan from the engine block

5. Remove the driveshaft heat shield.
6. Remove the A/C compressor without disconnecting the A/C hoses.
7. Remove the dipstick, then remove the dipstick tube.
8. Remove the Crankshaft Position (CKP) sensor cover (A), then disconnect the CKP sensor connector (B).
9. Remove the transmission mounting bolts.
10. Remove the oil pan bolts. Note the bolt locations by their size.
11. Insert a flat blade screwdriver where shown, and separate the oil pan from the engine block.
12. Remove the oil pan.

➡Lower the oil pan carefully not to damage the IMA motor rotor position sensor.

To install:

13. Remove all of the old liquid gasket from the oil pan mating surfaces, the bolts, and the bolt holes.
14. Clean and dry the oil pan mating surfaces and the O-ring groove.

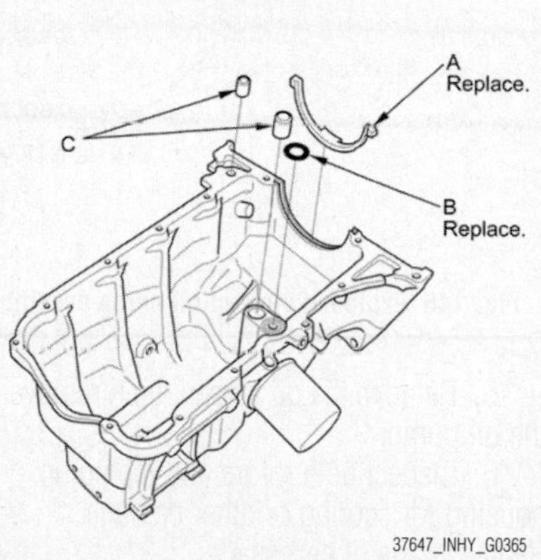
A
Replace.
C
B
Replace.
37647_INHY_G0365

Fig. 141 Install the new oil pan gasket (A), the new O-ring (B), and the dowel pins (C) on the oil pan

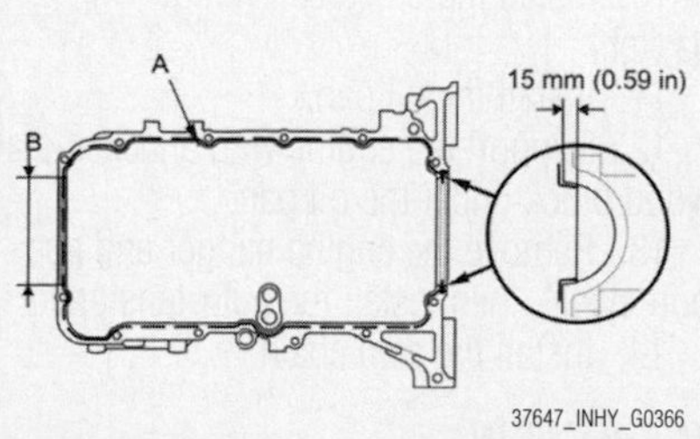
A
15 mm (0.59 in)
B
37647_INHY_G0366

Fig. 142 Apply a bead of liquid gasket along the broken line (A); and an extra bead of liquid gasket to the shaded area (B)

15. Install the new oil pan gasket, the new O-ring, and the dowel pins on the oil pan.
16. Apply liquid gasket (P/N 08717-0004, 08718-0003, or 08718-0009) to the engine block mating surface of the oil pan and to the inside edge of the bolt holes. Install the component within 5 minutes of applying the liquid gasket.
    a. Apply a bead of liquid gasket along the broken line.
    b. Apply an extra bead of liquid gasket to the shaded area.
    c. If you apply liquid gasket P/N 08718-0012, the component must be installed within 4 minutes.
    d. If too much time has passed after applying the liquid gasket, remove the old liquid gasket and residue, then reapply new liquid gasket.
17. Install the oil pan.
    a. Raise the oil pan carefully not to damage IMA motor rotor position sensor.
    b. Wait at least 30 minutes before filling the engine with oil.
    c. Do not run the engine for at least 3 hours after installing the oil pan.
    d. Make sure to install the bolts in the correct locations according to size.
18. Tighten the bolts in three steps in a crisscross pattern starting in the middle of the oil pan and working toward each end. Wipe off the excess liquid gasket on the each side of crankshaft pulley and the drive plate.
19. Install the transmission mounting bolts and tighten them to 47 ft. lbs. (64 Nm).
20. Connect the CKP sensor connector, then install the CKP sensor cover.
21. Install the dipstick tube with new O-ring, then install the dipstick.
22. If the engine is still in the vehicle, do steps 11 through 15.
23. Install the A/C compressor.
24. Install the driveshaft heat shield.
25. Install the drive belt.
26. Install the splash shields.
27. Refill the engine with engine oil.

OIL PUMP

REMOVAL & INSTALLATION

See Figures 143 through 146.

Special Tools Required*:

- Support Eyelet 07AAK-SNAA600
- Engine Support Hanger, A and Reds AAR-T1256

*Available through the Honda Tool and Equipment Program 888-424-6857

1. Remove the cam chain.

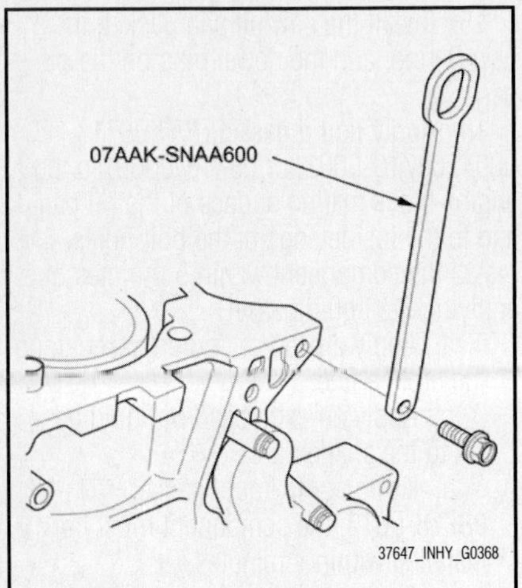
07AAK-SNAA600
37647_INHY_G0368

Fig. 143 Install the support eyelet

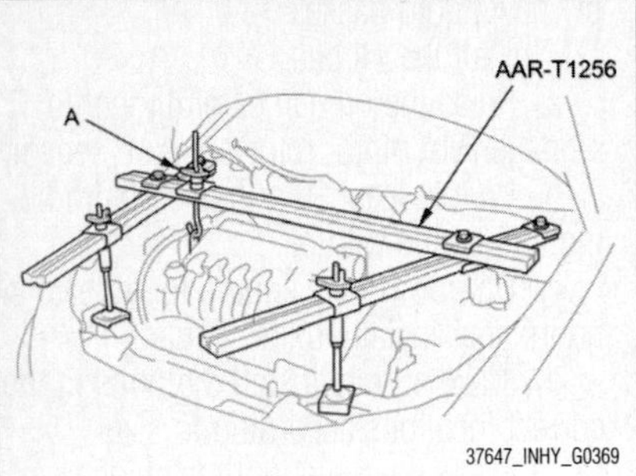
AAR-T1256
A
37647_INHY_G0369

Fig. 144 Install the engine support hanger, then attach the hook to the slotted hole in the support eyelet; tighten the wing nut (A) by hand to lift and support the engine

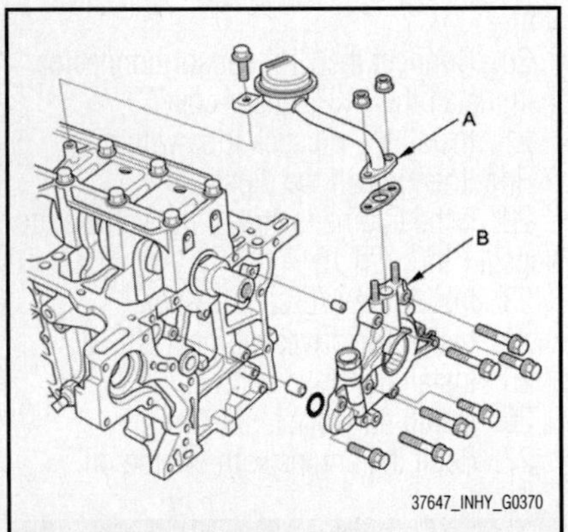
A
B
37647_INHY_G0370

Fig. 145 Remove the oil screen (A), then remove the oil pump (B)

2. Remove the auto-tensioner, then install the support eyelet.
3. Install the engine support hanger (AAR-T1256), then attach the hook to the slotted hole in the support eyelet. Tighten the wing nut by hand to lift and support the engine.
4. Remove the oil pan.

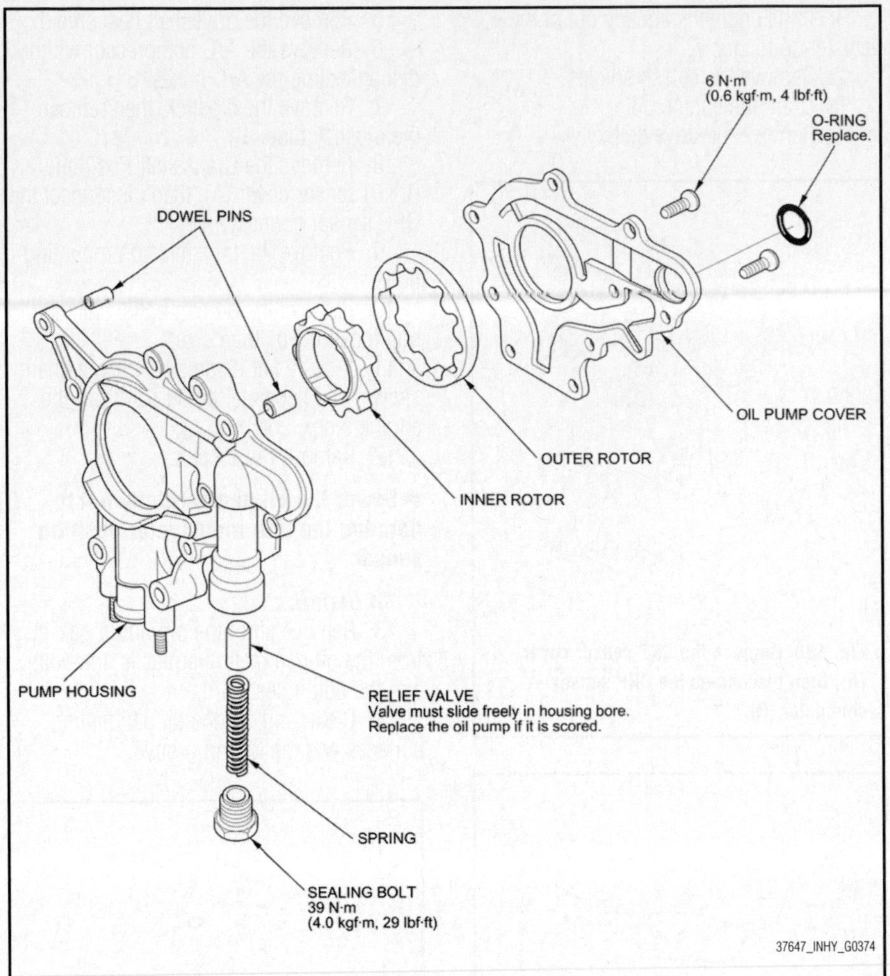
6 N·m
(0.6 kgf·m, 4 lbf·ft)
O-RING
Replace.
DOWEL PINS
OIL PUMP COVER
OUTER ROTOR
INNER ROTOR
PUMP HOUSING
RELIEF VALVE
Valve must slide freely in housing bore.
Replace the oil pump if it is scored.
SPRING
SEALING BOLT
39 N·m
(4.0 kgf·m, 29 lbf·ft)
37647_INHY_G0374

Fig. 146 Exploded view of oil pump assembly

5. Remove the oil screen, then remove the oil pump.
6. Inspect both rotors and the pump housing for scoring or other damage. Replace parts, if necessary.
7. Check that the oil pump turns freely.

To install:

8. Clean the O-ring groove and the mating surface of the engine block.
9. Install the oil pump with a new O-ring.
10. Install the oil screen with a new gasket.
11. Install the oil pan.
12. Support the engine with a jack and a wood block under the oil pan.
13. Remove the engine hanger and support eyelet, then install the auto-tensioner.
14. Install the cam chain.

INSPECTION

See Figures 147 through 149.

1. Remove the screws from the pump housing, then separate the housing and the cover.

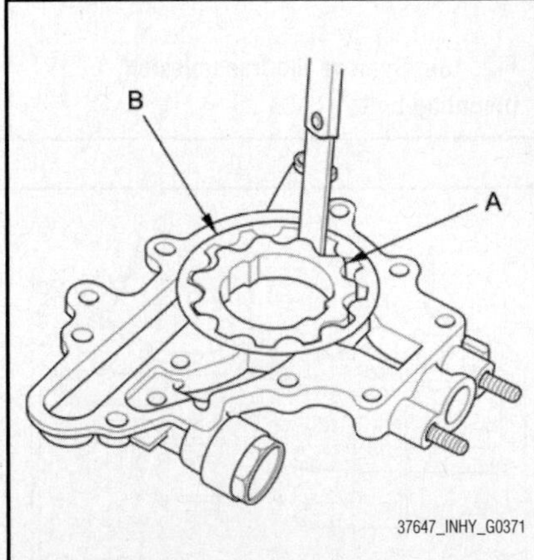
B
A
37647_INHY_G0371

Fig. 147 Check the inner-to-outer rotor radial clearance between the inner rotor (A) and the outer rotor (B)

2. Check the inner-to-outer rotor radial clearance between the inner rotor and the outer rotor. If the inner-to-outer rotor radial clearance exceeds the service limit, replace the oil pump.

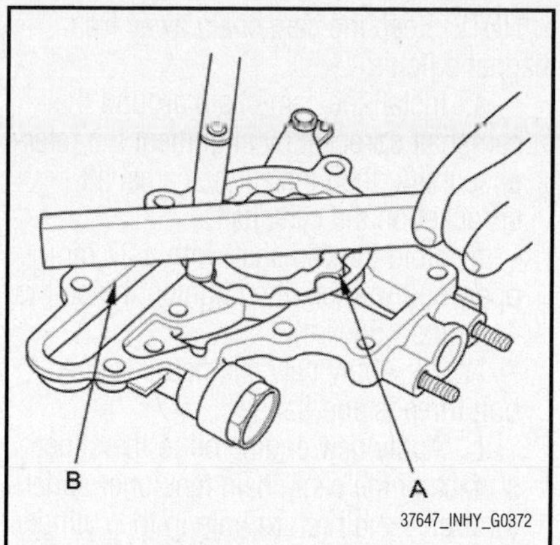
B
A
37647_INHY_G0372

Fig. 148 Check the pump housing-to-rotor axial clearance between the rotors (A) and the pump housing (B)

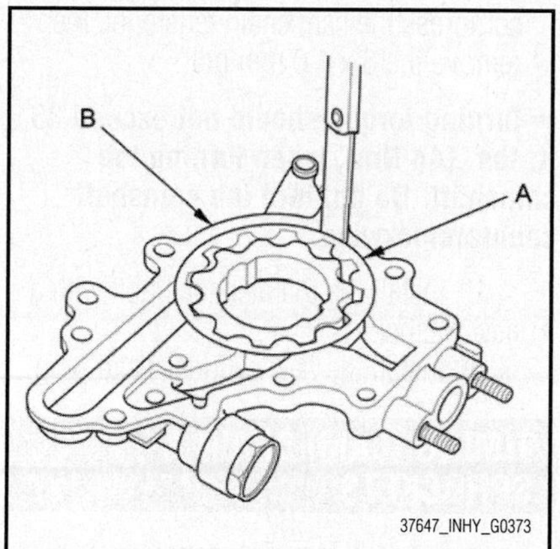
B
A
37647_INHY_G0373

Fig. 149 Check the pump housing-to-outer rotor radial clearance between the outer rotor (A) and the pump housing (B)

➡**Service Limit: 0.008 inches (0.20 mm)**

3. Check the pump housing-to-rotor axial clearance between the rotors and the pump housing. If the pump housing-to-rotor axial clearance exceeds the service limit, replace the oil pump.

➡**Service Limit: 0.006 inches (0.15 mm)**

4. Check the pump housing-to-outer rotor radial clearance between the outer rotor and the pump housing. If the pump housing-to-outer rotor radial clearance exceeds the service limit, replace the oil pump.

➡**Service Limit: 0.008 inches (0.20 mm)**

PISTON AND RING

POSITIONING

See Figures 150 and 151.

1. Install the rings as shown. The top ring has a R mark, and the second ring has a 2R mark. The manufacturing marks must face upward.
2. Rotate the rings in their grooves to make sure they do not bind.
3. Position the ring end gaps as shown.

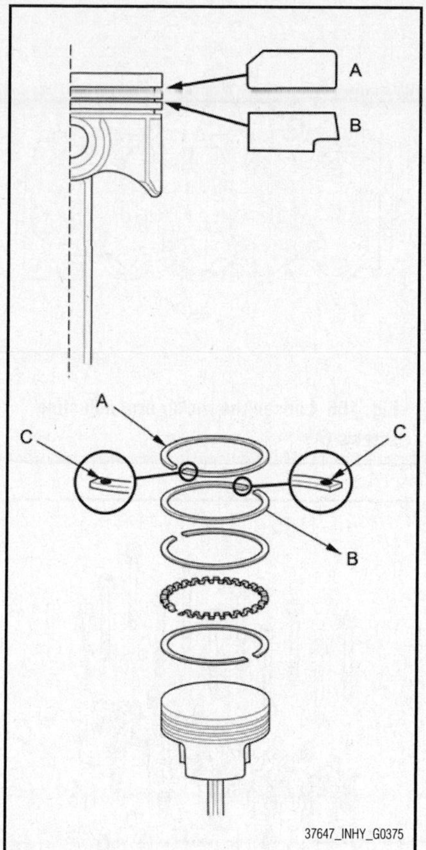
A
B
A
C
C
B
37647_INHY_G0375

Fig. 150 The top ring (A) has a R mark, and the second ring (B) has a 2R mark. The manufacturing marks (C) must face upward

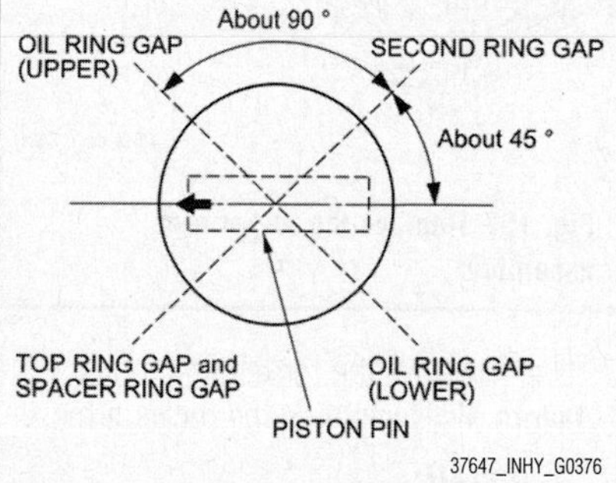
About 90 °
OIL RING GAP (UPPER)
SECOND RING GAP
About 45 °
TOP RING GAP and SPACER RING GAP
OIL RING GAP (LOWER)
PISTON PIN
37647_INHY_G0376

Fig. 151 Position the ring end gaps as shown

ROCKER ARMS

REMOVAL & INSTALLATION

See Figures 152 through 157.

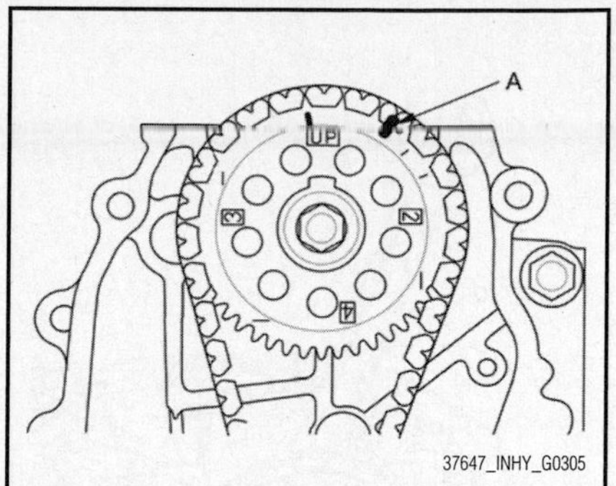
A
37647_INHY_G0305

Fig. 152 Make a reference mark (A) across the camshaft sprocket and cam chain

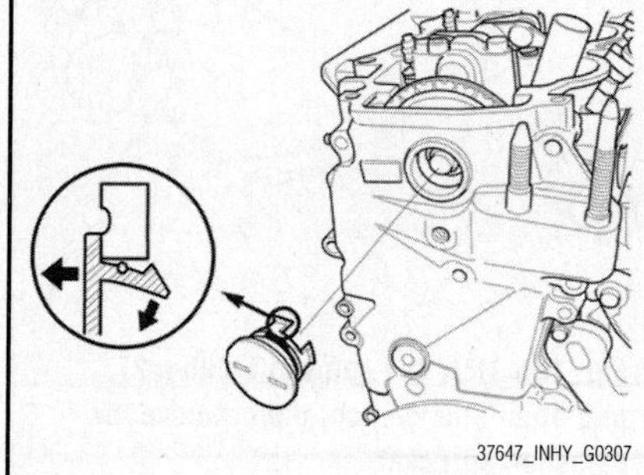
37647_INHY_G0307

Fig. 153 Remove the cylinder head plug

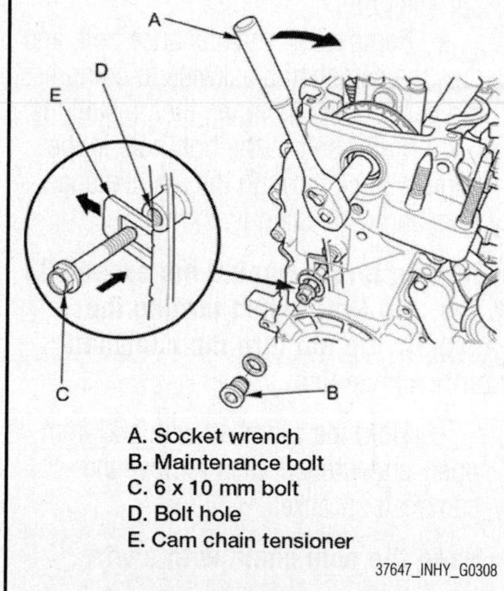
A
D
E
C
B
A. Socket wrench
B. Maintenance bolt
C. 6 x 10 mm bolt
D. Bolt hole
E. Cam chain tensioner
37647_INHY_G0308

Fig. 154 Hold the crankshaft pulley, and set the socket wrench on the camshaft sprocket bolt

1. Remove the camshaft sprocket.
   a. Make a reference mark across the camshaft sprocket and cam chain.
   b. Apply new engine oil to the slider surface of the cam chain tensioner slider through the oil return hole in the cylinder head.

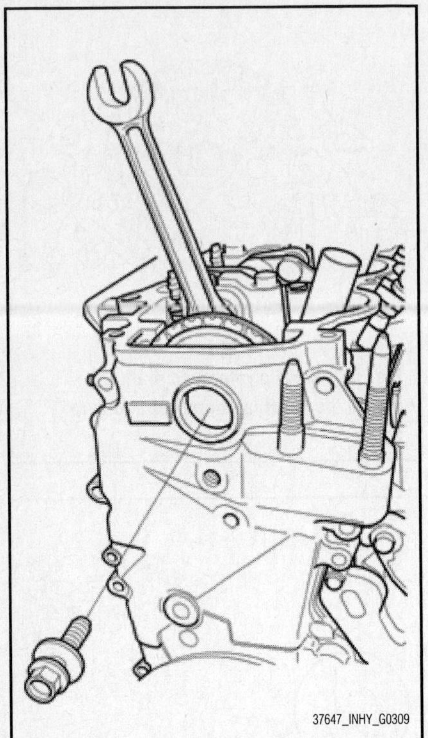
37647_INHY_G0309

Fig. 155 Hold the camshaft with a 27 mm open-end wrench, then remove the camshaft sprocket

c. Remove the cylinder head plug.

d. Hold the crankshaft pulley, and set the socket wrench on the camshaft sprocket bolt.

e. Remove the maintenance bolt, and turn the camshaft clockwise to compress the cam chain tensioner, then install the 6 x 1.0 mm bolt in the bolt hole in the engine block through the maintenance hole and cam chain tensioner.

➡**Turning torque should not exceed 33 ft. lbs. (44 Nm), when turning the camshaft. Do not turn the camshaft counterclockwise.**

f. Hold the camshaft with a 27 mm open-end wrench, then remove the camshaft sprocket.

➡**Hang the cam chain with a wire.**

2. Remove the rocker arm assembly.

a. Loosen the rocker arm adjusting screws.

b. Remove the camshaft holder bolts. To prevent damaging the camshaft, loosen the bolts in sequence two turns at a time, in a crisscross pattern.

c. Remove the rocker arm assembly.

d. Identify each part as it is removed so that each item can be reinstalled in their original location.

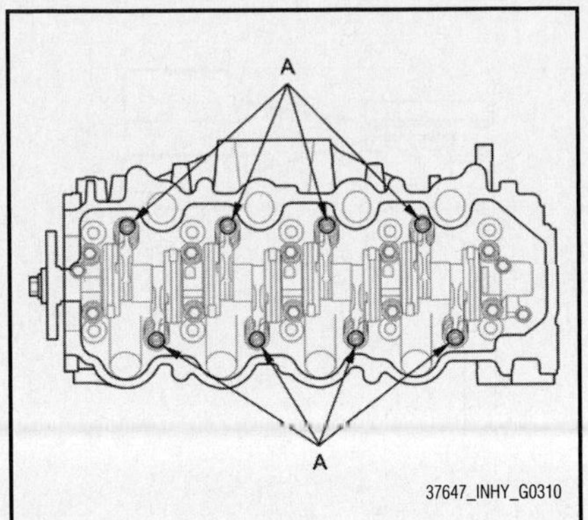
A
A
37647_INHY_G0310

Fig. 156 Loosen the rocker arm adjusting screws (A)

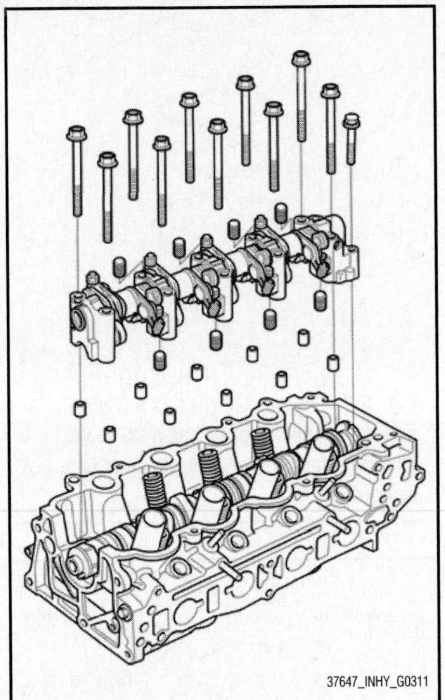
37647_INHY_G0311

Fig. 157 Remove the rocker arm assembly

e. Remove the rocker shaft bolts before disassembling the rocker arms.

To install:

3. Install the rocker arm assembly.

a. Reassemble the rocker arm assembly.

b. If reused, the rocker arms must be installed in their original location.

c. Prior to reassembling, clean all the parts in solvent, dry them, and apply new engine oil to any contact points.

d. Apply new engine oil to the threads of the rocker shaft bolts when installing them.

e. When replacing the rocker arm assembly, remove the fastening hardware from the new rocker arm assembly.

4. Install the camshaft sprocket.

NOTE: Keep the cam chain away from magnetic fields.

a. Install the cam chain around the camshaft sprocket by alignment the reference mark, then install the camshaft sprocket on the camshaft.

b. Hold the camshaft with a 27 mm open-end wrench, then tighten the bolt to 41 ft. lbs. (56 Nm).

NOTE: Apply new engine oil to the bolt threads and flange.

c. Apply new engine oil to the slider surface of the cam chain tensioner slider through the oil return hole in the cylinder head.

d. Hold the crankshaft pulley, and set the socket wrench on the camshaft sprocket bolt.

e. Turn the camshaft clockwise to compress the cam chain tensioner, then remove the 6 x 1.0 mm bolt.

➡**Turning torque should not exceed 33 ft. lbs. (44 Nm), when turning the camshaft. Do not turn the camshaft counterclockwise.**

f. Install the maintenance bolt with a new washer.

g. Install the new cylinder head plug.

TIMING (CAM) CHAIN & SPROCKETS

REMOVAL & INSTALLATION

See Figures 158 through 176.

➡**Keep the cam chain away from magnetic fields.**

1. Remove the cylinder head cover.

2. Set the No. 1 piston at Top Dead Center (TDC). The UP mark on the camshaft sprocket should be at the top, and

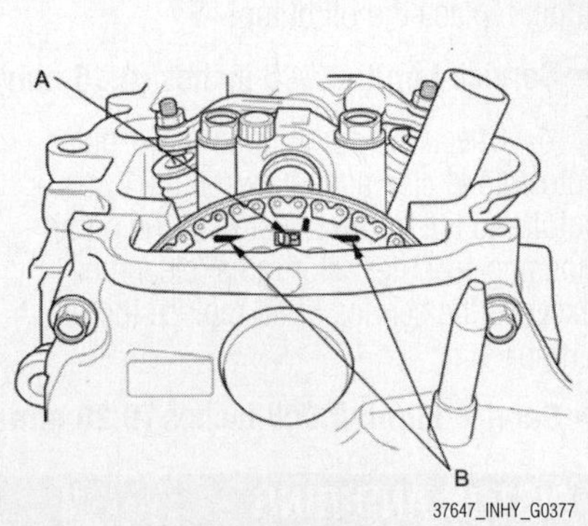
A
B
37647_INHY_G0377

Fig. 158 The UP mark (A) on the camshaft sprocket should be at the top, and the TDC grooves (B) on the camshaft sprocket should line up with the top edge of the head

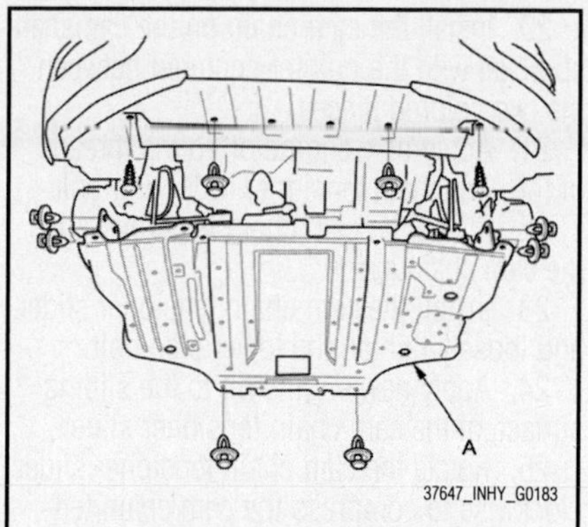
A
37647_INHY_G0183

Fig. 159 Remove the splash shields

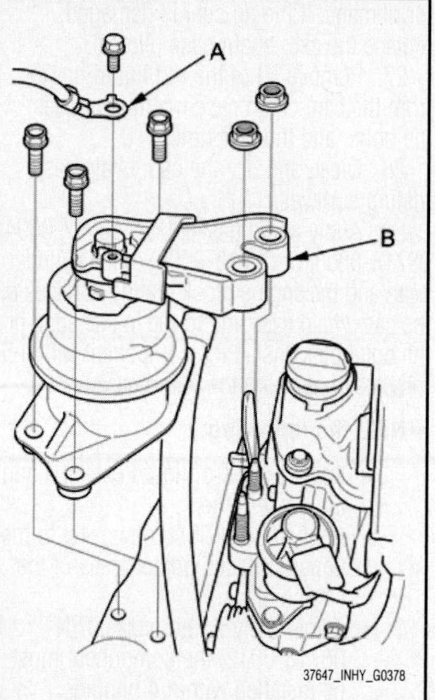
A
B
37647_INHY_G0378

Fig. 160 Disconnect the ground cable (A), then remove the side engine mount/bracket assembly (B)

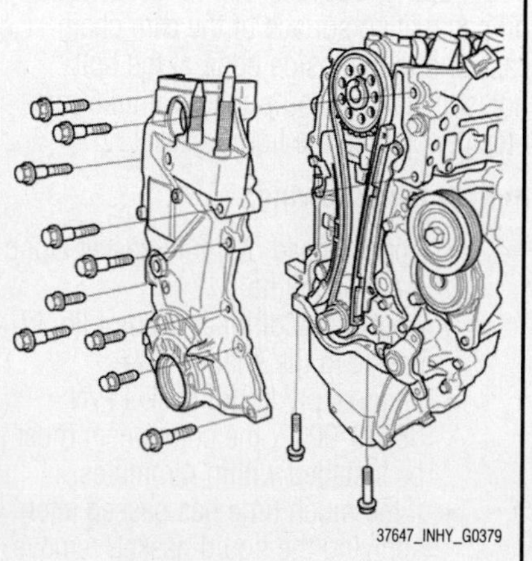
37647_INHY_G0379

Fig. 161 Remove the cam chain case

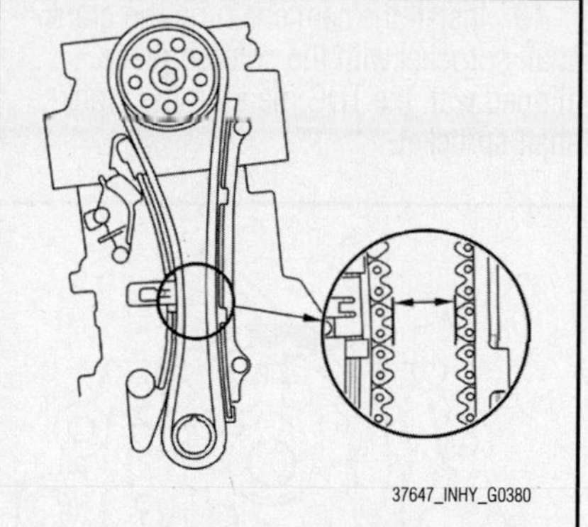
37647_INHY_G0380

Fig. 162 Measure the cam chain separation

the TDC grooves on the camshaft sprocket should line up with the top edge of the head.

3. Remove the front wheels.
4. Remove the splash shields.
5. Remove the drive belt.
6. Remove the water pump pulley.
7. Remove the crankshaft pulley.
8. Support the engine with a jack and a wood block under the oil pan.
9. Disconnect the ground cable, then remove the side engine mount/bracket assembly.
10. Remove the cam chain case.
11. Measure the cam chain separation. If the distance is less than the service limit, replace the cam chain and the cam chain tensioner.

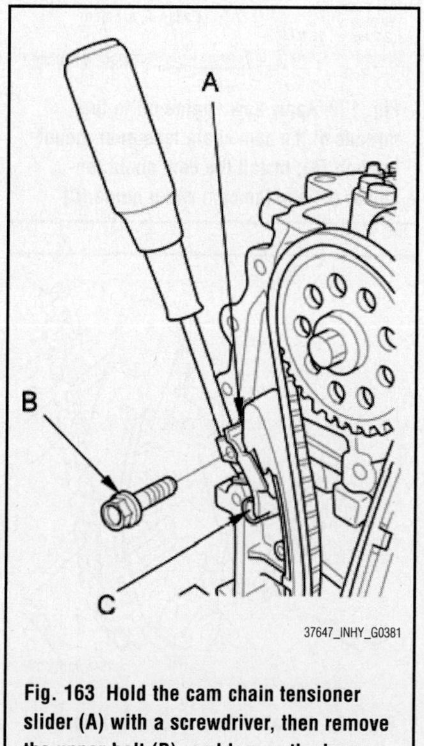
A
B
C
37647_INHY_G0381

Fig. 163 Hold the cam chain tensioner slider (A) with a screwdriver, then remove the upper bolt (B), and loosen the lower bolt (C)

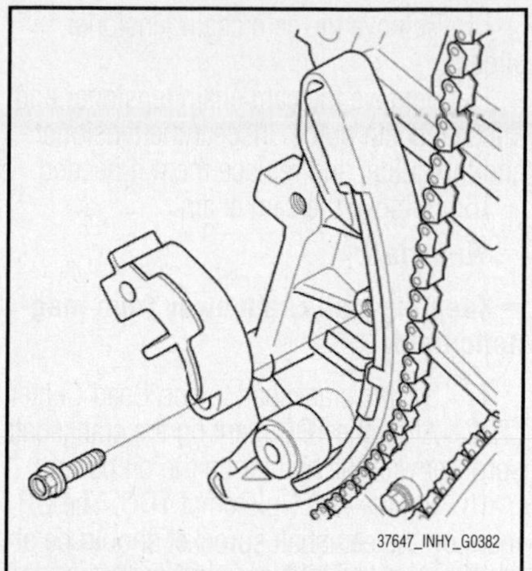
37647_INHY_G0382

Fig. 164 Remove the cam chain tensioner slider

➡Service Limit: 0.59 inches (15 mm)

12. Apply new engine oil to the sliding surface of the cam chain tensioner slider.
13. Hold the cam chain tensioner slider with a screwdriver, then remove the upper bolt, and loosen the lower bolt.

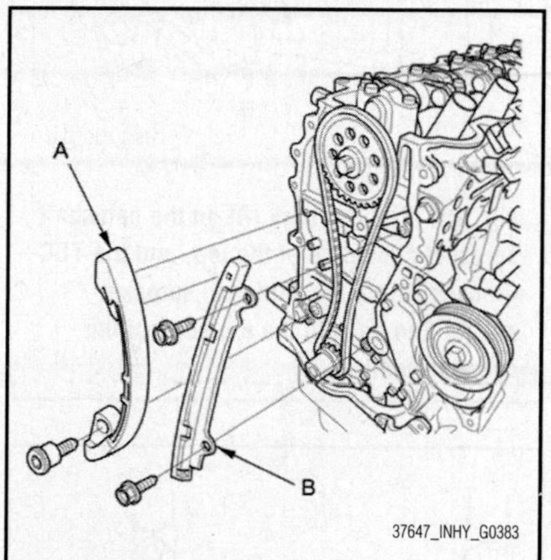
A
B
37647_INHY_G0383

Fig. 165 Remove the cam chain tensioner (A) and the cam chain guide (B)

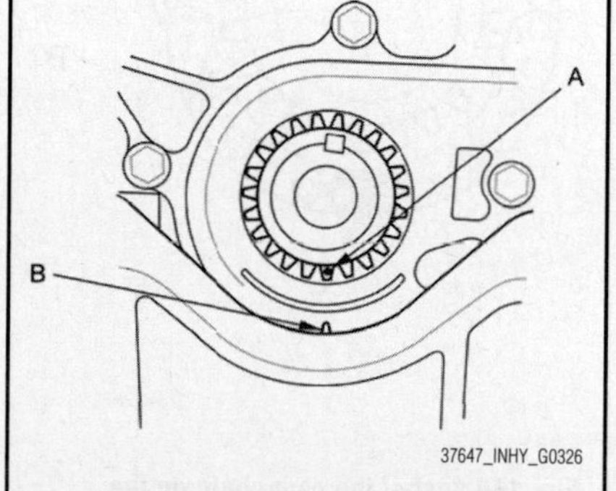
A
B
37647_INHY_G0326

Fig. 166 Align the TDC mark (A) on the crankshaft sprocket with the pointer (B) on the oil pump

14. Remove the cam chain tensioner slider.

15. Remove the cam chain tensioner and the cam chain guide. Inspect the tensioner and the guide, and replace them if needed.

16. Remove the cam chain.

To install:

➡Keep the cam chain away from magnetic fields.

17. Set the crankshaft to Top Dead Center (TDC). Align the TDC mark on the crankshaft sprocket with the pointer on the oil pump.

18. Set the No. 1 piston at TDC. The UP mark on the camshaft sprocket should be at the top, and the TDC grooves on the camshaft sprocket should line up with the top edge of the cylinder head.

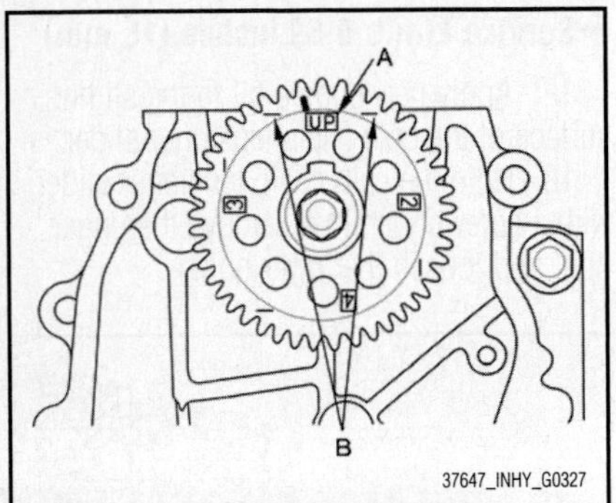
A
UP
B
37647_INHY_G0327

Fig. 167 The UP mark (A) on the camshaft sprocket should be at the top, and the TDC grooves (B) on the camshaft sprocket should line up with the top edge of the cylinder head

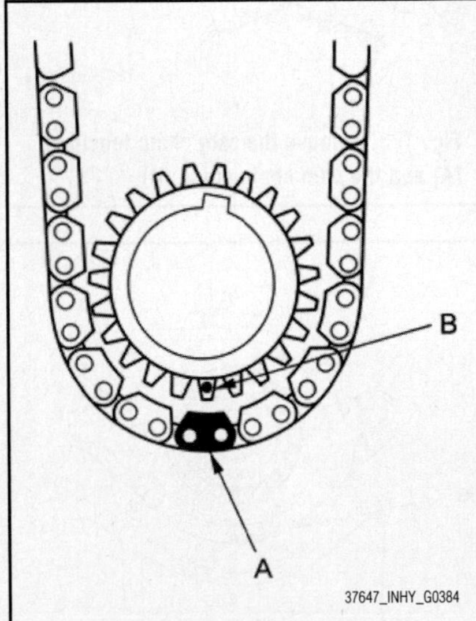
B
A
37647_INHY_G0384

Fig. 168 Install the cam chain on the crankshaft sprocket with the colored piece (A) aligned with the TDC mark (B) on the crankshaft sprocket

19. Install the cam chain on the crankshaft sprocket with the colored piece aligned with the TDC mark on the crankshaft sprocket.

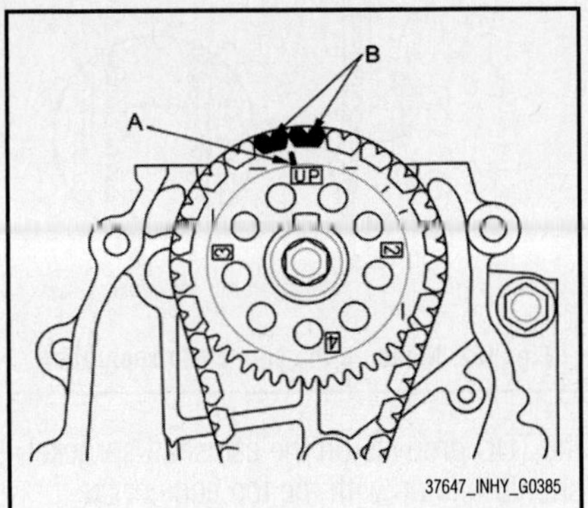
B
A
UP
37647_INHY_G0385

Fig. 169 Install the cam chain on the camshaft sprocket with the pointer (A) centered between the two colored pieces (B)

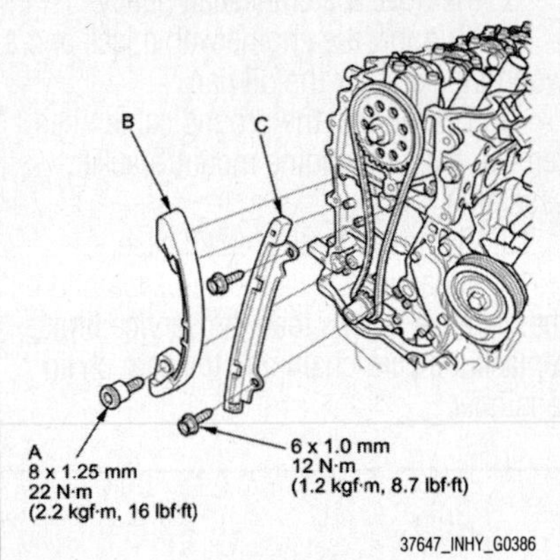
B
C
A
8 x 1.25 mm
22 N·m
(2.2 kgf·m, 16 lbf·ft)
6 x 1.0 mm
12 N·m
(1.2 kgf·m, 8.7 lbf·ft)
37647_INHY_G0386

Fig. 170 Apply new engine oil to the threads of the cam chain tensioner mounting bolt (A); install the cam chain tensioner (B) and the cam chain guide (C)

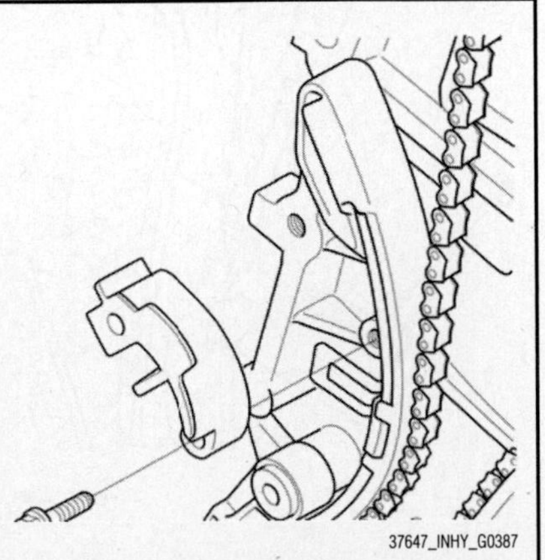
37647_INHY_G0387

Fig. 171 Install the cam chain tensioner slider, and loosely tighten the lower side bolt

20. Install the cam chain on the camshaft sprocket with the pointer centered between the two colored pieces.

21. Apply new engine oil to the threads of the cam chain tensioner mounting bolt.

22. Install the cam chain tensioner and the cam chain guide.

23. Install the cam chain tensioner slider, and loosely tighten the lower side bolt.

24. Apply new engine oil to the sliding surface of the cam chain tensioner slider.

25. Rotate the cam chain tensioner slider clockwise to compress the cam chain tensioner, and install the remaining bolt, then tighten the two bolts.

26. Check the cam chain case oil seal for damage. If the oil seal is damaged, replace the cam chain case oil seal.

27. Remove all of the old liquid gasket from the cam chain case mating surfaces, the bolts, and the bolt holes.

28. Clean and dry the cam chain case mating surfaces.

29. Apply liquid gasket (P/N 08717-0004, 08718-0003, or 08718-0009) to the cylinder head and the engine block mating surfaces of the cam chain case and to the inside edge of the bolt holes. Install the component within 5 minutes of applying the liquid gasket.

➡Note the following:

- Apply a bead of liquid gasket along the broken line.
- Apply a bead of liquid gasket to the upper surface contact areas of the engine block.
- If you apply liquid gasket P/N 08718-0012, the component must be installed within 4 minutes.
- If too much time has passed after applying the liquid gasket, remove the old liquid gasket and residue, then reapply new liquid gasket.

30. Apply liquid gasket (P/N 08717-0004, 08718-0003, or 08718-0009) to the oil pan mating surface of the cam chain case and to the inside edge of the bolt holes. Install the component within 5 minutes of applying the liquid gasket.

➡Note the following:

- Apply a bead of liquid gasket along the broken line.
- Apply an additional bead of liquid gasket to the shaded area.
- If you apply liquid gasket P/N 08718-0012, the component must be installed within 4 minutes.
- If too much time has passed after applying the liquid gasket, remove the old liquid gasket and residue, then reapply new liquid gasket.

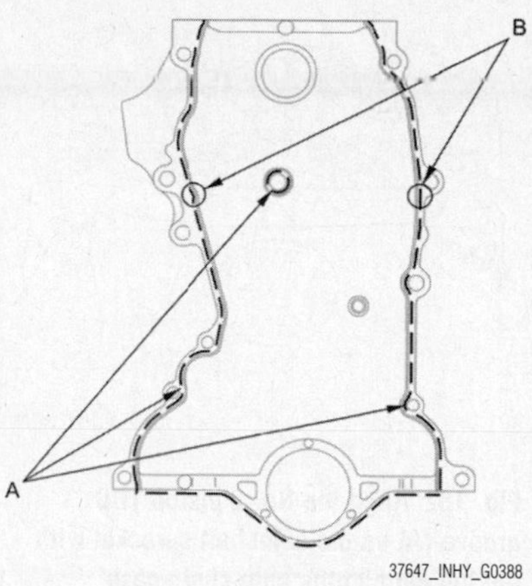
B
A
37647_INHY_G0388

Fig. 172 Apply a bead of liquid gasket along the broken line (A) and to the upper surface contact areas of the engine block (B)

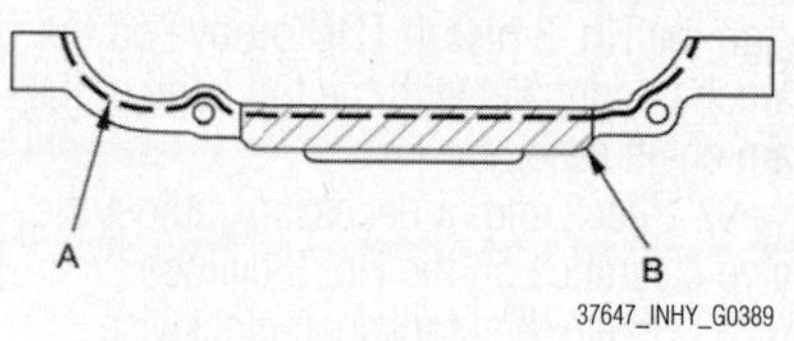
A
B
37647_INHY_G0389

Fig. 173 Apply a bead of liquid gasket along the broken line (A); and an additional bead to the shaded area (B)

31. Set the edge of the cam chain case on the edge of the oil pan, then install the cam chain case on the engine block.

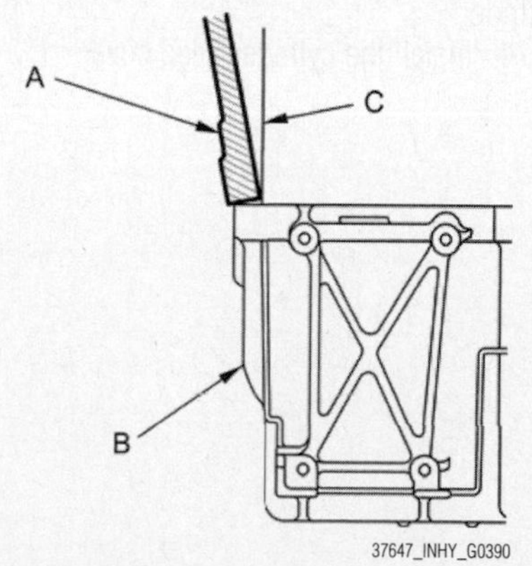
A
C
B
37647_INHY_G0390

Fig. 174 Set the edge of the cam chain case (A) on the edge of the oil pan (B), then install the cam chain case on the engine block (C)

➡Note the following:

- When installing the cam chain case, do not slide the bottom surface onto the oil pan mounting surface.
- Wait at least 30 minutes before filling the engine with oil.
- Do not run the engine for at least 3 hours after installing the cam chain case.

32. Tighten the cam chain case mounting bolts to 23 ft. lbs. (31 Nm). Wipe off the excess liquid gasket from the oil pan and the cam chain case mating area.
33. Install the side engine mount/bracket assembly, then tighten the new side engine mount/bracket assembly mounting bolts to 43 ft. lbs. (59 Nm).
34. Loosely tighten the new side engine mount/bracket assembly mounting nuts.
35. Connect the ground cable.
36. Remove the jack and the wood block.
37. Remove the air cleaner.
38. Loosen the transmission mount bracket mounting bolts and nuts.
39. Raise the vehicle on the lift to full height.
40. Loosen the torque rod mounting bolt and nut.
41. Lower the vehicle on the lift.
42. Tighten the side engine mount/bracket assembly mounting nuts to 36 ft. lbs. (49 Nm).
43. Tighten the transmission mount mounting bolts and nuts. Tighten the 10 x 1.25 mm through bolt to 40 ft. lbs. (54 Nm), then tighten the 12 x 1.25 mm bolt and nuts to 54 ft. lbs. (74 Nm).
44. Raise the vehicle on the lift to full height.
45. Tighten the torque rod mounting bolt

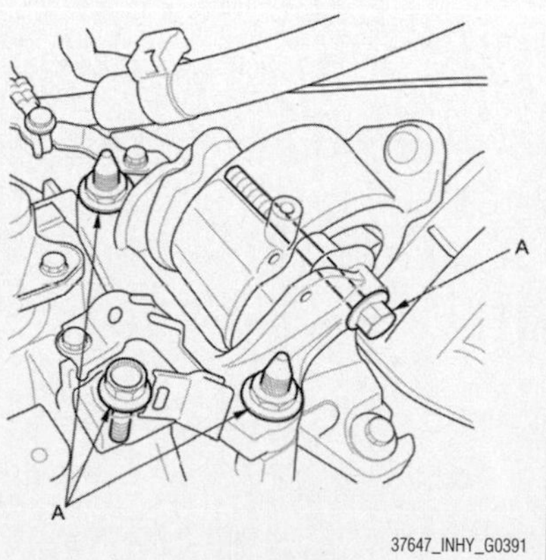
A
A
37647_INHY_G0391

Fig. 175 Loosen the transmission mount bracket mounting bolts and nuts (A)

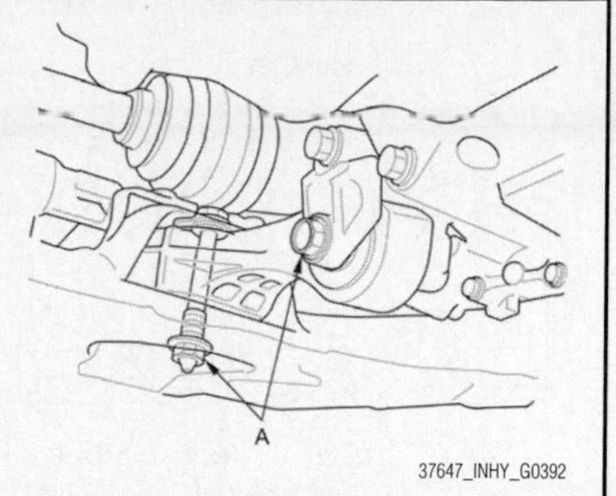
A
37647_INHY_G0392

Fig. 176 Loosen the torque rod mounting bolt and nut (A)

to 61 ft. lbs. (83 Nm),and then tighten the nut to 69 ft. lbs. (93 Nm).

46. Lower the vehicle on the lift.
47. Install the air cleaner.
48. Install the cylinder head cover.
49. Install the crankshaft pulley.
50. Install the water pump pulley.
51. Install the drive belt.
52. Install the splash shields.
53. Install the front wheel.
54. Do the Crankshaft Position (CKP) pattern clear/CKP pattern learn procedure.

VALVE LASH

ADJUSTMENT

See Figures 177 through 182.

➡Adjust the valves only when the cylinder head temperature is less than 100 °F (38 °C). Check the engine coolant temperature with the HDS if you are not sure.

1. Remove the cylinder head cover.
2. Set the No. 1 piston at top dead center (TDC). The UP mark on the camshaft sprocket should be at the top, and the TDC grooves on the camshaft sprocket should

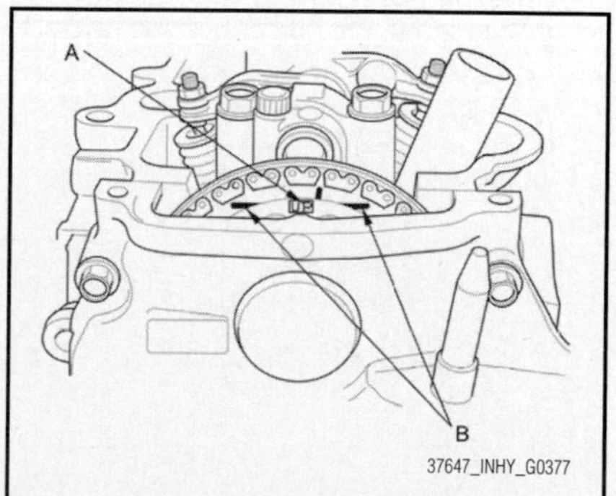
A
B
37647_INHY_G0377

Fig. 177 The UP mark (A) on the camshaft sprocket should be at the top, and the TDC grooves (B) on the camshaft sprocket should line up with the top edge of the cam chain case

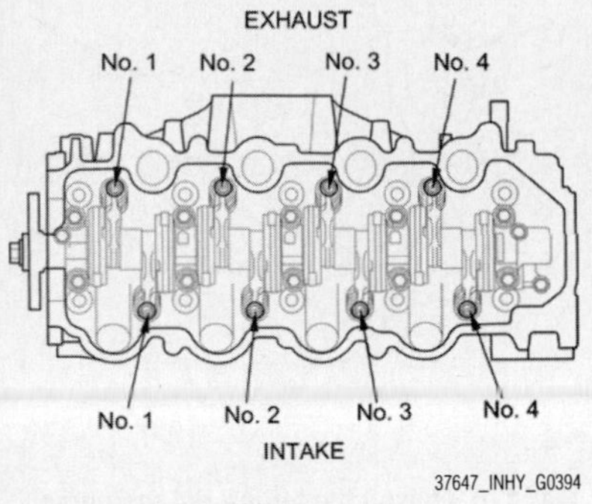
EXHAUST
No. 1
No. 2
No. 3
No. 4
No. 1
No. 2
No. 3
No. 4
INTAKE
37647_INHY_G0394

Fig. 178 Select the correct feeler gauge for the valves you are going to check

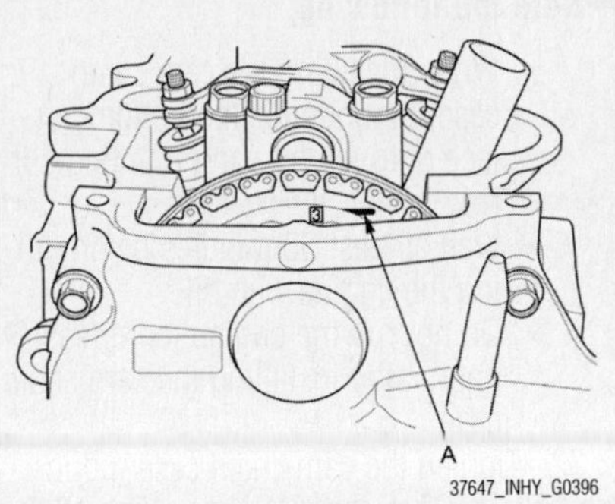
A
37647_INHY_G0396

Fig. 180 Align the No. 3 piston TDC groove (A) on the camshaft sprocket with the top edge of the cam chain case

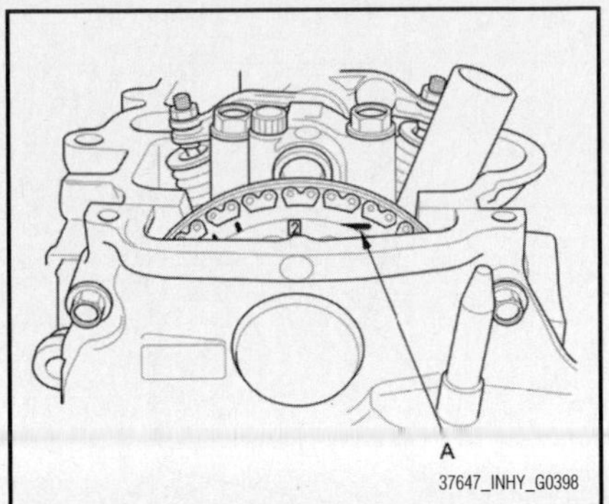
A
37647_INHY_G0398

Fig. 182 Align the No. 2 piston TDC groove (A) on the camshaft sprocket with the top edge of the cam chain case

line up with the top edge of the cam chain case.

3. Select the correct feeler gauge for the valves you are going to check.

➡Valve Clearance Specifications:

- Intake: 0.006–0.007 inches (0.15–0.19 mm)
- Exhaust: 0.009–0.011 inches (0.24–0.28 mm)

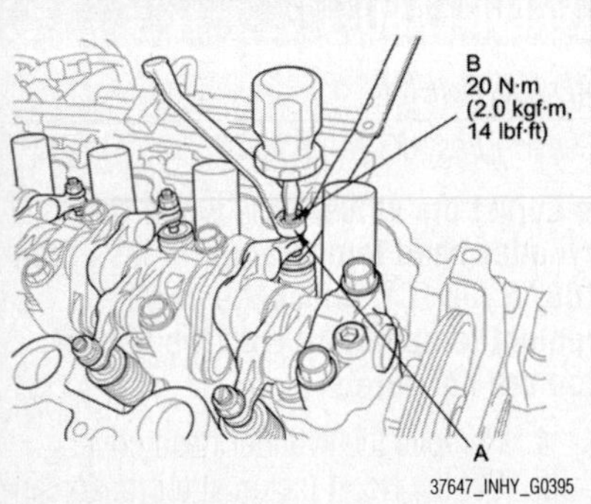
B
20 N·m
(2.0 kgf·m,
14 lbf·ft)
A
37647_INHY_G0395

Fig. 179 Loosen the locknut (A), and turn the adjusting screw (B) until the drag on the feeler gauge is correct

4. Insert the feeler gauge between the adjusting screw and the end of the valve stem on No. 1 cylinder and slide it back and forth; you should feel a slight amount of drag.

5. If you feel too much or too little drag, loosen the locknut and turn the adjusting screw until the drag on the feeler gauge is correct.

6. Tighten the locknut, and recheck the clearance. Repeat the adjustment if necessary.

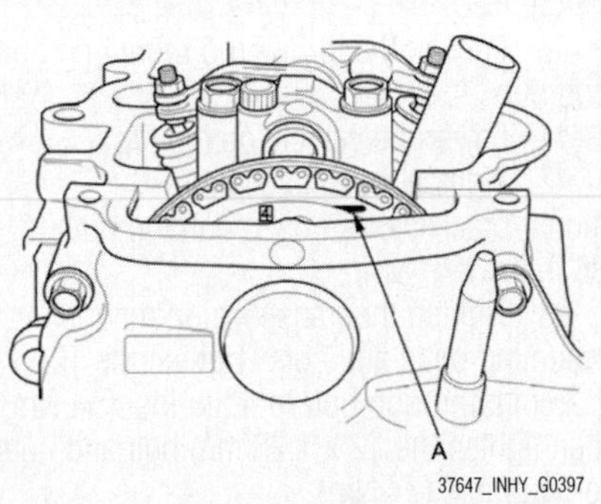
A
37647_INHY_G0397

Fig. 181 Align the No. 4 piston TDC groove (A) on the camshaft sprocket with the top edge of the cam chain case

7. Tighten the locknut to 14 ft. lbs. (20 Nm), and recheck the valve clearance. Repeat the adjustment if necessary.

8. Rotate the crankshaft clockwise. Align the No. 3 piston TDC groove on the camshaft sprocket with the top edge of the cam chain case.

9. Check and, if necessary, adjust the valve clearance on the No. 3 cylinder.

10. Rotate the crankshaft clockwise. Align the No. 4 piston TDC groove on the camshaft sprocket with the top edge of the cam chain case.

11. Check and, if necessary, adjust the valve clearance on the No. 4 cylinder.

12. Rotate the crankshaft clockwise. Align the No. 2 piston TDC groove on the camshaft sprocket with the top edge of the cam chain case.

13. Check and, if necessary, adjust the valve clearance on the No. 2 cylinder.

14. Install the cylinder head cover.

ENGINE PERFORMANCE & EMISSION CONTROLS

ACCELERATOR PEDAL POSITION (APP) SENSOR

LOCATION

The Accelerator Pedal Position (APP) sensor is an integral part of the accelerator pedal module.

REMOVAL & INSTALLATION

See Figure 183.

1. Remove the driver's dashboard lower cover.
2. Disconnect the APP sensor 6P connector.
3. Remove the accelerator pedal module.

➡The APP sensor is not available separately. Do not disassemble the accelerator pedal module. If the accelerator pedal module is dropped, replace it.

4. Install the parts in the reverse order of removal.

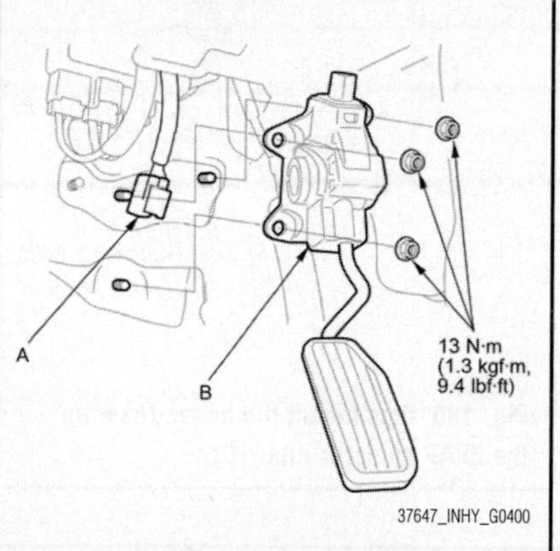
A
B
13 N·m
(1.3 kgf·m,
9.4 lbf·ft)
37647_INHY_G0400

Fig. 183 Disconnect the APP sensor 6P connector (A); remove the accelerator pedal module (B)

CAMSHAFT POSITION (CMP) SENSOR

LOCATION

Refer to the graphics in the Removal and Installation section for the location(s).

REMOVAL & INSTALLATION

See Figure 184.

1. Remove the air cleaner.
2. Remove the cowl cover and the under-cowl panel.
3. Disconnect the CMP sensor 3P connector.
4. Remove the CMP sensor.

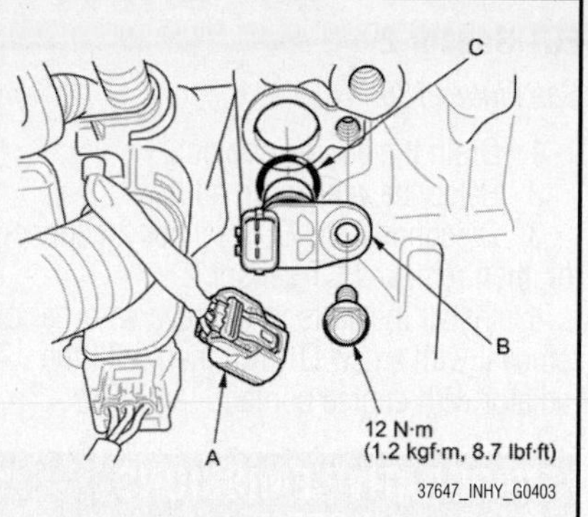
C
B
A
12 N·m
(1.2 kgf·m, 8.7 lbf·ft)
37647_INHY_G0403

Fig. 184 Disconnect the CMP sensor 3P connector (A); remove the CMP sensor (B) and the O-ring (C)

5. Install the sensor in the reverse order of removal with a new O-ring.

CRANKSHAFT POSITION (CKP) SENSOR

LOCATION

Refer to the graphics in the Removal and Installation section for the location(s).

REMOVAL & INSTALLATION

See Figures 185 and 186.

1. Raise the vehicle on a lift.

➡Make sure the vehicle is level, because engine oil will drip out when you remove the sensor.

2. Remove the engine undercover.
3. Remove the CKP sensor cover.
4. Disconnect the CKP sensor connector.
5. Remove the CKP sensor.

To install:

6. Install the parts in the reverse order of removal with a new O-ring.

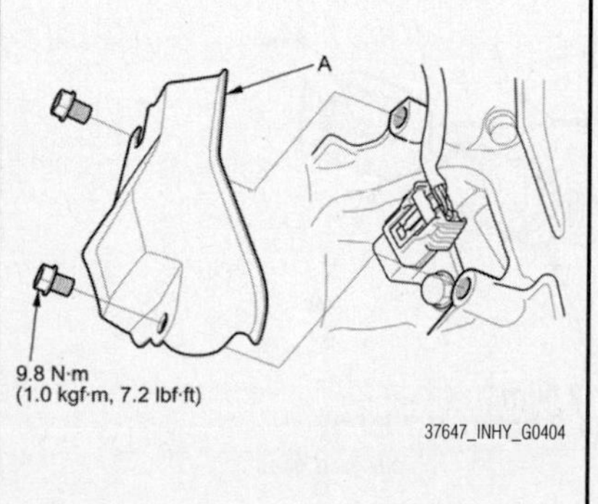
A
9.8 N·m
(1.0 kgf·m, 7.2 lbf·ft)
37647_INHY_G0404

Fig. 185 Remove the CKP sensor cover (A)

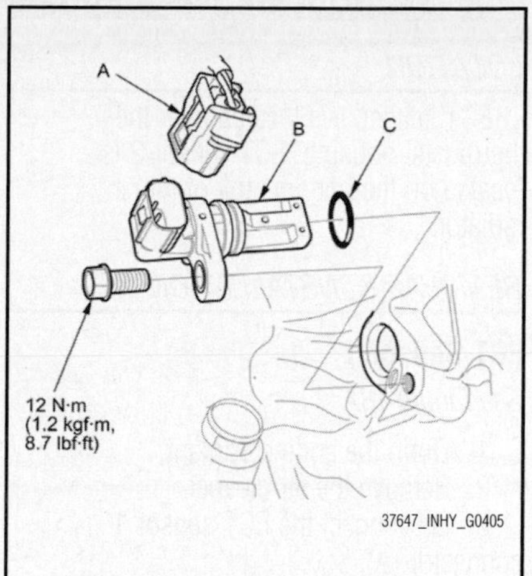
A
B
C
12 N·m
(1.2 kgf·m,
8.7 lbf·ft)
37647_INHY_G0405

Fig. 186 Disconnect the CKP sensor connector (A); remove the CKP sensor (B) and the O-ring (C)

7. Do the CKP pattern clear/CKP pattern learn procedure.
8. Check the engine oil level, and add more oil if needed.

CKP PATTERN CLEAR/CKP PATTERN LEARN PROCEDURE

Clear/Learn Procedure (with the HDS)

1. Connect the HDS to the Data Link Connector (DLC) located under the driver's side of the dashboard.
2. Turn the ignition switch to ON (II).
3. Make sure the HDS communicates with the PCM and all other vehicle systems. If it doesn't, go to the DLC circuit troubleshooting.
4. Select CRANK PATTERN in the ADJUSTMENT MENU with the HDS.
5. Select CRANK (CKP) PATTERN CLEAR, and clear the CKP pattern.
6. Select CRANK PATTERN LEARNING with the HDS, and follow the screen prompts.

Learn Procedure (without the HDS)

1. Start the engine. Hold the engine speed at 3,000 rpm without load (in P or N) until the radiator fan comes on.
2. Test-drive the vehicle on a level road: Decelerate (with the throttle fully closed) from an engine speed of 2,500 rpm down to 1,000 rpm with the transmission in L.
3. Repeat step 2 several times.
4. Turn the ignition switch to LOCK (0).
5. Turn the ignition switch to ON (II), and wait 30 seconds.

ENGINE COOLANT TEMPERATURE (ECT) SENSOR

LOCATION

ECT sensor 1 is located near the thermostat housing. ECT sensor 2 is located on the bottom tank of the radiator.

REMOVAL & INSTALLATION

ECT Sensor 1

See Figure 187.

1. Drain the engine coolant.
2. Remove the air cleaner.
3. Disconnect the ECT sensor 1 connector (A).
4. Remove ECT sensor 1.
5. Install the parts in the reverse order of removal with a new O-ring, then refill the radiator with engine coolant.

ECT Sensor 2

See Figure 188.

1. Drain the engine coolant.
2. Raise the vehicle on a lift.
3. Disconnect the ECT sensor 2 connector, then remove ECT sensor 2.
4. Install the parts in the reverse order of removal with a new O-ring, then refill the radiator with engine coolant.

EVAPORATIVE EMISSIONS (EVAP) CANISTER

LOCATION

Refer to the graphics in the Removal and Installation section for the location(s).

REMOVAL & INSTALLATION

See Figures 189 through 191.

1. Raise the vehicle on a lift.
2. Disconnect the hoses from the EVAP canister filter.
3. Remove the bolts, then remove the canister guard.
4. Disconnect the hoses.
5. Disconnect the FTP sensor 3P connector and the EVAP canister vent shut valve 2P connector, then remove the harness clip.
6. Remove the bolts, then remove the EVAP canister assembly.
7. Install the parts in the reverse order of removal.

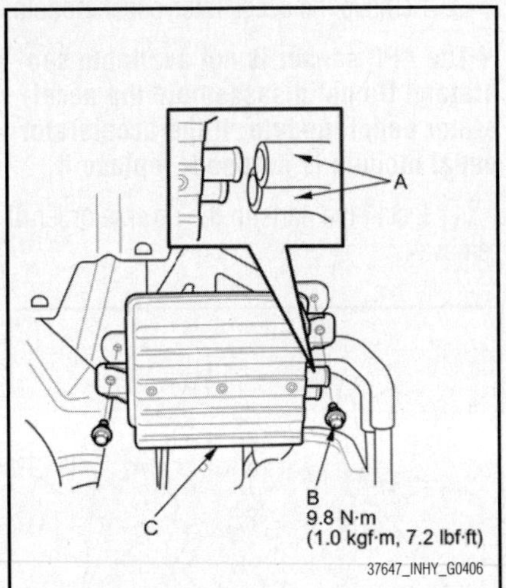
A
B
9.8 N·m
(1.0 kgf·m, 7.2 lbf·ft)
C
37647_INHY_G0406

Fig. 189 Disconnect the hoses (A) from the EVAP canister filter (C)

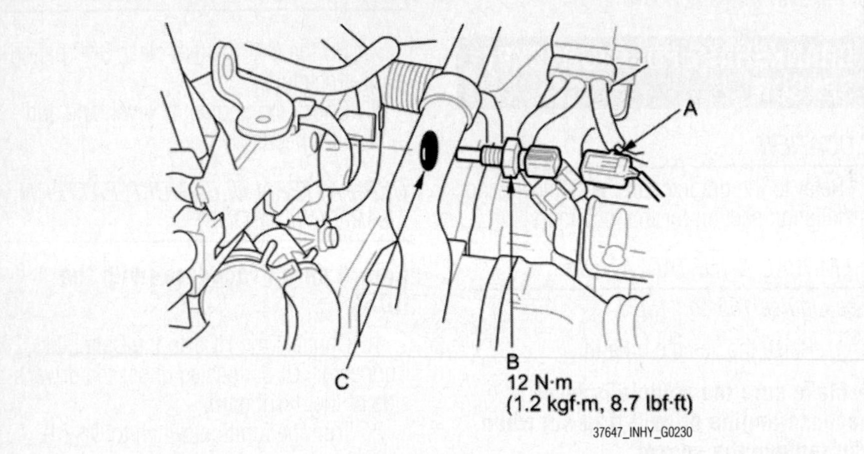
A
B
12 N·m
(1.2 kgf·m, 8.7 lbf·ft)
C
37647_INHY_G0230

Fig. 187 Disconnect the ECT sensor 1 connector (A); remove ECT sensor 1 (B) and O-ring (C)

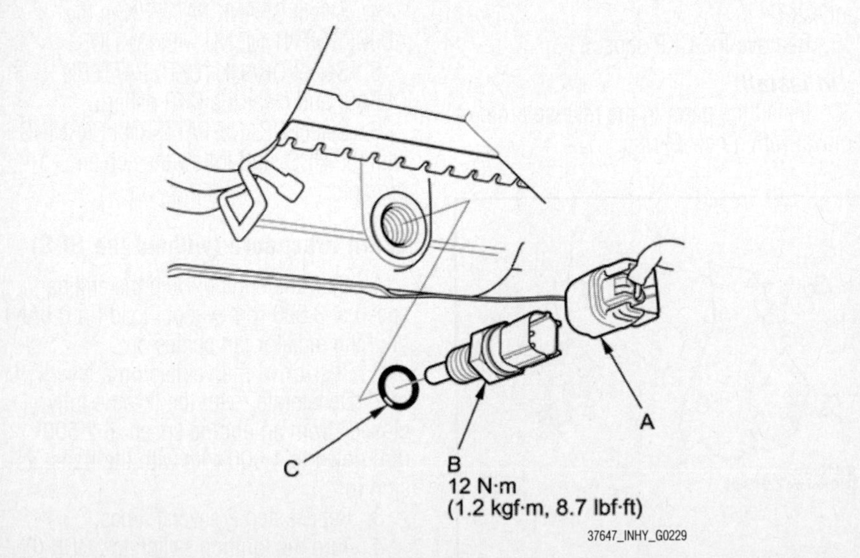
A
B
12 N·m
(1.2 kgf·m, 8.7 lbf·ft)
C
37647_INHY_G0229

Fig. 188 Disconnect the ECT sensor 2 connector (A), then remove ECT sensor 2 (B) and O-ring (C)

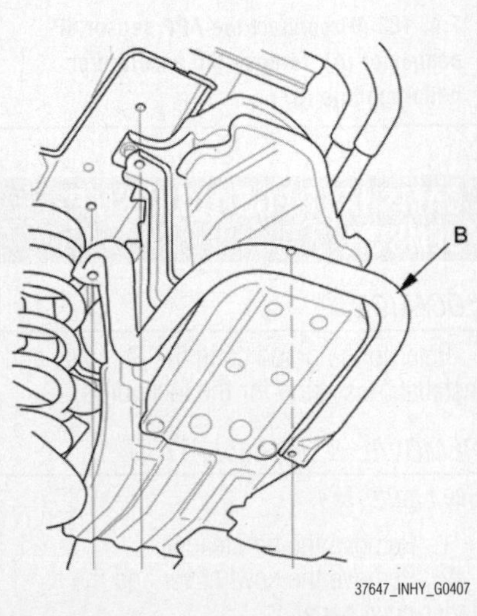
B
37647_INHY_G0407

Fig. 190 Remove the bolts (A), then remove the canister guard (B)

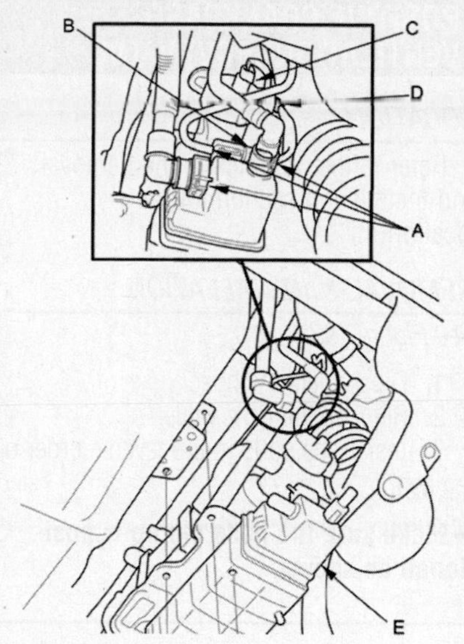
B
C
D
A
E

A. Hoses
B. FTP sensor 3P connector
C. EVAP canister vent shut valve 2P connector
D. Harness clip
E. EVAP canister assembly

37647_INHY_G0408

Fig. 191 Disconnect the hoses

EXHAUST GAS RECIRCULATION (EGR) VALVE

LOCATION

Refer to the graphics in the Removal and Installation section for the location(s).

REMOVAL & INSTALLATION

See Figure 192.

1. Disconnect the EGR valve 5P connector.

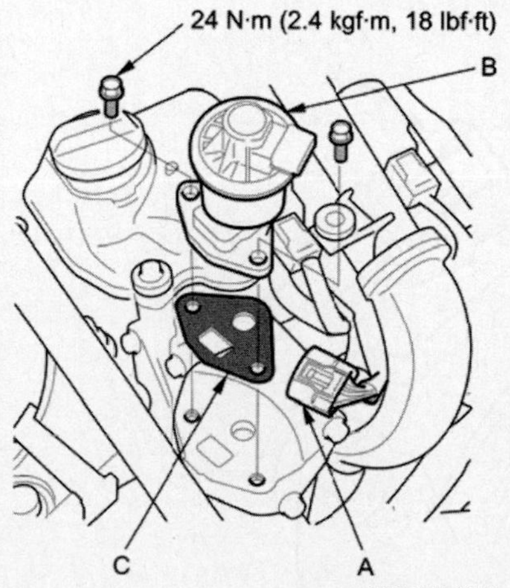
24 N·m (2.4 kgf·m, 18 lbf·ft)
B
C
A
37647_INHY_G0411

Fig. 192 Disconnect the EGR valve 5P connector (A); Remove the EGR valve (B) and the gasket (C)

2. Remove the EGR valve.
3. Install the parts in the reverse order of removal with a new gasket.

HEATED OXYGEN (HO2S) SENSOR

LOCATION

Refer to the graphics in the Removal and Installation section for the location(s).

REMOVAL & INSTALLATION

See Figures 193 and 194.

Special Tools Required: O2 Sensor Wrench, Snap-on S6176 or equivalent, commercially available

1. Remove the cowl cover and the under-cowl panel.
2. Disconnect the secondary HO2S connector.
3. Raise the vehicle on a lift.
4. Remove the wire clips, then remove the secondary HO2S.
5. Install the parts in the reverse order of removal.

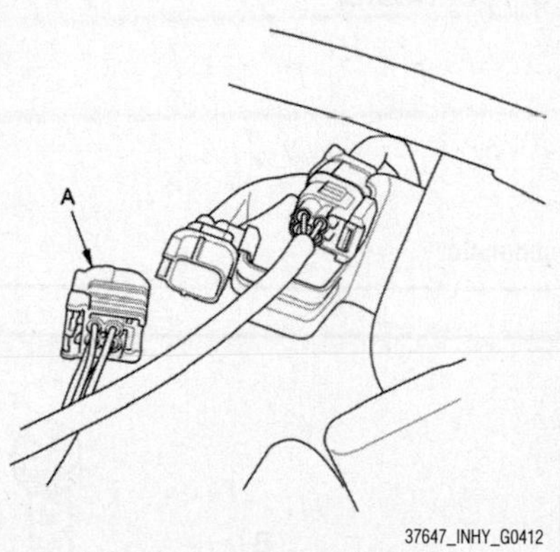
A
37647_INHY_G0412

Fig. 193 Disconnect the secondary HO2S connector (A)

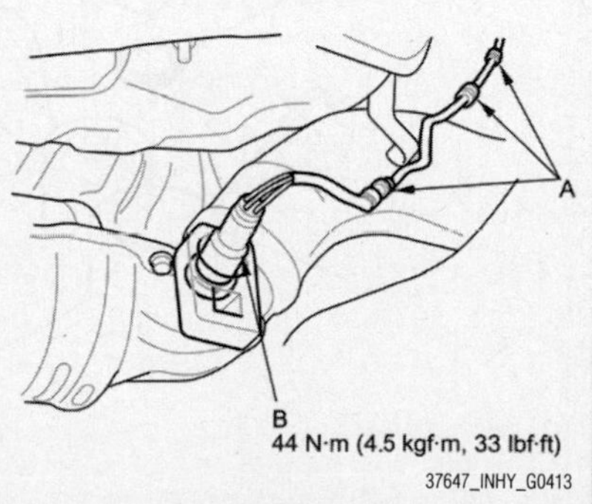
A
B
44 N·m (4.5 kgf·m, 33 lbf·ft)
37647_INHY_G0413

Fig. 194 Remove the wire clips (A), then remove the secondary HO2S (B)

INTAKE AIR TEMPERATURE (IAT) SENSOR

LOCATION

The Intake Air Temperature (IAT) sensor is an integral part of the Mass Air Flow (MAF)/Intake Air Temperature (IAT) sensor located on the engine air intake duct.

REMOVAL & INSTALLATION

Refer to the Mass Air Flow (MAF)/Intake Air Temperature (IAT) sensor section when servicing this component.

KNOCK SENSOR (KS)

LOCATION

Refer to the graphics in the Removal and Installation section for the location(s).

REMOVAL & INSTALLATION

See Figure 195.

1. Remove the engine oil dipstick.
2. Disconnect the knock sensor 1P connector.

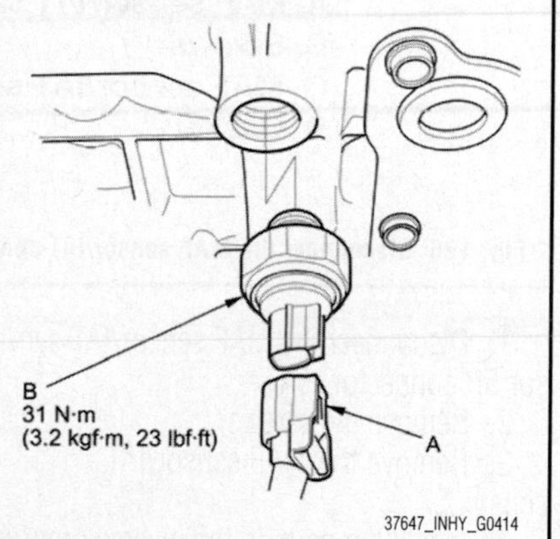
B
31 N·m
(3.2 kgf·m, 23 lbf·ft)
A
37647_INHY_G0414

Fig. 195 Disconnect the knock sensor 1P connector (A); remove the knock sensor (B)

3. Remove the knock sensor.
4. Install the parts in the reverse order of removal.

MASS AIR FLOW (MAF) SENSOR

LOCATION

The MAF sensor/IAT sensor is located on the air intake. Refer to the graphics in the Removal and Installation section for the location(s).

REMOVAL & INSTALLATION

See Figure 196.

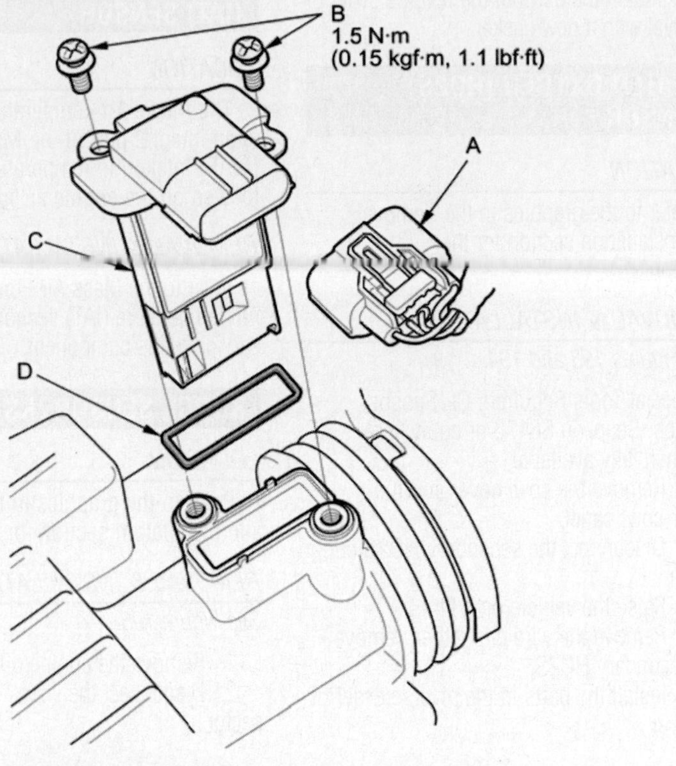
B
1.5 N·m
(0.15 kgf·m, 1.1 lbf·ft)
A
C
D

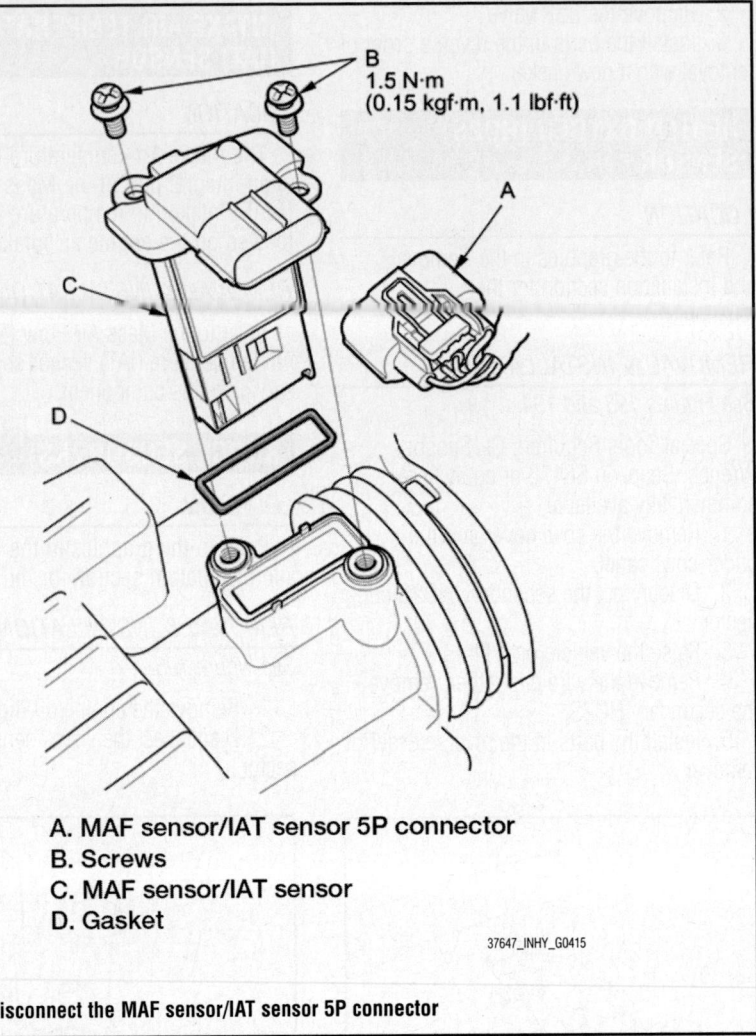

A. MAF sensor/IAT sensor 5P connector
B. Screws
C. MAF sensor/IAT sensor
D. Gasket

37647_INHY_G0415

Fig. 196 Disconnect the MAF sensor/IAT sensor 5P connector

1. Disconnect the MAF sensor/IAT sensor 5P connector.
2. Remove the screws.
3. Remove the MAF sensor/IAT sensor.
4. Install the parts in the reverse order of removal with a new gasket.

MANIFOLD ABSOLUTE PRESSURE (MAP) SENSOR

LOCATION

Refer to the graphics in the Removal and Installation section for the location(s).

REMOVAL & INSTALLATION

See Figure 197.

1. Remove the cowl cover and the under-cowl panel.
2. Disconnect the MAP sensor 3P connector.
3. Remove the MAP sensor.
4. Install the parts in the reverse order of removal with a new O-ring.

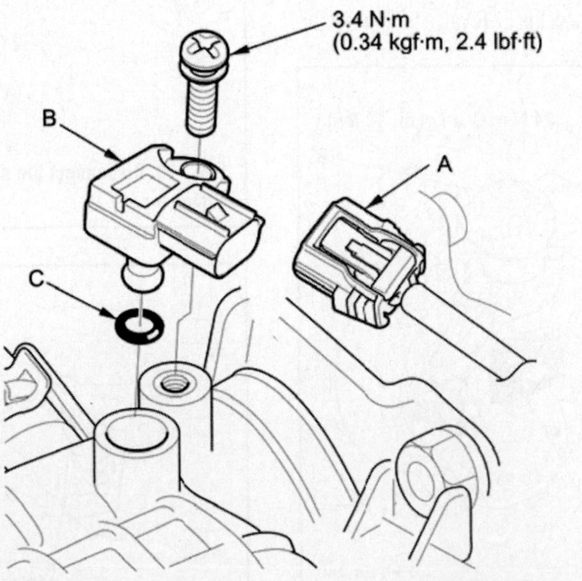
3.4 N·m
(0.34 kgf·m, 2.4 lbf·ft)
B
A
C
37647_INHY_G0416

Fig. 197 Disconnect the MAP sensor 3P connector (A); remove the MAP sensor (B)and the O-ring (C)

POSITIVE CRANKCASE VENTILATION (PCV) VALVE

LOCATION

Refer to the graphics in the Removal and Installation section for the location(s).

REMOVAL & INSTALLATION

See Figure 198.

1. Disconnect the hose.
2. Remove the PCV valve.
3. Install the parts in the reverse order of removal.

➡Make sure the hose clamp is positioned as shown.

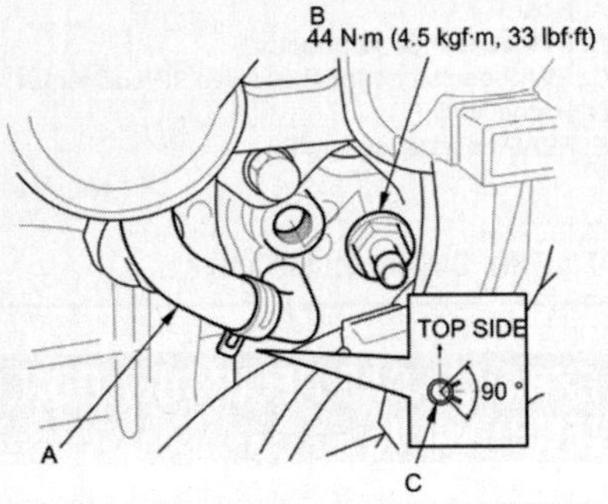
B
44 N·m (4.5 kgf·m, 33 lbf·ft)
TOP SIDE
90 °
A
C
37647_INHY_G0418

Fig. 198 Disconnect the hose (A); remove the PCV valve (B) and the clamp (C)

POWERTRAIN CONTROL MODULE (PCM)

LOCATION

The Powertrain Control Module (PCM) is located in the engine compartment on the left front.

REMOVAL & INSTALLATION

See Figures 199 through 201.

Special Tools Required:

- Honda Diagnostic System (HDS) tablet tester
- Honda Interface Module (HIM) and an iN workstation with the latest HDS software version
- HDS pocket tester
- GNA600 and an iN workstation with the latest HDS software version

➡Any one of the above updating tools can be used. Make sure the HDS/iN workstation has the latest HDS software version. The lifetime points of the Eco guide cannot be carried over to the replacement PCM.

1. Connect the HDS to the Data Link Connector (DLC) located under the driver's side of the dashboard.
2. Turn the ignition switch to ON (II).
3. Make sure the HDS communicates with the PCM and all other vehicle systems. If it doesn't, go to the DLC circuit troubleshooting. If you are returning from the DLC circuit troubleshooting, skip steps 4 through 7, 17 through 19, and 20 through 23, and do these procedures after replacing the PCM:
   - Replace the engine oil and the engine oil filter.
   - Clean the throttle body.
4. Select the PGM-FI system with the HDS.
5. Select the INSPECTION MENU with the HDS.
6. Select the ETCS TEST, then select the TP POSITION CHECK, and follow the screen prompts.

➡If the TP POSITION CHECK indicates FAILED, continue with this procedure.

7. Select the REPLACE PCM MENU, then select READ DATA, and follow the screen prompts.

➡Doing this step copies (READS) the engine oil life data from the original PCM so you can later download (WRITES) it into the new PCM. If the READ DATA indicates FAILED, continue with this procedure.

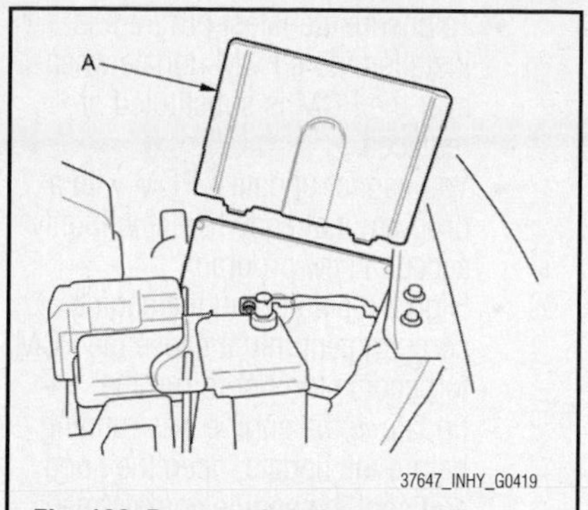
A
37647_INHY_G0419

Fig. 199 Remove the PCM cover (A)

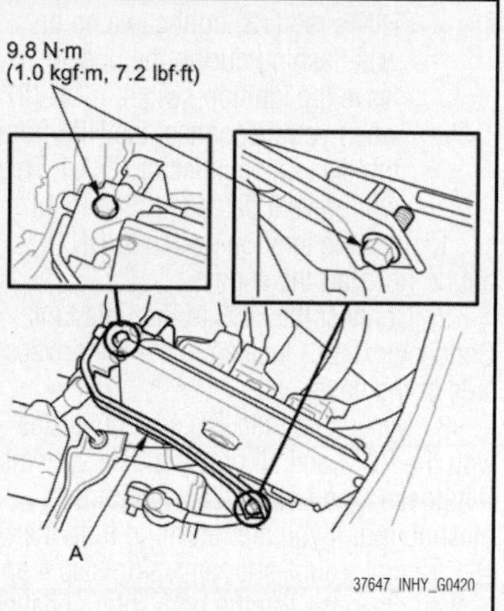
9.8 N·m
(1.0 kgf·m, 7.2 lbf·ft)
A
37647_INHY_G0420

Fig. 200 Remove the bracket (A)

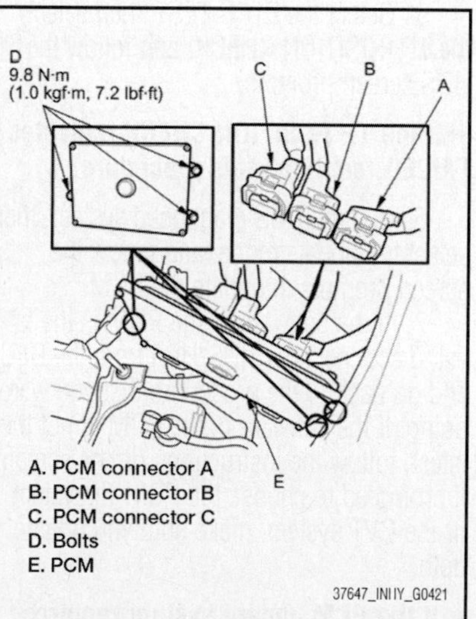
D
9.8 N·m
(1.0 kgf·m, 7.2 lbf·ft)
C
B
A
E
A. PCM connector A
B. PCM connector B
C. PCM connector C
D. Bolts
E. PCM
37647_INHY_G0421

Fig. 201 Remove the bolts

8. Jump the SCS line with the HDS.
9. Turn the ignition switch to LOCK (0).
10. Remove the PCM cover, then go to step 11 (KC model) or to step 12 (except KC model).
11. Remove the bracket.
12. Remove the bolts.
13. Disconnect PCM connectors A, B, and C, then remove the PCM.

➡PCM connectors A, B, and C have symbols (A=□, B=◎, C=○) embossed on them for identification.

To install:

14. Install all parts in the reverse order of removal.

➡If the IMA battery level indicator displays no level, start the engine, and hold it between 3,500 and 4,000 rpm without load (in P or N) until the level in the indicator is half full.

15. Turn the ignition switch to ON (II).
16. Manually input the VIN to the PCM with the HDS.

➡DTC P0630 VIN Not Programmed or Mismatch may be stored because the VIN has not been programmed into the PCM; ignore it, and continue this procedure.

17. If the READ DATA (engine oil life) failed in step 7, go to step 18.
18. Select the PGM-FI system with the HDS.
19. Select the REPLACE PCM MENU, then select WRITE DATA, and follow the screen prompts.

➡If the WRITE DATA indicates FAILED, continue with this procedure.

20. Select the IMMOBI system with the HDS.
21. Enter the immobilizer PCM code that you got from the iN, and use the PCM replacement procedure in the IMMOBI MENU of the HDS; it allows you to start the engine.
22. If the TP POSITION CHECK failed in step 6, clean the throttle body, then go to step 23.
23. If the READ DATA failed in step 7 or the WRITE DATA failed in step 19, replace the engine oil, and the engine oil filter, then go to step 24.
24. Select the PGM-FI system, and reset the PCM with the HDS.
25. Update the PCM if it does not have the latest software.
26. Do the PCM idle learn procedure.
27. Do the CKP pattern clear/CKP pattern learn procedure.
28. Do the start clutch control calibration procedure.

PCM IDLE LEARN PROCEDURE

The idle learn procedure must be done so the PCM can learn the engine idle characteristics.

Do the idle learn procedure whenever you do any of these actions:

- Replace the PCM.
- Reset the PCM.
- Update the PCM.
- Replace or clean the throttle body.
- Disassemble the engine or the transmission.

➡**Clearing DTCs with the HDS does not require you to do the idle learn procedure.**

1. Make sure all electrical items (A/C, navigation, lights, etc.) are off.
2. Reset the PCM with the HDS.
3. Turn the ignition switch to ON (II), and wait 2 seconds.
4. Start the engine. Hold the engine speed at 3,000 rpm without load (in P or N) until the radiator fan comes on, or until the engine coolant temperature reaches 194 °F (90 °C).
5. Let the engine idle for about 5 minutes with the throttle fully closed.

➡**If the radiator fan comes on, do not include its running time in the 5 minutes.**

PCM UPDATE

Special Tools Required:

- Honda Diagnostic System (HDS) tablet tester
- Honda Interface Module (HIM) and an iN workstation with the latest HDS software version
- HDS pocket tester
- GNA600 and an iN workstation with the latest HDS software version

➡**Any one of the above updating tools can be used.**

➡**Note the following:**

- Make sure the HDS/iN workstation has the latest HDS software version.
- Before you update the PCM, make sure the 12 volt battery in the vehicle is fully charged, and connect a jumper battery (not a battery charger) to maintain system voltage.
- Never turn the ignition switch to ACC (I) or to LOCK (0) during the update. If there is a problem with the update, leave the ignition switch to ON (II).
- To prevent PCM damage, do not operate anything electrical (headlights, navigation system, brakes, A/C, power windows, door locks, etc.) during the update.
- To ensure the latest program is installed, do a PCM update whenever the PCM is substituted or replaced.
- You cannot update a PCM with a program it already has. It will only accept a new program.
- High temperature in the engine compartment might cause the PCM to become too hot to run the update. If the engine was running before the update, open the hood and cool the engine compartment.
- If you need to diagnose the Honda interface module (HIM) because the HIM's red (#3) light came on or was flashing during the update, leave the ignition switch in ON (II) when you disconnect the HIM from the data link connector (DLC). This will prevent damage to the PCM.

1. Turn the ignition switch to ON (II), but do not start the engine.
2. Connect the HDS to the Data Link Connector (DLC) located under the driver's side of the dashboard.
3. Make sure the HDS communicates with the PCM and all other vehicle systems. If it doesn't, go to the DLC circuit troubleshooting. If you are returning from the DLC circuit troubleshooting, skip step 4 and 5, then clean the throttle body after updating the PCM.
4. Select the INSPECTION MENU with the HDS.
5. Select the ETCS TEST, then select the TP POSITION CHECK, and follow the HDS screen prompts.

➡**If the TP POSITION CHECK indicates FAILED, continue this procedure.**

6. Exit the HDS diagnostic system, then select the update mode, and follow the screen prompts to update the PCM.
7. If the software in the PCM is the latest, disconnect the HDS/HIM from the DLC, and go back to the procedure that you were doing. If the software in the PCM is not the latest, follow the instructions on the screen. If prompted to choose the PGM-FI system or the CVT system, make sure you update both.

➡**If the PCM update system requires you to cool the PCM, follow the instructions on screen. If you have a problem during the update procedure (programming takes over 15 minutes, status bar goes over 100 %, D or immobilizer indicator flashes, HDS tablet freezes, etc.), follow these steps to minimize the chance of damaging the PCM.**

a. Leave the ignition switch in the ON (II) position.
b. Connect a jumper battery (do not connect a battery charger).
c. Shut down the HDS.
d. Disconnect the HDS from the DLC.
e. Reboot the HDS.
f. Reconnect the HDS to the DLC, and try the update procedure again.

8. If the TP POSITION CHECK failed in step 5, clean the throttle body.
9. Do the PCM idle learn procedure.
10. Do the CKP pattern clear/CKP pattern learn procedure.
11. Select the AT SYSTEM, then reset the TCM with the HDS.

PCM IDLE LEARN PROCEDURE

The idle learn procedure must be done so the PCM can learn the engine idle characteristics.

Do the idle learn procedure whenever you do any of these actions:

- Replace the PCM.
- Reset the PCM.
- Update the PCM.
- Replace or clean the throttle body.
- Disassemble the engine or the transmission.

➡**Clearing DTCs with the HDS does not require you to do the idle learn procedure.**

1. Make sure all electrical items (A/C, navigation, lights, etc.) are off.
2. Reset the PCM with the HDS.
3. Turn the ignition switch to ON (II), and wait 2 seconds.
4. Start the engine. Hold the engine speed at 3,000 rpm without load (in P or N) until the radiator fan comes on, or until the engine coolant temperature reaches 194 °F (90 °C).
5. Let the engine idle for about 5 minutes with the throttle fully closed.

➡**If the radiator fan comes on, do not include its running time in the 5 minutes.**

THROTTLE POSITION (TP) SENSOR

LOCATION

See Figure 202.

REMOVAL & INSTALLATION

➡**The manufacturer does not provide a specific Removal and Installation procedure for this component. Refer to the graphic(s) when servicing this component.**

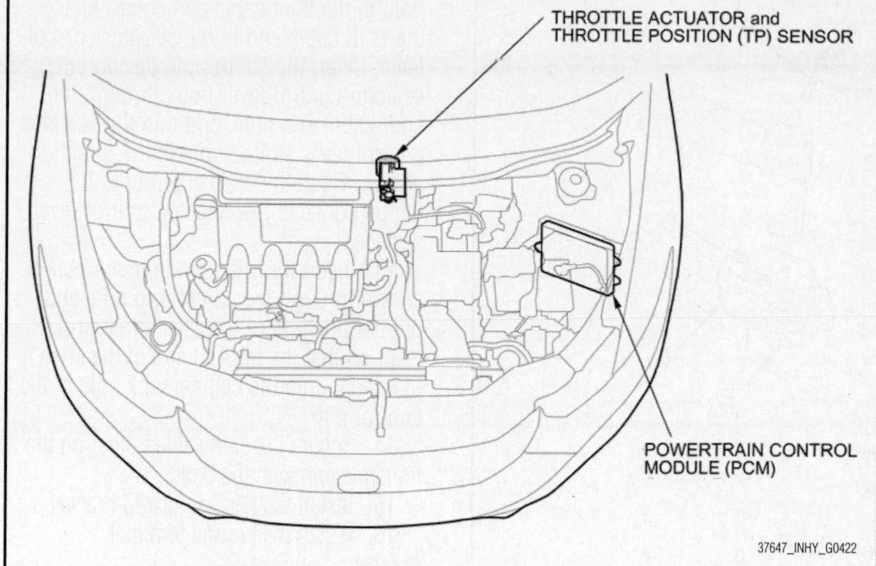
THROTTLE ACTUATOR and
THROTTLE POSITION (TP) SENSOR
POWERTRAIN CONTROL
MODULE (PCM)
37647_INHY_G0422

Fig. 202 Throttle Position (TP) sensor location

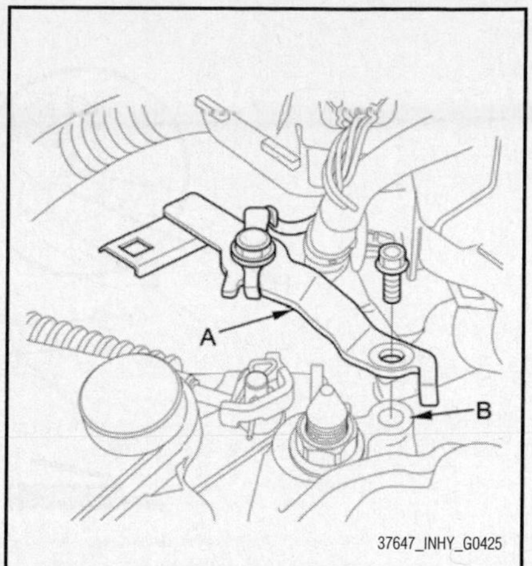
A
B
37647_INHY_G0425

Fig. 205 Remove the ground terminal bracket (A) from the transmission mount (B)

VEHICLE SPEED SENSOR (VSS)

REMOVAL & INSTALLATION

See Figures 203 through 208.

1. Remove the air cleaner.
2. Remove the heater hose clamps and the IMA motor power cable clamps from the bracket.
3. Disconnect the transmission range switch connector and the CVT output shaft (driven pulley) speed sensor connector.
4. Remove the harness clamp and the harness holder clamp from the harness clamp bracket.
5. Remove the ground terminal bracket from the transmission mount.
6. Remove the harness clamp bracket.
7. Remove the snap pin and the control pin from the control lever.
8. Remove the bolts securing the shift cable holder, then separate the shift cable from the control lever.
9. Disconnect the vehicle speed sensor connector, and remove the vehicle speed sensor.

To install:

10. Install a new O-ring on a new vehicle speed sensor, then install the vehicle speed sensor in the transmission housing.
11. Check the connector for rust, dirt, or oil, and clean or repair if necessary, then connect the connector securely.

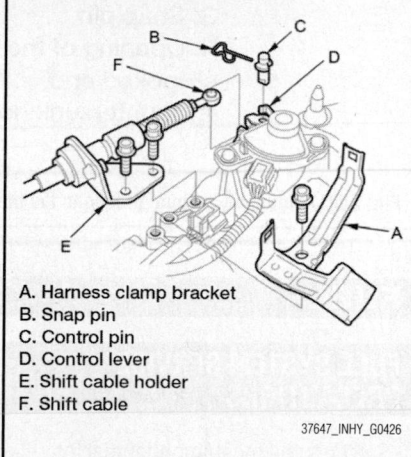

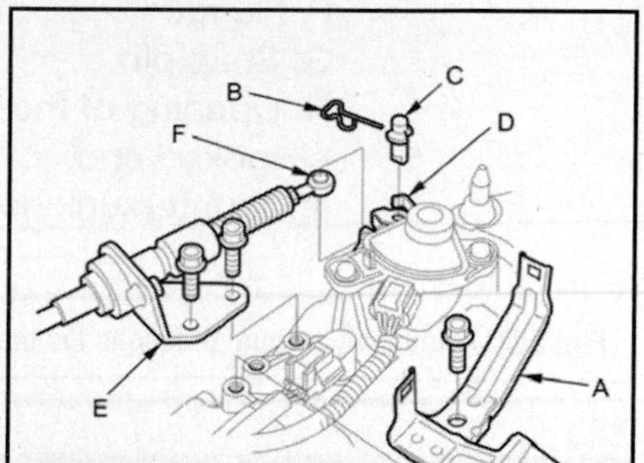
B
C
F
D
A
E
A. Harness clamp bracket
B. Snap pin
C. Control pin
D. Control lever
E. Shift cable holder
F. Shift cable
37647_INHY_G0426

Fig. 206 Remove the harness clamp bracket

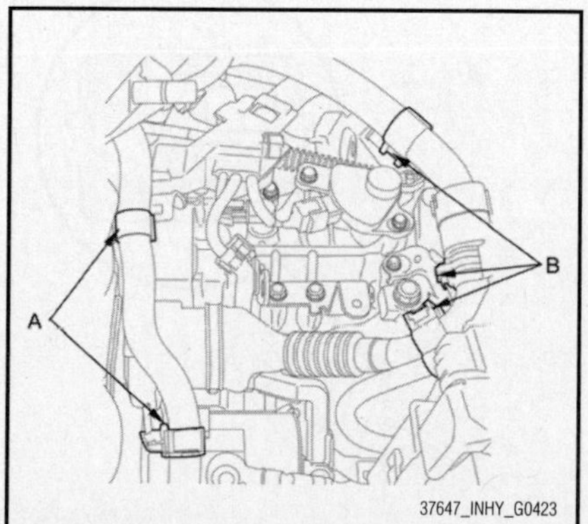
A
B
37647_INHY_G0423

Fig. 203 Remove the heater hose clamps (A) and the IMA motor power cable clamps (B) from the bracket

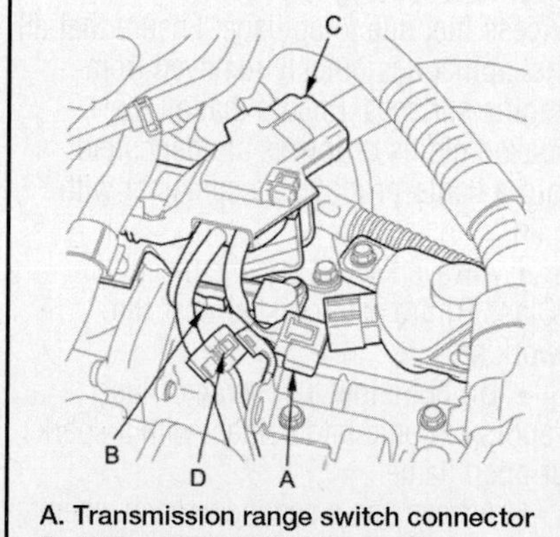
C
B
D
A
A. Transmission range switch connector
B. CVT output shaft (driven pulley) speed sensor connector
C. Harness clamp
D. Harness holder clamp
37647_INHY_G0424

Fig. 204 Disconnect the transmission range switch connector and the CVT output shaft (driven pulley) speed sensor connector

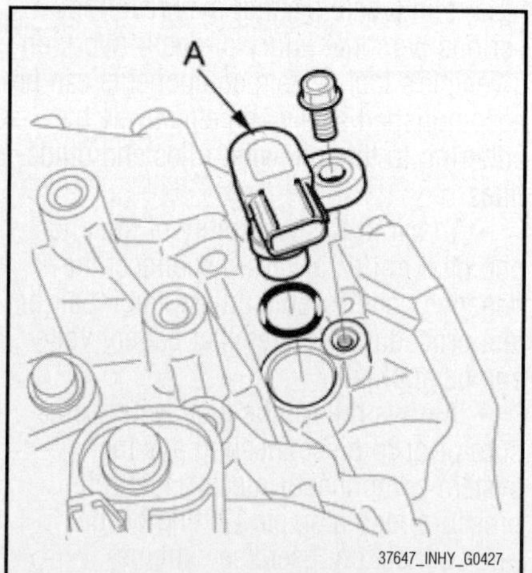
A
B
37647_INHY_G0427

Fig. 207 Disconnect the vehicle speed sensor connector, and remove the vehicle speed sensor (A) and the O-ring (B)

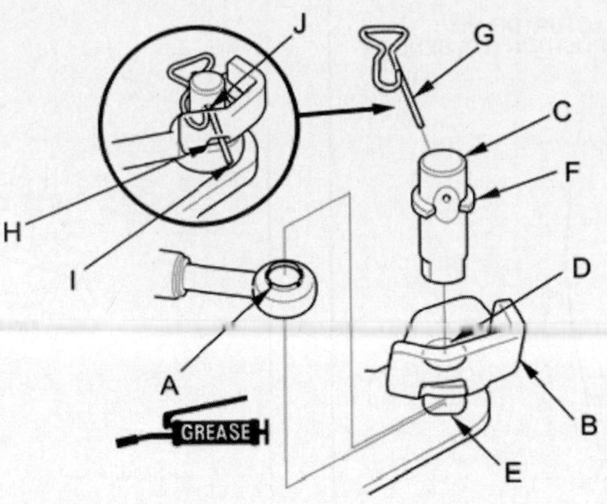
J
G
C
F
H
I
D
A
GREASE
B
E

A. Shift cable end collar
B. Selector control lever
C. Control pin
D. Selector control lever hole
E. Selector control lever slotted hole
F. Flange
G. Snap pin
H. Opening of the selector control lever
I. Hooked end
J. Countersunk hole

37647_INHY_G0428

Fig. 208 Apply molybdenum grease to the hole in the shift cable end collar

12. Apply molybdenum grease to the hole in the shift cable end collar. Attach the shift cable end to the selector control lever. Insert the control pin through the selector control lever hole, through the shift cable end hole, and into the selector control lever slotted hole in the direction shown. Push the control pin until its flange contacts the selector control lever surface.
13. Insert the snap pin in the direction shown through the control pin hole and out the opening of the selector control lever so that the hooked end of the snap pin snaps into the countersunk hole of the control pin.
14. Secure the shift cable holder on the transmission with the bolts.
15. Install the harness clamp bracket.
16. Install the ground terminal bracket.
17. Install the harness clamps and the harness hold clamp.
18. Connect the transmission range switch connector, and the CVT output shaft (driven pulley) speed sensor connector.
19. Install the heater hose clamps and IMA motor power cable clamp.
20. Install the air cleaner.

FUEL

GASOLINE FUEL INJECTION SYSTEM

FUEL SYSTEM SERVICE PRECAUTIONS

Safety is the most important factor when performing not only fuel system maintenance, but any type of maintenance. Failure to conduct maintenance and repairs in a safe manner may result in serious personal injury or death. Work on a vehicle's fuel system components can be accomplished safely and effectively by adhering to the following rules and guidelines.

- To avoid the possibility of fire and personal injury, always disconnect the negative battery cable unless the repair or test procedure requires that battery voltage be applied.
- Always relieve the fuel system pressure prior to disconnecting any fuel system component (injector, fuel rail, pressure regulator, etc.) fitting or fuel line connection. Exercise extreme caution whenever relieving fuel system pressure to avoid exposing skin, face and eyes to fuel spray. Please be advised that fuel under pressure may penetrate the skin or any part of the body that it contacts.
- Always place a shop towel or cloth around the fitting or connection prior to loosening to absorb any excess fuel due to spillage. Ensure that all fuel spillage is quickly removed from engine surfaces. Ensure that all fuel-soaked cloths or towels are deposited into a flame-proof waste container with a lid.
- Always keep a dry chemical (Class B) fire extinguisher near the work area.
- Do not allow fuel spray or fuel vapors to come into contact with a spark or open flame.
- Always use a second wrench when loosening or tightening fuel line connection fittings. This will prevent unnecessary stress and torsion on fuel piping. Always follow the proper torque specifications.
- Always replace worn fuel fitting O-rings with new ones. Do not substitute fuel hose where rigid pipe is installed.

FUEL SYSTEM PRESSURE

RELIEVING

See Figures 209 through 211.

Before disconnecting fuel lines or hoses, relieve pressure from the system by disabling

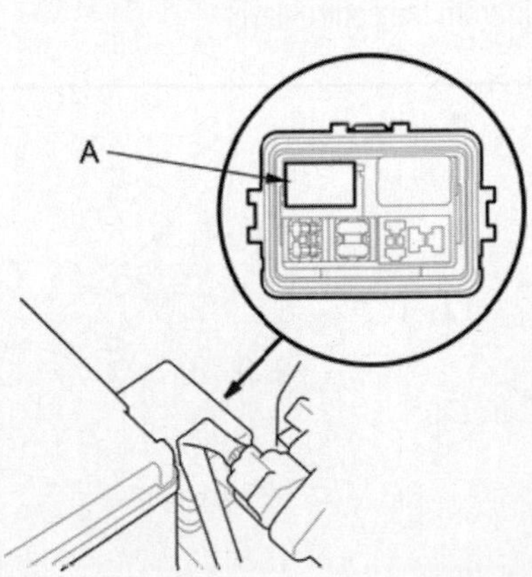
A

37647_INHY_G0430

Fig. 209 Remove PGM-FI main relay 2 (A) from the auxiliary under-hood fuse/relay box

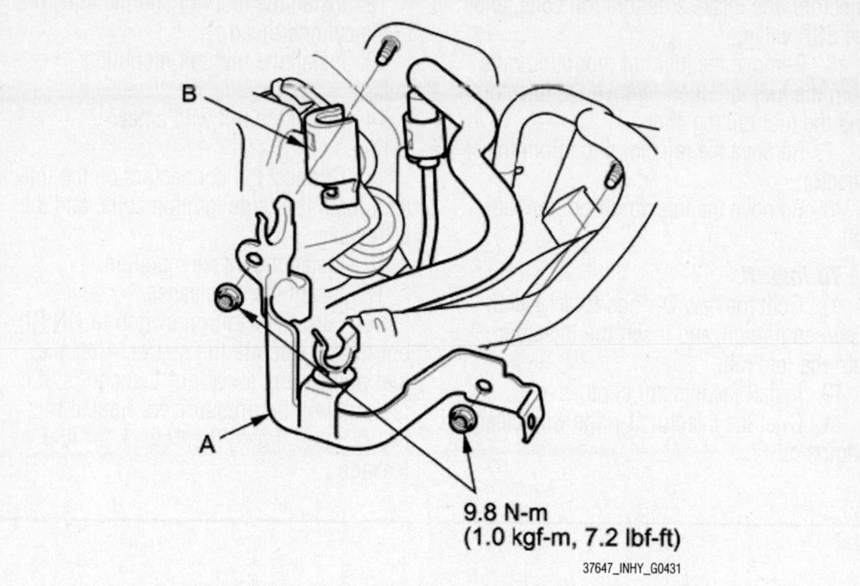
B
A
9.8 N-m
(1.0 kgf-m, 7.2 lbf-ft)
37647_INHY_G0431

Fig. 210 Remove the bracket (A) and the quick-connect fitting cover (B)

the fuel pump and then disconnecting the fuel tube/quick connect fitting in the engine compartment.

1. Remove PGM-FI main relay 2 from the auxiliary under-hood fuse/relay box (to the left of the 12 volt battery).
2. Start the engine, and let it idle until it stalls.

➡If any DTCs are stored, clear and ignore them.

3. Turn the ignition switch to LOCK (0).
4. Remove the fuel fill cap to relieve the pressure in the fuel tank.
5. Do the 12 volt battery terminal disconnection procedure.
6. Remove the bracket and the quick-connect fitting cover.
7. Check the fuel quick-connect fitting for dirt, and clean it if needed.
8. Place a rag or shop towel over the quick-connect fitting.
9. Disconnect the quick-connect fitting: Hold the connector with one hand, and squeeze the retainer tabs with the other hand to release them from the locking tabs. Pull the connector off.
   a. Be careful not to damage the fuel line or other parts.
   b. Do not use tools.
   c. If the connector does not move, keep the retainer tabs pressed down, and alternately pull and push the connector until it comes off easily.
   d. Do not remove the retainer from the line; once removed, the retainer must be replaced with a new one.
10. After disconnecting the quick-connect fitting, check it for dirt or damage.
11. Do the 12 volt battery terminal reconnection procedure.

FUEL FILTER

REMOVAL & INSTALLATION

See Figure 212.

The fuel filter should be replaced whenever the fuel pressure drops below the specified value, after making sure that the fuel pump and the fuel pressure regulator are OK.

1. Remove the fuel tank unit.
2. Remove the fuel filter set.

To install:

3. Check these items before installing the fuel tank unit:
   - When connecting the wire harness, make sure the connection is secure and the connectors are firmly locked into place.
   - When installing the fuel gauge sending unit, make sure the connection is secure and the connector is firmly locked into place. Be careful not to bend or twist it excessively.
4. Install the parts in the reverse order of removal with new O-rings and a new bracket. When installing the fuel tank unit, align the marks on the unit and the fuel tank.
   a. Coat the O-rings with clean engine oil; do not use any other oils or fluids.
   b. Do not pinch the O-rings during installation.

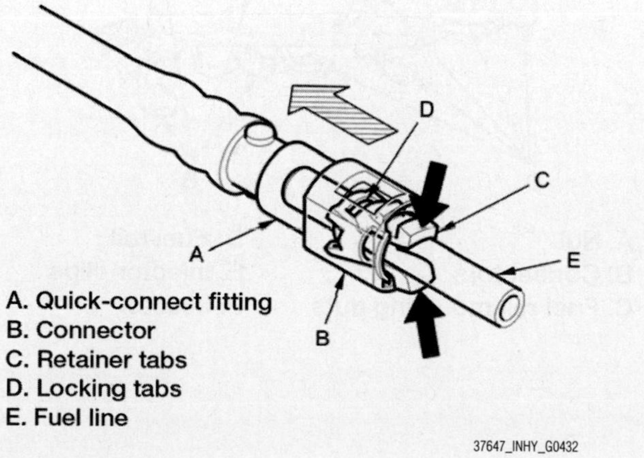
D
C
A
E
B
A. Quick-connect fitting
B. Connector
C. Retainer tabs
D. Locking tabs
E. Fuel line
37647_INHY_G0432

Fig. 211 Disconnect the quick-connect fitting

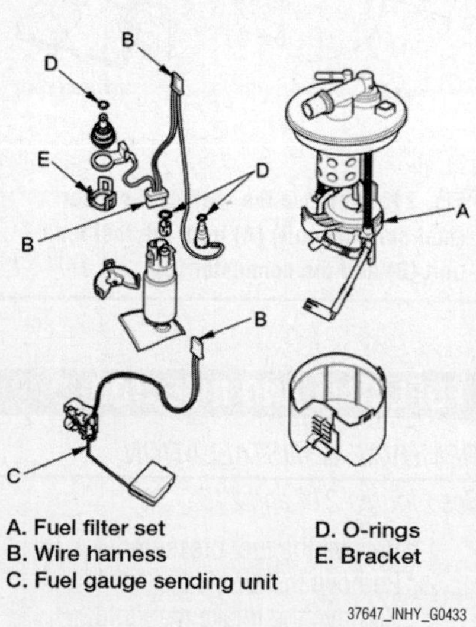
B
D
E
D
A
B
B
C
A. Fuel filter set
B. Wire harness
C. Fuel gauge sending unit
D. O-rings
E. Bracket
37647_INHY_G0433

Fig. 212 Remove the fuel filter set

c. Use all the new parts supplied in the fuel filter replacement kit.

FUEL LEVEL SENSOR/FUEL GAUGE SENDING UNIT

REMOVAL & INSTALLATION

See Figure 213.

1. Remove the fuel tank unit.
2. Remove the fuel level sensor (fuel sending unit) from the fuel tank unit.

To install:

3. Check these items before installing the fuel tank unit:
   - When connecting the wire harness, make sure the connection is secure and the connector is firmly locked into place.
   - When installing the fuel gauge sending unit, make sure the connection is secure. Be careful not to bend or twist it excessively.
4. Install the parts in the reverse order of removal. When installing the fuel tank unit, align the marks on the unit and the fuel tank.

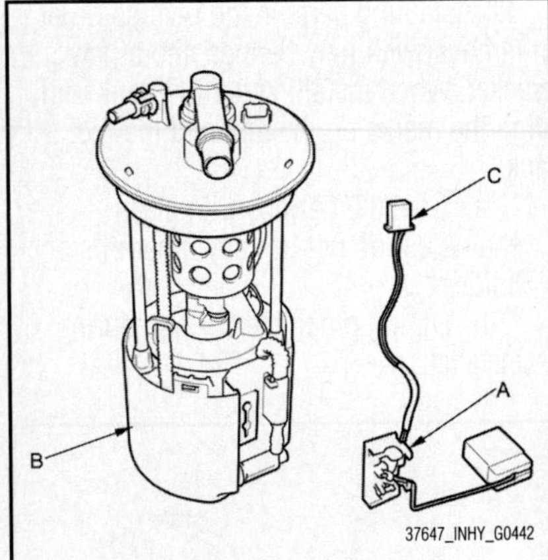

C
A
B
37647_INHY_G0442

Fig. 213 Remove the fuel level sensor (fuel sending unit) (A) from the fuel tank unit (B) and the connector (C)

FUEL RAIL AND INJECTOR

REMOVAL & INSTALLATION

See Figures 214 and 215.

1. Relieve the fuel pressure.
2. Remove the air cleaner.
3. Remove the intake manifold.
4. Remove the nut.
5. Disconnect the connectors from the injectors, the intake side ignition coils, and the EGR valve.
6. Remove the fuel rail mounting nuts from the fuel rail, then remove the injectors and the fuel rail together.
7. Remove the injector clips from the injector.
8. Remove the injectors from the fuel rail.

To install:

9. Coat the new O-rings (black) with clean engine oil, and insert the injectors into the fuel rail.
10. Install the injector clips.
11. Coat the injector O-rings with clean engine oil.
12. Install the fuel rail and the injectors in the cylinder head.
13. Install the fuel rail mounting nuts.
14. Install the nut with a new O-ring.
15. Connect the connectors on the injectors, the intake side ignition coils, and the EGR valve.
16. Install the intake manifold.
17. Install the air cleaner.
18. Turn the ignition switch to ON (II), but do not operate the starter. After the fuel pump runs for about 2 seconds, the fuel rail will be pressurized. Repeat this two or three times, then check for fuel leakage.

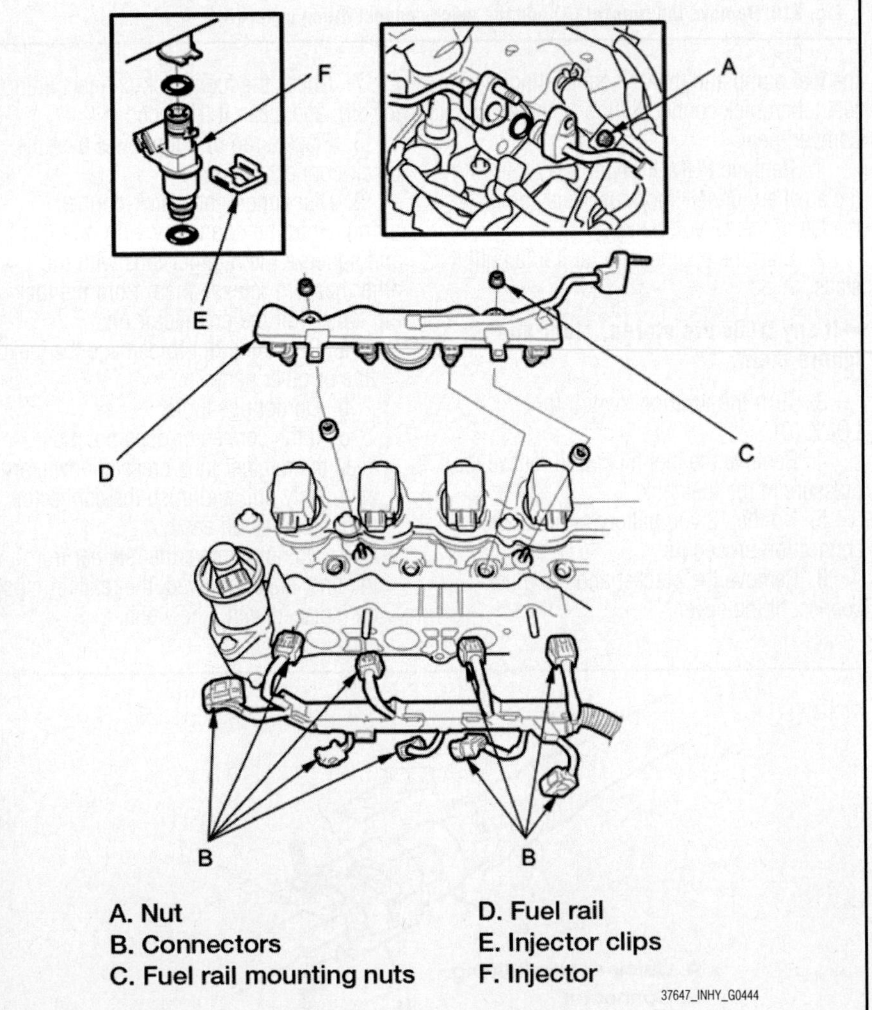

F
A
E
D
C
B
B
A. Nut
B. Connectors
C. Fuel rail mounting nuts
D. Fuel rail
E. Injector clips
F. Injector
37647_INHY_G0444

Fig. 214 Removing the fuel rail assembly and injectors

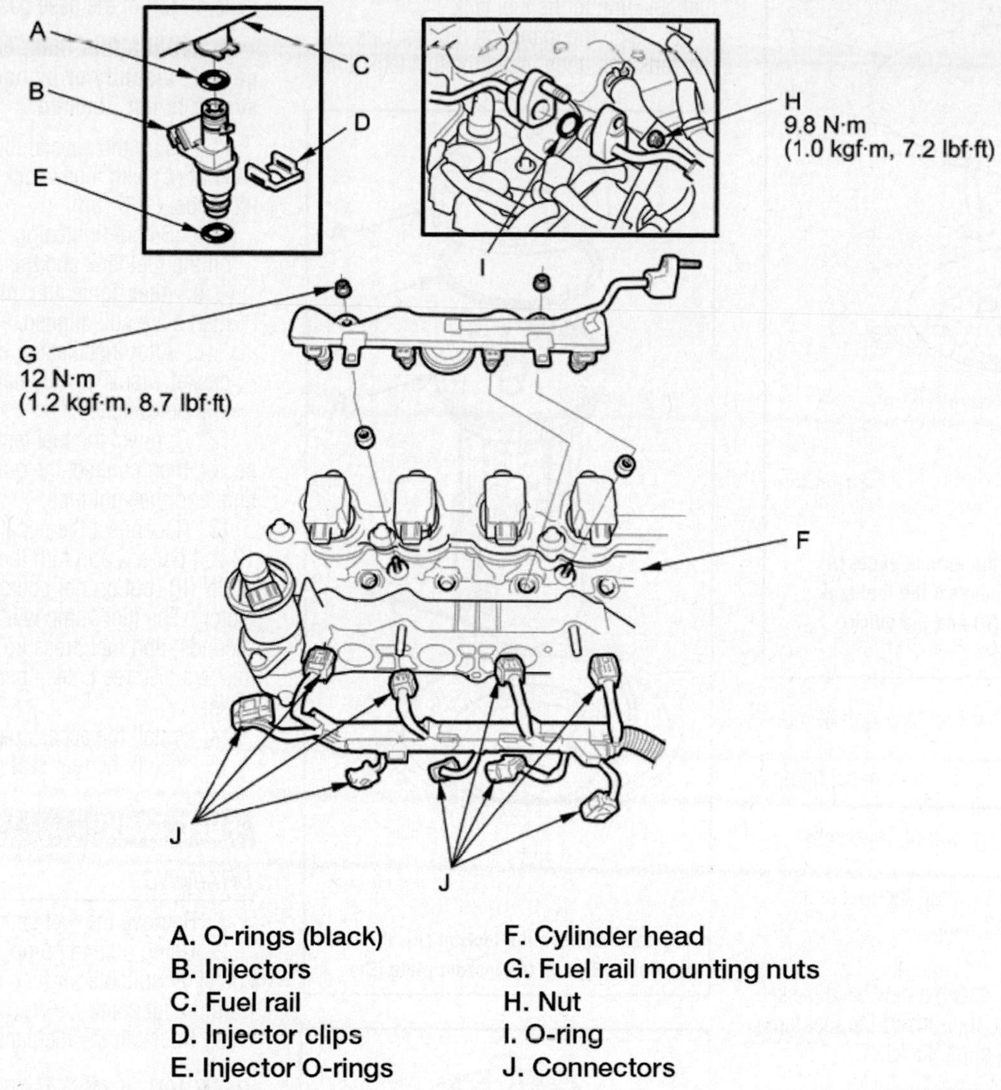
A
B
C
D
E
H
9.8 N·m
(1.0 kgf·m, 7.2 lbf·ft)
I
G
12 N·m
(1.2 kgf·m, 8.7 lbf·ft)
F
J
J
A. O-rings (black)
B. Injectors
C. Fuel rail
D. Injector clips
E. Injector O-rings
F. Cylinder head
G. Fuel rail mounting nuts
H. Nut
I. O-ring
J. Connectors
37647_INHY_G0445

Fig. 215 Installing the injectors and fuel rail

FUEL TANK UNIT

REMOVAL & INSTALLATION

See Figures 216 through 221.

Special Tools Required: Fuel Pump Module Locknut Wrench 07AAA-SNAA100

1. Relieve the fuel pressure.
2. Remove the rear seat cushion.
   a. Remove the bolt from the slit in the seat cushion.
   b. Pull the seat hook knobs while pushing down on the seat cushion to release the hooks.
   c. Pull up the seat cushion to release the seat belt buckles from their slots, and remove it.
3. Remove the access panel from the floor.

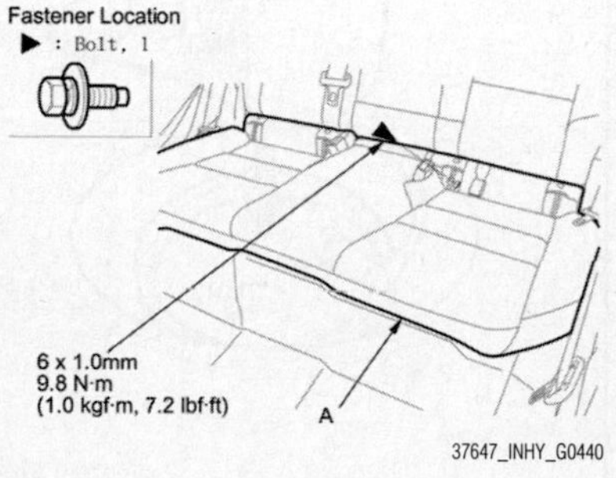
Fastener Location
▶ : Bolt, 1
6 x 1.0mm
9.8 N·m
(1.0 kgf·m, 7.2 lbf·ft)
A
37647_INHY_G0440

Fig. 216 Remove the bolt from the slit in the seat cushion (A)

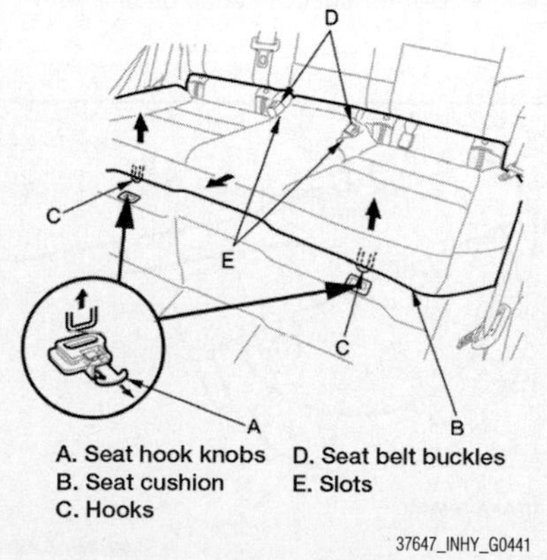
D
C
E
C
A
B
A. Seat hook knobs
B. Seat cushion
C. Hooks
D. Seat belt buckles
E. Slots
37647_INHY_G0441

Fig. 217 Pull the seat hook knobs while pushing down on the seat cushion to release the hooks

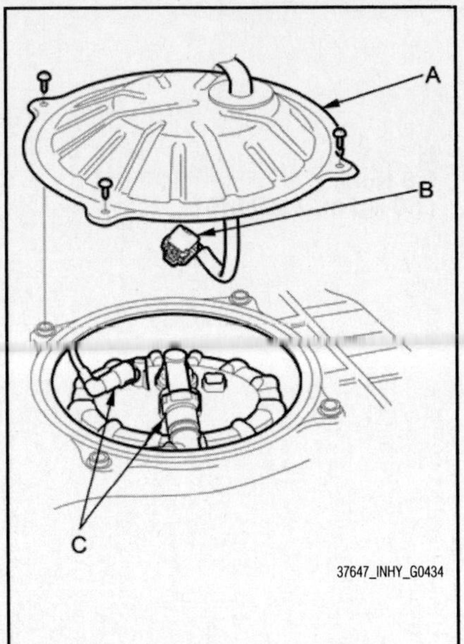
A
B
C
37647_INHY_G0434

Fig. 218 Remove the access panel (A) from the floor; disconnect the fuel tank unit 4P connector (B) and the quick-connect fittings (C)

4. Disconnect the fuel tank unit 4P connector.

5. Disconnect the quick-connect fittings from the fuel tank unit.

6. Using the special tool, loosen the locknut.

7. Remove the locknut, the fuel tank unit, and the locknut plate.

To install:

8. Temporarily attach a new base gasket to the fuel tank unit, then insert the fuel tank unit partially into the fuel tank.

➡Note the following during installation:

- Be careful not to damage the new base gasket.
- Be careful not to bend the fuel gauge sending unit.
- Do not coat the base gasket with oil.

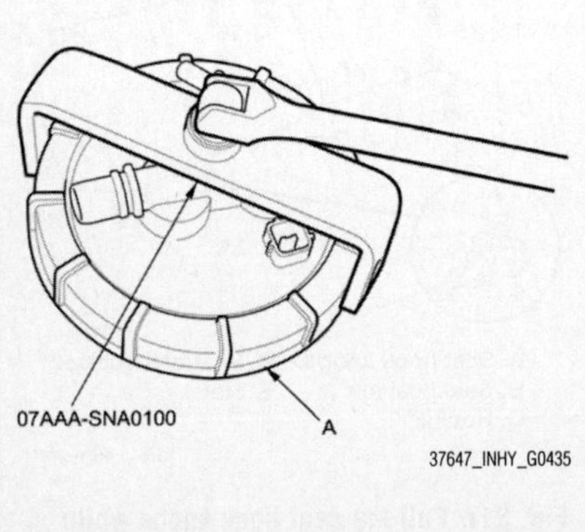
07AAA-SNA0100
A
37647_INHY_G0435

Fig. 219 Using the special tool, loosen the locknut (A)

9. Transfer the base gasket from the fuel tank unit to the fuel tank.

10. Align the marks on the fuel tank and fuel tank unit, then insert the fuel tank unit into the fuel tank until the fuel tank unit rests on top of the base gasket.

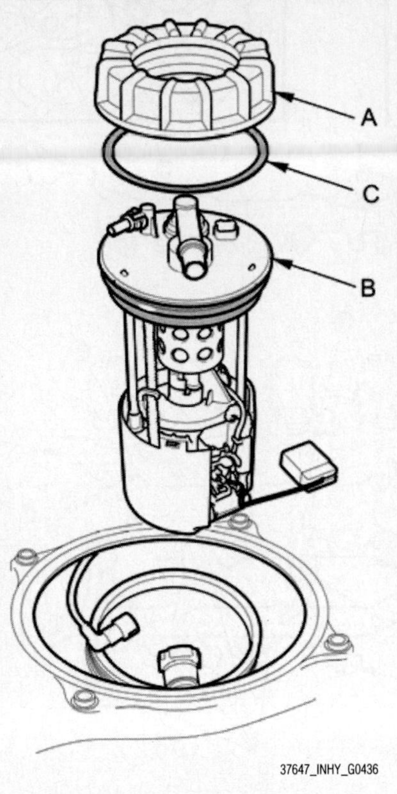
A
C
B
37647_INHY_G0436

Fig. 220 Remove the locknut (A), the fuel tank unit (B), and the locknut plate (C)

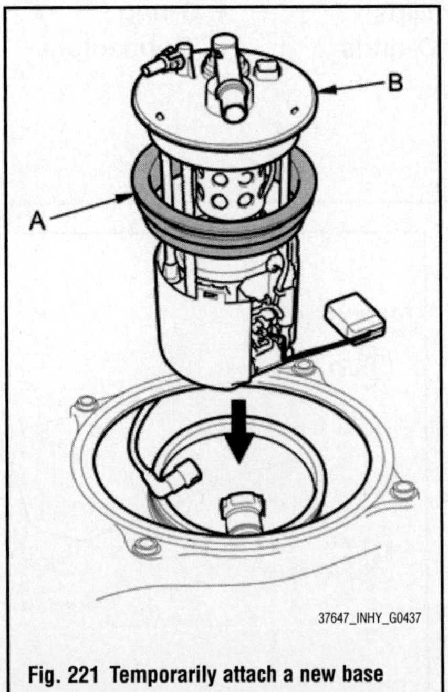
B
A
37647_INHY_G0437

Fig. 221 Temporarily attach a new base gasket (A) to the fuel tank unit (B), then insert the fuel tank unit partially into the fuel tank

➡To avoid a fuel leak, check the base gasket, visually or by hand, to make sure it is not pinched.

11. Using the special tool, tighten a new locknut (A) with a new locknut plate (B) to 92 ft. lbs. (125 Nm).

a. Before tightening , align the marks on the fuel tank and the locknut.

b. After tightening, make sure the marks are still aligned.

c. After installation, check the base gasket, visually or by hand, to make sure it is not pinched.

12. Connect the fuel tank unit 4P connector, then connect the quick-connect fitting, and the vent tube.

13. Reconnect the negative cable to the 12 volt battery, and turn the ignition switch to ON (II) (but do not operate the starter motor). The fuel pump will run for about 2 seconds, and fuel pressure rises. Repeat this two or three times, then check for fuel leaks.

14. Install the access panel.

15. Install the rear seat cushion.

FUEL TANK

DRAINING

1. Remove the fuel tank unit.

2. Using a hand pump, a hose, and a container suitable for fuel, draw the fuel from the fuel tank.

3. Reinstall the fuel tank unit.

REMOVAL & INSTALLATION

See Figures 222 through 225.

1. Drain the fuel tank until it is less than half full.

2. Reinstall the fuel tank unit without connecting the fuel tank unit 4P connectors, the quick-connect fitting or the vent tube.

3. Raise the vehicle on a lift.

4. Disconnect the quick-connect fittings.

5. Remove the hose from the clamp.

6. Remove the middle floor undercover.

7. Remove the trailing arm braces.

8. Place a jack or other support under the fuel tank.

9. Remove the strap bolts and the straps.

10. Remove the fuel tank.

11. Remove the fuel tank protectors.

To install:

12. Install the parts in the reverse order of removal.

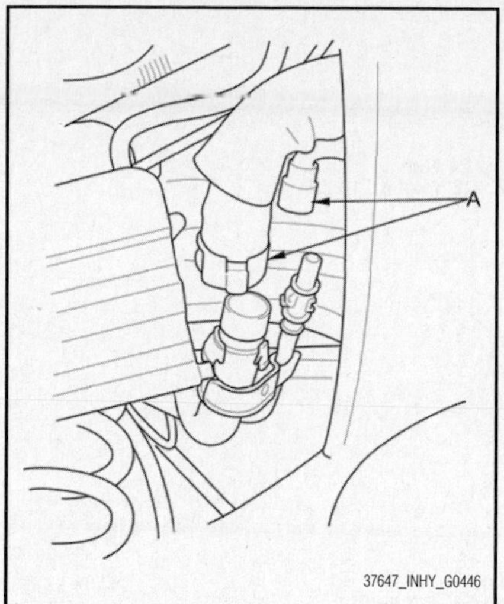
A
37647_INHY_G0446

Fig. 222 Disconnect the quick-connect fittings (A)

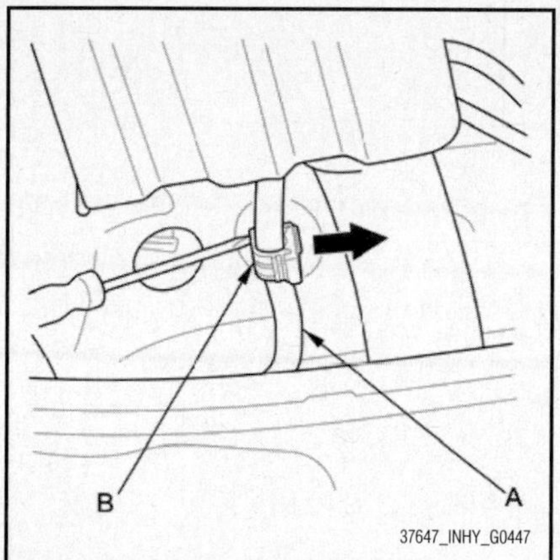
B
A
37647_INHY_G0447

Fig. 223 Remove the hose (A) from the clamp (B)

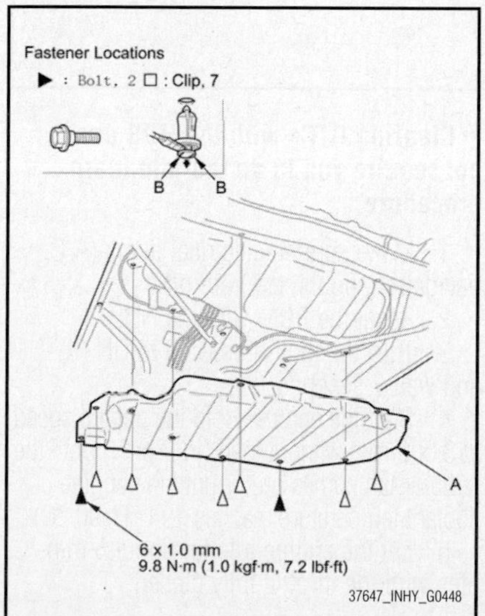
Fastener Locations
▶ : Bolt, 2 □ : Clip, 7
B
B
A
6 x 1.0 mm
9.8 N·m (1.0 kgf·m, 7.2 lbf·ft)
37647_INHY_G0448

Fig. 224 Remove the middle floor undercover (A)

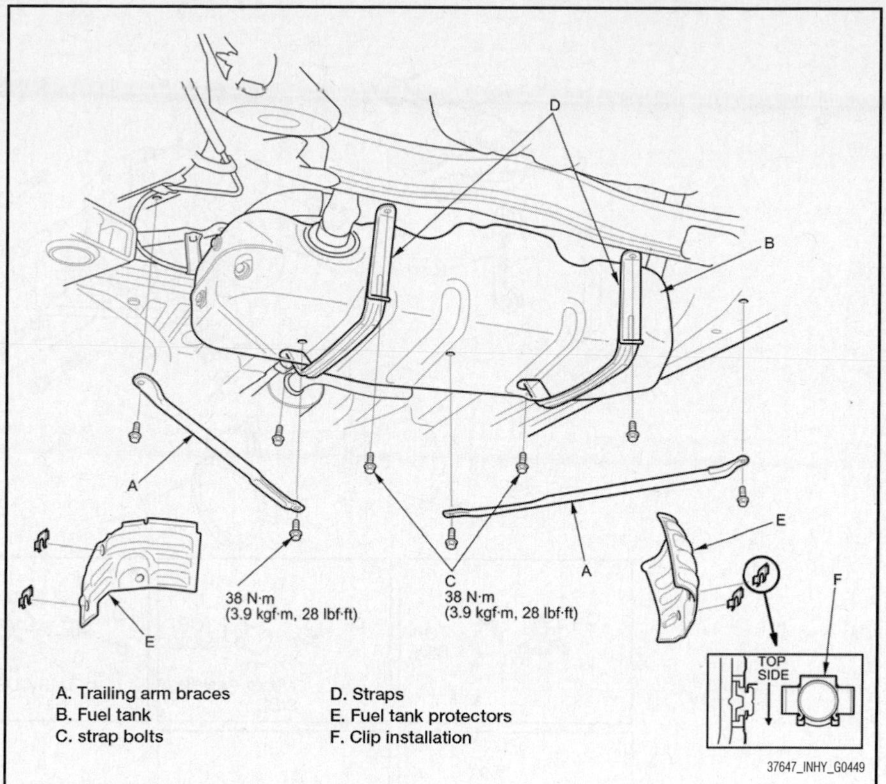
D
B
A
E
38 N·m
(3.9 kgf·m, 28 lbf·ft)
C
38 N·m
(3.9 kgf·m, 28 lbf·ft)
A
E
F
TOP
SIDE
A. Trailing arm braces
B. Fuel tank
C. strap bolts
D. Straps
E. Fuel tank protectors
F. Clip installation
37647_INHY_G0449

Fig. 225 Exploded view of fuel tank mounting assembly

➡When installing the fuel tank protectors, make sure to insert the clips in the direction shown.

IDLE SPEED

ADJUSTMENT

The idle speed is controlled by the powertrain Control Module (PCM). No adjustment is necessary or possible.

THROTTLE BODY

REMOVAL & INSTALLATION

See Figure 226.

⁂ CAUTION

Do not insert your fingers into the installed throttle body when you turn the ignition switch to ON (II) or while the ignition switch is in ON (II). If you do, you will seriously injure your fingers if the throttle valve is activated.

➡If you are replacing or cleaning the throttle body, start at step 1. If you are removing the throttle body, start at step 4.

1. Connect the HDS to the DLC while the engine is stopped.
2. Select the INSPECTION MENU on the HDS.
3. Do the TP POSITION CHECK in the ETCS TEST.
4. Turn the ignition switch to LOCK (0).
5. Remove the air cleaner.
6. Disconnect the MAP sensor 3P connector and the EVAP canister purge valve 2P connector, then remove the harness holder from the EVAP canister purge guard
7. Disconnect the throttle body connector, then disconnect the EVAP canister purge valve hose.
8. Disconnect the purge hoses.
9. Disconnect and plug the water bypass hoses.
10. Remove the throttle body.
11. Remove the purge pipe.
12. Install the parts in the reverse order of removal with a new gasket (M), then refill the radiator with engine coolant.

➡Note the following during installation:

- If you replace or clean the throttle body, go to step 13.
- If you did not replace or clean the throttle body, this procedure is complete.

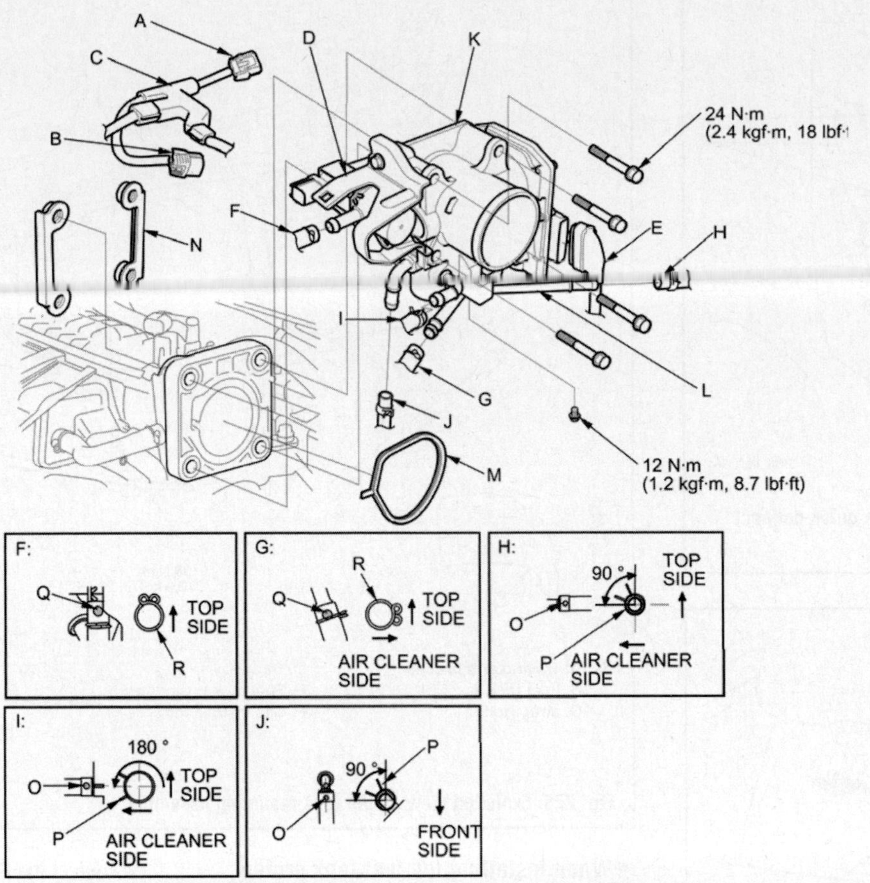
24 N·m
(2.4 kgf·m, 18 lbf·
12 N·m
(1.2 kgf·m, 8.7 lbf·ft)
F:
TOP SIDE
G:
TOP SIDE
AIR CLEANER SIDE
H:
90 °
TOP SIDE
AIR CLEANER SIDE
I:
180 °
TOP SIDE
AIR CLEANER SIDE
J:
90 °
FRONT SIDE

A. MAP sensor 3P connector
B. EVAP canister purge valve 2P connector
C. Harness holder
D. EVAP canister purge guard
E. Throttle body connector
F. EVAP canister purge valve hose
G. Purge hose
H. Purge hose
I. Water bypass hose
J. Water bypass hose
K. Throttle body
L. Purge pipe
M. Gasket
N. Plates
O. Marks
P. Clamp
Q. Mark
R. Clamps

37647_INHY_G0450

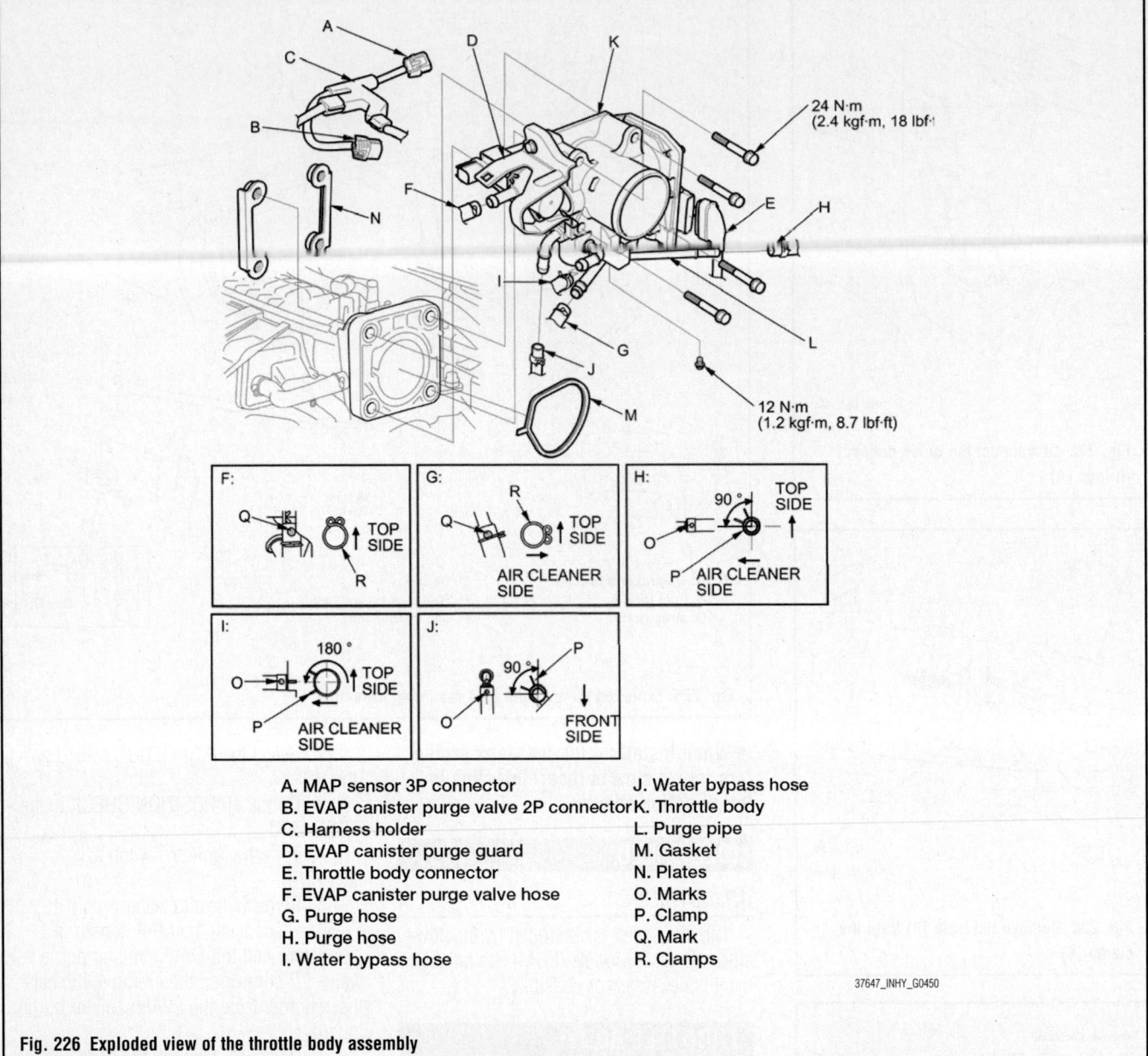

Fig. 226 Exploded view of the throttle body assembly

- Be careful not to drop or damage the plates.
- Align the marks on the hoses and the throttle body, then insert the hoses.
- Align the mark on the hoses, then insert the hoses.

13. Connect the HDS to the Data Link Connector (DLC) located under the driver's side of the dashboard.
14. Turn the ignition switch to ON (II).
15. Reset the PCM with the HDS.
16. Select the ETCS TEST in the INSPECTION MENU with the HDS.
17. Select the TP POSITION CHECK, then clear the Throttle Position (TP) learned value.
18. Turn the ignition switch to LOCK (0).
19. Turn the ignition switch to ON (II), and wait 2 seconds without pressing the accelerator pedal.
20. Do the PCM idle learn procedure.

PCM idle learn procedure

The idle learn procedure must be done so the PCM can learn the engine idle characteristics.

Do the idle learn procedure whenever you do any of these actions:

- Replace the PCM.
- Reset the PCM.
- Update the PCM.
- Replace or clean the throttle body.
- Disassemble the engine or the transmission.

➡Clearing DTCs with the HDS does not require you to do the idle learn procedure.

1. Make sure all electrical items (A/C, navigation, lights, etc.) are off.
2. Reset the PCM with the HDS.
3. Turn the ignition switch to ON (II), and wait 2 seconds.
4. Start the engine. Hold the engine speed at 3,000 rpm without load (in P or N) until the radiator fan comes on, or until the engine coolant temperature reaches 194 °F (90 °C).
5. Let the engine idle for about 5 minutes with the throttle fully closed.

➡If the radiator fan comes on, do not include its running time in the 5 minutes.

HEATING & AIR CONDITIONING SYSTEM

BLOWER MOTOR

REMOVAL & INSTALLATION

See Figures 227 through 231.

1. Remove the passenger's dashboard undercover and the glove box.

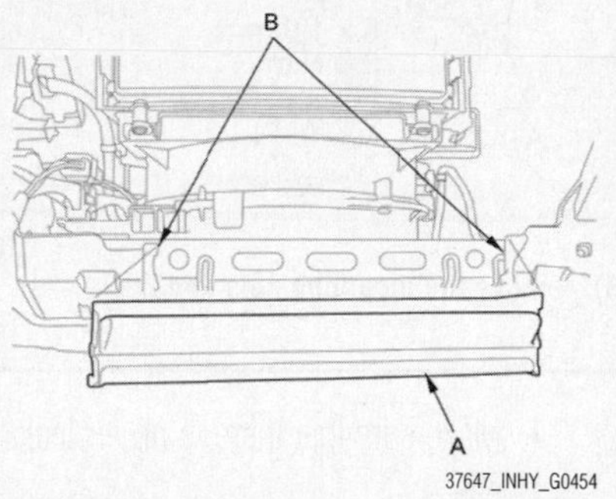
B
A
37647_INHY_G0454

Fig. 227 Cut the plastic cross brace (A) in the glove box opening in the area (B)

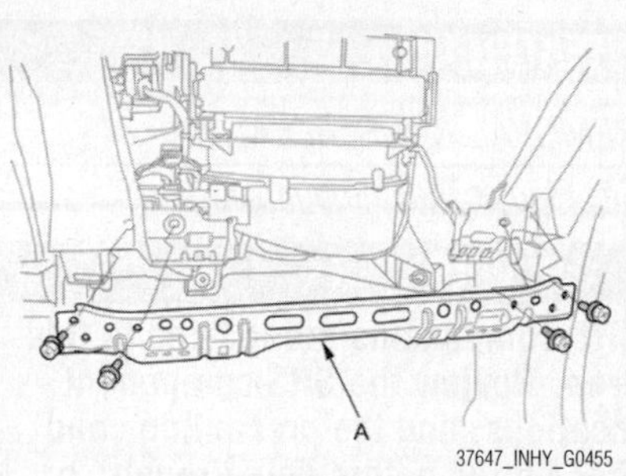
A
37647_INHY_G0455

Fig. 228 Remove the bolts and the glove box frame (A)

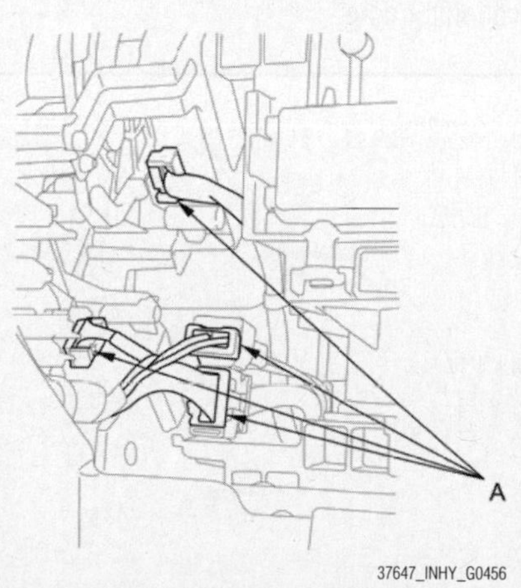
A
37647_INHY_G0456

Fig. 229 Disconnect the connectors (A) from the A/C wire harness and the mode control motor and air mix control motor

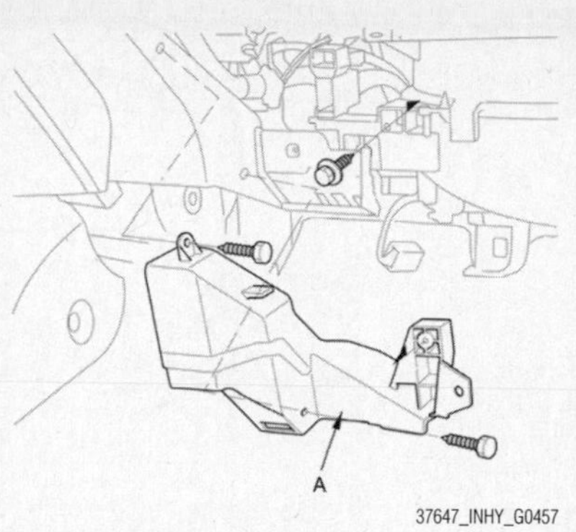
A
37647_INHY_G0457

Fig. 230 Remove the self-tapping screws and the passenger's heater duct (A)

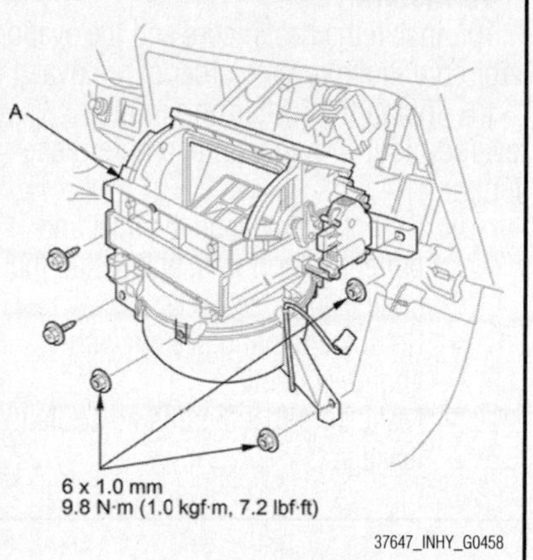
A
6 x 1.0 mm
9.8 N·m (1.0 kgf·m, 7.2 lbf·ft)
37647_INHY_G0458

Fig. 231 Remove the self-tapping screws, the mounting nuts, and the blower unit (A)

2. Cut the plastic cross brace in the glove box opening with diagonal cutters in the area, and discard it.
3. Remove the bolts and the glove box frame.
4. Disconnect the connectors from the A/C wire harness and the mode control motor and air mix control motor.
5. Remove the self-tapping screws and the passenger's heater duct.
6. Remove the self-tapping screws, the mounting nuts, and the blower unit.
7. Install the unit in the reverse order of removal. Make sure that there is no air leakage.

HEATER CORE

REMOVAL & INSTALLATION

See Figures 231 through 237.

⁂ WARNING

SRS components are located in this area. Review the SRS component locations, and the precautions and procedures before doing repairs or service.

1. Do the 12 volt battery terminal disconnection procedure.
2. Disconnect the A/C line from the evaporator core.
3. Drain the engine coolant from the radiator.
4. From the inlet heater hose, remove the clip. Slide the hose clamps back, then disconnect the inlet heater hose and the outlet heater hose from the heater unit. Note the layout of the hoses.

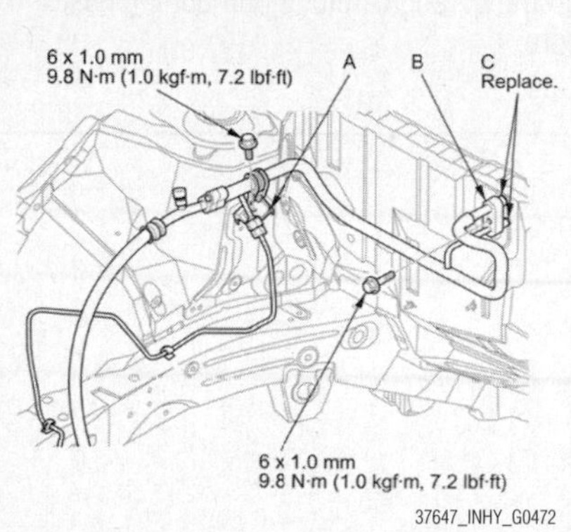
6 x 1.0 mm
9.8 N·m (1.0 kgf·m, 7.2 lbf·ft)
A
B
C
Replace.
6 x 1.0 mm
9.8 N·m (1.0 kgf·m, 7.2 lbf·ft)
37647_INHY_G0472

Fig. 232 Remove the bolts from the A/C line clamp (A); remove the bolt, then disconnect the A/C line (B) from the evaporator core and remove the O-rings (C)

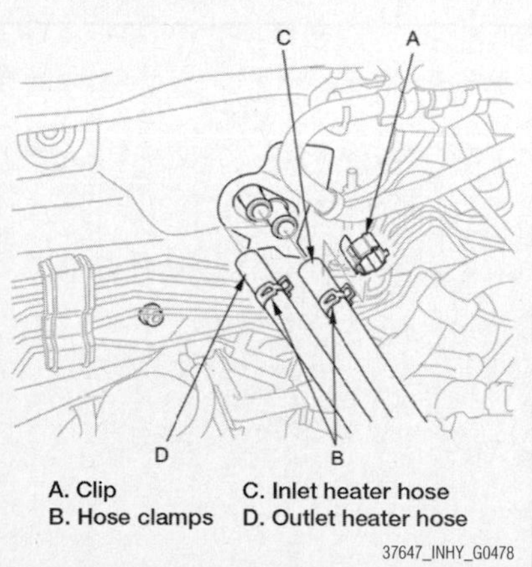
C
A
D
B
A. Clip
B. Hose clamps
C. Inlet heater hose
D. Outlet heater hose
37647_INHY_G0478

Fig. 233 Remove the clip; slide the hose clamps back, then disconnect the inlet heater hose and the outlet heater hose from the heater unit

➡Engine coolant will run out when the hoses are disconnected; drain it into a clean drip pan. Be sure not to let coolant spill on the electrical parts or the painted surfaces. If any coolant spills, rinse it off immediately.

5. Remove the mounting nut from the heater unit. Take care not to damage or bend the fuel lines or the brake lines, etc.
6. Remove the dashboard.
7. Disconnect these connectors: The mode control motor, the power transistor, the evaporator temperature sensor, and the air mix control motor.
8. Remove the duct and the drain hose. Remove the mounting nuts and the blower-heater unit.
9. Remove the self-tapping screws and the heater core cover, the cap, and the grommet, and carefully pull out the heater core.

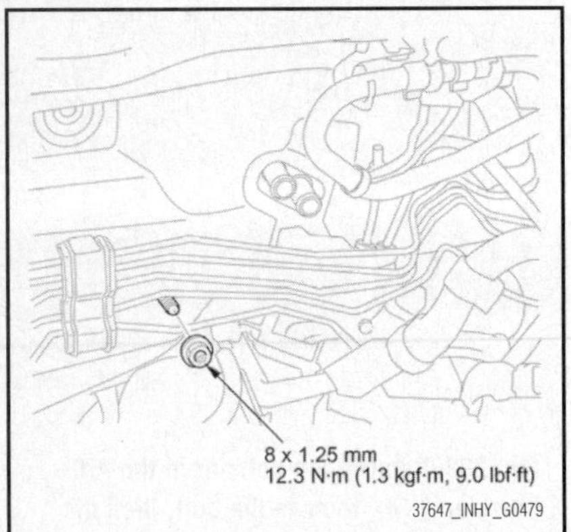
8 x 1.25 mm
12.3 N·m (1.3 kgf·m, 9.0 lbf·ft)
37647_INHY_G0479

Fig. 234 Remove the mounting nut from the heater unit

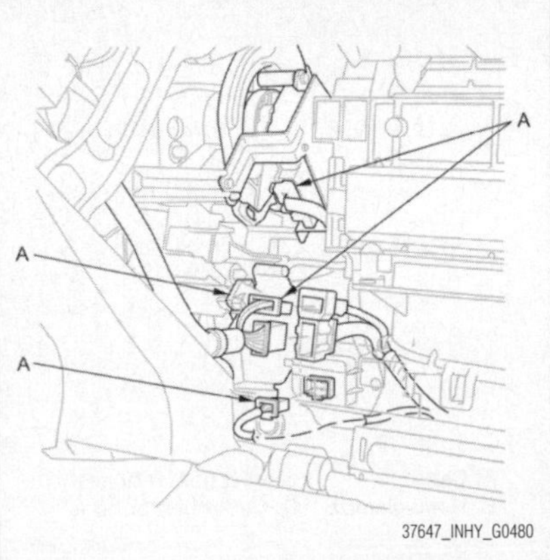
A
A
A
37647_INHY_G0480

Fig. 235 Disconnect these connectors (A): The mode control motor, the power transistor, the evaporator temperature sensor, and the air mix control motor

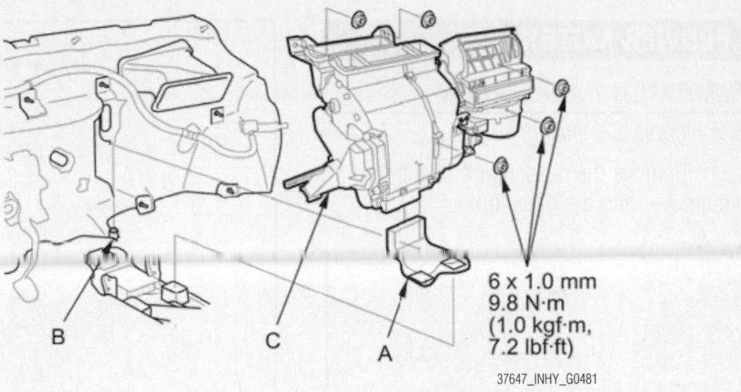
B
C
A
6 x 1.0 mm
9.8 N·m
(1.0 kgf·m,
7.2 lbf·ft)
37647_INHY_G0481

Fig. 236 Remove the duct (A) and the drain hose (B); remove the mounting nuts and the blower-heater unit (C)

To install:

10. Install the heater core and the evaporator core in the reverse order of removal.
11. Install the heater unit in the reverse order of removal, and note these items:
- Do not interchange the inlet and outlet heater hoses, and install the hose clamps securely.
- Refill the cooling system with engine coolant.
- Make sure that there is no coolant leakage.
- Make sure that there is no air leakage.
- Refer to the evaporator core replacement.

12. Do the 12 volt battery terminal reconnection procedure.

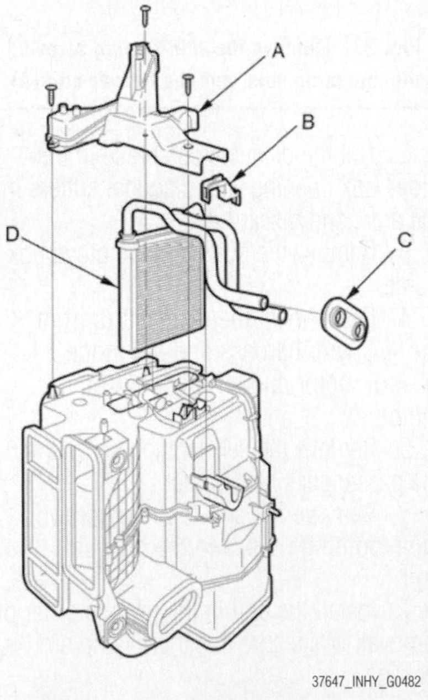
A
B
D
C
37647_INHY_G0482

Fig. 237 Remove the self-tapping screws and the heater core cover, the cap, and the grommet, and carefully pull out the heater core

HEATER UNIT

REMOVAL & INSTALLATION

See Figures 238 through 240.

⁂ WARNING

SRS components are located in this area. Review the SRS component locations, and the precautions and procedures before doing repairs or service.

1. Do the 12 volt battery terminal disconnection procedure.
2. Disconnect the A/C line from the evaporator core.

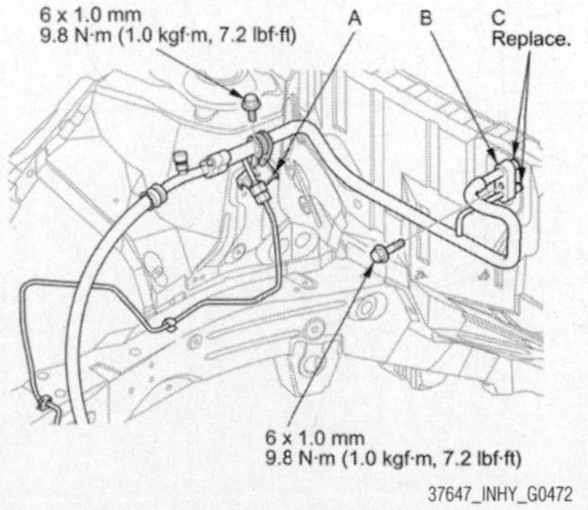
6 x 1.0 mm
9.8 N·m (1.0 kgf·m, 7.2 lbf·ft)
A
B
C
Replace.
6 x 1.0 mm
9.8 N·m (1.0 kgf·m, 7.2 lbf·ft)
37647_INHY_G0472

Fig. 238 Remove the bolts from the A/C line clamp (A); remove the bolt, then disconnect the A/C line (B) from the evaporator core and remove the O-rings (C)

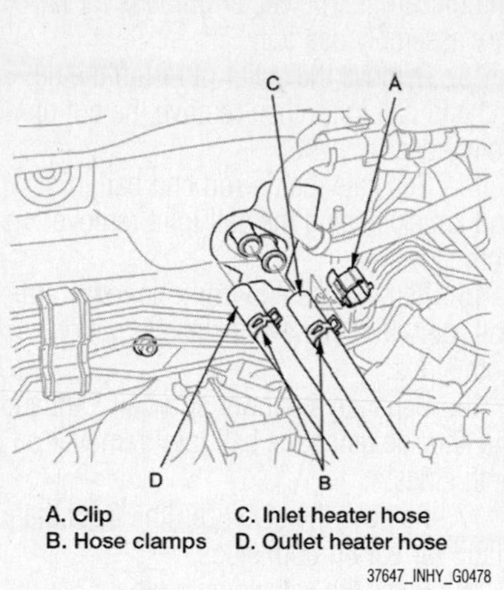
C
A
D
B
A. Clip
B. Hose clamps
C. Inlet heater hose
D. Outlet heater hose
37647_INHY_G0478

Fig. 239 Remove the clip; slide the hose clamps back, then disconnect the inlet heater hose and the outlet heater hose from the heater unit

3. Drain the engine coolant from the radiator.

4. From the inlet heater hose, remove the clip. Slide the hose clamps back, then disconnect the inlet heater hose and the outlet heater hose from the heater unit. Note the layout of the hoses.

➡Engine coolant will run out when the hoses are disconnected; drain it into a clean drip pan. Be sure not to let coolant spill on the electrical parts or the painted surfaces. If any coolant spills, rinse it off immediately.

5. Remove the mounting nut from the heater unit. Take care not to damage or bend the fuel lines or the brake lines, etc.

6. Remove the dashboard.

7. Disconnect these connectors: The mode control motor, the power transistor, the evaporator temperature sensor, and the air mix control motor.

8. Remove the duct and the drain hose. Remove the mounting nuts and the blower-heater unit.

To install:

9. Install the heater unit in the reverse order of removal, and note these items:

- Do not interchange the inlet and outlet heater hoses, and install the hose clamps securely.

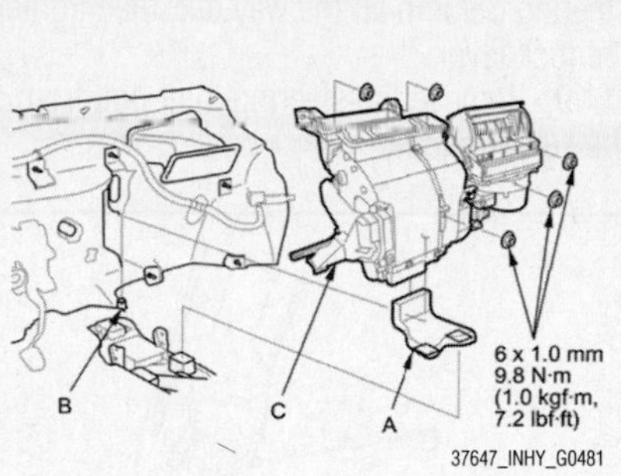
6 x 1.0 mm
9.8 N·m
(1.0 kgf·m,
7.2 lbf·ft)
B
C
A
37647_INHY_G0481

Fig. 240 Remove the duct (A) and the drain hose (B); remove the mounting nuts and the blower-heater unit (C)

- Refill the cooling system with engine coolant.
- Make sure that there is no coolant leakage.
- Make sure that there is no air leakage.
- Refer to the evaporator core replacement.

10. Do the 12 volt battery terminal reconnection procedure.

STEERING

POWER RACK & PINION STEERING GEAR

REMOVAL & INSTALLATION

See Figures 241 through 254.

Special Tools Required:

- Ball Joint Thread Protector, 10 mm 07AAF-SECA120
- Ball Joint Remover, 28 mm 07MAC-SL0A202
- Ball Joint Thread Protector, 14 mm 071AF-S3VA000

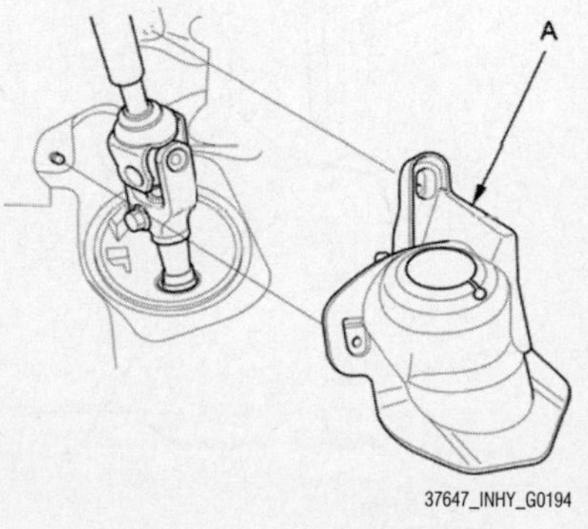
A
37647_INHY_G0194

Fig. 241 Remove the steering joint cover (A)

➡Note these items during removal:

- Use solvent and a brush, wash any oil and dirt off the end of the steering gearbox, but avoid any electrical parts. Blow dry with compressed air.
- Be sure to remove the steering wheel before disconnecting the steering joint or damage to the cable reel can occur.

1. Do the 12 volt battery terminal disconnection procedure.
2. Raise the vehicle on a lift.
3. Remove the front wheels.
4. Release the lock lever, and adjust the steering column to the full tilt down position, and to the full telescopic in position.
5. Tighten the lock lever.
6. Remove the steering joint cover.
7. Hold the lower slide shaft on the column with a piece of wire between the joint yoke of the lower slide shaft and joint yoke of the upper shaft to prevent the lower slide from pulling out.
8. Release the lock lever, and slide the

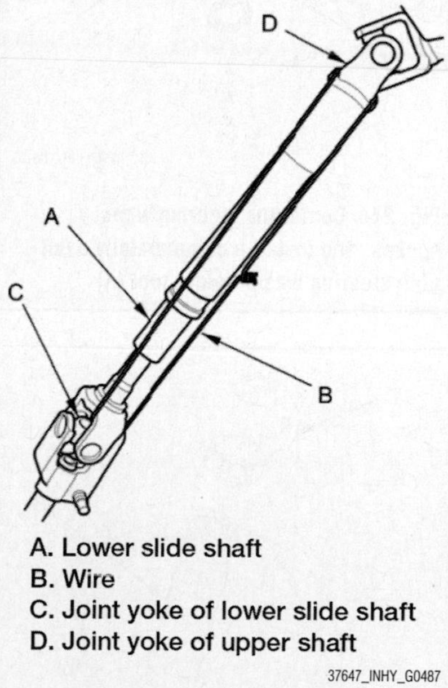
D
A
C
B
A. Lower slide shaft
B. Wire
C. Joint yoke of lower slide shaft
D. Joint yoke of upper shaft
37647_INHY_G0487

Fig. 242 Hold the lower slide shaft on the column with a piece of wire between the joint yoke of the lower slide shaft and joint yoke of the upper shaft to prevent the lower slide from pulling out

steering column all the way out, then tighten the lock lever.

9. Remove the steering joint bolt from the steering joint.

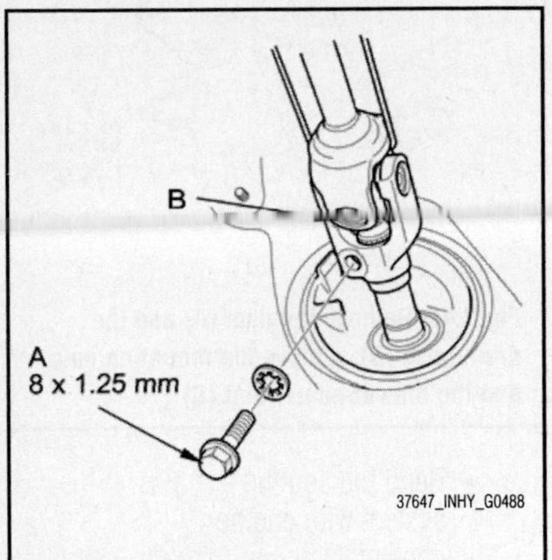
B
A
8 x 1.25 mm
37647_INHY_G0488

Fig. 243 Remove the steering joint bolt (A) from the steering joint (B)

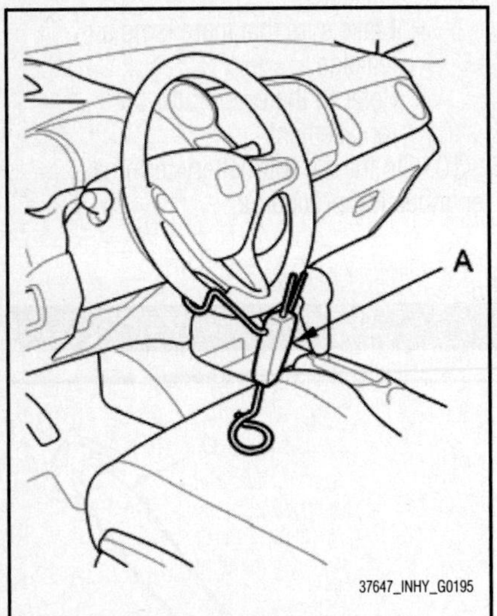
A
37647_INHY_G0195

Fig. 244 Center the steering wheel spokes, and install a commercially available steering wheel holder tool (A)

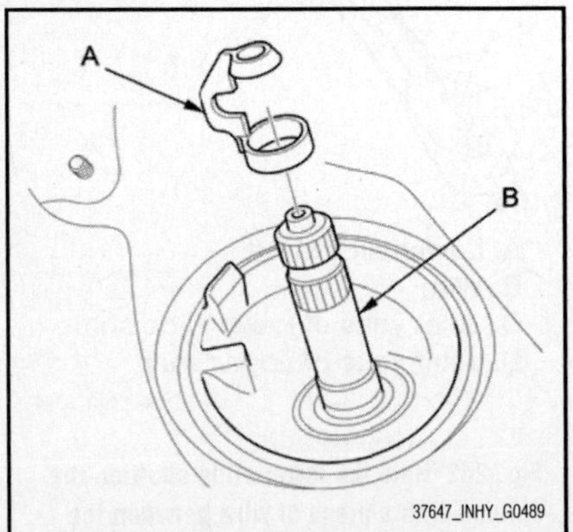
A
B
37647_INHY_G0489

Fig. 245 Remove the center guide (A) (if equipped) from the top of the pinion shaft (B)

10. Center the steering wheel spokes, and install a commercially available steering wheel holder tool.

11. Disconnect the steering joint by sliding the steering joint into the column.

12. Remove the center guide (if equipped) from the top of the pinion shaft and discard it. The center guide is for factory assembly use only.

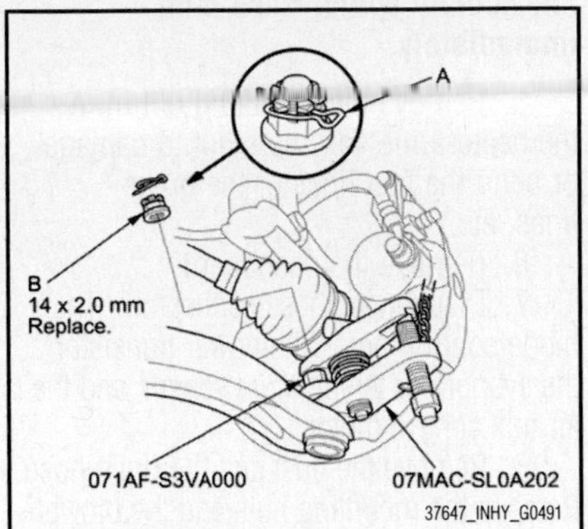
A
B
14 x 2.0 mm
Replace.
071AF-S3VA000
07MAC-SL0A202
37647_INHY_G0491

Fig. 246 Remove the clip (A) from the lower arm ball joint, then remove the castle nut (B) on both sides

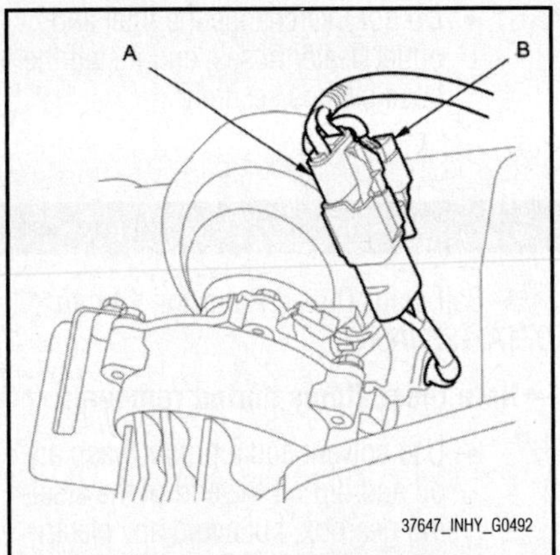
A
B
37647_INHY_G0492

Fig. 247 Disconnect the EPS motor 2P connector (A), and torque sensor 4P connector (B) from the torque sensor

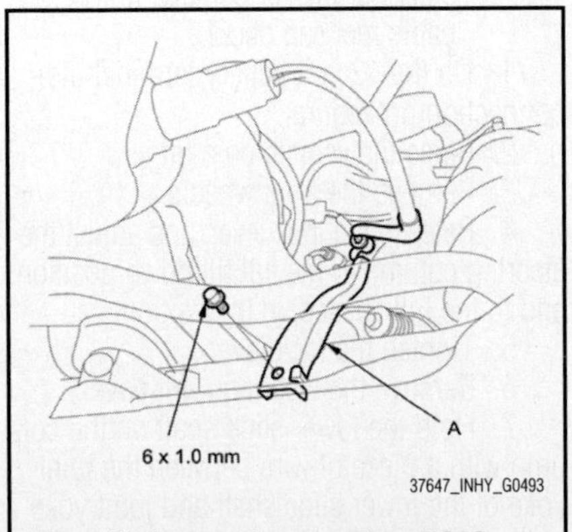
6 x 1.0 mm
A
37647_INHY_G0493

Fig. 248 Remove the secondary HO2S harness bracket (A) from the steering gearbox

13. Remove the cotter pin from the tie-rod end ball joint, then remove the nut on both sides.

14. Separate the tie-rod end ball joint and knuckle using the ball joint remover on both sides.

15. Remove the clip from the lower arm ball joint, then remove the castle nut on both sides.

16. Separate the lower arm ball joint and the knuckle using the ball joint remover on both sides.

17. Remove the stabilizer link from the stabilizer bar on both sides.

18. Raise the vehicle on a lift.

19. Remove the splash shield.

20. Disconnect the EPS motor 2P connector, and torque sensor 4P connector from the torque sensor.

21. Wrap the connectors with the vinyl tape to avoid contamination from grease or water.

22. Remove the secondary HO2S harness bracket from the steering gearbox. Do not disconnect secondary HO2S 4P connector and the secondary HO2S.

23. Place a jack under the engine oil pan.

24. Remove the torque rod.

25. Attach a transmission jack to the center of the front subframe, and support the front subframe securely by raising the transmission jack.

26. Remove the front subframe mounting bolts.

27. Lower the front subframe and the steering gearbox as an assembly by lowering the transmission jack slowly.

28. Remove the steering gearbox from the front subframe.

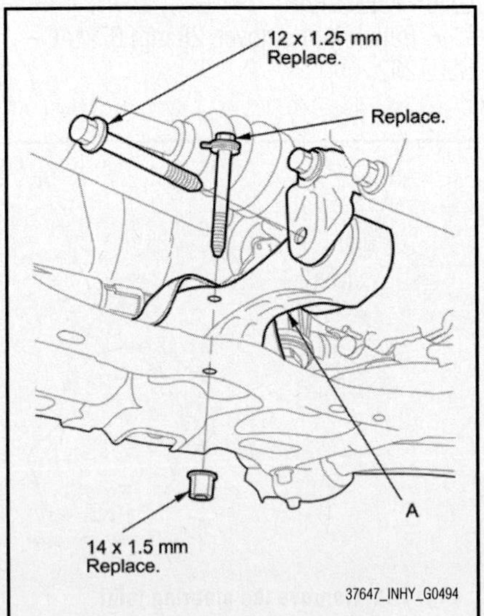
12 x 1.25 mm
Replace.
Replace.
A
14 x 1.5 mm
Replace.
37647_INHY_G0494

Fig. 249 Remove the torque rod (A)

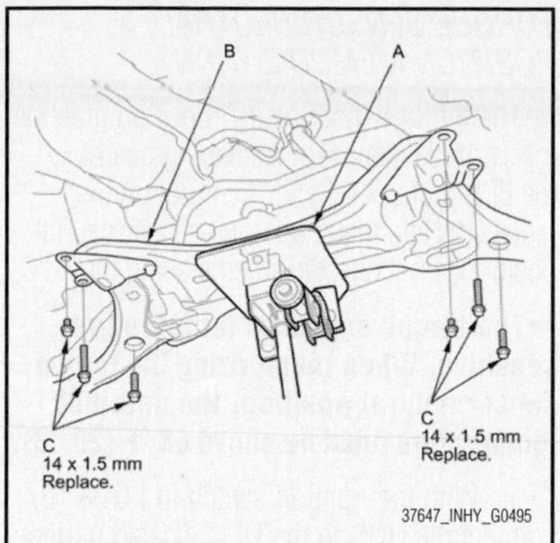
B
A
C
14 x 1.5 mm
Replace.
C
14 x 1.5 mm
Replace.
37647_INHY_G0495

Fig. 250 Attach a transmission jack (A) to the center of the front subframe (B); remove the front subframe mounting bolts (C)

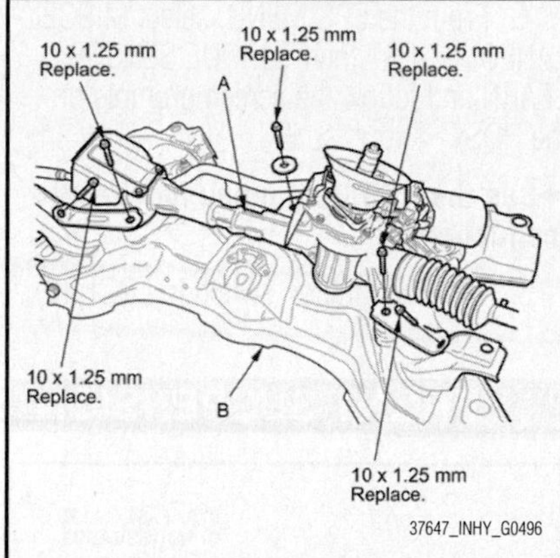
10 x 1.25 mm
Replace.
10 x 1.25 mm
Replace.
10 x 1.25 mm
Replace.
A
10 x 1.25 mm
Replace.
B
10 x 1.25 mm
Replace.
37647_INHY_G0496

Fig. 251 Remove the steering gearbox (A) from the front subframe (B)

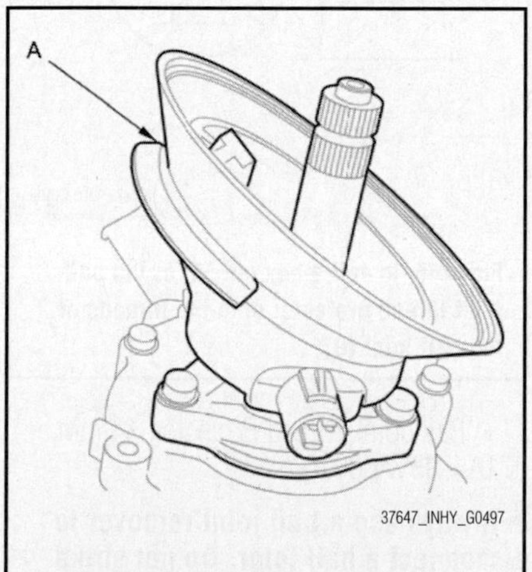
A
37647_INHY_G0497

Fig. 252 Remove the pinion shaft grommet (A) from the top of the torque sensor

29. Remove the pinion shaft grommet from the top of the torque sensor.

To install:

30. Before installing the steering gearbox, make sure that no grease is on the mating surface of the steering gearbox and the front subframe. To prevent the gearbox mounting bolts from loosening after the installation, remove any grease from the bolt holes.
31. Install the pinion shaft grommet. Align the ledge portion in the pinion shaft grommet with the lug portion on the torque sensor 3P connector. The grommet must not have a gap at the mating surface of the grommet and torque sensor.
32. Place the steering gearbox on the front subframe.
33. Loosely install the stiffener plates, and new gearbox mounting bolts, then tighten the bolts.
34. Turn the lip of the pinion shaft grommet to ease installation.
35. Set the front subframe with the steering gearbox on the transmission jack, and support it.
36. Carefully raise the front subframe with transmission jack, and pass the pinion shaft into the passenger's compartment.

➡Be sure that the pinion shaft grommet is securely in place. Make sure the lip of the pinion shaft grommet is not turned up. Incorrect installation can cause leakage of water or mud, or noise. Take care not to damage the lower arm ball joint boot with the edge of the knuckle, etc.

37. Loosely install new front subframe mounting bolts, then tighten the bolts to 69 ft. lbs. (93 Nm).

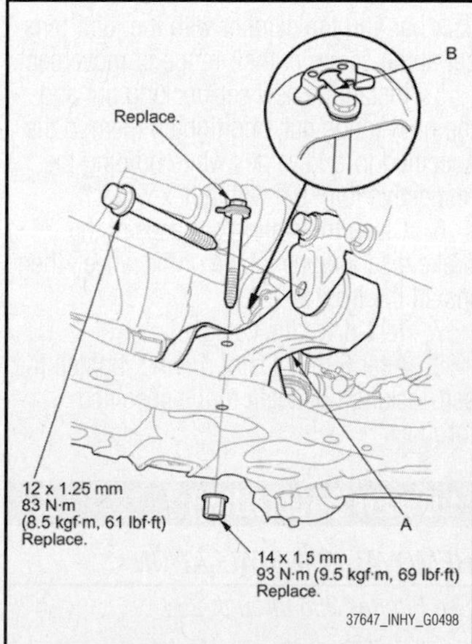
B
Replace.
12 x 1.25 mm
83 N·m
(8.5 kgf·m, 61 lbf·ft)
Replace.
A
14 x 1.5 mm
93 N·m (9.5 kgf·m, 69 lbf·ft)
Replace.
37647_INHY_G0498

Fig. 253 Install the torque rod (A) with new mounting bolts and new mounting nut; make sure the guide plate (B) of the flange bolt is securely installed

38. Install the torque rod with new mounting bolts and new mounting nut, and tighten them to the specified torque.

➡Make sure the guide plate of the flange bolt is securely installed.

39. Remove the jack.
40. Install the secondary HO2S harness bracket on the steering gearbox.
41. Remove the vinyl tape from the connectors.
42. Connect the EPS motor 2P connector and the torque sensor 4P connector to the torque sensor.
43. Install the splash shield.
44. Connect the stabilizer links.
45. Lower the vehicle.
46. Wipe off any grease contamination from the ball joint tapered section and threads. Reconnect the tie-rod end ball joints to the knuckles. Install the nuts, and tighten to 32 ft. lbs. (43 Nm) on both sides.
47. Install the new cotter pin.
48. Wipe off any grease contamination from the lower arm ball joint tapered section and threads. Then reconnect the lower arm to the knuckle. Install the new castle nut and tighten it on both sides.

➡Be careful not to damage the lower ball joint boot. Check the ball joint boot for deformation before connecting the knuckle. Torque the castle nut to the lower torque specification, then tighten it only far enough to align the slot with the joint pin clip hole. Do not align the castle nut by loosening it.

49. Install the clip.
50. Install the front wheels, then set the wheels in the straight ahead position.

➡Before installing the wheel, clean the mating surfaces of the brake disc and the inside of the wheel.

51. Cut the wire holding the steering shaft.
52. Slip the lower end of the steering joint onto the pinion shaft taking care to align the gap within the angle shown.
53. Remove the steering wheel holder tool.
54. Align the bolt hole on the steering joint with the groove around the pinion shaft, then loosely install the lower steering joint bolt. Be sure that the joint bolt is securely in the groove in the pinion shaft.
55. Pull on the steering joint to make sure that the steering joint is fully seated, then tighten the lower joint bolt to the specified torque.
56. Install the steering joint cover.
57. Install the steering wheel and the driver's airbag.

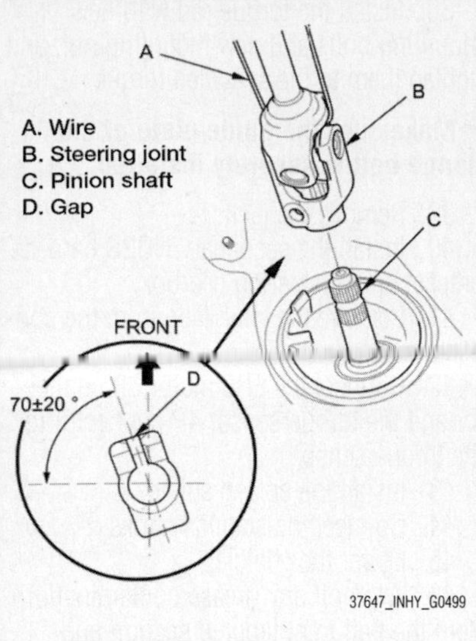
A
B
A. Wire
B. Steering joint
C. Pinion shaft
D. Gap
C
FRONT
70±20 °
D
37647_INHY_G0499

Fig. 254 Slip the lower end of the steering joint onto the pinion shaft taking care to align the gap within the angle shown

58. With the tires raised off the ground, check for the following symptoms by turning the steering wheel fully to the right and left several times.

59. Do the 12 volt battery terminal reconnection procedure, and do these tasks:

a. Turn the ignition switch to ON (II), and check that the SRS indicator comes on for about 6 seconds and then goes off.

b. Make sure the horn and turn signal switches work properly.

c. Make sure the steering wheel switches work properly.

60. Do the memorizing of the torque sensor neutral position.

61. After installation, do these checks:

a. Start the engine, allow it to idle, and turn the steering wheel to full left or full right.

b. Check that the EPS indicator does not come on.

c. Check the steering wheel spoke angle. If steering spoke angles to the right and left are not equal (steering wheel and rack are not centered), correct the engagement of the joint/pinion shaft serrations, then adjust the front toe by turning the tie-rod ends, if necessary.

62. Check the wheel alignment, and adjust it if necessary.

TORQUE SENSOR NEUTRAL POSITION MEMORIZATION

The torque sensor neutral position must be memorized whenever the steering gearbox, the EPS motor, or the EPS control unit is replaced. Note that the torque sensor neutral position is not affected when erasing the DTC.

➡The torque sensor is temperature sensitive. When memorizing the torque sensor neutral position, the ambient temperature must be above 68 °F (20 °C).

1. With the ignition switch in LOCK (0), connect the HDS to the DLC located under the driver's side of the dashboard.

2. Turn the ignition switch to ON (II).

3. Make sure the HDS communicates with the vehicle and the EPS control unit. If it doesn't, troubleshoot the DLC circuit.

4. From the EPS MENU, select MISCELLANEOUS TEST then TORQUE SENSOR LEARN and follow the screen prompts on the HDS.

➡See the HDS Help menu for specific instructions.

5. Turn the ignition switch to LOCK (0).

SUSPENSION

FRONT SUSPENSION

CONTROL LINKS

REMOVAL & INSTALLATION

Front Stabilizer Link

See Figure 255.

1. Raise the vehicle on a lift.

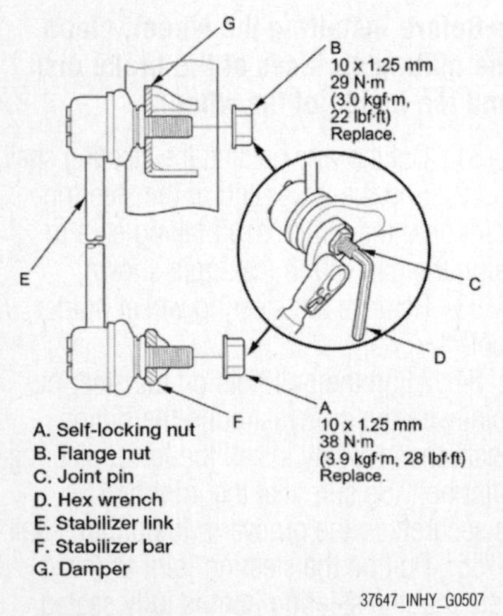
G
B
10 x 1.25 mm
29 N·m
(3.0 kgf·m,
22 lbf·ft)
Replace.
E
C
D
F
A
10 x 1.25 mm
38 N·m
(3.9 kgf·m, 28 lbf·ft)
Replace.
A. Self-locking nut
B. Flange nut
C. Joint pin
D. Hex wrench
E. Stabilizer link
F. Stabilizer bar
G. Damper
37647_INHY_G0507

Fig. 255 Remove the self-locking nut and the flange nut while holding the respective joint pin with a hex wrench, then remove the stabilizer link

2. Remove the front wheel.

3. Remove the self-locking nut and the flange nut while holding the respective joint pin with a hex wrench, then remove the stabilizer link.

To install:

4. Install the stabilizer link on the stabilizer bar and the damper with the joint pins set at the center of their range of movement.

5. Install the new self-locking nut and the new flange nut, and tighten them to the specified torque values while holding the respective joint pin with a hex wrench.

6. Clean the mating surfaces of the brake disc and the inside of the wheel, then install the front wheel.

7. Test-drive the vehicle.

8. After 5 minutes of driving, tighten the self-locking nut again to the specified torque.

LOWER BALL JOINT

REMOVAL & INSTALLATION

See Figures 256 through 258.

Special Tools Required:

- Ball Joint Thread Protector, 10 mm 07AAF-SECA120
- Ball Joint Remover, 28 mm 07MAC-SL0A202

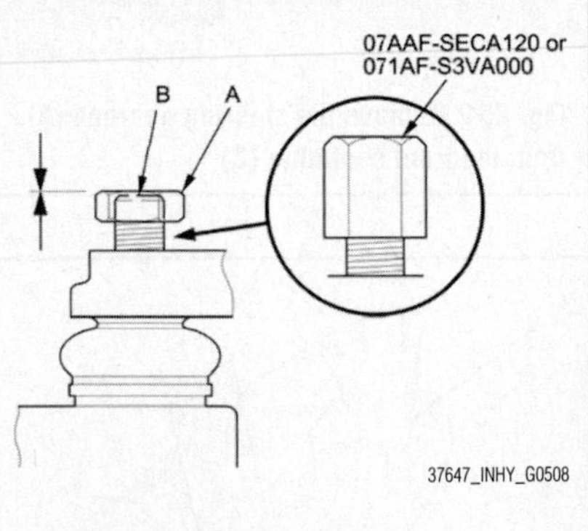
07AAF-SECA120 or
071AF-S3VA000
B
A
37647_INHY_G0508

Fig. 256 Install a hex nut (A) or the ball joint thread protector onto the threads of the ball joint (B)

- Ball Joint Thread Protector, 14 mm 071AF-S3VA000

➡Always use a ball joint remover to disconnect a ball joint. Do not strike the housing or any other part of the ball joint connection to disconnect it.

1. Install a hex nut or the ball joint thread protector onto the threads of the ball joint.

➡Using a hex nut, make sure the nut is flush with the ball joint pin end to prevent damage to the threaded end of the ball joint pin.

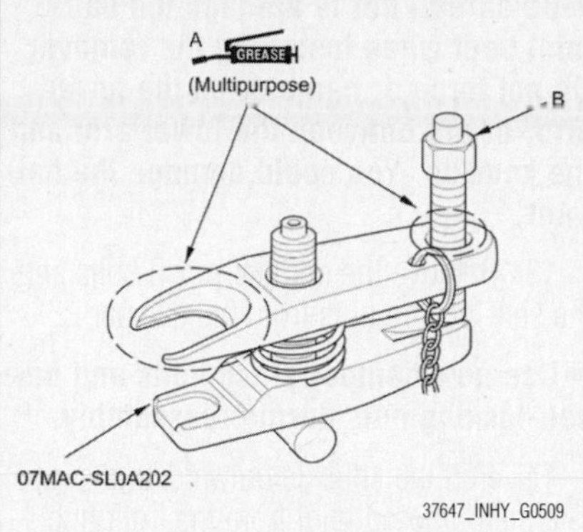
A
GREASE
(Multipurpose)
B
07MAC-SL0A202
37647_INHY_G0509

Fig. 257 Apply grease to the ball joint remover on the areas shown (A); this will ease installation of the tool and prevent damage to the pressure bolt (B) threads

2. Apply grease to the ball joint remover on the areas shown. This will ease installation of the tool and prevent damage to the pressure bolt threads.

3. Loosen the pressure bolt, and install the ball joint remover as shown. Insert the jaws carefully, making sure not to damage the ball joint boot. Adjust the jaw spacing by turning the adjusting bolt.

➡Fasten the safety chain securely to a suspension arm or the subframe. Do not fasten it to a brake line or wire harness.

4. After adjusting the adjusting bolt, make sure the head of the adjusting bolt is in the position shown to allow the jaw to pivot.

5. With a wrench, tighten the pressure bolt until the ball joint pin pops loose from the ball joint connecting hole. If necessary, apply penetrating type lubricant to loosen the ball joint pin.

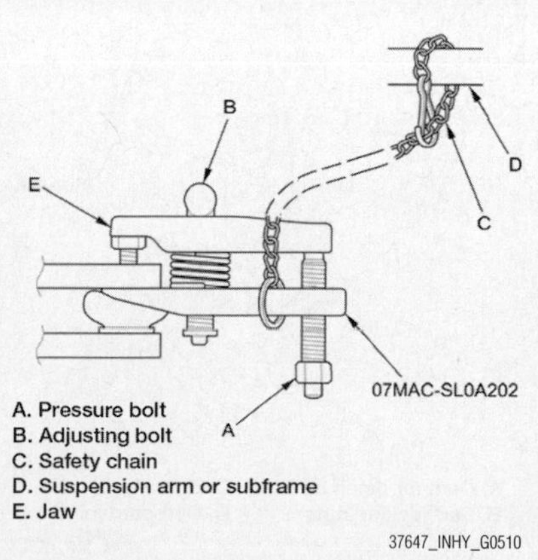
B
E
D
C
A
07MAC-SL0A202
A. Pressure bolt
B. Adjusting bolt
C. Safety chain
D. Suspension arm or subframe
E. Jaw
37647_INHY_G0510

Fig. 258 Loosen the pressure bolt, and install the ball joint remover as shown

➡Do not use pneumatic or electric tools on the pressure bolt.

6. Remove the ball joint remover, then remove the nut from the end of the ball joint pin, and pull the ball joint out of the ball joint connecting hole. Inspect the ball joint boot, and replace it if damaged.

LOWER CONTROL ARM

REMOVAL & INSTALLATION

See Figures 259 and 260.

Special Tools Required:

- Ball Joint Remover, 28 mm 07MAC-SL0A202
- Ball Joint Thread Protector, 14 mm 071AF-S3VA000

➡Do not remove the lower arm from both sides at the same time. The lower arm mounting bolts also secure the subframe to the vehicle.

1. Raise the vehicle on a lift.
2. Remove the front wheel.
3. Remove the splash shield.
4. Remove the lock pin from the lower arm ball joint, then remove the castle nut.

➡During installation, install the lock pin after tightening the new castle nut.

5. Disconnect the lower ball joint from the knuckle using the ball joint remover.

➡Be careful not to damage the ball joint boot when installing the remover. Do not force or hammer on the lower arm, or pry between the lower arm and the knuckle. You could damage the ball joint.

6. Remove the lower arm mounting bolts, then remove the lower arm from the front subframe.

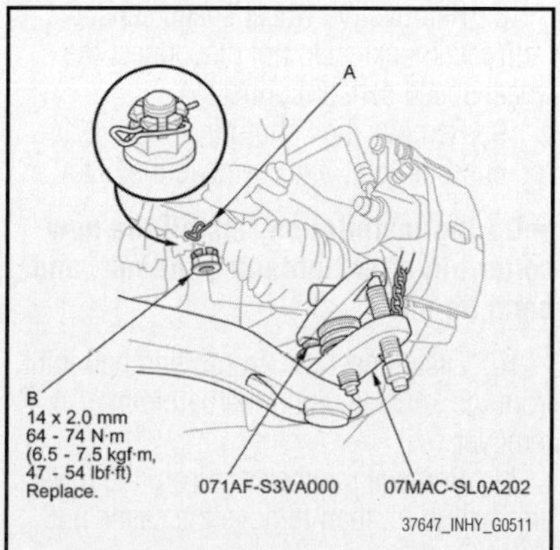
A
B
14 x 2.0 mm
64 - 74 N·m
(6.5 - 7.5 kgf·m,
47 - 54 lbf·ft)
Replace.
071AF-S3VA000
07MAC-SL0A202
37647_INHY_G0511

Fig. 259 Remove the lock pin (A) from the lower arm ball joint, then remove the castle nut (B)

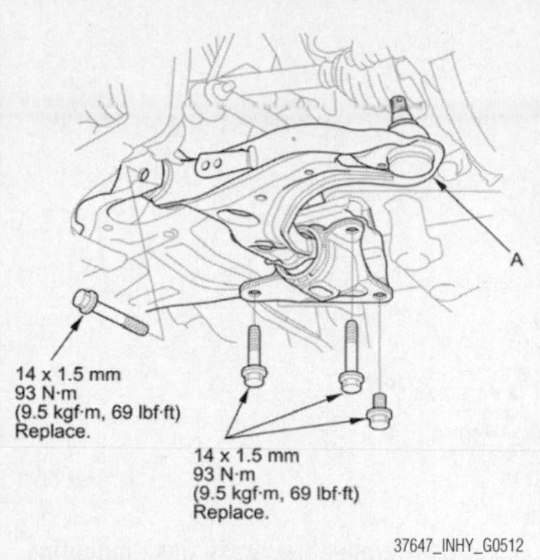
A
14 x 1.5 mm
93 N·m
(9.5 kgf·m, 69 lbf·ft)
Replace.
14 x 1.5 mm
93 N·m
(9.5 kgf·m, 69 lbf·ft)
Replace.
37647_INHY_G0512

Fig. 260 Remove the lower arm mounting bolts, then remove the lower arm (A) from the front subframe

➡Use new lower arm mounting bolts during reassembly.

To install:

7. Install the lower arm in the reverse order of removal, and note these items:
   - First install all the components, and lightly tighten the bolts and the nuts, then raise the suspension to load it with the vehicle's weight before fully tightening to the specified torque values. Do not place the jack against the ball joint of the lower arm.
   - Be careful not to damage the ball joint boot when connecting to the knuckle.
   - Before connecting the ball joint, degrease the threaded section and the tapered portion of the ball joint pin, the ball joint connecting hole, and the threaded section and the mating surfaces of the castle nut.
   - Torque the castle nut to the lower torque specification, then tighten it only far enough to align the slot with the ball joint pin hole. Do not align the castle nut by loosening it.
   - Before installing the wheel, clean the mating surfaces of the brake disc and the inside of the wheel.
8. Check the wheel alignment, and adjust it if necessary.

STEERING KNUCKLE

REMOVAL & INSTALLATION

See Figures 261 through 265.

Special Tools Required:

- Ball Joint Thread Protector, 10 mm 07AAF SECA120

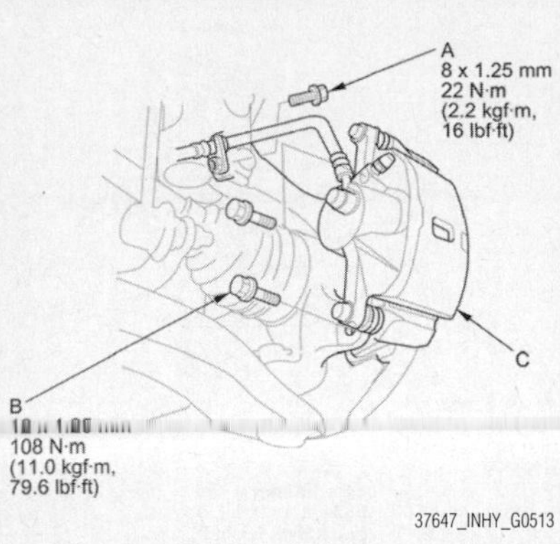

A
8 x 1.25 mm
22 N·m
(2.2 kgf·m,
16 lbf·ft)
C
B
108 N·m
(11.0 kgf·m,
79.6 lbf·ft)
37647_INHY_G0513

Fig. 261 Remove the brake hose mounting bolt (A), remove the brake caliper bracket mounting bolts (B), then remove the caliper assembly (C) from the knuckle

- Hub Dis/Assembly Pin, 40 mm 07GAF-SE00100
- Ball Joint Remover, 28 mm 07MAC-SL0A202
- Ball Joint Thread Protector, 14 mm 071AF-S3VA000
- Bearing Driver Attachment, 52 x 55 mm 07746-0010400
- Driver Handle, 15 x 135L 07749-0010000
- Support Base 07965-SD90100

1. Raise the vehicle on a lift.
2. Remove the wheel nuts and the front wheel.
3. Remove the brake hose mounting bolt from the damper bracket.
4. Remove the brake caliper bracket mounting bolts, then remove the caliper assembly from the knuckle. To prevent damage to the caliper assembly or the brake hose, use a short piece of wire to hang the caliper assembly from the undercarriage. Do not twist the brake hose excessively.
5. Pry up the stake on the spindle nut, then remove the nut.

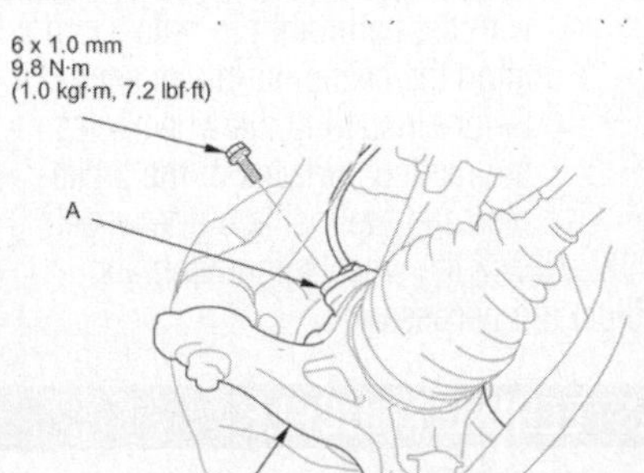

6 x 1.0 mm
9.8 N·m
(1.0 kgf·m, 7.2 lbf·ft)
A
B
37647_INHY_G0514

Fig. 262 Remove the wheel speed sensor (A) from the knuckle (B)

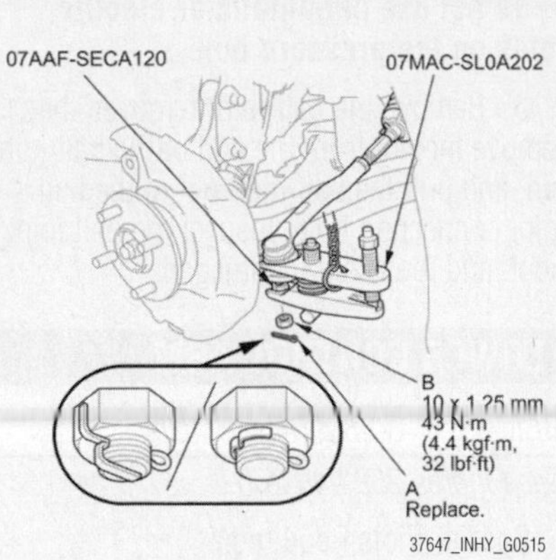

07AAF-SECA120
07MAC-SL0A202
B
10 x 1.25 mm
43 N·m
(4.4 kgf·m,
32 lbf·ft)
A
Replace.
37647_INHY_G0515

Fig. 263 Remove the cotter pin (A) from the tie-rod end ball joint, then remove the nut (B)

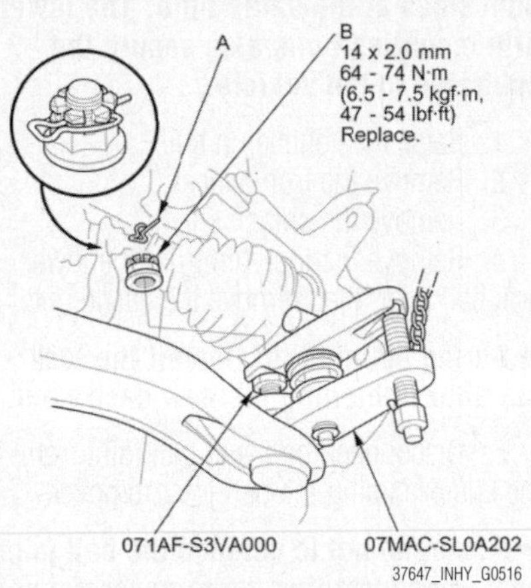

A
B
14 x 2.0 mm
64 - 74 N·m
(6.5 - 7.5 kgf·m,
47 - 54 lbf·ft)
Replace.
071AF-S3VA000
07MAC-SL0A202
37647_INHY_G0516

Fig. 264 Remove the lock pin (A) from the lower arm ball joint, then remove the castle nut (B)

6. Remove the front brake disc.
7. Check the front hub for damage and cracks.
8. Remove the wheel speed sensor from the knuckle. Do not disconnect the wheel speed sensor connector.
9. Remove the cotter pin from the tie-rod end ball joint, then remove the nut.

➡During installation, install the new cotter pin after tightening the nut, and bend its end.

10. Disconnect the tie-rod end ball joint from the knuckle using the ball joint remover.
11. Remove the lock pin from the lower arm ball joint, then remove the castle nut.

➡During installation, install the lock pin after tightening the new castle nut.

12. Disconnect the lower ball joint from the knuckle using the ball joint remover.

➡Be careful not to damage the ball joint boot when installing the remover. Do not force or hammer on the lower arm, or pry between the lower arm and the knuckle. You could damage the ball joint.

13. Remove the damper pinch bolts and the self-locking nuts from the damper.

➡Use new damper pinch bolts and new self locking nuts during reassembly.

14. Pull the knuckle outward, and separate the outboard joint from the front hub using a plastic hammer.

➡Do not pull the driveshaft end outward. The driveshaft inboard joint may come apart. During installation, apply grease to the mating surfaces of the wheel bearing and the driveshaft outboard joint.

To install:

15. Install the knuckle/hub in the reverse order of removal, and note these items:
    - First install all the components, and lightly tighten the bolts and the nuts, then raise the suspension to load it with the vehicle's weight before fully tightening to the specified torque values. Do not place the jack against the ball joint of the lower arm.
    - Be careful not to damage the ball joint boot when connecting the knuckle.
    - Before connecting the ball joint, degrease the threaded section and the tapered portion of the ball joint pin, the ball joint connecting hole, and the threaded section

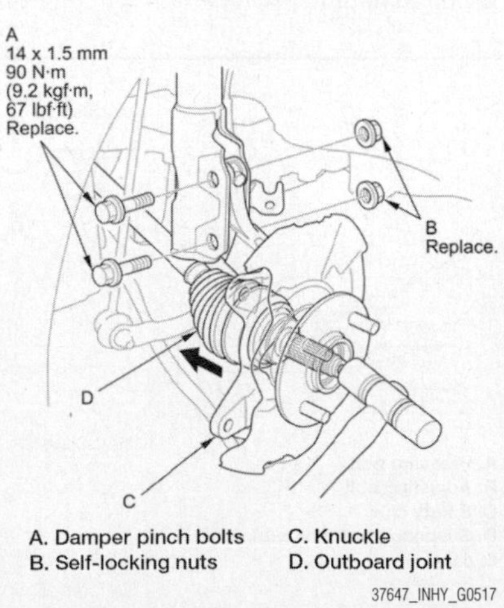

A
14 x 1.5 mm
90 N·m
(9.2 kgf·m,
67 lbf·ft)
Replace.
B
Replace.
D
C
A. Damper pinch bolts
B. Self-locking nuts
C. Knuckle
D. Outboard joint
37647_INHY_G0517

Fig. 265 Remove the damper pinch bolts and the self-locking nuts from the damper

and the mating surfaces of the castle nut.
- Torque the castle nut to the lower torque specification, then tighten it only far enough to align the slot with the ball joint pin hole. Do not align the castle nut by loosening it.
- Use a new spindle nut on reassembly.
- Before installing the spindle nut, apply a small amount of engine oil to the seating surface of the nut. After tightening, use a drift to stake the spindle nut shoulder against the driveshaft.
- Before installing the brake disc, clean the mating surfaces of the front hub and the inside of the brake disc.
- Before installing the wheel, clean the mating surfaces of the brake disc and the inside of the wheel.

16. Check the wheel alignment, and adjust it if necessary.

STRUT (DAMPER/SPRING)

REMOVAL & INSTALLATION

See Figures 262, 266 through 270.

1. Raise the vehicle on a lift.
2. Remove the front wheel.
3. Remove the wheel speed sensor from the knuckle. Do not disconnect the wheel speed sensor connector.
4. Disconnect the stabilizer link from the damper.
5. Remove the wheel speed sensor clip, the wire guide grommet, and the brake hose bracket from the damper. Do not disconnect the wheel speed sensor connector.

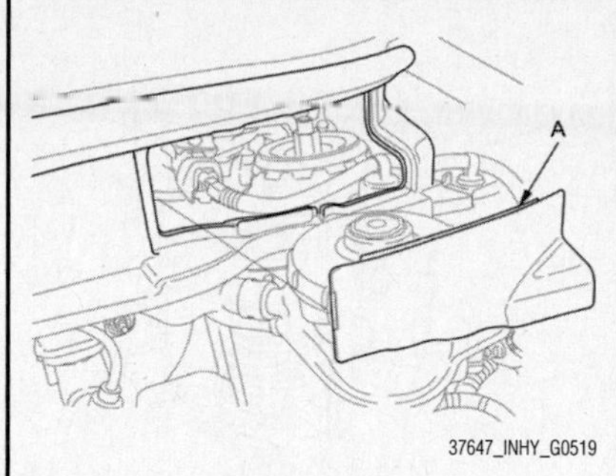
A
37647_INHY_G0519

Fig. 267 Remove the cowl lid from the cowl cover (A)

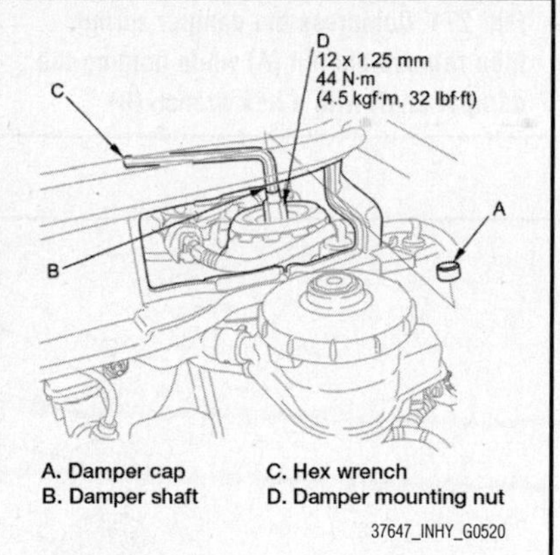
D
12 x 1.25 mm
44 N·m
(4.5 kgf·m, 32 lbf·ft)
C
A
B
A. Damper cap
B. Damper shaft
C. Hex wrench
D. Damper mounting nut
37647_INHY_G0520

Fig. 268 Remove the damper cap from the top of the damper

6. Remove the damper pinch bolts and the self-locking nuts from the damper.

➡**Do not allow the knuckle to rotate too far outward. This may allow the driveshaft inboard joint come apart. During installation, install new damper pinch bolts and new self-locking nuts, then tighten the damper pinch bolts to the specified torque value.**

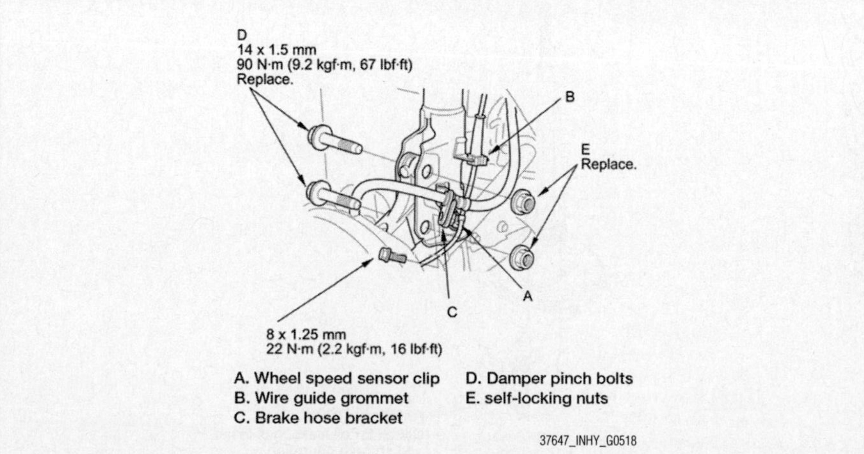
D
14 x 1.5 mm
90 N·m (9.2 kgf·m, 67 lbf·ft)
Replace.
B
E
Replace.
A
C
8 x 1.25 mm
22 N·m (2.2 kgf·m, 16 lbf·ft)
A. Wheel speed sensor clip
B. Wire guide grommet
C. Brake hose bracket
D. Damper pinch bolts
E. self-locking nuts
37647_INHY_G0518

Fig. 266 Remove the wheel speed sensor clip, the wire guide grommet, and the brake hose bracket from the damper

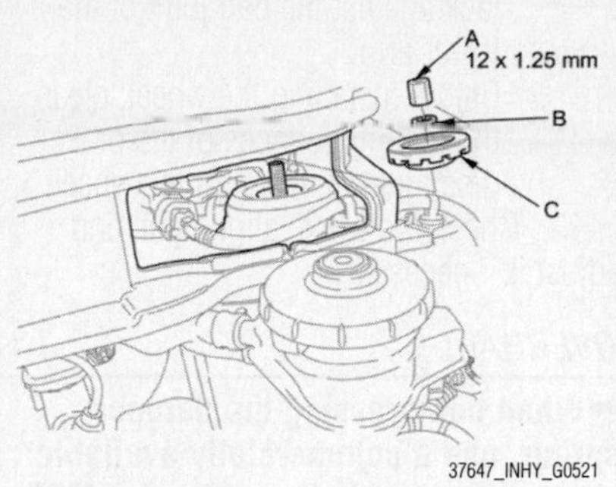
A
12 x 1.25 mm
B
C
37647_INHY_G0521

Fig. 269 Remove the damper mounting nut (A) and the wave washer (B), then remove the damper mounting base (C) from the top of the damper

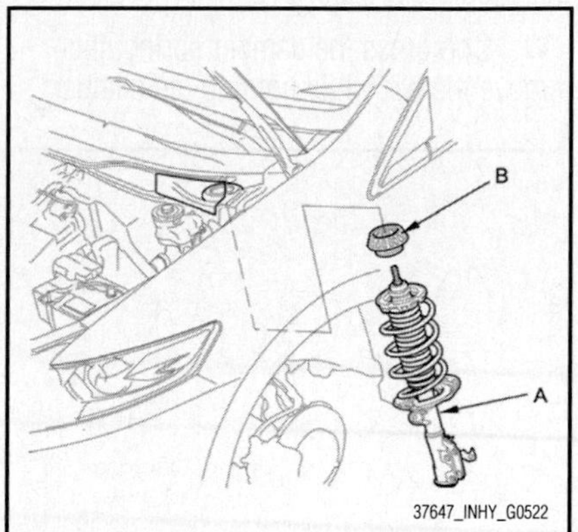
B
A
37647_INHY_G0522

Fig. 270 Remove the damper/spring (A) and the damper rubber mount (B)

7. Remove the cowl lid from the cowl cover.
8. Remove the damper cap from the top of the damper.
9. Hold the damper shaft using a hex wrench, and loosen the damper mounting nut.
10. Remove the damper mounting nut and the wave washer, then remove the damper mounting base from the top of the damper.
11. Remove the damper/spring and the damper rubber mount.

➡**Be careful not to damage the body.**

To install:

12. Install all of the removed parts in the reverse order of removal, and note these items:
    - First install all the components, and lightly tighten the bolts and the nuts, then raise the suspension to load it with the vehicle's weight before fully tightening to the specified torque values. Do not place the

jack against the ball joint of the lower arm.

- Before installing the wheel, clean the mating surfaces of the brake disc and the inside of the wheel.

13. Check the wheel alignment, and adjust it if necessary.

OVERHAUL

➡**When compressing the damper spring, use a commercially available strut spring compressor (Branick MST-580A or Model 7200, or equivalent) according to the manufacturer's instructions.**

Disassembly

See Figures 271 and 272.

1. Compress the damper spring, then remove the nut while holding the damper shaft with a hex wrench. Do not compress the damper spring more than necessary to remove the nut.

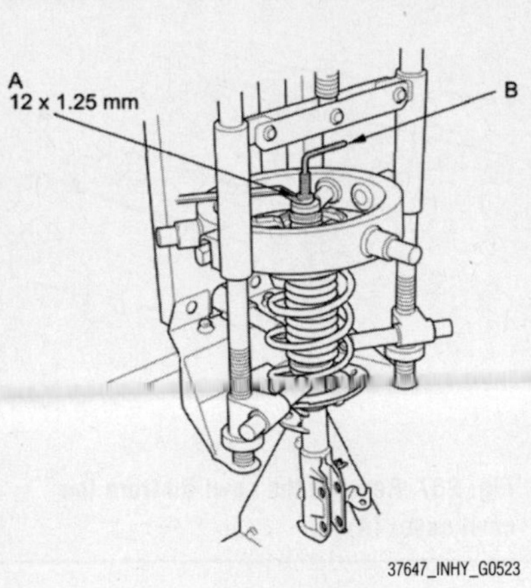
A
12 x 1.25 mm
B
37647_INHY_G0523

Fig. 271 Compress the damper spring, then remove the nut (A) while holding the damper shaft with a hex wrench (B)

2. Release the pressure from the strut spring compressor, then disassemble the damper as shown in the Exploded View.

Inspection

1. Install the nut on the damper shaft end, and set the socket wrench and T-handle on the nut.

2. Compress the damper unit by hand, and check for smooth operation through a full stroke, both compression and extension. The damper should extend smoothly and constantly when compression is released. If it does not, the gas is leaking and the damper should be replaced.

3. Check for oil leaks, abnormal noises, and binding during these tests.

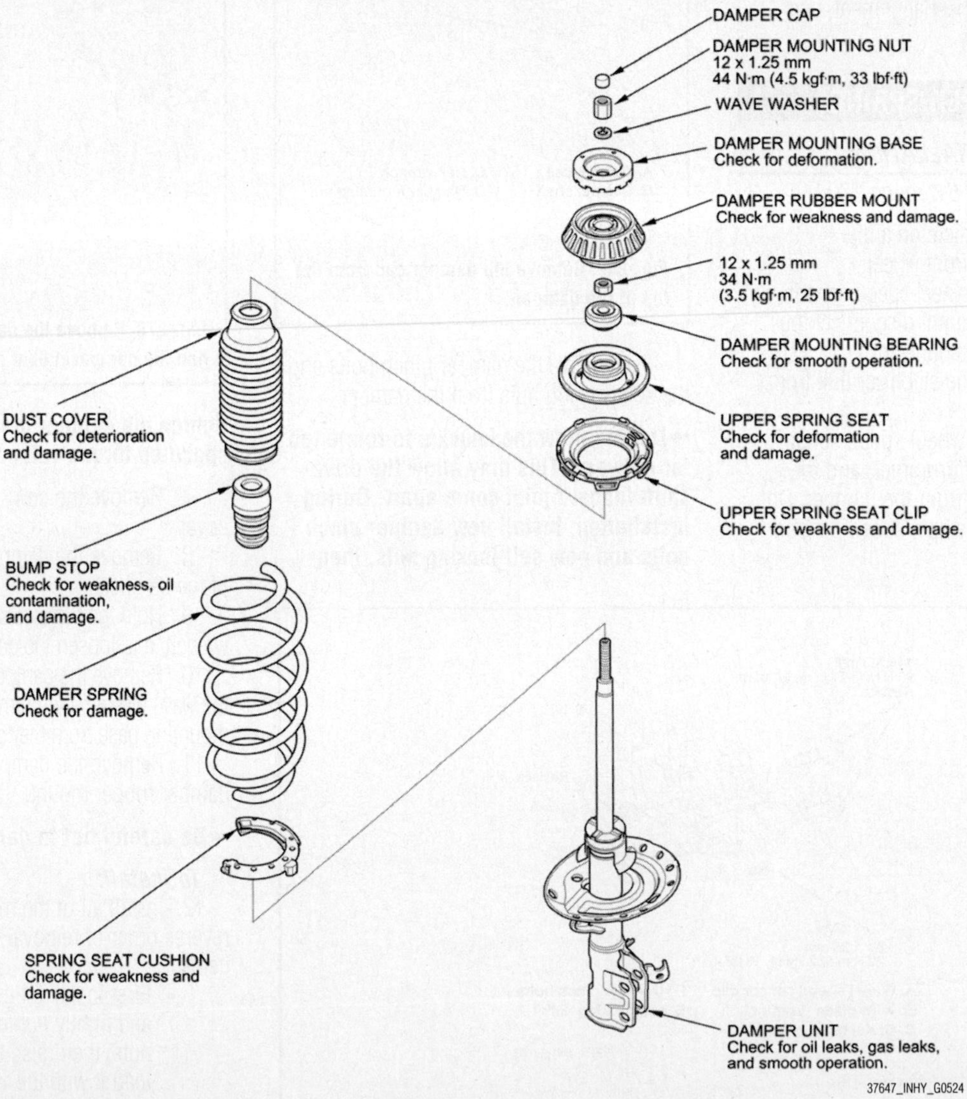
DAMPER CAP
DAMPER MOUNTING NUT
12 x 1.25 mm
44 N·m (4.5 kgf·m, 33 lbf·ft)
WAVE WASHER
DAMPER MOUNTING BASE
Check for deformation.
DAMPER RUBBER MOUNT
Check for weakness and damage.
12 x 1.25 mm
34 N·m
(3.5 kgf·m, 25 lbf·ft)
DAMPER MOUNTING BEARING
Check for smooth operation.
UPPER SPRING SEAT
Check for deformation and damage.
UPPER SPRING SEAT CLIP
Check for weakness and damage.
DUST COVER
Check for deterioration and damage.
BUMP STOP
Check for weakness, oil contamination, and damage.
DAMPER SPRING
Check for damage.
SPRING SEAT CUSHION
Check for weakness and damage.
DAMPER UNIT
Check for oil leaks, gas leaks, and smooth operation.
37647_INHY_G0524

Fig. 272 Exploded view of damper/spring assembly

Reassembly

See Figures 277 through 280.

1. Install the upper spring seat to the upper spring seat clip.

➡Make sure to securely set the upper spring seat to the hook on the spring seat clip.

2. Install the bump stop and the dust cover to the upper spring seat.

➡Push the bump stop and the dust cover into the dent of the upper spring seat securely.

3. Install the damper spring on the upper spring seat clip by aligning the upper end of the damper spring with the ledge portion of the upper spring seat clip.
4. Install the spring seat cushion on the damper spring.
5. Compress the damper spring.
6. Install all the parts except the nut and the damper bearing onto the damper unit by referring to the Exploded View.

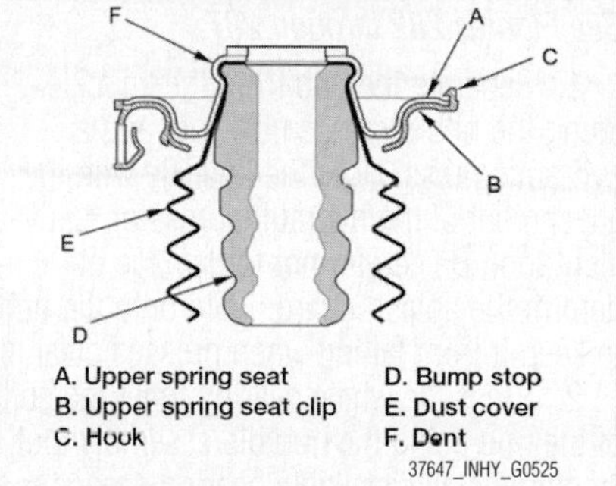
F
A
C
B
E
D
A. Upper spring seat
B. Upper spring seat clip
C. Hook
D. Bump stop
E. Dust cover
F. Dent
37647_INHY_G0525

Fig. 273 Install the upper spring seat to the upper spring seat clip

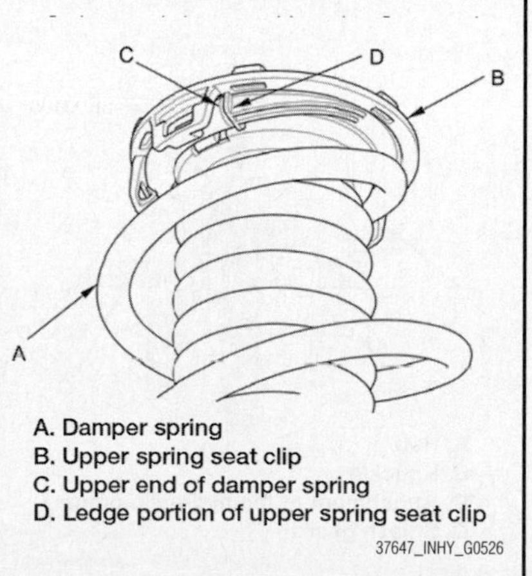
C
D
B
A
A. Damper spring
B. Upper spring seat clip
C. Upper end of damper spring
D. Ledge portion of upper spring seat clip
37647_INHY_G0526

Fig. 274 Install the damper spring on the upper spring seat clip by aligning the upper end of the damper spring with the ledge portion of the upper spring seat clip

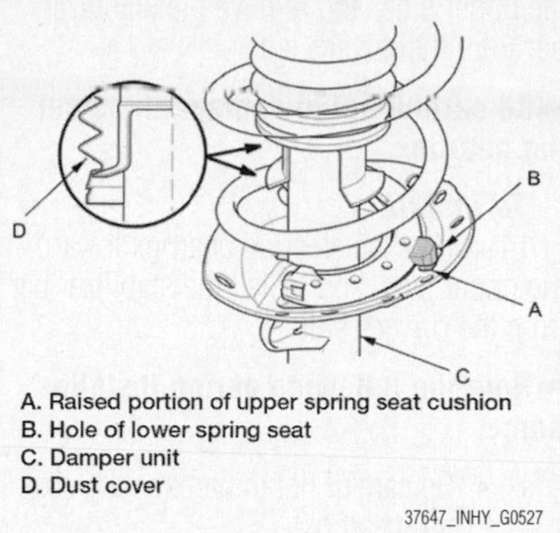
D
B
A
C
A. Raised portion of upper spring seat cushion
B. Hole of lower spring seat
C. Damper unit
D. Dust cover
37647_INHY_G0527

Fig. 275 Align the raised portion of the spring seat cushion and the hole of the lower spring seat on the damper unit

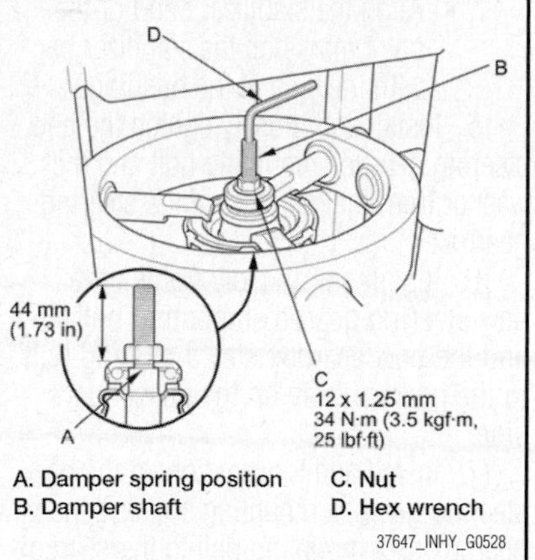
D
B
44 mm
(1.73 in)
A
C
12 x 1.25 mm
34 N·m (3.5 kgf·m,
25 lbf·ft)
A. Damper spring position
B. Damper shaft
C. Nut
D. Hex wrench
37647_INHY_G0528

Fig. 276 Compress the damper spring until the position of the damper shaft comes in contact with the damper bearing

7. Align the raised portion of the spring seat cushion and the hole of the lower spring seat on the damper unit.

➡After reassembling the damper/spring, install the dust cover into the damper unit.

8. Install the damper bearing onto the damper shaft.
9. Compress the damper spring until the position of the damper shaft comes in contact with the damper bearing. Do not excessively compress the damper spring.

➡Make sure the distance of the rod is 1.73 inches (44 mm) from the upper surface of the bearing.

10. Install the nut.
11. Hold the damper shaft with a hex wrench, and tighten the nut to 25 ft. lbs. (34 Nm).
12. Remove the damper/spring from the strut spring compressor.

STABILIZER BAR

REMOVAL & INSTALLATION

See Figures 277 through 281.

✲✲ WARNING

SRS components are located in this area. Review the SRS component locations, and the precautions and procedures before doing repairs or service.

1. Do the 12 volt battery terminal disconnection procedure.
2. Raise the vehicle on a lift.
3. Remove the front wheels.
4. Remove the steering wheel.
5. Disconnect the steering joint from the pinion shaft.

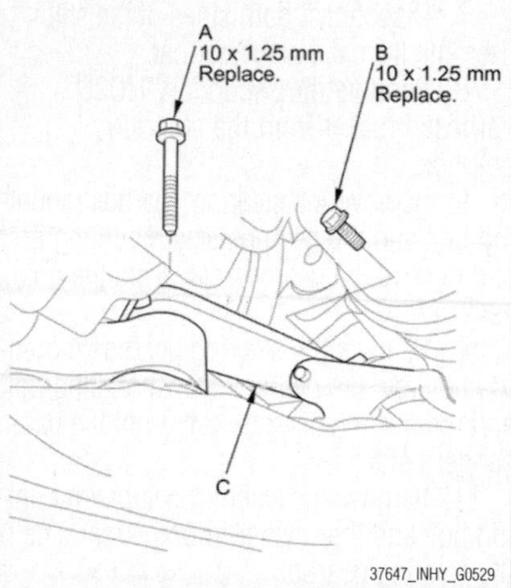
A
10 x 1.25 mm
Replace.
B
10 x 1.25 mm
Replace.
C
37647_INHY_G0529

Fig. 277 Remove the steering gearbox mounting bolt (A) and the gearbox stay mounting bolt (B), and remove the gearbox stay (C) from the driver's side

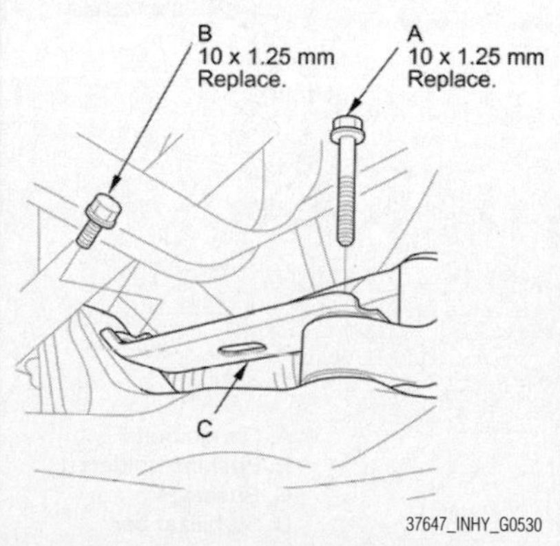
B
10 x 1.25 mm
Replace.
A
10 x 1.25 mm
Replace.
C
37647_INHY_G0530

Fig. 278 Remove the steering gearbox mounting bolt (A) and the gearbox stay mounting bolt (B), and remove the gearbox stay (C) from the passenger's side

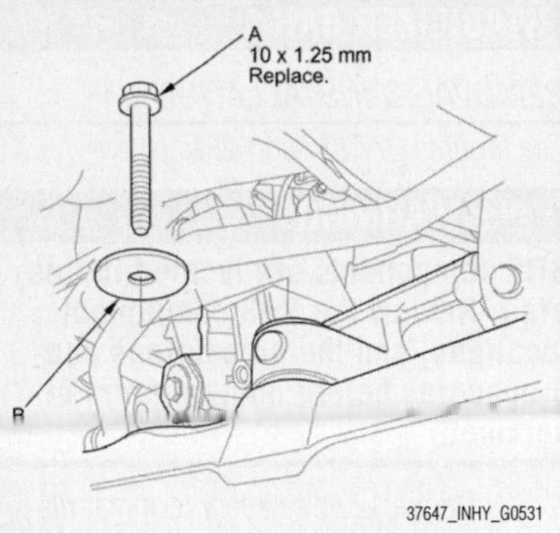
A
10 x 1.25 mm
Replace.
B
37647_INHY_G0531

Fig. 279 Remove the steering gearbox mounting bolt (A) and the washer (B) from the rear side of the steering gearbox

6. Disconnect both sides of the tie-rod end ball joint from the knuckle.

7. Disconnect both sides of the stabilizer link from the stabilizer bar.

8. Remove the secondary HO2S harness bracket from the steering gearbox.

9. Remove the steering gearbox mounting bolt and the gearbox stay mounting bolt, and remove the gearbox stay from the driver's side.

10. Remove the steering gearbox mounting bolt and the gearbox stay mounting bolt, and remove the gearbox stay from the passenger's side.

11. Remove the steering gearbox mounting bolt and the washer from the rear side of the steering gearbox.

12. Remove the flange bolts and the bushing holders, then remove the bushings.

13. Move the steering gearbox toward the upper side, and remove the stabilizer bar from the driver's side.

➡**Be careful not to damage the steering gearbox.**

To install:

14. Move the steering gearbox toward the upper side, and install the stabilizer bar from the driver's side.

➡**Note the following during installation:**

- Be careful not to damage the steering gearbox.
- Note the right and left direction of the stabilizer bar.
- Note the direction of installation for the bushings and the bushing holders.
- Align the stabilizer band or the paint marks on the stabilizer bar with the side of the bushings.

15. Install and loosely tighten the new steering gearbox mounting bolt and the washer from the rear side of the steering gearbox.

16. Install and loosely tighten the new steering gearbox mounting bolt and the new gearbox stay mounting bolt to the gearbox stay on the passenger's side.

17. Install and loosely tighten the new steering gearbox mounting bolt and the new gearbox stay mounting bolt to the gearbox stay on the driver's side.

18. Tighten the steering gearbox mounting bolts and the gearbox stay mounting bolts to 44 ft. lbs. (60 Nm).

19. Install the secondary HO2S harness bracket on the steering gearbox.

20. Connect both sides of the stabilizer link to the stabilizer bar.

21. Connect both sides of the tie-rod end ball joint to the knuckle.

22. Connect the steering joint.

23. Install the steering wheel.

24. Do the 12 volt battery terminal reconnection procedure, and do these tasks:

a. Turn the ignition switch to ON (II), and check that the SRS indicator should come on for about 6 seconds and then go off.

b. Make sure the horn and turn signal switches work properly.

c. Make sure the steering wheel switches work properly.

25. Clean the mating surfaces of the brake disc and the inside of the wheel, then install the front wheels.

26. Check the wheel alignment, and adjust it if necessary.

WHEEL HUB & BEARING

REMOVAL & INSTALLATION

See Figures 282 through 287.

1. Separate the hub from the knuckle using the hub dis/assembly tool and a hydraulic press. Hold the knuckle with the attachment of the hydraulic press or equivalent tool. Be careful not to damage or deform the splash guard. Hold onto the hub to keep it from falling when pressed clear.

2. Press the wheel bearing inner race off of the hub using the hub dis/assembly tool, a commercially available bearing separator, and a press.

3. Remove the splash guard and the snap ring from the knuckle.

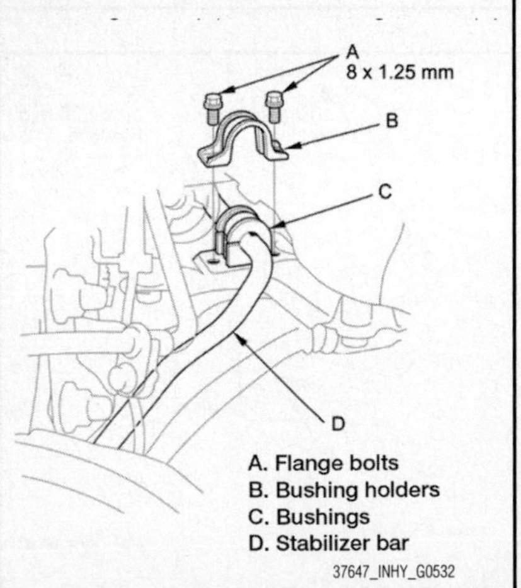
A
8 x 1.25 mm
B
C
D
A. Flange bolts
B. Bushing holders
C. Bushings
D. Stabilizer bar
37647_INHY_G0532

Fig. 280 Remove the flange bolts and the bushing holders, then remove the bushings

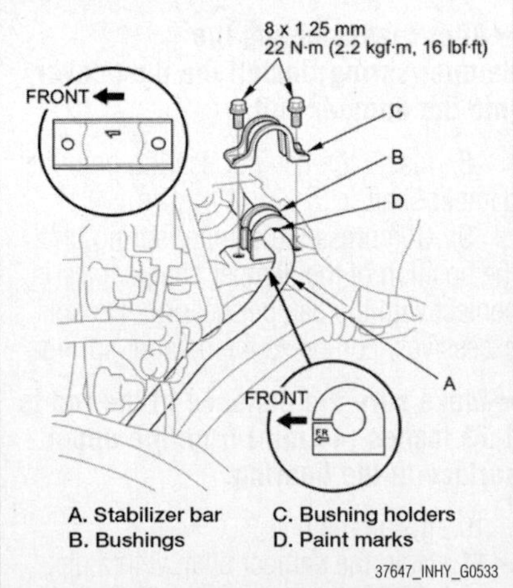
8 x 1.25 mm
22 N·m (2.2 kgf·m, 16 lbf·ft)
FRONT
C
B
D
A
FRONT
A. Stabilizer bar
B. Bushings
C. Bushing holders
D. Paint marks
37647_INHY_G0533

Fig. 281 Move the steering gearbox toward the upper side, and install the stabilizer bar from the driver's side

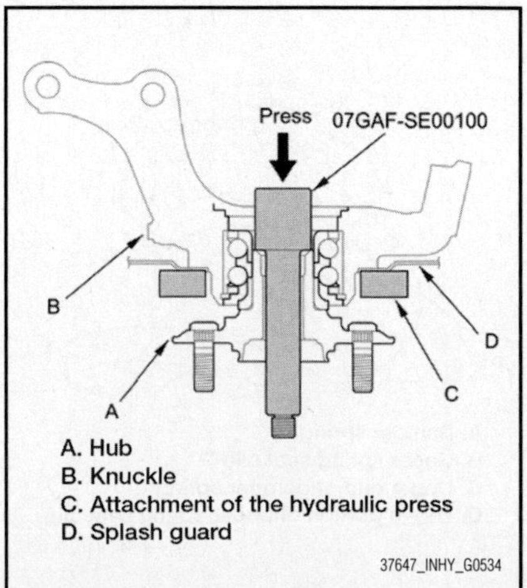
Press
07GAF-SE00100
B
D
A
C
A. Hub
B. Knuckle
C. Attachment of the hydraulic press
D. Splash guard
37647_INHY_G0534

Fig. 282 Separate the hub from the knuckle using the hub dis/assembly tool and a hydraulic press

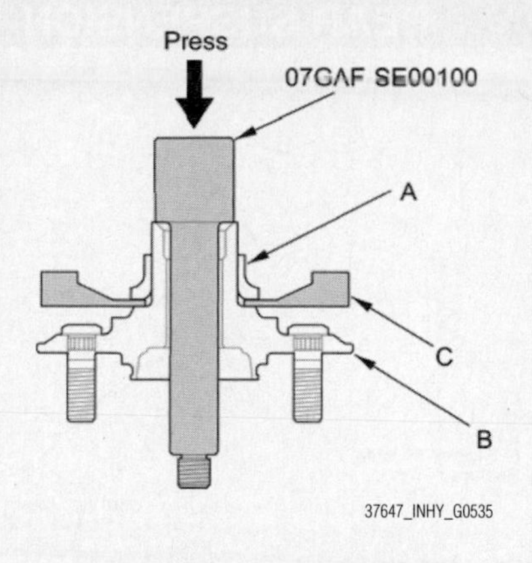
Press
07GAF SE00100
A
C
B
37647_INHY_G0535

Fig. 283 Press the wheel bearing inner race (A) off of the hub (B) using the hub dis/assembly tool, a bearing separator (C), and a press

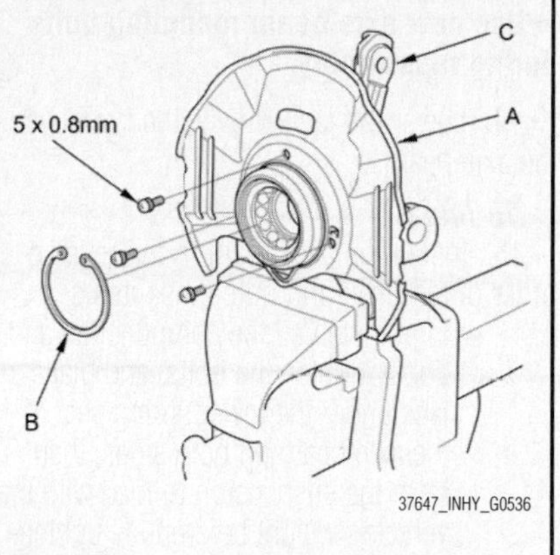
C
A
5 x 0.8mm
B
37647_INHY_G0536

Fig. 284 Remove the splash guard (A) and the snap ring (B) from the knuckle (C)

4. Press the wheel bearing out of the knuckle using the attachment, the driver handle, and a press.
5. Wash the knuckle and the hub thoroughly in high flash point solvent before reassembly.

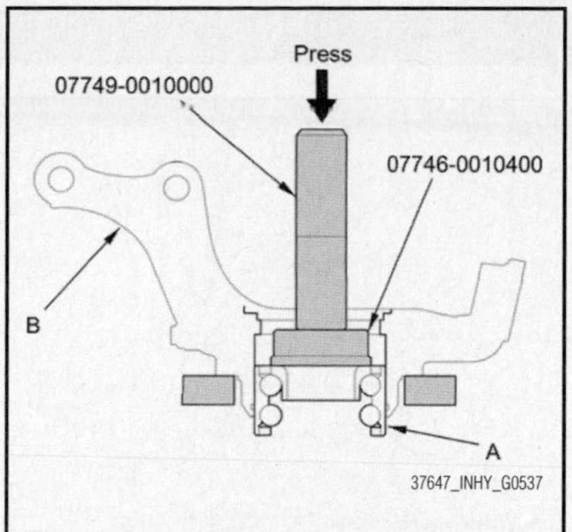
Press
07749-0010000
07746-0010400
B
A
37647_INHY_G0537

Fig. 285 Press the wheel bearing (A) out of the knuckle (B) using the attachment, the driver handle, and a press

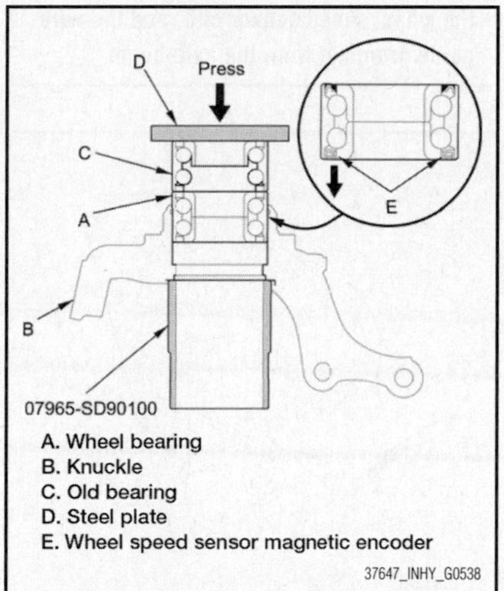
D
Press
C
A
E
B
07965-SD90100
A. Wheel bearing
B. Knuckle
C. Old bearing
D. Steel plate
E. Wheel speed sensor magnetic encoder
37647_INHY_G0538

Fig. 286 Press a new wheel bearing into the knuckle using the old bearing, a steel plate, the support base, and a press

6. Press a new wheel bearing into the knuckle using the old bearing, a steel plate, the support base, and a press.

➡**Note the following:**

- Install the wheel bearing with the wheel speed sensor magnetic encoder (brown color), toward the inside of the knuckle.
- Remove any oil, grease, dust, metal debris, and other foreign material from the magnetic encoder surface.
- Keep any magnetic tools away from the encoder surface.
- Be careful not to damage the encoder surface when you insert the wheel bearing.

7. Install the snap ring securely in the knuckle.
8. Install the splash guard, and tighten the screws.
9. Install the hub onto the knuckle using the attachment, the driver handle, the support base, and a hydraulic press. Be careful not to damage the splash guard.

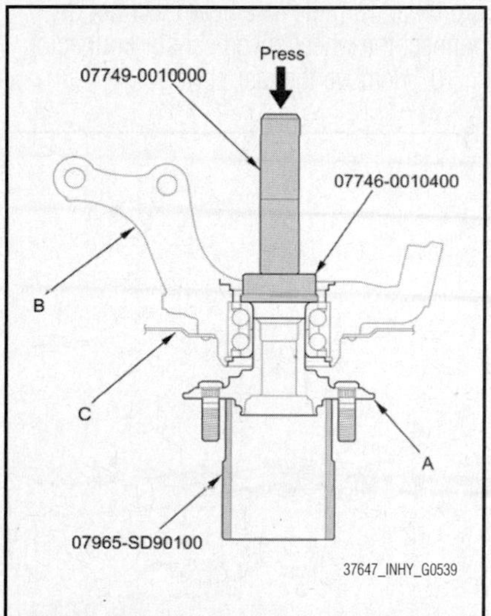
Press
07749-0010000
07746-0010400
B
C
A
07965-SD90100
37647_INHY_G0539

Fig. 287 Install the hub (A) onto the knuckle (B) using the attachment, the driver handle, the support base, and a hydraulic press

AXLE BEAM

REMOVAL & INSTALLATION

See Figures 288 through 292.

1. Raise the vehicle on a lift.
2. Remove the rear wheels.
3. Remove the rear suspension lower cover.
4. Remove the rear strake.
5. Remove the rear hub bearing unit.
6. Remove the parking brake cable from the axle beam.
7. Disconnect the brake line from the wheel cylinder, and plug the line with a shop towel.
8. Remove the backing plate with the brake shoes assembly from the spindle.
9. Remove the wheel speed sensor, the wheel speed sensor clip, and the wire guide grommet from the axle beam. Do not disconnect the wheel speed sensor connector.
10. Remove the rear spring.

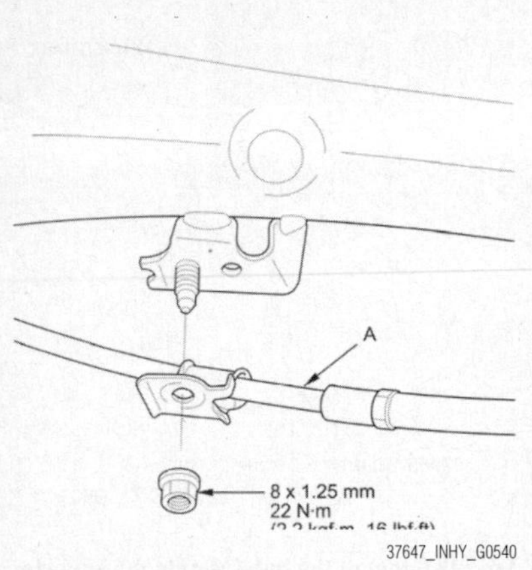
A
8 x 1.25 mm
22 N·m
37647_INHY_G0540

Fig. 288 Remove the parking brake cable (A) from the axle beam

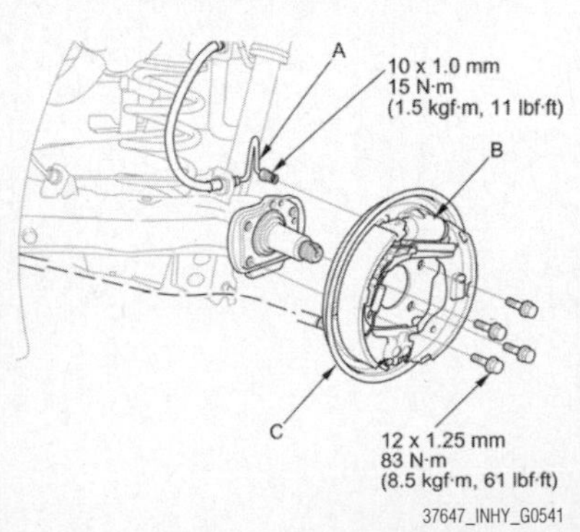
A
10 x 1.0 mm
15 N·m
(1.5 kgf·m, 11 lbf·ft)
B
C
12 x 1.25 mm
83 N·m
(8.5 kgf·m, 61 lbf·ft)
37647_INHY_G0541

Fig. 289 Disconnect the brake line (A) from the wheel cylinder (B); remove the backing plate (C) with the brake shoes assembly

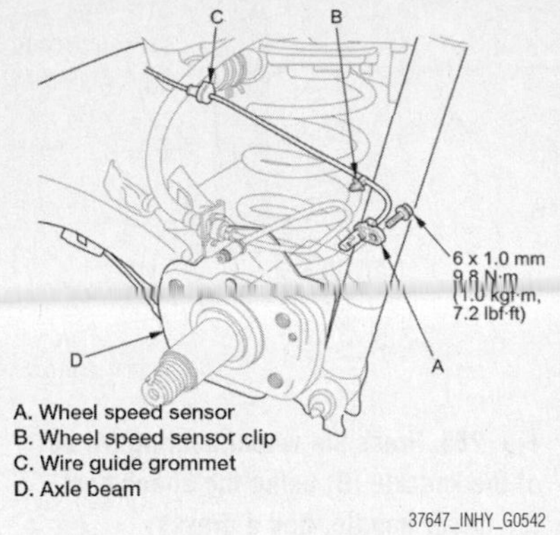
C
B
6 x 1.0 mm
9.8 N·m
(1.0 kgf·m,
7.2 lbf·ft)
D
A
A. Wheel speed sensor
B. Wheel speed sensor clip
C. Wire guide grommet
D. Axle beam
37647_INHY_G0542

Fig. 290 Remove the wheel speed sensor, the wheel speed sensor clip, and the wire guide grommet from the axle beam

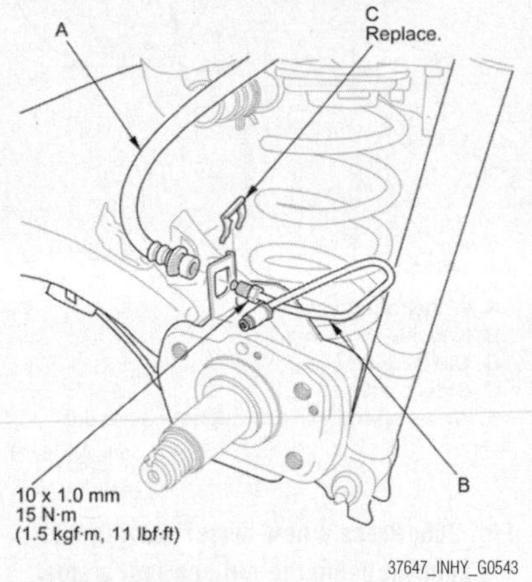
A
C
Replace.
10 x 1.0 mm
15 N·m
(1.5 kgf·m, 11 lbf·ft)
B
37647_INHY_G0543

Fig. 291 Disconnect the brake hose (A) from both sides of the brake line (B), then remove the brake line by removing the brake hose clip (C)

11. Disconnect the brake hose from both sides of the brake line, then remove the brake line by removing the brake hose clip.

➡Do not spill brake fluid on the vehicle; it may damage the paint; if brake fluid gets on the paint, wash it off immediately with water. Plug the end of a hose and joints to prevent spilling brake fluid. During installation, install new brake hose clips.

12. Place a floor jack under the lower spring seat on both sides of the axle beam, and support it by raising the floor jack. Do not place the floor jack under the center of the axle beam.
13. Remove the axle beam mounting bolt on both sides.

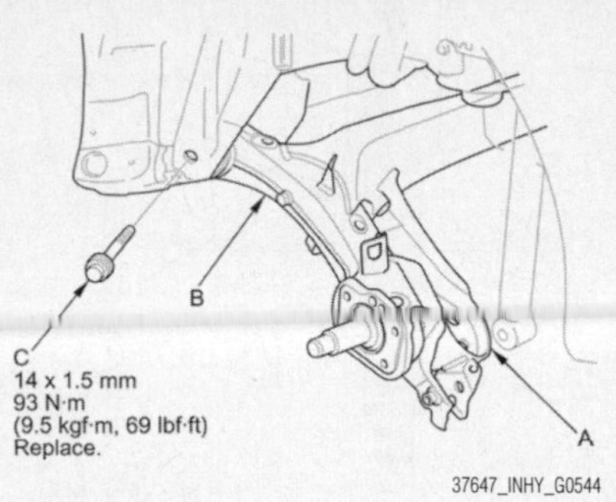
B
C
14 x 1.5 mm
93 N·m
(9.5 kgf·m, 69 lbf·ft)
Replace.
A
37647_INHY_G0544

Fig. 292 Place a floor jack under the lower spring seat (A) on both sides of the axle beam (B), and support it by raising the floor jack; remove the axle beam mounting bolt (C) on both sides

➡Use new axle beam mounting bolts during reassembly.

14. Lower the jack slowly, then remove the axle beam.

To install:

15. Install the axle beam in the reverse order of removal, and note these items:
    - First install all the components, and lightly tighten the bolts, and place a jack under the lower spring seat of the axle beam on both sides, then raise the suspension to load with the vehicle's weight before fully tightening to the specified torque values.
    - After installing the brake hose, the brake line, and the parking brake cable, check for interference and twisting of other parts.
    - Before installing the brake drum, clean the mating surfaces of the hub bearing unit and the inside of the brake drum.
    - After installing, fill the reservoir with new brake fluid, and bleed the brake system.
    - Check the brake hose and line joint for leaks, and tighten if necessary.
    - Before installing the wheel, clean the mating surfaces of the brake drum and the inside of the wheel.
16. Check the wheel alignment, and adjust it if necessary.

COIL SPRING

REMOVAL & INSTALLATION

See Figures 290, 293 through 297.

1. Raise the vehicle on a lift.
2. Remove the rear wheels.

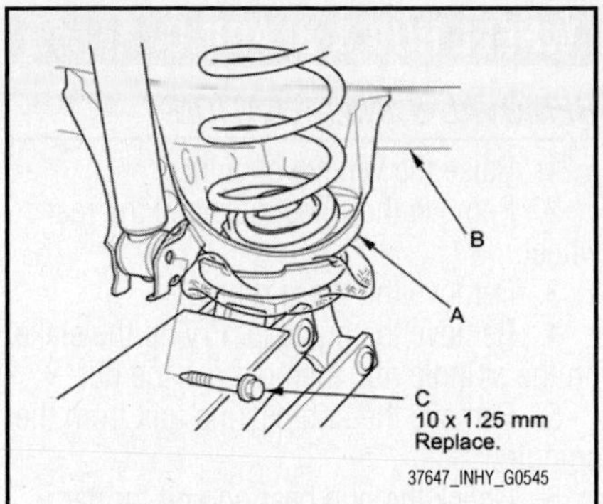
B
A
C
10 x 1.25 mm
Replace.
37647_INHY_G0545

Fig. 293 Position the floor jack under spring seat (A) on both sides of the axle beam (B); remove the damper mounting bolt (C) from both sides

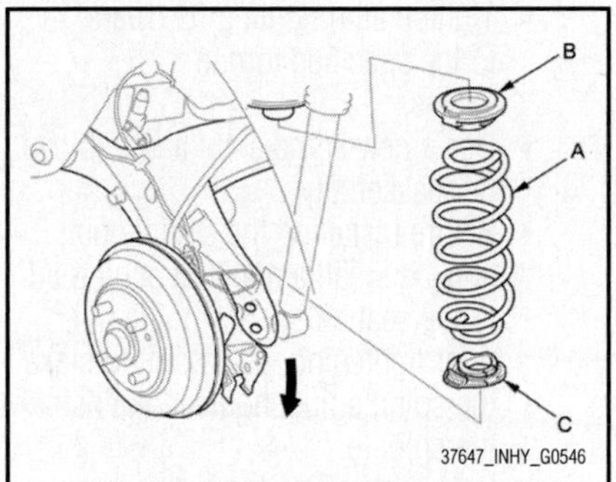
B
A
C
37647_INHY_G0546

Fig. 294 Remove the spring (A), the upper rubber mount (B), and the lower rubber mount (C)

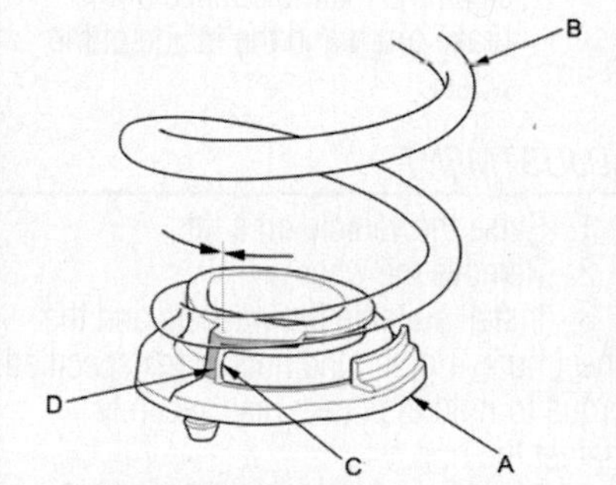
B
D
C
A

A. Upper rubber mount
B. spring
C. Upper end of spring
D. Ledge portion of upper rubber mount

37647_INHY_G0547

Fig. 295 Install the upper rubber mount on the spring by aligning the upper end of the spring with the ledge portion of the upper rubber mount

3. Remove the wheel speed sensor, the wheel speed sensor clip, and the wire guide grommet from both sides of the axle beam. Do not disconnect the wheel speed sensor connector.

4. Position the floor jack under spring seat on both sides of the axle beam. Raise the floor jack until the suspension begins to compress.

5. Remove the damper mounting bolt from both sides.

6. Lower the floor jack gradually.

7. Remove the spring, the upper rubber mount, and the lower rubber mount. Do not lower the jack more than necessary.

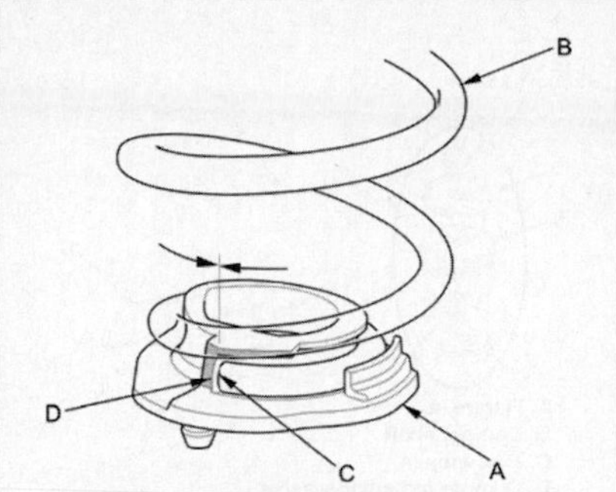
B
D
C
A

A. Lower rubber mount
B. Spring
C. Lower end of spring
D. Ledge portion of lower rubber mount

37647_INHY_G0548

Fig. 296 Install the lower rubber mount on the spring by aligning the lower end of the spring with the ledge portion of the lower rubber mount

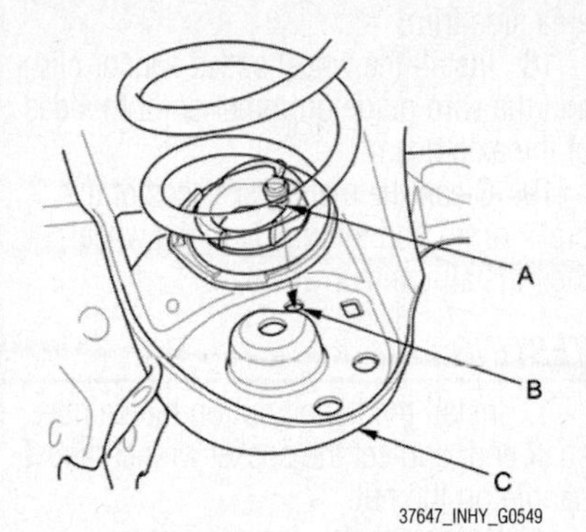
A
B
C
37647_INHY_G0549

Fig. 297 Install the tab (A) of the lower rubber mount into the groove (B) of the lower spring seat

To install:

8. Install the upper rubber mount on the spring by aligning the upper end of the spring with the ledge portion of the upper rubber mount.

9. Install the lower rubber mount on the spring by aligning the lower end of the spring with the ledge portion of the lower rubber mount.

10. Install the tab of the lower rubber mount into the groove of the lower spring seat.

➡Make sure that the tab of the lower rubber mount is properly installed into the axle beam. Make sure that the spring is installed correctly.

11. Position a floor jack under the lower spring seat on both sides of the axle beam.

12. Slowly raise the jacks until you can align the bolt hole with the holes in the axle beam and the damper, then loosely tighten the new damper mounting bolt on both sides.

13. Raise the rear suspension with the jacks until the vehicle just lifts off of the lift, then tighten the damper mounting bolts to the specified torque.

14. Install the wheel speed sensor, the wheel speed sensor clip, and the wire guide grommet on both sides of the axle beam.

15. Clean the mating surfaces of the brake drum and the inside of the wheel, then install the rear wheel.

SHOCK ABSORBER (DAMPER)

REMOVAL & INSTALLATION

See Figures 298 through 301.

1. Raise the vehicle on a lift.
2. Remove the rear wheels.
3. Remove the wheel speed sensor clip and the wire guide grommet from both sides of the axle beam.
4. Position a floor jack under lower spring seat on both sides of the axle beam. Raise the floor jack until the suspension begins to compress.
5. Remove the damper mounting bolt that connects the axle beam and the damper.
6. Remove the access panel on the cargo area side trim.
7. Remove the flange nut while holding the damper shaft with a hex wrench.
8. Remove the damper mounting washer and the damper rubber mount from the top of the damper.

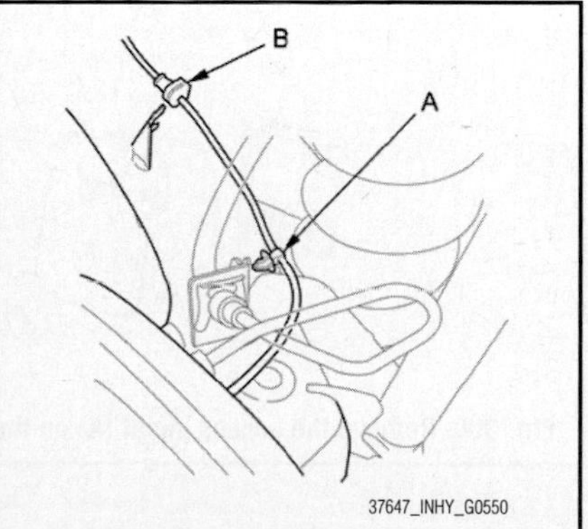
B
A
37647_INHY_G0550

Fig. 298 Remove the wheel speed sensor clip (A) and the wire guide grommet (B) from both sides of the axle beam

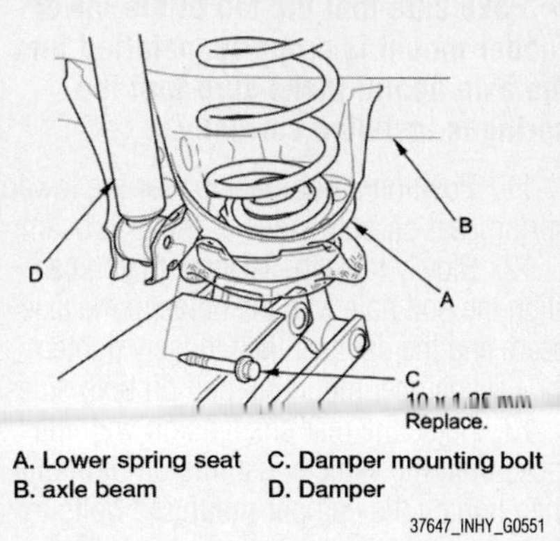
A
B
C
D
Replace.
A. Lower spring seat
B. axle beam
C. Damper mounting bolt
D. Damper
37647_INHY_G0551

Fig. 299 Position a floor jack under lower spring seat on both sides of the axle beam

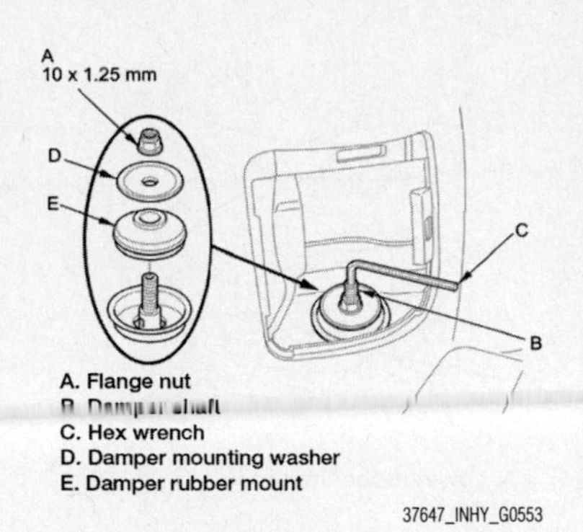
A
10 x 1.25 mm
D
E
C
B
A. Flange nut
C. Hex wrench
D. Damper mounting washer
E. Damper rubber mount
37647_INHY_G0553

Fig. 301 Remove the flange nut while holding the damper shaft with a hex wrench

9. Compress the damper assembly by hand, and remove it from the vehicle.

➡Be careful not to damage the body.

10. Remove the damper rubber mount.

To install:

11. Install the damper rubber mount onto the damper unit. Position the damper assembly between the body and the axle beam.
12. Position the floor jack under lower spring seat on both sides of the axle beam.
13. Slowly raise the jack until you can align the bolt hole with the holes in the axle beam and the damper, then loosely tighten the new damper mounting bolt on both sides.
14. Raise the rear suspension with the jack until the vehicle just lifts off of the lift, then tighten the damper mounting bolts to 40 ft. lbs. (54 Nm).
15. Install the damper rubber mount, the damper mounting washer, and the flange nut on the damper shaft.

➡During installation, note the direction of the damper rubber mount and the damper mounting washer.

16. Tighten the flange nut) to 22 ft. lbs. (29 Nm) while holding the damper shaft with a hex wrench.
17. Install the access panel on the cargo area side trim.
18. Install the wheel speed sensor clip and the wire guide grommet on both sides of the axle beam.
19. Clean the mating surfaces of the brake drum and the inside of the wheel, then install the rear wheel.

TESTING

1. Install the flange nut on the damper shaft end, and set the socket wrench and T-handle on the nut.
2. Compress the damper unit by hand, and check for smooth operation through a full stroke, both compression and extension. The damper should extend smoothly and constantly when compression is released. If it does not, the gas is leaking and the damper should be replaced.
3. Check for oil leaks, abnormal noises, and binding during these tests.

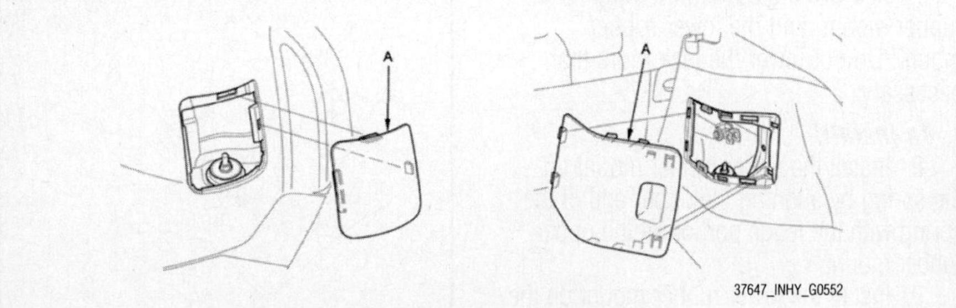
A
A
37647_INHY_G0552

Fig. 300 Remove the access panel (A) on the cargo area side trim

WHEEL HUB & BEARING

REMOVAL & INSTALLATION

1. Raise the vehicle on a lift.
2. Remove the wheel nuts and the rear wheel.
3. Remove the brake drum.
4. Remove the hub cap. Pry up the stake on the spindle nut, then remove the nut.
5. Remove the hub bearing unit from the spindle.
6. Check the hub bearing unit for damage and cracks.

To install:

7. Install the hub bearing unit in the reverse order of removal, and note these items:
   - Tighten all mounting hardware to the specified torque values.
   - Use a new spindle nut and hub cap on reassembly.
   - Before installing the spindle nut, apply a small amount of engine oil to the seating surface of the nut. After tightening, use a drift to stake the spindle nut shoulder against the spindle.
   - Before installing the brake drum, clean the mating surface of the hub bearing unit and the inside of the brake drum.
   - Before installing the wheel, clean the mating surface of the brake drum and the inside of the wheel.

ADJUSTMENT

1. Raise the vehicle on a lift.
2. Remove the wheels.
3. Install suitable flat washers and the wheel nuts. Tighten the nuts to the specified torque to hold the brake disc securely against the hub.
4. Attach the dial gauge. Place the dial gauge against the hub flange.
5. Measure the bearing end play while moving the brake disc or the brake drum inward and outward.

➡Wheel bearing end play: 0–0.002 inches (0–0.05 mm)

6. If the bearing end play measurement is more than the standard, replace the wheel bearing or the hub bearing unit.

HONDA

Odyssey

15

AUXILIARY HEATING & AIR CONDITIONING SYSTEM...15-68

Blower Motor 15-68
Removal & Installation........15-68
Heater Core 15-69
Removal & Installation........15-69

BRAKES 15-9

ANTI-LOCK BRAKE SYSTEM (ABS)........15-9
Wheel Speed Sensors 15-9
Removal & Installation........15-9
BLEEDING THE BRAKE SYSTEM........15-9
Bleeding Procedure........15-9
Bleeding Procedure 15-9
Bleeding the ABS System 15-9
FRONT DISC BRAKES........15-11
Brake Caliper........15-11
Removal & Installation........15-11
Disc Brake Pads........15-11
Removal & Installation........15-11
INFORMATION AND PRECAUTIONS 15-9
Anti-lock Systems........15-9
Disc and Drum Systems 15-9
PARKING BRAKE........15-16
Parking Brake Cables 15-16
Adjustment 15-16
Parking Brake Shoes 15-16
Adjustment 15-17
Removal & Installation........15-16
REAR DISC BRAKES 15-14
Brake Caliper........15-14
Removal & Installation........15-14
Disc Brake Pads........15-14
Removal & Installation........15-14

CHASSIS ELECTRICAL 15-19

AIR BAG (SUPPLEMENTAL RESTRAINT SYSTEM) 15-19
General Information........15-19
Arming the System 15-19
Clockspring Centering........15-19
Disarming the System........15-19
Service Precautions........15-19

DRIVE TRAIN 15-20

Front Halfshaft........15-20
CV-boots Inspection........15-22
Removal & Installation........15-22
Intermediate Shaft 15-24
Removal & Installation........15-24

ENGINE COOLING........15-25

Engine Cooling Fans & Shroud........15-25
Removal & Installation........15-25
Radiator........15-26
Removal & Installation........15-26
Thermostat 15-26
Removal & Installation........15-26
Water Pump 15-27
Removal & Installation........15-27

ENGINE ELECTRICAL 15-28

CHARGING SYSTEM 15-28
Alternator 15-28
Removal & Installation........15-28
IGNITION SYSTEM 15-28
Firing Order........15-28
Ignition Coil 15-28
Removal & Installation........15-28
Ignition Timing........15-28
Adjustment 15-28
Inspection 15-28
Spark Plugs........15-28
Removal & Installation........15-28
STARTING SYSTEM 15-29
Starter 15-29
Removal & Installation........15-29

ENGINE MECHANICAL......15-29

Accessory Drive Belts 15-29
Accessory Belt Routing.......15-29
Adjustment 15-30
Inspection 15-30
Removal & Installation........15-30
Camshaft and Valve Lifters........15-30
Inspection 15-30
Removal & Installation........15-30
Crankshaft Damper........15-32
Removal & Installation........15-32
Crankshaft Front Seal........15-33
Removal & Installation........15-33
Cylinder Head 15-33
Removal & Installation........15-33
Driveplate........15-35
Removal & Installation........15-35
Exhaust Manifold 15-35
Removal & Installation........15-35
Intake Manifold 15-35
Removal & Installation........15-35
Oil Pan 15-40
Removal & Installation........15-40
Oil Pump........15-41
Inspection 15-41
Removal & Installation........15-41
Piston and Ring........15-44
Positioning 15-44
Rear Main Seal........15-44
Removal & Installation........15-44
Rocker Arms/Shafts........15-45
Removal & Installation........15-45
Timing Belt & Sprockets 15-45
Removal & Installation........15-45
Timing Belt Front Cover........15-45
Removal & Installation........15-45
Valve Covers 15-45
Removal & Installation........15-45
Valve Lash........15-54
Adjustment 15-54

ENGINE PERFORMANCE & EMISSION CONTROLS 15-56

Accelerator Pedal Position (APP) Sensor........15-56
Location........15-56
Removal & Installation........15-56
Camshaft Position (CMP) Sensor 15-56
Location........15-56
Removal & Installation........15-56
Crankshaft Position (CKP) Sensor 15-56
CKP Pattern Clear/ckp Pattern Learn........15-56
Location........15-56
Removal & Installation........15-56

Engine Coolant
Temperature (ECT) Sensor....15-58
Location....15-58
Removal & Installation....15-58
Evaporative Emissions
(EVAP) Canister....15-57
Location....15-57
Removal & Installation....15-57
Exhaust Gas Recirculation
(EGR) Valvo....15-58
Location....15-58
Removal & Installation....15-58
Knock Sensor (KS)....15-59
Location....15-59
Removal & Installation....15-59
Malfunction Indicator Light
(MIL)....15-59
Reset Procedures....15-59
Manifold Absolute Pressure
(MAP) Sensor....15-59
Location....15-59
Removal & Installation....15-59
Mass Air Flow/Intake Air
Temperature (MAF/IAT)
Sensor....15-59
Location....15-59
Removal & Installation....15-59
Output (Countershaft) Speed
Sensor (OSS)....15-59
Location....15-59
Removal & Installation....15-59
Powertrain Control
Module (PCM)....15-60
Location....15-60
PCM IDLE Learn
Procedure....15-62
Removal & Installation....15-60
Variable Camshaft Timing
Oil Control Solenoid....15-62
Location....15-62
Removal & Installation....15-63

FUEL....15-63

GASOLINE FUEL INJECTION SYSTEM....15-63
Fuel Filter....15-64
Removal & Installation....15-64
Fuel Level Sending Unit....15-64
Removal & Installation....15-64
Fuel Pump....15-64
Removal & Installation....15-64
Fuel Rail & Injectors....15-64
Removal & Installation....15-64
Fuel System Service
Precautions....15-63
Fuel Tank....15-64
Draining....15-64
Removal & Installation....15-64
Idle Speed....15-65
Adjustment....15-65
Inspection....15-65
Relieving Fuel System
Pressure....15-63
Throttle Body....15-65
Removal & Installation....15-65

HEATING & AIR CONDITIONING SYSTEM...15-67

Blower Unit....15-67
Removal & Installation....15-67
Heater Unit & Heater
Core....15-67
Removal & Installation....15-67

SPECIFICATIONS AND MAINTENANCE CHARTS.....15-3

Brake Specifications....15-7
Camshaft and Bearing
Specifications....15-5
Capacities....15-4
Crankshaft and Connecting
Rod Specifications....15-5
Engine and Vehicle
Identification....15-3
Engine Tune-Up
Specifications....15-3
Fluid Specifications....15-4
General Engine
Specifications....15-3
Piston and Ring
Specifications....15-5
Maintenance Minder
Schedule....15-8
Tire, Wheel and Ball Joint
Specifications....15-7
Torque Specifications....15-6
Valve Specifications....15-4
Wheel Alignment....15-6

STEERING....15-71

Power Rack & Pinion
Steering Gear....15-71
Removal & Installation....15-71
Power Steering Pump....15-76
Bleeding....15-77
Removal & Installation....15-76

SUSPENSION....15-78

FRONT SUSPENSION....15-78
Lower Ball Joint....15-78
Removal & Installation....15-78
Lower Control Arm....15-78
Removal & Installation....15-78
MacPherson Strut....15-79
Overhaul....15-79
Removal & Installation....15-79
Stabilizer Bar....15-82
Removal & Installation....15-82
Steering Knuckle,
Hub & Wheel
Bearing....15-80
Removal & Installation....15-80
Wheel Bearings....15-83
Adjustment....15-83
Inspection....15-83
Removal & Installation....15-83

REAR SUSPENSION....15-84
Coil Spring....15-84
Removal & Installation....15-84
Control Arms/Links....15-84
Removal & Installation....15-84
Shock Absorber....15-85
Removal & Installation....15-85
Wheel Hub and Bearing
(Sealed Unit)....15-86
Adjustment....15-86
Removal & Installation....15-86

SPECIFICATIONS AND MAINTENANCE CHARTS

ENGINE AND VEHICLE IDENTIFICATION CHART

Engine Code							Model Year	
Code	Liters (cc)	Cu. In.	Cyl.	Fuel Sys.	Engine Type	Eng. Mfg.	Code ①	Year
J35A6	3.5 (3471)	212	6	SMFI	SOHC	Honda	9	2009
J35A7 ②	3.5 (3471)	212	6	SMFI	SOHC	Honda	A	2010

SOHC: Single Overhead Cam

SMFI: Sequential Multi-port Fuel Injection

① 10th position of VIN

② Variable cylinder management

37647_ODYS_C0001

GENERAL ENGINE SPECIFICATIONS

Year	Model	Engine Displacement Liters	Engine ID	Net Horsepower @ rpm	Net Torque @ rpm (ft. lbs.)	Bore x Stroke (in.)	Com-pression Ratio	Oil Pressure @ rpm
2009	Odyssey ①	3.5	J35A6	244@5750	240@5000	3.50x3.66	10.0:1	71@3000
	Odyssey ②	3.5	J35A7	244@5750	240@4500	3.50x3.66	10.0:1	71@3000
2010	Odyssey ①	3.5	J35A6	241@5700	242@4900	3.50x3.66	10.0:1	71@3000
	Odyssey ②	3.5	J35A7	244@5750	240@5000	3.50x3.66	10.5:1	71@3000

① LX and EX models

② EX-L and EX-L Touring models

37647_ODYS_C0002

ENGINE TUNE-UP SPECIFICATIONS

Year	Engine Displacement Liters	Engine ID	Spark Plug Gap (in.)	Ignition Timing (deg.)		Fuel Pump (psi)	Idle Speed (rpm)		Valve Clearance (in.)	
				MT	AT		MT	AT	In.	Ex.
2009	3.5	J35A6 J35A7	0.039-0.043	—	8-12B	55-63	—	600-700	0.008-0.009	0.011-0.013
2010	3.5	J35A6 J35A7	0.039-0.043	—	8-12B	55-63	—	600-700	0.008-0.009	0.011-0.013

NOTE: The Vehicle Emission Control Information label often reflects changes made during production and must be used if they differ from this chart.

NOTE: The fuel pressure readings are given with the vacuum hose connected to the regulator and the engine running

B: Before top dead center

37647_ODYS_C0003

CAPACITIES

Year	Model	Engine Displacement Liters	Engine ID	Engine Oil with Filter (qts.)	Transmission (qts.) 5-Spd	Transmission (qts.) Auto.	Fuel Tank (gal.)	Cooling System (qts.)
2009	Odyssey	3.5	J35A6/J35A7	4.5	—	6.6	21.0	7.4
2010	Odyssey	3.5	J35A6/J35A7	4.5	—	6.6	21.0	7.4

NOTE: All capacities are approximate. Add fluid gradually and check to be sure a proper fluid level is obtained.

37647_ODYS_C0004

FLUID SPECIFICATIONS

Year	Model	Engine Displ. Liters	Engine Oil	Man. Trans.	Auto. Trans.	Drive Axle Front	Drive Axle Rear	Power Steering Fluid	Brake Master Cylinder	Cooling System
2009	Odyssey	3.5	5W-20 Honda	N/A	Honda ATF-Z1	—	—	Honda PS Fluid	Honda DOT 3	①
2010	Odyssey	3.5	5W-20 Honda	N/A	Honda ATF-Z1	—	—	Honda PS Fluid	Honda DOT 3	①

DOT: Department Of Transpotation

① Honda Long Life Antifreeze/Coolant-Type2

37647_ODYS_C0005

VALVE SPECIFICATIONS

Year	Engine Displacement Liters	Engine ID	Seat Angle (deg.)	Face Angle (deg.)	Spring Test Pressure (lbs. @ in.)	Spring Installed Height (in.)	Stem-to-Guide Clearance (in.) Intake	Stem-to-Guide Clearance (in.) Exhaust	Stem Diameter (in.) Intake	Stem Diameter (in.) Exhaust
2009	3.5	J35A6	NA	NA	NA	①	0.0020-0.0040	0.0040-0.0060	0.2159-0.2163	0.2146-0.2150
	3.5	J35A7	NA	NA	NA	②	0.0020-0.0040	0.0040-0.0060	0.2159-0.2163	0.2146-0.2150
2010	3.5	J35A6	NA	NA	NA	①	0.0020-0.0040	0.0040-0.0060	0.2159-0.2163	0.2146-0.2150
	3.5	J35A7	NA	NA	NA	③	0.0020-0.0040	0.0040-0.0060	0.2159-0.2163	0.2146-0.2150

NA: Not Available

① Valve spring free length:
Intake: 2.029 in.
Exhaust: 2.010 in.

② Valve spring free length:
Intake: 2.069 in.
Exhaust: 2.069 in.

③ Valve spring free length:
Intake: 2.027 in.
Exhaust: 2.010 in.

37647_ODYS_C0007

CAMSHAFT AND BEARING SPECIFICATIONS

All measurements are given in inches.

Year	Engine Displacement Liters	Engine VIN	Journal Diameter	Brg. Oil Clearance	Shaft End-play	Runout	Journal Bore	Lobe Height	
								Intake	Exhaust
2009	3.5	J35A6	NA	0.0020-0.0035	0.0020-0.0080	0.0010	NA	①	1.4302
	3.5	J35A7	NA	0.0020-0.0035	0.0020-0.0080	0.0010	NA	②	③
2010	3.5	J35A6	NA	0.0020-0.0035	0.0020-0.0080	0.0010	NA	①	1.4302
	3.5	J35A7	NA	0.0020-0.0035	0.0020-0.0080	0.0010	NA	④	⑤

NA: Information not available

① Primary: 1.3796 in.
Mid: 1.4348 in.
Secondary: 1.3891 in.

② Front: 1.4232 in.
Rear: 1.3939 in.

③ Front: 1.4120 in.
Rear: 1.4581 in.

④ Intake Nos. 1-4: 1.3843 in.
Intake Nos. 5-6: 1.3841 in.

⑤ Intake Nos. 1-4: 1.4385 in.
Intake Nos. 5-6: 1.4378 in.

37647_ODYS_C0006

CRANKSHAFT AND CONNECTING ROD SPECIFICATIONS

All measurements are given in inches

Year	Engine Displacement Liters	Engine ID	Crankshaft				Connecting Rod		
			Main Brg. Journal Dia.	Main Brg. Oil Clearance	Shaft End-play	Thrust on No.	Journal Diameter	Oil Clearance	Side Clearance
2009	3.5	J35A6 J35A7	2.8337-2.8346	0.0008-0.0017	0.0040-0.0140	3	2.1644-2.1654	0.0008-0.0017	0.0060-0.0140
2010	3.5	J35A6 J35A7	2.8337-2.8346	0.0007-0.0017	0.0040-0.0140	3	2.1644-2.1654	0.0008-0.0017	0.0060-0.0140

37647_ODYS_C0008

PISTON AND RING SPECIFICATIONS

All measurements are given in inches

Year	Engine Displacement Liters	Engine ID	Piston Clearance	Ring Gap			Ring Side Clearance		
				Top Compression	Bottom Compression	Oil Control	Top Compression	Bottom Compression	Oil Control
2009	3.5	J35A6 J35A7	0.0006-0.0016	0.0080-0.0140	0.0160-0.0220	0.0080-0.0280	0.0022-0.0031	0.0012-0.0022	NA
2010	3.5	J35A6 J35A7	0.0006-0.0016	0.0080-0.0140	0.0160-0.0220	0.0080-0.0280	0.0022-0.0031	0.0012-0.0022	NA

NA: Not Applicable

37647_ODYS_C0009

TORQUE SPECIFICATIONS

All readings in ft. lbs.

Year	Engine Displacement Liters	Engine ID	Cylinder Head Bolts	Main Bearing Bolts	Rod Bearing Bolts	Crankshaft Damper Bolts	Flywheel Bolts	Manifold Intake	Manifold Exhaust	Spark Plugs	Oil Pan Drain Plug
2009	3.5	J35A6	①	②	③	④	54	16	23	13	29
		J35A7	①	②	③	④	54	16	23	13	29
2010	3.5	J35A6	①	②	③	④	54	16	23	13	29
		J35A7	①	②	③	④	54	16	23	13	29

NOTE: Dip main bearing bolts and crankshaft damper bolt in clean engine oil prior to tightening.

① 12-point head bolts
Step 1: Torque all bolts to 22 ft. lbs.
Step 2: Torque all bolts an additional 90 deg.
Step 3: Torque all bolts an additional 90 deg.
New Bolt Only: an additional 90 deg.

② Cap bolts: 54 ft. lbs.
Side bolts: 36 ft. lbs.

③ Step 1: 14 ft. lbs.
Step 2: 90 degrees

37647_ODYS_C0010

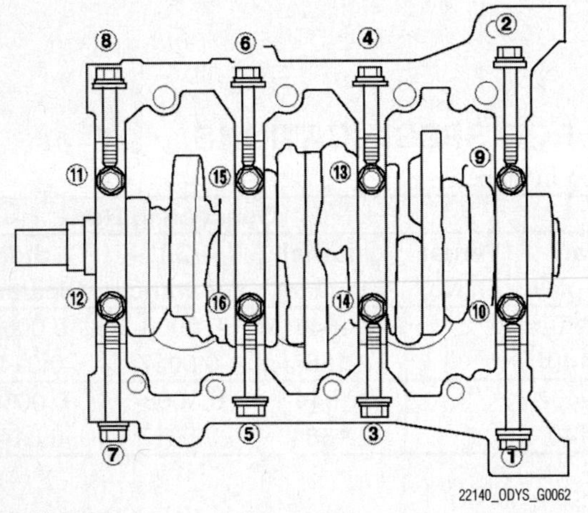
8
6
4
2
11
15
13
9
12
16
14
10
7
5
3
1
22140_ODYS_G0062

Fig. 1 Main Bearing Torque Sequence

WHEEL ALIGNMENT

Year	Model		Caster Range (+/-Deg.)	Caster Preferred Setting (Deg.)	Camber Range (+/-Deg.)	Camber Preferred Setting (Deg.)	Toe-in (in.)
2009	Odyssey	F	1.00	+2.53	0.50	0	0+/-0.08
		R	—	—	0.75	-0.50	0.08+/-0.08
2010	Odyssey	F	1.00	+2.53	0.50	0	0+/-0.08
		R	—	—	0.75	-0.50	0.08+/-0.08

37647_ODYS_C0011

TIRE, WHEEL AND BALL JOINT SPECIFICATIONS

Year	Model	OEM Tires		Tire Pressures (psi)		Wheel	Ball Joint	Lug Nut Torque
		Standard	Optional	Front	Rear	Size	Inspection	(ft. lbs.)
2009	Except Touring	P235/65R16	None	①	①	7.0	NS	94
	Touring	P235-710R460A	None	①	①	7.0	NS	94
2010	Except Touring	P235/65R16	None	①	①	7.0	NS	94
	Touring	P235-710R460A	None	①	①	7.0	NS	94

OEM: Original Equipment Manufacturer

PSI: Pounds Per Square Inch

STD: Standard

OPT: Optional

NS: Not specified by manufacturer

① See placard on vehicle

37647_ODYS_C0012

BRAKE SPECIFICATIONS

All measurements in inches unless noted

Year	Model		Brake Disc Original Thickness	Brake Disc Minimum Thickness	Brake Disc Maximum Runout	Minimum Lining Thickness Front	Minimum Lining Thickness Rear	Brake Caliper Bracket Bolts (ft. lbs.)	Brake Caliper Mounting Bolts (ft. lbs.)
2009	Odyssey	F	1.100	1.020	0.004	0.060	—	101	37
		R	0.440	0.350	0.004	—	0.060	65	16
2010	Odyssey	F	1.100	1.020	0.004	0.060	—	101	37
		R	0.440	0.350	0.004	—	0.060	65	16

F: Front

R: Rear

① Parking brake shoe

37647_ODYS_C0013

MAINTENANCE MINDER SCHEDULE

Honda Odyssey

The Odyssey displays engine oil life and maintenance service items in the information display to indicate when to perform maintenance service. If the engine oil life is 15% or less, based on the onboard computer's caluculations, you will see SERVICE DUE SOON in the information display every time the ignition key is turned to ON. The maintenance minder indicator will also come on and the maintenance code(s) for other scheduled maintenance items needing service will be displayed below the message.

Symbol	Item	Service
A	Engine oil ①	Change
B	Engine oil and filter	Change
	Tires	Rotate
	Brakes	Inspect
	Parking brake adjustment	Check
	Steering gear and linkage	Inspect
	Suspension components	Inspect
	Driveshaft boots	Inspect
	Brake hoses and lines	Inspect
	All fluid levels and condition	Inspect
	Exhaust system	Inspect
	Fuel lines and connections	Inspect
1	Tires	Rotate
2	Engine air filter ②	Replace
	Dust and pollen filter ③	Replace
	Accessory drive belt	Inspect
3	Transmission fluid ④	Replace
4	Spark plugs	Replace
	Timing belt ⑤	Replace
	Water pump	Inspect
	Valve clearance ⑥	Inspect
5	Engine coolant	Replace

① If the message SERVICE DUE NOW does not appear more than 12 months after the display is reset, change every year.
② If driven in dusty conditions, replace every 15,000 miles.
③ If driven in urban areas that have a high concentration of soot from industry and diesel, replace every 15,000 miles
④ If regularly driven in mountainous areas at very low speed or trailer towing, change the fluid every 30,000 miles.
⑤ If driven regularly in temperatures over 110 deg.F or below -20 deg.F, or towing a trailer, replace every 60,000 miles.
⑥ Adjust if necessary.

Additionally, replace the brake fluid every 3 years, and inspect the idle speed every 160,000 miles.

To reset the Engine Oil Life Display on LX, EX and EX-L models:

1. Turn the ignition switch to ON.
2. Press the SELECT/RESET button repeatedly until the engine oil life display or the service message is displayed.
3. Press the SELECT/RESET button for about 10 seconds. The engine oil life idicator and the codes will blink.
4. Press the SELECT/RESET knob for more than 5 seconds. The codes will disappear and the indicator will reset to 100.

To reset the Engine Oil Life Display on Touring models:

1. Turn the ignition switch to ON.
2. Press the SEL/RESET button on the steering wheel until the engine oil life is displayed.
3. Press the SEL/RESET button on the steering wheel for 10 seconds. The display will change to CUSTOM SETUP mode.
4. Press the SEL/RESET button on the steering wheel. The codes will disappear and the indicator will reset to 100.

37647_ODYS_C0014

BRAKES — INFORMATION AND PRECAUTIONS

ANTI-LOCK SYSTEMS

- Certain components within the ABS system are not intended to be serviced or repaired individually.
- Do not use rubber hoses or other parts not specifically specified for and ABS system. When using repair kits, replace all parts included in the kit. Partial or incorrect repair may lead to functional problems and require the replacement of components.
- Lubricate rubber parts with clean, fresh brake fluid to ease assembly. Do not use shop air to clean parts; damage to rubber components may result.
- Use only DOT 3 brake fluid from an unopened container.
- If any hydraulic component or line is removed or replaced, it may be necessary to bleed the entire system.
- A clean repair area is essential. Always clean the reservoir and cap thoroughly before removing the cap. The slightest amount of dirt in the fluid may plug an orifice and impair the system function. Perform repairs after components have been thoroughly cleaned; use only denatured alcohol to clean components. Do not allow ABS components to come into contact with any substance containing mineral oil; this includes used shop rags.
- The Anti-Lock control unit is a microprocessor similar to other computer units in the vehicle. Ensure that the ignition switch is **OFF**before removing or installing controller harnesses. Avoid static electricity discharge at or near the controller.
- If any arc welding is to be done on the vehicle, the control unit should be unplugged before welding operations begin.

DISC AND DRUM SYSTEMS

CAUTION

Dust and dirt accumulating on brake parts during normal use may contain asbestos fibers from production or aftermarket brake linings. Breathing excessive concentrations of asbestos fibers can cause serious bodily harm. Exercise care when servicing brake parts. Do not sand or grind brake lining unless equipment used is designed to contain the dust residue. Do not clean brake parts with compressed air or by dry brushing. Cleaning should be done by dampening the brake components with a fine mist of water, then wiping the brake components clean with a dampened cloth. Dispose of cloth and all residue containing asbestos fibers in an impermeable container with the appropriate label. Follow practices prescribed by the Occupational Safety and Health Administration (OSHA) and the Environmental Protection Agency (EPA) for the handling, processing, and disposing of dust or debris that may contain asbestos fibers.

BRAKES — BLEEDING THE BRAKE SYSTEM

BLEEDING PROCEDURE

BLEEDING PROCEDURE

See Figures 2 through 4.

➡Do not reuse the drained fluid. Use only clean Honda DOT 3 Brake Fluid from an unopened container. Using a non-Honda brake fluid can cause corrosion and shorten the life of the system.

Do not mix different brands of brake fluid; they may not be compatible.

Make sure no dirt or other foreign matter is allowed to contaminate the brake fluid.

WARNING

Do not spill brake fluid on the vehicle, it may damage the paint; if brake fluid does contact the paint, wash it off immediately with water.

The reservoir on the master cylinder must be at the MAX (upper) level mark at the start of the bleeding procedure and checked after bleeding each brake caliper. Add fluid as required.

1. Make sure the brake fluid level in the reservoir is at the MAX (upper) level line.
2. Attach a length of clear drain tube to the bleed screw.

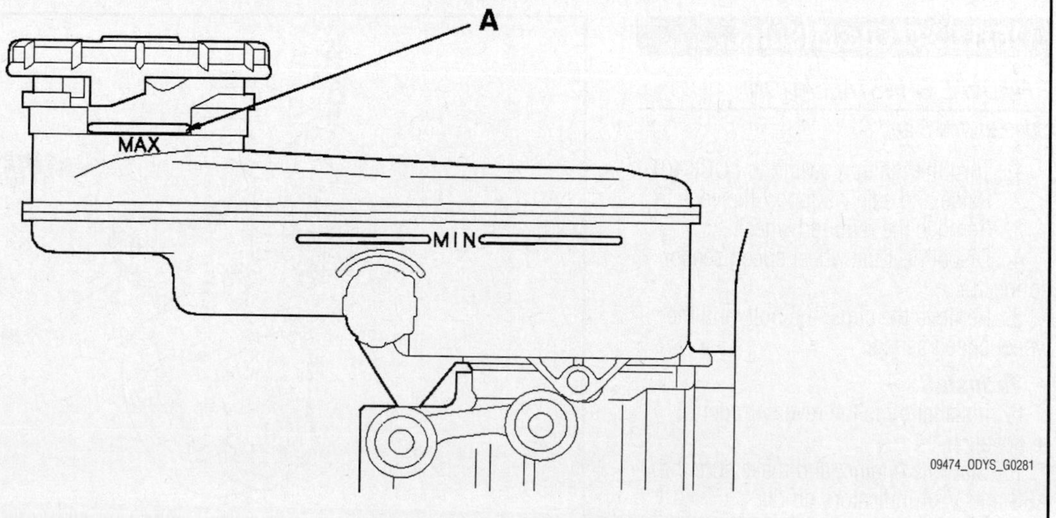
A
MAX
MIN
09474_ODYS_G0281

Fig. 2 The reservoir on the master cylinder must be at the MAX (A) level mark at the start of the bleeding procedure

3. Have someone slowly pump the brake pedal several times, then apply steady pressure.

4. Starting at the left-front, loosen the brake bleed screw to allow air to escape from the system. Then tighten the bleed screw securely.

5. Repeat the procedure for each caliper until no air bubbles are in the fluid. Bleed the calipers in the sequence shown.

6. Refill the master cylinder reservoir to the MAX (upper) level line.

BLEEDING THE ABS SYSTEM

The bleeding procedure for the ABS System is the same as the Conventional Bleeding Procedure. See "Bleeding the Brake System" in this section.

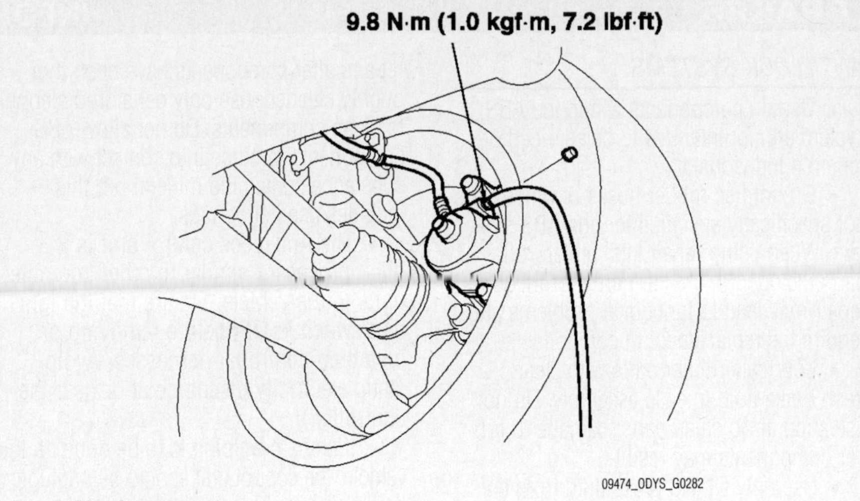

9.8 N·m (1.0 kgf·m, 7.2 lbf·ft)
09474_ODYS_G0282

Fig. 3 Bleed screw

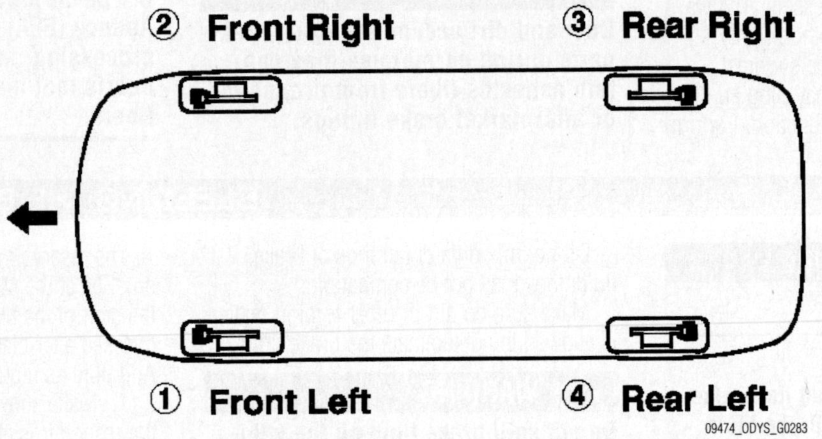

② Front Right
③ Rear Right
① Front Left
④ Rear Left
09474_ODYS_G0283

Fig. 4 Bleeding sequence

WHEEL SPEED SENSORS

REMOVAL & INSTALLATION

See Figures 5 and 6.

1. Turn the ignition switch to LOCK (0).
2. Raise and safely support the vehicle.
3. Remove the required wheel.
4. Disconnect the wheel speed sensor connector.
5. Remove the clips, the bolt, and the wheel speed sensor.

To install:

6. Installation is the reverse order of removal.
7. Start the engine, and make sure the ABS and VSA indicators go off.
8. Test-drive the vehicle, and make sure the ABS and VSA indicators do not come on.

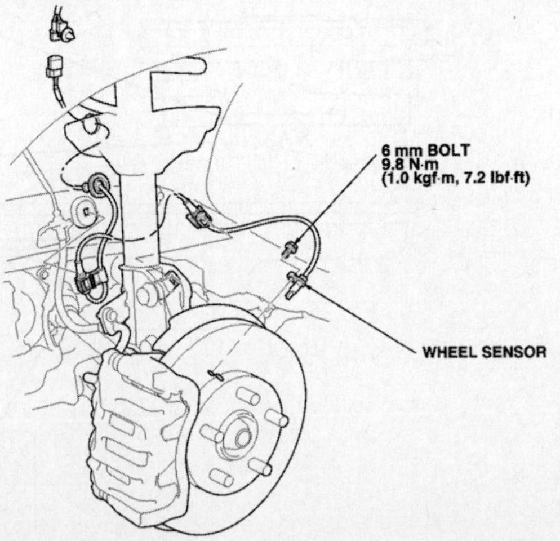

6 mm BOLT
9.8 N·m
(1.0 kgf·m, 7.2 lbf·ft)
WHEEL SENSOR
42050_ODYS_G0062

Fig. 5 View of the front wheel speed sensor

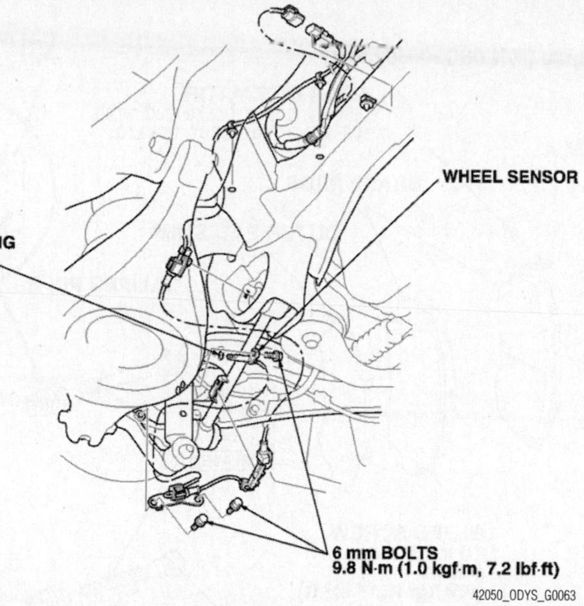
WHEEL SENSOR
O-RING
6 mm BOLTS
9.8 N·m (1.0 kgf·m, 7.2 lbf·ft)
42050_ODYS_G0063

Fig. 6 View of the rear wheel speed sensor

BRAKES

FRONT DISC BRAKES

BRAKE CALIPER

REMOVAL & INSTALLATION

See Figures 7 through 9.

CAUTION

Read and follow the "Precautions" in this section before beginning work.

1. Raise the front of the vehicle, and support it with safety stands in the proper locations.
2. Remove the front wheel.
3. Remove the brake hose mounting bolt.
4. Remove the brake hose-to-caliper union bolt and discard the washers. Plug the brake hose.
5. Remove the caliper mounting bolts.

To install:

6. Installation is the reverse of removal. Install the brake hose on the damper bracket with flange bolt first, then connect the brake hose to the caliper with the banjo bolt and new sealing washers. Observe the torques given in the accompanying illustrations.
7. Bleed the brakes.
8. Press the brake pedal several times to make sure the brakes work.

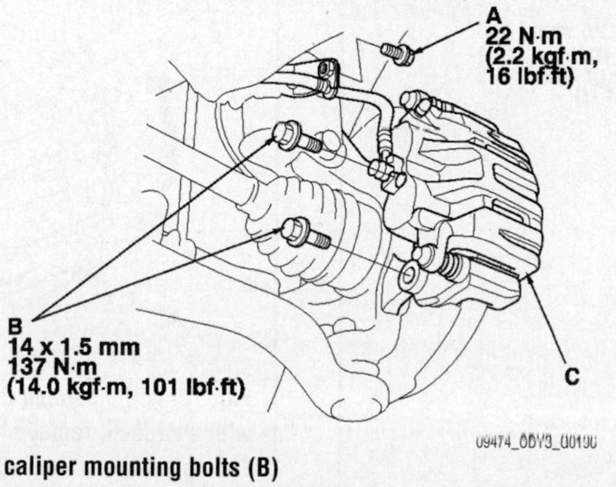
A
22 N·m
(2.2 kgf·m,
16 lbf·ft)
B
14 x 1.5 mm
137 N·m
(14.0 kgf·m, 101 lbf·ft)
C
09474_ODYS_G0190

Fig. 7 Remove the caliper mounting bolts (B)

DISC BRAKE PADS

REMOVAL & INSTALLATION

See Figures 10 through 14.

CAUTION

Read the "Precautions" in this section before beginning any repair work.

1. Raise the front of the vehicle, and support it with safety stands in the proper locations.
2. Remove the front wheel.
3. While holding the caliper pin with a wrench, remove the flange bolt. Be careful not to damage the pin boot. Then pivot the caliper up out of the way. Check the hose and pin boots for damage and deterioration.
4. Remove the pad shims and brake pads.
5. Remove the pad retainers from the caliper bracket.
6. Clean the caliper thoroughly; remove any rust, and check for grooves and cracks.
7. Check the brake disc for damage and cracks.

To install:

8. Apply a thin coat of M-77 assembly paste (P/N 08798-9010) to the retainers on their mating surfaces (indicated by the arrows) against the caliper bracket.
9. Install the pad retainers. Wipe excess paste off the retainers. Keep paste off the discs and pads.
10. Apply a thin coat of M-77 assembly paste (P/N 08798-9010) to the pad side of the shims, the back of the brake pads and the other areas indicated by the arrows. Wipe excess assembly paste off the shims and brake pads. Contaminated brake discs or brake pads reduce stopping ability. Keep grease and assembly paste off the brake discs and brake pads.

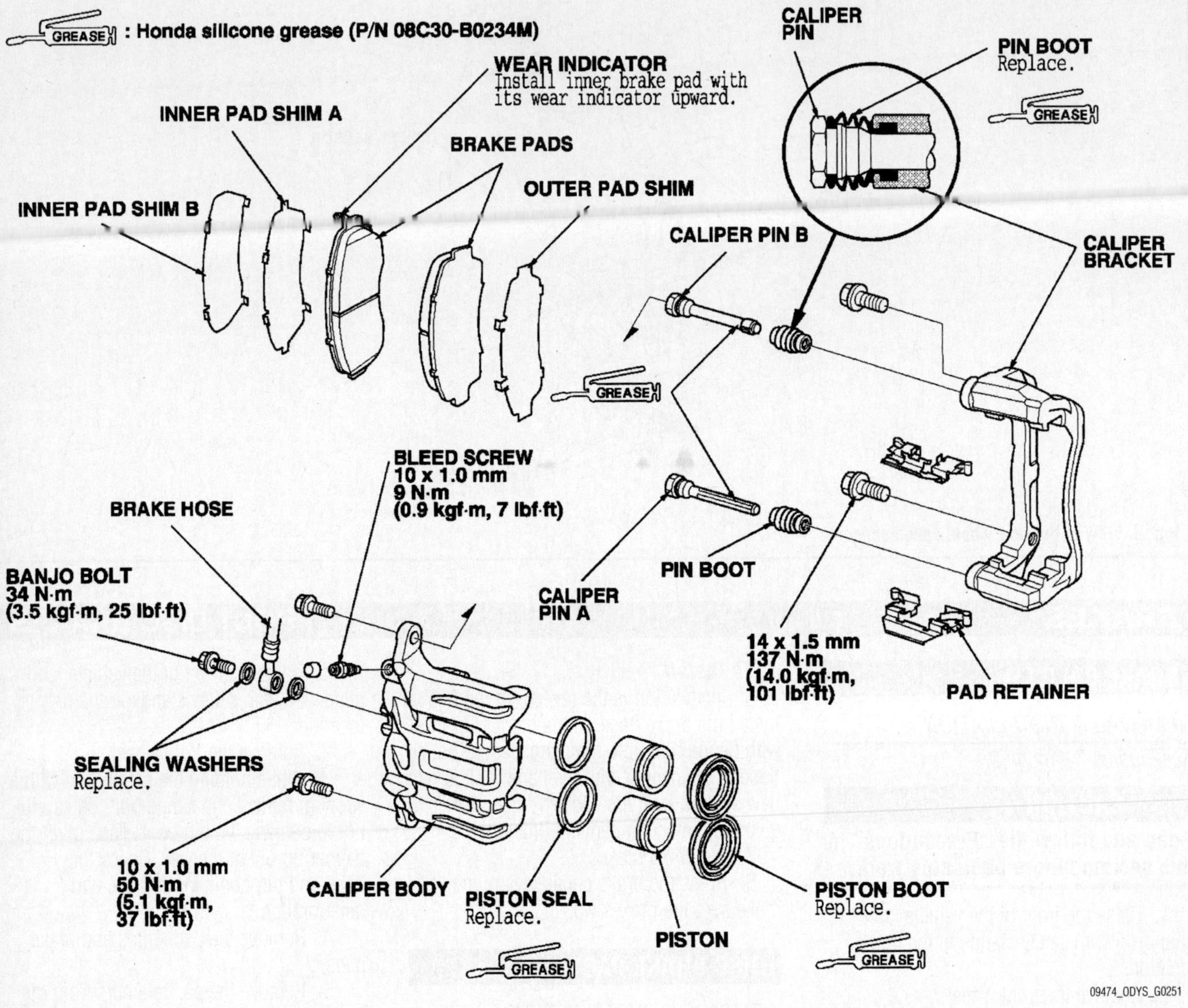
GREASE : Honda silicone grease (P/N 08C30-B0234M)
CALIPER PIN
PIN BOOT
Replace.
WEAR INDICATOR
Install inner brake pad with its wear indicator upward.
INNER PAD SHIM A
GREASE
BRAKE PADS
OUTER PAD SHIM
INNER PAD SHIM B
CALIPER PIN B
CALIPER BRACKET
GREASE
BLEED SCREW
10 x 1.0 mm
9 N·m
(0.9 kgf·m, 7 lbf·ft)
BRAKE HOSE
BANJO BOLT
34 N·m
(3.5 kgf·m, 25 lbf·ft)
PIN BOOT
CALIPER PIN A
14 x 1.5 mm
137 N·m
(14.0 kgf·m, 101 lbf·ft)
PAD RETAINER
SEALING WASHERS
Replace.
10 x 1.0 mm
50 N·m
(5.1 kgf·m, 37 lbf·ft)
CALIPER BODY
PISTON SEAL
Replace.
PISTON BOOT
Replace.
PISTON
GREASE
GREASE
09474_ODYS_G0251

Fig. 8 Exploded view of the front caliper

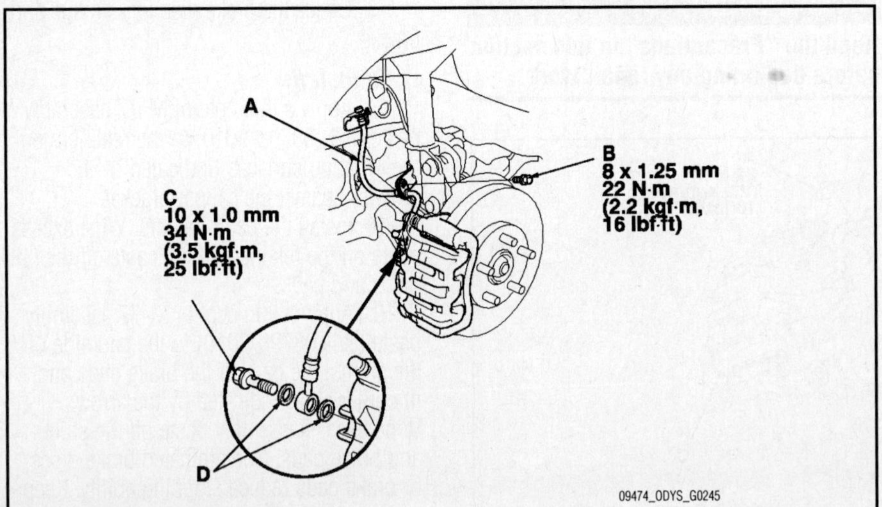
A
B
8 x 1.25 mm
22 N·m
(2.2 kgf·m, 16 lbf·ft)
C
10 x 1.0 mm
34 N·m
(3.5 kgf·m, 25 lbf·ft)
D
09474_ODYS_G0245

Fig. 9 Brake hose attachment

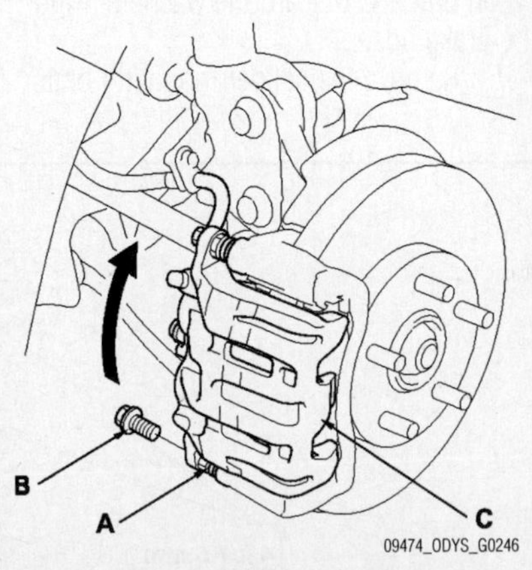
B
A
C
09474_ODYS_G0246

Fig. 10 While holding the caliper pin (A) with a wrench, remove the flange bolt (B). Then pivot the caliper (C) up out of the way

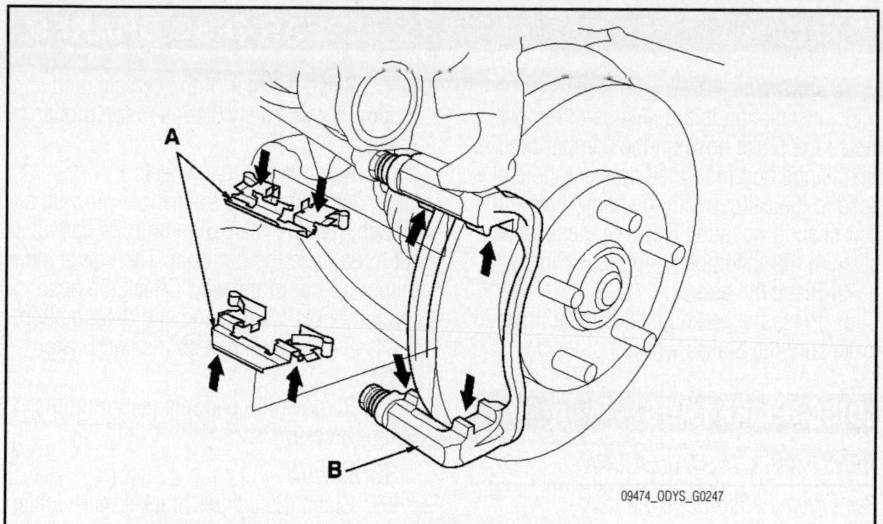
A
B
09474_ODYS_G0247

Fig. 11 Remove the pad retainers (A) from the caliper bracket (B)

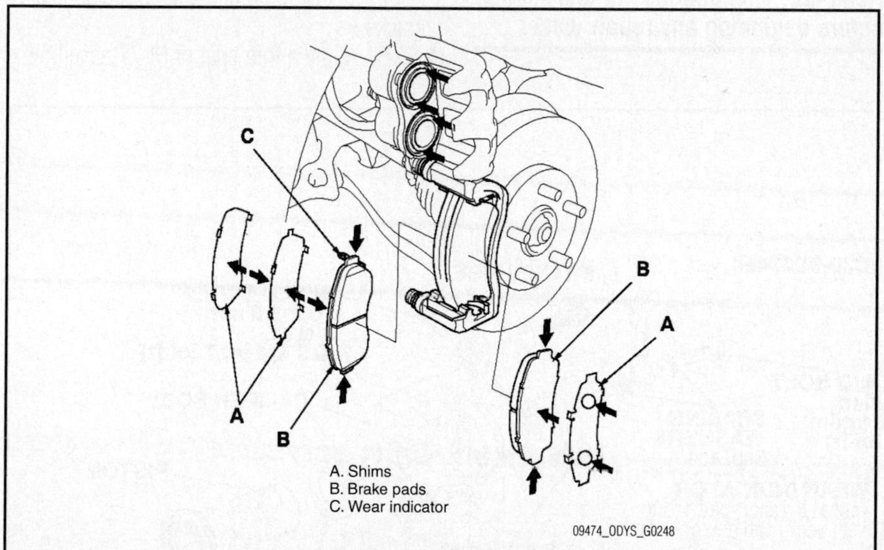
C
B
A
A
B
A. Shims
B. Brake pads
C. Wear indicator
09474_ODYS_G0248

Fig. 12 Showing the shims (A), brake pads (B), wear indicator (C)

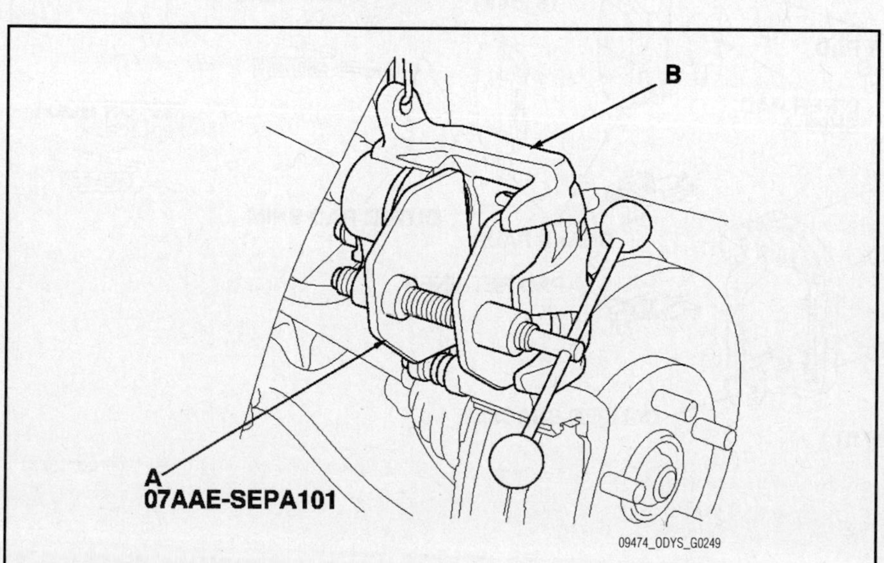
B
A
07AAE-SEPA101
09474_ODYS_G0249

Fig. 13 Caliper piston tool (A) and caliper (B)

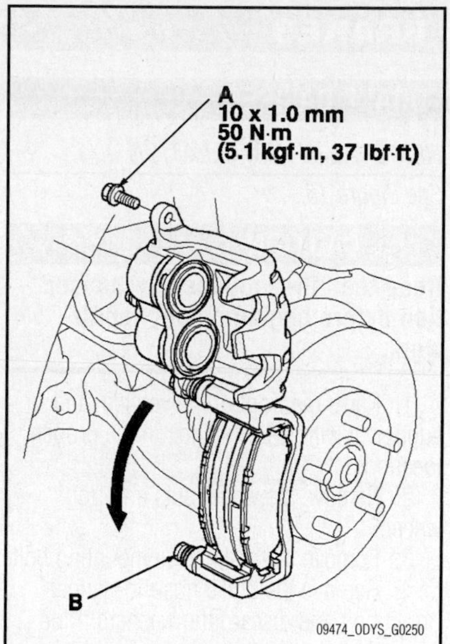
A
10 x 1.0 mm
50 N·m
(5.1 kgf·m, 37 lbf·ft)
B
09474_ODYS_G0250

Fig. 14 Pivot the caliper down into position and engage the lower retainer (B) with the retaining bolt (A)

11. Install the brake pads and pad shims correctly. Install the brake pad with the wear indicator on the inside. If you are reusing the brake pads, always reinstall the brake pads in their original positions to prevent a loss of braking efficiency.
12. Mount the special tool on the caliper.
13. Press in the piston with the special tool so the caliper will fit over the brake pads. Make sure the piston boot is in position to prevent damaging it when pivoting the caliper down.

➡Be careful when pressing in the piston, brake fluid might overflow from the master cylinder's reservoir.

14. Remove the special tool.
15. Pivot the caliper down into position. Install the flange bolt, and torque it to the specified torque while holding the caliper pin with a wrench. Be careful not to damage the pin boot.
16. Press the brake pedal several times to make sure the brakes work.

➡Engagement of the brakes may require a greater pedal stroke immediately after the brake pads have been replaced. Several applications of the brake pedal will restore the normal pedal stroke.

17. After installation, check for leaks at hose and line joints or connections, and retighten if necessary.
18. Install the front wheels, then test-drive the vehicle.

BRAKE CALIPER

REMOVAL & INSTALLATION

See Figure 15.

**** CAUTION**

Read the "Precautions" in this section before beginning any repair work.

1. Raise the front of the vehicle, and support it with safety stands in the proper locations.
2. Remove the wheel nuts and front wheel.
3. Remove the brake hose mounting bolt.
4. Remove the brake hose-to-caliper union bolt and discard the washers. Plug the brake hose.
5. Remove the caliper mounting bolts.

To install:

6. Installation is the reverse of removal. Install the brake hose on the damper bracket with flange bolt first, then connect the brake hose to the caliper with the banjo bolt and new sealing washers. Observe the torques given in the accompanying illustrations.
7. Bleed the brakes.
8. Press the brake pedal several times to make sure the brakes work.

DISC BRAKE PADS

REMOVAL & INSTALLATION

See Figures 16 through 19.

**** CAUTION**

Read the "Precautions" in this section before beginning any repair work.

1. Raise the rear of the vehicle, and support it with safety stands in the proper locations.
2. Remove the rear wheels.
3. While holding the caliper pin with a wrench, remove the flange bolt. Be careful not to damage the pin boot. Then pivot the caliper up out of the way. Check the hose and pin boots for damage and deterioration.
4. Remove the pad shims and brake pads.
5. Remove the pad retainers from the caliper bracket.

To install:

6. Clean the caliper bracket thoroughly; remove any rust, and check for grooves and cracks.
7. Check the brake disc for damage and cracks.
8. Apply a thin coat of M-77 assembly

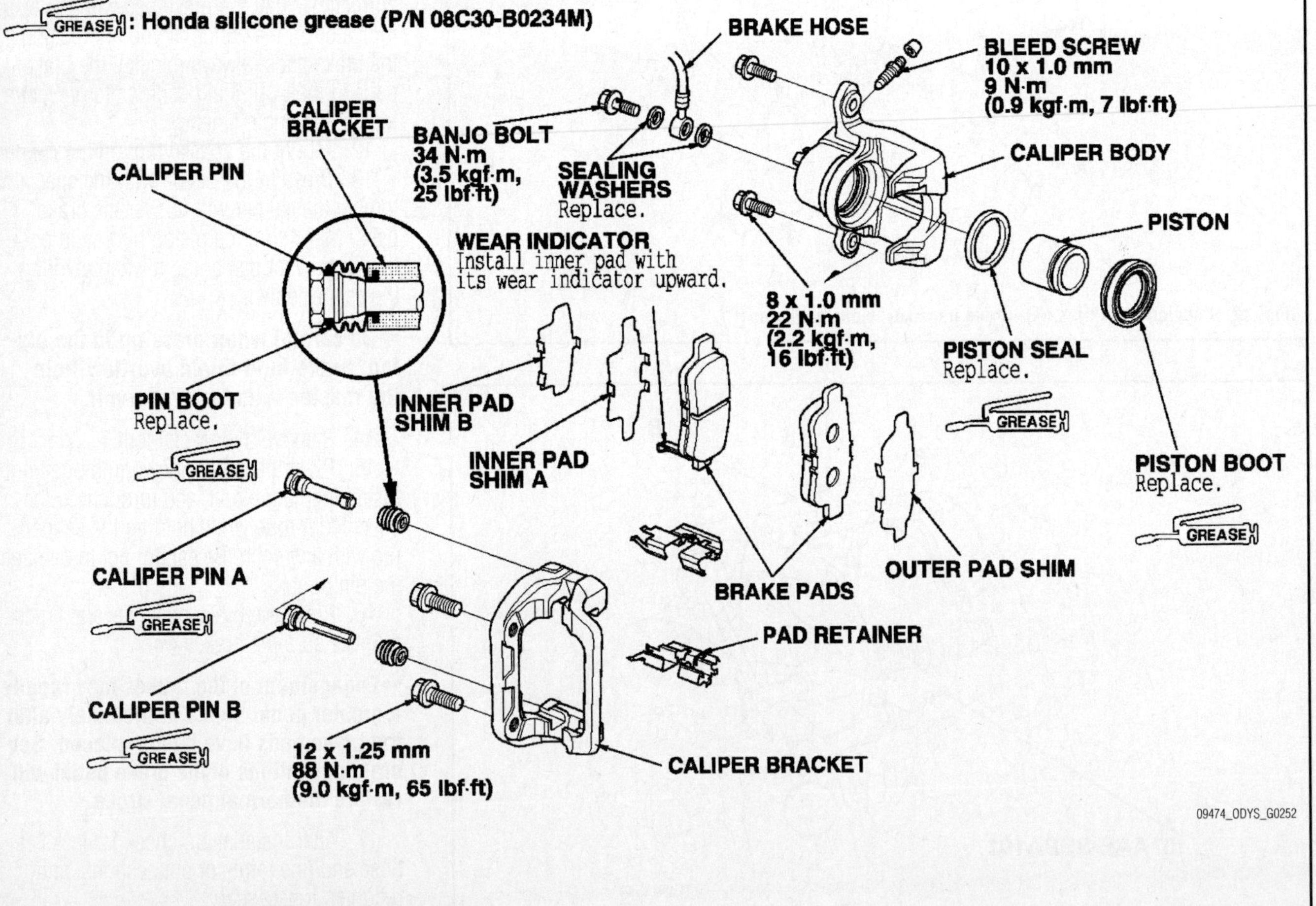

GREASE : Honda silicone grease (P/N 08C30-B0234M)
BRAKE HOSE
BLEED SCREW
10 x 1.0 mm
9 N·m
(0.9 kgf·m, 7 lbf·ft)
CALIPER BRACKET
BANJO BOLT
34 N·m
(3.5 kgf·m,
25 lbf·ft)
SEALING WASHERS
Replace.
CALIPER BODY
CALIPER PIN
PISTON
WEAR INDICATOR
Install inner pad with its wear indicator upward.
8 x 1.0 mm
22 N·m
(2.2 kgf·m,
16 lbf·ft)
PISTON SEAL
Replace.
PIN BOOT
Replace.
INNER PAD SHIM B
INNER PAD SHIM A
GREASE
PISTON BOOT
Replace.
CALIPER PIN A
OUTER PAD SHIM
BRAKE PADS
PAD RETAINER
CALIPER PIN B
12 x 1.25 mm
88 N·m
(9.0 kgf·m, 65 lbf·ft)
CALIPER BRACKET
09474_ODYS_G0252

Fig. 15 Exploded view of the rear caliper

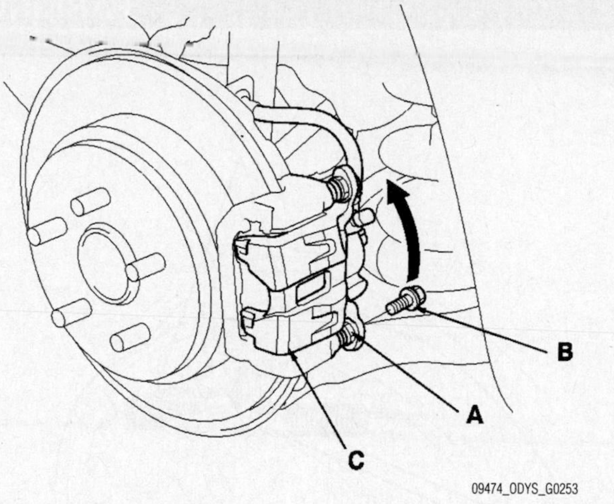
A
B
C
09474_ODYS_G0253

Fig. 16 While holding the caliper pin (A) with a wrench, remove the flange bolt (B). Then pivot the caliper (C) up out of the way

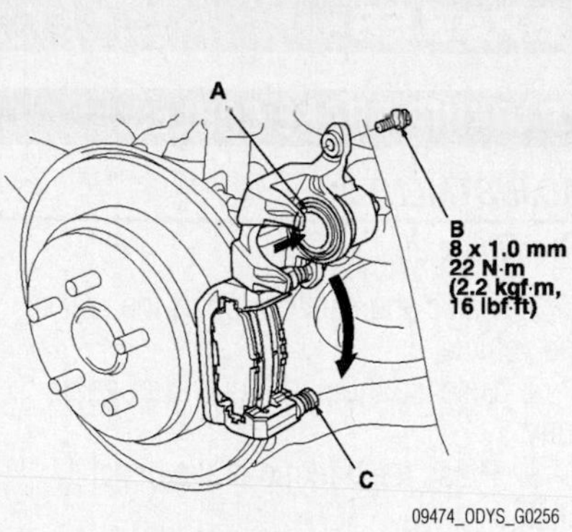
A
B
8 x 1.0 mm
22 N·m
(2.2 kgf·m,
16 lbf·ft)
C
09474_ODYS_G0256

Fig. 19 Push in the piston (A) so the caliper will fit over the brake pads. Install the flange bolt (B), and torque it to the specified torque while holding the caliper pin (C) with a wrench

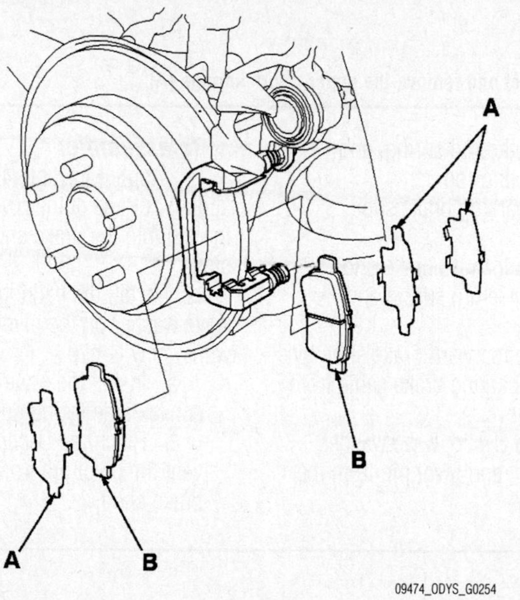
A
B
A
B
09474_ODYS_G0254

Fig. 17 Remove the pad shims (A) and brake pads (B)

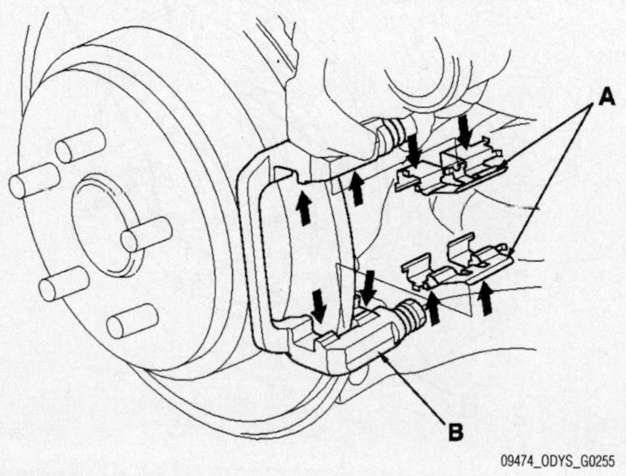
A
B
09474_ODYS_G0255

Fig. 18 Remove the pad retainers (A) from the caliper bracket (B)

paste (P/N 08798-9010) to the retainers on their mating surfaces (indicated by the arrow) against the caliper bracket.

9. Install the pad retainers. Wipe excess assembly paste off the retainers. Keep any assembly paste off the discs and pads.

10. Apply a thin coat of M-77 assembly paste (P/N 08798-9010) to the pad side of the shims, the back of the brake pads, and the other areas indicated by the arrows. Wipe excess assembly paste off the pad shims and brake pads. Contaminated brake discs or pads reduce stopping ability. Keep assembly paste off the brake discs and pads.

11. Install the brake pads and pad shims correctly. Install the brake pad with the wear indicator on the inside. If you are reusing the brake pads, always reinstall the brake pads in their original positions to prevent a momentary loss of braking efficiency.

12. Push in the piston so the caliper will fit over the brake pads. Make sure the piston boot is in position to prevent damaging it when pivoting the caliper down.

13. Pivot the caliper down into position. Install the flange bolt, and torque it to the specified torque while holding the caliper pin with a wrench. Be careful not to damage the pin boot.

14. Press the brake pedal several times to make sure the brakes work.

➡Brake engagement may require a greater pedal stroke immediately after the brake pads have been replaced. Several applications of the brake pedal will restore the normal pedal stroke.

15. After installation, check for leaks at hose and line joints or connections, and retighten if necessary.

PARKING BRAKE CABLES

ADJUSTMENT

See Figure 20.

1. Raise and safely support the rear of the vehicle.
2. Release the parking brake pedal fully.
3. Press the parking brake pedal 1 click.
4. Tighten the parking brake adjusting nut until the parking brakes drag slightly when the rear wheels are turned.
5. Release the parking brake pedal fully, and check that the parking brakes do not drag when the rear wheels are turned. Readjust if necessary.
6. Make sure the parking brake pedal is within the specified number of clicks (4 to 5 clicks).

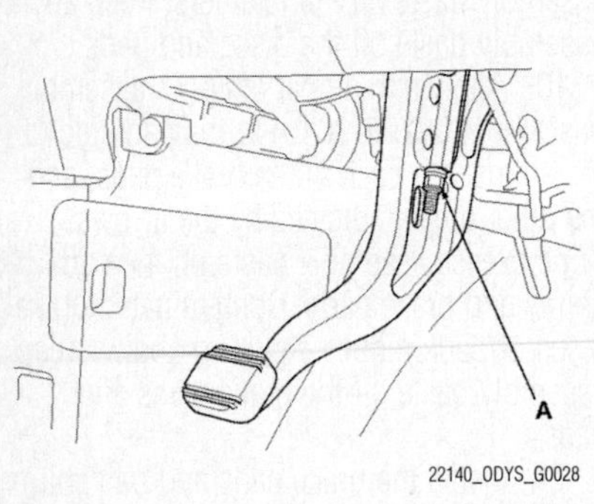
A
22140_ODYS_G0028

Fig. 20 Turn the adjusting nut (A) to make the adjustment

PARKING BRAKE SHOES

REMOVAL & INSTALLATION

See Figures 21 through 29.

⁂ CAUTION

Read the "Precautions" in this section before beginning any repair work.

1. Raise the rear of the vehicle, and support it with safety stands in the proper locations.
2. Remove the rear wheels.
3. Release the parking brake, and remove the rear brake caliper and brake disc.
4. Disconnect and remove the upper return springs.
5. Remove the tension pins by pushing and turning the retainer springs.

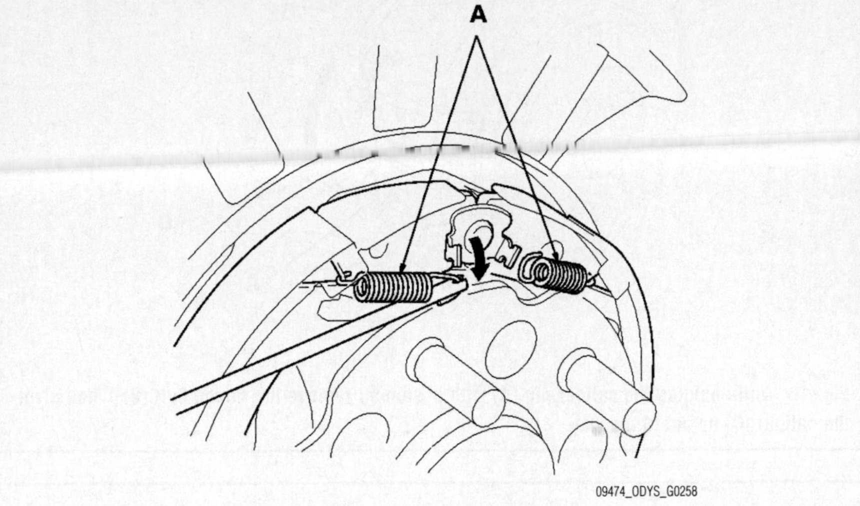
A
09474_ODYS_G0258

Fig. 21 Disconnect and remove the upper return springs (A)

6. Disconnect the rod spring, and remove the connecting rod.
7. Lower the parking brake shoe assembly.
8. Remove the forward brake shoe by removing the lower return spring and adjuster assembly.
9. Remove the rearward brake shoe by disconnecting the parking brake cable from the parking brake lever.
10. Remove the U-clip, wave washer, parking brake lever, and pivot pin from the brake shoe.

To assemble:

11. Apply Molykote 44 MA grease to the sliding surface of the pivot pin and insert the pin into the brake shoe from the rear side.
12. Install the parking brake lever and wave washer on the pivot pin, and secure with a new U-clip.
    a. Install the wave washer with its convex side facing out.
    b. Pinch the U-clip securely to prevent the pivot pin from coming out of the brake shoe.

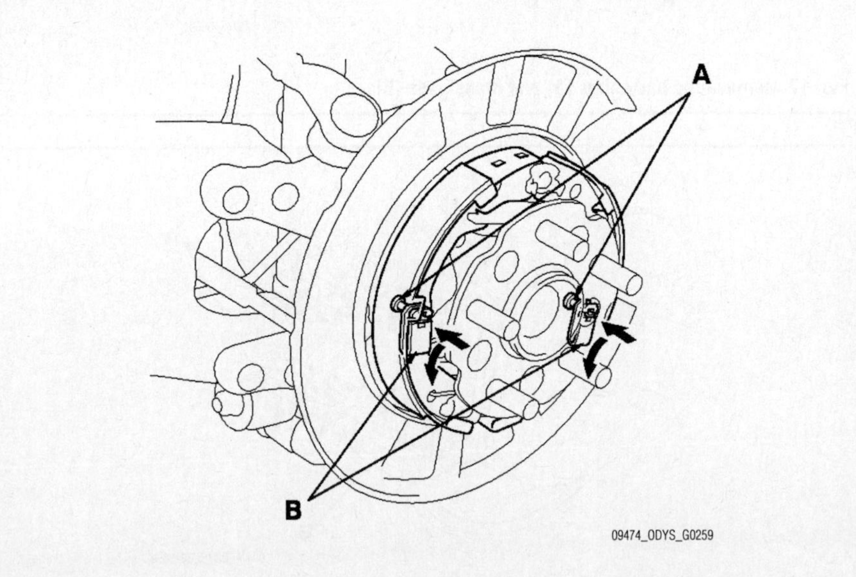
A
B
09474_ODYS_G0259

Fig. 22 Remove the tension pins (A) by pushing and turning the retainer springs (B)

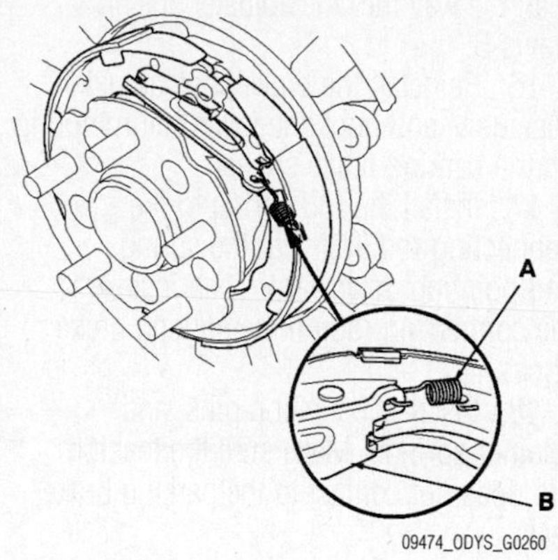
A
B
09474_ODYS_G0260

Fig. 23 Disconnect the rod spring (A), and remove the connecting rod (B)

13. Connect the parking brake cable to the parking brake lever. Apply silicone grease to the cable contact surface on the backing plate.

14. Apply Molykote® 44 MA grease to the sliding surfaces, the inner edges of the parking brake shoes, and the pivot of the parking brake lever as shown. Wipe off any excess. Keep grease off the brake linings.

15. Clean the threaded portions of clevis A, and coat the threads of clevis A

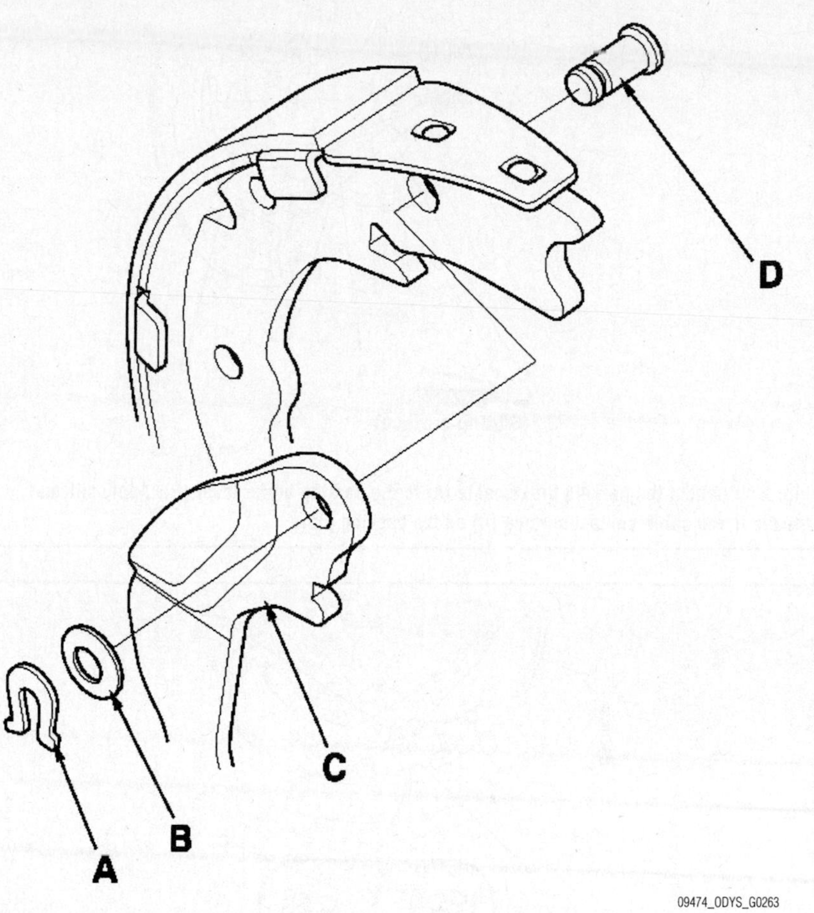
D
C
B
A
09474_ODYS_G0263

Fig. 25 Remove the U-clip, wave washer, parking brake lever, and pivot pin from the brake shoe

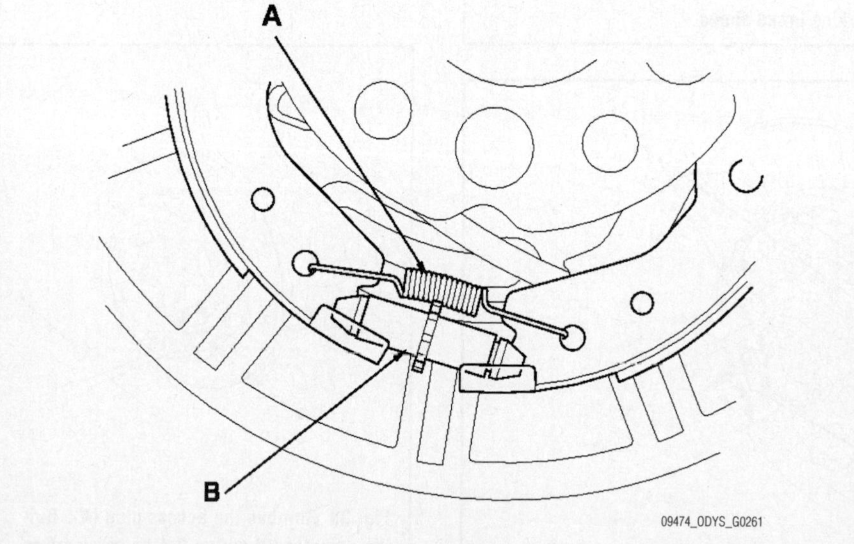
A
B
09474_ODYS_G0261

Fig. 24 Remove the forward brake shoe by removing the lower return spring (A) and adjuster assembly (B)

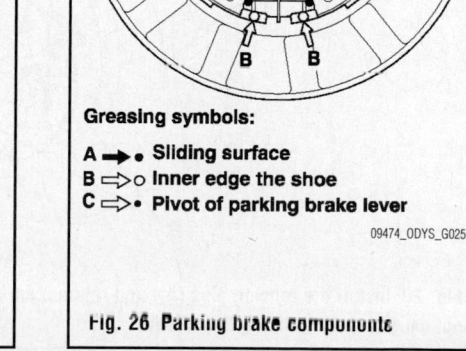

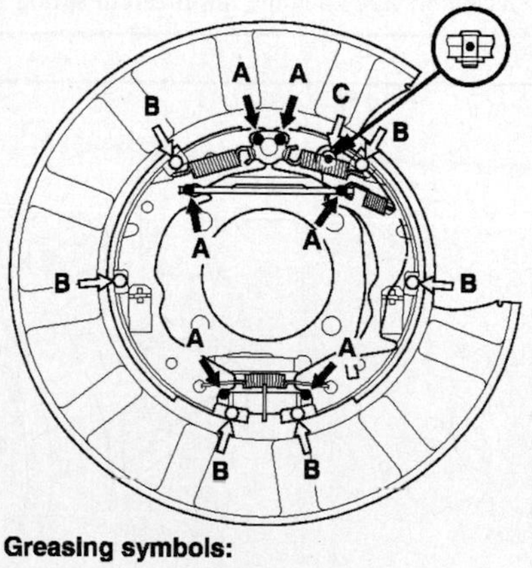
Greasing symbols:
A Sliding surface
B Inner edge the shoe
C Pivot of parking brake lever
09474_ODYS_G0257

Fig. 26 Parking brake components

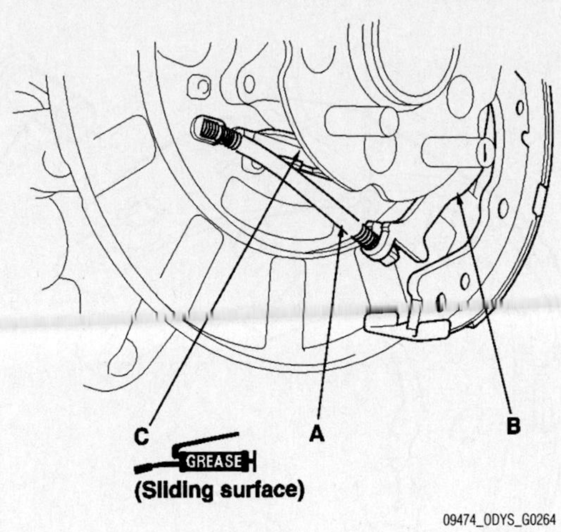
C
A
B
GREASE
(Sliding surface)
09474_ODYS_G0264

Fig. 27 Connect the parking brake cable (A) to the parking brake lever (B). Apply silicone grease to the cable contact surface (C) on the backing plate

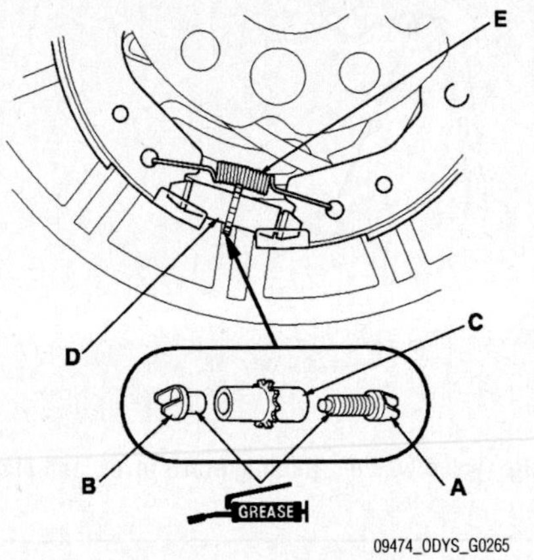
E
D
C
B
A
GREASE
09474_ODYS_G0265

Fig. 28 Clean the threaded portions of clevis A, and coat the threads of clevis A with grease. Clean the sliding surface of clevis B, and coat the sliding surface of clevis B with grease. Thread clevis A all the way into the adjuster. Install clevis B. Reinstall the brake shoe adjuster assembly and hook the lower return spring on the parking brake shoes

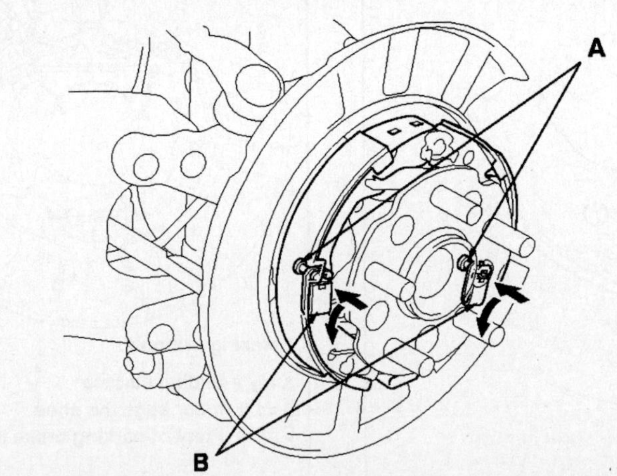
A
B
09474_ODYS_G0267

Fig. 29 Install the tension pins (A), and retainer springs (B). Make sure the tension pin does not contact to the parking brake lever

with grease. Clean the sliding surface of clevis B, and coat the sliding surface of clevis B with grease. Thread clevis A all the way into the adjuster. Install clevis B.

16. Reinstall the brake shoe adjuster assembly, and hook the lower return spring on the parking brake shoes.
17. Install the rod spring to the connecting rod first with the spring end pointing downward. Then install the connecting rod on the parking brake shoes.
18. Install the tension pins, and retainer springs. Make sure the tension pin does not contact to the parking brake lever.
19. Install the upper return springs.
20. Install the brake disc/drum and rear brake caliper.
21. Adjust the parking brake.

ADJUSTMENT

See Figure 30.

1. Raise the rear of the vehicle, and support it with safety stands in the proper locations.
2. Remove the rear wheels.
3. Release the parking brake, and back off the adjusting nut.
4. Remove the access plug.
5. Turn the adjuster with a flat-tip screwdriver until the shoes lock against the drum. Then back off ten clicks, and install the access plug.
6. Do the minor adjustment procedure. For additional information, see "Parking Brake Cable."
7. Install the rear wheels.

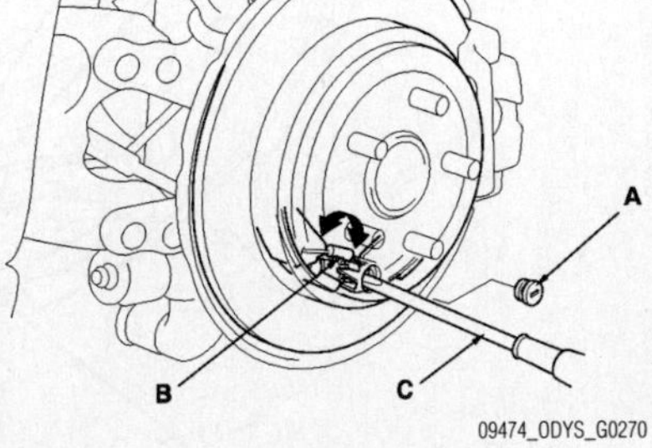
A
B
C
09474_ODYS_G0270

Fig. 30 Remove the access plug (A). Turn the adjuster (B) with a flat-tip screwdriver (C) until the shoes lock against the drum. Then back off ten clicks, and install the access plug

CHASSIS ELECTRICAL

AIR BAG (SUPPLEMENTAL RESTRAINT SYSTEM)

GENERAL INFORMATION

CAUTION

These vehicles are equipped with an air bag system. The system must be disarmed before performing service on, or around, system components, the steering column, instrument panel components, wiring and sensors. Failure to follow the safety precautions and the disarming procedure could result in accidental air bag deployment, possible injury and unnecessary system repairs.

SERVICE PRECAUTIONS

Disconnect and isolate the battery negative cable before beginning any airbag system component diagnosis, testing, removal, or installation procedures. Allow system capacitor to discharge for two minutes before beginning any component service. This will disable the airbag system. Failure to disable the airbag system may result in accidental airbag deployment, personal injury, or death.

DISARMING THE SYSTEM

Disconnect and isolate the negative battery cable. Wait 3 minutes for the system capacitor to discharge before performing any service.

ARMING THE SYSTEM

1. After performing service, connect the negative battery cable to re-arm the SRS system.

CLOCKSPRING CENTERING

1. First rotate the cable reel clockwise until it stops.
2. Then rotate it counterclockwise (three full turns) until the arrow mark on the cable reel label points straight up.
3. Position the two tabs of the turn signal canceling sleeve as shown, and install the steering wheel on to the steering column shaft, making sure the steering wheel hub engages the pins of the cable reel and tabs of the turn signal canceling sleeve. Do not tap on the steering wheel or steering column shaft when installing the steering wheel.

DRIVE TRAIN

FRONT HALFSHAFT

REMOVAL & INSTALLATION

See Figures 31 through 39.

1. Loosen the wheel nuts slightly.
2. Raise the front of the vehicle, and support it with safety stands in the proper locations.
3. Remove the wheel nuts and front wheels.
4. Lift up the locking tab on the spindle nut, then remove the nut.
5. Drain the automatic transmission fluid Reinstall the drain plug using a new washer.
6. Separate the front stabilizer link from the damper. See the procedure under Front Suspension.
7. Remove the lock pin from the lower arm ball joint castle nut, and remove the nut, then separate the ball joint from the lower arm with the ball joint thread protector and remover.

➡To avoid damaging the ball joint, install the ball joint thread protector onto the threads of the ball joint. Be careful not to damage the ball joint boot when installing the remover.

8. Pull the knuckle outward, and remove the outboard joint from the front wheel hub using a plastic hammer.
9. Remove exhaust pipe A.

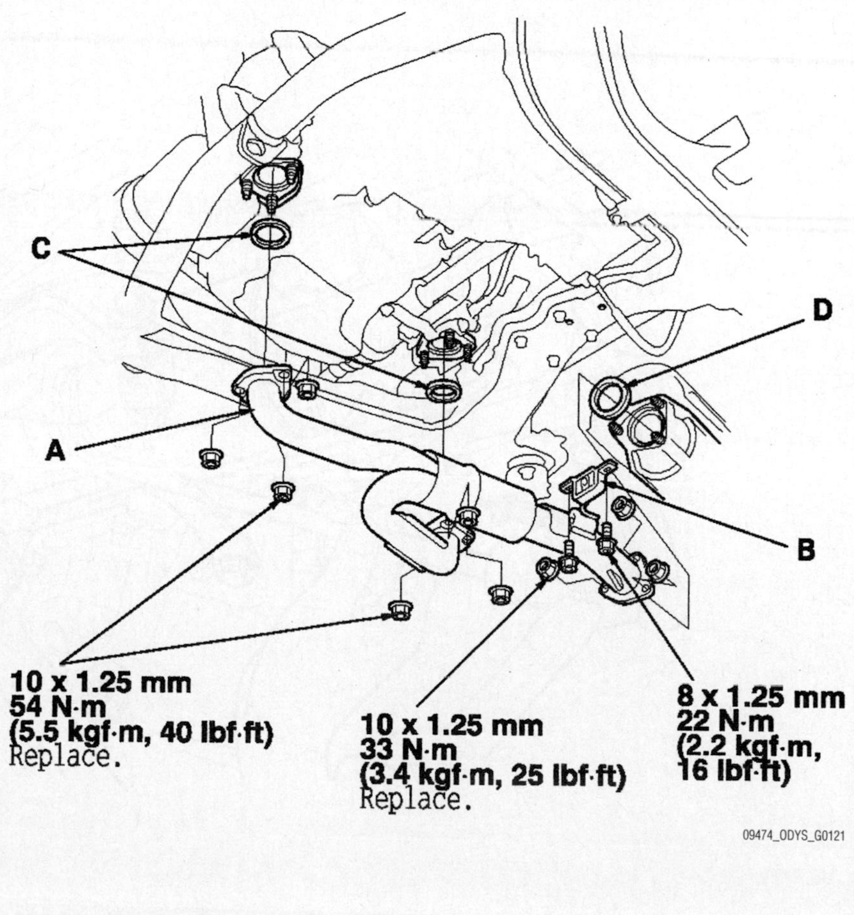
C
D
A
B
10 x 1.25 mm
54 N·m
(5.5 kgf·m, 40 lbf·ft)
Replace.
10 x 1.25 mm
33 N·m
(3.4 kgf·m, 25 lbf·ft)
Replace.
8 x 1.25 mm
22 N·m
(2.2 kgf·m,
16 lbf·ft)
09474_ODYS_G0121

Fig. 31 Remove exhaust pipe A

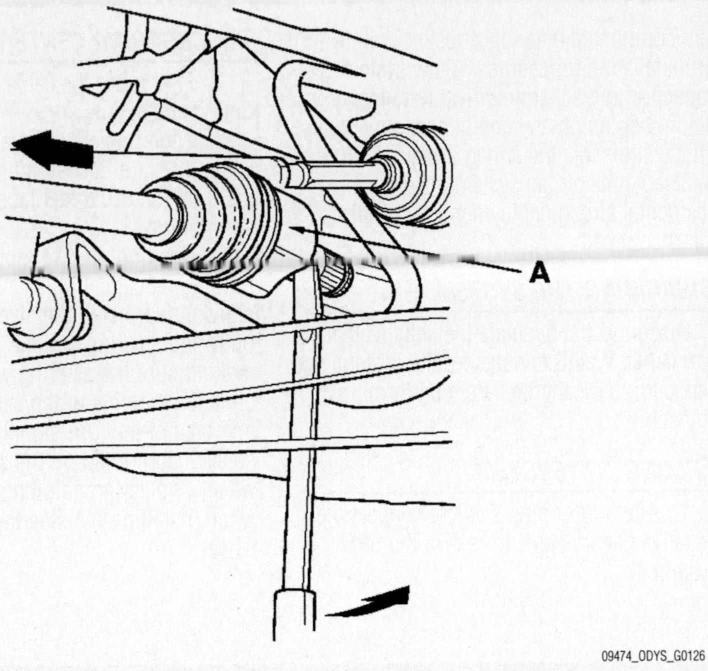
A
09474_ODYS_G0126

Fig. 32 Left halfshaft: Pry the inboard joint (A) from the transmission housing with a prybar. Remove the halfshaft as an assembly.

10. Left halfshaft: Pry the inboard joint from the transmission housing with a prybar. Remove the halfshaft as an assembly.

11. Right halfshaft: Drive the inboard joint off of the intermediate shaft with a drift and hammer. Remove the halfshaft as an assembly.

➡Do not pull on the halfshaft or the inboard joint may come apart. Pull the halfshaft straight out to avoid damaging the oil seal.

To install:

12. Install a new set ring in the set ring groove of the halfshaft (left halfshaft).

13. Apply 0.07–0.11 oz. of grease to the whole splined surface of the right halfshaft. After applying grease, remove the grease from the splined grooves at intervals of 2–3 splines and from the set ring groove so that air can bleed from the intermediate shaft.

14. Clean the areas where the halfshaft contacts the differential thoroughly with brake cleaner, and dry with compressed air. Do not wash the rubber parts with solvent. Insert the inboard end of the halfshaft into

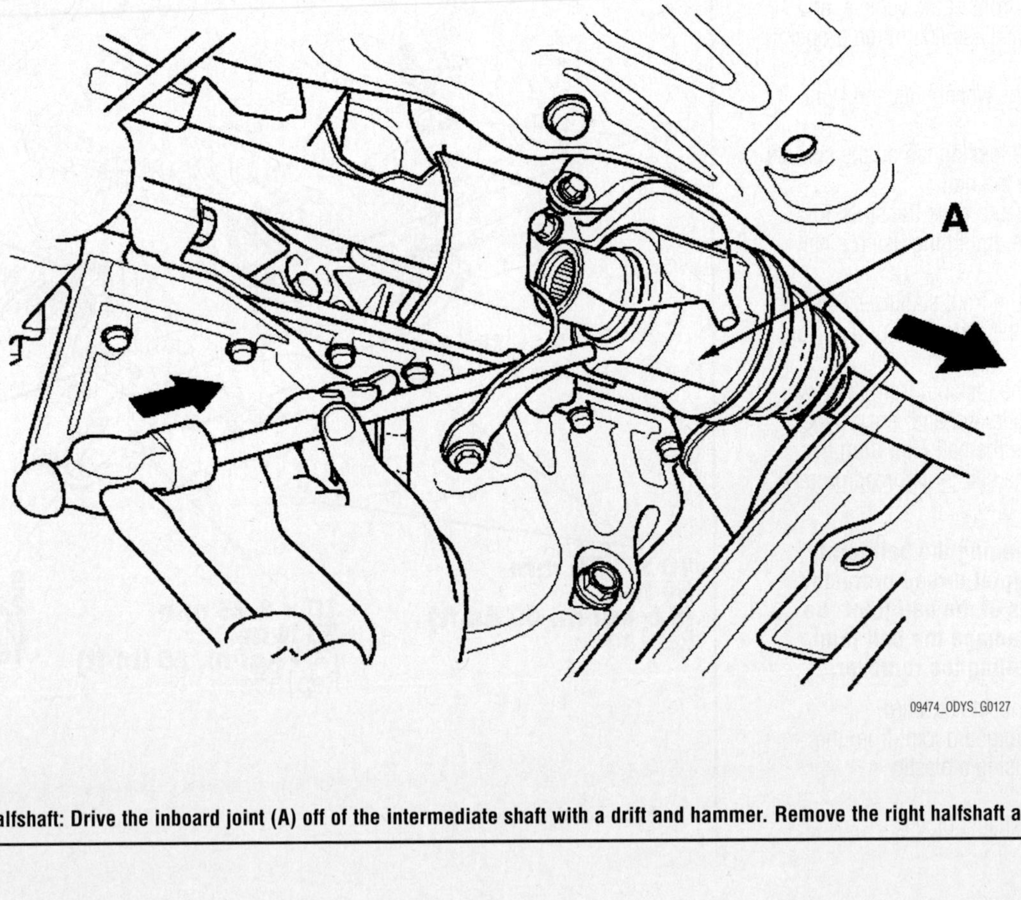
A
09474_ODYS_G0127

Fig. 33 Right halfshaft: Drive the inboard joint (A) off of the intermediate shaft with a drift and hammer. Remove the right halfshaft as an assembly.

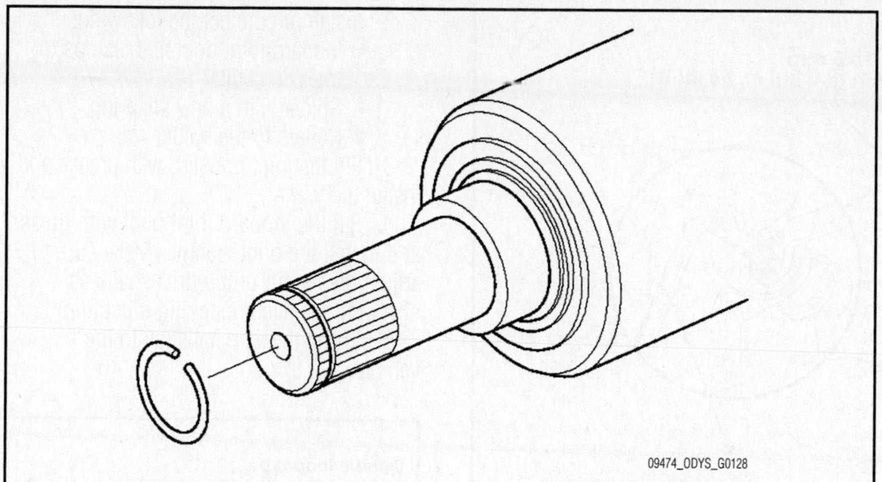
09474_ODYS_G0128

Fig. 34 Install a new set ring in the set ring groove of the halfshaft (left halfshaft)

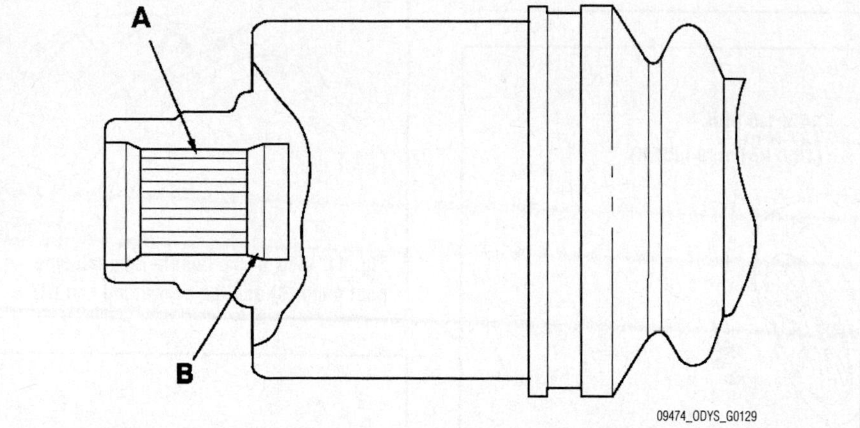
A
B
09474_ODYS_G0129

Fig. 35 Apply 0.07–0.11 oz. of grease to the whole splined surface (A) of the right halfshaft. After applying grease, remove the grease from the splined grooves at intervals of 2–3 splines and from the set ring groove (B) so that air can bleed from the intermediate shaft

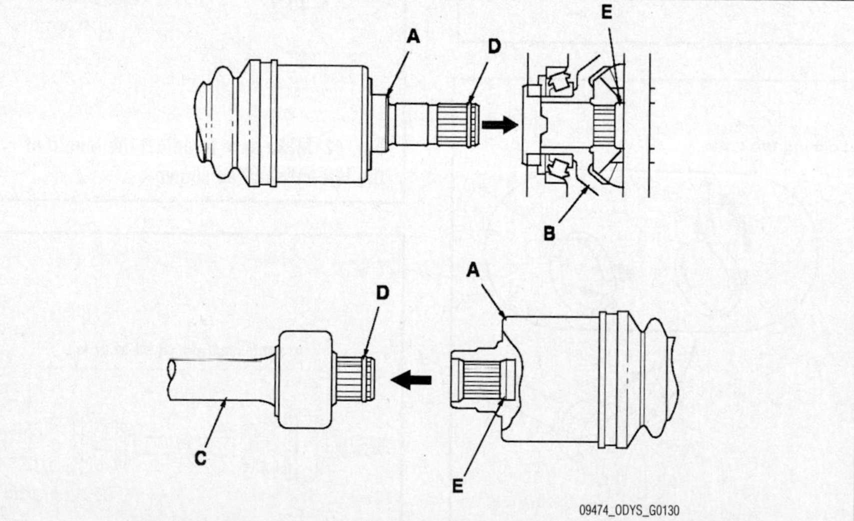
A
D
E
B
A
D
C
E
09474_ODYS_G0130

Fig. 36 Clean the areas where the halfshaft contacts the differential thoroughly with brake cleaner, and dry with compressed air. Do not wash the rubber parts with solvent. Insert the inboard end of the halfshaft into the differential or intermediate shaft until the new set ring locks in the groove.

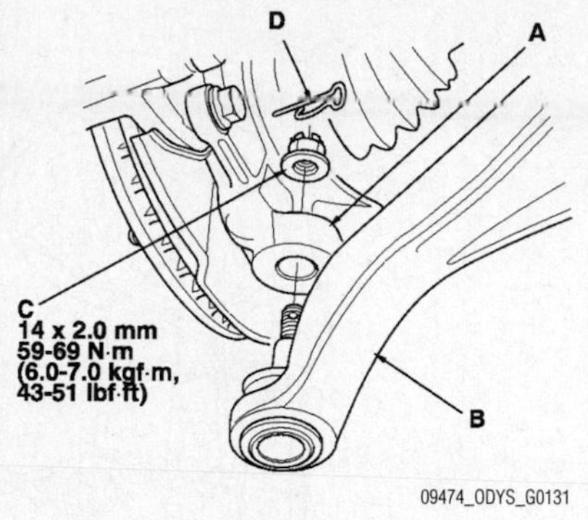
D
A
C
14 x 2.0 mm
59-69 N·m
(6.0-7.0 kgf·m,
43-51 lbf·ft)
B
09474_ODYS_G0131

Fig. 37 Clean off any grease contamination from the ball joint tapered section and threads, then install the knuckle (A) onto the lower arm (B). Torque the new castle nut (C) to the lower torque specification, then tighten it only far enough to align the slot with the ball joint pin (D) hole

the differential or intermediate shaft until the new set ring locks in the groove.

15. Install the outboard joint into the front hub.

16. Install exhaust pipe A.

17. Clean off any grease contamination from the ball joint tapered section and threads, then install the knuckle onto the lower arm. Torque the new castle nut to the lower torque specification, then tighten it only far enough to align the slot with the ball joint pin hole. Do not align the nut by loosening it.

➡Make sure the ball joint boot is not damaged or cracked.

18. Install the new lock pin into the ball joint pin hole.

19. Connect the front stabilizer link to the damper. Hold the stabilizer link ball joint pin with a hex wrench and tighten the new flange nut.

20. Install a new spindle nut, then tighten the nut. After tightening, use a drift to stake the spindle nut shoulder against the halfshaft.

21. Clean the mating surfaces of the brake disc and the front wheel, then install the front wheel with the wheel nuts.

22. Turn the front wheel by hand, and make sure there is no interference between the halfshaft and surrounding parts.

23. Refill the automatic transmission with recommended transmission fluid.

24. Check the front wheel alignment, and adjust it if necessary

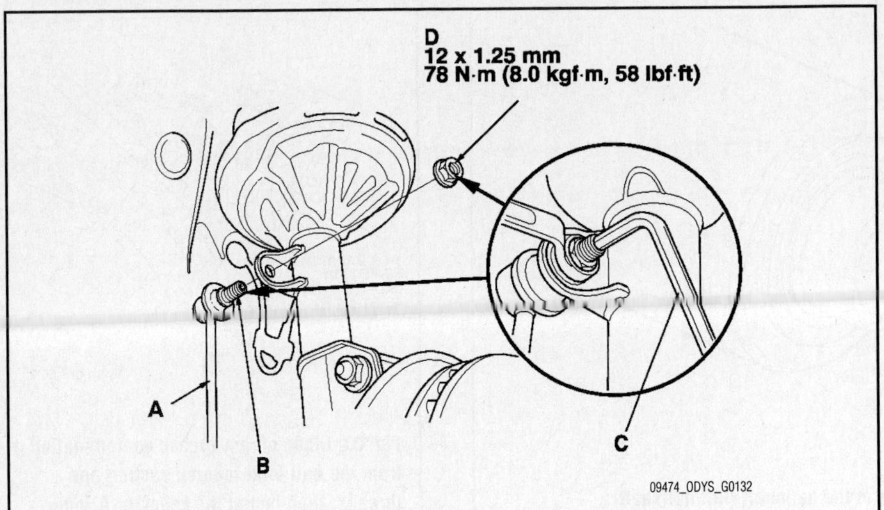
D
12 x 1.25 mm
78 N·m (8.0 kgf·m, 58 lbf·ft)
A
B
C
09474_ODYS_G0132

Fig. 38 Connect the front stabilizer link to the damper. Hold the stabilizer link ball joint pin with a hex wrench, and tighten the new flange nut.

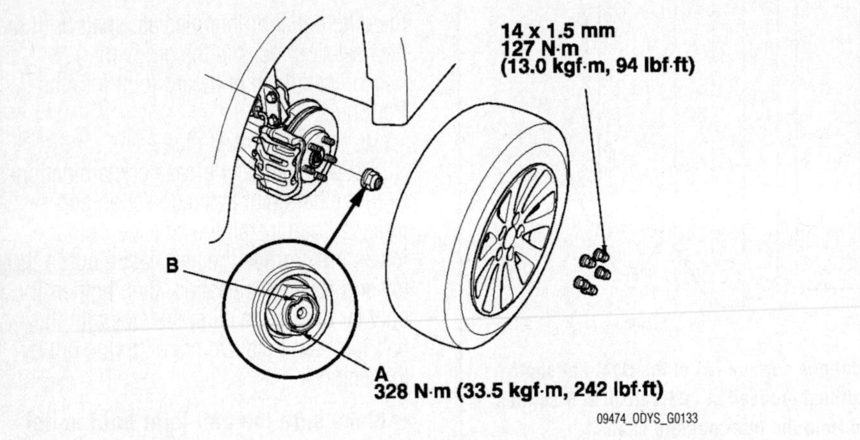
14 x 1.5 mm
127 N·m
(13.0 kgf·m, 94 lbf·ft)
B
A
328 N·m (33.5 kgf·m, 242 lbf·ft)
09474_ODYS_G0133

Fig. 39 Install a new spindle nut (A), then tighten the nut. After tightening, use a drift to stake the spindle nut shoulder (B) against the halfshaft

CV-BOOTS INSPECTION

Inboard Joint

See Figures 40 through 43.

1. Remove or disconnect the following:
   - Axle halfshaft from the vehicle.
   - Inboard joint boot clamps and push the boot back
   - Inboard joint housing from the axle
   - Rollers from the spider
   - Snapring and the spider from the axle shaft
   - Inboard joint boot

To install:

➡Use new circlips and boot clamps for assembly.

2. Install or connect the following:
   - Inboard joint boot and clamps to the axle shaft
   - Spider with a new snapring
   - Rollers to the spider
3. Fill the joint housing with grease and install it.
4. Fill the inboard joint boot with grease and install the boot clamps. Make sure to adjust the length of the driveshafts as shown in the accompanying illustration.
5. Install the axle halfshaft to the vehicle.

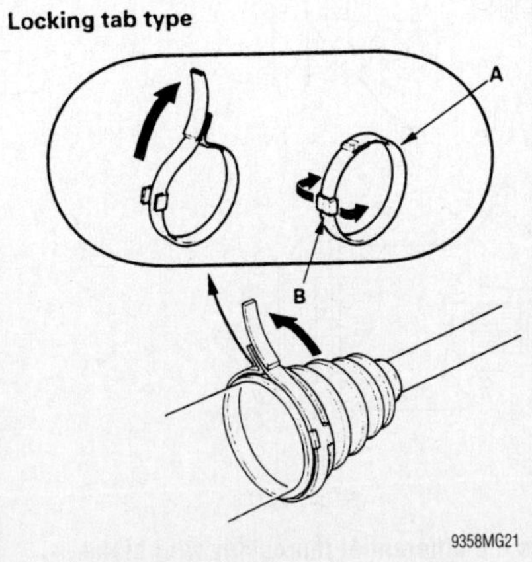
Locking tab type
A
B
9358MG21

Fig. 40 View of the locking tab type boot band (A) and the direction of rotation (B)

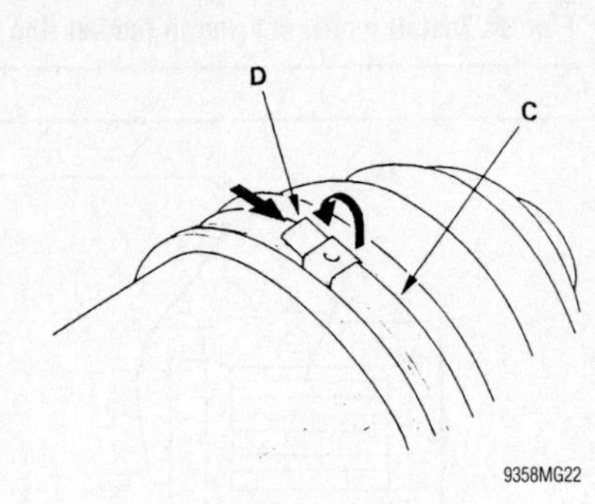
Double loop type
D
C
9358MG22

Fig. 41 View of the double loop tab type boot band (C) and the protruding end (D)

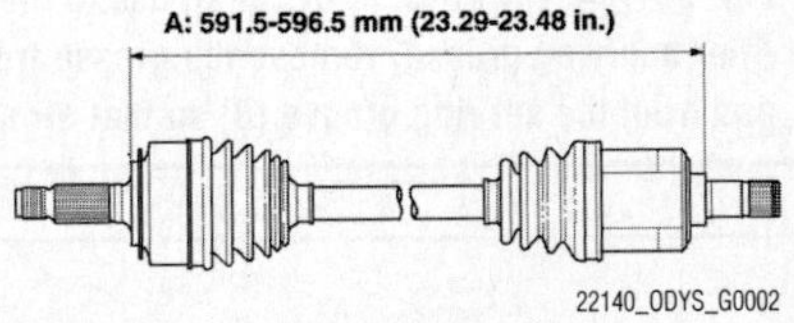
A: 591.5-596.5 mm (23.29-23.48 in.)
22140_ODYS_G0002

Fig. 42 Make sure to adjust the length of the left halfshaft as shown

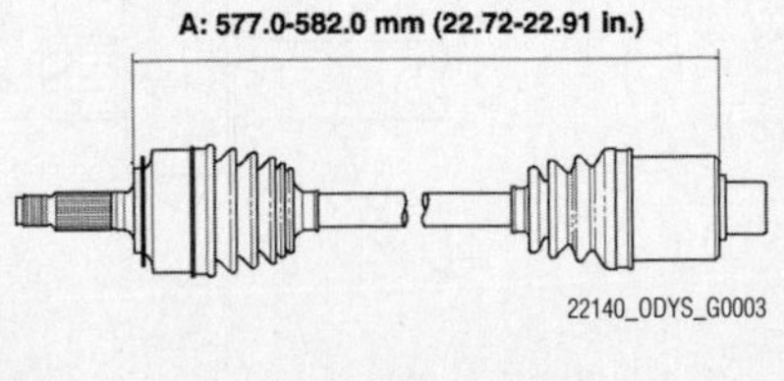
A: 577.0-582.0 mm (22.72-22.91 in.)
22140_ODYS_G0003

Fig. 43 Make sure to adjust the length of the right halfshaft as shown

Outboard Joint

See Figures 44 and 45.

1. Remove or disconnect the following:
   - Axle halfshaft from the vehicle and place it in a vise
   - Outboard joint boot clamps and push the boot back
   - Outboard joint by driving it off the axle shaft with a brass drift and hammer or tool 07XAC-001020A, if available
   - Outboard joint boot

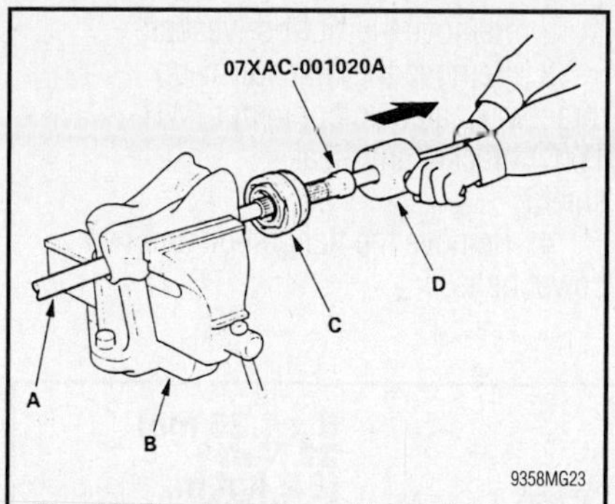
07XAC-001020A
A
B
C
D
9358MG23

Fig. 44 Removing the outboard joint using tool 07XAC-001020A

To install:

➡Use new circlips and boot clamps for assembly.

2. Install the outboard joint boot and clamps to the axle shaft.
3. Fill the outboard joint with grease. Install the outboard joint to the axle shaft. Tap the stub shaft with a brass hammer to seat the circlip.
4. Fill the outboard joint boot with grease and install the boot clamps.
5. Install the axle halfshaft to the vehicle.

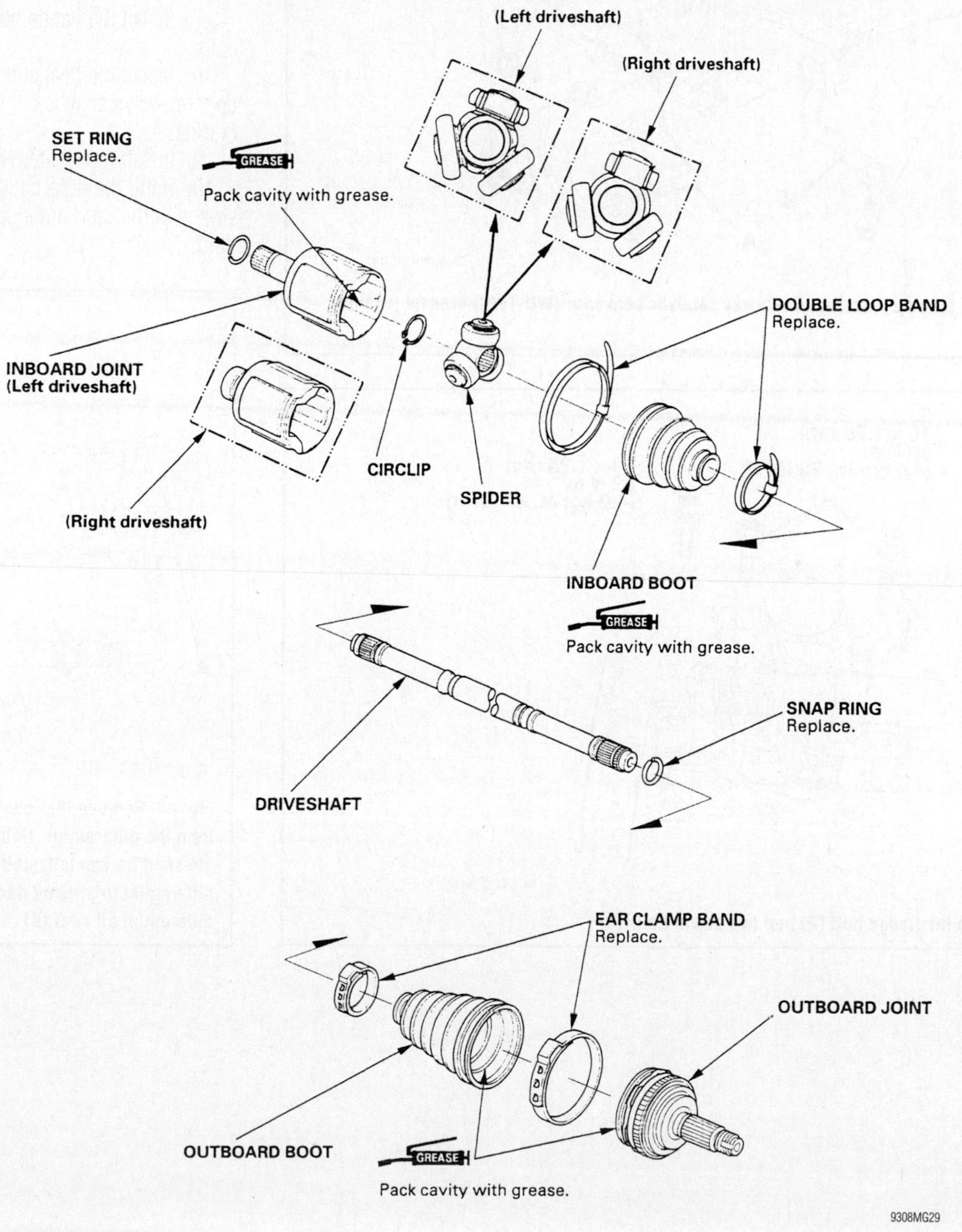
(Left driveshaft)
(Right driveshaft)
SET RING
Replace.
GREASE
Pack cavity with grease.
INBOARD JOINT
(Left driveshaft)
(Right driveshaft)
CIRCLIP
SPIDER
DOUBLE LOOP BAND
Replace.
INBOARD BOOT
GREASE
Pack cavity with grease.
SNAP RING
Replace.
DRIVESHAFT
EAR CLAMP BAND
Replace.
OUTBOARD JOINT
OUTBOARD BOOT
GREASE
Pack cavity with grease.
9308MG29

Fig. 45 Front axle exploded view

INTERMEDIATE SHAFT

REMOVAL & INSTALLATION

See Figures 46 through 48.

1. Drain the automatic transmission fluid Reinstall the drain plug, using a new washer.
2. Remove the right driveshaft.
3. Remove the rear warm-up three-way catalytic converter (WU-TWC) bracket and heat shield.
4. Remove the flange bolt and two dowel bolts.
5. Remove the intermediate shaft from the differential. Hold the intermediate shaft horizontally until it is clear of the differential to prevent damage to the differential oil seal.

To install:

6. Use brake cleaner to thoroughly clean the areas where the intermediate shaft contacts the transmission (differential), and dry them with compressed air. Do not wash the rubber parts with solvent. Insert the intermediate shaft assembly into the differential. Hold the intermediate shaft horizontally to prevent damage to the differential oil seal.
7. Install the flange bolt and two dowel bolts.
8. Install the heat shield and rear warm-up three-way catalytic converter (WU-TWC) bracket.
9. Install the right driveshaft.
10. Refill the automatic transmission with the recommended transmission fluid.

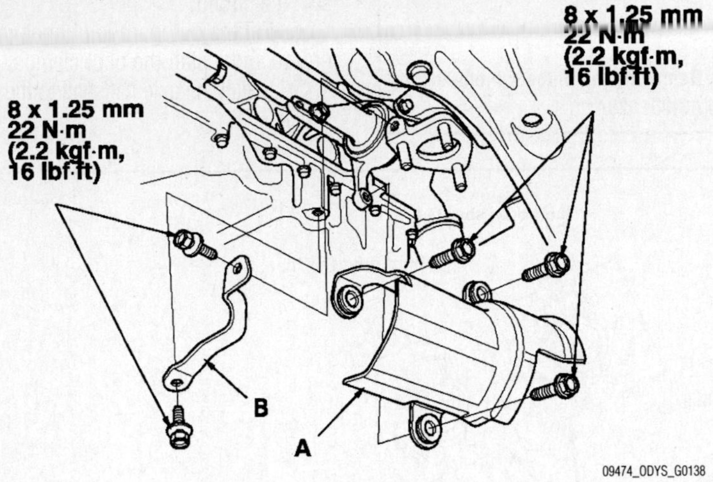
8 x 1.25 mm
22 N·m
(2.2 kgf·m,
16 lbf·ft)
8 x 1.25 mm
22 N·m
(2.2 kgf·m,
16 lbf·ft)
B
A
09474_ODYS_G0138

Fig. 46 Remove the rear warm-up three-way catalytic converter (WU-TWC) bracket (B) and heat shield (A)

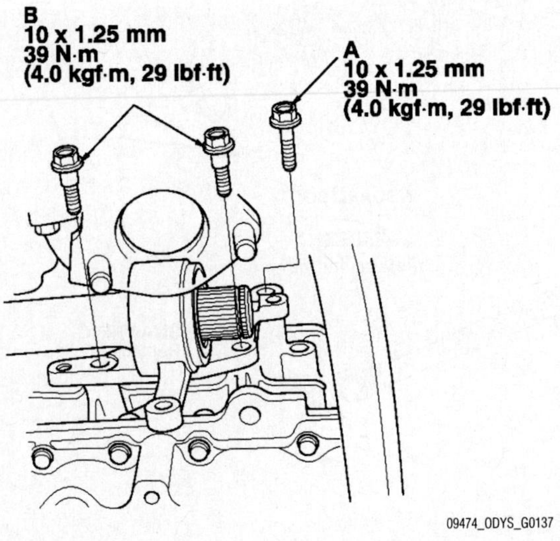
B
10 x 1.25 mm
39 N·m
(4.0 kgf·m, 29 lbf·ft)
A
10 x 1.25 mm
39 N·m
(4.0 kgf·m, 29 lbf·ft)
09474_ODYS_G0137

Fig. 47 Remove the flange bolt (A) and two dowel bolts (B)

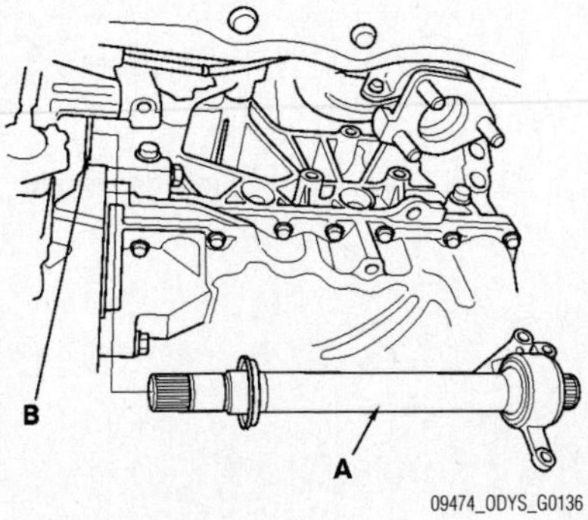
B
A
09474_ODYS_G0136

Fig. 48 Remove the intermediate shaft (A) from the differential. Hold the intermediate shaft horizontally until it is clear of the differential to prevent damage to the differential oil seal (B)

ENGINE COOLING

ENGINE COOLING FANS & SHROUD

REMOVAL & INSTALLATION

See Figures 49 through 51.

1. Do the battery terminal disconnect procedure, then wait at least 3 minutes before beginning work.
2. Remove the battery.
3. Remove the front bumper.
4. Disconnect the breather pipe, then remove the intake air duct.
5. Remove the battery base.
6. Disconnect the A/C compressor clutch connector and the fan motor connectors. Disconnect the hood switch connector.
7. If equipped with the J35A6 engine, disconnect the engine mount control solenoid valve connector and the vacuum hoses. Remove the ground cable and the harness clamp.
8. Remove the two bolts securing the coolant reservoir, then remove the coolant reservoir.
9. Remove the two bolts, and loosen the bolt securing the radiator fan shroud, then remove the radiator fan shroud assembly.
10. Remove the two bolts, and loosen the two bolts securing the A/C condenser fan shroud, then remove the A/C condenser fan shroud assembly from the battery side.

To install:

11. Install the A/C condenser fan shroud assembly from the battery side, then install the two bolts, and tighten the two bolts previously loosened.
12. Install the radiator fan shroud assembly, then install the two bolts, and tighten the bolt previously loosened.
13. If equipped with J35A6 engine, install the ground cable and the harness clamp. Connect the engine mount control solenoid valve connector and the vacuum hoses.
14. Connect the A/C compressor clutch connector and the fan motor connectors.

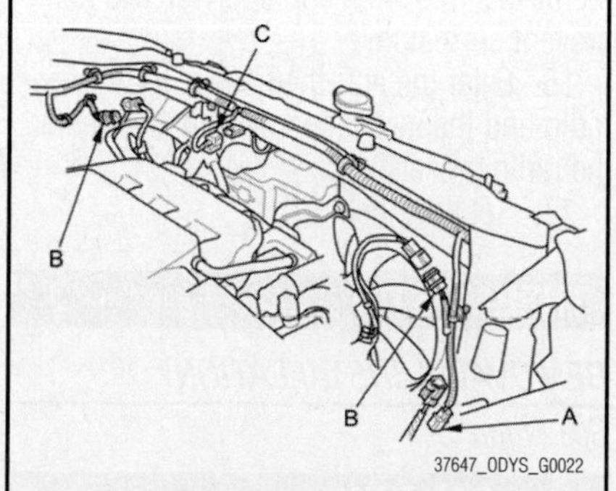
C
B
B
A
37647_ODYS_G0022

Fig. 49 Disconnect the A/C compressor clutch connector (A) and the fan motor connectors (B). Disconnect the hood switch connector (C).

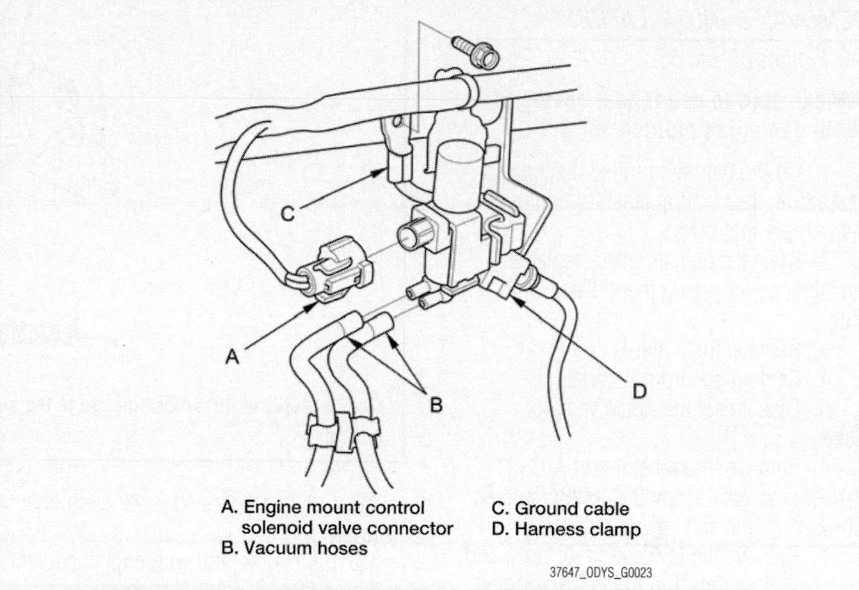
C
A
B
D
A. Engine mount control solenoid valve connector
B. Vacuum hoses
C. Ground cable
D. Harness clamp
37647_ODYS_G0023

Fig. 50 If equipped with the J35A6 engine, disconnect the engine mount control solenoid valve connector (A) and the vacuum hoses (B). Remove the ground cable (C) and the harness clamp (D).

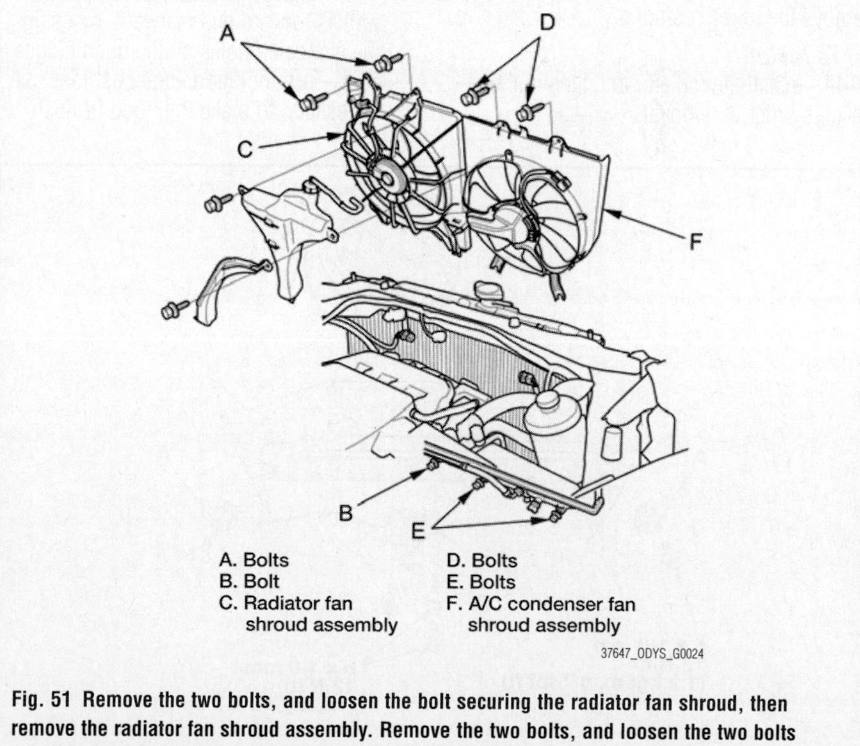
A
D
C
F
B
E
A. Bolts
B. Bolt
C. Radiator fan shroud assembly
D. Bolts
E. Bolts
F. A/C condenser fan shroud assembly
37647_ODYS_G0024

Fig. 51 Remove the two bolts, and loosen the bolt securing the radiator fan shroud, then remove the radiator fan shroud assembly. Remove the two bolts, and loosen the two bolts securing the A/C condenser fan shroud, then remove the A/C condenser fan shroud assembly from the battery side.

15. Connect the hood switch connector.
16. Install the battery base.
17. Install the intake air duct, then connect the breather pipe.
18. Install the front bumper.
19. Do the battery installation procedure.

RADIATOR

REMOVAL & INSTALLATION

See Figures 52 and 53.

➡**Make sure to use fender covers to avoid damaging painted surfaces.**

1. Do the battery terminal disconnect procedure, then wait at least 3 minutes before beginning work.
2. Drain the coolant into a sealable container by loosening the radiator drain plug.
3. Remove the battery.
4. Remove the front bumper.
5. Disconnect the upper radiator hose.
6. Remove the radiator and A/C condenser fans. See "Engine Cooling Fans & Shroud" in this section.
7. Remove the lower radiator hose and automatic transmission fluid (ATF) cooler hoses. Plug the ATF cooler hoses and lines.
8. Remove the radiator upper brackets/cushions.
9. Remove the engine hood latch.
10. Pull up the radiator assembly, then remove the lower cushions.

To install:

11. Install the radiator and fans in the reverse order of removal.
12. Set the upper and lower cushions securely.
13. Fill the radiator with engine coolant and bleed the air from the system.
14. Adjust the engine hood latch:
  a. Loosen each bolt slightly.
  b. Adjust the hood alignment in this sequence:
  c. Adjust the hood right and left, as well as forward and rearward, by using the elongated holes on the hood hinges.
  d. Turn the hood edge cushions, as necessary, to make the hood fit flush with the body at the front and side edges.
  e. Adjust the hood latch to obtain the proper height at the forward edge, and move the hood latch right or left until the strike is centered in the hood latch.
  f. Tighten each bolt securely.
  g. Check that the hood opens properly and locks securely.
  h. For some models, check that the security system operates properly with the hood opened and closed.
  i. Apply touch-up paint to the hinge mounting bolts and around the hinges.
  j. Apply multipurpose grease to the hood latch and hood hinge as indicated by the arrows.
15. Install the battery. Clean the battery posts and cable terminals with sandpaper, then assemble them and apply grease to prevent corrosion.
16. Enter the anti-theft codes for the radio and the navigation system, then enter the radio presets.
17. Set the clock.

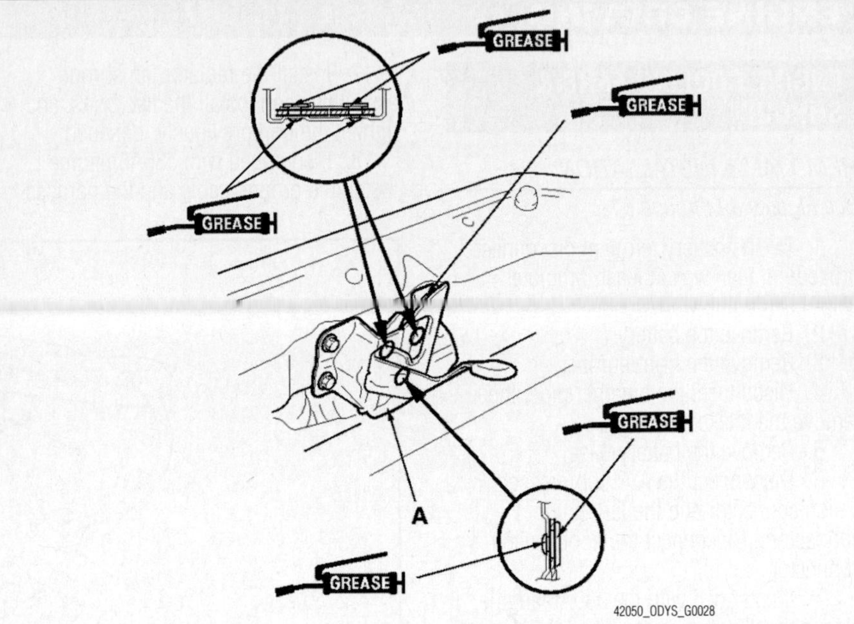
GREASE
GREASE
GREASE
GREASE
GREASE
A
42050_ODYS_G0028

Fig. 53 Apply multipurpose grease to the hood latch (A) and hood hinge (B) as indicated by the arrows

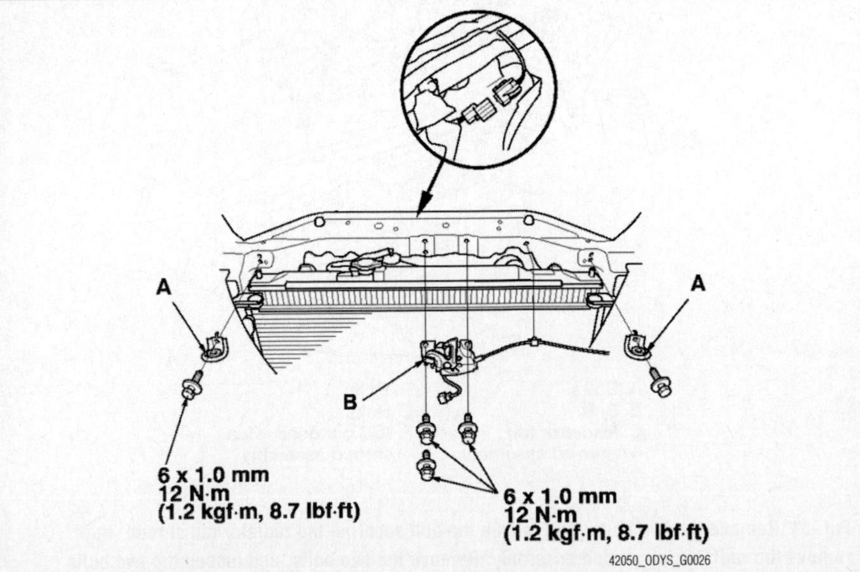
A
A
B
6 x 1.0 mm
12 N·m
(1.2 kgf·m, 8.7 lbf·ft)
6 x 1.0 mm
12 N·m
(1.2 kgf·m, 8.7 lbf·ft)
42050_ODYS_G0026

Fig. 52 Remove the upper brackets/cushions (A) and engine hood latch (B)

THERMOSTAT

REMOVAL & INSTALLATION

See Figure 54.

⁂ CAUTION

Never open, service or drain the radiator or cooling system when hot; serious burns can occur from the steam and hot coolant. Also, when

draining engine coolant, keep in mind that cats and dogs are attracted to ethylene glycol antifreeze and could drink any that is left in an uncovered container or in puddles on the ground. This will prove fatal in sufficient quantities. Always drain coolant into a sealable container. Coolant should be reused unless it is contaminated or is several years old.

1. Note the radio security code and station presets.
2. Disconnect the negative battery cable.
3. Drain the engine coolant into a sealable container.
4. Remove the fasteners from the thermostat housing, then remove the thermostat.

To install:

5. Install the thermostat using a new seal. If the thermostat has a small bleed hole, make sure the bleed hole is on the top.
6. Apply an anti-seize compound to the threads of the fasteners.
7. Reassemble in the reverse order of disassembly.
8. Set the heater to the full hot position. Set the heater to the full hot position.
9. Top off the cooling system and overflow reservoir with a 50/50 mixture of a recommended antifreeze and water solution.
10. Bleed the system to remove any air pockets as necessary. Simultaneously squeeze the upper and lower radiator hoses to help push any captured air pockets out of the system.
11. Inspect all coolant hoses and fittings to make sure they are properly installed and if previously opened, close the bleed valve.
12. Connect the negative battery cable.
13. Install the radiator cap loosely and start the engine. Allow the engine to run until the cooling fan has cycled two times, then turn the engine **OFF** and top off the cooling system as necessary.
14. Install the radiator cap and inspect for leaks.
15. Enter the radio security code.

WARNING

The manufacturer does not recommend using a coolant concentration of greater than 60 percent antifreeze.

➡When mixing a 50/50 solution of antifreeze and water, using distilled water may help to keep the cooling system from building up mineral deposits and internal blockage.

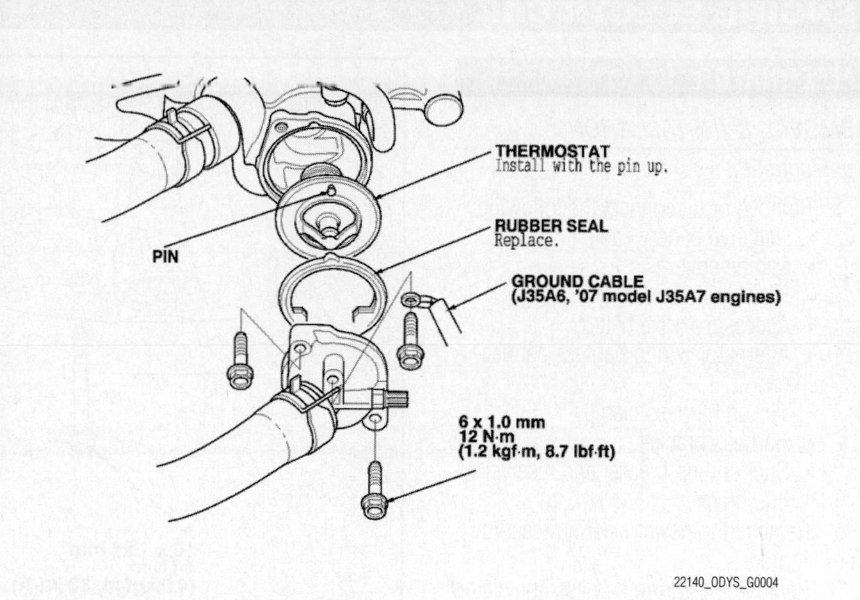
THERMOSTAT
Install with the pin up.
RUBBER SEAL
Replace.
PIN
GROUND CABLE
(J35A6, '07 model J35A7 engines)
6 x 1.0 mm
12 N·m
(1.2 kgf·m, 8.7 lbf·ft)
22140_ODYS_G0004

Fig. 54 Always install the thermostat with the small pin at the top. The rubber thermostat seal is installed around the thermostat

WATER PUMP

REMOVAL & INSTALLATION

See Figure 55.

1. Before servicing the vehicle, refer to the precautions in the beginning of this section.
2. Drain the cooling system.
3. Remove or disconnect the following:
   - Negative battery cable
   - Accessory drive belts
   - Front cover
   - Timing belt. Refer to the timing belt procedure.
   - Timing belt tensioner
   - Water pump

To install:

4. Install or connect the following:
   - Water pump. Use a new O-ring seal and tighten the bolts to 105 inch lbs. (12 Nm).
   - Timing belt tensioner
   - Timing belt
   - Front cover
   - Accessory drive belts
   - Negative battery cable
5. Refill the cooling system to the correct level.
6. Start the engine and check for leaks.

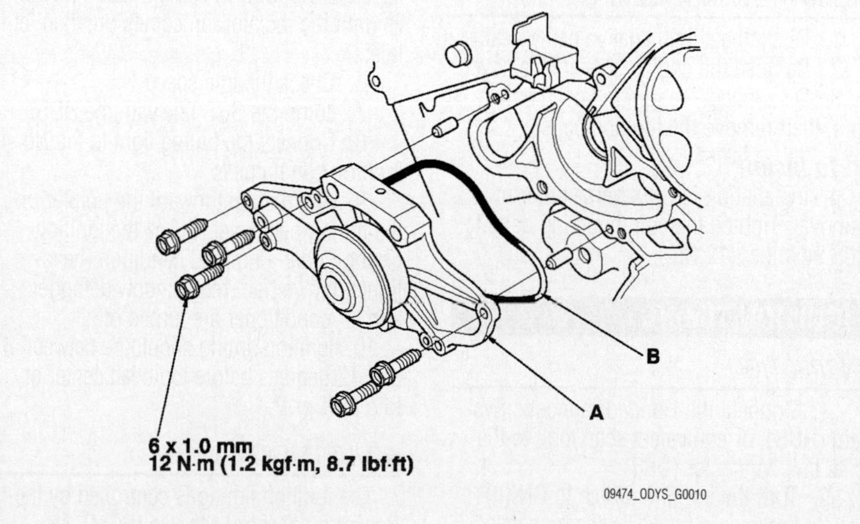
B
A
6 x 1.0 mm
12 N·m (1.2 kgf·m, 8.7 lbf·ft)
09474_ODYS_G0010

Fig. 55 Water pump (A) and gasket (B)

ENGINE ELECTRICAL — CHARGING SYSTEM

ALTERNATOR

REMOVAL & INSTALLATION

See Figure 56.

1. Remove or disconnect the following:
   - Negative battery cable
   - Engine cover
   - Front grille cover
   - Accessory drive belt
   - Alternator wiring harness connectors
   - Wiring harness clamp
   - Harness bracket
   - Suction line bracket and receiver line bracket
2. Remove the power steering reservoir from its bracket.
3. Remove the wiring harness line clamp below the reservoir location.
4. Remove the wiper washer reservoir.
5. Remove the two alternator retaining bolts and remove the alternator.

To install:

6. Install or connect the following:
   - Alternator; Tighten the 10mm bolt to 33 ft. lbs. (44 Nm) and the 8mm bolt to 16 ft. lbs. (22 Nm).
   - Wiper washer reservoir
   - Wiring harness clamp below the power steering reservoir location
   - Power steering fluid reservoir to the bracket
   - Suction line bracket and receiver line bracket
   - Harness bracket
   - Wiring harness clamp
   - Alternator wiring harness connectors
   - Accessory drive belt
   - Front grille cover
   - Engine cover
   - Negative battery cable

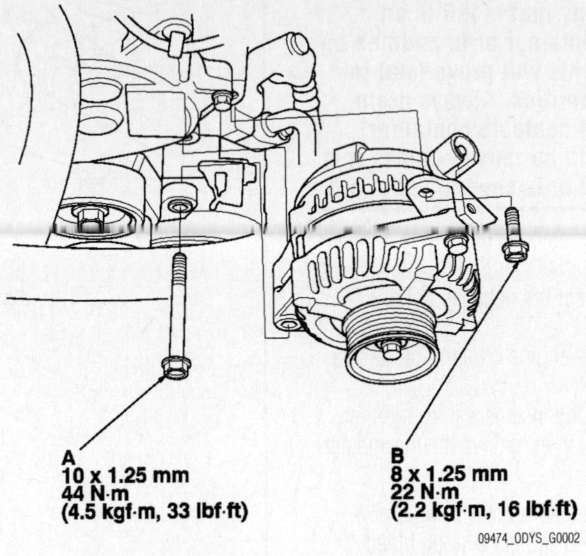
A
10 x 1.25 mm
44 N·m
(4.5 kgf·m, 33 lbf·ft)
B
8 x 1.25 mm
22 N·m
(2.2 kgf·m, 16 lbf·ft)
09474_ODYS_G0002

Fig. 56 Alternator mounting, showing the two different mounting bolts (A and B)

ENGINE ELECTRICAL — IGNITION SYSTEM

FIRING ORDER

Firing order is: 1–4–2–5–3–6

IGNITION COIL

REMOVAL & INSTALLATION

1. Disconnect the negative battery cable.
2. Remove the engine appearance cover.
3. Disconnect the ignition coil connectors, then remove the ignition coils.

To install:

4. Installation is the reverse order of removal. Tighten the coil mounting bolts to 105 inch lbs. (12 Nm).

IGNITION TIMING

INSPECTION

1. Connect the Honda Diagnostic System (HDS), or equivalent scan tool, to the data link connector (DLC).
2. Turn the ignition switch to ON (II).
3. Make sure the HDS communicates with the vehicle and the powertrain control module (PCM). If it does not communicate, troubleshoot the DLC circuit.
4. Check for DTCs. If a DTC is present, diagnose and repair the cause before continuing with this test.
5. Start the engine. Hold the engine speed at 3,000 RPM with no load (in N or P) until the radiator fan comes on, then let it idle.
6. Check the idle speed.
7. Jump the SCS line with the HDS.
8. Connect the timing light to the No. 1 ignition coil harness.
9. Aim the light toward the pointer on the timing belt cover. Check the ignition timing under a no load condition (headlights, blower fan, rear window defogger, and air conditioner are turned off).
10. Ignition timing should be between 8 and 12 degrees before top dead center at idle in **N** or **P**.

ADJUSTMENT

The ignition timing is controlled by the Powertrain Control Module (PCM). No adjustment is necessary.

SPARK PLUGS

REMOVAL & INSTALLATION

The Odyssey uses a coil over plug ignition system. Each of the spark plugs has its own ignition coil which mounts directly above the spark plug and eliminates the need for a distributor, distributor cap, rotor and spark plug wires. Because the ignition coils are placed above the spark plugs, the coils must be removed before the spark plugs can be accessed.

1. Disconnect the negative battery cable, note the radio security code and, if the vehicle has been run recently, allow the engine to thoroughly cool.
2. Remove the ignition coil for the spark plug that needs to be removed.
3. Using a spark plug socket equipped with a rubber insert to properly hold the plug, turn the spark plug counterclockwise to loosen and remove the spark plug from the threaded hole in the cylinder head.

⁂ WARNING

Avoid using a flexible extension on the spark plug socket. A flexible extension may allow a shear force to be applied to the plug. A shear force could break the plug off in the cylinder head, leading to costly repairs.

4. Inspect each ignition coil for damage, or deterioration. Make sure the coils are clean, and free of debris, such as engine oil. If a damaged coil is found, it should be replaced.

⁂ WARNING

The original equipment spark plugs installed in this engine are platinum tip spark plugs, and the plug gap must not be adjusted.

5. Using a wire feeler gauge, check, but **do not** adjust the spark plug gap. When using a gauge, the proper size should pass between the electrodes with a slight drag. The next larger size should not be able to pass, while the next smaller size should pass freely.

6. The spark plug gap should be 0.039–0.043 inches (1.0–1.1mm). If the spark plug gap exceeds 0.051inches (1.3mm) replace the spark plug.

To install:

7. Apply a light coating of an anti-seize compound to the spark plug threads.

8. Carefully thread the plug into the threaded spark plug hole by hand. If resistance is felt before the plug is almost completely threaded, back the plug out and begin threading again. In tight, hard to reach areas, a small piece of rubber hose pressed onto the spark plug can be used as a threading tool. The rubber hose will hold the plug and while twisting the end of the hose, and the hose will be flexible enough to twist before allowing the plug to cross thread.

⁂ WARNING

Do not use any excessive force when beginning to install the plugs. Always carefully thread the plug by hand or by using a rubber hose to prevent the possibility of cross threading and damaging the cylinder head threads.

9. Carefully tighten the spark plug to 13 ft. lbs. (18 Nm).

10. Install the ignition coils, then attach the electrical connector to each ignition coil.

11. Connect the negative battery cable and enter the radio security code.

ENGINE ELECTRICAL

STARTING SYSTEM

STARTER

REMOVAL & INSTALLATION

1. Make sure you have the anti-theft codes for the radio and navigation system.
2. Disconnect the battery cables.
3. Remove the battery and battery tray.
4. Remove the air intake assembly.
5. Remove the harness clamp, harness clamp bracket, and dipstick if necessary.
6. Disconnect the starter wiring connections.
7. Remove the mounting bolts and remove the starter motor.

To install:

8. Installation is the reverse order of removal. Tighten the starter mounting bolts to 33 ft. lbs. (44 Nm).

ENGINE MECHANICAL

ACCESSORY DRIVE BELTS

ACCESSORY BELT ROUTING

See Figure 57.

Refer to the accompanying illustration for belt routing.

INSPECTION

J35A6 Engine

See Figure 58.

1. Inspect the belt for cracks and damage. If the belt is cracked or damaged, replace it.
2. Check that the auto-tensioner indicator is within the standard range as shown. If it is out of the standard range, replace the drive belt.

J35A7 Engine

See Figure 59.

1. Inspect the belt for cracks and damage. If the belt is cracked or damaged, replace it.

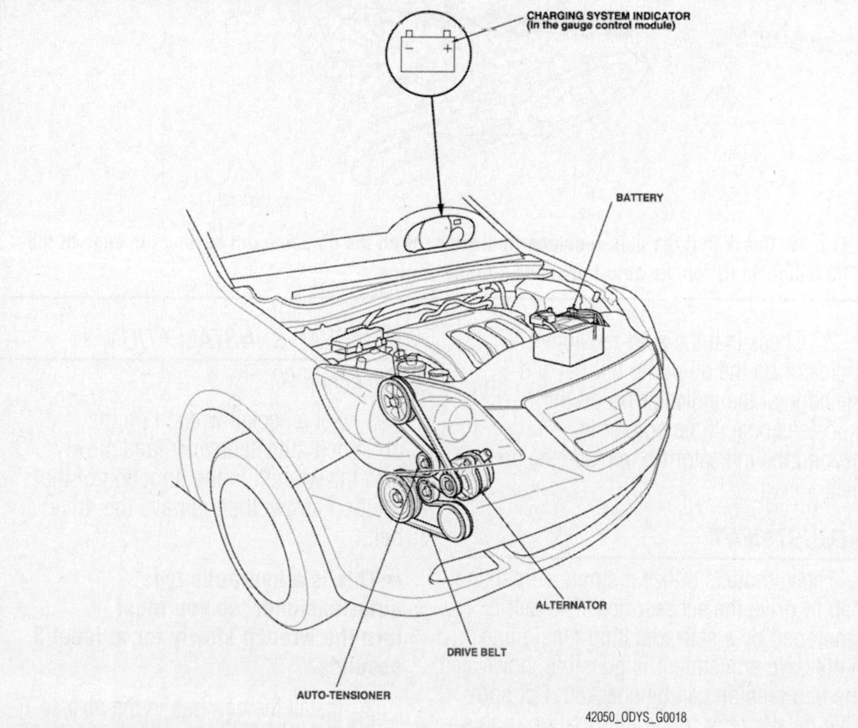
CHARGING SYSTEM INDICATOR
(In the gauge control module)
BATTERY
ALTERNATOR
DRIVE BELT
AUTO-TENSIONER
42050_ODYS_G0018

Fig. 57 Belt routing and related components—J35A6 engine shown; J35A7 engine the same routing

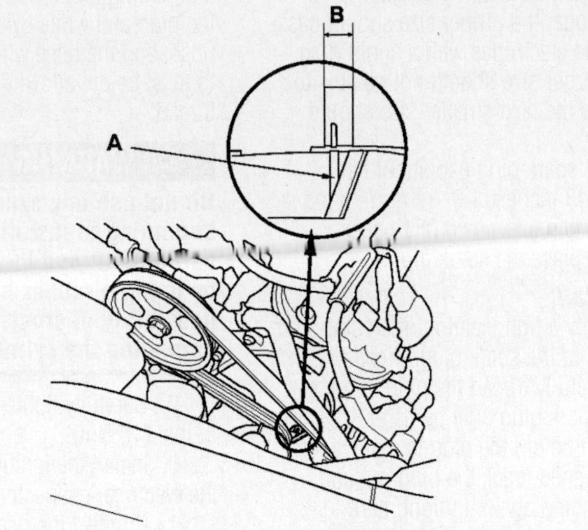
B
A
42050_ODYS_G0107

Fig. 58 Check that the auto-tensioner indicator (A) is within the standard range (B) as shown. If it is out of the standard range, replace the drive belt—J35A6 engine

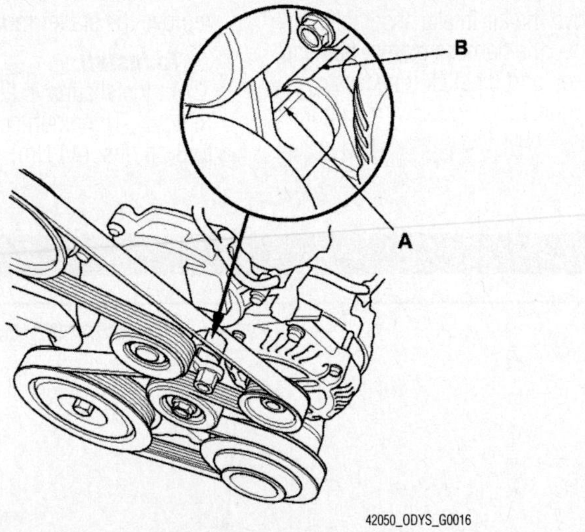
B
A
42050_ODYS_G0016

Fig. 59 Check that the auto-tensioner indicator (A) on the oil pump not beyond the edge of the indicator rib (B) on the auto-tensioner—J35A7 engine

2. Check that the auto-tensioner indicator on the oil pump not beyond the edge of the indicator rib on the auto-tensioner. If the pointer is beyond the indicator rib, replace the drive belt.

ADJUSTMENT

These models utilize a single serpentine belt to drive the accessories. The belt is tensioned by a self-adjusting tensioning pulley. No adjustment is possible, however the belt tension can be checked. For specific details refer to the information in this section.

REMOVAL & INSTALLATION

See Figure 60.

1. Set a socket wrench on the drive belt auto-tensioner, and slowly turn the wrench in the direction of the rotation arrow, then remove the drive belt.

➡This is a hydraulic type auto-tensioner, so you must turn the wrench slowly for at least 3 seconds.

2. Install the new belt in the reverse order of removal.

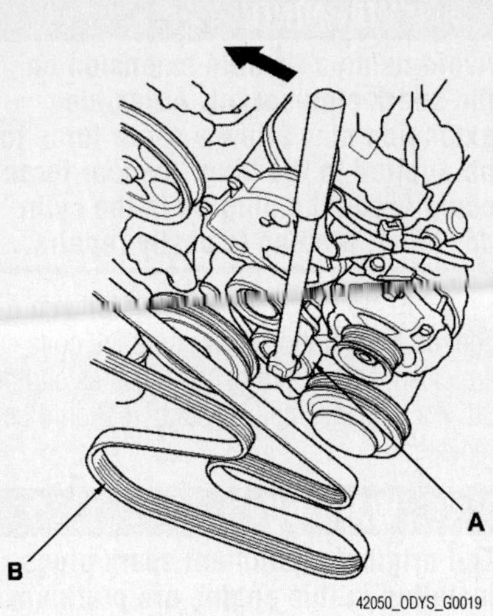
A
B
42050_ODYS_G0019

Fig. 60 Insert a hexagon socket wrench into the drive belt auto-tensioner (A), and slowly turn the wrench in the direction of the rotation arrow, then remove the drive belt (B)

CAMSHAFT AND VALVE LIFTERS

INSPECTION

➡Refer to the camshaft specification chart.

1. Remove the cylinder head.
2. Remove the rocker arm assembly.
3. Put the rocker shafts on the cylinder head, then tighten the bolts to 17 ft. lbs. (24 Nm).
4. Seat the camshaft by pushing it toward the rear of the cylinder head.
5. Zero the dial indicator against the end of the camshaft. Push the camshaft back and forth and read the end play. If the end play is beyond the service limit, replace the thrust cover and recheck. If it is still beyond the service limit, replace the camshaft.
6. Remove the camshaft thrust cover, then pull out the camshaft.
7. Wipe the camshaft clean, then inspect the lift ramps. Replace the camshaft if any lobes are pitted, scored, or excessively worn.
8. Measure the diameter of each camshaft journal.
9. Zero the gauge to the journal diameter.
10. Clean the camshaft bearing surfaces in the cylinder head. Measure the inside diameter of each camshaft bearing surface, and check for an out-of-round condition. If the camshaft-to-holder clearance is beyond the service limit and the camshaft has been replaced, replace the cylinder head.

11. If the camshaft-to-holder clearance is beyond the service limit and the camshaft has not been replaced, check total runout with the camshaft supported on V-blocks. If the total runout of the camshaft is within the service limit, replace the cylinder head. If the total runout is beyond the service limit, replace the camshaft and recheck the oil clearance. If the oil clearance is still out of tolerance, replace the cylinder head.
12. Measure cam lobe height.

REMOVAL & INSTALLATION

J35A6 Engine

Front

See Figure 61.

1. Make sure you have the anti-theft codes for the radio and navigation systems.
2. Remove the battery.
3. Drain the coolant.
4. Remove the upper radiator hose.
5. Remove the EGR valve.
6. Remove the timing belt.
7. Remove the rocker arm assembly.
8. Remove the camshaft pulley.
9. Remove the thrust cover.
10. Carefully remove the camshaft.

To install:

11. Installation is the reverse of removal. Always use new O-rings. Apply clean engine oil to the journals and lobes.

Rear

1. Relieve the fuel system pressure.
2. Remove the air cleaner assembly.
3. Remove the intake manifold.
4. Disconnect the fuel feed line.
5. Disconnect the brake lines at the master cylinder. Plug the openings.
6. Remove the timing belt.
7. Remove the rocker arm assembly.
8. Remove the camshaft pulley.
9. Remove the EVAP canister hose joint and bracket.
10. Remove the thrust cover and camshaft.

To install:

11. Installation is the reverse of removal. Always use new O-rings. Apply clean engine oil to the journals and lobes. Bleed the brakes. Perform the CKP pattern clear/learn procedure.

J35A7 Engine

Front

See Figure 62.

1. Make sure you have the anti-theft codes for the radio and navigation systems.
2. Remove the battery.
3. Drain the coolant.
4. Remove the upper radiator hose.
5. Remove the EGR valve.
6. Remove the timing belt.
7. Remove the rocker arm assembly.
8. Remove the camshaft pulley.
9. Remove the thrust cover.
10. Carefully remove the camshaft.

To install:

11. Installation is the reverse of removal. Always use new O-rings. Apply clean engine oil to the journals and lobes.

Rear

1. Relieve the fuel system pressure.
2. Remove the air cleaner assembly.
3. Remove the intake manifold.
4. Disconnect the fuel feed line.
5. Disconnect the brake lines at the master cylinder. Plug the openings.
6. Remove the timing belt.

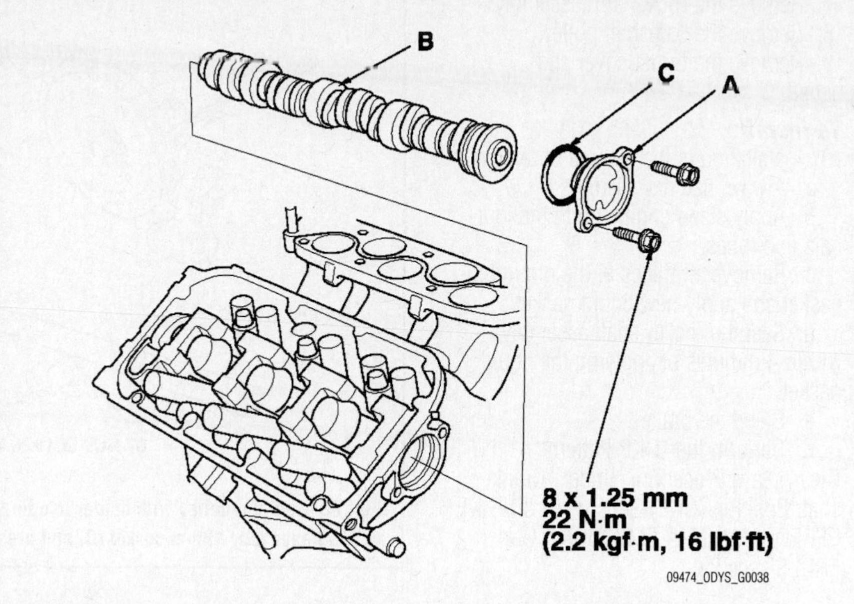
B
C
A
8 x 1.25 mm
22 N·m
(2.2 kgf·m, 16 lbf·ft)
09474_ODYS_G0038

Fig. 61 Remove the thrust cover (A) and then carefully remove the camshaft (B) and O-rings (C)

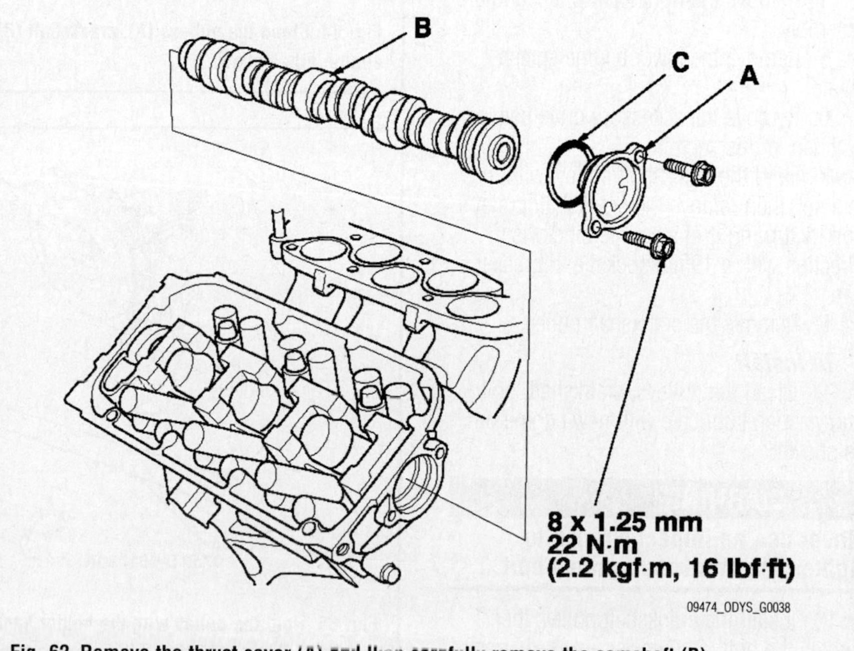
B
C
A
8 x 1.25 mm
22 N·m
(2.2 kgf·m, 16 lbf·ft)
09474_ODYS_G0038

Fig. 62 Remove the thrust cover (A) and then carefully remove the camshaft (B)

7. Remove the rocker arm assembly.
8. Remove the camshaft pulley.
9. Remove the thrust cover and camshaft.

To install:

10. Installation is the reverse of removal.
    a. Always use new O-rings.
    b. Apply clean engine oil to the journals and lobes.
    c. Remove all traces of the old liquid gasket and apply new liquid gasket.
    d. Seat the rocker shaft assembly within 4 minutes of applying the liquid gasket.
    e. Bleed the brakes.
    f. Perform the "CKP Pattern Clear/Learn Procedure" under "Crankshaft Position (CKP) Sensor" in "ENGINE PERFORMANCE & EMISSION CONTROLS" section.

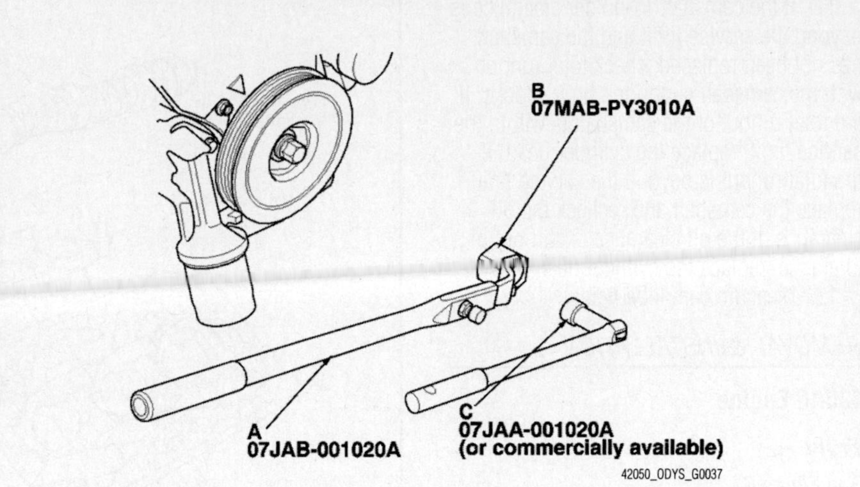
B
07MAB-PY3010A
A
07JAB-001020A
C
07JAA-001020A
(or commercially available)
42050_ODYS_G0037

Fig. 63 Hold the pulley with holder handle (A) and holder attachment (B). Remove the bolt with a heavy duty 19mm socket (C) and breaker bar

CRANKSHAFT DAMPER

REMOVAL & INSTALLATION

See Figures 63 through 66.

This procedure requires the use of the following special tools, or their equivalents:

- Holder handle (07JAB-001020A)
- Holder attachment, 50 mm, offset (07MAB-PY3010A)
- Socket, 19 mm (07JAA-001020A)

1. Note the radio security code and station presets.
2. Disconnect the negative battery cable.
3. Raise and safely support the vehicle.
4. Remove the right front wheel and tire assembly.
5. Remove the lower engine splash shield.
6. Remove the accessory drive belt, as outlined in this section.
7. Hold the crankshaft pulley, with the tool set, then remove the crankshaft pulley bolt by turning it in a counterclockwise direction with a 19mm socket and breaker bar.
8. Remove the crankshaft pulley.

To install:

9. Clean the pulleys, crankshaft, bolt and washer. Lubricate with new engine oil as shown.

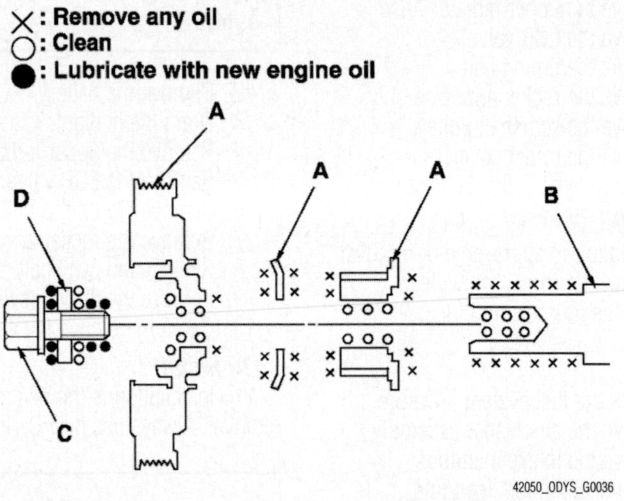
×: Remove any oil
○: Clean
●: Lubricate with new engine oil
A
A
A
B
D
C
42050_ODYS_G0036

Fig. 64 Clean the pulleys (A), crankshaft (B), bolt (C) and washer (D). Lubricate with new engine oil as shown

⁂ WARNING

Never use an impact wrench to tighten the crankshaft pulley bolt.

10. Install the crankshaft pulley, then tighten the bolt, as follows:

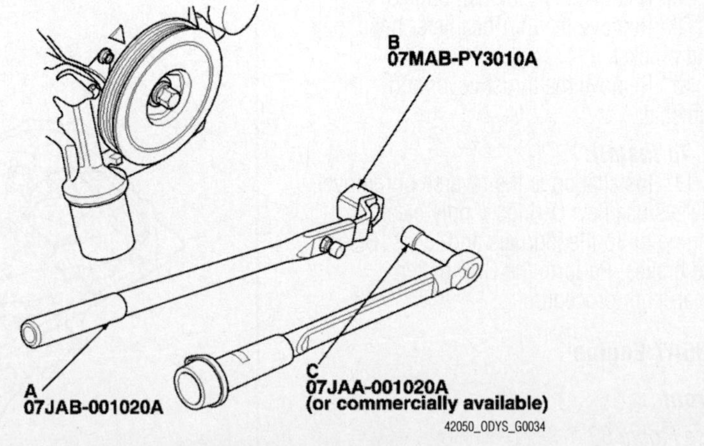
B
07MAB-PY3010A
A
07JAB-001020A
C
07JAA-001020A
(or commercially available)
42050_ODYS_G0034

Fig. 65 Hold the pulley with the holder handle (A) and holder attachment (B), then tighten the bolt to 47 ft. lbs. (65 Nm) with a torque wrench and heavy duty 19mm socket (C)

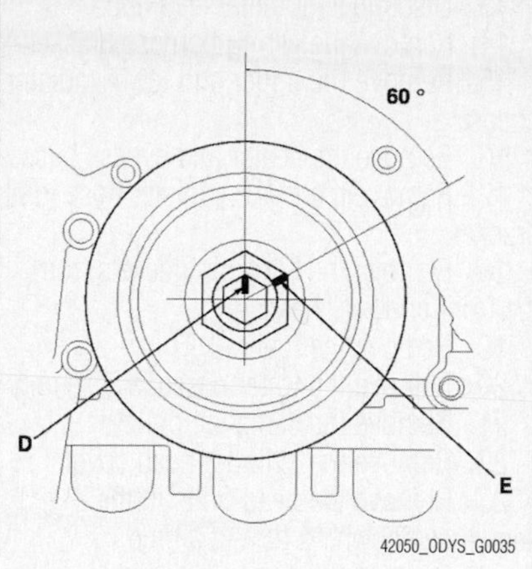
60 °
D
E
42050_ODYS_G0035

Fig. 66 Mark the bolt head (D) and crankshaft pulley (E) as shown, then tighten the bolt an additional 60 degrees. The mark on the bolt head should line up with the mark on the crankshaft pulley

a. Hold the pulley with the holder handle and holder attachment, then tighten the bolt to 47 ft. lbs. (65 Nm) with a torque wrench and heavy duty 19mm socket.

b. Mark the bolt head and crankshaft pulley as shown, then tighten the bolt an additional 60° (The mark on the bolt head should line up with the mark on the crankshaft pulley).

11. Install the drive belt, as outlined in this section.
12. Install the splash shield.
13. Install the right front tire/wheel.
14. Connect the negative battery cable, then enter the security code for the radio.

CRANKSHAFT FRONT SEAL

REMOVAL & INSTALLATION

1. Remove the Crankshaft Position (CKP) sensor.
2. Remove the timing belt.
3. Remove the timing belt drive sprocket.
4. Remove the oil seal.

To install:

5. Dry and clean the oil seal housing.
6. Apply a light coating of multi-purpose grease to the crankshaft and oil seal lip.
7. Using a seal driver, install the seal until it bottoms.
8. The remainder of installation is the reverse of removal.

CYLINDER HEAD

REMOVAL & INSTALLATION

J35A6 Engine

See Figures 67 and 68.

➡The engine coolant temperature must be below 100°F (38°C) before removing the head bolts.

1. Make sure that you have the anti-theft codes for the radio and navigation system.
2. Relieve the fuel system pressure.
3. Disconnect the battery ground.
4. Remove the accessory drive belt.
5. Drain the coolant.
6. Remove the power steering pump and the hose bracket.
7. Remove the alternator.
8. Remove the timing belt.
9. Remove the intake manifold.
10. Remove the ignition coils.
11. Tag and disconnect all wiring from the head.
12. Remove the front and rear warm-up converters.
13. Disconnect the fuel feed line.
14. Remove the wiring harness bracket.
15. Remove the upper and lower radiator hoses.
16. Remove the heater and bypass hoses.
17. Remove the EVAP canister hose joint bracket.
18. Remove the harness brackets from the front and rear heads.
19. Remove the fuel rail.
20. Remove the water passage assembly.
21. Remove the camshaft pulley.
22. Remove the cylinder head cover.
23. Remove the head bolts in the sequence shown ⅓ turn at a time.
24. Remove the head.

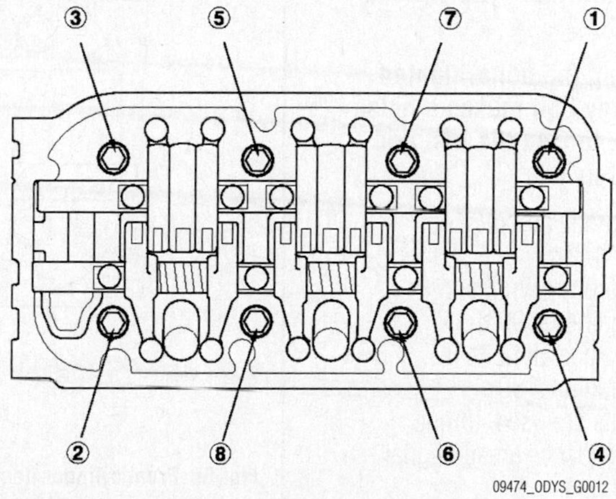
③ ⑤ ⑦ ①
② ⑧ ⑥ ④
09474_ODYS_G0012

Fig. 67 Cylinder head bolt removal sequence—J35A6 engine

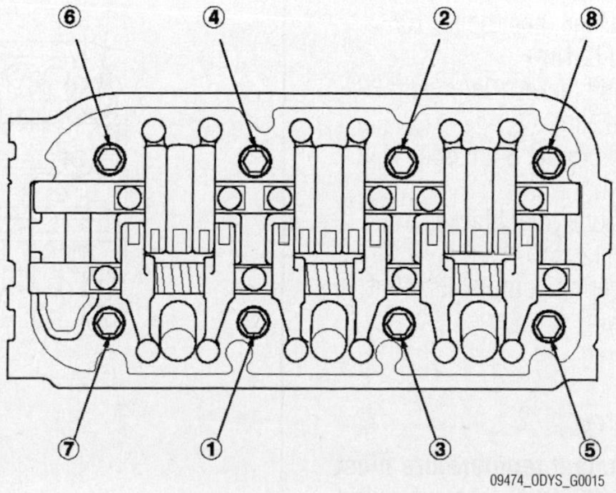
⑥ ④ ② ⑧
⑦ ① ③ ⑤
09474_ODYS_G0015

Fig. 68 Head bolt torque sequence—J35A6 engine

To install:

25. Clean all mating surfaces thoroughly.
26. Clean and install the oil control orifices with new O-rings.
27. Install the dowel pins and a new head gasket.
28. Set the crankshaft pulley to TDC by aligning the TDC mark with the pointer. See the "Timing Belt & Sprockets" in this section.
29. Set the camshaft pulley(s) to TDC by aligning the mark with the pointer. See the "Timing Belt & Sprockets" in this section..
30. Coat the threads and flanges of head bolts with clean engine oil.

CAUTION

There are 2 types of head bolts in service, 6-point and 12-point. Do not mix them on the same head.

➡There are 2 different tightening methods based on which type of head bolt is used.

➡When tightening the bolts, tighten them slowly. If any bolt makes a noise while tightening, loosen the bolt and retighten from Step 1.

a. Step 1: Torque all bolts in sequence to 22 ft. lbs.
b. Step 2: Torque all bolts in sequence an additional 90°
c. Step 3: Torque all bolts in sequence an additional 90°
- If new bolts are used, torque them in sequence an additional 90°

31. The remainder of installation is the reverse of removal. Note the following torques:
- Water passage 8mm bolts: 16 ft. lbs. (22 Nm)
- Water passage 6mm bolts: 105 inch lbs. (12 Nm)
- Front head bracket 10mm bolt: 33 ft. lbs. (44 Nm)
- Rear head bracket 8mm bolt: 16 ft. lbs. (22 Nm)
- EVAP canister joint bracket: 105 inch lbs. (12 Nm)
- Power steering pump bolts: 16 ft. lbs. (22 Nm)

J35A7 Engine

See Figures 69 through 72.

➡The engine coolant temperature must be below 100°F (38°C) before removing the head bolts.

1. Make sure that you have the anti-theft codes for the radio and navigation system.
2. Relieve the fuel system pressure.
3. Disconnect the battery ground.
4. Remove the accessory drive belt.
5. Drain the coolant.
6. Remove the power steering pump and the hose bracket.
7. Remove the alternator.
8. Remove the timing belt.
9. Remove the intake manifold.
10. Remove the ignition coils.
11. Tag and disconnect all wiring from the head.
12. Remove the front and rear warm-up converters.
13. Disconnect the fuel feed line.
14. Remove the wiring harness bracket.
15. Remove the upper and lower radiator hoses.
16. Remove the heater and bypass hoses.
17. Remove the EVAP canister hose joint bracket.
18. Remove the harness brackets from the front and rear heads.
19. Remove the fuel rail.
20. Remove the water passage assembly.
21. Remove the camshaft pulley.
22. Remove the cylinder head cover.
23. Remove the head bolts in the sequence shown ⅓ turn at a time.
24. Remove the head.

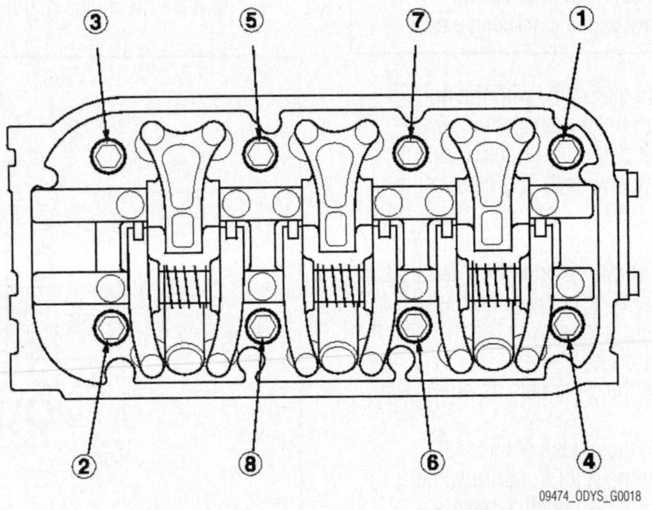

③
⑤
⑦
①
②
⑧
⑥
④
09474_ODYS_G0018

Fig. 69 Front cylinder head bolt removal sequence—J35A7 engine

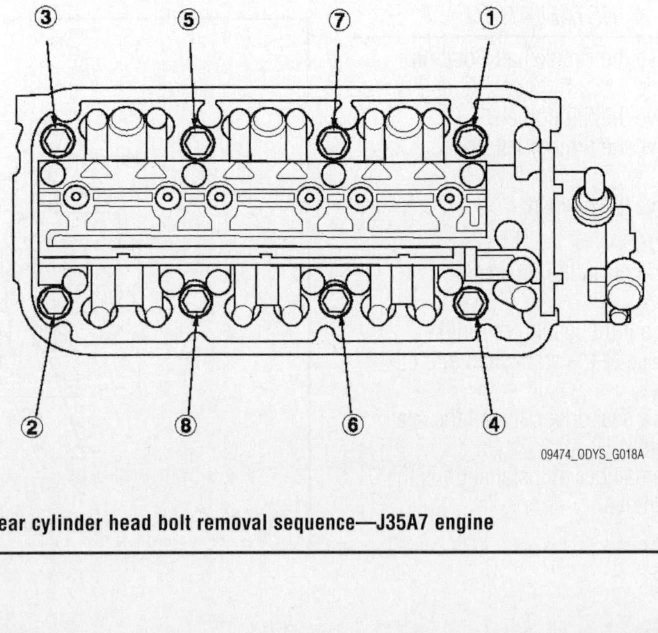

③
⑤
⑦
①
②
⑧
⑥
④
09474_ODYS_G018A

Fig. 70 Rear cylinder head bolt removal sequence—J35A7 engine

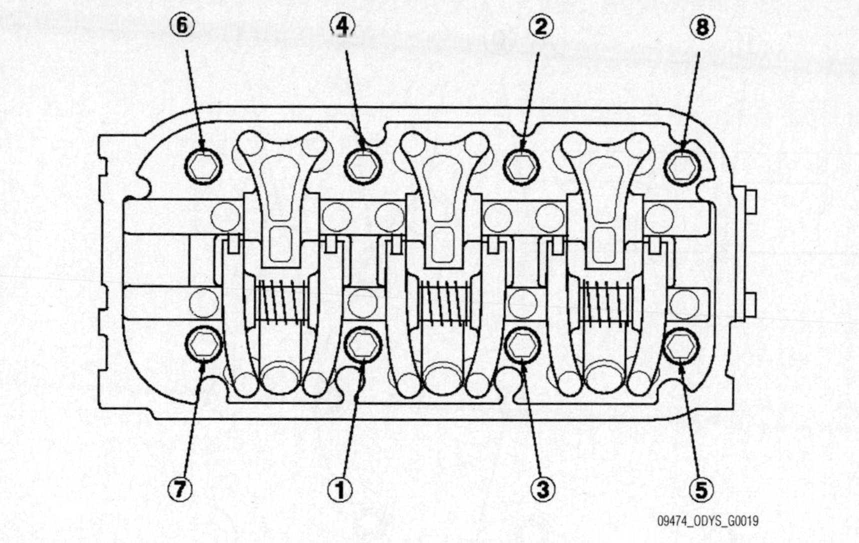
6
4
2
8
7
1
3
5
09474_ODYS_G0019

Fig. 71 Front head bolt torque sequence—J35A7 engine

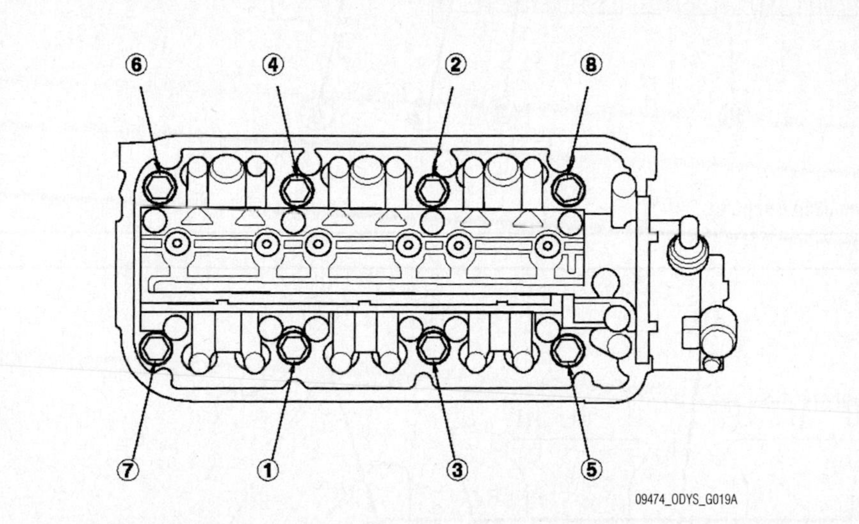
6
4
2
8
7
1
3
5
09474_ODYS_G019A

Fig. 72 Rear head bolt torque sequence—J35A7 engine

To install:

25. Clean all mating surfaces thoroughly.
26. Clean and install the oil control orifices with new O-rings.
27. Install the dowel pins and a new head gasket.
28. Set the crankshaft pulley to TDC by aligning the TDC mark with the pointer. See the "Timing Belt & Sprockets" in this section..
29. Set the camshaft pulley(s) to TDC by aligning the mark with the pointer. See the "Timing Belt & Sprockets" in this section..
30. Coat the threads and flanges of head bolts with clean engine oil.

➡When tightening the bolts, tighten them slowly. If any bolt makes a noise while tightening, loosen the bolt and retighten from Step 1.

31. Tighten the bolts, in sequence, in 3 steps. Perform each step twice.
    a. Step 1: 22 ft. lbs. (29 Nm)
    b. Step 2: Torque all bolts in sequence an additional 90°
    c. Step 3: Torque all bolts in sequence an additional 90°
32. The remainder of installation is the reverse of removal. Note the following torques:
    - Water passage 8mm bolts: 16 ft. lbs. (22 Nm)
    - Water passage 6mm bolts: 105 inch lbs. (12 Nm)
    - Front head bracket 10mm bolt: 33 ft. lbs. (44 Nm)
    - Rear head bracket 8mm bolt: 16 ft. lbs. (22 Nm)
    - EVAP canister joint bracket: 105 inch lbs. (12 Nm)
    - Power steering pump bolts: 16 ft. lbs. (22 Nm)

EXHAUST MANIFOLD

REMOVAL & INSTALLATION

1. Remove the engine cover.
2. Remove the front A/F sensor (Sensor 1) and the front secondary HO2S (Sensor 2).
3. Remove the exhaust pipe A mounting nuts (front WU-TWC side).
4. Remove the A/C condenser fan assemblies.
5. Carefully remove the front WU-TWC.

To install:

6. Carefully install the front WU-TWC with a new gasket and new self-locking nuts. Tighten the nuts in a crisscross pattern in two or three steps to 23 ft. lbs. (31 Nm).
7. The remainder of the installation is the reverse order of removal.

DRIVEPLATE

REMOVAL & INSTALLATION

1. Remove the transaxle assembly.
2. Remove the drive plate and the washer from the engine crankshaft.

To install:

3. Install the drive plate and the washer on the engine crankshaft, and tighten the eight bolts in a crisscross pattern to 54 ft. lbs. (74 Nm).
4. Install the transaxle assembly.

INTAKE MANIFOLD

REMOVAL & INSTALLATION

See Figures 73 through 78.

1. Remove the intake manifold cover.
2. Remove the air inlet duct.
3. Remove the PCV hose and brake booster hose.
4. Remove the EVAP canister hose and water bypass hoses. Plug the bypass hose.
5. Tag and disconnect all wiring connected to the manifold.
6. Remove the upper cover mounting bolts and nuts, in 2 equal steps, in the sequence shown. Remove the cover.
7. Remove the intake manifold mounting bolts and nuts, in 2 equal steps, in the sequence shown. Remove the manifold.

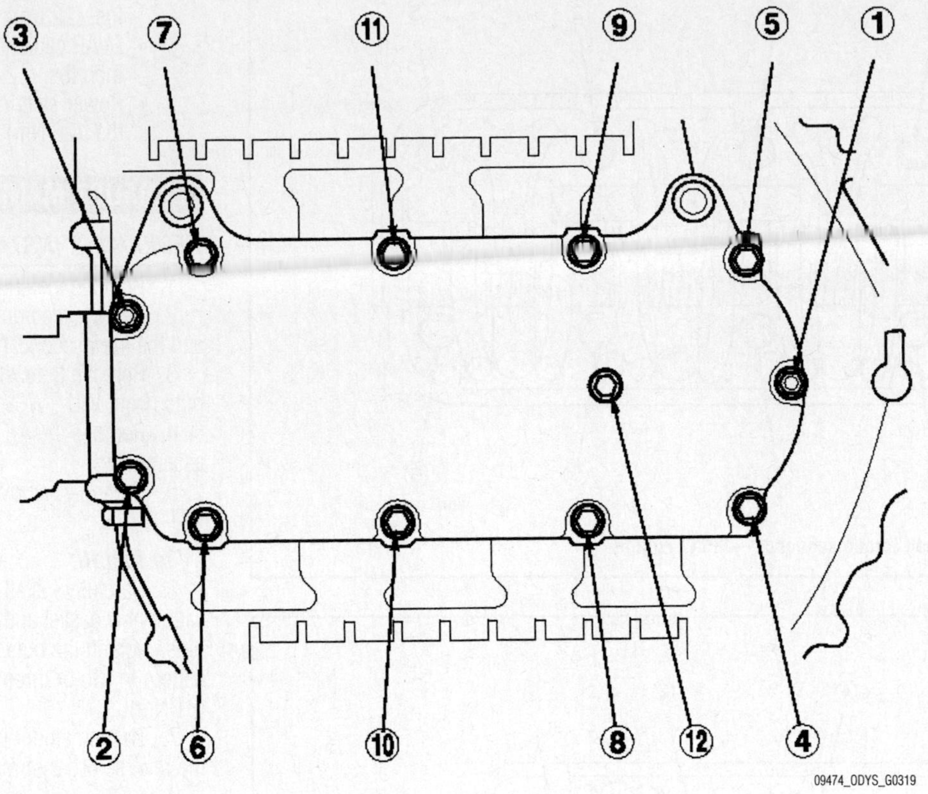

③
⑦
⑪
⑨
⑤
①
②
⑥
⑩
⑧
⑫
④
09474_ODYS_G0319

Fig. 73 Upper cover loosening sequence—J35A6 and J35A7 engines

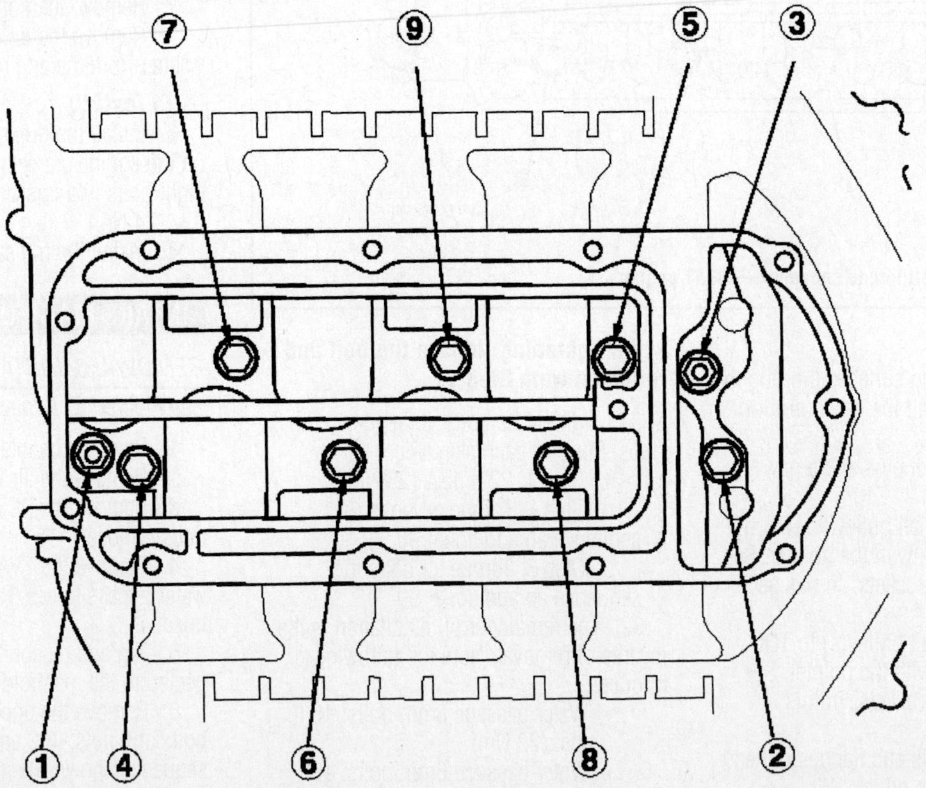

⑦
⑨
⑤
③
①
④
⑥
⑧
②
09474_ODYS_G0320

Fig. 74 Intake manifold loosening sequence—J35A6 and J35A7 engines

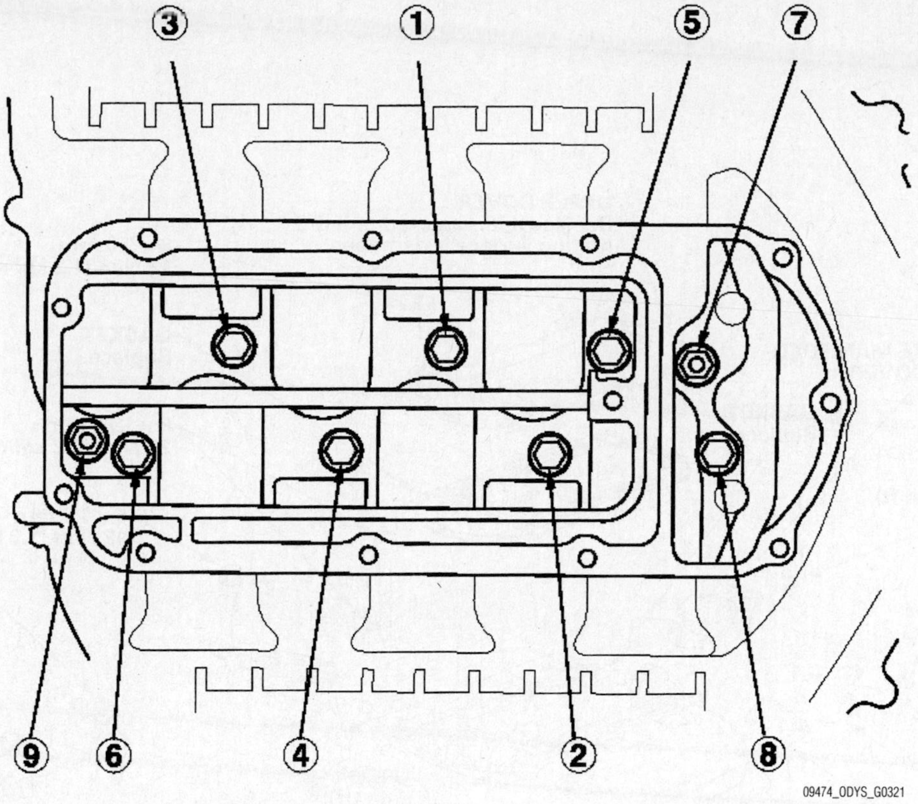
3
1
5
7
9
6
4
2
8
09474_ODYS_G0321

Fig. 75 Intake manifold torque sequence—J35A6 and J35A7 engines

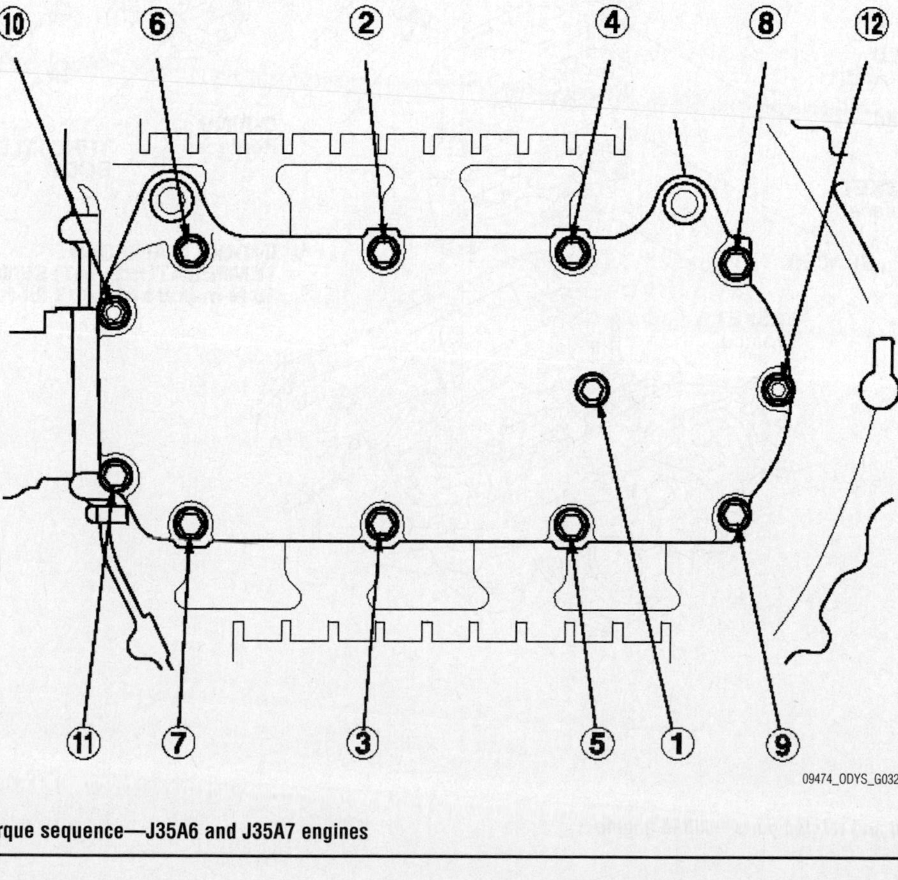
10
6
2
4
8
12
11
7
3
5
1
9
09474_ODYS_G0322

Fig. 76 Upper cover torque sequence—J35A6 and J35A7 engines

J35A6 engine:

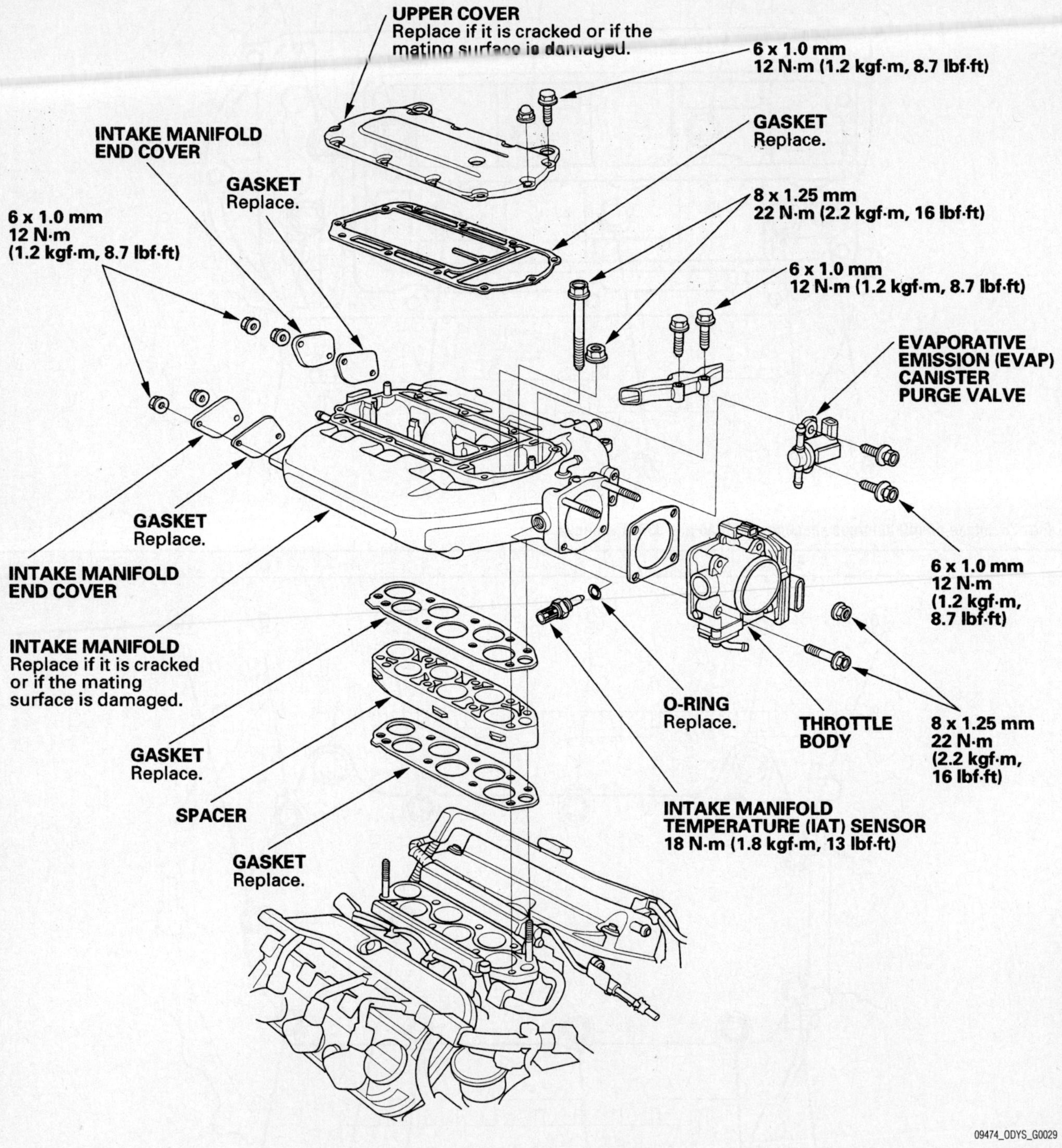
UPPER COVER
Replace if it is cracked or if the mating surface is damaged.
6 x 1.0 mm
12 N·m (1.2 kgf·m, 8.7 lbf·ft)
GASKET
Replace.
INTAKE MANIFOLD END COVER
GASKET
Replace.
6 x 1.0 mm
12 N·m
(1.2 kgf·m, 8.7 lbf·ft)
8 x 1.25 mm
22 N·m (2.2 kgf·m, 16 lbf·ft)
6 x 1.0 mm
12 N·m (1.2 kgf·m, 8.7 lbf·ft)
EVAPORATIVE EMISSION (EVAP) CANISTER PURGE VALVE
GASKET
Replace.
INTAKE MANIFOLD END COVER
6 x 1.0 mm
12 N·m
(1.2 kgf·m,
8.7 lbf·ft)
INTAKE MANIFOLD
Replace if it is cracked or if the mating surface is damaged.
O-RING
Replace.
THROTTLE BODY
8 x 1.25 mm
22 N·m
(2.2 kgf·m,
16 lbf·ft)
GASKET
Replace.
SPACER
INTAKE MANIFOLD TEMPERATURE (IAT) SENSOR
18 N·m (1.8 kgf·m, 13 lbf·ft)
GASKET
Replace.
09474_ODYS_G0029

Fig. 77 Intake manifold and related parts—J35A6 engine

J35A7 engine:

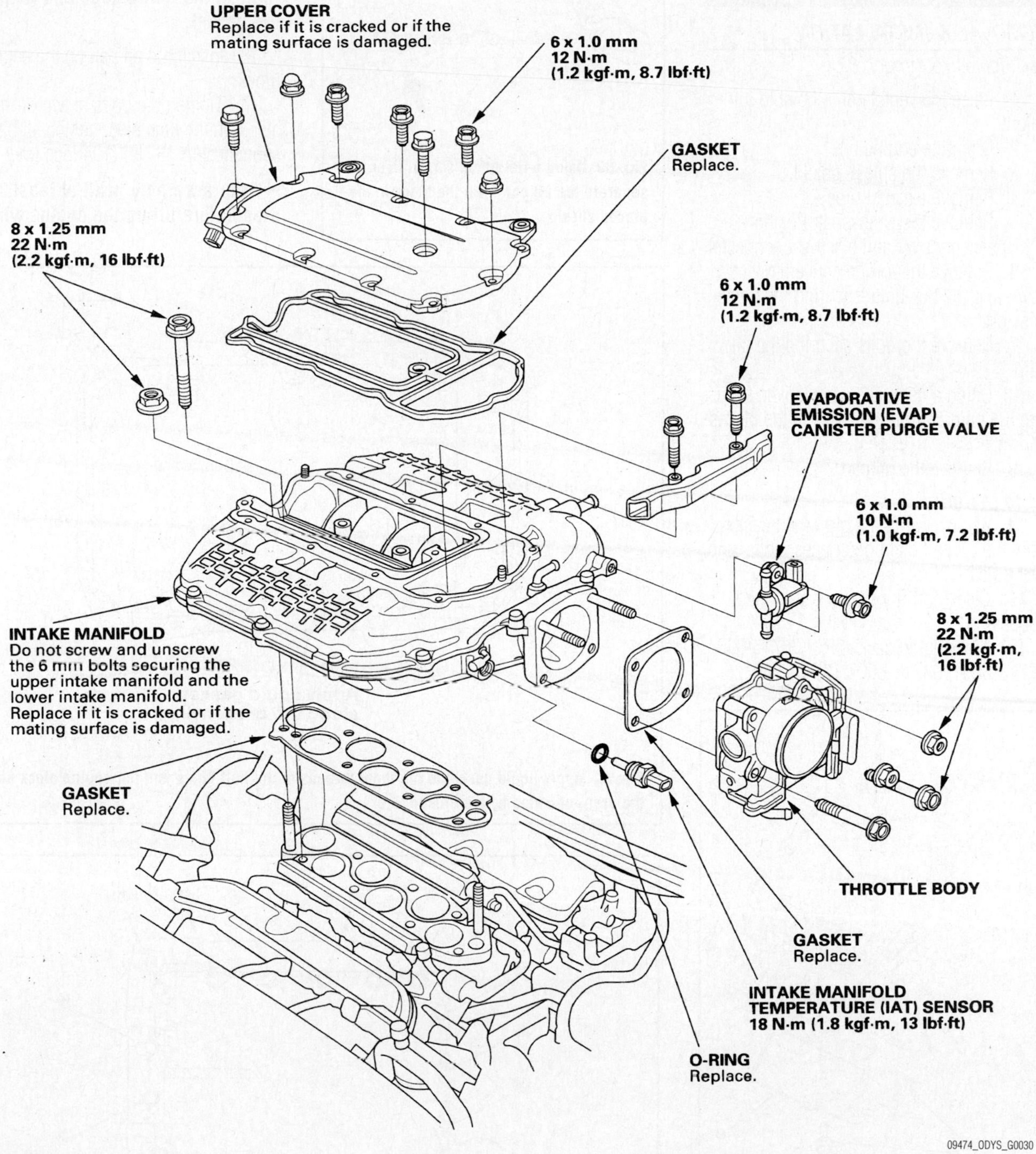
UPPER COVER
Replace if it is cracked or if the mating surface is damaged.
6 x 1.0 mm
12 N·m
(1.2 kgf·m, 8.7 lbf·ft)
GASKET
Replace.
8 x 1.25 mm
22 N·m
(2.2 kgf·m, 16 lbf·ft)
6 x 1.0 mm
12 N·m
(1.2 kgf·m, 8.7 lbf·ft)
EVAPORATIVE EMISSION (EVAP) CANISTER PURGE VALVE
6 x 1.0 mm
10 N·m
(1.0 kgf·m, 7.2 lbf·ft)
8 x 1.25 mm
22 N·m
(2.2 kgf·m, 16 lbf·ft)
INTAKE MANIFOLD
Do not screw and unscrew the 6 mm bolts securing the upper intake manifold and the lower intake manifold.
Replace if it is cracked or if the mating surface is damaged.
GASKET
Replace.
THROTTLE BODY
GASKET
Replace.
INTAKE MANIFOLD TEMPERATURE (IAT) SENSOR
18 N·m (1.8 kgf·m, 13 lbf·ft)
O-RING
Replace.
09474_ODYS_G0030

Fig. 78 Intake manifold and related parts—J35A7 engine

8. Installation is the reverse of removal. Tighten the manifold bolts and nuts, in the sequence shown, in 2 equal steps, to 16 ft. lbs. (22 Nm). Tighten the cover bolts and nuts, in the sequence shown, in 2 equal steps, to 105 inch lbs. (12 Nm).

OIL PAN

REMOVAL & INSTALLATION

See Figures 79 through 82.

1. Raise the vehicle on a hoist to full height.
2. Drain the engine oil.
3. Remove the splash shield.
4. Remove exhaust pipe A.
5. Remove the Crankshaft Position (CKP) sensor cover and electrical connector.
6. Remove the torque converter cover and the two bolts securing the transmission.
7. Remove the bolts securing the oil pan.
8. Using a flat blade screwdriver, separate the oil pan from the block in the places shown.
9. Remove the oil pan.

To install:

10. Remove all of the old liquid gasket from the oil pan mating surfaces, bolts, and bolt holes.
11. Clean and dry the oil pan mating surfaces.
12. Apply liquid gasket, P/N 08717-0004, 08718-0001, 08718-0002, 08718-0003, or 08718-009, evenly to the oil pan mating surface of the engine block.

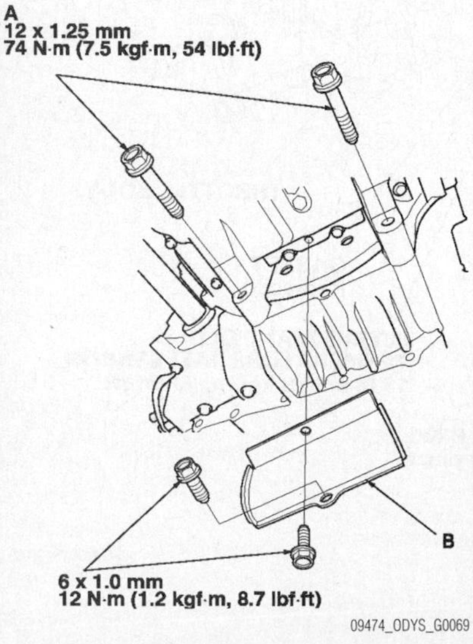
A
12 x 1.25 mm
74 N·m (7.5 kgf·m, 54 lbf·ft)
B
6 x 1.0 mm
12 N·m (1.2 kgf·m, 8.7 lbf·ft)
09474_ODYS_G0069

Fig. 79 Remove the torque converter cover (A) and the two bolts (B) securing the transmission

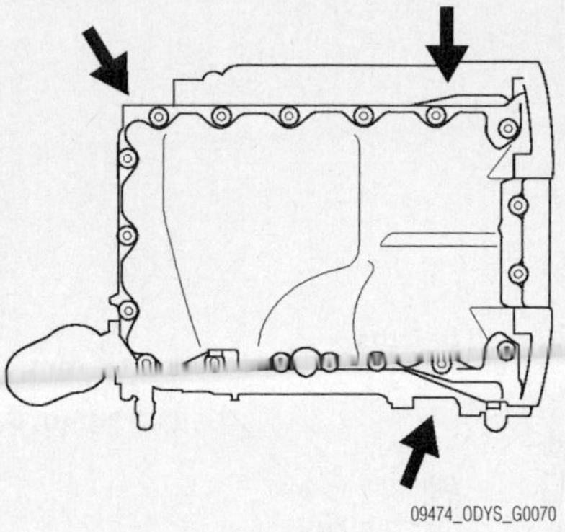
09474_ODYS_G0070

Fig. 80 Using a flat blade screwdriver, separate the oil pan from the block in the places shown

➡Do not install components if too much time has passed after applying the liquid gasket (for P/N 08718-0002, no more than 4 minutes, for all others, no more than 5 minutes). Instead, remove the old residue and reapply liquid gasket.

13. Install the oil pan on the engine block.
14. Tighten the bolts in two or three steps. In the final step, tighten all bolts, in sequence, to 12 Nm (104 inch lbs.).

➡After assembly, wait at least 30 minutes before filling the engine with oil.

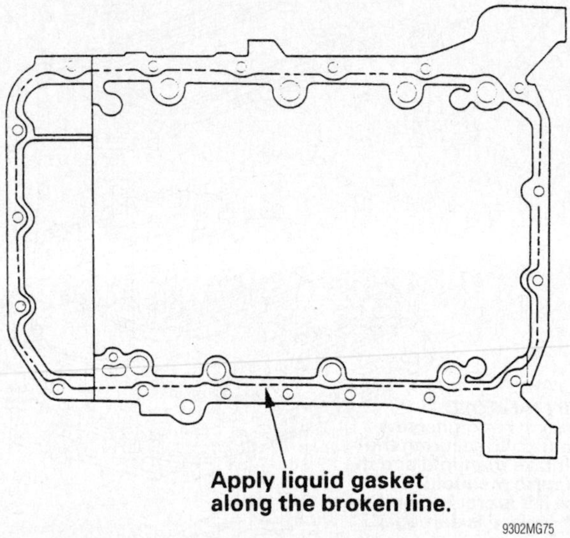
Apply liquid gasket along the broken line.
9302MG75

Fig. 81 Apply liquid gasket to the inner threads of the bolt holes and the engine block along the area indicated by the broken line

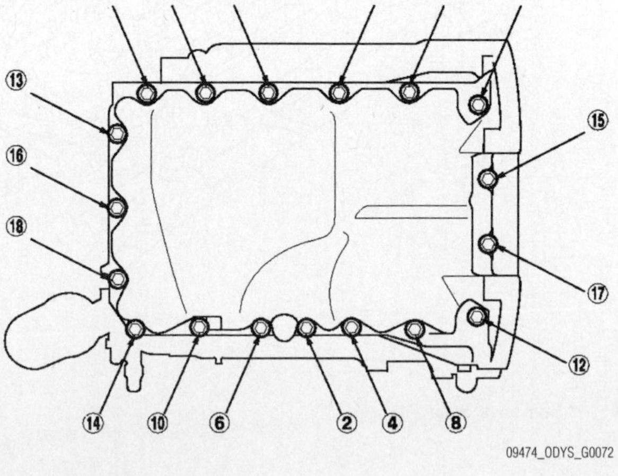
9
5
1
3
7
11
13
15
16
18
17
12
14
10
6
2
4
8
09474_ODYS_G0072

Fig. 82 Oil pan bolt tightening sequence

OIL PUMP

INSPECTION

See Figures 83 through 85.

1. Remove the screws from the pump housing, then separate the housing and cover.
2. Check the inner-to-outer rotor radial clearance between the inner rotor and outer rotor. If the inner-to-outer rotor clearance exceeds the service limit, replace the oil pump assembly.

Inner Rotor-to-Outer Rotor Radial Clearance:

- Standard (New): 0.04–0.16 mm (0.002–0.006 in.)
- Service Limit: 0.20 mm (0.008 in.)

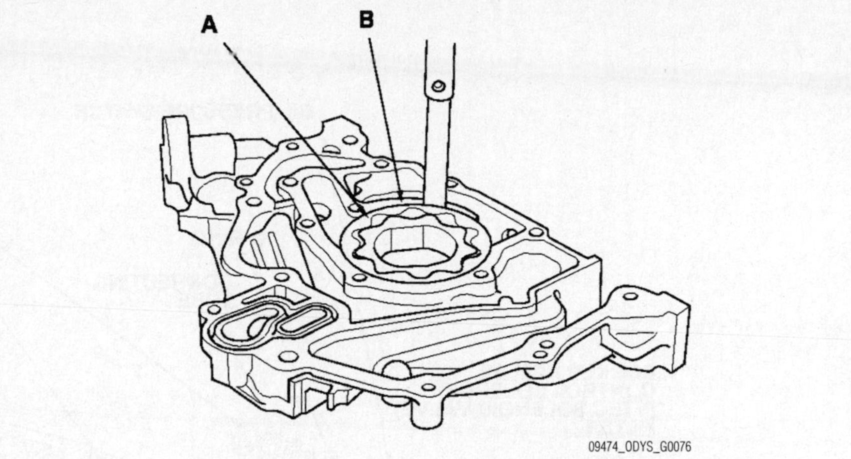
A
B
09474_ODYS_G0076

Fig. 85 Check the housing-to-outer rotor radial clearance between the outer rotor (A) and pump housing (B)

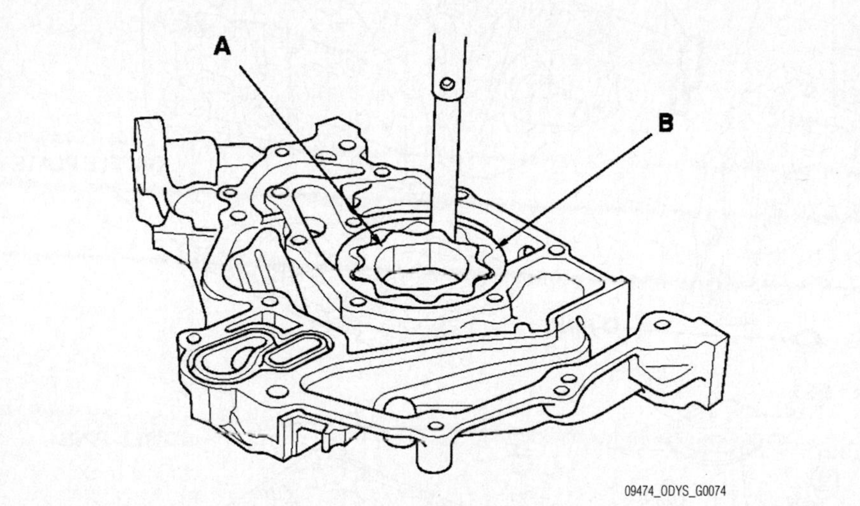
A
B
09474_ODYS_G0074

Fig. 83 Check the inner-to-outer rotor radial clearance between the inner rotor (A) and outer rotor (B)

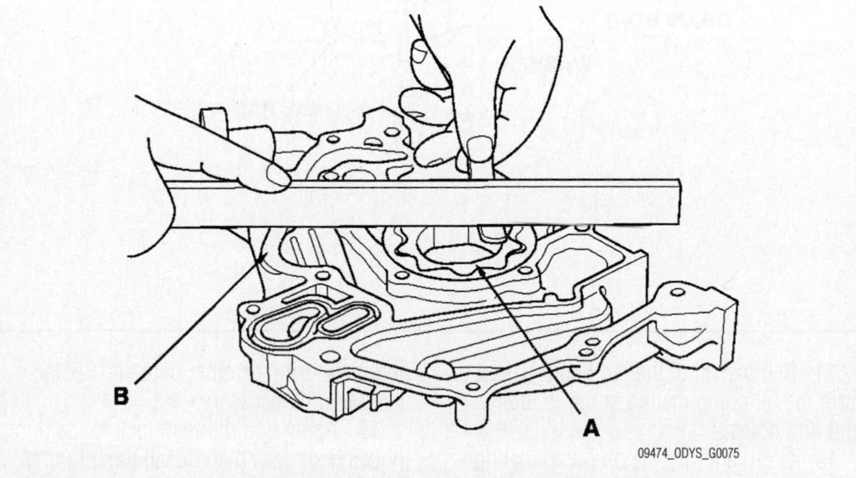
B
A
09474_ODYS_G0075

Fig. 84 Check the housing-to-rotor axial clearance between the rotors (A) and pump housing (B)

3. Check the housing-to-rotor axial clearance between the rotors and pump housing. If the housing-to-rotor axial clearance exceeds the service limit, replace the oil pump assembly.

Housing-to-Rotor Axial Clearance:

- Standard (New): 0.02–0.07 mm (0.001–0.003 in.)
- Service Limit: 0.12 mm (0.005 in.)

4. Check the housing-to-outer rotor radial clearance between the outer rotor and pump housing. If the housing-to-outer rotor radial clearance exceeds the service limit, replace the oil pump assembly.

Housing-to-Outer Rotor Radial Clearance:

- Standard (New): 0.10–0.19 mm (0.004–0.007 in.)
- Service Limit: 0.20 mm (0.008 in.)

REMOVAL & INSTALLATION

See Figures 86 through 88.

1. Drain the engine oil.
2. Turn the crankshaft so that the No. 1 piston is at top dead center.
3. Remove the timing belt.
4. Remove the idler pulley.
5. Remove the crankshaft position (CKP) sensor A/B.
6. Attach the chain hoist to the engine hanger on the power steering (P/S) pump bracket.
7. Remove the jack from under the oil pan.
8. Remove the rocker arm oil control solenoid (VTEC solenoid valve)/oil filter assembly (J35A6 engine) or oil filter base/oil filter assembly (J35A7 engine).
9. Remove the oil pan.
10. Remove the oil screen.
11. Remove the mounting bolts and the oil pump assembly.

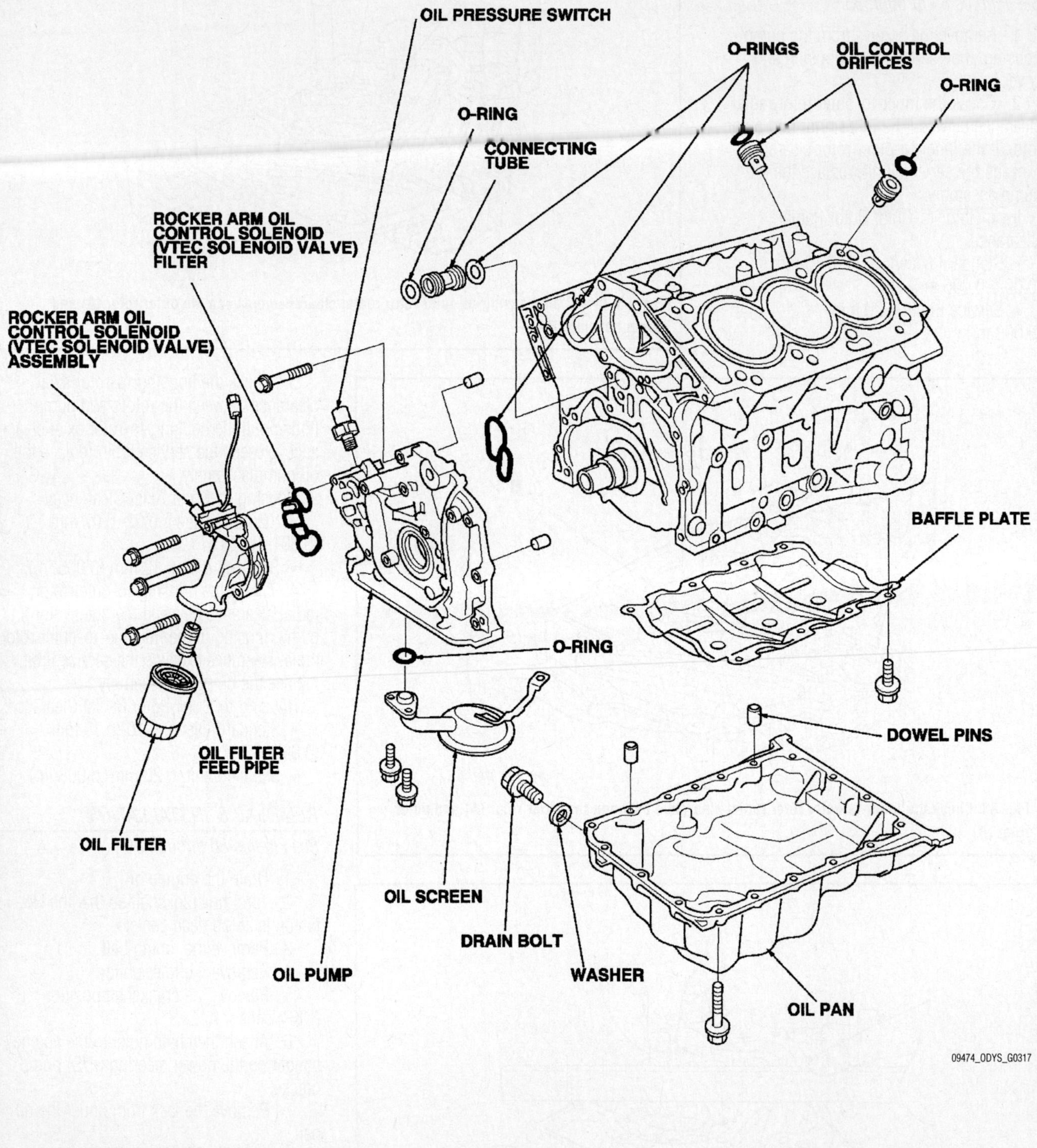
OIL PRESSURE SWITCH
O-RINGS
OIL CONTROL ORIFICES
O-RING
O-RING
CONNECTING TUBE
ROCKER ARM OIL CONTROL SOLENOID (VTEC SOLENOID VALVE) FILTER
ROCKER ARM OIL CONTROL SOLENOID (VTEC SOLENOID VALVE) ASSEMBLY
BAFFLE PLATE
O-RING
DOWEL PINS
OIL FILTER FEED PIPE
OIL FILTER
OIL SCREEN
DRAIN BOLT
OIL PUMP
WASHER
OIL PAN
09474_ODYS_G0317

Fig. 86 Oil pump and related parts—J35A6 engine

To install:

12. Remove the old oil seal from the oil pump.
13. Gently tap in the new oil seal until the oil seal driver bottoms on the pump.
14. Remove all of the old liquid gasket from the oil pump mating surfaces, bolts, and bolt holes.
15. Clean and dry the oil pump mating surfaces.
16. Inspect both rotors and pump housing for scoring or other damage. Replace the parts, if necessary.
17. Apply liquid thread lock to the pump housing screws, then install the oil pump cover.
18. Check that the oil pump turns freely.

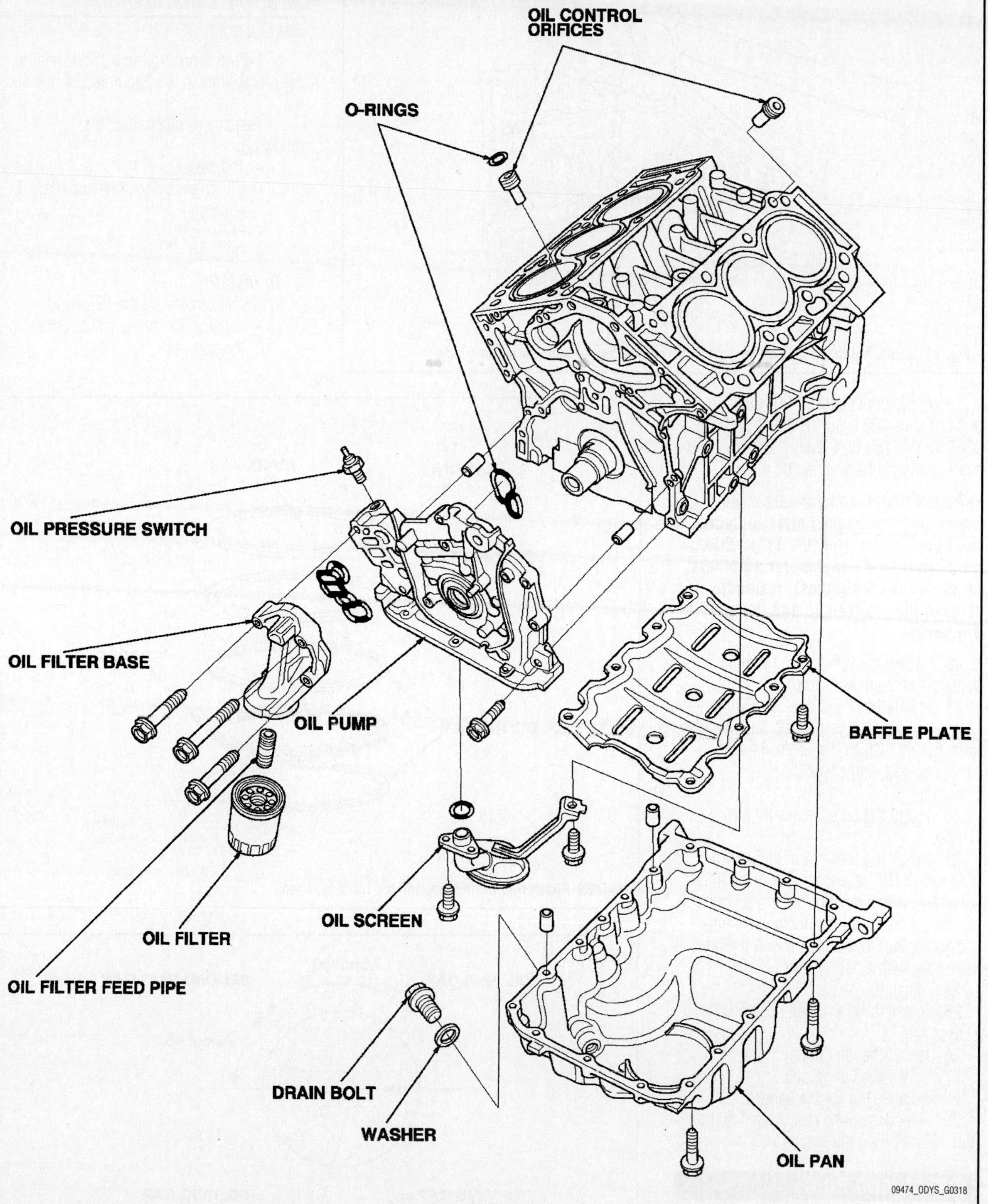
OIL CONTROL ORIFICES
O-RINGS
OIL PRESSURE SWITCH
OIL FILTER BASE
OIL PUMP
BAFFLE PLATE
OIL FILTER
OIL SCREEN
OIL FILTER FEED PIPE
DRAIN BOLT
WASHER
OIL PAN
09474_ODYS_G0318

Fig. 87 Oil pump and related parts—J35A7 engine

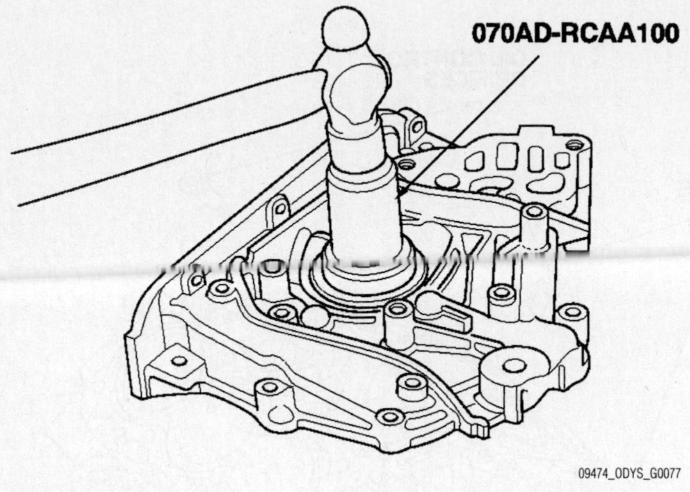
070AD-RCAA100
09474_ODYS_G0077

Fig. 88 Gently tap in the new oil seal until the oil seal driver bottoms on the pump

REAR MAIN SEAL

REMOVAL & INSTALLATION

See Figure 91.

1. Before servicing the vehicle, refer to the precautions in the beginning of this section.
2. Remove or disconnect the following:
   - Transaxle
   - Clutch pressure plate and disc, if equipped
   - Flywheel
   - Oil seal

To install:

3. Install or connect the following:
   - Oil seal. Drive the seal square into the seal case.

19. Apply liquid gasket, P/N 08717-0004, 08718-0001, 08718-0002, 08718-0003, or 08718-0009, evenly to the engine block mating surface of the oil pump.

➡Do not install components if too much time has passed after applying the liquid gasket (for P/N 08718-0002, no more than 4 minutes, for all others, no more than 5 minutes). Instead, remove the old residue and reapply liquid gasket.

20. Grease the lip of the oil seal, and apply oil to the new O-ring.
21. Install the dowel pins, then align the inner rotor with the crankshaft, and install the oil pump. Clean the excess grease off the crankshaft, and check the seal for distortion.
22. Install the oil screen with new O-ring.
23. Install the rocker arm oil control solenoid, VTEC solenoid valve or oil filter assembly, with a new rocker arm oil control solenoid filter (VTEC solenoid valve filter, J35A6 engine), or oil filter base/oil filter assembly, with a new O-ring.
24. Install the oil pan.
25. Install the crankshaft position (CKP) sensor A/B.
26. Install the idler pulley.
27. Install the timing belt.
28. Remove the engine hanger.
29. After assembly, wait at least 30 minutes before filling the engine with oil.

PISTON AND RING

POSITIONING

See Figures 89 and 90.

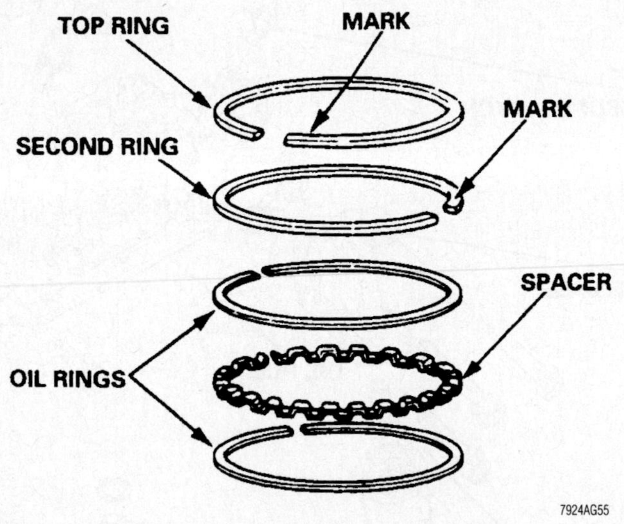
TOP RING
MARK
MARK
SECOND RING
SPACER
OIL RINGS
7924AG55

Fig. 89 Piston ring positioning and top mark location

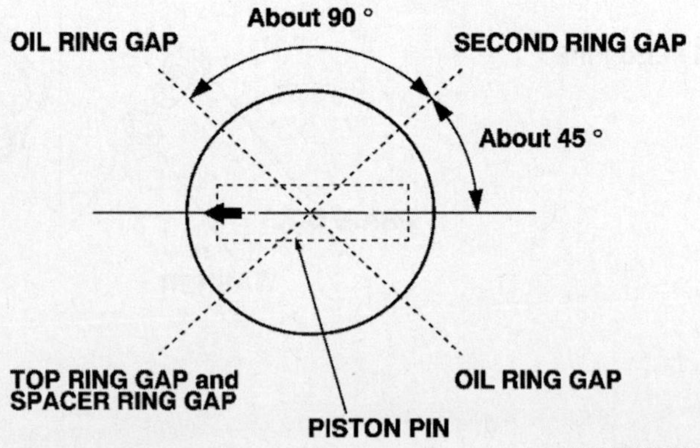
About 90 °
OIL RING GAP
SECOND RING GAP
About 45 °
TOP RING GAP and SPACER RING GAP
OIL RING GAP
PISTON PIN
22140_ODYS_G0005

Fig. 90 Ring end gap positioning

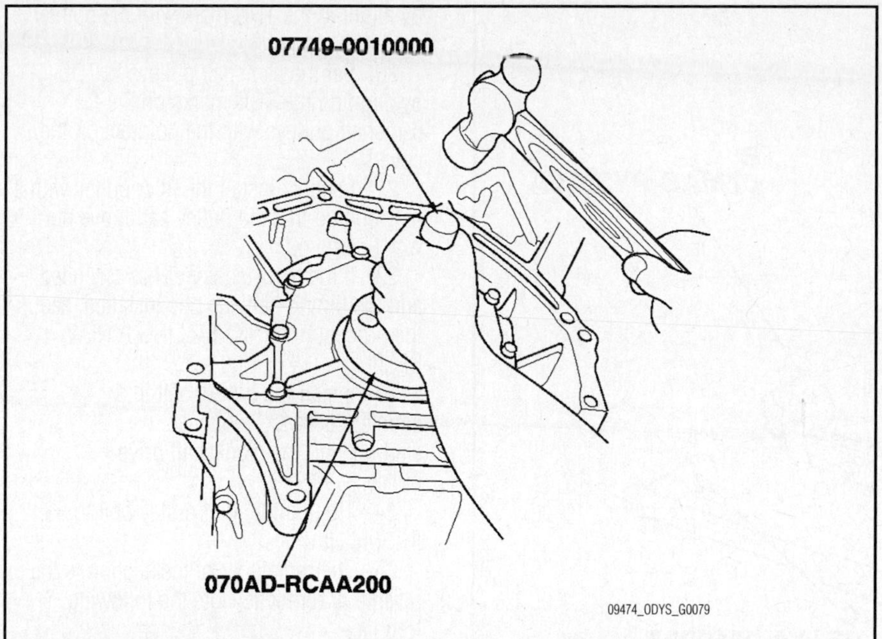
07749-0010000
070AD-RCAA200
09474_ODYS_G0079
Fig. 91 Rear seal installation

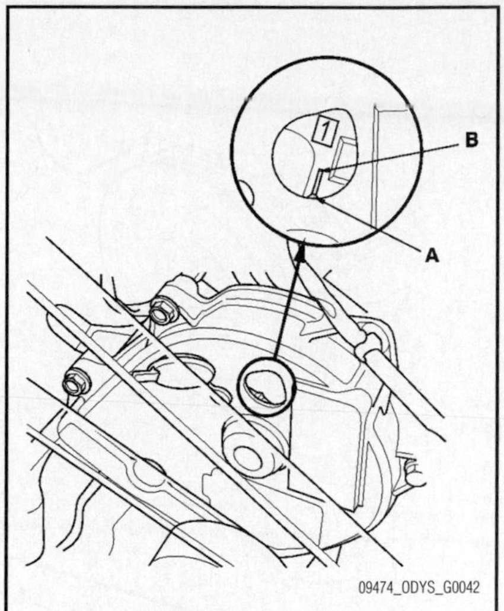
B
A
09474_ODYS_G0042
Fig. 93 Check that the TDC timing mark (A) on the front camshaft sprocket is aligned with the pointer (B) on the cover—J35A6 engine

- Flywheel. Tighten the bolts in a crossing pattern to 54 ft. lbs. (73 Nm).
- Clutch pressure plate and disc, if equipped
- Transaxle

4. Check the fluid levels.
5. Start the engine and check for leaks.

ROCKER ARMS/SHAFTS

REMOVAL & INSTALLATION

For additional information, see "Camshaft & Valve Lifters" in this section.

TIMING BELT FRONT COVER

REMOVAL & INSTALLATION

For additional information, see "Timing Belt & Sprockets" in this section.

TIMING BELT & SPROCKETS

REMOVAL & INSTALLATION

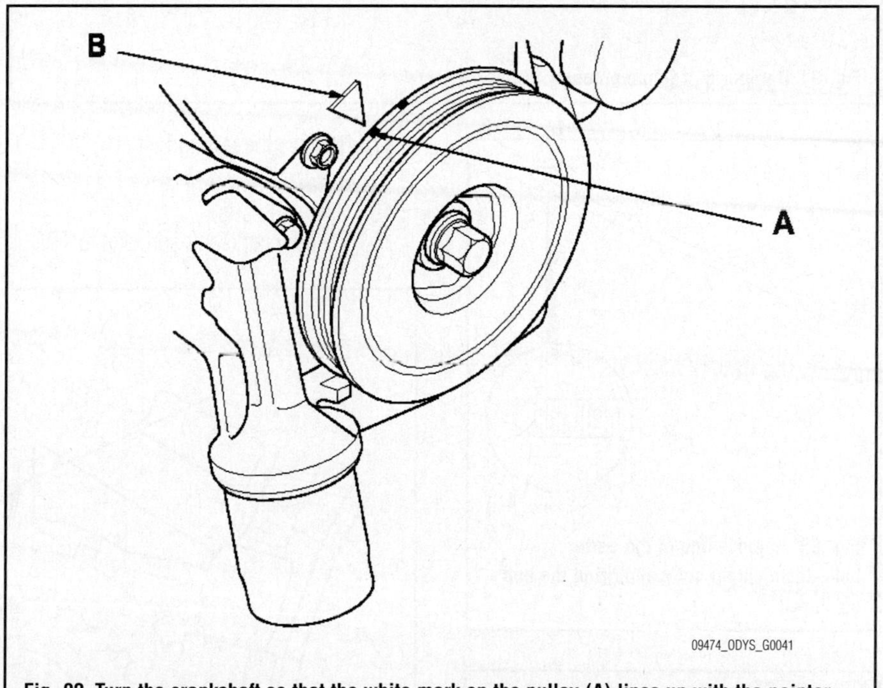
B
A
09474_ODYS_G0041
Fig. 92 Turn the crankshaft so that the white mark on the pulley (A) lines up with the pointer (B)—J35A6 engine

J35A6 Engine

Installing a used Belt

See Figures 92 through 98.

1. Remove the right front wheel.
2. Remove the splash shield.
3. Remove the accessory drive belt.
4. Turn the crankshaft so that the white mark on the pulley lines up with the pointer.
5. Check that the TDC timing mark on the front camshaft sprocket is aligned with the pointer on the cover.
6. Support the engine with a jack and a block of wood under the oil pan.
7. Remove the upper half of the side engine mount bracket.
8. Using the tools shown and a 19mm socket, remove the damper bolt.
9. Remove the damper.
10. Remove the front and rear upper belt covers.
11. Remove the lower cover.
12. Remove one of the battery hold-down clamp bolts and grind the end as shown.
13. Screw the battery clamp bolt in as shown to hold the timing belt adjuster in its current position. Tighten it by hand. Do not use a wrench.
14. Remove the timing belt guide plate.
15. Remove the lower side engine mount bracket.
16. Remove the idler pulley bolt and idler pulley. Discard the bolt.

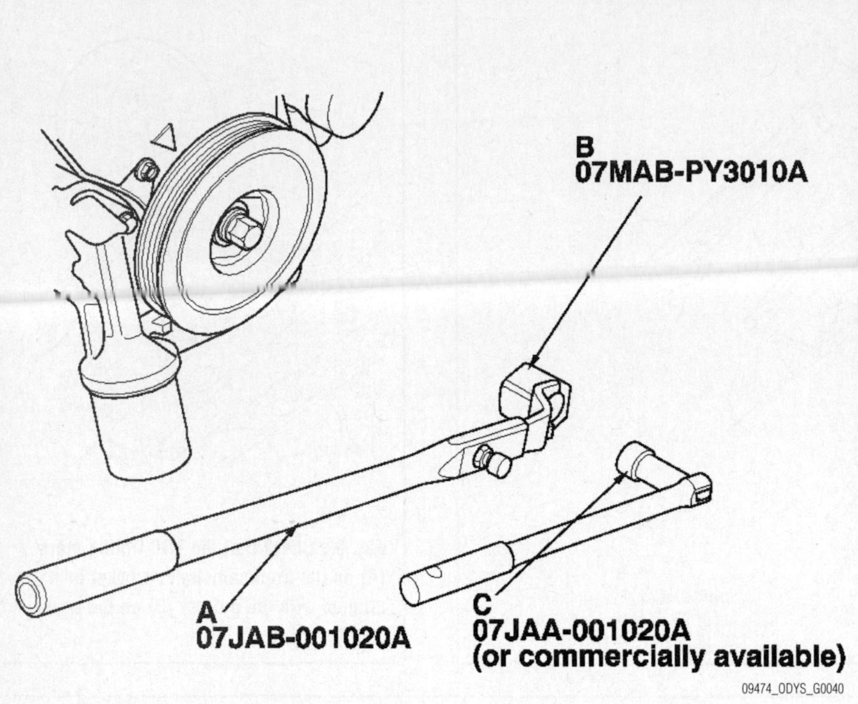
B
07MAB-PY3010A
A
07JAB-001020A
C
07JAA-001020A
(or commercially available)
09474_ODYS_G0040

Fig. 94 Damper bolt removal tools

17. Remove the timing belt.

To install:

18. Clean all parts.
19. Set the crankshaft sprocket to TDC by aligning the TDC mark with the pointer on the oil pump.
20. Set the camshaft pulleys to TDC by aligning the TDC marks on the camshaft pulleys with the pointers on the covers.
21. Loosely install the idler pulley with a new bolt so that the pulley can move but won't come off.
22. If the auto-tensioner has extended, and the timing belt can't be installed, see the procedure for INSTALLING A NEW BELT.
23. Install the timing belt in a counterclockwise sequence, starting with the crankshaft drive pulley.
24. Tighten the idler pulley bolt to 33 ft. lbs. (44 Nm).
25. The remainder of installation is the reverse of removal. Note the following torques:
    - Lower half engine mount bracket: 33 ft. lbs. (44 Nm)
    - Lower cover: 105 inch lbs. (12 Nm)
    - Upper covers: 105 inch lbs. (12 Nm)
    - Crankshaft damper: Tighten the bolt to 47 ft. lbs. (64 Nm). Mark the bolt head and pulley, then tighten the bolt an additional 60°.

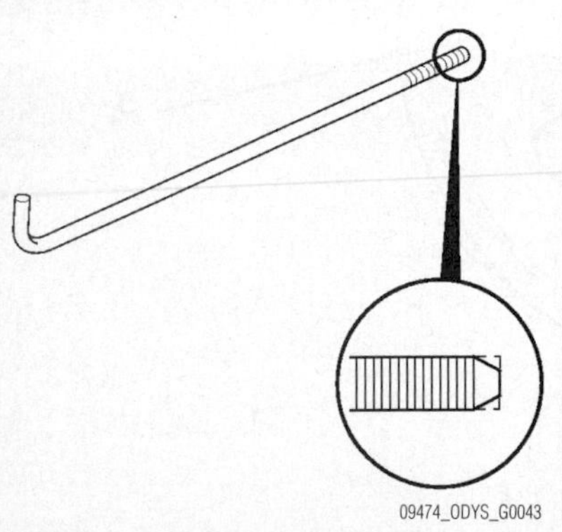
09474_ODYS_G0043

Fig. 95 Remove one of the battery hold-down clamp bolts and grind the end as shown

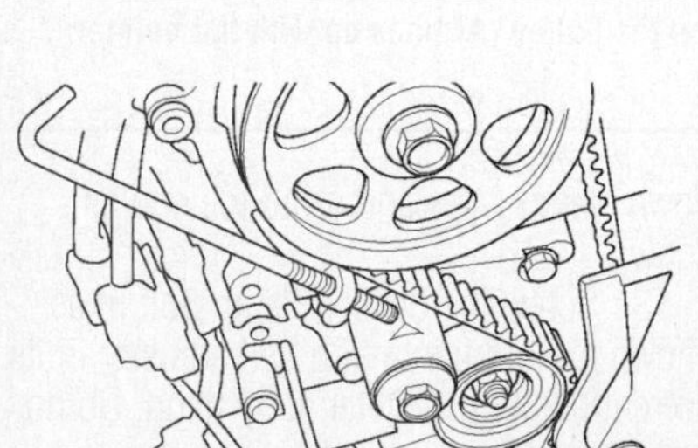
09474_ODYS_G0044

Fig. 96 Battery clamp bolt screwed into position—J35A6 engine

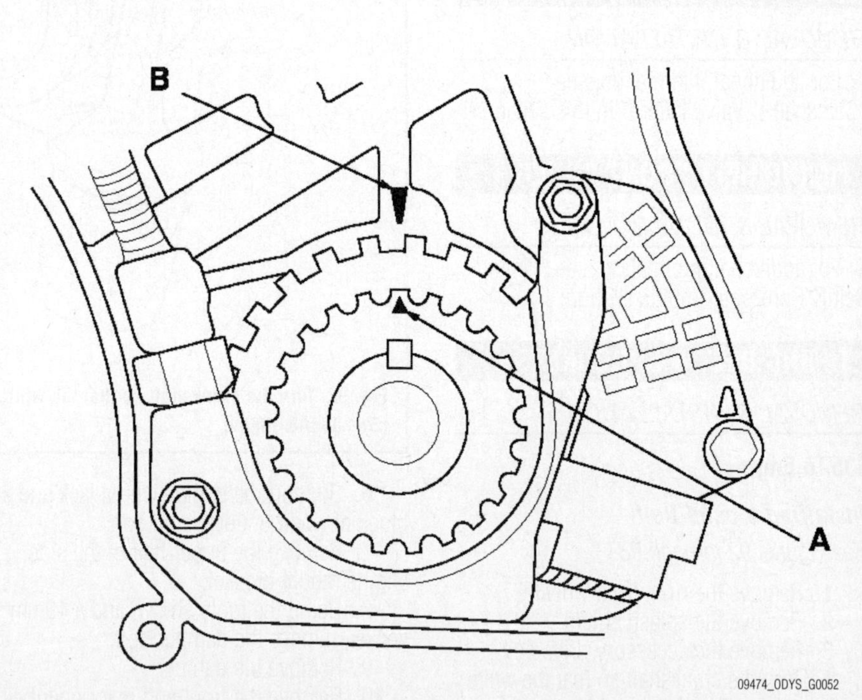
B
A
09474_ODYS_G0052

Fig. 97 Set the crankshaft sprocket to TDC by aligning the TDC mark (A) with the pointer (B) on the oil pump—J35A6 engine

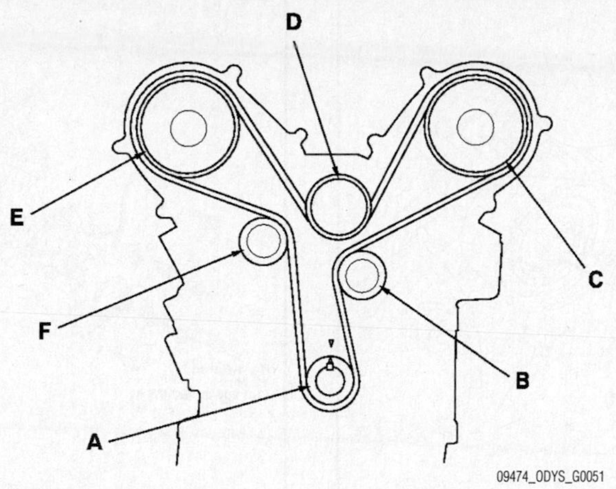
D
E
C
F
B
A
09474_ODYS_G0051

Fig. 98 Install the timing belt in a counterclockwise sequence (A through F) , starting with the crankshaft drive pulley (A)—J35A6 engine

13. Loosely install the idler pulley with a new idler pulley bolt so the pulley can move but does not come off.
14. Install the timing belt in a counterclockwise sequence starting with the drive pulley.
15. Tighten the idler pulley bolt.
16. Remove the pin from the auto-tensioner.
17. Remove the battery clamp bolt from the back cover.
18. Install the lower half of the side engine mount bracket.
19. Install the timing belt guide plate as shown.
20. Install the lower cover.
21. Install the front upper cover and rear upper cover.

Installing a New Belt

See Figures 99 through 113.

1. Remove the timing belt.
2. Clean the timing belt pulleys, timing belt guide plate, and the upper and lower covers.
3. Set the timing belt drive pulley to top dead center (TDC) by aligning the TDC mark on the tooth of the timing belt drive pulley with the pointer on the oil pump.
4. Set the camshaft pulleys to TDC by aligning the TDC marks on the camshaft pulleys with the pointers on the back covers.
5. Remove the battery clamp bolt from the back cover.
6. Remove the auto-tensioner.
7. Align the holes on the rod and housing of the auto-tensioner.
8. Use a hydraulic press to slowly compress the auto-tensioner. Insert a 2.0 mm (0.08 in.) pin through the housing and the rod.

➡The compression pressure should not exceed 9,800 N (2,200 lbs.).

9. Align the holes on the rod and housing of the auto-tensioner.
10. Use a hydraulic press to slowly compress the auto-tensioner. Insert a 2.0 mm (0.08 in.) pin through the housing and the rod.

➡The compression pressure should not exceed 9,800 N (2,200 lbs.).

11. Install the auto-tensioner.

➡Make sure the pin stays in place.

12. Screw the battery clamp bolt in as shown to hold the timing belt adjuster. Tighten it by hand; do not use a wrench.

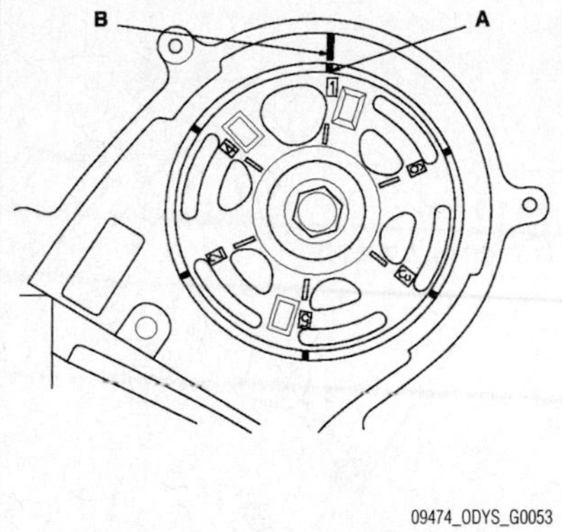
B
A
09474_ODYS_G0053

Fig. 100 Set the left camshaft pulley to TDC by aligning the TDC marks (A) on the camshaft pulleys with the pointers (B) on the back covers—J35A6 engine

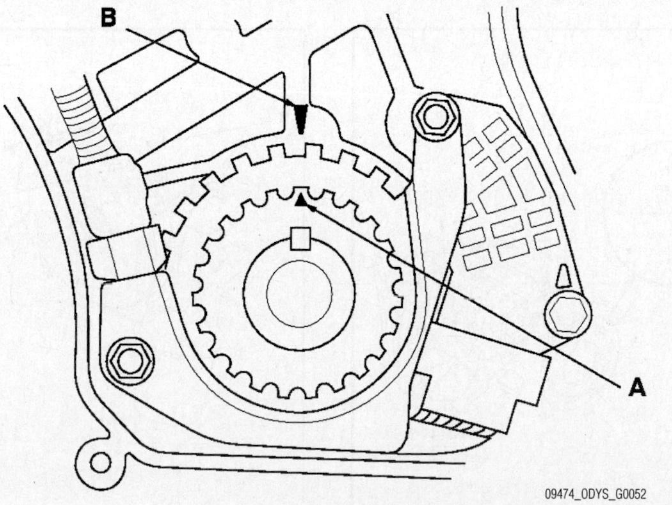
B
A
09474_ODYS_G0052

Fig. 99 Set the timing belt drive pulley to top dead center (TDC) by aligning the TDC mark (A) on the tooth of the timing belt drive pulley with the pointer (B) on the oil pump—J35A6 engine

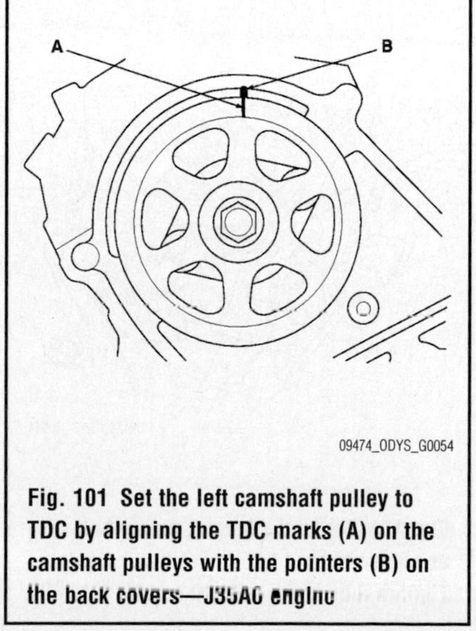
A
B
09474_ODYS_G0054

Fig. 101 Set the left camshaft pulley to TDC by aligning the TDC marks (A) on the camshaft pulleys with the pointers (B) on the back covers—J35A6 engine

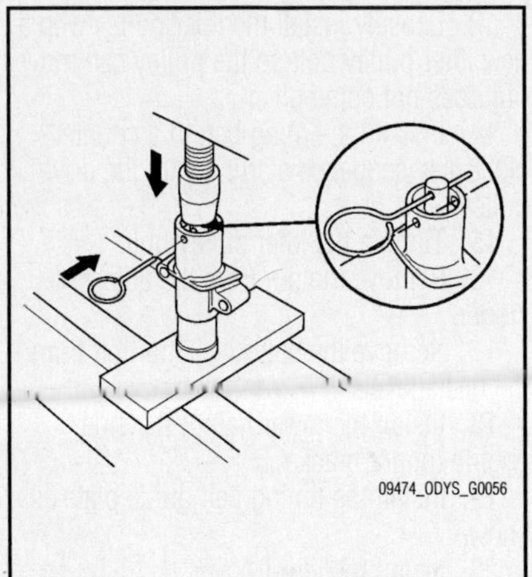
09474_ODYS_G0056

Fig. 102 Insert a 2.0 mm (0.08 in.) pin through the housing and the rod—J35A6 engine

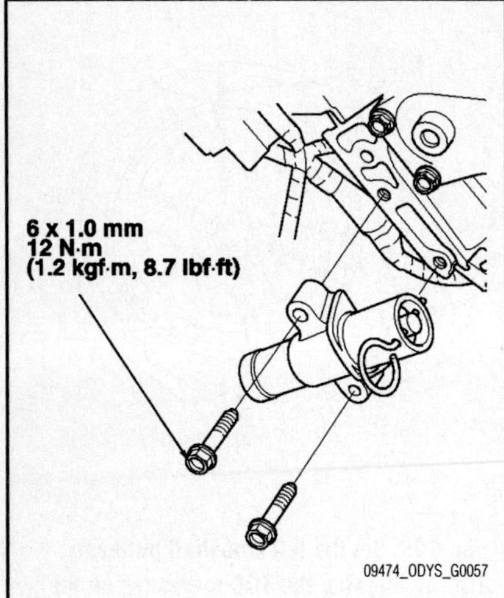
6 x 1.0 mm
12 N·m
(1.2 kgf·m, 8.7 lbf·ft)
09474_ODYS_G0057

Fig. 103 Install the auto-tensioner—J35A6 engine

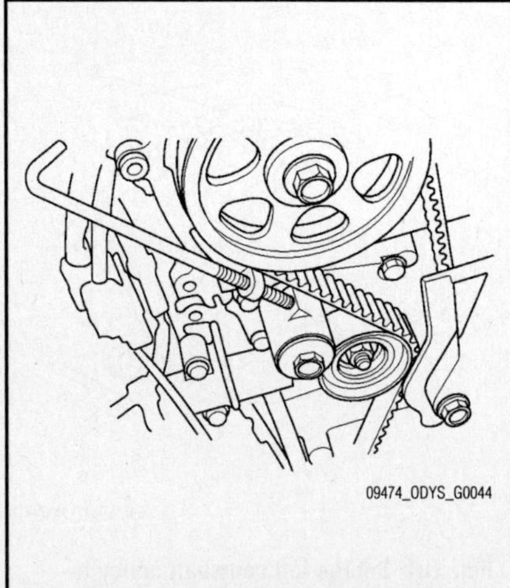
09474_ODYS_G0044

Fig. 104 Screw the battery clamp bolt in as shown to hold the timing belt adjuster. Tighten it by hand—J35A6 engine

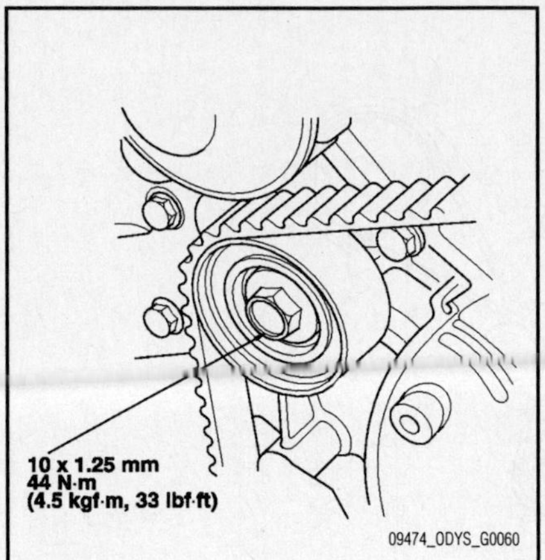
10 x 1.25 mm
44 N·m
(4.5 kgf·m, 33 lbf·ft)
09474_ODYS_G0060

Fig. 105 Tighten the idler pulley bolt—J35A6 engine

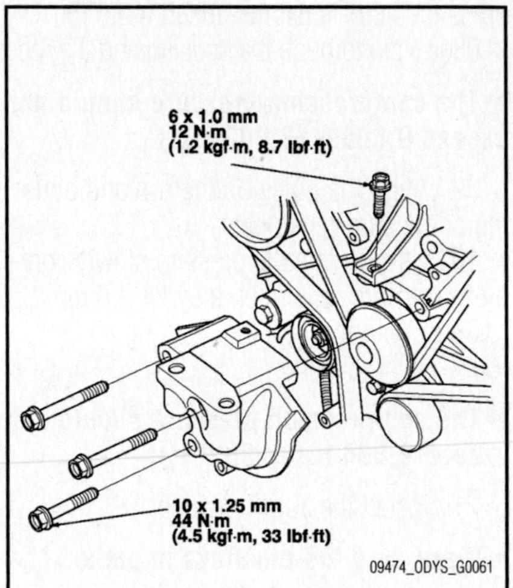
6 x 1.0 mm
12 N·m
(1.2 kgf·m, 8.7 lbf·ft)
10 x 1.25 mm
44 N·m
(4.5 kgf·m, 33 lbf·ft)
09474_ODYS_G0061

Fig. 106 Install the lower half of the side engine mount bracket—J35A6 engine

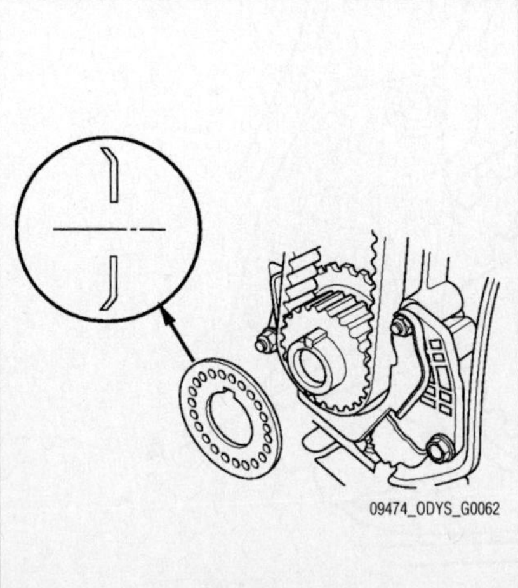
09474_ODYS_G0062

Fig. 107 Install the timing belt guide plate as shown—J35A6 engine

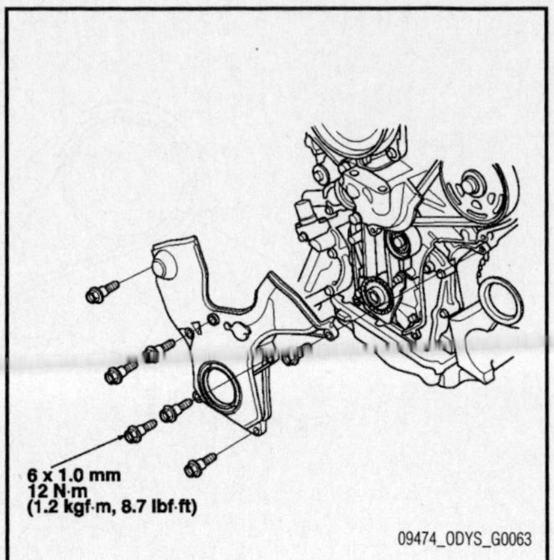
6 x 1.0 mm
12 N·m
(1.2 kgf·m, 8.7 lbf·ft)
09474_ODYS_G0063

Fig. 108 Install the lower cover—J35A6 engine

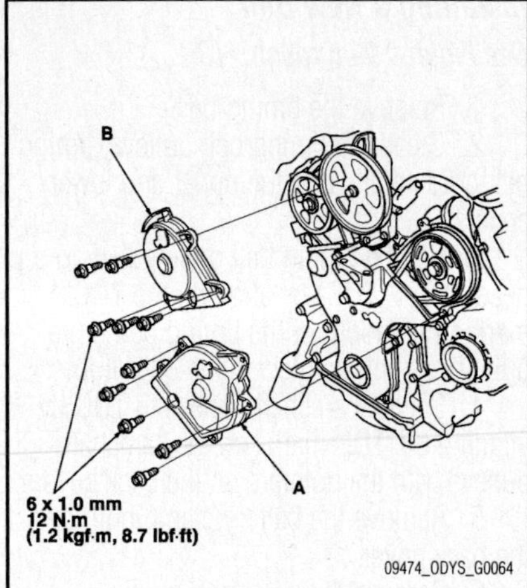
B
A
6 x 1.0 mm
12 N·m
(1.2 kgf·m, 8.7 lbf·ft)
09474_ODYS_G0064

Fig. 109 Install the front upper cover (A) and rear upper cover (B)—J35A6 engine

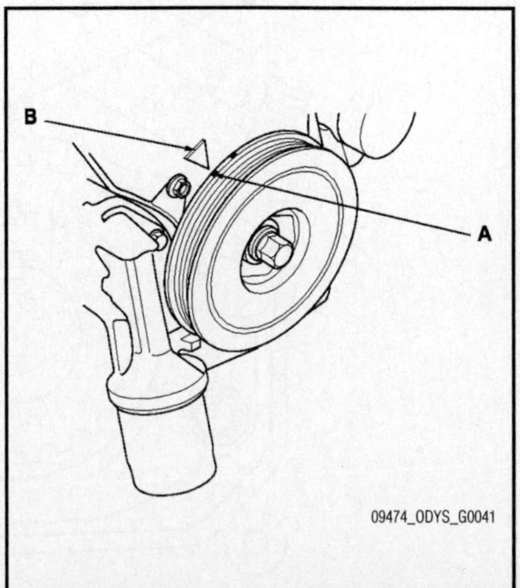
B
A
09474_ODYS_G0041

Fig. 110 Turn the crankshaft pulley so its white mark (A) lines up with the pointer (B)—J35A6 engine

22. Install the crankshaft pulley.
23. Rotate the crankshaft pulley six turns clockwise so the timing belt positions itself on the pulleys.
24. Turn the crankshaft pulley so its white mark lines up with the pointer.
25. Check the camshaft pulley marks, making sure the two marks are aligned.

➡If the marks are not aligned, rotate the crankshaft 360 degrees, and recheck the camshaft pulley mark.

a. If the camshaft pulley marks are at TDC, go to step 24.
b. If the camshaft pulley marks are not at TDC, remove the timing belt and repeat step 3 through 22.

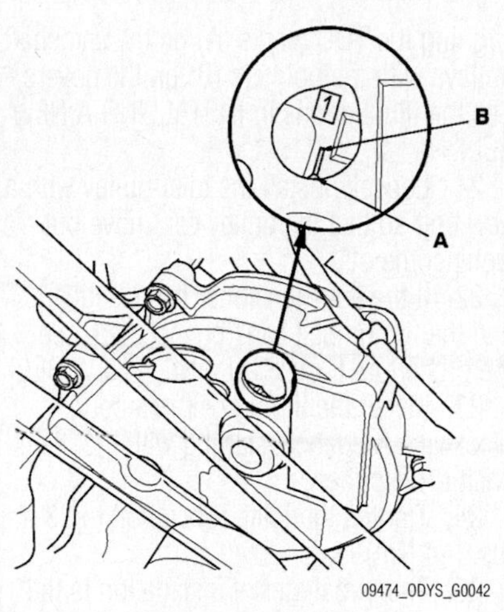
B
A
09474_ODYS_G0042

Fig. 111 Check the camshaft pulley marks, making sure the two marks (A and B) are aligned—J35A6 engine

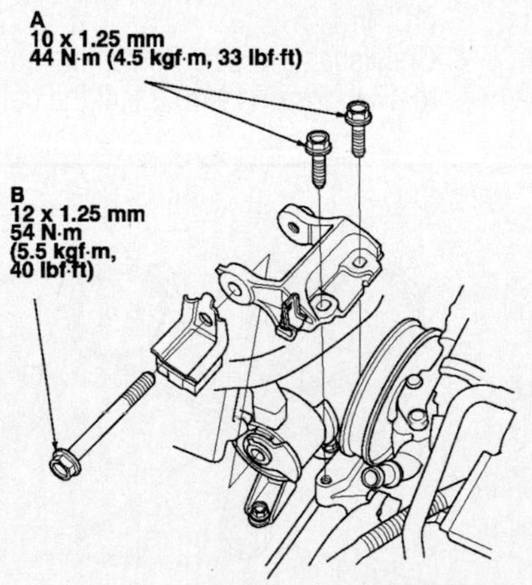
A
10 x 1.25 mm
44 N·m (4.5 kgf·m, 33 lbf·ft)
B
12 x 1.25 mm
54 N·m
(5.5 kgf·m,
40 lbf·ft)
09474_ODYS_G0067

Fig. 112 Install the upper half of the side engine mount bracket, and tighten the new mounting bolts (A), then the mass damper mounting bolt (B) —J35A6 engine

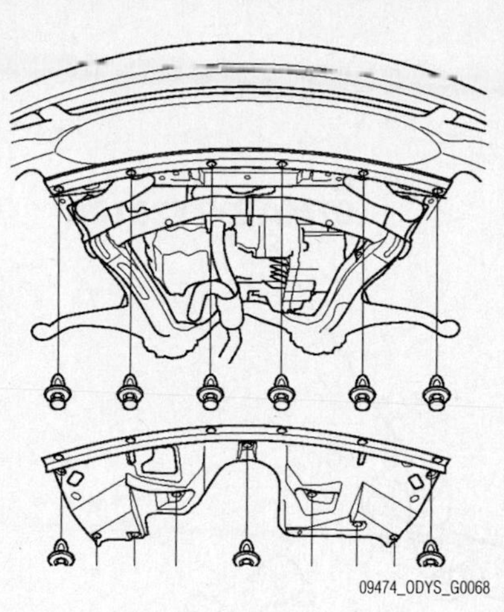
09474_ODYS_G0068

Fig. 113 Install the splash shield—J35A6 engine

26. Install the upper half of the side engine mount bracket, and tighten the new mounting bolts, then the mass damper mounting bolt.
27. Install the drive belt.
28. Install the splash shield.
29. Install the right front wheel.
30. Do the crankshaft position (CKP) pattern clear/CKP pattern learn procedure.

➡The ECT needs to be at 176 °F (80 °C) or higher.

a. Clear the CKP pattern while the engine is stopped.
b. Turn the ignition switch OFF.
c. Turn the ignition switch ON (II), and wait 30 seconds.
d. Test-drive the vehicle on a level road: decelerate (with the throttle fully closed) from an engine speed of 2,500 RPM to 1,000 RPM with the A/T in 2 position.
e. Stop the vehicle, but keep the engine running.
f. Check PULSER F/B LEARN in the DATA LIST with the HDS. If it is NOT COMPLETED, go to step 4. If it is COMPLETED, go to step 7.
g. Test-drive the vehicle on a level road: decelerate (with the throttle fully closed) from an engine speed of 5,000 RPM to 3,000 RPM with the A/T in 2 position.
h. Stop the vehicle, but keep the engine running.
i. Check the PULSER F/B LEARN (HIGH RPM) in the DATA LIST with the HDS. If it is NOT COMPLETED, go to step 7. If it is COMPLETED, go to step 10.
j. Turn the ignition switch OFF.
k. Turn the ignition switch ON (II), and wait 30 seconds. The CKP learning procedure is complete.

J35A7 Engine

Installing a used Belt

See Figures 114 through 121.

1. Remove the right front wheel.
2. Remove the splash shield.
3. Remove the accessory drive belt.
4. Turn the crankshaft so that the white mark on the pulley lines up with the pointer.
5. Check that the TDC timing mark on the front camshaft sprocket is aligned with the pointer on the cover.

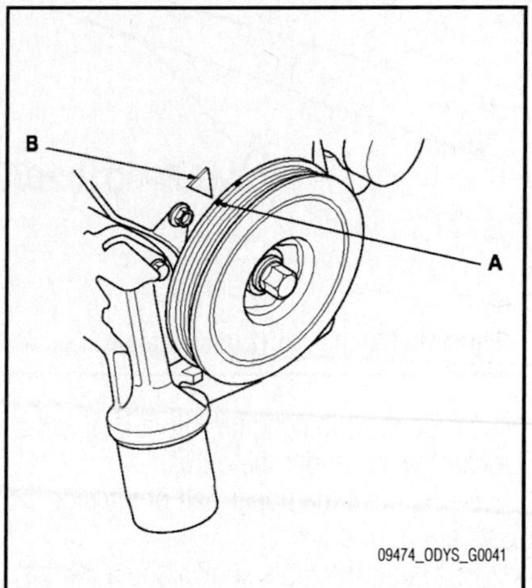
B
A
09474_ODYS_G0041

Fig. 114 Turn the crankshaft so that the white mark on the pulley (A) lines up with the pointer (B)—J35A7 engine

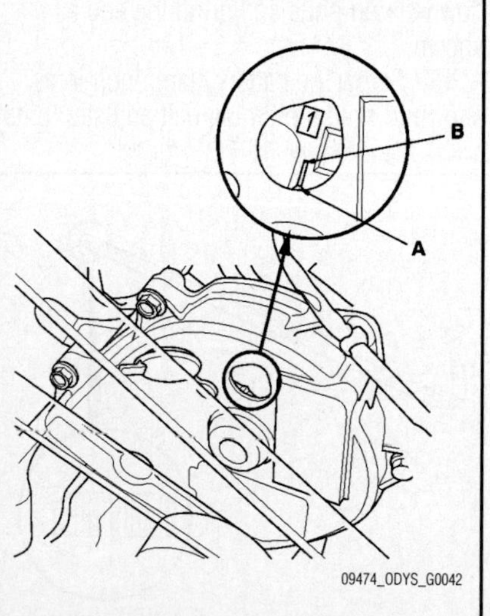
B
A
09474_ODYS_G0042

Fig. 115 Check that the TDC timing mark (A) on the front camshaft sprocket is aligned with the pointer (B) on the cover—J35A7 engine

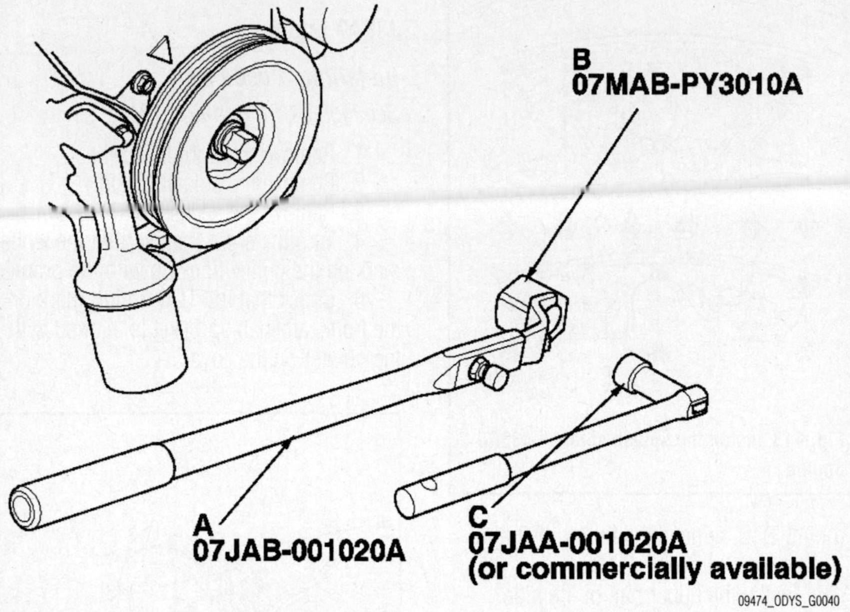
B
07MAB-PY3010A
A
07JAB-001020A
C
07JAA-001020A
(or commercially available)
09474_ODYS_G0040

Fig. 116 Damper bolt removal tools

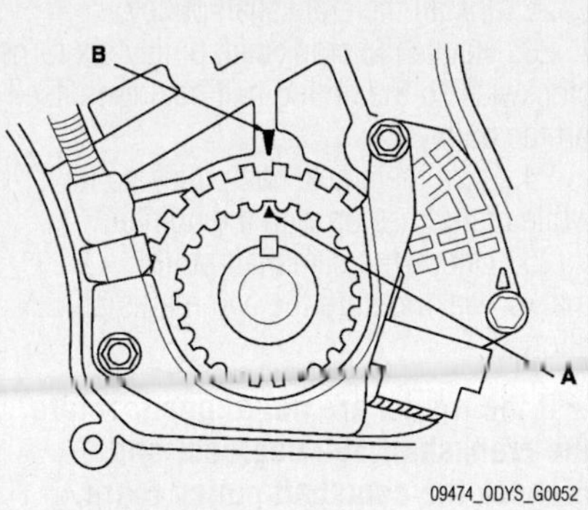
B
A
09474_ODYS_G0052

Fig. 119 Set the crankshaft sprocket to TDC by aligning the TDC mark (A) with the pointer (B) on the oil pump—J35A7 engine

6. Support the engine with a jack and a block of wood under the oil pan.
7. Remove the upper half of the side engine mount bracket.
8. Using the tools shown and a 19mm socket, remove the damper bolt.
9. Remove the damper.
10. Remove the front and rear upper belt covers.
11. Remove the lower cover.
12. Remove one of the battery hold-down clamp bolts and grind the end as shown.
13. Screw the battery clamp bolt in as shown to hold the timing belt adjuster in its current position. Tighten it by hand. Do not use a wrench.
14. Remove the timing belt guide plate.
15. Remove the lower side engine mount bracket.
16. Remove the idler pulley bolt and idler pulley. Discard the bolt.
17. Remove the timing belt.

To install:

18. Clean all parts.
19. Set the crankshaft sprocket to TDC by aligning the TDC mark with the pointer on the oil pump.
20. Set the camshaft pulleys to TDC by aligning the TDC marks (A) on the camshaft pulleys with the pointers (B) on the covers. See the illustrations in INSTALLING A NEW BELT.
21. Loosely install the idler pulley with a new bolt so that the pulley can move but won't come off.
22. If the auto-tensioner has extended, and the timing belt can't be installed, see the procedure for INSTALLING A NEW BELT.
23. Install the timing belt in a counterclockwise sequence, starting with the crankshaft drive pulley.
24. Tighten the idler pulley bolt to 33 ft. lbs. (44 Nm).
25. The remainder of installation is the reverse of removal. Note the following torques:

- Lower half engine mount bracket: 33 ft. lbs. (44 Nm)
- Lower cover: 105 inch lbs. (12 Nm)
- Upper covers: 105 inch lbs. (12 Nm)
- Crankshaft damper: Tighten the bolt to 47 ft. lbs. (64 Nm). Mark the bolt

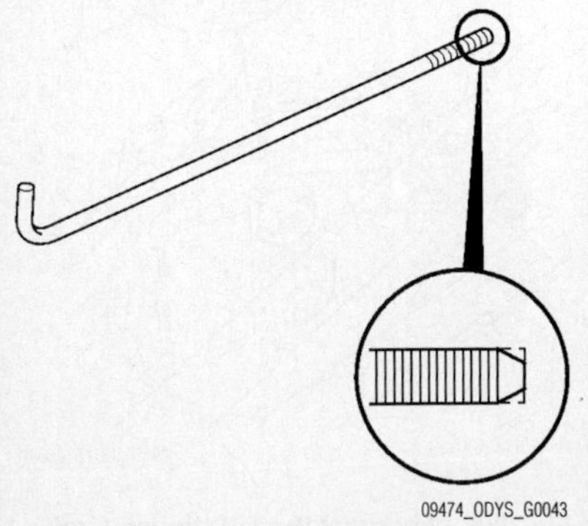
09474_ODYS_G0043

Fig. 117 Remove one of the battery hold-down clamp bolts and grind the end as shown

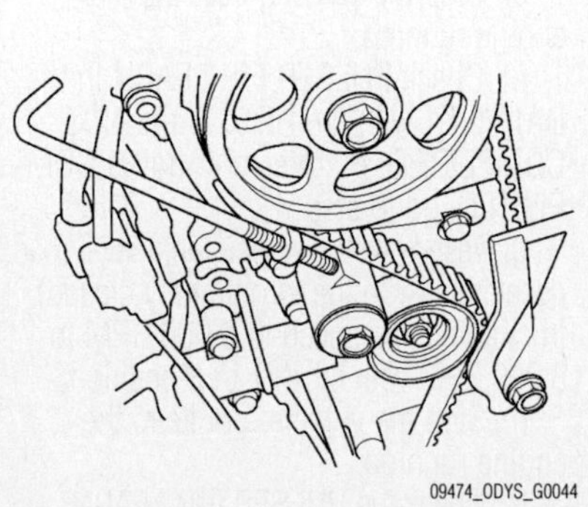
09474_ODYS_G0044

Fig. 118 Battery clamp bolt screwed into position

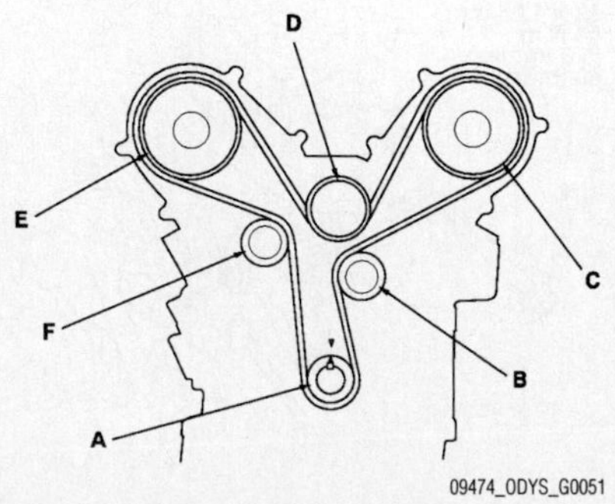
D
E
C
F
B
A
09474_ODYS_G0051

Fig. 120 Install the timing belt in a counterclockwise sequence (A through F), starting with the crankshaft drive pulley (A)—J35A7 engine

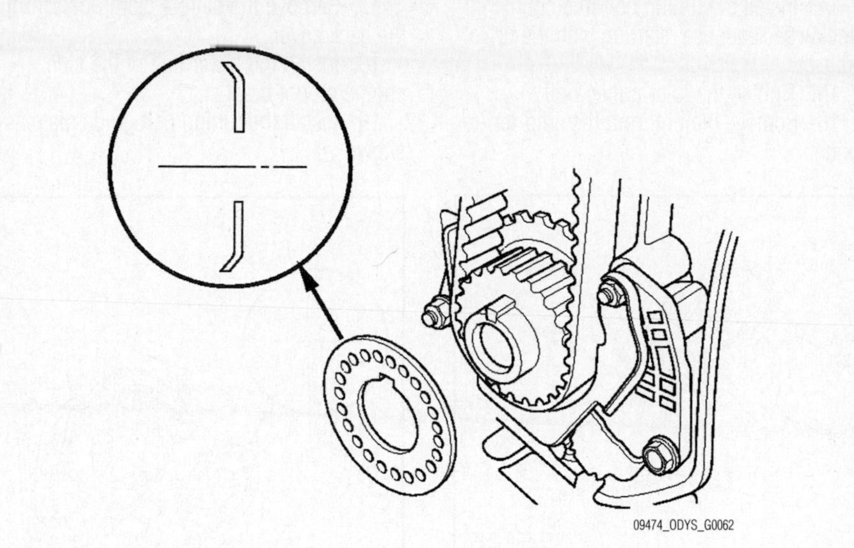

Fig. 121 Install the timing belt guide plate as shown—J35A7 engine

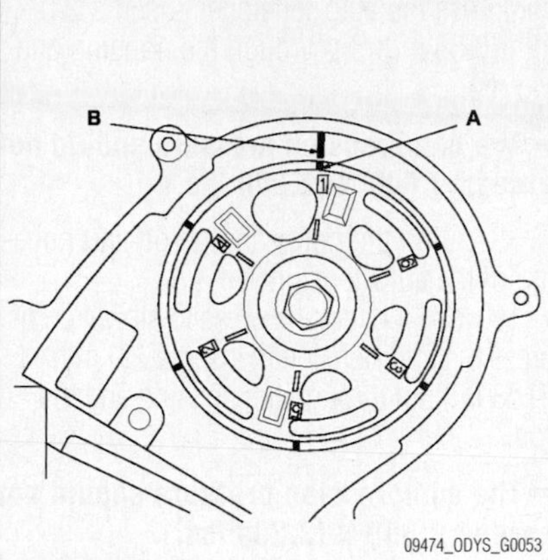
B
A

Fig. 123 Set the left camshaft pulley to TDC by aligning the TDC marks (A) on the camshaft pulleys with the pointers (B) on the back covers—J35A7 engine

head and pulley, then tighten the bolt an additional 60°.

26. Do the crankshaft position (CKP) pattern clear/CKP pattern learn procedure.

➡**The ECT needs to be at 176°F (80 °C) or higher.**

a. Clear the CKP pattern while the engine is stopped.
b. Turn the ignition switch OFF.
c. Turn the ignition switch ON (II), and wait 30 seconds.
d. Test-drive the vehicle on a level road: decelerate (with the throttle fully closed) from an engine speed of 2,500 RPM to 1,000 RPM with the A/T in 2 position.
e. Stop the vehicle, but keep the engine running.
f. Check PULSER F/B LEARN in the DATA LIST with the HDS. If it is NOT COMPLETED, go to step 4. If it is COMPLETED, go to step 7.
g. Test-drive the vehicle on a level road: decelerate (with the throttle fully closed) from an engine speed of 5,000 RPM to 3,000 RPM with the A/T in 2 position.
h. Stop the vehicle, but keep the engine running.
i. Check the PULSER F/B LEARN (HIGH RPM) in the DATA LIST with the HDS. If it is NOT COMPLETED, go to step 7. If it is COMPLETED, go to step 10.
j. Turn the ignition switch OFF.
k. Turn the ignition switch ON (II), and wait 30 seconds. The CKP learning procedure is complete.

Installing a New Belt

See Figures 122 through 138.

1. Remove the timing belt.
2. Clean the timing belt pulleys, timing belt guide plate, and the upper and lower covers.
3. Set the timing belt drive pulley to top dead center (TDC) by aligning the TDC mark on the tooth of the timing belt drive pulley with the pointer on the oil pump.
4. Set the camshaft pulleys to TDC by aligning the TDC marks on the camshaft pulleys with the pointers on the back covers.
5. Remove the battery clamp bolt from the back cover.
6. Remove the auto-tensioner.
7. Align the holes on the rod and housing of the auto-tensioner.
8. Use a hydraulic press to slowly

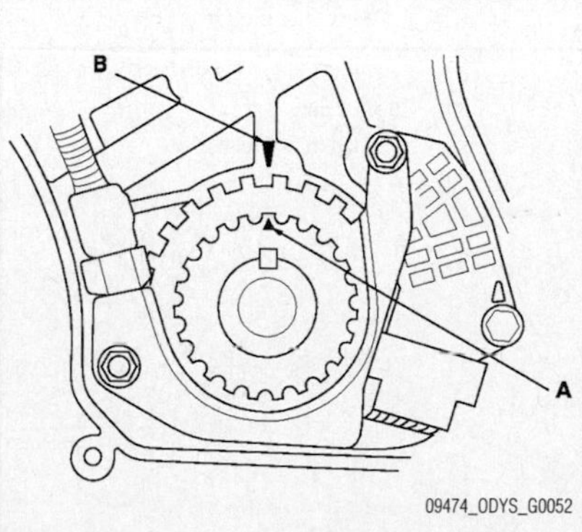
B
A

Fig. 122 Set the timing belt drive pulley to top dead center (TDC) by aligning the TDC mark (A) on the tooth of the timing belt drive pulley with the pointer (B) on the oil pump—J35A7 engine

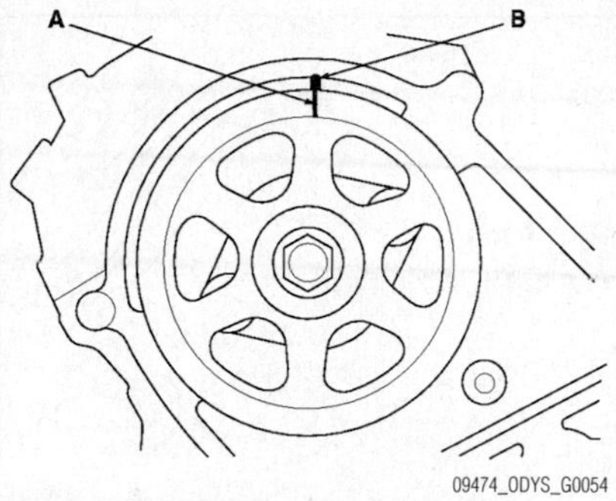
A
B

Fig. 124 Set the left camshaft pulley to TDC by aligning the TDC marks (A) on the camshaft pulleys with the pointers (B) on the back covers—J35A7 engine

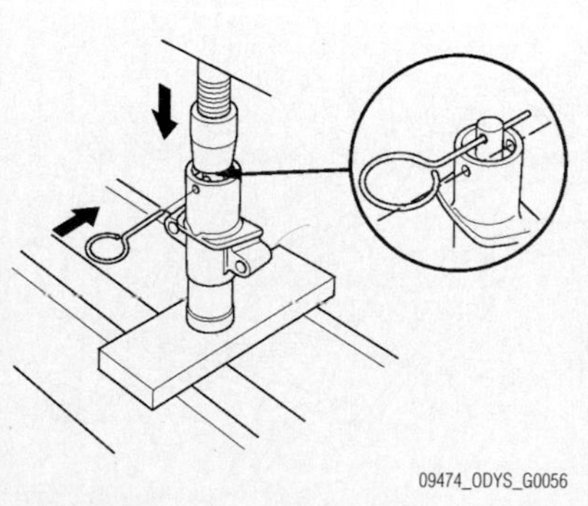

Fig. 125 Insert a 2.0 mm (0.08 in.) pin through the housing and the rod—J35A7 engine

compress the auto-tensioner. Insert a 2.0 mm (0.08 in.) pin through the housing and the rod.

➡**The compression pressure should not exceed 9,800 N (2,200 lbs.).**

9. Align the holes on the rod and housing of the auto-tensioner.
10. Use a hydraulic press to slowly compress the auto-tensioner. Insert a 2.0 mm (0.08 in.) pin through the housing and the rod.

➡**The compression pressure should not exceed 9,800 N (2,200 lbs.).**

11. Install the auto-tensioner.

➡**Make sure the pin stays in place.**

12. Screw the battery clamp bolt in as shown to hold the timing belt adjuster. Tighten it by hand; do not use a wrench.
13. Loosely install the idler pulley with a new idler pulley bolt so the pulley can move but does not come off.
14. Install the timing belt in a counterclockwise sequence, starting with the drive pulley.
15. Tighten the idler pulley bolt.
16. Remove the pin from the auto-tensioner.

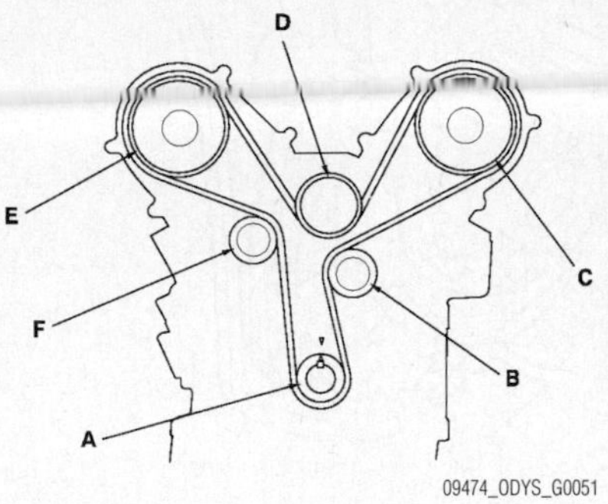
D
E
C
F
B
A
09474_ODYS_G0051

Fig. 128 Install the timing belt in a counterclockwise sequence (A through F), starting with the drive pulley (A)—J35A7 engine

17. Remove the battery clamp bolt from the back cover.
18. Install the lower half of the side engine mount bracket.
19. Install the timing belt guide plate as shown.

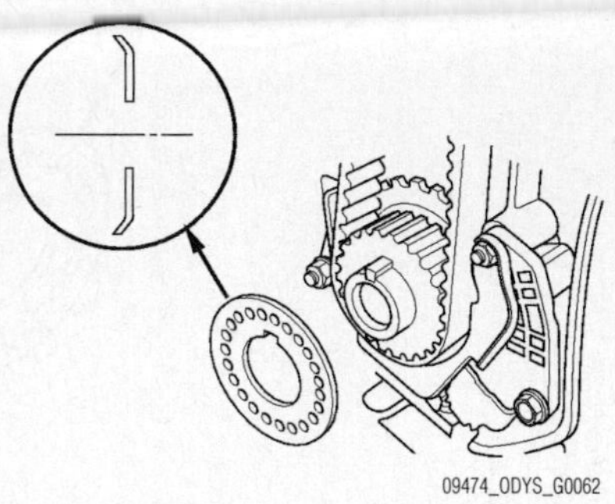
09474_ODYS_G0062

Fig. 131 Install the timing belt guide plate as shown—J35A7 engine

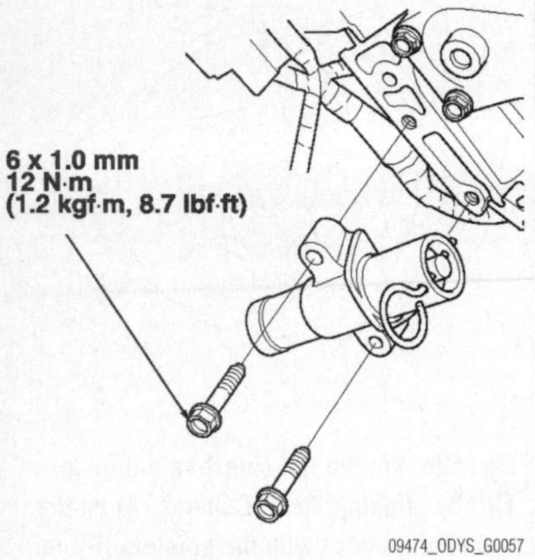
6 x 1.0 mm
12 N·m
(1.2 kgf·m, 8.7 lbf·ft)
09474_ODYS_G0057

Fig. 126 Install the auto-tensioner—J35A7 engine

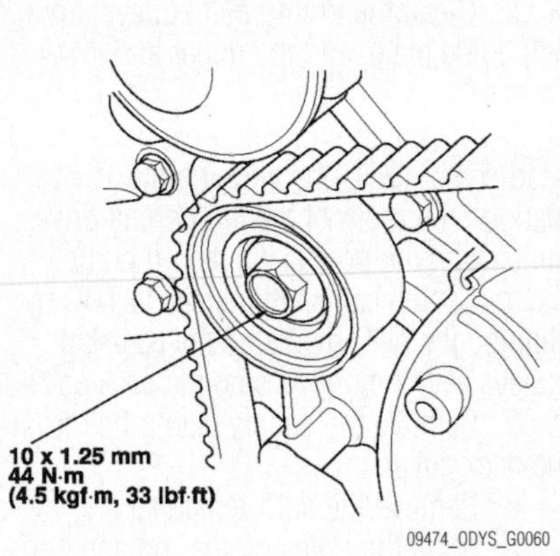
10 x 1.25 mm
44 N·m
(4.5 kgf·m, 33 lbf·ft)
09474_ODYS_G0060

Fig. 129 Tighten the idler pulley bolt—J35A7 engine

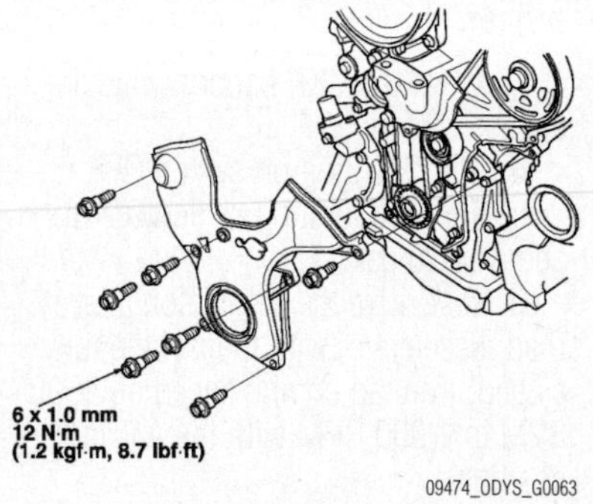
6 x 1.0 mm
12 N·m
(1.2 kgf·m, 8.7 lbf·ft)
09474_ODYS_G0063

Fig. 132 Install the lower cover—J35A7 engine

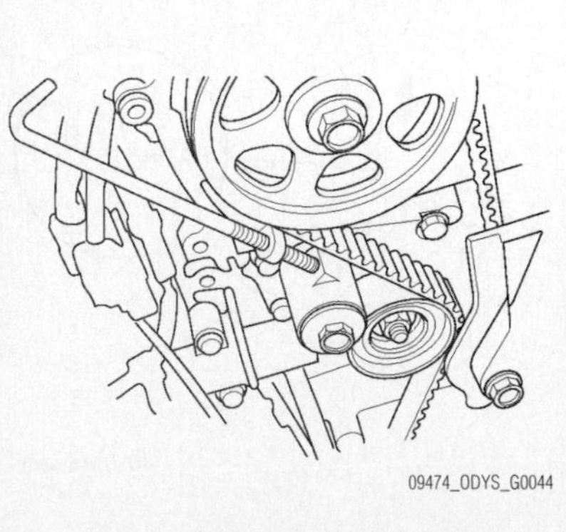
09474_ODYS_G0044

Fig. 127 Screw the battery clamp bolt in as shown to hold the timing belt adjuster. Tighten it by hand—J35A7 engine

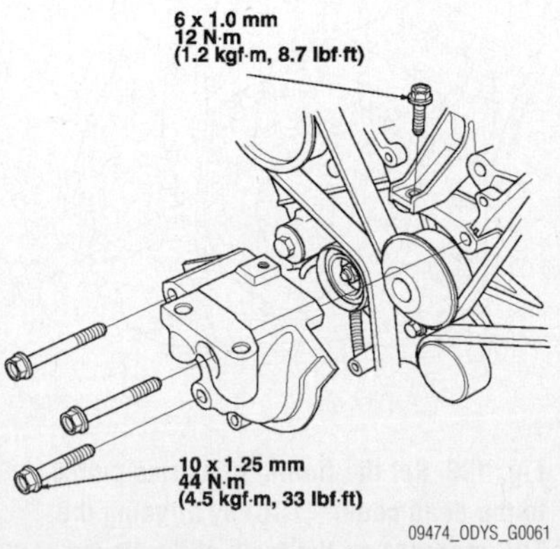
6 x 1.0 mm
12 N·m
(1.2 kgf·m, 8.7 lbf·ft)
10 x 1.25 mm
44 N·m
(4.5 kgf·m, 33 lbf·ft)
09474_ODYS_G0061

Fig. 130 Install the lower half of the side engine mount bracket—J35A7 engine

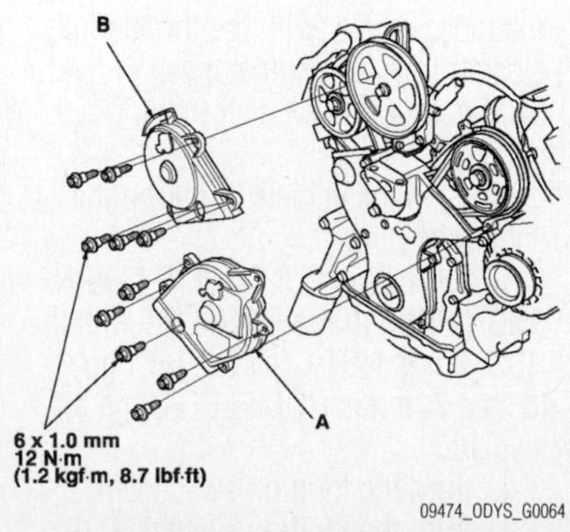
B
A
6 x 1.0 mm
12 N·m
(1.2 kgf·m, 8.7 lbf·ft)
09474_ODYS_G0064

Fig. 133 Install the front upper cover (A) and rear upper cover (B)—J35A7 engine

20. Install the lower cover.

21. Install the front upper cover and rear upper cover.

22. Install the crankshaft pulley.

23. Rotate the crankshaft pulley six turns clockwise so the timing belt positions itself on the pulleys.

24. Turn the crankshaft pulley so its white mark lines up with the pointer.

25. Check the camshaft pulley marks.

➡If the marks are not aligned, rotate the crankshaft 360 degrees, and recheck the camshaft pulley mark.

a. If the camshaft pulley marks are at TDC, go to step 24.

b. If the camshaft pulley marks are not at TDC, remove the timing belt and repeat step 3 through 22.

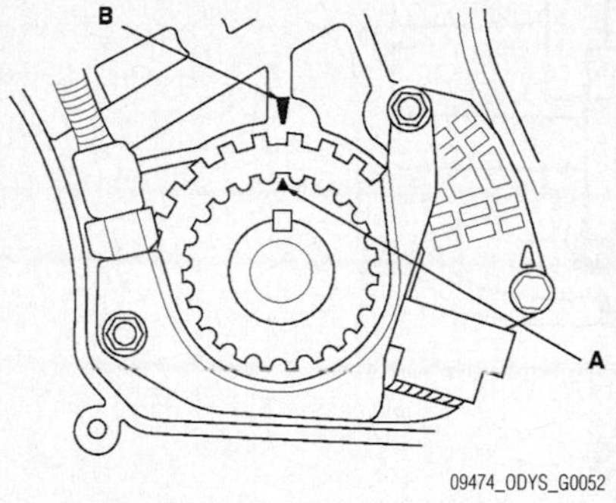
B
A
09474_ODYS_G0052

Fig. 134 Turn the crankshaft pulley so its white mark (A) lines up with the pointer (B) —J35A7 engine

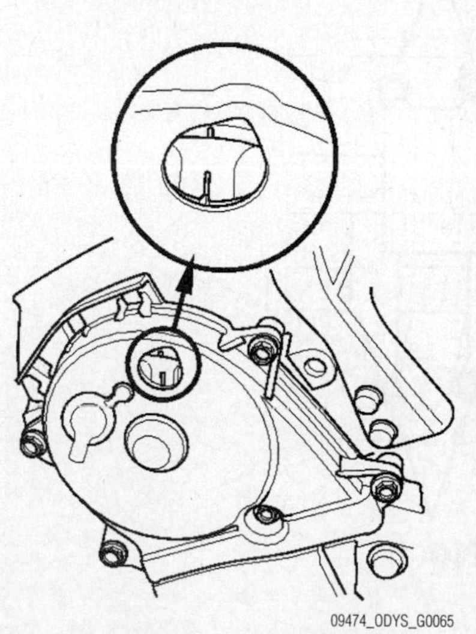
09474_ODYS_G0065

Fig. 135 Camshaft pulley marks—J35A7 engine rear head

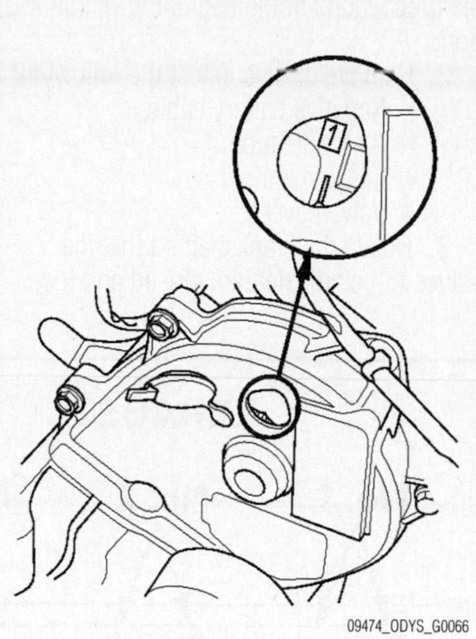
1
09474_ODYS_G0066

Fig. 136 Camshaft pulley marks—J35A7 engine front head

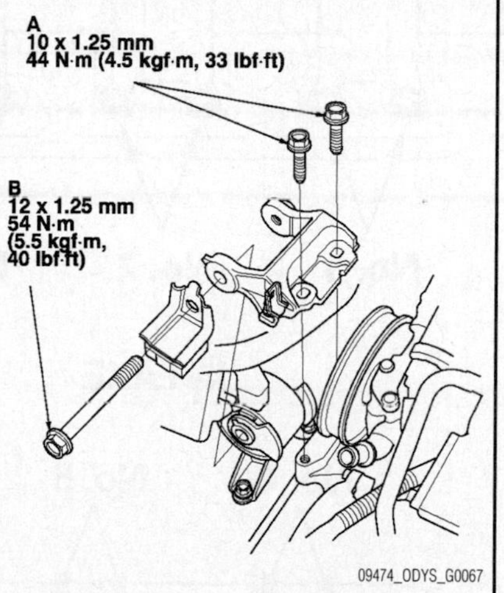
A
10 x 1.25 mm
44 N·m (4.5 kgf·m, 33 lbf·ft)
B
12 x 1.25 mm
54 N·m
(5.5 kgf·m,
40 lbf·ft)
09474_ODYS_G0067

Fig. 137 Install the upper half of the side engine mount bracket, and tighten the new mounting bolts (A), then the mass damper mounting bolt (B) —J35A7 engine

26. Install the upper half of the side engine mount bracket, and tighten the new mounting bolts, then the mass damper mounting bolt.

27. Install the drive belt.

28. Install the splash shield.

29. Install the right front wheel.

30. Do the crankshaft position (CKP) pattern clear/CKP pattern learn procedure.

➡The ECT needs to be at 176°F (80 °C) or higher.

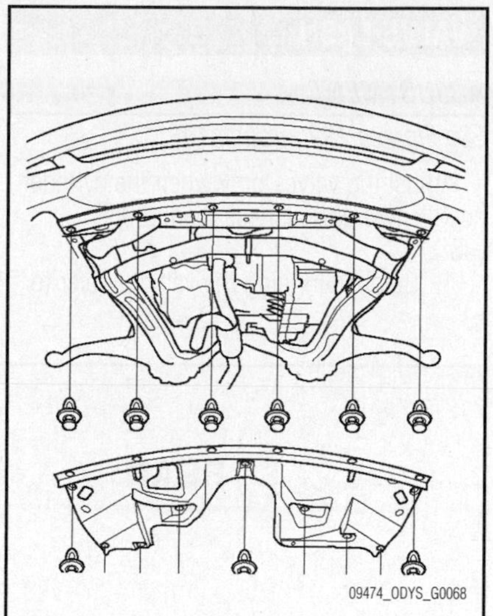
09474_ODYS_G0068

Fig. 138 Install the splash shield—J35A7 engine

a. Clear the CKP pattern while the engine is stopped.

b. Turn the ignition switch OFF.

c. Turn the ignition switch ON (II), and wait 30 seconds.

d. Test-drive the vehicle on a level road: decelerate (with the throttle fully closed) from an engine speed of 2,500 RPM to 1,000 RPM with the A/T in 2 position.

e. Stop the vehicle, but keep the engine running.

f. Check PULSER F/B LEARN in the DATA LIST with the HDS. If it is NOT COMPLETED, go to step 4. If it is COMPLETED, go to step 7.

g. Test-drive the vehicle on a level road: decelerate (with the throttle fully closed) from an engine speed of 5,000 RPM to 3,000 RPM with the A/T in 2 position.

h. Stop the vehicle, but keep the engine running.

i. Check the PULSER F/B LEARN (HIGH RPM) in the DATA LIST with the HDS. If it is NOT COMPLETED, go to step 7. If it is COMPLETED, go to step 10.

j. Turn the ignition switch OFF.

k. Turn the ignition switch ON (II), and wait 30 seconds. The CKP learning procedure is complete.

VALVE COVERS

REMOVAL & INSTALLATION

For additional information, refer to the following section, "Cylinder Head, Removal and Installation."

VALVE LASH

ADJUSTMENT

See Figures 139 through 143.

Adjust the valves only when the cylinder head temperature is less than 100°F (38°C).

1. Before servicing the vehicle, refer to the precautions in the beginning of this section.
2. Remove or disconnect the following:
   - Negative battery cable
   - Air intake tube
   - Intake manifold
   - Valve cover
3. Rotate the crankshaft so that the valves to be adjusted are closed and the rocker arm is contacting the camshaft lobe base circle.
4. Measure the valve clearance. If adjustment is necessary, loosen the locknut and turn the adjusting screw as necessary to achieve the correct valve clearance.
5. The correct valve clearance is:
   - Intake valves: 0.008–0.009 inches (0.20–0.24mm)

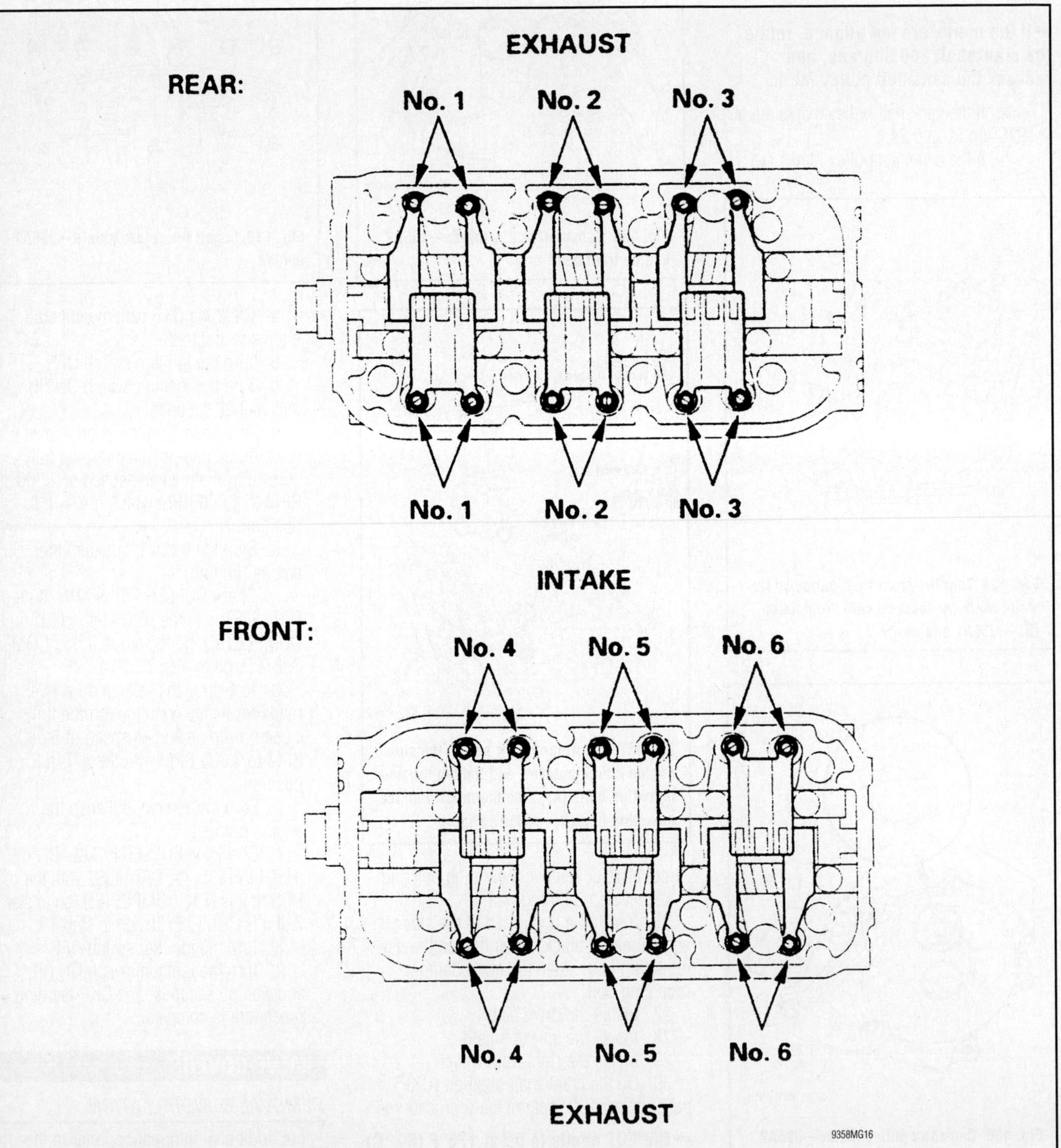
EXHAUST
REAR:
No. 1
No. 2
No. 3
No. 1
No. 2
No. 3
INTAKE
FRONT:
No. 4
No. 5
No. 6
No. 4
No. 5
No. 6
EXHAUST
9358MG16

Fig. 139 Adjusting screw locations—J35A6 engines

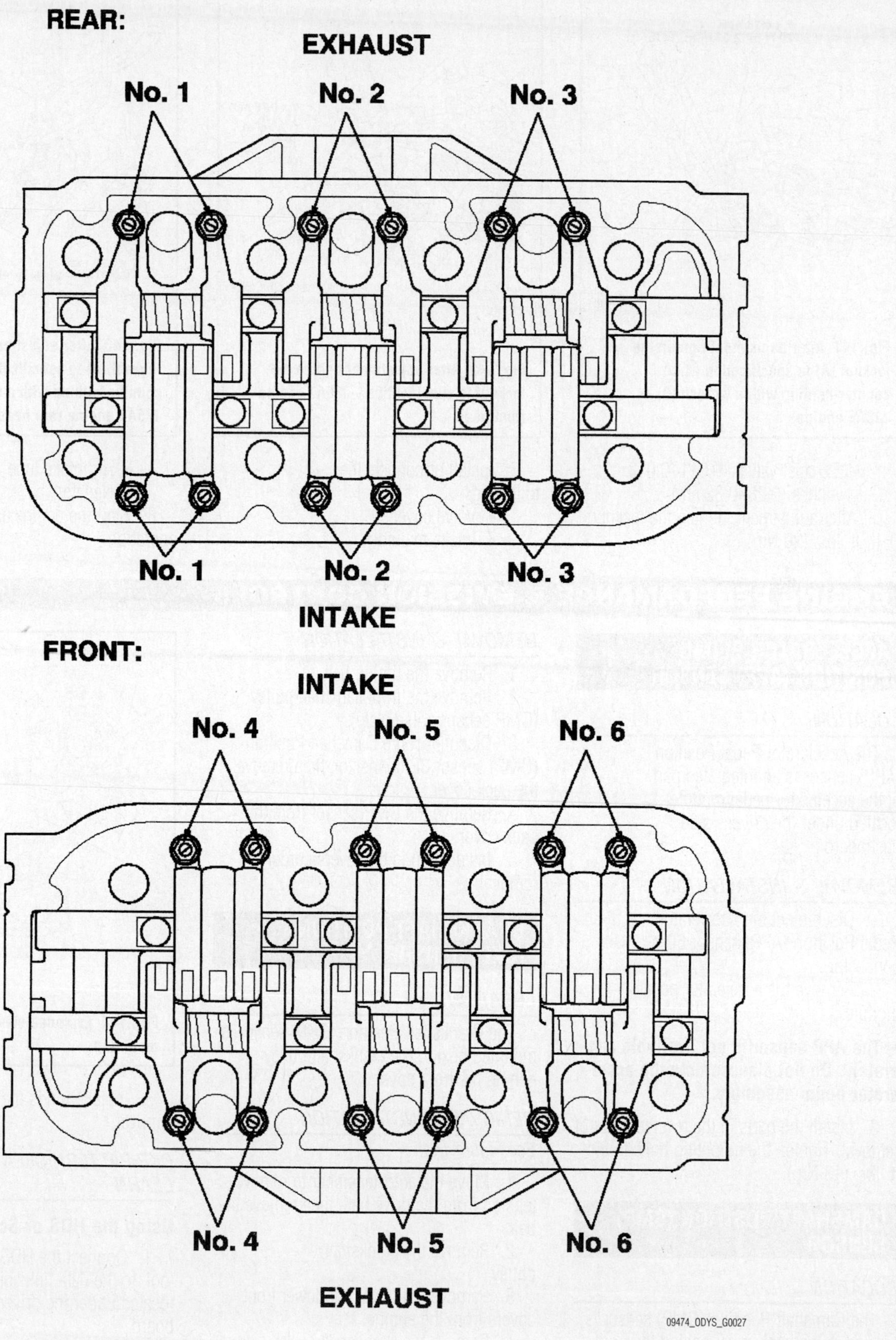
REAR:
EXHAUST
No. 1
No. 2
No. 3
No. 1
No. 2
No. 3
INTAKE
FRONT:
INTAKE
No. 4
No. 5
No. 6
No. 4
No. 5
No. 6
EXHAUST
09474_ODYS_G0027

Fig. 140 Adjusting screw locations—J35A7 engines

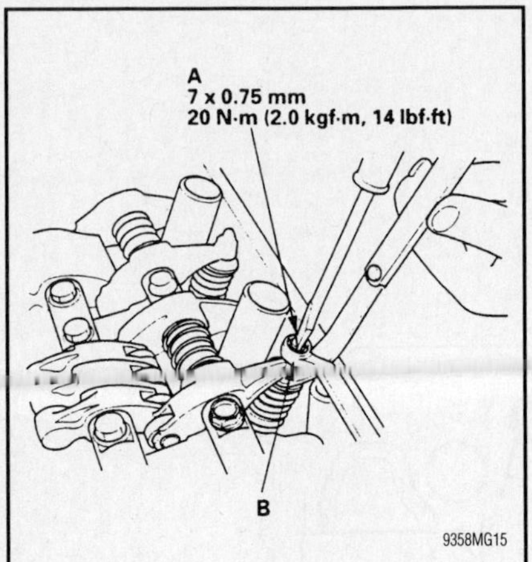
A
7 x 0.75 mm
20 N·m (2.0 kgf·m, 14 lbf·ft)
B
9358MG15

Fig. 141 After adjustment tighten the locknut (A) to specification while counter-holding with a wrench (B)—J35A6 engines

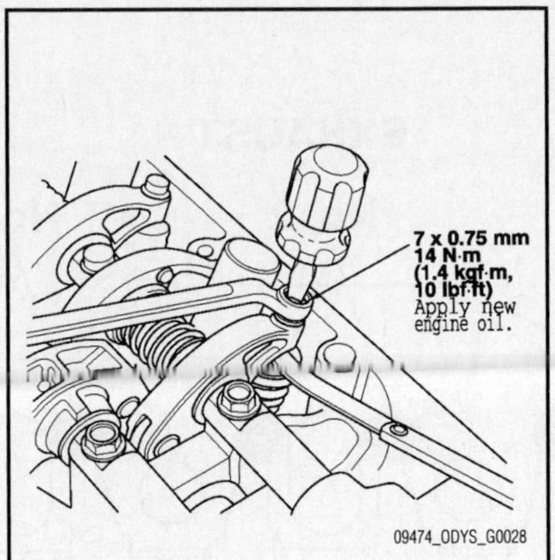
7 x 0.75 mm
14 N·m
(1.4 kgf·m,
10 lbf·ft)
Apply new
engine oil.
09474_ODYS_G0028

Fig. 142 After adjustment tighten the locknut to specification—J35A7 engine front head

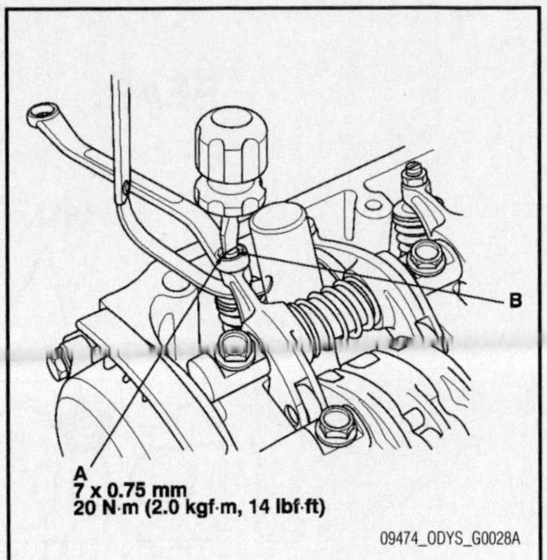
B
A
7 x 0.75 mm
20 N·m (2.0 kgf·m, 14 lbf·ft)
09474_ODYS_G0028A

Fig. 143 After adjustment tighten the locknut (A) to specification while counter-holding with a wrench (B)—J35A7 engine rear head

- Exhaust valves: 0.011–0.013 inches (0.28–0.32mm)

6. After adjustment, tighten the locknuts to 14 ft. lbs. (20 Nm).
7. Install or connect the following:
   - Valve cover
   - Intake manifold
   - Air intake tube
   - Negative battery cable
8. Start the engine and check for proper operation.

ENGINE PERFORMANCE & EMISSION CONTROLS

ACCELERATOR PEDAL POSITION (APP) SENSOR

LOCATION

The Accelerator Pedal Position (APP) sensor is an integrated part of the accelerator pedal module, located under the driver's side dashboard.

REMOVAL & INSTALLATION

1. Disconnect the Accelerator Pedal Position (APP) sensor 6P connector.
2. Remove the accelerator pedal module.

➡The APP sensor is not available separately. Do not disassemble the accelerator pedal assembly.

3. Install the parts in the reverse order of removal. Tighten the mounting bolts to 9.4 ft. lbs. (13 Nm).

CAMSHAFT POSITION (CMP) SENSOR

LOCATION

The Camshaft Position (CMP) sensor is located in the back cover behind the front camshaft pulley.

REMOVAL & INSTALLATION

1. Remove the timing belt.
2. Remove the front camshaft pulley (CMP sensor pulse plate).
3. Disconnect the Camshaft Position (CMP) sensor 3P connector, then remove the back cover.
4. Remove the CMP sensor from the back cover.
5. Installation is the reverse order of removal.

CRANKSHAFT POSITION (CKP) SENSOR

LOCATION

The Crankshaft Position (CKP) sensor is mounted next to the crankshaft pulley behind the front cover.

REMOVAL & INSTALLATION

See Figure 144.

1. Move the auto-tensioner to remove tension from the drive belt, then remove the belt.
2. Remove the crankshaft pulley.
3. Remove the upper and lower front covers from the engine.
4. Remove the Crankshaft Position (CKP) sensor from the oil pump.

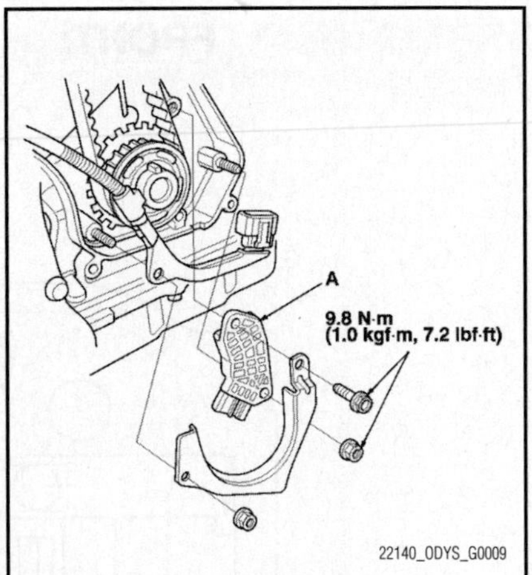
A
9.8 N·m
(1.0 kgf·m, 7.2 lbf·ft)
22140_ODYS_G0009

Fig. 144 Exploded view of the CKP sensor mounting

5. Installation is the reverse order of removal.

CKP PATTERN CLEAR/CKP PATTERN LEARN

Using the HDS or Scan Tool

1. Connect the HDS, or equivalent scan tool, to the data link connector (DLC) located under the driver's side of the dashboard.
2. Turn the ignition switch to ON (II).

3. Make sure the HDS communicates with the PCM and other vehicle systems.
4. Select CRANK PATTERN in the ADJUSTMENT MENU with the HDS and clear the CKP pattern.
5. Select CRANK PATTERN LEARNING with the HDS, and follow the screen prompts.

Without the HDS or Scan Tool

1. Start the engine. Hold the engine speed at 3,000 RPM without load (in P or N) until the radiator fan comes on.
2. Test-drive the vehicle on a level road: Decelerate (with the throttle fully closed) from an engine speed of 2,500 RPM down to 1,000 RPM with the transmission in 2. Repeat step several times.
3. Turn the ignition switch to LOCK (0).
4. Turn the ignition switch to ON (II), and wait 30 seconds.

EVAPORATIVE EMISSIONS (EVAP) CANISTER

LOCATION

See Figure 145.

The EVAP canister assembly is located under the vehicle, aft of the engine.

REMOVAL & INSTALLATION

See Figure 146.

1. Raise the vehicle on a lift.
2. Remove the EVAP canister cover.
3. Remove the hoses, the FTP sensor 3P connector, and the EVAP canister vent shut valve 2P connector.
4. Remove the bolts.
5. Remove the EVAP canister assembly.
6. Remove the EVAP canister bracket.

➡The canister filter remains on the bracket.

7. Install the parts in the reverse order of removal.

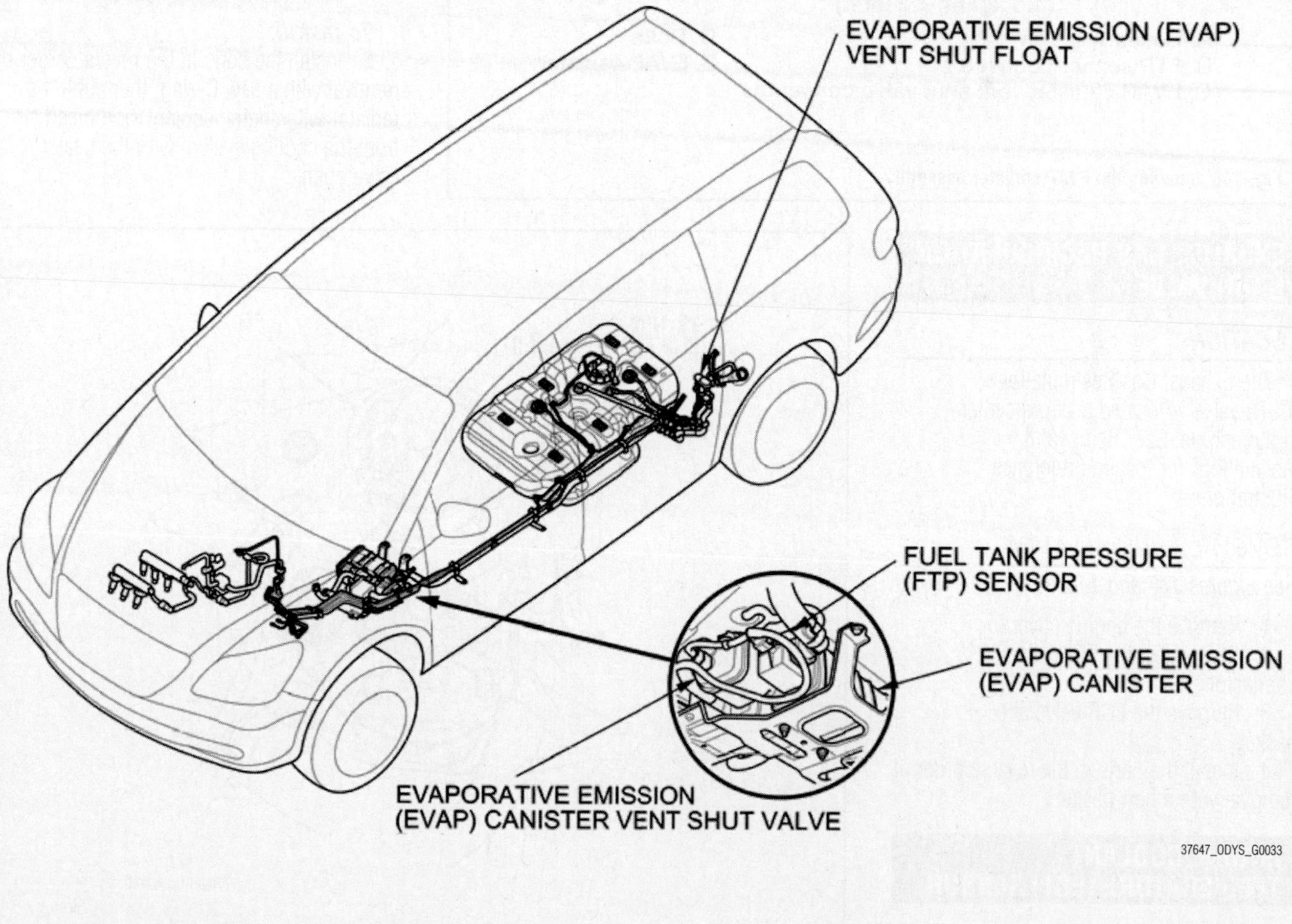
EVAPORATIVE EMISSION (EVAP) VENT SHUT FLOAT
FUEL TANK PRESSURE (FTP) SENSOR
EVAPORATIVE EMISSION (EVAP) CANISTER
EVAPORATIVE EMISSION (EVAP) CANISTER VENT SHUT VALVE
37647_ODYS_G0033

Fig. 145 Showing the EVAP system component locations

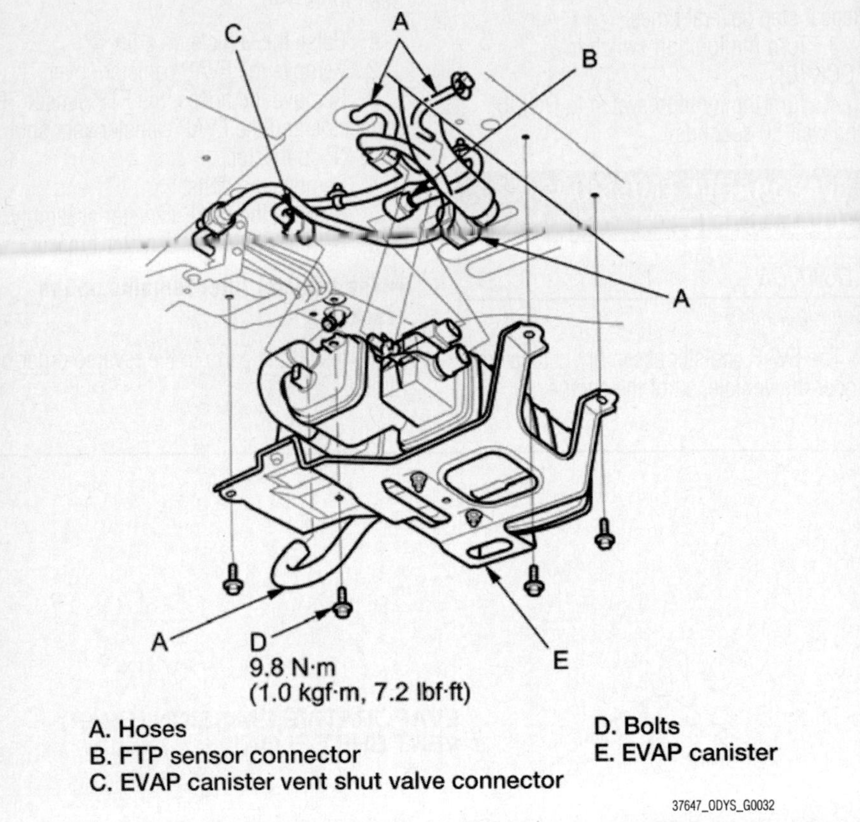
C
A
B
A
A
D
9.8 N·m
(1.0 kgf·m, 7.2 lbf·ft)
E
A. Hoses
B. FTP sensor connector
C. EVAP canister vent shut valve connector
D. Bolts
E. EVAP canister
37647_ODYS_G0032

Fig. 146 Showing the EVAP canister assembly

EXHAUST GAS RECIRCULATION (EGR) VALVE

LOCATION

The Exhaust Gas Recirculation (EGR) valve is located on front center of the engine. See "Removal & Installation" for location reference illustrations.

REMOVAL & INSTALLATION

See Figures 147 and 148.

1. Remove the engine cover.
2. Disconnect the EGR valve 6P connector.
3. Remove the EGR valve and gasket.
4. Install the parts in the reverse order of removal with a new gasket.

ENGINE COOLANT TEMPERATURE (ECT) SENSOR

LOCATION

There are two Engine Coolant Temperature (ECT) sensors located below the engine appearance cover.

REMOVAL & INSTALLATION

Sensor 1

1. Drain the engine coolant.
2. Remove the engine cover.
3. Disconnect the Engine Coolant Temperature (ECT) sensors 1 2P connector.
4. Remove the ECT sensor 1.

To install:

5. Install the parts in the reverse order of removal with a new O-ring, then refill the radiator with engine coolant, and bleed air from the cooling system with the heater valve open.

Sensor 2

1. Drain the engine coolant.
2. Remove the engine cover.
3. Disconnect the Engine Coolant Temperature (ECT) sensors 2 2P connector.
4. Remove the ECT sensor 2.

To install:

5. Install the parts in the reverse order of removal with a new O-ring, then refill the radiator with engine coolant, and bleed air from the cooling system with the heater valve open.

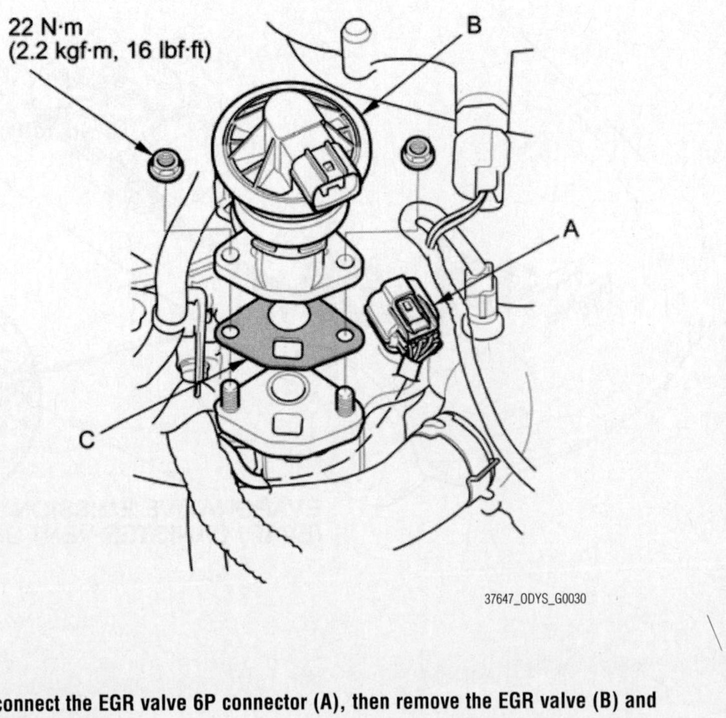
22 N·m
(2.2 kgf·m, 16 lbf·ft)
B
A
C
37647_ODYS_G0030

Fig. 147 Disconnect the EGR valve 6P connector (A), then remove the EGR valve (B) and gasket (C)—J35A6 engine

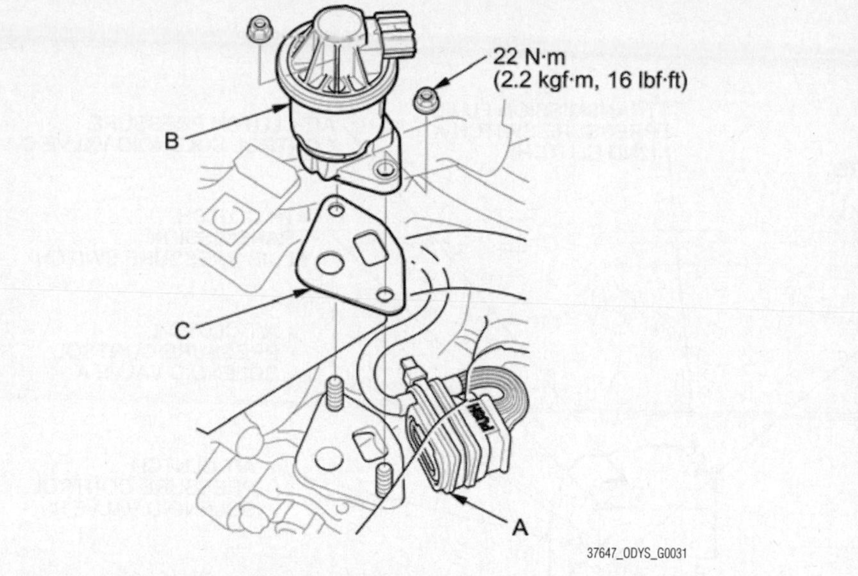
22 N·m
(2.2 kgf·m, 16 lbf·ft)
B
C
A
37647_ODYS_G0031

Fig. 148 Disconnect the EGR valve 6P connector (A), then remove the EGR valve (B) and gasket (C)—J35A7 engine

KNOCK SENSOR (KS)

LOCATION

The knock sensor is located in the top middle of the cylinder head underneath the intake manifold.

REMOVAL & INSTALLATION

1. Remove the intake manifold.
2. Remove the injector rails and the injector base.
3. Disconnect the knock sensor 1P connector, then remove the knock sensor.
4. Install the parts in the reverse order of removal.

MALFUNCTION INDICATOR LIGHT (MIL)

RESET PROCEDURES

The indicator is reset using the Honda Diagnostic System (HDS). Connect the HDS, or equivalent scan tool, to the Data Link Connector (DLC) port. Identify, troubleshoot and clear any Diagnostic Trouble Codes (DTC) to reset the MIL.

MASS AIR FLOW/INTAKE AIR TEMPERATURE (MAF/IAT) SENSOR

LOCATION

The Mass Air Flow/Intake Air Temperature (MAF/IAT) sensor is located on the air intake tube of the air intake assembly.

REMOVAL & INSTALLATION

See Figure 149.

1. Disconnect the MAF/IAT sensor 5P connector.
2. Remove the screws.
3. Remove the MAF/IAT sensor.
4. Install the parts in the reverse order of removal.

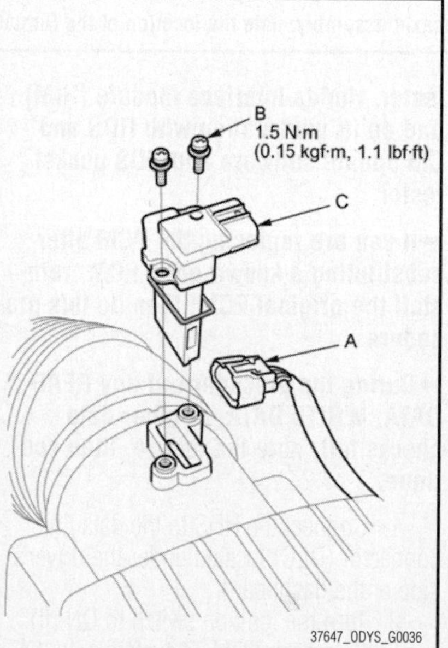
B
1.5 N·m
(0.15 kgf·m, 1.1 lbf·ft)
C
A
37647_ODYS_G0036

Fig. 149 Removing/installing the MAF/IAT sensor

MANIFOLD ABSOLUTE PRESSURE (MAP) SENSOR

LOCATION

The Manifold Absolute Pressure (MAP) sensor is located on the throttle bottle assembly.

REMOVAL & INSTALLATION

See Figure 150.

1. Remove the engine appearance cover.
2. Disconnect the Manifold Absolute Pressure (MAP) sensor 3P connector.
3. Remove the screw.
4. Remove the MAP sensor.
5. Install the parts in the reverse order of removal with a new O-ring.

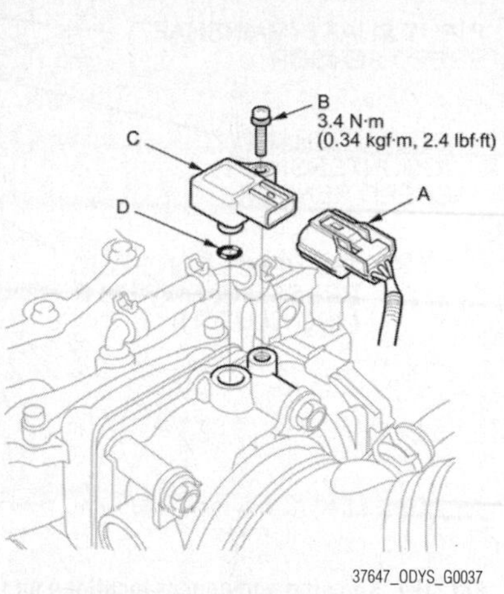
B
3.4 N·m
(0.34 kgf·m, 2.4 lbf·ft)
C
A
D
37647_ODYS_G0037

Fig. 150 Removing/installing the MAP sensor

OUTPUT (COUNTERSHAFT) SPEED SENSOR (OSS)

LOCATION

See Figure 151.

The Output Shaft Speed (OSS) sensor is located on the transaxle assembly, as shown in the illustration.

REMOVAL & INSTALLATION

1. Remove the engine splash shield.
2. Disconnect the output shaft (countershaft) speed sensor connector, and remove the output shaft (countershaft) speed sensor.

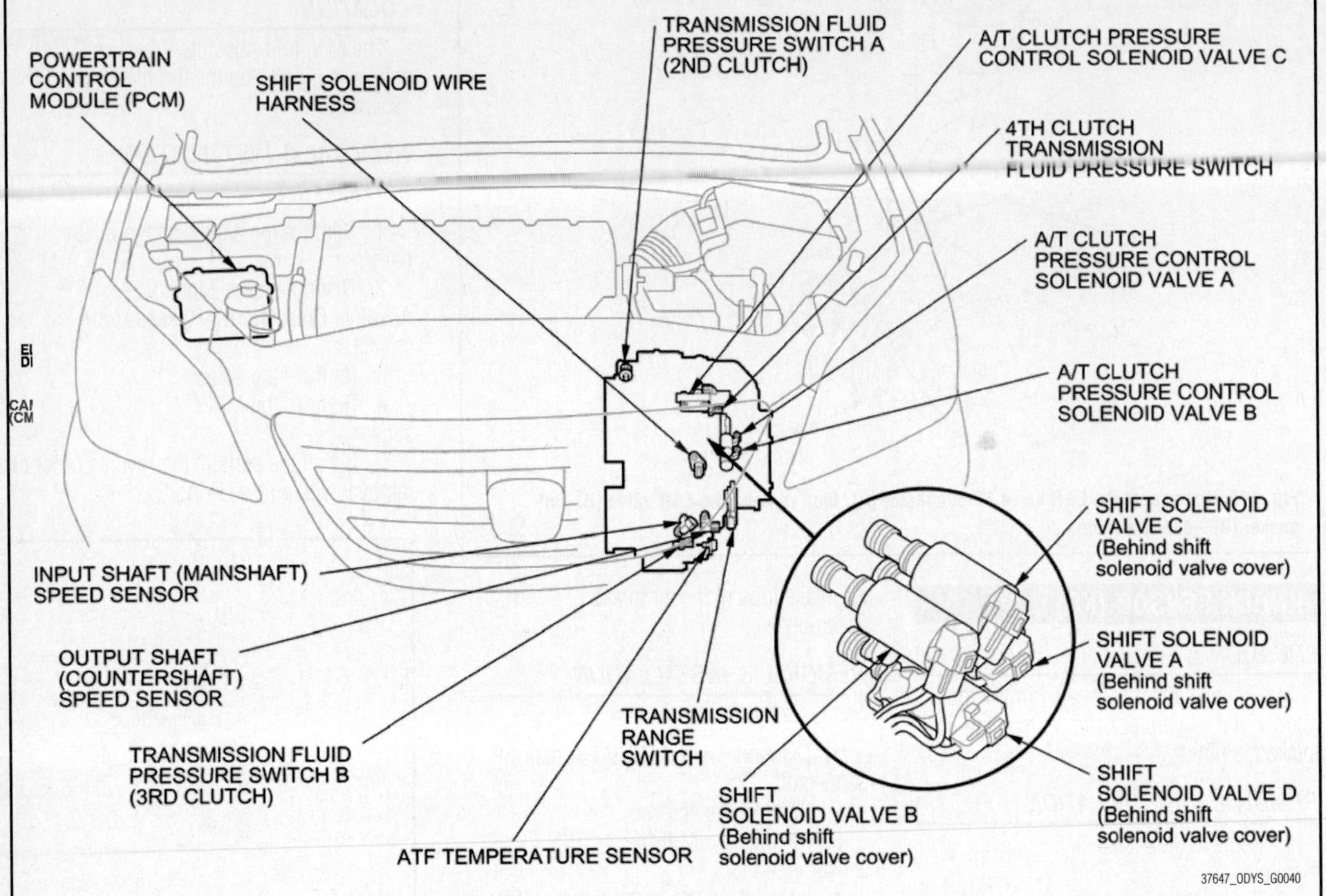
TRANSMISSION FLUID PRESSURE SWITCH A (2ND CLUTCH)
A/T CLUTCH PRESSURE CONTROL SOLENOID VALVE C
POWERTRAIN CONTROL MODULE (PCM)
SHIFT SOLENOID WIRE HARNESS
4TH CLUTCH TRANSMISSION FLUID PRESSURE SWITCH
A/T CLUTCH PRESSURE CONTROL SOLENOID VALVE A
A/T CLUTCH PRESSURE CONTROL SOLENOID VALVE B
SHIFT SOLENOID VALVE C (Behind shift solenoid valve cover)
INPUT SHAFT (MAINSHAFT) SPEED SENSOR
SHIFT SOLENOID VALVE A (Behind shift solenoid valve cover)
OUTPUT SHAFT (COUNTERSHAFT) SPEED SENSOR
TRANSMISSION RANGE SWITCH
TRANSMISSION FLUID PRESSURE SWITCH B (3RD CLUTCH)
SHIFT SOLENOID VALVE D (Behind shift solenoid valve cover)
SHIFT SOLENOID VALVE B (Behind shift solenoid valve cover)
ATF TEMPERATURE SENSOR
37647_ODYS_G0040

Fig. 151 Showing component locations on the transaxle assembly; note the location of the Output Shaft Speed Sensor

To install:

3. Install a new O-ring on a new output shaft (countershaft) speed sensor and apply clean ATF to the O-ring, then install the output shaft (countershaft) speed sensor in the transmission housing.
4. Check the connector for rust, dirt, or oil, then connect the connector securely.
5. Install the engine splash shield.

POWERTRAIN CONTROL MODULE (PCM)

LOCATION

See Figure 152.

REMOVAL & INSTALLATION

See Figure 153.

➡This procedure requires the use of a Honda diagnostic system (HDS) tablet tester, Honda interface module (HIM) and an iN workstation with HDS and CM update software and HDS pocket tester

➡If you are replacing the PCM after substituting a known-good PCM, reinstall the original PCM, then do this procedure.

➡During the procedure, if any READ DATA, WRITE DATA, or other data checks fail, note the failure, then continue.

1. Connect the HDS to the data link connector (DLC) located under the driver's side of the dashboard.
2. Turn the ignition switch to ON (II).
3. Select the PGM-FI system with the HDS.
4. Select the INSPECTION MENU with the HDS.
5. Select the ETCS TEST, then select the TP POSITION CHECK, and follow the screen prompts.

➡If the TP POSITION CHECK indicates FAILED, continue with this procedure.

6. Select the REPLACE PCM MENU, then select READ DATA and follow the screen prompts.

➡Doing this step copies (READS) the engine oil life data from the original PCM so you can later download (WRITES) it into the new PCM.

7. If READ DATA indicates FAILED, continue with this procedure.
8. Select the A/T system with the HDS.
9. Select the REPLACE TCM/PCM MENU, then select READ DATA and follow the screen prompts.

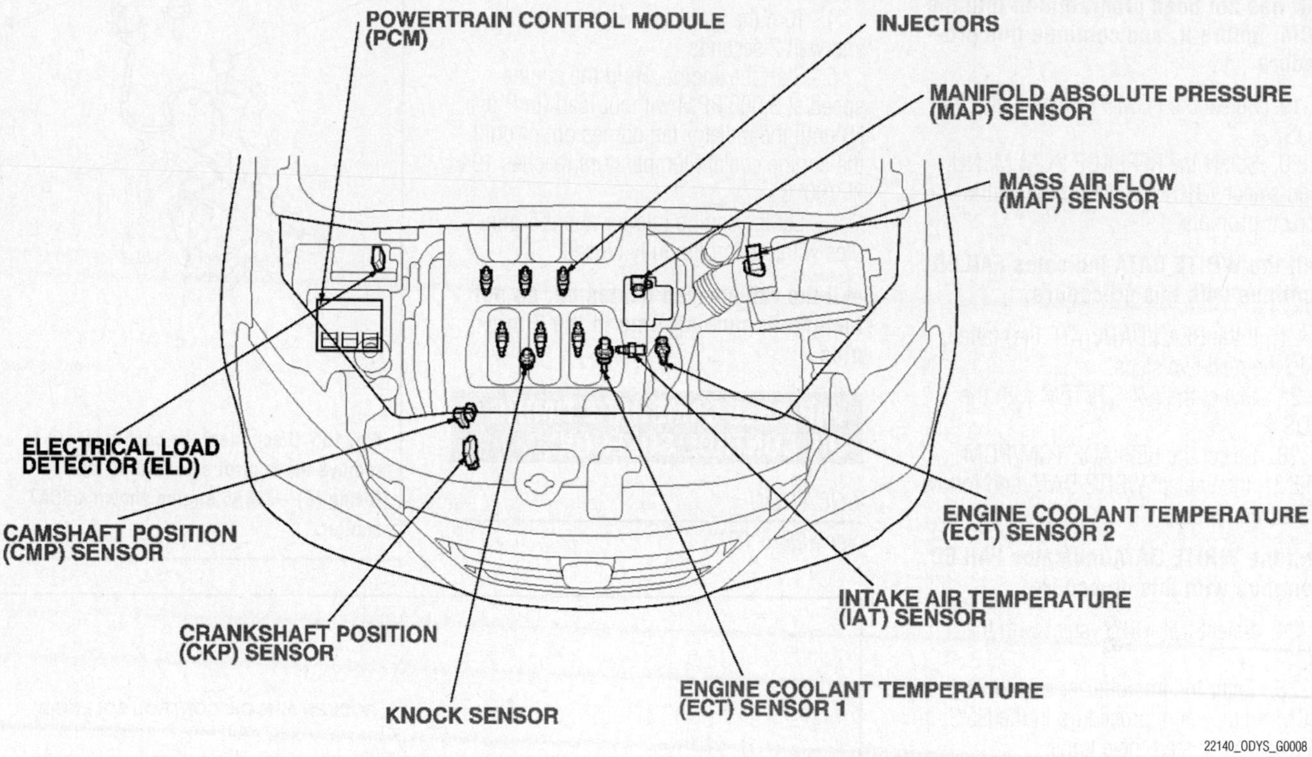
POWERTRAIN CONTROL MODULE (PCM)
INJECTORS
MANIFOLD ABSOLUTE PRESSURE (MAP) SENSOR
MASS AIR FLOW (MAF) SENSOR
ELECTRICAL LOAD DETECTOR (ELD)
CAMSHAFT POSITION (CMP) SENSOR
ENGINE COOLANT TEMPERATURE (ECT) SENSOR 2
INTAKE AIR TEMPERATURE (IAT) SENSOR
CRANKSHAFT POSITION (CKP) SENSOR
ENGINE COOLANT TEMPERATURE (ECT) SENSOR 1
KNOCK SENSOR
22140_ODYS_G0008

Fig. 152 The PCM is located in the engine compartment

➡Doing this step copies (READS) the ATF life data from the original PCM so you can later download (WRITES) it into the new PCM.

10. If READ DATA indicates FAILED, continue with this procedure.
11. Turn the ignition switch to LOCK (0).
12. Jump the SCS line with the HDS.
13. Remove the power steering reservoir from the body.
14. Remove the cover, then disconnect PCM connectors A, B, and C.

➡PCM connectors A, B, and C have symbols (A=square, B=triangle , C=circle) embossed on them for identification.

15. Remove the bolts, then remove the PCM.

To install:

16. Installation is the reverse order of removal.
17. Turn the ignition switch to ON (II).

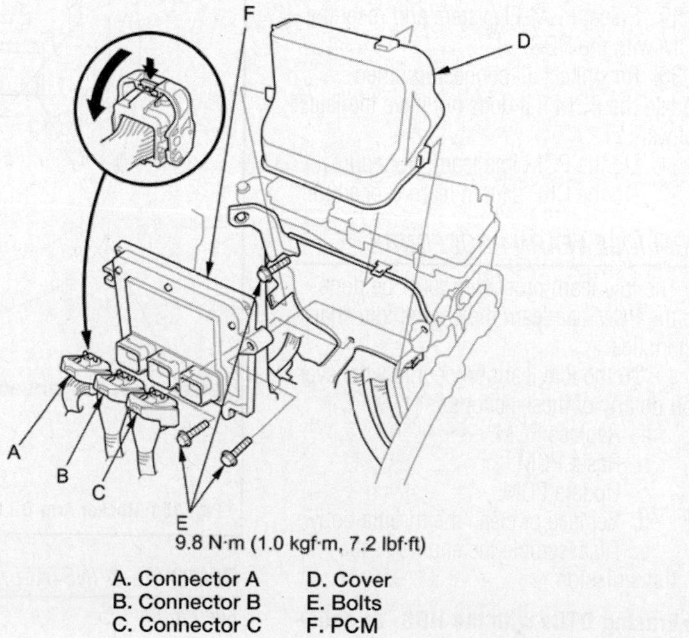
F
D
A
B
C
E
9.8 N·m (1.0 kgf·m, 7.2 lbf·ft)
A. Connector A
B. Connector B
C. Connector C
D. Cover
E. Bolts
F. PCM
37647_ODYS_G0038

Fig. 153 Removing the PCM

18. Manually input the VIN to the PCM with the HDS.

➡DTC P0630 "VIN Not Programmed or Mismatch" may be stored because the VIN has not been programmed into the PCM; ignore it, and continue this procedure.

19. Select the PGM-FI system with the HDS.
20. Select the REPLACE PCM MENU, then select WRITE DATA and follow the screen prompts.

➡If the WRITE DATA indicates FAILED, continue with this procedure.

21. If the READ DATA (ATF life) failed, skip the next two steps.
22. Select the A/T SYSTEM with the HDS.
23. Select the REPLACE TCM/PCM MENU, then select WRITE DATA and follow the screen prompts.

➡If the WRITE DATA indicates FAILED, continue with this procedure.

24. Select IMMOBI system with the HDS.
25. Enter the immobilizer code with the PCM replacement procedure in the HDS; it allows you to start the engine.
26. If the TP POSITION CHECK failed, clean the throttle body.
27. If the READ DATA or the WRITE DATA failed for engine oil life, replace the engine oil and engine oil filter.
28. If the READ DATA or the WRITE DATA failed for the A/T system, replace the ATF.
29. Select PGM-FI system and reset the PCM with the HDS.
30. Reconnect all connectors, then update the PCM if it does not have the latest software.
31. Do the PCM idle learn procedure.
32. Do the CKP pattern learn procedure.

PCM IDLE LEARN PROCEDURE

The idle learn procedure must be done so the PCM can learn the engine idle characteristics.

1. Do the idle learn procedure whenever you do any of these actions:
   a. Replace PCM.
   b. Reset PCM.
   c. Update PCM.
   d. Replace or clean the throttle body.
   e. Disassemble the engine or the transmission.

➡Erasing DTCs with the HDS, or equivalent scan tool, does not require you to do the idle learn procedure.

2. Make sure all electrical items (A/C, audio system, rear window defogger, lights, etc.) are off.
3. Reset the PCM with the HDS, or equivalent scan tool.
4. Turn the ignition switch to ON (II), and wait 2 seconds.
5. Start the engine. Hold the engine speed at 3,000 RPM without load (in P or N) until the radiator fan comes on, or until the engine coolant temperature reaches 194 °F (90°C).
6. Let the engine idle for about 5 minutes with the throttle fully closed.

➡If the radiator fan comes on, do not include its running time in the 5 minutes.

VARIABLE CAMSHAFT TIMING OIL CONTROL SOLENOID

LOCATION

See Figure 154.

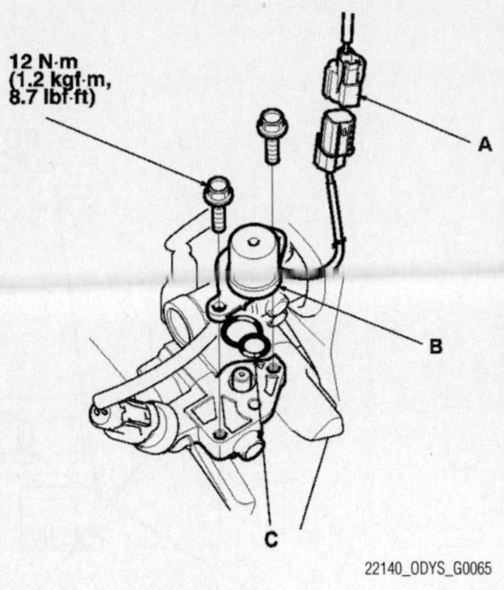
12 N·m
(1.2 kgf·m,
8.7 lbf·ft)
A
B
C
22140_ODYS_G0065

Fig. 155 Disconnect the connector (A) to remove the control solenoid (B) and O-ring (C)—J35A6 engine shown; J35A7 similar.

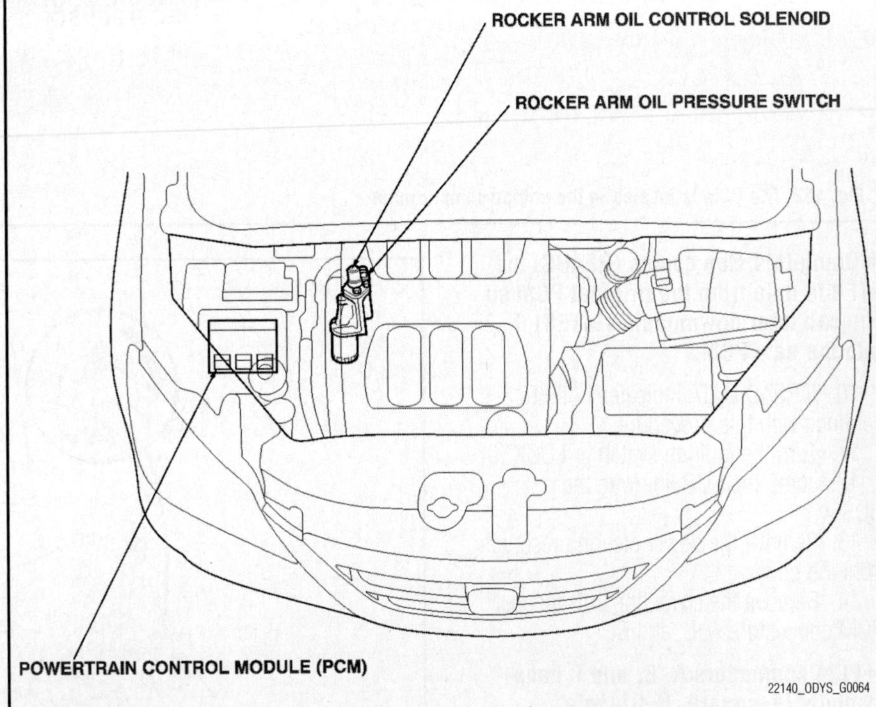
ROCKER ARM OIL CONTROL SOLENOID
ROCKER ARM OIL PRESSURE SWITCH
POWERTRAIN CONTROL MODULE (PCM)
22140_ODYS_G0064

Fig. 154 Rocker Arm Oil Control solenoid

REMOVAL & INSTALLATION

See Figure 155.

1. Disconnect the rocker arm oil control solenoid connector.
2. Remove the rocker arm oil control solenoid.
3. Install the parts in the reverse order of removal with a new O-ring.

FUEL SYSTEM SERVICE PRECAUTIONS

Safety is the most important factor when performing not only fuel system maintenance but any type of maintenance. Failure to conduct maintenance and repairs in a safe manner may result in serious personal injury or death. Maintenance and testing of the vehicle's fuel system components can be accomplished safely and effectively by adhering to the following rules and guidelines.

- To avoid the possibility of fire and personal injury, always disconnect the negative battery cable unless the repair or test procedure requires that battery voltage be applied.
- Always relieve the fuel system pressure prior to disconnecting any fuel system component (injector, fuel rail, pressure regulator, etc.), fitting or fuel line connection. Exercise extreme caution whenever relieving fuel system pressure to avoid exposing skin, face and eyes to fuel spray. Please be advised that fuel under pressure may penetrate the skin or any part of the body that it contacts.
- Always place a shop towel or cloth around the fitting or connection prior to loosening to absorb any excess fuel due to spillage. Ensure that all fuel spillage (should it occur) is quickly removed from engine surfaces. Ensure that all fuel soaked cloths or towels are deposited into a suitable waste container.
- Always keep a dry chemical (Class B) fire extinguisher near the work area.
- Do not allow fuel spray or fuel vapors to come into contact with a spark or open flame.
- Always use a back-up wrench when loosening and tightening fuel line connection fittings. This will prevent unnecessary stress and torsion to fuel line piping.
- Always replace worn fuel fitting O-rings with new Do not substitute fuel hose or equivalent where fuel pipe is installed.

Before servicing the vehicle, make sure to also refer to the precautions in the beginning of this section as well.

Do not insert your fingers into the installed throttle body when you turn the ignition switch to ON (II) or while the ignition switch is in ON (II). If you do, you will seriously injure your fingers if the throttle valve is activated.

RELIEVING FUEL SYSTEM PRESSURE

Before disconnecting fuel lines or hoses, relieve pressure from the system by stopping the fuel pump and then disconnecting the fuel tube/quick connect fitting in the engine compartment.

With the HDS or Scan Tool

See Figures 156 and 157.

1. Make sure you have the anti-theft codes for the radio and the navigation system (if equipped) then write down the radio station and XM radio channel presets.
2. Remove the fuel fill cap, and relieve the pressure in the fuel tank.
3. Turn the ignition switch ON (II).
4. From the INSPECTION MENU of the HDS, select Fuel Pump OFF, then start the engine and let it idle until it stalls.
5. Turn the ignition switch OFF.

➡Do not allow the engine to idle above 1,000 RPM or the PCM will continue to operate the fuel pump. A DTC or a Temporary DTC may be set during this procedure. Check for DTCs, and clear them as needed.

6. Turn the ignition switch OFF.
7. Disconnect the negative cable from the battery.
8. Remove the quick-connect fitting cover.
9. Check the fuel quick-connect fitting for dirt, and clean it if needed.
10. Place a rag or shop towel over the quick-connect fitting.

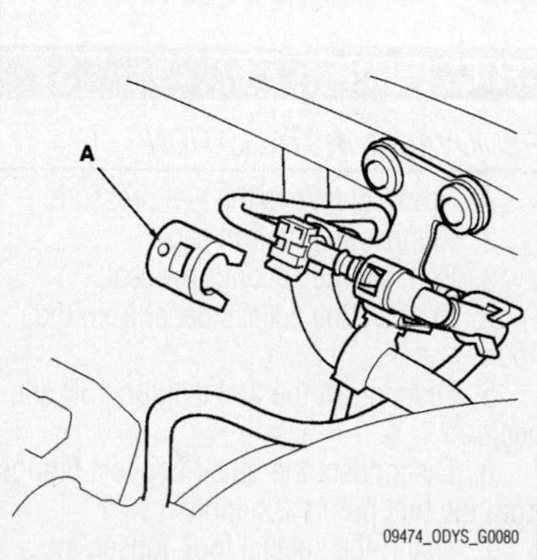
A
09474_ODYS_G0080

Fig. 156 Remove the quick-connect fitting cover (A)

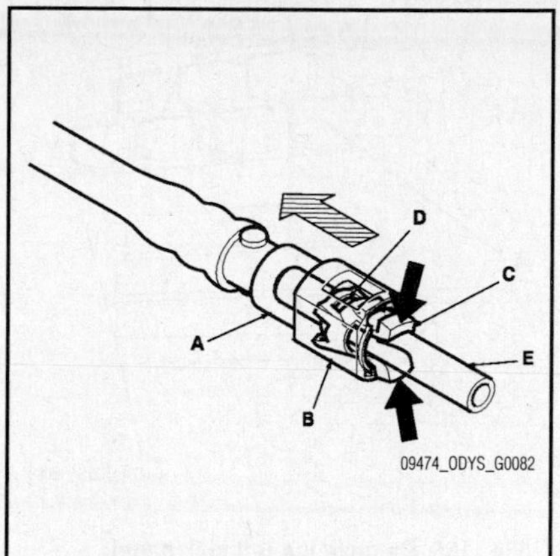
D
C
A
E
B
09474_ODYS_G0082

Fig. 157 Quick-connect fitting

11. Disconnect the quick-connect fitting as follows:
    a. Hold the connector with one hand and squeeze the retainer tabs with the other hand to release them from the locking tabs.
    b. Pull the connector off.

➡Be careful not to damage the line or other parts. Do not use tools. If the connector does not move, keep the retainer tabs pressed down, and alternately pull and push the connector until it comes off easily. Do not remove the retainer from the line; once removed, the retainer must be replaced with a new one.

12. After disconnecting the quick-connect fitting, check it for dirt or damage. Reconnect the negative cable to the battery. Enter the anti-theft codes for the radio and the navigation system, then enter the customer's radio station and XM radio channel presets. Set the clock.

Without the HDS or Scan Tool

See Figure 158.

1. Make sure you have the anti-theft codes for the radio and the navigation system (if equipped) then write down the radio station and XM radio channel presets.
2. Remove the left kick panel, then remove PGM-FI main relay 2 (FUEL PUMP) from the driver's under-dash fuse/relay box.
3. Start the engine, and let it idle until it stalls.

➡If any DTCs are stored, clear and ignore them.

4. Turn the ignition switch OFF.
5. Remove the fuel fill cap.

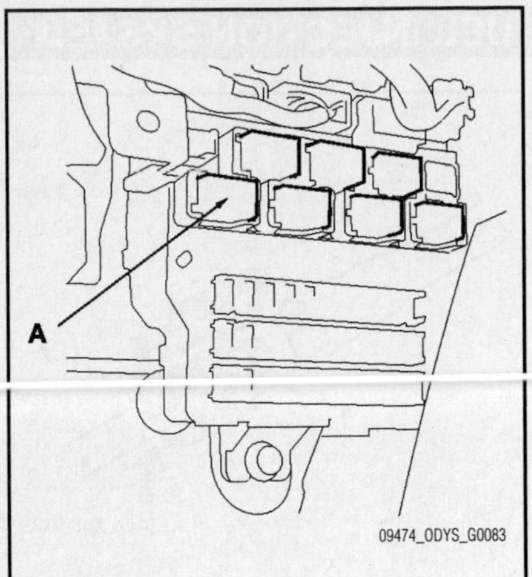
A
09474_ODYS_G0083

Fig. 158 Remove the left kick panel, then remove PGM-FI main relay 2 (FUEL PUMP) (A) from the driver's under-dash fuse/relay box

6. Disconnect the negative cable from the battery.
7. Remove the quick-connect fitting cover.
8. Check the fuel quick-connect fitting for dirt, and clean it if needed.
9. Place a rag or shop towel over the quick-connect fitting.
10. Disconnect the quick-connect fitting as follows:
   a. Hold the connector with one hand and squeeze the retainer tabs with the other hand to release them from the locking tabs.
   b. Pull the connector off.

➡Be careful not to damage the line or other parts. Do not use tools. If the connector does not move, keep the retainer tabs pressed down, and alternately pull and push the connector until it comes off easily. Do not remove the retainer from the line; once removed, the retainer must be replaced with a new one.

11. After disconnecting the quick-connect fitting, check it for dirt or damage.
12. Reconnect the negative cable to the battery. Enter the anti-theft codes for the radio and the navigation system, then enter the customer's radio station and XM radio channel presets. Set the clock.

FUEL FILTER

REMOVAL & INSTALLATION

1. Properly relieve the fuel system pressure.
2. Remove the fuel pump assembly. See Fuel Pump" in this section.
3. Remove the fuel filter set.
4. Installation is the reverse order of removal.

FUEL LEVEL SENDING UNIT

REMOVAL & INSTALLATION

See Figure 159.

1. Properly relieve the fuel system pressure.
2. Remove the fuel pump assembly. See "Fuel Pump" in this section.
3. Remove the fuel level sensor (fuel level sending unit) from the fuel pump assembly.
4. Installation is the reverse order of removal.

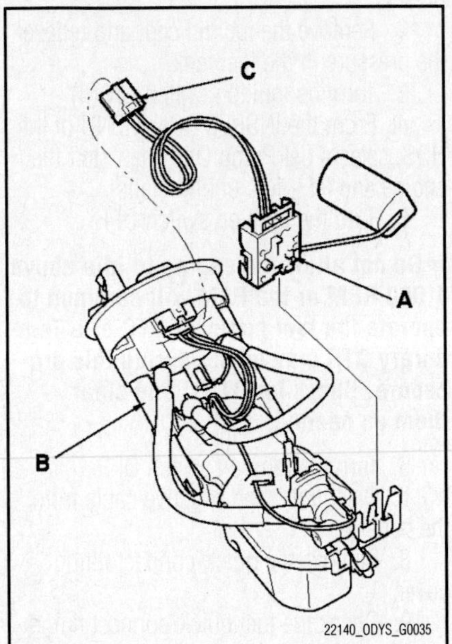
C
A
B
22140_ODYS_G0035

Fig. 159 Remove the fuel level sensor (A) from the fuel pump assembly (B)

FUEL PUMP

REMOVAL & INSTALLATION

1. Properly relieve the fuel pressure.
2. Remove the fuel fill cap.
3. Remove the second row seat.
4. Remove the access panel from the floor.
5. Disconnect the fuel tank unit 5P connector.
6. Disconnect the quick-connect fittings from the fuel pump assembly.
7. Using the special tool, loosen the locknut.
8. Remove the locknut and the fuel pump assembly.

To install:

9. Temporarily attach a new base gasket to the fuel pump assembly, then insert the fuel pump assembly partially into the fuel tank.
10. Transfer the base gasket from the fuel tank unit to the fuel tank.
11. Align the marks on the tank and the fuel tank unit, then insert the fuel tank unit into the fuel tank until the fuel tank unit rests on the base gasket.
12. Using the special tool, tighten the new locknut to 69 ft. lbs. (93 Nm).
13. Connect the electrical connector.
14. The remainder of the installation is the reverse order of removal.

FUEL TANK

DRAINING

1. Remove the fuel tank unit.
2. Using a hand pump, a hose, and a container suitable for fuel, draw the fuel from the fuel tank.
3. Reinstall the fuel tank unit.

REMOVAL & INSTALLATION

1. Properly relieve the fuel pressure.
2. Drain the fuel tank.
3. Raise and safely support the vehicle.
4. Disconnect the fuel pump sub-harness electrical connector.
5. Disconnect the fuel line quick-connect fitting.
6. Disconnect the fuel vapor hoses, and the filler neck hose and clamp. Gently twist the hoses as you pull to avoid damaging them.
7. Place a suitable jack or other support under the fuel tank.
8. Remove the strap bolts.
9. Remove the fuel tank. If it sticks to the undercoated mounts, carefully pry it off the mounts.

To install:

10. Installation is the reverse order of removal.
11. Tighten the fuel strap mounting bolts to 28 ft. lbs. (38 Nm).

FUEL RAIL & INJECTORS

REMOVAL & INSTALLATION

See Figure 160.

1. Properly relieve the fuel pressure.
2. Remove the intake manifold. See "Intake Manifold" in "ENGINE MECHANICAL" section.
3. Disconnect the connectors from the injectors.

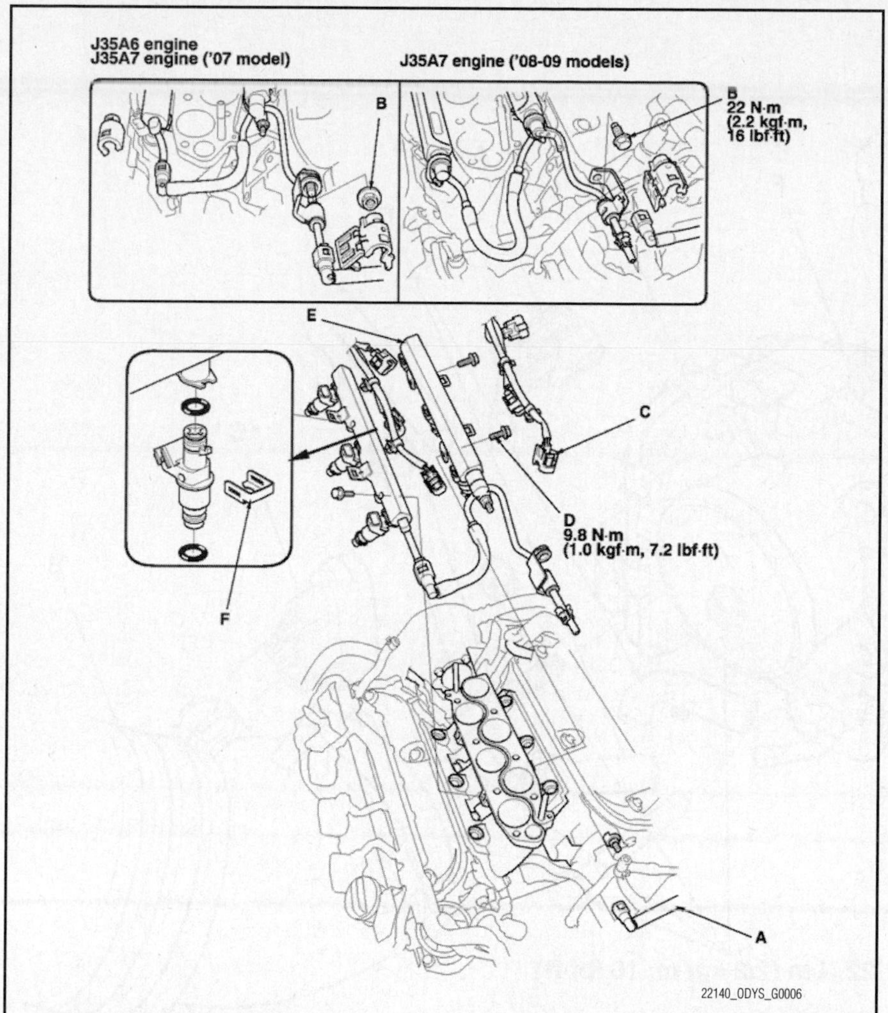
J35A6 engine
J35A7 engine ('07 model)
J35A7 engine ('08-09 models)
B
22 N·m
(2.2 kgf·m,
16 lbf·ft)
E
C
D
9.8 N·m
(1.0 kgf·m, 7.2 lbf·ft)
F
A
22140_ODYS_G0006
Fig. 160 Fuel rail and injectors

4. Disconnect the quick-connect fitting.
5. Remove the fuel rail mounting bolts from the fuel rail.
6. Remove the injector clip from the fuel rail.
7. Remove the injectors from the rails.

To install:

8. Coat the new O-ring with clean engine oil, and insert the injectors into the fuel rail.
9. Install the injector clip.
10. Coat the new injector O-ring with clean engine oil.
11. Install the injectors in the injector base.
12. Install the fuel rail mounting bolts.
13. Install the connectors on the injectors.
14. Connect the quick-connect fitting.
15. Turn the ignition switch ON (II), but do not operate the starter. After the fuel pump runs about 2 seconds, the fuel pressure in the fuel line rises. Repeat this step two or three times, then check for fuel leakage.
16. Install the intake manifold.

IDLE SPEED

INSPECTION

1. Before checking the idle speed, check these items:
   - The malfunction indicator lamp (MIL) has not been reported on, and there are no DTCs.
   - Ignition timing
   - Spark plugs
   - Air cleaner
   - PCV system
2. Apply the parking brake, and make sure the headlights are off.
3. Disconnect the evaporative emission (EVAP) canister purge valve connector.
4. Connect the HDS to the data link connector (DLC) located under the driver's side of the dashboard.
5. Turn the ignition switch to ON (II).
6. Make sure the HDS communicates with the PCM. If it doesn't, perform further DLC communication troubleshooting.
7. Start the engine. Hold the engine speed at 3,000 RPM without load (in P or N) until the radiator fan comes on, then let it idle.
8. Check the idle speed without load conditions: headlights, blower fan, radiator fan, and air conditioner off. Idle speed should be 650±50 RPM.
9. Let the engine idle for 1 minute with high electric load (A/C switch on, temperature set to max cool, blower fan on high, rear window defogger on, and headlights on high beam). Idle speed should be 700±50 RPM.
   a. If the idle speed is not within specification, do the "PCM Idle Learn Procedure". See "Powertrain Control Module (PCM)" in "ENGINE PERFORMANCE & EMISSION CONTROLS" section.
   b. If the idle speed is still not within specification, see "DIAGNOSTICS" section for any specific testing related to this component or for symptom troubleshooting.
10. Reconnect the EVAP canister purge valve connector.

ADJUSTMENT

The idle speed is controlled by the Powertrain Control Module (PCM). No adjustment is necessary or possible.

THROTTLE BODY

REMOVAL & INSTALLATION

See Figure 161.

✻✻ CAUTION

Do not insert your fingers into the installed throttle body when you turn the ignition switch ON (II) or while the ignition switch is ON (II). If you do, you will seriously injure your fingers. If the throttle valve is activated.

➡If you are replacing the throttle body, start at step 1. If you are removing the throttle body, start at step 4. This procedure requires the use of the HDS or equivalent scan tool.

1. Connect the HDS while the engine is stopped.
2. Select the INSPECTION MENU with the HDS.
3. Do the TP LEARNING CHECK in the ETCS TEST.
4. Disconnect the MAP sensor connector.
5. Remove the intake air duct.
6. Disconnect the throttle body connector.

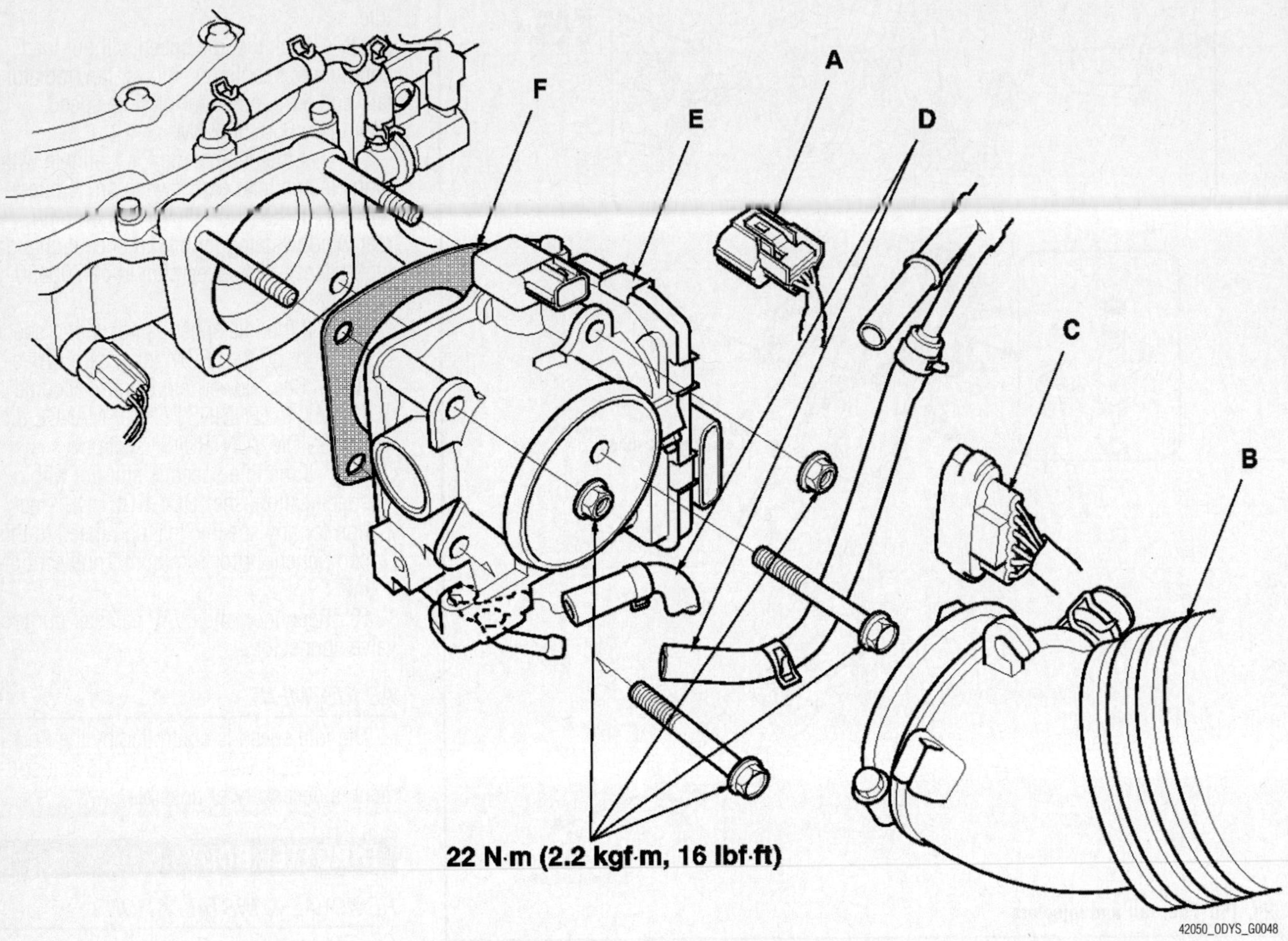
A
F
E
D
C
B
22 N·m (2.2 kgf·m, 16 lbf·ft)
42050_ODYS_G0048

Fig. 161 Exploded view of the throttle body and related components

7. Disconnect and plug the water bypass hoses.
8. Remove the throttle body.
9. Using a suitable plastic scraper, remove any gasket material from the throttle body and air intake plenum.

To install:

10. Install the parts in the reverse order of removal with a new gasket.

➡**The idle learn procedure must be done so the PCM can learn the engine idle characteristics. Perform the idle learn procedure whenever you do any of these actions:**

- Replace PCM
- Reset PCM
- Update PCM
- Clean or replace the throttle body

11. Do the PCM idle learn procedure after the throttle body has been replaced, as follows:

➡**Erasing DTCs with the HDS does not require you to do the idle learn procedure.**

a. Make sure all electrical items (A/C, audio, rear window defogger, lights, etc.) are off.
b. Reset the PCM with the HDS.
c. Turn the ignition switch ON (II), and wait 2 seconds.
d. Start the engine. Hold the engine speed at 3,000 RPM without load (in Park or neutral) until the radiator fan comes on, or until the engine coolant temperature reaches 194 °F (90 °C).
e. Let the engine idle for about 5 minutes with the throttle fully closed.

➡**If the radiator fan comes on, do not include its running time in the 5 minutes.**

12. Refill the radiator with engine coolant.

HEATING & AIR CONDITIONING SYSTEM

BLOWER UNIT

REMOVAL & INSTALLATION

See Figures 162 and 163.

1. Remove the glove box housing.
2. Remove the self-tapping screws, the bolts, and the glove box frame.
3. Cut the plastic cross brace in the glove box opening with diagonal cutters in the area shown, and discard it.
4. Disconnect the connectors from the front blower motor and the front power transistor, then remove the wire harness clips.
5. Disconnect the connector from the recirculation control motor, then remove the bolts, the mounting nuts and the blower unit.

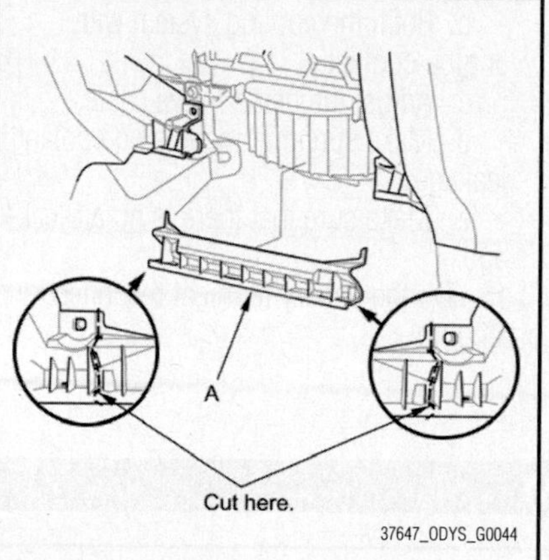
A
Cut here.
37647_ODYS_G0044

Fig. 162 Cut the plastic cross brace (A) in the glove box opening with diagonal cutters in the area shown, and discard it.

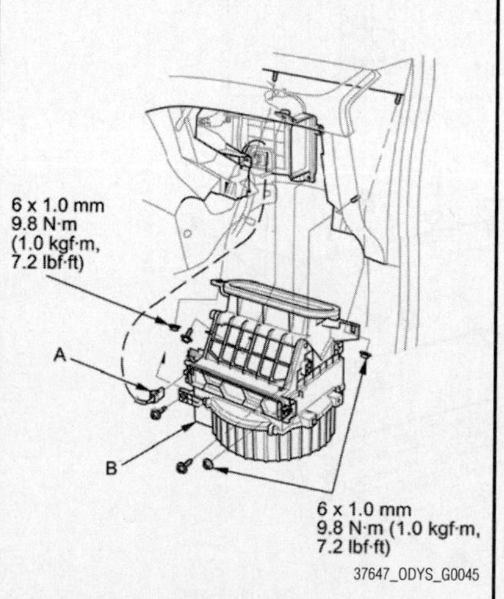
6 x 1.0 mm
9.8 N·m
(1.0 kgf·m,
7.2 lbf·ft)
A
B
6 x 1.0 mm
9.8 N·m (1.0 kgf·m,
7.2 lbf·ft)
37647_ODYS_G0045

Fig. 163 Disconnect the connector (A) from the recirculation control motor, then remove the bolts, the mounting nuts and the blower unit (B).

6. Install the unit in the reverse order of removal. Make sure that there is no air leakage.

HEATER UNIT & HEATER CORE

REMOVAL & INSTALLATION

See Figures 164 through 167.

⁂ WARNING

SRS components are located in this area. Review the SRS component locations and the precautions and procedures before doing repairs or service. See "AIR BAG (SUPPLEMENTAL RESTRAINT SYSTEM)" section.

1. Do the battery terminal disconnect procedure, then wait at least 3 minutes before beginning work.
2. Disconnect the suction and receiver lines from the front evaporator core.
3. From under the hood, open the cable clamp then disconnect the heater valve cable from the heater valve arm. Turn the heater valve arm to the fully opened position as shown.
4. When the engine is cool, drain the engine coolant from the radiator.
5. Slide the hose clamps back. Remove the nut and the water valve, then disconnect the inlet heater hose and the outlet heater hose from the heater unit.

➡Engine coolant will run out when the hoses are disconnected; drain it into a clean drip pan. Be sure not to let coolant spill on the electrical parts or the painted surfaces. If any coolant spills, rinse it off immediately.

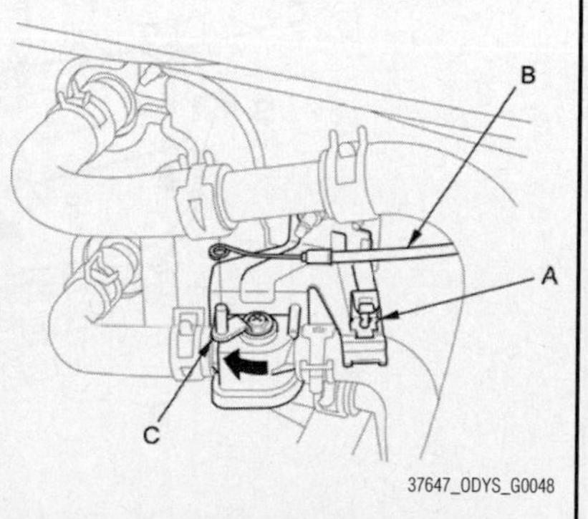
B
A
C
37647_ODYS_G0048

Fig. 164 From under the hood, open the cable clamp (A), then disconnect the heater valve cable (B) from the heater valve arm (C). Turn the heater valve arm to the fully opened position as shown.

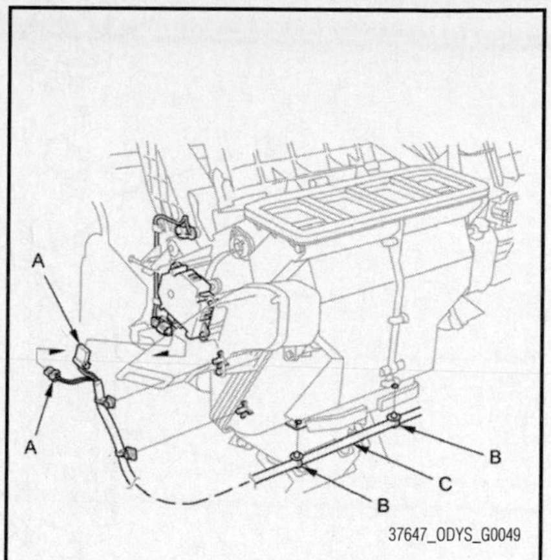
A
A
B
C
B
37647_ODYS_G0049

Fig. 165 Disconnect the connectors (A) from the evaporator sensor and the front air mix control motor, then remove the wire harness clips (B) and the wire harness (C).

6. Remove the mounting nut from the heater unit. Take care not to damage or bend the fuel lines and the brake lines, etc.
7. Remove the dashboard.
8. Disconnect the connectors from the front blower motor and the front power transistor, then remove the wire harness clips.
9. Disconnect the connectors from the front mode control motor and the recirculation control motor, then remove the wire harness clips and the connector clip.
10. Disconnect the connectors from the evaporator sensor and the front air mix control motor, then remove the wire harness clips and the wire harness.

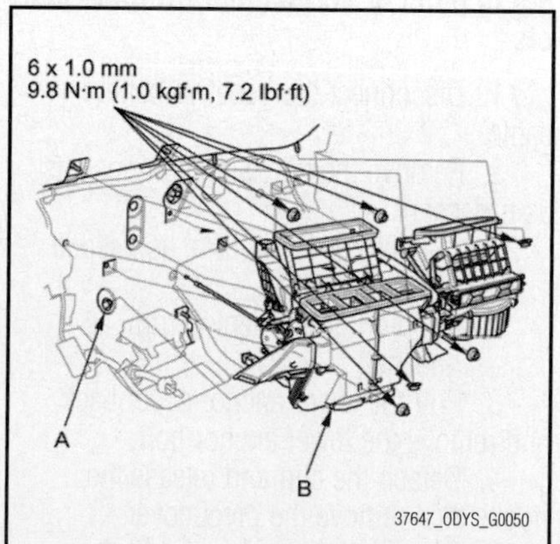
6 x 1.0 mm
9.8 N·m (1.0 kgf·m, 7.2 lbf·ft)
A
B
37647_ODYS_G0050

Fig. 166 Remove the drain hose (A), then remove the nuts and the blower-heater unit (B).

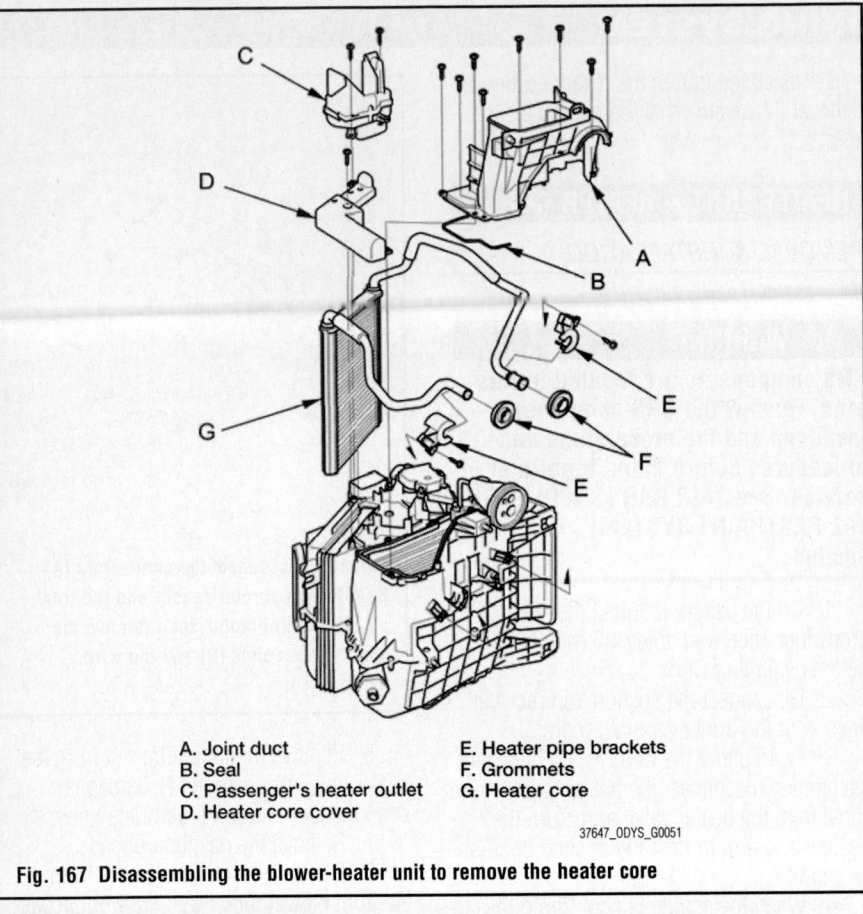

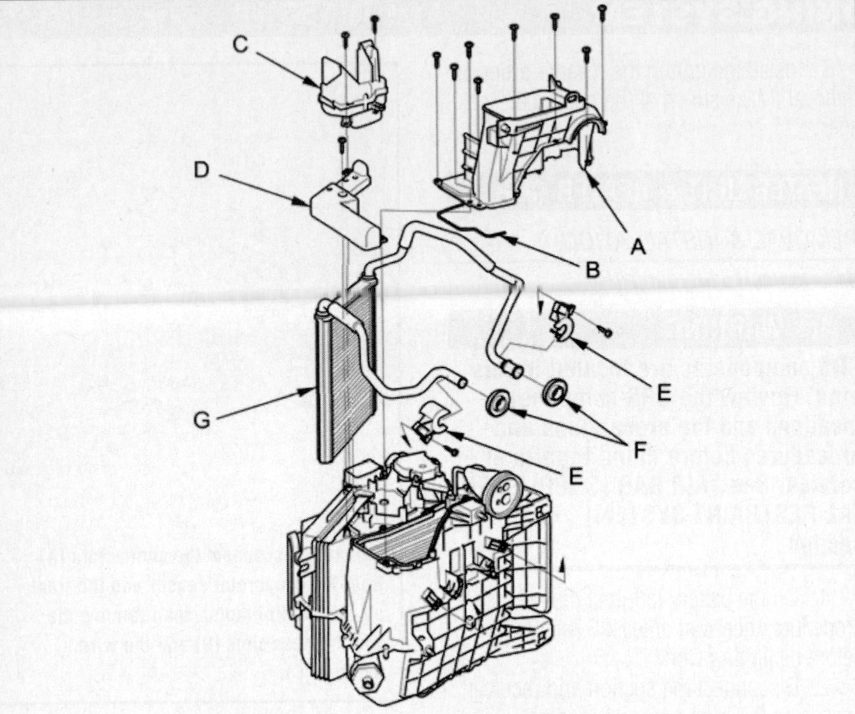
C
D
A
B
E
G
F
E
A. Joint duct
B. Seal
C. Passenger's heater outlet
D. Heater core cover
E. Heater pipe brackets
F. Grommets
G. Heater core
37647_ODYS_G0051

Fig. 167 Disassembling the blower-heater unit to remove the heater core

11. Remove the drain hose, then remove the nuts and the blower-heater unit.
12. Remove the self-tapping screws, the joint duct and seal. Remove the self-tapping screws, then remove the passenger's heater outlet, and the heater core cover. Remove the self-tapping screws, the heater pipe brackets, the grommets, and carefully pull out the heater core so you don't bend the inlet and outlet pipes.

To install:

13. Install the heater core in the reverse order of removal.
14. Install the heater unit in the reverse order of removal, and note these items:
    a. Do not interchange the inlet and outlet heater hoses, and install the hose clamps securely.
    b. Refill the cooling system with engine coolant.
    c. Adjust the heater valve cable.
    d. Make sure that there is no coolant leakage.
    e. Make sure that there is no air leakage.
15. Do the battery terminal reconnection procedure

AUXILIARY HEATING & AIR CONDITIONING SYSTEM

BLOWER MOTOR

REMOVAL & INSTALLATION

See Figures 168 through 170.

➡This procedure requires the use of KTC trim tool set (SOJATP2014, or equivalent). When removing trim components, wear gloves and be careful not to bend or scratch the trim and panels.

1. Disconnect the negative battery cable.
2. Remove the following components, as necessary:
    - Sliding door sill trim, as needed
    - C-pillar trim
    - Sliding door opening trim, as needed
3. Pull the lower anchor cover back, and remove the lower anchor bolt.
4. Detach the clip and release the hooks then remove the pivot cover.
5. Remove the lower anchor bolt.
6. Remove the rear trim panel, as follows:
    a. Remove the side bolts.

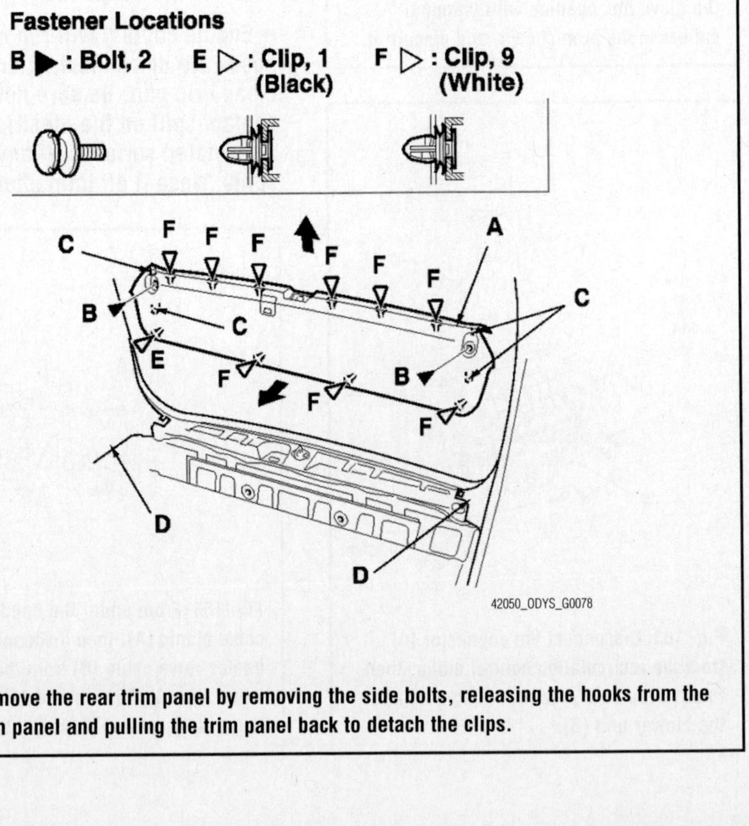
Fastener Locations
B ▶ : Bolt, 2
E ▷ : Clip, 1 (Black)
F ▷ : Clip, 9 (White)
A
B
C
D
E
F
42050_ODYS_G0078

Fig. 168 Remove the rear trim panel by removing the side bolts, releasing the hooks from the rear side trim panel and pulling the trim panel back to detach the clips.

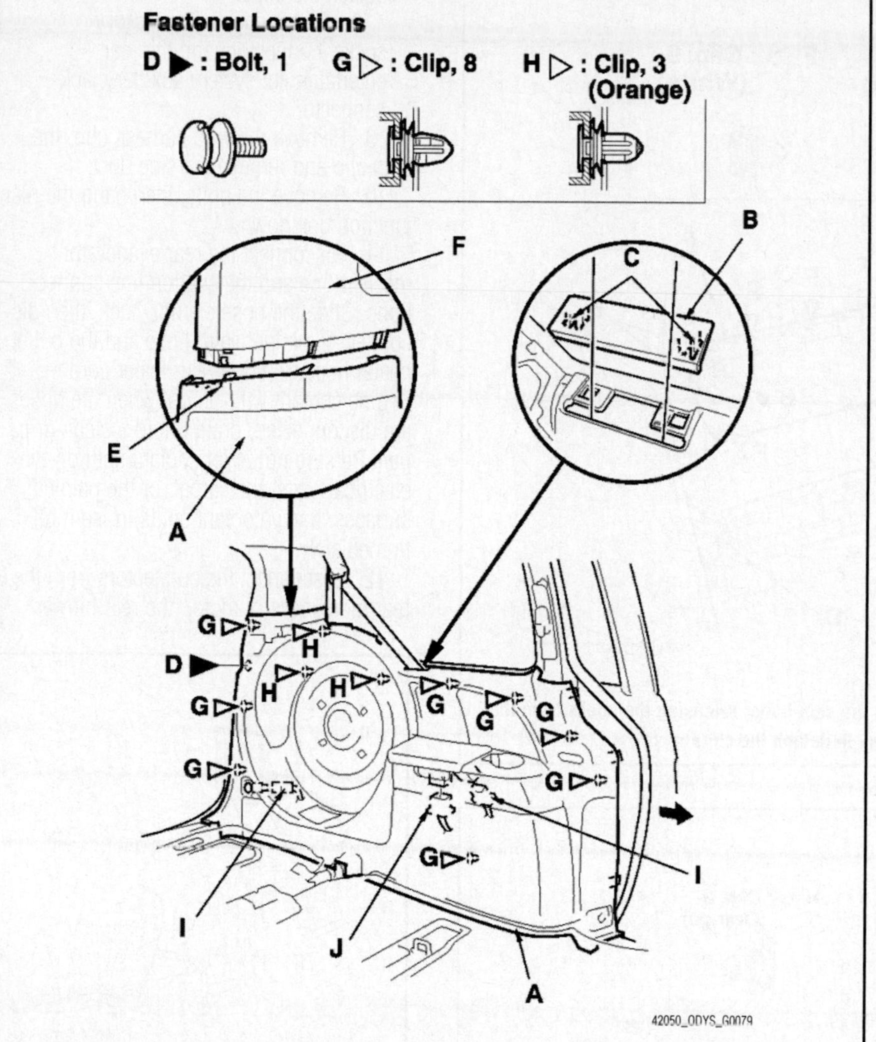

Fastener Locations
D ▶ : Bolt, 1
G ▷ : Clip, 8
H ▷ : Clip, 3 (Orange)
A
B
C
D
E
F
G
H
I
J
42050_ODYS_G0079

Fig. 169 Rear side trim panel and related components

b. Pull out both side edges of the trim panel by hand to release the hooks from the rear side trim panel.

c. Pull the trim panel back to detach the clips.

7. For the driver's side, remove the spare tire, as follows:

a. Place a hand in each grab handle on the spare tire cover, and pull back to release the clips. Remove the spare tire cover.

b. Remove the spare tire mounting bolt.

c. Remove the spare tire.

8. Remove the rear side trim panel, as follows:

a. Remove the cap by releasing the two hooks.

b. Remove the rear side bolt.

c. Pull out the rear upper edge of the trim panel by hand to release the hooks from the D-pillar trim.

d. Pull the trim panel back by hand to detach the clips.

e. Disconnect the accessory socket connectors and the rear entertainment system auxiliary jack connector.

9. Disconnect the connectors from the rear mode control motor and the rear blower motor, then remove the wire harness clips, the self-tapping screws and the rear blower motor.

10. Installation is the reverse order of removal.

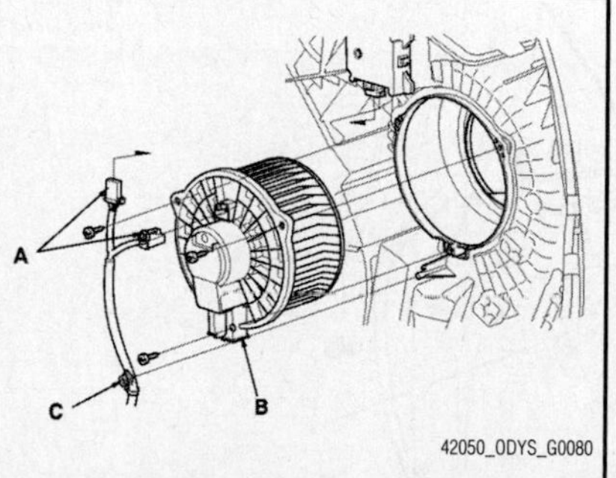

A
B
C
42050_ODYS_G0080

Fig. 170 Disconnect the connectors (A) from the rear mode control motor and the rear blower motor (B), then remove the wire harness clips (C), the self-tapping screws and the rear blower motor

HEATER CORE

REMOVAL & INSTALLATION

See Figures 171 through 176.

➡This procedure requires the use of KTC trim tool set (SOJATP2014, or equivalent). When removing trim components, wear gloves and be careful not to bend or scratch the trim and panels.

1. Disconnect the negative battery cable.

2. Remove the following components, as necessary:

- Sliding door sill trim, as needed
- C-pillar trim
- Sliding door opening trim, as needed

3. Pull the lower anchor cover back, and remove the lower anchor bolt.

4. Detach the clip and release the hooks then remove the pivot cover.

5. Remove the lower anchor bolt.

6. Remove the rear trim panel as follows:

a. Remove the side bolts.

b. Pull out both side edges of the trim panel by hand to release the hooks from the rear side trim panel.

c. Pull the trim panel back to detach the clips.

7. For the driver's side, remove the spare tire, as follows:

a. Place a hand in each grab handle on the spare tire cover, and pull back to release the clips. Remove the spare tire cover.

b. Remove the spare tire mounting bolt.

c. Remove the spare tire.

8. Remove the rear side trim panel, as follows:

a. Remove the cap by releasing the two hooks.

b. Remove the rear side bolt.

c. Pull out the rear upper edge of the trim panel by hand to release the hooks from the D-pillar trim.

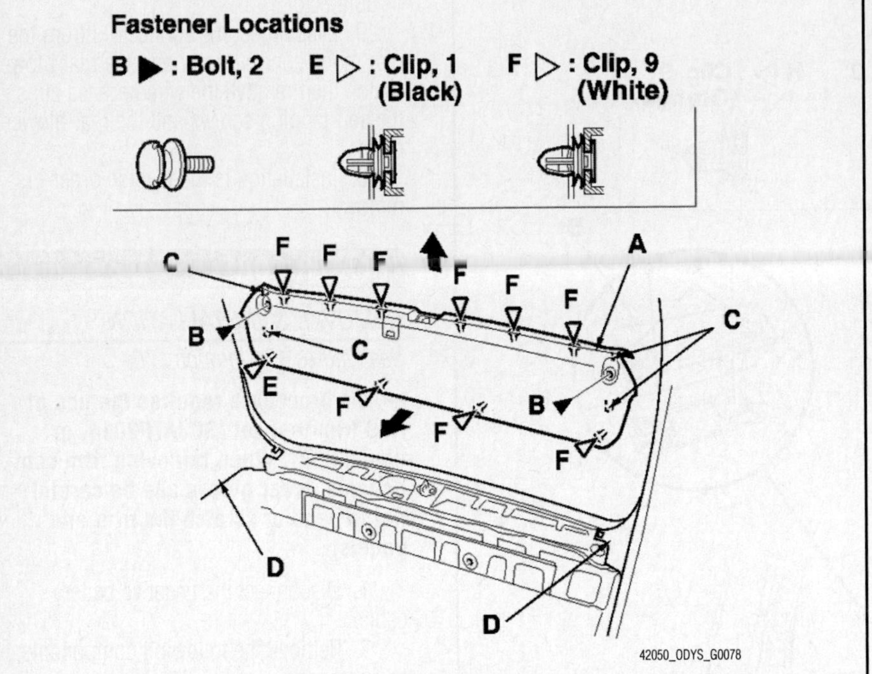
Fastener Locations
B ▶ : Bolt, 2
E ▷ : Clip, 1 (Black)
F ▷ : Clip, 9 (White)
A
B
C
D
E
F
42050_ODYS_G0078

Fig. 171 Remove the rear trim panel by removing the side bolts, releasing the hooks from the rear side trim panel and pulling the trim panel back to detach the clips

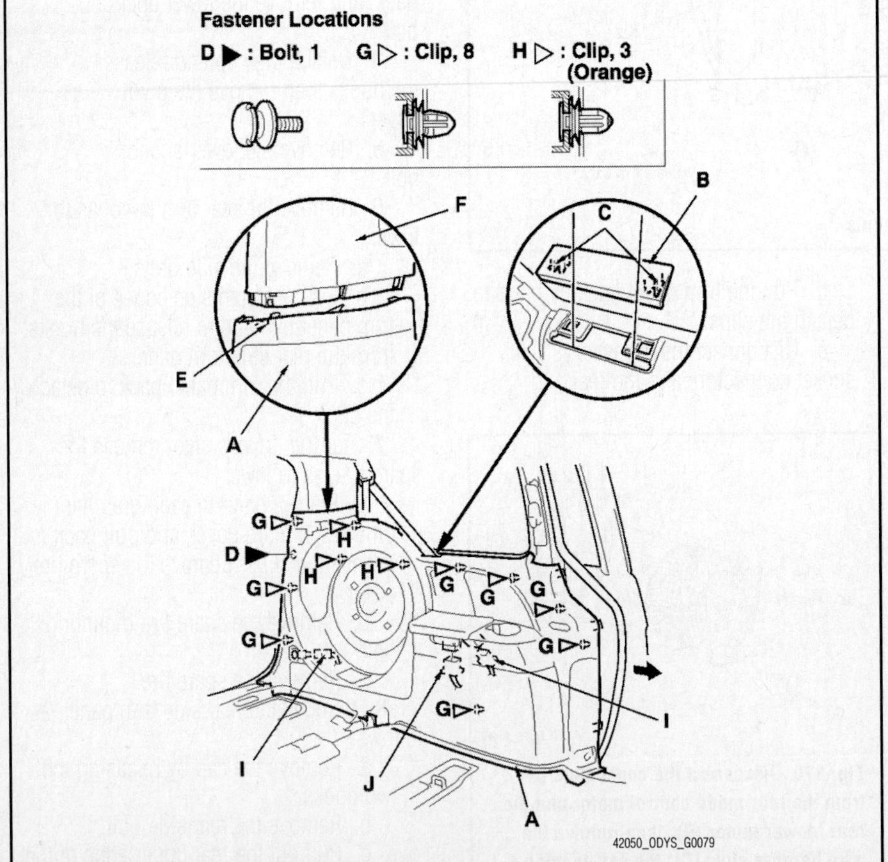
Fastener Locations
D ▶ : Bolt, 1
G ▷ : Clip, 8
H ▷ : Clip, 3 (Orange)
A
B
C
D
E
F
G
H
I
J
42050_ODYS_G0079

Fig. 172 Rear side trim panel and related components

d. Pull the trim panel back by hand to detach the clips.

e. Disconnect the accessory socket connectors and the rear entertainment system auxiliary jack connector.

9. Remove the wire harness clip, the duct clip and remove the side duct.

10. Remove the bolts, then hang the rear junction box down.

11. Disconnect the rear evaporator receiver line and the suction line connections. Slide the hose clamps back, then disconnect the inlet heater hose and the outlet heater hose from the rear heater core. Engine coolant will run out when the hoses are disconnected; drain it into a clean drip pan. Be sure not to let coolant spill on the electrical parts, the carpet, or the painted surfaces. If any coolant spills, rinse it off immediately.

12. Disconnect the connectors from the rear mode control motor, the rear blower

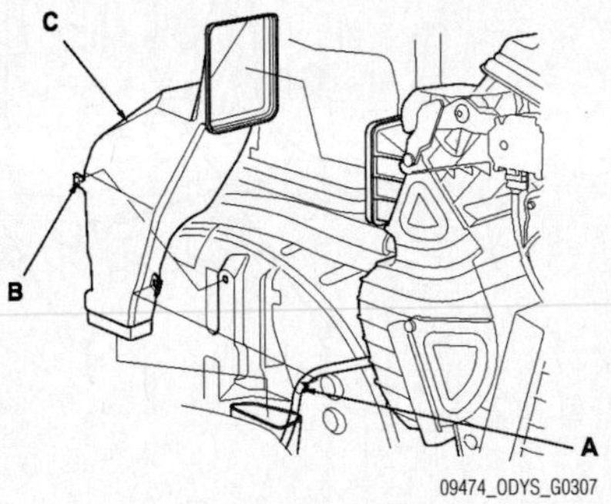
A
B
C
09474_ODYS_G0307

Fig. 173 Remove the wire harness clip (A), the duct clip (B) and remove the side duct (C)

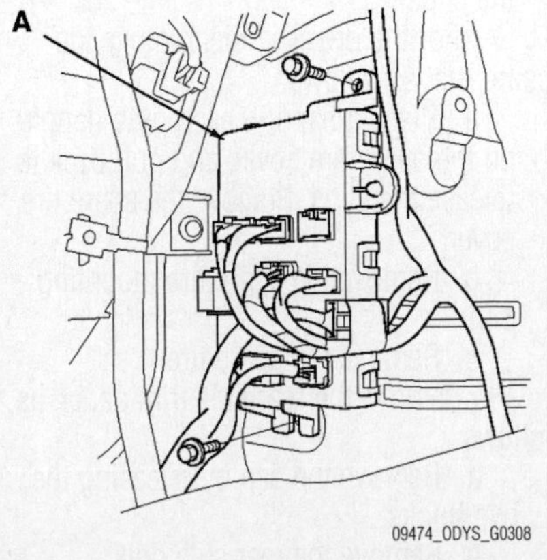
A
09474_ODYS_G0308

Fig. 174 Remove the bolts, then hang the rear junction box (A) down.

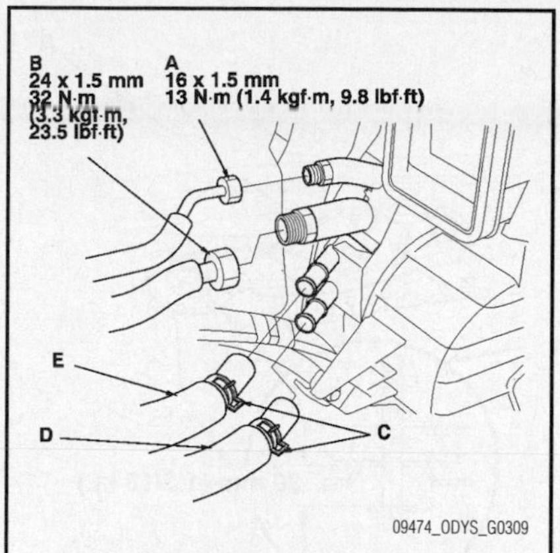
B
24 x 1.5 mm
32 N·m
(3.3 kgf·m,
23.5 lbf·ft)
A
16 x 1.5 mm
13 N·m (1.4 kgf·m, 9.8 lbf·ft)
E
D
C
09474_ODYS_G0309

Fig. 175 Disconnect the rear evaporator receiver line (A) and the suction line (B) connections. Slide the hose clamps (C) back, then disconnect the inlet heater hose (D) and the outlet heater hose (E) from the rear heater core

motor, rear in-car temperature sensor and rear climate control unit, then remove the wire harness clips, from the rear HVAC unit.

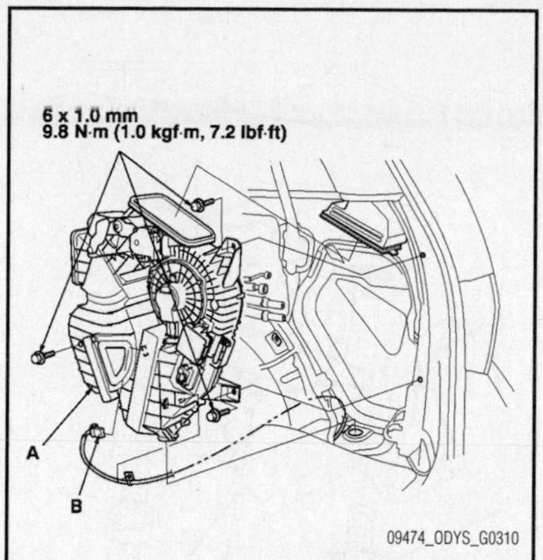
6 x 1.0 mm
9.8 N·m (1.0 kgf·m, 7.2 lbf·ft)
A
B
09474_ODYS_G0310

Fig. 176 Remove the bolts, then pull out the rear HVAC unit (A). Disconnect the rear air mix control motor connector (B), then remove the rear HVAC unit

13. Remove the bolts, then pull out the rear HVAC unit. Disconnect the rear air mix control motor connector, then remove the rear HVAC unit.

14. Remove the self-tapping screws, the clamps and the rear heater core.

To install:

15. Installation is the reverse order of removal.

- Before reassembly, make sure that the rear air mix control linkage and door move smoothly without binding.
- Before reassembly, make sure that the rear mode control linkage and door move smoothly without binding.
- Make sure no air is leaking from the right housing, the left housing and the lower housing fitting.
- After reassembly, make sure the rear air mix control motor runs smoothly.
- After reassembly, make sure the rear mode control motor runs smoothly.
- Make sure that there is no coolant leakage.

STEERING

POWER RACK & PINION STEERING GEAR

REMOVAL & INSTALLATION

See Figures 177 through 203.

1. Drain the power steering fluid.
2. Record the radio station preset.
3. Disconnect the battery negative cable, and wait at least 3 minutes before beginning work.

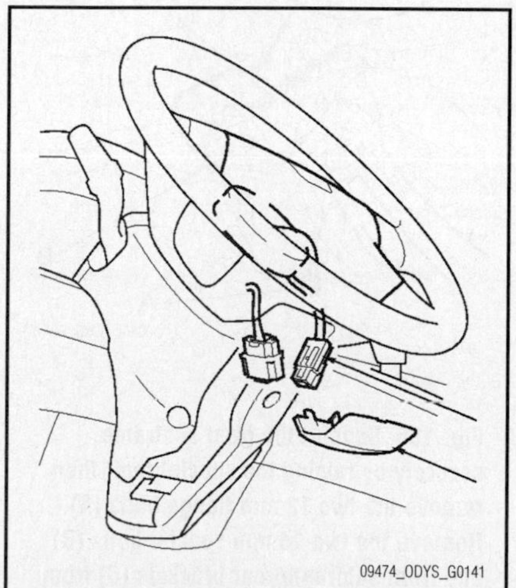
09474_ODYS_G0141

Fig. 177 Remove the access panel from the steering wheel, then disconnect the driver's air bag 4P connector from the cable reel

4. Remove the access panel from the steering wheel, then disconnect the driver's air bag 4P connector from the cable reel.
5. Using a Torx® T30 bit, remove the two Torx® bolts.
6. Disconnect the horn switch connector (1P), then remove the driver's air bag.
7. Align the front wheels straight ahead.
8. Disconnect the cable reel sub-harness connector from the cable reel.

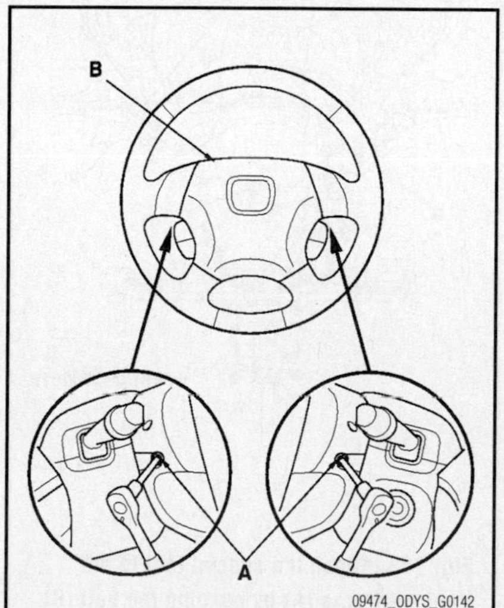
B
A
09474_ODYS_G0142

Fig. 178 Using a Torx® T30 bit, remove the two Torx® bolts (A)

9. Loosen the steering wheel nut three full turns.
10. Install a commercially available steering wheel puller on the steering wheel. Free the steering wheel from the steering column shaft by turning the pressure bolt of the puller.
11. Note these items when removing the steering wheel:
    a. Do not tap on the steering wheel or the steering column shaft when removing the steering wheel.
    b. If you thread the puller bolts into the wheel hub more than five threads, the bolts will hit the cable reel and damage it. To prevent this, install a pair

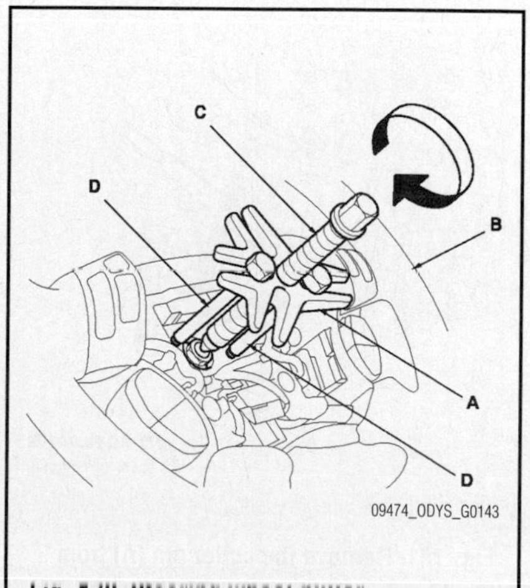
C
D
B
A
D
09474_ODYS_G0143

Fig. 179 Steering wheel puller

of jam nuts five threads up on each puller bolt.

12. Remove the steering wheel puller, then remove the steering wheel nut and steering wheel from the steering column.

13. Remove the steering joint cover.

14. Remove the steering joint bolts, disconnect the steering joint by moving the steering joint toward the column.

15. Remove the cotter pin from the 12 mm nut, and loosen the nut.

16. Separate the tie-rod ball joint and knuckle using the special tool.

17. Remove the 10 mm flange bolts of the exhaust rubber mount.

18. Remove the three self-locking nuts, and disconnect the three way catalytic converter (TWC) from the muffler.

19. Remove exhaust pipe A.

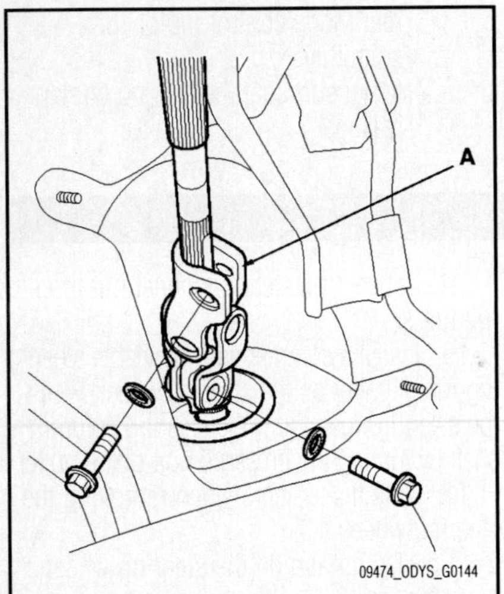
A
09474_ODYS_G0144

Fig. 180 Remove the steering joint bolts, disconnect the steering joint by moving the steering joint (A) toward the column

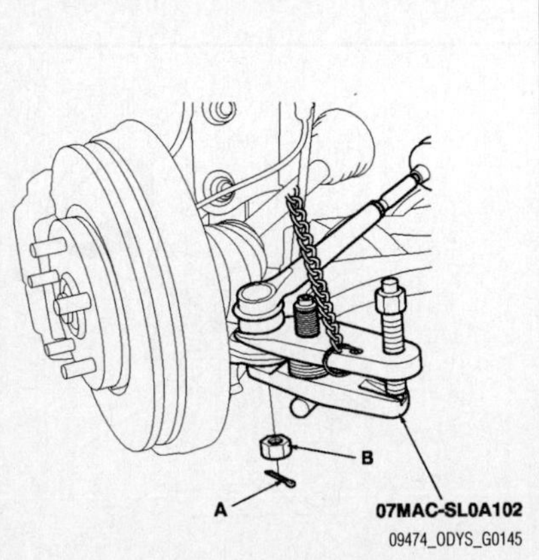
B
A
07MAC-SL0A102
09474_ODYS_G0145

Fig. 181 Remove the cotter pin (A) from the 12 mm nut (B), and loosen the nut

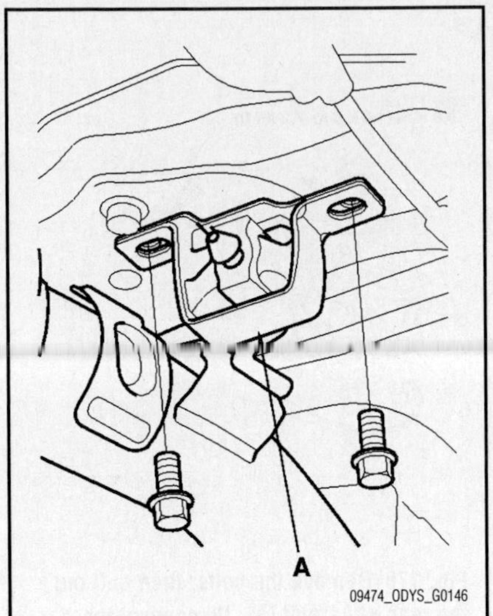
A
09474_ODYS_G0146

Fig. 182 Remove the 10 mm flange bolts of the exhaust rubber mount (A)

20. Disconnect the power steering pressure (PSP) switch connector.

21. Remove the front splash shield.

22. Attach the special tool to the front subframe by hanging the hook of the special tool over the front of the subframe, then tighten the special tool screw.

23. Raise the jack and line up the slots in the arms with the bolt holes on the corner of the jack base, then attach them with bolts securely.

24. Loosen the front subframe front bracket mounting bolts on the right and left of the vehicle so they are about 30 mm (1³⁄₁₆ in.) from the mounting surface.

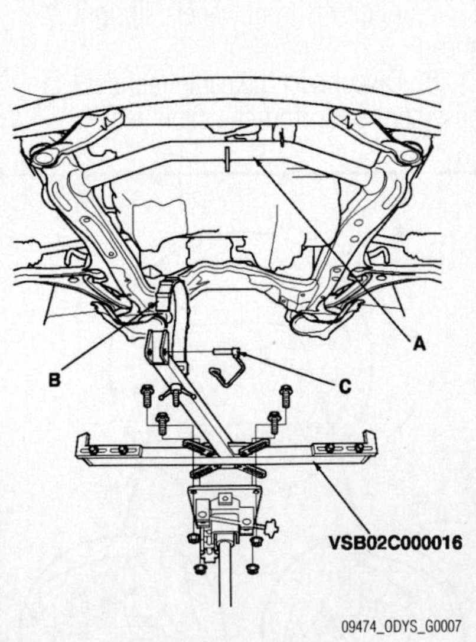
A
B
C
VSB02C000016
09474_ODYS_G0007

Fig. 183 Attach the special tool to the front subframe (A) by hanging the belt (B) of the special tool over the front of the subframe, secure it with the pin (C), then tighten the special tool screw

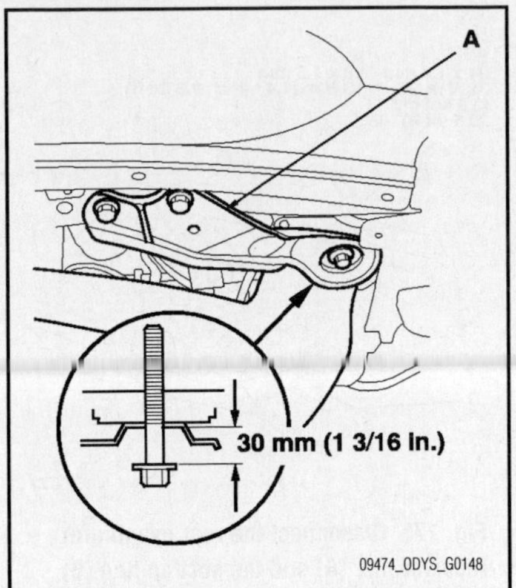
A
30 mm (1 3/16 in.)
09474_ODYS_G0148

Fig. 184 Loosen the front subframe front bracket (A) mounting bolts on the right and left of the vehicle so they are about 30 mm (1³⁄₁₆ in.) from the mounting surface

25. Support the front subframe securely by raising the special tool, then remove the two 12 mm flange bolts.

26. Remove the two 14 mm special bolts and front subframe rear brackets from the front subframe.

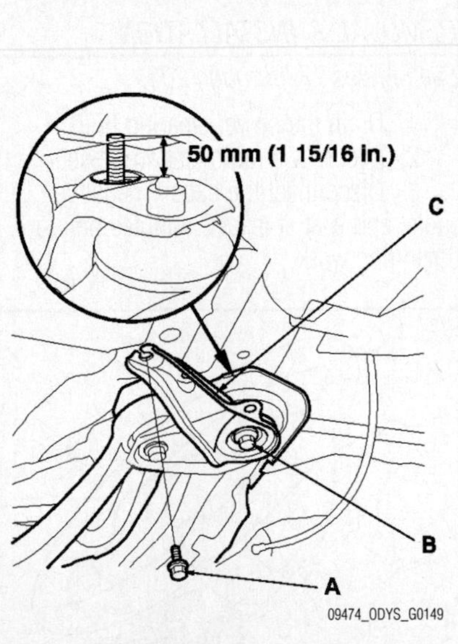
50 mm (1 15/16 in.)
C
B
A
09474_ODYS_G0149

Fig. 185 Support the front subframe securely by raising the special tool, then remove the two 12 mm flange bolts (A). Remove the two 14 mm special bolts (B) and front subframe rear brackets (C) from the front subframe. Lower the jack supporting the front subframe with the special tool slowly until the front subframe has dropped about 50 mm (1¹⁵⁄₁₆ in.)

27. Lower the jack supporting the front subframe with the special tool slowly until the front subframe has dropped about 50 mm (1 15/16 in.).

28. Remove the P/S line mounting brackets from the front subframe and gearbox mounting bracket.

29. Loosen the 16 mm flare nut, and disconnect the feed line from the gear box.

30. Loosen the adjustable hose clamp and disconnect the return hose.

31. Remove the P/S heat baffle plate.

32. Remove the two 10 mm flange bolts from the right side of the steering gearbox, then remove the mounting bracket and cushion.

33. Remove the two 10 mm flange bolts and nuts from the left side of the gearbox.

34. Loosen the steering gearbox bracket mounting bolts.

35. On EX-L, EX-L Touring models, do the following:

a. Remove the rear mount stop, then remove the rear mount bolt.

b. Remove the rear mount from the base bracket, and disconnect the connector.

c. Remove the base bracket from the front subframe.

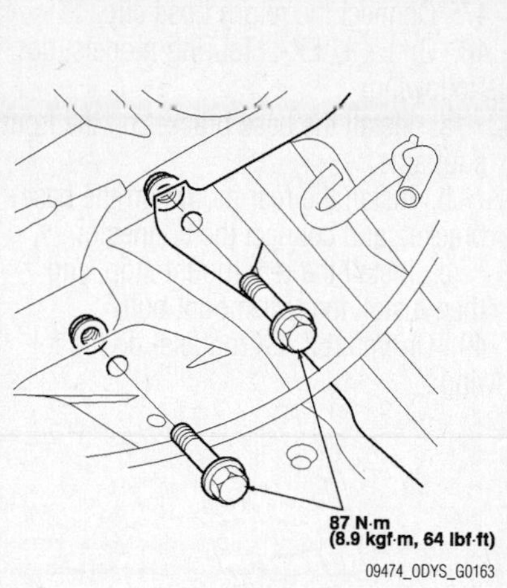
87 N·m
(8.9 kgf·m, 64 lbf·ft)
09474_ODYS_G0163

Fig. 187 Remove the two 10 mm flange bolts and nuts from the left side of the gearbox

36. On VAN, LX, EX models, do the following:

a. Remove the rear mount bracket bolt.

b. Remove the rear mount bracket from the front subframe.

37. Disconnect the return hose clip.

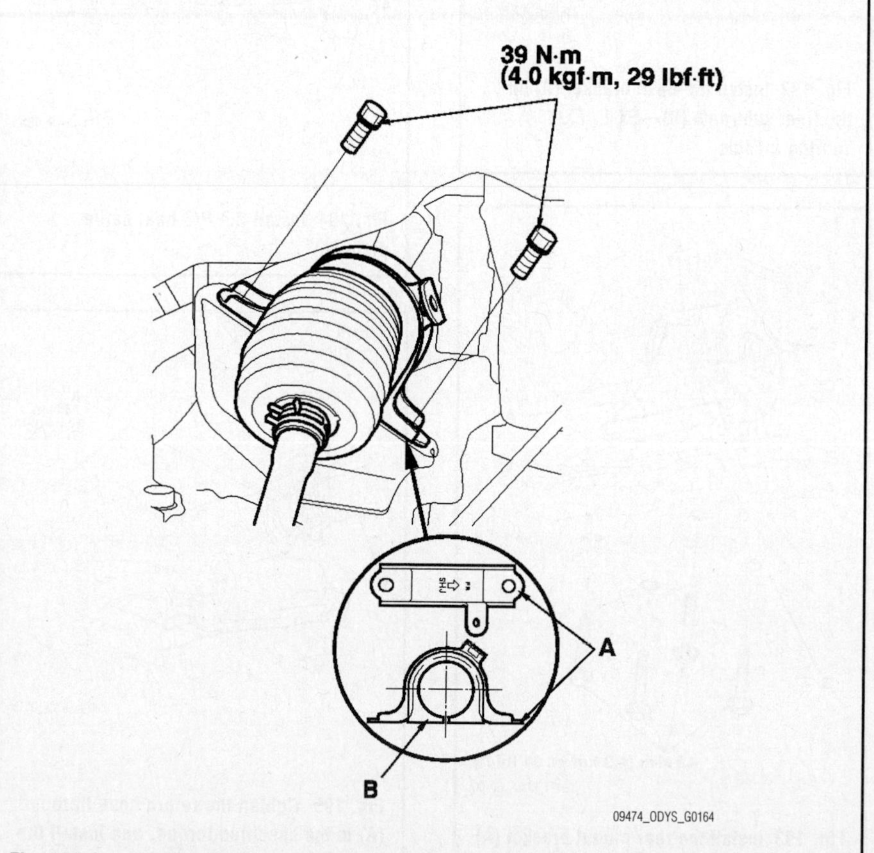
39 N·m
(4.0 kgf·m, 29 lbf·ft)
A
B
09474_ODYS_G0164

Fig. 186 Remove the two 10 mm flange bolts from the right side of the steering gearbox, then remove the mounting bracket (A) and cushion (B)

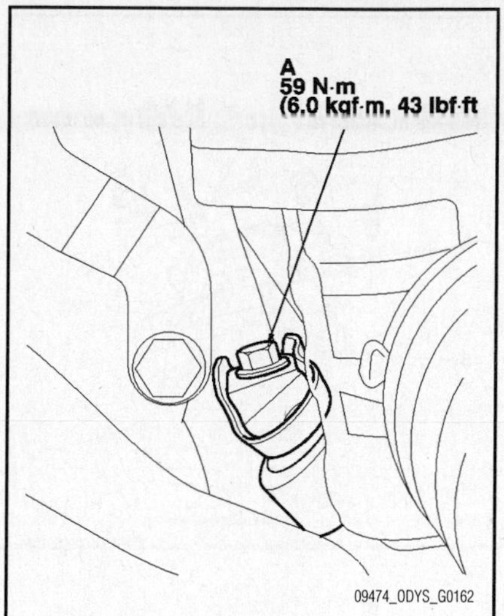
A
59 N·m
(6.0 kgf·m, 43 lbf·ft)
09474_ODYS_G0162

Fig. 188 Loosen the steering gearbox bracket mounting bolts (A)

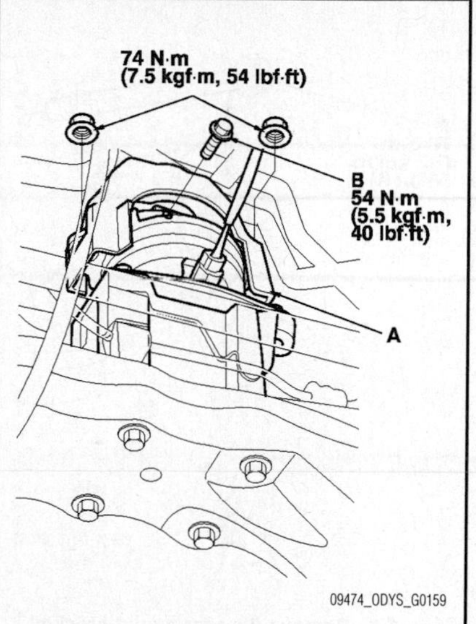
74 N·m
(7.5 kgf·m, 54 lbf·ft)
B
54 N·m
(5.5 kgf·m, 40 lbf·ft)
A
09474_ODYS_G0159

Fig. 189 Remove the rear mount stop (A), then remove the rear mount bolt (B)—EX-L, EX-L Touring models

38. Move the steering gearbox to the driver's side, and rotate it so the pinion shaft points toward the front of the vehicle.

39. Carefully move the steering gearbox as an assembly toward the left side of the vehicle until the pinion shaft clears the wheel well opening. Be careful not to damage the brake lines with the pinion shaft.

40. Remove the steering gearbox through the wheel well opening on the driver's side.

41. Remove the pinion shaft grommet from the steering joint cover B.

42. After removing the steering gearbox, make sure that no power steering fluid gets

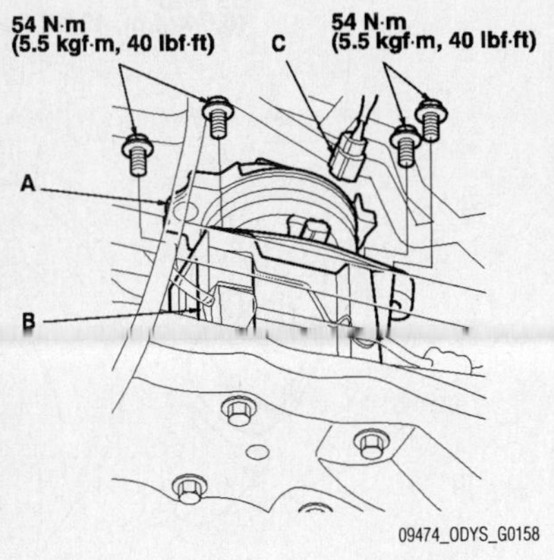
54 N·m
(5.5 kgf·m, 40 lbf·ft)
C
54 N·m
(5.5 kgf·m, 40 lbf·ft)
A
B
09474_ODYS_G0158

Fig. 190 Remove the rear mount (A) from the base bracket (B), and disconnect the connector (C)—EX-L, EX-L Touring models

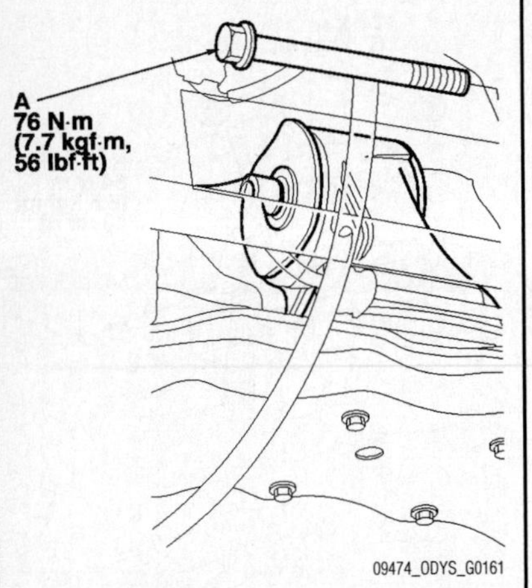
A
76 N·m
(7.7 kgf·m,
56 lbf·ft)
09474_ODYS_G0161

Fig. 191 Remove the rear mount bracket bolt (A)—Van, LX, EX models

on the gearbox mount cushions, gearbox housing, surface of the front subframe and stiffener. Wipe off any spilled fluid at once.

To install:

43. Install the pinion shaft grommet on the valve housing.
44. Slide the steering gearbox between the front subframe and body from the driver's side. Place the gearbox in position on the front suspension subframe.
45. Rotate the steering gearbox so the pinion shaft points upward.
46. Continue moving the gearbox toward the passenger's side until the steering gearbox is in position. Make sure the power steering return line and feed line are routed above the gearbox.
47. Connect the return hose clip.
48. On EX-L, EX-L Touring models, do the following:
    a. Install the base bracket on the front subframe.
    b. Install the rear mount on the base bracket, and connect the connector.
    c. Install the rear mount stop, and then install the rear mount bolt.
49. On Van, LX, EX models, do the following:

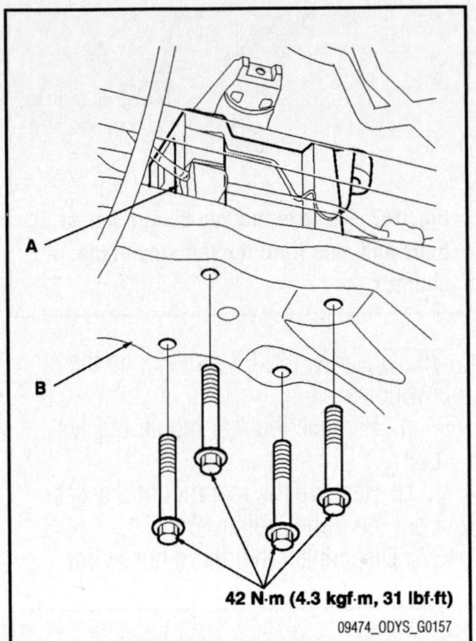
A
B
42 N·m (4.3 kgf·m, 31 lbf·ft)
09474_ODYS_G0157

Fig. 192 Install the base bracket (A) on the front subframe (B)—EX-L, EX-L Touring models

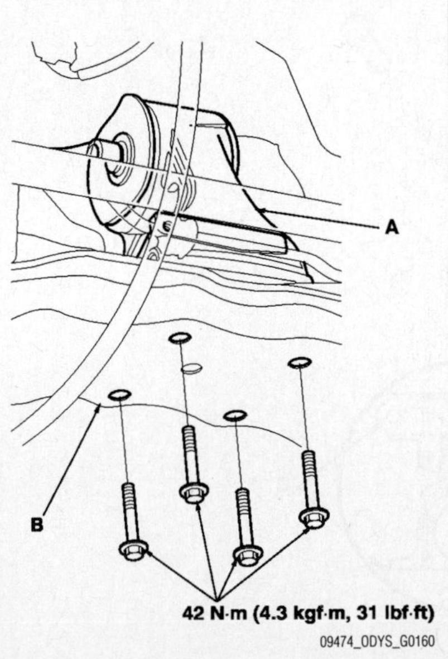
A
B
42 N·m (4.3 kgf·m, 31 lbf·ft)
09474_ODYS_G0160

Fig. 193 Install the rear mount bracket (A) on the front subframe (B)—Van, LX, EX models

    a. Install the rear mount bracket on the front subframe.
    b. Install the rear mount bracket bolt.
50. Install the steering gearbox bracket mounting bolts.
51. Install the two 12 mm flange bolts and nuts on the left side of the gearbox.
52. Install the two 10 mm flange bolts on the right side of the steering gearbox, then install the mounting bracket and cushion.
53. Install the P/S heat baffle plate.
54. Tighten the return hose flare nut to the specified torque, and install the adjustable hose clamp and the return hose.
55. Tighten the feed line flare nut to 37 Nm (27 ft. lbs.).
56. Install the P/S line mounting brackets on the front subframe and gearbox mounting bracket.

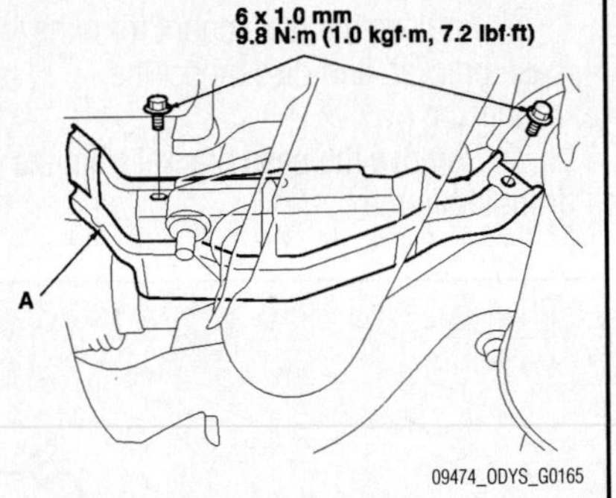
6 x 1.0 mm
9.8 N·m (1.0 kgf·m, 7.2 lbf·ft)
A
09474_ODYS_G0165

Fig. 194 Install the P/S heat baffle plate (A)

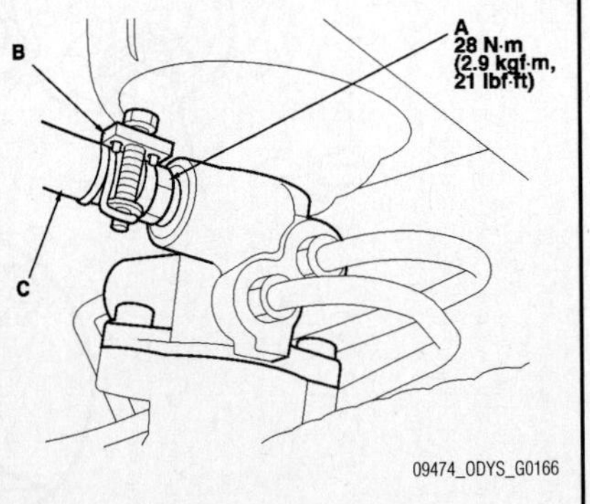
A
28 N·m
(2.9 kgf·m,
21 lbf·ft)
B
C
09474_ODYS_G0166

Fig. 195 Tighten the return hose flare nut (A) to the specified torque, and install the adjustable hose clamp (B) and the return hose (C)

57. Install the front subframe rear bracket with 12 mm flange bolts and 14 mm special bolts, and tighten to specified torque.

58. Install the front subframe front bracket with 12 mm flange bolts and 14 mm special bolts, and tighten to specified torque.

59. Install the front splash shield.

60. Connect the power steering pressure (PSP) switch connector.

61. Install exhaust pipe A using new gaskets and new self-locking nuts.

62. Connect the three way catalytic converter (TWC) to the muffler.

63. Install the new 10 mm self-locking nuts, and tighten them to 25 ft. lbs. (33 Nm).

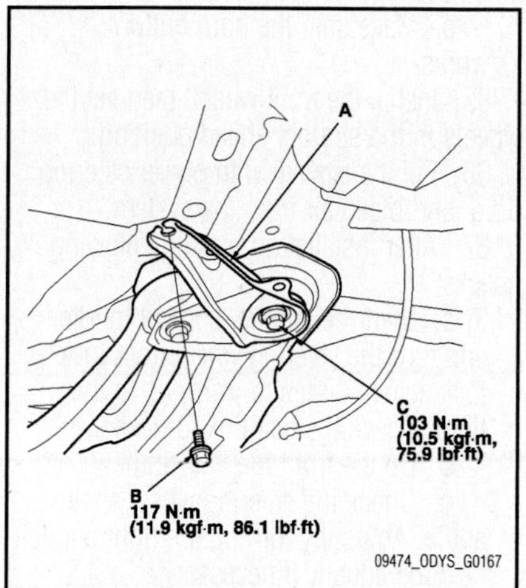
A
C
103 N·m
(10.5 kgf·m,
75.9 lbf·ft)
B
117 N·m
(11.9 kgf·m, 86.1 lbf·ft)
09474_ODYS_G0167

Fig. 196 Install the front subframe rear bracket (A) with 12 mm flange bolts (B) and 14 mm special bolts (C), and tighten to specified torque

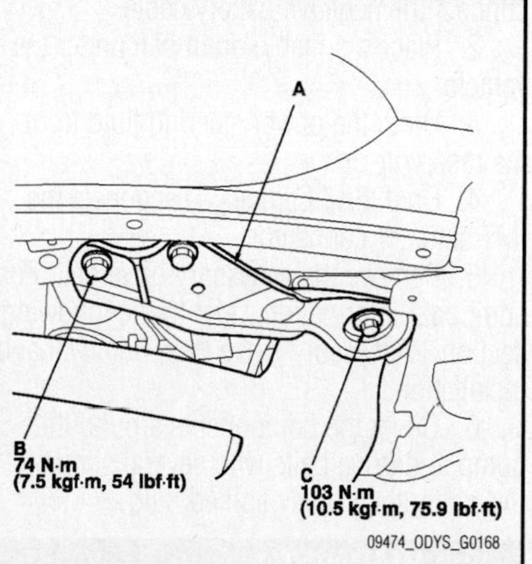
A
B
74 N·m
(7.5 kgf·m, 54 lbf·ft)
C
103 N·m
(10.5 kgf·m, 75.9 lbf·ft)
09474_ODYS_G0168

Fig. 197 Install the front subframe front bracket (A) with 12 mm flange bolts (B) and 14 mm special bolts (C), and tighten to specified torque

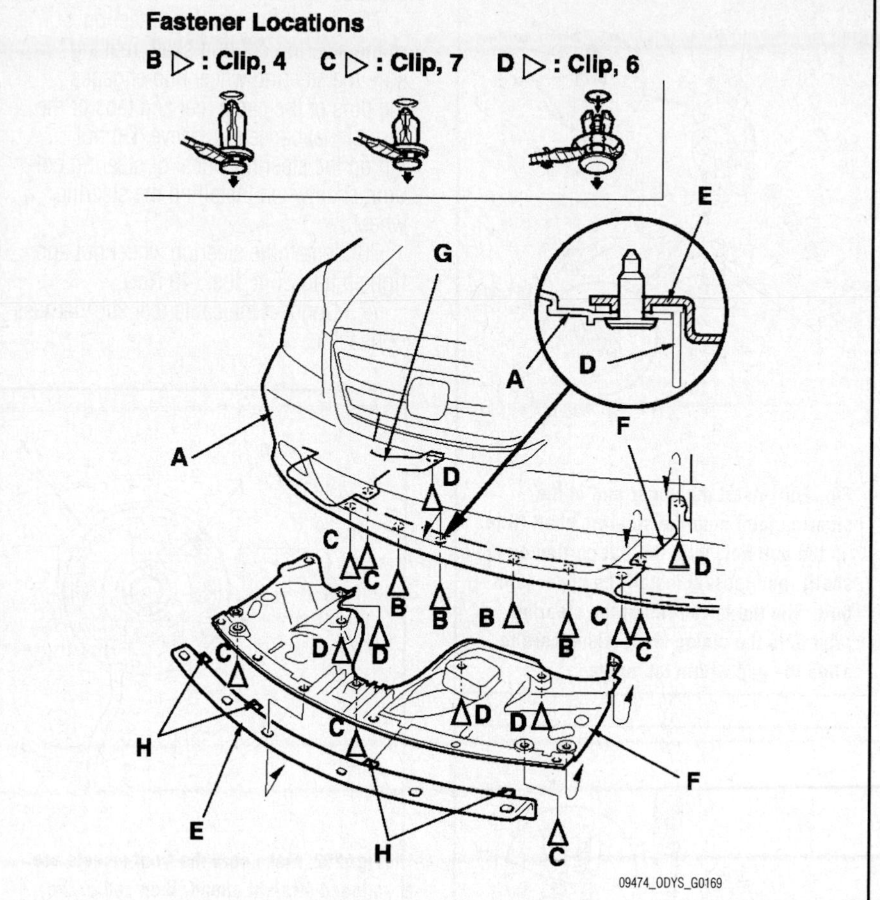
Fastener Locations
B ▷ : Clip, 4 C ▷ : Clip, 7 D ▷ : Clip, 6
E
G
A
D
A
F
D
C
C
B
B
B
C
D
D
D
C
B
C
D
D
C
H
E
H
F
C
09474_ODYS_G0169

Fig. 198 Front splash shield installation

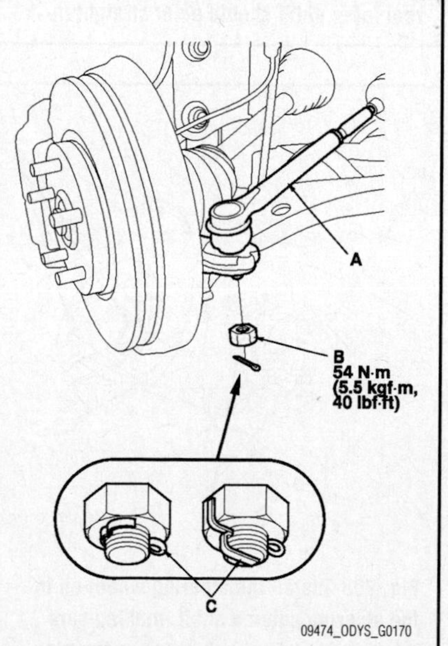
A
B
54 N·m
(5.5 kgf·m,
40 lbf·ft)
C
09474_ODYS_G0170

Fig. 199 Reconnect the tie-rod ends (A) to the steering knuckles. Install the 12 mm nut (B) and tighten it. Install a new cotter pin (C), and bend it as shown

64. Install the exhaust rubber mount on the frame. Torque to 16 ft. lbs. (22 Nm).

65. Wipe off any grease contamination from the ball joint tapered section and threads. Reconnect the tie-rod ends to the steering knuckles. Install the 12 mm nut and tighten it.

66. Install a new cotter pin and bend it over properly.

67. Center the steering rack within its stroke.

68. Insert the upper end of the steering joint onto the steering shaft (line up the bolt hole with the flat portion on the shaft), and loosely install the upper joint bolt.

69. Slip the lower end of the steering joint onto the pinion shaft taking care to align the gap within the angle.

70. Align the bolt hole on the steering joint with the groove around the pinion shaft, then loosely install the lower joint bolt.

71. Pull on the steering joint to make sure that the steering joint is fully seated, then tighten the lower joint bolt to the indicated torque.

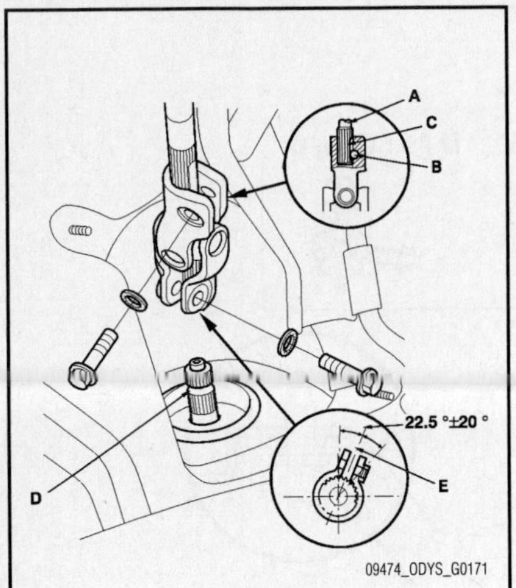
A
C
B
22.5 °±20 °
D
E
09474_ODYS_G0171

Fig. 200 Insert the upper end of the steering joint onto the steering shaft (line up the bolt hole with the flat portion on the shaft), and loosely install the upper joint bolt. Slip the lower end of the steering joint onto the pinion shaft taking care to align the gap within the angle

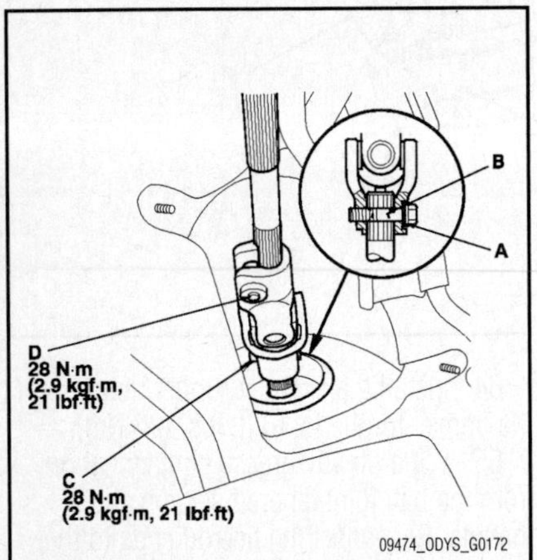
B
A
D
28 N·m
(2.9 kgf·m,
21 lbf·ft)
C
28 N·m
(2.9 kgf·m, 21 lbf·ft)
09474_ODYS_G0172

Fig. 201 Align the bolt hole on the steering joint with the groove around the pinion shaft, then loosely install the lower joint bolt. Pull on the steering joint to make sure that the steering joint is fully seated, then tighten the lower joint bolt to the indicated torque. Tighten the upper joint bolt to the indicated torque

72. Tighten the upper joint bolt to the indicated torque.

73. Reinstall the steering joint cover.

74. Before installing the steering wheel, make sure the front wheels are aligned straight ahead, then center the cable reel. Do this by first rotating the cable reel clockwise until it stops. Then rotate it counterclockwise about three full turns. The arrow mark on the cable reel label point should point straight up.

75. Install the steering wheel on to the steering column shaft, making sure the steering wheel hub engages the pins of the cable reel and tabs of the turn signal canceling sleeve. Do not tap on the steering wheel or steering column shaft when installing the steering wheel.

76. Install the steering wheel nut and tighten it to 36 ft. lbs. (49 Nm).

77. Connect the cable reel sub-harness connector.

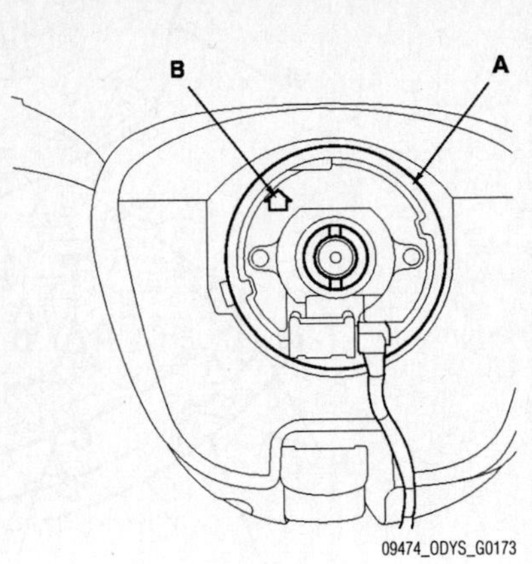
B
A
09474_ODYS_G0173

Fig. 202 Make sure the front wheels are aligned straight ahead, then center the cable reel (A). Do this by first rotating the cable reel clockwise until it stops. Then rotate it counterclockwise about three full turns. The arrow mark (B) on the cable reel label point should point straight up

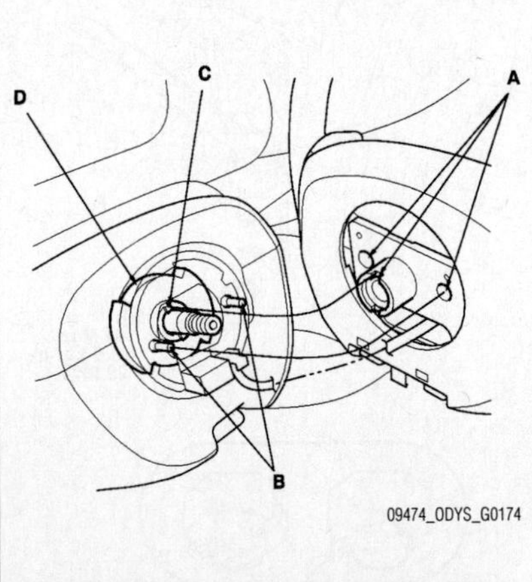
C
D
A
B
09474_ODYS_G0174

Fig. 203 Install the steering wheel on to the steering column shaft, making sure the steering wheel hub engages the pins of the cable reel and tabs of the turn signal canceling sleeve. Do not tap on the steering wheel or steering column shaft when installing the steering wheel

78. Enter the anti-theft codes for the radio and the navigation system, then enter the audio presets.

79. Without navigation: Reset the clock.

80. Check the cruise control, radio remote, navigation system, and turn signal canceling for proper operation

81. Place the new driver's air bag in the steering wheel, and secure it with new Torx® bolts. Tighten to 84 inch lbs. (10 Nm).

82. Connect the cable reel to the driver's air bag 4P connector, then install the access panel on the steering wheel.

83. Connect the battery negative cable.

84. After installing the air bag, confirm proper system operation:

   a. Turn the ignition switch ON (II); the SRS indicator should come on for about 6 seconds and then go off.

   b. Make sure the horn button works.

85. Install the front wheel, then set the wheels in the straight ahead position.

86. Fill the system with power steering fluid, and bleed air from the system.

87. After installation, do the following checks:

   a. Start the engine, allow it to idle, and turn the steering wheel from lock-to-lock several times to warm up to the fluid. Check the gearbox for leaks.

   b. Do the front toe inspection.

   c. Check the steering wheel spoke angle. Adjust by turning the right and left tie-rods equally if necessary.

POWER STEERING PUMP

REMOVAL & INSTALLATION

See Figure 204.

1. Note the radio security code and disconnect the negative battery cable.

2. Place a suitable drain pan under the vehicle.

3. Drain the power steering fluid from the reservoir.

4. On J35A7 Engines: Disconnect the IMT actuator connector .

5. Remove the accessory drive belt. For additional information, refer to the following section, "Accessory Drive Belt, Removal and Installation."

6. Cover the components around the pump and drive belts with several shop towels to absorb any spilled fluid.

WARNING

If any power steering fluid is spilled on a painted surface, wipe it off immediately.

7. Squeeze the power steering hose clamp at the pump, slide it up the hose and remove the hose from the pump.

8. Loosen the power steering pump outlet hose retaining bolts, and remove the hose fitting along with its O-ring from the pump.

WARNING

Protect all open power steering lines and fittings from debris and contaminants, otherwise internal damage may occur.

9. Wrap a clean cloth around the outlet hose fitting, and the port on the pump and hold in place with duct electrical tape.

10. Plug and cap the reservoir hose and metal hose fitting on the pump.

CAUTION

Do not turn the steering wheel with the pump removed.

11. Remove the power steering pump retainers, then remove the pump.

To install:

12. Install the pump and loosely install the lock and pivot bolts.

13. Install a new O-ring on the power steering hose outlet fitting, lubricate the O-ring with a light coating of Honda power steering fluid and install the fitting.

14. The balance of the installation is the reverse of the removal procedure.

15. Make note of the following points::

- Tighten the pump fasteners to 16 ft. lbs. (22 Nm)
- Make sure all fasteners, hose clamps and fittings are properly installed and tightened.

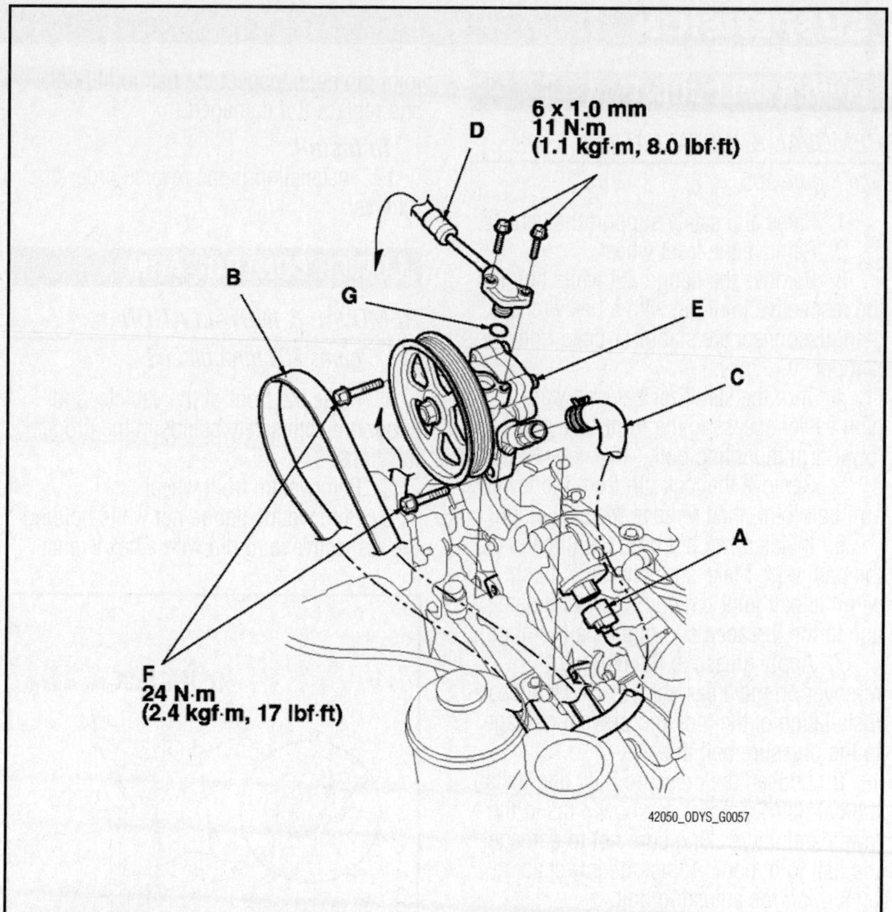
6 x 1.0 mm
11 N·m
(1.1 kgf·m, 8.0 lbf·ft)
D
B
G
E
C
A
F
24 N·m
(2.4 kgf·m, 17 lbf·ft)
42050_ODYS_G0057

Fig. 204 Exploded view of the IMT actuator connector, drive belt. power steering pump inlet hose, outlet hose, power steering pump, mounting bolts and nut

➡Manufacturer recommends, when topping off the system use only Genuine Honda Power Steerlng Fluid-V or S. Substituting another brand may damage internal components.

- Clean any spilled fluid before starting the vehicle.
- Bleed the system and top off as necessary.

BLEEDING

1. Fill the reservoir to the upper line with proper steering fluid.

2. Run the engine at idle and turn the steering wheel lock-to-lock several times to bleed air from the system and fill the rack valve body.

3. Recheck the fluid level and add more if necessary. Don't overfill the reservoir.

4. Check the power steering system for leaks.

LOWER BALL JOINT

REMOVAL & INSTALLATION

See Figure 205.

1. Raise and safely support the vehicle.
2. Remove the front wheel.
3. Remove the flange nut while holding the respective joint pin with a hex wrench, then disconnect the stabilizer links from the damper.
4. Turn the stabilizer bar backward to gain easier access to the front side of the lower arm mounting bolt.
5. Remove the lock pin from the lower arm ball joint, then remove the castle nut.
6. Install a hex nut onto the threads of the ball joint. Make sure the nut is flush with the ball joint pin end to prevent damage to the threaded end of the ball joint pin.
7. Apply grease to the ball joint remover on the areas shown. This will ease installation of the tool and prevent damage to the pressure bolt threads.
8. Loosen the pressure bolt, and install the ball joint remover as shown. Insert the jaws carefully, making sure not to damage the ball joint boot. Adjust the jaw spacing by turning the adjusting bolt.
9. After adjusting the adjusting bolt, make sure the head of the adjusting bolt is in the position shown to allow the jaw to pivot.
10. With a wrench, tighten the pressure bolt until the ball joint pin pops loose from the ball joint pin hole. If necessary, apply penetrating type lubricant to loosen the ball joint pin.
11. Remove the ball joint remover, then remove the nut from the end of the ball joint pin, and pull the ball joint out of the ball joint pin hole. Inspect the ball joint boot, and replace it if damaged.

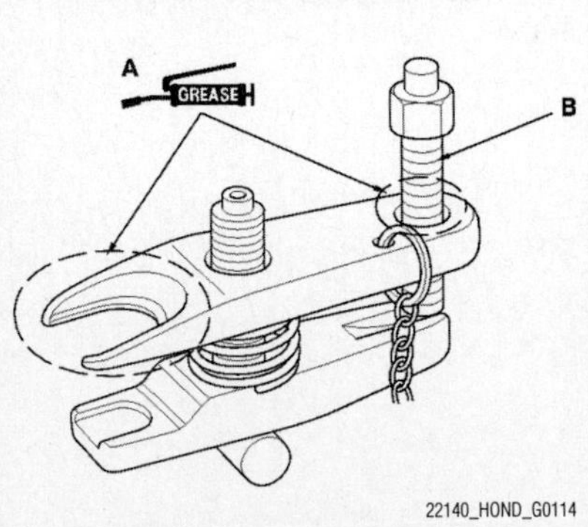
A
GREASE
B
22140_HOND_G0114

Fig. 205 Apply grease to the ball joint remover on the areas shown (A). This will ease installation of the tool and prevent damage to the pressure bolt (B) threads.

To install:

12. Installation is the reverse order of removal.

LOWER CONTROL ARM

REMOVAL & INSTALLATION

See Figures 206 through 208.

1. Raise the front of the vehicle, and support it with safety stands in the proper locations.
2. Remove the front wheel.
3. Remove the flange nut while holding the respective joint pin with a hex wrench, then disconnect the stabilizer links from the strut.

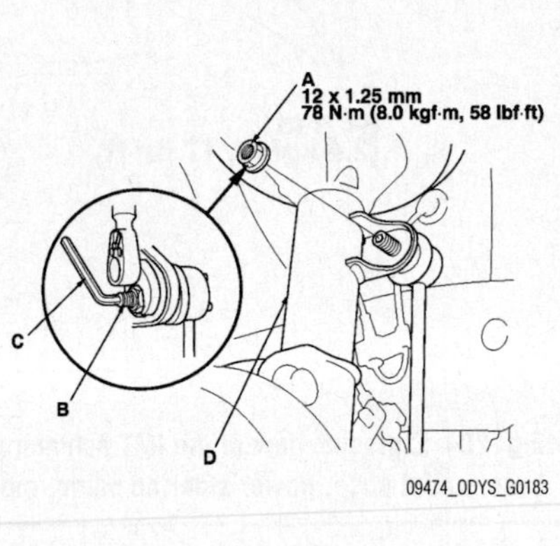
A
12 x 1.25 mm
78 N·m (8.0 kgf·m, 58 lbf·ft)
C
B
D
09474_ODYS_G0183

Fig. 206 Remove the flange nut while holding the respective joint pin with a hex wrench, then disconnect the stabilizer links from the strut

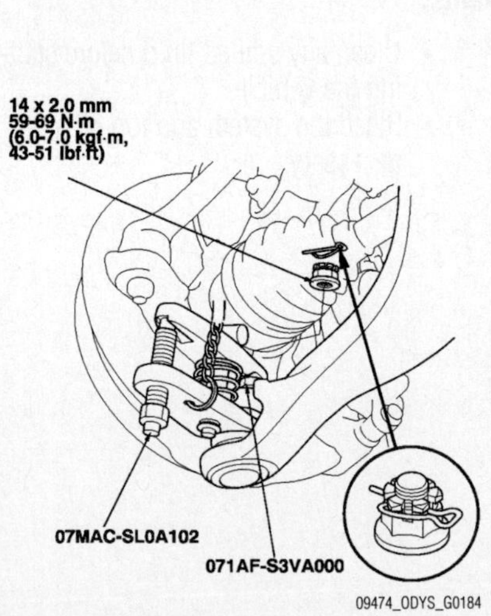
14 x 2.0 mm
59-69 N·m
(6.0-7.0 kgf·m,
43-51 lbf·ft)
07MAC-SL0A102
071AF-S3VA000
09474_ODYS_G0184

Fig. 207 Remove the lock pin from the lower arm ball joint castle nut, and remove the nut

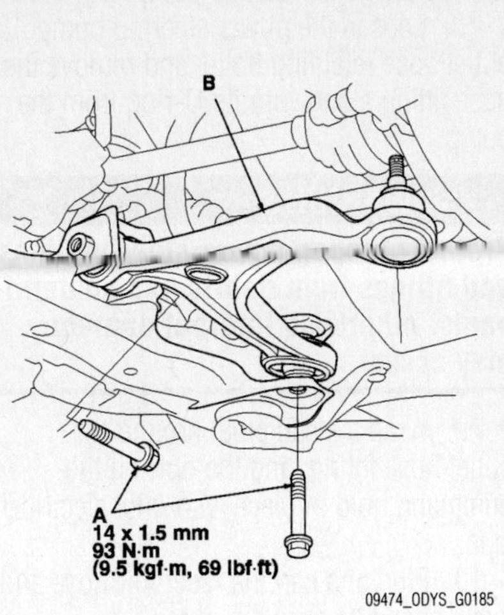
B
A
14 x 1.5 mm
93 N·m
(9.5 kgf·m, 69 lbf·ft)
09474_ODYS_G0185

Fig. 208 Remove the flange bolts (A), then remove the lower arm (B)

4. Turn the stabilizer bar backward to gain easier access to the front side of the lower arm mounting bolt.
5. Remove the lock pin from the lower arm ball joint castle nut, and remove the nut.
6. Remove the lower ball joint from the knuckle using the special tools.
7. Remove the flange bolts, then remove the lower arm.

To install:

8. Install the lower arm in the reverse order of removal, and note these items:
   - Be careful not to damage the ball joint boot when connecting the lower arm to the knuckle.
   - Place a jack under the lower arm, and raise the lower arm to load the suspension with the vehicle's weight.
   - Tighten all mounting hardware to the specified torque values.
   - Torque the castle nut to the lower torque specification, then tighten it only far enough to align the slot with the hole in the stud. Do not align the castle nut by loosening it.
   - Use a new lock pin on the castle nut.
   - Connect the struts to the links.
   - Before installing the wheel, clean the mating surface on the brake disc and the inside of the wheel.
   - Check the front wheel alignment, and adjust it if necessary.

MACPHERSON STRUT

REMOVAL & INSTALLATION

See Figures 209 through 211.

➡When compressing the damper spring, use a commercially available strut spring compressor (Branick MST-580A or Model 7200 or equivalent) according to the manufacturer's instructions.

1. Raise the vehicle, and support it with safety stands in the proper locations.
2. Remove the front wheel.
3. Disconnect the stabilizer link from the damper.
4. Remove the wheel sensor harness and the brake hose bracket from the damper. Do not disconnect the wheel sensor connector.
5. Remove the flange nuts and damper pinch bolts from the damper.
6. Remove the service caps, and remove the damper by removing the three flange nuts.
7. Remove the damper.

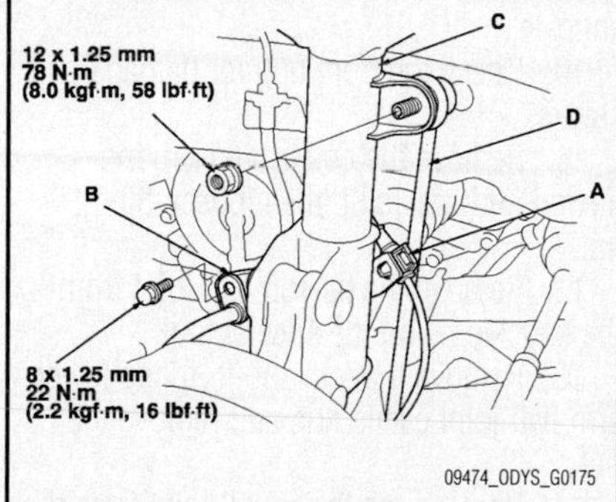
12 x 1.25 mm
78 N·m
(8.0 kgf·m, 58 lbf·ft)
C
D
B
A
8 x 1.25 mm
22 N·m
(2.2 kgf·m, 16 lbf·ft)
09474_ODYS_G0175

Fig. 209 Disconnect the stabilizer link from the damper. Remove the wheel sensor harness and the brake hose bracket from the damper

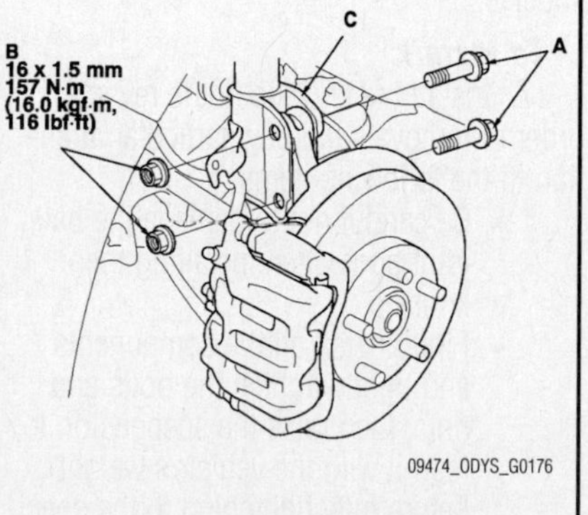
C
A
B
16 x 1.5 mm
157 N·m
(16.0 kgf·m,
116 lbf·ft)
09474_ODYS_G0176

Fig. 210 Remove the flange nuts (A) and damper pinch bolts (B) from the damper

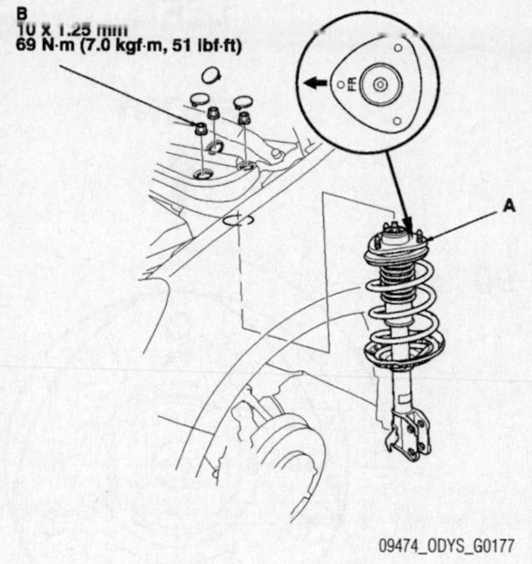
B
10 x 1.25 mm
69 N·m (7.0 kgf·m, 51 lbf·ft)
A
09474_ODYS_G0177

Fig. 211 Remove the service caps, and remove the damper by removing the three flange nuts (B)

➡Damper springs are different, left and right. Mark the springs "L" and "R" before you continue.

To install:

8. Install the damper onto the frame, then loosely install the three flange nuts.
9. Loosely install the damper pinch bolts and flange nuts to the damper.
10. Install the wheel sensor harness and the brake hose bracket to the damper.
11. Loosely install the stabilizer link to the damper.
12. Raise the front suspension with a floor jack to load the suspension with the vehicle's weight.
13. Tighten the damper pinch bolts and flange nuts to the specified torque value.
14. Tighten the flange nuts on top of the damper to the specified torque value.
15. Tighten the stabilizer link nuts to the specified torque value.
16. Install the service caps.
17. Install the front wheel.
18. Check the front wheel alignment, and adjust it if necessary.

OVERHAUL

See Figures 212 through 216.

➡When compressing the damper spring, use a commercially available strut spring compressor (Branick MST-580A or Model 7200 or equivalent) according to the manufacturer's instructions.

1. Compress the strut spring, then remove the self-locking nut while holding the strut rod with a hex wrench. Do not compress the spring more than necessary to remove the nut.
2. Release the pressure from the strut spring compressor, then disassemble the strut.
3. Reassemble all the parts, except for the spring.
4. Compress the strut assembly by hand, and check for smooth operation through a full stroke, both compression and extension. The strut should extend smoothly and constantly when compression is

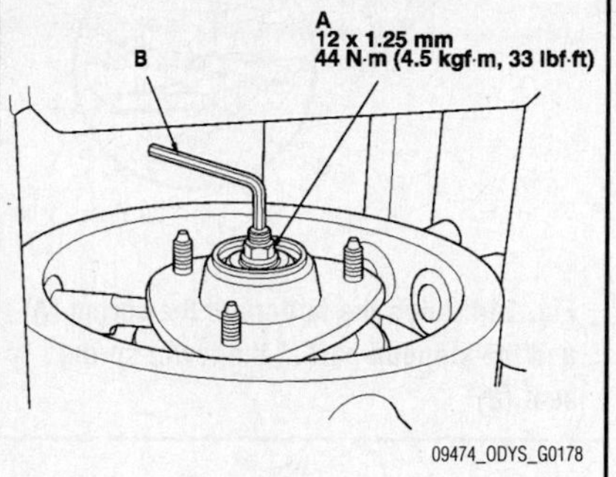
A
12 x 1.25 mm
44 N·m (4.5 kgf·m, 33 lbf·ft)
B
09474_ODYS_G0178

Fig. 212 Compress the strut spring, then remove the self-locking nut (A) while holding the strut rod with a hex wrench (B)

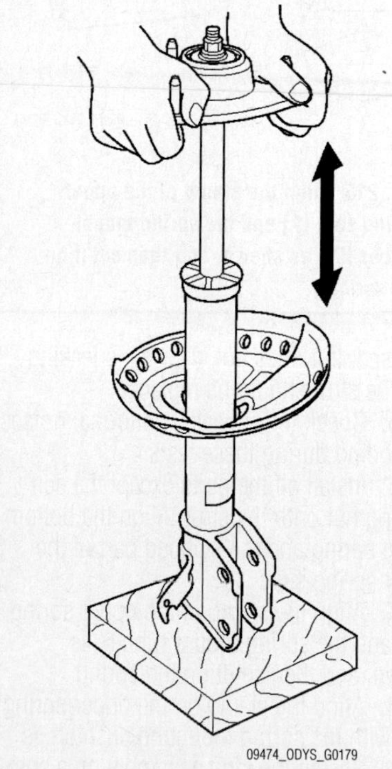
09474_ODYS_G0179

Fig. 213 Compress the strut assembly by hand

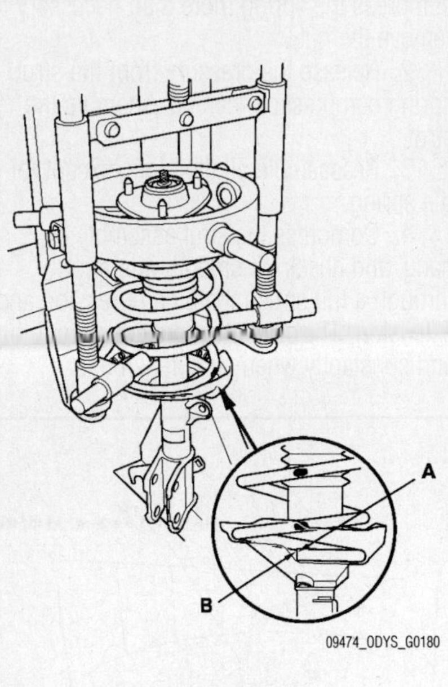
A
B
09474_ODYS_G0180

Fig. 214 Align the bottom of the spring (A) and the stepped part of the lower spring seat (B)

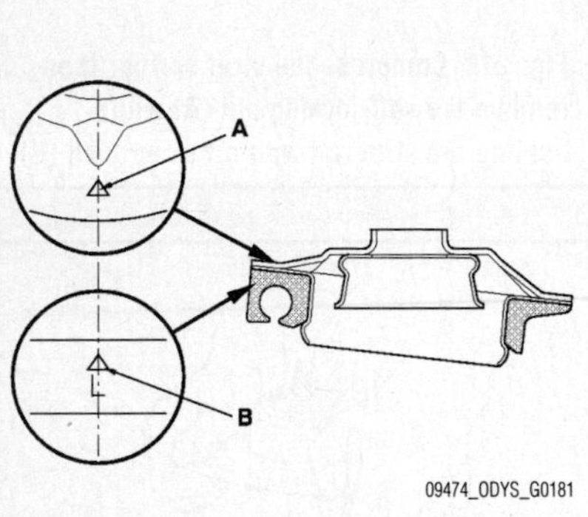
A
B
09474_ODYS_G0181

Fig. 215 Align the stamp of the upper spring seat (A) and the spring mount rubber (B), as shown, and then set it on the spring

released. If it does not, the gas is leaking and the strut should be replaced.

5. Check for oil leaks, abnormal noises, or binding during these tests
6. Install all the parts except the self-locking nut onto the strut. Align the bottom of the spring and the stepped part of the lower spring seat.
7. Align the stamp of the upper spring seat and the spring mount rubber, as shown, and then set it on the spring.
8. Align the stamp on the upper spring seat with the spring mounting cushion as shown. Install the strut assembly on a commercially available strut spring compressor.
9. Compress the strut spring.
10. Install a new self-locking nut.

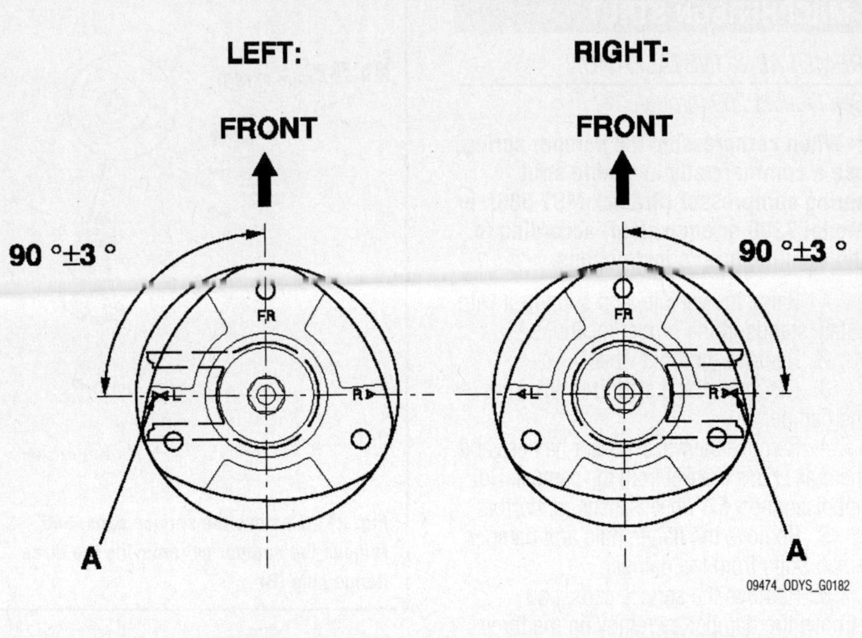
LEFT:
RIGHT:
FRONT
FRONT
90 °±3 °
90 °±3 °
FR
FR
L
R
L
R
A
A
09474_ODYS_G0182

Fig. 216 Align the stamp (A) on the upper spring seat with the spring mounting cushion as shown. Install the strut assembly on a commercially available strut spring compressor

11. Hold the strut rod using a hex wrench, and tighten the self-locking nut.

STEERING KNUCKLE, HUB & WHEEL BEARING

REMOVAL & INSTALLATION

See Figures 217 through 223.

1. Raise the front of the vehicle, and support it with safety stands in the proper locations.
2. Remove the wheel nuts and front wheel.
3. Remove the brake hose mounting bolt.
4. Remove the caliper mounting bolts, and hang the caliper assembly to one side.

CAUTION

To prevent damage to the caliper assembly or brake hose, use a short piece of wire to hang the caliper from the undercarriage.

5. Remove the wheel sensor from the knuckle. Do not disconnect the wheel sensor connector.
6. Raise the stake, and then remove the spindle nut.
7. Remove the 6mm brake disc retaining screws.
8. Screw two 8 x 1.25mm bolts into the brake disc to push it away from the hub. Turn each bolt two turns at a time to prevent cocking the disc excessively.
9. Remove the brake disc from the knuckle.
10. Check the front hub for damage and cracks.
11. Remove the cotter pin from the tie-rod end ball joint, then loosen the nut.
12. Remove the tie-rod ball joint from the knuckle using the special tool.
13. Remove the lock pin from the lower arm ball joint castle nut, and remove the nut.
14. Remove the lower ball joint from the knuckle using the special tools.
15. Remove the damper pinch bolts and flange nuts from the damper.
16. Pull the knuckle outward, and remove the driveshaft outboard joint from the knuckle by tapping the driveshaft end with a plastic hammer, then remove the knuckle.

To install:

17. Install the knuckle in the reverse order of removal, and pay particular attention to the following items:
   - Be careful not to damage the ball joint boot when installing the knuckle.
   - First, install all the components and lightly tighten the bolts and nuts, then raise the suspension to load it with the vehicle's weight before fully tightening to the specified torque values. Do not place the jack against the ball joint pin of the knuckle.

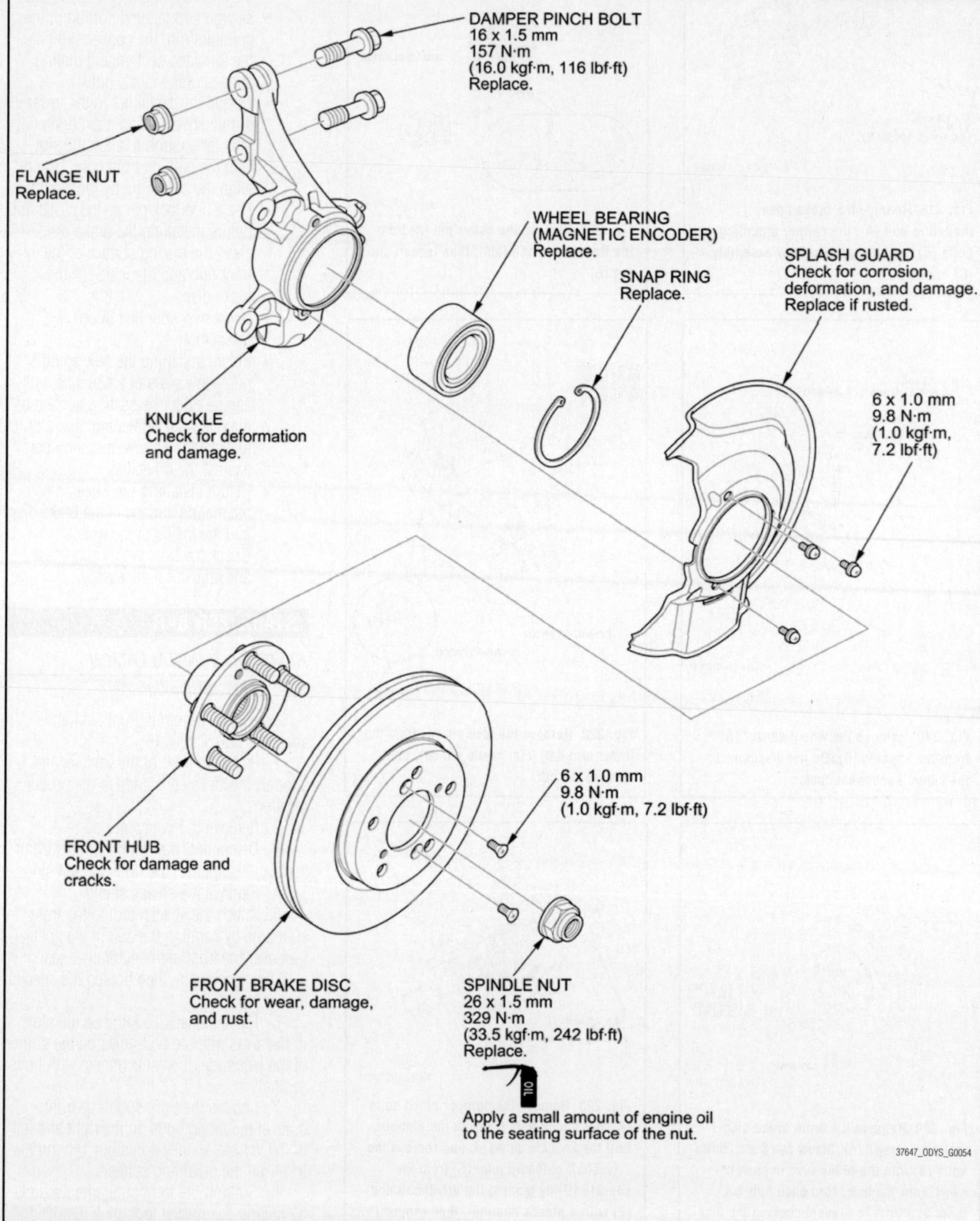
DAMPER PINCH BOLT
16 x 1.5 mm
157 N·m
(16.0 kgf·m, 116 lbf·ft)
Replace.
FLANGE NUT
Replace.
WHEEL BEARING
(MAGNETIC ENCODER)
Replace.
SNAP RING
Replace.
SPLASH GUARD
Check for corrosion,
deformation, and damage.
Replace if rusted.
KNUCKLE
Check for deformation
and damage.
6 x 1.0 mm
9.8 N·m
(1.0 kgf·m,
7.2 lbf·ft)
6 x 1.0 mm
9.8 N·m
(1.0 kgf·m, 7.2 lbf·ft)
FRONT HUB
Check for damage and
cracks.
FRONT BRAKE DISC
Check for wear, damage,
and rust.
SPINDLE NUT
26 x 1.5 mm
329 N·m
(33.5 kgf·m, 242 lbf·ft)
Replace.
OIL
Apply a small amount of engine oil
to the seating surface of the nut.
37647_ODYS_G0054

Fig. 217 Exploded view of the knuckle, hub and wheel bearing assembly

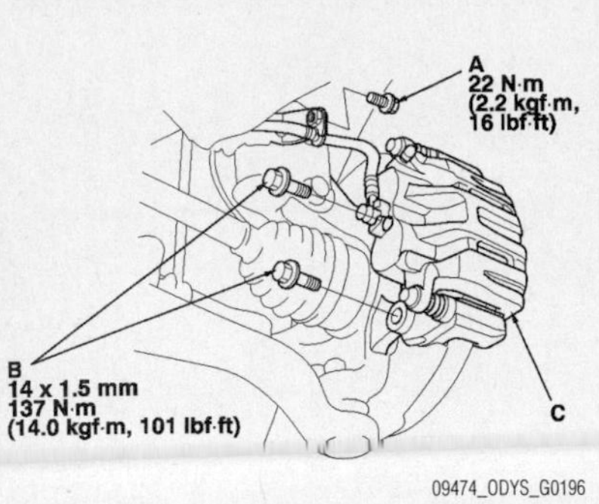
A
22 N·m
(2.2 kgf·m,
16 lbf·ft)
B
14 x 1.5 mm
137 N·m
(14.0 kgf·m, 101 lbf·ft)
C
09474_ODYS_G0196

Fig. 218 Remove the brake hose mounting bolt (A), the caliper mounting bolts (B), and hang the caliper assembly (C) to one side

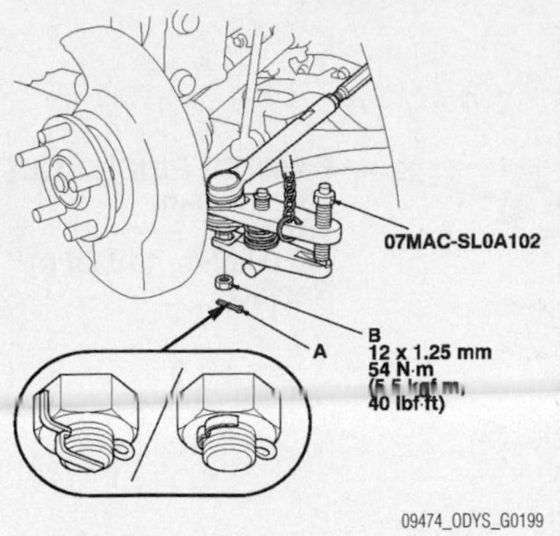
07MAC-SL0A102
A
B
12 x 1.25 mm
54 N·m
40 lbf·ft)
09474_ODYS_G0199

Fig. 221 Remove the cotter pin (A) from the tie-rod end ball joint, then loosen the nut (B)

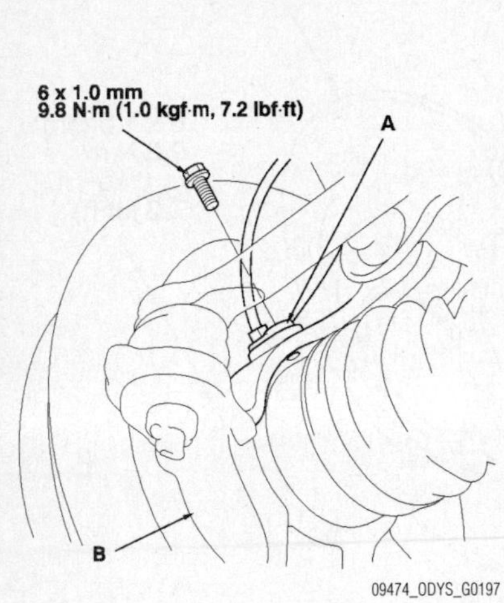
6 x 1.0 mm
9.8 N·m (1.0 kgf·m, 7.2 lbf·ft)
A
B
09474_ODYS_G0197

Fig. 219 Remove the wheel sensor (A) from the knuckle (B). Do not disconnect the wheel sensor connector

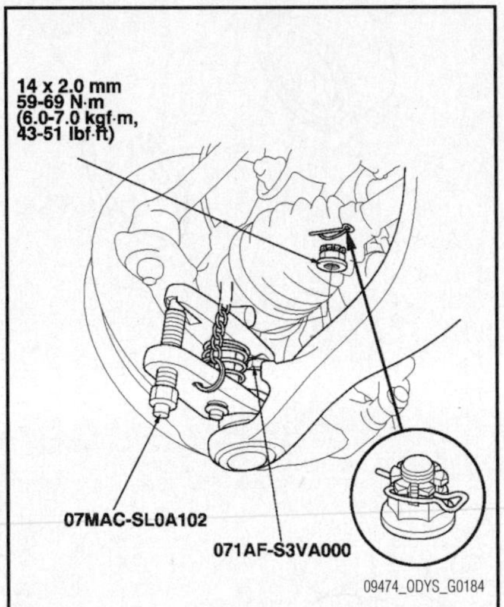
14 x 2.0 mm
59-69 N·m
(6.0-7.0 kgf·m,
43-51 lbf·ft)
07MAC-SL0A102
071AF-S3VA000
09474_ODYS_G0184

Fig. 222 Remove the lock pin (A) from the lower arm ball joint castle nut (B), and remove the nut

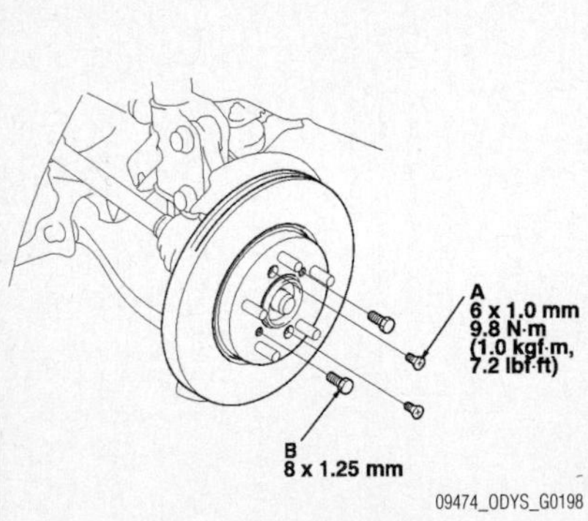
A
6 x 1.0 mm
9.8 N·m
(1.0 kgf·m,
7.2 lbf·ft)
B
8 x 1.25 mm
09474_ODYS_G0198

Fig. 220 Remove the 6mm brake disc retaining screws (A). Screw two 8 x 1.25mm bolts (B) into the brake disc to push it away from the hub. Turn each bolt two turns at a time to prevent cocking the disc excessively

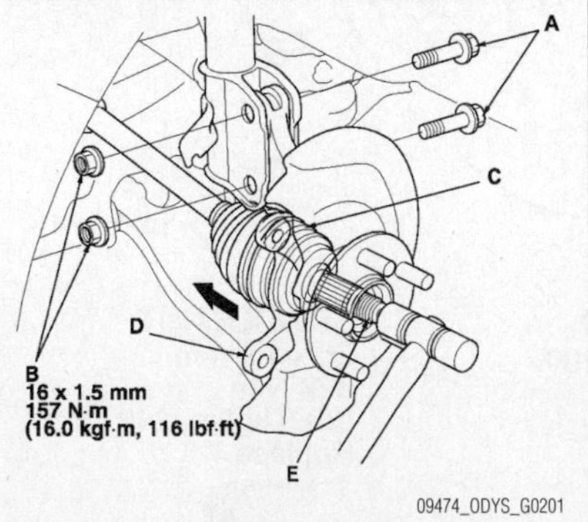
A
C
D
B
16 x 1.5 mm
157 N·m
(16.0 kgf·m, 116 lbf·ft)
E
09474_ODYS_G0201

Fig. 223 Remove the damper pinch bolts (A) and flange nuts (B) from the damper. Pull the knuckle outward, and remove the driveshaft outboard joint (C) from the knuckle (D) by tapping the driveshaft end (E) with a plastic hammer, then remove the knuckle

- Tighten all mounting hardware to the specified torque values.
- Before connecting the ball joint to the knuckle, degrease the threaded section and tapered portion of the ball joint pin, the connecting hole, the threaded section and mating surface of the castle nut.
- Torque the castle nut to the lower torque specification, then tighten it only far enough to align the slot with the ball joint pin hole. Do not align the castle nut by loosening it.
- Use a new lock pin on the castle nut.
- Before installing the brake disc, clean the mating surface of the front hub and the inside of the brake disc.
- Use a new spindle nut on reassembly.
- Before installing the new spindle nut, apply a small amount of engine oil to the seating surface of the nut. After tightening, use a drift to stake the spindle nut shoulder against the driveshaft.
- Before installing the wheel, clean the mating surface of the brake disc and the inside of the wheel.
- Check the front wheel alignment, and adjust it if necessary.

STABILIZER BAR

REMOVAL & INSTALLATION

See Figures 224 through 229.

Special tool required: front subframe adapter VSB02C000016

1. Raise the front of the vehicle, and support it with safety stands in the proper locations.
2. Remove the front wheels.
3. Disconnect the stabilizer links from the stabilizer bar on the right and left sides.
4. Remove the splash shield.
5. Attach the special tool to the front subframe by hanging the belt of the special tool over the front of the subframe, secure it with the special pin, then tighten the special tool screw.
6. Raise the jack and line up the slots in the arms with the bolt holes on the corner of the jack base, then attach them with bolts securely.
7. Loosen the front subframe front bracket mounting bolts on the right and left of the vehicle so they are about 14mm (9/16 in.) from the mounting surface.
8. Support the front subframe securely by raising the special tool, then remove the two 12mm flange bolts.

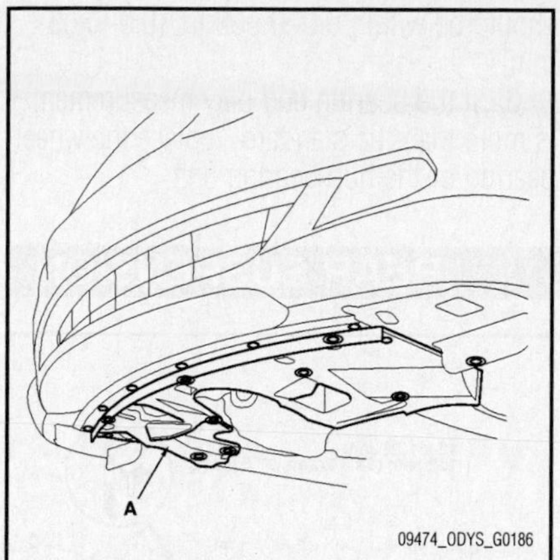
A
09474_ODYS_G0186

Fig. 224 Remove the splash shield (A)

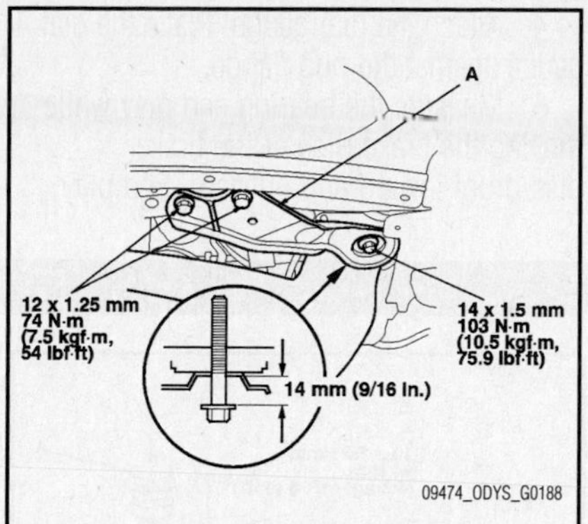
A
12 x 1.25 mm
74 N·m
(7.5 kgf·m,
54 lbf·ft)
14 x 1.5 mm
103 N·m
(10.5 kgf·m,
75.9 lbf·ft)
14 mm (9/16 in.)
09474_ODYS_G0188

Fig. 226 Loosen the front subframe front bracket (A) mounting bolts on the right and left of the vehicle so they are about 14mm (9⁄16 in.) from the mounting surface

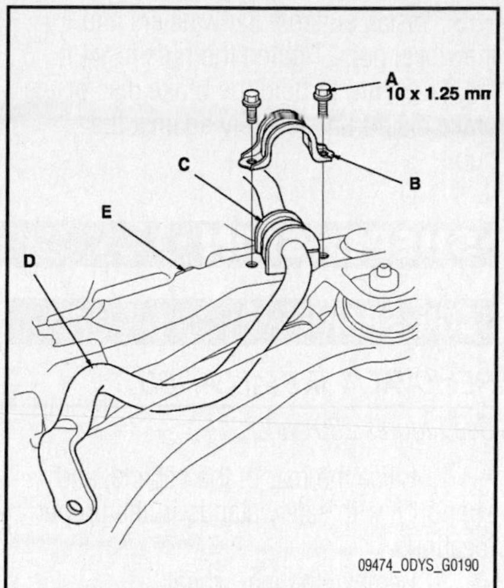
A
10 x 1.25 mm
C
B
E
D
09474_ODYS_G0190

Fig. 228 Remove the flange bolts and bushing holders, then remove the bushings and the stabilizer bar from the front subframe

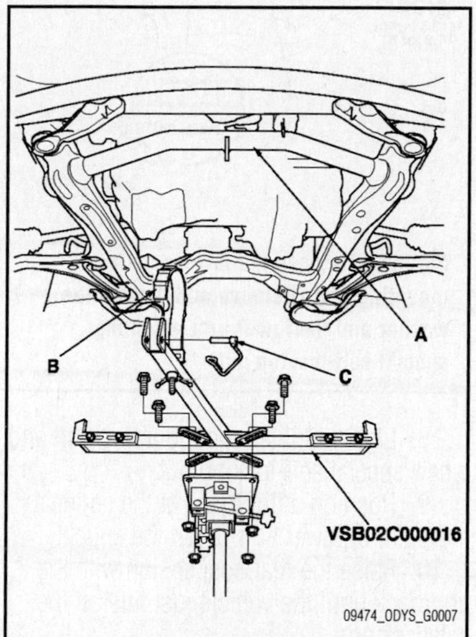
A
B
C
VSB02C000016
09474_ODYS_G0007

Fig. 225 Attach the special tool to the front subframe (A) by hanging the belt (B) of the special tool over the front of the subframe, secure it with the special pin (C), then tighten the special tool screw.

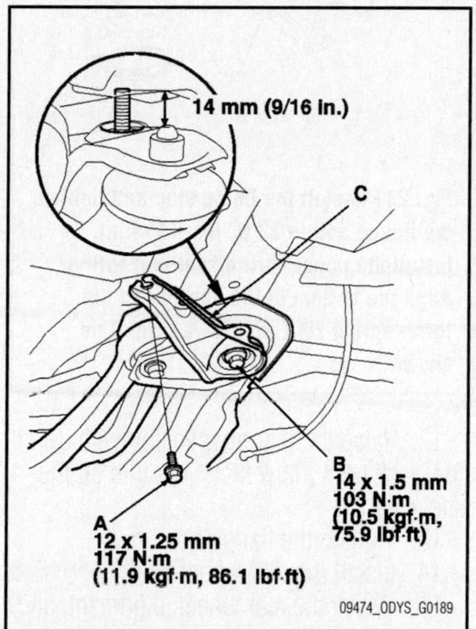
14 mm (9/16 in.)
C
B
14 x 1.5 mm
103 N·m
(10.5 kgf·m,
75.9 lbf·ft)
A
12 x 1.25 mm
117 N·m
(11.9 kgf·m, 86.1 lbf·ft)
09474_ODYS_G0189

Fig. 227 Support the front subframe securely by raising the special tool, then remove the two 12mm flange bolts (A). Loosen the two 14mm special bolts (B) so they are about 14mm (9⁄16 in.) from the mounting surface

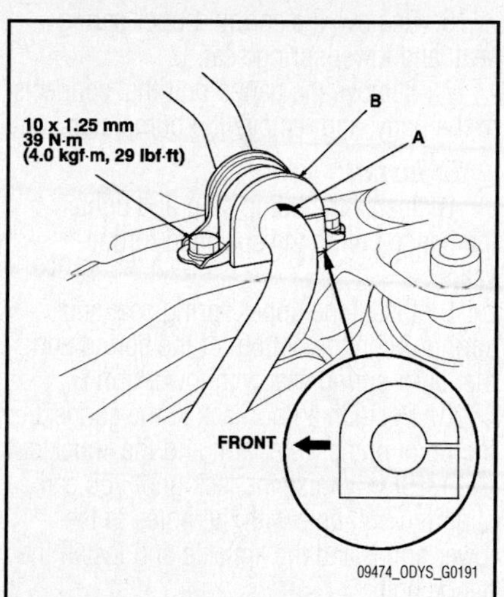
B
10 x 1.25 mm
39 N·m
(4.0 kgf·m, 29 lbf·ft)
A
FRONT
09474_ODYS_G0191

Fig. 229 Align the paint marks (A) on the stabilizer bar with the sides of the bushings (B)

9. Loosen the two 14mm special bolts so they are about 14mm (9⁄16 in.) from the mounting surface.

10. Lower the jack supporting the front subframe with the special tool slowly until the front subframe has dropped about 14mm (9⁄16 in.).

11. Remove the flange bolts and bushing holders, then remove the bushings and the stabilizer bar from the front subframe.

To install:

12. Install the stabilizer bar in the reverse order of removal, and note these items:

- Note the right and left direction of the stabilizer bar.
- Align the paint marks on the stabilizer bar with the sides of the bushings.
- Note the fore/aft direction of the bushing holders.
- Raise the front subframe up with the jack and special tool until it contacts the body frame, then tighten the mounting bolts to the specified torque.
- Refer to Stabilizer Link Removal/Installation to connect the stabilizer bar to the links.
- Do the front toe inspection, and adjust it if necessary.

WHEEL BEARINGS

REMOVAL & INSTALLATION

➡See "Steering Knuckle, Hub & Wheel Bearing".

ADJUSTMENT

The wheel bearings are not adjustable. If they are not within specification, the wheel bearings must be replaced.

INSPECTION

1. Raise the vehicle.
2. Remove the wheels.

3. Install suitable flat washers and the wheel nuts. Tighten the nuts to 94 ft. lbs. (127 Nm) to hold the brake disc or the brake disc/drum securely against the hub.

4. Attach the dial gauge. Place the dial gauge against the hub flange.

5. Measure the bearing end play while moving the brake disc or the brake disc/drum inward and outward. End play should be within 0.0–0.002 in. (0.0–0.05 mm).

6. If the bearing end play measurement is more than the standard, replace the wheel bearing or the hub bearing unit.

SUSPENSION

REAR SUSPENSION

COIL SPRING

REMOVAL & INSTALLATION

See Figures 230 and 231.

1. Raise the rear of the vehicle, and support it with safety stands in the proper locations.
2. Remove the rear wheel.
3. Position a floor jack at the connecting point of the lower arm B and the knuckle.
4. Remove the flange bolt that connects the lower arm B and the knuckle.
5. Lower the floor jack gradually.
6. Remove the spring, upper spring seat, and lower spring seat.
7. Remove the flange bolt that connects to the body, and remove the bump stop.

To install:

8. Install the bump stop and tighten the flange bolt to the specified torque value.
9. Install the upper spring seat and spring. Align the bottom of the spring and the lower spring seat with lower arm B.
10. Position a floor jack at the connecting point of the lower arm and the knuckle.
11. Slowly raise the jack until you can align the bolt hole with the holes in the lower arm B and the knuckle and install the flange bolt.

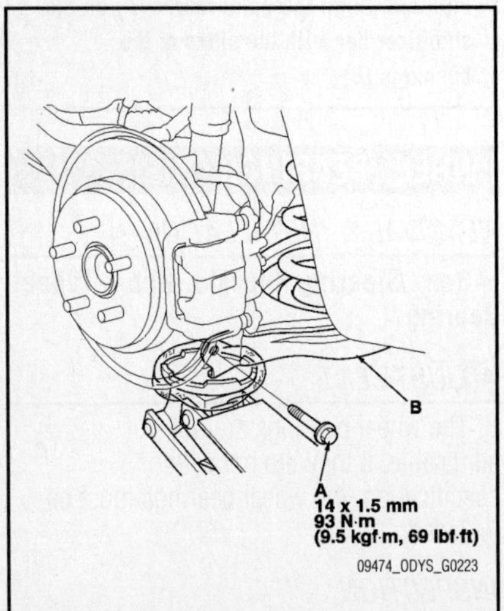
B
A
14 x 1.5 mm
93 N·m
(9.5 kgf·m, 69 lbf·ft)
09474_ODYS_G0223

Fig. 230 Remove the flange bolt (A) that connects the lower arm B and the knuckle

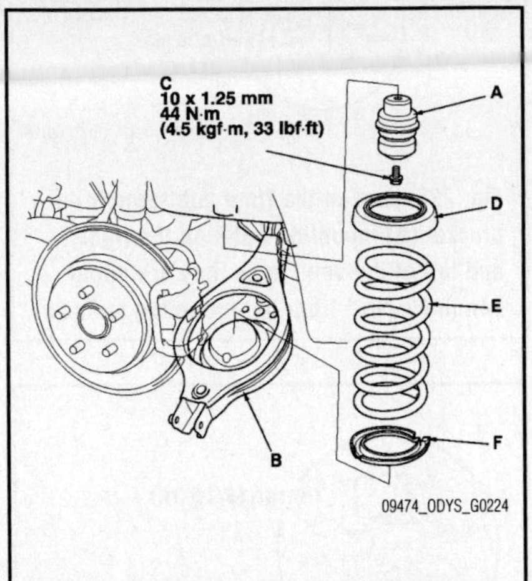
C
10 x 1.25 mm
44 N·m
(4.5 kgf·m, 33 lbf·ft)
A
D
E
F
B
09474_ODYS_G0224

Fig. 231 Install the bump stop and tighten the flange bolt to 33 ft. lbs. (44 Nm). Install the upper spring seat and spring. Align the bottom of the spring and the lower spring seat with lower arm B as shown

12. Raise the rear suspension with a floor jack until the vehicle just lifts off the safety stands.
13. Tighten the flange bolt.
14. Install the rear wheel.
15. Check the rear wheel alignment, and adjust it if necessary.

CONTROL ARMS/LINKS

REMOVAL & INSTALLATION

Lower Control Arm A

See Figure 232.

1. Raise the rear of the vehicle, and support it with safety stands in the proper locations.
2. Remove the rear wheel.
3. Remove the wheel sensor harness from the lower arm A.
4. Remove the lower arm A mounting bolt.
5. Remove and discard the washer and the lower arm mounting special self-locking nut.
6. Remove the lower arm A from the vehicle.

To install:

7. Install the lower arm A and a new washer.

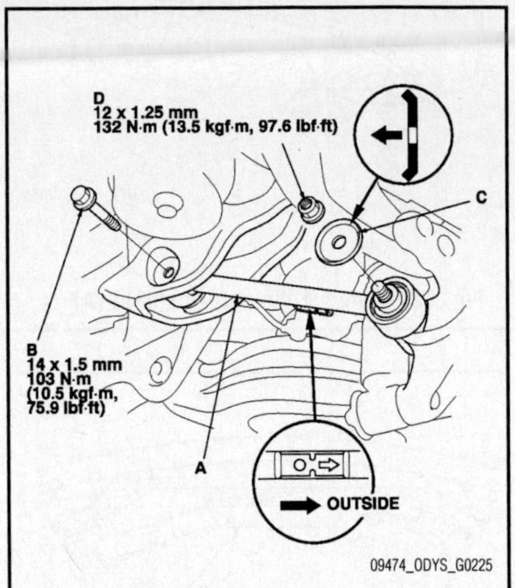
D
12 x 1.25 mm
132 N·m (13.5 kgf·m, 97.6 lbf·ft)
C
B
14 x 1.5 mm
103 N·m
(10.5 kgf·m,
75.9 lbf·ft)
A
OUTSIDE
09474_ODYS_G0225

Fig. 232 Remove the lower arm A mounting bolt. Remove and discard the washer and the lower arm mounting special self-locking nut

8. Lightly tighten the mounting bolt and a new special self-locking nut.
9. Position a floor jack at the connecting point of lower arm B and the knuckle.
10. Raise the rear suspension with the floor jack until the vehicle just lifts off the safety stands.
11. Tighten the lower arm bolt and nut to the specified torque value, and note these items:
    - Before installing the wheel, clean the mating surfaces on the brake disc/drum and the inside of the wheel.
    - Check the rear wheel alignment, and adjust it if necessary

Lower Control Arm B

See Figure 233.

1. Raise the rear of the vehicle, and support it with safety stands in the proper locations.
2. Remove the rear wheel.
3. Position a floor jack at the connecting point of lower arm B and the knuckle.
4. Remove the flange bolt that connects the lower arm B and the knuckle.
5. Lower the floor jack gradually.
6. Remove the spring, upper spring seat and lower spring seat.

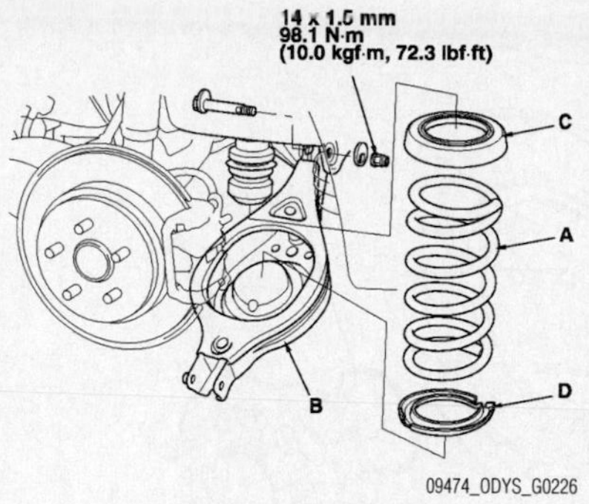
14 x 1.5 mm
98.1 N·m
(10.0 kgf·m, 72.3 lbf·ft)
C
A
B
D
09474_ODYS_G0226

Fig. 233 Install the spring and upper spring seat. Align the bottom of the spring and the lower spring seat with the lower arm B as shown

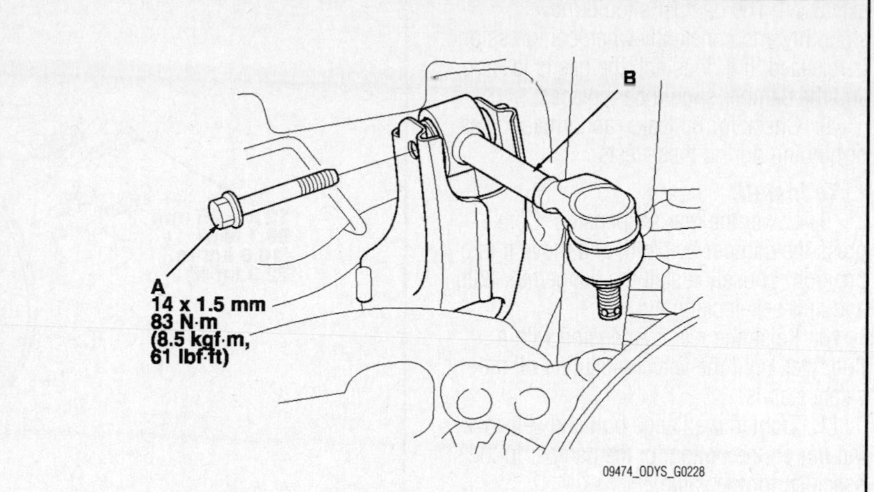
B
A
14 x 1.5 mm
83 N·m
(8.5 kgf·m,
61 lbf·ft)
09474_ODYS_G0228

Fig. 235 Remove the upper arm bolt (A). Remove the upper arm from the vehicle

7. Remove the self-locking nut and flange bolt.

To install:

8. Position the lower arm B, and loosely install the flange bolt and the self-locking nut.

9. Install the spring and upper spring seat. Align the bottom of the spring and the lower spring seat with the lower arm B as shown.

Upper Control Arm

See Figures 234 and 235.

1. Raise the rear of the vehicle, and support it with safety stands in the proper locations.
2. Remove the rear wheel.
3. Position a floor jack at the connecting point of the lower arm B and the knuckle.

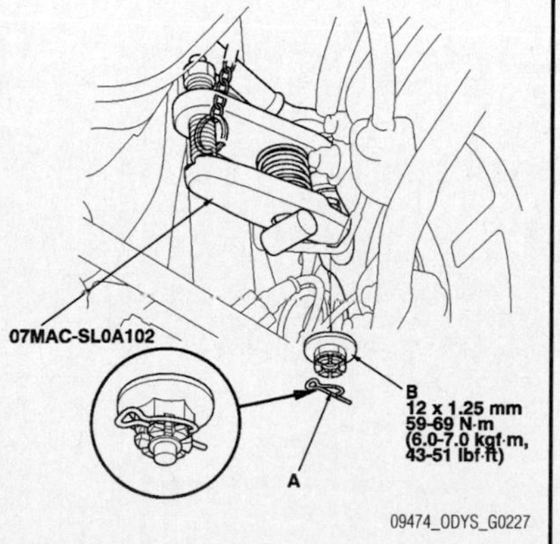
07MAC-SL0A102
B
12 x 1.25 mm
59-69 N·m
(6.0-7.0 kgf·m,
43-51 lbf·ft)
A
09474_ODYS_G0227

Fig. 234 Remove the lock pin (A) from the upper ball joint castle nut (B), and remove the nut

4. Remove the lock pin from the upper ball joint castle nut, and remove the nut.
5. Remove the upper ball joint from the knuckle using the special tool.
6. Remove the upper arm bolt. Remove the upper arm from the vehicle.

To install:

7. Install the upper arm.
8. Lightly tighten the upper arm bolt and castle nut.
9. Position a floor jack at the connecting point of the lower arm B and the knuckle.
10. Raise the rear suspension with the floor jack until the vehicle just lifts off the safety stands.
11. Tighten the upper arm bolt to the specified torque value, and note these items:
    - Be careful not to damage the ball joint boot when connecting the upper arm to the knuckle.
    - Before connecting the ball joint to the knuckle, degrease the threaded section and tapered portion of the ball joint pin, the connecting hole, and the threaded section and mating surface of the castle nut.
    - Torque the castle nut to the lower torque specification, then tighten it only far enough to align the slot with the hole in the stud. Do not align the castle nut by loosening it.
    - Use a new lock pin on the castle nut.
    - Before installing the wheel, clean the mating surfaces on the brake disc/drum and the inside of the wheel.
    - Check the rear wheel alignment, and adjust it if necessary.

SHOCK ABSORBER

REMOVAL & INSTALLATION

See Figure 236.

1. Raise the rear of the vehicle, and support it with safety stands in the proper locations.
2. Remove the rear wheel.
3. Position a floor jack at the connecting point of the lower control arm and knuckle. Raise the floor jack until the suspension begins to compress.
4. Remove the flange bolt and nut from the body.
5. Remove the self-locking nut from the knuckle.
6. Compress the damper by hand, and remove it from the vehicle.
7. Compress the damper assembly by hand, and check for smooth operation through a full stroke, both compression and

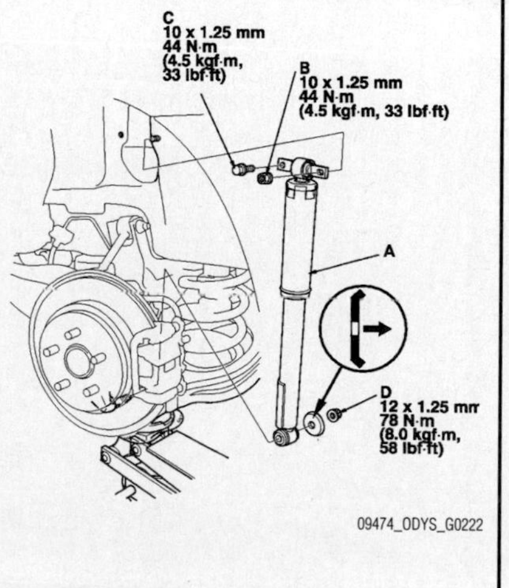
C
10 x 1.25 mm
44 N·m
(4.5 kgf·m,
33 lbf·ft)
B
10 x 1.25 mm
44 N·m
(4.5 kgf·m, 33 lbf·ft)
A
D
12 x 1.25 mm
78 N·m
(8.0 kgf·m,
58 lbf·ft)
09474_ODYS_G0222

Fig. 236 Rear shock absorber

extension. The damper should move smoothly and constantly when compression is released. If it does not, the gas is leaking and the damper should be replaced.

8. Check for oil leaks, abnormal noises, or binding during these tests.

To install:

9. Lower the rear suspension. Compress the damper by hand, and move it into position. Loosely install the flange nut, bolt, and new self-locking nut.

10. Raise the rear suspension with a floor jack until the vehicle just lifts off the safety stands.

11. Tighten the flange bolt and self-locking nut on the bottom of the damper to the specified torque value.

12. Install the rear wheel.

13. Check the rear wheel alignment, and adjust it if necessary.

WHEEL HUB AND BEARING (SEALED UNIT)

REMOVAL & INSTALLATION

See Figure 237.

1. Raise and safely support the vehicle.
2. Remove the rear wheel.
3. Remove the brake caliper bracket mounting bolts, then remove the caliper assembly from the knuckle.

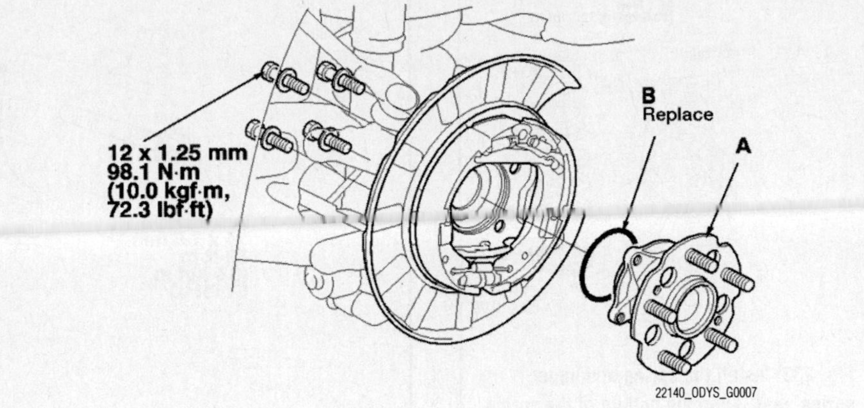
12 x 1.25 mm
98.1 N·m
(10.0 kgf·m,
72.3 lbf·ft)
B
Replace
A
22140_ODYS_G0007

Fig. 237 Install the hub bearing unit (A) with a new O-ring (B)

⁂ WARNING

To prevent damage to the caliper assembly or the brake hose, use a short piece of wire to hang the caliper from the undercarriage. Do not twist the brake hose excessively.

4. Remove the brake rotor.
5. Remove the hub bearing unit and the O-ring.

To install:

6. Installation is the reverse order of removal. Tighten the hub mounting bolts to 72 ft. lbs. (98 Nm).

ADJUSTMENT

The wheel bearings are not adjustable. If they are not within specification, the wheel bearings must be replaced.

HONDA

Pilot

16

AUXILIARY HEATING & AIR CONDITIONING SYSTEM...16-75

Blower Motor 16-75
Removal & Installation........16-75
Heater Core 16-75
Removal & Installation........16-75
HVAC Unit..........16-76
Removal & Installation........16-76

BRAKES 16-9

ANTI-LOCK BRAKE SYSTEM (ABS)..........16-10
Wheel Speed Sensors 16-10
Removal & Installation........16-10
BLEEDING THE BRAKE SYSTEM..........16-9
Bleeding Procedure..........16-9
Bleeding Procedure 16-9
Bleeding the ABS System 16-9
FRONT DISC BRAKES........16-11
Brake Caliper..........16-11
Removal & Installation........16-11
Disc Brake Pads..........16-12
Removal & Installation........16-12
INFORMATION AND PRECAUTIONS 16-9
Anti-lock Systems..........16-9
Disc and Drum Systems 16-9
PARKING BRAKE..........16-16
Parking Brake Cables..........16-16
Adjustment 16-16
Parking Brake Shoes..........16-16
Adjustment 16-17
Removal & Installation........16-16
REAR DISC BRAKES 16-13
Brake Caliper..........16-13
Removal & Installation........16-13
Disc Brake Pads..........16-15
Removal & Installation........16-15

CHASSIS ELECTRICAL 16-18

AIR BAG (SUPPLEMENTAL RESTRAINT SYSTEM) 16-18
General Information..........16-18
Arming the System 16-18
Disarming the System.........16-18
Service Precautions..........16-18

DRIVE TRAIN 16-18

Front Halfshaft..........16-18
Removal & Installation........16-18
Intermediate Shaft 16-19
Removal & Installation........16-19
Rear Driveshaft (Propeller Shaft) 16-20
Removal & Installation........16-20
Rear Halfshafts 16-20
Removal & Installation........16-20
Transfer Case Assembly 16-22
Removal & Installation........16-22

ENGINE COOLING..........16-23

Engine Fan 16-23
Removal & Installation........16-23
Radiator..........16-24
Removal & Installation........16-24
Thermostat 16-26
Removal & Installation........16-26
Water Pump 16-26
Removal & Installation........16-26

ENGINE ELECTRICAL 16-27

CHARGING SYSTEM 16-27
Alternator 16-27
Removal & Installation........16-27
IGNITION SYSTEM 16-28
Firing Order..........16-28
Ignition Coil 16-28
Removal & Installation........16-28
Ignition Timing..........16-28
Adjustment 16-28
Inspection 16-28
Spark Plugs..........16-28
Removal & Installation........16-28
STARTING SYSTEM 16-29
Starter 16-29
Removal & Installation........16-29
Solenoid or Relay Replacement..........16-29

ENGINE MECHANICAL......16-30

Accessory Drive Belts 16-30
Accessory Belt Routing.......16-30
Adjustment 16-30
Inspection 16-30
Removal & Installation........16-30
Camshaft and Valve Lifters......16-30
CKP Pattern Clear/CKP Pattern Learn Procedure...16-34
Inspection 16-30
Removal & Installation........16-31
Catalytic Converter..........16-34
Removal & Installation........16-34
Crankshaft Front Seal..........16-38
Removal & Installation........16-38
Crankshaft Pulley 16-37
Removal & Installation........16-37
Cylinder Head 16-39
CKP Pattern Clear/CKP Pattern Learn Procedure...16-43
Removal & Installation........16-39
Drive Plate..........16-43
Removal & Installation........16-43
Intake Manifold 16-44
Removal & Installation........16-44
Oil Pan 16-46
Removal & Installation........16-46
Oil Pump..........16-47
Inspection 16-49
Removal & Installation........16-47
Piston and Ring..........16-50
Positioning 16-50
Rear Main Seal..........16-51
Removal & Installation........16-51
Rocker Arms..........16-51
Removal & Installation........16-51
Timing Belt & Sprockets 16-53
CKP Pattern Clear/CKP Pattern Learn Procedure...16-56
Removal & Installation........16-53
Timing Belt Front Cover..........16-51
Removal & Installation........16-51
Valve Lash..........16-56
Adjustment 16-56

ENGINE PERFORMANCE & EMISSION CONTROLS 16-58

Accelerator Pedal Position (APP) Sensor 16-58
Location 16-58
Removal & Installation 16-58
Camshaft Position (CMP) Sensor 16-58
Location 16-58
Removal & Installation 16-58
Crankshaft Position (CKP) Sensor 16-58
CKP Pattern Clear/CKP Pattern Learn Procedure 16-58
Removal & Installation 16-58
Engine Coolant Temperature (ECT) Sensor 16-59
Location 16-59
Removal & Installation 16-59
Evaporative Emissions (EVAP) Canister 16-59
Location 16-59
Removal & Installation 16-59
Exhaust Gas Recirculation (EGR) Valve 16-60
Location 16-60
Removal & Installation 16-60
Heated Oxygen (HO2S) Sensor 16-60
Location 16-60
Removal & Installation 16-60
Intake Air Temperature (IAT) Sensor 16-61
Location 16-61
Knock Sensor (KS) 16-61
Location 16-61
Removal & Installation 16-61
Manifold Absolute Pressure (MAP) Sensor 16-62
Location 16-62
Removal & Installation 16-62
Mass Air Flow (MAF) Sensor/ Intake Air Temperature (IAT) Sensor (Hot Wire) 16-62
Location 16-62
Removal & Installation 16-62
Output Shaft Speed (OSS) Sensor 16-62
Location 16-62
Removal & Installation 16-62
Positive Crankcase Ventilation (PCV) Valve 16-63
Location 16-63
Removal & Installation 16-63
Powertrain Control Module (PCM) 16-63
CKP Pattern Clear/CKP Pattern Learn Procedure 16-65
Location 16-63
PCM Idle Learn Procedure 16-65
PCM Update 16-64
Removal & Installation 16-63

FUEL 16-65

GASOLINE FUEL INJECTION SYSTEM 16-65

Fuel Filter 16-66
Removal & Installation 16-67
Fuel Pump/Fuel Gauge Sending Unit 16-67
Removal & Installation 16-67
Fuel Rail and Injector 16-67
Removal & Installation 16-67
Fuel System Pressure 16-65
Relieving 16-65
Fuel System Service Precautions 16-65
Fuel Tank 16-67
Draining 16-67
Removal & Installation 16-68
Fuel Tank Unit 16-69
Removal & Installation 16-69
Idle Speed 16-70
Adjustment 16-70
Throttle Body 16- 71
Removal & Installation 16-71

HEATING & AIR CONDITIONING SYSTEM 16-71

Blower Motor 16-71
Removal & Installation 16-71
Heater Core 16-72
Removal and Installation 16-72
Heater Unit 16-73
Removal & Installation 16-73

SPECIFICATIONS AND MAINTENANCE CHARTS 16-3

Brake Specifications 16-7
Camshaft Specifications 16-5
Capacities 16-4
Crankshaft and Connecting Rod Specifications 16-5
Engine and Vehicle Identification Chart 16-3
Engine Tune-Up Specifications 16-3
Fluid Specifications 16-4
General Engine Specifications 16-3
Piston and Ring Specifications 16-5
Scheduled Maintenance Intervals 16-8
Tire, Wheel and Ball Joint Specifications 16-7
Torque Specifications 16-6
Valve Specifications 16-4
Wheel Alignment 16-6

STEERING 16-77

Power Rack & Pinion Steering Gear 16-77
Removal & Installation 16-77
Power Steering Pump 16-82
Removal & Installation 16-82

SUSPENSION 16-82

FRONT SUSPENSION 16-82

Control Links 16-82
Removal & Installation 16-82
Lower Ball Joint 16-82
Removal & Installation 16-82
Lower Control Arm 16-83
Removal & Installation 16-83
Stabilizer Bar 16-87
Removal & Installation 16-87
Steering Knuckle 16-84
Removal & Installation 16-84
Strut (Damper/Spring) 16-84
Overhaul 16-85
Removal & Installation 16-84
Wheel Hub & Bearing (sealed unit) 16-91
Adjustment 16-91
Removal & Installation 16-91

REAR SUSPENSION 16-92

Coil Spring 16-92
Removal & Installation 16-92
Control Arms/Links 16-92
Removal & Installation 16-92
Shock Absorber (Damper) 16-95
Removal & Installation 16-95
Testing 16-95
Stabilizer Bar 16-96
Removal & Installation 16-96
Wheel Hub & Bearing (sealed unit) 16-96
Adjustment 16-98
Removal & Installation 16-96

SPECIFICATIONS AND MAINTENANCE CHARTS

ENGINE AND VEHICLE IDENTIFICATION CHART

Engine Code							Model Year	
Code	Liters (cc)	Cu. In.	Cyl.	Fuel Sys.	Engine Type	Eng. Mfg.	Code ①	Year
J35Z4	3.5 (3471)	212	6	SMFI	SOHC	Honda	9	2009
							A	2010

SOHC: Single Overhead Cam

SMFI: Sequential Multi-port Fuel Injection

① 10th position of VIN

② Variable Cylinder Management

37647_PLOT_C0001

GENERAL ENGINE SPECIFICATIONS

Year	Model	Engine Displacement Liters (VIN)	Net Horsepower @ rpm	Net Torque @ rpm (ft. lbs.)	Bore x Stroke (in.)	Com-pression Ratio	Oil Pressure @ rpm
2009	Pilot	3.5 (J35Z4)	250@5700	253@4800	3.50x3.66	10.0:1	71@3000
2010	Pilot	3.5 (J35Z4)	250@5700	253@4800	3.50x3.66	10.0:1	71@3000

NA: Not Available

37647_PLOT_C0002

ENGINE TUNE-UP SPECIFICATIONS

Year	Engine Displacement Liters (VIN)	Spark Plug Gap (in.)	Ignition Timing (deg.)		Fuel Pump (psi)	Idle Speed (rpm)		Valve Clearance (in.)	
			MT	AT		MT	AT	In.	Ex.
2009	3.5 (J35Z4)	0.039-0.043	—	8-12B	57-64	—	650-750	0.008-0.009	0.011-0.013
2010	3.5 (J35Z4)	0.039-0.043	—	8-12B	57-64	—	650-750	0.008-0.009	0.011-0.013

NOTE: The Vehicle Emission Control Information label often reflects changes made during production and must be used if they differ from this chart.

NOTE: Pressure with fuel pressure gauge connected

B: Before top dead center

37647_PLOT_C0003

CAPACITIES

Year	Model	Engine Displacement Liters (VIN)	Engine Oil with Filter (qts.)	Transmission (pts.) 5-Spd	Transmission (pts.) Auto.	Transfer Case (pts.)	Drive Axle Front (pts.)	Drive Axle Rear (pts.)	Fuel Tank (gal.)	Cooling System (qts.)
2009	Pilot	3.5 (J35Z4)	4.5	—	3.6	0.9	—	5.6	21.0	8.0
2010	Pilot	3.5 (J35Z4)	4.5	—	3.6	0.9	—	5.6	21.0	8.0

NOTE: All capacities are approximate. Add fluid gradually and check to be sure a proper fluid level is obtained.

37647_PLOT_C0004

FLUID SPECIFICATIONS

Year	Model	Engine Displacement Liters (VIN)	Engine Oil	Auto. Trans.	Drive Axle	Power Steering Fluid	Brake Master Cylinder
2009	Pilot	3.5 (J35Z4)	①	②	③	④	⑤
2010	Pilot	3.5 (J35Z4)	①	②	③	④	⑤

DOT: Department Of Transportation

Note: If specification disagrees with specification in owners manual, use specification in owners manaual

① Honda Motor Oil: 5w-20

② Acura ATF-Z1 fluid

③ Transfer case: API classified GL4 or GL5 only. SAE 90 viscosity.
Rear differential: Honda VTM-4 differential fluid

④ Honda power steering fluid

⑤ Honda DOT 3 Brake Fluid

37647_PLOT_C0014

VALVE SPECIFICATIONS

Year	Engine Displacement Liters (VIN)	Seat Angle (deg.)	Face Angle (deg.)	Spring Test Pressure (lbs. @ in.)	Spring Installed Height (in.)	Stem-to-Guide Clearance (in.) Intake	Stem-to-Guide Clearance (in.) Exhaust	Stem Diameter (in.) Intake	Stem Diameter (in.) Exhaust
2009	3.5 (J35Z4)	NA	NA	NA	NA	0.0008-0.0018	0.0022-0.0032	0.2159-0.2163	0.2146-0.2150
2010	3.5 (J35Z4)	NA	NA	NA	NA	0.0008-0.0018	0.0022-0.0032	0.2159-0.2163	0.2146-0.2150

NA: Not Available

37647_PLOT_C0005

CAMSHAFT SPECIFICATIONS

All measurements in inches unless noted

Year	Model	Engine Displacement Liters (VIN)	Journal Dia.	Brg. Oil Clearance	Shaft End-play	Circle Runout	Lobe Height Intake	Lobe Height Exhaust
2009	Pilot	3.5 (J35Z4)	NA	0.00197-0.0035	0.0020-0.0079	NA	①	②
2010	Pilot	3.5 (J35Z4)	NA	0.00197-0.0035	0.0020-0.0079	NA	①	②

NA: Not Available

① Cylinders 1,2,3,4: 1.38433 inches

Cylinders 5, 6: 1.38405 inches

② Cylinders 1,2,3,4: 1.43846 inches

Cylinders 5, 6: 1.43748 inches

37647_PLOT_C0006

CRANKSHAFT AND CONNECTING ROD SPECIFICATIONS

All measurements are given in inches

	Engine	Crankshaft				Connecting Rod		
Year	Displacement Liters (VIN)	Main Brg. Journal Dia.	Main Brg. Oil Clearance	Shaft End-play	Thrust on No.	Journal Diameter	Oil Clearance	Side Clearance
2009	3.5 (J35Z4)	2.8337-2.8346	0.0008-0.0018	0.0039-0.0138	3	2.283	0.0008-0.0017	0.0059-0.0138
2010	3.5 (J35Z4)	2.8337-2.8346	0.0008-0.0018	0.0039-0.0138	3	2.283	0.0008-0.0017	0.0059-0.0138

37647_PLOT_C0007

PISTON AND RING SPECIFICATIONS

All measurements are given in inches

	Engine		Ring Gap			Ring Side Clearance		
Year	Displacement Liters (VIN)	Piston Clearance	Top Compression	Bottom Compression	Oil Control	Top Compression	Bottom Compression	Oil Control
2009	3.5 (J35Z4)	0.0006-0.0016	0.0079-0.0138	0.0157-0.0217	0.0079-0.0276	0.0022-0.0032	0.0012-0.0022	NA
2010	3.5 (J35Z4)	0.0006-0.0016	0.0079-0.0138	0.0157-0.0217	0.0079-0.0276	0.0022-0.0032	0.0012-0.0022	NA

NA: Not Available

37647_PLOT_C0008

TORQUE SPECIFICATIONS

All readings in ft. lbs.

Year	Engine Displacement Liters (VIN)	Cylinder Head Bolts	Main Bearing Bolts	Rod Bearing Bolts	Crankshaft Damper Bolts	Flywheel Bolts	Manifold Intake	Manifold Exhaust	Spark Plugs	Oil Pan Drain Plug
2009	3.5 (J35Z4)	①	②	③	④	54	16	23	16	29
2010	3.5 (J35Z4)	①	②	③	④	54	16	23	16	29

NOTE: Dip main bearing bolts and crankshaft damper bolt in clean engine oil prior to tightening.

① Step 1: 22 ft. lbs.
Step 2: plus 90 degrees
Step 3: plus 90 degrees
Step 4: If using a new bolt, plus 90 degrees

② Cap bolts: 54 ft. lbs.
Side bolts: 36 ft. lbs.

③ Step 1: 15 ft. lbs.
Step 2: 90 degrees

④ Step 1: 47 ft. lbs.
Step 2: plus 60 degrees

37647_PLOT_C0009

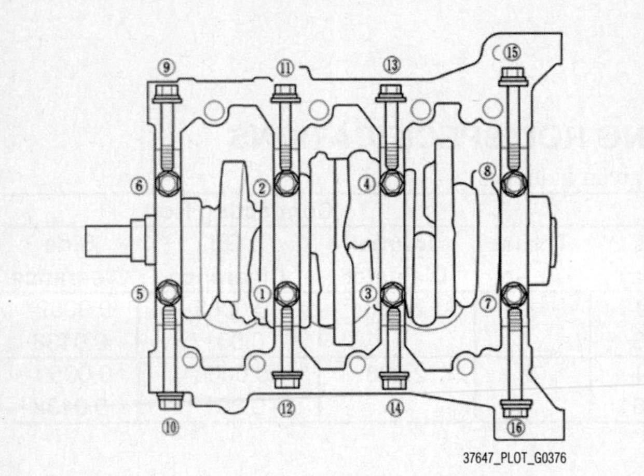
9
11
13
15
6
2
4
8
5
1
3
7
10
12
14
16
37647_PLOT_G0376

Fig. 1 Main bearing torque sequence for Honda J35Z4 engine

WHEEL ALIGNMENT

Year	Model		Caster Range (+/-Deg.)	Caster Preferred Setting (Deg.)	Camber Range (+/-Deg.)	Camber Preferred Setting (Deg.)	Toe-in (in.)
2009	Pilot	F	0.35	4.11	0.45	-0.30	0+/-0.08
		R	—	—	0	-0.30	0.08+/-0.08
2010	Pilot	F	0.35	4.11	0.45	-0.30	0+/-0.08
		R	—	—	0	-0.30	0.08+/-0.08

37647_PLOT_C0010

TIRE, WHEEL AND BALL JOINT SPECIFICATIONS

Year	Model	OEM Tires Standard	OEM Tires Optional	Tire Pressures (psi) Front	Tire Pressures (psi) Rear	Wheel Size	Ball Joint Inspection	Lug Nut (ft. lbs.)
2009	Pilot	P245/65R17	None	33	33	R17	NS	80
2010	Pilot	P245/65R17	None	33	33	R17	NS	80

OEM: Original Equipment Manufacturer

PSI: Pounds Per Square Inch

NS: Not specified by manufacturer

37647_PLOT_C0011

BRAKE SPECIFICATIONS

All measurements in inches unless noted

Year	Model		Brake Disc Original Thickness	Brake Disc Minimum Thickness	Brake Disc Maximum Runout	Brake Drum Diameter Original Inside Diameter	Brake Drum Diameter Max. Wear Limit	Brake Drum Diameter Maximum Machine Diameter	Minimum Lining Thickness Front	Minimum Lining Thickness Rear	Brake Caliper Bracket Bolts (ft. lbs.)	Brake Caliper Mounting Bolts (ft. lbs.)
2009	Pilot	F	1.100	1.024	0.0016	—	—	—	0.063	—	101	53
		R	0.430	0.354	0.0016	—	—	—	—	0.063	65	27
2010	Pilot	F	1.100	1.024	0.0016	—	—	—	0.063	—	101	53
		R	0.430	0.354	0.0016	—	—	—	—	0.063	65	27

F: Front

R: Rear

37647_PLOT_C0012

MAINTENANCE MINDER SCHEDULE

Honda Pilot

All Honda's displays engine oil life and maintenance service items in the information display to indicate when to perform maintenance service. If the engine oil life is 15% or less, based on the onboard computer's caluculations, you will see SERVICE DUE SOON in the information display every time the ignition key is turned to ON. The maintenance minder indicator will also come on and the maintenance code(s) for other scheduled maintenance items needing service will be displayed below the message.

Symbol	Item	Service
A	Engine oil ①	Change
B	Engine oil and filter	Change
	Fluid levels	Inspect
	Brakes	Inspect
	Parking brake adjustment	Check
	Steering gear and linkage	Inspect
	Suspension components	Inspect
	Driveshaft boots	Inspect
	Brake hoses and lines	Inspect
	Exhaust system	Inspect
	Fuel lines and connections	Inspect
1	Tires	Rotate
2	Engine air filter ②	Replace
	Dust and pollen filter ③	Replace
	Accessory drive belt	Inspect
3	Transmission fluid ④	Replace
	Transfer case fluid ④	Replace
4	Spark plugs	Replace
	Timing belt ⑤	Replace
	Water pump	Inspect
	Valve clearance ⑥	Inspect
5	Engine coolant	Replace
6	VTM-4 rear differential fluid	Replace

① If the message SERVICE DUE NOW does not appear more than 12 months after the display is reset, change every year.
② If driven in dusty conditions, replace every 15,000 miles.
③ If driven in urban areas that have a high concentration of soot from industry and diesel, replace every 15,000 miles
④ If regularly driven in mountainous areas at very low speed or trailer towing, change the fluid every 30,000 miles.
⑤ If driven regularly in temperatures over 110 deg.F or below -20 deg.F, or towing a trailer, replace every 60,000 miles.
⑥ Adjust if necessary.

Additionally, replace the brake fluid every 3 years, and inspect the idle speed every 160,000 miles.

To reset the Engine Oil Life Display:

1. Turn the ignition switch to ON.
2. Press the SELECT button repeatedly until the engine oil life display or the service message is displayed.
3. Press the RESET button for about 10 seconds. You will see a MAINT RESET message.
4. Select the appropriate answer, MAINT RESET >N (NO) or MAINT RESET > y (YES) by pressing the SELECT button repeatedly.
   >N or >Y is displayed on the outside temperature >N or >Y is displayed on the outside temperature display.
5. Select the MAINT RESET > Y (YES), and press and hold the RESET button again to reset the engine oil life to 100%.

37647_PLOT_C0013

BRAKES — INFORMATION AND PRECAUTIONS

ANTI-LOCK SYSTEMS

- Certain components within the ABS system are not intended to be serviced or repaired individually.
- Do not use rubber hoses or other parts not specifically specified for and ABS system. When using repair kits, replace all parts included in the kit. Partial or incorrect repair may lead to functional problems and require the replacement of components.
- Lubricate rubber parts with clean, fresh brake fluid to ease assembly. Do not use shop air to clean parts; damage to rubber components may result.
- Use only DOT 3 brake fluid from an unopened container.
- If any hydraulic component or line is removed or replaced, it may be necessary to bleed the entire system.
- A clean repair area is essential. Always clean the reservoir and cap thoroughly before removing the cap. The slightest amount of dirt in the fluid may plug an orifice and impair the system function. Perform repairs after components have been thoroughly cleaned; use only denatured alcohol to clean components. Do not allow ABS components to come into contact with any substance containing mineral oil; this includes used shop rags.
- The Anti-Lock control unit is a microprocessor similar to other computer units in the vehicle. Ensure that the ignition switch is **OFF** before removing or installing controller harnesses. Avoid static electricity discharge at or near the controller.
- If any arc welding is to be done on the vehicle, the control unit should be unplugged before welding operations begin.

DISC AND DRUM SYSTEMS

CAUTION

Dust and dirt accumulating on brake parts during normal use may contain asbestos fibers from production or aftermarket brake linings. Breathing excessive concentrations of asbestos fibers can cause serious bodily harm. Exercise care when servicing brake parts. Do not sand or grind brake lining unless equipment used is designed to contain the dust residue. Do not clean brake parts with compressed air or by dry brushing. Cleaning should be done by dampening the brake components with a fine mist of water, then wiping the brake components clean with a dampened cloth. Dispose of cloth and all residue containing asbestos fibers in an impermeable container with the appropriate label. Follow practices prescribed by the Occupational Safety and Health Administration (OSHA) and the Environmental Protection Agency (EPA) for the handling, processing, and disposing of dust or debris that may contain asbestos fibers.

BRAKES — BLEEDING THE BRAKE SYSTEM

BLEEDING PROCEDURE

BLEEDING PROCEDURE

See Figure 2.

➡Do not spill brake fluid on the vehicle; it may damage the paint. If brake fluid gets on the paint, wash it off immediately with water.

➡Note the following when bleeding the brake system:

- Do not reuse the drained fluid. Use only new Honda DOT 3 Brake Fluid from an unopened container. Using a non-Honda brake fluid can cause corrosion and shorten the life of the system.
- Make sure no dirt or other foreign matter is allowed to contaminate the brake fluid.
- The reservoir connected to the master cylinder must be at the MAX (upper) level mark at the start of the bleeding procedure, and checked after bleeding each wheel location. Add fluid as required.

1. Make sure the brake fluid level in the reservoir is at the MAX (upper) level line.
2. Have someone slowly pump the brake pedal several times, then apply steady pressure.
3. Start the bleeding at the driver's side of the front brake system.

➡Bleed the calipers in the sequence shown.

4. Attach a length of clear drain tube to the bleed screw, then loosen the bleed screw to allow air to escape from the system. Then tighten the bleed screw securely.
5. Refill the master cylinder reservoir to the MAX (upper) level line.
6. Repeat the procedure for each brake circuit until there are no air bubbles in the fluid.

BLEEDING THE ABS SYSTEM

Bleeding the ABS brake system is done in the same manner as bleeding a conventional brake system.

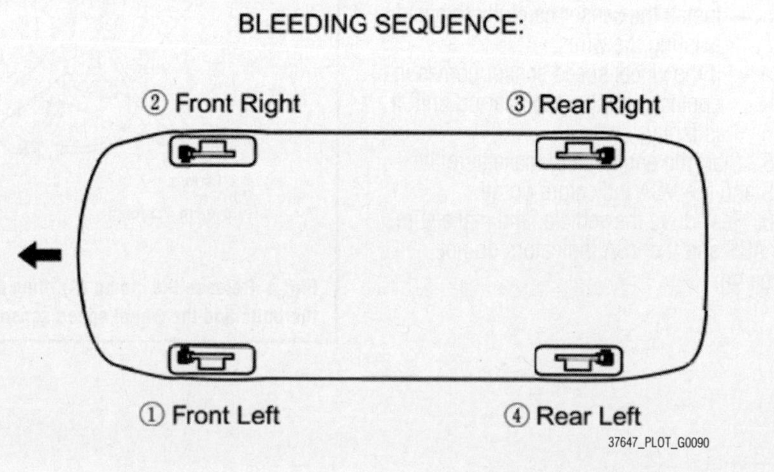
BLEEDING SEQUENCE:
② Front Right
③ Rear Right
① Front Left
④ Rear Left
37647_PLOT_G0090

Fig. 2 Brake bleeding sequence

WHEEL SPEED SENSORS

REMOVAL & INSTALLATION

Front

See Figure 3.

1. Turn the ignition switch to LOCK (0).
2. Release the clamp, then disconnect the wheel speed sensor connector.
3. Remove the bolts and the wheel speed sensor.
4. Install the wheel speed sensor in the reverse order of removal, and note these items:
    - Install the sensor carefully to avoid twisting the wires.
    - If the wheel speed sensor comes in contact with the hub bearing unit, it is faulty.
5. Start the engine, and make sure the ABS and the VSA indicators go off.
6. Test-drive the vehicle, and make sure the ABS and the VSA indicators do not come on.

Rear

See Figure 4.

1. Turn the ignition switch to LOCK (0).
2. Release the clamp, then disconnect the wheel speed sensor connector.
3. Remove the bolts and the wheel speed sensor.
4. Install the wheel speed sensor in the reverse order of removal, and note these items:
    - Install the sensor carefully to avoid twisting the wires.
    - If the wheel speed sensor comes in contact with the hub bearing unit, it is faulty.
5. Start the engine, and make sure the ABS and the VSA indicators go off.
6. Test-drive the vehicle, and make sure the ABS and the VSA indicators do not come on.

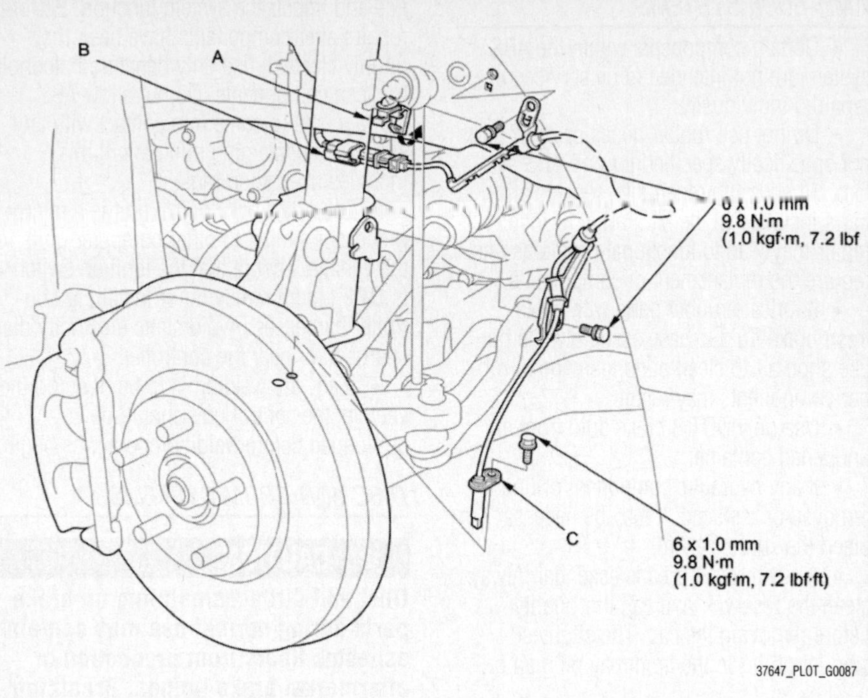
B
A
9.8 N·m
(1.0 kgf·m, 7.2 lbf
C
6 x 1.0 mm
9.8 N·m
(1.0 kgf·m, 7.2 lbf·ft)
37647_PLOT_G0087

Fig. 3 Release the clamp (A), then disconnect the wheel speed sensor connector (B); remove the bolts and the wheel speed sensor (C)

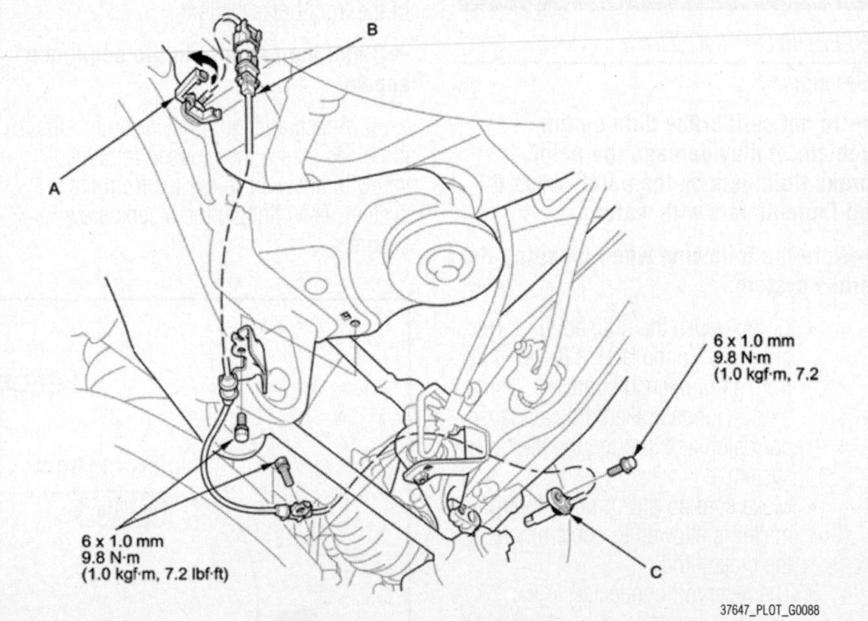
B
A
6 x 1.0 mm
9.8 N·m
(1.0 kgf·m, 7.2
6 x 1.0 mm
9.8 N·m
(1.0 kgf·m, 7.2 lbf·ft)
C
37647_PLOT_G0088

Fig. 4 Release the clamp (A), then disconnect the wheel speed sensor connector (B); remove the bolts and the wheel speed sensor (C)

BRAKE CALIPER

REMOVAL & INSTALLATION

See Figures 5 and 6.

➡Keep any grease off the brake disc and brake pads.

1. Raise the front of the vehicle, and support it with safety stands in the proper locations.
2. Remove the front wheel.
3. Remove the brake hose mounting bolt.
4. Remove the brake caliper bracket

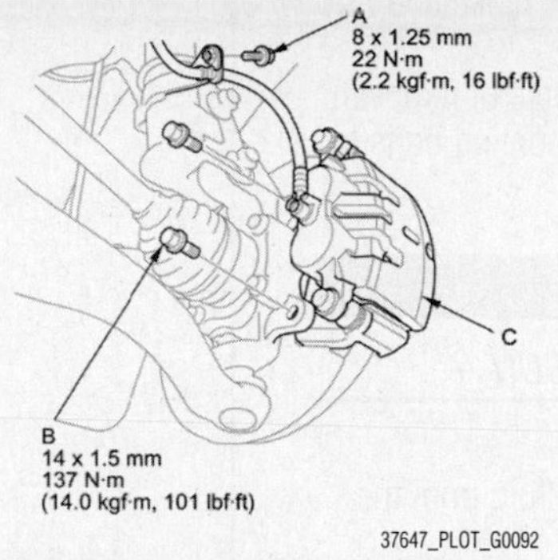
A
8 x 1.25 mm
22 N·m
(2.2 kgf·m, 16 lbf·ft)
C
B
14 x 1.5 mm
137 N·m
(14.0 kgf·m, 101 lbf·ft)
37647_PLOT_G0092

Fig. 5 Remove the brake hose mounting bolt (A), the brake caliper bracket mounting bolts (B), then remove the caliper assembly (C)

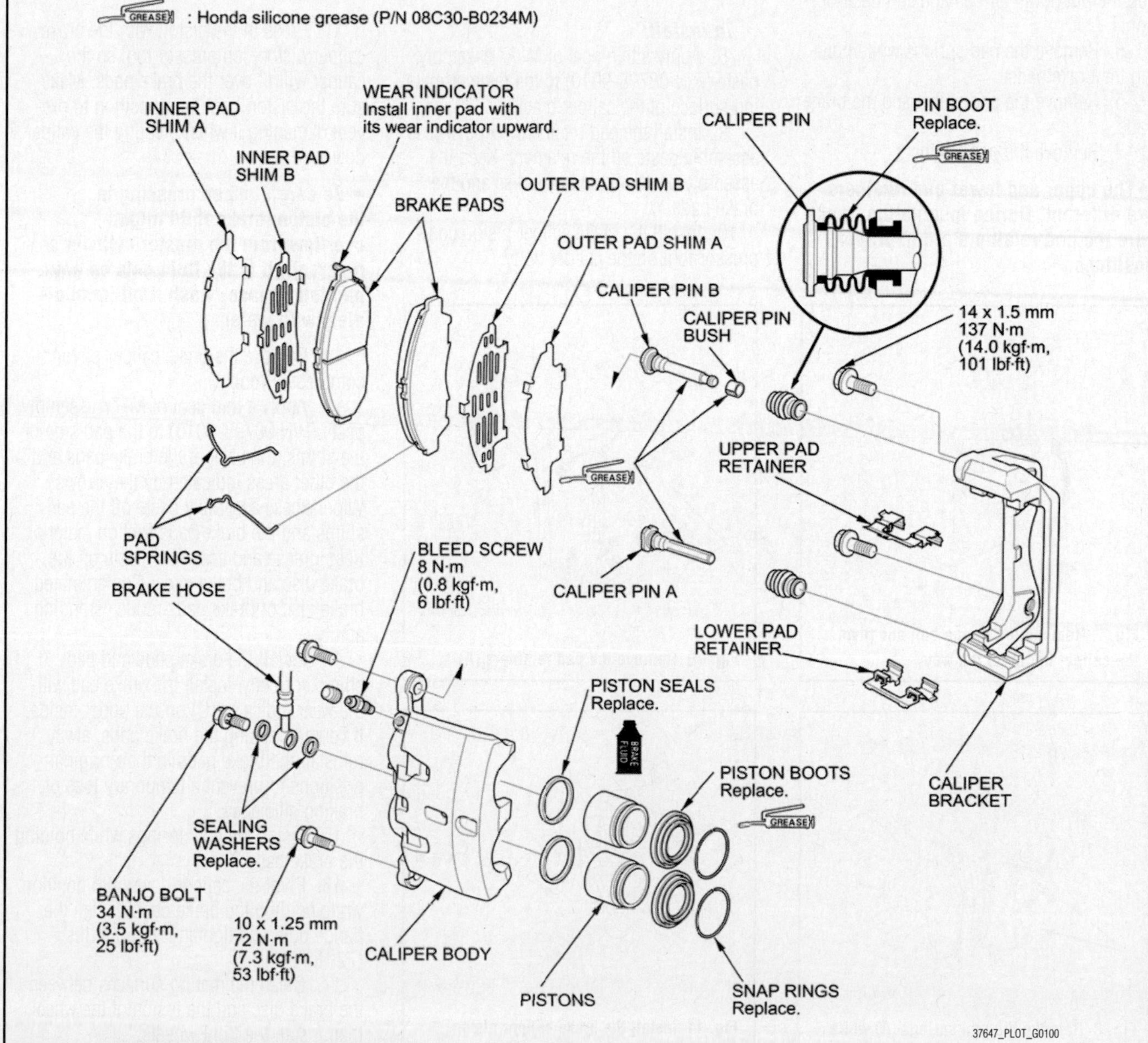
GREASE : Honda silicone grease (P/N 08C30-B0234M)
INNER PAD SHIM A
INNER PAD SHIM B
WEAR INDICATOR Install inner pad with its wear indicator upward.
BRAKE PADS
OUTER PAD SHIM B
OUTER PAD SHIM A
CALIPER PIN
PIN BOOT Replace.
GREASE
CALIPER PIN B
CALIPER PIN BUSH
14 x 1.5 mm
137 N·m
(14.0 kgf·m,
101 lbf·ft)
UPPER PAD RETAINER
GREASE
PAD SPRINGS
BLEED SCREW
8 N·m
(0.8 kgf·m,
6 lbf·ft)
CALIPER PIN A
BRAKE HOSE
LOWER PAD RETAINER
PISTON SEALS Replace.
BRAKE FLUID
PISTON BOOTS Replace.
GREASE
CALIPER BRACKET
SEALING WASHERS Replace.
BANJO BOLT
34 N·m
(3.5 kgf·m,
25 lbf·ft)
10 x 1.25 mm
72 N·m
(7.3 kgf·m,
53 lbf·ft)
CALIPER BODY
PISTONS
SNAP RINGS Replace.
37647_PLOT_G0100

Fig. 6 Exploded view of the front caliper assembly

mounting bolts, then remove the caliper assembly from the knuckle.

5. Disconnect the flexible brake line from the caliper.

6. Installation is the reverse of removal.

7. Tighten the caliper mounting bolts to 101 ft. lbs. (137 Nm).

DISC BRAKE PADS

REMOVAL & INSTALLATION

See Figures 7 through 12.

1. Remove some brake fluid from the master cylinder.
2. Raise and support the vehicle.
3. Remove the front wheels.
4. Remove the flange bolt, and pivot the caliper up out of the way. Check the hose and the pin boots for damage and deterioration.
5. Remove the pad springs while holding the brake pads.
6. Remove the pad shims and the brake pads.
7. Remove the pad retainers.

➡The upper and lower pad retainers are different. During installation, make sure the pad retainers are in proper positions.

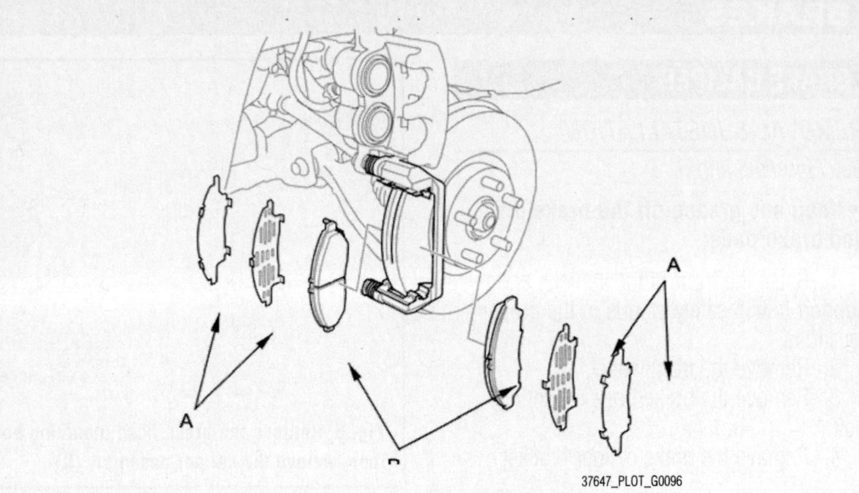
A
A
37647_PLOT_G0096

Fig. 9 Remove the pad shims (A) and the brake pads (B)

To install:

8. Apply a thin coat of M-77 assembly paste (P/N 08798-9010) to the retainer mating surface of the caliper bracket.
9. Install the pad retainers. Wipe excess assembly paste off the retainers. Keep the assembly paste off the brake disc and the brake pads.
10. Install the brake caliper piston compressor tool on the caliper body.
11. Press in the piston with the brake caliper piston compressor tool so the caliper will fit over the brake pads. Make sure the piston boot is in position to prevent damaging it when pivoting the caliper down.

➡Be careful when pressing in the piston; brake fluid might overflow from the master cylinder's reservoir. If brake fluid gets on any painted surface, wash it off immediately with water.

12. Remove the brake caliper piston compressor tool.
13. Apply a thin coat of M-77 assembly paste (P/N 08798-9010) to the pad side of the shims, the back of the brake pads and the other areas indicated by the arrows. Wipe excess assembly paste off the pad shims and the brake pads friction material. Keep grease and assembly paste off the brake disc and brake pads. Contaminated brake disc or brake pads reduce stopping ability.
14. Install the brake pads and pad shims correctly. Install the brake pad with the wear indicator (C) on the upper inside. If you are reusing the brake pads, always reinstall the brake pads in their original positions to prevent a temporary loss of braking efficiency.
15. Install the pad springs while holding the brake pads.
16. Pivot the caliper down into position. while holding the brake pads. Install the flange bolt, and tighten it to 53 ft. lbs. (72 Nm).
17. Clean the mating surfaces between the brake disc and the inside of the wheel, then install the front wheels.

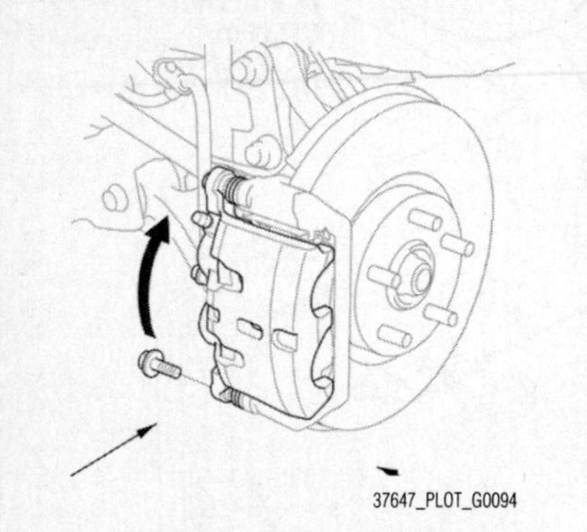
37647_PLOT_G0094

Fig. 7 Remove the flange bolt and pivot the caliper up out of the way

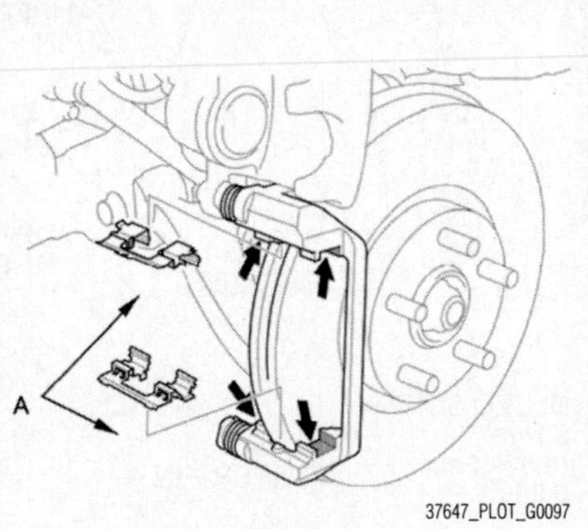
A
37647_PLOT_G0097

Fig. 10 Remove the pad retainers (A)

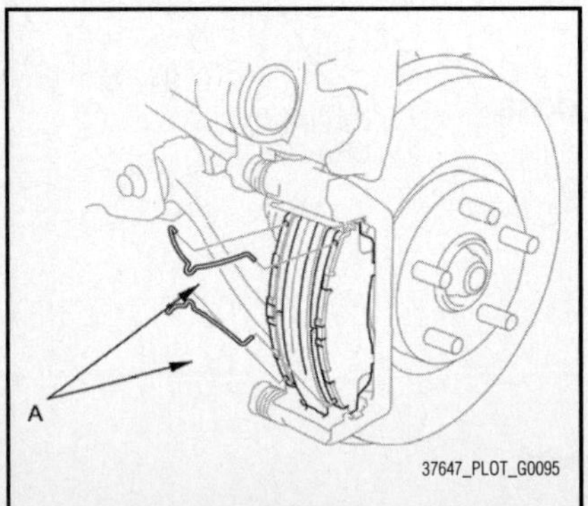
A
37647_PLOT_G0095

Fig. 8 Remove the pad springs (A) while holding the brake pads

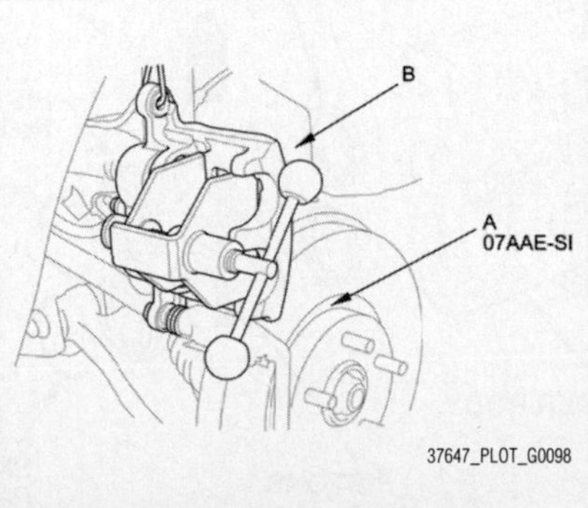
B
A
07AAE-SI
37647_PLOT_G0098

Fig. 11 Install the brake caliper piston compressor tool (A) on the caliper body (B)

18. Press the brake pedal several times to make sure the brakes work.

➡Engagement may require a greater pedal stroke immediately after the brake pads have been replaced as a set. Several applications of the brake pedal will restore the normal pedal stroke.

19. Add brake fluid as needed.
20. After installation, check for leaks at hose and line joints or connections, and retighten if necessary.
21. Test-drive the vehicle, then recheck for leaks.

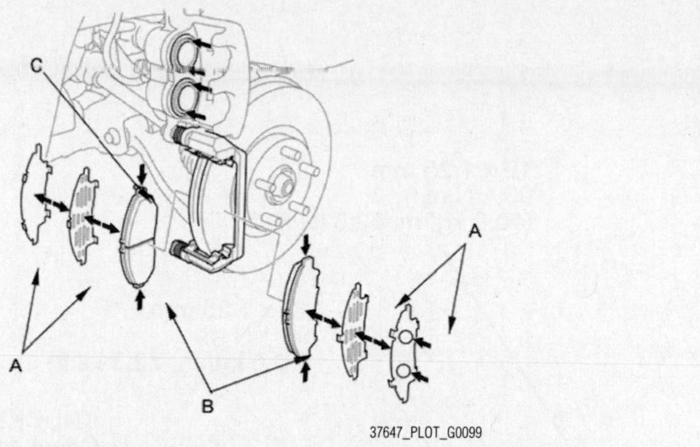
C
A
A
B
37647_PLOT_G0099

Fig. 12 Apply a thin coat of M-77 assembly paste (P/N 08798-9010) to the pad side of the shims (A), the back of the brake pads (B) and the other areas indicated by the arrows

BRAKES — REAR DISC BRAKES

BRAKE CALIPER

REMOVAL & INSTALLATION

See Figures 13 through 15.

➡Keep any grease off the brake disc/drum and brake pads.

1. Raise the rear of the vehicle, and support it with safety stands in the proper locations.
2. Remove the rear wheel.
3. Release the parking brake pedal fully.
4. Remove the brake caliper bracket mounting bolts, and remove the caliper assembly from the knuckle.
5. Disconnect the flexible brake line from the caliper.
6. Installation is the reverse of removal.
7. Tighten the caliper mounting bolts to 65 ft. lbs. (88 Nm).

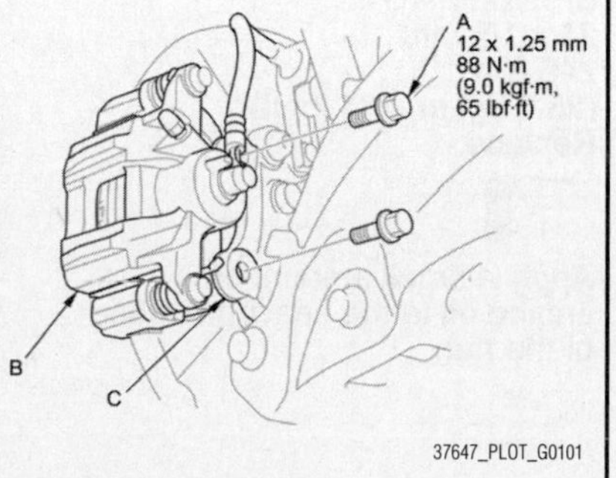
A
12 x 1.25 mm
88 N·m
(9.0 kgf·m,
65 lbf·ft)
B
C
37647_PLOT_G0101

Fig. 13 Remove the brake caliper bracket mounting bolts (A), and remove the caliper assembly (B) from the knuckle

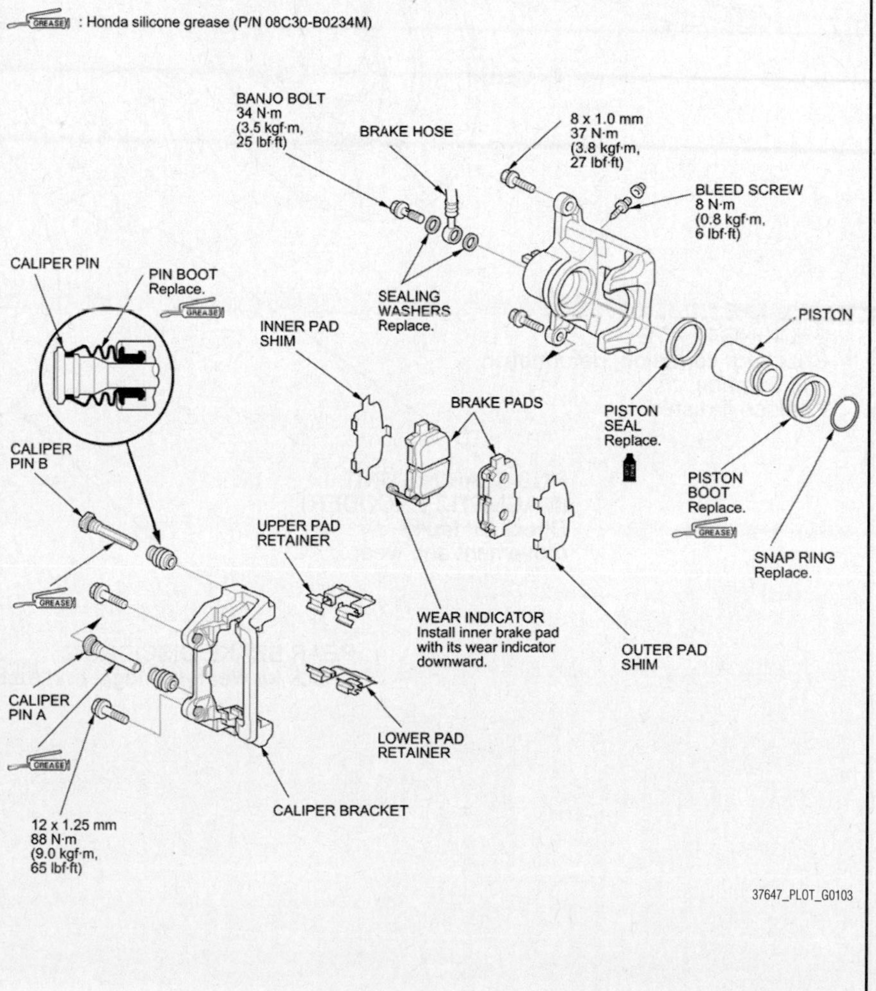
GREASE : Honda silicone grease (P/N 08C30-B0234M)
BANJO BOLT
34 N·m
(3.5 kgf·m,
25 lbf·ft)
BRAKE HOSE
8 x 1.0 mm
37 N·m
(3.8 kgf·m,
27 lbf·ft)
BLEED SCREW
8 N·m
(0.8 kgf·m,
6 lbf·ft)
CALIPER PIN
PIN BOOT
Replace.
SEALING
WASHERS
Replace.
INNER PAD
SHIM
PISTON
BRAKE PADS
PISTON
SEAL
Replace.
CALIPER
PIN B
PISTON
BOOT
Replace.
UPPER PAD
RETAINER
SNAP RING
Replace.
WEAR INDICATOR
Install inner brake pad
with its wear indicator
downward.
OUTER PAD
SHIM
CALIPER
PIN A
LOWER PAD
RETAINER
CALIPER BRACKET
12 x 1.25 mm
88 N·m
(9.0 kgf·m,
65 lbf·ft)
37647_PLOT_G0103

Fig. 14 Exploded view of the rear caliper assembly

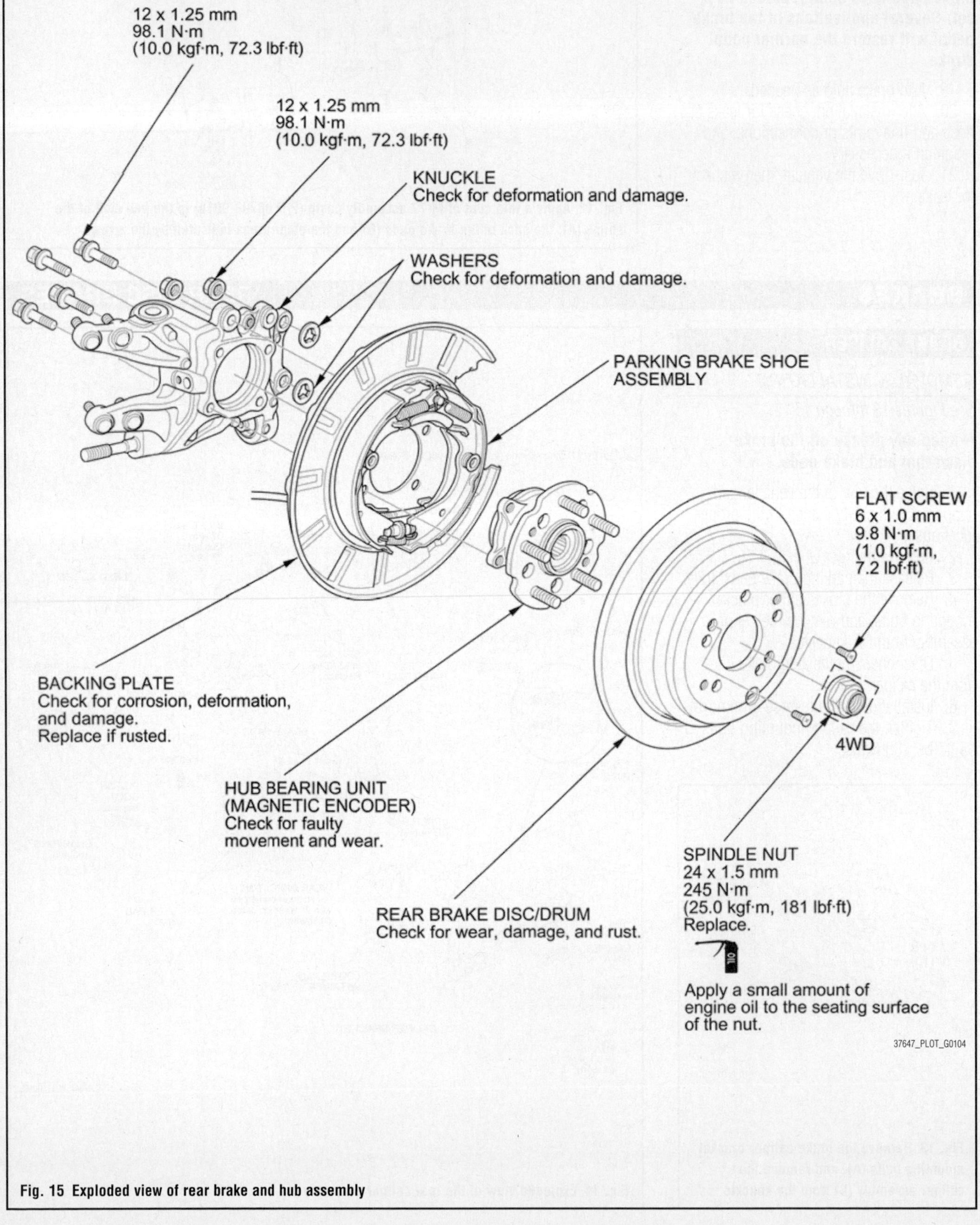
12 x 1.25 mm
98.1 N·m
(10.0 kgf·m, 72.3 lbf·ft)
12 x 1.25 mm
98.1 N·m
(10.0 kgf·m, 72.3 lbf·ft)
KNUCKLE
Check for deformation and damage.
WASHERS
Check for deformation and damage.
PARKING BRAKE SHOE ASSEMBLY
FLAT SCREW
6 x 1.0 mm
9.8 N·m
(1.0 kgf·m,
7.2 lbf·ft)
BACKING PLATE
Check for corrosion, deformation, and damage.
Replace if rusted.
4WD
HUB BEARING UNIT
(MAGNETIC ENCODER)
Check for faulty movement and wear.
SPINDLE NUT
24 x 1.5 mm
245 N·m
(25.0 kgf·m, 181 lbf·ft)
Replace.
OIL
Apply a small amount of engine oil to the seating surface of the nut.
REAR BRAKE DISC/DRUM
Check for wear, damage, and rust.
37647_PLOT_G0104

Fig. 15 Exploded view of rear brake and hub assembly

DISC BRAKE PADS

REMOVAL & INSTALLATION

See Figures 16 through 20.

1. Remove some brake fluid from the master cylinder.
2. Raise and support the vehicle.
3. Remove the rear wheels.
4. Remove the flange bolt and pivot the caliper up out of the way. Check the hose and pin boots for damage and deterioration.
5. Remove the pad shims and the brake pads.
6. Remove the upper and lower pad retainers.

➡The upper and lower pad retainers are different. During installation, make sure the pad retainers are in their proper positions.

7. Clean the caliper bracket thoroughly; remove any rust, and check for grooves and cracks. Verify that the caliper pins move in and out smoothly. Clean and lube if needed.
8. Apply a thin coat of M-77 assembly paste (P/N 08798-9010) to the retainer mating surface of the caliper bracket (indicated by the arrows).
9. Install the upper and lower pad retainers. Wipe excess assembly paste off the retainers. Keep the assembly paste off the brake discs/drum and brake pads.
10. Install the brake caliper piston compressor tool on the caliper body.
11. Press in the piston with the brake caliper piston compressor tool so the caliper will fit over the brake pads. Make sure the piston boot is in position to prevent damaging it when pivoting the caliper down.

➡Be careful when pressing in the piston; brake fluid might overflow from the master cylinder's reservoir. If brake fluid gets on any painted surface, wash it off immediately with water.

12. Remove the brake caliper piston compressor tool.
13. Apply a thin coat of M-77 assembly paste (P/N 08798-9010) to the pad side of the shims, the back of the brake pads, and the other areas indicated by the arrows. Wipe excess assembly paste from the pad shims and brake pads friction material. Keep grease and assembly paste away from the brake disc/drum and brake pads. Contaminated brake disc/drum or brake pads reduce stopping ability.
14. Install the brake pads and pad shims correctly. Install the brake pad with the wear indicator on the bottom inside. If you are reusing the brake pads, always reinstall the brake pads in their original positions to prevent a temporary loss of braking efficiency.
15. Pivot the caliper down into position. Install the flange bolt, and tighten it to 27 ft. lbs. (37 Nm).
16. Clean the mating surfaces between the brake disc/drum and the inside of the wheel, then install the rear wheels.
17. Press the brake pedal several times to make sure the brakes work.

➡Engagement may require a greater pedal stroke immediately after the brake pads have been replaced as a set. Several applications of the brake pedal will restore the normal pedal stroke.

18. Add brake fluid as needed.
19. After installation, check for leaks at hose and line joints or connections, and retighten if necessary.
20. Test-drive the vehicle, then check for leaks.

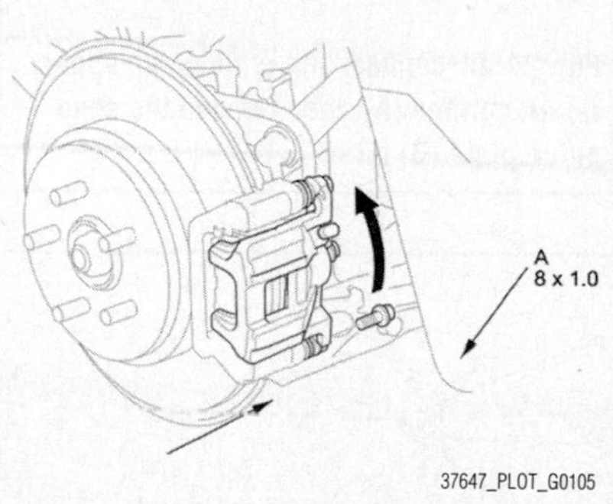
A
8 x 1.0
37647_PLOT_G0105

Fig. 16 Remove the flange bolt (A), and pivot the caliper up out of the way

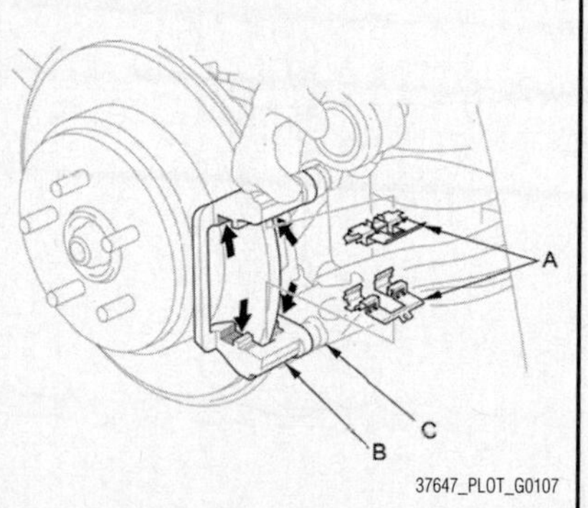
A
B
C
37647_PLOT_G0107

Fig. 18 Remove the upper and lower pad retainers (A)

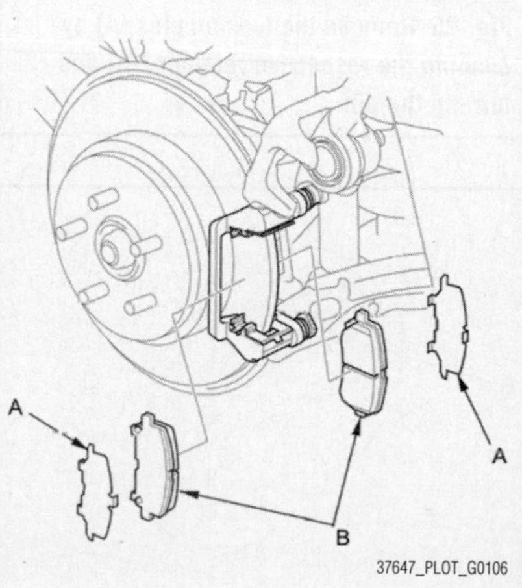
A
A
B
37647_PLOT_G0106

Fig. 17 Remove the pad shims (A) and the brake pads (B)

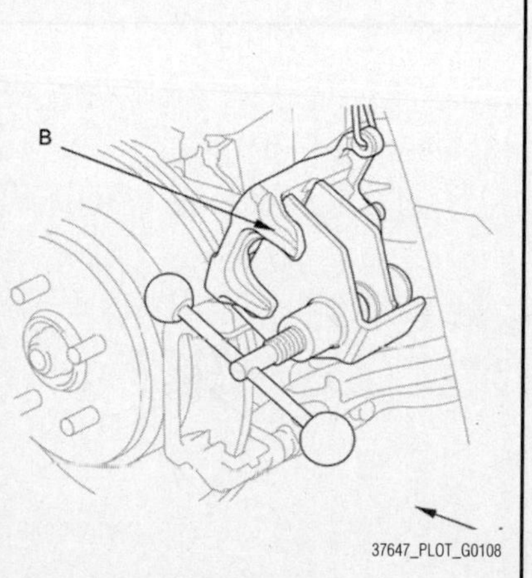
B
37647_PLOT_G0108

Fig. 19 Install the brake caliper piston compressor tool (A) on the caliper body (B)

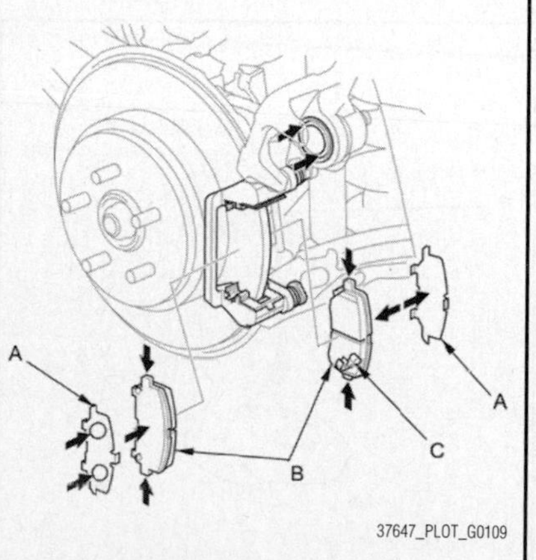
A
A
B
C
37647_PLOT_G0109

Fig. 20 Apply a thin coat of M-77 assembly paste to the pad side of the shims (A), the back of the brake pads (B), and the other areas indicated by the arrows

PARKING BRAKE CABLES

ADJUSTMENT

1. Press the parking brake pedal with enough force to fully apply the parking brake. The parking brake pedal should be locked within 8 to 10 clicks.
2. If the number of pedal clicks is not as specified, adjust the parking brake.

➡Minor parking brake pedal adjustments (1 to 2 clicks) can be made with the adjusting nut (see parking brake minor adjustment). If a larger adjustment is required, follow the major adjustment procedure using the adjuster nut at the parking brake drum (see parking brake major adjustment). After installing new parking brake shoes and/or new brake disc/drum, make sure you drive the vehicle for "break-in".

Minor Adjustment

See Figure 21.

1. Raise the rear of the vehicle, and support it with safety stands in the proper locations.
2. Release the parking brake pedal fully.
3. Press the parking brake pedal 1 click.
4. Tighten the parking brake adjusting nut until the parking brakes drag slightly when the rear wheels are turned.
5. Release the parking brake pedal fully, and check that the parking brakes do not drag when the rear wheels are turned. Readjust if necessary.
6. Make sure the parking brake pedal is fully applied within 8 to 10 clicks.

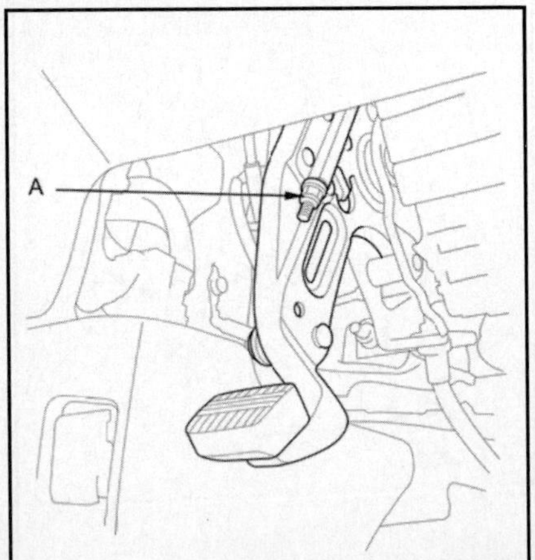
A
37647_PLOT_G0110

Fig. 21 Tighten the parking brake adjusting nut (A)

Major Adjustment (to be done when replacing parking brake shoes and after lining surface break-in)

See Figures 22 and 23.

1. Raise the rear of the vehicle, and support it with safety stands in the proper locations.
2. Release the parking brake pedal fully.
3. Loosen the parking brake adjusting nut.
4. Remove the rear wheels.
5. Remove the access plug.
6. Turn the ratchet teeth on the adjuster nut with a flat-tipped screwdriver until the shoes lock against the parking brake drum. Then back off the adjuster 10 clicks, and install the access plug.
7. Clean the mating surfaces between the brake disc/drum and the inside of the wheel, then install the rear wheels.

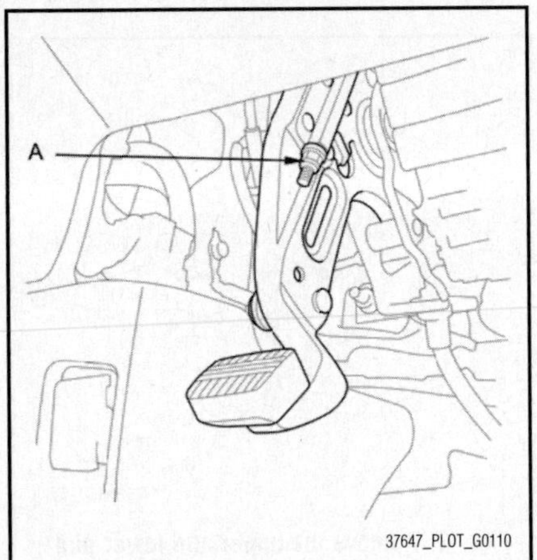
A
37647_PLOT_G0110

Fig. 22 Loosen the parking brake adjusting nut (A)

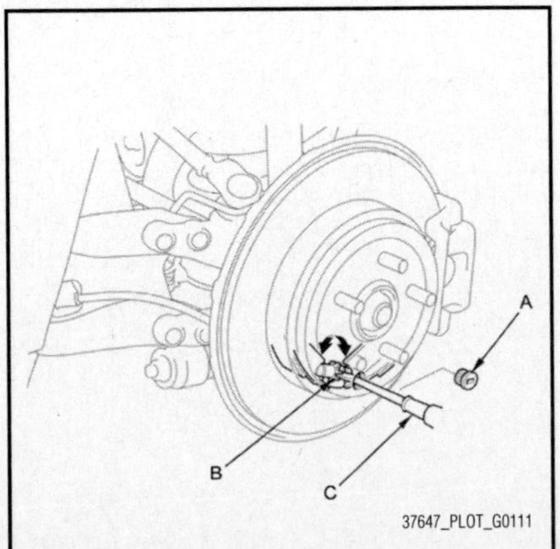
A
B
C
37647_PLOT_G0111

Fig. 23 Remove the access plug (A); turn the ratchet teeth on the adjuster nut (B) with a flat-tipped screwdriver (C)

8. Do the minor adjustment procedure.

PARKING BRAKE SHOES

REMOVAL & INSTALLATION

See Figures 24 through 30.

1. Raise and support the vehicle.
2. Remove the rear wheels.
3. Release the parking brake, and remove the rear brake disc/drum.

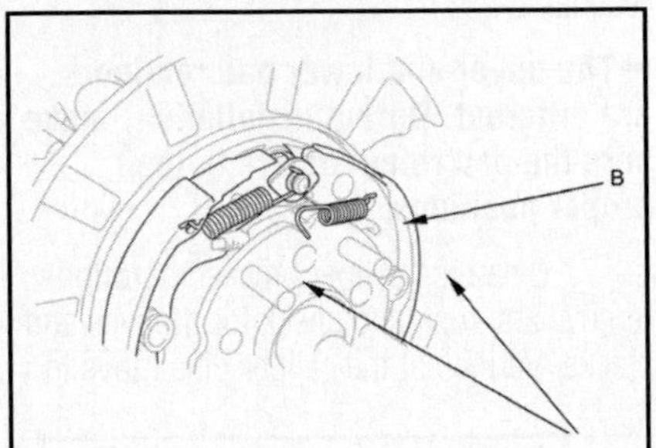
B
37647_PLOT_G0112

Fig. 24 Disconnect and remove the upper return springs (A), then remove the shoe guide plate (B)

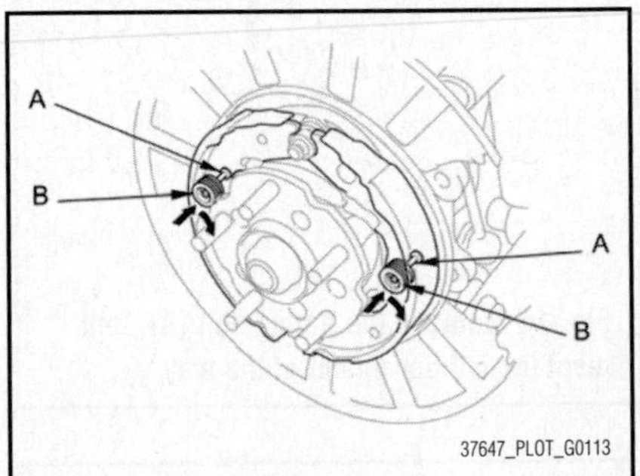
A
B
A
B
37647_PLOT_G0113

Fig. 25 Remove the tension pins (A) by pushing the respective retainer (B), and turning the pin

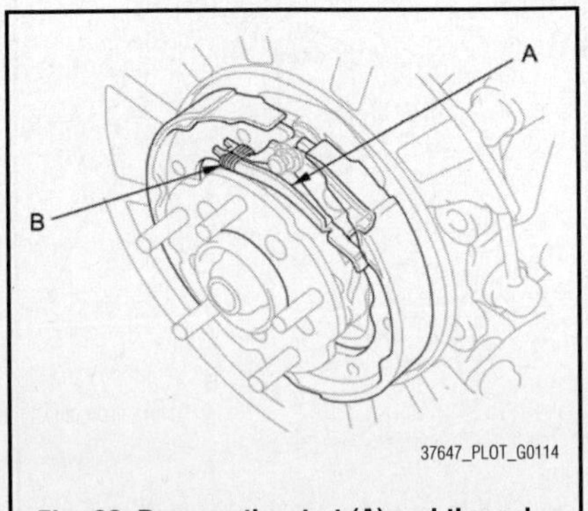
A
B
37647_PLOT_G0114

Fig. 26 Remove the strut (A) and the rod spring (B)

4. Disconnect and remove the upper return springs, then remove the shoe guide plate.

5. Remove the tension pins by pushing the respective retainer, and turning the pin.

6. Remove the strut and the rod spring.

7. Lower the parking brake shoe assembly.

8. Remove the forward brake shoe and the adjuster assembly by removing the lower return spring.

9. Remove the rearward brake shoe by disconnecting the parking brake cable from the parking brake lever.

10. Remove the U-clip, wave washer and parking brake lever from the brake shoe.

To install:

11. Apply Molykote 44MA to the sliding surface of the pivot pin of the rearward brake shoe.

12. Install the parking brake lever and the wave washer on the pivot pin, and secure with a new U-clip.

a. Install the wave washer with its convex side facing out.

b. Pinch the U-clip securely to prevent the parking brake lever from coming off the brake shoe.

13. Connect the parking brake cable to the parking brake lever. Apply Molykote 44MA to the cable contact surface on the backing plate.

14. Apply a thin coat of Molykote 44MA grease to the shoe ends and strut ends, sliding surfaces, and opposite edges of the parking brake shoe as shown. Wipe off any excess. Keep the grease off the brake linings.

15. Install the tension pin, the retainer spring, and the retainer on the rearward brake shoe. Make sure the tension pin does not contact the parking brake lever.

16. Install connecting rods A and B on the adjuster nut.

➡Clean the threaded portions of connecting rod A and the sliding surface of connecting rod B, then coat them with rubber grease. Shorten connecting rod A by fully turning the adjuster nut.

17. Position the brake shoe adjuster assembly on the parking brake shoes.

18. Hook the lower return spring on the parking brake shoes.

19. Install the rod spring to the strut first. Then install the strut on the parking brake shoes.

20. Install the tension pin, the retainer spring, and the retainer on the forward brake shoe.

21. Install the shoe guide plate.

22. Install the upper return springs.

23. Install the brake disc/drum and the rear brake caliper.

24. Do the major parking brake adjustment.

25. Clean the mating surfaces between the brake/drum and the inside of the wheel, then install the rear wheels.

ADJUSTMENT

Refer to Parking Brake Cable adjustment. They are the only adjustments possible.

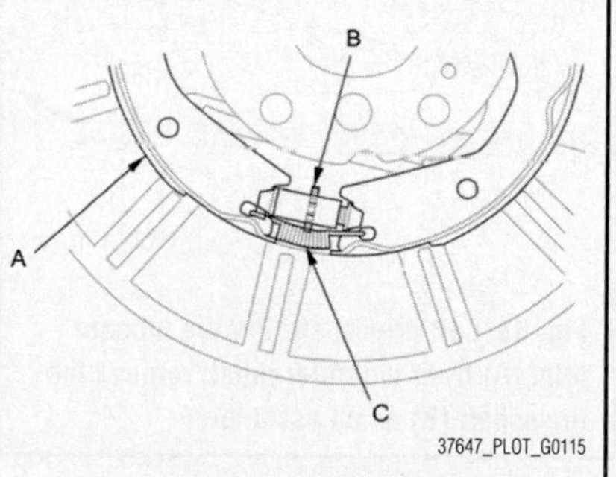
B
A
C
37647_PLOT_G0115

Fig. 27 Remove the forward brake shoe (A) and the adjuster assembly (B) by removing the lower return spring (C)

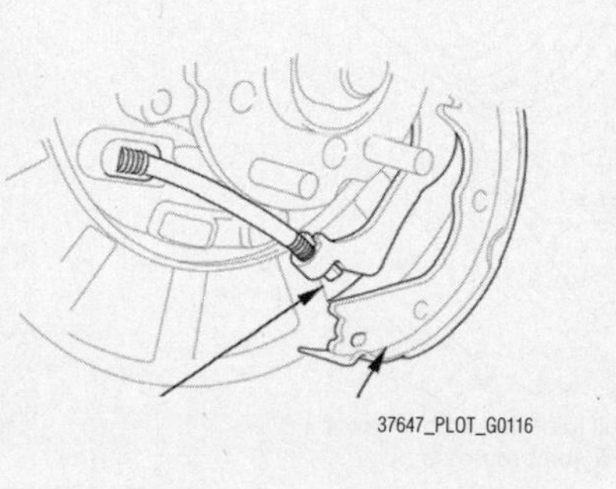
37647_PLOT_G0116

Fig. 28 Remove the rearward brake shoe by disconnecting the parking brake cable from the parking brake lever

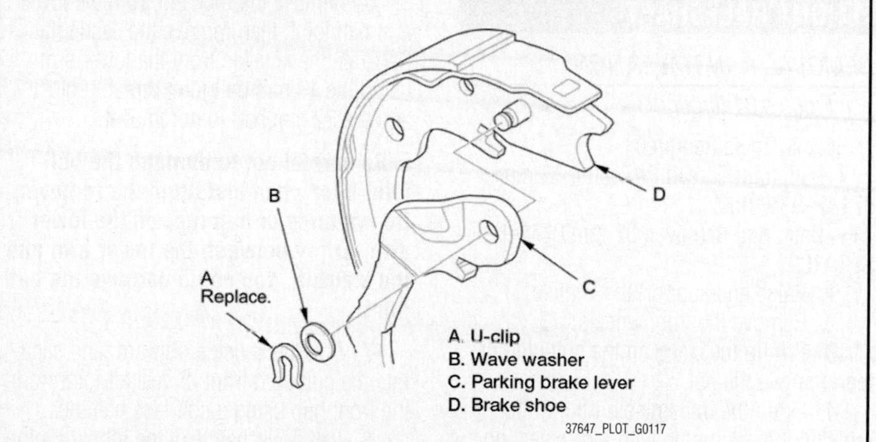
B
D
A
Replace.
C
A. U-clip
B. Wave washer
C. Parking brake lever
D. Brake shoe
37647_PLOT_G0117

Fig. 29 Remove the U-clip, wave washer and parking brake lever from the brake shoe

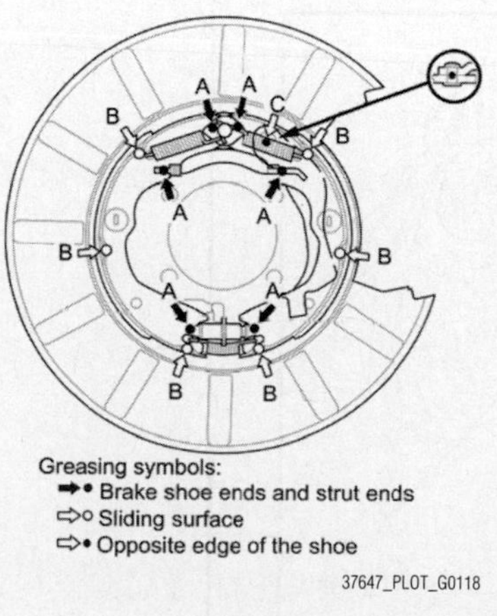
A A
B
C
B
A
A
B
B
A
A
B
B
Greasing symbols:
Brake shoe ends and strut ends
Sliding surface
Opposite edge of the shoe
37647_PLOT_G0118

Fig. 30 Apply a thin coat of Molykote 44MA grease to the shoe ends and strut ends (A), sliding surfaces (B), and opposite edges of the parking brake shoe (C) as shown

CHASSIS ELECTRICAL — AIR BAG (SUPPLEMENTAL RESTRAINT SYSTEM)

GENERAL INFORMATION

⁂ CAUTION

These vehicles are equipped with an air bag system. The system must be disarmed before performing service on, or around, system components, the steering column, instrument panel components, wiring and sensors. Failure to follow the safety precautions and the disarming procedure could result in accidental air bag deployment, possible injury and unnecessary system repairs.

SERVICE PRECAUTIONS

⁂ CAUTION

Disconnect and isolate the battery negative cable before beginning any airbag system component diagnosis, testing, removal, or installation procedures. Wait at least 90 seconds after the ignition switch is turned off and the negative (-) terminal cable is disconnected from the battery before starting the operation. The SRS is equipped with a backup power source, so if work is started within 90 seconds after disconnecting the negative (-) terminal cable from the battery, the SRS may be deployed. Failure to disable the airbag system may result in accidental airbag deployment, personal injury, or death.

DISARMING THE SYSTEM

1. Do the battery terminal disconnection procedure, then wait at least 3 minutes before starting work.

ARMING THE SYSTEM

1. Do the battery terminal reconnection procedure.
2. Clear any DTCs with the HDS.
3. After installing the airbag, confirm proper system operation, turn the ignition switch to ON (II); and check that the SRS indicator comes on for about 6 seconds and then goes off.

DRIVE TRAIN

FRONT HALFSHAFT

REMOVAL & INSTALLATION

See Figures 31 through 34.

Special Tools Required:

- Ball Joint Thread Protector, 14 mm 071AF-S3VA000
- Ball Joint Remover, 32 mm 07MAC-SL0A102

1. Raise and support the vehicle.
2. Remove the front wheels.
3. Pry up the stake on the spindle nut, then remove the nut.
4. Drain the transmission fluid, then reinstall the drain plug with a new sealing washer.
5. Remove the brake hose mounting bolt.
6. Remove the lock pin from the lower arm ball joint, then remove the castle nut. Separate the knuckle from the lower arm using the 14 mm ball joint thread protector and the 32 mm ball joint remover.

➡Be careful not to damage the ball joint boot when installing the remover. Do not force or hammer on the lower arm, or pry between the lower arm and the knuckle. You could damage the ball joint.

7. Pull the knuckle outward, and separate the outboard front driveshaft joint from the front hub using a soft face hammer.
8. Left driveshaft: Pry the inboard joint from the differential using a pry bar. Remove the driveshaft as an assembly.

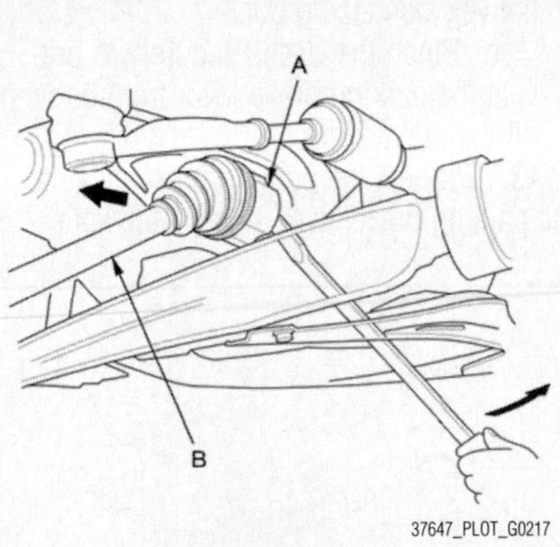
A
B
37647_PLOT_G0217

Fig. 33 Left driveshaft: Pry the inboard joint (A) from the differential; remove the driveshaft (B) as an assembly

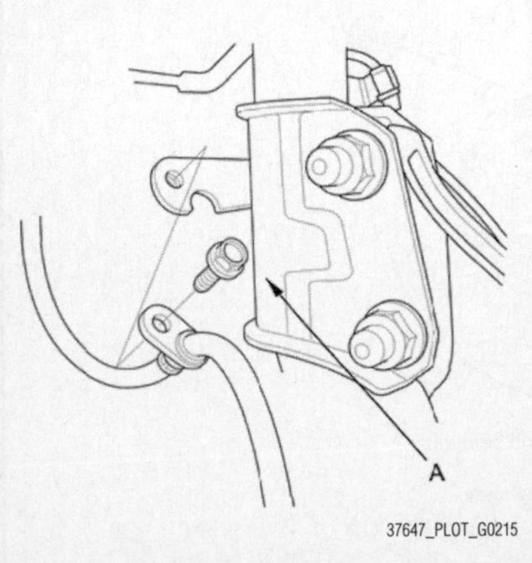
A
37647_PLOT_G0215

Fig. 31 Remove the brake hose mounting bolt (A)

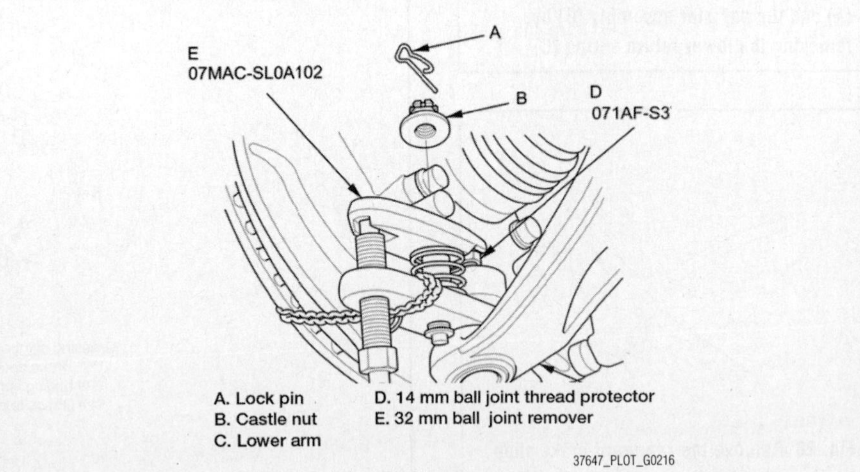
A
E
07MAC-SL0A102
B
D
071AF-S3
A. Lock pin
B. Castle nut
C. Lower arm
D. 14 mm ball joint thread protector
E. 32 mm ball joint remover
37647_PLOT_G0216

Fig. 32 Remove the lock pin from the lower arm ball joint, then remove the castle nut

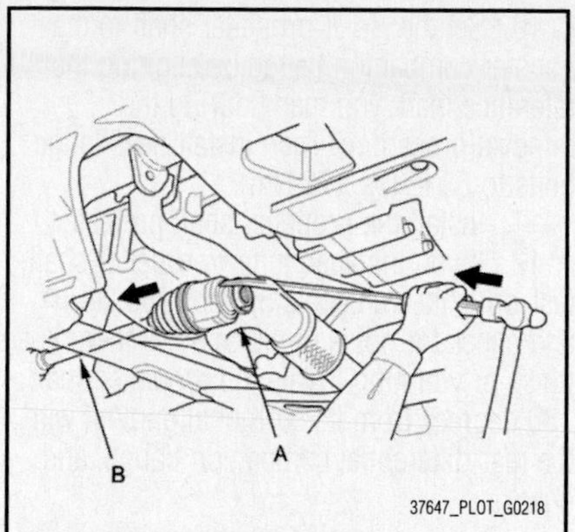
A
B
37647_PLOT_G0218

Fig. 34 Right driveshaft: Drive the inboard joint (A) off of the intermediate shaft; remove the driveshaft (B) as an assembly

➡**Do not pull on the driveshaft, or the inboard joint may come apart. Pull the inboard joint straight out to avoid damaging the oil seal. Be careful not to damage the oil seal or the end of the inboard joint using the pry bar.**

9. Right driveshaft: Drive the inboard joint off of the intermediate shaft using a drift and a hammer. Remove the driveshaft as an assembly.

➡**Do not pull on the driveshaft, or the inboard joint may come apart.**

10. Remove the set ring from the left driveshaft inboard joint.
11. Remove the set ring from the intermediate shaft.

To install:

➡**Before starting installation, make sure the mating surfaces of the joint and the splined section are clean.**

12. Apply about 0.18 oz (5 g) of Moly 60 paste (P/N 08734-0001) to the contact area (A) of the outboard joint and the front wheel bearing.

➡**The paste helps prevent noise and vibration.**

13. Install a new set ring into the set ring groove of the left driveshaft inboard joint.
14. Install a new set ring into the set ring groove of the intermediate shaft.
15. Apply 2.0–3.0 g (0.07–0.10 oz) of super high temp urea grease (P/N 08798-9002) to the whole splined surface of the right driveshaft. After applying grease, remove the grease from the splined grooves at intervals of 2–3 splines and from the set ring groove so that air can bleed from the intermediate shaft.
16. Clean the areas where the driveshaft contacts the differential thoroughly with solvent, and dry them with compressed air.

➡**Do not wash the rubber parts with solvent.**

17. Insert the inboard end of the driveshaft into the differential or the intermediate shaft until the set ring locks in the groove.

➡**Insert the driveshaft horizontally to prevent damaging the oil seal.**

18. Install the outboard joint into the front hub on the knuckle.
19. Wipe off any grease contamination from the ball joint tapered section and threads, then install the knuckle onto the lower arm. Be careful not to damage the ball joint boot. Wipe off the grease before tightening the nut at the ball joint. Torque a new castle nut to the lower torque specification, then tighten it only far enough to align the slot with the ball joint pin hole.

➡**Make sure the ball joint boot is not damaged or cracked. Do not align the nut by loosening it.**

20. Install the lock pin into the ball joint pin hole.
21. Install the brake hose mounting bolt.
22. Apply a small amount of engine oil to the seating surface of a new spindle nut.
23. Install the spindle nut, then tighten it. After tightening, use a drift to stake the spindle nut shoulder against the driveshaft.
24. Clean the mating surfaces of the brake disc and the wheel, then install the front wheels.
25. Turn the wheel by hand, and make sure there is no interference between the driveshaft and surrounding parts.
26. Lower the vehicle.
27. Refill the transmission with the recommended transmission fluid.
28. Check the wheel alignment, and adjust it if necessary.
29. Test-drive the vehicle.

INTERMEDIATE SHAFT

REMOVAL & INSTALLATION

See Figures 35 and 36.

1. Drain the transmission fluid, then reinstall the drain plug with a new sealing washer.
2. Remove the right front driveshaft.
3. Remove exhaust pipe A.
4. Remove the rear Warm Up Three Way Catalytic Converter (rear WU-TWC) bracket.
5. Remove the flange bolt, and the two dowel bolts.

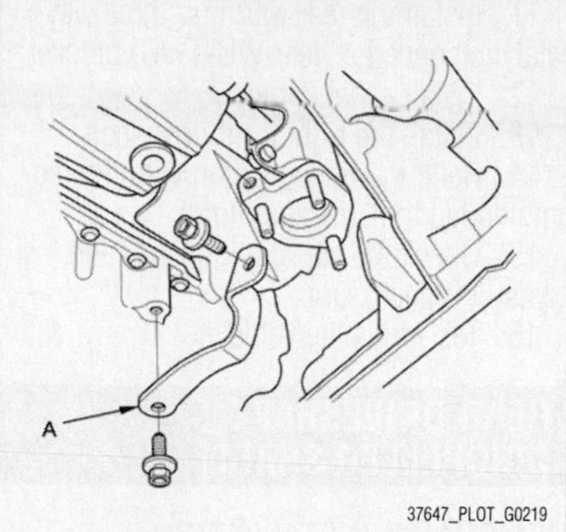
A
37647_PLOT_G0219

Fig. 35 Remove rear WU-TWC bracket (A)

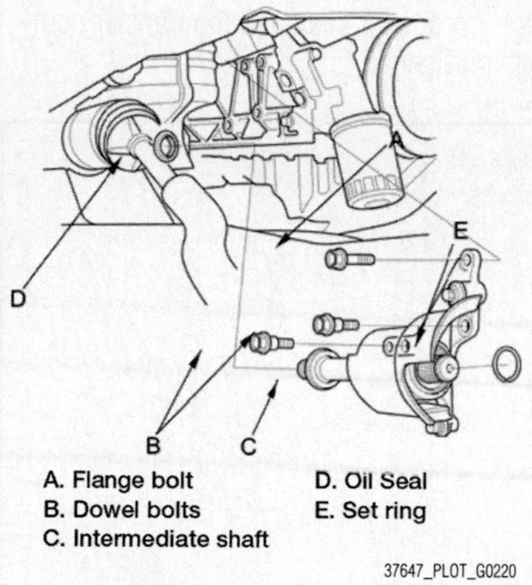
A
E
D
B
C
A. Flange bolt
B. Dowel bolts
C. Intermediate shaft
D. Oil Seal
E. Set ring
37647_PLOT_G0220

Fig. 36 Remove the flange bolt, and the two dowel bolts

6. Remove the intermediate shaft from the differential. Hold the intermediate shaft horizontal until it is clear of the differential to prevent damaging the oil seal, then remove the set ring from the intermediate shaft.

To install:

7. Clean the areas where the intermediate shaft contacts the differential thoroughly with solvent, and dry them with compressed air.

➡**Do not wash the rubber parts with solvent.**

8. Install a new set ring into the set ring groove of the intermediate shaft.
9. Insert the intermediate shaft into the differential correctly.

➡**Insert the intermediate shaft carefully to prevent damaging the oil seal.**

10. Install the flange bolt and the two dowel bolts.

11. Install the rear warm up three way catalytic converter (rear WU-TWC) bracket.
12. Install exhaust pipe A.
13. Install the right front driveshaft.
14. Refill the transmission with the recommended transmission fluid.
15. Check the wheel alignment, and adjust it if necessary.
16. Test-drive the vehicle.

REAR DRIVESHAFT (PROPELLER SHAFT)

REMOVAL & INSTALLATION

See Figures 37 through 40.

1. Raise and support the vehicle.
2. Remove the propeller shaft protector.
3. Make a reference mark across the No. 1 propeller shaft and the transfer companion flange.
4. Remove the flange bolts.
5. Remove the center support bearing mounting bolts.
6. Make a reference mark across the No. 2 propeller shaft and the rear differential companion flange.
7. Remove the flange bolts.

To install:

8. Set the No. 2 propeller shaft to the rear differential companion flange by aligning the reference mark you made during the removal procedure. Then install new flange bolts to 53 ft. lbs. (72 Nm).

➡ **When replacing the propeller shaft or the rear differential, align the factory reference marks.**

9. Install the center support bearing mounting bolts.
10. Set the No. 1 propeller shaft to the transfer companion flange by aligning the reference mark you made during the removal procedure. Then install new flange bolts to 53 ft. lbs. (72 Nm).
11. Install the propeller shaft protector.
12. If you installed a new propeller shaft, test-drive the vehicle at 55 mph (88 km/h) and check for noise or vibration. If there is a noise or vibration, rotate the propeller shaft 180 degrees from its current alignment with the rear differential companion flange, and check.

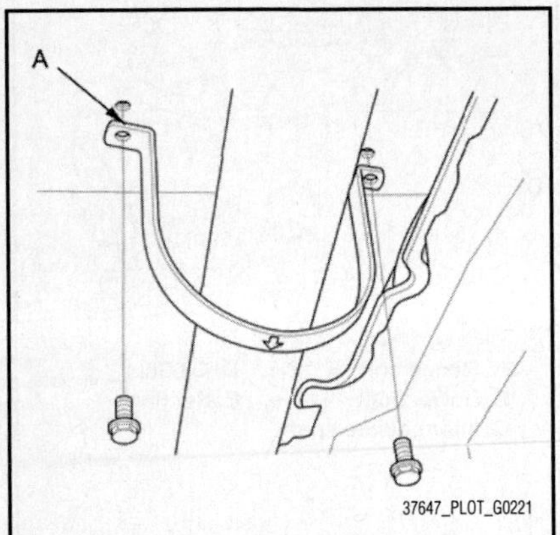
A
37647_PLOT_G0221

Fig. 37 Remove the propeller shaft protector (A)

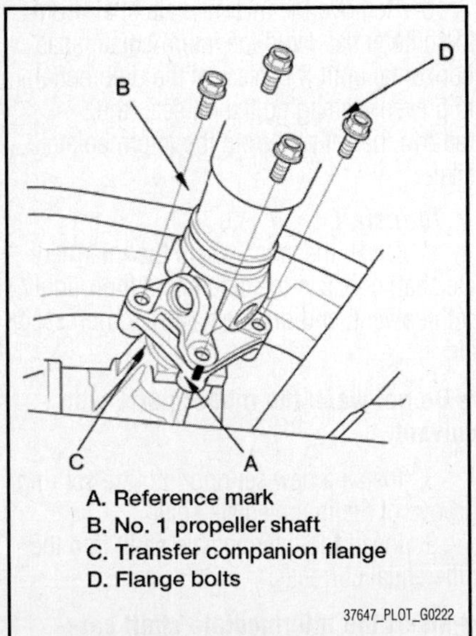
D
B
C
A
A. Reference mark
B. No. 1 propeller shaft
C. Transfer companion flange
D. Flange bolts
37647_PLOT_G0222

Fig. 38 Make a reference mark across the No. 1 propeller shaft and the transfer companion flange

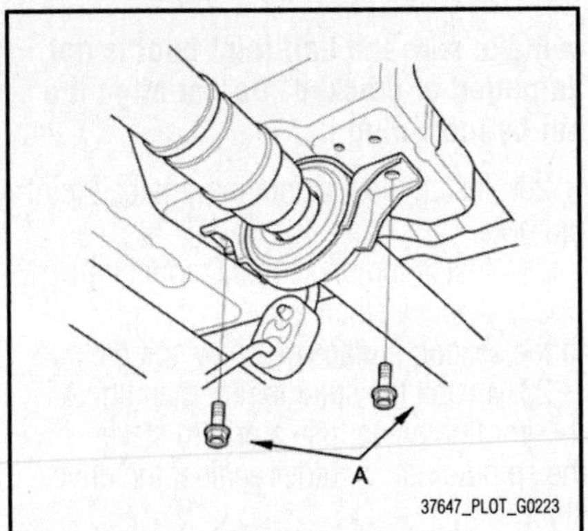
A
37647_PLOT_G0223

Fig. 39 Remove the center support bearing mounting bolts (A)

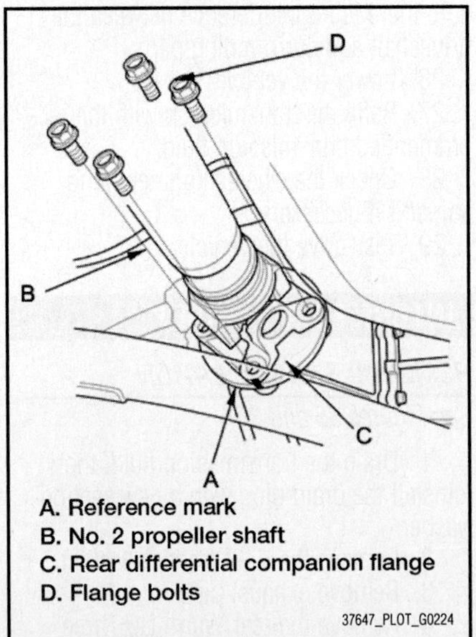
D
B
C
A
A. Reference mark
B. No. 2 propeller shaft
C. Rear differential companion flange
D. Flange bolts
37647_PLOT_G0224

Fig. 40 Make a reference mark across the No. 2 propeller shaft and the rear differential companion flange

REAR HALFSHAFTS

REMOVAL & INSTALLATION

See Figures 41 through 47.

Special Tools Required:
- Driveshaft Remover 07AAD-S9VA000
- Ball Joint Thread Protector, 12 mm 07AAF-SDAA100
- Ball Joint Remover, 28 mm 07MAC-SL0A202

➡ **Be careful not to damage the brake hose, sensor, and harness.**

1. Raise and support the vehicle.
2. Remove the rear wheels.
3. Pry up the stake on the spindle nut, then remove the nut.
4. Remove the rear wheel speed sensor and harness stay.
5. Remove the lock pin from the upper arm ball joint, then remove the castle nut. Separate the knuckle from the upper arm using the 12 mm ball joint thread protector and the 28 mm ball joint remover.

➡ **Be careful not to damage the ball joint boot when installing the remover. Do not force or hammer on the upper arm, or pry between the upper arm and the knuckle. You could damage the ball joint.**

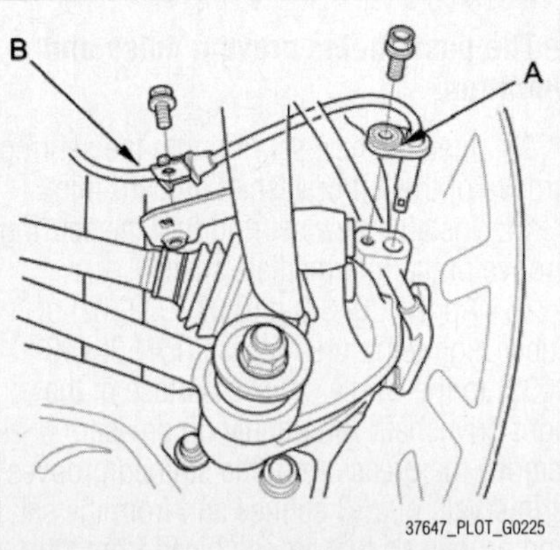
B
A
37647_PLOT_G0225

Fig. 41 Remove the rear wheel speed sensor (A) and harness stay (B)

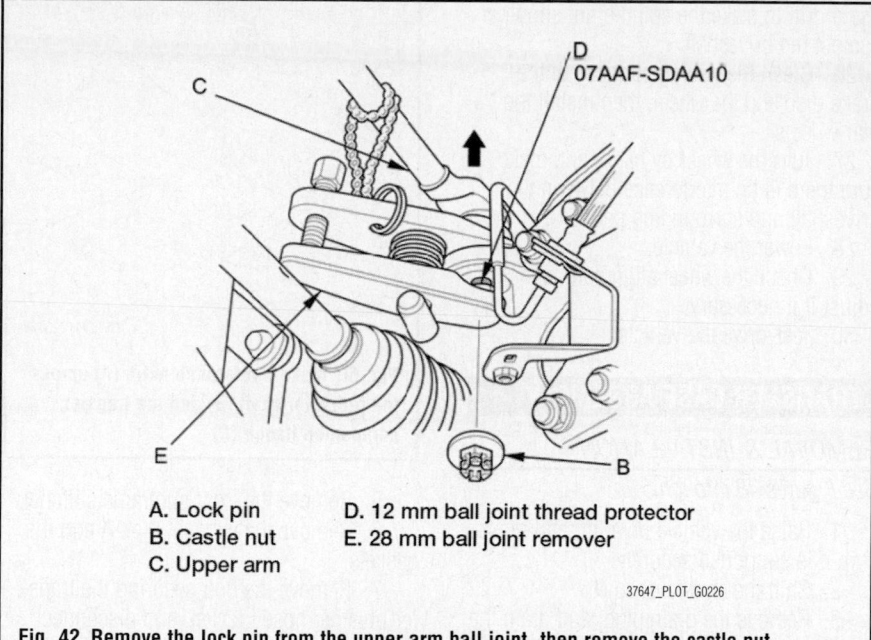

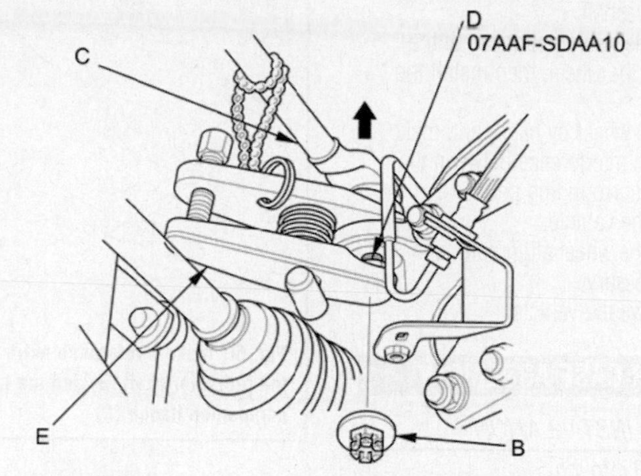
D
07AAF-SDAA10
C
E
B
A. Lock pin
B. Castle nut
C. Upper arm
D. 12 mm ball joint thread protector
E. 28 mm ball joint remover
37647_PLOT_G0226

Fig. 42 Remove the lock pin from the upper arm ball joint, then remove the castle nut

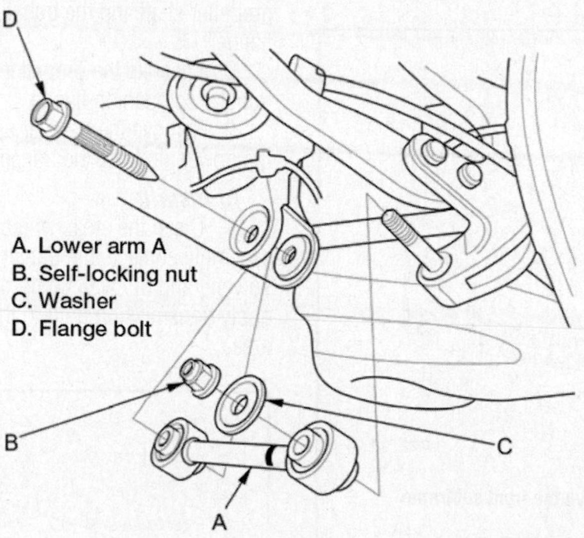
D
A. Lower arm A
B. Self-locking nut
C. Washer
D. Flange bolt
B
C
A
37647_PLOT_G0227

Fig. 43 Remove the self-locking nut, the washer, and the flange bolt, then remove lower arm A

6. Remove the self-locking nut, the washer, and the flange bolt, then remove lower arm A.
7. Place a transmission jack under lower arm B, and remove the flange bolt.
8. Pull the knuckle outward, and separate the rear driveshaft outboard joint from the rear hub using a soft face hammer.
9. Using the driveshaft remover pry out the inboard joint from the rear differential.

➡This is a prying tool, do not strike it with a hammer.

10. Remove the rear driveshaft.

➡When removing the outboard joint, continue supporting both the knuckle and the lower arm with the transmission jack. Make sure not to over extend the brake hose.

11. Remove the set ring from the rear differential.

To install:

➡Before starting installation, make sure the mating surfaces of the joint and the splined section are clean.

12. Apply 1.5–2.0 g (0.05–0.07 oz) of super high temp urea grease (P/N 08798-9002) to the whole splined surface. After applying grease, remove the grease from the splined grooves at intervals of 2–3 splines and from the set ring groove so that air can bleed from the differential.
13. Install a new set ring into the set ring groove of the differential.
14. Clean the areas where the driveshaft contacts the differential thoroughly with solvent, and dry them with compressed air.

➡Do not wash the rubber parts with solvent.

15. Make sure the inboard joint is installed all the way into the rear differential and to ensure the set ring is properly seated.

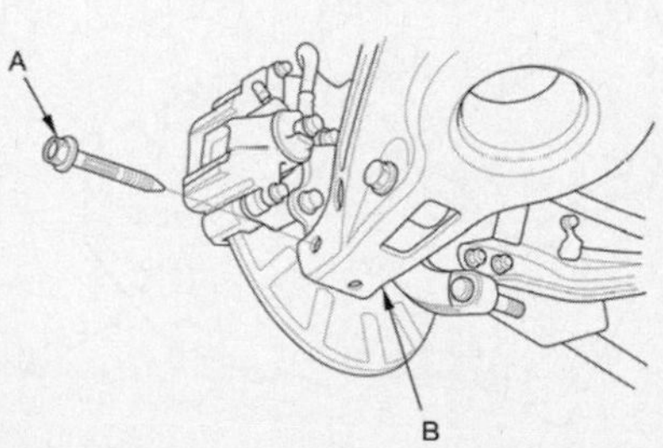
A
B
37647_PLOT_G0228

Fig. 44 Place a transmission jack under lower arm B, and remove the flange bolt (A)

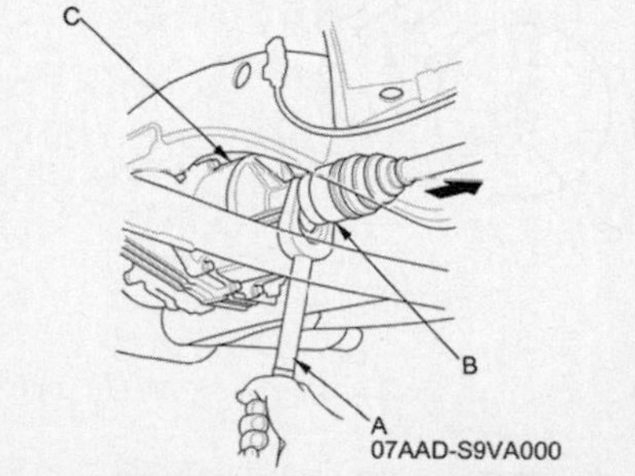
C
B
A
07AAD-S9VA000
37647_PLOT_G0229

Fig. 45 Using the driveshaft remover (A) pry out the inboard joint (B) from the rear differential (C)

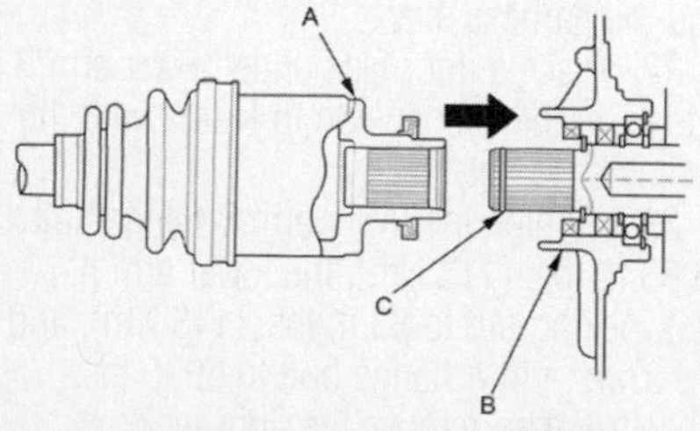
A
C
B
37647_PLOT_G0230

Fig. 46 Make sure the inboard joint (A) is installed all the way into the rear differential (B) and to ensure the set ring (C) is properly seated

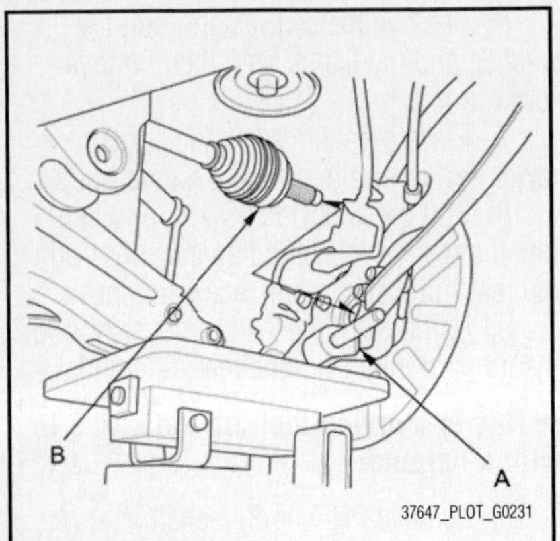
B
A
37647_PLOT_G0231

Fig. 47 Pull the knuckle (A) outward, and insert the rear driveshaft outboard joint (B) into the rear hub

16. Pull the knuckle outward, and insert the rear driveshaft outboard joint into the rear hub.

➡When installing the outboard joint continue supporting both the knuckle and the lower arm with the transmission jack. Make sure not to over extend the brake hose.

17. Loosely install a new flange bolt onto lower arm B.
18. Install lower arm A, then loosely install a new flange bolt and a new self-locking nut with the washer.

➡Position the paint mark on lower arm A toward the outside of the vehicle.

19. Wipe off any grease contamination from the ball joint tapered section and threads, then install the upper arm. Be careful not to damage the ball joint boot. Wipe off the grease before tightening the nut at the ball joint. Torque the castle nut to the lower torque specification, then tighten it only far enough to align the slot with the ball joint pin hole.
20. Install the lock pin into the ball joint pin hole as shown.
21. Install the rear wheel speed sensor and the harness stay.
22. Place a floor jack under lower arm B, and raise the suspension to load it with the vehicle's weight.
23. Tighten the lower arm B flange bolt to 83 ft. lbs. (113 Nm), the lower arm A self-locking nut to 83 ft. lbs. (113 Nm), and the lower arm A flange bolt to 69 ft. lbs. (93 Nm), then remove the floor jack.
24. Apply a small amount of engine oil to the seating surface of a new spindle nut.
25. Install the spindle nut, then tighten it to 181 ft. lbs. (245 Nm). After tightening, use a drift to stake the spindle nut shoulder against the driveshaft.
26. Clean the mating surfaces of the brake disc and the wheel, then install the rear wheels.
27. Turn the wheel by hand, and make sure there is no interference between the driveshaft and surrounding parts.
28. Lower the vehicle.
29. Check the wheel alignment, and adjust it if necessary.
30. Test-drive the vehicle.

TRANSFER CASE ASSEMBLY

REMOVAL & INSTALLATION

See Figures 48 through 51.

1. Raise the vehicle on a lift, and make sure it is supported securely.
2. Shift the shift lever to N.
3. Remove the drain plug, and drain the Automatic Transmission Fluid (ATF).
4. Reinstall the drain plug with a new sealing washer.

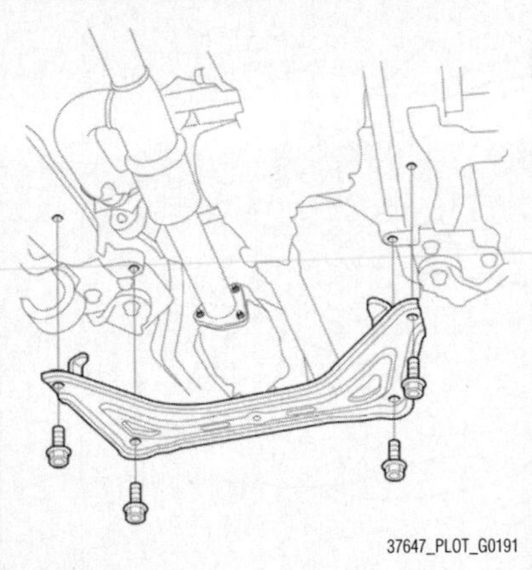
37647_PLOT_G0191

Fig. 48 Remove the front subframe stiffener

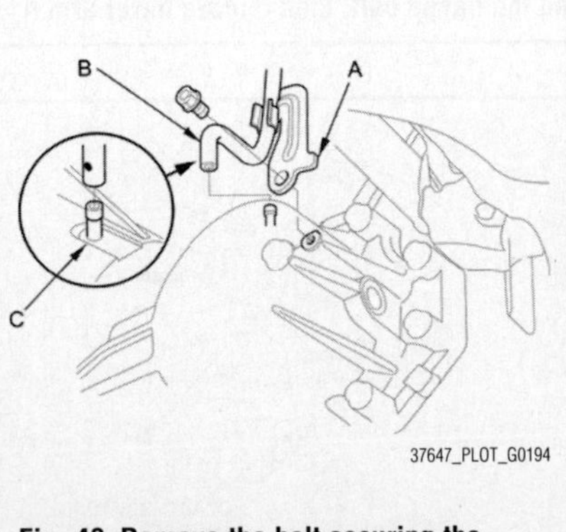
B
A
C
37647_PLOT_G0194

Fig. 49 Remove the bolt securing the transfer breather hose bracket (A), and disconnect the breather hose (B) from the breather pipe (C) on the transfer assembly

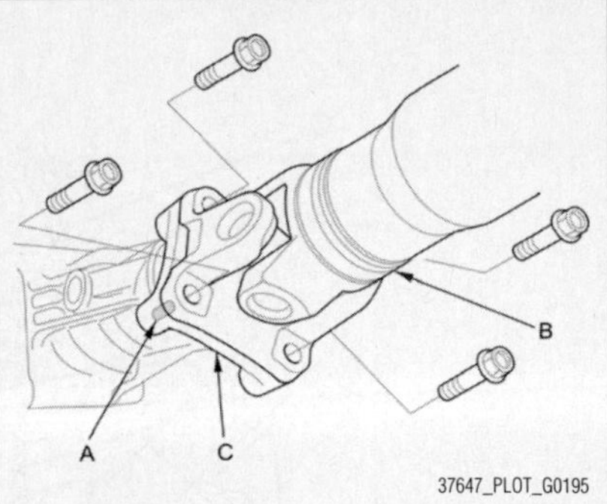
B
A
C
37647_PLOT_G0195

Fig. 50 Make a reference mark (A) across the propeller shaft (B) and the transfer companion flange (C)

5. Remove the front subframe stiffener.
6. Remove the exhaust pipe A and the gaskets.
7. Remove the bolt securing the transfer breather hose bracket, and disconnect the breather hose from the breather pipe on the transfer assembly.
8. Make a reference mark across the propeller shaft and the transfer companion flange.
9. Separate the propeller shaft from the transfer companion flange.
10. Remove the transfer assembly and the dowel pin from the transmission.

To install:

11. Clean the areas where the transfer assembly contacts the transmission with solvent, and dry with compressed air. Then apply transmission fluid to the seal contact area.

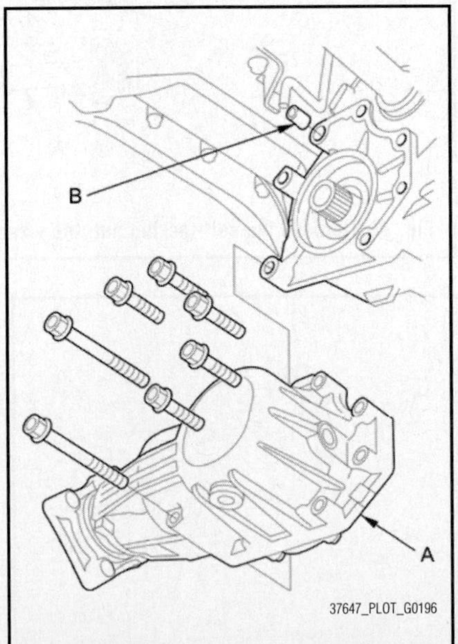
B
A
37647_PLOT_G0196

Fig. 51 Remove the transfer assembly (A) and the dowel pin (B) from the transmission

12. Install the 14 x 20 mm dowel pin in the transmission, and install the transfer assembly on the transmission. Tighten the mounting bolts to 38 ft. lbs. (51 Nm).

13. Install the propeller shaft to the transfer companion flange by aligning the reference mark. Make sure you use new mounting bolts. Tighten the mounting bolts to 53 ft. lbs. (72 Nm).

14. Secure the transfer breather hose bracket on the transfer assembly with the bolt, and connect the breather hose over the breather pipe with the dot mark facing out.

15. Install the exhaust pipe A with new self-locking nuts, its mount, and new gaskets.

16. Install the front subframe stiffener with new mounting bolts. Tighten the mounting bolts to 43 ft. lbs. (59 Nm).

17. Refill the transfer assembly with the transfer fluid (hypoid gear oil), if necessary.

18. Refill the transmission with ATF.

ENGINE COOLING

ENGINE FAN

REMOVAL & INSTALLATION

See Figures 52 through 60.

1. Remove the front grille.
2. Raise the vehicle on the lift.
3. Drain the engine coolant.
4. Remove the splash shield.
5. Remove the harness clamp from the radiator fan shroud.
6. Unclamp the Automatic Transmission Fluid (ATF) cooler hose clamps and loosen the A/C condenser fan shroud mounting bolts.
7. Lower the vehicle on the lift.
8. Disconnect the radiator fan motor connector and the radiator upper hose, then remove the harness clamps.
9. Disconnect the A/C condenser fan motor connector and the coolant reservoir hose, then remove the harness clamps.

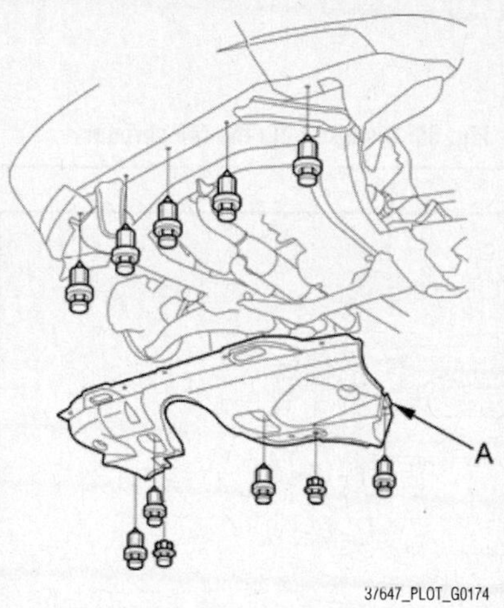
A
37647_PLOT_G0174

Fig. 53 Remove the splash shield

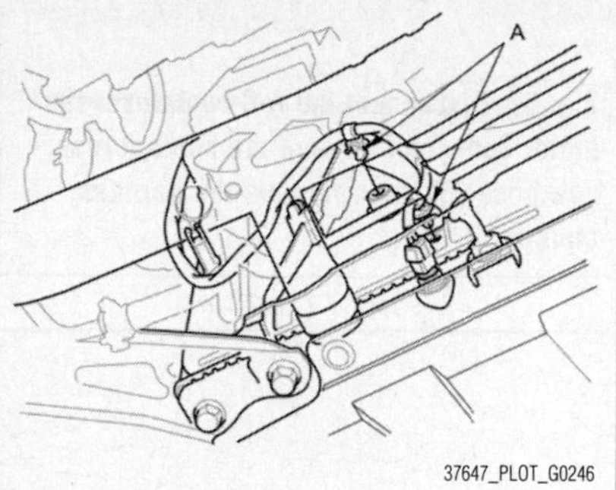
A
37647_PLOT_G0246

Fig. 54 Remove the harness clamp (A) from the radiator fan shroud

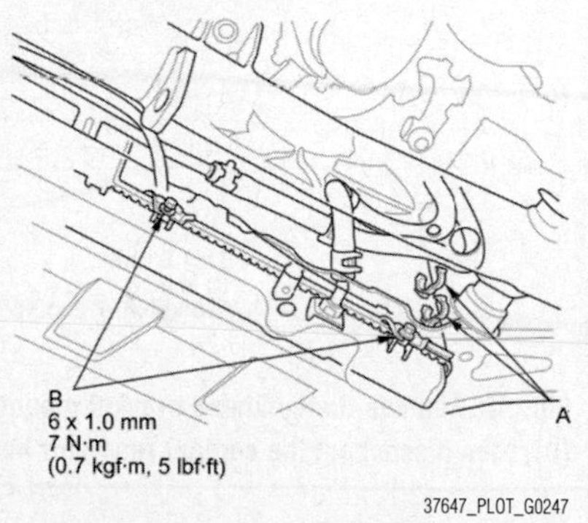
B
6 x 1.0 mm
7 N·m
(0.7 kgf·m, 5 lbf·ft)
A
37647_PLOT_G0247

Fig. 55 Unclamp the Automatic Transmission Fluid (ATF) cooler hose clamps (A) and loosen the A/C condenser fan shroud mounting bolts (B)

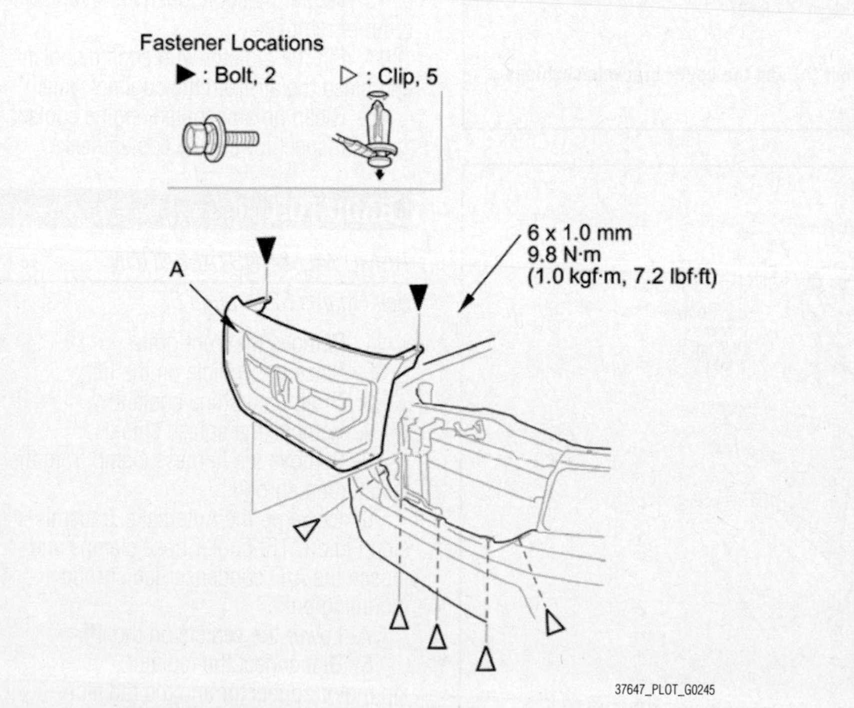
Fastener Locations
▶ : Bolt, 2
▷ : Clip, 5
A
6 x 1.0 mm
9.8 N·m
(1.0 kgf·m, 7.2 lbf·ft)
37647_PLOT_G0245

Fig. 52 Remove the front grille

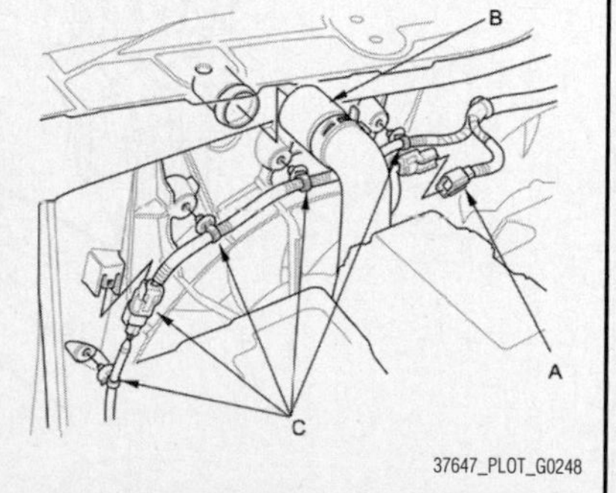
B
A
C
37647_PLOT_G0248

Fig. 56 Disconnect the radiator fan motor connector (A) and the radiator upper hose (B), then remove the harness clamps (C)

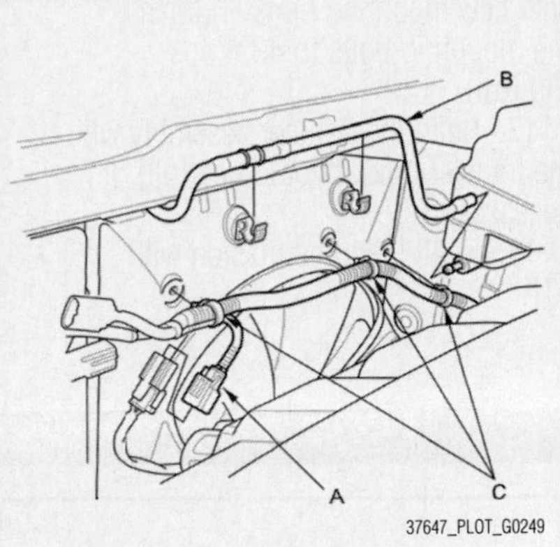
B
A
C
37647_PLOT_G0249

Fig. 57 Disconnect the A/C condenser fan motor connector (A) and the coolant reservoir hose (B), then remove the harness clamps (C)

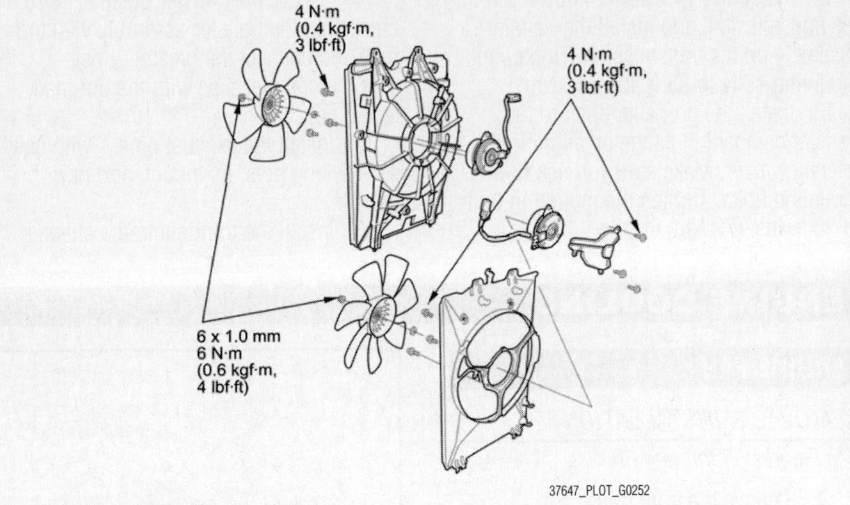
4 N·m
(0.4 kgf·m,
3 lbf·ft)
4 N·m
(0.4 kgf·m,
3 lbf·ft)
6 x 1.0 mm
6 N·m
(0.6 kgf·m,
4 lbf·ft)
37647_PLOT_G0252

Fig. 60 Disassemble the fan shrouds

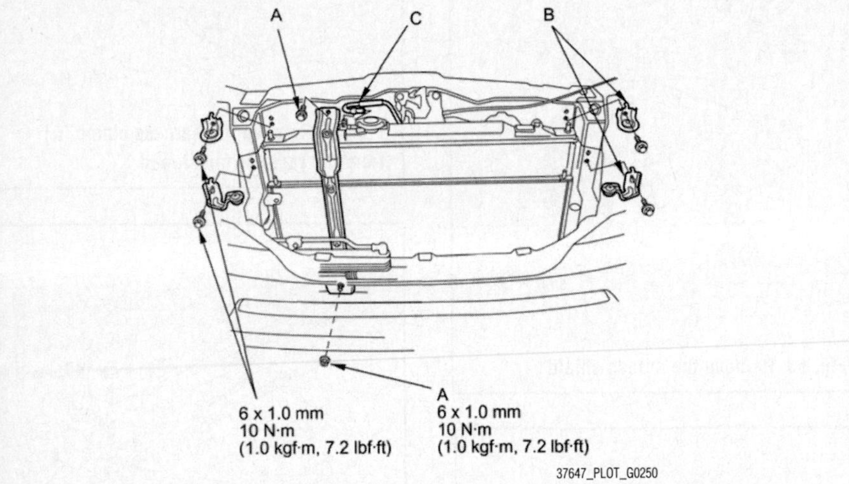
A
C
B
A
6 x 1.0 mm
10 N·m
(1.0 kgf·m, 7.2 lbf·ft)
6 x 1.0 mm
10 N·m
(1.0 kgf·m, 7.2 lbf·ft)
37647_PLOT_G0250

Fig. 58 Remove the bulkhead bracket mounting bolt/nut (A) and the upper brackets/cushions (B), then disconnect the coolant reservoir hose (C)

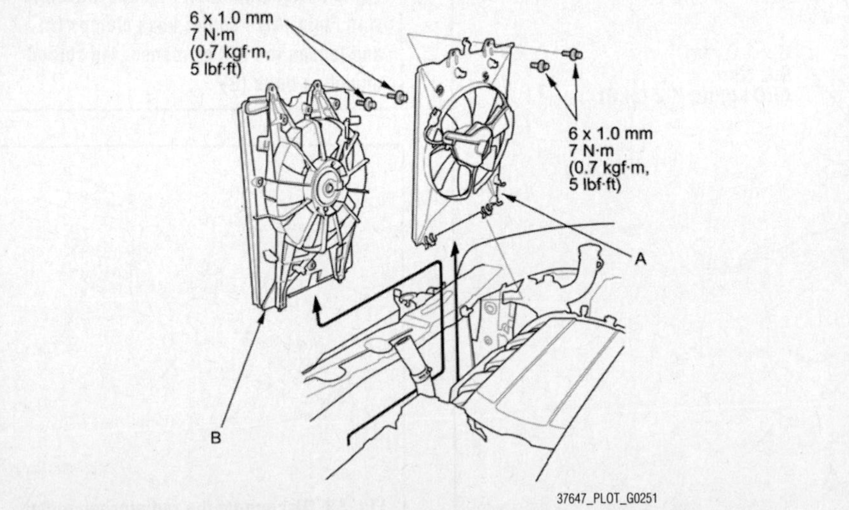
6 x 1.0 mm
7 N·m
(0.7 kgf·m,
5 lbf·ft)
6 x 1.0 mm
7 N·m
(0.7 kgf·m,
5 lbf·ft)
A
B
37647_PLOT_G0251

Fig. 59 Remove the A/C condenser fan shroud assembly (A), then remove the radiator fan shroud assembly (B)

10. Remove the bulkhead bracket mounting bolt/nut and the upper brackets/cushions, then disconnect the coolant reservoir hose.
11. Remove the A/C condenser fan shroud assembly, then remove the radiator fan shroud assembly.

➡Move the radiator fan shroud assembly toward the A/C compressor side of the vehicle to allow for enough space to lift it up and away from the A/C condenser fan shroud assembly.

12. Disassemble the fan shrouds.

To install:

13. Install the cooling fans in the reverse order of removal.
14. Fill the radiator with engine coolant, and bleed the air from the cooling system.
15. Clean up any spilled engine coolant.
16. Inspect for engine coolant leaks.

RADIATOR

REMOVAL & INSTALLATION

See Figures 57 through 67.

1. Remove the front grille.
2. Raise the vehicle on the lift.
3. Drain the engine coolant.
4. Remove the splash shield.
5. Remove the harness clamp from the radiator fan shroud.
6. Unclamp the Automatic Transmission Fluid (ATF) cooler hose clamps and loosen the A/C condenser fan shroud mounting bolts.
7. Lower the vehicle on the lift.
8. Disconnect the radiator fan motor connector and the radiator upper hose, then remove the harness clamps.

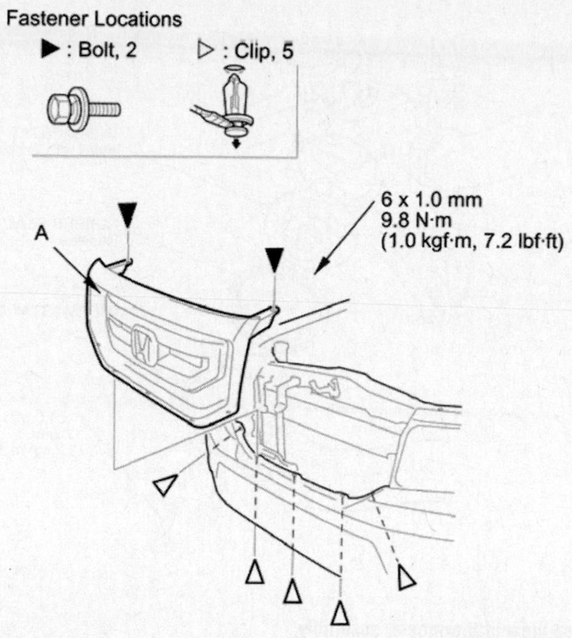
Fastener Locations
▶ : Bolt, 2
▷ : Clip, 5
A
6 x 1.0 mm
9.8 N·m
(1.0 kgf·m, 7.2 lbf·ft)
37647_PLOT_G0245

Fig. 61 Remove the front grille

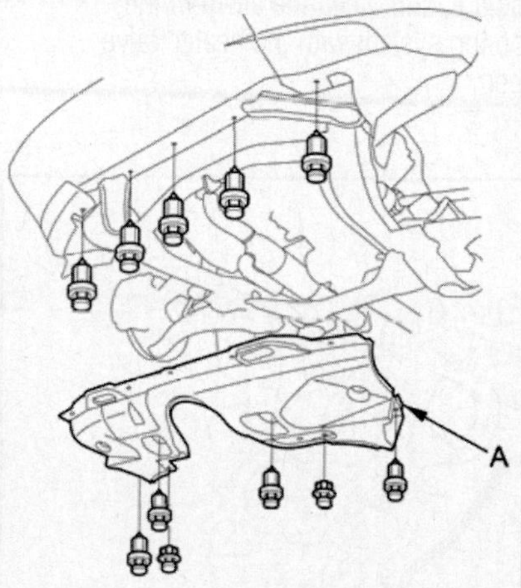
A
37647_PLOT_G0174

Fig. 62 Remove the splash shield

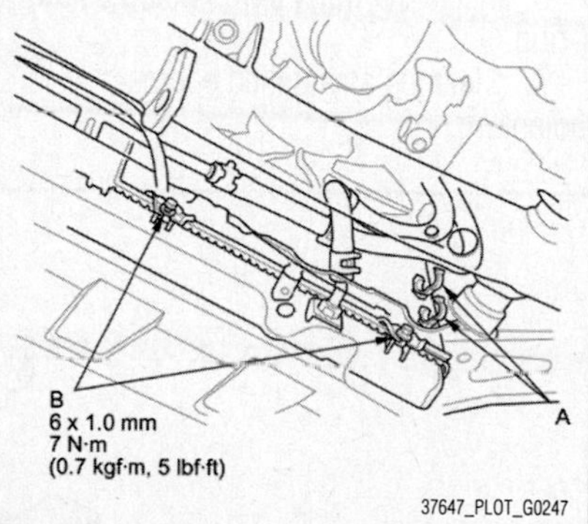
B
6 x 1.0 mm
7 N·m
(0.7 kgf·m, 5 lbf·ft)
A
37647_PLOT_G0247

Fig. 64 Unclamp the Automatic Transmission Fluid (ATF) cooler hose clamps (A) and loosen the A/C condenser fan shroud mounting bolts (B)

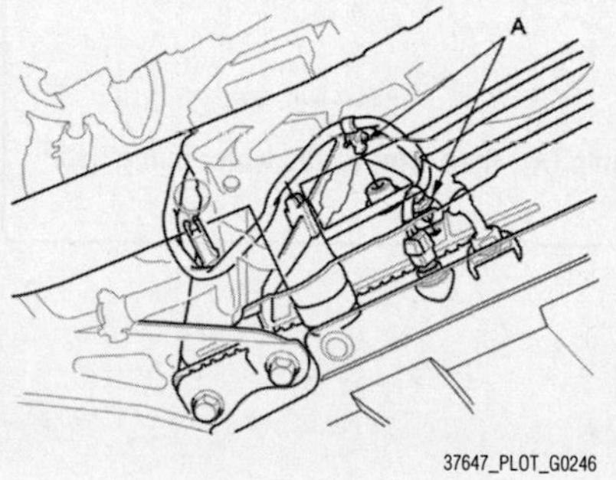
A
37647_PLOT_G0246

Fig. 63 Remove the harness clamp (A) from the radiator fan shroud

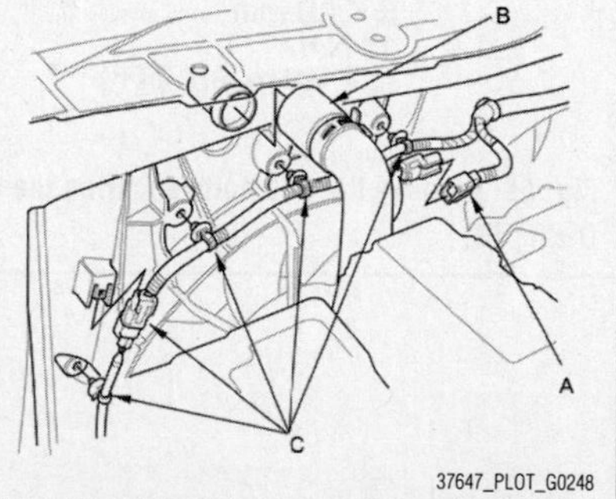
B
A
C
37647_PLOT_G0248

Fig. 65 Disconnect the radiator fan motor connector (A) and the radiator upper hose (B), then remove the harness clamps (C)

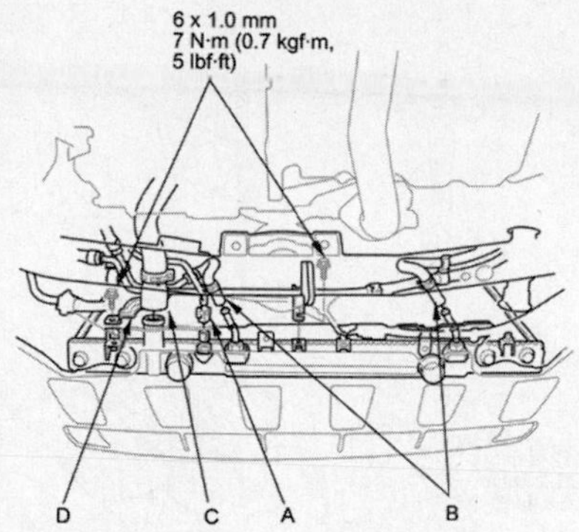
6 x 1.0 mm
7 N·m (0.7 kgf·m,
5 lbf·ft)
D
C
A
B
A. ECT sensor 2 connector
B. ATF cooler hoses
C. Lower radiator hose
D. ATF cooler hose brackets
37647_PLOT_G0253

Fig. 66 Disconnect the Engine Coolant Temperature (ECT) sensor 2 connector, the ATF cooler hoses, the lower radiator hose and remove the ATF cooler hose brackets

9. Disconnect the A/C condenser fan motor connector and the coolant reservoir hose, then remove the harness clamps.
10. Remove the bulkhead bracket mounting bolt/nut and the upper brackets/cushions, then disconnect the coolant reservoir hose.
11. Remove the A/C condenser fan shroud assembly, then remove the radiator fan shroud assembly.

➡Move the radiator fan shroud assembly toward the A/C compressor side of the vehicle to allow for enough space to lift it up and away from the A/C condenser fan shroud assembly.

12. Disassemble the fan shrouds.
13. Raise the vehicle on the lift.
14. Disconnect the Engine Coolant Temperature (ECT) sensor 2 connector, the ATF cooler hoses, the lower radiator hose and remove the ATF cooler hose brackets.
15. Lower the vehicle on the lift.
16. Pull up the radiator assembly, then remove the lower mounting cushions.
17. Remove the related parts from the radiator.

To install:

18. Install the radiator in the reverse order of removal. Make sure the upper and lower mounting cushions are set securely.
19. Fill the radiator with engine coolant, and bleed the air from the cooling system.
20. Clean up any spilled engine coolant.
21. Inspect for engine coolant leaks.

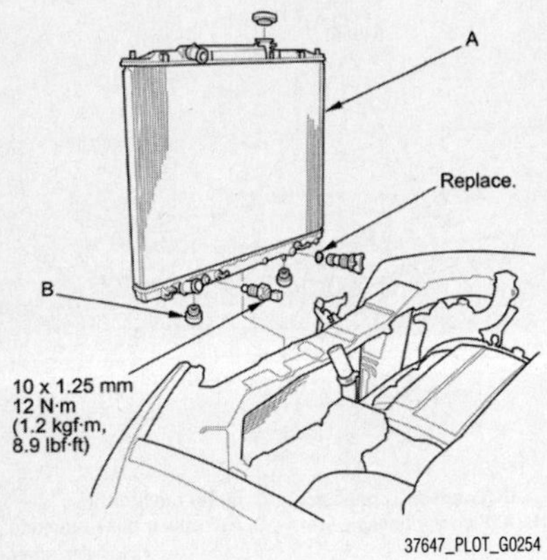
A
Replace.
B
10 x 1.25 mm
12 N·m
(1.2 kgf·m,
8.9 lbf·ft)
37647_PLOT_G0254

Fig. 67 Pull up the radiator assembly (A), then remove the lower mounting cushions (B)

THERMOSTAT

REMOVAL & INSTALLATION

See Figure 68.

1. Drain the engine coolant.
2. Remove the air intake duct.
3. Do the battery removal procedure.
4. Remove the battery base.
5. Remove the thermostat cover, then remove the thermostat.
6. Install the new thermostat with a new rubber seal, then install the thermostat cover.
7. Install the battery base.
8. Do the battery installation procedure.
9. Install the air intake duct.
10. Refill the radiator with engine coolant, and bleed the air from the cooling system.
11. Clean up any spilled engine coolant.
12. Inspect for engine coolant leaks.

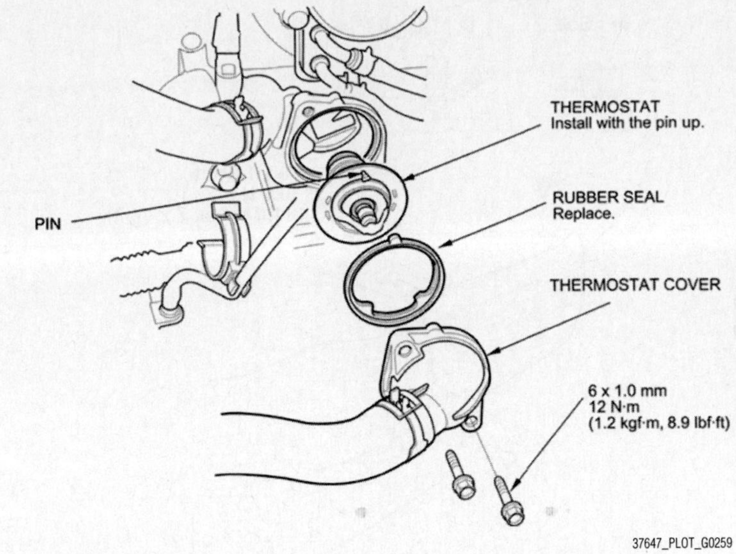
THERMOSTAT
Install with the pin up.
PIN
RUBBER SEAL
Replace.
THERMOSTAT COVER
6 x 1.0 mm
12 N·m
(1.2 kgf·m, 8.9 lbf·ft)
37647_PLOT_G0259

Fig. 68 Exploded view of thermostat assembly

WATER PUMP

REMOVAL & INSTALLATION

See Figure 69.

1. Drain the engine coolant.
2. Remove the timing belt.
3. Remove the timing belt adjuster.
4. Remove the five bolts securing the water pump, then remove the water pump.
5. Inspect and clean the O-ring groove and the mating surface of the engine block.
6. Install the water pump using a new O-ring.
7. Clean up any spilled engine coolant.
8. Install the timing belt adjuster.
9. Install the timing belt.
10. Refill the radiator with engine coolant, and bleed the air from the cooling system with the heater valve open.

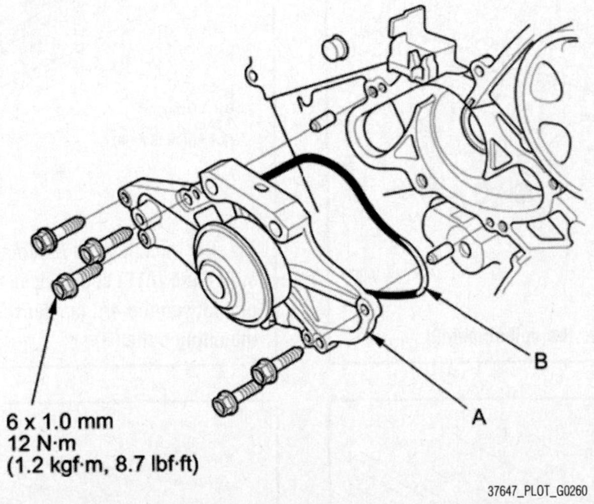
B
A
6 x 1.0 mm
12 N·m
(1.2 kgf·m, 8.7 lbf·ft)
37647_PLOT_G0260

Fig. 69 Remove the five bolts securing the water pump (A), then remove the water pump and O-ring (B)

ALTERNATOR

REMOVAL & INSTALLATION

See Figures 70 through 74.

1. Remove the air intake duct.
2. Do the battery terminal disconnection procedure.
3. Remove the engine cover.
4. Remove the coolant reservoir and the Power Steering (P/S) fluid reservoir, then remove the holder bracket.
5. Remove the accessory drive belt.
6. Disconnect the alternator connector and the positive alternator cable from the alternator.
7. Disconnect the A/C compressor clutch connector from the A/C compressor.

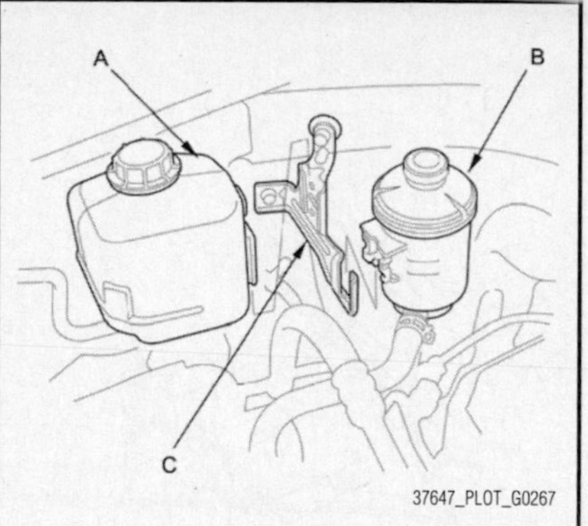
A
B
C
37647_PLOT_G0267

Fig. 72 Remove the coolant reservoir (A) and the power steering (P/S) fluid reservoir (B), then remove the holder bracket (C)

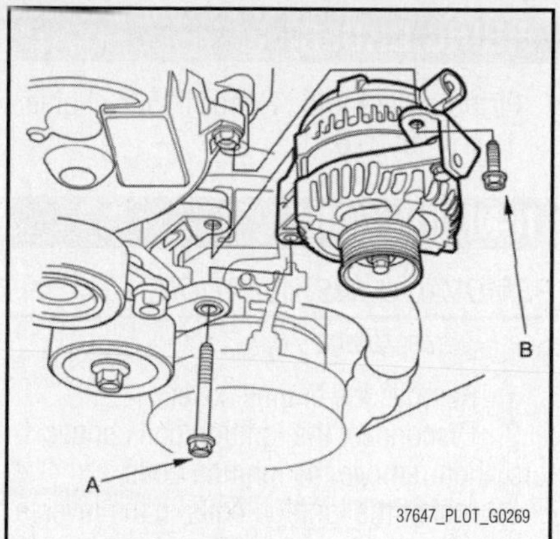
A
B
37647_PLOT_G0269

Fig. 74 Remove the mounting bolt (A) and the alternator bracket mounting bolt (B), then remove the alternator

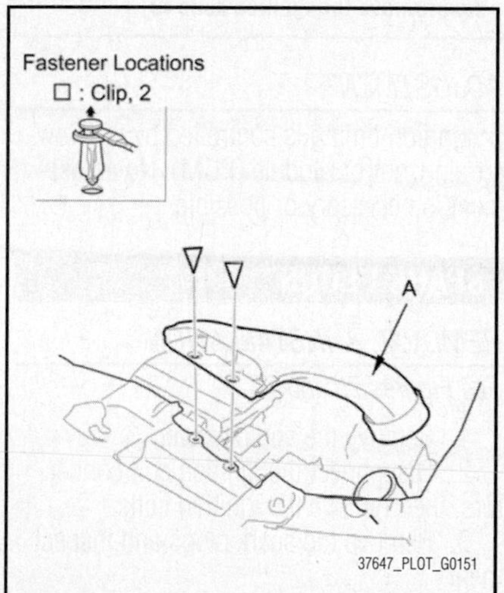
Fastener Locations
□ : Clip, 2
A
37647_PLOT_G0151

Fig. 70 Remove the clips, then remove the air intake duct (A)

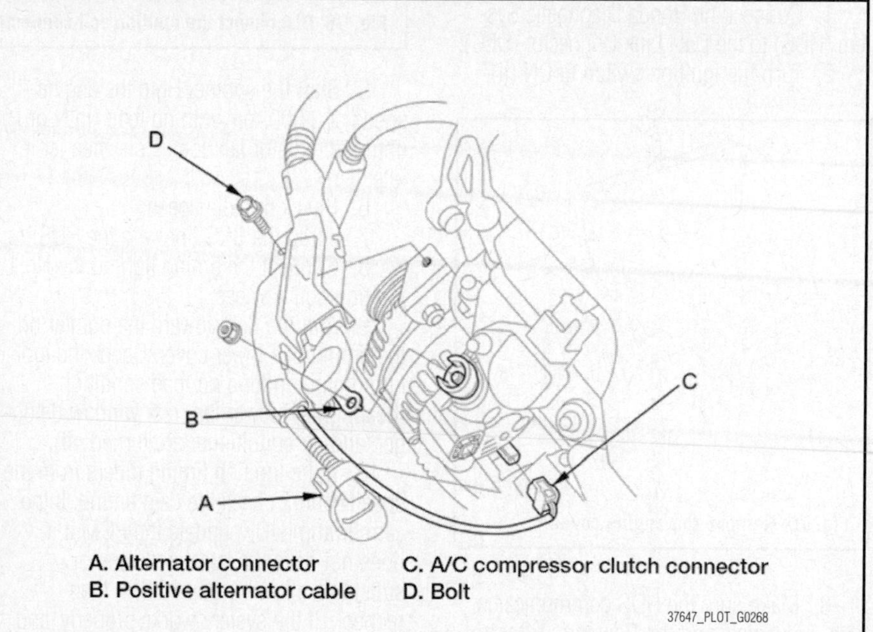
D
B
A
C
A. Alternator connector
B. Positive alternator cable
C. A/C compressor clutch connector
D. Bolt
37647_PLOT_G0268

Fig. 73 Disconnect the alternator connector and the positive alternator cable from the alternator

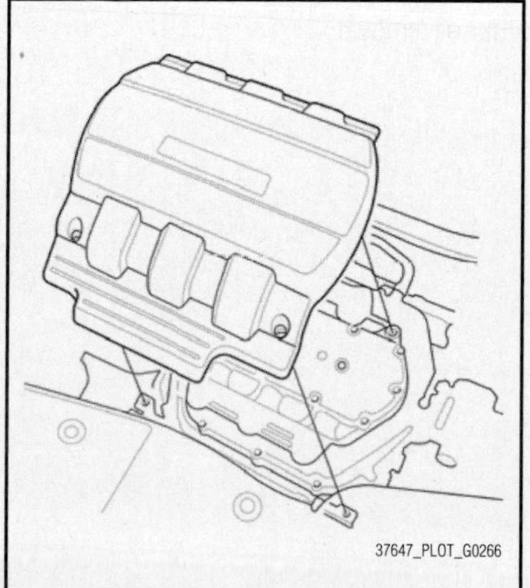
37647_PLOT_G0266

Fig. 71 Remove the engine cover

8. Remove the bolt securing the harness holder.
9. Remove the mounting bolt and the alternator bracket mounting bolt, then remove the alternator.

To install:

10. Install the alternator, then tighten the mounting bolt to 33 ft. lbs. (45 Nm), and the alternator bracket mounting bolt to 16 ft. lbs. (22 Nm).
11. Install the bolt securing the harness holder.
12. Connect the A/C compressor clutch connector to the A/C compressor.
13. Connect the alternator connector and the positive alternator cable to the alternator. Make sure the crimped it side of the ring terminal faces away from the alternator when you connect it.
14. Install the drive belt.
15. Install the holder bracket, then install the coolant reservoir and the Power Steering (P/S) fluid reservoir.
16. Install the engine cover.
17. Do the battery terminal reconnection procedure.
18. Install the air intake duct.

FIRING ORDER

Firing order for the Honda J35Z4 engine is: 1–4–2–5–3–6.

IGNITION COIL

REMOVAL & INSTALLATION

See Figures 75 and 76.

1. Remove the engine cover.
2. Disconnect the ignition coil connectors, then remove the ignition coils.
3. Install the ignition coils in the reverse order of removal.

IGNITION TIMING

INSPECTION

1. Connect the Honda Diagnostic System (HDS) to the Data Link Connector (DLC).
2. Turn the ignition switch to ON (II).

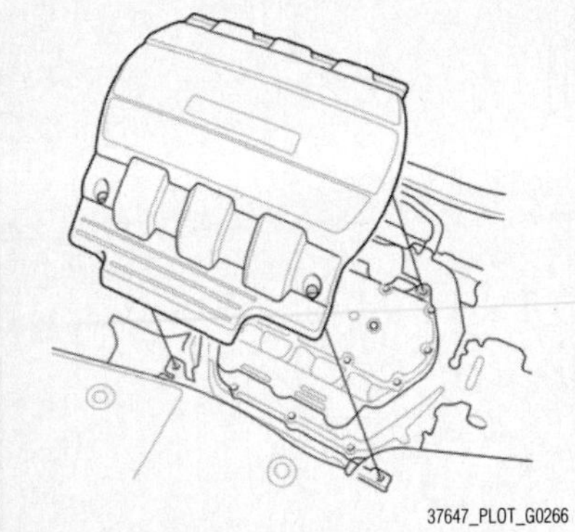
37647_PLOT_G0266

Fig. 75 Remove the engine cover

3. Make sure the HDS communicates with the vehicle and the Powertrain Control Module (PCM). If it does not communicate, troubleshoot the DLC circuit.
4. Check for DTCs. If a DTC is present, diagnose and repair the cause before continuing with this test.

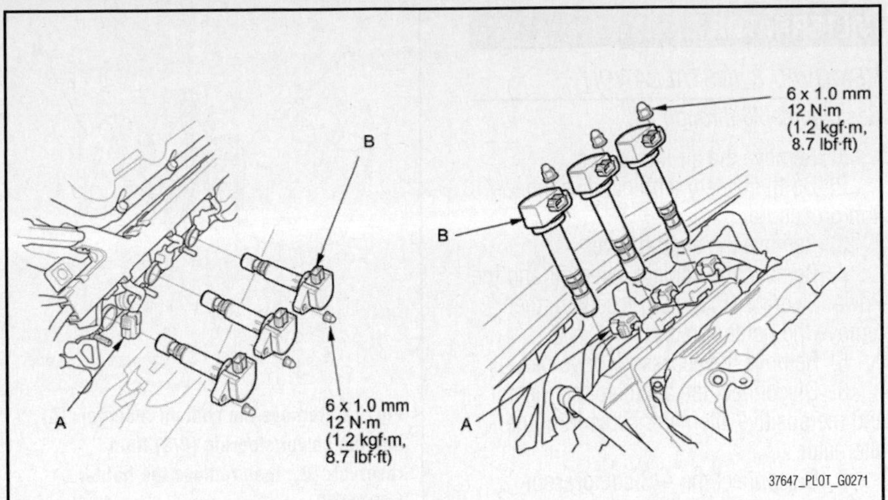
B
6 x 1.0 mm
12 N·m
(1.2 kgf·m,
8.7 lbf·ft)
A
B
6 x 1.0 mm
12 N·m
(1.2 kgf·m,
8.7 lbf·ft)
A
37647_PLOT_G0271

Fig. 76 Disconnect the ignition coil connectors (A), then remove the ignition coils (B)

5. Start the engine. Hold the engine speed at 3,000 rpm with no load (in N or P) until the radiator fan comes on, then let it idle.
6. Check the idle speed.
7. Jump the SCS line with the HDS.
8. Connect the timing light to the No.1 ignition coil harness.
9. Aim the light toward the pointer on the timing belt lower cover. Check the ignition timing under a no load condition (headlights, blower fan, rear window defogger, and air conditioner are turned off).
10. If the ignition timing differs from the specification, check the cam timing. If the cam timing is OK, update the PCM if it does not have the latest software, or substitute a known-good PCM, then recheck. If the system works properly, and the PCM was substituted, replace the original PCM.
11. Disconnect the HDS and the timing light.

ADJUSTMENT

Ignition timing is controlled by the Powertrain Control Module (PCM). No adjustment is necessary or possible.

SPARK PLUGS

REMOVAL & INSTALLATION

See Figures 75 and 76.

1. Remove the engine cover.
2. Disconnect the ignition coil connectors, then remove the ignition coils.
3. Remove the spark plugs and inspect them.
4. Apply a small amount of anti-seize compound to the plug threads, and screw the plugs into the cylinder head, finger-tight. Torque them to 16 ft. lbs. (22 Nm).
5. Install the ignition coils in the reverse order of removal.

STARTER

REMOVAL & INSTALLATION

See Figures 77 through 79.

1. Remove the air intake duct.
2. Do the battery removal procedure.
3. Remove the battery base.
4. Remove the harness clamp, then disconnect the positive starter cable and the S terminal connector from the starter.
5. Remove the lower radiator hose bracket and the dipstick.
6. Remove the two bolts holding the starter, then remove the starter.

To install:

7. Install the starter, then tighten the mounting bolts to 33 ft. lbs. (44 Nm).

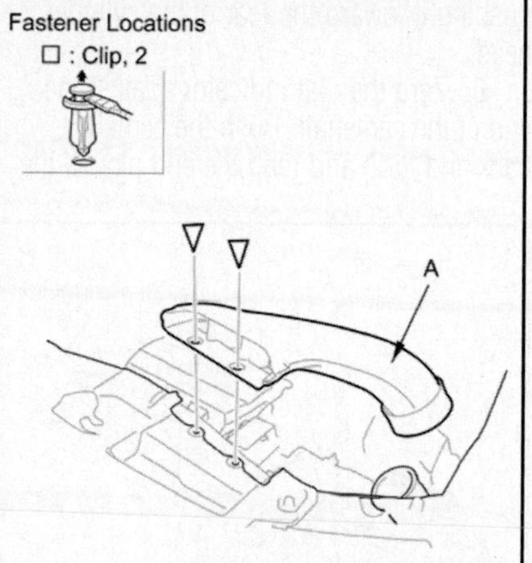
Fastener Locations
□ : Clip, 2
A
37647_PLOT_G0151

Fig. 77 Remove the air intake duct

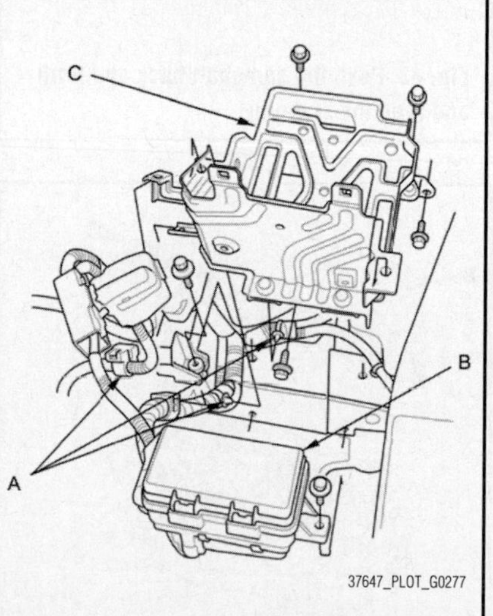
C
B
A
37647_PLOT_G0277

Fig. 78 Remove the harness clamps (A), and the auxiliary under-hood fuse/relay box (B), then remove the battery base (C)

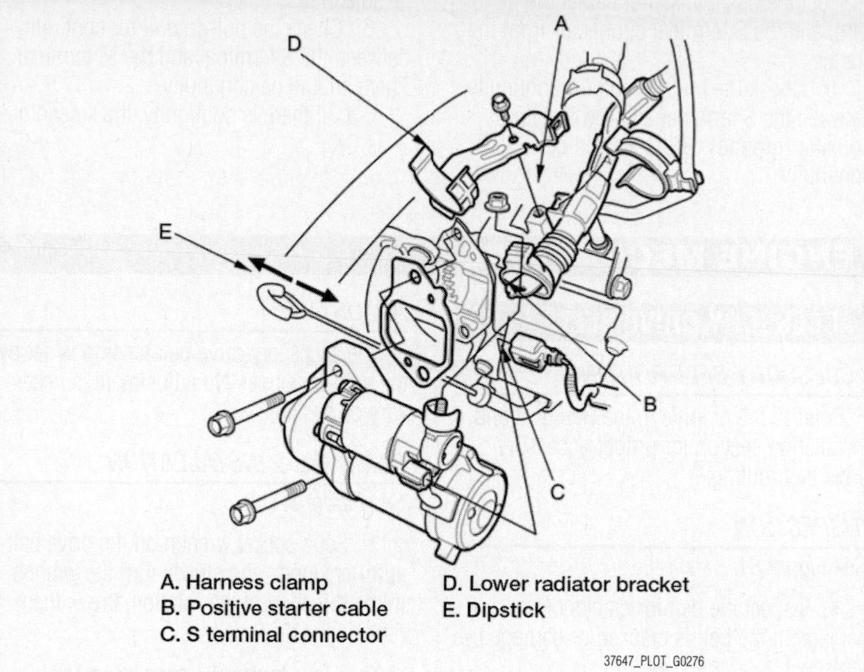
A
D
E
B
C
A. Harness clamp
B. Positive starter cable
C. S terminal connector
D. Lower radiator bracket
E. Dipstick
37647_PLOT_G0276

Fig. 79 Remove the harness clamp, then disconnect the positive starter cable and the S terminal connector from the starter

➡**Always use a new gasket.**

8. Install the lower radiator hose bracket and the dipstick.
9. Connect the positive starter cable and the S terminal connector to the starter, then install the harness clamp. Make sure the crimped side of the ring terminal faces away from the starter when you connect it.
10. Install the battery base.
11. Do the battery installation procedure.
12. Install the air intake duct.
13. Start the engine to make sure the starter works properly.

SOLENOID OR RELAY REPLACEMENT

See Figures 77,78 and 80.

1. Remove the air intake duct.
2. Do the battery removal procedure.

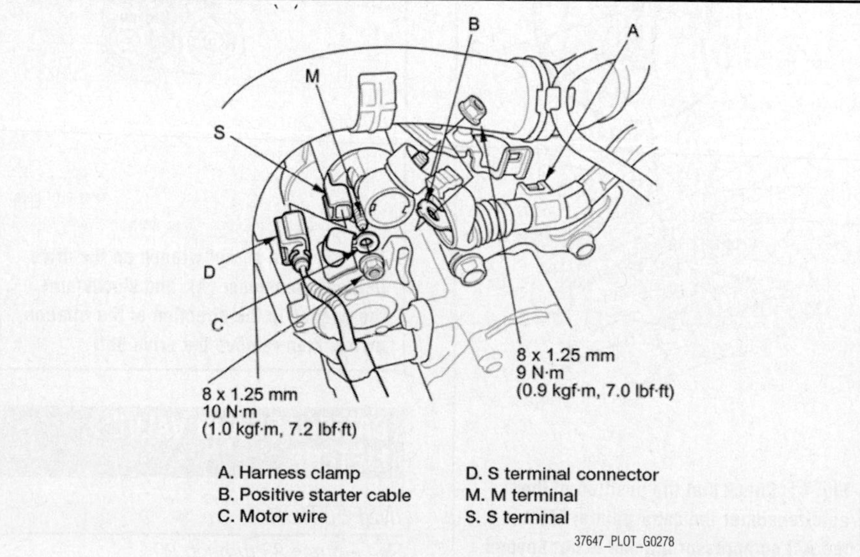
B
A
M
S
D
C
8 x 1.25 mm
10 N·m
(1.0 kgf·m, 7.2 lbf·ft)
8 x 1.25 mm
9 N·m
(0.9 kgf·m, 7.0 lbf·ft)
A. Harness clamp
B. Positive starter cable
C. Motor wire
D. S terminal connector
M. M terminal
S. S terminal
37647_PLOT_G0278

Fig. 80 Remove the harness clamp, then disconnect the positive starter cable, the motor wire, and the S terminal connector from the starter

3. Remove the battery base.
4. Remove the harness clamp, then disconnect the positive starter cable, the motor wire, and the S terminal connector from the starter.
5. Check the hold-in coil for continuity between the S terminal and the armature housing (ground). There should be continuity.
a. If there is continuity, go to step 6.
b. If there is no continuity, replace the solenoid.
6. Check the pull-in coil for continuity between the S terminal and the M terminal. There should be continuity.
a. If there is continuity, the solenoid is OK.
b. If there is no continuity, replace the solenoid.

To install:

7. Install the wire and the connector in the reverse order of removal.
8. Install the battery base.
9. Do the battery installation procedure.
10. Install the air intake duct.

ENGINE MECHANICAL

ACCESSORY DRIVE BELTS

ACCESSORY BELT ROUTING

Refer to the graphic in the Removal and Installation section for proper accessory drive belt routing.

INSPECTION

See Figure 81.

1. Inspect the belt for cracks or damage. If the belt is cracked or damaged, replace it.
2. Remove the Power Steering (P/S) fluid reservoir.
3. Check that the position of the auto-tensioner indicator pointer on the A/C compressor bracket is not beyond the edge of the indicator on the auto-tensioner. If the pointer is beyond the indicator, replace the drive belt.

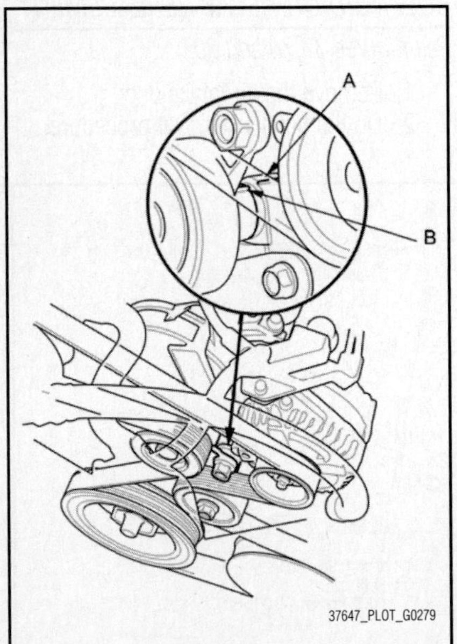
A
B
37647_PLOT_G0279

Fig. 81 Check that the position of the auto-tensioner indicator pointer (A) on the A/C compressor bracket is not beyond the edge of the indicator (B) on the auto-tensioner

ADJUSTMENT

The accessory drive belt tension is set by the auto-tensioner. No adjustment is necessary or possible.

REMOVAL & INSTALLATION

See Figure 82.

1. Set a socket wrench on the drive belt auto-tensioner, and slowly turn the wrench in the direction of the rotation arrow, then remove the drive belt.

➡This is a hydraulic type auto-tensioner, so you must turn the wrench slowly for at least 3 seconds.

2. Install the new belt in the reverse order of removal.

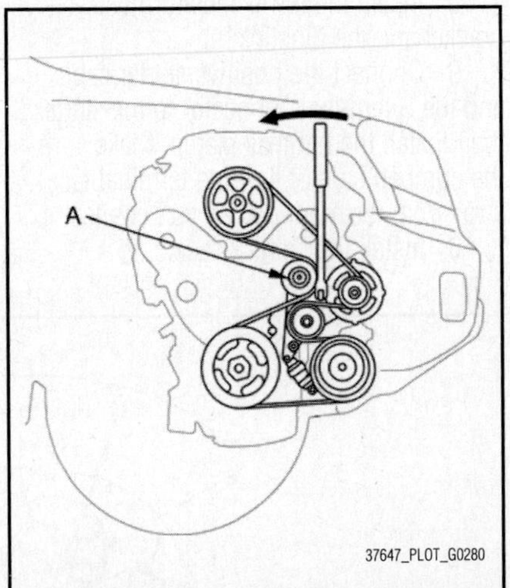
A
37647_PLOT_G0280

Fig. 82 Set a socket wrench on the drive belt auto-tensioner (A), and slowly turn the wrench in the direction of the rotation arrow, then remove the drive belt

CAMSHAFT AND VALVE LIFTERS

INSPECTION

See Figures 83 through 89.

1. Remove the cylinder head.
2. Remove the rocker arm assembly.
3. Front cylinder head: Put the rocker shafts bridge and the rocker shaft holder on the front cylinder head, then tighten the bolts to 16 ft. lbs. (22 Nm).
4. Rear cylinder head: Put the rocker shaft bridge and the rocker shaft holder on the rear cylinder head, then tighten the bolts to 16 ft. lbs. (22 Nm).
5. Seat the camshaft by pushing it toward the rear of the cylinder head.
6. Zero the dial indicator against the end of the camshaft. Push the camshaft back and forth and read the end play. If the

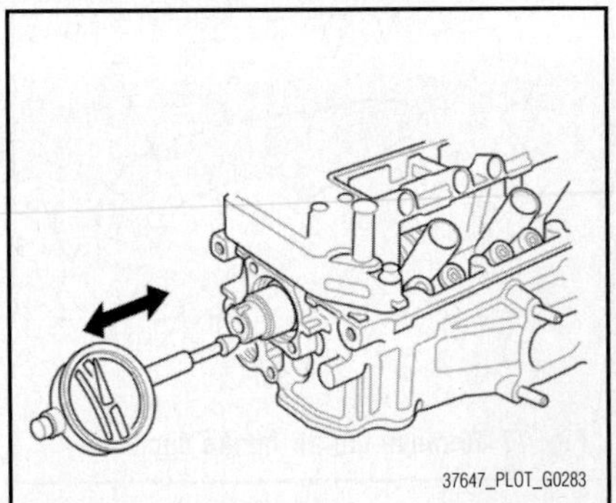
37647_PLOT_G0283

Fig. 83 Push the camshaft back and forth and read the end play

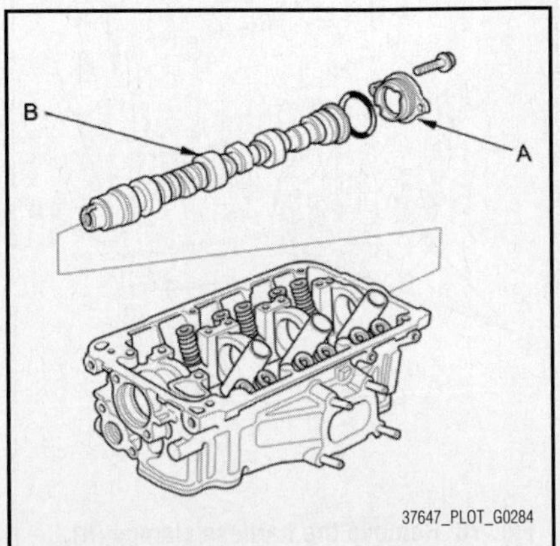
B
A
37647_PLOT_G0284

Fig. 84 Remove the camshaft thrust cover (A), then pull out the camshaft (B)

end play is beyond the service limit of 0.008 inches (0.20 mm), replace the thrust cover and recheck. If it is still beyond the service limit, replace the camshaft.

7. Remove the camshaft thrust cover, then pull out the camshaft.

8. Wipe the camshaft clean, then inspect the lift ramps. Replace the camshaft if any lobes are pitted, scored, or excessively worn.

9. Measure the diameter of each camshaft journal.

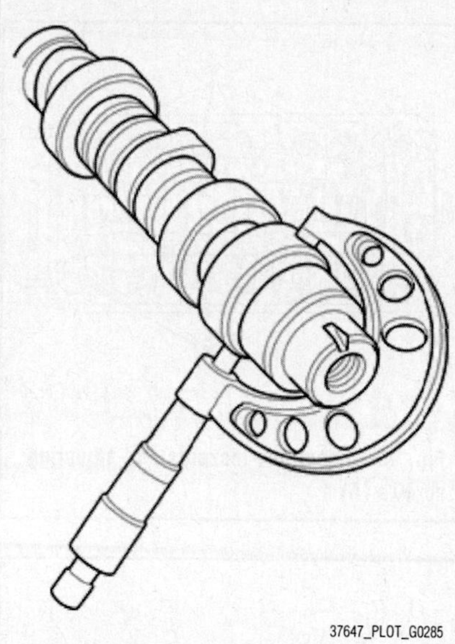
37647_PLOT_G0285

Fig. 85 Measure the diameter of each camshaft journal

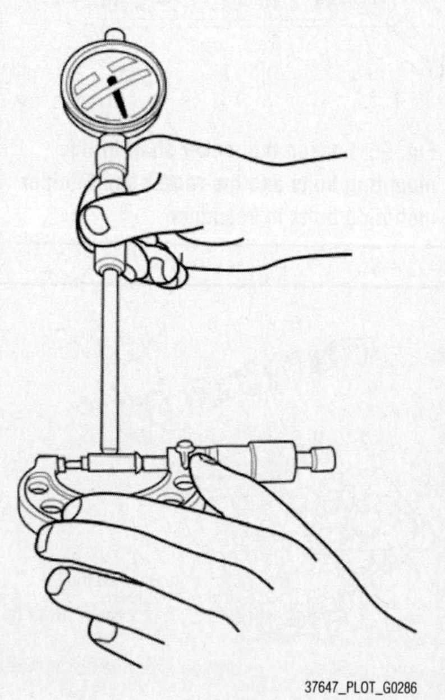
37647_PLOT_G0286

Fig. 86 Zero the gauge to the journal diameter

10. Zero the gauge to the journal diameter.

11. Clean the camshaft bearing surfaces in the cylinder head. Measure the inside diameter of each camshaft bearing surface, and check for an out-of-round condition.

a. If the camshaft-to-holder clearance is within the service limit, go to step 13. Standard (New): 0.0020–0.0035 inches (0.050–0.089 mm).

b. If the camshaft-to-holder clearance is beyond the service limit of 0.006 inches (0.15 mm) and the camshaft has been replaced, replace the cylinder head.

c. If the camshaft-to-holder clearance is beyond the service limit and the camshaft has not been replaced, go to step 12.

12. Check the total runout with the camshaft supported on V-blocks.

a. If the total runout of the camshaft is within the service limit, replace the cylinder head. Standard (New): 0.001 inches (0.03 mm) max.

b. If the total runout is beyond the service limit of 0.002 inches (0.04 mm), replace the camshaft and recheck the oil clearance. If the oil clearance is still out of tolerance, replace the cylinder head.

13. Measure the cam lobe height.

➡When measuring the No. 1, No. 2, No. 3, and No. 4 cylinders intake cam lobe height of the camshaft, measure the secondary cam lobes.

When measuring the No. 1, No. 2, No. 3, and No. 4 cylinders exhaust cam lobe height of the camshaft, measure the primary cam lobes.

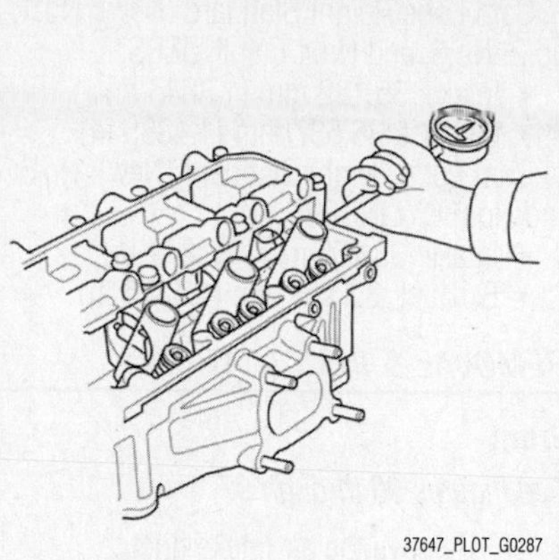
37647_PLOT_G0287

Fig. 87 Measure the inside diameter of each camshaft bearing surface, and check for an out-of-round condition

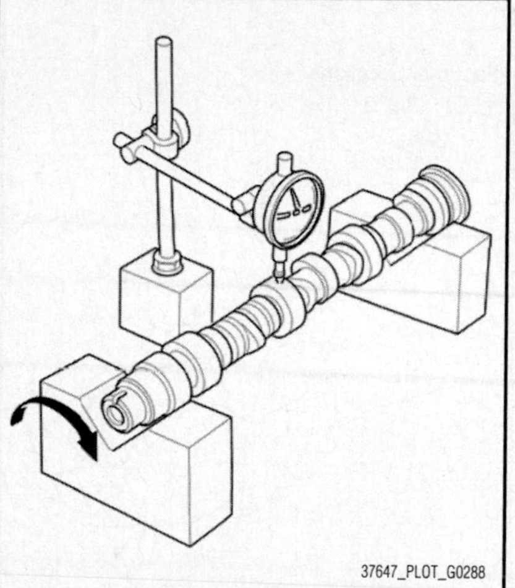
37647_PLOT_G0288

Fig. 88 Check the total runout with the camshaft supported on V-blocks

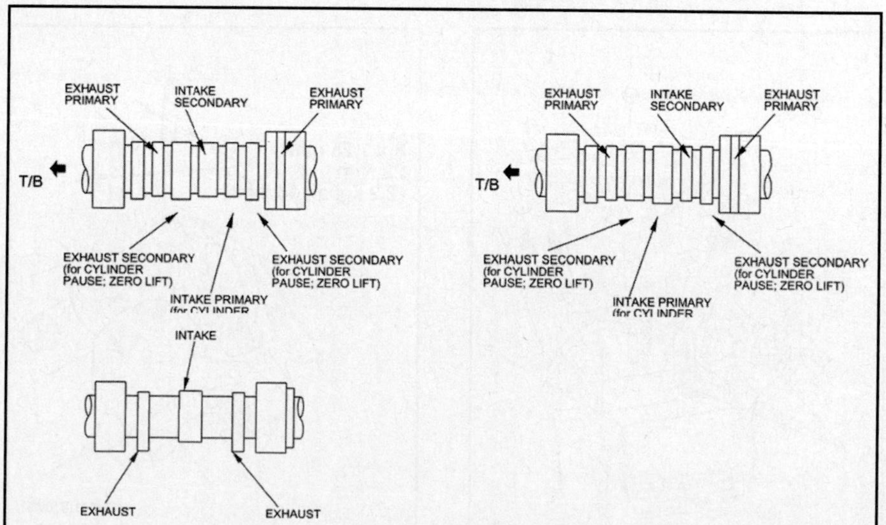
EXHAUST PRIMARY
INTAKE SECONDARY
EXHAUST PRIMARY
T/B
EXHAUST SECONDARY (for CYLINDER PAUSE; ZERO LIFT)
EXHAUST SECONDARY (for CYLINDER PAUSE; ZERO LIFT)
INTAKE PRIMARY
INTAKE
EXHAUST
EXHAUST
EXHAUST PRIMARY
INTAKE SECONDARY
EXHAUST PRIMARY
T/B
EXHAUST SECONDARY (for CYLINDER PAUSE; ZERO LIFT)
EXHAUST SECONDARY (for CYLINDER PAUSE; ZERO LIFT)
INTAKE PRIMARY
37647_PLOT_G0289

Fig. 89 Measure the cam lobe height

Cam Lobe Height Standard (New): No.1, No.2, No.3, and No.4 CYLINDERS:

- Intake: 35.162 mm (1.3843 in)
- Exhaust: 36.537 mm (1.4385 in)

Cam Lobe Height Standard (New): No.5, and No.6 CYLINDERS:

- Intake: 35.155 mm (1.3841 in)
- Exhaust: 36.512 mm (1.4375 in)

REMOVAL & INSTALLATION

Front

See Figures 90 through 97.

1. Remove the air intake duct.
2. Do the battery removal procedure.
3. Remove the battery base.
4. Drain the engine coolant.
5. Disconnect the radiator hoses.

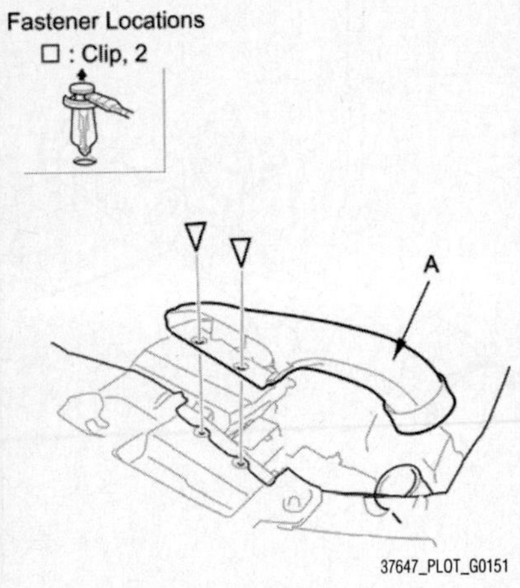
Fastener Locations
□ : Clip, 2
A
37647_PLOT_G0151

Fig. 90 Remove the air intake duct

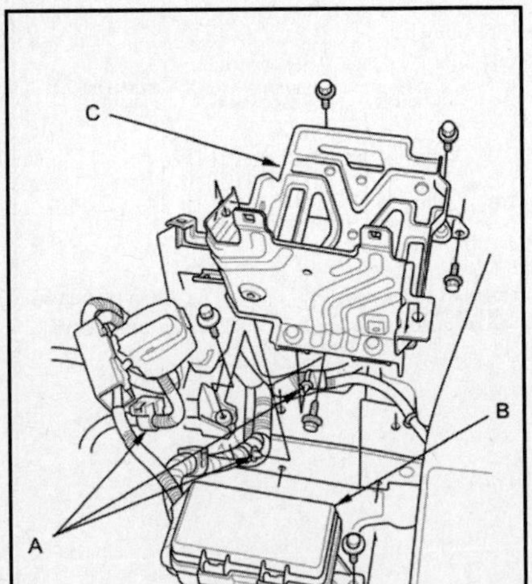
C
B
A
37647_PLOT_G0277

Fig. 91 Remove the battery base

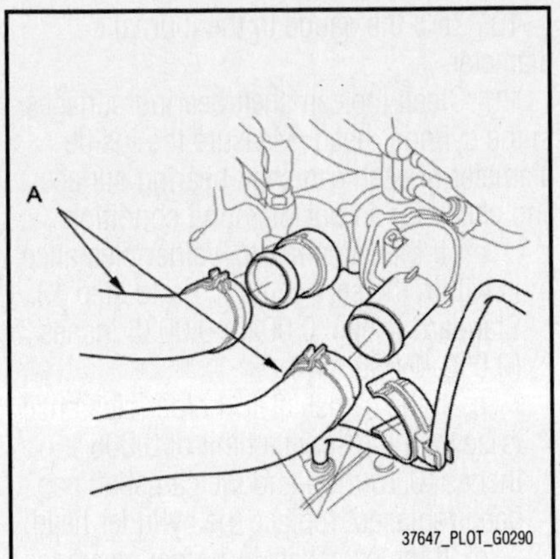
A
37647_PLOT_G0290

Fig. 92 Disconnect the radiator hoses (A)

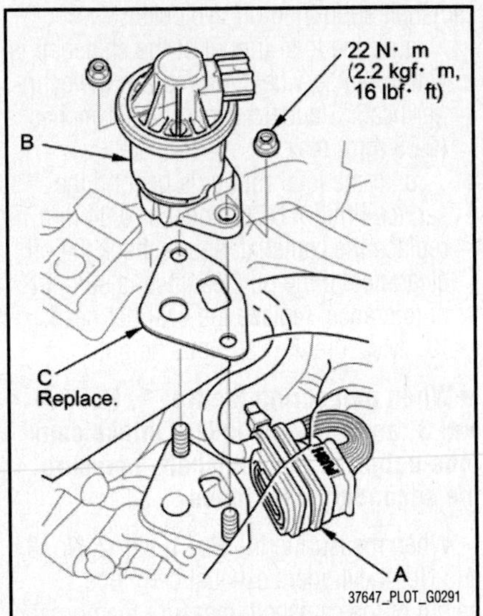
22 N· m
(2.2 kgf· m,
16 lbf· ft)
B
C
Replace.
A
37647_PLOT_G0291

Fig. 93 Disconnect the EGR 5P valve connector (A); remove the EGR valve (B)

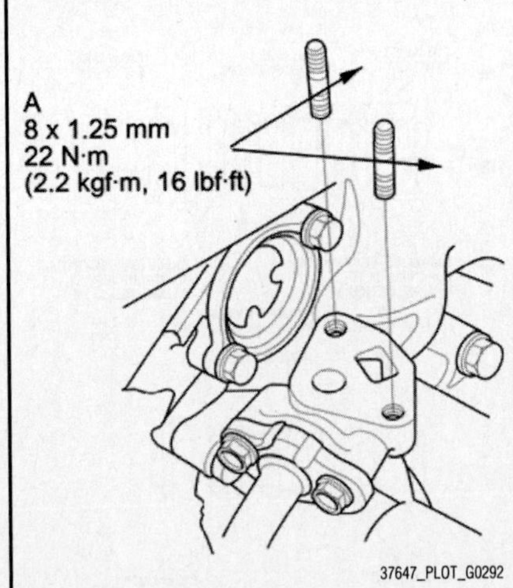
A
8 x 1.25 mm
22 N·m
(2.2 kgf·m, 16 lbf·ft)
37647_PLOT_G0292

Fig. 94 Remove the EGR valve stud bolts (A)

6. Remove the Exhaust Gas Recirculation (EGR) valve.
7. Remove the EGR valve stud bolts.
8. Remove the timing belt.
9. Remove the rocker arm assembly.
   a. Loosen the locknuts and adjusting screws.
   b. Remove the rocker shaft bridge mounting bolts, the rocker shaft holder mounting bolts, and the rocker arm assembly.
   c. Loosen the rocker shaft bridge mounting bolts and the rocker shaft

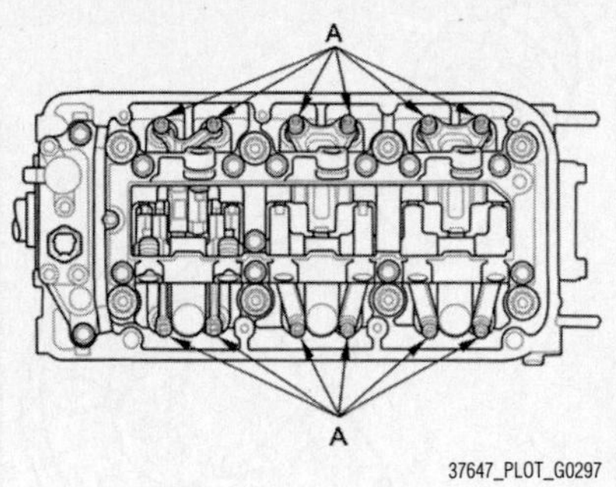
A
A
37647_PLOT_G0297

Fig. 95 Loosen the locknuts and adjusting screws (A)

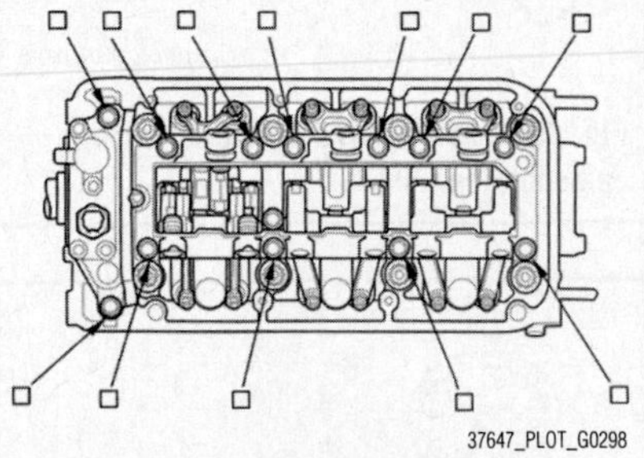
37647_PLOT_G0298

Fig. 96 Loosen the rocker shaft bridge mounting bolts and the rocker shaft holder mounting bolts in sequence

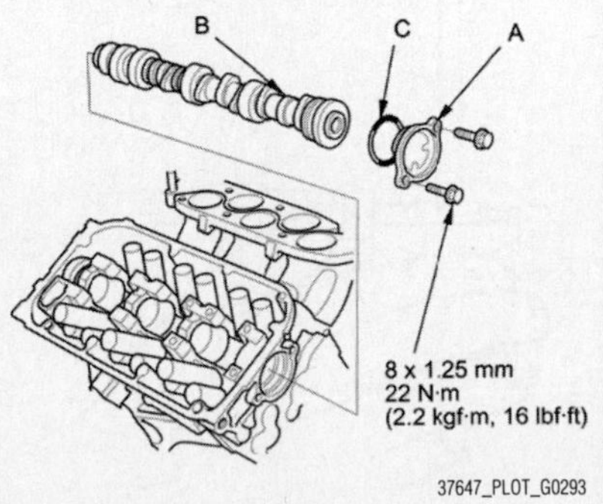
B
C
A
8 x 1.25 mm
22 N·m
(2.2 kgf·m, 16 lbf·ft)
37647_PLOT_G0293

Fig. 97 Remove the thrust cover (A), then remove the camshaft (B) and O-ring (C)

holder mounting bolts in sequence two turns at a time, starting at the ends in a crisscross pattern working toward the middle to prevent damaging the valves or the rocker arm assembly.

d. When removing the rocker arm assembly, do not remove the rocker shaft bridge mounting bolts and the rocker shaft holder mounting bolts. The bolts will keep the rocker arms on the shafts.

10. Remove the front camshaft pulley.
11. Remove the thrust cover, then remove the camshaft.

To install:

12. Install the camshaft in the reverse order of removal using a new O-ring. Apply new engine oil to the journals and cam lobes.
13. Apply new engine oil to the threads of the camshaft pulley mounting bolt, then install the front camshaft pulley.
14. Install the rocker arm assembly, then tighten the mounting bolts.
15. Install the timing belt.
16. Adjust the valve clearance.
17. Install the EGR valve stud bolts.
18. Install the EGR valve.
19. Connect the radiator hoses.
20. Install the battery base.
21. Do the battery installation procedure.
22. Install the air intake duct.
23. Fill the radiator with engine coolant and bleed the air from the cooling system with the heater valve open.
24. Do the Crankshaft Position (CKP) pattern clear/CKP pattern lean procedure.

Rear

See Figures 90, 98 through 103.

1. Relieve the fuel pressure.
2. Drain the engine coolant.
3. Remove the intake air duct.
4. Remove the air cleaner assembly.
5. Remove the quick-connect fitting cover, then disconnect the fuel feed hose.
6. Disconnect the heater hoses and the Evaporative Emission (EVAP) canister hose, then remove the purge joint bracket.
7. Remove the timing belt.
8. Remove the rocker arm assembly.

a. Loosen the locknuts and the adjusting screws (A).

b. Remove the rocker shaft bridge mounting bolts, the rocker shaft holder mounting bolts, and the rocker arm assembly.

c. Loosen the rocker shaft bridge mounting bolts and the rocker shaft holder mounting bolts in sequence two

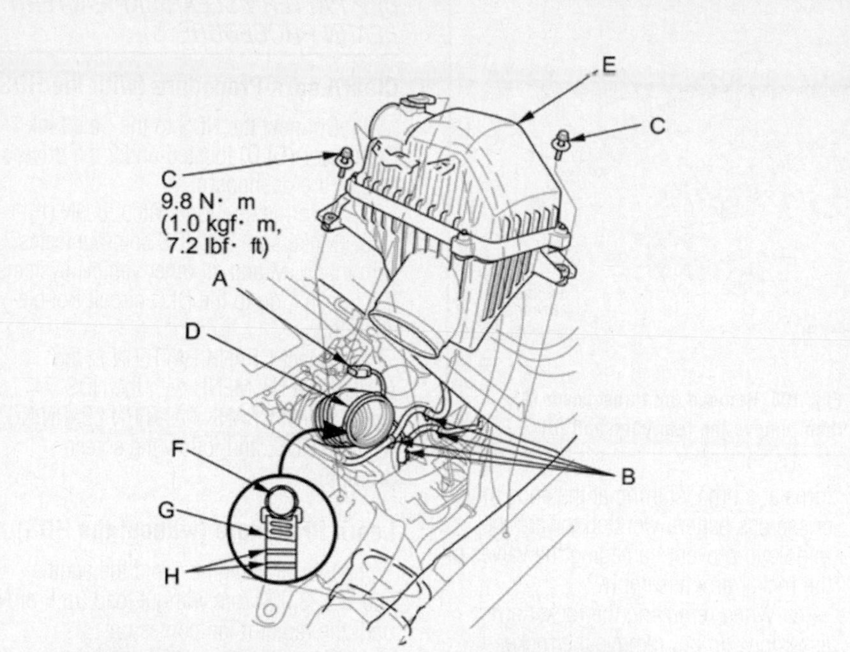
E
C
C
9.8 N· m
(1.0 kgf· m,
7.2 lbf· ft)
A
D
F
B
G
H

A. AMF sensor/IAT sensor connector
B. Harness clamps
C. Bolts
D. Band
E. Air cleaner
F. Screw of the hose band
G. Edge of the hose band
H. Paint marks

37647_PLOT_G0281

Fig. 98 Remove the air cleaner assembly

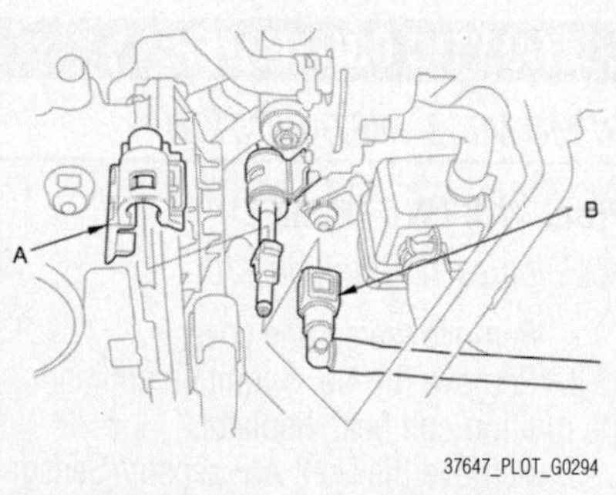
A
B

37647_PLOT_G0294

Fig. 99 Remove the quick-connect fitting cover (A), then disconnect the fuel feed hose (B)

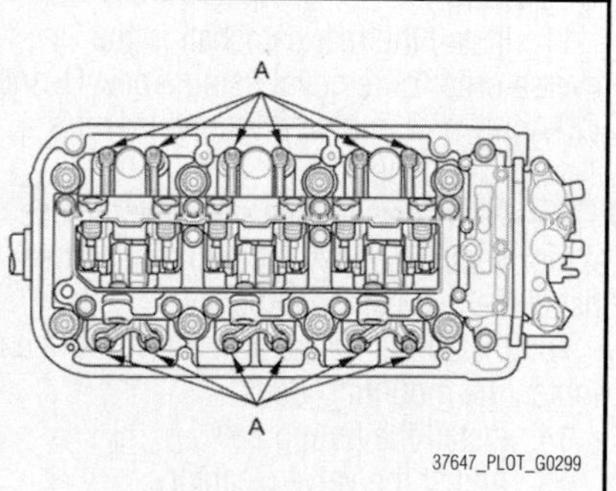
A
A

37647_PLOT_G0299

Fig. 101 Loosen the locknuts and the adjusting screws (A)

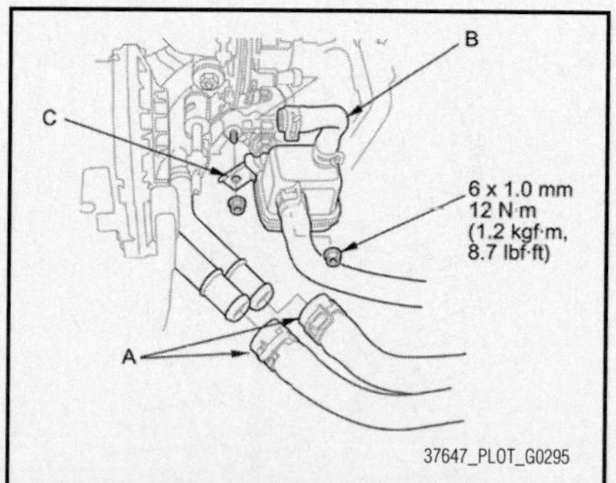
B
C
6 x 1.0 mm
12 N·m
(1.2 kgf·m,
8.7 lbf·ft)
A

37647_PLOT_G0295

Fig. 100 Disconnect the heater hoses (A) and the EVAP canister hose (B), then remove the purge joint bracket (C)

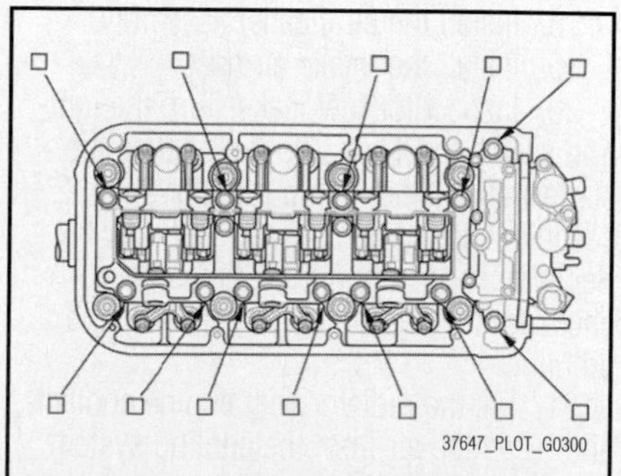

37647_PLOT_G0300

Fig. 102 Loosen the rocker shaft bridge mounting bolts and the rocker shaft holder mounting bolts in sequence

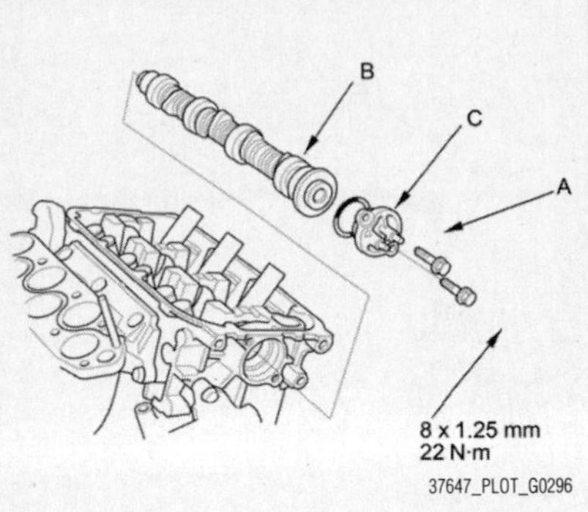
B
C
A
8 x 1.25 mm
22 N·m
37647_PLOT_G0296

Fig. 103 Remove the thrust cover (A), then remove the rear camshaft (B)

turns at a time, starting at the ends in a crisscross pattern working toward the middle to prevent damaging the valves or the rocker arm assembly.

d. When removing the rocker arm assembly, do not remove the rocker shaft bridge mounting bolts and the rocker shaft holder mounting bolts. The bolts will keep the rocker arms on the shafts.

9. Remove the rear camshaft pulley.

10. Remove the thrust cover, then remove the rear camshaft.

To install:

11. Install the rear camshaft in the reverse order of removal using a new O-ring (C). Apply new engine oil to the journals and the cam lobes.

12. Apply new engine oil to the threads of the camshaft pulley mounting bolt, then install the rear camshaft pulley.

13. Install the rocker arm assembly, then tighten the mounting bolts.

14. Install the timing belt.

15. Adjust the valve clearance.

16. Install the heater hoses and the purge joint bracket.

17. Connect the fuel feed hose, then install the quick-connect fitting cover.

18. Install the air cleaner assembly.

19. Install the intake air duct.

20. Inspect for fuel leaks. Turn the ignition switch to ON (II) (do not operate the starter) so the fuel pump runs for about 2 seconds and pressurizes the fuel line. Repeat this operation three times, then check for fuel leakage at any point in the fuel line.

21. Fill the radiator with engine coolant, and bleed the air from the cooling system with the heater valve open.

22. Do the CKP pattern clear/CKP pattern learn procedure.

CKP PATTERN CLEAR/CKP PATTERN LEARN PROCEDURE

Clear/Learn Procedure (with the HDS)

1. Connect the HDS to the Data Link Connector (DLC) located under the driver's side of the dashboard.

2. Turn the ignition switch to ON (II).

3. Make sure the HDS communicates with the PCM and all other vehicle systems. If it doesn't, go to the DLC circuit troubleshooting.

4. Select CRANK PATTERN in the ADJUSTMENT MENU with the HDS.

5. Select CRANK PATTERN LEARNING with the HDS, and follow the screen prompts.

Learn Procedure (without the HDS)

1. Start the engine. Hold the engine speed at 3,000 rpm without load (in P or N) until the radiator fan comes on.

2. Test-drive the vehicle on a level road: Decelerate (with the throttle fully closed) from an engine speed of 2,500 rpm down to 1,000 rpm with the transmission in 2.

3. Repeat step 2 several times.

4. Turn the ignition switch to LOCK (0).

5. Turn the ignition switch to ON (II), and wait for 30 seconds. The CKP pattern learn procedure is complete.

CATALYTIC CONVERTER

REMOVAL & INSTALLATION

Front WU-TWC (Bank 2)

See Figures 104 through 110.

1. Remove the engine cover.

2. Remove the No. 5 ignition coil and the ignition coil heat insulator.

3. Remove the front A/F sensor (Sensor 1) and front secondary HO2S (Sensor 2).

4. Remove the exhaust pipe A mounting nuts (front WU-TWC side).

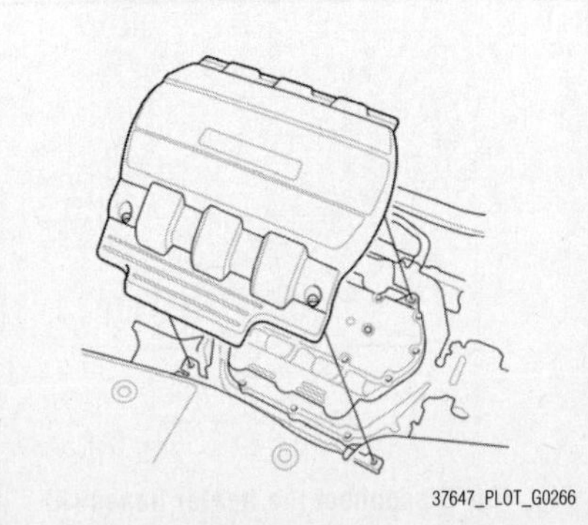
37647_PLOT_G0266

Fig. 104 Remove the engine cover

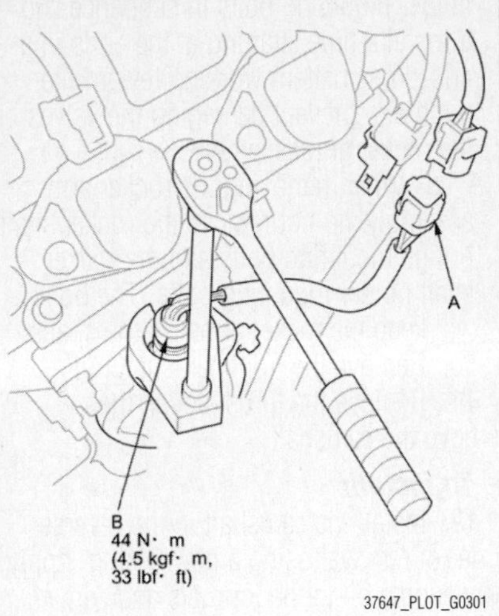
A
B
44 N· m
(4.5 kgf· m,
33 lbf· ft)
37647_PLOT_G0301

Fig. 105 Remove the front A/F sensor (Sensor 1)

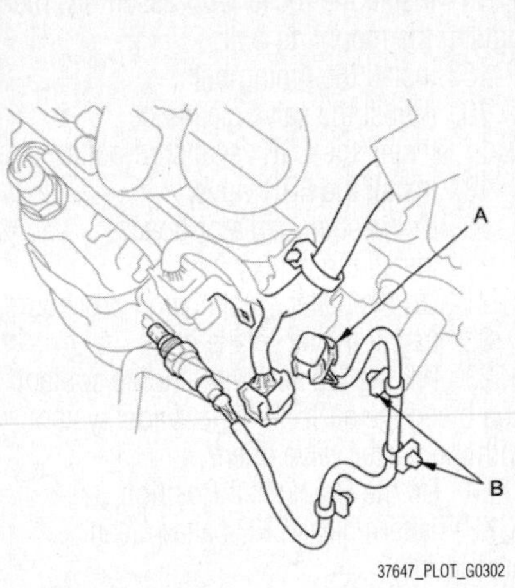
A
B
37647_PLOT_G0302

Fig. 106 Remove the front secondary HO2S (Sensor 2)

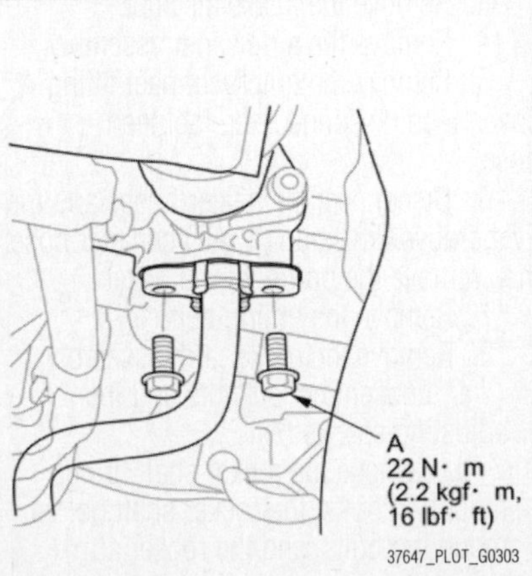
A
22 N· m
(2.2 kgf· m,
16 lbf· ft)
37647_PLOT_G0303

Fig. 107 Remove the EGR pipe mounting bolts (A)

5. Remove the EGR pipe.
   a. Remove the EGR pipe mounting bolts.
   b. Raise the vehicle on a lift.
   c. Remove the splash shield.
   d. Remove the EGR pipe mounting nuts and the EGR pipe.
6. Remove the A/C condenser fan assembly. Refer to Engine Fan Removal and Installation.

➡Place an appropriate size piece of cardboard in front of the radiator. The cardboard protects the radiator from damage when removing the warm up TWC.

7. Remove the front WU-TWC

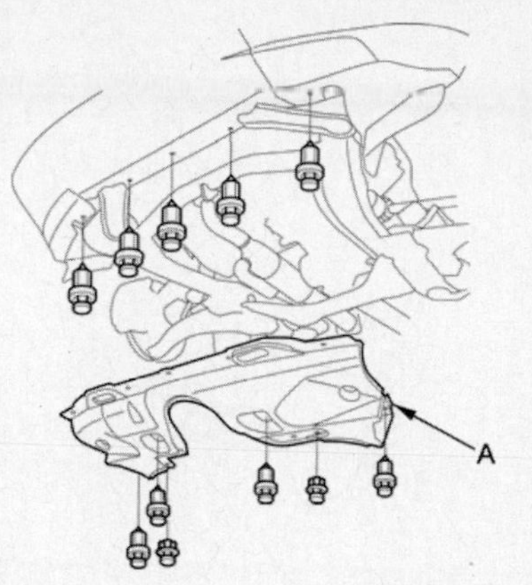
A
37647_PLOT_G0174
Fig. 108 Remove the splash shield

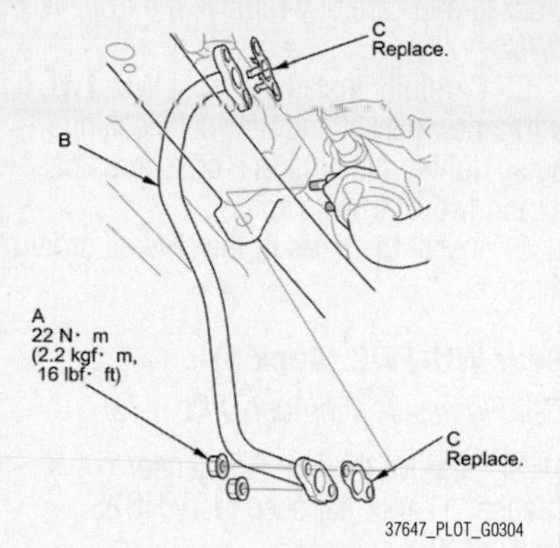
C
Replace.
B
A
22 N· m
(2.2 kgf· m,
16 lbf· ft)
C
Replace.
37647_PLOT_G0304
Fig. 109 Remove the EGR pipe mounting nuts (A) and the EGR pipe (B)

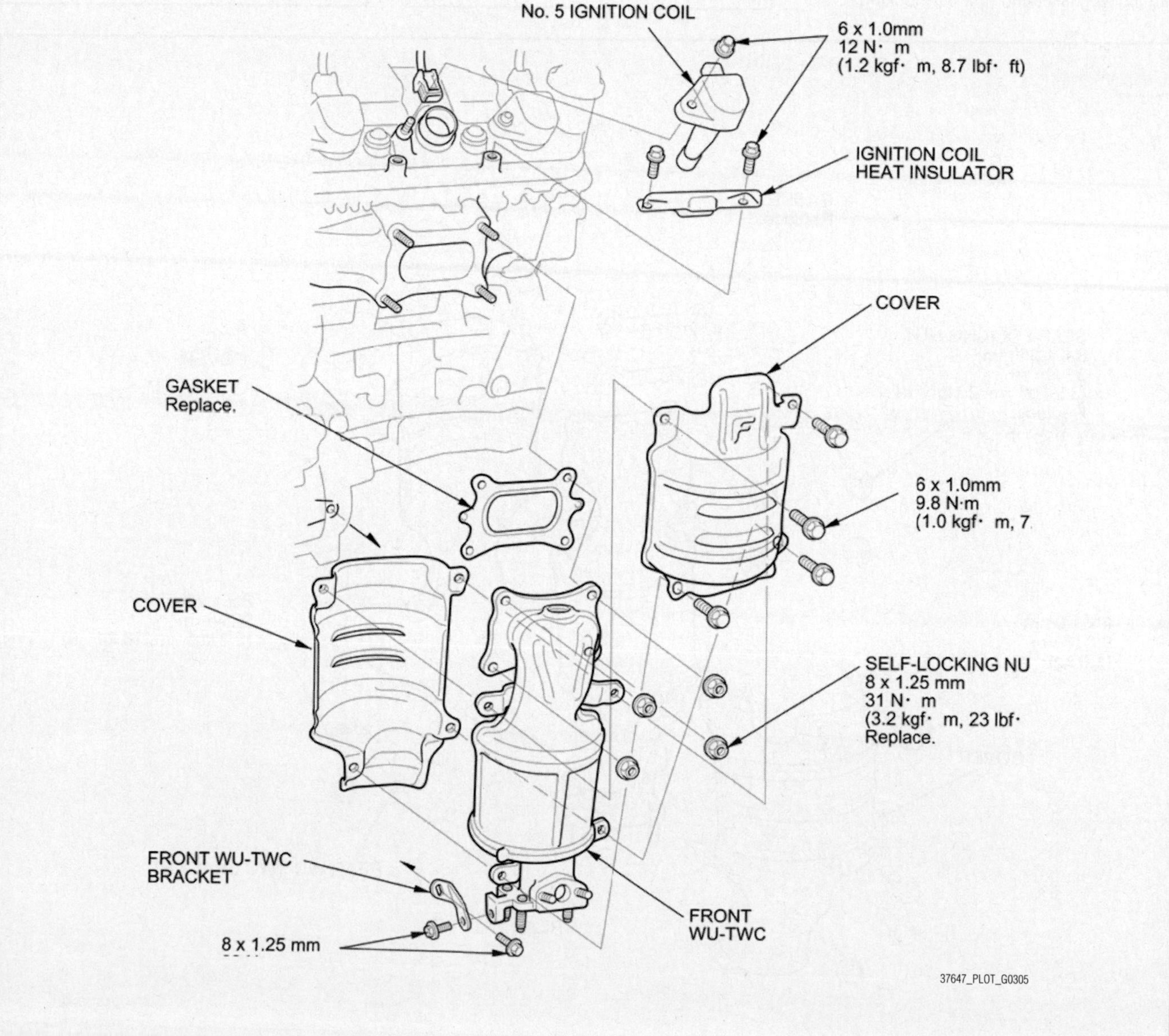
No. 5 IGNITION COIL
6 x 1.0mm
12 N· m
(1.2 kgf· m, 8.7 lbf· ft)
IGNITION COIL
HEAT INSULATOR
COVER
GASKET
Replace.
6 x 1.0mm
9.8 N·m
(1.0 kgf· m, 7.
COVER
SELF-LOCKING NU
8 x 1.25 mm
31 N· m
(3.2 kgf· m, 23 lbf·
Replace.
FRONT WU-TWC
BRACKET
FRONT
WU-TWC
8 x 1.25 mm
37647_PLOT_G0305
Fig. 110 Exploded view of front WU-TWC assembly

bracket, and carefully remove the front WU-TWC.

8. Carefully install the front WU-TWC with a new gasket and new self-locking nuts. Tighten the nuts in a crisscross pattern in two or three steps.

9. Install the parts in the reverse order of removal.

Rear WU-TWC (Bank 1)

See Figures 111 through 113.

1. Remove the rear A/F sensor (Sensor 1) and rear secondary HO2S (Sensor 2).
2. Remove exhaust pipe A, then remove the rear WU-TWC bracket.
3. Remove the intermediate shaft.
4. Carefully remove the rear WU-TWC.
5. Carefully install the rear WU-TWC with a new gasket and new self-locking

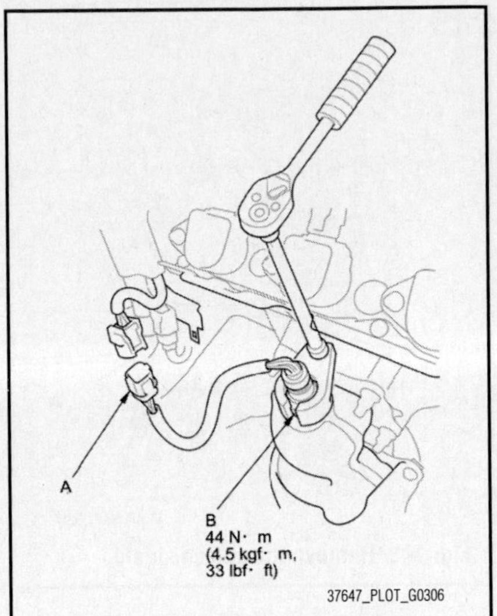
A
B
44 N· m
(4.5 kgf· m,
33 lbf· ft)
37647_PLOT_G0306

Fig. 111 Remove the rear A/F sensor (Sensor 1)

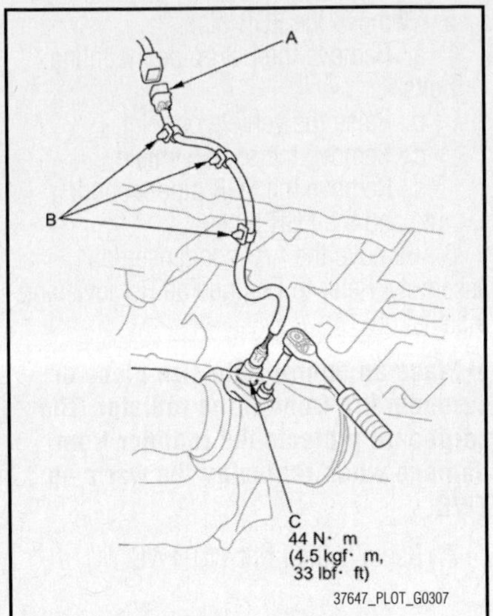
A
B
C
44 N· m
(4.5 kgf· m,
33 lbf· ft)
37647_PLOT_G0307

Fig. 112 Remove the rear secondary HO2S (Sensor 2)

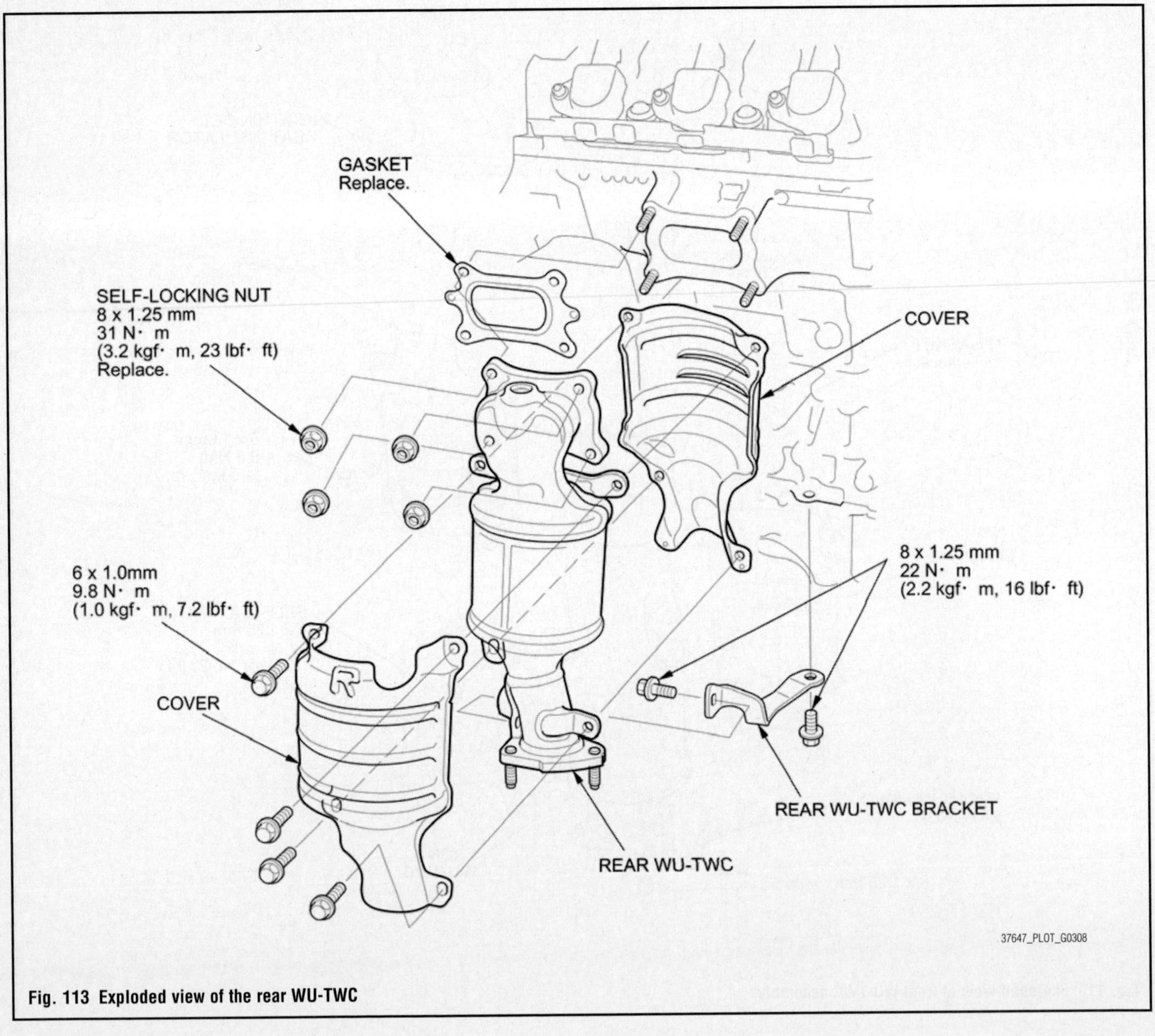
GASKET
Replace.
SELF-LOCKING NUT
8 x 1.25 mm
31 N· m
(3.2 kgf· m, 23 lbf· ft)
Replace.
COVER
6 x 1.0mm
9.8 N· m
(1.0 kgf· m, 7.2 lbf· ft)
8 x 1.25 mm
22 N· m
(2.2 kgf· m, 16 lbf· ft)
COVER
REAR WU-TWC BRACKET
REAR WU-TWC
37647_PLOT_G0308

Fig. 113 Exploded view of the rear WU-TWC

nuts. Tighten the nuts in a crisscross pattern in two or three steps.

6. Install the parts in the reverse order of removal.

Under-Floor TWC

See Figure 114.

1. Raise the vehicle on a lift.
2. Remove the exhaust pipe hangers.
3. Remove the under-floor TWC.
4. Remove the converter covers.
5. Install the parts in the reverse order of removal with new gaskets and new self-locking nuts.

CRANKSHAFT PULLEY

REMOVAL & INSTALLATION

See Figures 115 through 118.

Special Tools Required:

- Handle, 6-25-660L 07JAB-001020B
- Holder Attachment, 50 mm, Offset 07MAB-PY3010A
- Socket Wrench, 19 x 90L 07JAA-001020A or equivalent

1. Raise the vehicle on the lift.
2. Remove the right front wheel.
3. Remove the splash shield.

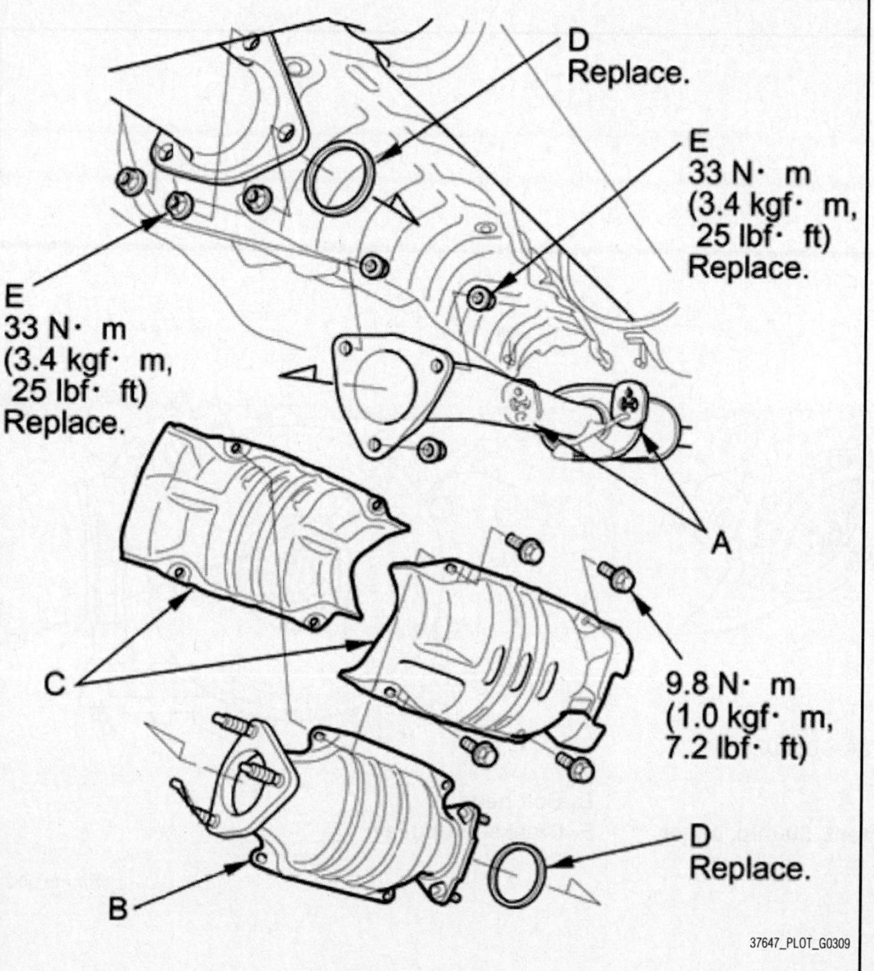
D
Replace.
E
33 N· m
(3.4 kgf· m,
25 lbf· ft)
Replace.
E
33 N· m
(3.4 kgf· m,
25 lbf· ft)
Replace.
A
C
9.8 N· m
(1.0 kgf· m,
7.2 lbf· ft)
D
Replace.
B
37647_PLOT_G0309

Fig. 114 Exploded view of the under-floor TWC assembly

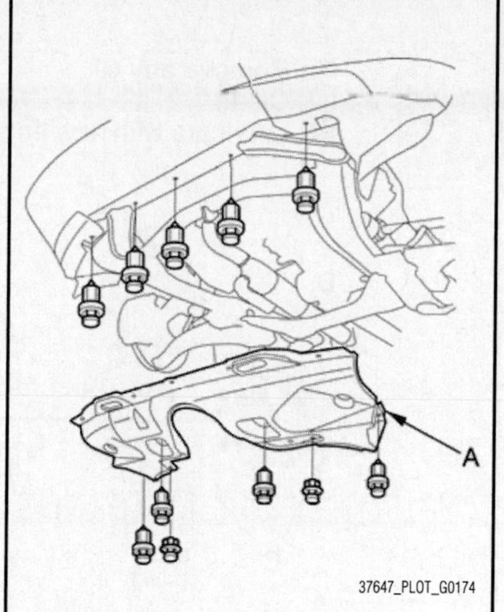
A
37647_PLOT_G0174

Fig. 115 Remove the splash shield

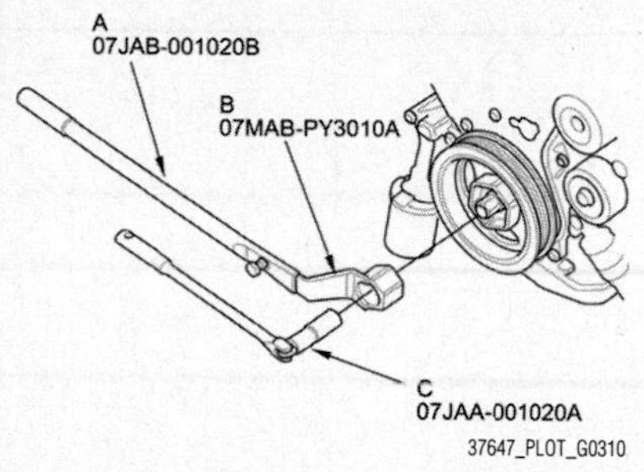
A
07JAB-001020B
B
07MAB-PY3010A
C
07JAA-001020A
37647_PLOT_G0310

Fig. 116 Hold the pulley with the holder (A) and the holder attachment, 50mm, offset (B); remove the bolt with a heavy duty socket, 19 mm (C) and a breaker bar

4. Remove the drive belt.
5. Hold the pulley with the holder and the holder attachment, 50mm, offset.
6. Remove the bolt with a heavy duty socket, 19 mm and a breaker bar, then remove the crankshaft pulley.

To install:

7. Remove any oil and clean the pulleys, the crankshaft, the bolt, and the washer. Lubricate using new engine oil as shown.
8. Install the crankshaft pulley, and tighten the bolt. Do not use an impact wrench.
   a. Hold the pulley with the handle and the holder attachment, 50mm, offset. Tighten the bolt to 47 ft. lbs. (64 Nm) with a torque wrench and a socket, 19 mm.
   b. Mark the bolt head and the crankshaft pulley as shown, then tighten the

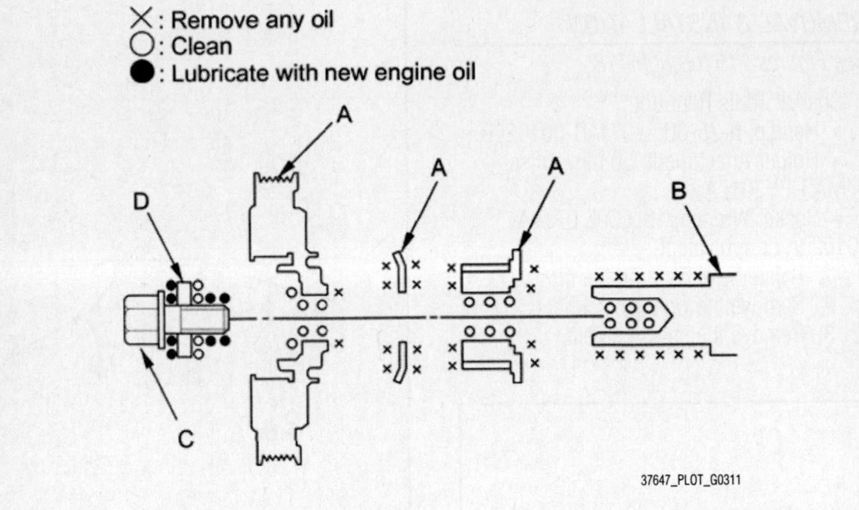

×: Remove any oil
○: Clean
●: Lubricate with new engine oil
A
A
A
B
D
C
37647_PLOT_G0311

Fig. 117 Remove any oil and clean the pulleys, the crankshaft, the bolt, and the washer; lubricate using new engine oil as shown

bolt an additional 60 degrees (The mark on the bolt head lines up with the mark on the crankshaft pulley).

9. Install the drive belt.
10. Install the splash shield.
11. Install the right front wheel.

CRANKSHAFT FRONT SEAL

REMOVAL & INSTALLATION

Special Tools Required: Oil Seal Driver, 64 mm 070AD-RCAA100

1. Remove the timing belt, the timing belt stopper, and the timing belt drive pulley.
2. Remove the pulley end crankshaft oil seal.
3. Clean and dry the crankshaft oil seal housing.
4. Apply a light coat of new engine oil to the crankshaft and to the lip of the crankshaft oil seal.

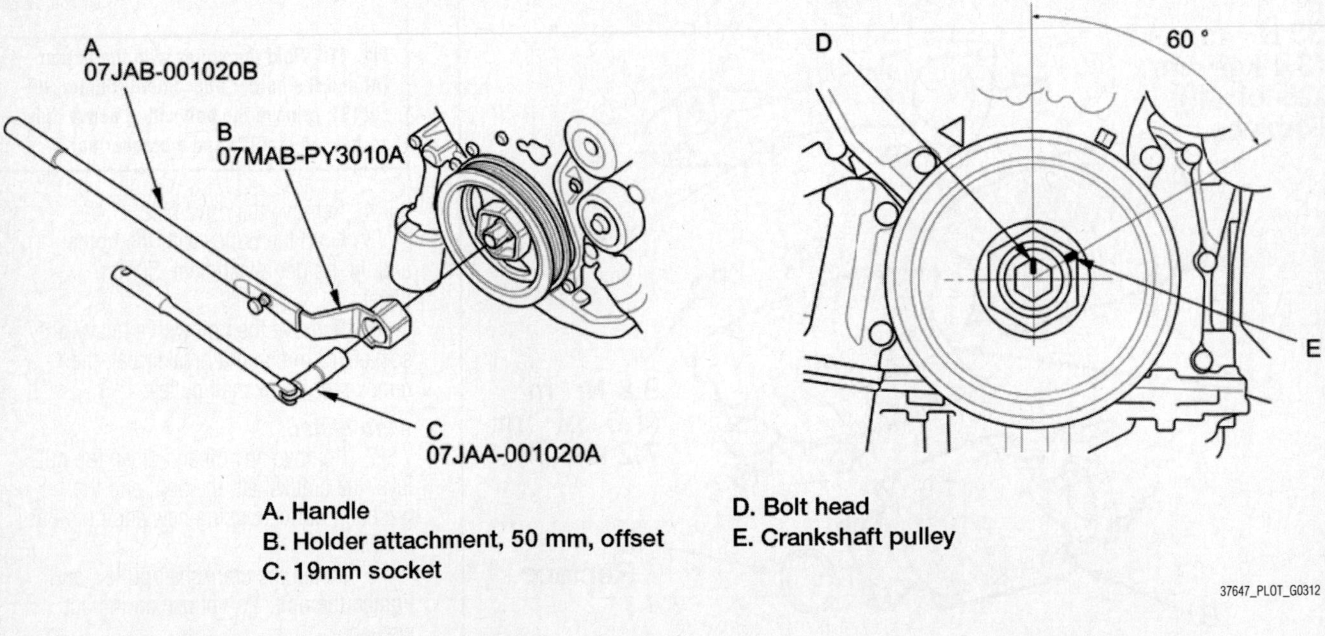

A
07JAB-001020B
B
07MAB-PY3010A
C
07JAA-001020A
D
60 °
E
A. Handle
B. Holder attachment, 50 mm, offset
C. 19mm socket
D. Bolt head
E. Crankshaft pulley
37647_PLOT_G0312

Fig. 118 Mark the bolt head and the crankshaft pulley as shown, then tighten the bolt an additional 60 degrees

5. Apply a light coat of new engine oil around the crankshaft oil seal, then using the oil seal driver, 64 mm, drive in the crankshaft oil seal until the driver bottoms against the oil pump. When the seal is in place, clean any excess grease off the crankshaft, and check that the oil seal lip is not distorted.

6. Install the timing belt drive pulley, the timing belt stopper, and the timing belt.

CYLINDER HEAD

REMOVAL & INSTALLATION

See Figures 119 through 131.

➡Note the following when removing the cylinder heads:

- Use fender covers to avoid damaging painted surfaces.
- To avoid damaging the wiring and terminals, unplug the wiring connectors carefully while holding the connector portion.
- Connect the Honda Diagnostic System (HDS) to the Data Link Connector (DLC), and monitor the Engine Coolant Temperature (ECT) sensor 1. To avoid damaging the cylinder head, wait until the ECT drops below 100°F (38°C) before loosening the cylinder head bolts.
- Mark all wiring and hoses to avoid misconnection. Also, be sure that they do not contact any other wiring or hoses, or interfere with any other parts.

1. Relieve the fuel pressure.
2. Remove the air intake duct.
3. Do the battery removal procedure.

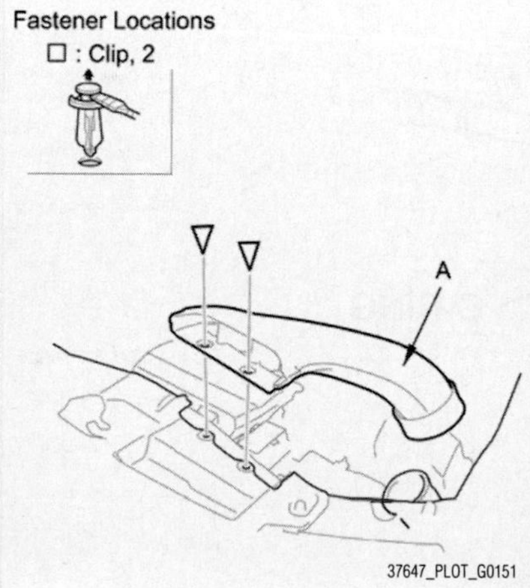
Fastener Locations
□ : Clip, 2
A
37647_PLOT_G0151
Fig. 119 Remove the air intake duct

4. Drain the engine coolant.
5. Remove the drive belt.
6. Remove the Power Steering (P/S) pump and the bolt securing the P/S hose bracket.
7. Remove the alternator.
8. Remove the timing belt.
9. Remove the intake manifold.
10. Remove the six ignition coils.
11. Disconnect the following engine wire harness connectors, and remove the wire harness clamps from the cylinder head:

- Six injector connectors
- Engine Coolant Temperature (ECT) sensor 1 connector
- Front rocker arm oil pressure switch connector
- Rear rocker arm oil pressure switch connector
- Camshaft Position (CMP) sensor connector
- Front Air Fuel Ratio (A/F) sensor 1 connector
- Rear Air Fuel Ratio (A/F) sensor 1connector
- Front secondary Heated Oxygen Sensor 2 (secondary HO2S) connector

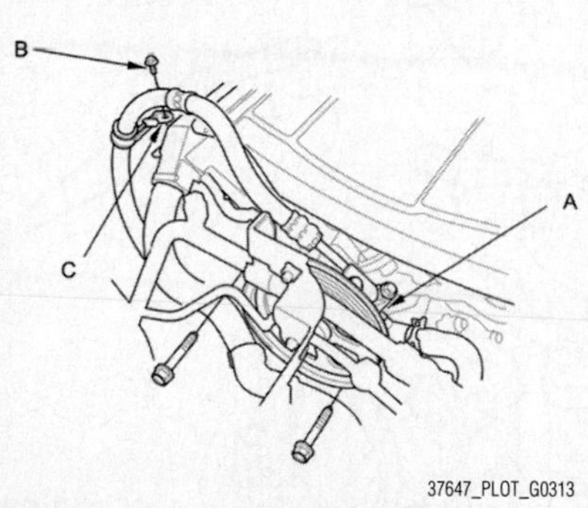
B
A
C
37647_PLOT_G0313
Fig. 120 Remove the P/S pump (A) and the bolt (B) securing the P/S hose bracket (C)

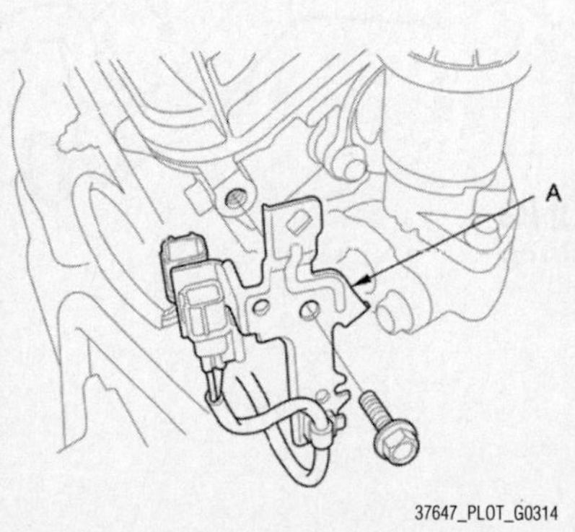
A
37647_PLOT_G0314
Fig. 121 Remove the connector bracket (A) from the front cylinder head

- Rear secondary Heated Oxygen Sensor 2 (secondary HO2S) connector
- Rocker arm oil control solenoid A (BANK 1) connector
- Rocker arm oil control solenoid A (BANK 2) connector
- Rocker arm oil control solenoid B (BANK 1) connector
- Knock sensor connector

12. Remove the front warm up three way catalytic converter (front WU-TWC) and the rear warm up three way catalytic converter (rear WU-TWC).
13. Remove the quick-connect fitting cover, then disconnect the fuel feed hose.
14. Remove the connector bracket from the front cylinder head.
15. Remove the harness clamp bracket (A) from the rear cylinder head.
16. Remove the injector bases.
17. Remove the water passage.
18. Remove the camshaft pulleys and the back covers.
19. Remove the cylinder head covers.
20. Remove the cylinder head bolts. To prevent warping, loosen the bolts in sequence ⅓ turn at a time; repeat the sequence until all the bolts are loosened.
21. Remove the cylinder heads.

To install:

22. Clean the cylinder head and the engine block surface.
23. Clean and install the oil control orifices using the new O-rings.
24. Install the dowel pins and the new cylinder head gaskets.
25. Clean the timing belt pulleys, the timing belt guide plate, and the upper and lower covers.
26. Set the timing belt drive

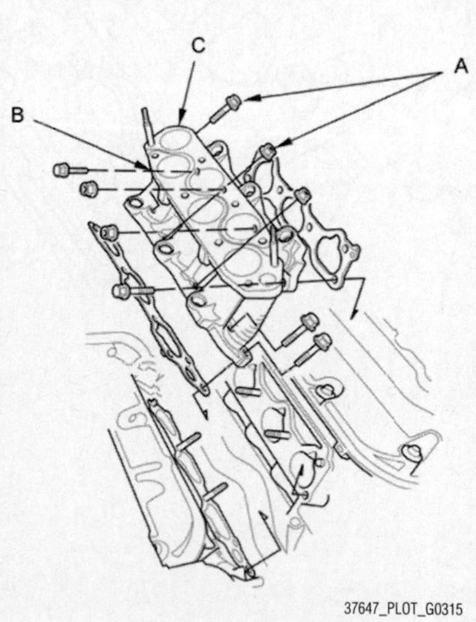
C
A
B
37647_PLOT_G0315
Fig. 122 Remove the injector bases

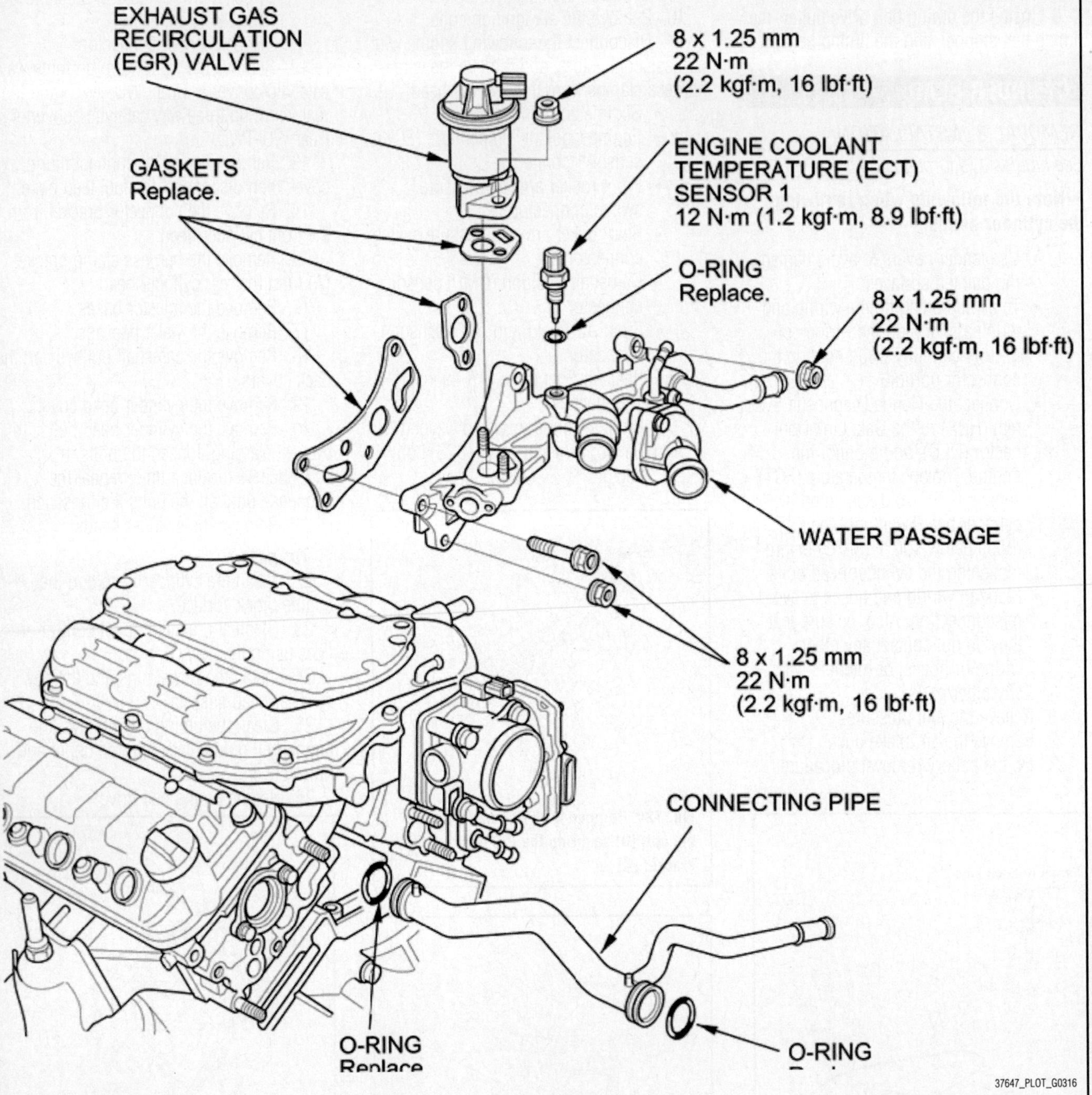
EXHAUST GAS
RECIRCULATION
(EGR) VALVE
8 x 1.25 mm
22 N·m
(2.2 kgf·m, 16 lbf·ft)
GASKETS
Replace.
ENGINE COOLANT
TEMPERATURE (ECT)
SENSOR 1
12 N·m (1.2 kgf·m, 8.9 lbf·ft)
O-RING
Replace.
8 x 1.25 mm
22 N·m
(2.2 kgf·m, 16 lbf·ft)
WATER PASSAGE
8 x 1.25 mm
22 N·m
(2.2 kgf·m, 16 lbf·ft)
CONNECTING PIPE
O-RING
O-RING
37647_PLOT_G0316

Fig. 123 Remove the water passage

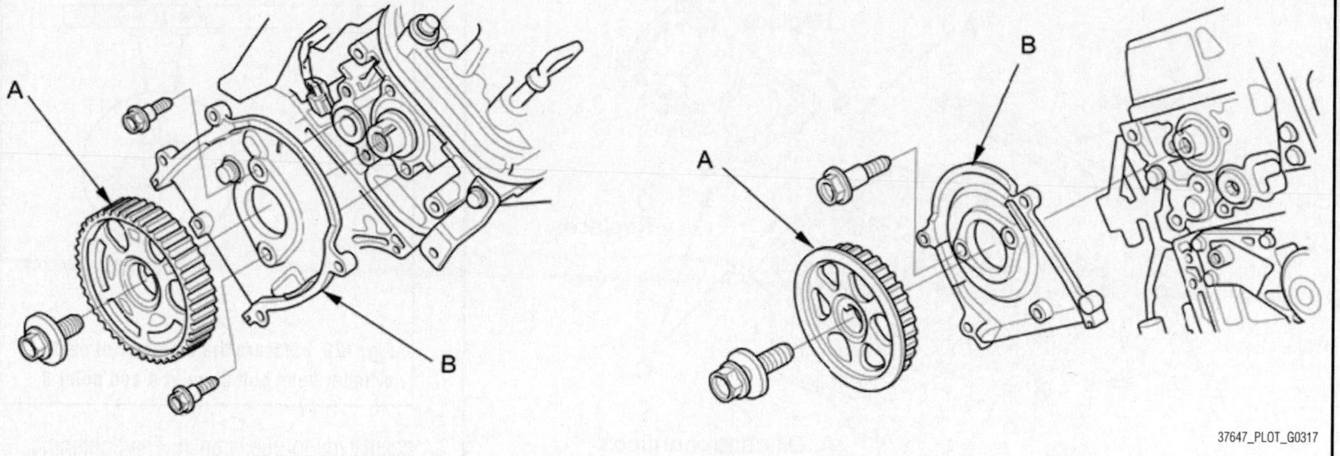
A
B
A
B
37647_PLOT_G0317

Fig. 124 Remove the camshaft pulleys (A) and the back covers (B)

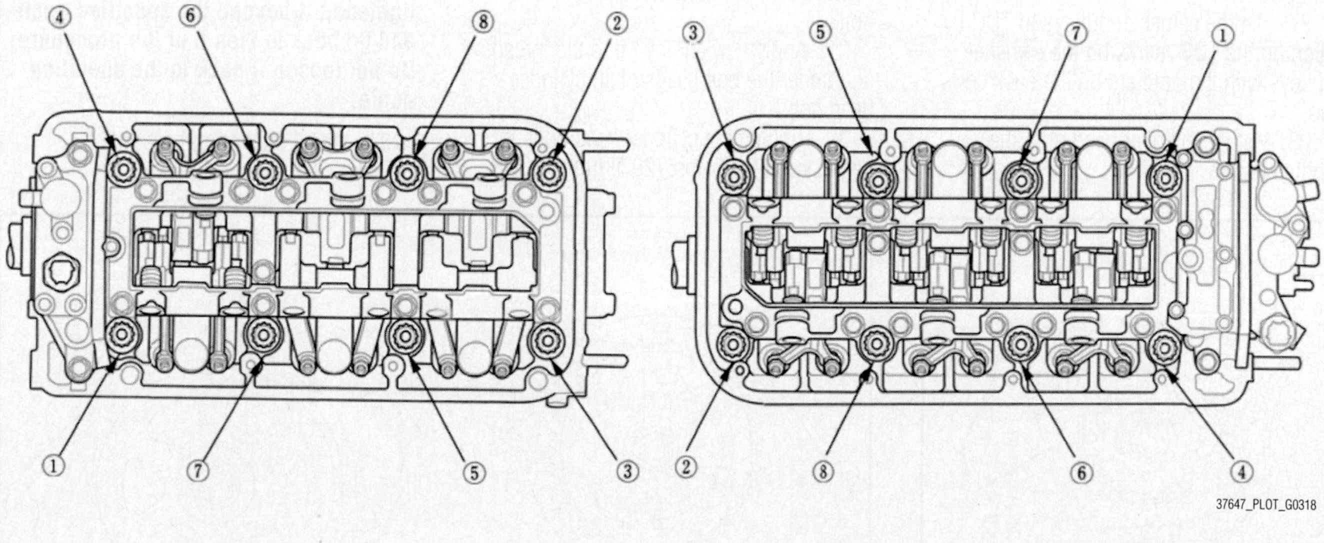
4
6
8
2
3
5
7
1
1
7
5
3
2
8
6
4
37647_PLOT_G0318

Fig. 125 Remove the cylinder head bolts

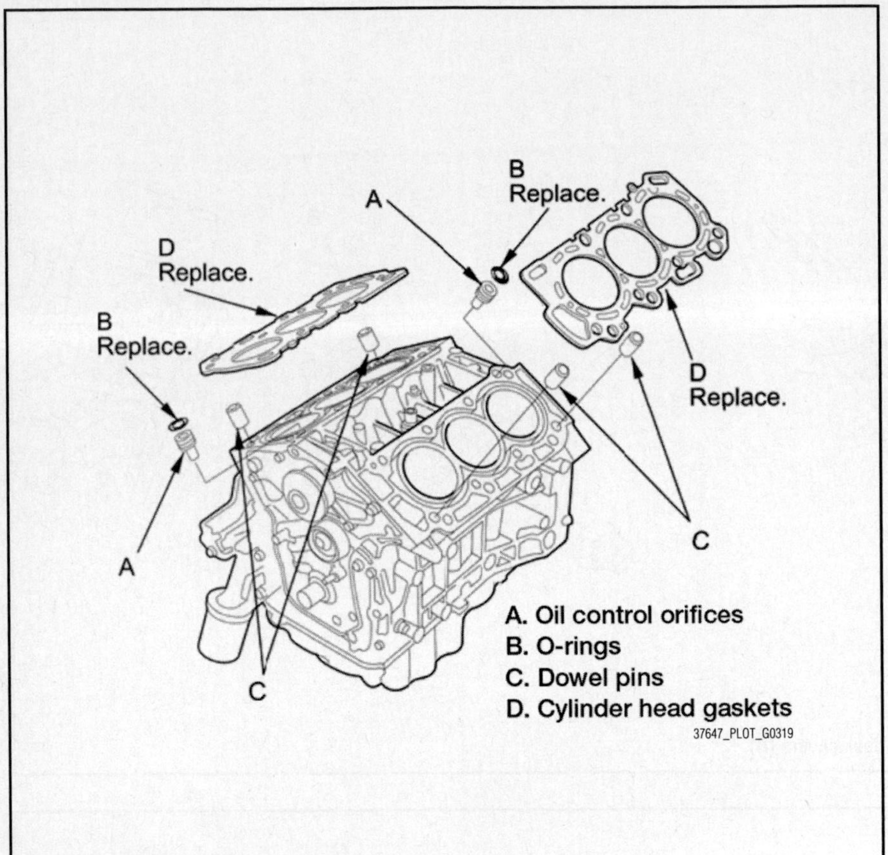
B
Replace.
A
D
Replace.
B
Replace.
D
Replace.
A
C
C
A. Oil control orifices
B. O-rings
C. Dowel pins
D. Cylinder head gaskets
37647_PLOT_G0319

Fig. 126 Clean and install the oil control orifices using the new O-rings

pulley to Top Dead Center (TDC) by aligning the TDC mark on the tooth of the timing belt drive pulley with the pointer on the oil pump.

27. Set the camshaft pulleys to TDC by aligning the TDC marks on the camshaft pulleys with the pointers on the back covers.

28. Put the cylinder head onto the engine block.

29. Measure the diameter of each cylinder head bolt at point A and point B.

30. If either diameter is less than 0.42 inches (10.6 mm), replace the cylinder head bolt.

31. Apply new engine oil to the threads and under the bolt heads of all cylinder head bolts.

32. Tighten the cylinder head bolts in sequence to 22 ft. lbs. (29 Nm) using a

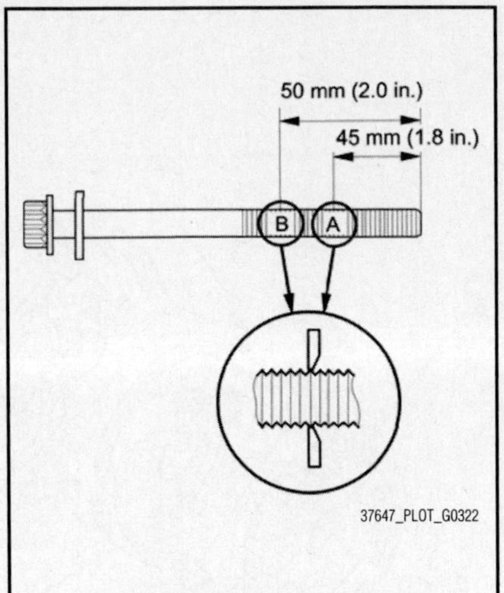
50 mm (2.0 in.)
45 mm (1.8 in.)
B
A
37647_PLOT_G0322

Fig. 129 Measure the diameter of each cylinder head bolt at point A and point B

beam-type torque wrench. The tightening sequence starts in the middle and uses a crisscross pattern working out to the ends. When using a preset click-type torque wrench, be sure to torque slowly and do not over tighten. If a bolt makes any noise while you are torquing it, loosen the bolt and retighten it from the first step.

33. After torquing, tighten all cylinder head bolts in two steps (90° per step) using the sequence in step 11. If you are using a new cylinder head bolt, tighten the bolt an extra 90°.

➡Remove the cylinder head bolt if you tightened it beyond the specified angle, and go back to step 8 of the procedure. Do not loosen it back to the specified angle.

34. Install the timing belt.

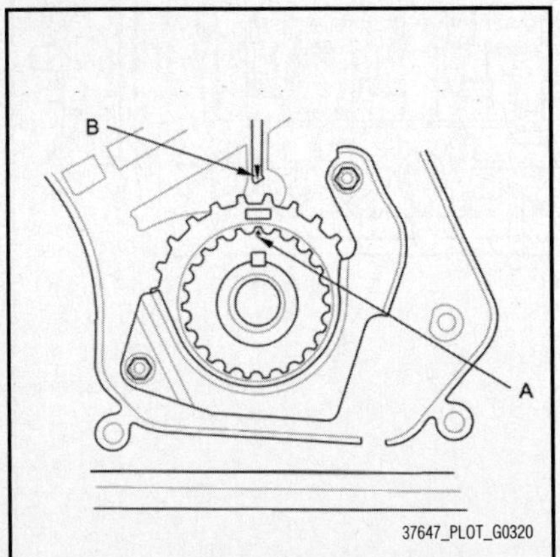
B
A
37647_PLOT_G0320

Fig. 127 Set the timing belt drive pulley to TDC by aligning the TDC mark (A) on the tooth of the timing belt drive pulley with the pointer (B) on the oil pump

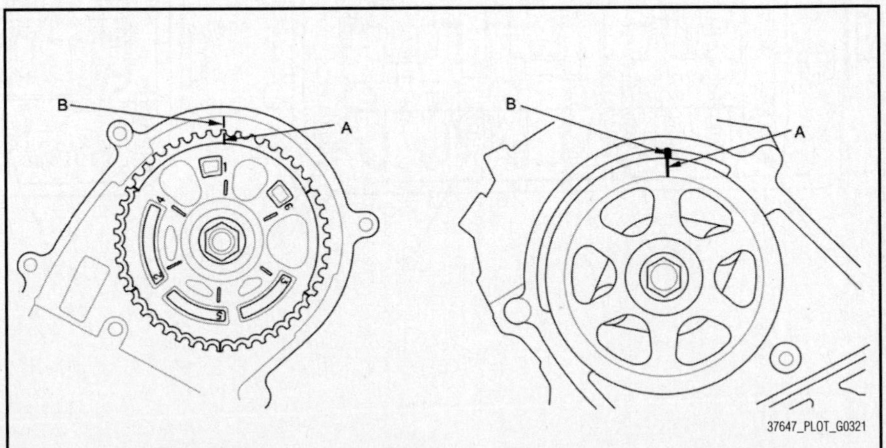
B
A
B
A
37647_PLOT_G0321

Fig. 128 Set the camshaft pulleys to TDC by aligning the TDC marks (A) on the camshaft pulleys with the pointers (B) on the back covers

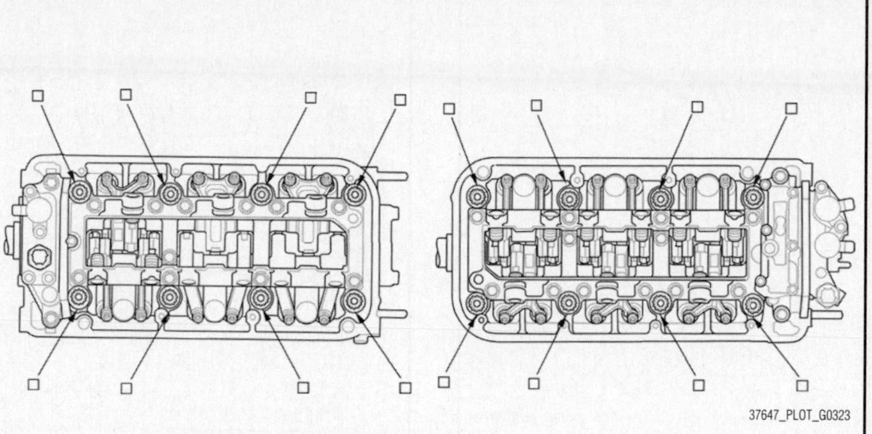

Fig. 130 Tighten the cylinder head bolts in sequence

35. Adjust the valve clearance.
36. Install the cylinder head covers.
37. Install the water passage.
38. Install the injector bases.
39. Install the connector bracket to the front cylinder head.
40. Install the harness clamp bracket to the rear cylinder head.
41. Connect the fuel feed hose, then install the quick-connect fitting cover.

Install the and the rear Warm Up Three Way Catalytic Converter (rear WU-TWC).

42. Connect the following engine wire harness connectors, and install the wire harness clamps to the cylinder head:
    - Six injector connectors
    - Engine Coolant Temperature (ECT) sensor 1 connector
    - Front rocker arm oil pressure switch connector
    - Rear rocker arm oil pressure switch connector
    - Camshaft Position (CMP) sensor connector
    - Front Air Fuel Ratio (A/F) (sensor 1) connector
    - Rear Air Fuel Ratio (A/F) (sensor 1) connector
    - Front secondary Heated Oxygen Sensor 2 (secondary HO2S) connector
    - Rear secondary Heated Oxygen Sensor 2 (secondary HO2S) connector
    - Rocker arm oil control solenoid A (BANK 1) connector
    - Rocker arm oil control solenoid A (BANK 2) connector
    - Rocker arm oil control solenoid B (BANK 1) connector
    - Knock sensor connector

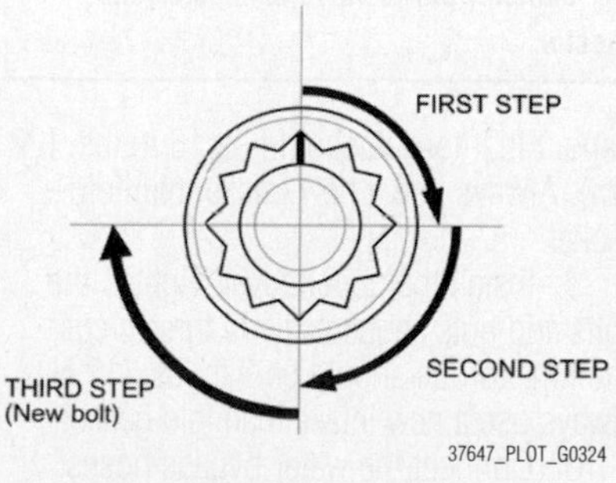
FIRST STEP
SECOND STEP
THIRD STEP
(New bolt)
37647_PLOT_G0324

Fig. 131 After torquing, tighten all cylinder head bolts in two steps (90 ° per step) using the sequence

43. Install the six ignition coils.
44. Install the intake manifold.
45. Install the alternator.
46. Install the Power Steering (P/S) pump and tighten the bolt securing the P/S hose bracket.
47. Install the drive belt.
48. Do the battery installation procedure.
49. Install the air intake duct.
50. After installation, check that all tubes, hoses, and connectors are installed correctly.
51. Inspect for fuel leaks. Turn the ignition switch to ON (II) (do not operate the starter) so the fuel pump runs for about 2 seconds and pressurizes the fuel line. Repeat this operation three times, then check for fuel leakage at any point in the fuel line.
52. Refill the radiator with engine coolant, and bleed the air from the cooling system with the heater valve open.
53. Check for fluid leaks.
54. Do the Powertrain Control Module (PCM) idle learn procedure.
55. Do the Crankshaft Position (CKP) pattern clear/CKP pattern learn procedure.
56. Inspect the idle speed.
57. Inspect the ignition timing.

CKP PATTERN CLEAR/CKP PATTERN LEARN PROCEDURE

Clear/Learn Procedure (with the HDS)

1. Connect the HDS to the Data Link Connector (DLC) located under the driver's side of the dashboard.
2. Turn the ignition switch to ON (II).
3. Make sure the HDS communicates with the PCM and all other vehicle systems. If it doesn't, go to the DLC circuit troubleshooting.
4. Select CRANK PATTERN in the ADJUSTMENT MENU with the HDS.
5. Select CRANK PATTERN LEARNING with the HDS, and follow the screen prompts.

Learn Procedure (without the HDS)

1. Start the engine. Hold the engine speed at 3,000 rpm without load (in P or N) until the radiator fan comes on.
2. Test-drive the vehicle on a level road: Decelerate (with the throttle fully closed) from an engine speed of 2,500 rpm down to 1,000 rpm with the transmission in 2.
3. Repeat step 2 several times.
4. Turn the ignition switch to LOCK (0).
5. Turn the ignition switch to ON (II), and wait for 30 seconds. The CKP pattern learn procedure is complete.

DRIVE PLATE

REMOVAL & INSTALLATION

See Figure 132.

1. Remove the transmission assembly.
2. Remove the drive plate and the washer from the engine crankshaft.

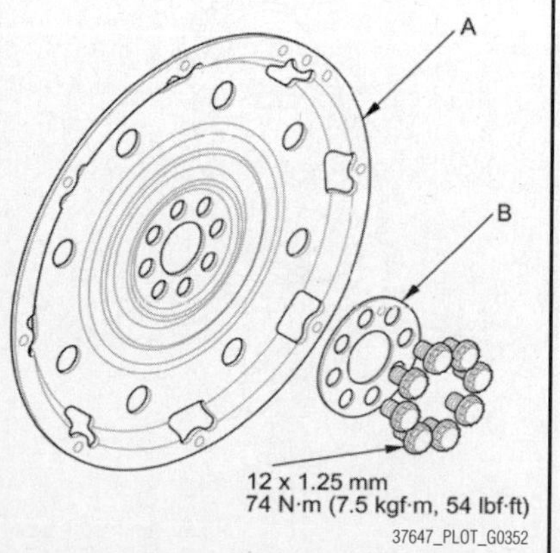
A
B
12 x 1.25 mm
74 N·m (7.5 kgf·m, 54 lbf·ft)
37647_PLOT_G0352

Fig. 132 Remove the drive plate (A) and the washer (B) from the engine crankshaft

3. Install the drive plate and the washer on the engine crankshaft, and tighten the eight bolts in a crisscross pattern in at least two steps.

4. Install the transmission assembly.

INTAKE MANIFOLD

REMOVAL & INSTALLATION

See Figures 133 through 142.

1. Remove the engine cover.
2. Disconnect the breather pipe, then remove the intake air duct.
3. Disconnect the Positive Crankcase Ventilation (PCV) hose, the brake booster vacuum hose, and the Intake Manifold Tuning (IMT) actuator connector.
4. Disconnect the Evaporative Emission (EVAP) canister hose, the EVAP canister purge valve connector, the throttle actuator connector, and the Manifold Absolute Pressure (MAP) sensor connector.
5. Disconnect and plug the water bypass hoses.

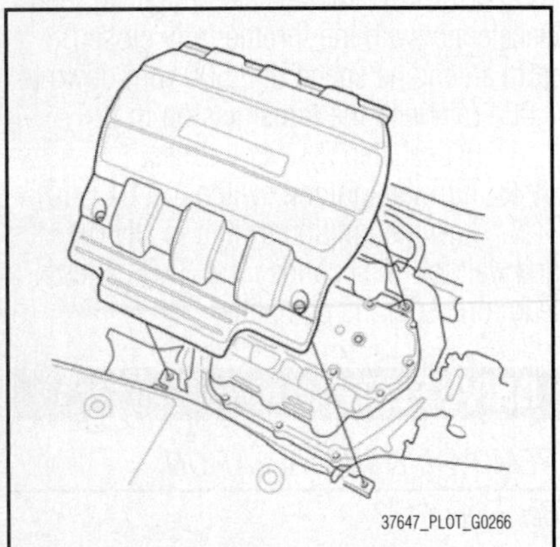
37647_PLOT_G0266

Fig. 133 Remove the engine cover

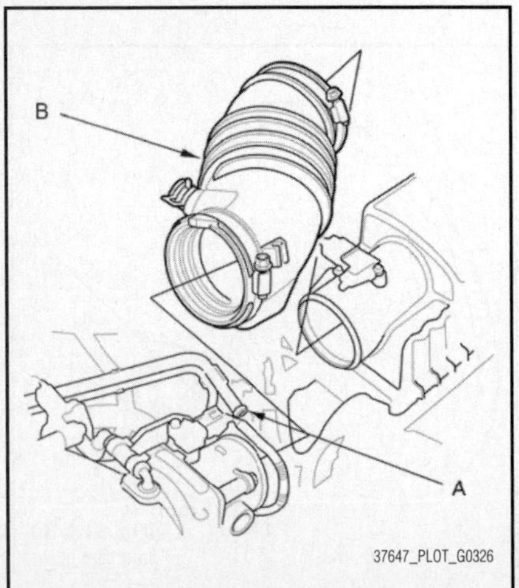
B
A
37647_PLOT_G0326

Fig. 134 Disconnect the breather pipe (A), then remove the intake air duct (B)

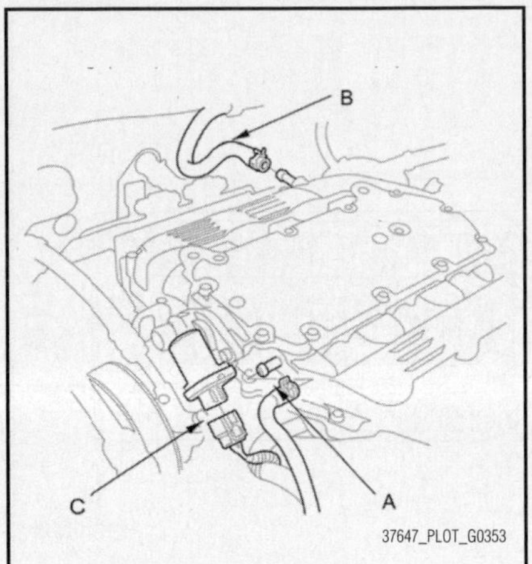
B
C
A
37647_PLOT_G0353

Fig. 135 Disconnect the PCV hose (A), the brake booster vacuum hose (B), and the IMT actuator connector (C)

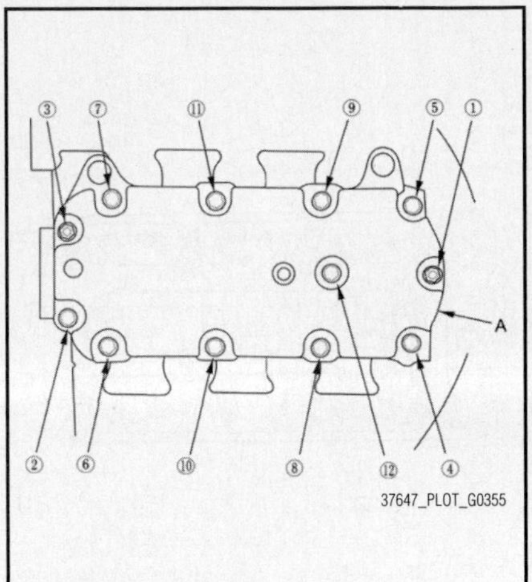
A
37647_PLOT_G0355

Fig. 137 Remove the upper cover mounting bolts and nuts sequentially in three steps, then remove the upper cover (A)

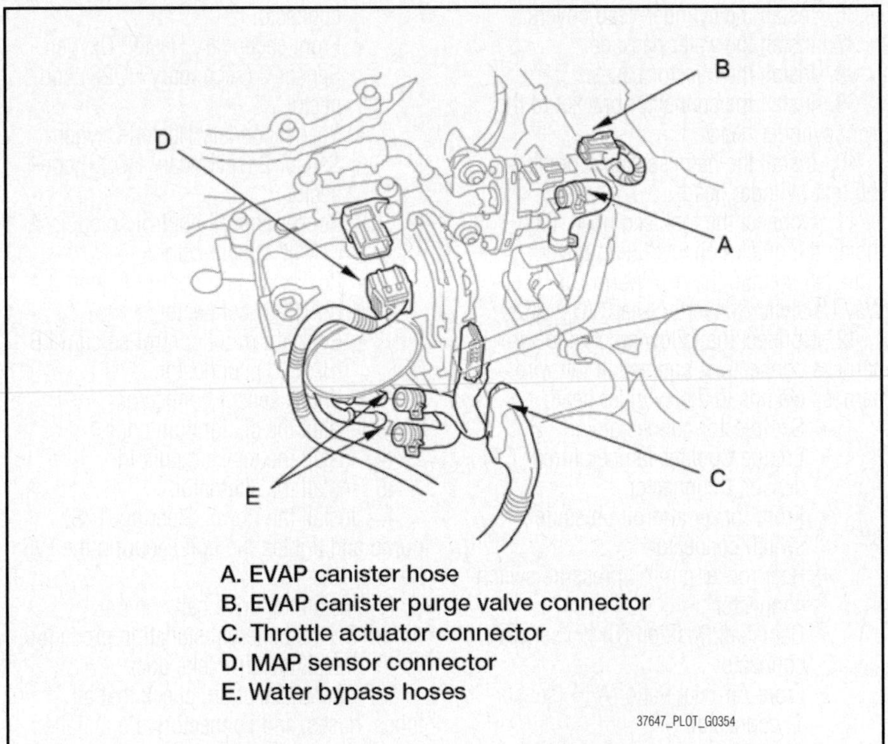
B
D
A
C
E
A. EVAP canister hose
B. EVAP canister purge valve connector
C. Throttle actuator connector
D. MAP sensor connector
E. Water bypass hoses
37647_PLOT_G0354

Fig. 136 Disconnect the EVAP canister hose, the EVAP canister purge valve connector, the throttle actuator connector, and the MAP sensor connector

6. Remove the upper cover mounting bolts and nuts sequentially in three steps, then remove the upper cover.

➡Refer to the Exploded View as needed during this procedure.

7. Remove the intake manifold mounting bolts and nuts sequentially in three steps, then remove the intake manifold.

To install:

8. Install the intake manifold. Tighten the bolts and nuts sequentially in three steps. Final torque should be 16 ft. lbs. (22 Nm). Always use a new intake manifold gasket.
9. Install the upper cover. Tighten the bolts and nuts sequentially in three steps. The final torque should be 9 ft. lbs. (12 Nm). Always use a new intake manifold gasket.
10. Connect the water bypass hoses.
11. Connect the MAP sensor connector, the throttle actuator connector, the EVAP canister purge valve connector, and the EVAP canister hoses.

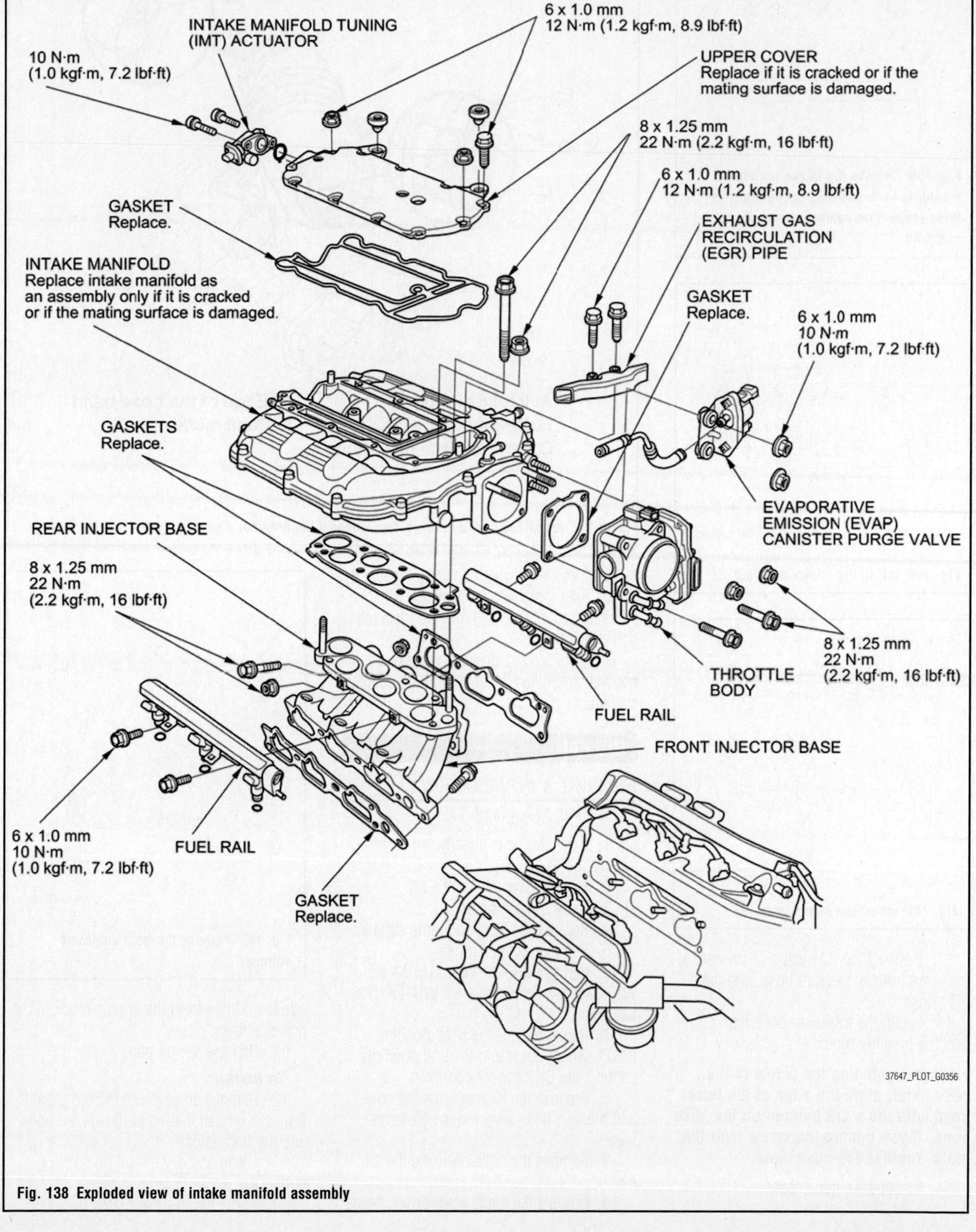
6 x 1.0 mm
12 N·m (1.2 kgf·m, 8.9 lbf·ft)
INTAKE MANIFOLD TUNING
(IMT) ACTUATOR
10 N·m
(1.0 kgf·m, 7.2 lbf·ft)
UPPER COVER
Replace if it is cracked or if the
mating surface is damaged.
8 x 1.25 mm
22 N·m (2.2 kgf·m, 16 lbf·ft)
6 x 1.0 mm
12 N·m (1.2 kgf·m, 8.9 lbf·ft)
GASKET
Replace.
EXHAUST GAS
RECIRCULATION
(EGR) PIPE
INTAKE MANIFOLD
Replace intake manifold as
an assembly only if it is cracked
or if the mating surface is damaged.
GASKET
Replace.
6 x 1.0 mm
10 N·m
(1.0 kgf·m, 7.2 lbf·ft)
GASKETS
Replace.
EVAPORATIVE
EMISSION (EVAP)
CANISTER PURGE VALVE
REAR INJECTOR BASE
8 x 1.25 mm
22 N·m
(2.2 kgf·m, 16 lbf·ft)
8 x 1.25 mm
22 N·m
(2.2 kgf·m, 16 lbf·ft)
THROTTLE
BODY
FUEL RAIL
FRONT INJECTOR BASE
6 x 1.0 mm
10 N·m
(1.0 kgf·m, 7.2 lbf·ft)
FUEL RAIL
GASKET
Replace.
37647_PLOT_G0356

Fig. 138 Exploded view of intake manifold assembly

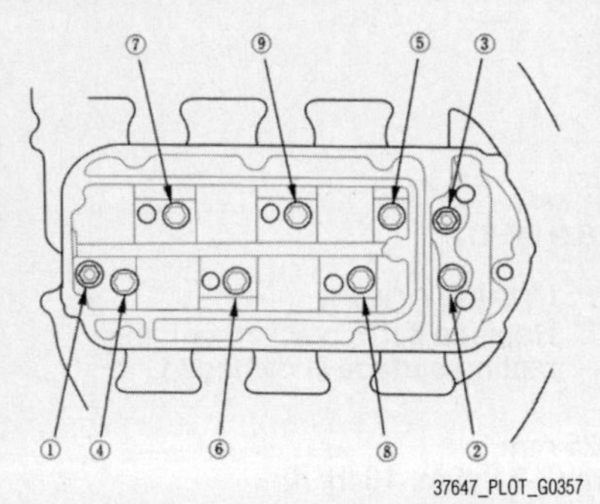

37647_PLOT_G0357

Fig. 139 Remove the intake manifold mounting bolts and nuts sequentially in three steps, then remove the intake manifold

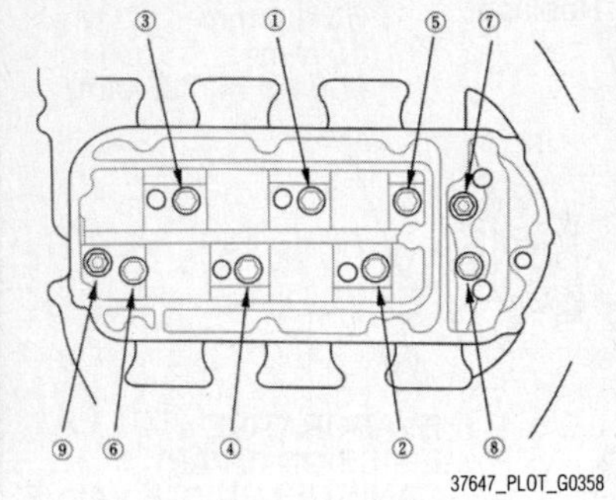

37647_PLOT_G0358

Fig. 140 Install the intake manifold

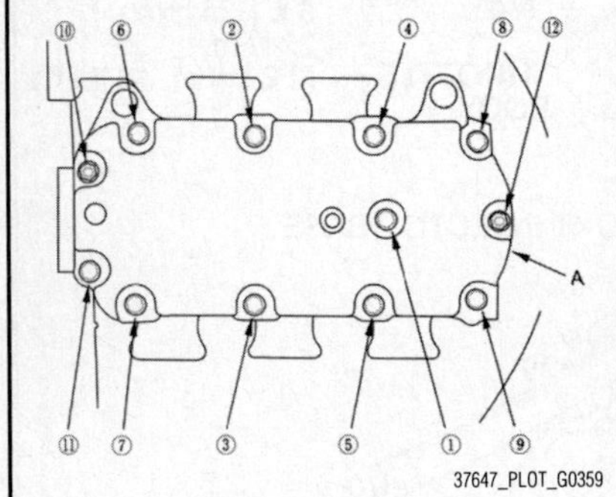

A
37647_PLOT_G0359

Fig. 141 Install the upper cover (A)

12. Connect the IMT actuator connector, the brake booster vacuum hose, and the PCV hose.

13. Install the intake air duct, then connect the breather pipe.

➡When tightening the screw of the hose band, align the edge of the hose band with the mark painted on the hose band. If you tighten the screw over the mark, replace the hose band.

14. Install the engine cover.

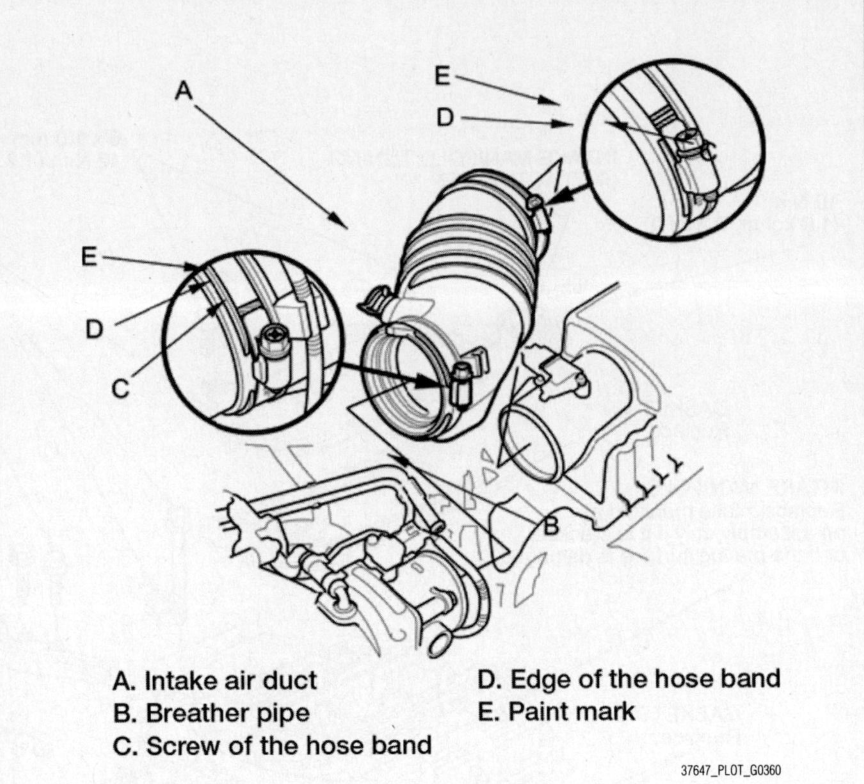

A
B
C
D
E
A. Intake air duct
B. Breather pipe
C. Screw of the hose band
D. Edge of the hose band
E. Paint mark
37647_PLOT_G0360

Fig. 142 Install the intake air duct, then connect the breather pipe

15. Clean up any spilled engine coolant.

16. After installation, check that all tubes, hoses, and connectors are installed correctly.

17. Refill the radiator with engine coolant, and bleed the air from the cooling system.

OIL PAN

REMOVAL & INSTALLATION

See Figures 143 through 149.

1. If the engine is already out of the vehicle, go to step 6.
2. Raise the vehicle on the lift.
3. Drain the engine oil.
4. Remove the front subframe stiffener.
5. Remove exhaust pipe A.
6. Remove the rear Warm Up Three Way Catalytic Converter (rear WU-TWC) bracket.
7. Remove the Crankshaft Position (CKP) sensor cover and the bolt, then disconnect the CKP sensor connector.
8. Remove the torque converter cover and the four bolts securing the transmission.
9. Remove the bolts securing the oil pan.
10. Using a flat blade screwdriver, separate the oil pan from the engine block in the places shown.

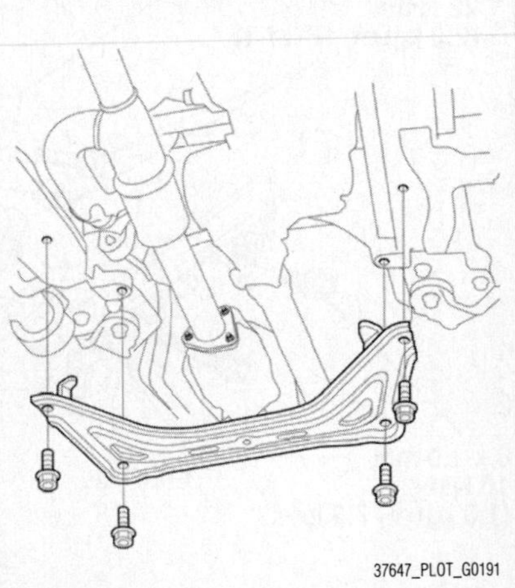

37647_PLOT_G0191

Fig. 143 Remove the front subframe stiffener

11. Remove the oil pan.

To install:

12. Remove all of the old liquid gasket from the oil pan mating surfaces, the bolts, and the bolt holes.
13. Clean and dry the oil pan mating surfaces.
14. Apply liquid gasket (P/N 08717-

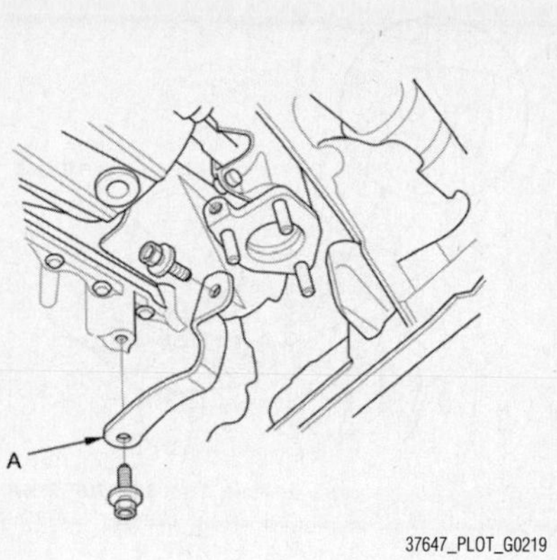
A
37647_PLOT_G0219

Fig. 144 Remove rear WU-TWC bracket (A)

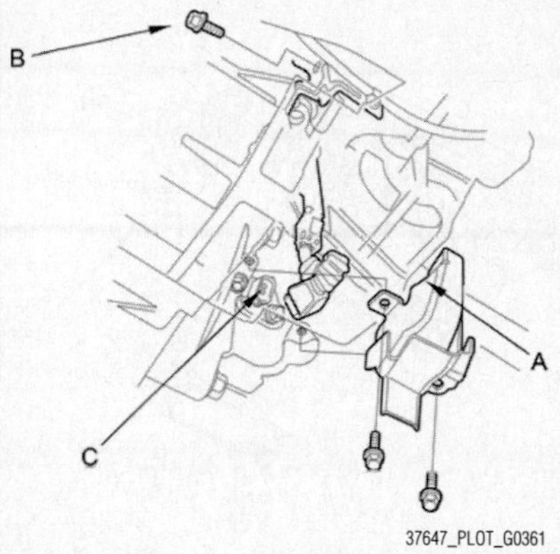
B
A
C
37647_PLOT_G0361

Fig. 145 Remove the CKP sensor cover (A) and the bolt (B), then disconnect the CKP sensor connector (C)

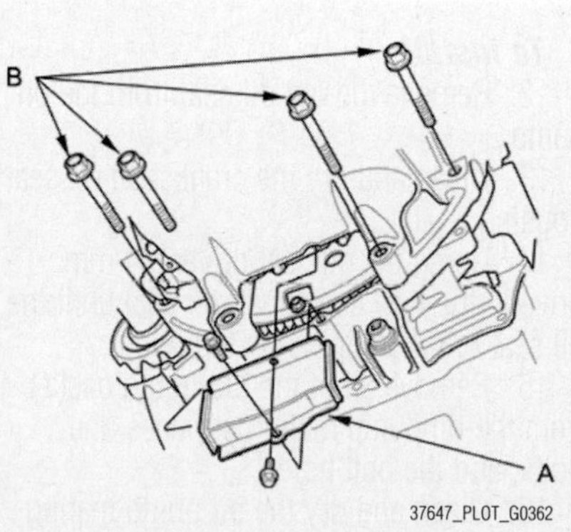
B
A
37647_PLOT_G0362

Fig. 146 Remove the torque converter cover (A) and the four bolts (B) securing the transmission

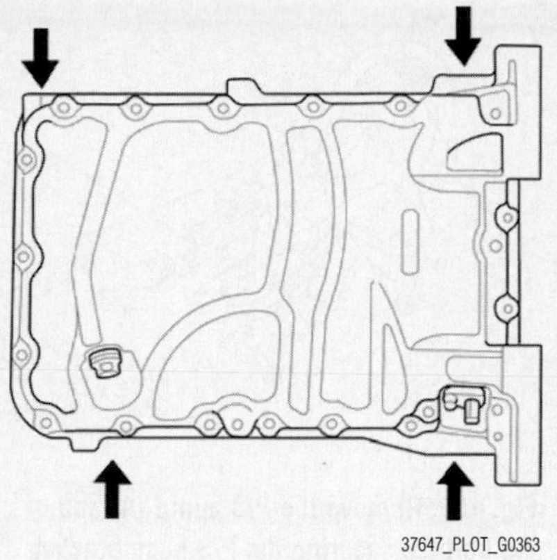
37647_PLOT_G0363

Fig. 147 Using a flat blade screwdriver, separate the oil pan from the engine block in the places shown

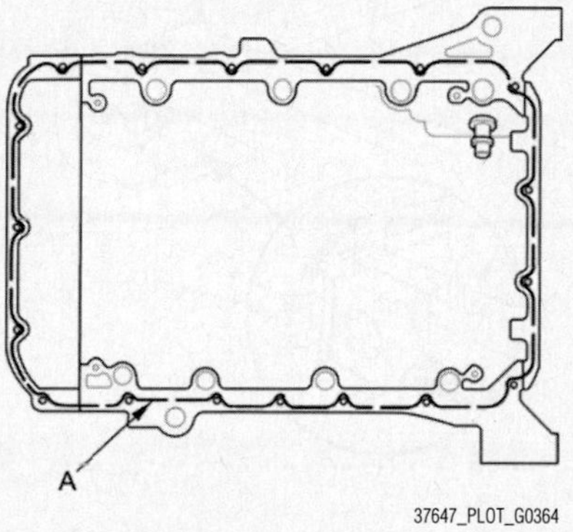
A
37647_PLOT_G0364

Fig. 148 Apply a 0.10 inches (2.5 mm) diameter bead of liquid gasket along the broken line (A)

0004, 08718-0003, 08718-0004, or 08718-0009) to the oil pan mating surface of the engine block and to the inside edge of the threaded bolt holes. Install the component within 5 minutes of applying the liquid gasket.

➡Note the following during oil pan installation:

- Apply a 0.10 inches (2.5 mm) diameter bead of liquid gasket along the broken line.
- If you apply liquid gasket P/N 08718-0012, the component must be installed within 4 minutes.
- If too much time has passed after applying the liquid gasket, remove the old liquid gasket and residue, then reapply the new liquid gasket.

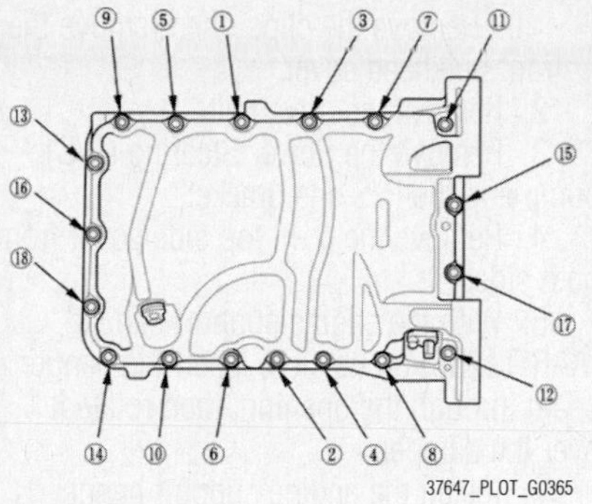
9
5
1
3
7
11
13
15
16
17
18
12
14
10
6
2
4
8
37647_PLOT_G0365

Fig. 149 Tighten the bolts in three steps in sequence

15. Install the oil pan on the engine block.
16. Tighten the bolts in three steps. In the final step, tighten all bolts, in sequence, to 9 ft. lbs. (12 Nm).

➡Wait at least 30 minutes before filling the engine with oil. Do not run the engine for at least 3 hours after installing the oil pan.

17. Tighten the four bolts securing the transmission to 54 ft. lbs. (74 Nm), then install the torque converter cover.
18. Connect the Crankshaft Position (CKP) sensor connector, then install the CKP sensor cover and the bolt.
19. Install the rear Warm Up Three Way Catalytic Converter (rear WU-TWC) bracket. Tighten the bolts to 16 ft. lbs. (22 Nm).
20. If the engine is still in the vehicle, do the following steps.
21. Install exhaust pipe A using new gaskets and new self-locking nuts.
22. Install the front subframe stiffener. Tighten the mounting bolts to 43 ft. lbs. (59 Nm).
23. Refill the engine with the recommended engine oil.

OIL PUMP

REMOVAL & INSTALLATION

See Figures 150 through 156.

Special Tools Required*:

- Engine Support Hanger, A and Reds AAR-T1256
- Oil Seal Driver, 64 mm 07OAD-RCAA100

*Available through the Honda Tool and Equipment Program 888-424-6857

1. Remove the bulkhead cover.

a. Remove the clips, then remove the air intake duct.

b. Remove the clips, then remove the front bulkhead cover.

2. Remove the drive belt.
3. Remove the Power Steering (P/S) pump and the P/S line bracket.
4. Remove the cowl top side cover from both sides.
5. With the engine support hanger (AAR-T1256) on its side, insert the hanger beam through the opening, then rotate it over the damper.
6. Install the engine support hanger (AAR-T1256) onto the vehicle as shown, and attach the hook to the engine hanger. Tighten the wing nut by hand, and lift and support the engine/transmission.

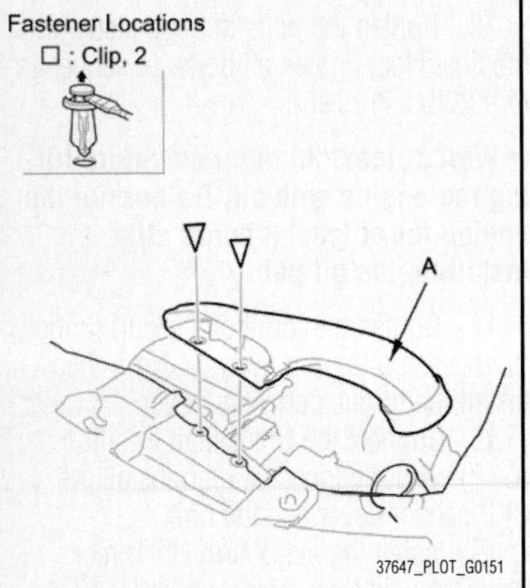
Fastener Locations
□ : Clip, 2
A
37647_PLOT_G0151

Fig. 150 Remove the clips, then remove the air intake duct (A)

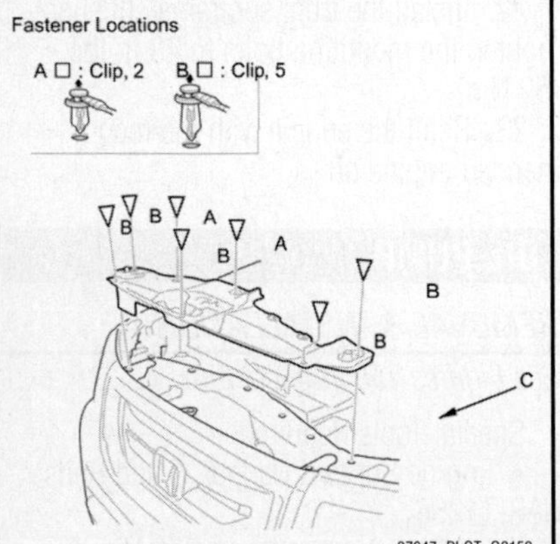
Fastener Locations
A □ : Clip, 2
B □ : Clip, 5
A
B
C
37647_PLOT_G0152

Fig. 151 Remove the clips (A, B), then remove the front bulkhead cover (C)

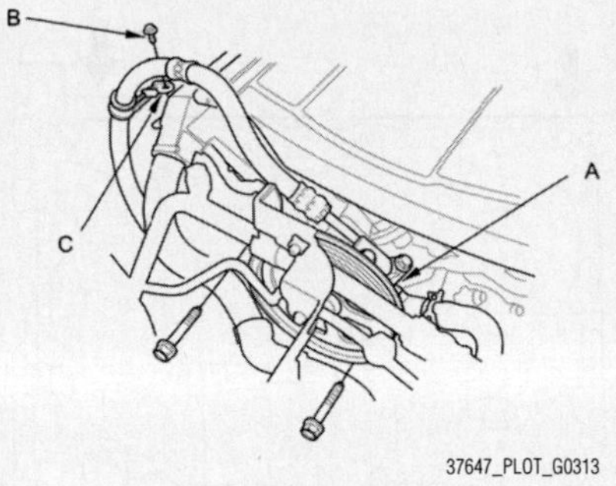
B
A
C
37647_PLOT_G0313

Fig. 152 Remove the P/S pump (A) and the bolt (B) securing the P/S hose bracket (C)

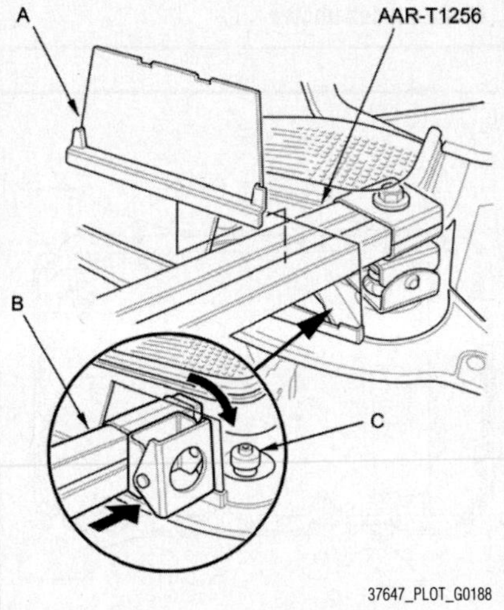
A
AAR-T1256
B
C
37647_PLOT_G0188

Fig. 153 Remove the cowl top side cover (A) from both sides; insert the hanger beam (B) through the opening then rotate it over the damper (C)

➡Note the following:

- Be careful when working around the windshield.
- Be careful not to damage the hood opener cable when installing the engine support hanger at the front bulkhead.
- AAR-T1256 two sets required for stacking additional cross section bar.

7. Remove the timing belt.
8. Remove the oil filter base/oil filter assembly.
9. Remove the oil pan.
10. Remove the oil strainer.
11. Remove the mounting bolts, then remove the oil pump assembly.

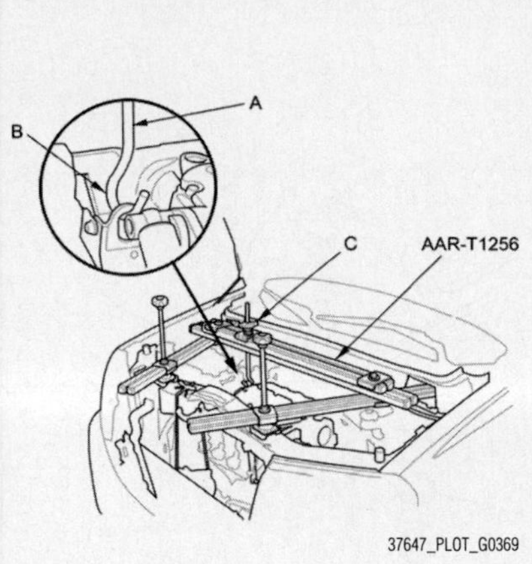
A
B
C
AAR-T1256
37647_PLOT_G0369

Fig. 154 Install the engine support hanger and attach the hook (A) to the engine hanger (B); tighten the wing nut (C) by hand

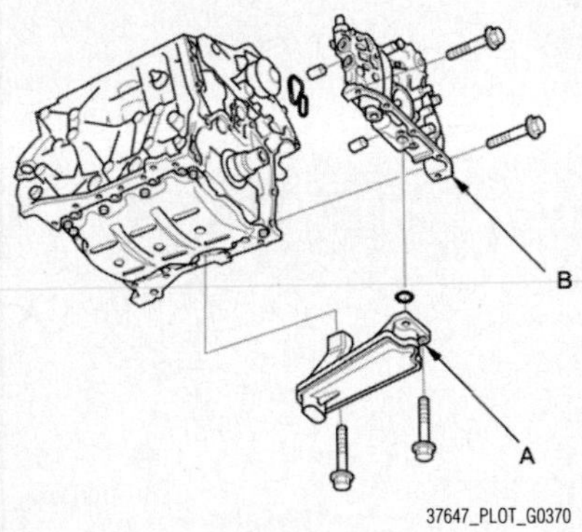
B
A
37647_PLOT_G0370

Fig. 155 Remove the oil strainer (A); remove the mounting bolts, then remove the oil pump assembly (B)

To install:

12. Remove the old oil seal from the oil pump.
13. Clean and dry the crankshaft oil seal housing.
14. Using the oil seal driver, 64 mm, drive in the new crankshaft oil seal until the oil seal driver bottoms on the pump.
15. Remove all of the old liquid gasket from the oil pump mating surfaces, the bolts, and the bolt holes.
16. Clean and dry the oil pump mating surfaces.
17. Apply liquid gasket (P/N 08717-0004, 08718-0003, 08718-0004, or 08718-0009) to the engine block mating surface of the oil

pump and to the inside edge of the threaded bolt holes. Install the component within 5 minutes of applying the liquid gasket.

➡Note the following:

- Apply a 0.10 inches (2.5 mm) diameter bead of liquid gasket along the broken line.
- If you apply liquid gasket P/N 08718-0012, the component must be installed within 4 minutes.
- If too much time has passed after applying the liquid gasket, remove the old liquid gasket and residue, then reapply the new liquid gasket.

18. Apply a light coat of new engine oil to the lip of the crankshaft oil seal, and apply new engine oil to the new O-ring.

19. Install the dowel pins, then align the inner rotor with the crankshaft, and install the oil pump.

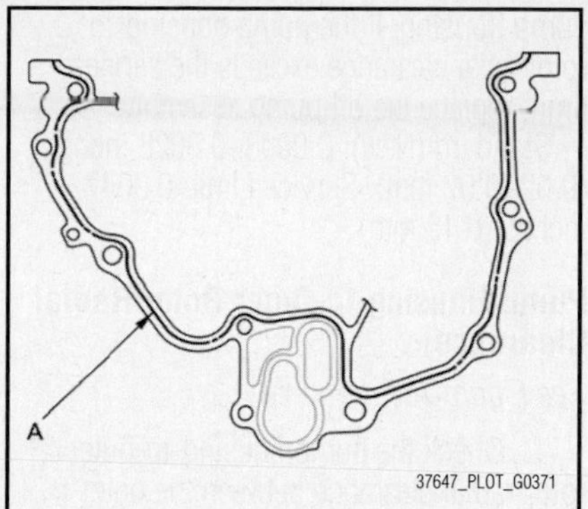
A
37647_PLOT_G0371

Fig. 156 Apply a 0.10 inches (2.5 mm) diameter bead of liquid gasket along the broken line (A)

➡Wait at least 30 minutes before filling the engine with oil. Do not run the engine for at least 3 hours after installing the oil pump.

20. Clean the excess grease off the crankshaft, and check the seal for distortion.

21. Install the oil strainer with a new O-ring.

22. Install the oil pan.

23. Install the oil filter base/oil filter assembly with a new O-ring.

24. Install the timing belt.

25. Remove the engine support hanger from the vehicle.

26. Install the both sides cowl top side lid on the cowl cover.

27. Install the Power Steering (P/S) pump and the P/S line bracket.

28. Install the drive belt.

29. Install the bulkhead cover.

30. Refill the engine with the recommended engine oil.

INSPECTION

See Figure 157.

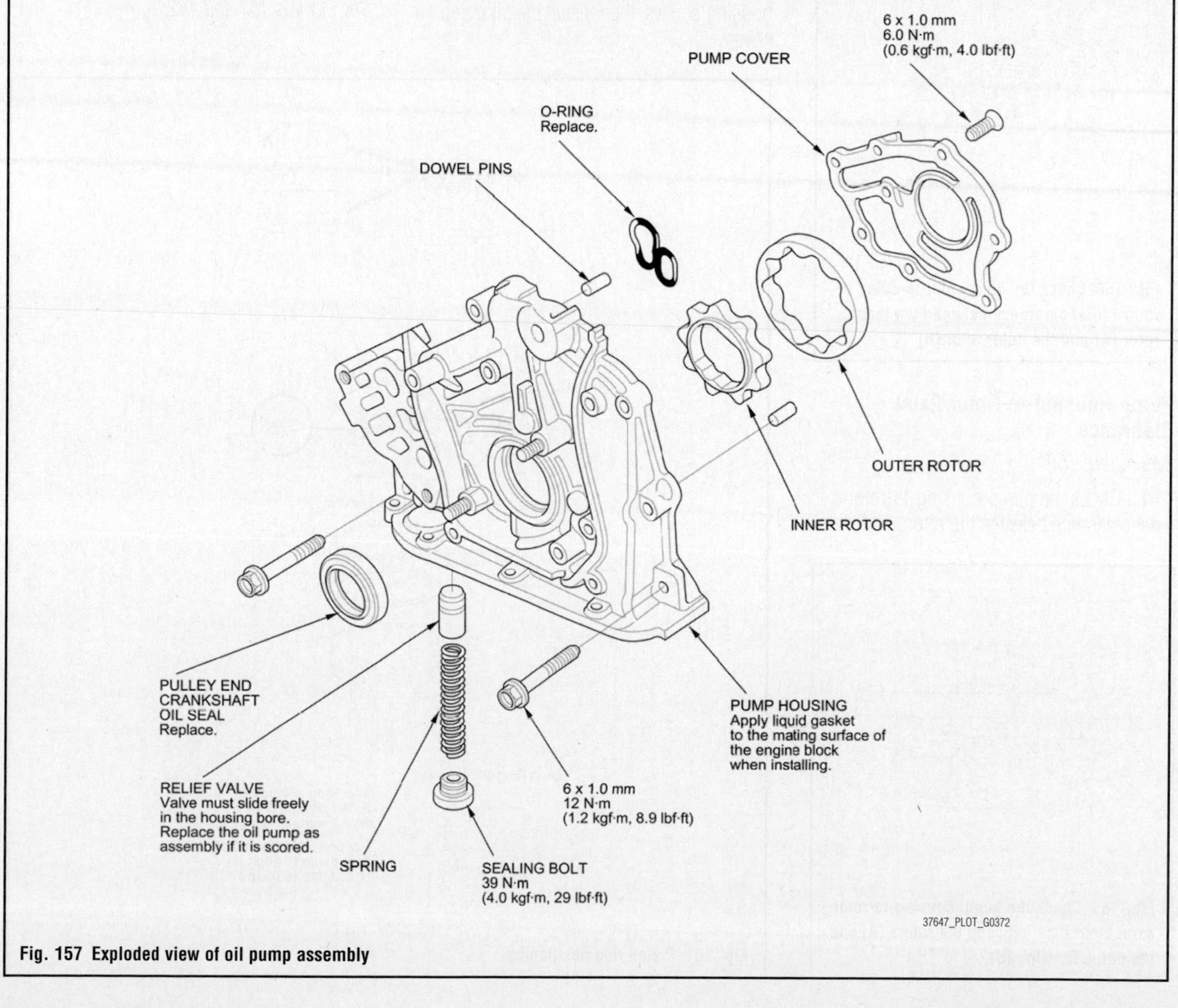
6 x 1.0 mm
6.0 N·m
(0.6 kgf·m, 4.0 lbf·ft)
PUMP COVER
O-RING
Replace.
DOWEL PINS
OUTER ROTOR
INNER ROTOR
PULLEY END
CRANKSHAFT
OIL SEAL
Replace.
PUMP HOUSING
Apply liquid gasket
to the mating surface of
the engine block
when installing.
RELIEF VALVE
Valve must slide freely
in the housing bore.
Replace the oil pump as
assembly if it is scored.
6 x 1.0 mm
12 N·m
(1.2 kgf·m, 8.9 lbf·ft)
SPRING
SEALING BOLT
39 N·m
(4.0 kgf·m, 29 lbf·ft)
37647_PLOT_G0372

Fig. 157 Exploded view of oil pump assembly

➡**Refer to the Exploded View as needed during this procedure.**

1. Remove the screws from the pump housing, then separate the pump housing and the pump cover.

Inner Rotor-to-Outer Rotor Radial Clearance

See Figure 158.

1. Check the inner rotor-to-outer rotor radial clearance between the inner rotor and the outer rotor. If the inner rotor-to-outer rotor clearance exceeds the service limit, replace the oil pump assembly.

Standard (New): 0.0016–0.0063 inches (0.04–0.16 mm); Service Limit: 0.0079 inches (0.20 mm).

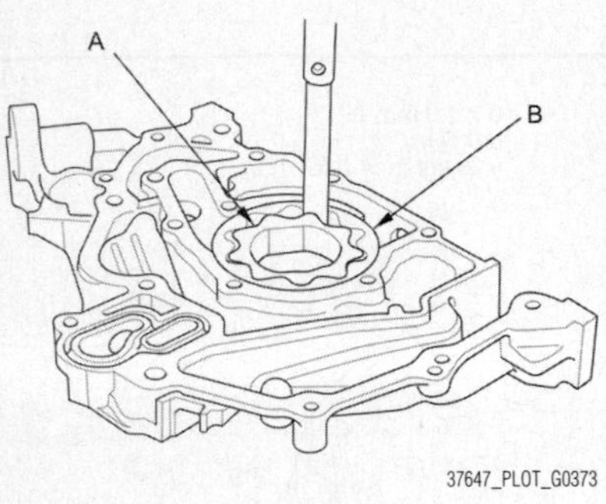
A
B
37647_PLOT_G0373

Fig. 158 Check the inner rotor-to-outer rotor radial clearance between the inner rotor (A) and the outer rotor (B)

Pump Housing-to-Rotor Axial Clearance

See Figure 159.

1. Check the pump housing-to-rotor axial clearance between the rotors and the pump housing. If the pump housing-to-rotor axial clearance exceeds the service limit, replace the oil pump assembly.

Standard (New): 0.0008–0.0028 inches (0.02–0.07 mm); Service Limit: 0.0047 inches (0.12 mm).

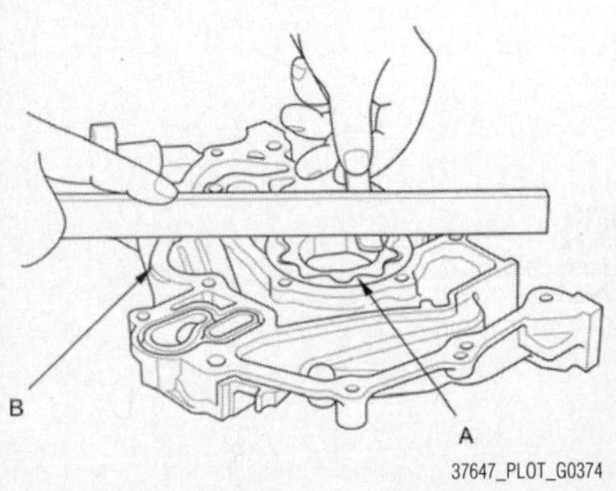
B
A
37647_PLOT_G0374

Fig. 159 Check the pump housing-to-rotor axial clearance between the rotors (A) and the pump housing (B)

Pump Housing-to-Outer Rotor Radial Clearance

See Figure 160.

1. Check the pump housing-to-outer rotor radial clearance between the outer rotor and the pump housing. If the pump housing-to-outer rotor radial clearance exceeds the service limit, replace the oil pump assembly.

Standard (New): 0.0039–0.0075 inches (0.10–0.19 mm); Service Limit: 0.0079 inches (0.20 mm).

2. Inspect both rotors and the pump housing for scoring or other damage. Replace the parts, if necessary.

3. Apply liquid thread lock to the pump housing screws, then install the oil pump cover.

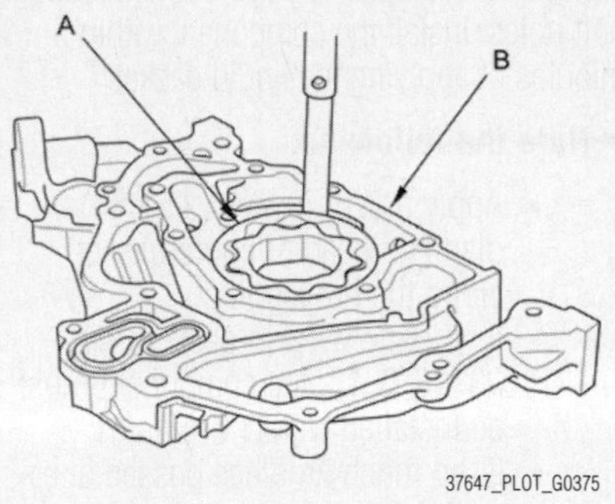
A
B
37647_PLOT_G0375

Fig. 160 Check the pump housing-to-outer rotor radial clearance between the outer rotor (A) and the pump housing (B)

4. Check that the oil pump turns freely.

PISTON AND RING

POSITIONING

See Figures 161 and 162.

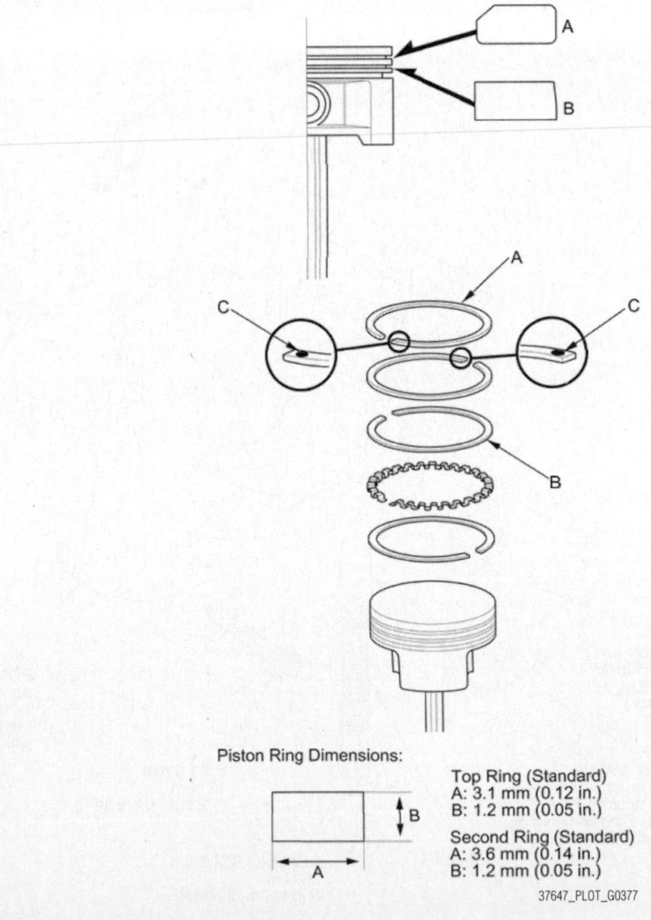
A
B
A
C
C
B
Piston Ring Dimensions:
B
A
Top Ring (Standard)
A: 3.1 mm (0.12 in.)
B: 1.2 mm (0.05 in.)
Second Ring (Standard)
A: 3.6 mm (0.14 in.)
B: 1.2 mm (0.05 in.)
37647_PLOT_G0377

Fig. 161 Piston ring positioning

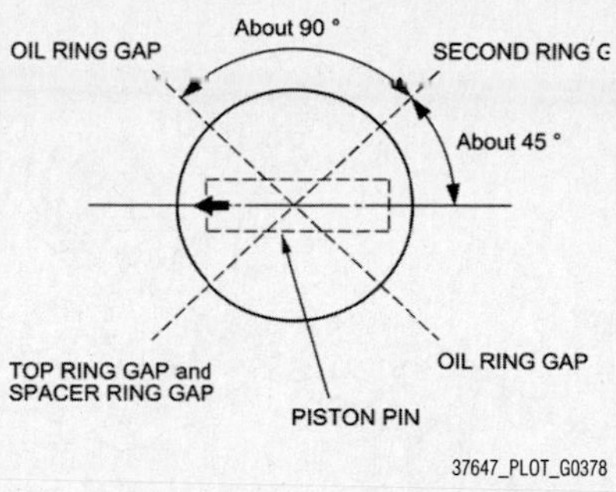
About 90 °
OIL RING GAP
SECOND RING G
About 45 °
TOP RING GAP and SPACER RING GAP
OIL RING GAP
PISTON PIN
37647_PLOT_G0378

Fig. 162 Piston ring end gap positioning

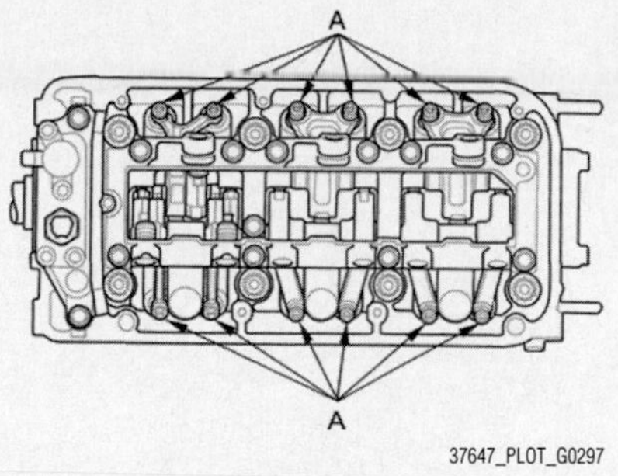
A
A
37647_PLOT_G0297

Fig. 163 Loosen the locknuts and adjusting screws (A)

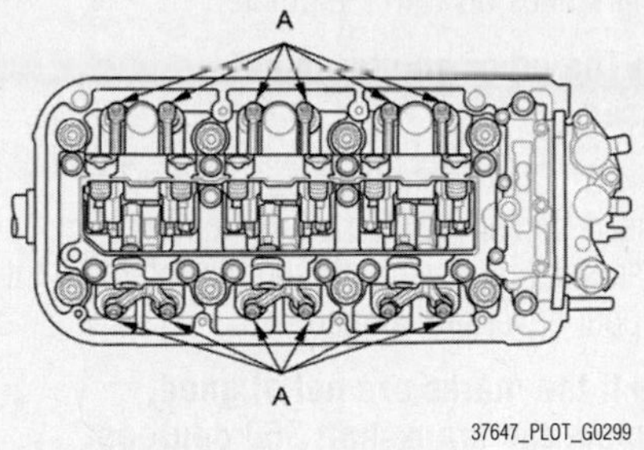
A
A
37647_PLOT_G0299

Fig. 165 Loosen the locknuts and the adjusting screws (A)

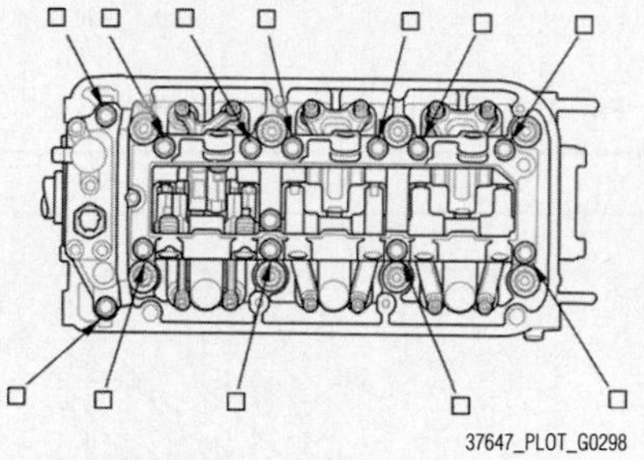
37647_PLOT_G0298

Fig. 164 Loosen the rocker shaft bridge mounting bolts and the rocker shaft holder mounting bolts in sequence

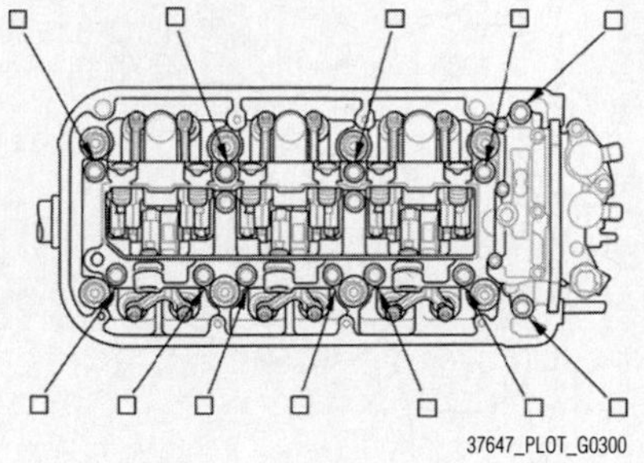
37647_PLOT_G0300

Fig. 166 Loosen the rocker shaft bridge mounting bolts and the rocker shaft holder mounting bolts in sequence

REAR MAIN SEAL

REMOVAL & INSTALLATION

Special Tools Required:

- Driver Handle, 15 x 135L 07749-0010000
- Oil Seal Driver Attachment, 106 mm 070AD-RCA0200

1. Remove the transmission and the drive plate.
2. Remove the transmission end crankshaft oil seal.
3. Clean and dry the crankshaft oil seal housing.
4. Apply a light coat of new engine oil to the lip of the crankshaft oil seal.
5. Using the driver handle, 15 x 135L and the oil seal driver attachment, 106 mm, drive in the new crankshaft oil seal until the oil seal driver attachment bottoms on the engine block end cover. Align the hole in the oil seal driver attachment with the pin on the crankshaft.
6. Clean the excess grease off the crankshaft, and check that the oil seal lip is not distorted.
7. Install the drive plate, and the transmission.

ROCKER ARMS

REMOVAL & INSTALLATION

Front

See Figures 163 and 164.

1. Remove the cylinder head cover.
   a. Remove the rocker shaft bridge mounting bolts, the rocker shaft holder mounting bolts, and the rocker arm assembly.
   b. Loosen the rocker shaft bridge mounting bolts and the rocker shaft holder mounting bolts in sequence two turns at a time, starting at the ends in a crisscross pattern working toward the middle to prevent damaging the valves or the rocker arm assembly.
   c. When removing the rocker arm assembly, do not remove the rocker shaft bridge mounting bolts and the rocker shaft holder mounting bolts. The bolts will keep the rocker arms on the shafts.

To install:

2. Installation is the reverse of removal.
3. Tighten the rocker shaft bridge mounting bolts and the rocker shaft holder mounting bolts in sequence two turns at a time, starting at the middle in a crisscross pattern working toward the ends to prevent damaging the valves or the rocker arm assembly.

Rear

See Figures 165 and 166.

1. Remove the cylinder head cover.
2. Loosen the locknuts and the adjusting screws.
3. Remove the rocker shaft bridge mounting bolts, the rocker shaft holder mounting bolts, and the rocker arm assembly.
   a. Loosen the rocker shaft bridge mounting bolts and the rocker shaft holder mounting bolts in sequence two turns at a time, starting at the ends in a crisscross pattern working toward the middle to prevent damaging the valves or the rocker arm assembly.
   b. When removing the rocker arm assembly, do not remove the rocker shaft bridge mounting bolts and the rocker shaft holder mounting bolts. The bolts will keep the rocker arms on the shafts.

To install:

4. Installation is the reverse of removal.
5. Tighten the rocker shaft bridge mounting bolts and the rocker shaft holder mounting bolts in sequence two turns at a time, starting at the middle in a crisscross pattern working toward the ends to prevent damaging the valves or the rocker arm assembly.

TIMING BELT FRONT COVER

REMOVAL & INSTALLATION

See Figures 167 through 173.

1. Remove the accessory drive belt.

2. Turn the crankshaft so the white mark lines up with the pointer.

➡The other pointer is not used.

3. Check that the No. 1 piston Top Dead Center (TDC) mark on the front camshaft pulley and the pointer on the front upper cover are aligned.

➡If the marks are not aligned, rotate the crankshaft 360 degrees, and recheck the camshaft pulley mark.

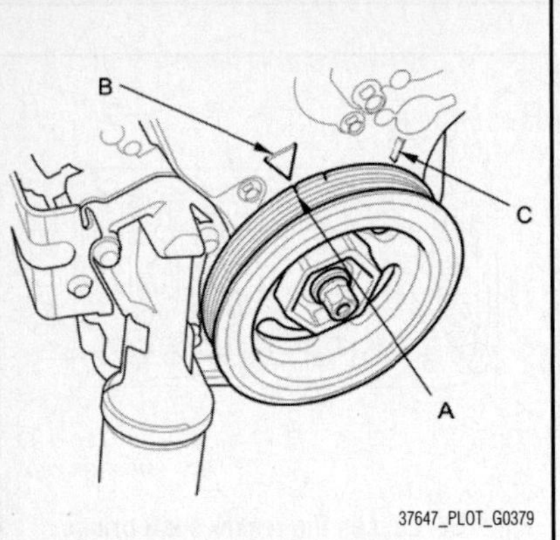
B
C
A
37647_PLOT_G0379

Fig. 167 Turn the crankshaft so the white mark (A) lines up with the pointer (B); pointer (C) is not used

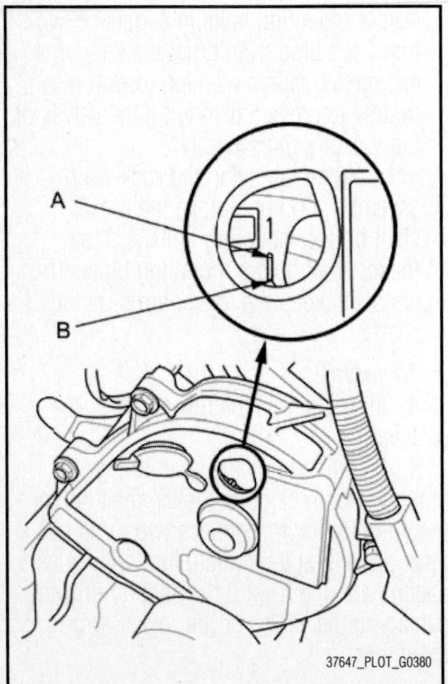
A
1
B
37647_PLOT_G0380

Fig. 168 Check that the No. 1 piston TDC mark (A) on the front camshaft pulley and the pointer (B) on the front upper cover are aligned

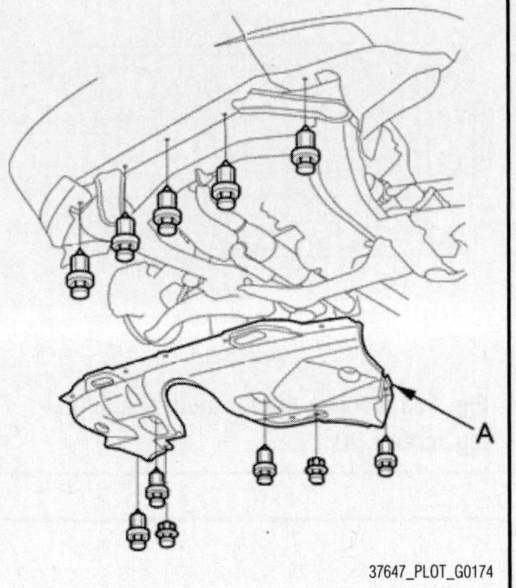
A
37647_PLOT_G0174

Fig. 169 Remove the splash shield

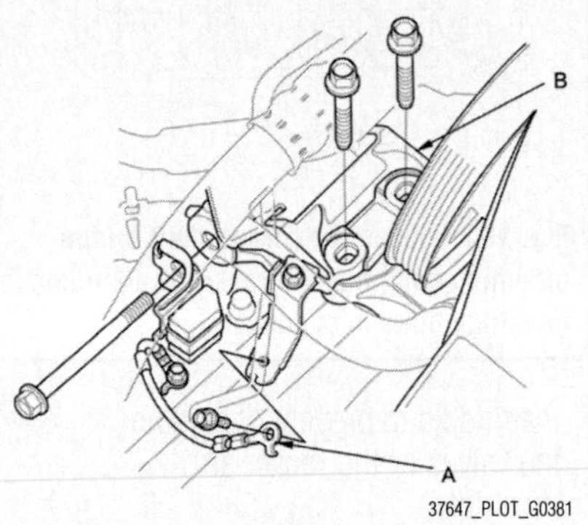
B
A
37647_PLOT_G0381

Fig. 170 Remove the ground cable (A), then remove the upper half of the side engine mount bracket (B)

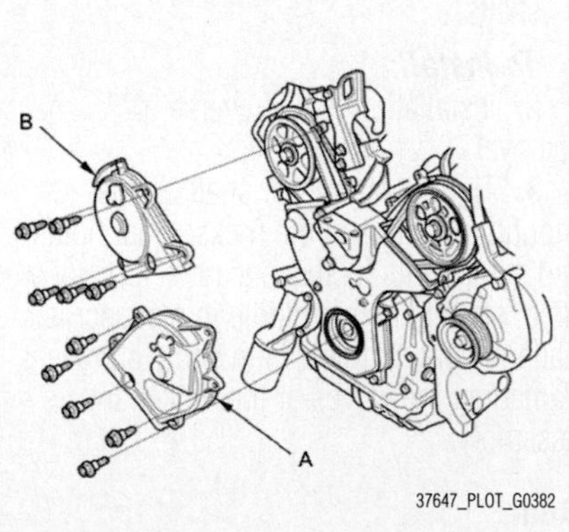
B
A
37647_PLOT_G0382

Fig. 171 Remove the front upper cover (A) and the rear upper cover (B)

4. Raise the vehicle on the lift, then remove the right front wheel.
5. Remove the splash shield.
6. Remove the drive belt auto-tensioner.
7. Lift and support the engine with a jack and a wood block under the oil pan.

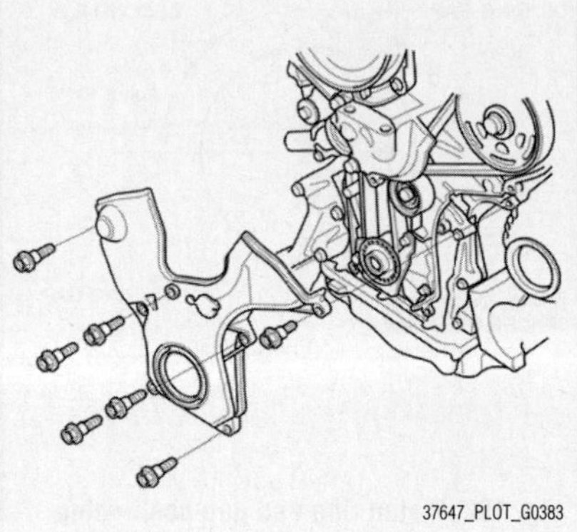
37647_PLOT_G0383

Fig. 172 Remove the lower cover

8. Remove the ground cable, then remove the upper half of the side engine mount bracket.
9. Remove the crankshaft pulley.
10. Remove the front upper cover and the rear upper cover.
11. Remove the lower cover.

To install:

12. Install the lower cover.
13. Install the front upper cover and the rear upper cover.
14. Install the crankshaft pulley.
15. Rotate the crankshaft pulley about six turns clockwise so the timing belt positions itself on the pulleys.
16. Turn the crankshaft pulley so its white mark lines up with the pointer.
17. Check the camshaft pulley marks.

➡If the marks are not aligned, rotate the crankshaft 360 degrees, and recheck the camshaft pulley mark.

a. If the camshaft pulley marks are at TDC, go to step 17.
b. If the camshaft pulley marks are not at TDC, remove the timing belt and repeat steps 2 through 16.

18. Install the upper half of the side engine mount bracket, and tighten the new mounting bolts to 33 ft. lbs. (44 Nm), then tighten the mass damper mounting bolt to 40 ft. lbs. (54 Nm).
19. Install the ground cable.
20. Install the drive belt auto-tensioner.
21. Install the splash shield.
22. Install the right front wheel.
23. Do the Crankshaft Position (CKP) pattern clear/CKP pattern learn procedure.

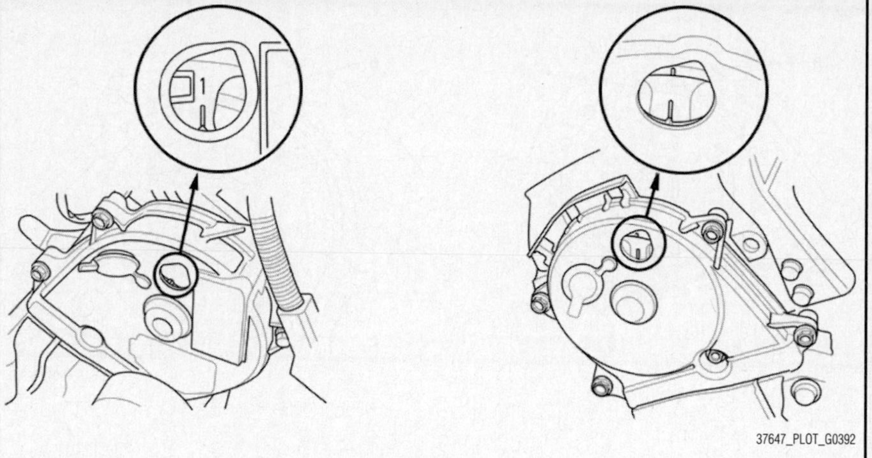
1
37647_PLOT_G0392

Fig. 173 Check the camshaft pulley marks

TIMING BELT & SPROCKETS

REMOVAL & INSTALLATION

See Figures 167 through 172, 174 through 178.

1. Turn the crankshaft so the white mark lines up with the pointer.

➡The other pointer is not used.

2. Check that the No. 1 piston Top Dead Center (TDC) mark on the front camshaft pulley and the pointer on the front upper cover are aligned.

➡If the marks are not aligned, rotate the crankshaft 360 degrees, and recheck the camshaft pulley mark.

3. Raise the vehicle on the lift, then remove the right front wheel.
4. Remove the splash shield.
5. Remove the drive belt auto-tensioner.
6. Lift and support the engine with a jack and a wood block under the oil pan.
7. Remove the ground cable, then remove the upper half of the side engine mount bracket.
8. Remove the crankshaft pulley.
9. Remove the front upper cover and the rear upper cover.
10. Remove the lower cover.
11. Remove one of the battery clamp bolts from the battery tray, and grind the end of it as shown.
12. Thread the battery clamp bolt in as shown to hold the timing belt adjuster in its current position. Tighten it by hand, do not use a wrench.
13. Remove the timing belt guide plate.
14. Remove the lower half of the side engine mount bracket.

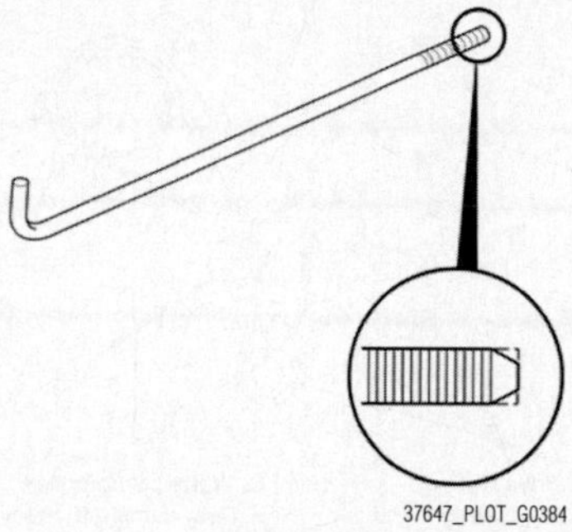
37647_PLOT_G0384

Fig. 174 Remove one of the battery clamp bolts from the battery tray, and grind the end of it as shown

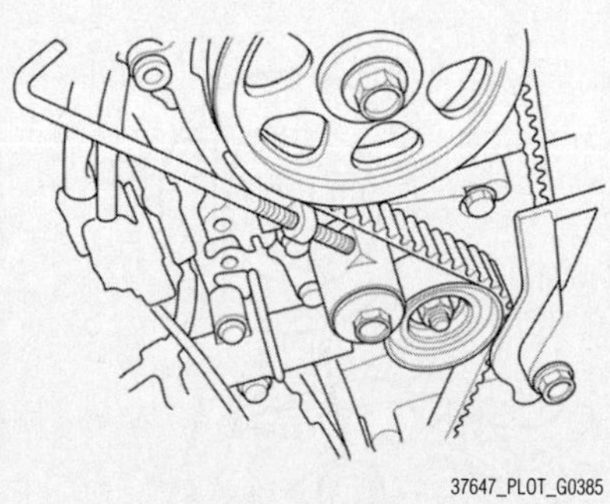
37647_PLOT_G0385

Fig. 175 Thread the battery clamp bolt in as shown to hold the timing belt adjuster in its current position

15. Remove the idler pulley bolt and the idler pulley, then remove the timing belt. Discard the idler pulley bolt.

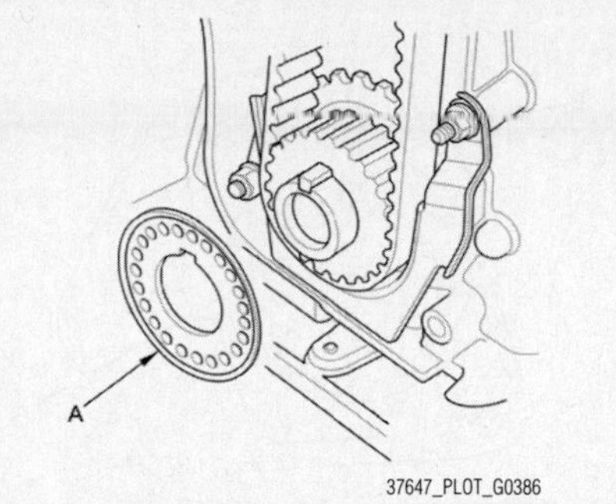
A
37647_PLOT_G0386

Fig. 176 Remove the timing belt guide plate (A)

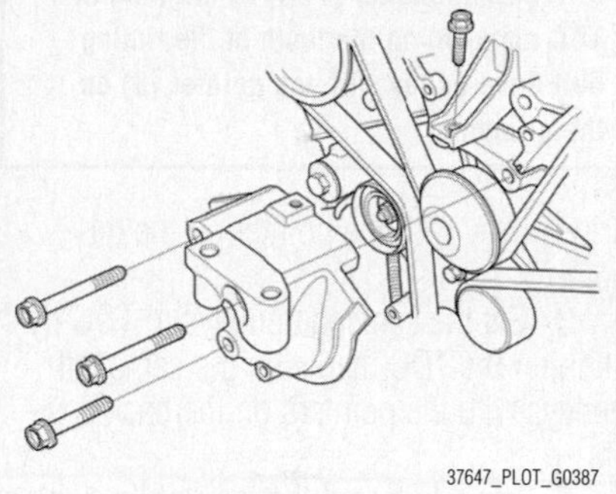
37647_PLOT_G0387

Fig. 177 Remove the lower half of the side engine mount bracket

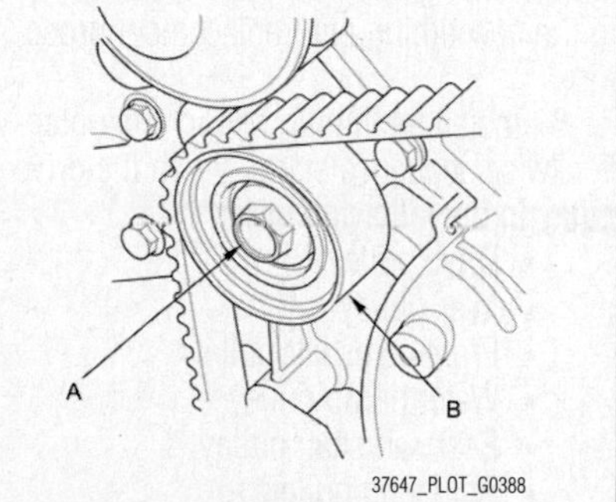
A
B
37647_PLOT_G0388

Fig. 178 Remove the idler pulley bolt (A) and the idler pulley (B), then remove the timing belt

To install:

Installing a Used Timing Belt

See Figures 179 through 182.

➡The following procedure is for installing a used timing belt. If you are installing a new belt, refer to the New Timing Belt replacement procedure.

1. Clean the timing belt pulleys, the timing belt guide plate, and the upper and lower covers.
2. Set the timing belt drive pulley to Top Dead Center (TDC) by aligning the TDC mark on the tooth of the timing belt

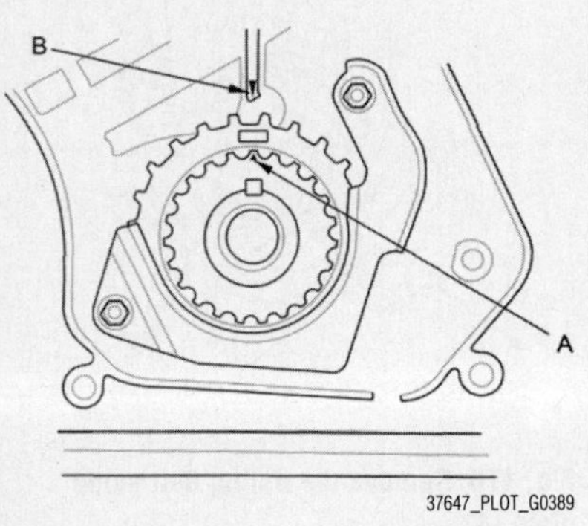
B
A
37647_PLOT_G0389

Fig. 179 Set the timing belt drive pulley to Top Dead Center (TDC) by aligning the TDC mark (A) on the tooth of the timing belt drive pulley with the pointer (B) on the oil pump

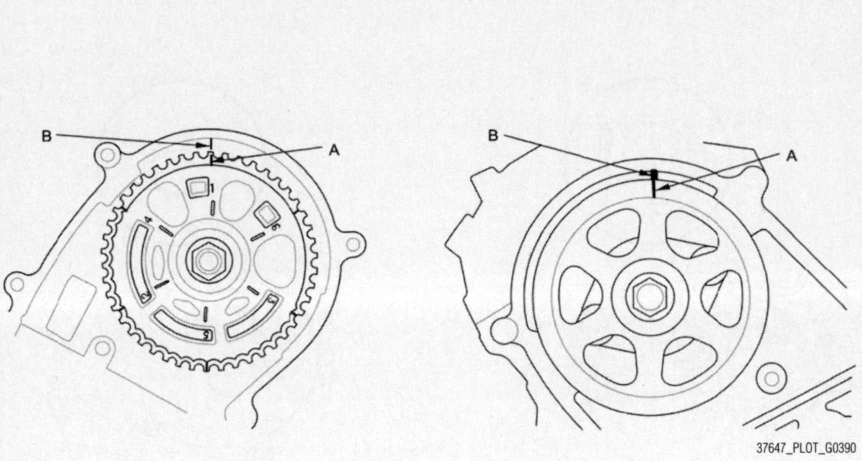
B
A
B
A
37647_PLOT_G0390

Fig. 180 Set the camshaft pulleys to TDC by aligning the TDC marks (A) on the camshaft pulleys with the pointers (B) on the back covers

drive pulley with the pointer on the oil pump.

3. Set the camshaft pulleys to TDC by aligning the TDC marks on the camshaft pulleys with the pointers on the back covers.

4. Loosely install the idler pulley with a new idler pulley bolt so the pulley can move but does not come off.

5. If the auto-tensioner has extended and the timing belt cannot be installed, do the new timing belt replacement procedure.

6. Install the timing belt in a counterclockwise sequence starting with the drive pulley in the following order:

- Drive pulley
- Idler pulley
- Front camshaft pulley
- Water pump pulley
- Rear camshaft pulley
- Adjusting pulley

➡Take care not to damage the timing belt during installation.

7. Tighten the idler pulley bolt to 33 ft. lbs. (44 Nm).

8. Remove the battery clamp bolt from the back cover.

9. Install the lower half of the side engine mount bracket. Tighten the three 10 x 1.25 mm bolts to 33 ft. lbs. (44 Nm).

10. Install the timing belt guide plate.

11. Install the lower cover.

12. Install the front upper cover and the rear upper cover.

13. Install the crankshaft pulley.

14. Rotate the crankshaft pulley about six turns clockwise so the timing belt positions itself on the pulleys.

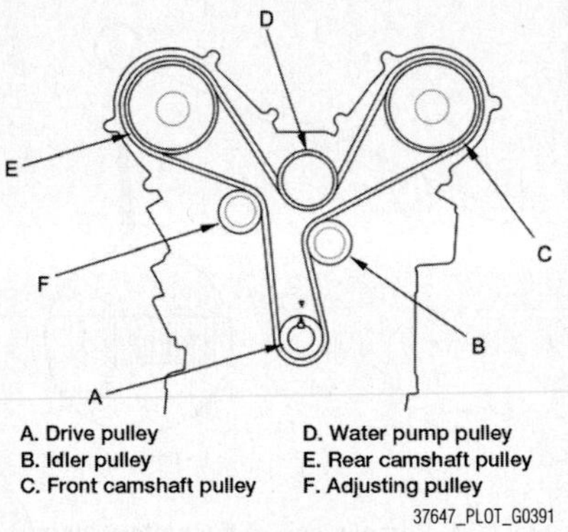
D
E
C
F
B
A
A. Drive pulley
B. Idler pulley
C. Front camshaft pulley
D. Water pump pulley
E. Rear camshaft pulley
F. Adjusting pulley
37647_PLOT_G0391

Fig. 181 Timing belt routing

15. Turn the crankshaft pulley so its white mark lines up with the pointer.

➡The other pointer is not used.

16. Check the camshaft pulley marks.

➡If the marks are not aligned, rotate the crankshaft 360 degrees, and recheck the camshaft pulley mark.

a. If the camshaft pulley marks are at TDC, go to step 17.

b. If the camshaft pulley marks are not at TDC, remove the timing belt and repeat steps 2 through 16.

17. Install the upper half of the side engine mount bracket, and tighten the new

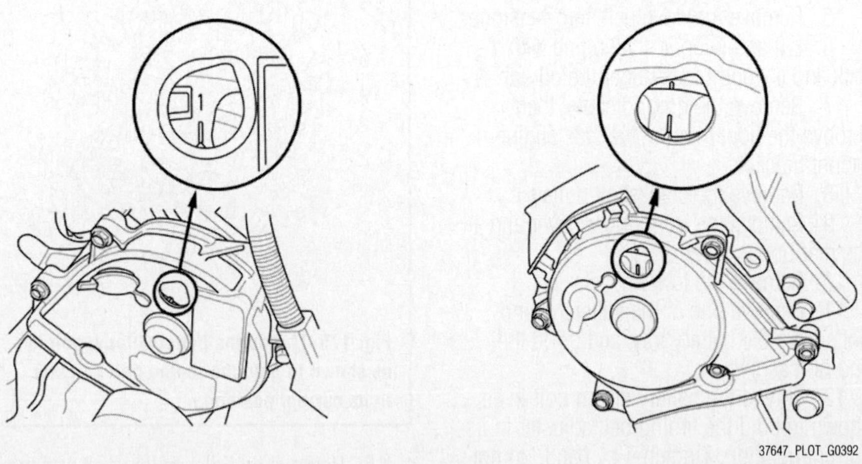
37647_PLOT_G0392

Fig. 182 Check the camshaft pulley marks

mounting bolts to 33 ft. lbs. (44 Nm), then tighten the mass damper mounting bolt to 40 ft. lbs. (54 Nm).

18. Install the ground cable.
19. Install the drive belt auto-tensioner.
20. Install the splash shield.
21. Install the right front wheel.
22. Do the Crankshaft Position (CKP) pattern clear/CKP pattern learn procedure.

Installing a New Timing Belt

See Figures 179 and 180, 183 through 188.

➡The following procedure is for installing a new timing belt. If you are installing a used belt, refer to the Used Timing Belt installation procedure.

1. Remove the timing belt.
2. Clean the timing belt pulleys, the timing belt guide plate, and the upper and lower covers.
3. Set the timing belt drive pulley to Top Dead Center (TDC) by aligning the TDC mark on the tooth of the timing belt drive pulley with the pointer on the oil pump.
4. Set the camshaft pulleys to TDC by aligning the TDC marks on the camshaft pulleys with the pointers on the back covers.
5. Remove the battery clamp bolt from the back cover.
6. Remove the auto-tensioner.
7. Align the holes on the rod and the housing of the auto-tensioner.
8. Use a hydraulic press to slowly compress the auto-tensioner. Insert a 0.08 inches (2.0 mm) pin through the housing and the rod.

➡The compression pressure should not exceed 2,200 lbs. (9,800 N).

9. Install the auto-tensioner.

➡Make sure the pin stays in place.

10. Thread the battery clamp bolt in as shown to hold the timing belt adjuster. Tighten it by hand, do not use a wrench.
11. Loosely install the idler pulley with a new idler pulley bolt so the pulley can move but does not come off.
12. Install the timing belt in a counterclockwise sequence starting with the drive pulley in the following order:

- Drive pulley
- Idler pulley
- Front camshaft pulley
- Water pump pulley
- Rear camshaft pulley
- Adjusting pulley

13. Tighten the idler pulley bolt to 33 ft. lbs. (44 Nm).
14. Remove the pin from the auto-tensioner.
15. Remove the battery clamp bolt from the back cover.
16. Install the lower half of the side engine mount bracket. Tighten the three 10 x 1.25 mm bolts to 33 ft. lbs. (44 Nm).
17. Install the timing belt guide plate.
18. Install the lower cover.
19. Install the front upper cover and the rear upper cover.
20. Install the crankshaft pulley.
21. Rotate the crankshaft pulley about six turns clockwise so the timing belt positions itself on the pulleys.
22. Turn the crankshaft pulley so its white mark lines up with the pointer.

➡The other pointer is not used.

23. Check the camshaft pulley marks.

➡If the marks are not aligned, rotate the crankshaft 360 degrees, and recheck the camshaft pulley mark.

a. If the camshaft pulley marks are at TDC, go to step 24.
b. If the camshaft pulley marks are not at TDC, remove the timing belt and repeat steps 3 through 22.

24. Install the upper half of the side engine mount bracket, and tighten the new mounting bolts to 33 ft. lbs. (44 Nm), then tighten the mass damper mounting bolt to 40 ft. lbs. (54 Nm).
25. Install the ground cable.
26. Install the drive belt auto-tensioner.
27. Install the splash shield.
28. Install the right front wheel.

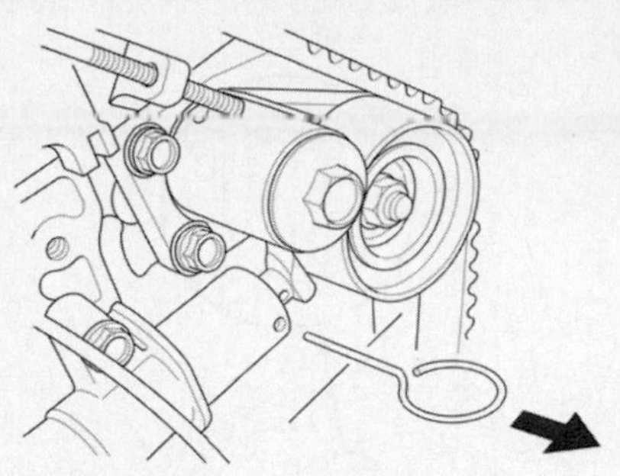
37647_PLOT_G0396

Fig. 187 Remove the pin from the auto-tensioner

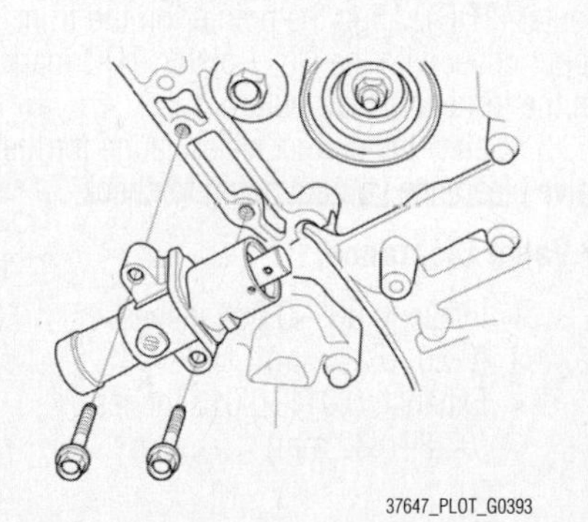
37647_PLOT_G0393

Fig. 183 Remove the auto-tensioner

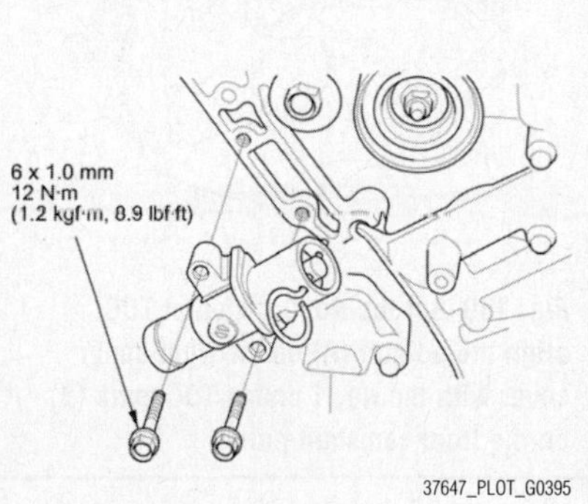
6 x 1.0 mm
12 N·m
(1.2 kgf·m, 8.9 lbf·ft)
37647_PLOT_G0395

Fig. 185 Install the auto-tensioner

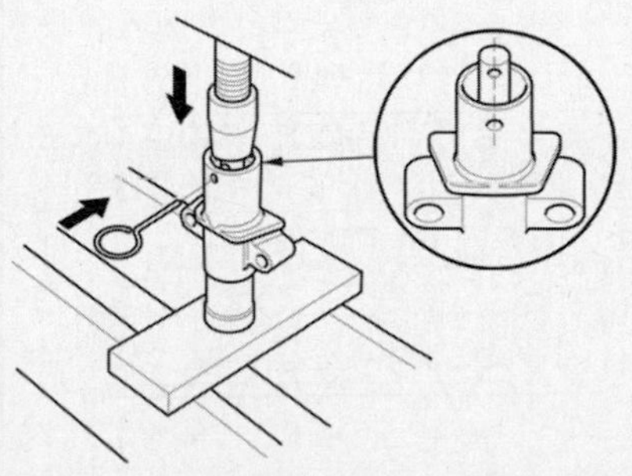
37647_PLOT_G0394

Fig. 184 Align the holes on the rod and the housing of the auto-tensioner

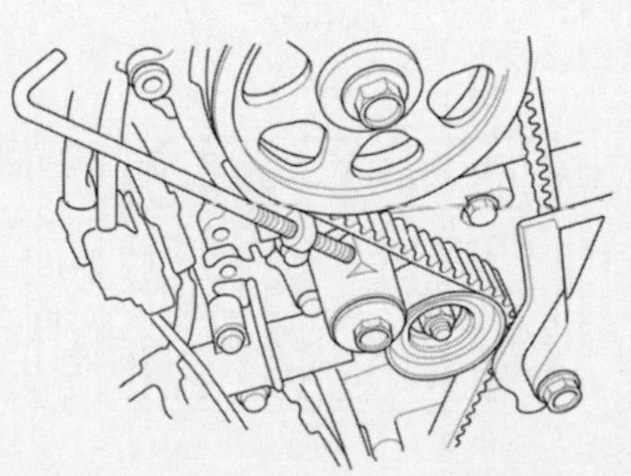
37647_PLOT_G0385

Fig. 186 Thread the battery clamp bolt in as shown to hold the timing belt adjuster

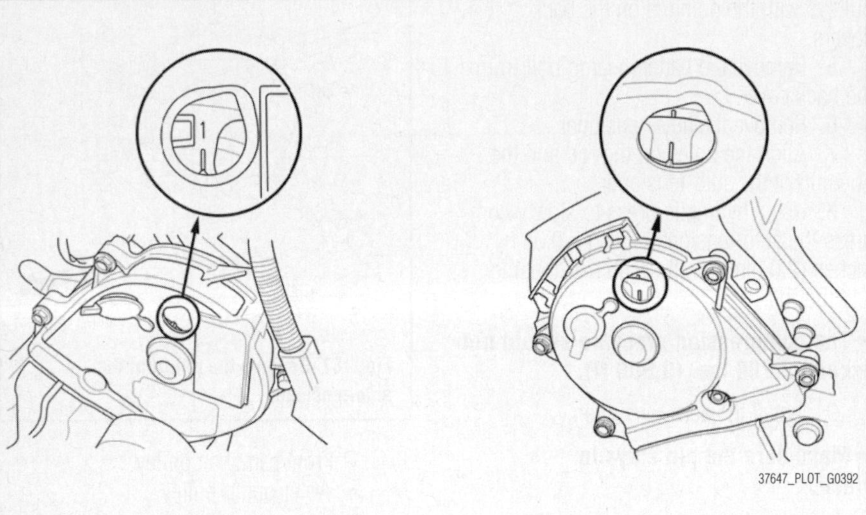
37647_PLOT_G0392

Fig. 188 Check the camshaft pulley marks

29. Do the Crankshaft Position (CKP) pattern clear/CKP pattern learn procedure.

CKP PATTERN CLEAR/CKP PATTERN LEARN PROCEDURE

Clear/Learn Procedure (with the HDS)

1. Connect the HDS to the Data Link Connector (DLC) located under the driver's side of the dashboard.
2. Turn the ignition switch to ON (II).
3. Make sure the HDS communicates with the PCM and all other vehicle systems. If it doesn't, go to the DLC circuit troubleshooting.
4. Select CRANK PATTERN in the ADJUSTMENT MENU with the HDS.
5. Select CRANK PATTERN LEARNING with the HDS, and follow the screen prompts.

Learn Procedure (without the HDS)

1. Start the engine. Hold the engine speed at 3,000 rpm without load (in P or N) until the radiator fan comes on.
2. Test-drive the vehicle on a level road: Decelerate (with the throttle fully closed) from an engine speed of 2,500 rpm down to 1,000 rpm with the transmission in 2.
3. Repeat step 2 several times.
4. Turn the ignition switch to LOCK (0).
5. Turn the ignition switch to ON (II), and wait for 30 seconds. The CKP pattern learn procedure is complete.

VALVE LASH

ADJUSTMENT

See Figures 189 through 195.

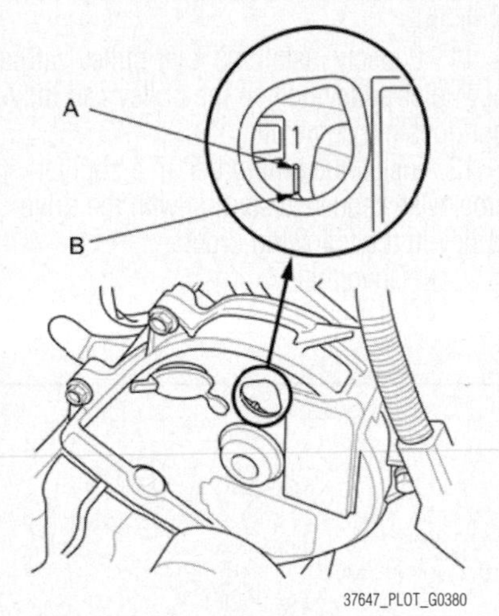
A
B
1
37647_PLOT_G0380

Fig. 189 Set the No. 1 piston at TDC; align the pointer (A) on the front upper cover with the No. 1 piston TDC mark (B) on the front camshaft pulley

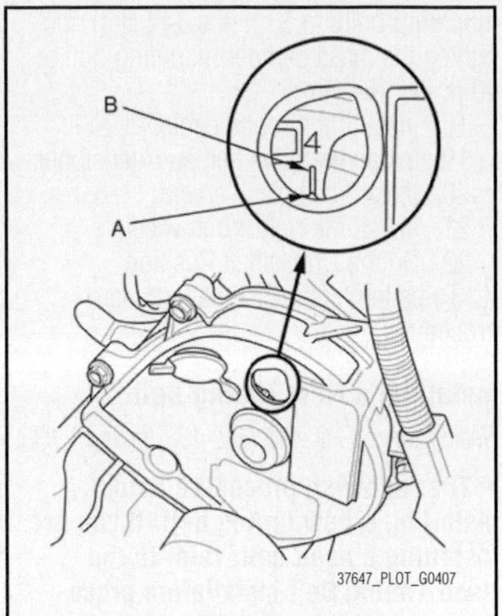
B
4
A
37647_PLOT_G0407

Fig. 191 Align the pointer (A) on the front upper cover with the No. 4 piston TDC mark (B) on the front camshaft pulley

➡Connect the Honda Diagnostic System (HDS) to the Data Link Connector (DLC), and monitor the Engine Coolant Temperature (ECT) sensor 1 with the HDS. Adjust the valve clearance only when the engine coolant temperature is less than 100°F (38°C).

1. Remove the cylinder head covers.
2. Set the No. 1 piston at Top Dead Center (TDC). Align the pointer on the front upper cover with the No. 1 piston TDC mark on the front camshaft pulley.
3. Select the correct feeler gauge for the valve clearance you are going to check.

➡Valve Clearance:

- Intake: 0.008–0.009 inches (0.20–0.24 mm)
- Exhaust: 0.011–0.013 inches (0.28–0.32 mm)

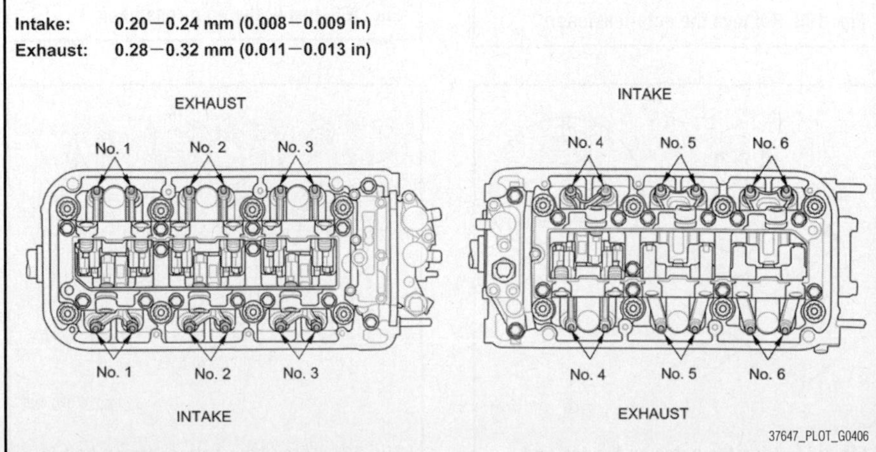
Intake: 0.20—0.24 mm (0.008—0.009 in)
Exhaust: 0.28—0.32 mm (0.011—0.013 in)
EXHAUST
No. 1
No. 2
No. 3
INTAKE
No. 4
No. 5
No. 6
No. 1
No. 2
No. 3
INTAKE
No. 4
No. 5
No. 6
EXHAUST
37647_PLOT_G0406

Fig. 190 Intake and exhaust valve locations

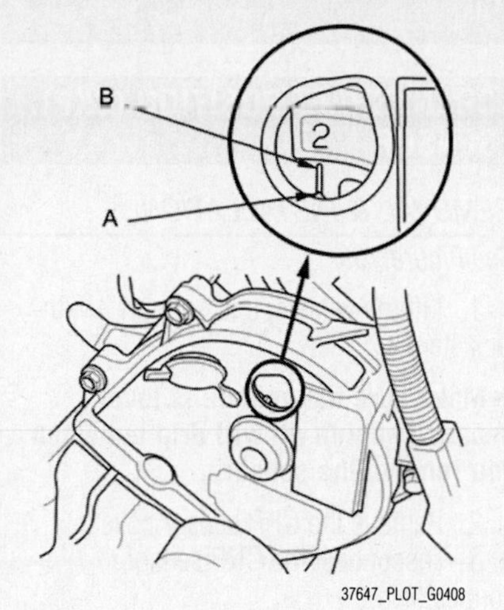
B
2
A
37647_PLOT_G0408

Fig. 192 Align the pointer (A) on the front upper cover with the No. 2 piston TDC mark (B) on the front camshaft pulley

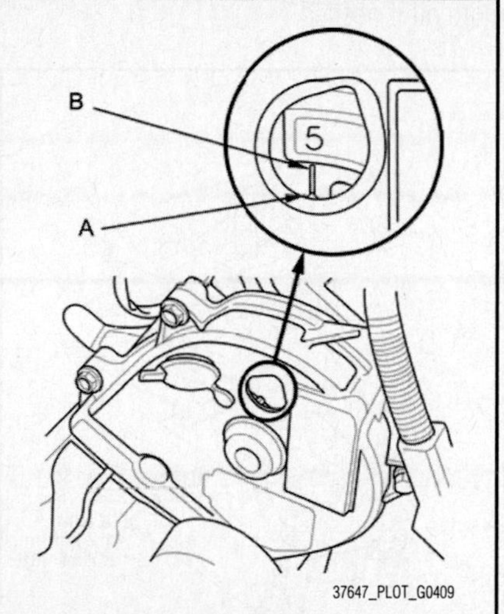
B
5
A
37647_PLOT_G0409

Fig. 193 Align the pointer (A) on the front upper cover with the No. 5 piston TDC mark (B) on the front camshaft pulley

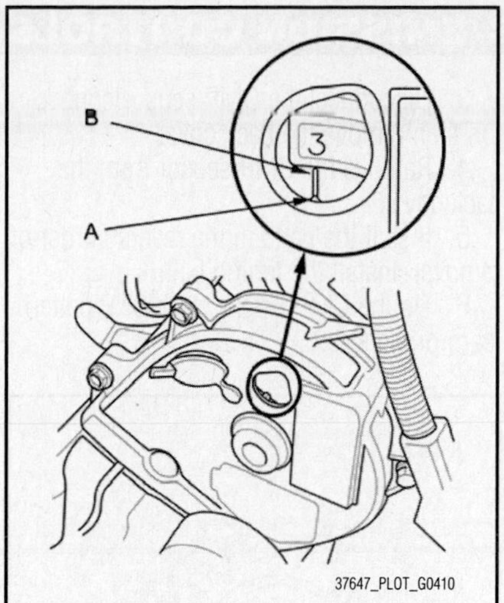
B
3
A
37647_PLOT_G0410

Fig. 194 Align the pointer (A) on the front upper cover with the No. 3 piston TDC mark (B) on the front camshaft pulley

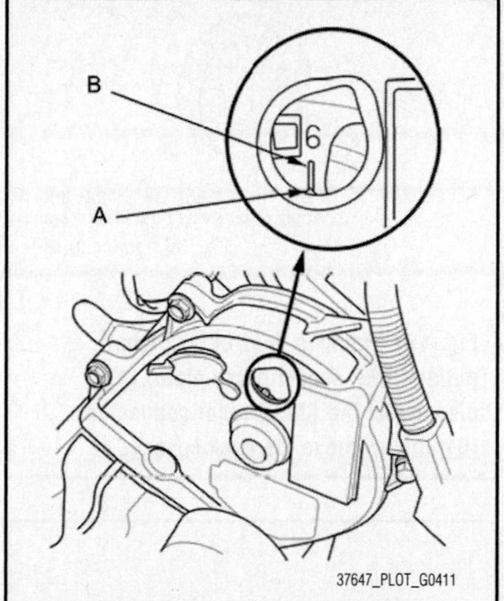
B
6
A
37647_PLOT_G0411

Fig. 195 Align the pointer (A) on the front upper cover with the No. 6 piston TDC mark (B) on the front camshaft pulley

4. Insert the feeler gauge between the adjusting screw and the end of the valve stem on the No. 1 cylinder and slide it back and forth; you should feel a slight amount of drag.
5. If you feel too much or too little drag, loosen the locknut, and turn the adjusting screw until the drag on the feeler gauge is correct.
6. While holding the adjusting screw with the screw driver, tighten the locknut, then recheck the clearance. Repeat the adjustment, if necessary.

➡**Specified Torque:**

- No. 1, No. 2,No. 3, and No. 4 cylinders: 14 ft. lbs. (20 Nm)
- No. 5 and No. 6 cylinders: 10 ft. lbs. (14 Nm)

7. Rotate the crankshaft clockwise. Align the pointer on the front upper cover with the No. 4 piston TDC mark on the front camshaft pulley.
8. Check and, if necessary, adjust the valve clearance on No. 4 cylinder.

Rotate the crankshaft clockwise. Align the pointer on the front upper cover with the No. 2 piston TDC mark on the front camshaft pulley.

9. Check and, if necessary, adjust the valve clearance on the No. 2 cylinder.
10. Rotate the crankshaft clockwise. Align the pointer on the front upper cover with the No. 5 piston TDC mark on the front camshaft pulley.
11. Check and, if necessary, adjust the valve clearance on No. 5 cylinder.
12. Rotate the crankshaft clockwise. Align the pointer on the front upper cover with the No. 3 piston TDC mark on the front camshaft pulley.
13. Check and, if necessary, adjust the valve clearance on No. 3 cylinder.
14. Rotate the crankshaft clockwise. Align the pointer on the front upper cover with the No. 6 piston TDC mark on the front camshaft pulley.
15. Check and, if necessary, adjust the valve clearance on the No. 6 cylinder.
16. Install the cylinder head covers.

ENGINE PERFORMANCE & EMISSION CONTROLS

ACCELERATOR PEDAL POSITION (APP) SENSOR

LOCATION

The Accelerator Pedal Position (APP) sensor is an integral part of the accelerator pedal module. The APP sensor is not available separately. Do not disassemble the accelerator pedal module.

REMOVAL & INSTALLATION

See Figure 196.

1. Disconnect the APP sensor 6P connector.
2. Remove the accelerator pedal module.

➡The APP sensor is not available separately. Do not disassemble the accelerator pedal module.

3. Install the parts in the reverse order of removal.

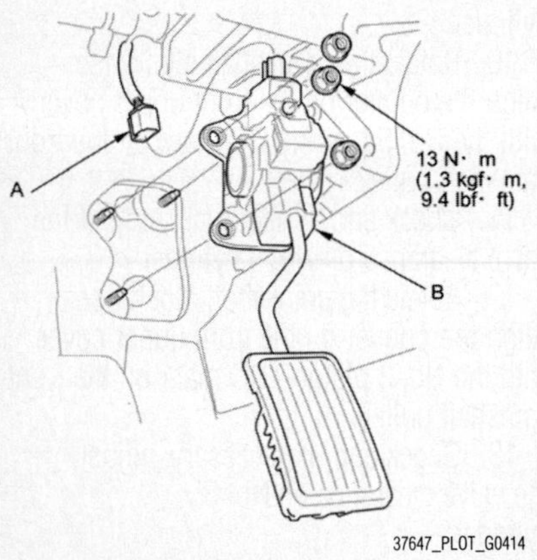
A
13 N· m
(1.3 kgf· m,
9.4 lbf· ft)
B
37647_PLOT_G0414

Fig. 196 Disconnect the APP sensor 6P connector (A); remove the accelerator pedal module (B)

CAMSHAFT POSITION (CMP) SENSOR

LOCATION

The Camshaft Position (CMP) sensor is located on the back cover for the front camshaft.

REMOVAL & INSTALLATION

See Figures 197 and 198.

1. Remove the timing belt.
2. Remove the front camshaft pulley (CMP sensor pulse plate).
3. Disconnect the CMP sensor connector, then remove the back cover.
4. Remove the CMP sensor from the back cover.
5. Install the parts in the reverse order of removal. Install the timing belt.
6. Do the CKP pattern clear/CKP pattern learn procedure.

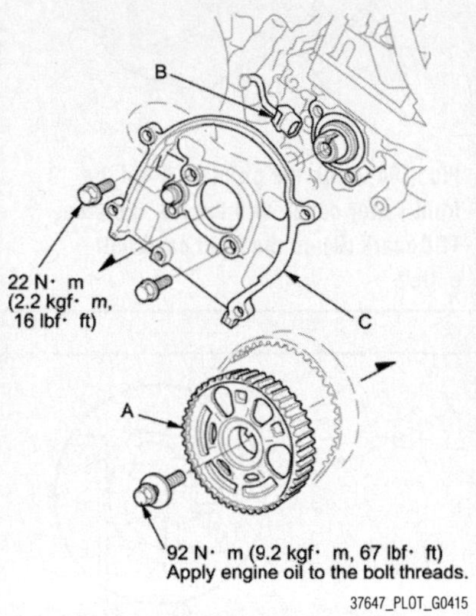
B
22 N· m
(2.2 kgf· m,
16 lbf· ft)
C
A
92 N· m (9.2 kgf· m, 67 lbf· ft)
Apply engine oil to the bolt threads.
37647_PLOT_G0415

Fig. 197 Remove the front camshaft pulley (CMP sensor pulse plate) (A); disconnect the CMP sensor connector (B), then remove the back cover (C)

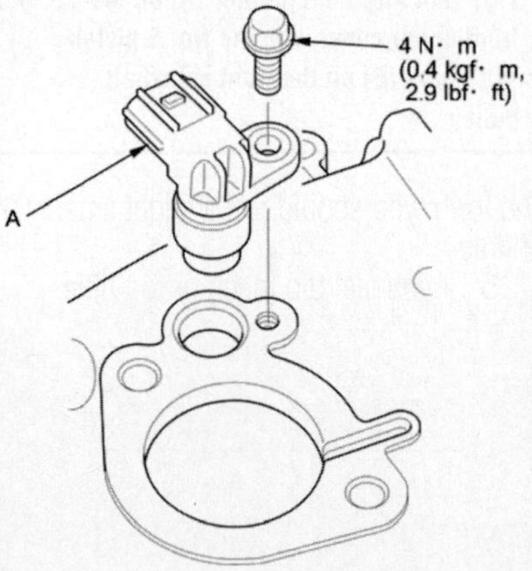
4 N· m
(0.4 kgf· m,
2.9 lbf· ft)
A
37647_PLOT_G0416

Fig. 198 Remove the CMP sensor (A) from the back cover

CRANKSHAFT POSITION (CKP) SENSOR

REMOVAL & INSTALLATION

See Figure 199.

1. Lift the vehicle, and support it with jack stands.

➡Make sure the vehicle is level, because engine oil will drip out when you remove the sensor.

2. Remove the CKP sensor cover.
3. Disconnect the CKP sensor connector.
4. Remove the CKP sensor.
5. Install the parts in the reverse order of removal with a new O-ring.
6. Do the CKP pattern clear/CKP pattern learn procedure.
7. Check the engine oil level, and add more oil if needed.

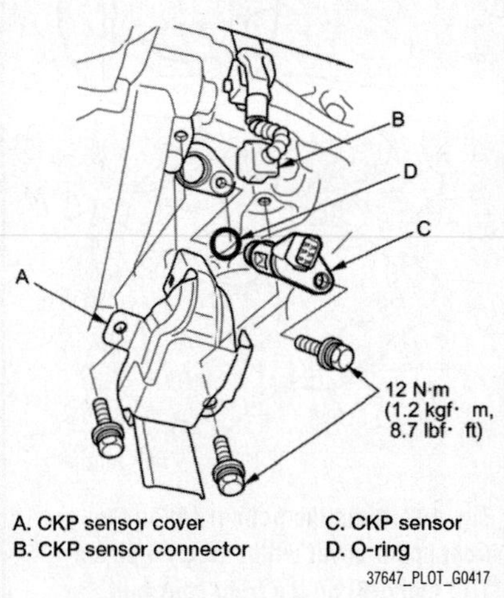
B
D
C
A
12 N·m
(1.2 kgf· m,
8.7 lbf· ft)
A. CKP sensor cover
B. CKP sensor connector
C. CKP sensor
D. O-ring
37647_PLOT_G0417

Fig. 199 Remove the CKP sensor cover

CKP PATTERN CLEAR/CKP PATTERN LEARN PROCEDURE

Clear/Learn Procedure (with the HDS)

1. Connect the HDS to the Data Link Connector (DLC) located under the driver's side of the dashboard.
2. Turn the ignition switch to ON (II).
3. Make sure the HDS communicates with the PCM and all other vehicle systems. If it doesn't, go to the DLC circuit troubleshooting.

4. Select CRANK PATTERN in the ADJUSTMENT MENU with the HDS.

5. Select CRANK PATTERN LEARNING with the HDS, and follow the screen prompts.

Learn Procedure (without the HDS)

1. Start the engine. Hold the engine speed at 3,000 rpm without load (in P or N) until the radiator fan comes on.

2. Test-drive the vehicle on a level road: Decelerate (with the throttle fully closed) from an engine speed of 2,500 rpm down to 1,000 rpm with the transmission in 2.

3. Repeat step 2 several times.

4. Turn the ignition switch to LOCK (0).

5. Turn the ignition switch to ON (II), and wait for 30 seconds. The CKP pattern learn procedure is complete.

ENGINE COOLANT TEMPERATURE (ECT) SENSOR

LOCATION

Refer to the graphics in the Removal and Installation section for the location(s).

REMOVAL & INSTALLATION

ECT Sensor 1

See Figures 200 and 201.

1. Drain the engine coolant.
2. Remove the engine cover.
3. Disconnect the ECT sensor 1 2P connector.
4. Remove ECT sensor 1.
5. Install the parts in the reverse order of removal with a new O-ring, then refill the radiator with engine coolant.

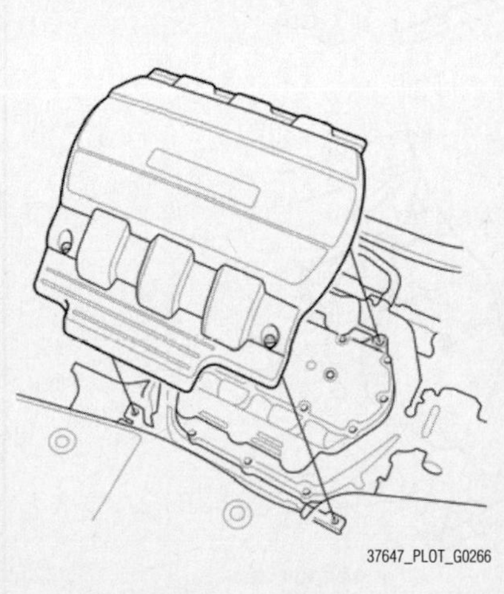
37647_PLOT_G0266

Fig. 200 Remove the engine cover

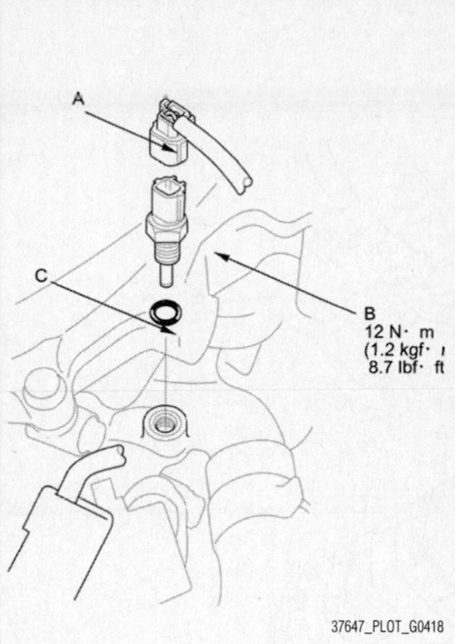
A
C
B
12 N· m
(1.2 kgf· m
8.7 lbf· ft
37647_PLOT_G0418

Fig. 201 Disconnect the ECT sensor 1 2P connector (A); remove ECT sensor 1 (B) and O-ring (C)

ECT Sensor 2

See Figures 202 and 203.

1. Remove the splash shield.
2. Drain the engine coolant.
3. Disconnect the ECT sensor 2 2P connector.
4. Remove ECT sensor 2.
5. Install the parts in the reverse order of removal with a new O-ring, then refill the radiator with engine coolant.

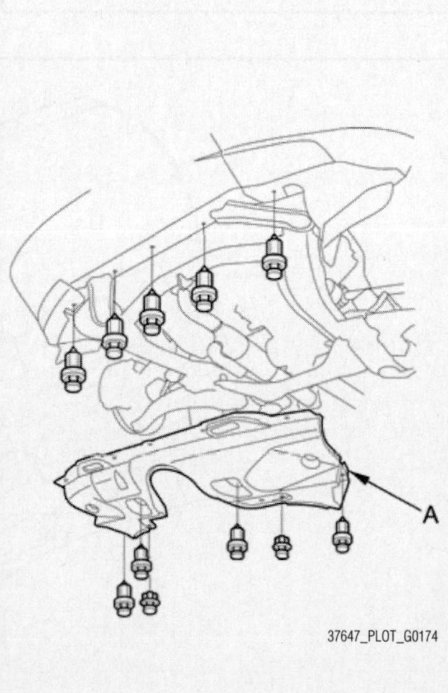
A
37647_PLOT_G0174

Fig. 202 Remove the splash shield

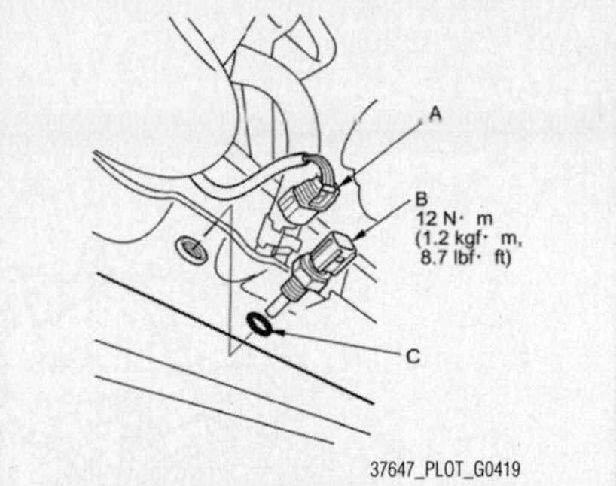
A
B
12 N· m
(1.2 kgf· m,
8.7 lbf· ft)
C
37647_PLOT_G0419

Fig. 203 Disconnect the ECT sensor 2 2P connector (A); Remove ECT sensor 2 (B) and the O-ring (C)

EVAPORATIVE EMISSIONS (EVAP) CANISTER

LOCATION

The EVAP canister is mounted on the vehicle frame. Refer to the graphics in the Removal and Installation section for the location(s).

REMOVAL & INSTALLATION

See Figures 204 and 205.

1. Raise the vehicle on a lift.
2. Remove the EVAP canister cover.
3. Disconnect the quick-connect fittings, the hoses, the EVAP canister vent shut valve connector, and the FTP sensor connector, then remove the harness clamps.
4. Remove the bolts.
5. Remove the EVAP canister assembly.

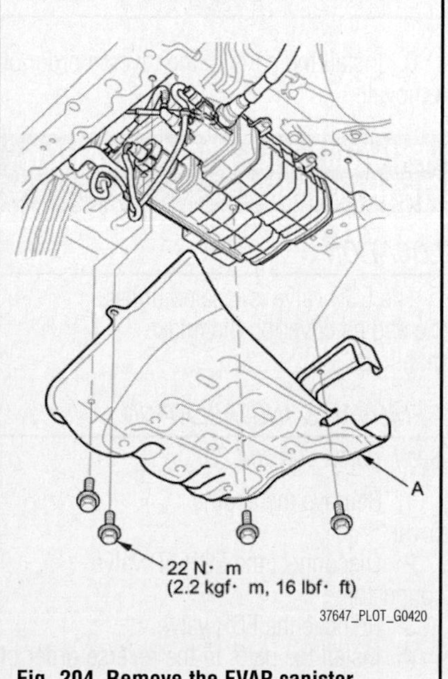
A
22 N· m
(2.2 kgf· m, 16 lbf· ft)
37647_PLOT_G0420

Fig. 204 Remove the EVAP canister cover (A)

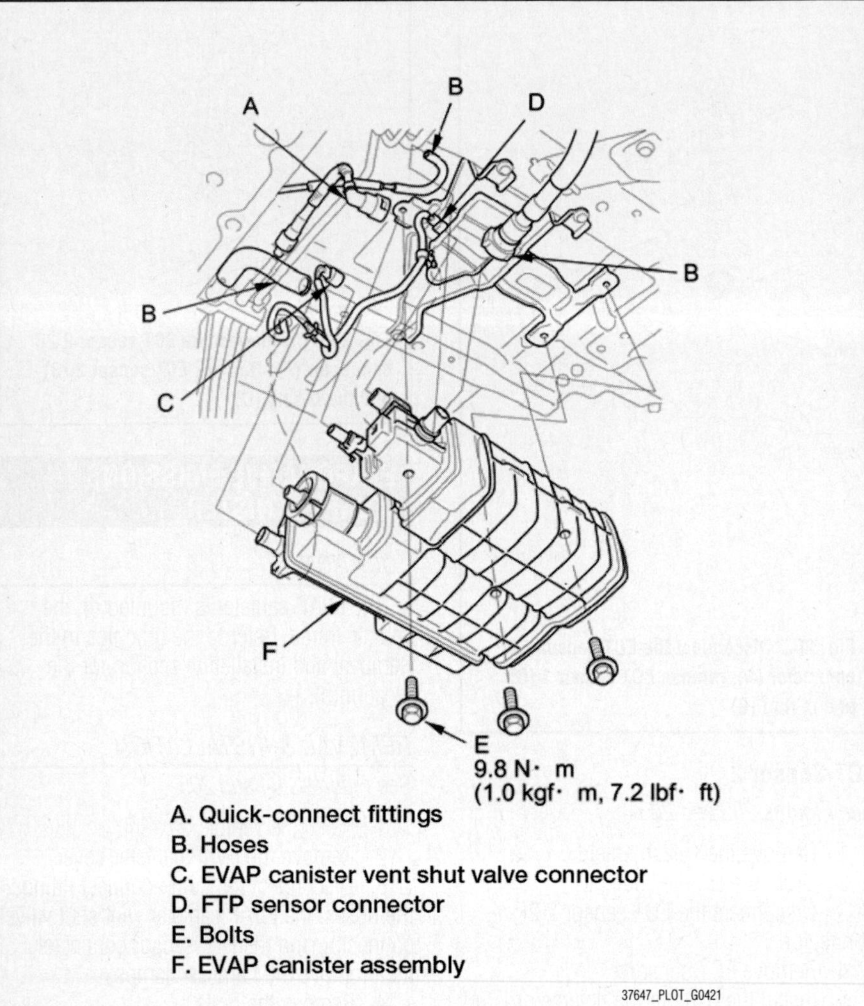
A
B
D
B
B
C
F
E
9.8 N· m
(1.0 kgf· m, 7.2 lbf· ft)
A. Quick-connect fittings
B. Hoses
C. EVAP canister vent shut valve connector
D. FTP sensor connector
E. Bolts
F. EVAP canister assembly
37647_PLOT_G0421

Fig. 205 Disconnect the quick-connect fittings, the hoses, the EVAP canister vent shut valve connector, and the FTP sensor connector, then remove the harness clamps

6. Install the parts in the reverse order of removal.

EXHAUST GAS RECIRCULATION (EGR) VALVE

LOCATION

The EGR valve is located under the engine cover on the intake manifold.

REMOVAL & INSTALLATION

See Figures 206 and 207.

1. Remove the engine cover.
2. Disconnect the EGR 5P valve connector.
3. Remove the EGR valve.
4. Install the parts in the reverse order of removal with a new gasket.

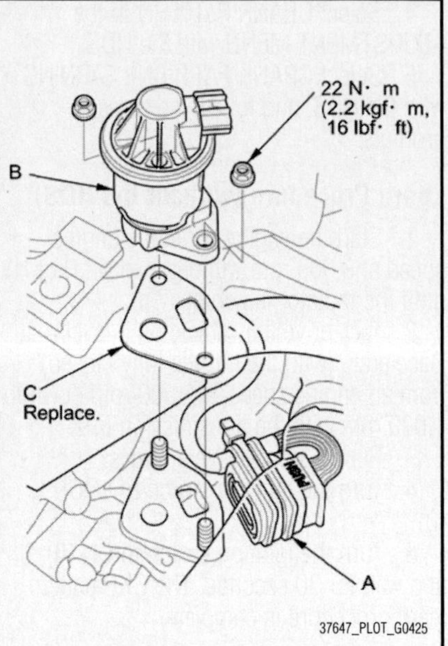
22 N· m
(2.2 kgf· m,
16 lbf· ft)
B
C
Replace.
A
37647_PLOT_G0425

Fig. 207 Disconnect the EGR 5P valve connector (A); remove the EGR valve (B) and the gasket (C)

HEATED OXYGEN (HO2S) SENSOR

LOCATION

See Figure 208.

Refer to the accompanying illustration.

REMOVAL & INSTALLATION

Special Tools Required: O2 Sensor Socket Wrench Snap-on® SWR2 or O2 Sensor Wrench Snap-on® YA8875, commercially available

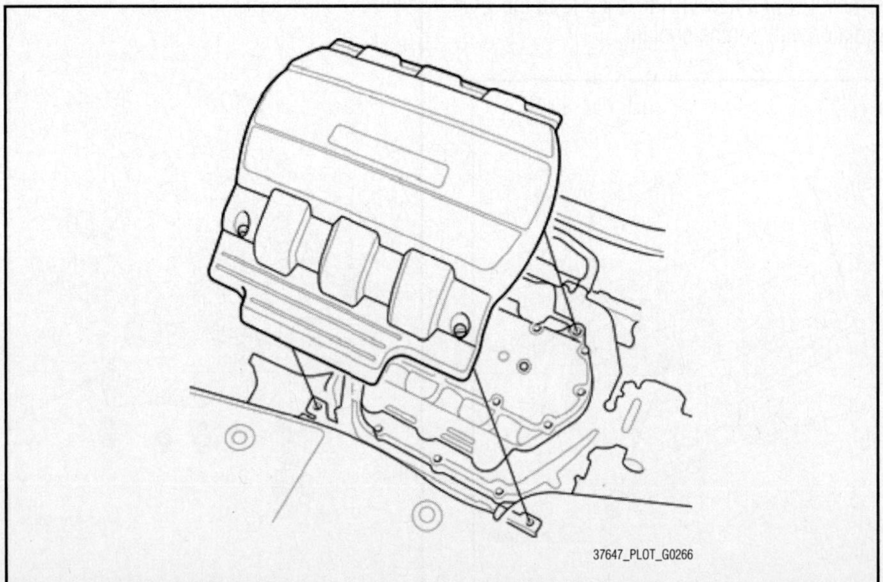
37647_PLOT_G0266

Fig. 206 Remove the engine cover

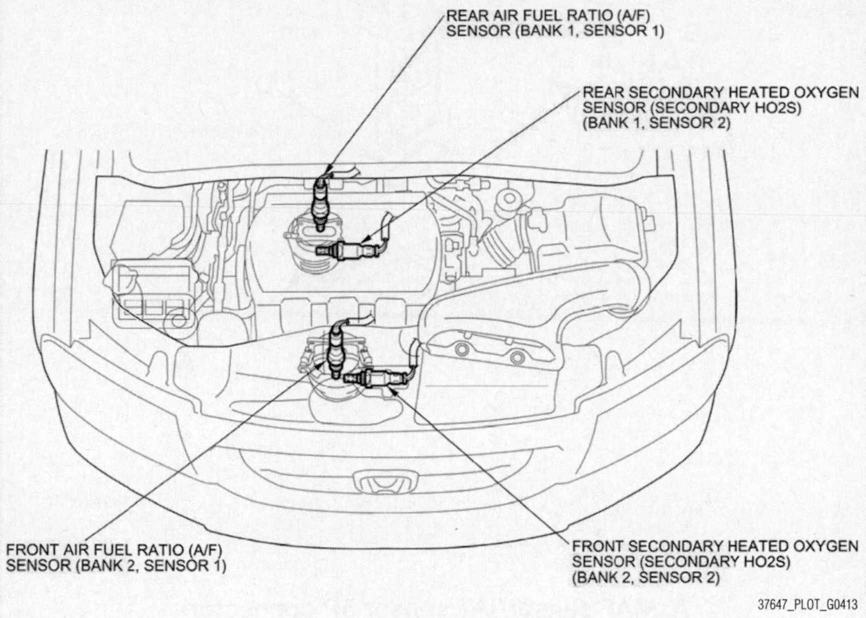
REAR AIR FUEL RATIO (A/F) SENSOR (BANK 1, SENSOR 1)
REAR SECONDARY HEATED OXYGEN SENSOR (SECONDARY HO2S) (BANK 1, SENSOR 2)
FRONT AIR FUEL RATIO (A/F) SENSOR (BANK 2, SENSOR 1)
FRONT SECONDARY HEATED OXYGEN SENSOR (SECONDARY HO2S) (BANK 2, SENSOR 2)
37647_PLOT_G0413

Fig. 208 Heated Oxygen Sensor locations

Front Bank (Bank2)

See Figures 209 and 210.

1. Disconnect the front secondary HO2S 4P connector, and remove the harness clamps.
2. Raise the vehicle on a lift.
3. Remove the harness clamp, then remove the front secondary HO2S.
4. Install the parts in the reverse order of removal.

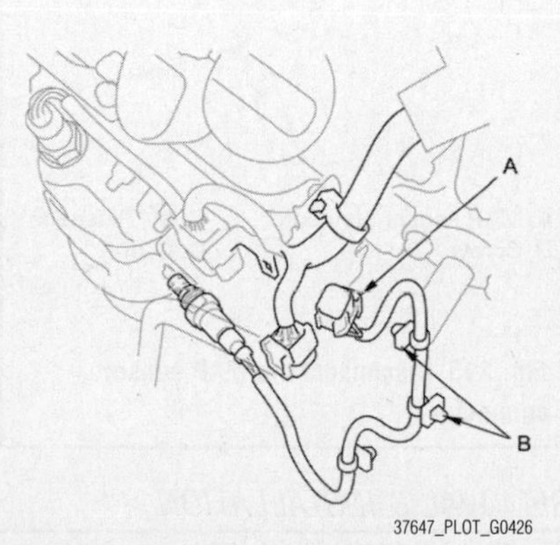
A
B
37647_PLOT_G0426

Fig. 209 Disconnect the front secondary HO2S 4P connector (A), and remove the harness clamps (B)

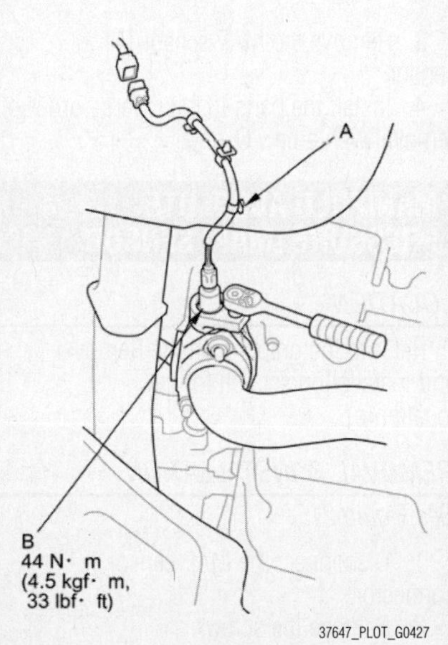
A
B
44 N· m
(4.5 kgf· m,
33 lbf· ft)
37647_PLOT_G0427

Fig. 210 Remove the harness clamp (A), then remove the front secondary HO2S (B)

Rear Bank (Bank 1)

See Figure 211.

1. Raise the vehicle on a lift.
2. Disconnect the rear secondary HO2S 4P connector, and remove the harness clamps.

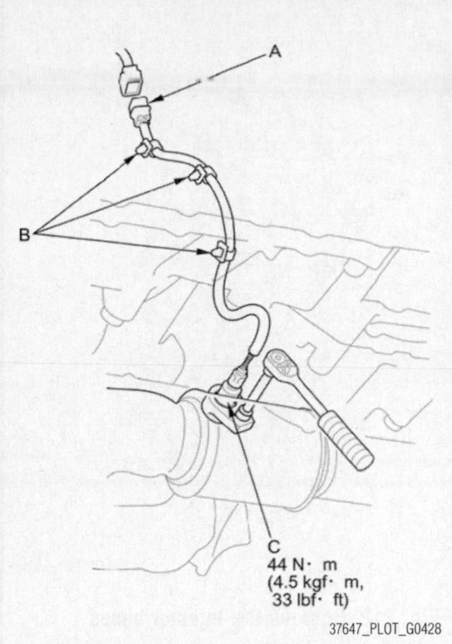
A
B
C
44 N· m
(4.5 kgf· m,
33 lbf· ft)
37647_PLOT_G0428

Fig. 211 Disconnect the rear secondary HO2S 4P connector (A), and remove the harness clamps (B); remove the rear secondary HO2S (C)

3. Remove the rear secondary HO2S.
4. Install the parts in the reverse order of removal.

INTAKE AIR TEMPERATURE (IAT) SENSOR

LOCATION

The Intake Air Temperature (IAT) sensor is an integral part of the Mass Air Flow (MAF) sensor/Intake Air Temperature (IAT) sensor which is located on the engine air take duct. Refer to the Mass Air Flow (MAF) sensor/Intake Air Temperature (IAT) sensor section when servicing this component.

KNOCK SENSOR (KS)

LOCATION

Refer to the graphics in the Removal and Installation section for the location(s).

REMOVAL & INSTALLATION

See Figures 212 and 213.

1. Remove the intake manifold.
2. Remove the injector base.
3. Disconnect the knock sensor connector.
4. Remove the knock sensor.
5. Install the parts in the reverse order of removal.

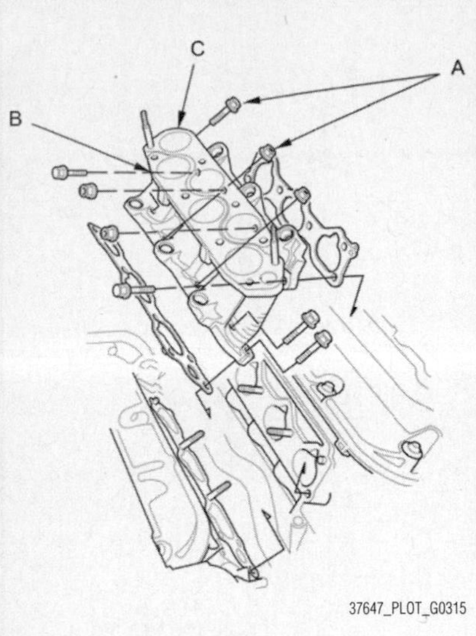
C
A
B
37647_PLOT_G0315

Fig. 212 Remove the injector bases

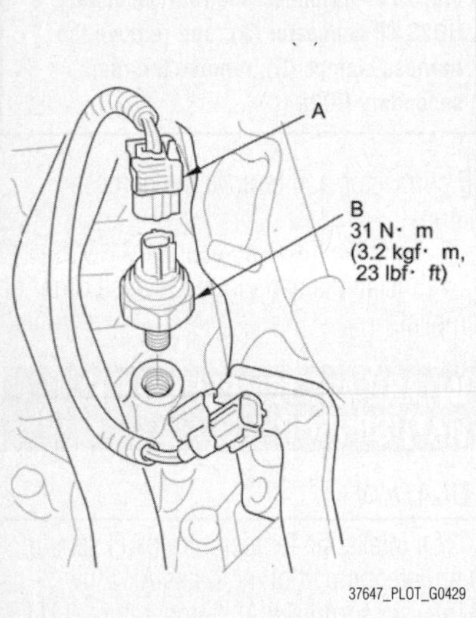
A
B
31 N· m
(3.2 kgf· m,
23 lbf· ft)
37647_PLOT_G0429

Fig. 213 Disconnect the knock sensor connector (A) and remove the knock sensor (B)

MASS AIR FLOW (MAF) SENSOR/INTAKE AIR TEMPERATURE (IAT) SENSOR (HOT WIRE)

LOCATION

The MAF sensor/IAT sensor is located on the engine air intake duct.

REMOVAL & INSTALLATION

See Figure 214.

1. Disconnect the MAF sensor/IAT sensor 5P connector.
2. Remove the screw.

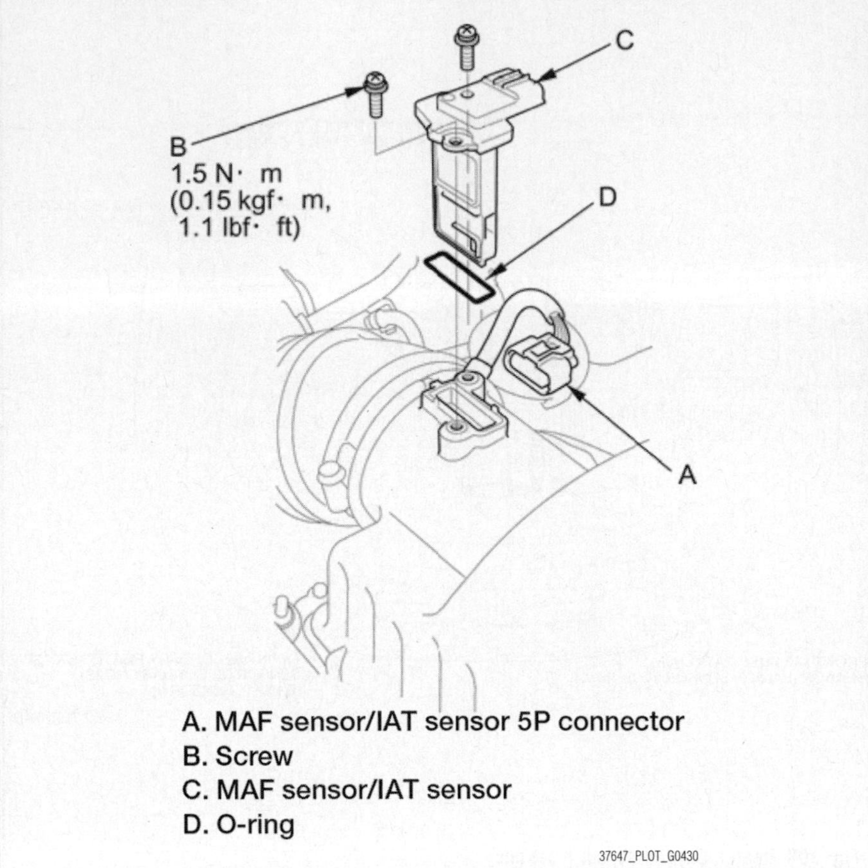
C
B
1.5 N· m
(0.15 kgf· m,
1.1 lbf· ft)
D
A
A. MAF sensor/IAT sensor 5P connector
B. Screw
C. MAF sensor/IAT sensor
D. O-ring
37647_PLOT_G0430

Fig. 214 Disconnect the MAF sensor/IAT sensor 5P connector

3. Remove the MAF sensor/IAT sensor.
4. Install the parts in the reverse order of removal with a new O-ring.

MANIFOLD ABSOLUTE PRESSURE (MAP) SENSOR

LOCATION

Refer to the graphics in the Removal and Installation section for the location(s).

REMOVAL & INSTALLATION

See Figure 215.

1. Disconnect the MAP sensor connector.
2. Remove the screw.
3. Remove the MAP sensor.
4. Install the parts in the reverse order of removal with a new O-ring.

OUTPUT SHAFT SPEED (OSS) SENSOR

LOCATION

The Output Shaft Speed (OSS) Sensor is located on the transaxle housing.

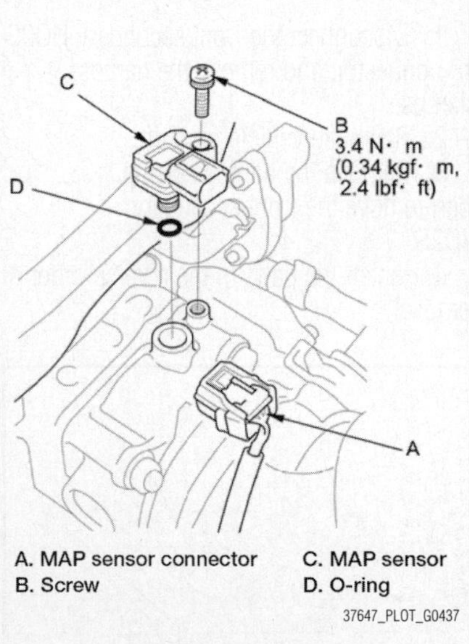
C
B
3.4 N· m
(0.34 kgf· m,
2.4 lbf· ft)
D
A
A. MAP sensor connector
B. Screw
C. MAP sensor
D. O-ring
37647_PLOT_G0437

Fig. 215 Disconnect the MAP sensor connector

REMOVAL & INSTALLATION

See Figures 216 through 218.

1. Raise the vehicle on a lift, or apply the parking brake, block both rear wheels, and raise the front of the

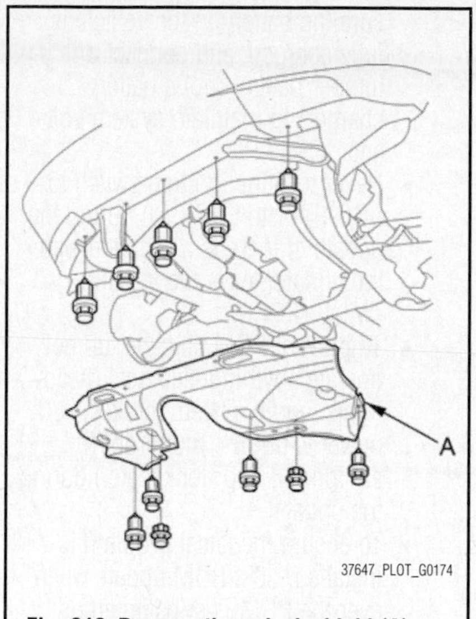
A
37647_PLOT_G0174

Fig. 216 Remove the splash shield (A)

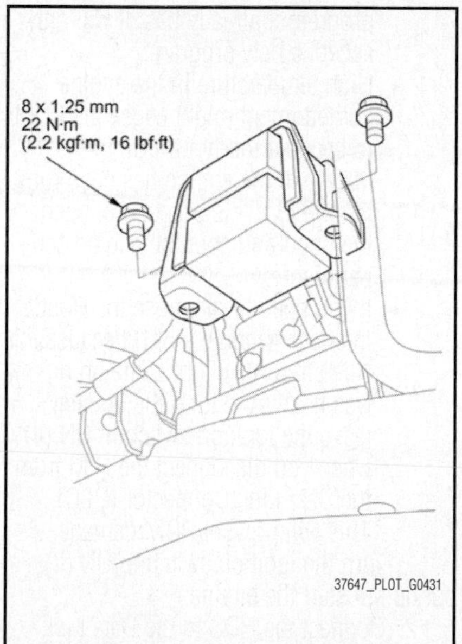
8 x 1.25 mm
22 N·m
(2.2 kgf·m, 16 lbf·ft)
37647_PLOT_G0431

Fig. 217 Remove the damper from the front subframe

vehicle. Make sure it is securely supported.

2. Remove the splash shield.
3. Remove the damper from the front subframe.
4. Disconnect the output shaft (countershaft) speed sensor connector, and remove the output shaft (countershaft) speed sensor and sensor washer.

To install:

5. Install the new O-ring on the new output shaft (countershaft) speed sensor, then install the output shaft (countershaft) speed sensor and sensor washer.
6. Check the connector for rust, dirt, or oil, and if clean necessary, then connect the connector securely.

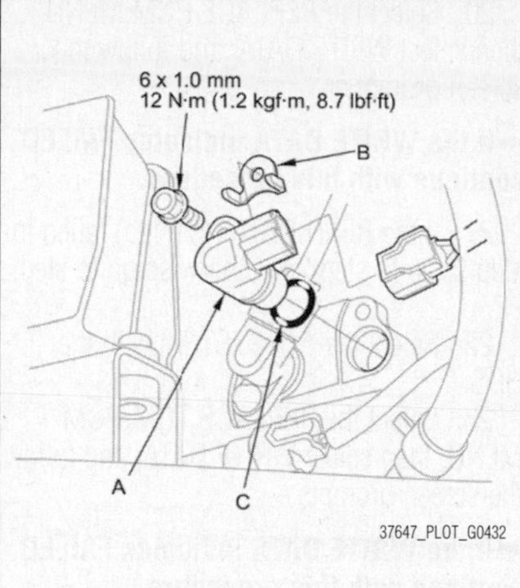
6 x 1.0 mm
12 N·m (1.2 kgf·m, 8.7 lbf·ft)
B
A
C
37647_PLOT_G0432

Fig. 218 Disconnect the output shaft speed sensor connector, and remove the output shaft speed sensor (A) and sensor washer (B) and O-ring (C)

7. Install the damper on the front subframe.
8. Install the splash shield.

POSITIVE CRANKCASE VENTILATION (PCV) VALVE

LOCATION

Refer to the graphics in the Removal and Installation section for the location(s).

REMOVAL & INSTALLATION

See Figure 219.

1. Remove the engine cover.
2. Remove the bolt.
3. Remove the PCV valve.

➡Do not to spill oil on the hot exhaust manifold.

4. Install the parts in the reverse order of removal.

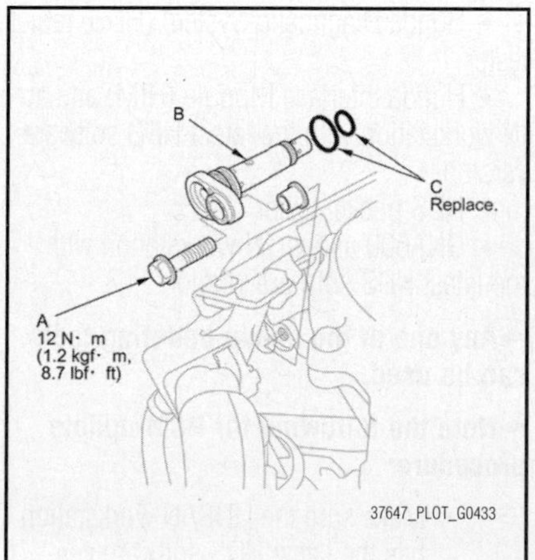
B
C
Replace.
A
12 N· m
(1.2 kgf· m,
8.7 lbf· ft)
37647_PLOT_G0433

Fig. 219 Remove the bolt (A), the PCV valve (B), and O-rings (C)

➡When installing a new PCV valve, make sure the new O-rings are in place.

POWERTRAIN CONTROL MODULE (PCM)

LOCATION

The Powertrain Control Module (PCM) is located in the engine compartment.

REMOVAL & INSTALLATION

See Figure 220.

Special Tools Required:

- Honda Diagnostic System (HDS) tablet tester
- Honda Interface Module (HIM) and an iN workstation with the latest HDS software version
- HDS pocket tester
- GNA600 and an iN workstation with the latest HDS software version

➡Any one of the above updating tools can be used.

1. Connect HDS to the Data Link Connector (DLC) located under the driver's side of the dashboard.
2. Turn the ignition switch to ON (II).
3. Make sure the HDS communicates with the PCM and other vehicle systems. If it doesn't, go to the DLC circuit troubleshooting. If you are returning from the DLC circuit troubleshooting, skip steps 5 through 9, 18 through 23, and 26 through 28, and do this after replacing the PCM:
   - Replace the engine oil and the engine oil filter.
   - Replace the ATF.
   - Clean the throttle body.
4. Select the PGM-FI system with the HDS.
5. Select the INSPECTION MENU with the HDS.
6. Select the ETCS TEST, then select the TP POSITION CHECK, and follow the screen prompts.

➡If the TP POSITION CHECK indicates FAILED, continue with this procedure.

7. Select the REPLACE PCM MENU, then select READ DATA, and follow the screen prompts.

➡Doing this step copies (READS) the engine oil life data from the original PCM so you can later download (WRITES) it into the new PCM. If the READ DATA indicates FAILED, continue with this procedure.

8. Select the A/T system with the HDS.

9. Select the REPLACE TCM/PCM MENU, then select READ DATA, and follow the screen prompts.

➡Doing this step copies (READS) the ATF life data from the original PCM so you can later download (WRITES) it into the new PCM. If READ DATA indicates FAILED, continue with this procedure.

10. Turn the ignition switch to LOCK (0).
11. Jump the SCS line with the HDS.
12. Remove the cover.
13. Disconnect PCM connectors A, B, and C.

➡PCM connectors A, B, and C have symbols (A=_, B=_, C=_) embossed on them for identification.

14. Remove the bolts and remove the PCM.
15. Install the PCM in the reverse order of removal.
16. Turn the ignition switch to ON (II).
17. Manually input the VIN to the PCM with the HDS.

➡DTC P0630 VIN Not Programmed or Mismatch may be stored because the VIN has not been programmed into the PCM; ignore it, and continue this procedure.

18. If the READ DATA (engine oil life) failed in step 7, go to step 21. Otherwise, go to step 19.
19. Select the PGM-FI system with the HDS.

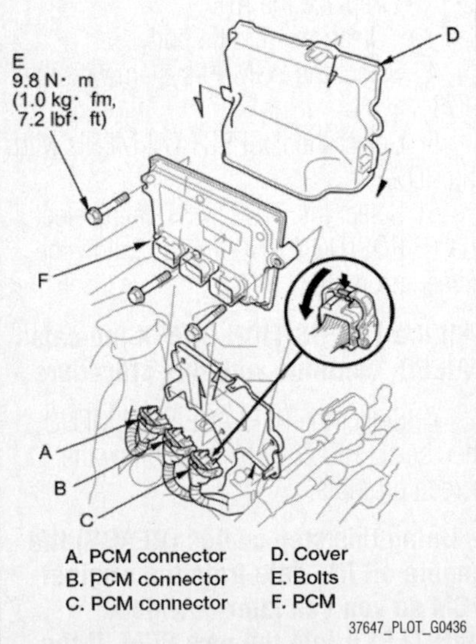
E
9.8 N· m
(1.0 kg· fm,
7.2 lbf· ft)
D
F
A
B
C
A. PCM connector
B. PCM connector
C. PCM connector
D. Cover
E. Bolts
F. PCM
37647_PLOT_G0436

Fig. 220 Exploded view of PCM assembly

20. Select the REPLACE PCM MENU, then select WRITE DATA, and follow the screen prompts.

➡If the WRITE DATA indicates FAILED, continue with this procedure.

21. If the READ DATA (ATF life) failed in step 9, go to step 24. Otherwise go to step 22.
22. Select the A/T SYSTEM with the HDS.
23. Select the REPLACE TCM/PCM MENU, then select WRITE DATA, and follow the screen prompts.

➡If the WRITE DATA indicates FAILED, continue with this procedure.

24. Select the IMMOBI SYSTEM with the HDS.
25. Enter the immobilizer PCM code that you got from the iN, and use the PCM replacement procedure in the IMMOBI MENU of the HDS; it allows you to start the engine.
26. If the TP POSITION CHECK failed in step 6, clean the throttle body, then go to step 27.
27. If the READ DATA failed in step 7 or the WRITE DATA failed in step 20, replace the engine oil and engine oil filter, then go to step 28.
28. If the READ DATA failed in step 9 or the WRITE DATA failed in step 23, replace the ATF, then go to step 29.
29. Select the PGM-FI system, and reset the PCM with the HDS.
30. Update the PCM if it does not have the latest software.
31. Do the PCM idle learn procedure.
32. Do the CKP pattern clear/CKP pattern learn procedure.

PCM UPDATE

Special Tools Required:

- Honda Diagnostic System (HDS) tablet tester
- Honda Interface Module (HIM) and an iN workstation with the latest HDS software version
- HDS pocket tester
- GNA600 and an iN workstation with the latest HDS software version

➡Any one of the above updating tools can be used.

➡Note the following for PCM update procedure:

- Make sure the HDS/iN workstation has the latest HDS software version.
- Before you update the PCM, make sure the battery in the vehicle is fully charged, and connect a jumper battery (not a battery charger) to maintain system voltage.
- Never turn the ignition switch to ACC (I) or to LOCK (0) during the update. If there is a problem with the update, leave the ignition switch ON (II).
- To prevent PCM damage, do not operate anything electrical (headlights, audio system, brakes, A/C, power windows, moonroof (if equipped), door locks, etc.) during the update.
- To ensure the latest program is installed, do a PCM update whenever the PCM is substituted or replaced.
- You cannot update a PCM with a program it already has. It will only accept a new program.
- High temperature in the engine compartment might cause the PCM to become too hot to run the update. If the engine has been running before this procedure, open the hood and cool the engine compartment.
- If you need to diagnose the Honda Interface Module (HIM) because the HIM's red (#3) light came on or was flashing during the update, leave the ignition switch in ON (II) when you disconnect the HIM from the Data Link Connector (DLC). This will prevent PCM damage.

1. Turn the ignition switch to ON (II), but do not start the engine.
2. Connect the HDS to the Data Link Connector (DLC) located under the driver's side of the dashboard.
3. Make sure the HDS communicates with the PCM and other vehicle systems. If it doesn't, go to the DLC circuit troubleshooting. If you are returning from DLC circuit troubleshooting, skip steps 4 and 5, and clean the throttle body after updating the PCM.
4. Select the INSPECTION MENU with HDS.
5. Select the ETCS TEST, then select the TP POSITION CHECK, and follow the HDS screen prompts.

➡If the TP POSITION CHECK indicates FAILED, continue this procedure.

6. Exit the HDS diagnostic system, then select the update mode, and follow the screen prompts to update the PCM.
7. If the software in the PCM is the lat-

est, disconnect the updating tool from the DLC, and go back to the procedure that you were doing. If the software in the PCM is not the latest, follow the instructions on the screen. If prompted to choose the PGM-FI system or the A/T system, make sure you update both.

➡If the PCM update system requires you to cool the PCM, follow the instructions on screen. If you run in to a problem during the update procedure (programming takes over 15 minutes, status bar goes over 100 %, D or immobilizer indicator flashes, HDS tablet freezes, etc.), follow these steps to minimize the chance of damaging the PCM.

- Leave the ignition switch is in ON (II).
- Connect a jumper battery (do not connect a battery charger).
- Shut down the HDS.
- Disconnect the HDS from the DLC.
- Reboot the HDS.
- Reconnect the HDS to the DLC, and try the update procedure again.

8. If the TP POSITION CHECK failed in step 5, clean the throttle body.
9. Do the PCM idle learn procedure.
10. Do the CKP pattern clear/CKP pattern learn procedure.

PCM IDLE LEARN PROCEDURE

The idle learn procedure must be done so the PCM can learn the engine idle characteristics.

Do the idle learn procedure whenever you do any of these actions:

- Replace the PCM.
- Reset the PCM.
- Update the PCM.
- Replace or clean the throttle body.
- Disassemble the engine or transmission.

➡Erasing DTCs with the HDS does not require you to do the idle learn procedure.

1. Make sure all electrical items (the A/C, the audio, the lights, etc.) are off.
2. Reset the PCM with the HDS.
3. Turn the ignition switch to ON (II), and wait 2 seconds.
4. Start the engine. Hold the engine speed at 3,000 rpm without load (in P or N) until the radiator fan comes on, or until the engine coolant temperature reaches 194°F (90°C).
5. Let the engine idle for about 5 minutes with the throttle fully closed.

➡If the radiator fan comes on, do not include its running time in the 5 minutes.

CKP PATTERN CLEAR/CKP PATTERN LEARN PROCEDURE

Clear/Learn Procedure (with the HDS)

1. Connect the HDS to the Data Link Connector (DLC) located under the driver's side of the dashboard.
2. Turn the ignition switch to ON (II).
3. Make sure the HDS communicates with the PCM and all other vehicle systems. If it doesn't, go to the DLC circuit troubleshooting.
4. Select CRANK PATTERN in the ADJUSTMENT MENU with the HDS.
5. Select CRANK PATTERN LEARNING with the HDS, and follow the screen prompts.

Learn Procedure (without the HDS)

1. Start the engine. Hold the engine speed at 3,000 rpm without load (in P or N) until the radiator fan comes on.
2. Test-drive the vehicle on a level road: Decelerate (with the throttle fully closed) from an engine speed of 2,500 rpm down to 1,000 rpm with the transmission in 2.
3. Repeat step 2 several times.
4. Turn the ignition switch to LOCK (0).
5. Turn the ignition switch to ON (II), and wait for 30 seconds. The CKP pattern learn procedure is complete.

FUEL — GASOLINE FUEL INJECTION SYSTEM

FUEL SYSTEM SERVICE PRECAUTIONS

Safety is the most important factor when performing not only fuel system maintenance, but any type of maintenance. Failure to conduct maintenance and repairs in a safe manner may result in serious personal injury or death. Work on a vehicle's fuel system components can be accomplished safely and effectively by adhering to the following rules and guidelines.

- To avoid the possibility of fire and personal injury, always disconnect the negative battery cable unless the repair or test procedure requires that battery voltage be applied.
- Always relieve the fuel system pressure prior to disconnecting any fuel system component (injector, fuel rail, pressure regulator, etc.) fitting or fuel line connection. Exercise extreme caution whenever relieving fuel system pressure to avoid exposing skin, face and eyes to fuel spray. Please be advised that fuel under pressure may penetrate the skin or any part of the body that it contacts.
- Always place a shop towel or cloth around the fitting or connection prior to loosening to absorb any excess fuel due to spillage. Ensure that all fuel spillage is quickly removed from engine surfaces. Ensure that all fuel-soaked cloths or towels are deposited into a flame-proof waste container with a lid.
- Always keep a dry chemical (Class B) fire extinguisher near the work area.
- Do not allow fuel spray or fuel vapors to come into contact with a spark or open flame.
- Always use a second wrench when loosening or tightening fuel line connection fittings. This will prevent unnecessary stress and torsion on fuel piping. Always follow the proper torque specifications.
- Always replace worn fuel fitting O-rings with new ones. Do not substitute fuel hose where rigid pipe is installed.

FUEL SYSTEM PRESSURE

RELIEVING

Before disconnecting fuel lines or hoses, relieve pressure from the system by disabling the fuel pump and disconnecting the fuel line/quick connect fitting in the engine compartment.

With the HDS

See Figures 221 and 222.

1. Connect the HDS to the Data Link Connector (DLC) located under the driver's side of the dashboard.
2. Turn the ignition switch to ON (II).
3. Make sure the HDS communicates with the PCM. If it doesn't, go to the DLC circuit troubleshooting.
4. Turn the ignition switch to LOCK (0).
5. Remove the fuel fill cap to relieve the pressure in the fuel tank.
6. Turn the ignition switch to ON (II).
7. From the INSPECTION MENU of the HDS, select Fuel Pump OFF, then start the engine, and let it idle until it stalls.

➡Do not allow the engine to idle above 1,000 rpm or the PCM will continue to operate the fuel pump. Pending or Confirmed DTC may be set during this pro-

cedure. Check for DTCs, and clear them as needed.

8. Turn the ignition switch to LOCK (0).
9. Do the battery terminal disconnection procedure.
10. Remove the quick-connect fitting cover.
11. Check the fuel quick-connect fitting for dirt, and clean it if needed.
12. Place a rag or a shop towel over the quick-connect fitting.
13. Disconnect the quick-connect fitting: Hold the connector with one hand, and squeeze the retainer tabs with the other hand to release them from the locking tabs. Pull the connector off.

➡Be careful not to damage the line or other parts. Do not use tools. If the connector does not move, keep the retainer tabs pressed down, and alternately pull and push the connector until it comes off easily. Do not remove the retainer from the line; once removed, the retainer must be replaced with a new one.

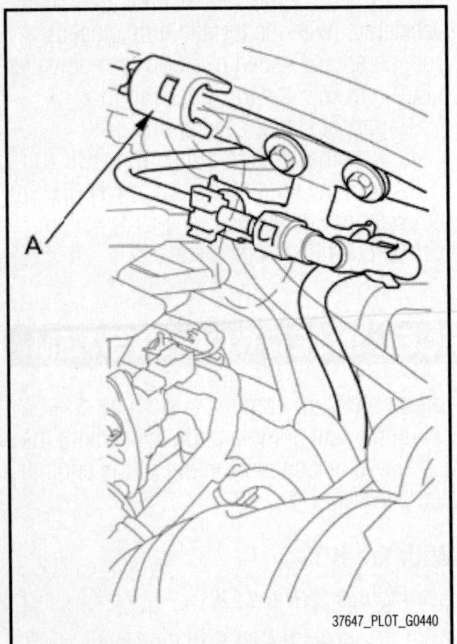
A
37647_PLOT_G0440
Fig. 221 Remove the quick-connect fitting cover (A)

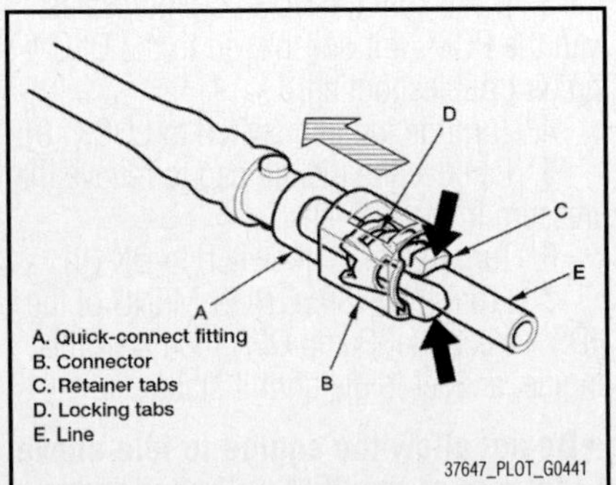
D
C
E
A
B
A. Quick-connect fitting
B. Connector
C. Retainer tabs
D. Locking tabs
E. Line
37647_PLOT_G0441
Fig. 222 Disconnect the quick-connect fitting

14. After disconnecting the quick-connect fitting, check it for dirt or damage.
15. Do the battery terminal reconnection procedure.

Without the HDS

See Figures 221 through 223.

1. Remove PGM-FI main relay 2 (FUEL PUMP) from the under-dash fuse/relay box.
2. Start the engine, and let it idle until it stalls.

➡If any DTCs are stored, clear and ignore them.

3. Turn the ignition switch to LOCK (0).
4. Remove the fuel fill cap to relieve the pressure in the fuel tank.
5. Do the battery terminal disconnection procedure.
6. Remove the quick-connect fitting cover.
7. Check the fuel quick-connect fitting for dirt, and clean it if needed.
8. Place a rag or a shop towel over the quick-connect fitting.
9. Disconnect the quick-connect fitting: Hold the connector with one hand, and squeeze the retainer tabs with the other hand to release them from the locking tabs. Pull the connector off.

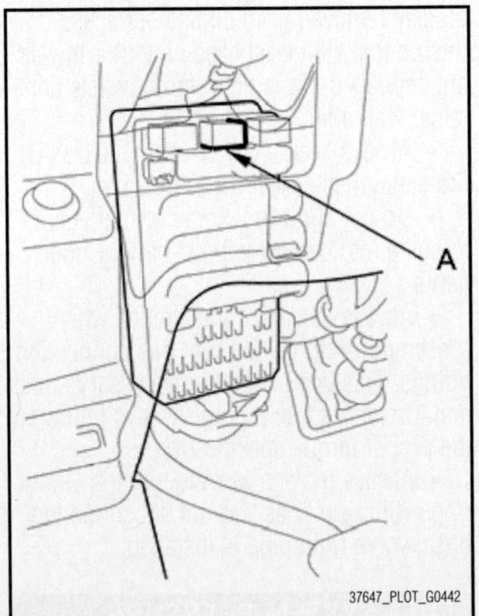
A
37647_PLOT_G0442
Fig. 223 Remove PGM-FI main relay 2 (FUEL PUMP) (A) from the under-dash fuse/relay box

➡Be careful not to damage the line or other parts. Do not use tools. If the connector does not move, keep the retainer tabs pressed down, and alternately pull and push the connector until it comes off easily. Do not remove the retainer from the line; once removed, the retainer must be replaced with a new one.

10. After disconnecting the quick-connect fitting, check it for dirt or damage.
11. Do the battery terminal reconnection procedure.

FUEL FILTER

REMOVAL & INSTALLATION

See Figure 224.

The fuel filter should be replaced whenever the fuel pressure drops below the specified value, after making sure that the fuel pump and the fuel pressure regulator are OK.

1. Remove the fuel tank unit.
2. Remove the fuel filter set.

To install:

3. Check these items before installing the fuel tank unit:
   a. When connecting the wire harness, make sure the connection is secure and the connectors are firmly locked into place.
   b. When installing the fuel gauge sending unit, make sure the connection

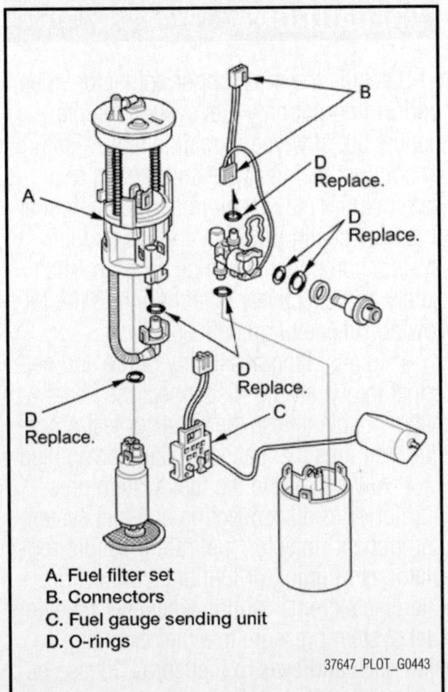
B
A
D
Replace.
D
Replace.
D
Replace.
D
Replace.
C
A. Fuel filter set
B. Connectors
C. Fuel gauge sending unit
D. O-rings
37647_PLOT_G0443
Fig. 224 Exploded view of the fuel tank unit

is secure and the connector is firmly locked into place. Be careful not to bend or twist it excessively.

4. Install the parts in the reverse order of removal with new O-rings. When installing the fuel tank unit, align the marks on the unit and the fuel tank.

➡Coat the O-rings with clean engine oil; do not use any other type oils or fluids. Do not pinch the O-rings during installation. Use all the new parts supplied in the fuel filter replacement kit.

FUEL PUMP/FUEL GAUGE SENDING UNIT

REMOVAL & INSTALLATION

See Figure 225.

1. Remove the fuel tank unit.
2. Remove the fuel level sensor (fuel sending unit) from the fuel tank unit.

To install:

3. Check these items before installing the fuel tank unit:
   a. When connecting the wire harness, make sure the connection is secure and the connector is firmly locked into place.
   b. When installing the fuel gauge sending unit, make sure the connection is secure. Be careful not to bend or twist it excessively.
4. Install the parts in the reverse order of removal. When installing the fuel tank unit, align the marks on the unit and the fuel tank.

FUEL RAIL AND INJECTOR

REMOVAL & INSTALLATION

See Figures 226 and 227.

1. Relieve the fuel pressure.
2. Remove the intake manifold. Refer to the Engine Mechanical section for Intake Manifold Removal and Installation.
3. Disconnect the quick-connect fitting.
4. Remove the fuel joint hose mounting bolt.
5. Disconnect the connectors from the injectors.
6. Remove the fuel rail mounting bolts from the fuel rails.
7. Remove the fuel rail and the injectors from the injector base.
8. Remove the injector clips from the fuel rail.
9. Remove the injectors from the fuel rails.

To install:

10. Coat the new O-rings (black) with clean engine oil, and insert the injectors into the fuel rails.
11. Install the injector clips.
12. Coat the new injector O-rings (brown) with clean engine oil.
13. Install the fuel rails and injectors in the injector base.
14. Install the fuel rail mounting bolts, and connect the connectors on the injectors.
15. Install the fuel joint hose mounting bolt.
16. Connect the quick-connect fitting.
17. Turn the ignition switch to ON (II), but do not operate the starter. After the fuel pump runs for about 2 seconds, the fuel rail will be pressurized. Repeat this two or three times, then check the fuel leakage.
18. Install the intake manifold with a new gasket.

FUEL TANK

DRAINING

1. Remove the fuel tank unit.
2. Using a hand pump, a hose, and a

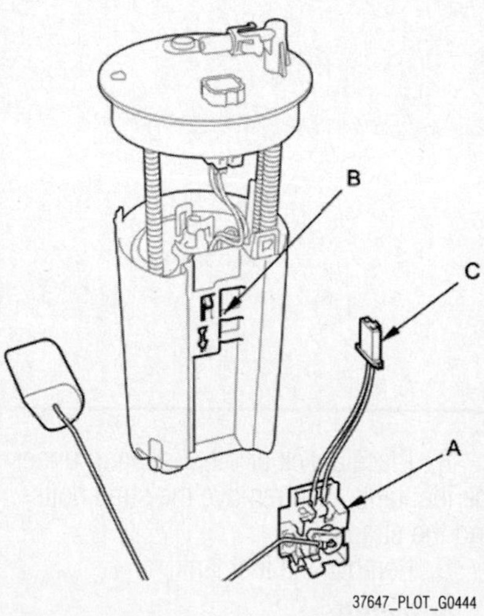
B
C
A
37647_PLOT_G0444

Fig. 225 Remove the fuel level sensor (fuel sending unit) (A) from the fuel tank unit (B)

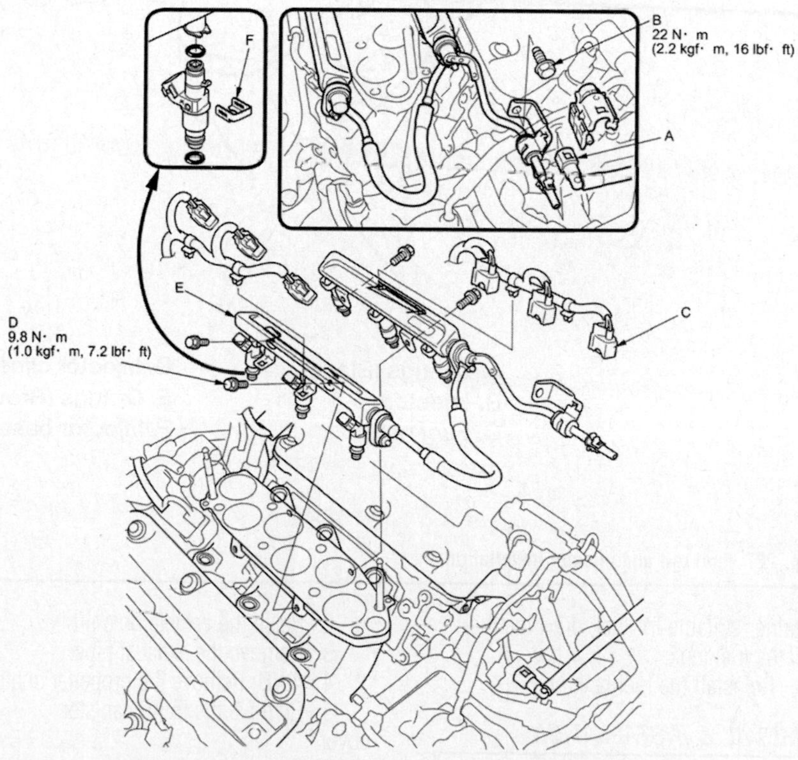
F
B
22 N· m
(2.2 kgf· m, 16 lbf· ft)
A
C
E
D
9.8 N· m
(1.0 kgf· m, 7.2 lbf· ft)
A. Quick-connect fitting
B. Fuel joint hose mounting bolt
C. Connectors
D. Fuel rail mounting bolts
E. Fuel rails
F. Injector clips
37647_PLOT_G0447

Fig. 226 Exploded view of the fuel rail assembly

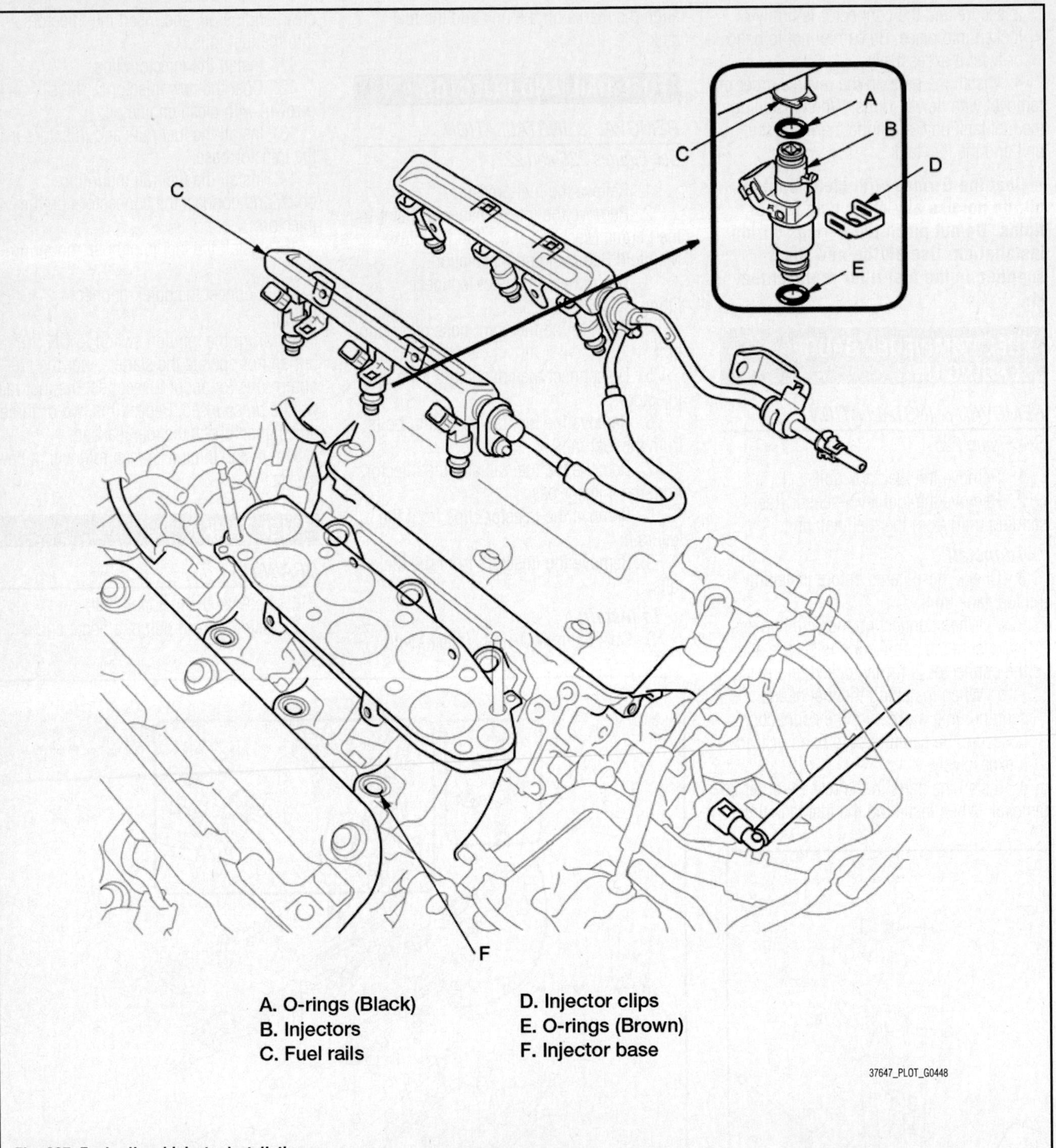
A
B
C
D
E
C
F
A. O-rings (Black)
B. Injectors
C. Fuel rails
D. Injector clips
E. O-rings (Brown)
F. Injector base
37647_PLOT_G0448

Fig. 227 Fuel rail and injector installation

container suitable for fuel, draw the fuel from the fuel tank.

3. Reinstall the fuel tank unit.

REMOVAL & INSTALLATION

See Figures 228 and 229.

1. Drain the fuel tank, then reinstall the fuel tank unit without connecting the fuel tank unit 4P connector and the fuel tank unit quick-connect fitting.
2. Raise the vehicle on a lift.
3. Remove the exhaust pipe.
4. 4WD: Remove the propeller shaft.
5. Remove the EVAP canister cover.
6. Remove the fuel tank protector.
7. Disconnect the quick-connector fittings and the hoses. Slide back the clamps, then twist the hoses as you pull to avoid damaging them.
8. Place a jack or other support under the fuel tank, then remove the strap bolts and the straps.
9. Remove the fuel tank.

To install:

10. Install the parts in the reverse order of removal.
    a. New fuel tanks have ring pull at the fuel vapor hose connector. When you connect the hose and confirm that the

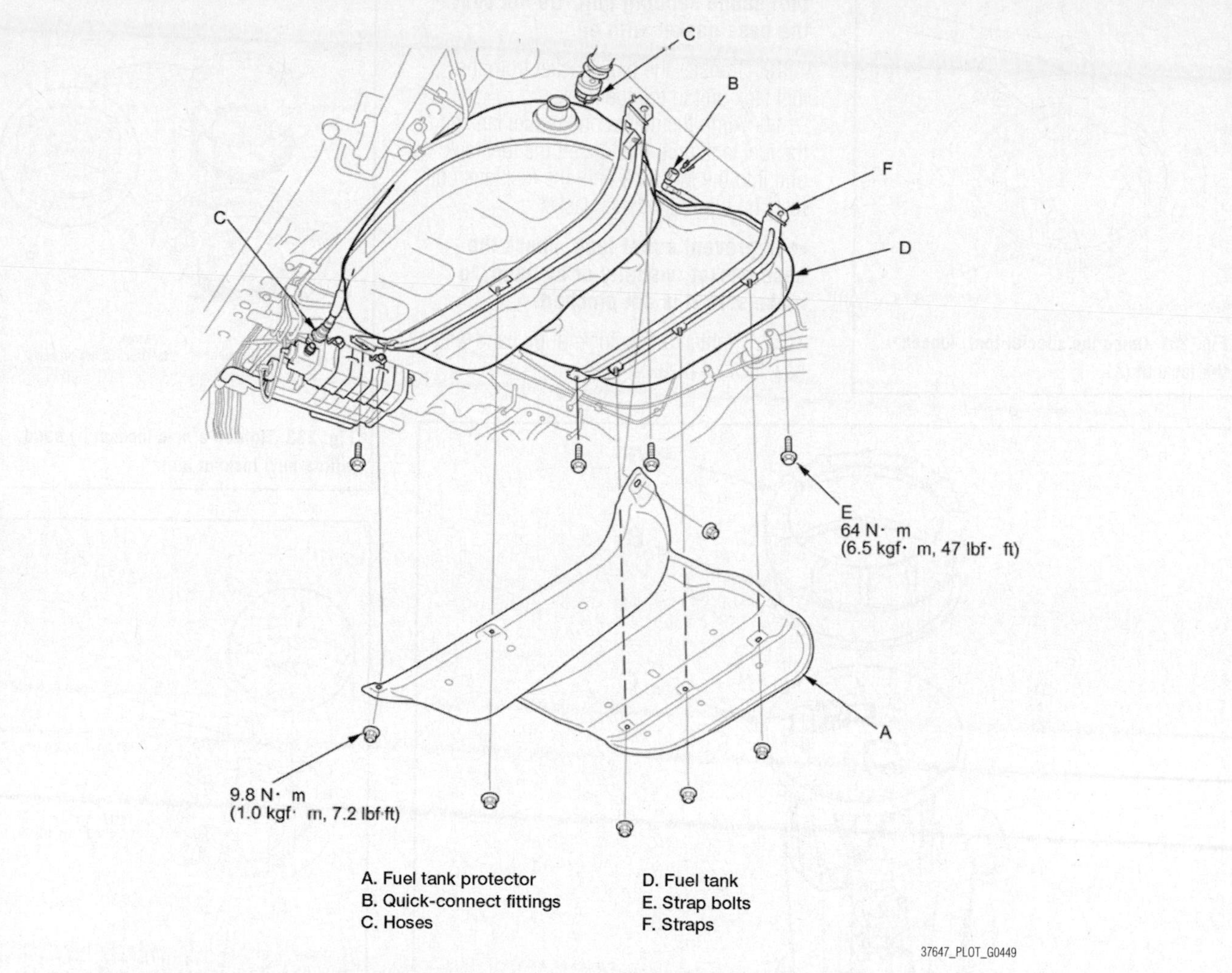

C
B
F
C
D
E
64 N· m
(6.5 kgf· m, 47 lbf· ft)
A
9.8 N· m
(1.0 kgf· m, 7.2 lbf·ft)
A. Fuel tank protector
B. Quick-connect fittings
C. Hoses
D. Fuel tank
E. Strap bolts
F. Straps
37647_PLOT_G0449

Fig. 228 Exploded view of the fuel tank assembly

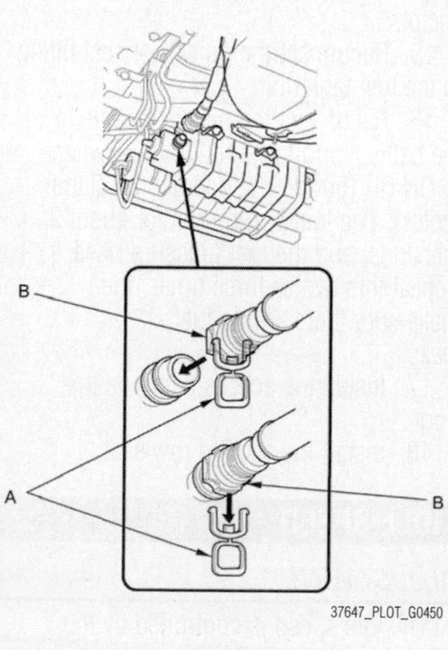

B
A
B
37647_PLOT_G0450

Fig. 229 Ring pull (A) at the fuel vapor hose connector (B)

connection is secure, remove the ring pull by pulling it down.

b. Before connecting the fuel fill pipe and the quick-connect fittings, check for dirt, and clean it if needed, taking care not to damage the fuel fill pipe and other parts.

FUEL TANK UNIT

REMOVAL & INSTALLATION

See Figures 230 through 234.

Special Tools Required: Fuel Sender Wrench 07AAA-S0XA100

1. Relieve the fuel pressure.
2. Remove the second row seat.
3. Pull the carpet out of way.
4. Remove the access panel from the floor.
5. Disconnect the fuel tank unit 4P connector.

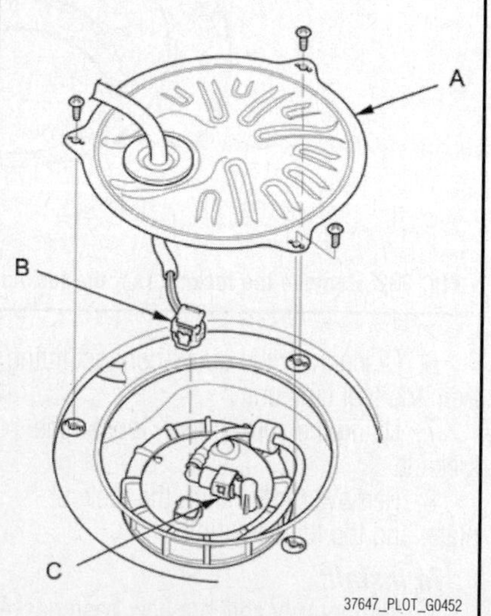

A
B
C
37647_PLOT_G0452

Fig. 230 Remove the access panel (A) from the floor; disconnect the fuel tank unit connector (B) and the quick-connect fitting (C)

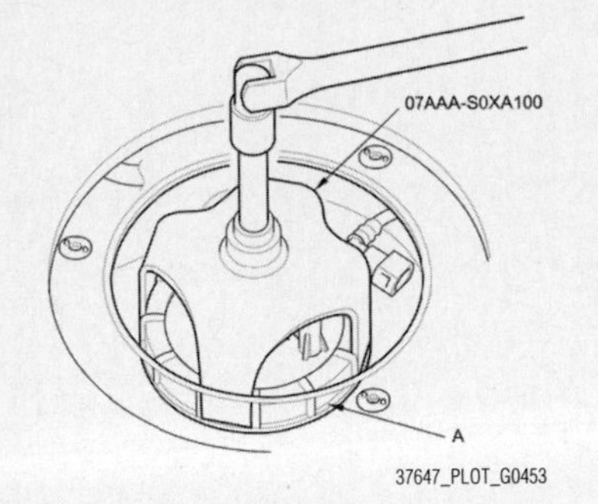
07AAA-S0XA100
A
37647_PLOT_G0453

Fig. 231 Using the special tool, loosen the locknut (A)

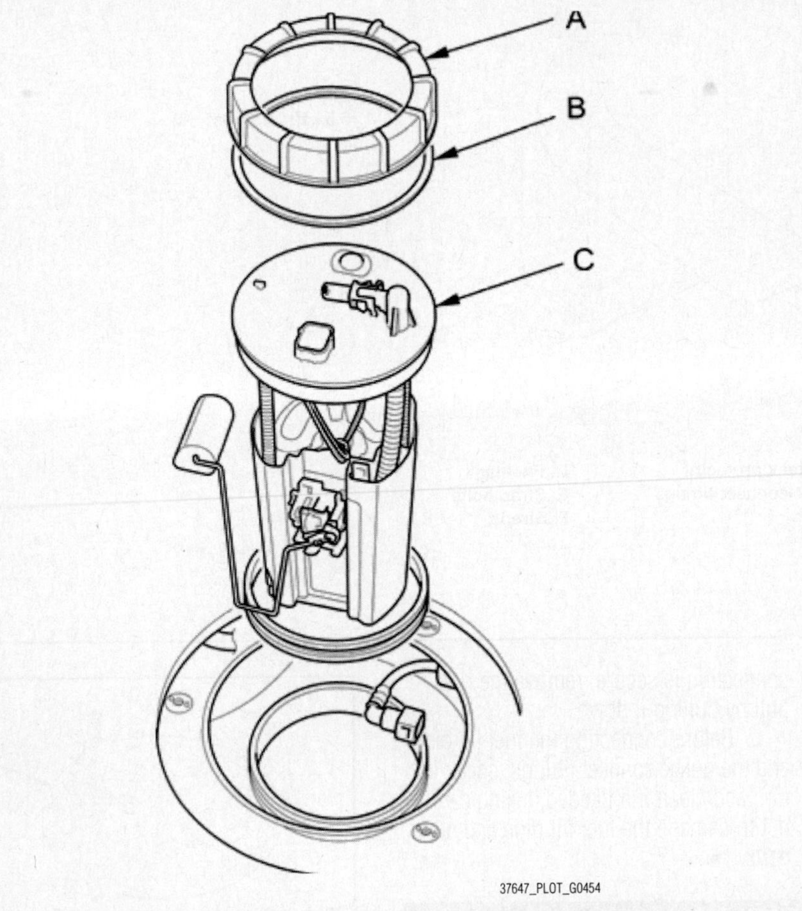
A
B
C
37647_PLOT_G0454

Fig. 232 Remove the locknut (A), the locknut plate (B), and the fuel tank unit (C)

6. Disconnect the quick-connect fitting from the fuel tank unit.
7. Using the special tool, loosen the locknut.
8. Remove the locknut, the locknut plate, and the fuel tank unit.

To install:

9. Temporarily attach a new base gasket to the fuel tank unit, then insert the fuel tank unit partially into the fuel tank.

➡Be careful not to damage the new base gasket. Be careful not to bend the fuel gauge sending unit. Do not coat the base gasket with oil.

10. Transfer the base gasket from the fuel tank unit to the fuel tank.
11. Align the marks on the fuel tank and the fuel tank unit, then insert the fuel tank unit into the fuel tank until the fuel tank unit rests on top of the base gasket.

➡To prevent a fuel leak, check the base gasket, visually or by hand, to make sure it is not pinched.

12. Tighten a new locknut by hand with a new locknut plate.

➡Before tightening, align the mark on the locknut with the start of the threads.

13. Using the special tool, tighten the locknut to the specified torque.

➡After tightening, make sure the marks are still aligned. After installation, check the base gasket, visually or by hand, to make sure it is not pinched.

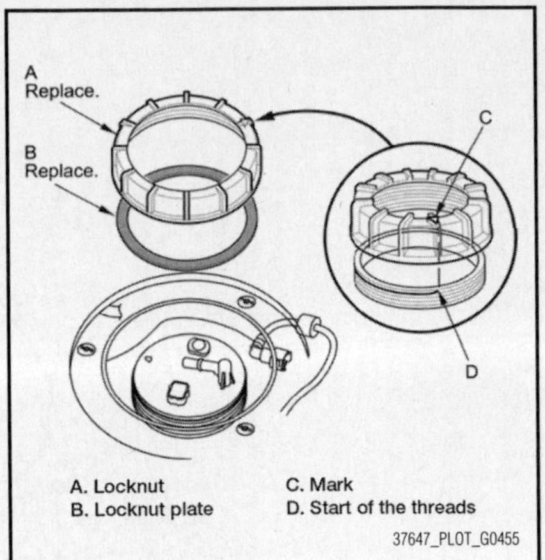
A
Replace.
B
Replace.
C
D
A. Locknut
B. Locknut plate
C. Mark
D. Start of the threads
37647_PLOT_G0455

Fig. 233 Tighten a new locknut by hand with a new locknut plate

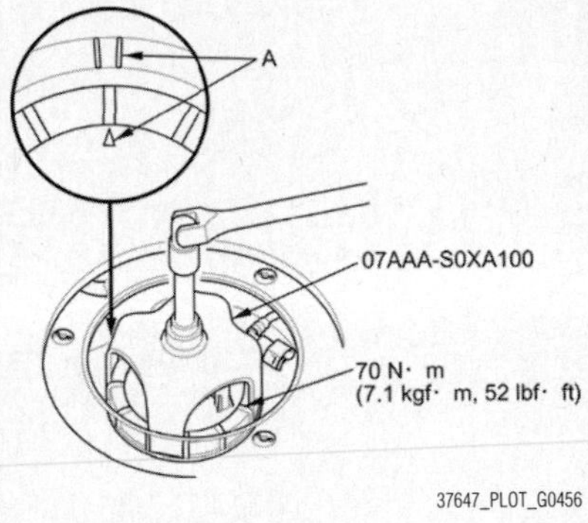
A
07AAA-S0XA100
70 N· m
(7.1 kgf· m, 52 lbf· ft)
37647_PLOT_G0456

Fig. 234 After tightening, make sure the marks (A) are still aligned

14. Connect the fuel tank unit 4P connector.
15. Reconnect the quick-connect fitting to the fuel tank unit.
16. Reconnect the negative cable to the battery, and turn the ignition switch to ON (II) (but do not operate the starter motor). The fuel pump runs for about 2 seconds, and the fuel pressure rises. Repeat this two or three times, then make sure there are no fuel leaks.
17. Install the access panel on the floor.
18. Install the second row seat.

IDLE SPEED

ADJUSTMENT

The idle speed is controlled by the PCM. No adjustment is necessary or possible.

THROTTLE BODY

REMOVAL & INSTALLATION

See Figure 235.

CAUTION

Do not insert your fingers into the installed throttle body when you turn the ignition switch to ON (II) or while the ignition switch is in ON (II). If you do, you will seriously injure your fingers if the throttle valve is activated.

➡If you are replacing the throttle body, begin at step 1. If you are removing the throttle body temporarily, begin at step 4.

1. Connect the HDS while the engine is stopped.
2. Select the INSPECTION MENU with the HDS.
3. Do the TP LEARNING CHECK in the ETCS TEST.
4. Turn the ignition switch to LOCK (0).
5. Disconnect the MAP sensor connector.
6. Remove the intake air duct.
7. Disconnect the throttle body connector.
8. Disconnect and plug the water bypass hoses.
9. Remove the throttle body.
10. Install the parts in the reverse order of removal with a new gasket.

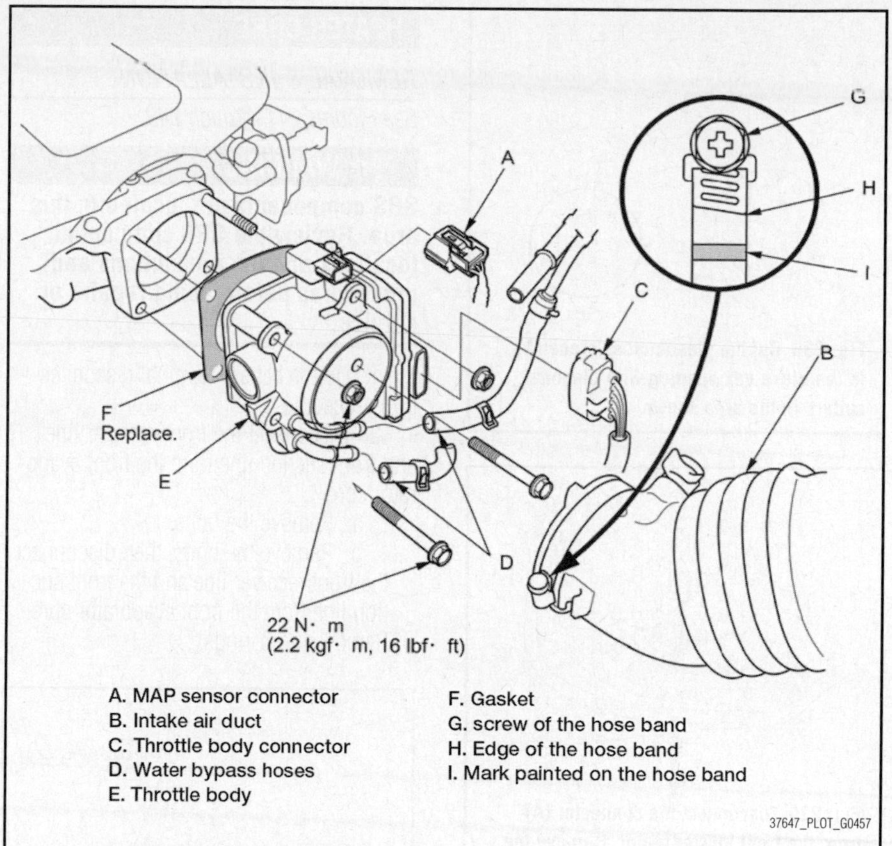

A
B
C
D
E
F
Replace.
G
H
I
22 N· m
(2.2 kgf· m, 16 lbf· ft)
A. MAP sensor connector
B. Intake air duct
C. Throttle body connector
D. Water bypass hoses
E. Throttle body
F. Gasket
G. screw of the hose band
H. Edge of the hose band
I. Mark painted on the hose band
37647_PLOT_G0457

Fig. 235 Exploded view of the throttle body assembly

➡When tightening the screw of the hose band, align the edge of the hose band with the mark painted on the hose band.

11. After installation, refill the radiator with engine coolant.
12. Do the PCM idle learn procedure.

HEATING & AIR CONDITIONING SYSTEM

BLOWER MOTOR

REMOVAL & INSTALLATION

See Figures 236 through 240.

1. Remove the glove box and passenger's center console trim.
2. Remove the wire harness clip, then remove the self-tapping screws and the passenger's heater duct.
3. Remove the harness clips, the bolts, and the glove box frame.
4. Cut the plastic cross brace in the glove box opening with diagonal cutters in the area shown, and discard it.
5. Disconnect the connector from the front blower motor. Remove the wire harness clips.
6. Disconnect the connector from the recirculation control motor. Remove the bolts, the self-tapping screws, the mounting nuts, and the blower unit.

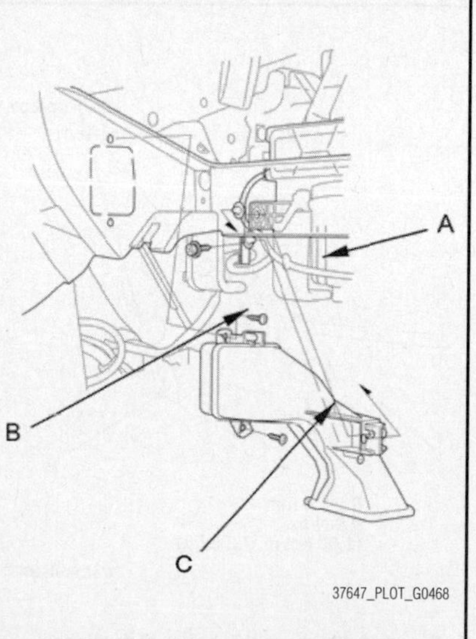

A
B
C
37647_PLOT_G0468

Fig. 236 Remove the wire harness clip (A), then remove the self-tapping screws (B) and the passenger's heater duct (C)

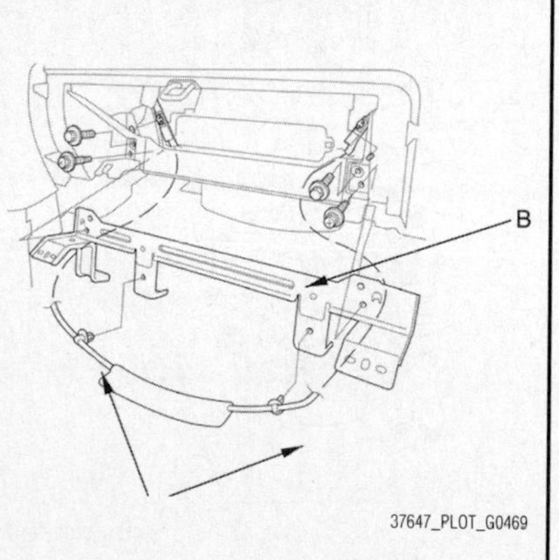

B
37647_PLOT_G0469

Fig. 237 Remove the harness clips (A), the bolts, and the glove box frame (B)

7. Install the unit in the reverse order of removal. Make sure that there is no air leakage.

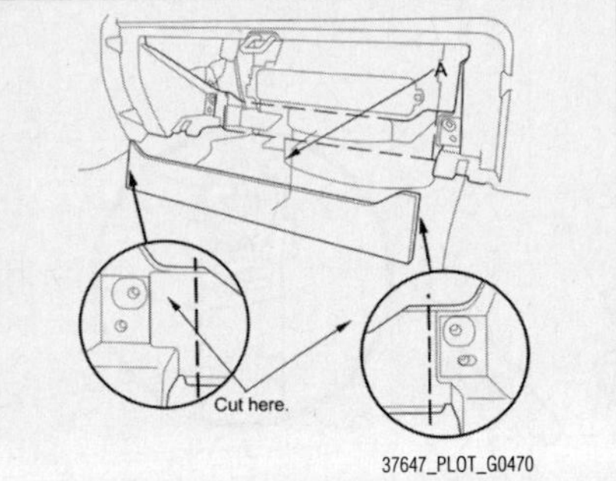
A
Cut here.
37647_PLOT_G0470

Fig. 238 Cut the plastic cross brace (A) in the glove box opening with diagonal cutters in the area shown

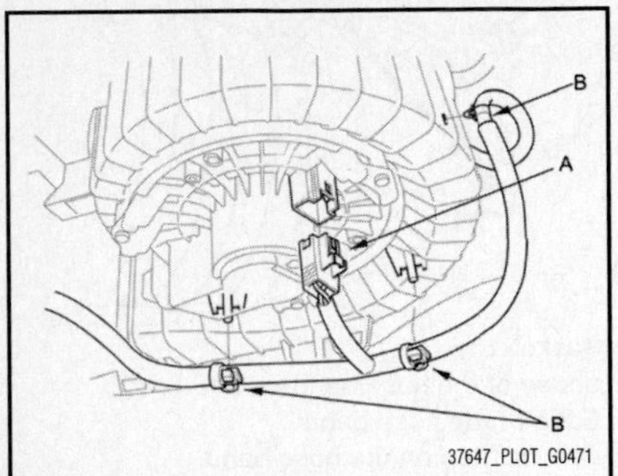
B
A
B
37647_PLOT_G0471

Fig. 239 Disconnect the connector (A) from the front blower motor. Remove the wire harness clips (B)

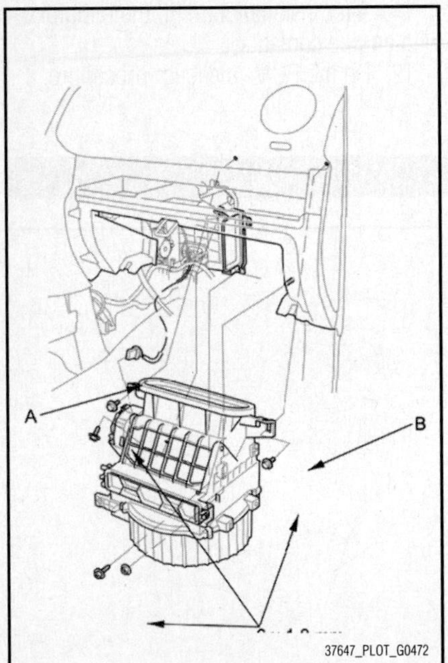
A
B
37647_PLOT_G0472

Fig. 240 Disconnect the connector (A) from the recirculation control motor. Remove the bolts, the self-tapping screws, the mounting nuts, and the blower unit (B)

HEATER CORE

REMOVAL & INSTALLATION

See Figures 241 through 249.

⁂ WARNING

SRS components are located in this area. Review the SRS component locations and the precautions and procedures before doing repairs or service.

1. Do the battery terminal disconnection procedure.
2. Disconnect the front receiver line and front suction line from the front evaporator core.
   a. Remove the bolts.
   b. Remove the bolts, then disconnect the front receiver line and the front suction line from the front evaporator core. Remove the O-rings.

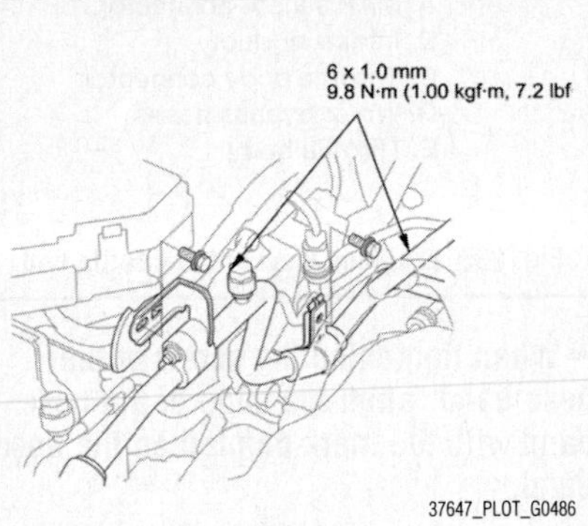
6 x 1.0 mm
9.8 N·m (1.00 kgf·m, 7.2 lbf
37647_PLOT_G0486

Fig. 241 Remove the bolts

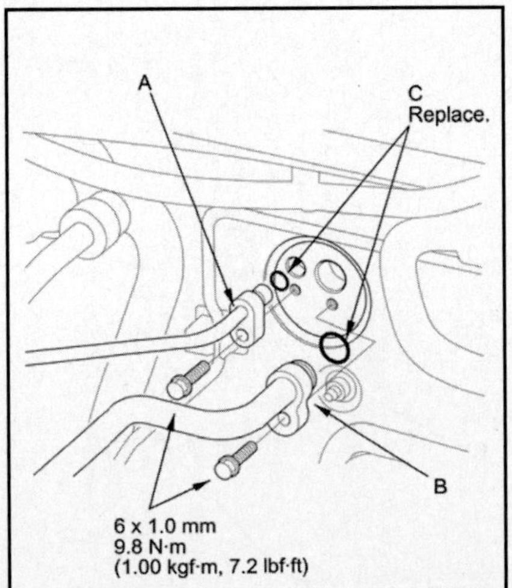
A
C
Replace.
B
6 x 1.0 mm
9.8 N·m
(1.00 kgf·m, 7.2 lbf·ft)
37647_PLOT_G0487

Fig. 242 Remove the bolts, then disconnect the front receiver line (A) and the front suction line (B) from the front evaporator core; remove the O-rings (C)

3. When the engine is cool, drain the engine coolant from the radiator.
4. From under the hood, slide the hose clamps back. Disconnect the inlet heater hose and the outlet heater hose from the heater unit. Note the layout of the hose.

➡Engine coolant will run out when the hoses are disconnected; drain it into a clean drip pan. Be sure not to let coolant spill on the electrical parts or the painted surfaces. If any coolant spills, rinse it off immediately.

5. Remove the mounting nuts from the heater unit. Take care not to damage or bend the fuel lines or the brake lines, etc.
6. Remove the dashboard.
7. Disconnect the connector from the front blower motor. Remove the wire harness clips.

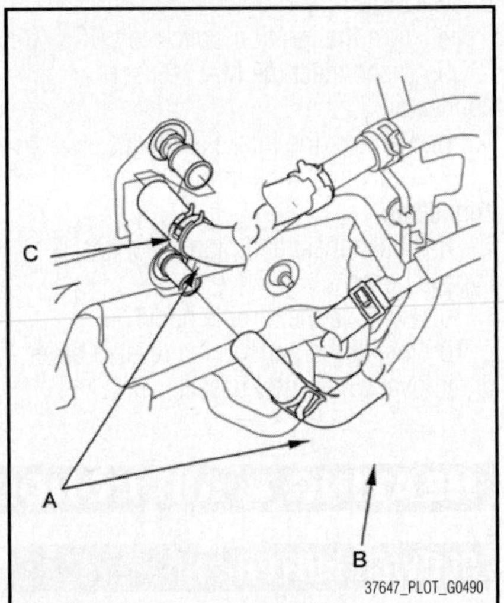
C
A
B
37647_PLOT_G0490

Fig. 243 Slide the hose clamps (A) back; disconnect the inlet heater hose (B) and the outlet heater hose (C) from the heater unit

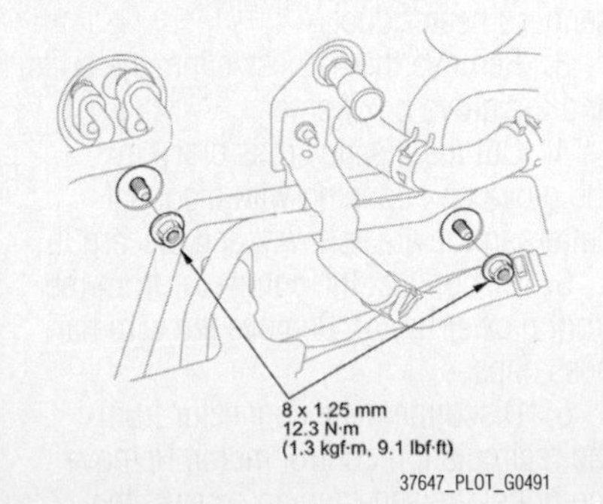
8 x 1.25 mm
12.3 N·m
(1.3 kgf·m, 9.1 lbf·ft)
37647_PLOT_G0491

Fig. 244 Remove the mounting nuts from the heater unit

8. Disconnect the connectors from the front mode control motor, the passenger's air mix control motor (with climate control), the recirculation control motor, and the front power transistor. Remove the wire harness clips.

9. Disconnect the connectors from the air mix control motor and the front evaporator temperature sensor. Remove the wire harness clips and the wire harness.

10. Turn over the carpet. Remove the clip and remove the rear heater duct.

11. Remove the mounting nuts and bolts. Slide the blower-heater unit back,

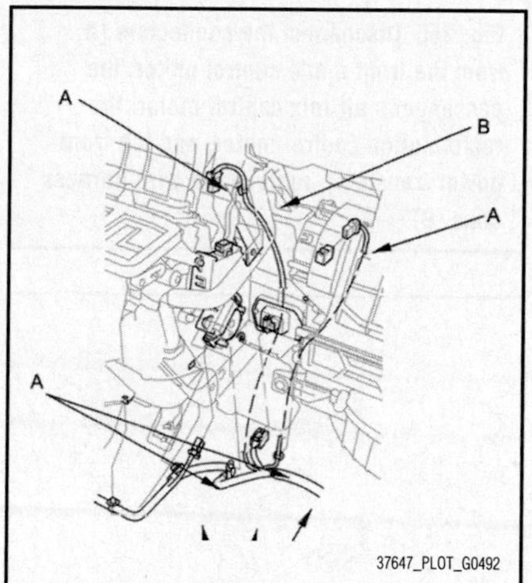
A
B
A
A
37647_PLOT_G0492

Fig. 245 Disconnect the connectors (A) from the front mode control motor, the passenger's air mix control motor, the recirculation control motor, and the front power transistor; remove the wire harness clips (B)

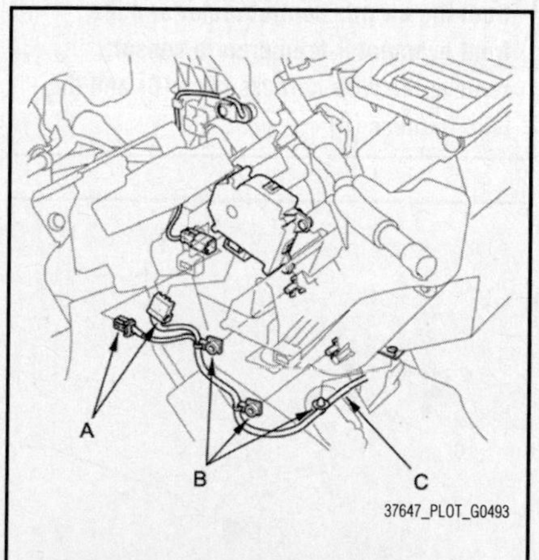
A
B
C
37647_PLOT_G0493

Fig. 246 Disconnect the connectors (A) from the air mix control motor and the front evaporator temperature sensor; remove the wire harness clips (B) and the wire harness (C)

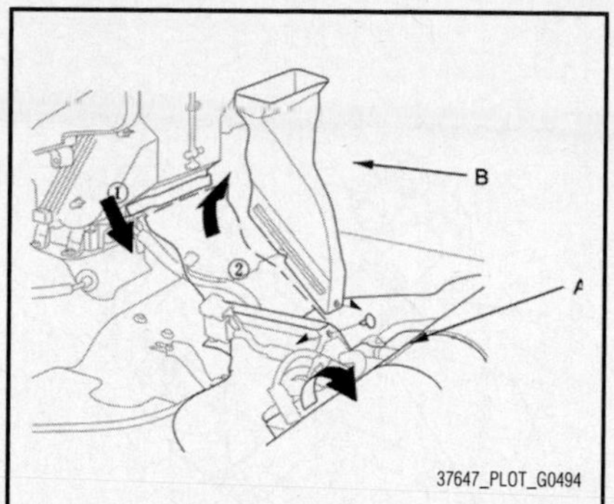
B
37647_PLOT_G0494

Fig. 247 Remove the clip (A) and remove the rear heater duct (B)

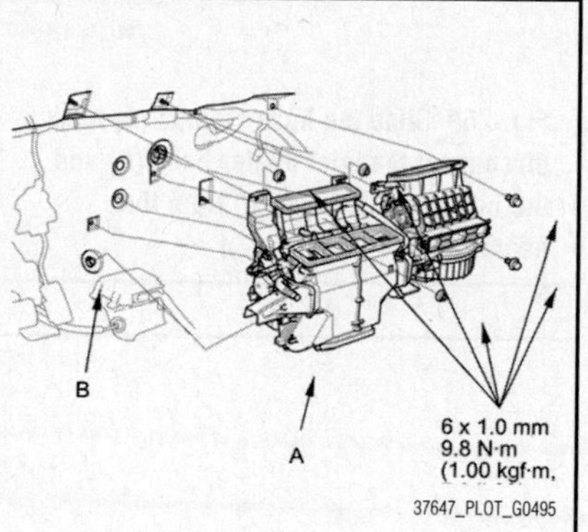
B
A
6 x 1.0 mm
9.8 N·m
(1.00 kgf·m,
37647_PLOT_G0495

Fig. 248 Remove the mounting nuts and bolts; slide the blower-heater unit (A) back, then remove the drain hose (B) and the blower-heater unit

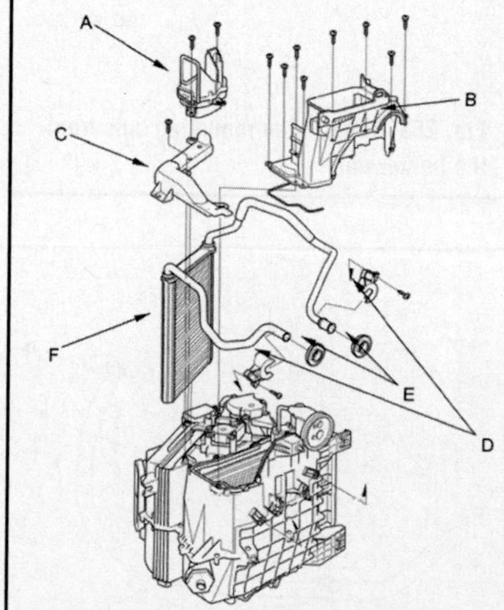
A
B
C
F
E
D

A. Passenger's heater duct
B. Expansion valve cover
C. Heater core cover
D. Heater pipe brackets
E. Grommets
F. Heater core

37647_PLOT_G0496

Fig. 249 Removing the heater core

then remove the drain hose and the blower-heater unit.

12. Remove the self-tapping screws and the passenger's heater duct. Remove the self-tapping screws and the expansion valve cover. Remove the self-tapping screw and the front heater core cover. Remove the self-tapping screws, the heater pipe brackets, the grommets, and carefully pull out the front heater core.

To install:

13. Install the front heater core, and the front evaporator core in the reverse order of removal.

14. Install the heater unit in the reverse order of removal, and note these items:

- Do not interchange the inlet and the outlet heater hoses.
- Install the inlet and the outlet hose clamps securely.
- Refill the cooling system with engine coolant.
- Make sure that there is no coolant leakage.
- Make sure that there is no air leakage.

15. Do the battery terminal reconnection procedure.

HEATER UNIT

REMOVAL & INSTALLATION

See Figures 250 through 258.

✻✻ WARNING

SRS components are located in this area. Review the SRS component locations and the precautions and procedures before doing repairs or service.

1. Do the battery terminal disconnection procedure.

2. Disconnect the front receiver line and front suction line from the front evaporator core.

a. Remove the bolts.

b. Remove the bolts, then disconnect the front receiver line and the front suction line from the front evaporator core. Remove the O-rings.

3. When the engine is cool, drain the engine coolant from the radiator.

4. From under the hood, slide the hose clamps back. Disconnect the inlet heater hose and the outlet heater hose from the heater unit. Note the layout of the hose.

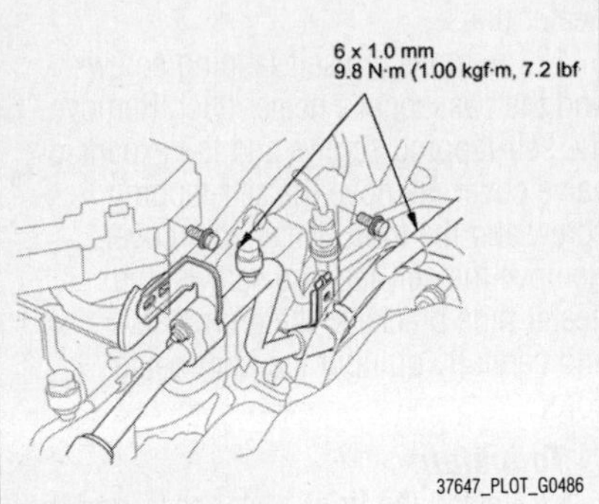
6 x 1.0 mm
9.8 N·m (1.00 kgf·m, 7.2 lbf
37647_PLOT_G0486

Fig. 250 Remove the bolts

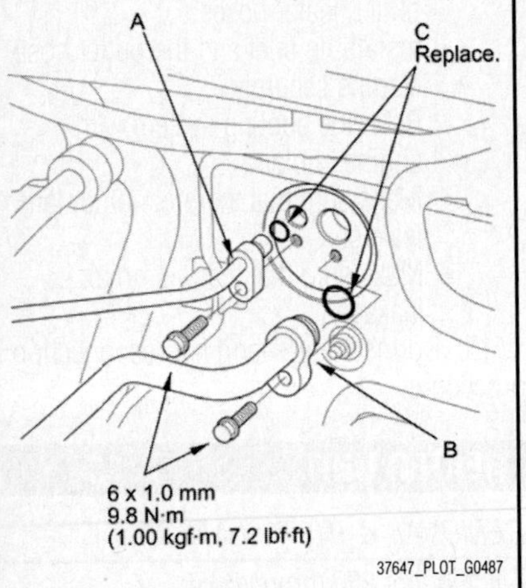
A
C
Replace.
B
6 x 1.0 mm
9.8 N·m
(1.00 kgf·m, 7.2 lbf·ft)
37647_PLOT_G0487

Fig. 251 Remove the bolts, then disconnect the front receiver line (A) and the front suction line (B) from the front evaporator core; remove the O-rings (C)

➡Engine coolant will run out when the hoses are disconnected; drain it into a clean drip pan. Be sure not to let coolant spill on the electrical parts or the painted surfaces. If any coolant spills, rinse it off immediately.

5. Remove the mounting nuts from the heater unit. Take care not to damage or bend the fuel lines or the brake lines, etc.

6. Remove the dashboard.

7. Disconnect the connector from the front blower motor. Remove the wire harness clips.

8. Disconnect the connectors from the front mode control motor, the passenger's air mix control motor (with climate control), the recirculation control motor, and the front

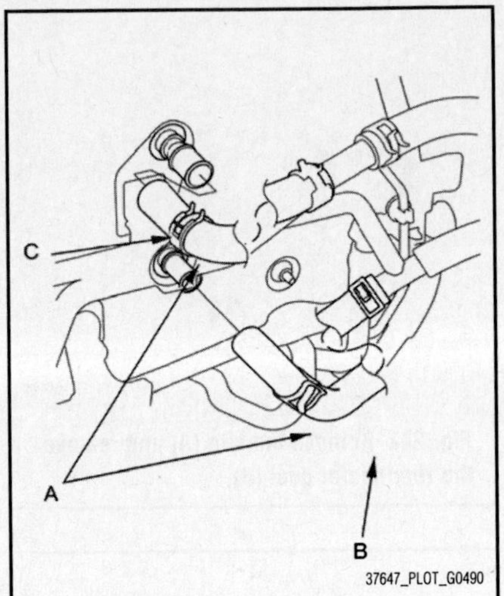
C
A
B
37647_PLOT_G0490

Fig. 252 Slide the hose clamps (A) back; disconnect the inlet heater hose (B) and the outlet heater hose (C) from the heater unit

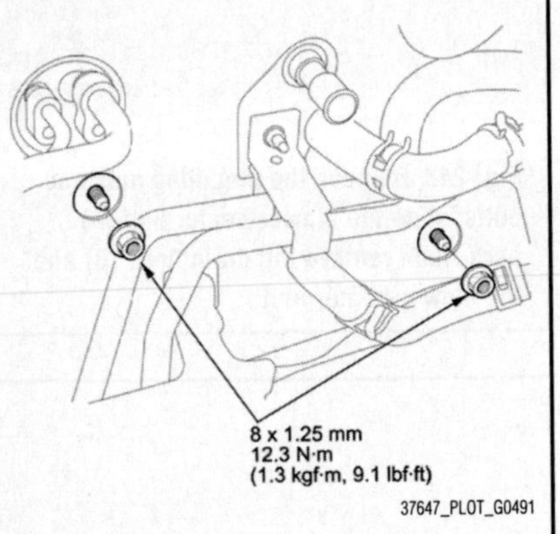
8 x 1.25 mm
12.3 N·m
(1.3 kgf·m, 9.1 lbf·ft)
37647_PLOT_G0491

Fig. 253 Remove the mounting nuts from the heater unit

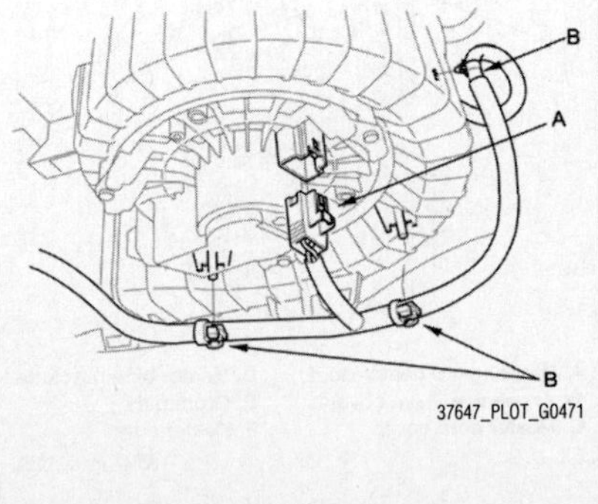
B
A
B
37647_PLOT_G0471

Fig. 254 Disconnect the connector (A) from the front blower motor. Remove the wire harness clips (B)

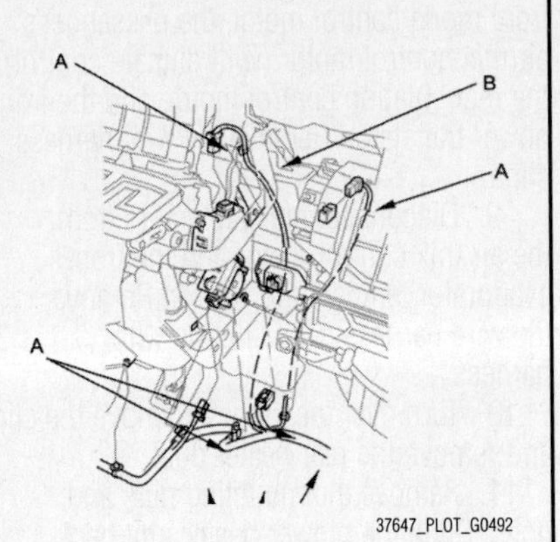
A
B
A
A
37647_PLOT_G0492

Fig. 255 Disconnect the connectors (A) from the front mode control motor, the passenger's air mix control motor, the recirculation control motor, and the front power transistor; remove the wire harness clips (B)

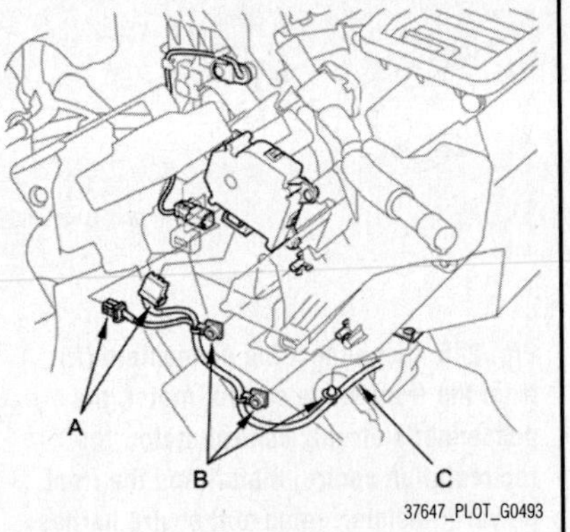
A
B
C
37647_PLOT_G0493

Fig. 256 Disconnect the connectors (A) from the air mix control motor and the front evaporator temperature sensor; remove the wire harness clips (B) and the wire harness (C)

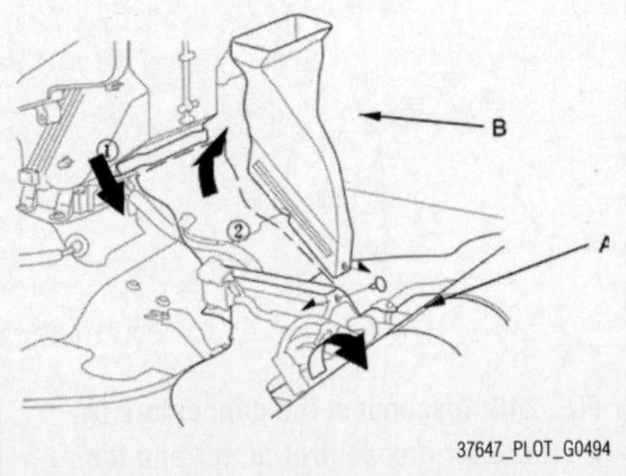
B
1
2
A
37647_PLOT_G0494

Fig. 257 Remove the clip (A) and remove the rear heater duct (B)

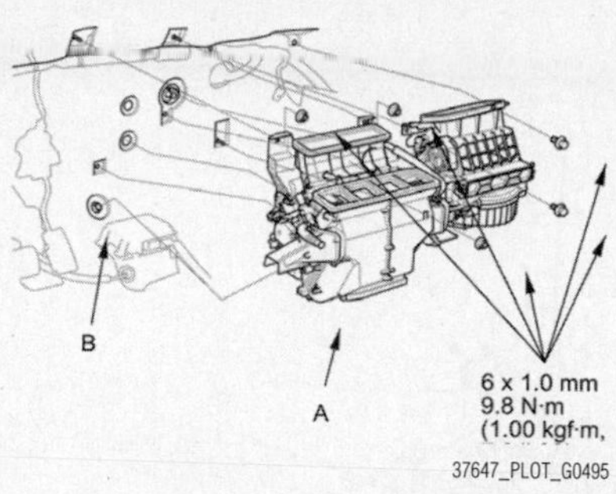
B
A
6 x 1.0 mm
9.8 N·m
(1.00 kgf·m,
37647_PLOT_G0495

Fig. 258 Remove the mounting nuts and bolts; slide the blower-heater unit (A) back, then remove the drain hose (B) and the blower-heater unit

power transistor. Remove the wire harness clips.

9. Disconnect the connectors from the air mix control motor and the front evaporator temperature sensor. Remove the wire harness clips and the wire harness.

10. Turn over the carpet. Remove the clip and remove the rear heater duct.

11. Remove the mounting nuts and bolts. Slide the blower-heater unit back, then remove the drain hose and the blower-heater unit.

To install:

12. Install the heater unit in the reverse order of removal, and note these items:

- Do not interchange the inlet and the outlet heater hoses.
- Install the inlet and the outlet hose clamps securely.
- Refill the cooling system with engine coolant.
- Make sure that there is no coolant leakage.
- Make sure that there is no air leakage.

13. Do the battery terminal reconnection procedure.

AUXILIARY HEATING & AIR CONDITIONING SYSTEM

BLOWER MOTOR

REMOVAL & INSTALLATION

See Figures 259 and 260.

⁂ WARNING

SRS components are located in this area. Review the SRS component locations and the precautions and procedures before doing repairs or service.

1. Remove the center console.

2. With navigation, disconnect the HandsFreeLink control unit connector and the USB adapter unit connectors.

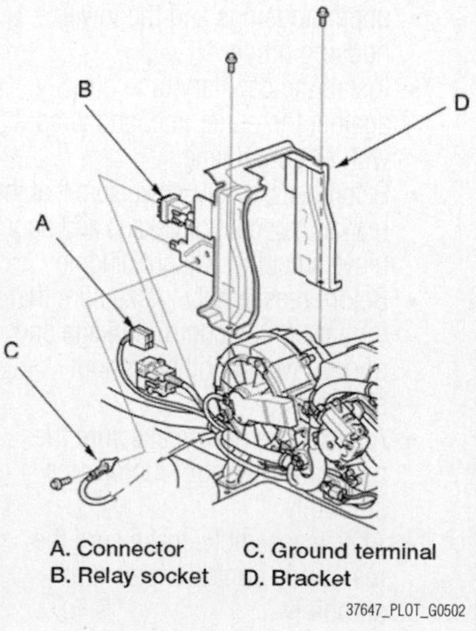
B
D
A
C
A. Connector
B. Relay socket
C. Ground terminal
D. Bracket
37647_PLOT_G0502

Fig. 259 Disconnect the connector and remove the relay socket

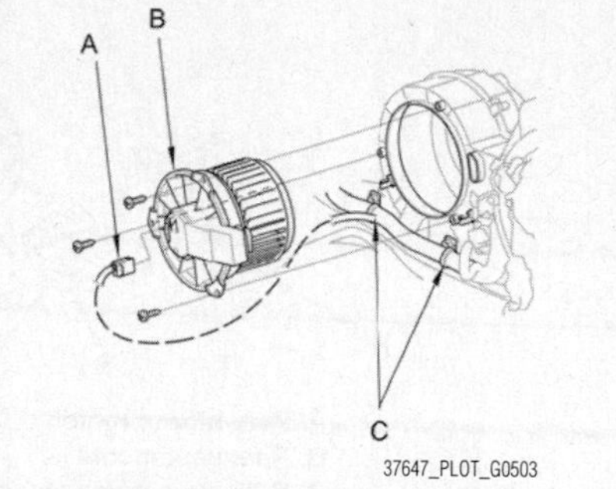
A
B
C
37647_PLOT_G0503

Fig. 260 Disconnect the 2P connector (A) from the rear blower motor (B), then remove the wire harness clips (C)

3. Disconnect the connector and remove the relay socket.

4. Remove the bolt and ground terminal, then remove the bolts and the bracket.

5. Disconnect the 2P connector from the rear blower motor, then remove the wire harness clips. Remove the self-tapping screws and the rear blower motor from the rear HVAC unit.

6. Install the motor in the reverse order of removal.

HEATER CORE

REMOVAL & INSTALLATION

See Figures 261 and 262.

⁂ WARNING

SRS components are located in this area. Review the SRS component locations and the precautions and procedures before doing repairs or service.

1. When the engine is cool, drain the engine coolant from the radiator.

2. Remove the center console.

3. Slide the hose clamps back, then disconnect the rear inlet heater hose and the rear outlet heater hose from the rear heater core. Note the orientation of the hose.

➡Engine coolant will run out when the hoses are disconnected; drain it into a clean drip pan. Be sure not to let coolant spill on the electrical parts o r the painted surfaces. If any coolant spills, rinse it off immediately.

4. Turn over the carpet. Remove the self-tapping screws and the clamps. Carefully pull out the rear heater core without bending lines.

To install:

5. Install the unit in the reverse order of removal, and note these items:

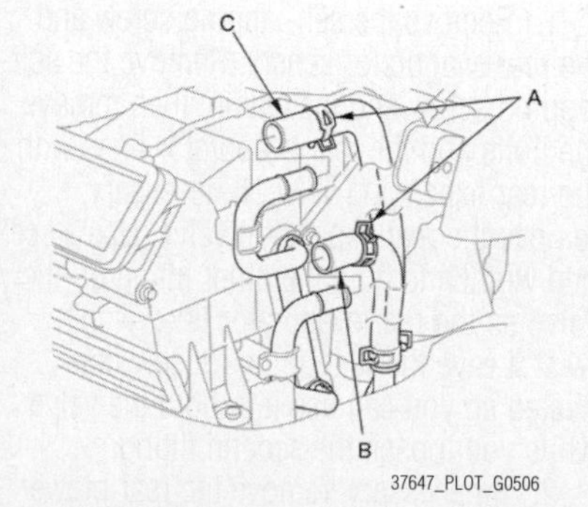
C
A
B
37647_PLOT_G0506

Fig. 261 Slide the hose clamps (A) back, then disconnect the rear inlet heater hose (B) and the rear outlet heater hose (C) from the rear heater core

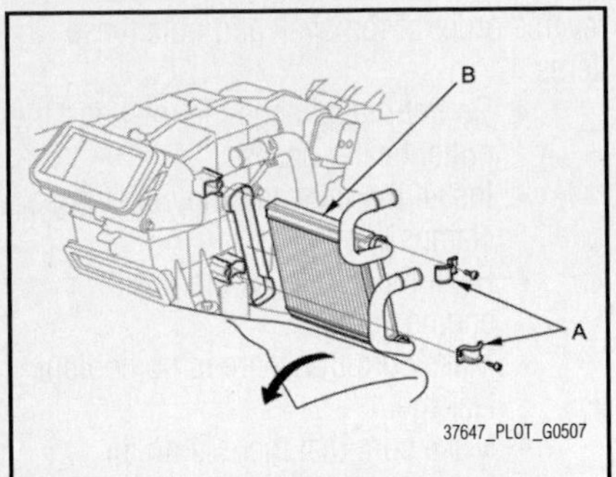
B
A
37647_PLOT_G0507

Fig. 262 Remove the self-tapping screws and the clamps (A); pull out the rear heater core (B) without bending lines

- Do not interchange the inlet and outlet heater hoses, and install the hose clamps securely.
- Refill the cooling system with engine coolant.
- Make sure that there is no coolant leakage.
- Make sure that there is no air leakage.

HVAC UNIT

REMOVAL & INSTALLATION

See Figure 263.

✻✻ WARNING

SRS components are located in this area. Review the SRS component locations and the precautions and procedures before doing repairs or service.

➡The rear blower motor, the rear heater core, the rear evaporator sensor, the rear power transistor, the rear air mix control motor, the rear mode control motor, and the rear expansion valve can be replaced without removing the rear HVAC unit.

1. Remove the self-tapping screw and the rear evaporator sensor. Remove the self-tapping screw and the clamp, then remove the bolts and the rear evaporator lines with the rear expansion valve. If necessary, remove the rear expansion valve. Use a second wrench to hold the other fitting on the valve so the rear evaporator lines won't twist. Leave the first fitting loosely connected so you can use it to hold the valve while you loosen the second fitting.
2. If necessary, remove the rear blower motor, the rear heater core, the rear power transistor, the rear air mix control motor, and the rear mode control motor.

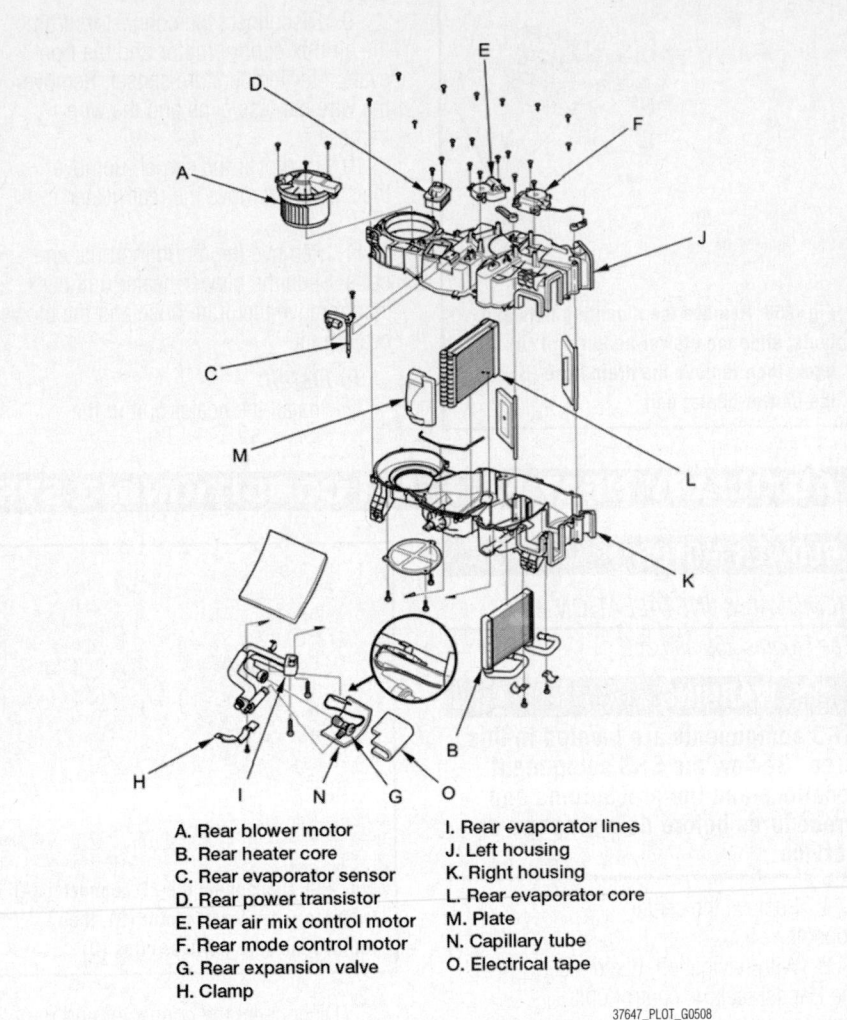

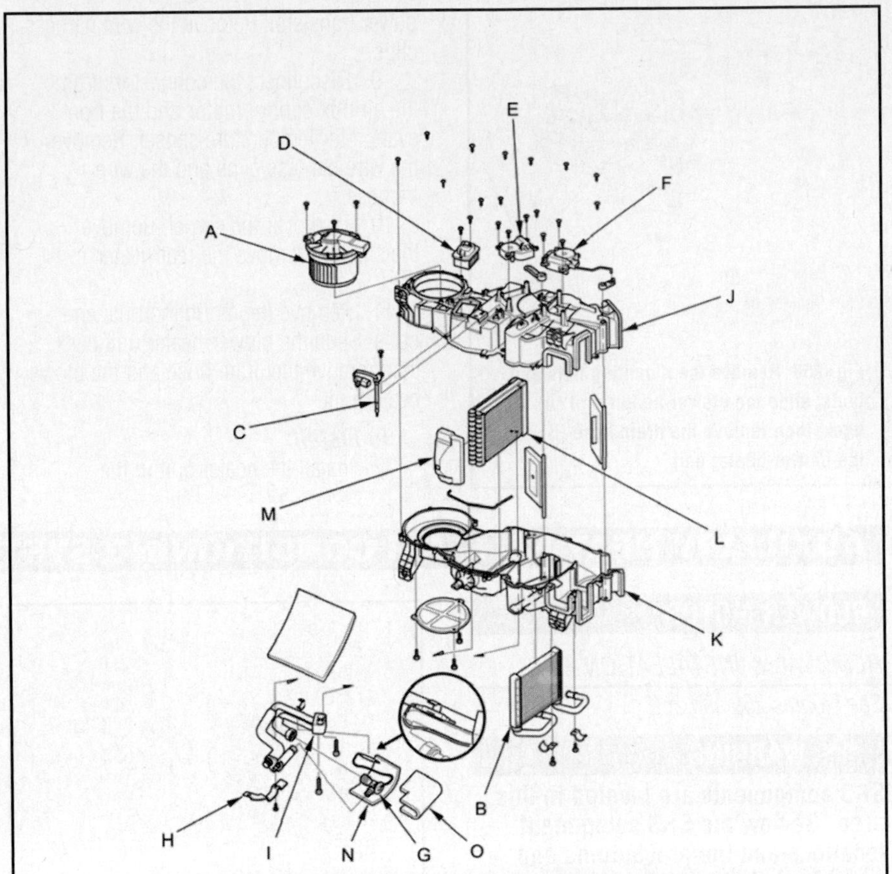
A
B
C
D
E
F
G
H
I
J
K
L
M
N
O
A. Rear blower motor
B. Rear heater core
C. Rear evaporator sensor
D. Rear power transistor
E. Rear air mix control motor
F. Rear mode control motor
G. Rear expansion valve
H. Clamp
I. Rear evaporator lines
J. Left housing
K. Right housing
L. Rear evaporator core
M. Plate
N. Capillary tube
O. Electrical tape
37647_PLOT_G0508

Fig. 263 Exploded view of the rear HVAC unit

3. Remove the self-tapping screws, and carefully separate the left housing from the right housing. Remove the rear evaporator core and plate.
4. Reassemble the unit in the reverse order of disassembly, and note these items:
   - Replace all O-rings with new ones at each fitting and apply a thin coat of refrigerant oil before installing them. Be sure to use the correct O-rings for HFC-134a (R-134a) to avoid leakage.
   - Immediately after using the oil, reinstall the cap on the container, and seal it to avoid moisture absorption.
   - Make sure no air is leaking from the left upper housing and the right upper housing fitting and from the upper housings and the lower housing fitting.
   - Install the capillary tube directly against the outlet line, and wrap it with electrical tape.
   - Before reassembly, make sure that the rear air mix control linkage and door move smoothly without binding.
   - Before reassembly, make sure that the rear mode control linkage and door move smoothly without binding.
   - After reassembly, make sure the rear air mix control motor runs smoothly.
   - After reassembly, make sure the rear mode control motor runs smoothly.
   - Make sure that there is no coolant leakage.

STEERING

POWER RACK & PINION STEERING GEAR

REMOVAL & INSTALLATION

See Figures 264 through 294.

Special Tools Required*:

- Ball Joint Thread Protector, 12 mm 07AAF-SDAA100
- Ball Joint Remover, 28 mm 07MAC-SL0A202
- Engine Support Hanger, A and Reds AAR-T1256
- 2008 V6 Attachment Arm SIL02C000033
- Subframe Adapter VSB02C000016
- Engine Hanger Balance Bar VSB02C000019

*Available through the Honda Tool and Equipment Program 888-424-6857

WARNING

SRS components are located in this area. Review the SRS component locations and the precautions and procedures before doing repairs or service.

Note these items during removal:

- Using clean solvent and a brush, wash any oil and dirt off the valve body unit, it's lines, and the end of the steering gearbox. Blow dry with compressed air.
- Be sure to remove the steering wheel before disconnecting the steering joint, or damage to the cable reel can occur.
- Lower the front subframe from the body, and remove the steering gearbox through the gap produced by lowering the front subframe.

1. Remove the hood support rod, then use it to prop the hood in the wide-open position.
2. Remove the engine cover.
3. Drain the power steering fluid.
4. Do the battery terminal disconnection procedure.
5. Raise and support the vehicle.
6. Remove the front wheels.
7. Remove the driver's airbag, and the steering wheel.
8. Remove the driver's dashboard lower cover.
9. Remove the footrest.
   a. Detach the hooks, then remove the footrest plate.
   b. Remove the nut with a 6 mm hexagonal wrench, release the clip, then remove the footrest.
10. Remove the center console trim.
11. Pull back the carpet.
12. Remove steering joint cover.
13. Release the lock lever, and adjust the steering column to the full tilt up position, and to the full telescope in position.
14. Tighten the lock lever.
15. Hold the lower slide shaft on the column with a piece of wire between the joint yoke of the lower slide shaft and joint yoke of the upper shaft to prevent the lower slide shaft from pulling out.
16. Release the lock lever, and adjust the steering column to the full telescopic out position, then tighten the lock lever.
17. Remove the steering joint bolt, and disconnect the steering joint by sliding the steering joint into the column.
18. Remove the center guide (if equipped) from the top of the pinion shaft,

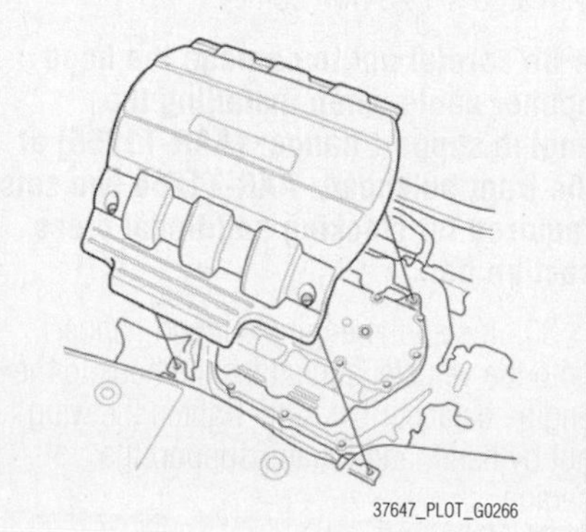
37647_PLOT_G0266

Fig. 264 Remove the engine cover

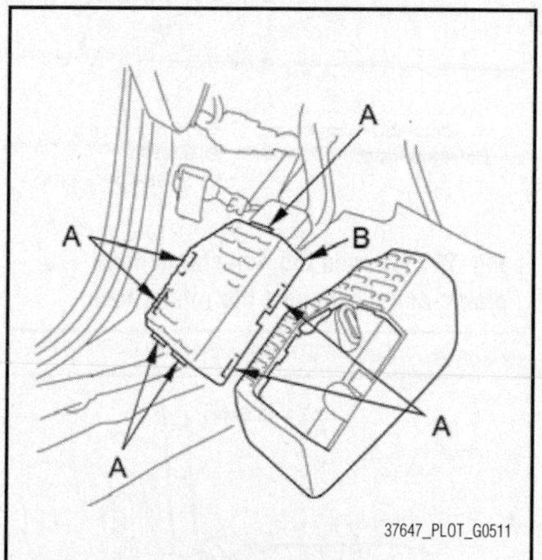
A
A
B
A
A
37647_PLOT_G0511

Fig. 265 Detach the hooks (A), then remove the footrest plate (B)

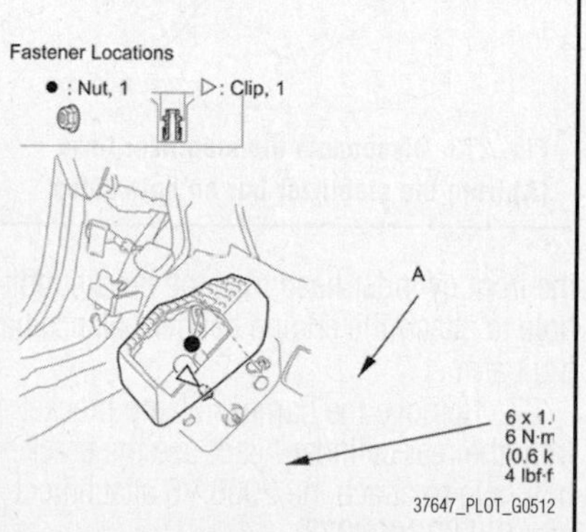
Fastener Locations
●: Nut, 1
▷: Clip, 1
A
37647_PLOT_G0512

Fig. 266 Remove the nut with a 6 mm hexagonal wrench, release the clip, then remove the footrest (A)

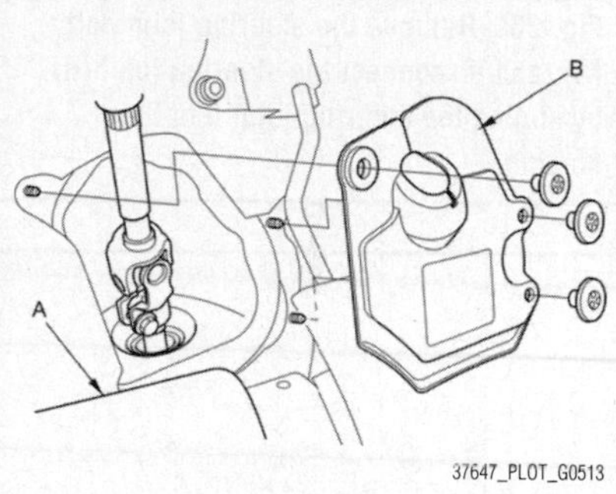
B
A
37647_PLOT_G0513

Fig. 267 Pull back the carpet (A), then remove steering joint cover (B)

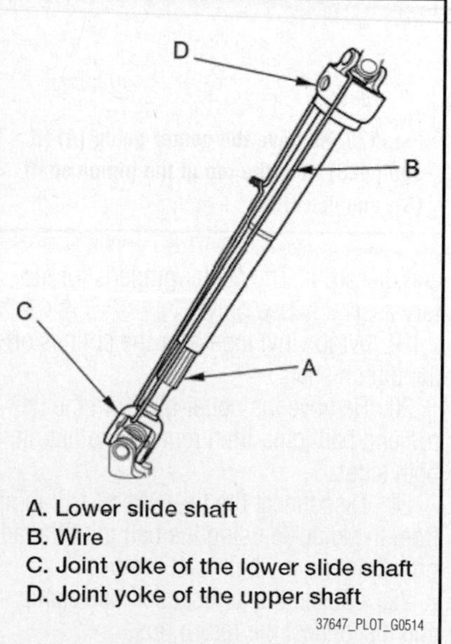
D
B
C
A
A. Lower slide shaft
B. Wire
C. Joint yoke of the lower slide shaft
D. Joint yoke of the upper shaft
37647_PLOT_G0514

Fig. 268 Hold the lower slide shaft on the column with a piece of wire between the joint yoke of the lower slide shaft and joint yoke of the upper shaft to prevent the lower slide shaft from pulling out

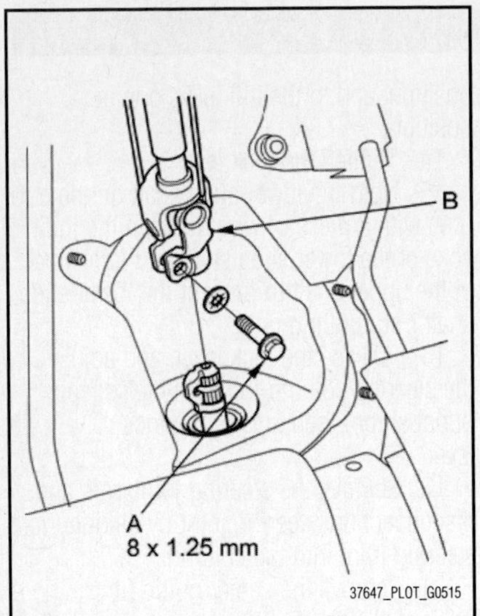
B
A
8 x 1.25 mm
37647_PLOT_G0515

Fig. 269 Remove the steering joint bolt (A), and disconnect the steering joint (B) by sliding the steering joint into the column

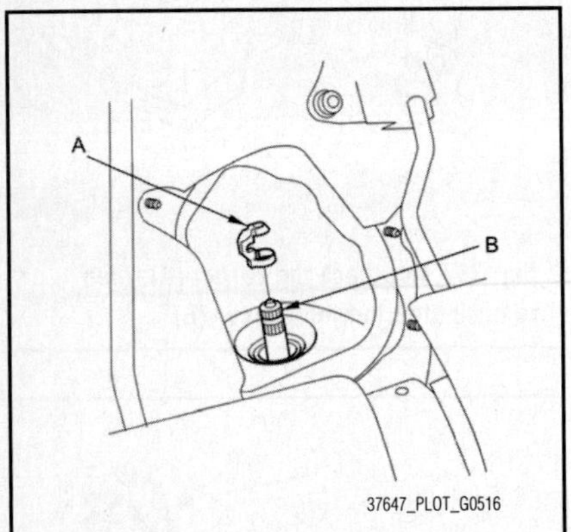
A
B
37647_PLOT_G0516

Fig. 270 Remove the center guide (A) (if equipped) from the top of the pinion shaft (B), and discard it

and discard it. The center guide is for factory assembly use only.

19. Wrap vinyl tape over the splines on the pinion shaft.

20. Remove the cotter pin from the tie-rod end ball joint, then remove the nut on both sides.

21. Disconnect the tie-rod end ball joint from the knuckle using the ball joint thread protector and the ball joint remover.

22. Loosen the adjustable hose clamp and disconnect the return hose.

23. Loosen the 16 mm flare nut and disconnect the inlet line.

24. Disconnect the stabilizer links from the stabilizer bar on both sides.

25. Remove the air cleaner assembly.

26. Remove the connector bracket from

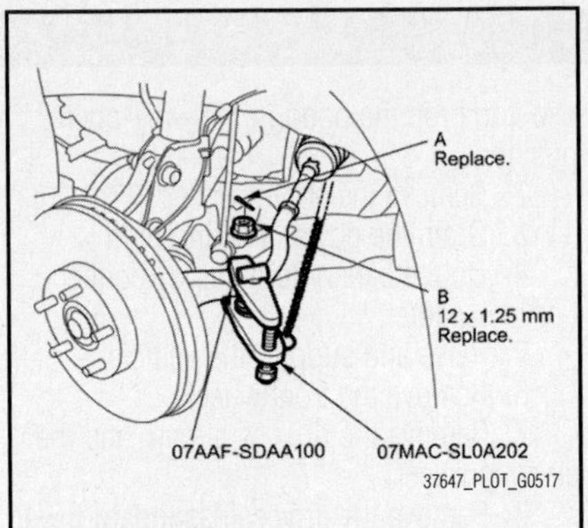
A
Replace.
B
12 x 1.25 mm
Replace.
07AAF-SDAA100
07MAC-SL0A202
37647_PLOT_G0517

Fig. 271 Remove the cotter pin (A) from the tie-rod end ball joint, then remove the nut (B) on both sides

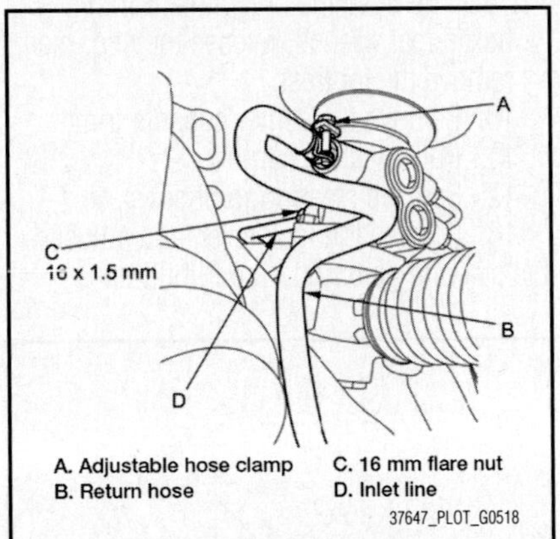
A
C
16 x 1.5 mm
B
D
A. Adjustable hose clamp
B. Return hose
C. 16 mm flare nut
D. Inlet line
37647_PLOT_G0518

Fig. 272 Loosen the adjustable hose clamp and disconnect the return hose

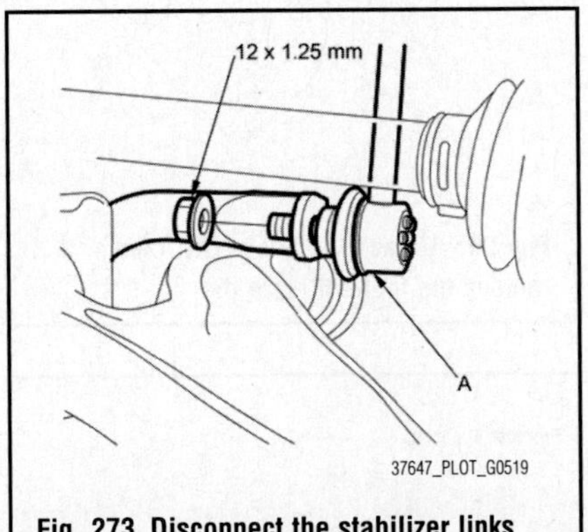
12 x 1.25 mm
A
37647_PLOT_G0519

Fig. 273 Disconnect the stabilizer links (A) from the stabilizer bar on both sides

the front cylinder head; use the bracket bolt hole to attach the engine hanger balance bar front arm.

27. Remove the harness clamp bracket from the rear cylinder head; use the bracket bolt hole to attach the 2008 V6 attachment arm (SIL02C000033).

28. Remove the cowl top side cover from both sides, then install the engine support hanger (AAR-T1256). With the engine sup-

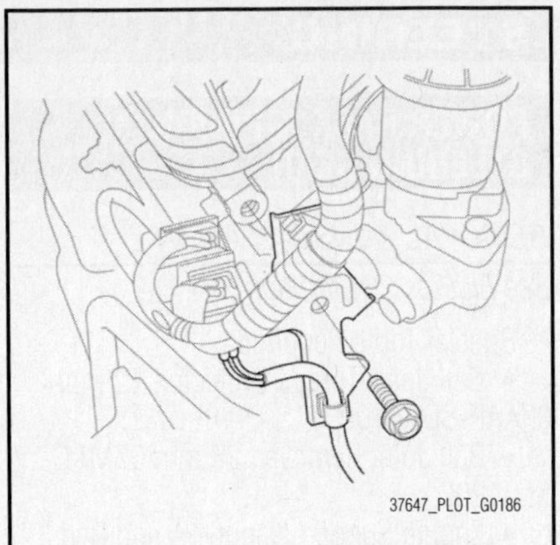
37647_PLOT_G0186

Fig. 274 Remove the connector bracket from the engine front cylinder head

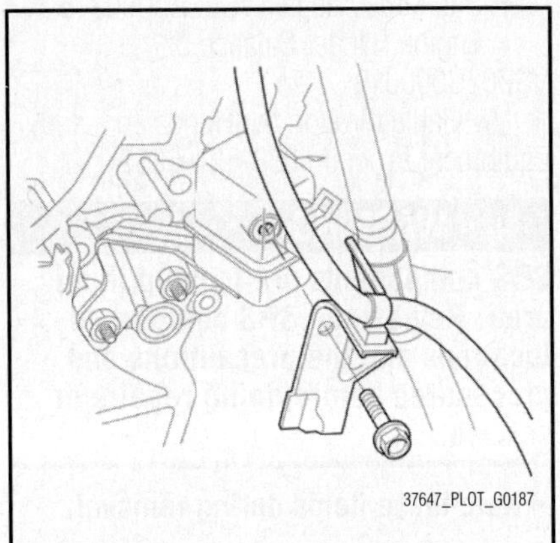
37647_PLOT_G0187

Fig. 275 Remove the harness clamp bracket (A) from the rear cylinder head

port hanger beam on its side, insert the hanger beam through the opening, then rotate it over the damper.

29. Install the engine hanger balance bar (VSB02C000019). Attach the front arm to the front cylinder head with several spacers and a 10 x 1.25 mm bolt. Remove the rear arm from the engine hanger balance bar, then install the 2008 V6 attachment arm (SIL02C000033). Attach the 2008 V6 attachment arm to the rear cylinder head with an 8 x 1.25 mm bolt.

➡Be careful not to damage the hood opener cable when installing the engine support hanger (AAR-T1256) at the front bulkhead. AAR-T1256 two sets required for stacking additional cross section bar.

30. Install the engine support hanger onto the vehicle, and attach the hook to the engine balancer bar slot. Tighten the wing nut by hand, and lift and support the engine.

31. Raise the vehicle.

32. Remove the splash shield.

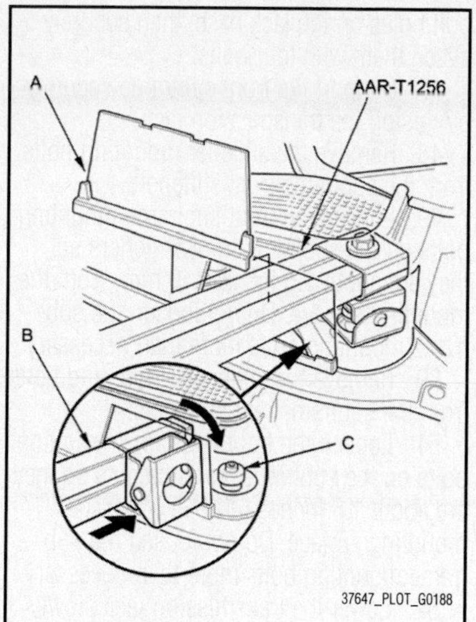
A
AAR-T1256
B
C
37647_PLOT_G0188

Fig. 276 Remove the cowl top side cover (A) from both sides, then install the engine support hanger; with the engine support hanger beam (B) on its side, insert the hanger beam through the opening, then rotate it over the damper (C)

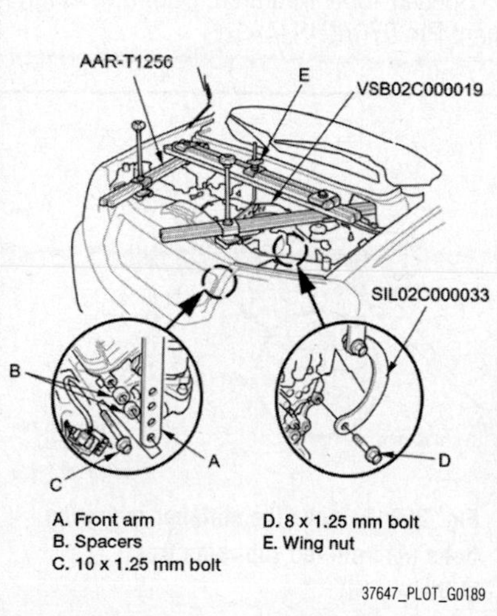
AAR-T1256
E
VSB02C000019
SIL02C000033
B
A
C
D
A. Front arm
B. Spacers
C. 10 x 1.25 mm bolt
D. 8 x 1.25 mm bolt
E. Wing nut
37647_PLOT_G0189

Fig. 277 Install the engine hanger balance bar

33. Remove the front subframe stiffener.
34. Remove exhaust pipe.
35. 4WD: Remove the propeller shaft and propeller shaft protectors.
36. Remove the inlet line clamp bolts.
37. Disconnect the front engine mount actuator connector and the harness clamps. Remove the front engine mount stop, then remove the front engine mount bolt.

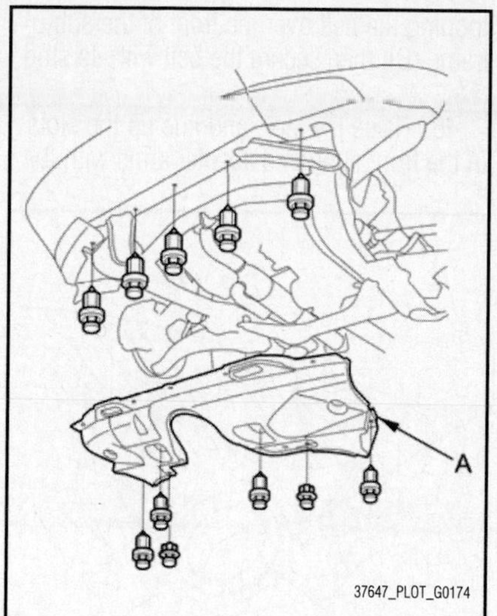
A
37647_PLOT_G0174

Fig. 278 Remove the splash shield (A)

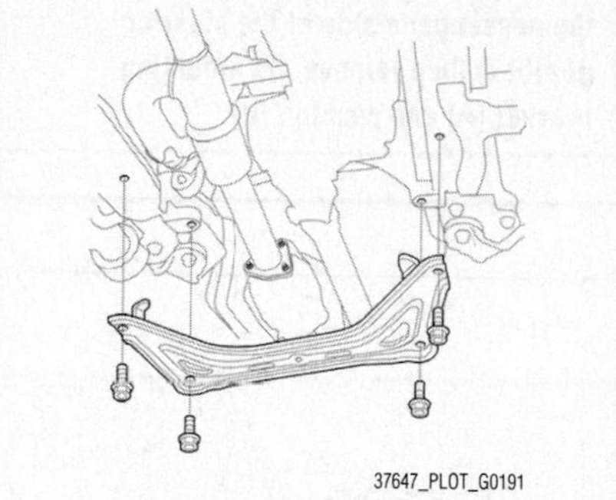
37647_PLOT_G0191

Fig. 279 Remove the front subframe stiffener (A)

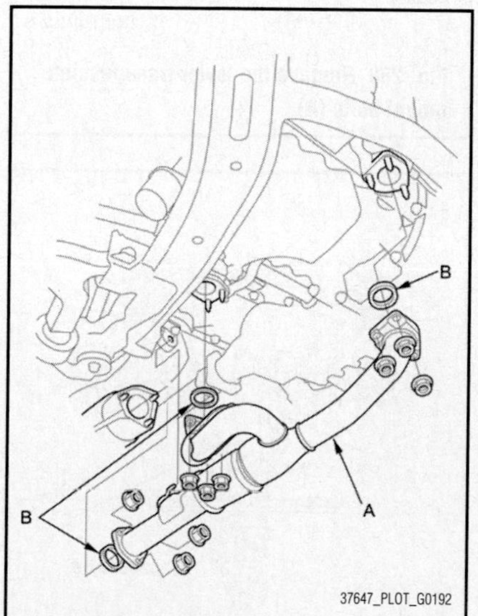
B
A
B
37647_PLOT_G0192

Fig. 280 Remove exhaust pipe

38. Remove the rear engine mount bolts, and disconnect the rear engine mount actuator connector.

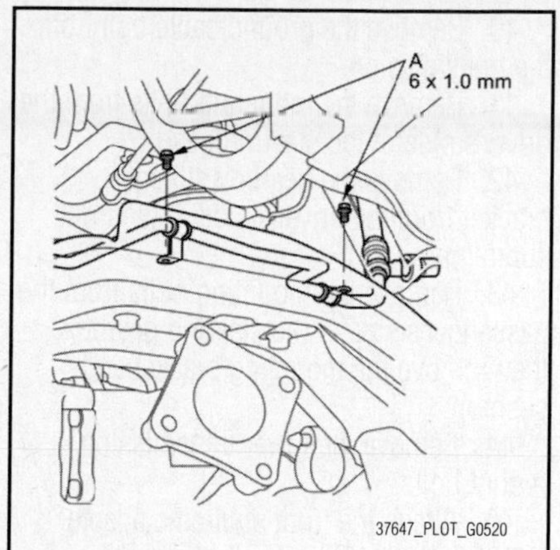
A
6 x 1.0 mm
37647_PLOT_G0520

Fig. 281 Remove the inlet line clamp bolts (A)

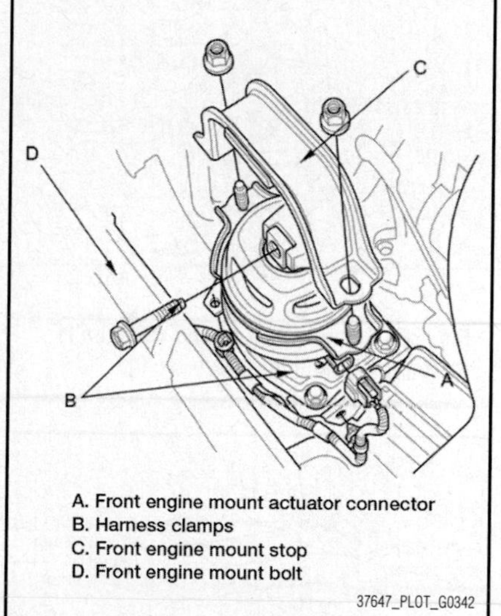
C
D
B
A
A. Front engine mount actuator connector
B. Harness clamps
C. Front engine mount stop
D. Front engine mount bolt
37647_PLOT_G0342

Fig. 282 Disconnect the front engine mount actuator connector and the harness clamps

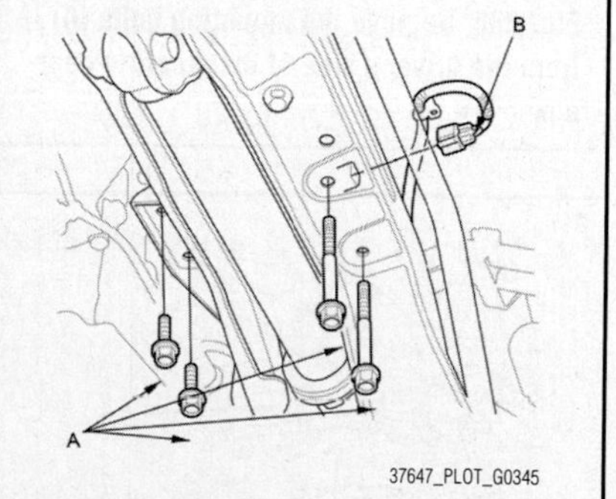
B
A
37647_PLOT_G0345

Fig. 283 Remove the rear engine mount bolts (A), and disconnect the rear engine mount actuator connector (B)

39. Remove the rear engine mount from the rear engine mount base bracket, and move it aside.

40. Remove the ground cable bolt from the transmission.

41. Remove the mounting bolts from the driver's side of the steering gearbox.

42. Remove the gearbox stiffener bracket from the driver's side of the front subframe.

43. Remove the mounting bolts from the passenger's side of the steering gearbox, then remove the mounting bracket and cushion.

44. Remove the lower transmission mount bolts.

45. Attach the front subframe adapter (VSB02C000016) to the subframe by looping the belt over the front of the subframe (B), then secure the belt with its stop (C).

46. Raise the jack, and line up the slots in the front subframe adapter arms with the bolt holes on the jack base, then securely attach them with four bolts.

47. Support the front subframe securely by raising the transmission jack.

48. Remove the stiffener mounting bolts from the subframe front stiffeners.

49. Loosen the front subframe mounting bolts on the subframe front stiffeners so they are about 1.6 inches (40 mm) from the mounting surface. Do not loosen the subframe mounting bolts more than necessary.

50. Remove the stiffeners mounting bolts from the subframe rear stiffeners.

51. Loosen the front subframe mounting bolts on the subframe rear stiffeners so they are about 1.6 inches (40 mm) from the mounting surface. Do not loosen the subframe mounting bolts more than necessary.

52. Lower the transmission jack slowly until the front subframe has dropped about 1.6 inches (40 mm).

53. Carefully move the steering gearbox toward the driver's side until the pinion shaft clears the fenderwell opening on the body.

54. Remove the steering gearbox through the fenderwell opening on the driver's side.

To install:

Special Tools Required: Subframe Alignment Pin 070AG-SJAA10S

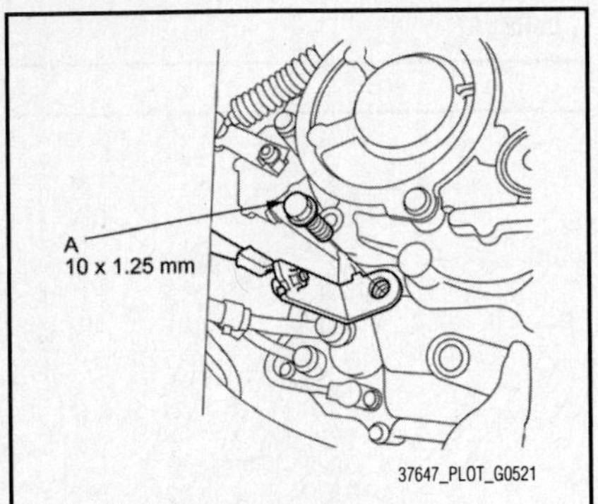
A
10 x 1.25 mm
37647_PLOT_G0521

Fig. 284 Remove the ground cable bolt (A) from the transmission

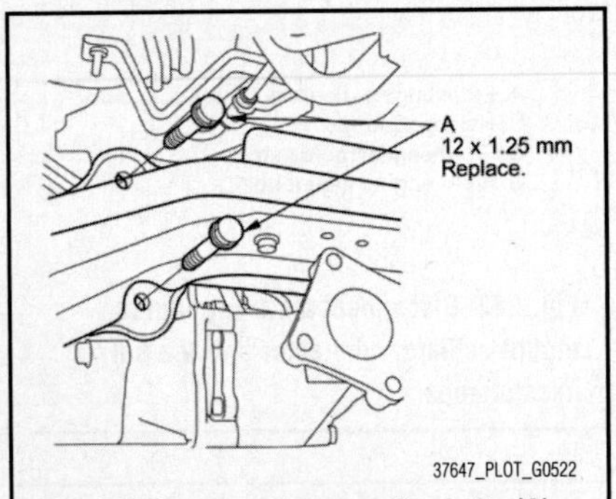
A
12 x 1.25 mm
Replace.
37647_PLOT_G0522

Fig. 285 Remove the mounting bolts (A) from the driver's side of the steering gearbox

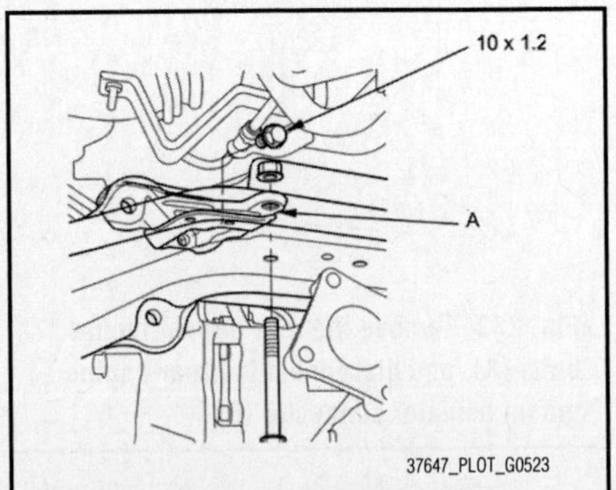
10 x 1.2
A
37647_PLOT_G0523

Fig. 286 Remove the gearbox stiffener bracket (A) from the driver's side of the front subframe

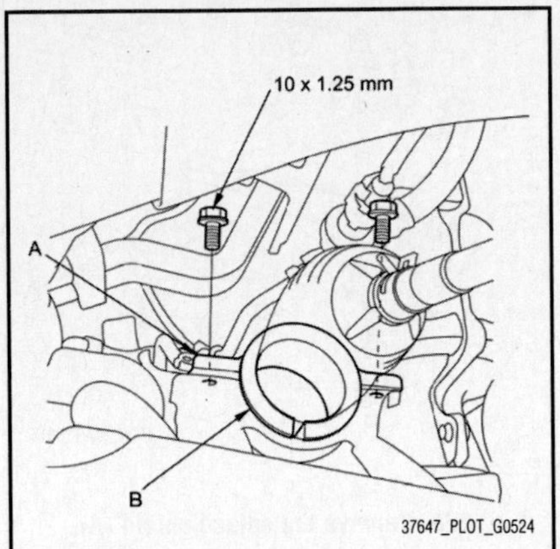
10 x 1.25 mm
A
B
37647_PLOT_G0524

Fig. 287 Remove the mounting bolts from the passenger's side of the steering gearbox, then remove the mounting bracket (A) and cushion (B)

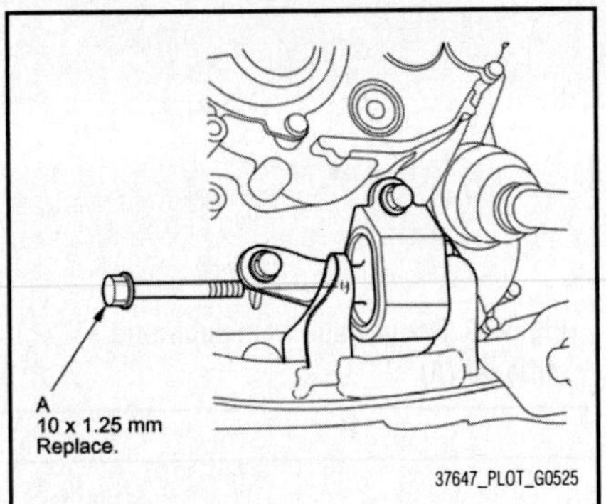
A
10 x 1.25 mm
Replace.
37647_PLOT_G0525

Fig. 288 Remove the lower transmission mount bolts (A)

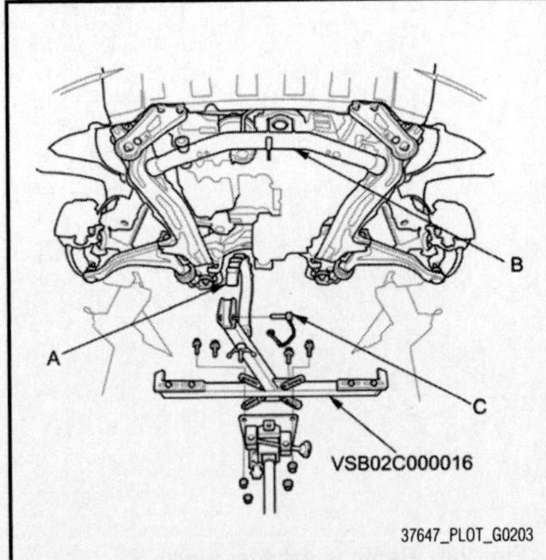
B
A
C
VSB02C000016
37647_PLOT_G0203

Fig. 289 Attach the front subframe adapter to the subframe by looping the belt (A) over the front of the subframe, then secure the belt with its stop

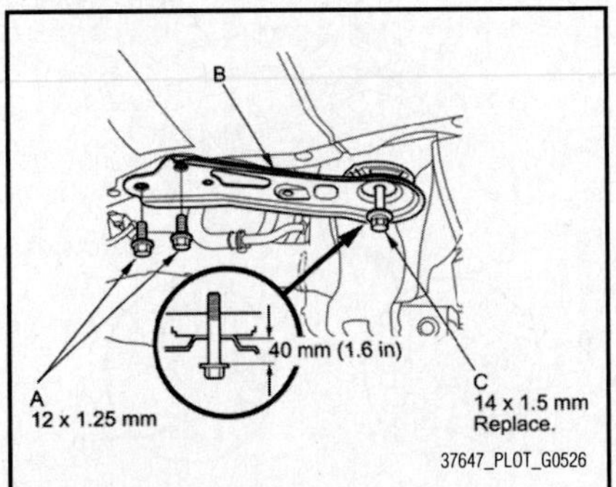
B
40 mm (1.6 in)
A
12 x 1.25 mm
C
14 x 1.5 mm
Replace.
37647_PLOT_G0526

Fig. 290 Remove the stiffener mounting bolts (A) from the subframe front stiffeners (B)

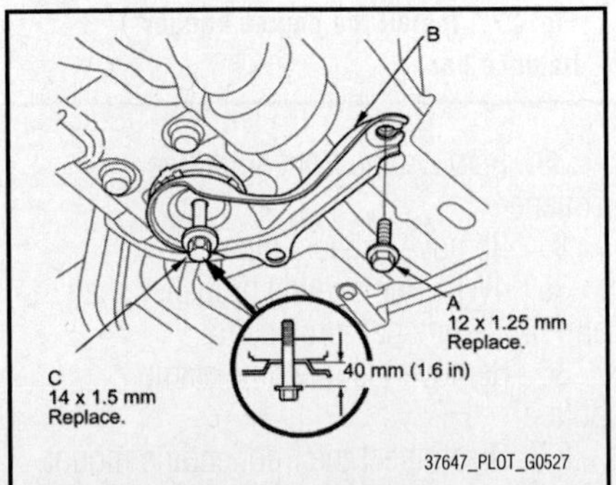
B
A
12 x 1.25 mm
Replace.
40 mm (1.6 in)
C
14 x 1.5 mm
Replace.
37647_PLOT_G0527

Fig. 291 Remove the stiffeners mounting bolts (A) from the subframe rear stiffeners (B)

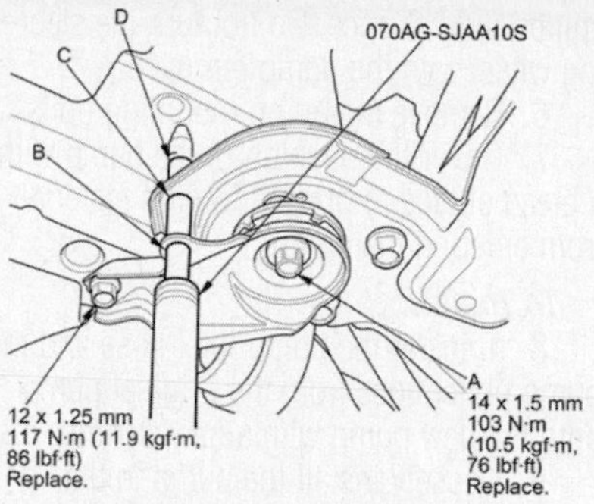
D
C
B
070AG-SJAA10S
12 x 1.25 mm
117 N·m (11.9 kgf·m,
86 lbf·ft)
Replace.
A
14 x 1.5 mm
103 N·m
(10.5 kgf·m,
76 lbf·ft)
Replace.
A. Left rear subframe mounting bolt
B. Positioning hole on the rear stiffener
C. Positioning hole on the subframe
D. Positioning hole on the body
37647_PLOT_G0212

Fig. 292 Loosely tighten the left rear subframe mounting bolt; insert the 15.7 mm side of the subframe alignment pin through the positioning hole on the rear stiffener, through the positioning hole on the subframe, and into the positioning hole on the body

55. Slide the steering gearbox) between the front subframe and body from the driver's side.
56. Carefully move the steering gearbox toward the passenger's side until the pinion shaft clears the fenderwell opening on the body.
57. Continue moving the gearbox toward the passenger's side until the steering gearbox is in position.
58. Loosely tighten the new left rear subframe mounting bolt and the new stiffener mounting bolt; insert the 15.7 mm side of the subframe alignment pin (070AG-SJAA10S) through the positioning hole on the rear stiffener, through the positioning hole on the subframe, and into the positioning hole on the body, then tighten the right rear subframe mounting bolt.
59. Loosely tighten the new right rear subframe mounting bolt and the new stiffener mounting bolt with the same procedure as the left rear using the subframe alignment pin.
60. Loosely tighten the right and left front new subframe mounting bolts and the stiffener mounting bolts.
61. Using the subframe alignment pin, tighten the rear-driver's side subframe mounting bolt to 76 ft. lbs. (103 Nm).
62. Using the subframe alignment pin, tighten the rear-passenger's side subframe mounting bolt to 76 ft. lbs. (103 Nm).
63. Tighten the front subframe mounting bolts to 76 ft. lbs. (103 Nm)..

➡Check all of the front subframe mounting bolts, and retighten if necessary.

64. Tighten the front and rear stiffener mounting bolts to 86 ft. lbs. (117 Nm).

➡Before tightening the stiffener mounting bolts, check that the positioning holes and slot are aligned using the subframe alignment pin.

65. Loosely tighten new lower transmission mount bolt.
66. Lower the transmission jack supporting the front subframe.
67. Install the gearbox stiffener bracket on the driver's side of the front subframe, and tighten the bolts and nut to 28 ft. lbs. (38 Nm).
68. Loosely install the new mounting bolts on the driver's side of the steering gearbox.
69. Position the cutout on the mounting cushion at the bottom, and install it on the passenger's side of the steering gearbox.
70. Install the gearbox mounting bracket over the mounting cushion, and loosely install the mounting bolts.
71. Tighten the mounting bolts on both sides of the steering gearbox to 37 ft. lbs. (50 Nm) alternately in two steps.
72. Install the ground cable bolt to the transmission.
73. Install the new rear engine mount bolts and the rear engine mount bolts to the rear engine mount base bracket. Tighten the mounting bolts to 31 ft. lbs. (42 Nm).
74. Connect the rear engine mount actuator connector.
75. Tighten the front engine mount bolt to 40 ft. lbs. (54 Nm), then install the front engine mount stop with the new mounting nuts. Tighten the nuts to 54 ft. lbs. (74 Nm). Connect the front engine mount actuator connector and the harness clamps.
76. Install the inlet line clamp bolts.
77. Install exhaust pipe A.
78. 4WD: Install the propeller shaft and propeller shaft protectors.
79. Install the front subframe stiffener with new mounting bolts, and tighten to 43 ft. lbs. (59 Nm).
80. Install the front splash shield.
81. Lower the vehicle.
82. Remove the engine support hanger, then install the cowl top side lid on both sides.
83. Tighten the lower transmission mount bolt to the specified torque.
84. Connect the stabilizer links to the stabilizer bar. Tighten the mounting nuts to 58 ft. lbs. (78 Nm).
85. Connect the return hose securely, and tighten the adjustable hose clamp.
86. Connect the inlet line, and tighten the 16 mm flare nut to 31 ft. lbs. (42 Nm).
87. On both sides, wipe off any grease contamination from the ball joint tapered section and threads. Reconnect the tie-rod ball joint to the knuckle. Install the new nut, and tighten it to 44 ft. lbs. (60 Nm).
88. Install a new cotter pin, and bend it.
89. Install the front wheels, then set the wheels in the straight ahead position.

➡Before installing the wheel, clean the mating surfaces of the brake disc and the inside of the wheel.

90. Center the steering rack within its stroke.
91. Cut the wire from the steering joint.
92. Slip the lower end of the steering joint onto the pinion shaft taking care to align the gap within the angle.
93. Align the bolt hole on the steering joint with the groove around the pinion shaft, then loosely install the joint bolt. Be sure that the joint bolt is securely in the groove in the pinion shaft.
94. Pull on the steering joint to make sure that the steering joint is fully seated, then tighten the joint bolt to 21 ft. lbs. (28 Nm).
95. Install the steering joint cover.
96. Set the carpet.
97. Install the center console trim.
98. Install the footrest.
99. Install the dashboard driver's lower cover.

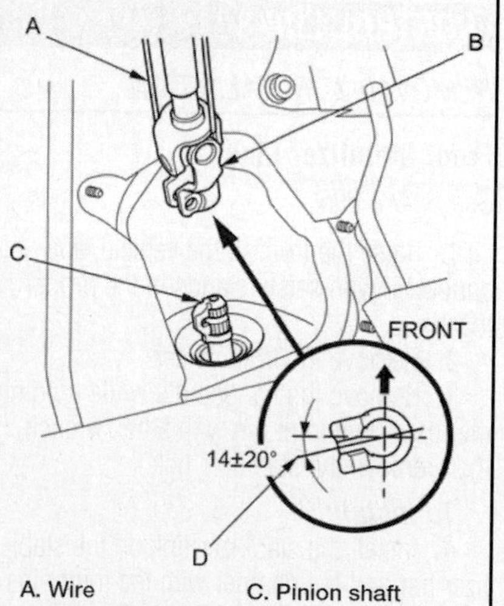
A
B
C
FRONT
14±20°
D
A. Wire
B. Steering joint
C. Pinion shaft
D. Gap
37647_PLOT_G0533

Fig. 293 Cut the wire, and slip the lower end of the steering joint onto the pinion shaft

100. Install the steering wheel, and the driver's airbag.

101. Do the battery terminal reconnection procedure, and do these tasks:

a. Turn the ignition switch to ON (II).The SRS indicator should come on for about 6 seconds, and then go off.

b. Make sure the horn and turn signal switches work properly.

c. Make sure the steering wheel switches work properly.

102. Fill the system with power steering fluid, and bleed the air from the system.

103. Install the engine cover.

104. Reinstall the support strut to the proper locations on the hood.

105. After installation, check these items.

a. Start the engine, allow it to idle, and turn the steering wheel from lock to lock several times to warm up the fluid. Check the gearbox for leaks.

b. Check the steering wheel spoke angle. If steering spoke angles to the right and left are not equal (steering wheel and rack are not centered), correct the engagement of the joint/pinion shaft serrations.

c. Set the steering column to the center tilt position, and to the center telescopic position, then do the front toe inspection.

POWER STEERING PUMP

REMOVAL & INSTALLATION

See Figure 294.

1. Place a suitable container under the vehicle to catch any spilled fluid.
2. Drain the power steering fluid from the reservoir.
3. Remove the engine cover.
4. Remove the drive belt from the pump pulley.
5. Cover the auto-tensioner, the alternator, and the A/C compressor with several shop towels to protect them from spilled power steering fluid. Disconnect the pump inlet hose and the pump outlet hose from the pump, and plug them. Take care not to spill the fluid on the vehicle. Wipe off any spilled fluid at once. Do not turn the steering wheel with the pump removed.
6. Remove the pump mounting bolts.
7. Cover the opening of the pump with a piece of tape to prevent foreign material from entering the pump.

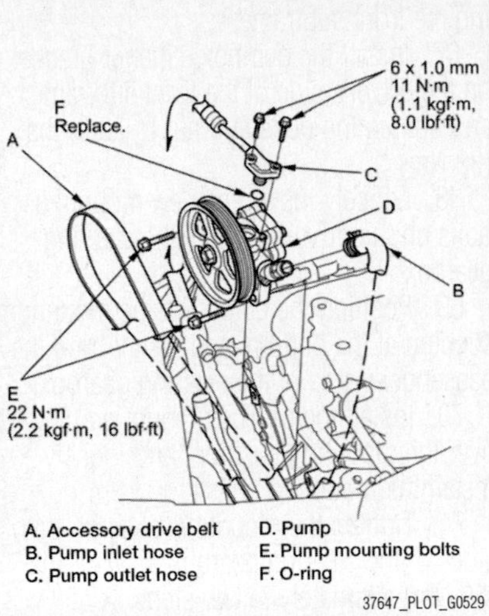
6 x 1.0 mm
11 N·m
(1.1 kgf·m,
8.0 lbf·ft)
F
Replace.
A
C
D
B
E
22 N·m
(2.2 kgf·m, 16 lbf·ft)
A. Accessory drive belt
B. Pump inlet hose
C. Pump outlet hose
D. Pump
E. Pump mounting bolts
F. O-ring
37647_PLOT_G0529

Fig. 294 Exploded view of the P/S pump assembly

To install:

8. Transfer the pump inlet hose and the pump outlet hose from the original pump onto the new pump with a new O-ring.
9. Loosely install the pump in the pump bracket with the mounting bolts, then tighten the pump fittings to 8 ft. lbs. (11 Nm).
10. Tighten the pump mounting bolts to 16 ft. lbs. (22 Nm).
11. Install the drive belt.
12. Note these items during drive belt installation:
    - Inspect the belt for wear and cracks. Replace the belt if necessary.
    - Make sure that the belt is properly positioned on the pulleys.
    - Do not get power steering fluid or grease on the auto-tensioner, alternator, A/C compressor, and drive belt or pulley faces. Clean off any fluid or grease before installation.
13. Install the engine cover.
14. Fill the reservoir to the upper level line.
15. Start the engine, and check for leaks.

SUSPENSION

FRONT SUSPENSION

CONTROL LINKS

REMOVAL & INSTALLATION

Front Stabilizer Link

See Figure 295.

1. Raise the front of the vehicle, and support it with safety stands in the proper locations.
2. Remove the front wheel.
3. Remove the flange nuts while holding the respective joint pin with a hex wrench, then remove the stabilizer link.

To install:

4. Install the stabilizer link on the stabilizer bar and the damper with the joint pins set at the center of their range of movement.
5. Install the flange nuts, and tighten them to the specified torque value while holding the respective joint pin with a hex wrench.
6. Clean the mating surfaces of the brake disc and the inside of the wheel, then install the front wheel.

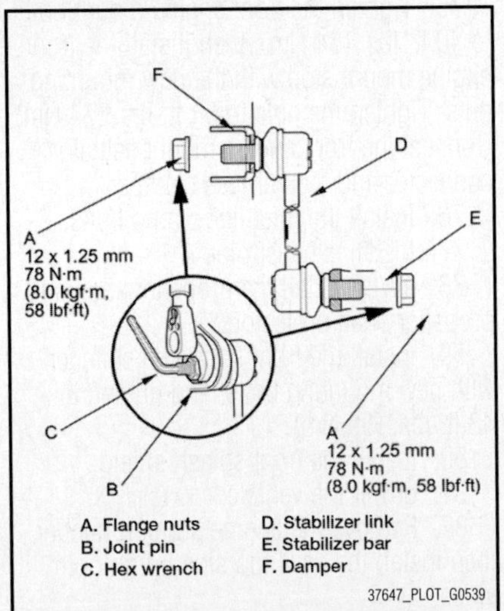
F
D
E
A
12 x 1.25 mm
78 N·m
(8.0 kgf·m,
58 lbf·ft)
C
B
A
12 x 1.25 mm
78 N·m
(8.0 kgf·m, 58 lbf·ft)
A. Flange nuts
B. Joint pin
C. Hex wrench
D. Stabilizer link
E. Stabilizer bar
F. Damper
37647_PLOT_G0539

Fig. 295 Remove the flange nuts while holding the respective joint pin with a hex wrench, then remove the stabilizer link

LOWER BALL JOINT

REMOVAL & INSTALLATION

See Figures 296 through 298.

Special Tools Required:

- Ball Joint Thread Protector, 12 mm 07AAF-SDAA100
- Ball Joint Remover, 32 mm 07MAC-SL0A102
- Ball Joint Remover, 28 mm 07MAC-SL0A202
- Ball Joint Thread Protector, 14 mm 071AF-S3VA000

➡Always use a ball joint remover to disconnect a ball joint. Do not strike the housing or any other part of the ball joint connection to disconnect it.

1. Install a hex nut or the ball joint

thread protector onto the threads of the ball joint.

➡**Using a hex nut, make sure the nut is flush with the ball joint pin end to prevent damage to the threaded end of the ball joint pin.**

2. Apply grease to the ball joint remover on the areas shown. This will ease the installation of the tool and prevent damage to the pressure bolt threads.

3. Loosen the pressure bolt, and install the ball joint remover as shown. Insert the jaws carefully, making sure not to damage the ball joint boot. Adjust the jaw spacing by turning the adjusting bolt.

➡**Fasten the safety chain securely to a suspension arm or the subframe. Do not fasten it to a brake line or wire harness.**

4. After adjusting the adjusting bolt, make sure the head of the adjusting bolt is in the position shown to allow the jaw to pivot.

5. With a wrench, tighten the pressure bolt until the ball joint pin pops loose from the ball joint connecting hole. If necessary, apply penetrating type lubricant to loosen the ball joint pin.

➡**Do not use pneumatic or electric tools on the pressure bolt.**

6. Remove the ball joint remover, then remove the nut or the ball joint thread protector from the end of the ball joint pin, and pull the ball joint out of the ball joint connecting hole. Inspect the ball joint boot, and replace it if damaged.

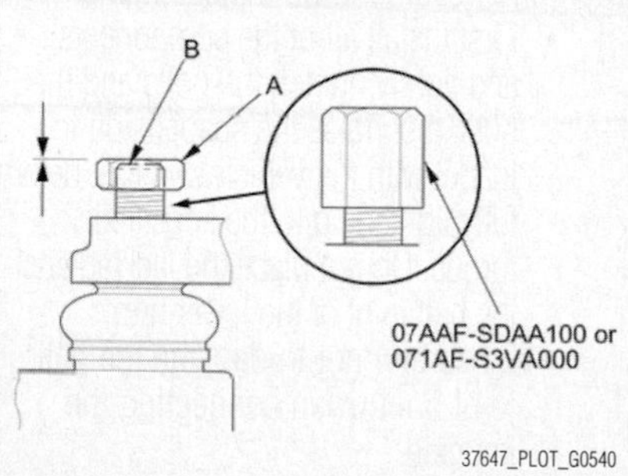
B
A
07AAF-SDAA100 or
071AF-S3VA000
37647_PLOT_G0540

Fig. 296 Install a hex nut (A) or the ball joint thread protector onto the threads of the ball joint (B)

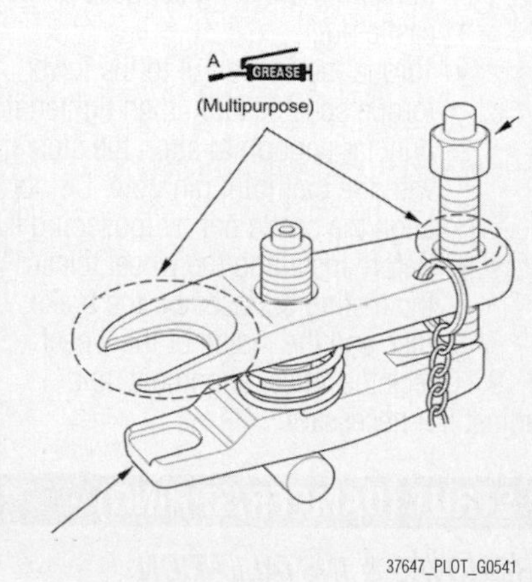
A GREASE
(Multipurpose)
37647_PLOT_G0541

Fig. 297 Apply grease to the ball joint remover on the areas shown (A)

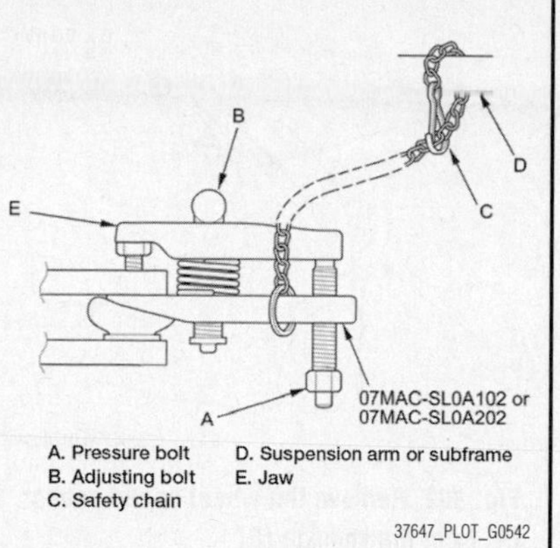
B
D
E
C
A
07MAC-SL0A102 or
07MAC-SL0A202
A. Pressure bolt
B. Adjusting bolt
C. Safety chain
D. Suspension arm or subframe
E. Jaw
37647_PLOT_G0542

Fig. 298 Loosen the pressure bolt, and install the ball joint remover as shown

LOWER CONTROL ARM

REMOVAL & INSTALLATION

See Figures 299 through 301.

Special Tools Required:

- Ball Joint Remover, 32 mm 07MAC-SL0A102
- Ball Joint Thread Protector, 14 mm 071AF-S3VA000

1. Raise the front of the vehicle, and support it with safety stands in the proper locations.
2. Remove the front wheel.

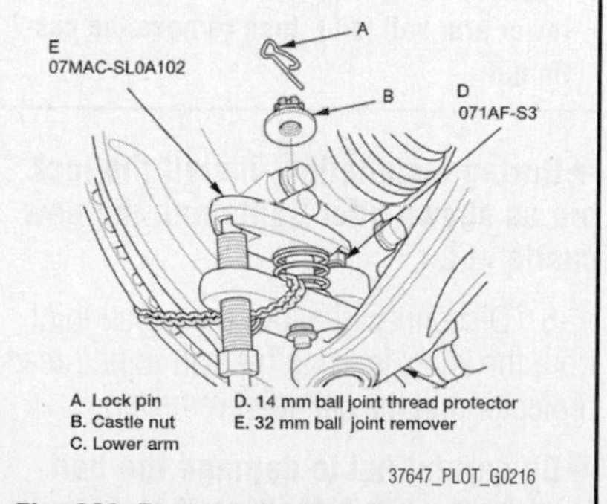
E
07MAC-SL0A102
A
B
D
071AF-S3
A. Lock pin
B. Castle nut
C. Lower arm
D. 14 mm ball joint thread protector
E. 32 mm ball joint remover
37647_PLOT_G0216

Fig. 299 Remove the lock pin from the lower arm ball joint, then remove the castle nut

3. Remove the lock pin from the lower arm ball joint, then remove the castle nut.

➡**During Installation, install the lock pin as shown after tightening the new castle nut.**

4. Disconnect the lower arm ball joint from the knuckle using the ball joint thread protector and the ball joint remover.

➡**Be careful not to damage the ball joint boot when installing the remover. Do not force or hammer on the lower arm, or pry between the lower arm and the knuckle. You could damage the ball joint.**

5. Remove the mounting bolt of the rear side on the stabilizer bar bushing holder.

➡**During installation, install the mounting bolt after tightening the lower arm mounting 14 mm bolts to the specified torque value.**

6. Remove the lower arm mounting 14 mm and 16 mm bolts, then remove the lower arm.

➡**Use new mounting bolts during reassembly.**

7. Remove the lower arm stops.

➡**During installation, align the slot on the lower arm stop with the lug portion on the front side lower arm bushing.**

To install:

8. Install the lower arm in the reverse order of removal, and note these items:

- First install all of the components, and lightly tighten the bolts and the nuts, then raise the suspension to load it with the vehicle's weight before fully tightening to the specified torque values. Do not place the jack against the ball joint of the lower arm.
- Be careful not to damage the ball joint boot when connecting the knuckle.
- Before connecting the ball joint, degrease the threaded section and the tapered portion of the ball joint pin, the ball joint connecting hole, and the threaded section and the mating surfaces of the castle nut.
- Torque the castle nut to the lower torque specification, then tighten it only far enough to align the slot with the ball joint pin hole. Do not align the castle nut by loosening it.
- Before installing the wheel, clean the mating surfaces of the brake disc and the inside of the wheel.

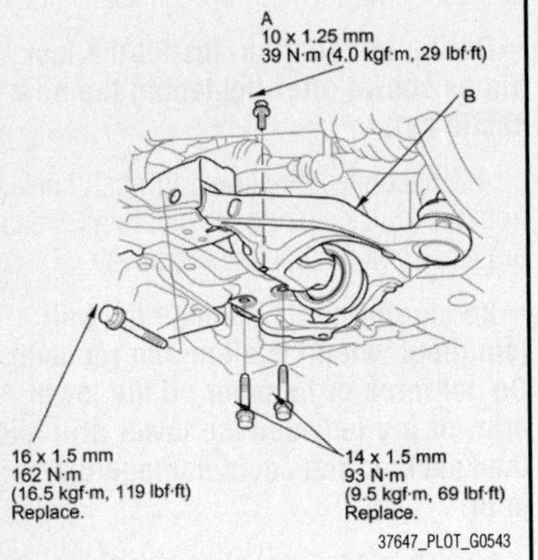

A
10 x 1.25 mm
39 N·m (4.0 kgf·m, 29 lbf·ft)
B
16 x 1.5 mm
162 N·m
(16.5 kgf·m, 119 lbf·ft)
Replace.
14 x 1.5 mm
93 N·m
(9.5 kgf·m, 69 lbf·ft)
Replace.
37647_PLOT_G0543

Fig. 300 Remove the mounting bolt (A) of the rear side on the stabilizer bar bushing holder; remove the lower arm mounting 14 mm and 16 mm bolts, then remove the lower arm (B)

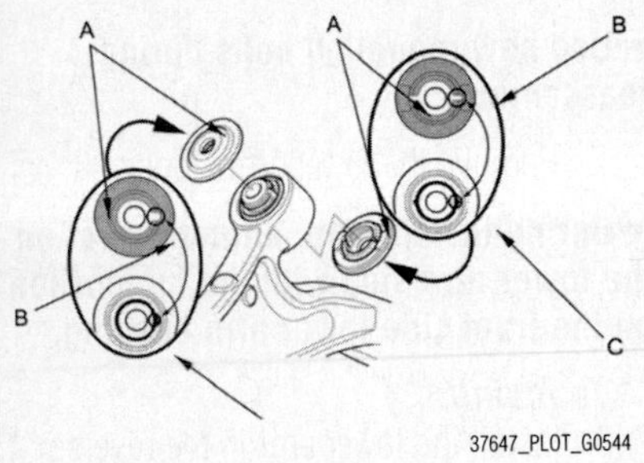

A
A
B
B
C
37647_PLOT_G0544

Fig. 301 Remove the lower arm stops (A)

9. Check the wheel alignment, and adjust it if necessary.

STEERING KNUCKLE

REMOVAL & INSTALLATION

See Figures 302 through 305.

1. Remove the hub bearing unit.
2. Remove the wheel speed sensor from the knuckle. Do not disconnect the wheel speed sensor connector.
3. Remove the cotter pin from the tie-rod end ball joint, then remove the nut.

➡During installation, install the new cotter pin after tightening the new nut, and bend its end as shown.

4. Disconnect the tie-rod end ball joint from the knuckle using the ball joint thread protector and the ball joint remover.
5. Remove the lock pin from the lower arm ball joint, then remove the castle nut.

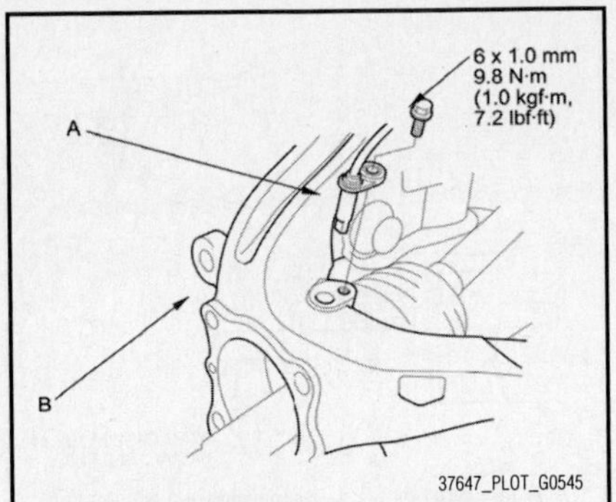

6 x 1.0 mm
9.8 N·m
(1.0 kgf·m,
7.2 lbf·ft)
A
B
37647_PLOT_G0545

Fig. 302 Remove the wheel speed sensor (A) from the knuckle (B)

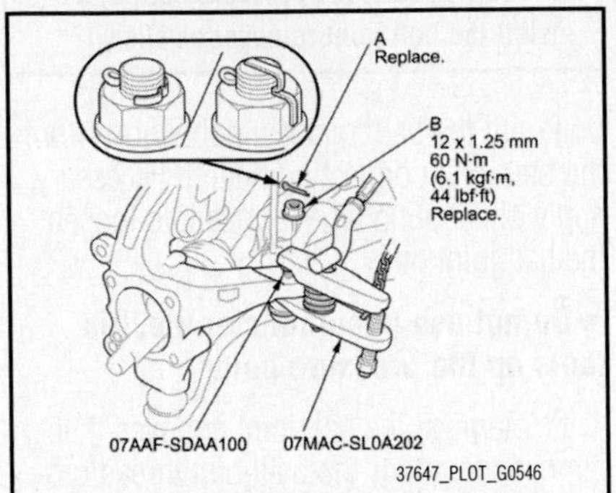

A
Replace.
B
12 x 1.25 mm
60 N·m
(6.1 kgf·m,
44 lbf·ft)
Replace.
07AAF-SDAA100
07MAC-SL0A202
37647_PLOT_G0546

Fig. 303 Remove the cotter pin (A) from the tie-rod end ball joint, then remove the nut (B)

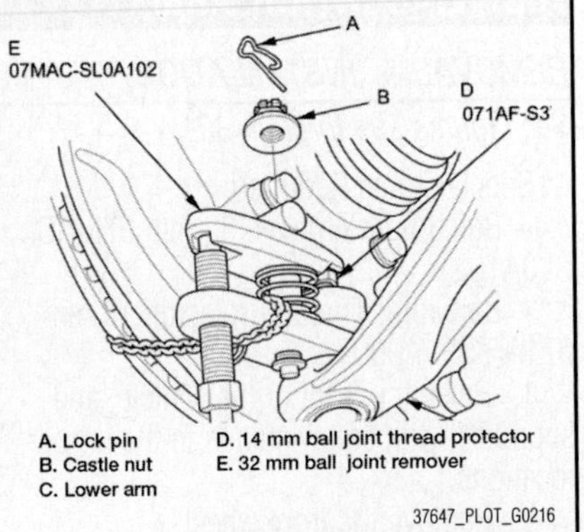

A
E
07MAC-SL0A102
B
D
071AF-S3
A. Lock pin
B. Castle nut
C. Lower arm
D. 14 mm ball joint thread protector
E. 32 mm ball joint remover
37647_PLOT_G0216

Fig. 304 Remove the lock pin from the lower arm ball joint, then remove the castle nut

➡During installation, install the lock pin as shown after tightening the new castle nut.

6. Disconnect the lower arm ball joint from the knuckle using the ball joint thread protector and the ball joint remover.

➡Be careful not to damage the ball joint boot when installing the remover. Do not force or hammer on the lower arm, or pry between the lower arm and the knuckle. You could damage the ball joint.

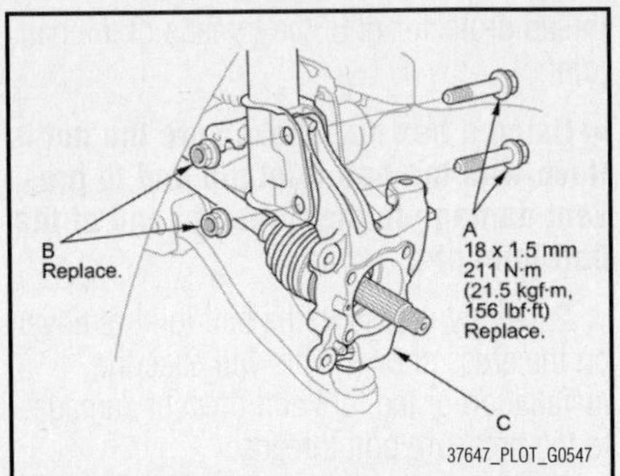

B
Replace.
A
18 x 1.5 mm
211 N·m
(21.5 kgf·m,
156 lbf·ft)
Replace.
C
37647_PLOT_G0547

Fig. 305 Remove the damper pinch bolts (A) and the flange nuts (B) from the damper, then remove the knuckle (C)

7. Remove the damper pinch bolts and the flange nuts from the damper, then remove the knuckle.

➡During installation, install new damper pinch bolts and new flange nuts.

To install:

8. Install the knuckle in the reverse order of removal, and note these items:
   - First install all of the components, and lightly tighten the bolts and the nuts, then raise the suspension to load it with the vehicle's weight before fully tightening to the specified torque. Do not place the jack against the ball joint of the lower arm.
   - Be careful not to damage the ball joint boot when connecting the knuckle.
   - Before connecting the ball joint to the knuckle, degrease the threaded section and the tapered portion of the ball joint pin, the ball joint connecting hole, and the threaded section and the mating surfaces of the castle nut.
   - Torque the castle nut to the lower torque specification, then tighten it only far enough to align the slot with the ball joint pin hole. Do not align the castle nut by loosening it.
   - Before installing the wheel, clean the mating surfaces on the brake disc and the inside of the wheel.
9. Check the wheel alignment, and adjust it if necessary.

STRUT (DAMPER/SPRING)

REMOVAL & INSTALLATION

See Figures 306 through 312.

1. Raise and support the vehicle.

2. Remove the front wheel.

3. Remove the wheel speed sensor harness and the brake hose from the damper. Do not disconnect the wheel speed sensor connector.

4. Remove the flange nut, while holding the joint pin with a hex wrench, and disconnect the stabilizer link from the damper.

5. Remove the damper pinch bolts and the flange nuts from the damper.

➡Do not allow the knuckle to rotate too far outward. This may allow the driveshaft inboard joint to come apart.

6. Remove the cowl top side cover, then remove the flange nuts from the top of the damper. Do not let the damper/spring drop down under its own weight.

7. Remove the damper/spring.

➡The left and right damper springs are different. Mark the springs L and R before you continue.

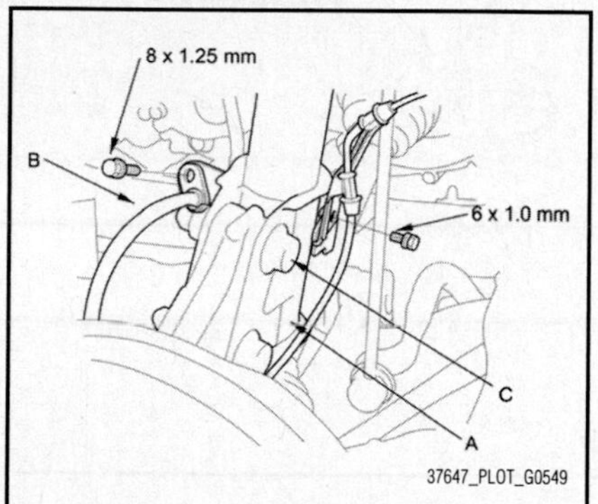
8 x 1.25 mm
B
6 x 1.0 mm
C
A
37647_PLOT_G0549

Fig. 306 Remove the wheel speed sensor harness (A) and the brake hose (B) from the damper (C)

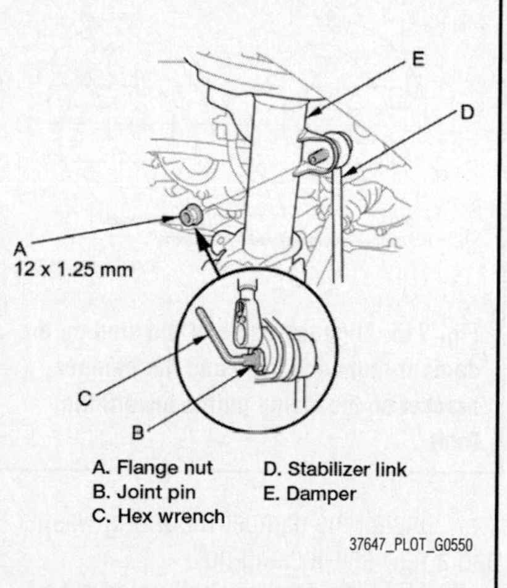
E
D
A
12 x 1.25 mm
C
B
A. Flange nut
B. Joint pin
C. Hex wrench
D. Stabilizer link
E. Damper
37647_PLOT_G0550

Fig. 307 Remove the flange nut, while holding the joint pin with a hex wrench, and disconnect the stabilizer link from the damper

To install:

8. Install the damper/spring on to the frame. Note the direction of the damper mounting base as shown.

9. Loosely install the new flange nuts to the top of the damper.

➡Install the cowl top side cover after tightening the flange nuts.

10. Loosely install new damper pinch bolts and new flange nuts to the damper.

11. Connect the stabilizer link to the damper, and loosely install the flange nut.

12. Tighten the flange nut to 58 ft. lbs. (78 Nm), while holding the joint pin with the hex wrench.

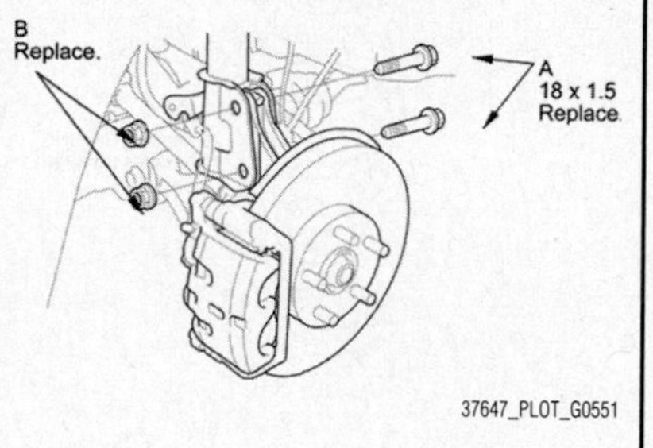
B
Replace.
A
18 x 1.5
Replace.
37647_PLOT_G0551

Fig. 308 Remove the damper pinch bolts (A) and the flange nuts (B) from the damper

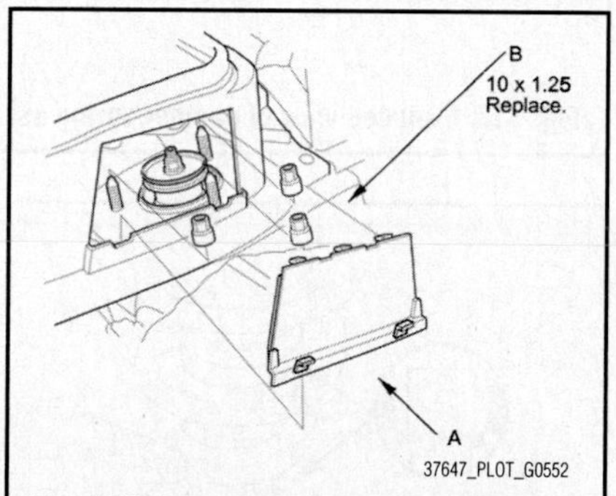
B
10 x 1.25
Replace.
A
37647_PLOT_G0552

Fig. 309 Remove the cowl top side cover (A), then remove the flange nuts (B) from the top of the damper

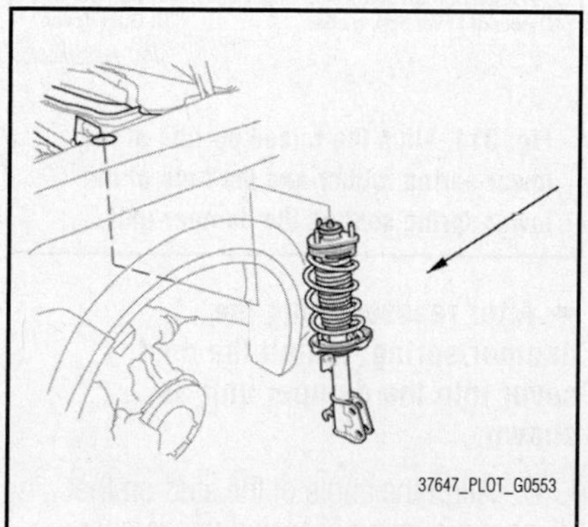
37647_PLOT_G0553

Fig. 310 Remove the damper/spring

13. Place a floor jack under the lower arm, and raise the suspension to load it with the vehicle's weight.

➡Do not place the jack against the ball joint of the lower arm.

14. Tighten the flange nuts on top of the damper to 43 ft. lbs. (59 Nm).

15. Tighten the damper pinch bolts and the flange nuts to 156 ft. lbs. (211 Nm).

16. Install the wheel speed sensor harness and the brake hose to the damper.

17. Clean the mating surfaces of the brake disc and the inside of the wheel, then install the front wheel.

18. Check the wheel alignment, and adjust it if necessary.

OVERHAUL

➡When compressing the damper spring, use a commercially available strut spring compressor (Branick MST-580A or Model 7200, or equivalent) according to the manufacturer's instructions.

Disassembly

See Figure 311.

1. Compress the damper spring, then remove the self-locking nut while holding the damper shaft with a hex wrench. Do not compress the damper spring more than necessary to remove the self-locking nut.

2. Release the pressure from the strut

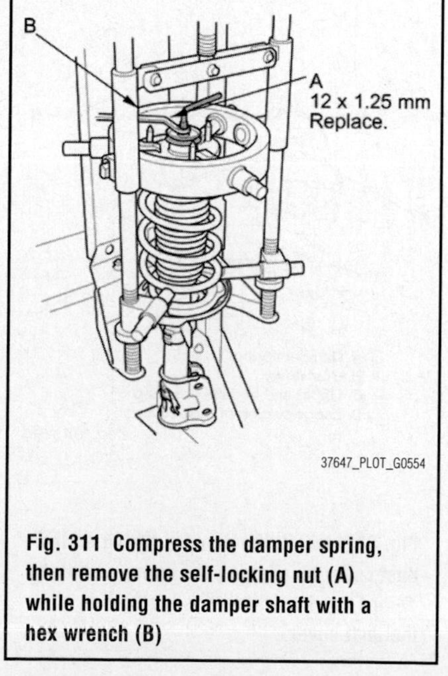
B
A
12 x 1.25 mm
Replace.
37647_PLOT_G0554

Fig. 311 Compress the damper spring, then remove the self-locking nut (A) while holding the damper shaft with a hex wrench (B)

spring compressor, then disassemble the damper as shown in the Exploded View.

Inspection

1. Reassemble the damper mounting base, the damper mounting washer, and the self-locking nut.

2. Compress the damper assembly by hand, and check for smooth operation through a full stroke, both compression and extension. The damper should extend smoothly and constantly when compression is released. If it does not, the gas is leaking and the damper should be replaced.

3. Check for oil leaks, abnormal noises, and binding during these tests.

Reassembly

See Figures 312 through 316.

1. Install the damper spring on the dust cover by aligning the upper end of the damper spring with the ledge portion of the dust cover.

2. Install the lower spring rubber on the damper spring.

3. Install all the parts except the damper mounting washer and the self-locking nut onto the damper unit by referring to the Exploded View.

4. Compress the damper spring using a strut spring compressor. Do not compress the spring excessively.

5. Align the raised portion of the lower spring rubber and the hole of the lower spring seat on the damper unit.

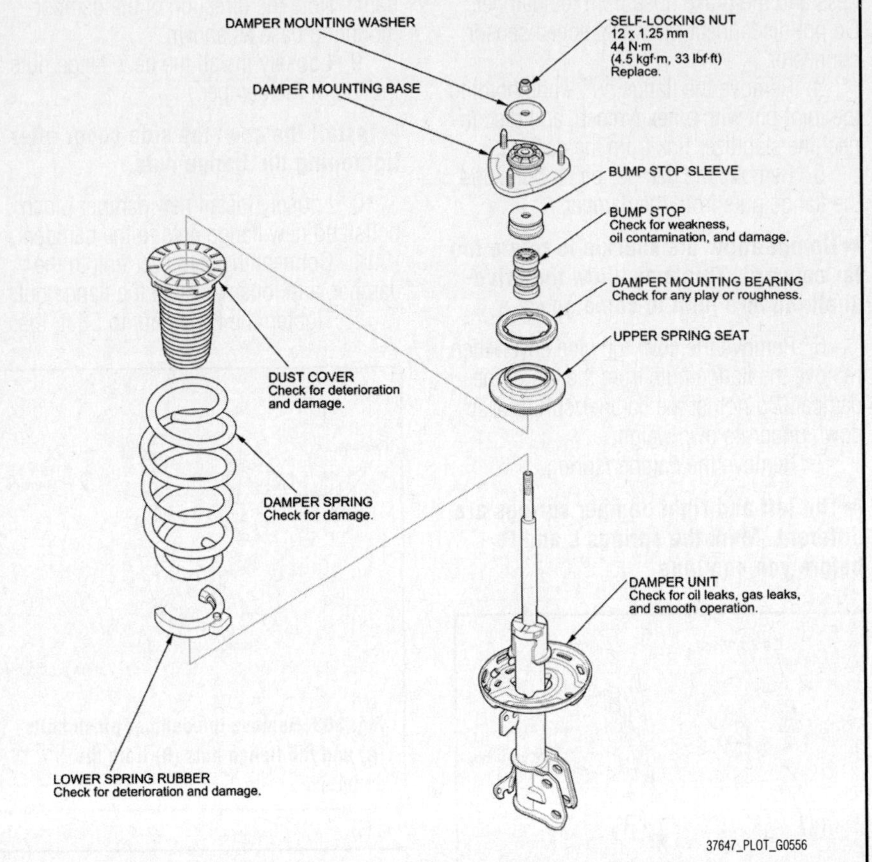
DAMPER MOUNTING WASHER
SELF-LOCKING NUT
12 x 1.25 mm
44 N·m
(4.5 kgf·m, 33 lbf·ft)
Replace.
DAMPER MOUNTING BASE
BUMP STOP SLEEVE
BUMP STOP
Check for weakness, oil contamination, and damage.
DAMPER MOUNTING BEARING
Check for any play or roughness.
UPPER SPRING SEAT
DUST COVER
Check for deterioration and damage.
DAMPER SPRING
Check for damage.
DAMPER UNIT
Check for oil leaks, gas leaks, and smooth operation.
LOWER SPRING RUBBER
Check for deterioration and damage.
37647_PLOT_G0556

Fig. 313 Exploded view of damper/spring assembly

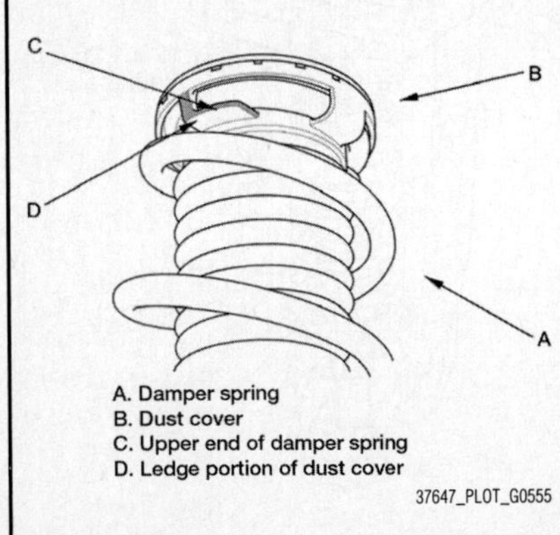
C
B
D
A
A. Damper spring
B. Dust cover
C. Upper end of damper spring
D. Ledge portion of dust cover
37647_PLOT_G0555

Fig. 312 Install the damper spring on the dust cover by aligning the upper end of the damper spring with the ledge portion of the dust cover

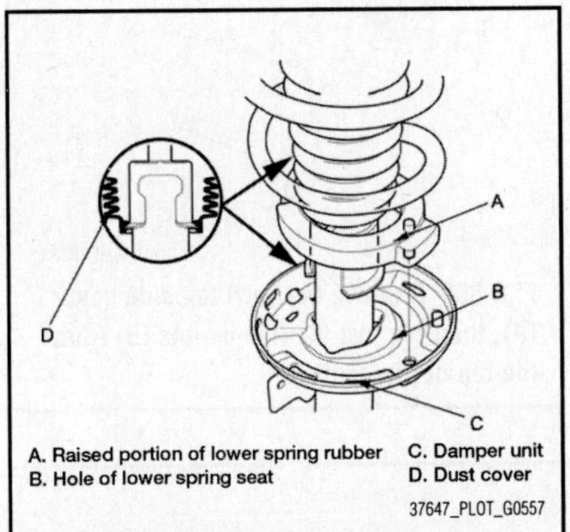
A
B
C
D
A. Raised portion of lower spring rubber
B. Hole of lower spring seat
C. Damper unit
D. Dust cover
37647_PLOT_G0557

Fig. 314 Align the raised portion of the lower spring rubber and the hole of the lower spring seat on the damper unit

➡ After reassembling the damper/spring, install the dust cover into the damper unit as shown.

6. Align the angle of the stud on the damper mounting base and the damper bracket so the stamp points toward the front.

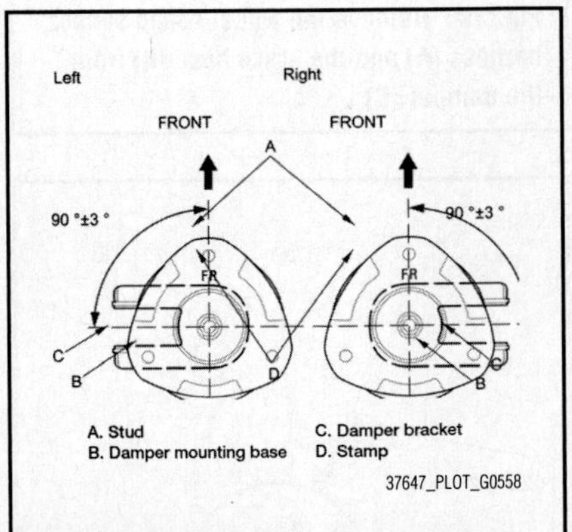
Left
Right
FRONT
FRONT
A
90 °±3 °
90 °±3 °
FR
FR
C
B
D
B
A. Stud
B. Damper mounting base
C. Damper bracket
D. Stamp
37647_PLOT_G0558

Fig. 315 Align the angle of the stud on the damper mounting base and the damper bracket so the stamp points toward the front

7. Install the damper mounting washer and a new self-locking nut.

8. Hold the damper shaft using a hex wrench and tighten the self-locking nut to the specified torque.

9. Remove the damper/spring from the strut spring compressor.

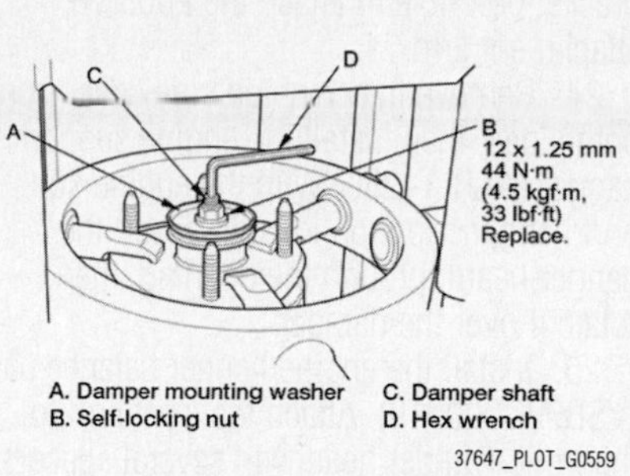
C
D
A
B
12 x 1.25 mm
44 N·m
(4.5 kgf·m,
33 lbf·ft)
Replace.
A. Damper mounting washer
B. Self-locking nut
C. Damper shaft
D. Hex wrench
37647_PLOT_G0559

Fig. 316 Install the damper mounting washer and a new self-locking nut

STABILIZER BAR

REMOVAL & INSTALLATION

See Figures 317 through 343.

Special Tools Required*:

- Ball Joint Thread Protector, 12 mm 07AAF-SDAA100
- Ball Joint Remover, 28 mm 07MAC-SL0A202
- Subframe Alignment Pin 070AG-SJAA10S
- 2008 V6 Attachment Arm SIL02C000033
- Engine Support Hanger, A and Reds AAR-T1256
- Engine Hanger Balance Bar VSB02C000019
- Subframe Adapter VSB02C000016

*Available through the Honda Tool and Equipment Program, 888-424-6857

WARNING

SRS components are located in this area. Review the SRS component locations and the precautions and procedures before doing repairs or service.

1. Note these items during replacement:
   - Be sure to remove the steering wheel before disconnecting the steering joint. Damage to the cable reel can occur.
   - Lower the front subframe from the body, and replace the front stabilizer bar through the gap created by lowering the front subframe.
2. Remove the support struts from the engine hood. Move the engine hood to a vertical position, then reinstall the support strut.
3. Remove the engine cover.
4. Do the battery terminal disconnection procedure.
5. Raise and support the vehicle.
6. Remove the front wheels.
7. Remove the driver's airbag and the steering wheel.
8. Remove the footrest.
   a. Detach the hooks, then remove the footrest plate.
   b. Remove the nut with a 6 mm hexagonal wrench, release the clip, then remove the footrest.
9. Remove the center console trim.
10. Pull back the carpet, then remove the steering joint cover.
11. Release the lock lever, and adjust the steering column to the full tilt up position, and to the full telescope in position.
12. Tighten the lock lever.
13. Hold the lower slide shaft on the column with a piece of wire between the joint yoke of the lower slide shaft and joint yoke of the upper shaft to prevent the slider shaft from pulling out.
14. Release the lock lever, and adjust the steering column to the full telescopic out position, then tighten the lock lever.
15. Remove the steering joint bolt, disconnect the steering joint by moving the steering joint toward the column.

➡If the center guide is in place and has not moved, leave it in place. If the center guide has moved or been removed, discard it.

16. Disconnect the Power Steering Pressure (PSP) switch connector).

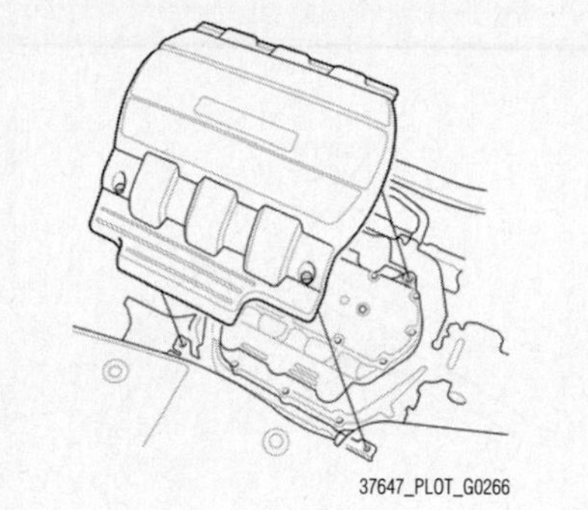
37647_PLOT_G0266

Fig. 317 Remove the engine cover

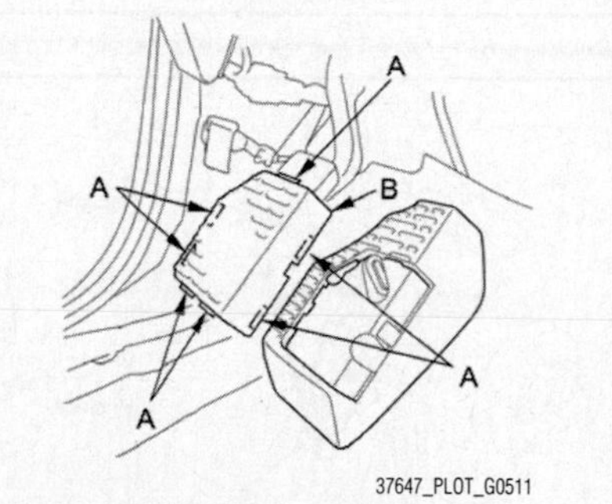
A
A
B
A
A
37647_PLOT_G0511

Fig. 318 Detach the hooks (A), then remove the footrest plate (B)

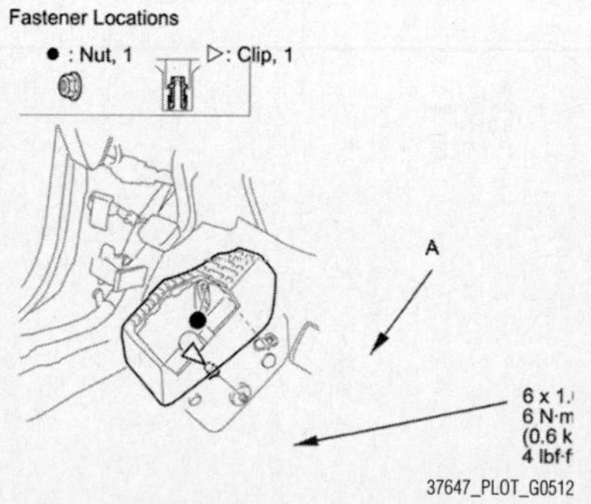
Fastener Locations
● : Nut, 1
▷: Clip, 1
A
6 x 1
6 N·m
(0.6 k
4 lbf·f
37647_PLOT_G0512

Fig. 319 Remove the nut with a 6 mm hexagonal wrench, release the clip, then remove the footrest (A)

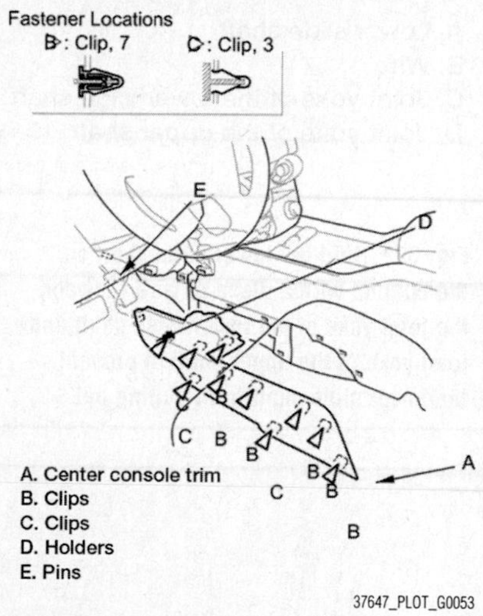
Fastener Locations
B: Clip, 7
C: Clip, 3
E
D
C
B
A
A. Center console trim
B. Clips
C. Clips
D. Holders
E. Pins
37647_PLOT_G0053

Fig. 320 Gently pull out the center console trim to detach the clips and release the holders from the pins, then remove the trim

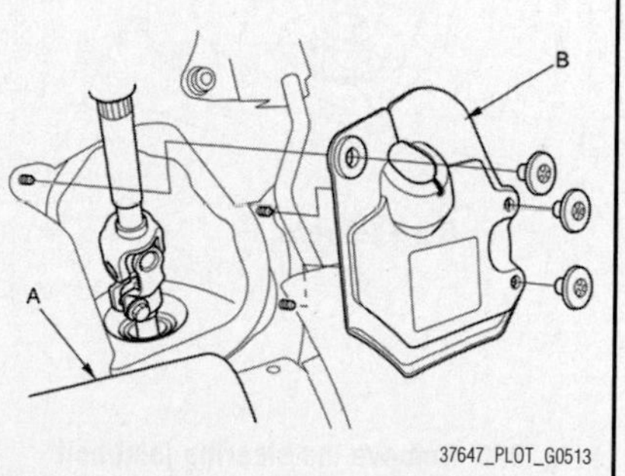
B
A
37647_PLOT_G0513

Fig. 321 Pull back the carpet (A), then remove steering joint cover (B)

17. Remove the cotter pin from the tie-rod end ball joint, then remove the nut.

➡During installation, install the new cotter pin after tightening the new nut, and bend its end.

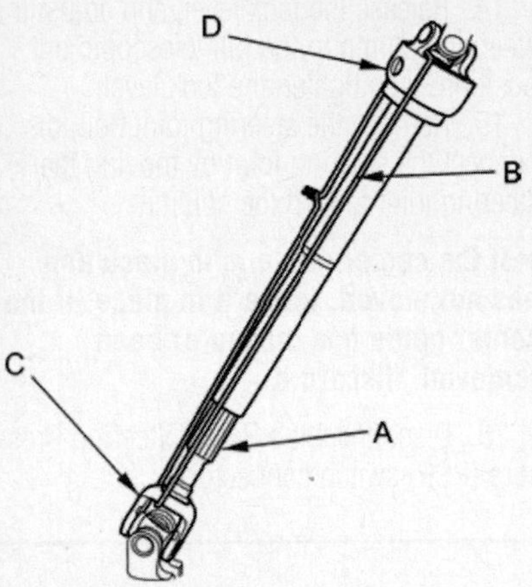
D
B
C
A

A. Lower slide shaft
B. Wire
C. Joint yoke of the lower slide shaft
D. Joint yoke of the upper shaft

37647_PLOT_G0514

Fig. 322 Hold the lower slide shaft on the column with a piece of wire between the joint yoke of the lower slide shaft and joint yoke of the upper shaft to prevent the lower slide shaft from pulling out

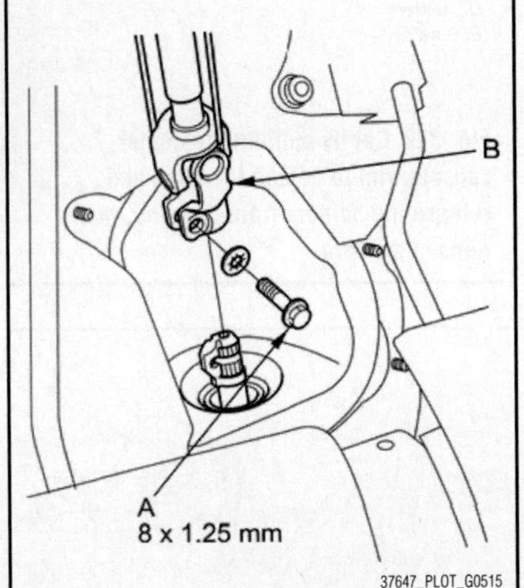
B
A
8 x 1.25 mm

37647_PLOT_G0515

Fig. 323 Remove the steering joint bolt (A), and disconnect the steering joint (B) by sliding the steering joint into the column

18. Disconnect the tie-rod end ball joint from the knuckle using the ball joint thread protector and the ball joint remover.

19. Disconnect both sides of the stabilizer link from the stabilizer bar.

20. Remove the power steering pump outlet hose mounting bolt.

21. Remove the air cleaner assembly.

22. Remove the connector bracket from the front cylinder head; use the bracket bolt hole to attach the engine hanger balance bar front arm.

23. Remove the harness clamp bracket from the engine rear cylinder head; use the bracket bolt hole to attach the 2008 V6 attachment arm.

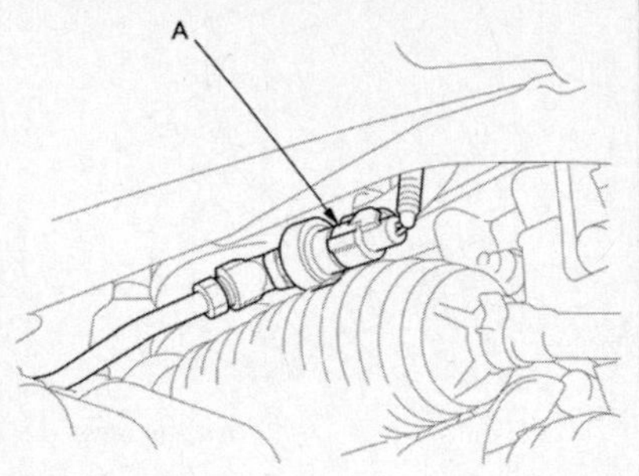
A

37647_PLOT_G0560

Fig. 324 Disconnect the PSP switch connector (A)

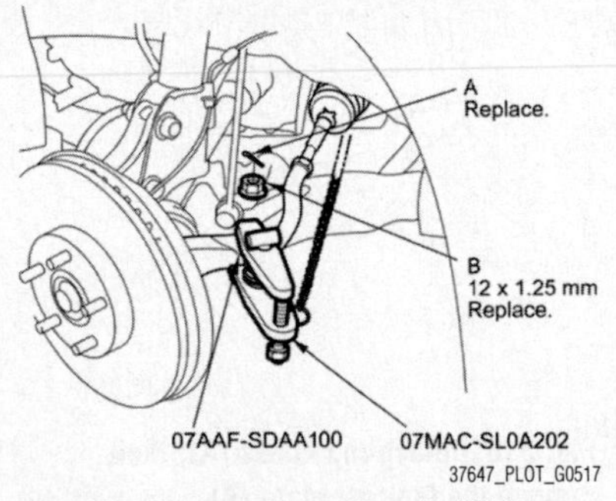
A
Replace.
B
12 x 1.25 mm
Replace.
07AAF-SDAA100
07MAC-SL0A202

37647_PLOT_G0517

Fig. 325 Remove the cotter pin (A) from the tie-rod end ball joint, then remove the nut (B)

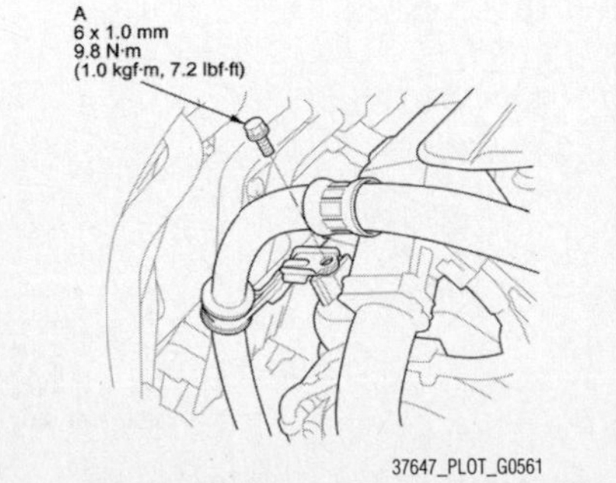
A
6 x 1.0 mm
9.8 N·m
(1.0 kgf·m, 7.2 lbf·ft)

37647_PLOT_G0561

Fig. 326 Remove the power steering pump outlet hose mounting bolt (A)

24. Remove the cowl top side cover from both sides, then install the engine support hanger (AAR-T1256).With the engine support hanger beam on its side, insert the hanger beam through the opening, then rotate it over the damper.

25. Install the engine hanger balance bar (VSB02C000019). Attach the front arm to the front cylinder head with several spacers and a 10 x 1.25 mm bolt. Remove the rear arm from the engine hanger balance bar, then install the 2008 V6 attachment arm (SIL02C000033). Attach the 2008 V6 attachment arm to the rear cylinder head with an 8 x 1.25 mm bolt.

➡Be careful not to damage the hood opener cable when installing the engine support hanger (AAR-T1256) at the front bulkhead. AAR-T1256 two sets required for stacking additional cross section bar.

26. Install the engine support hanger onto the vehicle, and attach the hook to the engine balancer bar slot. Tighten the wing nut by hand, and lift and support the engine.

27. Raise the vehicle.

28. Remove the splash shield.

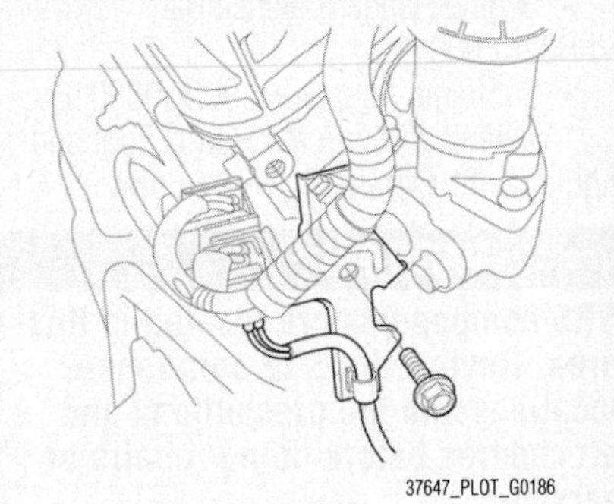

37647_PLOT_G0186

Fig. 327 Remove the connector bracket from the engine front cylinder head

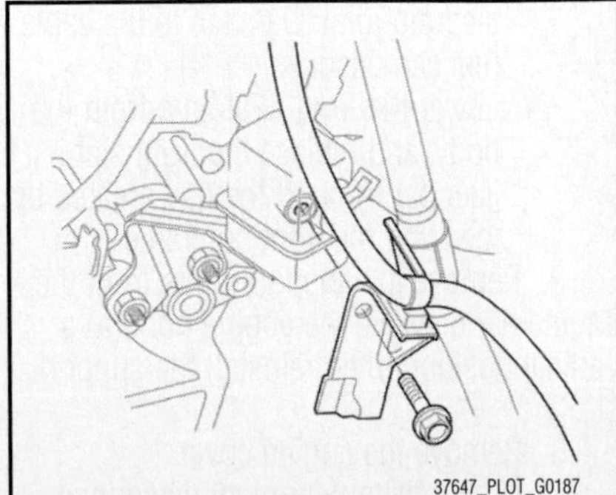

37647_PLOT_G0187

Fig. 328 Remove the harness clamp bracket from the engine rear cylinder head

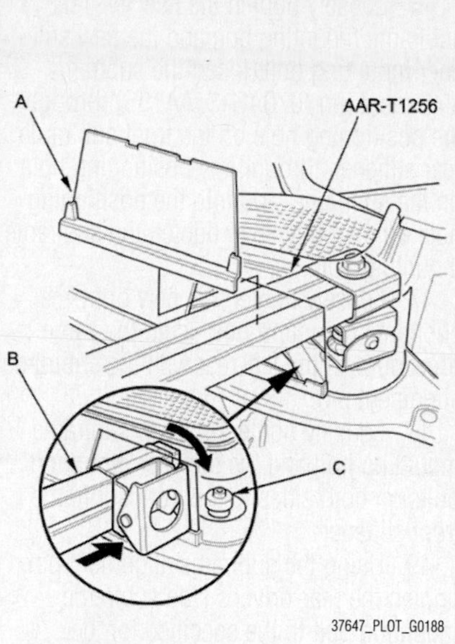
A
AAR-T1256
B
C
37647_PLOT_G0188

Fig. 329 Remove the cowl top side cover (A) from both sides, then install the engine support hanger; with the engine support hanger beam on its side, insert the hanger beam (B) through the opening, then rotate it over the damper (C)

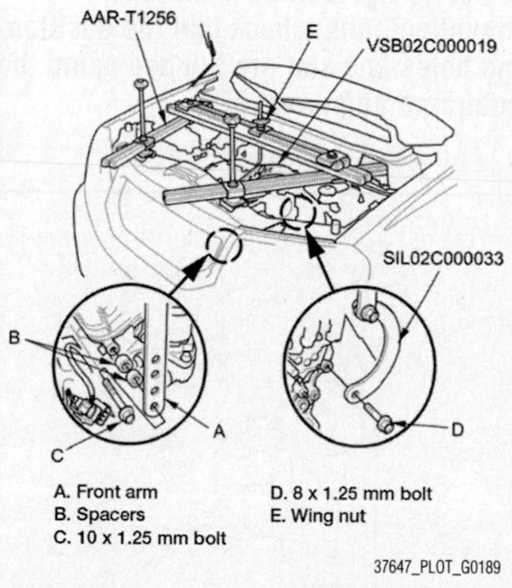
AAR-T1256
E
VSB02C000019
SIL02C000033
B
A
C
D
A. Front arm
B. Spacers
C. 10 x 1.25 mm bolt
D. 8 x 1.25 mm bolt
E. Wing nut
37647_PLOT_G0189

Fig. 330 Install the engine hanger balance bar and the engine support hanger

29. Remove the front subframe stiffener.
30. 4WD model: Remove the propeller shaft and propeller shaft protectors.
31. Disconnect the front engine mount actuator connector and the harness clamps. Remove the front engine mount stop, then remove the front engine mount bolt.
32. Remove the rear engine mount bolts and disconnect the rear engine mount actuator connector.

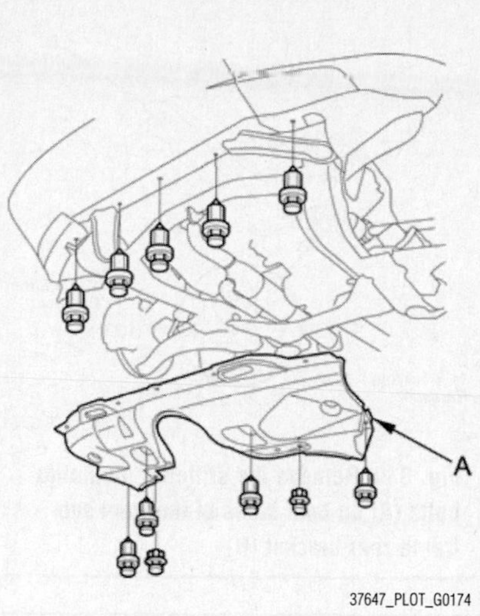
A
37647_PLOT_G0174

Fig. 331 Remove the splash shield

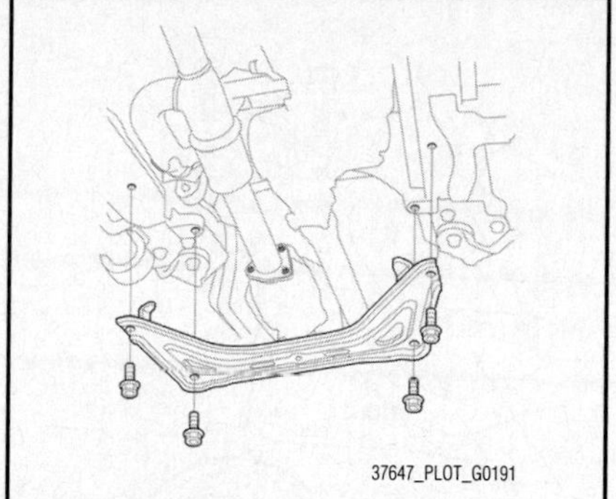
37647_PLOT_G0191

Fig. 332 Remove the front subframe stiffener (A)

33. Remove the rear engine mount from the rear engine mount base bracket, and move it aside.
34. Remove the ground cable bolt from the transmission.
35. Remove the lower transmission mount bolts.
36. Attach the front subframe adapter (VSB02C000016) to the subframe by looping the belt over the front of the subframe, then secure the belt with its stop.
37. Raise the jack and line up the slots in the front subframe adapter arms with the bolt holes on the jack base, then securely attach then with four bolts.
38. Remove the stiffener mounting bolts on both sides of the front subframe front stiffeners.
39. Loosen the front side subframe mounting bolts until there is 1.2 inches (30 mm) distance between the bolt seat and the mounting surface. Do not loosen the mounting bolts more than necessary.
40. Remove the stiffener mounting bolts on both sides of the front subframe rear bracket.
41. Loosen the rear side subframe mounting bolts until there is 1.2 inches (30 mm) distance between the bolt seat and the mounting surface. Do not loosen the mounting bolts more than necessary.

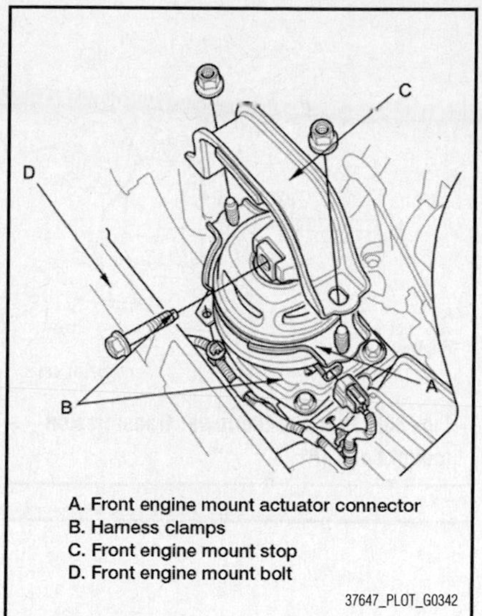
C
D
B
A
A. Front engine mount actuator connector
B. Harness clamps
C. Front engine mount stop
D. Front engine mount bolt
37647_PLOT_G0342

Fig. 333 Disconnect the front engine mount actuator connector and the harness clamps

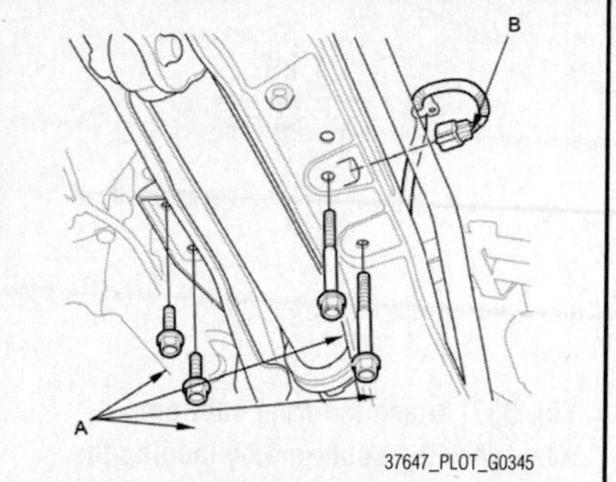
B
A
37647_PLOT_G0345

Fig. 334 Remove the rear engine mount bolts (A) and disconnect the rear engine mount actuator connector (B)

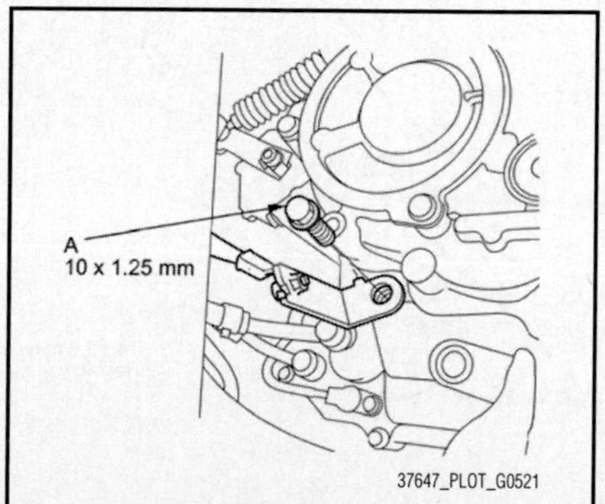
A
10 x 1.25 mm
37647_PLOT_G0521

Fig. 335 Remove the ground cable bolt (A) from the transmission

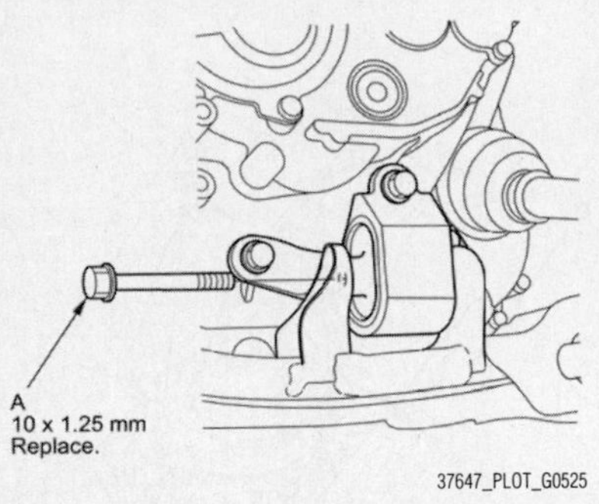
A
10 x 1.25 mm
Replace.
37647_PLOT_G0525

Fig. 336 Remove the lower transmission mount bolts (A)

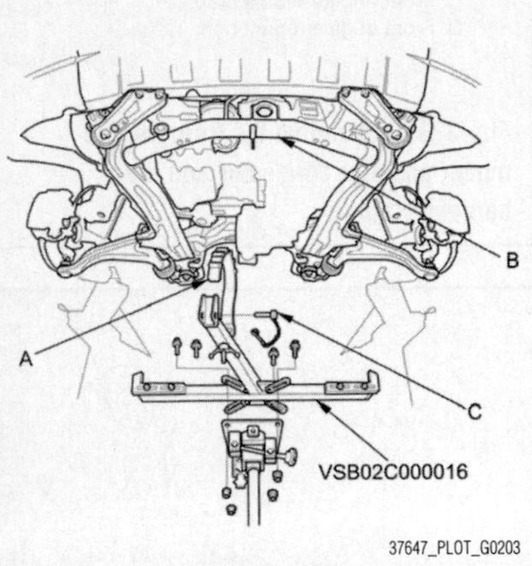
A
B
C
VSB02C000016
37647_PLOT_G0203

Fig. 337 Attach the front subframe adapter to the subframe by looping the belt (A) over the front of the subframe (B), then secure the belt with its stop (C)

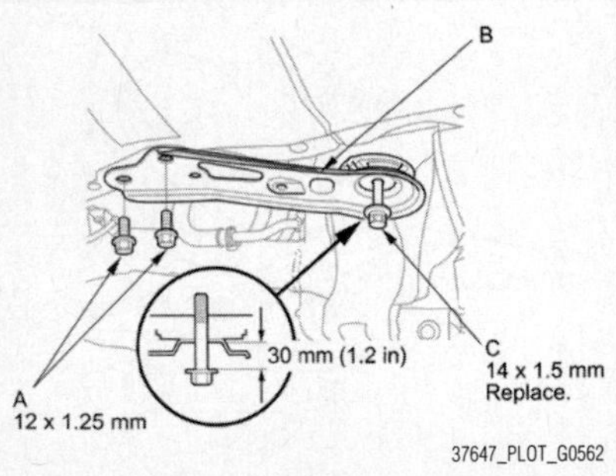
B
30 mm (1.2 in)
C
14 x 1.5 mm
Replace.
A
12 x 1.25 mm
37647_PLOT_G0562

Fig. 338 Remove the stiffener mounting bolts (A) from the subframe front stiffeners (B)

42. Lower the transmission jack slowly with the front subframe adapter until the front subframe has dropped about 1.2 inches (30 mm).

➡**Do not try to lower the front subframe beyond the loosened subframe mounting bolts clearance.**

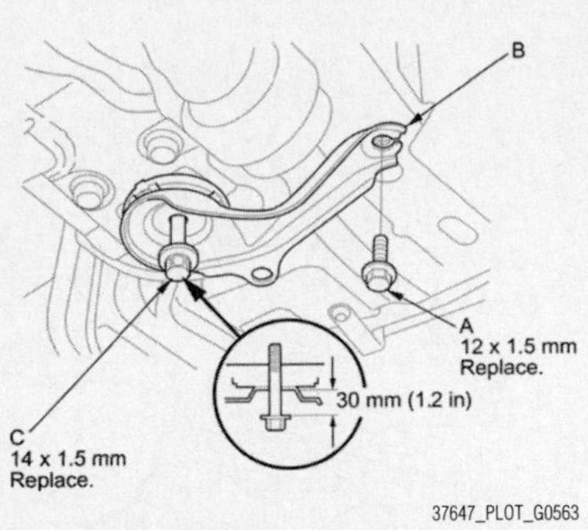
B
A
12 x 1.5 mm
Replace.
30 mm (1.2 in)
C
14 x 1.5 mm
Replace.
37647_PLOT_G0563

Fig. 339 Remove the stiffener mounting bolts (A) on both sides of the front subframe rear bracket (B)

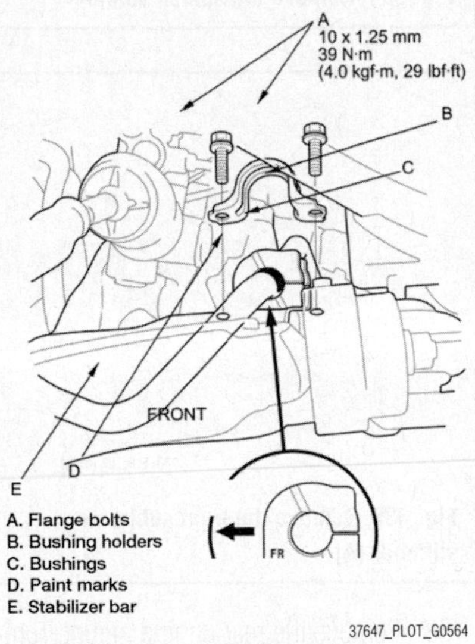
A
10 x 1.25 mm
39 N·m
(4.0 kgf·m, 29 lbf·ft)
B
C
FRONT
D
E
A. Flange bolts
B. Bushing holders
C. Bushings
D. Paint marks
E. Stabilizer bar
FR
37647_PLOT_G0564

Fig. 340 Remove the flange bolts and the bushing holders, then remove the bushings

43. Remove the flange bolts and the bushing holders, then remove the bushings.

➡**During installation, align the paint marks on the stabilizer bar with the side of the bushings.**

44. Move the stabilizer bar toward the passenger's side, and remove the stabilizer bar.

To install:

45. Install the stabilizer bar.

➡**Note the right and left direction of the stabilizer bar. Note the direction of installation for the bushings.**

46. Loosely tighten the new left-rear subframe mounting bolt and the new stiffener mounting bolt; insert the subframe alignment pin (070AG-SJAA10S) through the positioning hole on the front subframe rear stiffener, through the positioning hole on the subframe, and into the positioning hole on the body, then tighten the subframe mounting bolt.

47. Loosely tighten the new right-rear subframe mounting bolt using the same procedure as the left-rear with the subframe alignment pin.

48. Loosely tighten the new subframe mounting bolt and the stiffener mounting bolts on both sides to the front subframe front stiffener.

49. Using the subframe alignment pin, tighten the rear-driver's side subframe mounting bolt to the specified torque.

50. Using the subframe alignment pin, tighten the rear-passenger's side subframe mounting bolt to the specified torque.

51. Tighten the front subframe mounting bolts to the specified torque.

➡**Check all of the subframe mounting bolts, and retighten if necessary.**

52. Tighten the front and rear stiffener mounting bolts to the specified torque.

➡**Before tightening the stiffener mounting bolts, check that the positioning holes and slot are aligned using the subframe alignment pin.**

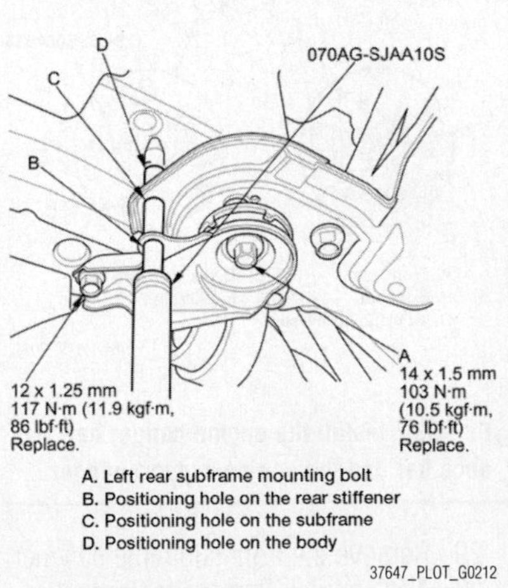
D
070AG-SJAA10S
C
B
A
14 x 1.5 mm
103 N·m
(10.5 kgf·m,
76 lbf·ft)
Replace.
12 x 1.25 mm
117 N·m (11.9 kgf·m,
86 lbf·ft)
Replace.
A. Left rear subframe mounting bolt
B. Positioning hole on the rear stiffener
C. Positioning hole on the subframe
D. Positioning hole on the body
37647_PLOT_G0212

Fig. 341 Loosely tighten the left rear subframe mounting bolt; insert the 15.7 mm side of the subframe alignment pin through the positioning hole on the rear stiffener, through the positioning hole on the subframe, and into the positioning hole on the body

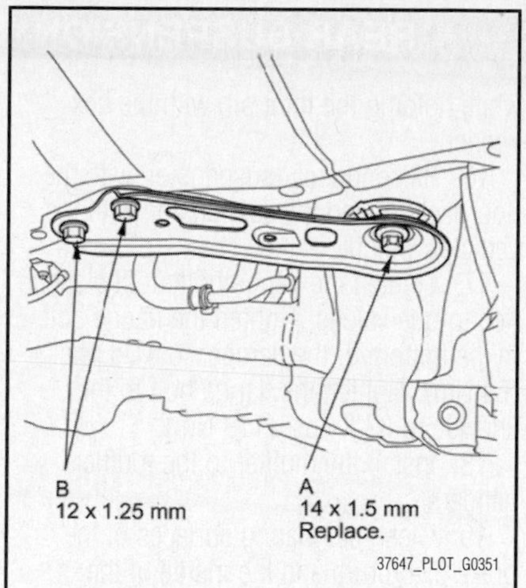
B
12 x 1.25 mm
A
14 x 1.5 mm
Replace.
37647_PLOT_G0351

Fig. 342 Loosely install the right and left front subframe mounting bolts (A) and stiffener mounting bolts (B)

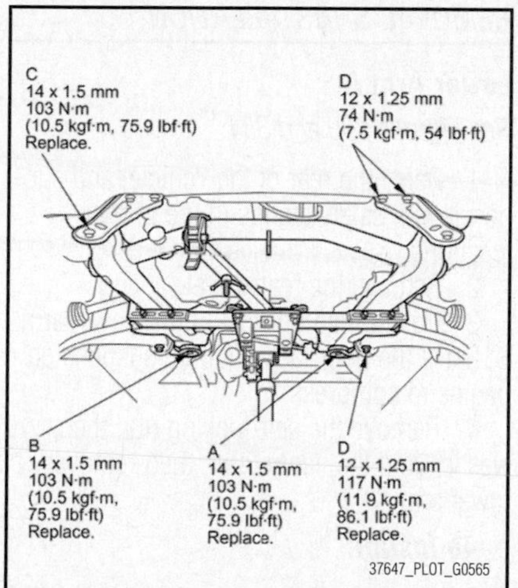
C
14 x 1.5 mm
103 N·m
(10.5 kgf·m, 75.9 lbf·ft)
Replace.
D
12 x 1.25 mm
74 N·m
(7.5 kgf·m, 54 lbf·ft)
B
14 x 1.5 mm
103 N·m
(10.5 kgf·m,
75.9 lbf·ft)
Replace.
A
14 x 1.5 mm
103 N·m
(10.5 kgf·m,
75.9 lbf·ft)
Replace.
D
12 x 1.25 mm
117 N·m
(11.9 kgf·m,
86.1 lbf·ft)
Replace.
37647_PLOT_G0565

Fig. 343 Torque specification for the subframe mounting bolts and the stiffener mounting bolts

53. Install all of the removed parts in the reverse order of removal, and note these items:
   - Refer to stabilizer link removal/installation to connect the stabilizer bar to the links.
   - If the center guide is in place, use it to determine the steering joint installation angle.
   - If the center guide is gone, check the steering joint installation angle.
   - Check the steering wheel installation.
   - When connecting the front and rear engine mount, first lightly tighten the mounting bolts, supporting the engine/transmission with a floor jack, then remove the jack, and tighten the bolts to the specified torque.
   - Before installing the wheel, clean the mating surfaces on the brake disc and inside of the wheel.

54. Do the battery terminal reconnection procedure, then turn the ignition switch to ON (II) and check that the SRS indicator should come on for about 6 seconds and then go off.

55. Check the wheel alignment, and adjust it if necessary.

WHEEL HUB & BEARING (SEALED UNIT)

REMOVAL & INSTALLATION

See Figures 344 and 345.

Special Tools Required:
- Ball Joint Thread Protector, 12 mm 07AAF-SDAA100
- Ball Joint Remover, 32 mm 07MAC-SL0A102
- Ball Joint Remover, 28 mm 07MAC-SL0A202
- Ball Joint Thread Protector, 14 mm 071AF-S3VA000

1. Raise and support the vehicle.
2. Remove the front wheel.
3. Remove the brake hose mounting bolt from the damper.
4. Remove the brake caliper bracket mounting bolts, then remove the caliper assembly from the knuckle. To prevent damage to the caliper assembly or the brake hose, use a short piece of wire to hang the caliper assembly from the undercarriage. Do not twist the brake hose excessively.
5. Pry up the stake on the spindle nut, then remove the nut.
6. Remove the front brake disc.
7. Remove the flange bolts, and remove the hub bearing unit by tapping the driveshaft end with a soft face hammer while drawing the hub bearing unit outward, then remove the splash guard.

➡Do not pull the driveshaft end outward. The driveshaft inboard joint may come apart. During installation, apply grease to the mating surfaces of the hub bearing unit and the driveshaft outboard joint.

8. Check the hub bearing unit for damage and cracks.

To install:

9. Install the hub bearing unit in the reverse order of removal, and note these items:
   - Use a new spindle nut on reassembly.
   - Before installing the spindle nut, apply a small amount of engine oil to the seating surface of the nut.

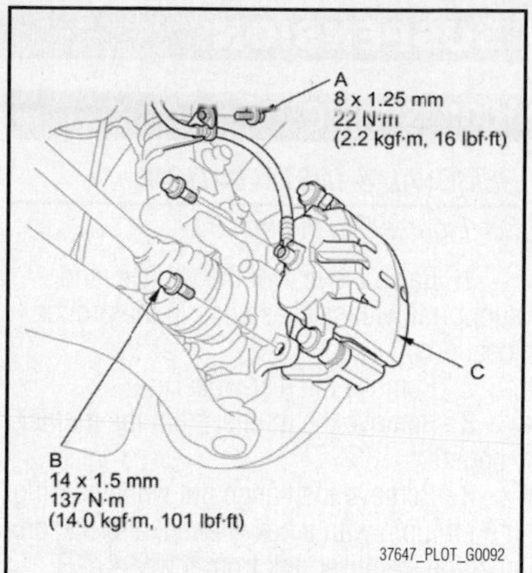
A
8 x 1.25 mm
22 N·m
(2.2 kgf·m, 16 lbf·ft)
C
B
14 x 1.5 mm
137 N·m
(14.0 kgf·m, 101 lbf·ft)
37647_PLOT_G0092

Fig. 344 Remove the brake hose mounting bolt (A), the brake caliper bracket mounting bolts (B), then remove the caliper assembly (C)

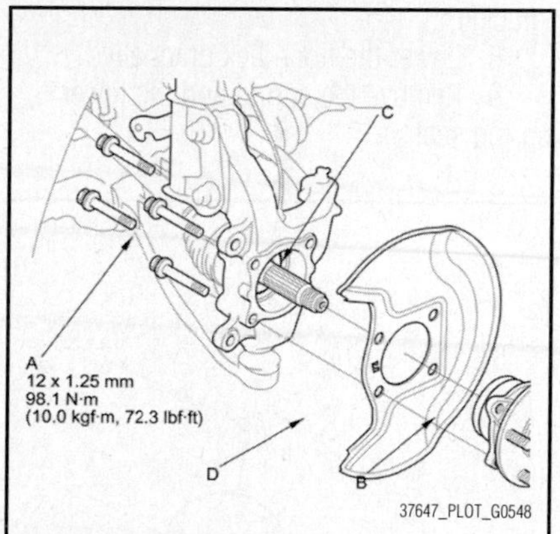
C
A
12 x 1.25 mm
98.1 N·m
(10.0 kgf·m, 72.3 lbf·ft)
D
B
37647_PLOT_G0548

Fig. 345 Remove the flange bolts, and remove the hub bearing unit by tapping the driveshaft end with a soft face hammer while drawing the hub bearing unit outward, then remove the splash guard

   After tightening, use a drift to stake the spindle nut shoulder against the driveshaft.
   - Before installing the brake disc, clean the mating surfaces on the hub bearing unit and the inside of the brake disc.
   - Before installing the wheel, clean the mating surfaces on the brake disc and the inside of the wheel.

10. Check the wheel alignment, and adjust it if necessary.

ADJUSTMENT

The hub bearing cannot be adjusted. If there is need for adjustment, replace the hub bearing.

COIL SPRING

REMOVAL & INSTALLATION

See Figures 346 through 349.

1. Raise the rear of the vehicle, and support it with safety stands in the proper locations.
2. Remove the rear wheel.
3. Remove the muffler from the muffler hangers.
4. Remove the flange nut while holding the joint pin with a hex wrench, and disconnect the stabilizer link from lower arm B.
5. Position a floor jack under lower arm B. Raise the floor jack until the suspension begins to compress.
6. Remove the flange bolt from the bottom of the damper.
7. Remove the flange bolt from the knuckle.
8. Lower the floor jack gradually.
9. Remove the spring and the lower spring seat.

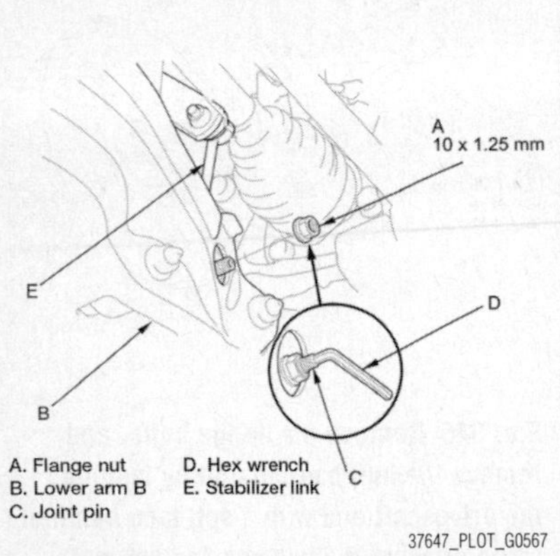
A
10 x 1.25 mm
E
D
B
C
A. Flange nut
B. Lower arm B
C. Joint pin
D. Hex wrench
E. Stabilizer link
37647_PLOT_G0567

Fig. 346 Remove the flange nut while holding the joint pin with a hex wrench, and disconnect the stabilizer link from lower arm B

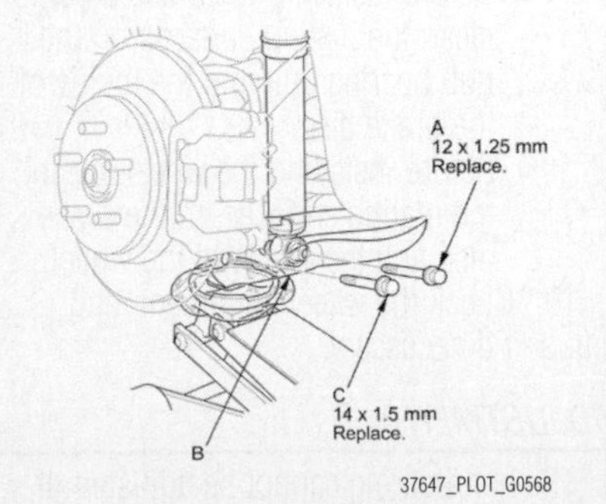
A
12 x 1.25 mm
Replace.
C
14 x 1.5 mm
Replace.
B
37647_PLOT_G0568

Fig. 347 Position a floor jack under lower arm B, remove the flange bolts (A, C)

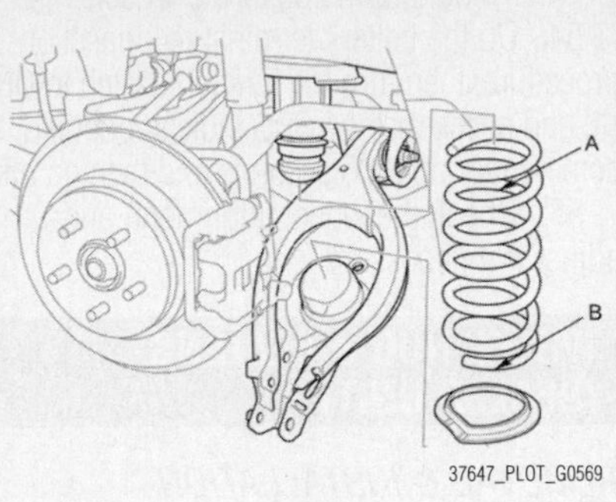
A
B
37647_PLOT_G0569

Fig. 348 Remove the spring (A) and the lower spring seat (B)

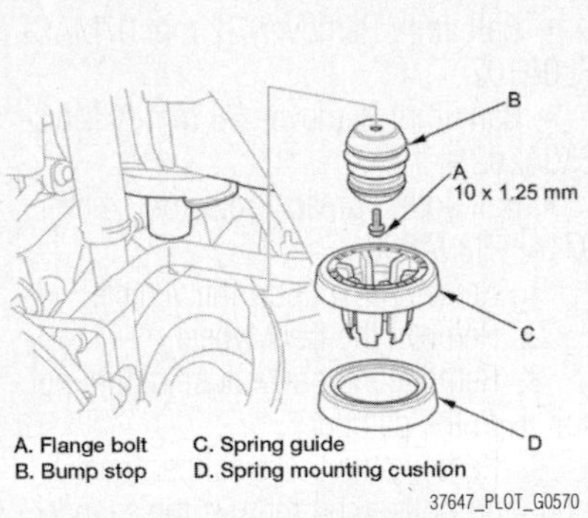
B
A
10 x 1.25 mm
C
D
A. Flange bolt
B. Bump stop
C. Spring guide
D. Spring mounting cushion
37647_PLOT_G0570

Fig. 349 Remove the flange bolt that connects the body, and remove the bump stop, the spring guide, and the spring mounting cushion if necessary

10. Remove the flange bolt that connects the body, and remove the bump stop, the spring guide, and the spring mounting cushion if necessary.

To install:

11. Install the bump stop, the spring guide, and the spring mounting cushion, then tighten the flange bolt to 43 ft. lbs. (59 Nm) if removed.
12. Install the spring and the lower spring seat. Align the bottom of the spring with the stepped part of the lower spring seat and lower arm B.
13. Position the floor jack under lower arm B. Raise the floor jack until the hole in lower arm B aligns with the hole in the damper, then loosely install the new flange bolt to the bottom of the damper.
14. Loosely install the new flange bolt to the knuckle.
15. Connect the stabilizer link to lower arm B, and install the new flange nut, and tighten the flange nut to 36 ft. lbs. (49 Nm), while holding the joint pin with the hex wrench.
16. Raise the rear suspension with the floor jack to load the suspension with the vehicle's weight.
17. Tighten the flange bolts to the specified torque values. Tighten the flange bolt in the bottom of the damper to 47 ft. lbs. (64 Nm). Tighten the flange bolt to the knuckle to 83 ft. lbs. (113 Nm).
18. Install the muffler to the muffler hangers.
19. Clean the mating surfaces of the brake disc/drum and the inside of the wheel, then install the rear wheel.
20. Check the wheel alignment, and adjust it if necessary.

CONTROL ARMS/LINKS

REMOVAL & INSTALLATION

Lower Arm A

See Figures 350 and 351.

1. Raise the rear of the vehicle, and support it with safety stands in the proper locations.
2. Remove the rear wheel.
3. Position a floor jack under lower arm B. Raise the floor jack until the suspension begins to compress.
4. Remove the self-locking nut, the washer, and the flange bolt, then remove lower arm A.

To install:

➡During installation, position the paint mark on lower arm A toward outside of the vehicle. Use a new self-locking nut and the new flange bolt, during assembly.

5. Install lower arm A in the reverse order of removal, and note these items:

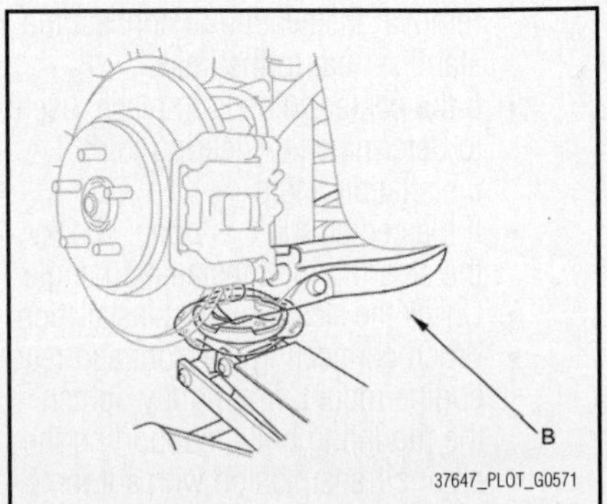
B
37647_PLOT_G0571

Fig. 350 Position a floor jack under lower arm B

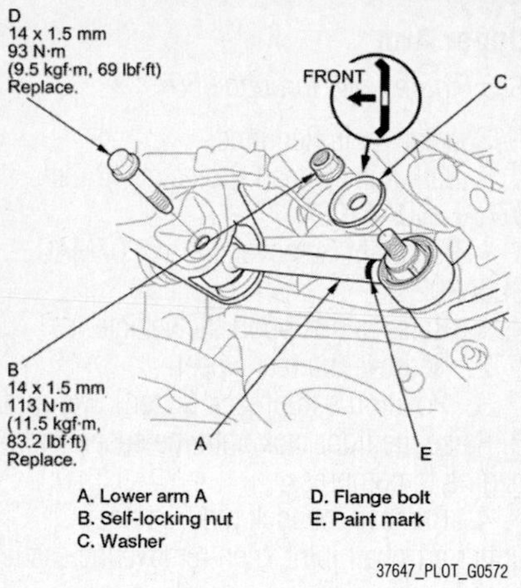
D
14 x 1.5 mm
93 N·m
(9.5 kgf·m, 69 lbf·ft)
Replace.
FRONT
C
B
14 x 1.5 mm
113 N·m
(11.5 kgf·m,
83.2 lbf·ft)
Replace.
A
E
A. Lower arm A
B. Self-locking nut
C. Washer
D. Flange bolt
E. Paint mark
37647_PLOT_G0572

Fig. 351 Remove the self-locking nut, the washer, and the flange bolt, then remove lower arm A

- First install all of the components, and lightly tighten the bolts and the nuts, then raise the suspension to load it with the vehicle's weight before fully tightening to the specified torque values.
- Before installing the wheel, clean the mating surfaces of the brake disc/drum and the inside of the wheel.

6. Check the wheel alignment, and adjust it if necessary.

Lower Arm B

See Figures 346 through 348, and 352.

1. Raise the rear of the vehicle, and support it with safety stands in the proper locations.
2. Remove the rear wheel.
3. Remove the muffler from the muffler hangers.
4. Remove the flange nut while holding the joint pin with a hex wrench, and disconnect the stabilizer link from lower arm B.
5. Position a floor jack under lower arm B. Raise the floor jack until the suspension begins to compress.
6. Remove the flange bolt from the bottom of the damper.
7. Remove the flange bolt from the knuckle.
8. Lower the floor jack gradually.
9. Remove the spring and the lower spring seat.
10. Mark the cam positions of the adjusting bolt and the adjusting cam plate with the frame.

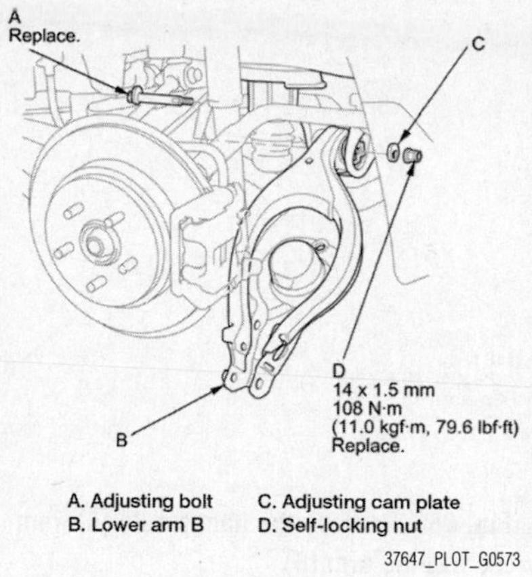
A
Replace.
C
D
14 x 1.5 mm
108 N·m
(11.0 kgf·m, 79.6 lbf·ft)
Replace.
B
A. Adjusting bolt
B. Lower arm B
C. Adjusting cam plate
D. Self-locking nut
37647_PLOT_G0573

Fig. 352 Mark the cam positions of the adjusting bolt and the adjusting cam plate with the frame

11. Remove the self-locking nut while holding the adjusting bolt, then remove the adjusting cam plate, the adjusting bolt, and lower arm B.

➡Use a new adjusting bolt and a new self- locking nut during reassembly.

To install:

12. Install lower arm B in the reverse order of removal, and note these items:
    - First install all of the components, and lightly tighten the bolts and nuts, then raise the suspension to load it with the vehicle's weight before fully tightening to the specified torque.
    - Align the cam positions of the adjusting bolt and the adjusting cam plate with the marked positions on the frame when tightening the self-locking nut.
    - Before installing the wheel, clean the mating surfaces on the brake disc/drum and the inside of the wheel.
13. Check the wheel alignment, and adjust it if necessary.

Stabilizer Link

See Figure 353.

1. Raise the rear of the vehicle, and support it with safety stands in the proper locations.
2. Remove the rear wheel.
3. Remove the self-locking nut and the flange nut while holding the respective joint pin with a hex wrench, then remove the stabilizer link.

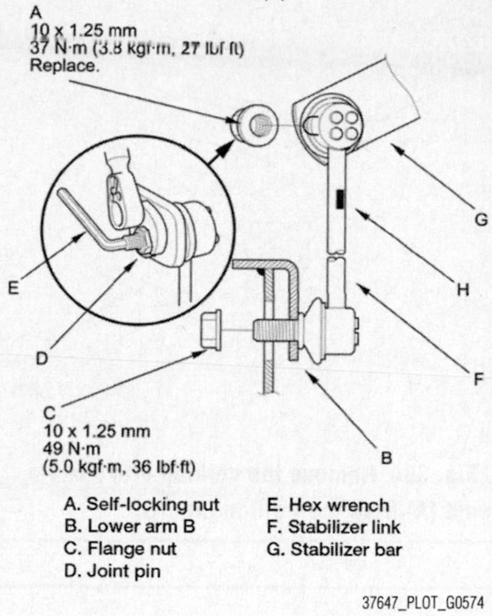
A
10 x 1.25 mm
37 N·m (3.8 kgf·m, 27 lbf·ft)
Replace.
G
E
H
D
F
C
10 x 1.25 mm
49 N·m
(5.0 kgf·m, 36 lbf·ft)
B
A. Self-locking nut
B. Lower arm B
C. Flange nut
D. Joint pin
E. Hex wrench
F. Stabilizer link
G. Stabilizer bar
37647_PLOT_G0574

Fig. 353 Remove the self-locking nut and the flange nut while holding the respective joint pin with a hex wrench, then remove the stabilizer link

To install:

4. Install the stabilizer link on the stabilizer bar and lower arm B with the joint pins set at the center of their range of movement.

➡The stabilizer link has a paint mark. The left stabilizer link is marked with yellow paint, and the right stabilizer link is marked with white paint. Install the end of the stabilizer link with the paint mark in the upper position.

5. Install a new self-locking nut and the new flange nut, and tighten them to the specified torque values while holding the respective joint pin with a hex wrench.
6. Clean the mating surfaces of the brake disc/drum and the inside of the wheel, then install the rear wheel.
7. Test-drive the vehicle.
8. After 5 minutes of driving, tighten the self-locking nut again to the specified torque value.

Trailing Arm

See Figures 354 through 357.

1. Raise the rear of the vehicle, and support it with safety stands in the proper locations.
2. Remove the rear wheel.
3. Remove the parking brake cable from the backing plate.
4. Remove the parking brake cable nut from the trailing arm.
5. Disconnect the brake line from the

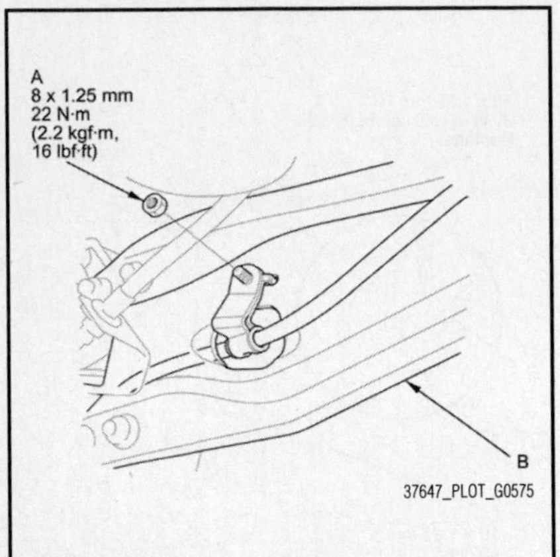
A
8 x 1.25 mm
22 N·m
(2.2 kgf·m,
16 lbf·ft)
B
37647_PLOT_G0575

Fig. 354 Remove the parking brake cable nut (A) from the trailing arm (B)

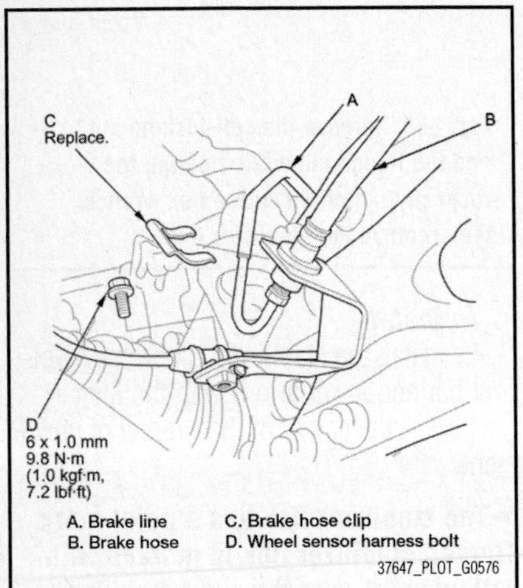
A
B
C
Replace.
D
6 x 1.0 mm
9.8 N·m
(1.0 kgf·m,
7.2 lbf·ft)
A. Brake line
B. Brake hose
C. Brake hose clip
D. Wheel sensor harness bolt
37647_PLOT_G0576

Fig. 355 Disconnect the brake line from the brake hose, then remove the brake hose clip

brake hose, then remove the brake hose clip.

➡Plug the end of a hose and joint to prevent spilling brake fluid. Use the new brake hose clip during reassembly.

6. Remove the wheel sensor harness bolt.

7. Remove the flange nuts from the trailing arm.

➡Use new flange nuts during reassembly.

8. Remove the flange bolts from the trailing arm, then remove the trailing arm.

➡Use new flange bolts during reassembly.

To install:

9. Install the trailing arm in the reverse order of removal, and note these items:

- First install all of the components, and lightly tighten the bolts and the nuts, then raise the suspension to load it with the vehicle's weight before fully tightening to the specified torque values.
- Do the parking brake adjustment.
- Fill the master cylinder reservoir to the MAX (upper) level line, and bleed the brake system. Check for a leak at the brake hose/line joint, and retighten it if necessary.
- Before installing the wheel, clean the mating surfaces of the brake disc/drum and the inside of the wheel.

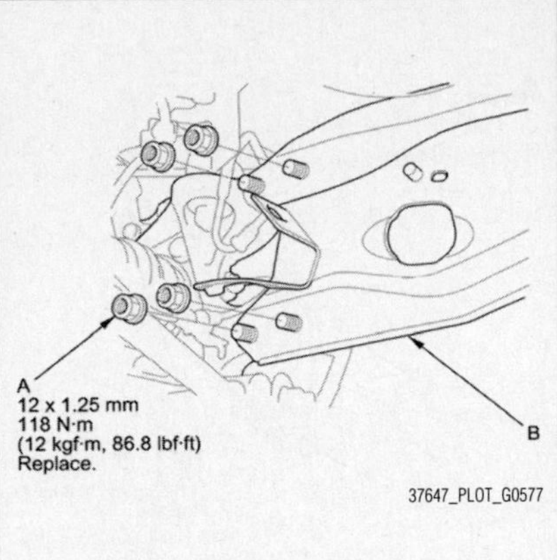
A
12 x 1.25 mm
118 N·m
(12 kgf·m, 86.8 lbf·ft)
Replace.
B
37647_PLOT_G0577

Fig. 356 Remove the flange nuts (A) from the trailing arm (B)

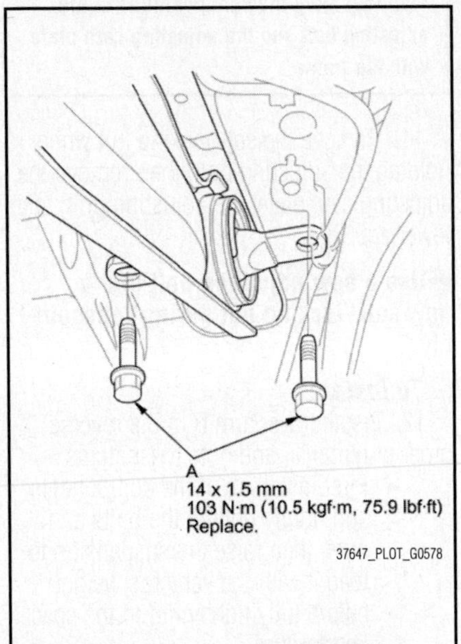
A
14 x 1.5 mm
103 N·m (10.5 kgf·m, 75.9 lbf·ft)
Replace.
37647_PLOT_G0578

Fig. 357 Remove the flange bolts (A) from the trailing arm, then remove the trailing arm

10. Check the wheel alignment, and adjust it if necessary.

Upper Arm

See Figures 358 through 360.

Special Tools Required:

- Ball Joint Thread Protector, 12 mm 07AAF-SDAA100
- Ball Joint Remover, 28 mm 07MAC-SL0A202

1. Raise and support the vehicle.
2. Remove the rear wheel.
3. Position a floor jack under lower arm B. Raise the floor jack until the suspension begins to compress.
4. Remove the lock pin from the upper arm ball joint, then remove the castle nut.

➡During installation, install the lock pin as shown after tightening the castle nut.

5. Disconnect the upper arm ball joint from the knuckle using the ball

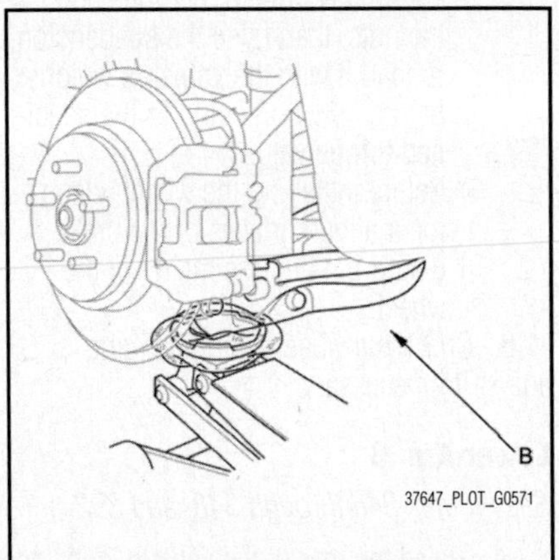
B
37647_PLOT_G0571

Fig. 358 Position a floor jack under lower arm B

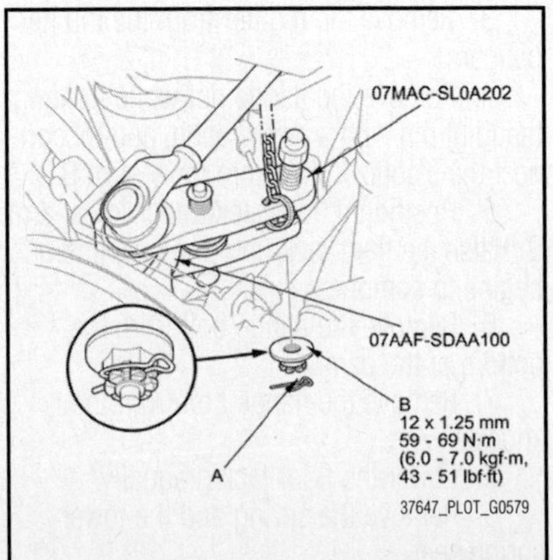
07MAC-SL0A202
07AAF-SDAA100
B
12 x 1.25 mm
59 - 69 N·m
(6.0 - 7.0 kgf·m,
43 - 51 lbf·ft)
A
37647_PLOT_G0579

Fig. 359 Remove the lock pin (A) from the upper arm ball joint, then remove the castle nut (B)

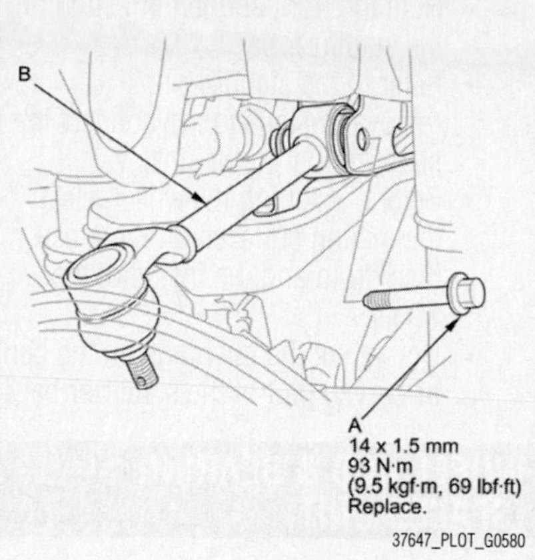
B
A
14 x 1.5 mm
93 N·m
(9.5 kgf·m, 69 lbf·ft)
Replace.
37647_PLOT_G0580

Fig. 360 Remove the upper arm mounting bolt (A), and remove the upper arm (B)

joint thread protector and the ball joint remover.

➡Be careful not to damage the ball joint boot when installing the remover.

6. Remove the upper arm mounting bolt and remove the upper arm.

➡Use the new mounting bolt during reassembly.

To install:

7. Install the upper arm in the reverse order of removal, and note these items:

- First install all of the components, and lightly tighten the bolts and the nuts, then raise the suspension to load it with the vehicle's weight before fully tightening to the specified torque.
- Be careful not to damage the ball joint boot when connecting the knuckle.
- Before connecting the ball joint to the knuckle, degrease the threaded section and the tapered portion of the ball joint pin, the ball joint connecting hole, the threaded section and the mating surfaces of the castle nut.
- Torque the castle nut to the lower torque specification, then tighten it only far enough to align the slot with the ball joint pin hole. Do not align the castle nut by loosening it.
- Before installing the wheel, clean the mating surfaces on the brake disc/drum and the inside of the wheel.

8. Check the wheel alignment, and adjust it if necessary.

SHOCK ABSORBER (DAMPER)

REMOVAL & INSTALLATION

See Figures 361 through 363.

1. Raise the rear of the vehicle, and support it with safety stands in the proper locations.
2. Remove the rear wheel.
3. Remove the flange nut while holding the joint pin with a hex wrench, and disconnect the stabilizer link from lower arm B.
4. Position a floor jack under lower arm B. Raise the floor jack until the suspension begins to compress.
5. Remove the flange bolt from the bottom of the damper.
6. Remove the flange bolt and the nut from the top of the damper.

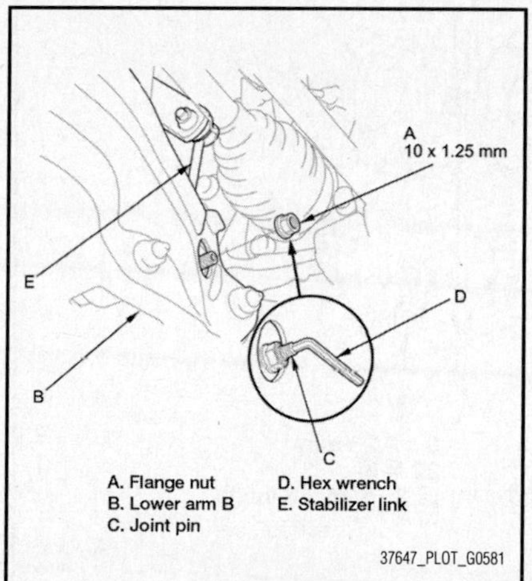
A
10 x 1.25 mm
E
D
B
C
A. Flange nut
B. Lower arm B
C. Joint pin
D. Hex wrench
E. Stabilizer link
37647_PLOT_G0581

Fig. 361 Remove the flange nut while holding the joint pin with a hex wrench, and disconnect the stabilizer link from lower arm B

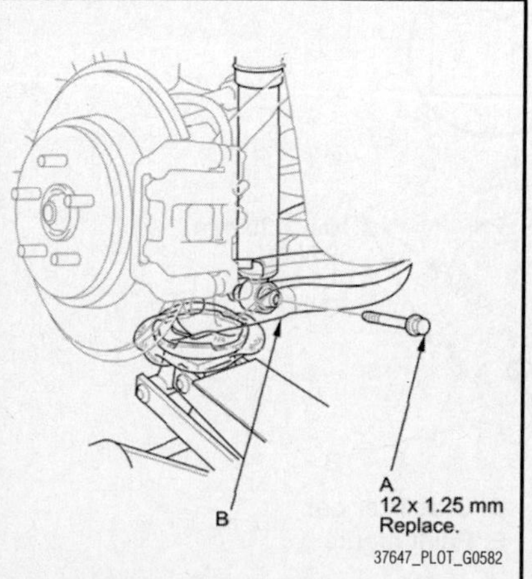
A
12 x 1.25 mm
Replace.
B
37647_PLOT_G0582

Fig. 362 Position a floor jack under lower arm B; remove the flange bolt (A) from the bottom of the damper

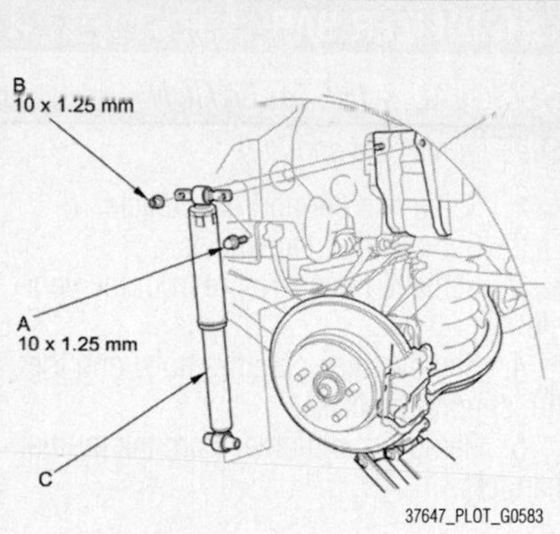
B
10 x 1.25 mm
A
10 x 1.25 mm
C
37647_PLOT_G0583

Fig. 363 Remove the flange bolt (A) and the nut (B) from the top of the damper; compress the damper unit (C)

7. Compress the damper unit by hand, and remove it from the vehicle.

To install:

8. Position the damper unit between the body and lower arm B.
9. Loosely install the new flange bolt and the nut to the top of the damper.
10. Raise the floor jack until the hole in lower arm B aligns with the hole in the damper, then loosely install the new flange bolt to the bottom of the damper.
11. Connect the stabilizer link to lower arm B, and install the new flange nut, and tighten the flange nut to 36 ft. lbs. (49 Nm), while holding the joint pin with the hex wrench.
12. Raise the rear suspension with the floor jack to load the suspension with the vehicle's weight.
13. Tighten the flange bolt and the nut on top of the damper to 33 ft. lbs. (44 Nm).
14. Tighten the flange bolt on bottom of the damper to 47 ft. lbs. (64 Nm).
15. Clean the mating surfaces of the brake disc/drum and the inside of the wheel, then install the rear wheel.
16. Check the wheel alignment, and adjust it if necessary.

TESTING

1. Compress the damper assembly by hand, and check for smooth operation through a full stroke, both compression and extension. The damper should extend smoothly and constantly when compression is released. If it does not, the gas is leaking and the damper should be replaced.
2. Check for oil leaks, abnormal noises, and binding during these tests.

STABILIZER BAR

REMOVAL & INSTALLATION

See Figures 364 and 365.

1. Raise and support the vehicle.
2. Remove the rear wheels.
3. Remove the spare tire from the vehicle.
4. Remove the spare tire hoist and the tire support bracket.
5. Remove the muffler from the muffler hangers.

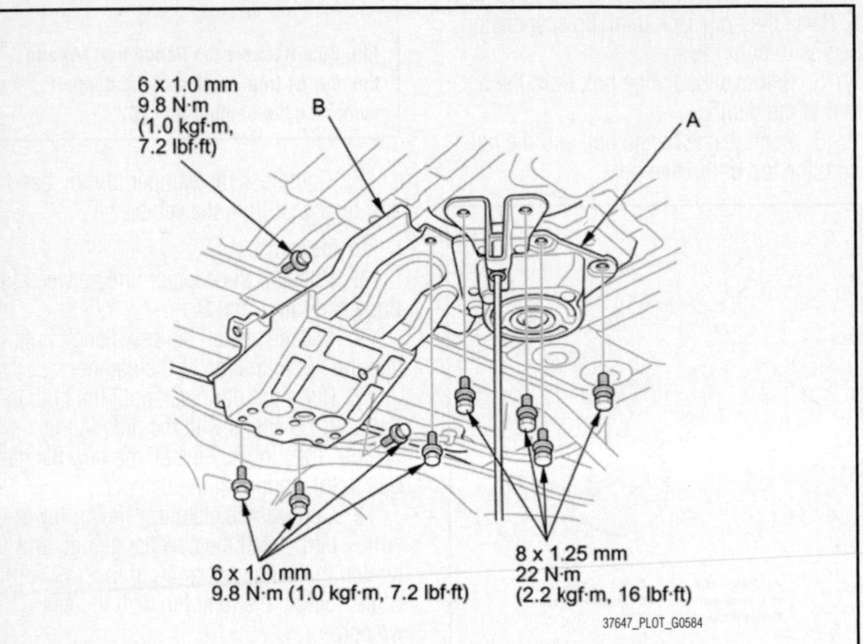
6 x 1.0 mm
9.8 N·m
(1.0 kgf·m,
7.2 lbf·ft)
B
A
6 x 1.0 mm
9.8 N·m (1.0 kgf·m, 7.2 lbf·ft)
8 x 1.25 mm
22 N·m
(2.2 kgf·m, 16 lbf·ft)
37647_PLOT_G0584

Fig. 364 Remove the spare tire hoist (A) and the tire support bracket (B)

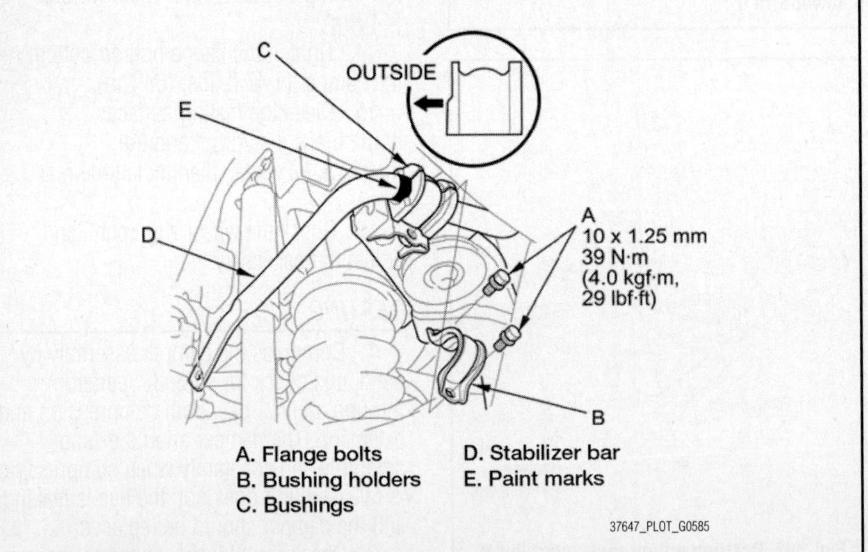
C
OUTSIDE
E
D
A
10 x 1.25 mm
39 N·m
(4.0 kgf·m,
29 lbf·ft)
B
A. Flange bolts
B. Bushing holders
C. Bushings
D. Stabilizer bar
E. Paint marks
37647_PLOT_G0585

Fig. 365 Remove the flange bolts and the bushing holders, then remove the bushings and the stabilizer bar

6. Disconnect both sides of the stabilizer link from the stabilizer bar.
7. Remove the flange bolts and the bushing holders, then remove the bushings and the stabilizer bar.

➡**During installation, align the paint marks on the stabilizer bar with the side of the bushings.**

To install:

8. Install the stabilizer bar in the reverse order of removal, and note these items:

- Note the right and left direction of the stabilizer bar.
- Refer to the stabilizer link removal/installation to connect the stabilizer bar to the links.
- Before installing the wheel, clean the mating surfaces on the brake disc/drum and the inside of the wheel.
- Do not set the bushings on the bent or curved part of the stabilizer bar.

WHEEL HUB & BEARING (SEALED UNIT)

REMOVAL & INSTALLATION

See Figures 366 through 370.

➡**Refer to the Exploded View as needed during the following procedure.**

1. Raise and support the vehicle.
2. Remove the wheel nuts (A) and the rear wheel.
3. Release the parking brake lever fully.
4. Remove the brake caliper bracket mounting bolts (A), then remove the caliper assembly (B) from the knuckle. To prevent damage to the caliper assembly or the brake hose, use a short piece of wire to hang the caliper assembly from the undercarriage. Do not twist the brake hose excessively.
5. Remove the two washers.

➡**During installation, make sure the washers are installed between the brake caliper bracket and the knuckle.**

6. 4WD model: Pry up the stake on the spindle nut, then remove the nut.
7. Remove the rear brake disc/drum.
8. 2WD model: Remove the flange bolts and remove the hub bearing unit.
9. 4WD model: Remove the flange bolts, and remove the hub bearing unit by tapping the driveshaft end with a soft face hammer while drawing the hub bearing unit outward.

➡**Do not pull the driveshaft end outward. The driveshaft inboard joint may come apart.**

10. Check the hub bearing unit for damage and cracks.

To install:

11. Install the hub bearing unit in the reverse order of removal, and note these items:

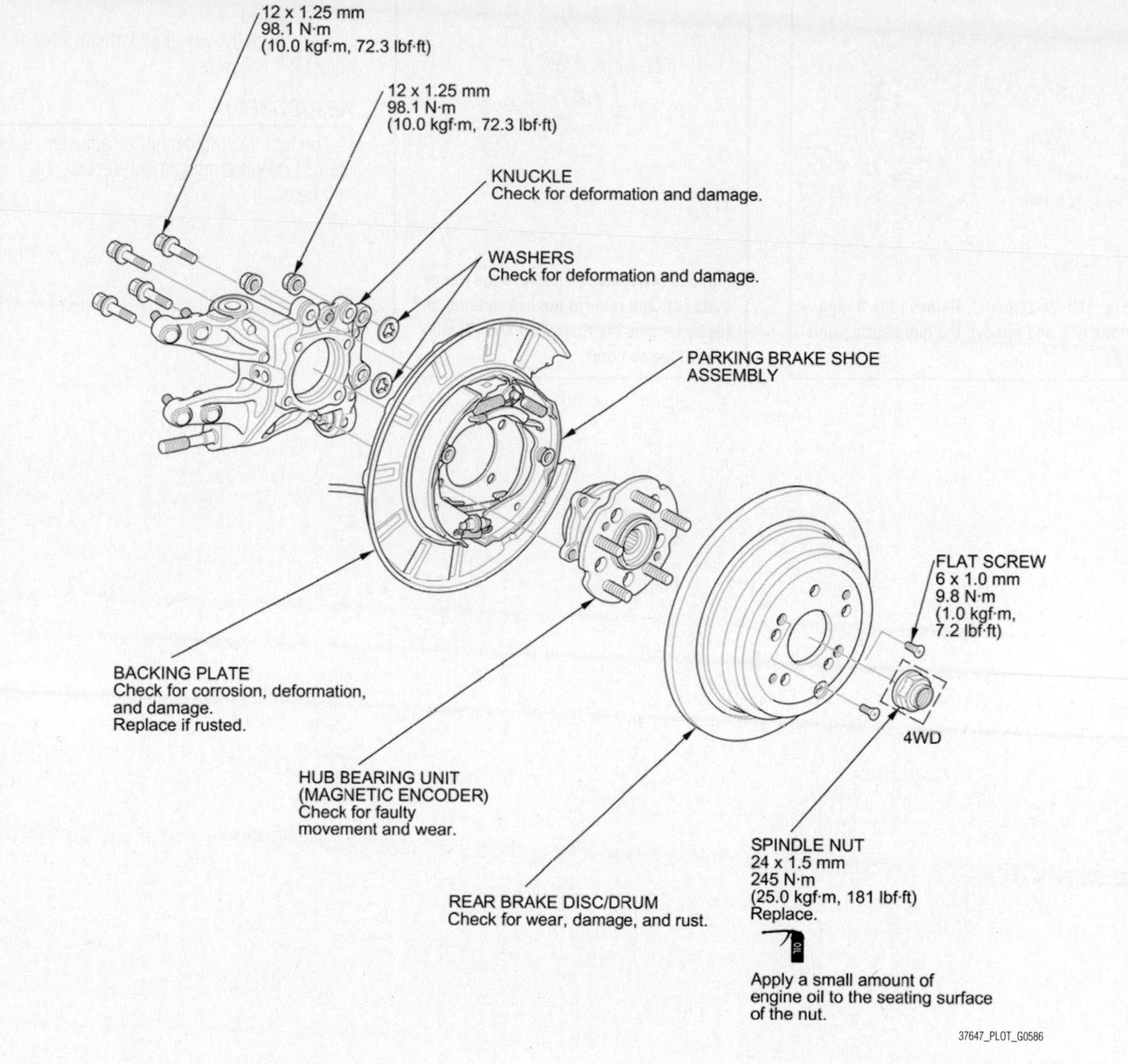
12 x 1.25 mm
98.1 N·m
(10.0 kgf·m, 72.3 lbf·ft)
12 x 1.25 mm
98.1 N·m
(10.0 kgf·m, 72.3 lbf·ft)
KNUCKLE
Check for deformation and damage.
WASHERS
Check for deformation and damage.
PARKING BRAKE SHOE ASSEMBLY
FLAT SCREW
6 x 1.0 mm
9.8 N·m
(1.0 kgf·m,
7.2 lbf·ft)
BACKING PLATE
Check for corrosion, deformation, and damage.
Replace if rusted.
4WD
HUB BEARING UNIT
(MAGNETIC ENCODER)
Check for faulty movement and wear.
SPINDLE NUT
24 x 1.5 mm
245 N·m
(25.0 kgf·m, 181 lbf·ft)
Replace.
REAR BRAKE DISC/DRUM
Check for wear, damage, and rust.
OIL
Apply a small amount of engine oil to the seating surface of the nut.
37647_PLOT_G0586

Fig. 360 Exploded view of rear wheel hub assembly

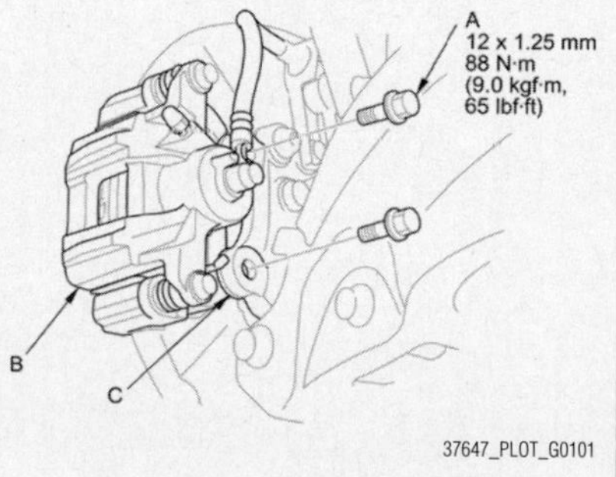
A
12 x 1.25 mm
88 N·m
(9.0 kgf·m,
65 lbf·ft)
B
C
37647_PLOT_G0101

Fig. 367 Remove the brake caliper bracket mounting bolts (A), and remove the caliper assembly (B) from the knuckle

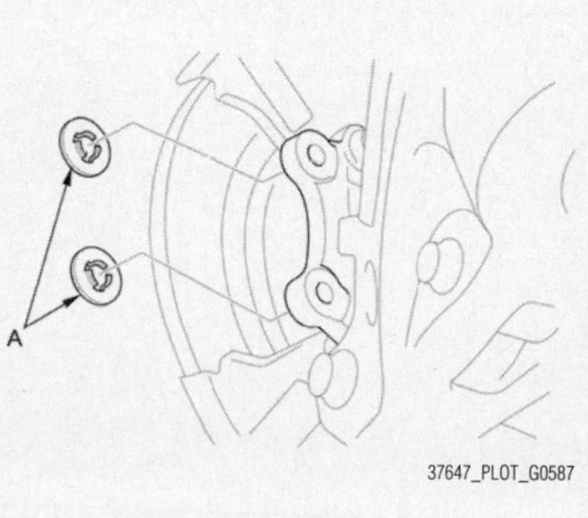
A
37647_PLOT_G0587

Fig. 368 Remove the two washers (A)

- 4WD model: Use a new spindle nut on reassembly.
- 4WD model: Before installing the spindle nut, apply a small amount of engine oil to the seating surface on the nut. After tightening, use a drift to stake the spindle nut shoulder against the driveshaft.
- Before installing the brake disc/drum, clean the mating surfaces of the hub bearing unit and the inside of the brake disc/drum.
- Before installing the wheel, clean

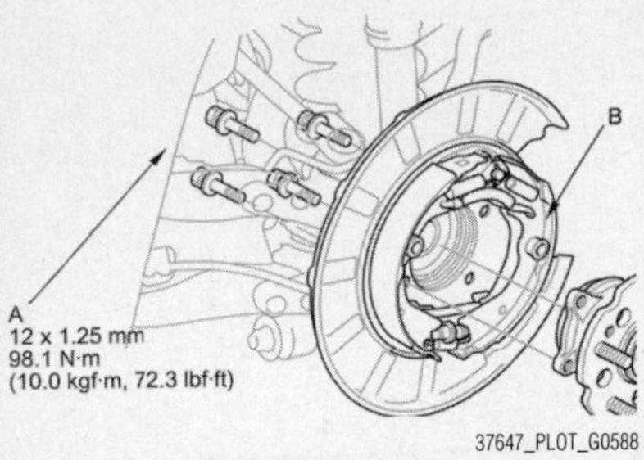
B
A
12 x 1.25 mm
98.1 N·m
(10.0 kgf·m, 72.3 lbf·ft)
37647_PLOT_G0588

Fig. 369 2WD model: Remove the flange bolts (A), and remove the hub bearing unit (B)

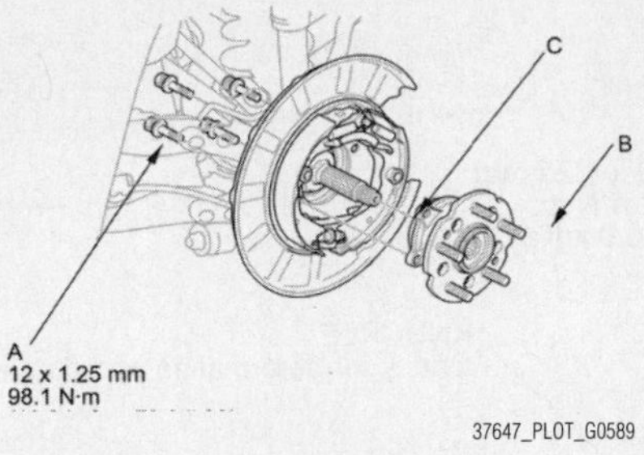
C
B
A
12 x 1.25 mm
98.1 N·m
37647_PLOT_G0589

Fig. 370 4WD model: Remove the flange bolts (A), and remove the hub bearing unit (B) by tapping the driveshaft end (C) with a soft face hammer

the mating surfaces on the brake disc/drum and the inside of the wheel.

12. Check the wheel alignment, and adjust it if necessary.

ADJUSTMENT

The hub bearing cannot be adjusted. If there is need for adjustment, replace the hub bearing.

HONDA

Ridgeline

17

BRAKES 17-10

ANTI-LOCK BRAKE SYSTEM (ABS) 17-11
Wheel Speed Sensors 17-11
Removal & Installation........17-11
BLEEDING THE BRAKE SYSTEM 17-10
Bleeding Procedure..............17-10
Bleeding Procedure 17-10
Bleeding the ABS System ...17-10
FRONT DISC BRAKES........17-12
Brake Caliper.........................17-12
Removal & Installation........17-12
Disc Brake Pads.....................17-12
Removal & Installation........17-12
INFORMATION AND PRECAUTIONS...............17-10
Anti-lock Systems..............17-10
Disc and Drum Systems 17-10
PARKING BRAKE.............17-14
Parking Brake Shoes 17-14
Adjustment 17-15
Removal & Installation........17-14
REAR DISC BRAKES 17-13
Brake Caliper..........................17-13
Removal & Installation........17-13
Disc Brake Pads.....................17-13
Removal & Installation........17-13

CHASSIS ELECTRICAL 17-15

AIR BAG (SUPPLEMENTAL RESTRAINT SYSTEM) 17-15
General Information................17-15
Arming the System 17-15
Clockspring Centering........17-15
Disarming the System.........17-15
Service Precautions............17-15

DRIVE TRAIN 17-16

Front Halfshaft (Driveshaft) 17-16
Removal & Installation........17-16
Intermediate Shaft 17-16
Removal & Installation........17-16
Rear Differential Assembly 17-16
Removal & Installation........17-16
Rear Halfshaft (Driveshaft).......17-18
Removal & Installation........17-18
Transfer Case Assembly 17-19
Removal & Installation........17-19

ENGINE COOLING...........17-19

Engine Fan 17-19
Removal & Installation........17-19
Radiator & Fan.......................17-19
Removal & Installation........17-19
Thermostat 17-19
Removal & Installation........17-19
Water Pump 17-20
Removal & Installation........17-20

ENGINE ELECTRICAL 17-21

CHARGING SYSTEM 17-21
Alternator 17-21
Removal & Installation........17-21
IGNITION SYSTEM 17-22
Firing Order...........................17-22
Ignition Coil 17-22
Removal & Installation........17-22
Ignition Timing.......................17-22
Adjustment 17-22
Inspection..........................17-22
Spark Plugs...........................17-22
Removal & Installation........17-22
STARTING SYSTEM 17-23
Starter 17-23
Removal & Installation........17-23

ENGINE MECHANICAL......17-23

Accessory Drive Belts 17-23
Accessory Belt Routing............................17-23
Adjustment 17-23
Inspection..........................17-23
Removal & Installation........17-23
Camshaft & Valve Lifters.........17-23
Inspection..........................17-23
Removal & Installation........17-23
Crankshaft Damper.................17-25
Removal & Installation........17-25
Crankshaft Front Seal.............17-26
Removal & Installation........17-26
Cylinder Head 17-26
Removal & Installation........17-26
Flexplate (Drive Plate)............17-29
Removal & Installation........17-29
Intake Manifold 17-29
Removal & Installation........17-29
Oil Pan 17-29
Removal & Installation........17-29
Oil Pump...............................17-31
Inspection..........................17-32
Removal & Installation........17-31
Piston and Ring......................17-33
Positioning 17-33
Rear Main Seal......................17-33
Removal & Installation........17-33
Rocker Arms/Shafts...............17-33
Removal & Installation........17-33
Timing Belt & Sprockets 17-35
Removal & Installation........17-35
Timing Belt Front Cover 17-33
Removal & Installation........17-33
Timing Belt Rear Cover 17-38
Removal & Installation........17-38
Valve Lash.............................17-39
Adjustment 17-39

ENGINE PERFORMANCE & EMISSION CONTROLS 17-40

Accelerator Pedal Position (APP) Sensor 17-40
Location.............................17-40
Removal & Installation........17-40
Camshaft Position (CMP) Sensor 17-40
Location.............................17-40
Removal & Installation........17-40
Crankshaft Position (CKP) Sensor 17-40
Location.............................17-40
Removal & Installation........17-40
Engine Coolant Temperature (ECT) Sensor 17-40
Location.............................17-40
Removal & Installation........17-40

Evaporative Emissions (EVAP) Canister 17-41
Location 17-41
Removal & Installation 17-42
Heated Oxygen (HO2S) Sensor 17-42
Location 17-42
Removal & Installation 17-42
Intake Air Temperature (IAT) Sensor 17-42
Removal & Installation 17-42
Knock Sensor (KS) 17-42
Location 17-42
Removal & Installation 17-43
Malfunction Indicator Light (MIL) 17-43
Reset Procedures 17-43
Manifold Absolute Pressure (MAP) Sensor 17-43
Location 17-43
Removal & Installation 17-43
Testing 17-43
Mass Air Flow/Intake Air Temperature (MAF/IAT) Sensor 17-43
Location 17-43
Removal & Installation 17-43
Output Shaft (Countershaft) Speed (OSS) Sensor 17-44
Location 17-44
Removal & Installation 17-44
Positive Crankcase Ventilation (PCV) Valve 17-44
Location 17-44
Removal & Installation 17-44
Powertrain Control Module (PCM) 17-45
Location 17-45
Removal & Installation 17-45

FUEL 17-46

GASOLINE FUEL INJECTION SYSTEM 17-46
Fuel Filter 17-46
Removal & Installation 17-46
Fuel Level Sending Unit 17-47
Removal & Installation 17-47
Fuel Pressure Regulator 17-47
Removal & Installation 17-47
Fuel Pump 17-47
Removal & Installation 17-47
Fuel Rail & Injectors 17-48
Removal & Installation 17-48
Fuel System Service Precautions 17-46
Fuel Tank 17-49
Draining 17-49
Removal & Installation 17-49
Idle Speed 17-49
Adjustment 17-50
Inspection 17-49
Relieving Fuel System Pressure 17-46
Throttle Body 17-51
Removal & Installation 17-51

HEATING & AIR CONDITIONING SYSTEM ...17-52

Blower Motor 17-52
Removal & Installation 17-52
Heater Core 17-52
Removal & Installation 17-52

SPECIFICATIONS AND MAINTENANCE CHARTS 17-3

Brake Specifications 17-7
Camshaft and Bearing Specifications 17-5
Capacities 17-4
Crankshaft and Connecting Rod Specifications 17-5
Engine and Vehicle Identification 17-3
Engine Tune-Up Specifications ...17-3
Fluid Specifications 17-4
General Engine Specifications ...17-3
Piston and Ring Specifications 17-5
Maintenance Minder Schedule 17-8,9
Tire, Wheel and Ball Joint Specifications 17-7
Torque Specifications 17-6
Valve Specifications 17-4
Wheel Alignment 17-6

STEERING 17-53

Power Rack & Pinion Steering Gear 17-53
Removal & Installation 17-53
Power Steering Pump 17-56
Bleeding 17-56
Removal & Installation 17-56

SUSPENSION 17-57

FRONT SUSPENSION 17-57
Lower Ball Joint 17-57
Removal & Installation 17-57
Lower Control Arm 17-57
Removal & Installation 17-57
MacPherson Strut 17-57
Overhaul 17-58
Removal & Installation 17-57
Stabilizer Bar 17-61
Removal & Installation 17-61
Steering Knuckle 17-59
Removal & Installation 17-59
Wheel Hub and Bearing 17-62
Removal & Installation 17-62

REAR SUSPENSION 17-62
Control Arms/Links 17-62
Removal & Installation 17-62
MacPherson Struts 17-63
Overhaul 17-63
Removal & Installation 17-63
Rear Knuckle 17-63
Removal & Installation 17-63
Stabilizer Bar 17-66
Removal & Installation 17-66
Wheel Hub and Bearing 17-68
Removal & Installation 17-68

SPECIFICATIONS AND MAINTENANCE CHARTS

ENGINE AND VEHICLE IDENTIFICATION CHART

Engine Code						
Code	Liters (cc)	Cu. In.	Cyl.	Fuel Sys.	Engine Type	Eng. Mfg.
J35Z5	3.5 (3471)	222	6	SMFI	SOHC	Honda

Model Year	
Code ①	Year
9	2009
A	2010

SOHC: Single Overhead Cam

SMFI: Sequential Multi-port Fuel Injection

① 10th position of VIN

37647_RIDG_C0001

GENERAL ENGINE SPECIFICATIONS

Year	Model	Engine Displacement Liters	Engine ID	Net Horsepower @ rpm	Net Torque @ rpm (ft. lbs.)	Bore x Stroke (in.)	Compression Ratio	Oil Pressure @ rpm
2009	Ridgeline	3.5	J35Z5	255@5750	252@4500	3.50x3.66	10:01	71@3000 ①
2010	Ridgeline	3.5	J35Z5	247@5750	245@4500	3.50x3.66	10:01	71@3000 ①

① At idle: 10 psi

37647_RIDG_C0002

ENGINE TUNE-UP SPECIFICATIONS

Year	Engine Displacement Liters	Engine ID	Spark Plug Gap (in.)	Ignition Timing (deg. BTDC)	Fuel Pump (psi)	Idle Speed (rpm)	Valve Clearance (in.)	
							In.	Ex.
2009	3.5	J35Z5	0.039-0.043	8-12	57-64	680-780	0.008-0.009	0.011-0.013
2010	3.5	J35Z5	0.039-0.043	8-12	57-64	680-780	0.008-0.009	0.011-0.013

NOTE: The Vehicle Emission Control Information label often reflects changes made during production and must be used if they differ from this chart.

BTDC: Before top dead center

37647_RIDG_C0003

CAPACITIES

Year	Model	Engine Displacement Liters	Engine ID	Engine Oil with Filter (qts.)	Transmission (pts.) *	Transfer Case (pts.)	Rear Drive Axle (pts.)	Fuel Tank (gal.)	Cooling System (qts.)
2009	Ridgeline	3.5	J35Z5	4.5	6.6	0.9	5.58	22.0	①
2010	Ridgeline	3.5	J35Z5	4.5	6.6	0.9	5.58	22.0	①

NOTE: All capacities are approximate. Add fluid gradually and check to be sure a proper fluid level is obtained.

* Fluid change

① Total capacity: 8.56 qts.
At coolant change: 6.26 qts.

37647_RIDG_C0004

FLUID SPECIFICATIONS

Year	Model	Engine Displacement Liters	Engine Oil	Auto. Trans.	Drive Axle	Power Steering Fluid	Brake Master Cylinder
2009	Ridgeline	3.5	①	Honda ATF-Z1	Honda VTM-4	Honda P/S Fluid	DOT 3
2010	Ridgeline	3.5	①	Honda ATF-Z1	Honda VTM-4	Honda P/S Fluid	DOT 3

DOT: Department Of Transportation

① See oil filler cap.

37647_RIDG_C0013

VALVE SPECIFICATIONS

Year	Engine Displacement Liters	Engine ID	Seat Angle (deg.)	Face Angle (deg.)	Spring Test Pressure (lbs. @ in.)	Spring Free Length (in.)	Stem-to-Guide Clearance (in.)		Stem Diameter (in.)	
							Intake	Exhaust	Intake	Exhaust
2009	3.5	J35Z5	45	45	NA	①	0.0008-0.0018	0.0022-0.0031	0.2159-0.2163	0.2146-0.2150
2010	3.5	J35Z5	45	45	NA	①	0.0008-0.0018	0.0022-0.0031	0.2159-0.2163	0.2146-0.2150

NA: Information not available

① Intake: 1.9713 in.
Exhaust: 2.1060 in.

37647_RIDG_C0005

CAMSHAFT AND BEARING SPECIFICATIONS

All measurements are given in inches.

Year	Engine Displacement Liters	Engine ID	Journal Diameter	Brg. Oil Clearance	Shaft End-play	Runout	Journal Bore	Lobe Lift Intake	Lobe Lift Exhaust
2009	3.5	J35Z5	NA	0.0002-0.0035	0.0002-0.0008	0.001	NA	①	1.4375
2010	3.5	J35Z5	NA	0.0002-0.0035	0.0002-0.0008	0.001	NA	①	1.4375

NA: Information not available

① Primary: 1.4024

Secondary 1.3504

37647_RIDG_C0007

CRANKSHAFT AND CONNECTING ROD SPECIFICATIONS

All measurements are given in inches

Year	Engine Displacement Liters	Engine ID	Crankshaft Main Brg. Journal Dia.	Crankshaft Main Brg. Oil Clearance	Crankshaft Shaft End-play	Crankshaft Thrust on No.	Connecting Rod Journal Diameter	Connecting Rod Oil Clearance	Connecting Rod Side Clearance
2009	3.5	J35Z5	2.8337-2.8346	0.0007-0.0018	0.0040-0.0140	3	2.1644-2.1654	0.0002-0.0006	0.0060-0.0140
2010	3.5	J35Z5	2.8337-2.8346	0.0007-0.0018	0.0040-0.0140	3	2.1644-2.1654	0.0002-0.0006	0.0060-0.0140

NA: Information not available

37647_RIDG_C0006

PISTON AND RING SPECIFICATIONS

All measurements are given in inches

Year	Engine Displacement Liters	Engine ID	Piston Clearance	Ring Gap Top Compression	Ring Gap Bottom Compression	Ring Gap Oil Control	Ring Side Clearance Top Compression	Ring Side Clearance Bottom Compression	Ring Side Clearance Oil Control
2009	3.5	J35Z5	0.0006-0.0016	0.0080-0.0140	0.0160-0.0220	0.0080-0.0280	0.0022-0.0031	0.0012-0.0022	snug
2010	3.5	J35Z5	0.0006-0.0016	0.0080-0.0140	0.0160-0.0220	0.0080-0.0280	0.0022-0.0031	0.0012-0.0022	snug

37647_RIDG_C0008

TORQUE SPECIFICATIONS

All readings in ft. lbs.

Year	Engine Displacement Liters	Engine ID	Cylinder Head Bolts	Main Bearing Bolts	Rod Bearing Bolts	Crankshaft Damper Bolts	Flywheel Bolts	Manifold Intake	Manifold Exhaust	Spark Plugs	Oil Pan Drain Plug
2009	3.5	J35Z5	①	②	③	④	54	16	NA	13	29
2010	3.5	J35Z5	①	②	③	④	54	16	NA	13	29

① Step 1: 22 ft. lbs.
Step 2: plus 90 degrees
Step 3: plus 90 degrees

② Cap bolts: 54 ft. lbs.
Side bolts: 36 ft. lbs.

③ Step 1: 12 ft. lbs.
Step 2: plus 90 degrees

④ Step 1: 47 ft. lbs.
Step 2: plus 60 degrees

37647_RIDG_C0009

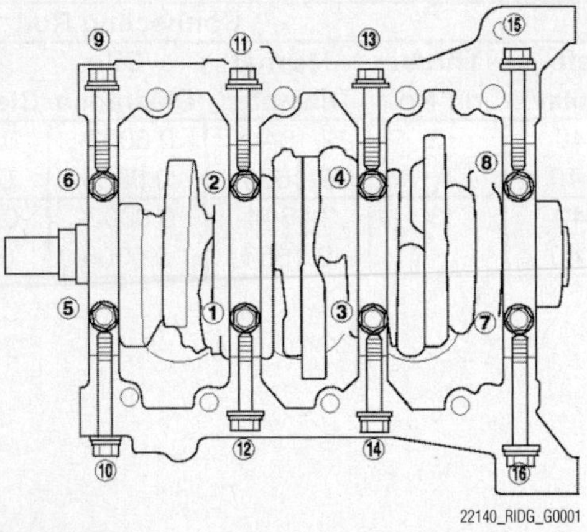
9
11
13
15
6
2
4
8
5
1
3
7
12
14
10
16
22140_RIDG_G0001

Fig. 1 Main bearing torque sequence—3.5L engine

WHEEL ALIGNMENT

Year	Model		Caster Range (+/-Deg.)	Caster Preferred Setting (Deg.)	Camber Range (+/-Deg.)	Camber Preferred Setting (Deg.)	Toe-in (in.)
2009	Ridgeline	F	1.00	+1.70	1.00	-0.50	0 +/- 0.08
		R	—	—	0.75	-0.50	0 +/- 0.08
2010	Ridgeline	F	1.00	+1.70	1.00	-0.50	0 +/- 0.08
		R	—	—	0.75	-0.50	0 +/- 0.08

37647_RIDG_C0011

TIRE, WHEEL AND BALL JOINT SPECIFICATIONS

		OEM Tires		Tire Pressures (psi)				Lug Nut
Year	Model	Standard	Optional	Front	Rear	Wheel Size	Ball Joint Inspection	Torque (ft. lbs.)
2009	Ridgeline	①	None	32	②	7.5	NA	94
2010	Ridgeline	①	None	32	②	7.5	NA	94

NA: Information not available

OEM: Original Equipment Manufacturer

PSI: Pounds Per Square Inch

① P245/65R17 (except RTL model)

P245/60R18 (RTL model)

② Consult door jamb label or Owner's Manual.

37647_RIDG_C0010

BRAKE SPECIFICATIONS

All measurements in inches unless noted

			Brake Disc				Brake Caliper	
Year	Model		Original Thickness	Minimum Thickness	Maximum Runout	Minimum Lining Thickness	Bracket Bolts (ft. lbs.)	Mounting Bolts (ft. lbs.)
2009	Ridgeline	F	1.10-1.11	1.020	0.0016	0.040	101	53
		R	0.43-0.44	0.350	0.0016	0.040	80	16
2010	Ridgeline	F	1.10-1.11	1.020	0.0016	0.040	101	53
		R	0.43-0.44	0.350	0.0016	0.040	80	16

F: Front

R: Rear

37647_RIDG_C0012

MAINTENANCE MINDER SCHEDULE

Honda Ridgeline

The Ridgeline displays engine oil life and maintenance service items in the information display to indicate when to perform maintenance service. If the engine oil life is 15% or less, based on the onboard computer's caluculations, you will see SERVICE DUE SOON in the information display every time the ignition key is turned to ON. The maintenance minder indicator will also come on and the maintenance code(s) for other scheduled maintenance items needing service will be displayed below the message.

Symbol	Item	Service
A	Engine oil ①	Change
B	Engine oil and filter	Change
	Brakes	Inspect
	Parking brake adjustment	Check
	Steering gear, boots and linkage	Inspect
	Suspension components	Inspect
	Driveshaft boots	Inspect
	Brake hoses and lines, including VSA	Inspect
	Exhaust system	Inspect
	All fluid levels and fluid condition	Inspect
	Exhaust system components	Inspect
	Fuel lines and connections	Inspect
1	Tires: condition and pressures	Rotate
2	Engine air filter ②	Replace
	Dust and pollen filter ③	Replace
	Accessory drive belt	Inspect
3	Transmission fluid	Replace
	Transfer case fluid	Replace
4	Spark plugs	Replace
	Timing belt ④	Replace
	Water pump	Inspect
	Valve clearance ⑤	Inspect
5	Engine coolant	Replace
6	Rear differential fluid ⑥	Replace

① If the message SERVICE DUE NOW does not appear more than 12 months after the display is reset, change every year.

② If driven in dusty conditions, replace every 15,000 miles.

③ If driven in urban areas that have a high concentration of soot from industry and diesel, replace every 15,000 miles

④ If driven regularly in temperatures over 110 deg.F or below -20 deg.F, or towing a trailer, replace every 60,000 miles.

⑤ Adjust if necessary.

⑥ Driving in mountainous areas at very low vehicle speeds or trailer towing results in higher level of mechanical (shear) stres to fluid. This requires differential fluid changes more frequently than recommended by the maintenance minder. If the vehicle regularly driven under these conditions, changed have the differential fluid at 7,500 miles, then every 15,000 miles.

Additionally, replace the brake fluid every 3 years, and inspect the idle speed every 160,000 miles.

To reset the Engine Oil Life Display:

1. Turn the ignition switch to ON.
2. Press the SELECT button repeatedly until the engine oil life display or the service message is displayed.
3. Press the TRIP/RESET button for about 10 seconds. The engine oil life indicator and maintenance item code(s) will blink.

NOTE: If you are resetting the display when the engine oil life is more than 15 %, make sure any maintenance item(s) requiring service are done before resetting the display.

4. Press TRIP/RESET button for another 5 seconds. The maintenance item code(s) will disappear, and the engine oil life will reset to “100”.

37647_RIDG_C0014

MAINTENANCE MINDER SCHEDULE

Honda Ridgeline (cont.)

To reset Individual Maintenance Item:

1. Connect the Honda Diagnostic System (HDS), or equivalent scan tool, to the data link connector (DLC).
2. Turn the ignition switch to ON (II).
3. Make sure the HDS communicates with the vehicle and the powertrain control module (PCM). If it doesn't communicate, troubleshoot the DLC circuit.
4. Select BODY ELECTRICAL with the HDS.
5. Select ADJUSTMENT in the GAUGE MENU with the HDS.
6. Select RESET in the MAINTENANCE MINDER with the HDS.
7. Select the individual maintenance item you wish to reset.

37647_RIDG_C0015

BRAKES — INFORMATION AND PRECAUTIONS

ANTI-LOCK SYSTEMS

- Certain components within the ABS system are not intended to be serviced or repaired individually.
- Do not use rubber hoses or other parts not specifically specified for and ABS system. When using repair kits, replace all parts included in the kit. Partial or incorrect repair may lead to functional problems and require the replacement of components.
- Lubricate rubber parts with clean, fresh brake fluid to ease assembly. Do not use shop air to clean parts; damage to rubber components may result.
- Use only DOT 3 brake fluid from an unopened container.
- If any hydraulic component or line is removed or replaced, it may be necessary to bleed the entire system.
- A clean repair area is essential. Always clean the reservoir and cap thoroughly before removing the cap. The slightest amount of dirt in the fluid may plug an orifice and impair the system function. Perform repairs after components have been thoroughly cleaned; use only denatured alcohol to clean components. Do not allow ABS components to come into contact with any substance containing mineral oil; this includes used shop rags.
- The Anti-Lock control unit is a microprocessor similar to other computer units in the vehicle. Ensure that the ignition switch is **OFF** before removing or installing controller harnesses. Avoid static electricity discharge at or near the controller.
- If any arc welding is to be done on the vehicle, the control unit should be unplugged before welding operations begin.

DISC AND DRUM SYSTEMS

CAUTION

Dust and dirt accumulating on brake parts during normal use may contain asbestos fibers from production or aftermarket brake linings. Breathing excessive concentrations of asbestos fibers can cause serious bodily harm. Exercise care when servicing brake parts. Do not sand or grind brake lining unless equipment used is designed to contain the dust residue. Do not clean brake parts with compressed air or by dry brushing. Cleaning should be done by dampening the brake components with a fine mist of water, then wiping the brake components clean with a dampened cloth. Dispose of cloth and all residue containing asbestos fibers in an impermeable container with the appropriate label. Follow practices prescribed by the Occupational Safety and Health Administration (OSHA) and the Environmental Protection Agency (EPA) for the handling, processing, and disposing of dust or debris that may contain asbestos fibers.

BRAKES — BLEEDING THE BRAKE SYSTEM

BLEEDING PROCEDURE

BLEEDING PROCEDURE

See Figure 2.

When bleeding the brake system, observe the following:

- Do not reuse the drained fluid. Use only clean Honda DOT 3 Brake Fluid from an unopened container. Using a non-Honda brake fluid can cause corrosion and shorten the life of the system. Do not mix different brands of brake fluid; they may not be compatible.
- Make sure no dirt or other foreign matter is allowed to contaminate the brake fluid.
- Do not spill brake fluid on the vehicle, it may damage the paint; if brake fluid does contact the paint, wash it off immediately with water.

1. The reservoir on the master cylinder must be at the MAX (upper) level mark at the start of the bleeding procedure and checked after bleeding each brake caliper. Add fluid as required.
2. Make sure the brake fluid level in the reservoir is at the MAX (upper) level line.
3. Slide a piece of clear plastic hose over the bleed screw, and submerge the other end in a container of new brake fluid.
4. Have someone slowly pump the brake pedal several times, then apply steady pressure.
5. Loosen the left-front brake bleed screw to allow air to escape from the system. Then tighten the bleed screw securely.
6. Repeat the procedure for each caliper until no air bubbles are in the fluid. Bleed the calipers in the sequence shown.
7. Refill the master cylinder reservoir to the MAX (upper) level line.

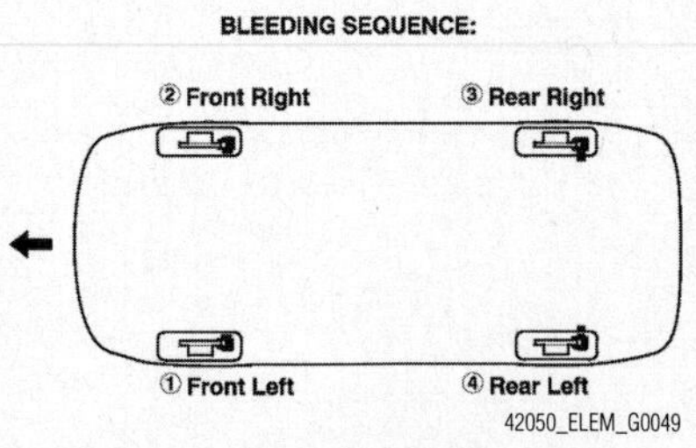
BLEEDING SEQUENCE:
② Front Right
③ Rear Right
① Front Left
④ Rear Left
42050_ELEM_G0049

Fig. 2 Proper brake bleeding sequence

BLEEDING THE ABS SYSTEM

The bleeding procedure for the ABS System is the same as the conventional bleeding procedure. Refer to "BLEEDING THE BRAKE SYSTEM" in this section.

WHEEL SPEED SENSORS

REMOVAL & INSTALLATION

See Figures 3 and 4.

⁂ CAUTION

Before beginning work, review the "Precautions" in this section.

1. Raise and safely support the vehicle.
2. Remove the front or rear wheel(s), as applicable.
3. Detach the speed sensor electrical connector, unfasten the retainers, then remove the speed sensor.

To install:

4. Install the speed sensor and tighten the retaining bolts to 7.2 ft. lbs. (9.8 Nm).
5. Attach the speed sensor connector.
6. Install the wheel(s).
7. Lower the vehicle.

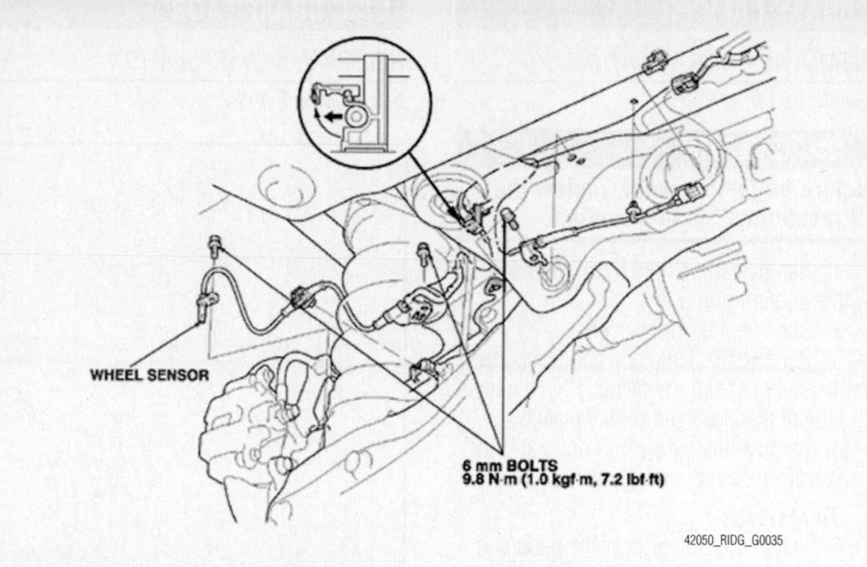
WHEEL SENSOR
6 mm BOLTS
9.8 N·m (1.0 kgf·m, 7.2 lbf·ft)
42050_RIDG_G0035

Fig. 4 View of the rear wheel speed sensor

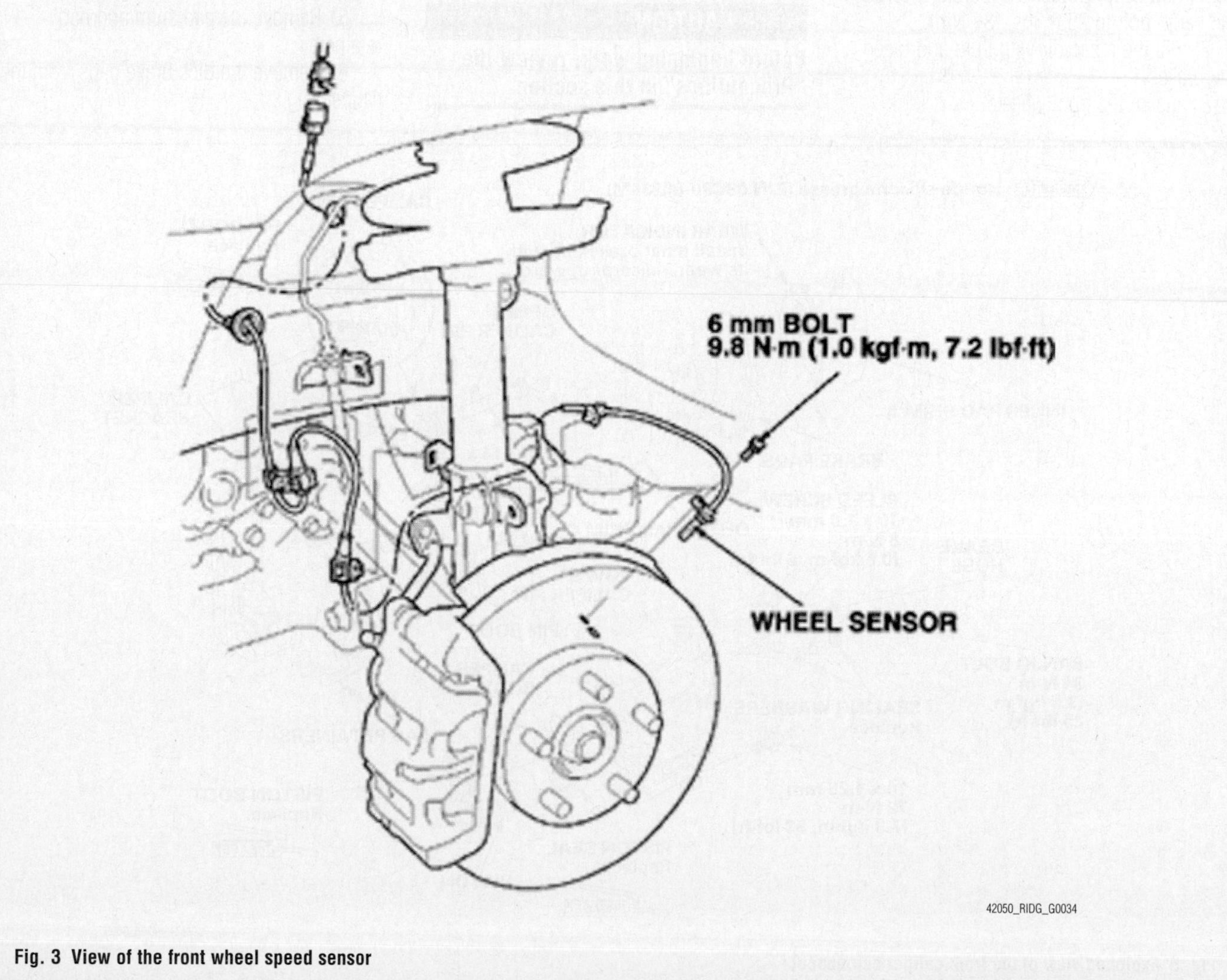
6 mm BOLT
9.8 N·m (1.0 kgf·m, 7.2 lbf·ft)
WHEEL SENSOR
42050_RIDG_G0034

Fig. 3 View of the front wheel speed sensor

BRAKE CALIPER

REMOVAL & INSTALLATION

See Figure 5.

CAUTION

Before beginning work, review the "Precautions" in this section.

1. Remove some fluid from the reservoir with a suction pump.
2. Remove the front wheels.
3. Remove the banjo bolt and disconnect the brake hose from the caliper. Plug the hose to prevent fluid loss and contamination.
4. Remove the caliper pin bolts and the caliper from its mounting bracket.

To install:

5. Install the caliper over the pads and onto its mounting bracket. Torque both caliper pin bolts to 53 ft. lbs. (72 Nm).
6. Install the brake hose to the caliper using new sealing washers. Carefully torque the banjo bolt to 25 ft. lbs. (34 Nm).
7. Fill the reservoir with fluid and bleed the brakes.
8. Install the front wheels

DISC BRAKE PADS

REMOVAL & INSTALLATION

See Figures 6 and 7.

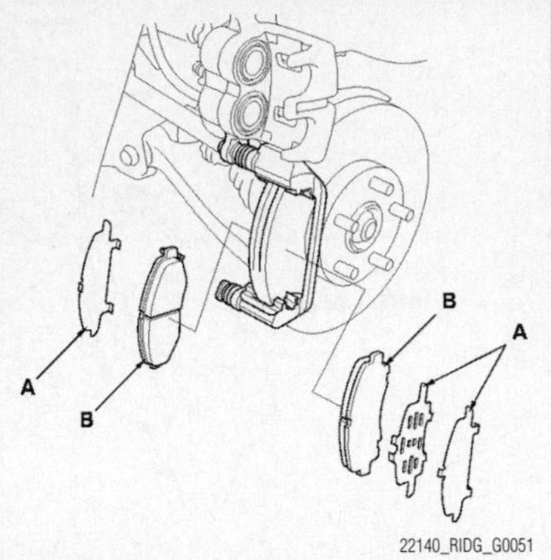
A
B
B
A
22140_RIDG_G0051

Fig. 6 Front pad shims (A) and the brake pads (B)

CAUTION

Before beginning work, review the "Precautions" in this section.

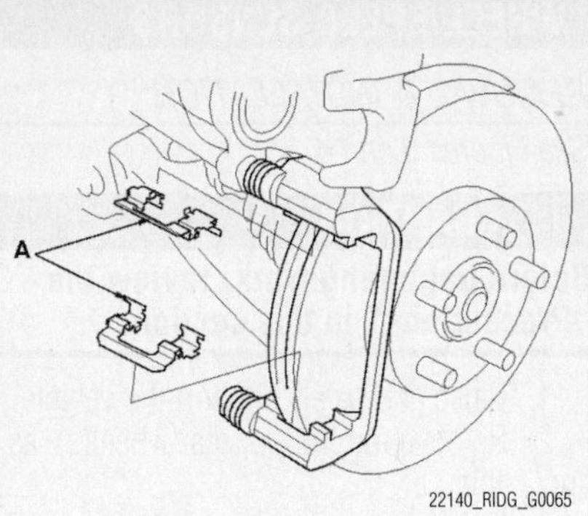
A
22140_RIDG_G0065

Fig. 7 Front pad retainers (A)

1. Remove a small amount of brake fluid from the reservoir using a suction pump.
2. Remove the front wheels.
3. Remove the lower caliper retaining bolt and pivot the caliper upward, off of the pads.
4. Remove the pad springs while holding the pads.
5. Remove the pad shim and pad retainers.
6. Remove the disc brake pads from the caliper.

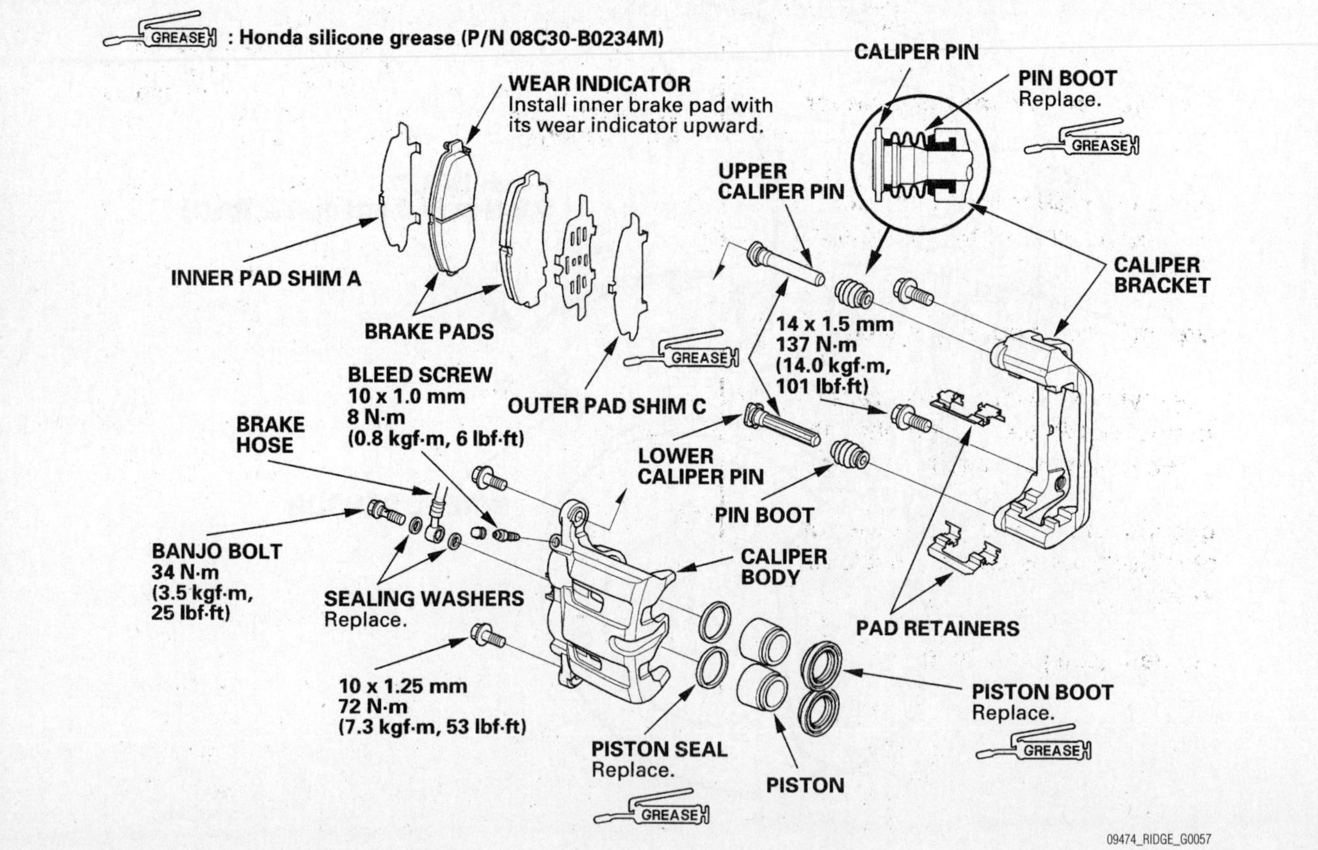
GREASE : Honda silicone grease (P/N 08C30-B0234M)
CALIPER PIN
PIN BOOT
Replace.
GREASE
WEAR INDICATOR
Install inner brake pad with its wear indicator upward.
UPPER CALIPER PIN
CALIPER BRACKET
INNER PAD SHIM A
BRAKE PADS
14 x 1.5 mm
137 N·m
(14.0 kgf·m, 101 lbf·ft)
GREASE
BLEED SCREW
10 x 1.0 mm
8 N·m
(0.8 kgf·m, 6 lbf·ft)
OUTER PAD SHIM C
BRAKE HOSE
LOWER CALIPER PIN
PIN BOOT
BANJO BOLT
34 N·m
(3.5 kgf·m, 25 lbf·ft)
SEALING WASHERS
Replace.
CALIPER BODY
PAD RETAINERS
10 x 1.25 mm
72 N·m
(7.3 kgf·m, 53 lbf·ft)
PISTON BOOT
Replace.
GREASE
PISTON SEAL
Replace.
GREASE
PISTON
09474_RIDGE_G0057

Fig. 5 Exploded view of the front caliper components

To install:

7. Clean the caliper thoroughly; remove any rust from the lip of the rotor. Check the brake rotor for grooves or cracks. If any heavy scoring is present, the rotor must be replaced.

8. Install the pad retainers. Apply molybdenum brake grease to both surfaces of the shims and the back of the disc brake pads.

9. Install the pads and shims. The pad with the wear indicator goes in the inboard position.

10. Install the pad springs while holding the pads.

➡Push in the caliper piston so the caliper will fit over the pads. This is most easily accomplished with a pad spreader or large C-clamp.

11. Install the caliper down into position and tighten the mounting bolt to 53 ft. lbs. (72 Nm).

12. Install the wheels.

13. Add brake fluid to the master cylinder reservoir and install the cap.

14. Depress the brake pedal several times and make sure that the movement feels normal. The first brake pedal application may result in a very long pedal action due to the pistons being retracted. Always make several brake applications before starting the vehicle. Bleed the system if necessary.

BRAKES — REAR DISC BRAKES

BRAKE CALIPER

REMOVAL & INSTALLATION

See Figure 8.

⁂ CAUTION

Before beginning work, review the "Precautions" in this section.

1. Remove some fluid from the reservoir with a suction pump.
2. Remove the rear wheels.
3. Remove the banjo bolt and disconnect the brake hose from the caliper. Plug the hose to prevent fluid loss and contamination.
4. Remove the caliper mounting bolts and the caliper from its mounting bracket.

To Install:

5. Install the caliper over the pads and onto its mounting bracket. Tighten the caliper bolts to 80 ft. lbs. (108 Nm).
6. Install the brake hose with new sealing washers. Tighten the banjo bolt to 25 ft. lbs. (34 Nm).
7. Fill the reservoir with fluid and bleed the brake system. Adjust the parking brake if necessary.
8. Install the rear wheels.

DISC BRAKE PADS

REMOVAL & INSTALLATION

See Figures 9 and 10.

⁂ CAUTION

Before beginning work, review the "Precautions" in this section.

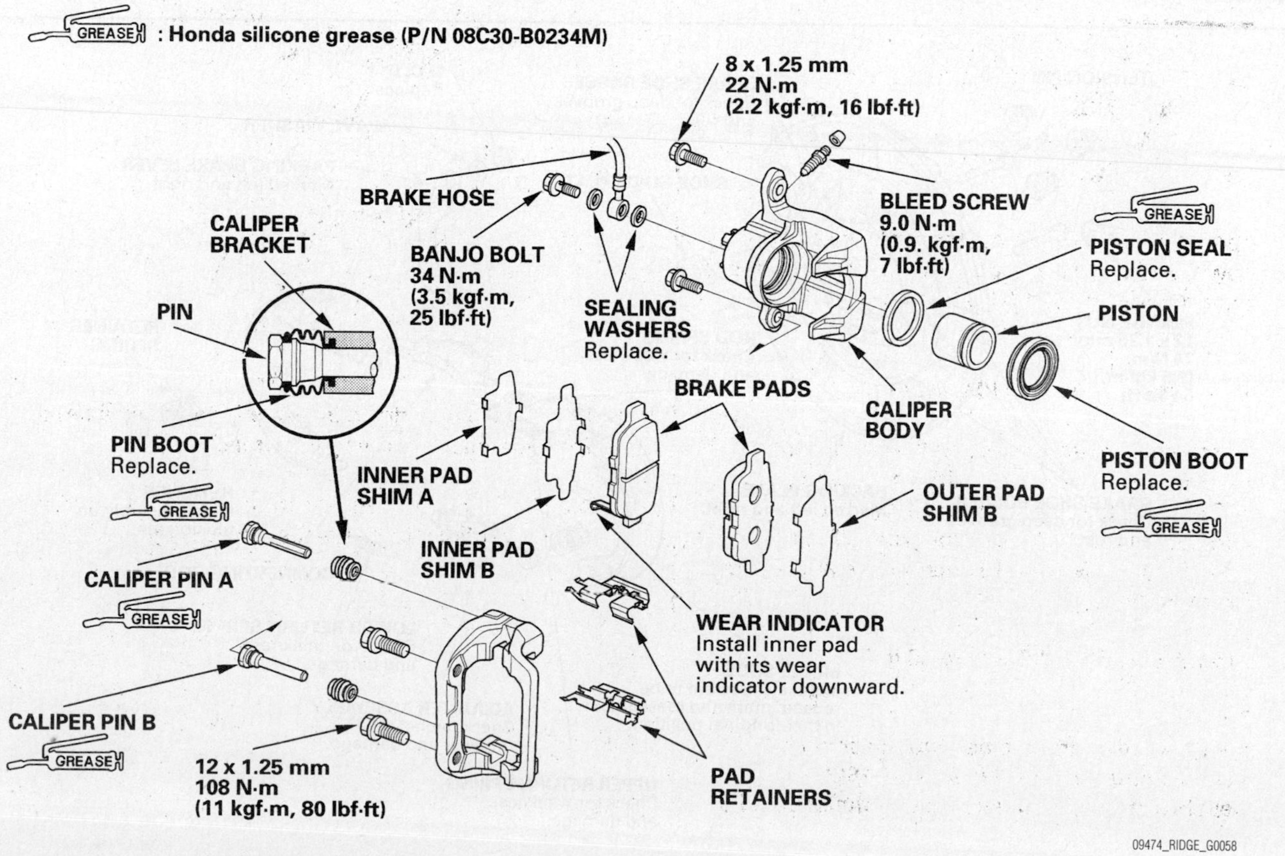

GREASE : Honda silicone grease (P/N 08C30-B0234M)
8 x 1.25 mm
22 N·m
(2.2 kgf·m, 16 lbf·ft)
BRAKE HOSE
BLEED SCREW
9.0 N·m
(0.9. kgf·m,
7 lbf·ft)
GREASE
PISTON SEAL
Replace.
CALIPER
BRACKET
BANJO BOLT
34 N·m
(3.5 kgf·m,
25 lbf·ft)
PIN
SEALING
WASHERS
Replace.
PISTON
BRAKE PADS
CALIPER
BODY
PIN BOOT
Replace.
INNER PAD
SHIM A
PISTON BOOT
Replace.
OUTER PAD
SHIM B
GREASE
GREASE
INNER PAD
SHIM B
CALIPER PIN A
GREASE
WEAR INDICATOR
Install inner pad
with its wear
indicator downward.
CALIPER PIN B
GREASE
12 x 1.25 mm
108 N·m
(11 kgf·m, 80 lbf·ft)
PAD
RETAINERS
09474_RIDGE_G0058

Fig. 8 Exploded view of the rear caliper components

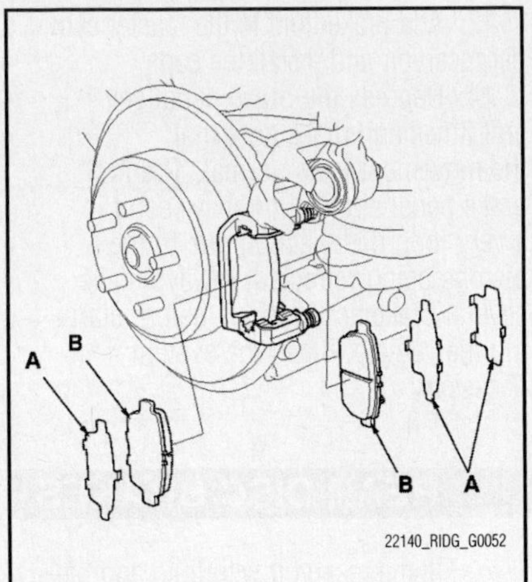
A
B
B
A
22140_RIDG_G0052
Fig. 9 Rear pad shims (A) and the brake pads (B)

1. Remove a small amount of brake fluid from the reservoir using a suction pump.

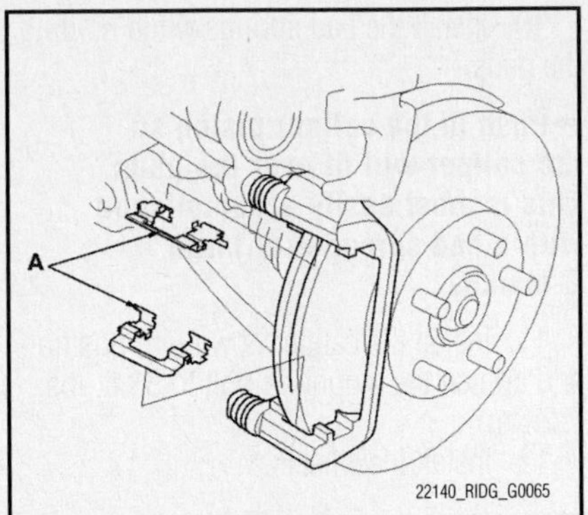
A
22140_RIDG_G0065
Fig. 10 Rear pad retainers (A)

2. Remove the lower caliper pin bolt and pivot the caliper upward.
3. Remove the pads, shims, and pad retainers.

To install:

4. Clean the caliper thoroughly; remove any dirt or dust. Check the brake rotor for grooves or cracks and machine or replace, as necessary.
5. Install the pad retainers. Apply molybdenum brake grease to both surfaces of the shims and the back of the disc brake pads.
6. Install the pads and shims. The wear retainer on the inboard pad faces down.
7. Use a suitable tool to push caliper piston into its bore and enable the caliper to fit over the pads. Lubricate the piston boot with silicon grease. Avoid twisting the boot.
8. Rotate the caliper down and tighten the mounting bolts to 80 ft. lbs. (108 Nm)
9. Install the rear wheels
10. Add brake fluid to the master cylinder reservoir. Depress the brake pedal several times to seat the pads. Bleed the brakes if necessary.

PARKING BRAKE SHOES

REMOVAL & INSTALLATION

See Figure 11.

CAUTION

Before beginning work, review the "Precautions" in this section.

1. Remove the rear wheels.
2. Release the parking brake tension.
3. Remove the caliper and support it out of the way.

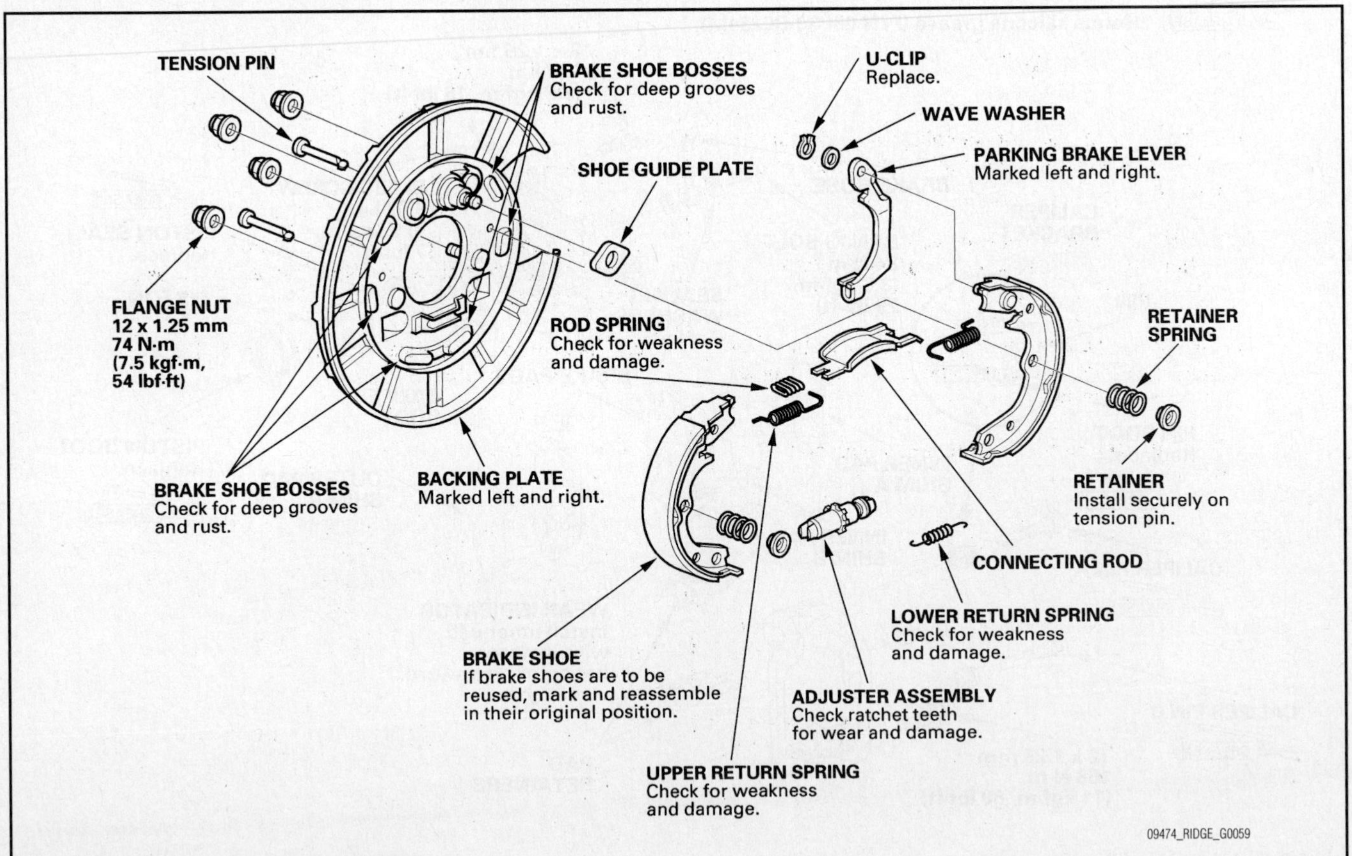
TENSION PIN
BRAKE SHOE BOSSES
Check for deep grooves
and rust.
U-CLIP
Replace.
WAVE WASHER
PARKING BRAKE LEVER
Marked left and right.
SHOE GUIDE PLATE
FLANGE NUT
12 x 1.25 mm
74 N·m
(7.5 kgf·m,
54 lbf·ft)
ROD SPRING
Check for weakness
and damage.
RETAINER
SPRING
BRAKE SHOE BOSSES
Check for deep grooves
and rust.
BACKING PLATE
Marked left and right.
RETAINER
Install securely on
tension pin.
CONNECTING ROD
LOWER RETURN SPRING
Check for weakness
and damage.
BRAKE SHOE
If brake shoes are to be
reused, mark and reassemble
in their original position.
ADJUSTER ASSEMBLY
Check ratchet teeth
for wear and damage.
UPPER RETURN SPRING
Check for weakness
and damage.
09474_RIDGE_G0059
Fig. 11 Parking brake shoes and related parts

4. Remove the rotor/drum assembly.
5. Remove the upper return springs.
6. Remove the hold-down pins and retainers.
7. Remove the connecting rod.
8. Remove the lower return spring and adjuster.
9. Remove the forward shoe.
10. Disconnect the cable and remove the rear shoe.

To install:

11. Installation is the reverse of removal.
12. Clean the backing plate thoroughly and apply a suitable brake grease to all mounting points and the adjuster threads.
13. Adjust the parking brake shoes.
14. While driving the vehicle safely, pull the parking brake release lever.
15. Press the parking brake pedal 2–4 clicks.
16. Drive the vehicle for one-quarter mile at no more that 30 mph.
17. Stop the vehicle and release the parking brake for 10 minutes to allow the drums to cool.
18. Repeat the procedure 3 more times.
19. Recheck the adjustment.

ADJUSTMENT

See Figures 12 and 13.

⁂ CAUTION

Before beginning work, review the "Precautions" in this section.

1. With the parking brake released, back off the adjusting nut at the pedal.
2. Remove the access plug from the drum.
3. Using an adjusting spoon, turn up the ratchet teeth on the adjuster until the shoes lock the drum. Then, back off 10 clicks and install the plug.
4. Press the parking brake pedal with about 66 lbs. of force. The pedal should travel 10–12 clicks. If it travels more than that.
5. Turn the adjusting nut at the pedal until the brake shoes drag slightly when the rear wheels are turned.
6. Back off the adjusting nut in half-turn increments until the proper pressure gives the proper number of clicks.

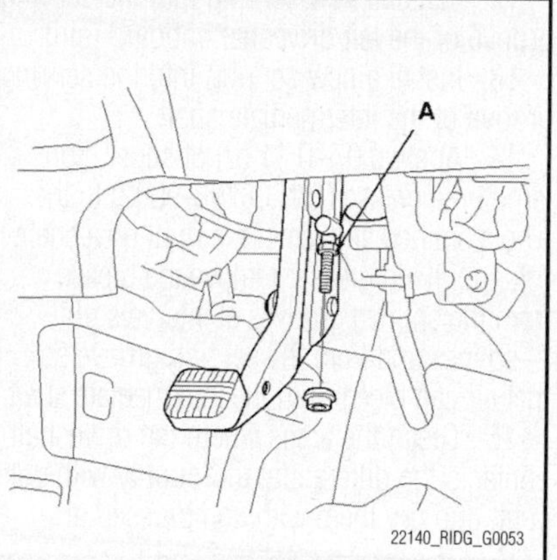
A
22140_RIDG_G0053

Fig. 12 With the parking brake released, back off the adjusting nut (A) at the pedal

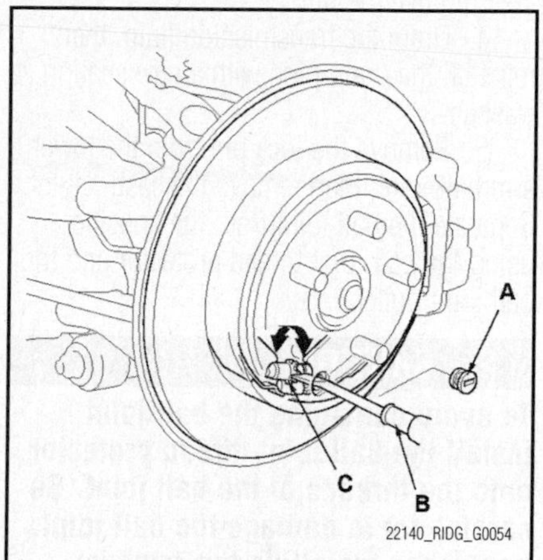
A
C
B
22140_RIDG_G0054

Fig. 13 Remove the access plug (A) from the drum. Using an adjusting spoon (B), turn up the ratchet teeth on the adjuster (C) until the shoes lock the drum. Then, back off 10 clicks and install the plug

CHASSIS ELECTRICAL

AIR BAG (SUPPLEMENTAL RESTRAINT SYSTEM)

GENERAL INFORMATION

⁂ CAUTION

These vehicles are equipped with an air bag system. The system must be disarmed before performing service on, or around, system components, the steering column, instrument panel components, wiring and sensors. Failure to follow the safety precautions and the disarming procedure could result in accidental air bag deployment, possible injury and unnecessary system repairs.

SERVICE PRECAUTIONS

⁂ CAUTION

Disconnect and isolate the battery negative cable before beginning any airbag system component diagnosis, testing, removal, or installation procedures. Wait at least 90 seconds after the ignition switch is turned off and the negative (-) terminal cable is disconnected from the battery before starting the operation. The SRS is equipped with a backup power source, so if work is started within 90 seconds after disconnecting the negative (-) terminal cable from the battery, the SRS may be deployed. Failure to disable the airbag system may result in accidental airbag deployment, personal injury, or death.

DISARMING THE SYSTEM

Disconnect and isolate the negative battery cable. Wait 3 minutes for the system capacitor to discharge before performing any service.

ARMING THE SYSTEM

To arm the system, connect the negative battery cable.

CLOCKSPRING CENTERING

See Figure 14.

1. Only used cable reel need be centered, new replacement cable reels come centered.

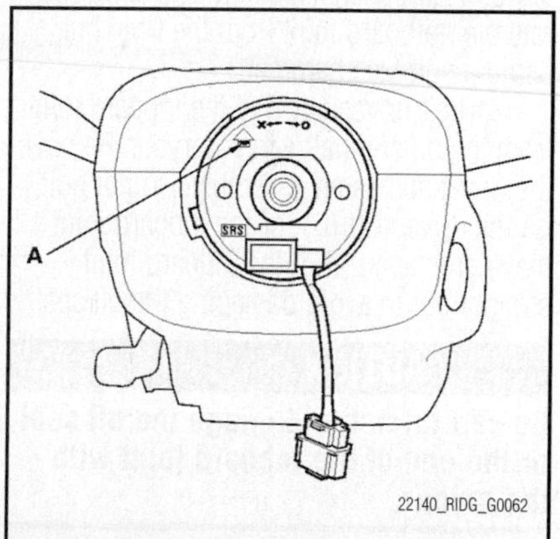
A
SRS
22140_RIDG_G0062

Fig. 14 Rotate the cable reel clockwise until it stops. Then rotate it counterclockwise (three full turns) until the arrow mark (A) on the cable reel label points straight up

2. Rotate the cable reel clockwise until it stops.
3. Then rotate it counterclockwise (three full turns) until the arrow mark on the cable reel label points straight up.
4. Install the cable reel.

DRIVE TRAIN

FRONT HALFSHAFT (DRIVESHAFT)

REMOVAL & INSTALLATION

1. Raise and support the vehicle.
2. Remove the front wheels.
3. Pry up the stake on the spindle nut, then remove the nut.
4. Drain the transmission fluid, then reinstall the drain plug with a new sealing washer.
5. Remove the lock pin from the lower arm ball joint, then remove the castle nut. Separate the ball joint from the knuckle, using the ball joint thread protector and the ball joint puller.

CAUTION

To avoid damaging the ball joint, install the ball joint thread protector onto the threads of the ball joint. Be careful not to damage the ball joint boot when installing the remover.

CAUTION

Do not force or hammer on the lower arm, or pry between the lower arm and the knuckle. You could damage the ball joint.

6. Pull the knuckle outward, and separate the outboard joint from the front hub using a soft face hammer.
7. Left driveshaft: Pry the inboard joint from the differential using a prybar. Remove the driveshaft as an assembly. Do not pull on the driveshaft (B), or the inboard joint may come apart. Pull the inboard joint straight out to avoid damaging the oil seal.

CAUTION

Be careful not to damage the oil seal or the end of the inboard joint with the prybar.

8. Right driveshaft: Drive the inboard joint off of the intermediate shaft using a drift punch and a hammer. Remove the driveshaft as an assembly. Do not pull on the driveshaft (B), or the inboard joint may come apart.

CAUTION

Be careful not to damage the end of the inboard joint with the drift.

9. Remove the set ring from the left driveshaft inboard joint.
10. Remove the set ring from the intermediate shaft.

To install:

➡Before starting installation, make sure the mating surfaces of the joint and the splined section are clean.

11. Apply about 0.18 oz. moly 60 paste, (P/N 08734-0001) to the contact area of the outboard joint and the front wheel bearing.

➡The paste helps to prevent noise and vibration.

12. Install a new set ring into the set ring groove of the left driveshaft inboard joint.
13. Install a new set ring into the set ring groove of the intermediate shaft.
14. Apply 0.07–0.11 oz) of super high temp urea grease (P/N 08798-9002) to the whole splined surface of the right driveshaft. After applying grease, remove the grease from the splined grooves at intervals of 2–splines and from the set ring groove so that air can bleed from the intermediate shaft.
15. Clean the areas where the driveshaft contacts the differential thoroughly with solvent, and dry them with compressed air.

CAUTION

Do not wash the rubber parts with solvent.

16. Insert the inboard end of the driveshaft into the differential or intermediate shaft until the set ring locks in the groove.

➡Insert the driveshaft horizontally to prevent damaging the oil seal.

17. Install the outboard driveshaft CV joint end into the front hub on the knuckle.
18. Wipe off any grease contamination from the ball joint tapered section and threads, then install the knuckle onto the lower arm. Be careful not to damage the ball joint boot.
19. Install a new castle nut and torque the castle nut to the lower torque specification, then tighten it only far enough to align the slot with the ball joint pin hole. Tightening range is 65–72 ft. lbs. (88–98 Nm). Install the lock pin into the pin hole.

➡Do not align the nut by loosening it.

20. Apply a small amount of engine oil to the seating surface of a new spindle nut.
21. Install the spindle nut, then tighten it to 242 ft. lbs. (328 Nm). After tightening, use a drift to stake the spindle nut shoulder against the driveshaft.
22. Clean the mating surfaces of the brake disc and the wheel, then install the front wheels.
23. Turn the front wheel by hand, and make sure there is no interference between the driveshaft and surrounding parts.
24. Lower the vehicle.
25. Refill the transmission with the recommended transmission fluid.
26. Check the wheel alignment, and adjust it if necessary.
27. Test-drive the vehicle.

INTERMEDIATE SHAFT

REMOVAL & INSTALLATION

See Figure 15.

1. Drain the transaxle.
2. Remove the right driveshaft.
3. Remove the subframe stiffener.
4. Remove the exhaust pipe and bracket.
5. Remove the heat shield.
6. Remove the intermediate shaft from the differential.

➡Hold the intermediate shaft horizontally to avoid damage to the seal.

To install:

7. Installation is the reverse of removal, observing the following torques:
   a. Torque the heat shield bolts to 29 ft. lbs. (39 Nm).

REAR DIFFERENTIAL ASSEMBLY

REMOVAL & INSTALLATION

See Figure 16.

1. Raise and support the vehicle.
2. Drain the differential fluid.
3. Make a reference mark across the propeller shaft, and the rear differential companion flange. Separate the propeller shaft from the rear differential.

➡Suspend the propeller shaft with an appropriate size wire or nylon strap.

4. Remove the left rear driveshaft.
5. Use a prybar or fork tool to disconnect the right rear driveshaft inboard joint from the differential.
6. Suspend the right rear driveshaft with an appropriate size wire or nylon strap.
7. Place a transmission jack under the rear differential.
8. Disconnect the 6P connector and the 2P connector connectors, then remove the mounting bolts and stopper.
9. Lower the rear differential a little on the transmission jack, then separate the right rear driveshaft inboard joint from the rear differential.

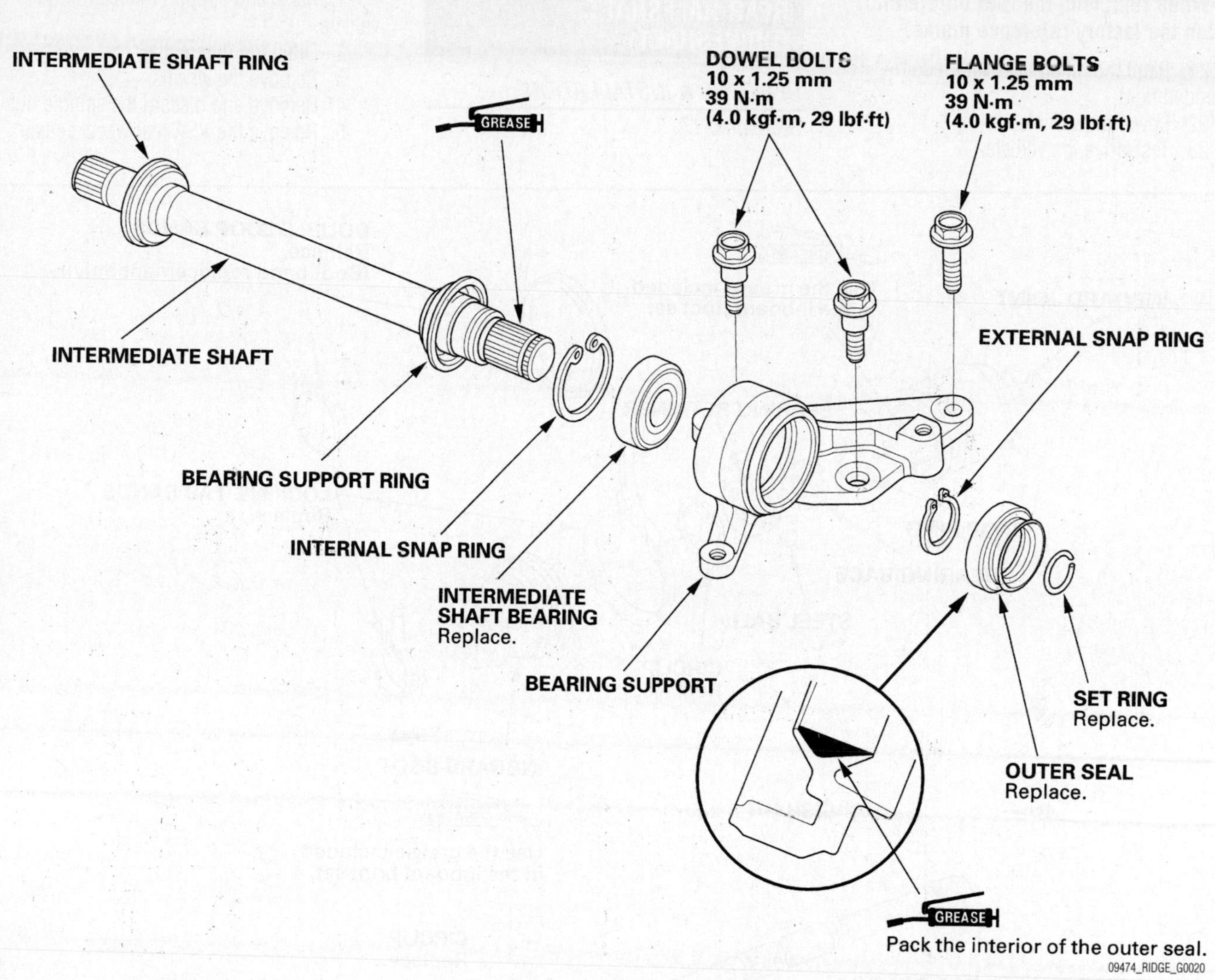
INTERMEDIATE SHAFT RING
GREASE
DOWEL BOLTS
10 x 1.25 mm
39 N·m
(4.0 kgf·m, 29 lbf·ft)
FLANGE BOLTS
10 x 1.25 mm
39 N·m
(4.0 kgf·m, 29 lbf·ft)
INTERMEDIATE SHAFT
EXTERNAL SNAP RING
BEARING SUPPORT RING
INTERNAL SNAP RING
INTERMEDIATE
SHAFT BEARING
Replace.
BEARING SUPPORT
SET RING
Replace.
OUTER SEAL
Replace.
GREASE
Pack the interior of the outer seal.
09474_RIDGE_G0020

Fig. 15 Intermediate shaft exploded view

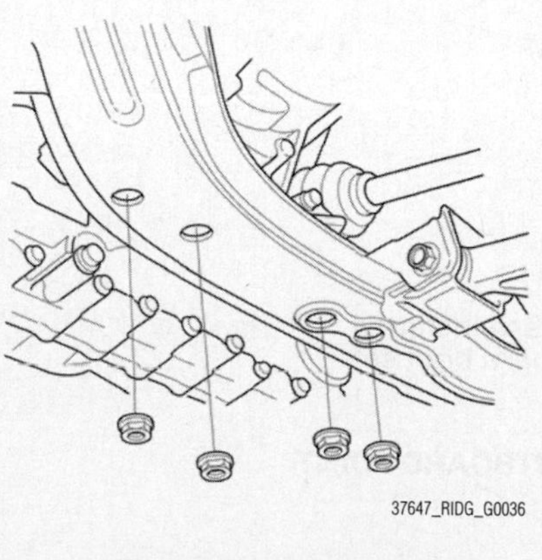
37647_RIDG_G0036

Fig. 16 Place a transmission jack under the rear differential. Disconnect the 6P connector and the 2P connector connectors, then remove the mounting bolts and stopper.

10. Disconnect the breather hose from the rear differential and remove the set rings.

11. Lower the rear differential on the transmission jack.

To install:

12. Place the rear differential on a transmission jack.

13. Install new set rings into the groove of the rear differential.

14. Raise the rear differential a little on the transmission jack, then connect the breather tube to the rear differential.

15. Apply 0.05–0.07 oz. of super high temp urea grease (P/N 08798-9002) to the splines of the right rear driveshaft inboard joint, then install the right rear driveshaft inboard joint to the rear differential.

16. Raise the rear differential to the mounting level, then install new mounting bolts.

17. Install the mounting bolt and the stopper.

18. Connect the 6P connector and 2P connector.

19. Install the left rear driveshaft.

20. Attach the propeller shaft to the rear differential by aligning the reference marks you made during the removal procedure. Then install new flange bolts to the specified torque:

- Differential mounting (14 x 1.5 mm) horizontal bolts: 63 ft. lbs. (85 Nm)
- Differential mounting (10 x 1.25 mm) vertical bolt: 41 ft. lbs. (55 Nm)
- Propeller shaft flange bolts: 53 ft. lbs. (72 Nm)

➡When replacing the rear differential, align the factory reference marks.

21. Refill the differential with recommended fluid.
22. Lower the vehicle.
23. Test-drive the vehicle.

REAR HALFSHAFT (DRIVESHAFT)

REMOVAL & INSTALLATION

See Figure 17.

1. Raise and support the rear of the vehicle.
2. Drain the differential.
3. Remove the wheels.
4. Remove and discard the spindle nut.
5. Remove the VSA rear wheel sensor.

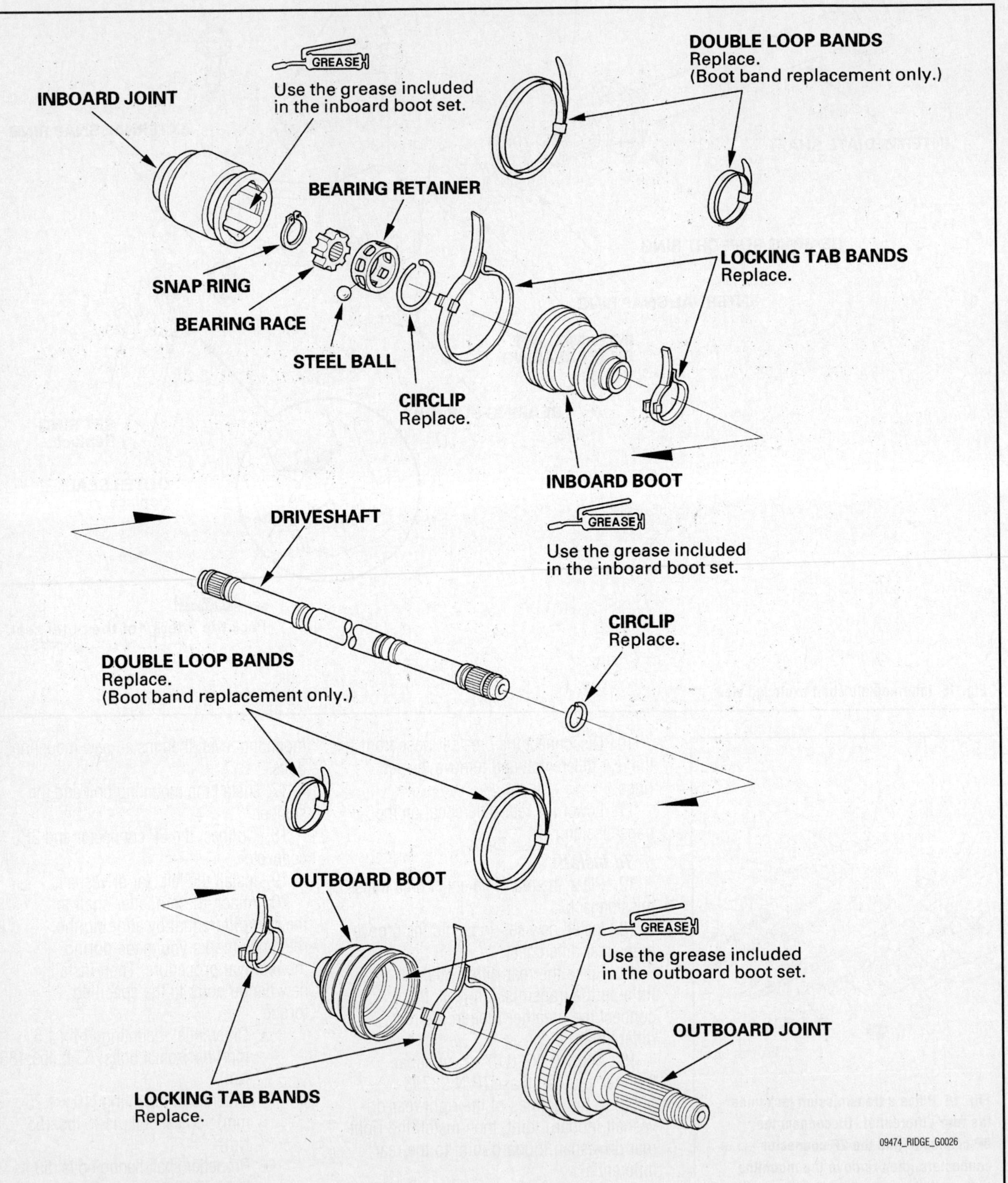
GREASE
Use the grease included in the inboard boot set.
INBOARD JOINT
DOUBLE LOOP BANDS
Replace.
(Boot band replacement only.)
BEARING RETAINER
LOCKING TAB BANDS
Replace.
SNAP RING
BEARING RACE
STEEL BALL
CIRCLIP
Replace.
INBOARD BOOT
DRIVESHAFT
GREASE
Use the grease included in the inboard boot set.
CIRCLIP
Replace.
DOUBLE LOOP BANDS
Replace.
(Boot band replacement only.)
OUTBOARD BOOT
GREASE
Use the grease included in the outboard boot set.
OUTBOARD JOINT
LOCKING TAB BANDS
Replace.
09474_RIDGE_G0026

Fig. 17 Rear halfshaft exploded view

6. Separate the upper ball joint from the uppor arm.
7. Remove the lower track rod.
8. Separate the lower arm from the knuckle.
9. Pull the knuckle outwards and separate the halfshaft from the hub.
10. Pry the halfshaft from the differential.
11. Installation is the reverse of removal.
   a. Use new snaprings and cotter pins.
   b. Always advance castellated nuts to align cotter pin holes.
   c. Observe the following torques:
   - Lower arm-to-knuckle bolt: 105 ft. lbs. (142 Nm)
   - Track rod inner end: 69 ft. lbs. (93 Nm)
   - Track rod outer end: 74 ft. lbs. (101 Nm)
   - Upper arm ball stud nut: 40 ft. lbs. (54 Nm)
   - Halfshaft end nut: 181 ft. lbs. (245 Nm)

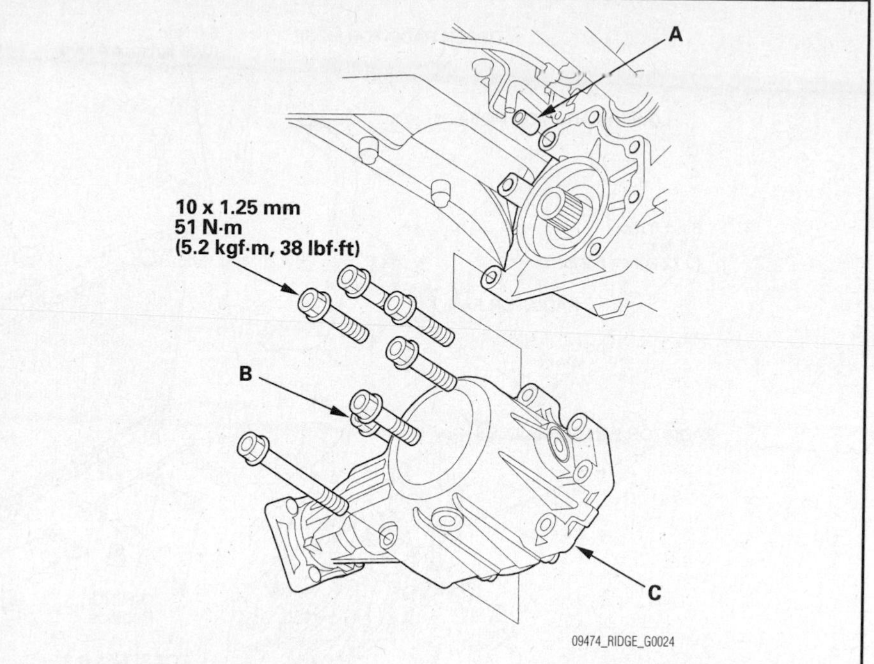
A
10 x 1.25 mm
51 N·m
(5.2 kgf·m, 38 lbf·ft)
B
C
09474_RIDGE_G0024
Fig. 18 Transfer case installation

TRANSFER CASE ASSEMBLY

REMOVAL & INSTALLATION

See Figure 18.

1. Raise and support the vehicle.
2. Place the transaxle in **N**.
3. Drain the transaxle. When the fluid is drained, install the drain plug, using a new washer and torque to 36 ft. lbs. (49 Nm).
4. Remove the subframe stiffener.
5. Remove the exhaust pipe and bracket.
6. Disconnect the transfer case breather.
7. Matchmark the driveshaft and disconnect it from the transfer case.
8. Remove the bolts and remove the transfer case.
9. Installation is the reverse of removal. Observe the following torques:
   - Transfer case bolts: 38 ft. lbs. (51 Nm)
   - Driveshaft bolts: 53 ft. lbs. (72 Nm)
   - Exhaust pipe-to-converter: 25 ft. lbs. (33 Nm)
   - Exhaust pipe hanger: 16 ft. lbs. (22 Nm)
   - Exhaust pipe-to-manifold: 40 ft. lbs. (54 Nm)
   - Stiffener (new bolts): 40 ft. lbs. (54 Nm)

ENGINE COOLING

ENGINE FAN

REMOVAL & INSTALLATION

➡See "Radiator & Fan".

RADIATOR & FAN

REMOVAL & INSTALLATION

See Figure 19.

1. Drain the engine coolant.
2. Remove the front bulkhead cover.
3. Disconnect the fan motor connectors and the engine coolant temperature sensor 2 connector, then remove the harness connector, the harness clamps, and the coolant reservoir hose.
4. Remove the upper radiator hose and the lower radiator hose.
5. Raise the vehicle on the lift to full height.
6. Remove the splash shield.
7. Remove the automatic transmission fluid (ATF) cooler hoses from the radiator, then plug the line and the hose.
8. Unclamp the ATF cooler hose clamp on the ATF cooler line.
9. Lower the vehicle on the lift.
10. Remove the upper bracket and cushions, then pull up the radiator.
11. Remove the fan shroud assemblies and other parts from the radiator.

To install:

12. Install the radiator in the reverse order of removal.
13. Make sure the upper and lower cushions are set securely.
14. Fill the radiator with engine coolant, then bleed the air from the cooling system.

THERMOSTAT

REMOVAL & INSTALLATION

See Figure 20.

1. Remove the breather pipe, then remove the air intake duct.
2. Drain the engine coolant.
3. Remove the ground cable and thermostat cover, then remove the thermostat.

To install:

4. Install the thermostat with a new rubber seal.
5. Refill the radiator with engine coolant, then bleed air from the cooling system.
6. Clean up any spilled engine coolant.

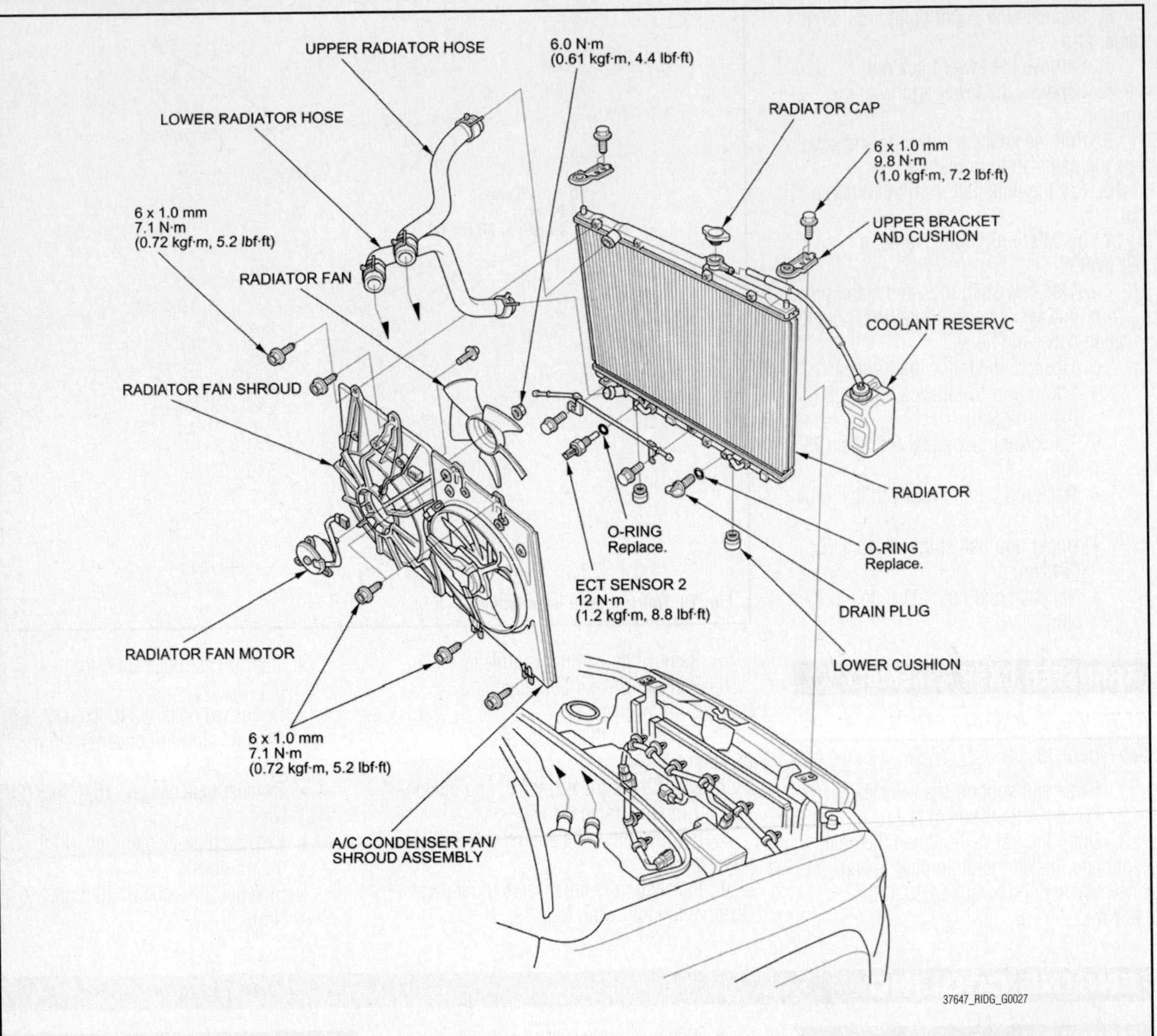
UPPER RADIATOR HOSE
6.0 N·m
(0.61 kgf·m, 4.4 lbf·ft)
LOWER RADIATOR HOSE
RADIATOR CAP
6 x 1.0 mm
9.8 N·m
(1.0 kgf·m, 7.2 lbf·ft)
6 x 1.0 mm
7.1 N·m
(0.72 kgf·m, 5.2 lbf·ft)
UPPER BRACKET
AND CUSHION
RADIATOR FAN
COOLANT RESERVC
RADIATOR FAN SHROUD
RADIATOR
O-RING
Replace.
O-RING
Replace.
ECT SENSOR 2
12 N·m
(1.2 kgf·m, 8.8 lbf·ft)
DRAIN PLUG
RADIATOR FAN MOTOR
LOWER CUSHION
6 x 1.0 mm
7.1 N·m
(0.72 kgf·m, 5.2 lbf·ft)
A/C CONDENSER FAN/
SHROUD ASSEMBLY
37647_RIDG_G0027

Fig. 19 Exploded view of the radiator and cooling fan assembly

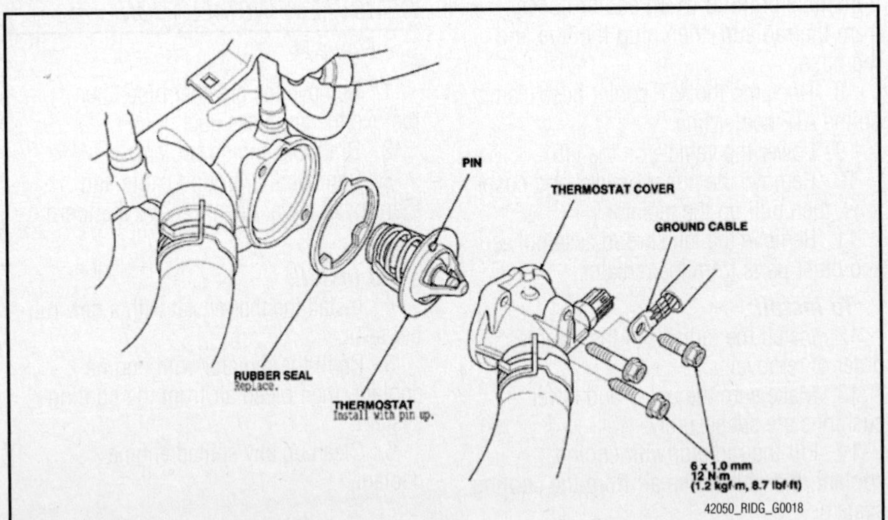
PIN
THERMOSTAT COVER
GROUND CABLE
RUBBER SEAL
Replace.
THERMOSTAT
Install with pin up.
6 x 1.0 mm
12 N·m
(1.2 kgf·m, 8.7 lbf·ft)
42050_RIDG_G0018

Fig. 20 Exploded view of the thermostat mounting

WATER PUMP

REMOVAL & INSTALLATION

See Figure 21.

1. Drain the cooling system.
2. Make sure the anti-theft codes for the audio system and navigation system (if equipped) are available.
3. Disconnect the negative battery cable.
4. Remove the accessory drive belts.
5. Remove the front cover.
6. Remove the timing belt and the belt tensioner. See "Timing Belt & Sprockets" in "ENGINE MECHANICAL" section.
7. Remove the water pump.

To install:

8. Install the water pump. Use a new O-ring seal and tighten the bolts to 105 inch lbs. (12 Nm).
9. Install the timing belt tensioner.
10. Install the timing belt.
11. Install the front cover.
12. Install the accessory drive belts
13. Connect the negative battery cable.
14. Enter the anti-theft codes for the audio system and the navigation system (if equipped).
15. Set the clock on vehicles without navigation.
16. Perform the power window control unit reset procedure.
17. Fill and bleed the cooling system.
18. Start the engine and check for leaks.

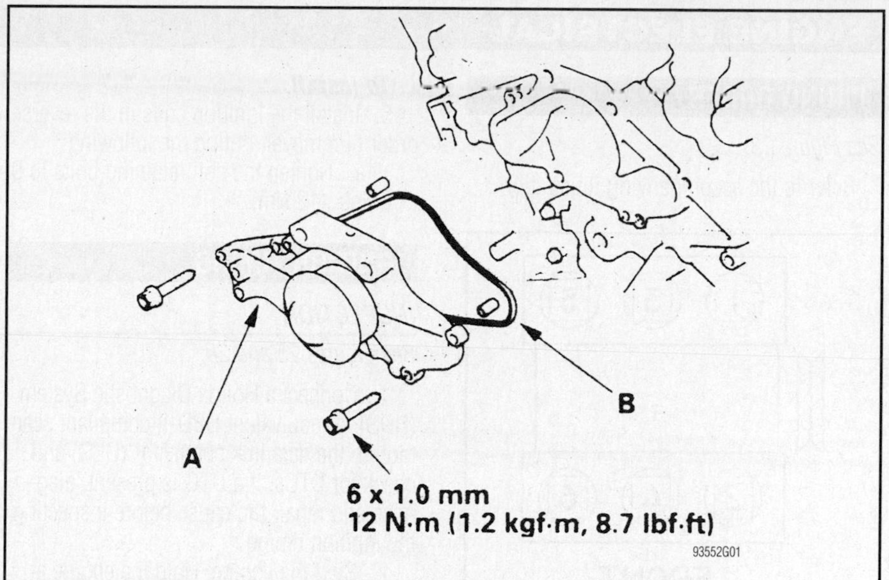
A
B
6 x 1.0 mm
12 N·m (1.2 kgf·m, 8.7 lbf·ft)
93552G01

Fig. 21 Exploded view of the water pump mounting. A is the pump; B is the seal

ENGINE ELECTRICAL

CHARGING SYSTEM

ALTERNATOR

REMOVAL & INSTALLATION

See Figure 22.

1. Make sure you have the anti-theft codes for the radio and navigation systems.
2. Make sure the anti-theft codes for the audio system and navigation system (if equipped) are available.
3. Disconnect the negative battery cable.
4. Disconnect the positive battery cable.
5. Remove the intake manifold cover.
6. Remove the accessory drive belt.
7. Remove the alternator wiring harness connectors.
8. Remove the alternator mounting bolts.
9. Remove the wiring harness clamp.
10. Remove the alternator.

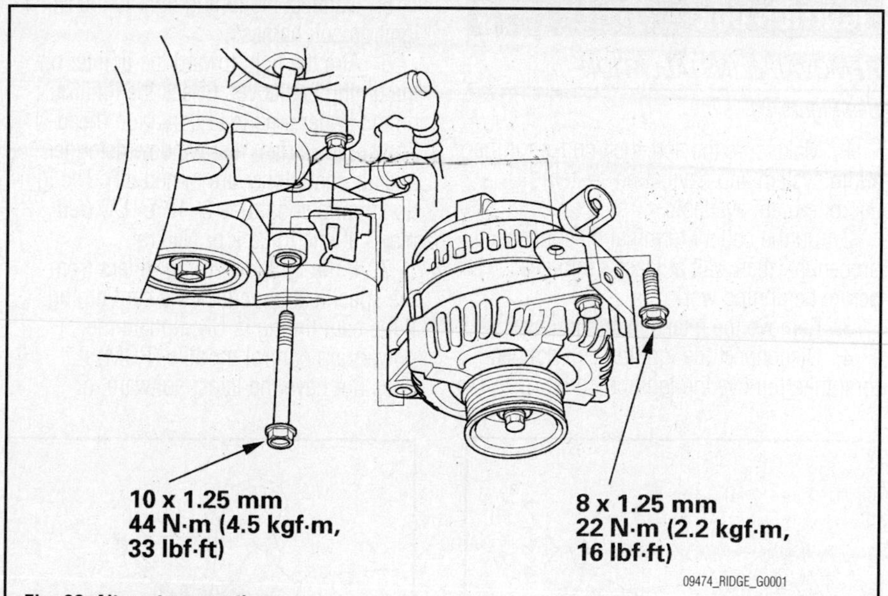
10 x 1.25 mm
44 N·m (4.5 kgf·m,
33 lbf·ft)
8 x 1.25 mm
22 N·m (2.2 kgf·m,
16 lbf·ft)
09474_RIDGE_G0001

Fig. 22 Alternator mounting

To install:

11. Install the alternator.
12. Install the wiring harness clamp. Tighten the bolt to 105 inch lbs. (12 Nm).
13. Install the alternator mounting bolts. Tighten the lower 10mm bolt to 33 ft. lbs. (44 Nm) and the upper 8mm bolt to 16 ft. lbs. (22 Nm).
14. Install the alternator wiring harness connectors. Tighten the battery terminal nut to 105 inch lbs. (12 Nm).
15. Install the accessory drive belt.
16. Install the intake manifold cover.
17. Connect the positive cable.
18. Connect the negative battery cable.
19. Enter the anti-theft codes for the audio system and the navigation system (if equipped).
20. Set the clock on vehicles without navigation.
21. Perform the power window control unit reset procedure.

ENGINE ELECTRICAL

FIRING ORDER

See Figure 23.

Refer to the accompanying illustration

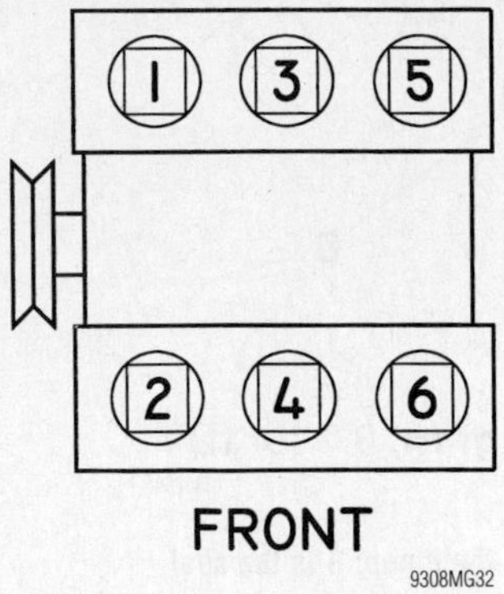
1
3
5
2
4
6
FRONT
9308MG32

Fig. 23 3.5L Engine—Firing Order: 1-4-2-5-3-6

IGNITION COIL

REMOVAL & INSTALLATION

See Figure 24.

1. Make sure the anti-theft codes for the audio system and navigation system (if equipped) are available.
2. Do the battery terminal disconnect procedure, then wait at least 3 minutes before beginning work.
3. Remove the intake manifold cover.
4. Disconnect the ignition coil connectors, then remove the ignition coils.

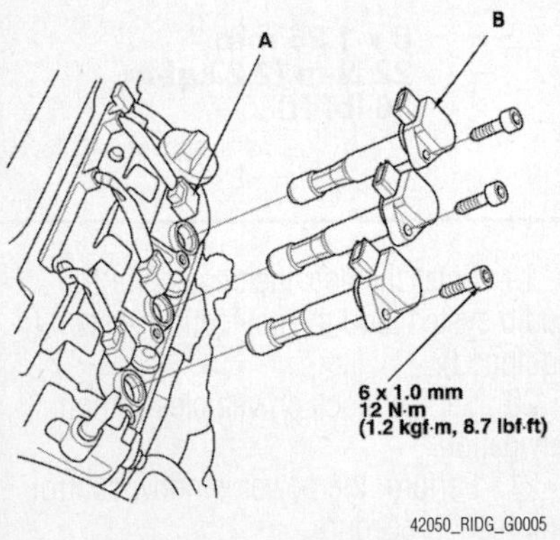
A
B
6 x 1.0 mm
12 N·m
(1.2 kgf·m, 8.7 lbf·ft)
42050_RIDG_G0005

Fig. 24 View of the ignition coil connectors (A) and ignition coils (B)—front bank shown, rear bank similar

To install:

5. Install the ignition coils in the reverse order of removal, noting the following:
   a. Tighten the coil retaining bolts to 9 ft. lbs. (12 Nm).

IGNITION TIMING

INSPECTION

See Figures 25 and 26.

1. Connect a Honda Diagnostic System (HDS), or equivalent OBD-II compliant scan tool, to the data link connector (DLC) and check for DTCs. If a DTC is present, diagnose and repair the cause before inspecting the ignition timing.
2. Start the engine. Hold the engine at 3,000 rpm with no load (in Neutral) until the radiator fan comes on, then let it idle.
3. Check the idle speed, as outlined in the Fuel System Section.
4. Select "SCS" mode using the HDS.
5. Connect the timing light to the No. 1 ignition coil harness.
6. Aim the light toward the pointer on the timing belt cover. Check the ignition timing under a no load condition (headlights, blower fan, rear window defogger, and air conditioner are turned off). The ignition timing should be 8–12°BTDC (red mark) at idle in Park or Neutral.
7. If the ignition timing differs from the specification, check the cam timing. If the cam timing is OK, update the powertrain control module (PCM) if it does not have the latest software, or substitute a known-good PCM, then recheck. If the system works properly, and the PCM was substituted, replace the original PCM.
8. Disconnect the HDS and the timing light.

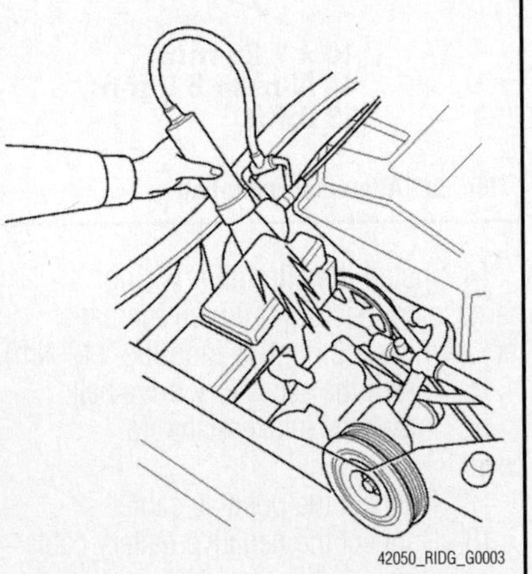
42050_RIDG_G0003

Fig. 25 Connect the timing light to the No. 1 ignition coil harness

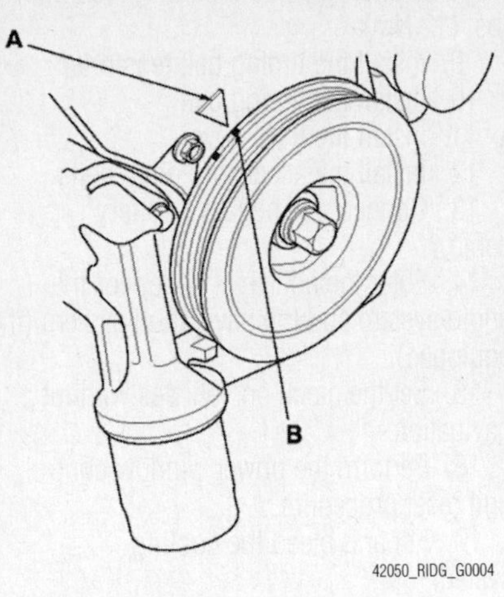
A
B
42050_RIDG_G0004

Fig. 26 Aim the light toward the pointer (A) on the timing belt cover. The timing mark is red (B)

ADJUSTMENT

The ignition timing is controlled by the Powertrain Control Module (PCM). No adjustment is necessary or possible.

SPARK PLUGS

REMOVAL & INSTALLATION

1. Make sure the anti-theft codes for the audio system and navigation system (if equipped) are available.
2. Disconnect the negative battery cable.
3. Remove the ignition coils.
4. Remove the spark plug.
5. Inspect the spark plug.

To install:

6. Install the spark plug and tighten to 13 ft. lbs. (18 Nm).
7. Install the ignition coils.
8. Connect the negative battery cable.
9. Enter the anti-theft codes for the audio system and the navigation system (if equipped).
10. Set the clock on vehicles without navigation.
11. Perform the power window control unit reset procedure.

ENGINE ELECTRICAL

STARTER

REMOVAL & INSTALLATION

See Figure 27.

1. Do the battery terminal disconnect procedure, then wait at least 3 minutes before beginning work.
2. Remove the automatic transmission fluid (ATF) dipstick and the radiator hose bracket.
3. Remove the harness clamp.
4. Disconnect the starter cable from the B terminal, then disconnect the BLK/WHT harness connector from the S terminal.
5. Remove the two bolts holding the starter.

To install:

6. Install the starter using a new gasket, then install the harness clamp, and connect the B terminal and the BLK/WHT harness connector. Make sure the crimped side of the B terminal faces away from the starter when you connect it.
7. Install the ATF dipstick and the radiator hose bracket.
8. Do the battery terminal reconnection procedure.
9. Start the engine to make sure the starter works properly.

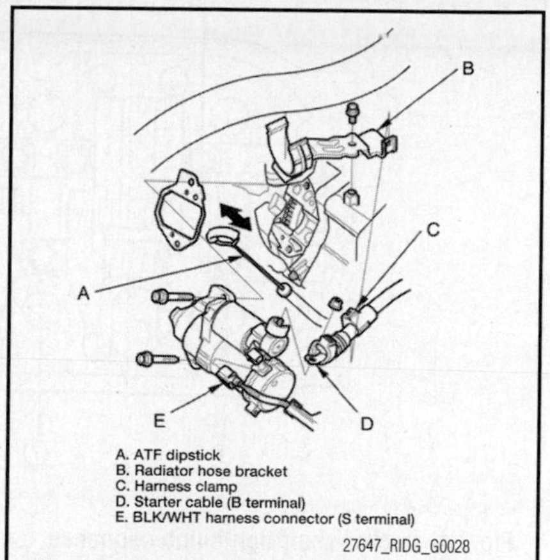
B
C
A
E
D
A. ATF dipstick
B. Radiator hose bracket
C. Harness clamp
D. Starter cable (B terminal)
E. BLK/WHT harness connector (S terminal)
27647_RIDG_G0028

Fig. 27 Removing the starter

ENGINE MECHANICAL

ACCESSORY DRIVE BELTS

ACCESSORY BELT ROUTING

See Figure 28.

Refer to the accompanying illustration.

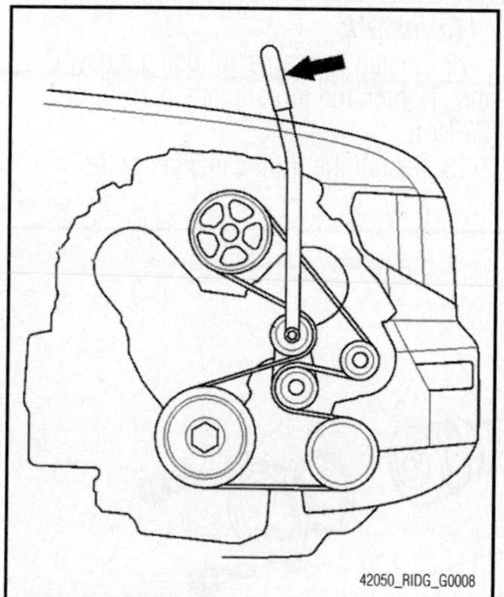
42050_RIDG_G0008

Fig. 28 Drive belt routing

INSPECTION

See Figure 29.

1. Inspect the belt for cracks or damage. If the belt is cracked or damaged, replace it.
2. Check that the auto-tensioner indicator (A) is within the standard range (B) as shown. If it is out of the standard range, replace the drive belt.

ADJUSTMENT

Belt tension is maintained by an automatic tensioner. No adjustments are necessary or possible.

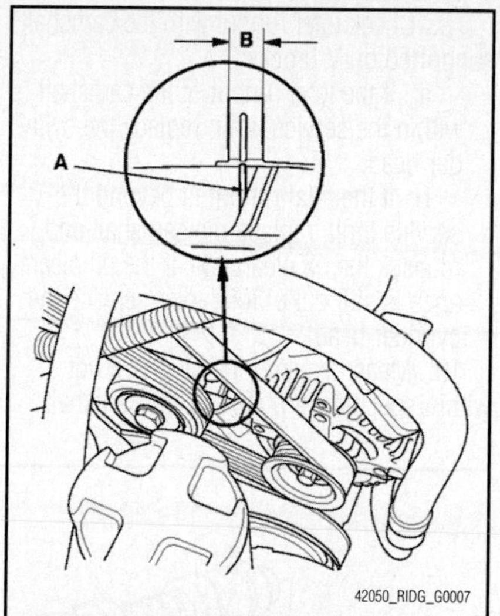
B
A
42050_RIDG_G0007

Fig. 29 Check that the auto-tensioner indicator (A) is within the standard range (B) as shown. If it is out of the standard range, replace the drive belt

REMOVAL & INSTALLATION

See Figure 30.

1. Move the auto-tensioner (A) to relieve tension from the drive belt, then remove the drive belt.
2. Install the new belt in the reverse order of removal.

CAMSHAFT & VALVE LIFTERS

INSPECTION

See Figures 31 and 32.

✲✲ CAUTION

Before beginning work, review the "Precautions" in this section.

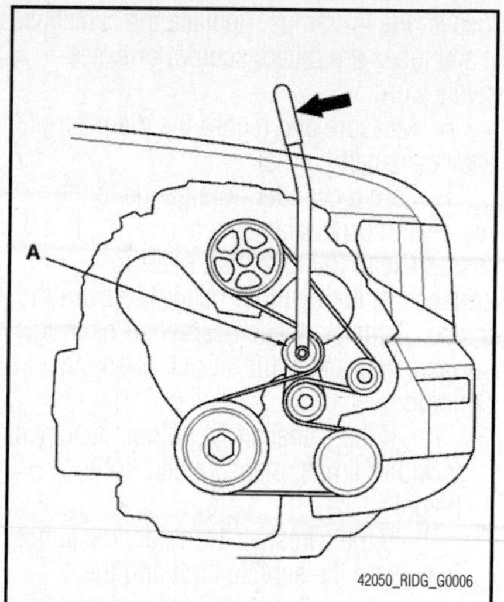
A
42050_RIDG_G0006

Fig. 30 Move the auto-tensioner (A) to relieve tension from the drive belt, then remove the drive belt

1. With the cylinder head and rocker arms removed, install the rocker shafts on the cylinder head, then torque the bolts in sequence to 17 ft. lbs. (24 Nm).

➡Apply new engine oil to the threads and flange of the exhaust rocker shaft mounting bolts.

2. Seat the camshaft by pushing it toward the rear of the cylinder head.
3. Zero the dial indicator against the end of the camshaft. Push the camshaft back and forth and read the end play. If the end play is beyond the service limit, replace the thrust cover and recheck. If it is still beyond the service limit, replace the camshaft.

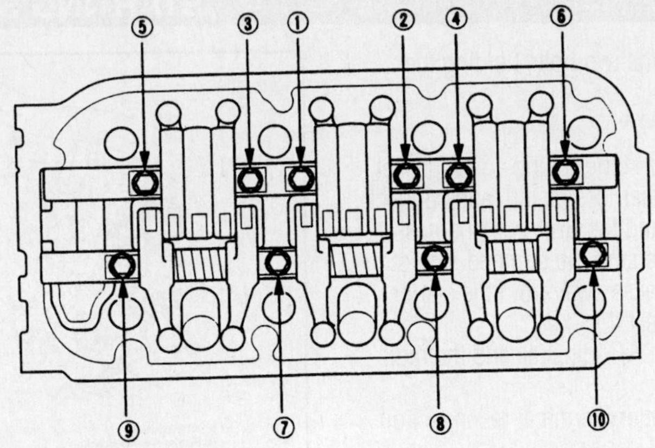

⑤
③
①
②
④
⑥
⑨
⑦
⑧
⑩
42356-HPIL-G15

Fig. 31 Rocker shaft tightening sequence

4. Remove the camshaft thrust cover, then pull out the camshaft.
5. Wipe the camshaft clean, then inspect the lift ramps. Replace the camshaft if any lobes are pitted, scored, or excessively worn.
6. Measure and record the diameter of each camshaft journal.
7. Zero a cylinder bore gauge to the recorded journal diameter.
8. Clean the camshaft bearing surfaces in the cylinder head. Measure the inside diameter of each camshaft bearing surface, and check for an out-of-round condition.
   a. If the camshaft-to-holder clearance is within limits, measure cam lobe height.
   b. If the camshaft-to-holder clearance is beyond the service limit and the camshaft has been replaced, replace the cylinder head.
   c. If the camshaft-to-holder clearance is beyond the service limit and the camshaft has not been replaced, Check total runout with the camshaft supported on V-blocks.
9. Check total runout with the camshaft supported on V-blocks.
   a. If the total runout of the camshaft is within the service limit, replace the cylinder head.
   b. If the total runout is beyond the service limit, replace the camshaft and recheck the oil clearance. If the oil clearance is still out of tolerance, replace the cylinder head.
10. Measure cam lobe height. If not within specification, replace the camshaft.

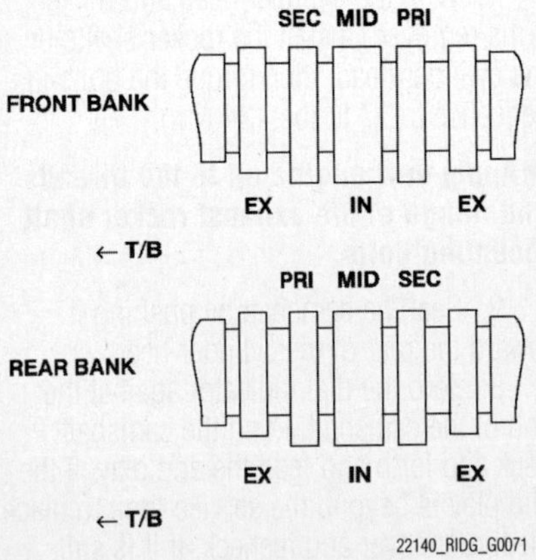

SEC MID PRI
FRONT BANK
EX IN EX
← T/B
PRI MID SEC
REAR BANK
EX IN EX
← T/B
22140_RIDG_G0071

Fig. 32 Camshaft lobe identification

REMOVAL & INSTALLATION

Front

See Figure 33.

⁂ CAUTION

Before beginning work, review the "Precautions" in this section.

1. Make sure the anti-theft codes for the audio system and navigation system (if equipped) are available.
2. Disconnect the negative battery cable.
3. Disconnect the positive battery cable.
4. Remove the battery and battery box.
5. Drain the coolant.
6. Remove the upper radiator hoses.
7. Remove the Exhaust Gas Recirculation (EGR) valve.
8. Remove the timing belt. See "Timing Belt & Sprockets" in this section.
9. Remove the rocker arm assembly. See "Rocker Arms & Shafts" in this section.
10. Remove the front camshaft pulley.
11. Remove the thrust plate and camshaft.

To install:

12. Install the camshaft using a new O-ring. Tighten the thrust plate to 16 ft. lbs. (22 Nm).
13. Install the front camshaft pulley.

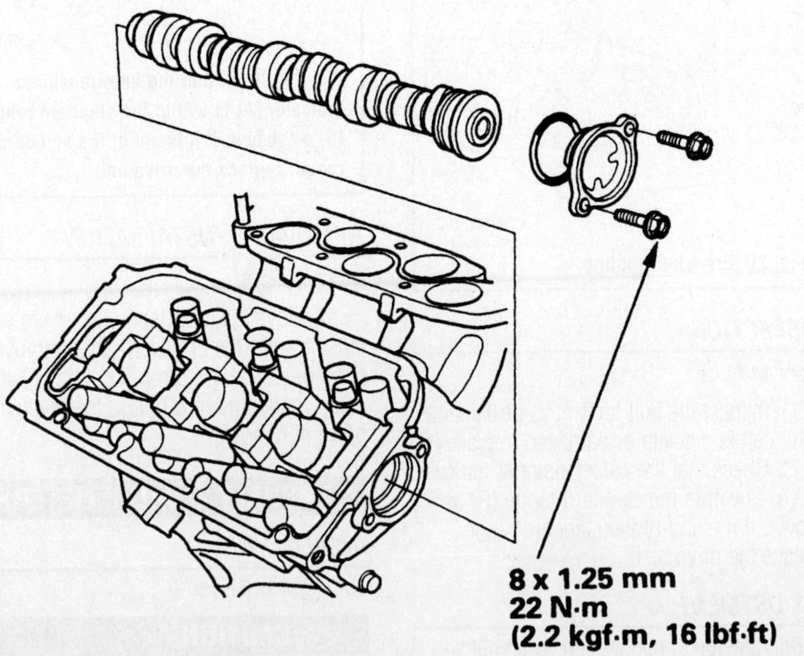

8 x 1.25 mm
22 N·m
(2.2 kgf·m, 16 lbf·ft)
42356-HPIL-G11

Fig. 33 Front camshaft assembly

14. Install the rocker arm assembly.
15. Install the timing belt.
16. Install the exhaust gas recirculation (EGR) valve.
17. Install the battery.
18. Fill the cooling system.
19. Connect the positive battery cable.
20. Connect the negative battery cable.
21. Enter the anti-theft codes for the audio system and the navigation system (if equipped).
22. Set the clock on vehicles without navigation.
23. Perform the power window control unit reset procedure.
24. Start the engine and check for leaks.

Rear

See Figure 34.

1. Drain the cooling system.
2. Relieve the fuel system pressure.
3. Make sure the anti-theft codes for the audio system and navigation system (if equipped) are available.
4. Disconnect the negative battery cable.
5. Remove the under-hood fuse box.
6. Remove the fuel feed hose.
7. Remove the nuts securing the fuel line.
8. Remove the brake lines from the master cylinder.
9. Remove the timing belt. See "Timing Belt & Sprockets" in this section.
10. Remove the rocker arm assembly. See "Rocker Arms & Shafts" in this section.
11. Remove the rear camshaft pulley.
12. Remove the thrust plate and camshaft.

To install:

13. Install the camshaft using a new O-ring. Tighten the thrust plate to 16 ft. lbs. (22 Nm).
14. Install the rear camshaft pulley.
15. Install the rocker arm assembly.
16. Install the timing belt.
17. Install the brake lines to the master cylinder.
18. Install the nuts securing the fuel line.
19. Install the fuel feed hose.
20. Install the under-hood fuse box.
21. Connect the negative battery cable.
22. Enter the anti-theft codes for the audio system and the navigation system (if equipped).
23. Set the clock on vehicles without navigation.
24. Perform the power window control unit reset procedure.

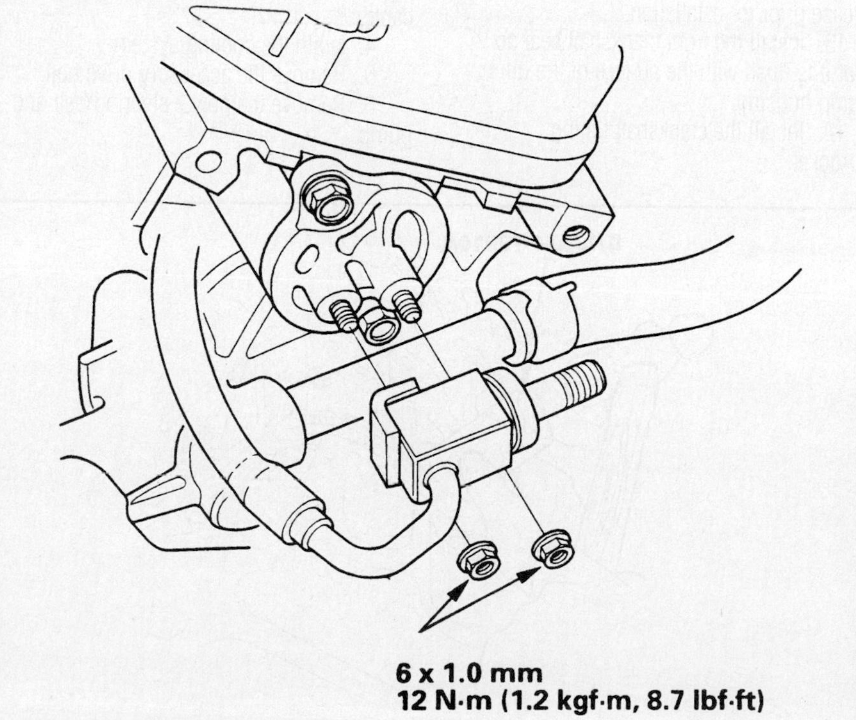
6 x 1.0 mm
12 N·m (1.2 kgf·m, 8.7 lbf·ft)
42356-HPIL-G12

Fig. 34 Remove the nuts attaching the fuel line when removing the rear camshaft

CRANKSHAFT DAMPER

REMOVAL & INSTALLATION

See Figures 35 through 37.

CAUTION

Before beginning work, review the "Precautions" in this section.

➡This procedure requires the use of the following special tools, or their equivalents:

- Holder handle 07JAB-001020A
- Holder attachment, 50 mm, offset 07MAB-PY3010A
- Socket, 19 mm 07JAA-001020A

1. Raise an safely support the vehicle.
2. Remove the right front wheel.
3. Remove the splash shield.
4. Remove the drive belt, as outlined in this section.
5. Hold the pulley with the holder handle and holder attachment.
6. Remove the bolt with a heavy duty 19 mm socket and breaker bar, then remove the crankshaft pulley.

To install:

7. Remove any oil or clean the pulley, crankshaft, bolt, and washer. Lubricate with new engine oil as shown.

WARNING

Never use an impact wrench to tighten the crankshaft pulley bolt.

8. Install the crankshaft pulley, and tighten the bolt, as follows:
    a. Hold the pulley with the holder handle and holder attachment, then tighten the bolt to 47 ft. lbs. (64 Nm) with a torque wrench and heavy duty 19 mm socket.

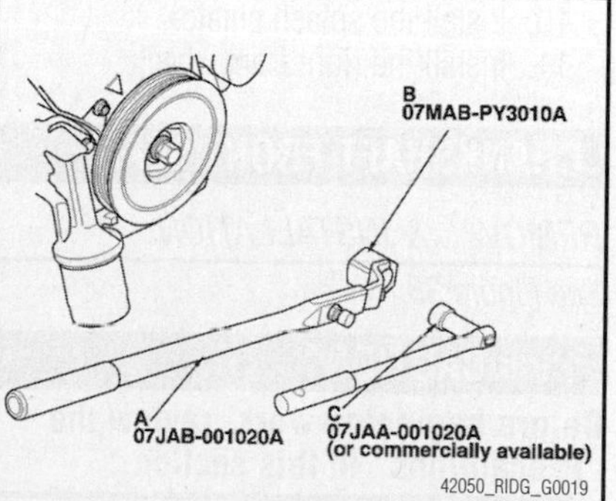
B
07MAB-PY3010A
A
07JAB-001020A
C
07JAA-001020A
(or commercially available)
42050_RIDG_G0019

Fig. 35 Hold the pulley with the holder handle (A) and holder attachment (B). Remove the bolt with a heavy duty 19 mm socket (C) and breaker bar, then remove the crankshaft pulley.

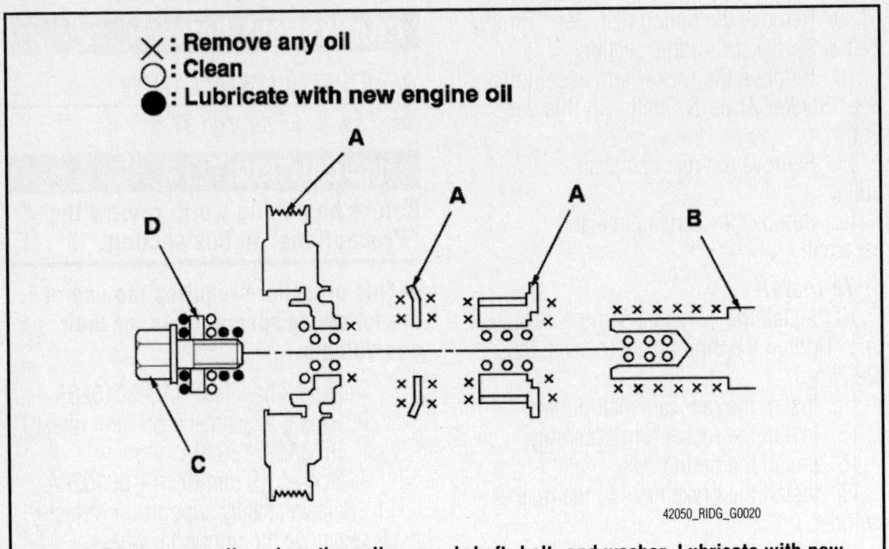
×: Remove any oil
○: Clean
●: Lubricate with new engine oil
A
A
A
B
D
C
42050_RIDG_G0020
Fig. 36 Remove any oil or clean the pulley, crankshaft, bolt, and washer. Lubricate with new engine oil as shown.

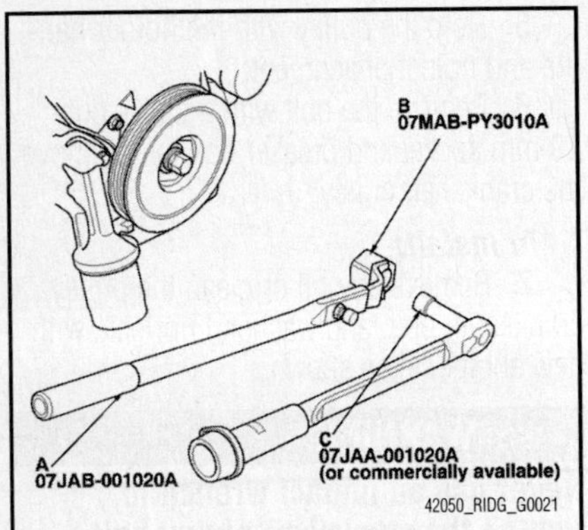
B
07MAB-PY3010A
A
07JAB-001020A
C
07JAA-001020A
(or commercially available)
42050_RIDG_G0021
Fig. 37 Crankshaft damper installation and tightening

b. Mark the bolt head and crankshaft pulley as shown, then tighten the bolt an additional 60° (The mark on the bolt head line up with the mark on the crankshaft pulley).

9. Install the drive belt.
10. Install the splash shield.
11. Install the right front wheel.

CRANKSHAFT FRONT SEAL

REMOVAL & INSTALLATION

See Figure 38.

⁂ CAUTION

Before beginning work, review the "Precautions" in this section.

1. Make sure the anti-theft codes for the audio system and navigation system (if equipped) are available.
2. Disconnect the negative battery cable.
3. Remove the accessory drive belts.
4. Remove the side engine mount.
5. Remove the valve cover.
6. Remove the crankshaft pulley.
7. Remove the front cover.
8. Remove the timing belt.
9. Remove the Top Dead Center (TDC) sensor, if equipped.
10. Remove the crankshaft timing sprocket.
11. Remove the front crankshaft seal.

To install:

12. Lubricate the crankshaft seal lip with grease prior to installation.
13. Install the front crankshaft seal so that it is flush with the surface of the oil pump housing.
14. Install the crankshaft timing sprocket.
15. Install the Top Dead Center (TDC) sensor, if equipped.
16. Install the timing belt.
17. Install the front cover.
18. Install the crankshaft pulley. Tighten the bolt to 181 ft. lbs. (245 Nm).
19. Install the valve cover.
20. Install the side engine mount.
21. Install the accessory drive belts.
22. Connect the negative battery cable.
23. Enter the anti-theft codes for the audio system and the navigation system (if equipped).
24. Set the clock on vehicles without navigation.
25. Perform the power window control unit reset procedure.
26. Check the engine oil level and add if necessary.
27. Start the engine and check for leaks.

CYLINDER HEAD

REMOVAL & INSTALLATION

See Figures 39 through 43.

⁂ CAUTION

Before beginning work, review the "Precautions" in this section.

1. Relieve the fuel system pressure.
2. Make sure the anti-theft codes for the audio system and navigation system (if equipped) are available.
3. Disconnect the negative battery cable.
4. Drain the cooling system.
5. Remove the accessory drive belt.
6. Remove the power steering belt and pump.

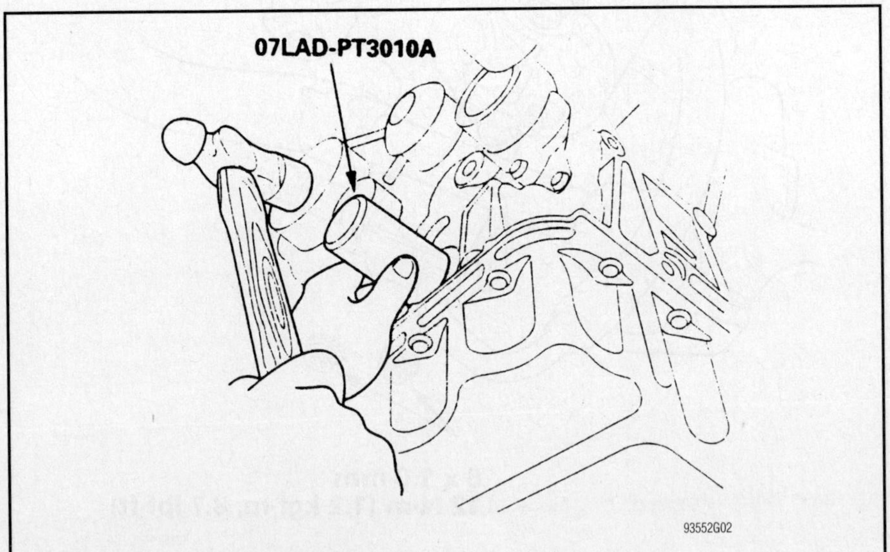
07LAD-PT3010A
93552G02
Fig. 38 Front crankshaft seal installation

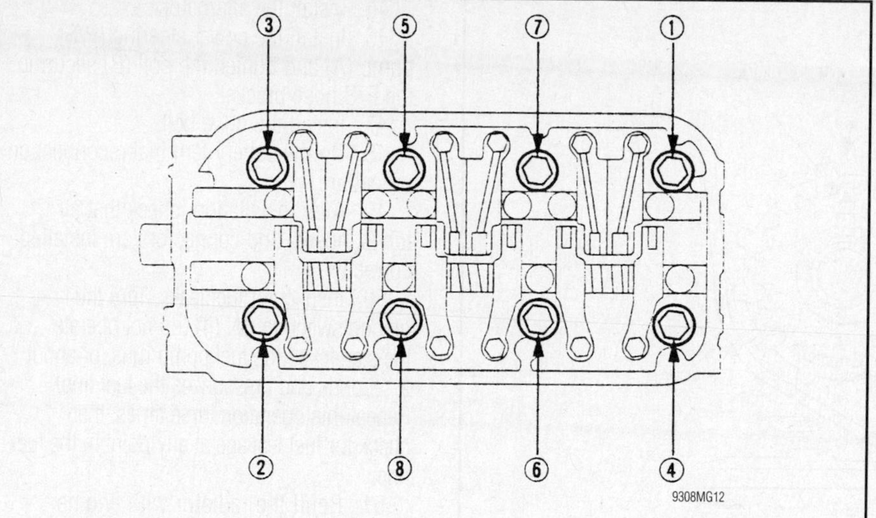
3
5
7
1
2
8
6
4
9308MG12

Fig. 39 Cylinder head bolt loosening sequence

7. Remove the power steering hose clamp.
8. Remove the alternator. See "Alternator" in "CHARGING SYSTEM" section.
9. Remove the intake manifold. See "Intake Manifold" in this section.
10. Remove the ignition coils.
11. Remove the timing belt. See "Timing Belt & Sprockets" in this section.
12. Remove the cylinder head covers.
13. Remove the following wiring and connectors from the engine:

➡**Mark all wiring connections for proper reinstallation.**

- Six injector connectors
- Engine coolant temperature (ECT) sensor 1 connector
- Oil pressure switch connector
- Camshaft position (CMP) sensor connector
- Rocker arm oil control solenoid connector
- Rocker arm oil pressure switch connector
- Rear air fuel ratio (A/F) sensor connector
- Rear secondary heated oxygen sensor (secondary HO2S) connector

14. Remove the catalytic converters.
15. Disconnect the fuel feed hose at the quick-connect fitting.
16. Remove the connector brackets from the front and rear cylinder heads.
17. Remove the fuel rails.
18. Remove the water passage.
19. Remove the front and rear camshaft pulleys and the front and rear back covers.
20. Remove the cylinder head bolts. To prevent warpage, loosen the bolts in sequence 1/3 turn at a time; repeat the sequence until all bolts are loosened.
21. Remove the cylinder heads.

To install:

22. Clean the cylinder head and the engine block surface.
23. Install a new coolant separator in the engine block whenever the engine block is replaced.
24. Clean and install the oil control orifices with new O-rings.
25. Install the dowel pins and the new cylinder head gaskets.
26. Clean the timing belt pulleys, the timing belt guide plate, and the upper and lower covers.
27. Set the timing belt drive pulley to top dead center (TDC) by aligning the TDC mark on the tooth of the timing belt drive pulley with the pointer on the oil pump.
28. Set the camshaft pulleys to TDC by aligning the TDC marks on the camshaft pulleys with the pointers on the back covers.
29. Install the cylinder heads on the engine block.
30. Measure the diameter of each cylinder head bolt at points shown.
If either diameter is less than 0.42 in. (10.6 mm), replace the cylinder head bolt. Apply new engine oil to the threads and under the bolt heads of all cylinder head bolts.
31. Tighten the cylinder head bolts in sequence to 22 ft. lbs. (29 Nm) using a beam-type torque wrench.

⁂ CAUTION

When using a preset-type torque wrench, be sure to torque slowly and do not over tighten. If a bolt makes any noise while you are torquing it, loosen the bolt and retighten it from the first step.

32. After torquing, tighten all cylinder head bolts in two steps (90° per step) using the proper sequence. If using a new cylinder head bolt, tighten the bolt an extra 90°.

➡**Remove the cylinder head bolt if it is tightened it beyond the specified angle, and go back to the start of the tightening sequence. Do not loosen it back to the specified angle.**

33. Install the timing belt. See "Timing Belt & Sprockets" in this section.
34. Adjust the valve clearance.

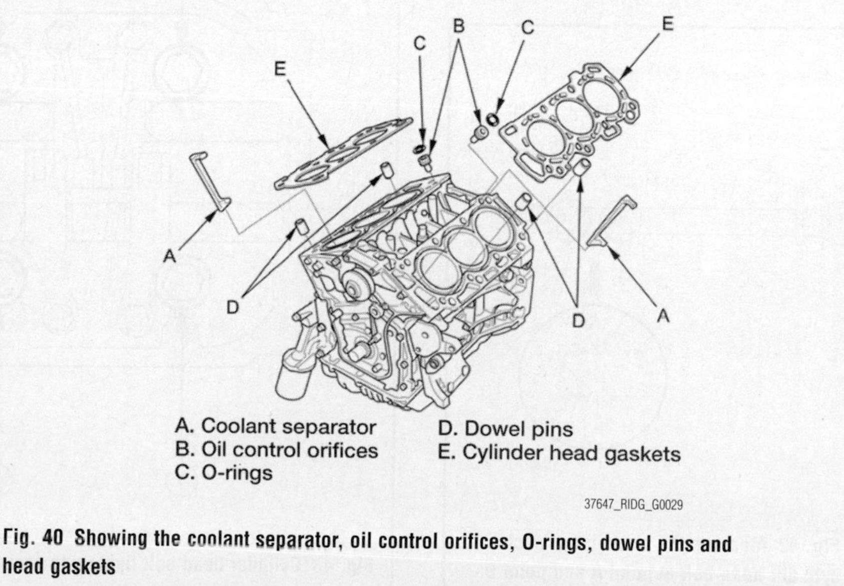
B
C
E
C
E
A
D
D
A
A. Coolant separator
B. Oil control orifices
C. O-rings
D. Dowel pins
E. Cylinder head gaskets
37647_RIDG_G0029

Fig. 40 Showing the coolant separator, oil control orifices, O-rings, dowel pins and head gaskets

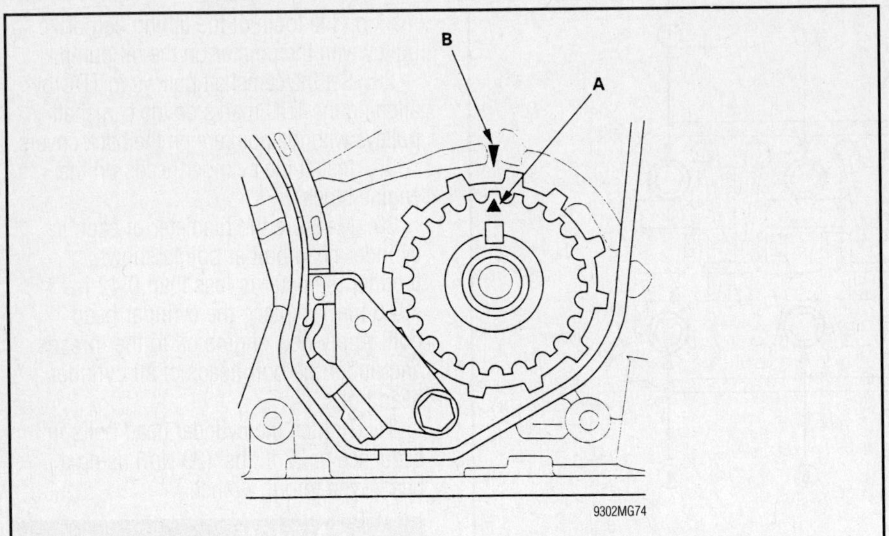

B
A
9302MG74

Fig. 41 Crankshaft timing belt sprocket TDC marks. Align sprocket mark (A) with pointer (B).

35. Install the water passage.
36. Install the fuel rails.
37. Install the connector bracket to the front cylinder head.
38. Install the harness clamp bracket to the rear cylinder head.
39. Connect the fuel feed hose, then install the quick-connect fitting cover.
40. Install the front warm up three way catalytic converter (front WU-TWC) and the rear warm up three way catalytic converter (rear WU-TWC).
41. Connect the following engine wire harness connectors, and install the wire harness clamps to the cylinder head:
- Six injector connectors
- Engine coolant temperature (ECT) sensor 1 connector
- Oil pressure switch connector
- Camshaft position (CMP) sensor connector
- Rocker arm oil control solenoid connector
- Rocker arm oil pressure switch connector
- Rear air fuel ratio (A/F) sensor connector
- Rear secondary heated oxygen sensor (secondary HO2S) connector

42. Install the cylinder head covers.
43. Install the six ignition coils.
44. Install the intake manifold.
45. Install the alternator.
46. Install the power steering (P/S) pump (A) and tighten the bolt (B) securing the P/S hose bracket.
47. Install the drive belt.
48. Do the battery terminal reconnection procedure.
49. After installation, check that all tubes, hoses, and connectors are installed correctly.
50. Inspect for fuel leaks. Turn the i gnition switch to ON (II) (do not operate the starter) so the fuel pump runs for about 2 seconds and pressurizes the fuel line. Repeat this operation three times, then check for fuel leakage at any point in the fuel line.
51. Refill the radiator with engine coolant, and bleed the air from the cooling system with the heater valve open.
52. Check for fluid leaks.
53. Do the "PCM Idle Learn Procedure". See "Powertrain Control Module (PCM)" in the "ENGINE PERFORMANCE & EMISSION CONTROLS" section.
54. Do the "CKP Pattern Clear/CKP Pattern Learn Procedure" in the "ENGINE PERFORMANCE & EMISSION CONTROLS" section.
55. Inspect the idle speed.
56. Inspect the ignition timing.

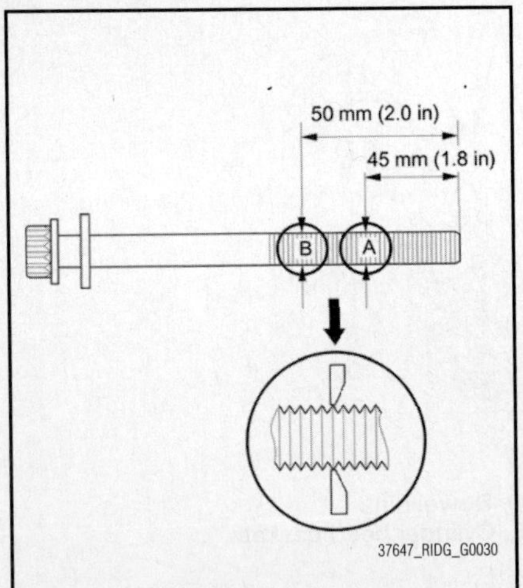

50 mm (2.0 in)
45 mm (1.8 in)
B
A
37647_RIDG_G0030

Fig. 42 Measure the diameter of each cylinder head bolt at point A and point B.

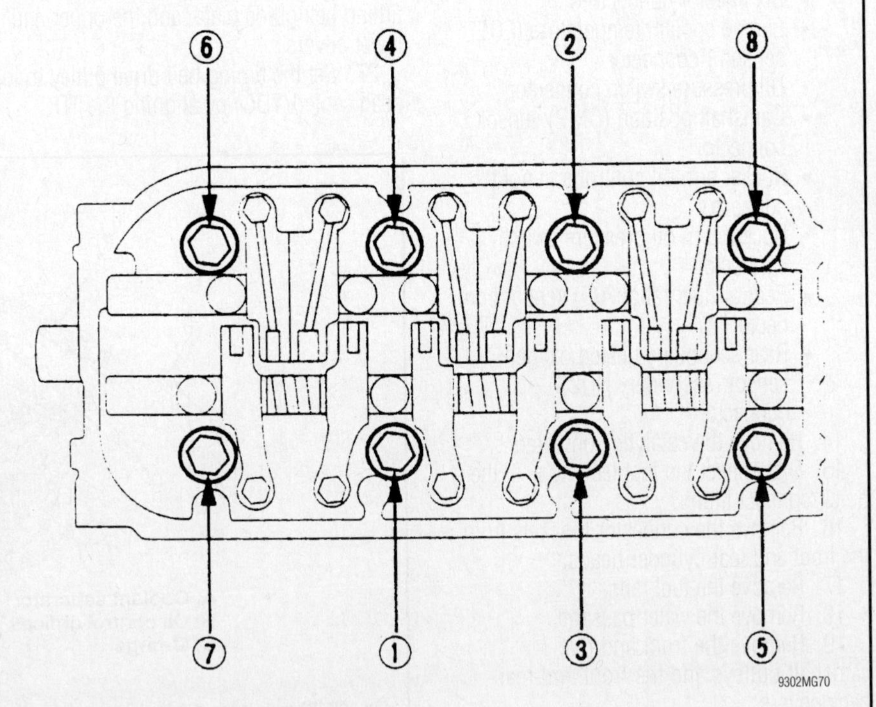

6
4
2
8
7
1
3
5
9302MG70

Fig. 43 Cylinder head bolt tightening sequence

FLEXPLATE (DRIVE PLATE)

REMOVAL & INSTALLATION

See Figure 44.

1. Remove the transmission.
2. Remove the drive plate and washer from the engine crankshaft.

To install:

3. Install the drive plate and washer on the engine crankshaft, and tighten the eight bolts in a crisscross pattern in two or more steps to 54 ft. lbs. (74 Nm).
4. Install the transmission.

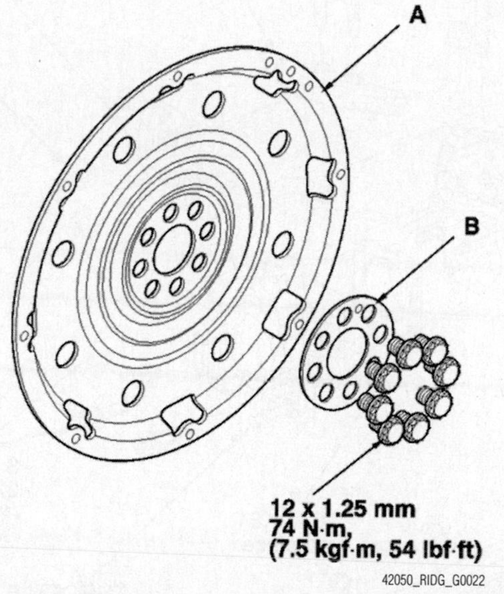
A
B
12 x 1.25 mm
74 N·m,
(7.5 kgf·m, 54 lbf·ft)
42050_RIDG_G0022

Fig. 44 Tighten the drive plate bolts, in several steps, in a crisscross pattern to specifications

INTAKE MANIFOLD

REMOVAL & INSTALLATION

See Figure 45.

⁂ CAUTION

Before beginning work, review the "Precautions" in this section.

1. Remove the engine cover.
2. Remove the breather pipe, then remove the air intake duct.
3. Mark and remove the engine wire harness connectors from the intake manifold:
   - Throttle actuator connector
   - Manifold absolute pressure (MAP) sensor connector
   - EVAP canister purge valve connector
   - Intake manifold tuning (IMT) valve actuator connector
4. Remove the positive crankcase ventilation (PCV) hose, the brake booster vacuum hose, and the evaporative emission (EVAP) canister purge hose.
5. Disconnect and plug the water bypass hoses.
6. Remove the upper cover mounting bolts and nuts sequentially, in three steps, then remove the upper cover.
7. Remove the intake manifold mounting bolts and nuts sequentially in three steps, then remove the intake manifold and the spacer.

To install:

8. Put a new gasket and the spacer on the injector base.
9. Install the intake manifold. Tighten the bolts and nuts sequentially in three steps, to a final torque of 16 ft. lbs. (22 Nm). Always use a new intake manifold gasket.
10. Install the upper cover. Tighten the bolts and nuts sequentially in three steps, to a final torque of 9 ft. lbs. (12 Nm). Always use a new gasket.
11. Install the water bypass hoses.
12. Install the PCV hose, the brake booster vacuum hose, and the EVAP canister purge hose.
13. Connect the engine wire harness connectors to the intake manifold:
    - Throttle actuator connector
    - MAP sensor connector
    - EVAP canister purge valve connector
    - IMT valve actuator connector
14. Install the air intake duct then install the breather pipe.
15. Clean up any spilled engine coolant.
16. After installation, check that all tubes, hoses and connectors are installed correctly.
17. Install the engine cover.
18. Refill the radiator with engine coolant, then bleed the air from the cooling system with the heater valve open.

OIL PAN

REMOVAL & INSTALLATION

See Figures 46 and 47.

⁂ CAUTION

Before beginning work, review the "Precautions" in this section.

1. Raise the vehicle on a hoist.
2. Drain the oil.
3. Remove the splash shield.
4. Remove the front subframe stiffener.
5. Remove the front exhaust pipe.
6. Remove the catalytic converter bracket.
7. Remove the torque converter cover.
8. Remove the 4 lower transaxle-to-engine bolts.
9. Remove the oil pan bolts.
10. Pry at the pry-points to break loose the oil pan.

To install:

11. Clean the pan and all mounting surfaces thoroughly.
12. Apply RTV gasket material to the oil pan flange as shown.
13. Position the pan on the block and install the bolts. Torque the bolts, in the sequence shown, in 2 even steps, to 104 inch lbs. (12 Nm).

➡Wait at least 30 minutes before filling with oil.

14. Torque the transaxle bolts to 54 ft. lbs. (74 Nm).
15. Install the cover.
16. Install the converter bracket.
17. Install the exhaust pipe using new gaskets and new self-locking nuts. Torque the flange-to-crossover nuts to 40 ft. lbs. (54 Nm), the hanger bolts to 16 ft. lbs. (22 Nm) and the flange-to-converter nuts to 25 ft. lbs. (33 Nm).
18. Install the stiffener. Torque the bolts to 40 ft. lbs. (54 Nm).
19. Install the splash shield.
20. Refill the engine oil.

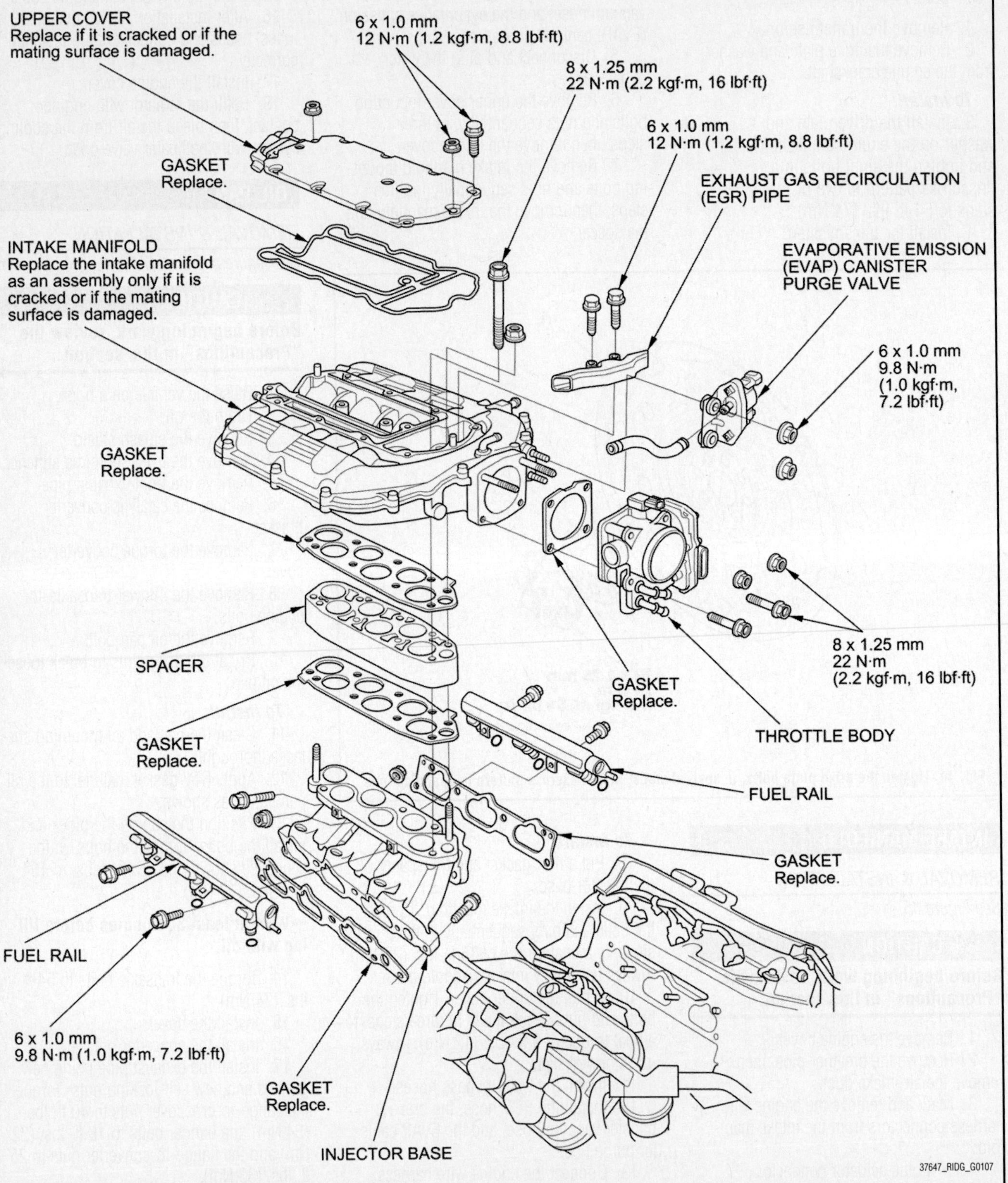
UPPER COVER
Replace if it is cracked or if the mating surface is damaged.
6 x 1.0 mm
12 N·m (1.2 kgf·m, 8.8 lbf·ft)
8 x 1.25 mm
22 N·m (2.2 kgf·m, 16 lbf·ft)
6 x 1.0 mm
12 N·m (1.2 kgf·m, 8.8 lbf·ft)
GASKET
Replace.
EXHAUST GAS RECIRCULATION (EGR) PIPE
INTAKE MANIFOLD
Replace the intake manifold as an assembly only if it is cracked or if the mating surface is damaged.
EVAPORATIVE EMISSION (EVAP) CANISTER PURGE VALVE
6 x 1.0 mm
9.8 N·m
(1.0 kgf·m,
7.2 lbf·ft)
GASKET
Replace.
SPACER
8 x 1.25 mm
22 N·m
(2.2 kgf·m, 16 lbf·ft)
GASKET
Replace.
THROTTLE BODY
GASKET
Replace.
FUEL RAIL
GASKET
Replace.
FUEL RAIL
6 x 1.0 mm
9.8 N·m (1.0 kgf·m, 7.2 lbf·ft)
GASKET
Replace.
INJECTOR BASE
37647_RIDG_G0107

Fig. 45 Exploded view of the intake manifold assembly

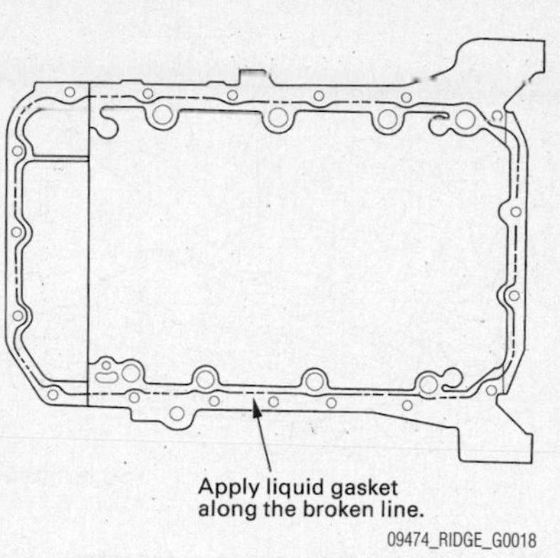
Apply liquid gasket
along the broken line.
09474_RIDGE_G0018

Fig. 46 Oil pan sealer application

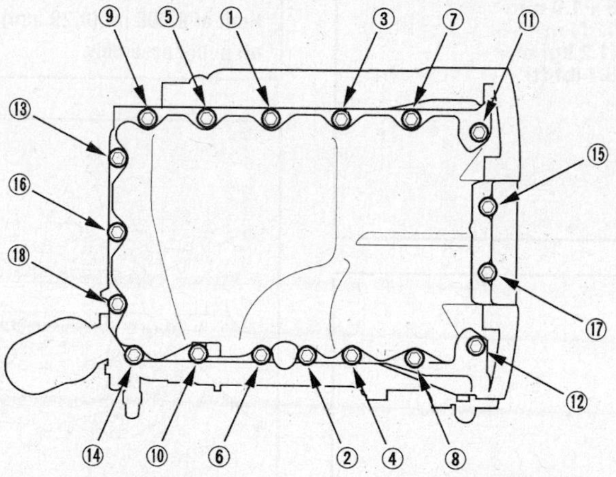
9
5
1
3
7
11
13
15
16
18
17
12
14
10
6
2
4
8
09474_RIDGE_G0019

Fig. 47 Oil pan bolt torque sequence

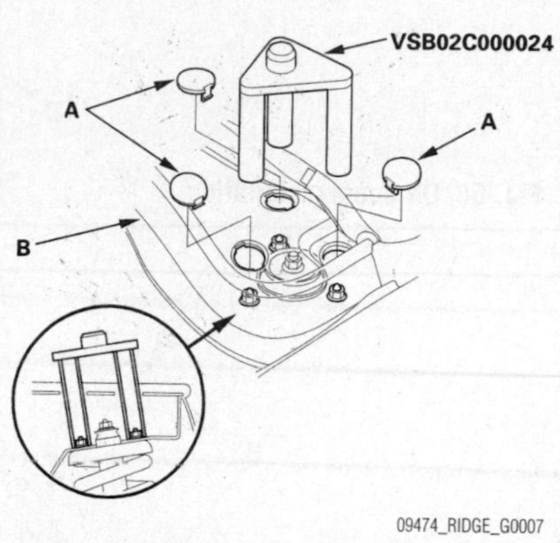
VSB02C000024
A
A
B
09474_RIDGE_G0007

Fig. 49 Engine hanger adapters installed

OIL PUMP

REMOVAL & INSTALLATION

See Figures 48 through 51.

⁂ CAUTION

Before beginning work, review the "Precautions" in this section.

1. Remove the bulkhead cover.
2. Remove the intake manifold cover.
3. Remove the drive belt.
4. Remove the power steering pump and the line bracket. Set the pump aside without disconnecting the lines.
5. Remove the caps from the front strut mounting nuts. Position the engine hanger adapters (VSB02C000024) with the FRONT mark facing forward, over the flange nuts.
6. Install the balancer bar (VSB02C000019):
   a. Attach the front arm to the front head with spacer and a 10x1.25mm bolt.
   b. Attach the rear arm to the rear head with an 8x1.25mm bolt.
7. Install the support hanger (AAR-T-12566):
   a. Attach the hook to the slotted hole in the balancer bar.
   b. Tighten the wing nut by hand to lift and support the engine/transaxle assembly.
8. Remove the timing belt.
9. Remove the crankshaft position sensor.
10. Remove the VTEC solenoid valve/oil filter assembly.
11. Remove the oil pan.
12. Remove the oil screen.
13. Remove the oil pump.

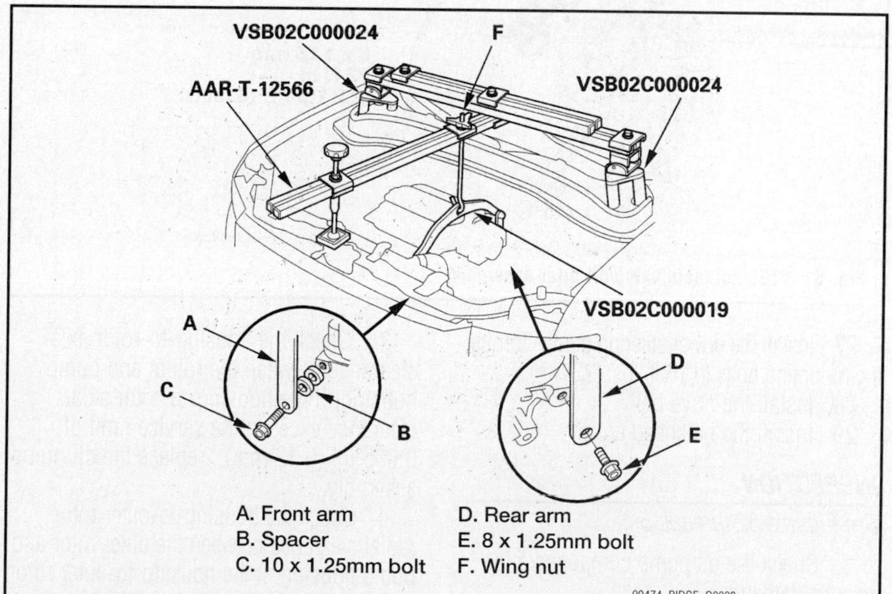
VSB02C000024
F
AAR-T-12566
VSB02C000024
VSB02C000019
A
C
B
D
E
A. Front arm
B. Spacer
C. 10 x 1.25mm bolt
D. Rear arm
E. 8 x 1.25mm bolt
F. Wing nut
09474_RIDGE_G0008

Fig. 48 Balancer bar installation and support hanger

To install:

14. Discard the oil seal.
15. Tap a new seal into place until it bottoms.
16. Thoroughly clean and dry all gasket surfaces.
17. Apply a bead of RTV gasket material to the pump mating surface of the block.

➡The oil pump must be installed within 4 minutes of applying the gasket material.

18. Apply all-purpose grease to the seal lip and clean engine oil to the new O-ring .
19. Install the dowel pins then align the inner rotor with the crankshaft and install the oil pump.
20. Install the screen with a new O-ring.
21. Install the VTEC solenoid valve/oil filter assembly, using a new filter.
22. Install the oil pan.
23. Install the CKP.
24. Install the timing belt.
25. Remove the engine hanger
26. Install the caps on the strut.

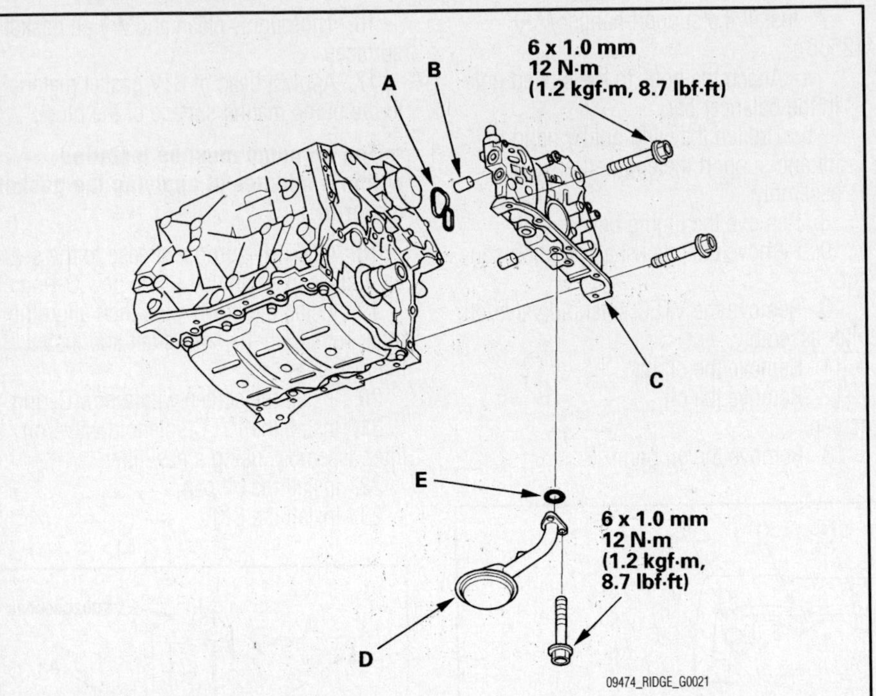

6 x 1.0 mm
12 N·m
(1.2 kgf·m, 8.7 lbf·ft)
A
B
C
E
6 x 1.0 mm
12 N·m
(1.2 kgf·m,
8.7 lbf·ft)
D
09474_RIDGE_G0021

Fig. 50 Oil pump installation

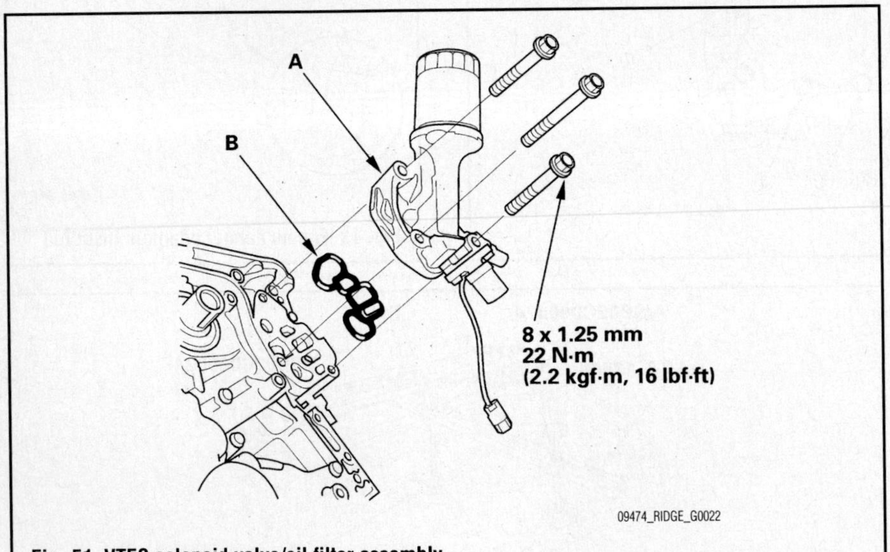

A
B
8 x 1.25 mm
22 N·m
(2.2 kgf·m, 16 lbf·ft)
09474_RIDGE_G0022

Fig. 51 VTEC solenoid valve/oil filter assembly

27. Install the power steering pump. Torque the mounting bolts to 16 ft. lbs. (22 Nm).
28. Install the drive belt.
29. Install the bulkhead cover.

INSPECTION

See Figures 52 through 54.

1. Check the oil pump components for wear or damage.
2. Check the inner-to-outer rotor radial clearance between the inner rotor and outer rotor. If the inner-to-outer rotor clearance exceeds the service limit of 0.008 in. (0.20 mm), replace the oil pump assembly.
3. Check the housing-to-rotor axial clearance between the rotors and pump housing. If the housing-to-rotor axial clearance exceeds the service limit of 0.005 in. (0.12 mm) , replace the oil pump assembly.
4. Check the housing-to-outer rotor radial clearance between the outer rotor and pump housing. If the housing-to-outer rotor radial clearance exceeds the service limit of 0.008 in. (0.20 mm), replace the oil pump assembly.
5. Inspect both rotors and pump housing for scoring or other damage. Replace the parts, if necessary.

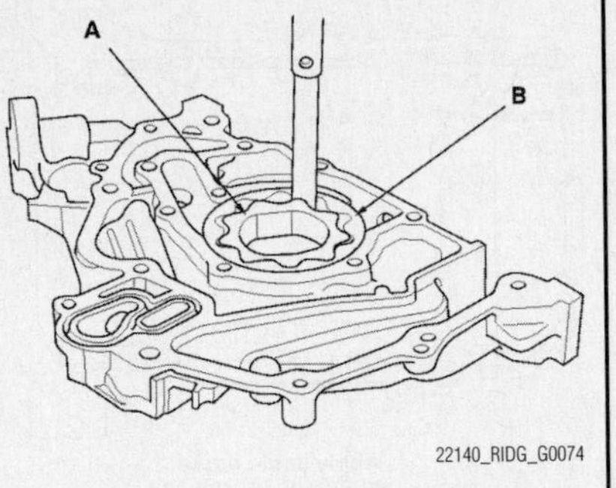

A
B
22140_RIDG_G0074

Fig. 52 Check the inner-to-outer rotor radial clearance between the inner rotor (A) and outer rotor (B). If the inner-to-outer rotor clearance exceeds the service limit of 0.008 in. (0.20 mm), replace the oil pump assembly.

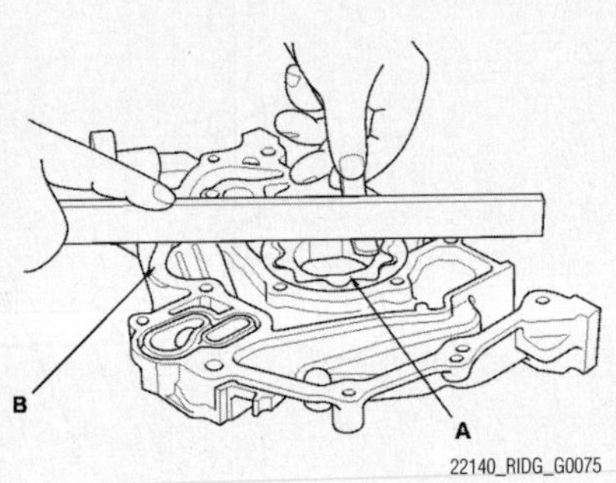

B
A
22140_RIDG_G0075

Fig. 53 Check the housing-to-rotor axial clearance between the rotors (A) and pump housing (B). If the housing-to-rotor axial clearance exceeds the service limit of 0.005 in. (0.12 mm) , replace the oil pump assembly

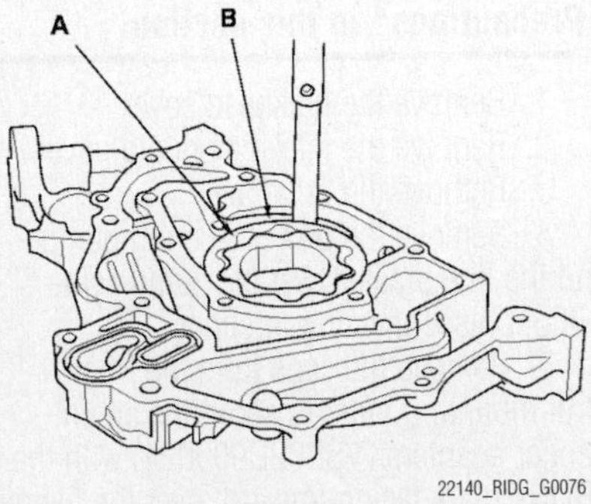

A
B
22140_RIDG_G0076

Fig. 54 Check the housing-to-outer rotor radial clearance between the outer rotor (A) and pump housing (B). If the housing-to-outer rotor radial clearance exceeds the service limit of 0.008 in. (0.20 mm), replace the oil pump assembly

6. Apply liquid thread lock to the pump housing screws, then install the oil pump cover.

7. Check that the oil pump turns freely.

PISTON AND RING

POSITIONING

See Figures 55 and 56.

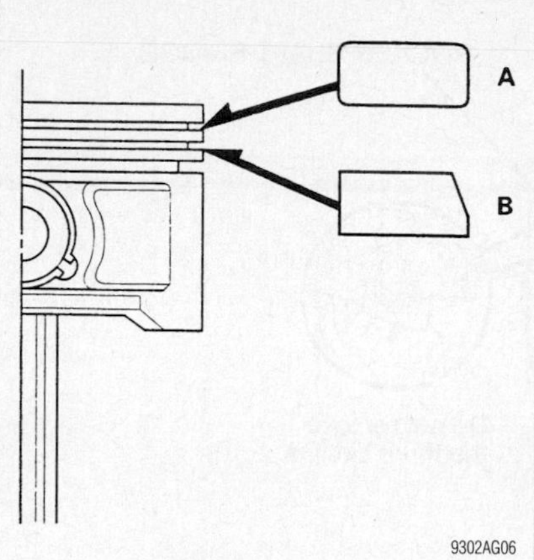
A
B
9302AG06

Fig. 55 Compression ring identification—3.5L engine

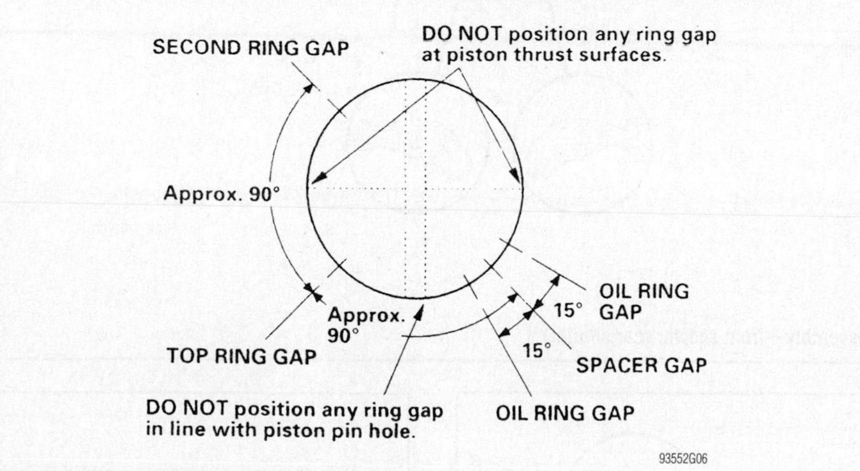
SECOND RING GAP
DO NOT position any ring gap at piston thrust surfaces.
Approx. 90°
Approx. 90°
15°
OIL RING GAP
15°
SPACER GAP
TOP RING GAP
DO NOT position any ring gap in line with piston pin hole.
OIL RING GAP
93552G06

Fig. 56 Ring end gap positioning—3.5L engine

REAR MAIN SEAL

REMOVAL & INSTALLATION

See Figure 57.

➡Oil seal Driver 07749-0010000 and Driver attachment, 106 mm 070AD-RCA0200 are required to perform this procedure.

⁂ CAUTION

Before beginning work, review the "Precautions" in this section.

1. Remove the transmission and the drive plate.

2. Remove the transmission end crankshaft oil seal.

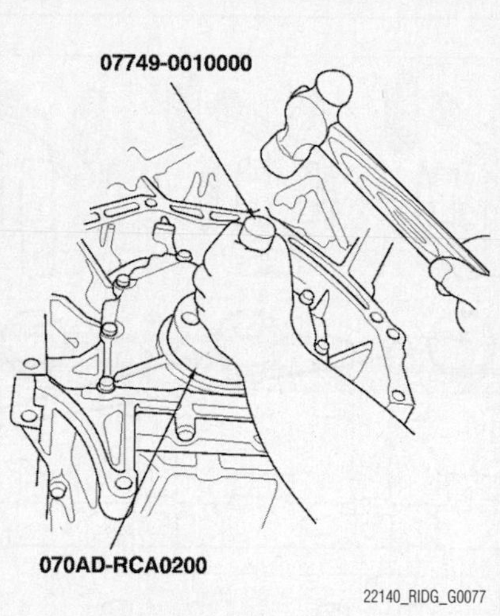
07749-0010000
070AD-RCA0200
22140_RIDG_G0077

Fig. 57 Using the special tools, drive in the crankshaft oil seal until the driver attachment bottoms against the engine block end cover

To install:

3. Clean and dry the crankshaft oil seal housing.

4. Apply a light coat of multipurpose grease to the crankshaft and to the lip of the seal.

5. Using the special tools, drive in the crankshaft oil seal until the driver attachment bottoms against the engine block end cover. Align the hole in the driver attachment with the pin on the crankshaft.

6. Clean any excess grease off the crankshaft, and check that the oil seal lip is not distorted.

7. Install the drive plate, and the transmission.

8. Check the fluid levels.

9. Start the engine and check for leaks.

ROCKER ARMS/SHAFTS

REMOVAL & INSTALLATION

See Figure 58.

➡Manufacturer does not provide a specific removal and installation procedure for this component. Use the illustration as a guide when servicing this component.

TIMING BELT FRONT COVER

REMOVAL & INSTALLATION

See Figures 59 through 62.

⁂ CAUTION

Before beginning work, review the "Precautions" in this section.

1. Turn the crankshaft so the white mark aligns with the pointer.

2. Make sure the number 1 piston is at Top Dead center (TDC).

3. Make sure the anti-theft codes for the audio system and navigation system (if equipped) are available.

4. Disconnect the negative battery cable

5. Remove the wheels and engine splash shield.

6. Remove the drive belts.

7. Support the engine with a block of wood and a jack under the oil pan.

8. Remove the ground cable bracket.

9. Remove the upper side engine mount.

10. Remove the dipstick tube.

11. Remove the crankshaft pulley using holder tool shown in the accompanying illustration and a breaker and socket, loosen the 19mm bolt and remove the pulley.

12. Remove the front upper cover, rear upper cover and the lower cover.

To install:

13. Install the lower timing cover and tighten the bolts to 9 ft. lbs. (12 Nm).

14. Install the front and rear upper timing covers and tighten the bolts to 9 ft. lbs. (12 Nm).

15. Install the crankshaft pulley and tighten the bolts to 181 ft. lbs. (245 Nm), using the holding tool to prevent the unit from turning.

16. Rotate the crankshaft pulley about 5 or 6 degrees clockwise so the belt positions itself on the pulleys.

17. Turn the crankshaft pulley so the white mark aligns with the pointer.

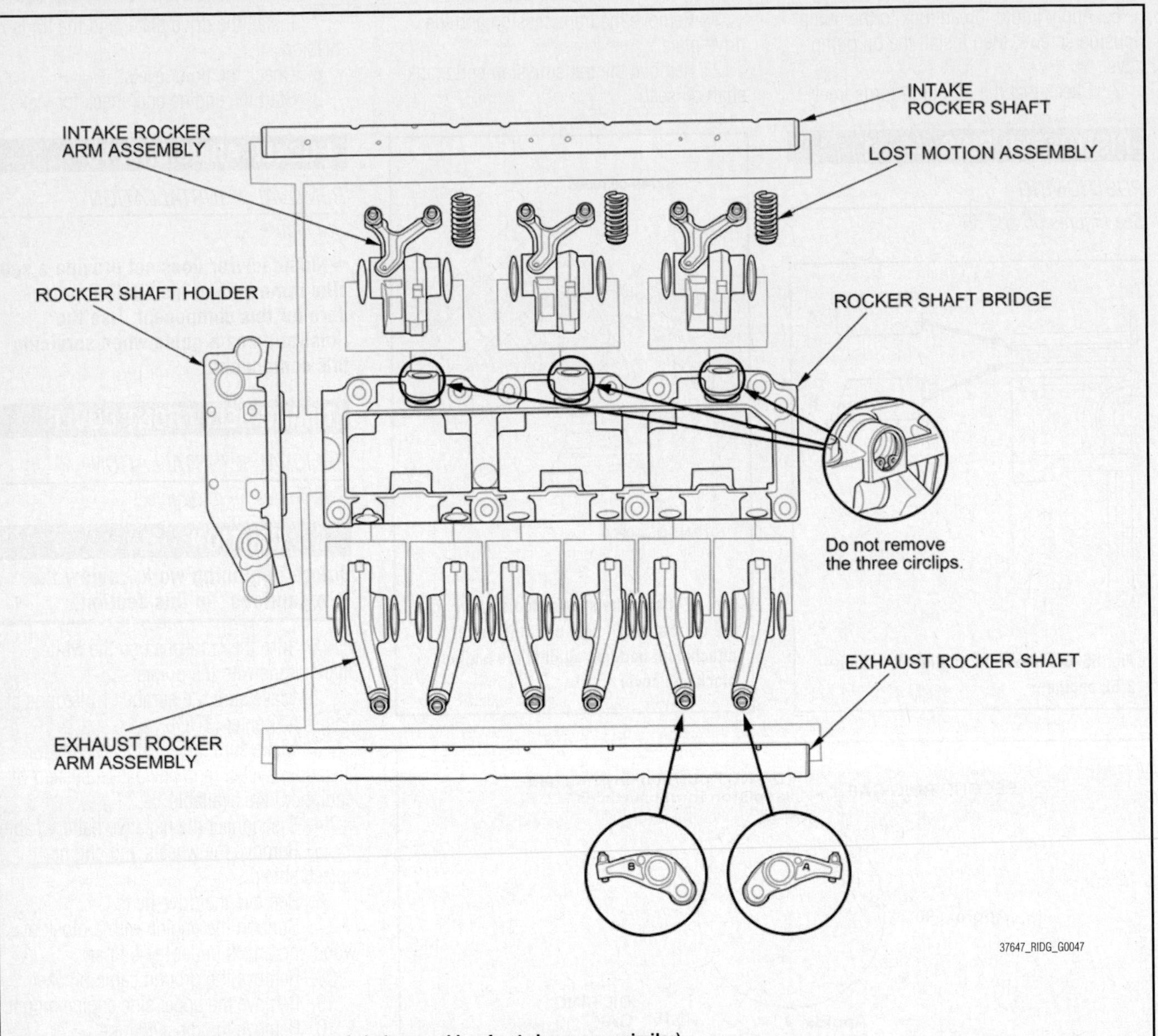
INTAKE ROCKER SHAFT
LOST MOTION ASSEMBLY
INTAKE ROCKER ARM ASSEMBLY
ROCKER SHAFT HOLDER
ROCKER SHAFT BRIDGE
Do not remove the three circlips.
EXHAUST ROCKER SHAFT
EXHAUST ROCKER ARM ASSEMBLY
B
A
37647_RIDG_G0047

Fig. 58 Exploded view of the rocker arm and shaft assembly—front shown; rear similar)

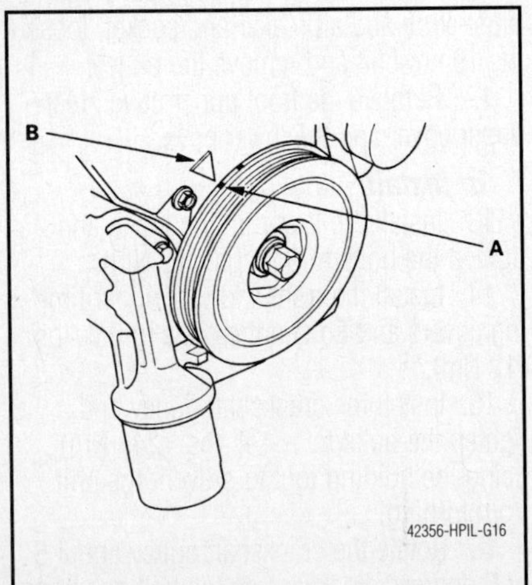
B
A
42356-HPIL-G16

Fig. 59 Turn the crankshaft so the white mark (A) aligns with the pointer (B)

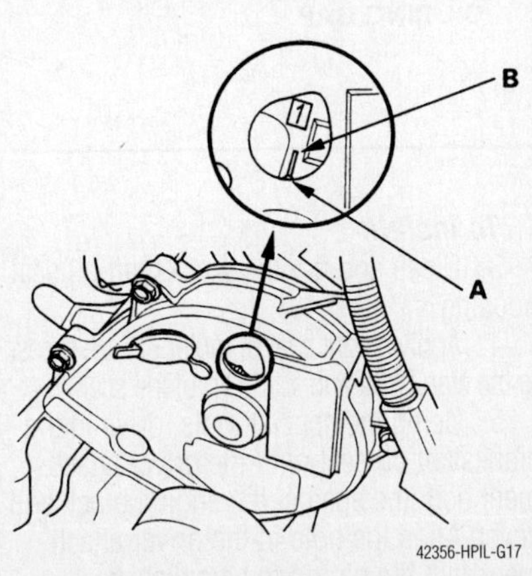
B
A
42356-HPIL-G17

Fig. 60 Make sure the number 1 piston is at top dead center (A) on the front camshaft pulley and pointer (B)

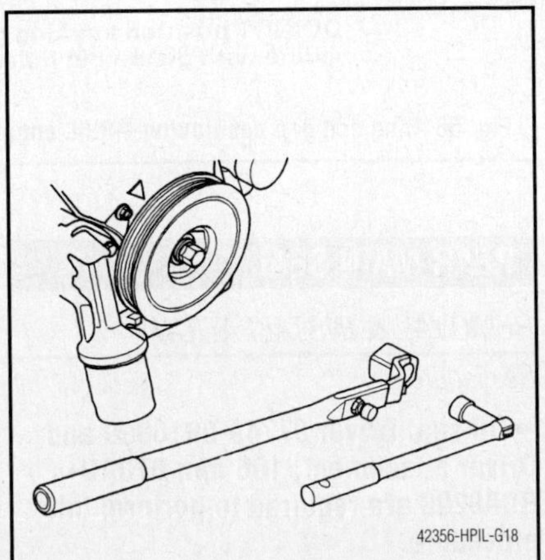
42356-HPIL-G18

Fig. 61 Remove the crankshaft pulley using holder tool shown

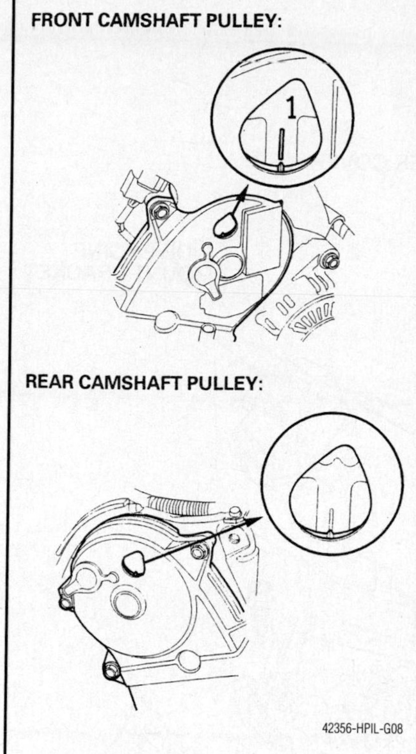
FRONT CAMSHAFT PULLEY:
1
REAR CAMSHAFT PULLEY:
42356-HPIL-G08

Fig. 62 Check that the camshaft pulley marks are aligned as shown

18. Check the camshaft pulley marks are aligned. If the marks are aligned, proceed to the next step. If the marks are not aligned, remove the timing belt and reinstall using the steps outlined before this step.
19. Install the drive belt.
20. Install the upper side mount and tighten the bolts in the sequence illustrated to the specifications in the illustration.
21. Using a suitable scan tool, perform the Powertrain Control Module (PCM) reset and the Crankshaft position (CKP) pattern clear/learn procedures, following the scan tool manufactures instructions.

TIMING BELT & SPROCKETS

REMOVAL & INSTALLATION

See Figures 63 through 75.

⁂ CAUTION

Before beginning work, review the "Precautions" in this section.

1. Turn the crankshaft so the white mark aligns with the pointer.
2. Make sure the number 1 piston is at Top Dead center (TDC).
3. Make sure the anti-theft codes for the audio system and navigation system (if equipped) are available.
4. Disconnect the negative battery cable
5. Remove the wheels and engine splash shield.
6. Remove the drive belts.
7. Support the engine with a block of wood and a jack under the oil pan.
8. Remove the ground cable bracket.
9. Remove the upper side engine mount.
10. Remove the dipstick tube.
11. Remove the crankshaft pulley using holder tool shown in the accompanying illustration and a breaker and socket, loosen the 19mm bolt and remove the pulley.
12. Remove the front upper cover, rear upper cover and the lower cover.
13. Remove the one of the battery clamp bolts and grind the end as shown in the illustration.
14. Screw the battery clamp bolt as shown in the illustration to hold the belt adjuster in position. Do not use a wrench, hand tighten only.
15. Remove the lower side engine mount.
16. Remove the idler pulley bolt and the pulley.
    - Timing belt

To install:

17. If installing a new belt, perform the following steps:
    a. Clean the pulleys, belt guide plate and the upper and lower covers.
    b. Set the timing belt drive pulley to TDC by aligning the TDC mark on the tooth of the belt drive pulley with the pointer on the oil pump.
    c. Set the camshaft pulleys to TDC by aligning the TDC marks on the camshaft pulleys with the pointers on the back covers.
    d. Remove the battery clamp bolt.
    e. Remove the belt tensioner.
    f. Align the holes on the rod and housing of the tensioner.
    g. Using a press or other suitable device, slowly compress the tensioner and insert a 0.08 inch (2mm) pin through the housing and rod.
    h. Install the tensioner making sure the pin is still installed.
    i. Apply thread locker to idler pulley bolt then hand tighten the bolt.
    j. Install the belt over the pulleys in this sequence; drive pulley, idler pulley, front camshaft pulley, water pump pulley, rear camshaft pulley and adjusting pulley.
    k. Tighten the idler pulley bolt to 33 ft. lbs. (44 Nm).
    l. Remove the pin from the tensioner.
18. Install the lower half of the side mount and tighten the 3 long bolts to 33 ft. lbs. (44 Nm) and the one short bolt to 9 ft. lbs. (12 Nm).
19. Install the timing belt guide plate as shown in the illustration.
20. Install the lower timing cover and tighten the bolts to 9 ft. lbs. (12 Nm).
21. Install the front and rear upper timing covers and tighten the bolts to 9 ft. lbs. (12 Nm).
22. Install the crankshaft pulley and tighten the bolts to 181 ft. lbs. (245 Nm), using the holding tool to prevent the unit from turning.
23. Rotate the crankshaft pulley about 5 or 6 degrees clockwise so the belt positions itself on the pulleys.
24. Turn the crankshaft pulley so the white mark aligns with the pointer.
25. Check the camshaft pulley marks are aligned. If the marks are aligned, proceed to the next step. If the marks are not aligned, remove the timing belt and reinstall using the steps outlined before this step.
26. Install the drive belt.
27. Install the upper side mount and tighten the bolts in the sequence illustrated to the specifications in the illustration.
28. Using a suitable scan tool, perform the following as found in "ENGINE PERFORMANCE & EMISSION CONTROLS" section":
    - PCM Reset Procedure
    - CKP Pattern Clear/Learn Procedures
29. If installing the old belt, perform the following steps:
    a. Clean the pulleys, belt guide plate and the upper and lower covers.
    b. Set the timing belt drive pulley to TDC by aligning the TDC mark on the tooth of the belt drive pulley with the pointer on the oil pump.
    c. Set the camshaft pulleys to TDC by aligning the TDC marks on the camshaft pulleys with the pointers on the back covers.
    d. Apply thread locker to idler pulley bolt then hand tighten the bolt.
30. If the tensioner was extended and the belt cannot be installed, perform the steps above for the new belt installation.
    a. Install the belt over the pulleys in this sequence; drive pulley, idler pulley, front camshaft pulley, water pump pulley, rear camshaft pulley and adjusting pulley.
    b. Tighten the idler pulley bolt to 33 ft. lbs. (44 Nm).
    c. Remove the battery clamp bolt.

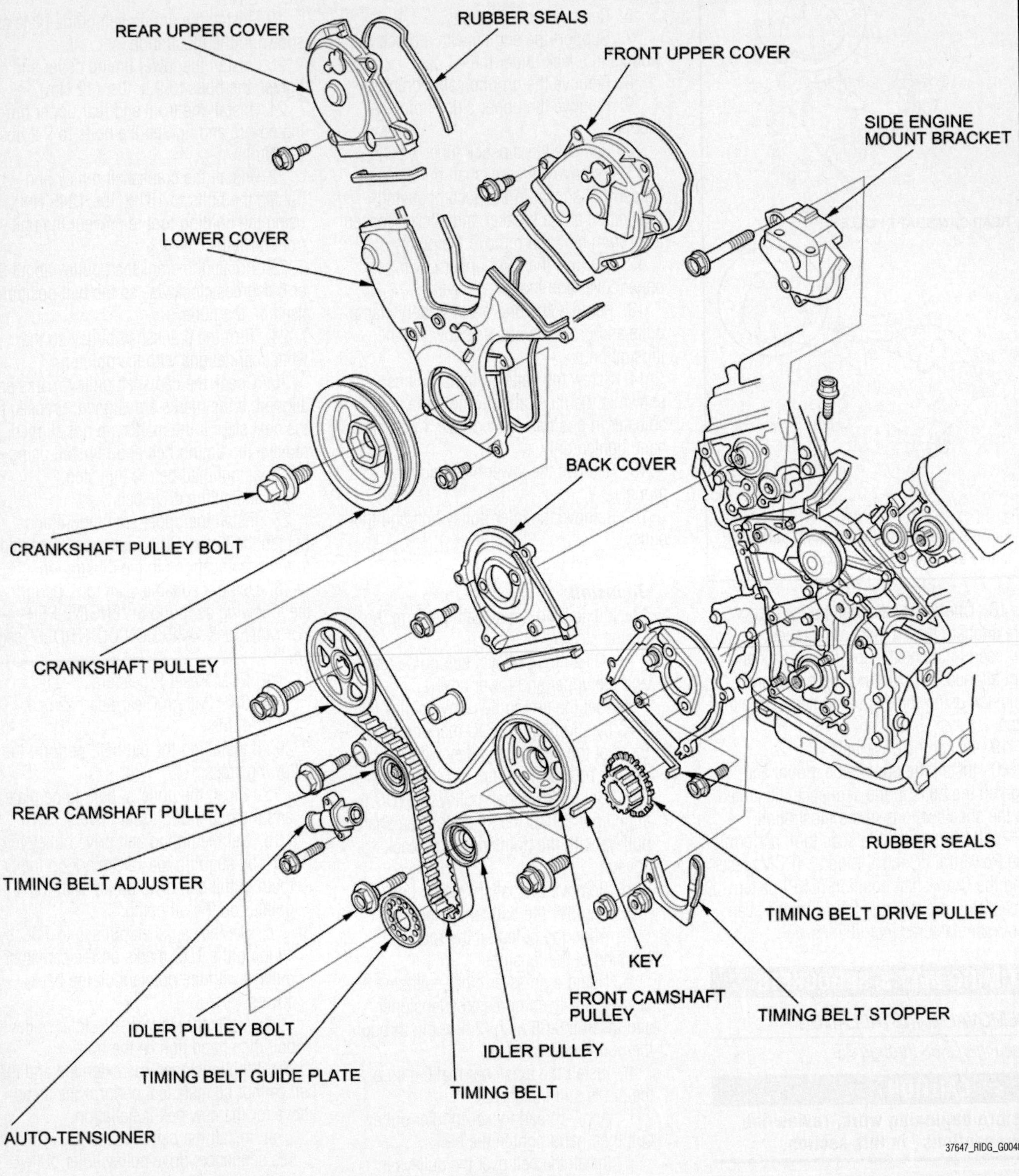
REAR UPPER COVER
RUBBER SEALS
FRONT UPPER COVER
SIDE ENGINE MOUNT BRACKET
LOWER COVER
BACK COVER
CRANKSHAFT PULLEY BOLT
CRANKSHAFT PULLEY
REAR CAMSHAFT PULLEY
RUBBER SEALS
TIMING BELT ADJUSTER
TIMING BELT DRIVE PULLEY
KEY
FRONT CAMSHAFT PULLEY
TIMING BELT STOPPER
IDLER PULLEY BOLT
IDLER PULLEY
TIMING BELT GUIDE PLATE
TIMING BELT
AUTO-TENSIONER
37647_RIDG_G0048

Fig. 63 Exploded view of the timing belt and cover assembly

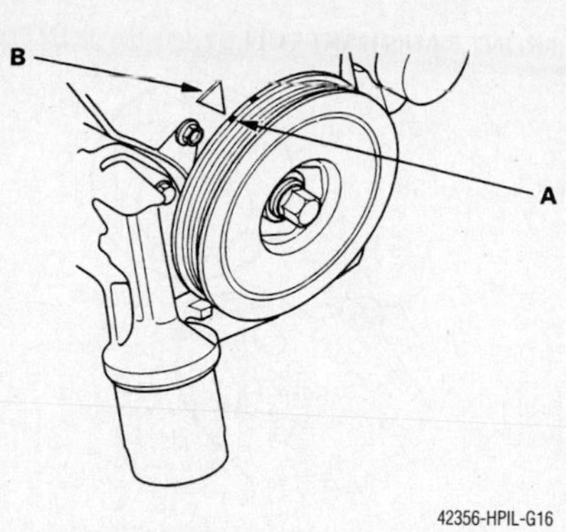
B
A
42356-HPIL-G16

Fig. 64 Turn the crankshaft so the white mark (A) aligns with the pointer (B)

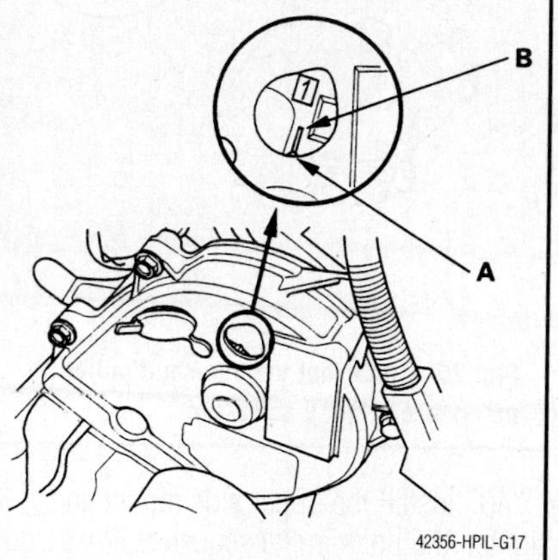
B
A
42356-HPIL-G17

Fig. 65 Make sure the number 1 piston is at top dead center (A) on the front camshaft pulley and pointer (B)

31. Install the lower half of the side mount and tighten the 3 long bolts to 33 ft. lbs. (44 Nm) and the one short bolt to 9 ft. lbs. (12 Nm).

32. Install the timing belt guide plate as shown in the illustration.

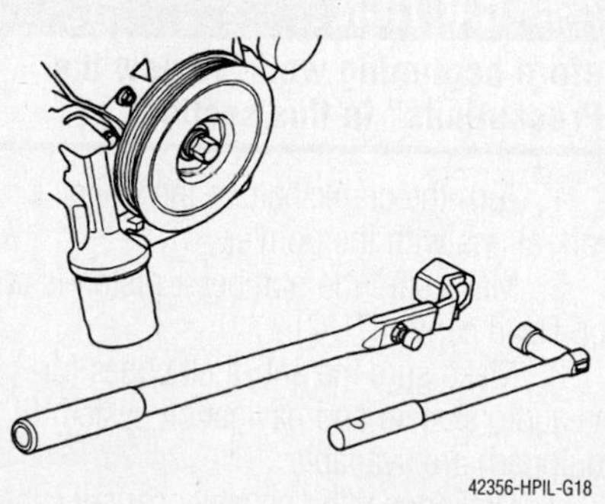
42356-HPIL-G18

Fig. 66 Remove the crankshaft pulley using holder tool shown

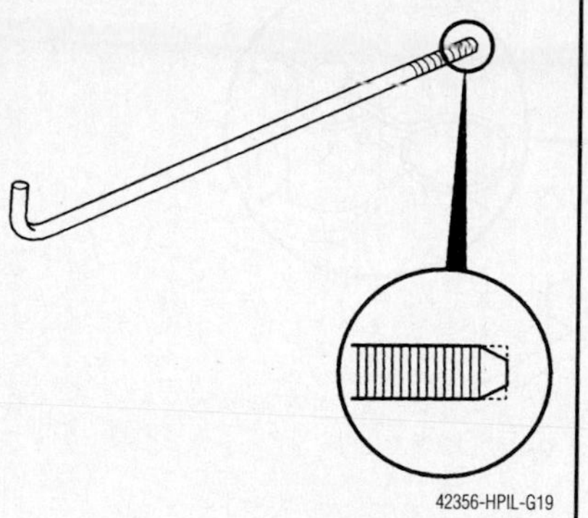
42356-HPIL-G19

Fig. 67 Remove a battery clamp bolt and grind the end as shown

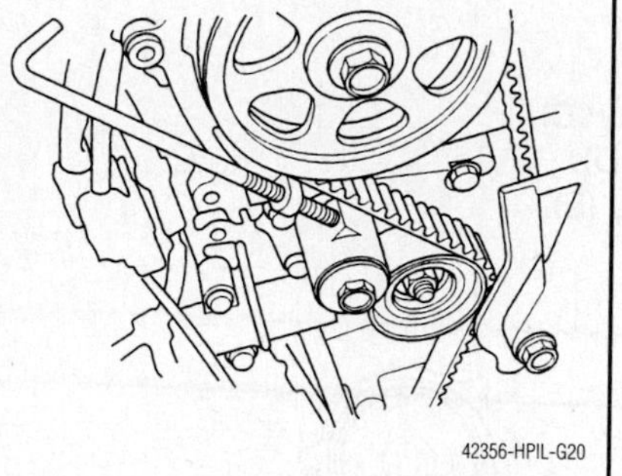
42356-HPIL-G20

Fig. 68 Install the battery clamp bolt as shown to hold the belt adjuster in position

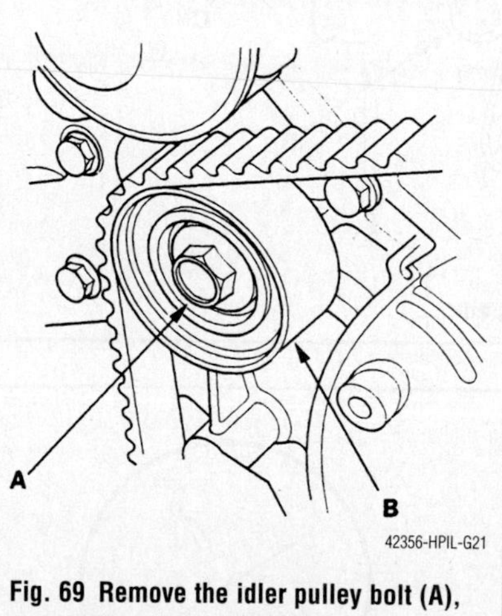
A
B
42356-HPIL-G21

Fig. 69 Remove the idler pulley bolt (A), pulley (B) and the timing belt

33. Install the lower timing cover and tighten the bolts to 9 ft. lbs. (12 Nm).

34. Install the front and rear upper timing covers and tighten the bolts to 9 ft. lbs. (12 Nm).

35. Install the crankshaft pulley and tighten the bolts to 181 ft. lbs. (245 Nm), using the holding tool to prevent the crankshaft from rotating.

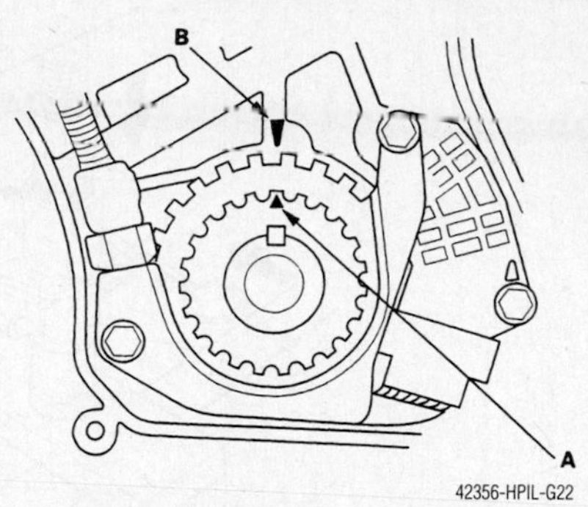
B
A
42356-HPIL-G22

Fig. 70 Set the timing belt pulley to TDC by aligning the TDC mark (A) on the tooth of the belt pulley with the pointer (B) on the oil pump

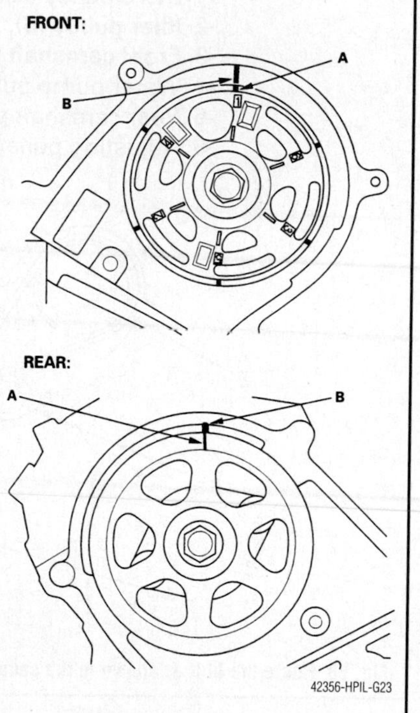
FRONT:
A
B
REAR:
A
B
42356-HPIL-G23

Fig. 71 Set the camshaft pulleys to TDC by aligning the TDC marks (A) on the camshaft pulleys with the pointers (B) on the back covers

36. Rotate the crankshaft pulley about 5 or 6 degrees clockwise so the belt positions itself on the pulleys.

37. Turn the crankshaft pulley so the white mark aligns with the pointer.

38. Check the camshaft pulley marks are aligned. If the marks are aligned, proceed to the next step. If the marks are not aligned, remove the timing belt and reinstall using the steps outlined before this step.

39. Install the drive belt.

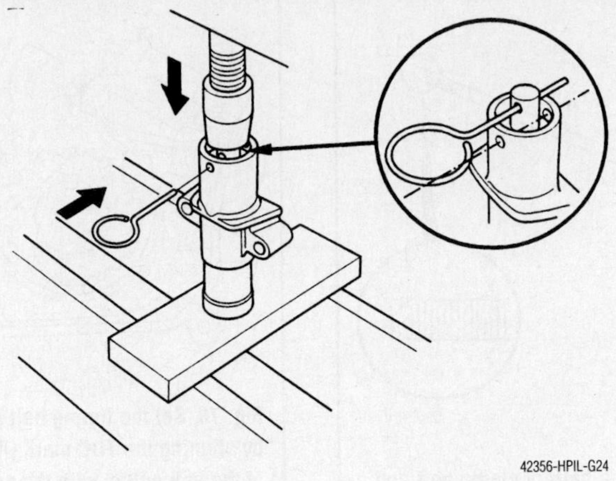
42356-HPIL-G24

Fig. 72 Insert a 0.08 inch (2mm) pin through the tensioner housing and rod

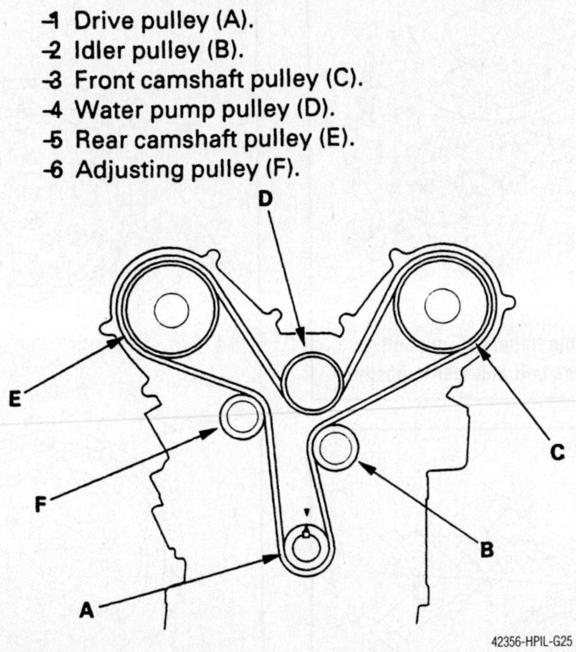
1 Drive pulley (A).
2 Idler pulley (B).
3 Front camshaft pulley (C).
4 Water pump pulley (D).
5 Rear camshaft pulley (E).
6 Adjusting pulley (F).
D
E
C
F
B
A
42356-HPIL-G25

Fig. 73 Route the belt as shown in the sequence listed

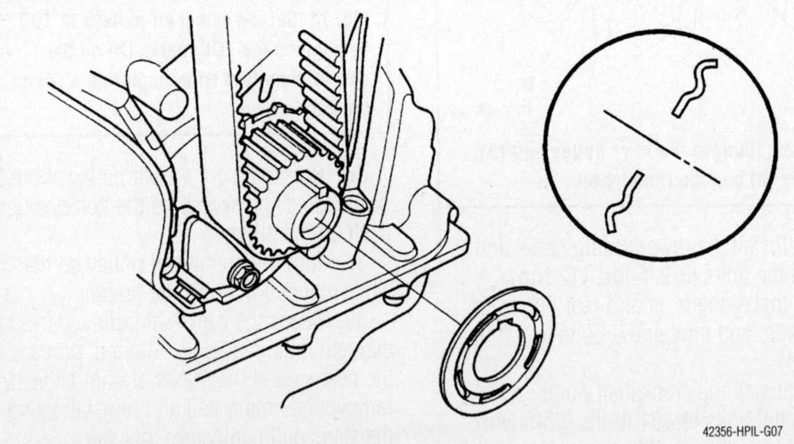
42356-HPIL-G07

Fig. 74 Install the timing belt guide plate as shown

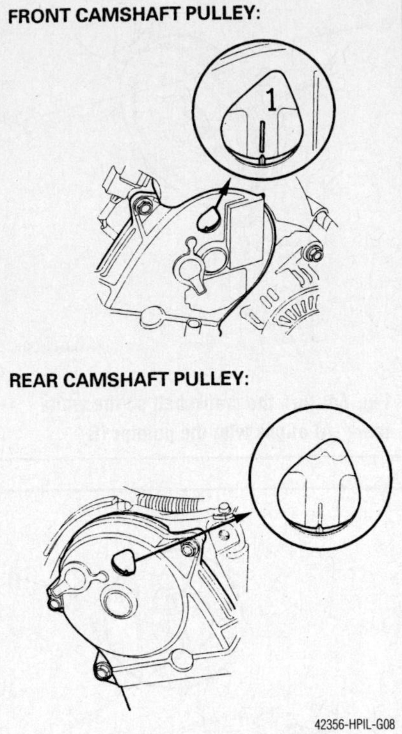
FRONT CAMSHAFT PULLEY:
1
REAR CAMSHAFT PULLEY:
42356-HPIL-G08

Fig. 75 Check that the camshaft pulley marks are aligned as shown

40. Install the upper side mount and tighten the bolts in the sequence illustrated to the specifications in the illustration.
41. Install the dipstick tube.
42. Using a suitable scan tool, perform the Powertrain Control Module (PCM) reset and the Crankshaft position (CKP) pattern clear/learn procedures, following the scan tool manufactures instructions.

TIMING BELT REAR COVER

REMOVAL & INSTALLATION

See Figures 76 through 78.

**** CAUTION**

Before beginning work, review the "Precautions" in this section.

1. Turn the crankshaft so the white mark aligns with the pointer.
2. Make sure the number 1 piston is at Top Dead center (TDC).
3. Make sure the anti-theft codes for the audio system and navigation system (if equipped) are available.
4. Disconnect the negative battery cable
5. Remove the wheels and engine splash shield.

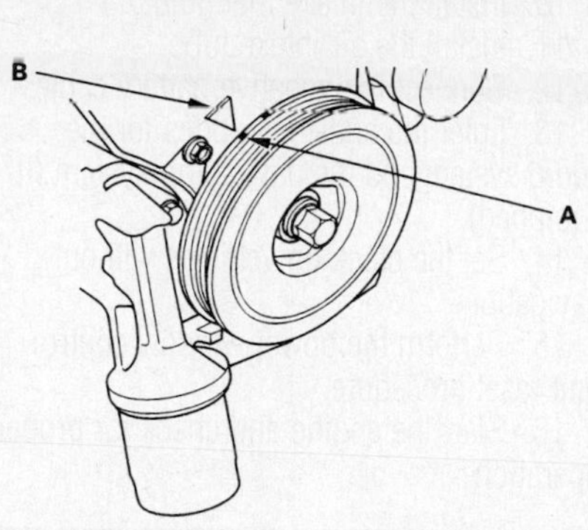
B
A
42356-HPIL-G16

Fig. 76 Turn the crankshaft so the white mark (A) aligns with the pointer (B)

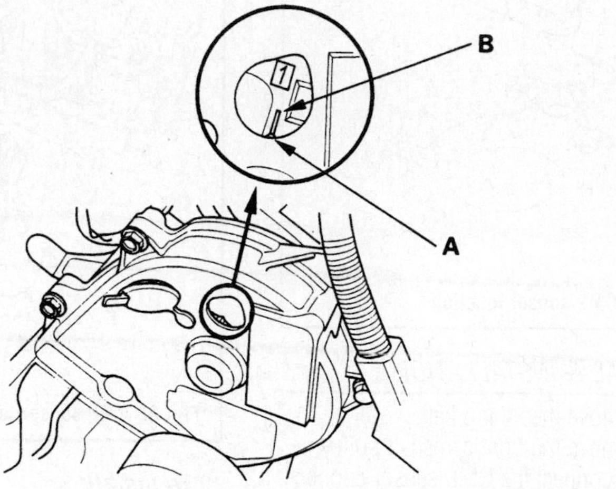
B
A
42356-HPIL-G17

Fig. 77 Make sure the number 1 piston is at top dead center (A) on the front camshaft pulley and pointer (B)

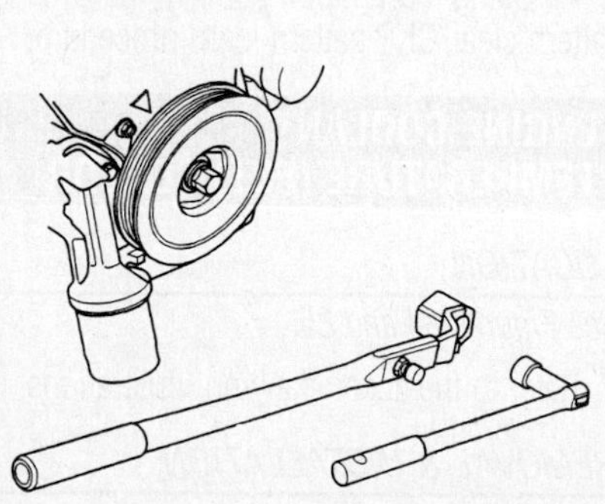
42356-HPIL-G18

Fig. 78 Remove the crankshaft pulley using holder tool shown

6. Remove the drive belts.
7. Support the engine with a block of wood and a jack under the oil pan.
8. Remove the ground cable bracket.
9. Remove the upper side engine mount.
10. Remove the dipstick tube.
11. Remove the crankshaft pulley using holder tool shown in the accompanying illustration and a breaker and socket, loosen the 19mm bolt and remove the pulley.
12. Remove the rear upper cover.

To install:

13. Install the rear upper timing covers and tighten the bolts to 9 ft. lbs. (12 Nm).
14. Install the crankshaft pulley and tighten the bolts to 181 ft. lbs. (245 Nm), using the holding tool to prevent the unit from turning.
15. Rotate the crankshaft pulley about 5 or 6 degrees clockwise so the belt positions itself on the pulleys.
16. Turn the crankshaft pulley so the white mark aligns with the pointer.
17. Check the camshaft pulley marks are aligned. If the marks are aligned, proceed to the next step. If the marks are not aligned, remove the timing belt and reinstall using the steps outlined before this step.
18. Install the drive belt.
19. Install the upper side mount and tighten the bolts in the sequence illustrated to the specifications in the illustration.
20. Using a suitable scan tool, perform the Powertrain Control Module (PCM) reset and the Crankshaft position (CKP) pattern clear/learn procedures, following the scan tool manufactures instructions.

VALVE LASH

ADJUSTMENT

See Figures 79 and 80.

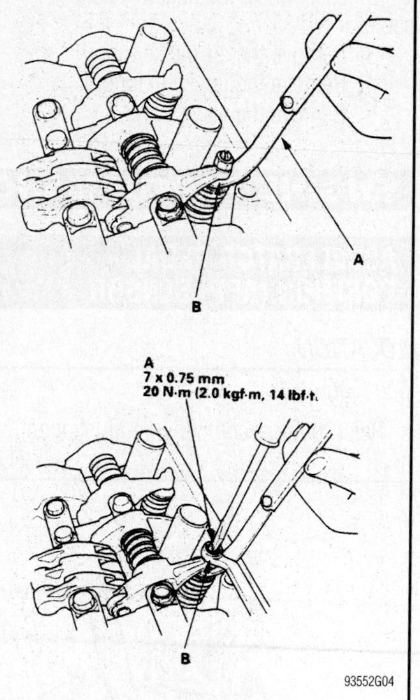
A
B
A
7 x 0.75 mm
20 N·m (2.0 kgf·m, 14 lbf·ft)
B
93552G04

Fig. 79 Inspect the valve clearance, adjust to specification and tighten the retainer to specification

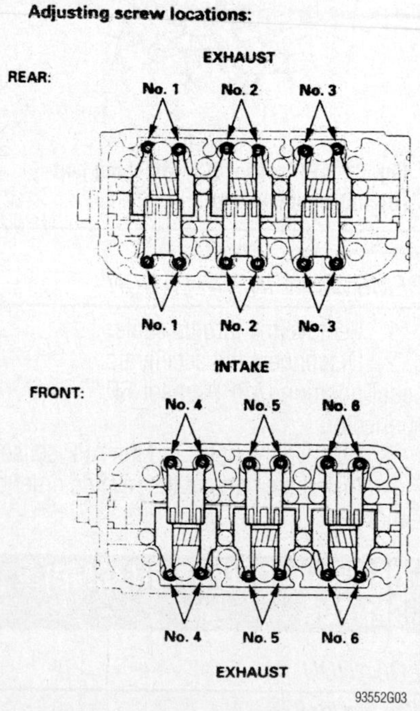
Adjusting screw locations:
REAR:
EXHAUST
No. 1
No. 2
No. 3
No. 1
No. 2
No. 3
INTAKE
FRONT:
No. 4
No. 5
No. 6
No. 4
No. 5
No. 6
EXHAUST
93552G03

Fig. 80 Valve adjusting retainer locations

➡Adjust the valves only when the cylinder head temperature is less than 100°F (38°C).

1. Make sure the anti-theft codes for the audio system and navigation system (if equipped) are available.
2. Disconnect the negative battery cable.
3. Remove the air intake tube.
4. Remove the intake manifold.
5. Remove the valve cover.
6. Rotate the crankshaft so that the valves to be adjusted are closed and the rocker arm is contacting the camshaft lobe base circle.
7. Measure the valve clearance.

If adjustment is necessary, loosen the locknut and turn the adjusting screw as necessary to achieve the correct valve clearance.

8. After adjustment, tighten the locknuts to 14 ft. lbs. (20 Nm).
9. Install the valve cover.
10. Install the intake manifold.
11. Install the air intake tube.
12. Connect the negative battery cable.
13. Enter the anti-theft codes for the audio system and the navigation system (if equipped).
14. Set the clock on vehicles without navigation.
15. Perform the power window control unit reset procedure.
16. Start the engine and check for proper operation.

ENGINE PERFORMANCE & EMISSION CONTROLS

ACCELERATOR PEDAL POSITION (APP) SENSOR

LOCATION

See Figure 81.

Refer to the accompanying illustration.

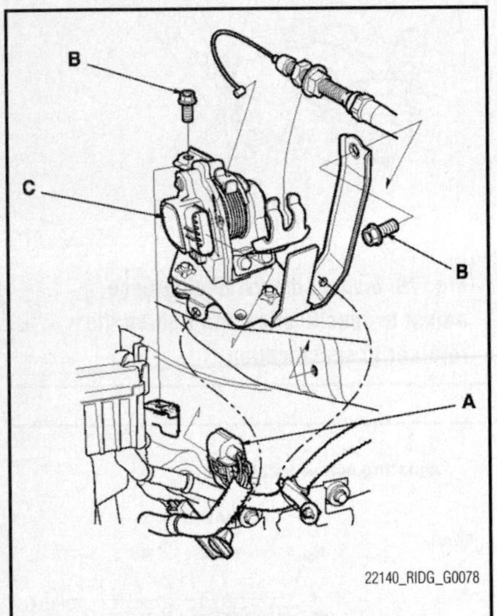
B
C
B
A
22140_RIDG_G0078

Fig. 81 APP sensor (C), attaching bolt (B) and connector (A)

REMOVAL & INSTALLATION

1. Remove the throttle cable.
2. Disconnect the accelerator pedal position (APP) sensor 6P connector.
3. Remove the bolts and the APP sensor.
4. Install the parts in the reverse order of removal.

CAMSHAFT POSITION (CMP) SENSOR

LOCATION

See Figure 82.

Refer to the accompanying illustration.

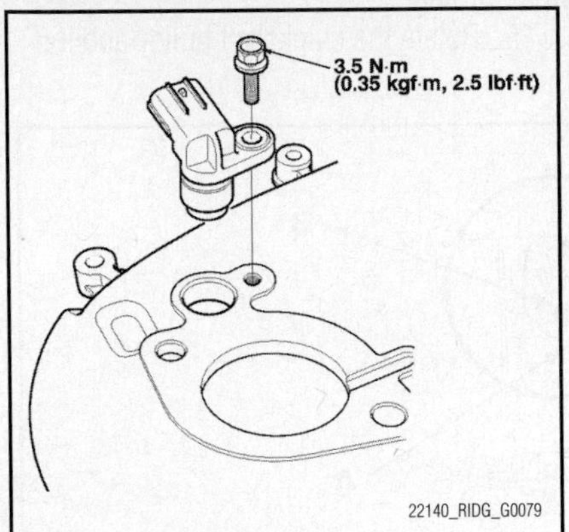
3.5 N·m
(0.35 kgf·m, 2.5 lbf·ft)
22140_RIDG_G0079

Fig. 82 CMP sensor location

REMOVAL & INSTALLATION

1. Remove the timing belt.
2. Remove the front camshaft pulley.
3. Disconnect the CMP sensor connector, then remove the back cover.
4. Remove the CMP sensor from the back cover.

To install:

5. Install the parts in the reverse order of removal.
6. Using a scan tool, perform the CKP pattern clear/CKP pattern learn procedure.

CRANKSHAFT POSITION (CKP) SENSOR

LOCATION

See Figure 83.

Refer to the accompanying illustration.

REMOVAL & INSTALLATION

1. Move the auto-tensioner to remove tension from the drive belt, then remove the belt.
2. Remove the crankshaft pulley.
3. Remove the upper and lower front covers from the engine.
4. Remove the CKP sensor from the oil pump.

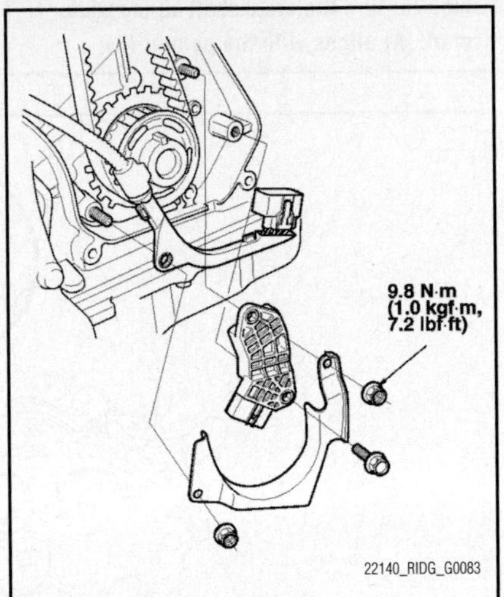
9.8 N·m
(1.0 kgf·m,
7.2 lbf·ft)
22140_RIDG_G0083

Fig. 83 CKP sensor location

To install:

5. Install the parts in the reverse order of removal.
6. Tighten sensor mounting bolt to 7 ft. lbs. (9 Nm).
7. Using a scan tool, perform the CKP pattern clear/CKP pattern learn procedure.

ENGINE COOLANT TEMPERATURE (ECT) SENSOR

LOCATION

See Figures 84 and 85.

Refer to the accompanying illustrations.

REMOVAL & INSTALLATION

ECT Sensor 1

See Figure 86.

1. Drain the engine coolant.
2. Remove the engine cover.
3. Disconnect the ECT sensor 1 2P connector.
4. Remove ECT sensor 1 and O-ring.
5. Install the parts in the reverse order of removal with a new O-ring, then refill the radiator with engine coolant.

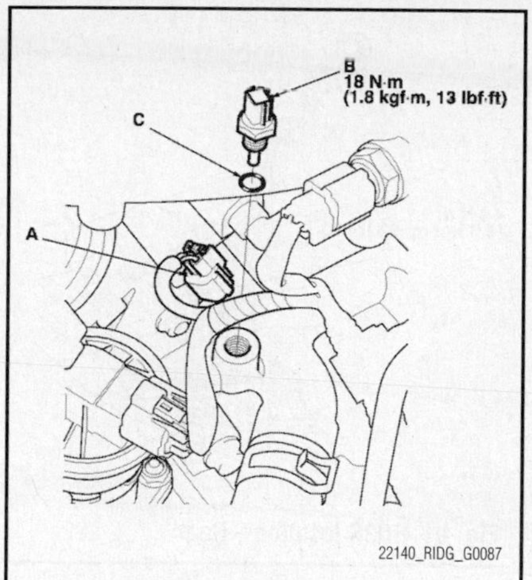
B
18 N·m
(1.8 kgf·m, 13 lbf·ft)
C
A
22140_RIDG_G0087

Fig. 84 ECT sensor 1 (B), O-ring (C) and connector (A) location

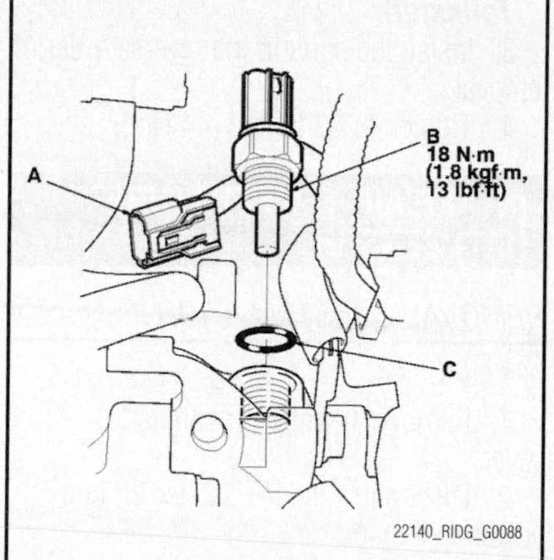
B
18 N·m
(1.8 kgf·m,
13 lbf·ft)
A
C
22140_RIDG_G0088

Fig. 85 ECT sensor 2 (B), O-ring (C) and connector (A) location

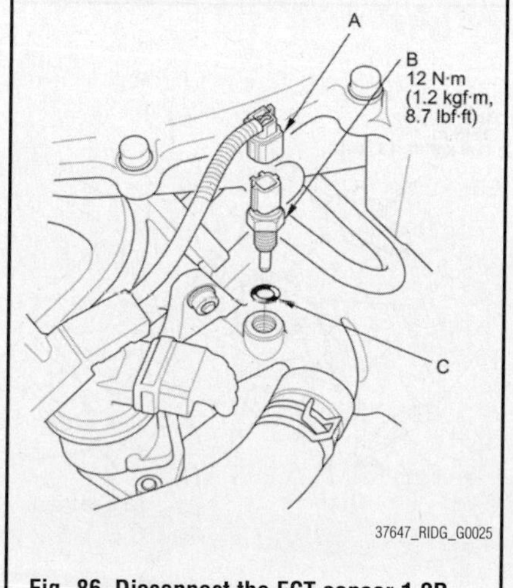
A
B
12 N·m
(1.2 kgf·m,
8.7 lbf·ft)
C
37647_RIDG_G0025

Fig. 86 Disconnect the ECT sensor 1 2P connector (A). Remove ECT sensor 1 (B) and O-ring (C).

ECT Sensor 2

Sco Figure 87.

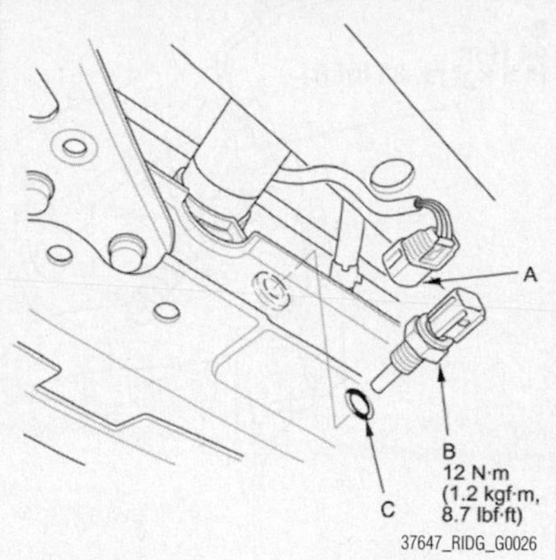
A
B
12 N·m
(1.2 kgf·m,
8.7 lbf·ft)
C
37647_RIDG_G0026

Fig. 87 Disconnect the ECT sensor 2 2P connector (A). Remove ECT sensor 1 (B) and O-ring (C).

1. Remove the front splash shield.
2. Drain the engine coolant.
3. Disconnect the ECT sensor 2 2P connector.
4. Remove ECT sensor 2 and O-ring.
5. Install the parts in the reverse order of removal with a new O-ring, then refill the radiator with engine coolant.

EVAPORATIVE EMISSIONS (EVAP) CANISTER

LOCATION

See Figure 88.

The Evaporative Emissions (EVAP) canister is located under the vehicle, just forward of the fuel tank, as shown.

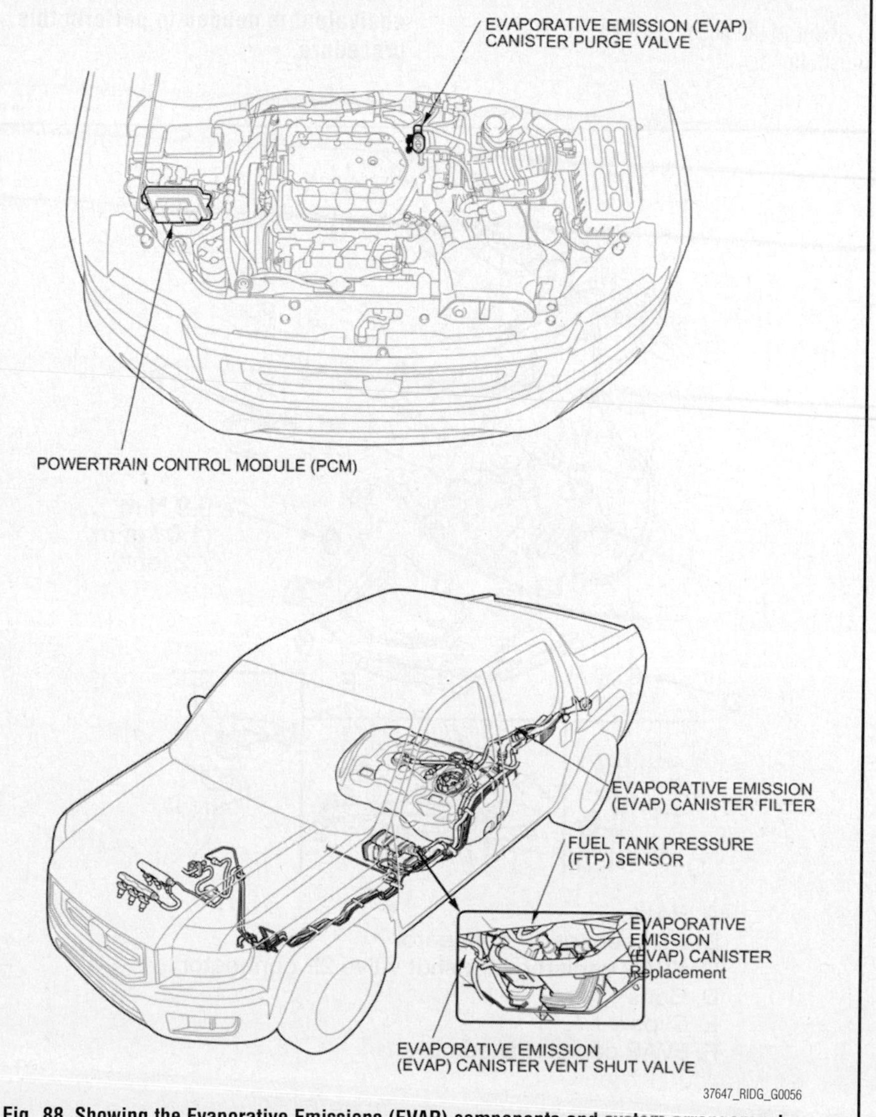
EVAPORATIVE EMISSION (EVAP)
CANISTER PURGE VALVE
POWERTRAIN CONTROL MODULE (PCM)
EVAPORATIVE EMISSION
(EVAP) CANISTER FILTER
FUEL TANK PRESSURE
(FTP) SENSOR
EVAPORATIVE
EMISSION
(EVAP) CANISTER
Replacement
EVAPORATIVE EMISSION
(EVAP) CANISTER VENT SHUT VALVE
37647_RIDG_G0056

Fig. 88 Showing the Evaporative Emissions (EVAP) components and system arrangement

REMOVAL & INSTALLATION

See Figure 89.

1. Raise the vehicle on a lift.
2. Remove the fuel tank guard.
3. Remove the hoses, the FTP sensor 3P connector, and the EVAP canister vent shut valve 2P connector. Remove the bolts and the clips. Remove the EVAP canister assembly.
4. Remove the EVAP canister bracket.

➡The canister filter remains on the bracket.

5. Installation is the reverse of the removal procedure.

HEATED OXYGEN (HO2S) SENSOR

LOCATION

See Figures 90 and 91.

Refer to the accompanying illustrations.

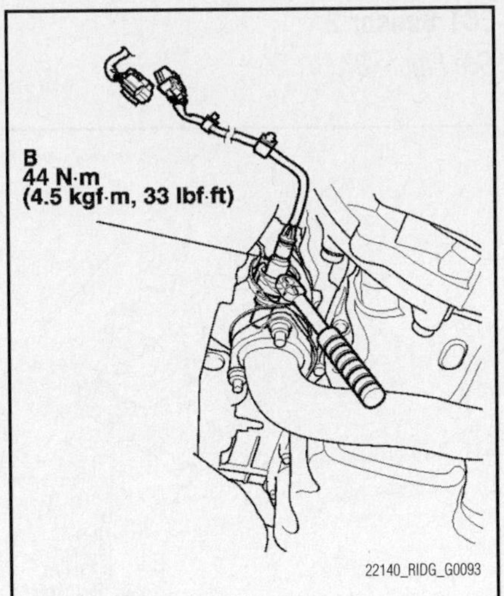
B
44 N·m
(4.5 kgf·m, 33 lbf·ft)
22140_RIDG_G0093
Fig. 90 HO2S location—Front

REMOVAL & INSTALLATION

➡O2 sensor socket wrench, Snap-on YA8875, SP Tools 93750, or equivalent is needed to perform this procedure.

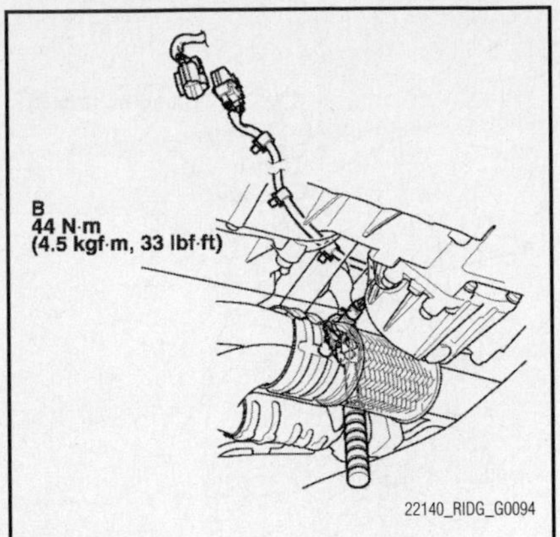
B
44 N·m
(4.5 kgf·m, 33 lbf·ft)
22140_RIDG_G0094
Fig. 91 HO2S location—Rear

1. Disconnect the oxygen sensor connector.
2. Remove the oxygen sensor using the special tool.

To install:

3. Install the parts in the reverse order of removal.
4. Tighten to 33 ft. lbs. (44 Nm).

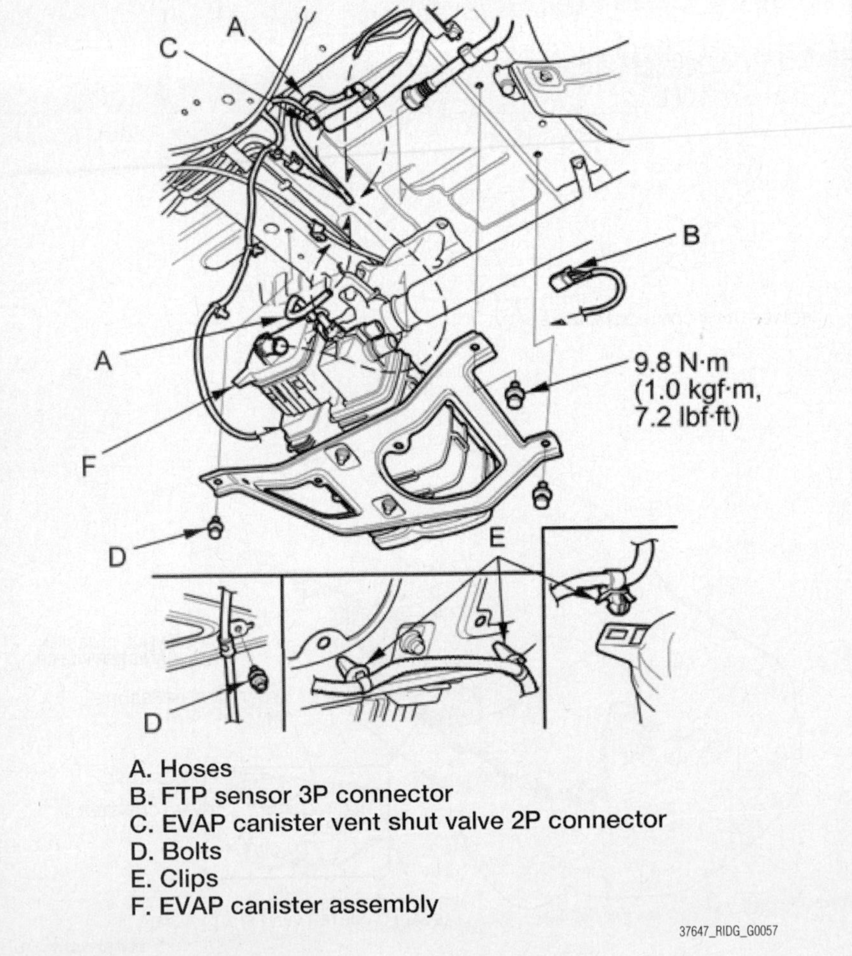
A
C
B
A
F
D
E
D
9.8 N·m
(1.0 kgf·m,
7.2 lbf·ft)
A. Hoses
B. FTP sensor 3P connector
C. EVAP canister vent shut valve 2P connector
D. Bolts
E. Clips
F. EVAP canister assembly
37647_RIDG_G0057
Fig. 89 Removing the EVAP canister

INTAKE AIR TEMPERATURE (IAT) SENSOR

REMOVAL & INSTALLATION

See Figure 92.

1. Remove the intake manifold cover.
2. Disconnect the IAT sensor 2P connector.
3. Remove the IAT sensor.

To install:

4. Install the parts in the reverse order of removal with a new O-ring.
5. Tighten to 13 ft. lbs. (18 Nm).

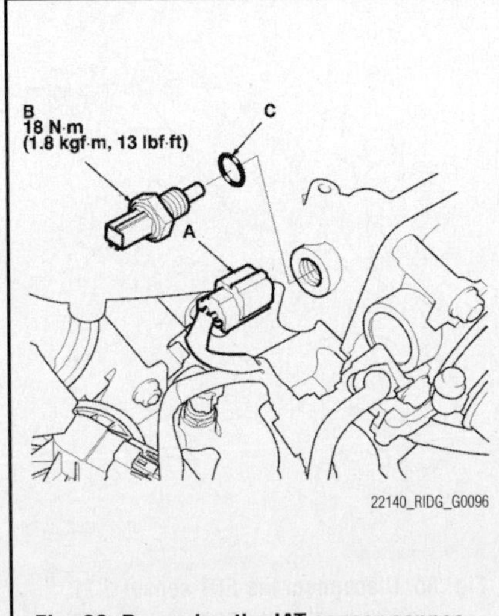
B
18 N·m
(1.8 kgf·m, 13 lbf·ft)
C
A
22140_RIDG_G0096
Fig. 92 Removing the IAT sensor connector (A), sensor (B) and O-ring (C)

KNOCK SENSOR (KS)

LOCATION

The Knock Sensor (KS) is located on top of the engine block, below the intake manifold.

REMOVAL & INSTALLATION

See Figure 93.

1. Remove the intake manifold. See "Intake Manifold" in "ENGINE MECHANICAL" section.
2. Disconnect the knock sensor 1P connector (A), then remove the knock sensor (B).

To install:

3. Install the parts in the reverse order of removal.
4. Tighten to 23ft. lbs. (31 Nm).

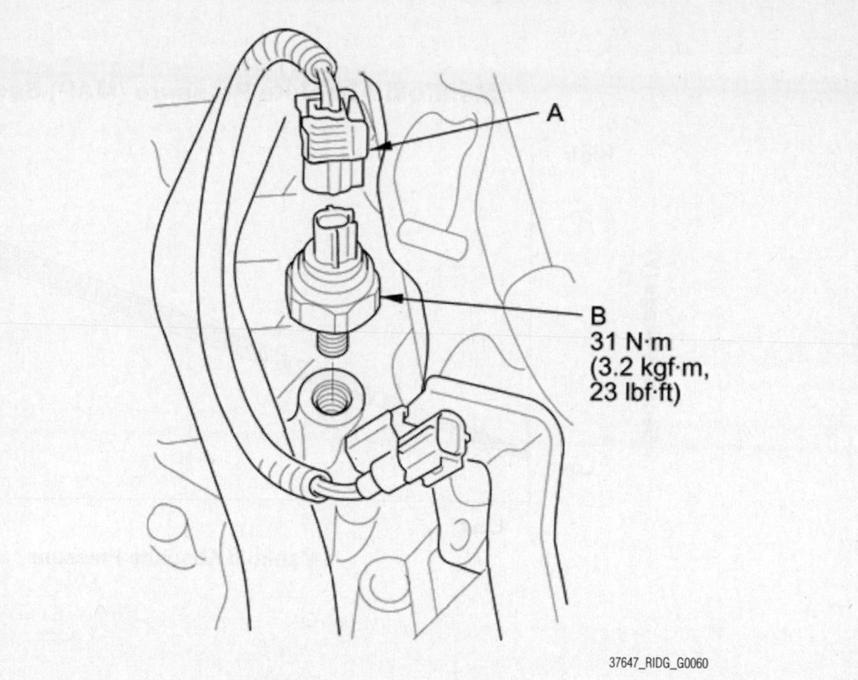
A
B
31 N·m
(3.2 kgf·m,
23 lbf·ft)
37647_RIDG_G0060

Fig. 93 KS sensor connector (A) and knock sensor (B)

MALFUNCTION INDICATOR LIGHT (MIL)

RESET PROCEDURES

1. Connect an HDS or equivalent OBD II scan tool to the diagnostic connector.
2. Turn the ignition switch to **ON**.
3. Turn the tester or scan tool **ON**.
4. Check whether any DTCs have been stored. Note them down if necessary.
5. Clear DTCs.
6. The MIL should turn **OFF**.

MANIFOLD ABSOLUTE PRESSURE (MAP) SENSOR

LOCATION

The Manifold Absolute Pressure (MAP) sensor is located on the inner end of the intake air duct.

REMOVAL & INSTALLATION

See Figure 94.

1. Remove the intake manifold cover.
2. Disconnect the MAP sensor connector.
3. Remove the screw.
4. Remove the MAP sensor.
5. Install the parts in the reverse order of removal with a new O-ring.

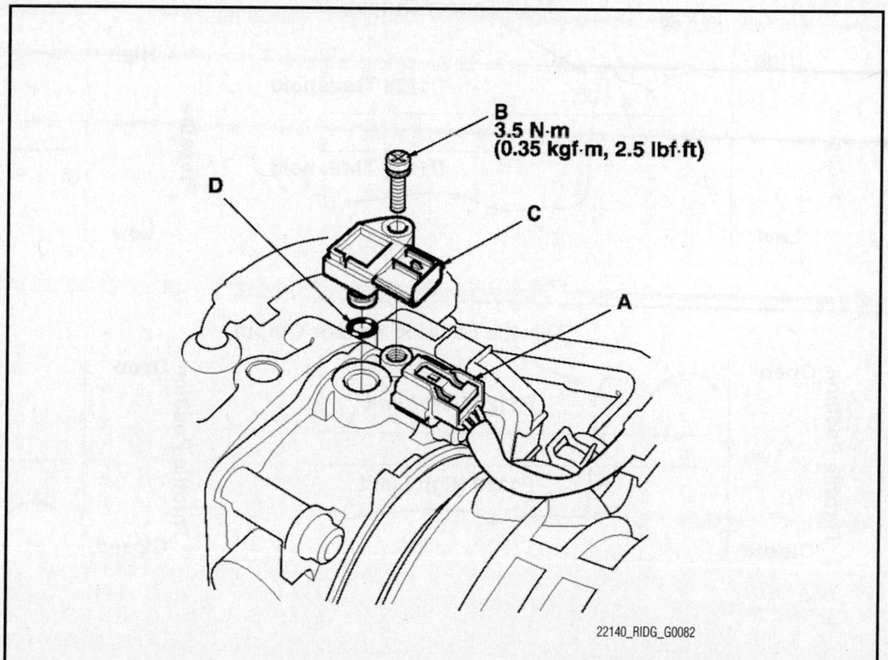
B
3.5 N·m
(0.35 kgf·m, 2.5 lbf·ft)
D
C
A
22140_RIDG_G0082

Fig. 94 MAP sensor (C), connector (A), and screw (B)

TESTING

See Figure 95.

1. Connect an HDS or equivalent scan tool to the data link connector (DLC) located under the driver's side of the dashboard.
2. Turn the ignition switch **ON**.
3. Check the MAP SENSOR in the DATA LIST.
4. If voltage is not as specified in the chart, inspect the connectors and wiring harness.
5. If no faults are found, replace the sensor.

MASS AIR FLOW/INTAKE AIR TEMPERATURE (MAF/IAT) SENSOR

LOCATION

The Mass Air Flow/Intake Air Temperature (MAF/IAT) sensor is located on the intake air duct, near the throttle body.

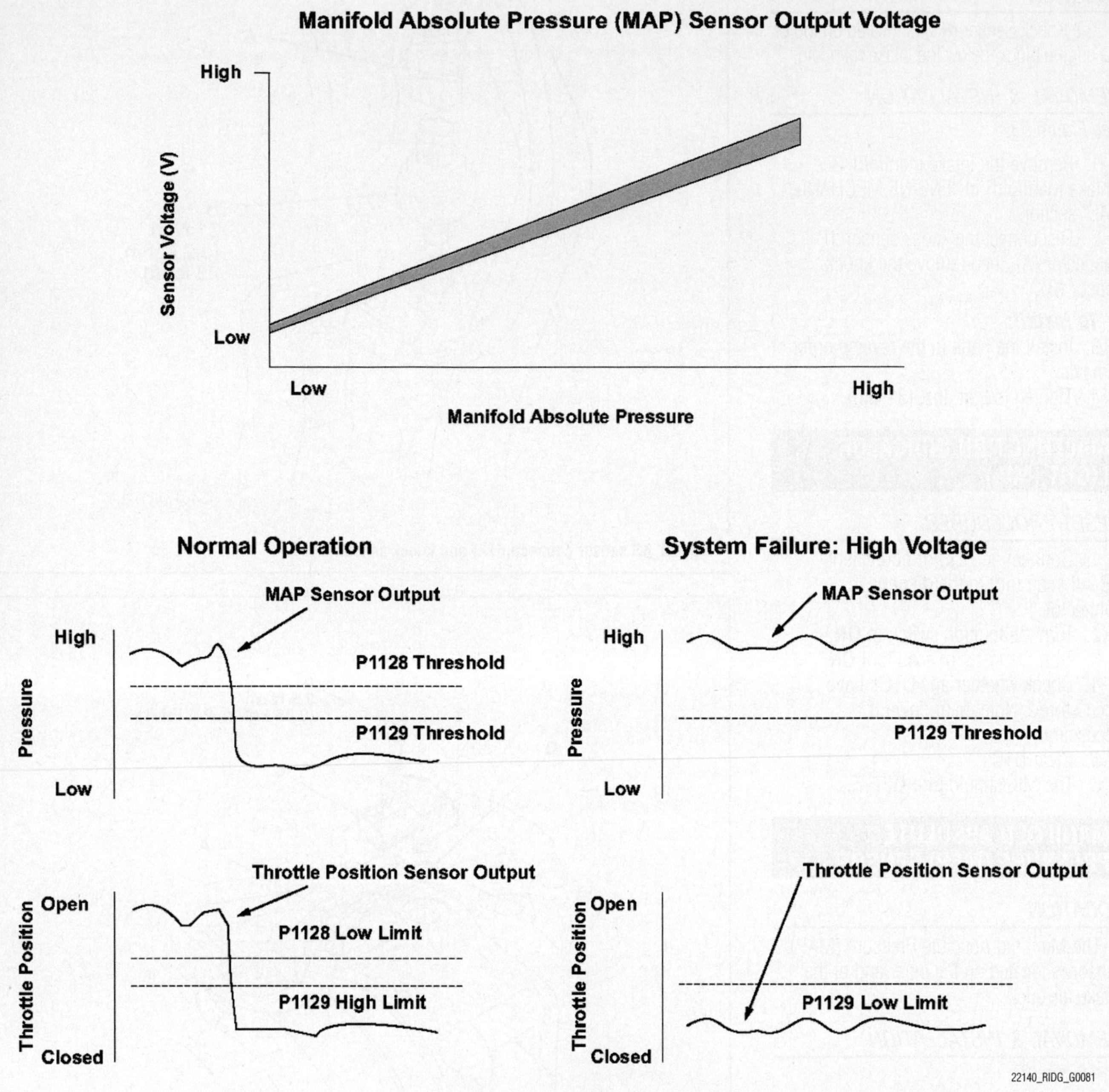
Manifold Absolute Pressure (MAP) Sensor Output Voltage
High
Sensor Voltage (V)
Low
Low
High
Manifold Absolute Pressure
Normal Operation
System Failure: High Voltage
MAP Sensor Output
High
P1128 Threshold
Pressure
P1129 Threshold
Low
MAP Sensor Output
High
Pressure
P1129 Threshold
Low
Throttle Position Sensor Output
Open
P1128 Low Limit
Throttle Position
P1129 High Limit
Closed
Throttle Position Sensor Output
Open
Throttle Position
P1129 Low Limit
Closed
22140_RIDG_G0081

Fig. 95 MAP sensor output voltage chart

REMOVAL & INSTALLATION

See Figure 96.

1. Disconnect the MAF/IAT sensor 5P connector.
2. Remove the screws.
3. Remove the MAF/IAT sensor and O-ring.
4. Install the parts in the reverse order of removal with a new O-ring.

OUTPUT SHAFT (COUNTERSHAFT) SPEED (OSS) SENSOR

LOCATION

The Output Shaft Speed (OSS) sensor is located on the side of the transmission housing. See "Removal & Installation" for figure reference.

REMOVAL & INSTALLATION

See Figure 97.

1. Remove the splash shield.
2. Disconnect the output shaft (countershaft) speed sensor connector, and remove the output shaft (countershaft) speed sensor, washer and O-ring.

To install:

3. Install a new O-ring on a new output shaft (countershaft) speed sensor, then install

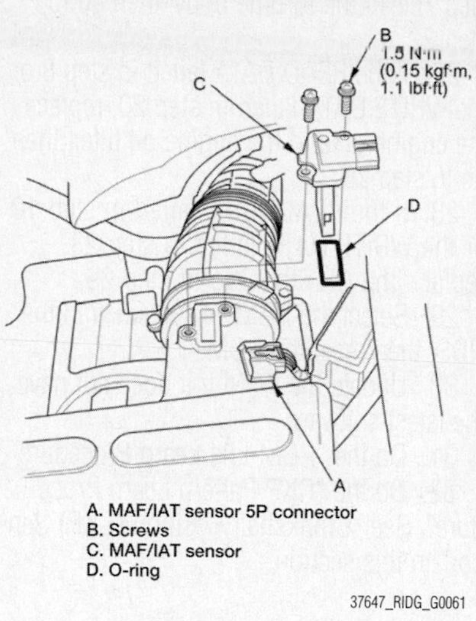
B
1.5 N·m
(0.15 kgf·m,
1.1 lbf·ft)
C
D
A
A. MAF/IAT sensor 5P connector
B. Screws
C. MAF/IAT sensor
D. O-ring
37647_RIDG_G0061

Fig. 96 Removing/installing the MAF/IAT sensor

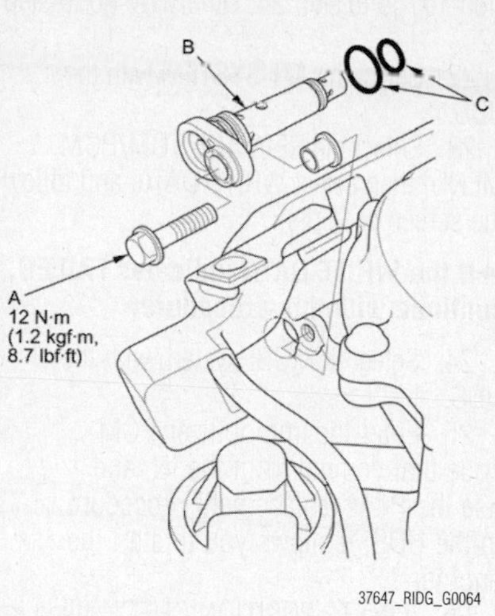
B
C
A
12 N·m
(1.2 kgf·m,
8.7 lbf·ft)
37647_RIDG_G0064

Fig. 98 Remove the bolt (A), PCV valve (B) and O-rings (C)

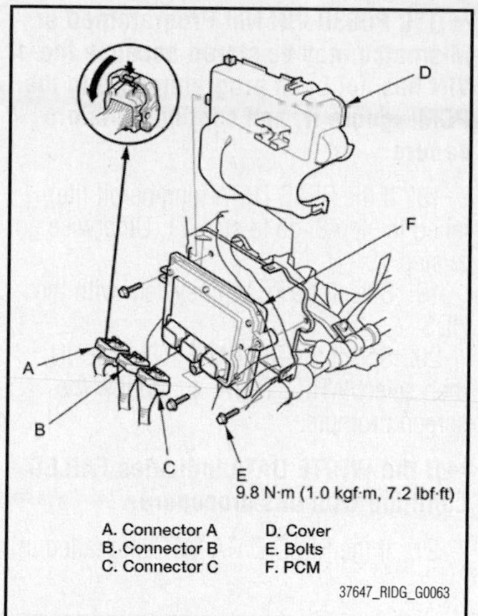
D
F
A
B
C
E
9.8 N·m (1.0 kgf·m, 7.2 lbf·ft)
A. Connector A
B. Connector B
C. Connector C
D. Cover
E. Bolts
F. PCM
37647_RIDG_G0063

Fig. 99 Removing the PCM

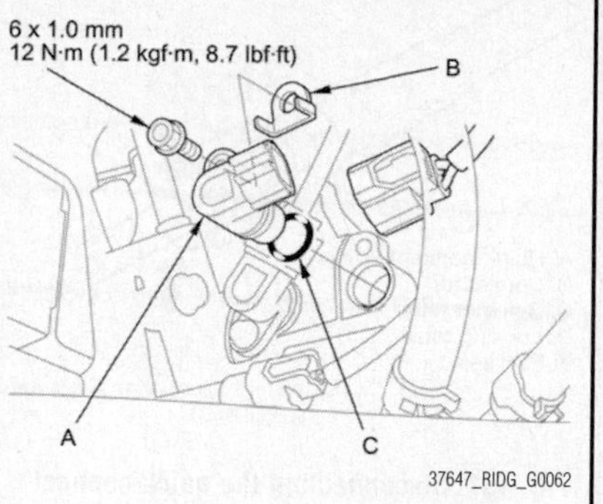
6 x 1.0 mm
12 N·m (1.2 kgf·m, 8.7 lbf·ft)
B
A
C
37647_RIDG_G0062

Fig. 97 Disconnect the output shaft (countershaft) speed sensor connector, and remove the output shaft (countershaft) speed sensor (A), washer (B) and O-ring (C).

the output shaft (countershaft) speed sensor and the washer in the transmission housing.

4. Check the connector for rust, dirt, or oil, and clean or repair if necessary, then connect the connector securely.
5. Install the splash shield.

POSITIVE CRANKCASE VENTILATION (PCV) VALVE

LOCATION

The Positive Crankcase Ventilation (PCV) valve is located at the left front corner of the engine, on the cylinder head.

REMOVAL & INSTALLATION

See Figure 98.

1. Remove the engine cover.
2. Remove the bolt.

➡Do not spill oil on the hot exhaust manifold.

3. Remove the PCV valve.

To install:

4. Install the parts in the reverse order of removal.
   a. When installing a new PCV valve, make sure the O-rings are in place.
   b. When installing a used PCV valve, use new O-rings.

POWERTRAIN CONTROL MODULE (PCM)

LOCATION

The Powertrain Control Module (PCM) is located in the forward right corner of the engine compartment.

REMOVAL & INSTALLATION

See Figure 99.

1. Connect the HDS, or equivalent scan tool, to the data link connector (DLC) located under the driver's side of the dashboard.
2. Turn the ignition switch to ON (II).
3. Make sure the HDS communicates with the PCM and other vehicle systems.
4. Select the PGM-FI system with the HDS.
5. Select the INSPECTION MENU with the HDS.
6. Select the ETCS TEST, then select the TP POSITION CHECK, and follow the screen prompts.

➡If the TP POSITION CHECK indicates FAILED, continue with this procedure.

7. Select the REPLACE PCM MENU, then select READ DATA, and follow the screen prompts.

➡Doing this step copies (READS) the engine oil life data from the original PCM so you can later download (WRITES) it into the new PCM. If READ DATA indicates FAILED, continue with this procedure.

8. Select the A/T system with the HDS.
9. Select the REPLACE TCM/PCM MENU, then select READ DATA, and follow the screen prompts.

➡Doing this step copies (READS) the ATF life data from the original PCM so you can later download (WRITES) it into the new PCM. If READ DATA indicates FAILED, continue with this procedure.

10. Turn the ignition switch to LOCK (0).
11. Jump the SCS line with the HDS.
12. Remove the cover.
13. Disconnect PCM connectors A, B, and C.

➡PCM connectors A, B, and C have symbols (A=square, B=triangle, C=circle) embossed on them for identification.

14. Remove the bolts, then remove the PCM assembly.

To install:

15. Install the PCM in the reverse order of removal.
16. Turn the ignition switch to ON (II).
17. Manually input the VIN to the PCM with the HDS.

➡DTC P0630 VIN Not Programmed or Mismatch may be stored because the VIN has not been programmed into the PCM; ignore it, and continue this procedure.

18. If the READ DATA (engine oil life) failed in step 8, go to step 21. Otherwise go to step 19.
19. Select the PGM-FI system with the HDS.
20. Select the REPLACE PCM MENU, then select WRITE DATA, and follow the screen prompts.

➡If the WRITE DATA indicates FAILED, continue with this procedure.

21. If the READ DATA (ATF life) failed in step 10, go to step 24. Otherwise go to step 22.
22. Select the A/T SYSTEM with the HDS.
23. Select the REPLACE TCM/PCM MENU, then select WRITE DATA, and follow the screen prompts.

➡If the WRITE DATA indicates FAILED, continue with this procedure.

24. Select IMMOBI system with the HDS.
25. Enter the immobilizer PCM code that you got from the iN, and use the PCM replacement procedure in the HDS; it allows you to start the engine.
26. If the TP POSITION CHECK failed in step 7clean the throttle body, then go to step 27.
27. If the READ DATA failed in step 8 or the WRITE DATA failed in step 20, replace the engine oil and the engine oil filter, then go to step 28.
28. If the READ DATA failed in step 10 or the WRITE DATA failed in step 23, replace the ATF, then go to step 29.
29. Select the PGM-FI system with the HDS, then reset the PCM.
30. Update the PCM if it does not have the latest software.
31. Do the "PCM Idle Learn Procedure".
32. Do the "CKP Pattern Learn Procedure". See "Crankshaft Position (CKP) Sensor" in this section.

FUEL — GASOLINE FUEL INJECTION SYSTEM

FUEL SYSTEM SERVICE PRECAUTIONS

Safety is the most important factor when performing not only fuel system maintenance but any type of maintenance. Failure to conduct maintenance and repairs in a safe manner may result in serious personal injury or death. Maintenance and testing of the vehicle's fuel system components can be accomplished safely and effectively by adhering to the following rules and guidelines.

- To avoid the possibility of fire and personal injury, always disconnect the negative battery cable unless the repair or test procedure requires that battery voltage be applied.
- Always relieve the fuel system pressure prior to disconnecting any fuel system component (injector, fuel rail, pressure regulator, etc.), fitting or fuel line connection. Exercise extreme caution whenever relieving fuel system pressure to avoid exposing skin, face and eyes to fuel spray. Please be advised that fuel under pressure may penetrate the skin or any part of the body that it contacts.
- Always place a shop towel or cloth around the fitting or connection prior to loosening to absorb any excess fuel due to spillage. Ensure that all fuel spillage (should it occur) is quickly removed from engine surfaces. Ensure that all fuel soaked cloths or towels are deposited into a suitable waste container.
- Always keep a dry chemical (Class B) fire extinguisher near the work area.
- Do not allow fuel spray or fuel vapors to come into contact with a spark or open flame.
- Always use a back-up wrench when loosening and tightening fuel line connection fittings. This will prevent unnecessary stress and torsion to fuel line piping.
- Always replace worn fuel fitting O-rings with new. Do not substitute fuel hose or equivalent where fuel pipe is installed.

Before servicing the vehicle, make sure to also refer to the precautions in the beginning of this section as well.

RELIEVING FUEL SYSTEM PRESSURE

See Figures 100 and 101.

1. Remove the left kick panel, then remove PGM-FI main relay 2 (FUEL PUMP) from the under-dash fuse/relay box.
2. Start the engine, and let it idle until it stalls.

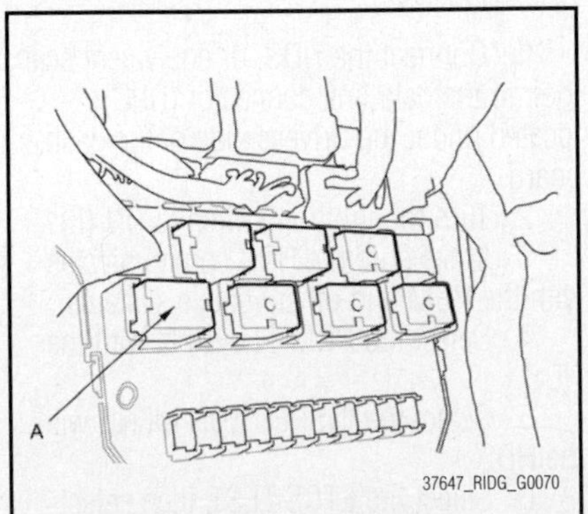
A
37647_RIDG_G0070

Fig. 100 Remove the left kick panel, then remove PGM-FI main relay 2 (FUEL PUMP) (A) from the under-dash fuse/relay box.

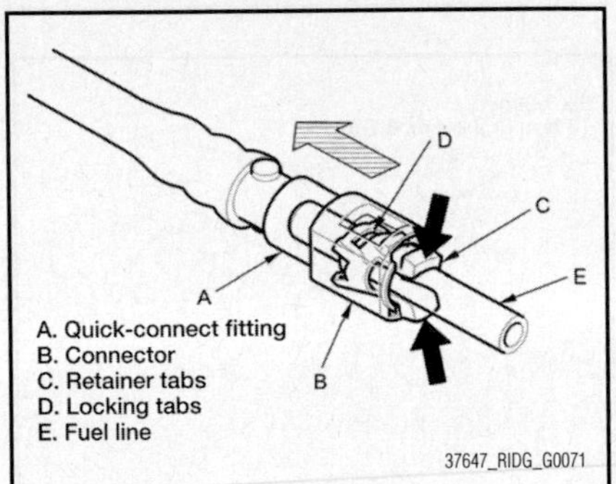
A
B
C
D
E
A. Quick-connect fitting
B. Connector
C. Retainer tabs
D. Locking tabs
E. Fuel line
37647_RIDG_G0071

Fig. 101 Disconnecting the quick-connect fuel line fitting

➡If any DTCs are stored, clear and ignore them.

3. Turn the ignition switch to LOCK (0).
4. Remove the fuel fill cap to relieve the pressure in the fuel tank.
5. Do the battery terminal disconnect procedure, then wait at least 3 minutes before beginning work.
6. Remove the quick-connect fitting cover.
7. Check the fuel quick-connect fitting for dirt, and clean it if needed.
8. Place a rag or shop towel over the quick-connect fitting.
9. Disconnect the quick-connect fitting: Hold the connector with one hand, and squeeze the retainer tabs with the other hand to release them from the locking tabs. Pull the connector off.

➡Be careful not to damage the line or other parts. Do not use tools. If the connector does not move, keep the retainer tabs pressed down, and alternately pull and push the connector until it comes off easily. Do not remove the

retainer from the line; once removed, the retainer must be replaced with a new one.

10. After disconnecting the quick-connect fitting, check it for dirt or damage.
11. Do the battery terminal reconnect procedure.

FUEL FILTER

REMOVAL & INSTALLATION

See Figures 102 through 104.

➡The fuel filter should be replaced whenever the fuel pressure drops below 57 psi.

1. Relieve the fuel system pressure.
2. Remove the rear back seat.
3. Remove the access panel on the floor.
4. Remove the fuel tank module lock ring using the special tool.
5. Remove the fuel filter.

To install:

6. Check these items before installing the fuel tank unit:
   a. When connecting the wire harness, make sure the connection is secure and the connectors are firmly locked into place.
   b. When installing the fuel gauge sending unit, make sure the connection is secure and the connector is firmly locked into place. Be careful not to bend or twist it excessively.
7. Install the parts in the reverse order of removal with new O-rings.

➡Coat the O-rings with clean engine oil.

8. When installing the fuel tank unit, align the marks on the unit and the fuel tank.

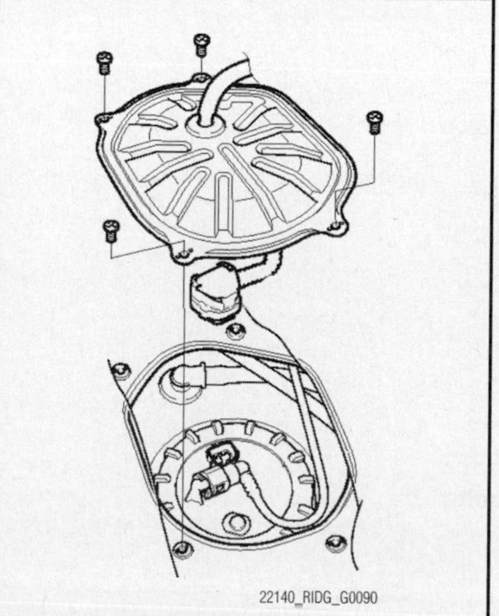
22140_RIDG_G0090

Fig. 102 Fuel pump access panel location

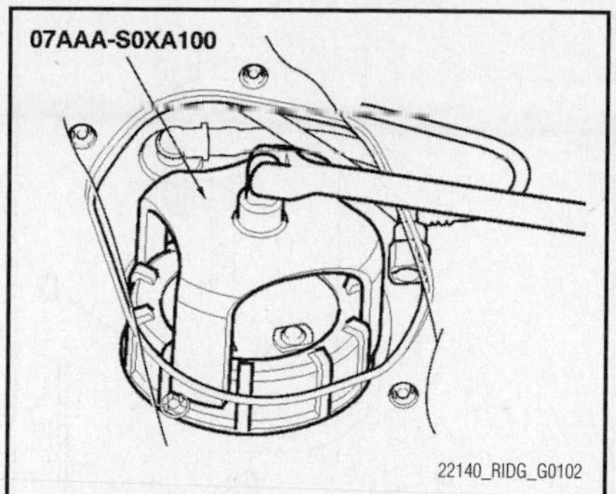
07AAA-S0XA100
22140_RIDG_G0102

Fig. 103 Fuel tank module lock ring special tool

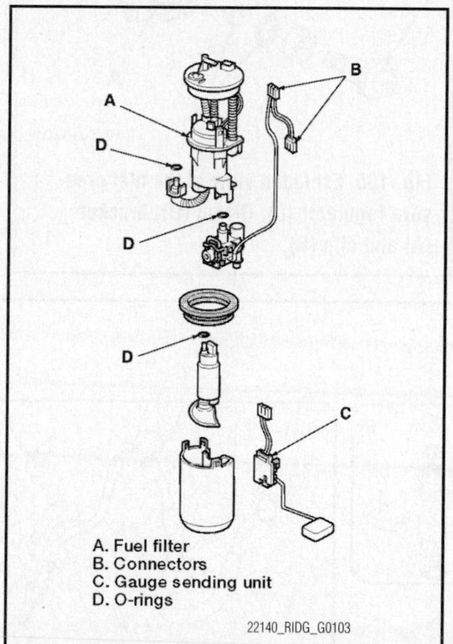
A
B
D
D
D
C
A. Fuel filter
B. Connectors
C. Gauge sending unit
D. O-rings
22140_RIDG_G0103

Fig. 104 Fuel tank module assembly

9. Torque the lock ring to 69 ft. lbs. (93 Nm).

FUEL LEVEL SENDING UNIT

REMOVAL & INSTALLATION

See Figures 103 and 104.

1. Relieve the fuel system pressure.
2. Remove the rear back seat.
3. Remove the access panel on the floor.
4. Remove the fuel tank module lockring using the special tool.
5. Remove the fuel level sending unit.

To install:

6. Check these items before installing the fuel tank unit:
   a. When connecting the wire harness, make sure the connection is secure and the connectors are firmly locked into place.
   b. When installing the fuel gauge sending unit, make sure the connection is secure and the connector is firmly locked into place. Be careful not to bend or twist it excessively.
7. Install the parts in the reverse order of removal with new O-rings.

➡Coat the O-rings with clean engine oil.

8. When installing the fuel tank unit, align the marks on the unit and the fuel tank.
9. Torque the lockring to 69 ft. lbs. (93 Nm).

FUEL PUMP

REMOVAL & INSTALLATION

See Figures 102 through 104.

1. Relieve the fuel system pressure.
2. Remove the rear back seat.
3. Remove the access panel on the floor.
4. Remove the fuel tank module lock ring using the special tool.
5. Disassemble the fuel tank module and remove the fuel pump.

To install:

6. Check these items before installing the fuel tank unit:
   a. When connecting the wire harness, make sure the connection is secure and the connectors are firmly locked into place.
   b. When installing the fuel gauge sending unit, make sure the connection is secure and the connector is firmly locked into place. Be careful not to bend or twist it excessively.
7. Install the parts in the reverse order of removal with new O-rings.

➡Coat the O-rings with clean engine oil.

8. When installing the fuel tank unit, align the marks on the unit and the fuel tank.
9. Torque the lock ring to 69 ft. lbs. (93 Nm).

FUEL PRESSURE REGULATOR

REMOVAL & INSTALLATION

See Figures 102, 103 and 105.

1. Relieve the fuel system pressure.
2. Remove the rear back seat.
3. Remove the access panel on the floor.
4. Remove the fuel tank module lock ring using the special tool.

5. Remove the bracket.
6. Remove the clip.
7. Remove the fuel pressure regulator.

To install:

8. Check these items before installing the fuel tank unit:
   a. When connecting the wire harness, make sure the connection is secure and the connectors are firmly locked into place.
   b. When installing the fuel gauge sending unit, make sure the connection is secure and the connector is firmly locked into place. Be careful not to bend or twist it excessively.
9. Install the parts in the reverse order of removal with new O-rings.

➡Coat the O-rings with clean engine oil.

10. When installing the fuel tank unit, align the marks on the unit and the fuel tank.
11. Torque the lock ring to 69 ft. lbs. (93 Nm).

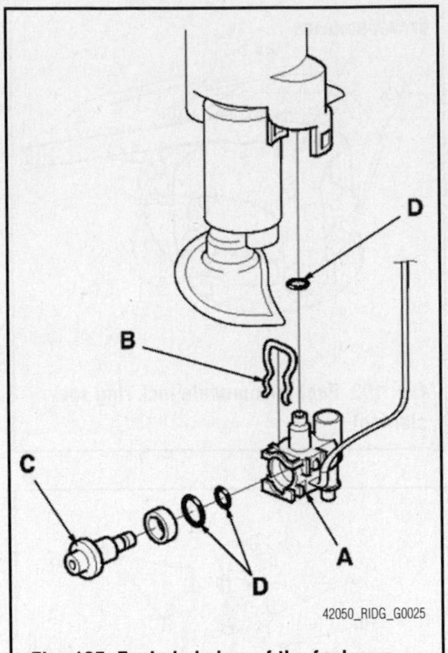

D
B
C
A
D
42050_RIDG_G0025

Fig. 105 Exploded view of the fuel pressure regulator (C), O-ring (D), bracket (A) and clip (B).

FUEL RAIL & INJECTORS

REMOVAL & INSTALLATION

See Figures 106 and 107.

1. Relieve the fuel pressure.
2. Remove the intake manifold.
3. Disconnect the quick-connect fitting.
4. Remove the fuel joint hose mounting bolt.
5. Disconnect the connectors from the injectors.
6. Remove the fuel rail mounting bolts from the fuel rails.
7. Remove the fuel rails and the injectors from the injector base.
8. Remove the injector clips from the fuel rails.

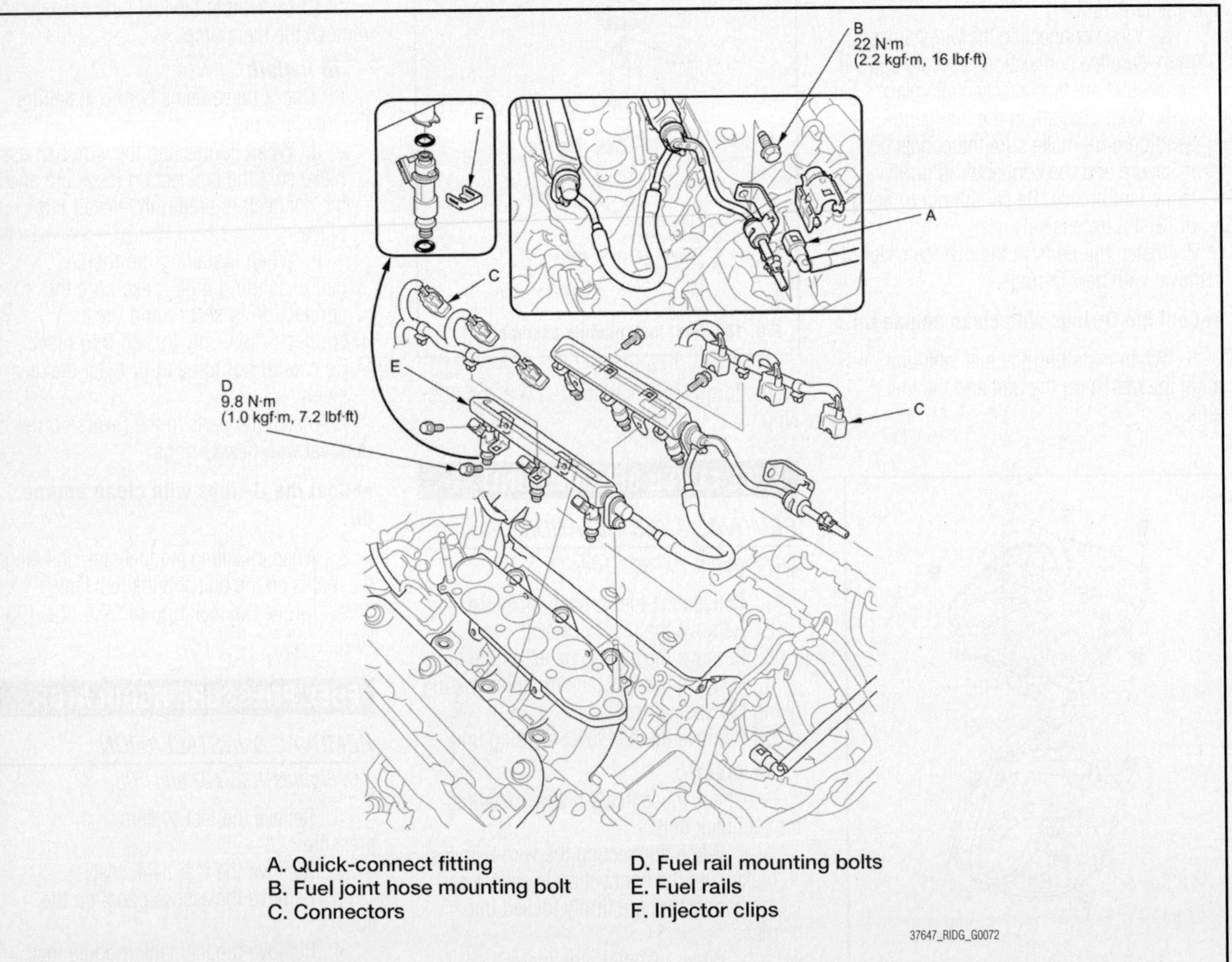

B
22 N·m
(2.2 kgf·m, 16 lbf·ft)
F
A
C
E
D
9.8 N·m
(1.0 kgf·m, 7.2 lbf·ft)
C
A. Quick-connect fitting
B. Fuel joint hose mounting bolt
C. Connectors
D. Fuel rail mounting bolts
E. Fuel rails
F. Injector clips
37647_RIDG_G0072

Fig. 106 Removing the fuel rail and injectors

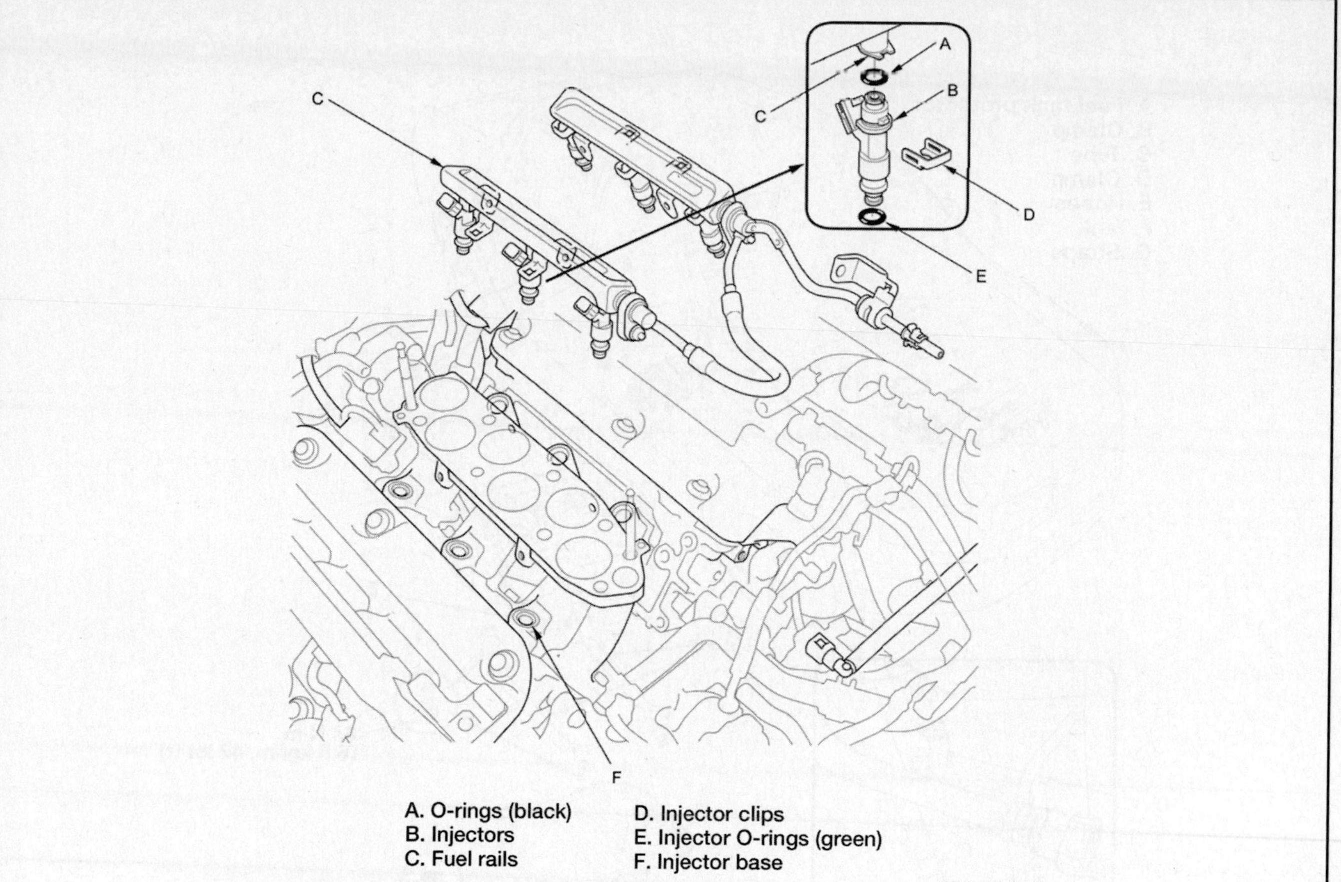
A
B
C
C
D
E
F
A. O-rings (black)
B. Injectors
C. Fuel rails
D. Injector clips
E. Injector O-rings (green)
F. Injector base
37647_RIDG_G0073

Fig. 107 Installing the fuel rail and injectors

9. Remove the injectors from the fuel rails.

To install:

10. Coat the new O-rings (black) with clean engine oil, and insert the injectors into the fuel rails.
11. Install the injector clips.
12. Coat the new injector O-rings (green) with clean engine oil.
13. Install the fuel rails and the injectors in the injector base.
14. Install the fuel rail mounting bolts, and connect the connectors to the injectors.
15. Install the fuel joint hose mounting bolt.
16. Connect the quick-connect fitting.
17. Turn the ignition switch to ON (II), but do not operate the starter. After the fuel pump runs for about 2 seconds, the fuel rail will be pressurized. Repeat this two or three times, then check for fuel leaks.
18. Install the intake manifold with a new gasket.

FUEL TANK

DRAINING

1. Remove the fuel tank unit. See "Fuel Level Sending Unit" in this section.
2. Using a hand pump, a hose, and a container suitable for fuel, draw the fuel from the fuel tank.
3. Reinstall the fuel tank unit.

REMOVAL & INSTALLATION

See Figure 108.

1. Drain the fuel tank.
2. Raise and safely support the vehicle securely on jackstands.
3. Remove the exhaust pipe.
4. Remove the propeller shaft, and support it with jackstands.
5. Remove the fuel tank protector.
6. Loosen the clamp, and disconnect the tube.
7. Open the clamp.
8. Disconnect the hoses. Slide back the clamps, then twist the hoses as you pull to avoid damaging them.
9. Place a jack or other support under the tank.
10. Remove the strap bolts and the straps.
11. Remove the fuel tank.

To install:

12. Install the parts in the reverse order of removal.
13. Tighten tank strap bolts to 47 ft. lbs. (64 Nm).
14. Tighten tank protector nuts to 7 ft. lbs. (10 Nm).

IDLE SPEED

INSPECTION

See Figure 109.

➡Before checking the idle speed, check these items:

- The malfunction indicator lamp (MIL) has not been reported on, and there are no DTCs.
- Ignition timing
- Spark plugs
- Air cleaner
- PCV system

1. Apply the parking brake.
2. Disconnect the Evaporative (EVAP) emission canister purge valve connector.
3. Connect the HDS to the Data Link Connector (DLC) located under the driver's side of the dashboard.
4. Start the engine. Hold the engine

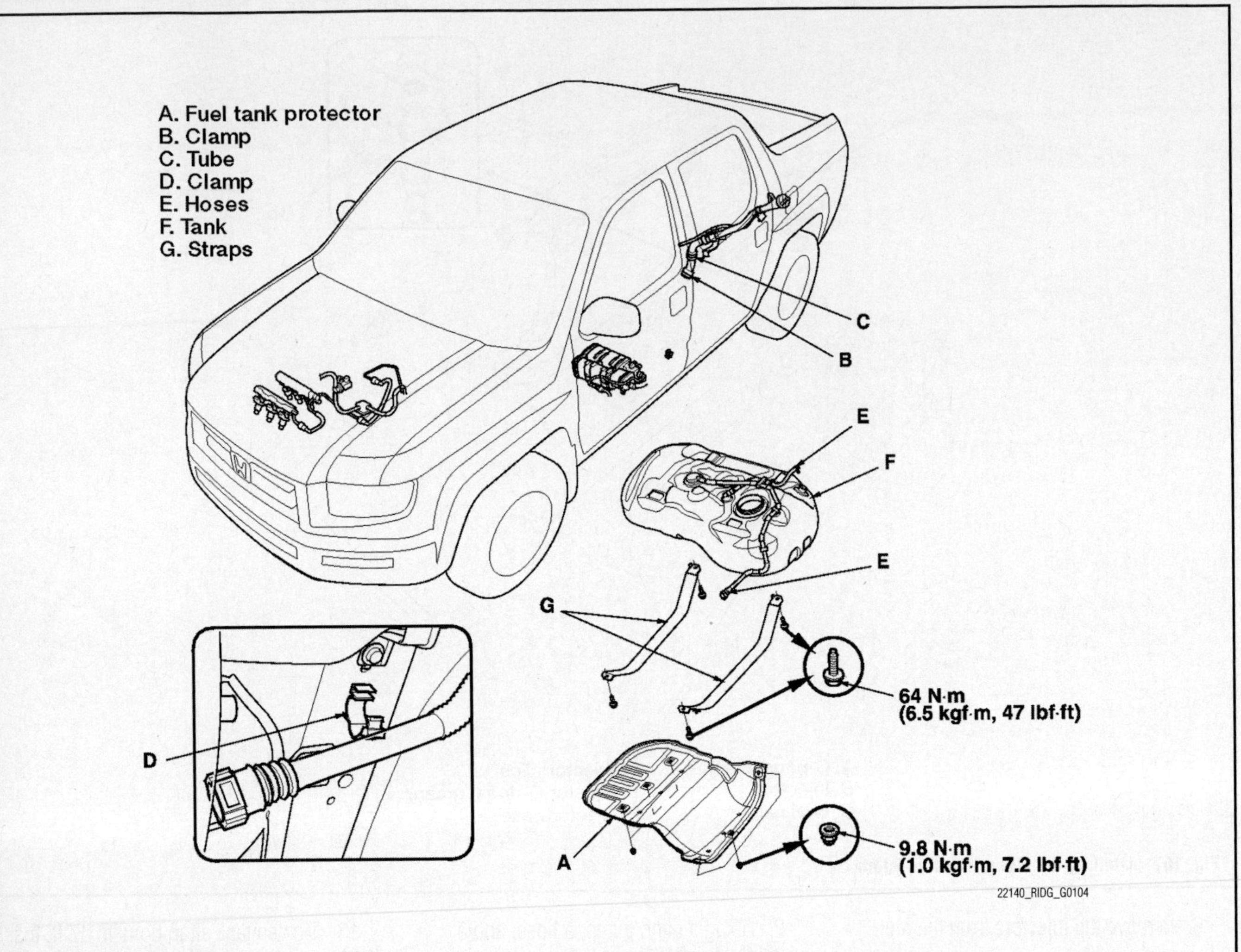
A. Fuel tank protector
B. Clamp
C. Tube
D. Clamp
E. Hoses
F. Tank
G. Straps
C
B
E
F
E
G
D
64 N·m
(6.5 kgf·m, 47 lbf·ft)
A
9.8 N·m
(1.0 kgf·m, 7.2 lbf·ft)
22140_RIDG_G0104

Fig. 108 Fuel tank assembly

speed at 3,000 rpm without load (in Park or neutral) until the radiator fan comes on, then let it idle.

5. Check the idle speed without load conditions: headlights, blower fan, radiator fan, and air conditioner off. The idle speed should be 680–780 rpm (in Park or Neutral).

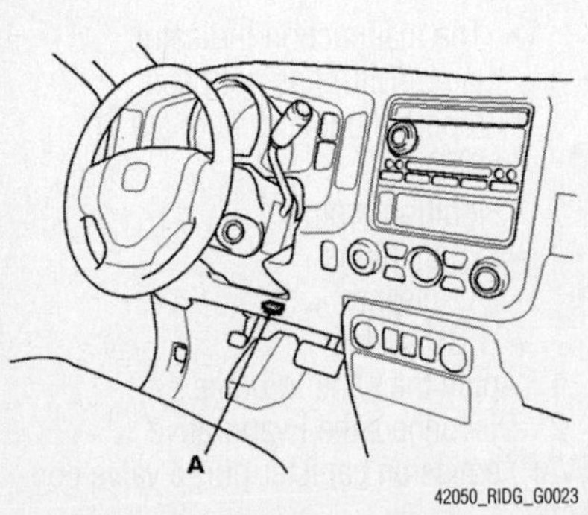
A
42050_RIDG_G0023

Fig. 109 Connect the scan tool to the DLC (A) which is found under the driver's side of the dashboard

6. Let the engine idle for 1 minute with high electric load (A/C switch ON, temperature set to Max Cool, blower fan on High, and headlights on high beam). The idle speed should be 680–780 rpm (in Park or Neutral).

➡**The idle learn procedure must be done so the PCM can learn the engine idle characteristics. Do the idle learn procedure whenever you do any of these actions:**

- Replace PCM.
- Reset PCM.
- Update PCM.
- Erasing DTCs with the HDS does not require you to do the idle learn procedure.
- Replace or clean the throttle body.

7. If the idle speed is not within specification, do the PCM idle learn procedure, as follows:

a. Make sure all electrical items (A/C, audio, lights, etc.) are off.

b. Reset the PCM with the HDS.

c. Turn the ignition switch **ON** (II), and wait 2 seconds.

d. Start the engine. Hold the engine speed at 3,000 rpm without load (in Park or neutral) until the radiator fan comes on, or until the engine coolant temperature reaches 194°F (90°C).

e. Let the engine idle for about 5 minutes with the throttle fully closed.

➡**If the radiator fan comes on, do not include its running time in the 5 minutes.**

8. Reconnect the EVAP canister purge valve connector.

ADJUSTMENT

Idle speed is maintained by the Powertrain Control Module (PCM). No adjustment is necessary or possible.

THROTTLE BODY

REMOVAL & INSTALLATION

See Figure 110.

⁂ CAUTION

Do not insert your fingers into the installed throttle body when you turn the ignition switch ON or while the ignition switch is ON. If you do, you will seriously injure your fingers if the throttle valve is activated.

1. Turn the ignition switch to LOCK (0).
2. Disconnect the MAP sensor connector.
3. Remove the intake air duct.
4. Disconnect the throttle body connector.
5. Disconnect and plug the water bypass hoses.
6. Remove the throttle body.

To install:

7. Install the parts in the reverse order of removal with a new gasket.
8. Tighten throttle body nuts to 16 ft. lbs. (22 Nm).

➡Do the PCM idle learn procedure after the throttle body has been replaced.

9. Refill the radiator with engine coolant.

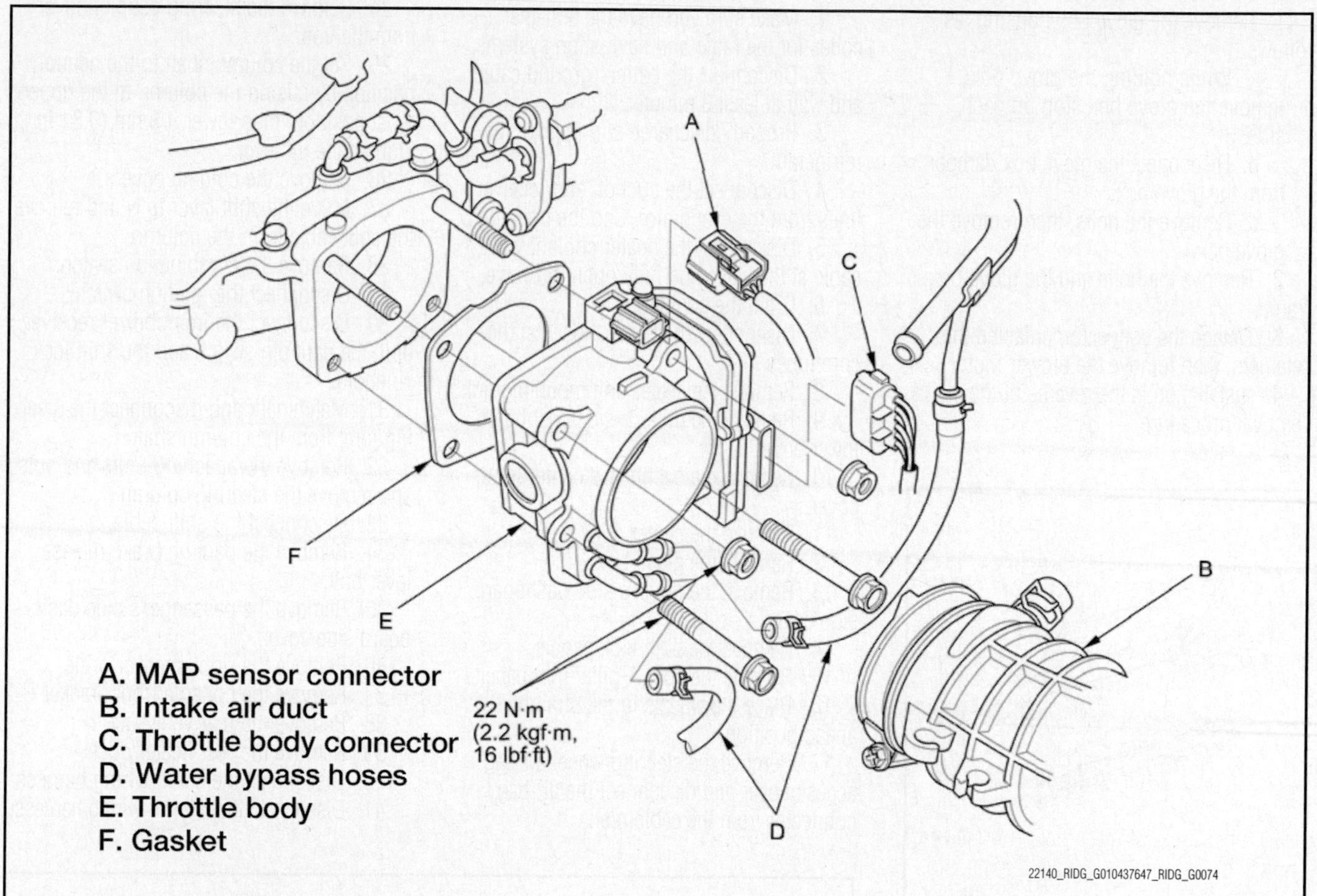
A
C
F
E
B
D
22 N·m
(2.2 kgf·m,
16 lbf·ft)
A. MAP sensor connector
B. Intake air duct
C. Throttle body connector
D. Water bypass hoses
E. Throttle body
F. Gasket
22140_RIDG_G010437647_RIDG_G0074

Fig. 110 Throttle body assembly exploded view

HEATING & AIR CONDITIONING SYSTEM

BLOWER MOTOR

REMOVAL & INSTALLATION

See Figures 111 and 112.

⁂ CAUTION

Before beginning work, review the "Precautions" in this section.

1. Remove the glove box housing, as follows:
   a. While holding the glove box, remove the glove box stop on each side.
   b. Disconnect the glove box damper from the glove box.
   c. Remove the bolts, then remove the glove box.
2. Remove the bolts and the glove box frame.
3. Detach the connector, unfasten the retainers, then remove the blower motor.
4. Installation is the reverse of the removal procedure.

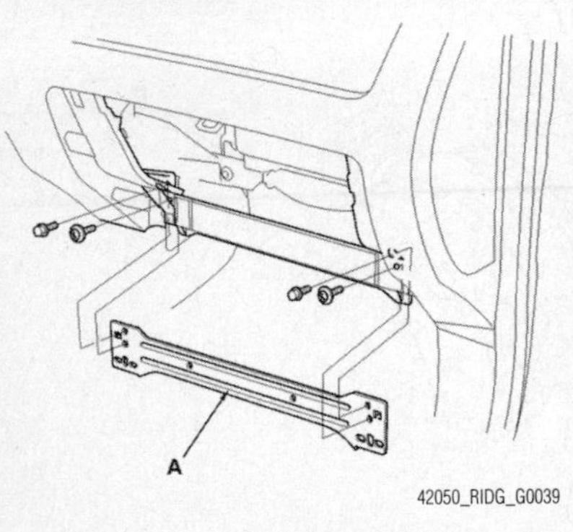
A
42050_RIDG_G0039

Fig. 111 Remove the bolts and the glove box frame (A)

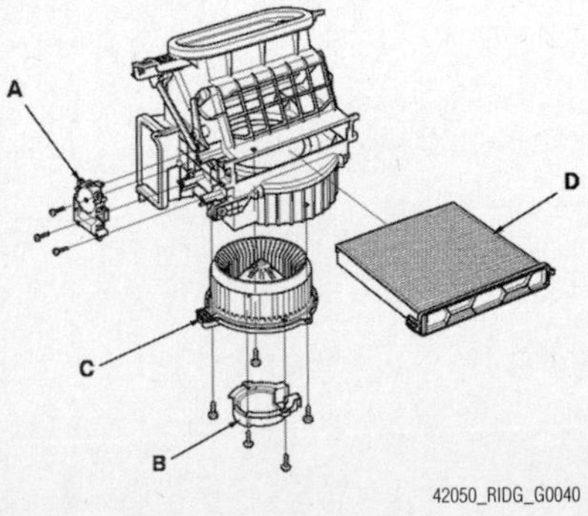
A
D
C
B
42050_RIDG_G0040

Fig. 112 Exploded view of the blower motor (C) and related components

HEATER CORE

REMOVAL & INSTALLATION

See Figures 113 through 115.

⁂ CAUTION

Before beginning work, review the "Precautions" in this section.

1. Make sure you have the anti-theft codes for the radio and navigation systems.
2. Disconnect the battery ground cable and wait at least 3 minutes.
3. Properly discharge and recover the refrigerant.
4. Disconnect the suction and receiver lines from the evaporator. Cap the openings.
5. Disconnect the heater coolant valve cable at the valve and fully open the valve.
6. Drain the coolant.
7. Disconnect the heater hoses at the core tubes.
8. Remove the heater unit mounting nut.
9. Remove the driver's side dashboard lower cover.
10. Remove the dashboard center lower cover.
11. Remove the center console.
12. Remove the glove box.
13. Remove the driver's side dashboard side cover.
14. Remove the both kick panels.
15. Remove the both A-pillar trim panels.
16. Place the wheels in the straight-ahead position.
17. Remove the steering wheel air bag access cover and disconnect the air bag connector from the cable reel.
18. Remove the 2 Torx® bolts, one each side of the steering wheel hub.
19. Disconnect the horn switch connector.
20. Remove the air bag.
21. Disconnect the cable reel harness.
22. Remove the steering wheel nut.
23. Using a puller, remove the steering wheel.
24. Remove the steering coupler cover from the floor.
25. Set the column shaft to the neutral position by raising the column to the uppermost position, then lower it 8mm (0.31 in.). Tighten the tilt lever.
26. Remove the column covers.
27. Move the shift lever to N and remove the shift cable from the column.
28. Remove the combination switch.
29. Disconnect the ignition switch.
30. Disconnect the immobilizer receiver unit, the park pin switch and the shift lock solenoid.
31. Matchmark and disconnect the steering joint from the column shaft.
32. Remove the attaching bolts and nuts and remove the steering column.
33. Disconnect the shift cable.
34. Remove the parking brake release lever bolt.
35. Remove the passenger's side dashboard side cover.
36. Remove the right speaker grille.
37. Remove the console front bracket.
38. Remove the rear vent duct.
39. Remove the rear heater duct.
40. Disconnect the cabin wiring harness.
41. Disconnect the interior wiring harness.

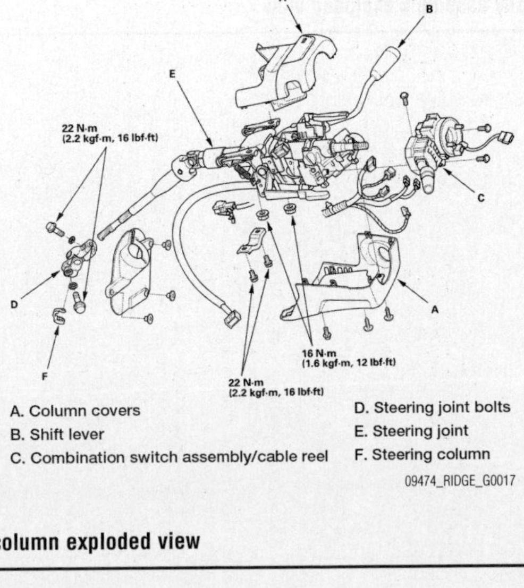
A
B
E
22 N·m (2.2 kgf·m, 16 lbf·ft)
C
D
A
16 N·m (1.6 kgf·m, 12 lbf·ft)
F
22 N·m (2.2 kgf·m, 16 lbf·ft)
A. Column covers
B. Shift lever
C. Combination switch assembly/cable reel
D. Steering joint bolts
E. Steering joint
F. Steering column
09474_RIDGE_G0017

Fig. 113 Steering column exploded view

42. Disconnect the left side wiring harness.
43. Disconnect the parking brake switch.
44. Disconnect the brake switch.
45. Disconnect the dashboard wiring harness.
46. Disconnect the under-dash fuse/relay box.

➡Lift the large harness connector locks before trying to remove the connectors from the fuse/relay box.

47. Disconnect the SRS control unit.
48. Disconnect the GPS antenna.
49. Disconnect the floor wiring harness, and remove the ground bolt.
50. From under the dash on the passenger's side, disconnect:
51. Disconnect the door wiring harness.
52. Disconnect the radio antenna.
53. Disconnect the cabin wiring harness.
54. Disconnect the right side harness.
55. Disconnect the front glass defogger.
56. Open the driver's door. See the illustration and completely loosen upper bolts A and B. Then, remove bolts C, D, E and F. Lift up on the dashboard (G) to release it from the guide pins (H&I).

➡Don't open the door fully with the driver's side upper dash bolts partially removed. Before removing the dashboard, open the driver's door to the half open position, then pull the upper dash bolts outward.

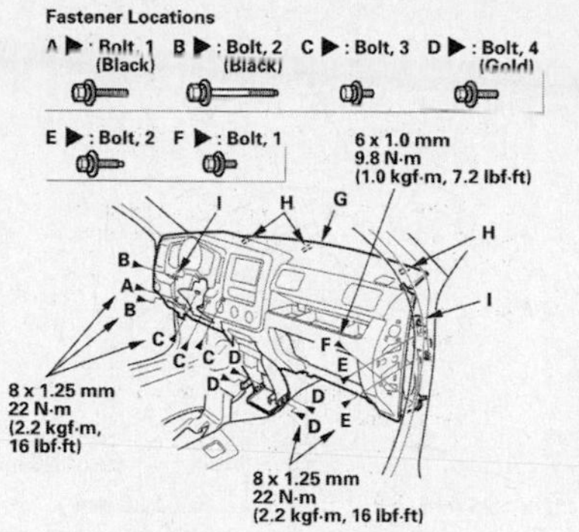
Fastener Locations
A ▶: Bolt, 1 (Black) B ▶: Bolt, 2 (Black) C ▶: Bolt, 3 D ▶: Bolt, 4 (Gold)
E ▶: Bolt, 2 F ▶: Bolt, 1
6 x 1.0 mm
9.8 N·m
(1.0 kgf·m, 7.2 lbf·ft)
8 x 1.25 mm
22 N·m
(2.2 kgf·m,
16 lbf·ft)
8 x 1.25 mm
22 N·m
(2.2 kgf·m, 16 lbf·ft)

A. Bolts
B. Bolts
C. Bolts
D. Bolts
E. Bolts
F. Bolts
G. Dashboard
H. Guide pins
I. Guide pins

09474_RIDGE_G0015

Fig. 114 Dashboard fastener locations

57. Disconnect the blower motor and power transistor wiring.
58. Disconnect the mode control motor and recirculation control motor.
59. Disconnect the evaporator sensor and air control mix motor.
60. Remove the mounting nuts and the blower/heater unit.
61. Remove the joint duct.
62. Remove the heater outlet.
63. Remove the heater core cover.

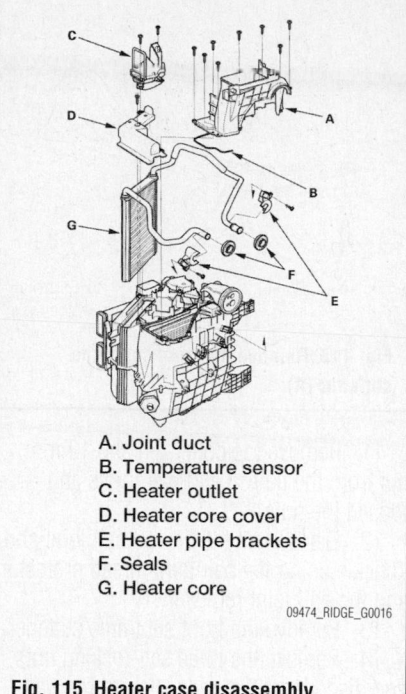

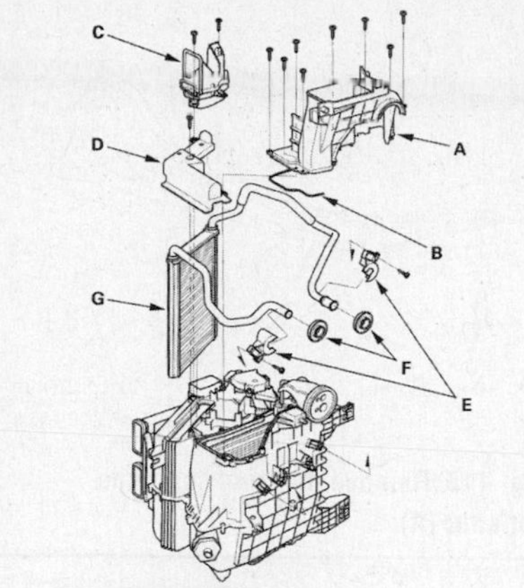

A. Joint duct
B. Temperature sensor
C. Heater outlet
D. Heater core cover
E. Heater pipe brackets
F. Seals
G. Heater core

09474_RIDGE_G0016

Fig. 115 Heater case disassembly

64. Remove the heater pipe brackets.
65. Remove the heater core.
66. Installation is the reverse of removal.
67. Evacuate, charge and leak test the A/C system.
68. Torque the steering wheel nut to 36 ft. lbs. (49 Nm).

STEERING

POWER RACK & PINION STEERING GEAR

REMOVAL & INSTALLATION

See Figures 116 through 126.

1. Drain the power steering fluid.
2. Do the battery terminal disconnect procedure, then wait at least 3 minutes before beginning work.
3. Remove the front wheels.
4. Remove the steering wheel.
5. Remove steering joint cover A at the floorboard.
6. Remove the steering joint bolts, disconnect the steering joint by moving the steering joint toward the column.
7. Remove the center guide (if equipped) from the top of the pinion shaft and discard it. The center guide is for factory assembly use only.
8. Disconnect the pump outlet hose from the power steering pump, and remove the clamp.
9. Remove the 10 mm flange bolts on the engine side mount bracket.

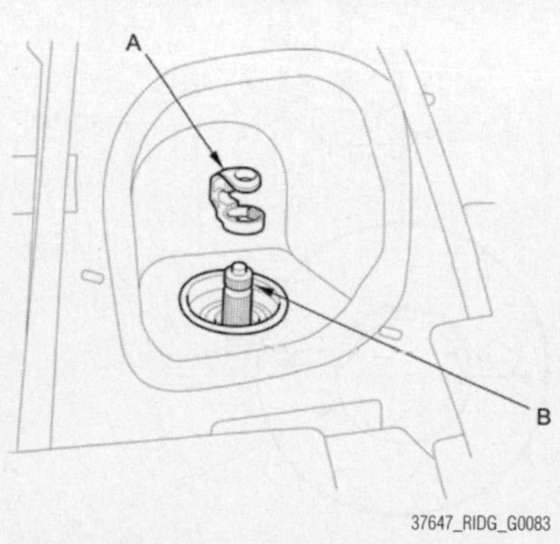
A
B

37647_RIDG_G0083

Fig. 116 Remove the center guide (A) (if equipped) from the top of the pinion shaft (B), and discard it. The center guide is for factory assembly use only.

10. Install a suitable engine support/lift assembly to the engine from the top in order to provide stability for the remainder of the removal procedure.

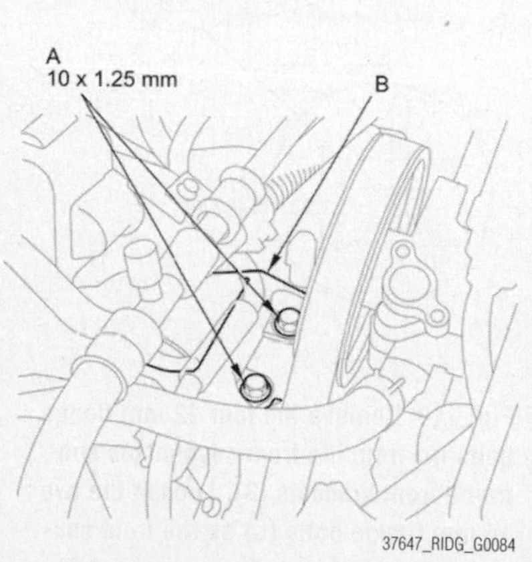
A
10 x 1.25 mm
B

37647_RIDG_G0084

Fig. 117 Remove the 10 mm flange bolts (A) on the engine side mount bracket (B).

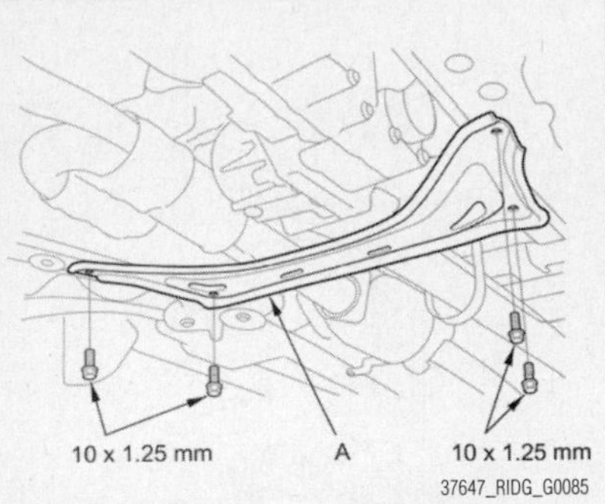
10 x 1.25 mm
A
10 x 1.25 mm
37647_RIDG_G0085

Fig. 118 Remove the front subframe stiffener (A).

11. Remove the cotter pin and 12 mm nut from the tie rod end ball joints and loosen the nuts.

12. Disconnect the tie-rod ball joint and knuckle using the ball joint thread protector and the ball joint remover.

13. Remove the front subframe stiffener.

14. Remove the three self-locking nuts and disconnect the under-floor three way catalytic converter (TWC) from the muffler.

15. Disconnect the power steering pressure (PSP) switch connector, near the lower end of the power steering hose, just above the joint boot.

16. Remove the propeller shaft U-clamp.

17. Remove the front splash shield.

18. Install a suitable front subframe adapter (EQ02BMDXSB0, or equivalent) on a transmission jack and raise the jack to vehicle height, then attach the front subframe adapter to the front subframe using the subframe stiffener mounting bolts and bolt holes.

19. Remove the four 12 mm flange bolts from the front suspension subframe front brackets.

20. Loosen the two 14 mm flange bolts on the front suspension subframe so they are about 1.2 in. (30 mm) from the mounting surface. Do not loosen the 14 mm flange bolts more than necessary.

21. Support the front subframe securely by raising the transmission jack. Remove the 10 mm flange bolts, nuts, and the subframe bolt retainers.

22. Remove the two 12 mm flange bolts, 14 mm special bolts and front subframe rear brackets on the right and left of the vehicle.

23. Lower the transmission jack slowly, until the front subframe has dropped about 2.0 in. (50 mm).

24. Remove the P/S line mounting brackets from the front subframe and gearbox mounting bracket.

25. Loosen the adjustable hose clamp and disconnect the return hose.

26. Remove the two 10 mm flange bolts from the right side of the steering gearbox, then remove the mounting bracket and cushion.

27. Remove the two 10 mm flange bolts from the left side of the gearbox.

28. Lower the transmission jack slowly until the front subframe has dropped 3.9 in. (100 mm) total.

29. Remove the gearbox stiffener bracket from the left side of the front subframe.

30. Loosen the 16 mm flare nut and disconnect the feed line.

31. Slide the steering gearbox between the body and front subframe toward the left, and remove the steering gearbox.

32. Remove the pinion shaft grommet from the frame.

33. After removing the steering gearbox, make sure that no power steering fluid gets on the gearbox mount cushions, gearbox housing, surface of the front subframe and stiffener. Wipe off any spilled fluid at once.

To install:

34. Before installing the steering gearbox, make sure that no power steering fluid is on the mating surface of the steering gearbox and the front subframe. To prevent the gearbox mounting bolts from loosening after the installation, remove any power steering fluid from the mount cushions and bolt holes.

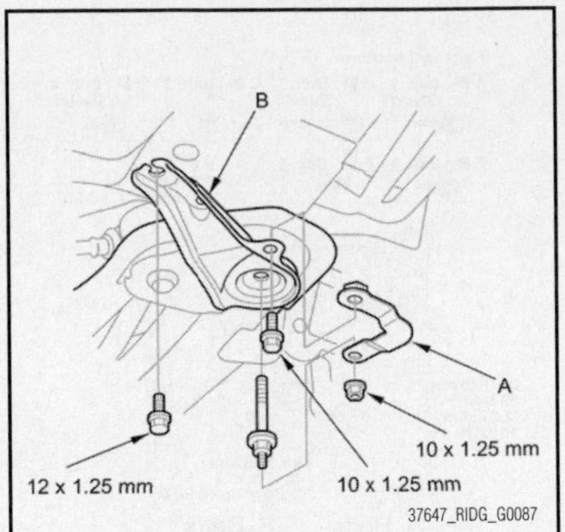
B
A
10 x 1.25 mm
12 x 1.25 mm
10 x 1.25 mm
37647_RIDG_G0087

Fig. 120 Support the front subframe securely by raising the transmission jack. Remove the 10 mm flange bolts, nuts, and the subframe bolt retainers (A). Remove the two 12 mm flange bolts, 14 mm special bolts and front subframe rear brackets (B) on the right and left of the vehicle.

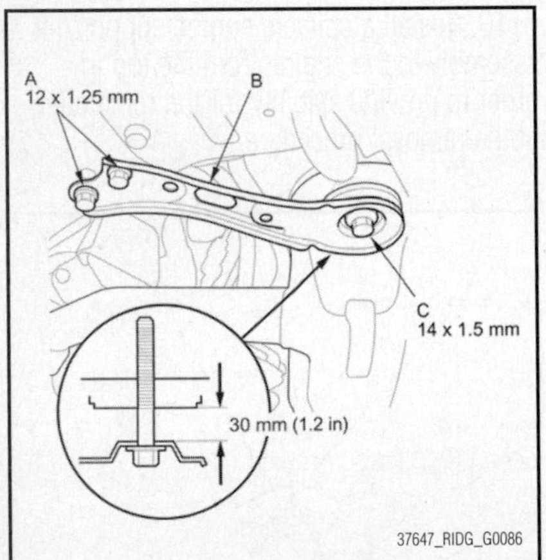
A
12 x 1.25 mm
B
C
14 x 1.5 mm
30 mm (1.2 in)
37647_RIDG_G0086

Fig. 119 Remove the four 12 mm flange bolts (A) from the front suspension subframe front brackets (B). Loosen the two 14 mm flange bolts (C) on the front suspension subframe so they are about 30 mm (1.2 in) from the mounting surface. Do not loosen the 14 mm flange bolts more than necessary.

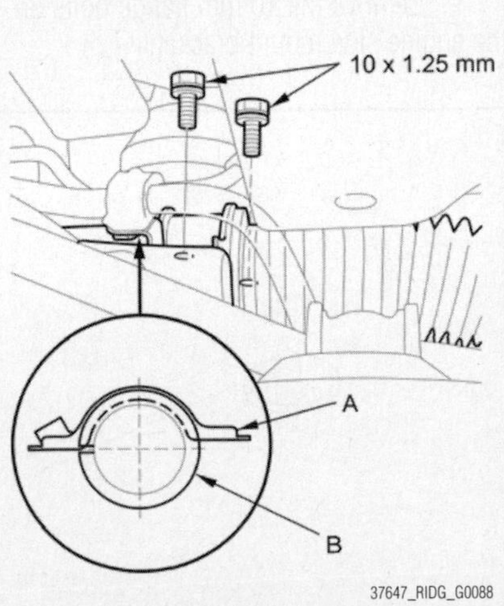
10 x 1.25 mm
A
B
37647_RIDG_G0088

Fig. 121 Remove the two 10 mm flange bolts from the right side of the steering gearbox, then remove the mounting bracket (A) and cushion (B).

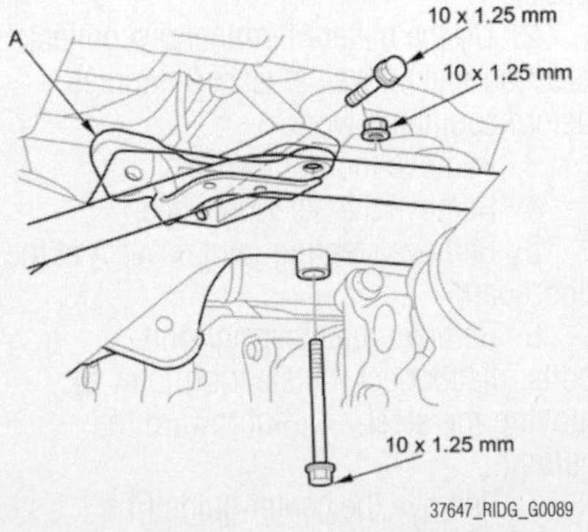
10 x 1.25 mm
A
10 x 1.25 mm
10 x 1.25 mm
37647_RIDG_G0089

Fig. 122 Remove the gearbox stiffener bracket (A) from the left side of the front subframe.

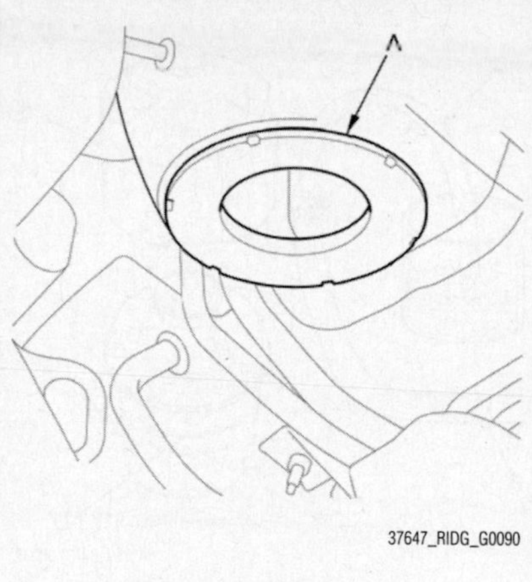
A
37647_RIDG_G0090

Fig. 123 Remove the pinion shaft grommet (A) from the frame.

35. Install the pinion shaft grommet on the valve housing.

36. Install the return line joint, if removed.

37. Slide the steering gearbox between the front subframe and body from the driver's side. Place the gearbox in position on the front subframe.

38. Connect the feed line and tighten the 16 mm flare nut to 31 ft. lbs. (42 Nm).

39. Install the gearbox stiffener bracket on the front subframe, and tighten the bolts and nut to 28 ft. lbs. (38 Nm).

40. Loosely install the new 10 mm flange bolts on the left side of the gearbox.

41. Install the mounting cushion on the right side of the gearbox.

42. Install the mounting bracket over the mounting cushion, then install the two 10 mm flange bolts. Tighten the four 10 mm flange bolts for the gearbox to 29 ft. lbs. (39 Nm).

43. Connect the return hose securely and tighten the adjustable hose clamp.

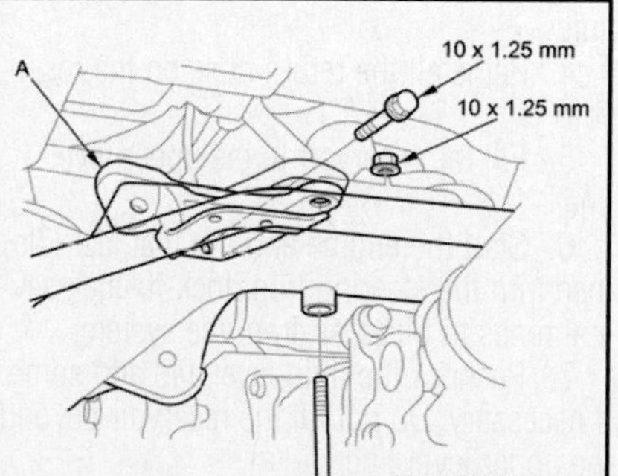
A
10 x 1.25 mm
10 x 1.25 mm
10 x 1.25 mm
37647_RIDG_G0089

Fig. 124 Install the gearbox stiffener bracket on the front subframe, and tighten the bolts and nut to 28 ft. lbs. (38 Nm).

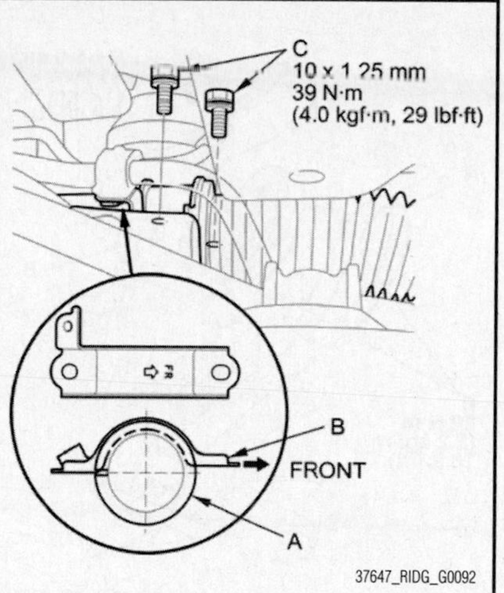
C
10 x 1.25 mm
39 N·m
(4.0 kgf·m, 29 lbf·ft)
B
FRONT
A
37647_RIDG_G0092

Fig. 125 Install the mounting cushion (A) on the right side of the gearbox. Install the mounting bracket (B) over the mounting cushion, then install the two 10 mm flange bolts (C). Tighten the four 10 mm flange bolts for the gearbox to 29 ft. lbs. (39 Nm).

44. Install the P/S line mounting brackets on the front suspension subframe and gearbox mounting bracket.

45. Carefully raise the front subframe with the front subframe adapter and the transmission jack or the powertrain lift until the subframe is in position.

46. Install the front subframe rear brackets. Tighten the 12 mm flange bolts and the new 14 mm special bolts on the right and the left of the vehicle to the indicated torque. Install the subframe bolt retainers. Tighten the two flange bolts and nuts to the indicated torque.

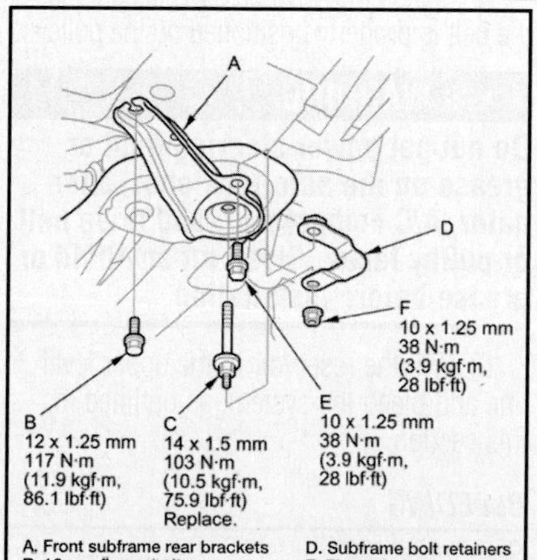
A
D
F
10 x 1.25 mm
38 N·m
(3.9 kgf·m,
28 lbf·ft)
B
12 x 1.25 mm
117 N·m
(11.9 kgf·m,
86.1 lbf·ft)
C
14 x 1.5 mm
103 N·m
(10.5 kgf·m,
75.9 lbf·ft)
Replace.
E
10 x 1.25 mm
38 N·m
(3.9 kgf·m,
28 lbf·ft)
A. Front subframe rear brackets
B. 12 mm flange bolts
C. 14 mm special bolts
D. Subframe bolt retainers
E. Flange bolts
F. Nuts
37647_RIDG_G0093

Fig. 126 Install the front subframe rear brackets. Tighten the 12 mm flange bolts and the new 14 mm special bolts on the right and the left of the vehicle to the indicated torque. Install the subframe bolt retainers. Tighten the two flange bolts and nuts to the indicated torque.

47. Install the front subframe front brackets with 12 mm flange bolts and the new 14 mm special bolts. Tighten the two inner 12 mm bolts to 54 ft. lbs. (74 Nm) and the single out 14 mm flange bolt to 76 ft. lbs. (103 Nm).

48. Lower the front subframe supporting the transmission jack or the powertrain lift.

49. Install the front splash shield.

50. Install the propeller shaft protector (U-clamp).

51. Connect the power steering pressure (PSP) switch connector.

52. Connect the under-floor three way catalytic converter (TWC) to the muffler. Install the new 10 mm self-locking nuts and tighten them to 25 ft. lbs. (33 Nm).

53. Install the front subframe stiffener and tighten to 40 ft. lbs. (54 Nm).

54. Install the new 10 mm flange bolts on the engine side mount bracket and torque to 33 ft. lbs. (44 Nm).

55. Remove the engine support hanger, the hanger balance bar, and the hanger adapter set.

56. Connect the pump outlet hose to the power steering pump, and install the feed hose clamp.

57. Wipe off any grease contamination from the ball joint tapered section and threads. Reconnect the tie-rod ends to the steering knuckles. Install the 12 mm nut and tighten it to 40 ft. lbs. (54 Nm).

58. Install a new cotter pin.

59. Center the steering rack within its stroke.

60. Insert the upper end of the steering joint onto the steering shaft (line up the bolt hole with the flat portion on the shaft), and loosely install the upper joint bolt.

61. Slip the lower end of the steering joint onto the pinion shaft taking care to align the gap within the angle.

62. Align the bolt hole on the steering joint with the groove around the pinion shaft, then loosely install the lower steering joint bolt. Be sure that the joint bolt is securely in the groove in the pinion shaft.

63. Pull on the steering joint to make sure that the steering joint is fully seated, then tighten the lower joint bolt to 16 ft. lbs. (22 Nm).

64. Tighten the upper steering joint bolt and tighten to 16 ft. lbs. (22 Nm).

65. Install steering joint cover A at the floorboard.

66. Install the front wheel, then set the wheels in the straight ahead position.

➡**Before installing the wheel, clean the mating surfaces of the brake disc and the inside of the wheel.**

67. Install the steering wheel.
68. Do the battery terminal reconnection procedure, and do these tasks:
    a. Turn the ignition switch to ON (II) and check that the SRS indicator comes on for about 6 seconds and then goes off.
    b. Make sure the horn and turn signal switches work properly.
    c. Make sure the steering wheel switches work properly.
    d. Make sure the steering wheel is centered.
69. Fill the system with power steering fluid, and bleed air from the system. See "Power Steering Pump" in this section.
70. After installation, do these checks:
    a. Start the engine, allow it to idle, and turn the steering wheel from lock to lock several times to warm up to the fluid.
    b. Check the gearbox for leaks.
    c. Do the front toe inspection.
    d. Check the steering wheel spoke angle.
71. If steering spoke angles to the right and left are not equal (steering wheel and rack are not centered), correct the engagement of the joint/pinion shaft serrations, then adjust the front toe by turning the tie-rod ends, if necessary.

POWER STEERING PUMP

REMOVAL & INSTALLATION

See Figure 127.

1. Place a suitable container under the vehicle.
2. Drain the power steering fluid from the reservoir.
3. Remove the engine cover.
4. Remove the drive belt from the pump pulley.
5. Cover the auto-tensioner, alternator, and A/C compressor with several shop towels to protect them from spilled power steering fluid.
6. Disconnect the pump inlet hose and pump outlet hose from the pump, and plug them. Take care not to spill the fluid on the body or parts. Wipe off any spilled fluid at once. Do not turn the steering wheel with the pump removed.

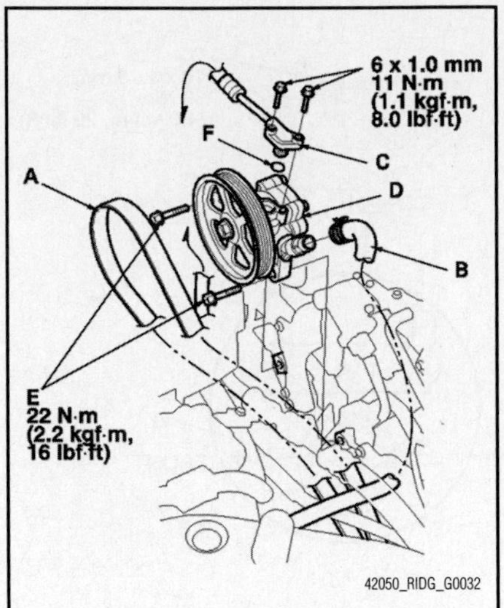
6 x 1.0 mm
11 N·m
(1.1 kgf·m,
8.0 lbf·ft)
F
C
A
D
B
E
22 N·m
(2.2 kgf·m,
16 lbf·ft)
42050_RIDG_G0032

Fig. 127 Exploded view of the power steering pump and related components

7. Remove the pump mounting bolts and remove the power steering pump.
8. Cover the opening of the pump with a piece of tape to prevent foreign material from entering the pump.

To install:

9. Connect the pump inlet hose and pump outlet hose onto the new pump with new O-ring.
10. Loosely install the pump in the pump bracket with the mounting bolts, then tighten the pump fittings securely.
11. Tighten the pump mounting bolts to 16 ft. lbs. (22 Nm).
12. Install the drive belt. Make sure that the belt is properly positioned on the pulleys.

WARNING

Do not get power steering fluid or grease on the auto-tensioner, alternator, A/C compressor, and drive belt or pulley faces. Clean off any fluid or grease before installation.

13. Fill the reservoir to the upper level line and bleed the system, as outlined in this section.

BLEEDING

See Figure 128.

Check the power steering fluid reservoir at regular intervals, and add the recommended fluid as necessary. Always use Honda Power Steering Fluid. Using any other type of power steering fluid or automatic transmission fluid can cause increased wear and poor steering in cold weather.

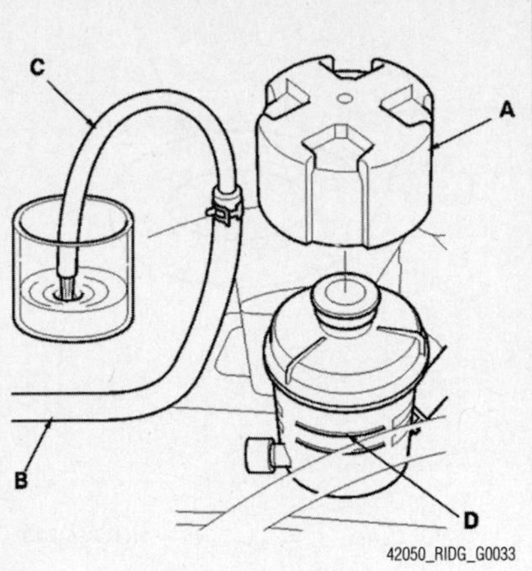
C
A
B
D
42050_RIDG_G0033

Fig. 128 Power steering filling and bleeding

➡**If the fluid is contaminated, the screen in the reservoir may be partially blocked. Replace the reservoir if necessary.**

The power steering fluid system capacity is 1.22 qt. at disassembly and reservoir capacity is 0.36 qt.

1. Pull the cover up, raise the reservoir, then disconnect the return hose to drain the reservoir. Take care not to spill the fluid on the body and parts. Wipe off any spilled fluid at once.

➡**Check the reservoir screen for any debris. If the reservoir screen is clogged, replace the reservoir.**

2. Connect a hose of suitable diameter to the disconnected return hose, and put the hose end in a suitable container.
3. Start the engine, let it run at idle, and turn the steering wheel from lock-to-lock several times. When fluid stops running out of the hose, shut off the engine. Discard the fluid.
4. Reinstall the return hose on the reservoir.
5. Fill the reservoir to the upper level line.
6. Start the engine and run it at fast idle, then turn the steering from lock-to-lock several times to bleed air from the system.
7. Recheck the fluid level and add some if necessary. Do not fill the reservoir beyond the upper level line.
8. If the fluid is contaminated, dark, or discolored, repeat the procedure as necessary.

LOWER BALL JOINT

REMOVAL & INSTALLATION

The lower ball joints are an integral part of the lower control arms and not serviced separately.

LOWER CONTROL ARM

REMOVAL & INSTALLATION

See Figures 129 and 130.

1. Raise and support the vehicle.
2. Remove the front wheels.
3. Remove the flange nut while holding the respective joint pin with a hex wrench, then disconnect the stabilizer links from the damper.
4. Turn the stabilizer bar backward to gain easier access to the front side of the lower arm mounting bolt.
5. Remove the lock pin from the lower arm ball joint, then remove the nut.

➡During installation, install the lock pin after tightening the new castle nut.

6. Disconnect the lower ball joint from the knuckle using the ball joint thread protector and the ball joint puller.
7. Remove the lower arm mounting bolt.

➡Use the new mounting bolt during reassembly.

8. Remove the lower arm mounting bolt, then remove the lower arm from the front suspension subframe.

NOTE: Use the new mounting bolt during reassembly.

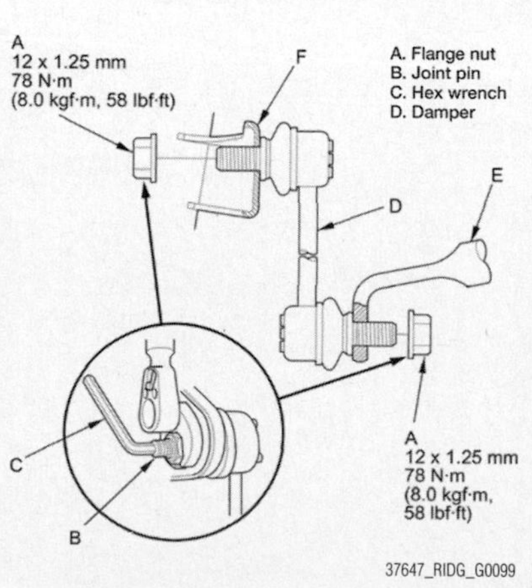
A
12 x 1.25 mm
78 N·m
(8.0 kgf·m, 58 lbf·ft)
F
A. Flange nut
B. Joint pin
C. Hex wrench
D. Damper
E
D
C
A
12 x 1.25 mm
78 N·m
(8.0 kgf·m,
58 lbf·ft)
B
37647_RIDG_G0099

Fig. 129 Remove the flange nut while holding the respective joint pin with a hex wrench, then disconnect the stabilizer links from the damper.

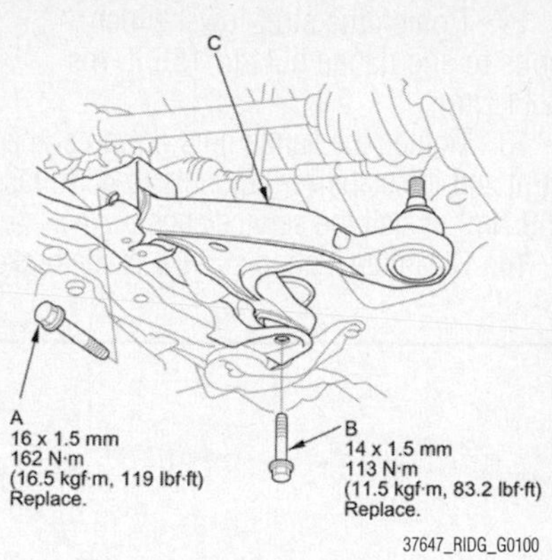
C
A
16 x 1.5 mm
162 N·m
(16.5 kgf·m, 119 lbf·ft)
Replace.
B
14 x 1.5 mm
113 N·m
(11.5 kgf·m, 83.2 lbf·ft)
Replace.
37647_RIDG_G0100

Fig. 130 Remove the lower arm mounting bolts (A and B), then remove the lower arm (C) from the front suspension sub-frame.

To install:

9. Install the lower arm in the reverse order of removal, and note these items:
   a. Be careful not to damage the ball joint boot when installing the knuckle.
   b. Before connecting the lower ball joint to the knuckle, degrease the threaded section and tapered portion of the ball joint pin, the lower arm connecting hole, the threaded section and mating surface of the castle nut.
   c. First install all the components and lightly tighten the bolts and nuts, then raise the suspension to load it with the vehicle's weight before fully tightening to the specified torque values shown in the figures.
   d. Torque the castle nut to the lower torque specification, then tighten it only far enough to align the slot with the ball joint pin hole. Do not align the castle nut by loosening it.
   e. Before installing the wheel, clean the mating surface of the brake disc and the inside of the wheel.
10. Check the wheel alignment, and adjust it if necessary.

MACPHERSON STRUT

REMOVAL & INSTALLATION

See Figure 131 and 132.

1. Raise and support the vehicle.
2. Remove the front wheel.
3. Disconnect the stabilizer link from the strut.
4. Remove the wheel speed sensor harness clips and the brake hose bracket from the strut. Do not disconnect the wheel speed sensor connector.

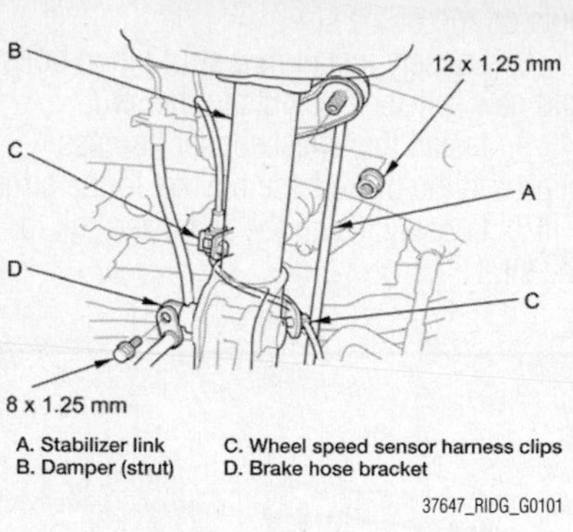
B
12 x 1.25 mm
C
A
D
C
8 x 1.25 mm
A. Stabilizer link
B. Damper (strut)
C. Wheel speed sensor harness clips
D. Brake hose bracket
37647_RIDG_G0101

Fig. 131 Disconnect the stabilizer link from the strut. Remove the wheel speed sensor harness clips and the brake hose bracket from the strut. Do not disconnect the wheel speed sensor connector.

5. Remove the lower pinch bolts and the self-locking nuts from the strut.

⁂ CAUTION

Do not allow the knuckle to rotate too far outward. This may allow the driveshaft inboard joint to come apart.

6. Remove the service caps and remove the three flange nuts from top of the strut, then remove the MacPherson strut assembly.

➡Strut springs are different, left and right. Mark the springs "L" and "R" before you continue.

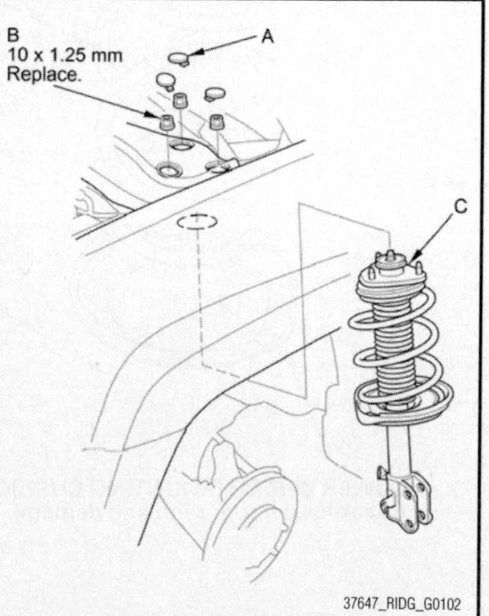
B
10 x 1.25 mm
Replace.
A
C
37647_RIDG_G0102

Fig. 132 Remove the service caps (A), and remove the three flange nuts (B) from top of the strut, then remove the MacPherson strut assembly (C).

To install:

7. Install the strut assembly onto the frame, then loosely install new three flange nuts on top of the strut.
8. Loosely install new strut pinch bolts and new self-locking nuts to the strut.
9. Install the wheel sensor harness clips and the brake hose bracket to the strut.
10. Loosely install the stabilizer link to the strut.
11. Raise the front suspension with a floor jack to load the suspension with the vehicle's weight.
12. Tighten the strut lower pinch bolts to and flange nuts to 156 ft. lbs. (211 Nm).
13. Tighten the flange nuts on top of the strut and the stabilizer link nuts to 43 ft. lbs. (59 Nm). Install the service caps.
14. Clean the mating surface of the brake disc and the inside of the wheel, then install the wheel.
15. Check the wheel alignment, and adjust it if necessary.

OVERHAUL

See Figures 133 through 136.

1. Remove the strut from the vehicle and install in a strut spring compressor. Com-

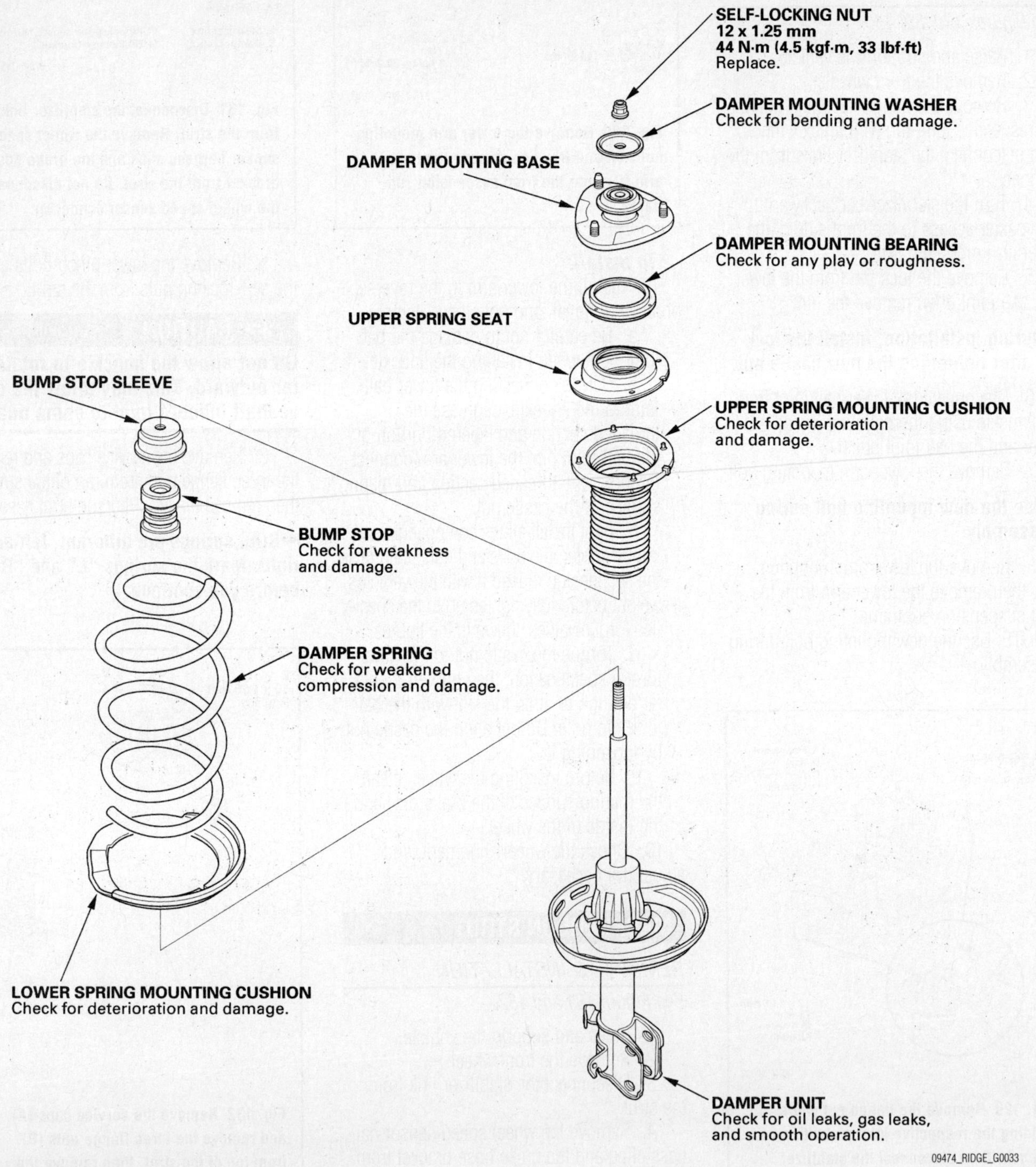
SELF-LOCKING NUT
12 x 1.25 mm
44 N·m (4.5 kgf·m, 33 lbf·ft)
Replace.
DAMPER MOUNTING WASHER
Check for bending and damage.
DAMPER MOUNTING BASE
DAMPER MOUNTING BEARING
Check for any play or roughness.
UPPER SPRING SEAT
BUMP STOP SLEEVE
UPPER SPRING MOUNTING CUSHION
Check for deterioration
and damage.
BUMP STOP
Check for weakness
and damage.
DAMPER SPRING
Check for weakened
compression and damage.
LOWER SPRING MOUNTING CUSHION
Check for deterioration and damage.
DAMPER UNIT
Check for oil leaks, gas leaks,
and smooth operation.
09474_RIDGE_G0033

Fig. 133 Exploded view of the front strut assembly

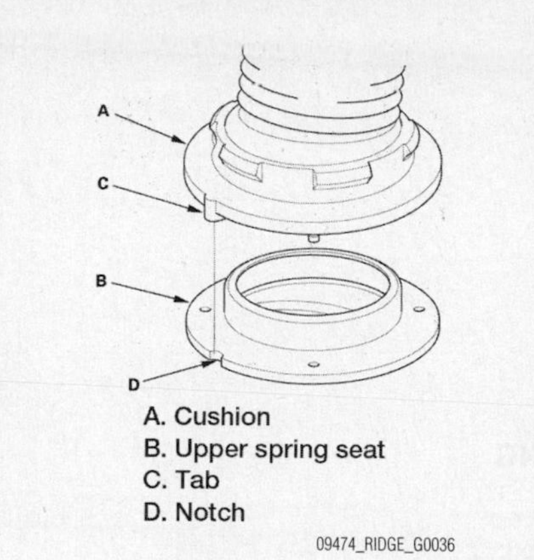
A
C
B
D
A. Cushion
B. Upper spring seat
C. Tab
D. Notch
09474_RIDGE_G0036

Fig. 134 Install the cushion on the upper spring seat by aligning the tab with the notch in the seat

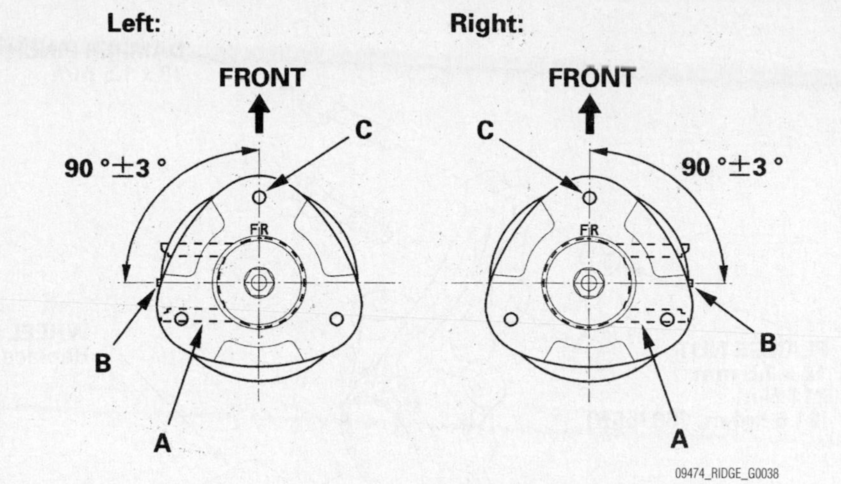
Left:
FRONT
C
90 °±3 °
FR
B
A
Right:
FRONT
C
90 °±3 °
FR
B
A
09474_RIDGE_G0038

Fig. 136 Align an angle of the bracket (A) and the tab portion (B) on the cushion. Position the angle of the tab portion and the stud bolt (C) near the FR stamp on the upper spring seat.

press the spring until the end of the spring comes away from the spring seat.

2. Remove the upper strut mount, spring seat and related components.

3. Remove the coil spring from the strut spring compressor.

To assemble:

➡Use a new self-locking nut.

4. Install the cushion on the upper spring seat by aligning the tab with the notch in the seat.

5. Compress the spring.

6. Align an angle of the bracket and the tab portion on the cushion. Position the angle of the tab portion and the stud bolt near the **FR**stamp on the upper spring seat.

7. Install the upper strut mounting components and tighten the nut to 33 ft. lbs. (44 Nm).

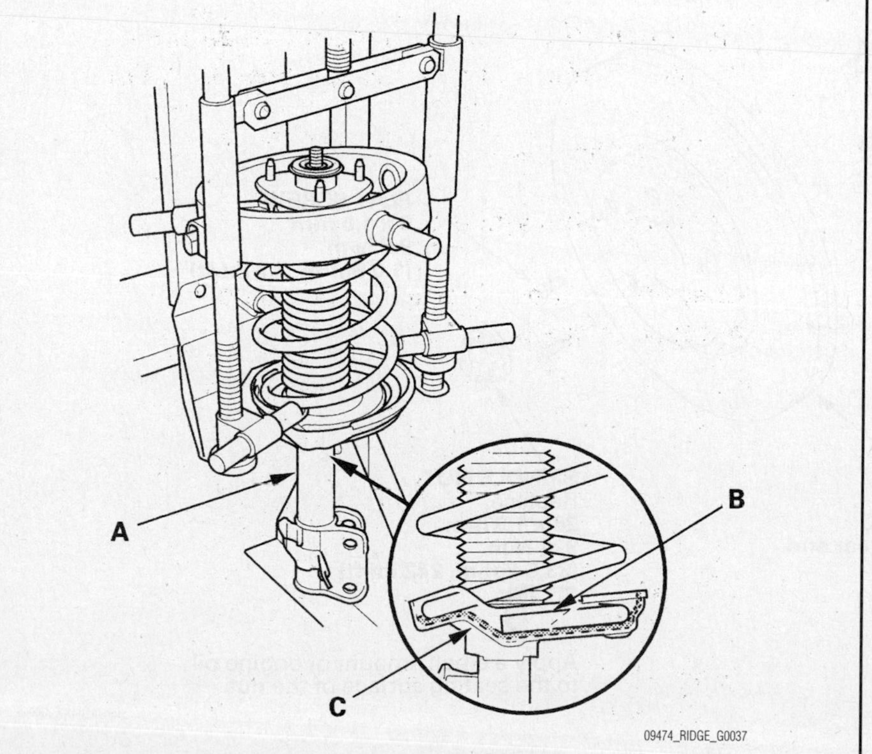
A
B
C
09474_RIDGE_G0037

Fig. 143 Align the bottom of the spring (B) and the stepped portion of the lower spring seat (C)

STEERING KNUCKLE

REMOVAL & INSTALLATION

See Figure 137.

1. Remove the wheel.
2. Remove the brake hose bracket bolt.
3. Remove the caliper bracket bolts and caliper and support it out of the way.
4. Remove the sensor from the knuckle.
5. Unstake and remove the halfshaft end nut.
6. Remove the brake rotor.

➡It may be necessary to push off the rotor by using two 8x1.25mm bolts in the holes provided.

7. Remove the nut and disconnect the tie rod end from the knuckle.
8. Remove the ball stud nut and separate the lower control arm from the knuckle.
9. Remove the strut-to-knuckle bolts/nuts.
10. Pull the knuckle outwards while taping the end of the halfshaft with a plastic hammer.
11. Installation is the reverse of removal, noting the following:
12. Use new bolts and nuts.
13. Install all fasteners loosely, then, final torque all fasteners with the suspension loaded (weight of the car). Never back off the ball stud nut to align the cotter pin holes.
14. Always tighten it to align.
15. Observe the following torque specifications:
    - Strut-to-knuckle: 156 ft. lbs. (211 Nm)
    - Lower arm ball stud nut: 69 ft. lbs. (93 Nm)

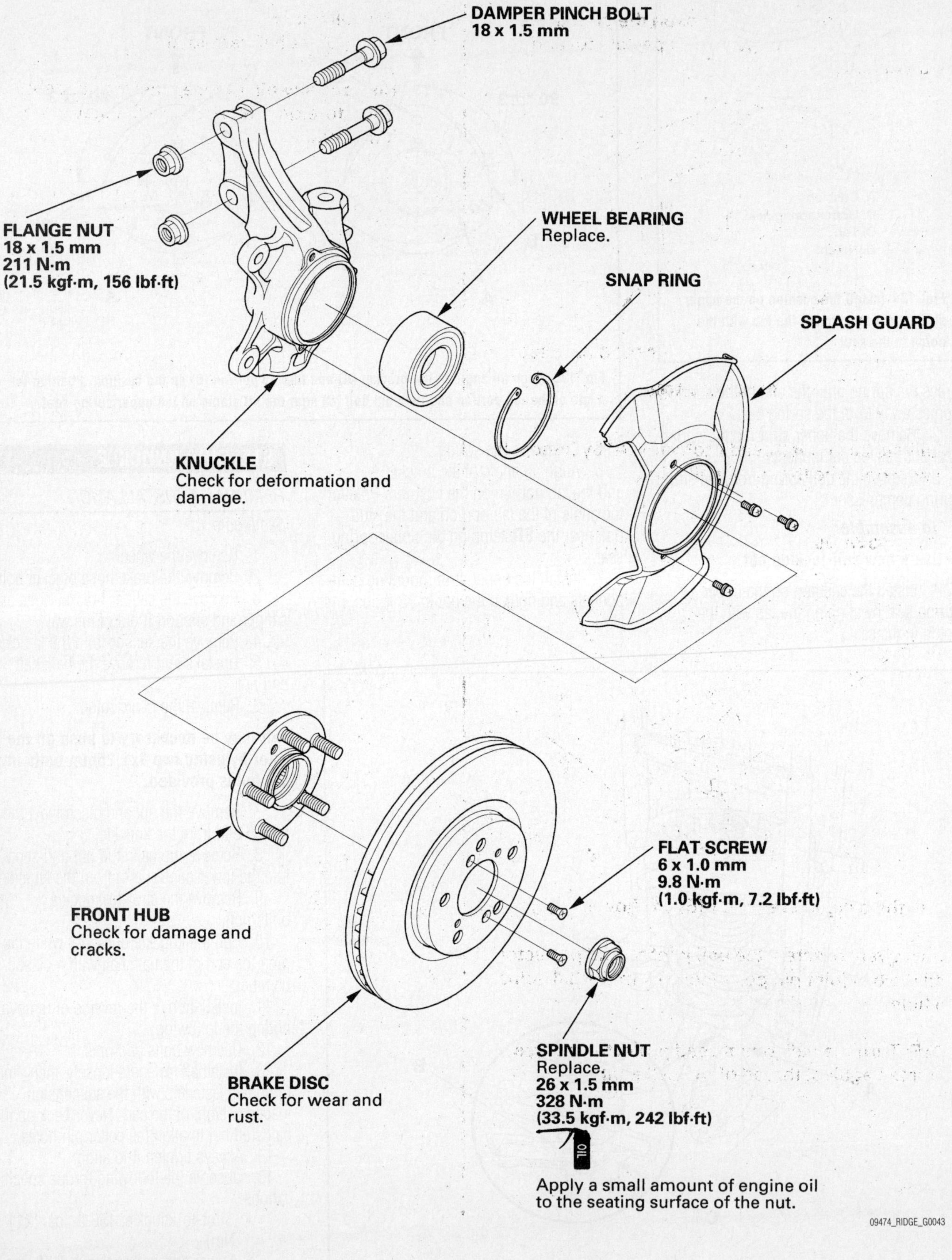
DAMPER PINCH BOLT
18 x 1.5 mm
FLANGE NUT
18 x 1.5 mm
211 N·m
(21.5 kgf·m, 156 lbf·ft)
WHEEL BEARING
Replace.
SNAP RING
SPLASH GUARD
KNUCKLE
Check for deformation and damage.
FLAT SCREW
6 x 1.0 mm
9.8 N·m
(1.0 kgf·m, 7.2 lbf·ft)
FRONT HUB
Check for damage and cracks.
SPINDLE NUT
Replace.
26 x 1.5 mm
328 N·m
(33.5 kgf·m, 242 lbf·ft)
BRAKE DISC
Check for wear and rust.
OIL
Apply a small amount of engine oil to the seating surface of the nut.
09474_RIDGE_G0043

Fig. 137 Hub/knuckle exploded view

- Tie rod end nut: 40 ft. lbs. (54 Nm)
- Halfshaft end nut: 242 ft. lbs. (328 Nm)
- Caliper bracket bolts: 101 ft. lbs. (137 Nm)

STABILIZER BAR

REMOVAL & INSTALLATION

See Figures 138 through 140.

1. Raise and support the vehicle.
2. Remove the front wheels.
3. Install a suitable engine balancer bar (VSB02C000019) and engine support hanger (AAR-T-1256) to the engine and put supporting tension on the engine.
4. Disconnect both sides of the stabilizer link from the stabilizer bar.
5. Remove the front splash shield.
6. Install and front subframe adapter on a transmission jack.
7. Raise the jack to vehicle height, then attach the front subframe adapter to the front subframe using the subframe stiffener mounting bolts and bolt holes.
8. Remove the four 12 mm flange bolts from the front suspension subframe front brackets.
9. Loosen the two 14 mm flange bolts on the front suspension subframe so they are about 14 mm (0.6 in) from the mounting surface. Do not loosen the 14 mm flange bolts more than necessary.
10. Support the front suspension subframe securely by raising the transmission jack, then remove the 10 mm flange bolts, the nuts, and the subframe bolt retainers.
11. Remove the two 12 mm flange bolts, the 14 mm special bolts and the front suspension subframe rear brackets on the right and left of the vehicle.
12. Lower the jack supporting the front suspension subframe with the special tool slowly until the front suspension subframe has dropped about 14 mm (0.6 in).
13. Remove the flange bolts and the bushing holders, then remove the bushings and the stabilizer bar from the front suspension subframe. Note the paint marks.

To install:

14. Install the stabilizer bar in the reverse order of removal, and note these items:
    a. Note the right and left direction of the stabilizer bar.
    b. Align the paint marks on the stabilizer bar with the sides of the bushings.

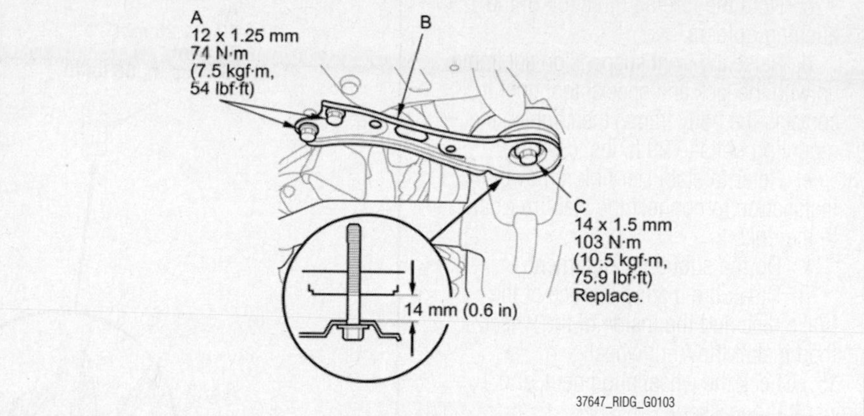
A
12 x 1.25 mm
74 N·m
(7.5 kgf·m,
54 lbf·ft)
B
C
14 x 1.5 mm
103 N·m
(10.5 kgf·m,
75.9 lbf·ft)
Replace.
14 mm (0.6 in)
37647_RIDG_G0103

Fig. 138 Remove the four 12 mm flange bolts (A) from the front suspension subframe front brackets (B). Loosen the two 14 mm flange bolts (C) on the front suspension subframe so they are about 14 mm (0.6 in) from the mounting surface. Do not loosen the 14 mm flange bolts more than necessary.

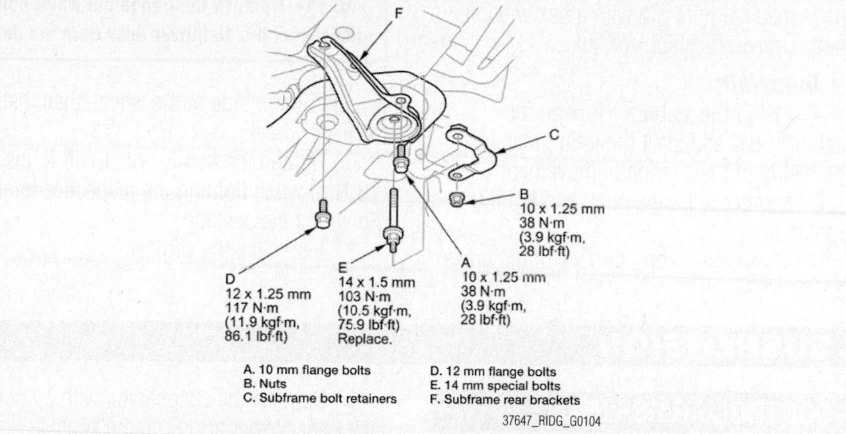
F
C
B
10 x 1.25 mm
38 N·m
(3.9 kgf·m,
28 lbf·ft)
A
10 x 1.25 mm
38 N·m
(3.9 kgf·m,
28 lbf·ft)
D
12 x 1.25 mm
117 N·m
(11.9 kgf·m,
86.1 lbf·ft)
E
14 x 1.5 mm
103 N·m
(10.5 kgf·m,
75.9 lbf·ft)
Replace.
A. 10 mm flange bolts
B. Nuts
C. Subframe bolt retainers
D. 12 mm flange bolts
E. 14 mm special bolts
F. Subframe rear brackets
37647_RIDG_G0104

Fig. 139 Support the front suspension subframe securely by raising the transmission jack, then remove the 10 mm flange bolts, the nuts, and the subframe bolt retainers. Remove the two 12 mm flange bolts, the 14 mm special bolts and the front suspension subframe rear brackets on the right and left of the vehicle.

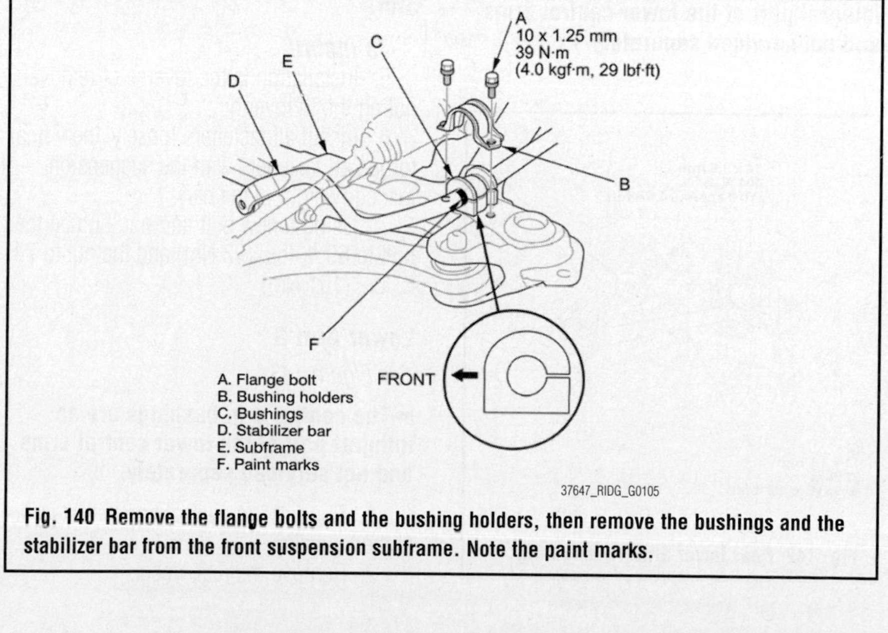
A
10 x 1.25 mm
39 N·m
(4.0 kgf·m, 29 lbf·ft)
C
E
D
B
F
FRONT
A. Flange bolt
B. Bushing holders
C. Bushings
D. Stabilizer bar
E. Subframe
F. Paint marks
37647_RIDG_G0105

Fig. 140 Remove the flange bolts and the bushing holders, then remove the bushings and the stabilizer bar from the front suspension subframe. Note the paint marks.

c. Note the fore/aft direction of the bushing holders.

d. Raise the front suspension subframe up with the jack and special tool until it contacts the body frame, then tighten the mounting bolts to 29 ft. lbs. (39 Nm).

e. Refer to stabilizer link removal/installation to connect the stabilizer bar to the links.

f. Do the subframe alignment.

g. Clean the mating surface of the brake disc and the inside of the wheel, then install the front wheel.

15. Check the wheel alignment, and adjust it if necessary.

Stabilizer Link

See Figure 141.

1. Raise and support the vehicle.
2. Remove the front wheel.
3. Remove the flange nuts while holding the respective joint pin with a hex wrench, then remove the stabilizer link.

To install:

4. Install the stabilizer link on the stabilizer bar and strut the joint pins set at the center of their range of movement.
5. Install the flange nuts, and lightly tighten them.
6. Clean the mating surface of the brake disc and the inside of the wheel, then install the front wheel.
7. Tighten the flange nuts to 58 ft. lbs. (78 Nm) while holding the respective joint pin with a hex wrench.
8. Test-drive the vehicle.

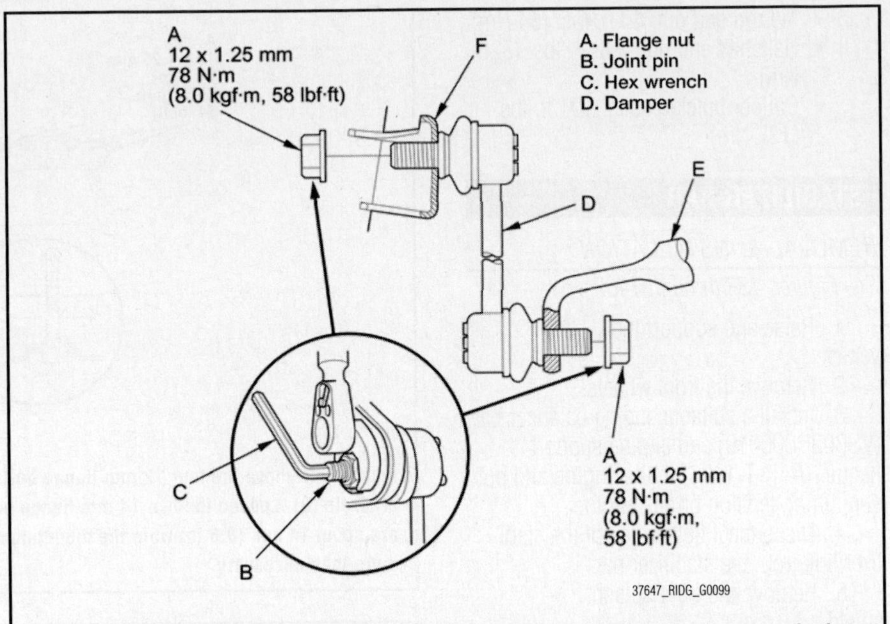
A
12 x 1.25 mm
78 N·m
(8.0 kgf·m, 58 lbf·ft)
F
A. Flange nut
B. Joint pin
C. Hex wrench
D. Damper
E
D
C
B
A
12 x 1.25 mm
78 N·m
(8.0 kgf·m,
58 lbf·ft)
37647_RIDG_G0099

Fig. 141 Remove the flange nut while holding the respective joint pin with a hex wrench, then disconnect the stabilizer links from the damper.

WHEEL HUB AND BEARING

REMOVAL & INSTALLATION

➡See "Steering Knuckle" in this section.

SUSPENSION

REAR SUSPENSION

CONTROL ARMS/LINKS

REMOVAL & INSTALLATION

Lower Arm A

See Figure 142.

➡The control arm bushings are an integral part of the lower control arms and not serviced separately.

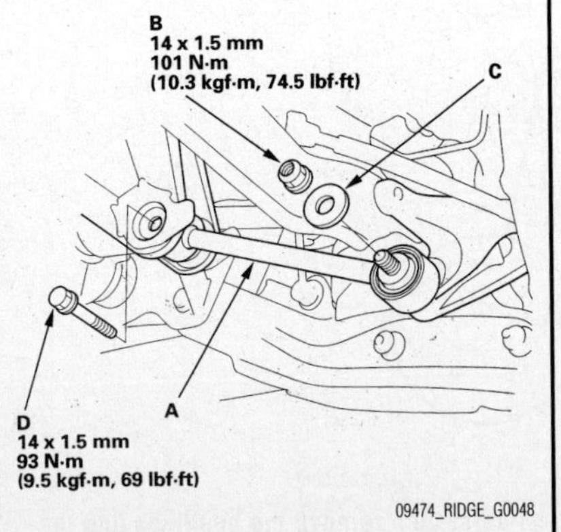
B
14 x 1.5 mm
101 N·m
(10.3 kgf·m, 74.5 lbf·ft)
C
D
14 x 1.5 mm
93 N·m
(9.5 kgf·m, 69 lbf·ft)
A
09474_RIDGE_G0048

Fig. 142 Rear lower arm (A) mounting

1. Support the suspension with a floor jack under lower arm B, at the knuckle.
2. Remove the rear wheel.
3. Remove the lower arm mounting bolts and nuts.
4. Remove the lower arm.

➡Raise the suspension to remove the arm.

To install:

5. Installation is the reverse of removal, noting the following:
6. Install all fasteners loosely, then, final torque all fasteners with the suspension loaded (weight of the car).
7. Using a new bolt and nut, tighten the bolt to 69 ft. lbs. (93 Nm) and the nut to 74 ft. lbs. (101 Nm).

Lower Arm B

See Figure 143.

➡The control arm bushings are an integral part of the lower control arms and not serviced separately.

1. Support the control arm with a jack at the knuckle.
2. Remove the rear wheel.
3. Remove the locknut.
4. Remove the bolt.
5. Remove the bolt.
6. Gradually lower the arm.
7. Matchmark the adjusting cam.
8. Remove the locknut, adjusting bolt and adjusting cam at the frame.
9. Remove the spring assembly.

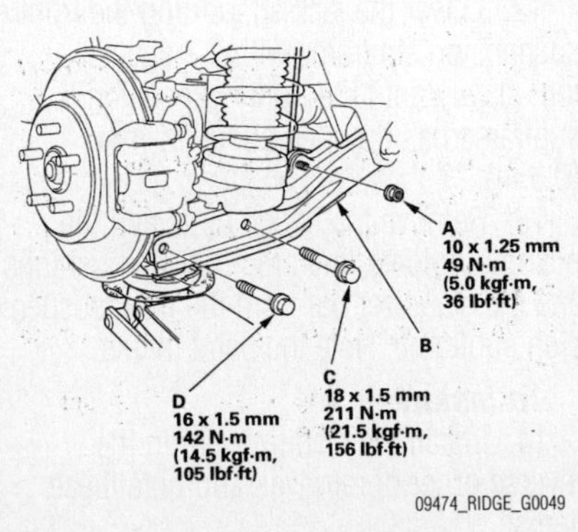
A
10 x 1.25 mm
49 N·m
(5.0 kgf·m,
36 lbf·ft)
B
C
18 x 1.5 mm
211 N·m
(21.5 kgf·m,
156 lbf·ft)
D
16 x 1.5 mm
142 N·m
(14.5 kgf·m,
105 lbf·ft)
09474_RIDGE_G0049

Fig. 143 Remove the locknut (A), remove the bolts (C and D) and gradually lower the arm B.

10. Remove the inner nuts and bolts and the arm.

To install:

11. Installation is the reverse of removal.
12. Install all fasteners loosely, then, final torque all fasteners with the suspension loaded (weight of the car).
13. Using new bolts and nuts, observe the following torques:
    - Adjusting cam nut: 61 ft. lbs. (83 Nm)
    - Nut A: 36 ft. lbs. (49 Nm)
    - Bolt C: 156 ft. lbs. (211 Nm)
    - Bolt D: 105 ft. lbs. (142 Nm)

Trailing Arm

See Figure 144.

1. Support control arm B with a jack at the knuckle.
2. Remove the caliper and support it out of the way.
3. Remove the rotor.
4. Disconnect the brake hose from the brake pipe and discard the clip. Plug the lines.
5. Remove the parking brake cable from the training arm.
6. Remove the trailing arm-to-knuckle bolts.
7. Remove the trailing arm mounting bolts.

To install:

8. Installation is the reverse of removal.
9. Install all fasteners loosely, then, final torque all fasteners with the suspension loaded (weight of the car).
10. Using new bolts and nuts and a new brake line clip, observe the following torque specifications:
    - Trailing arm-to-knuckle bolts: 47 ft. lbs. (64 Nm)
    - Trailing arm mounting bolts: 76 ft. lbs. (103 Nm).
11. Check the alignment.
12. Bleed the brakes.

Upper Control Arm

1. Support the control arm at the knuckle.
2. Remove the wheel.
3. Remove the upper ball joint from the knuckle.
4. Remove the upper arm bolt and the arm.

To install:

5. Installation is the reverse of removal.
6. Install all fasteners loosely, then, final torque all fasteners with the suspension loaded (weight of the car).
7. Using new bolts and nuts, and cotter pin, observe the following torque specifications:
    - Upper control arm bolt: 69 ft. lbs. (94 Nm)
    - Ball joint nut: 36–43 ft. lbs. (49–59 Nm).

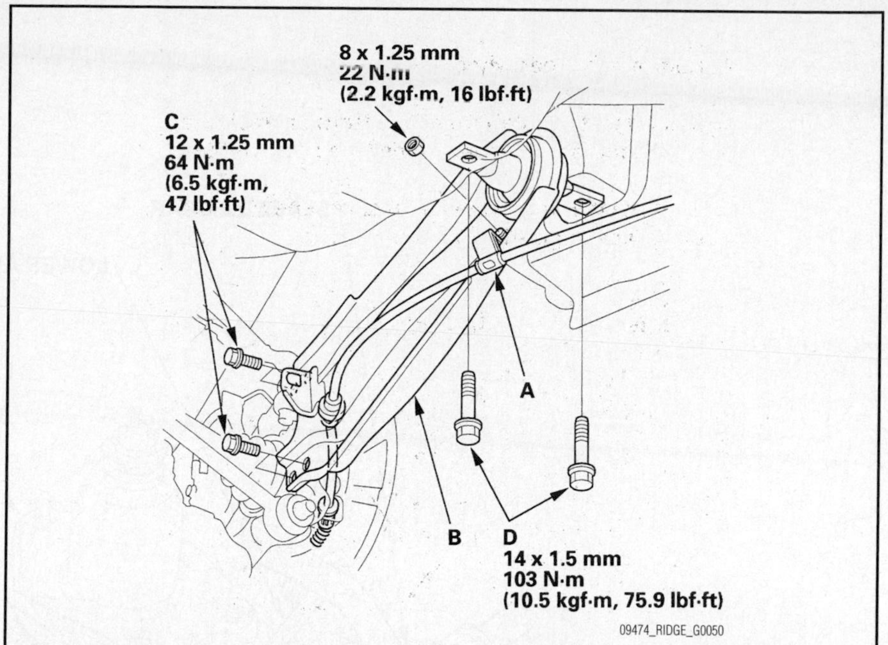
8 x 1.25 mm
22 N·m
(2.2 kgf·m, 16 lbf·ft)
C
12 x 1.25 mm
64 N·m
(6.5 kgf·m,
47 lbf·ft)
A
B
D
14 x 1.5 mm
103 N·m
(10.5 kgf·m, 75.9 lbf·ft)
09474_RIDGE_G0050

Fig. 144 Trailing arm mounting

MACPHERSON STRUTS

REMOVAL & INSTALLATION

See Figure 145.

1. Raise and support the rear end.
2. Remove the wheel(s).
3. Remove lower arm B. See "Lower Control Arm" in this section.
4. Remove the 3 bolts from the top of the strut.
5. Installation is the reverse of removal.
6. Torque the upper bolts to 25 ft. lbs. (34 Nm).

OVERHAUL

See Figures 146 and 147.

1. Place the strut in a spring compressor.
2. Compress the spring and remove the shaft nut.
3. Release the spring pressure

To assemble:

4. Compress the spring.
5. Assemble all parts except the shaft washer and nut.
6. Align the spring as shown.
7. Install the washer and NEW nut. Torque to 22 ft. lbs. (29 Nm).

REAR KNUCKLE

REMOVAL & INSTALLATION

See Figure 148.

1. Remove the wheel.
2. Remove and discard the brake hose clip.
3. Remove the brake hose bracket bolts.
4. Remove the caliper bracket mounting bolts and support the caliper out of the way.
5. Remove the sensor from the knuckle. Don't disconnect it.
6. Remove the halfshaft end nut.
7. Remove the rotor.
8. Remove the parking brake shoes and cable.
9. Disconnect the upper arm from the knuckle.
10. Remove lower arm A.
11. Support the lower arm B with a floor jack.
12. Disconnect lower arm B from the knuckle.
13. Pull outward on the knuckle while tapping the end of the halfshaft with a plastic hammer.
14. Installation is the reverse of removal.
15. Install all fasteners loosely, then, final torque all fasteners with the suspension loaded (weight of the car).
16. Use new bolts and nuts and a new brake line clip. See the relevant procedures for torque values.

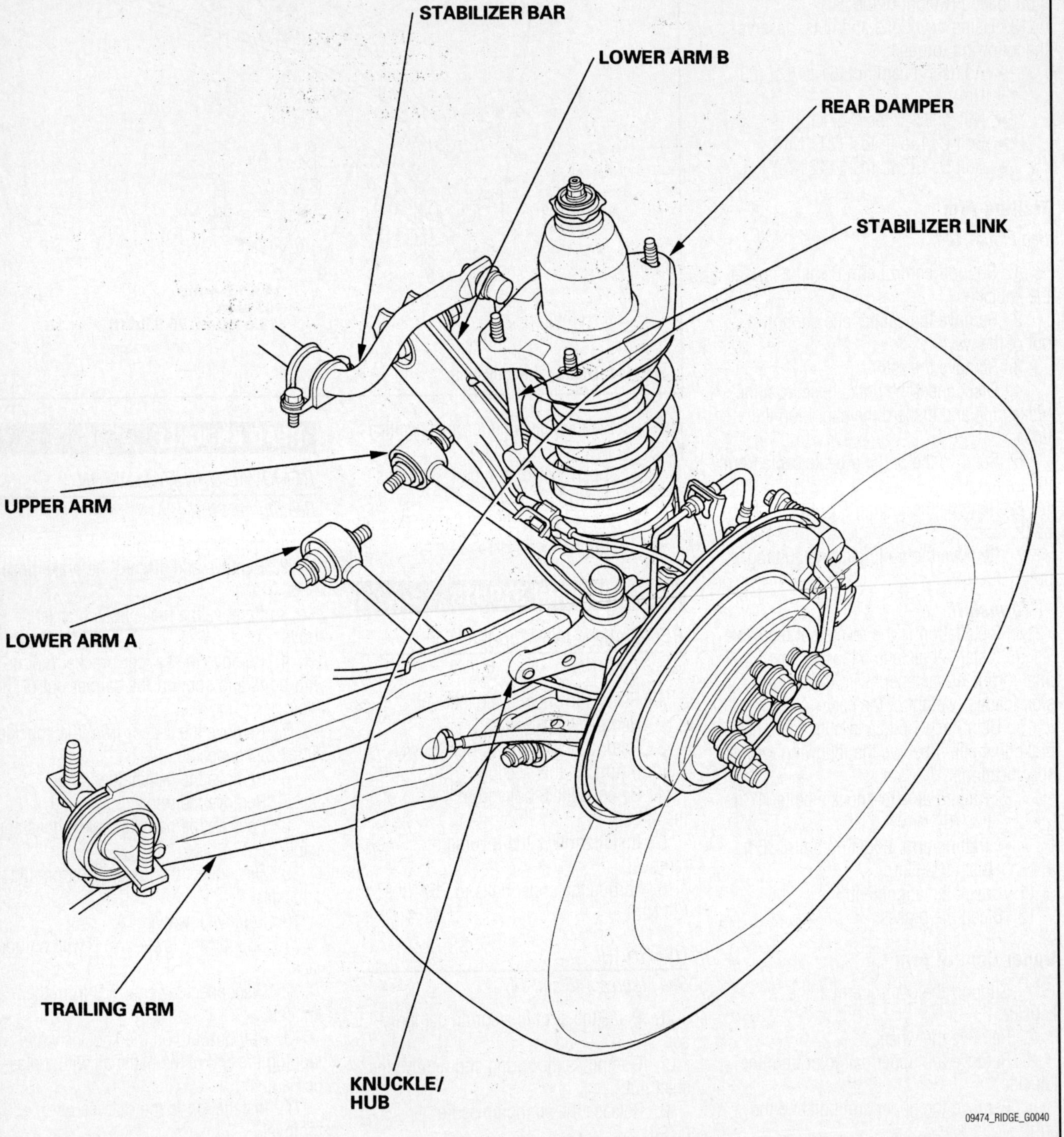
STABILIZER BAR
LOWER ARM B
REAR DAMPER
STABILIZER LINK
UPPER ARM
LOWER ARM A
TRAILING ARM
KNUCKLE/
HUB
09474_RIDGE_G0040

Fig. 145 Rear suspension components

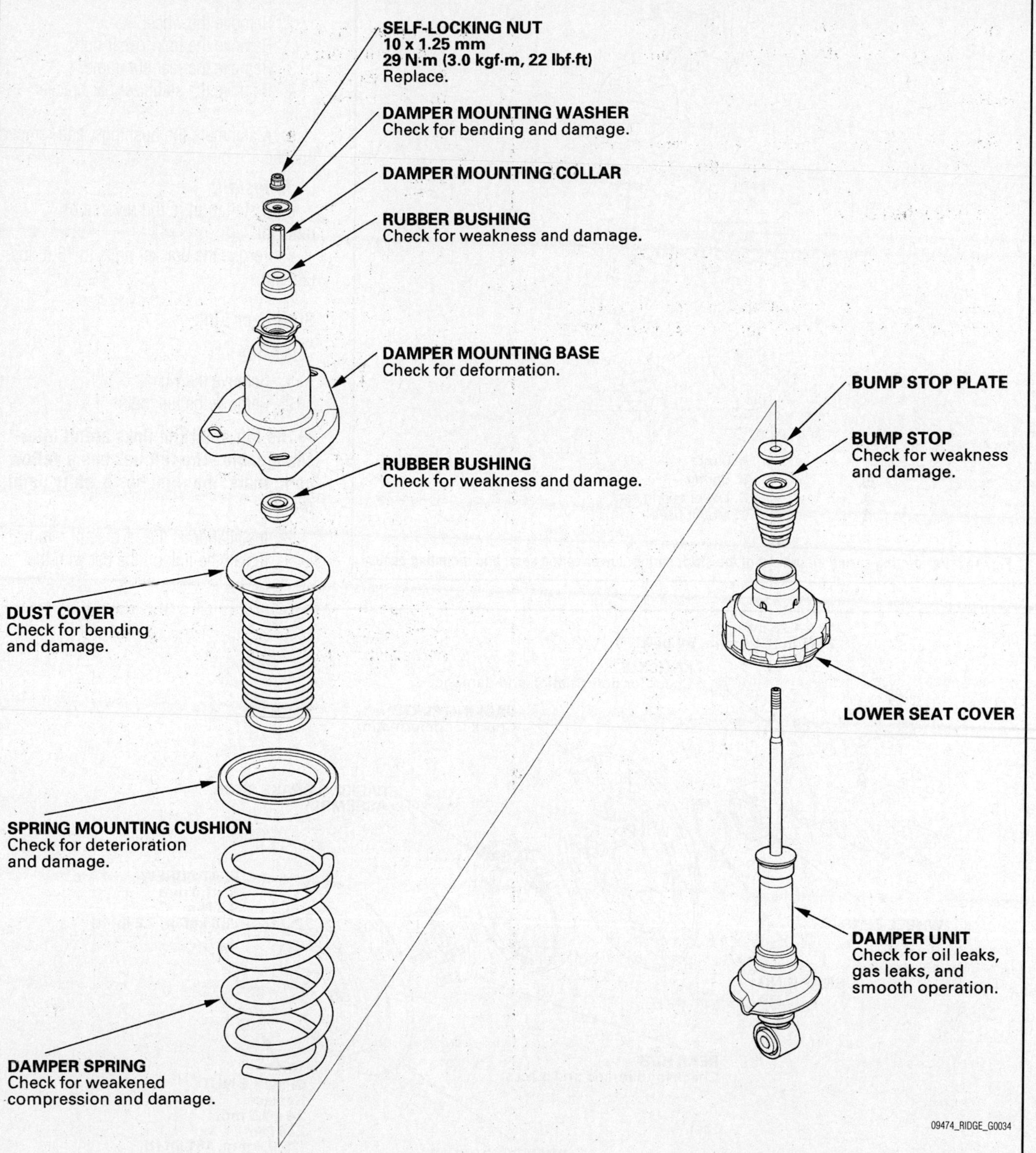
SELF-LOCKING NUT
10 x 1.25 mm
29 N·m (3.0 kgf·m, 22 lbf·ft)
Replace.
DAMPER MOUNTING WASHER
Check for bending and damage.
DAMPER MOUNTING COLLAR
RUBBER BUSHING
Check for weakness and damage.
DAMPER MOUNTING BASE
Check for deformation.
RUBBER BUSHING
Check for weakness and damage.
DUST COVER
Check for bending and damage.
SPRING MOUNTING CUSHION
Check for deterioration and damage.
DAMPER SPRING
Check for weakened compression and damage.
BUMP STOP PLATE
BUMP STOP
Check for weakness and damage.
LOWER SEAT COVER
DAMPER UNIT
Check for oil leaks, gas leaks, and smooth operation.
09474_RIDGE_G0034

Fig. 146 Rear strut exploded view

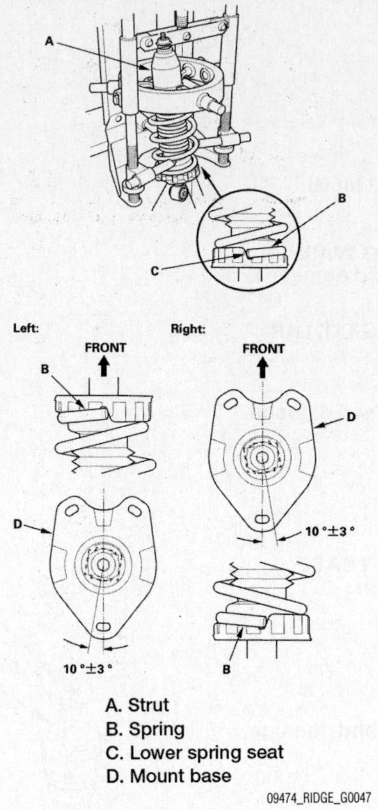
A
B
C
Left:
Right:
FRONT
FRONT
B
D
D
10 °±3 °
10 °±3 °
B
A. Strut
B. Spring
C. Lower spring seat
D. Mount base
09474_RIDGE_G0047

Fig. 147 Performing spring alignment of the strut, spring, lower spring seat, and mounting base.

STABILIZER BAR

REMOVAL & INSTALLATION

See Figure 149.

1. Remove the wheel.
2. Remove the link-to-bar nuts.
3. Remove the rear subframe.
4. Remove the stabilizer bar bracket bolts.
5. Matchmark the bushings, and remove the bar.

To install:

6. Installation is the reverse of removal.
7. Torque the bracket bolts to 16 ft. lbs. (22 Nm).

Stabilizer Link

See Figure 150.

1. Remove the wheel.
2. Remove the link nuts.

➡The left and right links aren't interchangeable. The left link has a yellow paint mark; the right has a white paint mark.

3. Installation is the reverse of removal.
4. Install the link on the bar with the

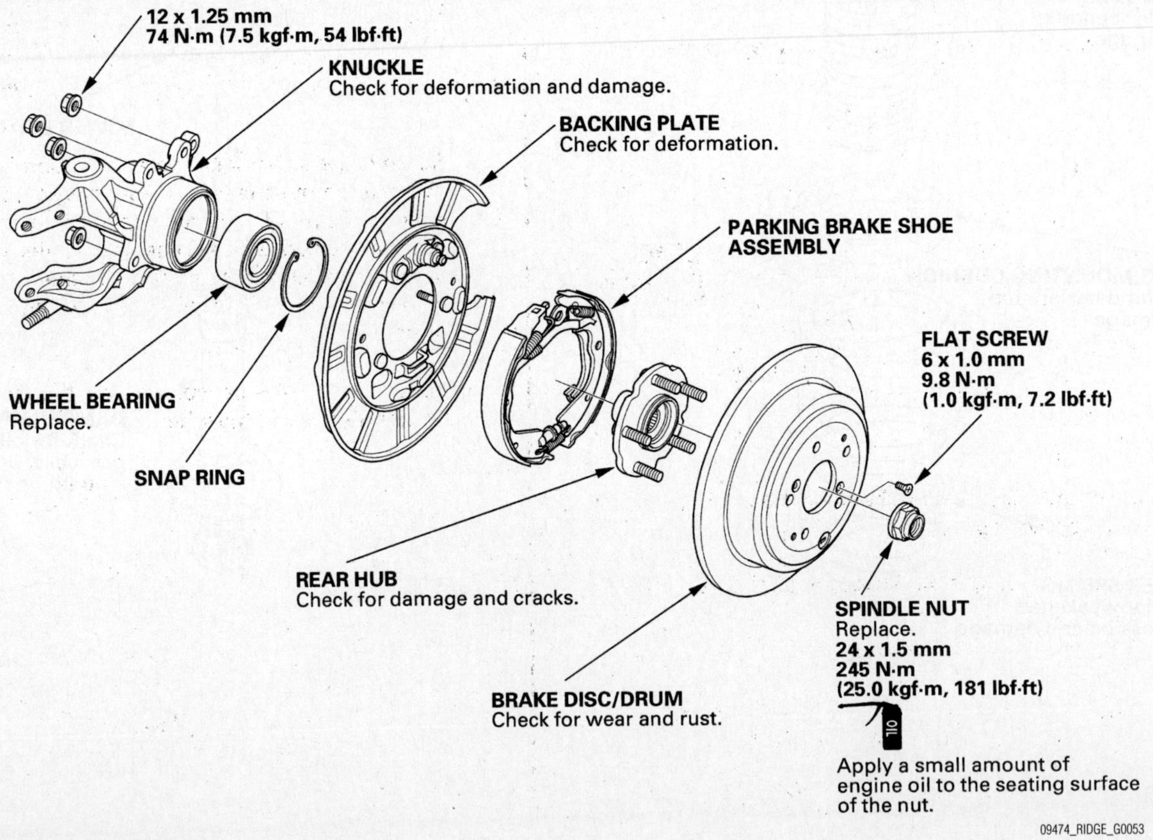
12 x 1.25 mm
74 N·m (7.5 kgf·m, 54 lbf·ft)
KNUCKLE
Check for deformation and damage.
BACKING PLATE
Check for deformation.
PARKING BRAKE SHOE ASSEMBLY
FLAT SCREW
6 x 1.0 mm
9.8 N·m
(1.0 kgf·m, 7.2 lbf·ft)
WHEEL BEARING
Replace.
SNAP RING
REAR HUB
Check for damage and cracks.
SPINDLE NUT
Replace.
24 x 1.5 mm
245 N·m
(25.0 kgf·m, 181 lbf·ft)
BRAKE DISC/DRUM
Check for wear and rust.
OIL
Apply a small amount of engine oil to the seating surface of the nut.
09474_RIDGE_G0053

Fig. 148 Rear knuckle/hub exploded view

Rear Subframe Torque

NOTE:
- When installing, align both installation reference holes in the subframe with the reference holes in the body using a screwdriver or tapered punch as a guide.
- After removing the subframe mounting bolts, be sure to replace them with new ones.

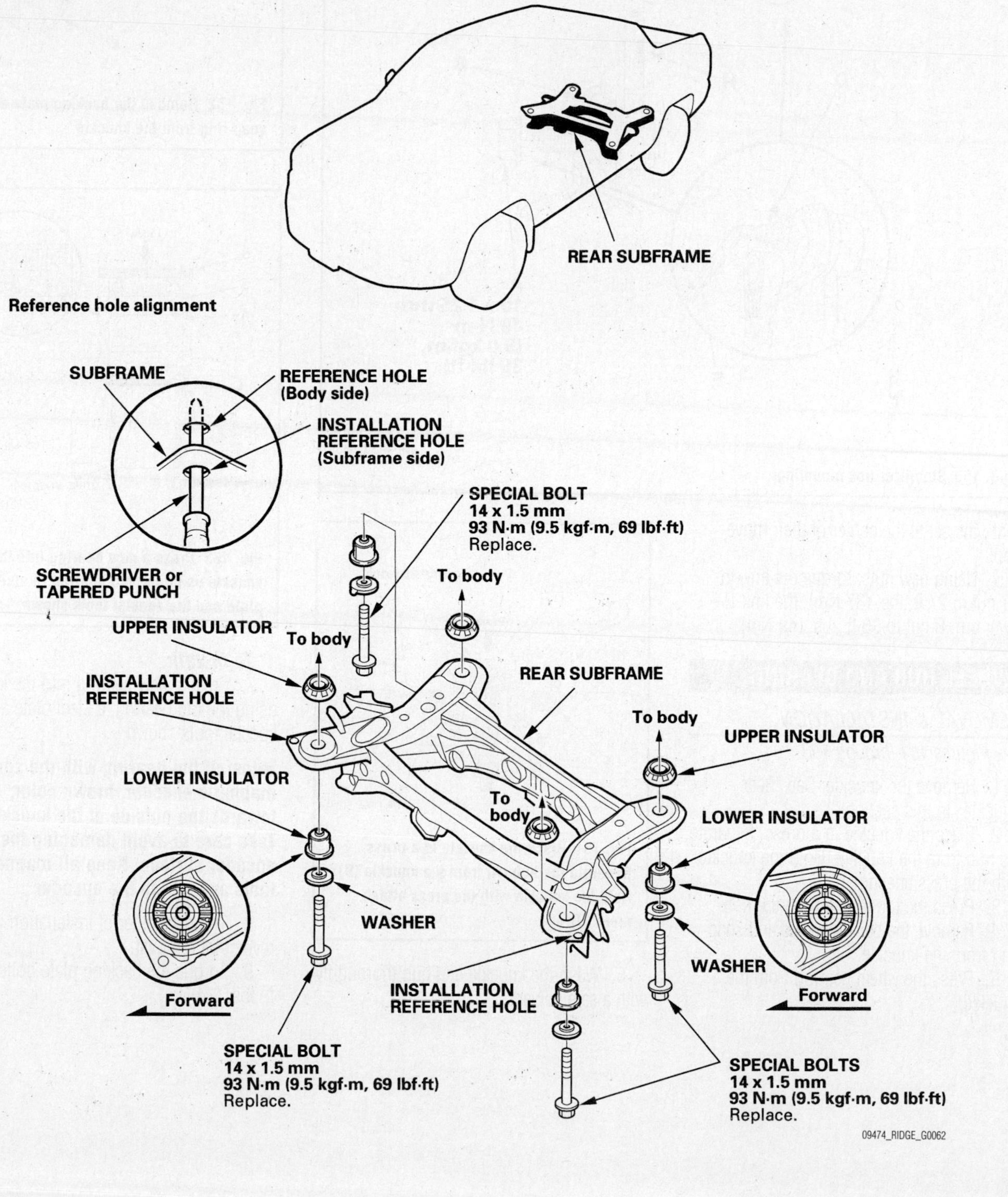
REAR SUBFRAME
Reference hole alignment
SUBFRAME
REFERENCE HOLE
(Body side)
INSTALLATION
REFERENCE HOLE
(Subframe side)
SCREWDRIVER or
TAPERED PUNCH
SPECIAL BOLT
14 x 1.5 mm
93 N·m (9.5 kgf·m, 69 lbf·ft)
Replace.
To body
UPPER INSULATOR
To body
INSTALLATION
REFERENCE HOLE
REAR SUBFRAME
To body
UPPER INSULATOR
LOWER INSULATOR
To
body
LOWER INSULATOR
WASHER
WASHER
Forward
INSTALLATION
REFERENCE HOLE
Forward
SPECIAL BOLT
14 x 1.5 mm
93 N·m (9.5 kgf·m, 69 lbf·ft)
Replace.
SPECIAL BOLTS
14 x 1.5 mm
93 N·m (9.5 kgf·m, 69 lbf·ft)
Replace.
09474_RIDGE_G0062

Fig. 149 Rear subframe mounting

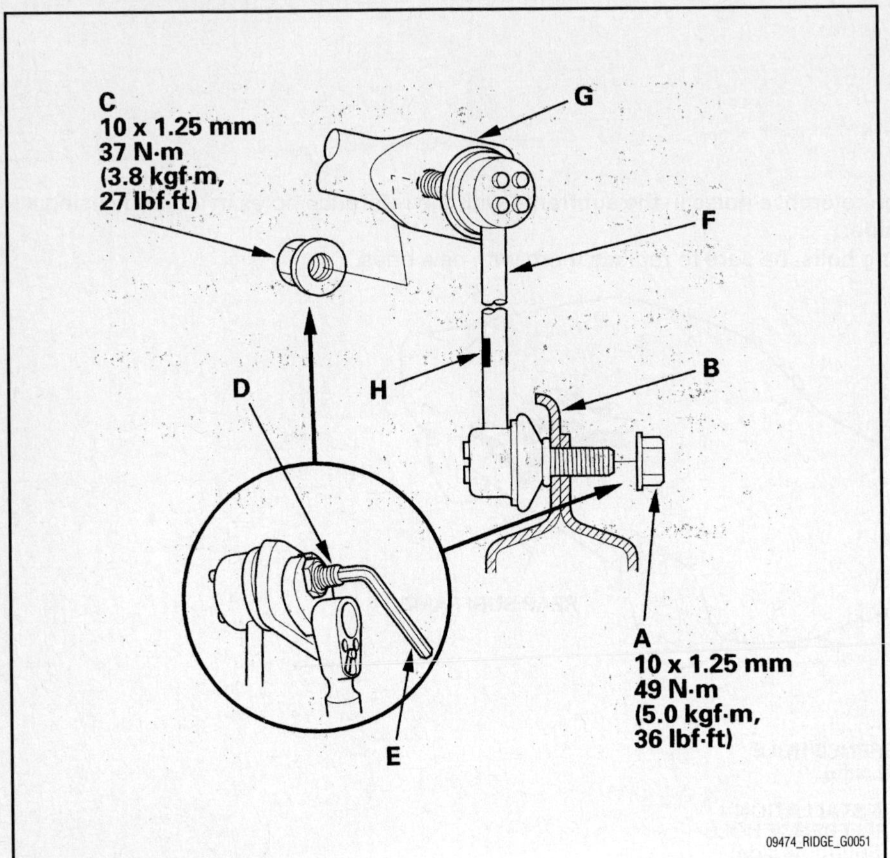
C
10 x 1.25 mm
37 N·m
(3.8 kgf·m,
27 lbf·ft)
G
F
H
D
B
A
10 x 1.25 mm
49 N·m
(5.0 kgf·m,
36 lbf·ft)
E
09474_RIDGE_G0051

Fig. 150 Stabilizer link mounting

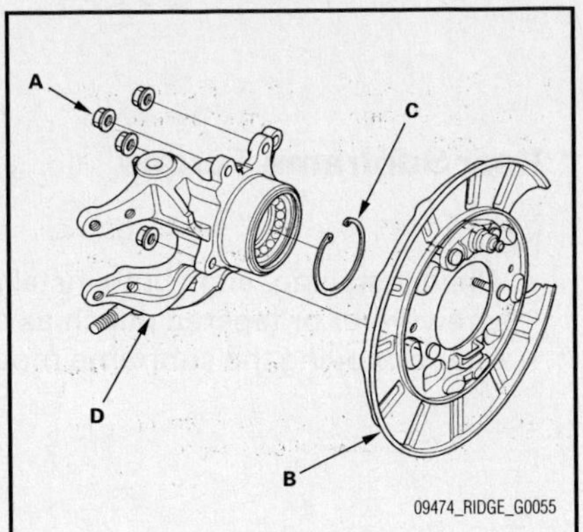
A
C
D
B
09474_RIDGE_G0055

Fig. 152 Remove the backing plate and snap ring from the knuckle

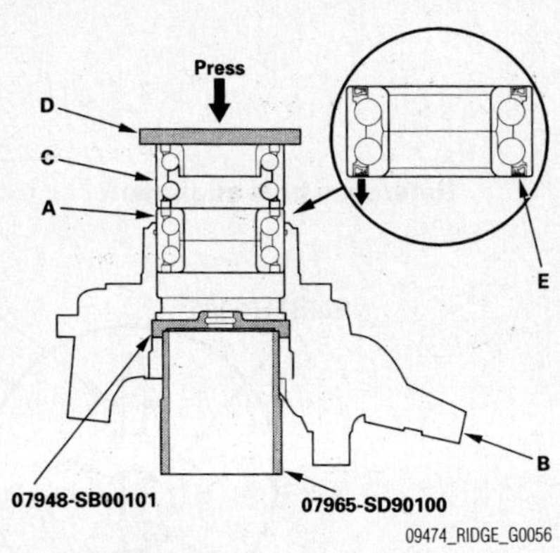
Press
D
C
A
E
B
07948-SB00101
07965-SD90100
09474_RIDGE_G0056

Fig. 153 Press a new bearing into the knuckle using the old bearing, a steel plate and the special tools shown.

joint pins set at the center of their movement.

5. Using new nuts, torque the link-to-bar nut to 27 ft. lbs. (37 Nm); the link-to-lower arm B nut to 36 ft. lbs. (49 Nm).

WHEEL HUB AND BEARING

REMOVAL & INSTALLATION

See Figures 151 through 153.

1. Remove the knuckle. See "Rear Knuckle" in this section.
2. Place the knuckle in a press. Separate the hub from the knuckle. Hold the knuckle with the press attachment.
3. Press the inner race from the hub.
4. Remove the backing plate and snap ring from the knuckle.
5. Press the wheel bearing from the knuckle.

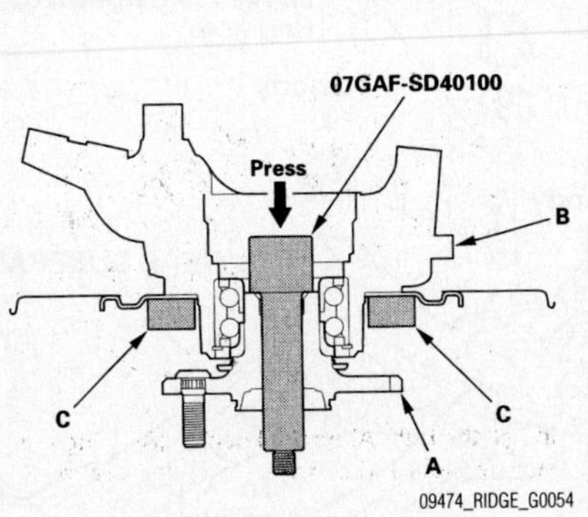
07GAF-SD40100
Press
B
C
C
A
09474_RIDGE_G0054

Fig. 151 Place the knuckle in a press. Separate the hub (A) from the knuckle (B). Hold the knuckle with the press attachment (C).

6. Wash the knuckle and hub thoroughly with a safe solvent.

To install:

7. Press a new bearing into the knuckle using the old bearing, a steel plate and the special tools shown.

➡Install the bearing with the sensor magnetic encoder, brown color, towards the outside of the knuckle. Take care to avoid damaging the encoder surface. Keep all magnetic tools away from the encoder.

8. The remainder of installation is the reverse of removal.
9. Torque the backing plate bolts to 54 ft. lbs. (74 Nm).

HONDA

S2000

18

BRAKES....................18-9

ANTI-LOCK BRAKE SYSTEM (ABS)..............18-10
Wheel Speed Sensors.............18-10
Removal & Installation........18-10
BLEEDING THE BRAKE SYSTEM..........................18-9
Bleeding Procedure..................18-9
Bleeding Procedure..............18-9
Bleeding the ABS System.....18-9
FRONT DISC BRAKES........18-10
Brake Caliper..........................18-10
Removal & Installation........18-10
Disc Brake Pads.....................18-10
Removal & Installation........18-10
INFORMATION AND PRECAUTIONS................18-9
Anti-lock Systems................18-9
Disc and Drum Systems.......18-9
PARKING BRAKE.............18-12
Parking Brake Cables.............18-12
Adjustment........................18-12
REAR DISC BRAKES.........18-11
Brake Caliper..........................18-11
Removal & Installation........18-11
Disc Brake Pads.....................18-11
Removal & Installation........18-11

CHASSIS ELECTRICAL.....18-13

AIR BAG (SUPPLEMENTAL RESTRAINT SYSTEM)......18-13
General Information................18-13
Arming the System.............18-13
Clockspring Centering........18-13
Disarming the System.........18-13
Service Precautions............18-13

DRIVETRAIN.................18-13

Clutch..................................18-13
Bleeding............................18-13
Removal & Installation........18-13
Front Driveshaft.....................18-13
Removal & Installation........18-13
Rear Halfshaft........................18-14
Removal & Installation........18-14
Rear Pinion Seal.....................18-14
Removal & Installation........18-14

ENGINE COOLING...........18-15

Thermostat.............................18-15
Removal & Installation........18-15
Water Pump............................18-15
Removal & Installation........18-15

ENGINE ELECTRICAL.......18-16

CHARGING SYSTEM.........18-16
Alternator..............................18-16
Removal & Installation........18-16
IGNITION SYSTEM...........18-16
Firing Order............................18-16
Ignition Coil...........................18-16
Removal & Installation........18-16
Ignition Timing.......................18-16
Adjustment........................18-16
Spark Plugs............................18-16
Removal & Installation........18-16
STARTING SYSTEM..........18-17
Starter....................................18-17
Removal & Installation........18-17

ENGINE MECHANICAL......18-17

Accessory Drive Belts.............18-17
Accessory Belt Routing.......18-17
Adjustment........................18-17
Removal & Installation........18-17
Camshaft and Valve Lifters......18-17
Removal & Installation........18-17
Cylinder Head.........................18-18
Removal & Installation........18-18
Exhaust Manifold....................18-19
Removal & Installation........18-19
Intake Manifold......................18-20
Removal & Installation........18-20
Oil Pan...................................18-21
Removal & Installation........18-21
Oil Pump................................18-21
Removal & Installation........18-21
Piston and Ring Positioning.........................18-22
Rear Main Seal.......................18-22
Removal & Installation........18-22
Timing Chain, Sprockets, Front Cover and Seal...........18-22
Removal & Installation........18-22
Valve Lash..............................18-24
Adjustment........................18-24

ENGINE PERFORMANCE & EMISSION CONTROL.......18-24

Camshaft Position (CMP) Sensor.................................18-24
Location............................18-24
Removal & Installation........18-24
Crankshaft Position (CKP) Sensor.................................18-24
Location............................18-24
Removal & Installation........18-24
Engine Coolant Temperature (ECT) Sensor....18-24
Location............................18-24
Removal & Installation........18-24
Heated Oxygen Sensor (HO2S)................................18-25
Location............................18-25
Removal & Installation........18-25
Intake Air Temperature (IAT) Sensor..........................18-25
Location............................18-25
Removal & Installation........18-25
Manifold Absolute Pressure (MAP) Sensor........18-25
Location............................18-25
Removal & Installation........18-25
Throttle Position Sensor (TPS).18-25
Location............................18-25
Removal & Installation........18-25
Vehicle Speed Sensor (VSS)...18-25
Location............................18-25
Removal & Installation........18-25

FUEL..........................18-26

GASOLINE FUEL INJECTION SYSTEM........18-26
Fuel Filter..............................18-26
Removal & Installation........18-26
Fuel Injectors.........................18-26
Removal & Installation........18-26

Fuel Pump....18-26
Removal & Installation....18-26
Fuel System Service Precautions....18-26
Idle Speed....18-26
Adjustment....18-26
Relieving Fuel System Pressure....18-26
Throttle Body....18-27
Removal & Installation....18-27

HEATING & AIR CONDITIONING SYSTEM...18-29

Blower Motor....18-29
Removal & Installation....18-29
Heater Core....18-29
Removal & Installation....18-29

SPECIFICATIONS AND MAINTENANCE CHARTS.....18-3

Brake Specifications....18-7
Camshaft Specifications....18-5
Capacities....18-4
Crankshaft and Connecting Rod Specifications....18-5
Engine and Vehicle Identification....18-3
Engine Tune-Up Specifications....18-3
Fluid Specifications....18-4
General Engine Specifications....18-3
Piston and Ring Specifications....18-5
Maintenance Minder Schedule....18-8
Tire, Wheel and Ball Joint Specifications....18-7
Torque Specifications....18-6
Valve Specifications....18-4
Wheel Alignment....18-6

STEERING....18-30

Power Steering Gear....18-30
Removal & Installation....18-30

SUSPENSION....18-31

FRONT SUSPENSION 18-31
Lower Ball Joint....18-31
Removal & Installation....18-31
Lower Control Arm....18-31
Removal & Installation....18-31
MacPherson Strut....18-31
Overhaul....18-31
Removal & Installation....18-32
Stabilizer Bar....18-31
Removal & Installation....18-32
Steering Knuckle....18-32
Removal & Installation....18-32
Upper Control Arm....18-32
Removal & Installation....18-32
Wheel Bearings....18-32
Adjustment....18-32
Removal & Installation....18-32

REAR SUSPENSION..........18-34
Lower Control Arm....18-34
Removal & Installation....18-34
Stabilizer Bar....18-34
Removal & Installation....18-34
Strut & Spring Assembly....18-34
Removal & Installation....18-34
Upper Control Arm....18-34
Removal & Installation....18-34
Wheel Bearings....18-34
Adjustment....18-36
Removal & Installation....18-34

SPECIFICATIONS AND MAINTENANCE CHARTS

ENGINE AND VEHICLE IDENTIFICATION

Engine						Model Year	
Code	Liters	Cu. In. (cc)	Cyl.	Fuel Sys.	Eng. Mfg.	Code	Year
F22C1	2.2	132.0 (2157)	4	PGM-FI	Honda	9	2009

PGM-FI: Programmed Fuel Injection

37647_S200_C0001

GENERAL ENGINE SPECIFICATIONS

Year	Model	Engine Displacement Liters	Engine ID/VIN	Net Horsepower @ rpm	Net Torque @ rpm (ft. lbs.)	Bore X Stroke (in.)	Com-pression Ratio	Oil Pressure @ rpm
2009	S2000	2.2	F22C1	237@7800	162@6800	3.43X3.57	11.1:1	85@3000

37647_S200_C0002

ENGINE TUNE-UP SPECIFICATIONS

Year	Engine Displacement Liters	Engine ID/VIN	Spark Plugs Gap (in.)	Ignition Timing (deg.)		Fuel Pump (psi)	Idle Speed (rpm)		Valve Clearance	
				MT	AT		MT	AT	In.	Ex.
2009	2.2	F22C1	0.039-0.043	3-7B	—	55-63	850-950	—	0.008-0.010	0.010-0.011

NOTE: The Vehicle Emission Control Information label often reflects specification changes made during production.
The label figures must be used if they differ from those in this chart

B: Before Top Dead Center

37647_S200_C0003

CAPACITIES

Year	Model	Engine Displacement Liters	Engine ID	Engine Oil with Filter	Transmission (pts.) 6-Spd	Transmission (pts.) Auto.	Drive Axle Front (pts.)	Drive Axle Rear (pts.)	Fuel Tank (gal.)	Cooling System (qts.)
2009	S2000	2.2	F22C1	5.1	3.1	—	—	—	13.2	6.9

NOTE: All capacities are approximate. Add fluid gradually and ensure a proper fluid level is obtained.

NOTE: Capacities given are service, not overhaul capacities

37647_S200_C0004

FLUID SPECIFICATIONS

Year	Model	Engine Displ. Liters	Engine Oil	Man. Trans.	Auto. Trans.	Drive Axle Front	Drive Axle Rear	Transfer Case	Power Steering Fluid	Brake Master Cylinder	Cooling System
2009	S2000	2.2	10W-30 Honda	Honda MTF	—	—	—	—	NA	Honda DOT 3	①

DOT: Department Of Transpotation

① Honda Long Life Antifreeze/Coolant-Type2

37647_S200_C0005

VALVE SPECIFICATIONS

Year	Engine Displacement Liters	Engine ID/VIN	Seat Angle (deg.)	Face Angle (deg.)	Spring Test Pressure (lbs. @ in.)	Spring Installed Height (in.)	Stem-to-Guide Clearance (in.) Intake	Stem-to-Guide Clearance (in.) Exhaust	Stem Diameter (in.) Intake	Stem Diameter (in.) Exhaust
2009	2.2	F22C1	45	45	NA	NA	0.0010-0.0020	0.0020-0.0030	0.2157-0.2161	0.2146-0.2150

NA: Information not available

37647_S200_C0007

CAMSHAFT AND BEARING SPECIFICATIONS

All measurements are given in inches.

Year	Engine Displacement Liters	Engine VIN	Journal Diameter	Brg. Oil Clearance	Shaft End-play	Runout	Journal Bore	Lobe Height	
								Intake	Exhaust
2009	2.2	F22C1	NA	0.0020-0.0040	0.0020-0.0060	0.0010	NA	①	②

NA: Information not available

① Primary: 1.3370 in.
Mid: 1.4340 in.
Secondary: 1.3370 in.

② Primary: 1.2902 in.
Mid: 1.3688 in.
Secondary: 1.2859 in.

37647_S200_C0006

CRANKSHAFT AND CONNECTING ROD SPECIFICATIONS

All measurements are given in inches.

Year	Engine Displacement Liters	Engine ID/VIN	Crankshaft				Connecting Rod		
			Main Brg. Journal Dia.	Main Brg. Oil Clearance	Shaft End-play	Thrust on No.	Journal Diameter	Oil Clearance	Side Clearance
2009	2.2	F22C1	2.1644-2.1654	0.0007-0.0016	0.0040-0.0140	4	1.8888-1.8898	0.0012-0.0021	0.0060-0.0120

37647_S200_C0008

PISTON AND RING SPECIFICATIONS

All measurements are given in inches.

Year	Engine Displacement Liters	Engine ID/VIN	Piston Clearance	Ring Gap			Ring Side Clearance		
				Top Compression	Bottom Compression	Oil Control	Top Compression	Bottom Compression	Oil Control
2009	2.2	F22C1	0.0002-0.0011	0.0100-0.0140	0.0240-0.0300	0.0080-0.0280	0.0018-0.0035	0.0016-0.0028	NA

NA: Information not available

37647_S200_C0009

TORQUE SPECIFICATIONS

All readings in ft. lbs.

Year	Engine Displacement Liters	Engine ID/VIN	Cylinder Head Bolts	Main Bearing Bolts	Rod Bearing Bolts	Crankshaft Damper Bolts	Flywheel Bolts	Manifold Intake	Manifold Exhaust	Spark Plugs	Oil Pan Drain Plug
2009	2.2	F22C1	①	②	③	192	94	16	23	13	29

① Step 1: 22 ft. lbs.
Step 2: Rotate 90 degrees
Step 3: Rotate 90 degrees
Step 4: If new bolt rotate additional 90 degrees

② Step 1: 22 ft. lbs.
Step 2: Rotate 60 degrees
Step 3: 8mm bolts to 16 ft. lbs.

③ Step 1: 18 ft. lbs.
Step 2: Rotate 90 degrees

37647_S200_C0010

WHEEL ALIGNMENT

Year	Model		Caster Range (+/-Deg.)	Caster Preferred Setting (Deg.)	Camber Range (+/-Deg.)	Camber Preferred Setting (Deg.)	Toe-in (in.)
2009	S2000	F	0.75	+6.00	0.50	-0.50	0.00+/-0.08
		R	—	—	0.50	-1.50	0.14+/-0.08

37647_S200_C0011

TIRE, WHEEL AND BALL JOINT SPECIFICATIONS

Year	Model	OEM Tires		Tire Pressures (psi)		Wheel	Ball Joint	Lug Nut
		Standard	Optional	Front	Rear	Size	Inspection	(ft. lbs.)
2009	S2000	①	None	32	32	NA	NA	80

OEM: Original Equipment Manufacturer

PSI: Pounds Per Square Inch

NA: Information not available

① Front: 215/45R17

Rear, Except CR: 245/40R17

Rear, CR Model: 255/40R17

37647_S200_C0012

BRAKE SPECIFICATIONS

All measurements in inches unless noted

Year	Model		Brake Disc			Minimum Lining Thickness		Brake Caliper	
			Original Thickness	Minimum Thickness	Maximum Runout	Front	Rear	Bracket Bolts (ft. lbs.)	Mounting Bolts (ft. lbs.)
2009	S2000	F	0.990	0.910	0.002	0.060	—	84	24
		R	0.476	0.390	0.002	—	0.060	41	17

NA: Information not available

F: Front

R: Rear

37647_S200_C0013

MAINTENANCE MINDER SCHEDULE

Honda S2000

All Honda's displays engine oil life and maintenance service items in the information display to indicate when to perform maintenance service. If the engine oil life is 15% or less, based on the onboard computer's caluculations, you will see SERVICE DUE SOON in the information display every time the ignition key is turned to ON. The maintenance minder indicator will also come on and the maintenance code(s) for other scheduled maintenance items needing service will be displayed below the message.

Symbol	Item	Service
A	Engine oil ①	Change
B	Engine oil and filter	Change
	Fluid levels	Inspect
	Brakes	Inspect
	Parking brake adjustment	Check
	Steering gear and linkage	Inspect
	Suspension components	Inspect
	Driveshaft boots	Inspect
	Brake hoses and lines	Inspect
	Exhaust system	Inspect
	Fuel lines and connections	Inspect
1	Tires	Rotate
2	Engine air filter ②	Replace
	Dust and pollen filter ③	Replace
	Accessory drive belt	Inspect
3	Transmission fluid ④	Replace
4	Spark plugs	Replace
	Water pump	Inspect
	Valve clearance ⑤	Inspect
5	Engine coolant	Replace
6	VTM-4 rear differential fluid	Replace

① If the message SERVICE DUE NOW does not appear more than 12 months after the display is reset, change every year.

② If driven in dusty conditions, replace every 15,000 miles.

③ If driven in urban areas that have a high concentration of soot from industry and diesel, replace every 15,000 miles

④ If regularly driven in mountainous areas at very low speed or trailer towing, change the fluid every 30,000 miles.

⑤ Adjust if necessary.

Additionally, replace the brake fluid every 3 years, and inspect the idle speed every 160,000 miles.

To reset the Engine Oil Life Display:

1. Turn the ignition switch to ON.
2. Press the SELECT button repeatedly until the engine oil life display or the service message is displayed.
3. Press the RESET button for about 10 seconds. You will see a MAINT RESET message.
4. Select the appropriate answer, MAINT RESET >N (NO) or MAINT RESET > y (YES) by pressing the SELECT button repeatedly.
   >N or >Y is displayed on the outside temperature >N or >Y is displayed on the outside temperature display.
5. Select the MAINT RESET > Y (YES), and press and hold the RESET button again to reset the engine oil life to 100%.

37647_S200_C0014

BRAKES — INFORMATION AND PRECAUTIONS

ANTI-LOCK SYSTEMS

- Certain components within the ABS system are not intended to be serviced or repaired individually.
- Do not use rubber hoses or other parts not specifically specified for and ABS system. When using repair kits, replace all parts included in the kit. Partial or incorrect repair may lead to functional problems and require the replacement of components.
- Lubricate rubber parts with clean, fresh brake fluid to ease assembly. Do not use shop air to clean parts; damage to rubber components may result.
- Use only DOT 3 brake fluid from an unopened container.
- If any hydraulic component or line is removed or replaced, it may be necessary to bleed the entire system.
- A clean repair area is essential. Always clean the reservoir and cap thoroughly before removing the cap. The slightest amount of dirt in the fluid may plug an orifice and impair the system function. Perform repairs after components have been thoroughly cleaned; use only denatured alcohol to clean components. Do not allow ABS components to come into contact with any substance containing mineral oil; this includes used shop rags.
- The Anti-Lock control unit is a microprocessor similar to other computer units in the vehicle. Ensure that the ignition switch is **OFF** before removing or installing controller harnesses. Avoid static electricity discharge at or near the controller.
- If any arc welding is to be done on the vehicle, the control unit should be unplugged before welding operations begin.

DISC AND DRUM SYSTEMS

⁂ CAUTION

Dust and dirt accumulating on brake parts during normal use may contain asbestos fibers from production or aftermarket brake linings. Breathing excessive concentrations of asbestos fibers can cause serious bodily harm. Exercise care when servicing brake parts. Do not sand or grind brake lining unless equipment used is designed to contain the dust residue. Do not clean brake parts with compressed air or by dry brushing. Cleaning should be done by dampening the brake components with a fine mist of water, then wiping the brake components clean with a dampened cloth. Dispose of cloth and all residue containing asbestos fibers in an impermeable container with the appropriate label. Follow practices prescribed by the Occupational Safety and Health Administration (OSHA) and the Environmental Protection Agency (EPA) for the handling, processing, and disposing of dust or debris that may contain asbestos fibers.

BRAKES — BLEEDING THE BRAKE SYSTEM

BLEEDING PROCEDURE

BLEEDING PROCEDURE

See Figure 2.

➡Do not reuse the drained fluid. Use only clean Honda DOT 3 Brake Fluid from an unopened container.

Using a non-Honda brake fluid can cause corrosion and shorten the life of the system.

➡Do not mix different brands of brake fluid; they may not be compatible.

➡Make sure no dirt or other foreign matter is allowed to contaminate the brake fluid.

⁂ WARNING

Do not spill brake fluid on the vehicle, it may damage the paint; if brake fluid does contact the paint, wash it off immediately with water.

➡The reservoir on the master cylinder must be at the MAX (upper) level mark at the start of the bleeding procedure and checked after bleeding each brake caliper. Add fluid as required.

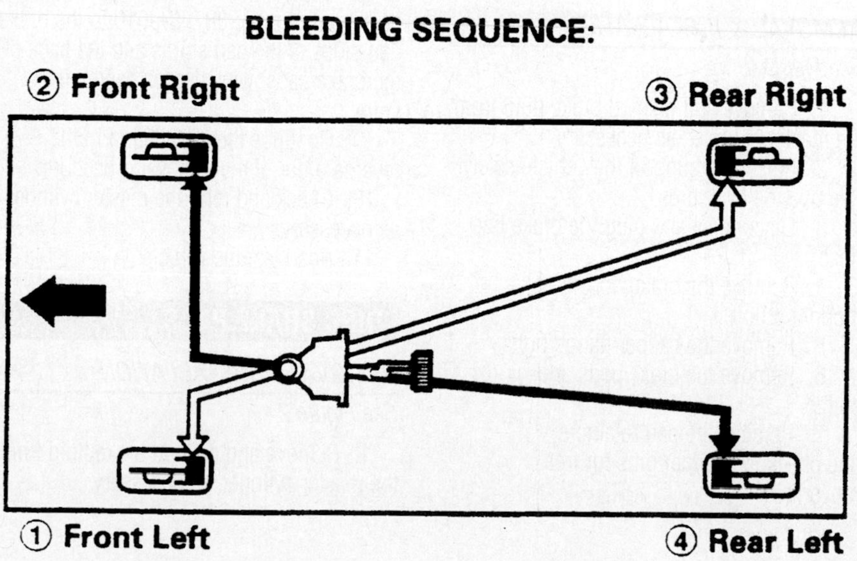
BLEEDING SEQUENCE:
② Front Right
③ Rear Right
① Front Left
④ Rear Left
91189G23

Fig. 1 Proper brake bleeding sequence

1. Make sure the brake fluid level in the reservoir is at the MAX (upper) level line.
2. Attach a length of clear drain tube to the bleed screw.
3. Have someone slowly pump the brake pedal several times, and then apply steady pressure.
4. Starting at the left-front, loosen the brake bleed screw to allow air to escape from the system. Then tighten the bleed screw securely.
5. Repeat the procedure for each wheel in the sequence shown following until air bubbles no longer appear in the fluid.
6. Refill the master cylinder reservoir to the MAX (upper) level line.

BLEEDING THE ABS SYSTEM

The bleeding procedure for the ABS System is the same as the Conventional Bleeding Procedure.

BRAKES — ANTI-LOCK BRAKE SYSTEM (ABS)

WHEEL SPEED SENSORS

REMOVAL & INSTALLATION

CAUTION

Vehicles equipped with air bag systems (SRS) have components and wiring in the same area as the front speed sensor wiring harnesses. The air bag system connectors are yellow. Do not use electrical test equipment on these circuits. Do not damage the SRS wiring while working on other wiring or components. Failure to observe correct procedures may cause the air bag system to inflate unexpectedly or render the system totally inoperative.

1. Raise and safely support the vehicle as necessary for access.
2. Make certain the ignition switch is **OFF**.
3. Detach the sensor harness connector.
4. Beginning at the connector end, remove grommets, clips or retainers as necessary to free the harness. Take careful note of the placement and routing of the harness; it must be reinstalled in the exact original position.
5. Remove the bolt holding the speed sensor to its mounting, and then remove the sensor. If it is stuck in place, gently tap on the side of the mounting flange with a hammer and small punch; do not tap on the sensor.

To install:

6. Place the sensor in position; install the retaining bolts loosely. Route the harness correctly. Avoid twisting or crimping the harness; use the white line on the wires as a guide.
7. Once the harness and sensor are correctly but loosely placed, tighten the sensor mounting bolts to the specifications shown in the accompanying illustrations.
8. Working from the sensor end to the connector, install each clip, retainer, bracket or grommet holding the sensor harness. The harness must not be twisted.
9. Attach the wiring connector.
10. Use the ABS checker to check for proper signal from the wheel speed sensor.
11. Carefully lower the vehicle to the ground.

BRAKES — FRONT DISC BRAKES

BRAKE CALIPER

REMOVAL & INSTALLATION

See Figure 2.

1. Remove and discard brake fluid from the master cylinder, as necessary.
2. Raise and support the vehicle safely. Remove the front tires.
3. Disconnect and plug the brake line hose.
4. Remove the brake hose bracket mounting bolt.
5. Remove the caliper flange bolts.
6. Remove the brake pads and shims.
7. Remove the pad retainers and check the caliper pins for free movement.

To install:

8. Apply a thin coat of M-77 assembly paste part number 08798-9010 to the brake pad sides of the pad shims and the back of the brake pads, wipe excess paste off the shim.
9. Continue the installation in the reverse order of the removal procedure.
10. Check and refill the master cylinder, as necessary.
11. Road test the vehicle.

DISC BRAKE PADS

REMOVAL & INSTALLATION

See Figure 2.

1. Remove and discard brake fluid from the master cylinder, as necessary.
2. Raise and support the vehicle safely. Remove the front tires.
3. Remove the caliper flange bolt.
4. Pivot the caliper assembly up and out of the way. Remove the brake pads and shims.
5. Remove the pad retainers and check the caliper pins for free movement.

To install:

6. Apply a thin coat of M-77 assembly paste part number 08798-9010 to the brake pad sides of the pad shims and the back of the brake pads, wipe excess paste off the shim.
7. Continue the installation in the reverse order of the removal procedure.
8. Check and refill the master cylinder, as necessary.

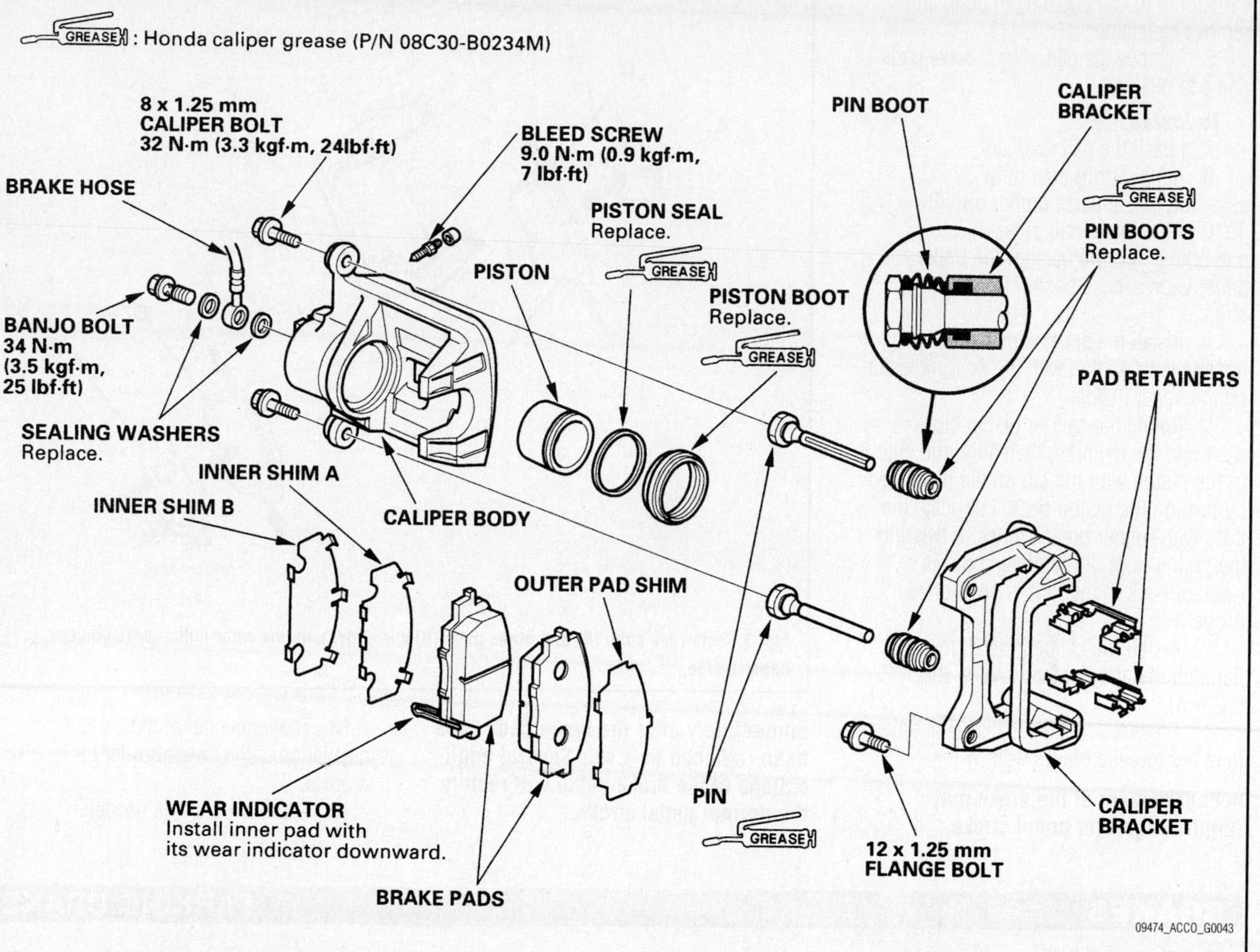
GREASE : Honda caliper grease (P/N 08C30-B0234M)
8 x 1.25 mm CALIPER BOLT 32 N·m (3.3 kgf·m, 24lbf·ft)
BRAKE HOSE
BLEED SCREW 9.0 N·m (0.9 kgf·m, 7 lbf·ft)
PISTON SEAL Replace.
GREASE
PISTON
PISTON BOOT Replace.
GREASE
PIN BOOT
CALIPER BRACKET
GREASE
PIN BOOTS Replace.
PAD RETAINERS
BANJO BOLT 34 N·m (3.5 kgf·m, 25 lbf·ft)
SEALING WASHERS Replace.
INNER SHIM A
INNER SHIM B
CALIPER BODY
OUTER PAD SHIM
WEAR INDICATOR Install inner pad with its wear indicator downward.
BRAKE PADS
PIN
GREASE
12 x 1.25 mm FLANGE BOLT
CALIPER BRACKET
09474_ACCO_G0043

Fig. 2 Exploded view of the front disc brakes

BRAKES

REAR DISC BRAKES

BRAKE CALIPER

REMOVAL & INSTALLATION

1. Remove and discard brake fluid from the master cylinder, as necessary.
2. Raise and support the vehicle safely. Remove the rear tires. Release the parking brake.
3. Disconnect and plug the brake line hose.
4. Remove the caliper bolts. Remove the caliper from the caliper bracket.
5. Remove the pad shims and brake pads.
6. Remove the pad retainers and check the caliper pins for free movement.

To install:

7. Apply a thin coat of M-77 assembly paste part number 08798-9010 to the brake pad sides of the pad shims and the back of the brake pads, wipe excess paste off the shim.
8. Install the brake pads and shims.
9. Rotate the caliper piston clockwise into the cylinder, then align the cutout in the piston with the tab on the inner pad by turning the piston back so the caliper can be installed on the brake pad. Lubricate the boot with rubber grease to avoid twisting the piston boot.
10. Continue the installation in the reverse order of the removal procedure.
11. After installation depress the brake pedal several times to be sure the brake works.

➡Engagement of the brake may require a greater pedal stroke effort immediately after the brake pads have been replaced as a set.

12. Check the parking brake adjustment.
13. Check and refill the master cylinder, as necessary.

DISC BRAKE PADS

REMOVAL & INSTALLATION

See Figure 3.

1. Remove some brake fluid from the master cylinder.
2. Raise and safely support the vehicle.
3. Remove the rear wheel.
4. Remove the brake hose mounting bolt.

5. Remove the flange bolts while holding respective caliper pin with a wrench. Be careful not to damage the pin boot, and remove the caliper.

6. Remove the pad shims, brake pads and pad retainers.

To install:

7. Install the pad retainers.

8. Apply a thin coat of M-77 assembly paste part number o8798-9010 to the brake pad sides of the pad shims and the back of the brake pads, wipe excess paste off the shim.

9. Install the brake pads and shims. Install the brake pad with the wear indicator on the upper inside.

10. Rotate the caliper piston clockwise into the cylinder, then align the cutout in the piston with the tab on the inner pad by turning the piston back. Lubricate the boot with rubber grease to avoid twisting the piston boot. If the piston boot is twisted, back it out so it is positioned properly.

11. Install the caliper. Install the flange bolts and tighten to 16 ft. lbs. (22 Nm).

12. Press the brake pedal several times to make sure the brakes work.

➡Engagement of the brake may require a greater pedal stroke immediately after the brake pads have been replaced as a set. Several applications of the brake pedal will restore the normal pedal stroke.

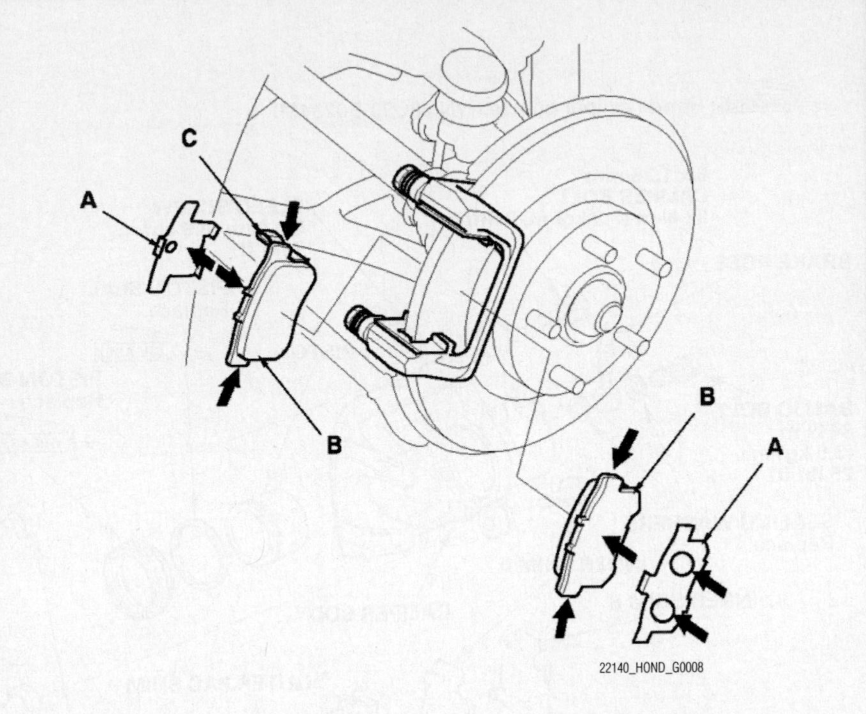
C
A
B
B
A
22140_HOND_G0008

Fig. 3 Install the shim (A) and brake pads (B) correctly, with the wear indicator (C) on the upper inside

13. The remainder of the installation is the reverse order of removal.

14. Add brake fluid as needed.

BRAKES — PARKING BRAKE

PARKING BRAKE CABLES

ADJUSTMENT

1. Raise and safely support the vehicle.
2. Remove the center console.
3. Release the parking brake lever fully.
4. Loosen the parking brake adjusting nut and remove the rear wheels.
5. Make sure the parking brake arm on the rear brake caliper contacts the brake caliper pin.

➡The parking brake arm will only contact the brake caliper pin when the parking brake adjusting nut is loosened.

6. Pull the parking brake lever up one click.
7. Tighten the adjusting nut until the parking brakes drag slightly when the rear wheels are turned.
8. Release the parking brake lever fully, and check that the parking brakes do not drag when the rear wheels are turned. Readjust if necessary.
9. Pull the parking brake all the way up, and make sure the parking brakes are fully applied.
10. Reinstall the center console.

CHASSIS ELECTRICAL

AIR BAG (SUPPLEMENTAL RESTRAINT SYSTEM)

GENERAL INFORMATION

CAUTION

These vehicles are equipped with an air bag system. The system must be disarmed before performing service on, or around, system components, the steering column, instrument panel components, wiring and sensors. Failure to follow the safety precautions and the disarming procedure could result in accidental air bag deployment, possible injury and unnecessary system repairs.

SERVICE PRECAUTIONS

CAUTION

Disconnect and isolate the battery negative cable before beginning any airbag system component diagnosis, testing, removal, or installation procedures. Allow system capacitor to discharge for two minutes before beginning any component service. This will disable the airbag system. Failure to disable the airbag system may result in accidental airbag deployment, personal injury, or death.

DISARMING THE SYSTEM

Disconnect and isolate the negative battery cable. Wait 3 minutes for the system capacitor to discharge before performing any service.

ARMING THE SYSTEM

Connect the negative battery cable.

CLOCKSPRING CENTERING

See Figure 4.

1. Rotate the cable reel clockwise until it stops.

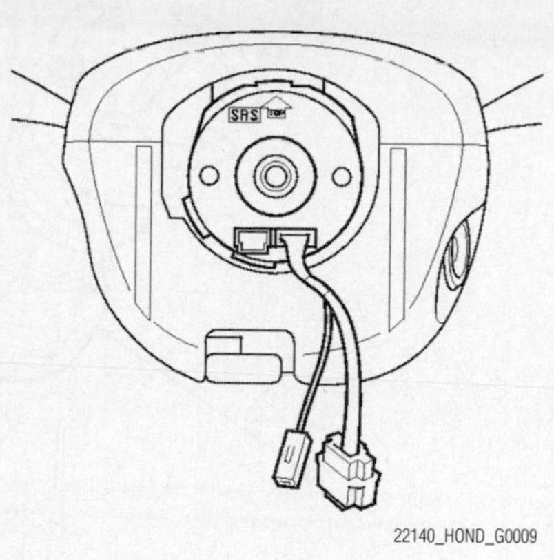
SRS
22140_HOND_G0009

Fig. 4 Ensure the arrow mark is facing straight up to center the clockspring

2. Then rotate it counterclockwise (about 2 ½–3 turns) until the arrow mark on the cable reel label points straight up.

DRIVETRAIN

CLUTCH

REMOVAL & INSTALLATION

1. Remove the transaxle assembly from the vehicle.
2. Install the ring gear holder, clutch alignment shaft and remover handle.
3. To prevent warping, unscrew the pressure plate mounting bolts in a crisscross pattern in several steps, then remove the pressure plate.
4. Remove the clutch disc, clutch alignment shaft, and remover handle.

To install:

5. Temporarily install the clutch disc onto the splines of the transmission mainshaft. Make sure the clutch disc slides freely on the mainshaft.
6. Install the ring gear holder.
7. Apply a light coat of super high temp urea grease (P/N 08798-9002) to the crankshaft pilot bushing.
8. Apply super high temp urea grease (P/N 08798-9002) to the splines of the clutch disc, then install the clutch disc using the clutch alignment shaft and remover handle.
9. Install the pressure plate and the mounting bolts finger-tight.
10. Torque the mounting bolts to 19 ft. lbs. (25 Nm) in a crisscross pattern. Tighten the bolts in several steps to prevent warping the diaphragm spring.
11. Remove the ring gear holder, clutch alignment shaft, and remover handle.

BLEEDING

1. Make sure the DOT 3 brake fluid level in the clutch reservoir is at the MAX (upper) level line

Attach one end of a clear tube to the bleeder screw, and put the other end to the container. Loosen the bleeder screw to allow air to escape from the system.

2. Make sure there is an adequate supply of fluid in the reservoir, then slowly push the clutch pedal all the way down. Before releasing the pedal, have an assistant temporarily tighten the bleeder screw. Loosen the bleeder screw, and push the clutch pedal down again. Repeat this step until no more bubbles appear at the clear tube.

WARNING

Make sure the fluid level on the reservoir does not go below MIN.

3. Tighten the bleeder screw securely.
4. Refill the brake fluid in the reservoir to the MAX (upper) level line.

FRONT DRIVESHAFT

REMOVAL & INSTALLATION

See Figures 5 and 6.

1. Raise and safely support the vehicle.
2. Remove the driveshaft protector.
3. Matchmark the driveshaft to the transmission companion flange.
4. Separate the driveshaft from the transmission.
5. Matchmark the driveshaft to the rear differential companion flange.
6. Separate the driveshaft form the rear differential, then remove the driveshaft.

To install:

7. Install the driveshaft onto the rear differential by aligning the matchmarks. Tighten the bolts to 36 ft. lbs. (49 Nm).

➡If the driveshaft is replaced, align the white marks on the new driveshaft with the white mark on the differential.

8. Install the driveshaft onto the transmission by aligning the matchmarks. Tighten the bolts to 36 ft. lbs. (49 Nm).

➡If the driveshaft is replaced, align the white marks on the new driveshaft with the white mark on the transmission.

9. Install the driveshaft protector, and tighten the bolts to 16 ft. lbs. (22 Nm).

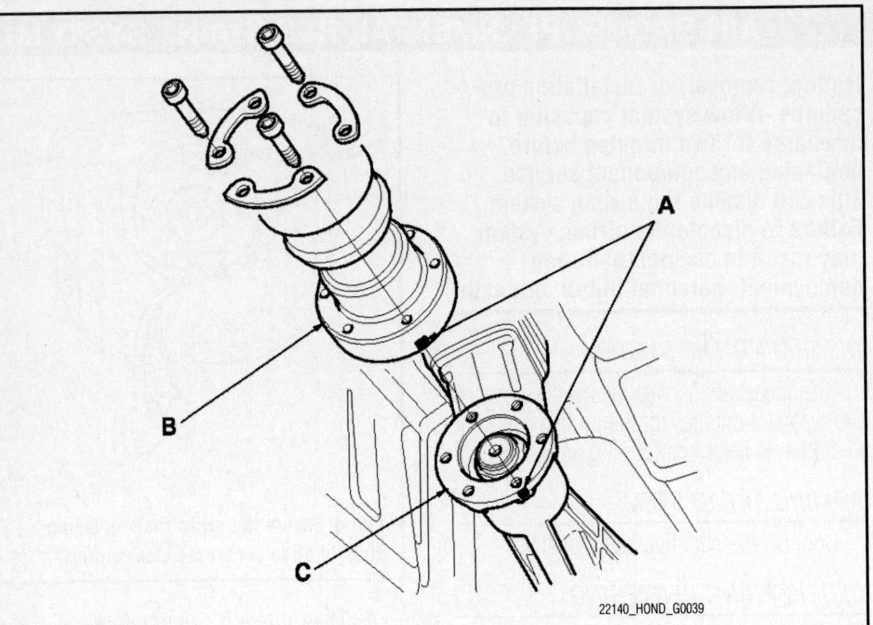
A
B
C
22140_HOND_G0039

Fig. 5 Make reference marks (A) across the propeller shaft (B) and the transmission companion flange (C)

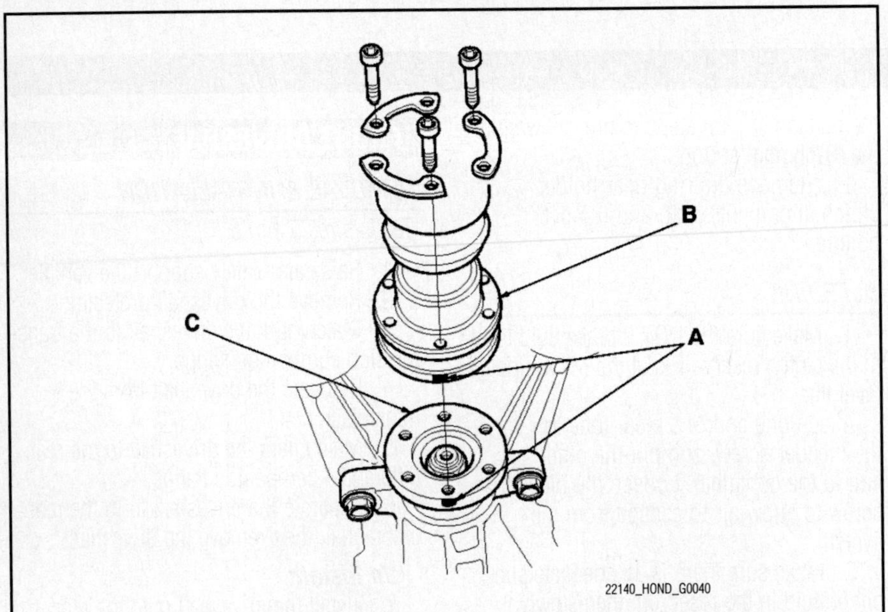
B
C
A
22140_HOND_G0040

Fig. 6 Make reference marks (A) across the propeller shaft (B) and the rear differential companion flange (C)

REAR HALFSHAFT

REMOVAL & INSTALLATION

1. Raise and safely support the vehicle.
2. Remove the rear tires.
3. Lift up the locking tab on the spindle nut. Remove the nut.
4. Remove the cotter pin from the lower arm ball joint castle nut. Remove the nut. Separate the ball joint from the lower arm.
5. Remove the wheel sensor harness from the upper arm.
6. Make reference marks across the inboard joint and the rear differential.
7. Remove the six inboard joint mounting bolts and nuts. Remove the inboard joint from the rear differential.
8. Pull the knuckle outward and remove the outboard joint from the wheel hub using a plastic hammer.
9. Remove the halfshaft.

To install:

10. Installation is the reverse of the removal procedure.
11. After the front tire is installed, turn the front wheel by hand and make sure there is no interference between the halfshaft and surrounding parts.
12. Check the rear wheel alignment. Adjust if necessary.

REAR PINION SEAL

REMOVAL & INSTALLATION

1. Raise and safely support the vehicle.
2. Remove the rear differential.
3. Mount the rear differential in a bench vise.
4. Remove the output shafts from the differential using pry bars.
5. Remove the ten mounting bolts in a crisscross pattern in several steps, then remove the differential case.
6. Make marks on the bearing cap, the adjustment screw, and the differential carrier.
7. Remove the lock plates and the bearing caps.
8. Remove the adjustment screws, the bearing outer races, and the Torsen LSD assembly.
9. Install the holder handle and the companion flange holder on the companion flange, then remove the locknut and the pinion washer.
10. Remove the oil seal from the differential carrier.

To install:

11. Using an oil seal driver tool, install the oil seal into differential carrier.
12. Apply molybdenum grease to the surface end of the companion flange, then install the companion flange, the drive pinion washer, and a new locknut.
13. Using a holder and companion flange holder, tighten the locknut to 14 ft. lbs. (20 Nm).
14. Rotate the drive pinion several times to ensure proper tapered roller bearing contact. Measure the drive pinion turning torque before tightening the locknut to the specified torque.
15. Tighten the locknut to 94 ft. lbs. (127 Nm), then remove the holder handle and the companion flange holder.
16. Rotate the drive pinion several times to assure proper tapered roller bearing contact. Measure the drive pinion turning torque. If the drive pinion turning torque exceeds the standard, replace the pinion spacer.
17. The remainder of the installation is the reverse order of removal.

ENGINE COOLING

THERMOSTAT

REMOVAL & INSTALLATION

See Figure 7.

1. Drain the engine coolant.
2. Remove the lower radiator hose and the thermostat cover, then remove the thermostat.

To install:

3. Install the thermostat with a new rubber seal.
4. Install the thermostat cover and the lower radiator hose.
5. Refill the radiator with engine coolant, then bleed air from the cooling system.

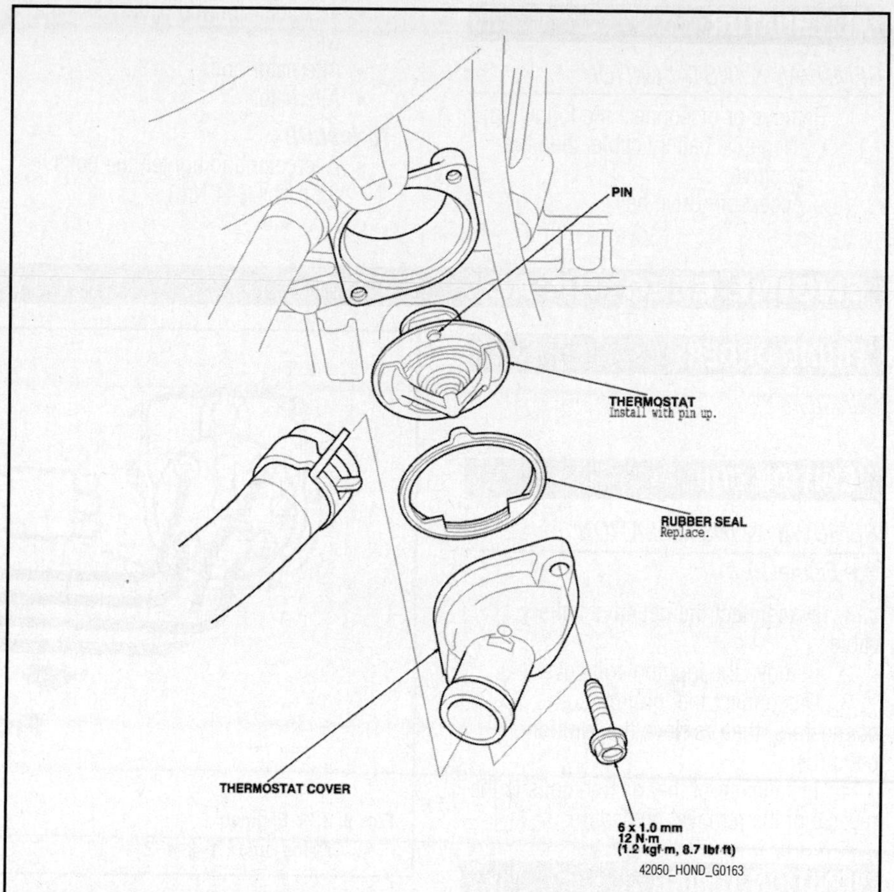
PIN
THERMOSTAT
Install with pin up.
RUBBER SEAL
Replace.
THERMOSTAT COVER
6 x 1.0 mm
12 N·m
(1.2 kgf·m, 8.7 lbf·ft)
42050_HOND_G0163

Fig. 7 Exploded view of the thermostat

WATER PUMP

REMOVAL & INSTALLATION

See Figure 8.

1. Before servicing the vehicle, refer to the precautions in the beginning of this section.
2. Drain the cooling system.
3. Remove the drive belt.
4. Remove the water pump pulley.

➡It may first be necessary to remove the power steering pump from its mounting, without disconnecting the fluid hoses.

5. Remove the water pump retaining bolts. Remove the water pump from the engine.

To install:

6. Clean the water pump and O-ring mating surfaces before installation.
7. Install the water pump to the engine using a new O-ring.
8. Continue the installation in the reverse order of the removal procedure.
9. Refill the radiator. Start the engine and check for leaks.

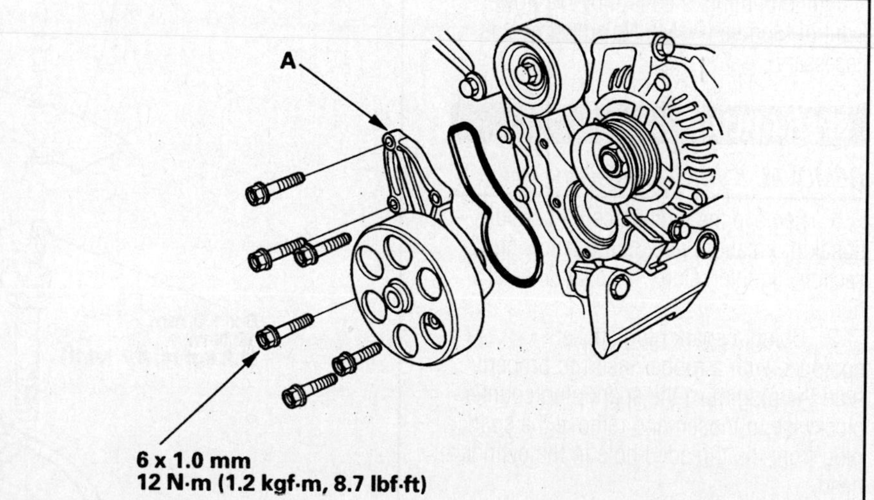
A
6 x 1.0 mm
12 N·m (1.2 kgf·m, 8.7 lbf·ft)
43256-ACCO-G02

Fig. 8 Water pump mounting

ENGINE ELECTRICAL — CHARGING SYSTEM

ALTERNATOR

REMOVAL & INSTALLATION

1. Remove or disconnect the following:
   - Negative battery cable, then the positive
   - Accessory drive belt
   - 4P connector and battery terminal wire
   - Alternator bolts
   - Alternator

To install:

- Alternator and tighten the bolts to 33 ft. lbs. (44 Nm)
- 4P connector and battery terminal wire. Tighten the battery terminal wire nut to 108 inch lbs. (12 Nm).
- Accessory drive belt
- Negative battery cable, then the positive

ENGINE ELECTRICAL — IGNITION SYSTEM

FIRING ORDER

See Figure 9.

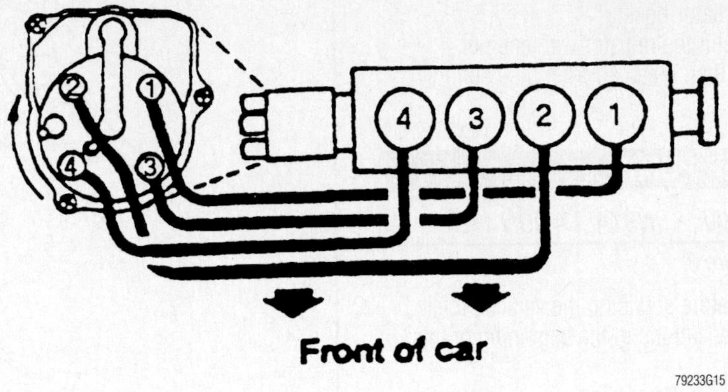

4
3
2
1
Front of car
79233G15

Fig. 9 2.2L Engines
Firing order: 1–3–4–2

IGNITION COIL

REMOVAL & INSTALLATION

See Figure 10.

1. Disconnect the negative battery cable.
2. Remove the ignition coil cover (A).
3. Disconnect the ignition coil connectors, then remove the ignition coils (B).
4. Installation of the ignition coils is the reverse of the removal procedure.

IGNITION TIMING

ADJUSTMENT

Ignition timing is control by the Power Control Module (PCM). No adjustment is necessary.

SPARK PLUGS

REMOVAL & INSTALLATION

1. Remove the ignition coil. For additional information, refer to the following section, "Ignition Coil, Removal & Installation."
2. Using a spark plug socket equipped with a rubber insert to properly hold the plug, turn the spark plug counterclockwise to loosen and remove the spark plug from the threaded hole in the cylinder head.

To install:

3. Apply a light coating of an anti-seize compound to the spark plug threads.
4. Carefully thread the plug into the threaded spark plug hole by hand. If resistance is felt before the plug is almost completely threaded, back the plug out and begin threading again.

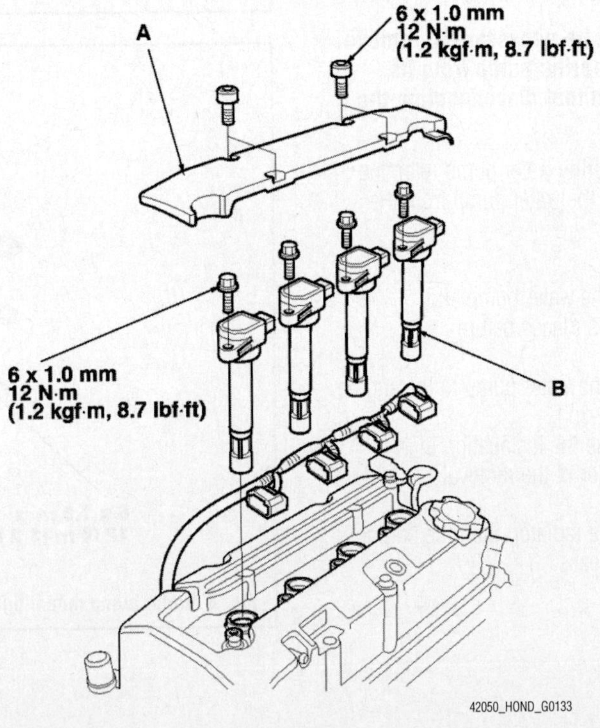

A
6 x 1.0 mm
12 N·m
(1.2 kgf·m, 8.7 lbf·ft)
6 x 1.0 mm
12 N·m
(1.2 kgf·m, 8.7 lbf·ft)
B
42050_HOND_G0133

Fig. 10 Remove the ignition coil cover (A), detach the connectors, then remove the ignition coils (B)

In tight, hard to reach areas, a small piece of rubber hose pressed onto the spark plug can be used as a threading tool. The rubber hose will hold the plug and while twisting the end of the hose, and the hose will be flexible enough to twist before allowing the plug to cross thread.

5. Carefully tighten the spark plug to 13 ft. lbs. (18 Nm).
6. The remainder of the installation is the reverse order of removal.

ENGINE ELECTRICAL

STARTING SYSTEM

STARTER

REMOVAL & INSTALLATION

1. Disconnect the negative battery cable. Disconnect the positive battery cable.
2. Disconnect the intake air temperature sensor connector and breather pipe. Remove the manifold absolute sensor harness from the holder. Remove the air cleaner housing cover and the air cleaner assembly. Remove the IAT sensor harness clamps and the intake air housing.
3. Remove the drive belt. Remove the alternator.
4. Disconnect the starter cable from the B terminal on the solenoid. Disconnect the BLK/WHT wire from the S terminal.
5. Remove the starter retaining bolts. Remove the starter from the vehicle.
6. Installation is the reverse order of removal.

ENGINE MECHANICAL

➡Disconnecting the negative battery cable may interfere with the functions of the on board computer systems and may require the computer to undergo a relearning process, once the negative battery cable is reconnected.

ACCESSORY DRIVE BELTS

ACCESSORY BELT ROUTING

See Figure 11.

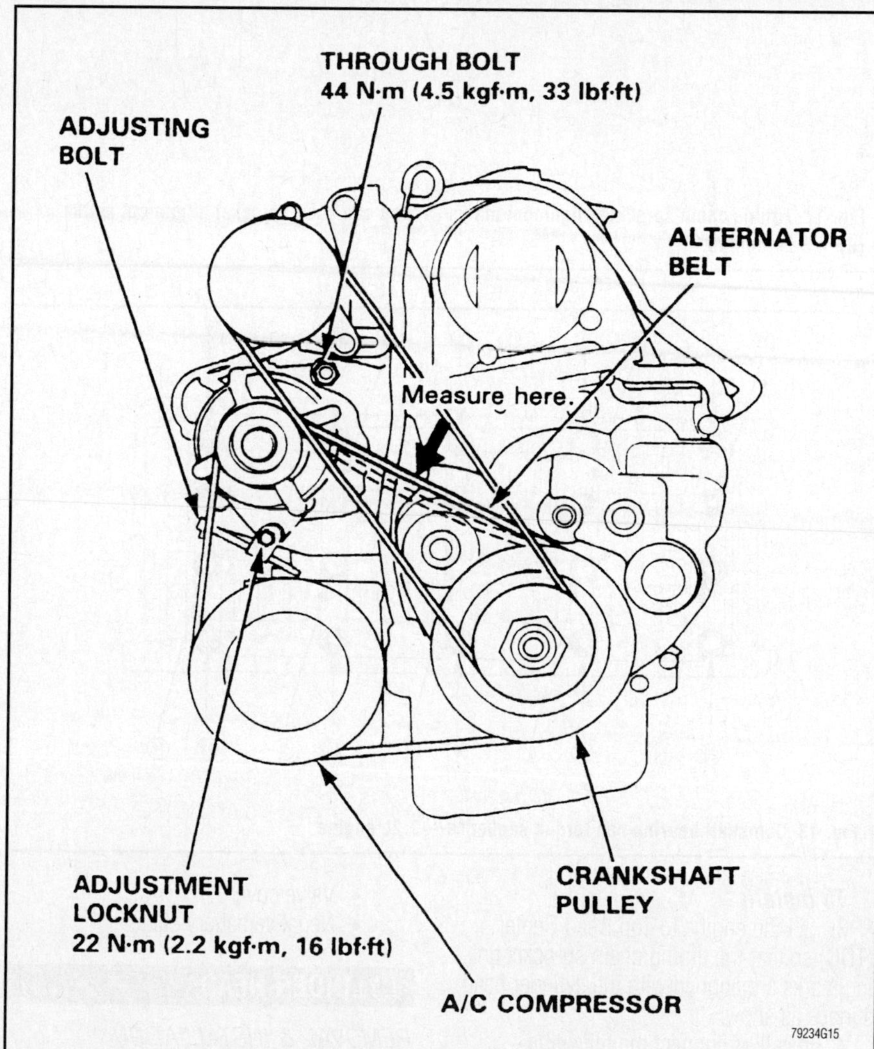
THROUGH BOLT
44 N·m (4.5 kgf·m, 33 lbf·ft)
ADJUSTING
BOLT
ALTERNATOR
BELT
Measure here.
ADJUSTMENT
LOCKNUT
22 N·m (2.2 kgf·m, 16 lbf·ft)
CRANKSHAFT
PULLEY
A/C COMPRESSOR
79234G15

Fig. 11 Accessory drive belt routing—2.2L engine

ADJUSTMENT

Belt tension is maintained by an automatic tensioner. No adjustment is necessary or possible.

REMOVAL & INSTALLATION

➡Refer to the Accessory Belt Routing Illustrations located earlier in this section for routing diagrams.

1. Place a long-handled, boxed-end wrench or a belt tension release tool on the drive belt auto-tensioner from above the engine. Slowly turn the wrench in the direction shown to release the tension, then remove the drive belt.

WARNING

This is a hydraulic type auto-tensioner; you must turn the wrench slowly.

2. Install the new belt in the reverse order of removal.

CAMSHAFT AND VALVE LIFTERS

REMOVAL & INSTALLATION

See Figures 12 and 13.

1. Loosen the valve adjustment screws so that all valves are closed and all rocker arms are loose.
2. Remove or disconnect the following:

- Negative battery cable
- Valve cover
- Camshaft bearing caps
- Camshafts

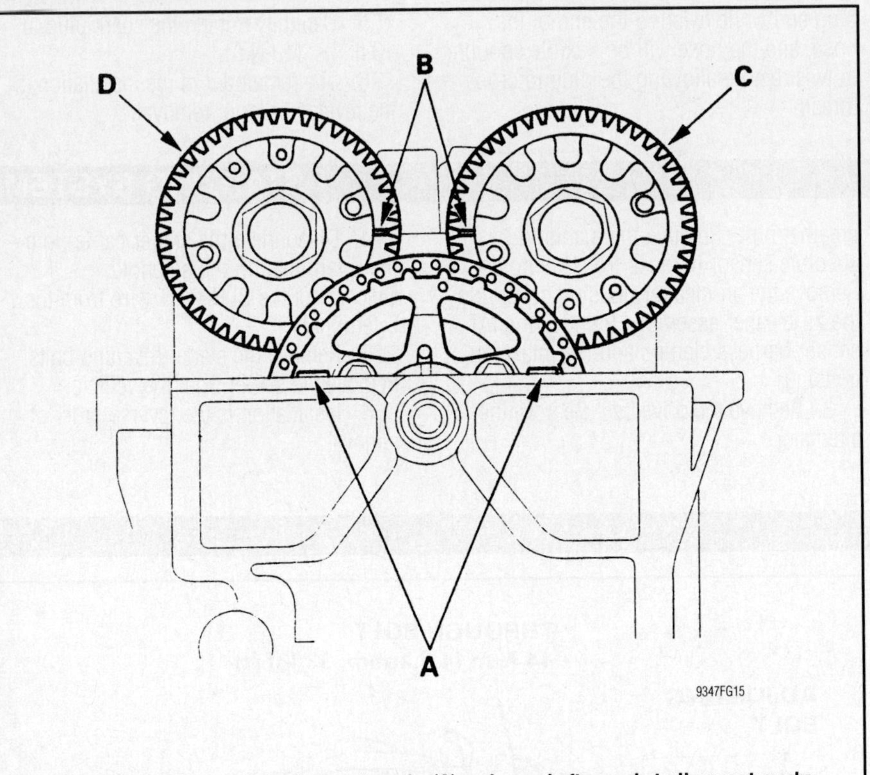
B
D
C
A
9347FG15

Fig. 12 Timing chain sprocket alignment marks (A) and camshaft sprocket alignment marks (B)— 2.2L engine

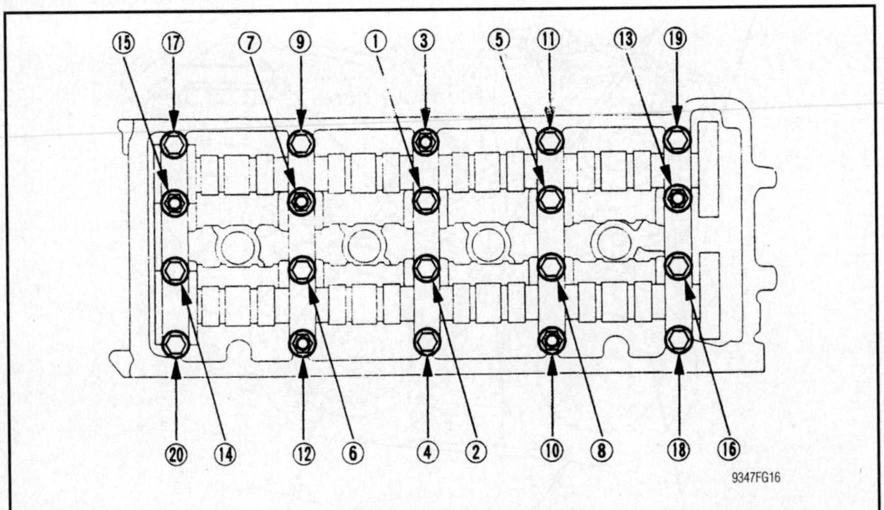
9347FG16

Fig. 13 Camshaft bearing cap torque sequence— 2.2L engine

To install:

3. Set the engine to Top Dead Center (TDC) so that the timing chain sprocket timing marks are aligned with the cylinder head surface as shown.
4. Install or connect the following:
   - Camshafts with the sprocket timing marks aligned as shown
   - Camshaft bearing caps and tighten the bolts in sequence to 16 ft. lbs. (22 Nm). Adjust the valve clearance.
   - Valve cover
   - Negative battery cable

CYLINDER HEAD

REMOVAL & INSTALLATION

See Figures 14 and 15.

➡Be sure the cylinder head is cool to the touch before beginning the removal procedure. The coolant temperature must be below 100°F (38°C).

1. Relieve the fuel system pressure. Disconnect the negative battery cable. Drain the engine coolant. Drain the engine oil.
2. Remove the air cleaner assembly. Remove the intake air cleaner housing. Remove the drive belt.
3. Remove the intake manifold cover, power brake booster hose and the quick-connect fitting cover, then disconnect the fuel feed hose.
4. Disconnect the evaporative emission canister hose. Remove the intake manifold bracket retaining bolt. Remove the water outlet cover.
5. Remove the engine wire harness connectors and the wire harness clamps from the cylinder head and intake manifold. Remove the four fuel injector connectors
6. Disconnect the engine coolant temperature sensor connector, throttle body connector, and the air/fuel sensor connector.
7. Disconnect the manifold absolute pressure connector, rocker arm oil control solenoid connector, rocker arm oil pressure switch connector and the crankshaft position sensor connector.
8. Remove the water bypass hose, EVAP hose and bracket and the intake manifold bracket.
9. Remove the four bolts retaining the exhaust manifold cover. Remove the heat shield retaining bolts. Remove the heat shield.
10. Remove the exhaust manifold cover. Remove the exhaust manifold retaining bolts and remove the exhaust manifold.
11. Remove the intake manifold bracket clips. Remove the intake manifold retaining bolts. Remove the intake manifold.
12. Remove the cylinder head cover retaining bolts. Remove the cylinder head cover.
13. Position the No. 1 piston at TDC. The TDC marks on the cam chain sprocket should align with the cylinder head surface.
14. Remove the end cover and nozzle from the cam chain auto tensioner. Thread a nut onto a 5x0.8 mm bolt at is at least 40 mm long. Thread the bolt into the maintenance hole in the cam chain auto tensioner.
15. Turn the bolt clockwise to compress the cam chain auto tensioner and lock it in place with the nut. Remove the cam chain auto tensioner.
16. Loosen the rocker arm adjusting screws. Remove the camshaft holders and camshafts.
17. Insert the bolts into the rocker shaft holder and remove the rocker arm assembly.
18. Remove the idler gear/cam chain sprocket assembly, idler gear collar and washer.

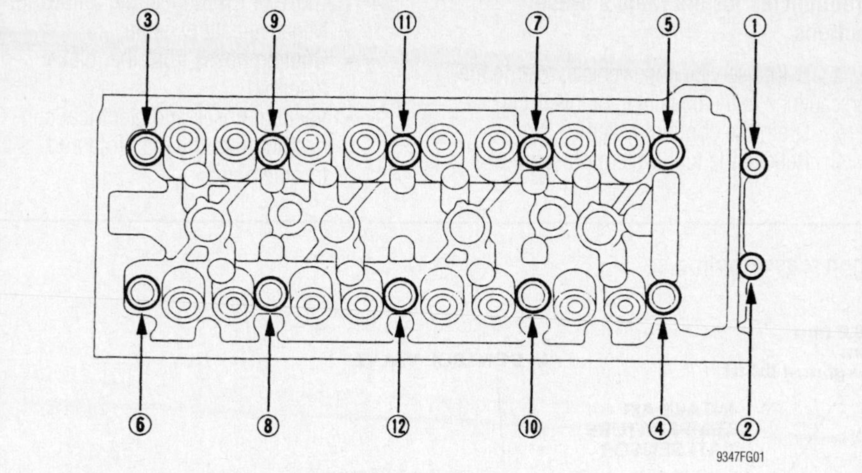
3
9
11
7
5
1
6
8
12
10
4
2
9347FG01

Fig. 14 Cylinder head loosening sequence— 2.2L engines

19. Remove the cylinder head retaining bolts. To prevent warpage, loosen the cylinder head bolts in a 3-step crisscross pattern in the reverse order of the tightening sequence.

20. Remove the cylinder head from the engine.

To install:

➡Use new O-ring, seals, and gaskets when installing the cylinder head and its components.

21. Be sure the cylinder head and the engine block surfaces are clean, level, and straight.

22. Be sure the cylinder head dowel pins and control orifice are aligned. Clean the oil control orifice and reinstall it with a new O-ring.

23. Apply liquid gasket, part number 08717-0004, 08718-0001, 08718-0003 or 08718-0009 to the cylinder head mating surface of the block and chain case within 5mm of the edge of the cylinder head gasket.

➡Do not install the parts if more than five minutes have elapsed since applying the liquid gasket. Instead, reapply the liquid gasket after removing the old residue.

24. Position the cylinder head on the engine.

25. Coat the threads of the cylinder head retaining bolts with clean engine oil. Tighten the cylinder head bolts sequentially to the proper torque.

26. Continue the installation in the reverse order of the removal procedure.

27. Reprogram the crankshaft position (CKP) pattern. Run the engine until the operating temperature reaches 176 degree. With the engine stopped clear the CKP pattern. Turn the ignition switch OFF. Turn the ignition switch ON and wait thirty seconds.

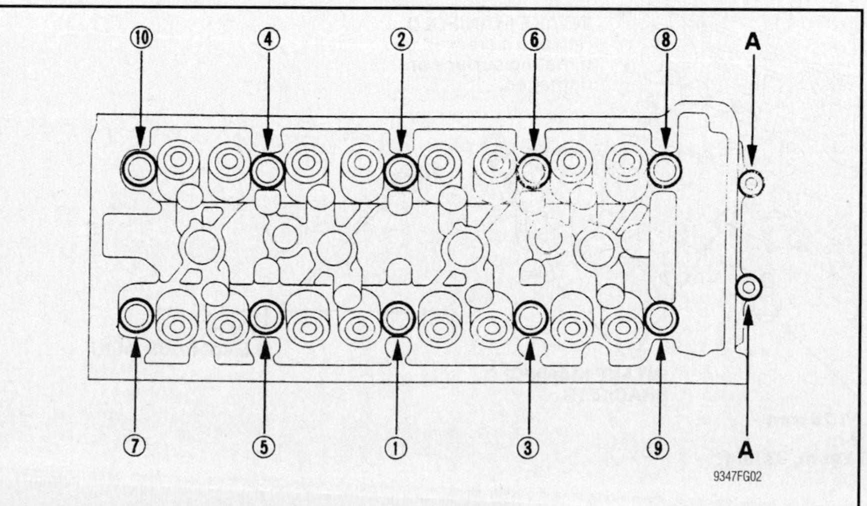
10
4
2
6
8
A
7
5
1
3
9
A
9347FG02

Fig. 15 Cylinder head torque sequence— 2.2L engines

28. Road test the vehicle on a level surface. Decelerate the engine speed of 2500 RPM to 1000 RPM. If equipped with automatic transaxle use two Drive positions. If equipped with manual transaxle use first gear.

29. Stop the vehicle, but keep the engine running.

30. Check PULSAR F/B LEARN in the data list with the HDS. If not complete repeat the procedure. If complete, road test the vehicle on a level surface. Decelerate the engine speed of 5000 RPM to 3000 RPM. If equipped with automatic transaxle use two Drive positions. If equipped with manual transaxle use first gear.

31. Stop the vehicle, but keep the engine running.

32. Check PULSAR F/B LEARN in the data list with the HDS. If not complete repeat the procedure.

33. If completed, turn the ignition switch OFF. Turn the ignition switch ON, wait thirty seconds. The learning procedure is now complete.

34. Enter the antitheft codes for the radio and the navigation system. Set the clock.

EXHAUST MANIFOLD

REMOVAL & INSTALLATION

1. Remove or disconnect the following:
   - Negative battery cable
2. Safely raise and support the vehicle.
3. Remove or disconnect the following:
   - Oxygen Sensor (O_2S) connector, if it is located in the manifold.
   - Exhaust manifold upper cover
   - Heat insulator from the manifold, if equipped with air conditioning.
   - Nuts attaching the exhaust manifold to the front exhaust pipe.
   - Pipe from the manifold and discard the gasket. Support the pipe with wire; do not allow it to hang by itself.
   - Exhaust manifold bracket(s) bolts and bracket(s).
   - Exhaust manifold attaching nuts, using a crisscross pattern (starting from the center).
   - Manifold and discard the gasket. Clean the manifold and cylinder head mating surfaces.
   - Lower manifold cover from the manifold, if equipped.
4. Installation is the reverse order of removal. Tighten the nuts to 23 ft. lbs. (31 Nm).

INTAKE MANIFOLD

REMOVAL & INSTALLATION

See Figure 16.

➡Make sure to acquire the anti-theft code from the radio and write down the frequencies for the radio's preset buttons.

1. Before servicing the vehicle, refer to the precautions in the beginning of this section.
2. Drain the cooling system.
3. Relieve the fuel system pressure.
4. Remove or disconnect the following:
   - Negative battery cable
   - Cooling hoses from the intake manifold
   - Vacuum hoses and electrical connectors from the manifold and throttle body

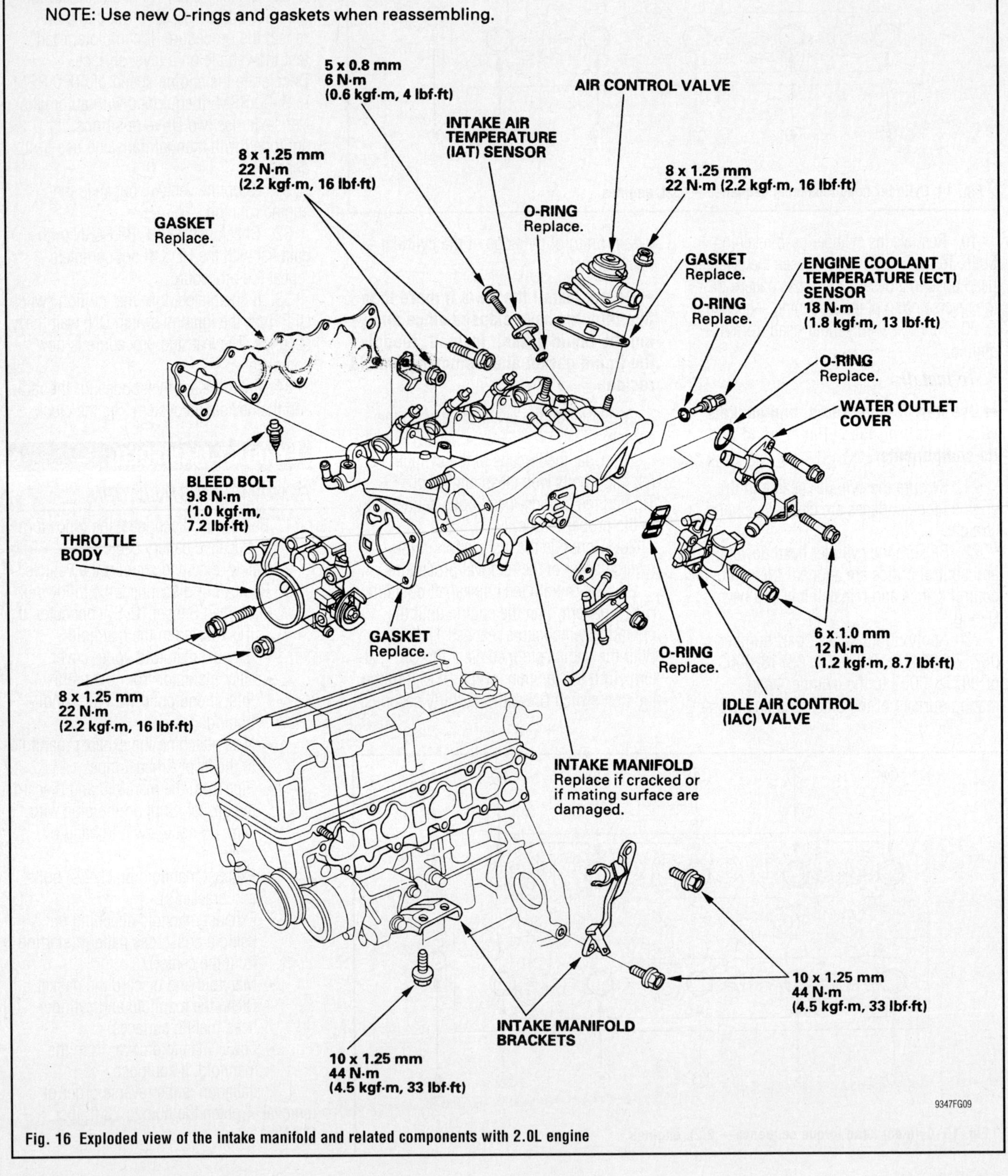
NOTE: Use new O-rings and gaskets when reassembling.
5 x 0.8 mm 6 N·m (0.6 kgf·m, 4 lbf·ft)
AIR CONTROL VALVE
INTAKE AIR TEMPERATURE (IAT) SENSOR
8 x 1.25 mm 22 N·m (2.2 kgf·m, 16 lbf·ft)
8 x 1.25 mm 22 N·m (2.2 kgf·m, 16 lbf·ft)
O-RING Replace.
GASKET Replace.
GASKET Replace.
ENGINE COOLANT TEMPERATURE (ECT) SENSOR 18 N·m (1.8 kgf·m, 13 lbf·ft)
O-RING Replace.
O-RING Replace.
WATER OUTLET COVER
BLEED BOLT 9.8 N·m (1.0 kgf·m, 7.2 lbf·ft)
THROTTLE BODY
GASKET Replace.
O-RING Replace.
6 x 1.0 mm 12 N·m (1.2 kgf·m, 8.7 lbf·ft)
IDLE AIR CONTROL (IAC) VALVE
8 x 1.25 mm 22 N·m (2.2 kgf·m, 16 lbf·ft)
INTAKE MANIFOLD Replace if cracked or if mating surface are damaged.
10 x 1.25 mm 44 N·m (4.5 kgf·m, 33 lbf·ft)
INTAKE MANIFOLD BRACKETS
10 x 1.25 mm 44 N·m (4.5 kgf·m, 33 lbf·ft)
9347FG09

Fig. 16 Exploded view of the intake manifold and related components with 2.0L engine

- Throttle cable from the throttle body
- Fuel rail and fuel injectors
- Intake manifold support brackets
- Intake manifold

To install:

➡Use new gaskets when installing the intake manifold. Use new O-rings when installing manifold sensors and components. Use new sealing washers when reconnecting the fuel lines.

5. Clean all gasket mating surfaces.
6. Torque the intake manifold retaining nuts to specification.
7. Continue the installation in the reverse order of the removal procedure.

OIL PAN

REMOVAL & INSTALLATION

See Figure 17.

1. Note the radio security code and the radio presets. Disconnect the negative battery cable.
2. Drain the engine oil. Raise and support the vehicle safely.
3. Remove the oil pan retaining bolts. Drive an oil pan seal cutter tool between the oil pan and engine block. Cut the oil pan seal by striking the side of the tool along the oil pan
4. Remove the oil pan from the vehicle.

To install:

5. Remove any old gasket from the mating surfaces. Be sure these surfaces are clean and dry.
6. Apply liquid gasket, part number 08717-004, 08718-0001, 08718-0003 or 08718-0009 evenly to the engine block mating surface of the oil pan.

➡Do not install the parts if more than four minutes have elapsed since applying the liquid gasket. Instead, reapply after removing the previous coating material.

7. Position the oil pan in place. Tighten the oil pan retaining bolts in two or three steps to specification and in the proper sequence.
8. Continue the installation in the reverse order of the removal procedure.
9. After assembly, wait at least thirty minutes before filling the engine with clean engine oil.
10. Do not run the engine for at least three hours after installing the oil pan.

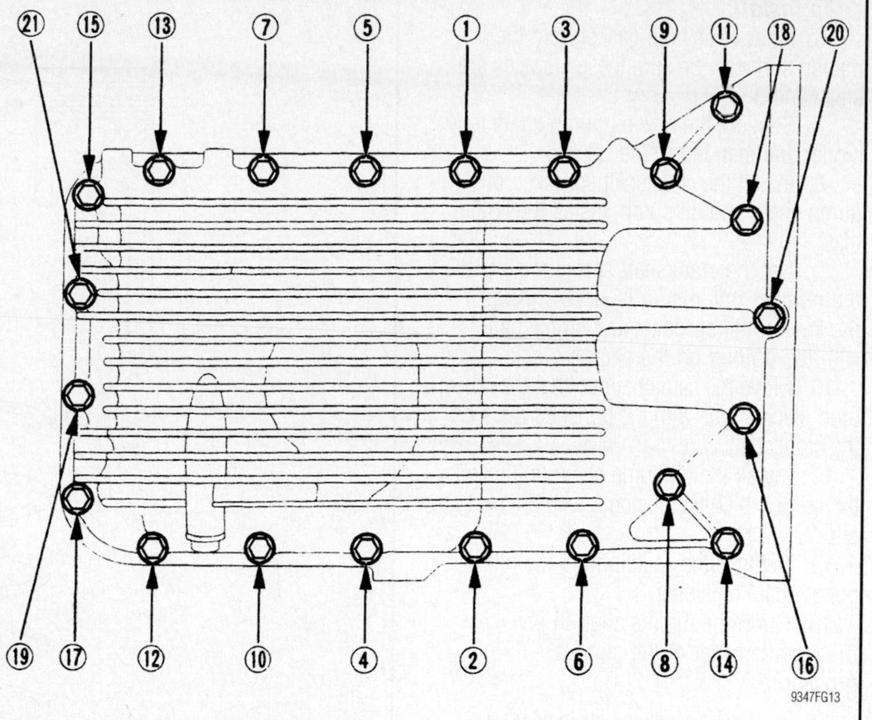
21
15
13
7
5
1
3
9
11
18
20
19
17
12
10
4
2
6
8
14
16
9347FG13

Fig. 17 Oil pan torque sequence—2.2L engines

OIL PUMP

REMOVAL & INSTALLATION

See Figure 18.

1. Note the radio security code and the radio presets. Disconnect the negative battery cable.
2. Remove the timing chain. For additional information, refer to the following section, "Timing Chain, Removal & Installation."
3. Remove the oil pan.
4. Remove the oil pump chain tensioner. Remove the baffle plate.
5. Remove the oil pump retaining bolts. Remove the oil pump, oil pump chain and crankshaft sprocket.

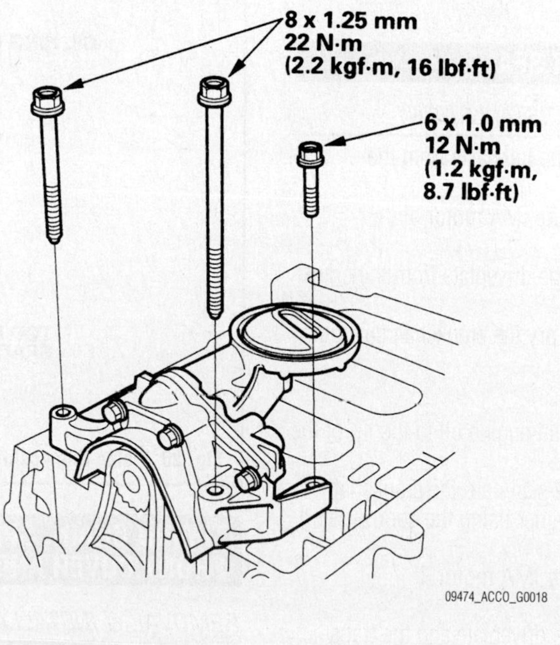
8 x 1.25 mm
22 N·m
(2.2 kgf·m, 16 lbf·ft)
6 x 1.0 mm
12 N·m
(1.2 kgf·m,
8.7 lbf·ft)
09474_ACCO_G0018

Fig. 18 Oil pump mounting—2.2L engines

To install:

6. Remove any old gasket from the mating surfaces. Be sure these surfaces are clean and dry.
7. Squeeze the new oil pump chain tensioner and then install the set clip.
8. Install the crankshaft sprocket, oil pump chain and oil pump. Install the baffle plate.
9. Set the crankshaft sprocket so that the number one piston is at TDC. Align the key on the sprocket and crankshaft with the pointer on the engine block.
10. Move the cam chain so that the colored piece aligns with the punched mark on the crankshaft.
11. Install the oil pump chain guide and the oil pump chain tensioner with the seat clip.
12. Remove the set clip from the oil pump chain tensioner.
13. Continue the installation in the reverse order of the removal procedure.

➡After assembly, wait at least thirty minutes before filling the engine with clean engine oil.

➡Do not run the engine for at least three hours after installing the oil pan.

14. Refill the engine with oil to the correct level.
15. Connect the negative battery cable.

PISTON AND RING

POSITIONING

See Figures 19 and 20.

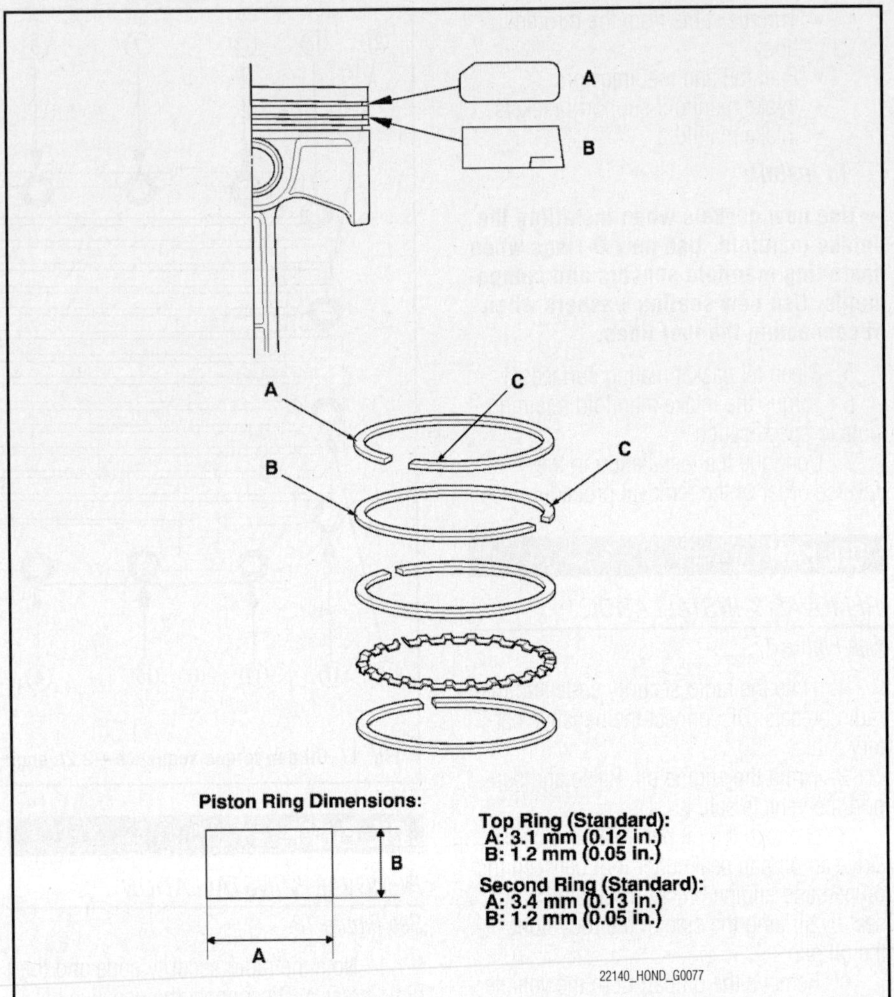
A
B
A
C
B
C
Piston Ring Dimensions:
B
A
Top Ring (Standard):
A: 3.1 mm (0.12 in.)
B: 1.2 mm (0.05 in.)
Second Ring (Standard):
A: 3.4 mm (0.13 in.)
B: 1.2 mm (0.05 in.)
22140_HOND_G0077

Fig. 19 Top ring (A), second ring (B) and the manufacturing marks (C) must face upward

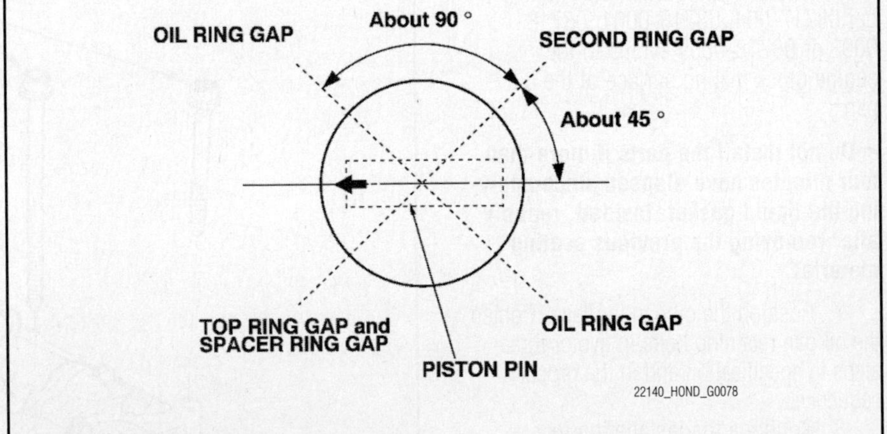
OIL RING GAP
About 90 °
SECOND RING GAP
About 45 °
TOP RING GAP and SPACER RING GAP
OIL RING GAP
PISTON PIN
22140_HOND_G0078

Fig. 20 Piston ring positioning

REAR MAIN SEAL

REMOVAL & INSTALLATION

1. Remove the transaxle from the vehicle.
2. Remove the IMA motor, if equipped.
3. Remove the driveplate from the crankshaft.
4. Carefully pry the crankshaft seal out of the retainer.

To install:

5. Apply clean engine oil to the lip of the new seal.
6. Install the seal onto the crankshaft and into the retainer using the appropriate seal driver.
7. Install the IMA motor, if equipped.
8. Install the driveplate and the transmission.

TIMING CHAIN, SPROCKETS, FRONT COVER AND SEAL

REMOVAL & INSTALLATION

See Figures 21 through 23.

➡Keep the cam chain away from magnetic fields.

1. Before servicing the vehicle, refer to the precautions in the beginning of this section.

2. Note the radio security code and the radio presets. Relieve the fuel system pressure.

3. Disconnect the negative battery cable. Disconnect the positive battery cable.

4. Drain the engine coolant. Drain the engine oil.

5. Loosen the water pump pulley bolts. Remove the drive belt.

6. Remove the cylinder head.

7. Remove the water bypass hose. Remove the water bypass tube retaining bolts and tube. Remove the water pump pulley. Remove the auto tensioner.

8. Remove the alternator. Remove the idler pulley. Remove the idler pulley base.

9. Remove the oil pan.

10. Remove the crankshaft pulley. Remove the chain case retaining bolts. Remove the chain case.

11. Remove the CKP pulse plate. Remove the oil pump chain guide. Remove the cam chain.

To install:

12. Set the crankshaft to TDC. Align the key on the sprocket and the crankshaft with the pointer on the engine block.

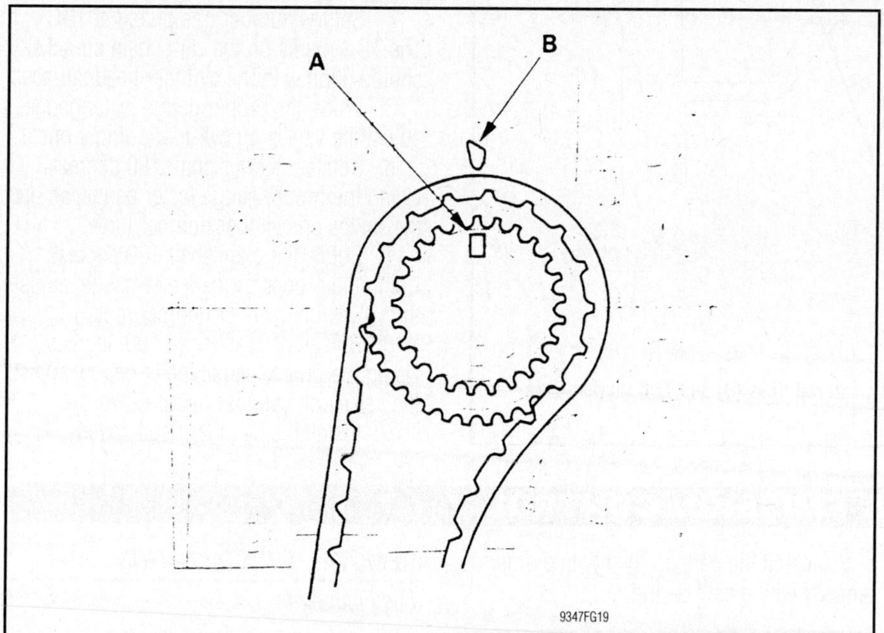
A
B
9347FG19

Fig. 21 Align the sprocket key (A) with the cylinder block pointer (B) to set the engine to TDC

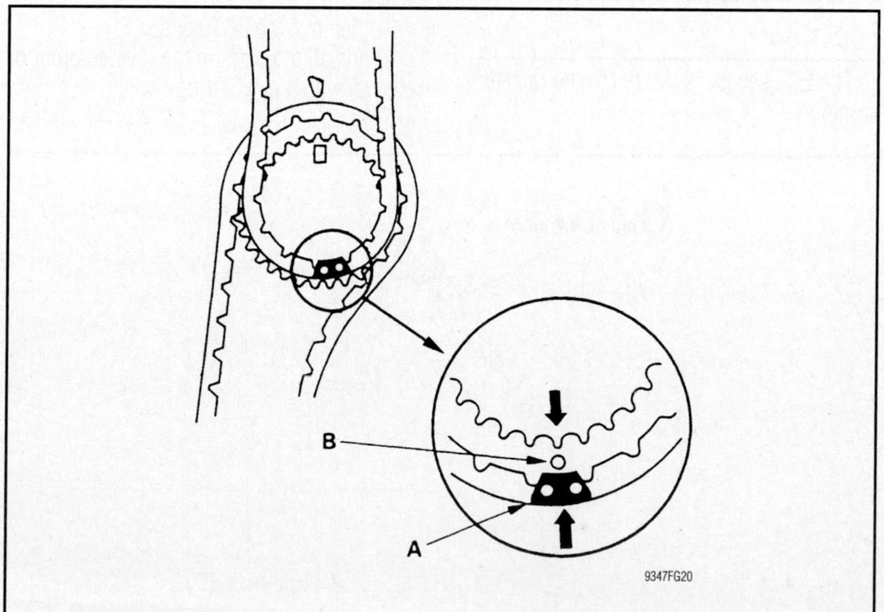
B
A
9347FG20

Fig. 22 Install the timing chain with the colored link (A) aligned with the crankshaft sprocket punch mark (B)

13. Install the cam chain with the colored piece aligned with the punch mark on the crankshaft sprocket.

14. Install the oil pump chain guide. Install the CKP pulse plate.

15. Replace the chain case oil seal.

16. Remove any old gasket from the mating surfaces. Be sure these surfaces are clean and dry.

17. Apply liquid gasket, part number 08717-004, 08718-0001, 08718-0003 or 08718-0009 evenly to the cylinder block mating surface of the chain case.

➡Do not install the parts if more than four minutes have elapsed since applying the liquid gasket. Instead, reapply after removing the previous coating material.

18. Install the dowel pins and the chain case using a new O-ring.

19. Continue the installation in the reverse order of the removal procedure.

20. After assembly, wait at least thirty minutes before filling the engine with clean engine oil.

21. Do not run the engine for at least three hours after installing the oil pan.

22. Reprogram the crankshaft position (CKP) pattern. Run the engine until the operating temperature reaches 176 degree. With the engine stopped clear the CKP pattern. Turn the ignition switch OFF. Turn the ignition switch ON and wait thirty seconds.

23. Road test the vehicle on a level surface. Decelerate the engine speed of 2500 RPM to 1000 RPM. If equipped with automatic transaxle use two Drive positions. If equipped with manual transaxle use first gear.

24. Stop the vehicle, but keep the engine running.

25. Check PULSAR F/B LEARN in the data list with the HDS. If not complete repeat the procedure. If complete, road test the vehicle on a level surface. Decelerate the engine speed of 5000 RPM to 3000 RPM. If equipped with automatic transaxle use two Drive positions. If equipped with manual transaxle use first gear.

26. Stop the vehicle, but keep the engine running.

27. Check PULSAR F/B LEARN in the data list with the HDS. If not complete repeat the procedure.

28. If completed, turn the ignition switch OFF. Turn the ignition switch ON, wait thirty seconds. The learning procedure is now complete.

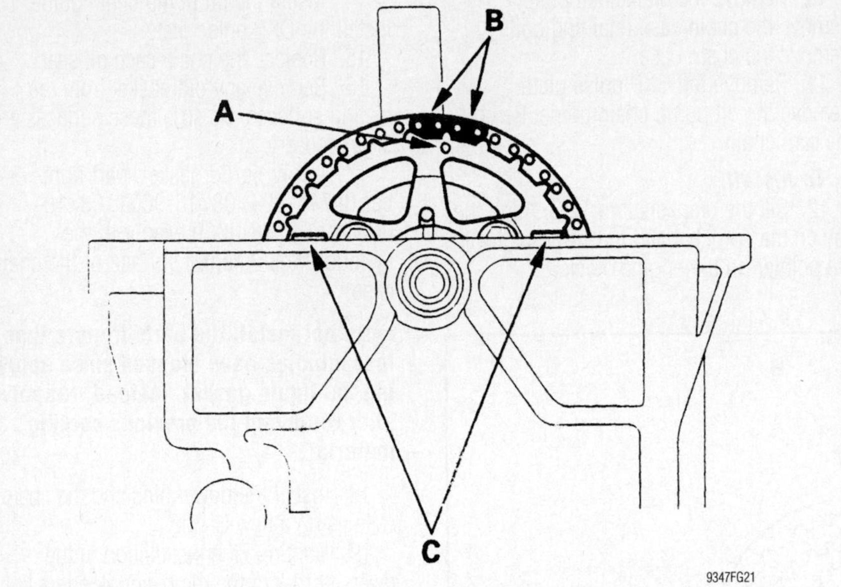
B
A
C
9347FG21

Fig. 23 Timing chain idler sprocket punch mark (A), colored links (B) and TDC marks (C) in proper alignment

VALVE LASH

ADJUSTMENT

➡Adjust valves only when the cylinder head temperature is less than 100 degrees F.

1. Before servicing the vehicle, refer to the precautions in the beginning of this section. Disconnect the negative battery cable.
2. Note the radio security code and the radio presets.
3. Remove the cylinder head cover retaining bolts. Remove the cylinder head cover from the engine.
4. Set the number one piston at TDC. The TDC marks on the cam chain sprocket should align with the cylinder head surface.
5. Using the proper gauge feeler gauge, adjust the valves on cylinder number one.
6. Rotate the crankshaft 180 degrees. Using the proper gauge feeler gauge, adjust the valves on cylinder number three.
7. Rotate the crankshaft 180 degrees. Using the proper gauge feeler gauge, adjust the valves on cylinder number four.
8. Rotate the crankshaft 180 degrees. Using the proper gauge feeler gauge, adjust the valves on cylinder number two.
9. Install the cylinder head cover.

ENGINE PERFORMANCE & EMISSION CONTROL

CAMSHAFT POSITION (CMP) SENSOR

LOCATION

The CMP sensor is located in the cylinder head.

REMOVAL & INSTALLATION

1. Disconnect the Camshaft Position (CMP) sensor 3P connector.
2. Remove the CMP sensor from the exhaust camshaft side of the cylinder head.
3. Install the parts in the reverse order of removal with a new O-ring

CRANKSHAFT POSITION (CKP) SENSOR

LOCATION

The CKP sensor is located in the engine block.

REMOVAL & INSTALLATION

1. Remove the Crankshaft Position (CKP) sensor.
2. Disconnect the CKP sensor 3P connector.
3. Install the parts in the reverse order of removal with a new O-ring.

ENGINE COOLANT TEMPERATURE (ECT) SENSOR

LOCATION

The ECT sensor is located in the radiator assembly.

REMOVAL & INSTALLATION

See Figure 24.

1. Drain the engine coolant.
2. Disconnect the ECT sensor 1 2P connector.
3. Remove the ECT sensor 1.
4. Install the parts in the reverse order of removal with a new O-ring.

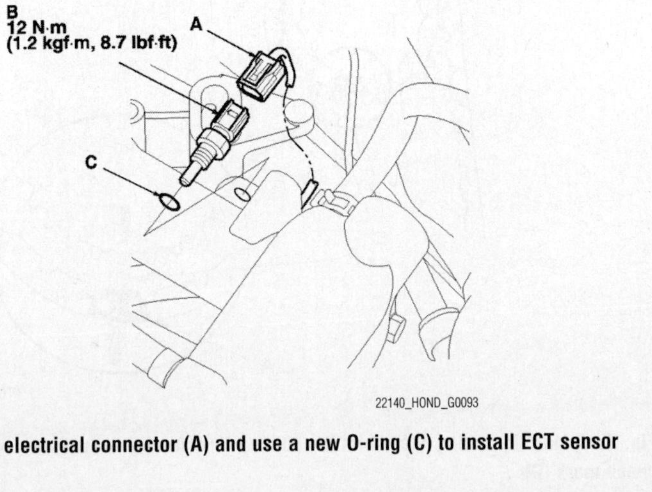
B
12 N·m
(1.2 kgf·m, 8.7 lbf·ft)
A
C
22140_HOND_G0093

Fig. 24 Connect the electrical connector (A) and use a new O-ring (C) to install ECT sensor (B)—2.2L Engine

5. Refill the radiator with engine coolant.

HEATED OXYGEN SENSOR (HO2S)

LOCATION

The HO2S sensors are located in the exhaust pipe assembly.

REMOVAL & INSTALLATION

See Figure 25.

1. Disconnect the secondary HO2S (Sensor 2) 4P connector.
2. Remove the secondary HO2S (Sensor 2).
3. Install the parts in the reverse order of removal.

INTAKE AIR TEMPERATURE (IAT) SENSOR

LOCATION

The IAT sensor is located in the throttle body assembly.

REMOVAL & INSTALLATION

See Figure 26.

1. Disconnect the IAT sensor 2P connector.
2. Remove the clip (B) and the IAT sensor.
3. Install the parts in the reverse order of removal.

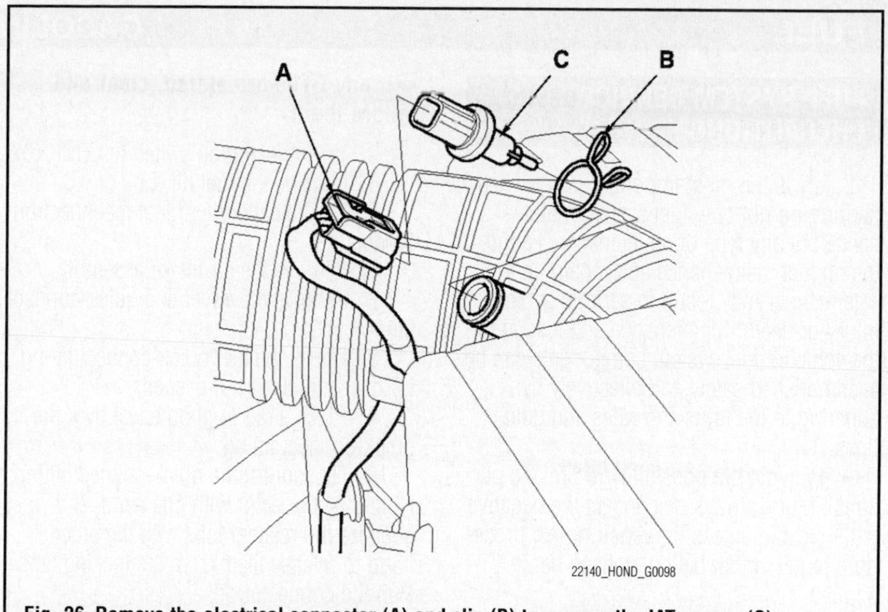
A
C
B
22140_HOND_G0098
Fig. 26 Remove the electrical connector (A) and clip (B) to remove the IAT sensor (C)

MANIFOLD ABSOLUTE PRESSURE (MAP) SENSOR

LOCATION

The MAP sensor is located on the air intake assembly.

REMOVAL & INSTALLATION

1. Disconnect the MAP sensor connector.
2. Remove the screw.
3. Remove the MAP sensor.
4. Install the parts in the reverse order of removal with a new O-ring.

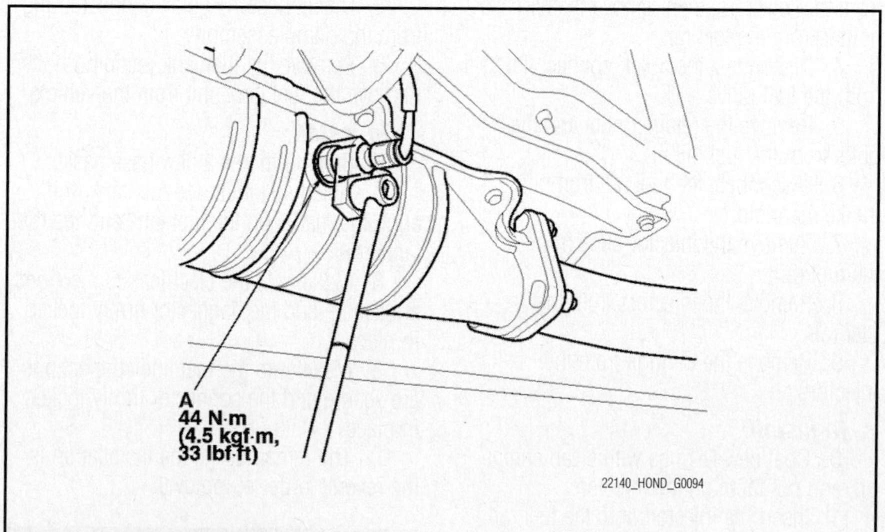
A
44 N·m
(4.5 kgf·m,
33 lbf·ft)
22140_HOND_G0094
Fig. 25 Installing the HO2S sensor

THROTTLE POSITION SENSOR (TPS)

LOCATION

The TPS is located under the dashboard as part of the throttle actuator module.

REMOVAL & INSTALLATION

1. Remove the passenger's dashboard lower cover.
2. Push the tab, and disconnect the throttle actuator control module 16P connector.
3. Remove the nuts and the throttle actuator control module.
4. Install the parts in the reverse order of removal.

VEHICLE SPEED SENSOR (VSS)

LOCATION

The Vehicle Speed Sensor (VSS) is located on the rear of the transaxle assembly.

REMOVAL & INSTALLATION

1. Disconnect the output shaft (countershaft) speed sensor 3P connector.
2. Remove the output shaft (countershaft) speed sensor.
3. Install the parts in the reverse order of removal with a new O-ring.

FUEL SYSTEM SERVICE PRECAUTIONS

Safety is the most important factor when performing not only fuel system maintenance but any type of maintenance. Failure to conduct maintenance and repairs in a safe manner may result in serious personal injury or death. Maintenance and testing of the vehicle's fuel system components can be accomplished safely and effectively by adhering to the following rules and guidelines.

- To avoid the possibility of fire and personal injury, always disconnect the negative battery cable unless the repair or test procedure requires that battery voltage be applied.
- Always relieve the fuel system pressure prior to disconnecting any fuel system component (injector, fuel rail, pressure regulator, etc.), fitting or fuel line connection. Exercise extreme caution whenever relieving fuel system pressure to avoid exposing skin, face and eyes to fuel spray. Please be advised that fuel under pressure may penetrate the skin or any part of the body that it contacts.
- Always place a shop towel or cloth around the fitting or connection prior to loosening to absorb any excess fuel due to spillage. Ensure that all fuel spillage (should it occur) is quickly removed from engine surfaces. Ensure that all fuel soaked cloths or towels are deposited into a suitable waste container.
- Always keep a dry chemical (Class B) fire extinguisher near the work area.
- Do not allow fuel spray or fuel vapors to come into contact with a spark or open flame.
- Always use a back-up wrench when loosening and tightening fuel line connection fittings. This will prevent unnecessary stress and torsion to fuel line piping.
- Always replace worn fuel fitting O-rings with new Do not substitute fuel hose or equivalent where fuel pipe is installed.

Before servicing the vehicle, make sure to also refer to the precautions in the beginning of this section as well.

RELIEVING FUEL SYSTEM PRESSURE

1. Remove PGM-FI main relay 2 from the auxiliary under-hood fuse/relay box.
2. Start the engine, and let it idle until it stalls.

➡**If any DTCs are stored, clear and ignore them.**

3. Turn the ignition switch to LOCK (0).
4. Remove the fuel fill cap.
5. Do the battery terminal reconnection procedure.
6. Remove the air intake assembly.
7. Remove the cover and quick-connect fitting cover.
8. Check the fuel quick-connect fitting for dirt, and clean it if needed.
9. Place a rag or shop towel over the quick-connect fitting.
10. Disconnect the quick-connect fitting: Hold the connector with one hand, and squeeze the retainer tabs with the other hand to release them from the locking tabs. Pull the connector off.

FUEL FILTER

REMOVAL & INSTALLATION

The fuel filter is an integrated part of the fuel pump assembly. For additional information, refer to the following section, "Fuel Pump, Removal & Installation."

FUEL INJECTORS

REMOVAL & INSTALLATION

1. Properly relieve the fuel system pressure.
2. Remove the fuel rail cover.
3. Disconnect the injector connectors from the injectors, then remove the harness holder from the fuel rail.
4. Disconnect the quick-connect fitting from the fuel rail.
5. Remove the retainer nuts and the bolts from the fuel rail.
6. Disconnect the fuel rail from the intake manifold.
7. Remove the injector clips from the fuel rail.
8. Remove the injectors from the fuel rail.
9. Remove the O-rings from the injectors.

To install:

10. Coat new O-rings with clean engine oil, and put them on the injectors.
11. Insert the injectors into the fuel rail.
12. Install the injector clips.
13. Install the injectors in the intake manifold.
14. Install and tighten the retainer nuts and bolts.
15. Connect the quick-connect fitting.
16. Connect the injector connectors and the harness holder.
17. Install the injector cover.
18. Turn the ignition switch to ON (II), but do not operate the starter. After the fuel pump runs for about 2 seconds, the fuel pressure in the fuel line rises. Repeat this 2 or 3 times, then check for fuel leakage.

FUEL PUMP

REMOVAL & INSTALLATION

CAUTION

The fuel injection system remains under pressure, even after the engine has been turned OFF. The fuel system pressure must be relieved before disconnecting any fuel lines. Failure to follow this procedure may result in fire, explosion, or personal injury.

1. Before servicing the vehicle, refer to the precautions in the beginning of this section.
2. Relieve the fuel system pressure.
3. Remove or disconnect the following:
   - Negative battery cable
   - Remove the fuel tank filler cap
   - Rear package tray
   - Fuel pump access panel
   - Fuel pump electrical harness

➡**Clean the fuel line fittings before disconnecting them.**

4. Disconnect the quick connect fitting from the pump assembly.
5. Remove the fuel tank retaining bolts. Remove the fuel tank unit from the vehicle.

To install:

6. Be sure to use a new base gasket.
7. When installing the fuel tank unit, align the marks on the fuel tank and the fuel tank unit.
8. Make sure the electrical connections are secure and the connector firmly locked in place.
9. Make sure the fuel line connections are secure and the connector firmly locked in place.
10. The remainder of the installation is the reverse order of removal.

IDLE SPEED

ADJUSTMENT

Idle speed is maintained by the Powertrain Control Module (PCM). No adjustment is necessary or possible.

THROTTLE BODY

REMOVAL & INSTALLATION

See Figures 27 through 31.

1. Refer to the accompanying illustration for Throttle Body replacement.

➡Do not adjust the throttle stop screw. The TP sensor is not removable.

2. After reassembly, adjust the cruise control actuator cable, as follows:
   a. Hold the cable sheath, removing all slack from the throttle cable.
   b. Turn the adjusting nut until it is 3 mm (1/8 in.) away from the throttle cable bracket. Tighten the locknut. The throttle cable deflection should now be 4-6 mm (0.16-0.24 in.).
   c. Remove the actuator cover, then disconnect the actuator cable end from the cruise control actuator.
   d. Turn the adjusting nut until it is 9 mm (0.35 in.) away from the actuator cable bracket when the throttle linkage starts open.
   e. Pull the cable so the adjusting nut touches the bracket, and tighten the locknut.
   f. Make sure the throttle linkage starts open when the actuator cable is pulled 9 mm (0.35 in.) from the starting point by rotating the actuator linkage.
3. Adjust the throttle cable as follows:
   a. Check cable free play at the throttle linkage. Cable free play should be 4-6 mm (3/16-1/4 in.).
   b. If free play is not within spec (4-6 mm, 3/16-1/4 in.) loosen the locknut, turn the adjusting nut until the free play is as specified, then retighten the locknut.
   c. With the cable properly adjusted, check the throttle valve to be sure it opens fully when you push the accelerator pedal to the floor. Also check the throttle valve to be sure it returns to the idle position whenever you release the accelerator pedal.

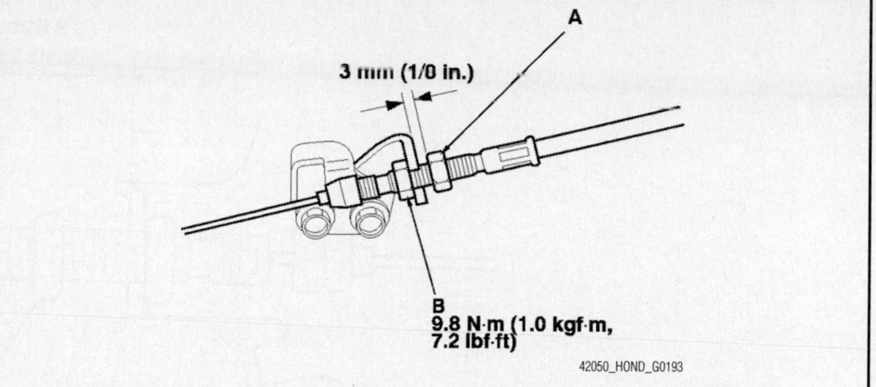
A
3 mm (1/8 in.)
B
9.8 N·m (1.0 kgf·m,
7.2 lbf·ft)
42050_HOND_G0193

Fig. 28 Turn the adjusting nut (A) until it is 3 mm (1/8 in.) away from the throttle cable bracket. Tighten the locknut (B). The throttle cable deflection should now be 4-6 mm (0.16-0.24 in.)

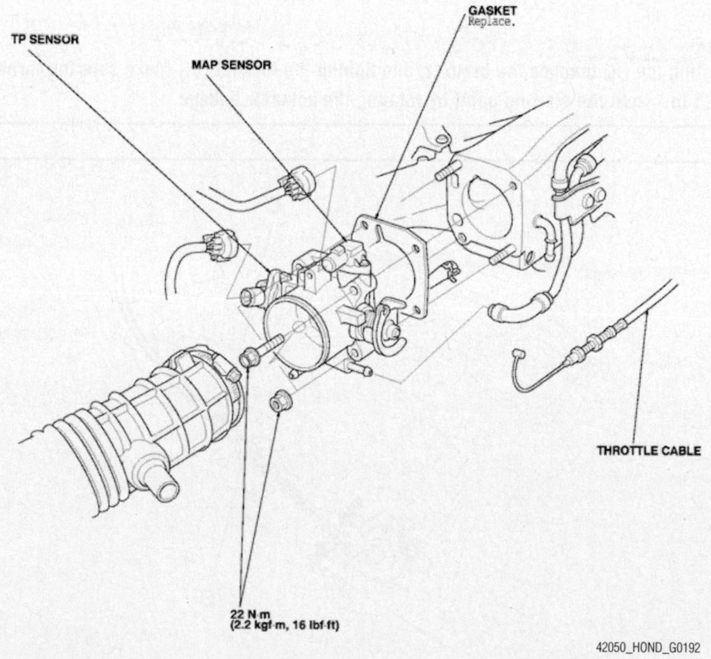
GASKET
Replace.
TP SENSOR
MAP SENSOR
THROTTLE CABLE
22 N·m
(2.2 kgf·m, 16 lbf·ft)
42050_HOND_G0192

Fig. 27 Throttle body removal and installation

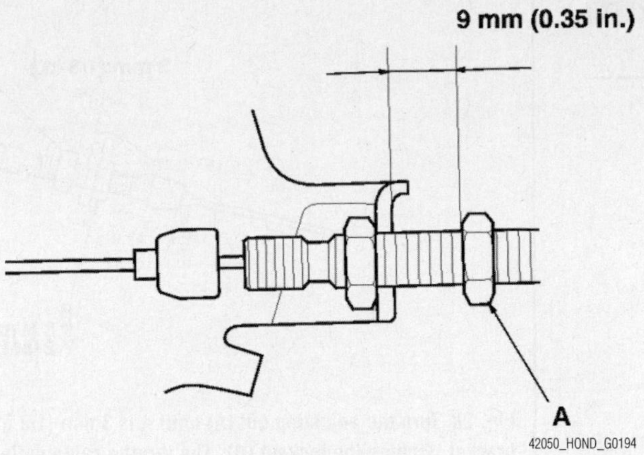
9 mm (0.35 in.)
A
42050_HOND_G0194

Fig. 29 Turn the adjusting nut (A) until it is 9 mm (0.35 in.) away from the actuator cable bracket when the throttle linkage starts open

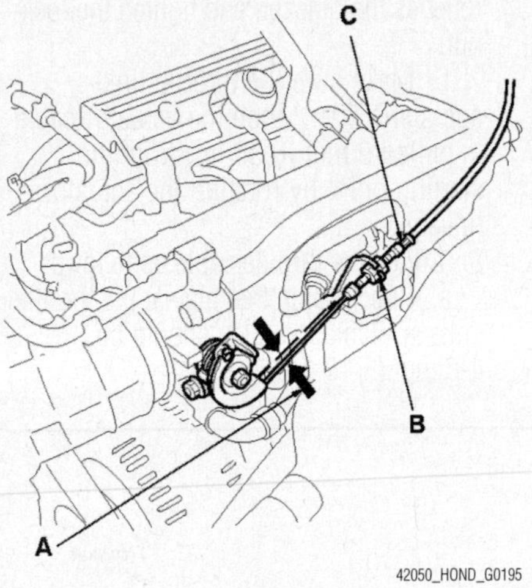
C
B
A
42050_HOND_G0195

Fig. 30 Pull the cable so the adjusting nut (A) touches the bracket, and tighten the locknut (B). Make sure the throttle linkage starts open when the actuator cable is pulled 9 mm (0.35 in.) from the starting point by rotating the actuator linkage.

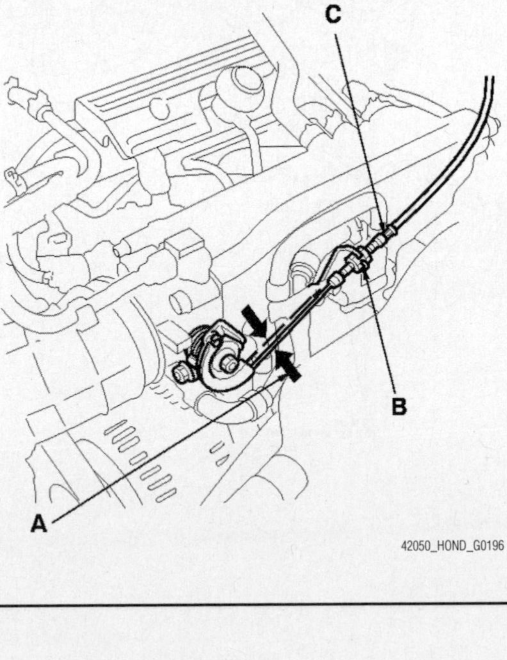
C
B
A
42050_HOND_G0196

Fig. 31 Throttle cable adjustment

HEATING & AIR CONDITIONING SYSTEM

BLOWER MOTOR

REMOVAL & INSTALLATION

1. Discharge and recover the A/C refrigerant.
2. Disconnect the battery cables and remove the battery.
3. Pull out the grommets of the A/C lines, then carefully separate the upper grommet from the lower grommet by releasing the lock tabs.
4. Remove the bolt, then disconnect the suction line and the receiver line from the blower/evaporator unit. Plug or cap the lines immediately after disconnecting them to avoid moisture and dust contamination.
5. Remove the passenger's dashboard lower cover and the right kick panel.
6. Disconnect the dashboard wire harness connector from the passenger's door wire harness connector, then remove it. Remove the wire harness connectors and the convertible top control unit from the steering hanger beam.
7. Disconnect the connectors from the blower motor, the power transistor, the evaporator sensor, and the recirculation control motor, then remove the wire harness clips. Remove the drain hose, the self-tapping screw, the mounting bolts, the mounting nuts, and the blower/evaporator unit.
8. Installation is the reverse order of removal.
9. Replace any O-ring for the A/C lines.
10. Recharge the A/C system refrigerant.

HEATER CORE

REMOVAL & INSTALLATION

See Figure 32.

➡Make sure to acquire the anti-theft code from the radio and write down the frequencies for the radio's preset buttons.

1. Disconnect the negative battery cable.

CAUTION

After disconnecting the negative battery cable, wait for at least 3 minutes for the air bag module to deplete its energy before working the on the instrument panel or steering wheel.

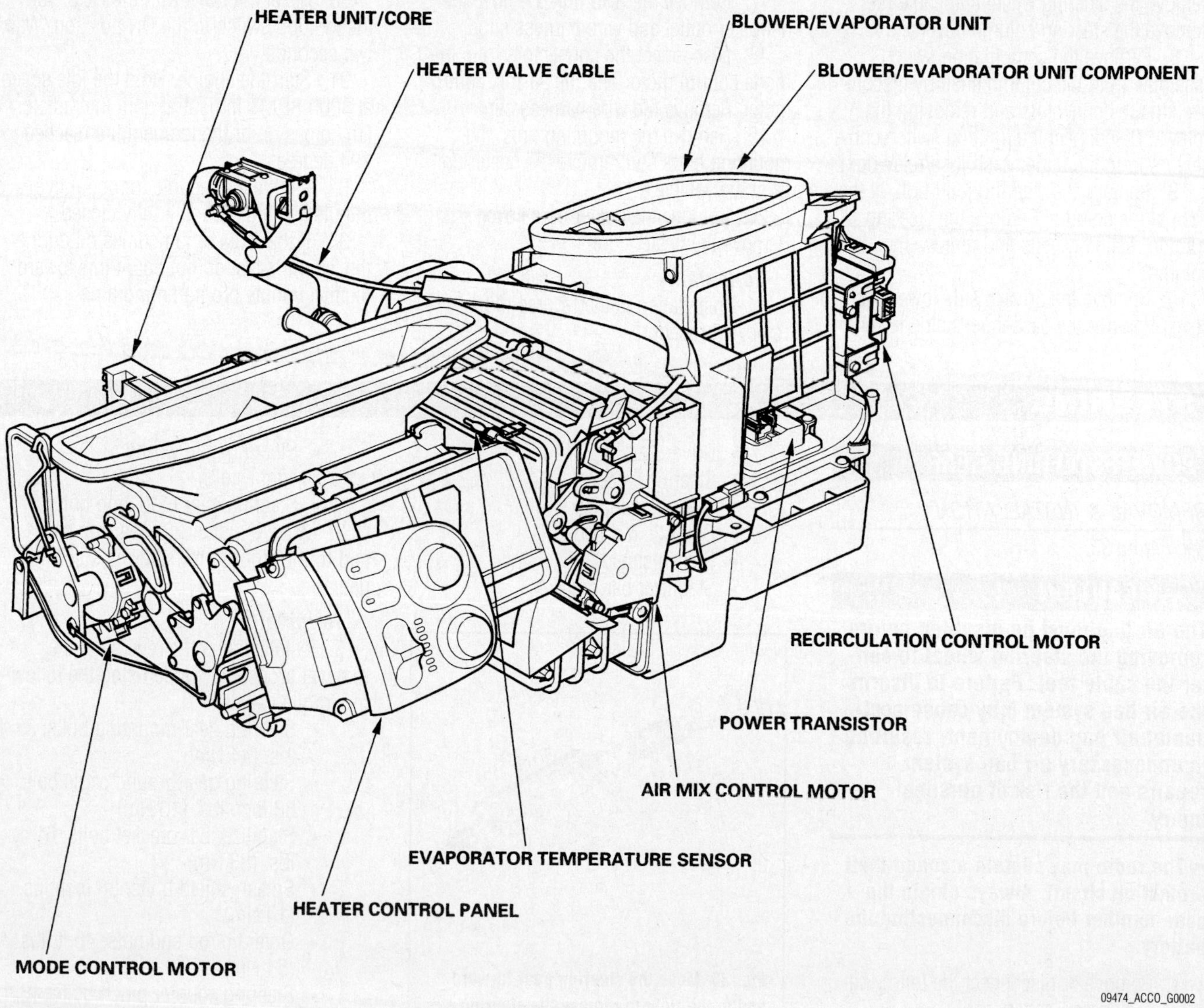
HEATER UNIT/CORE
HEATER VALVE CABLE
BLOWER/EVAPORATOR UNIT
BLOWER/EVAPORATOR UNIT COMPONENT
RECIRCULATION CONTROL MOTOR
POWER TRANSISTOR
AIR MIX CONTROL MOTOR
EVAPORATOR TEMPERATURE SENSOR
HEATER CONTROL PANEL
MODE CONTROL MOTOR
09474_ACCO_G0002

Fig. 32 View of the heater housing, evaporator housing and related components

2. Drain the engine coolant. Disconnect the heat shield from the exhaust manifold. Remove the battery.

3. Discharge the air conditioning system. Disconnect the suction and the receiver lines from the evaporator core.

4. From under the hood, open the cable clamp. Disconnect the heater control cable from the valve. Turn the valve arm to the fully open position.

5. Remove the mounting bolt from the heater valve. Disconnect the heater hoses from the heater unit. Remove the mounting nut from the heater unit. Disconnect the air conditioning lines from the evaporator housing.

6. Remove the radio panel retaining clips and remove the panel. Remove the radio retaining screws and pull the unit forward. Disconnect the antenna lead and electrical connectors and remove the radio.

7. Remove the driver's side air bag. Remove the steering wheel and cable reel. Remove the steering column covers.

8. Remove the combination switch assembly from the column shaft by disconnecting the connectors and removing the screws. Disconnect the ignition switch connectors from the under dash fuse/relay box.

9. Remove the steering joint bolt, at the base of the column. Remove the steering column retaining bolts and remove the column.

10. Remove the driver's side lower dash cover. Remove the passenger's side lower dash cover. Remove the drivers and passengers side front console cover.

11. Remove the passenger's side air bag assembly.

12. Remove the kick panels and A-pillar trim from both the drivers and passenger's side.

13. Disconnect all electrical connectors, air hoses and harness that interfere with the removal of the dashboard. Remove the ground bolts.

14. Open the driver's side door remove the bolts and clips after remove their protective caps. Lift upward on the dashboard/steering hanger beam to release if from the guide pins. Be sure all electrical connectors are unplugged.

15. Remove the blower/evaporator housing from the vehicle.

16. Remove the mounting bolts, the center brackets and the radio mounting brackets.

17. Remove the SRS unit. Remove the defroster outlet and wire harness clips.

18. Disconnect the connectors from the mode control motor and the air mix control motor. Remove the wire harness clip.

19. Remove the mounting nuts, the mounting bolts and remove the heater unit from the vehicle.

20. Remove the heater core cover. Remove the heater core.

To install:

21. Installation is the reverse of the removal procedure.

22. Evacuate, charge and leak test the air conditioning system refrigerant.

23. Operate the engine to normal operating temperatures; then, check the climate control operation and check for leaks.

24. Enter the antitheft codes for the radio and the navigation system. Set the clock.

25. On 2001–05 vehicles, reprogram the ECM engine idle characteristics. Be sure all electrical items are OFF.

26. Start the engine. Hold the idle speed at 3000 RPM's in neutral until the radiator fan comes on or the temperature reached 176 degrees.

27. Let the engine idle for about five minutes with the throttle fully closed.

28. If the radiator fan comes on during the five minutes, do not count this toward the five minute programming time.

29. On 2006 vehicles, reprogram the ECM engine idle characteristics. Be sure all electrical items are OFF.

30. Reset the ECM with the HDS. Turn the ignition switch to the ON position. Wait two seconds.

31. Start the engine. Hold the idle speed at 3000 RPM's in neutral until the radiator fan comes on or the temperature reached 194 degrees.

32. Let the engine idle for about five minutes with the throttle fully closed.

33. If the radiator fan comes on during the five minutes, do not count this toward the five minute programming time.

STEERING

POWER STEERING GEAR

REMOVAL & INSTALLATION

See Figure 33.

⁂ CAUTION

The air bag must be disabled before removing the steering wheel to center the cable reel. Failure to disarm the air bag system may cause accidental air bag deployment, resulting in unnecessary air bag system repairs and the risk of personal injury.

➡The radio may contain a coded theft protection circuit. Always obtain the code number before disconnecting the battery.

1. Remove or disconnect the following:
   - Negative battery cable
   - Front wheels
   - Driver's air bag
   - Steering wheel
   - Steering coupler
   - Outer tie rod ends
   - Splash shield
   - Stabilizer bar brackets
   - Steering gear wiring connectors
   - Steering gear mounting bolts

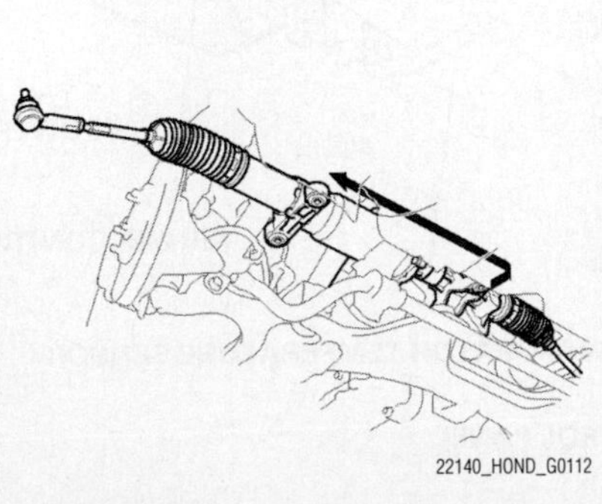
22140_HOND_G0112

Fig. 33 Move the steering gear forward and to the right to remove the steering gear Models

2. Move the steering gear forward and to the right to remove the steering gear.

To install:

3. Installation is the reverse of the removal procedure, while using the following torque values:
   - Steering gear mounting bolts: 33 ft. lbs. (44 Nm)
   - Steering gear ground cable bolt: 88 inch lbs. (10 Nm)
   - Stabilizer bar bracket bolts: 61 ft. lbs. (83 Nm)
   - Splash shield bolts: 88 inch lbs. (10 Nm)
   - Outer tie rod end nuts: 40 ft. lbs. (54 Nm)
   - Steering coupler pinch bolts: 16 ft. lbs. (22 Nm)

LOWER BALL JOINT

REMOVAL & INSTALLATION

See Figure 34.

1. Install a hex nut onto the threads of the ball joint. Make sure the nut is flush with the ball joint pin end to prevent damage to the threaded end of the ball joint pin.
2. Apply grease to the ball joint remover on the areas shown. This will ease installation of the tool and prevent damage to the pressure bolt threads.
3. Loosen the pressure bolt, and install the ball joint remover as shown. Insert the jaws carefully, making sure not to damage the ball joint boot. Adjust the jaw spacing by turning the adjusting bolt.
4. After adjusting the adjusting bolt, make sure the head of the adjusting bolt is in the position shown to allow the jaw to pivot.
5. With a wrench, tighten the pressure bolt until the ball joint pin pops loose from the ball joint pin hole. If necessary, apply penetrating type lubricant to loosen the ball joint pin.
6. Remove the ball joint remover, then remove the nut from the end of the ball joint pin, and pull the ball joint out of the ball joint pin hole. Inspect the ball joint boot, and replace it if damaged.

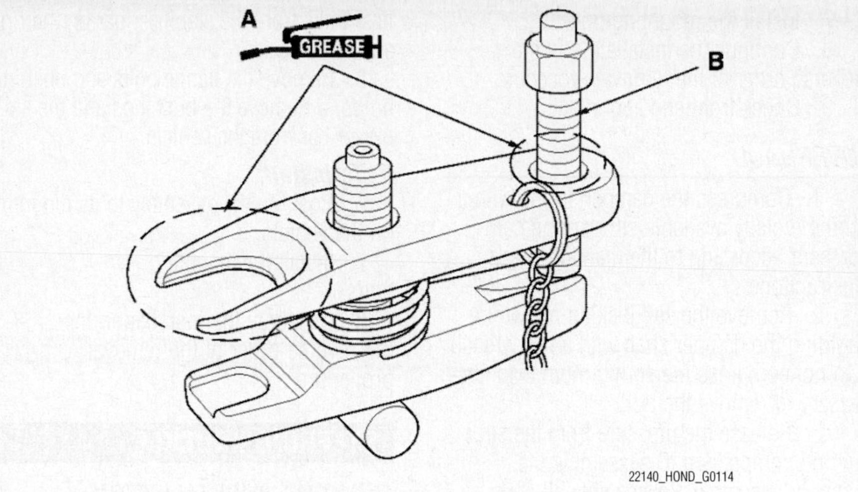
A
GREASE
B
22140_HOND_G0114

Fig. 34 Apply grease to the ball joint remover on the areas shown (A). This will ease installation of the tool and prevent damage to the pressure bolt (B) threads.

LOWER CONTROL ARM

REMOVAL & INSTALLATION

See Figure 35.

1. Raise and support the vehicle safely. Remove the front tires.
2. Remove the flange nut while holding the ball joint pin. Disconnect the stabilizer links from the lower arm.
3. Remove the flange bolt and nuts from the lower arm.
4. Remove the cotter pin from the lower arm ball joint. Remove the castle nut.
5. Remove the lower arm ball joint from the knuckle, using the ball joint removal tool.
6. Remove the self locking nut and self locking cam nut. Remove the cam plate adjusting bolt, cam collar, flange bolt and the lower arm.

To install:

7. Installation is in the reverse order of the removal procedure.
8. Be sure to use new flange bolts, castle nut and lock pin.
9. Check and adjust the wheel alignment, as necessary.

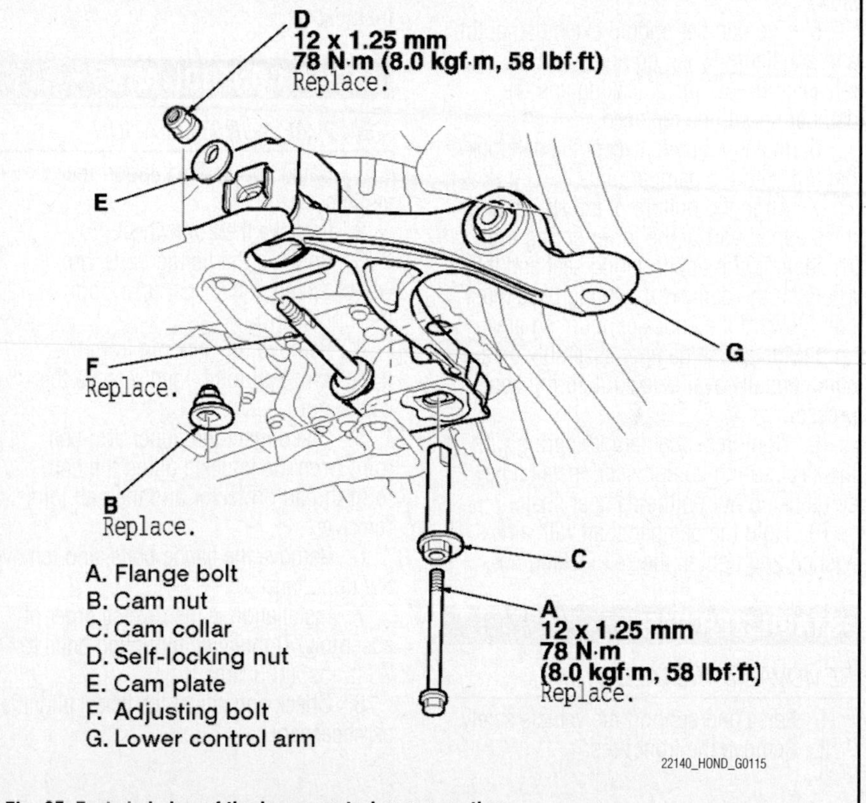
D
12 x 1.25 mm
78 N·m (8.0 kgf·m, 58 lbf·ft)
Replace.
E
G
F
Replace.
B
Replace.
C
A
12 x 1.25 mm
78 N·m
(8.0 kgf·m, 58 lbf·ft)
Replace.
A. Flange bolt
B. Cam nut
C. Cam collar
D. Self-locking nut
E. Cam plate
F. Adjusting bolt
G. Lower control arm
22140_HOND_G0115

Fig. 35 Exploded view of the lower control arm mounting

MACPHERSON STRUT

REMOVAL & INSTALLATION

1. Raise and support the vehicle safely. Remove the front tires.
2. Remove the flange bolt and brake hose mounting bracket from the damper.
3. Remove the cotter pin from the lower arm ball joint. Remove the castle nut. Remove the lower arm ball joint.
4. Remove the flange nuts from the top of the damper. Remove the flange bolt from the bottom of the damper.
5. Lower the lower arm and remove the damper assembly from the vehicle.

To install:

6. Position the assembly to its mounting on the vehicle.
7. Install the mounting bolts.
8. Continue the installation in the reverse order of the removal procedure.
9. Check front end alignment.

OVERHAUL

1. Compress the damper spring with a commercially available strut spring compressor according to the manufacturer's instructions.
2. Remove the self locking nut while holding the damper shaft with a hex wrench. Do not compress the spring more than necessary to remove the nut.
3. Release the pressure from the strut spring compressor. Disassemble the damper assembly. Reassemble all parts, except the spring.
4. Compress the damper assembly by hand and check for smooth operation thru full stroke, both compression and extension.
5. The damper should extend smoothly and constantly when compression is released. If not, gas is leaking and the damper should be replaced.
6. Install all parts except the self locking nut onto the damper unit.
7. Align the bottom of the spring and the stepped part of the lower spring seat. The hole in the upper spring seat and the arrow on the damper mounting base must point toward the knuckle mounting area.
8. Install the damper assembly on a commercially available strut spring compressor.
9. Compress the damper spring with the strut spring compressor. Install a new self locking nut on the damper shaft.
10. Hold the damper shaft with a hex wrench and tighten the self locking nut.

STABILIZER BAR

REMOVAL & INSTALLATION

1. Raise and support the vehicle safely.
2. Remove the front tires.
3. Remove the splash shield.
4. Remove the self locking nuts while holding the joint pin. Disconnect the stabilizer links from the stabilizer bar on the right and left sides.
5. Remove the flange bolts and bushing holders. Remove the bushings and the stabilizer bar from the vehicle.

To install:

6. Position the assembly to its mounting on the vehicle.
7. Install the mounting bolts.
8. Continue the installation in the reverse order of the removal procedure.

STEERING KNUCKLE

REMOVAL & INSTALLATION

The steering knuckle is removed with the wheel bearings as an assembly. Refer to the following section for additional information, "Wheel Bearings, Removal & Installation."

UPPER CONTROL ARM

REMOVAL & INSTALLATION

1. Raise and safely support the vehicle.
2. Remove the front wheel.
3. Remove the flange bolts and wheel speed sensor harness from the upper arm.
4. Remove the lock pin from the upper arm ball joint, then remove the castle nut.
5. Disconnect the upper arm ball joint from the knuckle using the ball joint thread protector and the ball joint remover.
6. Remove the flange bolts, and remove the upper arm.
7. Installation is the reverse order of assembly. Tighten the mounting bolts to 76 ft. lbs. (103 Nm).
8. Check and adjust the wheel alignment as necessary.

WHEEL BEARINGS

REMOVAL & INSTALLATION

See Figure 36.

➡A hydraulic press and several bearing drivers and attachments are needed to remove and install the hub and bearing.

1. Before servicing the vehicle, refer to the precautions in the beginning of this section.
2. Remove or disconnect the following:
   - Front wheel
   - Brake hose bracket mounting bolts
   - Brake caliper and caliper support
   - Wheel speed sensor
   - Brake rotor
   - Outer tie rod end
   - Upper and lower ball joints
   - Steering knuckle from the vehicle
   - Dust cover
   - Spindle nut
   - Wheel speed pulse ring
3. Mount the steering knuckle in a press and press the hub out of the wheel bearing.
4. Remove the splash guard and the wheel bearing snapring.
5. Press the wheel bearing out of the steering knuckle.

To install:

6. Installation is the reverse of the removal procedure, while using the following torque values:
   - Splash guard screws: 48 inch lbs. (5 Nm)
   - Spindle nut: 242 ft. lbs. (329 Nm)
   - Upper ball joint nut: 36–43 ft. lbs. (49–59 Nm)
   - Lower ball joint nut: 43–51 ft. lbs. (56–69 Nm)
   - Outer tie rod end nut: 40 ft. lbs. (54 Nm)
   - Brake caliper support bolts: 83 ft. lbs. (113 Nm)

ADJUSTMENT

The wheel bearings are not adjustable. If they are not within specification, the wheel bearings must be replaced.

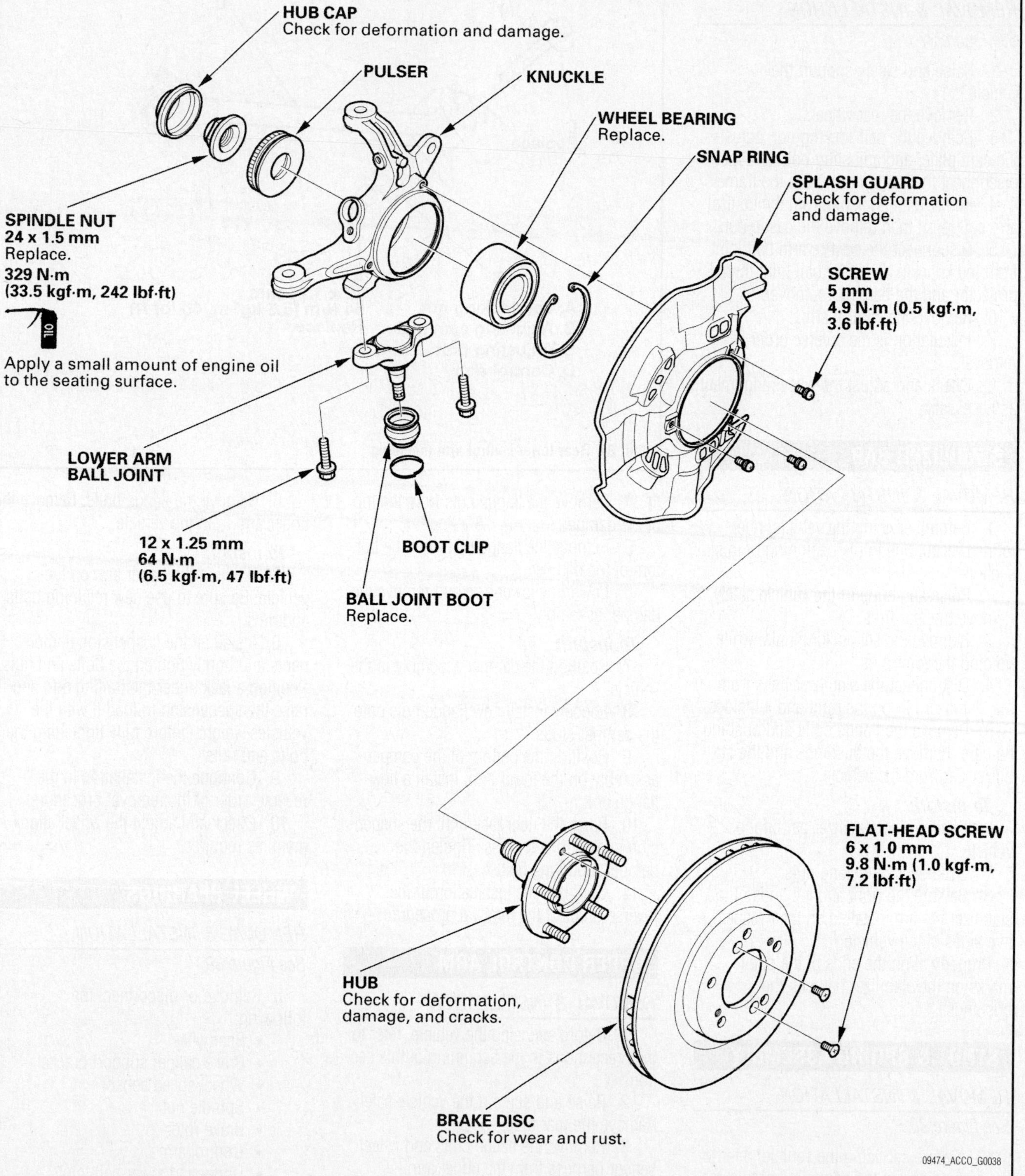
HUB CAP
Check for deformation and damage.
PULSER
KNUCKLE
WHEEL BEARING
Replace.
SNAP RING
SPLASH GUARD
Check for deformation and damage.
SPINDLE NUT
24 x 1.5 mm
Replace.
329 N·m
(33.5 kgf·m, 242 lbf·ft)
OIL
Apply a small amount of engine oil to the seating surface.
SCREW
5 mm
4.9 N·m (0.5 kgf·m, 3.6 lbf·ft)
LOWER ARM BALL JOINT
12 x 1.25 mm
64 N·m
(6.5 kgf·m, 47 lbf·ft)
BOOT CLIP
BALL JOINT BOOT
Replace.
FLAT-HEAD SCREW
6 x 1.0 mm
9.8 N·m (1.0 kgf·m, 7.2 lbf·ft)
HUB
Check for deformation, damage, and cracks.
BRAKE DISC
Check for wear and rust.
09474_ACCO_G0038

Fig. 36 Knuckle components

LOWER CONTROL ARM

REMOVAL & INSTALLATION

See Figure 37.

1. Raise and safely support the vehicle.
2. Remove the rear wheel.
3. Remove the self-locking nut, adjusting cam plate, and adjusting bolt, and then disconnect the control arm from the frame.
4. Remove the lock pin from the control arm ball joint, then remove the castle nut.
5. Disconnect the control arm ball joint from the knuckle using the ball joint thread protector and the ball joint remover.
6. Remove the control arm.
7. Installation is the reverse order of removal.
8. Check and adjust the wheel alignment as necessary.

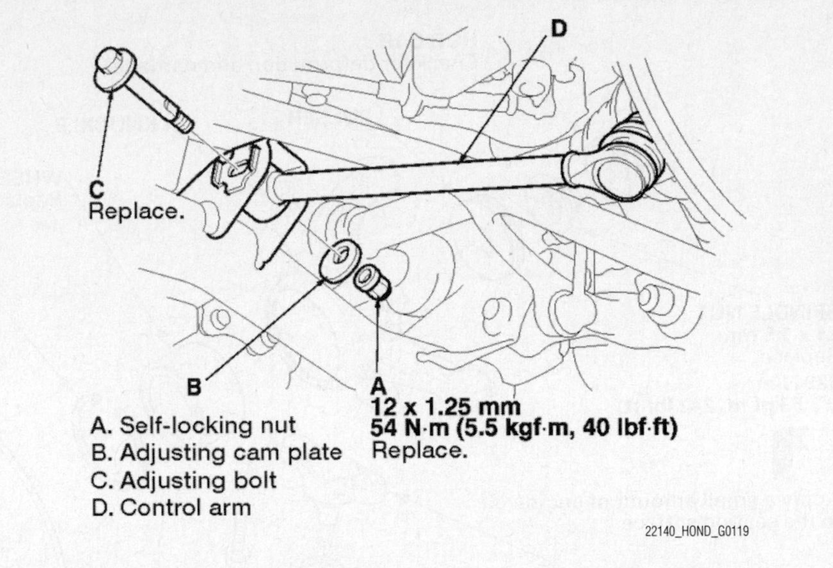
D
C
Replace.
B
A
12 x 1.25 mm
54 N·m (5.5 kgf·m, 40 lbf·ft)
Replace.
A. Self-locking nut
B. Adjusting cam plate
C. Adjusting bolt
D. Control arm
22140_HOND_G0119

Fig. 37 Rear lower control arm mounting

STABILIZER BAR

REMOVAL & INSTALLATION

1. Before servicing the vehicle, refer to the precautions in the beginning of this section.
2. Raise and support the vehicle safely. Remove the rear tires.
3. Remove the self locking nuts, while holding the joint pins.
4. Disconnect the stabilizer links from the stabilizer bar on the right and left sides.
5. Remove the flange bolts and bushing holders. Remove the bushings and the stabilizer bar from the vehicle.

To install:

6. Position the stabilizer bar on the vehicle.
7. Use new self locking nuts.
8. Be sure the right and left ends of the stabilizer bar are installed on their respective sides of the vehicle.

Properly align the ends of the paint marks on the stabilizer bar with the bushings

STRUT & SPRING ASSEMBLY

REMOVAL & INSTALLATION

See Figure 38.

1. Before servicing the vehicle, refer to the precautions in the beginning of this section.
2. Remove the spare tire from the trunk.
3. Raise and support the vehicle safely. Remove the rear tires.
4. Remove the flange nuts from the top of the damper.
5. Remove the flange bolt from the bottom of the damper.
6. Lower the lower arm and remove the damper assembly.

To install:

7. Position the damper assembly in the vehicle.
8. Loosely install new flange nuts onto the damper studs.
9. Position the bottom of the damper assembly on the lower arm. Install a new flange bolt.
10. Raise the floor jack until the suspension begins to compress. Tighten the damper mounting bolt.
11. Continue the installation in the reverse order of the removal procedure.

UPPER CONTROL ARM

REMOVAL & INSTALLATION

1. Before servicing the vehicle, refer to the precautions in the beginning of this section.
2. Raise and support the vehicle safely. Remove the rear tires.
3. Remove the flange bolts and wheel sensor harness from the upper arm.
4. Remove the lock pin from the upper arm ball joint, remove the castle nut.
5. Remove the upper arm ball joint from the knuckle using the ball joint removal tool.
6. Remove the flange bolts. Remove the upper arm from the vehicle.

To install:

7. Position the upper arm on the vehicle. Be sure to use new retaining bolts and nuts.
8. Install all the suspension components and lightly tighten the bolts and nuts. Position a jack under the trailing arm and raise the suspension to load it with the vehicles weight, before fully tightening the bolts and nuts.
9. Continue the installation in the reverse order of the removal procedure.
10. Check and adjust the wheel alignment, as required.

WHEEL BEARINGS

REMOVAL & INSTALLATION

See Figure 39.

1. Remove or disconnect the following:
   - Rear wheel
   - Brake caliper support bracket
   - Wheel speed sensor
   - Spindle nut
   - Brake rotor
   - Control arm
   - Upper and lower ball joints
   - Spindle from the vehicle
2. Mount the steering knuckle in a press and press the hub out of the wheel bearing.

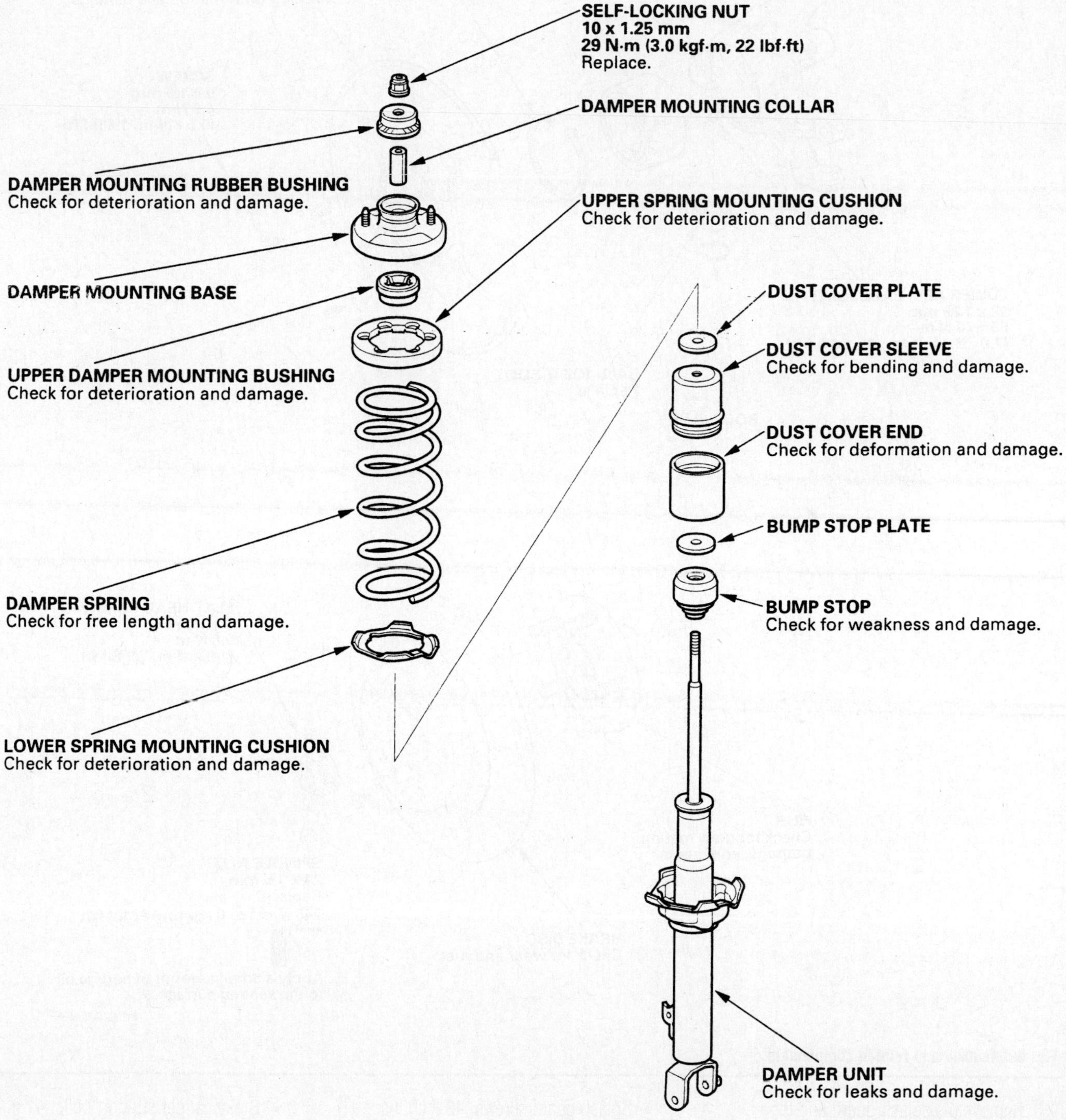
SELF-LOCKING NUT
10 x 1.25 mm
29 N·m (3.0 kgf·m, 22 lbf·ft)
Replace.
DAMPER MOUNTING COLLAR
DAMPER MOUNTING RUBBER BUSHING
Check for deterioration and damage.
UPPER SPRING MOUNTING CUSHION
Check for deterioration and damage.
DAMPER MOUNTING BASE
DUST COVER PLATE
DUST COVER SLEEVE
Check for bending and damage.
UPPER DAMPER MOUNTING BUSHING
Check for deterioration and damage.
DUST COVER END
Check for deformation and damage.
BUMP STOP PLATE
DAMPER SPRING
Check for free length and damage.
BUMP STOP
Check for weakness and damage.
LOWER SPRING MOUNTING CUSHION
Check for deterioration and damage.
DAMPER UNIT
Check for leaks and damage.
9347FG24

Fig. 38 Exploded view of the strut and spring assembly

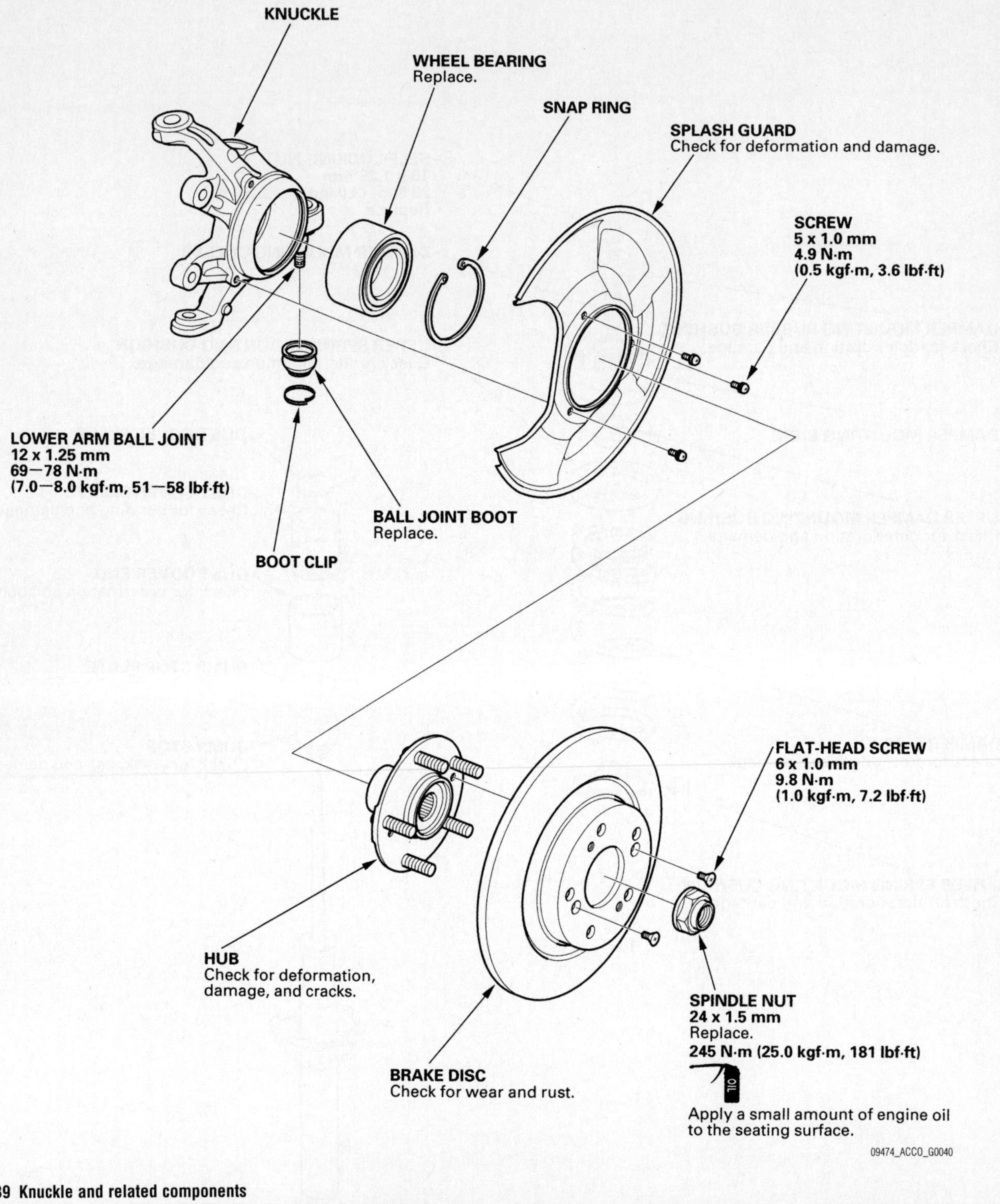
KNUCKLE
WHEEL BEARING
Replace.
SNAP RING
SPLASH GUARD
Check for deformation and damage.
SCREW
5 x 1.0 mm
4.9 N·m
(0.5 kgf·m, 3.6 lbf·ft)
LOWER ARM BALL JOINT
12 x 1.25 mm
69—78 N·m
(7.0—8.0 kgf·m, 51—58 lbf·ft)
BALL JOINT BOOT
Replace.
BOOT CLIP
FLAT-HEAD SCREW
6 x 1.0 mm
9.8 N·m
(1.0 kgf·m, 7.2 lbf·ft)
HUB
Check for deformation, damage, and cracks.
SPINDLE NUT
24 x 1.5 mm
Replace.
245 N·m (25.0 kgf·m, 181 lbf·ft)
OIL
Apply a small amount of engine oil to the seating surface.
BRAKE DISC
Check for wear and rust.
09474_ACCO_G0040

Fig. 39 Knuckle and related components

3. Remove the splash guard and the wheel bearing snapring.
4. Press the wheel bearing out of the steering knuckle.

To install:

5. Installation is the reverse of the removal procedure, while using the following torque values:

- Splash guard screws: 48 inch lbs. (5 Nm)
- Spindle nut: 181 ft. lbs. (245 Nm)
- Upper ball joint nut: 36–43 ft. lbs. (49–59 Nm)
- Lower ball joint nut: 51–58 ft. lbs. (68–78 Nm)
- Control arm ball joint nut: 36–43 ft. lbs. (49–59 Nm)
- Brake caliper support bolts: 41 ft. lbs. (55 Nm)

ADJUSTMENT

The wheel bearings are not adjustable. If they are not within specification, the wheel bearings must be replaced.

19

HONDA

Diagnostic Trouble Codes

DIAGNOSTIC TROUBLE CODES 19-1

OBD II Trouble Code List (P0XXX Codes) 19-2
OBD II Trouble Code List (P1XXX Codes) 19-50
OBD II Trouble Code List (P2XXX Codes) 19-65
OBD II Trouble Code List (P3XXX Codes) 19-79
OBD II Trouble Code List (U0XXX Codes) 19-80
OBD II Trouble Code List (U1XXX Codes) 19-86
OBD II Vehicle Applications 19-1
Honda 19-1

DIAGNOSTIC TROUBLE CODES

OBD II VEHICLE APPLICATIONS

HONDA

Accord
2009–2010
- 2.4L I4 .... K24Z3
- 3.5L V6.................... J35Z2

Civic
2009–2010
- 1.8L I4 .... R18A1
- 2.0L I4 .... K20Z3

Civic Hybrid
2009–2010
- 1.3L I4 .... LDA2

CR-V
2009–2010
- 2.4L I4 .... K24Z1, K24A6

Element
2009–2010
- 2.4L I4 .... K24A8

Fit
2009–2010
- 1.5L I4 .... CD3

Insight Hybrid
2010
- 1.3L I4 .... LDA3

Odyssey
2009–2010
- 3.5L V6............ J35A6, J35A7

Pilot
2009–2010
- 3.5L V6.................... J35Z4

Ridgeline
2009–2010
- 3.5L V6.................... J35A9

S2000
2009
- 2.2L I4 .... F22C1

OBD II Trouble Code List (P0XXX Codes)

DTC	Trouble Code Title, Conditions & Possible Causes
DTC: P0010 **1T PCM, MIL: Yes** **Year:** 2009, 2010 **Model:** Accord, Civic, CR-V, Element **Engine:** 2.0L L4, 2.4L L4 **Transmission:** All	**Variable Valve Timing Control (VTC) Oil Control Solenoid Valve Malfunction :** With the engine running one of the following conditions must be met: Condition 1: Output duty is 40% or more and the VTC current is 200 mA or less for at least 5 seconds. Condition 2: Output duty is 5% or less and the VTC current is 800 mA or more for at least 5 seconds. **NOTE: Before you troubleshoot, record all freeze data and any on-board snapshot, and review the general troubleshooting information.** **Possible Causes:** • Poor connections or loose terminals at the VTC oil control solenoid valve and the ECM/PCM • "Open" circuit between the VTC oil control solenoid valve and ground • "Open" or "Short" circuit between the ECM/PCM and the VTC oil control solenoid valve • Faulty VTC oil control solenoid valve
DTC: P0011 **2T PCM, MIL: Yes** **Year:** 2009, 2010 **Model:** Accord, Civic, CR-V, Element **Engine:** 2.0L L4, 2.4L L4, 3.5L V6 **Transmission:** All	**Variable Valve Timing Control (VTC) System Malfunction :** With the engine running and at operating temperature. The difference between the timing control command and the actual timing of the camshaft is +/- 50 degrees or more for at least 15 seconds. **NOTE: Before you troubleshoot, record all freeze data and any on-board snapshot, and review the general troubleshooting information.** **Possible Causes:** • Engine oil level low, If the level is OK, check the engine oil pressure • Poor connections or loose terminals at the VTC oil control solenoid valve and the ECM/PCM • Faulty VTC oil control solenoid valve or clogged VTC strainer • Faulty VTC actuator • Perform the ECM/PCM idle learn procedure and the CKP pattern clear/CKP pattern learn procedure
DTC: P0101 **2T PCM, MIL: Yes** **Year:** 2009, 2010 **Model:** Accord, Accord Crosstour, Civic, CR-V, Element, Fit, Odyssey, Pilot, Ridgeline **Engine:** 1.3L L4, 1.5L L4, 1.8L L4, 2.0L L4, 2.4L L4, 3.5L V6 **Transmission:** All	**Mass Airflow (MAF) Sensor Circuit Range/Performance Problem:** Elapsed time after engine start 5 seconds, engine coolant temperature 156°F (69°C), engine speed 650-2,100 rpm. No active DTCs P0102, P0103, P0107, P0108, P0112, P0113, P0117, P0118, P0171, P0172, P0174, P0175, P0300, P0301, P0302, P0303, P0304, P0305, P0306, P0335, P0339, P0340, P0341, P0344, P0401, P0404, P0443, P0496, P0497, P0506, P0507, P1128, P1129, P145C, P2413, P2646, P2647, P2648, P2649. The difference between the amount of intake air measured by the MAF sensor and the amount of intake air calculated from the MAP sensor output is out of the normal area for at least 10 seconds. **Possible Causes:** • Dirty air cleaner element • Faulty PCV valve or hose • Faulty EVAP canister purge valve • Vacuum leaks at the Throttle body, Intake manifold, Brake booster • Cracked or loose Air Intake Duct • Poor connections or loose terminals at the MAF sensor/IAT sensor and the PCM • Faulty MAF sensor/IAT sensor • Faulty PCM
DTC: P0102 **1T PCM, MIL: Yes** **Year:** 2009, 2010 **Model:** Accord, Accord Crosstour, Civic, CR-V, Element, Fit, Odyssey, Pilot, Ridgeline **Engine:** 1.3L L4, 1.5L L4, 1.8L L4, 2.0L L4, 2.4L L4, 3.5L V6 **Transmission:** All	**Mass Air Flow (MAF) Sensor Circuit Low Voltage:** The lower limit of the MAF sensor output is specified. If the output is below that limit, the PCM detects a malfunction and stores a DTC. Execution is continuous and the duration time is 2 seconds or more. The MAF sensor input voltage is 0.1 volt or less for at least 2 seconds. **NOTE: Before you troubleshoot, record all freeze data and any on-board snapshot.** **Possible Causes:** • Poor connections or loose terminals at the MAF sensor/IAT sensor and the PCM • Blown fuse • "Open" or "Short" in the wire between the MAF sensor and the fuse • "Open" or "Short" in the wire between the PCM and the MAF sensor • Faulty MAF sensor/IAT sensor • PCM may need to be updated with the latest software • Faulty PCM

DTC	Trouble Code Title, Conditions & Possible Causes
DTC: P0103 **1T PCM, MIL: Yes** **Year:** 2009, 2010 **Model:** Accord, Accord Crosstour, Civic, CR-V, Element, Odyssey, Pilot, Ridgeline **Engine:** 1.3L L4, 1.8L L4, 2.0L L4, 2.4L L4, 3.5L V6 **Transmission:** All	**Mass Airflow (MAF) Sensor Circuit High Voltage:** The upper limit of the MAF sensor output is specified. If the output is above that limit, the PCM detects a malfunction and stores a DTC. Execution is continuous and the duration time is 2 seconds or more. The MAF sensor input voltage is 4.89 V or more for at least 2 seconds. P0102 is not active. **NOTE: Before you troubleshoot, record all freeze data and any on-board snapshot.** **Possible Causes:** • Poor connections or loose terminals at the MAF sensor/IAT sensor and the PCM • "Short" in the wire between the PCM and the MAF sensor • Faulty MAF sensor/IAT sensor • Faulty PCM
DTC: P0107 **1T PCM, MIL: Yes** **Year:** 2009, 2010 **Model:** CR-V, Fit, Odyssey, Pilot, Ridgeline **Engine:** 1.5L L4, 2.4L L4, 3.5L V6 **Transmission:** All	**Manifold Absolute Pressure (MAP) Sensor Circuit Low Voltage (PGM-FI System):** The MAP sensor outputs low signal voltage at high vacuum (throttle valve closed) and high signal voltage at low vacuum (throttle valve wide open). If a signal voltage from the MAP sensor is a set value or less, the Powertrain Control Module (PCM) detects a malfunction and a DTC is stored. The execution time is continuous and the duration time is 2 seconds or more. The MAP sensor output voltage is 0.23 V or less for at least 2 seconds. DTC P0108 is not active. **NOTE: Before you troubleshoot, record all freeze data and any on-board snapshot.** **Possible Causes:** • Poor connections or loose terminals at the MAP sensor and the PCM • "Short" in the wire between the PCM and the MAP sensor • "Open" in the wire between the PCM and the MAP sensor • Faulty MAP sensor • Faulty PCM
DTC: P0107 **1T PCM, MIL: Yes** **Year:** 2009, 2010 **Model:** Accord, Accord Crosstour, Civic, Element **Engine:** 1.3L L4, 1.8L L4, 2.0L L4, 2.4L L4, 3.5L V6 **Transmission:** All	**Manifold Absolute Pressure (MAP) Sensor Circuit Low Voltage:** With the engine running the Map sensor output voltage is 0.24 volts or less for at least 2 seconds. **Possible Causes:** • Poor connections or loose wires at the MAP sensor and at the PCM • "Open" or "Short circuit between the Map sensor and PCM • Faulty MAP sensor • PCM needs to be updated with the latest software • Faulty PCM
DTC: P0108 **1T PCM, MIL: Yes** **Year:** 2009, 2010 **Model:** Accord, Accord Crosstour, Fit, Odyssey, Pilot, Ridgeline **Engine:** 1.5L L4, 2.4L L4, 3.5L V6 **Transmission:** All	**MAP Sensor Circuit High Voltage (A/T/System) (With Navigation):** The MAP sensor outputs low signal voltage at high vacuum (throttle valve closed) and high signal voltage at low vacuum (throttle valve wide open). If a signal voltage from the MAP sensor is a set value or more, the Powertrain Control Module (PCM) detects a malfunction and a DTC is stored. The execution time is continuous and the duration time is 2 seconds or more. DTC P0107 is not active. The MAP sensor output voltage is 4.49 V or more for at least 2 seconds. **NOTE: Before you troubleshoot, record all freeze data and any on-board snapshot.** **Possible Causes:** • Poor connections or loose terminals at the MAP sensor and the PCM • "Open" in the wire between the PCM and the MAP sensor • "Open" in the wire between the PCM and the MAP sensor • PCM may need to be updated with the latest software • Faulty PCM
DTC: P0108 **1T PCM, MIL: Yes** **Year:** 2009, 2010 **Model:** Civic, CR-V, Element **Engine:** 1.3L L4, 1.8L L4, 2.0L L4, 2.4L L4 **Transmission:** All	**Manifold Absolute Pressure (MAP) Sensor Circuit High Voltage:** With the engine running the Map sensor output voltage value is 4.49 V for at least 2 seconds. **NOTE: Before you troubleshoot, record all freeze data and any on-board snapshot, and review the general troubleshooting information.** **Possible Causes:** • Poor connections or loose wires at the MAP sensor and at the PCM • "Open" circuit between the MAP sensor and the PCM • Faulty MAP sensor • PCM needs to be updated with the latest software • Faulty PCM

DTC	Trouble Code Title, Conditions & Possible Causes
DTC: P0111 **2T PCM, MIL: Yes** **Year:** 2009, 2010 **Model:** Accord, Accord Crosstour, Civic, CR-V, Element, Fit, Odyssey, Pilot, Ridgeline **Engine:** 1.3L L4, 1.5L L4, 1.8L L4, 2.0L L4, 2.4L L4, 3.5L V6 **Transmission:** All	**Intake Air Temperature (IAT) Sensor Circuit Range/Performance:** The execution is once per driving cycle and the duration time is 10 seconds or more. Engine OFF time is 6 hours. DTCs P0112, P0113, P0116, P0117, P0118, P0125, P1116, P2183, P2184, P2185, P2610 are not active. A malfunction is detected if the following three conditions are present after the engine has stopped and the ignition switch has been turned to LOCK (0) for at least 6 hours before restarting the engine: * When the temperature difference between the IAT and ECT1 is 57°F (32°C) or more. * When the temperature difference between the IAT and ECT2 is 30°F (17°C) or more. * When the temperature difference between the ECT2 and ECT1 is 73°F (41°C) or more. **NOTE: Before you troubleshoot, record all freeze data and any on-board snapshot.** **Possible Causes:** • Poor connections or loose terminals at ECT sensor 1 and 2 and the MAF sensor/IAT sensor • Poor connections or loose terminals at the IAT sensor and the PCM • Faulty MAF sensor/IAT sensor • Faulty PCM
DTC: P0112 **1T PCM, MIL: Yes** **Year:** 2009, 2010 **Model:** Accord, Accord Crosstour, Fit, Pilot **Engine:** 1.5L L4, 2.4L L4, 3.5L V6 **Transmission:** All	**Intake Air Temperature (IAT) Sensor Circuit Low Voltage:** If the IAT sensor output voltage is excessively low, the Powertrain Control Module (PCM) detects a malfunction and a DTC is stored. The execution time is continuous and the duration time is 2 seconds or more. The IAT sensor output voltage is 0.08 V or less for at least 2 seconds. **NOTE: Before you troubleshoot, record all freeze data and any on-board snapshot.** **Possible Causes:** • Poor connections or loose terminals at the MAF sensor/IAT sensor and the PCM • "Short" in the wire between the MAF sensor/IAT sensor and the PCM • Faulty MAF sensor/IAT sensor • PCM may need to be updated with the latest software • Faulty PCM
DTC: P0112 **1T PCM, MIL: Yes** **Year:** 2009, 2010 **Model:** Civic, CR-V, Element, Odyssey **Engine:** 1.3L L4, 1.8L L4, 2.0L L4, 2.4L L4, 3.5L V6 **Transmission:** All	**Intake Air Temperature (IAT) Sensor Circuit Low Voltage:** With the ignition ON the IAT sensor output voltage is 0.08 V or less for at least 2 seconds. **Possible Causes:** • Poor connections or loose wires at the IAT sensor and at the PCM • "Short" circuit between the IAT sensor and the PCM • Faulty IAT sensor • Faulty PCM
DTC: P0112 **1T PCM, MIL: Yes** **Year:** 2009, 2010 **Model:** Ridgeline **Engine:** 3.5L V6 **Transmission:** All	**Intake Air Temperature (IAT) Sensor 1 Circuit Low Voltage:** With the ignition ON the IAT sensor 1 output voltage is 0.07 V or less for at least 5 seconds. **NOTE: Before you troubleshoot, record all freeze data and any on-board snapshot, and review the general troubleshooting information.** **Possible Causes:** • Poor connections or loose wires at the IAT sensor and at the PCM • "Short" circuit between the IAT sensor and the PCM • Faulty IAT sensor • PCM needs to be updated with the latest software • Faulty PCM
DTC: P0113 **1T PCM, MIL: Yes** **Year:** 2009, 2010 **Model:** Ridgeline **Engine:** 3.5L V6 **Transmission:** All	**Intake Air Temperature (IAT) Sensor 1 Circuit High Voltage:** With the ignition ON the IAT sensor 1 output voltage is 4.93 V or less for at least 5 seconds. **NOTE: Before you troubleshoot, record all freeze data and any on-board snapshot, and review the general troubleshooting information.** **Possible Causes:** • Poor connections or loose wires at the IAT sensor and at the PCM • "Open" circuit between the IAT sensor 1 and the PCM • Faulty IAT sensor 1 • PCM needs to be updated with the latest software • Faulty PCM

DTC	Trouble Code Title, Conditions & Possible Causes
DTC: P0113 **1T PCM, MIL: Yes** **Year:** 2009, 2010 **Model:** Element, Fit, Odyssey, Pilot **Engine:** 1.5L L4, 2.4L L4, 3.5L V6 **Transmission:** All	**Intake Air Temperature (IAT) Sensor Circuit High Voltage:** The IAT sensor resistance varies depending on temperature. The output voltage and the sensor resistance increase as the intake air temperature decreases. Conversely, the output voltage and the sensor resistance decrease as the intake air temperature increases. If the IAT sensor output voltage is excessively high, the Powertrain Control Module (PCM) detects a malfunction and a DTC is stored. The execution time is continuous and the duration time is 2 seconds or more. P0112 is not active. The IAT sensor output voltage is 4.92 V or more for at least 2 seconds. **NOTE: Before you troubleshoot, record all freeze data and any on-board snapshot.** **Possible Causes:** • Poor connections or loose terminals at the MAF sensor/IAT sensor and the PCM • "Open" in the wire between the PCM and the MAF sensor/IAT sensor • "Open" in the wire between the PCM and the MAF sensor/IAT sensor • Faulty MAF sensor/IAT sensor • PCM may need to be updated with the latest software • Faulty PCM
DTC: P0113 **1T PCM, MIL: Yes** **Year:** 2009, 2010 **Model:** Civic, CR-V **Engine:** 1.3L L4, 1.8L L4, 2.0L L4, 2.4L L4 **Transmission:** All	**Intake Air Temperature (IAT) Sensor Circuit High Voltage:** With ignition ON, the IAT sensor output voltage is 4.92 V or more for at least 2 seconds. **Possible Causes:** • Poor connections or loose wires at the IAT sensor and at the PCM • "Open" circuit between the PCM and the IAT sensor • Faulty IAT sensor • Faulty PCM
DTC: P0116 **2T PCM, MIL: Yes** **Year:** 2009, 2010 **Model:** Accord, Accord Crosstour, Civic, CR-V, Element, Fit, Odyssey, Pilot, Ridgeline **Engine:** 1.3L L4, 1.5L L4, 1.8L L4, 2.0L L4, 2.4L L4, 3.5L V6 **Transmission:** All	**Engine Coolant Temperature (ECT) Sensor 1 Circuit Range/Performance Problem :** The execution time is once per driving cycle and the duration time is 10 seconds or more. The following DTCs are not active P0101, P0102, P0103, P0107, P0108, P0117, P0118, P0134, P0135, P0154, P0155, P0171, P0172, P0174, P0175, P0300, P0301, P0302, P0303, P0304, P0305, P0306, P0335, P0339, P0340, P0344, P0401, P0404, P0443, P0496, P0497, P0506, P0507, P0627, P1077, P1078, P1128, P1129, P1172, P1174, P145C, P2195, P2197, P2227, P2228, P2229, P2237, P2238, P2240, P2241, P2243, P2245, P2247, P2249, P2251, P2252, P2254, P2255, P2413, P2610, P2646, P2647, P2648, P2649. Malfunction determination 1: With a completely cooled engine (one that has been off for at least 6 hours): When the change in coolant temperature after 42 minutes or more of running time or drive 5 miles (7 km) or more of driving time is 50°F (10°C) or less, a malfunction is detected. Malfunction determination 2: With a partially cooled engine (one that has been off for less than 6 hours): When the difference between the coolant temperature after 42 minutes or more of running time or drive 5 miles (7 km) or more of driving time the coolant temperature after the engine has been off for 150 minutes and then run for 10 seconds is 50°F (10°C) or less, a malfunction is detected. **NOTE: Before you troubleshoot, record all freeze data and any on-board snapshot.** **Possible Causes:** • Poor connections or loose terminals at ECT sensor 1 and the PCM • Faulty Replace ECT sensor (1) • Faulty PCM
DTC: P0117 **1T PCM, MIL: Yes** **Year:** 2009, 2010 **Model:** Accord, Accord Crosstour, Civic, CR-V, Element, Fit, Odyssey, Pilot, Ridgeline **Engine:** 1.3L L4, 1.5L L4, 1.8L L4, 2.0L L4, 2.4L L4, 3.5L V6 **Transmission:** All	**Engine Coolant Temperature (ECT) Sensor 1 Circuit Low Voltage:** If the ECT sensor 1 output voltage is less than a set value when the engine coolant temperature is high, the PCM detects a malfunction and a DTC is stored. The execution time is continuous and the duration time is 2 seconds or more. P0118 is not active. The ECT sensor 1 output voltage is 0.08 V or less for at least 2 seconds. **NOTE: Before you troubleshoot, record all freeze data and any on-board snapshot.** **Possible Causes:** • Poor connections or loose terminals at ECT sensor 1 and the PCM • "Short" in the wire between ECT sensor 1 and the PCM • Faulty ECT sensor (1) • PCM may need to be updated with the latest software • Faulty PCM

DTC	Trouble Code Title, Conditions & Possible Causes
DTC: P0118 **1T PCM, MIL: Yes** **Year:** 2009, 2010 **Model:** Accord, Accord Crosstour, Civic, CR-V, Element, Fit, Odyssey, Pilot, Ridgeline **Engine:** 1.3L L4, 1.5L L4, 1.8L L4, 2.0L L4, 2.4L L4, 3.5L V6 **Transmission:** All	**Engine Coolant Temperature (ECT) Sensor 1 Circuit High Voltage:** If the ECT sensor 1 output voltage is less than the set value when the engine coolant temperature is low, the PCM detects a malfunction and a DTC is stored. The execution time is continuous and the duration time is 2 seconds or more. DTC P0117 is not active. The ECT sensor 1 output voltage is 4.92 V or more for at least 2 seconds. **NOTE: Before you troubleshoot, record all freeze data and any on-board snapshot.** **Possible Causes:** • Poor connections or loose terminals at ECT sensor 1 and the PCM • "Open" in the wire between the PCM and ECT sensor 1 • "Open" in the wire between the PCM and ECT sensor 1 • Update the PCM if it does not have the latest software, • Faulty ECT sensor (1) • Faulty PCM
DTC: P0122 **1T PCM, MIL: Yes** **Year:** 2009, 2010 **Model:** Civic, CR-V, Element, Fit, Odyssey **Engine:** 1.3L L4, 1.5L L4, 1.8L L4, 2.0L L4, 2.4L L4, 3.5L V6 **Transmission:** All	**Throttle Position (TP) Sensor A Circuit Low Voltage:** With the ignition switch ON (II), the TP sensor output voltage is 0.3 V or less for at least 200 milliseconds. **NOTE: Before you troubleshoot, record all freeze data and any on-board snapshot, and review the general troubleshooting information.** **Possible Causes:** • Poor connections or loose terminals at the throttle body and the PCM • "Open" or "Short" circuit between the throttle body and the PCM • Faulty TP sensor A, replace throttle body • Update the PCM if it does not have the latest software, or substitute a known-good PCM • Faulty PCM
DTC: P0122 **1T PCM, MIL: Yes** **Year:** 2009, 2010 **Model:** Accord, Accord Crosstour, Pilot, Ridgeline **Engine:** 2.4L L4, 3.5L V6 **Transmission:** All	**Throttle Position (TP) Sensor A Circuit Low Voltage:** With the ignition switch ON (II), the TP sensor output voltage is 0.3 V or less for at least 200 milliseconds. **NOTE: Before you troubleshoot, record all freeze data and any on-board snapshot, and review the general troubleshooting information.** **Possible Causes:** • Poor connections or loose terminals at the throttle body and the PCM • "Open" or "Short" circuit between the throttle body and the PCM • Faulty throttle body • PCM may need to be updated with the latest software • Faulty PCM
DTC: P0123 **1T PCM, MIL: Yes** **Year:** 2009, 2010 **Model:** Accord, Accord Crosstour, Civic, CR-V, Element, Fit, Odyssey, Pilot, Ridgeline **Engine:** 1.3L L4, 1.5L L4, 1.8L L4, 2.0L L4, 2.4L L4, 3.5L V6 **Transmission:** All	**Throttle Position (TP) Sensor A Circuit High Voltage:** If the signal from TP sensor A is more than a fixed value for a set time, the PCM detects a TP sensor A malfunction and stores a DTC. The execution time is continuous and the duration time is 200 milliseconds or more. Ignition is running and DTCs P0122, P2101, P2118, P2135, P2176 are not active. The TP sensor A output voltage is 4.8 V or more for at least 200 milliseconds. **NOTE: Before you troubleshoot, record all freeze data and any on-board snapshot.** **Possible Causes:** • Poor connections or loose terminals at the throttle body and the PCM • "Open" in the wire between the throttle body and the PCM • "Open" in the wire between the throttle body and the PCM • Faulty throttle body • PCM may need to be updated with the latest software • Faulty PCM
DTC: P0125 **2T PCM, MIL: Yes** **Year:** 2009, 2010 **Model:** Accord, Accord Crosstour, Civic, CR-V, Element, Fit, Odyssey, Pilot, Ridgeline **Engine:** 1.3L L4, 1.5L L4, 1.8L L4, 2.0L L4, 2.4L L4, 3.5L V6 **Transmission:** All	**Engine Coolant Temperature (ECT) Sensor 1 Malfunction/Slow Response:** The execution time is once per driving cycle and the duration time is 20 minutes or less. The following DTCs are not active P0101, P0102, P0103, P0107, P0108, P0111, P0112, P0113, P0117, P0118, P0134, P0135, P0154, P0155, P0171, P0172, P0174, P0175, P0300, P0301, P0302, P0303, P0304, P0305, P0306, P0335, P0339, P0340, P0344, P0401, P0404, P0443, P0496, P0497, P0506, P0507, P0627, P1077, P1078, P1128, P1129, P1172, P1174, P145C, P2195, P2197, P2227, P2228, P2229, P2237, P2238, P2240, P2241, P2243, P2245, P2247, P2249, P2251, P2252, P2254, P2255, P2413, P2646, P2647, P2648, P2649. As the engine coolant warms, the ECT sensor 1 resistance decreases, and the PCM detects a low signal voltage. If the ECT sensor 1 output voltage does not reach a specified temperature at which closed-loop control for stoichiometric air/fuel ratio starts within a set time, depending on the initial coolant temperature after starting the engine, the PCM detects a malfunction. **NOTE: Before you troubleshoot, record all freeze data and any on-board snapshot.** **Possible Causes:** • Poor connections or loose terminals at ECT sensor 1, ECT sensor 2, and the PCM • Low coolant level • Faulty thermostat • Faulty ECT sensor (1) • Faulty PCM

DTC	Trouble Code Title, Conditions & Possible Causes
DTC: P0128 **2T PCM, MIL: Yes** **Year:** 2009, 2010 **Model:** Accord, Accord Crosstour **Engine:** 2.4L L4, 3.5L V6 **Transmission:** All	**Cooling System Malfunction:** With engine running the ECT sensor output (70 C) 158 F or less, an estimated engine coolant temperature is (75 C) 168 F or more. The difference between the estimated engine coolant temperature and the ECT sensor output is (15 C) 27 F or more. **NOTE: If the DTCs listed below are stored at the same time as DTC P0128, troubleshoot those DTCs first: P0107, P0108, P1128, P1129, P1106, P1107, P1108, P1259, P0401, P0116, P0117, P0118, P0112, P0113, P0335, P0336, P0300-P0306, P0505, P1519, then recheck for P0128.** **Possible Causes:** • Low coolant level • Faulty thermostat (Stuck Open) • Radiator fan runs constantly • PCM may need to be updated with the latest software • Faulty PCM
DTC: P0128 **2T PCM, MIL: Yes** **Year:** 2009, 2010 **Model:** Civic, CR-V, Element, Fit, Odyssey, Pilot, Ridgeline **Engine:** 1.3L L4, 1.5L L4, 1.8L L4, 2.0L L4, 2.4L L4, 3.5L V6 **Transmission:** All	**Cooling System Malfunction :** With the engine OFF for at least 6 hours, coolant temperature between 20 to 123°F (-7 to 50°C), the malfunction threshold are as follows. Malfunction determination 1: If the difference between the current measured coolant temperature at the radiator (ECT 2) and the initial coolant temperature at the radiator is at least 13°F (7°C) when the estimated coolant temperature at the engine (ECT 1) reaches 164°F (73°C), a malfunction is detected (Thermostat Stuck Open); or if the coolant temperature at the radiator (ECT 2) only reaches 68°F (20°C), a malfunction is detected. Malfunction determination 2: When the calculated engine coolant temperature (ECT 1) reaches 158°F (70°C) before the measured engine coolant temperature (ECT 1) reaches 158°F (70°C), a malfunction is detected. **Possible Causes:** • Radiator fan keep running, check the radiator fan circuit and the radiator fan relay • Faulty Thermostat • Low coolant level, and/or leaks • Poor connections or loose terminals at ECT sensor 1, ECT sensor 2, and the PCM • PCM may need to be updated with the latest software • Faulty PCM
DTC: P0133 **2T PCM, MIL: Yes** **Year:** 2009, 2010 **Model:** Accord, Civic, CR-V, Element, Fit **Engine:** 1.3L L4, 1.5L L4, 1.8L L4, 2.0L L4, 2.4L L4 **Transmission:** All	**Air/Fuel Ratio (A/F) Sensor (Sensor 1) Malfunction/Slow Response:** With the engine running and in closed loop, and the vehicle speed at least 33 mph. The total A/F sensor (SENSOR 1) output value for A/T models is 20 and for M/T models 38 or less in 0.8 seconds. **NOTE: Before you troubleshoot, record all freeze data and any on-board snapshot, and review the general troubleshooting information. If DTC P0139 is stored at the same time as DTC P0133, troubleshoot DTC P0139 first, then recheck for DTC P0133.** **Possible Causes:** • Poor connections or loose terminals at the A/F sensor (Sensor 1) and the ECM/PCM • Faulty A/F sensor (Sensor 1)
DTC: P0133 **2T O2S1, MIL: Yes** **Year:** 2009, 2010 **Model:** Accord, Accord Crosstour, Odyssey, Pilot, Ridgeline **Engine:** 3.5L V6 **Transmission:** All	**Rear Air/Fuel Ratio (A/F) Sensor (Bank 1, Sensor 1) Malfunction/Slow Response:** The execution time is once per drive cycle and the duration time is 6.8 seconds or more. Engine coolant temperature 156°F (69°C), Intake air temperature, Fuel trim, 0.73-1.47, Engine speed 1,250-2,200 rpm, Vehicle speed 33 mph or more. The total rear A/F sensor output value is 11 or less in 6.5 seconds. **NOTE: Before you troubleshoot, record all freeze data and any on-board snapshot.** Conditions for illuminating the MIL When a malfunction is detected during the first drive cycle, a Temporary DTC is stored in the PCM memory. If the malfunction recurs during the next (second) drive cycle, the MIL comes on and the DTC and the freeze frame data are stored. **Possible Causes:** • Poor connections or loose terminals at the A/F sensor (Sensor 1) and the PCM • Faulty A/F sensor (Sensor 1)
DTC: P0134 **2T PCM, MIL: Yes** **Year:** 2009, 2010 **Model:** Accord, Accord Crosstour, Odyssey, Pilot, Ridgeline **Engine:** 3.5L V6 **Transmission:** All	**Rear Air/Fuel Ratio (A/F) Sensor (Bank 1, Sensor 1) Heater System Malfunction:** The execution time is continuous and the duration time is 40 seconds or more. Battery voltage minimum 10.5 volts. Malfunction determination 1: The rear A/F sensor (bank 1, sensor 1) internal resistance value is 110 ohms or more for at least 40 seconds right after the engine starts. Malfunction determination 2: The rear A/F sensor (bank 1, sensor 1) internal resistance value is 110 ohms or more for at least 15 seconds. **NOTE: Before you troubleshoot, record all freeze data and any on-board snapshot.** The rear A/F sensor (bank 1, sensor 1) internal resistance value is 200 ohms or more for at least 1 second. **Possible Causes:** • Poor connections or loose terminals at the A/F sensor (Sensor 1), the relay and the PCM • "Open" circuit between the PCM and the A/F sensor • Faulty A/F Sensor (Bank 1, Sensor 1) • PCM needs to be updated with the latest software • Faulty PCM

DTC	Trouble Code Title, Conditions & Possible Causes
DTC: P0134 **2T PCM, MIL: Yes** **Year:** 2009, 2010 **Model:** Accord, Civic, CR-V, Element, Fit **Engine:** 1.3L L4, 1.5L L4, 1.8L L4, 2.0L L4, 2.4L L4 **Transmission:** All	**Air/Fuel Ratio (A/F) Sensor (Sensor 1) Heater System Malfunction :** With the engine running for 40 seconds or more and battery voltage above 10.5 V. Malfunction 1: From start up the A/F sensor (SENSOR 1) internal resistance value is 90 ohms or more for at least 40 seconds right after the engine starts. Malfunction 2: With the engine hot and at operating temperature the A/F sensor (SENSOR 1) internal resistance value is 250 ohms or more for at least 1.0 second. **NOTE: Before you troubleshoot, record all freeze data and any on-board snapshot, and review the general troubleshooting information. If DTC P0135 is stored at the same time as DTC P0134, troubleshoot DTC P0135 first, then recheck for DTC P0134.** **Possible Causes:** • Poor connections or loose terminals at the A/F sensor (Sensor 1), the A/F sensor relay, and the ECM/PCM • "Open" or "Short" between the A/F sensor (Sensor 1), the A/F sensor relay, and/or the ECM/PCM • Faulty A/F sensor relay • Faulty A/F sensor (Sensor 1)
DTC: P0135 **1T PCM, MIL: Yes** **Year:** 2009, 2010 **Model:** Accord, Accord Crosstour, Odyssey, Pilot, Ridgeline **Engine:** 3.5L V6 **Transmission:** All	**Rear Air/Fuel Ratio (A/F) Sensor (Bank 1, Sensor 1) Heater Circuit Malfunction:** The execution time is continuous and the duration time is 2 seconds or more. Engine running and battery at 10 volts or more. No return signal HIGH is detected when the PCM output duty is less than 2%. Return signal does not change when the PCM output duty is more than 20% and less than 80%. No return signal LOW is detected when the PCM output duty is more than 8%. **NOTE: Before you troubleshoot, record all freeze data and any on-board snapshot.** **Possible Causes:** • Poor connections or loose terminals at the A/F sensor (Sensor 1), the relay and the PCM • "Open" or Short" circuit between the A/F sensors, the relay box or the PCM • Faulty A/F sensor (Sensor 1) • Faulty PCM
DTC: P0135 **2T PCM, MIL: Yes** **Year:** 2009, 2010 **Model:** Accord, Civic, CR-V, Element, Fit **Engine:** 1.3L L4, 1.5L L4, 1.8L L4, 2.0L L4, 2.4L L4 **Transmission:** All	**Air/Fuel Ratio (A/F) Sensor (Sensor 1) Heater Circuit Malfunction :** With the engine running and the engine temperature at least 69 degrees the following malfunctions are detected. Malfunction 1: The heater current is 0.8 A or less for at least 4 seconds while the heater is active, and the heater current is 0.8 A or more for at least 4 seconds while the heater is not active. Malfunction 2: The heater current is 15.2 A or more for at least 0.6 second. **NOTE: Before you troubleshoot, record all freeze data and any on-board snapshot, and review the general troubleshooting information.** **Possible Causes:** • Blown fuse • Poor connections or loose wires at the A/F sensor (Sensor 1), the relay, and the ECM/PCM • "Open" or "Short" circuit between the A/F sensor (Sensor 1), A/F sensor relay and/or the ECM/PCM • Faulty A/F sensor relay • Faulty A/F sensor (Sensor 1) • Faulty ECM/PCM
DTC: P0137 **1T PCM, MIL: Yes** **Year:** 2009, 2010 **Model:** Accord, Civic, CR-V, Element, Fit **Engine:** 1.3L L4, 1.5L L4, 1.8L L4, 2.0L L4, 2.4L L4 **Transmission:** All	**Secondary HO2S (Sensor 2) Circuit Low Voltage :** With engine running the secondary HO2S (Sensor 2) output voltage remains in Zone 1 (0.29 V or below) for at least 72 seconds. **Possible Causes:** • Poor connections or loose wires at the secondary HO2S (Sensor 2) and the PCM • "Short" circuit between the secondary HO2S (Sensor 2) and the PCM • Faulty secondary HO2S (Sensor 2) • Faulty PCM
DTC: P0137 **1T PCM, MIL: Yes** **Year:** 2009, 2010 **Model:** Accord, Accord Crosstour, Odyssey, Pilot, Ridgeline **Engine:** 3.5L V6 **Transmission:** All	**Rear Secondary Heated Oxygen Sensor (Secondary HO2S (Bank 1, Sensor 2) Circuit Low Voltage:** The execution time is continuous and the duration time is 40 seconds or more. The system is in closed loop with the engine coolant temperature at 156°F (69°C) and the intake air temperature at -13°F (-25°C). The rear secondary HO2S output voltage is 0.05 V or less for at least 40 seconds. After current is applied to the rear secondary HO2S heater, if the rear secondary HO2S output continues low (lean) during feedback control, a malfunction is detected and a DTC is stored. **NOTE: Before you troubleshoot, record all freeze data and any on-board snapshot.** **Possible Causes:** • Poor connections or loose terminals at the secondary HO2S (Sensor 2) and the PCM • "Open" or "Short circuit between the PCM and the secondary HO2S (Sensor 2) • Faulty secondary HO2S (Sensor 2) • Faulty PCM

DTC	Trouble Code Title, Conditions & Possible Causes
DTC: P0138 **2T PCM, MIL: Yes** **Year:** 2009, 2010 **Model:** Accord, Civic, CR-V, Element, Fit **Engine:** 1.3L L4, 1.5L L4, 1.8L L4, 2.0L L4, 2.4L L4 **Transmission:** All	**Secondary HO2S (Sensor 2) Circuit High Voltage :** With engine running the Secondary HO2S (Sensor 2) remains in Zone 4 (0.80 V or above) for at least 72 seconds. **Possible Causes:** • Poor connections or loose wires at the secondary HO2S (Sensor 2) and at the PCM • "Open" circuit between the PCM and the secondary HO2S (Sensor 2) • Faulty secondary HO2S (Sensor 2) • Faulty PCM
DTC: P0138 **1T PCM, MIL: Yes** **Year:** 2009, 2010 **Model:** Accord, Accord Crosstour, Odyssey, Pilot, Ridgeline **Engine:** 3.5L V6 **Transmission:** All	**Rear Secondary Heated Oxygen Sensor (Secondary HO2S (Bank 1, Sensor 2) Circuit High Voltage:** The execution time is continuous and the duration time is 2 seconds or more. The rear secondary HO2S output voltage is 1.270 V or more for at least 5 seconds. After current is applied to the rear secondary HO2S heater, if the rear secondary HO2S output continues to exceed the upper limit used during feedback control, a malfunction is detected and a DTC is stored. **NOTE: Before you troubleshoot, record all freeze data and any on-board snapshot.** **Possible Causes:** • Poor connections or loose terminals at the secondary HO2S (Sensor 2) and the PCM • "Open" circuit between the PCM and the secondary HO2S (Sensor 2) • Faulty (Secondary HO2S (Bank 1, Sensor 2) • Faulty PCM
DTC: P0139 **2T PCM, MIL: Yes** **Year:** 2009, 2010 **Model:** Accord, Fit **Engine:** 1.5L L4, 2.4L L4 **Transmission:** All	**Secondary HO2S (Sensor 2) Slow Response :** Engine started, vehicle driven in closed loop in 4th or 6th gear at over 55 mph at steady speed, and the PCM detected the HO2S response time to switch between 300-600 mv was too slow, or that the rich to lean or lean to rich switch time was too slow. **Possible Causes:** • Poor connections or loose wires at the secondary HO2S (Sensor 2) and at the PCM • Faulty secondary HO2S (Sensor 2)
DTC: P0139 **2T PCM, MIL: Yes** **Year:** 2009, 2010 **Model:** Odyssey, Pilot **Engine:** 3.5L V6 **Transmission:** All	**Rear Secondary Heated Oxygen Sensor (Secondary HO2S (Bank 1, Sensor 2) Circuit Slow Response:** If the response time of the rear secondary HO2S becomes longer than the specified time after current to the secondary HO2S heater is applied, a malfunction is detected and a DTC is stored. The execution time is once per driving cycle and the duration time is 20.7 seconds or less. When a malfunction is detected during the first drive cycle, a Temporary DTC is stored in the PCM memory. If the malfunction recurs during the next (second) drive cycle, the MIL comes on and the DTC and the freeze frame data are stored. **NOTE: Before you troubleshoot, record all freeze data and any on-board snapshot.** **Possible Causes:** • Poor connections or loose terminals at the secondary HO2S (Sensor 2) and the PCM • Faulty (Secondary HO2S (Bank 1, Sensor 2) • Faulty PCM
DTC: P0139 **2T PCM, MIL: Yes** **Year:** 2009, 2010 **Model:** CR-V, Element **Engine:** 2.4L L4 **Transmission:** All	**Secondary HO2S (Sensor 2) Slow Response:** With the engine running for at least 2 minutes, coolant temperature at a minimum of 156 F (69 C) and the engine speed a maximum of 4,000 rpm, the secondary HO2S output voltage is 0.05 V or more for at least 10 seconds. **Possible Causes:** • Poor connections or loose terminals at secondary HO2S (Sensor 2) and the PCM • Faulty secondary HO2S (Sensor 2)
DTC: P0139 **2T PCM, MIL: Yes** **Year:** 2009, 2010 **Model:** Accord, Accord Crosstour, Ridgeline **Engine:** 3.5L V6 **Transmission:** All	**Rear Secondary HO2S (Bank 1, Sensor 2) Slow Response:** With the engine running in closed loop and driven at a steady speed of 35 mph (57 km/h) for at least 22.7 seconds. The rear secondary HO2S output drops to the response deterioration judgment threshold value and the response characteristics measurement is finished. OR. The voltage does not drop to the response deterioration judgment threshold value after the predetermined time has elapsed. (Elapsed time Minimum 0.51 seconds–Maximum 4.35 seconds) Before you troubleshoot, record all freeze data and any on-board snapshot, and review the general troubleshooting information. **Possible Causes:** • Poor connections or loose terminals at secondary HO2S (Sensor 2) and the ECM/PCM • Faulty secondary HO2S (Sensor 2)
DTC: P0141 **1T PCM, MIL: Yes** **Year:** 2009, 2010 **Model:** Accord, Accord Crosstour, Odyssey, Pilot, Ridgeline **Engine:** 3.5L V6 **Transmission:** All	**Rear Secondary Heated Oxygen Sensor (Bank 1 Sensor 2) Heater Circuit Malfunction:** The rear secondary HO2S heater output is 0.38 A or less, or 3.33 A or more, for at least 5 seconds when the heater is on. Engine running the battery voltage (IGP terminal of PCM) is 10.5-16 V, DTCs P0117, P0118 are not active. If the rear secondary HO2S heater draws more or less than a specified amperage, the PCM detects a malfunction and a DTC is stored. **NOTE: Before you troubleshoot, record all freeze data and any on-board snapshot.** **Possible Causes:** • Poor connections or loose terminals at the secondary HO2S (Sensor 2), the relay and the PCM • "Open" or "Short circuit between the PCM and the secondary HO2S (Sensor 2) • "Open or "Short circuit between the A/F sensors, the relay • Faulty (Secondary HO2S (Bank Sensor 2)

DTC	Trouble Code Title, Conditions & Possible Causes
DTC: P0141 **1T PCM, MIL: Yes** **Year:** 2009, 2010 **Model:** Accord, Civic, CR-V, Element, Fit **Engine:** 1.3L L4, 1.5L L4, 1.8L L4, 2.0L L4, 2.4L L4 **Transmission:** All	**Secondary HO2S (Sensor 2) Heater Circuit Malfunction :** With engine running the current is 0.5 A or less, or 3.6 A or more for at least 5 seconds when the heater is on. **Possible Causes:** • Poor connections or loose wires at the primary HO2S (Sensor 1) and the PCM • "Open" or "Short" circuit between the primary HO2S (Sensor 1) and the PCM • Faulty primary HO2S (Sensor 1) • Faulty PCM
DTC: P0153 **2T PCM, MIL: Yes** **Year:** 2009, 2010 **Model:** Accord, Accord Crosstour, Odyssey, Pilot, Ridgeline **Engine:** 3.5L V6 **Transmission:** All	**Front Air/Fuel Ratio (A/F) Sensor (Bank 2, Sensor 1) Malfunction/Slow Response:** The execution time is once per drive cycle and the duration time is 6.8 seconds or more. Engine coolant temperature 156°F (69°C), Intake air temperature, Fuel trim, 0.73-1.47, Engine speed 1,250-2,200 rpm, Vehicle speed 33 mph or more. The total front A/F sensor (bank 2, sensor 1) output value is 34 or less in 6.8 seconds. **Possible Causes:** • Poor connections or loose terminals at the A/F sensor (Sensor 1) and the PCM • Faulty A/F sensor (Sensor 1)
DTC: P0154 **2T PCM, MIL: Yes** **Year:** 2009, 2010 **Model:** Odyssey, Pilot, Ridgeline **Engine:** 3.5L V6 **Transmission:** All	**Front Air Fuel Ratio (A/F) Sensor (Bank 2, Sensor 1) Heater System Malfunction :** The execution time is continuous and the duration time is 40 seconds or more. Battery voltage minimum 10.5 volts. Malfunction determination 1: The front A/F sensor (bank 2, sensor 1) internal resistance value is 200 or more for at least 40 seconds right after the engine starts. Malfunction determination 2: The front A/F sensor (bank 2, sensor 1) internal resistance value is 270 or more for at least 1.0 second. **NOTE: Before you troubleshoot, record all freeze data and any on-board snapshot.** **Possible Causes:** • Poor connections or loose terminals at the A/F sensor (bank 2, Sensor 1), the under-hood fuse/relay box (PGM-FI subrelay) and the PCM • "Open" circuit between the PCM and the A/F sensor • Faulty A/F Sensor (Bank 2, Sensor 1) • PCM needs to be updated with the latest software • Faulty PCM
DTC: P0155 **2T PCM, MIL: Yes** **Year:** 2009, 2010 **Model:** Odyssey, Pilot, Ridgeline **Engine:** 3.5L V6 **Transmission:** All	**Front Air/Fuel Ratio (A/F) Sensor (Bank 2, Sensor 1) Heater Circuit Malfunction:** The execution time is continuous and the duration time is 2 seconds or more. Engine running and battery voltage at 10 volts or more. One of these 2 conditions must be met for at least 2 seconds: (1) No return signal HIGH is detected when the PCM output duty is less than 20%. (2) Return signal does not change when the PCM output duty is more than 20% and less than 80%. **NOTE: Before you troubleshoot, record all freeze data and any on-board snapshot.** **Possible Causes:** • Poor connections or loose terminals at the A/F sensor (Sensor 1), the relay and the PCM • "Open" or "Short" Circuit between the A/F sensors, the relay or PCM • Faulty (A/F) Sensor (Bank 2, Sensor 1) • Faulty PCM
DTC: P0157 **1T O2S HTR2, MIL: Yes** **Year:** 2009, 2010 **Model:** Accord, Accord Crosstour, Odyssey, Pilot, Ridgeline **Engine:** 3.5L V6 **Transmission:** All	**Front Secondary HO2S (Bank 2, Sensor 2) Circuit Low Voltage :** The execution time is continuous and the duration time is 40 seconds or more. The system is in closed loop with the engine coolant temperature at 156°F (69°C) and the intake air temperature at -13°F (-25°C). The rear secondary HO2S output voltage is 0.05 V or less for at least 40 seconds. After current is applied to the rear secondary HO2S heater, if the rear secondary HO2S output continues low (lean) during feedback control, a malfunction is detected and a DTC is stored. **NOTE: Before you troubleshoot, record all freeze data and any on-board snapshot.** **Possible Causes:** • Poor connections or loose terminals at the secondary HO2S (Sensor 2) and the PCM • "Open" or "Short circuit between the PCM and the secondary HO2S (Sensor 2) • Faulty secondary HO2S (Sensor 2) • Faulty PCM

DTC	Trouble Code Title, Conditions & Possible Causes
DTC: P0158 **2T PCM, MIL: Yes** **Year:** 2009, 2010 **Model:** Accord, Accord Crosstour, Odyssey, Pilot, Ridgeline **Engine:** 3.5L V6 **Transmission:** All	**Front Secondary HO2S (Bank 2, Sensor 2) Circuit High Voltage :** With the vehicle at operating temperature and driven between 1,000-3,000 rpm for at least 1 minute, the rear secondary HO2S output voltage is 1.270 V or more for at least 5 seconds. After current is applied to the rear secondary HO2S heater, if the rear secondary HO2S output continues to exceed the upper limit used during feedback control, a malfunction is detected and a DTC is stored. **NOTE: Before you troubleshoot, record all freeze data and any on-board snapshot.** **Possible Causes:** • Poor connections or loose terminals at the secondary HO2S (Sensor 2) and the PCM • "Open" circuit between the PCM and the secondary HO2S (Sensor 2) • Faulty (Secondary HO2S (Bank 1, Sensor 2) • Faulty PCM
DTC: P0159 **2T PCM, MIL: Yes** **Year:** 2009, 2010 **Model:** Accord, Accord Crosstour, Odyssey, Pilot, Ridgeline **Engine:** 3.5L V6 **Transmission:** All	**Front Secondary HO2S (Bank 2, Sensor 2) Slow Response :** The execution time is once per driving cycle and the duration time is 20.7 seconds or less. Engine coolant temperature is 156°F (69°C), intake air temperature -13°F (-25°C), engine speed 1,180-2,000 rpms and vehicle speed is 30 mph. When the front secondary HO2S output drops to the response deterioration judgment threshold value and the response characteristics measurement is finished.(0.77-4.58 seconds) **Possible Causes:** • Poor connections or loose terminals at the secondary HO2S (Sensor 2) and the PCM • Faulty Front Secondary HO2S (Bank 2, Sensor 2) • Faulty PCM
DTC: P0161 **1T PCM, MIL: Yes** **Year:** 2009, 2010 **Model:** Accord, Accord Crosstour, Odyssey, Pilot, Ridgeline **Engine:** 3.5L V6 **Transmission:** All	**Front Secondary HO2S (Bank 2, Sensor 2) Heater Circuit Malfunction :** The front secondary HO2S heater output is 0.38 A or less, or 3.33 A or more, for at least 5 seconds when the heater is on. Engine running the battery voltage (IGP terminal of PCM) is 10.5-16 V, DTCs P0117, P0118 are not active. If the front secondary HO2S heater draws more or less than a specified amperage, the PCM detects a malfunction and a DTC is stored. **NOTE: Before you troubleshoot, record all freeze data and any on-board snapshot.** **Possible Causes:** • Poor connections or loose terminals at the front secondary HO2S (Sensor 2), the relay and the PCM • "Open" or "Short circuit between the PCM and the front secondary HO2S (Sensor 2) • Faulty PCM
DTC: P0171 **2T PCM, MIL: Yes** **Year:** 2009, 2010 **Model:** Accord, Accord Crosstour, Odyssey, Pilot, Ridgeline **Engine:** 3.5L V6 **Transmission:** All	**Rear Bank (Bank 1) Fuel System Too Lean:** If long term fuel trim is higher than normal (too lean), a malfunction in the fuel metering components is detected and a DTC is stored. Long term fuel trim is higher than 1.188 (+18.8 %). Engine is at running temperature, vehicle speed at 550-4,000 rpms and the system is in closed loop. **NOTE: Before you troubleshoot, record all freeze data and any on-board snapshot.** **Possible Causes:** • Vacuum leaks • Clogged fuel filter • Faulty fuel pump or regulator • Faulty MAF sensor/IAT sensor • Faulty EVAP canister purge valve • Faulty Throttle body • Faulty fuel injectors
DTC: P0171 **2T PCM, MIL: Yes** **Year:** 2009, 2010 **Model:** Accord, Civic, CR-V, Element, Fit **Engine:** 1.3L L4, 1.5L L4, 1.8L L4, 2.0L L4, 2.4L L4 **Transmission:** All	**Fuel System Too Lean:** With the engine at operating temperature in closed loop, and the engine speed between 550-650 rpm. The Long term fuel trim is higher than 1.33 (+33%). If any related DTCs are listed first troubleshoot those DTCs first, then recheck for P0171. **NOTE: Before you troubleshoot, record all freeze data and any on-board snapshot, and review the general troubleshooting information.** **Possible Causes:** • Vacuum leaks • Improper valve clearances • Faulty injectors • Clogged fuel filter • Faulty fuel pump or regulator • Faulty EVAP canister purge valve • Faulty A/F sensor (Sensor 1)

DTC	Trouble Code Title, Conditions & Possible Causes
DTC: P0172 **2T PCM, MIL: Yes** **Year:** 2009, 2010 **Model:** Accord, Civic, CR-V, Element, Fit **Engine:** 1.3L L4, 1.5L L4, 1.8L L4, 2.0L L4, 2.4L L4 **Transmission:** All	**Fuel System Too Rich:** With the vehicle running, engine rpm between 550-650 and in closed loop. The long term fuel trim is lower than 0.785 (-215%). If any related DTCs are present, troubleshoot those DTCs first, then recheck P0172. **NOTE: Before you troubleshoot, record all freeze data and any on-board snapshot, and review the general troubleshooting information.** **Possible Causes:** • Improper fuel pressure • Improper valve clearances • Leaking injectors • MAF sensor/IAT sensor • Faulty EVAP canister purge valve • Faulty A/F sensor (Sensor 1)
DTC: P0172 **2T PCM, MIL: Yes** **Year:** 2009, 2010 **Model:** Accord, Accord Crosstour, Odyssey, Pilot, Ridgeline **Engine:** 3.5L V6 **Transmission:** All	**Rear Bank (Bank 1) Fuel System Too Rich:** The execution time is once per driving cycle and the duration time is 13 seconds or more. If long term fuel trim is lower than normal (too rich), a malfunction in the fuel metering components is detected and a DTC is stored. Engine is at running temperature, vehicle speed at 550-4,000 rpm and the system is in closed loop the long term fuel trim is lower than 0.820 (-18.0 %). **NOTE: Before you troubleshoot, record all freeze data and any on-board snapshot.** **Possible Causes:** • Faulty fuel pump or regulator • Engine valve clearance • Faulty coolant temp sensor • Faulty MAF sensor/IAT sensor • Faulty EVAP canister purge valve • Faulty Throttle body • Faulty fuel injectors
DTC: P0174 **1T PCM, MIL: Yes** **Year:** 2009, 2010 **Model:** Accord, Accord Crosstour, Odyssey, Pilot, Ridgeline **Engine:** 3.5L V6 **Transmission:** All	**Front Bank (Bank 2) Fuel System Too Lean :** If long term fuel trim is higher than normal (too lean), a malfunction in the fuel metering components is detected and a DTC is stored. Long term fuel trim is higher than 1.188 (+18.8 %). Engine is at running temperature, vehicle speed at 550-4,000 rpm and the system is in closed loop. **NOTE: Before you troubleshoot, record all freeze data and any on-board snapshot** **Possible Causes:** • Vacuum leaks • Clogged fuel filter • Faulty fuel pump or regulator • Faulty MAF sensor/IAT sensor • Faulty EVAP canister purge valve • Faulty Throttle body • Faulty fuel injectors
DTC: P0175 **2T PCM, MIL: Yes** **Year:** 2009, 2010 **Model:** Accord, Accord Crosstour, Odyssey, Pilot, Ridgeline **Engine:** 3.5L V6 **Transmission:** All	**Front Bank (Bank 2) Fuel System Too Rich :** The execution time is once per driving cycle and the duration time is 13 seconds or more. If long term fuel trim is lower than normal (too rich), a malfunction in the fuel metering components is detected and a DTC is stored. Engine is at running temperature, vehicle speed at 550-4,000 rpm and the system is in closed loop the long term fuel trim is lower than 0.820 (-18.0 %). **NOTE: Before you troubleshoot, record all freeze data and any on-board snapshot.** **Possible Causes:** • Faulty fuel pump or regulator • Engine valve clearance • Faulty coolant temp sensor • Faulty MAF sensor/IAT sensor • Faulty EVAP canister purge valve • Faulty Throttle body • Faulty fuel injectors
DTC: P0181 **2T PCM, MIL: Yes** **Year:** 2009, 2010 **Model:** Civic **Engine:** 1.8L L4 **Transmission:** All	**Fuel Temperature Sensor Circuit Range/Performance Problem :** With the engine OFF for a minimum of 6 hours, then start the engine and run for at least 10 seconds. When the difference between the fuel temperature sensor and the IAT is 45°F (25°C) or more a malfunction is detected. **Possible Causes:** • Poor connections or loose terminals at the fuel temperature sensor and the PCM • Faulty fuel temperature sensor

DTC	Trouble Code Title, Conditions & Possible Causes
DTC: P0182 **2T PCM, MIL: Yes** **Year:** 2009, 2010 **Model:** Civic **Engine:** 1.8L L4 **Transmission:** All	**Fuel Temperature Sensor Circuit Low Voltage :** With the ignition switch ON and P0183 not active, the fuel temperature output voltage is 0.08 V or less for at least 2 seconds. **Possible Causes:** • Poor connections or loose terminals at the fuel temperature sensor and the PCM • "Short" circuit between the PCM and the fuel temperature sensor • Faulty fuel temperature sensor • PCM may need to be updated with the latest software • Faulty PCM
DTC: P0183 **2T PCM, MIL: Yes** **Year:** 2009, 2010 **Model:** Civic **Engine:** 1.8L L4 **Transmission:** All	**Fuel Temperature Sensor Circuit High Voltage :** With the ignition switch ON and the Intake Air Temperature (IAT) a minimum of 22°F (30°C), the fuel temperature output voltage is 4.92 V or more for at least 2 seconds. **NOTE: Before you troubleshoot, record all freeze data and any on-board snapshot, and review the general troubleshooting information.** **Possible Causes:** • Poor connections or loose terminals at the fuel temperature sensor and the PCM • "Open" circuit between the PCM and the fuel temperature sensor • Faulty fuel temperature sensor • PCM may need to be updated with the latest software • Faulty PCM
DTC: P0191 **2T PCM, MIL: Yes** **Year:** 2009, 2010 **Model:** Civic **Engine:** 1.8L L4 **Transmission:** All	**Fuel Rail Pressure Sensor Circuit Range/Performance Problem :** With the engine running and the coolant temperature a minimum of 156°F (69°C), the 2 malfunction threshold modes are as follows. Mode 1: The difference between PF2A values when the MAP is maximum and when MAP is minimum, is 4 kPa (0.04 kgf/cm2, 0.5 psi) or less. Mode 2: PF2A is not between 204-326 kPa (2.1-3.3 kgf/cm2, 30-47 psi) for at least 3 seconds. (PF2S=PF2A-Map) **Possible Causes:** • Loose or damaged fuel pressure regulator vacuum line • Poor connections or loose terminals at the fuel rail pressure sensor and the PCM • Leaking fuel system • Faulty fuel rail pressure sensor • Faulty fuel pressure regulator • Faulty MAP sensor
DTC: P0192 **1T PCM, MIL: Yes** **Year:** 2009, 2010 **Model:** Civic **Engine:** 1.8L L4 **Transmission:** All	**Fuel Rail Pressure Sensor Circuit Low Voltage :** With the ignition switch ON and P0193 not active, the fuel rail pressure sensor output voltage is 0.23 V or less for at least 2 seconds. **Possible Causes:** • Poor connections or loose terminals at the fuel rail pressure sensor and the PCM • "Open" or "Short" circuit between the PCM and the fuel rail pressure sensor • Faulty fuel rail pressure sensor • PCM may need to be updated with the latest software • Faulty PCM
DTC: P0222 **1T PCM, MIL: Yes** **Year:** 2009, 2010 **Model:** Accord, Accord Crosstour, Civic, CR-V, Element, Fit, Odyssey, Pilot, Ridgeline **Engine:** 1.3L L4, 1.5L L4, 1.8L L4, 2.0L L4, 2.4L L4, 3.5L V6 **Transmission:** All	**Throttle Position (TP) Sensor B Circuit Low Voltage:** The execution time is continuous and the duration time is 200 milliseconds or more. Ignition on and DTCs P0223, P2101, P2118, P2135, P2176 are not active. The TP sensor B output voltage is 0.3 V or less for at least 200 milliseconds. If the signal from TP sensor B is less than a fixed value for a set time, the PCM detects a TP sensor B malfunction and stores a DTC. **NOTE: Before you troubleshoot, record all freeze data and any on-board snapshot.** **Possible Causes:** • Poor connections or loose terminals at the throttle body and the PCM • "Open" or "Short" circuit between the throttle body and the PCM • Faulty throttle body • PCM may need to be updated with the latest software • Faulty PCM
DTC: P0223 **1T PCM, MIL: Yes** **Year:** 2009, 2010 **Model:** Accord, Accord Crosstour, Civic, CR-V, Element, Fit, Odyssey, Pilot, Ridgeline **Engine:** 1.3L L4, 1.5L L4, 1.8L L4, 2.0L L4, 2.4L L4, 3.5L V6 **Transmission:** All	**Throttle Position (TP) Sensor B Circuit High Voltage :** The execution time is continuous and the duration time is 200 milliseconds or more. Ignition ON and DTCs P0222, P2101, P2118, P2135, P2176 are not active. The TP sensor B output voltage is 4.8 V or more for at least 200 milliseconds. If the signal from TP sensor B is more than a fixed value for a set time, the PCM detects a TP sensor B malfunction and stores a DTC. **NOTE: Before you troubleshoot, record all freeze data and any on-board snapshot.** **Possible Causes:** • Poor connections or loose terminals at the throttle body and the PCM • "Open" or "Short" between the throttle body and the PCM • PCM may need to be updated with the latest software • Faulty throttle body • Faulty PCM

DTC	Trouble Code Title, Conditions & Possible Causes
DTC: P0300 **2T PCM, MIL: Yes** **Year:** 2009, 2010 **Model:** Accord, Accord Crosstour, Civic, CR-V, Element, Fit, Odyssey, Pilot, Ridgeline **Engine:** 1.3L L4, 1.5L L4, 1.8L L4, 2.0L L4, 2.4L L4, 3.5L V6 **Transmission:** All	**Random Misfire Detected:** The execution time is once per driving cycle and the duration time is 200 revolutions for Type 1 and 1,000 revolutions for Type 2. Conditions for illuminating the MIL Type 1 and Type 2. * Misfire Type 1 (Severe) Per 200 revolutions 27-90 times. If a type 1 misfire (catalyst damaging) occurs once, the MIL blinks once per second, a Temporary DTC is stored, and the high rpm fuel injection stop system activates. The fuel injection stops, at high rpm only, on the cylinder that has the highest misfire counts. the MIL continues to blink, and the fuel injection stays off at high rpm, until the drive is completed. If a type 1 misfire occurs during a second drive cycle, the MIL and fuel injection behave the same and a DTC is stored. After a type 1 misfire has been detected during two drive cycles, the MIL comes on and stays on beginning with the third drive cycle, unless the Temporary DTC has been cleared by the PCM. Even if the MIL is on, it will start blinking if a type 1 misfire occurs. If the malfunction recurs during the next (second) drive cycle, the MIL comes on and the DTC and the freeze frame data are stored. * Misfire Type 2 (Light) Per 1,000 revolutions 55 times. If a type 2 misfire (emission-related but not severe enough to immediately damage the catalyst) occurs, a Temporary DTC is stored, but the MIL does come on or blink. If a type 2 misfire occurs during a second drive cycle, the MIL comes on and stays on unless the Temporary DTC has been cleared by the PCM. **NOTE: Before you troubleshoot, record all freeze data and any on-board snapshot.** **Possible Causes:** • Poor fuel quality • Clogged fuel filter • Faulty spark plugs • Faulty Fuel pump and/or regulator • Poor connections or loose terminals at the ignition coil, the injector, and the PCM • Check the CKP pattern learn procedure
DTC: P0301 **2T PCM, MIL: Yes** **Year:** 2009, 2010 **Model:** Accord, Accord Crosstour, Civic, CR-V, Element, Fit, Odyssey, Pilot, Ridgeline **Engine:** 1.3L L4, 1.5L L4, 1.8L L4, 2.0L L4, 2.4L L4, 3.5L V6 **Transmission:** All	**Cylinder 1 Misfire Detected :** The execution time is once per driving cycle and the duration time is 200 revolutions for Type 1 and 1,000 revolutions for Type 2. Conditions for illuminating the MIL Type 1 and Type 2. * Misfire Type 1 (Severe) Per 200 revolutions 27-90 times. If a type 1 misfire (catalyst damaging) occurs once, the MIL blinks once per second, a Temporary DTC is stored, and the high rpm fuel injection stop system activates. The fuel injection stops, at high rpm only, on the cylinder that has the highest misfire counts. the MIL continues to blink, and the fuel injection stays off at high rpm, until the drive is completed. If a type 1 misfire occurs during a second drive cycle, the MIL and fuel injection behave the same and a DTC is stored. After a type 1 misfire has been detected during two drive cycles, the MIL comes on and stays on beginning with the third drive cycle, unless the Temporary DTC has been cleared by the PCM. Even if the MIL is on, it will start blinking if a type 1 misfire occurs. If the malfunction recurs during the next (second) drive cycle, the MIL comes on and the DTC and the freeze frame data are stored. * Misfire Type 2 (Light) Per 1,000 revolutions 55 times. If a type 2 misfire (emission-related but not severe enough to immediately damage the catalyst) occurs, a Temporary DTC is stored, but the MIL does come on or blink. If a type 2 misfire occurs during a second drive cycle, the MIL comes on and stays on unless the Temporary DTC has been cleared by the PCM. **NOTE: Before you troubleshoot, record all freeze data and any on-board snapshot.** **Possible Causes:** • Faulty spark plug • Faulty ignition coil • Faulty fuel injector • Faulty fuel pump and/or regulator • Poor connections or loose terminals at the ignition coil, the injector, and the PCM • "Open" or "Short" circuit between the ignition coil and the under-hood fuse/relay box • "Open" or "Short" circuit between the PCM and the ignition coil • Internal engine problem • Incorrect PCM idle learn procedure • Incorrect CKP pattern learn procedure • MAF sensor

DTC	Trouble Code Title, Conditions & Possible Causes
DTC: P0302 **2T PCM, MIL: Yes** **Year:** 2009, 2010 **Model:** Accord, Accord Crosstour, Civic, CR-V, Element, Fit, Odyssey, Pilot, Ridgeline **Engine:** 1.3L L4, 1.5L L4, 1.8L L4, 2.0L L4, 2.4L L4, 3.5L V6 **Transmission:** All	**Cylinder 2 Micfire Detected :** The execution time is once per driving cycle and the duration time is 200 revolutions for Type 1 and 1,000 revolutions for Type 2. Conditions for illuminating the MIL Type 1 and Type 2. * Misfire Type 1 (Severe) Per 200 revolutions 27-90 times. If a type 1 misfire (catalyst damaging) occurs once, the MIL blinks once per second, a Temporary DTC is stored, and the high rpm fuel injection stop system activates. The fuel injection stops, at high rpm only, on the cylinder that has the highest misfire counts. the MIL continues to blink, and the fuel injection stays off at high rpm, until the drive is completed. If a type 1 misfire occurs during a second drive cycle, the MIL and fuel injection behave the same and a DTC is stored. After a type 1 misfire has been detected during two drive cycles, the MIL comes on and stays on beginning with the third drive cycle, unless the Temporary DTC has been cleared by the PCM. Even if the MIL is on, it will start blinking if a type 1 misfire occurs. If the malfunction recurs during the next (second) drive cycle, the MIL comes on and the DTC and the freeze frame data are stored. * Misfire Type 2 (Light) Per 1,000 revolutions 55 times. If a type 2 misfire (emission-related but not severe enough to immediately damage the catalyst) occurs, a Temporary DTC is stored, but the MIL does come on or blink. If a type 2 misfire occurs during a second drive cycle, the MIL comes on and stays on unless the Temporary DTC has been cleared by the PCM. **NOTE: Before you troubleshoot, record all freeze data and any on-board snapshot.** **Possible Causes:** • Faulty spark plug • Faulty ignition coil • Faulty fuel injector • Faulty fuel pump and/or regulator • Poor connections or loose terminals at the ignition coil, the injector, and the PCM • "Open" or "Short" circuit between the ignition coil and the under-hood fuse/relay box • "Open" or "Short" circuit between the PCM and the ignition coil • Internal engine problem • Incorrect PCM idle learn procedure • Incorrect CKP pattern learn procedure • MAF sensor
DTC: P0303 **2T PCM, MIL: Yes** **Year:** 2009, 2010 **Model:** Accord, Accord Crosstour, Civic, CR-V, Element, Fit, Odyssey, Pilot, Ridgeline **Engine:** 1.3L L4, 1.5L L4, 1.8L L4, 2.0L L4, 2.4L L4, 3.5L V6 **Transmission:** All	**Cylinder 3 Misfire Detected :** The execution time is once per driving cycle and the duration time is 200 revolutions for Type 1 and 1,000 revolutions for Type 2. Conditions for illuminating the MIL Type 1 and Type 2. * Misfire Type 1 (Severe) Per 200 revolutions 27-90 times. If a type 1 misfire (catalyst damaging) occurs once, the MIL blinks once per second, a Temporary DTC is stored, and the high rpm fuel injection stop system activates. The fuel injection stops, at high rpm only, on the cylinder that has the highest misfire counts. the MIL continues to blink, and the fuel injection stays off at high rpm, until the drive is completed. If a type 1 misfire occurs during a second drive cycle, the MIL and fuel injection behave the same and a DTC is stored. After a type 1 misfire has been detected during two drive cycles, the MIL comes on and stays on beginning with the third drive cycle, unless the Temporary DTC has been cleared by the PCM. Even if the MIL is on, it will start blinking if a type 1 misfire occurs. If the malfunction recurs during the next (second) drive cycle, the MIL comes on and the DTC and the freeze frame data are stored. * Misfire Type 2 (Light) Per 1,000 revolutions 55 times. If a type 2 misfire (emission-related but not severe enough to immediately damage the catalyst) occurs, a Temporary DTC is stored, but the MIL does come on or blink. If a type 2 misfire occurs during a second drive cycle, the MIL comes on and stays on unless the Temporary DTC has been cleared by the PCM. **NOTE: Before you troubleshoot, record all freeze data and any on-board snapshot.** **Possible Causes:** • Faulty spark plug • Faulty ignition coil • Faulty fuel injector • Faulty fuel pump and/or regulator • Poor connections or loose terminals at the ignition coil, the injector, and the PCM • "Open" or "Short" circuit between the ignition coil and the under-hood fuse/relay box • "Open" or "Short" circuit between the PCM and the ignition coil • Internal engine problem • Incorrect PCM idle learn procedure • Incorrect CKP pattern learn procedure • MAF sensor

DTC	Trouble Code Title, Conditions & Possible Causes
DTC: P0304 **2T PCM, MIL: Yes** **Year:** 2009, 2010 **Model:** Accord, Accord Crosstour, Civic, CR-V, Element, Fit, Odyssey, Pilot, Ridgeline **Engine:** 1.3L L4, 1.5L L4, 1.8L L4, 2.0L L4, 2.4L L4, 3.5L V6 **Transmission:** All	**Cylinder 4 Misfire Detected :** The execution time is once per driving cycle and the duration time is 200 revolutions for Type 1 and 1,000 revolutions for Type 2. Conditions for illuminating the MIL Type 1 and Type 2. * Misfire Type 1 (Severe) Per 200 revolutions 27-90 times. If a type 1 misfire (catalyst damaging) occurs once, the MIL blinks once per second, a Temporary DTC is stored, and the high rpm fuel injection stop system activates. The fuel injection stops, at high rpm only, on the cylinder that has the highest misfire counts. the MIL continues to blink, and the fuel injection stays off at high rpm, until the drive is completed. If a type 1 misfire occurs during a second drive cycle, the MIL and fuel injection behave the same and a DTC is stored. After a type 1 misfire has been detected during two drive cycles, the MIL comes on and stays on beginning with the third drive cycle, unless the Temporary DTC has been cleared by the PCM. Even if the MIL is on, it will start blinking if a type 1 misfire occurs. If the malfunction recurs during the next (second) drive cycle, the MIL comes on and the DTC and the freeze frame data are stored. * Misfire Type 2 (Light) Per 1,000 revolutions 55 times. If a type 2 misfire (emission-related but not severe enough to immediately damage the catalyst) occurs, a Temporary DTC is stored, but the MIL does come on or blink. If a type 2 misfire occurs during a second drive cycle, the MIL comes on and stays on unless the Temporary DTC has been cleared by the PCM. **NOTE: Before you troubleshoot, record all freeze data and any on-board snapshot.** **Possible Causes:** • Faulty spark plug • Faulty ignition coil • Faulty fuel injector • Faulty fuel pump and/or regulator • Poor connections or loose terminals at the ignition coil, the injector, and the PCM • "Open" or "Short" circuit between the ignition coil and the under-hood fuse/relay box • "Open" or "Short" circuit between the PCM and the ignition coil • Internal engine problem • Incorrect PCM idle learn procedure • Incorrect CKP pattern learn procedure • MAF sensor
DTC: P0305 **2T PCM, MIL: Yes** **Year:** 2009, 2010 **Model:** Accord, Accord Crosstour, Odyssey, Pilot, Ridgeline **Engine:** 3.5L V6 **Transmission:** All	**Cylinder 5 Misfire Detected :** The execution time is once per driving cycle and the duration time is 200 revolutions for Type 1 and 1,000 revolutions for Type 2. Conditions for illuminating the MIL Type 1 and Type 2. * Misfire Type 1 (Severe) Per 200 revolutions 27-90 times. If a type 1 misfire (catalyst damaging) occurs once, the MIL blinks once per second, a Temporary DTC is stored, and the high rpm fuel injection stop system activates. The fuel injection stops, at high rpm only, on the cylinder that has the highest misfire counts. the MIL continues to blink, and the fuel injection stays off at high rpm, until the drive is completed. If a type 1 misfire occurs during a second drive cycle, the MIL and fuel injection behave the same and a DTC is stored. After a type 1 misfire has been detected during two drive cycles, the MIL comes on and stays on beginning with the third drive cycle, unless the Temporary DTC has been cleared by the PCM. Even if the MIL is on, it will start blinking if a type 1 misfire occurs. If the malfunction recurs during the next (second) drive cycle, the MIL comes on and the DTC and the freeze frame data are stored. * Misfire Type 2 (Light) Per 1,000 revolutions 55 times. If a type 2 misfire (emission-related but not severe enough to immediately damage the catalyst) occurs, a Temporary DTC is stored, but the MIL does come on or blink. If a type 2 misfire occurs during a second drive cycle, the MIL comes on and stays on unless the Temporary DTC has been cleared by the PCM. **NOTE: Before you troubleshoot, record all freeze data and any on-board snapshot.** **Possible Causes:** • Faulty spark plug • Faulty ignition coil • Faulty fuel injector • Faulty fuel pump and/or regulator • Poor connections or loose terminals at the ignition coil, the injector, and the PCM • "Open" or "Short" circuit between the ignition coil and the under-hood fuse/relay box • "Open" or "Short" circuit between the PCM and the ignition coil • Internal engine problem • Incorrect PCM idle learn procedure • Incorrect CKP pattern learn procedure • MAF sensor

DTC	Trouble Code Title, Conditions & Possible Causes
DTC: P0306 **1T PCM, MIL: Yes** **Year:** 2009, 2010 **Model:** Accord, Accord Crosstour, Odyssey, Pilot, Ridgeline **Engine:** 3.5L V6 **Transmission:** All	**Cylinder 6 Misfire Detected:** The execution time is once per driving cycle and the duration time Is 200 revolutions for Type 1 and 1,000 revolutions for Type 2. Conditions for illuminating the MIL Type 1 and Type 2. * Misfire Type 1 (Severe) Per 200 revolutions 27-90 times. If a type 1 misfire (catalyst damaging) occurs once, the MIL blinks once per second, a Temporary DTC is stored, and the high rpm fuel injection stop system activates. The fuel injection stops, at high rpm only, on the cylinder that has the highest misfire counts. the MIL continues to blink, and the fuel injection stays off at high rpm, until the drive is completed. If a type 1 misfire occurs during a second drive cycle, the MIL and fuel injection behave the same and a DTC is stored. After a type 1 misfire has been detected during two drive cycles, the MIL comes on and stays on beginning with the third drive cycle, unless the Temporary DTC has been cleared by the PCM. Even if the MIL is on, it will start blinking if a type 1 misfire occurs. If the malfunction recurs during the next (second) drive cycle, the MIL comes on and the DTC and the freeze frame data are stored. * Misfire Type 2 (Light) Per 1,000 revolutions 55 times. If a type 2 misfire (emission-related but not severe enough to immediately damage the catalyst) occurs, a Temporary DTC is stored, but the MIL does come on or blink. If a type 2 misfire occurs during a second drive cycle, the MIL comes on and stays on unless the Temporary DTC has been cleared by the PCM. **NOTE: Before you troubleshoot, record all freeze data and any on-board snapshot** **Possible Causes:** • Faulty spark plug • Faulty ignition coil • Faulty fuel injector • Faulty fuel pump and/or regulator • Poor connections or loose terminals at the ignition coil, the injector, and the PCM • "Open" or "Short" circuit between the ignition coil and the under-hood fuse/relay box • "Open" or "Short" circuit between the PCM and the ignition coil • Internal engine problem • Incorrect PCM idle learn procedure • Incorrect CKP pattern learn procedure • MAF sensor
DTC: P0325 **1T PCM, MIL: Yes** **Year:** 2009, 2010 **Model:** Accord, Accord Crosstour, Civic, CR-V, Element, Fit, Odyssey, Pilot, Ridgeline **Engine:** 1.3L L4, 1.5L L4, 1.8L L4, 2.0L L4, 2.4L L4, 3.5L V6 **Transmission:** All	**Knock Sensor Circuit Malfunction :** The execution time is continuous and the duration time is 5 seconds or more. Engine running, speed 2,000 rpm or more, coolant temperature at 140°F (60°C) or more. No signals from the knock sensor are detected for at least 5 seconds. If the signals from the knock sensor do not vary for a set time, the PCM detects a malfunction and stores a DTC. **NOTE: Before you troubleshoot, record all freeze data and any on-board snapshot.** **Possible Causes:** • Poor connections or loose terminals at the knock sensor and the PCM • "Open" or "Short" circuit between the PCM and the knock sensor sub-harness • "Open" or "Short" circuit in the knock sensor subharness • Poor connections or loose terminals at the ignition coil, the injector, and the PCM • Faulty knock sensor • PCM may need to be updated with the latest software • Faulty PCM
DTC: P0335 **1T PCM, MIL: Yes** **Year:** 2009, 2010 **Model:** Odyssey **Engine:** 3.5L V6 **Transmission:** All	**Crankshaft Position (CKP) Sensor No Signal :** With the engine running with the speed over 750 rpm, and DTCs P0365, P0369 are not active. No input signals from the CKP sensor are detected while the signals from the CMP sensor are detected at least 125 times. **NOTE: Before you troubleshoot, record all freeze data and any on-board snapshot, and review the general troubleshooting information.** **Possible Causes:** • Poor connections or loose terminals at the CKP sensor and the PCM • "Open" circuit between the CKP sensor and PGM-FI main relay 1 • "Open" circuit between the CKP sensor and ground • "Open" or "Short" circuit between the PCM and the CKP sensor • Faulty CKP sensor • PCM may need to be updated with the latest software • Faulty PCM

DTC	Trouble Code Title, Conditions & Possible Causes
DTC: P0335 **1T PCM, MIL: Yes** **Year:** 2009, 2010 **Model:** Pilot, Ridgeline **Engine:** 3.5L V6 **Transmission:** All	**Crankshaft Position (CKP) Sensor No Signal:** With the engine running for 2 seconds or more at 750 rpm, no input signals from the CKP sensor are detected while the signals from the CMP sensor are detected for at least 125 times. **Possible Causes:** • Poor connections or loose terminals at the CKP sensor and the PCM • "Open" circuit between the CKP sensor and PGM-FI main relay 1 • "Open" in the wire between the CKP sensor and ground • "Open" or "Short" circuit between the CKP sensor and PCM • Faulty CKP sensor • PCM may need to be updated with the latest software • Faulty PCM
DTC: P0335 **1T PCM, TCIL: Yes** **Year:** 2009, 2010 **Model:** Accord, Accord Crosstour, CR-V, Element, Fit **Engine:** 1.5L L4, 2.4L L4, 3.5L V6 **Transmission:** All	**Crankshaft Position (CKP) Sensor A No Signal :** The execution is continuous and the duration is for 2 seconds or more (when the engine speed is 750 rpm). MIL OFF, D indicator blinks. No signals from the CKP sensor are input at least 63 times. If no pulsing signals from the CKP sensor are detected, a malfunction is detected and a DTC is stored. **Possible Causes:** • Poor connections or loose terminals at CKP sensor A and the PCM • "Open" or "Short" in the wire between the PCM and CKP sensor A • Faulty CKP sensor A • PCM may need to be updated with the latest software • Faulty PCM
DTC: P0335 **1T PCM, MIL: Yes** **Year:** 2009, 2010 **Model:** Accord, Civic **Engine:** 1.3L L4, 1.8L L4, 2.0L L4, 2.4L L4 **Transmission:** All	**Crankshaft Position (CKP) Sensor No Signal:** With the engine running the PCM has detected no CKP pulses over 75 times. **Possible Causes:** • Poor connections or loose wires at the CKP sensor and at the PCM • "Open" or "Short" circuit between the PCM and the CKP sensor • Faulty CKP sensor • Faulty PCM
DTC: P0339 **1T PCM, MIL: Yes** **Year:** 2009, 2010 **Model:** Ridgeline **Engine:** 3.5L V6 **Transmission:** All	**Crankshaft Position (CKP) Sensor A Intermittent Interruption:** With the engine running, other than 22 pulses are detected during intervals between reference pulses for each crankshaft revolution. This condition has been detected at least 30 times. DTCs P0335, P0340, P0385 and P0389 are not active. **NOTE: Before you troubleshoot, record all freeze data and any on-board snapshot, and review the general troubleshooting information.** **Possible Causes:** • Poor connections or loose terminals at CKP sensor A/B and the ECM/PCM • Poor connections at engine ground and/or body ground • Damaged CKP sensor A/B pulse plate on the timing belt drive pulley • Faulty CKP sensor
DTC: P0339 **1T PCM, MIL: Yes** **Year:** 2009, 2010 **Model:** Accord, Accord Crosstour, Civic, CR-V, Element, Fit, Odyssey, Pilot **Engine:** 1.3L L4, 1.5L L4, 1.8L L4, 2.0L L4, 2.4L L4, 3.5L V6 **Transmission:** All	**Crankshaft Position (CKP) Sensor Circuit Intermittent Interruption :** With the engine running other than 58 pulses are detected during intervals between reference pulses for each crank revolution. This condition has been detected 30 times. **NOTE: Before you troubleshoot, record all freeze data and any on-board snapshot, and review the general troubleshooting information.** **Possible Causes:** • Poor connections at the ECM/PCM, CKP sensor, engine ground, body ground • Damaged CKP sensor pulse plate • Faulty CKP sensor
DTC: P0340 **1T PCM, MIL: Yes** **Year:** 2009, 2010 **Model:** Accord, Civic, CR-V, Element **Engine:** 2.0L L4, 2.4L L4 **Transmission:** All	**Camshaft Position (CMP) Sensor A No Signal :** Engine running, NO signals from the CMP sensor A are detected 25 times or more in succession. **NOTE: Before you troubleshoot, record all freeze data and any on-board snapshot, and review the general troubleshooting information.** **Possible Causes:** • Poor connections or loose terminals at CMP sensor A and the ECM/PCM • "Open" circuit between the fuse in the under-dash fuse/relay box and CMP sensor A • "Open" in the wire between CMP sensor A and ground • "Open" or "Short" circuit between the ECM/PCM and CMP sensor A • Faulty CMP sensor A • ECM/PCM needs to be updated with the latest software • Faulty ECM/PCM

DTC	Trouble Code Title, Conditions & Possible Causes
DTC: P0340 **1T PCM, MIL: Yes** **Year:** 2009, 2010 **Model:** Accord, Accord Crosstour, Odyssey **Engine:** 3.5L V6 **Transmission:** All	**Camshaft Position (CMP) Sensor No Signal:** With the engine running no signals from CMP sensor A are detected while signals from the CKP sensor are detected 352 times in succession. The execution time is continuous and DTCs P0335, P0339, P0344 are not active. **NOTE: Before you troubleshoot, record all freeze data and any on-board snapshot.** **Possible Causes:** • Poor connections or loose terminals at the CMP sensor and the PCM • "Open" circuit between the CMP sensor and PGM-FI main relay • "Open" or "Short" circuit between the PCM and the CMP sensor • "Open" in the wire between the CMP sensor and ground • Faulty CMP sensor • PCM needs to be updated with the latest software • Faulty PCM
DTC: P0341 **2T PCM, MIL: Yes** **Year:** 2009, 2010 **Model:** Accord, Civic, CR-V, Element **Engine:** 2.0L L4, 2.4L L4 **Transmission:** All	**Camshaft Position (CMP) Sensor A and CKP Sensor Incorrect Phase Detected :** With the engine running and at operating temperature, and the vehicle speed between 19-38 mph (30-60 km/h) for at least 5 seconds. Malfunctions are as follows: Malfunction with VTC OFF: The gap between the CMP sensor A pulse and the middle of the CMP sensor A assembly is 10 degrees or more for at least 5 seconds. Malfunction with VTC ACTIVE: The timing of the camshaft is out of specified range (Other than when BTDC is between 10-100 degrees) for at least 5 seconds. **NOTE: Before you troubleshoot, record all freeze data and any on-board snapshot, and review the general troubleshooting information.** **Possible Causes:** • The camshaft timing needs to be reset • Damage or stretched cam chain • Faulty auto-tensioner • Poor connections or loose terminals at the VTC oil control solenoid valve and the ECM/PCM • Faulty VTC actuator • Faulty VTC oil control solenoid valve
DTC: P0344 **2T PCM, MIL: Yes** **Year:** 2009, 2010 **Model:** Accord, CR-V, Element **Engine:** 2.4L L4 **Transmission:** All	**Camshaft Position (CMP) Sensor A Circuit Intermittent Interruption:** With the engine running, more or less than two CMP sensor A pulses are detected during intervals between the CKP standard pulses. This condition occurs at least 30 times. **NOTE: Before you troubleshoot, record all freeze data and any on-board snapshot, and review the general troubleshooting information.** **Possible Causes:** • Poor connections or loose terminals at the engine and/or body ground • Poor connections or loose terminals at CMP sensor A and the ECM/PCM • Damaged CMP pulse plate A • Faulty CMP sensor A
DTC: P0344 **1T PCM, MIL: Yes** **Year:** 2009, 2010 **Model:** Accord, Accord Crosstour **Engine:** 3.5L V6 **Transmission:** All	**Camshaft Position (CMP) Sensor Circuit Intermittent Interruption:** Engine running, more or less that two CMP pulses are detected during intervals between the standard CKP pulses. This condition occurs at least 30 times. **NOTE: Before you troubleshoot, record all freeze data and any on-board snapshot, and review the general troubleshooting information.** **Possible Causes:** • Poor connections or loose terminals at the CMP sensor and the ECM/PCM • Poor connections or loose terminal at the engine and/or body ground • Damaged CMP sensor pulse projection on the front camshaft pulley • Faulty CMP sensor
DTC: P0344 **1T PCM, MIL: Yes** **Year:** 2009, 2010 **Model:** Odyssey **Engine:** 3.5L V6 **Transmission:** All	**Camshaft Position (CMP) Sensor Intermittent Interruption:** With the engine running more or less than two CMP sensor pulses are detected during intervals between the CKP standard pulses. This condition occurs at least 30 times. The execution time is continuous and DTCs P0335, P0339, P0340 are not active. **NOTE: Before you troubleshoot, record all freeze data and any on-board snapshot.** **Possible Causes:** • Damage to the CMP sensor pulse projection on the front camshaft pulley • Poor or loose connections at CMP sensor, PCM, engine ground, body ground • Faulty CMP sensor

DTC	Trouble Code Title, Conditions & Possible Causes
DTC: P0351 **1T PCM, MIL: Yes** **Year:** 2009, 2010 **Model:** Civic, Fit, Odyssey, Ridgeline **Engine:** 1.5L L4, 1.8L L4, 3.5L V6 **Transmission:** All	**Cylinder 1 Ignition Coil Circuit Malfunction:** With the engine running, the execution time is continuous and the duration time is 5 seconds or more. (when the engine speed is 700 rpm) The return signal does not change for at least 5 seconds when the ignition coil is triggered. **NOTE: Before you troubleshoot, record all freeze data and any on-board snapshot.** **Possible Causes:** • Blown fuse • Poor connections or loose terminals at the ignition coil and the PCM • "Open" or "Short" circuit between the ignition coils and the ignition coil relay • "Open" or "Short" circuit between the ignition coils and the PCM • "Open" in the wire between the ignition coil and ground • Faulty ignition coil relay • Faulty ignition coil
DTC: P0352 **1T PCM** **Year:** 2009, 2010 **Model:** Fit, Odyssey, Ridgeline **Engine:** 1.5L L4, 3.5L V6 **Transmission:** All	**Cylinder 2 Ignition Coil Circuit Malfunction:** With the engine running, the execution time is continuous and the duration time is 5 seconds or more. (when the engine speed is 700 rpm) The return signal does not change for at least 5 seconds when the ignition coil is triggered. **Possible Causes:** • Blown fuse • Poor connections or loose terminals at the ignition coil and the PCM • "Open" or "Short" circuit between the ignition coils and the ignition coil relay • "Open" or "Short" circuit between the ignition coils and the PCM • "Open" in the wire between the ignition coil and ground • Faulty ignition coil relay • Faulty ignition coil
DTC: P0353 **1T PCM, MIL: Yes** **Year:** 2009, 2010 **Model:** Fit, Odyssey, Ridgeline **Engine:** 1.5L L4, 3.5L V6 **Transmission:** All	**Cylinder 3 Ignition Coil Circuit Malfunction:** With the engine running, the execution time is continuous and the duration time is 5 seconds or more. (when the engine speed is 700 rpm) The return signal does not change for at least 5 seconds when the ignition coil is triggered. **NOTE: Before you troubleshoot, record all freeze data and any on-board snapshot** **Possible Causes:** • Blown fuse • Poor connections or loose terminals at the ignition coil and the PCM • "Open" or "Short" circuit between the ignition coils and the ignition coil relay • "Open" or "Short" circuit between the ignition coils and the PCM • "Open" in the wire between the ignition coil and ground • Faulty ignition coil relay • Faulty ignition coil
DTC: P0354 **1T PCM, MIL: Yes** **Year:** 2009, 2010 **Model:** Fit, Odyssey, Ridgeline **Engine:** 1.5L L4, 3.5L V6 **Transmission:** All	**Cylinder 4 Ignition Coil Circuit Malfunction:** With the engine running, the execution time is continuous and the duration time is 5 seconds or more. (when the engine speed is 700 rpm) The return signal does not change for at least 5 seconds when the ignition coil is triggered. **NOTE: Before you troubleshoot, record all freeze data and any on-board snapshot** **Possible Causes:** • Blown fuse • Poor connections or loose terminals at the ignition coil and the PCM • "Open" or "Short" circuit between the ignition coils and the ignition coil relay • "Open" or "Short" circuit between the ignition coils and the PCM • "Open" in the wire between the ignition coil and ground • Faulty ignition coil relay • Faulty ignition coil
DTC: P0355 **1T PCM, MIL: Yes** **Year:** 2009, 2010 **Model:** Odyssey, Ridgeline **Engine:** 3.5L V6 **Transmission:** All	**Cylinder 5 Ignition Coil Circuit Malfunction:** With the engine running, the execution time is continuous and the duration time is 5 seconds or more. (when the engine speed is 700 rpm) The return signal does not change for at least 5 seconds when the ignition coil is triggered. **NOTE: Before you troubleshoot, record all freeze data and any on-board snapshot** **Possible Causes:** • Blown fuse • Poor connections or loose terminals at the ignition coil and the PCM • "Open" or "Short" circuit between the ignition coils and the ignition coil relay • "Open" or "Short" circuit between the ignition coils and the PCM • "Open" in the wire between the ignition coil and ground • Faulty ignition coil relay • Faulty ignition coil

DTC	Trouble Code Title, Conditions & Possible Causes
DTC: P0356 **1T PCM, MIL: Yes** **Year:** 2009, 2010 **Model:** Odyssey, Ridgeline **Engine:** 3.5L V6 **Transmission:** All	**Cylinder 6 Ignition Coil Circuit Malfunction:** With the engine running, the execution time is continuous and the duration time is 5 seconds or more. (when the engine speed is 700 rpm) The return signal does not change for at least 5 seconds when the ignition coil is triggered. **NOTE: Before you troubleshoot, record all freeze data and any on-board snapshot** **Possible Causes:** • Blown fuse • Poor connections or loose terminals at the ignition coil and the PCM • "Open" or "Short" circuit between the ignition coils and the ignition coil relay • "Open" or "Short" circuit between the ignition coils and the PCM • "Open" in the wire between the ignition coil and ground • Faulty ignition coil relay • Faulty ignition coil
DTC: P0365 **1T PCM, MIL: Yes** **Year:** 2009, 2010 **Model:** Fit, Odyssey, Pilot, Ridgeline **Engine:** 1.5L L4, 3.5L V6 **Transmission:** All	**Camshaft Position (CMP) Sensor Circuit No Signal:** With the engine running at 750 rpm, DTCs P0335, P0339 and P0369 are not active. No input signals from the CMP sensor are detected while the signals from the CKP sensor are detected at least 300 times. **NOTE: Before you troubleshoot, record all freeze data and any on-board snapshot** **Possible Causes:** • Poor connections or loose terminals at the CMP sensor and the PCM • "Open" circuit between the CMP sensor and PGM-FI main relay 1 • "Open" or "Short" circuit between the CMP and the PCM • "Open" in the wire between the CMP sensor and ground • Faulty CMP sensor • PCM needs to be updated with the latest software • Faulty PCM
DTC: P0365 **1T PCM, MIL: Yes** **Year:** 2009, 2010 **Model:** Accord, Civic, CR-V, Element **Engine:** 1.3L L4, 1.8L L4, 2.0L L4, 2.4L L4 **Transmission:** All	**Camshaft Position (CMP) Sensor B No Signal:** With the engine running, NO CMP sensor B pulsing signals are detected for at least 50 times in succession. **Possible Causes:** • Poor connections or loose terminals at CMP sensor B and the ECM/PCM • "Open" circuit between CMP sensor B and PGM-FI main relay 1 • "Open" or "Short" circuit between the ECM/PCM and CMP sensor B • "Open" circuit between CMP sensor B and ground • Faulty CMP sensor B • PCM may need to be updated with the latest software • Faulty PCM
DTC: P0369 **1T PCM, MIL: Yes** **Year:** 2009, 2010 **Model:** Civic, Fit, Odyssey, Pilot, Ridgeline **Engine:** 1.3L L4, 1.5L L4, 1.8L L4, 3.5L V6 **Transmission:** All	**Camshaft Position (CMP) Sensor Circuit Intermittent Interruption:** Other than 30 times of CMP pulses are detected during two cycles of CKP. This condition has been detected at least 10 times. **NOTE: Before you troubleshoot, record all freeze data and any on-board snapshot.** **Possible Causes:** • Poor connections or loose terminals at the CMP and the PCM • Poor ground connections • Damaged front camshaft pulley (CMP pulse plate) • Faulty CMP sensor
DTC: P0369 **1T PCM, MIL: Yes** **Year:** 2009, 2010 **Model:** Accord, Civic, CR-V, Element **Engine:** 2.0L L4, 2.4L L4 **Transmission:** All	**Camshaft Position (CMP) Sensor B Intermittent Interruption:** With the engine running, more or less than four CMP B pulsing signals are detected during intervals between standard pulses at least 30 times in succession. **NOTE: Before you troubleshoot, record all freeze data and any on-board snapshot, and review the general troubleshooting information.** **Possible Causes:** • Poor connections or loose terminals at the engine and/or body ground • Poor connections or loose terminals at CMP sensor B and the ECM/PCM • Damaged CMP pulse plate B • Faulty CMP sensor B

DTC	Trouble Code Title, Conditions & Possible Causes
DTC: P0385 **1T PCM, MIL: Yes** **Year:** 2009, 2010 **Model:** Odyssey **Engine:** 3.5L V6 **Transmission:** All	**Crankshaft Position (CKP) Sensor B No Signal :** With the engine running the execution is continuous and P0385, P0389 are not active. No signals from CKP sensor A are detected while signals from CKP sensor B are detected 127 times in succession. If no pulsing signals are detected from CKP sensor A, a malfunction is detected and a DTC is stored. **NOTE: Before you troubleshoot, record all freeze data and any on-board snapshot.** **Possible Causes:** • Poor connections or loose terminals at CKP sensor A/B and the PCM • "Open" or "Short" in the wire between CKP sensor A/B and the under-hood fuse/relay box (PGM-FI main relay 1) • "Open" or "Short" in the wire between the PCM and CKP sensor A/B, • Faulty CKP sensor A/B • PCM may need to be updated with the latest software • Faulty PCM
DTC: P0389 **1T PCM, MIL: Yes** **Year:** 2009, 2010 **Model:** Odyssey **Engine:** 3.5L V6 **Transmission:** All	**Crankshaft Position (CKP) Sensor B Intermittent Interruption :** Engine running, other than 22 pulses are detected during intervals between reference pulses for each crankshaft revolution. This condition has been detected at least 30 times. An abnormal amount of pulsing signals are detected from CKP sensor B. **NOTE: Before you troubleshoot, record all freeze data and any on-board snapshot.** **Possible Causes:** • Poor connections or loose terminals at CKP sensor A/B and the PCM • Poor grounds at engine, body, CKP sensor A/B or PCM • Damaged CKP sensor A/B pulse plate/timing belt drive pulley • Faulty CKP sensor A/B • PCM may need to be updated with the latest software • Faulty PCM
DTC: P0400 **2T PCM, MIL: Yes** **Year:** 2009, 2010 **Model:** Accord, Accord Crosstour, Civic, Fit, Odyssey, Pilot, Ridgeline **Engine:** 1.3L L4, 1.5L L4, 1.8L L4, 2.0L L4, 3.5L V6 **Transmission:** All	**Exhaust Gas Recirculation (EGR) System Leak Detected:** With engine running, rpm 1,500-3,500 and system in closed loop the A/F sensor response parameter according to the opening and closing of the EGR is 14,746 (LT80A) or less. **NOTE: Before you troubleshoot, record all freeze data and any on-board snapshot.** **Possible Causes:** • Loose or damaged EGR pipe • Exhaust gas leakage between the EGR pipe and the EGR valve • Faulty EGR valve
DTC: P0401 **2T PCM, MIL: Yes** **Year:** 2009, 2010 **Model:** Accord, Accord Crosstour, Civic, CR-V, Fit, Odyssey, Pilot, Ridgeline **Engine:** 1.3L L4, 1.5L L4, 1.8L L4, 2.0L L4, 2.4L L4, 3.5L V6 **Transmission:** All	**Exhaust Gas Recirculation (EGR) Insufficient Flow:** Engine temperature is at a minimum of 156°F (69°C), engine speed is between 1,100-2,400 rpm with the map value of 14 kPa (4.0 in.Hg, 100 mmHg), throttle fully closed and battery 10.5 volt minimum. The execution time is once per driving cycle and the duration time is 3 seconds or more. The ratio of the current EGR flow to the normal EGR flow is 15% or less for at least 3 seconds. **NOTE: Before you troubleshoot, record all freeze data and any on-board snapshot.** **Possible Causes:** • Clogged intake manifold, and/or EGR ports • Poor connections or loose terminals at the EGR valve and the PCM • Faulty EGR valve • PCM may need to be updated with the latest software • Faulty PCM
DTC: P0404 **2T PCM, MIL: Yes** **Year:** 2009, 2010 **Model:** Accord, Accord Crosstour, Civic, CR-V, Fit, Odyssey, Pilot, Ridgeline **Engine:** 1.3L L4, 1.5L L4, 1.8L L4, 2.0L L4, 2.4L L4, 3.5L V6 **Transmission:** All	**Exhaust Gas Recirculation (EGR) Control Circuit Range/Performance Problem :** Vehicle at a speed between 15-75 mph, actual valve lift command at 0.012 in. (0.3 mm). The execution time is once per driving cycle and the duration time is 5 seconds or more. The difference between the command value of the amount of EGR valve lift in the PCM and the actual amount of valve lift is 0.041 in. (1.020 mm) or more for at least 5 seconds. **NOTE: Before you troubleshoot, record all freeze data and any on-board snapshot.** **Possible Causes:** • Carbon build-up on the EGR valve • Faulty EGR valve • Poor connections or loose terminals at the EGR valve and the PCM • "Short" or "Short" between the PCM and the EGR valve • "Open" in the wire between the EGR valve and ground • PCM may need to be updated with the latest software • Faulty PCM

DTC	Trouble Code Title, Conditions & Possible Causes
DTC: P0406 **1T PCM, MIL: Yes** **Year:** 2009, 2010 **Model:** Accord, Accord Crosstour, Civic, CR-V, Fit, Odyssey, Pilot, Ridgeline **Engine:** 1.3L L4, 1.5L L4, 1.8L L4, 2.0L L4, 2.4L L4, 3.5L V6 **Transmission:** All	**Exhaust Gas Recirculation (EGR) Valve Position Sensor Circuit High Voltage:** With the engine running the execution time is continuous and the duration time is 2 seconds or more. The EGR valve position sensor output voltage is 4.88 V or more for at least 2 seconds. **NOTE: Before you troubleshoot, record all freeze data and any on-board snapshot.** **Possible Causes:** • Poor connections or loose terminals at the EGR valve and the PCM • "Open" circuit between the EGR valve and the PCM • PCM may need to be updated with the latest software • Faulty EGR valve • Faulty PCM
DTC: P0420 **2T PCM, MIL: Yes** **Year:** 2009, 2010 **Model:** Accord, Civic, CR-V, Element, Fit **Engine:** 1.3L L4, 1.5L L4, 1.8L L4, 2.0L L4, 2.4L L4 **Transmission:** All	**Catalytic System Efficiency Below Threshold :** With the vehicle driven to a speed of 16-75 mph at less than 3500 rpm in closed loop for 2-5 minutes, ECT sensor more than 140°F, MAF sensor from 8-50 g/sec, engine load less than 99% (±8%), predicted Catalyst temperature over 750°F, and the PCM detected the catalyst oxygen storage capacity was below an acceptable threshold during the test. **NOTE: If some of the DTCs listed below are stored at the same time as DTC P0420, troubleshoot those DTCs first, then recheck for DTC P0420.** **Possible Causes:** • Air leaks in at the exhaust manifold or exhaust pipes • Poor fuel quality • Catalytic converter damaged or has failed (deteriorated) • Front HO2S is more aged than the rear HO2S (HO2S is lazy) • Faulty Secondary HO2S (Sensor 2) • PCM has failed
DTC: P0420 **2T PCM, MIL: Yes** **Year:** 2009, 2010 **Model:** Accord, Accord Crosstour, Odyssey, Pilot, Ridgeline **Engine:** 3.5L V6 **Transmission:** All	**Rear Bank Catalyst System Efficiency Below Threshold (Bank 1):** The execution time is once per driving cycle and the duration time is 102 seconds or more. The number of detections is 752 (CTAGLT67) or more. **NOTE: Before you troubleshoot, record all freeze data and any on-board snapshot.** **Possible Causes:** • Poor connections or loose terminals at the secondary HO2S (sensor 2) and the PCM • Faulty (Bank 1) WU-TWC
DTC: P0430 **2T PCM, MIL: Yes** **Year:** 2009, 2010 **Model:** Accord, Accord Crosstour, Odyssey, Pilot, Ridgeline **Engine:** 3.5L V6 **Transmission:** All	**Front Bank Catalyst System Efficiency Below Threshold (Bank 2) :** The execution time is once per driving cycle and the duration time is 102 seconds or more. The number of detections is 720 (CTAGLT68) or more. **NOTE: Before you troubleshoot, record all freeze data and any on-board snapshot.** **Possible Causes:** • Poor connections or loose terminals at the secondary HO2S (sensor 2) and the PCM • Faulty (Bank 2) WU-TWC
DTC: P0443 **1T PCM, MIL: Yes** **Year:** 2009, 2010 **Model:** Accord, Accord Crosstour, Civic, CR-V, Element, Fit, Odyssey, Pilot, Ridgeline **Engine:** 1.3L L4, 1.5L L4, 1.8L L4, 2.0L L4, 2.4L L4, 3.5L V6 **Transmission:** All	**Evaporative Emission (EVAP) Canister Purge Valve Circuit Malfunction:** With the engine running, EVAP canister purge valve output duty at 2% to 98% the execution time is continuous and the duration time is 5 seconds or more. When the return signal does not change according to the EVAP canister purge valve duty cycle for a set time, the PCM detects a malfunction. The return signal does not change according to the EVAP canister purge valve output for at least 5 seconds. **NOTE: Before you troubleshoot, record all freeze data and any on-board snapshot.** **Possible Causes:** • Poor connections or loose terminals at the EVAP canister purge valve and the PCM • "Open" or "short" between the EVAP canister purge valve and the PCM • "Open" or "short" between the EVAP canister purge valve and the under-dash fuse/relay box • EVAP canister purge valve • PCM may need to be updated with the latest software • Faulty PCM
DTC: P0451 **2T PCM, MIL: Yes** **Year:** 2009, 2010 **Model:** Accord, Accord Crosstour, Civic, CR-V, Element, Fit, Odyssey, Pilot, Ridgeline **Engine:** 1.3L L4, 1.5L L4, 1.8L L4, 2.0L L4, 2.4L L4, 3.5L V6 **Transmission:** All	**Fuel Tank Pressure (FTP) Sensor Circuit Range/Performance Problem:** Elapsed time after starting the engine is 2 seconds with the throttle fully closed. The execution time is once per driving cycle and the duration time is 20 seconds or more. The FTP sensor output fluctuates by 0.3 kPa (0.1 in.Hg, 2 mmHg) or more at least five times within 3 seconds. **NOTE: Before you troubleshoot, record all freeze data and any on-board snapshot.** **Possible Causes:** • Poor connections or loose terminals at the FTP sensor and the PCM • Faulty FTP sensor

DTC	Trouble Code Title, Conditions & Possible Causes
DTC: P0452 **2T PCM, MIL: Yes** **Year:** 2009, 2010 **Model:** Accord, Accord Crosstour, Civic, CR-V, Element, Fit, Odyssey, Pilot, Ridgeline **Engine:** 1.3L L4, 1.5L L4, 1.8L L4, 2.0L L4, 2.4L L4, 3.5L V6 **Transmission:** All	**Fuel Tank Pressure (FTP) Sensor Circuit Low Voltage:** Elapsed time after starting the engine is 2 seconds at idle. The execution time is once per driving cycle and the duration time is 3 seconds or more. The output from the fuel tank pressure sensor is less than -7 kPa (-2.1 in.Hg, -55 mmHg) for at least 3 seconds. **NOTE: Before you troubleshoot, record all freeze data and any on-board snapshot.** **Possible Causes:** • Poor connections or loose terminals at the FTP sensor and the PCM • "Open" or "Short" in the wire(s) between the PCM and the FTP sensor • Faulty FTP sensor
DTC: P0453 **2T PCM, MIL: Yes** **Year:** 2009, 2010 **Model:** Accord Crosstour, Civic, CR-V, Element, Fit, Odyssey, Pilot, Ridgeline **Engine:** 1.3L L4, 1.5L L4, 1.8L L4, 2.0L L4, 2.4L L4, 3.5L V6 **Transmission:** All	**Fuel Tank Pressure (FTP) Sensor Circuit High Voltage:** Elapsed time after starting the engine is 2 seconds at idle. The execution time is once per driving cycle and the duration time is 3 seconds or more. If the FTP sensor output voltage is higher than a target value within a set time after starting the engine in a cold condition. The output from the fuel tank pressure sensor is more than 8 kPa (2.2 in.Hg, 55 mmHg) for at least 3 seconds. **NOTE: Before you troubleshoot, record all freeze data and any on-board snapshot.** **Possible Causes:** • Poor connections or loose terminals at the FTP sensor and the PCM • "Open" in the wire between the PCM and the FTP sensor • Faulty FTP sensor • PCM may need to be updated with the latest software • Faulty PCM
DTC: P0455 **2T PCM, MIL: Yes** **Year:** 2009, 2010 **Model:** Ridgeline **Engine:** 3.5L V6 **Transmission:** All	**EVAP System Large Leak Detected:** After 6 hours, with initial engine coolant temperature between 40-95°F (5-35°C), Barometric pressure 76 kPa (569 mmHg, 22.5 in. Hg), battery voltage a minimum of 10.5 V, fuel trim 0.73-1.47 and fuel level is not full. The variation of pressure inside the fuel tank is -1 kPa (-7 mmHg, -0.3 in. Hg) or less. **Possible Causes:** • Faulty or loose fuel fill cap • Poor connection or damage at the fuel tank vapor control valve hose • Poor connections or loose terminals at the FTP sensor, the EVAP canister • purge valve, or the EVAP canister vent shut valve, and the PCM • Faulty EVAP canister vent shut valve • Faulty FTP sensor O-ring • Faulty fuel tank vapor control valve hose • Faulty EVAP canister vent shut valve case and O-ring • Faulty EVAP canister • Faulty fuel tank unit base gasket, and/or fuel tank
DTC: P0455 **2T PCM, MIL: Yes** **Year:** 2009, 2010 **Model:** Accord Crosstour, Civic, CR-V, Element, Fit, Odyssey, Pilot **Engine:** 1.3L L4, 1.5L L4, 1.8L L4, 2.0L L4, 2.4L L4, 3.5L V6 **Transmission:** All	**Evaporative Emission (EVAP) System Large Leak Detected:** The execution time is once per driving cycle and the duration time is 36 minutes and 37 seconds, or less. Here is an overview of the malfunction detection for the EONV method: 1: Judgment of detection of 0.09 inch leak as normal operation 2: Judgment of detection of 0.02 inch leak as normal operation 3: Detection of 0.02 inch leak 4: Detection of atmospheric pressure failure 5: Flickering of the FTP sensor The execution time is once per driving cycle and the duration time is 36 minutes and 37 seconds, or less. The variation of pressure inside the fuel tank is 0.03 kPa (0.009 in.Hg, 0.24 mmHg) or more. **NOTE: Before you troubleshoot, record all freeze data and any on-board snapshot.** **Possible Causes:** • Faulty or loose fuel fill cap • Poor connection or damage at the fuel tank vapor control valve hose • Poor connections or loose terminals at the FTP sensor, the EVAP canister • purge valve, or the EVAP canister vent shut valve, and the PCM • Faulty EVAP canister vent shut valve • Faulty FTP sensor O-ring • Faulty fuel tank vapor control valve hose • Faulty EVAP canister vent shut valve case and O-ring • Faulty EVAP canister • Faulty fuel tank unit base gasket, and/or fuel tank

DTC	Trouble Code Title, Conditions & Possible Causes
DTC: P0456 **2T PCM, MIL: Yes** **Year:** 2009, 2010 **Model:** Accord, Accord Crosstour, Civic, CR-V, Element, Fit, Odyssey, Pilot, Ridgeline **Engine:** 1.3L L4, 1.5L L4, 1.8L L4, 2.0L L4, 2.4L L4, 3.5L V6 **Transmission:** All	**EVAP System Very Small Leak Detected :** The execution time is once per driving cycle and the duration time is at least 16 minutes and 37 seconds, but not more than 36 minutes and 37 seconds. Here is an overview of the malfunction detection for the EONV method: 1: Judgment of detection of 0.09 inch leak as normal operation 2: Judgment of detection of 0.02 inch leak as normal operation 3: Detection of 0.02 inch leak 4: Detection of atmospheric pressure failure 5: Flickering of the FTP sensor The execution time is once per driving cycle and the duration time is 36 minutes and 37 seconds, or less. The variation of pressure inside the fuel tank is 0.03 kPa (0.009 in.Hg, 0.24 mmHg) or more. **NOTE: Before you troubleshoot, record all freeze data and any on-board snapshot.** **Possible Causes:** • Faulty or loose fuel fill cap • Poor connection or damage at the fuel tank vapor control valve hose • Poor connections or loose terminals at the FTP sensor, the EVAP canister • purge valve, or the EVAP canister vent shut valve, and the PCM • Faulty EVAP canister vent shut valve • Faulty FTP sensor O-ring • Faulty fuel tank vapor control valve hose • Faulty EVAP canister vent shut valve case and O-ring • Faulty EVAP canister • Faulty fuel tank unit base gasket, and/or fuel tank
DTC: P0457 **3T PCM, MIL: Yes** **Year:** 2009, 2010 **Model:** Accord, Accord Crosstour, Civic, CR-V, Element, Odyssey, Pilot, Ridgeline **Engine:** 1.3L L4, 1.8L L4, 2.0L L4, 2.4L L4, 3.5L V6 **Transmission:** All	**Evaporative Emission (EVAP) System Leak Detected/Fuel Fill Cap Loose or Missing:** Engine coolant temperature before EVAP purge control starts 140°F (60°C), 2 mph (2 km/h), Barometric pressure, 76 kPa (22.5 in. Hg, 569 mmHg), Battery voltage 10.5 volts or more, fuel trim 0.73-1.47, system is in close loop. The execution time is continuous and the duration time is 12 seconds or more. The output from the fuel cap monitor is 0.053 or less for at least 12 seconds (when there is no NG judgment history in this drive cycle). P0455 or P0456 are judged as NG. **NOTE: Before you troubleshoot, record all freeze data and any on-board snapshot.** **Possible Causes:** • Faulty fuel fill cap seal missing or damaged, fuel fill pipe damaged • Poor connections or loose terminals at the FTP sensor, the EVAP canister vent shut valve, and the PCM • Faulty routing of the EVAP canister vent tube • Faulty EVAP canister vent shut valve
DTC: P0461 **1T PCM** **Year:** 2009, 2010 **Model:** Civic, Fit **Engine:** 1.3L L4, 1.5L L4, 1.8L L4, 2.0L L4 **Transmission:** All	**Fuel Level Sensor (Fuel Gauge Sending Unit) Circuit Range/Performance Problem :** The execution time is every 125 miles (200 km).The change in the fuel level sensor output is 3.5 % or less. **NOTE: Because it requires 162 miles (260 km) of driving without refueling to complete this diagnosis, DTC P0461 cannot be duplicated during this troubleshooting.** **Possible Causes:** • Poor connections or loose terminals at the fuel gauge sending unit and the gauge control module • Faulty fuel gauge sending unit
DTC: P0461 **1T PCM** **Year:** 2009, 2010 **Model:** Accord, Accord Crosstour, CR-V, Element, Odyssey, Pilot, Ridgeline **Engine:** 2.4L L4, 3.5L V6 **Transmission:** All	**Fuel Level Sensor (Fuel Gauge Sending Unit) Range/Performance Problem :** The execution time is every 125 miles (200 km), DTCs P0462, P0463, U0028, U0155 are not active. If the powertrain control module (PCM) receives no change in the fuel level sensor output after driving for a specified number of miles, it detects a malfunction. The change in the fuel level sensor output is 3.5 % or less. **Possible Causes:** • Poor connections or loose terminals at the fuel gauge sending unit and the gauge control module • Faulty fuel gauge sending unit

DTC	Trouble Code Title, Conditions & Possible Causes
DTC: P0462 **1T PCM** **Year:** 2009, 2010 **Model:** Accord, Accord Crosstour, Civic, CR-V, Element, Fit, Odyssey, Pilot, Ridgeline **Engine:** 1.3L L4, 1.5L L4, 1.8L L4, 2.0L L4, 2.4L L4, 3.5L V6 **Transmission:** All	**Fuel Level Sensor (Fuel Gauge Sending Unit) Circuit Low Voltage:** The fuel level sensor (fuel gauge sending unit) output voltage is 0.10 V or less for at least 5 seconds. The execution time is continuous DTCs P0463, U0155 are not active. **NOTE: Before you troubleshoot, record all freeze data and any on-board snapshot.** **Possible Causes:** • Poor connections or loose terminals at the gauge control module, the fuel gauge sending unit, and the secondary fuel gauge sending unit • "Short" circuit between the gauge control module (signal line) and the fuel gauge sending unit • Faulty fuel gauge sending unit • Faulty gauge control module • PCM may need to be updated with the latest software • Faulty PCM
DTC: P0463 **1T PCM** **Year:** 2009, 2010 **Model:** Accord, Accord Crosstour, Civic, CR-V, Element, Fit, Odyssey, Pilot, Ridgeline **Engine:** 1.3L L4, 1.5L L4, 1.8L L4, 2.0L L4, 2.4L L4, 3.5L V6 **Transmission:** All	**Fuel Level Sensor (Fuel Gauge Sending Unit) Circuit High Voltage :** The execution time is every 125 miles (200 km), DTCs P0462, P0463, U0028, U0155 are not active. If the Powertrain Control Module (PCM) receives no change in the fuel level sensor output after driving for a specified number of miles, it detects a malfunction. The fuel level sensor output voltage is 4.92 V or more for at least 5 seconds. **Possible Causes:** • Poor connections or loose terminals at the gauge control module and the fuel gauge sending unit • "Open" in the wire between the gauge control module (GND line) and the fuel gauge sending unit • "Open" in the wire between the gauge control module (signal line) and the fuel gauge sending unit • Faulty fuel gauge sending unit • Faulty gauge control module • PCM may need to be updated with the latest software • Faulty PCM
DTC: P0496 **2T PCM, MIL: Yes** **Year:** 2009, 2010 **Model:** Accord, Accord Crosstour, Civic, CR-V, Element, Fit, Odyssey, Pilot, Ridgeline **Engine:** 1.3L L4, 1.5L L4, 1.8L L4, 2.0L L4, 2.4L L4, 3.5L V6 **Transmission:** All	**Evaporative Emission (EVAP) System High Purge Flow Detected:** Monitor execution is once per driving cycle. The output from the EVAP canister purge valve is 0.2 kPa (0.07 in.Hg, 2 mmHg) or more for at least 10 seconds. **NOTE: Before you troubleshoot, record all freeze data and any on-board snapshot.** **Possible Causes:** • Poor connections or loose terminals at the FTP sensor, the EVAP canister purge valve, the EVAP canister vent shut valve, and the PCM • Faulty EVAP canister purge valve
DTC: P0497 **2T PCM, MIL: Yes** **Year:** 2009, 2010 **Model:** Accord, Accord Crosstour, Civic, CR-V, Element, Fit, Odyssey, Pilot, Ridgeline **Engine:** 1.3L L4, 1.5L L4, 1.8L L4, 2.0L L4, 2.4L L4, 3.5L V6 **Transmission:** All	**Evaporative Emission (EVAP) System Low Purge Flow Detected:** The execution time is once per driving cycle and P145C is judged as NG. Enable conditions are low load duration time 10 seconds, Wait for 10 seconds after the ignition switch is turned to LOCK (0) and Engine coolant temperature before EVAP purge control starts is 140°F (60°C). The output from the fuel tank pressure sensor is 0.2 kPa (0.07 in.Hg, 2 mmHg) or less for at least 10 seconds. **NOTE: Before you troubleshoot, record all freeze data and any on-board snapshot.** **Possible Causes:** • Faulty or Loose fuel fill cap • Poor connections or loose terminals at the FTP sensor, the EVAP canister purge valve, the EVAP canister vent shut valve, and the PCM • Blockage in the vacuum hose between the EVAP canister purge valve and the EVAP canister • Faulty EVAP canister purge valve
DTC: P0498 **1T PCM, MIL: Yes** **Year:** 2009, 2010 **Model:** Accord, Accord Crosstour, Civic **Engine:** 1.3L L4, 1.8L L4, 2.0L L4, 2.4L L4, 3.5L V6 **Transmission:** All	**Evaporative Emission (EVAP) Canister Vent Shut Valve Control Circuit Low Voltage:** The execution time is continuous and the duration time is 5 seconds or more. If the return signal is OFF when the powertrain control module (PCM) outputs the ON signal to the EVAP canister vent shut valve, the PCM detects a malfunction. DTC P0499 is not active. The return signal is Low for at least 5 seconds when the PCM outputs the ON signal to the EVAP canister vent shut valve. **Possible Causes:** • Poor connections or loose terminals at the EVAP canister vent shut valve and the PCM • "Open" or "Short" in the wire between the EVAP canister vent shut valve and the PCM • "Open" in the wire between the EVAP canister vent shut valve and the under-hood fuse/relay box (PGM-FI sub-relay) • Faulty EVAP canister vent shut valve

DTC	Trouble Code Title, Conditions & Possible Causes
DTC: P0498 **1T PCM, MIL: Yes** **Year:** 2009, 2010 **Model:** CR-V, Element, Fit, Odyssey, Pilot, Ridgeline **Engine:** 1.5L L4, 2.4L L4, 3.5L V6 **Transmission:** All	**Evaporative Emission (EVAP) Canister Vent Shut Valve Control Circuit Low Voltage:** The execution time is continuous and the duration time is 5 seconds or more. If the return signal is OFF (Low) when the Powertrain Control Module (PCM) outputs the ON signal to the EVAP canister vent shut valve, the PCM detects a malfunction. The return signal is Low for at least 5 seconds when the PCM outputs the ON signal to the EVAP canister vent shut valve. **Possible Causes:** • Poor connections or loose terminals at the EVAP canister vent shut valve and the PCM • "Open" or "Short" in the wire between the EVAP canister vent shut valve and the PCM • "Open" in the wire between the EVAP canister vent shut valve and the A/F relay • Faulty EVAP canister vent shut valve
DTC: P0499 **1T PCM, MIL: Yes** **Year:** 2009, 2010 **Model:** Accord, Accord Crosstour, Civic, CR-V, Element, Fit, Odyssey, Pilot, Ridgeline **Engine:** 1.3L L4, 1.5L L4, 1.8L L4, 2.0L L4, 2.4L L4, 3.5L V6 **Transmission:** All	**Evaporative Emission (EVAP) Canister Vent Shut Valve Control Circuit High Voltage:** If the return signal is ON when the Powertrain Control Module (PCM) outputs the OFF signal to the EVAP canister vent shut valve, the PCM detects a malfunction. The execution time is continuous and the duration time is 5 seconds or more. DTC P0498 is not active. The return signal is ON for at least 5 seconds when the PCM outputs the Low signal to the EVAP canister vent shut valve. **NOTE: Before you troubleshoot, record all freeze data and any on-board snapshot.** **Possible Causes:** • Poor connections or loose terminals at the EVAP canister vent shut valve and the PCM • PCM may need to be updated with the latest software • Faulty PCM • Faulty EVAP canister vent shut valve
DTC: P0506 **2T PCM, MIL: Yes** **Year:** 2009, 2010 **Model:** Accord, Accord Crosstour, Civic, CR-V, Element, Fit, Odyssey, Pilot, Ridgeline **Engine:** 1.3L L4, 1.5L L4, 1.8L L4, 2.0L L4, 2.4L L4, 3.5L V6 **Transmission:** All	**Idle Control System RPM Lower Than Expected:** The execution time is once per driving cycle and the duration time is 20 seconds or more. Throttle fully closed, fuel trim 0.75-1.47, intake air temperature 19°F (-7°C) and battery voltage 10.5 or more. If the actual idle speed varies beyond a specified value from the target speed over a certain period of time, the PCM detects a malfunction. The actual idle speed is at least 100 rpm less than the target idle speed for at least 20 seconds. **NOTE: Before you troubleshoot, record all freeze data and any on-board snapshot.** **Possible Causes:** • Dirt, carbon, or damage in the throttle bore • Damaged air cleaner element • Incorrect DATA LIST parameter conditions • Poor connections or loose terminals at the throttle body and the PCM • Faulty throttle body • PCM may need to be updated with the latest software • Faulty PCM
DTC: P0507 **2T PCM, MIL: Yes** **Year:** 2009, 2010 **Model:** Accord, Accord Crosstour, Civic, CR-V, Element, Fit, Odyssey, Pilot, Ridgeline **Engine:** 1.3L L4, 1.5L L4, 1.8L L4, 2.0L L4, 2.4L L4, 3.5L V6 **Transmission:** All	**Idle Control System RPM Higher Than Expected:** Enable conditions are as follows: coolant temperature 156°F (69°C) minimum, intake air temperature 19°F (-7°C), fuel trim 0.73-1.47, throttle closed and battery voltage at least 10.5 volts. The execution time is once per driving cycle and the duration time is 20 seconds or more. If the actual idle speed varies beyond a specified value from the target speed over a certain period of time, the PCM detects a malfunction. The actual idle speed is at least 200 rpm greater than the target idle speed for at least 20 seconds. **NOTE: Before you troubleshoot, record all freeze data and any on-board snapshot.** **Possible Causes:** • Vacuum leaks • Faulty PCV valve • Dirty throttle bore • Faulty EVAP canister purge valve • Poor connections or loose terminals at the throttle body and the PCM • Improper PCM idle learn procedure • PCM may need to be updated with the latest software • Faulty PCM

DTC	Trouble Code Title, Conditions & Possible Causes
DTC: P050A **2T PCM, MIL: Yes** **Year:** 2009, 2010 **Model:** Accord, Accord Crosstour, Civic, CR-V, Element, Fit, Odyssey, Pilot, Ridgeline **Engine:** 1.3L L4, 1.5L L4, 1.8L L4, 2.0L L4, 2.4L L4, 3.5L V6 **Transmission:** All	**Cold Start Idle Air Control System Performance Problem:** The execution time is once per driving cycle and the duration time is 10 seconds or more. When the actual amount of air is less than the target amount, a malfunction is detected. The total airflow is decreased by a factor of 0.693 for at least 10 seconds. **NOTE: Before you troubleshoot, record all freeze data and any on-board snapshot.** **Possible Causes:** • Dirty air cleaner element • Damage air cleaner element or housing • Dirty or damaged throttle bore • Poor connections or loose terminals at the throttle body, the MAF sensor/IAT sensor • Faulty throttle body • Faulty MAF sensor/IAT sensor
DTC: P050B **2T PCM, MIL: Yes** **Year:** 2009, 2010 **Model:** Accord, Accord Crosstour, Civic, CR-V, Element, Fit, Pilot, Ridgeline **Engine:** 1.3L L4, 1.5L L4, 1.8L L4, 2.0L L4, 2.4L L4, 3.5L V6 **Transmission:** All	**Cold Start Ignition Timing Control System Performance Problem:** With the vehicle at idle and throttle position fully closed, the engine speed is 2,150 rpm or more for at least 3.5 seconds. **NOTE: Before you troubleshoot, record all freeze data and any on-board snapshot, and review the general troubleshooting information.** **Possible Causes:** • Poor connections or blockage at the air intake duct • Damaged air cleaner housing or dirty air cleaner • Damaged CKP sensor and/or the CKP sensor pulser plate • Faulty throttle body • Dirty or faulty MAF sensor/IAT sensor 1 • Faulty ECT SENSOR 1 and/or ECT SENSOR 2 • Check and repair any problems with the following items, Engine compression, VTEC system, Engine oil, A/C system, Power steering system • PCM may need to be updated with the latest software • Faulty PCM
DTC: P050B **2T PCM, MIL: Yes** **Year:** 2009, 2010 **Model:** Odyssey **Engine:** 3.5L V6 **Transmission:** All	**Cold Start Ignition Timing Performance Problem:** The execution time is once per driving cycle and the duration time is 3.5 seconds or more and the engine speed is 2,100 rpm or more for at least 3.5 seconds. When the actual engine speed is a specified value or more, and it continues for a specified time, a malfunction is detected. **NOTE: Before you troubleshoot, record all freeze data and any on-board snapshot.** **Possible Causes:** • Poor connections or blockage at the intake air duct • Damage at the air cleaner housing • Dirt or debris in the air cleaner element • Incorrect ignition timing • Faulty Throttle body • Faulty MAF sensor/IAT sensor • Poor connections or loose terminals at the CKP sensor, the throttle body, the MAF sensor/IAT sensor, ECT sensor 1, ECT sensor 2, and the PCM • Low engine coolant • Faulty ECT sensor 1, and/or ECT sensor 2
DTC: P0522 **2T PCM, MIL: Yes** **Year:** 2009, 2010 **Model:** Civic **Engine:** 1.3L L4, 1.8L L4, 2.0L L4 **Transmission:** All	**Rocker Arm Oil Pressure Sensor A Low Voltage:** With the ignition switch ON and the barometric pressure a minimum of 56 Kpa (17.3 in. Hg. 439 mmHg), the rocker arm oil pressure sensor A output voltage is 0.18 V or less for at least 2 seconds. **Possible Causes:** • Poor connections or loose terminals at rocker arm oil pressure sensor A and the PCM • "Open" or "Short" circuit between the PCM and rocker arm oil pressure sensor A • Faulty rocker arm oil pressure sensor A • PCM may need to be updated with the latest software • Faulty PCM
DTC: P0522 **1T PCM, MIL: Yes** **Year:** 2009, 2010 **Model:** Odyssey, Pilot **Engine:** 3.5L V6 **Transmission:** All	**Rocker Arm Oil Pressure Sensor Circuit Low Voltage :** With the engine running and the barometric pressure a minimum of 69 Kpa (20.5 in. Hg. 520 mmHg) or more, the rocker arm oil pressure sensor output voltage is 0.18 V or less for at least 2 seconds. **Possible Causes:** • Poor connections or loose terminals at rocker arm oil pressure sensor and the PCM • "Open" or "Short" circuit between the PCM and rocker arm oil pressure sensor • Faulty rocker arm oil pressure sensor • PCM may need to be updated with the latest software • Faulty PCM

DTC	Trouble Code Title, Conditions & Possible Causes
DTC: P0523 **1T PCM, MIL: Yes** **Year:** 2009, 2010 **Model:** Odyssey, Pilot **Engine:** 3.5L V6 **Transmission:** All	**Rocker Arm Oil Pressure Sensor Circuit High Voltage:** With the engine speed at a maximum of 6,000 rpm, engine coolant temperature a minimum of 104°F (40°C), the EOP sensor output voltage is 4.79 V or more for at least 2 seconds. **Possible Causes:** • Poor connections or loose terminals at rocker arm oil pressure sensor and the PCM • "Open" circuit between the PCM and rocker arm oil pressure sensor • Faulty rocker arm oil pressure sensor • PCM may need to be updated with the latest software • Faulty PCM
DTC: P0523 **2T PCM, MIL: Yes** **Year:** 2009, 2010 **Model:** Civic **Engine:** 1.3L L4, 1.8L L4, 2.0L L4 **Transmission:** All	**Rocker Arm Oil Pressure Sensor A High Voltage:** With the engine speed a maximum of 4,000 rpm and engine coolant a minimum of 140°F (60°C), the rocker arm oil pressure sensor A output voltage is 4.79 volts or more for at least 2 seconds. **Possible Causes:** • Poor connections or loose terminals at rocker arm oil pressure sensor A and the PCM • "Open" circuit between the PCM and rocker arm oil pressure sensor A • Faulty rocker arm oil pressure sensor A • PCM may need to be updated with the latest software • Faulty PCM
DTC: P0532 **2T PCM** **Year:** 2009, 2010 **Model:** Civic, CR-V **Engine:** 1.3L L4, 1.8L L4, 2.0L L4, 2.4L L4 **Transmission:** All	**A/C Pressure Sensor Circuit Low Voltage:** Ignition switch ON (II), DTC P0533 is not active. The A/C pressure sensor output voltage is 0.24 V for at least 10 seconds. **NOTE: Before you troubleshoot, record all freeze data and any on-board snapshot, and review the general troubleshooting information.** **Possible Causes:** • Poor connections or loose terminals at the A/C pressure sensor and the ECM/PCM • "Open" or "Short between the ECM/PCM and the A/C pressure sensor • Faulty A/C pressure sensor • PCM may need to be updated with the latest software • Faulty PCM
DTC: P0533 **1T PCM** **Year:** 2009, 2010 **Model:** Civic, CR-V **Engine:** 1.3L L4, 1.8L L4, 2.0L L4, 2.4L L4 **Transmission:** All	**A/C Pressure Sensor Circuit High Voltage:** With the ignition switch on, the A/C pressure sensor output voltage is 4.74 V or more for at least 10 seconds. DTC P0532 is not active. **NOTE: Before you troubleshoot, record all freeze data and any on-board snapshot, and review the general troubleshooting information.** **Possible Causes:** • Poor connections or loose terminals at the A/C pressure sensor and the PCM • "Open" circuit between the PCM and the A/C pressure sensor • Faulty A/C pressure sensor • PCM may need to be updated with the latest software • Faulty PCM
DTC: P0562 **1T PCM** **Year:** 2009, 2010 **Model:** Accord, Accord Crosstour, Civic, CR-V, Element, Fit, Odyssey, Pilot, Ridgeline **Engine:** 1.3L L4, 1.5L L4, 1.8L L4, 2.0L L4, 2.4L L4, 3.5L V6 **Transmission:** All	**Charging System Low Voltage:** The execution time is continuous and the duration time is 60 seconds or more, engine speed 550 rpm. When the IGP (power source) terminal voltage is a set value or less for a set time, the PCM detects a malfunction. The IGP terminal voltage is 11.0 V or less for at least 60 seconds. **NOTE: Before you troubleshoot, record all freeze data and any on-board snapshot. If any high current load accessories are installed, this DTC can be set.** **Possible Causes:** • Faulty battery, or connections • Faulty alternator • Poor connections or loose terminals at the alternator and the main under-hood fuse box
DTC: P0562/15 **1T PCM, MIL: Yes** **Year:** 2009, 2010 **Model:** Civic **Engine:** 1.3L L4 **Transmission:** All	**Motor Control Module (MCM) Power Source Circuit Unexpected Voltage (HYBRID MODEL):** With the ignition switch ON and the MCM power supply a minimum of 9.0 V, the VBU terminal voltage of the MCM is 4.0 V or less for at least 2 seconds. (IMA system indicator ON) WARNING: The IMA system uses high voltage (144 V), before servicing turn OFF the battery module switch. **Possible Causes:** • Blown back up (10 A) fuse • Poor connections or loose terminals at the BACK UP (10 A) fuse and the MCM • "Open" or "Short" to GND in the signal wire • MCM may need to be updated with the latest software • Faulty MCM

DTC	Trouble Code Title, Conditions & Possible Causes
DTC: P0563 **1T PCM, MIL: Yes** **Year:** 2009, 2010 **Model:** Odyssey, Pilot, Ridgeline **Engine:** 3.5L V6 **Transmission:** All	**Powertrain Control Module (PCM) Power Source Circuit Unexpected Voltage:** The PCM operates for at least 5 seconds after the ignition switch is turned to LOCK (0). Battery voltage 10.1 volts. (IGP terminal of PCM) **NOTE: Before you troubleshoot, record all freeze data and any on-board snapshot.** **Possible Causes:** • Faulty PGM-FI main relay 1 • Poor connections or loose terminals under-hood fuse/relay box (PGM-FI main relay 1) and the fuse in the under-hood fuse/relay box and the PCM • "Short" to power in the wire between the PCM and under-hood fuse/relay box (PGM-FI main relay 1) • PCM may need to be updated with the latest software • Faulty PCM
DTC: P0563 **1T PCM, MIL: Yes** **Year:** 2009, 2010 **Model:** Accord, Accord Crosstour, Civic, CR-V, Element, Fit **Engine:** 1.3L L4, 1.5L L4, 1.8L L4, 2.0L L4, 2.4L L4, 3.5L V6 **Transmission:** All	**Engine Control Module (ECM) Powertrain Control Module (PCM) Power Source Circuit Unexpected Voltage:** The ECM/PCM operates for at least 5 seconds after the ignition switch is turned to LOCK (0). Battery voltage 10.1 volts. (IGP terminal of PCM) **NOTE: Before you troubleshoot, record all freeze data and any on-board snapshot.** **Possible Causes:** • Faulty PGM-FI main relay 1 • Poor connections or loose terminals under-hood fuse/relay box (PGM-FI main relay 1) and the fuse • "Short" to power in the wire between the PCM (PGM-FI main relay 1) • ECM/PCM may need to be updated with the latest software • Faulty ECM/PCM
DTC: P0602 **T PCM, MIL: Yes** **Year:** 2009, 2010 **Model:** Civic **Engine:** 1.3L L4, 1.8L L4, 2.0L L4 **Transmission:** All	**PCM Programming Error:** With the ignition ON (II), a programming error was detected for at least 2 seconds. **NOTE: Do not turn the ignition switch to ACC (I) or to LOCK (0) while updating the PCM. If you turn the ignition switch to ACC (I) or to LOCK (0) before completion, the PCM can be damaged.** **Possible Causes:** • PCM needs to be updated with the latest software • Faulty PCM
DTC: P0602 **1T PCM, MIL: Yes** **Year:** 2009, 2010 **Model:** Accord, Accord Crosstour, CR-V, Element, Fit, Odyssey, Pilot, Ridgeline **Engine:** 1.5L L4, 2.4L L4, 3.5L V6 **Transmission:** All	**ECM/PCM Programming Error:** With ignition on the execution time is continuous and the duration time is 1 second or less. The ECM/PCM program update stops 1 second before it is finished. **NOTE: This DTC is indicated when a PCM update is not completed.** **WARNING: Do not turn the ignition switch to LOCK (0) or ACC (I) while updating the PCM. If you turn the ignition switch to LOCK (0) before completion, the ECM/PCM can be damaged.** **Possible Causes:** • ECM/PCM needs to be updated with the latest software • Faulty ECM/PCM
DTC: P0602/91 **1T PCM, MIL: Yes** **Year:** 2009, 2010 **Model:** Civic **Engine:** 1.3L L4 **Transmission:** All	**Motor Control Module (MCM) Programming Error:** With the ignition switch ON (II), the power is interrupted during reprogramming, and the MCM is not updated and programmed normally. (IMA system indicator is ON) **NOTE: Do not turn the ignition switch to ACC (I) or to LOCK (0) while updating the MCM. If you do, the MCM can be damaged.** **Possible Causes:** • Faulty MCM
DTC: P0603/60 **1T PCM, MIL: Yes** **Year:** 2009, 2010 **Model:** Civic **Engine:** 1.3L L4 **Transmission:** All	**Motor Control Module (MCM) Internal Circuit KAM Error (HYBRID MODELS):** Immediately after the ignition is turned ON (II), one of the following conditions occurs. Condition 1: The data input to or read out from the KAM is bad. Condition 2: The check sum of the data readout is abnormal three times. **Possible Causes:** • Poor connections or loose terminals at the MCM • MCM may need to be updated with the latest software • Faulty MCM
DTC: P0606 **1T PCM, MIL: Yes** **Year:** 2009, 2010 **Model:** Accord, Civic, CR-V, Element, Fit **Engine:** 1.3L L4, 1.5L L4, 1.8L L4, 2.0L L4, 2.4L L4 **Transmission:** All	**ECM/PCM Processor Malfunction:** After 30 seconds have elapsed since start-up, or after the engine speed exceeds 1,000 rpm once. No signal from the DKS CPU is detected or is abnormal for at least 5 seconds. **NOTE: Before you troubleshoot, record all freeze data and any on-board snapshot, and review the general troubleshooting information.** **Possible Causes:** • ECM/PCM may need to be updated with the latest software • Faulty ECM/PCM

DTC	Trouble Code Title, Conditions & Possible Causes
DTC: P060A **1T PCM, MIL: Yes** **Year:** 2009, 2010 **Model:** Accord, Accord Crosstour, CR-V, Fit **Engine:** 1.5L L4, 2.4L L4, 3.5L V6 **Transmission:** All	**Powertrain Control Module (PCM) Internal Control Module Malfunction:** With the ignition ON and battery voltage at a minimum of 10. 0 volts. The execution time is continuous and the duration time is 200 milliseconds or more. With keyless access system 110 milliseconds or more. One of these 2 symptoms occurs: (1) Symptom 1: The internal communication between the FI CPU and the A/T CPU is abnormal or the internal communication is interrupted for at least 200 milliseconds. (2) Symptom 2: The watchdog timer that monitors the A/T CPU detects an abnormality, and the FI CPU receives the A/T CPU check signal for at least 110 milliseconds. **Possible Causes:** • PCM needs to be updated with the latest software • Faulty PCM
DTC: P060A **1T PCM, MIL: Yes** **Year:** 2009, 2010 **Model:** Odyssey, Ridgeline **Engine:** 3.5L V6 **Transmission:** All	**Powertrain Control Module (PCM) (A/T System) Internal Control Module Malfunction:** (Symptom 1) The execution time is continuous and the duration time 500 milliseconds or more The serial communication between the FI CPU and A/T CPU is abnormal or the serial communication is interrupted for at least 500 milliseconds. (Symptom 2) The execution time is continuous and the duration time , 30 milliseconds or more The watchdog timer that monitors the A/T CPU detects an abnormality, and the FI CPU receives the A/T CPU check signal for at least 30 milliseconds. **NOTE: Before you troubleshoot, record all freeze data and any on-board snapshot.** **Possible Causes:** • PCM needs to be updated with the latest software • Faulty PCM
DTC: P060F **1T PCM, MIL: Yes** **Year:** 2009, 2010 **Model:** CR-V **Engine:** 2.4L L4 **Transmission:** All	**Powertrain Control Module (PCM) Internal Control Module Keep Alive Memory (KAM) Error :** With the ignition switch ON a malfunction is detected whenever the keep alive data retrieval and writing process is not completed normally. **NOTE: Before you troubleshoot, record all freeze data and any on-board snapshot with the HDS, and review General Troubleshooting Information.** **Possible Causes:** • Poor connections or loose terminals at the PCM • PCM may need to be updated with the latest software • Faulty PCM
DTC: P0615 **1T PCM** **Year:** 2009, 2010 **Model:** Pilot **Engine:** 3.5L V6 **Transmission:** All	**Starter Cut Relay STRLD Circuit Malfunction:** With the ignition switch ON (II). The diagnosis line (STRLD) input voltage is between 2.4 V to 2.6 V for at least 1 second. The execution time is continuous and the duration time is 1 seconds or more. **NOTE: Before you troubleshoot, record all freeze data and any on-board snapshot.** **Possible Causes:** • Blown fuse • Poor connections or loose terminals at starter cut relay 1, starter cut relay 2, and the PCM • "Open" circuit between the PCM and starter cut relay 1 • PCM needs to be updated with the latest software • Faulty PCM
DTC: P062F **1T PCM, TCIL: Yes** **Year:** 2009, 2010 **Model:** Accord, Accord Crosstour, Civic, Element, Fit, Odyssey, Pilot, Ridgeline **Engine:** 1.3L L4, 1.5L L4, 1.8L L4, 2.0L L4, 2.4L L4, 3.5L V6 **Transmission:** All	**ECM/PCM Internal Control Module Keep Alive Memory (KAM) Error :** A malfunction is detected whenever the keep alive data retrieval and writing process is not completed normally. **NOTE: Before you troubleshoot, record all freeze data and any on-board snapshot.** **Possible Causes:** • ECM/PCM needs to be updated with the latest software • Faulty ECM/PCM

DTC	Trouble Code Title, Conditions & Possible Causes
DTC: P0630 **1T PCM, MIL: Yes** **Year:** 2009, 2010 **Model:** Accord, Accord Crosstour, Civic, CR-V, Element, Fit, Odyssey, Pilot, Ridgeline **Engine:** 1.3L L4, 1.5L L4, 1.8L L4, 2.0L L4, 2.4L L4, 3.5L V6 **Transmission:** All	**VIN Not Programmed or Mismatch:** The VIN is not registered in the keep-alive memory in the PCM. **NOTE: Before you troubleshoot, record all freeze data and any on-board snapshot.** **Possible Causes:** • ECM/PCM needs to be updated with the latest software • Faulty ECM/PCM
DTC: P0641 **1T PCM, MIL: Yes** **Year:** 2009, 2010 **Model:** Accord, Accord Crosstour, Pilot **Engine:** 3.5L V6 **Transmission:** All	**Sensor Reference Voltage A Malfunction:** With the ignition ON the execution time is continuous and the sensor power voltage is 5.2 V or more, or 4.5 V or less, for at least 2.0 seconds. **NOTE: Before you troubleshoot, record all freeze data and any on-board snapshot. It may be possible to locate the fault by disconnecting one component at a time from the 5-volt reference circuit while viewing the 5-Volt Reference circuit parameter on the scan tool. The scan tool parameter would change from Fault to OK when the source of the fault is disconnected. If all 5-volt reference components have been disconnected and a Fault is still indicated, the fault may exist in the wiring harness.** **Possible Causes:** • Intermittent condition • "Open" or "Short" circuit in the following 5-volt reference circuits, APP sensor, Throttle body, Input shaft (mainshaft) speed sensor • PCM needs to be updated with the latest software • Faulty PCM
DTC: P0685 **2T PCM, MIL: Yes** **Year:** 2009, 2010 **Model:** Accord Crosstour, Odyssey, Pilot, Ridgeline **Engine:** 3.5L V6 **Transmission:** All	**Powertrain Control Module (PCM) Power Control Circuit/Internal Circuit Malfunction:** When the voltage to the PCM is turned off and the PCM shuts down without the normal shut down procedure, a malfunction in the PGM-FI main relay 1 control circuit is detected. **NOTE: Before you troubleshoot, record all freeze data and any on-board snapshot.** **Possible Causes:** • Loose terminals at the IGP line connectors • PCM needs to be updated with the latest software • Faulty PCM
DTC: P0685 **2T PCM, MIL: Yes** **Year:** 2009, 2010 **Model:** Accord, Civic, CR-V, Element, Fit **Engine:** 1.3L L4, 1.5L L4, 1.8L L4, 2.0L L4, 2.4L L4, 3.5L V6 **Transmission:** All	**ECM/PCM Power Control Circuit Malfunction:** When the voltage to the ECM/PCM is turned off and the ECM/PCM shuts down without the normal shut down procedure, a malfunction in the PGM-FI main relay 1 control circuit is detected. **NOTE: Before you troubleshoot, record all freeze data and any on-board snapshot.** **Possible Causes:** • ECM/PCM needs to be updated with the latest software • Faulty ECM/PCM
DTC: P0705 **1T PCM, MIL: Yes, TCIL: Yes** **Year:** 2009, 2010 **Model:** Element, Fit, Odyssey, Pilot, Ridgeline **Engine:** 1.5L L4, 2.4L L4, 3.5L V6 **Transmission:** All	**Short in Transmission Range Switch Circuit (Multiple Shift-position Input):** One of 3 conditions occurs: (1) The PCM detects the selected range switch input and another range switch input simultaneously for at least 1 seconds. (2) The PCM detects the P, R, or N range switch input and the FWD switch input simultaneously for at least 1 seconds (3) The PCM detects the D or D3 range switch input and the RVS switch input simultaneously for at least 1 second. **NOTE: Before you troubleshoot, record all freeze data and any on-board snapshot. This code is caused by an electrical circuit problem and cannot be caused by a mechanical problem in the transmission.** **Possible Causes:** • Intermittent "Short" in the wire between the transmission range switch and the PCM • Faulty transmission range switch • PCM needs to be updated with the latest software • Faulty PCM

DTC	Trouble Code Title, Conditions & Possible Causes
DTC: P0705 **1T PCM, MIL: Yes, TCIL: Yes** **Year:** 2009, 2010 **Model:** Accord, Accord Crosstour, Civic, CR-V **Engine:** 1.3L L4, 1.8L L4, 2.4L L4, 3.5L V6 **Transmission:** All	**Transmission Range Switch Circuit (Multiple Shift-position Input):** Malfunction 1: The PCM detects the selected range switch input and another range switch (except L2 switch) input simultaneously for at least 1 second. Malfunction 2: The PCM detects the P,R or N range switch input and the L2 switch input simultaneously for at least 1 second. **NOTE: Record all freeze data and review General Troubleshooting Information before you troubleshoot. This code is caused by an electrical circuit problem and cannot be caused by a mechanical problem in the transmission.** **Possible Causes:** • Intermittent "Short" in the wires between the transmission range switch and PCM • "Short" to ground in the wire between PCM connector terminal and the transmission range switch • "Open" in the wire between PCM connector terminals and ground • Faulty transmission range switch • PCM needs to be updated with the latest software • Faulty PCM
DTC: P0706 **2T PCM, MIL: Yes** **Year:** 2009, 2010 **Model:** Accord, Accord Crosstour, Civic, CR-V, Element, Fit, Odyssey, Pilot, Ridgeline **Engine:** 1.3L L4, 1.5L L4, 1.8L L4, 2.4L L4, 3.5L V6 **Transmission:** All	**Open in Transmission Range Switch Circuit:** No FWD position signal is detected when the vehicle speed changes from 6 mph (10 km/h) 25 mph (40 km/h) 6 mph (10 km/h) in D or D3. DTCs P0705, P0721, P0722 are not active. **NOTE: This code is caused by an electrical circuit problem and cannot be caused by a mechanical problem in the transmission.** **Possible Causes:** • Poor connections or loose terminals at the transmission range switch and the PCM • "Open" in the wire between the transmission range switch and ground • "Open" in the wire between the transmission range switch and PCM • Faulty transmission range switch • PCM needs to be updated with the latest software • Faulty PCM
DTC: P0711 **1T PCM, TCIL: Yes** **Year:** 2009, 2010 **Model:** Accord, Accord Crosstour, Civic, CR-V, Element, Fit, Odyssey, Pilot, Ridgeline **Engine:** 1.3L L4, 1.5L L4, 1.8L L4, 2.4L L4, 3.5L V6 **Transmission:** All	**Problem in ATF Temperature Sensor Circuit:** The ATF temperature sensor signal does not change. Stuck at low temperature or stuck at high temperature is detected. **NOTE: This code is caused by an electrical circuit problem and cannot be caused by a mechanical problem in the transmission.** **Possible Causes:** • Faulty ATF temperature sensor or temperature sensor/shift solenoid harness • Poor connections or loose terminals between the ATF temperature sensor and the PCM • PCM needs to be updated with the latest software • Faulty PCM
DTC: P0712 **1T PCM, TCIL: Yes** **Year:** 2009, 2010 **Model:** Accord, Accord Crosstour, Civic, CR-V, Element, Fit, Odyssey, Pilot, Ridgeline **Engine:** 1.3L L4, 1.5L L4, 1.8L L4, 2.4L L4, 3.5L V6 **Transmission:** All	**Short in ATF Temperature Sensor Circuit:** When the ATF temperature sensor signal voltage to the PCM is under the specification, indicating that the temperature is above the specification (a short to ground), a malfunction is detected. The ATF temperature sensor output voltage is less than 0.07 V for at least 10 seconds. **NOTE: This code is caused by an electrical circuit problem and cannot be caused by a mechanical problem in the transmission.** **Possible Causes:** • "Short" between the ATF temperature sensor and the PCM • Faulty ATF temperature sensor or temperature sensor/shift solenoid harness • PCM needs to be updated with the latest software • Faulty PCM
DTC: P0713 **1T PCM, TCIL: Yes** **Year:** 2009, 2010 **Model:** Accord, Accord Crosstour, Civic, CR-V, Element, Fit, Odyssey, Pilot, Ridgeline **Engine:** 1.3L L4, 1.5L L4, 1.8L L4, 2.0L L4, 2.4L L4, 3.5L V6 **Transmission:** All	**Open in ATF Temperature Sensor Circuit:** When the ATF temperature sensor signal voltage to the PCM is above the specification, indicating that the temperature is under the specification (open), a malfunction is detected. The ATF temperature sensor output voltage is 4.93 V or more for at least 10 seconds. **NOTE: This code is caused by an electrical circuit problem and cannot be caused by a mechanical problem in the transmission.** **Possible Causes:** • Poor connections or loose terminals at the ATF temperature sensor and the PCM • "Open" circuit between PCM connector terminal and the ATF temperature sensor • Faulty ATF temperature sensor or temperature sensor/shift solenoid harness • PCM needs to be updated with the latest software • Faulty PCM
DTC: P0714 **2T PCM, TCIL: Yes** **Year:** 2010 **Model:** Accord Crosstour **Engine:** 3.5L V6 **Transmission:** All	**ATF Sensor Intermittent Failure:** With the ignition OFF for at least 6 hours and batter voltage a minimum of 11 V, the temperature in the ATF temperature sensor deviates from the temperature in the ECT 2 sensor, lees than 11°F (-24°C) or greater than 9°F (32°C). Greater than 140°F (60°C) if equipped with a block heater. **NOTE: This code is caused by an electrical circuit problem and cannot be caused by a mechanical problem in the transmission.** **Possible Causes:** • Faulty ATF temperature sensor

DTC	Trouble Code Title, Conditions & Possible Causes
DTC: P0715 **1T ECM** **Year:** 2009, 2010 **Model:** Accord **Engine:** 3.5L V6 **Transmission:** All	**Input Shaft (Mainshaft) Speed Sensor Circuit Malfunction (M/T Model):** With the engine running and engine speed at a minimum of 1,000 rpm, No signal from the input shaft (Mainshaft) speed sensor is detected for at least 30 seconds. **NOTE: Before you troubleshoot, record all freeze data and any on-board snapshot, and review the general troubleshooting information.** **Possible Causes:** • Poor connections or loose terminals at the input shaft (mainshaft) speed sensor and the ECM • "Open" or "Short" between the ECM and the input shaft (mainshaft) speed sensor • Faulty input shaft (mainshaft) speed sensor • ECM needs to be updated with the latest software • Faulty ECM
DTC: P0716 **1T PCM, MIL: Yes, TCIL: Yes** **Year:** 2009, 2010 **Model:** Accord, Accord Crosstour, Civic, CR-V, Element, Fit, Odyssey, Pilot, Ridgeline **Engine:** 1.3L L4, 1.5L L4, 1.8L L4, 2.4L L4, 3.5L V6 **Transmission:** All	**Problem in Input Shaft (Mainshaft) Speed Sensor Circuit:** If no pulses occur with the input shaft (mainshaft) rotating, the PCM detects a malfunction that may be caused by an open, a temporary open, or a short to ground. The vehicle speed measured by the input shaft (mainshaft) speed sensor/(divided by) the vehicle speed measured by the output shaft (countershaft) speed sensor is less than 0.156 for at least 10 seconds. **NOTE: This code is caused by an electrical circuit problem and cannot be caused by a mechanical problem in the transmission.** **Possible Causes:** • Loose or poor connections at the PCM and input shaft (mainshaft) speed sensor connectors • Poor grounds • Faulty or improperly installed Input Shaft (Mainshaft) Speed Sensor • PCM needs to be updated with the latest software • Faulty PCM
DTC: P0717 **1T PCM, MIL: Yes, TCIL: Yes** **Year:** 2009, 2010 **Model:** Accord, Accord Crosstour, Civic, CR-V, Element, Fit, Odyssey, Pilot, Ridgeline **Engine:** 1.3L L4, 1.5L L4, 1.8L L4, 2.4L L4, 3.5L V6 **Transmission:** All	**Problem in Input Shaft (Mainshaft) Speed Sensor Circuit (No Signal Input):** If no pulses occur with the input shaft (mainshaft) rotating, the PCM detects a malfunction that may be caused by an open, a temporary open, or a short to ground. When the vehicle speed measured by the output shaft (countershaft) speed sensor is 13 mph (20 km/h) or more, the vehicle speed measured by the input shaft (mainshaft) speed sensor is 1 mph (2 km/h) or less for at least 10 seconds. **NOTE: This code is caused by an electrical circuit problem and cannot be caused by a mechanical problem in the transmission.** **Possible Causes:** • Loose or poor connections at the PCM and input shaft (mainshaft) speed sensor connectors • "Open" in the wires between PCM connector terminals and ground (G101), or repair poor ground • Faulty or improperly installed Input Shaft (Mainshaft) Speed Sensor • PCM needs to be updated with the latest software • Faulty PCM
DTC: P0718 **2T PCM, MIL: Yes, TCIL: Yes** **Year:** 2009, 2010 **Model:** Accord, Accord Crosstour, Civic, CR-V, Element, Fit, Odyssey, Pilot, Ridgeline **Engine:** 1.3L L4, 1.5L L4, 1.8L L4, 2.4L L4, 3.5L V6 **Transmission:** All	**Input Shaft (Mainshaft) Speed Sensor Intermittent Failure:** If no pulses occur with the input shaft (mainshaft) rotating, the PCM detects a malfunction that may be caused by an open, a temporary open, or a short to ground. The fluctuation of the vehicle speed measured by the input shaft (mainshaft) speed sensor in 10 milliseconds is 4 mph (6km/h) or more, and it fluctuates at least six times within 500 milliseconds. **NOTE: This code is caused by an electrical circuit problem and cannot be caused by a mechanical problem in the transmission.** **Possible Causes:** • Poor connections or loose terminals at the input shaft (mainshaft) speed sensor and the PCM • "Open" or "Short" in the wire between PCM connector terminal and the input shaft (mainshaft) speed sensor connector • "Open" in the wires between PCM connector terminals and ground • Faulty input shaft (mainshaft) speed sensor • PCM needs to be updated with the latest software • Faulty PCM
DTC: P0720 **2T PCM, MIL: Yes** **Year:** 2009, 2010 **Model:** Accord, CR-V, Element **Engine:** 2.4L L4, 3.5L V6 **Transmission:** All	**Output Shaft (Countershaft) Speed Sensor Circuit Malfunction :** With the engine speed at a minimum of 4,000 rpm and during fuel cut-off operation for deceleration. NO signal from the output shaft (Countershaft) speed sensor is detected for at least 5 seconds. **NOTE: Before you troubleshoot, record all freeze data and any on-board snapshot, and review the general troubleshooting information.** **Possible Causes:** • Poor connections or loose terminals at the output shaft (countershaft) speed sensor and the ECM • "Open" or "Short" circuit between the ECM and the output shaft (countershaft) speed sensor • Faulty output shaft (countershaft) speed sensor • ECM may need to be updated with the latest software • Faulty ECM

DTC	Trouble Code Title, Conditions & Possible Causes
DTC: P0721 **1T PCM, MIL: Yes, TCIL: Yes** **Year:** 2009, 2010 **Model:** Accord, Accord Crosstour, Civic, CR-V, Element, Fit, Odyssey, Pilot, Ridgeline **Engine:** 1.3L L4, 1.5L L4, 1.8L L4, 2.4L L4, 3.5L V6 **Transmission:** All	**Problem in Output Shaft (Countershaft) Speed Sensor Circuit:** If pulse dropouts occur with the output shaft (countershaft) rotating, the PCM detects a malfunction that may be caused by an open, a temporary open, or a short to ground. The vehicle speed measured by the input shaft (mainshaft) speed sensor/(divided by) the vehicle speed measured by the output shaft (countershaft) speed sensor is greater than 6.0 for at least 10 seconds. **NOTE: This code is caused by an electrical circuit problem and cannot be caused by a mechanical problem in the transmission.** **Possible Causes:** • Faulty or improperly installed output shaft (countershaft) speed sensor • "Open" in the wires between PCM connector terminals and ground • "Open" or "Short" in the wire between PCM connector and the output shaft (countershaft) speed sensor connector • PCM needs to be updated with the latest software • Faulty PCM
DTC: P0722 **1T PCM, MIL: Yes, TCIL: Yes** **Year:** 2009, 2010 **Model:** Accord, Accord Crosstour, Civic, CR-V, Element, Fit, Odyssey, Pilot, Ridgeline **Engine:** 1.3L L4, 1.5L L4, 1.8L L4, 2.4L L4, 3.5L V6 **Transmission:** All	**Problem in Output Shaft (Countershaft) Speed Sensor Circuit (No Signal Input):** If pulse dropouts occur with the output shaft (countershaft) rotating, the PCM detects a malfunction that may be caused by an open, a temporary open, or a short to ground. When the vehicle speed measured by the input shaft (mainshaft) speed sensor is 13 mph (20 km/h) or more, the vehicle speed measured by the output shaft (countershaft) speed sensor is 1 mph (2 km/h) or less for at least 10 seconds. **NOTE: This code is caused by an electrical circuit problem and cannot be caused by a mechanical problem in the transmission.** **Possible Causes:** • Faulty or improperly installed output shaft (countershaft) speed sensor • Loose or poor connections at the PCM and output shaft (countershaft) speed sensor connectors • "Open" in the wires between PCM connector terminals and ground or poor ground • "Open" or "Short" in the wire between PCM connector terminal and the output shaft (countershaft) speed sensor connector • PCM needs to be updated with the latest software • Faulty PCM
DTC: P0723 **2T PCM, MIL: Yes, TCIL: Yes** **Year:** 2009, 2010 **Model:** Accord, Accord Crosstour, Civic, CR-V, Element, Fit, Odyssey, Pilot, Ridgeline **Engine:** 1.3L L4, 1.5L L4, 1.8L L4, 2.4L L4, 3.5L V6 **Transmission:** All	**Output Shaft (Countershaft) Speed Sensor Intermittent Failure:** If pulse dropouts occur with the output shaft (countershaft) rotating, the PCM detects a malfunction that may be caused by an open, a temporary open, or a short to ground. Based on the fluctuation of the vehicle speed measured by the output shaft (countershaft) speed sensor, a malfunction is detected. The fluctuation of the vehicle speed measured by the output shaft (countershaft) speed sensor in 10 milliseconds is 4 mph (6km/h) or more, and it fluctuates at least six times within 500 milliseconds. **NOTE: This code is caused by an electrical circuit problem and cannot be caused by a mechanical problem in the transmission.** **Possible Causes:** • Faulty or improperly installed output shaft (countershaft) speed sensor • Poor connections and loose terminals at the output shaft (countershaft) speed sensor and the PCM • "Open" in the wires between PCM connector terminals and ground or poor ground • "Open" or "Short" in the wire between PCM connector terminal and the output shaft (countershaft) speed sensor connector • PCM needs to be updated with the latest software • Faulty PCM
DTC: P0729 **2T PCM, TCIL: Yes** **Year:** 2009, 2010 **Model:** Civic **Engine:** 1.3L L4, 1.8L L4 **Transmission:** All	**Problem in 6th Clutch and 6th Clutch Hydraulic Circuit:** With the vehicle running in drive and engine speed at a minimum of 7 mph, battery voltage 11 volts. The actual gear ratio must match one of these conditions for at least 12 seconds with the 6th gear shift command: • Actual gear ratio is greater than the 6th gear ratio by a factor of 1.25. • Actual gear ratio is less than the 6th gear ratio by a factor of 0.8. **NOTE: Before you troubleshoot, record all freeze data and any on-board snapshots.** **Possible Causes:** • Low or dirty transmission fluid • Faulty ATF pump or regulator valve • Faulty 6th clutch, replace the 3rd/6th clutch or the transmission

DTC	Trouble Code Title, Conditions & Possible Causes
DTC: P0731 **2T PCM, TCIL: Yes** **Year:** 2009, 2010 **Model:** Accord, Accord Crosstour, Civic, CR-V, Element, Fit, Odyssey, Ridgeline **Engine:** 1.5L L4, 1.8L L4, 2.4L L4, 3.5L V6 **Transmission:** All	**Problem in 1st Clutch and 1st Clutch Hydraulic Circuit (1st gear incorrect ratio):** The Powertrain Control Module (PCM) computes the ratio of the input shaft (mainshaft) speed to the output shaft (countershaft) speed. When the ratio is not the 1st gear ratio, it is detected as a malfunction of the hydraulic circuit or the 1st clutch. (Symptom 1) The actual gear ratio must match one of these conditions for at least 12 seconds with the 1st gear command: * Actual gear ratio is greater than the 1st gear ratio by a factor of 1.2 * Actual gear ratio is less than the 1st gear ratio by a factor of 0.75 (Symptom 2) The actual gear position is neutral for at least 3 seconds and then the gear up-shifted from 2nd to 3rd, even though 1st gear shift is commanded. **NOTE: Before you troubleshoot, record all freeze data and any on-board. snapshot,** **Possible Causes:** • Low or dirty transmission fluid • Faulty shift valves B and C • Faulty ATF pump and the regulator valve • Inspect the strainer for metal debris or excessive clutch material, if present • replace the transmission
DTC: P0732 **2T PCM, TCIL: Yes** **Year:** 2009, 2010 **Model:** Accord, Accord Crosstour, Civic, CR-V, Element, Fit, Odyssey, Pilot, Ridgeline **Engine:** 1.5L L4, 1.8L L4, 2.4L L4, 3.5L V6 **Transmission:** All	**Problem in 2nd Clutch and 2nd Clutch Hydraulic Circuit (2nd gear incorrect ratio):** The Powertrain Control Module (PCM) computes the ratio of the input shaft (mainshaft) speed to the output shaft (countershaft) speed. When the ratio is not the 2nd gear ratio, it is detected as a malfunction of the hydraulic circuit or the 2nd clutch. The actual gear ratio must match one of these conditions for at least 12 seconds with the 2nd gear command. * Actual gear ratio is greater than the 2nd gear ratio by a factor of 1.2. * Actual gear ratio is less than the 2nd gear ratio by a factor of 0.75. **NOTE: Before you troubleshoot, record all freeze data and any on-board snapshot.** **Possible Causes:** • Low or dirty transmission fluid • Faulty shift valves A, B and C • Faulty ATF pump and the regulator valve • Inspect the strainer for metal debris or excessive clutch material, if present • replace the transmission
DTC: P0733 **2T PCM, TCIL: Yes** **Year:** 2009, 2010 **Model:** Accord, Accord Crosstour, Civic, CR-V, Element, Fit, Odyssey, Pilot, Ridgeline **Engine:** 1.5L L4, 1.8L L4, 2.4L L4, 3.5L V6 **Transmission:** All	**Problem in 3rd Clutch and 3rd Clutch Hydraulic Circuit (3rd gear incorrect ratio):** The powertrain control module (PCM) computes the ratio of the input shaft (mainshaft) speed to the output shaft (countershaft) speed. When the ratio is not the 3rd gear ratio, it is detected as a malfunction of the hydraulic circuit or the 3rd clutch. The actual gear ratio must match one of these conditions for at least 12 seconds with the 3rd gear command. * Actual gear ratio is greater than the 3rd gear ratio by a factor of 1.2. * Actual gear ratio is less than the 3rd gear ratio by a factor of 0.75. **NOTE: Before you troubleshoot, record all freeze data and any on-board snapshot.** **Possible Causes:** • Low or dirty transmission fluid • Faulty shift valves A, B, and C are stuck • Faulty ATF pump and the regulator valve • Inspect the strainer for metal debris or excessive clutch material, if present • replace the transmission
DTC: P0734 **2T PCM, TCIL: Yes** **Year:** 2009, 2010 **Model:** Accord, Accord Crosstour, Civic, CR-V, Element, Fit, Odyssey, Pilot, Ridgeline **Engine:** 1.5L L4, 1.8L L4, 2.4L L4, 3.5L V6 **Transmission:** All	**Problem in 4th Clutch and 4th Clutch Hydraulic Circuit (4th gear incorrect ratio):** The Powertrain Control Module (PCM) computes the ratio of the input shaft (mainshaft) speed to the output shaft (countershaft) speed. When the ratio is not the 4th gear ratio, it is detected as a malfunction of the hydraulic circuit or the 4th clutch. The actual gear ratio must match one of these conditions for at least 12 seconds with the 4th gear command. * Actual gear ratio is greater than the 4th gear ratio by a factor of 1.2. * Actual gear ratio is less than the 4th gear ratio by a factor of 0.75. **NOTE: Before you troubleshoot, record all freeze data and any on-board snapshot.** **Possible Causes:** • Low or dirty transmission fluid • Faulty shift valves A, B, C, and D are stuck • Faulty ATF pump and the regulator valve • Inspect the strainer for metal debris or excessive clutch material, if present • replace the transmission

DTC	Trouble Code Title, Conditions & Possible Causes
DTC: P0735 **2T PCM, TCIL: Yes** **Year:** 2009, 2010 **Model:** Accord, Accord Crosstour, Civic, CR-V, Element, Fit, Odyssey, Pilot, Ridgeline **Engine:** 1.5L L4, 1.8L L4, 2.4L L4, 3.5L V6 **Transmission:** All	**Problem in 5th Clutch and 5th Clutch Hydraulic Circuit (5th gear incorrect ratio):** The Powertrain Control Module (PCM) computes the ratio of the input shaft (mainshaft) speed to the output shaft (countershaft) speed. When the ratio is not the 5th gear ratio, it is detected as a malfunction of the hydraulic circuit or the 5th clutch. The actual gear ratio must match one of these conditions for at least 12 seconds with the 5th gear command. * Actual gear ratio is greater than the 5th gear ratio by a factor of 1.2 * Actual gear ratio is less than the 5th gear ratio by a factor of 0.75 **NOTE: Before you troubleshoot, record all freeze data and any on-board snapshot** **Possible Causes:** • Low or dirty transmission fluid • Faulty shift valves A, B, C, and D are stuck • Faulty ATF pump and the regulator valve • Inspect the strainer for metal debris or excessive clutch material, if present • replace the transmission
DTC: P0741 **2T PCM, MIL: Yes** **Year:** 2009, 2010 **Model:** Accord, Accord Crosstour, Civic, CR-V, Element, Fit, Odyssey, Pilot, Ridgeline **Engine:** 1.3L L4, 1.5L L4, 1.8L L4, 2.0L L4, 2.4L L4, 3.5L V6 **Transmission:** All	**Torque Converter Clutch Hydraulic Circuit Stuck OFF:** If the ratio of engine speed and input shaft (mainshaft) speed is not about 1:1 while the PCM is issuing the command to turn shift solenoid valve D and A/T clutch pressure control solenoid valve C ON, the PCM detects a faulty lock-up control system. The ratio of the engine revolutions to the transmission input pulses does not reach about 100 % for at least 22 seconds. **NOTE: Before you troubleshoot, record all freeze data and any on-board snapshot.** **Possible Causes:** • Low or dirty transmission fluid • Inspect the strainer for metal debris or excessive clutch material, if present replace the transmission • Faulty shift solenoid valve D • Faulty torque converter clutch mechanism, torque converter clutch hydraulic circuit, lock-up shift valve, lock-up control valve, or replace the transmission
DTC: P0746 **2T PCM, MIL: Yes, TCIL: Yes** **Year:** 2009, 2010 **Model:** Accord, Accord Crosstour, Civic, Fit, Odyssey, Pilot, Ridgeline **Engine:** 1.5L L4, 1.8L L4, 3.5L V6 **Transmission:** All	**A/T Clutch Pressure Control Solenoid Valve A Stuck OFF:** When an improper gear ratio is output compared to the predetermined gear ratio, an A/T clutch pressure control solenoid valve A OFF failure is detected. One of these symptoms occur: * Transmission is held in 1st gear. * The engine speed flares when upshifting to 2nd-3rd. * The engine speed flares when upshifting to 3rd-4th or 5th if applicable. **NOTE: Before you troubleshoot, record all freeze data and any on-board snapshot.** **Possible Causes:** • Faulty hydraulic system related with shift valve A • Low or dirty transmission fluid • Inspect the strainer for metal debris or excessive clutch material, if present • replace the transmission
DTC: P0746 **2T PCM, MIL: Yes, TCIL: Yes** **Year:** 2009, 2010 **Model:** Civic **Engine:** 1.3L L4 **Transmission:** All	**CVT Drive Pulley Pressure Control Valve Stuck OFF:** With the vehicle at operating temperature, and driven for 15 seconds or more, both of the following judgment conditions must be met. Judgment A: The pulley ratio is 0.3-0.8 for at least 2.5 seconds. Judgment B: The pulley ratio is less than 3 for at least 10 seconds. **Possible Causes:** • Poor connections or loose terminals between the CVT drive pulley pressure control solenoid valve and the PCM • Faulty CVT drive pulley pressure control solenoid valve
DTC: P0747 **2T PCM, MIL: Yes, TCIL: Yes** **Year:** 2009, 2010 **Model:** Accord, Accord Crosstour, Civic, CR-V, Element, Fit, Pilot, Ridgeline **Engine:** 1.5L L4, 1.8L L4, 2.4L L4, 3.5L V6 **Transmission:** All	**A/T Clutch Pressure Control Solenoid Valve A Stuck ON:** When an improper gear ratio is output compared to the predetermined gear ratio, an A/T clutch pressure control solenoid valve A ON failure is detected. The execution time is continuous and the duration time is 20 seconds. One of these conditions occur: * The transmission is held in 2nd gear against the 2nd-3rd gear upshift command as long as 20 seconds though there is no record of being neutral when the 1st gear shift is commanded. * The transmission is held in 4th gear against the 4th-5th gear upshift command as long as 20 seconds though there is no record of being neutral when the 1st gear shift is commanded. **NOTE: Before you troubleshoot, record all freeze data and any on-board snapshot.** **Possible Causes:** • Low or dirty transmission fluid • Inspect the strainer for metal debris or excessive clutch material, if present • replace the transmission • Faulty hydraulic system related with shift valve A

DTC	Trouble Code Title, Conditions & Possible Causes
DTC: P0751 **2T PCM, MIL: Yes, TCIL: Yes** **Year:** 2009, 2010 **Model:** Accord, Accord Crosstour, Civic, Fit, Odyssey, Pilot, Ridgeline **Engine:** 1.5L L4, 1.8L L4, 2.4L L4, 3.5L V6 **Transmission:** All	**Shift Solenoid Valve A Stuck OFF:** When an improper gear ratio is output compared to the predetermined gear change mode, a shift solenoid valve A OFF failure is detected and a DTC is stored. The execution time is continuous and the duration time is 2 seconds or more. The transmission is held in 5th gear against the 3rd gear command for at least 2 seconds. **NOTE: Before you troubleshoot, record all freeze data and any on-board snapshot.** **Possible Causes:** • Low or dirty transmission fluid • Faulty shift solenoid valve A • Inspect the strainer for metal debris or excessive clutch material, if present • replace the transmission
DTC: P0752 **2T PCM, MIL: Yes, TCIL: Yes** **Year:** 2009, 2010 **Model:** Accord, Accord Crosstour, Civic, Element, Fit, Odyssey, Pilot, Ridgeline **Engine:** 1.5L L4, 1.8L L4, 2.4L L4, 3.5L V6 **Transmission:** All	**Shift Solenoid Valve A Stuck ON:** When the wrong transmission fluid switch is turned on for a given speed change mode, a shift solenoid valve turn-on malfunction is detected. The execution time is continuous depending on the driving pattern. The 3rd clutch transmission fluid switch is ON against the 4th-5th gear upshift command for at least 11 seconds. **NOTE: Before you troubleshoot, record all freeze data and any on-board snapshot.** **Possible Causes:** • Low or dirty transmission fluid • Faulty shift solenoid valve A • Inspect the strainer for metal debris or excessive clutch material, if present • replace the transmission
DTC: P0756 **2T PCM, MIL: Yes, TCIL: Yes** **Year:** 2009, 2010 **Model:** Accord, Accord Crosstour, Civic, CR-V, Element, Fit, Odyssey, Pilot, Ridgeline **Engine:** 1.5L L4, 1.8L L4, 2.4L L4, 3.5L V6 **Transmission:** All	**Shift Solenoid Valve B Stuck OFF:** When an improper gear ratio is output compared to the predetermined gear change mode, a shift solenoid valve B OFF failure is detected. The transmission is held in 4th gear against the 2nd gear command for at least 2 seconds. The execution time is continuous and the duration time is 2 seconds or more. **NOTE: Before you troubleshoot, record all freeze data and any on-board snapshot.** **Possible Causes:** • Low or dirty transmission fluid • Faulty shift valve B • Inspect the strainer for metal debris or excessive clutch material, if present • replace the transmission
DTC: P0757 **2T PCM, MIL: Yes, TCIL: Yes** **Year:** 2009, 2010 **Model:** Accord, Accord Crosstour, Civic, CR-V, Element, Fit, Odyssey, Pilot, Ridgeline **Engine:** 1.5L L4, 1.8L L4, 2.4L L4, 3.5L V6 **Transmission:** All	**Shift Solenoid Valve B Stuck ON:** When the wrong gear ratio is output for a given speed change mode, or when the wrong transmission fluid pressure switch is turned-on, a shift solenoid valve turn-on malfunction is detected. The execution time is continuous depending on the driving pattern. One of these conditions occur: * The 2nd clutch transmission fluid switch is ON against the 3rd-4th gear upshift command for at least 11 seconds. * After the 3rd-4th gear upshift command is output, it is neutral for at least 2 seconds when the 5th gear shift command is output, though there is no history of being neutral. **NOTE: Before you troubleshoot, record all freeze data and any on-board snapshot.** **Possible Causes:** • Low or dirty transmission fluid • Inspect the strainer for metal debris or excessive clutch material, if present • replace the transmission • Faulty shift solenoid valve B
DTC: P0761 **2T PCM, MIL: Yes, TCIL: Yes** **Year:** 2009, 2010 **Model:** Accord, Accord Crosstour, Civic, CR-V, Element, Fit, Pilot, Ridgeline **Engine:** 1.5L L4, 1.8L L4, 2.4L L4, 3.5L V6 **Transmission:** All	**Shift Solenoid Valve C Stuck OFF:** The execution time is continuous and the duration time is 20 seconds. When an improper gear ratio is output compared to the predetermined gear change mode, a shift solenoid valve C OFF failure is detected. One of these symptoms occurred when the actual gear position was neutral when the 1st gear shift is commanded: * The transmission is held in 2nd gear against the 2nd-3rd gear upshift command for at least 17 seconds. * The transmission is held in 4th gear against the 4th-5th gear upshift command for at least 17 seconds. **NOTE: Before you troubleshoot, record all freeze data and any on-board snapshot.** **Possible Causes:** • Low or dirty transmission fluid • Faulty shift solenoid valve C • Inspect the strainer for metal debris or excessive clutch material, if present • replace the transmission

DTC	Trouble Code Title, Conditions & Possible Causes
DTC: P0762 **2T PCM, MIL: Yes, TCIL: Yes** **Year:** 2009, 2010 **Model:** Accord, Accord Crosstour, Civic, Fit, Odyssey, Pilot, Ridgeline **Engine:** 1.5L L4, 1.8L L4, 3.5L V6 **Transmission:** All	**Shift Solenoid Valve C Stuck ON:** When an improper gear ratio is output compared to the predetermined gear change mode, a shift solenoid valve C ON failure is detected. The execution time is continuous and the duration time is 20 seconds. The transmission is held in 3rd gear against the 3rd-4th gear upshift command for as long as 20 seconds, without records that the gear change time was short when the 2nd-3rd gear upshift were commanded. **NOTE: Before you troubleshoot, record all freeze data and any on-board snapshot.** **Possible Causes:** • Low or dirty transmission fluid • Faulty shift solenoid valve C • Inspect the strainer for metal debris or excessive clutch material, if present replace the transmission
DTC: P0766 **2T PCM, MIL: Yes, TCIL: Yes** **Year:** 2009, 2010 **Model:** Civic, Fit, Odyssey **Engine:** 1.5L L4, 1.8L L4, 3.5L V6 **Transmission:** All	**Shift Solenoid Valve D Stuck OFF:** When an improper gear ratio is output compared to the predetermined gear change mode, a shift solenoid valve D OFF failure is detected. **NOTE: Before you troubleshoot, record all freeze data and any on-board snapshot.** **Possible Causes:** • Low or dirty transmission fluid • Faulty shift solenoid valve D • Inspect the strainer for metal debris or excessive clutch material, if present • replace the transmission
DTC: P0767 **2T PCM, MIL: Yes, TCIL: Yes** **Year:** 2009, 2010 **Model:** Civic, Fit **Engine:** 1.5L L4, 1.8L L4 **Transmission:** All	**Shift Solenoid Valve D Stuck ON:** When an improper gear ratio is output compared to the predetermined gear change mode, a shift solenoid valve D ON failure is detected. One of these conditions occur: * The actual gear position is neutral for at least 3 seconds when 1st gear in-gear is commanded, though there is no history of being neutral when reverse gear in-gear is commanded. * The actual gear position is neutral for at least 3 seconds, though 1st gear in-gear is commanded and reverse drive occurred during this driving cycle. **NOTE: Before you troubleshoot, record all freeze data and any on-board snapshot.** **Possible Causes:** • Low or dirty transmission fluid • Faulty shift solenoid valve D • Faulty ATF pump and the regulator valve • Inspect the strainer for metal debris or excessive clutch material, if present • replace the transmission
DTC: P0771 **2T PCM, MIL: Yes, TCIL: Yes** **Year:** 2009, 2010 **Model:** Accord, CR-V, Element **Engine:** 2.4L L4 **Transmission:** All	**Shift Solenoid Valve E Stuck OFF:** While driving the vehicle with the torque converter lock-up ON a malfunction is detected when shifting from 3rd gear into 4th gear. **NOTE: Before you troubleshoot, record all freeze data and any on-board snapshot with the HDS, and review General Troubleshooting Information.** **Possible Causes:** • Low or dirty transmission fluid • Strainer has metal debris or excessive clutch material, replace the transmission • Faulty shift solenoid valve E • Faulty transmission.
DTC: P0776 **2T PCM, MIL: Yes, TCIL: Yes** **Year:** 2009, 2010 **Model:** Accord, Accord Crosstour, Civic, CR-V, Element, Fit, Odyssey, Pilot, Ridgeline **Engine:** 1.5L L4, 1.8L L4, 2.4L L4, 3.5L V6 **Transmission:** All	**A/T Clutch Pressure Control Solenoid Valve B Stuck OFF:** When an improper gear ratio is output compared to the predetermined gear change mode, an A/T clutch pressure control solenoid valve B OFF failure is detected. The transmission is held in 3rd gear against the 3rd-4th gear upshift command for as long as 20 seconds, with records that the gear change time was short when the 2nd-3rd gear upshift was commanded. **NOTE: Before you troubleshoot, record all freeze data and any on-board snapshot.** **Possible Causes:** • Low or dirty transmission fluid • Faulty A/T clutch pressure control solenoid valve B • Inspect the strainer for metal debris or excessive clutch material, if present • replace the transmission

DTC	Trouble Code Title, Conditions & Possible Causes
DTC: P0777 **2T PCM, MIL: Yes, TCIL: Yes** **Year:** 2009, 2010 **Model:** Accord, Accord Crosstour, Civic, CR-V, Element, Fit, Odyssey, Pilot, Ridgeline **Engine:** 1.5L L4, 1.8L L4, 2.4L L4, 3.5L V6 **Transmission:** All	**A/T Clutch Pressure Control Solenoid Valve B Stuck ON:** When an improper gear ratio is output compared to the predetermined gear change mode, an A/T clutch pressure control solenoid valve B ON failure is detected. The engine speed flares during 2nd-3rd and 3rd-4th upshifts for at least 1 second. The execution time is continuous depending on the driving pattern. **NOTE: Before you troubleshoot, record all freeze data and any on-board snapshot.** **Possible Causes:** • Low or dirty transmission fluid • Faulty A/T clutch pressure control solenoid valve B • Inspect the strainer for metal debris or excessive clutch material, if present • replace the transmission
DTC: P0777 **2T PCM, MIL: Yes, TCIL: Yes** **Year:** 2009, 2010 **Model:** Civic **Engine:** 1.3L L4 **Transmission:** All	**CVT Driven Pulley Pressure Control Valve Stuck ON:** With the vehicle at operating temperature, and driven for 15 seconds or more, both of the following judgment conditions must be met. Judgment A: The pulley ratio is 0.3-0.8 for at least 2.5 seconds. Judgment B: The pulley ratio is less than 3 for at least 0.2 seconds. **Possible Causes:** • Poor connections or loose terminals between the CVT drive pulley pressure control solenoid valve and the PCM • Faulty CVT drive pulley pressure control solenoid valve
DTC: P0780 **2T PCM, MIL: Yes, TCIL: Yes** **Year:** 2009, 2010 **Model:** Accord, CR-V, Element **Engine:** 2.4L L4, 3.5L V6 **Transmission:** All	**Shift Control System:** This code is stored whenever DTCs P1730, P1731, P1732, P1733, and P1734 are detected. Refer to specific DTC information. Before you troubleshoot, record all freeze data and any on-board snapshot with the HDS, and review General Troubleshooting Information. **Possible Causes:** • Refer to specific DTC information.
DTC: P0796 **2T PCM, MIL: Yes, TCIL: Yes** **Year:** 2009, 2010 **Model:** Civic **Engine:** 1.3L L4 **Transmission:** All	**CVT Start Clutch Pressure Control Valve Stuck OFF:** With the vehicle at operating temperature, and driven for 15 seconds or more, the oil pressure of the start clutch is at specified value or more for at least 2 seconds. **Possible Causes:** • Poor connections or loose terminals between the CVT drive pulley pressure control solenoid valve and the PCM • Faulty CVT drive pulley pressure control solenoid valve
DTC: P0796 **2T PCM, MIL: Yes, TCIL: Yes** **Year:** 2009, 2010 **Model:** Accord, Accord Crosstour, Civic, CR-V, Element, Fit, Odyssey, Pilot, Ridgeline **Engine:** 1.5L L4, 1.8L L4, 2.4L L4, 3.5L V6 **Transmission:** All	**A/T Clutch Pressure Control Solenoid Valve C Stuck OFF:** When an improper gear ratio is output compared to the predetermined gear ratio, an A/T clutch pressure control solenoid valve C OFF failure is detected. The execution time is continuous depending on the driving pattern. **NOTE: Before you troubleshoot, record all freeze data and any on-board snapshot.** **Possible Causes:** • Low or dirty transmission fluid • Faulty A/T clutch pressure control solenoid valve C • Inspect the strainer for metal debris or excessive clutch material, if present • replace the transmission
DTC: P0797 **2T PCM, MIL: Yes, TCIL: Yes** **Year:** 2009, 2010 **Model:** Accord, Accord Crosstour, Civic, Element, Fit, Odyssey, Pilot, Ridgeline **Engine:** 1.5L L4, 1.8L L4, 2.4L L4, 3.5L V6 **Transmission:** All	**A/T Clutch Pressure Control Solenoid Valve C Stuck ON:** When the wrong transmission fluid pressure switch is turned on, an A/T clutch pressure control solenoid valve C turn-on malfunction is detected. The 2nd clutch transmission fluid switch is ON against the 3rd-4th gear upshift command for at least 11 seconds. The execution time is continuous and the duration time is 20 seconds. **NOTE: Before you troubleshoot, record all freeze data and any on-board snapshot,.** **Possible Causes:** • Low or dirty transmission fluid • Faulty hydraulic system related with shift valve C • Faulty A/T clutch pressure control solenoid valve C • Inspect the strainer for metal debris or excessive clutch material, if present • replace the transmission

DTC	Trouble Code Title, Conditions & Possible Causes
DTC: P0797 **T TCM, TCIL: Yes** **Year:** 2009, 2010 **Model:** CR-V **Engine:** 2.4L L4 **Transmission:** All	**Pressure Control Solenoid Valve C (Stuck On):** **NOTE: If DTC U1000 is displayed with this DTC, first perform the trouble diagnosis for DTC U1000.** • This malfunction will not be detected while the O/D OFF indicator lamp is indicating another self-diagnosis malfunction. • This is not caused by electrical malfunction (circuits open or shorted) but by mechanical malfunction such as control valve sticking, improper solenoid valve operation, etc. • The pressure control solenoid valve C is normally low, 3-port linear pressure control solenoid. • The pressure control solenoid valve C is activated to control the apply and release of the 2nd brake and 1st and reverse brake, and torque converter clutch. • Lock-up operation, however, is prohibited when A/T fluid temperature is too low. • When the accelerator pedal is depressed (less than 1/8) in lock-up condition, the engine speed should not change abruptly. If there is a big jump in engine speed, there is no lock-up. • This is an OBD-II self-diagnostic item. • Diagnostic trouble code "PC SOL C STC ON" with CONSULT-III or P0797 without CONSULT-III is detected when condition of pressure control solenoid valve C is different from monitor value, and relation between gear position and actual gear ratio or lock-up status is irregular. **Possible Causes:** • Pressure control solenoid valve C (On stick) • Hydraulic control circuit
DTC: P0812 **2T PCM, MIL: Yes, TCIL: Yes** **Year:** 2009, 2010 **Model:** Civic, Fit, Odyssey, Pilot, Ridgeline **Engine:** 1.5L L4, 1.8L L4, 3.5L V6 **Transmission:** All	**Open in Transmission Range Switch ATP RVS Switch Circuit:** If the R switch is OPEN with the shift lever in R, the PCM detects a switch OPEN failure. The RVS signal is detected but the R switch signal is not detected for at least 2 seconds. The execution time is continuous depending on the driving pattern. **NOTE: This code is caused by an electrical circuit problem and cannot be caused by a mechanical problem in the transmission.** **Possible Causes:** • Faulty transmission range switch • Poor connections and loose terminals at the transmission range switch and the PCM • "Open" in the wire between PCM and the transmission range switch • "Open" in the wire between transmission range switch and ground, or poor ground • PCM needs to be updated with the latest software • Faulty PCM
DTC: P0842 **1T PCM, MIL: Yes, TCIL: Yes** **Year:** 2009, 2010 **Model:** Accord, Accord Crosstour, Civic, CR-V, Element, Fit, Odyssey, Pilot, Ridgeline **Engine:** 1.5L L4, 1.8L L4, 2.4L L4, 3.5L V6 **Transmission:** All	**Short in 2nd Clutch Transmission Fluid Pressure Switch Circuit, or 2nd Clutch Transmission Fluid Pressure Switch Stuck ON:** The input signal from the 2nd clutch transmission fluid pressure switch to the PCM is low when driving in 1st gear, 3rd gear, or 5th gear. The execution time is continuous and the duration time is 2 seconds or more. **NOTE: This code is caused by an electrical circuit problem and cannot be caused by a mechanical problem in the transmission.** **Possible Causes:** • Faulty 2nd clutch transmission fluid pressure switch • Poor connections and loose terminals at the 2nd clutch transmission fluid pressure switch and the PCM • OP2SW wire for an intermittent short to ground between the 2nd clutch transmission fluid pressure switch and the PCM • "Short" in the wire between PCM connector terminal and the 2nd clutch transmission fluid pressure switch • PCM may need to be updated with the latest software • Faulty PCM
DTC: P0843 **1T PCM, MIL: Yes, TCIL: Yes** **Year:** 2009, 2010 **Model:** Accord, Accord Crosstour, Civic, CR-V, Element, Fit, Odyssey, Pilot, Ridgeline **Engine:** 1.5L L4, 1.8L L4, 2.4L L4, 3.5L V6 **Transmission:** All	**Open in 2nd Clutch Transmission Fluid Pressure Switch Circuit, or 2nd Clutch Transmission Fluid Pressure Switch Stuck OFF :** The input signal from the 2nd clutch transmission fluid pressure switch to the PCM is high when driving in 2nd gear. The execution time is continuous and the duration time is 2 seconds or more. **NOTE: Before you troubleshoot, record all freeze data and any on-board snapshot.** **Possible Causes:** • Faulty 2nd clutch transmission fluid pressure switch • Poor connections and loose terminals at the 2nd clutch transmission fluid pressure switch and the PCM. • "Open" circuit between PCM connector terminal and the 2nd clutch transmission fluid pressure switch • PCM needs to be updated with the latest software • Faulty PCM

DTC	Trouble Code Title, Conditions & Possible Causes
DTC: P0847 **1T PCM, MIL: Yes, TCIL: Yes** **Year:** 2009, 2010 **Model:** Accord, Accord Crosstour, Civic, CR-V, Element, Fit, Odyssey, Pilot, Ridgeline **Engine:** 1.5L L4, 1.8L L4, 2.4L L4, 3.5L V6 **Transmission:** All	**Short in 3rd Clutch Transmission Fluid Pressure Switch Circuit, or 3rd Clutch Transmission Fluid Pressure Switch Stuck ON:** The input signal from the 3rd clutch transmission fluid pressure switch to the PCM is low when driving in 1st gear, 2nd gear, 4th gear, or 5th gear. The execution time is continuous and the duration time is 2 seconds or more. **NOTE: This code is caused by an electrical circuit problem and cannot be caused by a mechanical problem in the transmission.** **Possible Causes:** • Faulty 3rd clutch transmission fluid pressure switch • Poor connections and loose terminals at the 3rd clutch transmission fluid pressure switch and the PCM • "Short" circuit between PCM connector terminal and the 3rd clutch transmission fluid pressure switch • PCM may need to be updated with the latest software • Faulty PCM
DTC: P0848 **1T PCM, MIL: Yes, TCIL: Yes** **Year:** 2009, 2010 **Model:** Accord, Accord Crosstour, Civic, CR-V, Element, Fit, Odyssey, Pilot, Ridgeline **Engine:** 1.5L L4, 1.8L L4, 2.4L L4, 3.5L V6 **Transmission:** All	**Open in 3rd Clutch Transmission Fluid Pressure Switch Circuit, or 3rd Clutch Transmission Fluid Pressure Switch Stuck OFF :** The input signal from the 3rd clutch transmission fluid pressure switch to the PCM is high when driving in 3rd gear. The execution time is continuous and the duration time is 2 seconds or more. **NOTE: Before you troubleshoot, record all freeze data and any on-board snapshot.** **Possible Causes:** • Faulty 3rd clutch transmission fluid pressure switch • Poor connections or loose terminals at the 3rd clutch transmission fluid pressure switch and the PCM • "Open" circuit between PCM connector terminal and the 3rd clutch transmission fluid pressure switch • PCM may need to be updated with the latest software • Faulty PCM
DTC: P0872 **1T PCM, MIL: Yes, TCIL: Yes** **Year:** 2009, 2010 **Model:** Accord, Accord Crosstour, Odyssey, Pilot, Ridgeline **Engine:** 2.4L L4, 3.5L V6 **Transmission:** All	**Short in 4th Clutch Transmission Fluid Pressure Switch Circuit, or 4th Clutch Transmission Fluid Pressure Switch Stuck ON :** The input signal from the 4th clutch transmission fluid pressure switch to the PCM is low when driving in 5th gear. The execution time is continuous and the duration time is 2 seconds or more. **NOTE: This code is caused by an electrical circuit problem and cannot be caused by a mechanical problem in the transmission.** **Possible Causes:** • Faulty 4th clutch transmission fluid pressure switch • Check the OP4SW wire for an intermittent short to ground between the 4th clutch transmission fluid pressure switch and the PCM
DTC: P0873 **1T PCM, MIL: Yes, TCIL: Yes** **Year:** 2009, 2010 **Model:** Accord, Accord Crosstour, Odyssey, Pilot, Ridgeline **Engine:** 2.4L L4, 3.5L V6 **Transmission:** All	**Open in 4th Clutch Transmission Fluid Pressure Switch Circuit, or 4th Clutch Transmission Fluid Pressure Switch Stuck OFF:** The input signal from the 4th clutch transmission fluid pressure switch to the PCM is high when driving in 4th gear. The execution time is continuous and the duration time is 2 seconds or more. **NOTE: Before you troubleshoot, record all freeze data and any on-board snapshot.** **Possible Causes:** • Faulty 4th clutch transmission fluid pressure switch • Poor connections and loose terminals at the 4th clutch transmission fluid pressure switch and the PCM • "Open" in the wire between PCM connector terminal and the 4th clutch transmission fluid pressure switch • PCM may need to be updated with the latest software • Faulty PCM
DTC: P0962 **1T PCM, MIL: Yes, TCIL: Yes** **Year:** 2009, 2010 **Model:** Accord, Accord Crosstour, Civic, CR-V, Element, Fit, Odyssey, Pilot, Ridgeline **Engine:** 1.5L L4, 1.8L L4, 2.4L L4, 3.5L V6 **Transmission:** All	**Problem in A/T Clutch Pressure Control Solenoid Valve A Circuit:** If the measured current for the PCM output duty cycle is not within a specified range (open or short), a malfunction is detected. The execution time is continuous and the duration time is 2 seconds or more. The measured current for the PCM command value is Duty 57-89 %, current less than 0.2, low input. **NOTE: This code is caused by an electrical circuit problem and cannot be caused by a mechanical problem in the transmission.** **Possible Causes:** • Poor connections and loose terminals at A/T clutch pressure control solenoid valve A and the PCM • Faulty A/T clutch pressure control solenoid valve A • "Open" circuit between A/T clutch pressure control solenoid valve A and ground, or poor ground • PCM may need to be updated with the latest software • Faulty PCM

DTC	Trouble Code Title, Conditions & Possible Causes
DTC: P0962 **1T PCM, MIL: Yes, TCIL: Yes** **Year:** 2009, 2010 **Model:** Civic **Engine:** 1.3L L4 **Transmission:** All	**CVT Drive Pulley Pressure Control Valve Circuit Low Voltage (Hybrid Models):** With the engine running for at least 1 second and battery voltage a minimum of 10.0 V, the measured current for the PCM's command value is as follows. Duty % 56.5-90.2, Current A less than 0.27, Failure mode (Low Input) **Possible Causes:** • Poor or loose terminals between the PCM and the CVT drive pulley pressure control solenoid valve • Faulty solenoid harness ("Open" or "Short") • Faulty CVT drive pulley pressure control solenoid valve • PCM may need to be updated with the latest software • Faulty PCM
DTC: P0963 **1T PCM, MIL: Yes, TCIL: Yes** **Year:** 2009, 2010 **Model:** Accord, Accord Crosstour, Civic, CR-V, Element, Fit, Odyssey, Pilot, Ridgeline **Engine:** 1.5L L4, 1.8L L4, 2.4L L4, 3.5L V6 **Transmission:** All	**Problem in A/T Clutch Pressure Control Solenoid Valve A:** Duty cycle is less than 13-27 %, current (A) 0.6-0.9, high input failure. The execution time is continuous and the duration time is 2 seconds or more. DTCs P0962, P0966, P0967, P0970, P0971 are not active. **NOTE: This code is caused by an electrical circuit problem and cannot be caused by a mechanical problem in the transmission.** **Possible Causes:** • Poor connections and loose terminals at A/T clutch pressure control solenoid valve A and the PCM • "Open" circuit between A/T clutch pressure control solenoid valve A and ground, or poor ground • Faulty A/T clutch pressure control solenoid valve A • PCM may need to be updated with the latest software • Faulty PCM
DTC: P0963 **1T PCM, MIL: Yes, TCIL: Yes** **Year:** 2009, 2010 **Model:** Civic **Engine:** 1.3L L4 **Transmission:** All	**CVT Drive Pulley Pressure Control Valve Circuit High Voltage:** With the engine running, battery voltage a minimum of 10.0 V, the measured current for the PCM's command is less than 0.57-0.89 (High input) for at least 1 second. **NOTE: This code is caused by an electrical circuit problem and cannot be caused by a mechanical problem in the transmission.** **Possible Causes:** • Poor connections or loose terminals between the PCM and the CVT drive pulley pressure control solenoid valve • Faulty solenoid wire harness (open condition) • Faulty CVT drive pulley pressure control solenoid valve • PCM may need to be updated with the latest software • Faulty PCM
DTC: P0966 **1T PCM, MIL: Yes, TCIL: Yes** **Year:** 2009, 2010 **Model:** Accord, Accord Crosstour, Civic, CR-V, Element, Fit, Odyssey, Pilot, Ridgeline **Engine:** 1.5L L4, 1.8L L4, 2.4L L4, 3.5L V6 **Transmission:** All	**Problem in A/T Clutch Pressure Control Solenoid Valve B Circuit:** The measured current is less than 0.2, duty cycle 57-89 %, low input. The execution time is continuous and the duration time is 1 seconds or more. **NOTE: This code is caused by an electrical circuit problem and cannot be caused by a mechanical problem in the transmission.** **Possible Causes:** • Poor connections and loose terminals at A/T clutch pressure control solenoid valve B and the PCM • "Open" or "Short" circuit between PCM connector terminal and A/T clutch pressure control solenoid valve B • "Open" circuit between A/T clutch pressure control solenoid valve B and ground, or poor ground • Faulty A/T clutch pressure control solenoid valve B • PCM may need to be updated with the latest software • Faulty PCM
DTC: P0966 **1T PCM, MIL: Yes, TCIL: Yes** **Year:** 2009, 2010 **Model:** Civic **Engine:** 1.8L L4 **Transmission:** All	**CVT Driven Pulley Pressure Control Valve Circuit Low Voltage:** With the engine running, battery voltage a minimum of 10.0 V, the measured current for the PCM's command is less than 0.19-0.27 (Low Input) for at least 1 second. **NOTE: This code is caused by an electrical circuit problem and cannot be caused by a mechanical problem in the transmission.** **Possible Causes:** • Poor connections or loose terminals between the PCM and the CVT drive pulley pressure control solenoid valve • Faulty solenoid wire harness (Short Condition) • Faulty CVT drive pulley pressure control solenoid valve • PCM may need to be updated with the latest software • Faulty PCM
DTC: P0967 **1T PCM, MIL: Yes, TCIL: Yes** **Year:** 2009, 2010 **Model:** Accord, Accord Crosstour, Civic, CR-V, Element, Fit, Odyssey, Pilot, Ridgeline **Engine:** 1.5L L4, 1.8L L4, 2.4L L4, 3.5L V6 **Transmission:** All	**Problem in A/T Clutch Pressure Control Solenoid Valve B:** The measured current for the PCMs command is 0.6-0.9, duty cycle 13-27%, high input failure. Engine running, DTCs P0962, P0963, P0966, P0970, P0971 are not active. The execution time is continuous and the duration time is 1 seconds or more. **NOTE: This code is caused by an electrical circuit problem and cannot be caused by a mechanical problem in the transmission.** **Possible Causes:** • Poor connections and loose terminals at A/T clutch pressure control solenoid valve B and the PCM • "Open" circuit between A/T clutch pressure control solenoid valve C and ground, or poor ground • Faulty A/T clutch pressure control solenoid valve B • PCM may need to be updated with the latest software • Faulty PCM

DTC	Trouble Code Title, Conditions & Possible Causes
DTC: P0967 **1T PCM, MIL: Yes, TCIL: Yes** **Year:** 2009, 2010 **Model:** Civic **Engine:** 1.3L L4 **Transmission:** All	**CVT Driven Pulley Pressure Control Valve Circuit High Voltage:** With the engine running, battery voltage a minimum of 10.0 V, the measured current for the PCM's command is less than 0.57-0.89 (High Input) for at least 1 second. **NOTE: This code is caused by an electrical circuit problem and cannot be caused by a mechanical problem in the transmission.** **Possible Causes:** • Poor connections or loose terminals between the PCM and the CVT drive pulley pressure control solenoid valve • Faulty solenoid wire harness • Faulty CVT drive pulley pressure control solenoid valve • PCM may need to be updated with the latest software • Faulty PCM
DTC: P0970 **1T PCM, MIL: Yes, TCIL: Yes** **Year:** 2009, 2010 **Model:** Accord, Accord Crosstour, Civic, CR-V, Element, Fit, Odyssey, Pilot, Ridgeline **Engine:** 1.5L L4, 1.8L L4, 2.4L L4, 3.5L V6 **Transmission:** All	**Problem in A/T Clutch Pressure Control Solenoid Valve C Circuit:** The measured current for the PCMs command value is 0.2-0.4, duty cycle 57-89%, low input failure. Engine is running with battery voltage at 11 volts, DTCs P0962, P0963, P0966, P0967, P0971 are not active. The execution time is continuous and the duration time is 1 seconds or more. **NOTE: This code is caused by an electrical circuit problem and cannot be caused by a mechanical problem in the transmission.** **Possible Causes:** • Poor connections and loose terminals at A/T clutch pressure control solenoid valve C and the PCM • "Open" or "Short" circuit between PCM connector terminal and A/T clutch pressure control solenoid valve C • "Open" circuit between A/T clutch pressure control solenoid valve C and ground, or poor ground • Faulty A/T clutch pressure control solenoid valve C • PCM needs to be updated with the latest software • Faulty PCM
DTC: P0970 **1T PCM, MIL: Yes, TCIL: Yes** **Year:** 2009, 2010 **Model:** Civic **Engine:** 1.3L L4 **Transmission:** All	**CVT Start Clutch Pressure Control Valve Circuit Low Voltage:** With the engine running, battery voltage a minimum of 10.0 V, the measured current for the PCM's command is less than 0.19-0.27 (Low Input) for at least 1 second. **NOTE: This code is caused by an electrical circuit problem and cannot be caused by a mechanical problem in the transmission.** **Possible Causes:** • Poor connections or loose terminals between the PCM and the CVT drive pulley pressure control solenoid valve • Faulty solenoid wire harness (Short Condition) • Faulty CVT drive pulley pressure control solenoid valve • PCM may need to be updated with the latest software • Faulty PCM
DTC: P0971 **1T PCM, MIL: Yes, TCIL: Yes** **Year:** 2009, 2010 **Model:** Accord, Accord Crosstour, Civic, CR-V, Element, Fit, Odyssey, Pilot, Ridgeline **Engine:** 1.5L L4, 1.8L L4, 2.4L L4, 3.5L V6 **Transmission:** All	**Problem in A/T Clutch Pressure Control Solenoid Valve C:** The measured current for the PCMs command value is 0.6-0.9, duty cycle 13-27, high input failure. The execution time is continuous and the duration time is 1 seconds or more. **NOTE: This code is caused by an electrical circuit problem and cannot be caused by a mechanical problem in the transmission.** **Possible Causes:** • Poor connections and loose terminals at A/T clutch pressure control solenoid valve C and the PCM • "Open" in the wire between A/T clutch pressure control solenoid valve C and ground,, or poor ground • Faulty A/T clutch pressure control solenoid valve C • PCM needs to be updated with the latest software • Faulty PCM
DTC: P0971 **1T PCM, MIL: Yes, TCIL: Yes** **Year:** 2009, 2010 **Model:** Civic **Engine:** 1.3L L4 **Transmission:** All	**CVT Start Clutch Pressure Control Valve Circuit High Voltage:** With the engine running, battery voltage a minimum of 10.0 V, the measured current for the PCM's command is less than 0.57-0.89 (High Input) for at least 1 second. **NOTE: This code is caused by an electrical circuit problem and cannot be caused by a mechanical problem in the transmission.** **Possible Causes:** • Poor connections or loose terminals between the PCM and the CVT drive pulley pressure control solenoid valve • Faulty solenoid wire harness (Open Condition) • Faulty CVT drive pulley pressure control solenoid valve • PCM may need to be updated with the latest software • Faulty PCM

DTC	Trouble Code Title, Conditions & Possible Causes
DTC: P0973 **1T PCM, MIL: Yes, TCIL: Yes** **Year:** 2009, 2010 **Model:** Accord, Accord Crosstour, Civic, CR-V, Element, Fit, Odyssey, Pilot, Ridgeline **Engine:** 1.5L L4, 1.8L L4, 2.4L L4, 3.5L V6 **Transmission:** All	**Short in Shift Solenoid Valve A Circuit:** The return signal does not match the command to turn ON shift solenoid valve A for at least 1 second. Tho execution time is continuous and the duration time is 2 seconds or more. DTCs P0974, P0982, P0983 are not active. **NOTE: This code is caused by an electrical circuit problem and cannot be caused by a mechanical problem in the transmission.** **Possible Causes:** • Blown fuse • Poor connections and loose terminals at shift solenoid valve A and the PCM • SHA wire for an intermittent "Short" to ground between shift solenoid valve A and the PCM • "Short" circuit between PCM connector terminal and the shift solenoid harness connector • "Open" circuit between PCM connector terminals and ground, or poor ground • Faulty shift solenoid valve A or the shift solenoid harness • PCM needs to be updated with the latest software • Faulty PCM
DTC: P0974 **1T PCM, MIL: Yes, TCIL: Yes** **Year:** 2009, 2010 **Model:** Accord, Accord Crosstour, Civic, CR-V, Element, Fit, Odyssey, Pilot, Ridgeline **Engine:** 1.5L L4, 1.8L L4, 2.4L L4, 3.5L V6 **Transmission:** All	**Open in Shift Solenoid Valve A Circuit:** The return signal does not match the command to turn OFF shift solenoid valve A for at least 1 second. The execution time is continuous and the duration time is 1 seconds or more. DTCs P0973, P0982, P0983 are not active. **NOTE: This code is caused by an electrical circuit problem and cannot be caused by a mechanical problem in the transmission.** **Possible Causes:** • Poor connections or loose terminals at shift solenoid valve A and the PCM • "Open" in the wire between PCM connector terminal and the shift solenoid harness connector • Faulty shift solenoid valve A or shift solenoid harness • PCM may need to be updated with the latest software • Faulty PCM
DTC: P0976 **1T PCM, MIL: Yes, TCIL: Yes** **Year:** 2009, 2010 **Model:** Accord, Accord Crosstour, Civic, CR-V, Element, Fit, Odyssey, Pilot, Ridgeline **Engine:** 1.5L L4, 1.8L L4, 2.4L L4, 3.5L V6 **Transmission:** All	**Short in Shift Solenoid Valve B Circuit:** The return signal does not match the command to turn ON shift solenoid valve B for at least 1 second. The execution time is continuous and the duration time is 2 seconds or more. DTCs P0977, P0979, P0980 are not active. **NOTE: This code is caused by an electrical circuit problem and cannot be caused by a mechanical problem in the transmission.** **Possible Causes:** • Blown fuse • Poor connections and loose terminals at shift solenoid valve B and the PCM • "Open" in the wires between PCM connector terminals and ground, or poor ground • SHB wire for an intermittent short to ground between shift solenoid valve B and the PCM • "Short" in the wire between PCM connector terminal and the shift solenoid harness connector • Faulty shift solenoid valve B or the shift solenoid harness • PCM may need to be updated with the latest software • Faulty PCM
DTC: P0977 **1T PCM, MIL: Yes, TCIL: Yes** **Year:** 2009, 2010 **Model:** Accord, Accord Crosstour, Civic, CR-V, Element, Fit, Odyssey, Pilot, Ridgeline **Engine:** 1.5L L4, 1.8L L4, 2.4L L4, 3.5L V6 **Transmission:** All	**Open in Shift Solenoid Valve B Circuit:** The return signal does not match the command to turn OFF shift solenoid valve B for at least 1 second. The execution time is continuous and the duration time is 1 seconds or more. Battery voltage minimum 11 volts, and DTCs P0976, P0979, P0980 are not active. **NOTE: This code is caused by an electrical circuit problem and cannot be caused by a mechanical problem in the transmission.** **Possible Causes:** • Poor connections or loose terminals at shift solenoid valve B and the PCM • "Open" circuit between PCM connector terminal and the shift solenoid harness connector • Faulty shift solenoid valve B or shift solenoid harness • PCM may need to be updated with the latest software • Faulty PCM
DTC: P0979 **1T PCM, MIL: Yes, TCIL: Yes** **Year:** 2009, 2010 **Model:** Accord, Accord Crosstour, Civic, CR-V, Element, Fit, Odyssey, Pilot, Ridgeline **Engine:** 1.5L L4, 1.8L L4, 2.4L L4, 3.5L V6 **Transmission:** All	**Short in Shift Solenoid Valve C Circuit:** The return signal does not match the command to turn ON shift solenoid valve C for at least 1 second. The execution time is continuous and the duration time is 1 seconds or more. Battery voltage is 11 volts and DTCs P0976, P0977, P0980 are not active. **NOTE: This code is caused by an electrical circuit problem and cannot be caused by a mechanical problem in the transmission.** **Possible Causes:** • Blown fuse • Poor connections and loose terminals at shift solenoid valve C and the PCM • SHC wire for an intermittent short to ground between shift solenoid valve C and the PCM • "Open circuit between PCM connector terminal and the under-dash fuse/relay box via the main relay • "Open" circuit between PCM connector terminals and ground, or poor ground • "Short" circuit between PCM connector terminal and the shift solenoid harness connector • Faulty shift solenoid valve C or the shift solenoid harness • PCM needs to be updated with the latest software • Faulty PCM

DTC	Trouble Code Title, Conditions & Possible Causes
DTC: P0980 **1T PCM, MIL: Yes, TCIL: Yes** **Year:** 2009, 2010 **Model:** Accord, Accord Crosstour, Civic, CR-V, Element, Fit, Odyssey, Pilot, Ridgeline **Engine:** 1.5L L4, 1.8L L4, 2.4L L4, 3.5L V6 **Transmission:** All	**Open in Shift Solenoid Valve C Circuit:** The return signal does not match the command to turn OFF shift solenoid valve C for at least 1 second. The execution time is continuous and the duration time is 1 seconds or more. Battery voltage is 11 V and DTCs P0976, P0977, P0979 are not active. **NOTE: This code is caused by an electrical circuit problem and cannot be caused by a mechanical problem in the transmission.** **Possible Causes:** • Poor connections or loose terminals at shift solenoid valve C and the PCM • "Open" circuit between PCM connector terminal and the shift solenoid harness connector • Faulty shift solenoid valve C or shift solenoid harness • PCM needs to be updated with the latest software • Faulty PCM
DTC: P0982 **1T PCM, MIL: Yes, TCIL: Yes** **Year:** 2009, 2010 **Model:** Accord, Accord Crosstour, Civic, CR-V, Element, Fit, Odyssey, Pilot, Ridgeline **Engine:** 1.5L L4, 1.8L L4, 2.4L L4, 3.5L V6 **Transmission:** All	**Short in Shift Solenoid Valve D Circuit:** The return signal does not match the command to turn ON shift solenoid valve D for at least 1 second. The execution time is continuous and the duration time is 1 seconds or more. Battery voltage is 11v, and DTCs P0973, P0974, P0983 are not active. **NOTE: This code is caused by an electrical circuit problem and cannot be caused by a mechanical problem in the transmission.** **Possible Causes:** • Blown Fuse • Poor connections and loose terminals at shift solenoid valve D and the PCM. If the PCM • SHD wire for an intermittent short to ground between shift solenoid valve D and the PCM • "Open" in the wire between PCM connector terminal and the under-dash fuse/relay box • "Short" in the wire between PCM connector terminal and the shift solenoid harness connector • Faulty shift solenoid valve D or the shift solenoid harness • PCM needs to be updated with the latest software • Faulty PCM
DTC: P0983 **1T PCM, MIL: Yes, TCIL: Yes** **Year:** 2009, 2010 **Model:** Accord, Accord Crosstour, Civic, CR-V, Element, Fit, Odyssey, Pilot, Ridgeline **Engine:** 1.5L L4, 1.8L L4, 2.4L L4, 3.5L V6 **Transmission:** All	**Open in Shift Solenoid Valve D Circuit:** The return signal does not match the command to turn OFF shift solenoid valve D for at least 1 second. The execution time is continuous and the duration time is seconds or more. Battery voltage is 11v and DTCs P0973, P0974, P0982 are not active. **NOTE: This code is caused by an electrical circuit problem and cannot be caused by a mechanical problem in the transmission.** **Possible Causes:** • Poor connections or loose terminals at shift solenoid valve D and the PCM • "Open" circuit between PCM connector terminal C9 and the shift solenoid harness connector • Faulty shift solenoid valve D or shift solenoid harness • PCM may need to be updated with the latest software • Faulty PCM
DTC: P0985 **1T PCM, MIL: Yes, TCIL: Yes** **Year:** 2009, 2010 **Model:** Accord, CR-V, Element **Engine:** 2.4L L4 **Transmission:** All	**Short in Shift Solenoid Valve E Circuit:** With the engine started and in park the return signal does not match the command to turn ON shift solenoid valve E for at least 1 second. **NOTE: Before you troubleshoot, record all freeze data and any on-board snapshot with the HDS, and review General Troubleshooting Information. This code is caused by an electrical circuit problem and cannot be caused by a mechanical problem in the transmission.** **Possible Causes:** • "Short" to body ground between PCM and the shift solenoid wire harness connector • "Short" between PCM and the shift solenoid wire harness connector • Faulty shift solenoid valve E or the shift solenoid wire harness
DTC: P0986 **1T PCM, MIL: Yes, TCIL: Yes** **Year:** 2009, 2010 **Model:** CR-V, Element **Engine:** 2.4L L4 **Transmission:** All	**Open in Shift Solenoid Valve E Circuit:** With the engine running and driven in 2nd gear in the D position, the return signal does not match the command to turn OFF shift solenoid E for at least 1 second. **NOTE: Before you troubleshoot, record all freeze data and any on-board snapshot with the HDS, and review General Troubleshooting Information. This code is caused by an electrical circuit problem and cannot be caused by a mechanical problem in the transmission.** **Possible Causes:** • Poor connections or loose terminals between shift solenoid valve E and the PCM • "Open" circuit between PCM connector and the shift solenoid wire harness connector • Faulty shift solenoid valve E, or shift solenoid wire harness • PCM may need to be updated with the latest software • Faulty PCM

DTC	Trouble Code Title, Conditions & Possible Causes
DTC: P0A14 **1T PCM** **Year:** 2009, 2010 **Model:** Accord, Accord Crosstour, Pilot **Engine:** 3.5L V6 **Transmission:** All	**Front Engine Mount Actuator Circuit Malfunction:** Engine running, and DTCs P0A15, P0AB6. P0AB7, P16C4 and P16C6 are not active. The current valve during one cycle of the front engine mount actuator, which outputs in 1 TDC duration as one cycle, is 7.5 A or more as a maximum current or 5.3 A or more as a minimum current, or the engine mounts cycle during start up and continue to cycle for at least 20 seconds. **NOTE: When testing the engine mount actuator, be sure to use a VOM capable of reading 0.01 ohms change of resistance. If the vehicle has been driven, allow the engine mount actuator to cool down for about 2 hours before testing.** **Possible Causes:** • Poor connections or loose terminals at the engine mount control unit and the engine mount actuator • "Open" or "Short" circuit between the engine mount control unit and the engine mount actuator • Faulty front engine mount • Faulty engine mount control unit
DTC: P0A15 **1T PCM** **Year:** 2009, 2010 **Model:** Accord, Accord Crosstour, Odyssey, Pilot **Engine:** 3.5L V6 **Transmission:** All	**Front Engine Mount Actuator Control Circuit Low Current:** With the engine running and the DTCs P0A14, P0AB6, P0AB7, P15AB, P16C4 and P16C6 are not active. The duty cycle of the front engine mount actuator, which outputs in 1 TDC duration as one cycle, reads 100% all the time during engine the mount operation and continues at least 20 seconds. **NOTE: When testing the engine mount actuator, be sure to use a VOM capable of reading 0.01 ohms change of resistance. If the vehicle has been driven, allow the engine mount actuator to cool down for about 2 hours before testing.** **Possible Causes:** • Poor connections or loose terminals at the engine mount control unit and the engine mount actuator, the engine mount control unit, and body ground • "Open" or "Short" to ground in the wire between the engine mount control unit and the engine mount actuator • Faulty front engine mount • Faulty engine mount control unit
DTC: P0A16 **1T PCM** **Year:** 2009, 2010 **Model:** Pilot **Engine:** 3.5L V6 **Transmission:** All	**Front Engine Mount Actuator Control Circuit High Current:** With the ignition switch ON and DTCs P15AF, P15B0, P15B1, P16C5, P16C9 not active. The minimum front ACM actuator current in one cycle is 6.6 A or more for at least 2.6 seconds, the front ACM actuator output cycle is one rotation of the crankshaft. **Possible Causes:** • Poor connections or loose terminals at the engine mount control unit and the engine mount actuator • "Open" or "Short" between the engine mount control unit and the engine mount actuator • Faulty front engine mount • Faulty engine mount control unit
DTC: P0A1F/112 **1T BCM, MIL: Yes** **Year:** 2009, 2010 **Model:** Civic **Engine:** 1.3L L4 **Transmission:** All	**F-CAN Malfunction (BCM Module-MCM) (HYBRID MODELS):** With the ignition ON, MCM supply-power voltage a minimum of 10.0 V, and the elapsed time after the ignition switch is tuned for 3 seconds. One of the following conditions occurs. (IMA system indicator is ON) Condition 1: No information is sent from the BCM for at least 0.5 seconds. Condition 2: The information sent from the BCM is abnormal at least 10 times. WARNING: The IMA system uses high voltage (144 V), before servicing turn OFF the battery module switch. **Possible Causes:** • Poor connections or loose terminals at the BCM module and the MCM • "Open" circuit between the BCM module and the MCM • BCM may need to be updated with the latest software • Faulty BCM
DTC: P0A3C/39 **1T , MIL: Yes** **Year:** 2009, 2010 **Model:** Civic **Engine:** 1.3L L4 **Transmission:** All	**Motor Control Module (MCM) Overheating (HYBRID MODELS):** After the ignition switch has been turned ON (II) for 2 seconds, the following conditions occurs. (IMA system indicator is ON) Condition 1: The MPI module temperature is 230°F (110°C) or more. Condition 2: The MCM internal temperature is 257°F (125°C) or more. WARNING: The IMA system uses high voltage (144 V), before servicing turn OFF the battery module switch. **Possible Causes:** • Blockage at the IPU module fan inlet duct • Poor connections or loose terminals at the MCM and the IPU module fan • Faulty IPU module fan • MCM may need to be updated with the latest software • Faulty MCM

DTC	Trouble Code Title, Conditions & Possible Causes
DTC: P0A3F/89 **1T PCM, MIL: Yes** **Year:** 2009, 2010 **Model:** Civic **Engine:** 1.3L L4 **Transmission:** All	**Motor Rotor Position Sensor Circuit Malfunction (HYBRID MODELS):** With the ignition switch ON and battery voltage at a minimum of 10.0 V, one of the following symptoms can occur. Symptom 1: An abnormal signal is output from the R/D converter for at least 0.4 seconds within 0.5 seconds. Symptom 2: The output voltage to the motor rotor position sensor is 0.1 V or less for at least 0.5 seconds. Symptom 3: The input voltage to the motor rotor position sensor is 1.2 V or less for at least 0.5 seconds. Symptom 4: The deviation of the IMA motor speed and the engine speed is 500 rpm or more for at least 2 seconds. (IMA system indicator is ON) WARNING: The IMA system uses high voltage (144 V), before servicing turn OFF the battery module switch. **Possible Causes:** • Poor connections or loose terminals at the motor rotor position sensor and the MCM • "Open" or "Short" circuit between the motor rotor position sensor and the MCM • Faulty motor rotor position sensor • MCM may need to be updated with the latest software • Faulty MCM
DTC: P0A5E/24 **1T PCM, MIL: Yes** **Year:** 2009, 2010 **Model:** Civic **Engine:** 1.3L L4 **Transmission:** All	**U Phase Motor Current Sensor Circuit Low Voltage (HYBRID MODELS):** With the ignition ON and the MCM power-supply voltage a minimum of 9.0 V, the MCM input voltage from the U phase motor current sensor is 0.156 V or less for at least o.5 seconds. P0A5F, P16C1, P16C2 and U1203 are not active. (IMA system indicator is ON) WARNING: The IMA system uses high voltage (144 V), before servicing turn OFF the battery module switch. **Possible Causes:** • Poor connections or loose terminals at the U phase motor current sensor and the MCM • "Open" or "Short" circuit between the U phase motor current sensor and the MCM • Faulty U phase motor current sensor • MCM may need to be updated with the latest software • Faulty MCM
DTC: P0A5F/25 **1T PCM, MIL: Yes** **Year:** 2009, 2010 **Model:** Civic **Engine:** 1.3L L4 **Transmission:** All	**U Phase Motor Current Sensor Circuit High Voltage (HYBRID MODELS):** With the ignition ON and MCM power-supply voltage a minimum of 9.0 V, the MCM input voltage from the U phase motor current sensor is 4.848 V or more for at least 0.5 seconds. P0A5E, P16C1, P16C2 and U1203 are not active. (IMA system indicator is ON) WARNING: The IMA system uses high voltage (144 V), before servicing turn OFF the battery module switch. **Possible Causes:** • Poor connections or loose terminals at the U phase motor current sensor and the MCM • "Open" circuit between the U phase motor current sensor and the MCM • Faulty U phase motor current sensor • MCM may need to be updated with the latest software • Faulty MCM
DTC: P0A61/26 **1T PCM, MIL: Yes** **Year:** 2009, 2010 **Model:** Civic **Engine:** 1.3L L4 **Transmission:** All	**V Phase Motor Current Sensor Circuit Low Voltage (HYBRID MODELS):** With the ignition switch ON and MCM power-supply voltage is a minimum of 9.0 V, the MCM input voltage from the V phase motor current sensor is 0.156 V or less for at least 0.5 seconds. (IMA system indicator is ON) WARNING: The IMA system uses high voltage (144 V), before servicing turn OFF the battery module switch. **Possible Causes:** • Poor connections or loose terminals at the V phase motor current sensor and the MCM • "Open" or "Short" circuit the V phase motor current sensor and the MCM • Faulty V phase motor current sensor • MCM may need to be updated with the latest software • Faulty MCM
DTC: P0A62/27 **1T PCM, MIL: Yes** **Year:** 2009, 2010 **Model:** Civic **Engine:** 1.3L L4 **Transmission:** All	**V Phase Motor Current Sensor Circuit High Voltage (HYBRID MODELS):** With the ignition ON and the MCM power-supply voltage a minimum of 9.0 V, the MCM input voltage from the V phase motor current sensor is 4.848 V or more for at least 0.5 seconds. (IMA system indicator is ON) WARNING: The IMA system uses high voltage (144 V), before servicing turn OFF the battery module switch. **Possible Causes:** • Poor connections or loose terminals at the V phase motor current sensor and the MCM • "Open" circuit between the V phase motor current sensor and the MCM • Faulty V phase motor current sensor • MCM may need to be updated with the latest software • Faulty MCM

DTC	Trouble Code Title, Conditions & Possible Causes
DTC: P0A64/28 **1T PCM, MIL: Yes** **Year:** 2009, 2010 **Model:** Civic **Engine:** 1.3L L4 **Transmission:** All	**W Phase Motor Current Sensor Circuit Low Voltage (HYBRID MODELS):** With the ignition ON and the MCM power-supply voltage a minimum of 9.0 V, the MCM input voltage from the W phase motor current sensor is 0.156 V or less for at least 0.5 seconds. (IMA system indicator is ON) WARNING: The IMA system uses high voltage (144 V), before servicing turn OFF the battery module switch. **Possible Causes:** • Poor connections or loose terminals at the W phase motor current sensor and the MCM • "Open" or "Short" circuit between the W phase motor current sensor and the MCM • "Short" circuit in the wire(s) between the U/V/W phase motor current sensors and the MCM • Faulty W phase motor current sensor • Faulty U phase motor current sensor • MCM may need to be updated with the latest software • Faulty MCM
DTC: P0A65/29 **1T PCM, MIL: Yes** **Year:** 2009, 2010 **Model:** Civic **Engine:** 1.3L L4 **Transmission:** All	**W Phase Motor Current Sensor Circuit High Voltage (HYBRID MODELS):** With the ignition ON and the MCM supply-power voltage a minimum of 9.0 V, the MCM input voltage from the W phase motor current sensor is 4.848 V or more for at least 0.5 seconds. (IMA system indicator is ON) WARNING: The IMA system uses high voltage (144 V), before servicing turn OFF the battery module switch. **Possible Causes:** • Poor connections or loose terminals at the W phase motor current sensor and the MCM • "Open" circuit between the W phase motor current sensor and the MCM • Faulty W phase motor current sensor • MCM may need to be updated with the latest software • Faulty MCM
DTC: P0A78/32 **1T , MIL: Yes** **Year:** 2009, 2010 **Model:** Civic **Engine:** 1.3L L4 **Transmission:** All	**Motor Control Module (MCM) Internal Circuit Malfunction (HYBRID MODELS):** With the ignition ON, battery voltage a minimum of 7.1 V. One of the following symptoms occurs. (IMA system indicator is ON) Symptom 1: Information is not sent from the motor control sub processor for at least 500 milliseconds, or information sent from the motor control subprocessor includes abnormal condition information for at least 0.5 second. Symptom 2: (one of the following conditions occurs) * The power supply circuit voltage (L=8V) is 4.30 V or more, or 2.47 V or less, for at least 40 milliseconds, or its deviation is 1.07 V or more for at least 40 milliseconds. * The constant current circuit detection voltage is 0.91 V or more, or 0.53 V or less, for at least 40 milliseconds, or its deviation is 1.07 V or more for at least 40 milliseconds. * The constant current circuit reference voltage is 0.84 V or more, or 0.6 V or less, for at least 40 milliseconds, or its deviation is 1.07 V or more for at least 40 milliseconds. * The platinum sensor disconnection threshold value is 4.66 V or more, or 1.8 V or less, for at least 0.8 second. Symptom 3: (one of the following conditions occur) * The power supply circuit voltage (+8V0 is 3.78 V or more, or 2.2 V or less, for at least 40 milliseconds, or its deviation is 1.07 V or more for at least 40 milliseconds. * The over current upper limit threshold voltage is 0.5 V or more, or 3.47 V or less, for at least 40 milliseconds, or its deviation is 1.07 V or more for at least 40 milliseconds. Symptom 4: The signal to the CAN controller and the return signal differ at least three times. Symptom 5: The over voltage notification is input three times in 25 seconds even though the control voltage is 235 V or less. **Possible Causes:** • Poor connections or loose terminals at the MCM • MCM may need to be updated with the latest software • Faulty MCM
DTC: P0AB6 **1T PCM** **Year:** 2009, 2010 **Model:** Accord, Accord Crosstour, Odyssey, Pilot **Engine:** 3.5L V6 **Transmission:** All	**Rear Engine Mount Actuator Circuit Malfunction:** Engine running, and DTCs P0A15, P0AB6. P0AB7, P16C4 and P16C6 are not active. The current valve during one cycle of the front engine mount actuator, which outputs in 1 TDC duration as one cycle, is 7.5 A or more as a maximum current or 5.3 A or more as a minimum current, or the engine mounts cycle during start up and continue to cycle for at least 20 seconds. **NOTE: When testing the engine mount actuator, be sure to use a VOM capable of reading 0.01 ohms change of resistance. If the vehicle has been driven, allow the engine mount actuator to cool down for about 2 hours before testing.** **Possible Causes:** • Poor connections or loose terminals at the engine mount control unit and the engine mount actuator • "Open" or "Short" circuit between the engine mount control unit and the engine mount actuator • Faulty rear engine mount • Faulty engine mount control unit

DTC	Trouble Code Title, Conditions & Possible Causes
DTC: P0AB7 **1T PCM** **Year:** 2009, 2010 **Model:** Accord, Accord Crosstour, Odyssey, Pilot **Engine:** 3.5L V6 **Transmission:** All	**Rear Engine Mount Actuator Control Circuit Low Current:** With the engine running and the DTCs P0A14, P0AB6, P0AB7, P15AB, P16C4 and P16C6 are not active. The duty cycle of the rear engine mount actuator, which outputs in 1 TDC duration as one cycle, reads 100% all the time during engine the mount operation and continues at least 20 seconds. **NOTE: When testing the engine mount actuator, be sure to use a VOM capable of reading 0.01 ohms change of resistance. If the vehicle has been driven, allow the engine mount actuator to cool down for about 2 hours before testing.** **Possible Causes:** • Poor connections or loose terminals at the engine mount control unit and the engine mount actuator • "Open" or "Short" circuit between the engine mount control unit and the engine mount actuator • Faulty rear engine mount • Faulty engine mount control unit
DTC: P0AB8 **1T PCM** **Year:** 2009, 2010 **Model:** Odyssey, Pilot **Engine:** 3.5L V6 **Transmission:** All	**Rear Engine Mount Actuator Control Circuit High Current:** With the ignition switch ON and DTCs P15AF, P15B0, P15B1, P16C5, P16C9 not active. The minimum front ACM actuator current in one cycle is 6.6 A or more for at least 2.6 seconds, the front ACM actuator output cycle is one rotation of the crankshaft. **Possible Causes:** • Poor connections or loose terminals at the engine mount control unit and the engine mount actuator • "Open" circuit between the engine mount control unit and the engine mount actuator • Faulty rear engine mount • Faulty engine mount control unit
DTC: P0AEE/109 **1T , MIL: Yes** **Year:** 2009, 2010 **Model:** Civic **Engine:** 1.3L L4 **Transmission:** All	**Motor Control Module (MCM) Internal Circuit Malfunction:** With the ignition ON, MCM supply-power voltage a minimum of 9 V, and the module temperature detection value at a minimum of 221°F (105°C), the MPI module temperature change is less than 39°F (4°C) within 10 minutes. (IMA system indicator is ON) **Possible Causes:** • Poor connections or loose terminals at the MCM • MCM may need to be updated with the latest software • Faulty MCM
DTC: P0AEF/110 **1T , MIL: Yes** **Year:** 2009, 2010 **Model:** Civic **Engine:** 1.3L L4 **Transmission:** All	**Motor Control Module (MCM) Internal Temperature Sensor Circuit Low Voltage (HYBRID MODELS):** With the ignition ON (II), the voltage change value of the MPI module temperature less than 1.79 V for at least 0.8 second. (IMA system indicator is ON) **Possible Causes:** • Poor connections or loose terminals at the MCM • MCM may need to be updated with the latest software • Faulty MCM
DTC: P0AF0/111 **1T , MIL: Yes** **Year:** 2009, 2010 **Model:** Civic **Engine:** 1.3L L4 **Transmission:** All	**Motor Control Module (MCM) Internal Temperature Sensor Circuit High Voltage (HYBRID MODELS):** With the ignition ON (II), the voltage change value of the MPI module temperature is more than 4.63 V for at least 0.8 second. (IMA system indicator is ON) **Possible Causes:** • Poor connections or loose terminals at the MCM • MCM may need to be updated with the latest software • Faulty MCM

OBD II Trouble Code List (P1XXX Codes)

DTC	Trouble Code Title, Conditions & Possible Causes
DTC: P1009 **1T PCM, MIL: Yes** **Year:** 2009, 2010 **Model:** Civic, CR-V, Element **Engine:** 2.0L L4, 2.4L L4 **Transmission:** All	**Variable Valve Timing Control (VTC) Advance Malfunction :** With the engine idling at operating temperature, the camshaft phase value is not 24.0 degrees or less within the monitored area (camshaft phase control directed value plus failure judgment value) after 3 seconds have passed, or when the camshaft phase value is 5.0 degrees for M/T and 20.0 degrees for A/T or less and continues for 0.5 seconds or more, even when the VTC does not actuate. **NOTE: Before you troubleshoot, record all freeze data and any on-board snapshot, and review the general troubleshooting information. If DTC P0341 is stored at the same time as DTC P1009, troubleshoot DTC P1009 first, then recheck for DTC P0341.** **Possible Causes:** • Clogged VTC strainer (A) • Clogged oil passages at the VTC system • Faulty VTC oil control solenoid valve • Faulty VTC actuator

DTC	Trouble Code Title, Conditions & Possible Causes
DTC: P1077 **2T PCM, MIL: Yes** **Year:** 2009, 2010 **Model:** Accord, Accord Crosstour **Engine:** 3.5L V6 **Transmission:** All	**Intake Manifold Runner Control (IMRC) Valve Stuck Open: Short Runner Position:** With the engine running the PCM sends a close (Long Runner) command, no long runner command is received for at least 1 second. **NOTE: If DTC P0651 is indicated at the same time as DTC P1077, troubleshoot DTC P0651 first, then recheck for P1077.** **Possible Causes:** • Poor connections or loose terminals at the IMT (IMRC) actuator and the PCM • "Open" or "Short" between the PCM and the IMT (IMRC) actuator • IMT (IMRC) actuator • Stuck valve, replace the intake manifold if necessary • PCM needs to be updated with the latest software • Faulty PCM
DTC: P1077 **2T PCM, MIL: Yes** **Year:** 2009, 2010 **Model:** Civic, Pilot, Ridgeline **Engine:** 1.8L L4, 2.0L L4, 3.5L V6 **Transmission:** All	**Intake Manifold Tuning (IMT) Valve Stuck in High RPM Position:** When the PCM sends a close (long runner) command, no long runner return signal is received for at least 7 seconds. Execution is once per drive cycle, engine speed, 3,600 rpm and intake air temperature a minimum of 5°F (-15°C). **NOTE: Before you troubleshoot, record all freeze data and any on-board snapshot.** **Possible Causes:** • Poor connections or loose terminals at the IMT actuator and the PCM • "Open" circuit between the PCM and the IMT actuator • Faulty IMT actuator • PCM may need to be updated with the latest software • Faulty PCM
DTC: P1078 **2T PCM, MIL: Yes** **Year:** 2009, 2010 **Model:** Accord, Accord Crosstour **Engine:** 3.5L V6 **Transmission:** All	**Intake Manifold Runner Control (IMRC) Valve Stuck Closed: Long Runner Position:** When the PCM sends an open (short runner) command, no short runner return signal is received for at least 1 second. The execution time is once per driving cycle. **NOTE: Before you troubleshoot, record all freeze data and any on-board snapshot.** **Possible Causes:** • Poor connections or loose terminals at the IMT (IMRC) actuator and the PCM • "Open" or "Short" circuit between the PCM and the IMT (IMRC) actuator • Faulty IMT (IMRC) actuator • Stuck valve, replace the intake manifold if necessary • PCM needs to be updated with the latest software • Faulty PCM
DTC: P1078 **2T PCM, MIL: Yes** **Year:** 2009, 2010 **Model:** Civic, Odyssey, Pilot, Ridgeline **Engine:** 1.8L L4, 2.0L L4, 3.5L V6 **Transmission:** All	**Intake Manifold Tuning (IMT) Valve Stuck in Low RPM Position:** When the PCM sends an open (short runner) command, no short runner return signal is received for at least 3 seconds. Intake air temperature 5°F (-15°C), engine speed is 3,800 rpm and battery voltage minimum of 10.5v. DTCs P0107, P0108, P0112, P0113, P0117, P0118, P0563, P1128, P1129, P2227, P2228, P2229 are not active. **NOTE:** **NOTE: Before you troubleshoot, record all freeze data and any on-board snapshot.** **Possible Causes:** • Poor connections or loose terminals at the IMT actuator and the PCM • "Open" or "Short" circuit between the PCM and the IMT actuator • Faulty IMT actuator • PCM needs to be updated with the latest software • Faulty PCM
DTC: P1109 **1T PCM, MIL: Yes** **Year:** 2009, 2010 **Model:** Accord, Civic, CR-V, Fit, Odyssey, Pilot, Ridgeline **Engine:** 1.3L L4, 1.5L L4, 1.8L L4, 2.0L L4, 2.4L L4, 3.5L V6 **Transmission:** All	**Barometric Pressure (BARO) Sensor Circuit Out of Range High:** The BARO sensor output voltage is between 3.59 V to 4.49 V for at least 2 seconds. The execution time is continuous and DTCs P2228, P2229 are not active. **NOTE: Before you troubleshoot, record all freeze data and any on-board snapshot.** **Possible Causes:** • PCM needs to be updated with the latest software • Faulty PCM

DTC	Trouble Code Title, Conditions & Possible Causes
DTC: P1116 **2T PCM, MIL: Yes** **Year:** 2009, 2010 **Model:** Accord, Accord Crosstour, Civic, CR-V, Element, Fit, Odyssey, Pilot, Ridgeline **Engine:** 1.3L L4, 1.5L L4, 1.8L L4, 2.0L L4, 2.4L L4, 3.5L V6 **Transmission:** All	**Engine Coolant Temperature (ECT) Sensor 1 Circuit Range/Performance Problem:** A malfunction is detected if the following 3 conditions are present after the engine has stopped and the ignition switch has been turned to LOCK (0) for at least 6 hours before restarting the engine: (1) When the temperature difference between the IAT and ECT1 is 57°F (32°C) or more. (2) When the temperature difference between the IAT and ECT2 is 30°F (17°C) or more. (3) When the temperature difference between the ECT2 and ECT1 is 73°F (41°C) or more. **NOTE: If DTC P0111 is stored at the same time as DTC P1116, troubleshoot DTC P0111 first, then recheck for DTC P1116.** **Possible Causes:** • Poor connections or loose terminals at ECT sensor 1 and ECT sensor 2 • Faulty ECT sensor 1 • Faulty ECT sensor 2
DTC: P1128 **2T PCM, MIL: Yes** **Year:** 2009, 2010 **Model:** Accord, Accord Crosstour, Civic, CR-V, Element, Fit, Odyssey, Pilot, Ridgeline **Engine:** 1.5L L4, 1.8L L4, 2.0L L4, 2.4L L4, 3.5L V6 **Transmission:** All	**Manifold Absolute Pressure (MAP) Sensor Signal Lower Than Expected:** The MAP sensor output is 33 kPa (9.8 in.Hg, 249 mmHg) or less for at least 2 seconds when atmospheric pressure is 52 kPa (15.4 in.Hg, 392 mmHg). The execution time is once per driving cycle. **NOTE: Before you troubleshoot, record all freeze data and any on-board snapshot.** **Possible Causes:** • Dirty air cleaner element • Poor connections or loose terminals at the MAP sensor and the PCM • Faulty MAP sensor
DTC: P1129 **2T PCM, MIL: Yes** **Year:** 2009, 2010 **Model:** Civic, CR-V, Element, Fit, Odyssey, Pilot, Ridgeline **Engine:** 1.5L L4, 1.8L L4, 2.0L L4, 2.4L L4, 3.5L V6 **Transmission:** All	**Manifold Absolute Pressure (MAP) Sensor Signal Higher Than Expected:** The MAP sensor output is 36 kPa (10.9 in.Hg, 277 mmHg) or more for at least 2 seconds. The execution time is once per driving cycle. **NOTE: Before you troubleshoot, record all freeze data and any on-board snapshot.** **Possible Causes:** • Vacuum leaks • Poor connections or loose terminals at the MAP sensor and the PCM • Faulty MAP sensor
DTC: P1129 **2T PCM, MIL: Yes** **Year:** 2009, 2010 **Model:** Accord, Accord Crosstour **Engine:** 2.4L L4, 3.5L V6 **Transmission:** All	**MAP Sensor Signal Higher Than Expected:** With the vehicle speed 1,750 or more and the throttle position is at 22° or more. The difference between the barometric pressure sensor measured during boost conditions and negative pressure (NO boost) is a specified output or less, an upward shift in the MAP sensor signal is detected. **NOTE: Before you troubleshoot, record all freeze data and any on-board snapshot, and review the general troubleshooting information.** **Possible Causes:** • Vacuum leaks • Poor connections or loose terminals at the MAP sensor and the PCM • Faulty MAP sensor
DTC: P1157 **1T PCM, MIL: Yes** **Year:** 2009, 2010 **Model:** Accord, Civic, CR-V, Element, Fit **Engine:** 1.3L L4, 1.5L L4, 1.8L L4, 2.0L L4, 2.4L L4 **Transmission:** All	**Air Fuel Ratio (A/F) Sensor (Sensor 1) AFS Line High Voltage:** With the engine running and battery voltage between 10.5-16.0 V, the A/F sensor (SENSOR 1) heater element resistance is 250 ohms or more for at least 5 seconds. **NOTE: Before you troubleshoot, record all freeze data and any on-board snapshot, and review the general troubleshooting information.** **Possible Causes:** • Poor connections or loose terminals at the A/F sensor (Sensor 1) and the ECM/PCM • "Open" circuit between the ECM/PCM and the A/F sensor (Sensor 1) • Faulty A/F sensor (Sensor 1) • PCM may need to be updated with the latest software • Faulty PCM
DTC: P1172 **2T PCM, MIL: Yes** **Year:** 2009, 2010 **Model:** Accord, Accord Crosstour, Odyssey, Pilot, Ridgeline **Engine:** 3.5L V6 **Transmission:** All	**Rear Air/Fuel Ratio (A/F) Sensor (Bank 1, Sensor 1) Circuit Out of Range High:** A malfunction is detected when the rear A/F sensor (bank 1, sensor 1) output voltage is 4.7 V or more. The execution time is continuous and the duration time is 7 seconds or more. **NOTE: Before you troubleshoot, record all freeze data and any on-board snapshot.** **Possible Causes:** • Poor connections or loose terminals at the A/F sensor (Sensor 1) and the PCM • Faulty A/F sensor (Sensor 1)

DTC	Trouble Code Title, Conditions & Possible Causes
DTC: P1172 **1T PCM, MIL: Yes** **Year:** 2009, 2010 **Model:** Accord, Civic, CR-V, Element, Fit **Engine:** 1.3L L4, 1.5L L4, 1.8L L4, 2.0L L4, 2.4L L4 **Transmission:** All	**Air/Fuel Ratio (A/F) Sensor (Sensor 1) Circuit Out of Range High:** A malfunction is detected when the rear A/F sensor (Sensor 1) output voltage is 4.9 V or more. The execution time is continuous and the duration time is 7 seconds or more. **NOTE: Before you troubleshoot, record all freeze data and any on-board snapshot.** **Possible Causes:** • Poor connections or loose terminals at the A/F sensor (Sensor 1) and the PCM • Faulty A/F sensor (Sensor 1)
DTC: P1174 **2T PCM, MIL: Yes** **Year:** 2009, 2010 **Model:** Accord, Accord Crosstour, Odyssey, Pilot, Ridgeline **Engine:** 3.5L V6 **Transmission:** All	**Front Air/Fuel Ratio (A/F) Sensor (Bank 2, Sensor 1) Circuit Out of Range High:** A malfunction is detected when the front A/F sensor (bank 2, sensor 1) output voltage is 4.7 V or more. The execution time is continuous and the duration time is 7 seconds or more. **NOTE: Before you troubleshoot, record all freeze data and any on-board snapshot.** **Possible Causes:** • Poor connections or loose terminals at the A/F sensor (Sensor 1) and the PCM • Faulty A/F sensor (Sensor 1)
DTC: P1187 **1T PCM** **Year:** 2009, 2010 **Model:** Civic **Engine:** 1.8L L4, 2.0L L4 **Transmission:** All	**Fuel Tank Temperature (FTT) Sensor Circuit Low Voltage :** With the ignition switch ON, the FTT output voltage is 0.08 V or less for at least 2 seconds. P01188 is not active. **Possible Causes:** • Poor connections or loose terminals at the FTT sensor and the PCM • short in the wire between the PCM and the FTT sensor • Faulty FTT sensor • PCM may need to be updated with the latest software • Faulty PCM
DTC: P1188 **1T PCM** **Year:** 2009, 2010 **Model:** Civic **Engine:** 1.8L L4, 2.0L L4 **Transmission:** All	**Fuel Tank Temperature (FTT) Sensor Circuit High Voltage :** With the ignition switch ON, the FTT output voltage is 4.92 V or more for at least 2 seconds. P01187 is not active. **Possible Causes:** • Poor connections or loose terminals at the FTT sensor and the PCM • "Open" circuit between the PCM and the FTT sensor • Faulty FTT sensor • ECM may need to be updated with the latest software • Faulty ECM
DTC: P1192 **1T PCM** **Year:** 2009, 2010 **Model:** Civic **Engine:** 1.8L L4, 2.0L L4 **Transmission:** All	**Fuel Tank Pressure (FTP) Sensor Circuit Low Voltage :** With the ignition switch ON, the FTP sensor output voltage is 0.23 V or less for at least 2 seconds. P1193 is not active. **NOTE: Before you troubleshoot, record all freeze data and any on-board snapshot, and review the general troubleshooting information.** **Possible Causes:** • Poor connections or loose terminals at the FTP sensor and the PCM • "Open" or "Short" circuit between the PCM and the FTP sensor • Faulty FTP sensor • PCM may need to be updated with the latest software • Faulty PCM
DTC: P1193 **1T PCM** **Year:** 2009, 2010 **Model:** Civic **Engine:** 1.8L L4, 2.0L L4 **Transmission:** All	**Fuel Tank Pressure (FTP) Sensor Circuit High Voltage:** With the ignition switch ON, the FTP sensor output voltage is 4.61 V or more for at least 2 seconds. P1193 is not active. **NOTE: Before you troubleshoot, record all freeze data and any on-board snapshot, and review the general troubleshooting information.** **Possible Causes:** • Poor connections or loose terminals at the FTP sensor and the PCM • "Open" circuit between the PCM and the FTP sensor • Faulty FTP sensor • PCM may need to be updated with the latest software • Faulty PCM

DTC	Trouble Code Title, Conditions & Possible Causes
DTC: P1220/34 **1T O2S4, MIL: Yes** **Year:** 2009, 2010 **Model:** Civic **Engine:** 1.3L L4 **Transmission:** All	**DC-DC Converter Lost Communication with Battery Condition Monitor (BCM) Module (Hybrid Models):** With the ignition switch ON and the BCM power supply voltage a minimum of 10.5 V, The serial communication data is not updated for at least 2 seconds. (IMA system indicator is ON) WARNING: The IMA system uses high voltage (144 V), before servicing turn OFF the battery module switch. **Possible Causes:** • Poor connections or loose terminals at the DC-DC converter and the BCM module • "Open" or "Short" circuit between the BCM module and the DC-DC converter • Faulty DC-DC converter • BCM may need to be updated with the latest software • Faulty BCM
DTC: P1221/35 **1T PCM, MIL: Yes** **Year:** 2009, 2010 **Model:** Civic **Engine:** 1.3L L4 **Transmission:** All	**Battery Condition Monitor (BCM) Module Lost Communication with DC-DC Converter (Hybrid Models):** With the ignition switch ON and the battery voltage a minimum of 8.4 V, the communication information from the DC-DC converter is not received or the received data is abnormal for at least 2 seconds. (IMA system indicator is ON) WARNING: The IMA system uses high voltage (144 V), before servicing turn OFF the battery module switch. **Possible Causes:** • Poor connections or loose terminals at the DC-DC converter and the BCM module • "Open" in the wire between the DC-DC converter and the No. 3 ALTERNATOR fuse in the under-dash fuse/relay box • "Open" or "Short" circuit between the BCM module and the DC-DC converter • Faulty DC-DC converter • BCM may need to be updated with the latest software • Faulty BCM
DTC: P1286 **1T PCM, MIL: Yes** **Year:** 2009, 2010 **Model:** Civic **Engine:** 1.3L L4 **Transmission:** All	**Rocker Arm Oil Pressure Sensor B Stuck Low:** With the engine running at operating temperature, and the vehicle speed is a minimum of 12 mph (18 km/h) or more with the engine speed between 900-3,000 rpm the following conditions occur for 2 seconds or more. Condition 1: rocker arm oil pressure sensor A is ON Condition 2: rocker arm oil pressure sensor B is ON Condition 3: rocker arm oil pressure sensor A output HIGH Condition 4: rocker arm oil pressure sensor B output LOW **Possible Causes:** • Poor connections or loose terminals at rocker arm oil pressure sensor A, or B • Low engine oil, or oil pressure • Faulty rocker arm oil pressure sensor A., or B • Faulty rocker arm oil control valve A, or B
DTC: P1287 **2T PCM, MIL: Yes** **Year:** 2009, 2010 **Model:** Civic **Engine:** 1.3L L4 **Transmission:** All	**Rocker Arm Oil Pressure Switch Circuit High Voltage:** With the engine running at a speed between 2,000-4,600 rpm in (P or N) for at least 5 seconds, the following conditions occur. * Rocker Arm Oil Control Solenoid 1 is OFF * Rocker Arm Oil Control Solenoid 2 is OFF * Rocker Arm Oil Pressure Switch is OFF * Rocker Arm Oil pressure Sensor Output HIGH 39 kPa or more. **Possible Causes:** • Poor connections or loose terminals at the rocker arm oil pressure control solenoid 1 and/or 2, rocker arm oil pressure switch, rocker arm oil pressure sensor and the PCM • "Open" between the rocker arm oil pressure switch and ground • "Open" circuit between the rocker arm oil pressure switch and the PCM • Faulty rocker arm oil pressure switch • PCM may need to be updated with the latest software • Faulty PCM
DTC: P1288 **2T PCM, MIL: Yes** **Year:** 2009, 2010 **Model:** Civic **Engine:** 1.3L L4 **Transmission:** All	**Rocker Arm Oil Pressure Switch Circuit Low Voltage:** With the engine running at a speed between 2,000-4,600 rpm in (P or N) for at least 5 seconds, the following conditions occur. * Rocker Arm Oil Control Solenoid 1 is ON. * Rocker Arm Oil Control Solenoid 2 is OFF. * Rocker Arm Oil Pressure Switch is ON. * Rocker Arm Oil pressure Sensor Output HIGH 39 kPa or less. **Possible Causes:** • Poor connections or loose terminals at the rocker arm oil pressure control solenoid 1 and/or 2, rocker arm oil pressure switch, rocker arm oil pressure sensor and the PCM • "Short" circuit between the rocker arm oil pressure switch and the PCM • Faulty rocker arm oil pressure switch • PCM may need to be updated with the latest software • Faulty PCM

DTC	Trouble Code Title, Conditions & Possible Causes
DTC: P1289 **2T PCM, MIL: Yes** **Year:** 2009, 2010 **Model:** Civic **Engine:** 1.3L L4 **Transmission:** All	**Rocker Arm Oil Pressure Sensor Stuck High:** With the engine running at a speed between 3,000-4,800 rpm in (P or N) for at least 5 seconds in high valve mode, the following conditions occur. * Rocker Arm Oil Control Solenoid 1 is ON * Rocker Arm Oil Control Solenoid 2 is OFF * Rocker Arm Oil Pressure Switch is OFF * Rocker Arm Oil pressure Sensor Output HIGH 39 kPa or more **Possible Causes:** • Poor connections or loose terminals at the rocker arm oil pressure control solenoid 1 and/or 2, rocker arm oil pressure switch, rocker arm oil pressure sensor and the PCM • Faulty rocker arm oil pressure sensor
DTC: P128A **2T PCM, MIL: Yes** **Year:** 2009, 2010 **Model:** Civic **Engine:** 1.3L L4 **Transmission:** All	**Valve Pause System (VPS) Stuck OFF:** With the engine running at a speed between 1,000-3,000 rpm in (P or N) for at least 5 seconds in the valve pause mode, the following conditions occur. * Rocker Arm Oil Control Solenoid 1 is ON * Rocker Arm Oil Control Solenoid 2 is ON * Rocker Arm Oil Pressure Switch is ON * Rocker Arm Oil pressure Sensor Output HIGH 39 kPa or more **Possible Causes:** • Low engine oil level • Faulty rocker arm oil pressure sensor • Faulty rocker arm oil control valve
DTC: P128B **2T PCM, MIL: Yes** **Year:** 2009, 2010 **Model:** Civic **Engine:** 1.3L L4 **Transmission:** All	**Valve Pause System (VPS) Malfunction:** With the engine running at a speed between 1,000-3,000 rpm in (P or N) for at least 5 seconds in the valve pause mode, the following conditions occur. * Rocker Arm Oil Control Solenoid 1 is ON * Rocker Arm Oil Control Solenoid 2 is ON * Rocker Arm Oil Pressure Switch is OFF * Rocker Arm Oil pressure Sensor Output HIGH 39 kPa or less **Possible Causes:** • Low oil level • Poor connections or loose terminals at the rocker arm oil pressure control solenoid 1 and/or 2, rocker arm oil pressure switch, rocker arm oil pressure sensor and the PCM • "Open" between the rocker arm oil pressure switch and ground • "Open" circuit between the rocker arm oil pressure switch and the PCM • Faulty rocker arm oil control valve • Faulty rocker arm oil pressure switch • PCM may need to be updated with the latest software • Faulty PCM
DTC: P1297 **1T PCM** **Year:** 2009, 2010 **Model:** Accord, Accord Crosstour, Civic, CR-V, Element, Fit, Odyssey, Pilot, Ridgeline **Engine:** 1.3L L4, 1.5L L4, 1.8L L4, 2.0L L4, 2.4L L4, 3.5L V6 **Transmission:** All	**Electrical Load Detector (ELD) Circuit Low Voltage:** The ELD output voltage is 0.27 V or less for at least 5 seconds. The execution time is continuous and DTC P1298 is not active. **NOTE: Before you troubleshoot, record all freeze data and any on-board snapshot.** **Possible Causes:** • Poor connections or loose terminals at the ELD and the PCM • "Short" circuit between the PCM and the ELD • Faulty left side engine compartment wire harness • PCM may need to be updated with the latest software • Faulty PCM
DTC: P1298 **1T PCM** **Year:** 2009, 2010 **Model:** Accord, Accord Crosstour, Civic, CR-V, Element, Fit, Odyssey, Pilot, Ridgeline **Engine:** 1.3L L4, 1.5L L4, 1.8L L4, 2.0L L4, 2.4L L4, 3.5L V6 **Transmission:** All	**Electrical Load Detector (ELD) Circuit High Voltage:** With the ignition switch ON (II) the ELD output voltage is 4.57 V or more for at least 5 seconds. The execution time is continuous and DTC P1297 is not active. **NOTE: Before you troubleshoot, record all freeze data and any on-board snapshot.** **Possible Causes:** • Blown fuse • Poor connections or loose terminals at the ELD and the PCM • "Open" circuit between the fuse in the under-dash fuse/relay box and the ELD • "Open" in the wire between the ELD and ground • Faulty left side engine compartment wire harness • PCM may need to be updated with the latest software • Faulty PCM

DTC	Trouble Code Title, Conditions & Possible Causes
DTC: P1437/41 **1T PCM, MIL: Yes** **Year:** 2009, 2010 **Model:** Civic **Engine:** 1.3L L4 **Transmission:** All	**Motor Power Inverter (MPI) Module Short Circuit (HYBRID MODELS):** With the ignition ON and the MCM supply-power voltage a minimum of 10.5 V, the MPI module transmits the shorting information for at least 2 seconds. (IMA system indicator is ON) WARNING: The IMA system uses high voltage (144 V), before servicing turn OFF the battery module switch. **Possible Causes:** • Poor connections or loose terminals at the MPI module and the MCM • "Open" or "Short" circuit between the No. 53 +B IMA (10 A) fuse in auxiliary fuse holder B, MCM relay 2 and the MPI module • Faulty MPI module
DTC: P1440/57 **1T PCM, MIL: Yes** **Year:** 2009, 2010 **Model:** Civic **Engine:** 1.3L L4 **Transmission:** All	**Motor Power Inverter (MPI) Module Output Circuit Malfunction (HYBRID MODELS):** With the ignition ON, MCM supply-power voltage a minimum of 7.0 V, IMA motor speed between 500-1500 rpm and IMA driving torque a minimum of 74 Ft lbs. (10 Nm). The following malfunctions are detected: Malfunction 1: The ratio of maximum to minimum for the three motor current sensor fluctuation values become 70% or less for at least 0.5 seconds. Malfunction 2: The deviation of the target torque and the actual torque for the motor control is at least 0.5 seconds and 30 Ft. lbs. (40 Nm). (IMA system indicator is ON) WARNING: The IMA system uses high voltage (144 V), before servicing turn OFF the battery module switch. **Possible Causes:** • Poor connections or loose terminals at the MPI module, the U/V/W phase motor current sensors, and the MCM • Poor motor power cable connection at the motor stator • "Open" or "Short" circuit between the MPI module and the MCM • Faulty MPI module • Faulty motor power cable • Faulty motor stator • Faulty motor rotor • MCM may need to be updated with the latest software • Faulty MCM
DTC: P1454 **2T PCM, MIL: Yes** **Year:** 2009, 2010 **Model:** Accord, Accord Crosstour, CR-V, Element, Fit, Odyssey, Pilot, Ridgeline **Engine:** 1.5L L4, 2.4L L4, 3.5L V6 **Transmission:** All	**Fuel Tank Pressure (FTP) Sensor Circuit Range/Performance Problem:** One of these 2 conditions is met. (1) The FTP sensor output fluctuates by 0.6 kPa (0.1 in.Hg, 5 mmHg) or more, or -0.6 kPa (-0.1 in.Hg, -5 mmHg) or less for at least 3 seconds. (2) The FTP sensor output value is -1.3 kPa (-0.3 in.Hg, -10 mmHg) or less for at least 3 seconds. DTCs P0452, P0453 are judged as OK. **NOTE: Before you troubleshoot, record all freeze data and any on-board snapshot.** **Possible Causes:** • Poor connections or loose terminals at the FTP sensor, the EVAP canister vent shut valve, and the PCM • Blockage in the EVAP canister, canister filter, vent hoses, and drain joint, • Blockage in the FTP sensor air tube or vent • Faulty FTP sensor • Faulty EVAP canister vent shut valve
DTC: P145C **2T PCM, MIL: Yes** **Year:** 2009, 2010 **Model:** Accord, Accord Crosstour, CR-V, Element, Fit, Odyssey, Ridgeline **Engine:** 1.5L L4, 2.4L L4, 3.5L V6 **Transmission:** All	**Evaporative Emission (EVAP) System Purge Flow Malfunction:** The pulses detected by the fuel tank pressure sensor are 1.0 % of the duty cycle or less for at least 31 seconds. The execution time is continuous. **NOTE: This DTC is representative of an EVAP system purge flow problem. If DTC P145C is indicated alone, troubleshoot P0496 and P0497 using the freeze data for P145C.** If any of the DTCs listed below are indicated at the same time as DTC P145C, troubleshoot those DTC first, then recheck for P145C: * P0496, P0497: EVAP system purge flow. **Possible Causes:** • Troubleshoot appropriate DTCs
DTC: P1549 **1T PCM** **Year:** 2009, 2010 **Model:** Accord, Accord Crosstour, Civic, CR-V, Element, Fit, Odyssey, Pilot, Ridgeline **Engine:** 1.3L L4, 1.5L L4, 1.8L L4, 2.0L L4, 2.4L L4, 3.5L V6 **Transmission:** All	**Charging System High Voltage:** The IGP terminal voltage is 16.0 volts or more for at least 60 seconds. The execution time is continuous. **NOTE: If a high voltage battery (24 V, etc.) is connected to the vehicle, this DTC can be stored.** **Possible Causes:** • Poor connections or loose terminals at the alternator and the main under-hood fuse box • Faulty alternator

DTC	Trouble Code Title, Conditions & Possible Causes
DTC: P1585/30 **1T PCM, MIL: Yes** **Year:** 2009, 2010 **Model:** Civic **Engine:** 1.3L L4 **Transmission:** All	**Motor Current Sensor Circuit Malfunction (HYBRID MODELS):** With the ignition ON and the MCM supply-power voltage a minimum of 9.0 V, the output electrical current value of the MCM internal summation circuit is -50 A or less or 50 A or more, for at least 0.5 seconds. (IMA system indicator is ON) WARNING: The IMA system uses high voltage (144 V), before servicing turn OFF the battery module switch. **Possible Causes:** • Poor connections or loose terminals at the U/V/W phase motor current sensors and the MCM • Faulty U/V/W phase motor current sensors • MCM may need to be updated with the latest software • Faulty MCM
DTC: P15A5/85 **1T PCM, MIL: Yes** **Year:** 2009, 2010 **Model:** Civic **Engine:** 1.3L L4 **Transmission:** All	**Motor Current Sensor Circuit Malfunction (HYBRID MODELS):** With the ignition ON and the MCM supply-power voltage a minimum of 10.0 V, the motor phase current is more than 270 A. (IMA system indicator is ON) WARNING: The IMA system uses high voltage (144 V), before servicing turn OFF the battery module switch. **Possible Causes:** • Poor connections or loose terminals at the motor power cable, the U/V/W phase motor current sensors, and the MCM • Perform the motor rotor position calibration • Faulty U/V/W phase motor current sensor • Faulty MPI module • Faulty motor power cable • MCM may need to be updated with the latest software • Faulty MCM
DTC: P15A6/86 **1T PCM, MIL: Yes** **Year:** 2009, 2010 **Model:** Civic **Engine:** 1.3L L4 **Transmission:** All	**U Phase Motor Current Sensor Circuit Malfunction (HYBRID MODELS):** With the ignition ON and the battery voltage a minimum of 6.0 V, the sensor output voltage when the U phase motor current sensor is 0 A, is 2.32 V or less, or 2.68 V or more for at least 0.5 seconds. (IMA system indicator is ON) WARNING: The IMA system uses high voltage (144 V), before servicing turn OFF the battery module switch. **Possible Causes:** • Poor connections or loose terminals at the U phase motor current sensor and the MCM • Faulty U phase motor current sensor • MCM may need to be updated with the latest software • Faulty MCM
DTC: P15A7/87 **1T PCM, MIL: Yes** **Year:** 2009, 2010 **Model:** Civic **Engine:** 1.3L L4 **Transmission:** All	**V Phase Motor Current Sensor Circuit Malfunction (HYBRID MODELS):** With the ignition ON, and the battery voltage a minimum of 6.0 V, the sensor output voltage when the V phase motor current sensor is 0 A, is 2.32 V or less or 2.68 V or more for at least 0.5 seconds. (IMA system indicator is ON) WARNING: The IMA system uses high voltage (144 V), before servicing turn OFF the battery module switch. **Possible Causes:** • Poor connections or loose terminals at the V phase motor current sensor and the MCM • Faulty V phase motor current sensor • MCM may need to be updated with the latest software • Faulty MCM
DTC: P15A8/88 **1T PCM, MIL: Yes** **Year:** 2009, 2010 **Model:** Civic **Engine:** 1.3L L4 **Transmission:** All	**W Phase Motor Current Sensor Circuit Malfunction (HYBRID MODELS):** With the ignition ON and the battery voltage a minimum of 6.0 V, the sensor output voltage when the W phase motor sensor is 0 A, is 2.32 V or less, or 2.68 V or more for at least 0.5 seconds. (IMA system indicator is ON) WARNING: The IMA system uses high voltage (144 V), before servicing turn OFF the battery module switch. **Possible Causes:** • Poor connections or loose terminals at the W phase motor current sensor and the MCM • Faulty W phase motor current sensor • MCM may need to be updated with the latest software • Faulty MCM
DTC: P15AA/93 **1T** **Year:** 2009, 2010 **Model:** Civic **Engine:** 1.3L L4 **Transmission:** All	**Motor Rotor Position Not Learned (HYBRID MODELS):** With the engine running between 500-1500 rpm, battery voltage a minimum of 10.0 V and the battery module voltage between 120.0-245.0 V. One of the following conditions occurs: Condition 1: Angle detection value is -30 degrees or less. Condition 2: Angle detection value is +30 degrees or more. Condition 3: Angle detection value is not stable for at least 15 seconds. (IMA system indicator is ON) WARNING: The IMA system uses high voltage (144 V), before servicing turn OFF the battery module switch. **Possible Causes:** • Poor connections at the motor rotor position sensor, motor rotor position rotor, and at the motor stator • Check the MOTOR ROTOR POSITION CALIBRATION in the DATA LIST in the PGM-FI SYSTEM with the HDS • Faulty motor rotor position sensor

DTC	Trouble Code Title, Conditions & Possible Causes
DTC: P15AB **1T PCM** **Year:** 2009, 2010 **Model:** Accord, Accord Crosstour, Odyssey, Pilot **Engine:** 3.5L V6 **Transmission:** All	**Engine Mount Control Unit Power Source Circuit Low Voltage:** With the ignition on (II), the IG1 terminal voltage is 4.0 V or less for at least 2.6 seconds. **NOTE: Before you troubleshoot, record all freeze data and any on-board snapshots, and review the general troubleshooting information.** **Possible Causes:** • Blown fuse • Poor connections or loose terminals at the engine mount control unit and the body ground • "Open" circuit between engine mount control unit and body ground • "Open" circuit between engine mount control unit and power source • Faulty engine mount control unit
DTC: P15AC **1T PCM** **Year:** 2009, 2010 **Model:** Pilot **Engine:** 3.5L V6 **Transmission:** All	**Engine Mount Control Unit Internal Circuit Malfunction:** With the ignition switch ON, P16C4 and P16C9 are not active. The ACM actuator driving power voltage is 24 V or less for at least 2.6 seconds. **NOTE: Before you troubleshoot, record all freeze data and any on-board snapshots, and review the general troubleshooting information.** **Possible Causes:** • Poor connections or loose terminals at the engine mount control unit and the engine mount actuator • Faulty engine mount control unit
DTC: P15AE **1T PCM** **Year:** 2009, 2010 **Model:** Accord, Accord Crosstour, Odyssey **Engine:** 3.5L V6 **Transmission:** All	**Cylinder Pause Signal 1 Malfunction:** With the engine running, the SCPA signal input value differs from the PCM output value (Received by CAN communication) for at least 2.6 seconds. **NOTE: Select the VTEC TEST in the PGM-FI INSPECTION menu, and do the VPS TEST of the 3 CYLINDER ACTIVATION TEST with the HDS.** **Possible Causes:** • Poor connections or loose terminals at the engine mount control unit and the PCM • "Open" or "Short" circuit between the engine mount control unit and PCM • Faulty engine mount control unit • PCM may need to be updated with the latest software • Faulty PCM
DTC: P15AE **1T PCM** **Year:** 2009, 2010 **Model:** Pilot **Engine:** 3.5L V6 **Transmission:** All	**Cylinder Pause Signal Malfunction:** With the ignition switch ON, P16C9 not active, the SCP signal input value differs from the PCM output value (received by the CAN communication) for at least 2.6 seconds. **NOTE: Before you troubleshoot, record all freeze data and any on-board snapshots, and review the general troubleshooting information.** **Possible Causes:** • Poor connections or loose terminals at the engine mount control unit and the rocker arm oil control solenoid • Open circuit between the PCM and the engine mount control unit • Faulty engine mount control unit
DTC: P15B0 **1T PCM** **Year:** 2009, 2010 **Model:** Accord, Accord Crosstour, Odyssey, Pilot **Engine:** 3.5L V6 **Transmission:** All	**Crankshaft Position (CKP) Sensor Signal Malfunction:** With the engine running at idle, the CKP signal is not input to the engine mount control unit for at least 20 seconds. **NOTE: If PGM-FI system's DTC P0335 and/or P0339 is stored at the same time as DTC P15B0 troubleshoot P0335 and/or P0339 first, then recheck for P15B0.** **Possible Causes:** • Poor connections or loose terminals at the engine mount control unit and the PCM • "Open" or "Short" to ground between engine mount control unit and the PCM • Faulty engine mount control unit • PCM may need to be updated with the latest software • Faulty PCM
DTC: P15BD **1T PCM** **Year:** 2009, 2010 **Model:** Accord, Accord Crosstour, Odyssey, Pilot **Engine:** 3.5L V6 **Transmission:** All	**Cylinder Pause Signal 2 Malfunction:** With the engine running, the SCPA signal input value differs from the PCM output value (Received by CAN communication) for at least 2.6 seconds. **NOTE: Select the VTEC TEST in the PGM-FI INSPECTION menu, and do the VPS TEST.** **Possible Causes:** • Poor connections or loose terminals at the engine mount control unit and the PCM • "Open" or "Short" circuit between the engine mount control unit and PCM • Faulty engine mount control unit • PCM may need to be updated with the latest software • Faulty PCM

DTC	Trouble Code Title, Conditions & Possible Causes
DTC: P15BE **1T PCM** **Year:** 2009, 2010 **Model:** Accord, Accord Crosstour, Pilot **Engine:** 2.4L L4, 3.5L V6 **Transmission:** All	**Camshaft Position (CMP) Sensor Signal Malfunction:** With the engine running at idle, the CMP signal is not input to the engine mount control unit for at least 20 seconds. **NOTE: If PGM-FI system's DTC P0365 and/or P0369 is stored at the same time as DTC P15BE, troubleshoot P0365 and/or P0369 first, then recheck for P15BE.** **Possible Causes:** • Poor connections or loose terminals at the engine mount control unit and the PCM • "Open" or "Short" to ground between the engine mount control unit and PCM • Faulty engine mount control unit • PCM may need to be updated with the latest software • Faulty PCM
DTC: P15BF **1T PCM** **Year:** 2009, 2010 **Model:** Accord, Accord Crosstour, Odyssey, Pilot **Engine:** 3.5L V6 **Transmission:** All	**Camshaft Position (CMP) Sensor Signal Intermittent Interruption:** With the engine running at idle, other than 10 CMP pulses are detected during two revolutions of the crankshaft. This condition has been detected for at least 127 times. **NOTE: If PGM-FI system's DTC P0365 and/or P0369 is stored at the same time as DTC P15BF, troubleshoot P0365 and/or P0369 first, then recheck for P15BF.** **Possible Causes:** • Poor connections or loose terminals at the engine mount control unit and the PCM • "Open" or "Short" circuit between the engine mount control unit and PCM • Faulty engine mount control unit • PCM may need to be updated with the latest software • Faulty PCM
DTC: P15C0 **1T PCM** **Year:** 2009, 2010 **Model:** Accord, Accord Crosstour, Odyssey, Pilot **Engine:** 3.5L V6 **Transmission:** All	**Crankshaft Position (CKP) Sensor Signal Intermittent Interruption:** With the engine running at idle, The number of CKP pulse is not 58 during intervals between reference pulses for each crank revolution. OR. No interval is detected while the CKP sensor pulse inputs 82. **Possible Causes:** • Poor connections or loose terminals at the engine mount control unit and the PCM • "Open" or "Short" to ground between the engine mount control unit and PCM • Faulty engine mount control unit • PCM may need to be updated with the latest software • Faulty PCM
DTC: P1658 **1T PCM, MIL: Yes** **Year:** 2009, 2010 **Model:** Accord, Accord Crosstour, Civic, CR-V, Fit, Odyssey, Pilot, Ridgeline **Engine:** 1.3L L4, 1.5L L4, 1.8L L4, 2.0L L4, 2.4L L4, 3.5L V6 **Transmission:** All	**Electronic Throttle Control System (ETCS) Control Relay ON Malfunction:** The communication signal is input from the throttle actuator control module for at least 2 seconds, after the throttle actuator control module relay is turned off. The execution time is once per driving cycle. DTCs P0122, P0123, P0222, P0223, P1659, P2101, P2118, P2122, P2123, P2127, P2128, P2135, P2138, P2176 are not active. **NOTE: Before you troubleshoot, record all freeze data and any on-board snapshot.** **Possible Causes:** • Poor connections or loose terminals at the under-hood fuse/relay box (ETCS control relay) and the PCM • "Short" in the wire between the PCM and the under-hood fuse/relay box (ETCS control relay) • "Short" to power in the wire between the PCM and the under-hood fuse/relay box (ETCS control relay) • Faulty relay control module (under-hood fuse/relay box) • PCM needs to be updated with the latest software • Faulty PCM
DTC: P1659 **1T PCM, MIL: Yes** **Year:** 2009, 2010 **Model:** Accord, Accord Crosstour, Civic, CR-V, Fit, Odyssey, Pilot, Ridgeline **Engine:** 1.3L L4, 1.5L L4, 1.8L L4, 2.0L L4, 2.4L L4, 3.5L V6 **Transmission:** All	**Electronic Throttle Control System (ETCS) Control Relay OFF Malfunction:** Ignition switch ON (II), battery voltage 10 volts minimum and DTCs P0122, P0123, P0222, P0223, P2101, P2118, P2122, P2123, P2127, P2128, P2135, P2138, P2176 are not active. The voltage is not applied from the throttle actuator controller relay for at least 200 milliseconds. The execution time is once per driving cycle. **NOTE: Before you troubleshoot, record all freeze data and any on-board snapshot.** **Possible Causes:** • Blown fuse • Poor connections or loose terminals at the under-hood fuse/relay box (ETCS control relay) and the PCM • "Open" or "Short" in the wire between the PCM and the under-hood fuse/relay box • Faulty right side engine compartment wire harness • Faulty relay control module (under-hood fuse/relay box) • PCM needs to be updated with the latest software • Faulty PCM

DTC	Trouble Code Title, Conditions & Possible Causes
DTC: P1683 **1T PCM, MIL: Yes** **Year:** 2009, 2010 **Model:** Accord, Accord Crosstour, Civic, CR-V, Element, Fit, Odyssey, Pilot, Ridgeline **Engine:** 1.3L L4, 1.5L L4, 1.8L L4, 2.0L L4, 2.4L L4, 3.5L V6 **Transmission:** All	**Throttle Valve Default Position Spring Performance Problem:** Ignition switch LOCK (0), battery voltage a minimum of 6 volts, engine coolant minimum 158°F (70°C) and DTCs P0117, P0118, P0122, P0123, P0222, P0223, P2101, P2118, P2122, P2123, P2127, P2128, P2135, P2138, P2176 not active. The throttle valve position is more than +5° from fully closed, or less than +3° from fully closed for at least 3 seconds. **NOTE: Before you troubleshoot, record all freeze data and any on-board snapshot.** **Possible Causes:** • Poor connections or loose terminals at the throttle body and the PCM • Faulty throttle body
DTC: P1684 **1T PCM, MIL: Yes** **Year:** 2009, 2010 **Model:** Accord, Accord Crosstour, Civic, CR-V, Element, Fit, Odyssey, Pilot, Ridgeline **Engine:** 1.3L L4, 1.5L L4, 1.8L L4, 2.0L L4, 2.4L L4, 3.5L V6 **Transmission:** All	**Throttle Valve Return Spring Performance Problem:** Ignition switch LOCK (0), coolant temperature minimum 158°F (70°C), battery voltage minimum 6 volts and DTCs P0117, P0118, P0122, P0123, P0222, P0223, P2101, P2118, P2122, P2123, P2127, P2128, P2135, P2138, P2176 not active. The throttle valve opening angle is 17° or more, or 11° or less, for at least 3 seconds. Before you troubleshoot, record all freeze data and any on-board snapshot. **Possible Causes:** • Poor connections or loose terminals at the throttle body and the PCM • Faulty throttle body
DTC: P16BB **1T PCM** **Year:** 2009, 2010 **Model:** Accord, Accord Crosstour, Civic, CR-V, Element, Fit, Odyssey, Pilot, Ridgeline **Engine:** 1.5L L4, 1.8L L4, 2.0L L4, 2.4L L4, 3.5L V6 **Transmission:** All	**Alternator B Terminal Circuit Low Voltage:** Engine speed 500-3,000 rpm, alternator control mode 14.5 volts. The IGP terminal voltage is 12.0 V or less, and the alternator power generation amount is within 1.0 % to 50.0 %, for at least 60 seconds. The execution time is continuous. **NOTE: Before you troubleshoot, record all freeze data and any on-board snapshot.** **Possible Causes:** • Faulty battery • Poor connections or loose terminals at the alternator and the main under-hood fuse box • "Open" circuit between the alternator and the main under-hood fuse box • Faulty alternator
DTC: P16BC **1T PCM** **Year:** 2009, 2010 **Model:** Accord, Accord Crosstour, CR-V, Element, Fit, Odyssey, Pilot, Ridgeline **Engine:** 1.5L L4, 2.4L L4, 3.5L V6 **Transmission:** All	**Alternator FR Terminal Circuit/IGP Circuit Low Voltage:** Engine speed 500-3,000 rpm, alternator control mode 14.5 volts minimum. The IGP terminal voltage is 12.0 V or less, and the alternator power generation amount is 0.5 % or less, for at least 60 seconds. The execution time is continuous. **NOTE: Before you troubleshoot, record all freeze data and any on-board snapshot.** **Possible Causes:** • Blown fuse • Poor connections or loose terminals at the alternator connector • Poor alternator ground • open in the wire between the alternator and the fuse in the under-dash fuse/relay box • Faulty alternator • PCM may need to be updated with the latest software • Faulty PCM
DTC: P16BD **1T PCM** **Year:** 2009, 2010 **Model:** Pilot **Engine:** 3.5L V6 **Transmission:** All	**Starter Cut Relay 2 Malfunction:** Engine running, starter switch OFF, DTC P16BE not active. The terminal voltage of the STRLD drops to 2.2 V for at least 5 seconds when the starter cut relay output is turned off. **NOTE: Before you troubleshoot, record all freeze data and any on-board snapshot.** **Possible Causes:** • Poor connections or loose terminals at starter cut relay 2 and the PCM • Faulty starter relay 2
DTC: P16BE **1T PCM** **Year:** 2009, 2010 **Model:** Pilot **Engine:** 3.5L V6 **Transmission:** All	**Starter Cut Relay 1 Malfunction:** Engine running, Starter switch OFF cut relay turn on switching failure, ON STRLD open circuit failure. One of these conditions occurs. (1) The terminal voltage of the STRLD exceeds 3.0 volts for at least 5 seconds when the starter cut relay output is turned off. (2) The terminal voltage of the STRLD is 2.4-2.6 volts for at least 0.8 second when the starter cut relay output is turned on. **NOTE: Before you troubleshoot, record all freeze data and any on-board snapshot.** **Possible Causes:** • Poor connections or loose terminals at starter cut relay 1 and the PCM • Faulty starter relay 1

DTC	Trouble Code Title, Conditions & Possible Causes
DTC: P16BF **1T PCM** **Year:** 2009, 2010 **Model:** Pilot **Engine:** 3.5L V6 **Transmission:** All	**Starter Cut Relay STRLY Circuit Malfunction:** With ignition switch ON (II) the signal of the on command from the PCM to the starter and the return signal from the starter cut relay do not coincide for at least 5 seconds. The execution time is continuous. **NOTE: Before you troubleshoot, record all freeze data and any on-board snapshot.** **Possible Causes:** • Poor connections or loose terminals at starter cut relay 1, starter cut relay 2, and the PCM • "Short" in the wire between the PCM and starter cut relay 1 or starter cut relay 2 • PCM needs to be updated with the latest software • Faulty PCM
DTC: P16C0 **1T PCM, MIL: Yes** **Year:** 2009, 2010 **Model:** Accord, Civic, CR-V, Odyssey, Pilot, Ridgeline **Engine:** 1.8L L4, 2.0L L4, 2.4L L4, 3.5L V6 **Transmission:** All	**PCM A/T Control System Incomplete Update:** The program rewriting process does not finish normally due to an irregular process, such as turning off the PCM power during the A/T CPU rewriting process, and then turning the ignition switch to ON (II) again. The execution time is continuous and the duration time is 1 seconds or more. **NOTE: This code is indicated when PCM updating is incomplete.** **Possible Causes:** • PCM may need to be updated with the latest software • Faulty PCM
DTC: P16C0/99 **1T PCM, MIL: Yes, TCIL: Yes** **Year:** 2009, 2010 **Model:** Civic **Engine:** 1.3L L4 **Transmission:** All	**Powertrain Control Module (PCM) CVT Control System Incomplete Update (Hybrid Models):** With the ignition switch ON, the program rewriting process does not finish normally due to an irregular process, such as turning off the power during the CVT CPU rewriting process, and the ignition switch is turned to ON, again. This code is indicated when PCM updating is incomplete. **NOTE: Do not turn the ignition switch to LOCK (0) or ACCESSORY (I) while updating the PCM. If you turn the ignition switch to LOCK (0) or ACCESSORY (I) before completion, the PCM can be damaged.** **Possible Causes:** • PCM may need to be updated with the latest software • Faulty PCM
DTC: P16C4 **1T PCM** **Year:** 2009, 2010 **Model:** Accord, Accord Crosstour, Odyssey, Pilot **Engine:** 3.5L V6 **Transmission:** All	**Engine Mount Actuator Control Power Circuit (Stuck Off):** With the ignition switch ON (II), the IGSOL voltage is 4.0 V or less for at least 2.6 seconds during the active engine mount (ACM) control relay is on. **NOTE: Before you troubleshoot, record all freeze data and any on-board snapshots, and review the general troubleshooting information.** **Possible Causes:** • Blown fuse • Poor connections or loose terminals at the engine mount control unit and the driver's under-dash fuse/relay box (active control engine mount (ACM) control relay) • "Open" or "Short" to ground between engine mount control unit and driver's under-dash fuse/relay box • Faulty (ACM) control relay in the driver's under-dash fuse/relay box • Faulty engine mount control unit
DTC: P16C5 **1T PCM** **Year:** 2009, 2010 **Model:** Accord, Accord Crosstour, Odyssey, Pilot **Engine:** 3.5L V6 **Transmission:** All	**Engine Mount Actuator Control Power Circuit (Stuck ON):** With the ignition ON (II), the IGSOL voltage is 8.0 V or more for at least 2.6 seconds during the active control engine mount (ACM) control relay is off. **NOTE: Before you troubleshoot, record all freeze data and any on-board snapshots, and review the general troubleshooting information.** **Possible Causes:** • "Short" to ground in the wire between engine mount control unit and driver's under-dash fuse/relay box • "Short" to power in the wire between engine mount control unit and driver's under-dash fuse/relay box • Faulty driver's under-dash fuse/relay box • Faulty engine mount control unit
DTC: P16C6 **1T PCM** **Year:** 2009, 2010 **Model:** Accord, Accord Crosstour, Odyssey, Pilot **Engine:** 3.5L V6 **Transmission:** All	**Engine Mount Actuator High Voltage During Function Test:** With the ignition switch ON (II), one of these 3 conditions occur: 1). The value during one cycle of the front engine mount actuator current is 7.5 A or more as a maximum current or 5.3 A or more as a minimum current. 2). The engine mount cycle occurs during current start up and continues. 3). The duty cycle of the front engine mount actuator reads 100% all the time during engine mount operation and continues at least 20 seconds. **NOTE: Before you troubleshoot, record all freeze data and any on-board snapshots, and review the general troubleshooting information.** **Possible Causes:** • Check for poor connections or loose terminals at the engine mount control unit, the engine mount actuator, and body ground.

DTC	Trouble Code Title, Conditions & Possible Causes
DTC: P16D5 **1T PCM, MIL: Yes** **Year:** 2009, 2010 **Model:** Civic **Engine:** 1.3L L4 **Transmission:** All	**F-CAN Malfunction (Internal Malfunction):** With the ignition switch ON, for 3 seconds and battery voltage at a minimum of 10.0 V, one of the following malfunction conditions exist. Condition 1: The PCM does not receive any signals via the F-CAN lines for at least 1 second. Condition 2: The information sent from the CVT CPU is abnormal at least 20 times. **Possible Causes:** • PCM may need to be updated with the latest software • Faulty PCM
DTC: P16D6 **1T PCM, MIL: Yes** **Year:** 2009, 2010 **Model:** Civic **Engine:** 1.3L L4 **Transmission:** All	**IMA-CAN Malfunction (Internal Malfunction) (HYBRID MODELS):** With the Ignition ON an minimum of 3 seconds, and battery voltage at a minimum of 10.0 V, one of the following conditions exist, Condition 1: The PCM does not receive any signals via the IMA-CAN lines. Condition 2: The information sent from the CVT CPU is abnormal at least 20 times. **Possible Causes:** • PCM may need to be updated with the latest software • Faulty PCM
DTC: P16D7/107 **1T PCM, TCIL: Yes** **Year:** 2009, 2010 **Model:** Civic **Engine:** 1.3L L4 **Transmission:** All	**Powertrain Control Module (PCM) Internal F-CAN Communication Circuit Malfunction:** With the ignition switch ON for at least 5 seconds and the battery voltage at a minimum of 10.0 V, no signal from the FI CPU via the F-CAN are received for at least 1.5 seconds. **Possible Causes:** • Poor connections or loose terminals between PCM and F-CAN circuit • PCM may need to be updated with the latest software • Faulty PCM
DTC: P16D8/120 **1T PCM, MIL: Yes, TCIL: Yes** **Year:** 2009, 2010 **Model:** Civic **Engine:** 1.3L L4 **Transmission:** All	**PCM Internal IMA-CAN Communication Circuit Malfunction (HYBRID MODELS):** With the ignition switch ON for at least 5 seconds and the battery voltage a minimum of 10.0 V, no signals from the FI CPU via the IMA-CAN are received for at least 1.5 seconds. **Possible Causes:** • Poor connections or loose terminals between PCM and F-CAN circuit • PCM may need to be updated with the latest software • Faulty PCM
DTC: P1717 **2T PCM, MIL: Yes, TCIL: Yes** **Year:** 2009, 2010 **Model:** Civic, CR-V, Element, Fit, Odyssey, Ridgeline **Engine:** 1.5L L4, 1.8L L4, 2.4L L4, 3.5L V6 **Transmission:** All	**Open in Transmission Range Switch ATPRVS Switch Circuit :** With the vehicle driven in reverse at a speed of 3 (5 km/h) mph or less for at least 2 seconds, no RVS signal is detected. **Possible Causes:** • Check for proper transmission range switch installation, adjust the shift cable if needed • "Open" circuit between PCM connector terminal and the transmission range switch • Faulty transmission range switch • PCM may need to be updated with the latest software • Faulty PCM
DTC: P1730 **2T PCM, MIL: Yes, TCIL: Yes** **Year:** 2009, 2010 **Model:** Accord, CR-V **Engine:** 2.4L L4 **Transmission:** All	**Problem in Shift Control System:** With the engine running and in Drive position allow trans mission to shift to 5th gear. Shift solenoid A or D stuck OFF, Shift solenoid B stuck ON, Shift Valves A, B or D stuck. **Possible Causes:** • Low or dirty transmission fluid. • Repair the hydraulic system related to shift valves A, B, and D, or replace the transmission
DTC: P1731 **2T PCM, MIL: Yes, TCIL: Yes** **Year:** 2009, 2010 **Model:** Accord, CR-V, Element **Engine:** 2.4L L4 **Transmission:** All	**Problem in Shift Control System:** With the engine running and in Drive position allow transmission to shift to 5th gear. Shift solenoid E stuck ON, Shift solenoid E stuck, A/T Clutch pressure control solenoid A stuck OFF. **NOTE: Before you troubleshoot, record all freeze data and any on-board snapshot with the HDS, and review General Troubleshooting Information.** **Possible Causes:** • Low or dirty transmission fluid • Faulty A/T clutch pressure control solenoid valve A • Repair the hydraulic system related to shift valve E, or replace the transmission

DTC	Trouble Code Title, Conditions & Possible Causes
DTC: P1732 **2T PCM, MIL: Yes, TCIL: Yes** **Year:** 2009, 2010 **Model:** Accord, CR-V **Engine:** 2.4L L4 **Transmission:** All	**Problem in Shift Control System: Shift Solenoid B or C Stuck ON:** With the engine running and in Drive position allow transmission to shift to 5th gear. Shift solenoid B or C stuck ON, Shift solenoid B or C Stuck. **NOTE: Before you troubleshoot, record all freeze data and any on-board snapshot with the HDS, and review General Troubleshooting Information.** **Possible Causes:** • Low or dirty transmission fluid • Faulty shift solenoid valve B or C Stuck ON • Faulty shift solenoid valve B or C Stuck • Faulty transmission
DTC: P1733 **2T PCM, MIL: Yes, TCIL: Yes** **Year:** 2009, 2010 **Model:** Accord, CR-V, Element **Engine:** 2.4L L4 **Transmission:** All	**Problem in Shift Control System:** With the engine running and in Drive position allow transmission to shift to 5th gear. Shift solenoid D stuck ON, Shift solenoid D Stuck. A/T clutch pressure control switch valve C Stuck OFF. **NOTE: Before you troubleshoot, record all freeze data and any on-board snapshot with the HDS, and review General Troubleshooting Information.** **Possible Causes:** • Low or dirty transmission fluid • Faulty shift solenoid valve D • Faulty A/T clutch pressure control switch valve C • Faulty transmission
DTC: P1734 **2T PCM, MIL: Yes, TCIL: Yes** **Year:** 2009, 2010 **Model:** Accord, CR-V, Element **Engine:** 2.4L L4 **Transmission:** All	**Problem in Shift Control System: :** With the engine running and in Drive position allow transmission to shift to 5th gear. Shift solenoid B or C stuck OFF, Shift solenoid B or C Stuck. **NOTE: Before you troubleshoot, record all freeze data and any on-board snapshot with the HDS, and review General Troubleshooting Information.** **Possible Causes:** • Low or dirty transmission fluid • Faulty shift solenoid valve B or C Stuck OFF • Faulty shift solenoid valve B or C Stuck • Faulty transmission
DTC: P1743 **2T PCM, TCIL: Yes** **Year:** 2009, 2010 **Model:** Accord, Accord Crosstour, Odyssey, Ridgeline **Engine:** 3.5L V6 **Transmission:** All	**Problem in Shift Control System; Shift Valve E Stuck OFF:** Vehicle speed is 5 mph (9 km/h), torque converter slip ratio 96-110 %. battery minimum of 11 volts. The transmission is neutral against the 5th gear shift command for at least 2 seconds, without records that the engine speed flares when upshifting to 3rd-4th. The execution time is continuous depending on the driving pattern. **NOTE: Before you troubleshoot, record all freeze data and any on-board snapshot,.** **Possible Causes:** • Low transmission fluid • Dirty transmission fluid • If the strainer has metal debris or excessive clutch material, replace the transmission • Faulty shift valve E in the main valve body, replace the main valve body • Faulty transmission
DTC: P1744 **2T PCM, TCIL: Yes** **Year:** 2009, 2010 **Model:** Accord, Accord Crosstour, Odyssey, Ridgeline **Engine:** 3.5L V6 **Transmission:** All	**Problem in Shift Control System; Shift Valve E Stuck ON:** The transmission is held in 5th gear against the 1st gear shift command for at least 2 seconds. After driving the vehicle in 3rd gear in D for at least 2 seconds. The execution time is continuous depending on the driving pattern. **NOTE: Before you troubleshoot, record all freeze data and any on-board snapshot.** **Possible Causes:** • Low or dirty transmission fluid • If the strainer has metal debris or excessive clutch material, replace the transmission • Faulty shift valve E in the main valve body, replace the main valve body • Faulty transmission
DTC: P1745 **2T PCM, TCIL: Yes** **Year:** 2009, 2010 **Model:** Accord, Accord Crosstour, Odyssey, Ridgeline **Engine:** 3.5L V6 **Transmission:** All	**Problem in Shift Control System; Servo Control Valve Stuck OFF or Servo Valve Stuck OFF:** The engine speed flares when upshifting to 2nd-3rd, the engine speed flares when upshifting to 3rd-4th, driving in 5th gear neutral condition. The execution time is continuous and the duration time is 1-2 seconds or more. **NOTE: Before you troubleshoot, record all freeze data and any on-board snapshot.** **Possible Causes:** • Low or dirty transmission fluid • If the strainer has metal debris or excessive clutch material, replace the transmission • Faulty servo control valve in the main valve body, servo valve in the regulator valve body, or replace the main valve body, regulator valve body • Faulty transmission

DTC	Trouble Code Title, Conditions & Possible Causes
DTC: P1746 **2T PCM, TCIL: Yes** **Year:** 2009, 2010 **Model:** Fit **Engine:** 1.5L L4 **Transmission:** All	**Hydraulic Control System (Cut Valve A Stuck OFF or Cut Valve B Stuck ON):** With the vehicle driven at a minimum of 5 mph (9km/h) and the accelerator position pedal a minimum of 2.0%, an improper gear ratio output compared to the predetermined gear change mode, the cut valve A OFF failure or the cut valve B ON failure is detected. (Poor shift quality) **NOTE: This code is stored simultaneously with DTC P1780, which is caused by a hydraulic control system problem.** **Possible Causes:** • Low or dirty transmission fluid • Inspect the strainer for metal debris or excessive clutch material, if present replace the transmission • Faulty cut valve A in the servo body, cut valve B in the main valve body, or replace the main valve body
DTC: P1747 **2T PCM, TCIL: Yes** **Year:** 2009, 2010 **Model:** Fit **Engine:** 1.5L L4 **Transmission:** All	**Hydraulic Control System (Cut Valve A Stuck ON or Cut Valve B Stuck OFF):** With the vehicle driven at a minimum of 5 mph (9km/h) and the accelerator position pedal a minimum of 2.0%, an improper gear ratio output compared to the predetermined gear change mode, the cut valve A ON failure or the cut valve B OFF failure is detected. (Poor shift quality) **NOTE: This code is stored simultaneously with DTC P1780, which is caused by a hydraulic control system problem.** **Possible Causes:** • Low or dirty transmission fluid • Inspect the strainer for metal debris or excessive clutch material, if present replace the transmission • Faulty cut valve A in the servo body, cut valve B in the main valve body, or replace the main valve body
DTC: P1780 **2T PCM, MIL: Yes, TCIL: Yes** **Year:** 2009, 2010 **Model:** Accord, Accord Crosstour, Fit, Odyssey, Ridgeline **Engine:** 1.5L L4, 3.5L V6 **Transmission:** All	**Problem in Shift Control System (Transmission is in Default Mode):** The A/T control switches to the default mode due to a mechanical malfunction. The execution time is continuous depending on the driving pattern. **NOTE: DTC P1780 means there is one or more A/T DTCs about the shift control system. Before you troubleshoot, record all freeze data and any on-board snapshot.** **Possible Causes:** • Poor connections and loose terminals at the PCM • PCM needs to be updated with the latest software • Faulty PCM
DTC: P1860/33 **1T PCM, TCIL: Yes** **Year:** 2009, 2010 **Model:** Civic **Engine:** 1.3L L4 **Transmission:** All	**Inhibitor Solenoid Circuit Low Voltage (Hybrid Models):** When the transmission gear selector is shifted into the reverse position by mistake during the forward operation, the inhibitor solenoid intercepts the activation of the reverse brake operation to prevent transmission damage. The return signal is inappropriate for the turn on command to the inhibitor solenoid. **NOTE: This code is caused by an electrical circuit problem and cannot be caused by a mechanical problem in the transmission.** **Possible Causes:** • Poor connections or loose terminals between the PCM and the inhibitor solenoid * • "Short" to body ground in the wire between PCM and solenoid wire harness • Faulty solenoid wire harness • Faulty inhibitor solenoid • PCM may need to be updated with the latest software • Faulty PCM
DTC: P1861/33 **1T PCM, TCIL: Yes** **Year:** 2009, 2010 **Model:** Civic **Engine:** 1.3L L4 **Transmission:** All	**Inhibitor Solenoid Circuit High Voltage:** When the transmission gear selector is shifted into the reverse position by mistake during the forward operation, the inhibitor solenoid intercepts the activation of the reverse brake operation to prevent transmission damage. The return signal is inappropriate for the turn on command to the inhibitor solenoid. **NOTE: This code is caused by an electrical circuit problem and cannot be caused by a mechanical problem in the transmission.** **Possible Causes:** • Poor connections or loose terminals between the PCM and the inhibitor solenoid * • "Open" to body ground in the wire between PCM and solenoid wire harness • "Open" circuit in the wire between PCM and solenoid wire harness • Faulty solenoid wire harness • Faulty inhibitor solenoid • PCM may need to be updated with the latest software • Faulty PCM
DTC: P1890/42 **1T PCM, TCIL: Yes** **Year:** 2009, 2010 **Model:** Civic **Engine:** 1.3L L4 **Transmission:** All	**Problem in CVT Speed Control System:** With the vehicle driven at 38 mph (60 km/h) for at least 1 minute, the difference between the target drive pulley speed and the actual drive pulley speed is 500 rpm or more for at least 18 seconds. **NOTE: Before you troubleshoot, record all freeze data and any on-board snapshot with the HDS, and review General Troubleshooting Information.** **Possible Causes:** • Faulty CVT fluid • Faulty transmission

DTC	Trouble Code Title, Conditions & Possible Causes
DTC: P1891/43 **1T PCM, TCIL: Yes** **Year:** 2009, 2010 **Model:** Civic **Engine:** 1.3L L4 **Transmission:** All	**Problem in Start Clutch Control System:** With the vehicle driven at 38 mph (60 km/h), the deviation in the vehicle speed measured by the vehicle speed sensor and vehicle speed measured by the driven pulley speed sensor is in correct. **Possible Causes:** • Poor connections or loose terminals between the CVT clutch pressure control solenoid valve and the PCM • Faulty start clutch • PCM may need to be updated with the latest software • Faulty PCM
DTC: P1898/100 **2T PCM, MIL: Yes, TCIL: Yes** **Year:** 2009, 2010 **Model:** Civic **Engine:** 1.3L L4 **Transmission:** All	**CVT Drive Pulley Pressure Control Valve Stuck ON or CVT Driven Pulley Pressure Control Valve Stuck OFF:** With the vehicle driven between 40-50 mph (65-80 km/h) on a flat road for at least 15 seconds, then stop the vehicle pressing the brake pedal for 5 seconds. Both of the following judgment malfunctions must be met. Judgment A: The pulley ratio is 2.2-2.7 for at least 0.4 second. Judgment B: The pulley ratio is 1.8 to 2.7 for at least 5 seconds. **Possible Causes:** • Poor connections or loose terminals between the CVT drive pulley pressure control solenoid valve, the CVT driven pulley pressure control solenoid valve, and the PCM • Faulty CVT drive pulley pressure control solenoid valve and CVT driven pulley pressure control solenoid valve • Faulty transmission
DTC: P1899/100 **2T PCM, MIL: Yes, TCIL: Yes** **Year:** 2009, 2010 **Model:** Civic **Engine:** 1.3L L4 **Transmission:** All	**CVT Drive Pulley Pressure Control Valve Stuck OFF or CVT Driven Pulley Pressure Control Valve Stuck ON:** With the vehicle driven between 40-50 mph (65-80 km/h) on a flat road for at least 15 seconds, then stop the vehicle pressing the brake pedal for 5 seconds. Both of the following judgment malfunctions must be met. Judgment A: The pulley ratio is 0.3-0.8 for at least 2.5 seconds. Judgment B: The pulley ratio is 3 for at least 12 seconds. **Possible Causes:** • Poor connections or loose terminals between the CVT drive pulley pressure control solenoid valve, the CVT driven pulley pressure control solenoid valve, and the PCM • Faulty CVT drive pulley pressure control solenoid valve and CVT driven pulley pressure control solenoid valve • Faulty transmission

OBD II Trouble Code List (P2XXX Codes)

DTC	Trouble Code Title, Conditions & Possible Causes
DTC: P2101 **1T PCM, MIL: Yes** **Year:** 2009, 2010 **Model:** Accord, Accord Crosstour, Civic, CR-V, Element, Fit, Odyssey, Pilot, Ridgeline **Engine:** 1.3L L4, 1.5L L4, 1.8L L4, 2.0L L4, 2.4L L4, 3.5L V6 **Transmission:** All	**Electronic Throttle Control System (ETCS) Malfunction (Includes Hybrid Models):** Ignition switch ON (II), battery voltage minimum of 6 volts and DTCs P0122, P0123, P0222, P0223, P2118, P2122, P2123, P2127, P2128,P2135, P2138, P2176 are not active. Difference between throttle valve target position and actual throttle valve position, 4-6° or more. The execution time is continuous and the duration time is 250-500 milliseconds or more. **NOTE: Before you troubleshoot, record all freeze data and any on-board snapshot.** **Possible Causes:** • Dirty throttle body • Poor connections or loose terminals at the throttle body and the PCM • "Open" circuit between the throttle body and the PCM • Faulty throttle body • PCM needs to be updated with the latest software • Faulty PCM
DTC: P2108 **1T PCM, MIL: Yes** **Year:** 2009, 2010 **Model:** Accord, Accord Crosstour, Element **Engine:** 2.4L L4, 3.5L V6 **Transmission:** All	**Throttle Actuator Control Module Problem:** With the ignition ON (II) one of these 4 conditions must be met for at least 200 milliseconds. (1) Data read from the ROM is abnormal. (2) Data read from the RAM is abnormal. (3) The A/D converter standard voltage is out of specified value. (4) The serial signals between the PCM and the throttle actuator control module do not agree. **NOTE: Before you troubleshoot, record all freeze data and any on-board snapshot.** **Possible Causes:** • Poor connections or loose terminals at the throttle body, the throttle actuator control module and the PCM • Faulty throttle actuator control module

DTC	Trouble Code Title, Conditions & Possible Causes
DTC: P2118 **1T PCM, MIL: Yes** **Year:** 2009, 2010 **Model:** Accord, Accord Crosstour, Civic, CR-V, Fit, Odyssey, Pilot, Ridgeline **Engine:** 1.3L L4, 1.5L L4, 1.8L L4, 2.0L L4, 2.4L L4, 3.5L V6 **Transmission:** All	**Throttle Actuator Current Range/Performance Problem (Includes Hybrid Models):** With ignition ON (II), battery voltage minimum of 6 volts and DTCs P0122, P0123, P0222, P0223, P2101, P2122, P2123, P2127, P2128, P2135, P2138, P2176 are not active. The current flow to the throttle actuator is 11 A or more for at least 200 milliseconds. The execution time is continuous. **NOTE: Before you troubleshoot, record all freeze data and any on-board snapshot.** **Possible Causes:** • Poor connections or loose terminals at the throttle body and the PCM • "Short" circuit between the PCM (ETCSM-line) and (ETCSM+line) • Faulty throttle body • Faulty throttle actuator control module • PCM may need to be updated with the latest software • Faulty PCM
DTC: P2122 **1T PCM, MIL: Yes** **Year:** 2009, 2010 **Model:** Accord, Accord Crosstour, Civic, CR-V, Element, Fit, Odyssey, Pilot, Ridgeline **Engine:** 1.3L L4, 1.5L L4, 1.8L L4, 2.0L L4, 2.4L L4, 3.5L V6 **Transmission:** All	**APP Sensor A or 1 (TP Sensor D) Circuit Low Voltage (Includes Hybrid Models):** With the ignition switch ON (II) the APP sensor A output voltage is 0.2 V or less for at least 200 milliseconds. The execution time is continuous. DTC P2123 is not active. **NOTE: Before you troubleshoot, record all freeze data and any on-board snapshot.** **Possible Causes:** • Poor connections or loose terminals at APP sensor A and the PCM • "Open" or "Short" circuit between the PCM and APP sensor A • Faulty APP sensor • Faulty accelerator pedal module • PCM needs to be updated with the latest software • Faulty PCM
DTC: P2123 **1T PCM, MIL: Yes** **Year:** 2009, 2010 **Model:** Accord, Accord Crosstour, Civic, CR-V, Element, Fit, Odyssey, Pilot, Ridgeline **Engine:** 1.3L L4, 1.5L L4, 1.8L L4, 2.0L L4, 2.4L L4, 3.5L V6 **Transmission:** All	**APP Sensor A or 1 (TP Sensor D) Circuit High Voltage (Includes Hybrid Models):** With engine running the APP sensor A output voltage is 4.9 V or more for at least 200 milliseconds. The execution time is continuous. DTC P2122 is not active. **Possible Causes:** • Poor connections or loose terminals at APP sensor A and the PCM • "Open" circuit between the PCM and APP sensor A • Faulty APP sensor • Faulty accelerator pedal module • PCM needs to be updated with the latest software • Faulty PCM
DTC: P2127 **1T PCM, MIL: Yes** **Year:** 2009, 2010 **Model:** Accord, Accord Crosstour, Civic, CR-V, Element, Fit, Odyssey, Pilot, Ridgeline **Engine:** 1.3L L4, 1.5L L4, 1.8L L4, 2.0L L4, 2.4L L4, 3.5L V6 **Transmission:** All	**APP Sensor B or 2 (Throttle Position (TP) Sensor E) Circuit Low Voltage (Includes Hybrid Models):** With the ignition ON (II) the APP sensor B output voltage is 0.2 V or less for at least 200 milliseconds. The execution time is continuous. DTC P2128 is not active. **NOTE: Before you troubleshoot, record all freeze data and any on-board snapshot.** **Possible Causes:** • Poor connections or loose terminals at APP sensor B and the PCM • "Open" or "Short" circuit between the PCM and APP sensor B • Faulty accelerator pedal module • Faulty APP sensor • PCM needs to be updated with the latest software • Faulty PCM
DTC: P2128 **1T PCM, MIL: Yes** **Year:** 2009, 2010 **Model:** Accord, Accord Crosstour, Civic, CR-V, Element, Fit, Odyssey, Pilot, Ridgeline **Engine:** 1.3L L4, 1.5L L4, 1.8L L4, 2.0L L4, 2.4L L4, 3.5L V6 **Transmission:** All	**APP Sensor B or 2 (Throttle Position (TP) Sensor E) Circuit High Voltage (Includes Hybrid Models):** With engine running the APP sensor B output voltage is 4 V or more for at least 200 milliseconds. The execution time is continuous. DTC P2127 is not active. **NOTE: Before you troubleshoot, record all freeze data and any on-board snapshot.** **Possible Causes:** • Poor connections or loose terminals APP sensor B and the PCM • "Open" circuit between the PCM and APP sensor B • Faulty accelerator pedal module • Faulty APP sensor • PCM may need to be updated with the latest software • Faulty PCM

DTC	Trouble Code Title, Conditions & Possible Causes
DTC: P2135 **1T PCM, MIL: Yes** **Year:** 2009, 2010 **Model:** Accord, Accord Crosstour, Civic, CR-V, Element, Fit, Odyssey, Pilot, Ridgeline **Engine:** 1.3L L4, 1.5L L4, 1.8L L4, 2.0L L4, 2.4L L4, 3.5L V6 **Transmission:** All	**Throttle Position (TP) Sensor A/B or 1/2 Incorrect Voltage Correlation (Includes Hybrid Models):** The difference between the throttle valve positions indicated by TP sensor A and TP sensor B exceeds the value as follows, for at least 200 milliseconds. Difference between TP sensor A and TP sensor B 1.8-14.7° or more, or 5° or less. **NOTE: Before you troubleshoot, record all freeze data and any on-board snapshot.** **Possible Causes:** • Poor connections or loose terminals at the throttle body and the PCM • "Short" circuit between the PCM (TPSA line) and the (TPSB line) • Faulty throttle body • PCM may need to be updated with the latest software • Faulty PCM
DTC: P2138 **1T PCM, MIL: Yes** **Year:** 2009, 2010 **Model:** Accord, Accord Crosstour, Civic, CR-V, Element, Fit, Odyssey, Pilot, Ridgeline **Engine:** 1.3L L4, 1.5L L4, 1.8L L4, 2.0L L4, 2.4L L4, 3.5L V6 **Transmission:** All	**APP Sensor A/B or 1/2 (Throttle Position (TP) Sensor D/E) Incorrect Voltage Correlation (Includes Hybrid Models):** With the ignition ON (II) one of these conditions must be met for at least 300 milliseconds: (1) APP sensor B voltage exceeds the range from 0V or less to 0.37V or more when the APP sensor A voltage is 0.37V. (2) APP sensor B voltage exceeds the range from 2.31V or less to 2.69V or more when the APP sensor A voltage is 5V. The execution time is continuous and DTCs P2122, P2123, P2127, P2128 are not active. **NOTE: Before you troubleshoot, record all freeze data and any on-board snapshot.** **Possible Causes:** • Poor connections or loose terminals at the APP sensor and the PCM • "Short" circuit between PCM (APSA line) and (APSB line) • Faulty accelerator pedal module • PCM needs to be updated with the latest software • Faulty PCM
DTC: P2176 **1T PCM, MIL: Yes** **Year:** 2009, 2010 **Model:** Accord, Accord Crosstour, Civic, CR-V, Element, Fit, Odyssey, Pilot, Ridgeline **Engine:** 1.3L L4, 1.5L L4, 1.8L L4, 2.0L L4, 2.4L L4, 3.5L V6 **Transmission:** All	**Throttle Actuator Control System Idle Position Not Learned (Includes Hybrid Models):** One of these condition must be met for at least 0.7 seconds. (1) The registration of the throttle valve fully closed position is not completed within a predetermined time after the ignition switch is turned ON. (2) The registered value of the throttle valve fully closed position is 0.74 volts TP1, 1.61 volts TP2 or more, or 0.49 volts TP1, 1.37 volts TP2 or less. The execution time is once per driving cycle and DTCs P0122, P0123, P0222, P0223, P2101, P2118, P2135 are not active. **NOTE: If DTC P2135 is stored at the same time as DTC P2176, troubleshoot DTC P2135 first, then recheck for DTC P2176.** **Possible Causes:** • Dirty throttle body • Poor connections or loose terminals at the throttle body and the PCM • "Open" circuit between the throttle body and the PCM • Faulty throttle body • PCM may need to be updated with the latest software • Faulty PCM
DTC: P2183 **2T PCM, MIL: Yes** **Year:** 2009, 2010 **Model:** Accord, Accord Crosstour, Civic, Element, Fit, Pilot, Ridgeline **Engine:** 1.3L L4, 1.5L L4, 1.8L L4, 2.0L L4, 2.4L L4, 3.5L V6 **Transmission:** All	**Engine Coolant Temperature (ECT) Sensor 2 Circuit Range/Performance Problem (Includes Hybrid Models):** A malfunction is detected if the following three conditions are present after the engine has stopped and the ignition switch has been turned to LOCK (0) for at least 6 hours before restarting the engine: (1) When the temperature difference between the IAT and ECT1 is 57°F (32°C) or more. (2) When the temperature difference between the IAT and ECT2 is 30°F (17°C) or more. (3) When the temperature difference between the ECT2 and ECT1 is 73°F (41°C) or more. The execution time is once per driving cycle and the duration time is 10 seconds or more. **NOTE: If DTC P0111 is stored at the same time as DTC P2183, troubleshoot DTC P0111 first, then recheck for DTC P2183.** **Possible Causes:** • Poor connections or loose terminals at ECT sensor 1, ECT sensor 2, and the PCM • Faulty ECT sensor 1 • Faulty ECT sensor 2
DTC: P2184 **1T PCM, MIL: Yes** **Year:** 2009, 2010 **Model:** Accord, Accord Crosstour, Civic, CR-V, Fit, Odyssey, Pilot, Ridgeline **Engine:** 1.3L L4, 1.5L L4, 1.8L L4, 2.0L L4, 2.4L L4, 3.5L V6 **Transmission:** All	**Engine Coolant Temperature (ECT) Sensor 2 Circuit Low Voltage (Includes Hybrid Models):** With ignition switch ON (II) the ECT sensor 2 output voltage is 0.08V or less for at least 2 seconds. The execution time is continuous. DTC P2185 is not active. **NOTE: Before you troubleshoot, record all freeze data and any on-board snapshot.** **Possible Causes:** • Poor connections or loose terminals at ECT sensor 2 and the PCM • "Short" in the wire between ECT sensor 2 and the PCM • Faulty ECT sensor 2 • PCM may need to be updated with the latest software • Faulty PCM

DTC	Trouble Code Title, Conditions & Possible Causes
DTC: P2185 **1T PCM, MIL: Yes** **Year:** 2009, 2010 **Model:** Accord, Accord Crosstour, Civic, CR-V, Element, Fit, Odyssey, Pilot, Ridgeline **Engine:** 1.3L L4, 1.5L L4, 1.8L L4, 2.0L L4, 2.4L L4, 3.5L V6 **Transmission:** All	**Engine Coolant Temperature (ECT) Sensor 2 Circuit High Voltage (Includes Hybrid Models):** With ignition ON (II) the ECT sensor 2 output voltage is 4.92 volts or more for at least 2 seconds. The execution time is continuous and DTC P2184 is not active. **NOTE: Before you troubleshoot, record all freeze data and any on-board snapshot.** **Possible Causes:** • Poor connections or loose terminals at ECT sensor 2 and the PCM • "Open" in the wire between the PCM and ECT sensor 2 • Faulty ECT sensor 2 • PCM may need to be updated with the latest software • Faulty PCM
DTC: P2195 **2T PCM, MIL: Yes** **Year:** 2009, 2010 **Model:** Accord, Accord Crosstour, Odyssey, Pilot, Ridgeline **Engine:** 3.5L V6 **Transmission:** All	**Rear Air/Fuel Ratio (A/F) Sensor (Bank 1, Sensor 1) Signal Stuck Lean:** The rear A/F sensor (bank 1, sensor 1) output voltage is 3.48 V or more for at least 7 seconds. The execution time is once per driving cycle and DTCs P0134, P0135, P0171, P0300, P0301, P0302, P0303, P0304, P0305, P0306, P0627, P1172, P2237, P2238, P2243, P2245, P2251, P2252 are not active. **NOTE: If DTC P2101, P2118, P2135, P2138, P2176, or a combination of P2122 and P2127, P2122 and P2138 or P2127 and P2138 is stored at the same time as DTC P2195, troubleshoot those DTCs first, then recheck for DTC P2195.** **Possible Causes:** • Dirty or Faulty MAF sensor/IAT sensor (If equipped) • Loose A/F sensor • Poor connections or loose terminals at the A/F sensor (Sensor 1) and the PCM • Faulty A/F sensor (Sensor 1) • PCM may need to be updated with the latest software • Faulty PCM
DTC: P2195 **1T PCM, MIL: Yes** **Year:** 2009, 2010 **Model:** Accord, Civic, CR-V, Element, Fit **Engine:** 1.3L L4, 1.5L L4, 1.8L L4, 2.0L L4, 2.4L L4 **Transmission:** All	**A/F Sensor (Sensor 1) Signal Stuck Lean (Includes Hybrid Models):** The A/F sensor (Sensor 1) output voltage is 2.5 V or more for at least 7 seconds. The execution time is once per driving cycle and DTCs P0134, P0135, P0171, P0300, P0301, P0302, P0303, P0304, P0305, P0306, P0627, P1172, P2237, P2238, P2243, P2245, P2251, P2252 are not active. **NOTE: If DTC P2101, P2118, P2135, P2138, P2176, or a combination of P2122 and P2127, P2122 and P2138 or P2127 and P2138 is stored at the same time as DTC P2195, troubleshoot those DTCs first, then recheck for DTC P2195.** **Possible Causes:** • Dirty or Faulty MAF sensor/IAT sensor (If equipped) • Loose A/F sensor • Poor connections or loose terminals at the A/F sensor (Sensor 1) and the PCM • Faulty A/F sensor (Sensor 1) • PCM may need to be updated with the latest software • Faulty PCM
DTC: P2197 **2T PCM, MIL: Yes** **Year:** 2009, 2010 **Model:** Accord, Accord Crosstour, Odyssey, Pilot, Ridgeline **Engine:** 3.5L V6 **Transmission:** All	**Front Air/Fuel Ratio (A/F) Sensor (Bank 2, Sensor 1) Signal Stuck Lean:** The front A/F sensor (bank 2, sensor 1) output voltage is 3.48 V or more for at least 7 seconds. The execution time is once per driving cycle and DTCs P0154, P0155, P0174, P0300, P0301, P0302, P0303, P0304, P0305, P0306, P0627, P1174, P2240, P2241, P2247, P2249, P2254, P2255 are not active. **NOTE: If DTC P2101, P2118, P2135, P2138, P2176, or a combination of P2122 and P2127, P2122 and P2138 or P2127 and P2138 is stored at the same time as DTC P2195, P2197, troubleshoot those DTCs first, then recheck for DTC P2197.** **Possible Causes:** • Loose A/F sensor • Poor connections or loose terminals at A/F sensor (Sensor 1) and the PCM • Faulty A/F sensor (Sensor 1) • PCM may need to be updated with the latest software • Faulty PCM
DTC: P2227 **2T PCM, MIL: Yes** **Year:** 2009, 2010 **Model:** Accord, Accord Crosstour, Civic, CR-V, Element, Fit, Odyssey, Pilot, Ridgeline **Engine:** 1.3L L4, 1.5L L4, 1.8L L4, 2.0L L4, 2.4L L4, 3.5L V6 **Transmission:** All	**Barometric Pressure (BARO) Sensor Circuit Range/Performance Problem:** Throttle position 17.0-33.0 degrees at 1,100-3,000 rpm. The difference between the BARO sensor output and the MAP sensor output is 26 kPa (7.5 in.Hg, 190 mmHg) or more for at least 2.5 seconds. The execution time is once per driving cycle and the duration time is 2.5 seconds or more. **NOTE: If DTC P0107, P0108, P1128, and/or P1129 are stored at the same time as DTC P2227, troubleshoot those DTCs first, then recheck for DTC P2227.** **Possible Causes:** • Dirty air cleaner element • Faulty BARO sensor • PCM needs to be updated with the latest software • Faulty PCM

DTC	Trouble Code Title, Conditions & Possible Causes
DTC: P2228 **1T PCM, MIL: Yes** **Year:** 2009, 2010 **Model:** Accord, Accord Crosstour, Civic, CR-V, Element, Fit, Odyssey, Pilot, Ridgeline **Engine:** 1.3L L4, 1.5L L4, 1.8L L4, 2.0L L4, 2.4L L4, 3.5L V6 **Transmission:** All	**Barometric Pressure (BARO) Sensor Circuit Low Voltage:** With ignition ON (II) the BARO sensor output voltage is 1.31 V or less for at least 2 seconds. The execution time is continuous and DTCs P1109, P2229 are not active. **NOTE: Before you troubleshoot, record all freeze data and any on-board snapshot.** **Possible Causes:** • Poor connections or loose terminals at the PCM • Faulty BARO sensor • PCM needs to be updated with the latest software • Faulty PCM
DTC: P2229 **1T PCM, MIL: Yes** **Year:** 2009, 2010 **Model:** Accord, Accord Crosstour, Civic, CR-V, Element, Fit, Odyssey, Pilot, Ridgeline **Engine:** 1.3L L4, 1.5L L4, 1.8L L4, 2.0L L4, 2.4L L4, 3.5L V6 **Transmission:** All	**Barometric Pressure (BARO) Sensor Circuit High Voltage:** With ignition switch ON (II) the BARO sensor output voltage is 4.49 V or more for at least 2 seconds. The execution time is continuous and DTCs P1109, P2228 are not active. **NOTE: Before you troubleshoot, record all freeze data and any on-board snapshot.** **Possible Causes:** • Faulty BARO sensor • PCM may need to be updated with the latest software • Faulty PCM
DTC: P2237 **1T PCM, MIL: Yes** **Year:** 2009, 2010 **Model:** Accord, Accord Crosstour, Odyssey, Pilot, Ridgeline **Engine:** 3.5L V6 **Transmission:** All	**Rear Air/Fuel Ratio (A/F) Sensor (Bank 1, Sensor 1) IP Circuit High Voltage:** With the engine running the IPB1 terminal voltage is 2 volts or less, or 5.6 volts or more, for at least 15 seconds. The execution time is continuous and DTCs P0135, P2195, P2238, P2243, P2245, P2251, P2252, P2627, P2628 are not active. **NOTE: Before you troubleshoot, record all freeze data and any on-board snapshot.** **Possible Causes:** • Poor connections or loose terminals at the A/F sensor (Sensor 1) and the PCM • "Open" circuit between the PCM and the A/F sensor (Sensor 1) • Faulty A/F sensor (Sensor 1) • PCM may need to be updated with the latest software • Faulty PCM
DTC: P2238 **1T PCM, MIL: Yes** **Year:** 2009, 2010 **Model:** Accord, Accord Crosstour, Odyssey, Pilot, Ridgeline **Engine:** 3.5L V6 **Transmission:** All	**Rear Air/Fuel Ratio (A/F) Sensor (Bank 1, Sensor 1) IP Circuit Low Voltage:** With engine running the IPB1 input terminal voltage is 1 volts or less for at least 5 seconds after the sensor becomes active and 85 seconds before the sensor becomes active. The execution time is continuous, DTCs P0135, P2195, P2237, P2243, P2245, P2251, P2252, P2627, P2628 are not active. **NOTE: Before you troubleshoot, record all freeze data and any on-board snapshot.** **Possible Causes:** • Poor connections or loose terminals at the A/F sensor (Sensor 1) and the PCM • "Short" circuit between the PCM and the A/F sensor (Sensor 1) • Faulty A/F sensor (Sensor 1) • PCM needs to be updated with the latest software • Faulty PCM
DTC: P2238 **1T PCM, MIL: Yes** **Year:** 2009, 2010 **Model:** Accord, Civic, CR-V, Element, Fit **Engine:** 1.3L L4, 1.5L L4, 1.8L L4, 2.0L L4, 2.4L L4 **Transmission:** All	**Air/Fuel Ratio (A/F) Sensor (Sensor 1) AFS+ Circuit Low Voltage:** With the engine running and battery voltage at a minimum of 10.5 V, the AFS+ voltage is 0.4 V or less for at least 4 seconds. **NOTE: Before you troubleshoot, record all freeze data and any on-board snapshot, and review the general troubleshooting information.** **Possible Causes:** • Poor connections or loose terminals at the A/F sensor (Sensor 1) and the PCM • "Short" circuit between the PCM and the A/F sensor (Sensor 1) • Faulty A/F sensor (Sensor 1) • PCM may need to be updated with the latest software • Faulty PCM
DTC: P2240 **1T PCM, MIL: Yes** **Year:** 2009, 2010 **Model:** Accord, Accord Crosstour, Odyssey, Pilot, Ridgeline **Engine:** 3.5L V6 **Transmission:** All	**Front Air/Fuel Ratio (A/F) Sensor (Bank 2, Sensor 1) IP Circuit High Voltage:** With the engine running the IPB2 terminal voltage is 5.6 volts or more, or 2 volts or less, for at least 15 seconds. The execution time is continuous and DTCs P0155, P2197, P2240, P2241, P2247, P2249, P2254, P2255, P2630, P2631 are not active. **NOTE: Before you troubleshoot, record all freeze data and any on-board snapshot.** **Possible Causes:** • Poor connections or loose terminals at the A/F sensor (Sensor 1) • "Open" circuit between the PCM and the A/F sensor (Sensor 1) • Faulty A/F sensor (Sensor 1) • PCM needs to be updated with the latest software • Faulty PCM

DTC	Trouble Code Title, Conditions & Possible Causes
DTC: P2241 **1T PCM, MIL: Yes** **Year:** 2009, 2010 **Model:** Accord, Accord Crosstour, Odyssey, Pilot, Ridgeline **Engine:** 3.5L V6 **Transmission:** All	**Front Air/Fuel Ratio (A/F) Sensor (Bank 2, Sensor 1) IP Circuit Low Voltage:** With engine running the IPB1 input terminal voltage is 1 volts or less for at least 5 seconds after the sensor becomes active and 85 seconds before the sensor becomes active. The execution time is continuous DTCs P0135, P2195, P2237, P2243, P2245, P2251, P2252, P2627, P2628 are not active. **NOTE: Before you troubleshoot, record all freeze data and any on-board snapshot.** **Possible Causes:** • Poor connections or loose terminals at the A/F sensor (Sensor 1) and the PCM • "Short" circuit between the PCM and the A/F sensor (Sensor 1) • Faulty A/F sensor (Sensor 1) • PCM needs to be updated with the latest software • Faulty PCM
DTC: P2243 **1T PCM, MIL: Yes** **Year:** 2009, 2010 **Model:** Accord, Accord Crosstour, Odyssey, Pilot, Ridgeline **Engine:** 3.5L V6 **Transmission:** All	**Rear Air/Fuel Ratio (A/F) Sensor (Bank 1, Sensor 1) VCENT Circuit High Voltage:** With engine running the VSB1 terminal voltage repeatedly fluctuates from a value above 4.8 volts to a value below 3.4 volts, at least 150 times. The execution time is continuous and DTCs P0135, P2195, P2237, P2238, P2245, P2251, P2252, P2627, P2628 are not active. **NOTE: Before you troubleshoot, record all freeze data and any on-board snapshot.** **Possible Causes:** • Poor connections or loose terminals at the A/F sensor (Sensor 1) and the PCM • "Open" circuit between the PCM and the A/F sensor (Sensor 1) • Faulty A/F sensor (Sensor 1) • PCM needs to be updated with the latest software • Faulty PCM
DTC: P2245 **1T PCM, MIL: Yes** **Year:** 2009, 2010 **Model:** Accord, Accord Crosstour, Odyssey, Pilot, Ridgeline **Engine:** 3.5L V6 **Transmission:** All	**Rear Air/Fuel Ratio (A/F) Sensor (Bank 1, Sensor 1) VCENT Circuit Low Voltage:** With engine running the IPB1 input terminal voltage is 1 volts or less for at least 5 seconds after the sensor becomes active, 85 seconds before the sensor becomes active. The execution time is continuous and DTCs P0135, P2195, P2237, P2238, P2243, P2251, P2252, P2627, P2628 are not active. **Possible Causes:** • Poor connections or loose terminals at the A/F sensor (Sensor 1) and the PCM • "Short" circuit between the PCM and the A/F sensor (Sensor 1) • Faulty A/F sensor (Sensor 1) • PCM needs to be updated with the latest software • Faulty PCM
DTC: P2247 **1T PCM, MIL: Yes** **Year:** 2009, 2010 **Model:** Accord, Accord Crosstour, Odyssey, Pilot, Ridgeline **Engine:** 3.5L V6 **Transmission:** All	**Front A/F Sensor (Bank 2, Sensor 1) VCENT Circuit High Voltage :** With engine running the VSB1 terminal voltage repeatedly fluctuates from a value above 4.8 volts to a value below 3.4 volts, at least 150 times. The execution time is continuous and DTCs P0135, P2195, P2237, P2238, P2245, P2251, P2252, P2627, P2628 are not active. **NOTE: Before you troubleshoot, record all freeze data and any on-board snapshot.** **Possible Causes:** • Poor connections or loose terminals at the A/F sensor (Sensor 1) and the PCM • "Open" circuit between the PCM and the A/F sensor (Sensor 1) • Faulty A/F sensor (Sensor 1) • PCM needs to be updated with the latest software • Faulty PCM
DTC: P2249 **1T PCM, MIL: Yes** **Year:** 2009, 2010 **Model:** Accord, Accord Crosstour, Odyssey, Pilot, Ridgeline **Engine:** 3.5L V6 **Transmission:** All	**Front A/F Sensor (Bank 2, Sensor 1) VCENT Circuit Low Voltage :** With the engine running the VSB2 terminal voltage between reads 0.3-1.5 volts. The execution time is continuous and DTCs P0155, P2197, P2240, P2241, P2247, P2254, P2255 are not active. The IPB2 input terminal voltage is 1.0 volts or less for at least 5 seconds, after the sensor becomes active, 85 seconds before the sensor becomes active. **NOTE: Before you troubleshoot, record all freeze data and any on-board snapshot.** **Possible Causes:** • Poor connections or loose terminals at the A/F sensor (Sensor 1) and the PCM • "Short" circuit between the PCM and the A/F sensor (Sensor 1) • Faulty A/F sensor (Sensor 1) • PCM needs to be updated with the latest software • Faulty PCM

DTC	Trouble Code Title, Conditions & Possible Causes
DTC: P2251 **1T PCM, MIL: Yes** **Year:** 2009, 2010 **Model:** Accord, Accord Crosstour, Odyssey, Pilot, Ridgeline **Engine:** 3.5L V6 **Transmission:** All	**Rear Air/Fuel Ratio (A/F) Sensor (Bank 1, Sensor 1) VS Circuit High Voltage:** With engine running the VSB1 terminal voltage is 6 V or more, and the PCM internal signal voltage is 4.6 V or more, for at least 5 seconds. The execution time is continuous and DTCs P0135, P2195, P2237, P2238, P2243, P2245, P2252, P2627, P2628 are not active. **NOTE: If DTC P2251 is stored at the same time as DTC P0134, troubleshoot DTC P2251 first, then recheck for P0134.** **Possible Causes:** • Poor connections or loose terminals at the A/F sensor (Sensor 1) and the PCM • "Open" circuit between the PCM and the A/F sensor (Sensor 1) • Faulty A/F sensor (Sensor 1) • PCM may need to be updated with the latest software • Faulty PCM
DTC: P2252 **1T PCM, MIL: Yes** **Year:** 2009, 2010 **Model:** Accord, Accord Crosstour, Odyssey, Pilot, Ridgeline **Engine:** 3.5L V6 **Transmission:** All	**Rear Air/Fuel Ratio (A/F) Sensor (Bank 1, Sensor 1) VS Circuit Low Voltage:** With engine running the VSB1 terminal voltage is 0.3 V or less, and the IPB1 terminal voltage is 1 V or more, for at least 5 seconds after the sensor becomes active, 85 seconds before the sensor becomes active. **Possible Causes:** • Poor connections or loose terminals at the A/F sensor (Sensor 1) and the PCM • "Short" circuit between the PCM and the A/F sensor (Sensor 1) • Faulty A/F sensor (Sensor 1) • PCM may need to be updated with the latest software • Faulty PCM
DTC: P2252 **1T PCM, MIL: Yes** **Year:** 2009, 2010 **Model:** Accord, Civic, CR-V, Element, Fit **Engine:** 1.3L L4, 1.5L L4, 1.8L L4, 2.0L L4, 2.4L L4 **Transmission:** All	**Air/Fuel Ratio (A/F) Sensor (Sensor 1) AFS- Circuit Low Voltage :** With the engine running the AFS-terminal voltage is 0.4 V or less for at least 5 seconds. **NOTE: Before you troubleshoot, record all freeze data and any on-board snapshot, and review the general troubleshooting information.** **Possible Causes:** • Poor connections or loose terminals at the A/F sensor (Sensor 1) and the ECM/PCM • "Short" in the wire between the ECM/PCM and the A/F sensor (Sensor 1) • Faulty A/F sensor (Sensor 1) • PCM may need to be updated with the latest software • Faulty ECM/PCM
DTC: P2254 **1T PCM, MIL: Yes** **Year:** 2009, 2010 **Model:** Accord, Accord Crosstour, Odyssey, Pilot, Ridgeline **Engine:** 3.5L V6 **Transmission:** All	**Front A/F Sensor (Bank 2, Sensor 1) VS Circuit High Voltage:** With engine running the VSB2 terminal voltage is 6 V or more, and the PCM internal signal voltage is 4.6 V or more, for at least 5 seconds. The execution time is continuous and DTCs P0155, P2197, P2240, P2241, P2247, P2249, P2255, P2630, P2631 are not active. **NOTE: If DTC P2254 is stored at the same time as DTC P0154, troubleshoot DTC P2254 first, then recheck for P0154.** **Possible Causes:** • Poor connections or loose terminals at the A/F sensor (Sensor 1) and the PCM • "Open" circuit between the PCM and the A/F sensor (Sensor 1) • Faulty A/F sensor (Sensor 1) • PCM may need to be updated with the latest software • Faulty PCM
DTC: P2255 **1T PCM, MIL: Yes** **Year:** 2009, 2010 **Model:** Accord, Accord Crosstour, Odyssey, Pilot, Ridgeline **Engine:** 3.5L V6 **Transmission:** All	**Front A/F Sensor (Bank 2, Sensor 1) VS Circuit Low Voltage :** With engine running the VSB2 terminal voltage is 0.3 V or less, and the IPB2 terminal voltage is 1 V or more, for at least 5 seconds after the sensor is active, 85 seconds before the sensor becomes active. **NOTE: Before you troubleshoot, record all freeze data and any on-board snapshot.** **Possible Causes:** • Poor connections or loose terminals at the A/F sensor (Sensor 1) and the PCM • "Short" circuit between the PCM and the A/F sensor (Sensor 1) • Faulty A/F sensor (Sensor 1) • PCM may need to be updated with the latest software • Faulty PCM
DTC: P2270 **2T PCM, MIL: Yes** **Year:** 2009, 2010 **Model:** Accord, Civic, CR-V, Element **Engine:** 1.3L L4, 1.8L L4, 2.0L L4, 2.4L L4 **Transmission:** All	**Secondary HO2S (Sensor 2) Circuit Signal Stuck Lean:** With the engine running, and in closed loop with a speed of 30 mph or more, the secondary HO2S output voltage is 0.650 V or less. **NOTE: Before you troubleshoot, record all freeze data and any on-board snapshot, and review the general troubleshooting information.** **Possible Causes:** • Poor connections or loose terminals at the secondary HO2S (Sensor 2) and the PCM • Faulty secondary HO2S (Sensor 2)

DTC	Trouble Code Title, Conditions & Possible Causes
DTC: P2270 **2T PCM, MIL: Yes** **Year:** 2009, 2010 **Model:** Accord, Accord Crosstour, Odyssey, Pilot, Ridgeline **Engine:** 3.5L V6 **Transmission:** All	**Rear Secondary Heated Oxygen Sensor (Secondary HO2S (Bank 1, Sensor 2) Circuit Signal Stuck Lean:** The rear secondary HO2S (bank 1, sensor 2) output voltage is 0.650 V or less. The execution time is once per driving cycle and the duration time is 36.3 seconds or less. **NOTE: Before you troubleshoot, record all freeze data and any on-board snapshot.** **Possible Causes:** • Poor connections or loose terminals at the secondary HO2S (Sensor 2) and the PCM • Faulty secondary HO2S (Sensor 2)
DTC: P2271 **2T PCM, MIL: Yes** **Year:** 2009, 2010 **Model:** Accord, Accord Crosstour, Odyssey, Pilot, Ridgeline **Engine:** 3.5L V6 **Transmission:** All	**Rear Secondary HO2S (Bank 1, Sensor 2) Circuit Signal Stuck Rich :** If, after current is applied to the rear secondary HO2S heater, the rear secondary HO2S does not fluctuate and the output is stuck within the specified area, a malfunction is detected. The rear secondary HO2S (bank 1, sensor 2) output voltage is 0.293 V or more. The execution time is once per driving cycle and the duration time is 36.3 seconds or more. **NOTE: Before you troubleshoot, record all freeze data and any on-board snapshot.** **Possible Causes:** • Poor connections or loose terminals at the secondary HO2S (Sensor 2) and the PCM • Faulty secondary HO2S (Sensor 2)
DTC: P2271 **2T PCM, MIL: Yes** **Year:** 2009, 2010 **Model:** Accord, Civic, CR-V, Element, Fit **Engine:** 1.3L L4, 1.5L L4, 1.8L L4, 2.0L L4, 2.4L L4 **Transmission:** All	**Secondary HO2S (Sensor 2) Circuit Signal Stuck Rich :** The secondary HO2S (Sensor 2) output voltage is 0.293 V or more. The execution time is once per driving cycle and the duration time is 24.2 seconds or more. **NOTE: Before you troubleshoot, record all freeze data and any on-board snapshot.** **Possible Causes:** • Poor connections or loose terminals at the secondary HO2S (Sensor 2) and the PCM • Faulty secondary HO2S (Sensor 2)
DTC: P2272 **2T PCM, MIL: Yes** **Year:** 2009, 2010 **Model:** Accord, Accord Crosstour, Odyssey, Pilot, Ridgeline **Engine:** 3.5L V6 **Transmission:** All	**Front Secondary HO2S (Bank 2, Sensor 2) Circuit Signal Stuck Lean :** If, after current is applied to the front secondary HO2S heater, the front secondary HO2S does not fluctuate and the output is stuck within the specified area, a malfunction is detected. The front secondary HO2S (bank 2, sensor 2) output voltage is 0.650 V or less. **Possible Causes:** • Poor connections or loose terminals at the secondary HO2S (Sensor 2) and the PCM • Faulty secondary HO2S (Sensor 2)
DTC: P2273 **2T PCM, MIL: Yes** **Year:** 2009, 2010 **Model:** Odyssey, Pilot, Ridgeline **Engine:** 3.5L V6 **Transmission:** All	**Front Secondary HO2S (Bank 2, Sensor 2) Circuit Signal Stuck Rich :** If, after current is applied to the front secondary HO2S heater, the front secondary HO2S does not fluctuate and the output is stuck within the specified area, a malfunction is detected. The front secondary HO2S (bank 2, sensor 2) output voltage is 0.293 V or more. **NOTE: Before you troubleshoot, record all freeze data and any on-board snapshot.** **Possible Causes:** • Poor connections or loose terminals at the secondary HO2S (Sensor 2) and the PCM • Faulty secondary HO2S (Sensor 2)
DTC: P2279 **2T PCM, MIL: Yes** **Year:** 2009, 2010 **Model:** Accord, Accord Crosstour, Civic, CR-V, Odyssey, Pilot, Ridgeline **Engine:** 1.3L L4, 2.4L L4, 3.5L V6 **Transmission:** All	**Intake Air System Leak:** With the engine running in closed loop for a minimum of 15 seconds. Either of these 2 conditions is met: (1) The estimated volume of intake air is 310 l/min (327.6 US qt/min, 272.8 Imp qt/min) or more when the MAP value is 35 kPa (10.3 in. Hg, 260 mmHg). (2) The estimated volume of intake air is 302 l/min (319.2 US qt/min, 265.8 Imp qt/min) or more when the MAP value is 62 kPa (18.2 in. Hg, 460 mmHg). The execution time is once per driving cycle and the duration time is 22 seconds or more. **NOTE: If DTC P0443 is stored at the same time as DTC P2279, troubleshoot DTC P0443 first, then recheck for DTC P2279.** **Possible Causes:** • Vacuum leaks at the PCV valve, the PCV hose, the purge (PCS) line, the throttle body, the intake manifold, and the brake booster hose • Incorrect camshaft timing

DTC	Trouble Code Title, Conditions & Possible Causes
DTC: P2413 **2T PCM, MIL: Yes** **Year:** 2009, 2010 **Model:** Accord, Accord Crosstour, Civic, CR-V, Fit, Odyssey, Pilot, Ridgeline **Engine:** 1.3L L4, 1.5L L4, 1.8L L4, 2.4L L4, 3.5L V6 **Transmission:** All	**Exhaust Gas Recirculation (EGR) System Malfunction:** With the engine running at 4,400 rpm, commanded EGR valve lift is 0.040 in. (1 mm). If the actual valve lift is 0.006 in. (0.15 mm) or less for at least 5 seconds, the valve is considered stuck closed. **NOTE: Before you troubleshoot, record all freeze data and any on-board snapshot.** **Possible Causes:** • Clogged intake manifold EGR port or EGR valve • Poor connections or loose terminals at the EGR valve and the PCM • "Open" in the wire between the EGR valve and ground • "Open" or "Short" circuit between the PCM and the EGR valve • Faulty EGR valve • PCM needs to be updated with the latest software • Faulty PCM
DTC: P2422 **2T PCM, MIL: Yes** **Year:** 2009, 2010 **Model:** Accord, Accord Crosstour, CR-V, Element, Fit, Odyssey, Pilot, Ridgeline **Engine:** 1.5L L4, 2.4L L4, 3.5L V6 **Transmission:** All	**EVAP Canister Vent Shut Valve Close Malfunction :** The output from the fuel tank pressure sensor is -4 kPa (-1.0 in. Hg, -25 mmHg), -2 kPa (-0.4 in. Hg, -10 mmHg) or less for at least 1.04-8 seconds. Elapsed time after the FTP sensor output exceeds the malfunction threshold 1.04 seconds. Excessive negative pressure is detected 8 seconds. **Possible Causes:** • Poor connections or loose terminals at the FTP sensor, the EVAP canister vent shut valve, and the PCM • Blockage in the EVAP canister, canister filter, vent hoses, and drain joint, • Blockage in the FTP sensor air tube or vent • Faulty FTP sensor • Faulty EVAP canister vent shut valve
DTC: P2446 **1T PCM, MIL: Yes** **Year:** 2009, 2010 **Model:** Odyssey **Engine:** 3.5L V6 **Transmission:** All	**Rocker Arm Oil Pressure Switch Circuit Low Voltage (J35A6 Engine):** With the engine running, engine speed a minimum of 4,400 rpm (Variable that is depending on the engine load). The rocker arm oil control solenoid is ON, the rocker arm oil pressure switch remains ON. The execution time is once per driving cycle and DTCs P0522, P0523, P2648, P2649 are not active. **NOTE: Before you troubleshoot, record all freeze data and any on-board snapshot.** **Possible Causes:** • Low engine oil or faulty oil pressure • Poor connections or loose terminals at the rocker arm oil pressure switch, the rocker arm oil control solenoid, and the PCM • "Short" circuit between the PCM and the rocker arm oil pressure switch • Faulty rocker arm oil pressure switch • Faulty rocker arm oil control solenoid • PCM may need to be updated with the latest software • Faulty PCM
DTC: P2552 **1T PCM, MIL: Yes** **Year:** 2009, 2010 **Model:** Element **Engine:** 2.4L L4 **Transmission:** All	**Throttle Actuator Control Module Relay Malfunction :** The serial signal is input from the throttle actuator control module for at least 2 seconds after the throttle actuator control module relay is turned OFF. **NOTE: Before you troubleshoot, record all freeze data and any on-board snapshot.** **Possible Causes:** • Poor connections or loose terminals at the throttle actuator control module relay, the throttle actuator control module, and the PCM • Faulty throttle actuator control module relay • "Short" circuit between the throttle actuator control module relay and the PCM • PCM may need to be updated with the latest software • Faulty PCM
DTC: P2610 **1T PCM, MIL: Yes** **Year:** 2009, 2010 **Model:** Accord, Accord Crosstour, Civic, CR-V, Element, Fit, Odyssey, Pilot, Ridgeline **Engine:** 1.3L L4, 1.5L L4, 1.8L L4, 2.0L L4, 2.4L L4, 3.5L V6 **Transmission:** All	**ECM/PCM Ignition Off Internal Timer Malfunction:** The access process to the ignition off timer fails, or a malfunction is found in the read data for at least 10 seconds. Ignition switch ON (II) when a battery is disconnected and connected again is excluded. **NOTE: Before you troubleshoot, record all freeze data and any on-board snapshot.** **Possible Causes:** • ECM/PCM needs to be updated with the latest software • Faulty ECM/PCM

DTC	Trouble Code Title, Conditions & Possible Causes
DTC: P2627 **1T PCM, MIL: Yes** **Year:** 2009, 2010 **Model:** Accord, Pilot, Ridgeline **Engine:** 3.5L V6 **Transmission:** All	**Rear Air/Fuel Ratio (A/F) Sensor (Bank 1, Sensor 1) LABEL Circuit Low Voltage:** With the engine running at 3,000 rpm for 2 minutes the VLBLB1 is 0.4 V or less for at least 5 seconds. **Possible Causes:** • Poor connections or loose terminals at the A/F sensor (Sensor 1) and the PCM • "Short" between the A/F sensor relay and each connector • Faulty A/F sensor (Sensor 1) • PCM needs to be updated with the latest software • Faulty PCM
DTC: P2627 **1T PCM, MIL: Yes** **Year:** 2009, 2010 **Model:** Odyssey **Engine:** 3.5L V6 **Transmission:** All	**VTEC System Stuck ON (J35A7 Engine):** With the engine running, engine speed between 1,500-3,000 rpm, the rocker arm oil control solenoid is ON, the EOP sensor output remains HIGH (39 kPa 11.6 in. Hg, 293 mmHg) or more. The execution time is once per driving cycle and DTCs P0522, P0523, P2227, P2228, P2229 are not active. **NOTE: Before you troubleshoot, record all freeze data and any on-board snapshot.** **Possible Causes:** • Low engine oil or faulty oil pressure • Poor connections or loose terminals at the rocker arm oil pressure switch, the rocker arm oil control solenoid, and the PCM • Faulty rocker arm oil pressure switch • Faulty rocker arm oil control solenoid
DTC: P2628 **1T PCM, MIL: Yes** **Year:** 2009, 2010 **Model:** Accord, Pilot, Ridgeline **Engine:** 3.5L V6 **Transmission:** All	**Rear Air/Fuel Ratio (A/F) Sensor (Bank 1, Sensor 1) LABEL Circuit High Voltage:** With the engine running at 3,000 rpm for 2 minutes the VLBLB1 is 4.69 V or more for at least 5 seconds. **Possible Causes:** • Poor connections or loose terminals at the A/F sensor (Sensor 1) and the PCM • "Open" between the A/F sensor relay and each connector • "Open" ground circuit • Faulty A/F sensor (Sensor 1) • PCM needs to be updated with the latest software • Faulty PCM
DTC: P2630 **1T PCM, MIL: Yes** **Year:** 2009, 2010 **Model:** Accord, Pilot, Ridgeline **Engine:** 3.5L V6 **Transmission:** All	**Front Air/Fuel Ratio (A/F) Sensor (Bank 2, Sensor 1) LABEL Circuit Low Voltage:** With the engine running at 3,000 rpm for 2 minutes the VLBLB2 is 0.3 V or less for at least 5 seconds. **Possible Causes:** • Poor connections or loose terminals at the A/F sensor (Sensor 1) and the PCM • "Short" between the A/F sensor relay and each connector • Faulty A/F sensor (Sensor 1) • PCM needs to be updated with the latest software • Faulty PCM
DTC: P2631 **1T PCM, MIL: Yes** **Year:** 2009, 2010 **Model:** Accord, Pilot, Ridgeline **Engine:** 3.5L V6 **Transmission:** All	**Front Air/Fuel Ratio (A/F) Sensor (Bank 2, Sensor 1) LABEL Circuit High Voltage:** With the engine running at 3,000 rpm for 2 minutes the VLBLB1 is 4.69 V or more for at least 5 seconds. **Possible Causes:** • Poor connections or loose terminals at the A/F sensor (Sensor 1) and the PCM • "Open" between the A/F sensor relay and each connector • "Open" ground circuit • Faulty A/F sensor (Sensor 1) • PCM needs to be updated with the latest software • Faulty PCM
DTC: P2646 **1T PCM, MIL: Yes** **Year:** 2009, 2010 **Model:** Civic **Engine:** 1.3L L4, 1.8L L4, 2.0L L4 **Transmission:** All	**VTEC System Stuck OFF :** With the engine running and at operating temperature, vehicle speed a minimum of 9 mph (14 km/h), the throttle plate anything but wide open, vehicle speed between 1,100-3,300 rpm, the rocker arm oil control solenoid is on and the EOP sensor remains (High). **Possible Causes:** • Low oil level • Poor connections or loose terminals at the rocker arm oil control solenoid, the rocker arm oil pressure sensor (EOP sensor), and the ECM/PCM • Faulty rocker arm Oil Pressure Sensor (EOP sensor) • Faulty rocker arm oil control valve • VTEC system oil line • Faulty rocker arm

DTC	Trouble Code Title, Conditions & Possible Causes
DTC: P2646 **1T PCM, MIL: Yes** **Year:** 2009, 2010 **Model:** Odyssey **Engine:** 3.5L V6 **Transmission:** All	**VTEC System Stuck OFF (J35A7 Engine):** With the engine running at idle, battery voltage a minimum of 10.5 V and DTCs P0522, P0523, P2227, P2228 and P2229 not active, the rocker arm oil control solenoid is OFF, the EOP sensor output remains LOW (39 kPa, 11.5 in. Hg, 292 mmHg) or less. **Possible Causes:** • Low engine oil or faulty oil pressure • Poor connections or loose terminals at the rocker arm oil pressure sensor, the rocker arm oil control solenoid, and the PCM • Faulty rocker arm oil pressure sensor • Faulty rocker arm oil control solenoid
DTC: P2646 **1T PCM, MIL: Yes** **Year:** 2009, 2010 **Model:** Accord, Accord Crosstour, CR-V, Element, Fit, Odyssey, Pilot, Ridgeline **Engine:** 1.5L L4, 2.4L L4, 3.5L V6 **Transmission:** All	**Rocker Arm Oil Pressure Switch (VTEC Oil Pressure Switch) Circuit Low Voltage :** With the engine running, engine speed a minimum of 4,800 rpm (Variable that is depending on the engine load). The rocker arm oil control solenoid is ON, the rocker arm oil pressure switch remains ON. The execution time is once per driving cycle and DTCs P2648, P2649 are not active. **NOTE: Before you troubleshoot, record all freeze data and any on-board snapshot.** **Possible Causes:** • Low engine oil or faulty oil pressure • Poor connections or loose terminals at the rocker arm oil pressure switch, the rocker arm oil control solenoid, and the PCM • "Short" circuit between the PCM and the rocker arm oil pressure switch • Faulty rocker arm oil pressure switch • Faulty rocker arm oil control solenoid • Faulty rocker arm
DTC: P2647 **1T PCM, MIL: Yes** **Year:** 2009, 2010 **Model:** Accord, Accord Crosstour, CR-V, Element, Fit, Pilot, Ridgeline **Engine:** 1.5L L4, 2.4L L4, 3.5L V6 **Transmission:** All	**Rocker Arm Oil Pressure Switch (VTEC Oil Pressure Switch) Circuit High Voltage :** When the rocker arm oil control solenoid is OFF, the rocker arm oil pressure switch remains OFF. (Low lift cam operation) **NOTE: Before you troubleshoot, record all freeze data and any on-board snapshot.** **Possible Causes:** • Low oil level • Low oil pressure • "Open" in the wire between the rocker arm oil pressure switch and ground • Poor connections or loose terminals at the rocker arm oil pressure switch, the rocker arm oil control solenoid, and the PCM • "Open" circuit between the PCM and the rocker arm oil pressure switch • Faulty rocker arm oil pressure switch • Faulty rocker arm oil control solenoid assembly • PCM needs to be updated with the latest software • Faulty PCM
DTC: P2647 **1T PCM, MIL: Yes** **Year:** 2009, 2010 **Model:** Odyssey **Engine:** 3.5L V6 **Transmission:** All	**Rocker Arm Oil Pressure Switch Circuit High Voltage (J35A6 Engine):** With the engine running at idle, and battery voltage a minimum of 10.5 V. The rocker arm oil control solenoid is OFF, the rocker arm oil pressure switch remains OFF. The execution time is once per driving cycle and DTCs P0522, P0523, P2648, P2649 are not active. **NOTE: Before you troubleshoot, record all freeze data and any on-board snapshot.** **Possible Causes:** • Low engine oil or faulty oil pressure • Poor connections or loose terminals at the rocker arm oil pressure switch, and the PCM • "Open" circuit between the PCM and the rocker arm oil pressure switch • Faulty rocker arm oil pressure switch • PCM may need to be updated with the latest software • Faulty PCM
DTC: P2647 **1T PCM, MIL: Yes** **Year:** 2009, 2010 **Model:** Civic **Engine:** 1.3L L4, 1.8L L4, 2.0L L4 **Transmission:** All	**VTEC System Stuck ON :** With the engine running for at least 2.5 seconds and battery voltage a minimum of 10.5 V, the rocker oil control solenoid is OFF and the EOP sensor remains (LOW). **Possible Causes:** • Low oil level • Low oil pressure • Poor connections or loose terminals at the rocker arm Oil Pressure Sensor (EOP sensor), the rocker arm oil control solenoid, and the ECM/PCM • Faulty rocker arm oil pressure sensor • Faulty rocker arm oil control valve

DTC	Trouble Code Title, Conditions & Possible Causes
DTC: P2648 **1T PCM, MIL: Yes** **Year:** 2009, 2010 **Model:** Odyssey **Engine:** 3.5L V6 **Transmission:** All	**Rocker Arm Oil Control Solenoid Circuit Low Voltage (J35A6 Engine):** With the vehicle driven in a lower gear at 4,400 rpm or more and battery voltage a minimum of 10.5 V, the return signal is OFF (LOW) for at least 2 seconds when the PCM outputs the ON (HIGH) signal to the rocker arm oil control solenoid. **Possible Causes:** • Poor connections or loose terminals at the rocker arm oil control solenoid, and the PCM • "Short" circuit between the PCM and the rocker arm oil control solenoid • Faulty rocker arm oil control solenoid • PCM may need to be updated with the latest software • Faulty PCM
DTC: P2648 **1T PCM, MIL: Yes** **Year:** 2009, 2010 **Model:** Accord, Accord Crosstour, Civic, CR-V, Element, Fit, Pilot, Ridgeline **Engine:** 1.3L L4, 1.5L L4, 1.8L L4, 2.0L L4, 2.4L L4, 3.5L V6 **Transmission:** All	**Rocker Arm Oil Control Solenoid (VTEC Solenoid Valve) Circuit Low Voltage :** The return signal is OFF (low) for at least 2 seconds when the PCM outputs the ON (high) signal to the rocker arm oil control solenoid. DTC P2649 is not active. **NOTE: Before you troubleshoot, record all freeze data and any on-board snapshot.** **Possible Causes:** • Poor connections or loose terminals at the rocker arm oil control solenoid and the PCM • "Short" circuit between the PCM and the rocker arm oil control solenoid • Faulty rocker arm oil control solenoid • PCM needs to be updated with the latest software • Faulty PCM
DTC: P2648 **1T PCM, MIL: Yes** **Year:** 2009, 2010 **Model:** Odyssey **Engine:** 3.5L V6 **Transmission:** All	**Rocker Arm Oil Control Solenoid A (Bank 1) Circuit Low Voltage (J35A7 Engine):** With the vehicle driven in a lower gear and engine speed at 3,300 rpm, battery voltage a minimum of 10.5 V, the return signal is OFF (LOW) for at least 1 second when the PCM outputs the ON (HIGH) signal to the rocker arm oil control solenoid A (Bank 1). **Possible Causes:** • Poor connections or loose terminals at rocker arm oil control solenoid A (Bank 1)and the PCM • "Short" circuit between rocker arm oil control solenoid A and the PCM • Faulty rear rocker arm oil control valve • PCM may need to be updated with the latest software • Faulty PCM
DTC: P2649 **1T PCM, MIL: Yes** **Year:** 2009, 2010 **Model:** Odyssey **Engine:** 3.5L V6 **Transmission:** All	**Rocker Arm Oil Control Solenoid Circuit High Voltage (J35A6 Engine):** With the engine running at idle for at least 7 seconds and battery voltage a minimum of 10.0 V. The return signal is ON (HIGH) for at least 2 seconds when the PCM outputs the OFF (LOW) signal to the rocker arm oil control solenoid. DTC P2648 is not active. **NOTE: Before you troubleshoot, record all freeze data and any on-board snapshot.** **Possible Causes:** • Poor connections or loose terminals at the rocker arm oil control solenoid and the PCM • "Open" circuit between the PCM and the rocker arm oil control solenoid • Faulty rocker arm oil control solenoid • PCM may need to be updated with the latest software • Faulty PCM
DTC: P2649 **1T PCM, MIL: Yes** **Year:** 2009, 2010 **Model:** Odyssey **Engine:** 3.5L V6 **Transmission:** All	**Rocker Arm Oil Control Solenoid A (Bank 1) Circuit High Voltage (J35A7 Engine):** The return signal is ON (High) for at least 2 seconds when the PCM outputs the OFF (Low) signal to the rocker arm oil control solenoid. With the engine running the execution time is continuous. DTC P2648 is not active. **NOTE: Before you troubleshoot, record all freeze data and any on-board snapshot.** **Possible Causes:** • Poor connections or loose terminals at the rocker arm oil control solenoid A (Bank 1) and the PCM • "Open" circuit between the PCM and the rocker arm oil control solenoid A (Bank 1) • Faulty rear rocker arm oil control solenoid • PCM may need to be updated with the latest software • Faulty PCM
DTC: P2649 **1T PCM, MIL: Yes** **Year:** 2009, 2010 **Model:** Accord, Accord Crosstour, Civic, CR-V, Element, Fit, Pilot, Ridgeline **Engine:** 1.3L L4, 1.5L L4, 1.8L L4, 2.0L L4, 2.4L L4, 3.5L V6 **Transmission:** All	**Rocker Arm Oil Control Solenoid (VTEC Solenoid Valve) Circuit High Voltage :** The return signal is ON (High) for at least 2 seconds when the PCM outputs the OFF (Low) signal to the rocker arm oil control solenoid. With the engine running the execution time is continuous. DTC P2648 is not active. **NOTE: Before you troubleshoot, record all freeze data and any on-board snapshot.** **Possible Causes:** • Poor connections or loose terminals at the rocker arm oil control solenoid and the PCM • "Open" circuit between the PCM and the rocker arm oil control solenoid • Faulty rocker arm oil control solenoid • PCM may need to be updated with the latest software • Faulty PCM

DTC	Trouble Code Title, Conditions & Possible Causes
DTC: P2651 **1T PCM, MIL: Yes** **Year:** 2009, 2010 **Model:** Civic **Engine:** 1.3L L4 **Transmission:** All	**Valve Pause System (VPS) Stuck OFF:** With the vehicle driven in a low gear at 4,800 rpm or more at around 7 mph for at least 5 seconds, The following conditions occur for at least 5 seconds. * Rocker arm oil control solenoid 1, ON * Rocker arm oil control solenoid 2, OFF * Rocker arm oil pressure switch, ON * Rocker arm oil pressure sensor, High 39 kPa or more **Possible Causes:** • Low oil level • Poor connections or loose terminals at the rocker arm oil pressure control solenoid 1 and/or 2, rocker arm oil pressure switch, rocker arm oil pressure sensor and the PCM • Faulty rocker arm oil control valve
DTC: P2652 **1T PCM, MIL: Yes** **Year:** 2009, 2010 **Model:** Civic **Engine:** 1.3L L4 **Transmission:** All	**Valve Pause System (VPS) Stuck ON:** With the engine running at a speed between 2,000-4,600 rpm in (P or N) for at least 5 seconds, the following conditions occur. * Rocker Arm Oil Control Solenoid 1 is OFF * Rocker Arm Oil Control Solenoid 2 is OFF * Rocker Arm Oil Pressure Switch is OFF * Rocker Arm Oil pressure Sensor Output HIGH 39 kPa or more **Possible Causes:** • Low oil level • Low oil pressure • Poor connections or loose terminals at the rocker arm oil pressure control solenoid 1 and/or 2, rocker arm oil pressure switch, rocker arm oil pressure sensor and the PCM • Faulty rocker arm oil control valve • Faulty rocker arm oil pressure switch
DTC: P2653 **2T PCM, MIL: Yes** **Year:** 2009, 2010 **Model:** Accord, Accord Crosstour, Odyssey, Pilot **Engine:** 3.5L V6 **Transmission:** All	**Rocker Arm Oil Control Solenoid B (Bank 1) Circuit Low Voltage:** With the engine running at about 3,000 rpm for 10 seconds, the return signal is OFF (Low) for at least 1.0 second when the PCM outputs the ON (High) signal to the rocker arm oil control solenoid B (Bank 1). **NOTE: Before you troubleshoot, record all freeze data and any on-board snapshot, and review the general troubleshooting information.** **Possible Causes:** • Poor connections or loose terminals at rocker arm oil control solenoid B (Bank 1) and the PCM • "Short" circuit between the PCM and rocker arm oil control solenoid B (Bank 1) • Faulty rear rocker arm oil control valve • PCM may need to be updated with the latest software • Faulty PCM
DTC: P2654 **2T PCM, MIL: Yes** **Year:** 2009, 2010 **Model:** Accord **Engine:** 2.4L L4 **Transmission:** All	**Rocker Arm Oil Control Solenoid B (Exhaust Valve Side) Circuit High Voltage:** With the engine running at idle for at least 10 seconds, the return signal is ON (High) for at least 2.0 seconds when the PCM outputs the OFF (Low) signal to the rocker arm oil control solenoid B. **NOTE: Before you troubleshoot, record all freeze data and any on-board snapshot, and review the general troubleshooting information.** **Possible Causes:** • Poor connections or loose terminals at rocker arm oil control solenoid B and the PCM • "Open" circuit between the rocker arm oil control solenoid B and ground • "Open" circuit between the PCM and the rocker arm oil control solenoid B • Faulty rocker arm oil control valve • PCM may need to be updated with the latest software • Faulty PCM
DTC: P2654 **1T PCM, MIL: Yes** **Year:** 2009, 2010 **Model:** Civic **Engine:** 1.3L L4 **Transmission:** All	**Rocker Arm Oil Control Solenoid 1 Circuit High Voltage:** With the ignition ON and the battery voltage a minimum of 10.0 V, the return signal is ON (HIGH) for at least 1.1 seconds when the PCM outputs the OFF (LOW) signal to the rocker arm oil control solenoid 1. **Possible Causes:** • Poor connections or loose terminals at the rocker arm oil pressure control solenoid 1 and the PCM • "Open" between the rocker arm oil control solenoid 1 and ground • "Open" circuit between the rocker arm oil control solenoid 1 and the PCM • Faulty rocker arm oil control valve • PCM may need to be updated with the latest software • Faulty PCM

DTC	Trouble Code Title, Conditions & Possible Causes
DTC: P2654 **2T PCM, MIL: Yes** **Year:** 2009, 2010 **Model:** Accord, Accord Crosstour, Odyssey, Pilot **Engine:** 3.5L V6 **Transmission:** All	**Rocker Arm Oil Control Solenoid B (Bank 1) Circuit High Voltage:** With the engine running at idle for at least 10 seconds, the return signal is ON (High) for at least 2.0 seconds when the PCM outputs the OFF (Low) signal to the rocker arm oil control solenoid B (Bank 1). **NOTE: Before you troubleshoot, record all freeze data and any on-board snapshot, and review the general troubleshooting information.** **Possible Causes:** • Poor connections or loose terminals at rocker arm oil control solenoid B (Bank 1)and the PCM • "Open" circuit between the rocker arm oil control solenoid B and ground • "Open" circuit between the PCM and the rocker arm oil control solenoid B • Faulty rocker arm oil control valve • PCM may need to be updated with the latest software • Faulty PCM
DTC: P2658 **2T PCM, MIL: Yes** **Year:** 2009, 2010 **Model:** Accord, Accord Crosstour, Odyssey, Pilot **Engine:** 3.5L V6 **Transmission:** All	**Rocker Arm Oil Control Solenoid A (Bank 2) Circuit Low Voltage:** With the engine running at about 3,000 rpm for 10 seconds, the return signal is OFF (Low) for at least 1.0 second when the PCM outputs the ON (High) signal to the rocker arm oil control solenoid A (Bank 2). **NOTE: Before you troubleshoot, record all freeze data and any on-board snapshot, and review the general troubleshooting information.** **Possible Causes:** • Poor connections or loose terminals at rocker arm oil control solenoid A (Bank 2) and the PCM • "Short" circuit between the PCM and rocker arm oil control solenoid A (Bank 2) • Faulty front rocker arm oil control valve • PCM may need to be updated with the latest software • Faulty PCM
DTC: P2659 **2T PCM, MIL: Yes** **Year:** 2009, 2010 **Model:** Accord, Accord Crosstour, Odyssey, Pilot **Engine:** 3.5L V6 **Transmission:** All	**Rocker Arm Oil Control Solenoid A (Bank 2) Circuit High Voltage:** With the engine running at idle for at least 10 seconds, the return signal is ON (High) for at least 2.0 seconds when the PCM outputs the OFF (Low) signal to the rocker arm oil control solenoid A (Bank 2). **NOTE: Before you troubleshoot, record all freeze data and any on-board snapshot, and review the general troubleshooting information** **Possible Causes:** • Poor connections or loose terminals at rocker arm oil control solenoid A (Bank 2) and the PCM • "Open" circuit between the rocker arm oil control solenoid A (Bank 2) and ground • "Open" circuit between the PCM and the rocker arm oil control solenoid A • Faulty rocker arm oil control valve • PCM may need to be updated with the latest software • Faulty PCM
DTC: P2769 **1T PCM, MIL: Yes, TCIL: Yes** **Year:** 2009 **Model:** Ridgeline **Engine:** 3.5L V6 **Transmission:** All	**Short in Torque Converter Clutch Solenoid Valve Circuit:** The return signal does not match the command to turn ON the torque converter clutch solenoid valve for at least 1 second. This code is caused by an electrical circuit problem and cannot be caused by a mechanical problem in the transmission. **NOTE: Before you troubleshoot, record all freeze data and any on-board snapshot, and review the general troubleshooting information.** **Possible Causes:** • Blown fuse • "Open" or "Short" in the wire between PCM connector terminal and the driver's under-dash fuse/relay box • "Short" to ground in the wire between PCM connector terminal and solenoid harness • Faulty torque converter clutch solenoid valve • PCM needs to be updated with the latest software • Faulty PCM
DTC: P2770 **1T PCM, MIL: Yes, TCIL: Yes** **Year:** 2009 **Model:** Ridgeline **Engine:** 3.5L V6 **Transmission:** All	**Open in Torque Converter Clutch Solenoid Valve Circuit:** With the engine running the return signal does not match the command to turn OFF the torque converter clutch solenoid valve for at least 1 second. This code is caused by an electrical circuit problem and cannot be caused by a mechanical problem in the transmission. **NOTE: Record all freeze data and review General Troubleshooting Information before you troubleshoot.** **Possible Causes:** • Poor connections or loose terminals the PCM • "Open" circuit between the PCM and the torque converter clutch solenoid • "Open" circuit in the torque converter clutch solenoid harness • Faulty torque converter clutch solenoid • PCM needs to be updated with the latest software • Faulty PCM

DTC	Trouble Code Title, Conditions & Possible Causes
DTC: P2A00 **2T PCM, MIL: Yes** **Year:** 2009, 2010 **Model:** Accord, Accord Crosstour, Odyssey, Pilot, Ridgeline **Engine:** 3.5L V6 **Transmission:** All	**Rear Air/Fuel Ratio (A/F) Sensor (Bank 1, Sensor 1) Circuit Range/Performance Problem:** The rear A/F sensor (bank 1, sensor 1) output voltage is 2.55 V or less, or 4.50 V or more. The execution time is once per driving cycle and the duration time is 5.3 seconds or more. **NOTE: Before you troubleshoot, record all freeze data and any on-board snapshot.** **Possible Causes:** • Poor connections or loose terminals at the A/F sensor (Sensor 1) and the PCM • Faulty A/F sensor (Sensor 1)
DTC: P2A00 **2T PCM, MIL: Yes** **Year:** 2009, 2010 **Model:** Accord, Civic, CR-V, Element, Fit **Engine:** 1.3L L4, 1.5L L4, 1.8L L4, 2.0L L4, 2.4L L4 **Transmission:** All	**Air/Fuel Ratio (A/F) Sensor (Sensor 1) Circuit Range/Performance Problem:** The A/F sensor (Sensor 1) output voltage is 3.0 V or less, or 4.8 V or more. The execution time is once per driving cycle and the duration time is 3.5 seconds or more. **NOTE: Before you troubleshoot, record all freeze data and any on-board snapshot.** **Possible Causes:** • Poor connections or loose terminals at the A/F sensor (Sensor 1) and the ECM/PCM • Faulty A/F sensor (Sensor 1)
DTC: P2A03 **2T PCM, MIL: Yes** **Year:** 2009, 2010 **Model:** Accord, Accord Crosstour, Odyssey, Pilot, Ridgeline **Engine:** 3.5L V6 **Transmission:** All	**Front Air/Fuel Ratio (A/F) Sensor (Bank 2, Sensor 1) Circuit Range/Performance Problem:** The front A/F sensor (bank 2, sensor 1) output voltage is 2.55 V or less, or 4.50 V or more. The execution time is once per driving cycle and the duration time is 5.3 seconds or more. **Possible Causes:** • Poor connections or loose terminals at the A/F sensor (Sensor 1) and the PCM • Faulty A/F sensor (Sensor 1)

OBD II Trouble Code List (P3XXX Codes)

DTC	Trouble Code Title, Conditions & Possible Causes
DTC: P3400 **1T PCM, MIL: Yes** **Year:** 2009, 2010 **Model:** Accord, Accord Crosstour, Odyssey, Pilot **Engine:** 3.5L V6 **Transmission:** All	**Valve Pulse System (VPS) Stuck Off (Bank 1):** With the engine running in closed loop, and in 5th gear at a steady speed of 12 mph or more for 5 seconds. When the PCM sends ON (Open) or OFF (Close) command to the rocker arm oil control solenoid A (Bank 1) the rear rocker arm oil pressure switch detects a malfunction. **NOTE: Before you troubleshoot, record all freeze data and any on-board snapshot, and review the general troubleshooting information.** **Possible Causes:** • Poor connections or loose terminals at rocker arm oil control solenoid A (Bank 1), rocker arm oil control solenoid B (Bank 1), the rear rocker arm oil pressure switch, and the PCM • Low or diluted engine oil • Clogged oil passage of the rear bank cylinder pause system • "Short" circuit between the PCM and the rear rocker arm oil pressure switch • Faulty rear rocker arm oil pressure switch • PCM may need to be updated with the latest software • Faulty PCM
DTC: P3400 **1T PCM, MIL: Yes** **Year:** 2009, 2010 **Model:** Civic **Engine:** 1.3L L4 **Transmission:** All	**Valve Pause System (VPS) Stuck OFF:** With the engine running at operating temperature, and the vehicle speed is a minimum of 12 mph (18 km/h) or more with the engine speed between 900-3,000 rpm the following conditions occur for 2 seconds or more. Condition 1: rocker arm oil pressure sensor A is ON Condition 2: rocker arm oil pressure sensor B is ON Condition 3: rocker arm oil pressure sensor A output HIGH Condition 4: rocker arm oil pressure sensor B output LOW **Possible Causes:** • Poor connections or loose terminals at rocker arm oil pressure sensor A, or B • Low engine oil, or oil pressure • Faulty rocker arm oil pressure sensor A., or B • Faulty rocker arm oil control valve A, or B

DTC	Trouble Code Title, Conditions & Possible Causes
DTC: P3497 **1T PCM, MIL: Yes** **Year:** 2009, 2010 **Model:** Accord, Accord Crosstour, Odyssey, Pilot **Engine:** 3.5L V6 **Transmission:** All	**Valve Pause System (VPS) Stuck Off (Bank 2):** With the engine running in closed loop, and in 5th gear at a steady speed of 68 mph or more for 5 seconds. When the PCM sends ON (Open) or OFF (Close) command to the rocker arm oil control solenoid A (Bank 2) the rear rocker arm oil pressure switch detects a malfunction. **Possible Causes:** • Poor connections or loose terminals at rocker arm oil control solenoid A (Bank 1), rocker arm oil control solenoid B (Bank 1), the rear rocker arm oil pressure switch, and the PCM • Low or diluted engine oil • Clogged oil passage of the rear bank cylinder pause system • "Short" circuit between the PCM and the rear rocker arm oil pressure switch • Faulty rear rocker arm oil pressure switch • PCM may need to be updated with the latest software • Faulty PCM

OBD II Trouble Code List (U0XXX Codes)

DTC	Trouble Code Title, Conditions & Possible Causes
DTC: U0028 **1T PCM, TCIL: Yes** **Year:** 2009, 2010 **Model:** Civic, CR-V, Element, Odyssey **Engine:** 1.3L L4, 1.8L L4, 2.0L L4, 2.4L L4, 3.5L V6 **Transmission:** All	**F-CAN Communication Circuit Error (F-CAN Bus OFF) :** When the powertrain control module (PCM) does not receive the signals via the CAN lines for a set time or more, the PCM detects a malfunction. The PCM does not receive any signals for at least 1 second. **NOTE: Before you troubleshoot, record all freeze data and any on-board snapshot.** **Possible Causes:** • PCM needs to be updated with the latest software • Faulty PCM
DTC: U0028/107 **1T PCM, MIL: Yes** **Year:** 2009, 2010 **Model:** Civic **Engine:** 1.3L L4 **Transmission:** All	**F-CAN Malfunction (BUS-OFF Motor Control Module (MCM) (HYBRID MODELS):** With the ignition switch ON (II) for at least 3 seconds and battery voltage a minimum of 10.0 V, the following conditions occur at the time within 2 seconds. (IMA system indicator is ON) WARNING: The IMA system uses high voltage (144 V), before servicing turn OFF the battery module switch. **Possible Causes:** • Poor connections or loose terminals at the PCM, the BCM module, and the MCM • PCM may need to be updated with the latest software • Faulty PCM • MCM may need to be updated with the latest software • Faulty MCM
DTC: U0029 **1T PCM** **Year:** 2009, 2010 **Model:** Accord, Accord Crosstour, CR-V, Fit, Odyssey, Ridgeline **Engine:** 1.5L L4, 2.4L L4, 3.5L V6 **Transmission:** All	**F-CAN A Malfunction (BUS-OFF (PCM)):** The PCM does not receive any signals via the F-CAN A lines for at least 1 second. When a malfunction is detected, the D indicator blinks, and a Pending DTC, a Confirmed DTC, and the freeze data are stored in the PCM memory. The MIL does not come on. **Possible Causes:** • Check battery and charging system condition • Loose or poor connections, or worn/shorted wires • "Short" in the F-CAN wires • Faulty gauge control module • PCM may need to be updated with the latest software • Faulty PCM

DTC	Trouble Code Title, Conditions & Possible Causes
DTC: U0037/98 **1T PCM, MIL: Yes** **Year:** 2009, 2010 **Model:** Civic **Engine:** 1.3L L4 **Transmission:** All	**IMA-CAN Malfunction (BUS-OFF Motor Control Module (MCM)) (HYBRID MODELS):** With the ignition switch ON (II) for at least 3 seconds and the battery voltage is a minimum of 10.0 V, the following conditions occur within 2 seconds. Condition 1: The information sent from the Powertrain Control Module (PCM) through the IMA-CAN is not received foe at least 1 second or the information sent from the PCM through the IMA-CAN is abnormal at least 10 times. Condition 2: The information sent from the Battery Condition Monitor (BCM) module through the IMA-CAN is not received for at least 0.5 second or the information sent from the BCM module through the IMA-CAN is abnormal at least 10 times. **Possible Causes:** • Poor connections or loose terminals at the BCM module, the A/C compressor driver, the PCM, and the MCM • "Open" circuit between the BCM module and the MCM • "Short" circuit between the MCM, the A/C compressor driver, the PCM, and BCM module • Faulty A/C compressor driver • PCM may need to be updated with the latest software • Faulty PCM • BCM may need to be updated with the latest software • Faulty BCM • MCM may need to be updated with the latest software • Faulty MCM
DTC: U0073 **1T PCM, MIL: Yes** **Year:** 2009, 2010 **Model:** Odyssey, Pilot, Ridgeline **Engine:** 3.5L V6 **Transmission:** All	**F-CAN Malfunction (BUS-OFF) :** The PCM does not receive any signals via the F-CAN lines for at least 1 second. **Possible Causes:** • PCM may need to be updated with the latest software • Faulty PCM
DTC: U0100 **1T PCM** **Year:** 2009, 2010 **Model:** Accord, Accord Crosstour **Engine:** 2.4L L4, 3.5L V6 **Transmission:** All	**Gauge Control Module Lost Communication With ECM/PCM:** F-CAN communication line malfunction. **NOTE: If you are troubleshooting multiple DTCs, be sure to follow the instructions in B-CAN System Diagnosis Test Mode A.** **Possible Causes:** • Faulty battery or charging system • Perform the gauge control module input test • Loose or poor connections at the gauge control module and the ECM/PCM • Check for faulty inputs • Gauge control module is faulty • Faulty ECM/PCM
DTC: U0100/102 **1T PCM, MIL: Yes** **Year:** 2009, 2010 **Model:** Civic **Engine:** 1.3L L4 **Transmission:** All	**F-CAN Malfunction (PCM (CVT System)-MCM) (HYBRID MODELS):** With the ignition switch ON (II) for at least 3 seconds, and battery voltage is a minimum of 10.0 V. One of the following conditions occurs. (IMA system indicator is ON) Condition 1: The information is not sent from the PCM through the F-CAN for at least 0.5 second and at least one other node is communicating after 2 seconds have passed. Condition 2: The information sent from the PCM through the F-CAN is abnormal at least 10 times and at least one other node is communicating after 2 seconds have passed. WARNING: The IMA system uses high voltage (144 V), before servicing turn OFF the battery module switch. **Possible Causes:** • Poor connections or loose terminals at the PCM, the BCM module, and the MCM • PCM may need to be updated with the latest software • Faulty PCM • MCM may need to be updated with the latest software • Faulty MCM
DTC: U0100/103 **1T , MIL: Yes** **Year:** 2009, 2010 **Model:** Civic **Engine:** 1.3L L4 **Transmission:** All	**F-CAN Malfunction (Powertrain Control Module (PCM)-Motor Control Module (MCM)) (HYBRID MODELS):** With the ignition ON and the MCM power-supply voltage a minimum of 10.2 V, no signal via the can lines are received for at least 0.5 seconds. **Possible Causes:** • Poor connections or loose terminals at the PCM and the MCM • To verify this is the appropriate troubleshooting procedure, jump the SCS with the HDS and read the flash code, or change the HDS setup to show Honda codes, and read the Honda code. • If DTC U0029 is indicated on the HDS, intermittent failure, this system is OK at this time. • If the DTC cannot be cleared, do the troubleshooting for DTC U1204/55.

DTC	Trouble Code Title, Conditions & Possible Causes
DTC: U0107 **1T PCM, MIL: Yes** **Year:** 2009, 2010 **Model:** Element **Engine:** 2.4L L4 **Transmission:** All	**Lost Communication With Throttle Actuator Control Module :** One of these 2 conditions must be met for at least 250 milliseconds: (1) No serial signals from the throttle actuator control module are detected. (2) The serial signals from the throttle actuator control module are abnormal. **NOTE: Before you troubleshoot, record all freeze data and any on-board snapshot.** **Possible Causes:** • Poor connections or loose terminals at the throttle body, the throttle actuator control module relay, the throttle actuator control module, and the PCM • "Open" in the wire between the PCM and ground • "Open" or "Short" circuit between the throttle actuator control module and the PCM • PCM may need to be updated with the latest software • Faulty PCM
DTC: U0110 **1T PCM, MIL: Yes** **Year:** 2009, 2010 **Model:** Civic **Engine:** 1.3L L4 **Transmission:** All	**F-CAN Malfunction (Powertrain Control Module (PCM)-Motor Control Module (MCM)):** With the ignition switch turned ON for a minimum of 3 seconds and battery voltage a minimum of 10.0 V, one of the following conditions must be met. Condition 1: The PCM does not receive any signals via the F-CAN lines for at least 1 second. Condition 2: The information sent from the MCM is abnormal at least 20 times. **Possible Causes:** • Poor connections or loose terminals at the Motor Control Module (MCM) and the PCM • "Open" circuit between the MCM and ground • Blown fuse • "Open" or "Short" circuit between MCM relay 1 and the MCM • Faulty MCM relay 1 • MCM may need to be updated with the latest software • Faulty MCM
DTC: U0111/100 **1T BCM** **Year:** 2009, 2010 **Model:** Civic **Engine:** 1.3L L4 **Transmission:** All	**F-CAN Malfunction (Battery Condition Monitor (BCM) Module- Motor Control Module (MCM) (HYBRID MODELS):** With the ignition switch ON (II) and the battery voltage at a minimum of 10.0 V. One of the following conditions occurs. (IMA system indicator is OFF) Condition 1: The information is not sent from the BCM module through the F-CAN for at least 0.5 second and at least one other node is communicating after 2 seconds have passed. Condition 2: The information is not sent from the BCM module is abnormal at least 10 times and at least one other node is communicating after 2 seconds have passed. WARNING: The IMA system uses high voltage (144 V), before servicing turn OFF the battery module switch. **Possible Causes:** • Poor connections or loose terminals at the BCM module and the MCM • "Open" circuit between the BCM module and the MCM • BCM may need to be updated with the latest software • Faulty BCM
DTC: U0114 **1T CCM, MIL: Yes** **Year:** 2009, 2010 **Model:** Pilot, Ridgeline **Engine:** 3.5L V6 **Transmission:** All	**FCAN Malfunction (VTM-4 Control Module To PCM):** DTC U0073 not set; and the PCM detected a problem in the FCAN communication circuit to the Vehicle Navigation Module (VTM) or the VSA control module. **Possible Causes:** • Check for loose connections at VTM-4 unit (intermittent fault) • Check for any Body Control Module trouble codes • FCAN communication circuit to the VTM-4 is open, shorted to ground or B+
DTC: U0121 **1T PCM, TCIL: Yes** **Year:** 2009, 2010 **Model:** Civic, Fit **Engine:** 1.3L L4, 1.5L L4, 1.8L L4, 2.0L L4 **Transmission:** All	**F-CAN Malfunction (Powertrain Control Module (PCM)-ABS):** With the ignition switch ON for at least 3 seconds and battery voltage at a minimum of 10.0 V, one of the following conditions must be met. Condition 1: The PCM does not receive any signals via the F-CAN lines for at least 1 second. Condition 2: The information sent from the ABS modulator-control unit is abnormal at least 20 times. **Possible Causes:** • Poor connections or loose terminals at the ABS modulator-control unit and the PCM • "Open" circuit between the PCM and the ABS modulator-control unit • Faulty ABS modulator-control unit

DTC	Trouble Code Title, Conditions & Possible Causes
DTC: U0122 **1T PCM, TCIL: Yes** **Year:** 2009, 2010 **Model:** Accord, Accord Crosstour **Engine:** 2.4L L4, 3.5L V6 **Transmission:** All	**Lost Communication with VSA Modulator-Control Unit:** No signals from the VSA modulator-control unit via the CAN lines are received for at least 1 second. The execution time is continuous and the duration time is 1 seconds or more. **NOTE: Before you troubleshoot, record all freeze data and any on-board snapshot.** **Possible Causes:** • Poor connections or loose terminals at the gauge control module, the VSA modulator-control unit, and the PCM • "Open circuit between the PCM and the VSA modulator-control • Perform DLC circuit troubleshooting • Faulty VSA modulator-control unit
DTC: U0122 **1T PCM, TCIL: Yes** **Year:** 2009, 2010 **Model:** Accord, Accord Crosstour, CR-V, Element, Odyssey, Pilot, Ridgeline **Engine:** 2.4L L4, 3.5L V6 **Transmission:** All	**F-CAN A Malfunction (Powertrain Control Module (PCM)-VSA Modulator-Control Unit) (PGM-FI System):** With the ignition switch ON for at least 3 seconds. One of these 2 conditions must be met: (1) The PCM does not receive any signals via the F-CAN A lines for at least 1 second. (2) The information sent from the VSA modulator-control unit is abnormal at least 20 times. **Possible Causes:** • Poor connections or loose terminals at the VSA modulator-control unit and the PCM • "Open" circuit between the PCM and the VSA modulator-control unit • VSA modulator-control unit needs to be updated with the latest software • Faulty VSA modulator-control unit
DTC: U0122 **1T PCM** **Year:** 2009, 2010 **Model:** Civic, Pilot **Engine:** 1.3L L4, 1.8L L4, 2.0L L4, 3.5L V6 **Transmission:** All	**Gauge Control Module Lost Communication With VSA Modulator-Control Unit (VSA message):** F-CAN communication line Malfunction. **NOTE: If you are troubleshooting multiple DTCs, be sure to follow the instructions in B-CAN System Diagnosis Test Mode A.** **Possible Causes:** • Faulty battery and/or charging system • Loose VSA ground or poor connections at the VSA modulator-control unit or gauge control module • Faulty inputs • Faulty gauge control module • Faulty ECM/PCM
DTC: U0122 **1T PCM** **Year:** 2009, 2010 **Model:** Fit **Engine:** 1.5L L4 **Transmission:** All	**Lost Communication with VSA Modulator-Control Unit (A/T System):** One of these 2 conditions must be met: (1) The PCM does not receive any signals via the F-CAN A lines for at least 1 second. (2) The information sent from the VSA modulator-control unit is abnormal at least 20 times. DTCs U0029, U0104, U0155 are not active. **NOTE: Before you troubleshoot, record all freeze data and any on-board snapshot. This code is caused by an electrical circuit problem and cannot be caused by a mechanical problem in the transmission.** **Possible Causes:** • Poor connections or loose terminals between the VSA modulator-control unit and the PCM • Check for Pending or Confirmed DTCs in the PGM-FI SYSTEM • PCM may need to be updated with the latest software • Faulty PCM
DTC: U0127 **1T BCM** **Year:** 2009, 2010 **Model:** Accord, Accord Crosstour, Pilot **Engine:** 2.4L L4, 3.5L V6 **Transmission:** All	**Gauge Control Module Lost Communication With the TPMS Control Unit (TPMS message):** Poor communication between the Gauge Control Module and TPMS Control Unit. **NOTE: If you are troubleshooting multiple DTCs, be sure to follow the instructions in B-CAN System Diagnosis Test Mode A. If the HDS does not communicate with the TPMS control unit, the TPMS indicator stay on, and no TPMS DTCs are stored, go to symptom troubleshooting for TPMS indicator does not go off, and no DTCs are stored.** **Possible Causes:** • Poor connections or loose terminals between gauge control module and TPMS control unit • "Open" circuit between gauge control module and TPMS control unit • Faulty TPMS control unit • Faulty gauge control module
DTC: U0151 **1T PCM** **Year:** 2009, 2010 **Model:** Accord, Accord Crosstour, Pilot **Engine:** 2.4L L4, 3.5L V6 **Transmission:** All	**Gauge Control Module Lost Communication With SRS Unit:** Poor communication between the gauge control module, and the SRS unit. **NOTE: If you are troubleshooting multiple DTCs, be sure to follow the instructions in B-CAN System Diagnosis Test Mode A. If the HDS does not communicate with the SRS unit, the SRS indicator stay on, go to symptom troubleshooting for SRS indicator stay on, but no DTCs are stored.** **Possible Causes:** • Loose or poor connections between the gauge control module, and the SRS unit. • "Open" circuit between the gauge control module, and the SRS unit • Faulty SRS unit • Faulty gauge control module

DTC	Trouble Code Title, Conditions & Possible Causes
DTC: U0155 **1T** **Year:** 2009, 2010 **Model:** Pilot **Engine:** 3.5L V6 **Transmission:** All	**Lost Communication With Gauge Control Module (VSP/NE Frame):** Poor communication between the Power Seat control Unit and Gauge Control Module. **NOTE: If you are troubleshooting multiple DTCs, be sure to follow the instructions in B-CAN System Diagnosis Test Mode A.** **Possible Causes:** • "Open" or poor connection in the wire between the gauge control module and the power seat control unit • Faulty power seat control unit • Faulty gauge control module
DTC: U0155 **2T PCM, MIL: Yes** **Year:** 2009, 2010 **Model:** Fit, Odyssey, Pilot **Engine:** 1.5L L4, 3.5L V6 **Transmission:** All	**F-CAN Malfunction (ECM/PCM-Gauge Control Module):** With the vehicle running, the ECM/PCM does not receive any signals from the gauge control module for at least 1 second. **NOTE: Before you troubleshoot, record all freeze data and any on-board snapshot, and review the general troubleshooting information.** **Possible Causes:** • Perform the gauge control module input test • Poor connections and loose terminals at the gauge control module and the ECM/PCM • "Open" circuit between the ECM/PCM and the gauge control module • ECM/PCM may need to be updated with the latest software • Faulty ECM/PCM
DTC: U0155 **1T** **Year:** 2009, 2010 **Model:** Accord, Accord Crosstour, Pilot **Engine:** 2.4L L4, 3.5L V6 **Transmission:** All	**Door Multiplex Control Unit Lost Communication With Gauge Control Module:** Poor communication between the Door Multiplex Control Unit and Gauge Control Module. **NOTE: If you are troubleshooting multiple DTCs, be sure to follow the instructions in B-CAN System Diagnosis Test Mode A.** **Possible Causes:** • Perform the gauge control module input test, and do all power, ground and communication input tests. If the tests prove OK, replace the gauge control module • Loose or poor connections at the gauge control module and the related units
DTC: U0155 **1T** **Year:** 2009, 2010 **Model:** Accord, Accord Crosstour, Pilot **Engine:** 2.4L L4, 3.5L V6 **Transmission:** All	**Immobilizer-Keyless Control Unit Lost Communication With Gauge Control Module:** Poor communication between the Immobilizer-Keyless Control Unit and Gauge Control Module. **NOTE: If you are troubleshooting multiple DTCs, be sure to follow the instructions in B-CAN System Diagnosis Test Mode A.** **Possible Causes:** • Perform the gauge control module input test, and do all power, ground, and communication input tests • Loose or poor connections at the gauge control module and the related units • Faulty gauge control module
DTC: U0155 **1T** **Year:** 2009, 2010 **Model:** Accord, Accord Crosstour **Engine:** 2.4L L4, 3.5L V6 **Transmission:** All	**Driver's MICU Lost Communication With Gauge Control Module:** Poor communication between the Driver's MICU and Gauge Control Module. **NOTE: If you are troubleshooting multiple DTCs, be sure to follow the instructions in B-CAN System Diagnosis Test Mode A.** **Possible Causes:** • Loose or poor connections at the gauge control module and the related units • Perform the gauge control module input test, and do all power, ground, and communication input tests. If the tests prove OK, replace the gauge control module
DTC: U0155 **1T** **Year:** 2009, 2010 **Model:** Accord, Accord Crosstour **Engine:** 2.4L L4, 3.5L V6 **Transmission:** All	**Passenger's MICU Lost Communication With Gauge Control Module:** Poor communication between the Passenger's MICU and Gauge Control Module. **NOTE: If you are troubleshooting multiple DTCs, be sure to follow the instructions in B-CAN System Diagnosis Test Mode A.** **Possible Causes:** • Loose or poor connections at the gauge control module and the related units • Perform the gauge control module input test, and do all power, ground, and communication input tests. If the tests prove OK, replace the gauge control module
DTC: U0155 **1T** **Year:** 2009, 2010 **Model:** Accord, Accord Crosstour, Pilot **Engine:** 2.4L L4, 3.5L V6 **Transmission:** All	**Climate Control Unit Lost Communication with Gauge Control Module:** Poor communication between the Climate Control Unit and Gauge Control Module. **NOTE: If you are troubleshooting multiple DTCs, be sure to follow the instructions in B-CAN System Diagnosis Test Mode A.** **Possible Causes:** • Loose wires or poor connections on the B-CAN lines between the gauge control module and the climate control unit • Perform the gauge control module input test • "Open" in the wire(s) between the climate control unit and the gauge control module • Faulty climate control unit

DTC	Trouble Code Title, Conditions & Possible Causes
DTC: U0155 **1T PCM, TCIL: Yes** **Year:** 2009, 2010 **Model:** CR-V, Element, Pilot, Ridgeline **Engine:** 2.4L L4, 3.5L V6 **Transmission:** All	**Lost Communication with Gauge Control Module :** No signals from the gauge control module via the CAN lines are received for at least 1 second. The execution time is continuous and the duration time is 1 second or more. **NOTE: Before you troubleshoot, record all freeze data and any on-board snapshot.** **Possible Causes:** • Poor connections and loose terminals at the gauge control module and the PCM • "Open" circuit between the PCM and the gauge control module • Faulty Gauge Control Module • PCM may need to be updated with the latest software • Faulty PCM
DTC: U0164 **1T BCM** **Year:** 2009, 2010 **Model:** Pilot **Engine:** 3.5L V6 **Transmission:** All	**Lost Communication With A/C:** A communication malfunction has been detected between the gauge control module and climate control unit or HVAC control unit. **NOTE: If you are troubleshooting multiple DTCs, be sure to follow the instructions in B-CAN System Diagnosis Test Mode A.** **Possible Causes:** • Loose or poor connections between the gauge control module and climate control unit or HVAC control unit • "Open" or poor connection in the wire between the SRS unit and the gauge control module • Faulty climate control unit or HVAC control unit
DTC: U0164 **1T** **Year:** 2009, 2010 **Model:** Accord, Accord Crosstour, Pilot **Engine:** 2.4L L4, 3.5L V6 **Transmission:** All	**Door Multiplex Control Unit Lost Communication With Climate Control Unit:** Poor communication between the door multiplex control unit and climate control unit. **NOTE: If you are troubleshooting multiple DTCs, be sure to follow the instructions in B-CAN System Diagnosis Test Mode A.** **Possible Causes:** • Perform the door multiplex control unit input test and check the power and ground. If OK, replace the driver's power window master switch • Loose or poor connections between the door multiplex control unit and climate control unit.
DTC: U0180 **1T BCM** **Year:** 2009, 2010 **Model:** Pilot **Engine:** 3.5L V6 **Transmission:** All	**Lost Communication With Auto Light Module (AUTOLT Frame):** A malfunction has been detected between the gauge control module and the automatic lighting control unit. **NOTE: If you are troubleshooting multiple DTCs, be sure to follow the instructions in B-CAN System Diagnosis Test Mode A.** **Possible Causes:** • Loose or poor connections between the gauge control module and the automatic lighting control unit • "Open" in the B-CAN wire • Check for lighting DTCs • Perform the power and ground input test for the automatic lighting control unit • Faulty automatic lighting control unit • Faulty gauge control module
DTC: U0199 **1T** **Year:** 2009, 2010 **Model:** Accord, Accord Crosstour, Pilot **Engine:** 2.4L L4, 3.5L V6 **Transmission:** All	**Driver's MICU Lost Communication With Door Multiplex Control Unit:** Poor communication between the Driver's MICU and Door Multiplex Control Unit. **NOTE: If you are troubleshooting multiple DTCs, be sure to follow the instructions in B-CAN System Diagnosis Test Mode A.** **Possible Causes:** • Perform the door multiplex control unit input test, and do all power, ground, and communication input tests. If the tests prove OK, replace the power window master switch • Loose or poor connections at the door multiplex control unit and the related units
DTC: U0199 **1T** **Year:** 2009, 2010 **Model:** Pilot **Engine:** 3.5L V6 **Transmission:** All	**Lost Communication With P/W (DRLockSW Frame):** Poor communication between the Door Multiplex Control Unit and Power Seat Control Unit. **NOTE: If you are troubleshooting multiple DTCs, be sure to follow the instructions in B-CAN System Diagnosis Test Mode A.** **Possible Causes:** • Loose or poor connections between the power seat control unit and the door multiplex control unit • Faulty power seat control unit • "Open" circuit between the door multiplex control unit and the power seat control unit • Faulty door multiplex control unit
DTC: U0199 **T BCM** **Year:** 2009, 2010 **Model:** Accord, Accord Crosstour, Pilot **Engine:** 2.4L L4, 3.5L V6 **Transmission:** All	**Immobilizer-keyless Control Unit Lost Communication With Door Multiplex Control Unit :** Poor communication between the Immobilizer-keyless Control Unit, and the Door Multiplex Control Unit. **NOTE: If you are troubleshooting multiple DTCs, be sure to follow the instructions in B-CAN System Diagnosis Test Mode A.** **Possible Causes:** • Perform the door multiplex control unit input test, and do all power, ground, and communication input tests. If the tests prove OK, replace the power window master switch • Loose or poor connections at the door multiplex control unit and the related units

DTC	Trouble Code Title, Conditions & Possible Causes
DTC: U0199 **1T** **Year:** 2009, 2010 **Model:** Pilot **Engine:** 3.5L V6 **Transmission:** All	**Lost Communication With Door Multiplex Control Unit (DRLOCK SW, KLDRLOCK Frame):** Poor communication between the Door Multiplex Control Unit Power Tailgate Control Unit. **NOTE: If you are troubleshooting multiple DTCs, be to follow the instructions in B-CAN System Diagnosis Test Mode A.** **Possible Causes:** • "Open" or high resistance between door power window master switch and power tailgate control unit connector • Perform the power window master switch input test, and do all power and ground input tests. If the tests prove OK, replace the power window master switch
DTC: U0230 **1T** **Year:** 2009, 2010 **Model:** Pilot **Engine:** 3.5L V6 **Transmission:** All	**Lost Communication With Power Tailgate Control Unit:** Poor communication between the Gauge Control Module and Power Tailgate Control Unit. **NOTE: If you are troubleshooting multiple DTCs, be sure to follow the instructions in B-CAN System Diagnosis Test Mode A.** **Possible Causes:** • Loose or poor connections between the gauge control module and the power tailgate control unit • "Open" circuit between the gauge control module connector and power tailgate control unit • Faulty power tailgate control unit • Faulty gauge control module
DTC: U0300 **1T PCM, MIL: Yes** **Year:** 2009, 2010 **Model:** Accord, Civic, CR-V, Fit, Odyssey, Pilot, Ridgeline **Engine:** 1.3L L4, 1.5L L4, 1.8L L4, 2.4L L4, 3.5L V6 **Transmission:** All	**PGM-FI System and A/T System Program Version Mismatch:** The rewriting of the FI CPU or the A/T CPU is done at the PCM, and different data of the serial communication data version is rewritten. The execution time is continuous and the duration time is 500 milliseconds or more. WARNING: Do not turn the ignition switch to LOCK (0) or ACC (I) while updating the PCM. If you turn the ignition switch to LOCK (0) before completion, the PCM will be damaged. **Possible Causes:** • PCM may need to be updated with the latest software • Faulty PCM

OBD II Trouble Code List (U1XXX Codes)

DTC	Trouble Code Title, Conditions & Possible Causes
DTC: U1028 **T BCM** **Year:** 2010 **Model:** Accord Crosstour **Engine:** 3.5L V6 **Transmission:** All	**Rear Window Wiper Motor (As) Signal Error:** A malfunction has been detected in the rear wiper motor circuit. **NOTE: If you are troubleshooting multiple DTCs, be sure to follow the instructions in B-CAN System Diagnosis Test Mode A.** **Possible Causes:** • Loose or poor connections • Blown fuse • "Open" or "Short" circuit between the under-hood fuse/relay box and the rear window wiper motor • "Open" in the wire between rear window wiper motor and body ground • Faulty rear window wiper motor • Faulty passenger's MICU; replace the passenger's under-dash fuse/relay box
DTC: U1101 **1T PCM** **Year:** 2009, 2010 **Model:** Accord, Accord Crosstour, Odyssey, Pilot **Engine:** 3.5L V6 **Transmission:** All	**F-CAN Malfunction Powertrain Control Module (PCM) Active Control Engine Mount (ACM):** With the battery voltage at a minimum of 10.0 V, and DTC U0029 not active. The PCM does not receive any signals via the F-CAN lines for at least 1 second. **NOTE: Before you troubleshoot, record all freeze data and any on-board snapshot, and review the general troubleshooting information.** **Possible Causes:** • Poor connections or loose terminals at the engine mount control unit and the PCM • "Open" circuit between the PCM and the engine mount control unit • Faulty engine mount control unit
DTC: U1202/64 **1T BCM, MIL: Yes** **Year:** 2009, 2010 **Model:** Civic **Engine:** 1.3L L4 **Transmission:** All	**IMA-CAN Malfunction (Battery Condition Monitor (BCM) Module-Motor Control Module (MCM) (HYBRID MODELS):** With the ignition switch on for at least 3 seconds and the battery voltage at a minimum of 10.0 V. One of the following conditions occurs. (IMA system indicator is ON) Condition 1: The information is not sent from the BCM module through the IMA-CAN for at least 0.5 second, and at least one other node is communicating after 2 seconds have passed. Condition 2: The information is not sent from the BCM module through the IMA-CAN is abnormal at least 10 times and at least one other node is communicating after 2 seconds have passed. WARNING: The IMA system uses high voltage (144 V), before servicing turn OFF the battery module switch. **Possible Causes:** • Poor connections or loose terminals at the PCM, the BCM module, the A/C compressor driver, and the MCM • "Open" circuit between the PCM, the BCM module, and the MCM • BCM may need to be updated with the latest software • Faulty BCM

DTC	Trouble Code Title, Conditions & Possible Causes
DTC: U1204/55 **1T , MIL: Yes** **Year:** 2009, 2010 **Model:** Civic **Engine:** 1.3L L4 **Transmission:** All	**IMA-CAN Malfunction (Powertrain Control Module (PCM)- Motor Control Module (MCM)) (HYBRID MODELS):** With the ignition ON (II) a minimum of 3 seconds and battery voltage at a minimum of 10.2 V, no signal from the PCM via the CAN lines are received for at least 1.0 second, or received data is abnormal at least 50 counts. (IMA system indicator is ON) **Possible Causes:** • Poor connections or loose terminals at the PCM and the MCM • "Open" or "Short" circuit between the PCM and the MCM • Update the PCM if it does not have the latest software, or substitute a known-good PCM • Update the MCM if it does not have the latest software, or substitute a known-good MCM
DTC: U1204/95 **1T PCM, MIL: Yes** **Year:** 2009, 2010 **Model:** Civic **Engine:** 1.3L L4 **Transmission:** All	**IMA-CAN Malfunction (Power Control Module (PCM) Motor Control Module (MCM)) (HYBRID MODELS):** With the ignition switch ON (II) for at least 3 seconds, and the battery voltage is a minimum of 10.0 V. (IMA system indicator is ON) One of the following conditions occurs. Condition 1: The information is not sent from the PCM through the IMA-CAN at least for 1 second and at least one other node is communicating after 2.0 seconds have passed. Condition 2: The information is not sent from the PCM through the IMA-CAN is abnormal at least 10 times and at least one other node is communicating after 2.0 seconds have passed. **Possible Causes:** • Poor connections or loose terminals at the PCM, the BCM module, the A/C compressor driver, and the MCM • "Open" Circuit between the PCM and the MCM • PCM may need to be updated with the latest software • Faulty PCM • MCM may need to be updated with the latest software • Faulty MCM
DTC: U1205 **1T PCM, MIL: Yes** **Year:** 2009, 2010 **Model:** Civic **Engine:** 1.3L L4 **Transmission:** All	**IMA-CAN Malfunction Powertrain Control Module (PCM)-Motor Control Module (MCM):** With the ignition switch ON for at least 3 seconds and the battery voltage is a minimum of 10.0 V, one of the following conditions must be met. Condition 1: The PCM does not receive any signals via the IMA-CAN lines for at least 1 second. Condition 2: The information sent from the MCM is abnormal at least 20 times. **Possible Causes:** • Poor connections or loose terminals at the MCM and the PCM • "Open" circuit between the PCM and the MCM • Faulty MCM
DTC: U1206/64 **1T PCM, MIL: Yes** **Year:** 2009, 2010 **Model:** Civic **Engine:** 1.3L L4 **Transmission:** All	**IMA-CAN Malfunction (BCM Module-MCM) (Hybrid Models):** With the ignition switch ON for a minimum of 3 seconds and the battery voltage a minimum of 10.0 V, one of the following conditions occur. (IMA system indicator is ON) Condition 1: The information is not sent from the BCM module through the IMA-CAN at least for 0.5 second and at least one other node is communicating after 2 seconds have passed. Condition 2: The information sent from the BCM module through the IMA-CAN is abnormal at least 10 times and at least one other node is communicating after 2 seconds have passed. WARNING: The IMA system uses high voltage (144 V), before servicing turn OFF the battery module switch. **Possible Causes:** • Poor connections or loose terminals at the PCM, the BCM module, the A/C compressor driver, and the MCM • "Open" circuit between the PCM, the BCM module, and the MCM • BCM may need to be updated with the latest software • Faulty BCM
DTC: U1207/80 **1T PCM, MIL: Yes** **Year:** 2009, 2010 **Model:** Civic **Engine:** 1.3L L4 **Transmission:** All	**IMA-CAN Malfunction (A/C Compressor Driver-BCM Module) (Hybrid Models):** With the ignition switch ON for at least 3 seconds and battery voltage a minimum of 10.0 V, one of the following conditions occur. (IMA system indicator is ON) Condition 1: The information is not sent from the A/C compressor driver through the IMA-CAN for at least 1.0 second and BUS-OFF is not detected after 2.0 seconds have passed. Condition 2: The information sent from the A/C compressor driver through the IMA-CAN is abnormal at least 10 times and when BUS-OFF is not detected after 2.0 seconds have passed. WARNING: The IMA system uses high voltage (144 V), before servicing turn OFF the battery module switch. **Possible Causes:** • Blown fuse • Poor connections or loose terminals at the BCM module, the A/C compressor driver, and the PCM • "Short" to ground in the wire between the A/C compressor driver and the fuse • "Open" circuit between the BCM module and the A/C compressor driver • "Open" circuit between the A/C compressor driver and Ground • Faulty A/C compressor driver

DTC	Trouble Code Title, Conditions & Possible Causes
DTC: U1280 **1T** **Year:** 2009, 2010 **Model:** Accord, Accord Crosstour, Pilot **Engine:** 2.4L L4, 3.5L V6 **Transmission:** All	**Communication Bus Line Error (BUS-OFF):** Communication Bus Line Error **NOTE: If you are troubleshooting multiple DTCs, be sure to follow the instructions in B-CAN System Diagnosis Test Mode A.** **Possible Causes:** • Check battery and charging system condition • Perform the following input test to help find the faulty unit: • Passenger's MICU input test • Door multiplex control unit (power window master switch) input test • Gauge control unit input test • Power control unit input test • Keyless access control unit input test • Remote slot control unit input test • Immobilizer-keyless control unit input test • Climate control unit power and ground circuit troubleshooting • Power seat control unit input test • HandsFreeLink control unit input test • AcuraLink control unit input test • Audio unit input test • Audio-navigation unit input test • Power tailgate control unit input test • BSI control unit input test • "Open" or "Short" between body ground and driver's under-dash fuse/relay box connector • Faulty driver's MICU
DTC: U1281 **1T BCM** **Year:** 2009, 2010 **Model:** Pilot **Engine:** 3.5L V6 **Transmission:** All	**Lost Communication with MICU (MICU frame):** A communication malfunction was detected between the gauge control module, the related units and the under-dash fuse/relay box. **NOTE: If you are troubleshooting multiple DTCs, be sure to follow the instructions in B-CAN System Diagnosis Test Mode A** **Possible Causes:** • Loose or poor connections between the gauge control module and the under-dash fuse/relay box • Perform the MICU input test and perform all power, ground and communication input tests • Faulty under-dash fuse/relay box • Faulty gauge control module
DTC: U1282 **1T PCM** **Year:** 2009, 2010 **Model:** Accord, Accord Crosstour, Pilot **Engine:** 2.4L L4, 3.5L V6 **Transmission:** All	**Door Multiplex Control Unit Lost Communication With Driver's MICU:** Poor communication between the Door Multiplex Control Unit, and the Driver's MICU. **NOTE: If you are troubleshooting multiple DTCs, be sure to follow the instructions in B-CAN System Diagnosis Test Mode A.** **Possible Causes:** • Perform the driver's MICU input test, and do all power, ground and communication input tests • Loose or poor connections at driver's under-dash fuse/relay box connector and the related units • Faulty driver's under-dash fuse/relay box
DTC: U1282 **1T PCM** **Year:** 2009, 2010 **Model:** Accord, Accord Crosstour, Pilot **Engine:** 2.4L L4, 3.5L V6 **Transmission:** All	**Immobilizer-keyless Control Unit Lost Communication With Driver's MICU:** Poor communication between the Immobilizer-keyless Control Unit, and the Driver's MICU. **NOTE: If you are troubleshooting multiple DTCs, be sure to follow the instructions in B-CAN System Diagnosis Test Mode A.** **Possible Causes:** • Perform the driver's MICU input test, and do all power, ground and communication input tests • Loose or poor connections at driver's under-dash fuse/relay box connector and the related units • Faulty driver's under-dash fuse/relay box
DTC: U1282 **1T PCM** **Year:** 2009, 2010 **Model:** Accord, Accord Crosstour **Engine:** 2.4L L4, 3.5L V6 **Transmission:** All	**Passenger's MICU Lost Communication With Driver's MICU:** Poor communication between the Passenger's MICU, and the Driver's MICU. **NOTE: If you are troubleshooting multiple DTCs, be sure to follow the instructions in B-CAN System Diagnosis Test Mode A.** **Possible Causes:** • Perform the driver's MICU input test, and do all power, ground and communication input tests • Loose or poor connections at driver's under-dash fuse/relay box connector and the related units • Faulty driver's under-dash fuse/relay box
DTC: U1282 **1T BCM** **Year:** 2009, 2010 **Model:** Pilot **Engine:** 3.5L V6 **Transmission:** All	**Lost Communication With MICU:** **NOTE: If you are troubleshooting multiple DTCs, be sure to follow the instructions in B-CAN System Diagnosis Test Mode A.** **Possible Causes:** • Loose or poor connections between the under-dash fuse/relay box and the automatic lighting control unit • Faulty automatic lighting control unit

DTC	Trouble Code Title, Conditions & Possible Causes
DTC: U1282 **1T PCM** **Year:** 2009, 2010 **Model:** Accord, Accord Crosstour, Pilot **Engine:** 2.4L L4, 3.5L V6 **Transmission:** All	**Lost Communication With Driver's MICU:** Poor communication between the power seat control unit, and the driver's MICU. **NOTE: If you are troubleshooting multiple DTCs, be sure to follow the instructions in B-CAN System Diagnosis Test Mode A.** **Possible Causes:** • Perform the power seat control unit input test • Loose or poor connections between the power seat control unit and the driver's MICU
DTC: U1282 **1T PCM** **Year:** 2009, 2010 **Model:** Accord, Accord Crosstour **Engine:** 2.4L L4, 3.5L V6 **Transmission:** All	**Keyless Access Control Unit Lost Communication With Driver's MICU:** Poor communication between the Keyless Access Control Unit, and the Driver's MICU. **NOTE: If you are troubleshooting multiple DTCs, be sure to follow the instructions in B-CAN System Diagnosis Test Mode A.** **Possible Causes:** • Perform the driver's MICU input test, and do all power, ground and communication input tests • Loose or poor connections at driver's under-dash fuse/relay box connector and the related units • Faulty driver's under-dash fuse/relay box
DTC: U1282 **1T** **Year:** 2009, 2010 **Model:** Accord, Accord Crosstour **Engine:** 2.4L L4, 3.5L V6 **Transmission:** All	**Gauge Control Module Lost Communication With Driver's MICU:** Poor communication between the Gauge Control Module and MICU. **NOTE: If you are troubleshooting multiple DTCs, be sure to follow the instructions in B-CAN System Diagnosis Test Mode A.** **Possible Causes:** • Perform the driver's MICU input test, and do all power, ground, and communication input tests. If the tests prove OK, replace the driver's under-dash fuse/relay box • Loose or poor connections at driver's under-dash fuse/relay box and the related units • Faulty gauge control module • Faulty MICU
DTC: U1283 **1T** **Year:** 2009, 2010 **Model:** Accord, Accord Crosstour **Engine:** 2.4L L4, 3.5L V6 **Transmission:** All	**Keyless Access Control Unit Lost Communication With Passenger's MICU:** Poor communication between the Keyless Access Control Unit and Passenger's MICU. **NOTE: If you are troubleshooting multiple DTCs, be sure to follow the instructions in B-CAN System Diagnosis Test Mode A.** **Possible Causes:** • Perform the passenger's MICU input test, and do all power, ground, and communication input tests. If the tests prove OK, replace the passenger's under-dash fuse/relay box • Loose or poor connections at passenger's under-dash fuse/relay box and the related units
DTC: U1283 **1T** **Year:** 2009, 2010 **Model:** Accord, Accord Crosstour **Engine:** 2.4L L4, 3.5L V6 **Transmission:** All	**Gauge Control Module Lost Communication With Passenger's MICU:** Poor communication between the Gauge Control Module and MICU. **NOTE: If you are troubleshooting multiple DTCs, be sure to follow the instructions in B-CAN System Diagnosis Test Mode A.** **Possible Causes:** • Perform the passenger's MICU input test, and do all power, ground, and communication input tests. If the tests prove OK, replace the passenger's under-dash fuse/relay box • Loose or poor connections at passenger's under-dash fuse/relay box and the related units
DTC: U1283 **1T** **Year:** 2009, 2010 **Model:** Accord, Accord Crosstour **Engine:** 2.4L L4, 3.5L V6 **Transmission:** All	**Lost Communication With Passenger's MICU:** Lost Communication With Passenger's MICU. **NOTE: If you are troubleshooting multiple DTCs, be sure to follow the instructions in B-CAN System Diagnosis Test Mode A.** **Possible Causes:** • Check the PCM for DTCs and troubleshoot PCM • Perform the passenger's MICU input test, and do all power, ground and communication input tests. If the tests prove OK, replace the passenger's under-dash fuse/relay box • Loose or poor connections at passenger's under-dash fuse/relay box and the related units
DTC: U1283 **1T** **Year:** 2009, 2010 **Model:** Accord, Accord Crosstour **Engine:** 2.4L L4, 3.5L V6 **Transmission:** All	**Driver's MICU Lost Communication With Passenger's MICU:** Poor communication between the Driver's MICU and Passenger's MICU. **NOTE: If you are troubleshooting multiple DTCs, be sure to follow the instructions in B-CAN System Diagnosis Test Mode A.** **Possible Causes:** • Perform the passenger's MICU input test, and do all power, ground, and communication input tests. If the tests prove OK, replace the driver's under-dash fuse/relay box and the related units • Loose or poor connections at driver's under-dash fuse/relay box and the related units
DTC: U1283 **1T** **Year:** 2009, 2010 **Model:** Accord, Accord Crosstour **Engine:** 2.4L L4, 3.5L V6 **Transmission:** All	**Door Multiplex Control Unit Lost Communication With Passenger's MICU:** Poor communication between the Door Multiplex Control Unit and Passenger's MICU. **NOTE: If you are troubleshooting multiple DTCs, be sure to follow the instructions in B-CAN System Diagnosis Test Mode A.** **Possible Causes:** • Perform the passenger's MICU input test, and do all power, ground and communication input tests. If the tests prove OK, replace the passenger's under-dash fuse/relay box • Loose or poor connections at passenger's under-dash fuse/relay box and the related units

DTC	Trouble Code Title, Conditions & Possible Causes
DTC: U1288 **1T BCM** **Year:** 2009, 2010 **Model:** Pilot **Engine:** 3.5L V6 **Transmission:** All	**Lost Communication with PARKSR:** A communication malfunction was detected between the gauge control module and the parking and back-up sensor control unit. **NOTE: If you are troubleshooting multiple DTCs, be sure to follow the instructions in B-CAN System Diagnosis Test Mode A** **Possible Causes:** • Loose or poor connections between the gauge control module and the parking and back-up sensor control unit • Faulty gauge control module
DTC: U128D **T PCM** **Year:** 2010 **Model:** Accord Crosstour **Engine:** 3.5L V6 **Transmission:** All	**Lost Communication With Gauge Control Module (VSP/NE Frame):** Poor communication between the gauge control module, and the power seat control unit. **NOTE: If you are troubleshooting multiple DTCs, be sure to follow the instructions in B-CAN System Diagnosis Test Mode A.** **Possible Causes:** • Loose or poor connections between the power seat control unit and the gauge control module • Perform the power seat control unit input test • Faulty gauge control module • Faulty power seat control unit

Commonly Used Abbreviations

2

2WD	Two Wheel Drive

4

4WD	Four Wheel Drive

A

A/C	Air Conditioning
ABDC	After Bottom Dead Center
ABS	Anti-lock Brakes
AC	Alternating Current
ACL	Air cleaner
ACT	Air Charge Temperature
AIR	Secondary Air Injection
ALCL	Assembly Line Communications Link
ALDL	Assembly Line Diagnostic Link
AT	Automatic Transaxle/Transmission
ATDC	After Top Dead Center
ATF	Automatic Transmission Fluid
ATS	Air Temperature Sensor
AWD	All Wheel Drive

B

BAP	Barometric Absolute Pressure
BARO	Barometric Pressure
BBDC	Before Bottom Dead Center
BCM	Body Control Module
BDC	Bottom Dead Center
BPT	Backpressure Transducer
BTDC	Before Top Dead Center
BVSV	Bimetallic Vacuum Switching Valve

C

CAC	Charge Air Cooler
CARB	California Air Resources Board
CAT	Catalytic Converter
CCC	Computer Command Control
CCCC	Computer Controlled Catalytic Converter
CCCI	Computer Controlled Coil Ignition
CCD	Computer Controlled Dwell
CDI	Capacitor Discharge Ignition
CEC	Computerized Engine Control
CFI	Continuous Fuel Injection
CIS	Continuous Injection System
CIS-E	Continuous Injection System - Electronic
CKP	Crankshaft Position
CL	Closed Loop
CMP	Camshaft Position
CPP	Clutch Pedal Position
CTOX	Continuous Trap Oxidizer System
CTP	Closed Throttle Position
CVC	Constant Vacuum Control
CYL	Cylinder

D

DBC	Dual Bed Catalyst
DC	Direct Current
DFI	Direct Fuel Injection
DIS	Distributorless Ignition System
DLC	Data Link Connector
DMM	Digital Multimeter
DOHC	Double Overhead Camshaft
DRB	Diagnostic Readout Box
DTC	Diagnostic Trouble Code
DTM	Diagnostic Test Mode
DVOM	Digital Volt/Ohmmeter

E

EBCM	Electronic Brake Control Module
ECM	Engine Control Module
ECT	Engine Coolant Temperature
ECU	Engine Control Unit or Electronic Control Unit
EDIS	Electronic Distributorless Ignition System
EEC	Electronic Engine Control
EEPROM	Electrically Erasable Programmable Read Only Memory
EFE	Early Fuel Evaporation
EGR	Exhaust Gas Recirculation
EGRT	Exhaust Gas Recirculation Temperature
EGRVC	EGR Valve Control
EPROM	Erasable Programmable Read Only Memory
EVAP	Evaporative Emissions
EVP	EGR Valve Position

F

FBC	Feedback Carburetor
FEEPROM	Flash Electrically Erasable Programmable Read Only Memory
FF	Flexible Fuel
FI	Fuel Injection
FT	Fuel Trim
FWD	Front Wheel Drive

G

GND	Ground

H

HAC	High Altitude Compensation
HEGO	Heated Exhaust Gas Oxygen sensor
HEI	High Energy Ignition
HO2 Sensor	Heated Oxygen Sensor

I

IAC	Idle Air Control
IAT	Intake Air Temperature
ICM	Ignition Control Module
IFI	Indirect Fuel Injection
IFS	Inertia Fuel Shutoff
ISC	Idle Speed Control
IVSV	Idle Vacuum Switching Valve

Commonly Used Abbreviations

K

KOEO	Key On, Engine Off
KOER	Key ON, Engine Running
KS	Knock Sensor

M

MAF	Mass Air Flow
MAP	Manifold Absolute Pressure
MAT	Manifold Air Temperature
MC	Mixture Control
MDP	Manifold Differential Pressure
MFI	Multiport Fuel Injection
MIL	Malfunction Indicator Lamp or Maintenance
MST	Manifold Surface Temperature
MVZ	Manifold Vacuum Zone

N

NVRAM	Nonvolatile Random Access Memory

O

O2 Sensor	Oxygen Sensor
OBD	On-Board Diagnostic
OC	Oxidation Catalyst
OHC	Overhead Camshaft
OL	Open Loop

P

P/S	Power Steering
PAIR	Pulsed Secondary Air Injection
PCM	Powertrain Control Module
PCS	Purge Control Solenoid
PCV	Positive Crankcase Ventilation
PIP	Profile Ignition Pick-up
PNP	Park/Neutral Position
PROM	Programmable Read Only Memory
PSP	Power Steering Pressure
PTO	Power Take-Off
PTOX	Periodic Trap Oxidizer System

R

RABS	Rear Anti-lock Brake System
RAM	Random Access Memory
ROM	Read Only Memory
RPM	Revolutions Per Minute
RWAL	Rear Wheel Anti-lock Brakes
RWD	Rear Wheel Drive

S

SBC	Single Bed Converter
SBEC	Single Board Engine Controller
SC	Supercharger
SCB	Supercharger Bypass
SFI	Sequential Multiport Fuel Injection
SIR	Supplemental Inflatible Restraint
SOHC	Single Overhead Camshaft
SPL	Smoke Puff Limiter
SPOUT	Spark Output
SRI	Service Reminder Indicator
SRS	Supplemental Restraint System
SRT	System Readiness Test
SSI	Solid State Ignition
ST	Scan Tool
STO	Self-Test Output

T

TAC	Thermostatic Air Clearner
TBI	Throttle Body Fuel Injection
TC	Turbocharger
TCC	Torque Converter Clutch
TCM	Transmission Control Module
TDC	Top Dead Center
TFI	Thick Film Ignition
TP	Throttle Position
TR Sensor	Transaxle/Transmission Range Sensor
TVV	Thermal Vacuum Valve
TWC	Three-way Catalytic Converter

V

VAF	Volume Air Flow, or Vane Air Flow
VAPS	Variable Assist Power Steering
VRV	Vacuum Regulator Valve
VSS	Vehicle Speed Sensor
VSV	Vacuum Switching Valve

W

WOT	Wide Open Throttle
WU-TWC	Warm Up Three-way Catalytic Converter

CHILTON LABOR GUIDE

Whether you are looking for labor times in print, or on CD-ROM, Chilton is your source! Chilton's editors have carefully crafted the latest edition of the famous Chilton Labor Guide to bring you the most accurate repair information available. Chilton's editors consider warranty times, component locations, component type, the environment in which technicians work, the training they receive, and the tools they use when calculating a labor time. To allow for vehicle age, operating conditions, and type of service, the Chilton Labor Guide provides standard and severe service times, plus OEM warranty times. Vehicle makes and models conform to current Automotive Aftermarket Industry Association (AAIA) standards.

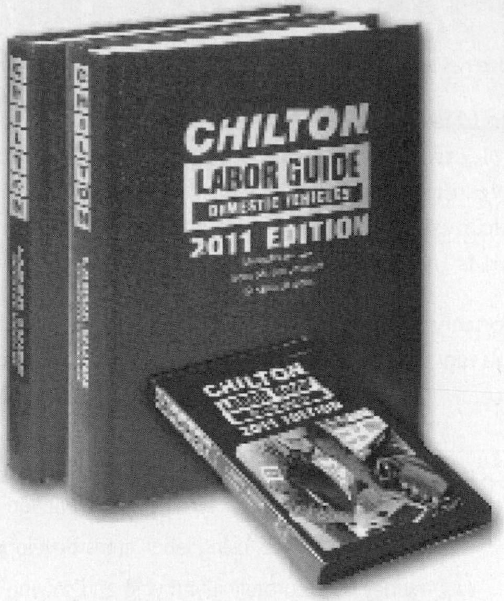
CHILTON
LABOR GUIDE
DOMESTIC VEHICLES
2011 EDITION

978-1-1115-4291-7 Chilton 2011 Labor Guide Manual Set (Domestic & Import)
978-1-1115-4294-8 Chilton 2011 Labor Guide CD-ROM (Domestic & Import)

CD-ROM FEATURES

- access labor times for 1981-2011 import and domestic vehicle models
- save time with automatically calculated labor charges, taxes, & parts as total job is estimated
- create professional estimates for your customer and worksheets for your technicians, printing them whenever needed
- keep track of customers, prior estimates, and your own parts or package jobs with less paper
- choose part names for estimates from an industry standard database to reduce typing
- estimate and track your work status with improved forms
- communicate easily with customers using re-designed printouts which show all labor and parts in an easy-to-read format.
- simplify adding parts to your estimate or work order with a helpful parts list
- locate information quick with a keyword search engine
- quickly locate work requests by day, week and month using the calendar feature

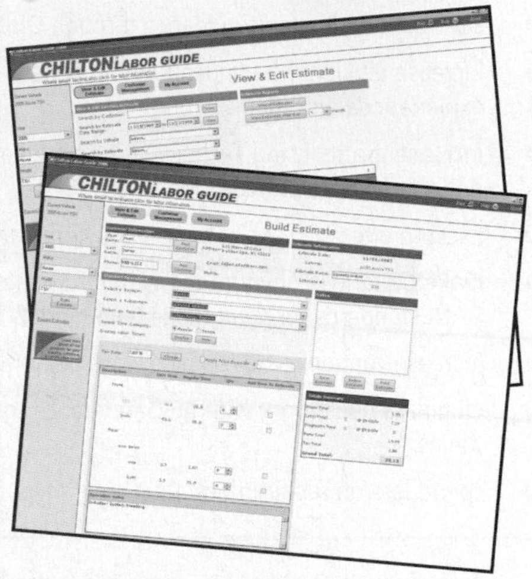
CHILTON LABOR GUIDE
View & Edit Estimate
CHILTON LABOR GUIDE
Build Estimate

Manual FEATURES

- more than 2,500 pages of updated Chilton labor times split into two volumes includes vehicle information from 1981 to 2011
- trusted by more service professionals than any other labor guide
- less flipping though pages with separate domestic and imported vehicle manuals
- convenient tabs display contents by manufacturer and model
- easy-to-find manufacturers are arranged alphabetically within each volume
- search using two-indexes - labor operations and systems - in each model group
- page numbers include manufacturer code so you know where you are in the book

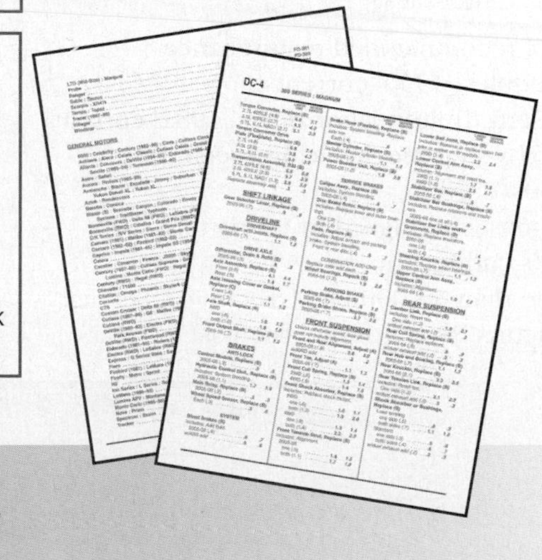
DC-4

Chilton's labor times are so trusted, even a competing publisher uses them!

CHILTONPRO.COM

Where smart technicians click for service information.

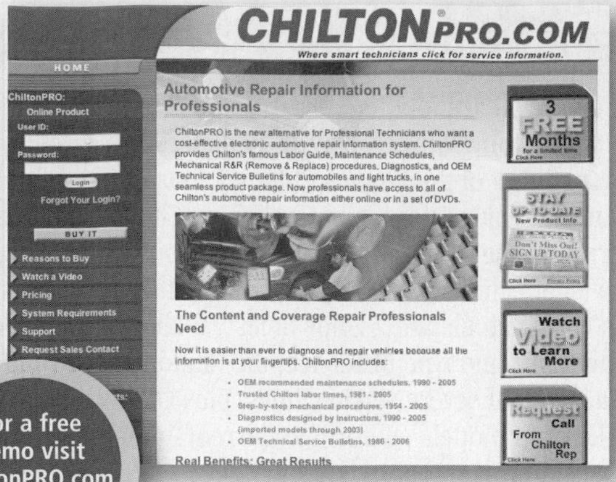
CHILTONPRO.COM
Where smart technicians click for service information.
Automotive Repair Information for Professionals
The Content and Coverage Repair Professionals Need
Real Benefits: Great Results

ChiltonPRO is the alternative for professional technicians who want a cost-effective electronic automotive repair system. It combines Chilton's famous automotive repair information into one solution covering more than 20 years of domestic and imported vehicles. The information is delivered online and is updated regularly throughout the year.

Online Monthly Payment
ISBN: 978-14180-3002-5

Online Annual Payment
ISBN: 978-14180-2876-3

For a free demo visit ChiltonPRO.com

ChiltonPRO FEATURES

- make repairs even easier with videos & animations which explain system operations & contribute to technician knowledge
- create better estimates using labor times developed with real-world factors
- save money by accurately identifying and solving engine performance problems
- save time with expert guidance through OBDII diagnostics
- increase efficiency by understanding system operation through detailed explanations and theory
- increase profits using Technical Service Bulletins (TSBs) to ensure that work is not going unperformed
- execute effective repairs by viewing cutaway diagrams and actual photos
- make better use of your time with information that can be found quicker using AAIA standards for year, make, and model
- increase confidence levels by always being able to print what you need
- eliminate guesswork with quick reference to critical specifications in helpful tables
- spend less on repair information

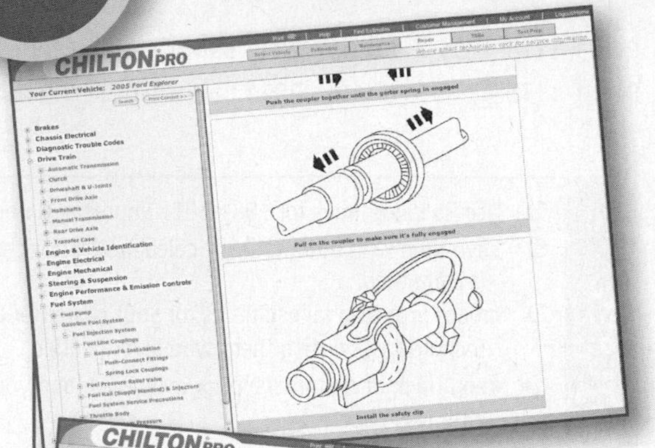
CHILTONPRO
Your Current Vehicle: 2005 Ford Explorer

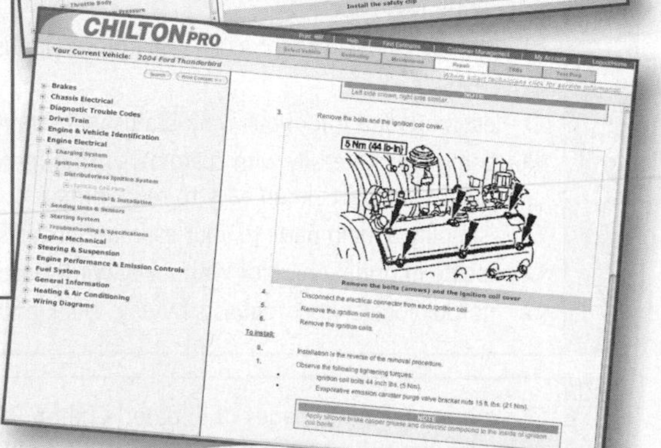
CHILTONPRO
Your Current Vehicle: 2004 Ford Thunderbird

Coverage Includes:

- OEM recommended maintenance schedules, 1990–current
- Trusted Chilton labor times, 1981–current
- Step-by-step mechanical procedures, 1950s–current
- Diagnostics designed by instructors, 1990–current
- More than 75,000 OEM Technical Service Bulletins issued during the past 20 years

System Requirements:
Web browser

- Internet Explorer 6.0 or above (recommended)
- Firefox 2 or 3, or Safari
- High-speed internet connection
- Adobe Flash Player
- Adobe Shockwave Player
- Windows XP or Vista

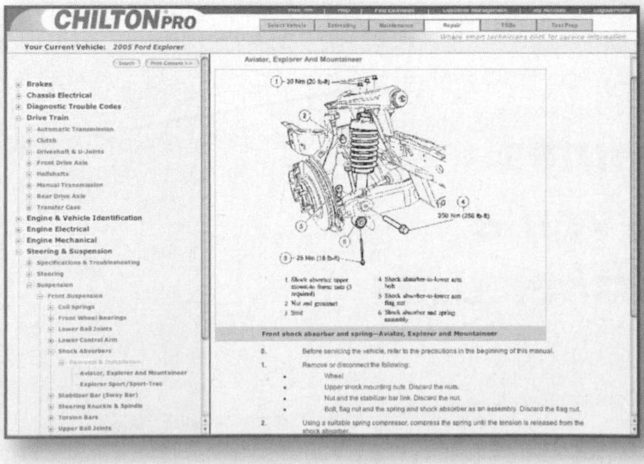
CHILTONPRO
Your Current Vehicle: 2005 Ford Explorer
Aviator, Explorer And Mountaineer
Front shock absorber and spring—Aviator, Explorer and Mountaineer

CHILTON
GENERAL MOTORS
SERVICE MANUAL
2010 EDITION

Chilton® 2010 Service Manuals

The Chilton 2010 Service Manuals now include even better graphics and expanded procedures! Chilton's editors have put together the most current automotive repair information available to assist users during daily repairs. These new manuals allow users to accurately and efficiently diagnose and repair late-model cars and trucks. Trust the step-by-step procedures and helpful illustrations that only Chilton can provide. The 2010 Service Manuals cover 2008 and 2009 models plus available 2010 models.

KEY FEATURES

- organized by vehicle manufacturer
- provides thousands of pages of expertly written content
- access new year, make, and model information without repeating previous edition's content
- comprehensive, technically detailed content, including exploded view illustrations, diagnostics and specification charts, arranged alphabetically by model group for quick, easy access

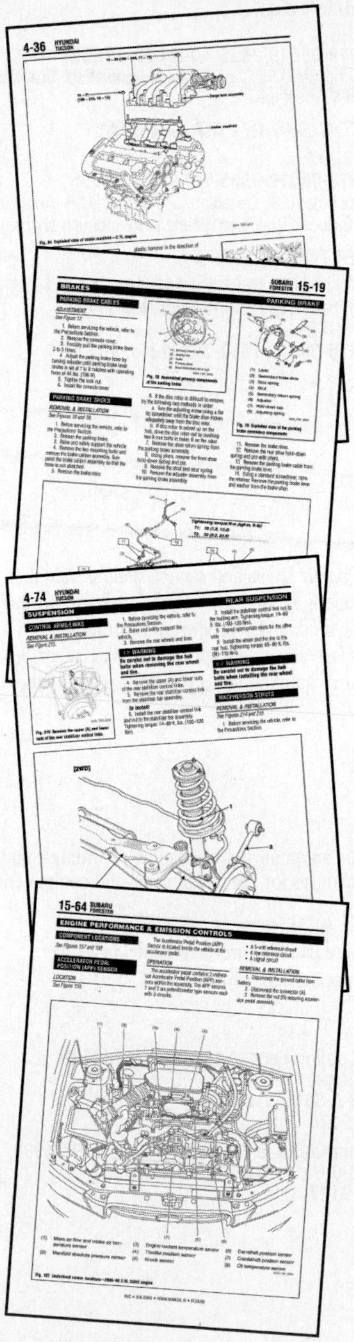

2010 EDITIONS

2010 Asian Service Manual Vol. 1*
ISBN 978-1-1110-3764-2
Part No. 163764

2010 Asian Service Manual Vol. 2*
ISBN 978-1-1110-3765-9
Part No. 163765

2010 Asian Service Manual Vol. 3*
ISBN 978-1-1110-3766-6
Part No. 163766

2010 Asian Service Manual Vol. 4*
ISBN 978-1-1110-3767-3
Part No. 163767

2010 Asian Service Manual Vol. 5*
ISBN 978-1-1110-3768-0
Part No. 163768

2010 European Service Manual*
ISBN 978-1-1110-3769-7
Part No. 163769

2010 Chrysler Service Manual, Volumes 1 & 2
ISBN 978-1-1110-3654-6
Part No. 163654

2010 Ford Service Manual, Vols. 1 & 2
ISBN 978-1-1110-3657-7
Part No. 163657

2010 General Motors Service Manuals, Vols. 1, 2, & 3
ISBN 978-1-111-03661-4
Part No. 163661

2008 EDITIONS

2008 Chrysler Service Manual, Vols. 1 & 2
ISBN 978-1-4283-2204-2
Part No. 142204

2008 Ford Service Manuals, Vols. 1 & 2
ISBN 978-1-4283-2208-0
Part No. 142208

2008 Edition General Motors Service Manuals, Vols. 1 & 2
ISBN 978-1-4283-2211-0
Part No. 142211

2008 Asian Service Manuals, Vols. 1-4
ISBN 978-1-4283-2214-1
Part No. 142214

2008 Asian Service Manual, Vol. 1
ISBN 978-1-4283-2215-8
Part No. 142215

2008 Asian Service Manual, Vol. 2
ISBN 978-1-4283-2216-5
Part No. 142216

2008 Asian Service Manual, Vol. 3
ISBN 978-1-4283-2217-2
Part No. 142217

2008 Asian Service Manual, Vol. 4
ISBN 978-1-4283-2218-9
Part No. 142218

2008 European Service Manual
ISBN 978-1-4283-2220-2
Part No. 142220

2006 EDITIONS

2006 DaimlerChrysler Diagnostic Service Manual
ISBN 978-1-4180-2118-4
Part No. 132118

2006 General Motors Diagnostic Service Manual
ISBN 978-1-4180-2120-7
Part No. 132120

2006 Asian Diagnostic Service Manual, Vol. 1
ISBN 978-1-4180-2913-5
Part No. 132913

2006 Asian Diagnostic Service Manual, Vol. 2
ISBN 978-1-4180-2914-2
Part No. 132914

2006 Asian Diagnostic Service Manual, Vol. 3
ISBN 978-1-4180-2915-9
Part No. 132915

2006 European Diagnostic Service Manual
ISBN 978-1-4180-2924-1
Part No. 132924

2006 DaimlerChrysler Mechanical Service Manual
ISBN 978-1-4180-0600-6
Part No. 130600

2006 Asian Mechanical Service Manual, Vol. 1
ISBN 978-1-4180-0947-2
Part No. 130947

2006 Asian Mechanical Service Manual, Vol. 2
ISBN 978-1-4180-0948-9
Part No. 130948

2006 Asian Mechanical Service Manual, Vol. 3
ISBN 978-1-4180-0949-6
Part No. 130949

2006 Asian Mechanical Service Manual, 3 Vol. Set
ISBN 978-1-4180-0603-7
Part No. 130603

2006 European Mechanical Service Manual
ISBN 978- 1-4180-0604-4
Part No. 130604

CHILTON
FORD
SERVICE MANUAL
2008 EDITION

*Available December 2010

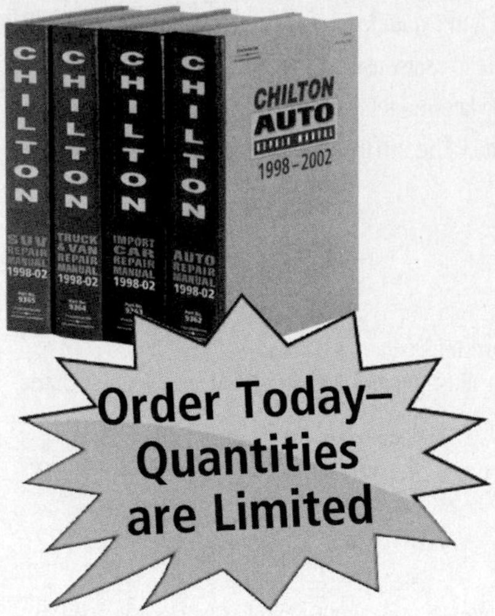
CHILTON
CHILTON
CHILTON
CHILTON
CHILTON AUTO
1998-2002
Order Today–
Quantities
are Limited

Chilton® Mechanical Service Manuals–Perennial Editions

These manuals contain repair and maintenance information for all major systems. Included are repair and overhaul procedures using thousands of illustrations.

CHILTON AUTO REPAIR MANUALS

1998-2002
ISBN 978-0-8019-9362-6/Part No. 9362
Covers all popular American and Canadian cars. An added feature includes scheduled maintenance interval charts.

1993-97
ISBN 978-0-8019-7919-4/Part No. 7919
Covers all popular American and Canadian cars.

1980-87
ISBN 978-0-8019-7670-4/Part No. 7670
Covers all popular American and Canadian cars.

CHILTON IMPORT AUTO REPAIR MANUALS

1998-2002
ISBN 978-0-8019-9363-3/Part No. 9363
Covers all popular Import cars. An added feature includes scheduled maintenance intervals charts.

1993-97
ISBN 978-0-8019-7920-0/Part No. 7920
Covers all popular Import cars.

1988-92
ISBN 978-0-8019-7907-1/Part No. 7907
Covers all popular Import cars.

1980-87
ISBN 978-0-8019-7672-8/Part No. 7672
Covers all popular Import cars.

CHILTON TRUCK AND VAN REPAIR MANUALS

1998-2002
ISBN 978-0-8019-9364-0/Part No. 9364
Covers popular U.S., Canadian, and Import Pick-Ups, Vans, and 4WDs. An added feature includes scheduled maintenance interval charts.

1993-97
ISBN 978-0-8019-7921-7/Part No. 7921
Covers popular U.S., Canadian, and Import Pick-Ups, Sport-Utilities, Vans, RVs and 4 wheel drives.

1991-95
ISBN 978-0-8019-7911-8/Part No. 7911
Covers popular U.S., Canadian, and Import Pick-Ups, Vans, RVs and 4 wheel drives.

1986-90
ISBN 978-08019-7902-6/Part No. 7902
Covers popular U.S., Canadian, and Import Pick-Us, Vans, RVs and 4 wheel drives.

1979-86
ISBN 978-08019-7655-1/Part No. 7655
Covers popular U.S., Canadian, and Import Pick-Ups, Vans, RVs and 4 wheel drives.

CHILTON SUV REPAIR MANUAL

1998-2002
ISBN 978-08019-9365-7/Part No. 9365
Covers popular U.S., Canadian, and import SUVs. An added feature includes scheduled maintenance intervals charts.

COLLECTOR'S SERIES

CHILTON AUTO REPAIR MANUAL 1964-1971
ISBN 978-08019-5974-5/Part No. 5974
1971-1978
ISBN 978-08019-7012-2/Part No. 7012

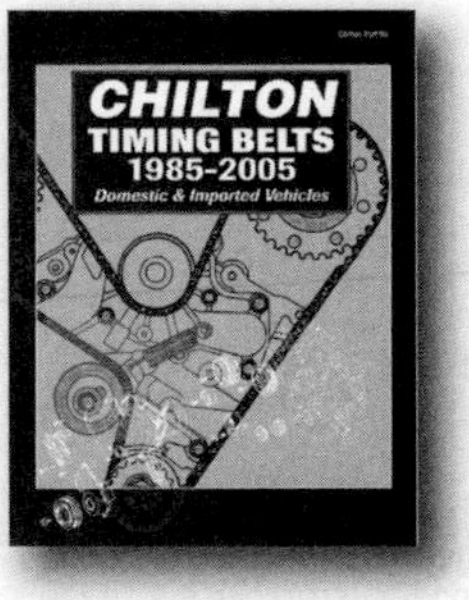
CHILTON
TIMING BELTS
1985-2005
Domestic & Imported Vehicles

ISBN 978-1-4018-9880-9
Part No. 129880
544 pp, 8" x 11", SC, ©2006

Chilton Timing Belts, 1985-2005

Timing belt procedures can represent increased profits for automotive repair shops and service stations, and this manual contains all the information automotive technicians need to properly service timing belts on domestic and imported cars, vans, and light trucks through 2005 models. Clear, straightforward procedures, illustrations, and specifications help to communicate 20 years of vehicle applications for fast, accurate inspection, replacement, and tensioning of timing belts. Users will learn how to perform key procedures quickly and safely, while learning the correct labor time to charge for the service.

ALSO AVAILABLE:

Quick-Reference Manuals

The Chilton Professional Series offers *Quick-Reference Manuals* for the automotive professional, providing complete coverage on repair and maintenance, adjustments, and diagnostic procedures for specific systems and components.

KEY FEATURES

- step-by-step procedures
- detailed illustrations and exploded views
- easy-to-use manufacturer and model indexing
- handy specifications or data charts

Heater Core Service 1990-2000,
ISBN 978-0-8019-9311-4
Part No. 9311

Brake Specifications and Service 1990-2000
ISBN 978-0-8019-9312-1
Part No. 9312

Electric Cooling Fans, Accessory Drive Belts & Water Pumps, 1995-1999,
ISBN 978-0-8019-9126-4
Part No. 9126

Powertrain Codes & Oxygen Sensors, 1990-1999,
ISBN 978-0-8019-9127-1
Part No. 9127

BRAKES
TEST H4
ENGINE REPAIR
TEST A1
GASOLINE ENGINES
TEST T1
Engine Machinist
Test M1
DIESEL ENGINES
TEST S2
PAINTING & REFINISHING
TEST B2
TRUCK EQUIPMENT INSTALLATION AND REPAIR
TEST E1

ASE Test Preparation for Transit Bus

153939	(H1) Compressed Natural Gas Engines	978-1-4354-3939-9
136570	(H2) Diesel Engines	978-1-4180-6570-6
155376	(H3) Drive Train	978-1-4354-5376-0
134998	(H4) Brakes	978-1-4180-4998-0
144011	(H5) Suspension & Steering	978-1-4283-4011-4
134999	(H6) Electrical/Electronic Systems	978-1-4180-4999-7
136571	(H7) Heating, Ventilation, & Air Conditioning	978-1-4180-6571-3
153938	(H8) Preventive Maintenance	978-1-4354-3938-2

ASE Test Preparation for Truck Equipment

153935	(E1) Truck Equipment Installation & Repair	978-1-4354-3935-1
153936	(E2) Electronic Systems Installation & Repair	978-1-4354-3936-8
153937	(E3) Auxilary Power Systems Installation & Repair	978-1-4354-3937-5

Online ASE Test Preparation

Covers the A1-A8, X1, P2, L1, C1, X1, and T1-T8 & B2-B6 Certification Exams

Visit www.techniciantestprep.com for a free demo.

131305	*Online (A1) Engine Repair	978-1-4180-1305-9
131306	*Online (A2) Automatic Transmissions & Transaxles	978-1-4180-1306-6
131307	*Online (A3) Manual Drive Trains & Axles	978-1-4180-1307-3
131308	*Online (A4) Suspension & Steering	978-1-4180-1308-0
131309	*Online (A5) Brakes	978-1-4180-1309-7
131310	*Online (A6) Electrical/Electronic Systems	978-1-4180-1310-3
131311	*Online (A7) Heating & Air Conditioning	978-1-4180-1311-0
131312	*Online (A8) Engine Performance	978-1-4180-1312-7
131313	*Online (X1) Exhaust Systems	978-1-4180-1313-4
131314	*Online (P2) Automobile Parts Specialist	978-1-4180-1314-1
131315	*Online (L1) Advanced Engine Performance	978-1-4180-1315-8
131316	*Online (C1) Service Consultant	978-1-4180-1316-5
127897	Online (T1) Gasoline Engines	978-1-4018-7897-9
127898	Online (T2) Diesel Engines	978-1-4018-7898-6
127900	Online (T3) Drive Train	978-1-4018-7900-6
127901	Online (T4) Brakes	978-1-4018-7901-3
127903	Online (T5) Suspension & Steering	978-1-4018-7903-7
131879	Online (T6) Electrical/Electronic Systems	978-1-4180-1879-5
131880	Online (T7) Heating, Ventilation, & Air Conditioning	978-1-4180-1880-1
127906	Online (T8) Preventive Maintenance	978-1-4018-7906-8
154748	Online (B2) Painting & Refinishing	978-1-4354-4748-6
154749	Online (B3) Non-Structural Analysis & Damage Repair	978-1-4354-4749-3
154750	Online (B4) Structural Analysis and Repair	978-1-4354-4750-9
154751	Online (B5) Mechanical & Electrical Components	978-1-4354-4751-6
154752	Online (B6) Damage Analysis & Estimating	978-1-4354-4752-3

* Switch between English & Spanish at the click of a button!

ASE Test Preparation in Spanish

Manuals

Reparación de Motores
Exámen A1
2a Edición
2a Edición
CERTIFICATION PARA TÉCNICO AUTOMOTRIZ

- Covers the A1-A8, L1, X1, P2, & B2-B6 Certification Exams in Spanish

Switch between English & Spanish at the click of a button!

- Covers the A1-A8, L1, X1, P2, & C1 exams online. Visit techniciantestprep.com

Online

21014	Spanish (A1) Engine Repair	978-1-4018-1014-6
21015	Spanish (A2) Transmissions and Transaxles	978-1-4018-1015-3
21016	Spanish (A3) Manual Drive Train and Axles	978-1-4018-1016-0
21017	Spanish (A4) Suspension and Steering	978-1-4018-1017-7
21018	Spanish (A5) Brakes	978-1-4018-1018-4
21019	Spanish (A6) Electrical/Electronic Systems	978-1-4018-1019-1
21020	Spanish (A7) Heating and Air Conditioning	978-1-4018-1020-7
21021	Spanish (A8) Engine Performance	978-1-4018-1021-4
21022	Spanish (L1) Advanced Engine Performance	978-1-4018-1022-1
21024	Spanish (X1) Exhaust Systems	978-1-4018-1024-5
21023	Spanish (P2) Parts Specialist	978-1-4018-1023-8
29255	Spanish (B2) Painting and Refinishin	978-1-4018-9255-5
22544	Spanish (B3) Non-Structural Analysis and Damage Repair	978-1-4018-2544-7
29131	Spanish (B4) Structural Analysis and Damage Repair	978-1-4018-9131-2
27759	Spanish (B5) Mechanical and Electrical Components	978-1-4018-7759-0
26573	Spanish (B6) Damage Analysis and Estimation	978-1-4018-6573-3

Complete Series

133954	ASE Manuals for Automotive (A1-A8, X1, P2, L1, C1)	978-1-4180-3954-7
136139	ASE Manuals for Automotive (A1-A8 & L1)	978-1-4180-6139-5
134197	ASE Manuals for Automotive (A1-A8, L1, & P2)	978-1-4180-4197-7
136237	ASE Manuals for Automotive (A1-A8)	978-1-4180-6237-8
136335	ASE Manuals for Automotive (A1-A8, L1, P2, & X1)	978-1-4180-6335-1
133447	Online ASE Manuals for Automotive (A1-A8, X1, P2, L1, C1)	978-1-4180-1344-8
134934	ASE Manuals for Medium/Heavy Duty Truck (T1-T8)	978-1-4180-4934-8
130611	Online ASE for Medium/Heavy Duty Truck (T1-T8)	978-1-4180-0611-2
125120	ASE Manuals for Collision (B2-B6)	978-1-4018-5120-0
24155	ASE Manuals for Collision in Spanish (B2-B6)	978-1-4018-4155-3
16283	ASE Manuals for Engine Machinist (M1-M3)	978-0-7668-6283-8

CSAT AUTO **Comprehensive Skill Assessment Tool**

CSAT-Automotive Series

The online *Comprehensive Skill Assessment Tool-Automotive Series* helps instructors and trainers implement the necessary training programs for individual areas needing improvement over various key automotive topics. As a true skill gap analysis tool, within each key topic, strategic learning areas are measured for knowledge of theory, hands-on application, and diagnostic skill. Areas of strength and areas needing improvement are identified. The combined phases of education and training, and post-assessment allow instructors to track skill level growth and target specific areas needing development.

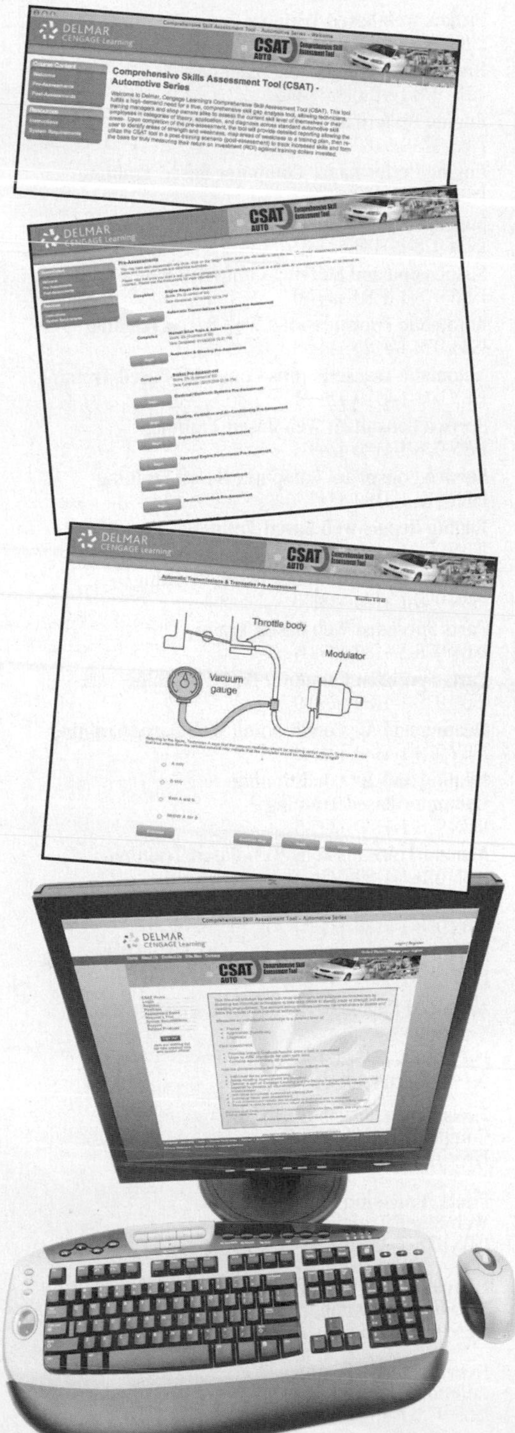
DELMAR CENGAGE Learning
CSAT AUTO Comprehensive Skill Assessment Tool
Comprehensive Skills Assessment Tool (CSAT) - Automotive Series
Pre-Assessments
Automatic Transmissions & Transaxles Pre-Assessment
Throttle body
Modulator
Vacuum gauge

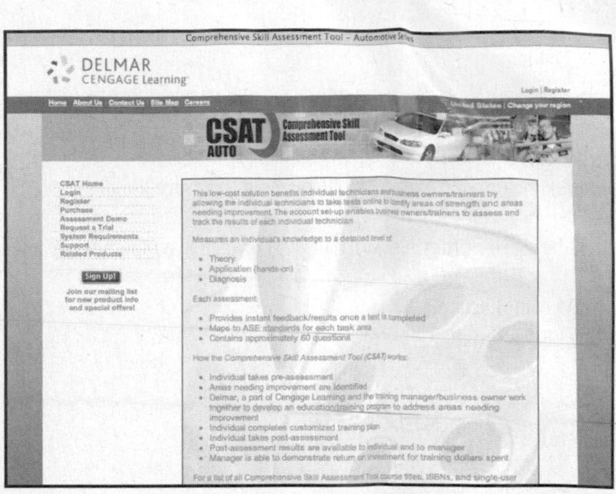
Comprehensive Skill Assessment Tool - Automotive Series
DELMAR CENGAGE Learning
CSAT AUTO Comprehensive Skill Assessment Tool

Courses Available in the CSAT Automotive Series

Parts Specialist
ISBN 978-1-4180-3225-8

Service Consultant
ISBN 978-1-4180-3223-4

Advanced Engine Performance
ISBN 978-1-4180-0073-8

Brakes
ISBN 978-1-4180-0069-1

Electrical/Electronic Systems
ISBN 978-1-4180-0070-7

Engine Performance
ISBN 978-1-4180-0072-1

Engine Repair
ISBN 978-1-4180-0065-3

Exhaust Systems
ISBN 978-1-4180-0074-5

Heating and Air Conditioning
ISBN 978-1-4180-0071-4

Manual Drive Train & Axles
ISBN 978-1-4180-0067-7

Suspension & Steering
ISBN 978-1-4180-0068-4

Transmissions & Transaxles
ISBN 978-1-4180-0066-0

All-in-One (contains questions from all eight core automotive areas in one product)
ISBN 978-1-4354-2825-6

FEATURES

- available tests include Engine Repair, Transmissions and Transaxles, Manual Drive Train and Axles, Suspension and Steering, Brakes, Electrical/Electronic Systems, Heating and Air Conditioning, Engine Performance, Advanced Engine Performance, and Exhaust Systems
- can be utilized by companies to measure the technical skill level of individuals against an "ideal" to identify areas of strength and creates a skill gap analysis to help users address areas needing improvement
- questions are written and reviewed by experts in the industry and offer users the opportunity to receive instant feedback
- account set-up that enables instructors and trainers to assess and track the results of individual students
- acts as a true return on investment (ROI) tool for companies to ensure they invest their training dollars in the most appropriate areas

Visit www.skillanalysis.com for a free demo!